American Men & Women of Science

1992-93 · 18th Edition

The 18th edition of *AMERICAN MEN & WOMEN OF SCIENCE* was prepared by the R.R. Bowker Database Publishing Group.

Stephen L. Torpie, Managing Editor
Judy Redel, Managing Editor, Research
Richard D. Lanam, Senior Editor
Tanya Hurst, Research Manager
Karen Hallard, Beth Tanis, Associate Editors

Peter Simon, Vice President, Database Publishing Group
Dean Hollister, Director, Database Planning
Edgar Adcock, Jr., Editorial Director, Directories

American Men & Women of Science

1992-93 • 18th Edition

A Biographical Directory of Today's Leaders in Physical, Biological and Related Sciences.

Volume 4 • J-L

R. R. BOWKER
New Providence, New Jersey

Published by R.R. Bowker, a division of Reed Publishing, (USA) Inc.

International Standard Book Number

Set:	0-8352-3074-0
Volume I:	0-8352-3075-9
Volume II:	0-8352-3076-7
Volume III:	0-8352-3077-5
Volume IV:	0-8352-3078-3
Volume V:	0-8352-3079-1
Volume VI:	0-8352-3080-5
Volume VII:	0-8352-3081-3
Volume VIII:	0-8352-3082-1

International Standard Serial Number: 0192-8570
Library of Congress Catalog Card Number: 6-7326
Printed and bound in the United States of America.

8 Volume Set

ISBN 0-8352-3074-0

9 780835 230742

Contents

Advisory Committee

Dr. Robert F. Barnes
    Executive Vice President
American Society of Agronomy

Dr. John Kistler Crum
    Executive Director
American Chemical Society

Dr. Charles Henderson Dickens
    Section Head, Survey & Analysis Section
Division of Science Resource Studies
National Science Foundation

Mr. Alan Edward Fechter
    Executive Director
Office of Scientific & Engineering Personnel
National Academy of Science

Dr. Oscar Nicolas Garcia
    Prof Electrical Engineering
Electrical Engineering & Computer Science Department
George Washington University

Dr. Charles George Groat
    Executive Director
American Geological Institute

Dr. Richard E. Hallgren
    Executive Director
American Meteorological Society

Dr. Michael J. Jackson
    Executive Director
Federation of American Societies for Experimental
Biology

Dr. William Howard Jaco
    Executive Director
American Mathematical Society

Dr. Shirley Mahaley Malcom
    Head, Directorate for Education and Human
    Resources Programs
American Association for the Advancement of Science

Mr. Daniel Melnick
    Sr Advisor Research Methodologies
Sciences Resources Directorate
National Science Foundation

Ms. Beverly Fearn Porter
    Division Manager
Education & Employment Statistics Division
American Institute of Physics

Dr. Terrence R. Russell
    Manager
Office of Professional Services
American Chemical Society

Dr. Irwin Walter Sandberg
    Holder, Cockrell Family Regent Chair
Department of Electrical & Computer Engineering
University of Texas

Dr. William Eldon Splinter
    Interim Vice Chancellor for Research,
    Dean, Graduate Studies
University of Nebraska

Ms. Betty M. Vetter
    Executive Director, Science Manpower Comission
Commission on Professionals in Science & Technology

Dr. Dael Lee Wolfe
    Professor Emeritus
Graduate School of Public Affairs
University of Washington

Preface

American Men and Women Of Science remains without peer as a chronicle of North American scientific endeavor and achievement. The present work is the eighteenth edition since it was first compiled as *American Men of Science* by J. Mckeen Cattell in 1906. In its eighty-six year history *American Men & Women of Science* has profiled the careers of over 300,000 scientists and engineers. Since the first edition, the number of American scientists and the fields they pursue have grown immensely. This edition alone lists full biographies for 122,817 engineers and scientists, 7021 of which are listed for the first time. Although the book has grown, our stated purpose is the same as when Dr. Cattell first undertook the task of producing a biographical directory of active American scientists. It was his intention to record educational, personal and career data which would make "a contribution to the organization of science in America" and "make men [and women] of science acquainted with one another and with one another's work." It is our hope that this edition will fulfill these goals.

The biographies of engineers and scientists constitute seven of the eight volumes and provide birthdates, birthplaces, field of specialty, education, honorary degrees, professional and concurrent experience, awards, memberships, research information and adresses for each entrant when applicable. The eighth volume, the discipline index, organizes biographees by field of activity. This index, adapted from the National Science Foundation's Taxonomy of Degree and Employment Specialties, classifies entrants by 171 subject specialties listed in the table of contents of Volume 8. For the first time, the index classifies scientists and engineers by state within each subject specialty, allowing the user to more easily locate a scientist in a given area. Also new to this edition is the inclusion of statistical information and recipients of theNobel Prizes, the Craaford Prize, the Charles Stark Draper Prize, and the National Medals of Science and Technology received since the last edition.

While the scientific fields covered by *American Men and Women Of Science* are comprehensive, no attempt has been made to include all American scientists. Entrants are meant to be limited to those who have made significant contributions in their field. The names of new entrants were submitted for consideration at the editors' request by current entrants and by leaders of academic, government and private research programs and associations. Those included met the following criteria:

1. Distinguished achievement, by reason of experience, training or accomplishment, including contributions to the literature, coupled with continuing activity in scientific work;

    or

2. Research activity of high quality in science as evidenced by publication in reputable scientific journals; or for those whose work cannot be published due to governmental or industrial security, research activity of high quality in science as evidenced by the judgement of the individual's peers;

    or

3. Attainment of a position of substantial responsibility requiring scientific training and experience.

This edition profiles living scientists in the physical and biological fields, as well as public health scientists, engineers, mathematicians, statisticians, and computer scientists. The information is collected by means of direct communication whenever possible. All entrants receive forms for corroboration and updating. New entrants receive questionaires and verification proofs before publication. The information submitted by entrants is included as completely as possible within the boundaries of editorial and space restrictions. If an entrant does not return the form and his or her current location can be verified in secondary sources, the full entry is repeated. References to the previous edition are given for those who do not return forms and cannot be located, but who are presumed to be still active in science or engineering. Entrants known to be deceased are noted as such and a reference to the previous edition is given. Scientists and engineers who are not citizens of the United States or Canada are included if a significant portion of their work was performed in North America.

The information in AMWS is also available on CD-ROM as part of *SciTech Reference Plus*. In adition to the convenience of searching scientists and engineers, *SciTech Reference Plus* also includes *The Directory of American Research & Technology*, *Corporate Technology Directory*, sci-tech and medical books and serials from *Books in Print* and *Bowker International Series*. *American Men and Women Of Science* is available for online searching through the subscription services of DIALOG Information Services, Inc. (3460 Hillview Ave, Palo Alto, CA 94304) and ORBIT Search Service (800 Westpark Dr, McLean, VA 22102). Both CD-Rom and the on-line subscription services allow all elements of an entry, including field of interest, experience, and location, to be accessed by key word. Tapes and mailing lists are also available through the Cahners Direct Mail (John Panza, List Manager, Bowker Files 245 W 17th St, New York, NY, 10011, Tel: 800-537-7930).

A project as large as publishing *American Men and Women Of Science* involves the efforts of a great many people. The editors take this opportunity to thank the eighteenth edition advisory committee for their guidance, encouragement and support. Appreciation is also expressed to the many scientific societies who provided their membership lists for the purpose of locating former entrants whose addresses had changed, and to the tens of thousands of scientists across the country who took time to provide us with biographical information. We also wish to thank Bruce Glaunert, Bonnie Walton, Val Lowman, Debbie Wilson, Mervaine Ricks and all those whose care and devotion to accurate research and editing assured successful production of this edition.

Comments, suggestions and nominations for the nineteenth edition are encouraged and should be directed to The Editors, *American Men and Women Of Science*, R.R. Bowker, 121 Chanlon Road, New Providence, New Jersey, 07974.

Edgar H. Adcock, Jr.
Editorial Director

Major Honors & Awards

Nobel Prizes
Nobel Foundation

The Nobel Prizes were established in 1900 (and first awarded in 1901) to recognize those people who "have conferred the greatest benefit on mankind."

1990 Recipients

Chemistry:
Elias James Corey
Awarded for his work in retrosynthetic analysis, the synthesizing of complex substances patterned after the molecular structures of natural compounds.

Physics:
Jerome Isaac Friedman
Henry Way Kendall
Richard Edward Taylor
Awarded for their breakthroughs in the understanding of matter.

Physiology or Medicine:
Joseph E. Murray
Edward Donnall Thomas
Awarded to Murray for his kidney transplantation achievements and to Thomas for bone marrow transplantation advances.

1991 Recipients

Chemistry:
Richard R. Ernst
Awarded for refinements in nuclear magnetic resonance spectroscopy.

Physics:
Pierre-Gilles de Gennes*
Awarded for his research on liquid crystals.

Physiology or Medicine:
Erwin Neher
Bert Sakmann*
Awarded for their discoveries in basic cell function and particularly for the development of the patch clamp technique.

Crafoord Prize
Royal Swedish Academy of Sciences
(Kungl. Vetenskapsakademien)

The Crafoord Prize was introduced in 1982 to award scientists in disciplines not covered by the Nobel Prize, namely mathematics, astronomy, geosciences and biosciences.

1990 Recipients

Paul Ralph Ehrlich
Edward Osborne Wilson
Awarded for their fundamental contributions to population biology and the conservation of biological diversity.

1991 Recipient

Allan Rex Sandage
Awarded for his fundamental contributions to extragalactic astronomy, including observational cosmology.

Charles Stark Draper Prize
National Academy of Engineering

The Draper Prize was introduced in 1989 to recognize engineering achievement. It is awarded biennially.

1991 Recipients

Hans Joachim Von Ohain
Frank Whittle
Awarded for their invention and development of the jet aircraft engine.

National Medal of Science

National Science Foundation

The National Medals of Science have been awarded by the President of the United States since 1962 to leading scientists in all fields.

1990 Recipients:

Baruj Benacerraf
Elkan Rogers Blout
Herbert Wayne Boyer
George Francis Carrier
Allan MacLeod Cormack
Mildred S. Dresselhaus
Karl August Folkers
Nick Holonyak Jr.
Leonid Hurwicz
Stephen Cole Kleene
Daniel Edward Koshland Jr.
Edward B. Lewis
John McCarthy
Edwin Mattison McMillan**
David G. Nathan
Robert Vivian Pound
Roger Randall Dougan Revelle**
John D. Roberts
Patrick Suppes
Edward Donnall Thomas

1991 Recipients

Mary Ellen Avery
Ronald Breslow
Alberto Pedro Calderon
Gertrude Belle Elion
George Harry Heilmeier
Dudley Robert Herschbach
George Evelyn Hutchinson**
Elvin Abraham Kabat
Robert Kates
Luna Bergere Leopold
Salvador Edward Luria**
Paul A. Marks
George Armitage Miller
Arthur Leonard Schawlow
Glenn Theodore Seaborg
Folke Skoog
H. Guyford Stever
Edward Carroll Stone Jr
Steven Weinberg
Paul Charles Zamecnik

National Medal of Technology

U.S. Department of Commerce,
Technology Administration

The National Medals of Technology, first awarded in 1985, are bestowed by the President of the United States to recognize individuals and companies for their development or commercialization of technology or for their contributions to the establishment of a technologically-trained workforce.

1990 Recipients

John Vincent Atanasoff
Marvin Camras
The du Pont Company
Donald Nelson Frey
Frederick W. Garry
Wilson Greatbatch
Jack St. Clair Kilby
John S. Mayo
Gordon Earle Moore
David B. Pall
Chauncey Starr

1991 Recipients

Stephen D. Bechtel Jr
C. Gordon Bell
Geoffrey Boothroyd
John Cocke
Peter Dewhurst
Carl Djerassi
James Duderstadt
Antonio L. Elias
Robert W. Galvin
David S. Hollingsworth
Grace Murray Hopper
F. Kenneth Iverson
Frederick M. Jones**
Robert Roland Lovell
Joseph A. Numero**
Charles Eli Reed
John Paul Stapp
David Walker Thompson

*These scientists' biographies do not appear in *American Men & Women of Science* because their work has been conducted exclusively outside the US and Canada.

**Deceased [Note that Frederick Jones died in 1961 and Joseph Numero in May 1991. Neither was ever listed in *American Men and Women of Science*.]

Statistics

Statistical distribution of entrants in *American Men & Women of Science* is illustrated on the following five pages. The regional scheme for geographical analysis is diagrammed in the map below. A table enumerating the geographic distribution can be found on page xvi, following the charts. The statistics are compiled by tallying all occurrences of a major index subject. Each scientist may choose to be indexed under as many as four categories; thus, the total number of subject references is greater than the number of entrants in *AMWS*.

All Disciplines

	Number	Percent
Northeast	58,325	34.99
Southeast	39,769	23.86
North Central	19,846	11.91
South Central	12,156	7.29
Mountain	11,029	6.62
Pacific	25,550	15.33
TOTAL	**166,675**	**100.00**

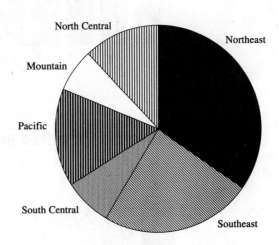

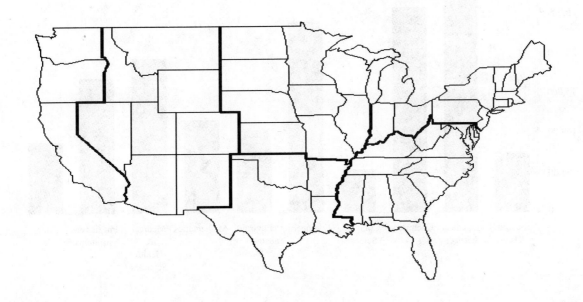

Age Distribution of American Men & Women of Science

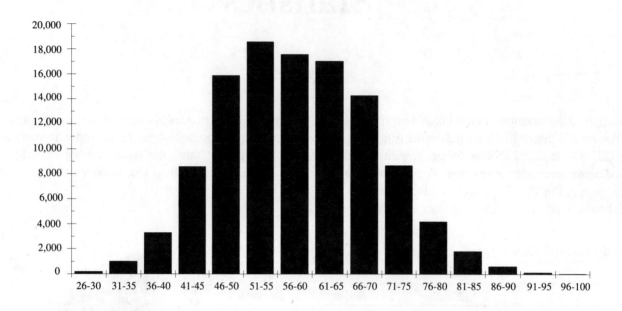

Number of Scientists in Each Discipline of Study

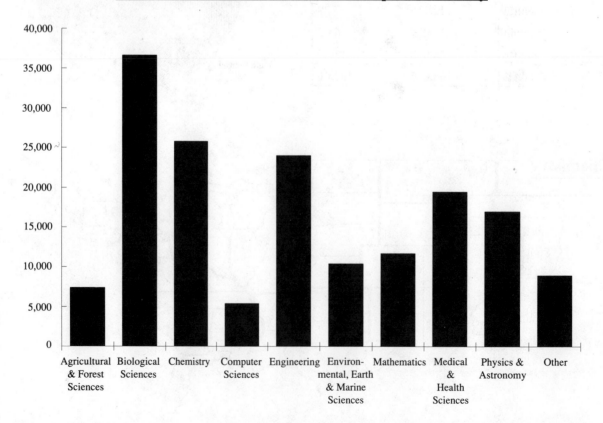

Agricultural & Forest Sciences

	Number	Percent
Northeast	1,574	21.39
Southeast	1,991	27.05
North Central	1,170	15.90
South Central	609	8.27
Mountain	719	9.77
Pacific	1,297	17.62
TOTAL	**7,360**	**100.00**

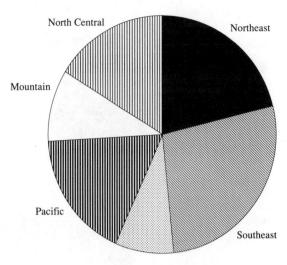

Biological Sciences

	Number	Percent
Northeast	12,162	33.23
Southeast	9,054	24.74
North Central	5,095	13.92
South Central	2,806	7.67
Mountain	2,038	5.57
Pacific	5,449	14.89
TOTAL	**36,604**	**100.00**

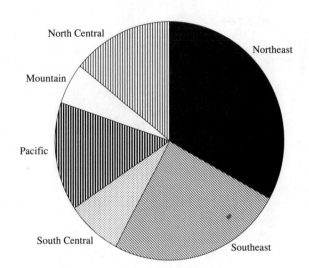

Chemistry

	Number	Percent
Northeast	10,343	40.15
Southeast	6,124	23.77
North Central	3,022	11.73
South Central	1,738	6.75
Mountain	1,300	5.05
Pacific	3,233	12.55
TOTAL	**25,760**	**100.00**

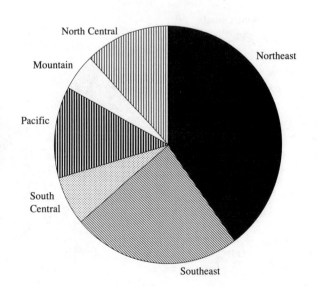

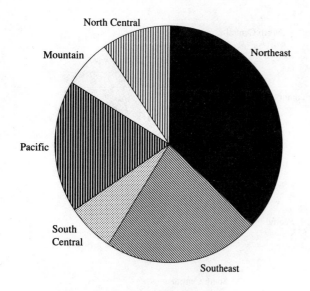

Computer Sciences

	Number	Percent
Northeast	1,987	36.76
Southeast	1,200·	22.20
North Central	511	9.45
South Central	360	6.66
Mountain	372	6.88
Pacific	976	18.05
TOTAL	**5,406**	**100.00**

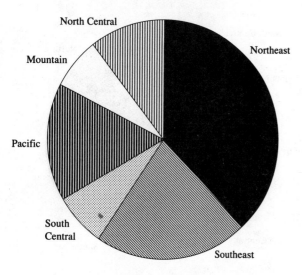

Engineering

	Number	Percent
Northeast	9,122	38.01
Southeast	5,202	21.68
North Central	2,510	10.46
South Central	1,710	7.13
Mountain	1,646	6.86
Pacific	3,807	15.86
TOTAL	**23,997**	**100.00**

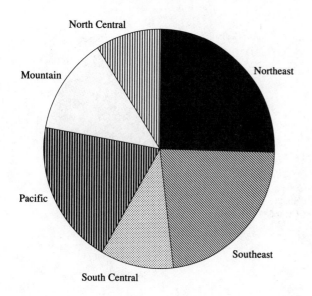

Environmental, Earth & Marine Sciences

	Number	Percent
Northeast	2,657	25.48
Southeast	2,361	22.64
North Central	953	9.14
South Central	1,075	10.31
Mountain	1,359	13.03
Pacific	2,022	19.39
TOTAL	**10,427**	**100.00**

Mathematics

	Number	Percent
Northeast	4,211	35.92
Southeast	2,609	22.26
North Central	1,511	12.89
South Central	884	7.54
Mountain	718	6.13
Pacific	1,789	15.26
TOTAL	**11,722**	**100.00**

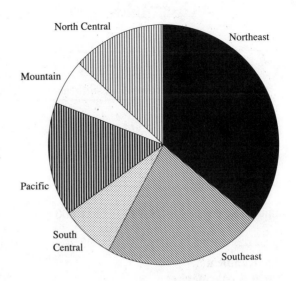

Medical & Health Sciences

	Number	Percent
Northeast	7,115	36.53
Southeast	5,004	25.69
North Central	2,577	13.23
South Central	1,516	7.78
Mountain	755	3.88
Pacific	2,509	12.88
TOTAL	**19,476**	**100.00**

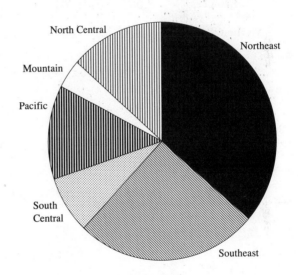

Physics & Astronomy

	Number	Percent
Northeast	5,961	35.12
Southeast	3,670	21.62
North Central	1,579	9.30
South Central	918	5.41
Mountain	1,607	9.47
Pacific	3,238	19.08
TOTAL	**16,973**	**100.00**

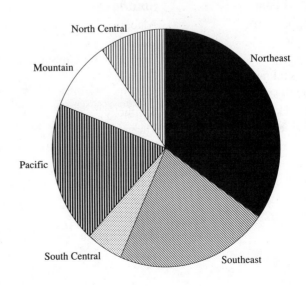

Geographic Distribution of Scientists by Discipline

	Northeast	Southeast	North Central	South Central	Mountain	Pacific	TOTAL
Agricultural & Forest Sciences	1,574	1,991	1,170	609	719	1,297	**7,360**
Biological Sciences	12,162	9,054	5,095	2,806	2,038	5,449	**36,604**
Chemistry	10,343	6,124	3,022	1,738	1,300	3,233	**25,760**
Computer Sciences	1,987	1,200	511	360	372	976	**5,406**
Engineering	9,122	5,202	2,510	1,710	1,646	3,807	**23,997**
Environmental, Earth & Marine Sciences	2,657	2,361	953	1,075	1,359	2,022	**10,427**
Mathematics	4,211	2,609	1,511	884	718	1,789	**11,722**
Medical & Health Sciences	7,115	5,004	2,577	1,516	755	2,509	**19,476**
Physics & Astronomy	5,961	3,670	1,579	918	1,607	3,238	**16,973**
Other Professional Fields	3,193	2,554	918	540	515	1,230	**8,950**
TOTAL	**58,325**	**39,769**	**19,846**	**12,156**	**11,029**	**25,550**	**166,675**

Geographic Definitions

Northeast
Connecticut
Indiana
Maine
Massachusetts
Michigan
New Hampshire
New Jersey
New York
Ohio
Pennsylvania
Rhode Island
Vermont

Southeast
Alabama
Delaware
District of Columbia
Florida
Georgia
Kentucky
Maryland
Mississippi
North Carolina
South Carolina
Tennessee
Virginia
West Virginia

North Central
Illinois
Iowa
Kansas
Minnesota
Missouri
Nebraska
North Dakota
South Dakota
Wisconsin

South Central
Arkansas
Louisiana
Texas
Oklahoma

Mountain
Arizona
Colorado
Idaho
Montana
Nevada
New Mexico
Utah
Wyoming

Pacific
Alaska
California
Hawaii
Oregon
Washington

Sample Entry

American Men & Women of Science (AMWS) is an extremely useful reference tool. The book is most often used in one of two ways: to find more information about a particular scientist or to locate a scientist in a specific field.

To locate information about an individual, the biographical section is most helpful. It encompasses the first seven volumes and lists scientists and engineers alphabetically by last name. The fictitious biographical listing shown below illustrates every type of information an entry may include.

The Discipline Index, volume 8, can be used to easily find a scientist in a specific subject specialty. This index is first classified by area of study, and within each specialty entrants are divided further by state of residence.

Name —

Date(s) of marriage —

Degrees Earned —

Professional Experience —

Current Position —

Birthplace & Date

Number of Children

Field of Specialty

Honorary Degrees

Concurrent Positions

Membership

Areas of research

Address

CARLETON, PHYLLIS B(ARBARA), b Glenham, SDak, April 1, 30. m 53, 69; c 2. ORGANIC CHEMISTRY. *Educ:* Univ Notre Dame, BSc, 52, MSc, 54, Vanderbilt Univ, PhD(chem), 57. *Hon Degrees:* DSc, Howard Univ, 79. *Prof Exp:* Res chemist, Acme Chem Corp, 54-59, sr res chemist, 59-60; from asst to assoc prof chem 60-63, prof chem, Kansas State Univ, 63-72; prof chem, Yale Univ, 73-89; CONSULT, CARLETON & ASSOCS, 89-. *Concurrent Pos:* Adj prof, Kansas State Univ 58-60; vis lect, Oxford Univ, 77, consult, Union Carbide, 74-80. *Honors & Awards:* Gold Medal, Am Chem Society, 81; *Mem:* AAAS, fel Am Chem Soc, Sigma Chi. *Res:* Organic synthesis, chemistry of natural products, water treatment and analysis. *Mailing Address:* Carleton & Assocs 21 E 34th St Boston MA 02108

Abbreviations

AAAS—American Association for the Advancement of Science
abnorm—abnormal
abstr—abstract
acad—academic, academy
acct—Account, accountant, accounting
acoust—acoustic(s), acoustical
ACTH—adrenocorticotrophic hormone
actg—acting
activ—activities, activity
addn—addition(s), additional
Add—Address
adj—adjunct, adjutant
adjust—adjustment
Adm—Admiral
admin—administration, administrative
adminr—administrator(s)
admis—admission(s)
adv—adviser(s), advisory
advan—advance(d), advancement
advert—advertisement, advertising
AEC—Atomic Energy Commission
aerodyn—aerodynamic
aeronaut—aeronautic(s), aeronautical
aerophys—aerophsical, aerophysics
aesthet—aesthetic
AFB—Air Force Base
affil—affiliate(s), affiliation
agr—agricultural, agriculture
agron—agronomic, agronomical, agronomy
agrost—agrostologic, agrostological, agrostology
agt—agent
AID—Agency for International Development
Ala—Alabama
allergol—allergological, allergology
alt—alternate
Alta—Alberta
Am—America, American
AMA—American Medical Association
anal—analysis, analytic, analytical
analog—analogue
anat—anatomic, anatomical, anatomy
anesthesiol—anesthesiology
angiol—angiology
Ann—Annal(s)
ann—annual
anthrop—anthropological, anthropology
anthropom—anthropometric, anthropometrical, anthropometry
antiq—antiquary, antiquities, antiquity
antiqn—antiquarian

apicult—apicultural, apiculture
APO—Army Post Office
app—appoint, appointed
appl—applied
appln—application
approx—approximate(ly)
Apr—April
apt—apartment(s)
aquacult—aquaculture
arbit—arbitration
arch—archives
archaeol—archaeological, archaeology
archit—architectural, architecture
Arg—Argentina, Argentine
Ariz—Arizona
Ark—Arkansas
artil—artillery
asn—association
assoc(s)—associate(s), associated
asst(s)—assistant(s), assistantship(s)
assyriol—Assyriology
astrodyn—astrodynamics
astron—astronomical, astronomy
astronaut—astonautical, astronautics
astronr—astronomer
astrophys—astrophysical, astrophysics
attend—attendant, attending
atty—attorney
audiol—audiology
Aug—August
auth—author
AV—audiovisual
Ave—Avenue
avicult—avicultural, aviculture

b—born
bact—bacterial, bacteriologic, bacteriological, bacteriology
BC—British Colombia
bd—board
behav—behavior(al)
Belg—Belgian, Belgium
Bibl—biblical
bibliog—bibliographic, bibliographical, bibliography
bibliogr—bibliographer
biochem—biochemical, biochemistry
biog—biographical, biography
biol—biological, biology
biomed—biomedical, biomedicine
biomet—biometric(s), biometrical, biometry
biophys—biophysical, biophysics

bk(s)—book(s)
bldg-building
Blvd—Boulevard
Bor—Borough
bot—botanical, botany
br—branch(es)
Brig—Brigadier
Brit—Britain, British
Bro(s)—Brother(s)
byrol—byrology
bull—Bulletin
bur—bureau
bus—business
BWI—British West Indies

c—children
Calif—California
Can—Canada, Canadian
cand—candidate
Capt—Captain
cardiol-cardiology
cardiovasc—cardiovascular
cartog—cartographic, cartographical, cartography
cartogr—cartographer
Cath—Catholic
CEngr—Corp of Engineers
cent—central
Cent Am—Central American
cert—certificate(s), certification, certified
chap—chapter
chem—chemical(s), chemistry
chemother—chemotherapy
chg—change
chmn—chairman
citricult—citriculture
class—classical
climat—climatological, climatology
clin(s)—clinic(s), clinical
cmndg—commanding
Co—County
co—Companies, Company
co-auth—coauthor
co-dir—co-director
co-ed—co-editor
co-educ—coeducation, coeducational
col(s)—college(s), collegiate, colonel
collab—collaboration, collaborative
collabr—collaborator
Colo—Colorado
com—commerce, commercial
Comdr—Commander

commun—communicable, communication(s)
comn(s)—commission(s), commissioned
comndg—commanding
comnr—commissioner
comp—comparitive
compos—composition
comput—computation, computer(s), computing
comt(s)—committee(s)
conchol—conchology
conf—conference
cong—congress, congressional
Conn—Connecticut
conserv—conservation, conservatory
consol—consolidated, consolidation
const—constitution, constitutional
construct—construction, constructive
consult(s)—consult, consultant(s), consultantship(s), consultation, consulting
contemp—contemporary
contrib—contribute, contributing, contribution(s)
contribr—contributor
conv—convention
coop—cooperating, cooperation, cooperative
coord—coordinate(d), coordinating, coordination
coordr—coordinator
corp—corporate, corporation(s)
corresp—correspondence, correspondent, corresponding
coun—council, counsel, counseling
counr—councilor, counselor
criminol—criminological, criminology
cryog—cryogenic(s)
crystallog—crystallographic, crystallographical, crystallography
crystallogr—crystallographer
Ct—Court
Ctr—Center
cult—cultural, culture
cur—curator
curric—curriculum
cybernet—cybernetic(s)
cytol—cytological, cytology
Czech—Czechoslovakia

DC—District of Columbia
Dec—December
Del—Delaware
deleg—delegate, delegation
delinq—delinquency, delinquent
dem—democrat(s), democratic
demog—demographic, demography
demogr—demographer
demonstr—demontrator
dendrol—dendrologic, dendrological, dendrology
dent—dental, dentistry
dep—deputy
dept—department
dermat—dermatologic, dermatological, dermatology
develop—developed, developing, development, developmental
diag—diagnosis, diagnostic
dialectol—dialectological, dialectology
dict—dictionaries, dictionary
Dig—Digest

dipl—diploma, diplomate
dir(s)—director(s), directories, directory
dis—disease(s), disorders
Diss Abst—Dissertation Abstracts
dist—district
distrib—distributed, distribution, distributive
distribr—distributor(s)
div—division, divisional, divorced
DNA—deoxyribonucleic acid
doc—document(s), documentary, documentation
Dom—Dominion
Dr—Drive
E—east
ecol—ecological, ecology
econ(s)—economic(s), economical, economy
economet—econometric(s)
ECT—electroconvulsive or electroshock therapy
ed—edition(s), editor(s), editorial
ed bd—editorial board
educ—education, educational
educr—educator(s)
EEG—electroencephalogram, electroencephalographic, electroencephalography
Egyptol—Egyptology
EKG—electrocardiogram
elec—elecvtric, electrical, electricity
electrochem-electrochemical, electrochemistry
electroph—electrophysical, electrophysics
elem—elementary
embryol—embryologic, embryological, embryology
emer—emeriti, emeritus
employ—employment
encour—encouragement
encycl—encyclopedia
endocrinol—endocrinologic, endocrinology
eng—engineering
Eng—England, English
engr(s)—engineer(s)
enol—enology
Ens—Ensign
entom—entomological, entomology
environ-environment(s), environmental
enzym—enzymology
epidemiol—epideiologic, epidemiological, epidemiology
equip—equipment
ERDA—Energy Research & Development Administration
ESEA—Elementary & Secondary Education Act
espec—especially
estab—established, establishment(s)
ethnog—ethnographic, ethnographical, ethnography
ethnogr—ethnographer
ethnol—ethnologic, ethnological, ethnology
Europ—European
eval—evaluation
Evangel—evangelical
eve—evening
exam—examination(s), examining
examr—examiner
except—exceptional
exec(s)—executive(s)

exeg—exegeses, exegesis, exegetic, exegetical
exhib(s)—exhibition(s), exhibit(s)
exp—experiment, experimental
exped(s)—expedition(s)
explor—exploration(s), exploratory
expos—exposition
exten—extension

fac—faculty
facil—facilities, facility
Feb—February
fed—federal
fedn—federation
fel(s)—fellow(s), fellowship(s)
fermentol—fermentology
fertil—fertility, fertilization
Fla—Florida
floricult—floricultural, floriculture
found—foundation
FPO—Fleet Post Office
Fr—French
Ft—Fort

Ga—Georgia
gastroenterol—gastroenterological, gastroenterology
gen—general
geneal—genealogical, genealogy
geod—geodesy, geodetic
geog—geographic, geographical, geography
geogr—geographer
geol—geologic, geological, geology
geom—geometric, geometrical, geometry
geomorphol—geomorphologic, geomorphology
geophys—geophysical, geophysics
Ger—German, Germanic, Germany
geriat—geriatric
geront—gerontological, gerontology
GES—Gesellschaft
glaciol—glaciology
gov—governing, governor(s)
govt—government, governmental
grad—graduate(d)
Gt Brit—Great Britain
guid—guidance
gym—gymnasium
gynec—gynecologic, gynecological, gynecology

handbk(s)—handbook(s)
helminth—helminthology
hemat—hematologic, hematological, hematology
herpet—herpetologic, herpetological, herpetology
HEW—Department of Health, Education & Welfare
Hisp—Hispanic, Hispania
hist—historic, historical, history
histol—histological, histology
HM—Her Majesty
hochsch—hochschule
homeop—homeopathic, homeopathy
hon(s)—honor(s), honorable, honorary
hort—horticultural, horticulture
hosp(s)—hospital(s), hospitalization
hq—headquarters

ABBREVIATIONS

HumRRO—Human Resources Research Office
husb—husbandry
Hwy—Highway
hydraul—hydraulic(s)
hydrodyn—hydrodynamic(s)
hydrol—hydrologic, hydrological, hydrologics
hyg—hygiene, hygienic(s)
hypn—hypnosis

ichthyol—ichthyological, ichthyology
Ill—Illinois
illum—illuminating, illumination
illus—illustrate, illustrated, illustration
illusr—illustrator
immunol—immunologic, immunological, immunology
Imp—Imperial
improv—improvement
Inc—Incorporated
in-chg—in charge
incl—include(s), including
Ind—Indiana
indust(s)—industrial, industries, industry
Inf—infantry
info—information
inorg—inorganic
ins—insurance
inst(s)—institute(s), institution(s)
instnl—institutional(ized)
instr(s)—instruct, instruction, instructor(s)
instrnl—instructional
int—international
intel—intellligence
introd—introduction
invert—invertebrate
invest(s)—investigation(s)
investr—investigator
irrig—irrigation
Ital—Italian

J—Journal
Jan—January
Jct—Junction
jour—journal, journalism
jr—junior
jurisp—jurisprudence
juv—juvenile

Kans—Kansas
Ky—Kentucky

La—Louisiana
lab(s)—laboratories, laboratory
lang—language(s)
laryngol—larygological, laryngology
lect—lecture(s)
lectr—lecturer(s)
legis—legislation, legislative, legislature
lett—letter(s)
lib—liberal
libr—libraries, library
librn—librarian
lic—license(d)
limnol—limnological, limnology
ling—linguistic(s), linguistical
lit—literary, literature
lithol—lithologic, lithological, lithology

Lt—Lieutenant
Ltd—Limited

m—married
mach—machine(s), machinery
mag—magazine(s)
maj—major
malacol—malacology
mammal—mammalogy
Man—Manitoba
Mar—March
Mariol—Mariology
Mass—Massachusetts
mat—material(s)
mat med—materia medica
math—mathematic(s), mathematical
Md—Maryland
mech—mechanic(s), mechanical
med—medical, medicinal, medicine
Mediter—Mediterranean
Mem—Memorial
mem—member(s), membership(s)
ment—mental(ly)
metab—metabolic, metabolism
metall—metallurgic, metallurgical, metallurgy
metallog—metallographic, metallography
metallogr—metallographer
metaphys—metaphysical, metaphysics
meteorol—meteorological, meteorology
metrol—metrological, metrology
metrop—metropolitan
Mex—Mexican, Mexico
mfg—manufacturing
mfr—manufacturer
mgr—manager
mgt—management
Mich—Michigan
microbiol—microbiological, microbiology
micros—microscopic, microscopical, microscopy
mid—middle
mil—military
mineral—mineralogical, mineralogy
Minn—Minnesota
Miss—Mississippi
mkt—market, marketing
Mo—Missouri
mod—modern
monogr—monograph
Mont—Montana
morphol—morphological, morphology
Mt—Mount
mult—multiple
munic—municipal, municipalities
mus—museum(s)
musicol—musicological, musicology
mycol—mycologic, mycology

N—north
NASA—National Aeronautics & Space Administration
nat—national, naturalized
NATO—North Atlantic Treaty Organization
navig—navigation(al)
NB—New Brunswick
NC—North Carolina
NDak—North Dakota
NDEA—National Defense Education Act
Nebr—Nebraska

nematol—nematological, nematology
nerv—nervous
Neth—Netherlands
neurol—neurological, neurology
neuropath—neuropathological, neuropathology
neuropsychiat—neuropsychiatric, neuropsychiatry
neurosurg—neurosurgical, neurosurgery
Nev—Nevada
New Eng—New England
New York—New York City
Nfld—Newfoundland
NH—New Hampshire
NIH—National Institute of Health
NIMH—National Institute of Mental Health
NJ—New Jersey
NMex—New Mexico
No—Number
nonres—nonresident
norm—normal
Norweg—Norwegian
Nov—November
NS—Nova Scotia
NSF—National Science Foundation
NSW—New South Wales
numis—numismatic(s)
nutrit—nutrition, nutritional
NY—New York State
NZ—New Zealand

observ—observatories, observatory
obstet—obstetric(s), obstetrical
occas—occasional(ly)
occup—occupation, occupational
oceanog—oceanographic, oceanographical, oceanography
oceanogr—oceanographer
Oct—October
odontol—odontology
OEEC—Organization for European Economic Cooperation
off—office, official
Okla—Oklahoma
olericult—olericulture
oncol—oncologic, oncology
Ont—Ontario
oper(s)—operation(s), operational, operative
ophthal—ophthalmologic, ophthalmological, ophthalmology
optom—optometric, optometrical, optometry
ord—ordnance
Ore—Oregon
org—organic
orgn—organization(s), organizational
orient—oriental
ornith—ornithological, ornithology
orthod—orthodontia, orthodontic(s)
orthop—orthopedic(s)
osteop—osteopathic, osteopathy
otol—otological, otology
otolaryngol—otolaryngological, otolaryngology
otorhinol—otorhinologic, otorhinology

Pa—Pennsylvania
Pac—Pacific
paleobot—paleobotanical, paleontology
paleont—paleontology

Pan-Am—Pan-American
parisitol—parasitology
partic—participant, participating
path—pathologic, pathological, pathology
pedag—pedagogic(s), pedagogical, pedagogy
pediat—pediatric(s)
PEI—Prince Edward Islands
penol—penological, penology
periodont—periodontal, periodontic(s)
petrog—petrographic, petrographical, petrography
petrogr—petrographer
petrol—petroleum, petrologic, petrological, petrology
pharm—pharmacy
pharmaceut—pharmaceutic(s), pharmaceutical(s)
pharmacog—pharmacognosy
pharamacol—pharmacologic, pharmacological, pharmacology
phenomenol—phenomenologic(al), phenomenology
philol—philological, philology
philos—philosophic, philosophical, philosophy
photog—photographic, photography
photogeog—photogeographic, photogeography
photogr—photographer(s)
photogram—photogrammetric, photogrammetry
photom—photometric, photometrical, photometry
phycol—phycology
phys—physical
physiog—physiographic, physiographical, physiography
physiol—physiological, phsysiology
Pkwy—Parkway
Pl—Place
polit—political, politics
polytech—polytechnic(s)
pomol—pomological, pomology
pontif—pontifical
pop—population
Port—Portugal, Portuguese
Pos:—Position
postgrad—postgraduate
PQ—Province of Quebec
PR—Puerto Rico
pract—practice
practr—practitioner
prehist—prehistoric, prehistory
prep—preparation, preparative, preparatory
pres—president
Presby—Presbyterian
preserv—preservation
prev—prevention, preventive
prin—principal
prob(s)—problem(s)
proc—proceedings
proctol—proctologic, proctological, proctology
prod—product(s), production, productive
prof—professional, professor, professorial
Prof Exp—Professional Experience
prog(s)—program(s), programmed, programming
proj—project(s), projection(al), projective

prom—promotion
protozool—protozoology
Prov—Province, Provincial
psychiat—psychiatric, psychiatry
psychoanal—psychoanalysis, psychoanalytic, psychoanalytical
psychol—psychological, psychology
psychomet—psychometric(s)
psychopath—psychopathologic, psycho pathology
psychophys—psychophysical, psychophysics
psychophysiol—psychophysiological, psychophysiology
psychosom—psychosomtic(s)
psychother—psychoterapeutic(s), psycho therapy
Pt—Point
pub—public
publ—publication(s), publish(ed), publisher, publishing
pvt—private

Qm—Quartermaster
Qm Gen—Quartermaster General
qual—qualitative, quality
quant—quantitative
quart—quarterly
Que—Quebec

radiol—radiological, radiology
RAF—Royal Air Force
RAFVR—Royal Air Force Volunteer Reserve
RAMC—Royal Army Medical Corps
RAMCR—Royal Army Medical Corps Reserve
RAOC—Royal Army Ornance Corps
RASC—Royal Army Service Corps
RASCR—Royal Army Service Corps Reserve
RCAF—Royal Canadian Air Force
RCAFR—Royal Canadian Air Force Reserve
RCAFVR—Royal Canadian Air Force Volunteer Reserve
RCAMC—Royal Canadian Army Medical Corps
RCAMCR—Royal Canadian Army Medical Corps Reserve
RCASC—Royal Canadian Army Service Corps
RCASCR—Royal Canadian Army Service Corps Reserve
RCEME—Royal Canadian Electrical & Mechanical Engineers
RCN—Royal Canadian Navy
RCNR—Royal Canadian Naval Reserve
RCNVR—Royal Canadian Naval Volunteer Reserve
Rd—Road
RD—Rural Delivery
rec—record(s), recording
redevelop—redevelopment
ref—reference(s)
refrig—refrigeration
regist—register(ed), registration
registr—registrar
regt—regiment(al)
rehab—rehabilitation
rel(s)—relation(s), relative
relig—religion, religious
REME—Royal Electrical & Mechanical

Engineers
rep—represent, representative
Repub—Republic
req—requirements
res—research, reserve
rev—review, revised, revision
RFD—Rural Free Delivery
rhet-rhetoric, rhetorical
RI—Rhode Island
Rm—Room
RM—Royal Marines
RN—Royal Navy
RNA—ribonucleic acid
RNR—Royal Naval Reserve
RNVR—Royal Naval Volunteer Reserve
roentgenol—roentgenologic, roentgenologi cal, roentgenology
RR—Railroad, Rural Route
Rte—Route
Russ—Russian
rwy—railway

S—south
SAfrica—South Africa
SAm—South America, South American
sanit—sanitary, sanitation
Sask—Saskatchewan
SC—South Carolina
Scand—Scandinavia(n)
sch(s)—school(s)
scholar—scholarship
sci—science(s), scientific
SDak—South Dakota
SEATO—Southeast Asia Treaty Organization
sec—secondary
sect—section
secy—secretary
seismog—seismograph, seismographic, seismography
seismogr—seismographer
seismol—seismological, seismology
sem—seminar, seminary
Sen—Senator, Senatorial
Sept—September
ser—serial, series
serol—serologic, serological, serology
serv—service(s), serving
silvicult—silvicultural, silviculture
soc(s)—societies, society
soc sci—social science
sociol—sociologic, sociological, sociology
Span—Spanish
spec—special
specif—specification(s)
spectrog—spectrograph, spectrographic, spectrography
spectrogr—spectrographer
spectrophotom—spectrophotometer, spectrophotometric, spectrophotometry
spectros—spectroscopic, spectroscopy
speleol—speleological, speleology
Sq—Square
sr—senior
St—Saint, Street(s)
sta(s)—station(s)
stand—standard(s), standardization
statist—statistical, statistics
Ste—Sainte
steril—sterility

ABBREVIATIONS

stomatol—stomatology
stratig—stratigraphic, stratigraphy
stratigr—stratigrapher
struct—structural, structure(s)
stud—student(ship)
subcomt—subcommittee
subj—subject
subsid—subsidiary
substa—substation
super—superior
suppl—supplement(s), supplemental, supplementary
supt—superintendent
supv—supervising, supervision
supvr—supervisor
supvry—supervisory
surg—surgery, surgical
surv—survey, surveying
survr—surveyor
Swed—Swedish
Switz—Switzerland
symp—symposia, symposium(s)
syphil—syphilology
syst(s)—system(s), systematic(s), systematical

taxon—taxonomic, taxonomy
tech—technical, technique(s)
technol—technologic(al), technology
tel—telegraph(y), telephone
temp—temporary
Tenn—Tennessee
Terr—Terrace
Tex—Texas
textbk(s)—textbook(s)
text ed—text edition
theol—theological, theology
theoret—theoretic(al)
ther—therapy
therapeut—therapeutic(s)
thermodyn—thermodynamic(s)
topog—topographic, topographical, topography
topogr—topographer
toxicol—toxicologic, toxicological,

toxicology
trans—transactions
transl—translated, translation(s)
translr—translator(s)
transp—transport, transportation
treas—treasurer, treasury
treat—treatment
trop—tropical
tuberc—tuberculosis
TV—television
Twp—Township

UAR—United Arab Republic
UK—United Kingdom
UN—United Nations
undergrad—undergraduate
unemploy—unemployment
UNESCO—United Nations Educational Scientific & Cultural Organization
UNICEF—United Nations International Childrens Fund
univ(s)—universities, university
UNRRA—United Nations Relief & Rehabilitation Administration
UNRWA—United Nations Relief & Works Agency
urol—urologic, urological, urology
US—United States
USAAF—US Army Air Force
USAAFR—US Army Air Force Reserve
USAF—US Air Force
USAFR—US Air Force Reserve
USAID—US Agency for International Development
USAR—US Army Reserve
USCG—US Coast Guard
USCGR—US Coast Guard Reserve
USDA—US Department of Agriculture
USMC—US Marine Corps
USMCR—US Marine Corps Reserve
USN—US Navy
USNAF—US Naval Air Force
USNAFR—US Naval Air Force Reserve
USNR—US Naval Reserve

USPHS—US Public Health Service
USPHSR—US Public Health Service Reserve
USSR—Union of Soviet Socialist Republics

Va—Virginia
var—various
veg—vegetable(s), vegetation
vent—ventilating, ventilation
vert—vertebrate
Vet—Veteran(s)
vet—veterinarian, veterinary
VI—Virgin Islands
vinicult—viniculture
virol—virological, virology
vis—visiting
voc—vocational
vocab—vocabulary
vol(s)—voluntary, volunteer(s), volume(s)
vpres—vice president
vs—versus
Vt—Vermont

W—west
Wash—Washington
WHO—World Health Organization
WI—West Indies
wid—widow, widowed, widower
Wis—Wisconsin
WVa—West Virginia
Wyo—Wyoming

Yearbk(s)—Yearbook(s)
YMCA—Young Men's Christian Association
YMHA—Young Men's Hebrew Association
Yr(s)—Year(s)
YT—Yukon Territory
YWCA—Young Women's Christian Association
YWHA—Young Women's Hebrew Association

zool—zoological, zoology

American Men & Women of Science

J

JA, WILLIAM YIN, b Mar 5, 36; US citizen. ANALYTICAL CHEMISTRY. *Educ:* Univ Calif, Berkeley, BS, 60. *Prof Exp:* Chemist, Qual Assurance Tech Agency, US Army Chem Corps, 60-62; anal chemist, Hyman (Julius) Labs, Inc, 62-64; anal chemist, Philadelphia Quartz Co, Calif, 64-66; RES ASSOC, RICHMOND RES CTR, STAUFFER CHEM CO, 66- *Mem:* Am Chem Soc; Am Soc Testing & Mat; Assoc Official Anal Chem. *Res:* Analytical methods development; trace analysis; separations and purification techniques, especially preparation of high-purity pesticide standards and metabolites by large-scale, high-speed column chromatography. *Mailing Add:* 145 Windward Ct ValleJo CA 94591

JAANUS, SIRET DESIREE, b Tallinn, Estonia; US citizen; m 73. PHARMACOLOGY, OCULAR PHARMACOLOGY. *Educ:* City Col New York, BS, 60; Hunter Col, MA, 66; State Univ NY Downstate Med Ctr, PhD(pharmacol), 70. *Prof Exp:* Res asst pharmacol, Albert Einstein Col Med, 60-64; res asst, State Univ Downstate Med Ctr, 64-66, NIH fel pharmacol, 66-70, path, 70-71; asst prof basic sci, State Univ NY Col Optom, 71-72, chmn dept, 72-73; chairperson, dept basic & visual sci, 78-81, PROF BASIC SCI, SOUTHERN CALIF COL OPTOM, 73- *Concurrent Pos:* Consult; assoc ed, Clin Ocular Pharmacol, 84, 89. *Honors & Awards:* Paul Yarwood Award, 79. *Mem:* Am Optom Asn; Am Soc Pharmacol & Exp Therapeut; Am Acad Optom. *Res:* Autonomic and endocrine pharmacology. *Mailing Add:* Dept Basic & Visual Sci Southern Calif Col Optom 2575 Yorba Linda Blvd Fullerton CA 92631

JABALPURWALA, KAIZER E, b Surat, India, May 4, 32; m 60; c 2. INORGANIC CHEMISTRY, PHYSICAL CHEMISTRY. *Educ:* Univ Bombay, BSc, 54, MSc, 56, PhD(coord chem), 60. *Prof Exp:* Res asst phys chem, Inst Sci, Univ Bombay, 58-61; res assoc inorg chem, Boston Univ, 61-64; chief chemist, Zinc Oxide Co Can Ltd, Hudson Bay Mining & Smelting Co, Montreal, 64-68, tech dir, 68-74; tech mgr, Zochem Ltd, 74; partner, G H Chem Ltd, 74-83; TECH DIR, JABALPUR INDUST, INC, 83- *Honors & Awards:* Gold Medalist, Univ Bombay, 61. *Mem:* AAAS; Am Chem Soc; Brit Chem Soc; Tech Asn Pulp & Paper Indust; Chem Inst Can; NY Acad Sci; Am Soc Testing & Mat. *Res:* Light scattering by colloid systems; solution stabilities of complex ions; electrophotography related to photoconductivity of zinc oxide; technology of zinc oxide. *Mailing Add:* 350 L'Esperance St Ste Lambert PQ J4P 1Y5 Can

JABARIN, SALEH ABD EL KARIM, b Haifa, Israel, Feb 7, 39; m 69; c 1. POLYMER CHEMISTRY. *Educ:* Dartmouth Col, BA, 66; Polytech Inst Brooklyn, MS, 68; Univ Mass, PhD(polymer sci & eng), 71. *Prof Exp:* SR SCIENTIST, OWENS-ILLINOIS TECH CTR, 71- *Mem:* Am Chem Soc; Soc Plastics Engrs. *Res:* Studies of thermal, mechanical and optical properties of polymers and polymer crystallization using light scattering, x-ray diffraction, infra-red dichroism and birefringence; molecular orientation and solution characterization. *Mailing Add:* Owens-Illinois One Seagate Toledo OH 43666

JABBAR, GINA MARIE, b Chicago, Ill. NEUROSCIENCES. *Educ:* Univ Bridgeport, Conn, BA, 80, MS, 82. *Prof Exp:* Lab scientist, Structure Probe Inc, 82-86; sr scientist, 86-87; RES COORDR, MOUNT SINAI MED CTR, NEUROPATH DIV, 87- *Res:* Electron optic and light microscopy methods and techniques. *Mailing Add:* 27 Koster Blvd Apt 3B Edison NJ 08837

JABBOUR, J T, b Tiptonville, Tenn, Aug 5, 27; c 5. PEDIATRIC NEUROLOGY. *Educ:* Univ Tenn, Martin, BS, 48; Univ Tenn, Memphis, MD, 51; Am Bd Pediat, dipl, 60. *Prof Exp:* Rotating intern, Baylor Univ Hosp, Dallas, 52-53; gen pract, Tenn, 53-56; resident, Col Med, Univ Tenn, Memphis, 56-58, res assoc pediat, 57; co-dir pediat neurol & seizure clin, Sch Med, Univ Okla, 59-61, asst prof pediat & neurol & assoc dir clin study ctr birth defects, 61-65; asst prof, 65-67, clin prof pediat & neurol & head child div child neurol, Univ Tenn, Memphis, 67-85; CONSULT, 85- *Concurrent Pos:* Fel neurol, Univ Minn, Minneapolis, 58-61; consult, Oklahoma City Speech & Hearing Ctr, 61-65; consult, Oklahoma City Children's Ctr, 61-65; chief pediat neurol, Child Develop Ctr, Med Units, Univ Tenn, Memphis, 65-68; consult, Ment Retardation Br, Bur Chronic Dis, USPHS, 66-69. *Mem:* AMA; Am Acad Pediat; Am Acad Neurol; Am Acad Cerebral Palsy; Child Neurol Soc. *Res:* Subacute sclerosing panencephalitis; behavioral neurology. *Mailing Add:* 848 Adams Ave Memphis TN 38103

JABBOUR, KAHTAN NICOLAS, b Safita, Syria, Aug 26, 34. ENGINEERING. *Educ:* Damascus Univ, cert math, 53 & 54; Sch Advan Eng, Beirut, BS, 57; Purdue Univ, MS, 60, PhD(struct), 62. *Prof Exp:* Field engr, Arabian Am Oil Co, Saudi Arabia, 57-58; design engr, El-Ghab Proj, Syria, 58-59; asst prof eng, Kans State Univ, 62-63; assoc prof eng sci, Tenn Technol Univ, 63-67; staff engr, Fairchild Hiller Corp, 67-69; aerospace engr, Goddard Space Flight Ctr, NASA, 69-73; sr mech engr, 73-80, SR PROJ MGR, US NUCLEAR REGULATORY COMN, 80- *Concurrent Pos:* NSF res grant, 65-66. *Mem:* Am Soc Eng Educ. *Res:* Structural mechanics and engineering; perforated plates. *Mailing Add:* 109 Lucas Lane Bethesda MD 20814

JABBUR, RAMZI JIBRAIL, b Beirut, Lebanon, Mar 9, 37; US citizen; div; c 1. INDUSTRIAL CHEMISTRY, ECONOMETRICS. *Educ:* Am Univ Beirut, BS, 58; Stanford Univ, MS, 60, PhD(high energy physics), 63. *Prof Exp:* Rockefeller Found fel, 57-58, Stanford Univ fel, 58-60; res assoc physics, Columbia Univ, 63-65; res physicist, Argonne Nat Lab, 65-67; asst prof physics, Univ Grad Prog, City Univ NY, 67-71; assoc dir, Mgt Sci Dept, BBDO Inc, 71-74; mgr corp strategic planning, W R Grace & Co, 74-78; MGT CONSULT, 78- *Concurrent Pos:* Dir, UBAF Arab Am Bank, 75-78. *Mem:* Am Phys Soc; NY Acad Sci; Inst Mgt Sci. *Res:* Management of major petrochemical projects; operations research and applications to managerial decisions; econometric statistical and dynamic models; strategic corp planning; investments, acquisitions and divestments. *Mailing Add:* 1380 Riverside Dr Apt 5H New York NY 10033

JABINE, THOMAS BOYD, b Brooklyn, NY, Jan 26, 25; m 50; c 4. APPLIED STATISTICS. *Educ:* Mass Inst Technol, BS & MS, 49. *Prof Exp:* Var pos, US Census Bur, 49-68, chief statist res div, 69-73; chief math statistician, Social Security Admin, 73-79; statist policy expert, Energy Info Admin, US Dept Energy, 79-80; STATIST CONSULT, 80- *Concurrent Pos:* Constituent mem, Inter-Am Statist Inst. *Mem:* Fel Am Statist Asn; Int Statist Inst. *Res:* Survey methodology; sampling; quality control. *Mailing Add:* 3231 Worthington St NW Washington DC 20015-2362

JABLON, SEYMOUR, b New York, NY, June 2, 18; m 41; c 2. BIOSTATISTICS, EPIDEMIOLOGY. *Educ:* City Col New York, BS, 39; Columbia Univ, MA, 40. *Prof Exp:* Prof assoc, Med Follow-up Agency, Nat Res Coun, 48-60; chief dept statist, Atomic Bomb Casualty Comn, 60-63; assoc dir, Med Follow-up Agency, Nat Res Coun, 63-68; chief dept statist, Atomic Bomb Casualty Comn, 68-71; assoc dir, 71-75, dir med follow-up agency, Nat Res Coun, 75-87; EXPERT, RADIATION EPIDEMIOL BR, NAT CANCER INST, 87- *Concurrent Pos:* Mem, Nat Comn Radiation Protection, 74-87. *Honors & Awards:* Order of Sacred Treasure, Govt of Japan, 87. *Mem:* Fel Am Statist Asn; Am Epidemiol Soc; Radiation Res Soc; Biomet Soc; Health Physics Soc; fel AAAS. *Res:* Late effects of radiation; epidemiology of cancer. *Mailing Add:* 6813 Persimmon Tree Rd Bethesda MD 20817

JABLONER, HAROLD, b New York, NY, Oct 25, 37; m 61; c 3. POLYMER CHEMISTRY, ORGANIC CHEMISTRY. *Educ:* City Col New York, BS, 57; Polytech Inst Brooklyn, PhD(chem), 63. *Prof Exp:* Res chemist, Hercules Powder Co, 63-68; sr res chemist, 68-71, res scientist, 71-78, RES ASSOC, HERCULES INC, 78- *Concurrent Pos:* Adj prof, Drexel Univ, 78- *Mem:* Am Chem Soc. *Res:* Solution properties and synthesis of macromolecules; hydrocarbon oxidation kinetics and mechanisms; thermally stable polymers; food; taste perception; polymer taste perception; solution properties of polymers; paper and synthetic pulp. *Mailing Add:* 527 Cherry St New Castle DE 19720

JABLONSKI, DANIEL GARY, b Washington, DC, Nov 15, 54; m 82; c 1. SUPERCONDUCTING ELECTRONICS. *Educ:* Mass Inst Technol, BS, 76, MS, 77; Cambridge Univ, PhD(physics), 82. *Prof Exp:* res physicist, Naval Surface Weapons Ctr, 81-86; RES STAFF MEM, SUPERCOMPUTING RES CTR, 86- *Concurrent Pos:* Adj prof, Capitol Inst Technol, 85- *Mem:* Am

Phys Soc; Inst Elec & Electronics Engrs. *Res:* Microwave properties of superconducting devices; properties of materials at millimeter wavelengths; chaotic behavior in nonlinear systems; supercomputing; theory of computation; computer science. *Mailing Add:* 12220 Somersworth Dr Silver Spring MD 20902

JABLONSKI, DAVID, b New York, NY, June 23, 53. EVOLUTIONARY BIOLOGY, PALEONTOLOGY. *Educ:* Columbia Univ, BA, 74; Yale Univ, MS, 76, PhD(geol), 79. *Prof Exp:* Asst res geol, Univ Calif, Santa Barbara, 79-80, Miller res fel paleobiol, dept paleont, Berkeley, 80-82; asst prof evolutionary biol, dept ecol & evolutionary biol, Univ Ariz, 82-85; ASSOC PROF PALEBIOL, DEPT GEOPHYS SCI, UNIV CHICAGO, 85- *Concurrent Pos:* Assoc ed, Paleobiol, 83-85 & 86-88, Evolution, 84-86. *Honors & Awards:* Paleont Soc Schuchert Award, 88. *Mem:* Paleont Soc; Soc Study Evolution; Int Paleont Union; Soc Syst Zool; Soc Econ Paleontologists & Mineralogists. *Res:* Evolutionary patterns and processes above the species level, in living and fossil organisms; marine inverts, particularly mollusks. *Mailing Add:* Dept Geophys Sci Univ Chicago Chicago IL 60637

JABLONSKI, FELIX JOSEPH, b Chicago, Ill, Jan 12, 25; m 56; c 3. MATHEMATICS, PHYSICS. *Educ:* Tex A&I Univ, BA, 55; US Navy Postgrad Sch, MS, 61. *Prof Exp:* Head, Naval Sci Opers Res Dept, US Naval Acad, 63-66; head opers res, LTV Electrosystems, 66-70; ASSOC DIV DIR, OPERS RES INC, 72- *Mem:* Sigma Xi. *Res:* Operations research and naval logistics. *Mailing Add:* 4113 Montpelier Rd Rockville MD 20853

JABLONSKI, FRANK EDWARD, b Brooklyn, NY, Feb 3, 15; m 49; c 3. ELECTROOPTICS. *Educ:* Fordham Univ, BS, 36; NY Univ, MS, 40; Harvard Univ, MS, 46. *Prof Exp:* Physicist bur ord, US Dept Navy, 40-43, active duty, US Navy, 43-46 physicist influence devices, US Naval Ord Lab, 46-57, ord engr, Spec Projs Off, 57-58; physicist, Nat Security Agency, 58-59; physicist, Goddard Space Flight Ctr & NASA Hqs, 59-61, physicist off long range plans & progs, 61-63; tech adv, Chief Naval Opers, 63-73, consult physicist, 73-87; RETIRED. *Concurrent Pos:* Mem, Polaris Re-entry Body Coord Comt, 57-58; res adv comt control, guid & navig, NASA, 59-60. *Mem:* Fel AAAS; Am Phys Soc; Wilderness Soc. *Res:* Degaussing of ships; proximity exploders and fuzes; warheads for missiles; communications; space electronics; sonar; infrared; lasers. *Mailing Add:* 9916 Julliard Dr Bethesda MD 20817

JABLONSKI, WERNER LOUIS, b Frankfurt, Ger, May 6, 24; m 54; c 5. ORGANIC CHEMISTRY. *Educ:* Univ Toronto, BA & MA, 49; McGill Univ, PhD(org chem), 53. *Prof Exp:* Chemist, Can Indust Ltd, 49-50; chemist, Dow Chem Co, 50-69 & US Plywood Champion Papers, Inc, 69-71; chemist, Foster Grant, Inc, 71-85; RETIRED. *Mem:* Soc Plastics Engrs; Am Chem Soc; Sigma Xi. *Res:* Polymers. *Mailing Add:* 404 Woodberry Dr Chesapeake VA 23320-5742

JACCARINO, VINCENT, b Brooklyn, NY, May 12, 24; m 65; c 2. SOLID STATE PHYSICS. *Educ:* Brooklyn Col, BS, 48; Mass Inst Technol, PhD(physics), 52. *Prof Exp:* Res assoc, Mass Inst Technol, 52-54; mem tech staff, Bell Tel Labs, 54-63, head solid state phys res dept, 63-66; chmn dept, 69-72, PROF PHYSICS, UNIV CALIF, SANTA BARBARA, 66-, DIR, QUANTUM INST, 85- *Concurrent Pos:* Guggenheim Found fel, 73-74, Lady Davis fel, 79, Yamada Found fel, Japan, 80; chmn, Int Conf Magnetism, 85-; USA chmn, magnetism sect, Int Union Pure & Appl Physics, 91. *Mem:* Fel Am Phys Soc. *Res:* Magnetic resonance in solids; magnetism; superionic conductors; critical phenomena. *Mailing Add:* Dept Physics Univ Calif Santa Barbara CA 93106

JACCHIA, LUIGI GIUSEPPE, b Trieste, Italy, June 4, 10; nat US. ASTRONOMY. *Educ:* Univ Bologna, PhD(physics), 32. *Prof Exp:* From asst to instr astron, Univ Bologna, 32-38; RES ASSOC, COL OBSERV, HARVARD UNIV, 39-; PHYSICIST, SMITHSONIAN ASTROPHYS OBSERV, 56- *Concurrent Pos:* Res assoc, Mass Inst Technol, 49-53. *Mem:* Am Astron Soc; Int Astron Union; Int Union Geod & Geophys. *Res:* Variable stars; meteors; upper atmosphere; artificial satellites. *Mailing Add:* Smithsonian Inst Astrophys Observ 60 Garden St Cambridge MA 02138

JACH, JOSEPH, b SAfrica, Dec 15, 29; m 64; c 2. PHYSICAL CHEMISTRY. *Educ:* Univ Cape Town, BSc, 50, MSc, 52; Oxford Univ, PhD, 55. *Prof Exp:* Lectr chem, Univ Cape Town, 53; res assoc, Brookhaven Nat Lab, 56-63; assoc prof eng, State Univ NY Stony Brook 63-91. *Concurrent Pos:* At Lawrance Radiation Lab, Calif, 69-70. *Res:* Solid state chemistry, particularly thermal decomposition of solids and Szillard-Challmers reactions and chemical reactivity at defect sites in solids. *Mailing Add:* 46 Bay Rd East Patchogue NY 11772

JACHE, ALBERT WILLIAM, b Manchester, NH, Nov 5, 24; m 48; c 4. INORGANIC CHEMISTRY. *Educ:* Univ NH, BS, 48, MS, 50; Univ Wash, PhD(chem), 52. *Prof Exp:* Sr chemist, Air Reduction Co, Inc, 52-53; res assoc physics, Duke Univ, 53-55; from asst prof to assoc prof chem, Agr & Mech Col Tex, 55-61; assoc res dir, Ozark Mahoning Co, 61-64; sr res assoc, Olin Mathieson Chem Corp, 64-67, sect mgr, 65-67; chmn dept chem, Marquette Univ, 67-72, prof, 67-90, dean, Grad Sch, 72-77, assoc acad vpres health sci, 74-77, assoc vpres acad affairs, 77-85, EMER PROF CHEM, MARQUETTE UNIV, 90- *Concurrent Pos:* Consult, Olin Corp, 67-75, Allied Chem Corp, 77-78; scientist in residence, Argonne Nat Lab, 85-86. *Honors & Awards:* Milwaukee Sect Award, Am Chem Soc. *Mem:* Fel AAAS; Am Chem Soc (chmn div Fluorine Chem, 82); NY Acad Sci; Sigma Xi; fel Am Inst Chem. *Res:* Fluorine chemistry; halogens; nonaqueous solvent systems; environmental problems. *Mailing Add:* 1618 Martha Washington Ave Milwaukee WI 53213

JACHENS, ROBERT C, b San Francisco, Calif, June 4, 39. MINING GEOPHYSICS. *Educ:* Santa Fe State Univ, BS, 62; Columbia Univ, MS, 68, PhD(geophys), 71. *Prof Exp:* Res assoc, Lamont Doherty, Geol Observ, Columbia Univ, 72-75; res geophysicist, 75-85, chief, 85-88, REG CRUSTEL STRUCT ANAL, CRUSTEL DYNAMICS SECT, US GEOL SURV OFF MINERAL RESOURCES & GEOPHYS BR, 88- *Mem:* Fel Geol Soc Am; AAAS; Am Geophys Union. *Res:* Solid earth geophysics & tectonics; mining geophysics. *Mailing Add:* US Geol Surv MS 989 345 Middlefield Rd Menlo Park CA 94025

JACHIMOWICZ, FELEK, b Poznan, Poland, July 2, 47; nat US; m 71; c 2. PHOTOCHEMISTRY, POLYMER CHEMISTRY. *Educ:* Univ Basel, Switz, dipl, 71, PhD(phys org chem), 75. *Prof Exp:* Fel phys org chem, Col Environ Sci & Forestry, State Univ NY, 78; SR RES CHEMIST, RES DIV, W R GRACE & CO, 78- *Mem:* Am Chem Soc. *Res:* Physical organic chemistry; synthetic chemistry; homogeneous catalysis; synthetic chemistry; radical ions chemistry; analytical chemistry; spectroscopy; polymer chemistry. *Mailing Add:* W R Grace & Co 7379 Rte 32 Columbia MD 21044

JACK, HULAN E, JR, b New York, NY, May 6, 35; m; c 4. PHYSICS. *Educ:* NY Univ, BS, 60, MS, 64, PhD(physics), 71. *Prof Exp:* Instr physics, NY Inst Technol, 61-66 & Eng Sch, Pratt Inst, 66-68; lectr, Wash Sq Col, NY Univ, 68-70; instr, Finch Col, 70-71; ASST PROF PHYSICS, KANS STATE UNIV, 71- *Mem:* AAAS; Am Asn Physics Teachers; Am Phys Soc; Asn Comput Mach. *Res:* Solid state and atomic physics. *Mailing Add:* 9832 57th Ave Apt 8D Flushing NY 11368

JACK, JOHN JAMES, b Trenton, NJ, Jan 11, 43; m 67; c 2. ANALYTICAL CHEMISTRY, PLASTICS MARKETING & PROCESSING. *Educ:* Princeton Univ, AB, 65; Mass Inst Technol, PhD(anal chem), 71. *Prof Exp:* RES CHEMIST & TECH MGT PLASTICS MKT, POLYMER PROD DEPT, E I DU PONT DE NEMOURS & CO, INC, 70- *Res:* Development of instrumental methods of analysis, especially spectroscopic, and application to industrial analytical problems; automation of laboratory testing and efficient use of newly developing mini- and micro-computers. *Mailing Add:* Polymer Prod Dept E I du Pont de Nemours & Co Inc 1007 Market St Wilmington DE 19898

JACK, ROBERT CECIL MILTON, b St Vincent, WI, Oct 10, 29; Brit citizen; m 59; c 2. BIOCHEMISTRY. *Educ:* McGill Univ, BSc, 56; Columbia Univ, PhD(plant biochem), 64. *Prof Exp:* Chemist, Cent Exp Sta, WI, 58-60; asst biochemist, Boyce Thompson Inst, 64-66, assoc biochemist, 67; assoc prof, 67-76, chmn, Dept Biol, 77-80, PROF BIOL, ARTS & SCI, ST JOHN'S UNIV, 76- *Mem:* Am Chem Soc; Am Soc Biochem Molecular Biol; Am Soc Microbiol; NY Acad Sci. *Res:* Lipid chemistry and metabolism; lipids; biological membranes; biomed applications of computers. *Mailing Add:* Dept Biol Sci St John's Univ Jamaica NY 11432

JACK, THOMAS RICHARD, b Toronto, Ont, Mar 4, 47; m; c 1. PETROLEUM MICROBIOLOGY, INORGANIC CHEMISTRY. *Educ:* Univ Toronto, BSc, 69, PhD(chem), 75. *Prof Exp:* Fel bioeng, Fac Eng Sci, Univ Western Ont, 75-76; vis asst prof chem, Scarborough Col, Univ Toronto, 76-77, asst prof, 77-79; petrol microbiologist, BC Res, 80-81; res scientist, 81-83, group leader, 83-87, MGR APPL SCI, NOVA HUSKY RES CORP, DIV NOVA CORP, 87- *Concurrent Pos:* Fel, Nat Res Coun Can, 75; indust assoc, Arctic Inst N Am, Univ Calgary, 81-83. *Mem:* Chem Inst Can; Am Soc Microbiol; Am Chem Soc. *Res:* Interface between inorganic chemistry and microbiology; biotechnology and inorganic chemistry in energy production. *Mailing Add:* Nova Husky Res Corp 2928 16th St NE Calgary AB T2E 7K7 Can

JACKANICZ, THEODORE MICHAEL, b Chicago, Ill, Oct 6, 38; div; c 1. REPRODUCTIVE ENDOCRINOLOGY. *Educ:* Northwestern Univ, BA, 59; Mich State Univ, PhD(biochem), 65. *Prof Exp:* Res fel endocrinol, Harvard Med Sch & Karolinska Inst, 65-68; proj specialist pop, Ford Found, 69-70; staff scientist 70-74, SCIENTIST REPRODUCTION, POP COUN, NY, 74- *Concurrent Pos:* NIH fel, Harvard Med Sch, 65-67 & Karolinska Inst, 67-68. *Mem:* Sigma Xi; AAAS; NY Acad Sci. *Res:* Reproductive endocrinology and contraceptive development. *Mailing Add:* Pop Coun Rockefeller Univ 66th St & York Ave New York NY 10021

JACKEL, LAWRENCE DAVID, b New York, NY, June 16, 48; m 69; c 2. EXPERIMENTAL SOLID STATE PHYSICS, MICROFABRICATION. *Educ:* Brandeis Univ, BA, 69; Cornell Univ, MA, 72, PhD(exp physics), 76. *Prof Exp:* Res asst, Sch Appl & Eng Physics, Cornell Univ, 71-75, res assoc, 75; mem tech staff exp solid state physics, 75-84, head, Device Struct Res Dept, 84-90, HEAD, ADAPTIVE SYSTS RES DEPT, AT&T BELL LABS, 90- *Honors & Awards:* Paul Rappaport Award, Inst Elec & Electronics Engrs Electron Device Soc, 85. *Mem:* Am Phys Soc; Inst Elec & Electronics Engrs. *Res:* Electronic neural networks. *Mailing Add:* 4D433 Bell Labs Holmdel NJ 07733

JACKEL, SIMON SAMUEL, b New York, NY, Nov 11, 17; m 54; c 2. FOOD SCIENCE, FOOD CHEMISTRY. *Educ:* City Col New York, BS, 38; Columbia Univ, AM, 47, PhD(biochem), 50. *Prof Exp:* Anal chemist, Plymouth Labs, 38-41; instr instrumentation, Air Corps Tech Training Sch, Ill, 41-43; instr chem & instrumentation, Army Air Forces Eng Officers Sch, Yale Univ, 43- 44; res chemist anal chem, Fleischmann Labs, 44-47; asst biochem, Columbia Univ, 47-50; head yeast dept, Fleischmann Labs, 50-54, head fermentation div, 54-59; vpres & dir res, Vico Prod Co, 59-61; dir res & develop lab, Qual Bakers Am Coop Inc, 61-76, vpres & dir res & develop, 76-84; CHMN BD, PLYMOUTH TECH SERV ASSOCS, 84- *Concurrent Pos:* Am Bakers Asn tech liaison comt, USDA, 70-86, chmn, 75-86; USPHS res grant, 47-50; pres, Plymouth Tech Serv, 66-; tech ed, Bakery Prod & Mkt Mag, 68-84; mem sci adv comt, Am Inst Baking, 70-; mem indust adv comt, NDak State Univ, 72-86; columnist, Cereal Foods World, 83- *Honors & Awards:* Charles N Frey Award, Am Asn Cereal Chemists, 81. *Mem:* AAAS; Am Chem Soc; Am Asn Cereal Chemists; Inst Food Technologists; fel Am Inst Chemists. *Res:* Fermentation; yeast metabolism; baking technology; nutrition; research management; new product development; regulatory affairs compliance; ingredient applications to food manufacturing; enzyme technology; foods. *Mailing Add:* 4523 Bardsdale Dr Palm Harbor FL 34685-2603

JACKELS, CHARLES FREDERICK, b St Paul, Minn, Nov 3, 46; m 70; c 1. QUANTUM CHEMISTRY. *Educ:* Univ Minn, Minneapolis, BChem, 68; Univ Wash, PhD(chem), 75. *Prof Exp:* Fel theoret chem, Battelle Mem Inst, 75-77; from asst prof to assoc prof, 77-90, PROF CHEM, WAKE FOREST UNIV, 90- *Mem:* Am Chem Soc. *Res:* Ab initio quantum chemical investigations of small molecules; potential energy surface calculations using self-consistent-field and configuration-interaction methods; applications to atmospheric chemistry. *Mailing Add:* Dept Chem Wake Forest Univ Winston-Salem NC 27109

JACKELS, SUSAN CAROL, b Wichita, Kans, July 12, 46; m 70. INORGANIC CHEMISTRY, BIO-INORGANIC CHEMISTRY. *Educ:* Carleton Col, BA, 68; Univ Wash, PhD(inorg chem), 73. *Prof Exp:* Res biochem, Univ Wash, 73-75 & inorg chem, Ohio State Univ, 75-77; asst prof, 77-83, ASSOC PROF CHEM, WAKE FOREST UNIV, 83- *Concurrent Pos:* Sabbatical leave, Mass Gen Hosp, Harvard Univ, 85-86. *Mem:* Am Chem Soc; Sigma Xi. *Res:* Design, synthesis and study of transition metal complexes relevant to biological systems; macrocyclic complexes; electrochemistry of transition metal complexes. *Mailing Add:* Dept Chem Wake Forest Univ Winston-Salem NC 27109

JACKISCH, PHILIP FREDERICK, b Oshkosh, Wis, June 25, 35; m 60; c 1. ORGANIC CHEMISTRY. *Educ:* Univ Wis, BS, 57; Univ Mich, PhD(chem), 65. *Prof Exp:* RES CHEMIST, RES LABS, ETHYL CORP, MICH & LA, 64- *Concurrent Pos:* Sci consult to wine industry, 70-81; Sci consult to wine indust, 70-81; tech ed, Am Wine Soc J, 73-81. *Mem:* Am Chem Soc. *Res:* Gasoline additives; lubricant additives; computer applications; enology; viticulture; sensory evaluation of foods; drugs, polymers; flame retardants. *Mailing Add:* Ethyl Tech Ctr PO Box 14799 Baton Rouge LA 70898

JACKIW, ROMAN WLADIMIR, b Lublinec, Poland, Nov 8, 39; US citizen; c 3. THEORETICAL PHYSICS, HIGH ENERGY PHYSICS. *Educ:* Swarthmore Col, AB, 61; Cornell Univ, PhD(physics), 66. *Prof Exp:* Soc Fels jr fel physics, Harvard Univ, 66-69; from asst prof to assoc prof, 69-77, PROF PHYSICS, MASS INST TECHNOL, 77- *Concurrent Pos:* Sloan Found res fel, 69-71; J S Guggenheim Mem Found fel, 77-78; consult, Los Alamos Nat Lab, 74-84; vis prof, Rockefeller Univ, 77-78, Univ Calif, Los Angeles & Santa Barbara, 80; group leader, Inst Theoret Physics, Univ Calif Santa Barbara & Columbia Univ, 89-90. *Mem:* Am Phys Soc; Am Acad Arts & Sci. *Res:* Theoretical and mathematical physics with specialization to particle, condensed matter and gravitational physics. *Mailing Add:* Ctr Theoret Physics Mass Inst Technol 6-320 Cambridge MA 02139

JACKLET, JON WILLIS, b Springfield Gardens, NY, Apr 16, 35; m 62; c 3. NEUROPHYSIOLOGY, ANIMAL BEHAVIOR. *Educ:* Univ Ore, BS, 62, MA, 64, PhD(biol), 66. *Prof Exp:* USPHS res fel neurophysiol, Calif Inst Technol, 67-68; From asst prof to assoc prog, 68-80, PROF BIOL SCI, STATE UNIV NY ALBANY, 81-. CHMN, DEPT BIOL SCI, 85- *Mem:* Soc Neurosci; Am Physiol Soc. *Res:* Neurophysiology of behavior; plasticity and specificity of neural organization; cellular aspects of circadian rhythms. *Mailing Add:* Dept Biol Sci State Univ NY Albany NY 12222

JACKMAN, DONALD COE, b Cleveland, Tenn, Nov 29, 40; m 65, 81; c 3. INORGANIC CHEMISTRY, ANALYTICAL CHEMISTRY. *Educ:* Maryville Col, BS, 62; Univ Tenn, Knoxville, PhD(inorg chem), 66. *Prof Exp:* From asst prof to assoc prof inorg & anal chem, 66-80, PROF CHEM, PFEIFFER COL, 80- *Concurrent Pos:* NSF res grant, 71; Celanese grant, 85-86; Dept Energy grant, 87 & 88. *Mem:* Am Chem Soc; IUPAC. *Res:* Electron exchange mechanisms in Cobalt-III and Chromium-II systems; immobilized liquid membrane; ruthenium polypyridyls. *Mailing Add:* Dept Chem Pfeiffer Col Misenheimer NC 28109

JACKMAN, LLOYD MILES, b Goolwa, SAustralia, Apr 1, 26; m 50; c 3. ORGANIC CHEMISTRY. *Educ:* Univ Adelaide, BSc, 45, Hons, 46, MSc, 48, PhD(chem), 51. *Prof Exp:* Beit fel, Univ London, 51-52; asst lectr chem, Imp Col, Univ London, 52-53, lectr, 53-61, reader org chem, Univ, 61-62; prof, Univ Melbourne, 62-67; PROF CHEM, PA STATE UNIV, UNIVERSITY PARK, 67- *Concurrent Pos:* Royal Commonwealth Soc bursary, 60; consult, Monsanto Co, UK & Australia, 60-67, Esso Res & Eng Co, 67-70 & Smith, Kline & French Labs, 67-; vis prof, Iowa State Univ, 62 & Univ Tenn, 65; NSF sr foreign fel, 65; Guggenheim Found fel, 73; Humboldt fel, 77-78; Wilsmore fel, 88. *Mem:* Am Chem Soc; The Chem Soc; fel Royal Australian Chem Inst; fel AAAS. *Res:* Applications of nuclear magnetic resonance spectroscopy in organic chemistry; structures and mechanisms of reactions of organic compounds of lithium. *Mailing Add:* 152 Davey Lab Pa State Univ University Park PA 16802

JACKMAN, THOMAS EDWARD, b Thamesville, Ont, Mar 16, 51; m 77; c 2. INTERFACE SCIENCE, ION-SOLID INTERACTIONS. *Educ:* Univ Guelph, BSc, 72, MSc, 74, PhD(physics), 79. *Prof Exp:* Guest scientist, Max Planck Inst, Stuttgart, 74-75; vis fel, Chalk River Nuclear Labs, Ont, 80, res officer, 80-86; RES OFFICER, NAT RES COUN, 86- *Concurrent Pos:* Res asst, Dept Physics, Univ Guelph, Ont, 76-; resident vis, Bell Labs, Murray Hill, NJ, 77-78; adj prof eng physics, MacMaster Univ, 88-; secy & treas, Surf Sci Div, Chem Inst Can & Can Asn Physicists, 87-90; mem, Grant Selection Comt, Nat Sci & Eng Res Coun Can, 90- *Honors & Awards:* Excellence Sci Award, 74; Sigma Xi. *Mem:* Chem Inst Can; Can Asn Physicists; Boehmische Phys Soc. *Res:* Fundamental interactions between MeV ion beams and solids and their application in material science; the growth and characterization of two-dimensional, multilayer semiconductor structures. *Mailing Add:* Microstruct Sci Inst Nat Res Coun Can Ottawa ON K1A 0R6 Can

JACKNOW, JOEL, b New York, NY, Dec 15, 37. PHYSICAL CHEMISTRY. *Educ:* City Col New York, BChE, 59; Univ Utah, PhD(phys chem), 63. *Prof Exp:* Res chemist fundamental res sect, Texaco Inc, 63-65; res assoc & fel chem, Polytech Inst Brooklyn, 65-66; res assoc med sch, Tufts Univ, 66-67; sr physiologist, Bioelectronics Br, Instrumentation Lab, NASA Electronics Res Ctr, 67-68; sr staff mem, Int Res & Technol, 68-71; phys sci adminr, Environ Protection Agency, Off Planning & Eval, 71-74; sr prog mgr, Environ Qual Systs Inc, 74-75; environ & energy consult, 75-79; proj mgr & sr scientist, Habitat Resources Prog, US Fish & Wildlife Serv, 79-86; PROG ANALYST & MGR, RES & DEVELOP SERV, FED AVIATION ADMIN, 86- *Concurrent Pos:* Mem, MBA prog, George Washington Univ, 71-74. *Mem:* Am Chem Soc; Wildlife Soc. *Res:* Sources of water publication; economic and technical analysis of environmental alternatives; research and progammatic strategic guidance; resource allocation; environmental contaminant impacts on fish and wildlife; airport safety. *Mailing Add:* 8110 Timber Valley Ct Dunn Loring VA 22027

JACKO, MICHAEL GEORGE, b Windsor, Ont, Oct 11, 38; m 63; c 1. PHYSICS, CHEMISTRY. *Educ:* Assumption Univ, BSc, 61; Univ Windsor, PhD(phys chem), 64. *Prof Exp:* Phys chemist, Imp Oil Res Dept, Ont, 64-66; phys chemist, 66-79, sr prin chemist, Bendix Advan Technol Ctr, 79-80, PRIN CHEMIST, BENDIX MAT CTR, BENDIX RES LABS, 81- *Concurrent Pos:* Res assoc chem, Univ Windsor, 73-74. *Mem:* Am Chem Soc; Soc Automotive Engrs; fel Am Chem Inst. *Res:* Gas kinetics; radical reactions; gas chromatography; petroleum products; polymers; thermal analysis; friction materials; functional fluids; brake wear debris studies; brake interace reactions. *Mailing Add:* 23721 Merrill Ave Southfield MI 48075

JACKOBS, JOHN JOSEPH, b Hibbing, Minn, Mar 25, 39; m 65; c 3. PHYSICAL CHEMISTRY, X-RAY CRYSTALLOGRAPHY. *Educ:* Wis State Univ, Superior, BA, 61; Iowa State Univ, MS, 64; Ariz State Univ, PhD(phys chem), 67. *Prof Exp:* Res asst chem, Case Western Reserve Univ, 67-69; asst prof physics, Heidelberg Col, 69-75; registr & dir, Comput Ctr, Coe Col, 75-87; programmer/analyst, Info Systs, Inc, 87-88; SR PROJ MGR, CARS INFO SYSTS CORP, 88- *Mem:* AAAS; Am Crystallog Asn; Sigma Xi. *Res:* X-ray diffraction studies of organic and small biological molecules. *Mailing Add:* CARS Info Systs Corp 4000 Executive Park Dr Cincinnati OH 45241

JACKOBS, JOSEPH ALDEN, b Shell Lake, Wis, Oct 23, 17; m 40; c 3. AGRONOMY. *Educ:* Univ Wis, BS, 40, MS, 43, PhD(agron), 47. *Prof Exp:* Asst agronomist, Irrig Exp Sta, State Col Wash, 46-51; assoc prof agron, 51-55, prof crop prod, Univ Ill, Urbana, 55-86; RETIRED. *Concurrent Pos:* Grass & fodder specialist, Int Coop Admin, India, 58-60, crop prod agronomist, US AID, 67-69, Int Soybean Agronomist, 79-86. *Mem:* Fel AAAS; fel Am Soc Agron; Sigma Xi. *Res:* Alfalfa management, cutting treatments; legume, grass and fertility interactions; seed rotting in sweetclover caused by Pythium; genetic shifts in forage species when grown outside region of adaptation; establishment of forage species; soybean production in the tropics; grassland ecology. *Mailing Add:* 12 Persimmon Circle Univ Ill Urbana IL 61801

JACKOVITZ, JOHN FRANKLIN, b Greensburg, Pa, Nov 9, 39; m 64; c 3. PHYSICAL INORGANIC CHEMISTRY. *Educ:* St Vincent Col, BSc, 61; Univ Notre Dame, PhD(chem), 65. *Prof Exp:* NSF vis scholar chem, Northwestern Univ, 66-67; sr scientist, 67-73, sr res scientist, 73-76, ADV SCIENTIST, WESTINGHOUSE RES & DEVELOP LABS, 76- *Concurrent Pos:* Res assoc, Univ Pittsburgh, 67-73. *Mem:* Soc Appl Spectros (pres, 81); Am Chem Soc; Electrochem Soc; Coblentz Soc. *Res:* Chelate chemistry, infrared and Raman spectra and force fields of inorganic molecules; uranium chemistry; electrod rxns. *Mailing Add:* Res & Develop Ctr Westinghouse Elec Co Pittsburgh PA 15235

JACKS, THOMAS JEROME, b Chicago, Ill, Jan 24, 38. BIOCHEMISTRY, PLANT PHYSIOLOGY. *Educ:* Western Reserve Univ, BS, 60, PhD, 65. *Prof Exp:* Nat Acad Sci res fel biochem, 65-67, RES LEADER, SOUTHERN REGIONAL RES CTR, USDA, 67- *Concurrent Pos:* Res staff, Tulane Univ, 74-; assoc ed, Am Oil Chemist's Soc, 75- *Mem:* Am Oil Chemist's Soc; Electron Micros Soc Am; Am Chem Soc; Am Soc Plant Physiol; AAAS. *Res:* Protein chemistry; enzymology; electron microscopy; oxygen radical chemistry. *Mailing Add:* PO Box 19687 Southern Regional Res Ctr New Orleans LA 70179

JACKS, THOMAS MAURO, b Harrisburg, Pa, Mar 13, 41; m 69; c 2. MEDICAL MICROBIOLOGY. *Educ:* Duquesne Univ, BS, 64; Pa State Univ, MS, 66, PhD(microbiol), 68. *Prof Exp:* Asst prof biol, State Univ NY Col New Paltz, 68-69; sr res microbiologist, Vet Microbiol Sect, Norwich Pharmacol Co, 69-76; SR RES FEL, MERCK RES LABS, RAHWAY, NJ, 76- *Concurrent Pos:* Instr, State Univ NY Agr & Tech Col Morrisville, 69-70 & 74-75; mem, NJ State Bd Vet Med Examrs, 84- *Mem:* NY Acad Sci; Am Soc Microbiol; Infectious Dis Soc Am. *Res:* Escherichia coli pathogenicity in man and animals; bovine antibodies against Escherichia coli; salmonellosis in cattle and swine; swine dysentery; pneumonic pasteurellosis in cattle and swine; endocrinology-tropic hormone secretion. *Mailing Add:* 842 Wallberg Ave Westfield NJ 07090

JACKSON, ALBERT S(MITH), b Sylvia, Kans, Feb 2, 27; m 78; c 5. COMPUTER SCIENCE, CONTROL SYSTEMS. *Educ:* Calif Inst Technol, BS, 51, MS, 52; Cornell Univ, PhD(elec eng), 56. *Prof Exp:* Engr, Bell Tel Labs, 52; instr, Cornell Univ, 52-55, asst prof, 56-59; mgr, Data Processing & Controls Dept, Thompson-Ramo-Wooldridge Prod Co, 59-61; pres, Control Technol, Inc, 61-65; chief scientist, Milgo Electronics Corp, 65-70; pres, Opto Logic Corp, 70-75; APPLN ENG MGR, MOTOROLA, INC, 75- *Concurrent Pos:* Consult, Gen Elec Co, 53-59, Gen Dynamics/Convair, 56-58 & Naval Res Lab, 57-59; chmn prof group Human Factors in Electronics, Inst Elec & Electronics Engrs, 63-64, educ coordr Region 6 & corresp mem Educ Activ Bd, 84-85; lectr, Univ Calif, Irving, 65- & Univ Calif, Los Angeles, 78-80. *Mem:* Inst Elec & Electronics Engrs. *Res:* Analog and digital computers; feedback control system theory; application of computers to control systems; human factors research; concurrent computer architecture; computer simulation, microprocessor-based system design. *Mailing Add:* Nine Aldergrove Irvine CA 92714

JACKSON, ANDREW, b Preston, Eng, Dec 5, 48; m 81; c 3. TRIBOLOGY. *Educ:* Imperial Col, London Univ, BSc, 70, PhD(mech eng), 74. *Prof Exp:* Res engr, 74-76, sr res engr, 76-80, assoc, 80-84, res assoc, 84-88, SR RES ASSOC, CENT RES LAB, MOBIL RES & DEVELOP CORP, PRINCETON, NJ, 88- *Honors & Awards:* Hunt Award, Am Soc Lubrication Engrs, 77 & Hodson Award, 82. *Mem:* Soc Tribologists & Lubrication Engrs, (dir, 87-); Am Soc Mech Engrs; Inst Mech Engrs. *Res:* Lubrication science (tribology); elastohydrodynamic lubrication; traction; rolling contact fatigue; internal combustion engine lubrication; synthetic lubricants. *Mailing Add:* Four Walking Purchase Dr Pennington NJ 08534-2917

JACKSON, ANDREW D, JR, b Orange, NJ, Dec 20, 41; m 66. THEORETICAL NUCLEAR PHYSICS. *Educ:* Princeton Univ, AB, 63, MA, 65, PhD(physics), 67. *Prof Exp:* Res assoc physics, Princeton Univ, 67; NATO res fel, Univ Sussex, 67-68; from asst prof to assoc prof, 68-76, PROF PHYSICS, STATE UNIV NY STONY BROOK, 76- *Concurrent Pos:* Alfred P Sloan Found fel, 71-72. *Mem:* Am Phys Soc. *Res:* Nucleon-nucleon interaction and nuclear structure calculations. *Mailing Add:* Dept Physics State Univ NY Stony Brook NY 11794

JACKSON, ANDREW OTIS, b Enterprise, Ala, Apr 14, 41. PLANT VIROLOGY. *Educ:* Okla State Univ, BS, 64, MS, 67; Univ Man, PhD(plant path), 70. *Prof Exp:* Fel plant virol, Dept Agr Biochem, Univ Ariz & Dept Plant Path, Univ Nebr, 70-73; asst prof, 73-77, ASSOC PROF PLANT VIROL, PURDUE UNIV, 77- *Mem:* Am Phytopathological Soc; AAAS; Sigma Xi; Soc Gen Microbiol. *Mailing Add:* Dept Plant Path Univ Calif Berkeley CA 94720

JACKSON, ANNE LOUISE, b Watertown, NY. IMMUNOLOGY. *Educ:* Cornell Univ, BS, 56; Univ Mich, MS, 57, PhD(microbiol), 63. *Prof Exp:* Res assoc biochem, Univ Mich, 61-63; asst prof microbiol, Sch Med & Dent, Georgetown Univ, 63-68; dir tech serv, Meloy Labs, 68-74; dir res immunol, Kent Labs, 74-77; mem staff, Dept Microbiol, Univ BC, 77-80; MGR TECH SERV, BECTON DICKINSON MONOCLONAL CTR, 80- *Concurrent Pos:* Guest worker, Lab Immunol, Nat Inst Allergy & Infectious Dis, NIH, 63-65; asst guest prof, Univ BC, 75- *Mem:* Am Asn Immunol; Fedn Am Socs Exp Biol; Can Asn Immunol; Am Fedn Clin Res. *Res:* Production and development of immunologic tests for in vitro diagnostics. *Mailing Add:* Becton Dickson 2350 Gume Dr San Jose CA 95131

JACKSON, BENJAMIN A, b Hillburn, NY, July 8, 29; m 55; c 4. TOXICOLOGY. *Educ:* NY State Col Teachers, Albany, BA, 50; Rensselaer Polytech Inst, MS, 51; NY Univ, PhD(biol), 57; Fairleigh Dickinson Univ, MBA, 78, Acad Toxicol Sci, dipl, 84, cert gen toxicol. *Prof Exp:* Res biologist, Am Cyanamid Co, 51-67; electron microscopist, Toxicol Dept, Sterling-Winthrop Res Inst, 67-69; sr res toxicologist, Reproductive Safety Eval Group, Toxicol Res, Lederle Labs, Am Cyanamid Co, 69-75, mgr Teratology & Mutagenicity, Toxicol Sect, 75-78; supvr, petitions reviewers, Ctr Food Safety & Appl Nutrit Food & Drug Admin, 78-80, chief, color & cosmetic eval br, Bur Foods, 80-85, dir path, Ctr Food Safety & Appl Nutrit, 85-90, prog mgr, Risk Assessment Res & Policy Develop, 87-90; SR SCI ADV, ENVIRON CORP, 90- *Concurrent Pos:* Res assoc, Cornell Med Col, 64-65; adj asst prof pharmacol & toxicol, Howard Univ, 80-85. *Mem:* Soc Toxicol; fel Acad Toxicol Sci; Am Col Toxicol; Soc Toxicol Path. *Res:* Experimental liver tumors; short term effects of drugs; mitotic activity; drug toxicity; quantitation of morphological changes; electron microscopy; teratology; mutagenicity; correlation between mutagenicity and carcinogenic potential of drugs; regulatory toxicology; color additive toxicology; cosmetic ingredients safety evaluation; mutagenicity testing, submitting pathology data; risk assessment. *Mailing Add:* Environ Corp 4350 N Fairfax Dr Arlington VA 22203

JACKSON, BENJAMIN T, b Jacksonville, Fla, Apr 28, 29; m 53; c 4. SURGERY, FETAL PHYSIOLOGY. *Educ:* Duke Univ, MD, 54. *Prof Exp:* Intern med, Duke Univ Hosp, 54-55; asst resident surg, Univ Minn Hosps, 57-58; resident, Med Col Va, 58-62, instr, 63-64; from asst prof to prof surg, Sch Med, Boston Univ, 64-80; PROF SURG, PROG MED, BROWN UNIV, 80- *Concurrent Pos:* USPHS res fel, Med Col Va, 58-62; Am Heart Asn advan res fel, 61-63; estab investr, Am Heart Asn, 63-68; asst chief surg serv, Boston Vet Admin Hosp, 74-80; chief surg serv, Providence Vet Admin Med Ctr, 80. *Mem:* Soc Gynec Invest; Am Fedn Clin Res; Am Soc Exp Path; Soc Univ Surg; Am Col Surg. *Res:* Fetal cardiovascular and endocrine physiology; pathophysiology of congenital cardiovascular anomalies; fetal hormonal responses in diabetic pregnancy. *Mailing Add:* 11 October Lane Weston MA 02193

JACKSON, BERNARD VERNON, b Peoria, Ill, Nov 7, 42; m 90; c 2. SOLAR PHYSICS, INTERPLANETARY MEDIUM PHYSICS. *Educ:* Univ Ill, Urbana, BS, 64; Ind Univ, Bloomington, MS, 67, PhD(astrophys),70. *Prof Exp:* Res asst, Ind Univ, Bloomington, 64-70; geophysicist, Univ Calif, Los Angeles, 70-73; sci programmer astrophys, Arecibo Observ, Cornell Univ, 73-75; res fel skylab, High Altitude Observ, Nat Ctr Atmospheric Res, Boulder, Colo, 75-77; res assoc astrophys, Commonwealth Sci & Indust Res Orgn, Sydney, Australia, 77-79; asst res physicist, 79-87, ASSOC RES PHYSICIST ASTROPHYS, DEPT ELEC ENG & COMPUT SCI, UNIV CALIF, SAN DIEGO, 88- *Honors & Awards:* Glacier named in honor, Jackson Glacier. *Mem:* Am Astron Soc; Am Geophys Union. *Res:* Solar and interplanetary physics; solar wind and its interaction with other material as this plasma flows outward from the sun. *Mailing Add:* CASS 0111 Univ Calif San Diego La Jolla CA 92093-0111

JACKSON, C(HARLES) IAN, b Keighley, Eng, Feb 11, 35; Can & UK citizen; div; c 1. SCIENCE ADMINISTRATION, SCIENCE POLICY. *Educ:* Univ London, BA, 56; McGill Univ, MSc, 59, PhD(geog), 61. *Prof Exp:* Lectr geog, London Sch Econ & Polit Sci, 59-69; head, Econ Geog Sect, Can Dept Energy, Mines & Resources, 69-71; dir priorities & planning, Can Ministry State Urban Affairs, 71-75; exec dir, Can Habitat & Energy Secretariat, 75, tech officer, 75-78; sr econ affairs officer, UN Econ Comn Europe, 78-81; Exec dir, Sigma Xi, 81-87; CONSULT, INST RES PUB POLICY, OTTAWA, 87-, GLOBAL WARMING, GOVT INDONESIA, 90- *Honors & Awards:* Darton Prize, Royal Meteorol Soc, 62; Evan Durbin Prize, Inst Econ Affairs, 66. *Mem:* Hakluyt Soc; Champlain Soc; Royal Geog Soc. *Res:* Environmental protection; energy use; fur-trade history; global warming. *Mailing Add:* 29 N Lake Dr Hamden CT 06517

JACKSON, CARL WAYNE, b Carbondale, Ill, Nov 27, 42; m 63. EXPERIMENTAL HEMATOLOGY, RADIATION BIOLOGY. *Educ:* Southern Ill Univ, BA, 63; Univ Tenn, Knoxville, PhD(radiation biol), 71. *Prof Exp:* Biol lab specialist physiol, Oak Ridge Nat Lab, 63-71; Nat Cancer Inst traineeship fel hemat, 71-73, res assoc, 73-75, asst mem, 76-80, assoc mem, 81-87, MEM, ST JUDE CHILDREN'S RES HOSP, 87- *Mem:* Am Soc Hemat; Int Soc Exp Hemat; Int Soc Thrombosis Haemostas. *Res:* Hemopoiesis; thrombopoiesis; cell kinetics; cell regulation and differentiation; radiation hematology; megakaryocyte differentiation; megakaryocytopoieses; micropectrophotometry; identification of megakaryocyte precursors. *Mailing Add:* St Jude Children's Res Hosp 332 N Lauderdale Memphis TN 38101

JACKSON, CARLTON DARNELL, b Wiggins, Miss, Dec 1, 38; div; c 2. BIOCHEMISTRY, ONCOLOGY. *Educ:* Miss Col, BS, 61; Univ Tenn, MS, 63, PhD(biochem), 67. *Prof Exp:* Res trainee biochem, St Jude Children's Res Hosp, 64-67; USPHS fel, Univ Miami, 67-69; asst prof biochem, Univ Tenn, 69-72; res chemist carcinogenesis, Nat Ctr Toxicol Res, Food & Drug Admin, 72-75, chief, Div Carcinogenic Res, 76-79; ASSOC PROF BIOCHEM & INTERDISCIPLINARY TOXICOL, MED SCH, UNIV ARK, LITTLE ROCK, 72- *Concurrent Pos:* Res biochemist carcinogenesis, Vet Admin Hosp, Memphis, 69-72; pharmacologist, Off Sci Intel, Nat Ctr Toxicol Res, Food & Drug Admin, Div Mollecular Biol, 81-84, Div Comp Toxicol, 84- *Mem:* Am Asn Cancer Res; Sigma Xi. *Res:* Mechanisms of chemical carcinogenesis; molecular mechanism of hormone action; protein and nucleic acid synthesis; molecular biology of cell division and differentiation. *Mailing Add:* Nat Ctr Toxicol Res Food & Drug Admin Jefferson AR 72079

JACKSON, CARMAULT B, JR, b Newton, Mass, Apr 19, 24; m 47; c 3. INTERNAL MEDICINE, HEMATOLOGY. *Educ:* Bucknell Univ, BS, 48; Univ Pa, MD, 52. *Prof Exp:* Mem, Proj Mercury, NASA Space Task Group, 58-61; pvt prac internal med, San Antonio, Tex, 61-76; assoc dir extramural progs, Univ Tex Syst, 77-79; exec vpres & adminr, Metrop Med Ctr, San Antonio, Tex, 79-83; med adv, baptist med ctr, San Antonio, Tex, 83-86; CONSULT, 86- *Concurrent Pos:* Mem bd, Tex Inst Med Assessment, 76-80, pres, 80-81. *Mem:* Inst Med Nat Acad Sci; Am Cancer Soc. *Res:* Cancer control information and technology dissemination. *Mailing Add:* Med Adv Serv 16902 Hidden Timber Wood San Antonio TX 78242

JACKSON, CRAIG MERTON, b Staples, Minn, Dec 2, 41; div; c 1. BIOCHEMISTRY. *Educ:* Wash State Univ, BS, 63; Univ Wash, PhD(biochem), 67. *Prof Exp:* Res assoc, Col Med, Univ Ariz, 67; Am Cancer Soc fel chem physics, Unilever Res Lab, Port Sunlight, Eng, 67-69; from asst prof to prof chem, Sch Med, Wash Univ, 69-83; SCI DIR, SOUTHEAST MICH REGION, AM RED CROSS, 83- *Concurrent Pos:* Res grants, Nat Heart, Lung & Blood Inst, res reviewer, Am Heart Asn, 71-74, estab investr, 74-79, chmn, Int Comt Thrombosis, 82-84; mem coun thrombosis, Am Heart Asn, 71-; vis prof, Kyushi Univ, Japan, 82; Am Red Cross estab investr, 83-; adj prof biochem, Sch Med, Wayne State Univ, 84-; mem adv comt, Div Blood Dis & Resources, Nat Heart, Lung & Blood Inst, 87-89, chmn, 89-90. *Mem:* Fel AAAS; Int Soc Thrombosis & Hemostasis; Nat Heart, Lung & Blood Inst; Am Chem Soc; Am Heart Asn; Sigma Xi. *Res:* Protein chemistry and enzymology of blood coagulation, lipid-protein interactions in blood coagulation; physical chemistry of lipids; plasma proteins. *Mailing Add:* 20630 Balfour Apt 3 Harper Woods MI 48225

JACKSON, CRAWFORD GARDNER, JR, b Birmingham, Ala, Jan 5, 31; div; c 1. VERTEBRATE BIOLOGY, PALEONTOLOGY. *Educ:* Emory Univ, AB, 52; Univ Fla, MS, 59, PhD(biol), 64; Nat Univ, MA, 85. *Prof Exp:* Instr biol, Armstrong Col, 52-53; asst prof, Univ S Ala, 65-67; from assoc prof to prof, Col Women, Miss State Univ, 67-74; managing ed, Ecol Monogr, 74-78; cur & actg dir, San Diego Natural Hist Mus, 78-80; PROF, DEPT NATURAL SCI, NAT UNIV, 80- *Concurrent Pos:* Vis lectr, San Diego State Col, 70-71; ed-in-chief, Herpetologica, 73-79; adj prof biol, San Diego State Univ, 74-79; res assoc, Smithsonian Inst, 74- & San Diego Natural Hist Mus, 75-; lectr, Sch Med & Continuing Educ, Univ Calif, San Diego, 77-83. *Mem:* Sigma Xi. *Res:* Ecology, morphology, ethology, pathobiology of amphibians and reptiles; paleobiology of reptiles; sexual orientation of vertebrates. *Mailing Add:* Natural Sci Nat Univ San Diego CA 92108

JACKSON, CURTIS M(AITLAND), b New York, NY, Apr 20, 33; m 57; c 2. METALLURGY. *Educ:* NY Univ, BMetE, 54; Ohio State Univ, MS, 59, PhD(metall), 66. *Prof Exp:* Prin metallurgist, Alloy Develop Div, Battelle Mem Inst, 54-61, proj leader, Specialty Alloys Div, 61-67, assoc chief, 67-77, ASSOC MGR PHYS & APPL METALL, COLUMBUS LABS, BATTELLE MEM INST, 77- *Concurrent Pos:* Chmn, N Cent US region, Am Inst Mining, Metall & Petrol Engrs, 65-66; dir, Wire J, 73-78 & Wire Found, 74-86. *Honors & Awards:* IR-100 Award, Indust Res Mag, 76; Mordica Mem Award, Wire Asn Int, 77, J Edward Donnellan Award, 78. *Mem:* Am Inst Mining, Metall & Petrol Engrs; Am Soc Metals; Am Vacuum Soc; Wire Asn Int (second vpres, 73-74, first vpres, 74-76, pres, 76-77); Sigma Xi. *Res:* Physical metallurgy; alloy development; nucleation and growth of thin films; technical economics; shape-memory alloys; metal failure analysis; electrical and electronic alloys; melting, casting and mechanical working of metals; wire technology. *Mailing Add:* 5088 Dalmeny Ct Columbus OH 43220-2693

JACKSON, CURTIS RUKES, b Kansas City, Mo, July 25, 27; m 51; c 6. RESEARCH MANAGEMENT, FOREIGN AGRICULTURAL DEVELOPMENT. *Educ:* Univ Miami, BS, 49; Fla State Univ, MS, 51; Univ Fla, PhD(plant path), 58. *Prof Exp:* Assoc dir, Agr Exp Sta, Univ Ga, 73-83; dir, Int Coop, Icrisat, India & Niger, 83-86; DIR, OFF RES & UNIV RELS AID, WASH, 86- *Concurrent Pos:* Consult foreign agr develop proj, Indust

& Govt, 70- *Honors & Awards:* Res Award, Nat Peanut Coun. *Res:* Diseases of peanuts; vegetables and ornamental plants; soil microbiology; tropical agricultural development. *Mailing Add:* Agency Int Develop Bur Sci & Technol Rm 309 SA-18 Washington DC 20523

JACKSON, DALE LATHAM, b Eng, May 20, 32; m 59; c 1. ZOOLOGY. *Educ:* Univ Durham, BSc, 55, PhD(entom), 59. *Prof Exp:* Asst prof zool, Univ Guelph, 59-61; from asst prof to assoc prof, 61-69, PROF BIOL, UNIV AKRON, 69-, HEAD DEPT, 68- *Mem:* Entom Soc Can; fel Royal Entom Soc; Soc Syst Zool; Sigma Xi. *Res:* Host relationships and taxonomy of proctotrupoidea experimental taxonomy. *Mailing Add:* 555 Royal Ave Akron OH 44303

JACKSON, DANIEL FRANCIS, b Pittsburgh, Pa, June 11, 25; m 51. LIMNOLOGY. *Educ:* Univ Pittsburgh, BS, 49, MS, 50; State Univ NY, PhD(water resources), 57. *Prof Exp:* Lectr biol, Univ Pittsburgh, 49-51; asst prof, Col of Steubenville, 51-52; hydrologist, US Army Corps Engrs, Pa, 52-53; asst col forestry, State Univ NY, 53-55; asst prof, Western Mich Univ, 55-59; from asst prof to assoc prof, Univ Louisville, 59-63; prof limnol, Syracuse Univ, 63-78; DIR & PROF ENVIRON & URBAN SYSTS, SCH TECHNOL, FLA INT UNIV, 78-; MEM STAFF, JACKSON & JACKSON ASSOC, 78- *Concurrent Pos:* Dir, C C Adams Ctr, 55-59; dir, Drinking Water Qual Res Ctr, Fla Int Univ, 76-78; dir & prof, Int Environ Studies, La State Univ, Baton Rouge, 82-86. *Mem:* Am Soc Limnol & Oceanog; Ecol Soc Am; Am Micros Soc; Am Fisheries Soc; Int Asn Theoret & Appl Limnol. *Res:* Limnology, primary productivity; environmental planning plankton; pollution, environmental toxicology. *Mailing Add:* 3323 Guilford Court Naples FL 33962

JACKSON, DARRYL DEAN, b Lexington, Okla, 1932; m 59; c 2. CHEMICAL INSTRUMENTATION. *Educ:* Univ Okla, BS, 55, MS, 56; Univ NMex, PhD(radiochem), 68. *Prof Exp:* STAFF MEM, LOS ALAMOS NAT LAB, 56- *Mem:* AAAS; Am Inst Chemists; Am Chem Soc; Sigma Xi. *Res:* Structure of optically active inorganic complexes; chemistry of solutions of carrier-free iodine-131; radiochemistry of transuranium elements and fission products; development of automated instruments for chemical analysis. *Mailing Add:* Box 1663 MS 740 Los Alamos NM 87544

JACKSON, DAVID ARCHER, b New York, NY, Apr 29, 42; m 66; c 1. MOLECULAR BIOLOGY, VIROLOGY. *Educ:* Harvard Univ, AB, 64; Stanford Univ, PhD(molecular biol), 69. *Prof Exp:* USPHS fel, Med Sch, Stanford Univ, 69-70, Nat Cystic Fibrosis Res Found basic sci fel, 70-72; asst prof microbiol, Univ Mich, Ann Arbor, 72-76, assoc prof, 77-81; vpres & sci dir, Genex Corp, Rockville, Md, 80-82, chmn sci adv bd, 77-85, sr vpres, 83-85, dir, 85; dir, Biotechnol Res, E I duPont, 86-89, dir virol res, 89-90, DIR VIROL RES, DUPONT MERCK PHARMACEUT CO, 91- *Concurrent Pos:* Consult, President's Comn for Study Ethical Probs in Med, Biomed, & Behav Res 81-; mem, Ad Hoc Comt on Nat Issues in Genetic Eng, NSF, 81-82; mem, Adv Panel Comp Assessment Develop Biotechnol, Off Technol Assessment, 81-83; adj prof appl molecular biol, Univ Md Baltimore County, 81-86; mem, bd dir, Indust Biotechnol Assoc, 86-89. *Mem:* Am Soc Microbiol; Genetic Soc Am. *Res:* Molecular biology; enzymes acting on nucleic acids; mammalian viruses. *Mailing Add:* DuPont Merck Pharmaceut Co Exp Sta PO Box 80328 Wilmington DE 19880-0328

JACKSON, DAVID DIETHER, b San Francisco, Calif, Sept 18, 43; m 68; c 2. GEOPHYSICS. *Educ:* Calif Inst Technol, BS, 65; Mass Inst Technol, PhD(geophys), 69. *Prof Exp:* Assoc prof, 69-80, PROF GEOPHYS, UNIV CALIF, LOS ANGELES, 80- *Concurrent Pos:* Mem, US Nat Comt Seismol, Panel on Crustal Movement Measurements & Comt Geodesy/Seismol, 78-81; sr resident res associateship, Nat Acad Sci/Nat Res Coun, 81; mem, Calif Earthquake Prediction Eval Coun, 84- *Mem:* AAAS; Am Geophys Union; Seismol Soc Am; Seismol Soc Japan. *Res:* Seismology; solid earth geophysics; geophysical inverse problems; applications of solid state physics to geophysics; earthquake prediction and control. *Mailing Add:* Dept Earth & Space Sci-1802 Univ Calif Los Angeles CA 90024-1567

JACKSON, DAVID PHILLIP, b Toronto, Ont, Oct 2, 40; m 75; c 2. MATHEMATICAL PHYSICS. *Educ:* Univ Toronto, BSc, 62, MA, 64, MASc, 66, PhD(eng physics), 68; Univ Ottawa, dipl, 85. *Prof Exp:* Mathematician, IBM, Toronto, 62-63; res asst, Inst Aerospace Studies, Toronto, 65-68; asst res officer, 68-70, assoc, 70-80, sr res officer, 81-85, mgr fusion progs, 85-86, DIR NAT FUSION PROG, ATOMIC ENERGY CAN LTD, 86- *Concurrent Pos:* Vis scientist, Max-Plank Inst Plasma Physics, 75-76; assoc prof eng physics, McMaster Univ, 78-81, mem staff, Inst Mat Res, 79-, mem staff, Inst Energy Studies, 80-, prof, 81-; vis prof, Bell Labs, Murray Hill, 81-87. *Mem:* Chem Inst Can; Can Asn Physicists. *Res:* Mathematical modeling of particle-surface and particle-solid interactions; the wall problem in fusion reactors. *Mailing Add:* Atomic Energy Can Ltd Chalk River ON K0J 1J0 Can

JACKSON, DONALD CARGILL, b Philadelphia, Pa, May 4, 37; m 66; c 2. PHYSIOLOGY. *Educ:* Geneva Col, BS, 59; Univ Pa, PhD(physiol), 63. *Prof Exp:* Res asst physiol, John B Pierce Lab, Conn, 61-63; res asst, Duke Univ, 63-65; res fel physiol, Sch Med, Univ Pa, 65-68, asst prof, 68-73; assoc prof, 73-79, PROF MED SCI, BROWN UNIV, 79- *Concurrent Pos:* Pa Plan scholar, 68-71. *Mem:* AAAS; Am Soc Zool; Am Physiol Soc. *Res:* Acid-base physiology in reptiles. *Mailing Add:* Div Biol & Med Brown Univ Box G Providence RI 02906

JACKSON, DUDLEY PENNINGTON, b Roanoke, Va, Apr 1, 24; m 48. MEDICINE, HEMATOLOGY. *Educ:* Johns Hopkins Univ, MD, 47. *Hon Degrees:* DSc, Randolph-Maron Col, Ashland, Va, 82. *Prof Exp:* Asst med, Sch Med, Johns Hopkins Univ, 48-49 & 51-52, USPHS res fel hemat, 52-54, from instr to prof, 54-72; chmn dept, 72-82, PROF MED, SCH MED, GEORGETOWN UNIV, 72-, PHYSICIAN HOSP, 72- *Concurrent Pos:* Intern, Osler Med Serv, Johns Hopkins Hosp, 47-48, asst resident physician, 48-49 & 51-52, asst physician, 52-53, physician, Hemat Clin, 53-72; Markle

scholar med sci, 54-59; mem coun on circulation & coun on thrombosis, Am Heart Asn. *Mem:* Am Physiol Soc; fel Am Col Physicians; Am Soc Hemat; Am Soc Clin Invest; Asn Am Physicians. *Res:* Blood coagulation. *Mailing Add:* Georgetown Univ Hosp 3800 Reservoir Rd NW Washington DC 20007

JACKSON, DURWARD P, b Hartsville, SC, Apr 12, 40; m 84. DATABASE MANAGEMENT. *Educ:* Univ Ariz, BS, 64; Univ Utah, MEA, 69; Golden Gate Univ, MBA, 78; Claremont Grad Sch, PhD(exec mgt), 83. *Prof Exp:* Consult, Touche Ross & Co, 69-71; owner & consult, Dataphase Inc, 71-75; data adminr, Air Force Flight Test Ctr, 76-81; PROF COMPUTER INFO SYSTS, CALIF STATE UNIV, LOS ANGELES, 81- *Concurrent Pos:* Consult, 81-; dir, Ctr Info Resource Mgt, 84-87. *Mem:* Asn Comput Mach; Data Processing Mgt Asn; Soc Info Mgt. *Res:* Techniques for enterprise-wide information management; integrated information structures. *Mailing Add:* Dept Info Systs Calif State Univ Los Angeles CA 90032

JACKSON, EARL GRAVES, b Springfield, Mass, Mar 27, 20; m 43; c 3. PHYSICAL CHEMISTRY. *Educ:* Am Int Col, BS, 42; Clark Univ, AM, 43; Rutgers Univ, PhD(chem), 51. *Prof Exp:* Chemist, Merck & Co, 46-48; lubricants specialist, Gen Elec Co, 51-55, supvr bearings, lubrication & seals develop, 55-60; sr phys chemist, Nat Res Corp, Cambridge, 60-62 & United Shoe Mach Corp, 62-69; dir chem physics lab, Res Div, USM Corp, Mass, 69-73, head adhesives sect, Lexington Res Lab, 73-79, mgr admin, Kendall Co, 80-83; RETIRED. *Concurrent Pos:* Mem subcomt lubrication & wear, Nat Adv Comt Aeronaut, 56-58. *Res:* Physical chemistry of polymeric soaps; properties and mechanisms of lubricating greases; high temperature lubrication and fatigue; structural sealants; shoe technology; pressure sensitive adhesives and tapes. *Mailing Add:* 201 Willow Oak Lane Hendersonville NC 28739

JACKSON, EARL ROGERS, b Madison, Ind, Aug 6, 30; m 51; c 1. FOOD CHEMISTRY. *Educ:* Hanover Col, AB, 52. *Prof Exp:* Chemist, Am Can Co, 52-57; chemist, Heekin Can Co, 58-64, asst res dir, 64-77, vpres res, 78-90, DIR RES, HEEKIN CAN, INC, 78-, SR VPRES RES, 90- *Mem:* Inst Food Technologists; Soc Soft Drink Technologists; Am Soc Testing & Mat. *Mailing Add:* Heekin Can Inc 8200 Broadwell Rd Cincinnati OH 45244

JACKSON, EDWARD MILTON, cell physiology, biochemistry, for more information see previous edition

JACKSON, EDWIN ATLEE, b Lyons, NY, Apr 18, 31; m 54; c 2. PHYSICS. *Educ:* Syracuse Univ, BS, 53, MS, 55, PhD(physics), 58. *Prof Exp:* Asst, Syracuse Univ, 54-57; asst & instr, Brandeis Univ, 57-58; Nat Acad Sci res assoc, Air Res & Develop Command, Air Force Cambridge Res Ctr, Mass, 58-59; staff mem, Proj Matterhorn, Princeton Univ, 59-61; asst prof, 61-63, assoc prof physics & mech eng, 63-77, PROF PHYSICS, UNIV ILL, URBANA, 77- *Concurrent Pos:* Vis sr physicist, Found Fundamental Res Matter, Inst Plasma-Physics, Netherlands, 67-68; vis mem staff, Los Alamos Sci Lab, 71; mem adv bd, Physica D: Nonlinear Phenomena, 80-; vis prof, Chalmers Univ, Sweden, 84, JIFT vis prof, Nagoya Univ, Japan, 84. *Honors & Awards:* Fel, Am Phys Soc. *Mem:* Am Phys Soc. *Res:* Nonlinear dynamics; plasma physics; kinetic theory of gases. *Mailing Add:* Dept Physics Univ Ill 1110 W Green St Urbana IL 61801

JACKSON, EDWIN KERRY, b Oct 21, 52; m; c 2. HIGH BLOOD PRESSURE. *Educ:* Univ Tex, Dallas, PhD(pharmacol), 79. *Prof Exp:* Assoc prof pharmacol, Sch Med, Vanderbilt Univ, 79-91; PROF PHARMACOL & MED, UNIV PITTSBURGH SCH MED, 91- *Concurrent Pos:* Estab investr, Am Heart Asn. *Honors & Awards:* Glaxo Cardiovasc Discovery Award. *Mem:* Am Soc Hypertension; Am Fedn Clin Res; Fedn Am Soc Exp Biol; Am Heart Asn. *Res:* Cardiovascular pharmacology. *Mailing Add:* Ctr Clin Pharmacol Univ Pittsburgh Sch Med Pittsburgh PA 15261

JACKSON, ELIZABETH BURGER, b Clay, WVa, Oct 20, 14; m 62. BIOLOGY, CHEMISTRY. *Educ:* Col William & Mary, BS, 34, MA, 35; Univ Va, DEd, 60. *Prof Exp:* Prof natural sci, 40-80, EMER PROF NATURAL SCI, LONGWOOD COL, 80- *Concurrent Pos:* Res grant, Univ Ctr, Univ Va, 58-59. *Mem:* AAAS; Nat Sci Teachers Asn; Sigma Xi. *Res:* Science education through commercial television; science for elementary schools. *Mailing Add:* c/o Brookview Lodge 2001 Cobb St Farmville VA 23901

JACKSON, ERNEST BAKER, b Bicknell, Utah, Mar 31, 14; m 43; c 8. AGRONOMY, BOTANY. *Educ:* Brigham Young Univ, BSc, 46; Univ Nebr, MSc, 54, PhD(agron), 56. *Prof Exp:* Pub sch teacher, Utah, 35-42, prin, 47-51; soils technologist, Bur Reclamation, Colo, 46-47; Fling Asst, Univ Nebr, 52-55, asst agronomist, 55-58; from asst agronomist to assoc agronomist, Univ Ariz, 58-71, prof agron & agronomist, Agr Exp Sta, 71-76, prof plant sci & res scientist, 76-83; RETIRED. *Mem:* Am Soc Agron. *Res:* Forage crops, cotton and small grains. *Mailing Add:* 1355 Gateway Yuma AZ 85364

JACKSON, ETHEL NOLAND, b Geneva, NY, Apr 27, 44; m 66; c 1. MOLECULAR BIOLOGY, RECOMBINANT DNA METHODOLOGY. *Educ:* Harvard Univ, BA, 66; Stanford Univ, PhD(biol sci), 73. *Prof Exp:* Fel biochem genetics, Dept Human Genetics, Sch Med, Univ Mich, 72-74, asst prof microbiol, 74-81; res dir, Molecular Genetics Dept, Genex Co, 81-85, exec dir res opers, 85; res supvr life sci, 87-90, RES MGR, CENT RES & DEVELOP DEPT, E I DU PONT DE NEMOURS & CO, 90- *Concurrent Pos:* Am Cancer Soc fel, 74. *Mem:* Am Soc Microbiol; Genetics Soc Am; AAAS; Am Women Sci. *Res:* Biological regulatory mechanisms and biochemical genetics of bacterial viruses; chromosome structure of viruses; viral DNA packaging; vector systems for gene cloning in gram-positive bacteria; protein secretion by prokaryotes; protein expression systems; plant molecular biology. *Mailing Add:* 298 Old Kennett Rd Kennett PA 19348

JACKSON, EUGENE BERNARD, b Frankfort, Ind, June 18, 15; m 41. INDUSTRIAL INFORMATION SYSTEMS, INDEXING & ABSTRACTING. *Educ:* Purdue Univ, BSc, 37; Univ Ill, BSLS, 38, MALS, 42. *Prof Exp:* Jr prof asst, Eng Libr, Univ Ill, 38-40, asst in chg newspaper div, Libr, 40-41; doc librn, Univ Ala, 41-42; prof asst, Detroit Pub Libr, 42-46; chief libr sect, Cent Air Doc Off, Army-Navy-Air Force, 46-49; chief res info div, Off Qm Gen, US Dept Army, 49; chief div res info, Nat Adv Comt Aeronaut, 49-56; head libr dept, Gen Motors Res Labs, 56-65; dir info retrieval & libr serv, IBM Corp, 65-71; prof libr sci, 71-85, EMER PROF LIBR & INFO SCI, UNIV TEX, AUSTIN, 85- *Concurrent Pos:* Chief, Wright Field Ref Libr, 46-49; US mem doc comt, Adv Group Aeronaut Res & Develop, NATO, Paris, 52-57; McBee lectr, Simmons Col, 55; mem, Kresge-Hooker Sci Libr Assocs, Wayne State Univ, 57-65, pres, 64-65; vis lectr, Simmons Col, 65; mem exec bd, US Nat Comt Int Fedn Doc, 65-70, chmn, 70-72; vpres, Eng Index, Inc, 68-69, pres, 69-73, secy, 75-77; vis lectr, Kans State Teachers Col, 70; Lincoln lectr, Bus Sch, Ariz State Univ, 70; proj leader, Centralized Processing Med Libr Res Proj, 71-72; ed, Spec Librarianship, New Reader, 80. *Mem:* Am Inst Aeronaut & Astronaut; Spec Libr Asn (pres, 61-62); Am Soc Info Sci; Am Rec Mgt Asn. *Res:* Analysis and use of scientific literature by and for physical scientists; application of conventional and nonconventional information retrieval procedures to scientific information problems; industrial information systems; indexing and abstracting of physical science literature; records management systems. *Mailing Add:* Univ Tex Grad Sch Libr Info Sci Austin TX 78712-1276

JACKSON, FRANCIS CHARLES, b Rutherford, NJ, Sept 2, 17; m 49; c 4. MEDICINE. *Educ:* Yale Univ, AB, 39; Univ Va, MD, 43. *Prof Exp:* Intern surg, New York Hosp, Cornell Med Ctr, 44, asst resident surgeon, 44-45 & 47-49; asst anat, Med Col, Cornell Univ, 46, instr surg, 50; chief surgeon & consult, Arabian-Am Oil Co, 51; asst chief surgeon, Vet Admin Ctr, Maine, 52; from asst prof to prof surg, Sch Med, Univ Pittsburgh, 53-70; dir surg serv, 70-72, dir emergency & disaster med serv staff, Vet Admin Cent Off, 72-75; interim med dir, South Plains Emergency Med Serv Syst, 79-87; chmn dept surg, 75-80, assoc dean clin educ, 80-82, PROF SURG, SCH MED, TEX TECH UNIV, 75- *Concurrent Pos:* Asst, Med Col, Cornell Univ, 45-49; chief resident surgeon, New York Hosp, Cornell Med Ctr, 50; chief surgeon, Vet Admin Hosp, 52-70; consult staff, Presby Univ Hosp, Pittsburgh, 59-70; mem ad hoc adv group emergency health serv, Bur Health Serv, USPHS, 67-75; chief surv teams, USPHS, 68-69; mem review comn, Vet Admin Res & Educ Trainee Progs, 68-70; consult & mem med surv team, Westinghouse Corp Proj New Generation Mil Hosps, 69-70; consult, Off Emergency Planning, 69-71; co-med dir & consult, Ctr Western Pa Health Res & Develop, Carnegie-Mellon Univ, 69-72; clin prof surg, Sch Med, Georgetown Univ, 70-75; mem surg adv group, Food & Drug Admin, 71-76; mem & vchmn comt emergency med serv, Nat Acad Sci-Nat Res Coun; mem, President's Comn Study Med Aspects of Los Angeles Earthquake, 71; clin prof surg, Sch Med, George Washington Univ, 71-75; mem ad hoc comt emergency med serv commun, Off Telecommun Policy, 73-74; mem interdept comt emergency med serv, Dept Health, Educ & Welfare, 73-75; chief surgeon, Lubbock Gen Hosp, 75-80. *Honors & Awards:* Billings Gold Medal, AMA, 66; Stitt Award, Asn Mil Surgeons US, 68; Distinguished Serv Citation, US Dept Defense, 69; Physician's Recognition Award, AMA, 70, 73, 76, 79. *Mem:* Fel Am Col Surg; Am Med Asn; Soc Surg Alimentary Tract; Soc Surg Chmn; Am Surg Asn. *Res:* Portal hypertension; cirrhosis; esophageal varices; vascular surgery; schistosomiasis; chemotherapy as an adjuvant to surgery in the control of cancer; percutaneous splenoportography; spleen pressure; portal hemodynamics in Wilson's and other diseases; emergency medical services; trauma; telecommunications; disaster medical care; mass casualty care. *Mailing Add:* Dean's Suite Tex Tech Univ Sch Med Lubbock TX 79430

JACKSON, FRANCIS J, b Providence, RI, May 23, 32; m 56, 83; c 3. PHYSICS, UNDERWATER ACOUSTICS. *Educ:* Providence Col, BS, 54; Brown Univ, ScM, 57, PhD(physics), 60. *Prof Exp:* Res assoc physics, Brown Univ, 59-60; sr scientist, 60-70, vpres, Phys Sci Div, 70-75, vpres Underwater Technol Div, 75-77, CORP VPRES, BOLT BERANEK & NEWMAN, INC, 77-; sr vpres, BBN Labs, Inc, 79-87, SR VPRES, BBN SYSTS & TECHNOL CORP, 87- *Concurrent Pos:* Adj prof, Cath Univ Am, 68-74. *Mem:* Acoust Soc Am; Inst Elec & Electronics Engrs. *Res:* Nonlinear acoustics; underwater acoustics and sonar; underwater sound propagation. *Mailing Add:* Bolt Beranek & Newman Inc 10 Moulton St Cambridge MA 02138

JACKSON, GARY LESLIE, b Minneapolis, Minn, Jan 5, 45; m 70; c 1. PLASMA PHYSICS, ELECTROMAGNETICS. *Educ:* Univ Idaho, BS, 67; Calif State Univ, Northridge, MS, 72; Univ Ariz, PhD(elec eng), 77. *Prof Exp:* SR ENGR, GEN ATOMICS CO, 77- *Mem:* Inst Elec & Electronics Engrs; Am Phys Soc. *Res:* Fusion plasmas. *Mailing Add:* 13-563 GA Tech PO Box 85608 San Diego CA 92138

JACKSON, GARY LOUCKS, b Skidmore, Mo, Nov 4, 38; m 70; c 1. ENDOCRINOLOGY, REPRODUCTIVE PHYSIOLOGY. *Educ:* Univ Mo, BS, 60, AM, 63; Univ Ill, Urbana, PhD(animal sci), 67. *Prof Exp:* NIH fel, 67-68, assoc prof, 68-77, PROF ENDOCRINOL, UNIV ILL, URBANA, 77- *Concurrent Pos:* NIH reprod biol study sect, 83-87; vis prof, Univ Bristol, 85; mem, Animal Reproduction Panel, USDA, 89-91. *Mem:* Soc Study Reproduction; Endocrine Soc; Brit Soc Study Fertil; Neurosci Soc. *Res:* Hypothalamic control of gonadotropin and prolactin secretion; biosynthesis of luteinizing hormone; biological rhythms. *Mailing Add:* Dept Vet Physiol & Pharmacol Univ of Ill Col of Vet Med Urbana IL 61801

JACKSON, GEORGE FREDERICK, III, b Brooklyn, NY, May 16, 43; m 66; c 2. PHYSICAL CHEMISTRY, ANALYTICAL CHEMISTRY. *Educ:* MacMurray Col, BA, 65; Northwestern Univ, PhD(chem), 69. *Prof Exp:* Asst prof chem, Lake Forest Col, 69-73; assoc prof, 73-80, PROF CHEM, UNIV TAMPA, 80-, DIV CHMN, 82- *Mem:* AAAS; Am Chem Soc. *Res:* Studies of atomic inversion; involving arsenic and phosphorus atoms; properties of the allenic bond; use of NMR shift reagents for determination of molecular structures. *Mailing Add:* Dept of Chem Univ of Tampa 401 W Kennedy Blvd Tampa FL 33606

JACKSON, GEORGE G, b Provo, Utah, Oct 5, 20; m 43; c 5. INTERNAL MEDICINE, MICROBIOLOGY. *Educ:* Brigham Young Univ, AB, 42; Univ Utah, MD, 45. *Prof Exp:* Intern, II & IV Med Serv, Boston City Hosp, Harvard Univ, 45-46, asst resident med, Hosp & teaching fel, Med Sch, 48-49, res fel, Med Sch, 49-50, asst med, Hosp & res fel, Thorndike Mem Lab, 49-51, Milton fel, Med Sch, 50-51; from asst prof to prof, 51-79, attend physician, 51-87, chief sect infectious dis, 59-87, Keeton prof med, 79-87, EMER PROF, UNIV ILL, CHICAGO, 87- *Concurrent Pos:* Consult, West Side Vet Admin Hosp, Chicago; spec fel, Virus Res Dept, Trop Inst, Hamburg, Ger, 68-69; vis prof, London Hosp Med Col, 77-78 & Max von Pettenkofer Inst, Univ Munich, Ger, 78-79; Alexander von Humboldt US Sr Scientist Award, 78; ed, J Infectious Dis, 79-84. *Honors & Awards:* Maxwell Finland Award, Infectious Dis Soc Am, 77; Ernst Jung Prize Med, 84. *Mem:* Asn Am Physicians; Am Col Physicians; emer mem Am Epidemiol Soc; Infectious Dis Soc Am (secy-treas, 67-72, pres, 74); Am Soc Clin Invest. *Res:* Infectious diseases; antibiotics; virology & microbiology. *Mailing Add:* Col Med Univ Ill 1853 W Polk St Chicago IL 60612

JACKSON, GEORGE JOHN, b Vienna, Austria, Dec 10, 31; nat US. FOOD MICROBIOLOGY, IMMUNOLOGY. *Educ:* Univ Chicago, AB, 51, MS, 54, PhD, 58. *Prof Exp:* Fel, La State Univ, 58; res assoc & instr, Univ Chicago, 58-59; USPHS res fel & guest investr, Rockefeller Inst, 59-63; fac mem, Rockefeller Univ, 63-72; head lab parasitol, 72-80, chief food & cosmetics, microbiol br, 80-87, CHIEF FOOD MICROBIOL METHODS DEVELOP BR, US FOOD & DRUG ADMIN, 88- *Concurrent Pos:* Consult, Pathway Labs, 58-59; guest researcher, Amazon Res Inst, Manaus, Brazil, 63; adj prof, Rockefeller Univ & Lehigh Univ, 72-76; ed, Exp Parasitol, 76-87; US deleg, Food Hyg Comt of Codex Alimentarius Comn of WHO/Food & Agr Orgn of UN, 80-88; mem, microbiol adv comt, Calif State Univ, Riverside, 85-; FDA Food microbiol liasion with int orgn, 87- *Mem:* Am Soc Parasitologists; Soc Protozoologists; Am Soc Trop Med & Hyg; Asn Off Anal Chem. *Res:* Immunity; invertebrate physiology; axenic culture; food parasitology; anisakiasis. *Mailing Add:* HFF-234 US Food & Drug Admin 200 C St SW Washington DC 20204

JACKSON, GEORGE RICHARD, b Chagrin Falls, Ohio, Sept 27, 20; m 43; c 3. CHEMISTRY. *Educ:* Baldwin-Wallace Col, BS, 42; Johns Hopkins Univ, MA, 43, PhD(org chem), 48. *Prof Exp:* From instr to asst prof chem, Western Reserve Univ, 47-53; dir res, H C Fisher Co, 53-55; pres, Cliffdale Prod Corp, 55-57, Top-Scor Prod Inc, 57-61 & SuCrest Corp, 61-69; pres, Southern Shortenings, 69-77; CONSULT, 78- *Mem:* Am Chem Soc; Am Oil Chem Soc; Am Asn Cereal Chem; Inst Food Technol. *Res:* Food chemistry; fats and oils; analytical chemistry. *Mailing Add:* 2828 N Atlantic Ave Apt 1903 Daytona Beach FL 32018

JACKSON, GRANT D, b Perryton, Tex, Apr 3, 45; m 65; c 2. SOIL & SOIL SCIENCE, AGRONOMY. *Educ:* Okla Panhandle State Col, BS, 68; Mont State Univ, MS, 70, PhD(crop & soil science), 74. *Prof Exp:* Soil scientist, US Dept Agr, 74-77; agronomist crop & soil sci, Univ Wyo, 77-84; exten specialist cropping systs, 77-79, AGRONOMIST CROP, MONT STATE UNIV, 84- *Mem:* Am Soc Agron; Soil Sci Soc Am. *Res:* Development of zero-till technology for dry land and irrigating cropping system. *Mailing Add:* Mont Res Ctr Mont State Univ PO Box 1474 Conrad MT 59425

JACKSON, HAROLD, b Preston, Lancashire, Eng, Aug 10, 37; m 61; c 3. FOOD SCIENCE. *Educ:* Univ Nottingham, BSc, 59, MSc, 61; Univ Alta, PhD(dairy sci), 63. *Prof Exp:* Asst prof dairy & food microbiol, 63-69, assoc prof food sci, 69-74, prof food sci & chmn dept, 74-82, PROF FOOD SCI, UNIV ALTA, 82- *Mem:* Am Soc Microbiol; Inst Food Technologists; Can Inst Food Sci & Technol; Brit Soc Appl Bact; Int Asn Milk Food & Environ Sanitarians. *Res:* Food microbiology; effects of environmental stress on microbial growth and activity; food-borne pathogens. *Mailing Add:* Dept Food Sci Univ Alta Edmonton AB T6G 2P5 Can

JACKSON, HAROLD E, JR, b Pittsburgh, Pa, Jan 5, 33; m 58; c 3. NUCLEAR PHYSICS. *Educ:* Princeton Univ, AB, 54; Cornell Univ, PhD(physics), 60. *Prof Exp:* Sr physicist, 59-81, res proj dir, 81-84, SR PHYSICIST, ARGONNE NAT LAB, 85- *Concurrent Pos:* Mem nuclear cross sect adv comt, AEC, 67-, secy, 71-73; chmn, US Nuclear Data Comt, 74-75; mem, nuclear data comt, Nuclear Energy Agency, 74-77; mem bd dirs, Los Alamos Meson Physics Fac Users Group, 79-80, chmn, 81-82; mem nat adv bd, Southeastern Univ Res Asn, 83-; chmn, CEBAF Users Group, 88-89; assoc ed, Phys Rev C, 91- *Mem:* Am Phys Soc; AAAS. *Res:* High energy physics; photoproduction of mesons; nuclear physics; study of nuclear structure with thermal and resonant neutron interactions; photonuclear interactions; medium energy physics; pion interactions with complex nuclei. *Mailing Add:* Physics Div Argonne Nat Lab Argonne IL 60439

JACKSON, HAROLD LEONARD, b Wichita, Kans, Mar 13, 23; m 52; c 4. POLYMER CHEMISTRY, FLUORINE CHEMISTRY. *Educ:* Munic Univ Wichita, AB, 43, MS, 46; Univ Ill, PhD(chem), 49. *Prof Exp:* Res chemist, Cent Res Dept, E I du Pont de Nemours & Co, 49-59, Org Chem Dept, 59-65, res assoc, Org Chem Dept, 66-77, res assoc, Petrolchem Dept, 78-90, RES FEL, DUPONT CHEMICALS, E I DUPONT DE NEMOURS & CO, INC, 90- *Concurrent Pos:* Vis prof, Univ Kans, 62-63. *Mem:* AAAS; Am Chem Soc; Soc Petroleum Engrs; Fedn of Socs of Coating Technol. *Res:* Organic solvents; polymer solvency; polymer characterization; fluorocarbon solvents and polymers; organic synthesis. *Mailing Add:* 102 Stratton Dr Canterbury Hills Hockessin DE 19707

JACKSON, HAROLD WOODWORTH, b Lawrence, Mass, Mar 14, 28; m 50; c 4. ANALYTICAL CHEMISTRY. *Educ:* Univ NH, BS, 49; Univ Conn, MS, 51, PhD(biochem, dairy technol), 55. *Prof Exp:* Instr dairy technol, Univ Conn, 52-53; prod leader cheese, Kraft Foods Res Lab, 54-55, res chemist, 55-59, group leader chromatog, 59-60; group leader chromatog & infrared spectros, Res & Develop Div, Fundamental Chem Sect, Nat Dairy Prod Corp, 60-69; group leader chromatog & instrumental anal, 69-75, sr group leader, Kraftco Corp, 75-78, SR GROUP LEADER, BASIC FLAVOR CHEM RES

& DEVELOP DIV, KRAFT, INC, 78- *Mem:* Am Oil Chemist's Soc; Am Chem Soc. *Res:* Isolation and identification of natural flavor compounds; analytical methodology on fatty acid derivatives; gas chromatography; nuclear magnetic resonance; basic flavor chemistry; mass spectrometry. *Mailing Add:* 56 Sunny Shores Dr Ormond Beach FL 32074

JACKSON, HERBERT LEWIS, b Sawyer, Kans, May 9, 21; m 46; c 2. NUCLEAR PHYSICS. *Educ:* Univ Wis, PhD(physics), 52. *Prof Exp:* Asst, Phys Inst, Univ Basel, 52-54; asst prof physics, Univ Nebr, 54-60; from asst prof to assoc prof, 60-68, PROF RADIOL, UNIV IOWA, 68- *Mem:* Am Phys Soc; Health Physics Soc; Am Asn Physicists in Med; Sigma Xi. *Res:* Scattering of protons and neutrons from light elements; radiological physics. *Mailing Add:* Dept Radiol Univ Iowa 614 Greenwood Dr Iowa City IA 52240

JACKSON, HERBERT WILLIAM, b Durham, NH, Jan 5, 11; m 87; c 1. POLLUTION BIOLOGY. *Educ:* Dartmouth Col, AB, 34; Cornell Univ, PhD(zool), 39. *Prof Exp:* Asst zool, Cornell Univ, 36-39; from instr to assoc prof biol, Va Polytech Inst, 39-51, tobacco specialist, Agr Exten Serv, 45; biologist, Training Sect, Robert A Taft Sanit Eng Ctr, USPHS, 51-67; chief biologist, Nat Training Ctr, Water Pollution Control Admin, US Dept Interior, 67-69; chief biologist, Environ Protection Agency, 69-73; consult water pollution ecol, 73; RETIRED. *Concurrent Pos:* Collabr, US Dept Interior, 47-51; Rockefeller Found fel, Mex, 49-50; consult biol & pollution ecol, Pan Am Health Orgn, 74-77. *Honors & Awards:* Bronze Medal for Meritorious Serv, US Environ Protection Agency, 73. *Mem:* Am Soc Limnol & Oceanog; Am Pub Health Asn; Sigma Xi; Am Fisheries Soc. *Res:* Biological and engineering aspects of water supply and pollution control; pollution ecology; marine and freshwater; plankton analysis techniques. *Mailing Add:* Old Piscataqua Rd Durham NH 03824

JACKSON, IVOR MICHAEL DAVID, b Glasgow, Scotland, Apr 17, 36; US citizen; m 72; c 2. NEUROENDOCRINOLOGY, COMPARATIVE ENDOCRINOLOGY. *Educ:* Univ Glasgow Med Sch, MB & ChB, 60; FACP. *Hon Degrees:* MA, Brown Univ, 85. *Prof Exp:* Res fel neuroendocrinol, Univ Conn, 71-72; res fel endocrinal, Tufts Univ, 72-73, from asst prof med to prof, Sch Med, 73-84; PROF MED, BROWN UNIV, 84-, DIR DIV ENDOCRINOL, RI HOSP, 84- *Concurrent Pos:* Prin investr, Diabetes, Digestive & Kidney Dis, Nat Inst Arthritis, 78-81 & 81-; mem, Spec Rev Comt, NIH, 81, Salk Inst Biol Studies, La Jolla, 84; co-organizer & co-chmn, Fedn Am Soc for Experimental Biol, 80, Wkshop on Hormone Standard, NIH, 82, Int Conf, NY Acad Sci, MD, 87; vis prof med, Univ Toronto, Can, 84, Univ Ottowa, 86. *Mem:* Endocrine Soc; fel Am Col Physicians; Am Thyroid Asn; Soc for Neurosci; Am Soc Clin Invest; Asn Am Physicians. *Res:* To determine the physiologic, cellular and molecular mechanisms regulating the biosynthesis and post-translational processing of proTRH, a 255 amino acid polyprotein which contains 5 copies of a thyrotrophin releasing hormone (TRH) progenitor sequence flanked by paired basic residues and the secretion of TRH in the various regions of the neuroendocrine system where TRH and its prohormone occur. *Mailing Add:* Div Endocrinology RI Hosp 593 Eddy St Providence RI 02903

JACKSON, JAMES EDWARD, b Rochester, NY, Jan 12, 25; m 47; c 3. MATHEMATICAL STATISTICS. *Educ:* Univ Rochester, AB, 47; Univ NC, MA, 49; Va Polytech Inst, PhD, 60. *Prof Exp:* Statistician, Eastman Kodak Co, 48-57; asst process engr, Hercules Powder Co, 57-58; asst prof, Va Polytech Inst, 58-59; statistician, Kodak Park, Eastman Kodak Co, 59-85; PVT CONSULT, 85- *Honors & Awards:* Brumbaugh Award, Am Soc Qual Control, 78. *Mem:* Psychomet Soc; fel Am Soc Qual Control; fel Am Statist Asn. *Res:* Development of statistical methods, particularly multivariate analysis; quality control; psychometric methods and their associated computer techniques for use in industrial problems. *Mailing Add:* 66 Kettering Dr Rochester NY 14612

JACKSON, JAMES F, NUCLEAR REACTOR ANALYSIS. *Prof Exp:* DEPT DIR, LOS ALAMOS NAT LAB, 86- *Mem:* Nat Acad Eng. *Mailing Add:* Los Alamos Nat Lab PO Box 1663 Los Alamos NM 87545

JACKSON, JAMES OLIVER, b New Iberia, La, July 16, 39; m 64; c 1. MEDICAL MICROBIOLOGY, IMMUNOLOGY. *Educ:* Univ Southwestern La, BS, 66, MS, 67; Univ Kans, PhD(microbiol), 70. *Prof Exp:* Asst prof microbiol, Southern Univ, 70-72; From asst prof to assoc prof, 72-80, PROF MICROBIOL, CALIF STATE POLYTECH UNIV, POMONA, 80-, COORDR MICROBIOL SECT, BIOL SCI DEPT, 84- *Concurrent Pos:* Ad hoc consult, NIH, 70- *Mem:* Am Soc Microbiol. *Res:* Metabolic changes in experimental Listeria monocytogenes infections; extracellular proteins produced by Listeria monocytogenes; pathogenic mechanisms of Chromobacterium violecium. *Mailing Add:* Dept Biol Sci Calif State Polytech Univ Pomona CA 91768

JACKSON, JASPER ANDREW, JR, b Washington, DC, Jan 26, 23; m 61; c 1. EXPERIMENTAL PHYSICS. *Educ:* Univ Okla, BS, 48, MS, 50, PhD(physics), 55. *Prof Exp:* Asst, Univ Okla, 49-51; mem staff, Los Alamos Sci Lab, 51-52; sr nuclear engr, Convair Div, Gen Dynamics Corp, 55; mem staff, Los Alamos Nat Lab, 56-86; RETIRED. *Mem:* Fel AAAS; Am Phys Soc; Soc Magnetics Resonance Med. *Res:* Nuclear magnetic resonance; infrared and Raman spectroscopy; dynamic nuclear polarization; mass spectrometry. *Mailing Add:* 1600 Conestoga Dr SE Albuquerque NM 87123

JACKSON, JEREMY BRADFORD COOK, b Louisville, Ky, Nov 13, 42. MARINE ECOLOGY, PALEOBIOLOGY. *Educ:* George Washington Univ, AB, 65, MA, 67; Yale Univ, MPhil, 70, PhD(geol), 71. *Prof Exp:* Asst prof, 71-77, assoc prof marine ecol, 77-81, PROF ECOL, JOHNS HOPKINS UNIV, 81- *Concurrent Pos:* Mem biol oceanog panel, NSF, 75 & 76; vis prof, Discovery Bay Marine Lab, Univ West Indies, 76 & 78, Marine Biol Lab, Woods Hole, 79-; res assoc, Smithsonian Inst, 76-; prin investr, NSF grants, 71- *Mem:* Am Soc Limnol & Oceanog; Brit Ecol Soc; Ecol Soc Am; Int Bryozoology Asn; Int Soc Reef Studies; Sigma Xi. *Res:* Population biology and evolution of marine invertebrates; adaptive significance of form in sessile organisms; competitive theory. *Mailing Add:* Smithsonian Topical Res Inst APO Miami Miami FL 34002-0011

JACKSON, JEROME ALAN, b Ft Benning, Ga, Feb 4, 43; m 65, 84; c 6. ORNITHOLOGY, ECOLOGY. *Educ:* Iowa State Univ, BS, 65; Univ Kans, PhD(zool), 70. *Prof Exp:* From asst prof to assoc prof, 70-79, PROF ZOOL, MISS STATE UNIV, 79- *Concurrent Pos:* Ed, Wilson Bull, 74-78, The Mississippi Kite, 76-, Inland Bird Banding, 78-81, J Field Ornithol, 80-85 & N Am Bird Bander, 81-83; regional ed, Am Birds, 78-; pres, Eco-Inventory Studies, Inc, 78-; leader endangered species recovery team, US Fish & Wildlife Serv, 75-82. *Honors & Awards:* Recognition for Contributions to Bird Conserv, Prov Assembly of Popular Power, Havana, Cuba. *Mem:* Fel Am Ornith Union; Wilson Ornith Soc (treas, 73-74, vpres, 79-, pres, 83-85); Sigma Xi; Cooper Ornith Soc; Am Soc Mammalogists; Herpetologists League; Asn Field Ornithologists; fel Explorer's Club. *Res:* Population dynamics and adaptation in hole-nesting birds; biology of endangered species; behavior of rat snakes, Elaphe. *Mailing Add:* Dept Biol Sci Miss State Univ Miss State MS 39762

JACKSON, JO-ANNE ALICE, b Washington, DC, July 30, 51. PHYSICAL CHEMISTRY. *Educ:* Am Univ, BS, 73, PhD(phys chem), 77. *Prof Exp:* Teaching asst gen, quant anal & phys chem, Lab Nursing Gen Chem, Am Univ, 73-75; chemist, Nat Bur Standards, 74-81; specialist, chem & mat export licensing, Int Trade Admin, 81-86, PHYS SCI FOREIGN TECH ANALYST, BUR EXPORT ADMIN, US DEPT COM, WASHINGTON, DC, 86- *Mem:* Am Chem Soc; Soc Appl Spectros; Fedn Orgs Prof Women. *Mailing Add:* 14711 Myer Terr Rockville MD 20853-2242

JACKSON, JOHN DAVID, b London, Ont, Jan 19, 25; nat US; m 49; c 4. THEORETICAL PHYSICS. *Educ:* Univ Western Ont, BSc, 46; Mass Inst Technol, PhD(physics), 49. *Hon Degrees:* DSc, Univ Western Ont, 89. *Prof Exp:* Res assoc physics, Mass Inst Technol, 49; from asst prof to assoc prof math, McGill Univ, 50-56; from assoc prof to prof physics, Univ Ill, 57-67; group leader theoret physics, Berkeley Lab, 74-78, chmn dept, 78-81, assoc dir & head physics div, 82-84, dep dir, SSC Cent Design Group, 85-87, PROF PHYSICS, UNIV CALIF, BERKELEY, 67- *Concurrent Pos:* Guggenheim fel, Princeton Univ, 56-57; consult, Argonne Nat Lab, 62-65, 75-78; Ford Found fel, Europ Orgn Nuclear Res, 63-64; assoc ed, Rev of Modern Physics, 68-72; vis fel, Clare Hall, Cambridge Univ, 70; consult, Stanford Linear Accelerator Ctr, 71-73 & Nat Accelerator Lab, 71-75; mem vis comt, Dept Physics, Mass Inst Technol, 73-76; ed, Annual Rev Nuclear & Particle Sci, 77-; chmn, Vis Comt for Fermilab, Univs Res Asn, 80-82; sr vis res fel, Jesus Col, Oxford, 88-89; sci assoc, Europ Orgn Nuclear Res, 76-77, 81, 84 & 88; mem, Prog Adv Comt, SSC Lab, 90- *Mem:* Nat Acad Sci; fel AAAS; Fel Am Phys Soc. *Res:* Theoretical physics of fundamental particles. *Mailing Add:* Dept Physics Univ Calif Berkeley CA 94720

JACKSON, JOHN EDWARD, agricultural biochemistry; deceased, see previous edition for last biography

JACKSON, JOHN ELWIN, JR, b Tuscaloosa, Ala, Oct 11, 48; m 71; c 3. FLUID-STRUCTURE INTERACTION, ANALYSIS OF METAL FLOW DURING FORMING. *Educ:* Univ Ala, BS, 71, MS, 73, PhD(eng mech), 77. *Prof Exp:* Mech engr, Tennessee Valley Authority, 76-78; asst prof, 78-82, ASSOC PROF ENG MECH, CLEMSON UNIV, 82- *Concurrent Pos:* Consult, finite element anal. *Mem:* Am Soc Mech Engrs; Am Acad Mech; Sigma Xi; Am Soc Metals. *Res:* Fluid-structure interaction including fluids in reactor containment vessels and underwater explosions on structures; improved design for nuclear power plant cable-tray hangers; finite element analysis of railroad track-structures; finite element analysis of nonlinear systems; finite element analysis of metal forming process. *Mailing Add:* 118 Valley View Dr Clemson SC 29631

JACKSON, JOHN ERIC, b Cincinnati, Ohio, July 25, 37; m 60; c 1. CHEMICAL ENGINEERING. *Educ:* Purdue Univ, BS, 58; Univ Mich, MSE, 59. *Prof Exp:* Res engr, Speedway Labs, Union Carbide Corp, 59-63, group leader welding & lasers, 63-66, suprv, 66-68, DIR TECHNOL, UNION CARBIDE COATING SERV CORP, 68- *Mem:* Am Inst Chem Engrs; Am Welding Soc. *Res:* Directing development of high performance materials for aircraft, nuclear, petroleum and related energy industries; wear and corrosion prevention; thermal barriers; composites; fossil fuel production and utilization; oil and gas extraction. *Mailing Add:* 60 Carnary Brownsburg IN 46112

JACKSON, JOHN FENWICK, b Kosciusko, Miss, Nov 19, 28; m 54; c 3. INTERNAL MEDICINE, GENETICS. *Educ:* Univ Miss, BA, 50; Tulane Univ, MD, 53; Am Bd Internal Med, dipl, 65; Am Bd Med Genetics, Clin Genetics, Clin Cytogenetics, dipl, 82. *Prof Exp:* Intern, Philadelphia Gen Hosp, 53-54; gen pract, Minter City, Miss, 54-56; resident res physician internal med, Med Ctr, Univ Miss, 58-60, chief resident physician, 60; from instr to asst prof internal med, Tulane Univ, 61-64; asst prof internal med & assoc prof prev med, 64-67, prof internal med, 80, chmn, 81, PROF PREV MED & ASSOC PROF INTERNAL MED, DEPT PREVENTIVE MED, SCH MED, UNIV MISS, 67- *Concurrent Pos:* Res fel cancer, Univ Miss, 60-61; trainee, Inst Med Genetics, Univ Uppsala, 62-63; vis physician, Charity Hosp La, New Orleans, 63-64; attend physician, Univ Miss Hosp, Jackson, 64-; consult, Vet Admin Hosp, Jackson, 64-; vis investr, Pop Genetics Lab, Univ Hawaii, 70-71. *Mem:* Am Fedn Clin; Sigma Xi; Am Soc Hemat; Soc Human Genetics. *Res:* Hematology; medical genetics; cytogenetic investigations in human disease; human linkage studies. *Mailing Add:* Dept Preventive Med Univ Miss Med Ctr Jackson MS 39216

JACKSON, JOHN MATHEWS, b Chicago, Ill, July 9, 08; m 31; c 7. FOOD SCIENCE. *Educ:* Univ Chicago, BS, 29, PhD(chem), 32. *Prof Exp:* Chemist, Thermal Eng Group, Am Can Co, Ill, 32-37, supvr, 38-41, asst chief packaging res, 42-49, asst mgr res, Pac Div, 49-51, mgr, Res Div Lab, 52-55, Ill, 55-57, sect mgr, 57-63; dir res, Green Giant Co, 63-67, dir packaging res, 68-73; CONSULT, 73- *Concurrent Pos:* Mem, Comt Foods, Subcomt Radiation Sterilization, Nat Res Coun, 55-57; pres, Res & Develop Assocs Food & Container Inst, 62-63; mem, Sci Res Comt, Nat Canners Asn, 65-72; mem, Comt Microbiol Food, Adv Bd Mil Personnel Supplies, Nat Acad Sci,

65-68, Container Comt, 68-71; consult, Process Rev Comt, Bur Foods, Food & Drug Admin, 74- *Mem:* Am Chem Soc; fel Inst Food Technologists (pres, 62-63). *Res:* Decomposition of organic compounds in electrical discharges; heat penetration in canned foods; sterilization of canned foods; packaging of frozen foods; aerosol packaging. *Mailing Add:* PO Box 87 Lakeside MI 49116

JACKSON, JOHN STERLING, b Bowling Green, Ky, Nov 3, 22; m 49; c 5. ECONOMIC SYSTEM MODELING. *Educ:* Calif Inst Technol, BS, 46, MS, 54. *Prof Exp:* Asst prof, 56-59, ASSOC PROF ELEC ENG, UNIV KY, 59- *Concurrent Pos:* Ed of proceedings, Carnahan Conf Crime Countermeasure, 67-, Carnahan Conf Electronic Prosthetics, 72-73 & Int Conf Crime Countermeasure, 73-, ed, Carnahan Conf on Harmonizing Technol with Soc, 85; consult, 77- *Honors & Awards:* Centennial Medal, IEEE. *Mem:* Inst Elec & Electronics Engrs; Am Soc Eng Educ; AAAS; Sigma Xi. *Res:* Social implications of technology; crime countermeasures. *Mailing Add:* 2176 Lakeside Dr Lexington KY 40502

JACKSON, JULIUS, b New York, NY, Apr 20, 16; m 41; c 1. INORGANIC CHEMISTRY. *Educ:* Polytech Inst Brooklyn, BS, 37, MS, 39, PhD(inorg chem), 41. *Prof Exp:* Res chemist, Otto H Henry, Brooklyn, 41-42; chemist-metallurgist, Nat Prod Refining Co, NJ, 42-44; res chemist, E I Du Pont De Nemours & Co, Inc, 44-65, sr res chemist, 65-66, res assoc, 66-80; CONSULT PIGMENTS, 80- *Concurrent Pos:* Eve instr, Polytech Inst Brooklyn, 42-47 & Brooklyn Col, 47. *Mem:* Am Chem Soc. *Res:* Inorganic chemistry; metallurgy of chromium alloys; nonaqueous solvents; chromium chemicals; selenium oxychloro compounds of pyridine and related compounds; inorganic pigments; phthalocyanine pigments; quinacridone pigments; flame retardants; light stabilizers. *Mailing Add:* 224 Charles St Westfield NJ 07090

JACKSON, KENNETH ARTHUR, materials science, for more information see previous edition

JACKSON, KENNETH LEE, b Berkeley, Calif, Jan 6, 26; m 48; c 4. PHYSIOLOGY. *Educ:* Univ Calif, Berkeley, AB, 49, PhD(physiol), 54. *Prof Exp:* Res asst physiol, Donner Lab, Univ Calif, Berkeley, 49-51; res physiologist, Off Naval Res Unit One, 51-53; sr investr, Biochem Br, US Naval Radiol Defense Lab, 54-60; head radiobiol group, Bioastronaut Sect, Boeing Co, 60-63; from asst prof to prof environ health, Univ Wash, 63-91, head div, 76-91, CHMN, RADIOL SCI GROUP, UNIV WASH, 67-, EMER PROF, ENVIRON HEALTH, 91- *Concurrent Pos:* Consult, Fred Hutchinson Cancer Res Ctr, 80-, DOE Pac Northwest Lab, 85-87; vis scientist, Inst Exp Gerontol, TNO, Rijswijk, Netherlands, 77. *Honors & Awards:* Sigma Xi. *Mem:* AAAS; Am Physiol Soc; Radiation Res Soc; Health Physics Soc. *Res:* Biochemical and physiological mechanisms in mammalian radiation biology; cell and intestinal physiology. *Mailing Add:* Radiol Sci SB-75 Univ Wash Seattle WA 98105

JACKSON, KENNETH RONALD, b Montreal, Que, Mar 3, 51; m 83. ORDINARY DIFFERENTIAL EQUATIONS, NUMERICAL LINEAR ALGEBRA. *Educ:* Univ Toronto, BSc, 73, MSc, 74, PhD(computer sci), 78. *Prof Exp:* Gibbs instr computer sci, Yale Univ, 78-81; asst prof, 81-86, assoc chair, 87-89, ASSOC PROF COMPUTER SCI, UNIV TORONTO, 86- *Mem:* Soc Indust & Appl Math; Asn Comput Mach; Sigma Xi. *Res:* Numerical analysis; mathematical software; scientific computing; numerical solution of ordinary differential equations. *Mailing Add:* Computer Sci Dept Univ Toronto Toronto ON M5S 1A4 Can

JACKSON, KERN CHANDLER, b Kansas City, Mo, Oct 13, 20; m 70; c 4. GEOLOGY. *Educ:* Mich Technol Univ, BS, 47, MS, 50; Univ Wis, PhD(geol), 51. *Prof Exp:* From instr to asst prof geol, Univ Maine, 50-52; from asst prof to assoc prof, 52-61, chmn dept, 55-59, PROF GEOL, UNIV ARK, FAYETTEVILLE, 61-, EMER PROF; RETIRED. *Mem:* AAAS; Am Mineral Soc; Geol Soc Am. *Res:* Petrology and petrography; petrography of Arkansas syenites and lamprophyres. *Mailing Add:* 235 Baxter Lane Fayetteville AR 72701

JACKSON, LARRY LAVERN, b Charlotte, Mich, June 2, 42; m 63; c 2. VETERINARY ANESTHESIOLOGY. *Educ:* Mich State Univ, BS, 64, DVM, 66; Iowa State Univ, MS, 71. *Prof Exp:* Vet, Miller Animal Clin, Lansing, Mich, 66-68; instr, 68-71, ASSOC PROF VET MED, IOWA STATE UNIV, 71- *Mem:* Am Vet Med Asn; Am Soc Vet Anesthesiol. *Res:* Cardio-pulmonary function of the equine and bovine species while under the influence of various anesthetic agents. *Mailing Add:* RR 4 Ames IA 50010

JACKSON, LARRY LEE, b Livingston, Mont, Oct 8, 40; m 62; c 4. BIOCHEMISTRY. *Educ:* Mont State Univ, BS, 62; NDak State Univ, PhD(biochem), 65. *Prof Exp:* From asst prof to assoc prof, 65-75, PROF CHEM, MONT STATE UNIV, 75- *Mem:* Int Soc Chem Ecol. *Res:* Lipid chemistry and biochemistry; surface lipids of insects and plants; hydrocarbon biosynthesis; microbial cell wall chemistry; insect pheromone biochemistry; insect sterols. *Mailing Add:* Dept Chem & Biochem Mont State Univ Bozeman MT 59717-0310

JACKSON, LARRY LYNN, b Defiance, Ohio, Aug 21, 40; m 62; c 3. POLYMER CHARACTERIZATION. *Educ:* Va Mil Inst, BS, 62; Ohio State Univ, PhD(org chem), 67. *Prof Exp:* RES MGR, DOW CHEM CO, 69- *Honors & Awards:* John C Vaaler, Chem Processing Mag, 72 & 78; IR-100, Indust Res Mag, 76. *Mem:* Am Chem Soc; Sigma Xi. *Res:* New product and process research with emphasis on advanced analytical instrumentation and technology. *Mailing Add:* 1320 Waldow Rd Midland MI 48640

JACKSON, LELAND BROOKS, b Atlanta, Ga, July 23, 40; m 68; c 1. ELECTRICAL ENGINEERING, ELECTRONICS ENGINEERING. *Educ:* Mass Inst Technol, SB & SM, 63; Stevens Inst Technol, ScD, 70. *Prof Exp:* Res eng radar, Sylvania Electronic Systs, 64-66; mem tech staff digital filters, Bell Tel Labs, 66-70; vpres instruments, Rockland Systs Corp, 70-74; assoc prof, 74-79, PROF ELEC ENG, UNIV RI, 79- *Concurrent Pos:* Res grants, NSF & Air Force Off Sci Res, 76, Off Naval Res, 77 & NSF, 83-

Honors & Awards: Tech Achievement Award, Acoust Speech & Signal Processing Soc. *Mem:* Fel Inst Elec & Electronics Engrs. *Res:* Optimum synthesis of digital filter structures; estimation of signal parameters and its application to speech analysis, sonar and bioengineering. *Mailing Add:* Dept Elec Eng Kelley Hall Univ RI Kingston RI 02881

JACKSON, LIONEL ERIC, JR, b San Mateo, Calif, Jan 20, 47; Can citizen; m 69. QUATERNARY GEOLOGY, ENVIRONMENTAL GEOLOGY. *Educ:* San Francisco State Univ, AB, 68; Stanford Univ, MS, 73; Univ Calgary, PhD(geol), 77. *Prof Exp:* Geologist, Hudson Bay Oil & Gas Ltd, 69-70; hydrologist, US Geol Surv, 71-73; GEOLOGIST, GEOL SURV CAN, 77- *Concurrent Pos:* Consult, Calif Dept Transp, 79; instr, Univ Calgary, 81- *Mem:* Fel Geol Asn Can; Am Quaternary Asn; Can Soc Petrol Geologists. *Res:* Quaternary geology and paleo-ecology of western interior plains and Rocky Mountain Foothills; natural hazards in northern Montane regions; quaternary geology of Yukon Territory. *Mailing Add:* 972 Porter St Coquitlam BC V3J 5C2 Can

JACKSON, LLOYD K, b Fairbury, Nebr, Aug 25, 22; m 43; c 1. MATHEMATICS. *Educ:* Univ Nebr, AB, 43, MA, 48; Univ Calif, Los Angeles, PhD, 50. *Prof Exp:* From asst prof to assoc prof, PROF MATH, UNIV NEBR, LINCOLN, 59- *Mem:* Am Math Soc; Math Asn Am; Sigma Xi. *Res:* Partial differential equations; function theory. *Mailing Add:* 4550 E Eden Dr Lincoln NE 68506

JACKSON, M(ELBOURNE) L(ESLIE), b Wisdom, Mont, Sept 27, 15; m 44; c 4. ENVIRONMENTAL ENGINEERING. *Educ:* Mont State Col, BS, 41; Univ Minn, PhD(chem eng), 48. *Hon Degrees:* Dr, Montana State Univ, 80. *Prof Exp:* Instr chem eng, Mont State Col, 42-44 & Univ Minn, 44-48; asst prof, Univ Colo, 48-50; head process develop br, US Naval Ord Test Sta, Calif, 50-53; prof chem eng & head dept, 53-65, dean, Grad Sch, 65-70, res prof chem eng, 71-80, dean eng, 73, 78-80 & 83, EMER PROF, UNIV IDAHO, 80- *Concurrent Pos:* Consult, indust firms, 53-; E L Phillips intern, univ admin, Pa State Univ, 63-64. *Mem:* Fel Am Inst Chem Engrs; Am Chem Soc. *Res:* Bioreactors for fermentation and waste water treatment. *Mailing Add:* Dept Chem Eng Univ Idaho Moscow ID 83843

JACKSON, MARGARET E, b Zanesville, Ohio, Sept 2, 28. INORGANIC CHEMISTRY, PHYSICAL CHEMISTRY. *Educ:* Muskingum Col, BS, 50; Auburn Univ, MS, 58, PhD(phys chem), 64. *Prof Exp:* Chemist, Dow Chem Co, Mich, 51-52; Atlantic Supply Co, Md, 53 & Calverton Chem Co, Md, 53-56; asst chem, Auburn Univ, 56-62, instr, 62-64; asst prof, Delta State Col, 64-65; asst prof chem, 65-77, ASSOC PROF CHEM, UNIV SALA, 77- *Mem:* Am Chem Soc; The Chem Soc. *Res:* The preparation and properties of organo-niobium compounds; determination of stability constants for metal diketo chelates and other coordination compounds. *Mailing Add:* Dept of Chem Univ of SAla Mobile AL 36608

JACKSON, MARION LEROY, b Reynolds, Nebr, Nov 30, 14; m 37; c 4. SOIL SCIENCE, TRACE ELEMENTS. *Educ:* Univ Nebr, BS, 36, MS, 37; Univ Wis, PhD(soil chem), 39. *Hon Degrees:* DSc, Univ Nebr, 74. *Prof Exp:* Land classification aide, Resettlement Admin, USDA, Nebr, 36-37; Alumni Res Found & univ fel, Univ Wis, 39-41, from instr to asst prof soils, 41-45; assoc prof agron, Exp Sta, Purdue Univ, 45-46; from assoc prof to prof soils, 46-74, chmn phys sci div, 52-55, FRANKLIN HIRAM KING PROF SOILS, UNIV WIS-MADISON, 74- *Concurrent Pos:* Vis prof, Cornell Univ, 59, Univ Calif, 60, Univ Wis, Egypt, 61, Va Polytech Inst, 62, Can Dept Agr, 64, Univ Thesaloniki, 66, Univ Tex, 67, Kyushu Univ, 69, Ohio Univ, 71 & distinguished vis prof, Univ Wash, 73; Nat Res Coun panel radioactive wastes, 76-77. *Honors & Awards:* Soil Sci Achievement Award, Am Soc Agron, 58; Career Award, Soil Sci Soc Am, 83. *Mem:* Nat Acad Sci; Clay Minerals Soc (vpres, 65, pres, 66); fel Soil Sci Soc Am (vpres, 67, pres, 68); fel Am Soc Agron; fel Am Mineral Soc; fel AAAS. *Res:* Crystal chemistry of soil colloids responsible for soil acidity; cation exchange, phosphate fixation and other chemical reactions; electron microscopy of soil minerals; world distribution of radioactive elements in global dust; soil influence on trace elements in the food chain and human health. *Mailing Add:* 309 Ozark Trail Madison WI 53705

JACKSON, MARION T, b Versailles, Ind, Aug 19, 33. PLANT ECOLOGY. *Educ:* Purdue Univ, BS, 61, PhD(plant ecol), 64. *Prof Exp:* From asst prof to assoc prof, 64-71, PROF LIFE SCI, IND STATE UNIV, TERRE HAUTE, 71- *Honors & Awards:* Oak Leaf Award, The Nature Conservancy. *Mem:* AAAS; Sigma Xi; Ecol Soc Am. *Res:* Forest ecology of Midwest; flowering phenology; biotic inventories of natural areas and national parks; ecological life histories; regional plant geography. *Mailing Add:* Dept of Life Sci Ind State Univ Terre Haute IN 47809

JACKSON, MARTIN PATRICK ARDEN, b Salisbury, Rhodesia, May 14, 47; Irish citizen; m 69; c 2. SEDIMENTARY TECTONICS, TECTONIC MODELING. *Educ:* Univ London, BSc, 68, BSc, 69; Univ Cape Town, PhD(geol), 76. *Prof Exp:* Proj geologist, Cominco, SAfrica, 70-72; res assoc, Precambrian Res Unit, Univ Cape Town, 72-76; lectr, geol dept, Univ Natal, 76-79; res sci assoc, 80-81, res scientist, 81-87, SR RES SCIENTIST, BUR ECON GEOL, UNIV TEX, AUSTIN, 87-, DIR, APPL GEODYNAMICS LAB, 88- *Concurrent Pos:* Prin investr, Dept Energy, Tex Bur Econ Geol, 80-; vis scientist, Univ Uppsala, Sweden, 84; mem, Int Union Geol Sci Comn Tectonics, 87- *Honors & Awards:* Sproule Award, Am Asn Petrol Geologists, 85, Matson Award, 90. *Mem:* Fel Geol Soc Am; Am Asn Petrol Geologists. *Res:* Diapirism and halokinesis; geodynamic centrifuge modeling; petroleum structural traps tectonic evolution of sedimentary basins; syndepositional deformation; geology of central Iran strain analysis; structural analysis of Precambrian gneisses and greenstone belts; metamorphic petrology and geochemistry; crustal evolution of Southern Africa. *Mailing Add:* Bur Econ Geol Univ Tex Austin Univ Station Box X Austin TX 78713-7508

JACKSON, MARVIN ALEXANDER, b Dawson, Ga, Oct 28, 27; m 57; c 2. MEDICINE, PATHOLOGY. *Educ:* Morehouse Col, BS, 47; Meharry Med Col, MD, 51; Univ Mich, MA, 56. *Prof Exp:* Intern, US Naval Hosp, St Albans, NY, 51-52, mem staff, 52-53; asst resident, Univ Mich Hosp, 53-54, teaching asst path, 54-55, instr, 55-56; from asst prof to assoc prof, 57-68, PROF PATH, COL MED, HOWARD UNIV, 68-, CHMN DEPT, 60- *Concurrent Pos:* Resident, Hosp Joint Dis, New York, 56-57; attend physician, Howard Univ Hosp, 57; consult physician, Vet Admin Hosp, Washington, DC, 60 & NIH, 68. *Mem:* Fel Col Am Path; Am Asn Anatomists; Am Asn Path & Bact; Int Acad Path; Tissue Cult Asn; Sigma Xi. *Mailing Add:* 1428 Iris St NW Washington DC 20012

JACKSON, MATTHEW PAUL, b Kansas City, Mo, July 17, 59. BACTERIAL PATHOGENESIS, MICROBIAL TOXINS. *Educ:* Univ Mo, BS, 80, MS, 82; Kans State Univ, PhD(microbiol), 85. *Prof Exp:* Postdoctoral microbiol, Uniformed Serv Univ Health Sci, 85- 89; ASST PROF MICROBIOL, SCH MED, WAYNE STATE UNIV, 89- *Mem:* Am Soc Microbiol; AAAS. *Res:* Genetic regulation and structure-function analyses of bacterial toxins; pathogenic mechanisms of diarrheagenic Escherichia coli with emphasis on the Shiga toxin family of cytotoxins. *Mailing Add:* Dept Immunol & Microbiol Med Sch Wayne State Univ 540 E Canfield Detroit MI 48201

JACKSON, MEL CLINTON, b Rochdale, Eng, Nov 3, 62; m 87; c 1. PROTEIN PURIFICATION, ENZYMOLOGY. *Educ:* Univ Keele, Staffordshire, Eng, BSc, 84, DPhil, 88. *Prof Exp:* POSTDOCTORAL RES FEL BIOCHEM, DEPT PHARMACOL, UNIV HAWAII, 88- *Mem:* MRNA and DNA sequencing. *Res:* Enzymology research on mammalian small intestinal peptidases and a human plasma dipeptidase; enzyme purification; structural and kinetic characterization; substrate specificity studies; immunology; immunohistochemical localization; molecular cloning. *Mailing Add:* Dept Pharmacol John A Burns Sch Med Univ Hawaii Leahi Hosp 3675 Kilauea Ave Honolulu HI 96816

JACKSON, MELVIN ROBERT, b Norwood, Pa, Nov 21, 43; m 68; c 1. METALLURGY, MATERIALS SCIENCE. *Educ:* Lehigh Univ, BS, 65, MS, 67, PhD(metall & mat sci), 71. *Prof Exp:* Metallurgist, Paul D Merios Res Labs, Int Nickel, 71-72; METALLURGIST, CORP RES & DEVELOP, GEN ELEC CO, 72- *Mem:* Am Inst Mining, Metall & Petrol Engrs; Sigma Xi. *Res:* Phase equilibria of high temperature metal and metal-ceramic systems, metallurgical coatings, tool materials. *Mailing Add:* 2208 Niskayuna Dr Schenectady NY 12309

JACKSON, MEYER B, b Iowa City, Iowa, Mar 24, 51; m 80; c 1. NEUROSCIENCE, DEVELOPMENTAL NEUROBIOLOGY. *Educ:* Brandeis Univ, BA, 73; Yale Univ, PhD(molecular biophys & biochem), 77. *Prof Exp:* NIH fel, 78-81; asst prof, 81-87, ASSOC PROF BIOL, UNIV CALIF, LOS ANGELES, 87- *Mem:* Biophys Soc; Soc Neurosci. *Res:* Physical and chemical factors involved in synaptic transmission and excitability. *Mailing Add:* Dept Biol Univ Calif 405 Hilgard Ave Los Angeles CA 90024

JACKSON, MICHAEL J, b Walton, Eng, Apr 12, 38; m 60. EPITHELIAL PHYSIOLOGY, MEMBRANE TRANSPORT. *Educ:* Univ London, BSc, 63; Univ Sheffield, PhD(physiol), 66. *Prof Exp:* Asst exp officer biochem, Vet Lab, Ministry Agr, Eng, 57-63; res asst physiol, Univ Sheffield, 63-65, from asst lectr to lectr, 65-67; from asst prof to prof physiol, George Washington Univ, 67-90, assoc dean Res & Sponsor Prog, Med Ctr, 85-89, dean res, 89-90; EXEC DIR, FEDN AM SOC EXP BIOL, 90- *Concurrent Pos:* USPHS Res Career Develop Award, 72-77; guest worker, Sect Gastroenterol, Nat Inst Arthritus, Metabolism & Digestive Dis, 75-76; consult, Vet Admin Merit Rev Bd Basic Sci, 78-80; assoc ed, Am J Physiol, Gastrointestinal & Liver Physics, 79-85. *Mem:* Brit Physiol Soc; Am Gastroenterol Asn; Am Physiol Soc; Nat Coun Univ Res Admin; Soc Res Admin. *Res:* Epithelial transport of drugs and electrolytes; gastro-intestinal physiology. *Mailing Add:* Exec Off Fedn Am Soc Exp Biol 9650 Rockville Pike Bethesda MD 20814

JACKSON, NOEL, b Northallerton, Eng, Dec 25, 31; m 56; c 2. PLANT PATHOLOGY. *Educ:* Univ Durham, BSc, 53, Hons, 54, PhD, 61. *Prof Exp:* Biologist, Sports Turf Res Inst, Eng, 58-65; from asst prof to assoc prof, 65-75, PROF PLANT PATH, UNIV RI, 75-, PROF ENTOM, 77- *Mem:* Brit Asn Appl Biol; Am Phytopath Soc; Trans Brit Mycol Soc; Brit Soc Plant Path; Int Turfgrass Soc. *Res:* Diseases of turf grasses and ornamentals. *Mailing Add:* Dept Plant Path-Entom 234 Woodward Hall Univ RI Kingston RI 02881

JACKSON, PETER RICHARD, b New York, NY, Dec 9, 48; m 76. PARASITOLOGY. *Educ:* Seton Hall Univ, BA, 70; Univ Tex, Austin, MA, 73; Rice Univ, PhD(biol), 76. *Prof Exp:* NIH assoc parasitol dept biol, Rice Univ, 76 & dept zool, Univ Mass, 76-78; Nat Res Coun assoc immunol, Walter Reed Army Inst Res, 78-87; res mgr, Am Inst Biol Sci, Washington, DC, 87-90; SCI REV OFFICER, NAT INST ALLERGY & INFECTIOUS DIS, NIH, 90- *Mem:* Am Soc Parasitologists; Soc Protozoologists; AAAS; Am Inst Biol Sci; Am Soc Zoologist; Sigma Xi. *Res:* Immunology of parasitic infections; biochemistry and physiology of parasites, especially parasitic protozoa. *Mailing Add:* NIAID/NIH Rm 3A07 Westwood Bldg 5333 Westbard Ave Bethesda MD 20892

JACKSON, PHILIP LARKIN, geophysics, geology; deceased, see previous edition for last biography

JACKSON, PRINCE A, JR, b Savannah, Ga, Mar 17, 25; m 50; c 4. PHYSICS, MATHEMATICS. *Educ:* Savannah State Col, BS, 49; NY Univ, MS, 50; Boston Col, PhD(sci, ed), 66. *Prof Exp:* Instr high sch, Ga, 50-55; instr physics, Savannah State Col, 55-61, asst prof math & physics, 61-64, assoc prof phys sci, 66-69, prof math & physics, 71-80, pres, 71-80. *Concurrent Pos:* Chmn div natural sci, Savannah State Col, 69-71. *Mem:* AAAS; Nat Sci Teachers Asn; Nat Inst Sci. *Res:* Improvement of science and mathematics education at all levels of instruction; differential equations of mathematical physics; pedagogical interrelationship between science and mathematics; philosophy of science. *Mailing Add:* Dept Math & Physics Savannah State Col Savannah GA 31404

JACKSON, RAY DEAN, b Shoshone, Idaho, Sept 28, 29; m 52, 68; c 7. SOIL PHYSICS. *Educ:* Utah State Univ, BS, 56; Iowa State Univ, MS, 57; Colo State Univ, PhD(soil sci), 60. *Prof Exp:* Soil scientist soil physics, Soil & Water Conserv Res Div, 57-60, soil scientist, US Water Conserv Lab, 60-62, PHYSICIST, US WATER CONSERV LAB, AGR RES SERV, USDA, 62- *Concurrent Pos:* Orgn Econ Coop & Develop Sci fel, Eng, 64; adj prof soil & water sci, Univ Ariz, Tuscon. *Honors & Awards:* Superior Serv Unit Award, USDA, 63. *Mem:* Fel AAAS; fel Soil Sci Soc Am; fel Am Soc Agron; Am Geophys Union. *Res:* Heat, water and water vapor transfer in soils; remote sensing of soil water and crop stress; agricultural remote sensing. *Mailing Add:* US Water Conserv Lab 4331 E Broadway Phoenix AZ 85040

JACKSON, RAY WELDON, b Toronto, Ont, Nov 11, 21; m 51; c 4. PHYSICS, SCIENCE POLICY. *Educ:* Univ Toronto, BASc, 44; McGill Univ, PhD(physics), 50. *Prof Exp:* Res asst physics, McGill Univ, 50-51; Am Coun Learned Soc fel, Yale Univ, 51-52, asst, 52-54; sr engr physics res, Sprague Elec Co, 54-56; assoc labs dir, RCA Victor Co, Ltd, 56-65; sci adv to sci secretariat, Privy Coun Off, Ottawa, 66-69; sci adv, Sci Coun Can, 69-86; RETIRED. *Concurrent Pos:* Vis prof, McMaster Univ, 64-65 & Carleton Univ, 78-80. *Mem:* AAAS; sr mem Inst Elec & Electronic Engrs; Can Asn Physicists. *Res:* Semiconductor physics; electronic circuitry; nuclear detectors; philosophy of science and technology; appropriate technology for development. *Mailing Add:* 208 Clemow Ottawa ON K1S 2B4 Can

JACKSON, RAYMOND CARL, b Medora, Ind, May 7, 28; m 47; c 2. CYTOGENETICS. *Educ:* Ind Univ, AB, 52, AM, 53; Purdue Univ, PhD, 55. *Prof Exp:* Asst, Purdue Univ, 53-55; from instr to asst prof biol, Univ NMex, 55-58; from asst prof to prof bot, Univ Kans, 58-71; PROF BIOL, TEX TECH UNIV, 71- *Mem:* AAAS; Bot Soc Am; Am Soc Plant Taxon; Int Asn Plant Taxon; Genetics Soc Am; Genetics Soc Can. *Res:* Evolutionary mechanisms in Haplopappus and Machaeranthera; cytogenetics of Haplopappus gracilis; genetics and cytogenetics of diploid species of Triticum; evolution and genetics of polyploids. *Mailing Add:* Dept of Biol Sci Tex Tech Univ Lubbock TX 79409

JACKSON, RICHARD H F, b Jan 29, 47; US citizen; m 81; c 2. PRECISION ENGINEERING, MECHANICAL METROLOGY. *Educ:* Johns Hopkins Univ, BA, 69; Southern Methodist Univ, MS, 70; George Wash Univ, DSc(opers res), 83. *Prof Exp:* Res asst, Computer Sci & Opers Res Ctr, Southern Methodist Univ, 69-70; opers res analyst, Appl Math Div, Nat Bur Standards, Wash, DC, 71-75, math consult, Boulder, Colo, 75-78, proj leader & sr mathematician, Ctr Appl Mat, Gaithersburg, 78-86, prog analyst, Nat Eng Lab, 86-87 & Off Dir, 87-88; prog mgr, Mfg Technol Ctr Prog, 88-89, DEP DIR, CTR MFG ENG, NAT INST STANDARDS & TECHNOL, 89- *Concurrent Pos:* Assoc prof, Opers Res Dept, George Wash Univ, Wash, DC; mem & chmn, numerous comts, Opers Res Soc Am, Math Prog Soc & Soc Indust & Appl Math. *Mem:* Opers Res Soc Am; Math Prog Soc; Soc Indust & Appl Math. *Res:* Author of numerous mathematical publications. *Mailing Add:* Nat Inst Standards & Technol Bldg 220 Rm B322 Gaithersburg MD 20899

JACKSON, RICHARD LEE, biochemistry, pharmacy, for more information see previous edition

JACKSON, RICHARD LEE, b Springfield, Ill, Dec 30, 39. BIOCHEMISTRY, PROTEIN CHEMISTRY. *Educ:* Univ Ill, BS, 63, PhD(microbiol), 67. *Prof Exp:* Res assoc protein chem, Biol Dept, Brookhaven Nat Lab, 67-69; Nat Heart & Lung Inst jr staff fel, 69-70; Nat Inst Arthritis & Metab Dis sr staff fel, 70-71; asst prof exp med, 71-73, assoc prof exp med & cell biol, Baylor Col Med, 73-77; PROF PHARMACOL & CELL BIOPHYS, UNIV CINCINNATI COL MED, 77- *Concurrent Pos:* Estab investr, Am Heart Asn, 72-77, mem coun arteriosclerosis. *Mem:* AAAS; Am Soc Biol Chem; Biophys Soc; Am Chem Soc; Am Heart Asn. *Res:* Structure and function of human plasma lipoproteins and their relationship to atherosclerosis. *Mailing Add:* Dir Dept Biochem Sci Merrell Dow Res Inst 2110 E Galbraith Rd Cincinnati OH 45215

JACKSON, RICHARD THOMAS, b Detroit, Mich, Jan 19, 30; m 55; c 2. PHYSIOLOGY. *Educ:* Univ Detroit, BS, 52, MS, 54; Fla State Univ, PhD(physiol), 60. *Prof Exp:* Asst physiol, Fla State Univ, 54-57, instr, 57-59; from asst prof to assoc prof, Loyola Univ, La, 59-63; res assoc, Lab Ophthal Res, 63-67; instr ophthal & physiol, 63-71, assoc prof, 71-76, dir otolaryngol lab, 67-, PROF SURG, EMORY UNIV, 76- *Concurrent Pos:* Consult, Comt Drugs, Am Acad Otolaryngol; consult, Yerkes Regional Primate Ctr, 82-; vis prof, Kagoshima Sch Med, Japan, 85; consult, Va Med Ctr, equilibrium tests, 88- *Honors & Awards:* Hon Award, Am Acad Otolaryngol. *Mem:* Asn Res Otolaryngol; Am Acad Otolaryngol. *Res:* Nasal and eustachian tube physiology; control of blood flow to the nose and ear; clinical and animal testing of drugs that effect blood flow; equilibrium testing, antivertigo drugs. *Mailing Add:* Dept Surg 441 WMB Emory Univ Atlanta GA 30322

JACKSON, RICHARD W, AMINO ACIDS, PROTEINS. *Educ:* Univ Ill, PhD(biochem), 25. *Prof Exp:* Chief fermentation div, US Dept Agr, Northern Regional Res Lab, 47-66; RETIRED. *Mailing Add:* 1319 N Institute Pl Peoria IL 61606

JACKSON, ROBERT BRUCE, JR, b Drakes Branch, Va, June 5, 29. MATHEMATICS. *Educ:* Davidson Col, BS, 50; Duke Univ, PhD(math), 57. *Prof Exp:* Teacher math, Battle Ground Acad, 50-51; asst, Duke Univ, 53-56; from asst prof to prof, 56-83, chmn dept, 78-83, VAIL PROF MATH, DAVIDSON COL, 83- *Mem:* Math Asn Am. *Res:* Probability. *Mailing Add:* Dept Math Davidson Col PO Box 1719 Davidson NC 28036

JACKSON, ROBERT C, CELLULAR BIOCHEMISTRY, EXOCYTOSIS. *Educ:* Harvard Univ, PhD(biochem), 76. *Prof Exp:* ASSOC PROF BIOCHEM, MED SCH, DARTMOUTH COL, 85- *Res:* Biochemistry of secretion. *Mailing Add:* Dept Biochem Med Sch Dartmouth Col Hanover NH 03756

JACKSON, ROBERT DEWEY, entomology, for more information see previous edition

JACKSON, ROBERT HENRY, nutrition, genetics, for more information see previous edition

JACKSON, ROBERT W, MICROBIOLOGY, IMMUNOLOGY. *Educ:* Purdue Univ, PhD(immunol), 63. *Prof Exp:* EXEC ASSOC DEAN, SCH MED, SOUTHERN ILL UNIV, 74- *Mailing Add:* PO Box 19230 Springfield IL 62794

JACKSON, ROSCOE GEORGE, II, b Eureka, Kans, May 14, 48. SEDIMENTOLOGY, FLUID DYNAMICS. *Educ:* Univ Kans, BS, 70; Univ Ill, Urbana-Champaign, MS, 73, PhD(geol), 75. *Prof Exp:* Instr, Northwestern Univ, Evanston, 74-75, asst prof geol, 75-80; mem fac, Dept Geol Sci, Univ Mich, Ann Arbor, 80-; AT JACKSON BROS CO, 80- *Mem:* Geol Soc Am; Int Asn Sedimentologists; Am Geophys Union; AAAS; Soc Econ Paleontologists & Mineralogists. *Res:* Ancient and modern alluvial sediments; mathematical models and mechanics of bedforms, sediment transport and fluid flow in modern sedimentary environments; flow structures of geophysical turbulent boundary layers. *Mailing Add:* Jackson Bros Co 514 N Main Eureka KS 67045

JACKSON, ROY, b Manchester, Eng, Oct 6, 31; m 57; c 2. TWO-PHASE FLOW. *Educ:* Cambridge Univ, BA, 54, MA, 58; Univ Edinburgh, DSc(chem eng), 68. *Prof Exp:* Tech officer, Imp Chem Industs Ltd, 55-61; lectr chem eng, Univ Edinburgh, 61-64, reader, 64-68; prof, Rice Univ, 68-77; prof, Univ Houston, 77-82; PROF CHEM ENG, PRINCETON UNIV, 82- *Concurrent Pos:* Consult, Shell Oil Co, 69- *Honors & Awards:* Alpha Chi Sigma Award, Am Inst Chem Engr, 80. *Mem:* Brit Inst Chem Engrs; Am Inst Chem Engrs. *Res:* Chemical reaction engineering; fluid-particle systems. *Mailing Add:* Dept Chem Eng Princeton Univ Princeton NJ 08544

JACKSON, ROY JOSEPH, b Cotton Port, La, Feb 8, 44. PHOTOCHEMISTRY. *Educ:* Southern Univ, Baton Rouge, BS, 65, MS, 69; Univ Calif, San Diego, PhD(chem), 75. *Prof Exp:* Instr chem, Southern Univ, Baton Rouge, 69-70; RES CHEMIST, SHELL DEVELOP CO, 75- *Mem:* Am Chem Soc. *Res:* Norish type II photoelimination, photocylization reactions, especially new aryl-alkyl systems; photocure of resins; development of new photocure systems. *Mailing Add:* 18707 Havant Crest Houston TX 77077

JACKSON, SHARON WESLEY, b Topeka, Kans, June 15, 36; m 57; c 3. GENETICS, ENVIRONMENTAL SCIENCES. *Educ:* Kans Wesleyan Univ, BA, 58; Univ Kans, MA, 60; NC State Univ, PhD(genetics), 67. *Prof Exp:* Teacher high sch, Kans, 60-62; instr biol, Kans Wesleyan Univ, 62-64, from asst prof to assoc prof, 67-71; prof environ studies & chmn dept, Sacramento State Col, 71, dir, Environ Studies Ctr, 74-76; CO-DIR, LAND INST, 76- *Mem:* AAAS; Int Asn Plant Taxon; Sigma Xi. *Res:* Development of perennial grain crops. *Mailing Add:* Land Inst 2440 E Waterwell Rd Salina KS 67401

JACKSON, SHIRLEY ANN, b Washington, DC, Aug 5, 46. THEORETICAL SOLID STATE PHYSICS. *Educ:* Mass Inst Technol, SB, 68, PhD(physics), 73. *Prof Exp:* Res assoc theoret physics, Fermi Nat Accelerator Lab, 73-74; vis sci assoc, Europ Orgn Nuclear Res, 74-75; res assoc theoret physics, Fermi Nat Accelerator Lab, 75-76; MEM TECH STAFF, BELL TEL LABS, 76- *Concurrent Pos:* Adv study fel, Ford Found, 71-73; fel, Martin Marietta Corp, 72-73; grant, Ford Found, 74-75; mem bd trustees, MIT Corp, 75-85 & Lincoln Univ, 80-; mem, Comt Educ & Employ Women Sci & Eng, Nat Acad Sci, 81-82. *Mem:* Am Phys Soc; AAAS; NY Acad Sci; Sigma Xi; Nat Inst Sci. *Res:* Landau theories of charge density waves in one and two dimensions; transport properties of random systems; correlation effects in electron-hole plasmas; channeling in metals and semiconductors. theory; two dimensional yang-mills gauge theories; neutrino reactions. *Mailing Add:* Bell Tel Labs 600 Mountain Ave Rm 1D-337 Murray Hill NJ 07974

JACKSON, SYDNEY VERN, nuclear chemistry, for more information see previous edition

JACKSON, THOMAS A J, b Sumter, SC, Dec 26, 42. ENVIRONMENTAL SAMPLE CONTROL. *Educ:* Seton Hall Univ, BS, 72; Fairleigh Dickinson Univ, MS, 77. *Prof Exp:* Chemist, Sherwin-Williams Co, 73-75 & Ethican Inc, 75-79; mgr, anal chem, Boyle-Midway, Inc, 79-84; SR RES SCIENTIST, NJ DEPT ENVIRON PROTECTION, 84- *Mem:* Fel Am Inst Chemists; Am Chem Soc. *Res:* Gel chromatography and high performance liquid chromatography methods development for environmental samples; environmental sample clean up and computer applications in gel chromatography and high performance liquid chromatography. *Mailing Add:* 11 Anita Dr Piscataway NJ 08854

JACKSON, THOMAS EDWIN, b Amarillo, Tex, May 7, 44; m 69. MEDICINAL CHEMISTRY. *Educ:* Rice Univ, BA, 66; Mass Inst Technol, PhD(org chem), 71. *Prof Exp:* Fel org chem, Univ SC, 71-73; sr scientist med chem, Sandoz Inc, 73-75; RES CHEMIST MED CHEM, BIOMED RES, ICI-AMERICAS, INC, 75- *Mem:* Am Chem Soc; The Chem Soc; NY Acad Sci; Sigma Xi. *Res:* Biochemical consideration in drug design; selectivity in organic synthesis; heterocyclic chemistry. *Mailing Add:* Biomed Res Dept ICI-Americas Inc Wilmington DE 19897

JACKSON, THOMAS GERALD, b Mt Sterling, Ala, Dec 26, 36; m 67. ORGANIC CHEMISTRY. *Educ:* Univ Southern Miss, BS, 59, MS, 61; Univ Tenn, Knoxville, PhD(org chem), 65. *Prof Exp:* Asst org chem, Univ Southern Miss, 59-61; PROF CHEM, UNIV S ALA, 65- *Concurrent Pos:* Consult, Res Prod, Inc, Ala, 71- *Mem:* Am Chem Soc; Sigma Xi; Royal Soc Chem; Nat Asn Prev Health Prog Advisors. *Res:* Synthesis of organic compounds of potential medicinal interest; metalation studies of nitrogen containing heterocycles; investigations of compounds containing the nitrogen-silicon bond. *Mailing Add:* Dept Chem Univ S Ala Mobile AL 36688

JACKSON, THOMAS LLOYD, soils, agronomy; deceased, see previous edition for last biography

JACKSON, TOGWELL ALEXANDER, b New York, NY, Nov 1, 39; m 67; c 2. BIOGEOCHEMISTRY OF AQUATIC ENVIRONMENTS. *Educ:* Columbia Univ, BA, 61; Univ Wis, MSc, 63; Univ Mo-Columbia, PhD(geol), 69. *Prof Exp:* Fel org geochem, Woods Hole Oceanog Inst, Dept of Chem, 68- 69; res assoc soil microbiol, Yale Univ Sch Forestry, 69-70; res assoc org geochem, Univ Calif, Santa Barbara, Dept Geol Sci, 70-72; res scientist biogeochem, Freshwater Inst, Winnipeg, 72-86, Nat Hydrol Res Inst, 86-90 & NAT WATER RES INST, 90- *Mem:* Sigma Xi; Int Asn Theoret & Appl Limnol; Rawson Acad Aquatic Sci. *Res:* Humic matter in recent lakes and streams, and its ecological role; paleobiological significance of organic matter in ancient, especially pre-Cambrian, sediments; biogeochemistry of toxic heavy metals, especially mercury in freshwater environments; forms speciation and bio-availability of metals; biochemical weathering of rocks; clay-organic interactions; microbial transformations of metals; microbial ecology (effects of metal species, clay humic matter). *Mailing Add:* Nat Water Res Inst PO Box 5050 Burlington ON L7R 4A6 Can

JACKSON, WARREN, JR, b Oak Park, Ill, May 8, 22; m 47; c 2. ELECTRICAL ENGINEERING. *Educ:* Purdue Univ, BS, 47; Case Inst Technol, MS, 54. *Prof Exp:* Radio engr, Police Dept, River Forest, Ill, 41-42 & Purdue Univ, 46-47; elec engr, Chem & Physics Res Div, Standard Oil Co, Ohio, 47-54, sr tech specialist, Process Eng Div, 54-61, sr proj leader, Mgt Sci Unit, 61-70, instrumentation supvr, res & develop, 70-74, sr process control specialist, res & develop lab, 74-82, sr res specialist, Alaska Iceberg Studies, res lab, 82-85; RETIRED. *Mem:* Inst Elec & Electronics Engrs; hon mem Soc Comput Simulation. *Res:* On line digital computer process control; electronic instrumentation for petroleum research; electronic computers; analog computer simulation of physical and business systems; iceberg drift in Alaskan waters. *Mailing Add:* 4871 Westbourne Rd Cleveland OH 44124-2361

JACKSON, WILLIAM ADDISON, b Castile, NY, Apr 24, 26; m 64. PLANT NUTRITION, SOIL FERTILITY. *Educ:* Cornell Univ, BS, 50; Purdue Univ, MS, 52; NC State Univ, PhD(soil sci), 57. *Prof Exp:* Res instr soil sci, NC State Univ, 52-57; Ford Found fel plant nutrit, Univ Mich, 57-58; from asst prof to prof, 58-72, WILLIAM NEAL REYNOLDS PROF SOIL SCI, NC STATE UNIV, 72-, ALUMNI DISTINGUISHED GRAD PROF, 84- *Concurrent Pos:* Mem comt post-doctoral fel eval, Div Biol & Agr, Nat Acad Sci-Nat Res Coun, 65-68; vis prof, Univ Ill, Urbana, 70-71. *Honors & Awards:* Co-recipient, Campbell Award, Am Inst Biol Sci, 64. *Mem:* Soil Sci Soc Am; Crop Sci Soc Am; Am Soc Agron; Am Soc Plant Pathologists; Japan Soc Soil Sci & Plant Nutrit; fel Australian Inst Biol Sci; fel Royal Irish Acad. *Res:* Absorption, assimilation and distribution of nitrogen by higher plants; effects of nitrogen assimilation on photosynthesis, respiration and mineral accumulation. *Mailing Add:* Dept of Soil Sci NC State Univ Raleigh NC 27607

JACKSON, WILLIAM BRUCE, b Milwaukee, Wis, Sept 10, 26; m 52; c 3. ANIMAL ECOLOGY. *Educ:* Univ Wis, BA, 48, MA, 49; Johns Hopkins Univ, ScD(hyg & vert ecol), 52. *Prof Exp:* Asst zool, Univ Wis, 47-49; asst vert ecol, Johns Hopkins Univ, 49-52; res assoc animal behav, Am Mus Natural Hist, 52; sr asst scientist ecol, USPHS, 52-55; biologist, Pac Sci Bd, 55-57; asst prof, 57-64, asst dean, Col Lib Arts, 64-69, asst dean grad sch, 69-70, dir, Environ Studies Ctr, 70-80, prof biol, 64-81, 81-84, dir, Ctr Environ Res & serv, 80-84, EMER PROF, BOWLING GREEN STATE UNIV, 85- *Concurrent Pos:* Collabr, US Fish & Wildlife Serv & Nat Pest Control Asn, 64-; consult, WHO & Food & Agr Orgn, UN, 69- & Ford Found, 75; consult ed, J Environ Educ, 73-77; chmn, Exec Comt, Ohio Biol Surv, 70-84; adj prof community med, Med Col Ohio, Toledo, 74-80; chmn, subcomt E 35.17 vert pesticides, Am Soc Testing & Mgt; pres, Bio Cenotics Inc, 85-; consult, 85- *Honors & Awards:* Envioron Qual Award, US Environ Protection Agency, Region V, 75. *Mem:* Fel AAAS; Animal Behav Soc; Am Inst Biol Sci; Am Ornithologists' Union; Ecol Soc Am; Sigma Xi; Am Soc Mammal. *Res:* Effects of insecticides on vertebrate populations; microclimatic factors in army ant behavior; ecology of small mammals and of arthropod disease vectors; economic and environmental biology; rodent and bird control methods; studies of anticoagulant resistance; environmental assessment. *Mailing Add:* Dept Biol Sci Bowling Green State Univ Bowling Green OH 43403

JACKSON, WILLIAM DAVID, b Edinburgh, Scotland, May 20, 27; div; c 2. ELECTRICAL POWER SYSTEMS, MAGNETOHYDRODYNAMICS. *Educ:* Glasgow Univ, BSc, 47, PhD(elec eng), 60. *Prof Exp:* Asst, Royal Col Sci & Technol, Glasgow Univ, 47-51; asst lectr elec eng, Col Sci & Technol, Univ Manchester Inst Sci & Technol, 51-54, lectr, 54-58; from asst prof to assoc prof, Mass Inst Technol, 58-66; prof, Univ Ill, Chicago Circle, 66-67; prin res scientist, Avco-Everett Res Lab, 67-72; prof elec eng, Univ Tenn Space Inst, 72-73; mgr thermal-mech energy conversion & storage, Elec Power Res Inst, Palo Alto, 73-74; mgr, MHD Prog, Off Coal Res & Energy Res & Develop Admin, Washington, DC, 74-75, dir MHD Div, 75-77; dir, Div Tech Anal & Spec Projs, Dept Energy, Washington, DC, 77-79; pres, Energy Consult Inc, 79-84; PRES, HMJ CORP, 82- *Concurrent Pos:* Vis lectr, Mass Inst Technol, 55-57, lectr, 68-72; Fulbright travel scholar, UK, 55-57; consult var industs & labs, 60-; vis prof, Tech Univ Berlin, 66, George Washington Univ, 86-87; mem int liaison group on magnetohydrodynamics, 66-, chmn, 68-73, secy, 84-; mem, US Steering Comt Eng Aspects Magnetohydrodynamics, 67-, prog chmn, 68-69, chmn, 69-70, secy, 71-; mem, US-Ger Natural Resources Panel Magnetohydrodynamic Power, 70-72; mem, Task Force Tech Aspects, Comt Conserv Energy, Nat Power Surv, Fed Power Comn, 73; chmn, Steering Comt, US-USSR Coop Prog on Magnethydrodynamic Power Generation, 73-79; prof lectr, George Washington Univ, 79-86, 87-; mem, Energy Resources Operating Bd, Am Soc Mech Engrs, 81-83; mem, Energy Develop Sub Comt, Inst Elec & Electronic Engrs, pres, 72-, chmn, 88-, Power Generation Comt, pres, 86-; prog chair, Intersoc Energy Conversion Eng Conf, 89- *Honors & Awards:* Energy Res & Develop Admin Spec Achievement Award, 76; SERI Award, 78. *Mem:* Fel Inst Elec & Electronics Engrs; Am Phys Soc; assoc fel Am Inst Aeronaut & Astronaut; fel Brit Inst Elec Engrs; fel Am Soc Mech Engrs. *Res:* Electrical power systems; magnetohydrodynamic power generation; analysis of energy systems, especially electrical aspects; development of technology base and engineering design data for first of a kind technologies; engineering of magnetohydrodynamic power systems; power electronics. *Mailing Add:* 3509 McKinley St NW Washington DC 20015

JACKSON, WILLIAM F, b Detroit, Mich, Nov 18, 52; m 76; c 3. VASCULAR SMOOTH MUSCLE, LOCAL CONTROL OF BLOOD FLOW. *Educ:* Mich State Univ, PhD(physiol), 79. *Prof Exp:* From asst prof to assoc prof physiol, Med Col Ga, 83-89; ASSOC PROF, WESTERN MICH UNIV, 89- *Mem:* Am Physiol Soc; Microcirculatory Soc Inc; Am Soc Zoologists; AAAS; Am Heart Asn. *Res:* Regulation of blood flow in the microcirculation; cardiovascular physiology. *Mailing Add:* Dept Biol Sci Western Mich Univ Kalamazoo MI 49008

JACKSON, WILLIAM G(ORDON), b Iron Mountain, Mich, Apr 22, 19; m 43; c 2. CHEMISTRY. *Educ:* Univ Mich, BS, 42; Univ Ill, MS, 43, PhD(org chem), 45. *Prof Exp:* Asst chem, Univ Ill, 42-43, spec asst, Nat Defense Res Comt Proj, 43-45; res chemist, Upjohn Co, 46-59; pres, Burdick & Jackson Labs, 59-77, consult, 78-84; RETIRED. *Mem:* Am Chem Soc; Sigma Xi. *Res:* Antibiotics, vitamin B12; natural product fractionation and structure; chromatography; countercurrent distribution; organic synthesis; laboratory automation apparatus; high-purity solvents. *Mailing Add:* 3840 Mariner's Way No 515 Cortez FL 34215

JACKSON, WILLIAM JAMES, b Houston, Tex, Aug 1, 40. NEUROPSYCHOLOGY. *Educ:* Univ Tex, El Paso, BA, 62; Tex Tech Col, PhD(psychol), 66. *Prof Exp:* Nat Acad Sci-Nat Res Coun fel psychol, Aeromed Res Lab, Holloman AFB, NMex, 66-68; res asst prof, Univ Houston, 68-69 & Univ S Fla, 69-71; asst prof, 71-77, ASSOC PROF PHYSIOL, MED COL GA, 77- *Mem:* Soc Neurosci. *Res:* Neural substrate of learning and motivation; neuropsycho-pharmacology. *Mailing Add:* Dept Physiol Med Col Ga Augusta GA 30902

JACKSON, WILLIAM MORGAN, b Birmingham, Ala, Sept 24, 36; m 59; c 2. PHOTOCHEMISTRY, CHEMICAL PHYSICS. *Educ:* Morehouse Col, BS, 56; Cath Univ Am, PhD(phys chem), 61. *Prof Exp:* Chemist res, Nat Bur Standards, 60-61; res scientist, Martin-Marietta Corp, 61-63; assoc, Nat Bur Standards, 63-64; asst, Goddard Space Flight Ctr, NASA, 64-67, sr chemist, 67-69; vis assoc prof physics, Univ Pittsburgh, 69-70; sr chemist, Goddard Space Flight Ctr, NASA, 70-74; PROF CHEM, UNIV CALIF, DAVIS, 85-, ASSOC DEAN, COL LETT & SCI, 90- *Concurrent Pos:* Mem, US comt, Int Comn Optics, 75-77, Int Astron Union Comn on Comets. *Honors & Awards:* Guggenheim Fel, 89; Miller Fel, 88. *Mem:* AAAS; Am Chem Soc; Am Phys Soc; Int Astron Union; Optical Soc Am; Sigma Xi. *Res:* Chemical kinetics; photochemistry; molecular beams; astrochemistry; mass spectroscopy; application of tunable lasers to problems in photochemistry and chemical kinetics; photochemistry of comets. *Mailing Add:* Dept Chem Univ Calif Davis CA 95616

JACKSON, WILLIAM MORRISON, b Colbert Co, Ala, Aug 2, 26; m 54; c 2. INORGANIC CHEMISTRY, PHYSICAL CHEMISTRY. *Educ:* Univ Ala, BS, 50; Univ Tenn, MS, 52 & 60, PhD(chem), 53; Am Bd Health Physics, cert, 80. *Prof Exp:* Chemist, Goodyear Atomic Corp, 53-55 & Union Carbide Corp, Oak Ridge Nat Lab, 55-61; group leader, Diamond Alkali Co, Ohio, 61-66; sr scientist, Oak Ridge Assoc Univs, Tenn, 66-68; tech mgr, Am Nuclear Corp, 68-70; syst chemist, ALA Power Co, 70-74, environ & health physics coordr, 74-83; sr proj mgr, Inst Nuclear Powers Opers, 83-89; RETIRED. *Concurrent Pos:* Tech Consult, Tiztek Corp, 91- *Mem:* Am Chem Soc; Health Physics Soc. *Res:* Radiochemistry; health physics. *Mailing Add:* 459 Bonner Rd Carrollton GA 30117

JACKSON, WILLIAM ROY, JR, b Port Lavaca, Tex, Nov 26, 36. NUCLEAR PHYSICS. *Educ:* Columbia Univ, BA, 59; Rice Univ, MA, 65, PhD(physics), 67. *Prof Exp:* Asst prof, 67-71, ASSOC PROF PHYSICS, SOUTHWEST TEX STATE UNIV, 71- *Mem:* Am Phys Soc. *Res:* Low energy experimental nuclear physics; reaction mechanisms. *Mailing Add:* Dept of Physics Southwest Tex State Univ San Marcos TX 78666

JACKSON, WILLIAM THOMAS, b Stockdale, Ohio, May 10, 23; m 49; c 3. PLANT PHYSIOLOGY. *Educ:* Ohio State Univ, BS, 47; Duke Univ, PhD(bot), 53. *Prof Exp:* Instr biol, WVa Univ, 49-50; asst bot & plant physiol, Duke Univ, 50-53; from instr to asst prof bot, Yale Univ, 53-59; from asst prof to prof biol, Dartmouth, 59-90; CONSULT, 90- *Concurrent Pos:* Mem fel panel, NIH, 63-66. *Mem:* Am Soc Plant Physiol; Bot Soc Am. *Res:* Mitosis and other events of cell cycle. *Mailing Add:* 8335 SW 72nd Ave Miami FL 33143

JACKSON, WINSTON JEROME, JR, b Asheville, NC, Feb 4, 26; m 52; c 2. POLYMER CHEMISTRY. *Educ:* Va Polytech Inst, BS, 49; Duke Univ, PhD(org chem), 52. *Prof Exp:* From res chemist to sr res chemist, Eastman Kodak Co, 52-57, res assoc, 58-72, sr res assoc, 73-80, res fel, Eastman Chemicals Div, 80-91; RETIRED. *Mem:* Am Chem Soc. *Res:* Organic synthesis; preparation, characterization and evaluation of new polymers and discovery of polymer-forming reactions; liquid crystalline polymers. *Mailing Add:* 4408 Greenspring Circle Kingsport TN 37664

JACO, CHARLES M, JR, b Montgomery Co, Miss, Jan 28, 24; m 46; c 1. INDUSTRIAL & MANUFACTURING ENGINEERING, SYSTEMS DESIGN & SYSTEMS SCIENCE. *Educ:* US Mil Acad, BS, 46; Univ Del, MChE, 57. *Prof Exp:* Proj & contract officer, Redstone Arsenol, 50-54; mem staff, Ord Off JTF Seven, Marshall Islands, 54-56; mil attache, US Embassy, Switz, 62-65; mgr corp develop, Res & Develop Div, Dravo Corp, 66-71; plant mgr, Midland Ross Corp, 71-72; pres, chief exec officer & gen mgr, Georgetown Ferreduction, 72-74; pres & chief oper officer, Midrex Corp, 74-76; PARTNER, JCI CONSULTS, 76- *Concurrent Pos:* Adj prof, Ballistic Res Lab Br, Univ Md, 60; mil attache & tech adv to ambassador, US Embassy, Switz, 63-66. *Mem:* Am Mgt Asn; Inst Mgt Consults; Am Inst Chem Eng;

Asn Iron & Steel Engrs. *Res:* Management of corporate organizations, including role of research, development and engineering; role of research, development and engineering to corporate growth and development; systems and multi-discipline engineering; co-author of two textbooks; process engineering. *Mailing Add:* 122 Pine Grove Lake Wylie SC 29710

JACO, WILLIAM HOWARD, b Grafton, WVa, July 14, 40; m 78; c 4. TOPOLOGY. *Educ:* Fairmont State Univ, BA, 62; Pa State Univ, MA, 64; Univ Wis, PhD(math), 68. *Prof Exp:* Proj mathematician underwater activities, Ord Res Lab, 61-64; instr math, Univ Mich, 68-70; from asst prof to prof math, Rice Univ, 70-82; prof math, Okla State Univ, 82-88, head dept, 82-87; EXEC DIR, AM MATH SOC, 88- *Concurrent Pos:* NSF fel math, Inst Advan Study, 71-72, 78-79. *Mem:* Am Math Soc; Math Asn Am; Nat Coun Teachers Math. *Res:* Geometric topology with particular interest in classical three-manifold topology; classification problems for three-manifolds and the structure of groups which are fundamental groups of three-manifolds. *Mailing Add:* Am Math Soc PO Box 6248 Providence RI 02940

JACOB, CHAIM O, b Carei, Romania, Sept 24, 51; Israeli citizen. IMMUNOGENETICS. *Educ:* Univ Tel-Aviv, Israel, MD, 76; Weizmann Inst Sci, Rehovot, Israel, PhD(immunol), 85. *Prof Exp:* Res fel immunogenetics, Stanford Univ Sch Med, 85-90; SR STAFF SCIENTIST, IMMUNOGENETICS, SYNTEX RES, 90- *Honors & Awards:* J F Kennedy Mem Prize, Pvt Int Orgn, 85; Res Presidential Award, Reticulo Endothelial Soc, 87. *Mem:* Am Asn Immunologists. *Res:* Immunogenetics basis of autoimmune diseases; genetic basis of lymphokine production and the interactions between cytokines and major histocompatibility complex genes; author of 45 publications. *Mailing Add:* Syntex Research R7-201 3401 Hillview Ave Palo Alto CA 94303

JACOB, FIELDEN EMMITT, b Columbia, Mo, July 20, 10; m 37; c 3. ANALYTICAL CHEMISTRY, PHYSICAL CHEMISTRY. *Educ:* Univ Mo, AB, 32, BS & MA, 35, PhD(chem), 39. *Prof Exp:* Instr chem, Univ Mo, 39-42; asst prof, Mont Sch Mines, 42-45; assoc prof, Kans State Teachers Col, Emporia, 45-47; assoc prof, 47-75, prof, 75-80, EMER PROF CHEM, DRAKE UNIV, 80- *Mem:* Am Asn Univ Profs; Am Chem Soc. *Res:* Colorimetric analysis; carotinoid pigments of egg yolks from hens on various diets; spectrophotometry; stability constants. *Mailing Add:* 1520 48th St Des Moines IA 50311

JACOB, GARY STEVEN, b St Louis, Mo, Mar 18, 47; m 75; c 2. SOLID STATE NUCLEAR MAGNETIC RESONANCE, GLYCOBIOLOGY. *Educ:* Univ Mo-St Louis, BS, 69; Univ Wis-Madison, PhD(biochem), 76. *Prof Exp:* Fel biophysics, Thomas J Watson Res Ctr, IBM Corp, 76-79; res specialist, Monsanto Co, St Louis, 79-86; res mgr, Oxford Proj, GD Searle, 86-90; MONSANTO FEL & HEAD GLYCOSCI, MONSANTO CO, 90- *Mem:* AAAS; Am Chem Soc; Sigma Xi; Am Soc Biochem & Molecular Biol. *Res:* Solid-state N-15 and C-13 nuclear magnetic resonance studies of bacterial metabolism, with emphasis on bacteria capable of degrading herbicides; cell-wall crosslinking in bacteria and nitrogen fixation; glycobiology; antivirals. *Mailing Add:* Monsanto Co 800 N Lindbergh Blvd St Louis MO 63167

JACOB, GEORGE KORATHU, b Calcutta, India, Aug 19, 59; m 89; c 1. PARALLEL CIRCUIT SIMULATION, PARALLEL COMPILERS. *Educ:* Indian Inst Technol, Kharagpur, BTech, 81; Pa State Univ, MS, 83; Univ Calif, Berkeley, PhD(elec eng & computer sci), 87. *Prof Exp:* Software developer, Berkeley, Calif, 87-89, SOFTWARE DEVELOPER, FRANZ INC, COLUMBUS, OHIO, 89- *Res:* Parallel processing applications, especially circuit simulation; compilers and programming environments for parallel processes. *Mailing Add:* Franz Inc 1995 University Ave Berkeley CA 94704

JACOB, HARRY S, b San Francisco, Calif, Apr 6, 33; m 54; c 3. INTERNAL MEDICINE, HEMATOLOGY. *Educ:* Reed Col, BA, 54; Harvard Univ, MD, 58. *Prof Exp:* Intern med, Boston City Hosp, 58-59, resident, 59-60; NIH fels hemat, Thorndike Mem Lab, Harvard Univ, 60-63, tutor med sci, Harvard Med Sch, 63-65; asst prof med, Sch Med, Tufts Univ, 65-68; assoc prof, 68-70, PROF MED & CHIEF SECT HEMAT, MED SCH, UNIV MINN, MINNEAPOLIS, 70- *Concurrent Pos:* NIH res grants, 65-; prof, Royal Postgrad Med Sch London, 66; prof, Med Sch, Univ Chicago, 71; prof med, Univ Man, 71; vice chmn Dept Med, Med Sch, Univ Minn, Minneapolis, 70, ed-in-chief, Journal Lab & Clin Med, 91- *Honors & Awards:* Conrad Elvehjem Mem Award, 71. *Mem:* Am Soc Clin Invest; Am Fedn Clin Res; Am Soc Hemat; Int Soc Hemat; Asn Am Physicians. *Res:* Red cell metabolism; hemoglobin function-structure relationships; granulocyte function; reticuloendothelial physiology. *Mailing Add:* Dept of Med Univ of Minn Med Ctr Minneapolis MN 55455-0132

JACOB, HENRY GEORGE, JR, b New Haven, Conn, June 11, 22; m 44; c 3. MATHEMATICS. *Educ:* Yale Univ, BE, 43, ME, 47, PhD(math), 53. *Prof Exp:* Asst instr calculus, Yale Univ, 50-53; from asst prof to assoc prof math, La State Univ, 53-62; prof, 62-87, EMER PROF MATH, UNIV MASS, AMHERST, 87- *Concurrent Pos:* asst prof math, Johns Hopkins Univ, 56-57. *Mem:* Am Math Soc; Math Asn Am. *Mailing Add:* Dept of Math Univ of Mass Amherst MA 01003

JACOB, HORACE S, b Varanasi, India, May 17, 31; US citizen; m 61; c 2. POULTRY SCIENCE, PHYSIOLOGICAL GENETICS. *Educ:* Univ Lucknow, BA, 54; Lucknow Christian Training Col, LT, 55; Sam Houston State Univ, BS, 58, MEd, 59; Pa State Univ, MS, 63; Tex A&M Univ, PhD(poultry sci & biol), 67. *Prof Exp:* Agr supvr & teacher, Ingraham Inst, 59-61; assoc prof biol & agr, Alcorn Agr & Mech Col, 67; ASST PROF BIOL, SOUTHWESTERN UNIV, 67- *Mem:* AAAS. *Res:* Population genetics through the aid of electrophoretic studies. *Mailing Add:* Dept Biol Southwestern Univ Box 770 Georgetown TX 78626

JACOB, JONAH HYE, b Calcutta, India, May 15, 43; US citizen; m 77; c 2. LASER PHYSICS, PLASMA PHYSICS. *Educ:* London Univ, BSc, 64; Yale Univ, PhD(plasma physics), 70. *Prof Exp:* Res assoc plasma physics, Yale Univ, 70-71; prin res scientist laser & plasma physics, Avco-Everett Res Lab, 71-82; PRES, SCI RES LAB, 83- *Mem:* Am Phys Soc; AAAS. *Res:* High power lasers; discharge physics; atomic physics; environmental physics. *Mailing Add:* Sci Res Lab 15 Ward St Somerville MA 02143

JACOB, K THOMAS, b India, May 13, 44; m 70; c 3. EXTRACTIVE METALLURGY, PROCESS METALLURGY. *Educ:* Mysore Univ, India, BSc, 64; Indian Inst Sci, BE, 67; Univ London, PhD(eng), 70; Imperial Col Sci & Technol, DIC, 70. *Hon Degrees:* DSc, Univ London, Eng, 86. *Prof Exp:* Sr res assoc, Univ Toronto, 71-76; spec lectr, 75-76, from asst prof to assoc prof metall & mat sci, 79-83; PROF & CHMN, DEPT METALL, INDIAN INST SCI, BANGALORE, INDIA, 83- *Concurrent Pos:* Consult, Lawrence Berkeley Lab, 77-79. *Honors & Awards:* Int Hoffmann Mem Prize, 82; Nat Metallurgist Award, Govt India, 86. *Mem:* Metall Soc; Am Ceramic Soc; Electrochem Soc; Can Inst Mining & Metall; Nat Inst Ceramic Engrs. *Res:* Physical chemistry of extractive metallurgy; interrelations between structure and thermodynamics properties of inorganic materials, alloys and melts; mathematical modelling; phase stability; chemical aspects of ceramic processing; transport properties of ceramics; irreversible thermodynamics; solid state sensors. *Mailing Add:* Indian Inst Sci Dept Metallurgy Bangalore 560012 India

JACOB, KLAUS H, Stuttgart, Ger, Aug 20, 36. SEISMOLOGY, TECTONICS. *Educ:* Univ Frankfurt, PhD(gophys), 68. *Prof Exp:* Res assoc, 68-83, SR SCIENTIST SEISMOL & TECTONICS, LAMONT-DOHERTY GEOL OBSERV, COLUMBIA UNIV, 83- *Concurrent Pos:* Assoc ed, Geophys Res Lett, Am Geophys Union, 83- *Mem:* Am Geophys Union; Seismol Soc Am; Am Geol Inst; Ger Geophys Soc. *Res:* Geophysics of active plate margins (subduction and continental collision zones) based on earthquake information; seismic and volcanic hazards; microearthquake studies in Alaska, Himalaya and Central America; earthquake prediction; earthquake engineering. *Mailing Add:* 348A Old Mill Rd Valley Cottage NY 10989

JACOB, LEONARD STEVEN, b Philadelphia, Pa, Mar 18, 49; m 69; c 2. CLINICAL INVESTIGATION. *Educ:* Philadelphia Col Pharm & Sci, BS, 70; Temple Univ, PhD(pharmacol), 75; Med Col Pa, MD, 78. *Prof Exp:* Group dir clin invest, 83-84, vpres clin res & develop NAm, 84-86, VPRES CLIN RES & DEVELOP WORLDWIDE, S K & F LABS, SMITH-KLINE BECKMAN CORP, 86- *Concurrent Pos:* Fel, Dept Pharmacol, Hahnemann Med Col, 70-71 & Temple Univ Sch Med, 73-75; house staff anesthesiol, Hosp Univ Pa, 78- *Mem:* AMA; AAAS; Sigma Xi; fel Am Col Clin Pharmacol; Pharmaceut Mfrs Asn. *Res:* Linking molecular biology to clinical trial design; pharmaceutical agents. *Mailing Add:* 5110 Campus Dr Plymouth Meeting PA 19462

JACOB, MARY, b Kerala, India, May 28, 33; US citizen. NUTRITIONAL STATUS ASSESSMENT OF ELDERLY, NURTIENT REQUIREMENT OF ELDERLY. *Educ:* Univ Madras, India, BS, 53, MS, 58; Univ London, MS, 63; Univ Ill, Urbana, PhD(nutrit biochem), 69. *Prof Exp:* Asst res nutritionist, Univ Calif, Los Angeles, 69-76; lectr nutrit, Western Australia Inst Technol, Perth, 76-77; asst prof, Ariz State Univ, Tempe, 77-80; PROF HUMAN NUTRIT, CALIF STATE UNIV, LONG BEACH, 80- *Concurrent Pos:* Res asst, Univ Bombay, India, 58-61; vis asst prof, Ariz State Univ, Tempe, 77-78. *Mem:* Am Inst Nutrit; Sigma Xi; Geront Soc Am; Am Col Nutrit; Inst Food Technol; NY Acad Sci. *Res:* Calcium, zinc metabolism and interdependence of these on vitamin D status; changes in body composition with age with main focus on elderly; effect of calorie restriction on lean body mass. *Mailing Add:* Dept Home Econ Calif State Univ 1250 Bellflower Blvd Long Beach CA 90840-0501

JACOB, PAUL B(ERNARD), JR, b Columbus, Miss, June 9, 22; m 46; c 2. ELECTRICAL ENGINEERING. *Educ:* Miss State Col, BS, 44; Northwestern Univ, MS, 48. *Prof Exp:* Jr engr, Tenn Eastman Corp, 44-46; from instr to assoc prof, 46-56, PROF ELEC ENG, MISS STATE UNIV, 56-, ASSOC HEAD DEPT, 61- *Mem:* Am Soc Eng Educ; Inst Elec & Electronics Engrs; Power Eng Soc. *Res:* High voltage engineering; electric power system analysis. *Mailing Add:* Box 5252 Mississippi State MS 39762

JACOB, PEYTON, III, b Ann Arbor, Mich, Sept 23, 47. DRUG METABOLISM, ORGANIC SYNTHESIS. *Educ:* Univ Calif, BS, 69; Purdue Univ, PhD(chem), 75. *Prof Exp:* NIH fel, 75-78, ASST RES CHEMIST, DIV CLIN PHARMACOL, SCH MED, UNIV CALIF, SAN FRANCISCO & SAN FRANCISCO GEN HOSP MED CTR, 78- *Mem:* Am Chem Soc; Sigma Xi. *Res:* Tobacco alkaloid metabolites; development of new analytical methodology for drugs and their metabolites in biologic fluids. *Mailing Add:* 3787 Highland Rd Lafayette CA 94549

JACOB, RICHARD JOHN, b Salt Lake City, Utah, Oct 9, 37; m 59; c 4. THEORETICAL PHYSICS. *Educ:* Univ Utah, BS, 58, PhD(physics), 63. *Prof Exp:* From asst prof to assoc prof, 63-78, chmn dept, 85-90, PROF PHYSICS, ARIZ STATE UNIV, 78- *Concurrent Pos:* Vis prof, Univ Karlsruhe, 70-71; Univ Kaiserlautern, 73, 78-79. *Mem:* Am Phys Soc; Am Asn Physics Teachers; Sigma Xi. *Res:* Theoretical elementary particle physics. *Mailing Add:* Dept Physics Ariz State Univ Tempe AZ 85281

JACOB, RICHARD L, b Ripon, Wis, July 6, 32; m 67; c 1. THEORETICAL PHYSICS, COMPUTER SCIENCES. *Educ:* Stanford Univ, BS, 55; Univ Wis, MS, 56, PhD(physics), 59. *Prof Exp:* Res assoc, Univ Wis, 59-60; asst prof physics, Tufts Univ, 60-65; assoc prof, Claremont Men's Col, 65-68; assoc prof, 68-82, PROF PHYSICS, CORNELL COL, 82-, ACAD COMP COORDR, 72- *Mem:* Am Phys Soc; Am Asn Physics Teachers. *Res:* Elementary particle physics; relativistic quantum mechanics; field theory; philosophy of physics. *Mailing Add:* Dept of Physics Cornell Col Mt Vernon IA 52314

JACOB, ROBERT ALLEN, b Chicago, Ill, Dec 16, 42; m 71. ANALYTICAL CHEMISTRY, CLINICAL CHEMISTRY. *Educ:* Ill Col, BA, 65; Southern Ill Univ, MA, 67, PhD, 70. *Prof Exp:* Res chemist, Human Nutrit Lab, 75-79, sr chemist, Midwest Res Inst, Kemo, 79-81, res chemist, Boston, 81-, AT NUTRIT RES CTR, USDA, SAN FRANCISCO. *Mem:* Am Chem Soc; Am Asn Clin Chem; Sigma Xi. *Res:* Trace metal analysis and metabolism; analytical chemistry of nutrients; clinical lab methods for assessing nutritional status; clinical chemistry. *Mailing Add:* Nutrit Res Ctr USDA PO Box 29997 Presidio San Francisco CA 94129

JACOB, ROBERT J(OSEPH) K(ASSEL), b Brooklyn, NY, Nov 11, 50; m 73; c 2. HUMAN-COMPUTER INTERACTION. *Educ:* Johns Hopkins Univ, BA, 72, MS, 74, PhD(elec eng), 76. *Prof Exp:* Res asst & instr, Johns Hopkins Univ, 72-76; COMPUTER SCIENTIST, HUMAN COMPUTER INTERACTION LAB, NAVAL RES LAB, WASHINGTON, DC, 77- *Concurrent Pos:* Vis asst prof, Math Dept, Towson State Univ, 76; full prof lectr, George Washington Univ, 78-; mem prog comt, Third Int Conf Software Eng & Knowledge Eng, 91 & var prog comts, Asn Comput Mach, 87-; vchair, Spec Interest Group Human-Computer Interaction, Asn Comput Mach, 90-; group leader, Input/Output Devices, George Washington Univ/NSF Workshop, 91. *Mem:* Sigma Xi; Asn Comput Mach; Inst Elec & Electronic Engrs; Human Factors Soc; Am Asn Artificial Intel; AAAS. *Res:* Human-computer interaction; eye movement-based interaction; formal specification of user-computer interfaces; visualization of multi-dimensional data; author of numerous publications on human-computer interaction. *Mailing Add:* 902 Portner Pl Alexandria VA 22314

JACOB, SAMSON T, RIBONUCLEIC ACID ENZYMOLOGY, GENE TRANSCRIPTION. *Educ:* Agra Univ, India, PhD(biochem), 64. *Prof Exp:* PROF PHARMACOL, COL MED, PENN STATE UNIV, 72- *Res:* Gene expression. *Mailing Add:* Dept Med Col Med Penn State Univ Hershey PA 17033

JACOB, STANLEY W, b Philadelphia, Pa, Jan 7, 24; m 64; c 2. MEDICINE, SURGERY. *Educ:* Ohio State Univ, BA, 45, MD, 48. *Prof Exp:* Instr surg, Harvard Med Sch, 57-59; asst prof, 59-65, ASSOC PROF SURG, MED SCH, UNIV ORE, 65-, GERLINGER ASSOC PROF, 81- *Mem:* Am Col Surg; Soc Univ Surg; NY Acad Sci. *Res:* Preservation and transplantation of tissues; biologic applications of dimethyl sulfoxide. *Mailing Add:* Ore Health Sci Univ 3181 SW Sam Jackson Park Rd Portland OR 97201

JACOB, THEODORE AUGUST, b Braddock, Pa, Aug 22, 19; m 44; c 3. ORGANIC CHEMISTRY, BIOCHEMISTRY. *Educ:* Col Wooster, BA, 41; Rensselaer Polytech Inst, MS, 43; Purdue Univ, PhD(org chem), 49. *Prof Exp:* Asst org chem, Rensselaer Polytech Inst, 41-43; org chemist, Standard Oil Co, NJ, 43-46; asst, Purdue Univ, 46-48; sr chemist, 49-57, group leader natural prod develop, 57-69, mgr animal drug metab, 69-76, sect dir, 76-80, DIR ANIMAL DRUG METAB & RADIOCHEM, MERCK, SHARP & DOHME RES LABS, 80- *Mem:* Am Chem Soc. *Res:* Isolation, purification and identification of biologically active products from plant, animal and fermentation sources; preparation and isolation of synthetic peptides, steroids and nucleotides; isolation and identification of drug metabolites. *Mailing Add:* 828 St Marks Ave Westfield NJ 07090

JACOBER, WILLIAM JOHN, b Newark, NJ, Feb 13, 17; m 42; c 4. CHEMISTRY. *Educ:* Union Col, NY, BS, 38; Brown Univ, PhD(chem), 42. *Prof Exp:* Res chemist, NY, E I du Pont de Nemours, 42-53, engr, 53-54, sr engr, 54-74, staff engr, 74-84; CONSULT, 84- *Mem:* Am Chem Soc; Sigma Xi. *Res:* Cellophane softeners and coatings; Mylar and other packaging films; separation of hydrogen isotopes; tritium control technology. *Mailing Add:* 1022 Hitchcock Dr Aiken SC 29801

JACOBI, GEORGE (THOMAS), b Mannheim, Ger, May 19, 22; nat US; m 55; c 1. ELECTRONICS ENGINEERING. *Educ:* Ohio State Univ, BEE, 47, MSc, 48. *Prof Exp:* Asst, Betatron Lab, Ohio State Univ, 48; engr, Res Lab, Gen Elec Co, NY, 48-50; mgr analog comput eng, Gen Eng Lab, 50-55, electronic recording mach acct systs lab, Comput Dept, Calif, 56-57, mgr spec comput eng, 57-59; dir comput & mgt sci res, ITT Res Inst, 59-77; dir bldg automotive syst, Johnson Controls, Inc, 77-84, vpres technol, 84-90; CONSULT, 90- *Honors & Awards:* Wiener Medal, Am Soc Cybernetics, 68. *Mem:* Sr mem Inst Elec & Electronics Engrs; Asn Comput Mach; NY Acad Sci; Sigma Xi. *Res:* Computer logic and storage devices; system theory; electronic component technology; engineering management. *Mailing Add:* 2375 N Wahl Ave Milwaukee WI 53211

JACOBI, PETER ALAN, b Abington, Pa, Sept 14, 45; m 75. SYNTHETIC ORGANIC CHEMISTRY. *Educ:* Univ NH, BS, 67; Princeton Univ, MS, 70, PhD(chem), 73. *Prof Exp:* ASST PROF CHEM, WESLEYAN UNIV, 75- *Concurrent Pos:* Corp appointee, Harvard Univ, 73-75; consult, Anderson Oil Co, 76- *Mem:* Am Chem Soc; Royal Soc Chem; Sigma Xi. *Res:* Mechanistic organic chemistry; chemistry of natural products. *Mailing Add:* Off of Chmn Dept Wesleyan Univ Middletown CT 06457

JACOBI, W(ILLIAM) M(ALLETT), b Elizabeth, NJ, Apr 27, 30; m 62; c 4. TECHNICAL MANAGEMENT. *Educ:* Syracuse Univ, BChE, 51; Univ Del, MChE, 53, PhD(chem eng), 55. *Prof Exp:* Supvry engr nuclear design, Bettis Atomic Power Div, Westinghouse Elec Corp, 55-61; consult, Nuclear Utilities Serv, Inc, 61-63; mgr adv reactor design, Westinghouse Astronuclear Lab, 63-66, mgr design & anal, systs & tech eng, 66-68, mgr mark 48 design eng, weapons dept, 68-70, eng mgr, Fast Flux Test Facility, Westinghouse Advan Reactors Div, 70-73; proj mgr, Clinch River Breeder Reactor Plant, systs eng mgr, Westinghouse Pressurized Water Reactor Systs Div, 78-79, gen mgr, Westinghouse Nuclear Technol div, 79-80, gen mgr, Westinghouse Nuclear Fuel Div, 81-84, vpres, Westinghouse Adv Power Systs, 84-86, pres Westinghouse Hanford Co, 87-88, vpres Westinghouse Govt Opers, 89-91; CONSULT, 91- *Mem:* Am Nuclear Soc. *Res:* Reactor physics; fluid flow and mechanical design. *Mailing Add:* 119 Mt Vernon Dr Monroeville PA 15146

JACOBOWITZ, DAVID, b Brooklyn, NY, July 15, 31; m 57; c 2. PHARMACOLOGY, BIOCHEMISTRY. *Educ:* City Col New York, BS, 53; Ohio State Univ, MS, 58, PhD(pharmacol), 62. *Prof Exp:* NIH fel pharmacol, Sch Med, Univ Pa, 62-63, instr, 63-65, assoc, 65-67, from asst prof to assoc prof, 67-71; HEAD, HISTOPHARMACOL SECT, LAB CLIN SCI, NIH, 71- *Concurrent Pos:* Pa Plan scholar, Sch Med, Univ Pa, 63-65; USPHS career develop award, 66; Lady Davis vis prof, Hebrew Univ Jerusalem, Israel, 81. *Mem:* Neurochem Soc; Am Soc Pharmacol & Exp Therapeut; Neurosci Soc; Am Col Neuropsychopharmacol & Psychoneuroendocrinol; Am Asn Anat. *Res:* Endocrine pharmacology; effect of stress on pituitary and hypothalamic metabolism and adrenocorticotropic hormone synthesis; cellular pharmacology; localization and mechanism of action of the autonomic neurotransmitters; histochemistry of catecholamines and acetylcholinesterase; immunohistochemistry of peptides; localization of neuromodulatory and neurotransmitter pathways in the brain; two-dimensional electrophoresis of brain proteins; insituhybridization histochemistry. *Mailing Add:* Lab of Clin Sci NIMH Bldg 10 Rm 3D-48 Bethesda MD 20892

JACOBOWITZ, RONALD, b New York, NY, Oct 18, 34; m 60; c 3. ALGEBRAIC NUMBER THEORY, BIOSTATISTICS EDUCATION. *Educ:* City Col New York, BA, 55; Univ Chicago, SM, 56; Princeton Univ, PhD(math), 60. *Prof Exp:* Instr math, Mass Inst Technol, 60-62; asst prof, Univ Ariz, 62-66; assoc prof, Univ Kans, 66-70; PROF MATH, ARIZ STATE UNIV, 70- *Concurrent Pos:* Vis statistician, NIH, 79-80. *Mem:* Math Asn Am; Am Statist Asn. *Res:* Algebra; biomedical statistics; theory of quadratic forms. *Mailing Add:* Dept Math Ariz State Univ Tempe AZ 85287

JACOBS, ABIGAIL CONWAY, b St Louis, Mo, Nov 11, 42; m 69; c 2. BIOCHEMISTRY, TOXICOLOGY. *Educ:* Univ Mich, Ann Arbor, BS, 64; Univ Calif, Berkeley, PhD(biochem), 68. *Prof Exp:* Am Cancer Soc fel immunochem, 68-70, res assoc biochem, Weizmann Inst Sci, Rehovot, Israel, 70-71; res assoc biochem, Queen's Univ, Belfast, 71-72; sr tech writer & researcher chem carcinogenesis, Tracor Jitco, 79-82; assoc proj mgr & sr scientist, Carltech Assocs, 82-89; SR BIOCHEMIST, TECHNOL RESOURCES, INC, 89- *Concurrent Pos:* Res assoc, Nuffield Found, 71-72. *Mem:* Sigma Xi; Am Chem Soc; NY Acad of Sci. *Res:* Biochemistry of the allergic response. *Mailing Add:* 9621 McAlpine Rd Silver Spring MD 20901

JACOBS, ALAN M(ARTIN), b New York, NY, Nov 14, 32; m 55, 78; c 4. NUCLEAR ENGINEERING, STATISTICAL MECHANICS. *Educ:* Cornell Univ, BEngPhys, 55; Pa State Univ, MS, 58, PhD(physics), 63. *Prof Exp:* Res assoc, Pa State Univ, University Park, 56-63, assoc prof nuclear eng, 63-68, prof, 68-80; prof & chmn nuclear eng sci, 80-82, PROF, UNIV FLA, 83- *Concurrent Pos:* Consult, Allis-Chalmers Mfg Co, 56-59, Westinghouse Astronuclear Lab, 61-62, Millitron, Inc & HRB-Singer, Inc, 63-, Combustion Eng Inc, 75- & Future-Tech Corp, 86- *Mem:* Am Nuclear Soc; Sigma Xi. *Res:* Many body problems, especially neutron transport and plasma physics; nuclear reactor theory; radiography. *Mailing Add:* Nuclear Eng Sci Dept Univ Fla Gainesville FL 32611

JACOBS, ALAN MARTIN, b New York, NY, Feb 17, 42; m 68; c 2. GEOLOGY. *Educ:* City Col New York, BS, 63; Ind Univ, MA, 65, PhD(geol), 67. *Prof Exp:* Teaching asst, Ind Univ, 63-64, teaching assoc, 64; asst geologist, Ill State Geol Surv, 67-74; from asst proj geologist to sr proj geologist, STS D'Appolonia Ltd, 74-85; mfr rep, Westinghouse Elec Corp, 81-88; PRES, GEOPROBE, ALAN M JACOBS, INC, 81-; GEOSCI MGR, IT CORP, 88-; LECTR, CIVIL ENG DEPT, CARNEGIE-MELLON UNIV, 87- *Honors & Awards:* Cert Merit, Am Inst Prof Geologists, 84. *Mem:* Geol Soc Am; Am Inst Prof Geologists; Explorers Club. *Res:* Engineering geology; glacial and quaternary geology; geomorphology; geologic factors in site selection; seismicity; age-dating; environmental geology; borehole camera surveys; hazardous waste chemicals management. *Mailing Add:* 323 Lime Oak Dr Pittsburgh PA 15235

JACOBS, ALLAN EDWARD, b Toronto, Ont, Aug 7, 38; m 62. THEORETICAL SOLID STATE PHYSICS. *Educ:* Univ Toronto, BASc, 60; Univ Waterloo, MSc, 62; Univ Ill, Urbana, PhD(physics), 68. *Prof Exp:* Res asst phys metall, Univ Toronto, 60-61; Nat Res Coun Can fel, Univ Hamburg, 68-69; from asst prof to assoc prof, 69-83, PROF PHYSICS, UNIV TORONTO, 83- *Concurrent Pos:* Nat Res Coun Can res grants, 69- *Mem:* Am Phys Soc; Can Asn Physicists. *Res:* Theory of inhomogeneous superconductors; superfluid helium; incommensurate systems; spin glasses; disordered. *Mailing Add:* Dept of Physics Univ of Toronto Toronto ON M5S 1A7 Can

JACOBS, ALLEN LEON, b New York, NY, May 22, 31; m 54; c 3. PHARMACEUTICAL CHEMISTRY. *Educ:* Columbia Univ, BS, 52, MS, 54, PhD(pharmaceut chem), 62. *Prof Exp:* Instr chem, Col Pharmacol, Columbia Univ, 60-62; res analalyst chemist, 62-63, MGR ANALYTICL RES, SANDOZ PHARMACEUT, INC, 63- *Mem:* Am Chem Soc; Am Pharmaceut Asn; Sigma Xi. *Res:* Plant biochemistry; analytical chemistry. *Mailing Add:* Five Alcor Rd Randolph NJ 07869

JACOBS, ALLEN WAYNE, b Quincy, Ill, May 12, 42. ANATOMY. *Educ:* Southern Ill Univ, BA, 65, MA, 67; Univ Iowa, PhD(anat), 71. *Prof Exp:* Asst physiol, Southern Ill Univ, 66-67; asst prof, 71-74, ASSOC PROF ANAT, MICH STATE UNIV, 74-, asst dean educ resources, 75-88, DIR SPORTS MED CLINIC, COL OSTEOP MED, 88- *Mem:* Am Inst Biol Sci; Asn Am Med Cols; Nat Sci Teachers Asn; Am Soc Performance Improvement. *Res:* Design and evaluation of anatomical instruction programs for undergraduate, graduate and medical students. *Mailing Add:* Dept Sports Med Kirksville Col Osteo 204 W Jefferson Kirksville MO 63501

JACOBS, BARBARA B, b Cambridge, Mass, July 23, 29; c 2. ENDOCRINOLOGY, IMMUNOLOGY. *Educ:* Mich State Univ, BS, 50, MS, 52; Ind Univ, PhD(zool), 56; Univ Hawaii, MPH, 86. *Prof Exp:* Res assoc cancer res, Med Ctr, Ind Univ, 56; Am Cancer Soc-NSF fel, Med Ctr, Univ

Colo, 56-58; USPHS fel, State Univ NY Downstate Med Ctr, 58-59; sr res scientist, Roswell Park Mem Inst, 59-63; assoc res scientist, 63-70, dir immunol, Am Med Ctr Denver, 70-77; from assoc prof to prof, Life Sci Ctr, Nova Univ, 78-80; clin prof, Sch Pub Health, Univ Hawaii, 87-88; CONSULT, 84- *Concurrent Pos:* Lect consult, State Univ NY Buffalo, 63; adj assoc prof, Sch Med, Univ Colo, 76-78. *Mem:* Am Pub Health Asn. *Res:* Endocrine and hormonally influenced neoplasms; growth of tumors in allogeneic hosts following passage in vitro; tumor-host immunologic interactions; epidemiology of goiter and malaria. *Mailing Add:* 2433 S Dahlia Lane Denver CO 80222

JACOBS, BARRY LEONARD, NEUROSCIENCE, NEUROPHYSIOLOGY. *Educ:* Univ Calif, Los Angeles, PhD(psychol & neurosci), 71. *Prof Exp:* PROF NEUROSCI, DEPT PSYCHOL, PRINCETON UNIV, 72- *Mailing Add:* Green Hall Princeton Univ Princeton NJ 08544

JACOBS, CARL HENRY, b Lewisburg, Pa, Jan 29, 48; m 70; c 2. ORTHOPEDICS, BIOMATERIALS. *Educ:* Univ Vt, BS, 70, MS, 73, PhD(mech eng), 74. *Prof Exp:* Asst prof, Sch Mech Eng, Ga Inst Technol, 74-79; group leader, Res & Develop, Howmedica Corp, Pfizer, Inc, 79-83; dir res labs & qual assurance, Zimmer Orthop Implant Div, Bristol-Myers, Inc, 83-89; prosthetics mfg, Orthomet Corp, 89-91; CONSULT, 91- *Concurrent Pos:* Clin asst prof, Dept Clin Rehab Med, Emory Univ, 77-79. *Mem:* Sigma Xi; Am Soc Mech Engrs; Soc Mfg Engrs; Orthop Res Soc; NY Acad Sci; Soc Biomat. *Res:* Development and design of orthopedic implants. *Mailing Add:* 2600 Monterey Ave S St Louis Park MN 55416

JACOBS, CHARLES WARREN, b Gainesville, Fla, Nov 2, 54; m 77; c 2. CELL DIVISION, CYTOSKELETON. *Educ:* Univ Miami, Fla, BS, 76; Univ Tex, Austin, PhD(microbiol), 83. *Prof Exp:* Fel cell biol, Univ Mich, 83-87, Ohio State Univ, 87-89; ASST PROF BIOL, ALBION COL, 89- *Concurrent Pos:* Instr phys sci, Eastern Mich Univ, 89. *Res:* Regulation of microtubules during the cell cycle; analyzing mutants of and molecular cloning the beta-tubulin gene of ustilago maydis. *Mailing Add:* Dept Biol Albion Col Albion MI 49224-1899

JACOBS, DAVID R, JR, b Brooklyn, NY, Apr 16, 45; div; c 3. CARDIOVASCULAR EPIDEMIOLOGY, PREVENTIVE CARDIOLOGY. *Educ:* Hofstra Univ, BA, 66; Johns Hopkins Univ, PhD(math statist), 71. *Prof Exp:* Asst prof, Towson State Col, 71; asst prof biostatist, Dept Social & Prev Med, Univ Md, 71-74; from asst prof to assoc prof biostatist, Lab Physiol Hyg, 74-81, assoc prof, div epidemiol, 82-88, PROF DIV EPIDEMIOL, SCH PUB HEALTH, UNIV MINN, MINNEAPOLIS, 88- *Concurrent Pos:* Fel, Coun Epidemiol, Am Heart Asn; Res Career Develop Award, 77-82. *Mem:* Am Statist Asn; Soc Epidemiol Res; Am Heart Asn; fel Am Col Epidemiol; Am Epidemiol Soc. *Res:* Carrying out of large-scale, long-term clinical trials; general research in cardiovascular epidemiology, including intervention methodologies for lowering the risk factors and relevant statistical techniques. *Mailing Add:* Div Epidemiol Sch Pub Health Univ Minn 1-210 Moos Tower 515 Delaware St SE Minneapolis MN 55455

JACOBS, DIANE MARGARET, b Port-of-Spain, Trinidad, Mar 24, 40; US citizen; m 85. IMMUNOLOGY. *Educ:* Radcliffe Col, AB, 61; Harvard Univ, PhD(bact), 67. *Prof Exp:* Instr immunol, Hadassah Med Sch, Hebrew Univ, Jerusalem, 67-68, lectr, 68-71; New York Cancer Res Inst fel, Dept Biol, Univ Calif, San Diego, 71-73; instr biol, 73-74; sr res assoc, Salk Inst Biol Sci, 74-76; from assoc prof to prof microbiol, State Univ NY, Buffalo, 80-89; PROF BIOL, ASSOC VCHANCELLOR RES & DEAN GRAD SCH, ECAROLINA UNIV, 89- *Concurrent Pos:* Consult, Hoffman-LaRoche, 75-76 & 78-79; mem cause and prev sci rev comt, Nat Cancer Inst, 77-81; prin investr on res grants from NIH, 74- & Am Cancer Soc, 77-80; mem, Spec Sci Rev Compt, NIH, 87. *Mem:* AAAS; NY Acad Sci; Am Asn Immunologists; Reticuloendothelial Soc; Am Soc Microbiol. *Res:* Immunomodulatory agents of bacterial origin, particularly lipopolysaccharide; mechanism of triggering lymphocytes; nature of interaction with lymphocytes; lymphocyte membrane determinants interacting with lipopolysaccharide; structural requirements for lipopolysaccharide biological activity. *Mailing Add:* Assoc VChancellor Res/ Dean Grad Sch ECarolina Univ 215 Brewster Greenville NC 27858-4353

JACOBS, DONALD THOMAS, b Detroit, Mich, Sept 16, 49; c 2. PHASE TRANSITIONS, CRITICAL PHENOMENA. *Educ:* Univ SFla, BA, 71, MA, 72; Univ Colo, PhD(physics), 76. *Prof Exp:* From asst prof to assoc prof, 76-87, PROF PHYSICS, COL WOOSTER, 87- *Concurrent Pos:* Vis scientist, Univ Md, 79; consult, The Col Wooster, 81-82. *Mem:* Am Phys Soc; Sigma Xi. *Res:* Experimental investigations of critical phenomena in binary fluid mixtures are conducted; coexistence curves, heat capacity, turbidity, dielectric constant, impurity and electric field effects are measured on various near-critical mixtures. *Mailing Add:* Physics Dept The Col Wooster Wooster OH 44691

JACOBS, EDWIN M, medical oncology, for more information see previous edition

JACOBS, ELLIOTT WARREN, b Brooklyn, NY, Feb 10, 50; m 71. APPLIED MATHEMATICS. *Educ:* State Univ New York, Stony Brook, BS, 71; Adelphi Univ, MS, 73, PhD(math), 76. *Prof Exp:* Asst prof math, Muskingum Col, 76-77 & Mt Union Col, 77-78; ASST PROF MATH, EMBRY-RIDDLE AERONAUT UNIV, 78- *Mem:* Am Math Soc; Math Asn Am. *Res:* Differential equations; nonstandard analysis. *Mailing Add:* Div Math & Phys Sci Embry-Riddle Aeronaut Univ Daytona Beach FL 32114

JACOBS, EMMETT S, b Selma, NC, Mar 17, 26; m 52; c 2. AIR POLLUTION. *Educ:* Univ NC, BS, 50; Lehigh Univ, MS, 55, PhD(anal chem), 58. *Prof Exp:* Chemist, Nitrogen Div, Allied Chem Corp, 50-52; instr anal chem, Lehigh Univ, 52-55; res chemist anal, Jackson Lab, Petrochem Dept, E I du Pont de Nemours & Co, Inc, 58-66, res supvr anal res, 66-69, supvr automotive emission studies, petrol Lab, 69-71, supvr anal & environ studies, 71-74, div head, Emissions & Eng Test Div, Petrol Lab, 74-78, div head, petrol additives & environ & mgr antiknocks tech serv, Petrol Lab, Petrochem Dept, 78-85; RETIRED. *Mem:* Am Chem Soc; Am Soc Testing & Mat. *Res:* Analytical chemistry, especially gas chromatography, electrochemistry, x-ray and infrared spectroscopy; atmospheric chemistry and analysis of automotive emissions; gasoline quality and volume demand; gasoline blinding; lead in gasoline environmental issues. *Mailing Add:* 33 Paxon Dr Wilmington DE 19803

JACOBS, FRANCIS ALBIN, b Minneapolis, Minn, Feb 23, 18; m 53; c 5. BIOCHEMISTRY, NUTRITION. *Educ:* Regis Col, BS, 39; St Louis Univ, PhD(biochem), 49. *Prof Exp:* Asst chem, Univ Denver, 39-41; chemist, Shattuck Chem Co, 41; biochemist, Off Sci Res & Develop, 42-45; Nat Cancer Inst fel chemotherapeut, 49 & 50; from instr to asst prof biochem, Sch Med, Univ Pittsburgh, 51-54; from asst prof to prof, 54-87, EMER PROF BIOCHEM, SCH MED, UNIV NDAK, 87- *Concurrent Pos:* Dir res participation for teacher training prog, Univ NDak, 59-63; mem adv comt sci & math, Dept Pub Instr, NDak, 59-80; mem rev & eval comt, NSF, 60-80. *Mem:* Fel AAAS; Am Soc Biochem & Molecular Biol; Am Chem Soc; Soc Exp Biol & Med; Am Inst Nutrit; Sigma Xi. *Res:* Antibiotics and antitumor agents from microorganisms; gastroenterology, intestinal transport of amino acids and lipids; audiovisual aids for teaching biochemisty and nutrition; trace metal nutrition and bioavailability of zinc and copper. *Mailing Add:* Dept Biochem & Molecular Biol Univ NDak Sch Med Grand Forks ND 58202

JACOBS, GEORGE JOSEPH, b New York, NY, Aug 30, 17; m 47; c 3. SPACE BIOLOGY, FISH & WILDLIFE SCIENCE. *Educ:* Univ Miss, AB, 40; WVa Univ, MSc, 46; George Washington Univ, PhD(zool), 55. *Prof Exp:* Biologist oceanog, US Navy Hydrographic Off, 49-50, biologist hematol, Naval Med Res Inst, 50-56; res biologist, Atomic Bomb Casualty Comn, Nat Acad Sci-Nat Res Coun, Japan, 56-59; space biologist, NASA, 59-62, chief phys biol, 62-71, chief ecol, 71-79; CONSULT LIFE SCI, 79- *Concurrent Pos:* Collabr, Brookhaven Nat Lab, 55; managing ed, J Am Soc Ichthyologists & Herpetologists, 66-72; exec-secy, NASA-USSR Comn Found Space Biol & Med, 67-72; hon res assoc, Smithsonian Inst, 71-; herpetol ed, J Am Soc Ichthyol & Herpetol, 78-82; consult, Am Physiol Soc, 80-86. *Honors & Awards:* Apollo Achievement Award, NASA, 69, Group Achievement Award, 72, 76 & 77. *Mem:* AAAS; Am Soc Ichthyologists & Herpetologists; Biophys Soc; Am Physiol Soc; Am Soc Gravitational & Space Biol; Sigma Xi. *Res:* Space biology; herpetology; environmental quality; ecology; remote sensing; comparative animal physiology. *Mailing Add:* Rd 2 Box 635 Montgomery PA 17752

JACOBS, GERALD DANIEL, b Perrysburg, Ohio, Jan 19, 35; m 58; c 2. PHYSICAL CHEMISTRY, SPECTROSCOPY. *Educ:* Bowling Green State Univ, BA, 57; Mich State Univ, PhD(microwave spectros), 61. *Prof Exp:* Res chemist, Chem Div, Union Carbide Corp, 63-64; assoc prof, 64-70, head dept, 70-89, PROF CHEM, NORTHERN MICH UNIV, 70- *Mem:* Am Chem Soc. *Res:* Microwave spectroscopy as applied to molecular structure determinations; determination of crystal structure by x-ray diffraction; physical chemistry of clathrates. *Mailing Add:* Dept of Chem Northern Mich Univ Marquette MI 49855

JACOBS, H KURT, b Sept 8, 43; m; c 2. CORONARY BLOOD FLOW, POSITIVE END EXPIRATORY PRESSURE. *Educ:* Univ Mo, PhD(physiol), 73. *Prof Exp:* ASSOC PROF SURG & PHYSIOL, HINES VET ADMIN HOSP, SCH MED, LOYOLA UNIV, 73- *Mem:* Am Physiol Soc; Am Asn Lab Animal Sci. *Res:* Cardiopulmonary interactions. *Mailing Add:* Dept Surg & Physiol Sch Med Loyola Univ 2160 S First Ave Maywood IL 60153

JACOBS, HAROLD ROBERT, b Portland, Ore, Nov 19, 36; m 61; c 3. MECHANICAL ENGINEERING, THERMAL SCIENCES. *Educ:* Univ Portland, BS, 58; Wash State Univ, MSME, 61; Ohio State Univ, PhD(mech eng), 65. *Prof Exp:* Res & develop engr, Gen Elec Co, 58-59 & Boeing Co, 61-62; mem tech staff, Aerospace Corp, 65-68; from asst prof to assoc prof mech eng, 67-74, chmn dept civil eng, 78-79, PROF MECH ENG, UNIV UTAH, 74-, ASSOC DEAN RES, COL ENG, 81- *Mem:* Assoc fel Am Inst Aeronaut & Astronaut; Am Soc Mech Engrs; Sigma Xi. *Res:* Heat transfer; fluid mechanics; geothermal energy; oil shale processing; direct contact processing; condensers; thermal stresses; fracture. *Mailing Add:* Dept Mech Eng Pa State Univ Main Campus University Park PA 16802

JACOBS, HARRY LEWIS, b Philadelphia, Pa, Apr 10, 25; m 50; c 4. PSYCHOPHYSIOLOGY, NUTRITION. *Educ:* Univ Del, BA, 50, MA, 51; Cornell Univ, PhD, 55. *Prof Exp:* Asst prof psychol, Bucknell Univ, 54-60; NIMH spec res fel physiol, Sch Med, Univ Rochester, 59-61, assoc prof psychol, Univ Ill, 61-67; assoc dir behav sci div, US Army Natick Labs, 66-68; RETIRED. *Concurrent Pos:* Vis lectr nutrit & food sci, Mass Inst Technol, 66-67; assoc prof physiol, Clark Univ, 66-69, prof, 70-76; affil scientist, Worcester Found Exp Biol, 71- *Mem:* AAAS; fel Am Psychol Asn; Am Physiol Soc; Am Inst Nutrit; Am Inst Biol Sci. *Res:* Appetite, hunger and food habits. *Mailing Add:* 63 Moore Rd Wayland MA 01778

JACOBS, HARVEY, b Cleveland, Ohio, Aug 10, 28; m 53; c 3. PHYSICAL CHEMISTRY, ORGANIC CHEMISTRY. *Educ:* Ohio State Univ, BS, 50; Temple Univ, MA, 54, PhD(chem), 56. *Prof Exp:* Res chemist, Anal Lab, Rohm & Haas Co, 56-71; toxicologist, Philadelphia Med Exam Off, 71-72; ANALYTICAL CHEMIST, THE GLIDDEN CO, 72- *Mem:* Am Chem Soc; Soc Appl Spectros. *Res:* Gas chromatography; nuclear magnetic resonance; mass spectrometry. *Mailing Add:* The Glidden Co PO Box 8827 Strongsville OH 44136

JACOBS, HYDE SPENCER, b Declo, Idaho, May 15, 26; m 50; c 5. AGRONOMY. *Educ:* Univ Idaho, BS, 52, MS, 54; Mich State Univ, PhD(soil chem), 57. *Prof Exp:* Instr soils, Mich State Univ, 53-57; from asst prof to assoc prof, 57-66, dir, Kans Water Resources Res Inst, 64-74, prof soils, Kans State Univ, 66-80, dir, Kans Evapotranspiration Lab, 68-79, head dept agron, 71-80, asst dir exten & dir agr progs, 81-87, ASST TO DEAN OF AGR, 87-, DIR, KANS WATER RESOURCE RES INST, KANS STATE UNIV, 88- *Concurrent Pos:* Consult, Earth Sci Curric Proj, Am Geol Inst, 64-65; NSF fel, Utah State Univ, 68-69; assoc ed, J Agron Educ, 70-74. *Mem:* Fel Am Soc Agron; Soil Sci Soc Am. *Res:* Agronomy, soils; irrigation; water resources. *Mailing Add:* 144 Waters Hall Kans State Univ Manhattan KS 66506

JACOBS, IRA, b Brooklyn, NY, Jan 3, 31; m 56; c 3. TELECOMMUNICATIONS, FIBER OPTICS. *Educ:* City Col New York, BS, 50; Purdue Univ, MS, 52, PhD(physics), 55. *Prof Exp:* Summer physicist, Signal Corps Eng Labs, 52; mem tech staff, Mil Res Div, Bell Labs, 55-60, supvr, 60-62, head electromagnetic res dept, 62-66, head mil commun res dept, 66-67, dir, Digital Transmission Lab, 71-76, Wideband Transmission Facil Lab, 76-85 & Transmission Technol Lab, Transmission Div, 85-87; PROF ELEC ENG, VA TECH, 87- *Mem:* AAAS; Am Phys Soc; Inst Elec & Electronics Engrs. *Res:* Study of radar and military communications systems; analysis of transmission systems performance; development and evaluation of pulse code modulation transmission systems; research and development of fiber optic communication technology and systems. *Mailing Add:* Dept Elec Eng Virginia Tech Blacksburg VA 24061-0111

JACOBS, IRWIN MARK, b New Bedford, Mass, Oct 18, 33; m 54; c 4. COMMUNICATION THEORY, COMPUTER SCIENCES. *Educ:* Cornell Univ, BEE, 56; Mass Inst Technol, MS, 57, ScD(elec eng), 59. *Prof Exp:* Res asst elec eng, Mass Inst Technol, 58-59, from asst prof to assoc prof, 59-66; from assoc prof to prof info & computer sci, Univ Calif, San Diego, 66-72; pres, Linkabit Corp, 68-85; PRES, CHMN & CHIEF EXEC OFFICER, QUALCOMM INC, 85- *Concurrent Pos:* Consult, Appl Res Lab, Sylvania Elec Prod, Inc, 59-, Lincoln Lab, Mass Inst Technol, 61-62, Indust Teaching, Minneapolis-Honeywell, Inc, 63 & Bolt Beranek & Newman, Inc, 65; NASA resident res fel, Jet Propulsion Lab, 64-65; chmn, Sci Adv Group, Defense Commun Agency & Eng Adv Coun, Univ Calif; mem gov bd, Inst Elec & Electronics Engrs Commun Soc. *Honors & Awards:* First Ann Excel Award, Am Electronics Asn, 89. *Mem:* Nat Acad Eng; fel Inst Elec & Electronics Engrs; Asn Comput Mach; Sigma Xi. *Res:* Information theory, coding theory and applications to digital communications; satellite multiple access, microprogrammed communications systems and packet switching; author of numerous technical publications. *Mailing Add:* Qualcomm Inc 10555 Sorrento Valley Rd San Diego CA 92121

JACOBS, ISRAEL S(AMSON), b Buffalo, NY, July 20, 25; m 50; c 2. SOLID STATE PHYSICS. *Educ:* Univ Mich, BS, 47; Univ Chicago, SM, 51, PhD(physics), 53. *Prof Exp:* Asst physics, Univ Chicago, 49-50; PHYSICIST, GEN ELEC RES & DEVELOP CTR, 54- *Concurrent Pos:* Mem adv comt, Conf Magnetism & Magnetic Mat, 59-, prog chmn, 61, pub chmn, 62-64, steering comt, 68-71; mem organizing comt, Int Cong Magnetism, 62-67 & 79-85, exec chmn, 66-67; res, High Magnetic Field Lab, Univ Grenoble, 65-66, Mass Inst Technol, 83-84, Solid State Chem Lab, Univ Bordeaux, 89; consult, US Energy Res & Develop Admin, 77, Nat Res Coun, 78 & 84 & 86; Coolidge fel, Gen Elec, 82; mem comt educ, Am Phys Soc, 79-82, comt opportunities, 80-87, chmn 82-84, panel pub affairs, 87-89. *Mem:* Fel Am Phys Soc; fel Inst Elec & Electronics Engrs; Magnetics Soc; fel AAAS; Mat Res Soc. *Res:* Magnetism and magnetic materials; high magnetic field phenomena; antiferromagnetism; low-dimensional magnetic model systems; industrial applications of magnetics research; microwave properties of composites. *Mailing Add:* Gen Elec Res Develop Ctr PO Box 8 Schenectady NY 12301

JACOBS, J(AMES) H(ARRISON), b St Charles, Mo, Apr 13, 16; m 42; c 2. METALLURGICAL ENGINEERING. *Educ:* Pa State Univ, BS, 36; Univ Mo, BS, 39, MS, 40. *Prof Exp:* Chemist, Western Elec Co, Ill, 36-39; metallurgist, US Bur Mines, 41-51; mgr chem eng develop, Tech Dept, Union Carbide Metals Co Div, 51-63, tech mgr nickel-cadmium battery develop, Consumer Prod Div, 63-72, TECH MGR CARBON/ZINC BATTERIES, BATTERY PROD DIV, UNION CARBIDE CORP, 72- *Mem:* Electrochem Soc; Am Inst Mining, Metall & Petrol Engrs. *Res:* Hydrometallurgy and electrometallurgy of non-ferrous metals. *Mailing Add:* 482 Edinborough Bay Village OH 44140

JACOBS, JACQUELINE E, b Wilkinsburg, Pa, June 17, 23; m 47; c 3. ENVIRONMENTAL BIOLOGY. *Educ:* Coker Col, AB, 44; Univ SC, MS, 61, PhD(biol), 68. *Hon Degrees:* LHD, Coker Col, 86. *Prof Exp:* Teacher, Moultrie High, Mt Pleasant, SC, 57-58 & Dreher High Sch, Columbia, 58-64; asst, Univ SC, 64-68; instr specialist, Instr TV, SC State Dept Educ, 68-71; instr, Spring Valley High Sch, 71-73; exec dir, SC Wildlife Fedn, 74-83; CONSULT, 83- *Concurrent Pos:* Bd trustees, Coker Col, 71-77; fel botany, W Gordon Belser, Univ SC, 64-65; grant, Belle W Baruch Found, Univ SC, 67-68; mem bd trustees, Coker Col, SC, 71-77; mem, Wildlife Adv Comt, Col Agr Sci, Clemson Univ, 76-82; mem, Harry Hampton Mem Wildlife Fund bd dir, 83-88; life mem, bd, SC Wildlife Fedn. *Honors & Awards:* SC Conserv Educ of Year Award, SC Wildlife Fedn; Conserv Award of Year, SC, Woodmen of the World, 75; F Bartow Culp Award Distinguished Serv, SC Wildlife Fedn, 75; Distinguished Serv Award, Nat Wildlife Fedn, 82. *Mem:* The Wildlife Soc; Sigma Xi; Bot Soc Am. *Res:* Freshwater algae of South Carolina; a curriculum guide for life science on educational TV. *Mailing Add:* Five Northlake Rd Columbia SC 29223

JACOBS, JEROME BARRY, b Worcester, Mass, Dec 15, 42; m; c 3. CANCER BIOLOGY, DIAGNOSTIC ELECTRON MICROSCOPY. *Educ:* Univ Vt, BS, 65, MS, 67; Clark Univ, PhD(cell biol), 71. *Prof Exp:* ASST CHIEF, DEPT LAB MED, ST VINCENT HOSP, 71- *Concurrent Pos:* Prof cell biol, Clark Univ; prof life sci, Worcester Polytech Inst; assoc prof path, Univ Mass Med Sch. *Mem:* Int Acad Path; Am Asn Pathologists; Am Soc Clin Pathologists; Am Soc Cell Biol; AAAS; Am Asn Cancer Res; Sigma Xi. *Res:* Renal diseases; bladder cancer. *Mailing Add:* Dept Path St Vincent Hosp Worcester MA 01604

JACOBS, JOHN ALLEN, b Cumberland, Ky, Aug 8, 39; m 64; c 2. ANIMAL SCIENCE. *Educ:* Univ Ky, BS, 63, MS, 65; Univ Wyo, PhD(animal sci), 70. *Prof Exp:* Supt beef, Stadler Packing Co, Columbus, 65-67; instr animal sci, Univ Wyo, 68-69; asst prof, 70-74, ASSOC PROF ANIMAL SCI, UNIV IDAHO, 74- *Mem:* Am Soc Animal Sci; Am Meat Sci Asn. *Res:* Physiology and biochemistry of domestic animals as related to meat quality and quantity. *Mailing Add:* Dept of Animal Sci Cal St Univ 6241 N Maple Ave Fresno CA 93740

JACOBS, JOHN EDWARD, b Kansas City, Mo, June 15, 20; m 46; c 6. BIOMEDICAL ENGINEERING, ELECTRICAL ENGINEERING. *Educ:* Northwestern Univ, BS, 47, MS, 48, PhD(elec eng), 50. *Hon Degrees:* ScD, Univ Strathclyde, 71. *Prof Exp:* Supvry res engr, X-Ray Dept, Gen Elec Co, 50-53, mgr adv develop lab, 53-59, elec engr res lab, 59-60; prof elec eng, 60-69, exec dir Biomed Eng Ctr, WALTER P MURPHY PROF ELEC ENG SCI, NORTHWESTERN UNIV, 69- *Concurrent Pos:* Lectr grad sch, Northwestern Univ, 52-; McKay vis prof, Univ Calif, 57-58; adj prof, Rensselaer Polytech Inst, 59-; pres, Biomed Eng Resource Corp, 69-; mem automated clin lab comt, Nat Inst Gen Med Sci, NIH, 70-, mem biomed training comt consult radiol sect, Army Res Off; chmn manpower comt, President's Adv Coun Mgt Improv, 71-; consult, Fed Coun Sci & Technol, 71- *Honors & Awards:* Coffin Award, Gen Elec Co, 53; Silver Medal, Am Soc Nondestructive Testing, 68. *Mem:* Nat Acad Eng; Biomed Eng Soc (treas, 68-); Instrument Soc Am. *Res:* Photoconduction electron optics; biomedicine; transducers. *Mailing Add:* 631 Milburn Evanston IL 60201

JACOBS, JOSEPH DONOVAN, b Motley, Minn, Dec 24, 08; m 37; c 1. CIVIL ENGINEERING. *Educ:* Univ Minn, BSCE, 34. *Prof Exp:* Civil engr & construct supvr, Walsh Construct Co, New York & San Francisco, 34-54; chief engr, Kaiser-Walsh-Perini-Raymond, Australia, 54-55; sr officer, Jacobs Assocs, 55-78; RETIRED. *Concurrent Pos:* Chmn US nat comt tunneling technol, Nat Acad Sci, 77. *Mem:* Nat Acad Eng; fel Am Soc Civil Engrs; Am Inst Mining, Metal & Petrol Engrs; Nat Soc Prof Engrs. *Mailing Add:* Jacobs Assocs 500 Sansome St San Francisco CA 94111

JACOBS, JOSEPH JOHN, b New York, NY, June 13, 16; m 42; c 3. CHEMICAL ENGINEERING. *Educ:* Polytech Inst Brooklyn, BChE, 37, MChE, 39, DChE, 42. *Prof Exp:* Res chem engr, Autoxygen, Inc, New York, 39-42; sr chem engr, Merck & Co, Rahway, NJ, 42-44; vpres & tech dir, Chemurgic Corp, Calif, 44-47; consult chem engr, 47-55; pres, Jacobs Eng Co, 55-74, CHMN BD & CHIEF EXEC OFFICER, JACOBS ENG GROUP INC, 74- *Mem:* AAAS; Am Chem Soc; Am Inst Chem Engrs. *Res:* Distillation; extraction; chlorination; plant design; unit processes; concentration of lactic acid; dehydration of caustic soda; continuous saponification; improved lubricating composition; deterioration of lubricating oils; economic surveys. *Mailing Add:* Jacobs Eng Group Inc 251 S Lake Ave Pasadena CA 91101

JACOBS, KENNETH CHARLES, b McAllen, Tex, Sept 17, 42; m 68. PHYSICS & ASTRONOMY, PHILOSOPHY OF SCIENCE. *Educ:* Mass Inst Technol, BS, 64; Calif Inst Technol, PhD(physics), 69. *Prof Exp:* Postdoctoral fel cosmology res, Physics Dept, Univ Md, College Park, 68-70; asst prof astron, Astron Dept, Univ Va, Charlottesville, 70-76; vis researcher astrophysics, Max Planck Inst, Munich, WGer, 77; mem tech staff econ, Bell Tel Labs, Murray Hill, NJ, 77-79, oper anal, AT&T Bell Labs, Whippany, NJ, 79-84; assoc prof & chair, Physics Dept, 84-90, SABBATICAL RESEARCHER, HOLLINS COL, ROANOKE, VA, 90- *Concurrent Pos:* Res aide, Lampf Accelerator, Los Alamos, NMex, 65; vis prof, Max Planck Inst, Munich, WGer, 72, Inst Astron, Cambridge Univ, Eng, 74-75, Kapteyn Lab, Univ Groningen, Neth, 75. *Mem:* Am Astron Soc; Int Astron Union; Sigma Xi. *Res:* Cosmological models; quantum cosmology; relativistic astrophysics; radio galaxies/quasars; tachyon theory; solar neutrino problem; celestial mechanics of Jovian Galilean moons; philosophy of science (mathematics); author of three publications. *Mailing Add:* Dept Physics Hollins Col Box 9661 Roanoke VA 24020

JACOBS, LAURENCE ALAN, b Mexico City, Mex, Dec 17, 49; div; c 2. THEORETICAL PHYSICS, ELEMENTARY PARTICLE PHYSICS. *Educ:* Nat Univ Mex, BS, 72; Mass Inst Technol, PhD(physics), 76. *Prof Exp:* Res assoc physics, Mass Inst Technol, 76-77; res assoc, Brookhaven Nat Lab, 77-79; asst prof physics, Nat Univ Mex, 79-81; asst prof, Inst Theoret Physics, Univ Calif, Santa Barbara, 81-83; mem staff, Bell Labs, 83-85; MEM STAFF, MASS INST TECHNOL, 85-; MEM STAFF, THINKING MACHINES, 90- *Concurrent Pos:* Sci adv physics, Nat Sci & Technol, Mex. *Honors & Awards:* Gunnar Kallén Award, 75. *Mem:* Am Phys Soc; Soc Mex Physics; Acad Invert Sci; NY Acad Sci. *Res:* Theoretical physics, field theory; mathematical physics; statistical mechanics; condensed-matter physics. *Mailing Add:* Ctr Theo Phys/6-403 MIT/77 Massachusetts Ave Cambridge MA 02139

JACOBS, LAURENCE STANTON, b Boston, Mass, Mar 24, 40; m 63; c 2. GENERAL ENDOCRINOLOGY, NEUROENDOCRINOLOGY. *Educ:* Harvard Univ, AB, 60; Univ Rochester, MD, 65. *Prof Exp:* Res asst, Res Inst Med & Chem, 60-61; intern & resident internal med, Wash Univ & Barnes Hosp, 65-67, fel endocrinol, 67-68 & 70-72, instr med, 71-72 & asst prof, 72-77; lieutenant comdr & med officer lab med, Ctrs Dis Control, USPHS, 68-70; assoc prof, Univ Rochester Med Ctr, 77-82, dir, Clin Res Ctr, 77-91, actg head gastroenterol, 83-88, PROF MED, UNIV ROCHESTER MED CTR, 82-, ASSOC DEAN STUDENT AFFAIRS, UNIV ROCHESTER SCH MED & DENT, 90- *Concurrent Pos:* Asst physician & consult clin chem, Barnes Hosp, St Louis, 71-77; sr assoc physician, Strong Mem Hosp, Rochester, NY, 77-82, sr physician, 82-; chmn, merit rev bd endocrinol, Vet Admin, 83-86; mem, Biomed Sci Study Sect, NIH, 87-91; pres, Asn Clin Res Ctr Prog Dirs, 87-89. *Mem:* AAAS; Am Soc Clin Invest; Endocrine Soc; Am Soc Biol Chemists; Int Soc Neuroendocrinol; Am Diabetes Asn; Am Fed Clin Res; Central Soc Clin Res. *Res:* Hypothalamic control and molecular mechanisms of secretion of growth hormone and prolactin; structure, composition and enzymatic activities of isolated adenohypophysial secretory granules. *Mailing Add:* Assoc Dean's Offices Univ Rochester Med Ctr 601 Elmwood Ave Box 601 Rochester NY 14642

JACOBS, LOIS JEAN, b Portage, Wis, Mar 10, 47. MEDICINE, GENETICS. *Educ:* Univ Wis-Madison, BS, 69, PhD(genetics), 77, MD, 87. *Prof Exp:* Res assoc genetics, Inst Med Res, 77-80; asst scientist dept med genetics, 81-83, first yr resident, Med Sch, Wis-Madison, 87-88; second yr resident, Bay State Med Ctr, Affil Tufts Univ Med Sch, 88-90; PRIVATE PRACT, 90- *Mem:* Environ Mutagen Soc; AMA. *Res:* Somatic cell genetics, specifically mutagenesis and carcinogenesis in diploid human cells. *Mailing Add:* 1339 Winnebago Ave Oshkosh WI 54901-5334

JACOBS, LOUIS JOHN, b Chicago, Ill, Mar 27, 43; m 65; c 2. FLUID MIXING TECHNOLOGY, PHASE SEPARATION TECHNOLOGY. *Educ:* Univ Wis-Madison, BS, 65; Washington Univ, St Louis, Mo, MS, 69. *Prof Exp:* Engr, Monsanto Co, 65-77, eng supt, 77-79, bus develop mgr, 79-80, mgr technol risk anal, Monsanto Co, 80-82; dir, Corp Eng Div, A E Staley Mfg Co, 82-88, vpres starch opers, 88-89; dir eng develop, Res & Develop Div, 89-91, DIR TECHNOL & ENG, KOCH INDUSTS INC, 91- *Concurrent Pos:* Chmn, Mixing Comt, Am Inst Chem Engrs, 75-76, St Louis Sect, 77; affil prof, Wash Univ, St Louis, 78-79. *Mem:* Am Inst Chem Engrs. *Res:* Author of various publications. *Mailing Add:* 8955 Boxthorn Ct Wichita KS 67226

JACOBS, LOYD DONALD, b Wolcott, Ind, Nov 22, 32; m 56; c 3. ACOUSTICS. *Educ:* Emporia State Univ, BA, 54; Univ Nebr, MS, 58. *Prof Exp:* Engr, Wichita, Kans, 57-62, sr engr, Seattle, Wash, 62-73, group engr, 74-82, PRIN ENGR, BOEING CO, SEATTLE, WASH, 82- *Mem:* Acoust Soc Am; Sigma Xi. *Res:* Control and reduction of aircraft noise; response of aircraft structure to noise; noise radiation from surfaces immersed in airflow. *Mailing Add:* 2004 128th Ave SE Bellevue WA 98005

JACOBS, MARC QUILLEN, b Chandler, Okla, June 28, 38. MATHEMATICS. *Educ:* Univ Okla, BS, 60, MA, 63, PhD(math), 66. *Prof Exp:* Jr mathematician, Int Bus Mach Corp, 60-61; instr math, Univ Okla, 65-66; res asst prof appl math, Brown Univ, 66-67; asst prof math, Rice Univ, 67-68 & appl math, Brown Univ, 68-71; assoc prof, 71-74, PROF MATH, UNIV MO-COLUMBIA, 74- *Mem:* Am Math Soc; Math Asn Am; Soc Indust & Appl Math; Sigma Xi. *Res:* Optimal control theory. *Mailing Add:* Dept Math Univ Mo Columbia MO 65211

JACOBS, MARK, b Princeton, NJ, May 19, 50; m 73; c 3. PLANT DEVELOPMENT, CELL RECEPTOR PROTEINS. *Educ:* Harvard Univ, BA, 71; Stanford Univ, PhD(biol), 76. *Prof Exp:* From asst prof to assoc prof, 81-89, chmn, 87-88 & 88-90, PROF BIOL, SWARTHMORE COL, 89- *Concurrent Pos:* NATO fel, 76-77; Guggenheim Fel, Cambridge Univ, England, 86-87; NAm ed, Plant Physiol & Biochem, 90- *Mem:* Am Soc Plant Physiologists; Sigma Xi. *Res:* Biochemical mode of action of the plant growth regulators; hormonal control of plant development. *Mailing Add:* Dept Biol Swarthmore Col Swarthmore PA 19081

JACOBS, MARTIN IRWIN, b Newark, NJ, Mar 29, 43; m 65; c 2. POLYMER FLAMMABILITY & SMOKE GENERATION, THERMOPLASTIC ELASTOMERS. *Educ:* Rensselaer Polytech Inst, BS, 64; Princeton Univ, MS, 67, PhD(polymer chem), 69. *Prof Exp:* Chemist, Allied chem Corp, 68-69; res scientist, Corp Res & Develop, Uniroyal Inc, 69-77, sr group leader, Chem Div, 77-78, res & develop mgr, 78-82, res & develop mgr, Eng Prod Div, 82-89, vpres res & develop, 89-90; MGR TECHNOL, BENTLEY-HARNS MFG CO, 90- *Mem:* Am Chem Soc; Soc Plastics Engrs. *Res:* Fire-resistant polymer formulations; lab flame-tests vs full-scale fire tests; suppression of smoke generation; thermoplastic elastomers: structure, properties and processing. *Mailing Add:* Bentley Harris Mfg Co 407 N Main St Lionville IN 46544

JACOBS, MARTIN JOHN, b Chicago, Ill, June 28, 44; m 66; c 5. SYNTHETIC ORGANIC CHEMISTRY. *Educ:* Ill Inst Technol, BS, 69; Colo State Univ, PhD(org chem), 75. *Prof Exp:* Res chemist, 75-82, SR RES SCIENTIST, INT MINERALS & CHEM CORP, 82- *Mem:* Am Chem Soc; Sigma Xi. *Res:* Synthesis of compounds as products for use in animal health, care and nutrition; industrial intermediates and pharmaceutical application. *Mailing Add:* PO Box 207 Terre Haute IN 47808

JACOBS, MARYCE MERCEDES, b El Paso, Tex, June 15, 44. TOXICOLOGY, CARCINOGENESIS. *Educ:* NMex State Univ, BS, 66; Univ Calif, Los Angeles, PhD(biochem), 70. *Prof Exp:* Res assoc, Univ Colo Med Ctr, 70-71; res assoc, M D Anderson Hosp & Tumor Inst, Univ Tex, 71-72, asst prof & asst biochemist, 72-77, co-chmn biochem, Grad Sch Biomed Sci, Houston, 73-75; from asst prof to assoc prof, Eppley Inst Res Cancer, Univ Nebr, 77-83, indust contract coordr, 79-83; biochem toxicologist, Metrek Div, Mitre Corp, 83-88; VPRES RES, AM INST CANCER RES, 88- *Concurrent Pos:* Consult, Nat Large Bowel Cancer Cadre, Nat Cancer Inst, 78-82; opponent, Univ Oulu, Finland, 80. *Mem:* Am Asn Cancer Res; NY Acad Sci; Sigma Xi; Soc Toxicol; Am Acad Clin Toxicol. *Res:* Inhibition of chemical carcinogenesis; dietary selenium inhibition of tumor induction by 1, 2-Dimethylhydrazine and other carcinogens in animals; determination of the mechanisms of inhibition in in-vivo and in-vitro systems. *Mailing Add:* 1800 Old Meadow Rd McLean VA 22102

JACOBS, MERLE EMMOR, b Hollsopple, Pa, Nov 30, 18; m 59. ZOOLOGY. *Educ:* Goshen Col, AB, 49; Univ Ind, PhD(zool), 53. *Prof Exp:* Instr zool, Goshen Col, 53-54; instr zool, Duke Univ, 54-57; asst prof biol, Bethany Col, WVa, 57-62; prof biol, EM Col, Va, 62-64; RES PROF ZOOL, GOSHEN COL, 64- *Concurrent Pos:* Grant, NIH, 59-85. *Honors & Awards:* Lalor Award, Sigma Xi. *Mem:* Sigma Xi; Am Men Sci; AAAS. *Res:* Behavior and biochemical genetics. *Mailing Add:* Dept of Biol Goshen Col Goshen IN 46526

JACOBS, MICHAEL MOISES, b Miami, Fla, June 21, 50; m 80; c 3. INFARED PHYSICS, ELECTRO-OPTIC SPACE SYSTEMS. *Educ:* Univ Miami, BS, 72, MS, 73, PhD(physics), 75. *Prof Exp:* Res physicist laser optics, Sci Applns Inc, Atlanta, Ga, 76; physicist & mem tech staff electro-optics,

Satellite Systs Div, Rockwell Int, Seal Beach, Calif, 77-80; PHYSICIST, SPACE SYSTS ANALYST & SYSTS ENG, ADVAN SYSTS TECHNOL DIV, AEROSPACE CORP, EL SEGUNDO, CALIF, 80- Concurrent Pos: Instr physics, Orange Coast Col, 76-79. Mem: Am Inst Physics; Optical Soc Am. Res: Electro-optic infrared & laser physics; acousto-optics; radiative transfer; atmospheric and oceanic optics; systems engineering; infrared focal plane device physics; space sensor mission planning/requirements; surveillance systems. Mailing Add: 2350 E El Segundo Blvd Aerospace Corp El Segundo CA 90245-4691

JACOBS, MORTON HOWARD, b Newark, NJ, June 28, 24; m 70; c 1. ANALYTICAL CHEMISTRY, TECHNICAL MANAGEMENT. Educ: Univ Pa, BA, 46. Prof Exp: Biochemist, Univ Hosp, Univ Mich, 50-52; chemist, Control Lab, 53-58, Res Dept, 58-60, PROJ LEADER SPECTROSCOPY, INSTRUMENTATION & ANALYTICAL RES DEPT, INT FLAVORS & FRAGRANCES INC, 60-, SR PROJ CHEMIST, CHROMATOGRAPHY & SPECTROSCOPY RES DEPT, 78-, SPEC PROJS, 90- Mem: Am Chem Soc; Int Soc Magnetic Resonance; Soc Appl Spectroscopy. Res: Application of chromatographic and spectroscopic techniques; magnetic resonance to structure elucidation of flavor and aroma chemicals, both natural and synthetic. Mailing Add: Int Flavors & Fragrances Inc 1515 Hwy 36 Union Beach NJ 07735

JACOBS, MYRON SAMUEL, b Jersey City, NJ, May 17, 22; c 2. PATHOLOGY. Educ: Univ Pa, BA, 45; NY Univ, MS, 51, PhD, 55. Prof Exp: Asst anat, Col Dent, NY Univ, 51-54; lectr histol & embryol, Queen's Univ, Ont, 54-56; instr histol, embryol & neuroanat, NY Med Col, 56-57, from asst prof to assoc prof anat, 57-65, dir cetacean brain lab, 62-65; res assoc comp path, Osborn Lab Marine Sci, NY Aquarium, 66-72; from assoc prof to prof path, 73-87, EMER PROF PATH, COL DENT, NY UNIV, 87- Mem: AAAS; Am Asn Anat; Am Soc Zool; Soc Neurosci; NY Acad Sci; Sigma Xi. Res: Skin grafts and thermal changes in transplantation, rat; alloxan diabetes, hamster; cetacean nervous system. Mailing Add: 4422 Langtry Dr Glen Arm MD 21057

JACOBS, NICHOLAS JOSEPH, b Oakland, Calif, Mar 29, 33; m 59; c 1. BACTERIOLOGY. Educ: Univ Ill, BS, 55; Cornell Univ, PhD, 60. Hon Degrees: MA, Dartmouth Col, 85. Prof Exp: Res assoc, Univ Ill, 59-61; from asst bacteriologist to assoc bacteriologist, Am Meat Inst Found, 61-64; from instr to assoc prof, 64-80, PROF MICROBIOL, DARTMOUTH MED SCH, 80- Mem: AAAS; Am Soc Microbiol; Sigma Xi. Res: Bacterial physiology; pathogenic bacteriology; heme synthesis in bacteria, plants, and animals. Mailing Add: Dept Microbiol & Physiol Dartmouth Med Sch Hanover NH 03755

JACOBS, PATRICIA ANN, b London, Eng, Oct 8, 34. GENETICS. Educ: Univ St Andrews, BSc, 56, DSc(cytogenetics), 66. Prof Exp: Scientist, Med Res Coun, 57-72; PROF ANAT, SCH MED, UNIV HAWAII, 73- Mem: Genetics Soc Gt Brit; Am Soc Human Genetics. Res: Human cytogenetics. Mailing Add: Opers Res Dept Naval Postgrad Sch Code 55 Monterey CA 93940

JACOBS, PATRICIA ANNE, b Chicago, Ill, May 27, 47. OPERATIONS RESEARCH. Educ: Northwestern Univ, BS, 69, MS, 71, PhD(appl math), 73. Prof Exp: Asst prof opers res, Stanford Univ, 72-78; assoc prof, 78-86, PROF OPERS RES, NAVAL POSTGRAD SCH, 86- Concurrent Pos: Assoc ed, Mgt Sci, 81- & Naval Res Logistics, 82- Mem: Int Statist Inst; Inst Math Statist. Res: Stochastic processes and their applications; stochastic modeling and statistical analysis in operations research and related fields. Mailing Add: Dept of Opers Res Naval Postgrad Sch Code OR Monterey CA 93943

JACOBS, PATRICK W M, b Durban, SAfrica, Sept 15, 23; m 50, 81; c 3. PHYSICAL CHEMISTRY, SOLID STATE CHEMISTRY. Educ: Univ Natal, BSc, 41, MSc, 43; Univ London, PhD(phys chem), 51, DSc(phys Chem), 63. Prof Exp: Lectr chem, Rhodes Univ, SAfrica, 46-48; Beit fel, Imp Col, London, 48-50, from asst lectr to sr lectr phys chem, 50-64, reader, 64-65; SR PROF CHEM, UNIV WESTERN ONT, 65- Concurrent Pos: Overseas fel, Churchill Col, Cambridge, Eng, 73-74; vis fel, Wolfson Col, Oxford, Eng, 82. Honors & Awards: Solid State Chem Award, Royal Soc Chem, 83. Mem: Fel Royal Soc Chem; Can Inst Chem; Can Asn Physicists. Res: Decomposition of solids; optical and electrical properties of solids; computer simulation of condensed matter. Mailing Add: Dept Chem Univ Western Ont London ON N6A 5B7 Can

JACOBS, R(OY) K(ENNETH), engineering mechanics, for more information see previous edition

JACOBS, RALPH R, b Niagara Falls, NY, Dec 31, 42; m 66; c 2. QUANTUM ELECTRONICS, ENGINEERING PHYSICS. Educ: NY Univ, BS, 64; Yale Univ, MS, 65, PhM, 67, PhD(physics), 69. Prof Exp: Lab instr physics, Yale Univ, 64; mem tech staff, GTE Labs, Inc, Bayside, NY, 69-72; sr physicist & proj mgr, laser-prog, Lawrence Livermore Lab, Univ Calif, 72-80; mgr res & advan develop, Spectra-Physics, 80-85, eng mgr, 85-89, dir, Corp Tech Develop, 89-90; MGR, NEW TECHNOL INITIATIVES, LAWRENCE LIVERMORE NAT LAB, 90- Mem: Sigma Xi; Am Phys Soc; Lasers & Electro-Optics Soc; Optical Soc Am; fel, Inst Elec & Electronics Engrs. Res: Basic and applied aspects of atomic, molecular and laser physics in gaseous, liquid and solid state media; quantum electronics; laser spectroscopy-linear and nonlinear (ultraviolet, visible, infrared); low and high pressure gas discharges; high resolution microwave spectroscopy; rotational, vibrational, and electronic relaxation in molecules; new laser research and development; heightened skills in management of technologists and strategic planning that emphasizes optimization of research and development, engineering, marketing, and manufacturing considerations for major program and commercial success. Mailing Add: Lawrence Livermore Nat Lab PO Box 5508 L-488 Livermore CA 94551

JACOBS, RICHARD L, b Elsberry, Mo, Dec 29, 30; m 54; c 3. ORTHOPEDIC SURGERY, BIOCHEMISTRY. Educ: State Univ Iowa, BA, 52, MD, 56, MS, 61. Prof Exp: Arthritis & Rheumatism Found fel, State Univ Iowa, 59-62, resident orthop surg, 62-65; from asst prof to prof orthop, Univ Ill Med Ctr, 66-74; PROF ORTHOP SURG & HEAD DIV, ALBANY MED CTR, 74- Concurrent Pos: Orthop Res & Educ Found res fel orthop, Mass Gen Hosp-Harvard Univ, 61-62; consult, Ill Div Serv Crippled Children, 65-74, Dixon State Sch, 66-74 & US Vet Admin Hosps, 68- Mem: AAAS; AMA; Am Chem Soc; Am Inst Chem; Am Acad Orthop Surg; Sigma Xi; Am Orthop Asn. Res: Vitamin metabolism; collagen chemistry; immunology. Mailing Add: Div Orthop Surg Albany Med Ctr Albany NY 12208

JACOBS, RICHARD LEE, b Perrysburg, Ohio, Aug 4, 31; m 56; c 4. ORGANIC CHEMISTRY. Educ: Bowling Green State Univ, BA, 53; Mich State Univ, MS, 55, PhD(org chem), 59. Prof Exp: Asst, Mich State Univ, 53-59; res chemist, Koppers Co, Inc, Pa, 59-62; sr chemist, Maumee Chem Co, 62-66; dir lab, Sherwin Williams Chems, 66-82,; tech & legal mgr, Prin Bus Enterprises, 82-88; ASSOC CHAIR, CHEM DEPT, BOWLING GREEN STATE UNIV. Mem: AAAS; Am Chem Soc; Soc Heterocyclic Chemists; Sigma Xi; fel Am Inst Chemists. Res: Organic sulfur compounds, especially thiacyclopropanes and thiophene; nitriles preparation and reactions; alkylation and reaction mechanisms of aromatics; nitrogen heterocycles. Mailing Add: 558 Clover Lane Perrysburg OH 43551

JACOBS, RICHARD LEWIS, b Pelham, Tenn, Mar 31, 27; m 59; c 4. CHEMOTHERAPY, IMMUNOLOGY. Educ: Tenn Polytech Inst, BS, 51; Tex A&M Univ, MS, 53, PhD(biochem, nutrit), 55. Prof Exp: Res asst biochem & nutrit, Tex A&M Univ, 51-55; mem staff, US Navy Med Res Inst, 55-60; sr scientist, Lab Parasite Chemother, Nat Inst Allergy & Infectious Dis, 60-76; sr scientist clin immunol, Parke Davis, div Warner-Lambert Co, 76-81; ASSOC PROF, VET MICROBIOL, UNIV MO, 81- Mem: AAAS; Am Soc Microbiol; Am Soc Trop Med & Hyg. Res: Animal growth-promoting effect of dietary antibiotics; animal nutrition; connective tissues; tissue homograft rejection; infectious diseases; metabolic pathways and mechanism of drug action in microorganisms; development of resistance to drugs in microorganisms; methods for determining drug levels in biological materials; chemotherapy of malaria; in vitro culture of malarial; parasites; immunology of malaria. Mailing Add: Rte 3 Decherd TN 37324

JACOBS, RICHARD M, b Wloclawek, Poland, Oct 31, 24; US citizen; m 50; c 1. ORTHODONTICS, ANATOMY. Educ: Maximilian Univ, Dr med dent, 48; NY Univ, DDS, 52; Univ Calif, Berkeley, MPH, 61; Med Col Va, PhD(anat), 64; Univ Ill, MS(orthod), 65. Prof Exp: Resident dent, NY State Dept Ment Hyg, 52-53; dentist, Nev State Dept Pub Health, 59-60; Nat Inst Dent Res fel, Med Col Va, 61-63 & Univ Ill, 63-65; assoc prof orthod & head dept, Fac Dent, Univ BC, 65-66; asst dean col dent, 66-67, actg head dept oral biol & curric coordr, 66-71, assoc dean col, 67-71, PROF ORTHOD, COL DENT, UNIV IOWA, 66- Concurrent Pos: Chmn, Grad Educ sect, Am Asn Dent Schs, 73-76. Mem: Fel Am Pub Health Asn; Am Asn Anat; Am Asn Orthod; fel Sigma Xi; Am Educ Res Asn; Asn Study Higher Educ. Res: Effects of spontaneous muscular activity on fetal development; Relation between knowledge and diagnostic reasoning; cost effectiveness of professional education; organizational behavior and control in formal organizations. Mailing Add: Col of Dent Univ of Iowa Iowa City IA 52242

JACOBS, ROBERT SAUL, MARINE PHARMACOLOGY, NEUROPHARMACOLOGY. Educ: Loyola Univ, PhD(pharmacol), 71. Prof Exp: PROF PHARMACOL, UNIV CALIF, SANTA BARBARA, 74- Mailing Add: Dept Biol Sci Univ Calif Santa Barbara CA 93106

JACOBS, ROSS, b Montreal, Que, Feb 27, 25; m 53; c 4. BIOCHEMISTRY. Educ: McGill Univ, BSc, 48, MSc, 52, PhD, 54. Prof Exp: Res biochemist, Ayerst Labs, Inc, 54-55, asst to dir pharmaceut develop labs, 55-57; res assoc med, Univ Southern Calif, 57-58; asst prof obstet & gynec, 58-64, ASST PROF BIOCHEM, COL MED, STATE UNIV NY UPSTATE MED CTR, 58- Concurrent Pos: Dir labs, Calif Found Med Res, 57-58. Mem: Biochem Soc; Endocrine Soc. Res: Biosynthesis and metabolism of steroid hormones; gonadotropins. Mailing Add: 465 S Collingwood Ave Syracuse NY 13206

JACOBS, S LAWRENCE, b New York, NY, Nov 27, 29; m 53; c 3. CLINICAL CHEMISTRY, ANALYTICAL TOXICOLOGY. Educ: Rensselaer Polytech Inst, BS, 51; Univ Ill, MS, 52, PhD(chem), 55. Prof Exp: Asst chem, Univ Ill, 52-55; res chemist, Northern Regional Lab, USDA, 55; chief spec projs div, Los Angeles Br Lab, Bio Sci Labs, 56-65; dir dept chem, 66-67, qual assurance officer, 68-71, asst to dir, 71-75, mgr prof rels & corp accounts, 76-77, dir, Dept Clin & Indust Toxicol, 78-81, dir, 81-84; dir, Pacific Toxicol Labs, 85-86; RETIRED. Mem: Am Chem Soc; Am Asn Clin Chem; Sigma Xi. Res: Clinical chemistry; adrenal hormones; bile pigments; enzymes; lipids; quality assurance; analytical toxicology. Mailing Add: 16055 Miami Way Pacific Palisades CA 90272-4232

JACOBS, SIGMUND JAMES, b Minneapolis, Minn, Mar 25, 12; m 44. PHYSICS, FLUID DYNAMICS. Educ: Univ Minn, BChE, 33, MS, 52; Univ Amsterdam, PhD(physics), 53. Prof Exp: Res assoc, Carnegie Inst Technol, 42-45; res assoc physics, Woods Hole Oceanog Inst, 46; res physicist, Naval Ord Lab, 46-50, chief detonation div, 50-57, sr scientist explosives res dept, 57-75; sr scientist Explosives Div, 75-80, CONSULT DETONATION PHYSICS, NAVAL SURFACE WEAPONS CTR, WHITE OAK LAB, 80- Honors & Awards: Meritorious Civilian Serv Award, US Navy, 51, 60; DuPont Medal, Soc Motion Picture & TV Eng, 64. Mem: Am Phys Soc; Am Chem Soc; Soc Motion Picture & TV Eng; Combustion Inst. Res: Deflagration of propellants and explosives; detonation of solid explosives; high pressure instrumentation; rapid expansion of gases; shock wave phenomena; electronic and photographic instrumentation; equation of fluid and solid states at high pressure. Mailing Add: 1208 Ruppert Rd Silver Spring MD 20903

JACOBS, STANLEY J, b Milwaukee, Wis, Feb 11, 36; m 65; c 2. GEOPHYSICS, FLUID MECHANICS. *Educ:* Northwestern Univ, BS, 59; Harvard Univ, AM, 60, PhD(appl math), 63. *Prof Exp:* Res fel atmospheric sci, Harvard Univ, 63-64; from asst prof to assoc prof, 64-74, PROF OCEANOG, UNIV MICH, ANN ARBOR, 74- *Mem:* Am Meteorol Soc. *Res:* Geophysical fluid mechanics. *Mailing Add:* Dept Atmospheric & Oceanic Sci Univ Mich Ann Arbor MI 48109

JACOBS, STANLEY S, b Rochester, NH, Apr 19, 40; m 79; c 2. ANTARCTIC OCEANOGRAPHY. *Educ:* Mass Inst Technol, BS, 62. *Prof Exp:* Res asst marine geophysics, Woods Hole Oceanog Inst, 61-62; from grad res asst to sr res asst, 62-72, res staff assoc, 72-73, SR STAFF ASSOC PHYS OCEANOG, LAMONT-DOHERTY GEOL OBSERV, COLUMBIA UNIV, 74- *Concurrent Pos:* Chief scientist oceanog res, NSF, Columbia Univ & US Coast Guard ships, 64-84; mem, Comt Polar Res & Panel Antartic Oceanog, Nat Acad Sci-Nat Res Coun, 71 & 84; prin investr, NAF, NASA & Dept Energy, 74- *Mem:* Am Geophys Union; Oceanog Soc. *Res:* Interactions between southern oceans and the Antarctic ice shelves; analyses of satellite sea ice observations; use of geochemical tracers in polar oceanography. *Mailing Add:* Lamont-Doherty Geol Observ Columbia Univ Palisades NY 10964

JACOBS, STEPHEN FRANK, b New York, NY, Oct 1, 28; m 63; c 3. PHYSICS. *Educ:* Antioch Col, BS, 51; Johns Hopkins Univ, PhD(physics), 56. *Prof Exp:* Engr, Perkin-Elmer Corp, 56-60; sr physicist, TRG, Inc, 60-65; PROF OPTICAL SCI, UNIV ARIZ, 65- *Mem:* Optical Soc Am; Am Phys Soc. *Res:* Infrared radiation detection; quantum electronics. *Mailing Add:* Box 7320 10100 E Catalina Hwy Tucson AZ 85749

JACOBS, THEODORE ALAN, b Atlanta, Ga, Oct 19, 27; m 61; c 1. CHEMICAL PHYSICS. *Educ:* Emory Univ, AB, 50; Univ Southern Calif, MSME, 54; Calif Inst Tech, PhD(chem physics), 60. *Prof Exp:* Designer, Douglas Aircraft Co, 51-52; res engr, G O Noville & Assocs, 53-55; res assoc & lectr mech eng, Univ Southern Calif, 55-57; sr res engr eng sci, Rocketdyne Div, NAm Aviation, Inc, 57-58; sr res fel, Calif Inst Tech, 60-61; head chem kinetics sect, Aerospace Corp, 61-67, head aerophys dept, 67-71; sr scientist & dir high energy laser technol, TRW, Inc, 71-76; supt, Optical Sci Div, Naval Res Lab, 76-78, dep asst secy navy, res eng & systs, 78-85; RETIRED. *Concurrent Pos:* Consult, Gen Appl Sci Labs, 58-60, Plasmadyne Corp, 60-61 & Vickers Div, Sperry Rand Corp, 61-62; pvt consult, 85- *Mem:* Fel Am Phys Soc; fel Optical Soc Am; Am Chem Soc; Am Defense Preparedness Asn; Sigma Xi. *Res:* High temperature chemical kinetics; chemical lasers; high energy lasers. *Mailing Add:* 4915 Loosestrife Ct Annandale VA 22003

JACOBS, THOMAS LLOYD, b Forest City, Iowa, Aug 18, 08; m 34; c 3. ORGANIC CHEMISTRY, POLYMER CHEMISTRY. *Educ:* Cornell Col, Iowa, AB, 30; Cornell Univ, PhD(org chem), 35. *Prof Exp:* Instr chem, Harvard Univ, 35-39; from instr to prof, 39-76, EMER PROF CHEM, UNIV CALIF, LOS ANGELES, 76- *Concurrent Pos:* Guggenheim fel, Univ Ill & Imp Col, London, 47-48; consult, Minn Mining & Mfg Co, 56-76. *Mem:* Am Chem Soc. *Res:* Chemistry of substituted acetylenes and allenes; heterocyclic compounds; synthesis of compounds for testing as antimalarials; stereochemistry of alicyclic compounds. *Mailing Add:* Dept Chem 4037 Young Hall Univ of Calif Los Angeles CA 90024-1569

JACOBS, VERNE LOUIS, b Los Angeles, Calif, Aug 30, 41; m 69; c 2. ATOMIC PHYSICS. *Educ:* Mass Inst Technol, BS, 64; Univ Calif, Berkeley, PhD(physics), 68. *Prof Exp:* Res fel appl math, Weizmann Inst Sci, 68-71; res fel atomic physics, Queen's Univ, Belfast, 71-72; Nat Res Coun assoc space physics, Goddard Space Flight Ctr, NASA, 72-74; scientist plasma physics, Sci Applns, Inc, 74-77; res physicist, Plasma Physics Div, 77-85, res physicist, E O Hulburt Ctr Space Res, 85-88, RES PHYSICIST, CONDENSED MATTER & RADIATION SCI DIV, NAVAL RES LAB, 88- *Mem:* Fel Am Phys Soc. *Res:* Atomic radiation processes in plasmas. *Mailing Add:* Condensed Matter & Radiation Sci Div Naval Res Lab Washington DC 20375

JACOBS, VIRGIL LEON, b Odin, Kans, Dec 27, 35; m 70. NEUROSCIENCE. *Educ:* St Benedict's Col, BS, 57; St Louis Univ, MS, 59; Univ Kans, PhD(anat), 65. *Prof Exp:* Instr anat, China Med Bd, Philippines, 62-63; fel neuroanat, Cent Inst Brain Res, Netherlands, 65-66; asst prof anat, Sch Med, Wayne State Univ, 66-69; asst prof & adv anat, AMA Educ Proj, Fac Med, Univ Saigon, 69-71; assoc prof anat, Sch Med, Univ Hawaii, 71-75; assoc prof anat, Tex A&M Univ, 75-78, res assoc, 78-80; ASSOC PROF ANAT, MOREHOUSE SCHOOL MED, ATLANTA, 80- *Concurrent Pos:* Exchange instr, Col Med, Univ Philippines, 62-63. *Mem:* Am Asn Anatomists; AAAS; Soc Neurosci; Sigma Xi. *Res:* Neuroanatomical pathways in reptiles; anatomical and behavioral studies on Flehmen in the Bovine; N Accumbens connections in the rat. *Mailing Add:* Morehouse Sch Med 830 Westview Dr SW Atlanta GA 30314

JACOBS, WILLIAM DONALD, b Birmingham, Ala, Apr 18, 28; m 63; c 3. ANALYTICAL CHEMISTRY. *Educ:* Col Charleston, BS, 51; Clemson Col, MS, 54; Univ Va, PhD(chem), 58. *Prof Exp:* Instr chem, Clemson Col, 52-54; instr chem, Gordon Mil Col, 54-55 & Univ Va, 57-58; asst prof, Univ Ga, 58-65; assoc prof, WGa Col, 65-68; assoc prof & chmn dept, 68-78, PROF CHEM, STILLMAN COL, 78-, CHMN, DIV MATH & SCI, 81- *Mem:* Am Chem Soc. *Res:* Spectrophotometric trace analysis; analytical chemistry of the platinum group elements. *Mailing Add:* 924 Cypress Garden's Rd Moncks Corner SC 29461

JACOBS, WILLIAM PAUL, b Boston, Mass, May 25, 19; m 49; c 2. PLANT DEVELOPMENT. *Educ:* Harvard Univ, AB, 42, MA, 45, PhD(biol), 46. *Prof Exp:* Fel histogenesis in vascular plants, Harvard Univ, 46-47, mem, Soc Fels, 47-48; mem fac, 48-89, EMER PROF BIOL, PRINCETON UNIV, 89- *Concurrent Pos:* Sheldon traveling fel, 45-46; Lalor fel, 50-51; sr fel, NSF, 56-57, sci fac fel, 62; vis prof, Univ Calif, Berkeley, 52, Zool Sta, Napoli, 57, Univ Oxford, 62, Univ Lausanne, 67, Univ Colo, 72, Univ Bristol, 80; mem adv panel, develop biol, NSF, 56, plant biol in space, NASA, 76-78 & US-Soviet space flights, 78; mem comt innovation in lab instr, Biol Sci Curriculum Study, Am Inst Biol Sci, 59-64; Guggenheim fel, 67. *Honors & Awards:* Morrison Award, NY Acad Sci, 51; Dimond Prize, Bot Soc Am, 75. *Mem:* Soc Develop Biol (secy, 58-60, pres, 60-61); Am Soc Plant Physiol; Bot Soc Am; Int Soc Plant Morphologists; Int Plant Growth Substances Asn; Int Phycol Soc. *Res:* Internal factors controlling cell and organ differentiation and longevity; hormone transport and polarity; gravitational effects on giant coenocytes. *Mailing Add:* Dept of Biol Princeton Univ Princeton NJ 08544

JACOBS, WILLIAM WESCOTT, b Madison, Wis, Sept 8, 43; m 77; c 2. INTERMEDIATE ENERGY NUCLEAR PHYSICS. *Educ:* Reed Col, Portland, BA, 65; Univ Wash, MS, 67, PhD(physics), 74. *Prof Exp:* Res assoc nuclear & atomic physics, Univ NC, Chapel Hill, 74-76; res assoc, 76-79, ASSOC RES SCIENTIST INTERMEDIATE ENERGY NUCLEAR PHYSICS, CYCLOTRON FACIL, IND UNIV, 79- *Mem:* Am Phys Soc; AAAS; Am Fedn Scientists. *Res:* Intermediate-energy nuclear physics; polarization effects in nuclear reactions and scattering; nuclear astrophysics; heavy-ion x-ray production; tests of fundamental symmetries. *Mailing Add:* Cyclotron Facil Milo B Sampson Ln Ind Univ Bloomington IN 47405

JACOBS, WILLIAM WOOD, JR, b Harrisburg, Pa, May 23, 47; m 69; c 3. ANIMAL BEHAVIOR, ETHOLOGY. *Educ:* Pa State Univ, BS, 69; Univ Chicago, MS, 71, PhD(biopsychol), 73. *Prof Exp:* Biologist rodenticides, US Environ Protection Agency, 74-75; fel olfaction & taste, Monell Chem Senses Ctr, 75-78; BIOLOGIST RODENTICIDES, US ENVIRON PROTECTION AGENCY, 78- *Concurrent Pos:* Fel, US Nat Inst Neurol Dis & Stroke, 75-78. *Res:* Investigations of taste, food selection, social and individual behavior of small mammals; use of containers in closed system transfer of pesticides; development of protective bait stations for rodenticides. *Mailing Add:* TS-767 Regist Div US Environ Protection Agency Washington DC 20460

JACOBS, WOODROW COOPER, oceanography, meteorology; deceased, see previous edition for last biography

JACOBSEN, BARRY JAMES, b Racine, Wis, Aug 6, 47; m 69; c 2. PLANT PATHOLOGY, ADMINISTRATION. *Educ:* Univ Wis-Madison, BSc, 69, MSc, 71; Univ Minn, PhD(plant path), 73. *Prof Exp:* From asst prof to prof plant path, Univ Ill, Urbana-Champaign, 73-87,proj leader, 79-87; PROF & HEAD DEPT, AUBURN UNIV, 87- *Concurrent Pos:* Vis prof, Friedrich Wilhelms Univ, WGer, 84-; consult plant path. *Honors & Awards:* Campbell Award, Am Phytopath Soc, 80; Merck Found Res Award. *Mem:* Sigma Xi; Am Phytopath Soc; Soc Nematologists; Organization Trop Am Nematologists. *Res:* Studies on generalized disease resistance in wheat and tomatoes; effects of pesticides on non-target plant pathogens and interactions of plant parasitic nematodes and fungi in plant diseases; chemical control of plant diseases; grain storage; development of disease management strategies. *Mailing Add:* Dept Plant Path Auburn Univ Auburn AL 36849-5409

JACOBSEN, CHRIS J, b Redwing, Minn, Oct 3, 60; m 82; c 2. X-RAY OPTICS. *Educ:* St Olaf Col, BA, 83; State Univ NY, Stony Brook, PhD(physics), 88. *Prof Exp:* Postdoctoral physics, Ctr X-Ray Optics, Lawrence Berkeley Lab, 88-89; postdoctoral physics, 89-91, ASST PROF PHYSICS, STATE UNIV NY, STONY BROOK, 91- *Concurrent Pos:* mem, Biophys Prog, State Univ NY, Stony Brook, 91- *Mem:* Optical Soc Am; Am Phys Soc. *Res:* X-ray microscopy using holography and zone plates; biological applications of x-ray microscopy; image processing; x-ray lithography; coherent x-ray sources. *Mailing Add:* Dept Physics State Univ NY-Stony Brook Stony Brook NY 11794-3800

JACOBSEN, DONALD WELDON, b Portland, Ore, Apr 26, 39; m 62; c 2. BIOCHEMISTRY, CHEMISTRY. *Educ:* Univ Pa, BA, 61; Ore State Univ, MS, 64, PhD(biochem, cell biol), 67. *Prof Exp:* Nat Insts Health fel biochem, 67-71, assoc biochem, Scripps Clin & Res Found, 71-84; PROF STAFF, CLEVELAND CLIN FOUND, 84- *Concurrent Pos:* Dernham fel, Am Cancer Soc, Calif, 72; mem, ad hoc adv comt on food hypersensitivity, Food & Drug Admin, 85-87; dir, Biochem Core Lab, Cleveland Clin, 87-; adj prof, Cleveland State Univ, 88-; chmn-elect, Cleveland Sect, Am Chem Soc, 90- *Mem:* AAAS; Am Chem Soc; Biophys Soc; Am Soc Microbiol; Am Soc Biochem & Molecular Biol; Am Soc Hemat. *Res:* Function of cobalamin-binding proteins; receptor-mediated transport of cobalamin-binding proteins; mechanisms of enzyme action, especially ribonucleotide reductase, methionine synthetase and other vitamin B12-dependent enzymes; cobalamin and cobinamide chemistry; synthesis and function of B12 coenzymes in mammalian cells; sulfur biochemistry of homocysteine as related to vascular disease; pathogenesis and prevention of sulfite hypersensitivity; angiotensin and the neurohormonal control of blood pressure. *Mailing Add:* Dept Lab Hemat L30 Cleveland Clin Found One Clin Ctr Cleveland OH 44195-5139

JACOBSEN, EDWARD HASTINGS, b Elizabeth, NJ, Jan 2, 26. PHYSICS, ELECTRON OPTICS. *Educ:* Mass Inst Technol, BS, 50, PhD(physics), 54. *Prof Exp:* Fulbright fel, Col France, 54-55; res physicist, Res Lab, Gen Elec Co, 55-61; Brit Dept Sci & Indust Res fel, Nottingham, 61; PROF PHYSICS, UNIV ROCHESTER, 62- *Concurrent Pos:* Vis scientist, Molecular Biol Lab, Harvard Med Sch, 65; vis prof, Mass Inst Technol, 67-68, vis scientist, Res Lab Electronics, 70-72. *Mem:* Am Phys Soc. *Res:* Magnetic resonance; x-ray and neutron diffraction; microwave ultrasonics; statistical mechanics; semiconductors; plasma physics; biophysics. *Mailing Add:* Dept Physics Univ Rochester Rochester NY 14627

JACOBSEN, FRED MARIUS, b Brooklyn, NY, May 19, 25; m 49; c 1. COMPUTER SCIENCE, APPLIED STATISTICS. *Educ:* Polytech Inst Brooklyn, BChE, 44; Iowa State Univ, PhD(chem eng), 54. *Prof Exp:* Minor shift foreman, US Army, Carbide & Carbon Chem Corp, 45; staff mem phys chem, Los Alamos Sci Lab, 46-50; asst chem engr, Inst Atomic Res, 51-54; chem engr statist, Am Oil Co, 55-56, group leader comput, 57-60, tech comput supvr, 61-69; dir comput serv, Standard Oil Co, Ind, 70-73, res supvr comput, Amoco Corp, 74-83,; CONSULT,83- *Concurrent Pos:* Consult, 83- *Mem:* Fel AAAS; Asn Comput Mach; Am Inst Chem Engrs; Am Phys Soc; Inst Math Statist; Am Chem Soc. *Res:* Scientific computing; engineering statistics; applied mathematics; systems design. *Mailing Add:* 828 Hawthorn Dr Naperville IL 60540-7425

JACOBSEN, FREDERICK MARIUS, b Ames, Iowa, Jan 19, 54; m 83. PSYCHOPHARMACOLOGY, BIOLOGICAL PSYCHIATRY. *Educ:* Cornell Univ, AB, 76; Univ Ill, MPH, 78, MD, 80; Am Bd Psychiat & Neurol, cert. 85. *Prof Exp:* Fel, Yale Univ Sch Med, 80-84, resident psychiat, 84; clin assoc, 84-87, med officer, 87, VIS SCIENTIST, NIMH, CLIN PSYCHOBIOL BR & LAB CLIN SCI, 88- *Concurrent Pos:* Asst clin prof psychiat, George Washington Univ, 85-; med dir, Transcult Mental Health Inst, 86-; med adv, Depressive & Manic Depressive Asn, 87-; consult, VOCA Corp, 87- *Mem:* Soc Biol Sci; AAAS; Math Asn Am; Am Psychiat Asn; AMA; NY Acad Sci. *Res:* Biological psychiatry; physiological bases of depression and manic-depressive illness; neuropsychiatric diagnosis; public health; manifestations of illness from a cultural perspective. *Mailing Add:* 1301 20th St NW Suite 711 Washington DC 20036

JACOBSEN, LYNN C, b Montevideo, Minn, Mar 12, 18; m 44; c 4. GEOLOGY. *Educ:* Univ Minn, BA, 41; Univ Okla, MS, 48; Pa State Col, PhD(mineral), 53. *Prof Exp:* Jr geologist, Panama Canal, 41-42; jr & asst geologist, US Geol Surv, 42-45; geologist, Union Oil Co, Calif, 45-47; instr geol, Univ Ky, 48-50 & 52-53, asst prof, 53-55; sr res geologist, Sohio Petrol Co, 55-59, asst to sr vpres, 59-70, mgr uranium opers, 70-78; CONSULT MINERAL INDUST, 78- *Concurrent Pos:* adj prof, NMex Inst Mines & Tech. *Mem:* AAAS; Geol Soc Am; Soc Econ Paleont & Mineral; Am Asn Petrol Geologists. *Res:* Sedimentary petrology; subsurface geology; fuel and mineral economics. *Mailing Add:* 510 Fruit NE Albuquerque NM 87102

JACOBSEN, NADINE KLECHA, b Milwaukee, Wis, Dec 13, 41; m 67. PHYSIOLOGICAL ECOLOGY. *Educ:* Drake Univ, BA, 64; Ore State Univ, MS, 66; Cornell Univ, PhD(wildlife ecol), 73. *Prof Exp:* Instr biol, Blue Mountain Community Col, 66-67; res assoc, Dept Radiation Biol, Cornell Univ, 68-69, fel, Dept Natural Resources, 73-74; ASSOC PROF WILDLIFE BIOL, UNIV CALIF, DAVIS, 74- *Mem:* Sigma Xi; Am Soc Mammal; Wildlife Soc; Am Soc Animal Scientists; Ecol Soc Am. *Res:* Ecological energetics of wildlife, particularly deer, over the annual cycle and between birth and weaning of young; how energy and nutrient metabolism are affected by environmental, behavioral and physiological states. *Mailing Add:* Div Wildlife & Fisheries Biol Univ Calif Davis CA 95616

JACOBSEN, NEIL SOREN, b Waterloo, Iowa, June 13, 30; m 54; c 3. ACADEMIC ADMINISTRATION. *Educ:* Univ Iowa, BA, 52; Univ Denver, MS, 56; Okla State Univ, PhD(physiol), 65. *Prof Exp:* Teacher high schs, Calif, 57-62; NIH fel, Univ Mass, 64-66; asst prof, NDak State Univ, 66-71, dir student acad affairs, 69-72, actg vpres acad affairs, 79-80, assoc prof zool, 71-86, dean univ studies, 72-86, assoc vpres acad affairs, 81-86; RETIRED. *Res:* Lipid metabolism of intestinal parasites; effect of pesticides on the cardiovascular system. *Mailing Add:* 5283 A Beach Dr State Univ Sta St Petersburg FL 33705

JACOBSEN, RICHARD T, b Pocatello, Idaho, Nov 12, 41; m 73; c 5. MECHANICAL ENGINEERING, THERMODYNAMICS. *Educ:* Univ Idaho, BS, 63, MS, 65; Wash State Univ, PhD(eng sci), 72. *Prof Exp:* From instr to prof eng, Univ Idaho, 64-80, prof mech eng & chmn dept, 80-85, assoc dir, ctr appl thermodyn studies, 75-86, assoc dean eng, 85-90, DIR, CTR APPL THERMODYN STUDIES, UNIV IDAHO, 86-, DEAN ENG, 90- *Concurrent Pos:* Mem correlating functions working panel, Int Union Pure & Appl Chem, 73-86; guest worker, Thermophysics Div, Nat Bur Standards, 79-86. *Mem:* Sigma Xi; fel Am Soc Mech Engrs; Soc Automotive Engrs; Am Soc Heating Refrig & Air Conditioning Engrs; Nat Soc Prof Engrs; Am Soc Eng Educ. *Res:* Thermodynamics; thermodynamic properties of fluids; thermodynamic system analysis. *Mailing Add:* Janssen Eng Bldg 125 Univ Idaho Moscow ID 83843

JACOBSEN, STEIN BJORNAR, b Baerum, Norway, Feb 12, 50. GEOCHEMISTRY. *Educ:* Univ Oslo, Norway, BS, 75, MS, 75; Calif Inst Technol, PhD(geochem), 80. *Prof Exp:* Fel geochem, Calif Inst Technol, 80-81; asst prof, 81-83, ASSOC PROF GEOCHEM, HARVARD UNIV, 84- *Mem:* Am Geophys Union; Geol Soc Am; Am Chem Soc; Meteoritical Soc. *Res:* Neodymium, strontium and lead isotope studies of mantle structure and differentiation, crustal evolution and chronology; petrological and geochemical studies of granulite and eclogite facies rocks, ophiolites, orogenic peridotites and chondritic meteorites. *Mailing Add:* Dept Earth & Planetary Sci Harvard Univ 20 Oxford St Cambridge MA 02138

JACOBSEN, STEPHEN C, ENGINEERING. *Prof Exp:* DIR, CTR ENG DESIGN. *Mailing Add:* 3176 Merrill Eng Bldg Ctr Eng Design Salt Lake City UT 84112

JACOBSEN, TERRY DALE, b Nampa, Idaho, Aug 17, 50; m 79. PLANT SYSTEMATICS. *Educ:* Col Idaho, BS, 72; Wash State Univ, MS, 75, PhD(bot), 78. *Prof Exp:* Lectr, Biol Prog, Wash State Univ, 77-78, instr bot, 78-79; asst to dir & res scientist, 79-81, ASST DIR & SR RES SCIENTIST, HUNT INST BOT DOC, CARNEGIE-MELLON UNIV, 81- *Concurrent Pos:* Adj res scientist, Carnegie Mus Natural Hist, 79-; ed, Bull Hunt Inst Bot Doc, 81-; adj assoc prof biol, Carnegie-Mellon Univ, 84- *Res:* Cytotaxonomy, anatomy and numerical analysis of the genus Allium L in North America. *Mailing Add:* Hunt Inst Bot Doc Carnegie-Mellon Univ 5000 Forbes Ave Pittsburgh PA 15213

JACOBSEN, WILLIAM E, b Brooklyn, NY, Dec 30, 12. NAVAL MACHINERY & SUBMARINE DESIGN. *Educ:* Polytech Inst Brooklyn, BEE, 34. *Prof Exp:* Test prog engr, Gen Elec Co, 34-37, mgr marine appln eng & design engr, 37-47, marine appln engr, 47-72; RETIRED. *Mem:* Fel Inst Elec & Electronics Engrs; Am Soc Naval Engrs; Soc Naval Archit & Marine Eng. *Res:* Marine power applications; electric propulsion drives. *Mailing Add:* Gen Elec Co One River Rd Schenectady NY 12305

JACOBS-LORENA, MARCELO, b Sao Paulo, Brazil, May 5, 42; m 70; c 1. MOLECULAR BIOLOGY, EMBRYOLOGY. *Educ:* Sao Paulo Univ, BS, 64; Osaka Univ, MS, 67; Mass Inst Technol, PhD(biol), 72. *Prof Exp:* Fel biol, Univ Geneva, 72-77; asst prof, 77-83, ASSOC PROF DEVELOP GENETICS, CASE WESTERN RESERVE UNIV, 83- *Concurrent Pos:* Fel, Europ Molecular Biol Orgn, 72-74; asst prof, Am Cancer Soc Inst grant, 77-78, NIH, 78-83, 85-90 & NSF, 83-86. *Mem:* Soc Develop Biol; Soc Cell Biol. *Res:* Drosophilia oogenesis and embryogenesis, Drosophila development and control of gene expression. *Mailing Add:* Dept Genetics Case Western Reserve Univ 2119 Abington Rd Cleveland OH 44106

JACOBSMEYER, VINCENT PAUL, b St Louis, Mo, July 7, 06. SOLID STATE PHYSICS. *Educ:* Gonzaga Univ, AB, 31; St Louis Univ, MA, 33, MS, 35, PhD(physics), 45; St Mary's Col, STL, 40. *Prof Exp:* Instr high sch, Marquette Univ, 35-36; from instr to prof, 44-74, EMER PROF PHYSICS, ST LOUIS UNIV, 74- *Mem:* Am Asn Physics Teachers. *Res:* Photoelectricity; photoconductivity; Hall effect and resistivity studies of semiconductors. *Mailing Add:* 3601 Lindell Blvd 233 N New Ballas Rd St Louis MO 63108

JACOBSOHN, GERT MAX, b Berlin, Ger, Aug 1, 29; nat US; m 59; c 4. BIOCHEMISTRY. *Educ:* Ill Col, AB, 52; Purdue Univ, MS, 55, PhD(biochem), 57. *Prof Exp:* Fel, Col Physicians & Surgeons, Columbia Univ, 57-60; asst mem endocrinol, Albert Einstein Med Ctr, 60-61; from asst prof to assoc prof, 62-77, PROF BIOL CHEM, HAHNEMANN UNIV, 77- *Mem:* AAAS; fel Am Cancer Soc; Endocrine Soc; Am Soc Biol Chem; Pan-Am Pigment Cell Soc. *Res:* Metabolic control of blood enzymes; membrane transfer phenomena; steroid interconversion and function in plants; regulation of steroid, methods of steroid assay and instrumentation; metabolism of catechol estrogens; melanin formation. *Mailing Add:* Dept Biol Chem Hahnemann Univ Philadelphia PA 19102

JACOBSOHN, MYRA K, b New York, NY, Feb 13, 39; c 4. MEMBRANE BIOCHEMISTRY. *Educ:* Columbia Univ, BA, 60; Univ Pa, MS, 62; Bryn Mawr Col, PhD(biol), 75. *Prof Exp:* Res assoc, Hahnemann Med Col, 71-74, fel biochem, 74-76; lectr, 74-76, from asst prof to assoc prof, 76-90, PROF BIOL, BEAVER COL, 90-; RES ASSOC, HAHNEMANN UNIV, 85- *Honors & Awards:* Lindback Award, 82. *Mem:* AAAS; Am Inst Biol Sci. *Res:* Enzymes; membrane structure; fungistatic substances. *Mailing Add:* Dept Biol Beaver Col Glenside PA 19038

JACOBSON, ABRAM ROBERT, PHYSICS. *Prof Exp:* US Army Corps Engrs, 44-45; eng asst, Harvard Univ, 48-49; staff mem, Earth Dams Sect, US Bur Reclamation, 50-53; staff mem, 53, assoc, 55, PARTNER, MUESER RUTLEDGE CONSULT ENGRS, 73- *Concurrent Pos:* In charge geotech studies, Reconstruct E Front, US Capitol Bldg, 55-60, US Navy Dry Dock 6, 56-60, landslide study, Pac Palisades, Los Angeles, 58-60, underpinning, House Rep Wing, 62-64, NY Waterfront Redevelop, 65-68 & Wash Metro Subway, 66-, Interceptor Sewer, Charleston, SC, 66-70, Nat Gallery Art Add, 70-72, Terminals Develop, Port Corp, 70-74, Battery Park City Develop, NY, 70-76, Locks C&E, Tenn & Tombigbee, 72-78, Dry Dock 4, 79-81, Nuclear Plant, Power Block Underpinning, Midland, Mich, 80-84, S Quadrangle Develop, Smithsonian Inst, 84-86, Little River Dam, Durham, NC, 82-88, Baldwin Bridge Replacement, 86-89, Schoharie Creek Bridge Failure Study, 87, US Navy Homefort Pier, Staten Island, 86-88, Conrail Tunnel Rehab, WPt, 85-88, Park Ave Tunnel Rehab, & NY Rock Slope Stabil, Westchester County, 88; consult mem adv bd, Wash Metro Subway, MBTA Boston Red Line Ext, Midland Nuclear Power Plant, Los Angeles Subway Construct Underpinning Mgt, Dallas Rapid Transit Proj, Utah Super Collider Proposal & Channel Tunnel Review Panel. *Honors & Awards:* Kapp Mem Lect, Am Soc Civil Engrs, 74 & 82; Crom Lectr, Univ Fla, 78; Haley Mem Lect, Soc Civil Engrs, 85. *Mem:* Am Soc Civil Engrs. *Res:* Author of 30 publications. *Mailing Add:* MS D466 LANL Los Alamos NM 87545

JACOBSON, ADA LEAH, b Boston, Mass, Oct 8, 33; m 58. PHYSICAL CHEMISTRY, BIOCHEMISTRY. *Educ:* Mass Inst Technol, BS, 54; Yale Univ, PhD(chem), 57. *Prof Exp:* Chemist, Shell Develop Co, Calif, 57-58; instr phys chem, Albertus Magnus Col, 58-59; instr biochem, Dartmouth Med Sch, 59-60; from instr to assoc prof chem, 60-86, ASSOC PROF BIOCHEM, UNIV CALGARY, 86- *Concurrent Pos:* Fel phys org chem, Yale Univ, 58-59; sr fel, Can Heart Found, 65- *Mem:* Am Chem Soc; Can Biochem Soc; Sigma Xi; Biophys Soc. *Res:* Physical chemistry of proteins, solutions and polymers. *Mailing Add:* 312 Superior Ave SW Calgary AB T3C 2J2 Can

JACOBSON, ALBERT H(ERMAN), JR, b St Paul, Minn, Oct 27, 17; m 60; c 2. INDUSTRIAL ENGINEERING, SYSTEMS ENGINEERING. *Educ:* Yale Univ, BS, 39; Mass Inst Technol, SM, 52; Univ Rochester, MS, 54; Stanford Univ, PhD(mgt eng), 76. *Prof Exp:* Personnel asst, Yale Univ, 39-40; indust engr, Radio Corp Am, NJ, 40-43; from chief engr to tech dir, Bur Ord, Navy Ord Div, Eastman Kodak Co, 46-57, from staff engr to mgr field opers, Space Satellite Prog, Apparatus Div, 57-59; assoc dean, Col Eng & Archit, Pa State Univ, 59-61; pres & gen mgr, Knapic Electro-Physics Co, Calif, 61-62; prof eng, 62-69, PROF INDUST & SYSTS ENG, SAN JOSE STATE UNIV, 69-, COFOUNDER & COORDR CYBERNETIC SYSTS GRAD PROG, 67- *Concurrent Pos:* Alfred P Sloan fel, Mass Inst Technol, 51-52; consult numerous indust firms, 63-; NSF sci fac fel, Stanford Univ, 65-66. *Mem:* AAAS; Am Soc Eng Educ; Am Inst Indust Engrs; Am Prod & Inventory Control Soc; Sigma Xi. *Res:* Infrared; guided missiles; satellites; engineering management; management development; cybernetic systems; transportation systems; information systems. *Mailing Add:* Sch Eng One Washington Sq San Jose CA 95192

JACOBSON, ALEXANDER DONALD, b New York, NY, Dec 1, 33; m 71; c 2. ARTIFICIAL INTELLIGENCE TECHNOLOGY, ELECTRICAL ENGINEERING. *Educ:* Univ Calif, Los Angeles, BS, 55, MS, 58; Calif Inst Technol, PhD(electro-magnetic theory), 64. *Prof Exp:* Mem tech staff optics res, Hughes Res Labs Div, Hughes Aircraft Co, 55-68, head unconventional imaging sect, 68- 72, assoc mgr explor studies dept, 72-76, prog mgr liquid crystal displays progs, Indust Prod Div, 76-77; consult, 77-79; CO-FOUNDER, CHMN & CHIEF EXEC OFFICER, INFERENCE CORP, 79- *Honors & Awards:* Rank Prize, 86. *Res:* Laser technology; electromagnetic theory; x-ray diffraction studies of crystals; display technology; computer systems; telecommunications; artificial intelligence computer software products; applications of artificial intelligence technology to the development of expert systems for industrial, commercial and military applications; development of computer software products that contain artificial intelligence technology for such purposes. *Mailing Add:* 12256 Canna Rd Los Angeles CA 90049

JACOBSON, ALLAN JOSEPH, b Newcastle, Eng, May 28, 44; m 68; c 1. INTERCALATION CHEMISTRY, SOLID STATE CHEMISTRY. *Educ:* Oxford Univ, BA, 65, MA, 69, DPhil, 69. *Prof Exp:* Lectr chem, Oxford Univ, 70-76; sr staff chemist, 76-80, RES ASSOC, EXXON RES & ENG CO, EXXON CORP, 80- *Mem:* Chem Soc; Am Chem Soc. *Res:* Synthetic and structural solid state inorganic chemistry; neutron x-ray and electron diffraction; intercalation chemistry. *Mailing Add:* Exxon Res & Eng Co Annandale NJ 08801

JACOBSON, ALLAN STANLEY (BUD), b Chattanooga, Tenn, June 18, 32; m 86. COMPUTER GRAPHICS, GAMMA RAY ASTRONOMY. *Educ:* Univ Calif Los Angeles, AB, 62; Univ Calif San Diego, MS, 64 & PhD(physics), 68. *Prof Exp:* Res asst, Univ Calif San Diego, 62-68, asst res physicist, 68-69; mem tech staff, 69-73, tech group supvr, 73-86, SR RES SCIENTIST, JET PROPULSION LAB, 81- *Concurrent Pos:* Guest lectr, Calif Inst Technol, 80-82; prin investr, NASA, 70-; consult, Ashton-Tate, 85-89; mem, Comput Soc, Inst Elec & Electronic Engrs. *Honors & Awards:* Bruno Rossi Prize, Am Astron Soc, 86. *Mem:* Fel Am Phys Soc; Am Astron Soc; Asn Comput Mach; Inst Elec & Electronics Engrs; Inst Elec & Electronics Engrs Computer Sci. *Res:* Gamma-ray astronomy; computer-aided science visualization; visual science data analysis. *Mailing Add:* Jet Propulsion Lab 4800 Oak Grove Dr 183-501 Pasadena CA 91109

JACOBSON, ALLEN F, b Omaha, Nebr, Oct 7, 26. MANUFACTURING ADMINISTRATION. *Educ:* Iowa State Univ, Ames, BS, 47. *Prof Exp:* Prod engr, Tape Lab, 3M, 47-50, tech asst to plant mgr, Hutchinson, 50-53, Bristol, 53-55, tape prod supt, Bristol, 55- 59, plant mgr, St Paul Tape Prod, 59-61, combined Tape & AC&S Opers, Bristol, 61-63, tape prod mgr, Tape & Allied Prod Group, 63, mfg mgr, 63-68, gen mgr, Indust Tape Div, 68-70, div vpres, 70-72, exec vpres & gen mgr, 3M Can Ltd, 73-75, vpres, Europ Opers, 75, Tape & Allied Prod Group, 75-81, exec vpres, Indust & Consumer Sector, pres, US Opers, 84-86, CHMN BD & CHIEF EXEC OFFICER, 3M, 86- *Concurrent Pos:* Dir, Mobil Corp, New York, NY, Northern States Power Co, Minneapolis, Minn, US West, Inc, Englewood, Colo, Valmont Industs, Inc, Valley, Nebr, Potlatch Corp, San Francisco, Calif, Sara Lee Corp, Chicago, Ill, Minn Bus Partnership, Minneapolis; mem bd dirs, Am Qual Found, Chamber Com US & Nat Legal Ctr Pub Interest; chmn, US Coun Int Bus, Emergency Comt Am Trade; mem, Bus Can, Washington, DC, Bus Roundtable, US-USSR Trade & Econ Coun, Inc & Bus Coun Sustainable Develop. *Mem:* Nat Acad Eng; hon mem Soc Mfg Engrs. *Mailing Add:* 3M Ctr Bldg 220-14W-04 St Paul MN 55144-1000

JACOBSON, ANN BEATRICE, b New York, NY, July 24, 38. MOLECULAR BIOLOGY. *Educ:* Univ Chicago, BS, 58, PhD(bot), 62; Purdue Univ, MS, 61. *Prof Exp:* Res assoc plant biochem, Univ Chicago, 63-67; biologist, Biol Div, Oak Ridge Nat Lab, 67-71; vis scientist, Max Planck Inst Biochem, Munich, 71-75; Europ Molecular Biol Orgn Fel, Dept Molecular Biol, Univ Geneva, Switz, 75-76; lectr, 76-80, res asst prof, 80-86, RES ASSOC PROF, DEPT MICROBIOL, STATE UNIV NY, STONY BROOK, 86- *Mem:* AAAS; Am Soc Microbiol; Am Women Sci. *Res:* RNA folding, computer modeling and electron microscopy. *Mailing Add:* Dept Microbiol State Univ NY Stony Brook NY 11794

JACOBSON, ANTONE GARDNER, b Salt Lake City, Utah, May 22, 29; m 63; c 2. DEVELOPMENTAL BIOLOGY. *Educ:* Harvard Univ, AB, 51; Stanford Univ, PhD(exp embryol, biol), 55. *Prof Exp:* From instr to assoc prof, 57-68, PROF ZOOL, UNIV TEX, AUSTIN, 68- *Mem:* AAAS; Am Soc Zool; Soc Develop Biol; Int Soc Develop Biol. *Res:* Embryonic induction, morphogenesis and development of the nervous system in embryos. *Mailing Add:* Dept Zool Univ Tex Austin TX 78712-1064

JACOBSON, ARNOLD P, b Rawlins, Wyo, July 21, 32; m 65. RADIATION BIOLOGY. *Educ:* Univ Wyo, BS, 58, MS, 60; Univ Mich, MPH, 62, PhD(radiation biol), 66. *Prof Exp:* Asst prof environ health, 65-79, assoc prof, 71-79, PROF ENVIRON & INDUST HEALTH, SCH PUB HEALTH, UNIV MICH, ANN ARBOR, 79- *Concurrent Pos:* Assoc res scientist, Inst Environ & Indust Health, 74-80. *Honors & Awards:* Marie Curie Gold Medal Award, Health Physics Soc. *Mem:* Am Radon Asn; Health Physics Soc. *Res:* Indoor radon risk, mitigation and measurement; low dose effects of radiations. *Mailing Add:* 9580 Stinchfield Woods Pickney MI 48169

JACOBSON, ARTHUR E, b New York, NY, May 2, 28; m 64; c 2. MEDICINAL CHEMISTRY. *Educ:* Fordham Univ, BS, 49; Rutgers Univ, MS(pharm chem), 52, MS(chem), 54, PhD(org & phys chem), 60. *Prof Exp:* Nat Insts Health, US Pub Health Serv fel & org chemist, Albert Einstein Col Med, 59-62; RES CHEMIST, NAT INST ARTHRITIS, METAB & DIGESTIVE DIS, NIH, 62- *Concurrent Pos:* Instr spectros, Found Adv Ed in Sci, Inc, 64-88 mem, Comt Probs Drug Dependence, 74-81; mem, Comt Probs Drug Dependence, 74-, chmn, drug testing prog, 77-; adj prof pharmacol & toxicol, Med Col Va, Va Commonwealth Univ, 84- *Honors & Awards:* J Michael Morrison Award, 90. *Mem:* AAAS; Am Chem Soc. *Res:* Synthesis of alkaloids and heterocycles; computer-assisted molecular modelling; spectroscopy; synthesis of affinity ligands for, and characterization of, opioid and phencyclidine receptors. *Mailing Add:* Nat Inst Diabetes & Digestive & Kidney Dis Bldg 8 Rm B1-22 Bethesda MD 20892

JACOBSON, BARRY MARTIN, mechanistic organic chemistry, pericyclic reactions, for more information see previous edition

JACOBSON, BARUCH S, b New York, NY, Nov 23, 25; m 51; c 4. BIOPHYSICS. *Educ:* Columbia Univ, AB, 51; Univ Calif, PhD(biophys), 56. *Prof Exp:* NSF fel, Donner Lab, Univ Calif, 56-58, biophysicist, 58; from instr to asst prof zool, Univ Tex, 58-61; asst prof radiol, Univ Minn, Minneapolis, 61-68; ASSOC PROF PHYSICS, CENT MICH UNIV, 68- *Concurrent Pos:* Vis res prof, Inst Ecol, Univ Calif, Davis, 75; res fel, Battelle Pac Northwest Labs, Richland, Wash, 82-83; fac res fels, Battelle Pac Northwest Labs, Richland, Wash, 85, 86 & 87. *Mem:* AAAS; Biophys Soc; Radiation Res Soc; Sigma Xi. *Res:* Cellular radiobiology; effects of ionizing radiation on cell reproduction; metabolic reversal of radiation damage; ultraviolet photobiology; environmental systems analysis. *Mailing Add:* Dept Physics Cent Mich Univ Mt Pleasant MI 48859

JACOBSON, BERNARD, b Cleveland, Ohio, Apr 7, 28; m 56; c 2. MATHEMATICS. *Educ:* Western Reserve Univ, BS, 51; Mich State Univ, MS, 52, PhD(math), 56. *Prof Exp:* From asst prof to assoc prof math, Franklin & Marshall Col, 56-61; assoc dir comt on undergrad prog in math, Math Asn Am, 62-63; assoc prof, 63-69, PROF MATH, FRANKLIN & MARSHALL COL, 69- *Mem:* Am Math Soc; Math Asn Am. *Res:* Number theory. *Mailing Add:* Dept of Math Franklin & Marshall Col Lancaster PA 17604

JACOBSON, BERTIL, b Stockholm, Sweden, Jan 21, 23; m 50; c 2. HUMAN IMPRINTING. *Educ:* Karolinska Inst, MD, 50. *Prof Exp:* Asst prof med electronics, Karolinska Inst, 57-60, prof med eng, 60-89; RETIRED. *Res:* Human imprinting; perinatal origin of self-destructive adult behavior. *Mailing Add:* Vastra Myrskaren 41 Huddinge Univ Hosp F60 Novum Balsta S-19800 Sweden

JACOBSON, BRUCE SHELL, b Los Angeles, Calif, Jan 11, 40. BIOLOGICAL CHEMISTRY. *Educ:* Calif State Col, Los Angeles, BA, 62, MA, 64; Univ Calif, Los Angeles, PhD(plant biochem), 70. *Prof Exp:* Res biologist, Univ Calif, Riverside, 70-71; res biochemist, Univ Calif, Davis, 71-73; res assoc, Harvard Univ, 73-77; from asst prof to assoc prof biochem, 77-86; PROF BIOCHEM, UNIV MASS, AMHERST, 86- *Concurrent Pos:* Estab investr, Am Heart Asn; NIH predoctoral fel, Univ Calif, Los Angeles; NIH postdoctoral fel, Harvard Univ; fel Maria Moor Cabot, Harvard Univ. *Mem:* Am Soc Biol Chem; Am Soc Cell Biologists. *Res:* Molecular and cellular biology of biological membranes; interaction of plasma membrane receptors with the extracellular matrix and the cytoskeleton; transcellular polarity of the plasma membrane in vascular endothelial cells; protein targetting the apical and basolateral plasma membrane domains during endo-lexocytosis, transcytosis, membrane protein synthesis and cell movement. *Mailing Add:* Dept Biochem Univ Mass Amherst MA 01003

JACOBSON, ELAINE LOUISE, b Miller, Kans, Mar 29, 45; m 67. BIOCHEMISTRY. *Educ:* Kans State Univ, BS, 67, PhD(biochem), 71. *Prof Exp:* Res fel biochem, Mayo Clin, Found & Grad Sch Med, 72-73; res assoc biochem, Dept Chem, NTex State Univ, 74-77; asst prof biol, Tex Woman's Univ, 77-81, assoc prof, 81-; ASSOC PROF MED & BIOCHEM, TEX COL OSTEOPATH MED. *Concurrent Pos:* NIH fel, Mayo Clin, Found & Grad Sch Med, 72-73, Extramanal Assoc, 79-80. *Mem:* Am Soc Biochem & Molecular Biol; Tissue Cult Asn. *Res:* Regulation of poly (adenosine diphosphate ribose) synthesis with a particular interest in events following DNA damage by carcinogens; niacin status and cancer prevention; poly(ADP-ribose) metabolism as a target for design of anti-cancer drugs. *Mailing Add:* Dept Med & Biochem Tex Col Osteopath Med 3500 Camp Bowie Ft Worth TX 76107

JACOBSON, EUGENE DONALD, b Bridgeport, Conn, Feb 19, 30; m 73; c 5. PHYSIOLOGY. *Educ:* Wesleyan Univ, BA, 51; Univ Vt, MD, 55; State Univ NY, MS, 60. *Prof Exp:* Intern, State Univ NY, 55-56, resident internal med, 57-60; assoc prof physiol, Sch Med, Univ Calif, Los Angeles, 64-66; prof & chmn dept, Sch Med, Univ Okla, 66-71; prof physiol & chmn dept, Univ Tex Med Sch, Houston, 71-77; vdean, Col Med, Univ Cincinnati, 77-85; dean, Sch Med, Univ Kans, 85-88; dean, 88-90, PROF MED & PHYSIOL, SCH MED, UNIV COLO, 90- *Concurrent Pos:* NIH spec fel, Univ Calif, Los Angeles, 64-66, NIH res career develop award, 66; consult, Gen Med Study Sect, NIH, 68-72, Vet Admin, 69-71 & Upjohn Co, Mich, 70-78; prof physiol, Baylor Col Med, 72-77; chairperson, Nat Comn Digestive Diseases, 77-79; mem, Nat Digestive Dis Adv Bd, NIH, 85-87. *Mem:* Am Physiol Soc; Soc Exp Biol & Med; Am Fedn Clin Res; Am Soc Clin Invest; Sigma Xi; Am Gastroenterol Asn (pres, 89-90); Asn Am Physicians. *Res:* Splanchnic circulation; gastrointestinal physiology. *Mailing Add:* 2000 E 12th Ave Box 27 Denver CO 80206

JACOBSON, FLORENCE DORFMAN, b Chicago, Ill, Mar 25, 18; m 42; c 2. MATHEMATICS. *Educ:* Univ Chicago, SB, 38, SM, 40. *Prof Exp:* Instr math, Univ NC, 42-43; lectr, 55-57, from asst prof to assoc prof, 58-66, PROF MATH, ALBERTUS MAGNUS COL, 66-, CHMN DEPT, 80- *Concurrent Pos:* Consult, Sch Math Study Group, 58-; mem adv panel, In-Serv Inst, NSF, 61-66, sci fac fel, 62-63. *Mem:* Am Math Soc; Math Asn Am. *Res:* Non-associative algebras; in-service and preservice education in mathematics for teachers from kindergarten through junior college. *Mailing Add:* Dept Math Albertus Magnus Col Prospect Ct Hamden CT 06511

JACOBSON, FRANK HENRY, b Providence, RI, Sept 11, 15; m 40; c 1. PHYSIOLOGY. *Educ:* Emory Univ, BA, 42; Univ Rochester, PhD(physiol), 51. *Prof Exp:* From instr to asst prof physiol, Jefferson Med Col, 51-59; physiologist, Aerospace Med Res Dept, US Naval Develop Ctr, 59-70, Appl Physiol Lab, Aircraft Crew Systs Directorate, 70-81; RETIRED. *Mem:* Am Physiol Soc; Aerospace Med Asn. *Res:* Temperature regulation of mammals; hypothalamus; acceleration; mechanism and pharmacology of cerebral concussion, sleep, arousal and affective illness. *Mailing Add:* 1035 Gravel Hill Rd Southampton PA 18966

JACOBSON, GAIL M, b Bartlesville, Okla, Feb 6, 38; m 63; c 3. BIOCHEMISTRY. *Educ:* Mt Holyoke Col, BA, 60, MA, 62; Cornell Univ, PhD(biochem), 66. *Prof Exp:* Res fel biochem, Calif Inst Technol, 66-68; asst mem, Okla Med Res Found, 68-76; LECTR, DEPT CHEM, CALIF POLYTECH STATE UNIV, 76- *Concurrent Pos:* Vis asst prof nutrit, Univ Okla, 68-75. *Mem:* Sigma Xi. *Res:* Enzymology; enzyme kinetics; coenzyme B12; enzymes; methylation of ribosomal RNA. *Mailing Add:* 156 Broad San Luis Obispo CA 93401

JACOBSON, GLEN ARTHUR, b Greensburg, Kans, Feb 10, 26; m 50; c 2. FOOD CHEMISTRY. *Educ:* Kans State Univ, BS, 48, MS, 49; Univ Ill, PhD(food technol, biochem), 53. *Prof Exp:* Res chemist, Res Labs, Swift & Co, 53-56; div head, Fat & Oil Res, Basic Res Dept, Campbell Soup Co, 56-70, DIV HEAD, LIPID CHEM RES, BASIC RES DEPT, CAMPBELL INST FOOD RES, 70- *Mem:* Am Chem Soc; Am Oil Chem Soc; Inst Food Technologists. *Res:* Fats and oil chemistry; biochemistry. *Mailing Add:* Basic Res Div Box 57L Campbell Soup Co Camden NJ 08101

JACOBSON, GUNNARD KENNETH, b Phoenix, Ariz, Feb 19, 47; m 71. FERMENTATION. *Educ:* Univ Ariz, BS, 69; Ore State Univ, PhD(microbiol), 73. *Prof Exp:* Fel, Univ Chicago, 72-75; res assoc, Argonne Nat Lab, 75-77; scientist, 77-82, sr scientist, molecular biol, 82-88, PROJ SCIENTIST, UNIVERSAL FOODS CORP, 88- *Mem:* Am Soc Microbiol; AAAS; Genetics Soc Am; Sigma Xi. *Res:* Genetics and molecular biology of industrial yeasts--new strain construction and improvement by classical genetics and recombinant DNA technology. *Mailing Add:* Universal Foods Corp 6143 N 60th St Milwaukee WI 53218

JACOBSON, HAROLD, b New York, NY, Jan 15, 29; m 52; c 3. PHYSICAL CHEMISTRY, PHYSICAL PHARMACY. *Educ:* City Col NY, BS, 50; Polytech Inst Brooklyn, PhD(phys chem), 59. *Prof Exp:* Chemist pharmaceut, Robin Pharmacal Co, 50-51; analyst, Nat Bur Standards, 51-54; sr res chemist electrochem devices, Nat Cash Register Co, 59-62; mem tech staff, Bell Tel Labs, 62-63; sr res scientist pharmaceut & biophys, Squibb Inst Med Res, 63-76, dept head methods develop & testing standards, 76-87, TECH DIR REGULATORY AFFAIRS, BRISTOL-MYERS SQUIBB, 87- *Concurrent Pos:* Adj prof, Middlesex Community Col, 67-70 & Fairleigh Dickinson Univ, 82-83. *Mem:* Am Pharmaceut Asn; Am Chem Soc; NY Acad Sci. *Res:* Ion-exchange membranes; electro- chemistry; solution chemistry; differential thermal analysis; ion-specific electrodes; drug dissolution; particle size and surface area measurements; particulate contamination of parenterals; x-ray diffraction; crystal polymorphism; high performance liquid chromatography. *Mailing Add:* Bristol-Myers Squibb Princeton NJ 08543-4000

JACOBSON, HAROLD GORDON, b Cincinnati, Ohio, Oct 12, 12; m 42; c 2. RADIOLOGY. *Educ:* Univ Cincinnati, BS, 34, BM, 36, MD, 37. *Prof Exp:* Asst radiol, Sch Med, Univ Tex, 41-42; instr, Sch Med, Yale Univ, 42; assoc to chief radiol serv & assoc radiologist, Vet Admin Hosp, Bronx, 46-50, chief radiol serv, 50-52; from asst clin prof to clin prof radiol, Col Med, NY Univ, 52-59, prof clin radiol, 59-64; RADIOLOGIST-IN-CHIEF, MONTEFIORE HOSP & MED CTR, 55-; PROF RADIOL, ALBERT EINSTEIN COL MED, 64-, CHMN DEPT, 72- *Concurrent Pos:* Dir dept radiol, Hosp Spec Surg, 53-54; consult, Vet Admin Hosp, Bronx, 57-; vis prof, Col Med, Univ Cincinnati, 59; chmn comt on affairs, Am Inst Radiol, Am Col Radiol, 71, co-chmn comt diag coding index & Thesaurus, 73-; Crookshank lectr, London, Eng, 74; Holmes lectr, Boston, 74; consult, Nat Bd Med Examiners, 75-; vis prof radiol, Inst Orthop, Univ London, Eng, 75- *Mem:* Radiol Soc NAm (1st vpres, 64-65, pres, 66-67); Am Roentgen Ray Soc; Int Skeletal Soc (pres, 74-75); AMA; Am Col Radiol. *Res:* Bone and joint radiology; neuroradiology; radiology of skeletal disorders. *Mailing Add:* 111 E 210th St Bronx NY 10467

JACOBSON, HARRY C, b Bozeman, Mont, July 13, 31; m 64; c 4. PHYSICS. *Educ:* Col Holy Cross, BS, 53; Yale Univ, MS, 61; PhD(physics), 65. *Prof Exp:* Physicist, Nuclear Div, Combustion Eng Inc, 56-60; PROF, DEPT PHYSICS & ASTRON, UNIV TENN, KNOXVILLE, 64-, PROF PHYSICS & ASTRON, 82- *Mem:* Am Phys Soc. *Res:* Theory of atomic and molecular structure; theory of spectral line shapes. *Mailing Add:* Dept of Physics & Astron Univ of Tenn Knoxville TN 37916

JACOBSON, HARRY R, b June 21, 47; US citizen; m. MEDICINE. *Educ:* Univ Ill, Champaign, BS, 69; Univ Ill, Chicago, MD, 72; Am Bd Internal Med, cert, 75. *Prof Exp:* Internship med, Johns Hopkins Hosp, Baltimore, Md, 72-73, med residency, 73-74; chief, Renal Sect, US Army Surg Res Ctr, Brooke Army Med Ctr, San Antonio, Tex, 76-78; nephrology fel, Univ Tex Health Sci Ctr, Dallas, 74-76, from asst prof to assoc prof internal med, 78-85; PROF MED & DIR, DIV NEPHROLOGY, DEPT MED, MED CTR, VANDERBILT UNIV, NASHVILLE, TENN, 85-; STAFF PHYSICIAN/ NEPHROLOGIST, VET ADMIN HOSP, NASHVILLE, 85- *Concurrent Pos:* NIH res career develop award, 78-83; assoc ed, News Physiol Sci, 88- & Kidney Int, 90-; mem, Nat Kidney & Urol Dis Adv Bd, NIH, 90- *Mem:* Am Fedn Clin Res; Am Soc Nephrology; Int Soc Nephrology; Am Physiol Soc; Am Soc Clin Invest; Asn Am Physicians. *Res:* Author of numerous publications. *Mailing Add:* Div Nephrol Dept Med Vanderbilt Univ Nashville TN 37232

JACOBSON, HERBERT (IRVING), b Chicago, Ill, Mar 17, 23; m 53; c 3. REPRODUCTIVE ENDOCRINOLOGY, BIOCHEMISTRY. *Educ:* Univ Chicago, SB, 48, SM, 49, PhD(chem), 57. *Prof Exp:* Res assoc & instr, Univ Chicago, 52-57, res assoc & asst prof, 57-59, asst prof, Ben May Lab Cancer Res, 59-64; assoc prof obstet & gynec, 64-73, RES PROF & CHIEF SECT REPROD STUDIES, DEPT OBSTET & GYNEC, ALBANY MED COL, 73-, RES PROF BIOCHEM, 75- *Concurrent Pos:* Res partic chem div, Oak Ridge Nat Lab, 61; USPHS spec fel, Max Planck Inst Biochem, Munich, 62-63; Nat Inst Arthritis & Metabolic Dis res career develop award, 66-70; vis prof, Max Planck Inst Cell Biol, 75-76. *Honors & Awards:* US Sr Scientist Award, Fed Repub Ger & Alexander von Humboldt Found, 75. *Mem:* Endocrine Soc; Soc Exp Biol & Med; NY Acad Sci; Soc Study Reprod. *Res:* Steroid biochemistry; mechanism of hormone action; regulation of hormone receptor synthesis; control processes in mammalian reproduction; response mechanisms in hormone-dependent cancer. *Mailing Add:* Dept Obstet & Gynec Albany Med Col Albany NY 12208

JACOBSON, HOMER, b Cleveland, Ohio, Nov 27, 22; m 57; c 3. PHYSICAL BIOCHEMISTRY. *Educ:* Calif Inst Technol, BS, 41; Columbia Univ, AM, 42, PhD(chem), 48. *Prof Exp:* Res scientist, Manhattan Proj, NY, 44-46; assoc chemist, Brookhaven Nat Lab, 47-49; instr chem, Hunter Col, 49-50; from instr to prof, 50-86, EMER PROF CHEM, BROOKLYN COL, 86- *Concurrent Pos:* Guggenheim fel, Calif Inst Technol, 59-60; NIH spec fel, 60-61. *Res:* Chemical mutagenesis in bacterial viruses; virus growth in continuous culture; steady-state kinetics; information theory in biological systems; configurational entropy of superhelices and catenanes. *Mailing Add:* Dept Chem Brooklyn Col Brooklyn NY 11210

JACOBSON, HOWARD NEWMAN, b St Paul, Minn, Aug 13, 23; m 61. PHYSIOLOGY. *Educ:* Northwestern Univ, BS, 47, BM, 50, MD, 51. *Prof Exp:* Intern med, Presby Hosp, Chicago, 50-51, resident obstet & gynec, 51-52; Asn Aid Crippled Children fel, 52-55; resident, Boston Lying-in-Hosp & Free Hosp for Women, 55-57; obstetrician-physiologist, PR Proj, Nat Inst Neurol Dis & Blindness, 58-60; asst obstet & gynec, Harvard Med Sch, 60-61, instr, 61-64, assoc, 64-65; actg assoc prof obstet & gynec, Sch Med & lectr pub health, Sch Pub Health, Univ Calif, San Francisco, 65-67, assoc coord allied health professions, 67-69; assoc prof obstet & gynec, Boston Hosp for Women & dir Macy prog, Harvard Med Sch, 69-74; prof community med, Rutgers Med Sch, Col Med & Dent NJ, 74-79; DIR, INST NUTRIT, UNIV NC, CHAPEL HILL, 79- *Concurrent Pos:* Res fel physiol, Harvard Med Sch, 52-54; mem adv panel maternal & child health, Children's Bur, 64-; NIH res career develop award, 64-65; mem food & nutrit bd, Nat Res Coun, 70-; chmn comt on maternal nutrit, Nat Acad Sci-Nat Res Coun, 71- *Mem:* Am Soc Clin Nutrit; assoc Am Physiol Soc. *Res:* Fetal physiology and responses of fetuses to asphyxia and hypoxia; autonomic nervous system in reproduction; maternity services, with emphasis on maternal nutrition; health manpower needs, nutrition monitoring. *Mailing Add:* Univ SFla 3500 E Fletcher Ave Tampa FL 33612

JACOBSON, IRA DAVID, b New York, NY, May 28, 42; m 54; c 2. FLIGHT MECHANICS, FLUID MECHANICS. *Educ:* New York Univ, BS, 63; Univ VA, MS, 67, PhD(aerospace eng), 70. *Prof Exp:* Aerospace engr, NASA, 63-67; res scientist & lectr, 70-73, from asst prof to assoc prof, 73-79, PROF AEROSPACE ENG, UNIV VA, 79-, DIR, CTR COMPUT AIDED ENG, 83- *Concurrent Pos:* Dir, Inst Comput Aided Eng, Ctr Innovative Technol. *Honors & Awards:* Atwood Award, Am Inst Aeronaut & Astronaut/Am Soc Eng Educr. *Mem:* Am Inst Aeronaut & Astronaut; Am Soc Eng Educr; Sigma Xi; Soc Mfg Engrs. *Res:* Flight mechanics, especially stability and control; fluid mechanics, especially the magnus effect; vehicle systems, especially vehicle ride quality; computer aided design and manufacturing. *Mailing Add:* Dept Mech & Aerospace Eng Univ Va Charlottesville VA 22901

JACOBSON, IRVEN ALLAN, JR, b Denver, Colo, Apr 19, 28; m 64; c 1. PHYSICAL CHEMISTRY, CHEMICAL KINETICS. *Educ:* Univ Colo, BA, 50. *Prof Exp:* Chemist, Laramie Energy Res Ctr, Bur Mines, US Dept Interior, 50-60, res chemist, 60-75, res chemist & proj leader, Laramie Energy Technol Ctr, US Dept Energy, 75-78, Chemical Engr 78-83; RETIRED. *Mem:* Fel AAAS; Am Chem Soc; Sigma Xi. *Res:* Directing research on the kinetics of oil shale thermal gasification and low temperature oxidation. *Mailing Add:* 2626 Park Ave Laramie WY 82070

JACOBSON, JAY STANLEY, b New York, NY, Oct 5, 34; m 71; c 1. PLANT PHYSIOLOGY, AIR POLLUTION. *Educ:* Cornell Univ, BS, 55; Columbia Univ, MA, 57, PhD(plant biochem), 60. *Prof Exp:* Assoc plant biochemist, 60-70, PLANT PHYSIOLOGIST, BOYCE THOMPSON INST, 70- *Concurrent Pos:* Adj prof, Dept of Natural Resources, Cornell Univ; consult to USDA & Environ Protection Agency. *Mem:* Am Soc Plant Physiol; Am Inst Biol Sci; Air Pollution Control Asn. *Res:* Effects of air pollutants and acid rain on growth, development, yield and quality of agricultural crops and forest trees. *Mailing Add:* Boyce Thompson Inst Tower Rd Cornell Univ Ithaca NY 14853

JACOBSON, JIMMY JOE, b Lonepine, Wyo, Feb 11, 37; m 73; c 3. GEOPHYSICS, EARTH SCIENCES. *Educ:* Univ Wyo, BS, 59; Colo Sch Mines, MSc, 64, DSc(geophys), 69. *Prof Exp:* From jr to sr geophysicist, Deco Electronics, 60-65; sr geophysicist, Westinghouse Geores Labs, 65-67; asst prof physics & geophys, Mont Col Mineral Sci & Technol, 70-74; mgr field opers, Geonomics, Inc, 74-77; sr res scientist geophys, Pac Northwest Div, Battelle Mem Inst, 77-80; MCR Geothermal, Denver, 80-83; CONSULT GEOPHYS, 83- *Concurrent Pos:* Geophysicist, Group Seven, Inc, 69-73. *Mem:* Sigma Xi. *Res:* Exploration of the earth and earth materials utilizing electrical and other geophysical tools. *Mailing Add:* 11009 W 65th Way Arvada CO 80004

JACOBSON, JOAN, b Hull, Iowa, Apr 26, 24; div. AUDIOLOGY, SPEECH PATHOLOGY. *Educ:* Morningside Col, BA, 44; Syracuse Univ, MA, 48, PhD(speech corrections), 58. *Prof Exp:* Instr speech path, Syracuse Univ, 46-51; therapist, Brookline Pub Schs, Mass, 51-57; res asst speech path, Syracuse Univ, 57-58; asst prof speech path & audiol & audiologist, Eastern Ill Univ, 58-62; assoc prof, 62-72, PROF SPEECH SCI, PATH & AUDIOL, ST CLOUD STATE UNIV, 72- *Concurrent Pos:* Therapist, Mass Gen Hosp, Boston, 51-57; mem prof adv comt, Minn Soc Crippled Children & Adults, 63- & Minn Easter Seal Soc, 64-; mem, Gov Adv Comt Serv Hearing Impaired, 64-66 & Model Presch Ctr Hearing-Impaired Children & Families, 69-72. *Mem:* Am Speech Lang Hearing Asn; Am Cleft Palate Asn; Am Acad Rehab Audiol. *Res:* Lip reading; cleft palate. *Mailing Add:* Dept of Commun Disorders St Cloud State Univ St Cloud MN 56301

JACOBSON, JOHN OBERT, b Alexandria, Minn, Oct 23, 39; c 1. MECHANICAL ENGINEERING. *Educ:* Wash State Univ, BS, 62; Univ Wash, MS, 65, PhD(bioeng), 72. *Prof Exp:* Safety engr, Boeing Co, 62-63; eng supvr, Lockheed Shipbuilding Co, 65-66; res scientist, Va Mason Med Ctr, 71-74 & Flow Res Inc, 74-75; res consult, Appl Physics Lab, 74; consult engr, Westinghouse Hausford, Olympic Engr, 76-78; asst prof, Cogswell Col, 79-82; assoc prof mech eng, Seattle Univ, 83-85; CONSULT ENGR, JACOBSON

ENGRS, 78- *Concurrent Pos:* Consult engr, Engineered Indust Systs, Inc, 66-86; comt mem, Nat Asn Fire Protection, 81-; rev comt mem, NIH, 86- *Mem:* Am Soc Mech Engrs; Human Factors Soc. *Res:* Ultrasonic imaging for medical diagnostics; laser development for military applications; material development for ultra high pressure applications; design and development of ultra high gravity centrifuge. *Mailing Add:* 5220 Roosevelt Way NE Seattle WA 98105

JACOBSON, KARL BRUCE, b Manning, Iowa, Mar 5, 28; m 51; c 4. BIOCHEMISTRY. *Educ:* St Bonaventure Col, BS, 48; Johns Hopkins Univ, PhD(biol), 56. *Prof Exp:* Am Cancer Soc fel chem, Calif Inst Technol, 56-58; BIOCHEMIST BIOL DIV, OAK RIDGE NAT LAB, 58-, PROF BIOMED SCI, UNIV TENN, 68- *Mem:* AAAS; Am Soc Biol Chem; Am Chem Soc. *Res:* New technologies for DNA sequencing; development and differentiation in terms of biochemical changes; relationship of structure of transfer RNA to its function; biochemical genetics in Drosophila; mechanism of toxicity of cadmium and other metals. *Mailing Add:* Biol Div Oak Ridge Nat Lab PO Box 2009 Oak Ridge TN 37831-8077

JACOBSON, KEITH HAZEN, b Yankton, SDak, Nov 4, 18. PHARMACOLOGY, TOXICOLOGY. *Educ:* Univ SDak, BA, 40, MA, 41; NY Univ, BS, 44; Univ Cincinnati, PhD(phys biochem), 49. *Prof Exp:* Chemist, Rustless Iron & Steel Corp, 41-43; biochemist, staff supv & coordr res, Res & Eng Div, US Army Chem Ctr, 49-51, indust toxicol res, Med Labs, Chem Corps, 51-56, coordr, Mil Chem Prog, 56-63, med res dir, Chem Warfare Labs, 57- 63, asst to dir, 57-63; supvr res pharmacol, Div Pharmacol, Food & Drug Admin, DC, 63-64, chief lab invests, Div Toxicol Eval, 64-68, chief toxicol res, Div Pharmacol & Toxicol, 68-69; assoc prof med, Lab Environ Med, Sch Med, Tulane Univ, 69-72; chief criteria develop br, Nat Inst Occup Safety & Health, 72-74, sr scientist, div criteria doc & standards develop, 74-81; RETIRED. *Concurrent Pos:* Consult toxicol & environ health, 81- *Mem:* Am Chem Soc; Am Indust Hyg Asn; AAAS; Am Acad Indust Hyg. *Res:* Occupational toxicology; pesticide and food additive toxicology. *Mailing Add:* 740 Beall Ave Rockville MD 20850-2107

JACOBSON, KENNETH ALAN, b Euclid, Ohio, July 18, 53; m 78; c 2. RECEPTOR PHARMACOLOGY, MEDICINAL CHEMISTRY. *Educ:* Reed Col, BA, 76; Univ Calif, San Diego, MS, 78, PhD(chem), 81. *Prof Exp:* Bantrell res fel, Dept Org Chem, Weizmann Inst Sci, Rehovot, Israel, 81-83; sr staff fel, 83-88, RES CHEMIST, NAT INST DIABETES, DIGESTIVE & KIDNEY DIS, NIH, BETHESDA, MD, 88- *Concurrent Pos:* Instr, Found Advan Educ Sci, Bethesda, Md, 84-85; scientific adv bd, Res Biochem, Inc, Natick, Mass, 90-91. *Mem:* Am Chem Soc; Soc Neurosci; Am Soc Pharmaceut & Exp Therapeut. *Res:* Developing novel ligands for neurotransmitter receptors; chemistry and pharmacology of purines; muscarinic acetylcholine receptors; neurochemistry and imaging. *Mailing Add:* Lab Chem Nat Inst Diabetes Digestive & Kidney Dis NIH Bldg 8A Rm B1A-17 Bethesda MD 20892-1008

JACOBSON, KENNETH ALLAN, b Milwaukee, Wis, Oct 29, 41; m 66; c 3. MEMBRANE BIOLOGY, CELL BIOLOGY. *Educ:* Univ Wis-Madison, BS, 64, MS, 66; State Univ NY, Buffalo, PhD(biophys), 72. *Prof Exp:* Physicist, Dow Corning Corp, 66-69; mem, Ctr Theoret Biol, State Univ NY, Buffalo, 72-80, asst prof biophys, 76-80, chmn grad group biomembranes, 77-80; pres, Fluorescence Unltd consults, 83-90; assoc prof anat & core mem, Cancer Res Ctr, 80-87, PROF CELL BIOL & ANAT, UNIV NC-CHAPEL HILL, 87- *Concurrent Pos:* Sr cancer res scientist, Roswell Park Mem Inst, 72-80; prin investr grants, NIH, 74- & Am Cancer Soc, 83-; estab investr, Am Heart Asn, 77-82; ed, Cell Physiol Sect, Am J Physiol, 80-84, Comments in Molecular & Cell Biophys, 86- *Mem:* Am Chem Soc; Biophys Soc; AAAS. *Res:* Development of fluorescence microscopy; use of digital image processing to measure the distribution of single classes of molecules in single, living cells; lateral diffusion in membranes; cell locomotion. *Mailing Add:* Dept Cell Biol & Anat Univ NC CB7090 108 Taylor Hall Chapel Hill NC 27599-9090

JACOBSON, LARRY A, b Madison, Wis, Oct 29, 40; m 65. NUCLEAR PHYSICS, GEOPHYSICS. *Educ:* Univ Wis, Madison, BS, 63; MS, 64, PhD(physics), 69. *Prof Exp:* Develop proj physicist, 69-74, sect mgr, Pulsed Neutron Sect, 74-76, dept mgr elec eng, 76-79, mgr sensor physics, 80-81, mgr nuclear logging, Schlumberger Well Serv Co, Schlumberger Ltd, 81-84; mgr nuclear res group, Austin Res Ctr, Gearhart Industs, Inc, 84-88; MGR NUCLEAR RES SECT, HALLIBURTON LOGGING SERV INC, 88- *Concurrent Pos:* Distinguished vis prof, Elec Eng Dept, Univ Houston, 80. *Mem:* AAAS; Am Phys Soc; Soc Prof Well Log Analysts; Soc Petrol Eng; Sigma Xi. *Res:* Negative ion source development; heavy ion nuclear elastic scattering; nuclear techniques in mineral and petroleum exploration and evaluation; data processing of nuclear data; computer simulations of nuclear processes. *Mailing Add:* 1610 Windsong Richmond TX 77469

JACOBSON, LEON ORRIS, b Sims, NDak, Dec 16, 11; m 38; c 2. MEDICINE. *Educ:* NDak State Univ, BS, 35, Univ Chicago, MD, 39; Am Bd Internal Med, dipl. *Hon Degrees:* DSc, NDak State Univ, 66, Acadia Univ, Can, 72. *Prof Exp:* Intern, Univ Clins, 39-40, asst resident med, 40-41, asst med, Univ, 41-42, from instr to prof, 42-65, assoc dean div biol sci, 45-51, head hemat sect, 51-61, chmn dept med, 61-65, dean div biol sci & Pritzker Sch Med, 66-75, dir, The Franklin McLean Mem Res Inst, 75-77, Joseph Regenstein prof, 65-75, EMER JOSEPH REGENSTEIN PROF BIOL & MED SCI, UNIV CHICAGO, 75- *Concurrent Pos:* From assoc dir to dir health, Plutonium Proj, Manhattan Dist, Univ Chicago, 43-46; consult, Div Biol & Med Res, Argonne Nat Lab; spec consult hemat study sect, USPHS, 49-53, mem comt radiation studies, 51-55, mem nat adv comt radiation, 61-66; mem comt cancer diag & ther, Panel Chemother, Nat Res Coun, 51-55; dir, Argonne Cancer Res Hosp, 51-67; R W Stewart Mem lectr, Pittsburgh Acad Med, 52; Janeway lectr, Am Radium Soc, 53; mem adv comt isotope distrib, USAEC, 52-56, mem adv comt to Yugoslavia, 60; Burrell O Ralston lectr, Univ Southern Calif, 57; Alice Messinger Band Mem lectr, Univ Md, 59; George Minot Mem lectr, AMA, 60; US rep, Int Conf Peaceful Uses of Atomic Energy, Geneva, 55 & 58; mem nat adv bd, Okla Med Res Found, 59-64; consult, Inst Cancer Res, Philadelphia, 59-70; mem expert adv panel radiation, WHO, 60-65 & adv comt biophys, Surgeon Gen Army, 61-70; Jacobaeus Mem lectr, Helsinki, 62; Malthe Found lectr, Oslo, 62; consult div radiol health, Univ Grant & Training Br, Dept Health Educ & Welfare, 62-65; mem sci adv bd, Inst Advan Learning Med Sci, City Hope Med Ctr, 62-; consult, Univ Calif, 62-; mem Morison panel, Wooldridge Comt, Off Sci & Technol, Exec Off President, 63-65; mem med & sci adv bd, Will Rogers Mem Fund, 63-66; mem erythropoietin comt, Nat Heart Inst, 63-67; mem bd sci counr, Nat Cancer Inst, 63-67; mem space biol adv subcomt, Space Sci & Appln Steering Comt, NASA, 68-70. *Honors & Awards:* Janeway Medal, Am Radium Soc, 53; Robert Roesler de Villiers Award, Leukemia Soc, 56; Ralston Award, Univ Southern Calif, 57; Borden Award, Asn Am Med Cols, 62; Mod Med Award, 63; Award, Am Nuclear Soc, 63; Citation, City of Hope Nat Med Ctr, 67; John Phillips Mem Award, 75; Theodore Roosevelt Rough Riders Award, State NDak, 76; Am Soc Contemp Med & Surg Award, 81. *Mem:* Inst Med-Nat Acad Sci; Am Asn Cancer Res; Am Nuclear Soc; Am Soc Clin Invest; master Am Col Physicians; Am Acad Arts & Sci; fel AAAS; AMA; Am Soc Exp Path; Am Soc Hemat. *Res:* Hematology; chemotherapeutic agents and radioisotopes on neoplastic diseases of the blood forming tissues; protection against radiation injury; radiobiology. *Mailing Add:* Div Biol Sci Univ Chicago 950 E 59th St Box 420 Chicago IL 60637

JACOBSON, LEWIS A, b Brooklyn, NY, Oct 10, 42; m 67; c 2. MOLECULAR BIOLOGY, BIOCHEMISTRY. *Educ:* Amherst Col, AB, 63; Univ Ill, MS, 65, PhD(biochem), 67. *Prof Exp:* From asst prof biophys to assoc prof biophys & microbiol, 67-76, ASSOC PROF BIOL SCI, UNIV PITTSBURGH, 76- *Mem:* AAAS; Am Soc Microbiol; Am Soc Biochem & Molecular Biol. *Res:* Regulation of gene expression in bacteria, biochemistry and genetics of lysosomal proteases; molecular mechanisms of aging. *Mailing Add:* Dept Biol Sci 304 Langley Hall Univ Pittsburgh Pittsburgh PA 15260

JACOBSON, LOUIS, b Chicago, Ill, Nov 21, 15; m 38; c 3. PLANT PHYSIOLOGY. *Educ:* Univ Calif, Los Angeles, AB, 36; Univ Calif, PhD(plant physiol), 43. *Prof Exp:* Analyst, Div Plant Nutrit, 38-42, radiochemist, Radiation Lab, 42-46, from asst prof & asst plant biochemist to assoc prof nutrit & assoc plant biochemist, 45-58, PROF PLANT NUTRIT & PLANT PHYSIOLOGIST EXP STA, DEPT SOILS & PLANT NUTRIT, UNIV CALIF, BERKELEY, 58- *Mem:* Am Chem Soc; Am Soc Plant Physiol; Bot Soc Am. *Res:* Absorption and accumulation of ions by plants; inorganic nutrition of plants; plant physiology. *Mailing Add:* 150 Poplar St Berkeley CA 94708

JACOBSON, MARCUS, b Houston, Tex, May 2, 30; m 65. ENGINEERING MECHANICS. *Educ:* Rice Univ, BA, 51, BSME, 52, MSME, 54; Univ Calif, Los Angeles, PhD(eng), 65. *Prof Exp:* Asst prof mech eng, Rice Univ, 52-62; design engr, Douglas Aircraft Co, 62; sr dynamics engr, Lockheed-Calif Co, 63-64; ENG SPECIALIST STRUCT DYNAMICS, AIRCRAFT DIV, NORTHROP CORP, 64- *Mem:* Am Inst Aeronaut & Astronaut. *Res:* Structural dynamics. *Mailing Add:* 5337 Holt Ave Los Angeles CA 90056

JACOBSON, MARCUS, b Cape Town, SAfrica, Apr 2, 30; US citizen; m 60; c 3. PHYSIOLOGY, BIOPHYSICS. *Educ:* Univ Cape Town, BSc, 51, MB, ChB, 56; Univ Edinburgh, PhD(physiol), 60. *Prof Exp:* Intern med, Groote Schuur Hosp, Cape Town, 56-57; lectr physiol, Med Sch, Univ Edinburgh, 60-65; guest investr biophys, Naval Med Res Inst, Bethesda, 65-66; from assoc prof to prof, Johns Hopkins Univ, 67-73; prof physiol, Sch Med, Univ Miami, 73-77; PROF & CHMN DEPT ANAT, SCH MED, UNIV UTAH, SALT LAKE CITY, 77- *Concurrent Pos:* Assoc ed, Brain Res, 71-74, J Neurosci Res, 74- & Exp Neurol, 81- *Mem:* Brit Physiol Soc; Am Physiol Soc; Soc Neurosci. *Res:* Development and growth of the nervous system. *Mailing Add:* Dept Anat Univ of Utah Sch Med Salt Lake City UT 84132

JACOBSON, MARTIN, b New York, NY, Nov 11, 19; m 42; c 1. PHYTOCHEMISTRY, PESTICIDES. *Educ:* City Univ NY, BS 40. *Prof Exp:* Lab aide chem, NIH, 41-42; res chemist, USDA, 42-86; RETIRED. *Concurrent Pos:* Res fel, Wash Acad Sci, 66; vis prof, dept chem, Univ Idaho, 71; agr consult, 86- *Honors & Awards:* Bronze Medals, Int Cong Pesticide Chem, 74; Hillebrand Prize, Wash Sect, Am Chem Soc, 64. *Mem:* Am Chem Soc; Entom Soc Am; AAAS. *Res:* Isolation, identification and synthesis of insect toxicants, repellents and attractants from plants and animals. *Mailing Add:* 1131 University Blvd W No 616 Silver Spring MD 20902-3308

JACOBSON, MARTIN MICHAEL, b New York, NY, Nov 24, 34; m 65. BIOCHEMICAL PHARMACOLOGY. *Educ:* City Col New York, BS, 57; Long Island Univ, MS, 68. *Prof Exp:* Res technician, Rockefeller Inst Med Res, 57-58; res biochemist, Wellcome Res Labs, Burroughs Wellcome & Co, 59-70; biochemist, Hoffmann- Laroche Inc, 70-73, sr scientist, 74, coordr exp therapeut, 74-78, res planning mgr exp therapeut, 78-80, asst dir exp therapeut, 80-81, asst dir res planning & develop, 81-83, asst dir pharm res & develop, 83-84, asst dir res qual assurance, 84-85, DIR RES PLANNING, SCI & FINANCIAL ADMIN, HOFFMANN-LAROCHE INC, 85- *Mem:* Am Soc Pharmacol & Exp Therapeut; NY Acad Sci; AAAS; Am Pharmaceut Asn; Acad Pharmaceut Sci; Sigma Xi; Drug Info Asn. *Res:* Biochemical, pharmacologic and toxicologic effects and metabolism of drugs, carcinogens and steroids; involvement of the various disciplines in drug development. *Mailing Add:* Res & Develop Hoffmann LaRoche Inc Bldg 76 Nutley NJ 07110

JACOBSON, MARVIN, b Chicago, Ill, Sept 27, 21; m 47; c 1. ORAL PATHOLOGY. *Educ:* Univ Ill, BS, 42, DDS, 46, MS, 48; Northwestern Univ , PhD(physiol), 50. *Prof Exp:* Periodontist, Brooklyn Vet Admin Hosp, 50-51, USAF Sch Aviation Med, 51-54; res fel path, Col Med, Northwestern Univ, 54-56; asst prof dent therapeut, 56-60, asst prof oral diag, 75-80, ASSOC PROF & HEAD, DEPT PERIO-PROSTHETICS, COL DENT, UNIV ILL, CHICAGO, 80- *Mem:* Am Acad Periodont; Am Dent Asn; Sigma Xi; Int Asn Dent Res; AMA. *Res:* Oral tissues, primarily the periondontal membranes in health and disease and its reattachment to the dentine of a tooth during healing stages; temporomandibular joint dysfunction syndrome. *Mailing Add:* 636 Church St Evanston IL 60201

JACOBSON, MELVIN JOSEPH, b Providence, RI, Nov 25, 28; m 86; c 2. APPLIED MATHEMATICS, WAVE PROPAGATION. *Educ:* Brown Univ, AB, 50; Carnegie Inst Technol, MS, 52, PhD(math), 54. *Prof Exp:* Res mathematician, United Aircraft Corp, 52; res assoc, Carnegie Inst Technol, 52-53, instr math, 53-54; mem tech staff, Bell Tel Labs, Inc, 54-56; from asst prof to assoc prof, 56-61; prof, 61-90, EMER PROF MATH, RENSSELAER POLYTECH INST, 91- *Concurrent Pos:* Prin investr, US Navy, Army, NASA & Indust, 57-; vis prof, Inst Marine Sci, Univ Miami, 63-64; adj prof, 69-72; consult indust, 65- *Mem:* Fel Acoust Soc Am; Soc Indust & Appl Math; Am Asn Univ Profs; Sigma Xi. *Res:* Theories of underwater acoustics, including ocean environmental effects; mathematical studies of electromagnetic wave propagation in the atmosphere; low frequency atmospheric acoustics. *Mailing Add:* Dept of Math Sci Rensselaer Polytech Inst Troy NY 12180-3590

JACOBSON, MICHAEL F, b Chicago, Ill, July 29, 43; m 89. SCIENCE POLICY, MOLECULAR BIOLOGY. *Educ:* Univ Chicago, BA, 65; Mass Inst Technol, PhD(microbiol), 69. *Prof Exp:* Res assoc, Salk Inst Biol Studies, 70-71; EXEC DIR, CTR SCI PUBLIC INTEREST, 71- *Concurrent Pos:* Tech consult, Ctr Study Responsive Law, 70-71; dir, Ctr Sci Pub Interest, 71-, Americans Concerned. About Corp Power, 79-81 & Nat Coalition Dis Prev & Environ Health, 79-82; co-founder, Ctr Study Commercialism, 90. *Res:* Impact of dietary and environmental factors on human health. *Mailing Add:* Ctr Sci Pub Interest 1875 Connecticut Ave NW Suite 300 Washington DC 20009-5728

JACOBSON, MICHAEL RAY, b Pittsburgh, Pa, Jan 11, 50; m 73; c 3. OPTICAL MEASUREMENTS, THIN FILM DEPOSITION. *Educ:* Harvard Col, AB, 71; Cornell Univ, MS, 75, PhD(astron), 77. *Prof Exp:* Res specialist, 77-86, assoc res scientist, Optical Sci Ctr, 86-90, ASSOC RES PROF OPTICAL SCI, OPTICAL SCI & MAT SCI & ENG DEPT, UNIV ARIZ, 90- *Concurrent Pos:* Consult, Energy Conversion Devices, 80-81 & Dow Chem, 89-; instr, Int Soc Optical Eng short courses, 84-; vis fac, Battelle Northwest Pac Lab, 90-; consult, Sci Applications Int Corp, 91- *Mem:* Optical Soc Am; Am Vacuum Soc; Int Soc Optical Eng; Am Meteorol Soc. *Res:* Optical materials, their deposition, characterization, and availability in thin film form; solar energy conversion materials; space applications; polymer films. *Mailing Add:* Ariz Mat Lab 4715 E Ft Lowell Rd Tucson AZ 85712

JACOBSON, MURRAY M, b Boston, Mass, Jan 2, 15; m 62; c 1. CHEMICAL ENGINEERING, MATERIALS ENGINEERING. *Educ:* Tufts Univ, BS, 35. *Prof Exp:* Chemist, Whiting Labs, Mass, 36-39; chief corrosion & lubrication sect, Watertown Arsenal Labs, 40-46, chem engr, 46-49, chief, Surface Chem Sect, 49-54, Phys Chem Br, 54-56, Chem Metall Lab, 56-59 & Mat Sci Lab, 59-64; dep chief, Mat Eng Div, Army Mat Res Agency, 64-66, chief prototypes lab, Army Mat & Mech Res Ctr, 66-69, chief mat test div, 69-74; PRES, JACON INDUSTS, 75- *Concurrent Pos:* Army liaison mem, Subgroup on Greases, Coord Res Coun, War Adv Comt & Adv Comt on Corrosion, Ord Dept, 42-45; liaison mem subpanel on chromium, Panel on Refractory Metals, Mat Adv Bd, 49 & Comt Coatings, 68-69. *Mem:* Am Chem Soc; Nat Asn Corrosion Engrs; Am Soc Metals. *Res:* Research management; corrosion; erosion; oxidation; wear; lubrication; protective coatings and treatments; antigalling coatings; titanium surface metallurgy. *Mailing Add:* Jacon Industs PO Box 231 Boston MA 02146-0002

JACOBSON, MYRON KENNETH, b Richland Center, Wis, Sept 20, 43; m 67. BIOCHEMISTRY. *Educ:* Univ Wis-Platteville, BS, 65; Kans State Univ, PhD(biochem), 70. *Prof Exp:* NIH fel biochem, Univ Utah, 70-72; from res asst to res assoc biochem, Mayo Clin & Found, 73-74; asst prof chem, NTex State Univ, 74-80, assoc prof chem & basic health sci, 80-; PROF BIOCHEM, TEX COL OSTEOPATH MED. *Mem:* AAAS; Am Soc Biol Chemists. *Res:* Nicotinamide nucleotide metabolism, chemical carcinogenesis and DNA repair mechanisms. *Mailing Add:* Dept Chem NTex State Univ Denton TX 76203

JACOBSON, NATHAN, b Warsaw, Poland, Sept 8, 10; US citizen; m 42; c 2. MATHEMATICS. *Educ:* Univ Ala, AB, 30; Princeton Univ, PhD(math), 34. *Hon Degrees:* DSc, Univ Chicago, 72. *Prof Exp:* Asst math, Inst Advan Study, Princeton Univ, 33-34, Procter fel, 34-35; lectr, Bryn Mawr Col, 35-36; Nat Res Found fel, Univ Chicago, 36-37; from instr to asst prof, Univ NC, 37-40; vis assoc prof, Johns Hopkins Univ, 40-41; assoc prof, Univ NC, 41-42, assoc ground sch instr, US Navy preflight sch, 42-43; assoc prof math, Johns Hopkins Univ, 43-47; from assoc prof to prof, 47-67, Henry Ford II prof, 67-81, EMER PROF MATH, YALE UNIV, 81- *Concurrent Pos:* Ed, Bull, Am Math Soc, 48-53; Guggenheim fel, 51-52; Fulbright res grant, Univ Paris, 51-52; vis prof, Univ Chicago, 64, Univ Tokyo, 65, Tata Inst Fundamental Res, India, 70, Univ Rome, 71, Hebrew Univ Jerusalem, 71, Rehovoth, 71 & Australian Nat Univ, 78, ETH Zurich, 81, Nanjing Univ, 83, Taiwan Nat Univ, 83, Wesleyan Univ, 88; vpres, IMU, 72-74. *Mem:* Nat Acad Sci; Am Math Soc (vpres, 57-58, pres, 71-73); French Math Soc; Math Soc Japan; Am Acad Arts & Sci; hon mem, London Math Soc, 72. *Res:* Topological algebra; structure theory of rings; non-associative algebra, especially Lie and Jordan algebras; Galois theory. *Mailing Add:* Dept Math Yale Univ Box 2155 Yale Sta New Haven CT 06520

JACOBSON, NORMAN LEONARD, b Eau Claire, Wis, Sept 11, 18; m 43; c 2. ANIMAL NUTRITION. *Educ:* Univ Wis, BS, 40; Iowa State Univ, MS, 41, PhD(nutrit), 47. *Prof Exp:* From asst prof to prof animal & dairy sci, 47-79, distinguished prof agr, 63-, assoc dean, Grad Col, 73-88, assoc vpres res, 79-88, ASSOC PROVOST RES, IOWA STATE UNIV, 63-, DEAN GRAD COL, 88- *Concurrent Pos:* Moorman travel fel nutrit, 66. *Honors & Awards:* Am Feed Mfgrs Award, 55; Borden Award, 60; Morrison Award, Am Soc Animal Sci, 70. *Mem:* AAAS; Am Dairy Sci Asn (vpres, 71, pres, 72); Am Inst Nutrit; Am Soc Animal Sci; Sigma Xi. *Res:* Animal nutrition, particularly nutrition and physiology of the ruminant. *Mailing Add:* 313 Kildee Hall Iowa State Univ Ames IA 50011

JACOBSON, RALPH ALLEN, b Jersey City, NJ, June 3, 40; m 63; c 3. MOLECULAR BIOLOGY, BIOCHEMISTRY. *Educ:* Montclair State Col, BA, 62: Cornell Univ, PhD(biochem), 66. *Prof Exp:* NIH res fel biol, Calif Inst Technol, 66-68; asst prof biochem, Univ Okla, 68-75; from asst prof to assoc prof, 75-83, PROF BIOCHEM, CALIF POLYTECH STATE UNIV, 83- *Concurrent Pos:* Vis scientist, W Alton Jones Cell Sci Ctr, NY, 83-84. *Mem:* AAAS; Sigma Xi; Am Chem Soc. *Res:* Molecular biology & biotechnology recombinant DNA. *Mailing Add:* Dept Chem Calif Polytech State Univ San Luis Obispo CA 93407

JACOBSON, RAYMOND E, b St Paul, Minn, May 25, 22; m 59, 86; c 3. ELECTRONICS, COMPUTER SCIENCE. *Educ:* Yale Univ, BE, 44; Harvard Univ, MBA, 48; Oxford Univ, BA, 50, MA, 54. *Prof Exp:* Asst to gen mgr, PRD Electronics Inc, 50-55; prod sales mgr, Curtiss Wright Corp, 55-57; from div sales mgr to dir mkt, TRW Comput Co, 57-60; vpres opers, Electro-Sci Investors, Inc, 60-63; pres, Maxson Electronics, 63-64; mgmt consult, 64-67; CHMN & PRES, ANDERSON JACOBSON, INC, 67- *Concurrent Pos:* Chmn, Staco, Inc & Gen Electronic Controls, 60-63 & Whitehall Electronics, Inc, 61-63; dir, Tamar Electronics, Inc, & Rawco Instruments, Inc, 60-63, Micro-Radionics, Inc, 65-67. *Mem:* Am Electronics Asn; Asn Am Rhodes Scholars; Sigma Xi. *Mailing Add:* 1540 Carnavon Way San Jose CA 95131-2477

JACOBSON, RICHARD MARTIN, b New York, NY, Dec 23, 47; m 75. SYNTHETIC ORGANIC CHEMISTRY. *Educ:* Case Western Reserve Univ, BS, 69; Columbia Univ, PhD(org chem), 73. *Prof Exp:* Asst prof chem, Ind Univ, Bloomington, 75-80; sr res chemist, 80-84, RES FEL, ROHM & HAAS RES LAB, 84- *Mem:* Am Chem Soc; Royal Soc Chem. *Res:* Design and synthesis of agricultural chemicals. *Mailing Add:* Rohm & Haas Co 727 Norristown Rd Spring House PA 19477

JACOBSON, ROBERT ANDREW, b Waterbury, Conn, Feb 16, 32; m 62; c 2. PHYSICAL CHEMISTRY, CRYSTALLOGRAPHY. *Educ:* Univ Conn, BA, 54; Univ Minn, PhD(phys chem), 59. *Prof Exp:* From instr to asst prof chem, Princeton Univ, 59-64; assoc prof, 64-69, PROF CHEM, IOWA STATE UNIV, 69- *Concurrent Pos:* Chemist, 64-69, sr chemist, Energy & Mineral Resources Res Inst & Ames Lab, 69- *Mem:* Am Chem Soc; Am Crystallog Asn. *Res:* Molecular structure of solids; x-ray and neutron diffraction. *Mailing Add:* Dept of Chem Iowa State Univ Ames IA 50011

JACOBSON, ROBERT LEROY, b Miles City, Mont, Mar 11, 32; m 53; c 4. CHEMICAL ENGINEERING. *Educ:* Mont State Univ, BS, 54, PhD(chem eng), 58. *Prof Exp:* Res engr petrol process develop, Chevron Res Co, 58-72, sr eng assoc, 72-77, group leader process develop, 77-84, group leader process res explor, Chevron Res Co, Standard Oil Co, Calif, 84-89. *Concurrent Pos:* Reforming process consult, 89- *Mem:* Am Inst Chem Engrs; Am Chem Soc; Catalysis Soc. *Res:* Petroleum process research and development; exploratory research. *Mailing Add:* Chevron Res Co 576 Standard Ave Richmond CA 94804

JACOBSON, STANLEY, b Chicago, Ill, Aug 24, 37; m 60; c 1. NEUROANATOMY. *Educ:* Univ Ill, BS, 59; Northwestern Univ, MS, 61, PhD(anat), 63. *Prof Exp:* Biologist, NIH, 63-65; biologist, Vet Admin Res Hosp, Chicago, 65-67; asst prof, 67-70, assoc prof, 70-80, PROF ANAT, SCH MED, TUFTS UNIV, 80- *Concurrent Pos:* Lectr, Sch Med, George Washington Univ, 64-65; assoc, Med Sch, Northwestern Univ, 65-67. *Mem:* AAAS; Am Asn Anatomists. *Res:* Structure of central nervous system in normal and diseased animals; connections between cerebral cortex and thalamus; degeneration of nerve fibers. *Mailing Add:* Dept Anat Tufts Univ Sch Med 136 Harrison Ave Boston MA 02111

JACOBSON, STEPHEN ERNEST, b State Center, Iowa, Apr 6, 45; m 75. INORGANIC CHEMISTRY, ORGANOMETALLIC CHEMISTRY. *Educ:* Iowa State Univ, BS, 67; Ohio State Univ, MS, 70, PhD(inorg chem), 72. *Prof Exp:* Res chemist inorg & catalysis, Allied-Signal Corp, 75-80; res assoc, Halcon Res, 80-86; RES ASSOC, E I DU PONT DE NEMOURS & CO, 87- *Concurrent Pos:* Fel, Univ Waterloo, 72-74, Univ Ala, 74-75. *Mem:* Am Chem Soc. *Res:* Process chemical research and development; new product research; organic synthesis; catalysis research. *Mailing Add:* Two Buchak Circle Princeton Jct NJ 08550

JACOBSON, STEPHEN RICHARD, b New York, NY, Sept 25, 44; m 81; c 1. GEOLOGY, PALEONTOLOGY. *Educ:* Dickinson Col, BS, 69; Harvard Univ, MA, 72; Ohio State Univ, PhD(geol), 78. *Prof Exp:* Instr geol, Lehman Col, 71-72; lectr geol, Hunter Col, 72-73; biostratigraphic supvr & geo chem coord, 82-86, GEOLOGIST, CHEVRON USA INC, 78-, GROUP LEADER ORG CHEM, RESERVOIR ARCHIT, 86- *Concurrent Pos:* Geologist, US Geol Surv, Coal Geol Lab, 75-76; group leader organic geochemistry, appln, Chevron USA, Inc, 86-; lectr geol, Calif State Univ Long Beach. *Mem:* Paleont Soc; Am Asn Stratig Palynologists; Palaeont Asn; Asn Am Petrol Geologists. *Res:* Stratigraphy; organic geochemistry; ordovician biostratigraphy, Cretaceous-Tertiary palynology, palynological techniques; petroleum source rock geochemistry; Ordovician and Permian acritarchs. *Mailing Add:* Chevron Oil Field Res Co PO Box 446 La Habra CA 90631

JACOBSON, STUART LEE, b Chicago, Ill, Apr 29, 34; m 59; c 2. CARDIAC PHYSIOLOGY, CARDIAC CELL CULTURE. *Educ:* Cornell Univ, BCE, 57; Univ Minn, Minneapolis, MS, 63, PhD(biophys), 68. *Prof Exp:* Proj engr, Aeromed Lab, Wright Air Develop Ctr, USAF, 57-59; asst prof, 68-75, ASSOC PROF BIOL, CARLETON UNIV, 75- *Mem:* Int Soc Heart Res; Biophys Soc; Soc Gen Physiologists. *Res:* Engineering of systems for maintaining life in sealed environments; instrumentation for ecological studies; sensory physiology; electrophysiology, culture and electrophysiology of myocardial cells. *Mailing Add:* Dept of Biol Carleton Univ Ottawa ON K1S 5B6 Can

JACOBSON, WILLARD JAMES, b Northfield, Wis, May 22, 22; m 46; c 3. SCIENCE EDUCATION, ENVIRONMENTAL SCIENCE. *Educ:* Univ Wis, River Falls, BS, 46; Columbia Univ, AM, 48, EdD, 52. *Prof Exp:* Teacher pub schs, Wis, 46-47; co-dir, Population Educ Proj, 72-80, chmn, dept sci educ, 65-73, dir, Citizens & Sci Educ Study, PROF NATURAL SCI, TEACHERS COL, COLUMBIA UNIV, 52- *Concurrent Pos:* Fulbright lectr, Univ London, 60; consult, Royal Afghan Ministry Educ, 54-56, Am Sch Guatemala, 58, 61 & 64, UNESCO, 65, 84, Nat Coun Sci Educ India, 69, Nat Textbook Comn Brazil, 71 & Ministry Educ, Jamaica, 79-80; chmn, educ adv comt, NY Acad Sci, 75-78, mem bd govs, 78-80; nat res coordr, Second Int Sci Study, 82- *Honors & Awards:* Robert Carleton Award, Nat Sci Teachers Asn. *Mem:* Fel AAAS (vpres, 68-69); Nat Asn Res Sci Teaching (pres, 68-69); fel NY Acad Sci; Asn Educ Teachers Sci (pres, 62-63). *Res:* Science for children and early adolescents; population education; science education survey. *Mailing Add:* 106 Morningside Dr New York NY 10027

JACOBUS, DAVID PENMAN, b Boston, Mass, Feb 26, 27; m 56; c 5. MEDICINE, RADIOBIOLOGY. *Educ:* Harvard Univ, BA, 49; Univ Pa, MD, 53. *Prof Exp:* Resident & researcher, Hosp Univ Pa, 53-57; chief dept radiobiol, Div Nuclear Med, Walter Reed Army Inst Res, 59-63, chief dept med chem, 63-65, dir, Div Med Chem, 65-69; vpres basic res, Merck Sharp & Dohme Res Labs, vpres, 74-78; OWNER, JACOBUS PHARMACEUT CO, INC, 78- *Concurrent Pos:* Mem revision comt, US Pharmacopeia, 70-; trustee, Cold Spring Harbor Lab, 70-; consult, St Luke's Hosp Ctr, New York, 71-; chmn Nat Cancer Inst Comt on Info Handling, 78- *Mem:* Am Chem Soc; Am Soc Info Sci; Asn Comput Mach; NY Acad Sci. *Res:* Information handling. *Mailing Add:* 37 Cleveland Lane Box 5290 Princeton NJ 08540

JACOBUS, OTHA JOHN, b Phoenix, Ariz, Dec 23, 39; m 62; c 2. ORGANIC CHEMISTRY. *Educ:* Southwestern at Memphis, BS, 62; Univ Tenn, Knoxville, PhD(org chem), 65. *Prof Exp:* NIH fel org chem, Princeton Univ, 66-67, instr, 67-69; from asst prof to prof org chem, Clemson Univ, 74-76; PROF ORG CHEM, TULANE UNIV, 78-, CHMN DEPT, 80- *Mem:* Am Chem Soc. *Res:* Stereochemistry; NMR spectroscopy; reaction mechanisms. *Mailing Add:* Dept of Chem Tulane Univ New Orleans LA 70118

JACOBUS, WILLIAM EDWARD, b Cleveland, Ohio, Nov 30, 42; m 66. BIOCHEMISTRY. *Educ:* Ohio Wesleyan Univ, BA, 64; Ohio State Univ, PhD(bchem), 69. *Prof Exp:* Fel biochem, Sch Med, Johns Hopkins Univ, 70-73, Heart Asn Md res fel & NIH res grant, Myocardial Infarction Res Unit, 71-73; asst prof zool, Univ Calif, Davis, 73-77; asst prof med & physiol chem, 77-82, ASSOC PROF MED, BIOL & CHEM, JOHNS HOPKINS SCH MED, 82- *Concurrent Pos:* Nat Heart & Lung Inst res grant, 74-77. *Mem:* AAAS; Am Chem Soc; Am Physiol Soc; Sigma Xi; Biophys Soc; Am Soc Biol Chem; Soc Magnetic Res Med; Int Soc Heart Res. *Res:* Mitochondrial oxidative phosphorylation and ion transport; mitochondrial compartmentation and enzymology; metabolic regulation; cardiac bioenergetics; tissue nuclear magnetic resonance. *Mailing Add:* 3902 Juniper Rd Baltimore MD 21218

JACOBY, ALEXANDER ROBB, b St Louis, Mo, Oct 8, 22; m 45. MATHEMATICS. *Educ:* Univ Chicago, SB, 41, SM, 42, PhD(math), 46. *Prof Exp:* Asst math, Univ Chicago, 43-45, instr, 45-47; asst prof, Univ Miami, 47-49 & Rutgers Univ, 49-57; with Gen Elec Corp, 57-61; PROF MATH, UNIV NH, 61- *Res:* Topology and algebra. *Mailing Add:* Dept of Math Univ of NH Durham NH 03824

JACOBY, HENRY I, b Scranton, Pa, Aug 26, 36; m 67; c 2. GASTROINTESTINAL PHARMACOLOGY, RENAL PHARMACOLOGY. *Educ:* Philadelphia Col Pharm & Sci, BSc, 58; Univ Mich, PhD(pharmacol), 63. *Prof Exp:* Res pharmacologist, Res Lab, Merck Sharpe & Dohme, 63-72; group leader, McNeil Labs Pharmaceut, 72-77, prin scientist, 78-79, res fel gen pharmacol, 79-87, RES FEL GEN PHARM, RW JOHNSON PHARMACEUT RES INST, 88- *Honors & Awards:* Philip B Hoffman Res Award, Johnson & Johnson, 79. *Mem:* Am Gastroent Asn; Am Soc Pharmacol & Exp Therapeut. *Res:* Neurokinin and opioid receptors; gastrointestinal pharmacology especially drug effecting motility; intestinal secretion. *Mailing Add:* R W Johnson Pharmaceut Res Inst McKean & Welsh Rds Spring House PA 19477-0776

JACOBY, JAY, b New York, NY, Dec 12, 17; m 42; c 3. ANESTHESIOLOGY. *Educ:* Univ Minn, BS, 39, MB, 41, MD, 42; Univ Chicago, PhD(anesthesiol), 47; Am Bd Anesthesiol, dipl. *Prof Exp:* Res assoc & instr anesthesia, Univ Chicago, 46-47; assoc prof surg & dir anesthesia, Ohio State Univ, 47-53, prof, 53-59; prof & chmn dept, Med Sch, Marquette Univ, 59-65; prof, 65-84, EMER PROF ANESTHESIOL & CHMN DEPT, JEFFERSON MED COL, 84- *Mem:* Am Soc Anesthesiol; Am Col Anesthesiol; Int Anesthesia Res Soc; AMA; Asn Univ Anesthetists. *Res:* Anesthetic and analgesic drugs; gas therapy. *Mailing Add:* Jefferson Med Col Thomas Jefferson Univ Philadelphia PA 19107

JACOBY, LAWRENCE JOHN, b Portland, Ore, May 19, 43; m 84; c 2. ORGANIC CHEMISTRY, ANALYTICAL CHEMISTRY. *Educ:* Ore State Univ, BS, 65; Colo State Univ, PhD(org chem), 69. *Prof Exp:* Asst prof org chem, Portland State Univ, 69-71; asst prof chem, Chemeketa Community Col, 71-76; anal chemist, Teledyne Wah Chang, Albany, 76-86; lab mgr, CH2M Hill, Corvallis, Ore, 86-88, client serv mgr, CH2M Hill, Redding, Calif, 88-90; QUAL ASSURANCE COORDR, COLUMBIA ANALYTIC SERV INC, KELSO, WASH, 90- *Mem:* Am Chem Soc; Asn Official Anal Chemists. *Res:* Valence tautomerism induced by electron transfer. *Mailing Add:* 104 Hackett Rd Longview WA 98632

JACOBY, ROBERT OTTINGER, b New York, NY, June 20, 39. COMPARATIVE PATHOLOGY. *Educ:* Cornell Univ, DVM, 63; Ohio State Univ, MSc, 68, PhD(path), 69. *Prof Exp:* Asst prof path, Ohio State Univ, 69; NIH fel, Univ Chicago, 69-71; From asst prof to assoc prof 71-87, CHMN, SECT COMPARATIVE MED & DIR, DIV ANIMAL CARE, 78-, PROF COMPARATIVE MED, YALE SCH MED, 87. *Honors & Awards:* Res Award, Am Asn Lab Animal Sci, 87. *Mem:* AAAS; Am Col Vet Pathologists; Int Acad Path; Am Vet Med Asn; Sigma Xi; Am Asn Pathologists. *Res:* Pathogenesis of infectious diseases; diseases of laboratory animals; animal models of human disease. *Mailing Add:* Comp Med 113 Yale Univ New Haven CT 06520

JACOBY, RONALD LEE, b Muskegon, Mich, Jan 30, 43; m 66; c 3. MEDICINAL CHEMISTRY, COMPUTER-ASSISTED INSTRUCTION. *Educ:* Ferris State Col, BSPharm, 66; Univ Conn, PhD(med chem), 71. *Prof Exp:* From asst prof to assoc prof, 70-80, PROF MED CHEM, FERRIS STATE COL, 80- *Res:* Computer-assisted instruction. *Mailing Add:* Sch Pharm Ferris State Univ Big Rapids MI 49307

JACOBY, RUSSELL STEPHEN, b Lehighton, Pa, June 23, 39; m 59; c 3. METAMORPHIC PETROLOGY, URANIUM GEOLOGY. *Educ:* Syracuse Univ, BS, 61, MS, 64; Queen's Univ, PhD(geol), 68. *Prof Exp:* Asst prof geol, Cent Mo State Univ, 68-71; asst prof, 71-73, ASSOC PROF GEOL, ST LAWRENCE UNIV, 73- *Mem:* Geol Soc Am; Geol Asn Can; Sigma Xi. *Res:* Structural geology and metamorphic petrology of precambrian metamorphic shield areas; uranium exploration. *Mailing Add:* PO Box 584 Canton NY 13617

JACOBY, WILLIAM R(ICHARD), b Columbia, NJ, Sept 8, 26; m 49; c 3. CERAMIC ENGINEERING. *Educ:* Dickinson Col, BS, 50; Rutgers Univ, MS, 53, PhD(ceramic eng), 56. *Prof Exp:* Asst instr ceramic eng, Rutgers Univ, 54-56; res assoc, Knolls Atomic Power Lab, Gen Elec Co, 57-59; sr ceramist, Reactor Mat Lab, Minn Mining & Mfg Co, 59-62, supvr, 62-63, res specialist, Nuclear Prod Dept, 63-64; sr engr, Atomic Power Div, Westinghouse Elec Corp, 65-67, mgr ceramic develop, Advan Reactors Div, 67-68, mgr plutonium fuel facil, 68-69, mgr, Cheswick Opers, 74-81; MGR DECONTAMINATION & DECOMISSIONING ENG, WEST VALLEY NUCLEAR SERV CO, INC, 81- *Concurrent Pos:* Mem staff, Argonne Nat Lab, 65- *Mem:* Fel Am Ceramic Soc; Am Nuclear Soc. *Res:* Materials, fabrication processes and product development in areas of fuel and control materials for thermal and fast reactors. *Mailing Add:* West Valley Nuclear Serv Co PO Box 191 West Valley NY 14171

JACOLEV, LEON, b Libau, Latvia, Dec 16, 14; nat US; m 44; c 2. CHEMICAL ENGINEERING. *Educ:* Northeastern Univ, BS, 39; Univ Pittsburgh, MS, 45. *Prof Exp:* Chemist, Eastern Gas & Fuel Assocs, Mass, 36-39; rating examr, Tech Personnel Div, US Civil Serv Comn, 40-41; asst chem engr, High Explosives Res Div, US Bur Mines, 41-43; chem engr, Nat Defense Res Comt, Off Sci Res & Develop, 43-45; tech & res div, Texaco Inc, 45-53; consult chem engr & tech dir, 53-55, CONSULT CHEM ENGR, TECH DIR & PRES, ASSOC TECH SERV, INC, 55- *Concurrent Pos:* Consult, Engrs Joint Coun, NY, 59; Am Inst Chem Eng rep, Mendeleev Cong Gen & Appl Chem, Moscow, 59; dir, Eng Socs Libr, NY; co-auth, German-English Sci Dict, 78; translr-ed over 20 Russ bks & monographs in sci & eng. *Mem:* AAAS; Am Chem Soc; Am Inst Chem Engrs; Geosci Info Soc; Spec Libr Asn. *Res:* Synthesis gas generation by partial combustion; petroleum refining and petrochemicals; documentation; Russian chemical and geological sciences and technology; retrieval, translation and dissemination of scientific information; patentee in synthesis gas generation. *Mailing Add:* 30 The Fairway Upper Montclair NJ 07043

JACOVIDES, LINOS J, b Paphos, Cyprus, May 10, 40; US citizen; m; c 3. ELECTRICAL ENGINEERING. *Educ:* Glasgow Univ, BSc, 61, MSc, 63; Univ London, PhD, 65. *Prof Exp:* Sr res engr, Defense Res Labs, Gen Motors Corp, 65-67, sr res engr, Res Labs, Warren, Mich, 67-75, dept res engr, asst dept head, Elec Eng Dept, 85-87, prin res engr, Elec Eng Dept, 87-88, HEAD, ELEC & ELECTRONIC ENG DEPT, RES LABS, GEN MOTORS CORP, 88- *Mem:* Fel Inst Elec & Electronics Engrs; Brit Inst Elec Engrs; Soc Automotive Engrs. *Res:* Electromagnetics and electromagnetic energy conversion; high performance electric drive systems; automotive electrical systems; electric vehicles and locomotive electric drives; electromechanical devices; automotive powertrain and chassis control systems. *Mailing Add:* Res Labs Gen Motors Tech Ctr Warren MI 48090

JACOX, ADA K, MEDICINE. *Educ:* Columbia Univ, BS; Wayne State Univ, MS, Case Western Reserve Univ, PhD. *Prof Exp:* Staff nurse, Kalamazoo State Hosp, Mich, 56-57; pvt duty nurse pediat & neurol, Columbia-Presby Hosp, NY, 58-59; staff nurse acute & chronic pediat, Children's Hosp Mich, Detroit, 59-60; assoc dir nursing & dir nursing educ, Plymouth State Home & Training Sch, Northville, Mich, 61-62, dir nursing, 62-64; psychiat nurse, Johnson County Mental Health Ctr, Kans City, 68-69; from assoc prof to prof, Col Nursing, Univ Iowa, 69-76; assoc dean res & doctoral prog & prof nursing, Sch Nursing, Univ Colo, 76-79; prof nursing, Sch Nursing, Univ Md, 80-90, dir, Ctr Nursing & Health Serv Res, 80-90; PROF INDEPENDANCE FOUND CHAIR & HEALTH POLICY, SCH NURSING, JOHNS HOPKINS UNIV, 90- *Concurrent Pos:* Carver fel, Univ Iowa, 72; asst prof, Dept Nursing Educ, Univ Kans, 68-69; actg dean advan studies, Grad Col, Univ Iowa, 75-76; prin investr, Dept Health, Educ & Welfare, 72-79; res consult, Mental Health Inst, Mt Pleasant, Iowa, 70-71, Iowa Vet Home, Marshalltown, 72-76, Vet Admin Hosp, Iowa City, 74-76 & Clin Ctr, NIH, 75-79. *Honors & Awards:* Shirley Titus Award, Am Nurses Asn, 88. *Mem:* Inst Med-Nat Acad Sci; Am Acad Nurses; Am Nurses Asn. *Res:* Author of 6 books and over 20 journal articles. *Mailing Add:* Sch Nursing Johns Hopkins Univ 484 Houck Baltimore MD 21205

JACOX, MARILYN ESTHER, b Utica, NY, Apr 26, 29. MOLECULAR SPECTROSCOPY, PHOTOCHEMISTRY. *Educ:* Syracuse Univ, BA, 51; Cornell Univ, PhD, 56. *Prof Exp:* Res assoc phys chem, Univ NC, 56-58; fel spectros of solids, Mellon Inst Sci Res, 58-62; PHYS CHEMIST, NAT INST STANDARDS & TECHNOL, 62- *Concurrent Pos:* Chief, Photochem Sect, Nat Bur Standards, 73-74; chief, Environ Chem Processes Sect, Nat Bur Standards, 74-78. *Honors & Awards:* Outstanding Alumnus Award, Utica Col, 63; Award Phys Sci, Washington Acad Sci, 68; Gold Medal Award for

Distinguished Serv, US Dept Commerce, 70; Federal Woman's Award, Fed Woman's Award Bd of Trustees, 73; Samuel Wesley Stratton Award, Nat Bur Standards, 73; Ellis R Lippincott Award, Coblentz Soc, 89. *Mem:* Fel Am Phys Soc; Am Chem Soc; Int-Am Photochem Soc; AAAS; Sigma Xi. *Res:* Chemistry of free radicals and molecular ions; molecular spectroscopy. *Mailing Add:* Nat Inst Standards & Technol Gaithersburg MD 20899

JACOX, RALPH FRANKLIN, b Alfred, NY, Oct 30, 13; m 40; c 3. INTERNAL MEDICINE. *Educ:* Alfred Univ, BS, 35; Univ Rochester, MD, 38; Am Bd Internal Med, dipl. *Prof Exp:* Intern med, Strong Mem Hosp, Univ Rochester, 38-39, asst resident, 39-40, chief resident med, 41-42; asst bact, Sch Hyg, Johns Hopkins Univ, 40-41; from instr to assoc prof, 46-63, PROF MED, SCH MED & DENT, UNIV ROCHESTER, 63-, DIR ARTHRITIS CLIN, 58- *Concurrent Pos:* Consult, Genesee Hosp, Rochester & Rochester Gen Hosp; trustee, Alfred Univ, 64. *Mem:* Am Fedn Clin Res; fel Am Col Physicians; Am Soc Clin Invest. *Res:* Cationic detergent fractionation of plasma proteins; glucuronidase studies of hemolytic streptococci and human joint fluid; factors concerned in coagulation of blood; bactericidal activity of human serum for B subtilis; bacteriology; hematology; a new method for production of nonspecific capsular swelling of the pneumococcus; protein chemistry and immunology. *Mailing Add:* Dept Med Univ Rochester Med Ctr Rochester NY 14642

JACQUES, FELIX ANTHONY, b Kansas City, Kans, July 17, 24. BIOLOGY. *Educ:* Univ Iowa, BA, 50, MS, 53; St Louis Univ, PhD(biol), 60. *Prof Exp:* Instr biol, Langston Univ, 53-55; lectr zool, Southern Ill Univ, 59-61; asst prof physiol & evolution, Webster Col, 61-62; asst prof, 62-64, ASSOC PROF PHYSIOL, ST BONAVENTURE UNIV, 64- *Concurrent Pos:* Nat Heart Inst fel, 60-62, res grant, 65-68. *Mem:* AAAS; Am Soc Zool; assoc Am Physiol Soc; Am Inst Biol Sci; Sigma Xi. *Res:* Comparative and environmental physiology; environmental effects on blood and other tissues. *Mailing Add:* PO Box 134 176 Portville Ceres Rd RT417 Poriville NY 14770

JACQUET, HERVE, b France, Aug 4, 39; m 69; c 1. NUMBER THEORY. *Educ:* Univ Paris, Licence, 61, PhD(math), 67. *Prof Exp:* Researcher, Nat Ctr Sci Res, France, 63-65; assoc prof math, Univ Md, 69-70; prof math, Grad Sch City Univ New York, 70-74; PROF MATH, COLUMBIA UNIV, 74- *Concurrent Pos:* Mem, Inst Advan Studies, 67-69. *Honors & Awards:* Prix Petit d'Ormoy, Acad Sci, Paris, 79. *Mem:* Am Math Soc. *Res:* Automorphic L-functions. *Mailing Add:* Columbia Univ New York NY 10027

JACQUET, YASUKO F, b San Francisco, Calif; c 1. BEHAVIORAL NEUROPHARMACOLOGY, OPIATE NEUROPHARMACOLOGY. *Educ:* Tsuda Col, Tokyo, Japan, dipl, 51; State Univ Iowa, Iowa City, BA, 53; Ind Univ, Bloomington, PhD(exp psychol), 62. *Prof Exp:* Postdoctoral fel, Yale Univ Sch Med, 62-64; staff fel, NIMH, Bethesda, Md, 64-65; vis scientist, Nat Inst Med Res, London, UK, 77, Uppsala Univ, Sweden, 79, Mass Gen Hosp, Boston, 82 & Columbia Univ Med Ctr, New York, 83-85; RES SCIENTIST, NY STATE DEPT MENT HYG, NEW YORK, 66- *Mem:* Am Soc Pharmacol & Exp Therapeut; Soc Neurosci; Europ Behav Pharmacol Soc; Am Pain Soc; Int Narcotics Res Conf; Asn Res Nervous & Ment Dis. *Res:* Central nervous system sites involved in opiate effects; physiological actions of endogenous opiate peptides; neurotransmitters mediating opiate effects. *Mailing Add:* Nathan Kline Inst Psychiat Res Orangeburg NY 10962

JACQUEZ, JOHN ALFRED, b Pfastatt, Alsace, France, June 26, 22; US citizen; m 48; c 4. COMPARTMENTAL ANALYSIS, EPIDEMIOLOGICAL MODELING. *Educ:* Cornell Univ, MD, 47. *Prof Exp:* Res fel, Sloan-Kettering Inst, 47-50, from asst to assoc exp chemother, 50-60, assoc mem, 60-62; from instr to assoc prof biol, Sloan-Kettering Div, Cornell Univ, 52-63; from assoc prof to prof physiol, Med Sch, 62-90, from assoc prof to prof biostatist, Sch Pub Health, 62-90, actg chmn physiol, Med Sch, 85-87, EMER PROF PHYSIOL & BIOSTATIST, UNIV MICH, 90- *Concurrent Pos:* Vis investr path & bact, Rockefeller Inst, 47-48; consult, Rand Corp, 59-64; mem sci adv comt, Ore Regional Primate Res Ctr, 65-68; mem comput res study sect, NIH, 66-70, chmn, 68-70, mem chem & biol info handling rev comt, 75-79 & biotech res adv comt, 77-79; assoc ed, Math Biosci, 67-75, ed, 75- *Mem:* AAAS; Am Physiol Soc; Biophys Soc; Soc Indust & Appl Math; Soc Math Biol (pres, 85-87); Sigma Xi. *Res:* Active transport; structure and function of membranes; mathematical modeling of physiological systems; compartmental systems; logical and probabilistic structure of the diagnostic process; respiratory physiology; epidemiological modeling. *Mailing Add:* 490 Huntington Dr Ann Arbor MI 48104

JACQUIN, ARNAUD ERIC, b Reims, France, Feb 18, 64. IMAGE & VIDEO CODING, FRACTAL CODING. *Educ:* Ecole Supérieure d'Electricitd, EngrESE, 86; Ga Inst Tech, MS, 87, PhD(math), 89. *Prof Exp:* MEM TECH STAFF SIGNAL PROCESSING RES, SIGNAL PROCESSING DEPT, AT&T BELL LABS, MURRAY HILL, NJ, 90- *Mem:* Inst Elec & Electronics Engrs. *Res:* Image processing; image and video coding. *Mailing Add:* AT&T Bell Labs Rm 2C-479 600 Mountain Ave Murray Hill NJ 07974-2070

JACQUOT, RAYMOND G, b Casper, Wyo, Nov 16, 38; div; c 1. MECHANICAL ENGINEERING, ELECTRICAL ENGINEERING. *Educ:* Univ Wyo, BS, 60, MS, 62; Purdue Univ, PhD(mech eng), 69. *Prof Exp:* Supply instr mech eng, Univ Wyo, 60-62, instr, 62-64; instr, Purdue Univ, 64-65; from asst prof to assoc prof, 69-77, PROF ELEC ENG, UNIV WYO, 77- *Concurrent Pos:* Vis prof elec & comput eng, Univ Calif, Davis, 84. *Mem:* Am Soc Mech Engrs; Am Soc Eng Educ; Inst Elec & Electronics Engrs. *Res:* Vibration of elastic systems; simulation of large scale systems by digital computer; digital control and signal processing; nonlinear system analysis; digital filtering. *Mailing Add:* Dept of Elec Eng Univ of Wyo Laramie WY 82071

JADUSZLIWER, BERNARDO, b Buenos Aires, Argentina, Oct 17, 43; m 68; c 1. ELECTRON SCATTERING, ATOMIC FREQUENCY STANDARDS. *Educ:* Univ Buenos Aires, Lic en physics, 68; Univ Toronto, MSc, 70, PhD(physics), 73. *Prof Exp:* Fel physics, Univ Toronto, 73-74; assoc res scientist,NY Univ, 74-78, from asst res prof to assoc res prof physics, 81-855; mem tech staff, 85-87, RES SCIENTIST, AEROSPACE CORP, 87- *Concurrent Pos:* Tech adv, Physics Today Buyers Guide, 83-; adj assoc prof, Univ Southern Calif, 85- *Mem:* Am Phys Soc; AAAS. *Res:* Low energy electron positron scattering on ground and excited state atoms and molecules; measurement of atomic and molecular polarizabilities; atom photon interaction; atomic frequency standards. *Mailing Add:* Aerospace Corp M2-253 PO Box 92957 Los Angeles CA 90009

JAECKER, JOHN ALVIN, b Troy, NY, Feb 21, 45; m 68; c 2. CATALYSIS. *Educ:* Hope Col, BA, 68; Purdue Univ, MA & PhD(inorg chem), 73. *Prof Exp:* Sr chemist, 73-84, RES ADV, ATLANTIC RICHFIELD CO, 84- *Concurrent Pos:* Purchasing agent. *Mem:* Am Chem Soc; Catalysis Soc; Am Soc Testing & Mat. *Res:* Analytical chemistry of petroleum catalysts and products; studies of catalysts. *Mailing Add:* 1248 Vally Vista Dr Fullerton CA 92631

JAECKS, DUANE H, b Wausau, Wis, Sept 24, 35; m 59; c 3. ATOMIC PHYSICS, TIME RESOLVED SPECTROSCOPY. *Educ:* Univ Wis, BS, 58; Miami Univ, MA, 60; Univ Wash, Seattle, PhD(physics), 64. *Prof Exp:* From asst prof to assoc prof, 66-74, PROF PHYSICS, UNIV NEBR-LINCOLN, 74- *Mem:* Fel Am Phys Soc. *Res:* Basic atomic collisions research. *Mailing Add:* Dept Physics Univ Nebr Lincoln NE 68588

JAEGER, CHARLES WAYNE, b Kissimmee, Fla, Sept 8, 43; m 65; c 2. INDUSTRIAL ORGANIC CHEMISTRY. *Educ:* Fla State Univ, BS, 66; Purdue Univ, PhD(org chem), 71. *Prof Exp:* Res fel org chem, Ga Inst Technol, 71-73; res chemist, Dyes & Chem Div, Crompton & Knowles Corp, 73-80; sr res scientist, 80-83, mgr Ink Jet Technol Group, 84-85, bus develop & spec proj mgr, Periphal Div, 85-86, PRIN SCIENTIST, TEKTRONIX, INC, 83- *Concurrent Pos:* Gen chmn, Soc Photographic Scientists & Engrs Fifth Int Congress on Adv in Non-Impact Printing Technol. *Mem:* Am Chem Soc; Am Asn Textile Chem & Colorists; Soc Info Display; Soc Photog Scientists & Engrs. *Res:* Industrial research and process development of disperse dyes, heat transfer printing dyes and fluorescent whitening agents primarily for textile applications; imaging research on hardcopy devices; ink-jet ink chemistry; ink-jet technology development. *Mailing Add:* Tektronix Inc PO Box 1000 MS 63-424 Wilsonville OR 97070

JAEGER, DAVID ALLEN, b San Diego, Calif, June 3, 44; m 68; c 1. ORGANIC CHEMISTRY. *Educ:* Stanford Univ, BS, 65; Univ Calif, Los Angeles, PhD(org chem), 70. *Prof Exp:* NSF fel chem, Stanford Univ, 70-71; from asst prof to assoc prof, 71-82, PROF CHEM, UNIV WYO, 82- *Mem:* Am Chem Soc. *Res:* Chemistry; micellar catalysis; surfactant chemistry. *Mailing Add:* Dept Chem Univ Wyo Box 3838 Laramie WY 82071-3838

JAEGER, HERBERT KARL, b Harpolingen-Saeckingen, Ger, June 29, 31; m 59; c 1. ORGANIC CHEMISTRY. *Educ:* Univ Basel, PhD(org chem), 58. *Prof Exp:* Res fel chem, Univ Basel, 58-59 & Univ Calif, Los Angeles, 60-61; res assoc, 61-65, sect head, 65-70, res mgr org chem, 70-78, group mgr, 78-85, DIR, FINE CHEM PROD, UPJOHN CO, 85- *Mem:* Am Chem Soc. *Res:* Isolation and chemistry of cardiac glycosides; microbial production and chemistry of carotenoids; process research and development of steroidal and other chemical products. *Mailing Add:* 4901 Gulf Shore Blvd N No 203 Naples FL 33940

JAEGER, JAMES J, CARDIOVASCULAR PHYSIOLOGY, PULMONARY PHYSIOLOGY. *Educ:* Rutgers Univ, PhD(physiol), 73. *Prof Exp:* Proj mgr, Walter Reed Army Inst Res, 73-85, CHIEF PHYSIOL BR, US ARMY MED RES INST CHEM DEFENSE, 85- *Mailing Add:* H A US Army Med R&D Command Ft Detrick Frederick MD 21701

JAEGER, KLAUS BRUNO, b Lübeck, Ger, May 8, 38; US citizen; m 69; c 1. TECHNICAL MANAGEMENT. *Educ:* Syracuse Univ, BS, 65, MS, 68, PhD(physics), 70. *Prof Exp:* Fel exp high energy physics, Argonne Nat Lab, 70-71, asst physicist, 71-75, physicist, 75-80; physicist, Brookhaven Nat Lab, 80-81; mgr, 85, PHYSICIST, LOCKHEED CO, 82- *Concurrent Pos:* Assoc group leader bubble chamber, Argonne Nat Lab, 71-77, proj mgr 12ft solenoid magnet, 77-80, dep group leader for measuring & testing superconducting magnets, 80-81, metrologist, 82-85. *Mem:* Am Phys Soc; Am Inst Physics. *Res:* Design and implementation of new particle detection techniques such as solenoidal magnets with shower counters; inclusive hadron physics at high and medium energies; colliding electron positron beams; design and implementation of measuring techniques for superconducting dipole and quadrapole magnets; primary standards development. *Mailing Add:* Lockheed Space & Missiles Co Inc PO Box 3504 Bldg 195A 0/48-75 Sunnyvale CA 94089

JAEGER, MARC JULES, b Berne, Switz, Apr 4, 29; m 60, 73; c 2. RESPIRATORY PHYSIOLOGY. *Educ:* Univ Berne, Baccalaureat, 48, MD, 54. *Prof Exp:* Res assoc med, Univ Berne, 57-61; res assoc physiol, Col Med, Univ Fla, 61-63; res asst med, Univ Berne, 63-65; asst prof physiol, Univ Fribourg, 65-70; PROF PHYSIOL & DENT, COL MED, UNIV FLA, 70- *Concurrent Pos:* Vis prof, Yale Univ, 68, Dept Med, McGill Univ, Montreal, 80-81. *Mem:* Swiss Med Asn; Swiss Asn Physiol & Pharmacol; Swiss Soc Advan Sci; Int Union Physiol Sci; Sigma Xi. *Res:* Mechanics of breathing; fluid dynamics; mechanical and analog computer modelling; environmental physiology, especially diving, smoking and air pollution. *Mailing Add:* 519 NW 19th St Gainesville FL 32603

JAEGER, RALPH R, b Cincinnati, Ohio, Jan 30, 40. INORGANIC CHEMISTRY. *Educ:* Univ Cincinnati, BS, 62; Purdue Univ, PhD(inorg chem), 67. *Prof Exp:* SR RES CHEMIST, MOUND LAB, MONSANTO RES CORP, 67- *Mem:* Am Chem Soc; Am Inst Chemists. *Res:* Chemical vapor deposition of refractory metals; high temperature chemistry of plutonium; fuel forms containing plutonium. *Mailing Add:* 5092 Benner Rd Miamisburg OH 45342

JAEGER, RICHARD CHARLES, b New York, NY, Sept 2, 44; m 64; c 2. MICROELECTRONICS, INTEGRATED CIRCUIT DESIGN. *Educ:* Univ Fla, BS, 66, ME, 66, PhD(elec eng), 69. *Prof Exp:* Staff eng, IBM Corp, 69-72, adv eng, 72-74, res staff mem, 74-76, adv eng, 76-79; assoc prof, 79-82, alumni prof, 83-88, PROF, AUBURN UNIV, 90-; DIR, ALA MICROELECTRONICS SCI & TECHNOL CTR, 84- *Concurrent Pos:* Prog comt mem, Int Solid-State Circuits conf, 78-92; mem, gov bd, comput soc, Inst Elec & Electronics Engrs, 85- 86, secy, Solid State Circuits Coun, 85-87, vpres, 87-89, pres, 90-91. *Honors & Awards:* Outstanding Contrib Award, Computer Soc, Inst Elec & Electronics Engrs. *Mem:* Fel, Elec & Electronics Engrs. *Res:* Microelectronic circuit, device and process design; electronic packaging and cooling; wafer-scale integration; low temperature semiconductor electronics. *Mailing Add:* Elec Eng Dept Auburn Univ 200 Broun Hall Auburn AL 36849

JAEGER, ROBERT GORDON, b Baltimore, Md, Dec 16, 37; m 64. POPULATION ECOLOGY. *Educ:* Univ Md, BS, 60, PhD(ecol), 69; Univ Calif, Berkeley, MA, 63. *Prof Exp:* Fac res asst ecol, Univ Md, 69-70, instr zool, 70-71; res assoc zool, Univ Wis, Madison, 71-74; asst prof zool, State Univ NY, Albany, 74-80, adj asst prof, 80-81; from asst prof to assoc prof, 81-86, PROF BIOL, UNIV SOUTHWESTERN LA, 86- *Concurrent Pos:* Ed, J Herpetologica, 82- *Honors & Awards:* Stoye Award, Am Soc Ichthyol & Herpet, 69. *Mem:* Ecol Soc Am; Animal Behav Soc; Soc Study Evolution; Am Inst Biol Sci; Am Soc Ichthyol & Herpet; Sigma Xi. *Res:* Competitive exclusion and environmental pressures in the distributions of salamander species; comparative phototactic responses of anuran species in relation to their natural habitats. *Mailing Add:* Dept Biol Univ Southwestern La Lafayette LA 70504-2451

JAEGER, RUDOLPH JOHN, b Weehawken, NJ, Jan 17, 44; m 66, 87; c 3. BIOCHEMISTRY. *Educ:* Renesselaer Polytech Inst, BS, 66; Johns Hopkins Univ, PhD(biochem toxicol), 71. *Prof Exp:* Res assoc toxicol, Harvard Sch Pub Health, 71-73, asst prof, 73-78, assoc prof, 78-79; assoc prof, 79-83, RES PROF ENVIRON MED, SCH MED, NY UNIV, 85- *Concurrent Pos:* Toxicol consult, Polaroid Corp, 77-, Southern Calif Edison, 84-87, Esselte Letraset Mfg, 85-, Goodwin, Proctor & Hoar, 87-, Womble, Carlyle, Sandridge & Rice, 88-90, Lockheed Aeronaut Systs, 87-89, Sidley & Austin, 89-, Polymerics Inc, 89-; tech support & prod eval, AT&T Bell Labs, 83-85; hazard commun training progs, Bell Commun Res, 88-89; mem, Toxicol Info Prog Comt, NAS, 90- *Honors & Awards:* Leslie Silverman Award, Am Indust Hyg Asn, 80. *Mem:* Soc Toxicol; AAAS; NY Acad Sci; Am Indust Hygiene Asn; Sigma Xi; Am Soc Testing Mat; Am Chem Soc; Am Col Toxicol; Am Asn Path; Am Acad Clin Toxicol. *Res:* Inhalation toxicology of plastics monomers, pulmonary toxicology of combustion products. *Mailing Add:* Environ Med 263 Center Ave Westwood NJ 07675

JAEHNING, JUDITH A, b Yakima, Wash, Oct 14, 50. BIOCHEMISTRY. *Educ:* Univ Wash, Seattle, BS, 72; Washington Univ, St Louis, PhD(biol chem), 77. *Prof Exp:* Fel biochem, Univ Calif, Berkeley, 77-78 & Stanford Univ, 78-80; asst prof biochem, Univ Ill, Urbana, 81-85; asst prof, 85-88, ASST PROF BIOL, IND UNIV, 88- *Mem:* Am Soc Microbiogists; Am Soc Biochem & Molecular Biol; Genetics Soc Am. *Res:* Regulation and mechanisms of eukaryotic transcription; nuclear and mitochondrial RNA polymerase; saccharomyces cerevisiae. *Mailing Add:* Dept Biol Ind Univ Bloomington IN 47405

JAENIKE, JOHN ROBERT, b Osawatomie, Kans, Feb 13, 26; m 47; c 3. MEDICINE. *Educ:* Univ Rochester, MD, 48. *Prof Exp:* From instr to assoc prof, 52-68, asst dean, 70-73, assoc dean, 73-75, PROF MED, SCH MED & DENT, UNIV ROCHESTER, 68-, CHIEF MED, HIGHLAND HOSP, 80- *Concurrent Pos:* Sr investr, Nat Heart Inst, 58-60; Fogarty fel award, Sabbatical to Eng, 75-76. *Mem:* Am Fedn Clin Res; Am Soc Clin Invest; Asn Am Physicians; Am Physiol Soc. *Res:* Clinical medicine; medical education. *Mailing Add:* Biol Dept Univ Rochester Rochester NY 14627

JAENISCH, RUDOLF, Ger citizen. RETROVIRUSES, MAMMALIAN DEVELOPMENT. *Educ:* Univ Munich, MD, 67. *Prof Exp:* Postdoc fel, Max-Planck Inst, 67-69, Princeton Univ, 70-72; asst prof, Salk Inst, San Diego, 73-77; prof, Henrich-Pette Inst, Hamburg, Ger, 77-84; PROF, WHITEHEAD INST & DEPT BIOL, MASS INST TECHNOL, 84- *Res:* Control of mammalian development using retroviruses and transgenic technology; generation of insertional mutations which affect mouse development; interaction of retroviruses with embryos. *Mailing Add:* Whitehead Inst for Biomed Res Nine Cambridge Ctr Cambridge MA 02142

JAESCHKE, WALTER HENRY, b Milwaukee, Wis, Nov 25, 09; m 38; c 1. PATHOLOGY. *Educ:* Univ Wis, BS, 32, MD, 34; Am Bd Path, dipl, 41. *Prof Exp:* Intern, Hosp Div, Med Col Va, 34-35; resident med, Wis Gen Hosp, 35-37, resident path, 37-38, asst, 38-39, instr clin path, 39-41; pathologist & dir labs, St Agnes Hosp, Wis, 41-42; from asst prof to prof clin path, 42-64, prof surg path & path, 64-80, DIR, LAB SURG PATH, GEN HOSP, UNIV WIS-MADISON, 58-, EMER PROF PATH & LAB MED, 80- *Concurrent Pos:* Consult surg pathologist, 80- *Mem:* Fel Am Soc Clin Path; fel AMA. *Res:* Teaching methods in basic and systemic pathology for medical students and surgical pathology for hospital staff. *Mailing Add:* Clin SoL Ctr E5/3 600 Highland Ave Madison WI 53792

JAFEK, BRUCE WILLIAM, b Chicago, Ill, Mar 4, 41; m 62; c 6. HEAD & NECK SURGERY, FACIAL PLASTIC SURGERY. *Educ:* Coe Col, Iowa, BS, 62; Univ Calif, Los Angeles, MD, 66. *Prof Exp:* Intern gen surg, New Haven Hosp, 66-67; resident gen surg, Univ Calif, Los Angeles, 67-68, res otolaryngol, head & neck surg, 68-71; instr otolaryngol, Johns Hopkins Univ, 71-73; asst prof otolaryngol, Univ Pa, 73-76; PROF OTOLARYNGOL & CHMN NECK-HEAD SURG, UNIV COLO, 76- *Honors & Awards:* Fowler Award, Triologic Soc, 83. *Mem:* Am Acad Facial Plastic Surg; Am Acad Otolaryngol-Head & Neck Surg; Am Col Surgeons; Asn Acad Dir Otolaryngol; Soc Univ Otolaryngologists; Triological Soc. *Res:* Chemosensation of the head and neck region; processing of biopsies from dysfunctional patients to correlate ultrastructural changes. *Mailing Add:* Dept Otolaryngol (Box B- 205) Univ Colo Health Sci Ctr Denver CO 80262

JAFFA, ROBERT E, b Berkeley, Calif, Nov 11, 35. MATHEMATICS. *Educ:* Univ Calif, BA, 57, MA, 60, PhD(math), 64. *Prof Exp:* Res mathematician, Univ Calif, 64-65; from asst prof to assoc prof, 65-76, prof math, Sacramento State Univ, 76-; AT DEPT MATH, CALIF STATE UNIV. *Mem:* Am Math Soc. *Res:* Abstract algebra; functional analysis. *Mailing Add:* Dept Math Calif State Univ 6000 J St Sacramento CA 95819

JAFFE, ANNETTE BRONKESH, b Munich, Ger, July 25, 46; US citizen; m 70; c 2. PHYSICAL ORGANIC CHEMISTRY, PHYSICAL CHEMISTRY. *Educ:* Douglass Col, Rutgers Univ, BA, 68; Yale Univ, MPhil, 70, PhD(chem), 72. *Prof Exp:* Mem res staff, Res Labs, IBM Corp, 74-90; PRIN SCIENTIST, IMAGING PRODS, APPLE COMPUTER, INC, SANTA CLARA, 90- *Mem:* Am Chem Soc; sr mem Soc Photog Scientists & Engrs; Soc Imaging Sci & Technol. *Res:* Solid state chemistry; electro-organic chemistry in aprotic solvents; reaction mechanisms; ink chemistry; electrophotography; contact electrification; color science; non-impact printing. *Mailing Add:* Apple Computer Inc 3535 Monroe St MS 69D Santa Clara CA 95051

JAFFE, ARTHUR MICHAEL, b New York, NY, Dec 22, 37; m 71. MATHEMATICAL PHYSICS. *Educ:* Princeton Univ, AB, 59, PhD(physics), 65; Cambridge Univ, BA, 61. *Hon Degrees:* MA, Harvard Univ, 70. *Prof Exp:* Res assoc physics, Princeton Univ, 65-66; actg asst prof math, Stanford Univ, 66-67; visitor natural sci, Inst Advan Study, 67; from asst prof to prof math physics, 67-85, LANDON T CLAY PROF MATH & THEORET SCI, HARVARD UNIV, 85- *Concurrent Pos:* Nat Acad Sci-Nat Res Coun, Air Force Off Sci Res fel, 65-67; res assoc, Stanford Linear Accelerator Ctr, 66-67; Alfred P Sloan Found fel, 68-70; John S Guggenheim Found fel, 77; ed, Comn Math Physics, 75-79, chief ed, 79-; ed, Progress in Physics, 80-, Selecta Mathematica Sovietica, 81- *Honors & Awards:* NY Acad Sci award, 79; Dannie Heineman Prize, Am Inst Physics & Phys Soc, 80. *Mem:* Am Math Soc; Am Phys Soc; NY Acad Sci; AAAS; Am Acad Arts Sci; Int Asn Math Physics. *Res:* Mathematics and theoretical physics. *Mailing Add:* Dept Physics Harvard Univ One Oxford St Cambridge MA 02138

JAFFE, BERNARD MORDECAI, b New York, NY, Mar 7, 17; m 62; c 1. PHYSICS, OPTICS. *Educ:* City Col New York, BS, 36; NY Univ, PhD(photoconductivity), 62. *Prof Exp:* Engr, Amperex Electronic Prod Co, 40-41; physicist, Signal Corps Labs, 41-44; engr, Lloyd Rogers Co, 44-45 & Airborne Instruments Lab, 45-46; tutor physics, City Col New York, 46-49 & 52-56; instr, Stevens Inst Technol, 56-59; from asst prof to prof, 59-83, EMER PROF PHYSICS, ADELPHI UNIV, 83- *Mem:* Am Phys Soc; Optical Soc Am; Am Asn Physics Teachers. *Res:* Photoconductivity; photovoltaic effect; persistent internal polarization; physical optics; coherence of light. *Mailing Add:* Dept Physics Adelphi Univ Garden City NY 11530

JAFFE, DONALD, b New York, NY, May 17, 31; m 53; c 5. MATERIALS SCIENCE, ELECTRONICS ENGINEERING. *Educ:* Mass Inst Technol, BS, 52, MS, 53; Carnegie-Mellon Univ, PhD(metall eng), 63. *Prof Exp:* Engr, Gen Elec Co, 53-58 & Westinghouse Elec Co, 58-65; mem tech staff magnetic mat, 65-68, supvr thin film mat, 68-72, supvr encapsulation mat, 72-77, supvr integrated circuit technol, 77-81, head, Film Technol Dept, 81-84 & Film Circuits Design & Technol Dept, 84-87, HEAD, ANALYTICAL TECHNOL DEPT, AT&T BELL LABS, 87- *Concurrent Pos:* Instr eve sch, Carnegie-Mellon Univ. *Mem:* Inst Elec & Electronics Engrs; Int Soc Hybrid Microelectronics; Am Vacuum Soc. *Res:* Technical management involving structure, properties, design and assembly techniques relative to materials and components used for microelectronics applications. *Mailing Add:* AT&T Bell Labs Inc 555 Union Blvd Allentown PA 18103

JAFFE, EDWARD E, b Poland, Sept 22, 28; nat US; m 54; c 3. ORGANIC CHEMISTRY. *Educ:* City Col NY, BS, 52; NY Univ, MS, 54, PhD(org chem), 57. *Prof Exp:* Tech asst res, Mt Sinai Hosp, NY, 51-54; from res chemist to sr res chemist, 57-65, res assoc, 65-73, res supvr, 73-75, tech supt, 75-78, res mgr, 78-80, res fel, E I du Pont de Nemours & Co, Inc, 80-84; distinguished res fel, 84-87, dir res, 87-88, VPRES RES, CIBA-GEIGY CORP, 88- *Mem:* Am Chem Soc; Sigma Xi. *Res:* Heterocyclic chemistry and the characterization of colored organic compounds; varied organic syntheses in the fields of organophosphorus chemistry, pigments, polymers; organic microchemistry; holder of over 100 United States and International patents. *Mailing Add:* Three Crenshaw Dr Wilmington DE 19810

JAFFE, EILEEN KAREN, b New York, NY, May 7, 54; m 83; c 1. ENZYMOLOGY, MAGNETIC RESONANCE. *Educ:* State Univ NY Col Cortland, BS, 75; Univ Pa, PhD(biochem), 79. *Prof Exp:* NIH res fel chem, Harvard Univ, 79-81; asst prof, Haverford Col, 81-83; res asst prof biochem, Jefferson Med Col, Thomas Jefferson Univ, 83-84; res asst prof biochem, Sch Dent Med, Univ Pa, 84-90, res assoc prof, 90-91; MEM, INST CANCER RES, FOX CHASE CANCER CTR, 91- *Concurrent Pos:* NSF fel, Sch Med, Univ Pa, 75-78; NIH prin investr, Haverford Col, Jefferson Med Col, Univ Pa & Inst Cancer Res, 81- *Mem:* Am Chem Soc; AAAS; Sigma Xi; Am Asn Women Sci; Am Soc Biochemists & Molecular Biologists. *Res:* The chemical mechanisms of enzyme catalyzed reactions; the role of metal ions; the binding of metals to adenosinetriphosphate; the role of zinc in the porphobilinogen synthase catalyzed reaction; protein structure. *Mailing Add:* Inst Cancer Res Fox Chase Cancer Ctr 7701 Burholme Ave Philadelphia PA 19111

JAFFE, ELAINE SARKIN, b New York, NY, Aug 27, 43; m 67; c 2. HEMATOPATHOLOGY, IMMUNOPATHOLOGY. *Educ:* Cornell Univ, AB, 65; Univ Pa, MD, 69. *Prof Exp:* Sr investr, Lab Path, 74-80, CHIEF, HEMATOPATH SECT, LAB PATH, NAT CANCER INST, NIH, 80-, DEP CHIEF, 82-; CLIN PROF PATH, SCH MED, GEORGE WASHINGTON UNIV, 85- *Concurrent Pos:* Chief, Path Anat Br, Clin Ctr, NIH, 82-; co-chair, Expert Panel Cytochem, ICSH; mem sci comt, Histiocyte Soc, 87-91; mem women's comt, Am Asn Pathologists, 90- *Honors & Awards:* Pritzker Mem Lectr, Acad Med, Toronto, Can, 89. *Mem:* Am Soc Hemat; US-Can Acad Path; Am Asn Pathologists; Soc Hematopath; Histiocyte Soc. *Res:* Analysis of human malignant lymphomas to determine relationship to normal immune system; immunologic characterization of lymphomas with delineation of new clinicopathologic entities. *Mailing Add:* Lab Path Nat Cancer Inst NIH Bldg 10 Rm 2N202 Bethesda MD 20892

JAFFE, ERIC ALLEN, b New York, NY, Apr 7, 42; m 71; c 2. HEMATOLOGY, ONCOLOGY. *Educ:* Downstate Med Ctr, State Univ NY, MD, 66. *Prof Exp:* Intern internal med, Kings County Hosp, Brooklyn, 66-67, resident, 67-68; resident, New York Hosp, 68-69; guest investr, Rockefeller Univ, 69; sr resident, New York Hosp, 70, fel hemat, 70-72; from instr to asst prof, 71-77, assoc prof, 77-82, PROF MED, CORNELL UNIV MED COL, 82- *Concurrent Pos:* Res fel, Nat Hemophilia Found, 74-76; career scientist, Health Res Coun City New York, 74-75, Irma T Hirschl Career Scientist Award, 76-81 & NIH Res Career Develop Award, 76-81. *Honors & Awards:* Passano Found Young Scientist Award, 77. *Mem:* Asn Am Physicians; Am Soc Clin Invest; Am Soc Cell Biol; Am Soc Hematology; Am Asn Pathologists. *Res:* Role of endothelial cells in coagulation and the relationship of enothelial cells to platelets, white cells, and atherosclerosis. *Mailing Add:* Cornell Univ Med Col 1300 York Ave New York NY 10021

JAFFE, ERNST RICHARD, b Chicago, Ill, Jan 4, 25; m 50; c 2. MEDICINE, HEMATOLOGY. *Educ:* Univ Chicago, BS, 45, MD & MS, 48; Am Bd Internal Med, dipl & Cert hemat. *Hon Degrees:* LHD, Yeshiva Univ, 87. *Prof Exp:* Asst path, Univ Chicago, 46 & 47-48, asst physiol, 47; intern, Med Serv, Presby Hosp, NY, 48-49, asst resident, 49-51 & 53-55; from instr to assoc prof, 56-69, actg dean, 72-74, & 83-84, head Div Hemat, 70-82, assoc dean fac, 76-83, PROF MED, ALBERT EINSTEIN COL MED, 69-, SR ASSOC DEAN, 74-; DISTINGUISHED PROF MED, BELFER INST ADVAN BIOMED STUDIES, 84- *Concurrent Pos:* Clin asst vis physician, Med Serv, Bronx Munic Hosp Ctr, 55-56, from asst vis physician to assoc vis physician, 55-63, attend physician, 63-; Nat Found Infantile Paralysis fel, 55-57; career scientist, Health Res Coun, NY 61-71; assoc vis physician, Lincoln Hosp, NY, 61-73; co-ed, Sem in Hemat, 68-; study Sect, NIH Hemat, 72-79, chmn, 79-81; adv coun Nat Diabetes, Digestive & Kidney Dis, NIH 84-87; Nat Bd of Govs, Am Red Cross, 84-, chmn, blood serv comt, 88- *Mem:* Am Soc Hemat (pres, 83); Am Fedn Clin Res; Am Physiol Soc; Asn Am Physicians; Am Med Writers' Asn; Am Soc Clin Invest. *Res:* Internal medicine and hematology; metabolism of the mammalian erythrocyte and alterations occurring with aging of the erythrocyte. *Mailing Add:* Nine Orchard Pl Tenafly NJ 07670

JAFFE, FRED, b Cleveland, Ohio, Apr 5, 30; m 60; c 3. ORGANIC CHEMISTRY. *Educ:* Western Reserve Univ, BS, 52; Cornell Univ, PhD, 57. *Prof Exp:* Sloan fel, Cornell Univ, 57-58; res chemist, Washington Res Ctr, W R Grace & Co, Md, 58-63, sr res chemist, 63-65, coordr oxidation chem, 64-65; res chemist, Miami Valley Labs, Procter & Gamble Co, 65-68; sr res chemist, 68-75, RES CHEMIST, EASTERN RES CTR, AZKO CHEM CO, 75- *Mem:* AAAS; Am Chem Soc; Royal Soc Chem. *Res:* Synthesis, stabilization and modification of formaldehyde polymers and copolymers; synthesis of para-bridged benzenes, liquid phase autoxidation, oxidation of decalin; hydroperoxides; carbanion oxidations; organometallic alkali chemistry; phosphorus and silicon; lubricant and hydraulic base stocks and additives; flame retardants. *Mailing Add:* Akzo Chem Inc Livingstone Ave Dobbs Ferry NY 10522- 3401

JAFFE, HANS, b Marburg, Ger, Apr 17, 19; US citizen; div; c 3. PHYSICAL CHEMISTRY, THEORETICAL CHEMISTRY. *Educ:* State Univ Iowa, BS, 41; Purdue Univ, MS, 42; Univ NC, PhD(chem), 52. *Prof Exp:* Phys chemist, USPHS, 46-54; from asst prof to assoc prof, 54-62, head dept, 66-71, PROF CHEM, UNIV CINCINNATI, 62- *Concurrent Pos:* Sect ed, Chem Abstr, 58-61; mem adv panel, Nat Bur Standards, 59-65; Fulbright fel, France, 61-62. *Mem:* Am Chem Soc; Am Phys Soc. *Res:* Quantum chemistry; basicities of weak bases; spectroscopy; excited state chemistry; physical organic chemistry; Hammett equation. *Mailing Add:* Dept Chem Univ Cincinnati Cincinnati OH 45221

JAFFE, HAROLD, b Chicago, Ill, May 8, 30; m 51; c 3. NUCLEAR SCIENCE. *Educ:* Univ Ill, BS, 51, PhD(nuclear chem), 54. *Prof Exp:* Asst res chemist, Union Oil Co, 54-55; sr chemist, Tracerlab, Inc, 55-56; prin nuclear chemist, Aerojet-Gen Nucleonics Div, Aerojet-Gen Corp, 56-57, prog mgr gas-cooled reactor exp, 57-59, mgr fuel develop dept, 60-62, asst mgr nuclear tech div, 62-64, mgr appl sci div, 64-66, asst to vpres, Nuclear Div, 66-69; mgr San Ramon Plant & asst to pres, 69-70; chief isotope power systs proj br, Space Nuclear Systs Div, US Atomic Energy Comn, 70-72, mgr Isotope Flight Systs, 72-75; tech asst to asst admin nuclear energy, Energy Res & Develop Agency, 75-76; dep dir, Off Int Tech Coop, 76-81, DIR, OFF INT RES & DEVELOP POLICY, DEPT ENERGY, 81- *Concurrent Pos:* Asst prof, John F Kennedy Univ, 66-70; mem bd dirs, Idaho Nuclear Corp, 68-70. *Mem:* AAAS; Am Nuclear Soc; Am Chem Soc; fel Am Inst Chemists. *Mailing Add:* 10702 Great Arbor Dr Potomac MD 20854

JAFFE, HOWARD WILLIAM, b New York, NY, Feb 16, 19; m 50; c 3. GEOLOGY. *Educ:* Brooklyn Col, BA, 42; Univ Geneve, DSc, 72. *Prof Exp:* Sr engr aid, US Geol Surv, 42-44; petrographer, US Bur Mines, 44-51; geologist, US Geol Surv, 51-58; res sect leader mineral & geochem, Union Carbide Nuclear Co Div, Union Carbide Corp, 58-65; assoc prof, 65-70, PROF GEOL, UNIV MASS, AMHERST, 70- *Concurrent Pos:* Vis prof, Univ Geneva, 71-72 & 78-79. *Mem:* Fel Am Mineral Soc; fel Geol Soc Am; Geochem Soc; Am Geophys Union; Mineral Soc Gt Brit & Ireland; Sigma Xi. *Res:* Optical properties and crystal chemistry of rockforming minerals; petrography of igneous and metamorphic rocks; physical and chemical mineralogy of ores; geochronology; geochemistry of minor elements; Precambrian geology of the Hudson highlands and Adirondacks in New York. *Mailing Add:* RR 1 540 Range Rd Underhill VT 05489

JAFFE, ISRAELI AARON, b New York, NY, Dec 21, 27; m 52; c 3. INTERNAL MEDICINE. *Educ:* NY Univ, BS, 46; Columbia Univ, MD, 50; Am Bd Internal Med, dipl. *Prof Exp:* Clin assoc, NIH, 53-55; CLIN ASSOC INSTR MED, COLUMBIA UNIV, 57-, PROF CLIN MED, COL PHYSICIANS & SURGEONS, 78- *Concurrent Pos:* Asst prof instr med, New York Med Col, 60-63, prof instr med & dir rheumatic dis serv, 63-78. *Mem:* Am Col Physicians; AMA; Am Rheumatism Asn. *Res:* Rheumatic diseases. *Mailing Add:* 161 Ft Washington Ave New York NY 10032

JAFFE, JAMES MARK, b New York, NY, Apr 11, 43; m 64; c 2. PHARMACEUTICS. *Educ:* Univ Pittsburgh, BS, 68, BA, 69, MS, 70, PhD(pharmaceut), 72. *Prof Exp:* Asst pharmaceut, Sch Pharm, Univ Pittsburgh, 68-70, from instr to asst prof pharmaceut, 70-75, assoc prof, 75-80. *Concurrent Pos:* Reviewer, J Pharmaceut Sci, 74- & Am J Hosp Pharm. *Mem:* Am Pharmaceut Asn; Acad Pharmaceut Sci; Sigma Xi. *Res:* Physiological and formulation factors that influence the absorption and excretion of drugs. *Mailing Add:* 60 W Main St Brookside NJ 07926

JAFFE, JONAH, b New York, NY, Oct 5, 29; m 51; c 2. PHARMACY, TECHNICAL MANAGEMENT. *Educ:* Columbia Univ, BS, 53; Univ Md, MS, 55, PhD(pharmaceut chem), 56. *Prof Exp:* Sr res chemist, E R Squibb, 56-62; dir, res & develop, Organon, US, 62-63, dir, res, develop & prod, 63-65; tech dir, Whitehall Labs Div, Am Home Prod, 65-70, asst vpres, 70-76; vpres res & develop, Johnson & Johnson, 76-85, vpres sci affairs, 85-88; vpres, Res & Develop Labs, McNeil Consumer Prod Co, 88-; RETIRED. *Mem:* Am Pharmaceut Asn; Sigma Xi; AAAS; NY Acad Sci. *Res:* Management of research and development; quality control; medical research and regulatory affairs; manufacture of proprietary and ethical drugs and toiletries. *Mailing Add:* Four Partridge Ct Cherry Hill NJ 08003

JAFFE, JULIAN JOSEPH, b New York, NY, Feb 17, 26; m 53; c 4. PHARMACOLOGY. *Educ:* Univ Conn, BA, 49; Harvard Univ, MA, 51, PhD(biol), 55. *Prof Exp:* Asst, Harvard Univ, 52-53; instr biol, Brown Univ, 54-55, USPHS fel, 55-56; from instr to asst prof pharmacol, Sch Med, Yale Univ, 56-61; assoc prof, 61-67, PROF PHARMACOL, COL MED, UNIV VT, 67- *Concurrent Pos:* Wellcome res travel grants, Oxford Univ, 58 & Nuffield Inst Comp Med, London, 67, guest worker, Nuffield Inst Comp Med, 67-68; mem, Panel Parasitic Dis, US-Japan Coop Med Sci Prog, 71-76 & Trop Med Parasitol Study Sect, 77-81; mem spec working group filariasis, WHO, 77, 80, 83 & 85. *Mem:* Am Soc Pharmacol & Exp Therapeut; Am Soc Parasitol; Am Soc Trop Med & Hyg; AAAS; Am Asn Univ Prof. *Res:* Biochemical pharmacology; biochemistry of parasites. *Mailing Add:* Dept of Pharmacol Univ of Vt Col of Med Burlington VT 05405

JAFFE, LAURINDA A, b Pasadena, Calif, Jan 9, 52. PHYSIOLOGY OF FERTILIZATION. *Educ:* Purdue Univ, BS, 73; Univ Calif, Los Angeles, PhD(biol), 77. *Prof Exp:* NIH fel electrophysiol fertil, Marine Biol Lab, Woods Hole, Mass, 78-79; NSF fel electrophysiol fertil, Univ Calif, San Diego, 79-81; asst prof, 81-86, ASSOC PROF PHYSIOL, UNIV CONN, 86- *Concurrent Pos:* Instr embryol, Marine Biol Lab, Woods Hole, 83-87; vis fac, Univ Wash, 85 & 88; assoc ed, Develop Biol, 85- *Mem:* Am Soc Cell Biol; Biophys Soc; Soc Develop Biol. *Res:* Physiology of fertilization. *Mailing Add:* Dept Physiol Univ Conn Health Ctr Rm L 5004 Farmington CT 06032

JAFFE, LEONARD, b Feb 1, 26. NUCLEAR SAFETY. *Prof Exp:* Aeronaut Res Scientist, Lewis Res Lab, Nat Adv Comt Aeronaut, 48-51, Instruments Res Div, 51-57, Chief, Data Systs Br, 57-59; chief, commun satellite prog, headquarters, NASA, 59-61, dir, commun systs, 61-63, dir, commun & navigation prog, 63-66, dir, space appln prog, 66-69, dep assoc admin space & sci appln, 69-71, space appln, 71-77, Dep Assoc Admin Space and Terrestrial Appln, Headquarters, 77-78, specialist to chief eng, 78-81; head, Tech Assessment Task Force, President's Comn Accident at Three Mile Island, 79; vpres, systs group, prog mgt & prod assurance, Comput Sci Corp, 81-82, pres, Systs Div, 82-84, VPRES, SYSTS GROUP, PROG MGT & PROD ASSURANCE, COMPUT SCI CORP, 84- *Concurrent Pos:* Chmn NASA Space Appln Adv Comt, 86- *Honors & Awards:* Inst Elec Electronics Engrs Award, 67; Am Inst Aeronaut & Astronaut, 78; Lloyd V Berkener Space Ulitization Award, Am Acad Sci; William Pecora Award, Dept Comn & NASA, 81. *Mem:* Fel Inst Elec Electronics Engrs; fel Am Inst Aeronaut & Astronaut; fel Am Astronaut Soc; Int Acad Astronaut; Int Astronaut Fedn, (pres, 74-76). *Mailing Add:* Comput Sci Corp 3160 Fairview Park Dr Falls Church VA 22042

JAFFE, LEONARD DAVID, b New York, NY, June 25, 19; m 45; c 4. SOLAR ENERGY, PLANETARY EXPLORATION. *Educ:* Mass Inst Technol, SB, 39, SM, 40; Harvard Univ, ScD(phys metall), 47. *Prof Exp:* Metallurgist, Watertown Arsenal, 40-42, metallurgist & supvr res on phase transformation in steel, 42-46, chief phys metall sect, 46-54; mgr, Mat Res Sect, Jet Propulsion Lab, 54-64, res specialist, Space Sci Div, 64, proj scientist surveyor proj, 65-68, mem tech staff, Space Sci & Energy Conversion Div, 68-81, SYST ENGR SOLAR THERMAL POWER SYSTS, JET PROPULSION LAB, CALIF INST TECHNOL, 81- *Mem:* AAAS; Am Geophys Union; Am Inst Aeronaut & Astronaut; Am Astron Soc; Am Inst Mining, Metall & Petrol Eng. *Res:* Solar thermal energy conversion; exploration of planetary surfaces; properties of the lunar surface. *Mailing Add:* Jet Propulsion Lab 4800 Oak Grove Dr Pasadena CA 91103

JAFFE, LIONEL F, b New York, NY, Dec 28, 27; m 49; c 3. DEVELOPMENTAL PHYSIOLOGY, BIOPHYSICS. *Educ:* Harvard Univ, SB, 48; Calif Inst Technol, PhD(embryol), 54. *Prof Exp:* Fel, Nat Res Coun, Hopkins Marine Sta, Stanford Univ, 53-54, NSF, 54-55; fel marine biol, Scripps Inst, Calif, 55-56; asst prof biol, Brandeis Univ, 56-60; from asst prof to assoc prof, Univ Pa, 60-67; prof biol, Purdue Univ, 67-84; DIR NAT VIBRATING PROBE FACIL, 82-; ADJ PROF BOT, UNIV MASS AMHERST, 90- *Mem:* AAAS; Biophys Soc; Am Soc Cell Biol; Dev Biol Soc; Soc Gen Physiol. *Res:* Development and nature of morphogenetic polarity; cellular tropisms; bioelectric and ionic aspects of development; calcium waves and gradients. *Mailing Add:* Marine Biol Lab Woods Hole MA 02543

JAFFE, MARVIN RICHARD, b New York, NY, May 23, 38. TEACHING CHEMISTRY FOR HEALTH SCIENCES, CONSUMER SCIENCES. *Educ:* Brooklyn Col, BS, 60, MA, 65; Fordham Univ, PhD(anal chem), 70. *Prof Exp:* Qual control chemist, Schrafft's, New York, 60-62, prod develop chemist, 62-63, supvr prod develop, 63-65; instr chem, Bronx Community Col, 65-68; res asst, Fordham Univ, 68-70; guest jr res assoc, Brookhaven Nat Lab, 69-70; from asst prof to assoc prof, 71-84, PROF, SCI DEPT, MANHATTAN COMMUNITY COL, 85- *Mem:* Sigma Xi; Nat Sci Teachers Asn; Am Chem Soc. *Res:* Consumer science; science education of non-scientists; kinetics and mechanisms of beta-diketones. *Mailing Add:* Dept Sci Manhattan Community Col 199 Chambers St New York NY 10007

JAFFE, MICHAEL, b New York, NY, May 10, 42; m; c 2. POLYMER PHYSICS, PHYSICAL CHEMISTRY. *Educ:* Cornell Univ, BA, 63; Rensselaer Polytech Inst, PhD(chem), 67. *Prof Exp:* Res chemist, Celanese Res Co, 67-71, sr res chemist, 71-73, res assoc, 73-74, res supvr, 74-78; group leader, Res & Develop Lab, Fiber Industs, Inc, 78-80; res supvr, Celanese Res Co, 81-82, res mgr, 82-88, SR RES ASSOC & RES FEL, HOECHST CELANESE RES DIV, 88- *Mem:* Am Chem Soc; Am Phys Soc; NAm Thermal Anal Soc. *Res:* Morphology of crystalline high polymers; transition behavior of polymers; structure-property relationships of polymers and related materials. *Mailing Add:* 86 Morris Ave Hoechst Celanese Res Div Summit NJ 07901

JAFFE, MIRIAM WALTHER, b Clinton, Ind, Feb 6, 22; m 49; c 3. ASTRONOMY. *Educ:* Ind Univ, AB, 43; Univ Va, MA, 45; Radcliffe Col & Harvard Univ, PhD(astron), 48. *Prof Exp:* Instr astron, Wellesley Col, 48-49 & Univ Southern Calif, 49-51; asst prof astron, Haverford Col, 65-66; asst prof astron, Purdue Univ, 68-85; RETIRED. *Mem:* Am Astron Soc. *Res:* Photographic photometry; classification of stellar spectra. *Mailing Add:* 59 Cumloden Dr Falmouth MA 02540-1609

JAFFE, MORDECAI J, b New York, NY, July 7, 33; m 61; c 3. PLANT PHYSIOLOGY. *Educ:* City Col New York, BS, 58; Cornell Univ, PhD(veg physiol), 64. *Prof Exp:* Lectr & res assoc biol, Yale Univ, 64-67; from asst prof to prof, plant physiol, Ohio Univ, 67-80; BABCOCK PROF BOTANY, WAKE FOREST UNIV, 80- *Concurrent Pos:* NSF res grant, 67-74 & 75-86; NASA res grant, 77-84; Bard res grant, 80-83; vis prof, Hebrew Univ, Jerusalem, 75, Lady Davis vis prof, 84; vis scientist, Boyce Thompson Inst Cornell Univ, 80; permanent vis prof, Hebrew Univ, Jerusalem, 84- *Mem:* AAAS; Am Soc Plant Physiologists; Soc Develop Biol; Phytochem Soc NAm; Japanese Soc Plant Physiol. *Res:* Sensory physiology and biochemistry; rapid movements in plants; biochemistry of touch mediated processes in plants; physiology of stress in plants; flowering mechanisms in plants. *Mailing Add:* Biol Dept Wake Forest Univ Winston-Salem NC 27109

JAFFE, MORRY, b New York, NY, Oct 10, 40; m 90. EXPERIMENTAL PHYSICS. *Educ:* City Col New York, BS, 62; Boston Univ, MA, 64; City Univ New York, PhD(physics), 71. *Prof Exp:* Asst, Boston Univ, 62-64; lectr, 64-65, res asst, City Col New York, 65-69 & 70-71; lab asst, New York Dept Air Resources, 73-78; DATA PROCESSING CONSULT, GROUP 88, NEW YORK, 84- *Honors & Awards:* Cert Appreciation, US Environ Protection Agency, 78. *Mailing Add:* 65 Park Terr E New York NY 10034

JAFFE, PHILIP MONLANE, b Bronx, NY, Aug 14, 27; m 50; c 3. INORGANIC CHEMISTRY. *Educ:* City Col NY, BS, 48; Polytech Inst Brooklyn, MS, 53, PhD, 62. *Prof Exp:* Chemist, City Chem Corp, NY, 48-51; res chemist, Westinghouse Elec Co, 53-63; sr staff scientist, Aerospace Res Ctr, Gen Precision, Inc, 63-66; sr res chemist, Zenith Radio Corp, 66-72; PROF CHEM, OAKTON COMMUNITY COL, 70- *Concurrent Pos:* Dean, Div Sci Allied Health, 85-91. *Mem:* Fel AAAS; Am Chem Soc; fel Am Inst Chem. *Res:* Inorganic phosphors and preparations; semiconductors; photoconductors; technical and educational writing. *Mailing Add:* Dept Chem Oakton Community Col Des Plaines IL 60016

JAFFE, RANDAL CRAIG, b St Louis, Mo, Dec 18, 47; m 70; c 3. ENDOCRINOLOGY, BIOCHEMISTRY. *Educ:* Univ Southern Calif, BS, 68; Univ Calif, Davis, PhD(biochem), 72. *Prof Exp:* Fel endocrinol, Sch Med, Vanderbilt Univ, 72-73; fel cell biol, Baylor Col Med, 73-75; asst prof, 75-83, ASSOC PROF PHYSIOL, MED CTR, UNIV ILL, 83- *Concurrent Pos:* NIH fel, 74-75; vis fel med, Mass Gen Hosp, 85-86; vis assoc prof med & physiol, Harvard Univ, 85-86. *Mem:* AAAS; Endocrine Soc; Am Soc Biol Chem. *Res:* Mechanism of hormone action; comparative endocrinology; hormonal control of development. *Mailing Add:* Dept Physiol & Biophys Univ Ill Col Med PO Box 6998 Chicago IL 60680

JAFFE, ROBERT B, b Detroit, Mich, Feb 18, 33; m 54; c 2. ENDOCRINOLOGY, OBSTETRICS & GYNECOLOGY. *Educ:* Univ Mich, MD, 57; Univ Colo, MS, 66; Am Bd Obstet & Gynec, dipl, 67 & Reproductive Endocrin cert. *Prof Exp:* Lab asst biochem, Univ Mich, 53-54; rotating intern, Univ Colo, 57-58, resident obstet & gynec, 59-63; from asst prof to prof obstet & gynec, Med Ctr, Univ Mich, Ann Arbor, 64-73; PROF OBSTET, GYNEC & REPROD SCI, CHMN DEPT & DIR, REPROD ENDOCRINOL CTR, UNIV CALIF, SAN FRANCISCO, 73- *Concurrent Pos:* USPHS postdoctoral fel endocrinol, Dept Internal Med, Med Ctr, Univ Colo, 58-59; NIH postdoctoral fel reprod endocrinol, Hormone Lab, Karolinska Sjukhuset, Sweden, 63-64; Josiah Macy, Jr Found fac fel, 66-69 & 80-81; chmn, Div Endocrinol & Fertil, Am Col Obstetricians & Gynecologists; mem, Human Embryol & Develop Study Sect, NIH, Sci Adv Bd, Nat Inst Child Health & Human Develop Coun, Reprod Biol Study Sect, NIH, Med Adv Bd, Nat Pituitary Agency, Steering Comt, Perinatal Res Soc, Sci Adv Bd, Prog Appl Res Fertil Regulation & Biomed Comt, Projs Pop Action; vis prof, Univ Mich, Univ Mo, Cornell Univ, Univ Wash, Univ Ore, Yale Univ, Univ Wis, Univ Va & Wash Univ; Fred Gellert chair reprod med & biol, 90. *Mem:* Inst Med-Nat Acad Sci; Endocrine Soc; Am Fedn Clin Res; Soc Gynec Invest (pres); fel Am Col Obstet & Gynec; Am Gynec Soc; Asn Am Physicians; Sigma Xi. *Res:* Endocrinology; gynecology; author of numerous technical publications. *Mailing Add:* Dept Obstet Gynec & Reprod Sci Univ Calif Med Sch San Francisco CA 94143

JAFFE, ROBERT LOREN, b Bath, Maine, May 23, 46. QUANTUM CHROMODYNAMICS, THEORY OF PARTICLES AND FIELDS. *Educ:* Princeton Univ, AB, 68; Stanford Univ, MS, 71, PhD(physics), 72. *Prof Exp:* Res assoc, 72-74, from asst prof to assoc prof, 74-83, PROF PHYSICS, MASS INST TECHNOL, 83- *Concurrent Pos:* Sloan fel, 75-77; vis scientist, Stanford Linear Accelerator Ctr, 76; sr vis fel, dept theoret physics & St Catherine's Col, Oxford, 79; sci assoc, Europ Coun Nuclear Res, 79; vis lectr, Beijing Univ, China, 81; consult, Los Alamos Nat Lab, 85- *Mem:* Am Phys Soc. *Mailing Add:* Dept Physics 6-306 Mass Inst Technol Cambridge MA 02139

JAFFE, RUSSELL M, b Albany, NY, Jan 1, 47. IMMUNOLOGY. *Educ:* Boston Univ, BA, MD & PhD(biochem & med sci), 72. *Prof Exp:* Intern, Univ Hosp, Boston Univ Med Col, 72-73; resident, Clin Ctr, NIH, 73-75, sr staff physician, 75-79, collab investr, Nat Heart Lung & Blood Inst & LEA, 76-79; FEL, HEALTH STUDIES COL, 79-, DIR, SERV PHYS LAB, 86- & PRINCETON BIO CTR, 89- *Mem:* AAAS; NY Acad Sci; AMA; fel Am Col Path. *Res:* Connective tissue; biology of mental function; coagulation; atherosclerosis; clinical biochemistry; biochemical immunology. *Mailing Add:* Princeton Bio Ctr 862 Rte 518 Skillman NJ 08558

JAFFE, SIGMUND, b New Haven, Conn, Mar 1, 21; m 46; c 2. PHYSICAL CHEMISTRY. *Educ:* Wesleyan Univ, AB, 49; Iowa State Univ, PhD(chem), 53. *Prof Exp:* Asst phys chem, Ames Lab, Atomic Energy Comn, 49-53; sr engr, Air Reduction Res Lab, 53-58; from asst prof to assoc prof, 58-64, PROF CHEM, CALIF STATE UNIV, LOS ANGELES, 64- *Concurrent Pos:* NIH fel, Weizmann Inst, 64-65; res grant, Environ Protection Agency-US Pub Health, 66; res fel, Weizmann Inst Sci, 71-72; res scientist, Jet Propulsion Lab, Calif, 72-; vis prof, Queen Mary Col, Univ London, 78-79. *Mem:* Am Chem Soc. *Res:* Chemical kinetics; photo-chemical reactions of stratosphere and atmosphere including the reactions of Cl, NO2 and O3. *Mailing Add:* Dept Chem Calif State Univ Los Angeles CA 90032

JAFFE, SOL SAMSON, b Ft Wayne, Ind, Feb 7, 20; m 52; c 2. TEACHING OF MATH & SCIENCE. *Educ:* Drew Univ, AB, 41; Stevens Inst Technol, MS, 45. *Prof Exp:* Chemist, Fidelity Chem Prod Corp Div, Maas & Waldstein Co, NJ, 42-46; res labs, Thomas A Edison, Inc, 46-51, res chemist, 51-57, Thomas A Edison Res Labs Div, McGraw Edison Co, 57-64; sr res chemist, Res & Develop Lab, Alkaline Battery Div, Gulton Industs, 64-65; chief chemist, Bright Star Industs, 65-70; formulating chemist, Nitine, Inc, 70-72; electrochem engr, Mallory Battery Co Div, P R Mallory & Co, Inc, 72-79; sect head electrochem eng, Battery Technol Co Div, Duracell Int Inc, 79-83, staff scientist, Duracell Prod Technol, 83-87; INSTR MATH & SCI, BLOOMFIELD COL, 87- *Mem:* Electrochem Soc. *Res:* Leclanche, air-zinc, mercury-zinc, divalent silver-zinc, alkaline and fuel cells; nickel-iron, nickel-cadmium, silver-cadmium and lead-acid storage batteries; hermetically sealed cells; charge control electrodes; state-of-charge indicators; electrochemical devices; metal finishing; electroless plating. *Mailing Add:* Seven Nance Rd West Orange NJ 07052

JAFFE, WALTER JOSEPH, astronomy, for more information see previous edition

JAFFE, WERNER G, b Frankfurt, Ger, Oct 27, 14; m 46; c 6. NUTRITION, BIOCHEMISTRY. *Educ:* Univ Zurich, PhD(chem), 39; Cent Univ Venezuela, DrSc(biochem), 50. *Hon Degrees:* Hon Prof, Cent Univ & Univ Simon Boliver. *Prof Exp:* Asst prof org chem, Sch Pharm, 47-50, assoc prof biochem, Sch Sci, 50-58, PROF BIOCHEM, SCH SCI, CENT UNIV VENEZUELA, 58-; RES ASSOC, NAT NUTRIT INST, 50-; DIR POSTGRAD COURSE NUTRIT PLANNING, 74- *Concurrent Pos:* Ed, Arch Latinoamerican Nutrit. *Honors & Awards:* Nat Sci Award, Ministry of Educ, 58; Gold Medal Sci Res, Cent Univ Venezuela, 60; Nat Sci Award of Venezuela, 78. *Mem:* AAAS; Venezuelan Asn Advan Sci (pres, 56); Venezuelan Chem Soc (pres, 53); cor mem Peruvian Chem Soc; cor mem Mex Chem Soc; Am Chem Soc. *Res:* Enzymology; toxicology; lectins, enzyme inhibition; toxic food constituents. *Mailing Add:* Apartado 21201 Caracas 1020-A Venezuela

JAFFE, WILLIAM J(ULIAN), b Passaic, NJ, Mar 22, 10. INDUSTRIAL ENGINEERING. *Educ:* NY Univ, BS, 30, Engr ScD(indust eng), 53; Columbia Univ, MA, 31, MS, 41. *Prof Exp:* Naval architect, Philadelphia Navy Yard, US Dept Navy, 41-45; instr math, 46, from instr to asst prof indust eng, 46-50, from asst exec assoc to exec assoc, 49-60, from assoc prof to prof eng, 50-73, distinguished prof, 73-75, EMER DISTINGUISHED PROF ENG, NJ INST TECHNOL, 75- *Concurrent Pos:* Adj assoc prof col eng, NY Univ, 53-54; mem, Inst Bus Admin & Mgt, Japan, Comn on Manpower, Israel & Clark Bd Int Mgt, Comt Int Orgn Sci, Bd Standards Rev, Am Nat Standards Inst, Standardization Bd, Am Soc Mech Engrs. *Honors & Awards:* Centennial Medal, Am Soc Mech Engrs. *Mem:* Fel AAAS; fel Am Soc Mech Engrs; fel Am Inst Indust Engrs; fel Soc Advan Mgt; assoc fel NY Acad Med; Am Math Soc. *Res:* Industrial and management engineering. *Mailing Add:* 1175 York Ave Apt 9E New York NY 10021

JAFFEE, OSCAR CHARLES, b New York, NY, Sept 26, 16; m 62; c 1. EXPERIMENTAL EMBRYOLOGY. *Educ:* NY Univ, BA, 46, MS, 48, Ind Univ, PhD(zool), 52. *Prof Exp:* Instr anat, Sch Med, Univ Ark, 52-56; embryologist, Chronic Dis Res Inst, State Univ NY Buffalo, 58-60; from lectr to prof biol, Univ Dayton, 60-89; RETIRED. *Concurrent Pos:* Mem staff, Mt Desert Island Biol Lab; mem, Am Heart Asn Coun; vis prof, dept anat & embryol, Hadassah Med Sch, Hebrew Univ Jerusalem, Israel, 85, vis prof, Zool Dept, 87. *Mem:* Am Asn Anat; Teratology Soc; Int Soc Stereology; Int Soc Biorheol. *Res:* Cardiovascular embryology; teratology; physiology of the embryonic heart. *Mailing Add:* 300 College Park Ave Dayton OH 45469

JAFFEE, ROBERT I(SAAC), b Chicago, Ill, July 11, 17; m 45; c 2. METALLURGY. *Educ:* Ill Inst Technol, BS, 39; Harvard Univ, SM, 40; Univ Md, PhD(metall) 43. *Prof Exp:* Lectr, Univ Md, 42; res metallurgist, Leeds & Northrup Co, Pa, 42-43 & Univ Calif, 44; res engr, Battelle-Columbus, 43-44, asst supvr, 45-50, div chief, 50-60, tech mgr, 60-65, chief mat scientist, 65-75; tech mgr, 75-80, SR TECH ADV, ELEC POWER RES INST, 80- *Concurrent Pos:* Consult, President's Sci Adv Comt, 66; chmn adv comt mat, NASA, 66-71, mem mat & struct comt, 71-74; panel magnetohydrodynamics, Off Sci & Technol, 68-69; Nat Mat Adv Bd, Nat Acad Sci-Nat Acad Eng, 71-74; consult prof, Stanford Univ, 74- *Honors & Awards:* Bronze Medal, Am Ord Asn, 66; Horace Gillett Mem Lectr, Am Soc Testing Mat, 75; Edward DeMille Campbell Lectr, Am Soc Metals, 77; James Douglas Gold Medal, Am Inst Mining, Metall and Petrol Engrs, 83; Distinguished Lectr Mat & Soc, ASM/TMS, 85. *Mem:* Nat Acad Eng; fel Metall Soc; The Metall Soc of the Am Inst Mining, Metall & Petrol Engrs (pres, 78); fel & hon mem Am Soc Metals; fel Brit Soc Metals. *Res:* Physical metallurgy in titanium and steel; high purity steels; stress corrosion of titanium; protective coatings. *Mailing Add:* 3851 May Ct Palo Alto CA 94304

JAFFEY, ARTHUR HAROLD, b Chicago, Ill, Dec 25, 14; m 45; c 2. NUCLEAR CHEMISTRY. *Educ:* Univ Chicago, BS, 36, PhD(phys chem), 41. *Prof Exp:* Instr chem lab, Univ Chicago, 37-40, res assoc detection poisonous gases, Off Sci Res & Develop Proj, 41-42 & Metall Lab, Manhattan Dist, 42-46; res assoc, 46-50, GROUP LEADER & SR CHEMIST, ARGONNE NAT LAB, 50- *Concurrent Pos:* Jr chemist, Rock Island Arsenal, US War Dept, Ill, 41; res assoc, Off Sci Res & Develop, Columbia Univ, 42. *Mem:* Am Phys Soc; Sigma Xi. *Res:* Measurement of disintegration properties of radioactive isotopes; measurement of thermal neutron reactions with heavy elements; statistical analysis of nuclear measurement data; development of superconducting heavy ion linear accelerator. *Mailing Add:* 5711 S Maryland Ave Chicago IL 60637

JAGADEESH, GOWRA G, b Karnatoka, India, Sept 18, 49; m 78; c 2. AUTONOMIC CARDIOVASCULAR, DRUG RECEPTOR-EFFECTOR COUPLING. *Educ:* Bangalore Univ, India, BPharm, 70; All-India Inst Med Sci, New Delhi, India, MS, 71; Banoras Hindu Univ, India, PhD(pharmacol), 80. *Prof Exp:* Dept physiol, Univ Sask, Can, 72-85; PHARMACOLOGIST, CARDIO-RENAL DIV, FOOD & DRUG ADMN, ROCKVILLE, MD 90- *Concurrent Pos:* Res, Inst Hist Med & Med Res, NDelhi, India, 75; from asst prof to assoc prof pharmacol, Banaras Hindu Univ, IT, Varanasi, India, 75-86; vis res assoc, Northeastern Univ Col Pharm, Boston, 86-89, staff sci, 89-90. *Mem:* Am Col Clin Pharmacol. *Res:* Alpha-adrenergic receptor (subtypes) coupling to G-protein and its modulation by protein kinase C, agonist (desensitization) and ions is investigated in vascular smooth muscle, liver and heart diseases; the reciprocal regulation of alpha-1 and beta-2 adrenergic receptor coupling events in liver, heart and lungs. *Mailing Add:* 8700 Cathedral Way Gaithersburg MD 20879

JAGANNATHAN, KANNAN, b Madras, India, Nov 4, 54. THEORETICAL HIGH ENERGY PHYSICS. *Educ:* Univ Madras, India, BSc, 73; Indian Inst Technol, Madras, MSc, 75; Univ Rochester, PhD(physics), 81. *Prof Exp:* Res assoc physics, Univ Rochester, 80-81; asst prof, 81-87, ASSOC PROF PHYSICS, AMHERST COL, 87- *Mem:* Am Phys Soc; Am Asn Physics Teachers; Am Math Asn. *Res:* Theoretical high energy physics; elementary particle physics; foundations of quantum mechanics. *Mailing Add:* Amherst Col Box 2244 Amherst MA 01002

JAGANNATHAN, SINGANALLUR N, b Coonoor, India, Mar 10, 34; m 68; c 2. NUTRITIONAL BIOCHEMISTRY, BIOCHEMICAL PATHOLOGY. *Educ:* Univ Bombay, BSc, 54, MSc, 59, PhD(biochem), 62. *Prof Exp:* Res asst to sr res officer, Indian Coun Med Res, Nat Inst Nutrit, 56-70; res scientist, lipid, lipoprotein & biochem, Col Med, Univ Iowa, 70-74; asst prof path & biochem, 74-77, ASSOC PROF PATH & BIOCHEM, DEPT PATH, SCH MED, WVA UNIV, MORGANTOWN, 77-, ASSOC PROF NUTRIT, SCH DENT, 79- *Concurrent Pos:* Nat Res Coun Can fel nutrit biochem, Queen's Univ, Kingston, Ont, 62-64; Brit Coun fel, 64; fel, Coun Arteriosclerosis, Am Heart Asn, 72. *Mem:* Am Inst Nutrit; Am Soc Clin Nutrit; fel Am Heart Asn. *Res:* Lipid and proteoglycans in relation to human atherosclerosis; platelet function; human nutrition; iron metabolism. *Mailing Add:* Dept Path Sch Med WVa Univ Health Sci Ctr Morgantown WV 26506

JAGEL, KENNETH I(RWIN), JR, b Jamaica, NY, Feb 2, 27; m 51; c 2. CHEMICAL ENGINEERING. *Educ:* Columbia Univ, BSc, 51, MSc, 53, DEngSc, 61. *Prof Exp:* Asst chem eng, Columbia Univ, 51-53, res asst, 55-57; res engr, Socony Mobil Oil Co, Inc, 57-59, sr res engr, 59-65, group leader, Mobil Res & Develop Corp, 65-68, eng assoc, 68-71; mgr mfg, Engelhard Minerals & Chem Corp, 71-72, mgr process eng & develop, 72-73, dir prod assurance, 73-81, dir qual & mfg serv, Engelhard Corp, 81-82, dir prod develop, 82-85, exec engr, 85-89; PRES, TRAILTREE ASSOCS, 89- *Mem:* AAAS; fel Am Inst Chem; Am Chem Soc; Am Inst Chem Engrs; Am Soc Testing & Mat. *Res:* Catalyst manufacture; auto-exhaust emission control catalysts; cracking catalysts; oil shale retorting processes; distillation equipment. *Mailing Add:* Box 112 Stanton NJ 08885

JAGENDORF, ANDRE TRIDON, b New York, NY, Oct 21, 26; m 52; c 3. PLANT PHYSIOLOGY, BIOCHEMISTRY. *Educ:* Cornell Univ, AB, 48; Yale Univ, PhD(plant sci), 51. *Prof Exp:* Res assoc bot, Univ Calif, Los Angeles, 51-53; from asst prof to prof biol, Johns Hopkins Univ, 53-66, with McCollum-Pratt Inst, 53-66; PROF PLANT PHYSIOL, CORNELL UNIV, 66- *Concurrent Pos:* Merck fel, 51-53, Weizmann fel, 62; Liberty Hyde Bailey prof, Cornell Univ, 81. *Honors & Awards:* Kettering Res Award, 63, Kettering Award in Photosyn, Am Soc Plant Physiologists, 78. *Mem:* Nat Acad Sci; Am Soc Photobiol; Am Soc Cell Biol; Am Soc Plant Physiol (pres, 67-68); fel AAAS; Am Acad Arts & Sci. *Res:* Photosynthetic phosphorylation; biochemistry; chloroplast biogenesis. *Mailing Add:* Plant Biol Div Plant Sci Bldg Cornell Univ Ithaca NY 14853

JAGERMAN, DAVID LEWIS, b Aug 27, 23; m 51; c 3. MATHEMATICS, ELECTRICAL ENGINEERING. *Educ:* Cooper Union Univ, BEE, 49; NY Univ, MS, 54, PhD(math), 62. *Prof Exp:* Sr engr analog comput, Reeves Instrument Corp, NY, 51-55; staff scientist math guided missiles, Stavid Eng, NJ, 55-57 & 57-59; design specialist trajectories, Gen Dynamics/Convair, Calif, 57; staff mem math control systs, Syst Develop Corp, NJ, 59-64; mem tech staff, 64-86, DISTINGUISHED MEM TECH STAFF, BELL LABS, 86- *Concurrent Pos:* Assoc prof math, Fairleigh Dickinson Univ, 59-66; prof, Stevens Inst Technol, 67-72. *Mem:* Am Math Soc. *Res:* Diophantine analysis and numerical quadrature theory with application to the mathematical properties of pseudo-random numbers; information theory; telephone traffic theory. *Mailing Add:* 32 Mendell Ave Cranford NJ 07016

JAGGARD, DWIGHT LINCOLN, b Oceanside, NY, Apr 14, 48; m 68; c 2. ELECTROMAGNETISM, OPTICS. *Educ:* Univ Wis-Madison, BSEE, 71, MSEE, 72; Calif Inst Technol, PhD(elec eng & appl physics), 76. *Hon Degrees:* MA, Univ Pa, 82. *Prof Exp:* Res asst geophys & elec eng, Univ Wis-Madison, 68-72; engr imaging radar & DFB lasers, Jet Propulsion Lab, Calif Inst Technol, 73-76, res fel, 76-78; asst prof optics & electromagnetics, Dept Elec Eng, Univ Utah, 78-80; asst prof, dept elec eng & sci, 80-82, assoc prof, dept elec eng, 82-85, PROF, MOORE SCH ELEC ENG, UNIV PA, 88- *Concurrent Pos:* Consult, Jet Propulsion Lab, 76-78, Environ Studies Lab, 78-81, Naval Res Lab, 82-85, Sohio, 85, IDA, 89-, Johns Hopkins Appl Physics Lab, 91-; forensic consult, 83-; dir, ExMSE grad prog, technol & leadership, Univ PA, 88-90. *Mem:* Fel Inst Elec & Electronics Engrs; Int Union Radio Sci; Optics Soc Am; Sigma Xi. *Res:* Electromagnetic chirality, fractal electrodynamics; inverse scattering and remote sensing; imaging and classification; light scattering from aerosols; fractal graphas; chaos and fractals. *Mailing Add:* Complex Media Lab Moore Sch Elec Eng 6314 Univ Pa Philadelphia PA 19104-6390

JAGGER, JOHN, b New Haven, Conn, Feb 22, 24; m 56; c 2. RADIATION BIOLOGY, PHOTOBIOLOGY. *Educ:* Yale Univ, BS, 49, MS, 53, PhD(biophys), 54. *Prof Exp:* Asst, Mem Hosp, NY, 50-51; asst, Yale Univ, 53-54; Nat Found Infantile Paralysis fel, Radium Inst, France, 54-55; biophysicist, Biol Div, Oak Ridge Nat Lab, 56-65; from assoc prof to prof biol, Southwest Ctr Advan Studies, 65-69; prof biol, Univ Tex, Dallas, 69-81, prof gen studies & biol, 81-86; RETIRED. *Concurrent Pos:* Lectr, Univ Tenn, 61-65; vis prof, Pa State Univ, 63 & Univ Kyoto, Japan, 79; consult, Aerojet Med & Biol Systs, 71-73; mem, US Nat Comt Photobiol, Nat Res Coun, 75-80, pres, 78-80. *Mem:* Fel AAAS; Am Soc Photobiol (pres, 83-84). *Res:* Effects of radiations on large molecules and cells; effects of ultraviolet on bacteria; photoprotection; effects of near ultraviolet on cell growth and membrane function; social impact of science. *Mailing Add:* 7532 Mason Dells Dr Dallas TX 75230

JAGIELLO, GEORGIANA MARY, b Boston, Mass, Aug 2, 27; m 57. GENETICS. *Educ:* Boston Univ, AB, 49; Tufts Univ, MD, 55. *Prof Exp:* Exchange fel surg, St Bartholomew's Hosp, London, 54; intern med, Res & Educ Hosps, Univ Ill, 55-56; resident, New Eng Med Ctr, Boston, 56-57; res fel endocrinol, Scripps Clin, La Jolla, 57-58; res fel, New Eng Med Ctr, Boston, 58-60; USPHS res fel cytogenetics, Guy's Hosp, London, 60-61; asst prof, Sch Med, Univ Ill, Chicago Circle, 61-66; sr lectr cytogenetics, Guy's Hosp, London, 66-69; res prof pediat, Sch Med, Univ Ill, Chicago Circle, 69-70; PROF OBSTET, GYNEC & HUMAN GENETICS, COL PHYSICIANS & SURGEONS, COLUMBIA UNIV, 70- *Concurrent Pos:* Mem, Inst Advan Study, Univ Ill, 64; NIH career develop award, 65; consult, Guy's Hosp, London, 66-69. *Mem:* Endocrine Soc; Teratology Soc; Environ Mutagen Soc; Am Soc Cell Biol; Soc Study Reproduction. *Res:* Mammalian meiosis; reproductive endocrinology. *Mailing Add:* Col Physicians & Surgeons 630 W 168th St New York NY 10032

JAGLAN, PREM S, b India, Sept 17, 29; US citizen; m 51; c 3. HIGH-PERFORMANCE LIQUID CHROMATOGRAPHY, GAS-LIQUID CHROMATOGRAPHY. *Educ:* Univ Calif, Riverside, PhD(toxicol), 69. *Prof Exp:* SR RES SCIENTIST IV BIOCHEM, UPJOHN CO, 69- *Mem:* Am Chem Soc. *Res:* Separation science and identification of metabolites using chemical and instrumental methods. *Mailing Add:* Upjohn Co 301 Henrietta St Kalamazoo MI 49001

JAHAN-PARWAR, BEHRUS, b Ghoochan, Iran, May 26, 38, nat US; m 66; c 2. NEUROBIOLOGY. *Educ:* Univ Gottingen, MD, 64, DMSc(physiol), 65. *Prof Exp:* Res assoc neurophysiol, Dept Physiol, Univ Gottingen, 64-66; asst res neurophysiologist, Mental Health Res Inst, Univ Mich, 66-68; from asst prof to assoc prof physiol, Dept Biol, Clark Univ, 68-73; SR SCIENTIST NEUROBIOL, WORCESTER FOUND EXP BIOL, 73- *Concurrent Pos:* Prin investr NIH grants, 69-73, 74-77, 75-83 & 78-81; NIH res career develop award, 70-73; Grass Found grant, 75-78; NSF grant, 78-82. *Mem:* Soc Neurosci; Am Physiolog Soc; Am Soc Zoologist; Europ Chemoreception Res Orgn. *Res:* The elucidation of the principles of neuronal organization underlying processing of sensory information and generation and modification of behavior; neural mechanisms of chemoreception, learning and rhythmic behaviors such as feeding and locomotion. *Mailing Add:* Wadsworth Ctr Labs & Res NY State Dept Health Empire State Plaza Albany NY 12201

JAHIEL, RENE, b France, Mar 29, 28; nat US; m 55; c 3. HEALTH SERVICES RESEARCH, MICROBIOLOGY. *Educ:* NY Univ, BA, 46; State Univ NY, MD, 50; Columbia Univ, PhD(microbiol), 57. *Prof Exp:* Intern, Montefiore Hosp, NY, 50-51; res, Mt Sinai Hosp, 51-52, fel, 52-55; asst prof microbiol, Sch Med, Univ Colo, 57-59; asst prof path, Columbia Univ, 59-61; asst prof pub health, Med Col, Cornell Univ, 61-67; from res assoc prof prev med to res prof prev med, NY Univ, 67-76, res prof med, 76-88; CONSULT, HEALTH SERV RES, 89- *Concurrent Pos:* Exp immunologist, Nat Jewish Hosp, Denver, Colo, 57-59; asst attend pathologist, Mt Sinai Hosp, New York, 59-61; career scientist, Health Res Coun New York, 62-66; prin investr, USPHS grants on several health res serv, 68-82; physician, Asn Children with Retarded mental Develop, 88- *Mem:* NY Acad Sci; Soc Social Study Sci; Tissue Cult Asn; Am Pub Health Asn; Asn Health Serv Res. *Res:* Immunopathology; autoantibodies; tissue culture virology; tissue culture; interferon; community medicine; sociology of knowledge; health serv res. *Mailing Add:* 60 E 8th St New York NY 10003-6519

JAHN, EDWIN CORNELIUS, b Oneonta, NY, Sept 6, 02; m 27, 70; c 2. ORGANIC CHEMISTRY. *Educ:* NY State Col Forestry, BS, 25, MS, 26; McGill Univ, PhD(org chem), 29. *Hon Degrees:* DSc, Syracuse Univ, 72. *Prof Exp:* Asst, NY State Col Forestry, 25-26 & McGill Univ, 26-29; fel, Am-Scand Found, 29-30; from assoc prof to prof forestry, Univ Idaho, 30-38; prof forest chem, State Univ NY Col Environ Sci & Forestry, 38-72, dir res, 49-52, assoc dean phys sci, 52-66, exec dean, 66-67, dean, 67-69; EXEC SECY, EMPIRE STATE PAPER RES ASSOCS, INC, 70- *Concurrent Pos:* Consult, USDA, Sweden, 43-44; sr econ analyst, US Dept Com, 45-46; tech attache to US State Dept for Sweden, Finland, Norway & Denmark, 45-46; mem wood chem comt, Food & Agr Orgn, UN, 47-64. *Honors & Awards:* Tech Asn Pulp & Paper Indust Award, 70. *Mem:* Am Chem Soc; fel Soc Am Foresters; fel Tech Asn Pulp & Paper Indust; AAAS. *Res:* Cellulose, wood and polymer chemistry. *Mailing Add:* 109 Hillcrest Rd Syracuse NY 13224

JAHN, ERNESTO, b Yumbel, Chile, Mar 29, 42; m 72. ANIMAL SCIENCE, RUMINANT NUTRITION. *Educ:* Univ Concepcion, MgAgr, 65; Va Polytech Inst & State Univ, MS, 69, PhD(ruminant nutrit), 74. *Prof Exp:* Agr engr ruminants, Inst Land & Cattle Invests, Chile, 65-72; res asst animal sci, Va Polytech Inst & State Univ, 72-74; SR RES BIOLOGIST ANIMAL SCI, RES LABS, MERCK & CO, INC, 74- *Mem:* Latin Am Soc Animal Prod; Agron Soc Chile. *Mailing Add:* Casilla 426 Chillian Chile

JAHN, J RUSSELL, b Spirit Lake, Iowa, Dec 2, 26; m 50; c 3. ANIMAL SCIENCE. *Educ:* SDak State Univ, BS, 59, MS, 60, PhD(animal sci), 63. *Prof Exp:* Assoc prof, Univ Wis-Platteville, 62-66, prof animal sci & head, Dept Agr Sci, 66-; RETIRED. *Mem:* Am Soc Animal Sci. *Res:* Artificial insemination of beef cattle. *Mailing Add:* 6700 Francis Sites Spirit Lake IA 51360

JAHN, LAURENCE R, b Jefferson, Wis, June 24, 26; m 47; c 2. MIGRATORY BIRDS, AQUATIC AREAS. *Educ:* Univ Wis-Madison, BS, 49, MS, 58, PhD(wildlife ecol & zool), 65. *Prof Exp:* Biologist migratory bird populations & habitats, Wis Dept Natural Resources, 49-59; vpres res mgt & res, 59-87, pres, 87-91, CHMN BD, WILDLIFE MGT INST, 91- *Concurrent Pos:* Assoc mem, Int Asn of Fish & Wildlife Agencies, 59; chmn, Nat Watershed Cong, 74-90; mem, adv comt marine fisheries, US Dept Com, 82-84, adv comt wildlife, Dept State, 72-77 & adv comt fish, wildlife & parks, US Dept Interior, 75-79; chmn, chief of engrs environ adv bd, US Dept Army, 83-85; chmn, Nat Resources Coun Am, 83-85; mem, US Implementations Bd, NAm Waterfowl Mgt Plan, 88-91, chmn, 90-91. *Mem:* Fel AAAS; Wildlife Soc (pres, 79-80); Am Fisheries Soc; Am Water Resources Asn; Soil & Water Conserv Soc Am; Am Inst Biol Sci; Am Wildlife Found (secy, 78-91). *Res:* Wildlife populations and habitats, particularly projects designed to provide information for strengthening management programs, especially guidelines for avoiding and minimizing adverse impacts on wildlife as development proceeds. *Mailing Add:* 2435 Riviera Dr Vienna VA 22181

JAHN, LAWRENCE A, b Cudahy, Wis, Dec 2, 41; m 66; c 2. AQUATIC BIOLOGY, FISHERIES MANAGEMENT. *Educ:* Univ Wis, BS, 63; Mont State Univ, MS, 66, PhD(zool), 68. *Prof Exp:* From asst prof to assoc prof, 68-81, PROF BIOL SCI, WESTERN ILL UNIV, 81- *Concurrent Pos:* Dir, Inst Environ Mgt, 88- *Mem:* Am Fisheries Soc; Am Soc Ichthyol & Herpet. *Res:* Fish management, ecology and life histories; aquarium management. *Mailing Add:* Dept of Biol Sci Western Ill Univ Macomb IL 61455

JAHN, ROBERT G(EORGE), b Kearny, NJ, Apr 1, 30; m 53; c 4. ENGINEERING PHYSICS. *Educ:* Princeton Univ, BS, 51, MA, 53, PhD(physics), 55. *Hon Degrees:* DSc, Andhra Univ, Vishakhapatnam, India, 86. *Prof Exp:* Asst prof physics. Lehigh Univ, 55-58; asst prof jet propulsion, Calif Inst Technol, 58-61; from asst prof to assoc prof, 62-67, dean, Sch Eng & Appl , 71-86, PROF AEROSPACE SCI, PRINCETON UNIV, 67-, DIR GRAD STUDIES, DEPT AEROSPACE & MECH SCI, 88- *Concurrent Pos:* Mem res & technol adv subcomt electrophys, NASA, 68-71, mem res & technol adv comt space propulsion & power, 71-72, mem res & technol adv coun comt space propulsion & power, 76-77, mem nat adv coun space systs & technol adv comt, 78-; mem bd trustees, Assoc Univs, Inc, Washington, DC, 71-74, chmn bd, 77-79, trustee rep, 74-86, mem nominating comt, 78-79; mem bd dirs, John E Fetzer Inst, 83-, vchmn, 83, nominating comt, 83-; mem, Soc Sci Explor, 83-, counr, 84-, prog comt, 91-; mem bd dirs, Hercules Inc, 85-, technol comt, 87-, strategic comt, 87-, compensation comt, 87-, nominating comt, 88-; mem bd dirs, Roy F Weston, Inc, 88-, Audit & Compensation Comt, 88- *Honors & Awards:* Curtis W McGraw Res Award, Am Soc Eng Educ, 69. *Mem:* Am Inst Aeronaut & Astronaut; Am Phys Soc. *Res:* Plasma propulsion; high temperature gasdynamics and fluid mechanics; shock tubes; plasmajets; ionization phenomena; electromagnetic wave propagation in ionized gases; engineering anomalies and man/machine interactions. *Mailing Add:* D-334 EQ Seas Princeton Univ Princeton NJ 08544-5263

JAHNGEN, EDWIN GEORG EMIL, JR, b Pittsburgh, Pa, Jan 8, 46. ORGANIC CHEMISTRY, BIORGANIC CHEMISTRY. *Educ:* Bates Col, BSc, 68; Univ Vt, PhD(org chem), 74. *Prof Exp:* Chemist, Water Improv Comn, Maine, 67-68; chemist org synthesis, Polaroid Corp, 68-70; res assoc nat prod, Univ BC, 74-76; groupleader, New Eng Nuclear Corp, 76-78; asst prof org chem, Wilkes Col, 78-; AT DEPT CHEM , UNIV LOWELL. *Concurrent Pos:* NIH fel, Nat Res Coun, Univ BC, 74-76; consult Med Sch, Univ Conn, 78- *Mem:* Am Chem Soc; Sigma Xi. *Res:* Studies of bio-organic systems; synthesis and modification of exogenous; endogenous drugs and hormones. *Mailing Add:* Dept Chem Univ Lowell Lowell MA 01854

JAHNS, HANS O(TTO), b Kamen, Ger, Sept 4, 31; m 59; c 5. PETROLEUM ENGINEERING, ARCTIC ENGINEERING. *Educ:* Clausthal Tech Univ, dipl, 55, dipl, 56, Dr(Ing), 61. *Prof Exp:* Res asst petrol eng, Inst Drilling & Petrol Prod, Clausthal Tech Univ, 56-59; reservoir engr, Reservoir Lab, Wintershall AG, Ger, 59-62; res engr prod res, Jersey Prod Res Co, Standard Oil Co NJ, Okla, 62-65, res engr, 65-68, res assoc, 68-73, res adv, 73-77, sr res adv, 77-80, res scientist, 80-82, SR RES SCIENTIST, EXXON PROD RES CO, 82- *Concurrent Pos:* Mem permafrost comt, Nat Res Coun, 75-80, mem polar res bd, 76-84; mem adv bd, Geophys Inst, Univ Alaska, 78-87; mem, polar adv comt, NSF, 85-87. *Mem:* Soc Petrol Engrs; AAAS; Am Petrol Inst. *Res:* Petroleum reservoir description; oceanography; arctic research; arctic engineering; sea ice mechanics; expert system, production geophysics; oil spill cleanup technology. *Mailing Add:* c/o Exxon Prod Res Co PO Box 2189 Houston TX 77252-2189

JAHNS, MONROE FRANK, b Seguin, Tex, May 16, 28; m 54; c 2. PHYSICS. *Educ:* Tex A&M Univ, BS, 49; Univ Tex, Austin, MA, 64, PhD(physics), 66. *Prof Exp:* Instr physics, Unit Tex, Austin, 65-66; asst prof, Sam Houston State Col, 66-67; advan sr fel med physics, Univ Tex M D Anderson Hosp & Tumor Inst, Houston, 67-68, assoc physicist & assoc prof biophysics, 68-88; RETIRED. *Mem:* Am Phys Soc; Soc Nuclear Med; Am Asn Physicists Med. *Mailing Add:* 6319 Vanderbilt St Houston TX 77005

JAHODA, FRANZ C, b Vienna, Austria, Sept 16, 30; nat US; m 55; c 2. PLASMA DIAGNOSTICS, FIBER OPTIC SENSORS. *Educ:* Swarthmore Col, BA, 51; Cornell Univ, PhD, 57. *Prof Exp:* MEM STAFF, LOS ALAMOS NAT LAB, 57-, GROUP LEADER, 74- *Concurrent Pos:* Mem staff, Culham Lab, Abingdon, Eng, 64-65; mem staff, Inst voor Plasmafysica, Jutphaas, Neth, 72-73. *Mem:* Optical Soc Am; Inst Elec & Electronics Engrs. *Res:* Plasma physics; spectroscopy; lasers; optical diagnostics. *Mailing Add:* 819 Bishops Lodge Rd Santa Fe NM 87501

JAHODA, GERALD, b Vienna, Austria, Oct 22, 25; US citizen; m 52. INFORMATION SCIENCE. *Educ:* NY Univ, AB, 47; Columbia Univ, MS, 52, DLS, 60. *Prof Exp:* Instr & chem librn, Univ Wis, 52-53; group leader, Colgate-Palmolive Co, 53-57; sect head, Esso Res & Eng Co, 57-63; PROF INFO SCI & LIBRARIANSHIP, FLA STATE UNIV, 63- *Concurrent Pos:* Instr, Polytech Inst Brooklyn, 53-54 & Rutgers Univ, 61-62; Air Force Sci Res grant, 65-66; mem, Sci Inst Pub Info; US Off Educ grant, 74-; NSF grant, Exxon Educ Found, 81-82; Am Coun Blind. *Mem:* Am Libr Asn; Am Soc Info Sci. *Res:* Information needs of scientists; organization of the literature of science and technology; information service for the blind. *Mailing Add:* Libr Sch Fla State Univ Tallahassee FL 32306

JAHODA, JOHN C, b Dalhart, Tex, Feb 9, 44. ECOLOGY, MAMMALOGY. *Educ:* Univ Conn, BA, 66; Okla State Univ, PhD(zool), 69. *Prof Exp:* Asst prof biol, State Univ NY Col Geneseo, 69-70; from asst prof to assoc prof, 70-81, PROF BIOL, BRIDGEWATER STATE COL, 81- *Concurrent Pos:* Consult, Raytheon Serv Co, 70-; mem, Corp Bermuda Biosta Res. *Mem:* Am Soc Mammal; Ecol Soc Am; Sigma Xi. *Res:* Ecology and ethology of mammals; salt marsh ecology; pollution of salt marshes. *Mailing Add:* Dept of Biol Sci Bridgewater State Col Bridgewater MA 02324

JAHSMAN, WILLIAM EDWARD, b Detroit, Mich, May 13, 26; m 49; c 3. MECHANICAL ENGINEERING. *Educ:* Cornell Univ, BEng, 51; Stanford Univ, MS, 53, PhD(eng mech), 54. *Prof Exp:* Mech specialist, Knolls Atomic Power Lab, Gen Elec, 54-56; consult scientist missiles & spacecraft, Lockheed Missiles & Space Co, 56-67; prof mech eng & chmn dept, Univ Colo, Boulder, 67-80; WITH INTEL CORP, 80- *Concurrent Pos:* Liaison scientist, Off Naval Res, 69-70; prin investr NSF grant, 70-76. *Mem:* Am Soc Mech Engrs; Am Soc Eng Educ; Am Acad Mech. *Res:* Wave propagation in inelastic materials; dynamic mechanical properties of coal and oil shale. *Mailing Add:* Intel Corp Chandler AZ 85226

JAICKS, FREDERICK G, ENGINEERING. *Prof Exp:* Chmn, Inland Steel, Chicago; RETIRED. *Mem:* Nat Acad Eng. *Mailing Add:* Meadow Lane Lakeside MI 49116

JAIKRISHNAN, KADAMBI RAJGOPAL, b Bangalore, India. FLUID DYNAMICS, MECHANICAL ENGINEERING. *Educ:* Univ Jodhpur, BE, 64; Univ Pittsburgh, MS, 70, PhD(mech eng), 76. *Prof Exp:* Planning engr mech eng, APEE Corp, Hyderabad, India, 65-67; grad student asst mech eng, Univ Pittsburgh, 68-71; engr heat transfer & fluid dynamics, Westinghouse Res & Develop Ctr, 71-75, sr eng Fluid Dynamics, 75- *Concurrent Pos:* Lectr mech eng dept, Univ Pittsburgh, 77. *Mem:* Am Soc Mech Engrs. *Res:* Compressible fluid dynamics; transonic flow of steam in blade passages of turbomachinery; laser doppler anemometry; flow visualization systems; flutter of blades; unstalled flutter. *Mailing Add:* 4767 Hillary Lane Cleveland OH 44143

JAIN, ANANT VIR, b Sardhana, India, March 15, 40; US citizen; m 71; c 2. TOXICOLOGY. *Educ:* Agra Univ, India, BS, 59, MS, 62; Purdue Univ, PhD(chem), 72. *Prof Exp:* Lectr chem, DAV Col, India, 62-64; res asst, Purdue Univ, 64-66, chem analyst, 66-72, res assoc, 72-74; anal chemist, 74-81, assoc anal toxicologist, 81-90, SR ANALYST TOXICOLOGIST & SUPVR TOXICOL LAB, UNIV GA, 90- *Concurrent Pos:* Pres, Southeast Regional Sect Assoc, Off Analytical Chemists, 86-87; mem, AOAC Comt, Regional Sect, 85-90. *Honors & Awards:* First Place Cert, Am Oil Chemists' Soc, 81. *Mem:* Am Chem Soc; Soc Toxicol; Asn Off Anal Chemists. *Res:* Analytical methods for the detection of poisons, drugs,and metals from biological and agricultural materials; analytical chemistry and toxicology of poisons; mycotoxin decontamination; diagnostic laboratory service. *Mailing Add:* Diag Asst Lab Col Vet Med Univ Ga Athens GA 30601

JAIN, ANIL KUMAR, b India, Jan 21, 46. ELECTRICAL ENGINEERING. *Educ:* Indian Inst Technol, Kharagpur, BTech Hons, 67; Univ Rochester, MS, 69, PhD(elec eng), 71. *Prof Exp:* Fel, Univ Southern Calif, 70-71, asst prof, 71-74; assoc prof elec eng, State Univ NY, Buffalo, 74-78; actg assoc prof, 78-79, PROF ELEC ENG, UNIV CALIF, DAVIS, 79- *Concurrent Pos:* Proj dir, NASA res grant, 75-76, Naval Undersea Ctr, San Diego, 76-77, Army Res Off grant, 76-78, Naval Ocean Syst Ctr grant, San Diego, 78-79, Army Res Off, 78-85, Off Naval Res, 82-84 & Univ Micro Projs, 82-86; topical ed, J Optical Soc Am. *Honors & Awards:* Image Coding Achievement Award, Int Picture Coding Symp, Tokyo, 77; Donald G Fink Prize Award, Inst Elec & Electronics Engrs, 83. *Mem:* Sr mem Inst Elec & Electronics Engrs; Optical Soc Am. *Res:* Digital image processing; signal processing; pattern recognition; communication theory; systems theory; computer applications; real time systems. *Mailing Add:* 52 Mahavir Nagar Near Tilak Nagar Kanadia Marg Indore MP 452001 India

JAIN, ANRUDH KUMAR, b India, Oct 23, 41; m 71; c 2. POPULATION STUDIES, PUBLIC HEALTH & EPIDEMIOLOGY. *Educ:* Agra Univ, BSc, 58; Delhi Univ, MA, 60; Univ Mich, Ann Arbor, MA, 65, PhD(sociol), 68. *Prof Exp:* Res asst pop studies & asst prof sociol, Univ Mich, Ann Arbor, 68-70; prog assoc family planning, Ford Found, India, 70-71; staff assoc family planning, Pop Coun, India, 71-73; asst dir biostatist, 73-76, assoc, 76-79, SR ASSOC, DEMOGRAPHIC IMPACTS DEVELOP PROG, POP COUN, 80-, DEP DIR, INT PROGS, 81- *Concurrent Pos:* Vis scholar, Univ Mich, 79-80. *Mem:* Int Union Sci Study Pop; Pop Asn Am; AAAS; Indian Asn Study Pop. *Res:* Consequences of population growth and determinants of fertility, fecundability, lactation and postpartum amenorrhea; assessing quality of family planning and health services. *Mailing Add:* Population Coun One Dag Hammarskjold Plaza New York NY 10017

JAIN, ARIDAMAN KUMAR, b Delhi, India, Apr 14, 38; m 63; c 2. APPLIED STATISTICS. *Educ:* Delhi Univ, BSc, 57; Purdue Univ, PhD(statist & indust eng), 68. *Prof Exp:* Statistician, SQC Units Indian Statist Inst, Baroda & Bombay, 60-61; Tata Oil Mills, Bombay, 61-63; mem tech staff, Bell Labs, 67-83, distinguished mem tech staff 84-86, DIST MGR, BELL COMMUN RES, 87- *Mem:* Am Statist Asn; Oper Res Soc Am; Am Soc Qual Control. *Res:* Data analysis; design of experiments; statistical modelling; monte-carlo simulation; survey sampling; quality assurance. *Mailing Add:* 15 Emory Dr Lincroft NJ 07738

JAIN, DULI CHANDRA, b Mungaoli, India, Feb 11, 29; m 51; c 2. MOLECULAR SPECTROSCOPY, DATABASE SYSTEMS. *Educ:* Banaras Hindu Univ, BS, 49; Univ Calcutta, MS, 51, DPhil(sci), 63; City Univ New York, MS, 74. *Prof Exp:* Lectr physics, Holkar Sci Col, India, 54-60, asst prof, 60-62; res fel, Saha Inst Nuclear Physics, India, 62-64; asst res scientist chem, NY Univ, 64-68; lectr, 68-70, from asst prof to assoc prof, 71-85, PROF PHYSICS, YORK COL, NY, 86- *Concurrent Pos:* Adj assoc prof comput sci, Queens Col, 78-; vis prof admin comput systs, Hofstra Univ, Hempstead, NY, 80-81. *Mem:* Am Phys Soc; Sigma Xi. *Res:* Intensity distribution in molecular band systems; potential energy curves and vibrational wave functions of diatomic molecules; programming systems for computers; quantum chemical study of molecular complexes. *Mailing Add:* York Col City Univ NY 94-20 Guy R Brewer Blvd Jamaica NY 11451

JAIN, HIMANSHU, b Mainpuri, India, Jan 20, 55; m 90; c 1. ELECTRONIC & SUPERCONDUCTING CERAMIC MATERIALS, INORGANIC GLASSES. *Educ:* Kanpur Univ, BS, 70; Banaras Univ, MS, 72; Indian Inst Technol, Kanpur, MTech, 74; Columbia Univ, EngScD(metall & mat sci), 79. *Prof Exp:* Asst ceramics, Argonne Nat Lab, Ill, 80-82; assoc metallurgist nuclear waste, Brookhaven Nat Lab, Upton, NY, 82-85; asst prof ceramics, 85-87, ASSOC PROF MAT SCI & ENG, LEHIGH UNIV, 88- *Concurrent Pos:* Vis scientist, Inst Physics, Univ Dortmund, Germany, 84 & Advan Ctr Mat Sci, Indian Inst Technol, 85; guest lectr, Krumb Sch Mines, Columbia Univ, 85; consult, MIT Sch Chem Eng Pract, Brookhaven Nat Lab, Upton, NY, 85. *Mem:* Am Ceramic Soc; Am Inst Mining Metall & Petrol Engrs; Ceramic Educ Coun. *Res:* Electrical relaxation, conductivity and dielectric properties of amorphous and crystalline ceramics; surface conduction and diffusion; effect of radiation on transport properties; diffusion and nuclear spin relaxation in glasses; corrosion in nuclear waste environment; sintering of ceramics. *Mailing Add:* Dept Mat Sci & Eng Lehigh Univ 264 Whitaker Lab 5 Bethlehem PA 18015-3195

JAIN, JAIN S, sanitary & environmental engineering, for more information see previous edition

JAIN, KAILASH CHANDRA, b Indore, Jan 1, 43; US citizen; m 69; c 3. FABRICATION METHODS, MICROSTRUCTURES. *Educ:* Banaras Hindu Univ, India, BEng, 64; State Univ NY, Stony Brook, MS, 69, PhD(mat sci), 72; Hofstra Univ, NY, MBA, 78. *Prof Exp:* Sci officer, Alloy Dept, Atomic Energy Comn, India, 64-66; teaching asst & lab instr variety, Dept Mat Sci, State Univ NY, Stony Brook, 66-72, postdoctoral res, Dept Earth & Space Sci, 72-73; sr process engr integrated circuits mfg, Gen Instrument Corp, NY, 73-76; device engr integrated circuits mfg, RCA Corp, WPalm Beach, Fla, 77-79; STAFF RES ENGR ELECTRONICS, GEN MOTORS RES LABS, WARREN, MICH, 79- *Concurrent Pos:* Consult, 70-76; recruiter, Electronics Dept, Gen Motors Res Lab, 80-86. *Mem:* Sr mem Inst Elec & Electronics Engrs. *Res:* Develop new materials and processes to realize novel sensors, power devices, and integrated circuits; propose improved fabrication methods and device structures; develop processes to improve current product; granted six patents. *Mailing Add:* E-3 Dept RSB 1B-500 Gen Motors Res Lab Warren MI 48090-9055

JAIN, MAHAVIR, b Barther, India, Jan 1, 41; US citizen; m 66. EXPERIMENTAL NUCLEAR PHYSICS. *Educ:* Agra Univ, BSc, 57; Univ Delhi, MSc, 59; Univ Md, College Park, PhD(physics), 69. *Prof Exp:* Fel nuclear physics, Univ Man, 69-71; res assoc nuclear physics, Tex A&M Univ, 71-78; MEM STAFF, LOS ALAMOS NAT LAB, 78- *Concurrent Pos:* Guest scientist, Los Alamos Sci Lab, 74-78. *Honors & Awards:* Res Publ Award, Naval Res Lab, 75. *Mem:* Am Phys Soc. *Res:* Direct interactions, especially quasi-free scattering from nucleons and clusters; excited states, breakup and polarization in three nucleon systems, neutron-proton scattering; polarization and pion production at LAMPF energies and transport calculations; diagnostics and simulations. *Mailing Add:* 1368 35th St Los Alamos NM 87544

JAIN, MAHENDRA KUMAR, b Ujjain, India, Oct 12, 38; m 74; c 1. BIOPHYSICS, NEUROSCIENCES. *Educ:* Vikram Univ, India, MSc, 59; Weizmann Inst Sci, PhD(chem), 67. *Prof Exp:* Lectr chem, Educ Dept, Govt Madhya Pradesh, India, 59-64 & Punjabi Univ, 64-65; res assoc biochem, Ind Univ, Bloomington, 67-73; from asst prof to assoc prof, 73-81, PROF BIOCHEM, UNIV DEL, 81- *Mem:* Biophys Soc; Fed Am Soc Exp Biol. *Res:* Membrane structure and function; mode of action of phospholipases on bilayers; reconstitution; effect of drugs on the phase properties of membrane; inhibitors of phospholipase. *Mailing Add:* Dept Chem Univ Del Newark DE 19711

JAIN, MAHENDRA KUMAR, b Muzaffarnagar, India, Jan 4, 29; m 49; c 4. MATHEMATICAL ANALYSIS. *Educ:* Univ Lucknow, BS, 48, MS, 51, PhD(math), 55. *Prof Exp:* Lectr math, M J Inter Col Asara, 51-52; Vidyant Col, Univ Lucknow, 52-55; H D Jain Col, Magadh Univ, 55-59; res instr, WVa Univ, 67-69, asst prof, 69-70; assoc prof, 75-83, PROF MATH, UNIV TENN, MARTIN, 83- *Concurrent Pos:* Asst prof, Bihar Inst Technol, Sindri, 59-72; Agency Int Develop fel, Univ Wis, 63-64. *Mem:* Am Math Soc. *Res:* Complex variables; integral transforms. *Mailing Add:* Math Dept Univ Tenn Martin TN 38238

JAIN, NARESH C, b Meerut, India, Dec 30, 32; m. TOXICOLOGY, ANALYTICAL CHEMISTRY. *Educ:* Univ Lucknow, BS, 51, MS, 54; Univ Calif, Berkeley, PhD(criminol/toxicol), 65. *Prof Exp:* Sci Officer toxicol, Govt Brit Guiana, 59-62; res toxicologist, Univ Calif, Berkeley, 63-66; assoc dir toxicol, Sch Med, Ind Univ, 66-71; assoc prof pharmacol & toxicol, 71-78, assoc prof community med & pub health, 72-78, PROF PHARMACOL & TOXICOL & PROF COMMUNITY MED & PUB HEALTH, SCH MED, UNIV SOUTHERN CALIF, 78-; DIR TOXICOL, NAT TOXICOL LABS, 87- *Concurrent Pos:* Mem clin toxicol devices panel, Food & Drug Admin, 75-; consult, US Navy, 83-, USAF, 85- *Mem:* Soc Toxicol; Am Acad Forensic Sci; Int Asn Toxicol. *Res:* Toxicology both clinical and forensic; application of instrumentation in the detection of drugs from biological fluids; drug metabolism; interaction of drugs, marijuana and alcohol; laboratory services to drug abuse and overdose patients; environmental monitoring of toxic wastes, herbicides, and pesticides; expert witness in toxicology; interpretation of alcohol & drug levels for impairment; environmental toxicology. *Mailing Add:* Nat Toxicol Labs 5451 Rockledge Dr Buena Park CA 90621

JAIN, NEMICHAND B, b Akola, India, July 1, 51; m 78; c 2. INDUSTRIAL PHARMACY, PHYSICAL PHARMACY. *Educ:* Nagpur Univ, India, BSc, 71; Univ Bombay, India, Bsc, 74; Univ Kans, MS, 76, PhD(pharmaceut), 78. *Prof Exp:* Res asst, Univ Kans, 74-78; res pharmacist, Wyeth Labs, subsid Am Home Prod Corp, 78-80; res investr, E R Squibb & Sons, 80-82, lab supvr, 82-85, RES GROUP LEADER & SECT HEAD, BRISTOL-MYERS SQUIBB CO, 85- *Mem:* Am Pharmaceut Asn; Acad Pharmaceut Sci; Controlled Release Soc; AAAS. *Res:* Development of pharmaceutical dosage forms; controlled drug delivery; in-vitro-in-vivo evaluation of dosage forms; drug stability degradation mechanisms; physio-chemical evaluation of drug entities. *Mailing Add:* 47 Wynwood Ave Monmouth Jct NJ 08852

JAIN, PIYARE LAL, b Punjab, India, Dec 11, 21; US citizen; m 66; c 2. PHYSICS. *Educ:* Punjab Univ, India, BA, 44, MA, 48; Mich State Univ, PhD(physics), 54. *Prof Exp:* Asst physics, Mich State Univ, 51-53; res assoc chem, Univ Minn, 53-54; from instr to assoc prof physics, 54-67, PROF PHYSICS, STATE UNIV NY BUFFALO, 67- *Concurrent Pos:* Res assoc, Univ Chicago, 59-60; mem staff, Lawrence Radiation Lab, Univ Calif, 61-62; vis prof, Bristol Univ, 61-62; Fulbright prof, Univ Rajasthan, India, 65-66. *Mem:* Fel Am Phys Soc. *Res:* Solid state, electron and nuclear magnetic resonance; nuclear physics; cosmic radiation and high energy physics; radiation physics; heavy ion physics. *Mailing Add:* Dept Physics State Univ NY Buffalo NY 14260

JAIN, RAKESH KUMAR, b Lalitpur, India, Dec 18, 50. TUMOR PHYSIOLOGY, MICROCIRCULATION. *Educ:* Indian Inst Technol, Kanpur, BTech, 72; Univ Del, Newark, MChE, 74, PhD(chem eng), 75. *Prof Exp:* Asst prof chem & biomed engr, Columbia Univ, 76-78; from asst prof to prof chem & biomed eng, Carnegie Mellon Univ, 78-91; ANDREW WORK COOK PROF TUMOR BIOL, HARVARD MED SCH, 91-; DIR, EDWIN L STEELE LAB TUMOR BIO, MASS GEN HOSP, 91-; PROF, DIV HEALTH SCI & TECHNOL, MASS INST TECHNOL, 91- *Concurrent Pos:* Adj prof neurosurg, Univ Pittsburgh Sch Med, 78-80; consult, Pathophysiol Lab, Nat Cancer Inst, 76-84 & Hybritech, 87-; vis prof, dept chem eng, dept chem eng, Mass Inst Technol, 83, dept bioeng, Univ Calif, San Diego & dept radiol, Stanford Univ Med Sch, 84; chair, Nat Prog Comt Life Sci, Am Inst Chem Engrs, 81-84, mem comt, Am Microcirculation Soc, 86-88 & meeeting prog comt, Biomed Eng Soc, 86-89; co-chair, Conf Thermal Characteristics Tumors, NY Acad Sci, 79-; assoc mem, Ctr Fluorescence Res Biol, 85-; mem, Pittsburg Cancer Inst, 86-; mem bd dirs, Int Inst Microcirculation, 87-,; John Simon Guggenheim Mem Found fel, 83-84; NIH res career develop award, 80-85. *Honors & Awards:* Int Inst Microcirculation Res Award, 84. *Mem:* Am Asn Cancer Res; AAAS; Am Inst Chem Engrs; Am Microcirculation Soc; Biomed Eng Soc; Int Inst Microcirculation; NY Acad Sci. *Res:* Develop a quantitative understanding of physiological events in the tumor microcirculation to improve cancer detection and treatment; transport of molecules in tumors; blood flow and microcirculatory hemodynamics in tumors; physiological studies in tissues isolated tumors; heat transfer and temperature distribution in tumors; rheology of malignant and non-malignant cells; interaction of cells with vasculature. *Mailing Add:* Harvard Med Sch 25 Shattuck St Boston MA 02115

JAIN, RAVINDER KUMAR, b Punjab, India, Oct 12, 35; US citizen. ENVIRONMENTAL ENGINEERING, INTELLIGENT SYSTEMS. *Educ:* Calif State Univ, Sacramento, BS, 61, MS, 68; Tex Tech Univ, PhD(civil eng), 71; Harvard Univ, MPA, 80. *Prof Exp:* Civil engr design, Spink Eng Corp, 61-64; assoc engr water resources, Calif Dept of Water Resources, 64-68; civil engr, Develop & Resources Corp, 68-69; chief Environ Div Environ Res, US Army Corps Engrs, Construct Eng Res Lab, 71-89; DIR, ARMY ENVIRON POLICY INST, 90- *Concurrent Pos:* Adj prof, Univ Ill, Urbana-Champaign, 75-; exec & prof develop fel, Harvard Univ, 79-80; res affil, Mass Inst Technol, 84-; Churchill Col fel, Cambridge Univ, Eng, 86. *Honors & Awards:* Sustained Super Performance Awards, US Army Corps Engr, 72, 73 & 74; Res & Develop Award, US Army, 76, Commendations for Exemplary Performance & Except Mgt Res Prog, 82-90. *Mem:* Fel Am Soc Civil Engrs; Soc Am Mil Engrs. *Res:* Environmental impact analysis, environ quality management related to solid waste, air, water, noise pollution and hazardous waste; management of research and development organizations; environmental policy development; computer systems and artificial intelligence; author/co-author six books. *Mailing Add:* Army Environ Policy Inst Champaign IL 61824-0798

JAIN, SUBODH K, b Nanauta, India, Dec 11, 34; m 57; c 3. POPULATION BIOLOGY, ECONOMIC BOTANY. *Educ:* Univ Delhi, BSc, 54; Indian Agr Res Inst, New Delhi, IARI, 56; Univ Calif, Davis, PhD(genetics), 60. *Prof Exp:* Pool off genetics, Coun Sci & Indust Res, New Delhi, 61-63; asst res geneticist, 63-67, assoc biologist, 67-72, PROF BIOL, UNIV CALIF, DAVIS, 72- *Concurrent Pos:* NSF res grant, 69-71, 76-78 & 79-82, SOHIO grant, 81-85, SLOAN grant, 87; Guggenheim Found fel & sr fel, Coun Sci &

Indust Res Orgn, Australia, 71-72; Indo-US fel, 78-79; consult & vis prof, Hyderabad, India, 79, 82, Mendoza, Argentina, 82, Piracicaba, Brazil, 83, Plant Breeding Inst, Wageningen, 84; assoc ed, Evolution, 81-83; mem, Nat Acad Sci Panel Amaranth, 82-85, Orgn Int Symp in pop biol, 78, 83, 86, 87, 88. *Mem:* Soc Study Evolution; Am Soc Naturalists; Bot Soc Am; Am Inst Biol Sci. *Res:* Population genetics; plant breeding; plant evolution; dynamics of grassland communities; genetics and ecology of avena, bromus, trifolium species; analysis of life histories and relative fitnesses; germplasm resources in crop breeding; development of new crops; conservation of rare and endangered plants. *Mailing Add:* Dept of Agronomy & Range Sci Univ of Calif Davis CA 95616

JAIN, SURENDER K, b Amritsar, India, Nov 16, 38; m 63; c 2. RING THEORY, LINEAR ALGEBRA. *Educ:* Panjab Univ, India, BA Hons, 57, MA, 59; Univ Delhi, India, PhD(ring theory), 63. *Prof Exp:* Res mathematician & lectr, Univ Calif, Riverside, 63-65; reader math, Univ Delhi, India, 65-69; PROF MATH, OHIO UNIV, 69- *Concurrent Pos:* Vis prof, Univ Frankfurt, Univ Chicago, McMaster Univ, Can, Kuwait Univ, Riyad Univ, Saudi Arabia, Ohio State Univ, NC State Univ & Univ Calif, Santa Barbara, 63-90. *Mem:* Am Math Soc; Math Asn Am; Soc Indust & Appl Math. *Res:* Noncommutative ring theory and applied linear algebra; author of 60 research publications and 6 books. *Mailing Add:* Three Ransom Rd Ohio Univ Athens OH 45701

JAIN, SUSHIL C, b Lucknow, India, June 14, 39; US citizen; m 69; c 2. PRODUCTIVITY IMPROVEMENT, QUALITY CONTROL. *Educ:* St John's Col, Agra, India, BS, 57; Indian Inst Technol, Kharagpur, BSEE, 61; Purdue Univ, Indiana, MSIE, 64. *Prof Exp:* Elec engr, Gwalior Rayons, India, 61-63; indust engr, Ford Motor Co, 64-65 & Safran Printing Co, 66-68; sr indust engr, Edwards Bros, Inc, 71-73; staff engr, Alco Gravure, Inc, 73-78; mgr, indust eng, Unified Data Prod, 79-81; dir, indust eng, Universal Folding Box Inc, 81-89; PROCESS ENGR, SEALED AIR CORP, 90- *Concurrent Pos:* Pres, Jain Consult, 89. *Mem:* Sr mem Inst Indust Engrs. *Mailing Add:* 60 Winthrop Rd Hillsdale NJ 07642

JAIN, SUSHIL KUMAR, b Nabha, Punjab, India, Mar 31, 50; m 80; c 2. HEMATOLOGY, NUTRITION. *Educ:* Punjab Univ, Chandigarh, BS, 70; Postgrad Inst Med Educ & Res, Chandigarh, MS, 72, PhD(biochem), 76. *Prof Exp:* Tutor biochem, Postgrad Inst Med & Res, 76-77; fel pharmacol & nutrit, Univ Southern Calif, Los Angeles, 77-79; fel hemat, Sch Med, Univ Calif, San Francisco, 79-81; from instr to asst prof, 81-91, PROF PEDIAT & BIOCHEM, LA STATE UNIV MED CTR, SHREVEPORT, CHIEF, SECT PEDIAT RES, 87- *Concurrent Pos:* Prin investr, NIH res grant, 85-88, Nat Am Diabetes Asn, 87- *Honors & Awards:* Founder's Award & Ross Award, Southern Soc Pediat Res; Beecham Award. *Mem:* NY Acad Sci; Am Soc Biol Chemists; Am Soc Hemat; Soc Pediat Res; Am Inst Nutrit; Am Fed Clin Res. *Res:* Mechanisms of reduced red blood cell life span in sickle cell disease, newborn red cells, copper deficiency, iron deficiency; red cell aging, membrane lipid peroxidation; hyperlipidemia and lecithin-cholesterol acyltranferase deficiency. *Mailing Add:* Dept Pediat La State Univ Med Ctr 1501 Kings Hwy Shreveport LA 71130

JAIN, VIJAY KUMAR, b Gwalior, India, Nov 15, 37; m 57; c 3. ELECTRICAL ENGINEERING. *Educ:* Univ Rajasthan, BE, 56; Univ Roorkee, ME dipl, 57; Mich State Univ, PhD(elec eng), 64. *Prof Exp:* Asst prof elec eng, Birla Eng Col, India, 57-61; asst prof, Mich State Univ, 64; asst prof, Birla Inst Technol & Sci, India, 65-66, assoc prof, 66-68; assoc prof, Univ SFla, 68-74, prof, 74-79; prof, Ga Inst Technol, 79-80; PROF ELEC ENG, UNIV SFLA, 80- *Concurrent Pos:* Consult, Honeywell, Sperry, Vet Admin Hosp & A C Nielsen; prof, Bell Lab, 82-84. *Mem:* Inst Elec & Electronics Engrs. *Res:* Communication Electronics; computer networking; digital signal-processing; pattern recognition; speech signals analysis; VLSI and microprocessors; system identification. *Mailing Add:* Dept Elec Eng Univ SFla Tampa FL 33620

JAIN, VINOD KUMAR, US citizen; m 67; c 2. POLYMER TRIBOLOGY, MACHINE DESIGN, ROBOTICS & FORGING. *Educ:* Univ Roorkee, India, BE, 64, ME, 70; Iowa State Univ, PhD(mech eng), 80. *Prof Exp:* Lectr, Mech Eng Dept, Univ Roorkee, India, 64-75; from instr to assoc prof, 79-89, PROF MECH ENG DEPT, UNIV DAYTON, 89- *Mem:* Am Soc Mech Engrs. *Res:* Friction and wear of polymers; characterization of surface topography; fatigue of polymeric composites; lubrication technology; metal processing sciences; forging. *Mailing Add:* Mech Eng Dept Univ Dayton Dayton OH 45469-0210

JAINCHILL, JEROME, b New York, NY, Jan 27, 32; m 64; c 3. ENVIRONMENT, BIOCHEMISTRY. *Educ:* NY Univ, BA, 53, MS, 60, PhD(genetics), 63. *Prof Exp:* Res assoc radiobiol, Sloan-Kettering Inst, NY, 63-65; res assoc biochem carcinogens, Dept Environ Med, NY Univ Med Ctr, 65-67; res biochemist, Endo Labs, 67-77; res biochemist, Cornell Univ, 77-80; res biochemist, North Star Res, 80-81; SCI EDUC, NEW YORK BD EDUC, 81- *Mem:* AAAS; NY Acad Sci; Am Chem Soc. *Res:* Biochemistry of carcinogenic agents on DNA; drug metabolism; pharmacokinetics; retrovirus; leukemia. *Mailing Add:* 2362 Garfield St North Bellmore NY 11710

JAISINGHANI, RAJAN A, b Karachi, Pakistan, Jan 21, 45; c 2. PRODUCT & BUSINESS DEVELOPMENT. *Educ:* Banaras Hindu Univ, India, BS, 69; Univ Wis, MS, 73. *Prof Exp:* Engr, Fiebing Chem Co, 71-73; res asst, Univ Wis, 71-73; mgr res, Nelson Indust, 74-82; mgr res & develop, Am Filtrona Corp, 82-90; PRES, PROD DEVELOP ASSISTANCE, INC, 90- *Mem:* Am Inst Chem Eng; Soc Automotive Eng; Am Asn Aerosol Res; Int Asn Colloid Scientists; Am Chem Soc; Am Inst Chem Engrs; Filtration Soc; Int Asn Colloid Scientists. *Res:* Air and liquid filtration; colloid and aerosols; electrically simulated filtration; capillarity and other surface phenomena; fluid flow coalescence; research management and planning. *Mailing Add:* Prod Develop Assistance Inc 4200 Northwich Rd Midlothian VA 23112

JAKAB, GEORGE JOSEPH, b Budapest, Hungary, April 7, 39; m 63; c 2. PULMONARY IMMUNOLOGY, DISEASES & TOXICOLOGY. *Educ:* Univ Wis-Madison, BS, 65, MS, 67, PhD(med microbiol), 70. *Prof Exp:* Fel, Univ Vt, 70-72, res assoc, 72-77; assoc prof, 77-86, PROF, SCH HYG & PUB HEALTH, JOHNS HOPKINS UNIV, 86-, ASSOC DEPT CHMN, 90- *Concurrent Pos:* Res career develop award, Nat Heart, Lung & Blood Inst, 77. *Mem:* Infectious Dis Soc Am; Am Thoracic Soc; Reticuloendothelial Soc; Am Soc Microbiol; Soc Toxicol. *Res:* Pulmonary defense mechanisms against infectious agents; interaction of infectious agents and environmental contaminants in the genesis and exacerbation of acute and chronic lung disease. *Mailing Add:* Dept Environ Health Sci Sch Hyg Pub Health Johns Hopkins Univ Baltimore MD 21205

JAKACKY, JOHN M, b Hartford, Conn, July 22, 56. ATOM MOLECULE COLLISION, ACOUSTIC. *Educ:* Univ Conn, BS, 78, MS, 79, PhD(physics), 84. *Prof Exp:* Analyst, 84-87, SR ANALYST, SONALYSTS INC, 87- *Mem:* Am Phys Soc. *Mailing Add:* 71 Gill St Colchester CT 06415

JAKES, KAREN SORKIN, b Washington, DC, June 18, 47; m 70; c 2. MOLECULAR BIOLOGY. *Educ:* Brown Univ, BSc, 69; Yale Univ, PhD(molecular biophys & biochem), 74. *Prof Exp:* Asst res genetics, 71-75, RES ASSOC GENETICS, ROCKEFELLER UNIV, 76- *Mem:* AAAS. *Res:* Mechanism of action and synthesis of colicin E3 and its immunity protein; export of colicins E1, E2 and E3; replication of bacteriophage fluid. *Mailing Add:* Dept Genetics 1230 York Ave New York NY 10021

JAKES, W(ILLIAM) C(HESTER), b Milwaukee, Wis, May 15, 22; m 48; c 2. ELECTRICAL ENGINEERING. *Educ:* Northwestern Univ, BS, 44, MS, 47, PhD(elec eng), 49. *Hon Degrees:* PhD, Iowa Wesleyan Univ, 61. *Prof Exp:* Mem tech staff, Bell Tel Labs, 49-62, head mobile radio res, 62-71, dir radio transmission lab, 71-84, dir Transmission Terminals & Radio Lab, 84-87; RETIRED. *Concurrent Pos:* Mem sci adv bd, Voice of Am, 60-62. *Honors & Awards:* Co-winner, Alexander Graham Bell Medal, Inst Elec & Electronics Engrs, 87. *Mem:* Fel Inst Elec & Electronics Engrs. *Res:* Microwave propagation and antennas; satellite communication; microwave transmission systems development. *Mailing Add:* 58 Wildrose Dr Andover MA 01810

JAKLEVIC, JOSEPH MICHAEL, b Kansas City, Kans, Jan 16, 41; m 66; c 2. PHYSICS, ENVIRONMENTAL SCIENCES. *Educ:* Rockhurst Col, AB, 62; Univ Notre Dame, PhD(physics), 66. *Prof Exp:* Fel nuclear physics, Univ Notre Dame, 66-67; fel, 67-69, staff scientist eng, 69-78, SR STAFF SCIENTIST, DEPT INST SCI, LAWRENCE BERKELEY LAB, 78- *Mem:* Am Phys Soc; Mat Res Soc; Air Pollution Control Asn; Mat Res Soc. *Res:* Application of nuclear and atomic physics principles and techniques to problems of environmental sampling and analysis; x-ray and atomic physics techniques. *Mailing Add:* Lawrence Berkeley Lab No 1 Cyclotron Rd Berkeley CA 94720

JAKLEVIC, ROBERT C, b Kansas City, Kans, July 27, 34; m 62; c 2. EXPERIMENTAL SOLID STATE PHYSICS. *Educ:* Rockhurst Col, BS, 56; Univ Notre Dame, PhD(physics), 60. *Prof Exp:* Fel, Univ Notre Dame, 61-62; STAFF SCIENTIST SOLID STATE PHYSICS, FORD SCI LABS, 62- *Honors & Awards:* Tech Achievement Award, Ford Motor Co, 90. *Mem:* Fel Am Phys Soc; Sigma Xi. *Res:* Superconductivity; Josephson tunneling; normal metal tunneling; photoelectric effect in metals; thin film technology; tunneling in semiconductors; organic conductors; surface science; scanning tunneling spectroscopy; nanoscale devices. *Mailing Add:* 31345 Old Cannon Rd Birmingham MI 48010

JAKOB, FREDI, b Horstein, Ger, Jan 11, 34; US citizen; m 57; c 4. ANALYTICAL CHEMISTRY. *Educ:* City Col NY, BS, 55; Rutgers Univ, PhD(anal Chem), 61. *Prof Exp:* Instr chem, Rutgers Univ, 60-61; from asst prof to assoc prof, 61-69, chmn dept, 65-68, PROF CHEM, CALIF STATE UNIV, SACRAMENTO, 69- *Concurrent Pos:* NSF grants, 61-; consult, St Bd Equalization, 62-78 & consult chemist, Anal Assocs Inc; vis assoc prof, Univ Wis, Madison, 68-69; vis prof, Victoria Univ, Wellington, NZ, 71, Univ Wollongong, Australia, 82. *Mem:* Am Chem Soc. *Res:* Theory and application of separation methods and chemical instrumentation; laboratory applications of computers. *Mailing Add:* Dept Chem Sacramento State Col 6000 Jay St Sacramento CA 95819

JAKOB, KARL MICHAEL, b Berlin, Ger, Nov 5, 21; nat US; m 54; c 2. NUCLEIC ACIDS BIOLOGY, CELL BIOLOGY. *Educ:* Univ Ill, BS, 43, MS, 48; Univ Calif, PhD(cytogenetics, bot), 52. *Prof Exp:* Plant breeder, Marshall Farm Serv, Ill, 43-45; asst bot & cytol, Univ Ill & Univ Calif, 47-51; res assoc plant genetics, 53-68, sr scientist, 69-78, ASSOC PROF, WEIZMANN INST SCI, ISRAEL, 79- *Concurrent Pos:* Vis sr lectr, Univ Bar Ilan, Israel, 62-72; vis investr, Biol Div, Oak Ridge Nat Lab, 63-64. *Mem:* Sigma Xi; Int Soc Plant & Molecular Biol. *Res:* Biochemistry of the cell division cycle of eukaryotes; plant RNA metabolism; use of antisense RNA probes to locate transcriptional activity by insitu hybridization; molecular biology of chromatin during DNA replication in vivo. *Mailing Add:* Dept Plant Genetics Weizmann Inst Sci Rehovot 76100 Israel

JAKOBIEC, FREDERICK ALBERT, OPHTHALMOLOGY, PATHOLOGY. *Educ:* Harvard Univ, MD. *Prof Exp:* CHMN, MANHATTAN EYE, EAR, & THROAT HOSP, 80- *Res:* Tumor surgery. *Mailing Add:* Manhattan Eye Ear & Throat Hosp 210 E 64th St New York NY 10021

JAKOBSEN, ROBERT JOHN, b Chicago, Ill, Jan 29, 29; m 52. VIBRATIONAL SPECTROSCOPY. *Educ:* Col of Emporia, BS, 51. *Prof Exp:* Asst phys chem, Kans State Col, 51-55 & Univ Kans, 55-56; prin chemist, Battelle Columbus Labs, 56-64, sr chemist, 64-76, res leader, 76-86; tech dir, Mattson Inst Spectros Res, 86-88; PRES, IR-ACTS, 88- *Concurrent Pos:* Adj prof chem, Kans State Univ,88- *Honors & Awards:* Coblentz Soc Williams-Wright Award, 84. *Mem:* Soc Appl Spectros; Coblentz Soc; Sigma Xi; NY Acad Sci. *Res:* Molecular spectroscopy, mainly infrared and Raman; application of molecular spectroscopy to structure, especially the structure of proteins. *Mailing Add:* IR-ACTS 326 Walhalla Rd Columbus OH 43202

JAKOBSON, MARK JOHN, b Carlyle, Mont, May 4, 23; m 45; c 2. NUCLEAR PHYSICS. *Educ:* Univ Mont, AB, 44, MA, 47; Univ Calif, PhD(physics), 51. *Prof Exp:* Asst physics, Univ Calif, 47-49, physicist, Radiation Lab, 50-52; instr physics, Univ Wash, 52-53; from asst prof to assoc prof, 53-58, chmn dept astron & physics, 68-73, PROF PHYSICS, UNIV MONT, 58- *Mem:* Fel Am Phys Soc. *Res:* Photonuclear reactions; accelerator design; pion interactions. *Mailing Add:* 3000 Queen St Missoula MT 59801

JAKOBSSON, ERIC GUNNAR, SR, b New York, NY, Nov 18, 38; m 63; c 6. BIOPHYSICS, PHYSIOLOGY. *Educ:* Columbia Univ, BA, 59, BS, 60; Dartmouth Col, PhD(physics), 69. *Prof Exp:* Process engr cryog, Air Prod & Chem, 60-62; develop engr, Malaker Corp, 62-65; fel, Case Western Reserve Univ, 69-71; res assoc, 71-72, asst prof, 72-78, ASSOC PROF PHYSIOL & BIOPHYS, UNIV ILL, URBANA, 78-, ASSOC PROF BIOENG, 81- *Concurrent Pos:* Fel, NSF, 70-71; vis assoc prof physiol, Duke Univ, 79. *Mem:* Biophys Soc; AAAS; NY Acad Sci; Am Phys Soc. *Res:* Mechanisms and significance of ion movement across biological membranes; osmoregulation of animal cells; physics of ion movement; rhythmic and repetitive electrical activity in nerve. *Mailing Add:* Div Biophysics Univ Ill 156 Davenport Urbana IL 61801

JAKOBY, WILLIAM BERNARD, b Breslau, Ger, Nov 17, 28; nat US. BIOCHEMISTRY, MICROBIOLOGY. *Educ:* Brooklyn Col, BS, 50; Yale Univ, PhD(microbiol), 54. *Prof Exp:* Fel pharmacol, NY Univ-Bellevue Med Ctr, 53-54, fel biochem, 54-55; sr investr, Nat Inst Arthritis, Metab & Digestive Dis, 55-68, CHIEF SECT ENZYMES & INTERMEDIARY METAB, NAT INST DIABETES & DIGESTIVE & KIDNEY DIS, 68-, CHIEF, LAB BIOCHEM & METAB, 84- *Concurrent Pos:* Mem bd dirs, Found Advan Educ in Sci, 68-87; mem adv bd, John F Fogarty Int Ctr Advan Study Health Sci, NIH, 70-73; consult, Molecular Biol Panel, NSF, 70-73 & 76; mem, Enzyme Comn, Int Union Biochem, 69-71, Comn Biochem Nomenclature, 74-80, life sci res off adv comt, Fedn Am Soc Exp Biol, 83-; ed in chief, Anal Biochem, 86-; assoc ed Hepatol, 80-85 & Protein Expression & Purification, 90-92. *Mem:* Am Soc Biol Chemists. *Res:* Enzymology, detoxication. *Mailing Add:* NIH Bldg 10 Rm 9N119 Bethesda MD 20892

JAKOI, EMMA RAFF, b Cornwall, Ont, May 10, 46; US citizen; m 71. CELL BIOLOGY, MOLECULAR BIOLOGY. *Educ:* Wash State Univ, BS, 68; Duke Univ, PhD(physiol, pharmacol), 73. *Prof Exp:* Asst prof anat, Med Ctr, Duke Univ, 77-89; ASSOC PROF, DEPT NEUROL, MED COL VA, 89- *Concurrent Pos:* Res assoc anat, Med Ctr, Duke Univ, 73-74; USPHS instnl res fel, 74-75, USPHS fel, 75-77. *Mem:* Am Soc Cell Biol; Biophys Soc; Sigma Xi. *Res:* Biochemical and morphological studies of ligatin, a membrane bound baseplate for cell surface proteins involved in intercellular adhesion during development of embryonic chick neural retina and in degradation of glycoproteins and glycolipids in suckling rat ileal epithelial cells. *Mailing Add:* Med Col Va Box 577 Sanger Hall Rm 6013 11th & Marshall St Richmond VA 23298

JAKOWATZ, CHARLES V, b Kansas City, Kans, Feb 6, 20; m 47; c 2. ELECTRICAL ENGINEERING. *Educ:* Kans State Univ, BS, 44, MS, 47; Univ Ill, PhD(elec eng), 53. *Prof Exp:* Instr math, Kans State Univ, 45-46, asst prof mech eng, 46-48; asst prof elec eng, Univ Ill, 48-53; commun engr, Res Lab, Gen Elec Co, 53-63, liaison scientist, 63-65; prof elec eng & dean eng, 65-69, PROF ELEC ENGRS, WICHITA STATE UNIV, 70- *Concurrent Pos:* Adj prof, Rensselaer Polytech Inst, 58-63. *Mem:* Math Asn Am; Am Soc Eng Educ; sr mem Inst Elec & Electronics Engrs. *Res:* Cognitive processes; communication and information theory; network synthesis. *Mailing Add:* 533 N Broadmoor Wichita State Univ Wichita KS 67206

JAKOWSKA, SOPHIE, b Warsaw, Poland, Feb 12, 22; nat US; m 41; c 3. PATHOBIOLOGY, ENVIRONMENTAL EDUCATION. *Educ:* Lycee Warsaw, Poland, cert, 39; Univ Rome, cert, 42; Fordham Univ, MS, 45, PhD(biol), 47. *Prof Exp:* Instr bact, Col Mt St Vincent, 46; asst chemother div, Sloan-Kettering Inst Cancer Res, 47-48; from asst prof to assoc prof, Col Mt St Vincent, 48-58; asst to vpres med affairs, Nat Cystic Fibrosis Res Found, 61-62; head dept path, Food & Drug Res Labs, Inc, 64-67; tech adv & res coordr biol, Santo Domingo Univ, 67-68; spec proj dir, Nat Cystic Fibrosis Res Found, 68-69; biologist, Food & Drug Admin, 69-71; prof biol sci, Col Staten Island, City Univ NY, 70-78; RETIRED. *Concurrent Pos:* Collabr, NY Aquarium, NY Zool Soc, 48-59, res assoc exp biol, Dept Marine Biochem & Ecol, 59-62; collabr, Brookhaven Nat Lab, 52-62; vis prof grad sch, St Louis Univ, 57; res assoc dept labs, Beth Israel Hosp, 59- & Inst Crippled & Disabled, 62; NSF biol teacher inst lectr, Iona Col, 63-; consult, Inst Marine Biol, Santo Domingo Univ, 63-, res coordr & hon prof fac sci, 68-; consult, Animal Med Ctr, 64-68; pvt consult, 66-; consult, Span Dept, Grolier, Inc, NY, 68-75; reviewer proposals & projs, Comn Educ, Int Union Conserv Nature; sci consult, 77-; mem Int Union Conserv Nature & Natural Resources Working Group on Ethics, 84-, liaison World Coun Churches, 88, adv bd, Global Harmony Found, 90- *Honors & Awards:* Tree Learning Award, Int Union Conserv of Nature & Natural Resources, 88; Liga Ochrony Przyrody gold medal, League Protection Nature, 89. *Mem:* Fel AAAS; Am Micros Soc; Soc Protozool; Am Soc Ichthyologists & Herpetologists; Am Inst Biol Sci; fel NY Acad Sci; Sigma Xi. *Res:* Plant and animal cytology; comparative pathology and hematology; experimental biology; parasitology; radiobiology; biochemical ecology; mucous secretions; conservation and religious environmental education; writing books for children and new readers in Spanish on conservation and environmental education, e.g., on crocodiles, parrots, etc; author of numerous scientific papers and books; continuing work and education of environmental conservation. *Mailing Add:* Arz Merino 154 Z-1 Santo Domingo Dominican Republic

JAKUBIEC, ROBERT JOSEPH, b Detroit, Mich, June 19, 41; m 64; c 2. ANALYTICAL CHEMISTRY. *Educ:* Univ Detroit, BS, 63; Wayne State Univ, PhD(anal chem), 68. *Prof Exp:* Chemist, US Food & Drug Admin, 63-65; sr chemist, Corn Prod Co, 68-69; sr chemist, Armak Co, Div Akzona, 69-70, sect mgr anal chem, 70-76; VPRES & LAB DIR, ENVIRO-TEST/PERRY CHICAGO DAIRY LABS, INC, 76- *Concurrent Pos:* Guest lectr, Northwestern Univ, 71- & Roosevelt Univ, 71-; instr, Chicago Gas Chromatog Sch, 73- *Mem:* Am Chem Soc; Am Oil Chem Soc; Am Soc Testing & Mat; Water Pollution Control Fedn; Am Asn Cereal Chemists; Asn Official Anal Chemists. *Res:* General analytical methods development; gas chromatography; thin layer chromatography; atomic absorption spectroscopy; ultraviolet and visible spectroscopy; residue analysis; general instrumentation; high pressure liquid chromatography; ion chromatography. *Mailing Add:* 5551 Lyman Downers Grove IL 60516

JAKUBOWSKI, GERALD S, b Toledo, Ohio, Nov 22, 49; m 72; c 2. ENGINEERING EDUCATION ADMINISTRATION, LASER DOPPLER VELOCIMETRY. *Educ:* Univ Toledo, BSME, 74, MSME, 76, PhD(eng sci), 78. *Prof Exp:* Grad & admin asst, Col Eng, Univ Toledo, 74-78, from asst prof to assoc prof mech eng, 78-86, asst dean eng, 86-88; assoc dean eng, Memphis State Univ, 88-89, interim dean, 89-90; DEAN, COL SCI & ENG, LOYOLA MARYMOUNT UNIV, 90- *Concurrent Pos:* Fel, NASA-Lewis Res Ctr, 84-85; chair, New Eng Educ Comt, Am Soc Eng Educ, 85-86; mem, Student Activ Comt, Soc Automotive Engrs, 86-, eng educ bd, 90- *Honors & Awards:* Ralph R Teetor Award, Soc Automotive Engrs, 85. *Mem:* Am Soc Eng Educ; Soc Automotive Engrs; Am Soc Mech Engrs; Am Inst Aeronaut & Astronaut. *Res:* Thermodynamics; fluid mechanics; heat transfer and energy; pump cavitation; ice melting; laser Doppler velocimetry. *Mailing Add:* Col Sci & Eng Loyola Marymount Univ Loyola Blvd at W 80th St Los Angeles CA 90045

JAKUBOWSKI, HIERONIM ZBIGNIEW, b Szczecinek, Poland, Sept 30, 46; m; c 2. PROTEIN SYNTHESIS REGULATION & ACCURACY, MOLECULAR MECHANISMS OF CELLULAR DEFENSES AGAINST STRESS. *Educ:* Poznan Univ, MSc, 69; Agr Univ, Poznan, PhD(biochem), 74; Inst Biochem & Biophysics, Warsaw, DrHabil, 78. *Prof Exp:* Res asst biochem, Akademia Rolniczaiw Poznaniu, 69-73, sr res asst, 73-75, adj, 75-78, adj habil, 79-87; adj asst prof, 84-91, ADJ ASSOC PROF MICROBIOL & MOLECULAR GENETICS, NJ MED SCH, NEWARK, 91- *Concurrent Pos:* Vis scientist, Univ NMex, Albuquerque, 75-76, Imp Col, London, Eng, 80 & Hanover Med Sch, Ger, 82; dep chmn, Dept Biochem, Agr Univ Poznan, 81. *Honors & Awards:* J Parnas Award, Polish Biochem Soc, 84. *Mem:* Am Soc Microbiol; Polish Inst Arts & Sci Am. *Res:* Mechanisms which maintain high degree of accuracy in the transmission and flow of information from gene to finished protein product; molecular mechanisms of cellular defenses against stress. *Mailing Add:* 347 Maple St Kearny NJ 07032

JAKUS, KARL, b Gyor, Hungary, Mar 21, 38; US citizen; m; c 2. MECHANICAL ENGINEERING, CERAMICS ENGINEERING. *Educ:* Univ Wis, Madison, BS, 63; Univ Calif, Berkeley, MS, 65, PhD(aerosci), 68. *Prof Exp:* Asst prof mech eng, Johns Hopkins Univ, 68-70; asst prof, 70-76, assoc prof, 77-83, PROF MECH ENG, UNIV MASS, AMHERST, 84- *Concurrent Pos:* Consult govt labs and indust. *Mem:* Am Ceramics Soc. *Res:* Mechanical behavior of ceramics. *Mailing Add:* Dept of Mech Eng Univ of Mass Amherst MA 01003

JAKUS, MARIE A, b Cleveland, Ohio, Oct 12, 14. BIOLOGY. *Educ:* Oberlin Col, AB, 37; Mass Inst Technol, PhD(biol), 45. *Prof Exp:* Asst zool, Wash Univ, 38-41; asst biol, Mass Inst Technol, 41-45, res assoc, 45-51; assoc, Retina Found, 51-61; res prog coordr, Extramural Progs, Nat Inst Neurol Dis & Blindness, NIH, 61-62, exec secy, Visual Sci B Study Sect, Div Res Grants, 62-77; RETIRED. *Concurrent Pos:* Rockefeller fel, Karolinska Inst, Sweden, 47-48. *Mem:* Electron Micros Soc Am; Asn Res Vision & Opthal. *Res:* Electron microscopy; biological electron microscopy; fine structure and properties of trichocysts, striated muscle, actin, myosin and actomyosin, paramyosin, cornea, sclera and lens. *Mailing Add:* 2370 Opalo Way San Diego CA 92111-5913

JAKWAY, GEORGE ELMER, b Twin Falls, Idaho, July 3, 31; div. VERTEBRATE PALEONTOLOGY. *Educ:* Idaho State Col, BA, 53; Univ Kans, MS, 58; Univ Nebr, PhD, 63. *Prof Exp:* Asst vert paleont, State Mus, Univ Nebr, 57-60, asst instr zool, Univ, 60-61, asst prof, 61-68, ASSOC PROF ZOOL, CALIF STATE UNIV, LOS ANGELES, 68- *Concurrent Pos:* Consult, Idaho State Univ Mus, 62; res & field assoc, State Mus, Univ Nebr, 64-; res assoc, Los Angeles County Mus, 65- & George Page Mus, 80- *Mem:* AAAS; Soc Vert Paleont; Soc Study Evolution; Am Soc Mammal. *Res:* Pleistocene paleomammalogy, speciation and ecology. *Mailing Add:* Dept of Biol Calif State Univ Los Angeles CA 90032

JAKWAY, JACQUELINE SINKS, b San Juan, PR, Dec 13, 28; div. ANATOMY. *Educ:* Park Col, AB, 50; Univ Kans, PhD(anat), 58. *Prof Exp:* Asst histochem, Sch Med, Univ Kans, 50-52, asst anat, Univ, 52-57; asst animal path & hyg, Col Agr, Univ Nebr, 58, res assoc animal husb, 59, res assoc animal path & hyg, 59-61; from instr to asst prof anat, Sch Dent, Univ Southern Calif, 61-67; ASST PROF ANAT, STATE UNIV NY DOWNSTATE MED CTR, 67- *Concurrent Pos:* Nat Cancer Inst fel, 59-61. *Mem:* Fel AAAS; NY Acad Sci; Soc Neurosci. *Res:* Comparative neuroanatomy; animal behavior. *Mailing Add:* Dept Anat State Univ NY Downstate Med Ctr Brooklyn NY 11203

JALAL, SYED M, b Ranchi, India, Dec 2, 38; nat US; m 66. CYTOGENETICS. *Educ:* Univ Bihar, BSc, 59; Univ Wis, MS, 62, PhD(cytogenetics), 65; Am Bd Med Genetics, dipl, 85. *Prof Exp:* From asst prof to assoc prof, 64-77, prof biol, Univ N Dak, 77-88; human cytogenetics consult, 80-88; DIR CYTOGENETICS, GENETIC SCREENING & COUN SERV, TEX, 88- *Concurrent Pos:* Vis prof, Univ Tex Cancer Ctr, Houston, 74, 79; consult clin cytogenetics, Dept Pediat & Path, 80- *Mem:* Am Genetic Asn; Am Inst Biol Sci; Environ Mutagen Soc; Am Soc Human Genetics; Sigma Xi. *Res:* Neonatal human cytogenetics; high resolution banded chromosome analysis; environmental mutagenesis, particularly of pesticides. *Mailing Add:* PO Box 2467 Denton TX 76202-8467

JALAN, VINOD MOTILAL, b Laxmangarh, India, May 2, 43; m 66; c 2. ELECTROCHEMISTRY, CATALYSIS. *Educ:* Bombay Univ, India, BChemEng, 67; Univ Fla, ME, 69, PhD(chem eng), 73. *Prof Exp:* Res engr, Power Systs Div, United Technol, 73-78; prin scientist, Stonehart Assocs, 78-79; vpres, Giner Inc, 79-86; PRES, ELECTROCHEM INC, 86- *Concurrent Pos:* Res assoc chem eng, Univ Fla, 67-73. *Honors & Awards:* H B Kapadia Matriculation Award, 60. *Mem:* Am Chem Soc; Am Inst Chem Engrs; Electrochem Soc; Catalysis Soc; Mats Res Soc; Indian Inst Chem Engrs. *Res:* Fuel cells, batteries, electrocatalysts, gas diffusion electrodes, coal gas clean-up, desulfurization, steam reforming, methanol. *Mailing Add:* Electrochem Inc 400 W Cummings Pk Woburn MA 01801

JALBERT, JEFFREY SCOTT, b Bridgeport, Conn, Jan 9, 40; m 66; c 2. NUCLEAR PHYSICS. *Educ:* Fairfield Univ, BS, 61; Va Polytech Inst, PhD(physics), 67. *Prof Exp:* Asst prof physics, Hollins Col, 66-67; assoc prof, 67-75, dir comput ctr, 76-84, PROF PHYSICS, DENISON UNIV, 75-; PRES, JCC CO, 84- *Mem:* Am Phys Soc; Am Math Soc; Sigma Xi. *Res:* Siting of power plants. *Mailing Add:* Dept Physics Denison Univ Granville OH 43023

JALIFE, JOSE, b Mex City, Mex, Mar 7, 47; m 71. CARDIAC ELECTROPHYSIOLOGY, ARRHYTHMIAS. *Educ:* Nat Univ Mex, BA, 65, MD, 72. *Hon Degrees:* Dr, Univ Buenos Aires, Arg, 85. *Prof Exp:* Fel pharmacol, Inst Cardiol, Mex, 68-70, instr, Univ Mex, 72-73; fel pharmacol, Upstate Med Ctr, State Univ NY, Syracuse, 73-75 & cardiac elec, Masonic Med Res Lab, Utica, 75-77; from asst prof to prof, 80-81, PROF & CHMN PHARMACOL, HEALTH SCI CTR, STATE UNIV NY, SYRACUSE, 88- *Concurrent Pos:* Res scientist cardiac elec, Masonic Med Res Lab, Utica, NY, 77-81; estab investr, Am Heart Asn, 82-87; fel, Cardiovasc Sect, Am Physiol Soc, 85; fac exchange scholar, State Univ NY, 87-; pres, res award, Health Sci Ctr, State Univ NY, 90. *Honors & Awards:* Young Investr Award, Am Col Cardiol, 79; Dr Harold Lamport Award, Am Physiol Soc, 80; Develop Achievement Award, Am Heart Assn. *Mem:* Am Physiol Soc; Cardiovasc Sec Am Physiol Soc; NY Acad Sci; AAAS; Electrophysiol Soc; Am Heart Asn; Biophys Soc; hon mem Arg Soc Cardiol. *Res:* Theoretical and experimental work related to three major areas of experimental cardiology; cellular mechanism of cardiac arrhythmias; mechanism of pacemaker synchronization in heart cells; nervous control of heart rate and atrioventricular conduction. *Mailing Add:* Health Sci Ctr State Univ NY 766 Irving Ave Syracuse NY 13210

JALIL, MAZHAR, b India, Nov 2, 38; US citizen; m 70; c 3. ACAROLOGY, BACTERIOLOGY. *Educ:* Univ Agra, BSc, 52, MSc, 54; Univ Nottingham, MSc, 63; Univ Waterloo, PhD(biol), 67; Am Registry Prof Entomologist cert, 71. *Prof Exp:* Farm supt, R A K Agr Inst, Sehore, India, 55-56; teacher & lectr agr, Govt Col, Sehore, 56-60; instr zool, Univ Nottingham, 62-64; instr biol, Univ Waterloo, 64-67; res assoc acarology, Univ Ky, 67-69; ENTOMOLOGIST & MICROBIOLOGIST, OHIO DEPT HEALTH, 69- *Concurrent Pos:* Lord Belper fel, 62-63, teaching fel, 64-67, Ontario Grad fel, 65-67; consult, United Nations Develop Prog, NIH, Govt Pakistan, 80-81; chmn, Sci Adv Comt, City Hall, Columbus, Ohio, 90-; mem, Columbus Comn Ethics & Values, 88-; bd trustees, Islamic Ctr, Columbus, Ohio, 88- *Mem:* Acarological Soc Am; Entom Soc Am; Royal Agr Soc Eng. *Res:* Bionomics and ecology of oribatid mites; genetic control of mites and insects; biology, ecology and reproductive physiology of mosquitoes; diagnosis of streptococcal infection. *Mailing Add:* Ohio Dept Health PO Box 2568 Columbus OH 43216-2568

JALUFKA, NELSON WAYNE, b Austwell, Tex, Dec 2, 32; m 62; c 2. ATOMIC PHYSICS, PLASMA PHYSICS. *Educ:* Lamar Univ, BS, 62; Col William & Mary, MA, 67; Univ Colo, Boulder, PhD(physics), 72. *Prof Exp:* RES SCIENTIST ATOMIC & PLASMA PHYSICS, LANGLEY RES CTR, NASA, 62- *Mem:* Am Phys Soc. *Res:* Nuclear pumped lasers, experimental; solar pumped lasers, experimental; basic atomic processes in plasmas. *Mailing Add:* 505 Bookenbridge Rd Yorktown VA 23692

JALURIA, YOGESH, b Nabha, Punjab, India, Sept 8, 49; m 75; c 3. NATURAL CONNECTION FLOWS & HEAT TRANSFER. *Educ:* Indian Inst Technol, Delhi, BS, 70; Cornell Univ, MS, 72, PhD(mech eng), 74. *Prof Exp:* Asst & fel, Cornell Univ, 70-74; mem res staff thermal eng, Bell Tel Syst, Princeton, NJ, 74-76; asst prof mech eng, Indian Inst Technol, Kanpur, 76-80; from asst prof to assoc prof, 80-85, PROF MECH ENG, RUTGERS UNIV, 85- *Concurrent Pos:* Consult, Steel Authority India, Ltd, 79-80, SRI Int & other co, 82-; prin invest, NSF, 82- & Dept Com, 83- *Honors & Awards:* Young Scientist Medal, Indian Nat Sci Acad, 79. *Mem:* Am Soc Mech Engrs; Combustion Inst; Am Phys Soc. *Res:* Natural convection flows, cooling of electronic equipment, enclosure fires, environmental heat transfer, solar ponds, and numerical simulation of manufacturing processes; computational heat transfer and thermal stratification; heat transfer; fire; computer methods; manufacturing processes; combustion and fire modeling. *Mailing Add:* Mech Eng Dept Rutgers Univ New Brunswick NJ 08903

JAMASBI, ROUDABEH J, US citizen. CLINICAL MICROBIOLOGY, CANCER IMMUNOLOGY. *Educ:* Univ Tehran, BS, 66; Antaeus Res Inst, MT, 69; Univ Ark, MS, 70, PhD(microbiol & immunol), 74. *Prof Exp:* Investr, Oak Ridge Nat Lab; cancer immunologist, Oak Ridge Nat Lab, 78-80; prog dir immunol, Antaeus Res Inst, 80-81; asst prof microbiol & immunol, 81-83, ASSOC PROF CLIN MICROBIOL & TUMOR IMMUNOL, BOWLING GREEN STATE UNIV, 84- *Concurrent Pos:* Assoc mem, Antaeus Res Inst, 81-; vis investr, Oak Ridge Nat Lab, 82- *Mem:* Am Asn Cancer Res; Am Soc Microbiol; Am Acad Microbiol; Am Asn Immunologists; Am Soc Clin Pathologists; Am Asn Blood Bank. *Res:* Immunological characterization of respiratory and digestive tract carcinomas; production of monoclonal antibodies; demonstration of cellular heterogeneity; isolation and characterization of radiation and drug resistance phenotypes. *Mailing Add:* Dept Med Technol & Biol Sci Bowling Green State Univ Bowling Green OH 43403

JAMBOR, PAUL EMIL, b Olomouc, Czechoslovakia, March 29, 37. RINGS & MODULES. *Educ:* Inst Advan Technol, Prague, Dipl Ing, 62; Columbia Univ, MA, 70; Charles Univ, Prague, PhD(math), 73. *Prof Exp:* Assoc prof math, Charles Univ, Prague, 71-76; vis position, Math Inst, Tubingen, 76-77; lectr, Univ Mich, 77-80; assoc prof, 81-87, PROF MATH, UNIV NC, 88- *Concurrent Pos:* Assoc ed, Math Rev, 77-80. *Mem:* Am Math Soc; Math Asn Am. *Res:* Homological properties and structure theory of associative unitary rings; rings with no superdecomposable modules. *Mailing Add:* Univ NC 4402 Jason Ct Wilmington NC 28406

JAMERSON, FRANK EDWARD, b Lowell, Mass, Nov 5, 27; m 50; c 5. PHYSICS. *Educ:* Mass Inst Technol, BS, 48; Univ Notre Dame, PhD(physics), 52. *Prof Exp:* Physicist atomics br, US Naval Res Lab, Washington, DC, 51-52, head neutron physics sect reactors br, 54-57; sr scientist atomic power div, Westinghouse Elec Corp, 53; sr res physicist, Nuclear Power Eng Dept, Gen Motors Corp, 57-61 & Physics Dept, 61-63, supvry res physicist & supv phys electronics group, Physics Dept, 63-69, head, Physics Dept, 69-85, Electrochem Dept, 85-87, CORP ASSOC, GEN MOTORS CORP, 70-, MGR, DIV & STAFF CONTACTS, 87- *Concurrent Pos:* Mem Nat Acad Sci-Nat Bur Standards eval panel, Off Air & Water Measurement, 71-72, chmn, 73-77; mem Nat Acad Sci-Nat Bur Standards eval panel, Inst Mat Res, 74-78 & panel Nat Measurement Lab, 78-80; chmn adv comt corp assoc, Am Inst Physics, 79-81, chmn comt pub policy, 85-88; mem, comt educ, Am Phys Soc, 83-85 & Nat Mat Adv Bd, 89-92; asst prog mgr, US Advan Battery Consortium. *Mem:* AAAS; fel Am Phys Soc; Soc Automotive Engrs; Sigma Xi; AAAS. *Res:* Plasma physics; nuclear reactor physics; energy conversion; research management solid state physics; surface physics; chemical physics; electro optical physics; metal physics; electrochemistry. *Mailing Add:* W3-evp Eng W 30200 Mound Rd Box 9010 Warren MI 48090-9010

JAMES, ALTON EVERETTE, JR, b Oxford, NC, Aug 22, 38; m 60; c 3. RADIOLOGY, NUCLEAR MEDICINE. *Educ:* Univ NC, AB, 59; Duke Univ, MD, 63; Johns Hopkins Univ, MS, 71; Am Bd Radiol, dipl, 69; Am Bd Nuclear Med, cert, 72, Vanderbilt Law Sch, 77-79, Harvard Bus Sch, 79. *Prof Exp:* Intern med, Univ Fla, 63-64; resident radiol, Mass Gen Hosp, 66-68; chief res & fel, Harvard Med Sch, 68-69; from asst prof to assoc prof radiol sci, Med Sch, Johns Hopkins Univ, 69-75, dir res radiol, 73-75; PROF RADIOL & RADIOL SCI & CHMN DEPT, VANDERBILT UNIV, 75-, PROF MED ADMIN & LECTR LEGAL MED, 79-, SR RES ASSOC, INST PUB POLICY, 80-, PROF BIOMED ENG, 81- *Concurrent Pos:* Nat Acad Sci-Nat Res Coun James Picker fel, Sch Hyg & Pub Health, Johns Hopkins Univ, 69-71; consult, Walter Reed Army Hosp, 73-75, Armed Forces Radiobiol Res Inst, 73-, Nat Zool Park, Smithsonian Inst, 73-, Nat Naval Med Ctr, 74-75 & Nuffield Inst Comp Zool, London, 74; hon fel, Royal Soc Med, 74; hon res fel, Univ Col, London, 74. *Honors & Awards:* Gold Medal, Soc Nuclear Med; Silver Medal, Am Roentgen Ray Soc; Bronze Medal, Soc Nuclear Med. *Mem:* AAAS; Am Soc Clin Invest; Radiol Soc NAm: Soc Chmn Acad Radiol Depts; Am Roentgen Ray Soc; Am Inst Ultrasound Med (treas, 78-81); Sigma Xi. *Res:* Cerebrospinal fluid physiology; avian respiration; computerized axial tomography; ultrasonography; medical jurisprudence; paleoradiology; nuclear magnetic resonance; positron emission tomography; evaluation of authenticity of paintings; xerography; medical jurisprudence; author or coauthor of 18 texts over 500 publications. *Mailing Add:* 519 Bellemeade Blvd Nashville TN 37205

JAMES, BELA MICHAEL, b Wichita Falls, Tex, Jan 20, 40; m 68; c 2. BIOLOGICAL OCEANOGRAPHY, FATE & EFFECTS OF OIL SPILLS. *Educ:* Tarleton State Col, BS, 63; Tex A&M Univ, MS, 66, PhD(oceanog), 72. *Prof Exp:* Res asst oceanog, Tex A&M Univ, 68-70, res scientist, 70-73; exec vpres & chief researcher, Tereco Corp, 73-83; SR SCIENTIST & OFF MGR, CONTINENTAL SHELF ASSOCS, 83- *Res:* Marine ecology; taxonomy and ecology of euphausiacean crustaceans and palaeotaxodont mollusks; deep-sea oceanography; water quality and pollution control; oil spill contigency planning; fate and effect of oil. *Mailing Add:* Continental Shelf Assocs 7607 Eastmark Dr Suite 250 College Station TX 77840

JAMES, BRIAN ROBERT, b Birmingham, Eng, Apr 21, 36; m 62; c 4. HOMOGENEOUS CATALYSIS, BIOINORGANIC CHEMISTRY. *Educ:* Oxford Univ, BA, 57, MA, DPhil(chem), 60. *Prof Exp:* Fel inorg reaction mechanisms, Univ BC, 60-62; sr sci officer, UK Atomic Energy Auth, 62-64; from asst prof to assoc prof, 64-74, PROF INORG CHEM, UNIV BC, 74- *Concurrent Pos:* Mem Nat Res Coun chem grants selection comt, 74-77; ed, Catalysis by Metal Complexes, 75- & Can J Chem, 78-88; vis prof, Univ Pisa, 79, Univ Venice, 83, Univ Amsterdam, 90, Australian Nat Univ, 91. *Honors & Awards:* Noranda Award, Chem Inst Can, 75, Can Catal Award, 90; Jacob Biely Award, 86. *Mem:* Fel Chem Soc; fel Chem Inst Can; NY Acad Sci; Am Chem Soc. *Res:* Synthesis, homogeneous catalytic properties of, and mechanistic studies on, coordination compounds, organometallics, and bioinorganic model systems; author of one book and 230 publications in journals. *Mailing Add:* Dept Chem Univ BC Vancouver BC V6T 1Y6 Can

JAMES, CHARLES FRANKLIN, JR, b Des Arc, Mo, July 16, 31; m; c 2. INDUSTRIAL ENGINEERING. *Educ:* Purdue Univ, BSc, 58, MSc, 60, PhD(indust eng), 63. *Prof Exp:* Sr indust engr, McDonnell Aircraft Co, 63; asst prof indust eng, Univ RI, 63-66; assoc prof, Univ Mass, 66-67; prof indust eng & chmn dept, Univ RI, 67-83; DEAN, COL ENG & APPL SCI, UNIV WIS-MILWAUKEE, 84- *Concurrent Pos:* Labor arbitrator, Am Arbit Asn & Fed Mediation & Conciliation Serv; indust consult, US & foreign indust & govt agencies, 65-90; US Dept Transp res grant, 72-82; vis fac mem, Massey Univ, NZ, 79; C Paul Stocker distinguished vis prof eng, Ohio Univ, 82-83; mem, Accreditation Processes Comt, Am Soc Eng Educ, 86 & NSF Panel Eval Grad Fel Applications, Nat Res Coun. *Mem:* Am Inst Indust Engrs; Am Soc Mech Engrs; Soc Mfg Engrs; Am Foundrymen's Soc; Am Soc Eng Educ; Am Arbit Soc; Nat Soc Prof Engrs; Sigma Xi. *Res:* Materials processing; robotics; highway safety. *Mailing Add:* Col Eng & Appl Sci Univ Wis PO Box 784 Milwaukee WI 53201

JAMES, CHARLES WILLIAM, b Dade City, Fla, Aug 13, 29; m 60; c 3. SYSTEMATIC BOTANY. *Educ:* Univ Fla, BS, 50, MS, 52; Duke Univ, PhD(bot), 55. *Prof Exp:* Instr bot, Univ Tenn, 55-56; res botanist herbarium, Harvard Univ, 56-57; from asst prof to assoc prof bot, 57-70, asst to dean, 63-70, PROF BOT & ASSOC DEAN COL ARTS & SCI, UNIV GA, 70- *Mem:* Am Soc Plant Taxon; Int Asn Plant Taxon. *Res:* Taxonomy of seed plants primarily of the southeastern United States. *Mailing Add:* 1175 Whit Davis Rd Athens GA 30605

JAMES, CHRISTOPHER ROBERT, b Vancouver, BC, Nov 15, 35; m 56; c 5. PLASMA PHYSICS, ELECTROMAGNETICS. *Educ:* Univ BC, BASc, 60, MASc, 61, PhD(elec Eng), 64. *Prof Exp:* Nat Res Coun-NATO fel, Oxford Univ, 64-65; from asst prof to assoc prof, 65-71, chmn dept, 74-87, PROF PLASMAS, UNIV ALTA, 71-, VPRES RES, 87- *Concurrent Pos:* Nat Res Coun grant, 65-81; mem, Dept External Affairs Negotiating Team, 78-, bd examiners, Asn Prof Engrs, Geologists & Geophysicists Alta, 74-; dir negotiated develop grant, Nat Res Coun, 71-75. *Mem:* Can Asn Physicists; Eng Inst Can; Eng Inst Can; Am Phys Soc; AAAS. *Res:* Nonlinear laser heating of plasmas; laser-plasma interaction studies. *Mailing Add:* VPres Res Univ Alta 3-5 Univ Hall Edmonton AB T6G 2J9 Can

JAMES, DANIEL SHAW, b Institute, WVa, May 23, 33; m 59; c 3. ORGANIC CHEMISTRY, TECHNICAL MANAGEMENT. *Educ:* Univ Ill, BA; Ill Inst Technol, PhD(org chem), 63. *Prof Exp:* Chemist org chem, Julian Labs, Ill, 58; NIH fel, Mass Inst Technol, 62-63; from res chemist to sr res chemist, E I Du Pont De Nemours & Co, Inc, 63-72, res supvr, 72-75, res assoc, 75-79; owner, Gemini Cosmetics Inc, 79-87, consult Gemini Consult Serv, 88-90; group leader, res & develop, Helene Curtis, Inc, 87-88; DIR LABS, NIACET CORP, 90- *Concurrent Pos:* Pilot, first lieutenant, US Air Force, 55-58. *Mem:* AAAS; Am Chem Soc; Sigma Xi. *Res:* Chemistry of heterocyclic compounds; studies of aromatic substitution reactions; organic syntheses; chemistry involving new and unusual ring closures; dye research and development for textile and paper products; permanent waving research and development; alkylation reaction studies of organic carboxylic acids and their salts. *Mailing Add:* 6873 Plaza Dr Apt C Niagara Falls NY 14304-2918

JAMES, DAVID EUGENE, b Washington, Iowa, June 19, 45; m 77; c 2. ORGANIC CHEMISTRY. *Educ:* Cornell Col, BA, 67; Univ Iowa, PhD(org chem), 75. *Prof Exp:* Instr chem, Linn Mar Community Sch Dist, 67-71; RES ASSOC, AMOCO CHEM CO, 75- *Mem:* Sigma Xi; Am Chem Soc. *Res:* Liquid chromatographic separations of industrially important compounds; photochemistry of aromatic hydrocarbons; homogeneous catalysis using transition metals; oxidation of aromatic hydrocarbons; condensation polymerization. *Mailing Add:* 1133 Woodland Hills Rd Batavia IL 60510

JAMES, DAVID EVAN, b Bellingham, Wash, Dec 14, 39; m 77; c 2. SEISMOLOGY. *Educ:* Stanford Univ, BS, 62, MS, 63, PhD(geophysics), 67. *Prof Exp:* Fel geophys, 66-68, assoc staff mem, 68-70, STAFF MEM GEOPHYS, DEPT TERRESTRIAL MAGNETISM, CARNEGIE INST, 70- *Concurrent Pos:* Ed, US Nat Report to the Int Union Geodesy & Geophysics, 79- 83. *Mem:* Am Geophys Union; Seismol Soc Am; Soc Explor Geophys. *Res:* Seismic studies of continental lithosphere and subduction zones; evolution of central Andean volcanic arc; isotope and trace element geochemistry of igneous rocks of volcanic arcs; precise hypocenter determinations; paleomagnetism. *Mailing Add:* Dept of Terrestrial Magnetism 5241 Broad Branch Rd NW Washington DC 20015

JAMES, DAVID F, b Belleville, Ont, July 9, 39. FLUID MECHANICS, BIOMEDICAL ENGINEERING. *Educ:* Queen's Univ, Ont, BSc, 62; Calif Inst Technol, MS, 63, PhD(mech eng), 67; Univ Cambridge, MA, 74. *Prof Exp:* Asst prof, 67-71, assoc prof, 71-79, PROF MECH ENG, UNIV TORONTO, 79- *Res:* Flow of dilute polymer solutions; rheology of non-Newtonian fluids; fluid mechanics of physiological systems. *Mailing Add:* 52 Beverley St Toronto ON M5T 1X9 Can

JAMES, DAVID WINSTON, b Logan, Utah, Apr 10, 29; m 52; c 6. AGRONOMY. *Educ:* Utah State Univ, BS, 56, MS, 57; Ore State Univ, PhD(soil chem), 62. *Prof Exp:* Instr soil chem, Ore State Univ, 60-62; from asst soil scientist to assoc soil scientist, Wash State Univ, 62-69; assoc prof soils & biometeorol, 69-75, PROF, DEPT SOILS SCI & BIOMETEROL, UTAH STATE UNIV, 75- *Concurrent Pos:* Tech adv, On-Farm Water Management Res in Develop Countries, Latin Am, USAID contract & Utah State Univ, 75-; res dir, Agr Res & Develop Prog for Utah State Univ, Bolivia, 77-80; tech adv & prog leader irrigated agr, Bangladesh, India, Nepal, Peru, Egypt, Dominican Republic, & Ecuador. *Honors & Awards:* Sigma Xi. *Mem:* Am Soc Agron; Soil Sci Soc Am; Soil Conserv Soc Am. *Res:* Chemistry of plant nutrients in soils and the interactions between plant nutrients, soil moisture and other factors of plant growth; modeling of crop yield responses to soil fertility and soil moisture. *Mailing Add:* Dept Plants Soils & Biometeorol Utah State Univ Logan UT 84322-4830

JAMES, DEAN B, b Ames, Iowa, June 14, 34; m 60; c 2. PHYSICAL INORGANIC CHEMISTRY. *Educ:* Iowa State Univ, BS, 56, PhD, 60. *Prof Exp:* Res asst, Ames Lab, Atomic Energy Comn, 52-60; staff mem, Los Alamos Sci Lab, 60-66; group leader rare-earth res, Mich Chem Corp, 66-68; fel scientist res & develop, Nuclear Mat & Equip Corp, Atlantic Richfield Co, Apollo, 68-71; prin engr, Nuclear Energy Group, Gen Elec Co, 72-75, mgr, Safeguards Audits, 75-85, advan tech, 85-88; ULTRAPURE WATER TECH, 88- *Res:* Ion exchange; waste treatment; process development; technical management; nuclear materials safeguards systems. *Mailing Add:* 20518 Deer Park Ct Saratoga CA 95070

JAMES, DONALD GORDON, b Auckland, NZ, Mar 18, 38; m 67; c 1. MATHEMATICS. *Educ:* New Zealand Univ, BSc, 59, MSc, 60; Mass Inst Technol, PhD(math), 63. *Prof Exp:* Lectr math, Univ Auckland, 64-65; from asst prof to assoc prof, 66-76, PROF MATH, PA STATE UNIV, 76- *Concurrent Pos:* Fel Alexander von Humboldt Stiftung, Ger, 69-70. *Mem:* Am Math Soc; London Math Soc. *Res:* Algebra and number theory, particularly quadratic and hermitian forms, orthogonal and unitary groups. *Mailing Add:* Dept Math Pa State Univ University Park PA 16802

JAMES, DOUGLAS GARFIELD LIMBREY, b London, Eng, Oct 31, 24; m 59; c 3. CHEMISTRY. *Educ:* Cambridge Univ, BA, 48, MA & PhD(chem), 55. *Prof Exp:* Lectr chem, Univ St Andrews, 54-59; from asst prof to assoc prof chem, Univ BC, prof, 68-; CONSULT. *Concurrent Pos:* Vis fel chem, Aberdeen Univ, 65-66. *Mem:* Fel Chem Inst Can; fel Royal Soc Chem. *Res:* Chemical kinetics; addition of free radicals to unsaturated molecules. *Mailing Add:* Dept of Chem Univ of BC Vancouver BC V6T 1Z1 Can

JAMES, EDWARD, JR, b El Paso, Tex, July 14, 17; m 40; c 1. PHYSICAL CHEMISTRY, EXPLOSIVES. *Educ:* Univ Mich, BS, 37. *Prof Exp:* Chemist, Sherwin Williams Co, 37-46; chemist, Los Alamos Sci Labs, 46-49, sect leader, 49-60; sect leader, Lawrence Livermore Nat Lab, Univ Calif, 60-63, asst div leader, 63-80; CONSULT, 80- *Concurrent Pos:* Mem, Sci Adv Bd, USAF, 85. *Mem:* Am Chem Soc; AAAS. *Res:* Resin bonded pigments for textiles; emulsion paints; polyester resins; plastic bonded explosives; explosives, polymer synthesis and manufacture; detonation hydrodynamics. *Mailing Add:* 1085 Peary Ct Livermore CA 94550

JAMES, FLOYD LAMB, organic chemistry, for more information see previous edition

JAMES, FRANCES CREWS, b Philadelphia, Pa, Sept 29, 30; c 3. ECOLOGY. *Educ:* Mt Holyoke Col, AB, 52; La State Univ, MS, 56; Univ Ark, PhD(zool), 70. *Prof Exp:* Instr zool & bot, Univ Ark, 60-70, res assoc, Mus, 71-73; asst prog dir, Ecol Prog, NSF, 73-76, assoc prog dir, 76-77; assoc prof, 77-84, PROF, DEPT BIOL SCI, FLA STATE UNIV, TALLAHASSEE, 84- *Concurrent Pos:* Res assoc, Smithsonian Inst, 75- *Honors & Awards:* E P Edwards Prize, Wilson Ornith Soc. *Mem:* Ecol Soc Am; Am Ornithologists Union (pres, 84-86); Soc Syst Zool; AAAS; Cooper Ornith Soc. *Res:* Geographic variation in vertebrates; analysis of avian communities; habitat selection in birds; thermal behavioral ecology of lizards; avian systematics; allometry. *Mailing Add:* Dept of Biol Sci Fla State Univ Tallahassee FL 32306-2043

JAMES, FRANKLIN WARD, b Montrose, Miss, Sept 2, 22; m 58; c 2. ANALYTICAL CHEMISTRY. *Educ:* Miss Col, BS, 47; Univ NC, PhD(chem), 52. *Prof Exp:* From assoc prof to prof chem, Millsaps Col, 51-58; sr chemist, Res & Tech Dept, Texaco, Inc, 58-61; prof chem, Mercer Univ, 61-70, chmn dept, 61-80, Fuller E Callaway prof, 70-88, EMER CALLAWAY PROF CHEM, MERCER UNIV, 88- *Mem:* Am Chem Soc. *Res:* Standard electrode potentials of electrodes in aqueous glycerol solutions. *Mailing Add:* Dept Chem Mercer Univ Macon GA 31207

JAMES, GARTH A, b Malad City, Idaho, Aug 1, 26; m 47; c 7. ENDODONTICS. *Educ:* Utah State Univ, BS, 48, MS, 51; Univ Nebr, DDS, 60. *Prof Exp:* Teacher pub sch, Idaho, 47-49; res technician, Naval Biol Lab, Univ Calif, 52; instr bact & pub health, Utah State Agr Col & bacteriologist, Exp Sta, 52-56; res assoc bact, Col Dent, Univ Nebr-Lincoln, 56-60, from assoc prof to prof endodontics, 60-88, chmn dept, 70-88; RETIRED. *Concurrent Pos:* Dir bact, St Elizabeth Hosp, 56-60. *Mem:* Am Dent Asn; Am Asn Endodont; Am Soc Microbiol; fel Am Col Dentists; fel Int Col Dentists. *Mailing Add:* Dept Endodont Col Dent Univ Nebr Lincoln NE 68508

JAMES, GEORGE ELLERT, b Douglas, Alaska, Apr 26, 17; m 53. ELECTRONICS ENGINEERING. *Educ:* Univ Wash, BS, 40; George Washington Univ, MS, 62, DSc(eng sci), 69. *Prof Exp:* Electronic develop engr, Gen Elec Co, 40-45; chief engr, Gen Commun Co, 46-47; asst proj engr, Hughes Aircraft Co, 47-48; chief engr, Lab for Electronics, Inc, 48-56; dir, Boston Div, Ramo-Wooldridge Corp, 56-57; tech staff mem, Inst Defense Anal, 58-70; vpres, Adcole Corp, 70-71; consult scientist, Missile Systs Div, Bedford Lab, Raytheon Co, 71-72; vpres, Adcole Corp, 72-90; RETIRED. *Mem:* Sr mem Inst Elec & Electronics Engrs. *Res:* Electromagnetic field theory; electronic circuit design; radar and control systems; applied mathematics; operations analysis; computer software development. *Mailing Add:* 14 Temple St Apt 3-B Framingham MA 01701

JAMES, GEORGE WATSON, III, b Richmond, Va, July 3, 18; m 43; c 3. MEDICINE. *Educ:* Washington & Lee Univ, AB, 40; Med Col Va, MD, 43. *Prof Exp:* USPHS fel, 48-49; from asst prof to assoc prof, 49-65, chmn, Div Hemat, 57-83, PROF MED, MED COL VA 65- *Concurrent Pos:* Markle scholar, Med Col Va, 49-54; consult, McGurie Vet Admin Hosp, 48-, Keecoughtan Vet Hosp 52-80. *Mem:* AAAS; Am Soc Clin Invest; Am Fedn Clin Res; Am Clin & Climat Asn; Am Soc Clin Nutrit. *Res:* Clinical investigations; bile pigment metabolism; red cell survival with N-15 label; leukemia and lymphoma chemotherapy; hematology. *Mailing Add:* Dept Med Med Col Va PO Box 113 MCV Sta Richmond VA 23298

JAMES, GIDEON T, b Kansas City, Mo, July 10, 27; m 50; c 3. VERTEBRATE PALEONTOLOGY. *Educ:* Univ Houston, BS, 53, MS, 56; Univ Calif, Berkeley, PhD(paleont), 61. *Prof Exp:* Instr paleont, Univ Calif, Berkeley, 61-62, res paleontologist, 62-63; asst prof geol, Univ Ariz, 63-64, res assoc physics, 64-65; asst prof earth sci, 65-71, PROF EARTH SCI, ETEX STATE UNIV, 71-, VPRES PLANNING & INSTNL ADVAN, 74- *Concurrent Pos:* Dir, NSF Earth Sci Inst Secondary Sch Sci Teachers, Univ Calif, Berkeley, 62-65; adj prof, Southern Methodist Univ, 71- *Mem:* Soc Vert Paleont; Soc Econ Paleont & Mineral; Paleont Soc; NY Acad Sci; Int Paleont Union. *Res:* Vertebrate paleontology and biostratigraphy of Tertiary deposits of North America; geochemical dating of Tertiary rocks and Tertiary geochronology; histologic and ultrafine anatomy of fossil vertebrate tissues. *Mailing Add:* Dept Archit Univ Tex Arlington Arlington TX 76019

JAMES, GORDON THOMAS, b Ft Scott, Kans, Mar 7, 40. PROTEIN CHEMISTRY, CELL CULTURE. *Educ:* Univ Calif, Riverside, PhD(biochem), 71. *Prof Exp:* Asst prof biochem, Dept Surg, Health Sci Ctr, Univ Colo, 76-86; sr biochemist, Electropore Co, Boulder, Co, 86-89; ANALYTICAL CHEMIST, NAT JEWISH HOSP, DENVER, 89- *Mem:* Am Soc Biochem & Molecular Biol. *Res:* Protein chemistry; pharmacokinetics of tuberculosis drugs. *Mailing Add:* Nat Jewish Hosp 1400 Jackson St Rm K-427 Denver CO 80206

JAMES, HAROLD LEE, b Taylorsville, NC, Oct 31, 39; m 65; c 1. MOLECULAR BIOLOGY & BIOCHEMISTRY, GENERAL PHYSIOLOGY. *Educ:* ETenn State Univ, BS, 62; Univ Tenn, Memphis, PhD(biochem), 68. *Prof Exp:* Res technician, Med Units, Univ Tenn, 62-63, res instr biochem, 68; res instr med & biochem, Sch Med, Temple Univ, 68-70; res scientist, Blood Res Lab, Am Nat Red Cross, 70-72; asst prof biochem, Univ Tenn Ctr for Health Sci, 72-75; res asst prof med, Sch Med, Pulmonary Div, Temple Univ, 76-80, res assoc prof med, Sch Med, Pulmonary Div, 80-83; ASSOC PROF BIOCHEM, HEALTH CTR, UNIV TEX-TYLER, 83- *Concurrent Pos:* Res assoc, Lab Hemat, St Jude Children's Res Hosp, 72-75. *Mem:* Sigma Xi; Am Physiol Soc; Int Soc Thrombosis & Haemostasis; Am Heart Asn. *Res:* Biochemistry and physiology of plasma and platelet fibrinogens; mechanism of interaction of alpha-1-antitrypsin with elastase; lung physiology of alpha-1-antitrypsin; animal models of emphysema; molecular biology of genetic variants of factors VIII and X; structure-function conclates of factors VIII and X; molecular. *Mailing Add:* Health Ctr Univ Tex-Tyler PO Box 2003 Tyler TX 75710

JAMES, HAROLD LLOYD, b Nanaimo, BC, June 11, 12; US citizen; m 36; c 4. MINERALOGY-PETROLOGY. *Educ:* State Col Wash, BS, 38; Princeton Univ, PhD(geol), 45. *Prof Exp:* Field asst, 38-40, from geologist to chief geologist, 40-71, RES GEOLOGIST, US GEOL SURV, 71- *Concurrent Pos:* Instr, Princeton Univ, 42; vis lectr, Northwestern Univ, Ill, 53-54; prof, Univ Minn, 61-65. *Honors & Awards:* Distinguished Serv Award, US Dept Int, 66; Penrose Medal, Soc Econ Geologists, 76. *Mem:* Nat Acad Sci; fel Geol Soc Am; Soc Econ Geol; Geochem Soc; Mineral Soc Am; Geol Asn Can. *Res:* Iron formations and iron ores; Precambrian history and time classification; Precambrian geology of southwestern Montana; geology of the Lake Superior region. *Mailing Add:* 1320 Lakeway Dr No 121 Bellingham WA 98226-2005

JAMES, HELEN JANE, b Nebraska City, Nebr, June 15, 43. ANALYTICAL CHEMISTRY. *Educ:* Univ Nebr, BS, 65, PhD(anal chem), 70. *Prof Exp:* Fel, Univ Ariz, 70-71; asst prof, 71-75, assoc prof chem, 75-80, PROF CHEM, WEBER STATE COL, 80- *Mem:* Sigma Xi; Am Chem Soc. *Res:* Development and application of ion selective electrodes; the use of coated wire electrodes containing liquid membranes. *Mailing Add:* Dept of Chem Weber State Col Ogden UT 84408

JAMES, HERBERT I, b St Thomas, VI, Mar 30, 33; US citizen; m 62; c 2. PHYSICAL CHEMISTRY. *Educ:* Hampton Inst, BS, 55; Clark Univ, MA, 58, PhD(chem), 65. *Prof Exp:* Teacher, Elec Storage Battery Co, 65-76; scientist, US, 76-84, MGR PERSONNEL, RES CTR, XEROX, ONT, 84- *Concurrent Pos:* Mgr affirmative action, Webster Res Ctr, Xerox. *Honors & Awards:* Commendation Award, President of US. *Mem:* Electrochem Soc. *Res:* Diffusion and sedimentation studies of macromolecules; nuclear and radiochemistry; electrochemistry. *Mailing Add:* 49 Cumberland Dr Mississauga ON L5G 3N1 Can

JAMES, HUGO A, b Bridgeport, Conn, May 24, 30. PARASITOLOGY, HELMINTHOLOGY. *Educ:* Univ Bridgeport, BA, 57, MS, 58; Univ Va, MA, 61; Iowa State Univ, PhD(parasitol), 68. *Prof Exp:* From instr to assoc prof, Univ Bridgeport, 58-73, prof biol, 73-85, chmn biol & dir, Div Biol & Health Technologies, 85-90. *Concurrent Pos:* NSF res grant, Univ Va, 69. *Mem:* Am Soc Parasitol; Am Micros Soc (treas, 79-81). *Res:* Host-parasite interrelationships of helminths, specifically the Cestoda; zoonotic associations, particularly aspects of taxonomy, morphology, pathology and evolution. *Mailing Add:* Seven Franklin St Trumbull CT 06611

JAMES, JACK N, b Dallas, Tex, Nov 22, 20; m 44; c 4. ELECTRICAL ENGINEERING. *Educ:* Southern Methodist Univ, BS, 42; Union Col, MS, 49. *Prof Exp:* Test engr, Gen Elec Co, 42-43 & 46-49; res engr, Radio Corp Am, 49-50; res engr, Jet Propulsion Lab, 50- 54, eng group supvr, 54-56, sect mgr, 56-58, div mgr, 58-60, dep prog mgr, 60-61, proj mgr, 61-65, dep asst lab dir, lunar & planetary projs, 65-67, asst lab dir, Tech Divs, 67-76, asst lab dir, Tech & Space Propulsion Develop, Jet Propulsion Lab, 76-; AT DEPT ELEC ENG, CALIF INST TECHNOL. *Honors & Awards:* Pub Serv Award, NASA, 63; Hill Award, Am Inst Aeronaut & Astronaut, 63; Except Sci Achievement Medal, 65; Stuart Ballantine Medal, Franklin Inst, 67. *Mem:* Am Inst Aeronaut & Astronaut; Inst Elec & Electronics Engrs. *Res:* Management of Mariner II to Venus and Mariner IV to Mars projects; guidance systems for Corporal and Sergeant missiles. *Mailing Add:* 1345 El Vago LaCanada Flintridge CA 91011

JAMES, JEFFREY, b Savannah, Ga, Aug 27, 44. ANALYTICAL CHEMISTRY, INORGANIC CHEMISTRY. *Educ:* Savannah State Col, BS, 66; Tuskegee Inst, MS, 70; Howard Univ, PhD(inorg chem), 73. *Prof Exp:* From asst prof to assoc prof, 72-84, PROF, SAVANNAH STATE COL, 84- *Concurrent Pos:* Res chemist, Agronne Nat Lab, 65, 69 & 83, Eli Lily & Co, 69, Savannah River Lab, 75 & 80, Lawrence Livermore Lab, 78. *Mem:* Am Chem Soc; AAAS. *Res:* Kinetic study of metalloporphyrins and oxidation of dithionite by manganese; hematoporphyrins in basic solution; characterization of mercury; electrodes; solubility products and thermodynamic functions for the Lanthanon fluride-water system. *Mailing Add:* 4625 Oakview Dr Savannah GA 31405

JAMES, JESSE, b Haynesville, La, Jan 26, 37; m 59; c 5. BIOCHEMISTRY. *Educ:* Tex Southern Univ, BSc, 61, MSc, 62; Univ Tex, PhD(biochem), 65. *Prof Exp:* Assoc prof, 65-73, chmn dept, 73-76, PROF CHEM, KNOXVILLE COL, 73- *Concurrent Pos:* Consult, Union Carbide Corp, Tenn, 66-70; res chemist, Nat Bur Standards, 70- *Res:* Kinetics and mechanisms of enzyme-catalyzed reactions; standardization of reference materials for clinical chemistry. *Mailing Add:* Dept Chem Knoxville Col 901 Col St Knoxville TN 37921

JAMES, JOHN CARY, b Ceredo, WVa, May 8, 26; m 58; c 1. ORGANIC CHEMISTRY. *Educ:* WVa Wesleyan Col, BS, 49; Univ Del, PhD(org chem), 60. *Prof Exp:* Teacher, Callao High Sch, Peru, 50-53; sr res chemist, Boston Labs, Monsanto Res Corp, 59-66; sr chemist, Northrop Carolina, Inc, 66-67; exec secy med chem fel rev comt, Div Res Grants, NIH, 67-70, chief sci eval sect, Res Anal & Eval Br, 70-71; chief res anal & eval br, 71-84, ASST DIR SPEC PROJS, DIV RES GRANTS, NIH, 84- *Mem:* AAAS. *Res:* Synthesis of anti-oxidants; research on jet fuels and jet engine lubricants; antiradiation drug research; high temperature explosives; health sciences administration; research analysis and information science; electronic communications. *Mailing Add:* 4874 Chevy Chase Dr Chevy Chase MD 20815

JAMES, KAREN K(ANKE), b Vinton, Iowa, June 2, 44; m 64; c 3. IMMUNOLOGY. *Educ:* Ohio State Univ, BS, 67, MS, 72; Rush Med Univ, PhD(immunol), 80. *Prof Exp:* Med technologist, Riverside Methodist Hosp, 67-71; supv clin immunol, Ohio State Univ Hosp, 73-76; instr, Rush Presby St Luke's Med Ctr, 76-80, dir immunol, 80-82, asst prof immunol, Rush Univ, 80-87; clin asst prof med lab sci, Univ Ill, Chicago, 77-87; ASSOC DIR LABS, CENT DUPAGE HOSP, 82-; ASSOC PROF PATH, LOYOLA UNIV MED CTR, CHICAGO, 86- *Concurrent Pos:* Clin instr allied med professions, Ohio State Univ, 73-77; chmn, immunol comt, Bd Registry, 77-85; consult, Smith Kline Biosci Labs, 84-86; acting ed, Lab Med, 85-86; chmn, lab mgt comt, Bd Registry, 86- *Mem:* Am Soc Clin Path; Am Asn Immunol. *Res:* Biologic response modifying properties of C-reactive protein; natural killer cells; cellular immunology. *Mailing Add:* Cent DuPage Hosp 25 N Winfield Rd Winfield IL 60190

JAMES, KENNETH EUGENE, b Los Angeles, Calif, Sept 14, 42; m; c 2. BIOMETRICS, BIOSTATISTICS. *Educ:* Walla Walla Col, Col Place, Wash, BS, 65; Univ Minn, MS, 67, PhD(biomet), 69. *Prof Exp:* Biometrician, Alaska Dept Fish & Game, 69-71; staff biostatistician, Vet Admin Coop Studies Analgesic Study, Palo Alto, Calif, 71-74; chief, Hines, Ill, 74-78, CHIEF, VET ADMIN COOP STUDIES PROG COORD CTR, PALO ALTO, CALIF, 78- *Concurrent Pos:* Lectr, Dept Preventive Med, Stanford Univ, 71-75, res assoc, Dept Anesthesia, Med Ctr, 71-74; asst prof, Dept Pharmacol, Stritch Sch Med, Loyola Univ, Chicago, 77-79; consult, Task Force Drug Develop, Muscular Dystrophy Asn, 77-, Chief Western Res & Develop Off, Vet Admin Med Ctr, Livermore, Calif, 78-81; consult prof, Dept Health Res & Policy, Stanford Univ, Palo Alto, 90- *Mem:* Am Statist Asn; Biomet Soc; Int Soc Clin Biostatist; Soc Controlled Clin Trials. *Res:* Administration and conduct of multi-center trials, including the design, conduct and analyses of such trials. *Mailing Add:* Coop Studies Prog Coord Ctr, 151-K Vet Admin Med Ctr 3801 Miranda Ave Palo Alto CA 94304

JAMES, L(AURENCE) ALLAN, b Glendale, Calif, Mar 18, 49. FLUVIAL GEOMORPHOLOGY, MODELING WATER & SEDIMENT YIELDS IN WATERSHEDS. *Educ:* Univ Calif, Berkeley, BA, 78; Univ Wis-Madison, MS, 81, MS, 83, PhD(geog & geol), 88. *Prof Exp:* Lectr phys geog, Geog Dept, Univ Wis-Madison, 86, Univ Oregon, 87, Univ Ga, 87-88; ASST PROF PHYS GEOG, GEOG DEPT, UNIV SC, 88- *Concurrent Pos:* Prin investr, res & prod scholarship, Univ SC, 89, NSF, 89-90 & SE Regional Climate Ctr, 90-91; travel grant, Nat Res Coun/Am Geophys Union, 91. *Mem:* Asn Am Geogr; Geol Soc Am; Am Geophys Union; Am Soc Photogram & Remote Sensing; Am Quaternary Asn; Int Asn Geomorphologists. *Res:* Water and sediment yields from fluvial systems; historical anthropogenic sedimentation; modeling watershed processes with digital geographic information techniques; updating the extent and behavior of the voluminous hydraulic gold mining sediment in the northern Sierra Nevada of California. *Mailing Add:* Geog Dept Univ SC Columbia SC 29208

JAMES, LARRY GEORGE, b Bellingham, Wash, May 1, 47; m 68; c 4. AGRICULTURAL ENGINEERING, IRRIGATION ENGINEERING. *Educ:* Wash State Univ, BS, 70; Univ Minn, PhD(agr eng), 75. *Prof Exp:* Asst prof agr eng, Cornell Univ, 75-77; from asst prof to assoc prof, 77-88, PROF & CHAIR, AGR ENG, WASH STATE UNIV, 88- *Mem:* Am Soc Agr Engrs; Am Soc Engr Educ; Am Soc Civil Eng. *Res:* Plant water requirements; energy requirements for irrigation; sprinkler irrigation; infiltration; water resources management. *Mailing Add:* Dept of Agr Eng Wash State Univ Pullman WA 99164-6120

JAMES, LAURENCE BERESFORD, b Hollywood, Calif, Aug 20, 16; m 39; c 4. GEOLOGY. *Educ:* Stanford Univ, AB, 40. *Prof Exp:* Mining geologist, Consol Coppermines Corp, Nev, 40-41; mining engr, Anaconda Copper Co, Mont, 41; eng geologist, Calif State Dept Water Resources, 46-56, chief eng geologist, 56-76; CONSULT GEOLOGIST, 76- *Concurrent Pos:* Mem, nat comt rock mech, Nat Acad Sci, 67; consult geologist, United Nations, 69-; mem, US Comt Large Dams; consult, US Army Corps Engrs, US Bur Reclamation & World Bank. *Mem:* Fel Geol Soc Am; Seismol Soc Am; fel Am Soc Civil Eng; Asn Eng Geol; Soc Am Mil Engrs. *Res:* Engineering and groundwater geology. *Mailing Add:* 120 Grey Canyon Dr Folsom CA 95630

JAMES, LAYLIN KNOX, JR, b Pittsburgh, Pa, Sept 17, 27; m 52; c 4. SURFACE CHEMISTRY. *Educ:* Univ Mich, BS, 50, MS, 52; Univ Ill, PhD(chem), 58. *Prof Exp:* Chemist, Shell Chem Co, div Shell Oil Co, 52-54; asst, Univ Ill, 54-56 & Wash State Univ, 57-58; res chemist, Procter & Gamble Co, 58-59; from asst prof to assoc prof, Lafayette Col, 59-77, actg dept head, 70-71, dept head, 79-85, prof chem, 77-90, EMER PROF CHEM, LAFAYETTE COL, 90- *Concurrent Pos:* Chemist, US Naval Res Lab, 63; guest prof, Hohenheim Univ, Stuttgart, WGer, 82. *Mem:* AAAS; Am Chem Soc. *Res:* Surface chemistry of proteins and lipoproteins. *Mailing Add:* Dept of Chem Lafayette Col Easton PA 18042

JAMES, LEE MORTON, b New York, NY, Dec 14, 16; m 46. FORESTRY. *Educ:* Pa State Col, BS, 37; Univ Mich, MF, 43, PhD(forest econ), 45. *Prof Exp:* Instr forestry, Pa State Col, 37-38; unit supvr, New Eng Forest Emergency Proj, US Forest Serv, 38-40, forester, Appalachian Forest Exp Sta, 40-41 & 43-46, forest economist in charge unit resource anal, Div Forest Econ, Southern Forest Exp Sta, 46-51; assoc prof, 51-58, chmn dept, 66-78,

PROF FORESTRY, MICH STATE UNIV, 58- Concurrent Pos: Consult, Resources for the Future, Inc, Forest Indust Coun, US Dept Interior, US Dept Com, Pub Land Law Rev Comt & President's Coun Environ Qual. Mem: Soc Am Foresters. Res: Forest resource and forest industry analysis; timber products marketing; forest policy. Mailing Add: Dept Forestry Mich State Univ East Lansing MI 48824

JAMES, MARLYNN REES, b Spanish Fork, Utah, Nov 20, 33; m 61; c 6. PHYSICAL CHEMISTRY. Educ: Brigham Young Univ, BS, 58, MS, 61; Univ Utah, PhD(theoret gas chromatography), 65. Prof Exp: Res asst chem, Purdue Univ, 64-66; PROF CHEM, UNIV NORTHERN COLO, 66- Mem: Am Chem Soc; Nat Sci Supvrs Asn; Nat Sci Teachers Asn. Res: Chemical education and curriculum development involving computers. Mailing Add: Dept Chem Univ Northern Colo Greeley CO 80639

JAMES, MARY FRANCES, b Clarksburg, WVa, Apr 21, 13. MEDICAL EDUCATION, LABORATORY MEDICINE. Educ: Randolph-Macon Woman's Col, AB, 35; Duke Univ Hosp, Cert Med Technol, 37; Univ Ala, MS, 61. Prof Exp: Res technician cellular physiol, Univ Hosp, Duke Univ, 36-37, head technician out-patient lab, Sch Med, 37-45; instr bact & clin path, Med Col Ala & Jefferson-Hillman Hosp, 45-54; teaching supvr, Div Med Tech, Med Ctr, Univ Miss, 55-63, instr, Sch Med, 55-59, asst prof, Clin Lab Sci, 59-63; teaching supvr, Med Ctr, Univ Hosp Sch Med Technol & asst prof path, Col Med, Dept Med Technol, 63-66, from assoc prof to prof, 68-78, chmn dept, 66-78, EMER PROF MED TECHNOL, COL ALLIED HEALTH PROFESSIONS, UNIV KY, 78- Concurrent Pos: Mem, Bd Sch Med Technol (ASCP), 53-59; mem, Coun Med Technol Educ, 64-69. Mem: Am Soc Med Technol; Am Soc Allied Health Professions; Sigma Xi. Res: Immunology; responses to penicillin; medical technology. Mailing Add: 3051 Rio Dosa Dr No 302 Lexington KY 40509-1548

JAMES, MERLIN LEHN, b Lincoln, Nebr, Sept 27, 22; m 44; c 3. ENGINEERING MECHANICS. Educ: Univ Nebr, BSME, 49, MSEM, 60. Prof Exp: Sales engr, Phillips Petrol Co, 49-51; salesman, Harrington Co, 51-57; from instr to assoc prof eng mech, 57-76, MEM FAC ENG MECH, UNIV NEBR, LINCOLN, 76- Concurrent Pos: NSF grant analog comput, 69. Mem: Sigma Xi. Res: Analog computation; damping characteristics of structural members. Mailing Add: 4620 Mohawk Lincoln NE 68510

JAMES, MICHAEL ROYSTON, b London, Eng, Sept 11, 50; US citizen; m 72; c 2. MATERIALS SCIENCE. Educ: Tulane Univ, BS, 72; Northwestern Univ, PhD(mat sci), 77. Prof Exp: Fel, Lab Metal Physics, State Univ Groningen, Neth, 77-78; consult, Am Anal Corp, 78; MEM TECH STAFF, ROCKWELL INT SCI CTR, ROCKWELL INT CORP, 78- Mem: Soc Exp Mech; Metall Soc; Am Soc Metals. Res: Nondestructive testing and component life prediction especially with residual stress measurement and its influence on metal fatigue; microstructural phenomena influencing microcrack initiation. Mailing Add: 419 Thunderhead St Thousand Oaks CA 91360

JAMES, ODETTE BRICMONT, b San Jose, Calif, Feb 7, 42; m 80; c 1. LUNAR PETROLOGY, IGNEOUS PETROLOGY. Educ: Stanford Univ, BS, 63, PhD(geol), 67. Prof Exp: GEOLOGIST, US GEOL SURV, 67- Concurrent Pos: Mem, Lunar Sample Anal Planning Team, NASA, 72-74, 80-82, chmn, 81-82, prin investr, 75-; mem, Lunar Planet Geosci Review Panel, 85-87. Mem: Mineral Soc Am (treas, 81-84); Geol Soc Am; Am Geophys Union; Meteoritical Soc. Res: Petrology of lunar highland rocks; lunar highland breccias; igneous petrology; shock metamorphism. Mailing Add: US Geol Surv 959 Nat Ctr Reston VA 22092

JAMES, PHILIP BENJAMIN, b Kansas City, Mo, Mar 18, 40; m 65; c 3. PLANETARY ATMOSPHERES. Educ: Carnegie-Mellon Univ, BS, 61; Univ Wis, Madison, MS, 63, PhD(physics), 66. Prof Exp: Off Naval Res res assoc physics, Univ Ill, Urbana, 66-68; from asst prof to prof physics, Univ Mo, St Louis, 68-90; CHMN DEPT PHYSICS & ASTRON, UNIV TOLEDO, 90- Concurrent Pos: Nat Res Coun assoc, Jet Propulsion Lab, Calif Inst Technol, 77-78; mem, Viking Mars Proj, 77-78; ed, Am J Physics, 82-83; adj scientist, Lowell Observ, 84-; prin investr, Hubble Space Telescope Observing Prog. Mem: Am Geophys Union; Am Astron Soc; fel Am Phys Soc. Res: Studies relevant to meteorology of and condensate cycles on Mars, includes analyses of spacecraft date, astronomical observations, and modeling; chemical physics. Mailing Add: Dept Physics & Astron Univ of Toledo Toledo OH 43606

JAMES, PHILIP NICKERSON, b Boston, Mass, Aug 16, 32; m 54; c 2. DATA PROCESSING. Educ: Mass Inst Technol, SB, 54; Univ Ill, PhD(org chem), 57. Prof Exp: Instr chem, Univ Calif, Berkeley, 57-58, asst prof, 58-59; res chemist, Lederle Labs, Am Cyanamid Co, 59-60, col rels rep, 60-62; photog chemist, Systs Res Div, Technicolor Corp, 62-63, proj leader, 63, sr res chemist, 63-64, staff asst to dir, 64-66; asst vchancellor grad studies & res, Univ Calif, San Diego, 66-69, exec asst to chancellor, 69-74; dir, Univ Southern Calif, Idyllwild Campus, 74-77; dir admin & planning, Deluxe Gen Inc, 20th Century-Fox, 77-78; dir mgt systs, Teledyne Systs Co, 78-79; sr res engr, Electronics Div, 79-80, DIR STRATEGY PLANNING DATA PROCESSING, NORTHROP CORP, 80- Mem: Am Chem Soc; Fedn Am Scientists; AAAS; Sigma Xi; Asn Comput Mach. Res: Biologically interesting compounds; structure; synthesis; chemical mechanisms; photographic chemistry; computer assisted solutions to synthetic problems; higher education. Mailing Add: 11400 Edenberg Ave Northridge CA 91326

JAMES, RALPH BOYD, b Nashville, Tenn, Nov 1, 53; c 1. CONDENSED MATTER PHYSICS. Educ: Univ Tenn, BS, 76; Ga Inst Technol, MS, 77, Calif Inst Technol, MS, 78, PhD(appl physics), 80. Prof Exp: Res fel, Calif Inst Technol, 80-81; Eugene P Wigner fel, Oak Ridge Nat Lab, 81-84; SR MEM TECH STAFF, SANDIA NAT LABS, 84- Mem: Am Phys Soc; Inst Elec Electronics Engrs; Sigma Xi; Mat Res Soc; Am Vacuum Soc; Soc Photo-Optical Instrumentation Engrs. Res: Semiconductor physics and non-linear optics. Mailing Add: 5420 Lenore Ave Livermore CA 94550

JAMES, RALPH L, b Portland, Ore, Apr 12, 41; m 69; c 2. MATHEMATICS, NUMERICAL ANALYSIS. Educ: Univ Wash, BS, 63; Ore State Univ, MS, 65, PhD, 70. Prof Exp: Vis asst prof math, Col of Idaho, 65-68; asst prof, 70-74, ASSOC PROF MATH, CALIF STATE COL, STANISLAUS, 74- Mem: Am Math Soc. Res: Functional analysis; ordered topological vector spaces; positive operators; approximation theory. Mailing Add: Dept of Math Calif State Col Stanislaus Turlock CA 95380

JAMES, RICHARD STEPHEN, b Hamilton, Ont, Feb 20, 40; m 64; c 2. GEOCHEMISTRY, PETROLOGY. Educ: McMaster Univ, BSc, 62, MSc, 64; Victoria Univ Manchester, PhD(geol), 67. Prof Exp: Fel, Univ Toronto, 67-69, lectr geol, 69-70; asst prof geol, 70-80, ASSOC PROF GEOL, LAURENTIAN UNIV, 80- Mem: Mineral Soc Am; Mineral Asn Can; Mineral Soc Gt Brit & Ireland. Res: Igneous and metamorphic petrology, application of experimental phase equilibria data to natural systems. Mailing Add: Dept Geol Laurentian Univ Ramsey Lake Rd Sudbury ON P3E 2C6 Can

JAMES, ROBERT CLARKE, b Bloomington, Ind, July 30, 18; m 45; c 4. MATHEMATICAL ANALYSIS. Educ: Univ Calif, Los Angeles, BA, 40; Calif Inst Technol, PhD(math), 47. Hon Degrees: DSc, Kent State Univ, 87. Prof Exp: Benjamin Pierce instr math, Harvard Univ, 46-47; from instr to asst prof, Univ Calif, 47-51; assoc prof, Haverford Col, 51-57; prof & chmn dept, Harvey Mudd Col, 57-67; prof, State Univ NY Albany, 67-68; prof, 68-81, EMER PROF MATH, CLAREMONT GRAD SCH, 81- Concurrent Pos: Mem, Inst Advan Study, Princeton Univ, 62-63, Jerusalem, 76-77 & Mittag-Leffler Inst, Sweden, 78-79. Mem: Am Math Soc; Soc Indust & Appl Math; Math Asn Am; AAAS; Fedn Am Scientists. Res: Normed vector spaces. Mailing Add: 14385 Clear Creek Pl Grass Valley CA 95949-8765

JAMES, RONALD VALDEMAR, b Oakland, Calif, Apr 27, 43; m 64; c 2. SOIL CHEMISTRY. Educ: Univ Calif, Davis, BS, 64; Univ Colo, Boulder, MS, 67, PhD(inorg chem), 69. Prof Exp: RES CHEMIST, US GEOL SURV, 68- Mem: Am Chem Soc; AAAS. Res: Chemistry and transport of solutes in the unsaturated zone and ground water; mathematical modeling; fate of pollutants in environmental waters; kinetics and mechanisms of inorganic reactions; ion exchange. Mailing Add: 595 Morey Dr Menlo Park CA 94025

JAMES, SHERMAN ATHONIA, b Hartsville, SC, Oct 25, 43; m 65, 90; c 2. BEHAVIORAL STRESS, PSYCHOPHYSIOLOGY. Educ: Talladega Col, AB, 64; Wash Univ, St Louis, PhD(psychol), 73. Prof Exp: From asst prof to prof epidemiol, Univ NC, Chapel Hill, 73-89; PROF EPIDEMIOL, UNIV MICH, ANN ARBOR, 89- Concurrent Pos: Consult, NIMH, 79-83 & NIH, 85-; vis prof, dept prev med, Fed Univ Bahia, Salvador, Brazil, 86-; fel coun epidemiol, Am Heart Asn. Mem: Fel Acad Behav Med Res; fel Am Epidemiol Soc; Am Heart Asn; Soc Behav Med; Am Col Epidemiol. Res: Psychosocial factors and cardiovascular disease risk in Black populations. Mailing Add: Dept Epidemiol Sch Pub Health Univ Mich Ann Arbor MI 48109

JAMES, STANLEY D, b Cardiff, UK, Aug 25, 32; m 61; c 2. ELECTROCHEMISTRY. Educ: Univ Wales, BSc, 53, PhD(phys chem), 59. Prof Exp: Vis scientist phys chem, NIH, 59-60; res fel phys & inorg chem, Univ Melbourne, 61-63; asst chemist electrochem, Brookhaven Nat Lab, 63-65, assoc chemist, 65-67; chemist, Electrochem Br, US Naval Surface Weapons Ctr, 67-; CHEMIST, DEPT NAVY. Mem: Electrochem Soc; Inst Elec & Electronics Engrs. Res: Ion exchange membranes; electrokinetics; electrode kinetics; fused salt electrochemistry. Mailing Add: Dept Electrochem Naval Surface Warfare Ctr Silver Spring MD 20910

JAMES, STEPHANIE LYNN, b Little Rock, Ark, m 82; c 2. IMMUNOPARASITOLOGY. Educ: Hendrix Col, BA, 72; Vanderbilt Univ, PhD(microbiol), 76. Prof Exp: Res fel, dept med, Harvard Med Sch, 77-79 & Lab Parasitic Dis, NIH, Nat Inst Allergy & Infectious Dis, 79-83; from asst res prof to assoc res prof med & microbiol, George Washington Univ, 83-87; parasitol prog officer, 87-90, CHIEF, PARASITOL & TROP DIS BR, NAT INST ALLERGY & INFECTIOUS DIS, NIH, 91- Concurrent Pos: Vis lectr, Univ de Sao Paulo, Brasil, 82; prin investr, NIH, NSF, WHO & Clark Found grants, 81-88; assoc ed, J Immunol, 85-88; sr investr, Biomed Res Inst, 88; travel award, Am Asn Immunologists, 80 & 83. Mem: Am Soc Trop Med Hyg; Am Asn Immunologists. Res: Parasite immunology, particularly in schistosomiasis, concentrating on the areas of immunopathology and vaccine production; elucidation of the roles of eosinophils and macrophages as effector cells of protective immunity which led to development of experimental vaccine based on cell-mediated immune mechanisms. Mailing Add: NIH Westwood Bldg Rm 737 Bethesda MD 20892

JAMES, STEPHEN P, b Columbus, Ohio, May 25, 47; m; c 2. EXPERIMENTAL BIOLOGY. Educ: Cornell Univ, BA, 69; Johns Hopkins Univ, MD, 73; Am Bd Internal Med, dipl, 76, dipl gastroenterol, 79. Prof Exp: Intern, Dept Med, Johns Hopkins Hosp, Baltimore, Md, 73-74, asst resident, 74-75, resident, 75-76; fel, Gastroenterol Div, Univ Md, Baltimore, 76-77; clin assoc, Liver Dis Sect, Nat Inst Arthritis, Diabetes, Digestive & Kidney Dis, NIH, 77-80, expert, Immunophysiol Sect, Metabol Br, Div Cancer Biol & Diag, Nat Cancer Inst, 80-82, sr clin investr, Mucosal Immunity Sect, Lab Clin Invest, Nat Inst Allergy & Infectious Dis, 82-91; HEAD, DIV GASTROENTEROL, UNIV MD, 91- Concurrent Pos: Chmn clin res comt, Nat Inst Allergy & Infectious Dis, 88-89; assoc ed, J Immunol, 88-; mem grants rev comt, Crohn's & Colitis Found Am, 88- Mem: Am Col Physicians; Am Fedn Clin Res; Am Asn Study Liver Dis; Am Asn Immunologists; Am Gastroenterol Asn; Soc Mucosal Immunol; AAAS; Int Asn Study Liver. Res: Regulatory functions of CD4 T cells; role of T cells in host defense and disease at mucosal surfaces; inflammatory bowel disease; gastrointestinal disease in immunodeficient patients; immune mechanisms in chronic liver disease; primary biliary cirrhosis. Mailing Add: Div Gastroenterol Rm N3W62 22 S Green St Bethesda MD 21201

JAMES, TED RALPH, b La Crosse, Wis, June 26, 36; m 62; c 2. ECOLOGY, VERTEBRATE ZOOLOGY. *Educ:* Wis State Univ, La Crosse, BS, 58; Univ NDak, MST, 63, PhD(ecol), 67. *Prof Exp:* Teacher, La Crosse City Sch Syst, Wis, 58-61 & 62-63; from instr to asst prof biol, Univ NDak, 66-68; from asst prof to assoc prof, 68-76, chmn dept, 73-85, PROF BIOL, UNIV TENN, MARTIN, 77-, ASSOC VCHANCELLOR, ACAD AFFAIRS, 87- *Mem:* Am Soc Mammalogists; Wildlife Soc; Sigma Xi. *Res:* Vertebrate ecology and mammalogy, especially population dynamics and radio telemetric evaluation of movement and habitat utilization in game species. *Mailing Add:* Dept Biol Sci Univ Tenn Martin TN 38238

JAMES, THOMAS LARRY, b North Platte, Nebr, Sept 8, 44; div; c 2. BIOPHYSICAL CHEMISTRY. *Educ:* Univ NMex, BS, 65; Univ Wis-Madison, PhD(anal chem), 69. *Prof Exp:* NIH trainee biochem, Univ Wis-Madison, 65-66, NIH fel anal chem, 66-69; res chemist, Tech Ctr, Celanese Chem Co, 69-71; NIH fel biophys, Johnson Res Found, Univ Pa, 71-73; from asst prof to assoc prof phys chem, 80-83, PROF PHARM CHEM, CHEM, RADIOL, UNIV CALIF, SAN FRANCISCO, 83- *Mem:* Am Chem Soc; Int Soc Magnetic Resonance; Soc Magnetic Resonance Med; Am Biophys Soc. *Res:* Nuclear magnetic resonance applications to biochemical and biological systems; nucleic acids; proteins. *Mailing Add:* Dept Pharmaceut Chem Univ Calif San Francisco CA 94143

JAMES, THOMAS NAUM, b Amory, Miss, Oct 24, 25; m 48; c 3. CARDIOVASCULAR DISEASES. *Educ:* Tulane Univ, BS, 46, MD, 49; Am Bd Internal Med, dipl, 57, cert cardiovasc dis, 60; Am Col Chest Physicians, dipl. *Prof Exp:* Intern & resident med & cardiol, Henry Ford Hosp, 49-53; cardiologist, Ochsner Clin, New Orleans, La, 55-59; chmn sect cardiovasc res, Henry Ford Hosp, 59-68; sr scientist & dir res, Cardiovasc Res & Training Ctr, 68-70, prof path, 68-73, dir cardiovasc res & training ctr, 70-77, prof med, Med Ctr, Univ Ala, Birmingham, 68-88, chmn, Dept Med, 73-88, Mary Gertrude Waters prof cardiol, 77-88; STAFF, UNIV TEX MED BR, UNIV TEX, GALVESTON, 88- *Concurrent Pos:* From instr to asst prof, Tulane Univ, 55-59; vis physician, Charity Hosp, New Orleans, 55-59; secy, Cardiac Electrophysiol Group, 64-65, pres, 65-66. *Mem:* Am Heart Asn; fel Am Col Physicians; fel Am Col Cardiol (vpres, 70-71); fel Am Col Chest Physicians; Soc Exp Biol & Med; Sigma Xi. *Res:* Anatomy, pathology, physiology and pharmacology of the heart, particularly coronary arteries and conduction system. *Mailing Add:* Off Pres Univ Tex Med Br Galveston TX 77550-2774

JAMES, THOMAS RAY, b Dayton, Ohio, Mar 23, 46; m 68; c 1. MATHEMATICS. *Educ:* Otterbein Col, BA, 68; Ohio Univ, MS, 71, PhD(math), 74. *Prof Exp:* Teaching asst, Ohio Univ, 70-74; instr math & physics, Sewickley Acad, 74-75; asst prof math, Lake Erie Col, 75-79; ASST PROF MATH, OTTERBEIN COL, 79- *Mem:* Am Math Soc; Asn Comput Mach; Math Asn Am; Inst Elec & Electronics Engrs. *Res:* Point set topology. *Mailing Add:* Dept Math Otterbein Col Westerville OH 43081

JAMES, THOMAS WILLIAM, b Haugen, Wis, July 30, 18; m 49; c 3. ZOOLOGY. *Educ:* Univ Minn, BA, 48; Univ Mo, MA, 51; Univ Calif, PhD(zool), 54. *Prof Exp:* From instr to assoc prof zool, 53-66, chmn dept, 66-68, PROF CELL BIOL, UNIV CALIF, LOS ANGELES, 66- *Concurrent Pos:* Lalor fel, 55; USPHS spec res fel, 62-63. *Mem:* Biophys Soc; Am Soc Protozool; Am Soc Zool; Int Soc Cell Biol; Sigma Xi. *Res:* Surface chemistry of biological molecules; electron microscopy of cellular components; nuclear physiology and synchronization of cell division; rotary dispersion of biological molecules; cell growth kinetics; yeast mtDNA; chloroplast DNA; evolution in a chemostat. *Mailing Add:* 2951 Mandeville Canyon Rd Los Angeles CA 90049

JAMES, V(IRGIL) EUGENE, b Braxton Co, WVa, May 6, 29; m 55; c 1. CHEMICAL ENGINEERING. *Educ:* WVa Univ, BS, 51, MS, 56, PhD(chem eng), 58. *Prof Exp:* Chem engr, nitrogen div, Allied Chem & Dye Corp, Va, 53-54; chem engr, Bur Mines, US Dept Interior, WVa, 54-58; res engr, Film Dept, Yerkes Res Lab, 58-61, res supvr, 61-65, process supvr, 65-67, SR SUPVR, TEXTILE FIBERS DEPT, E I DU PONT DE NEMOURS & CO, INC, 67- *Mem:* AAAS; assoc mem Am Inst Chem Engrs. *Res:* Film forming and polymer research; coal gasification research; synthetic fibers. *Mailing Add:* 214 Masters Ct Chattanooga TN 37343

JAMES, W(ILBUR) GERALD, b Parmele, NC, Oct 27, 22; m 44; c 2. SYSTEMS ENGINEERING, ELECTRONIC WARFARE ENGINEER. *Educ:* NC State Col, BS, 47; Univ Tenn, MS, 51. *Prof Exp:* Develop engr instrumentation dept, Oak Ridge Nat Lab, 47-51, group leader circuitry develop, 51-53; chief radio frequency sect, US Dept of Defense, Washington, DC, 53-54; mgr components develop sect, ACF Electronics Div, ACF Indust Inc, 54-56; proj mgr opers res sect, Am Mach & Foundry Co, Va, 56-58, mgr mil systs sect, 58-59, asst dept mgr electronics dept, 59-61; mgr appl sci lab, Melpar, Inc, 61-65; mem tech dirs staff, John Hopkins Univ, 65-66, head, Electronic Countermeasures Proj Off, Appl Physics Lab, 66-71, head, Oper Eval Group & Rep on Dir of Navy Labs Advan Tech Objectives Working Group Electronic Warfare, 71-75; asst dir, Systs Develop Dept, SRI Int, 75-80, sr staff engr, Ctr for Systs Develop, 80-84, asst dir, 84-91; RETIRED. *Concurrent Pos:* Lectr appl electronics counter measures, George Washington Univ, Sch Continuing Educ, 79-81. *Mem:* Inst Elec & Electronics Engrs. *Res:* Design and analysis for electronic warfare equipment and systems; threat studies and systems design for electronic warfare threat simulator; system design and analysis for military range instrumentation systems; electronic warfare systems and test and evaluation. *Mailing Add:* 7312 Redd Rd Falls Church VA 22043

JAMES, WILLIAM HOLDEN, b Ellijay, Ga, Apr 14, 09; m 47; c 1. FOOD SCIENCE. *Educ:* Emory Univ, BS, 28, MS, 29; Pa State Univ, PhD(org chem), 43. *Prof Exp:* Asst chem, Pa State Univ, 29-34; res chemist, Socony-Vacuum Oil Co, NJ, 34-41; asst chem, Pa State Univ, 41-43, instr org res, 43-44, animal nutrit, 44-45, asst prof, 45-48; assoc prof agr biochem, 48-62, prof, 62-77, EMER PROF FOOD SCI, LA STATE UNIV, 77- *Mem:* Fel AAAS; Am Chem Soc; Am Inst Nutrit; Inst Food Technologists; fel Royal Soc Health. *Res:* Energy metabolism; nutritional utilization of carotene and vitamin A; metabolic patterns in preadolescent children; use of electronic computers in searching out nutritional interrelationships; processing effects on rice quality. *Mailing Add:* Dept Food Sci La State Univ Baton Rouge LA 70803-4202

JAMES, WILLIAM JOSEPH, b Providence, RI, Sept 17, 22; m 42; c 2. SOLID STATE CHEMISTRY, ELECTROCHEMISTRY. *Educ:* Tufts Univ, BS, 49; Iowa State Univ, MS, 52, PhD(chem), 53. *Prof Exp:* Asst physics, Pa State Univ, 52-53; from assoc prof to prof, Univ Mo-Rolla, 53-84, dir grad ctr mat res, 64-75, assoc dir, 75-76, dir, 82-83, EMER PROF CHEM, UNIV MO-ROLLA, 84- *Concurrent Pos:* Fulbright res fel, 61-62; pres & founder, Mead Technol Inc, Rolla, Mo, 76-; pres, Incubator Technol Inc, Rolla, Mo, 84-86; vpres & bd dir, Filterteck Inc, 82-; bd dir, Brewer Sci, 81-; bd dir, APR Inc, Redwood, Ca, 87-; sr investr Grad Ctr Mat Res, 64- *Honors & Awards:* Thomas Jefferson Award, 89. *Mem:* Electrochem Soc; fel Am Inst Chemists; Am Chem Soc; Am Crystallog Asn; Mat Res Soc. *Res:* Lattice imperfections; magnetic and crystal structure determinations by x-ray and neutron diffraction; electrochemical kinetics and corrosion science; plasma polymerization. *Mailing Add:* Grad Ctr Mat Res Univ Mo Rolla MO 65401

JAMESON, A KEITH, b Provo, Utah, June 11, 33; m 63; c 2. PHYSICAL CHEMISTRY. *Educ:* Brigham Young Univ, BS, 56, BSE & MS, 57; Univ Ill, Urbana, PhD(phys chem), 63. *Prof Exp:* Res chemist, Esso Res & Eng Co, 62-65; vis assoc prof chem, Ateneo de Manila Univ, 65-67; vis asst prof, Univ Ill, Urbana, 67-68; asst prof, 68-73, assoc prof chem, 73-80, PROF CHEM, LOYOLA UNIV CHICAGO, 80- *Mem:* Am Chem Soc; Am Phys Soc. *Res:* Nuclear magnetic resonance; intermolecular interactions and spectroscopic observables; energy and environmental chemistry. *Mailing Add:* Dept of Chem Loyola Univ Chicago IL 60611

JAMESON, ARTHUR GREGORY, b Branford, Conn, Mar 26, 15; m 50; c 2. MEDICINE. *Educ:* Harvard Col, AB, 37; Mass Inst Technol, MS, 40; Columbia Univ, MD, 50. *Prof Exp:* Asst prof med, Col Med, State Univ NY Downstate Med Ctr, 57-59; asst prof pediat, 59-66, assoc clin prof med, 66-72, PROF CLIN MED, COL PHYSICIANS & SURGEONS, COLUMBIA UNIV, 72-; ACTG DIR, DEPT MED, ROOSEVELT HOSP, NEW YORK, 78- *Concurrent Pos:* Dir cardiovasc lab, Columbia-Presby Med Ctr, 59-66; consult, Brooklyn Hosp; dir cardiol, Dept Med, Roosevelt Hosp, 66-78. *Mem:* Fel Am Col Physicians; fel Am Col Cardiol. *Res:* Cardiovascular hemodynamics. *Mailing Add:* 17 E 89th St New York NY 10128

JAMESON, CHARLES WILLIAM, b LaPlata, Md, Feb 3, 48; m 69; c 1. ORGANIC CHEMISTRY, CHEMICAL CARCINOGENESIS. *Educ:* Mt St Mary's Col, BS, 70; Univ Md, PhD(org chem), 76. *Prof Exp:* Fac grad asst, Univ Md, 75-76; chemist bioassay, Tracor Jitco Inc, Tracor, Inc, 76-78, sr chemist, 78-79; expert chem, 79-80, prog leader chem, Nat Toxicol Prog, 80-90, SR CHEMIST, OFF OF DIR, NIEHS, 90- *Concurrent Pos:* Sr chemist Bioassay Prog & consult, Chem Selection Group, Nat Cancer Inst, 76-80; mem, WHO task group Environ Health Criteria Partially Halogenated Chlorofluorocarbons. *Res:* Structure activity relationships; toxicokinetics; leukemia. *Mailing Add:* PO Box 12233 Nat Toxicol Prog Research Triangle Park NC 27709

JAMESON, DAVID LEE, b Ranger, Tex, June 3, 27; m 49; c 4. EVOLUTION. *Educ:* Southern Methodist Univ, BS, 48; Univ Tex, MA, 49, PhD(zool), 52. *Prof Exp:* Asst prof biol, Pacific Univ, 52-53; from instr to asst prof, Univ Ore, 53-57; from asst prof to prof zool, San Diego State Col, 57-67; dir coastal ctr, 72-76, PROF BIOL, UNIV HOUSTON, 67- *Concurrent Pos:* Assoc dean grad sch, Univ Houston, 71-72, dean, 72-74; managing ed, Copeia, Am Soc Ichthyologists & Herpetologists; Nat Acad Sci Exchange scholar, Bulgarian Acad Sci, 77 & USSR Acad Sci, 78; sr res fel, Calif Acad Sci, 87. *Mem:* Am Soc Mammal; Ecol Soc Am; Soc Study Evolution (secy, 68-73); Am Soc Ichthyologists & Herpetologists; Am Inst Biol Sci. *Res:* Genetics; amphibians; population genetics; mitochondrial DNA evolution. *Mailing Add:* Dept Herpet Cal Acad Sci Golden Gate Park San Francisco CA 94118

JAMESON, DONALD ALBERT, b Pueblo, Colo, Nov 18, 29; m 54; c 5. COMPUTER SCIENCES. *Educ:* Colo State Univ, BS, 50; Mont State Col, MS, 52; Tex A&M Univ, PhD(range mgt), 58. *Prof Exp:* From assoc prof to prof range sci, Colo State Univ, 68-89, assoc dean, Col Forestry & Natural Resources, 74-78; range scientist, Rocky Mountain Forest & Range Exp Sta, 56-68, PROG ANALYST, USDA FOREST SERV, 89- *Mem:* Soc Range Mgt; Am Soc Photogram & Remote Sensing. *Res:* Ecology of native plants; quantitative natural resource management. *Mailing Add:* 12561 Quincy Adams Ct Herndon VA 22701

JAMESON, DOROTHEA, b Newton, Mass, Nov 16, 20; m 48. NEUROSCIENCE. *Educ:* Wellesley Col, BA, 42. *Hon Degrees:* MA, Univ Pa, 72; DSc, State Univ NY, 89. *Prof Exp:* Res asst, Harvard Univ, 41-47; res psychologist, Eastman Kodak Co, 47-57; res scientist, NY Univ, 57-62; res assoc, 62-68, res prof, 68-72, prof psychol, 72-74, UNIV PROF PSYCHOL & VISUAL SCI, UNIV PA, 75- *Concurrent Pos:* Prin investr res grant, NIH & NSF, 57-; mem, Nat Acad Sci-Nat Res Coun Vision Comt, 70-72 & 76-; vis prof, Univ Rochester, 74-75 & Columbia Univ, 74-76; mem, Visual Sci B Study Sect, NIH, 75-78 & Comn Human Resources, Nat Acad Sci-Nat Res Coun, 77-80; fel Ctr Adv Study Behav Sci, 81-82, nat adv eye coun, NIH, 85- *Honors & Awards:* Warren Medal, Soc Exp Psychologists, 71; Distinguished Sci Contrib Award, Am Psychol Asn, 72; Helmholtz Award, Cognitive Neurosci Asn, 87; Godlove Award for Res in Color Vision, Inter-Soc Color Coun, 73; Tillyer Medal, Optical Soc Am, 82; Judd Medal, Asn Int de la Couleur, 85. *Mem:* Nat Acad Sci; Am Acad Arts & Sci; Soc Neurosci; Optical Soc Am; Asn Res Vision & Ophthal; Sigma Xi; Int Res Group Color Vision Deficiencies; Int Brain Res Org. *Res:* Visual mechanisms; human perception. *Mailing Add:* Dept Psychol Univ Pa 3815 Walnut St Philadelphia PA 19104-6196

JAMESON, EVERETT WILLIAMS, JR, b Buffalo, NY, May 2, 21; m 69; c 5. VERTEBRATE ZOOLOGY, MEDICAL ENTOMOLOGY. *Educ:* Cornell Univ, BS, 43, PhD(vert zool), 48; Univ Kans, MA, 46. *Prof Exp:* Field observer, Hastings Reservation, Calif, 42; lab asst zool, Univ Kans, 45-46 & Cornell Univ, 46-48; from instr to prof zool, 48-88, from asst zoologist to assoc zoologist, exp sta, 48-65, vchmn dept, 69-74, EMER PROF ZOOL, UNIV CALIF, DAVIS, 88- *Concurrent Pos:* Guggenheim fel, 58-59. *Mem:* Am Soc Mammalogists; assoc Am Soc Ichthyologists & Herpetologists. *Res:* Population investigations of small mammals; food habits of vertebrates; fat and reproductive cycles of reptiles; oxygen consumption of reptiles; ecological, zoogeographic and taxonomic investigations of fleas and mites in North America and the Far East. *Mailing Add:* Dept Zool Univ Calif Davis CA 95616

JAMESON, JAMES LARRY, b Ft Benning, Ga, June 21, 54; m 84. GLYCOPROTEIN HORMONES, ENDOCRINOLOGY. *Educ:* Univ NC, BS, 76, MD, 81, PhD(biochem), 81. *Prof Exp:* Resident internal med, 81-83, fel endocrinol, 83-85, INSTR MED, LAB MOLECULAR ENDOCRINOL, MASS GEN HOSP, HARVARD MED SCH, 85- *Mem:* AMA. *Res:* Regulation of glycoprotein hormone gene expression in eukaryotic cell line sand pituitary tumors. *Mailing Add:* Thyroid Unit Mass Gen Hosp Bulfinch Basement Boston MA 02114

JAMESON, PATRICIA MADOLINE, b Rhinelander, Wis, Mar 17, 39. MICROBIOLOGY, VIROLOGY. *Educ:* Carroll Col, Wis, BS, 61; Ind Univ, MS, 63, PhD(microbiol), 65. *Prof Exp:* Microbiologist viruses, US Army Biol Labs, Ft Detrick, 65-69; from instr to assoc prof microbiol, Med Col Wis, 69-89; asst prof, Booth Libr, Eastern Ill Univ, 89-91; REF LIBRN, APPL SCI & TECHNOL, UNIV WIS, MILWAUKEE, 91- *Mem:* AAAS; Am Soc Microbiol; Sigma Xi. *Res:* Arboviruses; comparison of neuraminidases of neurotropic and nonneurotropic influenza virus strains, especially with respect to substrate specificity; interferon, especially standards, assay and inducers; feline leukemia virus. *Mailing Add:* Golda Meir Libr PO Box 604 Milwaukee WI 53201

JAMESON, ROBERT A, b Schenectady, NY, May 3, 37; m 59; c 2. PARTICLE ACCELERATOR PHYSICS & ENGINEERING. *Educ:* Univ Nebr, BS, 58; Univ Colo, MS, 62, PhD(elec eng), 65; Univ NMex, MMgt, 77. *Prof Exp:* From asst group leader to assoc group leader, Los Alamos Nat Lab, 63-71, group leader, Accelerator Systs, Group MP-9, MP-Div, Lampf, 72-80, from dep div leader to div leader, Accelerator Technol Div, 78-87, STAFF MEM, LOS ALAMOS NAT LAB, 88- *Concurrent Pos:* Vis prof, Ministry Educ, Japan, 88-89. *Mem:* Fel Am Phys Soc. *Res:* Application of automatic control theory to high power microwave systems; systems analysis and development of particle accelerator control and rf-accelerator systems; particle accelerator beam dynamics; electrical engineering. *Mailing Add:* AT-DO MS H811 Los Alamos Nat Lab Los Alamos NM 87545

JAMESON, WILLIAM J, JR, b Billings, Mont, June 8, 30; m 53; c 2. NUMERICAL MATHEMATICS, SYSTEMS ANALYSIS. *Educ:* Univ Mont, BA, 52; Univ Tex, MA, 54; Iowa State Univ, PhD(math), 62. *Prof Exp:* Physicist, Lockheed Missiles & Space Co, 58-59; fel & teaching asst appl math, Iowa State Univ, 59-62; mathematician, Collins Radio Co, Spectra Assocs, Inc, 62-72, vpres, 72-85; dir telecommunications, State Mich, 85-87; ELEC ENG DEPT, MONT STATE UNIV, 87- *Concurrent Pos:* Part-time asst prof math, Iowa State Univ, 62-67, assoc prof, 68-72, 80-81; corresp consult, Nat Acad Sci-Nat Acad Eng Comt Sci & Tech Commun, 67-69; mem, Comn Nat Info Syst Math, 68-70; mem, Pub Info Comt, Fedn Info Processing Socs, 69-72; chmn, Math Sect, Res Div, Am Defense Preparedness Asn, 69-78; mem, Comt on Commun & Info Policy, Inst Elec & Electronics Engrs, 85-88. *Mem:* Soc Indust & Appl Math (secy, 64-69, vpres, 69-74); Inst Elec & Electronics Engrs. *Res:* Numerical analysis and computation telecommunications; systems analysis. *Mailing Add:* PO Box 5279 Bozeman MT 59717

JAMIESON, ALEXANDER MACRAE, b Glasgow, Scotland, Sept 19, 44; m 71; c 3. CHEMICAL PHYSICS, POLYMER SCIENCE. *Educ:* Univ Glasgow, BS, 66; Oxford Univ, PhD(chem physics), 69. *Prof Exp:* From res assoc to sr res assoc, 72-74, from asst prof to assoc prof, 74-82, PROF MOLECULAR SCI, DEPT MACROMOLECULAR SCI, CASE WESTERN RESERVE UNIV, 82- *Mem:* Am Phys Soc; Am Chem Soc; Soc Rheology. *Res:* Physical characterization of polymer materials; hydrodynamic properties of macromolecules; rheological properties of polymer solutions; structure and function of polysaccharides and proteoglycans; quasielastic laser light scattering. *Mailing Add:* 801 Olin Bldg Case Western Reserve Univ Cleveland OH 44106

JAMIESON, DEREK MAITLAND, b Dundee, Scotland, Nov 27, 30; Can citizen; m 55, 83. MATHEMATICS, STATISTICS. *Educ:* St Andrews Univ, BSc, 51, Hons, 53. *Prof Exp:* Statistician, Can Industs, Ltd, 53-57; sci off math & statist, Defence Res Bd Can, 57-60; mem tech staff, Mitre Corp, 60-65; chief indust models div, Nat Energy Bd, 65-66; planning exec, Simpac Div, Treas Bd, 66-68; DIR INST ANAL & PLANNING, 68-, ADJ PROF MATH, UNIV GUELPH, 82- *Concurrent Pos:* Res dir, Comn Future Develop Univs On, 84. *Mem:* Fel AAAS; Inst Mgt Sci; Can Opers Res Soc; fel Royal Statist Soc; Opers Res Soc Am. *Res:* Computer aided analysis and study of large systems. *Mailing Add:* Univ of Guelph Guelph ON N1G 2W1 Can

JAMIESON, GLEN STEWART, b Montreal, Que. FISHERIES MANAGEMENT, INVERTEBRATE ECOLOGY. *Educ:* McGill Univ, BSc, 67; Univ BC, MSc, 70, PhD(zool), 73. *Prof Exp:* Fel, Dalhousie Univ, 74-75; sr marine biologist, Appl Marine Res Ltd, Halifax, 75-77; res scientist, Fisheries Res Br, Halifax, 77-81, res scientist, 81-82, HEAD SHELLFISH, FISHERIES RES BR, FISHERIES & OCEANS, NANAIMO, CAN, 82- *Concurrent Pos:* Head herring, Fisheries Res Br, Nanaimo, 82-84. *Res:* Fisheries management, emphasizing invertebrates; spatial and temporal distributions; stock assessment methodology; predator-prey interactions. *Mailing Add:* Pacific Biol Sta Wanaimo BC V9R 5K6 Can

JAMIESON, GRAHAM ARCHIBALD, b Wellington, NZ, Aug 14, 29; m 60; c 1. BIOCHEMISTRY. *Educ:* Univ Otago, NZ, MSc, 51; Univ London, PhD(org chem), 54, DSc(biochem), 72. *Prof Exp:* Res assoc org chem, Royal Inst Technol, Sweden, 55-56 & Med Col, Cornell Univ, 56-57; vis scientist, NIH, 57-61; res biochemist, Am Nat Red Cross, 61-64, asst dir res, 65-69, res dir, Blood Prog, Blood Res Lab, 69-79, assoc dir, Blood Serv, 79-84, SR SCIENTIST, AM NAT RED CROSS, 84- *Concurrent Pos:* Adj prof, Sch Med & Dent, Georgetown Univ, 74-; mem exec comt, Thrombosis Coun, Am Heart Asn; mem, Blood Res Study Sect, NIH; mem adv comt, Res Blood Prod & Preserv, Letterman Army Inst Res; ed, Thrombosis & Haemostasis Int J Haematol, Am Soc Biol Chemists; John Edmund fel, Univ Otago. *Honors & Awards:* Winzler Mem Lectr, Univ Fla, 75. *Mem:* AAAS; Am Soc Biol Chem; Int Soc Thrombosis & Haemostasis; Am Chem Soc; Soc Exp Biol & Med. *Res:* Platelet receptor function and membrane biochemistry. *Mailing Add:* Cell Biol Lab Am Nat Red Cross Rockville MD 20855

JAMIESON, J(OHN) A(NTHONY), b Barnet, Eng, Mar 16, 29; nat US; m 56; c 3. INFRARED PHYSICS, ENGINEERING. *Educ:* Univ London, BSc, 52; Stanford Univ, MS, 55, PhD(elec eng), 57. *Prof Exp:* Head detector systs anal, Avionics Div, Aerojet-Gen Corp, 56-59; sr scientist, Aeronutronic Div, Ford Motor Co, 59-62; mgr res, Astrionics Div, Aerojet-Gen Corp, 62-66, mgr electronic systs div, 66-70; asst dir & chief, Optics Div, US Army Advan Ballistic Missile Defense Agency, 70-73; PRES, JAMIESON SCI & ENG INC, 73- *Mem:* Optical Soc Am; sr mem Inst Elec & Electronics Engrs. *Res:* Information theory; applied infrared physics; noise analysis. *Mailing Add:* Suite 549 W 7315 Wisconsin Ave Bethesda MD 20814

JAMIESON, JAMES C, b Aberdeen, Scotland, May 15, 39. GLYCOPROTEIN BIOSYNTHESIS. *Educ:* Aberdeen Univ, Scotland, PhD(biochem), 67. *Prof Exp:* PROF CHEM, UNIV MAN, 78- *Mem:* Am Asn Biol Syst; Can Biochem Soc; Chem Inst Can; Royal Inst Chem; Soc Complex Carbohydrates. *Mailing Add:* Univ Man Winnipeg MB R3T 2N2 Can

JAMIESON, JAMES DOUGLAS, b Armstrong, BC, Jan 22, 34; m 64; c 2. CELL BIOLOGY. *Educ:* Univ BC, MD, 60; Rockefeller Univ, PhD(cell biol), 66. *Prof Exp:* Res assoc cell biol, Rockefeller Univ, 66-67, from asst prof to assoc prof, 67-73; assoc prof, 73-75, PROF CELL BIOL, YALE UNIV MED SCH, 75-, CHMN DEPT. *Mem:* Am Soc Cell Biol; Am Soc Biol Chemists. *Res:* Intracellular transport of secretory proteins; membrane formation and function; cell-hormone interactions; immunocytochemistry; pathophysiology of vascular smooth muscle; cytodifferentiation of glandular epithelia. *Mailing Add:* Sect Cell Biol Yale Univ Med Sch 333 Cedar St PO Box 3333 New Haven CT 06510

JAMIESON, LEAH H, b Trenton, NJ, Aug 27, 49. SPEECH PROCESSING, PARALLEL PROCESSING. *Educ:* Mass Inst Technol, SB, 72; Princeton Univ, MA & MSE, 74, PhD(elec eng & comput sci), 77. *Prof Exp:* Researcher ins comput systs, Prudential Ins Co, 69 & info systs, Mass Inst Technol, 70; res teaching asst elec eng & comput sci, Princeton Univ, 72-76; asst prof, 76-82, ASSOC PROF ELEC ENG, PURDUE UNIV, 82- *Concurrent Pos:* Int assoc chairperson, Distrib Comput Systs, Inst Elec & Electronics Engrs, 81-82; co-organizer, Workshop Algorithmically Specialized Comput Orgns, 82 & Taxon Parallel Algorithms Workshop, 83; assoc ed, Jour Parallel & Distrib Comput. *Mem:* Inst Elec & Electronics Engrs; Asn Comput Mach. *Res:* Computer analysis and recognition of speech; design of parallel processing algorithms for digital speech, signal and image processing, including parallel languages and modelling of parallel and distributed computing. *Mailing Add:* Sch Elec Eng Purdue Univ West Lafayette IN 47907

JAMIESON, NORMAN CLARK, b Edinburgh, Scotland, Nov 21, 35; US citizen; m 64; c 1. ORGANIC CHEMISTRY. *Educ:* Univ Edinburgh, BSc, 58; Univ Alta, MSc, 61; Univ Adelaide, PhD(org chem), 66. *Prof Exp:* Fel org chem, Rensselaer Polytech Inst, 66-67; sr res scientist, Merck & Co, Inc, 67-70; sr chemist, 70-77, res assoc, 77-80, dir res & develop, Sci Prods Div, 80-86, ASST DIR CORP ANALYST RES, MALLINCKRODT, INC, 86- *Mem:* Am Chem Soc; Royal Soc Chem. *Res:* Photochemistry; carbohydrates. *Mailing Add:* Mallinckrodt Specialty Chem Co PO Box 5439 St Louis MO 63147-0339

JAMIESON, WILLIAM DAVID, b Toronto, Ont, Aug 6, 29; m 51; c 1. ANALYTICAL MASS SPECTROMETRY, QUALITY ASSURANCE. *Educ:* Dalhousie Univ, BSc & dipl chem eng, 50, MSc, 51; Cambridge Univ, PhD(phys chem), 54. *Prof Exp:* Prin res officer, Atlantic Res Lab, Nat Res Coun Can, 54-90, asst to dir, 64-75, head marine anal chem, 75-90; SR SCIENTIST, FENWICK LABS LTD, 90- *Concurrent Pos:* Coordr, Atlantic Prov Interuniv Comt Sci, 63-65; head clean-up technol coord, Oper Oil, 70; mgr, Marine Anal Chem Standards Prog, 79-90; chair, Group Experts Standards & Ref Mat, Int Oceanog Comn, 87-, Standards Comt, Can Asn Environ Anal Chem, 89- *Honors & Awards:* Caledon Award, 91. *Mem:* Fel Chem Inst Can; Am Soc Mass Spectrometry; Spectros Soc Can; Marine Technol Soc; Can Soc Mass Spectrometry. *Res:* Mass spectrometry; instrumentation development; analytical chemistry; kinetics of gas phase ion reactions; oil pollution clean-up technology; marine analytical chemistry; development of analytical chemistry reference materials and standards; quality assurance in analytical chemistry. *Mailing Add:* 30 Colindale Halifax NS B3P 2A4 Can

JAMIESON-MAGATHAN, ESTHER R, b Kirland Lake, Ont, Can, Feb 6, 37. CARBONATE SEDIMENTATION, APPLIED PALEONTOLOGY. *Educ:* Queens Univ, Kingston, Can, BA, 60, BS, 61, Reading Univ, PhD(geol), 67. *Prof Exp:* Res geologist, Sask Dept Mineral Resources, Can, 67-70; sr staff assoc, Union Oil Co, 70-74; CONSULT GEOLOGIST, ESTHER MAGATHAN ASSOC, 75- *Honors & Awards:* A E Levorsen, Am Asn Petrol Geologist. *Mem:* Fel Geol Soc Am; Sigma Xi; Can Soc Petrol Geologist; AAAS; Soc Econ Paleont & Mineralogist. *Mailing Add:* 114 Cove Creek Lane Houston TX 77042

JAMISON, HOMER CLAUDE, b Marion, NC, Apr 14, 21; c 3. EPIDEMIOLOGY, DENTISTRY. *Educ:* Western Carolina Teachers Col, AB, 42; Emory Univ, DDS, 50; Univ Mich, MPH, 57, DrPH(epidemiol), 61; Am Bd Dent Pub Health, dipl. *Prof Exp:* Pub health dentist, NC State Bd Health, 51-54; dent officer, Mecklenburg Health Dept, 54-56; pub health dentist, Mich Dept Health, 57-58; from asst prof to prof dent, Med Ctr, Univ Ala, Birmingham, 60-68, dir grad prog, 62-63, dir comput res lab, 63-64; prof dent, Sch Dent, Univ Mo-Kansas City, 68-72; prof, 72-86, PROF EMER, SCH DENT, UNIV ALA, BIRMINGHAM, 86- *Concurrent Pos:* Consult, Div Radiol Health, USPHS, 64-66; mem bd dirs, Jefferson County Anti-Tuberc Asn, 64-68. *Mem:* Fel Am Pub Health Asn; Am Dent Asn; Biomet Soc; Am Statist Asn. *Res:* Clinical studies of potential prophylactic and therapeutic agents; applications and uses of computers in health research; patterns and trends in oral health and diseases. *Mailing Add:* 3586 Rockhill Rd Birmingham AL 35223-1402

JAMISON, JOEL DEXTER, b Roanoke, Va, Nov 22, 32; m 59; c 3. ORGANIC CHEMISTRY. *Educ:* Col William & Mary, BS, 55; Northwestern Univ, PhD(org chem), 60. *Prof Exp:* Res chemist, 60-72, res scientist, 72-79, TECH DIV MGR, HERCULES INC, 79- *Mem:* Am Chem Soc. *Res:* Molecular structure elucidation; synthesis and investigation of condensation reactions in strong acid; synthesis of biologically active organic compounds for screening as pesticides; synthesis of lubrication base stocks. *Mailing Add:* 26-5 Walnut Hill Ct Hockessin DE 19707-9802

JAMISON, KING W, JR, b Meridian, Miss, Aug 8, 31; m 53; c 4. MATHEMATICS. *Educ:* Union Univ, Tenn, BS, 52; George Peabody Col, MA, 53, PhD(math educ), 62. *Prof Exp:* Lectr math, Vanderbilt Univ, 61-62; from asst prof to assoc prof, 62-72, PROF MATH, MIDDLE TENN STATE UNIV, 72- *Res:* Mathematics education, especially the relationship of mathematical symbols to English words; variable base abacus as a visual aid. *Mailing Add:* Dept Math Middle Tenn State Univ Box 163 Murfreesboro TN 37132

JAMISON, RICHARD MELVIN, b Rayne, La, Oct 28, 38; div; c 3. VIROLOGY. *Educ:* Univ Southwestern La, BS, 58; Baylor Univ, MS, 62, PhD(virol), 66; Am Bd Med Microbiol, dipl, 76. *Prof Exp:* Res assoc biol div, Oak Ridge Nat Lab, 65-67; asst prof path, Univ Colo, Denver, 67-70; assoc prof, 70-78, dir, Diag Virol Lab, 79-85, PROF MICROBIOL & IMMUNOL, LA STATE UNIV, SHREVEPORT, 78-, PROF PEDIAT, 87- *Concurrent Pos:* Vis prof microbiol, Fac Med, Al Fetah Univ, Tripoli, Libya, 81-82; trustee, Am Bd Med Microbiol, 79; Russel V Lee Award Lectureship, 82. *Mem:* Am Soc Microbiol; Sigma Xi; Am Fedn Clin Res; fel Am Acad Microbiol. *Res:* Replication of picornaviruses; viral oncology; rapid diagnosis of viral infections. *Mailing Add:* Dept Pediat & Microbiol & Immunol La State Univ Sch of Med Shreveport LA 71130

JAMISON, ROBERT EDWARD, b Tampa, Fla, Dec 21, 48; m 78; c 2. ABSTRACT CONVEXITY. *Educ:* Clemson Univ, BS, 70; Univ Wash, MS, 73, PhD(math), 74. *Prof Exp:* Asst prof, La State Univ, 74-79; assoc prof, 79-83, PROF MATH SCI, CLEMSON UNIV, 83- *Concurrent Pos:* Vis asst, Inst Appl Math, Univ Bonn, 75-76; Alexander von Humboldt fel, Univ Erlangen, 76-77; vis prof, Tech Univ Darmstadt, 79; Humboldt fel, Univ Freiburg, 84; vis prof, Univ Berne, 86. *Mem:* Math Asn Am. *Res:* Combinatorial problems of a geometric nature, primarily those concerned with the theory of convex sets. *Mailing Add:* Math Sci Dept Clemson Univ Clemson SC 29631

JAMISON, RONALD D, b 1931. MATHEMATICS. *Educ:* Univ Utah, PhD(math), 65. *Prof Exp:* PROF MATH, BRIGHAM YOUNG UNIV, 63- *Mem:* Am Math Soc. *Res:* Differential equations & applied math. *Mailing Add:* Bringham Young Univ 330-A MSC B TMCB Provo UT 84602

JAMISON, WILLIAM H, b Burlington, Iowa, May 4, 32; m 62. MATHEMATICS. *Educ:* Mont State Col, BS, 59, MS, 61. *Prof Exp:* Instr math, Mont State Col, 59-62; assoc prof, 62-68, PROF MATH & CHMN DIV NATURAL SCI & MATH, ROCKY MOUNTAIN COL, 68- *Mem:* Math Asn Am; Am Math Soc; Am Asn Physics Teachers. *Res:* Boolean algebra; logic; fossil fuel utilization. *Mailing Add:* Div of Natural Sci & Math Rocky Mountain Col Billings MT 59102

JAMMU, K S, b India, Jan 1, 35; m 59; c 2. PHYSICS. *Educ:* Aligarh Muslim Univ, India, MSc, 57; Univ Toronto, MA, 60, PhD(physics), 65. *Prof Exp:* Lectr physics, Khalsa Col, Amritsar, India, 57-59; asst prof, Mem Univ, 65-67; from asst prof to assoc prof, St Dunstan's Univ, 67-69; assoc prof, 69-80, PROF PHYSICS, UNIV PRINCE EDWARD ISLAND, 80- *Mem:* Am Asn Physics Teachers; Can Asn Physicists. *Res:* Molecular physics; spectroscopy. *Mailing Add:* Dept of Physics Univ of Prince Edward Island Charlottetown PE C1A 4P3 Can

JAMNBACK, HUGO ANDREW, JR, b Fitchburg, Mass, Sept 18, 26; m 53; c 3. MEDICAL ENTOMOLOGY. *Educ:* Boston Univ, BA, 48; Univ Mass, MS, 51, PhD, 53; London Sch Hyg & Trop Med, Dipl, 66. *Prof Exp:* Scientist entom, NY State Mus & Sci Serv, 53-59, sr scientist, 59-67, assoc scientist, 67-71, dir, NY State Sci Serv, 71-81; RETIRED. *Concurrent Pos:* Consult, WHO, 67-; sr res assoc, Col Environ Sci & Forestry, Syracuse Univ, 73- *Mem:* Entom Soc Am; Am Mosquito Control Asn. *Res:* Taxonomy, biology and control of biting flies. *Mailing Add:* RD 1 Singer Rd East Berne NY 12059

JAMPEL, ROBERT STEVEN, b New York, NY, Nov 3, 26; m 52; c 4. OPHTHALMOLOGY. *Educ:* Columbia Univ, AB, 47, MD, 50; Univ Mich, MS, 57, PhD(anat), 58. *Prof Exp:* Clin instr ophthal, Univ Mich, 56-57, instr neurol, 57-58; asst prof, State Univ NY Downstate Med Ctr, 58-62; assoc ophthal, Columbia Univ, 62-70; PROF OPHTHAL & CHMN DEPT & DIR, KRESGE EYE INST, WAYNE STATE UNIV, 70- *Concurrent Pos:* Chief dept ophthal, Hutzel Hosp & Detroit Med Ctr, 88. *Mem:* Asn Res Vision & Ophthal; Am Acad Ophthal & Otolaryngol; Am Acad Neurol. *Res:* Physiology of the ocular muscles. *Mailing Add:* Kresge Eye Inst 4717 St Antoine Detroit MI 48201

JAMPLIS, ROBERT W, b Chicago, Ill, Apr 1, 20; m; c 2. MEDICAL FOUNDATION EXECUTIVE. *Educ:* Univ Chicago, BS, 41, MD, 44; Univ Minn, MS, 51; Am Bd Surg, dipl, 52; Am Bd Thoracic Surg, dipl, 53. *Prof Exp:* Fel thoracic surg, Mayo Clin, 47-52; CHIEF THORACIC SURG, PALO ALTO MED CLIN, 54-; CLIN PROF SURG, MED SCH, STANFORD UNIV, 58- *Concurrent Pos:* Pres & chief exec officer, Palo Alto Med Found; exec dir, Palo Alto Clin, 66-82; nat bd dir, Am Can Soc. *Honors & Awards:* Nat Div Award, Am Cancer Soc, 79; Russel V Lee Award Lectureship, 82 & Yater Award, 85, Am Group Pract Asn. *Mem:* Nat Acad Sci; Sigma Xi; Am Asn Thoracic Surg; Am Col Chest Physicians; Am Col Surgeons; Soc Thoracic Surgeons. *Mailing Add:* Palo Alto Med Found 300 Homer Ave Palo Alto CA 94301

JAMPOLSKY, ARTHUR, b Bismarck, NDak, Apr 24, 19; m 57; c 3. OPHTHALMOLOGY. *Educ:* Univ Calif, AB, 40; Stanford Univ, MD, 44; Am Bd Ophthal, dipl, 50. *Prof Exp:* Chief strabismus clin, 50-60, DIR SMITH-KETTLEWELL INST VISUAL SCI, PRESBY MED CTR, 60- *Concurrent Pos:* Mem comt on vision, Armed Forces-Nat Res Coun, 58-, exec coun, comt on vision, 60-64; vis sci study sect, NIH, 67-71, chmn, 70-71; regional consult ophthal, Oak Knoll Naval Hosp, Oakland & Travis AFB; consult, Letterman Gen Hosp, San Francisco & Calif State Bd Health; spec consult, Nat Inst Neurol Dis & Blindness. *Mem:* Am Optom Asn; Am Acad Ophthal & Otolaryngol; Am Ophthal Soc; Am Asn Ophthal; fel Am Col Surg. *Res:* Binocular vision; strabismus; physiological optics. *Mailing Add:* Smith-Kettlewell Inst Visual Sci 2232 Webster St San Francisco CA 94115

JAMRICH, JOHN XAVIER, b Muskegon Heights, Mich, June 12, 20; m 44; c 3. ACADEMIC ADMINISTRATION, STATISTICS. *Educ:* Univ Chicago, BS, 43; Marquette Univ, MS, 48; Northwestern Univ, PhD(admin), 51. *Hon Degrees:* LLD, Northern Mich Univ, 68, Grand Valley State Col, 85. *Prof Exp:* Instr math, Marquette Univ, 46-48; asst inst, Univ Wis, 48-49; asst dean of men, Northwestern Univ, 49-51; dean of students, Coe Col, 51-55; dean of fac, Doane Col, 55-57; prof & dir, Ctr Study Higher Educ, Mich State Univ, 57-63; assoc dean, Col Educ, 63-68; pres, Northern Mich Univ, 68-83; CONSULT, 83- *Concurrent Pos:* From asst dir to assoc dir, Legis Surv Higher Educ in Mich, 57-61; dir, Surv Higher Educ Grand Rapids, 59, Saginaw Valley, 62 & Study of Capital Outlay Needs for Ohio's State Insts Higher Educ, 62-63; accreditation examr & consult, NCent Asn Cols & Sec Schs, 62-; consult, Ford Found on Univ Nigeria, 64, State Bd Regents of Ohio, 65, Study of Capital Outlay Needs for Va State Comn Higher Educ, 65 & Facil Study for SC Comn Hihgher Educ, 66. *Mem:* AAAS; Am Math Soc; Am Educ Res Asn. *Res:* Educational statistics in connection with administration of the university. *Mailing Add:* Box 3129 Venice FL 34293

JAMSHIDI, MOHAMMAD MO, b Shiraz, Iran, May 10, 44; US citizen; m 74; c 2. ROBOTICS, COMPUTER-AIDED DESIGN. *Educ:* Ore State Univ, BSEE, 67; Univ Ill, Urbana-Champaign, MS, 69, PhD(elec eng), 71. *Prof Exp:* Res assoc, Univ Ill, 70-71; from asst prof to assoc prof, Dept Elec Eng, Pahlavi Univ, Iran, 71-75, prof, 77-79; scientist, Int Bus Mach Res Ctr, Yorktown Heights, 75-77; PROF ELEC ENG, UNIV NMEX, 80-, DIR, COMPUTER AIDED DESIGN LAB, 84-, AT&T PROF MFG ENG, 89- *Concurrent Pos:* Ed, Inst Elec & Electronics Engrs Control Systs Mag, 80-84; Am Soc Mech Engrs Ser Robotics & Mfg, 86- & Int J Computers & Elec Eng, 89-; adv engr, Info Prod Div, Int Bus Mach, 82-83; consult, USAF Phillips Lab, 84-, Oak Ridge Nat Lab, 88- & Los Alamos Nat Lab, 90-; hon prof, Nanjing Aeronaut Inst, People's Repub China, 86; vis prof, George Washington Univ & Nat Inst Standards & Technol, 87-88 & Univ Va, 88. *Honors & Awards:* Centennial Medal, Inst Elec & Electronics Engrs. *Mem:* Fel Inst Elec & Electronics Engrs; Inst Elec & Electronics Engrs Control Systs Soc; Soc Photo-Optical Instrumentation Engrs; Am Soc Mech Engrs. *Res:* Intelligent control systems including fuzzy logic, neural network, and expert systems; robotics control; adaptive control of nuclear reactors; computer-aided design of control systems. *Mailing Add:* 13407 Deer Trail Ct NE Albuquerque NM 87111

JAN, KUNG-MING, CARDIOLOGY, CARDIOVASCULAR PHYSIOLOGY. *Educ:* Nat Taiwan Univ, Taipei, MD, 67; Columbia Univ, PhD(physiol), 71. *Prof Exp:* ASSOC PROF PHYSIOL & MED, COL PHYSICIANS & SURGEONS, COLUMBIA UNIV, 68-; ASSOC ATTEND PHYSICIAN, PRESBY HOSP, 78- *Mailing Add:* Dept Med Columbia Presby Allen Pavilion 5141 New York NY 10034

JANAK, JAMES FRANCIS, b Yonkers, NY, Dec 5, 38; m 65; c 2. ELECTRICAL ENGINEERING, PHYSICS. *Educ:* Mass Inst Technol, SB, 60, SM, 62, ScD(elec eng), 64. *Prof Exp:* Instr elec eng, Mass Inst Technol, 62-64, asst prof, 64-65; MEM RES STAFF THEORET PHYSICS, THOMAS J WATSON RES CTR, IBM CORP, 65- *Concurrent Pos:* Ford Found fel, 64-65; adj assoc prof math, Pace Univ, 78-83, adj prof physics, 83- *Mem:* Am Phys Soc. *Res:* Solid state physics. *Mailing Add:* T J Watson Res Ctr IBM Corp Box 218 Yorktown NY 10598

JANAKIDEVI, K, GROWTH CONTROL MECHANISMS, GENE REGULATION. *Educ:* Ofmania Univ, India, PhD(protozool), 57. *Prof Exp:* ASSOC PROF MOLECULAR PATH, ALBANY MED SCH, 70- *Mailing Add:* 191 Ormond St Albany NY 12208

JANATA, JIRI, b Podebrady, Czech, July 12, 39; Brit citizen; m 62; c 2. ELECTROANALYTICAL, SOLID STATE DEVICES. *Educ:* Charles Univ, Pargue, MSc, 61, PhD(anal chem), 65. *Prof Exp:* Res fel, Univ Mich, 66-68; sr chemist, Imperial Chem Indust, 68-76; PROF BIOENG, UNIV UTAH, 76- *Concurrent Pos:* Prin investr, NSF, NIH & Dept Defense, 76-; consult, Johnson & Johnson, 79-; adj prof, Dept Chem Univ Utah, 80- *Mem:* Royal Soc Chem; Electrochem Soc; Am Chem Soc. *Res:* Electroanalytical chemistry: solid state chemically inactive devices and in flow through electrochemical detectors. *Mailing Add:* Dept Mat Sci Univ Utah Salt Lake City UT 84112

JANATOVA, JARMILA, b Pisek, Czech, Jan 9, 29; US citizen; m 62; c 2. BIOCHEMISTRY. *Educ:* Charles Univ, MSc, 61; Czech Acad Sci, PhD(biochem), 65. *Prof Exp:* Res scientist, Inst Org Chem & Biochem, Czech Acad Sci, 65; sr lectr, Dept Phys Chem, Charles Univ, Prague, Czech, 65-66; post doctoral res assoc, Biophysics Div, Inst Sci & Technol, Univ MIch, 66-67; post doctoral res assoc, Dept Biol, Univ Utah, 73-75, 76-77 & 77-79; sr exp officer, Dept Biochem, Univ Liverpool, Eng, 75-76; res instr, 79-81, res asst prof, 81-85, RES ASSOC PROF, DEPT PATHOL, UNIV UTAH, 85-, ADJ ASSOC PROF, DEPT BIOENG, 89- *Concurrent Pos:* Sr lectr, Charles Univ, 65-66; postdoctoral res assoc, Univ Mich, 66-67, dept biol, Univ Utah, 73-75, dept mat sci & eng, 76-77, dept pathol, 77-79; sr acad vis, MRC Immunochem Unit, Dept Biochem, Univ Oxford, 86-87; acad mem, Ctr Biopolymers, Univ Utah, 86- *Mem:* Sigma Xi; AAAS; Am Chem Soc; Am Soc Biochem & Molecular Biol; Am Asn Immunol; NY Acad Sci. *Res:* Biochemistry of complement proteins; protein chemistry structure/function of proteins from the complement system; isolation and characterization of proteins; biocompatibility of biomedical polymers. *Mailing Add:* Sch Med Univ Utah 50 N Medical Dr Salt Lake City UT 84132

JANAUER, GILBERT E, b Vienna, Austria, Feb 26, 31; m 58; c 1. PHYSICAL ANALYTICAL CHEMISTRY. *Educ:* Univ Vienna, PhD(chem), 62. *Prof Exp:* Jr chemist, Oemvag, Austria, 58-60; res asst, Anal Inst, Univ Vienna, 60-61, instr anal chem, 61-62; res assoc chem, Clarkson Tech, 63-64; from asst prof to assoc prof, 64-80, PROF CHEM, STATE UNIV NY, BINGHAMTON, 81- *Concurrent Pos:* Speaker, Gordon Res Conf Ion Exchange, 69, 75, 77 & vchmn, 79; vis prof, Graz Inst Technol, Austria, 71-72; NSF fac adv, 73 & 76. *Mem:* AAAS; Am Chem Soc; Sigma Xi. *Res:* Ion exhange equilibria and kinetics in aqueous and aqueous-organic solvents; separation methods; trace preconcentration and analysis; reactive ion exchange; chemical disinfection. *Mailing Add:* Dept of Chem State Univ of NY Binghamton NY 13901

JANCA, FRANK CHARLES, b Chicago, Ill, Oct 27, 46. GENETICS, DROSOPHILA MUTAGENESIS. *Educ:* Western Mich Univ, BA, 68; MA, 72; La State Univ, PhD(zool), 78. *Prof Exp:* Fel genetics, Univ Alta, 78-81; res assoc reproductive toxicol, SDak State Univ, 83-86; vis res scientist, 88-90, ADJ ASST PROF, WESTERN MICH UNIV, 90- *Mem:* Genetics Soc Am; Environ Mutagen Soc; Sigma Xi. *Res:* Mutagen testing and mutagenesis; bacteria; maize; Drosophila and mammalian systems; reproductive toxicology using mouse and rat testes as test system and flow cytometry as research tool. *Mailing Add:* Rt One 38660 76th Ave Decatur MI 49045

JANCARIK, JIRI, b Brno, Czech, Oct 9, 41; m 63; c 2. PLASMA PHYSICS. *Educ:* Charles Univ, Prague, RNDr(exp physics), 63; Czech Acad Sci, CSc(plasma physics), 68. *Prof Exp:* Fel electron beam & plasma physics, Inst Plasma Physics, Czech Acad Sci, 63-68; res officer plasma turbulence, Culham Lab, UK Atomic Energy Authority, 68-72; res assoc beam-plasma interactions, Eng Dept, Univ Oxford, 69-72; res scientist, Fusion Res Ctr, Univ Tex, Austin, 72-81; PHYSICIST, LAWRENCE LIVERMORE NAT LAB, UNIV CALIF, LIVERMORE, 81- *Mem:* Am Phys Soc. *Res:* Plasma heating and containment for thermonuclear applications; study of plasma waves; turbulence using x-ray, magnetic and electromagnetic diagnostics; computer simulation of relativistic beams, plasma turbulence; collective ion accelerators; laser isotope separation. *Mailing Add:* Lawrence Livermore Nat Lab PO Box 808 L-470 Livermore CA 94550

JANDA, JOHN MICHAEL, b Burbank, Calif, Nov 4, 49; m 79; c 3. INFECTIOUS DISEASES, MICROBIAL PATHOGENICITY & VIRULENCE. *Educ:* Loyola Univ, BS, 71; Calif State Univ, Los Angeles, MS, 75; Univ Calif, Los Angeles, PhD(microbiol & immunol), 79. *Prof Exp:* Postdoctoral fel, Dept Microbiol, Mt Sinai Hosp, 79-81, asst dir/prof, 81-84, assoc dir/prof, 84-86; res microbiologist, 86-90, RES SCIENTIST, MICROBIOL DIS LAB, 90- *Concurrent Pos:* Mem, subcomt Facultatively Anaerobic Gram-Negative Rods, Am Soc Microbiol, 87- *Honors & Awards:* Aeromonas Jandaei named in honor, 91. *Mem:* Am Soc Microbiol. *Res:* Microbial pathogenesis; clinical microbiology; diagnostic microbiology; Aeromonas pathogenesis and taxonomy; cellular replication and invasion by inertia bacteria; Vibrio infections; toxigenic bacteria. *Mailing Add:* Microbial Dis Lab Calif Dept Health Serv 2151 Berkeley Way Berkeley CA 94704-1011

JANDA, KENNETH CARL, b Denver, Colo, Nov 28, 50; m 71; c 2. MOLECULAR SPECTROSCOPY, SURFACE CHEMISTRY. *Educ:* Hope Col, AB, 73; Harvard Univ, AM, 75, PhD(phys chem), 77. *Prof Exp:* Fel, Univ Chicago, 77-78; res instr physics, A A Noyes, 78-80; asst prof, Calif Inst Technol, 80-85; PROF DEPT CHEM, UNIV PITTSBURGH, 86- *Concurrent Pos:* A P Sloan fel, 81-84; Dreyfus fel, 83-85; Fulbright fel, 85. *Mem:* Am Chem Soc; Am Phys Soc. *Res:* Spectroscopy of weakly bound molecules in molecular beams and on solid surfaces; dynamics of energy transfer from strong to weak bonds; dynamics of molecular processes on surfaces, applications of lasers in physical chemistry. *Mailing Add:* Dept Chem Univ Pittsburgh Chem Bldg Pittsburgh PA 15260

JANDACEK, RONALD JAMES, b Chattanooga, Tenn, Dec 26, 42. LIPID NUTRITION, FAT DIGESTION & ABSORPTION. *Educ:* Rice Univ, BA, 64; Univ Tex, PhD(chem), 68. *Prof Exp:* RES CHEMIST, MIAMI VALLEY LABS, PROCTER & GAMBLE, CINCINNATI, OHIO, 68- *Mem:* Am Chem Soc; Am Oil Chemists Soc; Am Inst Nutrit. *Res:* Physical and biological properties of lipids, including phase behavior, digestion and intestinal absorption. *Mailing Add:* Procter & Gamble Miami Valley Labs PO Box 398707 Cincinnati OH 45239-8707

JANDE, SOHAN SINGH, anatomy, histology; deceased, see previous edition for last biography

JANDL, JAMES HARRIMAN, b Racine, Wis, Oct 30, 25; m 50; c 5. MEDICINE. *Educ:* Franklin & Marshall Col, BS, 45; Harvard Med Sch, MD, 49. *Prof Exp:* Res fel med, Harvard Med Sch, 52-55, instr med, 55-57, assoc, 57-59, from asst prof to assoc prof, 59-68, dir, Harvard Med Unit, 68-70, GEORGE RICHARDS MINOT PROF MED, HARVARD MED SCH, BOSTON CITY HOSP, 68-, HEAD, DEPT HEMAT, HARVARD MED SCH, 73- *Mem:* Am Soc Clin Invest; Am Fedn Clin Res; Asn Am Physicians; Am Clin & Climat Asn; Am Soc Hemat. *Res:* Hematology; mechanisms of the anemias, especially the hemolytic anemias; immune hematology; functions of the reticuloendothelial system. *Mailing Add:* Harvard Med Sch 25 Shattuck St Boston MA 02115

JANDORF, BERNARD JOSEPH, b Berlin, Ger, May 19, 15; nat US; m 46; c 1. BIOCHEMISTRY. *Educ:* Cambridge Univ, BA, 38; Harvard Univ, AM, 40, PhD(biochem), 42. *Prof Exp:* Asst, Lilly Res Labs, Woods Hole, 41; Commonwealth Fund sr fel, Thorndike Mem Lab, Boston City Hosp, 42-44; res biochemist, Chem Corps, Med Labs, US Army Chem Center, Md, 44-49, chief enzyme chem br, 49-56, biochem res div, Chem Warfare Labs, 56-62; dep dir res directorate weapons systs, Edgewood Arsenal, 62-65, chief chem res div, 65-74, sr scientist, Frederick Cancer Res Ctr, 75-89; RETIRED. *Concurrent Pos:* Fel, Harvard Univ, 42-44; lectr, Sch Hyg & Pub Health, Johns Hopkins Univ, 46-49; assoc prof, Univ Md, 59-62. *Mem:* AAAS; Am Chem Soc; Am Soc Biol Chem:. *Res:* Biological oxidations in mammalian tissues; intermediary carbohydrate metabolism; enzyme isolations; action of toxic agents on enzymes. *Mailing Add:* PO Box 20924 Baltimore MD 21209

JANE, JOHN ANTHONY, b Chicago, Ill, Sept 21, 31; m 60; c 4. NEUROSURGERY. *Educ:* Univ Chicago, BA, 51, MD, 56, PhD(biol, psychol), 67. *Prof Exp:* From instr to assoc prof neurosurg, Sch Med, Case Western Reserve, 65-69; PROF NEUROSURG, SCH MED, UNIV VA, 69-, CHMN DEPT, 80- *Honors & Awards:* Herbert Olivecrona Lectr. *Mem:* Am Asn Anat; Am Physiol Soc. *Res:* Head injury; neuroplasticity; craniofacial surgery. *Mailing Add:* Dept Neurosurg Sch Med Univ Va Charlottesville VA 22908

JANECKE, JOACHIM WILHELM, b Heidelberg, Ger, Feb 5, 29; m 54; c 2. NUCLEAR PHYSICS, NUCLEAR STRUCTURE. *Educ:* Univ Heidelberg, Dipl Physics, 52, Dr rer nat(physics), 55. *Prof Exp:* Res asst, Max Planck Inst Nuclear Res, 55-60; res assoc, Univ Mich, 60-62; res assoc, Nuclear Res Ctr, Karlsruhe, Ger, 62-65; assoc prof, 65-69, PROF PHYSICS, UNIV MICH, ANN ARBOR, 69- *Concurrent Pos:* vis prof, MPI Heidelberg, UVI Groningen, Tel-Aviv Univ, 72, 79, 80, 86 & 88. *Mem:* Fel Am Phys Soc; Sigma Xi. *Res:* Nuclear physics; nuclear astrophysics; nuclear reactions; accelerators; nuclear structure and masses; cosmo-chronology. *Mailing Add:* Dept Physics Univ Mich Ann Arbor MI 48109

JANES, DONALD LUCIAN, b Fresno, Calif, July 1, 39. SOLID STATE CHEMISTRY. *Educ:* Grinnell Col, AB, 61; Purdue Univ, PhD(inorg chem), 66. *Prof Exp:* Sr chemist, Cent Res Lab, Minn Mining & Mfg Co, 65-71, res specialist, 71-73, supvr, Magnetic Audio-Video Prod Div, 73-79, mgr, 79-81, mgr, Info Storage Lab, 81-85, lab mgr, Appl Res Lab, 85-90, DIR, CORP ANALYSIS LAB, 3M CO, 90- *Mem:* Am Chem Soc. *Res:* Preparation and properties of magnetic materials. *Mailing Add:* 3M Corp Analysis Lab Bldg 201-1S-07 3M Ctr St Paul MN 55144-1000

JANES, DONALD WALLACE, b Kansas City, Mo, June 12, 29; m 53; c 4. BACTERIOLOGY, BIOLOGY. *Educ:* Baker Univ, AB, 51; Univ Kans, MA, 56; Kans State Univ, PhD(zool), 62. *Prof Exp:* Instr biol, Washburn Univ, 57-60; asst prof, Parsons Col, 61-62; from asst prof to assoc prof, 62-78, assoc vpres acad affairs & dean grad sch, 68-78, PROF BIOL, UNIV SOUTHERN COLO, 78- *Concurrent Pos:* Fulbright fel, 56-57; intern Acad Admin, Am Coun Educ, 68; consult & examr, NCent Asn Cols & Sec Schs, 72- *Mem:* AAAS; Am Soc Zoologists; Am Soc Microbiol; Am Soc Mammal. *Res:* Problems of vertebrate distribution and histology; reproduction; vertebrate fauna of Colorado; chemistry and biosynthesis of bacterial pigments, particularly pigments of Serratia marcescens. *Mailing Add:* Dept Biol Univ Southern Colo Pueblo CO 81001-4901

JANES, GEORGE SARGENT, b Brooklyn, NY, Apr 12, 27; m 52; c 5. PHYSICS. *Educ:* Cornell Univ, AB, 49; Mass Inst Technol, PhD(physics), 53. *Prof Exp:* Mem res staff, Nuclear Sci Div Indust Coop, Mass Inst Technol, 53-56; prin res scientist, 56-77, vpres isotope res, 77-83, DIR DYE LASER TECHNOL, AVCO-EVERETT RES LAB DIV, AVCO CORP, 83- *Concurrent Pos:* Mem adv comt, Regional Laser Ctr, Mass Inst Technol, 81- *Mem:* Fel Am Phys Soc; assoc fel Am Inst Aeronaut & Astronaut; Sigma Xi. *Res:* Meson physics and cosmic rays; magnetohydrodynamics; high temperature gas physics; plasma physics; ionization phenomena; lasers. *Mailing Add:* 34 Conant Rd Lincoln MA 01773

JANEWAY, CHARLES ALDERSON, JR, b Boston, Mass, Feb 5, 43; m 77; c 3. IMMUNE RECOGNITION, IMMUNOGENICITY. *Educ:* Harvard Col, BA, 63; Harvard Univ, MD, 69. *Hon Degrees:* Dr, Univ Cracow, Poland, 91. *Prof Exp:* Res assoc, NIH, 70-75; Moseley fel, Univ Uppsala, 75-77; from asst prof to assoc prof, 77-83, PROF PATH & BIOL, SCH MED, YALE UNIV, 83-, PROF IMMUNOBIOL, 88- *Concurrent Pos:* Investr, Howard Hughes Med Inst, 77-; lectr biol, Yale Univ, 79-; mem, Immunobiol Study Sect, NIH, 81-85, 88-92. *Mem:* AAAS; Am Asn Immunologists; Am Soc Microbiol. *Res:* Molecular basis of specific immune recognition and the genes that control it, focusing primarily on T cells that activate all immune responses. *Mailing Add:* Dept Immunobiol Sch Med Yale Univ New Haven CT 06510

JANEWAY, RICHARD, b Los Angeles, Calif, Feb 12, 33; m 55; c 3. MEDICINE, NEUROLOGY. *Educ:* Colgate Univ, AB, 54; Univ Pa, MD, 58. *Prof Exp:* Intern, Hosp Univ Pa, 58-59; resident neurol, NC Baptist Hosp, 63-66; from instr to assoc prof neurol, Bowman Gray Sch Med, Wake Forest Univ, 66-71, actg chmn, Dept Neurol, 69-70, prog dir, Cerebral Vascular Res Ctr, 69-71, dean, 71-85, vpres, Health Affairs, 83-90, PROF NEUROL, BOWMAN GRAY SCH MED, WAKE FOREST UNIV, 71-, EXEC DEAN, 85-, EXEC VPRES, HEALTH AFFAIRS, 90- *Concurrent Pos:* Prog admin, Cerebral Vascular Res Ctr, 66-69; mem spec task force, Joint Coun Subcomt, Cerebrovascular Dis, 68; coun cerebrovascular dis, Am Heart Asn,

69; mem spec procedures & equip study group, Joint Comt for Stroke Facil, 70; mem nat adv coun regional med prog, Dept Health, Educ & Welfare; consult, US-Egypt Collab Prog on Stroke; mem, Coun Deans, Asn Am Med Col, 71-, chmn, 82-85, exec coun, 77-86; Markle Scholar, 68-73. *Mem:* Inst Med-Nat Acad Sci; fel Am Heart Asn; fel Am Col Physicians; Soc Med Adminr; Am Col Physician Exec; AAAS; Sigma Xi; Am Med Asn; Am Clin & Climatological Asn. *Res:* Neurology; cerebrovascular disease. *Mailing Add:* Bowman Gray Sch Med Wake Forest Univ Winston-Salem NC 27103

JANG, SEI JOO, b Andong, Korea, Dec 30, 47; m 83; c 2. RELAXOR MATERIALS, OPTICS. *Educ:* Sogang Univ, BS, 73; Boston Col, MS, 76; Pa State Univ, PhD(solid state sci) , 79. *Prof Exp:* Serv engr med equip, Siemens Elec Eng Co, 73-74; teaching asst physics, Boston Col, 74-76; res asst mat, Pa State Univ, 76-79; sr res staff optics, AT&T Eng Res Ctr, 79-83; res assoc elec mat, Mat Res Labs, 83-87; ASSOC PROF SOLID STATE SCI, PA STATE UNIV, 87- *Concurrent Pos:* Secy, JBS Consult Inc, 84-; pres, Matronix Inc, 85-; sr res assoc, Mat Res Lab, Pa State Univ, 87- *Mem:* Am Ceramic Soc; Nat Inst Ceramic Engrs; Optical Soc Am. *Res:* Relaxor materials; microwave measurements and materials; optics and electro-optics materials; electrostrictive abd piezoelectric materials for actuator, transducers and motors. *Mailing Add:* 259 Mat Res Lab Pa State Univ University Park PA 16802

JANGAARD, NORMAN OLAF, b Seattle, Wash, Oct 11, 41; m 63; c 2. BIOCHEMISTRY. *Educ:* San Diego State Univ, BS, 62; Univ Calif, Los Angeles, PhD(biochem), 66; Univ Denver, JD, 76. *Prof Exp:* Lab technician, Scripps Inst Oceanog, Univ Calif, San Diego; biochemist, Pfizer, Inc, 66-68 & Shell Develop Co, 68-72; dir res, Adolph Coors Co, 72-74, dir qual assurance, 74-78, vpres qual assurance & res & develop, 78-80, vpres eng & res & develop, 80-81, vpres qual assurance, regulatory affairs res & develop, 81-83, vpres prod, 83-84; PRES, COORS BIOTECH INC, 85- *Mem:* Am Soc Brewing Chemists; Am Chem Soc; Inst Food Technologists; Am Asn Cereal Chemists; Master Brewer's Asn Am. *Res:* Fermentation and yeast physiology; microbiological control; brewing and malting technology; packaging materials; breeding and growing of hops, barley and rice; waste treatment technology; vitamin production; food ingredient technology. *Mailing Add:* 11297 Ranch Pl Denver CO 80234

JANGHORBANI, MORTEZA, b Isfahan, Iran, Sept 29, 43; US citizen; m 69; c 1. STABLE ISOTOPES, NEUTRON ACTIVATION. *Educ:* Am Univ Beirut, Lebanon, BS, 66; Oregon State Univ, MS, 68, PhD(chem), 72. *Prof Exp:* Assoc vis asst prof chem, Univ Ky, 72-73; res chemist, Univ Marburg, Ger, 73-75; group leader, Environ Trace Substances Res Ctr, Univ Mo, 75-77; prin res scientist, Mass Inst Technol, 77-; ASSOC PROF PATH, MALLORY INST PATH, BOSTON UNIV SCH MED. *Mem:* Am Chem Soc; AAAS. *Res:* Trace element research in relation to biology and human nutrition; analytical chemistry of trace elements. *Mailing Add:* Univ Chicago 5841 S Maryland Box 223 Chicago IL 60637

JANICK, JULES, b New York, NY, Mar 16, 31; m 52; c 2. PLANT BREEDING, TISSUE CULTURE. *Educ:* Cornell Univ, BS, 51; Purdue Univ, MS, 52, PhD(plant genetics & breeding), 54. *Hon Degrees:* DS, Univ Bologna, Italy. *Prof Exp:* From instr to prof, 54-88, DISTINGUISHED PROF HORT, PURDUE UNIV, 88- *Concurrent Pos:* Hon res assoc bot, Univ Col, Univ London, 63 & 85, Univ Pisa, 85; horticulturist, Agr Univ Minas Gerais, 63-65; vis colleague, Univ Hawaii, 69; consult, World Bank, Indonesia, 73, AID Portugal, 83, 86 & 87, Morocco, 88, China, 88, Italy, 89 & 90, Equador, 90. *Honors & Awards:* Paul Howe Shepard Award, Am Pomol Soc, 60 & 70; Marion W Meadows Award, Am Soc Hort Sci, 71, Kenneth Post Award, 81, Stark Award, 78 & 82, N F Childers Award, 82. *Mem:* Am Pomol Soc; fel Am Soc Hort Sci (pres, 86-87); fel Portuguese Hort Asn; fel AAAS; Sigma Xi; Int Soc Hort Sci. *Res:* Genetics and breeding of horticultural crops; tissue culture. *Mailing Add:* Dept Hort Purdue Univ West Lafayette IN 47907

JANICKI, BERNARD WILLIAM, b Wilmington, Del, Oct 14, 31; m 54; c 5. IMMUNOLOGY, MICROBIOLOGY. *Educ:* Univ Del, BA, 53, MA, 55; George Washington Univ, PhD(microbiol), 60. *Prof Exp:* Microbiologist, Tuberc Res Lab, Vet Admin Hosp, DC, 55-60, chief, 60-63, chief microbiol res lab, 63-72, chief pulmonary immunol res lab, 72-74; chief, Immunol Br, 74-77, Immunol & Biochem Br, 78-83, HEALTH SCI ADMINR, NAT INST ALLERGY & INFECTIOUS DIS, 74-, DEP DIR, IMMUNOL, ALLERGIC & IMMUNOL DIS PROG, 83- *Concurrent Pos:* Lectr microbiol, Univ Md, 69-79; consult, Nat Inst Allergy & Infectious Dis, 69-75; mem US tuberc panel, US-Japan Coop Med Sci Prog, 69-75; spec lectr med, George Washington Univ, 74-79. *Mem:* Am Soc Microbiol; Soc Exp Biol & Med; NY Acad Sci; Am Asn Immunol; Am Thoracic Soc. *Res:* Immunity and hypersensitivity in infectious diseases. *Mailing Add:* Dana-Farber Cancer Inst 44 Binney St Rm 1826 Boston MA 02115

JANICKI, CASIMIR A, b Milwaukee, Wis, Sept 20, 34; m 59; c 2. ANALYTICAL CHEMISTRY. *Educ:* LaSalle Col, BA, 56; Marquette Univ, MS, 58; Loyola Univ, PhD(anal chem), 64. *Prof Exp:* Anal chemist, Smith, Kline & French Labs, 57-60; sr anal chemist, 63-66, group leader, 66-74, sect head, McNeil labs, Ft Washington, 74-80, sect head, 74-81, TECH DIR ANAL QUAL CONTROL, MCNEIL PHARMACEUT, SPRING HOUSE, PA, 82- *Mem:* Am Asn Pharm Sci. *Res:* Pharmaceutical analytical chemistry, including thin layer, chromatography, ultra violet visible and infrared spectrometry, separation techniques including high performance liquid and gas liquid chromatography; kinetics and drug stability; robotics in pharmaceutical analysis, laboratory computers, validation of computer systems in GMP area; FTIR in QC analysis. *Mailing Add:* 2888 Hickory Hill Dr RD 1 Norristown PA 19403

JANICZEK, PAUL MICHAEL, b Hazleton, Pa, Oct 5, 37. ASTRONOMY, NAVIGATION. *Educ:* King's Col, BA, 60; Georgetown Univ, MA, 65, PhD(astron), 70. *Prof Exp:* Supvr qual control, Lansdale Div, Philco Corp, 60-61; programmer analyst sci satellites, Fed Syst Div, IBM Corp, 61-66; astronr, 67-84, ed, Navig, 78-85, chief ephemerides div, 84-90, DIV ASTRON APPLNS DEPT, US NAVAL OBSERV, 90- *Concurrent Pos:* Instr, Maryland Col Art & Design, 81-83. *Mem:* Am Astron Soc; AAAS; Inst Navigation; Sigma Xi; Int Astron Union. *Res:* Dynamical astronomy; celestial navigation. *Mailing Add:* US Naval Observ Washington DC 20392-5100

JANIK, BOREK, b Brno, Czech, Oct 29, 33; m 65; c 2. CLINICAL CHEMISTRY, BIOTECHNOLOGY. *Educ:* Purkyne Univ, Brno, MS, 56; Czech Acad Sci, Brno, PhD(chem, biophys), 64; Purkyne Univ, RNDr, 66. *Prof Exp:* Res assoc org chem, Lachema, Pure Chem Corp, Czech, 56-60; fel, Inst Biophys, Czech Acad Sci, Brno, 60-64, res scientist electrochem & biophys, 64-66 & 67-68; fel chem, Univ Mich, Ann Arbor, 66-67, res assoc electrochem, 68-69; sr res scientist phys biochem, Molec Biol, Miles Lab, Inc, 69-74, mem staff & mgr res & develop, Res Prod Div, 75-78, mgr res & develop, Ames Div, 78-79; dir clin res & develop, Gelman Sci, 79-82, dir lab prod develop, 82-85, dir lab technol, 85-90; DIR TECH AFFAIRS, BIOPORE, CAAN, FRANCE, 91-; OWNER, MORAX, CHELSEA, MICH, 91- *Mem:* Electrophoresic Soc; Am Asn Clin Chem. *Res:* Test and instrument programs in clinical chemistry and biotechnology (proteins, nucleic acids, immunochemistry, enzymology and hematology) utilizing separation technologies; binding and transfer of nucleic acids and proteins to binding membranes; author of over 70 technical publications. *Mailing Add:* Morax 13805 Waterloo Rd Chelsea MI 48118

JANIK, GERALD S, b Niagara Falls, NY, July 2, 40; m 67; c 3. ENVIRONMENTAL SCIENCES, ANALYTICAL CHEMISTRY. *Educ:* Niagara Univ, BS, 61; Purdue Univ, MS, 64; Tex A&M Univ, PhD(phy chem), 66. *Prof Exp:* Sr engr, Bell Aerospace Co, 66-72; engr, 73-76, res engr, 76-81, RES DIR, NY STATE ELEC & GAS CORP, 81- *Concurrent Pos:* Fossil Fuel Comn, Empire State Elec Energy Res Corp, 76-; reviewer, NY State Energy Res & Develop Authority, 81-; task force, 86-88, div comt, Elec Power Res Inst, 91-94. *Res:* Energy production and pollution control measures. *Mailing Add:* 817 Catalina Blvd Endwell NY 13760

JANIS, ALLEN IRA, b Chicago, Ill, Sept 11, 30; m 53; c 2. PHYSICS. *Educ:* Northwestern Univ, BS, 51; Syracuse Univ, PhD(physics), 57. *Prof Exp:* From instr to assoc prof, 57-68, PROF PHYSICS, UNIV PITTSBURGH, 68-, ASSOC DIR, PHILOS SCI CTR, 75- *Concurrent Pos:* Sr res assoc, Philos Sci Ctr, Univ Pittsburgh, 67-75. *Mem:* AAAS; Am Phys Soc; Am Asn Physics Teachers; Philos Sci Asn. *Res:* General relativity. *Mailing Add:* Dept of Physics Univ of Pittsburgh Pittsburgh PA 15260

JANIS, CHRISTINE MARIE, b London, Eng, Oct 18, 50; m 91. MAMMALIAN PALEOBIOLOGY & SYSTEMATICS. *Educ:* Univ Cambridge, UK, BA, 73; Harvard Univ, PhD(biol), 79. *Prof Exp:* Res fel, Newnham Col, Univ Cambridge, UK, 79-83; asst prof, 83-89, ASSOC PROF BIOL, DIV BIOL & MED, BROWN UNIV, 89- *Concurrent Pos:* Officer, Harvard Univ, 84- *Honors & Awards:* G G Simpson Prize in Paleont, 85. *Mem:* Soc Vert Paleont; Paleont Soc; Am Soc Mammalogists; Am Soc Zoologists; Int Soc Cryptozool. *Res:* Paleoecology and patterns of evolutionary diversification in ungulates, hoofed mammals, in relation to environmental change; combining data from living and fossil taxa. *Mailing Add:* Div Biol & Med Brown Univ Providence RI 02912

JANIS, F TIMOTHY, b Chicago, Ill, Apr 11, 40; m 62; c 3. THEORETICAL CHEMISTRY, COMPUTER SCIENCE. *Educ:* Wichita State Univ, BS, 62, MS, 63; Ill Inst Technol, PhD(chem), 68. *Prof Exp:* Res assoc chem, Argonne Nat Lab, 66-68; instr chem & data processing, Col DuPage, 68-69; from asst prof to assoc prof chem, Ill Benedictine Col, 69-74; assoc prof & asst acad dean/registr, Franklin Col, 74-77; admin mgr, Indianapolis Ctr Advan Res, 77-78, actg dir, Indust Liaison Off, 78-80, dir, Bus Develop Div, 80-83, dir prog develop, 83-84, technol transfer, 84-90; dir, ARAC, 91; PRES, J-TECH & ASSOCS, INC, 90- *Concurrent Pos:* Fel, Argonne Nat Lab, 68, consult, 68- *Mem:* Am Chem Soc. *Res:* Technology transfer methodology and delivery; information engineering; ab-initio caculations on molecules and entrepreneurial development. *Mailing Add:* ARAC 611 N Capitol Indianapolis IN 46204

JANIS, RONALD ALLEN, b Sask, Can, Oct 11, 43; m 68; c 2. CALCIUM CHANNEL MODULATORS, POTASSIUM CHANNEL LIGANDS. *Educ:* Univ BC, BSP, 66, MSP, 68; State Univ NY Buffalo, PhD(biochem pharmacol), 72. *Prof Exp:* Fel pharmacol, Univ Alta, 72-74; asst prof physiol, Northwestern Univ, 74-80, assoc prof, 80; prin res scientist, 84, HEAD & PRIN STAFF SCIENTIST, MILES INST PRECLIN PHRMACOL, 80-; ASSOC PROF MED, UNIV CONN HEALTH CTR, 81-, ASSOC CLIN PROF, 89-; PRIN STAFF SCIENTIST, MILES INC, 84- *Concurrent Pos:* Prin investr, NIH grants, 74-83; vis scientist, Bayer AG, Inst Pharmacol, 80; asst prof med, Univ Conn Health Ctr, 80-81. *Mem:* Am Physiol Soc; Am Soc Pharmacol & Exp Therapeut; Biophys Soc; British Pharmacol Soc; Soc Gen Physiologists; Am Soc Neurochem. *Res:* Mechanisms of action of drugs acting on calcium and potassium channels and development of such drugs; drug development; binding studies; calcium channel activators and antagonists. *Mailing Add:* Miles Inst Preclin Pharmacol 400 Morgan Lane West Haven CT 06516

JANISCHEWSKYJ, W, b Prague, Czech, Jan 21, 25; Can citizen; m 51; c 2. ELECTRICAL ENGINEERING. *Educ:* Univ Toronto, BASc, 52, MASc, 54. *Prof Exp:* Demonstr elec eng, Univ Toronto, 52-54, instr, 54-55; elec engr, Aluminum Labs, Ltd, 55-59; lectr elec eng, 59-62, from asst prof to assoc prof, 62-70, asst head Elec Eng Dept, 65-70, assoc dean fac, Appl Sci & Eng, 78-82, PROF ELEC ENG, UNIV TORONTO, 70- *Concurrent Pos:* Nat Res Coun Can res grant, 61-; consult, Elec Eng Consociates, 68- *Mem:* Fel Inst Elec & Electronics Engrs; Can Elec Asn. *Res:* Distribution of mechanical stress in composite transmission-line conductors; extra high voltage transmission of electric power; radio interference caused by high voltage corona; fault behavior of complex electric power systems; methods of testing underground cable; lightning studies; microgap discharges; television interference. *Mailing Add:* Dept Elec Eng Univ Toronto Toronto ON M5S 1A1 Can

JANKE, NORMAN C, b Milwaukee, Wis, Sept 5, 23; m 52; c 1. ENVIRONMENTAL GEOLOGY, MINING-PETROGRAPHY. *Educ:* Univ Chicago, MS, 52; Univ Calif, Los Angeles, PhD(geol), 63. *Prof Exp:* Consult geologist, Geo-Sci Inc, Tex, 53; instr geol, Fresno State Col, 55; instr geol & math, 56-60, from asst prof to prof, 60-83, head dept, 68-74, EMER PROF GEOL, CALIF STATE UNIV, SACRAMENTO, 83; CONSULT GEOLOGIST, NORMAN JANKE ASSOC. *Concurrent Pos:* Trustee bd mem, Moss Landing Marine Lab, 67-71; consult mining, eng & forensic geol, fault & seismic risk, petrography. *Mem:* NY Acad Sci; Sigma Xi; Asn Eng Geologists; Soc Econ Paleont & Mineral; Am Military Eng; Am Soc Appl Technol; Am Soc Testing Mat; Soc Explosive Engrs; Soc Mining Engrs; Am Inst Mining, Metall & Petrol Engrs. *Res:* Slumping and land sliding mechanisms; effects of shape upon settling velocity and sieving; photogrammetric uses of ordinary camera equipment; particle size and shape analysis, sieving and settling methods; swelling clays genesis and effects. *Mailing Add:* 2670 Fair Oaks Blvd Sacramento CA 95864

JANKE, ROBERT A, b Detroit, Mich, Aug 19, 22; m 44; c 4. PLANT ECOLOGY, PHYSICS. *Educ:* Univ Mich, AB, 44, MS, 52; Mich Technol Univ, BS, 48; Univ Colo, PhD(ecol), 68. *Prof Exp:* Teacher pub sch, 44; from instr to assoc prof physics, 44-62, assoc prof, 62-84, EMER PROF BIOL, MICH TECHNOL UNIV, 84- *Concurrent Pos:* NSF sci fac fel, 63-65. *Mem:* AAAS; Ecol Soc Am; Sigma Xi; Bot Soc Am; George Wright Soc. *Res:* Fire ecology; boreal forest ecology; vascular flora inventory of Isle Royale National Park; physical ecology. *Mailing Add:* Dept Biol Mich Technol Univ Houghton MI 49931

JANKE, WILFRED EDWIN, b Morris, Man, Dec 24, 32; m 58; c 4. SOIL SCIENCE. *Educ:* Univ Man, BSA, 55, MSc, 57; Univ Wis-Madison, PhD(soils, geol), 62. *Prof Exp:* Pedologist, Soil Surv Div, Can Dept Agr, 57-59, res scientist, Res Sta, 62-63; dir, Soil Testing Lab, Univ Man, 63-66; res agronomist & mkt coordr, fertilizer mkt div, Sherritt Gordon Mines Ltd, 66-78; fertilize prod mgr, Federated Cooperatives Ltd, 78-81, mkt res sr analyst, Potash Corp, Saskatoon, 81-83; res & develop specialist, BASF Can, Inc 83-88. *Mem:* Am Soc Agron; Can Soc Soil Sci; Agr Inst Can; Int Soc Soil Sci. *Res:* Soil fertility, nutrient requirements of various crops under various soil and climatic conditions; fertilizer research, development of new fertilizer products, determining agronomic uses and effectiveness; evaluation of pesticide products. *Mailing Add:* 14 DeGeer Cres Saskatoon SK S7H 4P7 Can

JANKOWSKI, CHRISTOPHER K, b Warsaw, Poland, July 31, 40; m 66; c 1. ORGANIC CHEMISTRY. *Educ:* Univ Warsaw, MSc, 63; Univ Montreal, PhD(chem), 68; Univ Paris, Doct Etat(phys), 85. *Prof Exp:* Asst org chem, Univ Warsaw, 63-64; fel, Univ Montreal, 67-68, asst prof, 68-69; from asst prof to assoc prof, 69-78, PROF ORG CHEM, UNIV MONCTON, 78-, DEAN, FAC OF RES & GRAD STUDENTS, 85- *Concurrent Pos:* Res fel chem, Syntex, SA, Mex, 75; res fel, Nuclear Studies Ctr, Saclay, France, 75-79; chmn dept chem, Univ Moncton, 78-81; adj prof, Univ Nacional Autonome, Mex, 88-, Univ New Brunswick, Can, 90-, Univ de Paris VI, France, 89- *Mem:* Fel Chem Inst Can; Fr-Can Asn Advan Sci. *Res:* Synthesis of organic compounds with physiological activity; organic application of mass spectrometry and nuclear magnetic resonance; natural products; alkaloids, carbohydrates. *Mailing Add:* Univ Moncton Moncton NB E1A 3E9 Can

JANKOWSKI, CONRAD M, b Chicago, Ill, Feb 25, 28; m 53. ENVIRONMENTAL CHEMISTRY, HIGH TEMPERATURE REACTION. *Educ:* Mich State Univ, BS, 51, MS, 53; State Univ Iowa, PhD(anal chem), 60. *Prof Exp:* Chief anal chemist, Rayovac Corp, 53-55; group leader instrumentation res, Cent Sci Co, 55-58; asst prof anal chem, 60-63, ASSOC PROF ANALYTICAL CHEM & CHEM OCEANOG, NORTHEASTERN UNIV, 63- *Concurrent Pos:* Indust consult; vis prof, Trent Polytechnic Nottingham, Eng. *Mem:* AAAS; Am Chem Soc; fel Royal Soc Chem; fel Am Inst Chemists. *Res:* Electroanalytical chemistry; high temperature reactions; chemical instrumentation; air and water pollution measurements. *Mailing Add:* Dept Chem Northeastern Univ Boston MA 02115

JANKOWSKI, FRANCIS JAMES, b Amsterdam, NY, Nov 22, 22; m 46; c 1. DESIGN ENGINEERING, NUCLEAR ENGINEERING. *Educ:* Union Col, NY, BScCE, 43; Univ Cincinnati, MSE, 47, ScD(physics), 49. *Prof Exp:* Res engr nuclear, Battelle Mem Inst, 49-50; adv scientist, Westinghouse Elec Corp, 50-55; consult, Battelle Mem Inst, 55-59; prof nuclear & mech, Rutgers Univ, 59-69; chmn dept eng, 69-74, prof systs eng, Wright State Univ, 69-84; RETIRED. *Concurrent Pos:* Consult, Englehard Industs, NJ, 59-62, United Nuclear Corp, 59-65, Picatinny Arsenal, US Army, 64-69; Westinghouse Elec Corp, 60- & US Dept Energy, 73-78; sabbatical leave, Foreign Technol Div, US Air Force, 79-80. *Mem:* Sigma Xi; Am Nuclear Soc; Am Phys Soc; Am Soc Eng Educ. *Res:* Principles and methodologies of engineering design process, with emphasis on incorporating human factors variables, life cycle costs, and systems approach. *Mailing Add:* 5800 Mahogany Pl NE Albuquerque NM 87111

JANKOWSKI, STANLEY JOHN, b Detroit, Mich, Dec 19, 28; m 54; c 4. ANALYTICAL CHEMISTRY. *Educ:* Washington & Jefferson Col, BA, 53; Univ Pittsburgh, PhD(anal chem), 60. *Prof Exp:* Supvr, Neville Chem Co, 58-60; anal chemist, Celanese Corp Am, 60-62, sr anal chemist, 62-66; sr res chemist, Atlas Chem Indust Inc, 66-70, res supvr, 70-81, RES SPECIALIST, ICI AMERICAS INC, 81- *Mem:* Am Chem Soc; Am Indust Hyg Asn. *Res:* Instrumental methods of analysis; chromatographic methods of analysis; drugs; organic chemicals; industrial hygiene analysis. *Mailing Add:* 28 Amarante Laguna Niguel CA 92677-8929

JANKUS, EDWARD FRANCIS, b Chicago, Ill, Mar 17, 30; m 55; c 2. PHYSIOLOGY. *Educ:* Univ Minn, BS, 57, DVM, 59, PhD(vet physiol), 66. *Prof Exp:* Vet pvt pract, 59-61; from instr to assoc prof vet physiol & pharmacol, Univ Minn, St Paul, 61-89; RETIRED. *Mem:* NY Acad Sci; Am Vet Med Asn; Am Animal Health Asn. *Res:* Comparative cardiovascular physiology. *Mailing Add:* Dept Vet Biol Univ Minn St Paul MN 55108

JANKUS, VYTAUTAS ZACHARY, b Girvalakis, Lithuania, Sept 6, 19; US citizen; wid; c 1. NUCLEAR SCIENCE. *Educ:* Univ Vilnius, Lithuania, dipl, 43; Stanford Univ, PhD(physics), 56. *Prof Exp:* Instr physics, Seattle Univ, 48-51; asst, Stanford Univ, 51-55; assoc physicist, Mat Sci Div, Argonne Nat Lab, 55-75; sr physicist, 75-81; RETIRED. *Mem:* Am Asn Physics Teachers; Am Phys Soc; Sigma Xi. *Res:* Electron scattering; neutron thermalization; reactor safety; performance. fuel element. *Mailing Add:* 801 McCarthy Rd Lemont IL 60439-4044

JANKY, DOUGLAS MICHAEL, b Hastings, Nebr, Nov 3, 46; m 70; c 3. FOOD SCIENCE & TECHNOLOGY, POULTRY SCIENCE. *Educ:* Univ Nebr, BS, 69, MS, 71, PhD(food sci & technol), 74. *Prof Exp:* From asst prof to assoc prof, 74-83, PROF, POULTRY PROD TECHNOL, UNIV FLA, GAINESVILLE, 84- *Concurrent Pos:* Consult qual control, USDA. *Honors & Awards:* Poultry & Egg Inst of Am Res Award, 83. *Mem:* Poultry Sci Asn; Inst Food Sci & Technol. *Res:* Poultry products technology; color and pigmentation of poultry products, meat tenderness and acceptability; mechanically deboned poultry; meat and emulsions. *Mailing Add:* Dept Poultry Sci Univ Fla Gainesville FL 32611

JANNA, WILLIAM SIED, b Toledo, Ohio, Mar 23, 49; m 75; c 1. SPRAY RESEARCH. *Educ:* Univ Toledo, BSME, 71, MSME, 73, PhD(transport phenomena), 76. *Prof Exp:* From asst prof to assoc prof mech eng, Univ New Orleans, 76-87, chmn dept, 78-83; chmn, 87-91, PROF, DEPT MECH ENG, MEMPHIS STATE UNIV, 87- *Mem:* Am Soc Mech Engrs; Am Soc Eng Educ. *Res:* Windmill economics; droplet sizes of airless sprays; heat transfer from high pressure sprays; heat transfer from fluid flow in a tube to a cooled isothermal wall; economics of pipeline sizing. *Mailing Add:* 3126 Autumn Gold Lane Memphis TN 38119

JANNASCH, HOLGER WINDEKILDE, b Holzminden, Ger, May 23, 27; m 56; c 1. MICROBIOLOGY. *Educ:* Univ Gottingen, PhD(microbiol), 55. *Prof Exp:* Asst scientist microbiol, Max Planck Soc, 56-60; asst prof, 61-63, PVT DOCENT, UNIV GÖTTINGEN, 63-; SR SCIENTIST, WOODS HOLE OCEANOG INST, 63- *Concurrent Pos:* Fel, Scripps Inst Oceanog, Univ Calif, San Diego, 57-58; fel, Univ Wis, 58-59; mem, Marine Microbiol Panel, Off Naval Res, 65-70; dir microbiol ecol course & mem corp, Marine Biol Lab, Woods Hole, Mass, 71-80; comt Environ Microbiol, Am Soc Microbiol, 71-73 & 79-81; mem panel water criteria, Nat Acad Sci & NSF, 75-78; trustee, Marine Biol Lab, Woods Hole, 80-88; mem panel, Comt Probs Environ & Global Sulfur Transformations, 84-; mem panel, Comt Ocean Res & Hydrothermal Emanations Plate Boundaries, 87-; mem, Space Sci Bd Nat Res Coun Comt Planetary Biol & Chem Evolution, 87- *Honors & Awards:* Henry Bryant Bigelow Medal Oceanog, 80; Fisher Award, in Environ & Appl Microbiol, 82. *Mem:* Am Soc Microbiol; Am Soc Limnol & Oceanog; Int Asn Theoret & Appl Limnol; fel AAAS; Gottingen Acad Sci; Am Acad Arts & Sci; Am Chem Soc. *Res:* Physiology and ecology of freshwater and marine bacteria; deep sea microbiology; growth of microorganisms at extreme temperatures and pressures; deep sea hydrothermal vents; author or co-author of 140 scientific publications in microbiology. *Mailing Add:* Woods Hole Oceanog Inst Woods Hole MA 02543

JANNETT, FREDERICK JOSEPH, JR, b Newark, NJ, Mar 6, 46; m 72. MAMMALOGY. *Educ:* Cornell Univ, BS, 67, PhD(ecol & evolutionary biol), 77; Tulane Univ, MS, 69. *Prof Exp:* Vis fel, Cornell Univ, 77-78, postdoctoral fel, 78-81; lectr mammal, 81; HEAD & CUR, DEPT BIOL, SCI MUS MINN, ST PAUL, 82- *Concurrent Pos:* Exchange scientist, Acad Sci, USSR, 84, vis scientist, 88; postdoctoral assoc, Bell Mus Natural Hist, Univ Minn, Minneapolis, 85- *Honors & Awards:* A B Howell Award, Am Soc Mammalogists, 77. *Res:* Variation, social dynamics and demography of microtine rodents. *Mailing Add:* Sci Mus Minn 30 E Tenth St St Paul MN 55101

JANNETTA, PETER JOSEPH, b Philadelphia, Pa, Apr 5, 32; m 54; c 6. SURGERY. *Educ:* Univ Pa, AB, 53, MD, 57; Am Bd Surg, dipl, 64; Am Bd Neurol Surg, dipl, 69. *Prof Exp:* From asst instr to instr surg, Sch Med, Univ Pa, 58-63, instr pharmacol, 60-63; assoc surg & neurosurg, Univ Calif, Los Angeles, 63-66; assoc prof surg & chmn div neurosurg, Med Ctr, La State Univ, 66-71; PROF NEUROL SURG & CHMN DEPT, SCH MED, UNIV PITTSBURGH, 71- *Concurrent Pos:* NIH training grant, Univ Pa, 60-63, res grants, Med Ctr, La State Univ, 67-70; develop training grant, 68-71. *Mem:* Fel Am Col Surg; Soc Neurol Surgeons; Am Asn Neurol Surgeons; Cong Neurol Surgeons; Neurosurg Soc Am. *Res:* Pheochromocytoma; catechol amine determinations; single unit recording in the vestibular system; mesoscopic central nervous system anatomy and pathology; trigeminal nerve function; trigeminal neuralgia; cranial nerve dysfunction syndromes-etiology and treatment; spinal cord injury. *Mailing Add:* Dept Neurol Surg Univ Pittsburgh Sch Med 3550 Terrace St Pittsburgh PA 15261

JANNEY, CLINTON DALES, b Dover, NJ, Mar 10, 20; m 43. RADIOLOGICAL PHYSICS. *Educ:* Univ Ill, BS, 41; Univ Calif, PhD(physics), 45. *Prof Exp:* Physicist, Manhattan Proj, Univ Calif, 42-46 & 47; from asst prof to assoc prof physiol & med physics, Col Med, Univ Iowa, 47-53; sr physicist, Southwest Res Inst, 53-54; assoc cancer res scientist physics, Roswell Park Mem Inst, 54-59; assoc prof, Univ Vt, 59-70, prof radiol physics, 70-82; RETIRED. *Concurrent Pos:* Am Cancer Soc fel, Nat Res Coun, 46-47. *Mem:* AAAS; Radiol Soc NAm; Am Asn Physicists in Med; Am Phys Soc; Biophys Soc; Sigma Xi. *Res:* Medical radiologic physics. *Mailing Add:* 51 Oak St Hyannis MA 02601

JANNEY, DONALD HERBERT, b Kansas City, Mo, Nov 26, 31; div; c 2. PHYSICS. *Educ:* Univ Ill, BS, 52; Stanford Univ, MS, 53, PhD(appl physics), 57. *Prof Exp:* Asst, Los Alamos Sci Lab, 52; asst, Microwave Lab, Stanford Univ, 53-56; staff mem, 56-65, alternate group leader, 65-74, group leader, 74-81, STAFF MEM, LOS ALAMOS NAT LAB, 81- *Mem:* AAAS; Am Phys Soc; sr mem Inst Elec & Electronics Engrs; Sigma Xi. *Res:* Gamma ray measurements; flash radiography; image processing; image analysis; non-destructive evaluation; intelligence analyst. *Mailing Add:* 229 Barranca Rd Los Alamos NM 87544-2409

JANNEY, GARETH MAYNARD, b Toledo, Ohio, Feb 19, 34; m 60; c 1. OPTICS. *Educ:* Columbia Univ, AB, 55; Georgetown Univ, MS, 62, PhD(physics), 65. *Prof Exp:* Physicist, US Army Night Vision Lab, US Army Electronics Command, 65-69; from mem tech staff lasers to sr staff physicist, 69-75, asst dept mgr lasers, Hughes Res Labs, 75-79, PROJ MGR, SPACE SENSORS DIV, HUGHES AIRCRAFT CO, 79- *Mem:* Am Optical Soc; Inst Elec & Electronics Engrs. *Res:* Diatomic molecular spectroscopy, gas laser research, laser mode control and diffractionoptics for high energy lasers, tunable electro-optical infrared filters. *Mailing Add:* 332 16th St Manhattan Beach CA 90266

JANOFF, AARON, experimental pathology; deceased, see previous edition for last biography

JANOS, DAVID PAUL, b Chicago, Ill, Nov 24, 47. TROPICAL PLANT ECOLOGY, MYCORRHIZAE. *Educ:* Carleton Col, BA, 69; Univ Mich, Ann Arbor, MS, 71, PhD(bot), 75. *Prof Exp:* Herbarium asst trop bot, Field Mus Natural Hist, 70; fel mycorrhizae, 76-79, asst prof, 79-84, ASSOC PROF BIOL, UNIV MIAMI, 84- *Concurrent Pos:* Field sta mgr, Orgn Trop Studies, Inc, 75; mem, Nat Acad Sci-Nat Res Coun Comt Selected Biol Problems Humid Trop, 80-81. *Mem:* Mycological Soc Am; Asn Trop Biol; Sigma Xi; Ecol Soc Am; Orgn Trop Studies (secy, 82-83). *Res:* The evolutionary ecology of mutualistic associations, and the influences of mutualistic root associations on plant community composition and dynamics, especially those of vesicular-arbuscular mycorrhizae in the tropics. *Mailing Add:* Dept Biol Univ Miami PO Box 249118 Coral Gables FL 33124-9118

JANOS, LUDVIK, b Brno, Czech, Oct 3, 22. MATHEMATICS. *Educ:* Charles Univ, Prague, Dr rer nat(math), 50. *Prof Exp:* Mathematician, Res Inst, Prague, 50-63; vis assoc prof math, George Washington Univ, 63-65; assoc prof, Dalhousie Univ, 65-66; vis assoc prof, Univ Fla, 66-69, assoc prof, 69-74; vis prof, Univ Mont, Missoula, 74-75; assoc prof math, Wash State Univ, 75-77; assoc ed, Math Rev, Univ Mich, 77-80; res prof, Univ Md, 80-86; ASSOC PROF MATH, CALIF STATE UNIV, LONG BEACH, 86- *Mem:* Am Math Soc. *Res:* Functional analysis applied to the theory of differential equations; theory of fixed points; general topology; mathematical statistics; algebraic topology applied to digital geometry and pattern recognition; mathematical logic; algebraic topology applied to dynamical systems; Ergodic theory; partial differential theory; dimension theory. *Mailing Add:* Dept Math Univ Calif Los Angeles CA 90024

JANOS, WILLIAM AUGUSTUS, b Easton, Pa, Nov 9, 26; m 59. PHYSICS, INFORMATION SCIENCE. *Educ:* Rutgers Univ, BS, 51; Univ Calif, Berkeley, MA, 54, PhD(physics), 58. *Prof Exp:* Res physicist, Convair Astronaut Div, Gen Dynamics Corp, 51-60; staff physicist, Res Div & Advan Develop Lab, Raytheon Co, 60-63; sr tech specialist, NAm Space & Info Systs Div, 63-66; prin scientist, Philco-Ford Aeronutronic Appl Res Lab, 66-67; sr staff physicist, Missile Syst Div Labs, Raytheon Co, 67; sr scientist, Technol Serv Corp, 67-74; sr staff engr, McDonnell Douglas Astronautics Co, 74-78; prin electronics engr, Interstate Electronics Corp, 78-84; STAFF CONSULT, AEROJET GEN CORP, 84-; STAFF SPECIALIST, ROCKWELL INST, 84- *Concurrent Pos:* USAEC del, Int Conf Controlled Thermonuclear Fusion, 61; US del, Plasma Physics Symp, Int Union Pure & Appl Chem, USSR, 65. *Mem:* Sigma Xi; Am Phys Soc; Inst Elec & Electronics Engrs; Am Asn Advan Sci. *Res:* Statistical physics and electromagnetics of Boltzmann and Fokker-Planck equations; Wiener-Hopf integral equations of statistical communications and information theory; systems analysis; phenomenology hydrodynamics; analytical modeling of physical systems, sensors and signal processing; mathematical physics; statistical optics. *Mailing Add:* 8381 Snowbird Dr Huntington Beach CA 92646

JANOTA, HARVEY FRANKLIN, b Gonzales, Tex, Nov 30, 35. ANALYTICAL CHEMISTRY. *Educ:* Tex Lutheran Col, BS, 57; Univ Tex, PhD(chem), 63. *Prof Exp:* Instr chem, Tex Lutheran Col, 60-63; asst prof, Muhlenberg Col, 63-68; assoc prof, 68-74, PROF CHEM, CALIF STATE UNIV, FULLERTON, 74- *Mem:* Am Chem Soc. *Res:* Spectrophotometry of the platinum elements; instrumental methods of analysis; infrared determination of minerals. *Mailing Add:* Dept Chem Calif State Univ 800 N State Col Blvd Fullerton CA 92634

JANOVY, JOHN, JR, b Houma, La, Dec 27, 37; m 61; c 3. ZOOLOGY. *Educ:* Univ Okla, BS, 59, MS, 62, PhD(zool), 65. *Prof Exp:* Trainee, Rutgers Univ, 65-66; assoc prof, 66-74, prof zool, 74-, asst dean arts & sci, 66-, AT DEPT LIFE SCI, UNIV NEBR, LINCOLN. *Mem:* Am Soc Trop Med & Hyg; Am Soc Parasitol. *Res:* Epidemiology of parasitic protozoa; comparative metabolism and evolution and of parasitic flagellates. *Mailing Add:* Dept Life Sci Univ Nebr Lincoln NE 68588

JANOWITZ, GERALD S(AUL), b Bronx, NY, Apr 5, 43; m 68; c 1. FLUID MECHANICS. *Educ:* Polytech Inst Brooklyn, BS, 63; Johns Hopkins Univ, MS, 65, PhD(mech), 67. *Prof Exp:* Fel mech, Johns Hopkins Univ, 67-68; asst prof fluid mech, Case Western Reserve Univ, 68-75; assoc prof oceanog, 75-80, PROF MARINE SCI, NC STATE UNIV, 80- *Mem:* Am Geophys Union. *Res:* Geophysical fluid mechanics; motion of bodies through stratified fluids; flows in lakes, ocean basins, and the coastal boundary layers. *Mailing Add:* Dept Marine Earth & Atmospheric Sci NC State Univ Raleigh NC 27695-8208

JANOWITZ, HENRY DAVID, b Paterson, NJ, Mar 23, 15; m 42; c 2. GASTROENTEROLOGY. *Educ:* Columbia Univ, AB, 35, MD, 39, Univ Ill, MS, 49. *Prof Exp:* Intern, Mt Sinai Hosp, 39-41, fel path, 46, resident med, 47-48, asst physiol, Univ Ill, 48-49; asst gastroenterol, 50-54, chief gastro-intestinal clin, 56-62, dir NIH training prog gastroenterol, 59-75, head div gastroenterol, 56-83, EMER PROF, MT SINAI SCH MED, 83- *Concurrent Pos:* Hon lectr, Guy's Hosp, London, Eng, 56; McArthur lectr, Univ Edinburgh, 56; ed, Am J Digestive Dis, 56-65; asst clin prof med, Columbia Univ, 62-66; ed sect alimentary canal, Handbook Physiol, 65; Comfort Mem lectr, Mayo Found, 65; mem, Am Bd Gastroenterol, 65, chmn gastroenterol res group steering comt, 65; mem prog proj comt, Nat Inst Arthritis & Metab Dis, 65, chmn, 70; clin prof, Mt Sinai Sch Med, 66-85; consult, Bronx Vet Admin Hosp, Horton Hosp, Middletown, NY & Englewood Hosp, NJ; pvt pract. *Honors & Awards:* Friedenwald Metal, Am Gastroenterol Asn, 84. *Mem:* Am Physiol Soc; Soc Exp Biol & Med; Am Soc Clin Investrs; Am Gastroenterol Asn (pres, 72); Asn Am Physicians. *Res:* Gastrointestinal physiology; clinical investigation in gastroenterology, especially the application of physiological methods to the study of intestinal function and disease; research in imflammatory bowel diseases. *Mailing Add:* Div Gastroenterol Mt Sinai Sch Med 11 E 100th St New York NY 10029

JANOWITZ, MELVIN FIVA, b Minneapolis, Minn, May 8, 29; m; c 3. ALGEBRA, CLUSTER ANALYSIS. *Educ:* Univ Minn, BA, 50; Wayne State Univ, PhD(math), 63. *Prof Exp:* Asst prof math, Univ NMex, 63-66; assoc prof, Western Mich Univ, 66-67; assoc prof, 67-70, asst dean, Natural Sci & Math, 79-83, PROF MATH, UNIV MASS, AMHERST, 70- *Mem:* Am Math Soc; Math Asn Am; Classification Soc. *Res:* Lattice theory; mathematical models for ordinal cluster analysis; ordinal models for semiorders, internal orders and social choice functions; connections between percentile based cluster techniques and probabilistic metric spheres. *Mailing Add:* Dept of Math Univ of Mass Amherst MA 01003

JANOWSKY, DAVID STEFFAN, b San Diego, Calif, June 24, 39; m 62; c 4. PSYCHOPHARMACOLOGY. *Educ:* Univ Calif, San Francisco, BS, 61, MD, 64. *Prof Exp:* Asst prof, Dept Psychiat, Sch Med, Univ Calif, Los Angeles, 69-70; head physician, Crisis Clin, Psychiat Emergency Serv, Dept Psychiat, Harbor Gen Hosp, Calif, 69-70; asst prof pharmacol, Sch Med, Vanderbilt Univ, 70-73, asst prof psychiat, 70-72, assoc prof, 72-73; chief, Vet Admin Liaison Serv, 73-74, assoc prof, 73-76, PROF, DEPT PSYCHIAT, SCH MED, UNIV CALIF, SAN DIEGO, 76- *Concurrent Pos:* Chief psychiat serv, Univ Hosp, Univ Calif, San Diego, 74-78; prin investr, Mental Health Clin Res Ctr, NIMH, Univ Calif, Calif, San Diego. *Mem:* Am Col Neuropsychopharmacol; Am Psychiat Asn; Psychiat Res Soc; Soc Neurosci; Col Int Neuropsychopharmacol. *Res:* Effects of adrenergic-cholinergic balance in the affective disorders, using cholinesterase inhibitors and psychostimulant challenges as investigative probes, and correlating these results with pre-clinical animal models. *Mailing Add:* Psychiat Dept CB No 7160 Med Sch Wing B/U NC Chapel Hill NC 27599

JANS, JAMES PATRICK, b Detroit, Mich, Apr 6, 27; m 50; c 2. MATHEMATICS. *Educ:* Univ Mich, AB, 49, MA, 50, PhD(math), 55. *Prof Exp:* Jr instr math, Univ Mich, 53-54; instr, Yale Univ, 54-56; asst prof, Ohio State Univ, 56-57; asst prof, 57-64, PROF MATH, UNIV WASH, 64- *Mem:* Am Math Soc. *Res:* Algebra; structure of rings; homological and topological algebra. *Mailing Add:* Dept Math Univ Wash Seattle WA 98105

JANSEN, BERNARD JOSEPH, b Rockville, Minn, Aug 10, 27; m 55; c 4. MATHEMATICS, SOFTWARE. *Educ:* St John's Univ, Minn, BA, 50; St Louis Univ, MA, 52. *Prof Exp:* Instr math, St John's Univ, Minn, 54-56; comput analyst prog, Unisys, 55-66, mgr Titan III software, 66-69, avionics software, 69-75, systs & software, Int Systs Div, 76-77, planning, control & change proposals, Int Telecommun Div, 77-82, underseas proj mgr, 82-85, mgr syst scheduling & planning, Comput Systs Div, Unisys Defense Systs, 85-88; RETIRED. *Concurrent Pos:* Mem adv panel, Spaceborne Digital Comput Systs, NASA, 68-80; instr math, Univ Wis, River Falls, 90. *Mem:* Math Asn Am; Sigma Xi; Performance Measurement Asn. *Res:* Technical management of and application of computers to systems and software in the underseas, avionics, aerospace, command and control, and telecommunications fields. *Mailing Add:* 1859 Hillcrest St Paul MN 55116-1934

JANSEN, FRANK, b Emmeloord, Neth, Feb 9, 46; m 73; c 3. THIN FILM TECHNOLOGY, ELECTRONIC FABRICATION PROCESSES. *Educ:* Tech Univ Delft, ingenieur, 73; Case Western Reserve Univ, PhD(physics), 77. *Prof Exp:* Tech Specialist, Webster Res Ctr, Xerox Corp, 77-80, proj mgr & mem res staff, 80-86, prin scientist, 86-90; MGR, THIN FILM TECHNOLOGIES, BOC GROUP, 90- *Concurrent Pos:* Lectr, Am Vacuum Soc, 88-; chmn, Thin Film Div, Am Vacuum Soc, 89, prog chmn, 90. *Mem:* Am Vacuum Soc. *Res:* Thin film physics and technology including optical, electrical and mechanical properties of amorphous semiconductors, surface morphology and defect structures, novel deposition methods, diamond films; glow discharge deposition processes and materials. *Mailing Add:* BOC Group 100 Mountain Ave Murray Hill NJ 07974

JANSEN, GEORGE, JR, b Aloha, Ore, Nov 15, 34; m 56; c 4. CHEMICAL ENGINEERING. *Educ:* Ore State Univ, BS & BA, 55; Mass Inst Technol, SM, 57, ScD(chem eng), 59. *Prof Exp:* Chem engr, Hanford Labs, Gen Elec Co, Wash, 59-62, sr engr, 62-65; sr develop engr, Battelle-Northwest, 65-68, res assoc, 68-75; sr engr, Exxon Nuclear Co, Inc, 75-81; ANALYSIS ENGR, BATTELLE MEM INST, 81- *Mem:* AAAS; Am Chem Soc; Am Inst Chem Engrs; Am Nuclear Soc. *Res:* Ion exchange; heat transfer; process development in nuclear fuel processing; solvent extraction; radioactive waste disposal; risk analysis; centrifuge enrichment. *Mailing Add:* 18365 SW Sandra Lanew Aloha OR 97006

JANSEN, GEORGE JAMES, b Canton, Ohio, Apr 22, 25; m 53, 71; c 2. MINERALOGY. *Educ:* Univ Notre Dame, BS, 51; Bryn Mawr Col, MA, 52. *Prof Exp:* Hydrol field asst, US Geol Surv, 51, geologist, 52-57; prin geologist, Battelle Mem Inst, 57; supvr mineral & metallog, Res Ctr, Repub Steel Corp, Ohio, 57-69; mineralogist, Climax Molybdenum Lab, 69-76; mineralogist, Com Test & Eng Co, 76-78; VPRES PRIN INVEST, ROCKY MOUNTAIN CONSULT PETROG, INC, 78- *Mem:* Soc Econ Geologists; Asn Petrol Geochemical Explorationists; Soc Org Petrol. *Res:* Mineralogy of base metals; quantitative metallography; reflected light optics; coal petrography. *Mailing Add:* 12870 W 15th Dr Golden CO 80401

JANSEN, GUSTAV RICHARD, b Staten Island, NY, May 19, 30; m 53; c 4. NUTRITION. *Educ:* Cornell Univ, BA, 50, PhD(biochem), 58. *Prof Exp:* Jr & assoc chemist, Am Cyanamid Co, 53-54; asst biochem, Cornell Univ, 54-58; res chemist, E I du Pont de Nemours & Co, 58-62; res fel, Merck Inst Therapeut Res, 62-69; prof & dept head, 69-90, EMER PROF FOOD SCI & NUTRIT, COLO STATE UNIV, 90- *Concurrent Pos:* Prog mgr, USDA, competition grants prog human nutrit, 81-82; mem, exec comt, Inst Food Technol, 89-91, Human Nutrit Bd Sci Counselors, USDA, 86- *Honors & Awards:* Cert of Merit, USDA, 83; Babcock-Hart Award, Inst Food Technologists, 85. *Mem:* Am Inst Nutrit; Am Soc Biochem & Molecular Biol; AAAS; Inst Food Technologists; Sigma Xi. *Res:* Protein nutrition; processed weaning foods; nutrition during lactation; nutrition education. *Mailing Add:* 1804 Seminole Dr Ft Collins CO 80525

JANSEN, HENRICUS CORNELIS, b Bergen op Zoom, Holland, Aug 3, 42; US citizen; m 77. RANGE MANAGEMENT. *Educ:* Univ Calif, Berkeley, BS, 69, PhD(natural res sci & range mgt), 74. *Prof Exp:* Res forester range mgt, Pac Southwest Forest & Range Exp Sta, US Forest Serv, 72-76; from asst prof to assoc prof, 76-86, PROF RANGE MGT, CALIF STATE UNIV, CHICO, 86- *Concurrent Pos:* Consult range conservationist, Soil Conserv Serv, 79 & Bur Land Mgt, 81; botanist, Fish & Wildlife Serv, 80; range mgt expert, UN Food & Agr Orgn, 84-85. *Mem:* Soc Range Mgt; AAAS. *Res:* Computerized planning method including documentation for the management of federal grazing lands; grazing management of arid lands in North Africa. *Mailing Add:* Dept Agr Calif State Univ First & Normal Sts Chico CA 95929

JANSEN, IVAN JOHN, b Newton, Iowa, Mar 17, 41; m 60; c 5. AGRONOMY, SOIL SCIENCE. *Educ:* Iowa State Univ, BS, 63; Cornell Univ, MS, 71, PhD(soil sci), 72. *Prof Exp:* Soil scientist, Soil Conserv Serv, USDA, 63-67; soil technologist, dept agron, Cornell Univ, 67-68; soil scientist, Soil Conserv Serv, USDA, 72-74; from asst prof to assoc prof, 74-85, PROF PEDOLOGY, UNIV ILL, URBANA, 85- *Mem:* Am Soc Agron; Soil Sci Soc Am; Soil Conserv Soc Am; Am Soc Adv Sci; Am Soc Surface Mining & Reclamation; fel AAAS. *Res:* Characterizing geographic bodies of soil; reclamation of surface-minded lands for rowcrop production; conceptual foundation of pedology. *Mailing Add:* Dept Agron Univ Ill Urbana Campus 1102 S Goodwin Ave Urbana IL 61801

JANSEN, MICHAEL, b Bucharest, Romania, Jan 27, 56; m 84; c 2. QUANTUM ELECTRONICS, SEMICONDUCTOR LASERS. *Educ:* Univ Calif Los Angeles, BS, 78, MS, 79, PhD(quantum electronics), 84. *Prof Exp:* Teaching asst & assoc optics, quantum electronics & electronics, UCLA, 80-83, res assoc, 83-84; SR SCIENTIST, SPACE & TECHNOL GROUP, ADVAN TECHNOL DIV, RES CTR, TRW, 84- *Concurrent Pos:* Consult, Monosolar Industs, 80-81; lectr, Elec Eng Dept, Univ Calif Los Angeles, 85-86. *Mem:* Inst Elec & Electronics Engrs. *Res:* Design, development, and characterization of diode lasers and integrated optics; development of monolithic two-dimensional, coherent and incoherent surface-emitting arrays; monolithic components for optical integration; large optical cavity and evanescently-coupled and diffraction-coupled laser arrays; unstable resonators, amplifiers, and LEDs. *Mailing Add:* TRW D1/2519 One Space Park Redondo Beach CA 90278

JANSEN, ROBERT BRUCE, b Spokane, Wash, Dec 14, 22. CIVIL ENGINEERING. *Educ:* Univ Denver, BSCE, 49; Univ SCalif, MSCE, 55. *Prof Exp:* Chief, Calif Div Dam Safety, 65-68; chief oper, Calif Dept Water Resources, 68-71, dept dir, 71-75, chief design & cons, 75-77; asst comnr, US Bus Reclamation, 77-80; CONSULT CIVIL ENG, 80- *Concurrent Pos:* Chmn, US Comn on Large Dams, 79-81. *Mem:* Nat Acad Eng. *Mailing Add:* 509 Briar Rd Bellingham WA 98225

JANSING, JO ANN, b Louisville, Ky, Mar 23, 38. ANALYTICAL CHEMISTRY. *Educ:* Ursuline Col, Ky, BA, 65; Fordham Univ, MS, 67, PhD(anal & phys chem), 70. *Prof Exp:* Teacher high sch, Ky, 62-65; instr chem, Mt St Agnes Col, 69-70; from asst prof to assoc prof, 70-81, chair, Div Natural Sci, 79-86, PROF CHEM, IND UNIV SOUTHEAST, 81-, COORDR CHEM, 76-79, 89- *Mem:* Am Chem Soc; Sigma Xi. *Res:* X-ray crystallographic structure studies of organic molecules. *Mailing Add:* Ind Univ SE PO Box 679 New Albany IN 47150

JANSKI, ALVIN MICHAEL, b Braham, Minn, May 27, 49; m 71; c 2. BIOCHEMISTRY. *Educ:* St Cloud State Univ, Minn, BA, 71; NDak State Univ, PhD(biochem), 75. *Prof Exp:* Res assoc, Dept Biochem & Biophys, Iowa State Univ, 75-78, NIH fel, 78; sr staff fel, Lab Metab, Nat Inst Alcohol Abuse & Alcoholism, 78-81; mgr, Biochem Res Sect, Int Minerals & Chem Corp, 81-86; DIR LIFE SCI, PITMAN-MOORE INC, 86- *Concurrent Pos:* Prin investr, Int Minerals & Chem Corp, Northbrook, Ill, 81- *Mem:* Am Soc Biol Chemists; NY Acad Sci; Am Chem Soc; AAAS; Endocrine Soc. *Res:* Protein vaccines by recombinant DNA technology; metabolic pathways through intracellular compartmentation of metabolites and enzymes and hormone-depedent phosphorylation of enzymes; in vitro study of growth. *Mailing Add:* Pitman-Moore Inc Box 207 Terre Haute IN 47808

JANSON, BLAIR F, b East Trumbull, Ohio, Jan 6, 18; m 44; c 3. PLANT PATHOLOGY. *Educ:* Ohio State Univ, BS, 40, MS, 47, PhD, 50. *Prof Exp:* Asst exten plant pathologist, Ohio State Univ, 46, asst instr bot, 47-50, prof plant path, 62-80, exten plant pathologist, 50-80, EMER PROF PLANT PATH, OHIO STATE UNIV, 80- *Mem:* Am Phytopath Soc. *Res:* Ornamental, fruit, cereal and forage crop diseases. *Mailing Add:* 266 Canyon Dr Columbus OH 43214

JANSONS, VILMA KARINA, b Riga, Latvia; US citizen. MICROBIOLOGY. *Educ:* Brooklyn Col, BA, 61; Rutgers Univ, New Brunswick, PhD(microbiol), 67. *Prof Exp:* Lectr biol, Princeton Univ, 68-70, mem res staff biochem sci, 70-72; asst prof microbiol, 72-77, ASSOC PROF MICROBIOL, NJ MED SCH, UNIV MED & DENT NJ, 77- *Mem:* AAAS; Am Soc Microbiol; Am Soc Cell Biol. *Res:* Surface properties of normal and malignant cells; biochemistry of morphogenesis. *Mailing Add:* NJ Med Sch Univ Med & Dent NJ 100 Bergen St Newark NJ 07103

JANSSEN, ALLEN S, civil engineering; deceased, see previous edition for last biography

JANSSEN, FRANK WALTER, b St Paul, Minn, Sept 10, 26; m 52; c 2. DRUG METABOLISM. *Educ:* Col St Thomas, BS, 50; Iowa State Univ, MS, 52. *Prof Exp:* Asst scientist biochem, Hormel Inst, Univ Minn, 52-61; res scientist protein chem, 61-66, sr res scientist & group leader drug metab, 66-78, supvr pharmacokinetic eval unit, 78-85, MGR DRUG DIPOSITION SECT, WYETH LABS, INC, 85- *Mem:* Sigma Xi. *Res:* Drug disposition; biotransformation and pharmacokinetics of drugs. *Mailing Add:* 309 Westbrook Dr Westchester PA 19382

JANSSEN, JERRY FREDERICK, b Mason City, Iowa, Mar 22, 36; m 59; c 2. ORGANIC CHEMISTRY. *Educ:* Iowa State Teachers Col, BA, 57, MA, 59; Mich State Univ, PhD(chem), 67; JD, Suffolk Univ, 83. *Prof Exp:* Instr sci & math, Mason City Jr Col, Iowa, 59-61; asst instr, Mich State Univ, 61-63; asst prof chem, Antioch Col, 66-69; from asst prof to assoc prof, Eisenhower Col, 69-74; sr environ engr, GTE Sylvania, Inc, Seneca Falls, NY, 74-77, CHEM PATENT AGENT, GTE SERV CORP, 77-; PARKE-DAVIS DIV, WARNER LAMBERT. *Concurrent Pos:* Consult, Vernay Labs, Ohio, 68-69 & Sylvania Elec Prod, Inc, NY, 69- *Mem:* Am Chem Soc; fel Am Inst Chem; Am Bar Asn; Am Intellectual Property Asn. *Res:* Organic reaction mechanisms; molecular photochemistry; rearrangement reactions of aromatic compounds. *Mailing Add:* Abbott Lab Dept 377 One Abbott Park Rd Abbott Park IL 60064

JANSSEN, MICHAEL ALLEN, b Boise, Idaho, Sept 30, 37; m; c 2. RADIO ASTRONOMY, PLANETARY SCIENCES. *Educ:* Univ Calif, AB, 63, PhD(atmospheric & space sci), 72. *Prof Exp:* Physicist, Lawrence Radiation Lab, 63-67; Nat Res Coun resident res assoc planetary radio astron, 72-74, sr scientist, 74-76, MEM TECH STAFF, JET PROPULSION LAB, CALIF INST TECHNOL, 76- *Concurrent Pos:* Prin investr, Microwave Atmospheric Exp, Venus Orbiting Imaging Radar Mission, partic scientist, cosmic background explorer. *Mem:* Int Union Radio Sci; Am Astron Soc; Am Geophys Union; Int Astron Union. *Res:* Development of radio interferometric techniques at millimeter wavelengths; investigation of the atmospheres of Venus and the outer planets by microwave techniques; spacecraft microwave radiometry; cosmic microwave background; microwave remote sensing. *Mailing Add:* MS 169-506 Jet Propulsion Lab 4800 Oak Grove Dr Pasadena CA 91109

JANSSEN, RICHARD WILLIAM, b Weehawken, NJ, June 22, 40; m 68; c 2. PHARMACEUTICAL CHEMISTRY, PHYSICAL PHARMACY. *Educ:* Ferris State Col, BS, 62; Rutgers Univ, MS, 66, PhD(pharmaceut chem), 69. *Prof Exp:* Sr anal chemist, Smith, Kline & French Labs, 69-71; sr res scientist, Lescarden Ltd, 71-75; group leader, William H Rorer Inc, 75-87; SECT HEAD, E R SQUIBB, 87- *Concurrent Pos:* Mem Dissolution Comt, Pharmaceut Mfr Asn. *Mem:* Am Asn Pharmaceut Scientists; Am Pharmaceut Asn; Am Chem Soc. *Res:* Research and development documentation; analytical methods development; pharmaceutics research and development. *Mailing Add:* Bristol-Myers Squibb Pharmaceut Res Inst One Squibb Dr Box 191 New Brunswick NJ 08903-0191

JANSSEN, ROBERT (JAMES) J, b Geneva, Ill, Feb 28, 31; m 57; c 2. VIROLOGY, IMMUNOLOGY. *Educ:* Cornell Col, BA, 53; State Univ Iowa, MS, 55, PhD(bact), 57. *Prof Exp:* Med bacteriologist virol, Biol Labs, US Army Chem Corps, Md, 57-61; asst prof, 61-67, assoc prof, 67-80, PROF MICROBIOL & MED TECHNOL, UNIV ARIZ, 80- *Mem:* Sigma Xi; Am Soc Microbiol. *Res:* Smallpox, influenza, enteroviruses, arborviruses; combined infections with two or more microbial agents; aerobiology studies with viral agents; serological techniques; effects of certain drugs on viral infections. *Mailing Add:* Dept Microbiol Immunol Univ Ariz Bldg 90 Tucson AZ 85721

JANSSON, BIRGER, b Stockholm, Sweden, Sept 4, 21; m 54; c 4. BIOMATHEMATICS, EPIDEMIOLOGY. *Educ:* Univ Stockholm, FilKand, 46, FilLic, 65, FilDr(math statist), 66. *Hon Degrees:* Docent, Univ Stockholm, 67. *Prof Exp:* Head res math sci, Res Inst Nat Defense, Stockholm, Sweden, 48-73; biomathematician, Nat Large Bowel Cancer Proj, 73-84, PROF BIOMATH, UNIV TEX M D ANDERSON CANCER CTR, 73- *Concurrent Pos:* Consult mathematician, Swed Money Lottery, Stockholm, 66-73; consult biomath, Tumor Biol, Karolinska Inst, 66-73; Eleanor Roosevelt fel, Int Union Against Cancer, 70; vis assoc prof, Univ Tex, Houston, 70-71; assoc ed, Math Biosci, 74-87; adj prof, Rice Univ, 74-; pres, Texas Swed Cult Found, 81-; vchmn, Swed Coun Am, 87- *Mem:* AAAS; Soc Environ Geochem & Health; Int Soc Prev Oncol; Am Soc Prev Oncol; European Inst Ecol & Cancer. *Res:* Cancer, epidemiology and prevention, especially cancer of the colon, rectum, stomach and breast; mathematical models of the cell cycle and their use for finding rational protocols for cancer treatment. *Mailing Add:* Dept Biomath MD Anderson Cancer Ctr PO Box 237 Houston TX 77030

JANSSON, DAVID GUILD, b Quincy, Mass. TECHNOLOGICAL INNOVATION. *Educ:* Mass Inst Technol, SB, 68, SM, 70, ScD, 73. *Prof Exp:* Scientific officer, Off Naval Res, 72-75; from asst prof to assoc prof, Dept Mech Eng, Mass Inst Technol, 75-83, dir, Innovation Ctr, 79-86; assoc prof mech eng & dir, Inst Innovation & Design Eng, Tex A&M Univ, 86-91; PRES, SUGARTREE TECHNOL, BOULDER, COLO, 91- *Res:* Development of methodology for understanding and teaching the technological innovation process, including specific applications to energy conservation, unique electronic display devices and many other areas of innovation technology. *Mailing Add:* Sugartree Technol 1035 Pearl St Boulder CO 80302

JANSSON, PETER ALLAN, b Teaneck, NJ, May 20, 42; div; c 2. OPTICAL PHYSICS, DIGITAL IMAGE PROCESSING. *Educ:* Stevens Inst Technol, BS, 64; Fla State Univ, PhD(physics), 68. *Prof Exp:* Infrared physics, Fla State Univ, 67-68; res physicist, 68-71, E I Du Pont de Nemours & Co, Inc, sr res physicist, 71-76, res assoc, 76-80, sr res assoc, 80-90, RES FEL & TECH LEADER, DIGITAL IMAGE PROCESSING GROUP, EXP STA, E I DU PONT DE NEMOURS & CO, INC, 80- *Mem:* Int Neural Network Soc; fel Optical Soc Am; Soc Photog Instrumentation Engrs. *Res:* Infrared and optical physics; digital image processing; optical information processing; molecular spectroscopy; super resolving method of deconvolution; artificial neural networks. *Mailing Add:* Rm 212 Bldg 357 Du Pont Exp Sta Wilmington DE 19898

JANTZ, O K, b Newton, Kans, June 16, 34; m 57; c 3. ENTOMOLOGY. *Educ:* Kans State Univ, BS, 57, MS, 62; Ore State Univ, PhD(entom), 65. *Prof Exp:* Lab asst entom, Kans State Univ, 52-53; field aide, Agr Mkt Serv, USDA, Kans, 53-55, biol aide, 55-57, entomologist, 57; res asst entom, Kans State Univ, 60-63; res asst, Ore State Univ, 63-65; entomologist, Agr Res Serv, USDA, Mich, 65-67; regional tech specialist, Dow Chem, Mich, 67-68, mgr field res sta, Ill, 68-71, develop specialist, 71-73; mgr res & develop agr & spec prods, Dow Chem Pac Ltd, Hong Kong, 73-77, mgr, tech serv & develop plant prod, Agr Prod Dept, 77-81, dir, Agr Prod Develop & Regist, 77-84, bus mgr, Agr Herbicide, 84, dir res & develop, NAm Agr Prod, Dow Chem USA, 85-89, GLOBAL DIR RES & DEVELOP OPERS, DOW ELANCO, 89- *Mem:* Entom Soc Am; Weed Sci Soc Am. *Res:* Field development of agricultural chemicals; forest insects; stored grain pests; field crop insects. *Mailing Add:* Dow Elanco 9002 Purdue Rd Indianapolis IN 46268-1189

JANUARY, LEWIS EDWARD, b Haswell, Colo, Nov 14, 10; m 41; c 2. CARDIOLOGY. *Educ:* Colo Col, BA, 33; Univ Colo, MD, 37. *Hon Degrees:* DSc, Colo Col, 66. *Prof Exp:* Jr intern, Univ Hosp, 37-38, from asst resident to resident internal med, 38-41, asst physician, 41-42, assoc chmn dept med, 73-80, from asst prof to prof, 46-80, EMER PROF MED, COL MED, UNIV IOWA, 80- *Concurrent Pos:* Chmn coun clin cardiol, Am Heart Asn, 61-63, fel coun clin cardiol; vchmn sect cardiovascular dis, AMA, 70-73; mem, Int-Soc Comn Heart Dis Resources; vis prof, Ain Shams Univ, Cairo, Egypt, 72; mem bd dirs, Joffrey Ballet, 74-; mem exec comt, Int Soc & Fedn Cardiol, 76-78. *Honors & Awards:* Honor Achievement Award, Angiol Res Found, 65; Distinguished Serv Citation, Coun Clin Cardiol, Am Heart Asn, 67, Gold Heart Award, Am Heart Asn, 69; Distinguished Serv Citation, Int-Soc Comn Heart Dis Resources, 71; Silver & Gold Award, Univ Colo, 71; Helen B Taussig Award, 72; Int Achievement Award, Am Heart Asn, 77, Citation Distinguished Serv to Int Cardiol, 78; Angel Award, Int soc Performing Arts Adminr Found, 90. *Mem:* Am Heart Asn (vpres, 63-65, pres, 66-67); fel Am Col Cardiol; Asn Univ Cardiol; Inter-Am Soc Cardiol; Int Cardiol Fedn (vpres, 70-76). *Res:* Diabetes insipidus; mercurial diuretics; electrocardiography; hypertension; heart diseases. *Mailing Add:* 3324 Hanover Ct Iowa City IA 52245

JANUS, ALAN ROBERT, b Utica, NY, Dec 27, 37; m 59, 70. SOLID STATE SCIENCE. *Educ:* Utica Col, BA, 59; Syracuse Univ, PhD(inorg chem), 64. *Prof Exp:* Lab technician qual control, Utica Drop Forge & Tool Co, 56-59; res chemist organometallic, Solvay Process Div, Allied Chem Corp, 60; sr chemist thin films, Sprague Elec Co, Mass, 63-66, assoc prog mgr ceramic develop, 66-68; asst prof, Roanoke Col, 68-70; mgr thin film eng, Electronic Mat Div, Bell & Howell Res Labs, 70-71, dir eng, 71-74; sr scientist, Hughes Aircraft, Calif, 74-80; WITH BOURNE, INC, 80- *Concurrent Pos:* Res technician, Metals Div, Kelsey Hayes Co, 58-59; consult, Am Safety Razor Div, Philip Morris Co, 68-70, Electron Tube Div, Int Tel & Tel, 68-70, Bell & Howell, 73-74, Nat Micrometrics, 74- & Optifilm, 75- *Mem:* Am Vacuum Soc. *Res:* Thin film preparation and evaluation; magnetic susceptibilities; organometallic compound preparation and evaluation; coordination chemistry; chrome photoplates; III-IV compounds; ferrites; microanalytical services; surface acoustic wave device development; microwave hybrid device development. *Mailing Add:* Bourne Inc MS-94 1200 Columbia Ave Riverside CA 92507

JANUSEK, LINDA WITEK, b La Salle, Ill, Jan 11, 52; m 75; c 2. NEONATAL ENDOCRINOLOGY, ADRENAL CORTEX. *Educ:* Bradley Univ, BS, 74; Univ Ill, PhD(physiol), 78. *Prof Exp:* Asst prof, 78-84, ASSOC PROF PHYSIOL & NURSING, LOYOLA UNIV CHICAGO, 84- *Honors & Awards:* Young Investr Award, Circulatory Shock Soc, 80. *Mem:* Am Physiol Soc; Circulatory Shock Soc; Fedn Am Soc Exp Biol. *Res:* Metabolic and hormonal responses of the neonate (rat model) to a septic insult. *Mailing Add:* Dept Physiol Loyola Univ 2160 S First Ave Maywood IL 60153

JANUSZ, GERALD JOSEPH, b Aug 20, 40; US citizen; m 61; c 2. MATHEMATICS. *Educ:* Marquette Univ, BS, 62; Univ Wis, MS, 63; Univ Ore, PhD(math), 65. *Prof Exp:* Mem, Inst Advan Study, 65-66; instr math, Univ Chicago, 66-68; from asst prof to assoc prof, 68-73; PROF MATH, UNIV ILL, URBANA, 73- *Concurrent Pos:* Exec ed, J Math Rev, 90- *Mem:* Am Math Soc; Math Asn Am. *Res:* Representations of finite groups; finite dimensional algebras. *Mailing Add:* Dept of Math 273 Altgeld Hall Univ of Ill Urbana IL 61801

JANUSZ, MICHAEL JOHN, b Pawtucket, RI, Feb 18, 54; m 85. INFLAMMATION IMMUNOBIOLOGY. *Educ:* RI Col, BA, 76; Smith Col, AM, 79; Univ NC, PhD(microbiol & immunol), 85. *Prof Exp:* Postdoctoral fel immunol, Brigham & Women's Hosp & Harvard Med Sch, 85-88, instr & res immunobiologist, 88-89; SR RES IMMUNOBIOLOGIST, MARION MERRELL DOW RES INST, 89- *Mem:* Am Asn Immunologists. *Res:* Role of proteinases in connective tissue matrix turnover; immunobiology of inflammatory diseases. *Mailing Add:* Marion Merrell Dow Res Inst 2110 E Galbraith Rd Cincinnati OH 45215

JANUTOLO, DELANO BLAKE, b Bluefield, WVa, July 7, 52; c 2. PLANT PATHOLOGY, MYCOLOGY. *Educ:* WVa Univ, BS, 73; Va Polytech Inst & State Univ, PhD(plant path), 77. *Prof Exp:* PROF BIOL & DEAN, SCH THEORET & APPL SCI, ANDERSON UNIV, 77- *Mem:* Am Phytopath Soc; Sigma Xi. *Res:* Evaluation and testing of fungicides; systemic fungicides. *Mailing Add:* Dept of Biol Anderson Univ Anderson IN 46012

JANZ, GEORGE JOHN, b Russia, Aug 24, 17; nat US; m 51; c 4. PHYSICAL CHEMISTRY. *Educ:* Univ Man, BSc, 40; Univ Toronto, MA, 41, PhD(chem), 43; Univ London, DSc, 54. *Prof Exp:* Res chemist, Can Indust Ltd, 43-46; hon res assoc, Univ Col, London, 46-49; asst prof chem, Pa State Col, 49-50; from asst prof to assoc prof, 50-53, chmn dept chem, 62-72, prof, 53-80, WM WEIGHTMAN WALKER PROF CHEM, RENSSELAER POLYTECH INST, 80-, DIR, MOLTEN SALTS DATA CTR, 68- *Concurrent Pos:* Vis prof, Rockefeller Univ, 72-73. *Mem:* Am Chem Soc; Royal Soc Chem; Electrochem Soc; Faraday Soc; NY Acad Sci. *Res:* Electrochemistry; physical properties ties and spectroscopy of inorganic compounds in the molten state; aqueous and nonaqueous electrolytes; electrolysis of molten salts; thermodynamics and reaction energies of cyanogen-like compounds. *Mailing Add:* 401 Winter St Ext Troy NY 12181

JANZEN, ALEXANDER FRANK, b Einlage, Ukraine, Apr 19, 40; Can citizen; m 67; c 3. INORGANIC CHEMISTRY, ORGANOMETALLIC CHEMISTRY. *Educ:* McMaster Univ, BSc, 63; Western Ont Univ, PhD(chem), 66. *Prof Exp:* Fel chem, Univ London, 66-67; from asst prof to assoc prof, 67-78, PROF CHEM, UNIV MAN, 78- *Concurrent Pos:* Vis scientist, Max Planck Inst Exp Med, Ger, 73. *Mem:* Chem Inst Can; Am Chem Soc. *Res:* Synthesis of inorganic fluorine compounds and study of dynamic properties. *Mailing Add:* Dept Chem Univ Man Winnipeg MB R3T 2N2 Can

JANZEN, DANIEL HUNT, b Milwaukee, Wis, Jan 18, 39; c 2. ECOLOGY, EVOLUTION. *Educ:* Univ Calif, Berkeley, PhD(entom), 65. *Prof Exp:* Asst prof biol, Univ Kans, 65-68; from asst prof to assoc prof, Univ Chicago, 68-72; from assoc prof to prof, Univ Mich, 72-76; PROF BIOL, UNIV PA, 76- *Concurrent Pos:* Adv, Orgn Trop Studies, Costa Rica & Costa Rican Nat Park Serv. *Honors & Awards:* Gleason Award, Am Bot Soc, 75; Crawford Prize, Royal Swed Acad Sci, 84. *Mem:* Soc Study Evolution; Ecol Soc Am; Am Soc Naturalists; Brit Ecol Soc; Asn Trop Biol. *Res:* Interactions of plants and animals, with emphasis on tropical field systems. *Mailing Add:* Dept of Biol Univ of Pa Philadelphia PA 19104

JANZEN, EDWARD GEORGE, b Manitoba, Man, May 23, 32; m 52; c 2. PHYSICAL ORGANIC CHEMISTRY. *Educ:* Univ Man, BSc, 57, MSc, 60; Iowa State Univ, PhD(org chem), 63. *Prof Exp:* Fel dept chem, Iowa State Univ, 63-64; from asst prof to prof spectros, Univ Ga, 64-75; prof & chmn dept chem, 76-86, PROF & DIR MAGNETIC RESONANCE IMAGING, UNIV GUELPH, 86- *Concurrent Pos:* Vis prof & scientist, Okla Med Res Found, 81 & IBM Instruments, Inc, San Jose, 82. *Honors & Awards:* Fulmer Award, Iowa State Univ; Syntex Award, Can Soc Chem - Chem Inst Can. *Mem:* Fel Can Inst Chem; Soc Free Radical Res; Sigma Xi. *Res:* Physical organic, biochemical and biomedical topics in electron spin resonance spectroscopy; spin trapping techniques, development and practice. *Mailing Add:* Dept Clin Studies & Biomed Sci-Ont Vet Col Univ Guelph Guelph ON N1G 2W1 Can

JANZEN, JAY, b Chickasha, Okla, Mar 24, 40; m 62; c 2. PHYSICAL CHEMISTRY. *Educ:* Univ Kans, BS, 62; Iowa State Univ, PhD(phys chem), 68. *Prof Exp:* SR RES CHEMIST, PHILLIPS PETROL CO, 68- *Mem:* Am Chem Soc; Sigma Xi. *Res:* Carbon black; reinforcement of elastopolymers; physics and chemistry of carbon surfaces; statistical morphology of particulate materials and composite media; random geometry; automatic image analysis; colloid physics. *Mailing Add:* 2727 SE Evergreen Dr Bartlesville OK 74006

JANZOW, EDWARD F(RANK), b St Louis, Mo, Mar 19, 41; m 67; c 1. NUCLEAR ENGINEERING, MECHANICAL ENGINEERING. *Educ:* Washington Univ, BS, 63; Univ Mo, MS, 64; Univ Ill, Urbana, PhD(nuclear eng), 70; Univ Dayton, MBA, 81. *Prof Exp:* NASA traineeship, Univ Mo-Columbia, 63-64; NSF traineeship, Univ Ill, Urbana, 64-68, asst, Off Water Resources, 68-69, nuclear eng prog, 69-70; sr res engr, Monsanto Res Corp, 71-72, engr group leader, 72-75, supvr design & develop eng, 75-76, mgr eng design & develop, 76-79, mgr mfg & qual assurance, 79-81, mgr opers, 81-84; PRES, FRONTIER TECHNOL CORP, 85 - *Concurrent Pos:* Mem comt, Sealed Radioactive Sources, Am Nat Standard Inst, 74 - *Mem:* Am Nuclear Soc; Am Soc Testing & Mat. *Res:* Nuclear radiation and heat sources; research, development and design relating to such sources and techniques, apparatus and facilities for their fabrication; development and design of radioisotope shipping containers. *Mailing Add:* 2671 Crone Rd Xenia OH 45385

JAOUNI, KATHERINE COOK, b Alexandria, Va, Nov 8, 29; m 64; c 1. MICROBIOLOGY. *Educ:* Col William & Mary, BS, 49; George Washington Univ, MS, 52, PhD(microbiol), 57. *Prof Exp:* Bacteriologist, Alexandria Health Dept, 49-52; parasitologist, Trop Dis Lab, 52-57, virologist, Infectious Dis Lab, 57-78, RES MICROBIOLOGIST, NAT INST ALLERGY & INFECTIOUS DIS, 78- *Concurrent Pos:* Researcher, Pasteur Inst & St Vincent de Paul Hosp, France, 59-61 & Max Planck Inst, Tuebingen, Ger, 61-62; pres, Grad Women in Sci, Inc, 81-82. *Mem:* Am Soc Trop Med & Hyg; Am Soc Microbiol; Sigma Xi; AAAS. *Res:* Tissue culture of protozoa and mode of action of drugs against toxoplasma; characterization and antigenic analysis of respiratory viruses; oncogenic virology; viruses of protozoa; mode of action of drugs against protozoa and viruses; science administration. *Mailing Add:* Nat Inst Allergy & Infectious Dis NIH Bethesda MD 20892

JAOUNI, TAYSIR M, b Jerusalem, Palestine, Aug 29, 24; US citizen; m 64; c 1. ORGANIC CHEMISTRY. *Educ:* Univ Calif, Berkeley, BA, 50, MA, 51, BSc, 60; Univ Colo, MSc, 63. *Prof Exp:* RES CHEMIST, LAB CHEM, NAT HEART & LUNG INST, 63- *Res:* Synthesis of diribonucleoside phosphates; RNA codewords and protein synthesis; GC/MS. *Mailing Add:* Bldg 10 Rm 7N323 Nat Heart & Lung Inst Bethesda MD 20205

JAPAR, STEVEN MARTIN, New York, NY, Nov 11, 44; m 84; c 2. PHOTOCHEMISTRY, CHEMICAL KINETICS. *Educ:* City Col New York, BS, 65; Case Inst Technol, PhD(phys chem), 69. *Prof Exp:* Fel, Div Physics, Nat Res Coun Can, 69-71; fel, Chem Dept, Univ Calif, Riverside, 71-72; instr, Chem Dept, Drexel Univ, 72-73; sr res scientist, 73-80, PRIN RES SCIENTIST ASSOC, RES STAFF, FORD MOTOR CO, 81- *Concurrent Pos:* Instr, Natural Sci Dept, Univ Mich, Dearborn, 75-76. *Honors & Awards:* Arch T Colwell Award, Soc of Automotive Engrs, 81. *Mem:* Am Chem Soc; Air Pollution Control Asn; Sigma Xi; Optical Soc Am; Inter-Am Photochem Soc. *Res:* Chemistry and physics of gas phase aerosols generated from combustion sources; photochemistry, spectroscopy and chemical kinetics of species important in atmospheric chemistry; development of analytical methods for the measurement of such species. *Mailing Add:* 4518 Whisper Way Troy MI 48098

JAQUES, LOUIS BARKER, b Toronto, Ont, July 10, 11; m 37; c 1. PHYSIOLOGY, PHARMACOLOGY. *Educ:* Univ Toronto, BA, 33, MA, 35, PhD(physiol), 41; Univ Sask, DSc, 74. *Prof Exp:* Asst physiol, Univ Toronto, 34-42, lectr & res assoc, 43-44, asst prof, 44-46; prof physiol & pharmacol & head dept, 46-71, Lindsay res prof, 71-79, EMER PROF, COL MED, RES ASSOC COL DENT, UNIV SASKATOON, 79- *Concurrent Pos:* Claude Bernard vis prof, Univ Montreal, 48; mem, Int Conf Thrombosis & Embolism, Univ Basel, 54; mem adv comt, Med Div, Nat Res Coun Can, 52-55 & 59-61, mem Int comt nomenclature of blood clotting factors, 54-66; chmn subcomt hemostasis, 62-65; chmn, Can Nat Comt, Int Union Physiol Sci, 62-64; mem Can nat comt, Int Coun Sci Unions, 63-64. *Mem:* Am Physiol Soc; Am Soc Pharmacol & Exp Therapeut; fel NY Acad Sci; fel Royal Soc Can; Pharmacol Soc Can. *Res:* Pharmacology of blood coagulation; anticoagulants; hemorrhage and thrombosis. *Mailing Add:* Col Dent Univ of Sask Saskatoon AB S7N 0W0 Can

JAQUES, ROBERT PAUL, b Caledonia, Ont, Jan 1, 31; m 54; c 3. INSECT PATHOLOGY. *Educ:* Univ Toronto, BSA, 52, MSA, 54; Cornell Univ, PhD(insect ecol), 60. *Prof Exp:* Res scientist, Kentville, NS, 54-67, RES SCIENTIST INSECT PATH, RES STA, CAN DEPT AGR, HARROW, ON, 67- *Concurrent Pos:* Res assoc biol, Acadia Univ, 62-65; assoc prof, 65-67; assoc fac, Univ Guelph, 76-84. *Mem:* Entom Soc Can; Soc Invert Path; Entom Soc Am. *Res:* Factors affecting development of disease in populations of insects; persistence of insect viruses in the environment; microbial control of insects. *Mailing Add:* Res Sta Can Dept Agr Harrow ON N0R 1G0 Can

JAQUES, WILLIAM EVERETT, b Newbury, Mass, July 11, 17; m 68; c 7. PATHOLOGY. *Educ:* McGill Univ, MD & CM, 42. *Prof Exp:* Intern, Bridgeport Hosp, 42-43; asst resident path, Mass Mem Hosp, 47-49; instr, Harvard Med Sch, 49-53; assoc prof, La State Univ, 53-57, prof & chmn dept, Sch Med, Univ Okla, 57-66; prof path & chmn dept, Sch Med, Univ Ark, Little Rock, 66-74; prof path, Okla Col Osteop Med & Surg, Tulsa, Okla, 74-81; clin prof path, Univ SFla, 84-89; dean med sci, 82-84, PROF PATH, AM UNIV CARIBBEAN, 81- *Concurrent Pos:* Resident, Children's Med Ctr, Boston, 49-50, assoc pathologist, Peter Bent Brigham Hosp, 51-53; res fel, Children Med Ctr, Boston, 50-51; vis prof, Nat Defense Med Ctr, Taiwan, 65-66; consult, Vet Admin, Oklahoma City, Muskogee & Little Rock; dir path, Nat Ctr Toxicol Res, Jefferson, Ark, 71-74; clin prof path, Univ Okla, 74-81. *Mem:* Am Soc Exp Path; Am Soc Clin Path; Am Asn Path & Bact; Int Acad Path; Am Col Angiol. *Res:* Pathological physiology; lesser circulation; embolism; mitral stenosis. *Mailing Add:* 509 S Park Blvd Venice FL 34285

JAQUISS, DONALD B G, organic chemistry; deceased, see previous edition for last biography

JAQUITH, RICHARD HERBERT, b Newton, Mass, Mar 31, 19; m 42; c 5. INORGANIC CHEMISTRY. *Educ:* Univ Mass BS, 40, MS, 42; Mich State Univ, PhD(inorg chem), 55. *Prof Exp:* Instr chem, Univ Conn, 42-44 & 46-47; asst prof, Colby Col, 47-54; from asst prof to assoc prof, 54-65, PROF CHEM, UNIV MD, COLLEGE PARK, 65-, ASST VCHANCELLOR ACAD AFFAIRS, 73- *Mem:* Sigma Xi. *Res:* Nonaqueous inorganic solvents; rare earth compounds. *Mailing Add:* 5807 Cherrywood Terr No 201 Greenbelt MD 20770

JARAMILLO, JORGE, b Chinchina, Colombia, Jan 7, 34; m 61; c 4. PHARMACOLOGY. *Educ:* Univ Caldas, MD, 58; Tulane Univ, MS, 62, PhD(pharmacol), 66. *Prof Exp:* Instr, Tulane Univ, 66-67; asst prof, Univ Conn, 67-68; sr pharmacologist, 68-70, res assoc pharmacol, Ayerst Res Labs, 70-77, sr res assoc, 77-84; coordr pharmacol, 84-88, DIR PHARMACOL, BIO-MEGA INC, 88- *Mem:* Pharmacol Soc Can; Soc Toxicol Can; Am Soc Pharmacol Exp Therapeut. *Res:* Cardiovascular. *Mailing Add:* Dept Pharmacol Bio-Mega Inc 2100 Cunard St Laval Montreal PQ H7S 2G5 Can

JARBOE, CHARLES HARRY, b Louisville, Ky, Oct 3, 28; div; c 6. TOXICOKINETICS. *Educ:* Univ Louisville, BSc, 51, PhD(chem), 56. *Prof Exp:* Chemist, E I du Pont de Nemours & Co, 51-53; asst, AEC, Univ Louisville, 56, res asst prof org chem, 56-58, res assoc pharmacol, Sch Med, 58-62, assoc prof, Health Sci Ctr, 62-72, prof pharmacol & dir, Therapeut & Toxicol Lab, 72-89, EMER PROF PHARMACL, SCH MED, UNIV LOUISVILLE, 89- *Concurrent Pos:* Consult, Brown & Williamson Tobacco Corp, 57-58, chief scientist, 58-61, consult, 62-64; spec fel, Nat Heart Inst, 61-62; asst dean planning & proj coordr, Univ Louisville, 65-67, actg chmn dept pharmacol, 68; vis scientist, Sci Div, Abbott Labs, 70-71; consult, Am Horse Shows Asn, 73; vis prof, Med Col Va, Va Commonwealth Univ, 74; Ky Med Assistance Prog Formulary Subcomt & Pest Control Adv Bd, Ky Environ Qual Comn, 78-; vis prof, Med Sch, Auckland Univ, 79, Univ Utah, 80, King Faisal Univ Col Med, Saudi Arabia, 82-84 & US Naval Regional Med Ctr, Portsmouth, Va, 82. *Mem:* Am Chem Soc; The Chem Soc; Am Acad Clin Toxicol; Am Soc Pharmacol & Exp Therapeut; NY Acad Sci; Am Col Clin Pharmacol. *Res:* Kinetic aspects of drug action; human pharmacokinetics; human toxicokinetics. *Mailing Add:* Dept Pharmacol & Toxicol Sch Med Univ Louisville Louisville KY 40292

JARBOE, THOMAS RICHARD, b Paxton, Ill, Aug 23, 45; m 70; c 5. PLASMA PHYSICS. *Educ:* Univ Ill, BS, 67; Univ Calif, Berkeley, PhD(plasma physics), 75. *Prof Exp:* Physicist optics, Naval Weapons Ctr, China Lake, Calif, 68; physicist fusion res, Los Alamos Nat Lab, 74-80, group leader, 80-89; PROF, UNIV WASH, 89- *Mem:* Fel Am Phys Soc. *Res:* Relaxation processes during the interaction of plasma and magnetic field in toroidal geometry; goal is to understand helicity conservation during plasma relaxation processes to understand relaxation in general. *Mailing Add:* 6508 NE 192nd Pl Seattle WA 98155-3458

JARCHO, LEONARD WALLENSTEIN, b New York, NY, Aug 12, 16; m 56; c 2. MEDICINE. *Educ:* Harvard Univ, AB, 36; Columbia Univ, MA, 37, MD, 41. *Prof Exp:* Instr physiol, Col Physicians & Surgeons, Columbia Univ, 41; Denison fel, Johns Hopkins Univ, 46-47, asst med, 48-51, instr, 51-52; from asst prof med to prof neurol, 53-86, chmn neurol div, Sch Med, 59-65, chmn , 65-81, neurologist-in-chief, Med Ctr, 65-81, EMER PROF, SCH MED, UTAH UNIV, 86- *Concurrent Pos:* Archbold fel med, Johns Hopkins Univ, 48-50, asst physician, Outpatient Dept, 48-52, Nat Found Infantile Paralysis fel, 50-52; asst chief med, Vet Admin Hosp, Salt Lake City, Utah, 53-57, chief med serv, 57, neurol serv, 59; spec clin trainee, NIH, Nat Hosp, London, 58 & Mass Gen Hosp, Boston, 58-59; neurologist-in-chief, Salt Lake County Gen Hosp, 59-65. *Mem:* Am Physiol Soc; Am Neurol Asn; Am Asn Res Nerv & Ment Dis; Am Acad Neurol; Am Fedn Clin Res; Sigma Xi. *Res:* Neurophysiology; neuromuscular disease. *Mailing Add:* 1497 Devonshire Dr Salt Lake City UT 84108

JARCHO, SAUL, b New York, NY, Oct 25, 06; m 48; c 2. HISTORY OF MEDICINE, MEDICAL CARTOGRAPHY. *Educ:* Harvard Univ, BA, 25; Columbia Univ, MA, 26, MD, 30. *Prof Exp:* consult, Nat Lib Med, 46-88; MED HISTORIAN, 38- *Concurrent Pos:* Consult to Surgeon Gen Army; consult to armed forces, Soc Med; ed, Bull NY Acad Med, 67-77; ed, Trans & Studies, Col Physicians Philadelphia, 79-83. *Honors & Awards:* William Welch Medal, Am Asn History Med, 63; Jacobi Medal, Mt Sinai Hosp, NY, 70; NY Acad Med Medal, 79. *Mem:* AMA; fel Am Col Physicians; Am Asn Path & Bact; Am Pub Health Asn; Am Asn Hist Med; Am Asn Advan Sci; NY Acad Sci; NY Acad Med; Am Soc Trop Med; Am Soc Parasitologists. *Res:* Cardiology; paleopathology; history of medicine; pathology; medical cartography. *Mailing Add:* 11 W 69th St New York NY 10023

JARDETZKY, OLEG, b Belgrade, Yugoslavia, Feb 11, 29; nat US; m 52, 65; c 3. MOLECULAR BIOLOGY, PHARMACOLOGY. *Educ:* Macalester Col, BA, 50; Univ Minn, MD, 54, PhD(chem physiol), 56. *Hon Degrees:* DSc, Macalester Col, 74; LLD, Calif Western Univ, 78. *Prof Exp:* Res asst physiol, Univ Minn, 50-54; Nat Res Coun fel chem, Calif Inst Technol, 56-57; assoc pharmacol, Harvard Med Sch, 57-59, asst prof, 59-66; dir dept biophys & pharmacol, Merck Sharp & Dohme Res Labs, NJ, 66-68, exec dir basic med sci, Merck Inst Therapeut Res, 68-69; actg chmn, 73-74, PROF PHARMACOL, SCH MED, STANFORD UNIV, 69-, DIR, MAGNETIC RESONANCE LAB, 75- *Concurrent Pos:* Irvine McQuarrie scholar award, 54; consult, Mass Gen Hosp, 61-67; vis prof, State Univ NY Buffalo, 63, State Univ NY Albany, 70; consult coun drugs, AMA, 64; Japan Chem Soc lectr, Univ Tokyo, 65; vis scientist, Cambridge Univ, 65-66; Chem Students Asn lectr, Univ Amsterdam, 66; basic sci lectr, Med Ctr, Univ Calif, San Francisco, 68; USSR Acad Sci lectr, 70; chmn, Nat Comn, 10th Int Conf on Magnetic Resonance Biol Syst, Stanford Univ, 82; adv bd mem, 22nd Cong Appl Mgt Planning & Eng Resources Eval, 82; plenary lectr, 58th Ann meeting Japanese Biochem Soc, 85. *Mem:* AAAS; Am Chem Soc; Am Soc Biol Chemists; Biophys Soc; Int Soc Magnetic Resonance; Soc Magnetic Resonance Med; fel Japan Soc Prom Sci. *Res:* Molecular mechanisms of protein function; biological applications of nuclear magnetic resonance. *Mailing Add:* Stanford Magnetic Resonance Lab Stanford Univ Stanford CA 94305-5055

JARDINE, D(ONALD) A(NDREW), b Kingston, Ont, July 23, 30; div; c 2. COMPUTER SCIENCE. *Educ:* Queen's Univ Ont, BSc, 52, MSc, 54; Univ Del, PhD, 57. *Prof Exp:* Res engr, Du Pont Can Ltd, 56-68, res assoc, 68-70; assoc prof, 70-73, head dept, 73-78, PROF COMPUT SCI, QUEEN'S UNIV ONT, 73- *Concurrent Pos:* Pres, Common Comput Users Group, 66-68. *Mem:* Asn Comput Mach. *Res:* Data base management systems; data description languages. *Mailing Add:* Dept Comput Sci Queens Univ Kingston ON K7L 3N6 Can

JARDINE, IAN, b Glasgow, Scotland, Sept 17, 48. MASS SPECTROMETRY. *Educ:* Univ Glasgow, BSc, 70, PhD(chem), 73. *Prof Exp:* Fel pharmacol, Med Sch, Johns Hopkins Univ, 73-76; asst prof med chem & pharmacog, Sch Pharm & Pharmacol Sci, Purdue Univ, West Lafayette, 76-79; from assoc prof to prof pharmacol, Mayo Med Sch, 79-86; consult pharmacol, Mayo Clin, 79-86; DIR ANALYTICAL BIOCHEM, FINNEGAN MAT, 88- *Mem:* Am Chem Soc; Am Soc Mass Spectrometry; AAAS. *Res:* Development of mass spectrometric methods for biochemical and pharmacological analysis. *Mailing Add:* Finnegan Mat 355 River Oaks Pkwy San Jose CA 95134-1991

JARDINE, JOHN MCNAIR, b Moncton, NB, June 25, 19; m 45; c 2. ATOMIC ENERGY, MATHEMATICAL STATISTICS. *Educ:* Mt Allison Univ, BSc, 40; McGill Univ, MSc, 48. *Prof Exp:* Res chemist, Refining Div, Eldorado Mining & Refining Ltd, 48-51, chief analyst, 51-53, chief chemist, 53-62, supt metall lab, Res & Develop Div, 62-69, res supt, Eldorado Nuclear Ltd, 69-72; scientific adv mining, Atomic Energy Control Bd, Can, 72-81; RETIRED. *Honors & Awards:* Can Forces Decoration, Can Govt, 60. *Mem:* Fel Chem Inst Can; Can Soc Chem Eng. *Res:* Analytical chemistry in the Canadian uranium industry; solvent extraction of uranium and thorium; separation of copper, cobalt and nickel by solvent extraction; development of process for production of hafnium free zirconium metal from zircon sands; transportation of radioactive materials; uranium mining and milling. *Mailing Add:* 467 Broadview Ave Ottawa ON K2A 2L2 Can

JARED, ALVA HARDEN, b Roseville, Ill, Jan 15, 34; m 55; c 2. WOOD TECHNOLOGY, DRAFTING & DESIGN. *Educ:* Western Ill State Col, BS, 55; Ball State Teachers Col, MAE, 56; Ariz State Univ, EdD(indust educ), 68. *Prof Exp:* Res asst, Ariz State Univ, 65-66; PROF INDUST EDUC, UNIV WIS-PLATTEVILLE, 56-; DEPT CHMN, INDUST STUDIES DEPT, 65- *Concurrent Pos:* Consult energy mgt & construct. *Mem:* Int Technol Educ Asn; Nat Asn Indust Technol; Int Coun Indust Teacher Educators. *Res:* Technology management; supervision and training of workers; construction management; energy management; technology education; author of several publications. *Mailing Add:* 945 St James Circle Platteville WI 53818

JAREM, JOHN, b Jarembina, Czech, July 4, 21; US citizen; c 4. ELECTRICAL ENGINEERING. *Educ:* Polytech Inst Brooklyn, BEE, 47, MEE, 50; Univ Pa, MS, 57, PhD(plasma physics), 60. *Prof Exp:* Electronic res engr, Tele-Register Corp, NY, 47; asst prof elec eng & math, US Naval Postgrad Sch, 48-51; math specialist, Lockheed Aircraft Corp, Calif, 51-54; mem systs eng tech staff, Radio Corp Am, NJ, 54-59, engr, 59-62; sr staff mem, Inst Defense Anal, 62-63; staff engr dir & systs engr, Radio Corp Am, 63-64; head dept elec eng, 64-68, PROF ELEC ENG, DREXEL UNIV, 64- *Concurrent Pos:* Consult, Radio Corp Am, 64-65, Inst Defense Anal, 64- & Aero Chem, 66- *Mem:* Sr mem Inst Elec & Electronics Engrs; Am Phys Soc. *Res:* Systems engineering; applied mathematics; plasma physics. *Mailing Add:* Dept Elec Eng 32nd & Chestnut Sts Philadelphia PA 19104

JARETT, LEONARD, b Lubbock, Tex, Aug 25, 36; m 62; c 3. CLINICAL PATHOLOGY, BIOCHEMISTRY. *Educ:* Rice Univ, BA, 58; Wash Univ, MD, 62. *Hon Degrees:* MA, Univ Pa, 82. *Prof Exp:* Intern path, Barnes Hosp, St Louis, Mo, 62-63, resident, 63-64; res assoc, Sect Cellular Physiol, Lab Biochem, Nat Heart Inst, 64-66; from instr path to assoc prof path & med, Wash Univ, 66-75, head div lab med, 69-75, prof path & med, Sch Med, 75-80; PROF & CHMN DEPT PATH & LAB MED, SCH MED, UNIV PA, 80- *Concurrent Pos:* Dir labs & head div lab med, Barnes Hosp, 69-80; mem sci adv bd, St Jude Children's Res Hosp, 80-83 & metab study sect, 83-87; mem adv bd, Juv Diabetes Found, 81-84. *Honors & Awards:* David Rumbough Award, Juv Diabetes Found, 80; Cotlove Award, Acad Clin Lab Physicians & Scientists, 85. *Mem:* Endocrine Soc; Acad Clin Lab Physicians & Scientists; Am Soc Biol Chemists; Am Fedn Clin Res; fel Am Soc Clin Path; Am Asn Physicians; Am Soc Clin Invest. *Res:* Biochemical and ultrastructural techniques in the study of signal transduction by insulin. *Mailing Add:* Dept Path & Lab Med Box 671 HUP 3400 Spruce Street Philadelphia PA 19104

JARGIELLO, PATRICIA, b Erie, Pa, July 9, 44. MOLECULAR GENETICS. *Educ:* Mercyhurst Col, BA, 66; Univ Pittsburgh, PhD(microbiol), 73. *Prof Exp:* Lab asst, Fed Bur Invest, 66-67; teaching asst biol, Univ Pittsburgh, 67-68; instr, Univ Parana, Brazil, 72-73; fel pediat, 73-76, instr, 74-76, RES ASST PROF MICROBIOL, MED SCH, UNIV PITTSBURGH, 77- *Concurrent Pos:* Vis fel cytogenetics, Southbury Training Sch Hosp, 74. *Mem:* Am Soc Human Genetics; Genetics Soc Am; Tissue Cult Asn; Am Soc Microbiol; AAAS. *Res:* Regulation of genes coding for enzymes involved in deoxynucleoside catabolism and deoxyribose utilization in salmonella; regulation of globin chain synthesis and hemoglobin formation in human fibroblast-rabbit erythroblast heterokaryons; ribose metabolism in hepatoma; metaphase chromosome isolation. *Mailing Add:* c/o Jarrett Lakewood Psychiat Hosp RD 2 Box 34A Canonsburg PA 15317

JARGON, JERRY ROBERT, b Beckemeyer, Ill, Aug 2, 39; m 63; c 3. PETROLEUM ENGINEERING, CHEMICAL ENGINEERING. *Educ:* Univ Ill, BS, 63; Univ Denver, MS, 67. *Prof Exp:* Assoc engr res, Chicago Bridge & Iron Co, 62; assoc engr, 63-66, engr, 66-72, adv engr prod res, 72-77, sr engr petrol technol, 77-80, adv sr engr, 80-85, sr staff engr, 85-87, sr tech consult, 87-88, MGR, RESERVOIR MGT DEPT, MARATHON OIL CO, 88- *Concurrent Pos:* Lectr continuing educ courses, Soc Petrol Engrs, 73-74, prog comt, Rocky Mt Region, 74-75, formation evalu comt, Nat Meeting, 75, mem, Monogr Rev Comt Gas Well Performance, 77-80; tech ed, J Petrol Tech, 85-86; ed, Gaswell Testing, 85-; eng adv bd, Univ Denver, 89- *Mem:* Soc Petrol Engrs; assoc Inst Mech Engrs; Opers Res Soc Am. *Res:* Reservoir modeling and engineering; pressure transient testing in wells; multiphase flow in wells and pipelines. *Mailing Add:* Marathon Oil Co PO Box 269 Littleton CO 80160

JARIWALA, SHARAD LALLUBHAI, b Bombay, India, Oct 15, 40; m 69; c 2. CHEMICAL ENGINEERING. *Educ:* Univ Bombay, BSChE, 62; Johns Hopkins Univ, PhD(chem eng), 66. *Prof Exp:* Res engr, Tenneco Chem, Inc, 66-70; sr scientist, Fermentation Res & Develop, 70-75, head fermentation prod prod, 75-76, mgr, 77-79, group mgr, 80-83, dir, 83-84, VPRES FERMENTATION OPER, UPJOHN CO, 84- *Mem:* Am Inst Chem Engrs. *Res:* Developing new technology for the separation and recovery of antibiotics from fermentation broths; separations technology; reaction engineering in fixed and fluidized beds; liquid phase oxidations. *Mailing Add:* Upjohn Co 1800-91-2 7000 Portage Rd Portage MI 49001

JARIWALLA, RAXIT JAYANTILAL, b Bombay, India, Nov 18, 49; US citizen; m. VIROLOGY, GENETICS. *Educ:* Bombay Univ, BSc, 71; Med Col Wisc, MS, 74, PhD(microbiol), 76. *Prof Exp:* Fel, div biophys, Sch Hyg & Pub Health, Johns Hopkins Univ, 76-79, instr, div biophys, 79-82; res scientist, 82-84, SR SCIENTIST, LINUS PAULING INST SCI MED, 84-, HEAD VIRAL CARCINOGENESIS & IMMUNOL PROG, 86- *Concurrent Pos:* Guest lectr, virol course, Dept Med Microbiol, Stanford Univ, 85-86; prin investr, NCI sponsored res grant, Linus Pauling Inst Sci Med, 86-; co-organizer, XVI Int Herpesvirus Workshop, Asilomar, Calif, 91. *Mem:* Am Soc Microbiol; AAAS; Am Soc Virol; NY Acad Sci. *Res:* Mechanisms of cancer induction by human herpesviruses: elucidation of the genes and biochemical mechanisms involved in cellular immortalization and tumorigenic transformation by herpes simplex-2 and cytomegaloviruses; role of herpesvirus as a cofactor in AIDS: activation of latent HIV infection by viral cofactors and suppression of HIV replication by antioxidants and reducing agents; control of cell proliferation, tumor growth and blood lipids by dietary phytate (inositol hexaphosphate). *Mailing Add:* Linus Pauling Inst 440 Page Mill Rd Palo Alto CA 94306

JARKE, FRANK HENRY, b Bloomington, Ill, Mar 28, 46; m 71. PHYSICAL CHEMISTRY, ANALYTICAL CHEMISTRY. *Educ:* Southern Ill Univ, BA, 69; Ill Inst Technol, MS, 74. *Prof Exp:* Asst chemist, ITT Res Inst, 69-73, assoc chem, 73-78, res chemist odor sci, 78-81; mgr anal serv, chem waste mgt, Riverdale, Ill, 81-83; asst mgr, Environ Waste Mgt, Oak Brook, 83-87, MGR, QUAL PROG, EML WASTE MGT, GENEVA, ILL, 87- *Mem:* Am Chem Soc; AAAS; Soc Appl Spectros; Am Soc Heating, Refrig & Air-Conditioning Engrs; NY Acad Sci. *Res:* Fundamental and applied research of odors and air pollution; development and use of both subjective and objective methods using humans as detectors. *Mailing Add:* Waste Mgt Inc 2100 Cleanwater Rd Geneva IL 60134

JARMAKANI, JAY M, US citizen. PEDIATRIC CARDIOLOGY. *Educ:* Damascus Univ, BCP, 56, MD, 62. *Prof Exp:* Pediat resident, Buffalo Children's Hosp & Children's Hosp Philadelphia, 63-65; fel pediat cardiol, Children's Hosp Pittsburgh, 65-66; fel, Med Ctr, Duke Univ, 66-68, asst prof pediat, 68-73; assoc prof, 73-78, DIR CARDIOPULMONARY LAB & PROF PEDIAT, MED CTR, UNIV CALIF, LOS ANGELES, 78- *Mem:* Soc Pediat Res; fel Am Col Cardiol; Am Heart Asn; Am Physiol Soc; Int Soc Heart Res. *Res:* Developmental myocardial function with emphasis on congenital heart disease and the effect of hypoxia on cardiac cell function. *Mailing Add:* Dept Pediat Univ Calif Med Ctr Los Angeles CA 90024

JARMIE, NELSON, b Santa Monica, Calif, Mar 24, 28; m 52, 89; c 2. NUCLEAR PHYSICS. *Educ:* Calif Inst Technol, BS, 48; Univ Calif, PhD(physics), 53. *Prof Exp:* Res physicist, Radiation Lab, Univ Calif, 50-53; prof physics, Los Alamos Grad Ctr, Univ NMex, 57-75; MEM STAFF, LOS ALAMOS NAT LAB, 53- *Concurrent Pos:* Vis asst prof, Univ Calif, 59-60; partic, Vis Scientist Prog, 65-85. *Honors & Awards:* Distinguished Performance Award, 85, Los Alamos Nat Lab. *Mem:* Fel AAAS; fel Am Phys Soc; Am Asn Physics Teachers; Asn Appl Psychophysiol & Biofeedback. *Res:* Light-nuclei energy levels; 3-body breakup, nucleon-nuclear scattering, astrophysical reactions; kinematic codes, straggling calculations and infrared laser diagnostics; fundamental properties of antimatter and gravitational acceleration of anti-protons; nuclear physics, particle physics and astrophysics. *Mailing Add:* Los Alamos Nat Lab Mail Stop D449 Los Alamos NM 87545

JARNAGIN, RICHARD CALVIN, b Dallas, Tex, Aug 26, 30; m 52; c 2. PHYSICAL CHEMISTRY. *Educ:* Southern Methodist Univ, BS, 52; Yale Univ, PhD, 58. *Prof Exp:* Res chemist, Wright Air Develop Ctr, US Air Force, 53-55; from instr to assoc prof, 58-68, PROF CHEM, UNIV NC, CHAPEL HILL, 68- *Concurrent Pos:* Guggenheim fel, 67-68 & NSF fel, Sandia Nat Labs, 78-79. *Mem:* Am Chem Soc; fel Am Phys Soc. *Res:* Electrical and optical properties of molecular systems; photo conduction in organic solids and liquids; kinetics of excited molecular states; catalytic properties of solid oxides and stabilization of oxide films. *Mailing Add:* 609 Caswell Rd Chapel Hill NC 27514

JAROLMEN, HOWARD, b New York, NY, Oct 19, 37; m 66; c 2. MEDICAL MICROBIOLOGY, GENETICS. *Educ:* Alfred Univ, BA, 58; Hahnemann Med Col, MS, 60, PhD(microbiol), 64. *Prof Exp:* NIH fel genetics, Cornell Univ, 64-67; res bacteriologist, Am Cyanamid Co, 67-70, group leader bact chemother, 70-74, head dept microbiol & chemother, 74-76, group leader, 76-80, PRIN RES MICROBIOLOGIST, FERMENTATION PROCESS RES & DEVELOP DEPT, LEDERLE LABS DIV, AM CYANAMID CO, 80- *Mem:* Am Soc Microbiol; Soc Indust Microbiol. *Res:* In vitro and vivo studies of transferable resistance amongst the Enterobacteriaceae; veterinary microbiology; prophylaxis and therapy of experimental infections; bacterial mutagenicity testing; antibiotic discoveries, discovery of antimycobacterials and antiparasitics; strain and media improvement for antibiotic-producing cultures. *Mailing Add:* Lederle Labs Pearl River NY 10965

JARON, DOV, b Tel-Aviv, Israel, Oct 29, 35; US citizen; m 79; c 2. BIOMEDICAL ENGINEERING. *Educ:* Univ Denver, BS, 61; Univ Pa, PhD(biomed eng), 67. *Prof Exp:* Sr res assoc, Maimonides Med Ctr, 67-70; dir surg res lab, Sinai Hosp, Detroit, 70-73; from assoc prof toprof biomed eng, Univ RI, 73-80; PROF & DIR BIOMED ENG & SCI INST, DREXEL UNIV, 80- *Concurrent Pos:* Consult circulatory syst devices panel, Food & Drug Admin, 76-79; chmn, Sixth Ann New Eng Bioeng Conf, 78. *Mem:* Biomed Eng Soc; AAAS; NY Acad Sci; fel Inst Elec & Electronics Engrs; Am Soc Artificial Internal Organs; Eng Med & Biol Soc (vpres, 84-85, pres, 86-87); Int Soc Artificial Organs. *Res:* Cardiovascular dynamics; control and optimization of assisted circulation; cardiovascular modeling and assessment of function; biomedical instrumentation; computer applications to health care. *Mailing Add:* Biomed Engr & Sci Inst Drexel Univ 32nd & Chestnut Sts Philadelphia PA 19104

JAROS, STANLEY E(DWARD), b Syracuse, NY, Mar 23, 19; m 42, 59; c 6. CHEMICAL ENGINEERING. *Educ:* Syracuse Univ, BChE, 40, MChE, 42. *Prof Exp:* Chem engr, Exxon Res & Eng Co, 42-55, asst dir, Chem Develop Div, 55-57, dir, 57-61, assoc dir, Process Eng Div, 61-65, res coordr, Chem Planning Staff, 65-66, res coordr, New Projs Develop, 66-72, res coordr, Corp Res Feasibility Unit, 72-78; consult, 78-82; RETIRED. *Mem:* Am Chem Soc; Am Inst Chem Engrs. *Res:* Translating research results to commercial projects in the process industries. *Mailing Add:* 1199 Monticello Rd Lafayette CA 94549

JAROWSKI, CHARLES I, b Baltimore, Md, July 29, 17; m 45; c 3. PHARMACEUTICAL CHEMISTRY. *Educ:* Univ Md, BS, 38, PhD(pharmaceut chem), 43. *Prof Exp:* Fel, Univ Ill, 42-44; res chemist, Wyeth Inc, Pa, 44-46; chief chemist, Vick Chem Co, NY, 46-48; res chemist, Chas Pfizer & Co, Inc, 48-50, mgr pharmaceut res & develop, 50-60, dir, 60-69; from assoc prof to prof pharmaceut, St Johns Univ, 69-88, chmn, Dept Allied Health & Indust Sci, 78-88; RETIRED. *Mem:* Am Pharmaceut Asn. *Res:* Synthesis of antibacterial agents, antioxidants, and medicinals; antioxidant for food and drug industry; germicidal steam aerosolic compounds; antibiotic derivatives; drug detoxification; scientific nutrition; drug delivery systems. *Mailing Add:* 67 Harbor Lane Massapequa Park NY 11762

JARRELL, JOSEPH ANDY, b Bad Kissengr, Ger, Mar 5, 50. INSTRUMENTATION, BIOPHYSICS. *Educ:* Mass Inst Technol, BS, 71, PhD(physics), 79. *Prof Exp:* Res assoc, Mass Inst Technol, 78-85; DIR, WATERS CHROMATOGRAPHY, DIV MILLTORE, 86- *Mem:* Am Phys Soc; Am Vacuum Soc. *Mailing Add:* Rm 26-457 Mass Inst Technol Cambridge MA 02139

JARRELL, WESLEY MICHAEL, b Forest Grove, Ore, May 23, 48; m 72; c 2. PLANT-SOIL RELATIONSHIPS, ECOSYSTEM SCIENCE. *Educ:* Stanford Univ, AB, 70; Ore State Univ, MS, 74, PhD(soil sci), 77. *Prof Exp:* From asst prof to assoc prof soil & plant relationships, Univ Calif, Riverside, 76-88, dir, Dry Lands Res Inst, 85-88; assoc prof, 88-91, PROF ECOSYST SCI, ORE GRAD INST SCI & TECHNOL, 91- *Concurrent Pos:* Consult, Allergan Pharmaceut, 80-85, Benchmark, 84 & Jardinier, 87- *Honors & Awards:* Alex B Laurie Award, Am Soc Hort Sci, 79. *Mem:* Soil Sci Soc Am; Am Soc Agron; Am Soc Hort Sci; Ecol Soc Am; AAAS. *Res:* Relationships between soil conditions and plant growth, water quality and air quality; restoration of disturbed landscapes to stable ecosystems; nutrient and hydrologic cycles, particularly the interactions between hydrology and elemental transport. *Mailing Add:* Ore Grad Inst 19600 NW von Neumann Dr Beaverton OR 97006-1999

JARRET, RONALD MARCEL, b Woonsocket, RI, Dec 22, 60; m 82; c 2. PHYSICAL ORGANIC. *Educ:* Rhode Island Col, BA, 82, BS, 82; Yale Univ, PhD(chem), 87. *Prof Exp:* ASST PROF ORG CHEM, COL HOLY CROSS, 86- *Concurrent Pos:* Prin investr, NSF grants, 88-92. *Mem:* Am Chem Soc. *Res:* Generate novel carbocations and use spectroscopic methods for structure identification. *Mailing Add:* Dept Chem Holy Cross Col Worcester MA 01610

JARRETT, HARRY WELLINGTON, III, b Charleston, SC, June 19, 50; m 77; c 3. PLANT BIOCHEMISTRY, PROTEIN CHEMISTRY. *Educ:* Univ SC, BS, 72; Univ NC, PhD(biochem), 76. *Prof Exp:* Postdoctoral fel biochem, Mayo Clin, 76-77; postdoctoral fel chem, Univ Calif, San Diego, 77-80; asst biochemist biochem, Univ Ga, 80-82; from asst prof to assoc prof biol, Ind Univ-Purdue Univ, Indianapolis, 82-89; ASSOC PROF BIOCHEM, UNIV TENN, MEMPHIS, 89- *Mem:* Am Soc Biochem & Molecular Biol; AAAS. *Res:* Calmodulin and Ca2-dependent metabolic regulation in plants are main areas; development of new supports for affinity, ion exchange, gel filtration and DNA high pressure liquid chromatography. *Mailing Add:* Dept Biochem Univ Tenn-Memphis 800 Madison Ave Memphis TN 38163

JARRETT, HOWARD STARKE, JR, b Charleston, WVa, Oct 24, 27; m 51; c 4. SOLID STATE PHYSICS. *Educ:* Rensselaer Polytech Inst, BS, 47, MS, 48; Mass Inst Technol, PhD(physics), 51. *Prof Exp:* Res physicist, Cent Res Dept, E I du Pont de Nemours & Co, Inc, 51-55 & 80-90, res supvr, 55-80; RETIRED. *Concurrent Pos:* Prog co-chmn, Conf Magnetism & Magnetic Mat, 65, conf chmn, 69, chmn adv comn, 70. *Mem:* Fel Am Phys Soc. *Mailing Add:* 805 Sycamore Lane Wilmington DE 19807

JARRETT, JEFFREY E, b Bronx, NY, Dec 13, 40; m 64; c 3. MANAGEMENT SCIENCE. *Educ:* Univ Mich, BBA, 62; NY Univ, MBA, 63, PhD(statist/opers res), 67. *Prof Exp:* Statistician, Columbia Rec Div, CBS, 62-63; instr statist/opers res, Univ Scranton, 65-66; prof statist/qual control, Wayne State Univ, 66-71; res analyst, Social Security Admin, 74-75; PROF MGT SCI/STATIST, UNIV RI, 71- *Concurrent Pos:* Sears Found fed fac fel, Div Health Ins, Soc Sec Admin, 74-75; consult, Abt Assocs, 79, RI Dept Health, 83, RI Pub Utilities Comn, 84, Tex Instrument, New Eng Tel, 88, Fed Paperboard, 86 & Eastern Utilities, 87; prof, Ohio Savings & Loan Acad, 81-91; mem, Prog Community Pharm Mgt, W M S Apple Found, 91; outstanding res scholar, URICBA, 91-92. *Honors & Awards:* Am Statist Asn. *Mem:* Decision Sci Inst; Int Inst Forecasters. *Res:* Applying statistics to business decision making, forecasting, quality control, and other managerial problems. *Mailing Add:* 133 Terre Mar Dr North Kingstown RI 02852

JARRETT, NOEL, b Long Eaton, Eng, Nov 17, 21; nat US; m 49; c 4. METALLURGICAL PROCESS ENGINEERING. *Educ:* Univ Pittsburgh, BS, 48; Univ Mich, MS, 51. *Prof Exp:* Sales engr indust oil sales, Freedom-Valvoline Oil Co, 49-50; res engr smelting, Aluminum Co Am 51-55, sect head chem eng, 55-59, asst chief process metall div, 60-69, mgr, 69-73, dir smelting res & develop, 73-81, tech dir chem engr res & develop, Alcoa Labs, 81-87; RETIRED. *Concurrent Pos:* Mem, numerous Nat Res Coun & Nat Mat Adv Bd comts, 81-90; consult, Noel Jarrett Assoc, 87-; Krumb lectr, Am Inst Mining, Metall & Petrol Engrs, 87- *Mem:* Nat Acad Eng; Am Inst Chem Engrs; Am Inst Mining, Metall & Petrol Engrs; fel Am Soc Metals. *Res:* Electrochemical cell development; optimization of Hall-Heroult Process; coker reactor; cell development of Alcoa Smelting Process; pollution control by scrubbing of chlorine from furnace effluent; high purity Al via crystallization; materials science engineering. *Mailing Add:* 149 Jefferson Ave Lower Burrell PA 15068-3127

JARRETT, STEVEN MICHAEL, b New York, NY, Mar 17, 36; m 61; c 4. PHYSICS. *Educ:* City Col New York, BS, 56; Univ Mich, MS, 58, PhD(physics), 63. *Prof Exp:* Res assoc physics, Univ Mich, 62-63; sr scientist, TRG, Inc, Control Data Corp, NY, 63-66; sr res, Coherent Radiation Labs, Calif, 66-71; pres, Quantum Systs Corp, 71-75; ENG PROJ MGR, SPECTRA PHYSICS, 75- *Mem:* Am Phys Soc; Optical Soc Am. *Res:* Lasers, especially gas, solid state and dye lasers; optics; spectroscopy. *Mailing Add:* 474 Los Ninos Way Los Altos CA 94022

JARROLL, EDWARD LEE, JR, b Huntington, WVa, Jan 4, 48; m 75; c 1. BIOLOGY, PARASITOLOGY. *Educ:* WVa Univ, AB, 69, MS, 71, PhD(biol), 77. *Prof Exp:* From instr to asst prof biol, Salem Col, 73-77; microbiologist, WVa Dept Health, 77; FEL MICROBIOL, HEALTH SCI CTR, UNIV ORE, 77-; Sr res assoc, Cornell Univ, 80-82; asst prof biol, West Chester Univ, 84; ASSOC PROF BIOL, CLEVELAND STATE UNIV, 85- *Mem:* Am Soc Trop Med & Hyg; Am Soc Parasitologists; AAAS; Sigma Xi. *Res:* Giardia and Trichomonas culture; physiology, immunology, and epidemiology; efficacy of disinfectants on Giardia cyst viability; helminth population biology. *Mailing Add:* Dept of Biol 1983 24th St Cleveland State Univ Cleveland OH 44115

JARUZELSKI, JOHN JANUSZ, b Poland, Oct 4, 26; nat US; m 56; c 2. INDUSTRIAL ORGANIC CHEMISTRY & MANUFACTURING LUBRICATING OILS COMPONENTS. *Educ:* Alliance Col, BS, 51; Pa State Univ, PhD(chem), 54. *Prof Exp:* Res chemist, Pittsburgh Plate Glass Co, 54-56; fel, Mellon Inst, 56-59; res chemist prod develop div, US Steel Corp, 59-60; sr chemist, Esso Res & Eng Co, 60-74, SR STAFF CHEMIST & RES ASSOC PARAMINS, TECHNOL DIV, EXXON CHEM CO, 74- *Mem:* Am Chem Soc. *Res:* Substitution reactions of aromatic hydrocarbons, especially chloroalkylations; esterification and polyesterification of alcohols and phenols; epoxydation and epoxy resins; thermosetting resins and reinforced plastics; chemistry of lubricating oils additives and antiwear chemicals; antioxidants. *Mailing Add:* 475 Channing Ave Westfield NJ 07090

JARVI, ESA TERO, b Turku, Finland, May 12, 54; US citizen; m 83. PHARMACEUTICAL CHEMISTRY, ORGANIC CHEMISTRY. *Educ:* Ohio State Univ, BS, 75; Univ Wis, Madison, PhD(org chem), 80. *Prof Exp:* SR RES CHEMIST, MERRELL-DOW PHARMACEUT, 82- *Mem:* Am Chem Soc. *Res:* Synthesis of new potential drugs; enzyme inhibitors. *Mailing Add:* 3953 Saint Johns Terr Cincinnati OH 45236

JARVIK, JONATHAN WALLACE, b Charleston, SC, Mar 18, 45; m; c 3. ORGANELLE MORPHOGENESIS. *Educ:* Columbia Col, BA, 67; Mass Inst Technol, PhD(biol), 75. *Prof Exp:* Helen Hay Whitney fel, Yale Univ, 75-78; asst prof, 78-83, ASSOC PROF BIOL SCI, CARNEGIE-MELLON UNIV, 84- *Mem:* Genetics Soc Am; Am Soc Microbiol; Am Soc Cell Biol; Soc Protozoologists. *Res:* Genetic biochemical and ultrastructural analysis of eucaryotic flagellar morphogenesis. *Mailing Add:* Dept Biol Sci Carnegie-Mellon Univ 4400 Fifth Ave Pittsburgh PA 15213

JARVIK, LISSY F, b The Hague, Neth; nat US; m 54; c 2. HUMAN GENETICS, GERIATRIC PSYCHIATRY. *Educ:* Hunter Col, AB, 46; Columbia Univ, MA, 47, PhD(phychol), 50; Western Reserve Univ, MD, 54; Am Bd Pediat, dipl. *Prof Exp:* Asst psychiat, Columbia Univ, 46-48, res assoc, 48-50; asst, Sch Med, Western Reserve Univ, 53; intern, Mt Sinai Hosp, New York, 54-55; sr res scientist med genetics, Radiation Safety Officer, NY State Psychiat Inst, 55-62, assoc res scientist & assoc attend psychiatrist, 63-72; PROF PSYCHIAT, UNIV CALIF, LOS ANGELES, 72-, CHIEF SECT NEUROPSYCHOGERIAT, 83- *Concurrent Pos:* NSF traveling fel, Int Cong Human Genetics, Denmark, 56; fel, Vanderbilt Clin, 57-58; resident pediat, Columbia-Presby Med Ctr, 55-56, from asst clin prof to assoc clin prof psychiat, Columbia Univ, 62-72; res psychiatrist, NY State Psychiat Inst, 69-72; vis assoc prof, Univ Calif, Los Angeles, 70-71; chief psychogenetics unit, Vet Admin Hosp Brentwood, Los Angeles, 70-; mem joint psychotomimetic adv comt, Nat Inst Ment Health-Food & Drug Admin, 70-72; tech comt res & develop, White House Conf Aging, 71-73; vis McCleod prof, Univ Adelaide, SAustralia, 81; distinguished physician, Vet Admin, 87-; co-ed, Alzheimer Dis & Assoc Disorders, Int J, 87-, mem, Nat Adv Mental Health Coun, 84-87, action comt, White House conf aging, 86, Nat Inst Mental Health, Dept Health & Human Serv coun, Alzheimer's Dis, Workshop, epidemiol, 87; testimony, Joint House Subcomt hearings, Alzheimer's Dis, 83; bd dir & med & sci adv coun, Alzheimer's Dis & Rel Disorders Asn, 80-; vis lectr geriat & geront, Am Asn Med Col, 83-84; Nat Sci Adv Coun, Am Fed Aging Res, 86-; Brookdale Nat Fel, med adv bd, 87-; selection comt, Merck fel clin geriat pharmacol, Am Fed Aging Res, 87-89; bd mem, Am Aging Asn, 91-; found fel, Ctr Advan Study Behav Sci, Stanford, 88-89. *Honors & Awards:* Jack Weinberg Award Geriat Psychiat, Am Psychiat Asn, 86; Robert W Kleemeier Award Outstanding Res Aging, Geront Soc Am, 86; Edward B Allen Award, Am Geriat Soc, 86; Irving S Wright Award Distinction, Am Fedn Aging Res, 88; Founder's Award, Am Asn Geriat Psychiat, 90. *Mem:* Fel Am Psychol Asn; Am Soc Human Genetics; Soc Study Social Biol; Am Psychopath Asn; Am Psychiat Asn; Am Asn Geriat Psychiat; Am Geriat Soc; Geront Soc Am; Behav Genetics Asn; Int Asn Geront; World Psychiat Asn. *Res:* Normal pathological mental changes with aging, particularly dementia of the Alzheimer type and depression; geriatric psychopathology; drug treatment and psychotherapy; basic biological mechanisms in Alzheimer's disease, especially microtubules; family studies. *Mailing Add:* Dept Psychiat Univ Calif Los Angeles CA 90024-1759

JARVIK, MURRAY ELIAS, b New York, NY, June 1, 23; m 54; c 2. PHARMACOLOGY. *Educ:* City Col New York, BS, 44; Univ Calif, Los Angeles, MA, 45; Univ Calif, Berkeley, MD, 51, PhD, 52. *Prof Exp:* Res technician phys chem, Rockefeller Inst, 43-44; asst exp psychol, Univ Calif, Los Angeles, 44-45; asst physiol psychol, Univ Calif, Berkeley, 45-46, comp psychol, 47 & pharmacol, Med Sch, 51; res assoc comp physiol psychol, Yerkes Labs, Fla, 51-53; lectr physiol psychol, Columbia Univ, 53-55; vis asst prof, Univ Calif, 55; res assoc psychopharmacol, Long Island Biol Asn, NY, 55-56; from asst prof to prof pharmacol, Albert Einstein Col Med, 56-72, prof psychiat, 69-72; PROF PSYCHIAT & PHARMACOL, UNIV CALIF, LOS ANGELES, 72- *Concurrent Pos:* Res assoc, Mt Sinai Hosp, NY, 53-55; adj asst prof physiol psychol, Grad Div, NY Univ, 57; managing ed, Psychopharmacologia, 65; mem psychopharmacology study sect, NIMH, 65-70; adv comt abuse of stimulant & depressant drugs, Bur Drug Abuse Control, Food & Drug Admin, 66-68; investr, VA Med Res, 71; chief, Psychopharmacol Unit, Vet Admin Hosp, West Los Angeles, 72- *Honors & Awards:* Career Develop Scientist Award, NIMH, 71. *Mem:* Am Soc Pharmacol & Exp Therapeut; fel Am Psychol Asn; fel NY Acad Sci; Am Col Neuropsychopharmacol; Int Brain Res Orgn; fel, CASBS Ctr Advon Study Behav Sci. *Res:* Effects of drugs upon learning and retention; neurophysiological basis of learning; localization of drug effects in the central nervous system; psychopharmacology; primate behavior; techniques for chronic implantation of arterial catheters; smoking behavior nicotine addiction. *Mailing Add:* Dept of Psychiat Neuropsychiat Inst Hosp Univ Calif Ctr for Health Sci Los Angeles CA 90024

JARVINEN, RICHARD DALVIN, b Virginia, Minn, Dec 5, 38; m 61; c 2. MATHEMATICS. *Educ:* St John's Univ, Minn, BA, 60; Vanderbilt Univ, MAT, 61; Syracuse Univ, PhD(math), 71. *Prof Exp:* Analyst missile simulations, Remington Rand Univac, 61-62; asst prof math, Carleton Col, 67-72; from assoc prof to prof math & statist, St Mary's Col, 72-90; PROF MATH & STATIST, WINONA STATE UNIV, 90- *Concurrent Pos:* Researcher math & statist, St Mary's Col, 75- *Mem:* Math Asn Am; Sigma Xi. *Res:* Bases in topological linear spaces; applications of undergraduate mathematics; computer generated movies and slides for learning mathematics. *Mailing Add:* Dept Math & Statistics Winona State Univ Gildemeister Hall Winona MN 55987

JARVIS, BRUCE B, b Van Wert, Ohio, Sept 30, 42; m 63; c 3. ORGANIC CHEMISTRY. *Educ:* Ohio Wesleyan Univ, BA, 63; Univ Colo, PhD(chem), 66. *Prof Exp:* Instr chem, Northwestern Univ, 66-67; asst prof, 67-71, assoc prof, 71-79, PROF CHEM, UNIV MD, COLLEGE PARK, 79- *Mem:* AAAS; Am Chem Soc; Sigma Xi. *Res:* Natural product chemistry; nucleophilic displacements; sulfur chemistry and molecular rearrangements; mycotoxins. *Mailing Add:* Dept of Chem Univ of Md College Park MD 20742

JARVIS, CHRISTINE WOODRUFF, b Raleigh, NC, June 19, 49; m 71; c 1. TEXTILE SCIENCE, PHYSICAL CHEMISTRY. *Educ:* Univ NC, Chapel Hill, 71; Mass Inst Technol, PhD(phys chem), 76. *Prof Exp:* Res assoc textiles, Clemson Univ, 76, instr chem, 76-78, from asst prof to prof textiles, 78-89, J E SIRRINE PROF TEXTILES, CLEMSON UNIV, 89- *Concurrent Pos:* Res assoc, Nat Bur Standards, 85. *Mem:* Am Chem Soc; Am Asn Textile Technologists; Sigma Xi; Am Asn Textile Chemists & Colorists; Tech Asn Pulp & Paper Indust. *Res:* Fiber physics; nonwovens; chemical kinetics of polymer flammability; cotton dust analysis; apparel manufacturing. *Mailing Add:* Sch Textiles Sirrine Hall Clemson Univ Clemson SC 29634-1307

JARVIS, FLOYD ELDRIDGE, JR, b Richmond, Va, Aug 15, 21; m 53; c 1. GENETICS. *Educ:* Univ Richmond, AB, 47; Va Polytech Inst, PhD(biol), 56. *Prof Exp:* From assoc prof to prof biol, Radford Univ, 55-87; RETIRED. *Mem:* Entom Soc Am. *Res:* Inheritance of insecticidal resistance; residual effectiveness of insecticide formulations. *Mailing Add:* 103 Dogwood Lane Radford VA 24141

JARVIS, JACK REYNOLDS, b Menomonie, Wis, Oct 31, 15; m 45; c 2. MEDICINE. *Educ:* Birmingham-South Col, BS, 34; Vanderbilt Univ, MD, 38; Am Bd Psychiat & Neurol, dipl, 45. *Prof Exp:* Assoc prof psychiat, Med Col Ala, 48-61; ASST PROF PSYCHIAT, SCH MED, EMORY UNIV, 61- *Concurrent Pos:* Chief psychiat serv, Vet Admin Hosp, Birmingham, Ala, 55-61; area chief psychiat, Vet Admin Ga, 61-65; staff physician, Regional Off, Vet Admin, 65-; staff physician, Vet Admin Hosp, 66- *Mem:* Am Psychiat Asn. *Res:* Psychiatry. *Mailing Add:* Vet Admin Hosp 1670 Clairmont Rd Decatur GA 30033

JARVIS, JAMES GORDON, b Aultsville, Ont, July 13, 24; nat US; m 47; c 2. PHYSICS. *Educ:* Queen's Univ, Can, BSc, 45; Univ Rochester, MS, 54. *Prof Exp:* Instr, Queen's Univ, Can, 45-46; res assoc, Photomat Div, Eastman Kodak Co, 46-69, sr lab head, Res Labs, 69-84; RETIRED. *Mem:* Optical Soc Am; Am Soc Photog Sci & Eng. *Res:* Colorimetry; physiological optics; solid state physics; electrophotography. *Mailing Add:* 846 Dewitt Rd Webster NY 14580

JARVIS, JOHN FREDERICK, b Montreal, Can, June 22, 41; US citizen; m 63; c 5. PHYSICS. *Educ:* Univ Fla, BS, 62; Duke Univ, PhD(physics), 67. *Prof Exp:* Res assoc physics, Duke Univ, 67-68; mem tech staff systs res, 68-82, HEAD, ROBOTICS SYSTS RES DEPT, AT&T BELL LABS, 82- *Concurrent Pos:* Vis lectr, Princeton Univ, 78-79; vis astronr, KH Peak Nat Observ, 79-81; adj prof, EE Stevens Inst Technol, 91. *Mem:* Inst Elec & Electronics Engrs; AAAS; Am Astron Soc. *Res:* Computer graphics, computer vision, automated inspection, pattern recognition; robotics; astronomy. *Mailing Add:* 96 Forrest Ave Fair Haven NJ 07704

JARVIS, JOHN J, b Donnelson, Tenn, Aug 7, 41; m 63. TRANSPORTATION & DISTRIBUTION, LOGISTICS. *Educ:* Univ Ala, BSIE, 63, MSIE, 65; Johns Hopkins Univ, PhD(opers res), 68. *Prof Exp:* Numerical analyst, NASA, 63; res asst, Univ Ala, 63-65; res assoc, Johns Hopkins Univ, 65-68; from assoc prof to prof indust & systs eng, 68-91, DIR, GA INST TECHNOL, 91- *Concurrent Pos:* Consult, Southern Rwy Syst, 69-75, Comput Aided Planning & Scheduling, Inc, 78-, Environ Protection Agency, 78-, Sohio, 82, Coca Cola, 84-85 & Sears, 84-85. *Mem:* Opers Res Soc Am; Inst Mgt Sci; Soc Indust & Appl Math; Am Inst Indust Engrs. *Res:* Modeling and methodology in operations research and network theory-analysis; transportation, distribution and logistics systems analysis. *Mailing Add:* Sch Indust Systs Eng Ga Tech 765 Ferst Dr Atlanta GA 30332-0205

JARVIS, LACTANCE AUBREY, b Homer, Mich, Sept 7, 21; m 43; c 2. ORGANIC POLYMER CHEMISTRY. *Educ:* Mich State Univ, BS, 43. *Prof Exp:* Chem engr, Firestone Rubber & Tire Co, 43-49; chemist rubber & plastics, Wyandotte Chem Corp, 49-55; chem engr, Whirlpool Corp, 55-60; dir res, Modern Plastics Corp, Mich, 60-66; mgr mat res, Clark Equip Co, 66-82; CONSULT, 82- *Mem:* Am Chem Soc; Soc Plastics Engrs; Sigma Xi. *Res:* Development and application of plastic materials and processes. *Mailing Add:* PO Box 64 Buchanan MI 49107

JARVIS, NELDON LYNN, b Salt Lake City, Utah, Nov 16, 35; m 53; c 4. SURFACE CHEMISTRY. *Educ:* Brigham Young Univ, BS, 52; Kans State Univ, PhD(agron), 58. *Prof Exp:* Nat Acad Sci-Nat Res Coun res assoc, US Naval Res Lab, 57-59, phys chemist, 59-69, HEAD SURFACE CHEM BR, CHEM DIV, US NAVAL RES LAB, 69- *Concurrent Pos:* Lectr, Am Univ, 64 & 69- *Mem:* Sigma Xi. *Res:* Adsorption-desorption phenomena at solid-liquid and liquid-air interfaces; wetting and spreading phenomena; tribology; surface chemistry of ocean; surface analysis. *Mailing Add:* 120 S Union Ave Havre de Grace MD 21078

JARVIS, RICHARD S, b Nottingham, Eng, Feb 13, 49. HYDROLOGY. *Educ:* Cambridge Univ, BA, 70, PhD(geog), 75. *Prof Exp:* Lectr geog, Durham Univ, 73-74; ASSOC PROF GEOG, STATE UNIV NY BUFFALO, 74- *Mem:* Am Geophys Union; Asn Am Geogrs; Inst Brit Geogrs. *Res:* Network analysis of hydrologic systems; fluvial geomorphology; digitized data systems in hydrology and geomorphology; computer applications in geography; biogeography. *Mailing Add:* State Univ NY State Univ Plaza Rm 5301 Albany NY 12246

JARVIS, ROGER GEORGE, b Hugglescote, Eng, Apr 26, 28; Can citizen; m 54; c 2. ENVIRONMENTAL SCIENCES. *Educ:* Oxford Univ, BA, 49, MA & DPhil(physics), 53. *Prof Exp:* Fel nuclear physics, Nat Res Coun, Chalk River, 53-54; res fel, Imp Chem Indust, Univ Liverpool, 55; mem, Atomic Power Div, Gen Elec Co, Eng, 56; sr res off, waste mgt technol, Atomic Energy Can, Ltd, 56-89; RETIRED. *Mem:* Can Appl Math Soc; Math Asn Am. *Res:* Operations research, mainly in nuclear energy; risk analysis; mathematical modeling. *Mailing Add:* Box 1570 Deep River ON K0J 1P0 Can

JARVIS, SIMON MICHAEL, b London, Eng. MEMBRANE TRANSPORT. *Educ:* Univ Nottingham, UK, BSc Hons, 77; Cambridge Univ, UK, PhD(physiol), 80. *Prof Exp:* Res fel cancer res, Cancer Res Unit, Univ Alta, Edmonton, Can, 80-82, from asst prof to assoc prof physiol, 82-86; LECTR BIOCHEM, BIOL LAB, UNIV KENT, CANTERBURY, 86- *Mem:* Am Physiol Soc; Biochem Soc. *Res:* Nucleoside and nucleobase transport in mammalian cells; facilitated-diffusion and sodium-dependent systems; comparison of sugar and nucleoside carriers; dopamine uptake by the central nervous system; physiological actions of adenosine. *Mailing Add:* Biol Lab Univ Kent Canterbury Kent CT2 7NJ England

JARVIS, WILLIAM ROBERT, b Olney, Eng, Nov 15, 27; m 52; c 1. PLANT PATHOLOGY, MYCOLOGY. *Educ:* Univ Sheffield, BSc, 51; Univ London, PhD(plant Path), 53, DIC, 53, cBiol, 85. *Prof Exp:* Prin sci officer, Scottish Hort Res Inst, 53-74; asst specialist, Univ Calif, 63-64; scientist, Dept Sci & Indust Res, New Zealand, 69-70; HEAD, PLANT PATH SECT, CAN DEPT AGR, 74- *Concurrent Pos:* Assoc ed, Can J Plant Path, sr ed, Plant Dis; chmn, Biol Control Comt, chmn Publ Comt, Can Pythopath Soc, 85-; ed Hort Res. *Honors & Awards:* Bailey Award, Can Phytopath Soc, 85. *Mem:* Am Phytopath Soc; Brit Fedn Plant Pathologists; Brit Mycol Soc; Can Phytopathol Soc. *Res:* Biology of botrytis species; powdery mildews; biological control; diseases of field and greenhouse vegetables, small berry fruits and ornamental bulb crops. *Mailing Add:* Agr Can Res Sta Harrow ON N0R 1G0 Can

JARVIS, WILLIAM TYLER, b Takoma Park, Md, Oct 19, 35; m 62; c 2. PUBLIC HEALTH. *Educ:* Univ Minn, Duluth, BS, 61; Kent State Univ, Ohio, MA, 68; Univ Ore, PhD(health educ), 73. *Prof Exp:* Instr, Parkersburg Jr Acad, WVa, 61-62, Mt Vernon Acad, Ohio, 62-68; asst prof health & phys educ, Loma Linda Univ, Calif, 68-71; fel health educ, Univ Ore, 71-73; from asst prof to prof prev & community dent, 73-82, prof pub health sci & chmn dept, 82-86, ASSOC PROF PREV MED, LOMA LINDA UNIV, 86- , PROF DEPT PUB HEALTH & PREV MED, 89- *Concurrent Pos:* Ed newslett, Nat Coun Against Health Fraud, 78-; mem bd sci adv, Am Coun Sci & Health, 78-; mem nat comt unproven methods cancer mgt, Am Cancer Soc, 86- *Mem:* Nat Coun Against Health Fraud (pres, 77-); Am Pub Health Asn; Am Asn Health Phys Educ & Recreation; Am Cancer Soc. *Res:* Consumer health education; health fraud, misinformation and quackery. *Mailing Add:* Loma Linda Univ Loma Linda CA 92354

JARZEN, DAVID MACARTHUR, b Cleveland, Ohio, Oct 19, 41; m 62; c 2. PALYNOLOGY, PALEOBOTANY. *Educ:* Kent State Univ, BS, 67, MA, 69; Univ Toronto, PhD(geol), 73. *Prof Exp:* Palynologist & cur fossil plants, Nat Mus Can, 73-89, RES SCIENTIST, CAN MUS NATURE, 89- *Concurrent Pos:* Vis scholar, Univ Queensland, Brisbane, Australia, 87-88; hon mem, St John's Col, Univ Queensland, Brisbane, Australia. *Mem:* Can Asn Palynologists (pres, 80); Am Asn Stratig Palynologists; Asn Trop Biol; Int Asn Angiosperm Paleobot; Int Fedn Palynology Soc (secy/treas, 84-88); Palynology & Paleobot Asn Australasia; Sigma Xi; Int Asn Palaeobot; Soc Preserv Natural Hist Col. *Res:* Palynological investigations of terminal cretaceous and lower Tertiary floras, to discover the paleoenvironmental setting based on the botanical affinities of the fossil pollen and spores; cretaceous spore pollen floras from Australasian (Gondwanan) sediments with an emphasis on evolutionary trends. *Mailing Add:* Earth Sci Div Can Mus Nature PO Box 3443 Sta D Ottawa ON K1P 6P4 Can

JARZYNSKI, JACEK, b Warsaw, Poland, Mar 28, 35; US citizen. PHYSICS. *Educ:* Imp Col, Univ London, BS, 57, PhD(physics), 61. *Prof Exp:* Fel phys acoustics, Cath Univ Am, 61-62; tech officer optics, Imp Chem Industs, Ltd, Gt Brit, 63; assoc prof physics, Am Univ, 63-71; RES PHYSICIST, NAVAL RES LAB, 71- *Concurrent Pos:* Consult, Naval Ord Lab, 67-70. *Mem:* Acoust Soc Am; Am Phys Soc; Sigma Xi. *Res:* Parametric underwater acoustic arrays and sound propagation; development of ultrasonic methods for study of materials; measurement of thermodynamic properties of metals and alloys and comparison with pseudopotential theory. *Mailing Add:* Georgia Inst Technol Sch Mech Eng 1146 Roxboro Rd NE Atlanta GA 30332

JASANOFF, SHEILA SEN, b Calcutta, India, Feb 15, 44; US citizen; m 68; c 2. LAW & SCIENCE, ENVIRONMENTAL POLICY. *Educ:* Harvard Col, AB, 64, PhD(ling), 73, JD, 76; Univ Bonn, WGer, MA, 66. *Prof Exp:* Assoc, Bracken, Selig & Baram, 76-78; sr res assoc sci policy & law, Prog Sci, Technol & Soc, Cornell Univ, 78-84, assoc prof, 84-89, dir, 88-91, PROF SCI POLICY & LAW, PROG SCI, TECHNOL & SOC, CORNELL UNIV, 90-, CHAIR, DEPT SCI & TECHNOL STUDIES, 91- *Concurrent Pos:* Consult, Orgn Econ Coop & Develop, 80-89 & Off Technol Assessment, 83-87; mem, Nat Conf Lawyers & Scientists, AAAS & Am Bar Asn, 85-91, Comt Govt-Indust Collab Biomed Res, Inst Med, 88-89 & Adv Comt, NSF, 90-; contrib ed, Sci, Technol & Human Values, 88-; vis prof, Yale Univ, 90-91. *Mem:* Fel AAAS; Soc Social Studies Sci. *Res:* Comparative studies of US and European health, safety and environmental regulations; US science policy; law, science and technology; risk management of chemicals and biotechnology. *Mailing Add:* Dept Sci & Technol Studies Cornell Univ 632 Clark Hall Ithaca NY 14853

JASCH, LAURA GWENDOLYN, developmental biology; deceased, see previous edition for last biography

JASELSKIS, BRUNO, b Suraitciai, Lithuania, Mar 9, 24; nat US; m 55; c 6. ANALYTICAL CHEMISTRY, INORGANIC CHEMISTRY. *Educ:* Union Univ, NY, BS, 52; Iowa State Univ, MS, 54, PhD, 55. *Prof Exp:* Instr chem, Univ Mich, 56-59, asst prof, 59-62; from asst prof to assoc prof, 62-69, PROF CHEM, LOYOLA UNIV CHICAGO, 69- *Mem:* Am Chem Soc; AAAS; Sigma Xi. *Res:* Complex ions and their application to analytical problems; solution chemistry of Xenon compounds: determination of micro amounts of various substances. *Mailing Add:* Dept of Chem Loyola Univ Chicago IL 60626-5385

JASHNANI, INDRU, b Ghotki, Pakistan, Nov 2, 44. CHEMICAL ENGINEERING. *Educ:* Indian Inst Technol, Bombay, BTech, 67; Univ Cincinnati, PhD(chem eng), 71. *Prof Exp:* Fel chem & nuclear eng, Univ Cincinnati, 71-72; sr engr, APT, Inc, 72-74; staff mem, Arthur D Little, Inc, 74-77; SR STAFF ENGR, MARTIN MARIETTA CORP, 77-; PRES, ENG & COMPUTER SERV, INC, 90- *Mem:* Am Inst Chem Engrs; Am Chem Soc; Air Pollution Control Asn. *Res:* Environmental control, air, water and solid, for process industries and utility boilers. *Mailing Add:* 3891 Woodville Lane Ellicott City MD 21043

JASIN, HUGO E, b Buenos Aires, Arg, Jan 22, 33; US citizen; m 66; c 3. INTERNAL MEDICINE, IMMUNOLOGY. *Educ:* Univ Buenos Aires, MD, 56. *Prof Exp:* Fel internal med, Univ Tex Southwestern Med Sch, 59-62 & 64-65; Nuffield fel, Med Res Coun Rheumatism Res Unit, Eng, 62-64; from instr to ssoc prof, 65-78, PROF INTERNAL MED, UNIV TEX SOUTHWESTERN MED SCH, 78- *Concurrent Pos:* Arthritis Found fel, 70-72; USPHS career develop award, 73-77; mem, Gen Med Study Sect, USPHS, 74-78. *Mem:* Fel Am Col Physicians; Am Asn Immunologists; Am Rheumatism Asn; Am Soc Clin Invest; Am Asn Physicians. *Res:* Immunological mechanisms in rheumatic diseases and chronic inflammation. *Mailing Add:* Univ of Tex Southwestern Med Sch 5323 Harry Hines Blvd Dallas TX 75235-8577

JASINSKI, DONALD ROBERT, b Chicago, Ill, Aug 27, 38; m 64; c 4. CLINICAL PHARMACOLOGY. *Educ:* Loyola Univ, Ill, 56-59; Univ Ill, MD, 63. *Prof Exp:* Intern, Res & Educ Hosps, Univ Ill, Chicago, 63-64, fel neuropharmacol, 64-65; staff physician, 65-67, chief opiate unit, 67-68, chief clin pharmacol sect, 69-77, DIR, ADDICTION RES CTR, NAT INST DRUG ABUSE, 77- *Concurrent Pos:* Assoc mem grad fac, Dept Pharmacol, Col Med, Univ Ky; clin asst prof pharmacol, Univ Ill; clin prof pharmacol & toxicol & mem grad fac, Univ Louisville, Ky. *Mem:* AAAS; Am Soc Clin Pharmacol & Therapeut; Am Soc Pharmacol & Exp Therapeut; Soc Neurosci; Int Brain Res Orgn; Sigma Xi. *Res:* Neuropharmacology; Psychopharmacology. *Mailing Add:* c/o Baltimore City Hosp Nat Inst Drug Abuse Addiction Res Ctr 4940 Eastern Ave Baltimore MD 21224

JASINSKI, JERRY PETER, b Newport, NH, July 28, 40; m 66; c 3. X-RAY CRYSTALLOGRAPHY, BIOCHEMISTRY. *Educ:* Univ NH, BA, 64, MST, 68; Worcester Polytech Inst, MNS, 68; Univ Wyo, PhD(chem), 74. *Prof Exp:* Teacher, high schs, NY, NH & Vt, 64-70 & 75-78; teaching & res assoc chem, Univ Wyo, 70-73; Assoc Western Univs fel, Los Alamos Sci Lab, 73-74; res assoc, Univ Va, 74-75; from asst prof to assoc prof, 78-89, coordr phys sci, 81-83, PROF CHEM, KEENE STATE COL, NH, 89- *Concurrent Pos:* Consult, US Army Mat & Mech Res Ctr, Watertown, Mass, 83- *Mem:* Sigma Xi; Am Chem Soc; AAAS; Am Asn Physics Teachers; Am Crystallog Asn; Health Physics Soc. *Res:* Experimental and theoretical molecular electronic spectroscopy; solid state and coordination chemistry; x-ray crystallography; bioinorganic chemistry; industrial chemistry; chemical design. *Mailing Add:* 12 Orchard Lane Springfield VT 05156

JASKOSKI, BENEDICT JACOB, b Velva, NDak, July 25, 15; m 56; c 1. PARASITOLOGY. *Educ:* Jamestown Col, AB, 39; Univ Notre Dame, MS, 42, PhD(zool), 50. *Hon Degrees:* DSc, Ill Col Podiatry, 64. *Prof Exp:* Prin pub sch, NDak, 39-40; asst instr biol, Univ Notre Dame, 49-50; asst prof, Creighton Univ, 50-54; from assoc prof to prof, 54-86, EMER PROF BIOL, LOYOLA UNIV CHICAGO, 86- *Concurrent Pos:* Fel trop med, La State Univ, 61; assoc Am Univ Beirut, 66; USPHS res award; Am Cancer Soc inst grant partic. *Mem:* Fel AAAS; Am Soc Parasitologists; Am Soc Zoologists; Am Micros Soc; fel Am Pub Health Asn. *Res:* Parasites of captive animals; nematode parasites; biochemistry and physiology of parasitic nematodes; human parasitology; culture of metazoan parasites. *Mailing Add:* Dept Biol Damen Hall Loyola Univ 6525 N Sheraton Rd Chicago IL 60626

JASMIN, GAETAN, b Montreal, Que, Nov 24, 24; m 52; c 3. PATHOLOGY. *Educ:* St Laurent Col, BA, 45; Univ Montreal, MD, 51, PhD(exp med), 56; CSPQ, 68; FRCP(C), 78. *Prof Exp:* From asst prof exp path to assoc prof path, 56-67, chmn dept, 70-82, PROF PATH, UNIV MONTREAL,. *Concurrent Pos:* Med res assoc, Nat Res Coun Can, 58-70; ed, Revue Canadienne de Biologie, 60-70, Methods & Achievements in Exp Path, 66- *Mem:* Soc Exp Biol & Med; Am Physiol Soc; Histochem Soc; Can Soc Clin Invest; Int Acad Path. *Res:* Endocrinology; muscle diseases and cancer. *Mailing Add:* Dept Path Fac Med Univ Montreal Box 6128 Sta A Montreal PQ H3C 3J7 Can

JASNY, GEORGE R, b Katowice, Poland, June 6, 24; US citizen; m 51; c 2. CHEMICAL ENGINEERING. *Educ:* Univ Wash, Seattle, BS, 49; Mass Inst Technol, ScM, 52. *Prof Exp:* Engr, Y-12 plant, Union Carbide Corp, 50-56, eng dept head, 56-62, tech div head, 62-65, chief engr, 65-71, dir eng, 71-80, vpres eng & comput sci, Nuclear Div, 80-84; vpres eng & comput sci, Martin Marietta Energy Systs, 84-89, vpres tech opers, 89; RETIRED. *Mem:* Nat Acad Eng; AAAS; Sci Res Soc NAm; Fel Am Inst Chem Eng; Nat Soc Prof Engrs; Am Soc Eng Mgt. *Res:* Solvent extractions; enriched uranium scrap processing; plant design; quality control; uranium enrichment. *Mailing Add:* 106 Dixie Lane Oak Ridge TN 37830

JASON, ANDREW JOHN, b Detroit, Mich, Jan 27, 38; m 62; c 1. MOLECULAR PHYSICS, SURFACE PHYSICS. *Educ:* Mass Inst Technol, SB, 59; Univ Chicago, MS, 60, PhD(physics), 67. *Prof Exp:* Res assoc physics, Univ Chicago, 67-68; asst prof, 68-73, ASSOC PROF PHYSICS, UNIV ALA, 73- *Mem:* AAAS; Am Phys Soc. *Res:* Field ionization; atomic physics; high field studies of atoms and molecules; kinetics of evaporation; ion optics; mass spectrometry; optical properties of surfaces. *Mailing Add:* 300 Rim Rd Los Alamos NM 87544

JASON, EMIL FRED, b Edwardsville, Ill, Aug 7, 27; m 55; c 2. CHEMISTRY. *Educ:* Lincoln Univ, Mo, BS, 49; Washington Univ, St Louis, MA, 55, PhD(chem), 57. *Prof Exp:* Teacher pub sch, Ethiopia, 49-51; fel, Wash Univ, St Louis, 57-58; asst proj chemist, Stand Oil Co, Ind, 58-60, proj chemist, 60; asst prof chem, Lincoln Univ, Mo, 60-71; PROF CHEM & ASST VPRES, SOUTHERN ILL UNIV, 71- *Mem:* Am Chem Soc. *Res:* Organic syntheses; oxidation; free radical reactions. *Mailing Add:* Rte 8 Box 183 Edwardsville IL 62025

JASON, MARK EDWARD, b Grand Rapids, Mich, Oct 18, 49; m 73; c 2. PHYSICAL ORGANIC, REACTION MECHANISMS. *Educ:* Univ Mich, BS, 71; Yale Univ, PhD(chem), 76. *Prof Exp:* Post doctoral assoc, Northwestern Univ, 76, NSF post doctoral assoc, 76-77; asst prof org chem, Amherst Col, 77-83; ASSOC FEL, MONSANTO, 83- *Concurrent Pos:* Vis prof, Cornell Univ, 81. *Mem:* Am Chem Soc. *Res:* Physical organic chemistry; chemistry of small ring compounds; chelation chemistry. *Mailing Add:* Monsanto 800 N Lindbergh Blvd St Louis MO 63167

JASPER, DONALD EDWARD, b La Grande, Ore, Dec 30, 18; m 43; c 2. VETERINARY MEDICINE, CLINICAL PATHOLOGY. *Educ:* State Col Wash, BS, 40, DVM, 42; Iowa State Col, MS, 44; Univ Minn, PhD(vet med), 47. *Prof Exp:* Asst clinician, Iowa State Col, 42-44; from asst prof to assoc prof, Univ Calif, Davis, 47-54, prof clin path, Sch Vet Med, 54-89; RETIRED. *Concurrent Pos:* Dean sch vet med & asst dir exp sta, Univ Calif, Davis, 54-62; sr NIH fel, 68; Fulbright Hays sr res scholar, 75; Fulbright Hays Distinguished Prof, 78. *Honors & Awards:* Borden Award, 67. *Mem:* US Animal Health Asn; Am Vet Med Asn; Am Soc Microbiol. *Res:* Bovine mastitis; mycoplasma infections. *Mailing Add:* 1826 Alameda Davis CA 95616

JASPER, DONALD K, b Miami, Fla. BIOLOGY. *Educ:* Howard Univ, BS, 52; Univ York, PhD(cell ultrastruct & physiol), 69. *Prof Exp:* Electron microscopist, Rockefeller Inst Med Res, 56-60; res asst cytol, Columbia Univ, 60-66; res fel cell biol, Univ York, 66-69; asst prof, 69-75, assoc dean, grad studies, 86-88, ASSOC PROF CELL BIOL, TECHNOL, 75-,. *Concurrent Pos:* Fac res fel, Argonne Nat Lab. *Mem:* AAAS; Am Soc Cell Biol; Electron Microscopy Soc Am; Am Inst Biol Sci; Fedn Am Soc Exp Biol. *Res:* Cellular ultrastructure as related to function, especially as mucosal epithelial and muscle cell structure and function. *Mailing Add:* Dept Biol Ill Inst Technol Chicago IL 60616

JASPER, HERBERT HENRY, b La Grande, Ore, July 27, 06; nat Can; m 40, 83; c 2. NEUROPHYSIOLOGY. *Educ:* Reed Col, BA, 27; Univ Ore, MA, 29; Univ Iowa, PhD(psychol), 31; Univ Paris, Dr es Sc(physiol), 35; McGill Univ, MDCM, 43. *Hon Degrees:* Dr, Univ Bordeaux, 49, Univ Aix-Marseille, 60, McGill Univ, 71, Univ Western Ont, 77 & Queens Univ, 79 & Mem Univ, 83. *Prof Exp:* Instr psychol, Univ Ore, 27-29; instr, Univ Iowa, 29-31; asst prof, Brown Univ, 33-38; asst prof neurol & neurosurg, McGill Univ, 38-46; prof exp neurol & dir neurophysiol & EEG labs, Montreal Neurol Inst, 46-64; res prof neurophysiol, Labs Neurol Sci, 65-76, & dir med res coun group neurol sci, Dept Physiol, 67-76, EMER PROF NEUROPHYSIOL, UNIV MONTREAL, 76- *Concurrent Pos:* First pres, Int Fedn Socs EEG & Clin Neurophysiol, 47-49; founding ed-in-chief & publisher Int J EEG & Clin Neurophysiol, 49-62; founding hon exec secy, Int Brain Res Orgn. *Honors & Awards:* Ralph Gerard Prize, Soc Neurosci, 81; Karl Lashley Prize, Am Philos Soc, 82; McLaughlin Medal, Royal Soc Can, 85; Officer Order Can, 72. *Mem:* Am Physiol Soc; Am Neurol Asn; Am EEG Soc (pres, 46-48); fel Royal Soc Can; Int Brain Res Orgn (exec secy, 61-63, hon exec secy, 71-72). *Res:* Brain research; behavioral sciences; neurology; electrical activity of the brain in man and experimental animals in relation to neuro chemistry; states of consciousness; epilepsy; sensori-motor functions and mechanisms of learning and memory. *Mailing Add:* 4501 Sherbrooke W No 1F Westmount PQ H3Z 1E7 Can

JASPER, MARTIN THEOPHILUS, b Hazlehurst, Miss, Mar 19, 34; m 63; c 5. MECHANICAL & CHEMICAL ENGINEERING. *Educ:* Miss State Univ, BS, 55, MS, 62; Univ Ala, PhD(mech eng), 67. *Prof Exp:* Engr, Am Cast Iron Pipe Co, 55-56; plant metallurgist, Vickers, Inc, 57-59; design engr, Missile Div, Chrysler Corp, 59-60; from instr to assoc prof, 60-75, PROF MECH ENG, MISS STATE UNIV, 75- *Mem:* Am Soc Eng Educ; Am Soc Mech Engrs; Soc Mfg Engrs; NY Acad Sci. *Res:* Parametric analysis, modeling and optimization of thermal and hydrodynamic systems. *Mailing Add:* Dept Mech Eng Miss State Univ Box 155 Mississippi State MS 39762

JASPER, ROBERT LAWRENCE, b Windsor, Ky, Apr 24, 18; m 44, 62; c 1. ENDOCRINOLOGY, TOXICOLOGY. *Educ:* Berea Col, AB, 49; Univ Ky, MS, 50; Purdue Univ, PhD, 55. *Prof Exp:* Instr physiol, Univ Ky, 50-51; sect chief environ physiol, Army Med Res Lab, Ky, 54-58; co-dir endocrinol, Endocrine Consult Lab, 58-62; pharmacologist-in-chg, Pharm Lab, Agr Res Serv, USDA, Md, 62-64, asst chief staff officer pharmacol, Pesticides Regulation Div, 64-67, pharmacologist-in-chg, Pharm Lab, 67-70, head safety & biol sect, Chem & Biol Invest Br, Tech Serv Div, Off Pesticides Progs, 70-73, ASST BR CHIEF, CHEM & BIOL INVEST BR, TECH SERV DIV, OFF PESTICIDES PROGS, ENVIRON PROTECTION AGENCY, 73- *Mem:* Am Inst Chemists; Am Chem Soc; Am Soc Zoologists; Am Asn Lab Animal Sci; NY Acad Sci. *Res:* Effect of low environmental temperature on fat metabolism; tissue and steroid metabolism; development and clinical application of hormone assays; pesticides toxicology. *Mailing Add:* 1731 Maple Ave Hanover MD 21076

JASPER, SAMUEL JACOB, b Lancaster, Ohio, Nov 1, 21; m 44; c 3. MATHEMATICS. *Educ:* Univ Ohio, AB, 43; Ohio State Univ, MA, 46; Univ Ky, PhD(math), 48. *Prof Exp:* Instr math, Univ Ohio, 43-44 & Univ Ky, 46-48; asst prof, Kent State Univ, 48-51 & ETenn State Col, 51-54; from asst prof to assoc prof, 54-67, PROF MATH, OHIO UNIV, 67- *Concurrent Pos:* Asst dean col arts & sci, Ohio Univ, 58-63, dir hon col, 63-66, chmn dept math, 67-68. *Mem:* Math Asn Am. *Res:* Differential geometry; homogeneous functions; calculus of variations. *Mailing Add:* Dept of Math Ohio Univ Athens OH 45701

JASPERSE, CRAIG PETER, b Sheboygan, Wis, Feb 2, 60; m 91. ORGANOLANTHANIDES, RADICALS. *Educ:* Calvin Col, BS, 82; Univ Wis-Madison, PhD(org chem), 87. *Prof Exp:* Postdoctoral, Univ Pittsburgh, 87-89; ASST PROF ORG CHEM, UNIV NDAK, 89- *Mem:* Am Chem Soc. *Res:* Organic chemistry, synthesis; ketyl radical anions, preparation and rearrangements; mechanism and application of SMI2 as a one-electron reducing agent; preparation and rearrangements of alkoxy radicals. *Mailing Add:* Chem Dept Univ NDak Grand Forks ND 58202-7185

JASPERSE, JOHN R, b Seattle, Wash, May 8, 35; m 58; c 2. PLASMA PHYSICS, SPACE PHYSICS. *Educ:* Harvard Univ, BA, 57; Northeastern Univ, MS, 63, PhD(physics), 66. *Prof Exp:* Res physicist, Arthur D Little, Inc, 59-65; RES PHYSICIST, AIR FORCE GEOPHYSICS LAB, 65- *Concurrent Pos:* Lectr, Northeastern Univ, 68-71; vis scientist, Mass Inst Technol, 79- *Honors & Awards:* Marcus D O'Day Mem Award; Guenter Loeser Mem Award. *Mem:* Am Geophys Union; Sigma Xi; Am Phys Soc. *Res:* Quantum theory of atoms and molecules; scattering theory; three-body problem; electromagnetic theory; plasma theory; space physics. *Mailing Add:* 198 Conant Rd Weston MA 02193

JASPERSON, STEPHEN NEWELL, b Wisconsin Rapids, Wis, May 10, 41; m 65; c 2. SOLID STATE PHYSICS. *Educ:* Univ Wis, BS, 63; Princeton Univ, MA, 65, PhD(physics), 68. *Prof Exp:* Res assoc physics, Princeton Univ, 67-68 & Univ Ill, 68-70; asst prof, 70-74, ASSOC PROF PHYSICS, WORCESTER POLYTECH INST, 74-, DEPT HEAD, 84- *Concurrent Pos:* Res scientist, Physics Br, Naval Weapons Ctr, 71-78; vis scientist, Nat Magnet Lab, Mass Inst Technol, 82-86. *Mem:* AAAS; Am Phys Soc. *Res:* Optical properties of metals and semiconductors; modulation spectroscopy techniques such as electroreflectance, polarization modulation and magnetoreflectance. *Mailing Add:* Dept Physics Worcester Polytech Inst 100 Institute Rd Worcester MA 01609

JASS, HERMAN EARL, b Chicago, Ill, Mar 30, 18; m 47; c 2. BIOCHEMISTRY, PHARMACOLOGY. *Educ:* Univ Ill, BS, 39; Northwestern Univ, MS, 50, PhD(chem), 53. *Prof Exp:* Org chemist, Gas Res Dept, People's Gas Co, Ill, 40-41; chief biochemist, Helene Curtis Indust, Inc, 42-51; group leader biochem, Armour & Co, 53-55; assoc res dir, Revlon, Inc, 55-64; vpres res, Carter Prod Div, Carter-Wallace, Inc, 64-76; TECH MGT CONSULT, 76- *Concurrent Pos:* Guest lectr, Columbia Col Pharm, 64 & 65. *Honors & Awards:* CIBS Sci Award, Cosmetic Toiletry & Fragrance Asn, 78. *Mem:* AAAS; Am Chem Soc; Soc Cosmetic Chem; Dermal Clin Eval Soc; Am Soc Consult Pharmacists. *Res:* Proprietary drugs and toiletries; biochemistry and physiology of the skin; regulation and safety of cosmetics & drugs. *Mailing Add:* 29 Platz Dr Skillman NJ 08558

JASSBY, DANIEL LEWIS, b Montreal, Que, Jan 27, 42; US citizen. CONTROLLED NUCLEAR FUSION. *Educ:* McGill Univ, BSc, 62; Univ BC, MS, 64; Princeton Univ, PhD(plasma physics), 70. *Prof Exp:* Asst prof elec sci, Univ Calif, Los Angeles, 70-73; res staff, 73-76, res physicist, 76-80, PRIN RES PHYSICIST, PRINCETON PLASMA PHYSICS LAB, 80- *Mem:* fel Am Phys Soc; Inst Elec & Electronics Engrs; Am Nuclear Soc. *Res:* Production, measurement and application of fusion neutrons; heating of toroidal plasmas; design of magnetic confinement fusion devices. *Mailing Add:* Princeton Plasma Physics Lab PO Box 451 Princeton NJ 08543

JASTAK, J THEODORE, b Astoria, NY, Dec 1, 36; m 62; c 3. ORAL & MAXILLOFACIAL SURGERY, DENTAL ANESTHESIOLOGY. *Educ:* Seton Hall Univ, DDS, 62; Univ Rochester, PhD(path), 67. *Prof Exp:* Resident oral maxillofacial surg, Henry Ford Hosp, Detroit, 67-69; assoc prof, 69-80, PROF ORAL SURG, SCH DENT, ORE HEALTH SCI UNIV, 80-, CHMN HOSP DENT, 80- *Concurrent Pos:* Vis asst prof, Dent Sch, Univ Detroit, 68-69. *Mem:* AAAS; Am Col Oral Maxillofacial Surg; Am Dent Soc Anesthesiol; Int Asn Dent Res. *Res:* Anesthesia and pain control for dental outpatients. *Mailing Add:* 3181 Sam Jackson Park Rd Portland OR 97201

JASTRAM, PHILIP SHELDON, b Providence, RI, Feb 28, 20; m 47; c 2. PHYSICS, ELECTROMAGNETISM. *Educ:* Harvard Univ, SB, 43; Univ Mich, PhD(physics), 48. *Prof Exp:* Res assoc, Radio Res Lab, Harvard Univ, 43-45; instr, Univ Mich, 48-49, res physicist, US Navy Proj, 49; asst prof physics, Wash Univ, St Louis, 49-55; from asst prof to assoc prof, 55-64, prof, 64-89, EMER PROF PHYSICS, OHIO STATE UNIV, 89- *Concurrent Pos:* Orgn European Econ Coop fel, Copenhagen, 61; consult, US Agency Int Develop, India, 64, 65 & 68. *Honors & Awards:* Nemzer Award, 82; Jefferson Award, 82. *Mem:* AAAS; fel Am Phys Soc; Inst Elec & Electronics Engrs; Am Asn Physics Teachers; Fedn Am Sci. *Res:* Nuclear structure and spectroscopy; angular and polarization correlation measurements; nuclear orientation at low temperatures; Mossbauer scattering; double inner bvemsstrahlung. *Mailing Add:* 115 W Royal Forest Blvd Columbus OH 43214-2026

JASTROW, ROBERT, b New York, NY, Sept 7, 25. EARTH SCIENCES. *Educ:* Columbia Univ, BA, 44, MA, 45, PhD(physics), 48. *Hon Degrees:* DSc, Manhattan Col, 80. *Prof Exp:* Fel, Univ Leiden, 48-49; mem, Inst Advan Study, Princeton Univ, 49-50, 53; mem fac, Univ Calif, Berkeley, 50-53; asst prof physics, Yale Univ, 53-54; consult nuclear physics, US Naval Res Lab, 54-58; head theoret div, Goddard Space Flight Ctr, NASA, 58-61, dir inst space studies, 61-81; PROF PHYSICS & EARTH SCI, DARTMOUTH COL, 74- *Concurrent Pos:* Chmn lunar explor comt, NASA, 59-60, mem comt, 60-62; adj prof, Columbia Univ, 61- & Dartmouth Col, 74-; ed, J Atmospheric Sci, Am Meteorol Soc, 62-74. *Honors & Awards:* Arthur S Fleming Award, 65; NASA Medal, 68. *Mem:* Fel AAAS; fel Am Geophys Union; fel Am Phys Soc; Am Astron Soc; Am Meteorol Soc. *Res:* Nuclear physics; physics of atmosphere, the moon and terrestrial planets. *Mailing Add:* Dept Physics & Earth Sci Dartmouth Col Hanover NH 03755

JASWAL, SITARAM SINGH, b Bham, India, Sept 15, 37. PHYSICS, MATHEMATICS. *Educ:* Univ Panjab, India, BSc, 58, MSc, 59; Mich State Univ, PhD(physics), 64. *Prof Exp:* Asst geophys, Oil & Natural Gas Comn, India, 59-60; asst physics, Univ Alta, 60-61 & Mich State Univ, 61-64; fel, Univ Pa, 64-66; from asst prof to assoc prof, 66-74, PROF PHYSICS, UNIV NEBR, LINCOLN, 74- *Concurrent Pos:* Vis scientist, Max Planck Inst Solid State Res, 74-75; Fulbright fel, Tech Univ Vienna, Austria, 86-87. *Mem:* Fel Am Phys Soc. *Res:* Electronic structure and properties of metallic glasses and magnetic materials. *Mailing Add:* Behlen Lab Physics Univ Nebr Lincoln NE 68588

JASZBERENYI, JOSEPH C, b Budafok, Hungary, Aug 19, 48. CHEMISTRY OF ANTIBIOTICS, HETEROCYCLIC CHEMISTRY. *Educ:* Kossuth L Univ, Debrecen, Hungary, dipl chem, 72, Dr rer nat, 75. *Prof Exp:* Asst lectr chem, Kossuth L Univ Debrecen, 72-75, lectr, 76-77; postdoctoral, Oxford Univ, Eng, 75-76; res fel, 77-83, SR RES FEL ANTIBIOTICS, RES GROUP ANTIBIOTICS, DEBRECEN, 84-; DIR RES, DEPT CHEM, TEX A&M UNIV, 90- *Concurrent Pos:* Postdoctoral org chem, Univ New Castle Upon Tyne, 79-80; reader, Kossuth L Univ, Debrecen, Hungary, 86; vis assoc prof org chem, Dept Chem, Tex A&M Univ, 88- *Res:* Development of new reagents for radical chemistry; new reactions applicable to the synthetic transformations of organic molecules; natural products. *Mailing Add:* Dept Chem Tex A&M Univ College Station TX 77843

JASZCZAK, RONALD JACK, b Chicago Heights, Ill, Aug 23, 42; m 67; c 2. MEDICAL PHYSICS, IMAGE PROCESSING PHYSICS. *Educ:* Univ Fla, BS, 64, PhD(physics), 68. *Prof Exp:* Postdoctoral fel, US AEC, Oak Ridge Nat Lab, 68-69, staff physicist, 69-71; prin res scientist, Nuclear Chicago Corp, Searle Diagnostics Inc, 71-73, sr prin res scientist, 73, res group leader, 73-77, chief scientist, 77-79; assoc prof, Dept Radiol, 79-89, ASSOC PROF BIOMED ENG, DUKE UNIV, 86-, PROF, DEPT RADIOL, 89- *Concurrent Pos:* NIH res fel, 80-82; prin investr, Nat Cancer Inst, 83- & Dept Energy, 89-; assoc ed, Inst Elec & Electronics Engrs Trans Med Imaging, 86- & J Nuclear Med Technol, 88- *Mem:* AAAS; Am Phys Soc; Inst Elec & Electronics Engrs; Am Asn Physicists Med; Soc Photo-Optical Instrumentation Engrs; Sigma Xi. *Res:* Nuclear medicine instrumentation; imaging systems for single photon emission computed tomography; nuclear radiation detectors; image reconstruction and restoration; Monte Carlo modeling; quantitative application of medical imaging. *Mailing Add:* Duke Univ Med Ctr Box 3949 Durham NC 27710

JATLOW, J(ACOB) L(AWRENCE), b Poland, Apr 7, 03; nat US; m 51; c 1. ELECTRONICS, COMMUNICATIONS ENGINEERING. *Educ:* Rensselaer Polytech Inst, EE, 24. *Prof Exp:* Develop engr, Conner Crouse Corp, 24-32; asst chief engr, F A D Andrea Radio Corp, 32-35; develop engr, Photo Positive Corp, 35-40; chief engr photochem res, Repub Eng Prod, Inc, 40-42; chief engr, Wire Transmission Div, Defense Commun Div, Int Tel & Tel, Inc, 42-54, assoc dir, Radio Transmission Lab, 54-60, dir, Systs Eng Lab, 60-64, tech dir commun div, 64-77; RETIRED. *Concurrent Pos:* Consult, ITT, 77-85. *Honors & Awards:* Region 1 Award, Inst Elec & Electronics Engrs, 83. *Mem:* Fel Inst Elec & Electronics Engrs; fel AAAS; NY Acad Sci. *Res:* Communications systems engineering; wire and radio transmission; switching systems; command and control systems; communication and electronic equipment; development of alternating current operated radio sets and power supplies; development of photographic emulsions and photoprocesses. *Mailing Add:* 166 E 61st St Apt 12-J New York NY 10021

JAUCHEM, JAMES ROBERT, b Washington, DC, Mar 22, 51; m 90. CARDIOVASCULAR PHYSIOLOGY, RADIOFREQUENCY RADIATION BIOEFFECTS. *Educ:* Heidelberg Col, BS, 73; Baylor Col Med, PhD(physiol), 78. *Prof Exp:* Res assoc, Microcirculatory Systs Res Group, Univ Miss, 77-79; postdoctoral fel, Dept Path, Health Sci Ctr, Univ Tex, San Antonio, 80-81; res scientist, Life Sci Div, Technol Inc, 81-84; Nat Res Coun, sr res assoc, Med Sci Div, Johnson Space Ctr, NASA, 84-85; res physiologist, Crew Tech Div, Sch Aerospace Med, USAF, 85-86; prin sci ed, Dept Info & Defense Progs, Tracor Inc, 86-87; RES PHYSIOLOGIST, RADIOFREQUENCY RADIATION BR, DIRECTED ENERGY DIV, OCCUP & ENVIRON HEALTH DIRECTORATE, ARMSTRONG LAB, USAF, 87- *Concurrent Pos:* Consult, Northrop Corp, 86. *Honors & Awards:* Award Excellence, Soc Tech Commun, 86. *Mem:* Am Physiol Soc; Soc Exp Biol & Med; Aerospace Med Asn; Bioelectromagnetics Soc. *Res:* Physiological effects of radiofrequency radiation; thermoregulation; cardiovascular pharmacology; circulatory shock; electromagnetic field bioeffects. *Mailing Add:* USAF Armstrong Lab-OEDR Brooks AFB TX 78235

JAUHAR, PREM P, b India, Sept 15, 39. GENETICS, PLANT BREEDING. *Educ:* Agra Univ, India, BS, 57, MS, 59; Indian Agr Res Inst, New Delhi, PhD, 65. *Prof Exp:* Asst cytogeneticist, Indian Agr Res Inst, 63-70, asst prof genetics, 70-72 & 75-76; sr sci officer cytology, Univ Col Wales, Welsh Plant Breeding Sta, UK, 72-75; res assoc agron, Univ Ky, 76-78; res geneticist bot & plant sci, Univ Calif, Riverside, 78-81; CYTOGENETICIST, DIV CYTOGENETICS & CYTOL, CITY HOPE NAT MED CTR, DUARTE, CALIF, 81- *Concurrent Pos:* Vis scientist, Welsh Plant Breeding Sta, UK, 72-75; post grad fac, Indian Agr Res Inst, New Delhi, 60-70; Genetics Soc Am travel award, 78. *Mem:* Genetics Soc Am; Crop Sci Soc Am; Am Genetic Asn; Indian Soc Genetics & Plant Breeding. *Res:* Regulatory mechanism that controls chromosome pairing in the polyploid species of Festuca; breeding work on Panicum and Pennisetum; tropical and temperature herbage crops; polyploidy and mutation breeding techniques. *Mailing Add:* 1460 E 1220 Logan VT 84321

JAUMOT, FRANK EDWARD, JR, b Charleston, WVa, Aug 3, 23; m 47; c 2. SOLID STATE PHYSICS. *Educ:* Western Md Col, BS, 47, DSc, 66; Univ Pa, PhD(physics), 51. *Prof Exp:* Instr physics, Univ Pa, 51-52; chief physics, Metals Sect, Labs Res & Develop, Franklin Inst, 52-56; dir res & eng semiconductors, Delco Radio Div, Gen Motors Corp, 56-70, dir res & eng, 70-79, dir advan eng, Delco Electronics Div, 79-83, dir, Automotive Elec Bus Unit, 83-84; PRES, JAUMOT CONSULTING, INC, 84- *Concurrent Pos:* Instr asst, Univ Pa, 52-54, vis asst prof, 54-56. *Mem:* Am Phys Soc; Am Inst Aeronaut & Astronaut; Inst Elec & Electronics Engrs; Soc Auto Engrs; Am Asn Physics Teachers; Sigma Xi. *Res:* Order-disorder phenomena and other cooperative phenomena; diffusion in metals; thermoelectricity; semiconductors. *Mailing Add:* 7549 Maholo Hui Dr Bay St Louis MS 39530-4229

JAUSSI, AUGUST WILHELM, b Paris, Idaho, Aug 26, 25; m 55; c 6. PHYSIOLOGY. *Educ:* Univ Idaho, BS, 53; Brigham Young Univ, MS, 55; Okla State Univ, PhD, 60. *Prof Exp:* From instr to asst prof physiol, Okla State Univ, 56-62; from asst prof to assoc prof, 62-72, prof physiol, 72-77, PROF ZOOL, BRIGHAM YOUNG UNIV, 77- *Res:* Environmental effects on physiological activity. *Mailing Add:* Dept Zool Brigham Young Univ Provo UT 84602

JAVADPOUR, NASSER, b March 23, 37; c 3. UROLOGIC ONCOLOGY. *Educ:* Shiraz Med Ctr, MD, 62. *Prof Exp:* Asst prof surgery, Chicago Med Sch, 69-72; Staff Urologist , NIH, 72-83; PROF & DIR, UROL ONCOL SECT, UNIV MD SCH MED, 83- *Concurrent Pos:* Clin consult, Nat Naval Hosp, 72-, Walter Reed Army Hosp, 72-; consult, Nat Bladder Cancer Proj, 81-84, Nat Prostatic Proj, 77-85. *Mem:* Am Urol Asn; Am Asn Cancer Res; Am Col Surgeons; Am Soc Clin Oncol; Soc Univ Urol. *Res:* Tumor markers. *Mailing Add:* Univ Md Hosp Urol 22 S Green St Baltimore MD 21201

JAVAHER, JAMES N, b Tehran, Iran, Jan 1, 24; US citizen; m 63; c 4. MATHEMATICS. *Educ:* Univ Tehran, LLB, 46; San Jose State Col, BA, 50; Univ Calif, Berkeley, MA, 52; Stanford Univ, BS, 54; Univ Paris, ScD(math, physics), 60. *Prof Exp:* Instr math & eng, Sacramento City Col, 55-59 & 60-61; asst prof math & physics, 61-62, assoc prof, 62-66, PROF MATH, CALIF STATE COL, STANISLAUS, 66- *Concurrent Pos:* Chmn dept math, Stanislaus State Col, 61-70. *Mem:* Am Math Soc; Math Asn Am; Am Asn Physics Teachers. *Res:* Foundations of mathematics; applied mathematics. *Mailing Add:* Dept Math Calif State Col Stanislaus 801 W Monte Vista Ave Turlock CA 95380

JAVAID, JAVAID IQBAL, b Lahore, Pakistan, Oct 1, 42. BIOCHEMISTRY, PSYCHOPHARMACOLOGY. *Educ:* Univ Panjab, W Pakistan, BS, Hons, 64, MS, 65; State Univ NY, Buffalo, PhD(biochem), 72. *Prof Exp:* Res asst, W Regional Labs, Lahore, 65-67; res fel, 72-74, RES SCIENTIST, ILL STATE PSYCHIAT INST, 74- *Mem:* AAAS. *Mailing Add:* Ill State Psychiat Inst 1601 W Taylor Chicago IL 60612

JAVAN, ALI, b Tehran, Iran, Dec 27, 26; nat US; m 62; c 2. PHYSICS. *Educ:* Columbia Univ, PhD(physics), 54. *Prof Exp:* Res assoc physics, Columbia Univ, 54-59; mem tech staff, Bell Tel Labs, Inc, 59-62; from assoc prof to prof, 62-78, FRANCIS WRIGHT PROF PHYSICS, MASS INST TECHNOL, 78-; FOUNDER & CHIEF SCIENTIST, LASER SCI, INC, 81- *Concurrent Pos:* Sr US Scientist Award, Humboldt Found, 80. *Honors & Awards:* Stuart Ballentine Medal, 62; Hertz Found Award, 66; Sepas Medal, Govt Iran, 71; Frederic Ives Medal, Optical Soc Am, 75. *Mem:* Fel Nat Acad Sci; fel Am Phys Soc; fel Optical Soc Am; Sigma Xi; fel Am Acad Arts & Sci; Soc Found Third World Acad Sci. *Res:* Atomic spectroscopy and physics of quantum electronics. *Mailing Add:* Dept Physics Mass Inst Technol Cambridge MA 02139

JAVEL, ERIC, b Elizabeth, NJ, Mar 2, 47; m 68; c 3. NEUROPHYSIOLOGY. *Educ:* Johns Hopkins Univ, BA, 68; Univ Pittsburgh, PhD(bioacoust), 72. *Prof Exp:* DIR, CTR HEARING RES, BOYS TOWN NAT INST COMMUN DIS IN CHILDREN, 76- *Concurrent Pos:* NIH fel, Dept Neurophysiol, Med Sch, Univ Wis, 73-75; prof otolaryngol & physiol, Sch Med, Creighton Univ; mem comt hearing & bioacoustics, Nat Acad Sci; assoc prof psychol, Univ Neb; vis res fel, Univ Melbourne, Australia, 84-85. *Mem:* Acoust Soc Am; Soc Neurosci; Asn Res Otolaryngol; Sigma Xi. *Res:* Stimulus coding in auditory nuclei; computer applications in physiology; developmental neurophysiology; neurosensory prostheses. *Mailing Add:* Ten Upchurch Circle Durham NC 27705

JAVICK, RICHARD ANTHONY, b Plains, Pa, Aug 29, 32; m 61; c 10. ANALYTICAL CHEMISTRY, ENVIRONMENTAL CHEMISTRY. *Educ:* King's Col, Pa, BS, 54; Pa State Univ, PhD(chem), 58. *Prof Exp:* Res chemist, E I du Pont de Nemours & Co, Del, 58-59; asst prof chem, King's Col, Pa, 59-61 & State assoc prof, 62-69; res assoc, Pa Univ, 61-62; sr res chemist, 69-80, RES ASSOC, CHEM RES & DEVELOP CTR, FMC CORP, 80- *Concurrent Pos:* Chmn dept chem, King's Col, Pa, 67-69. *Mem:* Am Chem Soc; Chem Mfg Asn. *Res:* Application of instrumental methods to analytical investigations; electrochemical kinetics; polymer chemistry, synthesis and applications thereof; applications of analytical methods to wastewater analysis; air monitoring for pesticides and pesticide/herbicide residue analysis. *Mailing Add:* Chem Res & Develop Ctr FMC Corp Box 8 Princeton NJ 08543

JAVID, MANSOUR, b Hamadan, Iran, Mar 25, 19; nat US. ELECTRICAL ENGINEERING. *Educ:* Univ Birmingham, BSc, 41; McGill Univ, MEng, 50, PhD, 56. *Prof Exp:* Consult engr, 45-49; lectr elec eng, McGill Univ, 50-52; asst prof, Ill Inst Technol, 53-55; from asst prof to assoc prof, 55-70, prof elec eng, City Col New York, 70-; CHMN, ELEC ENG, MANHATTAN COL. *Concurrent Pos:* Mem, Nat Electronic Conf, 54. *Mem:* Inst Elec & Electronics Engrs. *Res:* Field and circuit analysis; computer aided design. *Mailing Add:* 2150 Westchester Ave Bronx NY 10462

JAVID, MANUCHER J, b Tehran, Iran, Jan 11, 22; nat US; m 51; c 4. NEUROSURGERY. *Educ:* Univ Ill, MD, 46; Am Bd Neurol Surg, dipl, 55. *Prof Exp:* Intern, Augustana Hosp, Ill, 46-47, resident gen surg, 47-48, resident neurosurg, 48-49; resident, New Eng Ctr Hosp, 50 & Mass Gen Hosp, 51-53; from instr to assoc prof, 53-62, chmn dept, 63, PROF NEUROL SURG, SCH MED, UNIV WIS-MADISON, 62- *Concurrent Pos:* Fel neurosurg, Lahey Clin, Mss, 49; fel neuropath, Ill Neuropsychiat Inst, 49; res fel, Mass Gen Hosp, 50; teaching fel, Harvard Med Sch, 52; pres, Int Intradiscal Ther Soc, 91-92. *Mem:* AAAS; AMA; fel Am Col Surg; Am Asn Neurol Surg; Soc Neurol Surgeons; Am Orthopsychiat Asn. *Res:* Intracranial pressure, cerebrovascular diseases, intracranial neoplasms and chemonucleolysis. *Mailing Add:* Dept Neurosurg Univ Wis Med Ctr Madison WI 53792

JAVIDI, BAHRAM, b Tehran, Iran, Feb 14, 59; US citizen. OPTICAL COMPUTING & PROCESSING, SIGNAL & IMAGE PROCESSING. *Educ:* George Washington Univ, BS, 80; Pa State Univ, MS, 82, PhD(elec eng), 86. *Prof Exp:* Prof elec eng, Mich State Univ, 86-88; PROF ELEC ENG, UNIV CONN, 88- *Concurrent Pos:* Prin investr, Inst Elec & Electronics Engrs & Eng Found, 87-88, USAF, 88-89, USAF & US Army, 89-91, NSF presidential young investr, 90-95; ed, spec issue J Optical Eng, 88, guest ed, 89 & 92; conf chmn, Prog Nonlinear Optical Processing, Inst Elec & Electronics Engrs, Lasers & Electr-optics Soc, 91; consult, US Army & USAF, 90-; reviewer & panelist, Nat Res Coun, 90. *Mem:* Optical Soc Am; Optical Eng Soc; Inst Elec & Electronics Engrs. *Res:* optical image processing; pattern recognition; neural networks; associative processing; nonlinear signal processing; holography; applications of spatial light modulators to information processing; communication systems. *Mailing Add:* Elec Eng Dept U-157 Univ Conn 260 Glenbrook Rd Storrs CT 06269-3157

JAVITT, NORMAN B, b New York, NY, Mar 9, 28; m 55; c 4. MEDICINE, PHYSIOLOGY. *Educ:* Syracuse Univ, AB, 47; Univ NC, PhD(physiol), 51; Duke Univ, MD, 54; Am Bd Internal Med, dipl, 62. *Prof Exp:* Intern med, Mt Sinai Hosp, New York, 54-55, asst resident, 57-58; Am Heart Asn adv fel, Col Physicians & Surgeons, Columbia Univ, 58-59; chief resident med, Mt Sinai Hosp, 60, res assoc, 61-62; from instr to asst prof, Sch Med, NY Univ, 62-68; assoc prof 68-72, PROF MED, MED COL, CORNELL UNIV, 72-, HEAD, DIV GASTROENTEROL, 70-; PROF MED & PEDIAT & DIR DIV HEPATIC DIS, NY UNIV MED CTR. *Concurrent Pos:* USPHS spec fel, Mt Sinai Hosp, 61-62; career investr, Health Res Coun City NY, 62-68. *Mem:* Am Fedn Clin Res; fel Am Col Physicians; Am Soc Clin Invest; Am Gastroenterol Asn; Am Asn Study Liver Dis. *Res:* Biochemical and physiological investigations related to human liver disease. *Mailing Add:* Dept Med & Pediat NY Univ Med Ctr 550 First Ave New York NY 10016

JAWA, MANJIT S, b Patiala, India, Aug 5, 34; m 64; c 2. APPLIED MATHEMATICS, CONTINUUM MECHANICS. *Educ:* Indian Inst Technol, PhD(appl math), 67. *Prof Exp:* Res asst statist, Panjab Govt, India, 56-58; lectr math, Panjab Univ Cols, India, 58-63 & Indian Inst Technol, 66-67; asst prof appl math & eng mech, Univ Mo-Rolla, 67-70; assoc prof, Hartwick Col, 70-71; PROF MATH, FAYETTEVILLE STATE UNIV, 71- *Res:* Exact numerical analysis of fluid dynamics, heat transfer and magnetohydrodynamics problems on digital computers. *Mailing Add:* Dept of Math Fayetteville State Univ Fayetteville NC 28301

JAWAD, MAAN HAMID, b Baghdad, Iraq, Dec 2, 43; US citizen; m 68; c 2. STRUCTURAL ENGINEERING, ENGINEERING MECHANICS. *Educ:* Al-Hikma Univ, BSc, 64; Univ Kans, MS, 65; Iowa State Univ, PhD(struct eng), 68. *Prof Exp:* Bridge engr, Iowa State Hwy Comn, 67-68; design engr, 68-70, staff consult, 70-77, mgr eng design, 77-87, ASST CHIEF ENG, NOOTER CORP, 87- *Mem:* Am Soc Civil Engrs; Am Soc Mech Engrs. *Res:* Pressure vessels area, mainly layered vessels, expansion joints, and high pressure gaskets. *Mailing Add:* Nooter Corp PO Box 451 St Louis MO 63166

JAWED, INAM, b Sagar, India, Sept 27, 47; US citizen; m; c 2. PHYSICAL & CEMENT CHEMISTRY, MATERIALS SCIENCE. *Educ:* Karachi Univ, BS, 66, MS, 68; Oxford Univ, PhD(chem), 71. *Prof Exp:* Asst prof phys chem, Peshawar Univ, 72-74; vis prof cement chem, Tokyo Inst Technol, 74-76; res scientist, 76-78, sr scientist, 78-80, head anal chem dept, 80-83, mgr, Res & Eng, Martin Marietta Labs, 83-87, PROG MGR, STRATEGIC HWY RES PROG, NAT RES COUN, 87- *Concurrent Pos:* Unesco fel, 74; Brit Coun fel, 68. *Honors & Awards:* UNESCO Fel, 74; Fel, Am Ceramic Soc, 87. *Mem:* Am Chem Soc; Am Ceramic Soc; Royal Inst Chem Brit; Am Concrete Inst; Am Soc Testing & Mat. *Res:* Materials science of cements, concrete, ceramics, and composite materials; technical management; kinetics of formation and hydration of silicates, aluminates, and ferrites; processing and properties of structural and electronic ceramics and composite materials. *Mailing Add:* Strategic Hwy Res Program 818 Connecticut Ave NW Washington DC 20006

JAWEED, MAZHER, EXERCISE, NERVE REGENERATION. *Educ:* Osmania Univ, BS, 62; WVa Univ, MS, 66; Thomas Jefferson Univ, PhD(pharmacol), 88. *Prof Exp:* Biochemist rehab med, 70-72, res assoc, 72-78, res asst, 78-82, RES ASST PROF PHARMACOL, THOMAS JEFFERSON UNIV, 79-, RES ASSOC PROF REHAB MED, 82- *Concurrent Pos:* Ed, J Archives Phys Med & Rehab, 88-; prin investr, Nat Inst Disability & Rehab, 88-; mem, Am Cong Rehab Med. *Mem:* AAAS; NY Acad Sci; Sigma Xi. *Res:* Nerve and muscle interactions as affected by drugs; evaluations of effects by electrophysiological, histochemical and immunological procedures. *Mailing Add:* Dept Rehab Jefferson Med Col Thomas Jefferson Univ Philadelphia PA 19107

JAWETZ, ERNEST, b Vienna, Austria, June 9, 16; nat US; m 54; c 4. MICROBIOLOGY, MEDICINE. *Educ:* Univ Vienna, 37; Univ NH, MA, 40; Univ Calif, PhD(microbiol), 42; Stanford Univ, MD, 46. *Prof Exp:* Lectr bact, Univ Calif, 42-44; sr asst surgeon, NIH, 46-48; from asst prof to assoc prof bact, 48-53, chmn dept microbiol, 62-78, PROF MICROBIOL & MED & LECTR PEDIAT, SCH MED, UNIV CALIF, 54- *Concurrent Pos:*

Almroth Wright lectr, London, 52 & 58; vis prof, Univ Shiraz, Iran, 77. *Honors & Awards:* Florey Mem lectr, Adelaide, 81. *Mem:* Am Soc Clin Invest; Am Soc Microbiol; Soc Exp Biol & Med; Am Asn Immunol; Am Acad Microbiol. *Res:* Clinical bacteriology; antibiotics; chemotherapy; infectious diseases; virology. *Mailing Add:* Dept of Microbiol & Med Univ of Calif Med Ctr San Francisco CA 94143

JAWOROWSKI, ANDRZEJ EDWARD, b Lublin, Poland, Dec 28, 42; m 65; c 1. SOLID STATE PHYSICS. *Educ:* Univ Warsaw, MSc, 66, PhD(physics), 74. *Prof Exp:* Instr physics, Univ Warsaw, 66-68, lectr physics, 68-74, asst prof physics, 74-78; sr res assoc, State Univ NY-Albany, 78-83; assoc prof physics, semiconductor group leader, Wright State Univ, Dayton, Ohio, 83-89; PRES, MICRONETICS, DAYTON, 90- *Concurrent Pos:* Res assoc, Radiation Physics Lab, Solid State Div, Inst Nuclear Res, Swierk, 67-74; prog head, Inst Physics, Polish Acad Sci, Warsaw, 76-77; consult, Mobil Solar Energy Co, Waltham, 80-85, Univ Energy Systs, Dayton, 85-88; rev, NSF, 87- *Honors & Awards:* Prize of Ministry Sci, Schs Acad Rank & Technol, Warsaw, Poland, 75. *Mem:* Europ Phys Soc; Am Phys Soc; Polish Phys Soc; Electrochem Soc; Mat Res Soc. *Res:* Physics of electronic materials; defects in semiconductors; radiation effects and damage; deep levels spectroscopy; hydrogen in solids; real-time measurements. *Mailing Add:* Micronetics PO Box 31467 Dayton OH 45431-0467

JAWOROWSKI, JAN W, b Augustow, Poland, Mar 2, 28; m 54; c 1. TOPOLOGY. *Educ:* Univ Warsaw, Magister, 52; Polish Acad Sci, PhD(math), 55. *Prof Exp:* Asst math, Univ Warsaw, 50-52, from adj to docent, 52-63; extraordinary prof, Math Inst, Polish Acad Sci, 63-64; assoc prof, Cornell Univ, 64-65; PROF MATH, IND UNIV, 65- *Concurrent Pos:* Fel, Polish Acad Sci, 57-58; NSF grant, Inst Advan Study, 60-61. *Mem:* Am Math Soc; Polish Math Soc. *Res:* Algebraic and geometric topology. *Mailing Add:* Prof Off Ind Univ Bloomington IN 47405

JAWORSKI, CASIMIR A, b South Bend, Ind, Aug 1, 30; c 2. SOIL SCIENCE, AGRONOMY. *Educ:* Purdue Univ, BSc, 52, MSc, 57; Rutgers Univ, PhD(soil sci), 62. *Prof Exp:* Resident hall counr, Purdue Univ, 54-56, res asst agron, 55-57, fac adv resident halls, 56-57; res fel soil, Rutgers Univ, 57-60; SOIL SCIENTIST, S ATLANTIC AREA, AGR RES SERV, USDA, 62- *Mem:* Am Soc Hort Sci; Int Soc Hort Sci; Am Inst Biol Sci; Sigma Xi. *Res:* Develop edible (conela) and industrial rapeseed cultural pract; Develop rapeseed harvesting systems; screen cruciferae; germplasm and accessions for adaptation to southeast US. *Mailing Add:* ARS USDA PO Box 748 Tifton GA 31793-0748

JAWORSKI, ERNEST GEORGE, b Minneapolis, Minn, Jan 10, 26; m 50; c 3. BIOLOGICAL CHEMISTRY, MOLECULAR BIOLOGY. *Educ:* Univ Minn, BChem, 48; Ore State Col, MS, 50, PhD(biochem), 52. *Prof Exp:* Asst chem, Ore State Col, 48-49, asst biochem, 49-52; res biochemist, 52-54, res group leader, 54-60, scientist, 60-62, sr scientist, 62-70, DISTINGUISHED SCI FEL, MONSANTO CO, 70- *Concurrent Pos:* Mem, Frasch Found Awards Comt, Am Chem Soc, 69-; chmn, Gordon Conf Plant Cell & Tissue Culture, 73-75, trustee, Gordon Res Conf, Inc, 75-81; mem ed bd, J Am Soc Plant Physiologists, 73-83; mem panel, Int Cell Res Orgn UNESCO, 77-; chmn bd trustees, Gordon Res Conf Inc, 78-79; Nat Res Coun, 85- *Honors & Awards:* David Rivette Mem lectr Commonwealth Sci & Indust Res, Australia. *Mem:* Fel AAAS; Am Chem Soc; Sigma Xi; Am Soc Plant Physiologists; Weed Sci Soc Am. *Res:* Plant growth regulation, hormones and metabolism; plant chemotherapeutic investigations; mechanism of action of herbicides; radioisotope techniques; biosynthesis of chitin; plant cell and tissue culture; plant organogenesis; cell biology; molecular biology; genetic engineering, biotechnology. *Mailing Add:* 11 Clerbrook Lane St Louis MO 63124

JAWORSKI, JAN GUY, b Woonsocket, RI, Dec 7, 46; m 69; c 2. BIOCHEMISTRY. *Educ:* Col of the Holy Cross, BA, 68; Purdue Univ, PhD(biochem), 72. *Prof Exp:* Res biochemist, Dept Biochem & Biophys, Univ Calif, Davis, 72-74; ASST PROF CHEM, MIAMI UNIV, 74- *Mem:* AAAS; Sigma Xi. *Res:* Metabolism of prostaglandins, long chain fatty acids and lipids. *Mailing Add:* Dept of Chem Miami Univ Oxford OH 45056

JAY, JAMES MONROE, b Ben Hill Co, Ga, Sept 12, 27; m 59; c 3. BACTERIOLOGY, MICROBIAL ECOLOGY. *Educ:* Paine Col, AB, 50; Ohio State Univ, MSc, 53, PhD(bact), 56. *Prof Exp:* Asst, Ohio State Univ, 53-55, Agr Exp Sta, 55-56, res assoc, 56-57; from asst prof to prof bact, Southern Univ, 57-61; from asst prof to assoc prof, 61-69, PROF BIOL SCI, WAYNE STATE UNIV, 69- *Concurrent Pos:* Mem, Govt Univ Indust Res Round Table, Nat Acad Sci, 84-87, coun Int Exchange of Scholars (Fulbright Prog), 85-88, Nat Adv Comt on Microbiol Criteria for Foods, US Dept Agr, 87-91; distinguished fac fel, 87; chmn, Food Microbiol Div, Am Soc Microbiologists, Food Microbiol Div, Inst Food Technologists, 90-91. *Honors & Awards:* Probus Award, 69. *Mem:* AAAS; Am Soc Microbiol; Inst Food Technologists; Soc Appl Bact; Sigma Xi; Int Asn Milk, Food & Environ Sanitarians. *Res:* Biochemistry and rapid techniques for measuring meat spoilage; rapid determination of microorganisms in foods; microbial ecology; limulus lysate test; lipopolysaccharides in foods; selective culture media for listeria; periplasmic binding proteins. *Mailing Add:* Dept Biol Sci Wayne State Univ Detroit MI 48202

JAYACHANDRAN, TOKE, b Madras, India; US citizen. STATISTICS, MATHEMATICS. *Educ:* V R Col, Nellore, India, BA, 51; Univ Wyo, MS, 62; Case Inst Technol, PhD(math statist), 67. *Prof Exp:* Res asst statist, Univ Wyo, 61-62; grad asst math, Case Inst Technol, 62-67; asst prof math, 67-70, ASSOC PROF MATH, NAVAL POSTGRAD SCH, 70- *Concurrent Pos:* Consult, Litton Sci Support Labs, Ft Ord, 68-72; BDM Corp, 72-75 & Sci Appln Inc, Monterey, 78-; opers analyst, Off Naval Res, Arlington, 75-77. *Mem:* Sigma Xi; Am Statist Asn; Am Math Soc. *Res:* Design of experiments, prediction intervals, reliability and life testing. *Mailing Add:* 22483 Estoque Pl Salinas CA 93908

JAYADEV, T S, b Bangalore, India; US citizen. ELECTRICAL ENGINEERING, PHYSICS. *Educ:* Univ Mysore, BSE, 58; Ill Inst Technol, MS, 62; Univ Notre Dame, PhD(elec eng), 68. *Prof Exp:* Asst prof elec eng, Karnatak Univ, India, 62-64, prof, 64-65; from asst prof to assoc prof, Univ Wis-Milwaukee, 68-76, prof elec eng, 76-78, mem lab surface studies, 68-78; mgr, Thermoelec, Energy Conversion Devices, 80-; SR SCIENTIST, LOCKHEED PALO ALTO RES LAB. *Concurrent Pos:* Sr to prin scientist, Solar Energy Res Inst, 78- *Mem:* Inst Elec & Electronics Engrs; Am Phys Soc; Am Vacuum Soc; Int Solar Energy Soc. *Res:* Infrared and visible sensors; sensor signal processing; solar, wind, geothermal energy conversion systems; solid state energy conversion; electromechanical conversion systems; electrical properties of thin films; thin film devices; device physics; surface physics. *Mailing Add:* Lockheed Palo Alto Res Lab 3251 Hanover St Palo Alto CA 94304

JAYANT, NUGGEHALLY S, b Bangalore, India, Jan 9, 46. COMMUNICATIONS SCIENCE, SPEECH PROCESSING. *Educ:* Univ Mysore, BSc, 62; Indian Inst Sci, Bangalore, BE, 65, PhD(elec commun), 70. *Prof Exp:* Res assoc commun, Stanford Univ, 67-68; mem tech staff speech & acoust res, 68-86, HEAD, SIGNAL PROCESSING RES DEPT, BEL TEL LABS, 86- *Concurrent Pos:* Fel Coun Sci & Indust Res, India, 66-67; vis scientist, Indian Inst Sci, Bangalore, 72, 75; vis prof, Univ Calif, 83. *Mem:* Fel Inst Elec & Electronics Engrs. *Res:* Speech communication and information systems; image processing. *Mailing Add:* Bell Tel Labs Rm 2D 539 600 Mountain Ave Murry Hill NJ 07974

JAYARAMAN, AIYASAMI, b Madras, India, Dec 5, 26; m 45; c 2. HIGH PRESSURE PHYSICS. *Educ:* Univ Madras, BSc, 46, MSc, 54, PhD(solid state physics), 60. *Prof Exp:* Res asst physics, Raman Res Inst, India, 49-54, asst prof, 54-60; asst res geophysicist, Inst Geophys, Univ Calif, Los Angeles, 60-63; MEM TECH STAFF, BELL LABS, NJ, 63- *Concurrent Pos:* Guggenheim fel, 70-71; vis prof, Indian Inst Sci, Bangalore, 70-71; vis scientist, Nat Aeronaut Lab, India, 70-71; vis prof, Max Planck Inst, 79-80; US sr scientist award, Alexander Von Humboldt Found, 78. *Mem:* Fel Am Phys Soc; fel Indian Acad Sci (treas, 56-60); Sigma Xi. *Res:* Optical, x-ray crystallography and luminescence; phase transitions in solids at high pressures; transport properties in semiconductors, magnetic and superconducting properties of metals and alloys. *Mailing Add:* AT&T Bell Labs Rm ID230 Murray Hill NJ 07974

JAYARAMAN, H, b Gudiattam, India, Dec 21, 36; US citizen; m 69; c 1. ORGANIC CHEMISTRY. *Educ:* Univ Madras, BSc, 56, MA, 58, PhD(chem), 63. *Prof Exp:* Lectr, Madras Christian Col, India, 57-65; Fulbright-Hays res fel, Univ Kans, 65-66; res assoc, Pa State Univ, 66-68; Pool off, Madras Christian Col, India, 69-70; res assoc, Univ Kans, 70-72 & Univ Pa, 72-73; info scientist, 73-80, sr info scientist, 80-86, SUPVR, TECH PLANNING & INTELLIGENCE DIV, RES & DEVELOP, PHILLIPS PETROL CO, OKLA, 86- *Mem:* Am Chem Soc. *Res:* Computerized retrieval and dissemination of technical information; writing on topics of value to technology planning and administrative divisions of corporations. *Mailing Add:* 111 PLB Phillips Petrol Co Bartlesville OK 74004

JAYARAMAN, NARAYANAN, b Tamilnadu, India, June 30, 48; m 77. CREEP-FATIGUE-ENVIRONMENT. *Educ:* Indian Inst Sci, Bangalore, India, BE, 70, ME, 72, PhD(metal), 79. *Prof Exp:* Res fel metal, Indian Inst Sci, 72-77; scientist, Nat Aeronaut Lab, India, 77-79; fel, 79-80, vis asst prof, 80-81, asst prof, 81-85, ASSOC PROF METAL, UNIV CINCINNATI, 85- *Mem:* Am Soc Metals; Metal Soc; AAAS; Sigma Xi. *Res:* Fracture and fatigue behavior of ni-base superalloys in relationship with their microstructures; stress generation due to oxidation of metals and alloys; life prediction models for high temperature materials. *Mailing Add:* Rm 498 Dept Mat Sci & Metall Eng M L 12 Rhodes Hall Univ Cincinnati Cincinnati OH 45221

JAYAS, DIGVIR SINGH, b Mant, Uttar Pradesh, India, Jan 10, 58; Can citizen; m 82; c 2. GRAIN DRYING & STORAGE, MODELLING OF BIOLOGICAL SYSTEMS. *Educ:* G B Pant Univ, Pantnagar, BTech, 80; Univ Man, Winnepeg, MSc, 82; Univ Sask, Saskatoon, PhD(agr eng), 87. *Prof Exp:* Pool scientist agr eng, G B Pant Univ, 82; res assoc, Univ Sask, 82-85; asst prof, 85-89, ASSOC PROF AGR ENG, UNIV MAN, 89- *Concurrent Pos:* Ed, Can Soc Agr Eng Newslett. *Mem:* Am Soc Agr Engrs; Can Soc Agr Eng; Can Inst Food Sci & Technol; Inst Food Technologists. *Res:* Physical and thermal properties of agricultural products; mathematic modelling of biological systems in relation to biotic and abiotic variables; controlled-atmosphere storage of agricultural products; instrumentation; sterilization of canned foods. *Mailing Add:* Dept Agr Eng Univ Man Winnepeg MB R3T 2N2 Can

JAYASWAL, RADHESHYAM K, b Ramganj, India, July 6, 49; m 77; c 2. MOLECULAR BIOLOGY, MICROBIOLOGY. *Educ:* Bombay Univ, India, BSc, 73, MSc, 80; Purdue Univ, PhD(molecular genetics), 85; Bhabha Inst, India, dipl anal methods, 74. *Prof Exp:* Sci asst molecular biol, Tata Inst Fundamental Res, Bombay, 74-80; res asst molecular biol, Dept Hort, Purdue Univ, 80-85, res assoc, Dept Biol, 85-88; ASST PROF MICROBIOL GENETICS, DEPT BIOL SCI, ILL STATE UNIV, NORMAL, 88- *Concurrent Pos:* NIH award, 90-92; Am Heart Asn Award, Am Heart Asn-IA, 91-93. *Mem:* Am Soc Microbiol; Am Phytopath Soc; Am Heart Asn. *Res:* Investigate the possibility of using pseudomonas cepacia as a biocontrol agent after genetic manipulations of antifungal genes; gene cloning, sequencing, and promoter modifications to enhance production of antifungal compound and field testing to determine the efficacy of genetically modified strains. *Mailing Add:* 1418 Godfrey Dr Normal IL 61761-6901

JAYAWEERA, KOLF, b Kalutara, Ceylon, Dec 2, 38; m 65; c 3. CLOUD PHYSICS. *Educ:* Univ Ceylon, BS, 60; Univ London, PhD(physics), 65, Imp Col, Univ London, DIC(cloud physics), 65. *Prof Exp:* Asst lectr physics, Univ Ceylon, 60-62, lectr, 65-67; scientist, Commonwealth Sci & Indust Res Orgn res fel, Sydney, 67-70; asst prof, 70-74, assoc prof, 74-81, PROF GEOPHYS, UNIV ALASKA, 81-, DEAN COL NAT SCI, 85- *Concurrent Pos:* Res grants, NSF, Geophys Inst, Univ Alaska, 71-84, Nat Oceanog & Atmospheric Admin, 73-75 & Off Naval Res, 72-75; prog assoc meteorol, NSF, 78-79 & Air Force Off Sci Res, 79-84. *Mem:* AAAS; fel Royal Meteorol Soc; Am Meteorol Soc; Am Geophys Union. *Res:* Nucleation, growth and aerodynamics of ice crystals in clouds; weather modification; satellite meteorology and sea ice; atmosphere turbulence. *Mailing Add:* Univ Alaska Col Natural Sci PO Box 84104 Fairbanks AK 99708

JAYCOX, ELBERT RALPH, b Miami, Ariz, Oct 13, 23; m 47; c 4. ENTOMOLOGY. *Educ:* Univ Calif, BS, 49, MS, 51, PhD(entom, apicult), 56. *Prof Exp:* Supvr apiary inspection, Calif Dept Agr, 53-58; entomologist apicult, wild bee pollination invests, USDA, Utah State Univ, 58-63; assoc prof hort, Univ Ill, Urbana, 63-69, prof hort & entom, 69-80, prof entom, 80-81; ADJ PROF ENTOM, NMEX STATE UNIV, 81- *Concurrent Pos:* Vis prof, Univ Bern, Switz, 73-74; assoc ed, J Apicult Res. *Honors & Awards:* Outstanding Serv Beekeeping, Western Apicult Soc. *Mem:* Sigma Xi; Entom Soc Am; Int Bee Res Asn. *Res:* Honey bee diseases and parasites; pesticides and bees; bee behavior and biology; pollination; taxonomy of Anthidium. *Mailing Add:* 6100 Shadow Hills Rd Las Cruces NM 88001

JAYE, MURRAY JOSEPH, b New York, NY, Aug 17, 37; m 60; c 2. FOOD SCIENCE, MICROBIOLOGY. *Educ:* Univ Ga, BS, 59; Univ Ill, Urbana, MS, 61, PhD(food sci), 64. *Prof Exp:* Res scientist, Hercules Inc, 64-67; sr scientist, Frito-Lay Inc, 67-68, sect mgr, 69-70, prin scientist, 71-73, mgr corp develop, 74-75; MGR NEW FOOD PROD, CLOROX CO, 75-; TECH DIR, FAIRMONT FOODS CO, 80- *Mem:* Am Chem Soc; Inst Food Technologists; Am Mgt Asn. *Res:* New food products research; flavor chemistry and utilization; starch and hydrocolloid chemistry and utilization; food systems development; research administration. *Mailing Add:* 2517 Via Verde Walnut Creek CA 94598

JAYE, SEYMOUR, b Chicago, Ill, Oct 1, 31; m 58; c 3. ENGINEERING PHYSICS, NUCLEAR ENGINEERING. *Educ:* Univ Ill, BS, 54, MS, 55. *Prof Exp:* Asst radiant heating, Univ Ill, 54-55; assoc physicist nuclear reactor design, Oak Ridge Nat Lab, 55, physicist, 56-60; group leader nuclear design & reactor physics, high temperature gas-cooled reactor, Gen Atomic Div, Gen Dynamics Corp, 60-66, mgr off high temperature gas-cooled reactor planning & asst dept chmn nuclear anal & reactor physics, 66-70, mgr nuclear fuel mkt, Gulf Gen Atomic, 70-71; mgr fuel studies, S M Stoller Corp, 71-73, dir, vpres & gen mgr, 74-82, pres, 83-85, chief exec officer & chmn, 85-89; SR ADV, RCG HAGLER BAILEY INC, 89- *Concurrent Pos:* Lectr, Univ Tenn, 60. *Mem:* Am Nuclear Soc. *Res:* Nuclear design of power reactors; nuclear reactor fuel; reactor physics. *Mailing Add:* RCG Hagler Bailey Inc PO Drawer O Boulder CO 80306-1906

JAYME, DAVID WOODWARD, b Peterson, NJ, Dec 30, 50; m 73; c 4. MEMBRANE TRANSPORT, SOMATIC CELL GENETICS. *Educ:* Brigham Young Univ, BS, 74, MS, 75; Univ Mich, PhD(biol chem), 79. *Prof Exp:* Fel human genetics, Yale Univ, 79-81; asst prof pharmacol, Med Col Va, Va Commonwealth Univ, 81-; MRG CELL BIOL RES & DEVELOP, GIBCO LABS-LIFE TECHNOL INC. *Mem:* NY Acad Sci. *Res:* Membrane transport of ions and metabolities in mammalian cells; isolation and characterization of transport and regulatory mutants; Mechanism of drug therapy, toxicity and resistance. *Mailing Add:* Gibco/Life Technol Inc 2086 Grand Island Blvd Grand Island NY 14072

JAYNE, BENJAMIN A, b Enid, Okla, Oct 10, 28; m 50; c 3. FOREST PRODUCTS, THEORETICAL ECOLOGY. *Educ:* Univ Idaho, BS, 52; Yale Univ, MF, 53, DFor, 55. *Prof Exp:* From instr to asst prof wood technol, Yale Univ, 55-58; assoc wood technologist, Wash State Univ, 59-61; NSF sr fel physics & phys chem, Univ Calif, San Diego, 61-62; prof wood technol, NC State Univ, 62-66; prof wood physics, Univ Wash, 66-76, dir, Ctr Quantitative Sci in Forestry, Fisheries & Wildlife, 71-76; dean, sch forestry & environ studies, duke univ, 76-88; MAURICE K GODDARD PROF FORESTRY & ENVIRON RESOURCE CONSERV, PA STATE UNIV, 88- *Concurrent Pos:* Assoc dean, Col Forest Resources, Univ Wash, 68-71; ed, Wood & Fiber; fel, Int Acad Wood Sci, Inst Wood Sci. *Mem:* Forest Prod Res Soc; Soc Wood Sci & Technol (pres elect, 60); AAAS; Soc Am Foresters. *Res:* Theoretical and experimental physics of wood and wood composites; mass and energy transport in ecosystems; development of a general theoretical framework for the physical properties of wood and fiber composite materials; application of mathematical modeling to management and policy analysis of natural resource system; global forest policy analysis. *Mailing Add:* 113 Ferguson Pa State Univ Univ Park PA 16802

JAYNE, JACK EDGAR, b Spokane, Wash, Dec 18, 25; m 47; c 4. CHEMISTRY. *Educ:* Univ Wis, BS, 47, MS, 48; Lawrence Univ, MS, 50, PhD, 53. *Prof Exp:* Sr res scientist, Kimberly-Clark Corp, 52-74; corp environ dir, Green Bay Packaging, Inc, 74-88; RETIRED. *Mem:* Tech Asn Pulp & Paper Indust. *Res:* Environmental research on effluents and emissions from pulp and paper manufacture. *Mailing Add:* N4218 Gonnering Ct Kaukauna WI 54130-9377

JAYNE, JERROLD CLARENCE, b Stevens Point, Wis, Feb 8, 31; m 60; c 2. ANALYTICAL CHEMISTRY, INORGANIC CHEMISTRY. *Educ:* Univ Wis, BS, 52, PhD(anal chem), 63. *Prof Exp:* From asst prof to assoc prof, 63-74, PROF CHEM, SAN FRANCISCO STATE UNIV, 75- *Concurrent Pos:* Partic, Water Chem Prog, Univ Wis, 70-72. *Mem:* Am Chem Soc; Sigma Xi. *Res:* Coordination chemistry; water chemistry. *Mailing Add:* 2351 Evergreen Dr San Bruno CA 94066

JAYNE, THEODORE D, b Painesville, Ohio, Dec 3, 29; m 59; c 3. SURFACE SCIENCES, PHYSICAL CHEMISTRY. *Educ:* Univ Chicago, AB, 50. *Prof Exp:* Head, Mat & Instrument Sect, Rand Develop Corp, 50-64; lab dir, Gen Tech Serv Inc, 64-72; TECH DIR, T JAYNE CO, 69- *Concurrent Pos:* Chief metallurgist, Rand Develop Corp, 50-64; prin investr, Gen Tech Serv Inc, 65-72 & T Jayne Co, 69-; consult, T Jayne Co, 69-, tech dir, 84- *Res:* Materials; instrumental techniques, light, x-ray, electron, stm microscopies, inertial sensors, stress sensors; vacuum techniques; metrology; instrument design; industrial processes; system design. *Mailing Add:* 10234 Johnnycake Painesville OH 44077-2055

JAYNES, EDWIN THOMPSON, b Waterloo, Iowa, July 5, 22. THEORETICAL PHYSICS. *Educ:* Univ Iowa, BA, 42; Princeton Univ, MA, 48, PhD(physics), 50. *Prof Exp:* Proj engr, Sperry Gyroscope Co, 42-44; actg asst prof, Stanford Univ, 50-55, assoc prof physics, 55-60; from assoc prof to prof, 60-75, WAYMAN CROW PROF PHYSICS, WASHINGTON UNIV, 75- *Concurrent Pos:* Vis fel, St John's Col, Cambridge, Eng, 83-84. *Mem:* AAAS; Am Phys Soc; Am Asn Physics Teachers. *Res:* Electromagnetic theory; statistical mechanics. *Mailing Add:* Washington Univ Campus Box 1105 One Brooking Dr St Louis MO 63130

JAYNES, HUGH OLIVER, b Greeneville, Tenn, Aug 14, 31; m 53; c 2. FOOD SCIENCE. *Educ:* Univ Tenn, BS, 53, MS, 54; Univ Ill, PhD(food sci), 70. *Prof Exp:* Bacteriologist, Res & Develop Ctr, Pet, Inc, 56-63, sect leader chem, 63-67; res fel food sci, Univ Ill, 67-70; assoc prof, 70-79, PROF FOOD TECHNOL & SCI, UNIV TENN, KNOXVILLE, 79-, DEPT HEAD, 85- *Concurrent Pos:* Vis prof, Univ Alexandria, Egypt, 80, coordr, Int Agr Progs, 82-85. *Mem:* Inst Food Technologists; Am Dairy Sci Asn; Sigma Xi. *Res:* Applied research in food color, food chemistry and food product development. *Mailing Add:* Food Technol & Sci Dept Univ Tenn PO Box 1071 Knoxville TN 37901

JAYNES, JOHN ALVA, b Bonham, Tex, Sept 27, 29; m 55; c 2. FOOD SCIENCE. *Educ:* Sam Houston State Teachers Col, BS, 51; Tex Tech Col, BS, 56, MS, 57; Mich State Univ, PhD(dairy), 60. *Prof Exp:* Proj leader food res, 60-63, assoc dir res, 63-67, prod mgr, Canned Milk Prod, 67-73, pres beverage prod, 73-75, vpres foods div, 75-76, pres refrigerated prod, 77-79, vpres, 80-81, VPRES OPERS GROCERY PROD, BORDEN INC, 81- *Mem:* Am Dairy Sci Asn. *Res:* Canned sterile milk and milk based drinks. *Mailing Add:* 2776 W Dublin Worthington OH 43223

JAYNES, RICHARD ANDRUS, b New Iberia, La, May 27, 35; m 59; c 3. PLANT BREEDING. *Educ:* Wesleyan Univ, BA, 57; Yale Univ, MS, 59, PhD(bot), 61. *Prof Exp:* From asst geneticist to geneticist, Conn Agr Exp Sta, 61-80, horticulturist, 80-84; CONSULT, 84- *Concurrent Pos:* Owner, Broken Arrow Nursery. *Mem:* Am Soc Hort Sci; Int Plant Propagators Soc. *Res:* Development of hybrid chestnut trees resistant to the chestnut blight fungus; biological control of the chestnut blight fungus; breeding improved woody ornamentals, especially laurel (Kalmia); vegetative propagation of woody plants. *Mailing Add:* 13 Broken Arrow Rd Hamden CT 06518

JEAN, GEORGE NOEL, b New York, NY, Aug 2, 29. ORGANIC CHEMISTRY. *Educ:* Fordham Univ, BS, 49, MS, 51, PhD(chem), 57. *Prof Exp:* Dye chemist, J P Stevens & Co, Inc, NY, 48; asst phys sci, Med Labs, Army Chem Ctr, Md, 53-55; asst chem, Fordham Univ, 55-56; patent chemist, Patent Dept, Legal Div, Chas Pfizer & Co, Inc, 56-71, patent chemist, 71-76, SR PATENT CHEMIST, PATENT DEPT, LEGAL DIV, PFIZER INC, 76- *Mem:* AAAS; Am Chem Soc; Am Inst Chemists. *Res:* Synthesis and stereochemistry of biaromatic heterocycles; analytical detection of mercaptans; dye chemistry; chemical pharmaceutical patents. *Mailing Add:* 6739 Ingram St Flushing NY 11375

JEAN-BAPTISTE, EMILE, b Port-au-Prince, Haiti, Mar 15, 47; m 68; c 2. MEDICINE, PHYSIOLOGY. *Educ:* Fordham Univ, BS, 71, MS, 72, PhD(physiol), 76; Med Col, Cornell Univ, MD, 83. *Prof Exp:* Res assoc metab & endocrinol, Rockefeller Univ, 76-80, internship/residency internal med, 83-84, asst prof, lab cell biochem & pharmacol, 80-85, ADJ FAC, LAB CELL BIOCHEM & PHARMACOL, ROCKEFELLER UNIV, 85-, ADJ FAC, LAB BACT & IMMUNOL, 89-; PVT MED PRACT, 85- *Concurrent Pos:* Adj asst prof, Col New Rochelle, 75-76; consult-tranlr, Fr ed, The Med Lett, 77-87; fel, Rockefeller Univ, 76-80, NIH, 78-80; asst prof, City Univ NY, 75-80, prof, 87-; prof, Hofstra Univ, 89-; med epidemiologist, New York City Health Dept, 89- *Mem:* Am Soc Zoologists; Sigma Xi; AAAS; NY Acad Sci; Am Soc Pharmacol & Exp Therapeut; AMA; Am Acad Pain Mgt; Nat Coun Int Health. *Res:* Hormonal regulation of lipolysis in adipose tissue; steroidogenesis in adrenal cortex; magnesium flux in plasma membranes; ACTH and glucagon analogs, endorphins, enkephelins, naloxone, luteinizing-hormone releasing hormone and their mechanism of action through the cyclic adenosine monophosphate system; endocrine pharmacology; biochemistry; public health; endocrinology; pharmacology. *Mailing Add:* 1015 Kings Pkwy Baldwin NY 11510

JEANES, JACK KENNETH, b McKinney, Tex, July 2, 23; m 47; c 4. CHEMISTRY. *Educ:* NTex State Col, BS, 47, MS, 48; Oak Ridge Inst Nuclear Studies, cert, 52; Univ Tex, Austin, PhD(biophys), 57. *Prof Exp:* Assoc prof chem, Southwestern State Col, Okla, 48-51; res assoc, Tex Res Found, 51-54; res asst biophys, Southwestern Med Sch, Univ Tex, 54-56, from instr to asst prof, 56-60; assoc prof chem & chmn dept, Univ Dallas, 60-69; PRES & CHIEF EXEC OFF, INDUST RI CHEM LAB, INC, 69- *Concurrent Pos:* Chem consult, 57-; NIH grant geront, 57-60. *Mem:* Am Chem Soc. *Res:* Biophysics; organic chemistry. *Mailing Add:* 1003 Sierra Pl Richardson TX 75080

JEANLOZ, RAYMOND, b Winchester, Mass, Aug 18, 52. MINERAL PHYSICS, PLANETARY INTERIORS. *Educ:* Amherst Col, BA, 75; Calif Inst Technol, PhD(geol & geophysics), 79. *Prof Exp:* Asst prof, Harvard Univ, 79-81; from asst prof to assoc prof, 82-85, PROF GEOL & GEOPHYS, UNIV CALIF, BERKELEY, 85- *Concurrent Pos:* Mem, Mat Res Lab, Harvard Univ, 79-81; A P Sloan Found fel, 81-85; assoc fac, Lawrence Berkeley Lab, 84-; Fairchild scholar, Calif Inst Technol, 88. *Honors & Awards:* J B Macelwane Award, Am Geophys Union, 84, First Birch Lectr, 88; Pres Young Investr Award, 84; Mineral Soc Am Award, 88; MacArthur Found Award, 88; Eyring Lectr, Ariz State Univ, 89; Hudnall Lectr, Univ Chicago, 90. *Mem:* fel AAAS; fel Am Geophys Union; Geol Soc Am; Mineral Soc Am; Mat Res Soc. *Res:* Experimental and theoretical study of minerals and other materials at high pressures, with particular application to the state of planetary interiors. *Mailing Add:* Dept Geol & Geophysics Univ Calif Berkeley CA 94720

JEANLOZ, ROGER WILLIAM, b Berne, Switz, Nov 3, 17; nat US; m 45; c 5. BIOCHEMISTRY. *Educ:* Univ Geneva, ChE, 41, PhD(org chem), 43. *Hon Degrees:* AM, Harvard Univ, 61; DSc, Univ Paris, 80. *Prof Exp:* Instr chem, Univ Geneva, 41-44; assoc chem, Univ Montreal, 47; sr mem & head biochem lab, Worcester Found Exp Biol, 49-51; assoc biochemist, Mass Gen Hosp, 51-61; res assoc, 51-57, assoc org chem, Dept Med, 57-60, from asst prof to assoc prof, 60-69, PROF BIOL CHEM, HARVARD MED SCH, 69-; biochemist, 61-88, EMER BIOCHEMIST, MASS GEN HOSP, 88- *Concurrent Pos:* Swiss Found fel, Univ Basel, 43-45; NIH sr res fel, 48; lectr Swiss-Am Found Sci exchange, 53-54; NSF sr fel, 59-60; guest prof, Univ Cologne, 59-60, Univ Freiburg, 60, Univ Tokyo & Univ Kyoto, 75, Univ Geneva, 76, Univ Saar, 83, Univ Kiel, 84; tutor, Harvard Univ, 61; mem study sect physiol chem, NIH, 64-68 & 69-70; Nat Acad Sci & Acad Sci USSR exchange fel, 70; mem physiol chem B res study comn, Am Heart Asn, 72-75; lectr, Grenoble, 72 & Lille, 73; Guggenheim Found fel, 76-77. *Honors & Awards:* Fr Soc Biol Chem Medal, 60; Liege Univ Medal, 64; Hudson Prize, Am Chem Soc, 73; Alexander von Humboldt Sr Scientist Award, 83. *Mem:* Am Chem Soc; Am Soc Biol Chemists; Royal Soc Chem; Swiss Chem Soc; Biochem Soc; Soc Clin Biol (France). *Res:* Chemistry of carbohydrates; amino sugars; mucopolysaccharides; glycolipids; glycoproteins; bacterial cell walls; deoxysugars; ribose derivatives; glycogen; steroids; metabolism of corticosteroids. *Mailing Add:* Mass Gen Hosp Boston MA 02114

JEANMAIRE, ROBERT L, b Rockford, Ill, Feb 28, 20; m 58; c 3. SCIENCE EDUCATION. *Educ:* Univ Ill, BS, 50, MS, 52; Rensselaer Polytech Inst, MS, 65. *Prof Exp:* Teacher, Melvin Sibley High Sch, 50-52, W Sr High Sch, 52-60 & Auburn High Sch, 60-64; instr physics & math, San Joaquin Delta Col, 64-65; assoc prof, 65-85, PROF PHYSICS, CARTHAGE COL, 85- *Concurrent Pos:* Writer and teacher oper jet engine control, Woodward Governor Co, 57-58. *Mem:* Am Asn Physics Teachers. *Res:* Teaching general physics using a computer. *Mailing Add:* Dept Physics Carthage Col Kenosha WI 53141

JEARLD, AMBROSE, JR, b Annapolis, Md, Mar 6, 44; m 76; c 1. FISHERIES BIOLOGY, FISHERIES RESEARCH. *Educ:* Univ Md, Eastern Shore, BS, 65; Okla State Univ, MS, 70, PhD(zool), 75. *Prof Exp:* Chemist, Publickers Indust Inc, 65-67; biol asst med res, US Army Edgewood Arsenal, 69-71; asst prof biol & anat, Lincoln Univ, 75-77; asst prof animal behav & ecol, Howard Univ, 77-78; SUPVRY RES FISHERY BIOLOGIST, NORTHEAST FISHERIES CTR, WOODS HOLE LAB, US DEPT OF COM, 78- *Concurrent Pos:* Fel, Nat Sci, Okla State Univ, 73; fac mem, Sandy Hook Lab, Dept Com, 77-78; mem, Annapolis Environ Comn, 77-78. *Mem:* Sigma Xi; Animal Behav Soc; Am Fisheries Soc; Int Asn Fish Ethologists. *Res:* Animal behavior with emphasis on behavioral ecology in an aquatic environment; aging and growth problems and their influence on conservation and management of fishery resources in the northeast Atlantic. *Mailing Add:* 135 Tanglewood Dr East Falmouth MA 02536

JEBE, EMIL H, b Clutier, Iowa, Feb 26, 09; m 41. APPLIED STATISTICS, DESIGN OF EXPERIMENTS. *Educ:* Iowa State Univ, BS, 38, MS, 41; NC State Univ, PhD(exp statist), 50. *Prof Exp:* Agr statistician, USDA, 38-40 & 46-49, supvr, USDA-Works Progress Admin & agr statistician, Pilot Res Survey, 40-41; assoc prof statist, Iowa State Univ, 49-59; res mathematician, Infrared & Optics Div, Willow Run Labs, Univ Mich, Ann Arbor, 59-72; res mathematician & consult statistician, 73-79, EMER STATISTICIAN, ENVIRON RES INST MICH, 79- *Concurrent Pos:* Consult Statistician, 74-; chmn, Scio Twpship, Zoning Bd Appeals, Am Soc Testing & Mat, 80-84, Am Statist Asn Youden Award Comt. *Mem:* Fel Am Statist Asn; Int Biomet Soc; Int Asn Statist Phys Sci & Eng; Int Asn Survey Statist; hon fel Am Soc Testing & Mat; NY Acad Sci; sr mem Am Soc Qual Control. *Res:* Application of sampling theory to the design and analysis of sample surveys and experiments; computer programing of least squares and ANOVA; systems analysis and operations research; property assessment analyses; author of over seventy publications. *Mailing Add:* 2650 Laurentide Dr Ann Arbor MI 48103-2116

JEBSEN, ROBERT H, b New York, NY, Sept 5, 31; m 51; c 3. PHYSICAL MEDICINE & REHABILITATION. *Educ:* Brooklyn Col, BA, 53; State Univ NY Downstate Med Ctr, MD, 56; Ohio State Univ, MMS, 60. *Prof Exp:* Intern, Harrisburg Hosp, Pa, 56-57; resident phys & rehab med, Ohio State Univ, Hosp, 57-60; chief phys med & rehab serv, Carswell AFB Hosp, Fort Worth, Tex, 60-62; dir rehab ctr & muscular dystrophy clin, St Luke's Hosp, Cedar Rapids, Iowa, 62-63; from asst prof to assoc prof phys med & rehab, Univ Wash, 63-68; prof phys med & rehab & dir, 68-74, CLIN PROF PHYS MED & REHAB, UNIV CINCINNATI, 74- *Concurrent Pos:* Attend physician, Iowa City Vet Admin Hosp, 62-63; consult, Knoxville Vet Admin Hosp, Iowa, 62-63. *Mem:* Am Acad Phys Med & Rehab; Am Asn Electromyog & Electrodiag (pres, 74-75); Am Cong Rehab Med; Asn Acad Physiatrists. *Res:* Neuromuscular electrodiagnosis; orthotics; objective measurements of physical function. *Mailing Add:* Univ Cincinnati 2900 Losantiville Ave Univ Cincinnati Cincinnati OH 45213

JECH, THOMAS J, b Prague, Czech, Jan 29, 44; US citizen; m 65; c 2. LOGIC, TOPOLOGY. *Educ:* Charles Univ, Prague, PhD(math), 66. *Prof Exp:* Jr fel math, Univ Bristol, 68-69; assoc prof math, State Univ NY, Buffalo, 69-74; PROF MATH, PA STATE UNIV, 74- *Concurrent Pos:* Vis assoc prof, Univ Calif, Los Angeles, 70-71 & Princeton Univ, 72; vis prof, Stanford Univ, 74, Univ Calif, Los Angeles, 81 & Univ Hawaii, 84; ed, Proceedings Am Math Soc, 80-; guest prof, Beijing Normal Univ, 85- *Mem:* Inst Advan Study; Am Math Soc; Asn Symbolic Logic. *Res:* Set theory. *Mailing Add:* Math Dept Pa State Univ State Col PA 16802

JECK, RICHARD KAHR, b Iola, Kans, Oct 6, 38; m 63; c 2. CLOUD PHYSICS, AEROSOL PHYSICS. *Educ:* Rockhurst Col, Kansas City, BA, 60; St Louis Univ, MS, 63, PhD(physics), 68. *Prof Exp:* Fel, Nat Acad Sci, Nat Res Coun, US Naval Res Lab, 68-70; res & develop physicist, Bruker Physik, Ger, 70-71; staff scientist, Smithsonian Radiation Biol Lab, 71-73; RES PHYSICIST, US NAVAL RES LAB, 73- *Concurrent Pos:* Res assoc prof, US Naval Acad, Annapolis, Md, 80-81. *Mem:* Am Meteorol Soc; Am Inst Aeronaut & Astronaut. *Res:* Airborne measurments of cloud characteristics related to aircraft icing; shipboard, airborne, and island based measurements of particulate aerosol size distributions in the maritime environment. *Mailing Add:* Fed Aviation Admin Tech Ctr Atlantic City Airport Atlantic City NJ 08405

JEDLINSKI, HENRYK, plant pathology; deceased, see previous edition for last biography

JEDRUCH, JACEK, b Warsaw, Poland, Feb 22, 27; US citizen; m 72. NUCLEAR REACTOR DEVELOPMENT. *Educ:* Northeastern Univ, BS, 56; Mass Inst Technol, MS, 58; Pa State Univ, PhD(nuclear eng), 66. *Prof Exp:* Eng trainee, H B Smith Co, 53-56; res asst, Columbia Nat Co, 57; assoc scientist, Atomic Power Dept, Westinghouse Elec Co, 57-62, sr scientist, Advan Reactor Div, 66-69, fel scientist, Astronuclear Lab, 69-72, Advan Energy Syst Dept, 72-74, Fusion Power Syst Dept, 74-80 & Nuclear Fuels Div, 80-82; prin engr, Nuclear Eng Dept, 85-90, SR PRIN ENGR, APPL PHYSICS DEPT, EBASCO SERV, INC, 90- *Concurrent Pos:* Mem standards comt, Am Nuclear Soc, 69-81; freelance tech writer, 82-84; consult comput methods, 85; adj assoc prof, Appl Physics & Nuclear Eng Dept, Columbia Univ, NY, 89. *Mem:* Am Soc Mech Engrs; Am Nuclear Soc. *Res:* Development and design of nuclear power sources for electric power generation, space and surface propulsion; fission and fusion technology, safety, economics of fuel cycle of water, gas and liquid metal cooled reactors; development of computing methods for the above; development of nuclear analysis methods. *Mailing Add:* Appl Physics Dept EBASCO Serv Inc Two World Trade Ctr 89N New York NY 10048

JEDYNAK, LEO, b Flint, Mich, Sept 15, 28; m 54; c 4. ELECTRICAL ENGINEERING. *Educ:* Mich State Univ, BSc, 54; Mass Inst Technol, MSc, 56, ScD(elec eng), 62. *Prof Exp:* Prog engr, Gen Elec Co, 54; res asst elec eng, Mass Inst Technol, 54-56; instr, Mich State Univ, 56-57; teaching asst, Mass Inst Technol, 57-58, instr, 58-62; from asst prof to assoc prof, Univ Wis-Madison, 62-76, prof elec eng, 76-80; sr vpres corp res & develop, Oak Indust Inc, 80-88; VPRES OPERS RES & DEVELOP, CUE PAGING CORP, 88- *Concurrent Pos:* Dir corp res, Oak Electro/netics Corp, 69-71, sci & eng consult, 71-; mem bd dirs, Oak Industs Inc, 70- *Mem:* AAAS; Inst Elec & Electronics Engrs; Sigma Xi. *Res:* Insulation of high voltages in high vacuum; electric switches, contacts and relays; real time applications of microcomputer systems. *Mailing Add:* 25931 Serenata Dr Mission Viejo CA 92691-5729

JEE, WEBSTER SHEW SHUN, b Oakland, Calif, June 25, 25; m 51; c 1. ANATOMY. *Educ:* Univ Calif, BA, 49, MA, 51; Univ Utah, PhD(anat), 59. *Prof Exp:* Asst zool, Univ Calif, 49-51; asst anat, 52-58, actg bone group leader, Radiol Div, 56-58, instr, 59-60, asst res prof, 60-61, assoc prof, 63-67, dir training prog mineralized tissues, 64-74, actg chmn anat, 73-77, actg dir div radiobiol, 73-79, PROF ANAT, COL MED, UNIV UTAH, 67-, BONE GROUP LEADER RADIOBIOL DIV, 58- *Concurrent Pos:* Spec consult, Int Atomic Energy Agency, 60 & 64, Proctor & Gamble, 75-78, Upjohn Co, 78-87, Colgate-Palmolive, 79-81, Monsanto, 83-, Eli Lilly, 84-86 & Sch Dent, China Med Col, Tarchung, Repub of China; mem staff, Radiol Health Res Activ, 63-; mem training comt, Nat Inst Dent Res, 66-70, chmn, 68-70; mem sci comt 33, Nat Coun Radiation Protection & Measurements, 69-86; assoc ed, Anat Rec, 69-; consult ed var jour, 70-; assoc ed, Calcified Tissue Res, 77-78; mem, Comt Animal Models for Res on Aging, Nat Res Coun, 78-81; mem peer rev comt musculoskeletal physiol, Am Inst Biol Sci & NASA, 78- *Mem:* Radiation Res Soc; Reticuloendothelial Soc; Orthop Res Soc; Am Asn Anat; Int Asn Dent Res; Geront Soc; Am Soc Bone & Mineral Res; Am Soc Gravitational & Space Biol. *Res:* Physiology and metabolism of bone and teeth; radiation biology. *Mailing Add:* Div Radiobiol Bldg 586 Univ Utah Sch Med Salt Lake City UT 84112

JEEJEEBHOY, KHURSHEED NOWROJEE, b Rangoon, Burma, Aug 26, 35; Can citizen; m 61; c 3. GASTROENTEROLOGY, EXERCISE PHYSIOLOGY. *Educ:* Madras, India, MB & BS, 59; FRCP, 61; London Univ, PhD(clin gastrointestinal res) 63; FRCP(E), 66; FRCP(C), 69; FRCP, 75. *Prof Exp:* Tutor gastroenterol, Postgrad Med Sch London, 61-63; staff radiation, Bhabbha Atomic Res Ctr, Bombay, 63, in-chg, 63-65; head radiation, Radiation Med Ctr, Bombay, 65-67; from asst prof to assoc prof, 68-75, PROF GASTROENTEROL, UNIV TORONTO, 75- *Concurrent Pos:* Prin investr, three grants, Med Res Coun Can, 68-; spec lectr, Dept Nutrit, Univ Toronto, 74-76, hon lectr, Dept Nutrit & Food Sci, 76-79; prof, Dept Nutrit & Food Sci, Univ Toronto, 79-81, Dept Nutrit Sci, 81-, Dept Physiol, 84-; vis prof, Santa Clara Valley Med Ctr, Stanford Univ, 91. *Mem:* Am Soc Clin Invest; Am Soc Clin Nutrit; Am Gastroenterol Asn; Can Asn Gastroenterol; Can Soc Clin Invest; Nutrit Soc Can; Am Asn Physicians. *Res:* Nutritional support of patients with gastrointestinal disease; effect of nutrition on muscle performance; interaction of nutrition and sepsis; long term support of patients with a short bowel. *Mailing Add:* Rm 6352 Med Sci Bldg Univ Toronto Toronto ON M5S 1A8 Can

JEEVANANDAM, MALAYAPPA, b Tirumangalam, India, June 14, 31; m; c 2. SURGICAL NUTRITION. *Educ:* Columbia Univ, PhD(chem), 65. *Prof Exp:* Res assoc, Col Physicians & Surgeons, Columbia Univ, 71-81; assoc lab mem, Sloan-Kettering Cancer Ctr, 81-86; DIR RES, TRAUMA CTR, ST JOSEPH'S HOSP, PHOENIX, AZ, 86- *Mem:* Inst Nutrit; Am Soc Clin Nutritionists; NY Acad Sci. *Res:* Nutrition of cancer; metabolism and nutrition of trauma victims. *Mailing Add:* Dir Res Trauma Ctr St Josephs Hosp & Med Ctr 350 W Thomas Rd Phoenix AZ 85013

JEFCOATE, COLIN R, b Chesham, Bucks, Eng, Sept 28, 42. PHARMACOLOGY. *Educ:* Oxford Univ, Eng, BS, 63, PhD(chem), 66. *Prof Exp:* NATO fel, Basel Univ, Switz, 66-67; NIH & NATO fel, Cornell Univ, Ithaca, 67-69; MRC fel, Edinburgh Univ, Scotland, 69-72; res assoc, Dept Biochem, 72-73, from asst prof to assoc prof, Dept Pharmacol, Med Sch, 73-82, PROF, DEPT PHARMACOL, MED SCH, UNIV WIS-MADISON, 82-, DIR, ENVIRON TOXICOL CTR, 83- *Concurrent Pos:* Mem, NATO Sci Comt Conf on Catalysis, Italy, 72; mem, four study sects, NIH, 76- *Res:* Author of numerous publications. *Mailing Add:* Dept Pharmacol Med Sch Univ Wis 1300 University Ave Madison WI 53706

JEFFAY, HENRY, b Brooklyn, NY, Feb 9, 27; m 57; c 5. BIOCHEMISTRY. *Educ:* Univ Wis, BS, 48, MS, 49, PhD(biochem), 53. *Prof Exp:* Instr biochem, Sch Med, Univ PR, 53-55; res assoc, 55-56, from asst prof to assoc prof, 56-68, asst dean fac affairs, 70-72, assoc dean basic sci, 72-74, dean, 76-79, PROF BIOCHEM, UNIV ILL COL MED, 68- *Concurrent Pos:* Consult, Vet Admin Hosp, Chicago & Norwegian Am Hosp; consult, Roosevelt Mem Hosp, dir med educ; dir basic sci, Rockford Sch Med, Univ Ill, 74-76, actg dean basic sci, 76-79. *Mem:* AAAS; Am Chem Soc; Am Soc Biol Chemists; Int Asn Dent Res. *Res:* Protein metabolism; metabolism of oral tissue; obesity. *Mailing Add:* Dept Biochem Univ Ill Col Med M/C 536 Chicago IL 60612

JEFFCOAT, MARJORIE K, b Boston, Mass, June 14, 51; m 73. PERIODONTOLOGY, RADIOLOGY. *Educ:* Mass Inst Technol, SB, 72; Harvard Sch Dent Med, DMD, 76, cert periodont, 78. *Prof Exp:* Res fel, 75-78, from instr to asst prof, 78-85, ASSOC PROF PERIODONT, HARVARD SCH DENT MED,85-; PROF & CHMN, DEPT PERIODONTOL, UNIV ALA BIRMINGHAM, 88- *Concurrent Pos:* Consult, Brigham & Womens Hosp, 80- & Children's Hosp Med Ctr, 81. *Mem:* Am Dent Asn; Am Acad Periodont; Int Asn Dent Res. *Res:* Bone resorption and diagnosis periodontal disease utilizing the following approaches, bone scanning, radiolabeled microsphere measurements of blood flow, studies of chemotherapeutic agents for treatment of periodontal disease, studies of the effects of local factors on periodontal disease. *Mailing Add:* Sch Dent Univ Ala Birmingham UAB Station Birmingham AL 35294

JEFFERIES, JOHN TREVOR, b Kellerberrin, Western Australia, Apr 2, 25; m 49; c 3. ASTROPHYSICS. *Educ:* Western Australia Univ, BSc, 46, DSc(physics), 61; Cambridge Univ, MA, 49. *Prof Exp:* Res off solar physics, Commonwealth Sci & Indust Res Orgn, NSW, 49-56, prin res off astrophys, 59-60; res assoc, Harvard Col Observ, 56-57, res staff, High Altitude Observ, Colo, 57-58 & Sacramento Peak Observ, 58-59; consult to dir, Nat Bur Standards, Colo, 60-62; fel, Joint Inst Lab Astrophys, 62-64; dir, 83-87, ASTRONR, NAT OPTICAL ASTRON OBSERV, 87-; PROF ASTROPHYS, UNIV HAWAII, 64- *Concurrent Pos:* Adj prof, Univ Colo, 61-64; res assoc, High Altitude Observ, 61-64; prof, Col France, 70 & 77; Guggenheim fel, 70-71; dir, Inst Astron, Univ Hawaii, 67-83. *Mem:* Am Astron Soc; Int Astron Union; Royal Astron Soc. *Res:* Solar physics; radiative transfer; spectral line information; analysis of stellar spectra. *Mailing Add:* Nat Solar Observ 950 N Cherry Tucson AZ 85719

JEFFERIES, MICHAEL JOHN, b London, Eng, Feb 2, 41; m 69; c 2. TECHNICAL MANAGEMENT. *Educ:* Univ Nottingham, BSc, 63, PhD(elec eng), 67. *Prof Exp:* Elec engr, Gen Elec Co, 67-76, managerial res & develop positions, 76-80, res & develop mgr, Eng Physics Labs, Corp Res & Develop, 80-87, gen mgr technol, Gen Elec Motor Bus, 87-90; MEM FAC, PURDUE UNIV, 90- *Mem:* Fel Inst Elec & Electronics Engrs; Brit Inst Elec Engrs; Inst Elec Engrs UK. *Res:* Automation and controls: computer-aided design/computer-aided manufacturing, computers, controls, engineering analysis. *Mailing Add:* 6728 Sweetwood Ct Ft Wayne IN 46804-8127

JEFFERIES, STEVEN, b Abington, Pa, Sept 27, 51; m 73. SYNTHETIC BIOPOLYMERS. *Educ:* Johns Hopkins Univ, BA, 73; Rutgers, MS, 77, Univ Md Dent Sch, DDS(dent), 80. *Prof Exp:* Res asst, environ health, Dept Environ Health, Johns Hopkins Sch Hygiene & Health, 73-75; res intern, chem biochem eng, Dept Chem & Biochem Eng, Rutgers Univ, 75-77; resident, gen dent, USPHS Hosp, New Orleans, 80-81; staff dentist, Municipal Health Servs Prog, Albert Witzke Med Ctr, 81-83; gen dent, Steven R Jefferies, DDS, Pa, 83-86; clin res dentist, 86-89. DIR CLIN RES, LD CAULK DIV, DENTSPLY INT INC, MILFORD, DEL, 90- *Concurrent Pos:* Consult, Johns Hopkins Sch Med, 77-78, Stacogen Corp, 86-; ed, Caulk Dent Educ Bull, 86- *Mem:* Am Asn Dent Res; Int Asn Dent Res; AAAS; Am Dent Asn. *Res:* Natural and synthetic biopolmers, controlled drug release technology, applied connective tissue research, clinical applications of collagen-based biomaterials, wound healing. *Mailing Add:* 715 N Shore Dr Milford DE 19963

JEFFERS, THOMAS KIRK, b Syracuse, NY, Apr 30, 41; m 69; c 2. PARASITOLOGY, POULTRY SCIENCE. *Educ:* Cornell Univ, BS, 63; Univ Wis, PhD(zool & poultry sci), 69. *Prof Exp:* Geneticist, Animal Res Inst, Can Dept Agr, 68-69; dept head parasitol, Hess & Clark Div, Rhodia, Inc, 69-74; sr parasitologist, 74-83, head, 83-86, DIR ANIMAL SCI DISCOVERY RES, LILLY RES LABS, 86- *Honors & Awards:* P P Levine Award, Am Asn Avian Pathologists, 74. *Mem:* Am Soc Parasitologists; Soc Protozoologists; Sigma Xi; Am Asn Avian Pathologists; Poultry Sci Asn; World's Poultry Sci Asn. *Res:* Avian coccidiosis; anticoccidial chemotherapy; intraspecific variation in the coccidia; anticoccidial drug resistance; host response to coccidia. *Mailing Add:* Lilly Res Labs PO Box 708 Greenfield IN 46140

JEFFERS, WILLIAM ALLEN, JR, b Philadelphia, Pa, May 4, 36; m 58; c 3. LOW TEMPERATURE PHYSICS. *Educ:* Amherst Col, AB, 57; Mass Inst Technol, PhD(physics), 62. *Prof Exp:* Sr physicist, Battelle Mem Inst, 62-66; asst prof, 66-76, dean col, 78-87, ASSOC PROF PHYSICS, LAFAYETTE COL, 76- *Concurrent Pos:* Dean studies, Lafayette Col, 72-75. *Mem:* Am Phys Soc; Am Asn Physics Teachers; Sigma Xi. *Res:* Ultrasonic absorption in liquid helium; superconductivity; transport properties in metals. *Mailing Add:* Dept of Physics Lafayette Col Easton PA 18042-1782

JEFFERSON, CAROL ANNETTE, b Minneapolis, Minn, July 4, 48; m 75; c 2. PLANT ECOLOGY. *Educ:* St Olaf Col, BA, 70; Ore State Univ, PhD(bot), 74. *Prof Exp:* Asst prof biol, Eckerd Col, 74-76; asst prof, 76-81, ASSOC PROF BIOL, WINONA STATE UNIV, 81- *Mem:* AAAS; Ecol Soc Am. *Res:* Great Lakes sand vegetation; driftless area-relict communities; flood plain vegetation; wetland ecotones. *Mailing Add:* Dept of Biol Winona State Univ Winona MN 55987

JEFFERSON, DAVID KENOSS, b Pasadena, Calif, Dec 21, 38; m 67; c 2. COMPUTER SCIENCE. *Educ:* Calif Inst Technol, BS, 60; Columbia Univ, AM, 62; Univ Mich, PhD(comput sci), 69. *Prof Exp:* Vis prof comput sci, Naval Postgrad Sch, 72-73; mathematician, Naval Weapons Lab, 60-72, res mathematician, 73-75; proj leader info syst design, David W Taylor Naval Ship Res & Develop Ctr, 75-82; mgr, Database Archit Group, Nat Bur Standards, 82-87; CHIEF INFO SYSTS ENG DIV, NAT INST STANDARDS & TECHNOL, 87- *Concurrent Pos:* Lectr, Univ Md, 77-82. *Mem:* Sigma Xi; Asn Comput Mach; Inst Elec & Electronics Engrs; Sr Exec Serv. *Res:* Standards and guides for data dictionary systems, database languages, data interchange, graphics, data administration, database design, hypertext, object-oriented databases and knowledge-based systems; development and administration of conformance tests and procedures. *Mailing Add:* Nat Inst Standards & Technol Gaithersburg MD 20899

JEFFERSON, DONALD EARL, b Homeland, Fla, Sept 27, 27; m 51; c 2. SCIENCE POLICY, PHYSICAL OCEANOGRAPHY. *Educ:* Morehouse Col, BS, 48; Howard Univ, MS, 50. *Prof Exp:* Instr physics, Va Union Univ, 49-51; physicist, US Naval Ord Lab, 51-52, elec engr, 54-72, elec engr, Naval Surface Weapon Ctr, 72-81, sci adv to commander second fleet, Naval Assistance Prog, 81-82, Naval Surface Weapon Ctr, 83-85; RETIRED. *Concurrent Pos:* Vpres eng, Copycomposer Corp, 69-71. *Mem:* Am Defense Preparedness Asn. *Res:* Review of naval operational systems for modification or replacement as needed; design and modification of instrumentation for measuring ocean currents; statistical analysis and prediction of system and environment interactions; underwater acoustics. *Mailing Add:* 13321 Bea Kay Dr Silver Spring MD 20904

JEFFERSON, EDWARD G, b London, Eng, July 15, 21; m 53; c 3. RESEARCH ADMINISTRATION. *Educ:* Univ London, PhD. *Prof Exp:* Supvr, Du Pont, 51-60, mgr res, Plastics Dept, 60-64, asst dir, 64-66, dir Flurocarbons Div, 66-69, asst gen mgr, Plastics Dept, 69-70, asst gen mgr, Explosives Dept, 70-72, asst gen mgr, Polymer Intermediates Dept, 72, vpres & gen mgr, Film Dept, 72- 73, dir, sr vpres & mem exec comt, 73-80, pres & chief oper officer, 80-81, chmn & chief exec off, Du Pont, 81-86; RETIRED. *Concurrent Pos:* Mem bd dirs, Du Pont Co, 86-; dir, Chem Banking Corp & Am Tel & Tel Co; mem, President's Export Coun & US Coun Int Bus. *Honors & Awards:* Warren K Lewis Lectr, Mass Inst Technol; Chem Indust Medal, Soc Chem Indust. *Mem:* Nat Acad Eng; Am Philos Soc; Am Acad Arts & Sci; Am Inst Chem Engrs; Am Chem Soc. *Mailing Add:* 1007 Market St Wilmington DE 19898

JEFFERSON, JAMES WALTER, b Mineola, NY, Aug 14, 37; m 65; c 3. PSYCHOPHARMACOLOGY. *Educ:* Bucknell Univ, BS, 58; Univ Wis, MD, 64. *Prof Exp:* Intern, St Lukes Hosp, NY, 64-65; resident internal med, Univ Wis, 65-67; fel, Univ Chicago, 67-68; resident psychiat, 71-74, from asst prof to assoc prof, 74-81, PROF PSYCHIAT, UNIV WIS-MADISON, 81- *Concurrent Pos:* Staff psychiatrist, Vet Admin Hosp, Wis, 74-81; dir, Lithium Info Ctr, Madison, 74-; dir, Ctr Affective Disorders, Madison, 84- *Mem:* Am Psychiat Asn; Am Psychopath Asn; Int Neuropsychopharmacologium Soc. *Res:* Clinical psychopharmacology; compliation and dissertation of information through interactive computer programs in psychiatry; neuropsychiatric aspects of medical disorders. *Mailing Add:* Dept Psychiat Clin Sci Ctr Univ Wis 600 Highland Ave Madison WI 53792

JEFFERSON, LEONARD SHELTON, b Maysville, Ky, Jan 14, 39. PHYSIOLOGY. *Educ:* Eastern Ky Univ, BS, 61; Vanderbilt Univ, PhD(physiol), 66. *Prof Exp:* Vis scientist, Cambridge Univ, 66-67; res assoc physiol, Col Med, Vanderbilt Univ, 67; from instr to assoc prof, 67-75, PROF PHYSIOL, COL MED, PA STATE UNIV, 75- *Concurrent Pos:* USPHS fel, 66-67. *Honors & Awards:* Lilly Award, Am Diabetes Asn, 79. *Mem:* Am Soc Biol Chemists; Biochem Soc; Am Physiol Soc; Am Diabetes Asn. *Res:* Regulation of skeletal muscle and hepatic carbohydrate and protein metabolism by hormones and other factors, especially mechanism of action of insulin and growth hormone. *Mailing Add:* Milton S Hershey Med Ctr Pa State Univ PO Box 850 Hershey PA 17033

JEFFERSON, MARGARET CORREAN, b Eau Claire, Wis, Aug 22, 47. GENETICS. *Educ:* Univ Dubuque, BS, 69; Univ Colo, MA, 71; Univ Ariz, PhD(genetics), 77. *Prof Exp:* Asst prof, 77-81, assoc prof biol, 81-85, PROF BIOL, CALIF STATE UNIV, LOS ANGELES, 85- *Concurrent Pos:* Consult, Compton Sickle Cell Educ & Detection Ctr, 77-; prin invester biomed res support grants, NIH, 77- & res apprenticeships minority high sch students, NSF, 81-82. *Mem:* AAAS; Am Genetics Asn; Genetics Soc Am; Soc Study Evolution. *Res:* Ecological and behavioral genetics of desert-adapted Drosophila; specifically, pheromonal regulation of reproductive strategies in desert-adapted Drosophila; genetics of learning behavior; cytogenetics of Cycads; eye pigmentation systems. *Mailing Add:* Dept of Biol Calif State Univ 5151 State Univ Dr Los Angeles CA 90032-8201

JEFFERSON, ROLAND NEWTON, b Washington, DC, Nov 7, 11; m 46; c 3. ENTOMOLOGY. *Educ:* Va Polytech Inst, BS, 34, MS, 36; Iowa State Col, PhD(entom), 42. *Prof Exp:* Asst entomologist, Va Exp Sta, 36-39; instr entom, Va Polytech Inst, 39-40; asst entomologist, Va Exp Sta, 41-42; from asst prof & asst entomologist to prof & entomologist, Exp Sta, Univ Calif, Los Angeles, 46-60; prof entom, Univ Calif, Riverside, 60-76, emer prof entom & entomologist, Exp Sta, 76-77; RETIRED. *Mem:* Entom Soc Am. *Res:* Insect morphology; insects affecting floricultural crops. *Mailing Add:* 5499 Grassy Trail Dr Riverside CA 92504

JEFFERSON, THOMAS BRADLEY, b Urich, Mo, Nov 25, 24; m 46; c 3. MECHANICAL ENGINEERING. *Educ:* Kans State Col, BS, 49; Univ Nebr, MS, 50; Purdue Univ, PhD, 55. *Prof Exp:* Instr mech eng, Univ Nebr, 49-52; from instr to asst prof, Purdue Univ, 52-58; prof & head dept, Univ Ark, Fayetteville, 58-68, assoc dean eng, 68-69; dean, Sch Eng & Technol, 69-78, PROF MECH ENG, SOUTHERN ILL UNIV, 78- *Concurrent Pos:* Consult, Allison Div, Gen Motors Corp, 56-57, Martin Marietta Aerospace, Denver, 58-68. *Mem:* Am Soc Mech Engrs; Am Soc Eng Educ. *Res:* Heat transfer. *Mailing Add:* Sch of Eng & Technol Southern Ill Univ Carbondale IL 62901

JEFFERSON, THOMAS HUTTON, JR, b Mineola, NY, June 6, 41. MASS STORAGE, OPERATING SYSTEMS. *Educ:* Rensselaer Polytech Inst, BS, 63; NC State Univ, MAM, 65; Univ Colo, Boulder, PhD(appl math), 69. *Prof Exp:* DISTINGUISHED MEM TECH STAFF, SANDIA LABS, LIVERMORE, 69- *Res:* Computer software libraries; computer mass storage; nonlinear parameter determination. *Mailing Add:* Computer Opers & Systs Div 2913 Sandia Nat Labs Livermore CA 94551-0969

JEFFERSON, WILLIAM EMMETT, JR, biochemistry; deceased, see previous edition for last biography

JEFFERTS, KEITH BARTLETT, b Raymond, Wash, May 10, 31; m 53; c 4. ATOMIC PHYSICS, FISHERIES MANAGEMENT. *Educ:* Univ Wash, PhD(physics), 62. *Prof Exp:* Mem tech staff physics, Bell Tel Labs, 63-75; PRES, NORTHWEST MARINE TECHNOL, 72- *Mem:* Am Phys Soc. *Res:* Structure of simple molecules; molecular astrophysics; application of physical techniques to problems of fishery management. *Mailing Add:* PO Box 363 Shaw Island WA 98286

JEFFERY, DUANE ELDRO, b Delta, Utah, Sept 28, 37; m 61; c 3. GENETICS, EVOLUTIONARY BIOLOGY. *Educ:* Utah State Univ, BS, 62, MS, 63; Univ Calif, Berkeley, MA, 66, PhD(zool, genetics), 72. *Prof Exp:* Asst prof, 69-77, ASSOC PROF ZOOL, BRIGHAM YOUNG UNIV, 77- *Concurrent Pos:* Vis colleague genetics, Univ Hawaii, 74-75. *Mem:* Soc Study Evolution; Genetics Soc Am; Am Soc Human Genetics; AAAS. *Res:* Developmental and evolutionary genetics in Drosophila populations; human transmission genetics; cytogenetics. *Mailing Add:* Dept of Zool Brigham Young Univ Provo UT 84602

JEFFERY, GEOFFREY MARRON, b Dundee, NY, May 13, 19; m 41; c 4. MALARIOLOGY. *Educ:* Hobart Col, BA, 40; Syracuse Univ, MA, 42; Johns Hopkins Univ, ScD(parasitol), 44; Yale Univ, MPH, 61. *Prof Exp:* Biol aide, Tenn Valley Authority, Wilson Dam, Alta, 44; asst sanitarian, USPHS, 44-45, asst sanitarian, Commun Dis Ctr, Ga, 45-46; from asst sanitarian to sr asst scientist, Sch Trop Med, PR, 46-47; from sr asst scientist to scientist, Malaria Res Lab, Lab Trop Dis, NIH, Ga, 48-54, from scientist to sr scientist, SC, 54-60, sci dir, 60-63, asst chief, Lab Parasite Chemother, Nat Inst Allergy & Infectious Dis, 63-66, actg chief, 66, chief, 67-68; chief, Cent Am Res Sta, Ctr Dis Control, 69-74, asst dir, 74-75, dir, Vector Biol & Control Div, Bur Trop Dis, 75-81, asst dir, Div Parasitic Dis, Bur Trop Dis, 81-85; RETIRED. *Concurrent Pos:* Asst prof, Univ Bridgeport, 47-48; mem expert panel malaria, WHO, 63-, scientific group on chemother of malaria, Geneva, 67, parasitol of malaria, Teheran, 68 & Cent Am Malaria Assessment Mission, AID, 64; assoc mem comn malaria, Armed Forces Epidemiol Bd, 65-69, mem, 69-73; deleg, Int Cong Trop Med & Malaria, Lisbon, 58, Rio de Janeiro, 63, Teheran, 68, Int Cong Parasitol, Rome, 64, Wash, 70 & Latin Am Cong Parasitol, 73. *Honors & Awards:* Ashford Medal, Am Soc Trop Med & Hyg, 59. *Mem:* AAAS; Am Soc Parasitol; Am Soc Trop Med & Hyg (secy-treas, 61-67, vpres, 71 & pres, 75); Am Mosquito Control Asn; Royal Soc Trop Med & Hyg. *Res:* Malarias of man and lower animals; chemotherapy of malaria and parasitic infections; epidemiology of malaria and intestinal parasites; biology of human malarias; immunology and pathology of malaria; diagnosis of parasitic infections; drug resistant strains of malaria parasites; methodology of malaria eradication and control. *Mailing Add:* 1093 Blackshear Dr Atlanta GA 30304

JEFFERY, LARRY S, b Delta, Utah, June 21, 36; m 59; c 7. WEED SCIENCE, WEED BIOLOGY. *Educ:* Utah State Univ, BS, 62; NDak State Univ, PhD(plant sci), 66. *Prof Exp:* Asst prof weed sci, Univ Nebr Lincoln & Bogota, Colombia, 66-69; ASSOC PROF WEED SCI, UNIV TENN, 69- *Concurrent Pos:* Consult, Univ Wis & EMBRAPA-Ministry Agr Brazil, Lordrina, 76. *Mem:* Weed Sci Soc Am; Coun Agr Sci & Technol; Int Weed Sci Soc. *Res:* Weed control in economic crops; development of weed control systems in corn, soybeans, grain sorghum, tobacco, alfalfa, small grains and pastures. *Mailing Add:* Agron Brigham Young Univ Provo UT 84602

JEFFERY, LAWRENCE R, b Memphis, Tenn, June 30, 27; m 48; c 5. SYSTEMS DESIGN, SYSTEMS SCIENCE. *Educ:* Univ Chicago, MS, 53. *Prof Exp:* Instr electronic eng & math, Am TV Inst, Ill, 46-51; engr, Raytheon Mfg Co, Mass, 53-54; staff mem command & control systs, Lincoln Lab, Mass Inst Technol, 54-58, sect leader, 58; assoc dept head, Mitre Corp, Mass, 59-61, dept head, 61-63, assoc tech dir, 63-73, tech dir commun, 73-86; RETIRED. *Mem:* Sr mem Inst Elec & Electronics Engrs. *Res:* Design and evaluation of computer-based command; control and communication systems; military operations research; digital computer engineering. *Mailing Add:* 16 Sherwood Dr Hollis NH 03049

JEFFERY, RONDO NELDEN, b Provo, Utah, Apr 16, 40; m 65; c 7. SOLID STATE PHYSICS, ELECTRONICS PHYSICS. *Educ:* Brigham Young Univ, BS, 63, MS, 65; Univ Ill-Urbana, PhD(physics), 70. *Prof Exp:* Res assoc physics, Rensselaer Polytech Inst, 70-73; asst prof physics, Wayne State Univ, 73-80; vis assoc prof, Weber State Col, 80-83, assoc prof, 83-86, PROF, WEBER STATE UNIV, 86-; MEM TECH STAFF, TRW, 86- *Concurrent Pos:* Asst prof, Cottrell res grant, Res Corp, 75-78 & NSF res grant, 78-81 & CSIP grant, 87-89. *Mem:* Am Phys Soc; Am Asn Physics Teachers; Sigma Xi. *Res:* High pressure effects in solids; properties of point defects such as vacancies under high-pressure, high-temperature conditions using diffusion and positron annihilation techniques; developing microcomputer based physics laboratory experiments and electronics; physic education material. *Mailing Add:* Dept Physics Weber State Univ Ogden UT 84408-2508

JEFFERY, WILLIAM RICHARD, b Chicago, Ill, June 9, 44. DEVELOPMENTAL BIOLOGY, CELL BIOLOGY. *Educ:* Univ Ill, BS, 67; Univ Iowa, PhD(zool), 71. *Prof Exp:* Res asst biol, Univ Ill, 65-66; NIH fel zool, Univ Iowa, 67-71; Am Cancer Soc fel oncol, Univ Wis, 71-72; res assoc biochem, Sch Med, Tufts Univ, 72-74; asst prof biophys, Univ Houston, 74-77; from asst prof to prof zool, Univ Tex, 77-90; PROF ZOOL, UNIV CALIF, DAVIS, 90- *Concurrent Pos:* Corp mem, Marine Biol Lab, Woods Hole 75-, instr, 80-82, dir embryol, 83- *Mem:* AAAS; Am Soc Cell Biol; Asn Develop Biol. *Res:* Molecular and cellular mechanisms of cell development and differentiation. *Mailing Add:* Bodega Marine Lab Univ Calif PO Box 247 Bodega Bay CA 94923

JEFFERYS, WILLIAM H, III, b New Bedford, Mass, July 8, 40. ASTRONOMY. *Educ:* Wesleyan Univ, BA, 62; Yale Univ, MS, 64, PhD(astron), 65. *Prof Exp:* Instr astron, Wesleyan Univ, 64-65; from asst prof to assoc prof, 65-79, PROF ASTRON, UNIV TEX, AUSTIN, 79-, HARLAN J SMITH CENTENNIAL PROF ASTRON, 85- *Concurrent Pos:* Alfred P Sloan fel, 65-67. *Mem:* AAAS; Am Astron Soc; Royal Astron Soc; Int Astron Union. *Res:* Astrometry; celestial mechanics; dynamical astronomy. *Mailing Add:* Dept of Astron Univ of Tex Austin TX 78712

JEFFORDS, RUSSELL MACGREGOR, b Shinglehouse, Pa, May 11, 18; m 43; c 1. GEOLOGY. *Educ:* Syracuse Univ, AB, 39; Univ Kans, MA, 41, PhD(geol), 46. *Prof Exp:* Asst geologist, Kans Geol Surv, 39-42; geologist, US Geol Surv, 42-54 & Humble Oil & Refining Co, 54-64; res adv, 64-79, CONSULT, EXXON PROD RES CO, 79- *Concurrent Pos:* Instr, Brown Univ, 46-47; asst prof, Univ Tex, 47-48; assoc ed, Paleont Inst, Univ Kans, 69- *Mem:* Soc Tech Commun; Geol Soc Am; Paleont Soc; Soc Econ Paleontologists & Mineralogists; Am Asn Petrol Geologists. *Res:* Stratigraphic paleontology; ground-water hydrology and geochemistry; Paleozoic corals; crinoids; chitinozoans; editing. *Mailing Add:* 8002 Beverly Hill Houston TX 77063

JEFFREY, GEORGE ALAN, b Cardiff, Eng, July 29, 15; nat US; m 42; c 2. CRYSTALLOGRAPHY. *Educ:* Univ Birmingham, BSc, 36, PhD(chem), 39, DSc, 53. *Prof Exp:* X-ray crystallographer, Brit Rubber Producers Res Asn, 39-45; lectr inorg & phys chem, Univ Leeds, 45-53; prof chem & physics, 53-64, prof crystallog, 65-85, chmn dept, 69-85, EMER PROF CRYSTALLOG, UNIV PITTSBURGH, 85- *Concurrent Pos:* Vis prof crystallog, Univ Pittsburgh, 50-51; mem exec comt gov bd, Am Inst Physics, 71- Award, Am Chem Soc, 78. *Honors & Awards:* Hudson Award, Am Chem Soc, 80. *Mem:* Am Chem Soc; Am Crystallog Asn (treas, 54-58, pres, 63); The Chem Soc; Brit Inst Physics & Phys Soc. *Res:* Structure of hydrates and carbohydrates; biochemical crystallography; hydrogen bonding; molecular distortions in crystals. *Mailing Add:* Dept of Crystallog Univ of Pittsburgh Pittsburgh PA 15260

JEFFREY, JACKSON EUGENE, b Oakhurst, NJ, May 28, 31; m 54; c 3. BIOLOGICAL STRUCTURE. *Educ:* Col William & Mary, BS, 54; Va Polytech Univ, MS, 59; Med Col Va, PhD(anat), 63. *Prof Exp:* Asst prof, 62-65, ASSOC PROF BIOL, VA COMMONWEALTH UNIV, 65-, CHMN DEPT, 85- *Mem:* AAAS. *Res:* Fatty acid oxidation in cattle; effects of avitaminosis C on enzyme activity; interaction of various sex hormones on the reproductive system of the golden hamster; serum protein polymorphism in fish; age and growth of freshwater fish. *Mailing Add:* Dept Biol Va Commonwealth Univ S7N 0W0 Richmond VA 23284-2012

JEFFREY, JOHN J, b Worcester, Mass, May 3, 37; m 72. BIOLOGICAL CHEMISTRY, ENDOCRINOLOGY. *Educ:* Col of the Holy Cross, BS, 58; Georgetown Univ, PhD(chem), 65. *Prof Exp:* From instr to assoc prof, 67-81, PROF MED, SCH MED, WASHINGTON UNIV, 81- *Res:* Enzymatic mechanisms of collagen degradation; hormonal regulation of mammalian collagenase activity. *Mailing Add:* Dept Biochem & Med Washington Univ Med 660 S Euclid Ave St Louis MO 63110

JEFFREY, KENNETH ROBERT, b Toronto, Ont, May 7, 41; m 67; c 2. NUCLEAR MAGNETIC RESONANCE. *Educ:* Univ Toronto, BSc, 64, MA, 66, PhD(physics), 69. *Prof Exp:* From asst prof physics to assoc prof, 69-81, PROF PHYSICS, UNIV GUELPH, 81- *Mem:* Can Asn Physicists; Biophys Soc. *Res:* Nuclear magnetic resonance studies of molecular reorientation and phase transitions in solids and lyotropic liquid crystals; biophysical techniques (nuclear magnetic resonance, x-ray diffraction, calorimetry, dielectric relaxation), applied to model and biological membranes. *Mailing Add:* Dept Physics Univ Guelph Guelph ON N1G 2W1 Can

JEFFREYS, DONALD BEARSS, plant physiology, microbiology; deceased, see previous edition for last biography

JEFFRIES, CARSON DUNNING, b Lake Charles, La, Mar 20, 22; m 90; c 2. SOLID STATE PHYSICS. *Educ:* La State Univ, BS, 43; Stanford Univ, PhD(physics), 51. *Prof Exp:* Res assoc, Radio Res Lab, Harvard Univ, 43-45; instr physics, Phys Inst, Univ Zurich, 51; from instr to assoc prof, 52-63, PROF PHYSICS, UNIV CALIF, BERKELEY, 63-, FAC SR SCIENTIST, LAWRENCE BERKELEY LAB, 77- *Concurrent Pos:* NSF fels, 58 & 65-66; Fulbright res scholar, 59; Miller res prof, 61 & 83. *Mem:* Nat Acad Sci; fel Am Phys Soc; fel Am Acad Arts & Sci. *Res:* Nuclear and electronic paramagnetism and magnetic resonance; nuclear orientation; low temperature and solid state physics; paramagnetic relaxation; optical properties of semi-conductors; excitons; chaotic behavior of physical systems; superconductivity. *Mailing Add:* Dept of Physics LeConte Hall Univ of Calif Berkeley CA 94720

JEFFRIES, CHARLES DEAN, b Rome, Ga, Apr 9, 29; m 53. MICROBIOLOGY. *Educ:* NGa Col, BS, 50; Univ Tenn, MS, 55, PhD(bact), 58. *Prof Exp:* Technician, Div Labs, State Dept Pub Health, Ga, 50-51; from instr to assoc prof, 58-70, from actg dep chmn to assoc chmn dept, 70-75, asst dean curric affairs & dir grad progs, 75-80, ASSOC PROF DERMAT, SCH MED, WAYNE STATE UNIV, 68-, PROF IMMUNOL & MICROBIOL, 70- *Concurrent Pos:* Fulbright lectr, Cairo Univ, 65-66; mem bd exam basic sci, State Mich, 67-72, yvres, 71-72; guest res, Mycol Div, Ctr Dis Control, US Pub Health Serv, Atlanta, Ga, 80-81; microbiologist consult, Vet Admin Med Ctr, Allen Park, Mich, 89- *Mem:* Fel Am Acad Microbiol; Am Soc Microbiol; Soc Exp Biol & Med; Brit Soc Gen Microbiol; Int Soc Human & Animal Mycol. *Res:* Bacterial identification; medical mycology. *Mailing Add:* Dept Immunol & Microbiol Wayne State Univ Sch of Med Detroit MI 48201

JEFFRIES, GRAHAM HARRY, b Barmera, S Australia, May 31, 29; m 55; c 4. INTERNAL MEDICINE, GASTROENTEROLOGY. *Educ:* Univ NZ, BMedSc, 49, MB, ChB, 53; Oxford Univ, DPhil(physiol), 55. *Prof Exp:* Assoc prof med, Med Col, Cornell Univ, 64-69; PROF MED & CHMN DEPT, COL MED, MILTON S HERSHEY MED CTR, PA STATE UNIV, 69- *Mem:* Am Fedn Clin Res; Am Gastroenterol Asn; Am Soc Clin Invest; fel Am Col Physicians. *Res:* Gastric secretion; vitamin B-12 metabolism; intestinal absorption; liver disease. *Mailing Add:* Milton S Hershey Med Ctr Pa State Univ Hershey PA 17033

JEFFRIES, HARRY PERRY, b Newark, NJ, Apr 15, 29; m 51; c 5. ZOOLOGY. *Educ:* Univ RI, BS, 51, MS, 55; Rutgers Univ, PhD(zool), 59. *Prof Exp:* Asst biol oceanog, Univ RI, 51-55; pharmacologist, Ciba Pharmaceut Prod, Inc, NJ, 55-56; asst, Rutgers Univ, 56-59; from asst prof to assoc prof, 59-73, PROF BIOL OCEANOG, UNIV RI, 73- *Concurrent Pos:* Grants, Dept Energy, Nat Oceanog & Atmospheric Admin; Environ Protection Agency Sea Grant, Am Petroleum Inst; pres, Estuarine Res Fedn, 73-75, Nat Sci Found, Off Naval Res. *Mem:* Fel AAAS; Am Soc Limnol & Oceanog; Sigma Xi. *Res:* Comparative ecology of estuarine habitats; biological fertility of inshore marine areas and characterization of community structure; chemical homeostasis of marine organisms in relation to environmental stress; biochemical systematics. *Mailing Add:* PO Box 64 Kingston RI 02881

JEFFRIES, JAY B, b June 3, 47; US citizen. MOLECULAR PHYSICS. *Educ:* Univ Iowa, BA, 69; Univ Colo, PhD(physics), 80. *Prof Exp:* Assoc prof res, Univ Pittsburgh, 80-83; CHEM PHYSICIST, SRI INT, 83- *Mem:* Am Phys Soc; Am Chem Soc; Combustion Inst; Mat Res Soc. *Res:* Laser-based diagnostic measurements of reacting flows and plasmas with ultimate goal of understanding the fundamental chemical mechanism of the process. *Mailing Add:* Molecular Physics Lab SRI Int 333 Ravenswood Ave Menlo Park CA 94025

JEFFRIES, NEAL POWELL, b Indianapolis, Ind, Aug 25, 35; m 58; c 2. MECHANICAL ENGINEERING. *Educ:* Purdue Univ, BS, 57; Mass Inst Technol, MS, 58; Univ Cincinnati, PhD(mech eng), 69. *Prof Exp:* Res asst heat transfer, Stanford Univ, 61-63; engr, Gen Elec Co, Ohio, 63-65, proj mgr heat transfer, 65-67; res assoc mech eng, Univ Cincinnati, 67-69, asst prof, 69-74; mgr educ mech eng dept, Struct Dynamic Res Corp, 74-78; EXEC DIR, CTR MFG TECH, 78- *Concurrent Pos:* Lectr, Gen Elec Co, 63-74; consult, Struct Dynamics Res Corp, 68-74, Honeywell Res Lab & Am Laundry Mfg, 69-70, Vortex Corp, 70-, & Avco Electronics, 71-; US Navy grant, 70-71. *Mem:* Am Soc Mech Engrs; Soc Mfg Engrs; Am Soc Eng Educ; Robotics Int; Comput & Automated Syst Asn. *Res:* Heat transfer; fluid flow; thermodynamics; boiling phenomena; heat pipe; manufacturing engineering. *Mailing Add:* 3112 Cooper Rd Cincinnati OH 45241

JEFFRIES, QUENTIN RAY, b Terre Haute, Ind, Feb 28, 20; m 51; c 2. CHEMICAL ENGINEERING. *Educ:* Rose Polytech Inst, BS, 41; Univ Mich, MS, 47; Univ Ill, PhD(chem eng), 53. *Prof Exp:* Asst chem engr, Commercial Solvents Corp, 48-49, shift supvr, Penicillin Plant, 49-51; prin chem engr, Battelle Mem Inst, 53-56; prin chem engr, 56-59, tech develop engr, 59-65, chem engr res, eng dept, Int Minerals Corp, 65-; AT CORN SOLVENTS CORP. *Mem:* Am Chem Soc; Sigma Xi. *Res:* Gaseous diffusion. *Mailing Add:* Com Solvents Corp 1331 S First St Terre Haute IN 47802

JEFFRIES, ROBERT ALAN, b Indianapolis, Ind, Nov 11, 33; m 54; c 2. OPTICAL PHYSICS. *Educ:* Univ Okla, BS, 54, MS, 61, PhD(ionization kinetics), 65. *Prof Exp:* Proj engr, Pontiac Motor Div, Gen Motors Corp, 54-55; staff mem, Los Alamos Nat Lab, 57-76, group leader, 76-77, asst div leader, 77-79, prog mgr, 79-83, off leader, 83-86, prog dir, 86-88, off dir, 88-89. *Concurrent Pos:* Delegation, Nuclear Testing Talks, Geneva, 86-90; mem, US-Soviet Bilateral Consultative Comn. *Mem:* Am Phys Soc; Sigma Xi. *Res:* Ionization kinetics; shock hydrodynamics; laser produced plasmas; electro optical instrumentation; arms control and verification technology. *Mailing Add:* 160 La Cueva Los Alamos NM 87544

JEFFRIES, THOMAS WILLIAM, b New Orleans, La, Oct 31, 47; m 74; c 3. BIOCHEMISTRY. *Educ:* Calif State Univ, Long Beach, BS, 69, MS, 72; Rutgers Univ, PhD(microbiol), 75. *Prof Exp:* Asst microbiol, Calif State Univ, Long Beach, 69-71; res intern, Rutgers Univ, 72-75; staff mem microbiol, Lawrence Livermore Lab, Univ Calif, 75-77; res assoc chem eng & appl chem, Columbia Univ, 77-79; MICROBIOLOGIST, FOREST PROD LAB, USDA, 79-; ASSOC PROF DEPT BACT, UNIV WIS, 87- *Concurrent Pos:* USDA career develop award, 87. *Mem:* Am Soc Microbiol; Soc Indust Microbiol; AAAS; Am Chem Soc; Tech Pulp & Paper Indust. *Res:* Applied microbial ecology; polysaccharide biochemistry; biochemical engineering; biochemistry; biotechnology; enzymology; microbial photosynthesis; biofuels; environmental toxicology; lignin biodegradation; pentose fermentation; yeasts; fermentation, metabolic regulation. *Mailing Add:* Forest Products Lab One Gifford Pinchot Dr Madison WI 53705-2398

JEFFRIES, WILLIAM BOWMAN, b Chicago, Ill, Mar 5, 26; m; c 3. INVERTEBRATE ZOOLOGY. *Educ:* Univ Pittsburgh, BS, 49; Univ NC, MA, 52, PhD(zool), 55. *Prof Exp:* Asst zool, Univ NC, 50-55; Nat Cancer Inst fel, Ind Univ, 55-56; from instr to asst prof microanat, Med Col Ga, 56-59; from asst prof to prof 59-81, chmn dept, 65-68, 74-77, 83-86 & 89-90, CHARLES A DANA PROF BIOL, DICKINSON COL, 81- *Concurrent Pos:* NIH spec res fel, Vet Admin Hosp, Miami, 68-69; res assoc biochem, Sch Med, Univ Miami, 68-69; res assoc, Dept Zool, Field Mus Natural Hist, 77- *Mem:* AAAS; Am Soc Zool; Soc Protozool; Sigma Xi; Am Asn Univ Prof; Crustacean Soc. *Res:* Physiology; parasitology; protozoology; biology of the barnacle genus Octolasmis. *Mailing Add:* Dept Biol Dickinson Col Carlisle PA 17013

JEFFS, GEORGE W, b Stockton, Calif, Mar 9, 25; m; c 3. SPACE & COMMUNICATION TECHNOLOGY. *Educ:* Univ Wash, BS & MS. *Hon Degrees:* DEE, West Coast Univ, 84. *Prof Exp:* Mem, Aerophysics Lab, sect chief advan eng, sect chief systs eng, mgr corp tech develop & planning, vpres & prog mgr, Paraglider Prog, corp exec dir eng, Rockwell Int, 47-66; asst prog mgr & chief prog engr, 66-69, vpres & prog mgr, Apollo CSM Progs, 69-73, pres space div, 74-76; corp officer, Rockwell Int, 76-78, pres NAm Aerospace Opers, 78-86, pres & ctr dir, Strategic Defense Ctr, 86-91, dir vpres strategic defense & technol, 88-91; CONSULT, 91- *Concurrent Pos:* Mem, Adv Panel Ballistic Missile Defense, Cong Off Technol Assessment; US deleg, Prog Indust & Tech Coop Aerospace, China, 84; Jimmy Doolittle educ fel, 89. *Honors & Awards:* Presidential Medal of Freedom, 70; Golden Knight of Mgt Award, Nat Mgt Asn, 80; Astronaut Engr Award, Nat Space Club, 82; Von Karman Lectr, Am Inst Aeronaut & Astronaut, 83, Elmer Sperry Award, 86. *Mem:* Nat Acad Eng; fel Am Inst Aeronaut & Astronaut; fel Am Astronaut Soc; fel Inst Advan Eng. *Res:* Advanced space engines; major launch vehicle propulsion engines; solid rockets. *Mailing Add:* Strategic Defense Ctr Rockwell Int 2800 Westminster Blvd Seal Beach CA 90740-2089

JEFFS, PETER W, b Luton, Eng, Jan 9, 33; m 57; c 3. ORGANIC CHEMISTRY. *Educ:* Univ Natal, PhD(chem), 61. *Prof Exp:* Res asst, Akers Res Labs, Imp Chem Indust, Eng, 50-57; lectr org chem, Univ Natal, 60-62; from asst prof to assoc prof, 64-71, PROF ORG CHEM, DUKE UNIV, 71- *Mem:* Am Chem Soc; fel The Chem Soc; assoc Royal Inst Chem. *Res:* Chemistry of alkaloids, terpenes and mould metabolites; alkaloid biosynthesis; application of nuclear magnetic resonance to structure determination. *Mailing Add:* Glaxo Inc Five Moore Dr Research Triangle Park NC 22709

JEFIMENKO, OLEG D, b USSR, Oct 14, 22; m 45. ELECTROMAGNETIC THEORY, ELECTROSTATICS. *Educ:* Univ Göttingen, Ger, Vordiplom, 49; Lewis & Clark Col, BA, 52; Univ Ore, MA, 54, PhD(physics), 56. *Prof Exp:* Asst physics, Univ Ore, 52-55; from asst prof to assoc prof, 56-67, PROF PHYSICS, WVA UNIV, 67- *Mem:* Am Phys Soc; Am Asn Physics Teachers; Electrostatic Soc Am. *Res:* Electromagnetic theory; cosmical electrodynamics; electrostatics; electrets; atomic physics. *Mailing Add:* Dept of Physics WVa Univ Morgantown WV 26506

JEGASOTHY, BRIAN V, b Colombo, Sri Lanka, Mar 3, 43. DERMATOLOGY. *Educ:* Univ Sri Lanka, MD, 67. *Prof Exp:* PROF, DEPT DERMAT, UNIV PA, 82-, ACTG CHMN, 86- *Mem:* Am Acad Dermat; Soc Invest Dermat; Am Fedn Clin Res. *Mailing Add:* Dept Dermat Losthrop Hall 190 Univ Pittsburgh Sch Med Suite 145 Lothrop St Pittsburgh PA 15228

JEGLA, DOROTHY ELDREDGE, b Brooklyn, NY, Sept 19, 39; m 65; c 2. DEVELOPMENTAL BIOLOGY, PLANT BIOLOGY. *Educ:* Mt Holyoke Col, AB, 61; Yale Univ, MS, 64, PhD(biol), 85. *Prof Exp:* asst prof, 72-87, ASSOC PROF BIOL, KENYON COL, 87- *Concurrent Pos:* Plant tissue cult facil, Comprehensive Sch Improv Proj, NSF, 87; vis assoc prof biol, Rennsalaer Polytech Inst, 88. *Mem:* AAAS; Bot Soc Am; Int Soc Plant Molecular Biologists. *Res:* Organization and regulation of apical meristem development in the sunflower, Helianthus annus, by clonal analysis, grafting and sterile culture techniques. *Mailing Add:* Dept Biol Kenyon Col Gambier OH 43022

JEGLA, THOMAS CYRIL, b St Johns, Mich, July 5, 35; m 65; c 2. CELL BIOLOGY, NEUROSCIENCE. *Educ:* Mich State Univ, BS, 58; Univ Ill, MS, 60, PhD(zool), 64. *Prof Exp:* Asst prof biol, Univ Minn, 63-64 & Yale Univ, 64-66; from asst prof to assoc prof, 66-85, chmn dept, 76-79 & 84-87, PROF BIOL, KENYON COL, 85- *Concurrent Pos:* NSF res grants, 70, 73 & 85-; vis assoc prof, Yale Univ, 81-83; vis prof, Univ Bonn, 81, 83 & 84; NIH res grant, 85; vis res prof, Univ WFla, 88. *Mem:* AAAS; Am Soc Zool; Soc Europ Comp Endocrinol. *Res:* Molting physiology of arthropods; invertebrate biology; comparative endocrinology; biochemistry of steroid and peptide hormones. *Mailing Add:* Dept Biol Kenyon Col Gambier OH 43022

JEGLUM, JOHN KARL, b Medford, Wis, Dec 9, 38; m 64; c 2. FOREST ECOLOGY, SILVICULTURE. *Educ:* Univ Wis, BS, 60, MS, 62; Univ Sask, PhD(plant ecol), 68. *Prof Exp:* Asst prof bot, Eastern Ill Univ, 65-66; RES SCIENTIST FORESTRY ECOL, CAN FORESTRY SERV, 68- *Concurrent Pos:* Vis researcher, Dept Peatland Forestry, Univ Helsinki, 84-85. *Mem:* Can Bot Asn; Can Inst Forestry; Int Peat Soc. *Res:* Wetland classification and ecology; boreal vegetational ecology; autecology of black spruce; regeneration silviculture; strip cutting in black spruce; environmental impacts of harvesting; peatland forestry. *Mailing Add:* Great Lakes Forestry Ctr Box 490 Sault Ste Marie ON P6A 5M7 Can

JEHN, LAWRENCE A, b Dayton, Ohio, Aug 7, 21; m 44; c 9. COMPUTER SCIENCE. *Educ:* Dayton Univ, BS, 43; Univ Mich, ScM, 49, ABD, 50. *Prof Exp:* Instr math, Univ Dayton, 46-47, asst prof, 50-56; res assoc, Univ Mich, 56-57; assoc prof computer sci, Univ Dayton, 57-63, res mathematician, Res Inst, 63-86, chmn computer sci, 82-86, prof, 74-88, EMER PROF COMPUTER SCI, UNIV DAYTON, 88- *Concurrent Pos:* Consult, Univ Dayton, 51-56 & 57-63, assoc prof comput sci & mech eng, 68-74; chmn, Asn Comput Mach, Comput Sci Conf, 79 & 86, Nat Educ Comput Conf, Comput Sci & Comput Conf Comt, 86-88. *Mem:* Asn Comput Mach; Inst Elec & Electronics Engrs; Comput Soc. *Res:* Computer science education; numerical analysis and simulation. *Mailing Add:* Dept Comput Sci Univ Dayton 300 College Park Dayton OH 45469-2160

JEKEL, EUGENE CARL, b Holland, Mich, Dec 19, 30; m 60; c 2. INORGANIC CHEMISTRY. *Educ:* Hope Col, AB, 52; Purdue Univ, MS, 55, PhD(inorg chem), 64. *Prof Exp:* From instr to assoc prof, 55-69, chmn dept, 67-70 & 73-76, PROF CHEM, HOPE COL, 69-, CHIEF HEALTH PROFESSIONS ADV, 77- *Concurrent Pos:* Vis prof, Univ Calif, Berkeley, 70-71. *Mem:* Asn Am Med Cols; Nat Sci Teachers Asn; Am Chem Soc; Sigma Xi. *Res:* Thermodynamics of aqueous solutions at high temperature. *Mailing Add:* Dept Chem Hope Col Holland MI 49423

JEKEL, JAMES FRANKLIN, b St Louis, Mo, Oct 14, 34; m 58; c 4. EPIDEMIOLOGY, PUBLIC HEALTH. *Educ:* Wesleyan Univ, AB, 56; Wash Univ, MD, 60; Yale Univ, MPH, 65. *Prof Exp:* Res asst pub health, St Louis County Health Dept, 58; epidemiologist, Ctrs Dis Control, 62-67; ass prof, 67-71, assoc prof pub health, 71-80, PROF EPIDEMIOL & PUB HEALTH, YALE UNIV, 80-, CEA WINSLOW PROF PUBLIC HEALTH, 82- *Concurrent Pos:* Fulbright fel, 85-86. *Mem:* Am Pub Health Asn; fel Am Col Prev Med; fel Am Sci Affiliation. *Res:* Program evaluation, especially health programs for teenage mothers; cocaine abuse. *Mailing Add:* Dept of Epidemiol & Pub Health Yale Univ 60 College St New Haven CT 06510

JEKELI, CHRISTOPHER, b Marburg, WGermany, Dec 21, 53; m 84. GEODESY. *Educ:* McGill Univ, BA, 76; Ohio State Univ, MSc, 78, PhD(geod), 81. *Prof Exp:* Res assoc geod, Ohio State Univ, 77-81; GEODESIST, AIR FORCE GEOPHYS LAB, 81- *Honors & Awards:* Weikko A Herskanen Award, Ohio State Univ, 80. *Mem:* Am Geophys Union. *Res:* Physical geodesy: methods to analyze and improve knowledge of the earth's external gravity field and application of these methods to gravimetric data. *Mailing Add:* 82 Ash St Concord MA 01742

JEKELI, WALTER, physics, for more information see previous edition

JELEN, FREDERIC CHARLES, b Chelsea, Mass, Jan 17, 10; m 43; c 2. CHEMICAL ENGINEERING. *Educ:* Mass Inst Technol, SB, 31, SM, 32; Harvard Univ, AM, 34, PhD(phys chem), 35. *Prof Exp:* Chemist phosphates, Monsanto Co, 35-41; engr electrochem, Battelle Mem Inst, 42-43; chief engr silicates, Cowles Chem Co, 43-49; engr corrosion, Allied Chem Co, 49-61; prof chem eng, Lamar Univ, 61-80; prof, McNeese State Univ, 81-83; RETIRED. *Concurrent Pos:* Consult cost eng, 61-, PPG Industs, Inc, 62-75, Int Nickel, 63, Mobil Oil Corp, 63-, Mobil Chem Co, 64-71 & E I du Pont de Nemours & Co, Inc, 75. *Honors & Awards:* Diamond Qual Award, Asn Pushing Gravity Res, 85. *Mem:* Fel Am Asn Cost Engrs. *Res:* Cost engineering. *Mailing Add:* Two Hull Circle Austin TX 78746

JELENKO, CARL, III, surgery; deceased, see previous edition for last biography

JELINEK, ARTHUR GILBERT, b Milwaukee, Wis, May 6, 17; m 45; c 3. SYNTHETIC ORGANIC CHEMISTRY. *Educ:* Univ Wis, BS, 40, PhD(org chem), 44. *Prof Exp:* Res & control chemist, Fox River Paper Corp, Wis, 40-41; asst org chem, Univ Wis, 41-44; res chemist, Grasselli Chem Dept, E I du Pont de Nemours & Co Inc, 44-55, sr res chemist, Biochem Dept, 55-79; RETIRED. *Mem:* Emer mem Am Chem Soc. *Res:* Agricultural chemicals. *Mailing Add:* 2500 Lindell Rd Grendon Farm Wilmington DE 19808

JELINEK, BOHDAN, b Jimaramov, Czech, June 21, 10; nat US; m 65; c 1. BIOCHEMISTRY, CLINICAL MEDICINE. *Educ:* Brno Tech Univ, BS, 32, DSc(chem), 35; Masaryk Univ, BSc, 33, MD, 39. *Prof Exp:* French Govt Scholar, Inst Pasteur, Paris, France, 35-37; asst prof appl biochem, Brno Tech Univ, 38-39; res assoc, Sugar Res Inst, 40-45; Brit Res Coun scholar, Univ Birmingham, Eng, 45-46, Univ res fel, 48-49; prof fermentation chem & indust mycol & chmn dept, Univ Prague, 46-48; asst prof biochem, Univ Alta, 49-52; assoc prof, Univ Mo, 52-54; res assoc, Med Res Inst, City of Hope Med Res Ctr, 54-64; head biochemist & asst to dir res, Adolph's Food Prod Mfg Co, 57-70; RETIRED. *Concurrent Pos:* Consult, 70-; intern, Sante Fe Mem Hosp, Los Angeles, 72-73; med dir, Alcohol Prog, Los Angeles County Health Serv, 74-75; med consult, Dept Rehab, State Calif, 75- *Mem:* AAAS; Am Chem Soc; Inst Food Technologists; NY Acad Sci; Royal Soc Chem. *Res:* Chemistry and biochemistry of carbohydrates and their derivatives; fermentations, physical chemistry and enzymatic degradation of starch and its components; nutritional and industrial aspects of carbohydrates; dietetic foods; papain. *Mailing Add:* 231 N Primrose Ave Monrovia CA 91016

JELINEK, CHARLES FRANK, b Miles City, Mont, Feb 6, 17; m 44; c 2. ORGANIC CHEMISTRY. *Educ:* Mont State Col, BS, 38; Oxford Univ, BSc, 41; Univ Ill, PhD(org chem), 44. *Prof Exp:* Res chemist, Gen Aniline & Film Corp, 46-47, asst to dir res, 47-49, sales engr & asst dir cent sales develop dept, 49-50, sect leader appln res, 50-52, mgr surfactants res, 52-55, mgr process res & develop dept, 55-59, dir dyestuff & chem div, Cent Res Lab, 59-63; sr staff adv, Chem Staff, Esso Res & Eng Co, 63-66; coordr new ventures, Enjay, 66-71; coordr tech opportunities div, Dart Industs, Inc, 71-72; dir div chem technol, Bur Foods, Food & Drug Admin, 79-86, 72-75, dept assoc dir technol, 75-79, dept dir phys sci, 79-86; CONSULT, 86- *Mem:* Am Chem Soc; Asn Off Anal Chemists; Am Inst Chemists; Commercial Develop Asn. *Res:* Derivatives of acetylene; dyes; pigments; surfactants; polymers; solvents; coatings; chemical contaminants in foods. *Mailing Add:* 8229 Kay Court Annandale VA 22003

JELINEK, FREDERICK, b Prague, Czech, Nov 18, 32; nat US; m 61; c 2. ELECTRONICS ENGINEERING. *Educ:* Mass Inst Technol, SB, 56, SM, 58, PhD(elec eng), 62. *Prof Exp:* Instr elec eng, Mass Inst Technol, 59-62; from asst prof to prof, Cornell Univ, 62-74; SR MGR CONTINUOUS SPEECH RECOGNITION, T J WATSON RES CTR, IBM CORP, 72- *Concurrent Pos:* Vis lectr, Harvard Univ, 62; NSF grant, 64-66; vis scientist, IBM T J Watson Res Ctr, 68-69; NASA contracts, 66-72. *Mem:* Fel Inst Elec & Electronics Engrs. *Res:* Transmission of information; coding; data compression; speech recognition; information theory. *Mailing Add:* T J Watson Res Ctr IBM Corp Yorktown Heights NY 10598

JELINEK, ROBERT V(INCENT), b New York, NY, Mar 5, 26; m 55; c 3. CHEMICAL ENGINEERING. *Educ:* Columbia Univ, BS, 45, MS, 47, PhD(chem eng), 53. *Prof Exp:* Asst drafting, Columbia Univ, 43-45, instr chem eng, 49-51; chem engr, Develop Div, Standard Oil Develop Co, 51-53; asst prof chem eng, Columbia Univ, 53-54; from asst prof to prof, Syracuse Univ, 54-72, dir summer res prog high sch teachers, 64-69, asst to dean eng, 55-60, fac secy, 62-64, chmn eng fac, 64-65; prof & dean, Sch Environ & Resource Eng, 72-80, PROF DEPT PAPER SCI & ENG, STATE UNIV NY COL ENVIRON SCI & FORESTRY, 80- *Concurrent Pos:* Assoc, Danforth Found, 56-60; NSF res grant, 59-61; prog dir eng chem, NSF, 71-72. *Mem:* Am Chem Soc; Electrochem Soc; Nat Asn Corrosion Engrs; Am Inst Chem Engrs. *Res:* Reaction kinetics; corrosion; electrochemistry; adsorption; process design and computer simulation. *Mailing Add:* Col Environ Sci & Forestry State Univ NY 424 Walters Hall Syracuse NY 13210

JELINSKI, LYNN W, b Arlington, Va, Jan 19, 49. PROTEIN FOLDING, BIOLOGICAL RECOGNITION. *Educ:* Duke Univ, BS, 71; Univ Hawaii, PhD(chem), 76. *Prof Exp:* Fel chem, Johns Hopkins Univ, 76-77; fel, Nat Inst Health, 77-78, staff fel, biophysics, 78-80; mem tech staff chem, 80-84, head polymar chem, 84-85, HEAD BIOPHYS, AT&T BELL LAB, 85- *Res:* Nuclear magnetic resonances and imaging; protein folding, biological recognition. *Mailing Add:* AT&T Bell Lab 600 Mountain Ave Murray Hill NJ 07974

JELLARD, CHARLES H, b Abergavenny, Wales, Dec 25, 16; m 50; c 4. MEDICAL BACTERIOLOGY, PUBLIC HEALTH. *Educ:* Oxford Univ, BA & BM, BCh, 42; Univ London, dipl bact, 51; FRCPath, 65; Oxford Univ, DM, 75. *Hon Degrees:* MA, Cambridge Univ, 48. *Prof Exp:* Dir, Pub Health Lab Serv, Plymouth, UK, 53-68; assoc prof bacj & dep dir, Prov Lab Pub Health, Univ Alta, 68-82; RETIRED. *Concurrent Pos:* Hon consult bacteriologist, Plymouth Hosps, UK, 53-68. *Mem:* Path Soc Gt Brit & Ireland; Brit Soc Gen Microbiol. *Res:* Diagnostic medical bacteriology; epidemiology. *Mailing Add:* 12504 Lansdone Dr Edmonton AB T6H 4L5 Can

JELLIFFE, DERRICK BRIAN, b Rochester, UK, Jan 20, 21; m 42. NUTRITION, PEDIATRICS. *Educ:* Univ London, MB, BS, 43, MRCP, 44, DCH & MD, 45, DTH & H, 47; Am Bd Pediat, dipl, 58; FRCP, 61. *Prof Exp:* Lectr med, Univ Ibadan, 49-52; sr lectr pediat, Univ WI, 52-54; WHO vis prof, All-India Inst Hyg & Pub Health, Calcutta, 54-56; vis prof trop med, Med Sch, Tulane Univ, 56-59; prof pediat, Med Sch, Makerere Univ, Uganda, 59-66; prof community nutrit & dir, Caribbean Food & Nutrit Inst, Jamaica, 67-71; prof pub health, 71-77, PROF ECON PUB HEALTH, SCH PUB HEALTH, UNIV CALIF, LOS ANGELES, 77- *Concurrent Pos:* Fulbright fel, Med Sch, Tulane Univ, 51; vis prof, Sch Pub Health, Univ Calif, Berkeley, 63; chmn comt ecol prev young child malnutrit, Int Union Nutrit Sci, 70- *Honors & Awards:* Rosen von Rosenstein Medal, Swed Pediat Soc, 69. *Mem:* Fel Am Pub Health Asn; fel Am Acad Pediat; Am Inst Nutrit; Asn Pediat India (pres, 54); Asn Physicians E Africa (pres, 63). *Res:* Health and nutrition in young children in developing countries, particularly the epidemiology field assessment and evaluation of programs. *Mailing Add:* Univ Calif Sch Pub Health Los Angeles CA 90024

JELLIFFE, ROGER WOODHAM, b Cleveland, Ohio, Feb 18, 29; m 54; c 4. CARDIOLOGY, CLINICAL PHARMACOLOGY. *Educ:* Harvard Col, AB, 50; Columbia Univ, MD, 54; Am Bd Internal Med, dipl, 62; Am Bd Cardiovasc Dis, dipl, 65. *Prof Exp:* Intern med, Univ Hosps, Cleveland, Ohio, 54-55, asst resident, 55-56; Nat Found Infantile Paralysis fel exp med, Sch Med, Western Reserve Univ, 56-58; staff physician, Vet Admin Hosp, Cleveland, 58-60, resident med, 60-61; from instr to assoc prof, 61-76, PROF MED, SCH MED, UNIV SOUTHERN CALIF, 76- *Concurrent Pos:* Los Angeles County Heart Asn res fel, Sch Med, Univ Southern Calif, 61-63; NIH res grants digitalis & appl pharmacokinetics, 64- *Mem:* Fel Am Col Physicians; fel Am Heart Asn; Am Soc Clin Pharmacol & Therapeut. *Res:* Cardiovascular pharmacology; chemical measurements of digitalis glycosides and mathematical descriptions of the kinetics of digitalis, kanamycin, gentamicin, streptomycin, procainamide, lidocaine and other drugs in man; computer assistance for planning, monitoring and adjusting dosage regimens of the above drugs; methods for optimal study and control of pharmacokinetic systems. *Mailing Add:* Univ Southern Calif Sch Med 2025 Zonal Ave Los Angeles CA 90033

JELLINCK, PETER HARRY, b Paris, France, Feb 20, 28; m 54; c 3. BIOCHEMISTRY, ENDOCRINOLOGY. *Educ:* Cambridge Univ, BA, 48; Univ London, BSc, 50, MSc, 52, PhD(biochem), 54. *Prof Exp:* Can Nat Res Coun fel biochem, McGill Univ, 55-56; lectr chem, Norwood Tech Col, Eng, 56-57; lectr biochem, St Bartholomew's Hosp Med Col, London, 57-58 & Middlesex Hosp Med Sch, 58-59; from asst prof to prof, Univ BC, 60-67; head dept, 67-78, PROF BIOCHEM, QUEEN'S UNIV, ONT, 67- *Concurrent Pos:* Nat Cancer Inst Can-Med Res Coun Can res grant, 59-; vis prof, Rockefeller Univ, 78-79, 82- *Mem:* Am Asn Cancer Res; Brit Biochem Soc; Can Biochem Soc; Endocrine Soc. *Res:* Estrogen metabolism and action; hormonal carcinogenesis. *Mailing Add:* Dept of Biochem Queen's Univ Kingston ON K7L 3N6 Can

JELLINEK, HANS HELMUT GUNTER, b Free City of Danzig, Mar 25, 17; m 48; c 1. PHYSICAL CHEMISTRY. *Educ:* Univ London, DIC, 41, PhD(phys chem), 42; Cambridge Univ, PhD(colloid & phys chem), 45, ScD, 64. *Prof Exp:* Sect head phys chem, J Lyons & Co, Eng, 45-50; assoc prof, Univ Adelaide, 50-54; vis prof, Polytech Inst Brooklyn, 54-57; assoc prof, Univ Cincinnati, 57-59; prof chem, Univ Windsor, 59-64, head dept, 59-63; PROF CHEM, CLARKSON COL, 64- *Concurrent Pos:* Sci expert, US Dept Army, 54-63 & 70-; mem comt high polymer res, Nat Res Coun Can, 62-64. *Mem:* AAAS; Am Chem Soc; fel Am Inst Chem; fel Chem Inst Can; fel Royal Inst Chem; Sigma Xi. *Res:* Stability of high polymers; reaction kinetics; surface chemistry; energy production; adhesive and rheological properties of ice; author of over 160 publications. *Mailing Add:* Dept Chem Clarkson Univ Tech Potsdam NY 13676

JELLINEK, MAX, b 1929; m 65; c 3. ORGAN TRANSPLANTATION, SHOCK. *Educ:* St Louis Univ, PhD(biochem), 61. *Prof Exp:* PROF BIOCHEM, ST LOUIS UNIV, 62- *Mem:* Am Physiol Soc; Am Chem Soc. *Res:* Metabolism of ischimic hypoxic; organs and shock. *Mailing Add:* Dept Surg Sch Med St Louis Univ 1402 S Grand Blvd St Louis MO 63104

JELLING, MURRAY, b Brooklyn, NY, Jan 7, 18; m 41; c 1. PATENT LICENSING, EXPERT WITNESS. *Educ:* Brooklyn Col, BS, 37; Polytech Univ, Brooklyn, MS, 41, PhD(chem), 45. *Prof Exp:* Res chemist, Autoxygen, Inc, 38-41; res chemist, Nopco Chem Co, Inc, 41-43; res assoc, polymer, Polytech Univ, 43-45; res dir org chem, Maguire Industs, Inc, 45-47; pres, Cidex Corp, 48-57; CONSULT CHEMIST, 57- *Concurrent Pos:* Pres, Jonelle Indust Prod Inc, 67-72. *Mem:* Am Chem Soc; Sigma Xi; Soc Asphalt Technologists; Asn Asphalt Paving Technologists; Asn Consult Chemists & Chem Engrs; Transp Res Bd. *Res:* Industrial organic chemistry; polymers; textile maintenance (dry-cleaning and laundry) products; bituminous products; patents; licensing. *Mailing Add:* 21 Spring Hill Rd Roslyn Heights NY 11577

JELLINGER, THOMAS CHRISTIAN, b Seaton, Ill, Sept 17, 23; m 46; c 3. CONSTRUCTION ENGINEERING. *Educ:* Univ Ill, Urbana-Champaign, BS, 49; Iowa State Univ, MS, 63. *Prof Exp:* Engr, various consult firms, Cincinnati, 49-52 & Wagner, Inc, 52-57; proprietor eng & architect, Thomas C Jellinger, 57-60; prof-in-charge construct eng, Iowa State Univ, 60-; RETIRED. *Concurrent Pos:* Dir & vpres, Assoc Gen Contractor Educ & Res Found, 70; founding dir, Am Inst Constructors. *Mem:* Nat Soc Prof Engrs; Am Soc Eng Educ. *Res:* Construction management, techniques, and advanced methods of construction scheduling. *Mailing Add:* 23 Knighton Dr Bella Vista AR 72714

JELLISON, GERALD EARLE, JR, b Bangor, Maine, Mar 27, 46; m 70; c 2. OPTICS. *Educ:* Bowdoin Col, BA, 68; Brown Univ, ScM, 73, PhD(physics), 77. *Prof Exp:* Nat Res Coun fel, Naval Res Lab, 76-78; STAFF SCIENTIST, OAK RIDGE NAT LAB, 78- *Mem:* Am Phys Soc; Am Optical Soc; Mat Res Soc. *Res:* Physics of semiconductors as related to laser annealing mechanisms; measurement of optical properties of materials as a function of doping and temperature, as well as time-resolved optical measurements. *Mailing Add:* Bldg 2000 Solid State Div Oak Ridge Nat Lab Oak Ridge TN 37831-6056

JELLUM, MILTON DELBERT, b Starbuck, Minn, Oct 26, 34; m 57; c 2. AGRONOMY, PLANT BREEDING. *Educ:* Univ Minn, BS, 56; Univ Ill, MS, 58, PhD(agron), 61. *Prof Exp:* Asst agronomist, Ga Exp Sta, Univ Ga, 60-67, from assoc prof to prof agron, 67-85; RETIRED. *Mem:* AAAS; Am Soc Agron; Crop Sci Soc Am; Am Oil Chem Soc; Am Asn Cereal Chem. *Res:* Environmental and genetic study of oil content and fatty acid composition of corn grain oil; study of yield components of corn and corn breeding. *Mailing Add:* PO Box 187 Orchard Hill GA 30266

JEMAL, MOHAMMED, Ethiopian citizen; c 1. PHARMACOKINETICS, DRUG METABOLISM. *Educ:* Haile Sellassie Univ, BS, 70; Purdue Univ, PhD(pharm anal), 76. *Prof Exp:* Post doctoral res, Purdue Univ, 76-77; res investr anal res & develop, Squibb Inst med Res, 78-82, GROUP LEADER BIOANAL RES, BRISTOL-MYERS SQUIBB, 82- *Honors & Awards:* Haile Sellassie Medal Award. *Mem:* Am Chem Soc; Am Asn Pharmaceut Scientists. *Res:* Quantification of drugs and metabolites in body fluids for assessment of pharmacokinetics and safety. *Mailing Add:* Bristol-Myers Squibb New Brunswick NJ 08903

JEMIAN, WARTAN A(RMIN), b Lynn, Mass, Dec 31, 25; m 51; c 4. PHYSICAL METALLURGY. *Educ:* Univ Md, BS, 50; Rensselaer Polytech Inst, MS, 53, PhD(metall eng), 56. *Prof Exp:* Engr, Semiconductor Dept, Westinghouse Elec Corp, 55-57; sr fel & head power rectifiers fel, Mellon Inst, 57-62; dir res & develop, Rectifier-Capacitor Div, Fansteel Metall Corp, 62; assoc prof, 62-65, PROF MECH ENG, AUBURN UNIV, 65-, CHMN MAT ENG CURRICULUM COMT, 63-, PROF MAT ENG, 75- *Concurrent Pos:* Lectr & adj prof, Univ Pittsburgh, 56-62. *Mem:* Am Inst Mining, Metall & Petrol Engrs; Am Soc Metals; Am Soc Eng Educ; Biomat Res Soc; Int Asn Math & Comput Simulation; Sigma Xi. *Res:* Education; structure and properties of composite materials; computer analysis of materials; education in materials science and engineering. *Mailing Add:* 350 Singleton St Auburn AL 36830

JEMMERSON, RONALD RENOMER WEAVER, b Baltimore, Md, Mar 7, 51. IMMUNOLOGY, BIOCHEMISTRY. *Educ:* Western Md Col, BA, 73; Northwestern Univ, PhD(biochem), 78. *Prof Exp:* Res fel, Scripps Clin & Res Found, 78-81; res assoc, La Jolla Cancer Res Found, 81-84 & Scripps Clin & Res Found, 84-85; ASST PROF, DEPT MICROBIOL, UNIV MINN, 85- *Concurrent Pos:* Damon Runyon-Walter Winchell fel, 79-80. *Mem:* AAAS; Am Asn Immunologists. *Res:* Protein antigenicity; use of synthetic molecules as vaccines; antibody and B cell repertoires. *Mailing Add:* Dept Microbiol 1460 Mayo Box 196 Univ Minn Med Sch 420 Delaware St SE Minneapolis MN 55455

JEMSKI, JOSEPH VICTOR, b Blackstone, Mass, Mar 19, 20; m 43; c 1. MEDICAL MICROBIOLOGY. *Educ:* Fordham Univ, BS, 42; Univ Pa, PhD(med microbiol), 52. *Prof Exp:* Head bact dept, Maltine Co, 46-49; chief animal path unit, Ralph M Parsons Co, 52-55; chief animal path sect, Chem Corps, 55-59, chief test sphere br, US Army Biol Defense Res Labs, 59-72, sr investr, US Army Res Inst Infectious Dis, 72-83; RETIRED. *Concurrent Pos:* Comn Rickettsial Dis, Armed Forces Epidemiol Bd, 65-71; chmn Biol Safety Comt, Asn Lab Anal Sci, 65-71. *Honors & Awards:* Barnett L Cohen Award, Am Soc Microbiol. *Mem:* Am Soc Microbiol (pres, 74-75); Am Soc Microbiol; Am Asn Lab Animal Sci; fel Am Acad Microbiol. *Res:* Experimental aerosol induced diseases in laboratory animals; aerobiology; biological safety; immunogenesis and immunoprophylaxis of respiratory diseases. *Mailing Add:* 7922 Long Meadow Dr Frederick MD 21701

JEN, CHIH KUNG, b Chin Yuan, Shansi, China, Aug 15, 06; nat US; m 37; c 4. MICROWAVE PHYSICS. *Educ:* Mass Inst Technol, SB, 28; Univ Pa, SM, 29; Harvard Univ, PhD(physics), 31. *Prof Exp:* Asst physics, Harvard Univ, 30-32, instr, 32-33; prof, Shuntung Univ, China, 33-34; prof, Tsing Hua Univ, Peking, China, 34-37, dir radio res int, 37-45; res lectr physics, Harvard Univ, 46-50; vchmn res ctr, Johns Hopkins Univ, 58-74, William S Parsons vis prof chem physics, 66-67, physicist & prin staff mem, Appl Physics Lab, 50-76, consult, 78-87; RETIRED. *Concurrent Pos:* Fel, China Found, 31-32; hon prof, Tsinghua Univ, Beijing & Univ Sci & Technol, Hefei, People's Repub China, 78. *Mem:* Fel Am Phys Soc. *Res:* Ionosphere; quantum mechanics; electron tube phenomena; microwave spectroscopy. *Mailing Add:* 10203 Lariston Ln Silver Springs MD 20903

JEN, JOSEPH JWU-SHAN, b Sichuan, China, May 8, 39; US citizen; m 65; c 2. FOOD BIOCHEMISTRY, ACADEMIC ADMINISTRATION. *Educ:* Nat Taiwan Univ, BS, 60; Wash State Univ, MS, 64; Univ Calif, Berkeley, PhD(comp biochem), 69; Southern Ill Univ, MBA, 86. *Prof Exp:* From asst prof to prof food biochem, Clemson Univ, SC, 69-79; assoc prof, Mich State Univ, East Lansing, 79-80; mgr, food enzyme, Cambell Inst Res & Tech, Camden, NJ, mgr,vegetable biochem, 83-85, dir, biochem, 85-86; CHMN DIV FOOD SCI & TECHNOL, UNIV GA, 86- *Concurrent Pos:* Res food technologist, USDA, 75; vis prof, Nat Taiwan Univ, 76; chmn fruit & vegetables prod, Inst Food Technol, 88-89. *Mem:* Inst Food Technol; Am Chem Soc; Chinese Am Foods Soc (pres, 77-78). *Res:* Food enzymology; vegetable texture; pectin chemistry and function; fruit and vegetable shelf-life extension and quality measurements; value added product development; food biotechnology. *Mailing Add:* Food Sci & Technol Dept Univ Ga Athens GA 30602

JEN, KAI-LIN CATHERINE, b Taiwan, Repub China, July 18, 49; US citizen; m 70; c 2. OBESITY, TYPE II DIABETES. *Educ:* Wayne State Univ, PhD(nutrit), 77. *Prof Exp:* Asst res scientist, Univ Mich, 78-83; asst prof nutrit, 84-87, ASSOC PROF NUTRIT, WAYNE STATE UNIV, 87- *Mem:* Am Inst Nutrit; Am Physiol Soc; NAm Asn Study Obesity; AAAS. *Res:* Regulation of appetite and body weight; animal model of human gestational diabetes; exercise and obesity; lipid metabolism in obesity; type two diabetes and nutrition. *Mailing Add:* Dept Nutrit & Food Sci Wayne State Univ 160 Old Main Detroit MI 48202

JEN, PHILIP HUNGSUN, b Hunan, China, Jan 11, 44; US citizen; m 71. AUDITORY PHYSIOLOGY, NEUROETHOLOGY. *Educ:* Tunghai Univ BS, 67; Washington Univ, MA, 71, PhD(biol), 74. *Prof Exp:* Res assoc, Washington Univ, 74-75; from asst prof to assoc prof, 75-84, PROF NERVOUS SYST, UNIV MO, COLUMBIA, 84- *Concurrent Pos:* Vis prof, J W Goethe Univ, Frankfurt, 79; prin investr, NSF, 78- & NIH, 80-; guest lectr, Inst Acoust, Chinese Acad Sci, 80; NIH res career develop award, 80. *Mem:* Am Soc Zoologists; Acoust Soc Am; Soc Neurosci; AAAS; NY Acad Sci. *Res:* Neuroethological investigation of acoustic signal encoding, processing and control in the auditory system of echo-locating bats. *Mailing Add:* 208 Lefevre Hall Div Biol Sci Univ Mo Columbia MO 65211

JEN, SHEN, b Shanghai, China, Dec 8, 47; m 75. APPLIED PHYSICS. *Educ:* Nat Taiwan Univ, BS, 68; Harvard Univ, MS, 70, PhD(appl physics), 75. *Prof Exp:* Res assoc light scattering spectros, Dept Chem, State Univ NY Stony Brook, 75-76; Proj mgr, Xerox Corp, 77-78, mem res staff, 79-82; mem res staff, IBM, 82-84; mem tech staff, 84-87, SR MEM TECH STAFF, TEX INSTRUMENTS, 88- *Mem:* Soc Photog Scientists & Engrs; Inst Elec & Electronics Engrs. *Res:* Light scaterring spectroscopy; physics of liquid ceystals; electro-photography; magnetic storage technology; acoustic surface wave devices; microwave signal processing. *Mailing Add:* 1200 Stratford Dr Richardson TX 75080

JEN, YUN, b China, Oct 5, 27; nat US; m 51; c 2. ORGANIC CHEMISTRY. *Educ:* Shanghai Univ, BS, 48; Carnegie Inst Technol, MS, 49. *Prof Exp:* Res chemist, Am Cyanamid Co, 51-56; mgr eng, Anaheim Plant, Oronite Chem Co Div, Calif Chem Co, Standard Oil Co, Calif, 56-60; res engr, Gen Elec Co, 60-63; dir res & develop, Tenneco Chem Co, 63-75; mem staff, Chem Div, Union Camp Corp, 75-77; PRES, J J CHEM, INC, 77- *Mem:* Am Chem Soc. *Res:* Polymers; water soluble resins; alkyds; acrylics; polyesters; pulp and paper; naval stores products; ore benefication. *Mailing Add:* 16 Gale Break Circle Savannah GA 31406

JENA, PURUSOTTAM, b Orissa, India, Feb 5, 43; m 69; c 1. ATOMIC CLUSTERS, METAL DEFECTS. *Educ:* Utkal Univ, India, BSc, 64, MSc, 66; Univ Calif, Riverside, PhD(physics), 70. *Prof Exp:* Lectr physics, State Univ NY, Albany, 70-71; postdoctoral fel, Dlahousie Univ, 71-73; res assoc physics, Univ BC, Vancouver, 73-75; vis asst prof physics, Northwestern Univ, 75-77; vis scientist, Argonne Nat Lab, 77-78; assoc prof physics, Mich Inst Technol Univ, 78-80; PROF PHYSICS, VA COMMONWEALTH UNIV, 80- *Concurrent Pos:* Prog dir, Div Mats Res, Nat Sci Found, 86-87; prin investr, Nat Sci Found, 84-, Army Res Office, 85-, Dept Energy, 87-; consult, BDM Corp, 86- *Mem:* Am Phys Soc; Indian Phys Soc; Mat Res Soc. *Res:* Theoretical condensed matter physics; defects in metals; small atomic clusters; electronic structure and properties. *Mailing Add:* Physics Dept Va Commonwealth Univ Richmond VA 23284-2000

JENCKS, WILLIAM PLATT, b Bar Harbor, Maine, Aug 15, 27; m 50; c 2. BIOCHEMISTRY, ORGANIC CHEMISTRY. *Educ:* Harvard Univ, MD, 51. *Prof Exp:* Intern, Peter Bent Brigham Hosp, 51-52; res fel biochem, Res Lab, Mass Gen Hosp, 52-53; res fel pharmacol, Army Med Serv Grad Sch, Walter Reed Army Med Ctr, 53-55, chief, dept pharm, 54-55; res fel biochem, Res Lab, Mass Gen Hosp, 55-56; res fel, Harvard Univ, 56-57; from asst prof to assoc prof, 57-63, PROF BIOCHEM, BRANDEIS UNIV, 63- *Honors & Awards:* Eli Lilly Co Award, Am Chem Soc, 62. *Mem:* Nat Acad Sci; AAAS; Am Soc Biol Chem; Am Chem Soc; Am Acad Arts & Sci. *Res:* Mechanism and catalysis of carbonyl, acyl, phosphate transfer and other reactions; mechanism of enzyme action; intermolecular forces in aqueous solution; mechanism of coupled vectorial processes. *Mailing Add:* Grad Dept Biochem Brandeis Univ Waltham MA 02254

JENDEN, DONALD JAMES, b Horsham, Eng, Sept 1, 26; nat US; m 50; c 3. PHARMACOLOGY, ANALYTICAL CHEMISTRY. *Educ:* Univ London, BSc, 47, MB, BS, 50. *Hon Degrees:* Dr, Univ Uppsala, Sweden, 80. *Prof Exp:* Demonstr pharmacol, Univ London, 48-49; lectr pharmacol, Univ Calif, 50-51; actg chmn, Univ Calif, Los Angeles, 56-57, from asst prof to assoc prof, 52-67, prof & chmn, dept pharmacol, 68-89, PROF PHARMACOL & BIOMATH, UNIV CALIF, LOS ANGELES, 67- *Concurrent Pos:* Mem, Brain Res Inst, Univ Calif, Los Angeles, 61-; NSF sr fel, hon res assoc, Univ Col Univ Col, Univ London, 61-62; Wellcome vis prof, Univ Ala, Birmingham, 84- *Mem:* Soc Neurosci; Am Soc Pharmacol & Exp Therapeut; Am Physiol Soc; Am Soc Med Sch Pharmacol; fel Am Col Neuropsychopharmacol; AAAS; Am Chem Soc, div Med Chem; Am Soc Neurochem; NY Acad Sci; Physiol Soc London; Int Union Pharmacol. *Res:* Chemical and biochemical pharmacology; applications of mass spectrometry and stable isotopes in pharmacology and toxicology; cholinergic mechanisms; mathematical biology. *Mailing Add:* Dept Pharmacol Sch Med 23-273 Ctr Health Sci Univ Calif Los Angeles CA 90024-1735

JENDREK, EUGENE FRANCIS, JR, b Baltimore, Md, June 11, 49; m 73; c 2. X-RAY POWDER DIFFRACTION. *Educ:* Loyola Col, Baltimore, Md, BS, 71; Univ Conn, Storrs, MS, 73; Univ Md, College Park, PhD(chem), 79. *Prof Exp:* Sr analyst, Davison Chem Div, W R Grace & Co, 74-75; res specialist, Monsanto Res Group, 79-89, res chemist; ANAL SPECIALIST, EG&G MOUND, 89- *Mem:* Am Chem Soc; Am Crystallog Soc. *Res:* Powder X-ray diffraction computation; laboratory computer automation and data management. *Mailing Add:* EG&G Mound Appl Technol Box 3000 Miamisburg OH 45343

JENDRESEN, MALCOLM DAN, b Janesville, Wis, June 6, 33; m 54; c 1. BIOMATERIALS. *Educ:* Marquette Univ, DDS, 61; Univ Lund, Sweden, PhD(surface sci), 80. *Prof Exp:* Instr & res assoc dent mat, Marquette Univ, Wis, 61-64; chief restoration dent, USAF Sch Aerospace Med, 64-68; PROF BIOMAT SCI, SCH DENT, UNIV CALIF, SAN FRANCISCO, 68-, ASST DENT RES, 72- *Concurrent Pos:* Consult, USAF Sch Aerospace Med, 68-72, Vet Admin Hosp, San Francisco, 68-, WHO, 72-, Nordisk Inst Odontologisk Mat, 78-, Surg Gen, US Army, 79 & Dept Health & Human Serv, Pub Health Serv, Nat Inst Dent Res, NIH; vis prof, Univ Lund, Sweden, 78-79. *Mem:* Fel Int Asn Dent Res; fel Am Col Dentists; fel Int Col Dentists; fel AAAS; fel Sigma Xi. *Res:* General materials with emphasis on adhesion in biological environments; characterization of biofilm and the clinical adhesiveness of intact biological surfaces and subsequent adhesive events. *Mailing Add:* Biomat Sci 507-S Sch Dent Univ Calif San Francisco CA 94143

JENERICK, HOWARD PETER, b Cicero, Ill, May 20, 23; m 47; c 3. PHYSIOLOGY. *Educ:* Univ Chicago, PhB, 46, SB, 48, PhD(physiol), 51. *Prof Exp:* From instr biol to asst prof gen physiol, Mass Inst Technol, 51-58; exec secy, Res Training Br, Div Gen Med Sci, NIH, 58-60; assoc prof physiol, Emory Univ, 60-64; chief spec res resources br, Div Res Facil & Resources, NIH, 64-65, chief res grants br, 65-67, prog dir biophys sci, 67-72, spec asst to the dir, 72-76; chief off prog anal, Nat Inst Gen Med Sci, 76-90, DIR, EXTRAMURAL INVENTIONS OFF, NIH, 90- *Mem:* Biophys Soc; Am Physiol Soc. *Res:* Electrophysiology; scientific administration. *Mailing Add:* Off Dir NIH Bethesda MD 20894

JENG, DUEN-REN, b Taipei, Taiwan, China, Mar 1, 32; nat US; m 66; c 3. FLUID MECHANICS. *Educ:* Nat Univ Taiwan, BS, 55; Univ Ill, MS, 60, PhD(mech eng), 65. *Prof Exp:* Asst mech eng, Nat Univ Taiwan, 56-69 & Univ Ill, 60-64; asst prof, Univ Ala, 65-67; from asst prof to assoc prof, 65-77, PROF MECH ENG, UNIV TOLEDO, 77- *Mem:* Am Soc Mech Engrs; Sigma Xi; Am Inst Aeronaut & Astronaut. *Res:* Metal cutting; thermal contact resistance and transient heat transfer in laminar boundary layer; wind energy; non-newtonian flow; radiation. *Mailing Add:* Dept Mech Eng Col Eng 2801 W Bancroft Toledo OH 43606

JENG, RAYMOND ING-SONG, b Taipei, Taiwan, Jan 1, 40; m 71. HYDROLOGY, HYDRAULICS. *Educ:* Nat Taiwan Univ, BS, 62; Colo State Univ, MS, 65, PhD(civil eng), 68. *Prof Exp:* Res asst hydrol invest, Colo State Univ, 64-67; from asst prof to assoc prof, 68-79, PROF CIVIL ENG, CALIF STATE UNIV, LOS ANGELES, 79-, CHMN DEPT, 85- *Concurrent Pos:* Consult, Boise Cascade Property Inc, 70 & Los Angeles County Flood Control Dist, 71-78; vis prof, Nat Taiwan Univ, 80-83. *Mem:* Am Soc Civil Engrs; Am Geophys Union; Am Water Works Asn. *Res:* Hydrologic system analysis; statistical and stochastic hydrology. *Mailing Add:* Dept of Civil Eng 5151 State University Dr Los Angeles CA 90032

JENICEK, JOHN ANDREW, b Chicago, Ill, 22; m; c 1. ANESTHESIOLOGY. *Educ:* Univ Ill, MD, 46; Am Bd Anesthesiol, cert, 57. *Prof Exp:* Intern, St Mary Nazareth Hosp, Chicago, 46-47; resident anesthesiol, Brooke Army Hosp, 49-51; chief anesthesiol & oper serv, Tripler Army Hosp, 52-54, asst chief, 54-55; asst chief anesthesiol & oper serv, Walter Reed Army Hosp, 55-57; chief anesthesiol & oper serv, Brooke Army Hosp, 57-61; chief anesthesiol & oper serv, Walter Reed Army Hosp, 62-67; assoc prof, 67-72, PROF ANESTHESIOL, UNIV TEX MED BR, GALVESTON, 72- *Concurrent Pos:* Consult anesthesiol, Surgeon Gen, US Army, 62-67. *Mem:* AMA; Am Soc Anesthesiol; fel Am Col Anesthesiol. *Mailing Add:* 2802 Beluche Dr Galveston TX 77551

JENIKE, ANDREW W(ITOLD), b Warsaw, Poland, Apr 16, 14; nat US; m 43; c 2. MECHANICAL ENGINEERING. *Educ:* Warsaw Tech Univ, Dipl, 39; Univ London, PhD(struct eng), 48. *Hon Degrees:* DTech, Univ Bradford, Eng, 72. *Prof Exp:* Design & develop engr, Poland, Can & US, 39 & 48-51; res prof mech & mining eng & dir bulk solids flow proj, Eng Exp Sta, Univ Utah, 56-62; consult engr, 62-66; pres, Jenike & Johanson Inc, 66-79; consult engr, 80-85; RETIRED. *Concurrent Pos:* Alexander von Humboldt Found Sr scientist award, W Ger, 76. *Mem:* Am Soc Mech Engrs; Am Inst Mining, Metall & Petrol Engrs; Am Inst Mech Engrs. *Res:* Storage and flow of solids; flowability testing equipment. *Mailing Add:* Three Newcastle Dr Nashua NH 03060

JEN-JACOBSON, LINDA, b Kunming, China, Oct 29, 41; US citizen; m 67; c 2. BIOCHEMISTRY, BIOPHYSICS. *Educ:* Radcliffe Col, AB, 62; Univ Ill, MS, 65, PhD(biochem), 67. *Prof Exp:* Res assoc biol sci, Univ Pittsburgh, 67-69, lectr, 69-70, res assoc biophys, 70-80, grad fac, 84, RES ASST PROF BIOL SCI, UNIV PITTSBURGH, 81- *Res:* Physicochemical determinants of protein conformation; structure-function relationships in proteins; mechanisms of protein-nucleic acid interactions. *Mailing Add:* Dept of Biol Sci Univ of Pittsburgh Pittsburgh PA 15260

JENKIN, HOWARD M, b New York, NY, May 1, 25; div; c 3. MICROBIOLOGY. *Educ:* Univ Wis, BS, 49; Univ Chicago, PhD(microbiol), 60. *Prof Exp:* Nat Res Coun fel microbiol, Virus-Rickettsiae Div, Biol Labs, Ft Detrick, Md, 60-61, mem staff, Immunol Br, 61-62; res asst prof prev med, Sch Med, Univ Wash, 62-66; from assoc prof to prof microbiol, Med Sch, Univ Minn, Minneapolis & from assoc prof to prof, Hormel Inst, Grad Sch, 66-84, head sect, 66-84; RETIRED. *Concurrent Pos:* Head virol-tissue cult dept, US Naval Med Res Unit 2, Taiwan, 63-66. *Mem:* Am Soc Microbiol; Sigma Xi; Tissue Cult Asn; Soc Exp Biol & Med. *Res:* Comparative lipid biochemistry; biology and serology of members of Chlamydia; herpes virus; leptospires; treponema and arbovirus groups of microorganisms; tumor-lipid membrane studies. *Mailing Add:* 520 Palm Springs Blvd Apt 407 Indian Harbor Beach FL 32937

JENKINS, ALFRED MARTIN, b Boston, Mass, July 27, 17. ORGANIC CHEMISTRY. *Educ:* Tufts Univ, BS, 42; Boston Univ, AM, 47; Okla State Univ, PhD(chem), 52. *Prof Exp:* Metallurgist, Watertown Arsenal, 42; res chemist, E I du Pont de Nemours & Co, 52-60; PROF SCI, GLASSBORO STATE COL, 60- *Mem:* Am Chem Soc. *Res:* Cyclic polymerization of aldehydes. *Mailing Add:* 1212 N Main St Glassboro NJ 08028

JENKINS, ALVIN WILKINS, JR, b Raleigh, NC, Dec 30, 28; m 51; c 2. PLASMA PHYSICS, ASTROPHYSICS. *Educ:* NC State Col, BEE, 51, MS, 55; Univ Va, PhD, 58. *Prof Exp:* Sr physicist theoret physics, Ord Res Lab, Univ Va, 58-59; res physicist, Univ Res Inst, Denver, 59-61; assoc prof physics, Wichita State Univ, 61-66; assoc prof, 66-70, PROF PHYSICS, NC STATE UNIV, 70-, HEAD DEPT, 76- ACTG CHMN DEPT, 75- *Mem:* Am Phys Soc; Am Geophys Union; Am Astron Soc. *Res:* Atmospheric and magnetospheric physics; plasma physics. *Mailing Add:* Dept Physics NC State Univ Box 8202 Raleigh NC 27695

JENKINS, BURTON CHARLES, b New Westminster, BC, June 13, 20; m 47; c 4. CYTOGENETICS. *Educ:* Univ Alta, BSc, 41, MSc, 44; Univ Calif, PhD(genetics), 50. *Prof Exp:* Asst field crops, Univ Alta, 39-44; asst cerealist, Exp Sta, Swift Current, Sask, 44-48; chief asst, Lab Cereal Breeding, Exp Farm, Lethbridge, Alta, 48-50; assoc prof field husb, Univ Sask, 50-54; prof, Div Plant Sci & Rosner Chair Agron, Univ Man, 54-66; cytogeneticist & head basic res, World Seeds, Inc, 66-68; dir res, Jenkins Found Res, 68-80; sta head & sr plant breeder triticale, Dessert Seed Co, Inc, 80-85; res prof plant sci, Graceland Col, 68-85; RETIRED. *Concurrent Pos:* Mem, Wartime Bur Tech Personnel, 44-46; spec sci aide, Wheat Improv Prog, Rockefeller Found, Mex, 64. *Mem:* AAAS; Agr Inst Can; Am Soc Agron; Sigma Xi. *Res:* Crop improvement, with special reference to breeding grains; fundamental cytogenetic study in wheats and related species. *Mailing Add:* 418A Cayuga St Salinas CA 93901

JENKINS, CHARLES ROBERT, b Newton, Ill, Aug 17, 30; m 53; c 4. SANITARY ENGINEERING, AQUATIC BIOLOGY. *Educ:* Eastern Ill State Col, BS, 52; Univ Ill, MS, 59; Okla State Univ, PhD(zool), 64. *Prof Exp:* From asst prof to assoc prof sanit eng, 61-77, prof environ eng, 77-80, PROF CIVIL ENG, WVA UNIV, 80- *Mem:* Am Soc Limnol & Oceanog; Water Pollution Control Fedn; Am Water Works Asn; Sigma Xi. *Res:* Water pollution control; waste treatment. *Mailing Add:* 432 Wilburn Morgantown WV 26506

JENKINS, DALE WILSON, b Wapakoneta, Ohio, June 17, 18; m 42; c 5. ECOLOGY, ENVIRONMENTAL SCIENCES. *Educ:* Ohio State Univ, BSc, 38, MA, 39, PhD, 47. *Prof Exp:* Ecologist, Soil Conserv Serv, 35; instr, Ohio State Univ, 38-39, Univ Chicago, 39-40, Univ Ill, 40-41 & Univ Minn, 41-42; agr specialist, Foreign Econ Admin, Washington, DC, 42-43; entomologist & chief animal ecol br, Army Med Labs, Md, 46-52, dep chief allied sci div, 54-56, chief entom div, 53-62; chief environ biol prog, NASA Hq, 62-66, asst dir biosci progs, 66-70; dir ecol prog, Smithsonian Inst, 70-74; dep dir, Ctr Human Ecol & Health, Pan Am Health Orgn, 75-78; ECOL & ENVIRON CONSULT, WHO WORLD HEALTH ORGN, USAID, WORLDBANK, UN DEVELOP PROG, INTER-AM DEVELOP BANK, 78- *Concurrent Pos:* Lectr, Sch Pub Health & Hyg, Johns Hopkins Univ, 50-; consult, USDA, Alaska, 47, USPHS, 48, Northern Insect Surv, Defence Res Bd Can, 49-50 & USAF, 59-; planning conf partic, Life Sci Prog, NASA, 60 & WHO, Bangkok, 60; mem, Armed Forces Pest Control Bd, 55-64 & Interdept Pest Control Comt, 58-64; chmn bd gov, Inst Lab Animal Resources, Nat Res Coun, 55-60, adv to UNESCO, 57-; US Dept State deleg, Int Conf Peaceful Uses Atomic Energy, Geneva, 55; Nat Acad Sci-Nat Res Coun deleg, Int Cong Entom, Montreal, 56. *Honors & Awards:* Distinguished Serv Award, Ohio State Univ. *Mem:* Fel AAAS; Ecol Soc Am; Lepidopterists Soc. *Res:* Ecology of plants and animals; radioisotope tracers; laboratory animals; epidemiology; environmental impacts. *Mailing Add:* 3028 Tanglewood Dr Sarasota FL 34239

JENKINS, DAVID A, b Seattle, Wash, Dec 28, 37. INTERMEDIATE ENERGY PHYSICS. *Educ:* Yale Univ, BE, 59; Univ Calif, Berkeley, MS, 61, PhD(physics), 64. *Prof Exp:* Res asst, Lawrence Berkeley Lab, 67; dir, NSF, 73-74; assoc prof, 67-73, PROF PHYSICS, VA TECH, 74- *Mem:* Sigma Xi; Am Phys Soc. *Res:* Intermediate energy physics; mesonic atoms; pion-nucleon scattering; pion production; photodisintegration of light nuclei. *Mailing Add:* Physics Dept Va Tech Blacksburg VA 24061-0435

JENKINS, DAVID ISAAC, b Shropshire, Eng, Oct 4, 35; m 60; c 2. SANITARY ENGINEERING. *Educ:* Univ Birmingham, BSc, 57; Univ Durham, PhD(sanit eng), 60. *Prof Exp:* Res chemist, 60-61, from asst prof to assoc prof, 63-74, PROF SANIT ENG, UNIV CALIF, BERKELEY, 74-, DIR, SANIT ENG RES LAB, 61- *Concurrent Pos:* Sabbatical leave, Dept Eng & Appl Physics, Harvard Univ, 69-70. *Mem:* Am Chem Soc; Water

Pollution Control Fedn; Asn Environ Eng Prof; fel Royal Inst Chem; Brit Inst Water Pollution Control. *Res:* Chemistry and biochemistry of processes and phenomena associated with the control of environment, especially the upgrading of water quality; biological waste treatment processes; activated sludge operation. *Mailing Add:* 11 Yale Circle Kensington CA 94708

JENKINS, DAVID JOHN ANTHONY, m; c 1. DIABETES, HYPERLIPIDEMIA. *Educ:* Oxford Univ, Eng, PhD(clin nutrit), 71, MD, 76, DSc, 86. *Prof Exp:* PROF MED & CLIN NUTRIT , DEPT NUTRIT SCI & DEPT MED, UNIV TORONTO & ST MICHAEL'S HOSP, 80- *Honors & Awards:* Borden Award, Can, 83; Goldsmith Award, Am Col Nutrit, 85. *Mem:* Am Inst Nutrit; Am Soc Clin Nutrit. *Mailing Add:* Fac Med Univ Toronto 150 College St Toronto ON M5S 1A8 Can

JENKINS, DAVID R(ICHARD), b Lima, Ohio, Oct 24, 24; m 47; c 3. STRUCTURAL MECHANICS, COMPOSITE MATERIALS. *Educ:* Case Inst Technol, BSc, 48; Ohio State Univ, MSc, 54; Univ Mich, PhD(eng mech), 62. *Prof Exp:* Stress analyst, Airplane Div, Curtiss-Wright Corp, 48, tech asst, Battelle Mem Inst, 48-50, prin mech engr, 50-55, proj leader, 55-58; instr eng mech, Univ Mich, 58-62, asst prof, 62-65; sr res fel, Tech Ctr, Owens-Corning Fiberglas Corp, 65-69; from assoc prof to prof & chmn engr mech & mat sci, Univ Cent Fla, 69-75, prof civil eng & environ sci & actg chmn, 75-76, actg chmn mech eng & aerospace sci, 81-82, chmn civil eng & environ sci, 84-90, PROF ENG, UNIV CENT FLA, 76- *Concurrent Pos:* Fac res grant, Univ Mich, 65; on sabbatical leave, mat res eng, Nat Bur Standards, Ctr for Bldg Technol, 78-80; sabbatical, Univ Aalborg, Denmark & Polytech, Wales, 90. *Mem:* AAAS; Am Soc Civil Engrs; Soc Exp Mech Engrs; Am Soc Metals; Soc Eng Sci; Am Soc Mech Engrs. *Res:* Structural testing, composite materials; high temperature structural behavior; crack propagation in steel shells; aircraft structural investigations; yielding and strain hardening in metallic materials; composite materials. *Mailing Add:* Univ Cent Fla Col Eng PO Box 25000 Orlando FL 32816

JENKINS, EDGAR WILLIAM, b Columbus, Ohio, Apr 29, 33; m 59; c 3. HIGH ENERGY PHYSICS. *Educ:* Harvard Univ, AB, 55; Columbia Univ, PhD(physics), 62. *Prof Exp:* From asst physicist to assoc physicist, Brookhaven Nat Lab, 60-64; from asst prof to assoc prof, 64-71, PROF PHYSICS, UNIV ARIZ, 71- *Mem:* Am Phys Soc; Am Asn Physics Teachers. *Res:* Interactions, properties and decays of elementary particles. *Mailing Add:* Dept of Physics Univ of Ariz Tucson AZ 85721

JENKINS, EDWARD BEYNON, b San Francisco, Calif, Mar 20, 39; m 63; c 2. ASTROPHYSICS. *Educ:* Univ Calif, Davis, BA, 62; Cornell Univ, PhD(physics), 66. *Prof Exp:* Res assoc astrophys, 66-67, res staff mem, 67-73, res astronomer, 73-79, SR RES ASTRONOMER, PRINCETON UNIV OBSERV, 79- *Concurrent Pos:* Prin Investr of a sounding rocket res prog, 80-; mem Comt on Space Astron & Astrophys, 86-88; mem, Sci definition teams, Space Telescope Imaging Spectrog & Far Ultraviolet Spectros Explorer, 87- *Mem:* Am Astron Soc; Int Astron Union. *Res:* Rocket and satellite ultraviolet astronomy; interstellar medium; image sensor development. *Mailing Add:* Princeton Univ Observ Princeton NJ 08544-1001

JENKINS, EDWARD FELIX, b Baltimore, Md, Aug, 17, 06. ASTRONOMY. *Educ:* Villanova Col, AB, 27; Cath Univ, MS, 33, PhD(chem), 39. *Hon Degrees:* DSc, Villanova Univ, 77. *Prof Exp:* From instr to prof chem, 35-60, head dept astron, 61-74, PROF ASTRON, VILLANOVA UNIV, 58- *Mem:* Am Chem Soc; Am Astron Soc; Royal Astron Soc Can; Sigma Xi. *Res:* Reagents for organic analysis; catalytic hydrogenation; history of chemistry and astronomy. *Mailing Add:* Dept of Astron Villanova Univ Villanova PA 19085

JENKINS, FARISH ALSTON, JR, b New York, May 19, 40; m 63; c 2. VERTEBRATE PALEONTOLOGY, ANATOMY. *Educ:* Princeton Univ, AB, 61; Yale Univ, MSc, 66, PhD(geol), 68; Harvard Univ, MA, 74. *Prof Exp:* From instr to asst prof anat, Col Physicians & Surgeons, Columbia Univ, 68-71; assoc prof biol & assoc cur vert paleont, 71-74, PROF BIOL, DEPT ORGANISMIC & EVOLUTIONARY BIOL & CUR VERT PALEONT, MUS COMP ZOOL, HARVARD UNIV, 74-, PROF ANAT, HARVARD-MASS INST TECHNOL DIV HEALTH SCI & TECHNOL, HARVARD MEDICAL SCH, 82- *Res:* Vertebrate anatomy and evolution, especially reptiles and mammals; biomechanics of musculoskeletal system. *Mailing Add:* Mus Comp Zool Labs Harvard Univ Cambridge MA 02138

JENKINS, FLOYD ALBERT, b Los Angeles, Calif, Aug 14, 16. COMPARATIVE ANATOMY, VERTEBRATE PALEONTOLOGY. *Educ:* St Louis Univ, AB, 40, MA, 42, MS, 43, PhD, 54; Alma Col, Calif, STL, 49. *Prof Exp:* Lab instr biol, St Louis Univ, 41-43; instr, 43-45 & 53-55, asst prof, 55-60, assoc prof, 60-77, PROF BIOL, LOYOLA MARYMOUNT UNIV, 77- *Mem:* Paleont Soc; Soc Study Evolution; Soc Vert Paleont. *Res:* Early evolution of mammals. *Mailing Add:* 7101 W 80th Los Angeles CA 90045

JENKINS, GEORGE ROBERT, b Denver, Colo, Dec 14, 14; m 41; c 2. GEOLOGY. *Educ:* Univ Colo, AB, 36; Univ Wis, PhM, 38. *Prof Exp:* Asst geog, Univ Colo, 36-37; asst, Univ Wis, 37-40; instr, Univ Mo, 40-42; officer in chg, US Weather Bur, Madison, Wis, 46-47; from asst prof to assoc prof geol, 48-63, from asst dir to assoc dir, Inst Res, 56-63, prof geol, 63-80, EMER DIR, INST RES, 80- *Mem:* AAAS; Am Meteorol Soc; Am Geophys Union. *Res:* Water resources. *Mailing Add:* 2121 Greenleaf St Allentown PA 18104

JENKINS, HOWARD BRYNER, b Arimo, Idaho, Jan 30, 28; m 51; c 3. MATHEMATICS. *Educ:* Mass Inst Technol, BS, 50; Univ Southern Calif, PhD(math), 58. *Prof Exp:* Lectr math, Univ Southern Calif, 54-57; instr, Calif Inst Technol, 57-58; temp mem, Inst Math Sci, NY Univ, 58-59, res assoc, 59-60; vis asst prof math, Stanford Univ, 60-61; asst prof, 61-65, ASSOC PROF MATH, UNIV MINN, MINNEAPOLIS, 65-, ASSOC HEAD SCH MATH, 71- *Concurrent Pos:* Vis assoc prof, Stanford Univ, 66-67. *Mem:* Am Math Soc. *Res:* Partial differential equations; variational problems; minimal surfaces. *Mailing Add:* Sch Math Inst Technol Univ Minn 206 Church St SE Minneapolis MN 55455

JENKINS, HOWARD JONES, b Oak Hill, Ohio, Sept 14, 16; m 51; c 4. PHARMACOLOGY. *Educ:* Ohio State Univ, PhD(pharmacol), 50. *Prof Exp:* Res pharmacologist, Armour Labs, 51-53; from asst prof to prof, Mass Col Pharm & Allied Health Sci, 53-84, dir, Div Pharmacol & Allied Sci, 64-73, emer prof, 84-; RETIRED. *Mem:* AAAS; Acad Pharmaceut Sci; Sigma Xi; Am Pharmaceut Asn. *Res:* Cardiovascular pharmacology; autonomic pharmacology; structure-activity relationships involved in antispasmodic and antihistaminic responses; analgetic potentiation; biological assay. *Mailing Add:* 24 Brooks Rd Wayland MA 01778

JENKINS, HUGHES BRANTLEY, JR, b Jacksonville, Fla, Oct 17, 27. THEORETICAL PHYSICS. *Educ:* Univ Ga, AB, 48, MS, 55; Univ Ky, PhD(physics), 63. *Prof Exp:* Instr math & physics, Univ Ga, 52-55; asst physicist, Oak Ridge Nat Lab, Union Carbide Corp, Tenn, 55-57, asst mathematician, 57-58; intr physics, Univ Ky, 58-62; assoc prof, Ga State Col, 62-68; assoc prof physics & astron, Valdosta State Col, 68-83; RETIRED. *Res:* Statistical mechanics and mathematical physics. *Mailing Add:* 810 Millpond Rd Valdosta GA 31602

JENKINS, JAMES ALLISTER, b Toronto, Ont, Sept 23, 23; nat US. MATHEMATICS. *Educ:* Toronto, BA, 44, MA, 45; Harvard Univ, PhD(math), 48. *Prof Exp:* Jewett fel, Harvard Univ, 48-49 & Inst Advan Study & Princeton Univ, 49-50; asst prof math, Johns Hopkins Univ, 50-54; from assoc prof to prof, Univ Notre Dame, 54-59; PROF MATH, WASHINGTON UNIV, 59- *Concurrent Pos:* Mem, Inst Advan Study, 57-59, 61-62, 73-74 & 80-81; Fulbright vis prof, Imp Col, Univ London, 62. *Mem:* Am Math Soc; Math Soc France; Ger Math Asn. *Res:* Geometrical and analytical function theories; topological theory functions. *Mailing Add:* Dept Math Washington Univ St Louis MO 63130

JENKINS, JAMES WILLIAM, b Jamestown, NY, May 5, 21; c 3. ORGANIC CHEMISTRY. *Educ:* Allegheny Col, BS, 44; Univ Buffalo, MS, 48, PhD(chem), 50. *Prof Exp:* Asst prof chem, Lafayette Col, 49-51; res anal chemist, Gen Aniline & Film Corp, 51-52; res anal chemist, Colgate Palmolive Co, 52-54, group leader anal sect, 54-58, sr group leader, 59-60, sect head, 60-63, res mgr, 63-64; dir, Pfizer Inc, 65-69, vpres res & Develop, consumer prod Div, 69-83; RETIRED. *Mem:* Am Chem Soc; Sigma Xi. *Res:* Product development; proprietary pharmaceuticals; toiletries and cosmetics; hair and skin research. *Mailing Add:* 135 Cheeskogili Way Louden TN 37774-2524

JENKINS, JEFF HARLIN, b Gamaliel, Ky, Mar 8, 37; m 59; c 2. PLANT PATHOLOGY. *Educ:* Western Ky Col, BS, 59; La State Univ, MS, 61, PhD(plant path), 63. *Prof Exp:* From asst prof to assoc prof bot, 63-74, PROF BIOL, WESTERN KY UNIV, 74- *Mem:* Am Phytopath Soc. *Res:* Taxonomic mycology; fusarium wilt of alfalfa; bacterial leaf spot of bell pepper. *Mailing Add:* Dept of Biol Western Ky Univ Bowling Green KY 42101

JENKINS, JIMMY RAYMOND, b Selma, NC, Mar 18, 43; m 65; c 2. BIOLOGICAL STRUCTURE. *Educ:* Elizabeth City State Univ, BS, 65; Purdue Univ, MS, 70, PhD(biol educ), 72. *Prof Exp:* Teacher high schs, Md, 65-69; fel biol, Purdue Univ, 69-70, teaching fel & res asst instrnl develop, 70-72; asst prof biol & asst acad dean, 72-74, ASSOC PROF BIOL & DEAN, ELIZABETH CITY STATE UNIV, 73-, V CHANCELLOR ACAD AFFAIRS, 77- *Concurrent Pos:* Adv coun mem, Albemarle Regional Planning & Develop Comn, 74-; proposal reviewer, NSF, 74; individualized instr, Region 15, Northeastern NC, 75; instrnl consult, Halifax County Schs, NC, 75-76; mem, Health Manpower Develop Corp. *Mem:* Nat Asn Res Sci Teaching. *Res:* Instructional development and design geared to biology and the facilitation of biological concepts. *Mailing Add:* Elizabeth City State Univ Parkview Dr Elizabeth City NC 27909

JENKINS, JOE WILEY, b Bronaugh, Mo, Oct 17, 41; m 75; c 5. NON-ABELIAN HARMONIC ANALYSIS. *Educ:* Univ Ill, Urbana, PhD(math), 68. *Prof Exp:* Asst prof to assoc prof, 68-80, PROF MATH, STATE UNIV NY ALBANY, 80- *Concurrent Pos:* Mem, Inst Advan Study, 71-72; vis prof, Univ Würzburg, 78-79. *Mem:* Am Math Soc. *Res:* Analysis on lie groups; representation theory of nilpotent lie groups. *Mailing Add:* Dept Math State Unv NY 1400 Washington Ave Albany NY 12222

JENKINS, JOHN BRUNER, b Springfield, Mass, July 20, 41; m 63. GENETICS. *Educ:* Utah State Univ, BS, 64, MS, 65; Univ Calif, Los Angeles, PhD(zool), 68. *Prof Exp:* Asst prof, 68-74, assoc prof, 74-80, PROF & CHMN BIOL, SWARTHMORE COL, 80- *Mem:* AAAS; Genetics Soc Am; Am Genetic Asn. *Res:* Chemical mutagenesis in Drosophila and its relation to genetic fine structure. *Mailing Add:* Dept of Biol Swarthmore Col Swarthmore PA 19081

JENKINS, JOHNIE NORTON, b Barton, Ark, Nov 3, 34; m 59; c 2. PLANT GENETICS, AGRONOMY. *Educ:* Univ Ark, BSA, 56; Purdue Univ, MS, 58, PhD(genetics), 60. *Prof Exp:* Res assoc agron, Univ Ill, 60-61; res geneticist, 61-80, DIR CROP SCI RES LAB, AGR RES SERV, USDA, 80- *Concurrent Pos:* Prof crop sci & mem grad fac, Miss State Univ, 64- *Honors & Awards:* Mobay Cotton Res Recognition Award. *Mem:* Am Soc Agron; Crop Sci Soc Am; Entom Soc Am; AAAS. *Res:* Host plant resistance to cotton insects and nematodes; investigations of basic causes of insect and nematode resistance in cotton plants and development of factors which will confer resistance. *Mailing Add:* Crop Sci Res Lab PO Box 5367 Mississippi State MS 39762

JENKINS, KENNETH DUNNING, b New York, NY, Apr 8, 44; c 3. DEVELOPMENTAL BIOLOGY. *Educ:* Calif State Univ, Northridge, BA, 66; Univ Calif, Los Angleles, PhD(develop biol), 70. *Prof Exp:* From asst prof to assoc prof biol, 70-80, PROF BIOL, CALIF STATE UNIV, LONG BEACH, 80- *Concurrent Pos:* Dep assoc vpres acad affairs, Molecular Ecol Inst, 78-80, dir of inst, 82-; mem panel on fate & effects of drilling fluids on the marine environ, Nat Acad Sci, 82-83; sci adv bd, US Environ Protection Agency, 82- *Mem:* AAAS; Soc Develop Biol. *Res:* Molecular ecology; aquatic toxicology. *Mailing Add:* Dept Biol Calif State Univ Long Beach CA 90840

JENKINS, KENNETH JAMES WILLIAM, b Montreal, Que, Oct 1, 29; m 69; c 1. BIOCHEMISTRY, NUTRITION. *Educ:* McGill Univ, BSc, 51; Univ Sask, MSc, 53; Univ Wis, PhD(biochem), 58. *Prof Exp:* Head res & develop emergency rations, Defense Res Med Labs, Dept Nat Defense, Can, 53-54; asst biochem, Univ Wis, 54-58; asst prof biochem & nutrit, Ont Agr Col, Guelph, 58-65; res officer, 65-73, head, Trace Mineral Nutrit Sect, Trace Minerals & Pesticide Div, 73-80, PRIN SCIENTIST, ANIMAL RES CTR, CENT EXP FARM, CAN DEPT AGR, 80- *Honors & Awards:* Borden Award, Can, 74; Medal for Excellence in Nutrit, Can Packers, 84. *Mem:* Can Nutrit Soc; Can Biochem Soc; Can Soc Animal Sci; Am Dairy Sci Asn. *Res:* Nutritional requirements of animals; biochemical role of mineral elements; tocopherol and selenium metabolism; neonatal nutrition and metabolism. *Mailing Add:* Animal Res Ctr Can Dept of Agr Ottawa ON K1A 0C6 Can

JENKINS, LAWRENCE E, b Salt Lake City, Utah, Mar 12, 33; m. MECHANICAL ENGINEERING. *Educ:* Univ Utah, BEE & MEE. *Hon Degrees:* AA, Austin Community Col, 89. *Prof Exp:* Vpres & gen mgr, Austin Div, Lockheed Missiles & Space Co; RETIRED. *Concurrent Pos:* Dir bd, Austin Symphony, 83- *Honors & Awards:* Significant Achievement Award, USAF, 77. *Mem:* Nat Acad Eng. *Res:* Education. *Mailing Add:* 3101 Sweet Gum Cove Austin TX 78735

JENKINS, LEONARD CECIL, b Vancouver, BC, June 23, 26; m 66. ANESTHESIOLOGY, PHARMACOLOGY. *Educ:* Univ BC, BA, 48; McGill Univ, MD, CM, 52; FRCP(C), 59. *Prof Exp:* McLaughlin travel fel, 58-59; clin instr, 59-61, clin asst prof, 61-67, assoc prof, 67-70, ASSOC PROF PHARMACOL, FAC MED, UNIV BC, 68-, PROF ANESTHESIA & HEAD DEPT, 70-; DIR ANESTHESIA, VANCOUVER GEN HOSP, 70- *Concurrent Pos:* Med Res Coun res grant, 66-68. *Mem:* Am Soc Anesthesiol; Can Anesthetists Soc. *Res:* Anesthesia and the central nervous system; mechanisms of anesthesia; shock. *Mailing Add:* 7084 Balaclaua St Vancouver BC V6N 1M5 Can

JENKINS, LESLIE HUGH, b Bryson City, NC, Sept 26, 24; m 51; c 2. SURFACE PHYSICS. *Educ:* Univ NC, BS, 49, PhD(phys chem), 54. *Prof Exp:* Group leader, Va-Carolina Chem Corp, 54-56; mem res staff, 56-73, SECT HEAD, OAK RIDGE NAT LAB, 73- *Mem:* AAAS; Am Phys Soc; Sigma Xi. *Res:* Surface physics; secondary electron emission and Auger spectroscopy; low energy electron diffraction; particle-solid interactions at surfaces. *Mailing Add:* 817 Whirlaway Circle Knoxville TN 37923

JENKINS, MAMIE LEAH YOUNG, b Washington, DC, July 10, 40; m 73. NUTRITION, BIOCHEMISTRY. *Educ:* Howard Univ, BS, 62, MS, 65, PhD(nutrit), 80. *Prof Exp:* Chemist, Agr Res Serv, US Dept Agr, 64-67; RES CHEMIST, DIV NUTRIT, FOOD & DRUG ADMIN, BUR FOODS, 67- *Mem:* Animal Nutrit Res Coun; Am Chem Soc; Am Inst Nutrit. *Res:* Protein quality, amino acid fortification, amino acid derivatives and vitamins; emphasis on the metabolic role of lecithin as a dietary choline source, and its interrelationships with other nutrients. *Mailing Add:* 1658 Hobart St NW Washington DC 20009

JENKINS, MELVIN EARL, b Kansas City, Mo, June 24, 23; m 48; c 3. METABOLISM. *Educ:* Univ Kans, AB, 44, MD, 46. *Prof Exp:* From instr to assoc prof pediat, Col Med, Howard Univ, 50-69; prof, Col Med, Univ Nebr Med Ctr, Omaha, 69-73; prof chmn dept pediat & child health, 73-86, EMER PROF PEDIAT, COL MED, HOWARD UNIV, 86- *Mem:* Am Acad Pediat; Sigma Xi; Am Pediat Soc, Soc for Pediat Res. *Res:* Gonadal function; human growth and development; sickle cell hemoglobin; fetal and newborn physiology; steroid metabolism. *Mailing Add:* 10401 Grosvenor Pl Apt 504 Rockville MD 20852

JENKINS, PHILIP WINDER, b Birmingham, Ala, Nov 7, 33; m 56; c 4. ORGANIC CHEMISTRY. *Educ:* Univ Ill, BS, 55; Mass Inst Technol, PhD(chem), 59. *Prof Exp:* Asst, Mass Inst Technol, 55-59; res chemist, 59-61, sr res chemist, 61-65, res assoc, 65-68, lab head, 68-71, res assoc, Emulsion Res Div, Res Labs, 71-74, res assoc, Anal Sci Div, Res Labs, 74-80, TECH ASSOC, HEALTH SAFETY & HUMAN FACTORS LAB, EASTMAN KODAK CO, 78- *Concurrent Pos:* Inst fel, Mass Inst Technol, 59. *Mem:* Am Chem Soc; Royal Inst Chem; Soc Environ Toxicol & Chem; Sigma Xi; NY Acad Sci. *Res:* Cyclooctatetraene derivatives; proximity effects in medium ring compounds; gas chromatography; heterocycles; photographic sensitizing dyes; photochemistry; excited state energy processes; environmental fate and effects of chemicals. *Mailing Add:* 49 Dorvid Rd Rochester NY 14617

JENKINS, RICHARD LEOS, b Brookings, SDak, June 3, 03; m 27; c 3. PSYCHIATRY. *Educ:* Stanford Univ, BA, 25; Rush Med Col, MD, 30. *Prof Exp:* Intern med, surg, pediat & obstet, Res & Educ Hosps, Chicago, 29-30; instr physiol, Univ Chicago, 30-32; pediatrician, Inst Juvenile Res, Chicago, 32-36; Rockefeller Found fel psychiat, Henry Phipps Clin, Johns Hopkins Hosp, 36-37; psychiatrist, NY State Training Sch Boys, 38-41; psychiatrist, Mich Child Guid Inst, Ann Arbor, 41-43; psychiatrist, Inst Juvenile Res, Chicago, 43-44, actg dir, 44-46; psychiatrist, Health Serv, Univ Ill, 46-49; chief psychiat res, Cent Off Vet Admin, Washington DC, 49-55, dir psyhiat eval proj, 55-61; prof & chief, Child Psychiat Serv, Univ Iowa, 61-71, EMER PROF CHILD PSYCHIAT, UNIV IOWA, 71- *Concurrent Pos:* From asst prof to assoc prof criminol, social hyg & med jurisp, Col Med, Univ Ill, 43-46, actg head dept, 44-46, assoc prof psychiat, 45-49; mem comt area-wide planning facilities for ment retarded, Dept Health, Educ & Welfare, 62-64, mem comt ment retardation grant rev, 65-66; consult, Iowa Training Sch Boys, Eldora, 71-83 & Ment Health Inst, Mt Pleasant, 71-81. *Mem:* Fel Am Orthopsychiat Asn; Am Psychiat Asn; Am Acad Child Psychiat. *Res:* Diagnostic grouping and treatment in child psychiatry; juvenile delinquency. *Mailing Add:* Dept Psychiat Univ Iowa Iowa City IA 52242

JENKINS, ROBERT ALLAN, b Logan, Utah, Apr 1, 34; m 56; c 2. CELL BIOLOGY. *Educ:* Utah State Univ, BS, 57; Syracuse Univ, MS, 61; Iowa State Univ, PhD(cell biol), 64. *Prof Exp:* Teacher, Jr High Sch, Utah, 56-60, instr & assoc cell biol & electron micros, Iowa State Univ, 63-66; from asst prof to assoc prof zool, 66-74, PROF ZOOL, UNIV WYO, 74- *Mem:* AAAS; Am Soc Cell Biol; Soc Protozool. *Res:* Use of electron microscopy, cytochemistry and biochemical techniques for cytological studies of filamentous structures related to morphogenetic processes typical of dividing, regenerating and excysting ciliates. *Mailing Add:* Dept Zool Univ Wyo Laramie WY 82071

JENKINS, ROBERT BRIAN, b Oct 23, 55; m; c 2. CYTOGENETICS. *Educ:* Northwestern Univ, BA, 77; Univ Chicago, PhD(develop biol), 81, MD, 83. *Prof Exp:* Resident clin path, Mayo Grad Sch Med, 86, fel hematopath/cytogenetics, 87, instr, Lab Med, Mayo Med Sch, 87-89, sr assoc consult, Dept Lab Med & Path, Mayo Clin & Found, 87-90, ASSOC DIR, CYTOGENETICS LAB, MAYO CLIN, 87-, ASST PROF LAB MED, MAYO MED SCH, 89- *Concurrent Pos:* Assoc mem, Biochem & Molecular Biol Dept, Mayo Grad Sch, 90- & consult, Dept Lab Med & Path, Mayo Clin & Found, 90- *Mem:* AAAS; AMA; Am Soc Clin Pathologists; Col Am Pathologists; Am Soc Human Genetics; Am Asn Pathologists; Int Acad Path; Sigma Xi. *Mailing Add:* Cytogenetics Lab Mayo Clin Rochester MN 55905

JENKINS, ROBERT EDWARD, b Baltimore, Md, June 2, 38; m 70; c 2. ANALOG SUBTHRESHOLD VERY LARGE SCALE INTEGRATION CIRCUIT DESIGN, NEURAL NETWORKS. *Educ:* Univ Md, BS, 60, MS, 66. *Prof Exp:* PRIN ENGR, JOHNS HOPKINS APPL PHYSICS LAB, 61- *Concurrent Pos:* Vis scientist, Defense Mapping Agency, Wash, DC, 78-79; vis prof, Johns Hopkins Univ, 84-85, lectr elec eng, 84-91, prog coordr, Sch Continuing Prof Studies, 86-91. *Mem:* Inst Elec & Electronics Engrs. *Res:* Implementation of advanced sensors and neural networks using analog very large scale integration circuits silicon methods for innovative information processing. *Mailing Add:* 7987 Aladdin Dr Laurel MD 20723

JENKINS, ROBERT ELLSWORTH, JR, b Lewistown, Pa, Sept 30, 42; m 64; c 2. ECOLOGICAL CONSERVATION. *Educ:* Rutgers Univ, AB, 64; Harvard Univ, PhD(biol), 70. *Prof Exp:* VPRES SCI PROGS, NATURE CONSERV, 70- *Concurrent Pos:* Mem US comt, Conserv Ecosysts Sect, Int Biol Prog, 70-75, AAAS, 72-73, Rep biol sect comt, 71-74; mem, Fed Comt Res Natural Areas, 70-; res assoc, Smithsonian Inst, 71-72; assoc dir, Ctr Appl Res & Environ Sci, 71-73; mem, US Nat Comn, UNESCO, 74-76; mem bd, Rare Animal Relief Effort, 74-76; founder & nat dir, State Natural Heritage Progs, 75-; mem, US Man & the Biosphere Directorate, Proj 8, 78-; mem adv coun, Ctr Plant Conserv, 84-, Inst Conserv Biol, 84-; counr, Xerces Soc, 85-; sci adv, WNET Nature Ser, 86-; mem bd, Soc Cons Biol, 88-; mem bd gov, Am Inst Biol Sci, 70-84, coun, 84-89. *Honors & Awards:* Am Motors Prof Conserv Award, 78. *Mem:* Fel AAAS; Am Inst Biol Sci; Ecol Soc Am; Soc Study Evolution; Wildlife Soc. *Res:* Animal and plant ecology and evolution; human population and environment; applied research in land management, ecosystem preservation and restoration; ecological inventory and data banking. *Mailing Add:* The Nature Conservancy 1815 N Lynn St Arlington VA 22209-2016

JENKINS, ROBERT GEORGE, b Gwent, Wales, Sept 29, 44; m 69; c 2. COAL CONVERSION FUNDAMENTALS, MATERIALS CHARACTERIZATION. *Educ:* Univ Leeds, Eng, BSc, 67, PhD(fuel sci), 70. *Prof Exp:* Res assoc, Dept Mat Sci, Pa State Univ, 70-73; res fel chem, Imperial Col Sci & Technol, Eng, 73-75; sr res assoc, Dept Mat Sci & Eng, 75-78, from asst prof to prof fuel sci, Pa State Univ, 78-88; ASSOC DEAN RES & PROF CHEM ENG, MAT SCI & ENG, COL ENG, UNIV CINCINNATI, 88- *Mem:* Fel Inst Energy; Sigma Xi; Am Chem Soc; Combustion Inst. *Res:* Coal conversion chemistry; modification and characterization of carbons and zeolites as molecular sciences and absorbants. *Mailing Add:* Mail Location 18 Baldwin Hall Col Eng Univ Cincinnati Cincinnati OH 45221-0018

JENKINS, ROBERT M, b Kansas City, Mo, June 18, 23; m 56; c 4. FISH BIOLOGY. *Educ:* Univ Okla, BS, 48, MS, 49. *Prof Exp:* Regional fishery biologist, Okla Game & Fish Dept, 49-50; dir fishery res, Okla Fishery Res Lab, 52-57; asst exec vpres fish conserv, Sport Fishing Inst, 58-62; dir nat reservior res prog, US Fish & Wildlife Serv, 63-83; sr scientist, Aquatic Ecosystem Analysts, 83-88; CONSULT, 89- *Concurrent Pos:* Mem panel fishery experts, Food & Agr Orgn, 63- *Mem:* Am Fisheries Soc, (pres, 70-71); fel Am Inst Fish Res Biol. *Res:* Large reservoir fish production nationally as influenced by various environmental parameters. *Mailing Add:* PO Box 4188 Fayetteville AR 72702-4188

JENKINS, ROBERT WALLS, JR, b Richmond, Va, June 12, 36; m 58; c 3. RADIOCHEMISTRY, PLANT ECOLOGY. *Educ:* Va Mil Inst, BS, 58; Purdue Univ, MS, 61; Calif Western, PhD, 80. *Prof Exp:* Instr chem, Purdue Univ, 58-60; asst prof, Va Mil Inst, 60-61; chief nuclear chem div, Nuclear Defense Lab, 61-63; res scientist, Naval Res Lab, 63-65; res scientist, 65-67, chief radiochem group, 67-76, ASSOC PRIN SCIENTIST, PHILIP MORRIS INC RES CTR, 76- *Honors & Awards:* Philip Morris Award Distinguished Achievement Tobacco Sci, 75. *Mem:* Am Chem Soc; Am Nuclear Soc. *Res:* Radioisotopes; biosynthetic production of radiochemicals and their use in experimentation; gas radiochromatography; neutron activation analysis; smoke formation mechanisms; smoke aerosol generation; smoke chemistry; neutron radiography. *Mailing Add:* Philip Morris Inc Res Ctr PO Box 26583 Richmond VA 23261

JENKINS, RONALD LEE, b Atlanta, Ga, Oct 24, 52; m 78; c 2. COMPARATIVE ANIMAL PHYSIOLOGIST, CELLULAR BIOLOGY-ENZYMOLOGY. *Educ:* Carson-Newman Col, BS, 74; Auburn Univ, MS, 76, PhD(anat-physiol), 80. *Prof Exp:* Asst prof biol, La Col, 79-81; res chemist diabetes, Vet Admin Med Ctr, 81-88; asst prof endocrinol, Univ Ala, Birmingham, 85-88; ASSOC PROF BIOL, SAMFORD UNIV, BIRMINGHAM, 88- *Concurrent Pos:* Prin investr, res grants Am Diabetes Asn, 85-88; lectr, Ala Gov Sch, 86-; chmn, curric renewal, Samford Univ, 90-91. *Mem:* Am Physiol Soc; Am Diabetes Asn. *Res:* Metabolic dysfunction of the heart of diabetic animals and humans; changes in isoenzymes and substrate specificities for the enzymes of nuclectide catabolism. *Mailing Add:* Dept Biol Samford Univ 800 Lakeshore Dr Birmingham AL 35229

JENKINS, SAMUEL FOREST, JR, plant pathology; deceased, see previous edition for last biography

JENKINS, TERRY LLOYD, b Beresford, SDak, Nov 7, 35; m 57; c 6. MATHEMATICS. *Educ:* Univ SDak, BA, 57; Univ Iowa, MS, 59; Univ Nebr, PhD(math), 66. *Prof Exp:* Instr math, Univ SDak, 59-60; from instr to asst prof, Univ Nebr, 61-66; from asst prof to assoc prof, 66-74, PROF MATH, UNIV WYO, 74- *Mem:* Am Math Soc; Math Asn Am. *Res:* Ring theory; radicals of rings. *Mailing Add:* Divine Word Col Epworth IA 52045

JENKINS, THOMAS GORDON, b Ft Lewis, Wash, Jan 28, 47; m 68; c 2. SYSTEMS ANALYSIS. *Educ:* Univ Ark, BS, 72, MS, 74; Tex A&M Univ, PhD(animal breeding), 77. *Prof Exp:* Res assoc, Tex A&M Univ, 77-78; RES GENETICIST, US MEAT & ANIMAL RES CTR, AGR RES SERV, USDA, 78- *Concurrent Pos:* Consult, Wintock Int Livestock Res & Training Ctr, 76. *Mem:* Am Soc Animal Sci. *Res:* Development and validation of the impact of innovative technologies on the efficiency of production of beef cattle and sheep production systems. *Mailing Add:* US Meat & Animal Res Ctr PO Box 166 Clay Center NE 68933

JENKINS, THOMAS LLEWELLYN, b Cambridge, Mass, July 16, 27; m 51; c 4. ASTROPHYSICS. *Educ:* Pomona Col, BA, 50; Cornell Univ, PhD(physics), 56. *Prof Exp:* Physicist, Lawrence Radiation Lab, Univ Calif, 55-60; from asst prof to assoc prof, 60-68, PROF PHYSICS, CASE WESTERN RESERVE UNIV, 68- *Concurrent Pos:* Sci & eng res coun fel, Southampton Univ, UK, 83. *Mem:* AAAS; Am Phys Soc. *Res:* Neutrino induced reactions; low level counting; electron pair production; photoproduction of mesons; shock hydrodynamics; experimental elementary particle physics; gamma ray astronomy. *Mailing Add:* Dept Physics Case Western Reserve Univ Cleveland OH 44106

JENKINS, THOMAS WILLIAM, b Adrian, Mich, Jan 25, 22; m 48; c 3. ANATOMY, NEUROPATHOLOGY. *Educ:* Kent State Univ, BS, 47; Mich State Col, MS, 50, PhD(zool, anat), 54. *Prof Exp:* Asst biol, Kent State Univ, 41-43, 46-47; asst zool, Mich State Univ, 48-52, from instr to prof anat & path, 51-88; RETIRED. *Concurrent Pos:* NIH spec fel, Sch Med, Temple Univ, 62-63. *Mem:* Am Asn Anat; Am Asn Vet Anat; Am Acad Neurol; Soc Neurosci; Sigma Xi. *Res:* Functional anatomy of the nervous system. *Mailing Add:* 304 Wayland East Lansing MI 48823

JENKINS, VERNON KELLY, b Chattanooga, Tenn, Dec 29, 32; m 54; c 2. RADIOLOGICAL HEALTH, RADIOBIOLOGY. *Educ:* Carson-Newman Col, BS, 54; Univ Tenn, Knoxville, MS, 65, PhD(zool), 67. *Prof Exp:* Res assoc, Biol Div, Oak Ridge Nat Lab, 59-65, res scientist, 67-68; NIH fel exp biol, Baylor Col Med, 68-69, asst prof, 69-70; asst prof, 70-76, ASSOC PROF RADIATION BIOL, UNIV TEX MED BR GALVESTON, 76- *Mem:* Radiation Res Soc; Am Soc Exp Path; Soc Exp Hemat; Reticuloendothelial Soc; NY Acad Sci. *Res:* Effects of radiation on hemopoiesis in mammals, including effects on the immune mechanism; studies of the interrelationships among radiation, immunity, hemopoiesis and the carcinogenic process. *Mailing Add:* Radiation Ther D11 Univ Tex Med Br 310 Gail Boeden Bldg Galveston TX 77550

JENKINS, WILLIAM KENNETH, b Pittsburgh, Pa, Apr 12, 47; m 70. ELECTRICAL ENGINEERING. *Educ:* Lehigh Univ, BSEE, 69; Purdue Univ, MSEE, 71, PhD(elec eng), 74. *Prof Exp:* Res scientist assoc elec eng, Lockheed Missiles & Space Co, Inc, 74-77; from asst prof to assoc prof, 77-83, PROF ELEC ENG, UNIV ILL, URBANA-CHAMPAIGN, 83-, DIR, COORD SCI LAB, 86- *Concurrent Pos:* Consult, Ill State Water, 78, Siliconix, Inc, 78-80, AT&T Bell Labs, 83-84, Lockheed Missiles & Spare Co, 83-85. *Honors & Awards:* Cas Soc Distinguished Serv Award, Inst Elec & Electronics Engrs, 90. *Mem:* Fel Inst Elec & Electronics Engrs (secy & treas, 82-84, pres, 85). *Res:* Circuit and system theory; digital signal processing: digital filters, algorithms and structures; adaptive signal processing; computed imaging. *Mailing Add:* 1913 Moraine Dr Champaign IL 61921

JENKINS, WILLIAM L, b Johannesburg, SAfrica, Jan 29, 37; m 61; c 4. VETERINARY PHARMACOLOGY, VETERINARY PHYSIOLOGY. *Educ:* Univ Pretoria, BVSc, 58, M Med Vet, 68; Univ Mo, PhD(pharmacol), 70. *Prof Exp:* Asst pvt pract, 59-62; lectr vet med, Univ Pretoria, 62-66, sr lectr, 69-71, prof & head physiol & pharmacol, 71-75, prof & head vet physiol & pharmacol, 76-78; res assoc pharmacol, Univ Mo, 66-69; vis prof physiol & pharmacol, Tex A&M Univ, 75-76, prof vet physiol & pharmacol, 78-88; DEAN, SCH VET MED, LA STATE UNIV, 88- *Concurrent Pos:* Mem, FDA/CVM Adv Comt, 85-88, USP Convention Comt, Vet Med, 87-; mem, Subcomt Radiation Appln, Agr SAfrican Atomic Energy Bd, 75-78. *Mem:* Am Acad Vet Pharmacol & Therapeut; Am Col Vet Toxicologists; Am Soc Vet Physiologists & Pharmacologists; Am Vet Med Assoc. *Res:* Veterinary pharmacology and therapeutics including comparative pharmacokinetics; pathophysiology of stress in cattle and sheep; ruminant physiology and pharmacology. *Mailing Add:* Sch Vet Med La State Univ Baton Rouge LA 70803

JENKINS, WILLIAM ROBERT, b Hertford, NC, Sept 12, 27; m 51; c 3. NEMATOLOGY. *Educ:* Col William & Mary, BS, 50; Univ Va, MS, 52; Univ Md, PhD(hort, plant path), 54. *Prof Exp:* Asst biol, Univ Va, 50-51; asst plant path, Univ Md, 51-54, from instr to asst prof, 54-60; assoc res specialist, 60-63, res specialist, 63-65, res prof, 65-69, assoc dean, col, 74-77, PROF BIOL & CHMN DEPT, LIVINGSTON COL, RUTGERS UNIV, 69-, DEAN, COL, 77- *Mem:* Soc Nematologists. *Res:* Nematodes in relation to water pollution; transmission of human pathogens in nematodes borne by domestic water supplies; nematodes in soil. *Mailing Add:* Livingston Col Rutgers Univ New Brunswick NJ 08903

JENKINS, WILLIAM WESLEY, b Chicago, Ill, Oct 22, 17; m 52; c 3. RESEARCH ADMINISTRATION. *Educ:* DePauw Univ, AB, 39; Loyola Univ, Ill, MS, 42; Northwestern Univ, PhD(chem), 50. *Prof Exp:* Res chemist, G D Searle & Co, 39-42 & 49-53, res admin asst, 53-60, head new prod develop, 60-63, dir, 63-67, dir res serv, 67-69, dir develop, 69-71, dir prod affairs, 71-79, dir preclin oper, Searle Res & Develop Div, 79-83; RETIRED. *Concurrent Pos:* Standard Oil fel, 49. *Mem:* Am Chem Soc; Am Pharmaceut Asn; Acad Pharmaceut Sci; Sigma Xi. *Res:* Heterocycles; nitrogen compounds; steroids. *Mailing Add:* 623 Washington Ave Wilmette IL 60091

JENKINS, WILMER ATKINSON, II, b Chicago, Ill, Feb 10, 28; m 49; c 3. PHYSICAL CHEMISTRY, INORGANIC CHEMISTRY. *Educ:* Swarthmore Col, BA, 49; Calif Inst Technol, PhD(chem), 53. *Prof Exp:* Res chemist, E I du Pont de Nemours & Co, Inc, 52-57, res supvr, 57-60, asst dir res, 60-62, tech mgr, 62-66, asst plant mgr, 66-68, plants tech mgr, 68-70, dir res & develop, 70-76, dir, Polyester & Acrylics Div, 76-78, dir, Flexible Pkg Div, 78-88; CONSULT, 88- *Mem:* Am Chem Soc. *Res:* Extractive metallurgy; catalytic processes for polymer intermediates; explosives compositions; polymer chemistry. *Mailing Add:* 115 Locksley Rd Glen Mills PA 19342

JENKINS, WINBORNE TERRY, b Waupun, Wis, Mar 23, 32; div; c 3. BIOLOGICAL CHEMISTRY, ENZYMOLOGY. *Educ:* Cambridge Univ, BA, 53; Mass Inst Technol, PhD(biol), 57. *Prof Exp:* Instr, Mass Inst Technol, 57-58; instr biochem, Univ Calif, Berkeley, 58-66; instr chem, 66-68, PROF CHEM, IND UNIV, BLOOMINGTON, 68- *Concurrent Pos:* Spec res fel, NIH, 61-62, career develop award, 69-74. *Mem:* Am Soc Biochem & Molecular Biol; Protein Soc. *Res:* Intermediary metabolism of amino acids, especially the purification, characterization and general enzymological properties of transaminases; enzymology of calcium and magnesium; Fi-ATPases. *Mailing Add:* Dept of Chem Ind Univ Bloomington IN 47405

JENKINSON, MARION ANNE, b Lancaster, Ohio, Apr 10, 37; m 63; c 1. ORNITHOLOGY. *Educ:* Otterbein Col, BA & BS, 58; Univ Kans, MA, 63. *Prof Exp:* High sch teacher biol, Mifflin, Ohio, 58-60; teaching asst zool, Univ Kans, 60-63; ADJ CUR ORNITH, MUS NATURAL HIST, UNIV KANS, 70- *Concurrent Pos:* Assoc ed, Auk, Am Ornithologists' Union, 66-67, spec ed, Ornith Monographs, 73. *Mem:* Fel Am Ornithologists' Union (treas, 85-). *Res:* Museum collections; characterization of specimen data; world-wide surveys of museum holdings; surveys of Latin American libraries in natural history; financial management. *Mailing Add:* Mus Natural Hist Univ Kans Lawrence KS 66045-2454

JENKINSON, STEPHEN G, b Shreveport, La, Dec 9, 47. PULMONARY DISEASE. *Educ:* La State Univ, Shreveport, MD, 73. *Prof Exp:* CHIEF, PULMONARY DIS, AUDIE MURPHY VET ADMIN HOSP, 82- *Mailing Add:* Dept Med Univ Tex 7703 Floyd Curl Dr San Antonio TX 78284

JENKS, GLENN HERBERT, b Savanna, Ill, June 8, 16; m 45; c 2. PHYSICAL CHEMISTRY. *Educ:* Mich State Col, BS, 39; Northwestern Univ, PhD(phys chem), 45. *Prof Exp:* Res assoc, Metall Lab, Univ Chicago, 43; sr chemist, Oak Ridge Nat Lab, 43-82; RETIRED. *Mem:* AAAS; Sigma Xi; Am Chem Soc; Am Phys Soc. *Res:* Calorimetry; radiation chemistry; radiation effects in solids. *Mailing Add:* 369 East Dr Oak Ridge TN 37830

JENKS, RICHARD D, b Chicago, Ill, Nov 16, 37; m 60; c 3. MATHEMATICS. *Educ:* Univ Ill, BS, 60, PhD(math), 66. *Prof Exp:* Res asst, Coordinated Sci Lab, Univ Ill, 60-66; fel math, Brookhaven Nat Lab, 66-68; RES STAFF MEM MATH SCI, THOMAS J WATSON RES CTR, IBM CORP, 68-; PROF COMPUT SCI, NY UNIV, 80- *Concurrent Pos:* Adj assoc prof math, NY Univ, 69-72; vis lectr comput sci, Yale Univ, 64 & 74; vis prof, Univ Utah, 76; Nat lectr, Asn Comput Mach, 78-80. *Mem:* Asn Comput Mach. *Res:* Computer language and system design, translator writing systems, computer algebra, non-numerical computation; study of very high level languages and their compilation. *Mailing Add:* IBM Research Ctr Box 218 Yorktown Heights NY 10598

JENKS, WILLIAM FURNESS, b Philadelphia, Pa, June 28, 09; m 35; c 3. ECONOMIC GEOLOGY. *Educ:* Harvard Univ, AB, 32, PhD(struct geol), 36; Univ Wis, MA, 33. *Prof Exp:* Jr geologist, Tex Col, 36-38; geologist, Cerro de Pasco Copper Corp, Peru, 38-45; US Dept State vis prof, Univ San Agustin, Peru, 45-46; from asst prof to assoc prof geol, Univ Rochester, 46-55; head dept & dir univ mus, Univ Cincinnati, 55-68, prof geol, 55-79; RETIRED. *Concurrent Pos:* Lectr, Univ San Agustin, Peru, 43; hon prof, Univ San Agustin, Peru, 46; Fulbright lectr, Univ Tokyo, 62-63; consult geologist, Newburyport, Mass, 79-82; coordr, Merrimack Valley Coun Nuclear Weapons Freeze, 82-85. *Mem:* Fel Geol Soc Am; Soc Econ Geol; Am Asn Petrol Geol; Am Geophys Union. *Res:* Mineral deposits of South America; disseminated copper deposits; tertiary volcanic rocks of western North America; massive concordant sulfide ore deposits. *Mailing Add:* 19 Munroe St Newburyport MA 01950

JENNE, EVERETT A, b Beattie, Kans, Mar 2, 30; m 58, 85; c 3. GEOCHEMICAL MODELING, METAL BIOAVAILABILITY. *Educ:* Univ Nebr, BS, 52, MS, 53; Ore State Univ, PhD, 60. *Prof Exp:* Res fel soil chem & clay mineral, Ore State Soil Dept, 56-60; res fel rheology, Univ Calif, Berkeley, 60-62; soil scientist, US Geol Surv, Colo, 62-68, Calif, 68-79; RES SCIENTIST ENVIRON GEOCHEM, BATTELLE PAC NW LAB, RICHLAND, 80- *Concurrent Pos:* Mem ad hoc comt trace elements & uralithiasis incidence, Nat Acad Sci, 75-76; mem subcomt, 76-81, Geochem Environ Rel Health & Dis Comt, Nat Acad Sci, 81-83. *Mem:* Soil Sci Soc Am; Soc Environ Geochem & Health; Am Geophys Union; AAAS; Am Chem Soc; Mineral Soc Am. *Res:* Trace element geochemistry; trace element analyses and partitioning processes among solute, sediment and biota; adsorption phenomenon; colloid chemistry of metal oxides; mineral-water reactions of fossil and nuclear fuel wastes; bioavailability of trace elements; metal adsorption by oxides and sediments; watershed acidification modeling; water-sediment reactions twenty five to one hundred twenty degrees celsius; aquifer thermal energy storage. *Mailing Add:* Pac NW Lab 999 Battelle Blvd Richland WA 99352

JENNEMANN, VINCENT FRANCIS, b St Louis, Mo, Nov 27, 21; m 46; c 7. COMPUTER SCIENCE, EXPLORATION GEOPHYSICS. *Educ:* St Louis Univ, BS, 47, MS, 49; Univ Tulsa, PhD(earth sci), 72. *Prof Exp:* Instr math, St Louis Univ, 46-48; res computer, Seismog Dept, Sun Oil Co, 48-51; instr math, Lamar Col, 49-51; asst, Lamont Geol Observ, Columbia Univ, 51-54; res engr, Amoco Prod Co, Amoco Corp, 54-58, sr res engr, 58-64, sr res scientist, 64-66, comput analyst, 66-74, staff computer analyst, 74-84; RETIRED. *Mem:* Seismol Soc Am; Soc Explor Geophys; Am Geophys Union; Sigma Xi. *Res:* Various aspects of the metric system (SI). *Mailing Add:* 203 Sunset Dr Tulsa OK 74114-1239

JENNER, CHARLES EDWIN, b Indianola, Iowa, Nov 5, 19; m 42; c 2. BIOLOGY. *Educ:* Cent Col, Mo, AB, 41; Harvard Univ, MA, 49, PhD(biol), 51. *Prof Exp:* From asst prof to assoc prof & chmn dept, 50-60, PROF ZOOL, UNIV NC, CHAPEL HILL, 60- *Concurrent Pos:* Instr, Marine Biol Lab, Woods Hole, 52-54 & 61. *Mem:* AAAS; Ecol Soc Am; Am Soc Limnol & Oceanog; Am Soc Zool. *Res:* Aquatic ecology; animal photoperiodism. *Mailing Add:* Dept Biol Univ NC Chapel Hill NC 27515

JENNER, DAVID CHARLES, b Seattle, Wash, Oct 21, 43; m 69; c 2. ASTRONOMY, COMPUTER SCIENCE. *Educ:* Univ Wash, BS(physics) & BS(math), 66; Univ Wis-Madison, PhD(astron), 70. *Prof Exp:* Asst prof astron, NMex State Univ, Las Cruces, 70-72; adj asst prof astron, Univ Calif, Los Angeles, 72-78; RES ASSOC, DEPT ASTRON & DIR, MANASTASH RIDGE OBSERV, UNIV WASH, 78- *Mem:* Am Astron Soc; Int Astron Union. *Res:* Masses of galaxies; stellar populations in galaxies; the nuclei of active galaxies; planetary nebulae; instrumentation and observational techniques; software systems; hardware systems; laboratory data acquisition and instrument control. *Mailing Add:* 3153 NE 84th St Seattle WA 98115

JENNER, EDWARD L, b Pontiac, Mich, Mar 27, 18; m 42; c 3. AGRICULTURAL BIOCHEMISTRY. *Educ:* Lake Forest Col, AB, 39; Univ Mich, MS, 40, PhD(chem), 42. *Prof Exp:* Res chemist, Univ Mich, 41-45; res chemist, Exp Sta, E I du Pont de Nemours & Co, Inc, 45-82; RETIRED. *Concurrent Pos:* Res assoc cell physiol, Univ Calif, 62-63. *Mem:* Am Chem Soc; Sigma Xi; Fel AAAS. *Res:* Synthesis of nitramines; acid-catalyzed telomerizations; reactions of hydroxyl and amino radicals, halogen atoms and aliphatic free radicals; catalysis by soluble derivatives of transition metals; oxidative and photosynthetic phosphorylation; biochemistry of phytochrome; ozone damage to vegetation. *Mailing Add:* 107 Lands End Rd Wilmington DE 19807-2519

JENNESS, ROBERT, b Rochester, NH, Sept 21, 17; m 40; c 3. BIOCHEMISTRY. *Educ:* Univ NH, BS, 38; Univ Vt, MS, 40; Univ Minn, PhD(agr biochem), 44. *Prof Exp:* From instr to prof agr biochem, Univ Minn, St Paul, 40-66; prof biochem, 66-84; ADJ PROF CHEM, NMEX STATE UNIV, LAS CRUCES, 84- *Honors & Awards:* Borden Award, 53. *Mem:* AAAS; Am Soc Biol Chem; Univ Vt, MS, 40; Am Chem Soc; Am Soc Mammalogists; Sigma Xi. *Res:* Biosynthesis of ascorbate by mammals; chemistry of milk proteins and salts; comparative biochemistry of milks of various species. *Mailing Add:* 1837 Corte del Ranchero Alamogordo NM 88310

JENNESS, STUART EDWARD, b Ottawa, Ont, Aug 22, 25; m 49, 80; c 2. GEOLOGY. *Educ:* Queens Univ, Ont, 48; Univ Minn, MS, 50; Yale Univ, PhD(geol), 55. *Prof Exp:* Instr geol, Muhlenberg Col, 49-51; geologist, Nfld Geol Surv, 52-53; geologist, Can Geol Surv Can, 54-67; publ supvr, 67-85, ED CONSULT GEOL, CAN JOUR RES, NAT RES COUN CAN, 85- *Mem:* Geol Soc Am; Geol Asn Can; life mem Arctic Inst NAm; Asn Earth Sci Ed. *Mailing Add:* 9 2051 Jasmine Crescent Ottawa ON K1J 7W2 Can

JENNETT, JOSEPH CHARLES, b Dallas, Tex, June 11, 40; m 63; c 2. CIVIL & ENVIRONMENTAL ENGINEERING. *Educ:* Southern Methodist Univ, BSCE, 63, MSCE, 66; Univ NMex, PhD(sanit eng), 69; Am Acad Environ Engrs, Dipl, 78. *Prof Exp:* Engr, Southwestern Design Br, US Corp Engrs, 63; construct engr, Calif State Dept Water Resources, Orville, 63-64; consult engr, Pitotmeter Assocs, 65-66 & 69; from asst prof to assoc prof civil eng, Univ Mo-Rolla, 69-75; prof civil eng & chmn dept, Syracuse Univ, 75-81; ACAD DEAN ENG & PROF ENVIRON SYST ENG, CLEMSON UNIV, 81- *Concurrent Pos:* Chmn task force on toxic trace substances in water, 75 & comt of water treatment and water resources mgt, 76; mem, Prof Coord Comt, 77-; mem, Environ Eng Div, Res Coun, 78-; ed, E N Am Minerals Environ J, 78; vis res, Appl Geochem Res Group, Imp Col, UK, 77. *Mem:* Am Soc Civil Engrs; Am Acad Environ Engrs, (trustee, 88-91); Am Soc Eng Educ; Water Pollution Control Fedn; Nat Soc Prof Engrs; Am Asn Environ Eng Prof. *Res:* Urban and rural runoff pollutants; drying of digested sludge; industrial waste treatment techniques; effects of heavy metals on aquatic ecosystems and treatment devices; biological operations on domestic and industrial wastes; analysis and treatment of toxic metals and trace organics; urban and rural run-off quality. *Mailing Add:* Col Eng Clemson Univ 109 Riggs Hall Clemson SC 29634-0901

JENNEY, DAVID S, b Mattapoisett, Mass, May 18, 31; m 53; c 3. MECHANICS, AERODYNAMICS. *Educ:* Worcester Polytech Inst, BS, 53; Univ Conn, MS, 56; Rensselaer Polytech Inst, PhD(mech), 68. *Prof Exp:* Res engr, Res Labs, United Aircraft Corp, 53-58; supvr rotary wing aerodyn, 58-62; eng mgt pos, 62-77; DIR TECH ENG, SIKORSKY AIRCRAFT DIV, UNITED TECHNOL CORP, 77- *Mem:* Am Helicopter Soc. *Res:* Helicopter and vertical take-off aircraft aerodynamics and dynamics. *Mailing Add:* 109 Wilbrook Rd Stratford CT 06497

JENNEY, ELIZABETH HOLDEN, b Bennington, Vt, Nov 4, 12. PHARMACOLOGY. *Educ:* Mt Holyoke Col, AB, 34; Univ Ill, MS, 47. *Prof Exp:* Asst pharmacol, Sch Med, Boston Univ, 35-36; med technologist, Rutland Hosp, Vt, 37-41; med technologist, Cooly Dickinson Hosp, Northampton, Mass, 41-43; res assoc, Univ Ill Col Med, 48-54; instr, Sch Med, Emory Univ, 54-60; res scientist, Sect Pharmacol, Bur Res, NJ Neuropsychiat Inst, 60-73; PHARMACOLOGIST, BRAIN BIO CTR, 73- *Mem:* Sigma Xi; Am Soc Pharmacol & Exp Therapeut. *Res:* Neuropharmacology; psychopharmacology; schizophrenia. *Mailing Add:* Pennswood Village J-212 Newtown PA 18940

JENNI, DONALD ALISON, b Pueblo, Colo, June 20, 32; m; c 7. ETHOLOGY. *Educ:* Ore State Univ, BS, 53; Utah State Univ, MS, 56; Univ Fla, PhD(zool), 61. *Prof Exp:* Asst prof zool, Univ Fla, 61-62 & Eastern Ill Univ, 62-66; assoc prof zool, 66-71, chmn dept, 72-75 & 85-88, PROF ZOOL, UNIV MONT, 71-, ASSOC DEAN BIOL SCI, COL ARTS & SCI, 88- *Concurrent Pos:* NIH fel & res biologist, Univ Leiden, 64-66; vis prof, Cornell Univ, 75, Univ Wash, 79-80, Univ Melbourne & James Cooke Univ, 85; prin

investr, NSF, 70-76, 85, 91-93, BLM, 77-80, Campfire, 83. *Mem:* Animal Behav Soc; Am Ecol Soc; Am Ornith Union; Wilson Ornith Soc; Asn Trop Biol; Coop Ornith Soc. *Res:* Ethology and behavioral ecology, especially behavioral approach to classic ecological problems; adaptation and evolution of behavioral patterns including social organization in response to ecological pressures; evolution of mateship systems, especially non-monogamous systems; behavioral problems of territoriality. *Mailing Add:* Div Biol Sci Univ Mont Missoula MT 59812-1002

JENNINGS, ALBERT RAY, b Grosvenor, Tex, Nov 11, 26; m 42; c 2. GEOLOGY. *Educ:* Hardin-Simmons Univ, BA, 58; Tex A&M Univ, MS, 60, PhD(geol), 64. *Prof Exp:* Res asst hydrol, Tex Eng Exp Sta, Tex A&M Univ, 60-63; explor geologist, Mobil Oil Co, 64-68; from asst prof to prof geol, ECarolina Univ, 68-74, chmn dept, 70-74; from assoc prof to prof, head dept geol, Hardin-Simmons Univ, 76-88; RETIRED. *Concurrent Pos:* Consult geologist, 74- *Mem:* Am Inst Prof Geol; Am Asn Petrol Geol; fel Geol Soc Am. *Res:* Utilization of radioisotopes as ground-water tracers. *Mailing Add:* 241 Styers St Houston TX 77022

JENNINGS, ALFRED ROY, JR, b Duncan, Okla, Sept 27, 45; m 67; c 3. HYDRAULIC FRACTURING, FORMATION DAMAGE. *Educ:* Okla Univ, BS, 67, MS, 72. *Prof Exp:* Chemist, Halliburton Serv, 67-72, sr engr, 72-74, dev engr, 74-77, group leader, 77-79, sect supvr, 79-82; RES ASSOC, MOBIL RES & DEVELOP CORP, 82- *Concurrent Pos:* Mem comts, Am Petrol Inst, 77-83, chmn, work group, 81- *Mem:* Soc Petrol Engrs; Asn Inst Mech Engrs. *Res:* Formation fines and damage control; well stimulation. *Mailing Add:* Mobil Res & Develop Corp Dallas Res Lab PO Box 819047 Dallas TX 75381

JENNINGS, ALFRED S(TONEBRAKER), b St Louis, Mo, Sept 30, 25; m 49; c 4. CHEMICAL ENGINEERING. *Educ:* Washington Univ, St Louis, BS, 48, MS, 49, DSc(chem eng), 51. *Prof Exp:* Chem engr, Savannah River Lab, E I du Pont de Nemours & Co Inc, 51-57, res supvr, 57-68, res mgr, Separations Eng Div, 68-80, sr res assoc, 80-85; RETIRED. *Mem:* Am Chem Soc; Am Inst Chem Engrs. *Res:* Radiochemical separations and solvent extraction process development; isotope separation processes; high-level waste immobilization. *Mailing Add:* 1469 Canterbury Ct SE Aiken SC 29801

JENNINGS, ALLEN LEE, b Quincy, Ill, July 5, 43; m 67; c 3. PESTICIDES, TOXIC CHEMICALS. *Educ:* Western Ill Univ, BS, 65; Univ Ark, PhD(chem), 70. *Prof Exp:* Res assoc biochem, Iowa State Univ, 70-71; dir, Chem & Statist Policy Div, 85-87, RES ASSOC BIOCHEM, US ENVIRONMENTAL PROTECTION AGENCY, 71-; DIR, BIOL & ECON ANALYSIS DIV, 87- *Mem:* AAAS; Am Chem Soc. *Mailing Add:* 2306 S Dinwiddie St Arlington VA 22206

JENNINGS, BOJAN HAMLIN, b Waukegan, Ill, Apr 4, 20; m 42; c 3. ORGANIC CHEMISTRY. *Educ:* Bryn Mawr Col, AB, 42; Radcliffe Col, MA, 43, PhD(chem), 55. *Prof Exp:* Res chemist, Dewey & Almy Chem Co, 42-43; from instr to assoc prof, 43-62, PROF CHEM, WHEATON COL, MASS, 62-, CHMN DEPT, 68- *Mem:* Am Chem Soc; NY Acad Sci; Asn Women Sci. *Res:* Steroid chemistry; cancer research; physical organic chemistry; photochemistry. *Mailing Add:* 56 Winthrop St Taunton MA 02780

JENNINGS, BURGESS H(ILL), b Baltimore, Md, Sept 12, 03; m 25; c 1. MECHANICAL ENGINEERING. *Educ:* Johns Hopkins Univ, BE, 25; Lehigh Univ, MS, 28, MA, 35. *Prof Exp:* Mem fac, Lehigh Univ, 26-35, assoc prof, 35-40; prof, 40-73, chmn dept, 41-57, assoc dean eng, 62-70, EMER PROF MECH ENG, NORTHWESTERN UNIV, 73- *Concurrent Pos:* Consult, var US co & labs, 34-71; mem, Refrig Res Found, 44-57; ed, Lubricating Eng, Am Soc Lube Engrs, 44-51; vpres, Int Inst Refrig, 57-67. *Honors & Awards:* Off Sci Res & Develop Citation, 45; Richards Mem Award, 50; Am Soc Heat, Refrig & Air-Conditioning Engrs Plaques, 61 & 84; Worcester Reed Warner Medal, 72; F Paul Anderson Medal, 80. *Mem:* Nat Acad Eng; fel Am Soc Heat, Refrig & Air-Conditioning Engrs (treas & vpres, 46-48, pres, 49); fel & hon mem Am Soc Mech Engrs; Am Soc Lubrication Engrs (secy & vpres, 44-49); Am Soc Eng Educ. *Res:* Applied thermodynamics; refrigeration; environmental control; author & co-author of eleven engineering textbooks. *Mailing Add:* 4576 Springmoor Circle Raleigh NC 27615-5708

JENNINGS, BYRON KENT, b Musquodoboit, NS, Can, Mar 29, 51. THEORETICAL PHYSICS. *Educ:* Mt Allison Univ, BSc, 72; McMaster Univ, MSc, 73 & PhD (physics), 76. *Prof Exp:* Postdoctoral fel physics, State Univ NY, Stony Brook, 76-80; res fel physics, Univ Regensburg, 80; res fel physics, McGill Univ, 80-82; RES SCIENTIST, TRIUMF, 82- *Concurrent Pos:* Vis prof, Univ Toronto, 86; assoc ed, Can J Physics, 83-89; adj prof, Simon Fruser Univ, 87- *Mem:* Can Asn Physics. *Res:* Study of nucleon structure as it impacts on nuclear properties. *Mailing Add:* TRIUMF 4004 Wesbrook Mall Vancouver BC V6T 2A3 Can

JENNINGS, CARL ANTHONY, b Harrisburg, Ill, Dec 28, 44; m 65; c 2. ORGANIC CHEMISTRY, CHEMICAL MANUFACTURING MANAGEMENT. *Educ:* Southern Ill Univ, BS, 67, PhD(org chem), 71. *Prof Exp:* Res assoc org chem, Univ Ill, 71-72; from res chemist to mgr, Photog Emulsion Mfg, GAF Corp, 72-77; from asst to vpres, Ind Chem, 77-78; mgr develop chem, 78-80, mgr agr chem mfg, 80-85, DIR MFG & TECHNOL, BASF WYANDOTTE CORP, 85- *Mem:* Am Chem Soc; Am Inst Chem Engrs; Soc Chem Indust. *Res:* Agricultural chemicals; chemical manufacturing management; polyoxyalkylenes and organic oxide chemicals; urethanes; organic synthesis; photographic emulsion theory; organometallics. *Mailing Add:* 1912 Torreypines Pl Raleigh NC 27615

JENNINGS, CHARLES DAVID, b Newtonia, Mo, May 21, 39; c 2. OCEANOGRAPHY. *Educ:* Northwest Nazarene Col, BA, 61; Ore State Univ, MS, 66, PhD(oceanog), 68. *Prof Exp:* Instr physics, Ore Col Educ, 62-63; instr oceanog, World Campus Afloat, 66, asst prof, 68; oceanogr, US

Bur Com Fisheries, 68-70; asst prof oceanog, Ore Col Educ, 70-74, assoc prof physics, 74-78; PROF PHYSICS, WESTERN ORE STATE COL, 78- *Mem:* AAAS; Am Soc Limnol & Oceanog. *Res:* Radioactivity and trace elements in the marine environment; circulation of estuaries; marine radioecology. *Mailing Add:* 1519 NE 77th St Seattle WA 98115

JENNINGS, CHARLES WARREN, b Toledo, Ohio, Dec 3, 18; m 49; c 3. ELECTROCHEMISTRY. *Educ:* Univ Toledo, BEng, 40; Univ Calif, MS, 43; Duke Univ, PhD(chem), 51. *Prof Exp:* Res chemist, Dow Chem Co, Calif, 42-43; chemist, Nat Bur Standards, 46-47, res assoc, 47-48; res assoc, Res Proj, Duke Univ, 48-50; assoc prof chem, NC State Col, 50-57; mem staff, Sandia Corp, 57-88; RETIRED. *Res:* Physical properties of electrodeposited metals; electrochemistry of batteries and fused salt systems; chlorination of hydrocarbons; thermal batteries; printed circuit boards; adhesives. *Mailing Add:* 1209 Mesilla NE Albuquerque NM 87110

JENNINGS, DANIEL THOMAS, b Fulton, Ky, July 4, 35; m 55; c 2. FOREST ENTOMOLOGY, ARACHNOLOGY. *Educ:* Colo State Univ, BS, 60; Univ NMex, MS, 67, PhD(biol), 72. *Prof Exp:* Entomologist, Forest Serv, 62-65, res entomologist, NC Forest Exp Sta, 65-68, res entomologist, Rocky Mountain Forest & Range Exp Sta, 68-76, PRIN RES ENTOMOLOGIST, NORTHEASTERN FOREST EXP STA, FOREST SERV, USDA, 76- *Concurrent Pos:* Collabr, Environ Qual Inst, Biol Active Natural Prod Lab, Agr Res Serv, Beltsville, Md, 73-; adj asst prof biol, Univ NMex, 74-77; fac assoc, Univ Maine, 76- *Mem:* Entom Soc Am; Am Entom Soc; Am Arachnological Soc; Brit Arachnological Soc; Entom Soc Can; Sigma Xi. *Res:* Life histories and habits of forest insects, their biological control by natural enemies and pheromones; the arachnid fauna associated with forest trees. *Mailing Add:* Northeastern Forest Exp Sta Univ Maine USDA Bldg Orono ME 04469

JENNINGS, DAVID PHIPPS, b Columbia, Mo, Aug 3, 41; m 64; c 2. MEDICAL INFORMATICS, VETERINARY MEDICINE. *Educ:* Univ Mo, BS, 63, DVM, 65; Okla State Univ, PhD(physiol), 69. *Prof Exp:* NIH trainee, 65-66, fel, 66-68, from asst prof to assoc prof physiol, Okla State Univ, 68-77; PROF PHYSIOL, MISS STATE UNIV, 77- *Concurrent Pos:* NIH spec fel anat, Sch Med, Univ Calif, Los Angeles, 71-72; clin neurol trainee, Univ Ga, 78. *Mem:* AAAS; Am Vet Med Asn; Am Asn Vet Anatomists; Am Asn Vet Cols; Am Soc Vet Physiol & Pharmacol. *Res:* Central nervous system mechanisms for control of physiologic systems; medical informatics and computerized medical records. *Mailing Add:* Col of Vet Med Miss State Univ Mississippi State MS 39762

JENNINGS, DONALD B, b Windsor, Ont, July 20, 32; m 57; c 4. MEDICINE, PHYSIOLOGY. *Educ:* Queen's Univ, Ont, MD, CM, 57, MSc, 60, PhD(physiol), 62. *Prof Exp:* Jr intern, Montreal Gen Hosp, 57-58, jr asst res med, 58-59; res fel med & physiol, Cardiovasc Res Inst, Med Ctr, Univ Calif, San Francisco, 62-64; from asst prof to assoc prof, 64-74, PROF PHYSIOL, QUEEN'S UNIV, ONT, 74- *Concurrent Pos:* George Christian Hoffman fel path, 62-63; Can Heart Found sr res fel physiol, 64-69; assoc ed, Can J Physiol Pharmacol, 78; fel, Max Planck Inst Exp Med, 81; vis prof, Med Sch, Dartmouth Col, 81. *Mem:* Can Physiol Soc (secy, 75-78); Am Physiol Soc; Can Soc Clin Invest. *Res:* Humoral and nervous regulation of cardiovascular, respiratory, metabolic and erythropoietic adjustments to high carbon dioxide and low oxygen environments and anaemic anoxia; interaction of temperature regulation with the cardio-respiratory admustment to acute and chronic hypercapnia and hypoxia. *Mailing Add:* Dept Physiol Queen's Univ Kingston ON K7L 3N6 Can

JENNINGS, DONALD EDWARD, b New Rochelle, NY, May 30, 48; m 70. MOLECULAR SPECTROSCOPY. *Educ:* Northern Ariz Univ, BS, 70; Univ Tenn, PhD(physics), 74. *Prof Exp:* Res assoc physics, Univ Tenn, Knoxville, 74-75; Nat Acad Sci-Nat Res Coun res assoc, 76-77, SPACE SCIENTIST, GODDARD SPACE FLIGHT CTR, NASA, 77- *Res:* Molecular spectroscopy; fourier transform, tuneable diode laser, and grating spectroscopy; planetary infrared astronomy; radio astronomy of interstellar molecules. *Mailing Add:* NASA Goddard Space Flight Ctr Code 693 Greenbelt MD 20771

JENNINGS, FRANK LAMONT, b Minneapolis, Minn, Apr 25, 21; m 48; c 4. PATHOLOGY. *Educ:* Ind Univ, AB, 42, MD, 47. *Prof Exp:* asst path, Univ chicago, 48, atomic energy fel, 48-51, from instr path, 52-60; from assoc prof to prof path, Univ Tex Med Br, Galveston, 60-76, chmn dept, 63-76; PROF & CHMN DEPT PATH, SCH MED, WRIGHT STATE UNIV. *Concurrent Pos:* Mem staff, Armed Forces Inst Path, 55-57; bd gov, Col Am Path, 75-81. *Mem:* Am Soc Exp Path; Radiation Res Soc; Col Am Path; Am Soc Clin Path; Int Acad Path. *Res:* Radiation pathology and recovery; protein nutrition; tumor metabolism. *Mailing Add:* Wright State Univ Sch Med PO Box 927 Dayton OH 45401

JENNINGS, HARLEY YOUNG, JR, b Clio, Mich, Sept 29, 26; m 50; c 3. CHEMISTRY. *Educ:* Univ NC, BS, 48; Univ Mich, MS, 49, PhD(chem), 52. *Prof Exp:* Res chemist, Parker Pen Co, 50-52; res chemist, Chevron Res Co, 52-59, group supvr, 59-62, sr res chemist, 62-67, sr res assoc, 67-83, MGR, CHEVRON OIL FIELD RES CO, LA HABRA, 83- *Honors & Awards:* Lester C Uren Award, 83. *Mem:* AAAS; Am Chem Soc; Am Inst Mining, Metall & Petrol Engrs. *Res:* Surface energy relationships, contact angle and interfacial tension; capillarity; fluid flow and enhanced recovery of petroleum; colloid and surface chemistry; oil well stimulation and stimulation and damage prevention mechanisms; phase behavior and fluid analysis. *Mailing Add:* 2501 Terraza Pl Fullerton CA 92635

JENNINGS, LAURENCE DUANE, b New Haven, Conn, Nov 14, 29; m 51; c 3. SOLID STATE PHYSICS. *Educ:* Mass Inst Technol, SB, 50, PhD(physics), 55. *Prof Exp:* Asst prof chem, Iowa State Univ, 55-59; SOLID STATE PHYSICIST, US ARMY MAT TECH LAB, 59- *Mem:* Am Phys Soc; Am Crystallog Asn; Inst Elec & Electronics Engr. *Res:* Diffraction; equilibrium properties of solids. *Mailing Add:* SLC MT-OMM Army Mat Tech Lab Watertown MA 02172-0001

JENNINGS, LISA HELEN KYLE, b Kingsport, Tenn, Apr 1, 55; m 76; c 1. EXPERIMENTAL HEMATOLOGY, PLATELET MEMBRANE BIOCHEMISTRY. *Educ:* Univ Tenn, BA, 76, PhD(biochem), 83; Memphis State Univ, MS, 78. *Prof Exp:* Res fel, dept biochem, St Jude Children's Res Hosp, Memphis, Tenn, 83-84, Leon Journey fel, 84-85; ASST PROF HEMAT & ONCOL, DIV HEMAT & ONCOL, DEPT MED & ASST PROF, DEPT BIOCHEM, UNIV TENN, MEMPHIS, 85- *Concurrent Pos:* Young investr, Am Heart Asn, 85-87. *Mem:* AAAS; Soc Anal Cytol. *Res:* Structure and function of platelet membrane surface proteins and their role in thrombosis and hemostasis, particularly the mechanism by which platelet surface proteins mediate platelet aggregation. *Mailing Add:* Dept Med Div Hemat/Oncol Univ Tenn 956 Court Ave Rm H316 Memphis TN 38163

JENNINGS, MICHAEL LEON, b Cleveland, Ohio, June 10, 48; m 76. TRANSPORT PHYSIOLOGY, MEMBRANE BIOCHEMISTRY. *Educ:* Mass Inst Technol, SB, 70; Harvard Univ, PhD(biophysics), 76. *Prof Exp:* Fel, Max Planck Inst Biophysics, 77-78; ASST PROF PHYSIOL, COL MED, UNIV IOWA, 78- *Mem:* Biophys Soc; Am Physiol Soc; Sigma Xi. *Res:* Structure and function of biological ion transport proteins, especially the inorganic anion transport protein of the erythrocyte membrane. *Mailing Add:* Dept Physiol & Biophys Univ Iowa Iowa City IA 52242

JENNINGS, PAUL BERNARD, JR, b Medford, Mass, Oct 13, 38; m 64; c 4. VETERINARY SURGERY, COMPARATIVE MEDICINE. *Educ:* Tufts Univ, BS, 60; Univ Pa, VMD, 64, M Med Sci, 70; Am Col Vet Surgeons, dipl, 73. *Prof Exp:* US Army, 65-, res investr surg, Walter Reed Army Inst Res, 65-67, asst chief, Clin Invest Serv, Madigan Army Med Ctr, Tacoma, Wash, 71-77, staff officer animal med, US Army Health Serv Command, Directorate Vet Serv, Ft Sam Houston, 78-79, chief, Div Res Support, Letterman Army Inst Res, San Francisco, 79-81, commander, 167th med detachment, Stuttgart, Germany, dir animal medicine, US Army Health Serv Command, Ft Sam Houston, 84-86, CHIEF MIL DOG VET SERV, LACKLAND AFB, 86- *Concurrent Pos:* Consult to surgeon gen, US Army Med Dept, 72-; Am Cancer Soc grant, 69-70; ed consult, J Vet Surg, 77- *Honors & Awards:* Gold Medal-Sci Exhib, Am Acad Pediat, 76 & 78; Cert Merit-Sci Exhib, AMA, 77 & 78. *Mem:* AAAS; Am Col Vet Surgeons; Sigma Xi; Am Vet Med Asn; Asn Mil Surgeons US. *Res:* Shock, trauma & surgical infections; transplantation immunology; comparative medical aspects of human disease, including animal models. *Mailing Add:* DOD Mil Working Dog Agency Bldg 7595 Lackland AFB TX 78236-5000

JENNINGS, PAUL C(HRISTIAN), b Brigham City, Utah, May 21, 36; m 81; c 2. CIVIL ENGINEERING, APPLIED MECHANICS. *Educ:* Colo State Univ, BS, 58; Calif Inst Technol, MS, 60, PhD(civil eng), 63. *Prof Exp:* From instr to assoc prof mech, USAF Acad, 63-66; from asst prof to prof appl mech, 66-76, chmn div eng & appl sci, 85-89, PROF CIVIL ENG & APPL MECH, CALIF INST TECHNOL, 76-, VPRES & PROVOST, 89- *Concurrent Pos:* Mem, eng panel, Nat Acad Sci Comt on Alaskan Earthquake, 65-; Erskine fel, Univ Canterbury, 70, 85; pres, Earthquake Eng Res Inst, 80-82. *Honors & Awards:* Huber Res Prize, Am Soc Civil Engrs, 76. *Mem:* Nat Acad Eng; Am Soc Civil Engrs; Seismol Soc Am (pres, 81); Earthquake Eng Res Inst; Am Soc Eng Educ; Am Geophys Union. *Res:* Structural dynamics and engineering seismology, especially response of structures to earthquake motion; earthquake engineering. *Mailing Add:* Mail Code 206-31 Calif Inst of Technol Pasadena CA 91125

JENNINGS, PAUL HARRY, b Brockton, Mass, Jan 31, 38; m 60; c 3. PLANT PHYSIOLOGY. *Educ:* Univ Mass, Amherst, BVA, 60; NC State Univ, MS, 62, PhD(plant physiol), 65. *Prof Exp:* Asst res plant physiol, Univ Calif, Davis, 67-69; from asst prof to assoc prof plant physiol, Univ Mass, Amherst, 69-82, secy fac senate, 80-82; PROF & HEAD, DEPT HORT, KANS STATE UNIV, 82- *Mem:* Am Soc Plant Physiol; Am Soc Hort Sci; Crop Sci Soc Am; Sigma Xi. *Res:* Physiology of disease resistance; anabolic and catabolic pathways of carbohydrate metabolism as related to genetic potential, stage development and isozymic differences in plants; effects of low temperatures on germination and growth of crop plants susceptible to chilling injury. *Mailing Add:* Dept Hort Waters Hall Kans State Univ Manhattan KS 66506

JENNINGS, PAUL W, b Denver, Colo, Sept 24, 36; m 61; c 2. ORGANOMETALLIC CHEMISTRY, COMPLEX MATERIALS. *Educ:* Univ Colo, BA, 58, MS, 61; Univ Utah, PhD(org chem), 65. *Prof Exp:* Res fel chem, Calif Inst Technol, 64-66; from asst prof to assoc prof, 66-75, PROF CHEM, MONT STATE UNIV, 75- *Mem:* Am Chem Soc; Sigma Xi. *Res:* Organometallic chemistry; photochemical energy transfer; chemistry of bituminous materials. *Mailing Add:* 2024 Baxter Dr Bozeman MT 59715

JENNINGS, RICHARD LOUIS, b Newark, NJ, July 28, 33; m 56; c 2. CIVIL ENGINEERING, APPLIED MECHANICS. *Educ:* Univ Ohio, BS, 56, BSCE, 57; Univ Ill, MS, 58, PhD(civil eng), 64. *Prof Exp:* Asst prof, 63-67, ASSOC PROF CIVIL ENG, UNIV VA, 67- *Concurrent Pos:* Consult, Babcock & Wilcox Corp, Va. *Mem:* Am Soc Civil Engrs. *Res:* Earthquake and nuclear blast resistant design structures; mechanical vibrations of thin shells; structural design of large steerable radio telescopes; rehabilitation engineering; highway pavement analysis. *Mailing Add:* Dept of Civil Eng Thornton Hall Univ of Va Charlottesville VA 22903

JENNINGS, ROBERT BURGESS, b Baltimore, Md, Dec 14, 26; m 52; c 5. PATHOLOGY, EXPERIMENTAL PATHOLOGY. *Educ:* Northwestern Univ, BS, 47, MS & BM, 49, MD, 50; Am Bd Path, dipl, 54. *Prof Exp:* Intern, Passavant Mem Hosp, Chicago, Ill, 49-50, resident path, 50-51; from instr to prof, Med Sch, Northwestern Univ, Ill, 53-69, Magerstadt prof path & chmn dept, 69-75; prof, 75-80, chmn dept, 75-89, JAMES B DUKE PROF PATH, MED SCH, DUKE UNIV, 80- *Concurrent Pos:* Lab officer, US Navy, 51-53; attend physician, Vet Admin Res Hosp, Chicago, Ill, 55-69, consult physician, 69-75; pathologist, Community Hosp, Evanston, Ill, 57-67; Markle scholar, 58-63; vis scientist, Middlesex Hosp Med Sch, London, 61-62; mem path A study sect, USPHS, 60-65, mem, Cardiol Adv Comt, Nat Heart, Lung &

Blood Inst, NIH, 78-82; attend physician & chief labs, Passavant Mem Hosp, 69-72; attend staff, Northwestern Mem Hosp, 72-75. *Mem:* Am Asn Pathologists; Soc Exp Biol & Med; Int Soc Heart Res (pres, 78-80); Am Soc Cell Biol; Am Heart Asn. *Res:* Cardiovascular and renal disease; cell physiology; cell injury; electron microscopy; biology of experimental myocardial infarction; molecular mechanisms which cause the death of ischemic myocytes. *Mailing Add:* Dept Path Duke Univ Med Ctr Box 3712 Durham NC 27710

JENNINGS, VIVAN M, b Columbus Junction, Iowa, May 2, 36. AGRICULTURAL RESEARCH. *Educ:* Iowa State Univ, BS, MS, PhD(agron). *Prof Exp:* Prof plant path, seed & weed sci & exten specialist, Integrated Pest Mgt, Weed Control & Agron, Iowa State Univ; assoc dean & assoc dir, Iowa Exten Serv; DEP ADMIN AGR, EXTEN SERV, USDA, 85- *Concurrent Pos:* Interim assoc admin, Exten Serv, USDA, 89; interim sr exten adv, Polish Ministry Agr & Food Econ, 90. *Res:* Integrated pest management; pesticide impact assessment; pesticide applicator training; urban gardening; farm safety and farmers with disabilities along with more traditional programs of farm management, marketing, crop and livestock production systems and agricultural engineering. *Mailing Add:* Exten Serv USDA Rm 3851 S Bldg Washington DC 20250-0900

JENNINGS, WALTER GOODRICH, b Sioux, Iowa, Mar 2, 22; m 47; c 3. GAS CHROMATOGRAPHY. *Educ:* Univ Calif, BS, 50, MS, 52, PhD(agr chem), 54. *Prof Exp:* Instr dairy indust, Univ Calif, Davis, 54-59, from asst prof to assoc prof food sci, 59-65, from jr chemist to assoc chemist, 54-65, prof food sci & chem exp sta, 65-89; CONSULT, J & W SCI, 75- *Concurrent Pos:* NIH sr scientist award, Vienna, Austria, 67-68; spec award sr Am scientist, Alexander von Humboldt Found, 74-75; consult, several indust firms; ed, J High Resolution Chromatography & Chromotography Commun, J Food Chem & Chemi, Mikrobiologie, Technologie der Lebensmittel; Founder of J & W Sci, Inc, Rancho Cordova, Calif. *Honors & Awards:* Medal, Univ Bologna, 67; Medal, Fr Asn Agr Chemists, 71; Beckman Award in Gas Chromatog, 90. *Mem:* Am Chem Soc; hon mem Soc Flavor Chemists. *Res:* Isolation and characterization of trace volatiles; flavor chemistry; glass capillary gas chromatography; author of over 250 publications. *Mailing Add:* J & W Sci 91 Blue Ravine Rd Folsom CA 95630

JENNINGS, WILLIAM HARNEY, JR, b Ames, Iowa, Dec 6, 31; m 57; c 2. BIOPHYSICS. *Educ:* Duke Univ, BS, 54; George Washington Univ, MS, 59. *Prof Exp:* Physicist, Naval Med Res Inst, 54-58; res physicist, Nat Inst Arthritis, Metab & Digestive Dis, 61-88; RETIRED. *Mem:* AAAS; Chem Soc. *Res:* Laboratory computers, dedicated, multi-user and networks. *Mailing Add:* 5432 Carolina Pl Washington DC 20016

JENNISON, DWIGHT RICHARD, b Teaneck, NJ, June 11, 43; m 68; c 3. SURFACE SCIENCE, ELECTRONIC STRUCTURE. *Educ:* Rensselaer Polytech Inst, BS, 65, MS, 73, PhD(physics), 74. *Prof Exp:* Res asst physics, Univ NDak, 69-70; teaching asst, Rensselaer Polytech Inst, 71, NIH trainee, 71-74; res assoc, Dept Physics & Mat Res Lab, Univ Ill, Urbana, 74-75, res asst prof physics, 76, assoc, 74-76; mem tech staff, Solid State Theory Div, 77-80, SUPVR, CONDENSED MATTER THEORY DIV 1151, SANDIA NAT LAB, 81- *Concurrent Pos:* Vis scientist, Univ Liverpool, UK, 83 & Tech Univ Munich, Ger, 85. *Mem:* Am Phys Soc. *Res:* Electronic structure of solids, surfaces and molecules; theory of surface spectroscopies and stimulated desorption; theory of high-temperature superconducting compounds. *Mailing Add:* Sandia Nat Labs 1151 Albuquerque NM 87185

JENNRICH, ELLEN COUTLEE, b Kankakee, Ill, Dec 16, 39; m 71; c 2. ANIMAL BEHAVIOR, ECOLOGY. *Educ:* Wayne State Univ, BA, 60, MS, 62; Univ Calif, Los Angeles, PhD(zool), 66. *Prof Exp:* Lectr biol, Mt St Mary's Col, 63-64; lectr, Univ Calif, Riverside, 66-68; lectr zool, Univ Calif, Los Angeles, 68-72; SUBSTITUTE TEACHER SCI & BIOL, WESTLAKE SCH, LOS ANGELES, 85- *Res:* Population biology of starlings; comparative breeding behavior of goldfinches, fluctuations in population size, avian communication, maintenance and agonistic behavior. *Mailing Add:* 3400 Purdue Ave Los Angeles CA 90066

JENNRICH, ROBERT I, b Milwaukee, Wis, Feb 11, 32. STATISTICS. *Educ:* Univ Wis, BS, 54, MS, 56; Univ Calif, Los Angeles, PhD(math), 60. *Prof Exp:* Asst prof math, Univ Wis, 60-62; asst prof math & asst res statistician, 62-70, assoc prof math & biomath, 70-74, PROF MATH & BIOMATH, UNIV CALIF, LOS ANGELES, 74- *Mem:* Am Statist Asn; Inst Math Statist. *Res:* Computer algorithms for data analysis; non-linear least squares, methods and statistical properties; factor analysis, rotation and maximum likelihood algorithms; analysis of variance, properties of the mixed model; time series analysis. *Mailing Add:* Dept of Math Univ of Calif Los Angeles CA 90024

JENNY, HANS K, b Glarus, Switz, Sept 14, 19; nat US; m 49; c 3. ELECTRONIC & ELECTRICAL ENGINEERING. *Educ:* Swiss Fed Inst Technol, MSEE, 43. *Prof Exp:* Asst prof & res engr, Swiss Fed Inst Technol, 43-46; RETIRED. *Mem:* Fel Inst Elec & Electronics Engrs. *Res:* Parametric amplifiers; variable capacitance and tunnel amplifiers; phase shifters; microwave laser modulators and detectors; microwave devices and systems; engineering organizations; technical information, including communications, publications and information systems. *Mailing Add:* 210 Riveredge Dr RD 1 Leola PA 17540

JENNY, NEIL ALLAN, b Milwaukee, Wis, Sept 6, 36; m 60; c 5. PESTICIDE CHEMISTRY. *Educ:* Univ Wis, BS, 58; Univ Kans, PhD(med chem), 63. *Prof Exp:* Res chemist, Polymer Div, Morton Chem co, Ill, 63-66, res chemist, Org Div, 66-69, contract mfg coordr, Div Schering Agr, 77-79, supvr anal res, Nor-Am Agr Prod, 69-85, SR RES CHEMIST, SHEREX, DIV SCHERING AGR, BERLIN, 86- *Mem:* Am Chem Soc; Am Pharmaceut Asn; Tech Asn Pulp & Paper Indust. *Res:* Resistance factors of crops; pesticide residue analytical methods; pesticide metabolism; effect of pesticide residues on environment; retail pharmacy; pesticide formulation; quality control; production and contract manufacturing; residue chemistry; paper chemistry; surfactant chemistry. *Mailing Add:* PO Box 1018 2001 Afton Rd Janesville WI 53545

JENS, WAYNE H(ENRY), b Manitowoc, Wis, Dec 20, 21; m 46; c 4. MECHANICAL ENGINEERING & NUCLEAR ENGINEERING. *Educ:* Univ Wis, BS, 43; Purdue Univ, MS, 48, PhD(mech eng), 49. *Prof Exp:* Eng designer, NAm Aviation, Inc, 43-44; eng asst heat transfer, Purdue Univ, 46-49; head eng anal group, Argonne Nat Lab, 49-53; proj leader & mgr, Nuclear Develop Corp Am, 53-57; gen mgr, Atomic Power Develop Assocs, Inc, 57-71; mgr eng & construct, Detroit Edison Co, 76-78, asst vpres eng & construct, 78-80, vpres nuclear oper, 80-86; PRES, JENS & JENS INC, 88- *Concurrent Pos:* Mem bd trustees, Argonne Univ Asn, 77-; mem nuclear power div comn, Elec Power Res Inst, 78-; mem, Nuclear Training Accrediting Bd, Inst Nuclear Power Opers, 85-88. *Honors & Awards:* Gold Award, Eng Soc of Detroit, 78. *Mem:* Fel Am Nuclear Soc; Am Soc Mech Engrs. *Res:* Boiling heat transfer; nuclear fuel irradiation stability; reactor design; nuclear operations and training. *Mailing Add:* 1220 Wild Azalea Pt Seneca SC 29678

JENSEN, ADOLPH ROBERT, b Elmhurst, Ill, Apr 14,15; m 50; c 2. ANALYTICAL CHEMISTRY. *Educ:* Wheaton Col, BS, 37; Univ Ill, MS, 40, PhD(anal chem), 42. *Prof Exp:* Asst chem, Wheaton Col, 37-38; asst anal chem, Univ Ill, 38-42; asst chemist & head anal chem sect, Aircraft Engine Res Lab, Nat Adv Comt Aeronaut, 42-46; from asst prof to prof, 46-83, chmn dept, 56-71, EMER PROF CHEM, BALDWIN-WALLACE COL, 84- *Concurrent Pos:* Consult, Stouffer Frozen Foods, Solon, Ohio, 85. *Mem:* AAAS; Am Chem Soc; Sigma Xi. *Res:* Analytical chemistry of foods; analytical chemistry of fuels and lubricants; instrumental methods of analysis. *Mailing Add:* 25527 Butternut Rd North Olmsted OH 44070-4505

JENSEN, ALBERT CHRISTIAN, b New York, NY, Jan 26, 24; m; c 4. ECOLOGY, MARINE ENVIRONMENTAL SCIENCE. *Educ:* State Univ NY Syracuse, BS, 51, MS, 54. *Prof Exp:* Res biologist marine fisheries, US Fish & Wildlife Serv, Woods Hole, Mass, 54-65; managing ed marine sci, Marine Lab, Univ Miami, 65-67; asst dir coastal environ, NY State Dept Environ Conserv, 67-77; CONSULT COASTAL ENVIRON, ENVIRON ASSOCS, 77- *Concurrent Pos:* Adv, Atlantic States Marine Fisheries Comn, Washington, DC, 67-80 & US Deleg to Int Comn Northwest Atlantic Fisheries, 72-75; asst prof, Grad Dept Marine Sci, C W Post Col, 75-77; prof, Cent Fla Community Col, 78- *Honors & Awards:* George Washington Hon Medal, Freedoms Found, 73; Spec Sci Book Award, NY Acad Sci, 79. *Mem:* Nat Marine Educrs Asn; Am Inst Fishery Res Biologists; Fla Acad Sci. *Res:* Marine science education; coastal zone management; fisheries management. *Mailing Add:* Environ Assocs PO Box 223 Inglis FL 32649-0223

JENSEN, ALDON HOMAN, b Massena, Iowa, Dec 20, 22; m 48; c 3. ANIMAL NUTRITION. *Educ:* Univ Ill, BS, 49, MS, 50; Iowa State Col, PhD, 53. *Prof Exp:* Asst, Iowa State Col, 52-54, asst prof, 54-60; from asst prof to assoc prof, 60-68, PROF ANIMAL SCI, UNIV ILL, URBANA, 68- *Mem:* Am Soc Animal Sci; Animal Nutrit Res Coun; Sigma Xi. *Res:* Animal physiology and environment. *Mailing Add:* 1301 W Gregory Dr Urbana IL 61801

JENSEN, ARNOLD WILLIAM, b Racine, Wis, Apr 30, 28; m 60; c 1. ORGANIC CHEMISTRY, POLYMER CHEMISTRY. *Educ:* Dana Col, Nebr, BA, 50; Okla State Col, PhD(chem), 58. *Prof Exp:* Res chemist, Dow Chem Co, Tex, 52-53; RES ASSOC, TEXTILE FIBERS DEPT, E I DU PONT DE NEMOURS & CO, INC, 58- *Mem:* AAAS; Am Chem Soc; Sigma Xi. *Res:* Nuclear magnetic resonance; infrared; synthetic fibers. *Mailing Add:* 213 Camellia Dr Charlottesville VA 22903

JENSEN, ARTHUR SEIGFRIED, b Trenton, NJ, Dec 24, 17; m 41; c 3. ELECTRONIC PHYSICS. *Educ:* Univ Pa, BS, 38, MS, 39, PhD(physics), 41; Westinghouse Sch Appl Eng Sci, dipl(advan eng technol), 72; dipl(comput sci), 77. *Prof Exp:* Lab asst physics, Univ Pa, 38-39; res physicist, Naval Res Lab, Washington, DC, 41; res physicist labs, Radio Corp Am, 45-57; mgr spec electron devices, Electronic Tube Div, Appl Res Dept, 57-65, SR ADV PHYSICIST, DEFENSE & ELECTRONICS CTR, WESTINGHOUSE ELEC CORP, 65- *Concurrent Pos:* Instr physics, US Naval Acad, 41-46; Regist prof eng, Md, 66; vchmn, Md State Bd Prof Engrs, 79-86. *Mem:* AAAS; Am Phys Soc; Am Asn Physics Teachers; fel Inst Elec & Electronics Engrs; Nat Coun Eng Examrs; Sigma Xi; Soc Photo-Optical Instrumentation Engrs. *Res:* Solid state electro-optical imaging systems; imaging techniques and sensing devices; image quality and information theory; noise and image sensor detection limitations; electron optics and integrated circuits; infrared image sensors and systems; modeling solid state devices and systems. *Mailing Add:* 5602 Purlington Way Baltimore MD 21212

JENSEN, BARBARA LYNNE, US citizen; c 2. PHYSICS, MATHEMATICS. *Educ:* Univ Utah, BS, 64; Columbia Univ, MA, 72, PhD(physics), 73. *Prof Exp:* Res assoc physics, IBM Thomas J Watson Res Ctr, Yorktown Heights, NY, 70-73; instr, Univ Lowell, 74-77; ASST PROF PHYSICS, BOSTON UNIV, 78- *Mem:* AAAS; Am Phys Soc; Optical Soc Am. *Res:* Solid state physics; condensed matter physics; plasma physics; interaction of radiation and matter; optical and electronic properties of semiconductors and metals at high frequencies. *Mailing Add:* 1901 Stearns Hill Rd Waltham MA 02154

JENSEN, BETTY KLAINMINC, b Poland, Jun 20, 49; US citizen; m 71; c 4. ENERGY USE & THE ENVIRONMENT, FUELS TECHNOLOGY. *Educ:* Brooklyn Col, BS, 70; Columbia Univ, MS, 73, MPhil, 74, PhD (physics), 76; St John's Univ, MBA, 81. *Prof Exp:* Fac fel physics, Columbia Univ, 70-76; adj instr physics, City Univ New York, 73-76; sr physicist, 76-81, prin physicist, 81-84, mgr res & develop nuclear & environ sci, 84-88, MGR RES & DEVELOP FUELS & ENVIRON SERV, PUB SERV ELEC & GAS CO, 88- *Concurrent Pos:* Adv, Gas-Cooled Reactor Asn, 78-87, Off Technol Assessment, 86-87, Elec Power Res Inst & Princeton Plasma Physics Lab, 77-, Edison Elec Inst, 87- & Hazardous Substance Mgt Res Ctr, 88- *Mem:* Sigma Xi; AAAS; Inst Elec & Electronics Engrs; Air & Waste Mgt Asn. *Res:* Electric and gas utilities; commercialization aspects of new technology. *Mailing Add:* Pub Serv Elec & Gas 80 Park Plaza T16 Newark NJ 07101

JENSEN, BETTY KLAINMINC, b Poland, June 20, 49; US citizen; m 71; c 4. ENVIRONMENTAL POLICY, RISK COMMUNICATION. *Educ:* Brooklyn Col, BS, 70; Columbia Univ, NY, PhD(physics), 76; St Johns Univ, NY, MBA, 81. *Prof Exp:* Instr physics, City Univ NY, 73-76; sr physicist, Pub Serv Elec & Gas Co, 76-79, prin physicist, 79-84, Nuclear & Environ Prog mgr, 84-88, FUELS & ENVIRON SCI MGR, PUB SERV ELEC & GAS CO, 88- *Concurrent Pos:* Adv, Elec Power Res Inst & Princeton Plasma Physics Lab, 77-, Mass Inst Technol, 78-89, Gas Cooled Res Assocs, 79-88, Off Technol Assessment, 82-84 & NJ Inst Technol, 89- *Mem:* Air & Waste Mgt Asn; AAAS; Inst Elec & Electronics Engrs; Am Phys Soc; Bioelectromagnetics Soc. *Res:* Environmental impact of electric power generation, transmission and distribution; risk communication; commercialization of new technologies. *Mailing Add:* 630 Armstrong Ave Staten Island NY 10308

JENSEN, BRUCE A, b Spencer, Iowa, Aug 6, 30; m 51; c 2. MATHEMATICS. *Educ:* Dana Col, BA, 52; Univ Wis-Madison, MS, 55; Univ Nebr-Lincoln, PhD(math), 66. *Prof Exp:* Instr math & physics, Dana Col, 55-58, asst prof math, 58-59; from asst prof to assoc prof, Nebr Wesleyan Univ, 59-66; assoc prof, 66-73, PROF MATH, PORTLAND STATE UNIV, 73-, DEPT CHMN, 86- *Mem:* Am Math Soc; Math Asn Am. *Res:* Algebraic semigroups; finiteness conditions on infinite semigroups; extensions of semigroups; decompositions of semigroups. *Mailing Add:* Dept Math Prof/ Dept Head Portland State Univ PO Box 751 Portland OR 97207

JENSEN, BRUCE DAVID, b Chicago, Ill, Mar 22, 54; m 87; c 1. PHARMACEUTICAL RESEARCH & DEVELOPMENT, CLINICAL DIAGNOSTIC RESEARCH & DEVELOPMENT. *Educ:* Univ Calif, Berkeley, AB, 77; Univ Rochester, MS, 80, PhD(biophys), 84. *Prof Exp:* Res assoc, Smith Kline & French Labs, 83-87, assoc sr investr, 87-88; RES GROUP LEADER, ZYNAXIS CELL SCI, INC, 88- *Mem:* Biophys Soc; NY Acad Sci; Am Soc Cell Biol; Soc Anal Cytol. *Res:* Production of novel clinical diagnostic assays; novel drug delivery systems. *Mailing Add:* Dept Biol Zynaxis Cell Sci Inc 371 Phoenixville Pike Malvern PA 19355-9603

JENSEN, BRUCE L, b Three Rivers, Mich, Aug 6, 44; m 65; c 2. ORGANIC CHEMISTRY. *Educ:* Western Mich Univ, BS, 66, PhD(org chem), 70. *Prof Exp:* Nat Cancer Inst fel, Univ Mich, Ann Arbor, 70-72; instr chem, Univ Maine, Orono, 72-73; asst prof, 73-78, ASSOC PROF CHEM, UNIV MAINE, ORONO, 78- *Concurrent Pos:* Sabbatical leave, Univ Southern Calif, 83-84. *Mem:* Am Chem Soc. *Res:* Organic synthesis; infrared, nuclear magnetic resonance and mass spectroscopy; heterocycles; natural products; medicinal chemistry; halonium ion chemistry; steroids; antineoplastic drugs; antiarrhythmic drugs. *Mailing Add:* Dept Chem Univ Maine Orono ME 04469

JENSEN, CLAYTON EVERETT, b Hartford, Conn, Oct 23, 20; m 77; c 3. METEOROLOGY, COMPUTER SCIENCE. *Educ:* Trinity Col, Conn, BS, 44; Mass Inst Technol, SM, 51, PhD(meteorol), 60. *Prof Exp:* Chief eval & develop div, Hq, Air Weather Serv, 51-53, detachment comdr & staff weather officer, Air Force Cambridge Res Labs, 56-58, meteorol systs analyst, Strategic Air Command, 60-63; assoc prof math & dir comput ctr, Va Mil Inst, 63-65; chief supporting res group, Off Fed Coord Meteorol, Environ Sci Serv Admin, 65-69, chief fed plans & coord div, 69-71, chief environ monitoring div, Nat Oceanic & Atmospheric Admin, 71-73, dep assoc adminr, 73-75; CONSULT & WEATHER ANALYST, WINK TV, CBS, FT MYERS, 78- & US DEPT STATE, 80- *Concurrent Pos:* Lectr, Univ Omaha, 60-63; consult, Nat Environ Satellite Ctr, 64-65; chmn, Interdept Comt Appl Meteorol Res, 65-73 & Interdept Comt Meteorol Serv; observer, Interdept Comt Atmospheric Sci, Fed Coun Sci & Technol; fed coordr meteorol, Dept Com, 73-75; govt & acad consult. *Honors & Awards:* Gold Medal Award, Dept Comm, 72. *Mem:* Am Meteorol Soc; Sigma Xi. *Res:* General circulation of the atmospheres; cloud physics; instrumentation for atmospheric electricity and airborne measurement of liquid water; satellite meteorology; computer education; global environmental research. *Mailing Add:* 4419 SE 20th Pl Cape Coral FL 33904

JENSEN, CLYDE B, b Rigby, Idaho, Aug 14, 48; m 69; c 2. PHARMACOLOGY. *Educ:* Brigham Young Univ, BS, 70; Univ NDak, MS, 73, PhD(pharmacol), 74. *Prof Exp:* Asst prof pharmacol, Okla Col Osteop Med & Surg, 74-, asst dean student affairs, 77-; pres, WVa Sch Osteopath Med; UNIV HEALTH & SCI, COL MED. *Concurrent Pos:* Consult pharmacol, Nat Bd Examrs Osteop Physicians & Surgeons Inc, 75- *Mem:* Sigma Xi. *Res:* The effects of centrally-acting and ototoxic drugs on the vestibulo-ocular reflex arc. *Mailing Add:* Univ Health & Sci Col Med 2105 Independent Blvd Kansas City MO 64124

JENSEN, CRAIG LEEBENS, b Rochester, Minn, Dec 8, 50. METALLURGY. *Educ:* Univ Minn, BS, 73; Iowa State Univ, PhD(metall), 77. *Prof Exp:* Asst prof mat sci, Univ Minn, 77-81; SR SCI ASSOC, ALCOA TECH CO, 81- *Mem:* Am Soc Metals; Am Inst Mining, Metall & Petrol Engrs; Sigma Xi. *Res:* Transport properties of hydrogen in transition metals. *Mailing Add:* 102 Weir Dr Pittsburgh PA 15215

JENSEN, CREIGHTON RANDALL, b Harlan, Iowa, Dec 27, 29; div. SOIL PHYSICS. *Educ:* Calif State Polytech Col, BS, 56; Iowa State Univ, MS, 59, PhD(agron), 61. *Prof Exp:* Res asst soil physics, Iowa State Univ, 56-61; soil physicist, Univ Calif, Riverside, 62-63 & 64-67; DIR, JENSEN INSTRUMENTS, 68- *Mem:* Int Soc Soil Sci; Am Soc Agron; Soil Sci Soc Am. *Res:* Soil aeration. *Mailing Add:* 2021 S Seventh St Tacoma WA 98405-3014

JENSEN, CYNTHIA G, b Wheeling, WVa, Nov 7, 38; m 60; c 2. CELL BIOLOGY. *Educ:* Brown Univ, AB, 60; Univ Minn, PhD(zool), 66. *Prof Exp:* Res assoc biol, Univ Ore, 66-68; asst prof path, Univ Utah, 68-71; sr lectr anat, 72-87, ASSOC PROF ANAT, SCH MED, UNIV AUCKLAND, 88- *Concurrent Pos:* Vis scientist, NY State Dept Health, Albany, 84-85. *Mem:* NZ Soc Electron Micros (vpres, 83-85, pres, 85-87); Am Soc Cell Biol; Electron Micros Soc Am; Australia & NZ Soc Cell Biol (NZ secy/treas, 86-, vpres, 90-); Anat Soc Australia N; Asia Pac Orgn Cell Biol. *Res:* Ultrastructural studies of cell division; microtubule structure and organization; cells exposed to anti-tumor drugs; neural cytoskeleton. *Mailing Add:* Dept Anat Sch Med Univ Auckland Auckland New Zealand

JENSEN, DAVID, b San Francisco, Calif, Oct 14, 26; m 50; c 2. MEDICAL PHYSIOLOGY. *Educ:* Univ Calif, Berkeley, BA, 48, MA, 50, PhD(physiol), 54. *Prof Exp:* Asst res physiol chemist, Sch Med, Univ Calif, Los Angeles & Vet Admin Hosp, 55-56, Am Heart Asn estab investr & res assoc, Scripps Inst, Univ Calif, 56-57; asst prof physiol, Med Ctr, Univ Colo, Denver, 67-71; RETIRED. *Concurrent Pos:* Los Angeles Co Heart Asn estab investr, Univ Calif, Los Angeles, 55-56, Riverside Co Heart Asn fel, 56-58 & San Diego Co Heart Asn res fel, 58-60; Am Heart Asn advan res fel, Scripps Inst, Univ Calif, 60-62, estab investr, 62-67; elected to Royal Soc Med, London, 70; sci author, 71- *Mem:* AAAS; Soc Gen Physiol; Roy Soc Med, London. *Res:* Basic mechanisms of cardiac automatism using electrophysiological techniques as well as biochemical approach; comparative physiological studies on a variety of species; intrinsic cardiac rate regulation; neuroanatomy; neurophysiology. *Mailing Add:* 121 Arbor Dr Moab UT 84532

JENSEN, DAVID JAMES, b Racine, Wis, May 10, 35; m 56; c 5. ANALYTICAL CHEMISTRY. *Educ:* Univ Wis-Milwaukee, BS, 58; Purdue Univ, MS, 65, PhD(biochem), 67. *Prof Exp:* Instr chem, Univ Wis-Milwaukee, 57-61; instr anal chem & state chemist of Ind, Purdue Univ, 61-67; res chemist anal chem, 67-84, SR LAB SUPVR, DOW CHEM USA, 84- *Mem:* Am Chem Soc; Sigma Xi. *Res:* Studies on pesticide residues; analysis of pesticide formulations and associated analytical methods development; priority pollutants analysis by GC/MS, product analysis, and industrial quality assurance. *Mailing Add:* 2218 Cranbrook Dr Midland MI 48640-3218

JENSEN, DONALD RAY, b Nashville, Tenn, Apr 25, 32; m 64; c 4. MATHEMATICAL STATISTICS. *Educ:* Univ Tenn, BS, 55; Iowa State Univ, MS, 57, PhD(statist, soils), 62. *Prof Exp:* Asst prof statist, Ore State Univ, 62-65; from asst prof to assoc prof, 65-73, PROF STATIST, VA POLYTECH INST & STATE UNIV, 73- *Concurrent Pos:* NIH career develop award, 67-72. *Mem:* Biomet Soc; Am Statist Asn; Am Inst Math Statist; Soc Indust Appl Math. *Res:* Probability inequalities; multivariate statistical analysis; multivariate distributions; simultaneous statistical inference; large-sample theory. *Mailing Add:* Dept of Statist Va Polytech Inst & State Univ Blacksburg VA 24061

JENSEN, DONALD REED, b Pocatello, Idaho, May 4, 31; m 56; c 3. MAMMALIAN PHYSIOLOGY. *Educ:* Idaho State Univ, BS, 53; Univ Wash, BA, 54; Utah State Univ, MS, 61, PhD(physiol), 64. *Prof Exp:* NIH fel, Inst Physiol Chem, Univ Cologne, 64-66; asst prof, 66-69, asst to chmn dept biol sci, 68-78, PROF PHYSIOL, ILL STATE UNIV, 69- *Mem:* Fel AAAS; Am Soc Zool; Sigma Xi. *Res:* Toxic effect of gossypol and selected pesticides on physiological processes. *Mailing Add:* Dept Biol Ill State Univ Normal IL 61761

JENSEN, DOUGLAS ANDREW, b Muskegon, Mich, Oct 18, 40; m 65; c 2. ELEMENTARY PARTICLE PHYSICS. *Educ:* Kalamazoo Col, AB, 63; Univ Chicago, MS, 65, PhD(physics), 70. *Prof Exp:* NSF fel, Joseph Henry Labs, Princeton Univ, 70-71, asst prof physics, 71-77; ASSOC PROF PHYSICS, UNIV MASS, AMHERST, 77- *Mem:* Am Asn Physics Teachers; Am Phys Soc. *Res:* Elementary particle physics; weak interaction and symmetries; hadron production of strange and charmed particles. *Mailing Add:* Dept Physics & Astron Univ Mass Grad Res 930 C Amherst MA 01003

JENSEN, EDWIN HARRY, b Phillips, Wis, Aug 29, 22; m 47; c 2. FORAGE ALFALFA. *Educ:* Univ Wis, BS, 49, MS, 50 & PhD(agron & soil), 52. *Prof Exp:* Soil scientist, Soil Conserv Serv, USDA, 48-49; asst agronomist & asst prof agron, Univ Nev, 52-54; exten agronomist, Univ Minn, 54-56; assoc agronomist & assoc prof agron, 56-64, PROF AGRON & AGRONOMIST, UNIV NEV, RENO, 64- *Concurrent Pos:* Vis prof, People's Repub China, 83, Kyong Hee Univ, Scoul, Korea, 85 & Univ Seregia, Italy, 87. *Mem:* Am Soc Agron; Crop Sci Soc; Sigma Xi. *Res:* Forage crop management; forage quality; water use; alfalfa nodalation. *Mailing Add:* Dept Plant Sci Univ Nev Reno NV 89507

JENSEN, ELWOOD VERNON, b Fargo, NDak, Jan 13, 20; m 41, 83; c 2. ENDOCRINOLOGY. *Educ:* Wittenberg Col, AB, 40; Univ Chicago, PhD(org chem), 44. *Hon Degrees:* DSc, Wittenberg Univ, 63, Acadia Univ, 76, Med Col Ohio, 91. *Prof Exp:* Asst prof, Dept Surg, Univ Chicago, 47-51, from asst prof to assoc prof, Dept Biochem, 51-60, from asst prof to prof, Ben May Lab Cancer Res, 51-63, Am Cancer Soc-Charles Hayden Found res prof, Dept Physiol & Ben May Lab Cancer Res, 63-69, dir, Lab, 69-82, prof biophys, 73-82, prof physiol, 77-82, prof biochem & Chas B Huggins distinguished serv prof biol sci, 80-90, EMER PROF BIOL, UNIV CHICAGO, 90- *Concurrent Pos:* Guggenheim fel, Swiss Fed Inst Technol, 46-47; USPHS spec fel, 58; vis prof, Max Planck Inst, Munich, Germany, 58 & Kyoto Univ, 65; res dir, Ludwig Inst Cancer Res, Zurich, Switz, 83-87; scholar-in-residence, Fogarty Int Ctr, NIH, 88 & Med Col, Cornell Univ, 90-91. *Honors & Awards:* D R Edwards Medal, 70; La Madonnina Prize, 73; GHA Clowes Award, 75; Papanicolaou Award, 75; Prix Roussel, 76; Nat Award, Am Cancer Soc, 76; Amory Prize, 77; Gregory Pincus Mem Award, 78; Gairdner Award, 79; C F Kettering Prize, 80; Lucy Wortham James Award, 80; Nat Acad Clin Biochem Award, 81; Pharmacia Award, 82; Rolf Luft Medal, 83; Hubert Humphrey Award, 83. *Mem:* Nat Acad Sci; Am Acad Arts & Sci; Am Chem Soc; Am Soc Biol Chemists; Endocrine Soc (pres, 80-81); Am Asn Cancer Res; AAAS. *Res:* Steroid hormone receptors; breast cancer; proteins; organophosphorus chemistry. *Mailing Add:* NY Hosp Cornell Med Ctr 525 E 68th St Box 340 New York NY 10021-4873

JENSEN, EMRON ALFRED, b Richfield, Utah, Jan 5, 25; m 49; c 8. PARASITOLOGY, PROTOZOOLOGY. *Educ:* Utah State Univ, BS, 50, MS, 61, PhD(zool), 63. *Prof Exp:* Teacher high sch, Idaho, 50-52; technician, Am Cyanamid Co, 52-53; teacher elem sch, Utah, 54-59; lab instr zool, Utah State Univ, 59-63; from asst prof to prof, Weber State Col, 63-83, chmn dept, 70-83; RETIRED. *Res:* Parasite protozoa, particularly trichomonads. *Mailing Add:* Dept Zool Weber State Col Ogden UT 84408

JENSEN, ERIK HUGO, b Fredericia, Denmark, June 27, 24; nat US; m 49; c 3. PHARMACY. *Educ:* Royal Danish Sch Pharm, BSc, 45, MS, 48, PhD, 54. *Prof Exp:* Res assoc, Upjohn Co, 50-56; head pharmaceut res & develop dept, Ferrosan Inc, Malmo, Sweden, 56-57; res assoc pharm, Upjohn Co, 57-62, sect head qual control, 62-63, mgr, 63-66, asst dir qual control, 66-81, dir, 81-85, exec dir control develop & admin, 85-86; PRES, JENSEN ENTERPRISES, 86- *Honors & Awards:* W E Upjohn Award, 62. *Mem:* Am Chem Soc; Am Asn Pharmaceut Scientists. *Res:* Controlled release of pharmaceuticals; stability of pharmaceuticals; assays of pharmaceuticals; analytical applications of sodium borohydride; analytical chemistry; quality control procedures. *Mailing Add:* Jensen Enterprises 2125 Crosswind Dr Kalamazoo MI 49008-1734

JENSEN, ERLING, microbiology, for more information see previous edition

JENSEN, ERLING N, nuclear physics; deceased, see previous edition for last biography

JENSEN, GARY LEE, b Hyrum, Utah, Sept 5, 33; m 58; c 5. EXPERIMENTAL NUCLEAR PHYSICS. *Educ:* Utah State Univ, BS, 58; Univ Mich, MS, 60, PhD(physics), 64. *Prof Exp:* From asst prof to assoc prof, 66-83, PROF PHYSICS, BRIGHAM YOUNG UNIV, 83- *Mem:* Am Phys Soc. *Res:* Decay modes and branching ratios for the K-plus meson; low-energy nuclear physics; neutron-energy spectrometers. *Mailing Add:* 57 S Eastwood Dr Orem UT 84058

JENSEN, GARY RICHARD, b Miles City, Mont, Mar 19, 41; m 65; c 3. DIFFERENTIAL GEOMETRY. *Educ:* Mass Inst Technol, BS, 63; Univ Calif, Berkeley, PhD(math), 68. *Prof Exp:* Asst prof math, Carnegie-Mellon Univ, 68-69; fel, 69-70, from asst prof to assoc prof, 70-82, PROF MATH, WASHINGTON UNIV, ST LOUIS, 83-, CHMN DEPT, 90- *Concurrent Pos:* Vis assoc prof math, Univ Calif, Berkeley, 76-77 & Univ Nancy, France, 82-83. *Mem:* Am Math Soc; Math Asn Am. *Res:* Differential geometry, especially of submanifolds of homogeneous spaces. *Mailing Add:* Math Dept Box 1146 Washington Univ St Louis MO 63130

JENSEN, GORDON D, b Seattle, Wash, Jan 28, 26; m 57; c 3. PEDIATRICS, PSYCHIATRY. *Educ:* Yale Univ, MD, 49. *Prof Exp:* Asst prof pediat, Sch Med, Univ Wash, 57-60; res asst prof psychiat, 61-62, asst psychiat, 62-65, from asst prof to assoc prof, 65-69; PROF PSYCHIAT & PEDIAT, SCH MED, UNIV CALIF, DAVIS, 69- *Concurrent Pos:* Mem core staff, Regional Primate Res Ctr, Univ Wash, 67-69; sr consult child psychiat, Sacramento Med Ctr, 69-74. *Mem:* Soc Biol Psychiat; Animal Behav Soc; Am Acad Pediat; Am Col Psychiat; Psychiat Res Soc. *Res:* Primate behavior; sexuality; aging. *Mailing Add:* Div of Ment Health Univ of Calif Sch Med Davis CA 95616

JENSEN, HANNE MARGRETE, b Copenhagen, Denmark, Dec 9, 35; US citizen; m 57; c 4. PRECANCER. *Educ:* Univ Wash, MD, 61; Am Bd Path, cert anatomic & clin path, 68, cert blood banking, 79. *Prof Exp:* Fel exp path, Dept Path, Sch Med, Univ Wash, 65-67; asst prof, 69-79, ASSOC PROF PATH, DEPT PATH, SCH MED, UNIV CALIF, DAVIS, 79- *Concurrent Pos:* Mem Treatment Comt, Breast Cancer Task Force, Nat Cancer Inst, 77-81, prin investr, Contract Breast Cancer Task Force, 78-81. *Mem:* Am Asn Blood Banks; AAAS; Am Soc Clin Pathologists; Int Acad Path. *Res:* Assessment of precancer of breast parenchyma, using assays for angiogenesis factor; assays of breast fluids for angiogenesis factor; prediction of high cancer risk; morphologic studies of precancer of the prostate gland. *Mailing Add:* Dept Path Sch Med Univ Calif Davis CA 95616

JENSEN, HARBO PETER, b Boston, Mass, Mar 27, 48; m. POLYMER CHEMISTRY. *Educ:* Northeastern Univ, BA, 71; Mass Inst Technol, PhD(org chem), 74. *Prof Exp:* Polaroid Corp, 67-70; Chevron Res Co, Standard Oil Co Calif, 74-78, proj supvr, Huntington Beach Co, 78-80; pres, Timoc, 75-80; govt affairs coordr, Chevron USA, 80-81, foreign staff adv, 81-90, MGR, CHEVRON INST OIL CO, STANDARD OIL CO, CALIF, 90- *Concurrent Pos:* Pres & chmn, Cal Bionics, 81- *Mem:* AAAS; Am Chem Soc; Sigma Xi; Contact Lens Mfrs Asn. *Res:* Petroleum science and synthetic fuels; polymer science, especially hydrophilic polymers for soft contact lenses. *Mailing Add:* Chevron Corp 555 Market St San Francisco CA 94105

JENSEN, HAROLD JAMES, b Sunnyside, Wash, Sept 16, 21; m 46; c 3. NEMATOLOGY. *Educ:* Univ Calif, BS, 47, PhD(nematol), 50. *Prof Exp:* Instr & asst, 50-51, from asst prof to prof bot & nematologist & from asst nematologist to nematologist, Ore State Univ, 51-84; RETIRED. *Concurrent Pos:* Consult, Hawaiian Sugar Planters Asn, 58. *Mem:* Am Phytopath Soc; Soc Nematol (vpres, 70-71, pres, 71-72). *Res:* Identification, symptomatology and pathology of plant diseases caused by nematodes; nematological control techniques, taxonomy, and teaching; relationships of nematodes with other plant pathogens. *Mailing Add:* 23619 Harris Rd Philomath OR 97370

JENSEN, J(OHN) H(ENRY), JR, b Aurora, Ill, June 17, 16; m 48. CHEMICAL ENGINEERING. *Educ:* SDak Sch Mines & Tech, BS, 39; Iowa State Univ, MS, 42, PhD(chem eng), 48. *Prof Exp:* Instr, Iowa State Univ, 40-48; sr chem engr, Tenn Eastman Co, 48-81; RETIRED. *Mem:* Instrument Soc Am. *Res:* Production of acetic anhydride; application of a digital computer to a chemical manufacturing process; process control by analog instruments or by digital computer; analog computing; interactive computer graphics system. *Mailing Add:* 4560 Old Stage Rd Kingsport TN 37664

JENSEN, JAMES BURT, b Los Angeles, Calif, Mar 8, 43; m 65; c 6. MALARIAL HOST-PARASITE RELATIONSHIPS. *Educ:* Brigham Young Univ, BS, 70, MS, 72; Auburn Univ, PhD(zool & parasitol), 76. *Prof Exp:* Teaching fel malariology, Rockefeller Univ, 75-76, asst prof parasitol, 76-79; asst prof, 79-82, assoc prof, 82-87, PROF PARASITOL & MICROBIOL, MICH STATE UNIV, 87- *Concurrent Pos:* Vis asst prof, Med Col, Cornell Univ, 77-79, Med Sch, Univ Col, Ibadan, Nigeria, 78 & Med Sch Paulista, Sao Paulo, Brazil, 79; consult malaria, WHO, 77-78, Pan Am Health Orgn, Brazil, 79, US Naval Med Res Unit No 2, Jakarta & Dept State, USAID-NIH, Israel-Egypt, 83-84, USAID Malaria Immunity & Vaccination Res Prog, 88- *Mem:* Soc Protozoologists; Am Soc Parasitologists; Am Soc Trop Med & Hyg; Royal Soc Trop Med & Hyg; Am Micros Soc; AAAS. *Res:* Biology, biochemistry and host-parasite relationship of protozoan parasites, specifically malarial parasites; human immune response to falcparum malaria; biochemical pharmacology of antimalarial chemotherapeutics. *Mailing Add:* Dept Microbiol & Pub Health Mich State Univ East Lansing MI 48824

JENSEN, JAMES EJLER, nuclear engineering, management, for more information see previous edition

JENSEN, JAMES LE ROY, b Hopkins, Minn, July 2, 15; m 40; c 5. NUTRITION. *Educ:* Univ Minn, BS, 37, MS, 38, PhD(agr biochem), 41. *Prof Exp:* Asst dir biol dept, Distillation Prod, Inc, NY, 42-45; sales rep, Nutrena Mills, Inc, Minn, 47-52, asst div mgr, Nebr, 52-54, div sales mgr, 54-55, qual control mgr, Minn, 56-59; gen mgr, Beebe Labs, Inc, 59-60; tech dir & div sales mgr, Ulmer Pharmacal Co, 60-70; dist sales mgr, Physicians & Hosp Supply Co, 70-74, sales mgr, 74-77; RETIRED. *Concurrent Pos:* Nutrit officer, Sanit Corp, US Army, 45-47. *Mem:* Fel Am Chem Soc; Sigma Xi. *Res:* Role of vitamins A, C and E in nutrition and physiology; pharmaceuticals and biologicals; chemistry. *Mailing Add:* 4639 Williston Rd Minnetonka MN 55345

JENSEN, JAMES LESLIE, b Tulare, Calif, Oct 17, 39; m 60; c 2. BIOPHYSICAL ORGANIC CHEMISTRY. *Educ:* Westmont Col, BA, 61; Univ Calif, Santa Barbara, MA, 63; Univ Wash, PhD(org chem), 67. *Prof Exp:* Instr chem, Westmont Col, 62-64; instr, Univ Wash, 68; from asst prof to assoc prof, 68-76, assoc dean, Sch Natural Sci, 83-88, PROF CHEM CALIF STATE UNIV, LONG BEACH, 76- *Concurrent Pos:* Calif & Long Beach Heart Asn grants, 69-78; vis scientist biochem, Brandeis Univ, 74-75; NSF grants, 79; vis prof chem, Univ Calif, Irvine, 81-82; NIH grants, 87- *Mem:* AAAS; Am Chem Soc; Royal Soc Chem; Sigma Xi; IUPAC; Nat Asn Sci; Technol Soc. *Res:* Solution kinetics; deuterium isotope effects; acid catalysis; acidity functions; mechanisms of hydration and hydrolysis reactions; linear free energy relationships; analytical organic chemistry. *Mailing Add:* 3301 Huntley Dr Los Alamitos CA 90720

JENSEN, KEITH EDWIN, b Council Grove, Kans, Sept 6, 24; m 43; c 4. CANCER. *Educ:* Univ Kans, AB, 48, MA, 49; Jefferson Med Col, PhD, 51. *Prof Exp:* Asst bacteriologist, State Bd Health, Kans, 49; asst instr, Univ Kans, 49; asst, Jefferson Med Col, 49-51; res assoc epidemiol, Univ Mich, 51-55, asst prof, 55-56; dir, Int Influenza Ctr, USPHS, 56-58; mgr Respiratory Dis Sect, 58-61, asst dir, Biol Res, 61-65; dir virol, 65-68, dir virol & oncol, Med Prod Res & Develop, 68-72, exec dir cancer res, 72-80, SR SCI ADV, PFIZER INC, 80- *Mem:* Am Soc Microbiol; Am Acad Microbiol; Am Asn Immunol. *Res:* Epidemiology and immunology of mycoplasmal and viral respiratory diseases; viral oncology; interferon inducers; tumor immunology; cancer chemotherapy; chemical carcinogenesis; antimicrobiols; immunotherapeutics; rheumatology. *Mailing Add:* Evergreen Cloning Nurseries 30 Trumbull Rd Waterford CT 06385

JENSEN, KEITH FRANK, b Fontanelle, Iowa, Apr 9, 38; m 60; c 3. FORESTRY. *Educ:* Iowa State Univ, BS, 60, PhD(plant physio physiol, silvicult), 63. *Prof Exp:* PLANT PHYSIOLOGIST, DIS DIV, US FOREST SERV, 63- *Concurrent Pos:* Res fel, Univ Wis, 73-74. *Mem:* Bot Soc Am; Air Pollution Control Asn; Sigma Xi. *Res:* Effect of air pollution and environmental stresses on growth and development of forest trees. *Mailing Add:* Forest Insect & Dis Lab PO Box 365 Delaware OH 43015

JENSEN, LAWRENCE CRAIG-WINSTON, b New York, NY, Oct 5, 36; m 60; c 2. PLANT ANATOMY, CYTOLOGY. *Educ:* Brown Univ, BA, 60; Univ Minn, MSc, 62, PhD(bot), 66. *Prof Exp:* NIH fel, 66-68; asst prof biol, Univ Utah, 68-72; MEM FAC DEPT BOT, UNIV AUCKLAND, 72- *Mem:* Bot Soc Am; NZ Soc Electron Micros; Australian & NZ Soc Cell Biol. *Res:* Ultrastructure of the mitotic spindle. *Mailing Add:* Dept Bot Univ Auckland Auckland New Zealand

JENSEN, LEO STANLEY, b Bellingham, Wash, Feb 28, 25; m 54; c 4. ANIMAL NUTRITION. *Educ:* Wash State Univ, BS, 49; Cornell Univ, PhD(animal nutrit), 54. *Prof Exp:* Jr poultry scientist, Wash State Univ, 49-51, from asst prof to prof poultry sci, 54-73, chmn grad prog nutrit, 70-73; prof poultry sci, 73-84, D W BROOKS DISTINGUISHED PROF, UNIV GA, 84- *Concurrent Pos:* Oak Ridge Inst Nuclear Studies res partic, AEC, Univ Tenn, 64-65. *Honors & Awards:* AFMA Award, Poultry Sci Asn, 66; Merck Award, Poultry Sci Asn, 79. *Mem:* Poultry Sci Asn; Am Inst Nutrit; Soc Exp Biol & Med. *Res:* Vitamins, minerals, fatty acids and unidentified factors in poultry nutrition; nutritional factors affecting abdominal fat accumulation; amino acid requirements and interactions. *Mailing Add:* Dept of Poultry Sci Univ of Ga Athens GA 30602

JENSEN, LYLE HOWARD, b East Stanwood, Wash, Nov 24, 15; m 40; c 3. BIOPHYSICAL CHEMISTRY. *Educ:* Walla Walla Col, BA, 39; Univ Wash, PhD(phys chem), 43. *Prof Exp:* Res assoc, Univ Chicago, 43-44; assoc prof chem, Emmanuel Missionary Col, 44-46; res assoc, Ohio State Univ, 46-47; actg asst prof, 47-48, Anderson fel x-ray diffraction, 48-49, from instr to assoc prof anat, 49-61, PROF ANAT, UNIV WASH, 61- *Mem:* AAAS; Am Chem Soc; Am Crystallog Asn; Am Asn Anat; Sigma Xi. *Res:* Chemistry of heavy metals; low temperature thermodynamics of gases; molecular structure; x-ray diffraction studies of biologically important molecules. *Mailing Add:* Dept Biol Struct Univ Wash Seattle WA 98195

JENSEN, MARCUS MARTIN, b Mantua, Utah, May 26, 29; m 59; c 3. MEDICAL MICROBIOLOGY. *Educ:* Utah State Univ, BS, 52, MS, 54; Univ Calif, Los Angeles, PhD(med microbiol), 61. *Hon Degrees:* Dr, Utah State Univ, 91. *Prof Exp:* Res virologist, Res Serv, Vet Admin Ctr, 61-63; asst prof med microbiol, Sch Med, Univ Calif, Los Angeles, 63-69; assoc prof, 69-78, PROF MICROBIOL, BRIGHAM YOUNG UNIV, 78- *Concurrent Pos:* Assoc mem, Brain Res Inst, Med Sch, Univ Calif, Los Angeles, 68-69; pres, Robbins Aseptic Air Systs Inc, Calif, 68-69 & Jensen Res Labs, Utah, 69- *Honors & Awards:* George N Raines Award, Am Psychiat Asn, 62. *Mem:* AAAS; Am Soc Microbiol; Am Asn Avian Pathologists. *Res:* Natural resistance to infectious diseases, influence of emotional stress on suscepsusceptibility to viral infections; role of viruses in kidney diseases; methods of controlling the airborne spread of microorganisms in hospitals; development of vaccines for turkey diseases. *Mailing Add:* Dept Microbiol Brigham Young Univ Provo UT 84602

JENSEN, MARVIN E(LI), b Clay Co, Minn, Dec 23, 26; m 47; c 3. AGRICULTURAL ENGINEERING. *Educ:* NDak State Univ, BS, 51, MS, 52; Colo State Univ, PhD, 65. *Hon Degrees:* DSc, NDak State Univ, 88. *Prof Exp:* Asst, NDak State Univ, 51-52, instr & asst agr engr, 52-54, asst prof agr eng & asst agr engr, 54-55; agr eng, Agr Res Serv, USDA, 55-59, invests leader irrig, drainage & water storage facil, 59-61, invests leader water mgt, Northwest Br, 61-69, dir, Snake River Conserv Res Ctr, Sci & Educ Admin, 69-79, nat prog leader, Water Mgt, 79-87; DIR, COLO INST IRRIGATION MGT, 87- *Concurrent Pos:* Pres, Int Comn Irrig & Drainage, 84-87. *Honors & Awards:* Huber Res Prize, Am Soc Civil Engrs, 68, R J Tipton Award, 82; Hancor Soil & Water Eng Award, Am Soc Agr Engrs, 74, John Deere Medal Award, 82; Arid Lands Hydraul Eng Award, Am Soc Civil Engrs, 90. *Mem:* Nat Acad Eng; hon mem Am Soc Civil Engrs; AAAS; Am Soc Agron; Am Soc Agr Engrs (vpres, 83-86). *Res:* Irrigation engineering research; crop water requirement and irrigation scheduling; irrigation management. *Mailing Add:* 1207 Springwood Dr Ft Collins CO 80525

JENSEN, MEAD LEROY, b Salt Lake City, Utah, June 11, 25; m 47; c 5. ECONOMIC GEOLOGY. *Educ:* Univ Utah, BS, 48; Mass Inst Technol, PhD(geol), 51. *Prof Exp:* From instr to assoc prof geol, Yale Univ, 51-64, dir grad studies, 64-65; PROF GEOL & GEOPHYS, UNIV UTAH, 65- *Concurrent Pos:* Lectr, Andhra Univ, India, 55; sr scientist, Australian Acad Sci, 57; hon lectr, Sigma Xi, 67. *Honors & Awards:* Sr Scientist Award, Australian Acad Sci, 62. *Mem:* Fel Geol Soc Am; Am Inst Mining, Metall & Petrol Eng; Am Geophys Union; Soc Petrol Eng; Mineral Soc Am. *Res:* Isotopic and economic geology, metallic, nonmetallic and petroleum; exploration geology. *Mailing Add:* 1359 S Ambassador Way Salt Lake City UT 84108

JENSEN, NORMAN P, b Pontiac, Mich, Dec 12, 38; m 65; c 3. MEDICINAL CHEMISTRY. *Educ:* Univ Mich, BS, 61; Mass Inst Technol, PhD(org chem), 65. *Prof Exp:* Res chemist, Socony Mobil, 61; NIH fel org chem, Stanford Univ, 65-66; sr res chemist, 66-73, res fel, 73-78, sr res fel, 78-79, asst dir, 80-83, dir, Merck & Co, 83; dir chem, Ayerst Labs, 83-85, ASST VPRES, CHEM, WYETH-AYERST LABS, 85- *Mem:* Am Chem Soc. *Res:* Search for new drugs in the fields of cardiovascular, anti-inflammatory, metabolic and central nervous system diseases. *Mailing Add:* Wyeth-Ayerst Labs CN-8000 Princeton NJ 08543-8000

JENSEN, PAUL ALLEN, b Chicago, Ill, Aug 27, 36; m 63; c 4. OPERATIONS RESEARCH, ELECTRICAL ENGINEERING. *Educ:* Univ Ill, BS, 59; Univ Pittsburgh, MS, 63; Johns Hopkins Univ, PhD(opers res), 67. *Prof Exp:* Engr, Surface Div, Westinghouse Elec Corp, 59-63; from asst prof to assoc prof, 67-73, PROF MECH ENG, UNIV TEX, AUSTIN, 73- *Mem:* Opers Res Soc Am; Inst Mgt Sci; Inst Indust Eng. *Res:* Mathematical optimization theory and application; network flow techniques used for optimization; reliability engineering; transportation systems; water resources. *Mailing Add:* Dept of Mech Eng Univ of Tex Austin TX 78712

JENSEN, PAUL EDWARD T, b New Orleans, La, Apr 27, 26; m 53; c 3. SYSTEMS ANALYSIS, OPERATIONS RESEARCH. *Educ:* Tulane Univ, BS, 47, BBA, 49; Golden Gate Univ, MBA, 75. *Prof Exp:* Asst mgr, Atlantic Gulf Sugar Co, Cuba, 52-55; sr engr, Electronic Defense Labs, GTE Prod Corp, 55-59, develop engr, 59-60, supvr tech pub, 60-63, mgr tech pub, 63-64, eng specialist, 64-76, sr eng specialist, 76-82; SR STAFF ENG SYSTEM, ESL INC, 82- *Concurrent Pos:* Consult, Asn Continuing Educ, Stanford, Calif, 74-82; Stanford Univ, 77-79 & GTE Prod Corp, 80-82; lectr, Cogswell Col, San Francisco, 79-; lectr, Northwestern Polytech Univ, Fremont, CA, 88- *Mem:* Am Phys Soc; Inst Elec & Electronics Engrs; Soc Tech Commun; assoc fel Soc Tech Comm. *Res:* Systems analysis of tactical and strategic communications and electronics systems; electronic warfare vulnerability analysis. *Mailing Add:* 495 Java Dr PO Box 3510 Ms607 1191 Bruckner Circle Sunnyvale CA 94088

JENSEN, RANDOLPH A(UGUST), b Lyon Co, Minn, May 25, 19; m 42; c 3. CHEMICAL ENGINEERING, POLLUTION CONTROL. *Educ:* Univ Minn, BChE, 40; Univ Iowa, MSChE, 46. *Prof Exp:* Res chem engr, Cliffs Dow Chem Co, Mich, 40-42; proj engr, eng exp sta, Pa State Col, 42-43; res assoc, Inst Hydrol Res, Univ Iowa, 43-46; proj engr, US Govt Synthetic Rubber Labs, Ohio, 46-47; res engr, Battelle Mem Inst, 47-51; chem engr, Houston Plant, Rohm & Haas Co Inc, 51-62, chief chem engr, Louisville Plant, 62-71, pollution control mgr, 71-79; PRES, JENSEN CONSULT INC, 79- *Concurrent Pos:* Mem, Nat Adv Comt Aeronaut, 42-43. *Mem:* Am Inst Chem Engrs; Water Pollution Control Asn. *Res:* Fluid flow low and high velocity gas streams; chemical plant process improvement; air and water pollution control; solid waste disposal; numerous publications on air and water pollution control, electropolishing, heat transfer and crystallization. *Mailing Add:* Jensen Consult Inc PO Box 43079 Louisville KY 40243

JENSEN, REED JERRY, b Dec 16, 36; m 60; c 6. PHYSICAL CHEMISTRY. *Educ:* Brigham Young Univ, BA, 60, PhD(phys chem), 65. *Prof Exp:* Fel phys chem, Univ Calif, Berkeley, 65-66; staff mem, Los Alamos Sci Lab, 66-67; asst prof, Brigham Young Univ, 67-69; staff mem phys chem, Los Alamos Nat Lab, 69-72, group leader chem lasers, 72-76, alt div leader laser chem, 76-89, DEP ASSOC DIR, CHEM & MAT, LOS ALAMOS NAT LAB, 89- *Mem:* Am Chem Soc. *Res:* Research in lasers and applications to chemistry; chemical separations with lasers and modern methods; chemical process development for nuclear transmutation processes. *Mailing Add:* 121 La Vista Los Alamos NM 87544

JENSEN, RICHARD ARTHUR, b Ogden, Utah, Oct 24, 36. PHARMACOLOGY. *Educ:* Univ Ore, BS, 60; Univ Wash, MS, 63, PhD(pharmacol), 66. *Prof Exp:* Fel, Univ Calif, San Francisco, 66-68; asst prof pharmacol, 69-74; sr pharmacologist, 74-78, DIR CARDIOVASC PHARMACOL PROG, STANFORD RES INST, 78- *Mem:* Am Soc Pharmacol & Exp Therapeut. *Res:* Evaluation of the effects of drugs and other chemicals on cardiac electrical and mechanical activity; antiarrhythmic drug action, the electrophysiological action of drugs that induce cardiac arrhythmias; drug induced cardiac muscle dis; hemodynamic studies in conscious dogs using biotelemetry. *Mailing Add:* Stanford Res Inst Menlo Park CA 94025

JENSEN, RICHARD DONALD, b Hartington, Nebr, Oct 6, 36; m 57; c 4. VETERINARY PATHOLOGY. *Educ:* Iowa State Univ, DVM, 64; Univ Minn, St Paul, PhD(vet path), 70. *Prof Exp:* Res fel path, 70-76, DIR TOXICOL & PATH, MERCK INST THERAPEUT RES, MERCK & CO, INC, 77- *Mem:* Am Col Vet Path; Int Acad Path; Am Vet Med Asn. *Res:* Avian mycoplasma infection; toxicologic and pathologic evaluation of potential therapeutic agents. *Mailing Add:* 463 Ferry Rd Doylestorm PA 18901

JENSEN, RICHARD ERLING, b Des Moines, Iowa, Apr 3, 38; m 60; c 2. ANALYTICAL CHEMISTRY, TOXICOLOGY. *Educ:* Iowa State Univ, BS, 60; Univ Iowa, MS, 64, PhD(anal chem), 65. *Prof Exp:* Asst prof anal chem, Mankato State Col, 65-66; from asst prof to assoc prof, Gustavus Adolphus Col, 66-79; supvr, Alcohol Sect, Forensic Sci Lab, State of Minn, 79-80, asst dir, 80-84; DIR & PRES, FORENSIC ASSOCS, 84-; DIR FORENSIC TOXICOL, MEDTOX LABS. *Mem:* Am Chem Soc; Am Acad Sci; Sigma Xi. *Res:* Alcohol and drug analysis for evidential purposes; trace analysis of metals using spectrophotometry, fluorescence and atomic absorption. *Mailing Add:* Forensic Assocs 4690 IDS Ctr Minneapolis MN 55402-2207

JENSEN, RICHARD EUGENE, b Unity, Sask, June 30, 27; US citizen; m 63. PHYSICS. *Educ:* Univ Sask, BS, 49, MS, 52; Ariz State Univ, PhD(physics), 66. *Prof Exp:* Proj engr physics, Motorola Inc, 56-59 & Nuclear Corp Am, 59-63; RES PHYSICIST, NAVAL SURFACE WEAPONS CTR, 67- *Mem:* Am Phys Soc; Optical Soc Am; Inst Elec & Electronics Engrs. *Res:* Lasers and optical propagation. *Mailing Add:* 1939 N East Rd North East MD 21901

JENSEN, RICHARD GRANT, b Los Angeles, Calif, Apr 16, 36; m 61; c 4. BIOCHEMISTRY. *Educ:* Brigham Young Univ, BA, 61, PhD(biochem), 65. *Prof Exp:* Chas F Kettering res fel biochem, Chas F Kettering Res Lab, Ohio, 63-65; NIH fel, Lawrence Radiation Lab, Univ Calif, 65-67; from asst prof to assoc prof biochem, 67-79, assoc prof plant sci, 76-79, PROF BIOCHEM & PLANT SCI, UNIV ARIZ, 79- *Concurrent Pos:* Vis prof, Chem Inst Tech Univ Munich, Freising-Weihens Tephan, West Ger, 74-75; vis prof, Bot Inst, Univ Brone, Switz, 75; consult, Agr Div, Monsanto Co, 76; prog dir, Photosynthesis Prog, Competitive Res Grants Off, Sci & Educ Admin, US Dept Agr, 81. *Mem:* Am Soc Biol Chemists; Am Soc Plant Physiol; fel AAAS. *Res:* Cell biology and metabolism; photosynthesis; metabolic regulation in plant cells; carbon dioxide fixation. *Mailing Add:* Dept Biochem Univ Ariz Tucson AZ 85721

JENSEN, RICHARD HARVEY, b Estherville, Iowa, June 14, 41; m; c 2. ANATOMY, IMMUNOLOGY. *Educ:* Univ Northern Iowa, BA, 63; Univ Iowa, MA, 69, PhD(anat), 73. *Prof Exp:* Instr math & sci, Charles City High Sch, Iowa, 63-66; clin phys therapist, Univ Iowa, 67-68, from teaching asst to instr gross anat, 69-73; grant seed res, Univ Nebr Med Ctr, Omaha, 73-75, asst prof gross anat, 73-77; MEM STAFF PROG PHYS THER, MARQUETTE UNIV, 77- *Concurrent Pos:* Vis instr gross anat, Univ Miami, 72; consult design & orgn gross anat prog phys ther, Fla Int Univ, 73. *Mem:* Am Phys Ther Asn; Am Asn Anatomists; Am Col Sports Med. *Res:* Hematology, especially stimulation of bone marrow; biomechanics, with emphasis on kinetic and kinematic analysis of extremities. *Mailing Add:* Phys Ther Prog Marquette Univ Walter Schroeder Complex Milwaukee WI 53233

JENSEN, RICHARD JORG, b Erie Co, Ohio, Jan 17, 47; m 70. SYSTEMATIC BOTANY. *Educ:* Austin Peay State Univ, BS, 70, MS, 72; Miami Univ, PhD(bot), 75. *Prof Exp:* Asst prof biol, Wright State Univ, 75-79; PROF BIOL, ST MARYS COL, 79- *Concurrent Pos:* NSF res grant, 73, 78, 84, & 87; Sigma Xi grant in aid of res, 74; guest assoc prof biol, Univ Notre Dame, 81-; res corp grant, 84; elected fel, Ind Acad Sci, 86; sr res fel, APSU Ctr Field Biol, 86-87; dir, Greene-Nieuwland Herbarium, 89-; Lilly Found grant, 90. *Mem:* Torrey Bot Club; Int Asn Plant Taxon; Bot Soc Am; Sigma Xi; Am Soc Plant Taxonomists; Soc Syst Zool. *Res:* Systematic and taxonomic studies of Quercus, the oaks, emphasizing numerical taxonomic and morphometric approaches. *Mailing Add:* Dept Biol St Marys Col Notre Dame IN 46556

JENSEN, ROBERT ALAN, b Bainbridge, NY, Sept 25, 40; m 85. NEUROBIOLOGY, PSYCHOBIOLOGY. *Educ:* Col Wooster, Ohio, BA, 65; Kent State Univ, MA, 70; Northern Ill Univ, PhD(biopsychol), 76. *Prof Exp:* Res psychologist, Kent State Univ, 68-71; asst res psychobiologist, Univ Calif, Irvine, 76-81; asst prof, 81-83, ASSOC PROF, DEPT PSYCHOL, SOUTHERN ILL UNIV, CARBONDALE, 83-, ASSOC DEAN, COL

LIBERAL ARTS, 88-, ASSOC PROF, SCH MED, 89- *Concurrent Pos:* Fel, Univ Calif, Irvine, 75-78; managing ed, Behav & Neural Biol, 78-81; consult, G D Searle Co, Skokie, Ill, 83-85; prin investr res grant, R J Reynolds Tobacco Co, Office Naval Res. *Mem:* AAAS; Int Soc Develop Psychobiol; Soc Neurosci; Sigma Xi. *Res:* Neurobiological aspects of memory modulation; role of catecholamine and opioid systems in the modulation of learning and memory; electrophysical correlates of neural plasticity; neurological basis of smoking behavior. *Mailing Add:* Dept Psychol Southern Ill Univ Carbondale IL 62901

JENSEN, ROBERT GORDON, b Carthage, Mo, Jan 2, 26; m 47; c 2. BIOCHEMISTRY, NUTRITION. *Educ:* Univ Mo, BS, 50, MS, 51, PhD(dairy bact), 54. *Prof Exp:* From instr to asst prof dairy bact, Univ Mo, 54-56; from asst prof to prof dairy mfg, 56-70, prof nutrit sci, 70-90, EMER PROF, UNIV CONN, 91- *Mem:* AAAS; Am Oil Chem Soc; Am Dairy Sci Asn; Am Inst Nutrit. *Res:* Lipases, human milk lipids; bovine milk lipids. *Mailing Add:* Dept Nutrit Sci U-17 3624 Horsebarn Rd Ext Univ Conn Storrs CT 06269-4017

JENSEN, RONALD HARRY, b Chicago, Ill, Nov 25, 38; m 58; c 3. BIOPHYSICAL CHEMISTRY, CYTOCHEMISTRY. *Educ:* Lawrence Col, BS, 60; Calif Inst Technol, PhD(chem), 64. *Prof Exp:* Res fel biol, Calif Inst Technol, 64-67; res scientist molecular biol, Int Minerals & Chem Corp, 67-69; sr investr microbiol, Smith Kline & French Labs, 70-74; life scientist biol & med, 75-79, SECT LEADER CYTOCHEMISTRY, LAWRENCE LIVERMORE NAT LAB, 79- *Mem:* Soc Anal Cytol; AAAS. *Res:* Fluorescent probes of cellular structure and the use of flow and image cytometry of stained cells or chromosomes to study mutagenesis and carcinogenesis. *Mailing Add:* Biomed Div Lawrence Livermore Nat Lab PO Box 5507 Livermore CA 94550

JENSEN, ROY A, b Racine, Wis, Apr 8, 36; m 56; c 5. MICROBIOLOGY, BIOCHEMISTRY. *Educ:* Ripon Col, BA, 58; Univ Tex M D Anderson Hosp & Tumor Inst, PhD(biochem, genetics), 63. *Prof Exp:* Res instr, Sch Med, Univ Wash, 65; asst prof biol, State Univ NY Buffalo, 66-68; assoc prof microbiol, Baylor Col Med, 68-73; prof biol, Univ Tex M D Anderson Hosp & Tumor Inst Houston, 73-76; prof biol, State Univ NY, Binghamton, 76-86, dir Ctr Somatic-Cell Genetics & Biochem, 78-86; CONSULT, 86- *Concurrent Pos:* USPHS fel microbiol, Sch Med, Univ Wash, 64-66. *Mem:* Am Soc Microbiol; Tissue Cult Asn. *Res:* Biochemical genetics; gene-enzyme relationships; regulation of gene and enzyme activities; metabolic interlock; plant tissue culture. *Mailing Add:* Dept Microbiol & Cell Sci Univ Fla Gainesville FL 32666

JENSEN, RUE, b Vermillion, Utah, Oct 24, 11; m 42; c 2. VETERINARY PATHOLOGY. *Educ:* Utah State Univ, BS, 37, MS, 39; Colo State Univ, DVM, 42; Univ Minn, PhD, 53; Kasetsart Univ, Bangkok, DVSc, 65. *Prof Exp:* Instr vet sci, La State Univ, 42-43; from asst prof to assoc prof, 43-48, dir diag lab, 73-77, PROF PATH, COLO STATE UNIV, 48- *Concurrent Pos:* From asst pathologist exp sta to pathologist exp sta, Colo State Univ, 43-57, chief animal dis sect & dean col vet med & biomed sci, 57-66, dir agr exp sta, 66-69, vpres res, 66-73; consult, USDA, 57-, Agency Int Develop, Univ Teheran, 62 & Kasetsart Univ, Bangkok, 64; USDA del, USSR, 58; mem, Agr Res Inst; consult pathologist, Monfort Colo, Inc, 77- & Univ Wyo, 78- *Mem:* Soc Exp Biol & Med; Am Vet Med Asn; Am Col Vet Path; Int Acad Path; Sigma Xi. *Res:* Necrobacillosis of cattle; vibriosis of sheep; diseases of feedlot cattle; diseases of sheep. *Mailing Add:* 620 Matthews 102 Colo State Univ Ft Collins CO 80524

JENSEN, STANLEY GEORGE, plant pathology; deceased, see previous edition for last biography

JENSEN, SUSAN ELAINE, b Edmonton, Alta, Jan 30, 50; m 71. ANTIBIOTICS, STREPTOMYCES. *Educ:* Univ Alta, BSc, 70, PhD(microbiol), 75. *Prof Exp:* Teaching fel, Univ BC, 74-76; sessional lectr & res assoc, 77-81, ALTA HERITAGE FOUND MED RES SCHOLAR MICROBIOL, UNIV ALTA, 81- *Mem:* Am Soc Microbiol; Can Soc Microbiologists. *Res:* Biosynthesis of beta-lactam antibiotics by Streptomyces; cell-free enzymatic synthesis of unnatural beta-lactam antibiotics; isolation of genes coding for enzymes involved in antibiotic biosynthesis. *Mailing Add:* Dept Microbiol Univ Alta Edmonton AB T6G 2M7 Can

JENSEN, THOMAS E, b Waverly, Iowa, Sept 21, 32; m 56; c 2. 2. CELL BIOLOGY. *Educ:* Wartburg Col, BA, 58; SDak State Univ, MA, 62; Iowa State Univ, PhD(cytol), 65. *Prof Exp:* Res assoc, Iowa State Univ, 64-65; asst prof biol, Wayne State Univ, 65-70; assoc prof, 70-72, PROF BIOL, LEHMAN COL, 73- *Mem:* AAAS; Electron Micros Soc Am; Am Soc Cell Biol; Bot Soc Am; Sigma Xi. *Res:* Ultrastructure of cells. *Mailing Add:* Dept of Biol Sci Lehman Col City Univ New York Bronx NY 10468

JENSEN, THORKIL, b Vejle, Denmark, Jan 23, 19; nat US; m 43; c 1. MICROBIOLOGY. *Educ:* Gustavus Adolphus Col, BA, 41; Univ Minn, MS, 49, PhD(zool), 52. *Prof Exp:* Instr embryol & histol, Vet Sch, Univ Minn, 51-52; from asst prof to assoc prof microbiol, Sch Med, Univ Kans, 52-63, prof, 63-; RETIRED. *Concurrent Pos:* China Med Bd fel, 55; consult, St Mary's Hosp, Kansas City, 53-58, Vet Admin Hosp, Mo, 54-61, Midwest Res Inst, 61-63 & Baptist Mem Hosp, Kansas City, 65- *Mem:* Am Soc Parasitol; Am Trop Med & Hyg; Sigma Xi. *Res:* In vitro culture of some parasitic protozoa and helminths; possible host-parasite relationships between viruses and protozoa and helminths; biochemistry of excystation in acanthamoeba. *Mailing Add:* 7029 Glenwood Overland Park KS 66204

JENSEN, TIMOTHY B(ERG), b Willmar, Minn, Oct 25, 39; div; c 2. CHEMICAL ENGINEERING. *Educ:* Univ Minn, Minneapolis, BS, 61; Princeton Univ, PhD(chem eng), 65. *Prof Exp:* Sr chem engr, 64-68, res supvr, 68-73, res mgr, 73-74, tech mgr, 74-84, RES MGR, MINN MINING & MFG CO, 84- *Mem:* Am Soc Testing & Mat; Am Inst Chem Engrs. *Res:* Optimal control theory; reactor design; urethane chemistry; oriented polyester; packaging products. *Mailing Add:* 6221 Loch Moor Dr Edina MN 55439-1619

JENSEN, TORKIL HESSELBERG, b Kolding, Denmark, Apr 9, 32; m 56; c 3. PLASMA PHYSICS. *Educ:* Tech Univ Denmark, MS, 56. *Prof Exp:* Staff mem reactor & plasma physics, Danish Atomic Energy Comn, 56-60; MEM STAFF PLASMA PHYSICS, GEN ATOMIC CO, 64- *Mem:* Am Phys Soc. *Res:* Experimental plasma physics. *Mailing Add:* Gen Atomics PO Box 85608 San Diego CA 92138

JENSEN, WALLACE NORUP, b Moroni, Utah, Aug 31, 21; m 47; c 3. HEMATOLOGY. *Educ:* Univ Utah, BS, 42, MD, 45; Am Bd Internal Med, dipl. *Prof Exp:* Intern med, Johns Hopkins Hosp, 45-46; asst resident, Univ Utah Hosps, 48-49, resident, 49-50; fel hemat, Univ Utah, 50-52, sr Damon Runyon fel med, Med, Ctr, 50-53; asst prof med, Sch Med, Duke Univ, 53-55; asst prof to prof Univ Pittsburgh, 58-67, prof med & chmn dept, head div hemat, 55-69; prof med & chmn dept, Sch Med, George Washington Univ, 69-76, Eugene Meyer prof med, prof 71-75; PROG CHMN, DEPT MED, ALBANY MED COL, 76- *Concurrent Pos:* Chief hemat sect, Vet Admin Hosp, Durham, 53-55; mem study sect hemat, NIH, 57-61; mem study sect, Grad Training Grants Hemat, USPHS, 62-66; NIH spec fel, Nat Transfusion Ctr & Sch Advan Studies, France, 63-64; mem bd exam, Am Bd Internal Med, 70-78; nat consult internal med-hemat, Off Surg Gen, 70-; Asn Prof Med, Liaison Comt, NIH, 71. *Mem:* Am Soc Hemat; Am Col Physicians; Am Fedn Clin Res; Soc Nuclear Med; Int Soc Hemat. *Res:* Medicine. *Mailing Add:* 1321 NW 14th St Suite 401 Miami FL 33136

JENSEN, WILLIAM AUGUST, b Chicago, Ill, Aug 22, 27; m 48; c 2. BOTANY. *Educ:* Univ Chicago, PhB, 49, MS, 50, PhD(bot), 53. *Prof Exp:* USPHS fel, Calif Inst Technol, 53-55; NSF fel, Univ Brussels, 55-56; asst prof biol, Univ Va, 56-57; from asst prof to prof, Univ Calif, Berkeley, 57-84, chmn dept, 71-84; dean, 84-89, PROF, COL BIOL SCIS, OHIO STATE UNIV, 84- *Concurrent Pos:* Prog dir develop biol, NSF, 73-74. *Honors & Awards:* NY Bot Garden Award, Bot Res, 60. *Mem:* Bot Soc Am (vpres, 75-76); Soc Develop Biol (secy, 62-64). *Res:* Botanical histochemistry; botanical cytology; plant embryology. *Mailing Add:* 396 Pebble Creek Dr Dublin OH 43017

JENSEN, WILLIAM PHELPS, b Minneapolis, Minn, May 22, 37; m 62; c 3. CHEMISTRY. *Educ:* Univ Minn, BS, 59; Univ Iowa, MS, 62, PhD(inorg chem), 64. *Prof Exp:* Res chemist, Pittsburgh Plate Glass Co, 63-66; vis asst prof, La State Univ, 66-67; assoc prof, 67-77, PROF CHEM, SDAK STATE UNIV, 77- *Mem:* Am Chem Soc. *Res:* Chemistry of lanthanide and actinide elements; structure determination of complex compounds by x-ray diffraction. *Mailing Add:* Dept of Chem SDak State Univ Brookings SD 57007

JENSH, RONALD PAUL, b New York, NY, June 14, 38; m 62; c 2. RADIATION EMBRYOLOGY, BEHAVIORAL TERATOLOGY. *Educ:* Bucknell Univ, BA, 60, MA, 62; Jefferson Med Col, PhD(anat), 66. *Prof Exp:* Instr anat & res assoc radiol, 66-68, from asst prof to assoc prof radiol & anat, 68-82, PROF ANAT, JEFFERSON MED COL & THOMAS JEFFERSON UNIV, 82-, VCHMN ANAT, 84- *Concurrent Pos:* Investr, NIH grants, Stein Res Ctr & Dept Anat, Jefferson Med Col, 66- *Mem:* AAAS; Am Asn Anat; Teratology Soc (treas, 89-92); NY Acad Sci; Neurobehav Teratology Soc (pres, 85-86); Sigma Xi; Soc Exp Biol Med. *Res:* Teratology; embryology, statistical applications; behavioral toxicology; reproductive biology, developmental biology and radiobiology. *Mailing Add:* Dept Anat 561 JAH Thomas Jefferson Univ Philadelphia PA 19107-6799

JENSKI, LAURA JEAN, b Chicago, Ill, Feb 23, 52; m. MHC RESTRICTION, CYTOXIC T-LYMPHOCYTE. *Educ:* Northern Ill Univ, BS, 73, MS, 75; Univ NC, PhD(oncol), 79. *Prof Exp:* Res assoc, Childrens Hosp Res Found, 83-86; asst prof, 87-91, ASSOC PROF BIOL, IND UNIV-PURDUE UNIV INDIANAPOLIS, 91- *Concurrent Pos:* Grants, var corp & inst, 86-92. *Mem:* Am Asn Immunologists; Am Soc Cell Biol; AAAS; Asn Women in Sci. *Res:* T-lymphocyte activity and regulation; immunological effects of long chain omega-3 fatty acids. *Mailing Add:* Dept Biol Ind Univ-Purdue Univ Indianapolis 723 W Michigan Indianapolis IN 46202-5132

JENSON, A BENNETT, IMMUNOPATHOLOGY, IMMUNOVIROLOGY. *Educ:* Baylor Col Med, MD, 66. *Prof Exp:* ACTG CHMN, DEPT DENT, MED & GRAD PATH, GEORGETOWN UNIV, 80- *Mailing Add:* Dept Path Georgetown Univ Med Ctr Med Sch Washington DC 20057

JENSSEN, THOMAS ALAN, b South Bend, Ind, Mar 18, 39; m 62; c 3. ANIMAL BEHAVIOR, ECOLOGY. *Educ:* Univ Redlands, BS, 62; Southern Ill Univ, MA, 64; Univ Okla, PhD(zool), 69. *Prof Exp:* Nat Inst Ment Health assoc herpet, Harvard Univ, 69-71; asst prof, 71-77, ASSOC PROF BIOL, VA POLYTECH INST & STATE UNIV, 77- *Concurrent Pos:* Res asst, Med Ctr, Univ Okla, 69-70. *Mem:* Am Soc Ichthyol & Herpet; Animal Behav Soc; Ecol Soc Am; Soc Study Amphibians & Reptiles; Sigma Xi. *Res:* Behavior and ecology of various species of anurans and lizards, especially communicative value of anoline lizard displays. *Mailing Add:* Dept of Biol Va Polytech Inst & State Univ Blacksburg VA 24061

JENTOFT, JOYCE EILEEN, b Canton, Ohio, Mar 10, 45; m. STRUCTURE-FUNCTION RELATIONSHIPS, PHYSICAL BIOCHEMISTRY. *Educ:* Capital Univ, BS, 66; Univ Minn, PhD(inorg chem), 71. *Prof Exp:* Fel phys biochem, Univ Minn, 72; fel phys biochem, Case Western Reserve Univ, 77, immunol, 78, sr res assoc phys biochem, Dept Pediat, 79-81, asst prof, 81- 89, ASSOC PROF, DEPT BIOCHEM, SCH MED, CASE WESTERN RESERVE UNIV, 89- *Concurrent Pos:* Instr biochem, Case Western Univ, 79-81. *Mem:* Am Chem Soc; Biophys Soc; Am Soc Biochem & Molecular Biol. *Res:* Structure-function relationships in proteins and enzymes; protein-nucleic acid interactions; molecular virology (retroviruses); spectroscopy (fluorescence, CD, NMR). *Mailing Add:* Dept Biochem Sch Med Case Western Reserve Univ Cleveland OH 44106

JENTOFT, RALPH EUGENE, JR, b Tacoma, Wash, Nov 30, 18; m 54; c 2. PHYSICAL CHEMISTRY, ANALYTICAL CHEMISTRY. *Educ:* Univ Wash, BS, 41, PhD(chem), 52. *Prof Exp:* Chemist, Oceanog Surv Philippines, US Fish & Wildlife Serv, 47-48; res assoc oceanog, Office Naval Res, Univ Wash, 49-52; res chemist, Chevron Res Co, Standard Oil Co Calif, 52-60, sr res chemist, 60-64, sr res assoc phys & anal chem, 64-79; CONSULT, 80- *Mem:* AAAS; Am Chem Soc. *Res:* Phase studies and thermodynamic measurements in field of petroleum chemistry; separation and purification; trace analysis for hydrocarbons and petrochemicals; analytical separations; liquid chromatography and supercritical fluid chromatography. *Mailing Add:* 11601 Occidental Rd Sebastopol CA 95472

JENZANO, ANTHONY FRANCIS, b Philadelphia, Pa, May 20, 19; m 40; c 2. ASTRONOMY, PHYSICS. *Prof Exp:* Head technician, Fels Planetarium, Pa, 46-49; head technician, Univ NC, Chapel Hill, 49-51, mgr, 51-60, dir, Morehead Planetarium, 60-81; planetarium counr US & Can, Carl Zeiss Optical Co, 81-86; RETIRED. *Concurrent Pos:* Consult, London Planetarium, Eng, 55-57, Buhl Planetarium, Pa, 59, var proposed planetaria, 63-, Carl Zeiss Optical Co, 65- & Fernbank Sci Ctr, Ga, 66- *Mem:* Assoc Am Astron Soc; Am Asn Mus. *Res:* Initiation and direction of celestial training program for United States Mercury, Gemini, Apollo, Skylab and Apollo-Soyuz astronauts. *Mailing Add:* 37 Oakwood Dr Chapel Hill NC 27514

JEON, KWANG WU, b Korea, Nov 10, 34; m 58; c 2. CELL BIOLOGY, DEVELOPMENTAL BIOLOGY. *Educ:* Seoul Nat Univ, BS, 57, MS, 59; Univ London, PhD(cell physiol), 64. *Prof Exp:* Res fel electron micros, Middlesex Hosp, Univ London, 64-65; res asst prof cell physiol, State Univ NY Buffalo, 65-69; assoc prof, 70-75, PROF ZOOL, UNIV TENN, KNOXVILLE, 76- *Concurrent Pos:* Ed, Int Rev Cytol; Am Cancer Soc res grant, Univ Tenn, Knoxville, 71, Am Heart Asn & Nat Inst Child Health & Human Develop res grants, 71-73; Nat Inst Gen Med Sci res grant, 74-77; NSF res grant, 77-; Chancellor's res scholar, Univ Tenn. *Mem:* NY Acad Sci; Am Soc Cell Biol; Soc Develop Biol; Soc Protozoologists; fel AAAS; Sigma Xi. *Res:* Cell growth and division; nucleocytoplasmic interactions; cell organelle structure and function; symbiosis. *Mailing Add:* Dept of Zool Univ of Tenn Knoxville TN 37996-0810

JEONG, TUNG HON, b Kwangtung, China, Dec 19, 35; US citizen; m 63; c 3. NUCLEAR PHYSICS. *Educ:* Yale Univ, BS, 57; Univ Minn, PhD(physics), 62. *Prof Exp:* Res assoc physics, Univ Minn, 62-63; from asst prof to assoc prof, 63-78, PROF PHYSICS & CHMN DEPT, LAKE FOREST COL, 78- *Concurrent Pos:* Tech consult; fel, Optical Soc Am. *Honors & Awards:* Robert Millikin Medal, Am Asn Physics Teachers, 76. *Mem:* Am Asn Physics Teachers; Laser Inst Am; AAAS; Soc Photo-Optical Instrumentation Engrs; Am Phys Soc. *Res:* Precision proton-nuclear elastic scattering; linear proton accelerator injector; H-source for pre-injectors; optics; physics education; lasers and holography; non-destructive testing; laser applications and holography. *Mailing Add:* Dept Physics Lake Fores Col Lake Forest IL 60045

JEPPESEN, RANDOLPH H, PHYSICS & ASTRONOMY. *Educ:* Univ Mont, BA, 58; Univ Ill, MS, 60; NMex State Univ, PhD (physics), 80. *Prof Exp:* From instr to assoc prof, 61-81, chmn dept, 73-81, PROF PHYSICS & ASTRON, UNIV MONT, 81- *Concurrent Pos:* IBM res staff mem, Thomas J Watson Res Ctr, 60-61; co-prin investr, AEC grant, Dept Energy, 72-80; AWA fac partic grants, Los Alamos Nat Lab, 81-86, collabr, exps 665 & 770, 87. *Mem:* Am Phys Soc. *Mailing Add:* 1824 Dexon A Missoula MT 59801-8418

JEPPSON, LEE RALPH, b Brigham City, Utah, Feb 17, 10; m 36; c 6. ENTOMOLOGY. *Educ:* Brigham Young Univ, BS, 31; Utah State Univ, MS, 40; Univ Calif, PhD(entom), 43. *Prof Exp:* Teacher, Utah Pub Schs, 31-36; entomologist, Calif Conserv Co, 40-42; assoc, Exp Sta, 42-45, from jr entomologist to assoc entomologist, Res Ctr & Exp Sta, 45-60, lectr, Univ, 66-74, prof, 74-77, ENTOMOLOGIST, CITRUS RES CTR & AGR EXP STA, UNIV CALIF, RIVERSIDE, 60-, EMER PROF ENTOM, UNIV, 77- *Mem:* AAAS; Entom Soc Am; Acarol Soc Am; Sigma Xi. *Res:* Biology; ecology; morphology; taxonomy; host specificity and control of phytophagous mites. *Mailing Add:* Citrus Exp Sta Univ of Calif Riverside CA 92521

JEPPSON, ROLAND W, b Brigham City, Utah, Aug 30, 33; m 59; c 9. CIVIL ENGINEERING. *Educ:* Utah State Univ, BS, 58, MS, 60; Stanford Univ, PhD(civil eng), 67. *Prof Exp:* Res engr, Utah State Univ, 58-60; asst prof civil eng, Humboldt State Col, 60-64; res engr, summers, 61-64, head, Dept Civil & Environ Eng, 73-77, assoc prof, 66-71, PROF CIVIL ENG, UTAH STATE UNIV, 71- *Honors & Awards:* J C Stevens Award, Am Soc Civil Engrs, 68; Horton Award, Am Geophys Union, 76. *Mem:* Am Soc Civil Engrs; Am Soc Eng Educ; Am Geophys Union. *Res:* Numerical solutions to free surface fluid and porous media flow problems; water resource planning and design; pipeline hydraulics; open chemical hydraulics. *Mailing Add:* Dept Civil & Environ Eng Utah State Univ Logan UT 84321-4110

JEPSEN, DONALD WILLIAM, b Lincoln, Nebr, Jan 14, 32; m; c 1. SURFACE PHYSICS, STATISTICAL MECHANICS. *Educ:* Univ Rochester, BS, 53; Univ Wis, MS, 56, PhD(theoret chem), 59. *Prof Exp:* Gen Motors fel, Inst Fluid Dynamics & Appl Math, Univ Md, 59-60; STAFF MEM, IBM CORP RES CTR, 60- *Mem:* Am Phys Soc; Am Chem Soc; Sigma Xi. *Res:* Theoretical chemical physics; nonequilibrium properties of large systems; properties of solid surfaces and low energy electron diffraction. *Mailing Add:* 507 Woodland Hills Rd White Plains NY 10603

JEPSON, WILLIAM W, b Minneapolis, Minn, Apr 14, 26; m 51; c 4. MEDICINE, PSYCHIATRY. *Educ:* Swarthmore Col, BA, 47; Cornell Univ, MD, 50. *Prof Exp:* Res psychiat, Cincinnati Gen Hosp, 56; from instr to asst prof, 57-68, ASSOC PROF PSYCHIAT, MED SCH, UNIV MINN, MINNEAPOLIS, 68- *Concurrent Pos:* Chief psychiat, Hennepin County Med Ctr, 59-, prog dir, Ment Health Ctr, 60-; mem psychiat training rev comt, NIMH, 67-71. *Mem:* Am Psychiat Asn. *Mailing Add:* Hennepin Co Med Ctr Minneapolis MN 55415

JEREMIAH, LESTER EARL, b Walla Walla, Wash, Dec 9, 41; m 66; c 3. MEAT SCIENCES. *Educ:* Wash State Univ, BS, 65; Univ Mo, MS, 67; Tex A&M Univ, PhD(meat sci), 71. *Prof Exp:* Meat lab technician, Wash State Univ, 65, exten agent, 67-69; res asst, Univ Mo, 65-67; grad asst meat sci, Tex A&M Univ, 69-71; salesman real estate, David A Gamache Real Este Co, 72-73; co exten dir, Colo State Univ, 73-74; tech writer human nutrit, Agriserv Found, 74-75; RES SCIENTIST MEAT SCI, CAN DEPT AGR, 75- *Mem:* Am Soc Animal Sci; Inst Food Technologists; Am Meat Sci Asn; Can Meat Sci Asn. *Res:* Beef and pork tenderness, quality, preservation, retail case-life, and meat handling systems; frozen storage and display of meat; sensory evaluation and consumer acceptance. *Mailing Add:* Agr Can Res Br Lacombe Res Sta Lacombe AB T0C 1S0 Can

JEREMIAS, CHARLES GEORGE, b Marlborough, Mass, July 8, 20; m 80; c 2. ORGANIC CHEMISTRY, INORGANIC CHEMISTRY. *Educ:* Univ Ga, BS, 42; Tulane Univ, PhD(chem), 49. *Prof Exp:* Chemist, US Rubber Co, 42-45; res chemist, Tenn Eastman Co, 48-60; group leader res, Southern Dyestuff Co, Martin-Marietta Co, 60-62; assoc prof & actg head dept, 62-64, PROF CHEM & HEAD DEPT, NEWBERRY COL, 64- *Concurrent Pos:* Consult, Delta 2 Finishing Plant, J P Stevens Co, 65-70 & James Flett Orgn, Inc, 77-79. *Mem:* Am Chem Soc; Am Inst Chemists. *Res:* Organic intermediates for synthetic fibers, dyes and insecticides; sulfur dyes and intermediates. *Mailing Add:* 2103 Johnstone St Newberry SC 29108

JERGER, E(DWARD) W, b Milwaukee, Wis, Mar 13, 22; m 82; c 2. MECHANICAL ENGINEERING. *Educ:* Marquette Univ, BS, 46; Univ Wis, MS, 47; Iowa State Univ, PhD(theoret & appl mech), 51. *Prof Exp:* Dir process eng, Wis Malting Co, 46-48; asst prof mech eng, Iowa State Col, 48-55; assoc prof, 55-61, prof & head dept, 61-68, assoc dean eng, 68-82, PROF MECH ENG, UNIV NOTRE DAME, 82- *Mem:* Am Soc Mech Engrs; Am Soc Eng Educ; Nat Fire Protection Asn; Int Asn Arson Investr. *Res:* Thermal systems; fire protection engineering; protective construction; product liability. *Mailing Add:* Col Eng Univ Notre Dame Notre Dame IN 46556

JERINA, DONALD M, b Chicago, Ill, Jan 17, 40; m 64; c 2. ORGANIC CHEMISTRY, BIOCHEMISTRY. *Educ:* Knox Col, Ill, BA, 62; Northwestern Univ, PhD(org chem), 66. *Prof Exp:* Fel org chem & biochem, 66-68, sr fel, 69-70, res chemist, 70-73, CHIEF, OXIDATION MECHANISMS SECT, NAT INST ARTHRITIS, DIABETES, DIGESTIVE & KIDNEY DIS, NIH, 73- *Honors & Awards:* Hillebrand Prize, Am Chem Soc, 79; Brodie Award, Am Soc Pharmacol & Exp Therapeut, 82. *Mem:* AAAS; Am Chem Soc; Am Cancer Soc; Fedn Am Socs Exp Biol; Am Soc Biochem & Molecular Biol. *Res:* Synthesis of peptides and oligonucleotides on polymer supports; enzymes drug metabolism; microsomal hydroxylation; biochemical mechanisms; migration of ring substituents during aryl hydroxylation, particularly the NIH shift; chemistry and biochemistry of arene oxides; chemical carcinogenesis. *Mailing Add:* Lab Bioorganic Chem Bldg Eight Nat Inst Diabetes & Digestive & Kidney Dis Bethesda MD 20892

JERIS, JOHN S(TRATIS), b Boston, Mass, June 6, 30; m 58; c 2. ENVIRONMENTAL ENGINEERING, SCIENCE. *Educ:* Mass Inst Technol, BS, 53, MS, 54, ScD(sanit eng), 62. *Prof Exp:* Proj engr, Stearns & Wheler, NY, 56-59; res asst, Mass Inst Technol, 59-62; from asst prof to assoc prof, 62-71, dir environ eng & sci grad prog, 66-78, 86-, PROF CIVIL ENG, MANHATTAN COL, 71- *Concurrent Pos:* Vpres res & develop, Ecolotrol Inc, 70- *Honors & Awards:* Kenneth Allen Mem Award, NY Water Pollution Control Asn, 75; Thomas R Camp Medal, Water Pollution Control Fedn, 79. *Mem:* Sigma Xi; Am Water Works Asn; Water Pollution Control Fedn; Am Soc Civil Engrs; Asn Environ Eng Prof. *Res:* Biological waste treatment; use of biological fluid beds, transport of polychlorinated biphenyl through sediment; anaerobic and aerobic stabilization of sludges. *Mailing Add:* Dept Civil Eng Manhattan Col Bronx NY 10471

JERISON, HARRY JACOB, b Bialystok, Poland, Oct 13, 25; US citizen; m 50; c 3. NEUROBIOLOGY, MEDICAL PSYCHOLOGY. *Educ:* Univ Chicago, BS, 47, PhD(psychol), 54. *Prof Exp:* Res psychologist, AeroMed Lab, USAF, 49-57; assoc prof psychol, Antioch Col, 57-64, dir, Behav Res Lab, 57-69, prof psychol, 64-68, prof biol, 68-69; PROF BIOBEHAV SCI, DEPT PSYCHIAT, SCH MED & PROF, DEPT PSYCHOL, UNIV CALIF, LOS ANGELES, 69- *Concurrent Pos:* Fel, Ctr Advan Study Behav Sci, 67-68; hon res assoc, Dept Vert Paleont, Los Angeles County Mus, 70-; vis scientist, Med Res Coun, Appl Psychol Unit, Cambridge, Eng, 78-79; vis scholar, Rockefeller Found Bellagio Ctr, 83; vis prof anthrop, Univ Florence, Italy, 86-87; acad vis, Oxford Univ, 86; vis prof psychol, Univ Hawaii, 87; vis prof neurobiol, Max-Plank Inst fo Biologica Cybernetics, Tuebingen, Ger, 89. *Honors & Awards:* James Arthur lectr, Am Mus Nat Hist, 89. *Mem:* Psychonomic Soc; Am Psychol Asn; Int Soc Evolutionary Biol; Am Soc Naturalists; Soc Vert Paleont. *Res:* Paleoneurology; evolutionary biopsychology; evolution of specialized and generalized behavioral and cognitive capacities in vertebrates, and its relation to allometry and encephalization (brain/body relations) among living and fossil animals; quantitative neuroanatomy. *Mailing Add:* Dept Psychiat Univ Calif Los Angeles CA 90024

JERISON, MEYER, b Bialystok, Poland, Nov 28, 22; nat US; m 45; c 2. MATHEMATICS. *Educ:* City Col New York, BS, 43; Brown Univ, MS, 47; Univ Mich, PhD(math), 50. *Prof Exp:* Physicist, Nat Adv Comt Aeronaut, 44-46; res instr math, Univ Ill, 49-51; from asst prof to assoc prof math, 51-60, chmn div math sci & head dept math, 69-75, PROF MATH, PURDUE UNIV, WEST LAFAYETTE, 60- *Concurrent Pos:* Lectr, Case Inst Technol, 45-46; res engr, Lockheed Aircraft Corp, 52; mem, Inst Advan Study, 58-59; mem comt undergrad prof in math, Math Asn Am, 68-71, bd govs, 81-84; ed, Bull Am Math Soc, 80-85. *Mem:* Am Math Soc; Math Asn Am. *Res:* Linear topological spaces; spaces of continuous functions; group algebra. *Mailing Add:* Dept Math Purdue Univ West Lafayette IN 47907

JERKOFSKY, MARYANN, b Alameda, Calif, Feb 18, 43. VIROLOGY, CELL CULTURE. *Educ:* Univ Tex, BA, 65; Baylor Col Med, PhD(virol), 69. *Prof Exp:* Fel microbiol, Col Med, Pa State Univ, 69-72, res assoc, 72-73, instr, 73-74; res asst prof, Sch Med, Univ Miami, 74-75; asst prof, 76-81, ASSOC PROF MICROBIOL, UNIV MAINE, ORONO, 81- *Concurrent Pos:* Vis prof, Univ Amsterdam, Neth, 83 & Am Univ Les Cayes, Haiti, 88, 90. *Mem:* Am Soc Microbiol; Sigma Xi; Am Soc Virol. *Res:* Herpesviruses in vitro model for Reye's Syndrome; lipid metabolism modifications produced by herpes viruses; interaction between unrelated animal viruses. *Mailing Add:* Dept Biochem Microbiol & Molecular Biol Univ Maine Orono ME 04469

JERMANN, WILLIAM HOWARD, b Cleveland, Ohio, June 29, 35; m 63; c 3. ELECTRICAL ENGINEERING. *Educ:* Univ Detroit, BEE, 58, MA, 62; Univ Conn, PhD(elec eng), 67. *Prof Exp:* Jr engr, Toledo Edison Co, Ohio, 58; instr elec eng, Univ Detroit, 61-62; asst prof, US Coast Guard Acad, 62-67; from asst prof to assoc prof, 67-77, PROF, DEPT ELEC ENG, MEMPHIS STATE UNIV, 77- *Concurrent Pos:* NSF res grant, 69-70. *Mem:* Am Soc Eng Educ; Simulation Coun. *Res:* Hybrid Monte-Carlo solutions to partial differential equations; development of engineering concepts curriculum project. *Mailing Add:* Dept Elec Eng Memphis State Univ Memphis TN 38152

JERNE, NIELS KAJ, b London, Eng, Dec 23, 11. IMMUNOLOGY, EXPERIMENTAL THERAPY. *Educ:* Univ Leiden, PhD. *Prof Exp:* Dir, Basel Inst Immunol, 69-80; prof, Pasteur Inst, Paris, 81-82; RETIRED. *Honors & Awards:* Nobel Prize in Med, 84; Paul Ehrlich Prize, 82; Marcel Prize, 84. *Mem:* Nat Acad Sci; Royal Danish Acad Sci; Am Acad Arts & Sci. *Mailing Add:* Pasteur Inst Castillon-du-Gard France

JERNER, R CRAIG, b St Louis, Mo, Oct 12, 38; m 57; c 4. METALLURGICAL ENGINEERING, MATERIALS SCIENCE. *Educ:* Univ Wash, St Louis, BS, 61, MS, 62; Univ Denver, PhD(metall), 65. *Prof Exp:* Res assoc, Univ Denver, 61-64; from asst prof to assoc prof metall eng, Univ Okla, 65-76, asst dean grad col, 71-72; PRES, EMTEC CORP, 73- *Concurrent Pos:* Consult, var indust co; assoc staff mem, Transp Safety Inst, US Dept Transp, 73-78; adj prof metall eng, Univ Okla, 76- *Mem:* Microbeam Anal Soc; Am Acad Forensic Sci; Sigma Xi. *Res:* Application of scanning electron microscopy and energy dispersive x-ray spectroscopy to the analysis of metallic and non-metallic failures. *Mailing Add:* Jerner Eng 10725 Sandhill Rd No 106 Dallas TX 75238

JERNIGAN, HOWARD MAXWELL, JR, b Winston-Salem, NC, Apr 13, 43; m 68; c 1. BIOCHEMISTRY. *Educ:* WVa Univ, Morgantown, BS, 65; Univ NC, Chapel Hill, PhD(biochem), 70. *Prof Exp:* Fel, dept biochem, Univ Fla, 70-73; from asst prof to assoc prof, 73-90, PROF BIOCHEM, UNIV TENN, MEMPHIS, 90- *Concurrent Pos:* Res assoc, Lab Vision Res, Nat Eye Inst, NIH, 78-80. *Mem:* Am Soc Biochem & Molecular Biol; Am Chem Soc; Sigma Xi; Asn Res Vision & Ophthal; Am Asn Univ Profs. *Res:* Biochemistry of the eye; lens metabolism; cataract; oxidative damages to tissues; membrane transport; amino acids. *Mailing Add:* Univ Tenn Coleman Bldg Rm D-222 956 Court Ave Memphis TN 38163

JERNIGAN, ROBERT LEE, b Portales, NMex, May 4, 41; c 1. PHYSICAL CHEMISTRY, COMPUTER SIMULATIONS. *Educ:* Calif Inst Technol, BS, 63; Stanford Univ, PhD(phys chem), 67. *Prof Exp:* NIH fel, Univ Calif, San Diego, 68-70; sr staff fel chem, 70-75, theoret chemist, 75-89, DEP LAB CHIEF, NIH, 89- *Mem:* AAAS; Am Chem Soc; Biophys Soc; Protein Soc. *Res:* Protein, polypeptide and NA conformations; dimensional, electrical and optical properties; conformations of biopolymers; biophysics. *Mailing Add:* Lab Math Biol Bldg 10 Rm 4B-58 NIH Bethesda MD 20892

JERNOW, JANE L, b Shanghai, China; US citizen; m 65; c 1. ORGANIC CHEMISTRY. *Educ:* Univ Ill, Urbana, BS, 58, MS, 61; Pa State Univ, PhD(org chem), 63. *Prof Exp:* Res chemist, Sterling Winthrop Res Inst, 68-69; res assoc, State Univ NY Albany, 69-90; group leader, Hoffmann-La Roche Inc, 71-86; prog dir, Nat Acad Sci, Wash, DC, 86-89; CONSULT, 89- *Concurrent Pos:* NIH fel chem, Cornell Univ, 64-65. *Mem:* Am Chem Soc; Sigma Xi; Asn Women Sci. *Res:* Mechanism study in organic chemistry; synthetic and medicinal chemistry. *Mailing Add:* Hoffmann-LaRoche Inc 340 Kingsland St Nutley NJ 07110

JEROME, JOSEPH WALTER, b Philadelphia, Pa, June 7, 39; m 88; c 2. SEMICONDUCTOR MODELING, NONLINEAR SYSTEMS. *Educ:* St Joseph's Col, Pa, BS, 61; Purdue Univ, MS, 63, PhD(math), 66. *Prof Exp:* Asst prof math, Math Res Ctr, Univ Wis-Madison, 66-68; asst prof, Case Western Reserve Univ, 68-70; from asst prof to assoc prof, 70-76, PROF MATH, NORTHWESTERN UNIV, 76- *Concurrent Pos:* Vis sr fel, Oxford Univ, 74-75; sr fel, British Sci Coun, 74-75; vis prof, Univ Tex, 78-79; vis mem tech staff, Bell Labs, NJ, 81 & 82-83; vis scholar, Univ Chicago, 85. *Mem:* Am Math Soc; Soc Indust Appl Math. *Res:* Approximation of nonlinear partial differential equation models. *Mailing Add:* Dept of Math Northwestern Univ Evanston IL 60208

JEROME, NORGE WINIFRED, b Grenada, WI, Nov 3, 30; US citizen. NUTRITION, PUBLIC HEALTH. *Educ:* Howard Univ, BS, 60; Univ Wis-Madison, MS, 62, PhD(nutrit, anthrop), 67. *Prof Exp:* Instr foods & nutrit, Howard Univ, 62-63; res assoc nutrit & anthrop, Univ Wis-Madison, 66-67; asst prof nutrit, 69-70, asst prof human ecol, 70-72, assoc prof human ecol & community health, Col Health Sci, 72-78, dir, Educ Resources Ctr, Div Learning Resources, 74-77, PROF, DEPT COMMUNITY HEALTH, SCH MED, UNIV KANS MED CTR, KANSAS CITY, 78-, DIR, DEPT PREV MED, COMMUNITY NUTRIT DIV, 81- *Concurrent Pos:* Mem, Inst Res on Poverty, Univ Wis-Madison, 66-67; mem awards bd, Am Dietetic Asn, 68-71; assoc ed, J Nutrit Educ, 71-77, mem, Nat Adv Coun, 77-; mem nat adv panel, Children's Advert Rev Unit, 74-; mem nat adv coun, Children's TV Workshop, 74-75; chairperson comt nutrit anthrop, 74-77; mem, Food & Nutrit Coun, Am Pub Health Asn, 75-78; mem study panel 12, World Food & Nutrit Study, Nat Acad Sci, 76, Cancer & Nutrit Sci Review Comt, Diet, Nutrit & Cancer Prog, Nat Cancer Inst, NIH, 76-78 & Lipid Metab Adv Comt, Nat Heart, Lung & Blood Inst, NIH, 78-82, man-food systs interaction comt, Nat Res Coun, 80- *Mem:* Fel Am Anthrop Asn; Am Dietetic Asn; Am Pub Health Asn; Soc Med Anthrop; Am Inst Nutrit; Inst Food Technologists. *Res:* Dietary patterns of population groups; modernization, diet and health; compliance to medical regimen; consumer response to nutritional and health prescriptions. *Mailing Add:* US Agency Int Develop SA 18 Rm 411 Washington DC 20523

JEROSLOW, ROBERT G, operations research, mathematical logic; deceased, see previous edition for last biography

JERRARD, RICHARD PATTERSON, b Evanston, Ill, July 23, 25; m 51; c 3. FIXED POINTS, MULTIPLE-VALUED FUNCTIONS. *Educ:* Univ Wis, BS, 49, MS, 50; Univ Mich, PhD(math), 58. *Prof Exp:* Engr, Gen Elec Co, 50-54; instr math, Univ Mich, 56-57; mathematician, Bell Labs, 57-58; from asst prof to assoc prof, 58-69, PROF MATH, UNIV ILL, URBANA-CHAMPAIGN, 69- *Concurrent Pos:* Vis fel, Univ Warwick, 65- 66, 77, 85 & 90, Cambridge Univ, 72-73. *Mem:* Am Math Soc; Math Asn Am; Sigma Xi. *Mailing Add:* Dept Math Univ Ill Urbana-Champaign 355 Altgeld Hall 1409 W Green Urbana IL 61801

JERRELLS, THOMAS RAY, b Wickenburg, Ariz, Feb 28, 44; m 65; c 2. TUMOR IMMUNOLOGY, CELLULAR IMMUNOLOGY. *Educ:* Univ Ariz, BS, 72; Wash State Univ, MS, 74, PhD(microbiol), 76. *Prof Exp:* Tumor immunologist, Litton Bionetics, Inc, 76-78, head, Immunoregulation Sect, 78-80; MEM STAFF, DEPT RICKETTSIAL DIS, WALTER REED ARMY INST RES, 80- *Mem:* Am Soc Microbiol; Am Med Technologists. *Res:* Defining immunoregulatory cells involved in cell-mediated immune responses and role in the immunodepression associated with tumor burden. *Mailing Add:* Dept Path 101 Keller Bldg F-09 Univ Texas Galveston TX 77550

JERRI, ABDUL J, b Amarah, Iraq, July 20, 32; m; c 3. APPLIED MATHEMATICS. *Educ:* Univ Baghdad, BSc, 55; Ill Inst Technol, MSc, 60; Ore State Univ, PhD(math), 67. *Prof Exp:* Instr physics, Baquba Teacher Col, Iraq, 56-58; asst physicist, IIT Res Inst, 59-62; asst prof, 67-70, ASSOC PROF MATH, CLARKSON UNIV, 70- *Concurrent Pos:* Head dept, Am Univ Cairo, 72-73, vis assoc prof, 73-74; vis assoc prof, Am Univ Cairo, 86-88; assoc prof, Kuwait Univ, 78-79, 89-90. *Mem:* Am Math Soc; Soc Indust & Appl Math; Pattern Recognition Soc. *Res:* Sampling expansion; integral and discrete transforms; numerical method. *Mailing Add:* Dept Math & Comp Sci Clarkson Univ Potsdam NY 13699-5815

JERSEY, GEORGE CARL, b Highland Park, Mich, Aug 20, 40; m 58; c 2. VETERINARY PATHOLOGY. *Educ:* Eastern Mich Univ, BA, 64; Mich State Univ, BS, 65; DVM, 67, MS, 69, PhD(vet path), 73. *Prof Exp:* Upjohn fel, Dept Path, Mich State Univ, 67-68, instr, 68-70, instr clin path, 70-72; res specialist path, Toxicol Res Lab, Dow Chem Co, 72-80. *Mem:* Am Vet Med Asn. *Res:* Pathological and toxicological evaluation of industrial, agricultural and consumer chemicals; chemical products in laboratory animals to help establish safe production, handling, transportation and use of these materials. *Mailing Add:* 2716 Jeffrey Lane Midland MI 48640-0000

JERSILD, RALPH ALVIN, JR, b Janesville, Wis, Sept 29, 31; m 53; c 2. MICROSCOPIC ANATOMY, CELL BIOLOGY. *Educ:* St Olaf Col, BA, 53; Univ Ill, MS, 57, PhD(zool), 61. *Prof Exp:* From instr to assoc prof, 61-71, PROF ANAT, SCH MED, IND UNIV INDIANAPOLIS, 71- *Mem:* Am Asn Anat; Electron Micros Soc Am; Am Soc Cell Biol. *Res:* Electron microscopy; intestinal lipid absorption and transport; glycoprotein synthesis and transport; golgi apparatus; cell surface. *Mailing Add:* Dept Anat Ind Univ Med Ctr Indianapolis IN 46223

JERUCHIM, MICHEL CLAUDE, b Paris, France, Apr 4, 37; US citizen; m 69; c 2. COMPUTER SIMULATION OF COMMUNICATION SYSTEMS, INTERFERENCE ANALYSIS OF COMMUNICATIONS SYSTEMS. *Educ:* City Col NY, BEE, 61; Univ Pa, MSEE, 63, PhD(elec eng/commun), 67. *Prof Exp:* Commun engr, Space Systs Div, Gen Elec Co, 61-71, sr commun engr, 71-84, SR STAFF CONSULT COMMUN SCI, GE AEROSPACE, 84- *Concurrent Pos:* Secy, Inst Elec & Electronics Engrs Commun Soc, subcomt comput-aided modeling, anal & design commun systs, 84-86, vchmn, 86- *Mem:* Fel Inst Elec & Electronics Engrs. *Res:* Analysis and design of communications systems, especially satellite-based systems; developing computer-aided tools such as simulation, for doing the analysis and design. *Mailing Add:* Gen Elec Co PO Box 8048 Bldg 9 Rm 2123 Philadelphia PA 19101

JERVIS, HERBERT HUNTER, b Wilmington, Del, June 25, 42; m. MOLECULAR GENETICS. *Educ:* Springfield Col, BS, 64, MEd, 66; Fla State Univ, MS, 71, PhD(genetics), 73; St John's Univ, JD, 87. *Prof Exp:* Res assoc biochem, VA Polytech Inst & State Univ, 73-75; from asst prof to assoc prof biol, Adelphi Univ, 81-87; patent atty, Fitzpatrick, Cella, Haper & Scinto, 87-89; PATENT ATTY, SMITHKLINE BEECHAM, 90- *Mem:* Genetics Soc Am; Am Soc Microbiol; Patent & Trademark Off Soc; Sigma Xi. *Res:* Role of transfer RNAs in the development and differentiation of fungi, especially Neurospora and Allomyces. *Mailing Add:* Three Continental Dr PO Box 405 Valley Forge PA 19481

JERVIS, ROBERT ALFRED, b Wilmington, Del, May 15, 38; m 81; c 2. BOTANY, ECOLOGY. *Educ:* Dartmouth Col, BA, 60; Rutgers Univ, MS, 62, PhD(ecol), 64. *Prof Exp:* From asst prof to assoc prof biol, Emory & Henry Col, 64-68; prof biol, Goddard Col, 68-81 & Community Col Vt, 81-84; INSTR, HARWOOD UNION HIGH SCH, MORETOWN, 81- *Concurrent Pos:* Dir Goddard Col non-resident ecol study projs, Southeast & Southwest US, 71-72 & Northwest & Alaska, 76 & Bahamas, 78; dir, Goddard Col Raptor Rehab Ctr; dir, Summer Prog in Outdoor Educ, Goddar Col; adj prof biol, Johnson State Col, 84- *Mem:* Ecol Soc Am; Am Nature Study Soc; Am Littoral Soc. *Res:* Freshwater marsh vegetation and productivity; vegetation patterns in the south Appalachians; New England ecology; ornithology; travel study programs in ecology. *Mailing Add:* Harwood Union High Sch Rte One Moretown VT 05660

JERVIS, ROBERT E, b Toronto, Int, May 21, 27; m 50; c 2. RADIOCHEMISTRY, APPLICATIONS. *Educ:* Univ Toronto, BA, 49, MA, 50, PhD(phys chem), 52. *Prof Exp:* Assoc res officer, Atomic Energy Can, Ltd, Ont, 52-58; assoc prof, 58-67, assoc dean res eng, 74-78, res chmn, 81-85, PROF APPL CHEM, DEPT CHEM ENG, UNIV TORONTO, 66- *Concurrent Pos:* Vchmn, Can Sci Fairs Coun, 63-65; vis prof, Fac Sci, Univ Tokyo, 65-66, Energy Res Group, Cambridge Univ, 78 & Nat Univ Malaysia, 79; vchmn Nuclear Safety Comt, Atomic Energy Control Bd, Can Fed Govt, chmn, 88. *Honors & Awards:* Hevesy Medal. *Mem:* AAAS; Can Soc Forensic Sci; fel Chem Inst Can; Can Nuclear Asn; fel Indian Acad Forensic Sci; Fel, Royal Soc Can. *Res:* Radioactivation research, especially application of nuclear detection methods to crime detection and to environmental pollution problems from heavy metals, mercury, arsenic, cadmium and lead; radiochemical studies of nuclear power reactor safety. *Mailing Add:* Dept of Chem Eng Univ of Toronto Toronto ON M5S 1A1 Can

JESAITIS, ALGIRDAS JOSEPH, b Fed Repub Ger, Aug 21, 45; US citizen; m 79; c 3. CELL BIOLOGY, IMMUNOLOGY. *Educ:* Sch Eng & Sci, NY Univ, BS, 67; Calif Inst Technol, PhD(biophysics), 73. *Prof Exp:* Fel, Univ Freiburg, WGer, 73-75; fel, Univ Calif, San Diego, 75-79; fel, Scripps Clin & Res Found, 79-85, asst mem, 85-89; RES PROF, DEPT CHEM & BIOCHEM, MONTANA STATE UNIV, 89- *Mem:* Biophys Soc; AAAS; Am Soc Cell Biol; Protein Soc. *Res:* Biophysics and cell biology of sensory transduction mechanisms; role of cell membrane processes in inflammation, chemotaxis and mechanisms of host defense; structure of neutrophil cytochrome. *Mailing Add:* Dept Chem Montana State Univ Bozeman MT 59717

JESAITIS, RAYMOND G, b Vilnius, Lithuania, Jan 20, 43. PHYSICAL ORGANIC CHEMISTRY. *Educ:* Cooper Union, BChE, 63; Cornell Univ, PhD(org chem), 67. *Prof Exp:* Res fel chem, Univ Calif, Berkeley, 67-68; asst prof, State Univ NY Stony Brook, 68-74; from assoc prof to prof chem, 74-84, PROF COMPUT SCI & CHEM, STATE UNIV NY COL TECHNOL, UTICA, 84- *Mem:* Am Chem Soc; Royal Soc Chem. *Res:* Physical and theoretical organic chemistry, including molecular structure; molecular interactions; ecological systematics. *Mailing Add:* Dept Comput Sci State Univ NY Col Technol Utica NY 13504-3050

JESKA, EDWARD LAWRENCE, b Erie, Pa, Aug 6, 23; m 50; c 4. IMMUNOLOGY. *Educ:* Gannon Col, BA, 51; Marquette Univ, MS, 54; Univ Pa, PhD, 66. *Prof Exp:* Chief parasitologist, Pa Dept Health, 55-63; fel, Univ Pa, 65-67, res asst prof parasitol, Sch Vet Med, 67; from asst prof to assoc prof, 67-74, chmn dept immunobiol, 75-80, PROF VET PATH & VET MED, RES INST, IOWA STATE UNIV, 74- *Mem:* Reticuloendothelial Soc; Soc Exp Biol & Med; Am Asn Immunologists. *Res:* Characterization of parasitic nematode antigens involved in white cell reactions of vertebrate hosts; macrophage as effector mechanisms of resistance to infection. *Mailing Add:* 12800 Marion Lane W No 517W Hopkins MN 51900

JESKEY, HAROLD ALFRED, b St Louis, Mo, Aug 18, 12; m 39; c 2. ORGANIC CHEMISTRY. *Educ:* St Louis Col Pharm, BS, 33; Wash Univ, BA, 37; Univ Wis, PhD(org chem), 42. *Prof Exp:* Asst chemist, James F Ballard, Inc, 33-35; instr chem, St Louis Col Pharm, 35-38; asst, Univ Wis, 38-41; from instr to asst prof, Univ Tenn, 41-44; from asst prof to assoc prof, Southern Methodist Univ, 45-57, chmn dept, 62-72, prof, 57-79, emer prof chem, 79-87; prof biochem, Southwestern Med Sch, Univ Tex, 80-87; RETIRED. *Mem:* Am Chem Soc. *Res:* Organic synthesis; carbonation of phenols. *Mailing Add:* 2929 Fondren Dr Dallas TX 75205

JESMOK, GARY J, b Milwaukee, Wis, Dec 13, 47. INFLAMMATION. *Educ:* Med Col Wis, PhD(pharmacol), 76. *Prof Exp:* MGR, PHARMACOL SECT, TRAVENOL LABS, 80- *Mem:* Am Soc Pharmacol & Exp Therapeut; Am Heart Asn; AAAS; Soc Exp Biol & Med. *Mailing Add:* Dept Exp Therapeut Cutter Biologics Fourth & Parker Sts Berkeley CA 94710

JESPERSEN, JAMES, b Weldona, Colo, Nov 17, 34; c 3. RADIOPHYSICS, COMMUNICATION THEORY. *Educ:* Colo Univ, BA, 56, MS, 61. *Prof Exp:* Proj leader radio astron, Cent Radio Propagation Lab, 56-61, group leader satellite ionospheric scintillation studies, 64-66; exchange scientist theory of VLF radio propagation, Radio Res Lab, Slough, Eng, 62-63; consult time broadcast studies, Nat Bur Standards, 67-68, chief time & frequency, Dissemination Res Group, Exp & Theoret Studies Time Dissemination, 69-72; Dept Comm Sci fel & consult tele-commun, 72-73; CONSULT THEORET RADIO PROPAGATION STUDIES, NAT BUR STANDARDS, 74- *Concurrent Pos:* Consult, Inst-Range Instrumentation Group, 73-79, Korean Standards Res Inst, 77-78 & UN Develop Plan, 78-79. *Honors & Awards:* Bronze Plaque, Korean Standards Res Inst, 78. *Mem:* Sr mem Inst Elec & Electronics Engrs; Sigma Xi; Inst Navig. *Res:* Radio astronomy; ionospheric physics; radio propagation; communication and information theory; time dissemination and navigation systems; communication aids for the deaf. *Mailing Add:* 87 Camiino Bosque Four Mile Canyon Boulder CO 80302

JESPERSEN, NEIL DAVID, b Brooklyn, NY, Mar 5, 46; m 70; c 2. ANALYTICAL CHEMISTRY. *Educ:* Washington & Lee Univ, BS, 67; Pa State Univ, PhD(chem), 71. *Prof Exp:* Asst prof chem, Univ Tex, Austin, 71-77; asst prof, 77-80, ASSOC PROF CHEM, ST JOHN'S UNIV, NY, 80- *Mem:* Am Chem Soc; AAAS; Sigma Xi. *Res:* Thermometric titrimetry; clinical analysis; environmental mutagens. *Mailing Add:* Dept of Chem St John's Univ Jamaica NY 11439

JESPERSEN, NILS VIDAR, b Horten, Norway, Aug 1, 51; US citizen; m 78; c 2. COMMUNICATION SYSTEM DESIGN & ANALYSIS, ANTENNA ENGINEERING. *Educ:* Drexel Univ, BS, 75, MS, 76, PhD(elec eng), 91. *Prof Exp:* Develop engr, Am Electronic Labs, 76-78; sr engr, Aydin Monitor Systs, 78-81 & Teledynamics, Subsid United Tech, 81-82; com equip engr, 82-87, mgr antenna eng, 87-90, SR STAFF ENGR, ASTRO SPACE DIV, GEN ELEC, 90- *Mem:* Inst Elec & Electronics Engrs. *Res:* Space applications of phased array technology; optical and fiber optic approaches to microwave phased array beamsteering and beamforming; awarded one patent. *Mailing Add:* 40 Grove Rd Havant Hampshire P09 1AR England

JESS, EDWARD ORLAND, b Westbrook, Maine, Oct 9, 17; m 44. METEOROLOGY. *Educ:* Univ Southern Maine, BS, 39; NY Univ, MS, 48; Univ Stockholm, Fil Lic(meteorol), 58. *Prof Exp:* Comdr, 15th Weather Squadron, Far E Air Forces, Manila, USAF, 46, comdr, Weather Control Detachment, Berlin Airlift, 48-49, opers officer & chief tech serv, Rhein Main Weather Cent, Air Weather Serv, Ger, 49-52, chief extended forecasting, Global, Weather Cent, Strategic Air Command, Offutt AFB, Nebr, 52-55, opers officer, Joint Task Force 7 Weather Ctr, Oper Red Wing, Eniwetok, 55-56, chief support tech serv, 4th Weather Group, Air Weather Serv, Andrews AFB, Md, 58-64, comdr, Asian Weather Cent, Air Weather Serv, Tokyo, 64-66, comdr, 20th Weather Squadron, 67, dir, dir aerospace servs directorate, Hq Air Weather Serv, Scott AFB, 67-72; staff meteorologist & dir state air pollution episode, Control Ctr, Va Ste Air Pollution COnytol Bd, 72-82; RETIRED. *Mem:* Fel Am Meteorol Soc. *Res:* Role of the physical environment in the genesis, transport, intensification and dissipation of air pollution. *Mailing Add:* 4712 Monument Ave Richmond VA 23230-3727

JESSE, KENNETH EDWARD, b Chicago, Ill, Jan 3, 33; m 59; c 3. SOLID STATE PHYSICS. *Prof Exp:* Res physicist, Aerospace Res Lab, Wright-Patterson AFB, 66-67; PROF PHYSICS, ILL STATE UNIV, 67-, RADIATION SAFETY OFFICER, 71- *Mem:* Am Asn Physics Teachers; Sigma Xi. *Res:* Thermoelectrical and electrical properties of nonmetallic materials. *Mailing Add:* Dept of Physics Ill State Univ Normal IL 61761

JESSEN, CARL ROGER, b Fairmont, Minn, Jan 12, 33; m 55; c 3. RADIOLOGY, GENETICS. *Educ:* Univ Minn, BS, 54, DVM, 56, PhD(genetics), 69. *Prof Exp:* Pvt pract, 56-64; assoc prof vet clin sci, 74-77, RADIOLOGIST, DEPT CLIN SCI, COL VET MED, UNIV MINN, ST PAUL, 69-, PROF VET CLIN SCI, 77-, ASSOC DEAN VET MED SERV, 78- *Mem:* Am Vet Med Asn; Am Vet Radiol Soc; Genetics Soc Am. *Res:* Canine hip dysplasia; bone dysplasias in general. *Mailing Add:* 2161 Folwell St St Paul MN 55108

JESSEN, NICHOLAS C, SR, b Ger, Feb 13, 13. NUCLEAR ENGINEERING. *Prof Exp:* Dir technol, Babcock & Wilcox Co, 30-78; RETIRED. *Concurrent Pos:* Instr metall, Univ Akron, 38-42. *Mem:* Fel Am Soc Metals Int; Am Welding Soc. *Res:* Development of welding processes for heavy pressure vessels and tubular products for low and high alloy steels; developed original guidelines for welding of stainless steel. *Mailing Add:* 4307 Village Oaks Lane Dunwoody GA 30338

JESSEN, NICHOLAS C, JR, b Barberton, Ohio, Mar 29, 44. METALLURGICAL ENGINEERING. *Educ:* Univ Tenn, BS, 67, MS, 72. *Prof Exp:* SUPT, MARTIN MARIETTA CORP, 86- *Mem:* Fel Am Soc Metals; Nat Mgt Asn. *Mailing Add:* Martin Marietta Corp Y-12 Plant Bldg 9212 Oak Ridge TN 37830-8194

JESSER, WILLIAM AUGUSTUS, b Waynesboro, Va, Dec 20, 39; m 62; c 2. METAL PHYSICS, MATERIALS SCIENCE. *Educ:* Univ Va, BA, 62, MS, 64, PhD(physics), 66. *Prof Exp:* Lectr physics, Univ Witwatersrand, 66-68; from asst prof to assoc prof, 68-78, PROF MAT SCI, UNIV VA, 78- *Concurrent Pos:* Mem, Ctr Advan Studies, Univ Va, NSF, 68-70; vis prof, Nagoya Univ, Japan, 78, Univ Pretoria, 82, 87, Univ Witwatersrand, 83 & Hunan Univ, Changsha, 90; Thomas Goodwin Digges Chair, 89. *Honors & Awards:* Alan Talbott Gwathmey Award, 67. *Mem:* Am Soc Metals; Electron Micros Soc Am; Am Inst Mining, Metall & Petrol Engrs; Mat Res Soc. *Res:* Growth and properties of thin films; transmission electron microscopy and diffraction; surface and interface properties; radiation damage; electronic materials. *Mailing Add:* Dept Mat Sci Thornton Hall Univ Va Charlottesville VA 22901

JESSOP, ALAN MICHAEL, b Wellingborough, UK, Feb 4, 34; Brit & Can citizen; m 59; c 3. GEOTHERMICS, GEOTHERMAL ENERGY. *Educ:* Univ Nottingham, BSc, 55, PhD(mining), 58. *Prof Exp:* Res officer, Brit Cotton Indust Res Asn, 58-60; Nat Res Coun Can fel geophysics, Univ Western Ont, 60-62; sci officer, Dominion Observ, Dept Energy Mines & Resources, Ottawa, 62-65; res scientist, Earth Phys Br, 65-86; RES SCIENTIST, GEOL SURVEY CAN, CALGARY, 86- *Concurrent Pos:* Mem, Int Heat Flow Comn, 63-75, secy, 71-75. *Mem:* Geol Asn Can; Am Geophys Union. *Res:* Thermal state, thermal history, and hydrodynamics of sedimentary basins; energy content of deep groundwater; applications to formation and migration of hydrocarbons. *Mailing Add:* Geol Survey Can Dept Energy Mines & Resources 3303 33rd St NW Calgary AB T2L 2A7 Can

JESSOP, NANCY MEYER, b Pasadena, Calif, Dec 24, 26; m 47; c 2. ZOOLOGY. *Educ:* Univ Redlands, BA, 45; Univ Ore, MA, 47; Univ Calif, Berkeley, PhD(zool), 53. *Prof Exp:* Asst zool & biol, Univ Ore, 45-46; asst exp zool, Univ Calif, 46-48; teacher, Calif Pub Schs, 53-55; teacher biol, Oceanside-Carlsbad Col, 55-60; from asst prof to prof, US Int Univ, Calif Western Campus, 60-75, chmn dept, 67-73; chmn dept, 83-86, PROF LIFE SCI, PALOMAR COL, 75- *Mem:* Am Soc Zoologists; Animal Behavior Soc (secy, 72-75). *Res:* Peromyscus genetics; tissue reactions to deep freezing; evolution and ontogeny of behavior; sociobiology. *Mailing Add:* Dept Life Sci Palomar Col San Marcos CA 92069

JESSUP, GORDON L, JR, b Hampton, Va, Apr 29; 22; m 55, 83; c 5. BIOSTATISTICS. *Educ:* Univ Md, BS, 50, MS, 55; Ore State Univ, PhD(genetics, statist), 63. *Prof Exp:* Animal husbandman, Southwest Range Sheep Breeding Lab, USDA, Ft Wingate, NMex, 52-57 & 60-61; math statistician, Biomath Div, US Army Biol Labs, Ft Detrick, 61-67; math statistician, Div Biol Effects, Bur Radiol Health, Food & Drug Admin, HEW, 67-81; RETIRED. *Mem:* Am Statist Asn; Biomet Soc. *Res:* Biostatistics and epidemiology in radiological health. *Mailing Add:* 109 Bayside Blvd No 6 Betterton MD 21610

JESTER, GUY EARLSCORT, b Dyersburg, Tenn, Oct 20, 29; m 53; c 4. STRUCTURAL DYNAMICS, SOIL MECHANICS. *Educ:* US Mil Acad, BS, 51; Univ Ill, MS, 58, PhD(civil eng). 69. *Prof Exp:* Chief, Eng Br, Corps Engrs, Europ, US Army, 59-61, asst prof civil eng, US Mil Acad, 62-65, dep dir & actg dir, Dept Res & Mgt, Waterways Exp Sta, 65-67, div engr, Viet Nam, 68-69, asst to chief res & develop & chief of info systs, 68-71, dist engr, Corp Engrs, St Louis Dist, 71-73; VPRES CORP PLANNING & MKT & VPRES & DIR INT WASTE ENERGY SYSTS, J S ALBERICICONSTRUCT CO INC, 73- *Concurrent Pos:* Pres, Asn Improvement Mississippi River, 74-78; vchmn, Prof Code Comt Metrop St Louis, 74-; Bldg & Indust Develop Comn, 78- & Metrop Area Bldg Code Rev Comt. *Mem:* Am Soc Civil Engrs; fel Soc Am Mil Engrs; Sigma Xi. *Res:* Soil-structure interaction; soils; structure design under dynamic loading conditions. *Mailing Add:* 13093 Greenbough Dr St Louis MO 63146

JESTER, JAMES VINCENT, b Riverside, Calif, Sept 7, 50; m 77. EXPERIMENTAL PATHOLOGY, OPHTHALMOLOGY. *Educ:* Univ Southern Calif, BS, 72, PhD(exp path), 78. *Prof Exp:* Fel ophthal path, Estelle Doheny Eye Found, 78, INSTR, DEPT OPHTHAL & PATH & VIS PROF, DEPT BIOL, UNIV SOUTHERN CALIF, 81- *Concurrent Pos:* Prin investr, Fight for Sight-Grant-in-Aid, 81-82. *Mem:* Asn Res Vision & Ophthal; AAAS. *Res:* Ophthalmic experimental pathology with specific emphasis on elucidating the pathogenetic mechanism involved in corneal and lid margin disease using morphologic and biochemical techniques. *Mailing Add:* Dept Ophthal Georgetown Univ Sch Med 3900 Reservoir NW Washington DC 20007

JESTER, WILLIAM A, b Philadelphia, Pa, June 16, 34; m 67; c 2. CHEMICAL & NUCLEAR ENGINEERING, NUCLEAR CHEMISTRY. *Educ:* Drexel Inst, BS, 57; Pa State Univ, MS, 61, PhD(chem eng), 65. *Prof Exp:* From asst prof to assoc prof, 65-86, PROF NUCLEAR ENG, PA STATE UNIV, 86- *Concurrent Pos:* Consult. *Honors & Awards:* Joan Hodes Queneal Palladium Medal, Nat Audubon Soc & Am Asn Eng Soc, 85. *Mem:* Am Nuclear Soc; fel Am Inst Chem; Am Chem Soc; Sigma Xi; Am Soc Eng Educ; Am Nuclear Sci Teachers Asn. *Res:* Development of radio-nuclear techniques for the solution of scientific and engineering problems; the development of radiation monitoring instrumentaion and methods for testing such monitors. *Mailing Add:* Radiation Sci & Eng Ctr Pa State Univ University Park PA 16802

JESWIET, JACOB, b Neth, Feb 24, 46; Can citizen; m 75; c 3. MANUFACTURING AUTOMATION. *Educ:* Queen's Univ, BSc, 70, MSc, 74, PhD(mech eng), 81. *Prof Exp:* Design engr, DuPont Can, 70-71; design & maintenance engr, Celanese, 71-73; asst prof eng, Univ NB, 79-82; asst prof, 82-86, ASSOC PROF MECH ENG, QUEEN'S UNIV, 86- *Concurrent Pos:* Assoc prof mech eng, Queen's Univ, 86- *Mem:* Sr mem Soc Mfg Engrs; NAm Mfg Res Inst; Am Soc Mech Engrs. *Res:* Friction and temperature at metal forming inter-faces; manufacturing automation, fns and fnc with emphasis upon diagnostic and robotic use. *Mailing Add:* Queen's Univ McLaughlin Hall Kingston ON K7L 3N6

JETER, HEWITT WEBB, b Cincinnati, Ohio, Sept 9, 41; m 66; c 2. ENVIRONMENTAL RADIOCHEMISTRY, GEOPHYSICS. *Educ:* Yale Univ, BE, 63; Ore State Univ, PhD(oceanog), 72. *Prof Exp:* Scientist oceanog, 72-74, LAB MGR RADIOCHEM, TELEDYNE ISOTOPES, 74-, SR SCIENTIST GEOPHYS, 78- *Res:* Mathematical modeling geophysics and oceanography; development of radiochemical procedures. *Mailing Add:* Teledyne Isotopes 50 Van Buren Ave Westwood NJ 07675

JETER, JAMES ROLATER, JR, b Ennis, TX, Sept 4, 40; m 63; c 1. CELL DIFFERENTIATION, CELL PROLIFERATION. *Educ:* Univ Tex, San Antonio, PhD(anat), 73. *Prof Exp:* asst prof anat, 75-78, ASSOC PROF ANAT, NEUROSCI, CELL BIOL & HISTOL, TULANE UNIV, 78- *Concurrent Pos:* Consult, NIH; ed adv bd, Cell Biol, A Series of Monographs, 85-90; secy & treas, Int Cell Cycle Soc, 84-90, pres elect, 90-92. *Mem:* AAAS; Am Asn Anatomists; Am Asn Cancer Res; Am Heart Asn; Am Soc Cell Biol; Int Cell Cycle Soc; Sigma Xi. *Res:* Role of nuclear proteins & protein phonylation in controlling cell proliferation and differentiation. *Mailing Add:* Med Sch Tulane Univ New Orleans LA 70112

JETER, WAYBURN STEWART, b Cooper, Tex, Feb 16, 26; m 47; c 3. MEDICAL MICROBIOLOGY, IMMUNOLOGY. *Educ:* Univ Okla, BS, 48, MS, 49; Univ Wis, PhD(med microbiol), 50; Am Bd Med Microbiol, dipl. *Prof Exp:* Instr plant sci, Univ Okla, 48; asst med microbiol, Univ Wis, 48-50; instr bact, Col Med, Univ Iowa, 50-51, assoc, 51-52, from asst prof to assoc prof, 52-63; dir, Med Technol Prog, 75-77, head, dept microbiol, 68-83, PROF MICROBIOL, UNIV ARIZ, 63-, DIR LAB CELLULAR IMMUNOL, 76-, PROF PHARMACOL & TOXICOL, 84- *Mem:* AAAS; Am Soc Microbiol; Soc Exp Biol & Med; Am Asn Immunol; Sigma Xi; Am Acad Microbiol. *Res:* Hypersensitivity; complement; transfer factor; tissue transplantation; pathogenic bacteria. *Mailing Add:* Dept Pharmacol Toxicol Univ Ariz Tucson AZ 85721

JETT, JAMES HUBERT, b Washington, DC, Nov 27, 38; m 62; c 2. FLOW CYTOMETRY. *Educ:* Univ NMex, BS, 60, MS, 61; Univ Colo, PhD(nuclear physics), 69. *Prof Exp:* Fel, Physics Div, 69-71, mem staff, Exp Pathol Group, 71-86, DEP GROUP LEADER, CELL BIOL GROUP, LIFE SCI DIV, LOS ALAMOS NAT LAB, 86- *Concurrent Pos:* Adj assoc prof, dept cell biol, Univ NMex, Albuquerque, 85- *Mem:* Am Phys Soc; AAAS; Soc Anal Cytology (secy-treas, 85-). *Res:* Ziomedical instrumentation, development, application and data interpretation; interpretation of biological experiments and computer applications. *Mailing Add:* 545 Navajo Los Alamos NM 87544

JETTE, ARCHELLE NORMAN, b Portland, Ore, May 15, 34; m 72; c 1. PHYSICS SURFACES, STRUCTURE DEFECT CENTERS. *Educ:* Univ Calif, Riverside, AB, 61, MA, 63, PhD(physics), 65. *Prof Exp:* Res assoc fel physics, Columbia Univ, 65; instr, Whiting Sch Eng, 83-90, RES PHYSICIST, APPL PHYSICS LAB, JOHNS HOPKINS UNIV, 65- *Concurrent Pos:* Vis prof solid state physics, Cath Univ, Rio de Janeiro, Brazil, 72; vis scientist, Ctr Interdisciplinary Res, Univ Bielefeld, WGermany, 80. *Mem:* Am Phys Soc; Am Vacuum Soc. *Res:* Surface structure; defect centers in ionic crystals; atomic and molecular physics. *Mailing Add:* Appl Physics Lab Johns Hopkins Univ Laurel MD 20723-6099

JETTEN, ANTON MARINUS, b June 26, 46; m; c 2. DIFFERENTIATION, RETINOIDS. *Educ:* Univ Nijmegen, Neth, PhD(biochem), 73. *Prof Exp:* HEAD, CELL BIOL GROUP & LAB PULMONARY PATH, NAT INST ENVIRON HEALTH SCI, NIH, 82- *Mem:* Am Asn Cancer Res; Am Soc Cell Biol. *Res:* Understanding the molecular mechanisms that regulate the proliferation and differentiation of tracheo bronchial epidermal cells. *Mailing Add:* Nat Inst Environ Health Sci NIH Res Triangle Park NC 27709

JETT-TILTON, MARTI, b Springfield, Ohio, July 22, 41; m 71; c 2. SIGNAL TRANSDUCTION, PHOSPHOLIPID METABOLISM. *Educ:* Marion Col, BA, 62; Georgetown Univ PhD(biochem), 73. *Prof Exp:* Fel biochem, Blood Res Lab, Am Red Cross, 73-75; res scientist, 75-80; sr fel, Nat Res Coun, 81-82, STAFF SCIENTIST, WALTER REED ARMY INST RES, 82- *Concurrent Pos:* Adj prof, Catholic Univ, 80- *Mem:* NY Acad Sci; Am Chem Soc; AAAS; Am Tissue Cult Asn; Am Women Sci. *Res:* Signal transduction through phosphoinositide hydrolysis; multi-drug resistance and phospholipid metabolism; controlling disease states by manipulation of signal transduction pathways; mechanism of toxin action. *Mailing Add:* 3446 Oakwood Terr NW Washington DC 20010

JEUTTER, DEAN CURTIS, b Bradford, Pa, Dec 27, 44; m 67; c 1. ELECTRONICS & RADIO FREQUENCY ENGINEERING. *Educ:* Drexel Univ, BS, 67, MS, 69, PhD(biomed eng), 74. *Prof Exp:* Chief engr, Electronics Div, Ventron Corp, 69-70; res assoc biotelemetry, dept biomed eng, Drexel Univ, 70-74; fel reprod biol, dept obstet & gynec, Univ Pa, 74-76; asst prof biomed eng, 76-83, ASSOC PROF ELEC, COMPUT & BIOMED ENG, MARQUETTE, 83- *Concurrent Pos:* Adj asst prof physiol, dept biol, Drexel Univ, 75-76 & dept physiol, Med Col Wis, 77-; asst clin prof neurosurg, Med Col Wis, 78-; consult, Symbion Inc, 85-88. *Honors & Awards:* Earl W Hatz Mem Award, 84. *Mem:* Sr mem Inst Elec & Electronics Engrs; Sigma Xi; sr mem Biomed Eng Soc. *Res:* Transcutaneous data and powering; cochlear prostheses; totally implanted artificial heart; sensors; signal processing; regenerative electrical stimulation. *Mailing Add:* Col Eng Marquette Univ 1515 W Wisconsin Ave Milwaukee WI 53233

JEVNING, RON, b Winnemucca, Nev, Oct 20, 42. BIOPHYSICS. *Educ:* Stanford Univ, PhD, 71. *Prof Exp:* Asst prof, Dept Med, Med Ctr, Univ Calif, Irvine, 72-88, ASSOC PROF BIOMED SCI, 88- *Mem:* Am Physiol Soc; Am Soc Psychophysiol Res; Soc Neurosci; Soc Behav Med; Am Asn Scientists Practicing TM Tech. *Mailing Add:* 13197 E Gwyneth Tustin CA 92680

JEWELL, FREDERICK FORBES, SR, b Oil City, Pa, June 4, 28; m 51; c 5. HISTOPATHOLOGY. *Educ:* Mich State Col, BS, 51, MS, 52; Univ WVa, PhD, 55. *Prof Exp:* Asst plant path, Univ WVa, 52-55; prin plant pathologist forest tree dis, Southern Inst Forest Genetics, 55-67; assoc prof, 67-69, PROF FOREST PROTECTION, LA TECH UNIV, 69- *Mem:* Am Phytopath Soc. *Res:* Disease resistance in forest treas; rust-resistance in Southern pines; pathological anatomy; needle diseases. *Mailing Add:* La Tech Univ Sch of Forestry Ruston LA 71270

JEWELL, JACK LEE, b Jacksonville, Fla, Jan 7, 54; m 83; c 1. DIODE LASERS, MICRO-OPTICS. *Educ:* Univ Fla, BS, 75; Fla Inst Technol, MS, 77; Univ Ariz, MS, 81, PhD(optical sci), 84. *Prof Exp:* Mem tech staff, AT&T Bell Labs, 84-91; VPRES, PHOTONICS RES, INC, 91- *Concurrent Pos:* Distinguished lectr, Inst Elec & Electronics Engrs Lasers & Electro-Optics Soc, 91-92. *Mem:* Optical Soc Am; Am Phys Soc; Soc Photo-optical Instrumentation Engrs. *Res:* Vertical-cavity surface-emitting microlasers; decreasing electrical resistance; extending wavelength range; opto-electronic integration with transistors; micro-optic integration. *Mailing Add:* Photonics Res Inc 100 Technology Dr Broomfield CO 80021

JEWELL, NICHOLAS PATRICK, b Paisley, Scotland, Sept 3, 52; m 80. BIOSTATISTICS, TIME SERIES. *Educ:* Univ Edinburgh, BSc, 73, PhD(math), 76. *Prof Exp:* Harkness fel, Commonwealth Fund, NY, 76-78; res fel, Univ Edinburgh, UK, 78-79; asst prof statist, Princeton Univ, 79-81; from asst prof to assoc prof, 81-87, PROF BIOSTATIST, UNIV CALIF, BERKELEY, 87- *Mem:* Am Statist Asn; Inst Math Statist (treas, 85-); Biomet Soc. *Res:* Biostatistics; mathematical statistics; functional analysis; function theory. *Mailing Add:* Dept Environ Health Prog Biostatist Univ Calif Sch Pub Health Berkeley CA 94720

JEWELL, WILLIAM R, b Evanston, Ill, Oct 7, 35; m 60; c 4. SURGERY. *Educ:* Blackburn Col, BA, 57; Univ Ill, BS, 59, MD, 61. *Prof Exp:* Asst prof surg, Med Ctr, Univ Ky, 68-71; assoc prof med, 71-78, PROF MED & CHIEF GEN SURG, UNIV KANS MED CTR, 78- *Concurrent Pos:* Consult, US Vet Admin Hosp, Lexington, Ky, 68- *Honors & Awards:* Meade Johnson Sr Res Award Surg, 66, Health Sci Achievement Award, 71. *Res:* Carcinogenesis; protein metabolism in cancer bearing hosts; oncologic immunology; wound healing. *Mailing Add:* Dept Surg Univ Kans Med Ctr 39th & Rainbow Blvd Kansas City KS 66103

JEWELL, WILLIAM S(YLVESTER), b Detroit, Mich, July 2, 32; m 56; c 4. OPERATIONS RESEARCH, ACTUARIAL SCIENCE. *Educ:* Cornell Univ, BEngPhys, 54; Mass Inst Technol, SM, 55, ScD, 58. *Prof Exp:* Asst, Mass Inst Technol, 55-58; assoc dir, Mgt Sci Div, Broadview Res Corp, 58-60; from asst prof to assoc prof indust eng, 61-67, chmn dept, 67-69 & 76-80, PROF INDUST ENG & OPERS RES, UNIV CALIF, BERKELEY, 67- *Concurrent Pos:* Fulbright res scholar, France, 65; bd mem, Teknekron Indust Inc, Berkeley, 68-83; res scholar, Int Inst Appl Systs Anal, 74-75; guest prof, Fed Inst Technol, Switzer, 80-81. *Honors & Awards:* Halmstead Memorial Prize. *Mem:* Opers Res Soc Am; Inst Mgt Sci; Swiss Actuarial Asn; Sigma Xi; Int Asn Actuaries. *Res:* Operations research; prediction and estimation; reliability; risk theory. *Mailing Add:* Dept Indust Eng & Opers Res Univ Calif Berkeley CA 94720

JEWETT, DON L, b Eureka, Calif, Jan 28, 31; m 54; c 2. ORTHOPEDIC SURGERY, NEUROPHYSIOLOGY. *Educ:* San Francisco State Col, AB, Univ Calif, Berkeley, 54-56; Univ Calif, San Francisco, MD, 60; Oxford Univ, DPhil(physiol), 63. *Prof Exp:* NIH fel, Yale Univ, 63-64; asst prof physiol & neurosurg, 64-72, clin instr orthop surg, 72-75, ASSOC PROF ORTHOP SURG, MED SCH, UNIV CALIF, SAN FRANCISCO, 75- *Mem:* Am Physiol Soc; Soc Neurosci; Sigma Xi. *Res:* Central and peripheral nervous system physiology related to clinical conditions; bioengineering; averaged far field potentials. *Mailing Add:* Dept Orthop Surg Univ Calif Med Ctr San Francisco CA 94143

JEWETT, HUGH JUDGE, b Baltimore, Md, Sept 26, 03; m 41; c 1. UROLOGY. *Educ:* Johns Hopkins Univ, AB, 26, MD, 30. *Prof Exp:* Assoc prof, 51-66, PROF UROL, SCH MED, JOHNS HOPKINS UNIV, 66-, UROLOGIST, JOHNS HOPKINS HOSP, 36- *Concurrent Pos:* Emer ed, J Urol; ed-in-chief, Urol Surv; pres, Md Med Serv, 50-54; trustee, Am Urol Res Found, 52; counsr med & chirurgical fac, State Md, 53-55; chmn registry genito-urinary path, Armed Forces Inst Path, 58-64; mem bd gov, Am Col Surg, 58-64 & Am Bd Urol, 59-66. *Honors & Awards:* Barringer Medal, Am Asn Genito-Urinary Surg, 62; Ramon Guiteras Award, Am Urol Asn, 63; Ferdinand C Valentine Medal, NY Acad Med, 84; Keyes Gold Medal, Am Asn Geritourinary Surgeons, 85. *Mem:* Fel Am Col Surg; Am Urol Asn (pres, 65-66); Am Asn Genito-Urinary Surg (pres, 70-71); Clin Soc Genito-Urinary Surg (pres, 68-69); Int Soc Urol. *Res:* Clinical research on cancer of the bladder and prostate. *Mailing Add:* Dept Urol Johns Hopkins Univ Baltimore MD 21205

JEWETT, JOHN GIBSON, b Birmingham, Ala, Jan 21, 37; m 62. ORGANIC CHEMISTRY. *Educ:* Harvard Univ, AB, 58; Mass Inst Technol, PhD(org chem), 62. *Prof Exp:* Res assoc org chem, Ind Univ, 62-64; from asst prof to prof chem, Ohio Univ, 64-77; dean Col Arts & Sci, 77-89, PROF CHEM, UNIV VT, 77- *Mem:* AAAS; Am Chem Soc. *Res:* Reaction mechanisms; isotope effects; simple displacement reactions; highly strained ring systems; fragmentation reactions. *Mailing Add:* Chem Dept Univ Vt Burlington VT 05405-0125

JEWETT, ROBERT ELWIN, b Jackson, Mich, Feb 3, 34. PHARMACOLOGY. *Educ:* Univ Mich, MD, 58. *Prof Exp:* Intern, Toledo Gen Hosp, Ohio, 58-59; asst resident ophthal, Univ Hosp, Ann Arbor, Mich, 59-60; physician, Ciba Pharmaceut Co, 60-62; instr pharmacol, Med Sch, Univ Kans, 64-65; from instr to prof, Med Sch, Emory Univ, 65-74; prof pharmacol & dean, Col Med, ETenn State Univ, 74-76; assoc dean, Sch Med, Wright State Univ, 76-80, prof pharmacol, 76-82, sr assoc dean acad affairs, 80-82; dep vpres, Nat Bd Med Examr, 82-87, vpres eval progs, 87-88; ADJ PROF, OFF DEAN, COL MED, UNIV FLA, 89- *Concurrent Pos:* Nat Inst Neurol Dis & Blindness spec fels, 63-65. *Res:* Neuropharmacology; pharmacology of sleep, amines and behavior; educational evaluation. *Mailing Add:* 5055 SW Ninth Lane Gainesville FL 32607

JEWETT, SANDRA LYNNE, b Lone Pine, Calif, Nov 13, 45. BIOCHEMISTRY, BIO-ORGANIC CHEMISTRY. *Educ:* Univ Calif, Santa Barbara, BA, 67, PhD(chem), 71. *Prof Exp:* Res fel biochem, Stanford Univ, 71-73; res fel enzyme immunoassays, Syva Co, Palo Alto, Calif, 74-75; asst prof chem, Williams Col, 75-77; asst prof, 77-82, assoc prof chem, 82-88, PROF CHEM, CALIF STATE UNIV, NORTHRIDGE, 88- *Concurrent Pos:* NIH fel, 72-73. *Mem:* AAAS; Am Chem Soc; Sigma Xi. *Res:* Studies of erythrocyte superoxide dismutase; chemical studies of active site and intersubunit interactions; formation of and properties of metal deficient enzymes; reaction of copper-zinc dismutase with hydrogen peroxide; iron-catecholamine complexes. *Mailing Add:* Dept of Chem Calif State Univ Northridge CA 91330

JEWSBURY, WILBUR, b Jacksonville, Ill, Dec 13, 06; m 33; c 3. CHEMISTRY. *Educ:* Ill Col, BA, 27; Western Ill Univ, MS, 45; NDak State Univ, PhD(chem), 65. *Prof Exp:* Teacher pub schs, Ill, 28-55; instr chem, Western Ill Univ, 55 & State Col Wash, 55-58; from asst prof to prof, 58-77, EMER PROF CHEM, MANKATO STATE COL, 77- *Mem:* AAAS; Am Chem Soc. *Res:* Inorganic and organic chemistry; phase selenate; humic acids. *Mailing Add:* 202 Long St Mankato MN 56001

JEYAPALAN, KANDIAH, b Sri Lanka, June 24, 38; m 64; c 2. PHOTOGRAMMETRY. *Educ:* Univ Ceylon, BSc, 60; Univ London, MSc, 67, PhD(photogram), 72. *Prof Exp:* Asst supt, Surv Dept, Sri Lanka, 61-67, chief photogrammetrist, 67-69; res assoc, Dept Geodetic Sci, Ohio State Univ, 69-72; asst prof surv, geod & photogram, Calif State Univ, Fresno, 72-74, assoc prof, 74-78, prof, 78-79; PROF SURV, GEOD & PHOTOGRAM, IOWA STATE UNIV, 79- *Concurrent Pos:* Ceylon Govt scholar, 63; UN fel, UN Educ & Sci Orgn, 66; lectr, Dept Geod Sci, Ohio State Univ, 69-72; admin asst, Highway Dept, Columbus, Ohio, 72; sr lectr, Univ Dar-es-Salaam, Tanzania, 73 & 74; res civil engr, US Geol Surv, 77; vis prof, Naval Postgrad Sch, Montrey, Calif, 87; UN fel, 86. *Mem:* Am Soc Photogram; Am Congress Surv & Mapping; Sigma Xi. *Res:* Photogrammetry: development of analytical plotter, calibration of cameras, analytical triangulation, shortwave photogrammetry and digital terrain model; geodesy: electronic surveying, Doppler surveying and geoposition system; numerical cadastral survey. *Mailing Add:* Dept Civil Eng Univ Iowa Ames IA 50011

JEZAK, EDWARD V, b Czestochowa, Poland, Mar 29, 34; US citizen; m 62; c 2. PHYSICS. *Educ:* Harvard Univ, AB, 57; Univ Minn, PhD(physics), 62. *Prof Exp:* Asst prof physics, Boston Col, 62-68; ASSOC PROF MATH, ROYAL MIL COL CAN, 68- *Mem:* Am Phys Soc. *Res:* Nuclear theoretical physics; three body problem; molecular dynamics. *Mailing Add:* Dept Math Royal Mil Col Kingston Kingston ON K7K 5L0 Can

JEZESKI, JAMES JOHN, b Minneapolis, Minn, June 8, 18; m 43. MICROBIOLOGY, FOOD SCIENCE. *Educ:* Univ Minn, BS, 40, MS, 42, PhD(bact), 47. *Prof Exp:* Asst dairy bact, Univ Minn, St Paul, 41-43, from asst prof to prof, 48-69; prof bot & microbiol, Mont State Univ, 69-73; dir res & develop, Monarch Chem Div, H B Fuller Co, 73-78; mem staff exten food technol, Univ Fla, 78-84; RETIRED. *Mem:* Am Soc Microbiol; Am Dairy Sci Asn; Nat Environ Health Asn; Int Asn Milk, Food & Environ Sanit; Inst Food Technol. *Res:* Role of microorganisms in manufacturing and deterioration of foods; quality assurance and public health safety of foods. *Mailing Add:* 1704 NW 39th Terrace Gainesville FL 32605

JEZL, JAMES LOUIS, b Tobias, Nebr, Dec 12, 18; m; c 6. ORGANIC CHEMISTRY, RESEARCH ADMINISTRATION. *Educ:* Univ Nebr, AB, 41; Pa State Col, MS, 42; Univ Del, PhD(org chem), 49. *Prof Exp:* Supvry chemist, US Rubber Co, 42-43; jr anal chemist, Sun Oil Co, Ohio, 43-45, sr anal chemist, 45-47, develop chemist, Pa, 47-49, res chemist, 49-54, res group leader, 54-58, sect chief, 58-60, mgr res div, Avisun Corp, 60-68, dir res, 68-70; div dir, 70-76, mgr explor res, Res & Develop Dept, Naperville Tech Ctr, Amoca Chem Corp, 76-86; RETIRED. *Mem:* AAAS; Am Chem Soc. *Res:* Petrochemicals; polyolefins; petroleum processing. *Mailing Add:* 35 W 094 Army Trail St Charles IL 60174

JEZOREK, JOHN ROBERT, b Baltimore, Md, June 12, 42; m 67; c 4. ANALYTICAL CHEMISTRY, INORGANIC CHEMISTRY. *Educ:* Loyola Col, Md, BS, 64; Univ Del, PhD(anal chem), 69. *Prof Exp:* Res assoc, Univ Mich, 69-70, Univ Ariz, 77-78; from asst prof to assoc prof, 70-81, PROF ANALYTICAL CHEM, UNIV NC, GREENSBORO, 82- *Mem:* Am Chem Soc; Sigma Xi. *Res:* Liquid chromatography; design of novel LC stationary phases; surface modification chemistry. *Mailing Add:* Dept of Chem Univ of NC Greensboro NC 27412

JEZYK, PETER FRANKLIN, b Ware, Mass, Nov 7, 39; c 2. VETERINARY MEDICINE. *Educ:* Univ Mass, BS, 61, PhD(zool), 66; Univ Pa, VMD, 75. *Prof Exp:* NIH fel biol chem & res assoc, Univ Mich, 66-67; asst prof biochem, Med Col Va, 67-71; asst prof, 75-81, ASSOC PROF MED GENETICS, SCH VET MED, UNIV PA, 81- *Concurrent Pos:* Dir, Metab Screening Lab, Children's Hosp Philadelphia, 76- *Mem:* AAAS; Am Vet Med Asn. *Res:* Metabolic aspects of inherited disease in companion animals. *Mailing Add:* Dept Med Genetics Univ Pa Sch Ved Med 3850 Spruce St Philadelphia PA 19104-6010

JHA, MAHESH CHANDRA, b Bihar, India, March 13, 45; US citizen; m 64; c 2. COAL CONVERSION TECHNOLOGY, PROCESS DEVELOPMENT. *Educ:* Bihar Inst Technol, India, BScEng, 65; Mich Tech Univ, MS, 70; Iowa State Univ, Ames, PhD(metall & chem eng), 74. *Prof Exp:* Lectr extractive metall, Bihar Inst Technol, India, 65-66 & Univ Rorkee, India, 66-69; grad res asst, Inst Mineral Res, Mich Tech Univ, 69-70 & Ames Lab, Iowa State Univ, 70-73; res metallurgist, 73-75, group leader, 75-78, sect supvr, 78-84, mgr contract res & develop, 85-86, MGR ENERGY RES & DEVELOP, AMAX RES & DEVELOP CTR, 87- 85- *Honors & Awards:* Extractive Metall Technol Award, Metall Soc, 86. *Mem:* Am Inst Mining, Metall & Petrol Engrs; Am Inst Chem Engrs. *Res:* Improving the processes for extraction of non-ferrous metals such as nickel, cobalt, molybdenum, tnugsten, gold, silver from low-grade ores; waste streams; coal conversion and utilization tehnology. *Mailing Add:* Amax Res & Develop Ctr 5950 McIntyre St Golden CO 80403-7499

JHA, SHACHEENATHA, b Darbhanga, Bihar, India, Nov 15, 18; m 55; c 4. EXPERIMENTAL NUCLEAR PHYSICS. *Educ:* Patna Univ, BS, 39, MS, 41; Univ Edinburgh, PhD(nuclear physics), 50. *Prof Exp:* Res scholar physics, Patna Sci Col, 41-44, lectr, 44-46, asst prof, 51; Govt Bihar scholar nuclear physics, Univ Edinburgh, 46-51; res fel physics, Tata Inst Fundamental Res, India, 51-56, fel, 56-61; asst prof, Carnegie Inst Technol, 61-66; assoc prof, Case Western Reserve Univ, 66-69; PROF PHYSICS, UNIV CINCINNATI, 69- *Mem:* Fel Am Phys Soc; Am Asn Physics Teachers. *Res:* Nuclear spectroscopy and reaction; Mossbauer effect; molecular spectroscopy. *Mailing Add:* Dept of Physics Univ of Cincinnati Cincinnati OH 45221

JHAMANDAS, KHEM, b EAfrica, May 11, 39; m 71. PHARMACOLOGY. *Educ:* Univ London, BSc, 64; Univ Alta, MSc, 66, PhD(pharmacol), 69. *Prof Exp:* Med Res Coun Can fel pharmacol & therapeut, Univ Man, 69-70; asst prof, 70-75, PROF PHARMACOL & TOXICOL, FAC MED, QUEEN'S UNIV, ONT, 75- *Concurrent Pos:* Vis scientist, Mayo Clinic & Killam res fel, 80-81, Kyoto Univ, 86, Univ Melbourne, 87. *Mem:* Pharmacol Soc Can; Am Soc Pharmacol & Exp Therapeut; Soc Neurosci. *Res:* Neuropharmacology; action of drugs on transmitter substances in the central nervous system; mechanisms underlying drug dependence on opioids; neuropharmacology of enkephalins, endorphins and neuropeptides; excitotoxins. *Mailing Add:* Dept Pharmacol & Toxicol Queen's Univ Kingston ON K7L 3N6 Can

JHANWAR, SURESH CHANDRA, CYTOGENETICS. *Educ:* Univ Delhi, India, PhD(genetics), 76. *Prof Exp:* ASST PROF GENETICS & ASST ATTEND GENETICIST, SLOAN-KETTERING CANCER CTR, 76- *Mailing Add:* Sloan-Kettering Cancer Ctr Box 147 New York NY 10021

JHIRAD, DAVID JOHN, b India, May 29, 39; US citizen; m; c 3. PHYSICS. *Educ:* Delhi Univ, BSc, 58; Cambridge Univ, BA, 61, MA, 64; Harvard Univ, PhD(appl physics), 72. *Prof Exp:* Asst prof physics, Boston Univ & Univ Mass, 70-75; staff dir energy, Union Concerned Scientists, 75-78; sr res scientist energy, Jet Propulsion Lab, Calif Inst Technol, 78-80; dir, Int Energy Prog, Brookhaven Nat Lab, 80-84; CONSULT, 84- *Concurrent Pos:* Sr Energy Adv, US Agency for Int Develop, 84- *Mem:* Am Phys Soc; AAAS; NY Acad Sci; Int Solar Energy Soc; Scientists Inst Pub Info. *Res:* New power technology, energy technology and applications assessment and policy analysis, international energy and power investment planning and technology development; thermodynamics and statistical mechanics. *Mailing Add:* 3009 Daniel Lane NW Washington DC 20015-1435

JHON, MYUNG S, b Korea; US citizen. POLYMER ENGINEERING, TRIBOLOGY. *Educ:* Seoul Nat Univ, Seoul, Korea, BS, 67; Univ Chicago, PhD(physics), 74. *Prof Exp:* Res asst physics, James Franck Inst, Univ Chicago, 70-74; postdoctoral fel physics, Univ Toronto, 74-76; res specialist chem, Univ Minn, 76-80; from asst prof to assoc prof, 80-88, PROF CHEM ENG, CARNEGIE MELLON UNIV, 88- *Concurrent Pos:* Vis prof, Magnetics Recording Inst, Int Bus Mach, 85 & Univ Calif, Berkeley, 89; consult, UN Indust Develop Orgn, 86; sr vis prof, Naval Res Lab, Washington, DC, 86; vis scientist, Int Bus Mach Res Div, Almaden Res Ctr, 88. *Mem:* Am Inst Chem Engrs; Am Phys Soc; Am Chem Soc; Sigma Xi; NY Acad Sci. *Res:* Magnetic and magneto-optical recording; polymer and suspension rheology; interfacial dynamics; membrane science and technology; equilibrium and nonequilibrium statistical mechanics; chemical kinetics; fluid mechanics; turbulent drag reduction. *Mailing Add:* Dept Chem Eng Carnegie Mellon Univ Pittsburgh PA 15213

JI, CHUENG RYONG, b Seoul, Korea, Jan 7, 54; m 83; c 2. INTERSECTION BETWEEN PARTICLE & NUCLEAR THEORY. *Educ:* Seoul Nat Univ, BS, 76; Korea Advan Inst Sci & Technol, MS, 78, PhD(physics), 82. *Prof Exp:* Postdoctoral nuclear physics, Stanford Univ, 84-86; res assoc, Brooklyn Col, City Univ New York, 86-87; ASST PROF NUCLEAR PHYSICS, NC STATE UNIV, 89- *Concurrent Pos:* Vis scholar elem particle physics, Stanford Linear Accelerator Ctr, 82-86; vis asst prof nuclear physics, NC State Univ, 87-89; theory consult, Continuous Electron Beam Accelerator Facil, 90; prin investr, Dept Energy, 90- *Mem:* Am Phys Soc; Sigma Xi. *Res:* Quark and gluon structures of hadron; theory of strong interaction, quantum chromodynamics, based on the light cone formulation. *Mailing Add:* Dept Physics NC State Univ Raleigh NC 27695-8202

JI, GUANGDA WINSTON, b Tianjin, People's Repub China; c 1. III-V SEMICONDUCTOR DEVICES RESEARCH & DEVELOPMENT, REACTIVE ION ETCHING AND PLASMA ENHANCED CHEMICAL VAPOR DEPOSITION SPECIALIST. *Educ:* Univ Ill Urbana, MS, 81, PhD(physics), 86. *Prof Exp:* Res assoc elec eng, Coord Sci Lab, Univ Ill Urbana, 86-88; RES ASSOC ELEC ENG, ONT LASER & LIGHTWAVE RES CTR, 88- *Honors & Awards:* Chinese Award in Sci & Technol, 82. *Mem:* Am Phys Soc. *Res:* Semiconductor laser; optoelectronic devices; high-speed semiconductor electronic devices; device modeling and optical measurements for superlattice and heterojunction devices in III-V semiconductors; reactive ion etching and plasma enhanced chemical vapor deposition techniques and diagnosis; author of one publication. *Mailing Add:* 145 St George St Apt 301 Toronto ON M5R 2M1 Can

JI, INHAE, b Seoul, Korea, May 17, 38; m 65; c 2. HORMONES, RECEPTORS. *Educ:* Seoul Nat Univ, BS, 61, MS, 63; Univ Wyo, PhD(biochem), 77. *Prof Exp:* Postdoctoral biochem, Harvard Med Sch, 77-78; res asst biochem, 78-91, RES PROF MOLECULAR BIOL, UNIV WYO, 91- *Mem:* Endocrine Soc; Am Soc Cell Biologists. *Mailing Add:* Dept Molecular Biol Univ Wyo Laramie WY 82071-3944

JI, SUNGCHUL, b Sheenweejoo, Korea, Dec 17, 37; m 67; c 1. BIOPHYSICS, CELL PHYSIOLOGY. *Educ:* Univ Minn, Duluth, BA, 65; State Univ NY Albany, PhD(org chem), 70. *Prof Exp:* Asst prof chem, Mankato State Col, 68-70; NIH trainee & res asst prof, Inst Enzyme Res, Univ Wis-Madison, 70-74; res assoc, Johnson Res Found, Univ Pa, 74-76; res scientist, Max Planck Inst Systs Physiol, 76-; res asst prof, dept pharmacol, Univ NC, 79-; AT GRAD PROG TOXICOL, RUTGERS UNIV. *Mem:* Am Chem Soc; AAAS; Sigma Xi. *Res:* Anion radical chemistry; electron transfer reactions in organic solvents; energy-coupling mechanism in mitochondria; nicotinamide-adenine dinucleotide fluorescence photography; micro-light guide tissue photometry; flow-metabolism coupling in the liver; alcohol-induced liver injury; lobular oxygen gradient in the liver. *Mailing Add:* Grad Prog Toxicol Rutgers Univ PO Box 789 Piscataway NJ 08854

JI, TAE H(WA), b Andong, Korea, Apr 7, 41; US citizen; m 65; c 2. BIOCHEMISTRY, MOLECULAR BIOLOGY. *Educ:* Seoul Nat Univ, BS, 64; Univ Calif, San Diego, PhD(biol), 68. *Prof Exp:* Inst Biomed Res fel, AMA, 68-69; fel, Univ Minn, 69-70; from asst prof to assoc prof, 70-77, PROF BIOCHEM, UNIV WYO, 77- *Concurrent Pos:* Vis prof, Harvard Univ, 77-78; regent fel, Univ Calif, 64-65; fac res award, Am Cancer Soc, 83-88; mem, Physiol Chem Study Sect, NIH, 90-94. *Honors & Awards:* Burlington Award, 88. *Mem:* Am Soc Biochem & Molecular Biologists; Endocrine Soc; Soc Study Reproduction. *Res:* Structure function and gene expression of gonadotropin receptors; photoaffinity labeling. *Mailing Add:* Dept Molecular Biol Univ Wyo Laramie WY 82071-3944

JIANG, JACK BAU-CHIEN, b Sze-chuan, China, Nov 15, 47; m 73; c 2. ORGANIC CHEMISTRY, MEDICINAL CHEMISTRY. *Educ:* Nat Cheng Kung Univ, BS, 70; Mich State Univ, PhD(org chem), 75. *Prof Exp:* Res specialist, Univ Minn, 75-77; res chemist drug synthesis, Am Cyanamid Co, 77-79; scientist, 79-81, sr scientist, Ortho Pharm Corp, 81-84, DuPont Pharmaceuts, 84-87; group leader, Du Pont Pharmaceut, 87-89; DIR MED CHEM, SPHINX PHARMACEUT CORP, 89- *Mem:* Am Chem Soc. *Res:* Design and synthesis of medicinal agents; anticancer chemotherapy discovery and development. *Mailing Add:* 1604 Skye Dr Chapel Hill NC 27516-9016

JIANG, NAI-SIANG, b Nanking, China, June 6, 31; m 58; c 2. BIOCHEMISTRY. *Educ:* Nat Taiwan Univ, BS, 55; Emory Univ, MS, 59, PhD(biochem), 62. *Prof Exp:* Instr biochem, Emory Univ, 62-66; res assoc, Mayo Found, Mayo Clinic, 66-67, consult, Dept Endoctrine Res, 67-70, asst prof biochem, Mayo Grad Sch Med, 67-75, assoc prof biochem & lab med, 75-80, PROF LAB MED, MAYO MED SCH, UNIV MINN, 80-, HEAD SECT CLIN CHEM, 84- *Concurrent Pos:* Dir & consult, Endocrine Lab, Dept Lab Med, Mayo Clin & Found, 71- *Mem:* AAAS; Am Chem Soc; Sigma Xi. *Res:* Measurement of hormones in body fluid. *Mailing Add:* Dept Lab Med Mayo Clin FDN Rochester MN 55901

JIBSON, RANDALL W, b San Jose, Calif, Apr 17, 56; m 82; c 3. EARTHQUAKE-INDUCED GROUND FAILURE, SEISMIC ENGINEERING. *Educ:* San Diego State Univ, BS, 80; Stanford Univ, MS, 83, PhD(geol), 85. *Prof Exp:* Geologist, 83-88, SUPVRY GEOLOGIST & GEOMECH RES COORDR, US GEOL SURV, 88- *Concurrent Pos:* Res fel,

Japan Pub Works Res Inst, 87; mem, Landslide Comt, Asn Eng Geologists, 87- *Mem:* Geol Soc Am; Asn Eng Geologists. *Res:* Basic and applied research in the field of geologic hazards, specifically in earthquake effects, ground-failure processes and coastal erosion. *Mailing Add:* US Geol Surv MS 966 Denver Fed Ctr Box 25046 Denver CO 80225

JICHA, HENRY LOUIS, JR, b New York, NY, June 25, 28; m 51; c 3. ECONOMIC GEOLOGY. *Educ:* Columbia Univ, BA, 48, MA, 51, PhD(econ geol), 52. *Prof Exp:* Geologist, Mineral Deposits Br, US Geol Surv, Colo, 48-49 & Fla, 49; field asst, NMex Bur Mines & Mineral Resources, 50-51, econ geologist, 51-56; asst prof geol, Colo Sch Mines, 56-58; analyst mining & metal stocks, Baker, Weeks & Co, 58-61; ed-analyst, Value Line Invest Surv, Metals, Oils, Brewing, 61-62; mgr, New York Res, Courts & Co, 62-70; sr analyst, Newberger, Loeb & Co, 70-71; mgr res, Jesup & Lamont, 71-73; vpres & sr analyst, Prudential Bache Securities, Inc, New York, 74-83; dir res, Wood Gundy Corp, NY, 83-89; RETIRED. *Concurrent Pos:* Consult, Baumgartner Oil Co, Colo, 57-58. *Mem:* Sigma Xi. *Res:* Uranium deposits in Colorado, phosphate deposits in Florida; lead-zinc deposits in Europe; tertiary volcanics, lead-zinc-silver deposits, Mesa del Oro Quadrangle and manganese deposits in New Mexico. *Mailing Add:* 12 Western Dr Ardsley NY 10502

JILES, CHARLES WILLIAM, b Vienna, La, Aug 11, 27; m 50; c 4. ELECTRICAL ENGINEERING. *Educ:* La Polytech Inst, BS & BA, 49; Okla State Univ, MS, 50, PhD, 55. *Prof Exp:* Asst physics, La Polytech Inst, 47-49; res instr, Okla State Univ, 50-55; sr aerophysics engr, Convair Div, Gen Dynamics Corp, 55-58, proj aerophysics engr, 58-60, design specialist, 60; PROF ELEC ENG, UNIV TEX, ARLINGTON, 60- *Mem:* Inst Elec & Electronics Engrs; Am Astronaut Soc; Sigma Xi. *Res:* Application of matrix algebra and tensor analysis to electric circuits and machines; network analysis and synthesis; design of automatic control systems. *Mailing Add:* Box 19018 Arlington TX 76019

JILKA, ROBERT LAURENCE, b Salina, Kans, Nov 26, 48. MOLECULAR ENDOCRINOLOGY. *Educ:* Kans State Univ, BS, 70, MS, 72; St Louis Univ, PhD(biochem), 75. *Prof Exp:* Fel, Roche Inst Molecular Biol, 75-78; staff scientist, Calcium Res Lab, Kansas City, 78-, VET ADMIN MED CTR, INDIANAPOLIS. *Concurrent Pos:* Adj asst prof, Dept Biochem, Univ Kans Med Ctr, 79- *Mem:* Am Soc Bone & Mineral Res. *Res:* Biochemical changes caused by parathormone, vitamin D and calcitonin on bone in organ culture and partially purified bone cells in tissue culture; both normal and genetically defined osteopetrotic bone is studied. *Mailing Add:* Vet Admin Med Ctr 1481 W Tenth St Indianapolis IN 46202

JILLIE, DON W, b San Lois Opsipo, Calif, May 19, 48; m 72; c 2. MATERIALS SCIENCE ENGINEERING. *Educ:* Stanford Univ, BS, 70; State Univ NY Stony Brook, PhD(physics), 76. *Prof Exp:* Mem tech staff, Sperry Res Ctr, 76-83; staff process engr, 83-86, GROUP LEADER, INTEL CORP, 86- *Mem:* Am Phys Soc; Inst Elec & Electronics Engrs. *Res:* Process engineering in support of volume manufacturing of silicon MDs IC's; specializing in etch and resist clean; 4 patents. *Mailing Add:* Group Leader for Prod Etch Intel Corp Sc9-13 PO Box 58125 Santa Clara CA 95014

JIM, KAM FOOK, b Po On, China, Nov 13, 53; US citizen; m 81; c 2. CARDIOVASCULAR PHARMACOLOGY, BIOCHEMICAL PHARMACOLOGY. *Educ:* NY Univ, BA, 76; State Univ NY, Buffalo, PhD(pharmacol), 81. *Prof Exp:* Postdoctoral fel biochem, Case Western Res Univ, 80-81; res assoc biochem, Cornell Univ Med Col, 81-83; postdoctoral scientist pharmacol, SmithKline & French Lab, SmithKline Beecham Co, 83-86; consult pharmacol, Med Col Pa, 86-88; SR SCI WRITER CLIN COMMUN, WYETH-AYERST RES, 88- *Mem:* Am Fed Clin Pharmacol; NY Acad Sci; Am Soc Pharmacol & Exp Therapeut; AAAS. *Res:* Writing responsibilities include clinical trials of cardiovascular drugs, especially antianginal and antiarrhythmic compounds. *Mailing Add:* Dept Clin Commun B-2 Wyeth-Ayerst Res PO Box 8299 Philadelphia PA 19101-1245

JIMBOW, KOWICHI, b Nagoya City, Japan, June 4, 41; c 5. DERMATOLOGY, PATHOLOGY. *Educ:* Sapporo Med Col, MD, 66, PhD(med sci), 74. *Prof Exp:* Instr dermat, Mass Gen Hosp, Boston, 74-75; from asst prof to assoc prof, Sapporo Med Col, Japan, 75-87; PROF & DIR DERMAT & CUTANEOUS SCI, UNIV ALTA, EDMONTON, 87-, PROF PATH, 88- *Concurrent Pos:* Vis assoc prof dematopath, Dept Path, Univ Ark, 75-78; counr, Int Soc Pigment Cell Res, 84-87; mem organizing comt, 2nd Int Melanoma Conf, 86-; chmn & organizer, Third Meeting Pan Am Soc Pigment Cell Res, 90- *Honors & Awards:* Alfred-Marchionini Prize, Int Asn Dermat, 82; Seiji Mem Prize, Japanese Soc Dermat, 84. *Mem:* Am Soc Cancer Res; Am Soc Photobiol; Can Dermat Asn; fel Am Acad Dermat. *Mailing Add:* 260G Heritage Med Res Ctr Univ Alta Edmonton AB T6G 2S2 Can

JIMENEZ, AGNES E, b Farrell, Pa, Oct 21, 43. NEUROENDOCRINOLOGY. *Educ:* Univ Louisville, PhD(physiol), 76. *Prof Exp:* Asst prof, 77-85, ASSOC PROF PHYSIOL, SCH MED, UNIV LOUISVILLE, 85- *Mem:* Soc Neurosci; Am Physiol Soc; Sigma Xi. *Mailing Add:* Dept Physiol & Biophysics Sch Med Univ Louisville Health Sch Med Louisville KY 40292

JIMENEZ, SERGIO, b Cuzco, Peru, Feb 21, 42; m. BIOCHEMISTRY. *Educ:* Univ San Marcos, Lima, MD, 64. *Hon Degrees:* MS, Univ Pa. *Prof Exp:* Assoc prof med & rheumatology, Sch Med, 80-85, prof, 85-87, DIR, COLLAGEN RES, DEPT MED, UNIV PA, 73-; PROF MED, BIOCHEM & MOLECULAR BIOL, THOMAS JEFFERSON UNIV. *Concurrent Pos:* NIH Gen Med Study Sect, Arthritis Found Res Comt. *Res:* Biochemistry and molecular biology of inherited and acquired connective tissue diseases. *Mailing Add:* Jefferson Hall-M 46 Thomas Jefferson Univ Rm M26 1020 Locust St Philadelphia PA 19107

JIMENEZ-MARIN, DANIEL, anatomy, genetics, for more information see previous edition

JIMERSON, GEORGE DAVID, b Little Rock, Ark, May 12, 44; m 65; c 2. INORGANIC CHEMISTRY, ANALYTICAL CHEMISTRY. *Educ:* Ouachita Baptist Univ, BS, 66; Ind Univ, Bloomington, PhD(chem), 70. *Prof Exp:* Asst prof, 75, ASSOC PROF CHEM, ARK STATE UNIV, 75- *Concurrent Pos:* Prin investr res grant, Ark Educ Res & Develop Proj, 71-72; co-prin investr res contract, Ark Highway Dept, 72- *Honors & Awards:* Hon Sci Award, Bausch & Lomb, 62. *Mem:* Am Chem Soc; Sigma Xi. *Res:* Waste utilization and resource conservation, specifically the development and evaluation of a substitute for petroleum asphalt that can be produced from wood and other cellulosic wastes; preparation and identification of cyano-halo complexes of chromium III. *Mailing Add:* Dept Phys Sci Box 26 Ark State Univ State University AR 72467

JIMESON, ROBERT M(ACKAY), JR, b Charleroi, Pa, Jan 29, 21; m 46; c 4. FUEL TECHNOLOGY & PETROLEUM ENGINEERING, RESOURCE MANAGEMENT. *Educ:* Pa State Univ, BS, 42; George Wash Univ, MS, 65. *Prof Exp:* Engr, Glenn L Martin Co, 42-45; res assoc org synthesis, Mellon Inst Indust Res, 45-47; res assoc, Sales Admin, Union Carbide Corp, 47-49; chem engr, US Bur Mines, US Dept Interior, 49-59, phys sci adminstr, 59-64; phys sci adminstr, US Pub Health Serv, HEW, 64-70; asst adv environ qual, Fed Power Comn, 70-74; staff, Off Technol Assessment, US Cong, 74-76; mgr, Fossil Technol Overview, Dept Energy, 76-78; CONSULT, ENERGY FUELS, ENVIRON MGT, INT ACTIV & CHEM ENG, 78- *Concurrent Pos:* Lectr, McKeesport Ctr, Pa State Univ, 57-59 & George Washington Univ Grad Sch Eng, 77-79. *Mem:* Am Chem Soc; Am Inst Chem Engrs (treas, 64-65, pres, 74-75). *Res:* Engineering administration; processes for production of natural fuels, synthetic fuels and chemicals; formulation of plans and policies affecting federal program for prevention and control of air pollution. *Mailing Add:* 1501 Gingerwood Ct Vienna VA 22182-1437

JIN, RONG-SHENG, b Foochow, Fukien, China, Dec 4, 33; US citizen; m 62; c 3. PLANETARY MAGNETISM. *Educ:* Denison Univ, BS, 57; Ohio State Univ, PhD(physics), 65. *Prof Exp:* Instr physics, Denison Univ, 59-60; asst prof, Loyola Univ, Calif, 65-67; assoc scientist, Lockheed Missiles & Space Co, Calif, 67-69; ASSOC PROF SPACE SCI & PHYSICS, FLA INST TECHNOL, 69- *Mem:* Sigma Xi; Am Inst Physics; Am Geophys Soc. *Res:* Nuclear and space physics; planetary magnetism. *Mailing Add:* Dept Physics & Space Sci Fla Inst of Technol Melbourne FL 32901

JIN, SUNGHO, b Daejon, Korea, Nov 6, 45; US citizen; m 72; c 2. MATERIALS SCIENCE ENGINEERING. *Educ:* Seoul Nat Univ, BS, 69; Univ Calif, Berkeley, MS, 72, PhD(phys metall), 74. *Prof Exp:* Res staff, Univ Calif, Berkeley, 74-76; mem tech staff, 76-81, SUPVR, AT&T BELL LABS, MURRAY HILL, 81- *Mem:* Am Soc Metals; Am Inst Mining, Metall & Petroleum Engrs Metall Soc; Mat Res Soc. *Res:* New alloys and thin films with unique magnetic, mechanical, electrical or thermal properties useful for applications in electronics or telecommunications industry. *Mailing Add:* AT&T Bell Labs 600 Mountain Ave Rm-1A123 Murray Hill NJ 07974

JINDRAK, KAREL, b Merin, Czech, Mar 29, 26; m 51; c 1. PATHOLOGY. *Educ:* Charles Univ, Prague, MUC, 47, MUDr, 50; Charles Univ, Hradec Kralove, CSc, 65; Educ Coun Foreign Med Grad, cert, 68. *Prof Exp:* Intern med, Gen Hosp, Roznava, Czech, 51; pathologist & asst prof path, Med Fac, Charles Univ, Hradec Kralove, 56-65; pathologist & head dept path, Res Inst Pharm & Biochem, Prague, 66-67; pathologist, Dept Animal Sci, Univ Hawaii, 67-68; res pathologist, Mt Sinai Hosp, New York, 68-71; PATHOLOGIST, METHODIST HOSP, NEW YORK, 71-; CLIN ASST PROF PATH, STATE UNIV NY DOWNSTATE MED CTR, 72- *Concurrent Pos:* Ministry Health app head dept path, Czech Hosp, Haiphong, Vietnam, 58-60; consult, Med Fac, Charles Univ, Hradec Kralove, 61-64; Czech Ministry Health res grant, 63-65; NIH grant, Univ Hawaii, 66. *Honors & Awards:* Slovak Nat Coun Award, 51; Czech Ministry Health Award, 66. *Mem:* Am Soc Trop Med & Hyg; Czech Med Soc; Am Asn Pathologists; fel Col Am Path. *Res:* Pathology of infectious and parasitic diseases of man and animals; neuropathology; pathology of chronic drug toxicity; problems related to cerebral nematodiasis; mechanical effect of vocalization on brain and meninges. *Mailing Add:* Dept Path Methodist Hosp 506 Sixth St Brooklyn NY 11215

JIRGENSONS, ARNOLD, b Latvia, Dec 2, 06; nat US; m 42. POLYMER CHEMISTRY. *Educ:* Univ Latvia, Chem Eng, 32. *Prof Exp:* Instr chem, Univ Latvia, 32-44; res chemist, Zellwolle & Kunstseide Ring, Ger, 44-45; res chemist, Boston Blacking Co, Sweden, 47-50, B B Chem Co, Can 50-54; Endicott Johnson Corp, NY, 54-60 & Jersey State Chem Co, NJ, 60-61; res chemist, RA Chem Corp, 61-73, chief tech dir, 73-77; CONSULT, 77- *Mem:* Am Chem Soc. *Res:* Emulsion polymerization; water base coatings; new emulsion polymers for flame retardant textile coatings; new emulsion copolymers capable of self-crosslinking. *Mailing Add:* 13 Courtshire Dr Brick Town NJ 08723

JIRKOVSKY, IVO, b Prague, Czech, June 26, 35; Can citizen; m 65. MEDICINAL CHEMISTRY, RESEARCH ADMINISTRATION. *Educ:* Chem Univ, Prague Dipl chem eng, 58; Czech Acad Sci, PhD (org chem), 63. *Prof Exp:* Asst res chemist, Res Inst Pharm & Biochem, Prague, 58-60, assoc res chemist, 63-67; sr res chemist, Wyeth-Ayerst Res Labs, 68-73, sect head med chem, 73-77, sr res assoc, 77-84, assoc dir, 84-89, DIR CHEM, WYETH-AYERST RES, 89- *Concurrent Pos:* Fel, Univ NB, 66-67. *Mem:* Am Chem Soc; fel Chem Inst Can; AAAS. *Res:* Organic syntheses; alkaloids; heterocycles; physical organic chemistry; biochemistry; structure-activity relationships; antihypertensives; psychotherapeutics and cognition enhancers; hypoglycemic drugs; enzyme inhibitors; antiobesity and hypolipidemic agents; immunoregulation; atherosclerosis; bone metabolism. *Mailing Add:* Wyeth-Ayerst Res CN 8000 Princeton NJ 08543-8000

JIRMANUS, MUNIR N, b Jerusalem, Apr 23, 44; US citizen; m 68; c 2. CRYOGENICS. *Educ:* Am Univ, Beirut, Lebanon, BSc, 64; Tufts Univ, MSc, 66, PhD(physics), 73. *Prof Exp:* Lectr physics, Tufts Univ, 74-75; asst prof, Am Univ, Lebanon, 75-77; sr appl engr cryogenics, 78-88, TECH DIR, JANIS RES CO, 89- *Mem:* Am Phys Soc; Mat Res Soc; Am Chem Soc. *Res:* Design and testing of cryogenic equipment for low temperature physics research. *Mailing Add:* Janis Res Co PO Box 696 2 Jewel Dr Willmington MA 01887-0696

JIRSA, JAMES O, b Lincoln, Nebr, July 30, 38; m 65; c 2. EARTHQUAKE ENGINEERING, STRUCTURAL ENGINEERING. *Educ:* Univ Nebr, BS, 60; Univ Ill, MS, 62, PhD(civil eng), 63. *Prof Exp:* Asst prof civil eng, Univ Nebr, 64-65; from asst prof to assoc prof, Rice Univ, 65-71; from assoc prof to prof, 72-84, Ferguson prof civil eng, 84-88 JANET S COCKRELL CENTENNIAL CHAIR ENG, UNIV TEX, AUSTIN, 88- *Concurrent Pos:* Fulbright scholar, Inst Appl Res Reinforced Concrete, France, 63-64, Portland Cement Asn, 65 & H J Degenkolb Assocs, 80. *Honors & Awards:* Reese Award, Am Soc Civil Engr, 70; Wason Medal, Am Concrete Inst, 77; Reese Struct Award, Am Concrete Inst, 77, 79, Lindau Award, 86; Huber Res Prize, Am Soc Civil Engr, 78; Bloem Award, Am Cancer Inst, 90. *Mem:* Nat Acad Eng; Am Soc Civil Engrs; Am Concrete Inst; Earthquake Eng Res Inst; Int Asn Bridge & Struct Engrs. *Res:* Reinforced concrete behavior and design of reinforced concrete structures; earthquake engineering. *Mailing Add:* Ferguson Struct Eng Lab Univ Tex 10100 Burnet Rd Austin TX 78758

JISCHKE, MARTIN C(HARLES), b Chicago, Ill, Aug 7, 41; m 70. FLUID MECHANICS. *Educ:* Ill Inst Technol, BS, 63; Mass Inst Technol, SM, 64, PhD(aeronaut & astronaut), 68. *Prof Exp:* Asst aeronaut & astronaut, Mass Inst Technol, 66-68; from asst prof to assoc prof aerospace & mech eng, 68-75, prof, 75-81, DEAN ENG, UNIV OKLA, 81-; PROF AEROSPACE & MECH ENG, UNIV OKLA, 75- *Concurrent Pos:* On leave, White House fel, US Dept Transp, 75-76; prin investr, USAF, 77- & US Nuclear Regulatory Comn, 77- *Honors & Awards:* Ralph R Teetor Award, Soc Automotive Engrs, 70. *Mem:* Am Inst Aeronaut & Astronaut; Am Phys Soc; Soc Automotive Engrs; Sigma Xi. *Res:* Viscous flows; aerodynamics; geophysical; fluid dynamics; heat transfer. *Mailing Add:* Off Chancellor Univ Mo Rolla MO 65401

JIU, JAMES, b Oakland, Calif, July 7, 29; m 56; c 4. MICROBIAL BIOCHEMISTRY. *Educ:* Univ Calif, BS, 52, PhD, 55. *Prof Exp:* Res scientist, 55-80, HEAD, G D SEARLE & CO, 80- *Mem:* Am Chem Soc; Am Soc Pharmacog; Am Soc Microbiol. *Res:* Analytical chemistry; microbial metabolites; chemotherapy, microbial transformation; synthetic drugs; natural products. *Mailing Add:* G D Searle & Co 4901 Searle Pkwy Skokie IL 60077

JIUSTO, JAMES E, meteorology, atmospheric physics; deceased, see previous edition for last biography

JIZBA, ZDENEK VACLAV, b Prague, Czech, Feb 25, 27; nat US; m 60; c 3. EXPLORATION GEOLOGY. *Educ:* State Col Wash, BS, 49, MS, 50; Univ Wis, PhD, 53. *Prof Exp:* Res geologist, Chevron Co, 55-62, sr res geologist, 62-67, sr res assoc, Chevron Oil Field Res Co, Standard Oil Co, Calif, 67-86; RETIRED. *Res:* Mathematical geology; man-machine interaction to solve complex geological problems; computer applications in geology. *Mailing Add:* 1341 N Rebecca Dr La Habra CA 90631

JOACHIM, FRANK G, entomology; deceased, see previous edition for last biography

JOANNOPOULOS, JOHN DIMITRIS, b New York, NY, Apr 26, 47. SURFACES, AMORPHOUS MATERIALS. *Educ:* Univ Calif, Berkeley, BA, 68, PhD(physics), 74; Univ Calif, Davis, MA, 70. *Prof Exp:* From asst prof to assoc prof, 74-83, PROF PHYSICS, MASS INST TECHNOL, 83- *Concurrent Pos:* Fel, Alfred P Sloan Found, 76-80, John Simon Guggenheim Found, 81-82. *Honors & Awards:* Fel, Am Phys Soc, 83. *Mem:* Am Phys Soc; Am Vacuum Soc. *Res:* Theoretical condensed matter physics: including properties of crystalline solids, surfaces of solids, defects and amorphous solids. *Mailing Add:* Dept Physics 12-116 Mass Inst Technol Cambridge MA 02139

JOB, ROBERT CHARLES, b Honolulu, Hawaii, May 19, 43. INORGANIC CHEMISTRY. *Educ:* Univ Calif, Berkeley, BS, 67; Univ Mich, PhD(inorg chem), 71. *Prof Exp:* Assoc chem, Univ Calif, Santa Barbara, 71-74, res chemist, 74-75; ASST PROF CHEM, COLO STATE UNIV, 75- *Mem:* Sigma Xi. *Res:* Inorganic analogs of biological systems; organometallic chemistry of transition metals with Group IV-a prosthetics; asymmetric induction involving optically active transition metal systems; coordination chemistry. *Mailing Add:* Shell Development Co PO Box 1380 Houston TX 77251

JOBE, JOHN M, b Ponca City, Okla, June 9, 33; m 54; c 5. TOPOLOGY. *Educ:* Univ Tulsa, BS, 55; Okla State Univ, MS, 63, PhD(math), 66. *Prof Exp:* Teacher high sch, Okla, 55-62; from asst prof to assoc prof, 74-77, PROF MATH, OKLA STATE UNIV, 77- *Mem:* Math Asn Am; Am Math Soc. *Res:* Point set topology. *Mailing Add:* Dept Math Okla State Univ Stillwater OK 74078

JOBE, LOWELL A(RTHUR), b Lead, SDak, Aug 28, 14; m 42, 85; c 2. CHEMICAL ENGINEERING. *Educ:* SDak Sch Mines & Technol, BS, 38; Univ Iowa, MS, 39. *Prof Exp:* Asst metall, Univ Iowa, 38-39; chief chemist & chem engr, Graver Tank & Mfg Co, Inc, 39-47; from asst prof to assoc prof chem eng, Univ Idaho, 47-60; process control engr, Atomic Energy Div, Phillips Petrol Co, 60-66; sr process control engr, Idaho Nuclear Corp, 66-71; sr process control engr, Idaho Chem Prog, Allied Chem Corp, 71-77; mem staff, Exxon Nuclear, 77-80; instr process technol, Eastern Idaho Voc Tech Sch, 80-85; RETIRED. *Mem:* Instrument Soc Am; Am Inst Chem Engrs. *Res:* Automatic process control; industrial water and waste treatment; nuclear engineering. *Mailing Add:* 14469 N 55th E Idaho Falls ID 83401

JOBE, PHILLIP CARL, b Carlsbad, NMex, Jan 9, 40; m 59; c 2. NEUROPHARMACOLOGY. *Educ:* Univ NMex, BS, 63; Univ Ariz, PhD(pharmacol), 70. *Prof Exp:* Teaching asst, Univ Ariz, 60-63, assoc, 63-67; asst prof pharmacol, Univ Nebr, 69-70; asst prof, Northeast La Univ, 70-74, dir, Drug Abuse Ctr, 71-74; asst prof, 74-75, assoc prof, 75-80, prof pharmacol, therapeut & psychiat, Sch Med, La State Univ, Shreveport, 80-; AT COL MED, UNIV ILL. *Concurrent Pos:* Consult neuropharmacol & clin pharmacologist, Vet Admin Hosp, 74- *Mem:* Soc Neurosci; Sigma Xi. *Res:* Role of central nervous system neurotransmitters in the regulation of seizure intensity and susceptibility with special emphasis on the relative importance of discrete catecholaminergic neuron systems. *Mailing Add:* Col Med Univ Ill One Illini Dr Peoria IL 61656

JOBES, FORREST CROSSETT, JR, b Trenton, NJ, Nov 26, 35; m 58; c 1. PHYSICS. *Educ:* Oberlin Col, AB, 57; Yale Univ, MS, 58, PhD(physics), 62. *Prof Exp:* Asst physics, Yale Univ, 57-62; res physicist cent res div lab, Mobil Oil Co, 62-65, sr res physicist, Mobil Oil Corp, 65-71; MEM RES STAFF, PLASMA PHYSICS, PHYSICS LAB, PRINCETON UNIV, 71- *Mem:* Am Phys Soc; Sigma Xi. *Res:* Plasma and nuclear physics. *Mailing Add:* Eight MacKenzie Lane Plainsboro NJ 08536

JOBS, STEVEN P, b Feb 24, 55. ELECTRONICS. *Prof Exp:* Co-founder Apple Computer, 76, chmn, exec vpres & gen mgr, Macintosh Div; PRES & CHMN, NEXT COMPUTER, INC, 85- *Concurrent Pos:* Bd dirs, Pixar. *Honors & Awards:* Nat Technol Medal, Pres Reagan, 85; Jefferson Award, 87. *Mem:* Nat Acad Sci. *Res:* Co-designed Apple II; implementation of PostScript and LaserWriting which helped create the desktop publishing industry. *Mailing Add:* NEXT Computer Inc 3475 Deer Creek Rd Palo Alto CA 94304

JOBSIS, FRANS FREDERIK, b Batavia, Indonesia, Apr 1, 29; nat US; m 51; c 5. PHYSIOLOGY. *Educ:* Univ Md, BS, 51; Univ Mich, MS, 53, PhD(zool), 58. *Prof Exp:* Res fel biophys, Johnson Found, Univ Pa, 58-59, res assoc, 61-62, asst prof biophys & physiol, Univ, 62-64; fel biochem, Univ Amsterdam, 59-60; fel, Nobel Inst Neurophysiol, Sweden, 60-61; from asst prof to assoc prof, 64-69, PROF PHYSIOL, DUKE UNIV, 69- *Concurrent Pos:* Guggenheim fel, 71-72. *Mem:* Fel AAAS; Am Physiol Soc; Biophys Soc; Soc Neurosci. *Res:* Physiology, biochemistry and biophysics of muscle and nervous tissue; comparative physiology; physiology of behavior. *Mailing Add:* Dept of Physiol Duke Univ Durham NC 27710

JOBST, JOEL EDWARD, b South Milwaukee, Wis, May 13, 36; m 59; c 3. NUCLEAR PHYSICS. *Educ:* Marquette Univ, BS, 59; Univ Wis, MS, 61, PhD(physics), 66. *Prof Exp:* Sci specialist, 66-91, REMOTE SENSING LAB, EG&G, 91- *Mem:* Health Physics Soc; Am Nuclear Soc; Solar Energy Soc. *Res:* Nuclear research; detector technology; operation and development of particle accelerators and neutron generators; airborne remote sensing, including infrared scanner; preparation of terrestrial radiation maps from gamma data recorded on an aerial survey platform. *Mailing Add:* PO Box 1912 Las Vegas NV 89125

JOCHIM, KENNETH ERWIN, b St Louis, Mo, July 30, 11; m 37; c 2. PHYSIOLOGY. *Educ:* Univ Chicago, BS, 39, PhD, 41. *Prof Exp:* Res assoc cardiovasc dept, Michael Reese Hosp, Chicago, 31-42; from instr to asst prof physiol, Sch Med, St Louis Univ, 42-46; prof & chmn dept, Univ Kans, 46-61, asst dean, Sch Med, 52-57; sr res scientist biol sci dept, Defense Systs Div, Gen Motors Corp, Mich, 61-63; PROF PHYSIOL, UNIV MICH, ANN ARBOR, 63- *Concurrent Pos:* Fulbright res scholar, Univ Munich, 56-57. *Mem:* Soc Exp Biol & Med; Sigma Xi. *Res:* Coronary circulation; electrocardiography; cardiodynamics; peripheral circulatory dynamics. *Mailing Add:* 2066 Chaucer Dr Ann Arbor MI 48103

JOCHLE, WOLFGANG, b Munich, Ger, Oct 5, 27; m 64. VETERINARY SCIENCE, THERIOGENOLOGY. *Educ:* Univ Munich, DrMedVet, 53; Am Col Theriogenologists, dipl, 75. *Prof Exp:* Ger Res Asn fel endocrinol, Vet Fac, Univ Munich, 53-54, vet res scientist, Hormon-Chemie, 54-56; asst animal husb, Vet Fac, Free Univ Berlin, 56-59; vet res scientist, Schering AG, 59-63; res dir vet med, Fecunda AG, Switz, 64-65; dir inst vet sci, Syntex Corp, Mex, 66-68, Calif, 68-73, vpres int vet sect, Syntex Res Div, 73-75; PRES, WOLFGANG JÖCHLE ASSOCS, CONSULT VET SCIENTISTS & THERIOGENOLOGISTS, 75- *Concurrent Pos:* Vet res, Syntex Int, Mex, 66-68. *Mem:* Am Vet Med Asn; Asn Gnotobiotics; Am Soc Animal Sci; Soc Theriogenology; NY Acad Sci. *Res:* Interaction between environment and reproductive functions in animals; use of hormones as therapeutic and managerial tools in veterinary medicine and animal industry; comparative reproductive neuroendocrinology; endocrinology of parturition; new drug development in the animal health field; animal models for clinical conditions. *Mailing Add:* Ten Old Boonton Rd Denville Township NJ 07834

JOCHMAN, RICHARD LEE, b Appleton, Wis, Jan 10, 48; m 69; c 1. MEDICINAL CHEMISTRY, ORGANIC CHEMISTRY. *Educ:* St Norbert Col, BS, 70; Univ Kans, MS, 74, PhD(med chem), 78. *Prof Exp:* From instr to asst prof, 77-85, ASSOC PROF CHEM, COL ST BENEDICT, 85- *Mem:* AAAS; Am Chem Soc. *Res:* Synthesis of metabolically stable analogs of neuropeptides. *Mailing Add:* Dept Chem St Johns Univ Collegeville MN 56321

JOCHSBERGER, THEODORE, b New York, NY, Mar 6, 40; m 84; c 2. PHYSICAL ORGANIC CHEMISTRY. *Educ:* Hunter Col, AB, 61, MA, 63; City Univ New York, PhD(phys chem), 69; Brooklyn Col Pharm, BS, 77. *Prof Exp:* PROF PHARMACEUT, ARNOLD & MARIE SCHWARTZ COL PHARM & HEALTH SCI, LONG ISLAND UNIV, 68- *Mem:* NY Acad Sci; Am Chem Soc. *Res:* Kinetics and mechanisms of free radical reactions; polymers and polymerization mechanisms; metal-peroxide catalyzed reactions; biopharmaceutics. *Mailing Add:* 75 Widsor Rd Staten Island NY 10314

JOCKUSCH, CARL GROOS, JR, b San Antonio, Tex, July 13, 41; m 64; c 3. MATHEMATICAL LOGIC. *Educ:* Swarthmore Col, BA, 63; Mass Inst Technol, PhD(math), 66. *Prof Exp:* Instr math, Northeastern Univ, 66-67; from asst to assoc prof, 67-75, PROF MATH, UNIV ILL, URBANA-CHAMPAIGN, 75- *Concurrent Pos:* Ed, J Symbolic Logic, 74-75. *Mem:* Am Math Soc; Math Asn Am; Asn Symbolic Logic. *Res:* Recursion theory. *Mailing Add:* Dept of Math Univ of Ill Urbana-Champaign 1409 W Green St Urbana IL 61801

JOCOY, EDWARD HENRY, b Buffalo, NY, Oct 24, 33; m 68; c 1. ELECTRICAL ENGINEERING. *Educ:* Rensselaer Polytech Inst, BEE, 55; Univ Buffalo, MS, 59; Cornell Univ, PhD(elec eng), 69. *Prof Exp:* Electronics engr, 55-64, head radar & electronics sect, 64-65, 71-74, PRIN ENGR, CALSPAN CORP, 74- *Mem:* Inst Elec & Electronics Engrs; Sigma Xi. *Res:* Radar and communications; analytical and experimental research of radar and communications systems; mathematical modeling; signal processing. *Mailing Add:* 100 Wiltshire Rd Williamsville NY 14221

JODEIT, MAX A, JR, b Tulsa, Okla, Apr 14, 37; div; c 3. MATHEMATICS. *Educ:* Rice Univ, BA, 62, MA, 65, PhD(math), 67. *Prof Exp:* Instr math, Univ Chicago, 67-69, vis asst prof, 69-70, asst prof, 70-73; ASSOC PROF MATH, UNIV MINN, MINNEAPOLIS, 73- *Mem:* Am Math Soc; Math Asn Am; Soc Indust & Appl Math. *Res:* Mathematical analysis; singular integrals. *Mailing Add:* Univ Minn Sch Math 127 Vincent Hall Minneapolis MN 55455

JODRY, RICHARD L, b Toledo, Ohio, May 17, 22; m 45; c 7. EXPLORATION GEOLOGY. *Educ:* Mich State Univ, BS, 45, MS, 54. *Prof Exp:* Geologist, Magnolia Petrol Co, 45-47 & Ohio Oil Co, 47-50; chief geologist, Rex Oil & Gas Co, 50-55; from res geologist & group supvr to sr res geologist, Billings Res Group, Sun Oil Co, 55-70, chief geologist geothermal energy, 70-77; PRES, ENERGY & NATURAL RESOURCE CONSULTS, INC, 77- *Concurrent Pos:* Distinguished lectr, Am Asn Geologists, 70-72; mem, Bd Mineral Resources, Nat Res Coun, Nat Acad Sci, 75-78. *Mem:* Am Asn Petrol Geologists; Geol Soc Am; Soc Econ Paleontologists & Mineralogists; Soc Explor Geophys; Geothermal Resources Coun (vpres, 74-75). *Res:* Deposition of carbonate sediments; formation of carbonate rocks and their petrographic and petrophysical characteristics; unexplored basin evaluation; world hydrocarbon resource evaluation; coal and geothermal exploration and development. *Mailing Add:* 641 Strings San Antonio TX 78216

JOEBSTL, JOHANN ANTON, b Graz, Austria, July 17, 27; US citizen; m 57; c 1. ENERGY CONVERSION. *Prof Exp:* Res chemist, Eng Res & Develop Lab, 58-68, Mobility Equip Res & Develop Ctr, 68-76, br chief, 76-81, div chief, Electrochem Div, Mobility Equip Res & Develop Command, 81-85, TECH ADV, BELVOIR RES & DEVELOP CTR, US ARMY, 85- *Mem:* Am Chem Soc; Am Vacuum Soc. *Res:* Electrocatalysis; novel electrolytes; advanced fuel conditioning techniques; fundamental investigations in electrochemistry; fuel cells. *Mailing Add:* 6641 Wakefield Dr Alexandria VA 22307

JOEDICKE, INGO BERND, b Grossfurra, Germany, May 17, 48; US citizen; m 68; c 2. INORGANIC CHEMISTRY. *Educ:* Univ Wash, BS, 70; Ore State Univ, PhD(inorg chem), 76. *Prof Exp:* Res asst inorg chem, Ore State Univ, 71-76; SR INORG CHEMIST, GAF CORP, 76- *Mem:* Am Chem Soc; Sigma Xi. *Res:* Homogeneous catalysis of coordinated phosphorus ester autooxidation; high temperature chemistry of silicates and clays; silicate films and coatings. *Mailing Add:* PO Box 1418 Hagerstown MD 21740

JOEL, AMOS EDWARD, JR, b Philadelphia, Pa, Mar 12, 18; m; c 3. ELECTRICAL ENGINEERING. *Educ:* Mass Inst Technol, BS, 40, MS, 42. *Prof Exp:* Switching systs develop engr, 54-60, head, Electronic Switching Planning Dept, 60-61, dir, Switching Systs Develop Lab, 61-62 & Local Switching Lab, 62-67, switching consult, Bell Tel Labs, Inc, 67-82; SWITCHING EXEC CONSULT, BELL TEL LABS & VARIOUS FIRMS, 82- *Honors & Awards:* Outstanding Patent Award, NJ Coun Res & Develop, 72; Alexander Graham Bell Medal, Inst Elec & Electronics Engrs, 76; Columbian Award, City of Genoa, 84; ITU Award, ITU Geneva Switzerland, 83; Kyoto Prize, 89. *Mem:* Nat Acad Eng; Asn Comput Mach; fel Inst Elec & Electronics Engrs; AAAS. *Res:* Design of automatic telephone switching systems; communication privacy systems; design of research computer systems; relay and transistor switching circuits; design of automatic accounting systems; teaching telephone switching circuit design and system principles; electronic information processing systems. *Mailing Add:* Bell Tel Labs Holmdel NJ 07733

JOEL, CLIFFE DAVID, b Saskatoon, Sask, Aug 10, 32; US citizen; m 58; c 3. BIOCHEMISTRY. *Educ:* Pomona Col, AB, 53; Harvard Univ, MA, 55, PhD(biochem), 59. *Prof Exp:* Res fel biol chem, Harvard Med Sch, 59-60, from instr to asst prof, 60-68; chmn dept, 71-73, ASSOC PROF CHEM, LAWRENCE UNIV, 68- *Concurrent Pos:* NIH res fel, 59-60; biochemist, Mass Ment Health Ctr, 63-68; career develop award, Nat Inst Neurol Dis & Stroke, 68. *Mem:* AAAS; Am Chem Soc; Am Soc Neurochem; Int Soc Neurochem. *Res:* Chemistry and metabolism of lipids, especially polyunsaturated fatty acids; neurochemistry; chemistry of the eye. *Mailing Add:* Dept of Chem Lawrence Univ Box 599 Appleton WI 54912

JOEL, DARREL DEAN, b Woodlake, Minn, Apr 26, 33; m 65. EXPERIMENTAL PATHOLOGY, IMMUNOLOGY. *Educ:* Univ Minn, BS, 56, DVM, 58, PhD(vet path), 64. *Prof Exp:* Instr vet path, Univ Minn, 58-60, res fel exp path, 60-64; from asst scientist to assoc scientist, 64-72, scientist, 72-79, SR SCIENTIST, BROOKHAVEN NAT LAB, 79- *Concurrent Pos:* Assoc prof, State Univ NY Stony Brook, 72-85, res prof, 85- *Mem:* AAAS; Am Physiol Soc; Am Soc Vet Clin Pathologists; Conf Res Workers Animal Dis; NY Acad Sci. *Res:* Physiology of the lymphoid system; regulation of lymphocyte kinetics and its relationship to the immune response; experimental pathology; radiation biology. *Mailing Add:* Med Res Ctr Brookhaven Nat Lab Upton NY 11973

JOENK, RUDOLPH JOHN, JR, solid state physics, for more information see previous edition

JOENSUU, OIVA I, b Finland, May 6, 15; nat US; m 41; c 3. GEOLOGY. *Educ:* Univ Helsinki, MS, 46, Fil Lic, 68. *Prof Exp:* Spectrochemist, Geol Surv, Finland, 45-47, Dept of Geol, Univ Chicago, 48-57 & Vitro Chem Co, 57-60; from res asst prof to res assoc prof geochem, 60-76, assoc prof, 76-80, EMER ASSOC PROF MARINE GEOL & GEOPHYS, INST MARINE SCI, UNIV MIAMI, 80- *Mem:* Am Geochem Soc; Geol Soc Finland. *Res:* Geochemistry. *Mailing Add:* 5040 SW 103rd Pl Miami FL 33165

JOERN, ANTHONY, b Omaha, Neb, Sept 6, 48; m 79. POPULATION BIOLOGY, INSECT ECOLOGY. *Educ:* Univ Wis, BS, 70; Univ Tex, PhD(pop biol), 77. *Prof Exp:* PROF ECOL, UNIV NEBR, 78- *Mem:* Ecol Soc Am; Soc Study Evolution; Entom Soc Am; Orthopterists Soc; Sigma Xi; Asn Study Animal Behav; Brit Ecol Soc. *Res:* Factors responsible for resource use by assemblages of grasshoppers; factors influencing the population dynamics of grasshoppers; the evolution of diet by herbivores. *Mailing Add:* Sch Biol Sci Univ Nebr Lincoln NE 68588-0118

JOESTEN, MELVIN D, b Rochelle, Ill, Oct 27, 32; m 53; c 2. INORGANIC CHEMISTRY. *Educ:* Northern Ill Univ, BS, 54; Univ Ill, MS, 59, PhD(inorg chem), 62. *Prof Exp:* Teacher, Ill High Sch, 56-58; asst prof chem, Southern Ill Univ, 62-66; assoc prof, 66-75, chmn dept, 76-82, PROF CHEM, VANDERBILT UNIV, 75- *Concurrent Pos:* Vis prof, Univ NC, 82-83. *Honors & Awards:* Fulbright lectr, Trinity Col, Dublin, Ireland, 72-73. *Mem:* Am Chem Soc; Sigma Xi; AAAS. *Res:* Hydrogen bonding; bioinorganic and coordination chemistry; medicinal uses of coordination compounds. *Mailing Add:* Dept Chem Vanderbilt Univ Nashville TN 37235

JOESTEN, RAYMOND, b San Francisco, Calif, Sept 12, 44; m 67; c 2. METAMORPHIC PETROLOGY. *Educ:* San Jose State Col, BS, 66; Calif Inst Technol, PhD, 74. *Prof Exp:* Instr, 71-83, head dept, 83-88, PROF, GEOL & GEOPHYS, UNIV CONN, 88- *Concurrent Pos:* Vis scholar, Dept Mineral & Petrol, Cambridge Univ, 79; vis assoc prof, dept Earth & Planetary Sci, Johns Hopkins Univ, 87. *Mem:* Fel Geol Soc Am; Mineral Soc Am; Am Geophys Union; Geochem Soc. *Res:* Analysis of mass transport in metamorphic rocks through study of natural systems and modelling using methods of non-equilibrium thermodynamics. *Mailing Add:* Dept Geol & Geophysics U-45 Univ Conn 354 Mansfield Rd Storrs CT 06269-2045

JOFFE, ANATOLE, b Belg, Sept 1, 32; c 2. MATHEMATICS. *Educ:* Univ Brussels, Lic Sc & advan teaching degree agr, 54, Lic Sc, 55; Cornell Univ, PhD(sci math), 59. *Prof Exp:* Asst prof math, McGill Univ, 60-61; from asst prof to assoc prof, 61-73, PROF MATH & DIR, MATH RES CTR, UNIV MONTREAL, 73- *Concurrent Pos:* Mem, Comt Aid Nat Res Coun, 74-77, Comt Basic Sci Coun Univ, 74- *Mem:* Am Math Soc; Inst Math Statist; Math Soc Can. *Res:* Theory of pure and applied probability; Galton-Watson process; some of independent random variables index by a tree and applications to biology. *Mailing Add:* Math Res Ctr Univ Montreal Montreal PQ H3C 3J4 Can

JOFFE, FREDERICK M, b Chicago, Ill, Oct 26, 36; m 59; c 4. BIOCHEMISTRY, FOOD TECHNOLOGY. *Educ:* Mich State Univ, BS, 58, MS, 59; Rutgers Univ, PhD(food sci), 61. *Prof Exp:* Basic develop scientist, Foods Div, Procter & Gamble Co, 62-63, process develop group leader, Folger Coffee Co, 63-64, prod res group leader, 64-68, head prod res & prof serv, 68-70, head shampoo prod develop, Procter & Gamble, 70-72, assoc dir toilet goods prod develop, Procter & Gamble, 72-76, ASSOC DIR PAPER PROD DEVELOP, PROCTER & GAMBLE INT, 77- *Res:* Kinetics of enzyme activity; autoxidation of lipids; instant coffee processes; extraction; spray and freeze drying; sensory perception effects on food acceptability; products research; process development and packaging management. *Mailing Add:* 368 Oliver Rd Cincinnati OH 45215

JOFFE, JOSEPH, b Moscow, Russia, Oct 14, 09; nat US; m 31; c 3. PHYSICAL CHEMISTRY. *Educ:* Columbia Univ, AB, 29, BS, 30, MA, 31, PhD(chem), 33. *Prof Exp:* Asst physics, Columbia Univ, 31, asst chem, Univ Exten, 32-33; from instr to assoc prof chem eng, Newark Col Eng, 32-40, from prof to distinguished prof chem eng, 40-75, res dir, Res Found, 59-61, chmn dept, 63-75, EMER PROF CHEM ENG, NJ INST TECHNOL, 75- *Concurrent Pos:* Sr asst, Div War Res, S A M Labs, Manhattan Proj, Columbia Univ, 43; develop phys chemist, Fed Tel & Radio Corp, 44; chem engr, Exxon Res & Eng Co (summers),50-74 & 77. *Honors & Awards:* Cullimore Medal, NJ Inst Technol, 76. *Mem:* Am Chem Soc; Am Inst Chem Engrs; Am Soc Eng Educ; Sigma Xi. *Res:* Absorption spectroscopy; selenium rectifiers; thermodynamics of gases and gas mixtures; combustion of carbon; flow of gases in pipelines; thermal cracking of hydrocarbons; chemical reaction kinetics; equations of state; vapor-liquid equilibria. *Mailing Add:* 77 Parker Ave Maplewood NJ 07040

JOFFE, MORRIS H, chemistry, for more information see previous edition

JOFFE, STEPHEN N, b Springs, SAfrica, Jan 11, 43; c 2. LASER SURGERY, GASTROENTEROLOGY. *Educ:* Univ Witwatersrand, SAfrica, MD, 67. *Prof Exp:* PROF SURG, UNIV CINCINNATI, 80- *Concurrent Pos:* Healthcare consult, lasers med & surg. *Res:* Laser surgery and general surgery; gastroenteroloy. *Mailing Add:* Dept Surg Univ Cincinnati Med Ctr 231 Bethesda Ave Cincinnati OH 45267-0558

JOFFEE, IRVING BRIAN, b Rochester, NY, Sept 9, 46; m 68; c 3. ORGANIC CHEMISTRY, SURFACE CHEMISTRY. *Educ:* Mass Inst Technol, SB, 68; Brandeis Univ, MA, 71, PhD(org chem), 73. *Prof Exp:* Fel, Hebrew Univ, Israel, 73; sr chemist res & develop, Dead Sea Bromine Co, Ltd, Israel, 74-75; res chemist, 75-83, mgr anal res, 83-87, ASSOC DIR, RES & DEVELOP, PALL CORP, 87- *Mem:* Am Chem Soc. *Res:* Polymer modification; membrane technology; filtration technology; filing and prosecution of patents. *Mailing Add:* 19 Clearview St Huntington NY 11743

JOFTES, DAVID LION, b Brooklyn, NY, Apr 30, 24; m 48; c 1. PHYSIOLOGY, DEVELOPMENTAL BIOLOGY. *Educ:* Tufts Col, BS, 44, MS, 47; Boston Univ, PhD(physiol), 51. *Prof Exp:* Asst physiol, Tufts Col, 47; res assoc radiation, Boston Univ, 50-52; res assoc, Col Med, Univ Ill, 52-53; radiobiologist, USAF Atomic Warfare Directorate, 53-54 & Cancer Res Inst, 54-67; biomed sci adminr, Ment Retardation Prog, Nat Inst Child Health & Human Develop, 68-74; chief, Nat Organ Site Progs Br, 74-76, chief, Rev & Referral Br, 76-78, CHIEF, CONTRACTS REV BR, NAT CANCER INST, 78- *Concurrent Pos:* Collab scientist, Brookhaven Nat Lab, 58-61; res assoc path, Harvard Med Sch, 61-66; adj prof, Boston Univ, 66-67. *Mem:* Am Soc Cell Biologists; Tissue Cult Asn; Soc Develop Biol; AAAS; Soc Nuclear Med (founding pres, 64). *Res:* Cancer; radiation effects; radioautography; circulation physiology. *Mailing Add:* 2133 NW 12th St Delray Beach FL 33445

JOH, TONG HYUB, BIOCHEMISTRY, MOLECULAR BIOLOGY. *Educ:* NY Univ, PhD(biochem), 71. *Prof Exp:* PROF NEUROBIOL, CORNELL UNIV MED COL, 72- *Mailing Add:* Cornell Univ Med Col Burke Rehab Ctr 785 Mamaronect Ave White Plains NY 10605

JOHAM, HOWARD ERNEST, b Los Angeles, Calif, Oct 12, 19; m 42; c 2. PLANT PHYSIOLOGY. *Educ:* Univ Calif, BA, 41; Agr & Mech Col, Tex, MS, 43; Iowa State Col, PhD(plant physiol), 50. *Prof Exp:* Jr plant physiologist, USDA, Calif, 43-44; instr bot, 46-47, from asst prof to prof plant physiol, 47-75, sect leader, 59-75, prof plant sci & head dept, 74-80, EMER PRO PLANT SCI, TEX A&M UNIV, 80- *Concurrent Pos:* Mem, Nat Cotton Task Force, 70-72. *Mem:* Am Soc Agron; Am Soc Plant Physiol; Scand Soc Plant Physiol. *Res:* Plant nutrition; role of calcium in translocation of carbohydrates; cation interactions in cotton nutrition. *Mailing Add:* 9633 E SH21 Bryan TX 77803

JOHANNES, ROBERT, b Philadelphia, Pa, Jan 16, 27; m 61; c 3. PHYSICS. *Educ:* Dickinson Col, BS, 50; Lehigh Univ, MS, 52, PhD(physics), 61. *Prof Exp:* Asst physics, Lehigh Univ, 52-58, res asst, 58-60; proj scientist res lab, Philco Corp, Ford Motor Co, 60-64, res specialist, Appl Res Lab, 64-66; sr scientist, Westinghouse Res Lab, 66-70; sr scientist, Superior Electronics Res Lab, Que, 70-72; res physicist, 72-77, PRIN SCIENTIST, CALSPAN CORP, 77- *Mem:* Am Phys Soc. *Res:* Electro-optics; ferroelectrics; infrared spectroscopy; optical data processing; optical properties; transition metal oxides; lasers; optics; system analysis. *Mailing Add:* 1217 Edgewood Ave Las Cruces NM 88005

JOHANNES, ROBERT EARL, b North Battleford, Sask, Sept 26, 36; m 59; c 1. MARINE ECOLOGY. *Educ:* Univ BC, BSc, 58, MSc, 59; Univ Hawaii, PhD(zool), 63. *Prof Exp:* Res asst, Univ BC, 58-59; res asst, Univ Hawaii, 60-63; res assoc, Marine Inst, Univ Ga, 63-65, res assoc, Dept Zool, 65-66, from asst prof to assoc prof, 66-77; vis assoc researcher, Hawaii Inst Marine Biol, 77-79; SR PRIN RES SCIENTIST, COMMONWEALTH SCI & INDUST RES ORGN DEPT, FISHERIES DIV, 79- *Concurrent Pos:* Guggenheim fel, 74-75. *Mem:* Am Soc Limnol & Oceanog. *Res:* Fisheries, ethnobiology and pollution in tropical marine communities; marine environmental impacts pulp mills; biology of krill. *Mailing Add:* Commonwealth Sci & Indust Res Orgn Fisheries Div PO Box 1538 Hobart 7001 Australia

JOHANNES, VIRGIL IVANCICH, b Omaha, Nebr, Feb 7, 30; m 62; c 1. DIGITAL TELECOMMUNICATIONS SYSTEMS & HARDWARE, HIGH-SPEED DIGITAL CIRCUITS. *Educ:* City Col New York, BS, 53; Columbia Univ, MS, 54, ScD(comm), 61. *Prof Exp:* Lectr elec eng, City Col New York, 53-58; prof & chmn, Elec Eng Dept, Fairleigh Dickinson Univ, 62-63; dept head, AT&T Bell Labs, 63-89; PRES, VIRGIL I JOHANNES INC, 89- *Concurrent Pos:* Adj assoc prof, Columbia Univ, 64-68; vchmn, Study Group XVIII, Int Consultative Comt Tel & Tel, 78- *Mem:* Fel Inst Elec & Electronics Engrs. *Res:* High speed digital transmission systems on copper, optical fiber and satellite media (system concepts and detailed implementation); international standards for digital telecommunications. *Mailing Add:* Two Cardinal Rd Holmdel NJ 07733

JOHANNESEN, ROLF BRADFORD, inorganic chemistry, for more information see previous edition

JOHANNESSEN, CARL L, b Santa Ana, Calif, July 28, 24; m; c 1. BIOGEOGRAPHY, CULTURAL GEOGRAPHY. *Educ:* Univ Calif, Berkeley, BA, 50, MA, 53, PhD(geog), 59. *Prof Exp:* Instr geog, Univ Calif, Davis, 59; PROF GEOG, UNIV ORE, 59- *Concurrent Pos:* Agr Develop Coun grant, Costa Rica, 65, Brazil, 79, Guggenheim Found fel, 65-66; pres, Neopropagations, Inc, 69-78; mem, Conf Latin Am Geogr, Chair, 84-86. *Mem:* AAAS; Asn Am Geogr; Am Geog Soc; Soc Econ Bot; Sigma Xi; Soc Ethnobiol. *Res:* Ways in which man has modified plants and animals in the domestication process and the distributions of domestic and wild biota; Latin America, Himalayas and India in pre-Columbian times. *Mailing Add:* Dept Geog Univ Ore Eugene OR 97403

JOHANNESSEN, GEORGE ANDREW, b Seattle, Wash, Jan 10, 19; m 49; c 4. HORTICULTURE, PLANT BREEDING. *Educ:* Rutgers Univ, BS, 41; Purdue Univ, MS, 48; Cornell Univ, PhD(veg crops, plant breeding, physiol), 50. *Prof Exp:* Asst soil technologist, Va Truck Exp Sta, 46; asst hort, Purdue Univ, 46-48; asst hort, NY Exp Sta, Cornell Univ, Geneva, 48-50; assoc prof veg crops & pomol, Cornell Univ, 50-53; western area agronomist, Am Can Co, 53-60; head plant breeding dept, Pineapple Res Inst Hawaii, 60-64; dir raw prod res, Calif Canners & Growers, 64-67; dir, Calif Tomato Res Inst, 68-72; mgr, Calif Processing Tomato Adv Bd, 72-78; dir, 78-90, EMER DIR, CALIF TOMATO RES INST, 90- *Concurrent Pos:* Affil mem grad fac, Univ Hawaii, 60-64; vis assoc prof, Cornell Univ, 63-64; consult tomato & pineapple prod, Agency Int Develop, Africa, 68; mem gov bd, Agr Res Inst, Washington, DC, 71-73; consult, Food & Agr Orgn, UN, Ivory Coast, Africa, 80. *Mem:* Fel Am Soc Hort Sci; Am Path Soc; Sigma Xi; Inst Food Technologists. *Res:* Vegetable crops; physiology; tomato fruit cracking;

histology of tomato fruit skin; fruit and vegetable crop production; post-harvest handling and storage of fruit and vegetable crops; tomato and pineapple breeding; research administration. *Mailing Add:* 333 Hartford Rd Danville CA 94526

JOHANNESSEN, JACK, b Alameda, Calif, June 22, 15; m 34; c 2. ELECTRICAL ENGINEERING. *Educ:* Calif Inst Technol, BS, 38. *Prof Exp:* Asst recorder geophys surv party, Tex Co, Calif, 38-39; elec distrib engr, Imp Irrig Dist, 39, elec inspector, 39-40; elec engr, Basic Magnesium Co, Nev, 42; elec engr, Imp Irrig Dist, Calif, 42, supt generation, 42-45; consult elec engr, 45-50; gen supvr, NAm Aviation, 50-62; consult engr & property develop, 62-65; res engr, Saturn V, Boeing Co, 65-67; design specialist, Lockheed, 67-80; RETIRED. *Concurrent Pos:* Consult engr, 80- *Mem:* Assoc Inst Elec & Electronics Engrs. *Res:* Design of remote control of central stations and of machine tools; basic principles of electrical engineering; electronics; electro-mechanical devices. *Mailing Add:* 76632 Morocco Rd Palm Desert CA 92260

JOHANNESSEN, PAUL ROMBERG, b Oslo, Norway, Aug 12, 26; nat US; m 50; c 2. SOLID STATE ELECTRONICS. *Educ:* Mass Inst Technol, SB & SM, 53, ScD, 58. *Prof Exp:* Res engr, Electronic Systs Lab, Mass Inst Technol, 53-56, res asst & instr, 56-58, asst prof, 58-59; sr scientist, Sylvania Elec Prod Inc, 59-69; vpres, Symbionics, 69-70; PRES, MEGAPULSE, INC, 70- *Mem:* Sr mem Inst Elec & Electronics Engrs; Sigma Xi. *Res:* Solid state power sources; automatic controls; nonlinear circuits; electronics. *Mailing Add:* 40 Tyler Rd Lexington MA 02173

JOHANNINGSMEIER, ARTHUR GEORGE, b Lafayette, Ind, Nov 5, 30; m 56; c 2. ECOLOGY. *Educ:* Purdue Univ, BS, 56, MS, 62, PhD, 66. *Prof Exp:* Teacher, High Sch, Mich, 56-58; instr biol, bot & zool, Purdue Univ, 58-62, teaching asst biol & zool, 62-64; asst prof biol, Boston Univ, 64-71; NSF fac fel, Grasslands IBP, Colo State Univ, 71-72; CHMN SCI DEPT, CUSHING ACAD, 72- *Concurrent Pos:* Consult water qual, New Eng Interstate Water Pollution Control Comn, Boston, 75-78. *Mem:* Ecol Soc Am; Am Inst Biol Scientists; Sigma Xi; Am Soc Mammal. *Res:* Food and energy relationships of small mammals in natural communities; development of field methods for the study of small mammal movements and physiology; water quality assessment. *Mailing Add:* Dept Sci Cushing Acad Ashburnham MA 01430

JOHANNSEN, CHRISTIAN JAKOB, b Randolph, Nebr, July 24, 37; m 59; c 2. SOILS & SOIL SCIENCE, ENVIRONMENTAL SCIENCES. *Educ:* Univ Nebr, Lincoln, BS, 59, MS, 61; Purdue Univ, PhD(soil physics, agron), 69. *Prof Exp:* Area agronomist, Chevron Chem Co, 61-62; exten agronomist, Univ Mo-Columbia, 67-85; exten agronomist, Purdue Univ, 63-65, res asst soil physics, 65-66, res agronomist, Lab Appln Remote Sensing, 66-69, prog leader, 69-72, Agr Data Network, 85-86, DIR, LAB APPLNS REMOTE SENSING, PURDUE UNIV, 86-, DIR, NATURAL RESOURCES RES INST, 87- *Mem:* Fel Am Soc Agron; fel Soil Sci Soc Am; fel Soil Conserv Soc Am (pres, 82-83); Int Soil Sci Soc; Am Soc Photogram; Sigma Xi. *Res:* Developing natural resources data and information; emphasis on remote sensing and geographic information systems. *Mailing Add:* Natural Resources Res Inst Purdue Univ ENTM Hall West Lafayette IN 47907

JOHANNSEN, FREDERICK RICHARD, b St Louis, Mo, Feb 17, 46; m; c 2. TOXICOLOGY, OCCUPATIONAL TOXICOLOGY. *Educ:* William Jewell Col, AB, 68; Univ Mo, MS, 70, PhD(toxicol), 73; Am Bd Toxicol, dipl, 81. *Prof Exp:* Grad res asst toxicol, Toxicol Lab, Univ Mo, 68-72, res assoc, 72-73; sr toxicologist, Monsanto Co, 73-78, toxicologist specialist, 78-79, group leader, Environ Health Lab, 79-80, toxicol mgr, 79-86, DIR TOXICOL, MONSANTO CO, 86- *Concurrent Pos:* Lectr, Am Indust Hyg Asn, 79-; dir, Toxicol Lab Accreditation BRD Inc, 82- *Mem:* Soc Toxicol; Am Indust Hyg Asn; Am Chem Soc; Entom Soc Am; Am Asn Lab Animal Sci. *Res:* Toxicology and risk assessment for use in support of environmental and occupational safety. *Mailing Add:* Monsanto Co 800 N Lindbergh Blvd St Louis MO 63167

JOHANSEN, ELMER L, b Lake Forest, Ill, June 28, 30; m 58; c 4. ELECTRICAL ENGINEERING. *Educ:* Harvard Univ, BA, 52; Univ Mich, Ann Arbor, MSEE, 54, PhD(elec eng), 64. *Prof Exp:* Sr engr, Cook Res Labs, 56-58, res asst radar systs, Univ Mich, Ann Arbor, 58-60, res assoc, 60-63, assoc res engr, 63-65, res engr, 65-77, lectr elec eng, 66-70; RES ENGR, ENVIRON RES INST, MICH, 73- *Honors & Awards:* Barry Carleton Award, Inst Elec & Electronics Engrs Group on Aerospace & Electronic Systs, 73. *Mem:* Inst Elec & Electronics Engrs; Sigma Xi. *Res:* Radar systems; electromagnetic scattering properties of radar targets; radar systems engineering; synthetic aperture radar; radar cross-section measurements; radar propagation; radar data analysis; millimeter-wave radar. *Mailing Add:* Environ Res Inst of Mich Box 8618 Ann Arbor MI 48107

JOHANSEN, ERLING, b Overhalla, Norway, Apr 8, 23; nat US; m 52; c 3. DENTISTRY, ORAL PATHOLOGY. *Educ:* Tufts Col, DMD, 49; Univ Rochester, PhD, 55. *Hon Degrees:* PhD, Univ Athens, 81. *Prof Exp:* Asst, Dent Sch, Tufts Col, 46-49; instr histol, Eastman Sch Dent Hyg, 52-64; from asst prof to prof dent res, Sch Med & Dent, Univ Rochester, 55-66, Margaret & Cy Welcher prof, 66-80, chmn dept, 55-80, prof clin dent, 74-80; PROF & DEAN DENT SCI, SCH DENT MED, TUFTS UNIV, 80- *Concurrent Pos:* Consult, Nat Inst Dent Res; consult, Bur Environ Health, mem clin fel rev panel & anat & path fel comt, USPHS; lectr, XIVth World Dent Cong, Paris, France, First Pan-Pac Cong Dent Res, Tokyo, Japan & Asian Pac Regional Orgn Cong, Bangkok, Thailand; spec consult, Comt Asn Role & Function, mem comt advan educ, Task Force on Advan Educ & Exec Comt, chmn sect advan educ & vpres, Advan Educ Prog, Am Asn Dent Schs; hon guest prof, Kanagawa Dent Sch, Japan, 69; int lectr & adv, Pan-Am Health Orgn, WHO, Colombia, Peru & Chile, 73; int lectr, Venezuela, 74; ed, J Dent Educ; USPHS grants; consult, King Abdulaziz Univ, Sch Dent, Jeddah, Saudi Arabia; travelling scholar, Int Col Dentists, Asian Pac Countries, 70; hon prof, Yonsei Univ Col Dent, Seoul, Korea, Peruvian Univ, Lima, Peru, 73; merit award, Rochester Acad Med, 80; mem, Coun Dent Res, Am Dent Asn, 83-87, chmn, 86-87. *Mem:* Fel AAAS; Am Dent Asn; Norweg Dent Asn; Int Asn Dent Res; Sigma Xi; Am Asn Dent Schs; fel Am Col Dentists; fel Int Col Dentists; hon mem Korean Dent Asn; hon mem Pedodontic Soc Peru; hon mem Am Acad Dent Sci. *Res:* Experimental dental caries; electron microscopy; mineralized tissues; graduate education. *Mailing Add:* Sch Dent Med Tufts Univ One Kneeland St Boston MA 02111

JOHANSEN, HANS WILLIAM, b Worcester, Mass, June 11, 32; m 82; c 2. MARINE PHYCOLOGY. *Educ:* San Jose State Col, BA, 55; San Francisco State Col, MA, 61; Univ Calif, Berkeley, PhD(phycol), 66. *Prof Exp:* Teacher, San Mateo High Sch, 56-60; USPHS fel, 66-68; asst prof bot, 68-72, assoc prof, 72-81, PROF BIOL, DEPT BIOL, CLARK UNIV, 81- *Mem:* Phycol Soc Am; Int Phycol Soc; Sigma Xi. *Res:* Systematics, structure, reproduction and morphogenesis of Corallinaceae; ecology of marine benthic algae. *Mailing Add:* Dept Biol Clark Univ Worcester MA 01610

JOHANSEN, NILS IVAR, b Oslo, Norway, Dec 25, 41. GEOTECHNICAL ENGINEERING, PERMAFROST ENGINEERING. *Educ:* Purdue Univ, BSCE, 66 MSCE, 67, PhD(civil eng & eng geol), 71. *Prof Exp:* Hwy engr, Ind Dept Highways, 67-71; asst prof, 71-76, ASSOC PROF ENG, UNIV ALASKA, FAIRBANKS, 76-, HEAD, MINING TECHNOL, 90- *Concurrent Pos:* Consult, Geotech Eng, 73-; vis prof, Univ Mo, Rolla, 81-82; vis assoc prof & acad skills coordr, Univ Southern Ind, 88-89; chair, Eng Sect, Ind Acad Sci, 91. *Mem:* Am Soc Civil Engrs; Soc Mining Engrs; Sigma Xi; Nat Asn Develop Educ. *Res:* Geotechnical engineering and permafrost engineering; resource development in arctic and subarctic regions; infrastructure related to resource development; engineering education. *Mailing Add:* Mining & Geol Eng Univ Alaska Fairbanks AK 99775

JOHANSEN, PETER HERMAN, zoology; deceased, see previous edition for last biography

JOHANSEN, ROBERT H, b Grafton, NDak, July 26, 22; m 48; c 4. HORTICULTURE. *Educ:* NDak State Univ, BS, 49, MS, 56; La State Univ, PhD(hort), 64. *Prof Exp:* From asst horticulturist to assoc horticulturist, 53-65, HORTICULTURIST, NDAK STATE UNIV, 65-, PROF HORT & FORESTRY, 73- *Res:* Potato breeding. *Mailing Add:* Dept Hort & Forestry NDak State Univ State Univ Sta Fargo ND 58105

JOHANSON, CHRIS ELLYN, b Tacoma, Wash, June 18, 45; m 72. PSYCHOPHARMACOLOGY. *Educ:* Univ Ill, Chicago, BS, 68; Univ Chicago, PhD(psychol), 72. *Prof Exp:* USPHS trainee psychol, 68-72, RES ASSOC, DEPT PSYCHIAT, UNIV CHICAGO, 72- *Concurrent Pos:* Consult behav res, Behav Res & Action Social Sci, 72-74, Schering Labs, 72-76 & Merrell-Nat Labs, 74-76. *Mem:* Am Psychol Asn; Behav Pharmacol Soc; Int Asn Study Drug Dependence. *Res:* The effects in the rhesus monkey of a variety of environmental and pharmacological variables on drug self-administration and effect of chronic drug administration on behavior. *Mailing Add:* Dept Psychiat Uniformed Serv Univ Health Sci 4301 Jones Bridge Rd Bethesda MD 20814

JOHANSON, JERRY RAY, b Salt Lake City, Utah, Aug 29, 37; m 57; c 5. MECHANICAL ENGINEERING, APPLIED MECHANICS. *Educ:* Univ Utah, BS, 59, PhD(mech eng), 62. *Prof Exp:* Res engr, Appl Res Lab, US Steel Corp, 62-65, sr res engr, 65-66; vpres, Jenike & Johanson, Inc, 72-80, pres, 80-85; PRES, J R JOHANSON, INC, 85- *Honors & Awards:* Henry Hess Award, Am Soc Mech Engrs, 66; Neal Rice Award, Int Briquetting Asn. *Mem:* Am Soc Mech Engrs; Int Briquetting Asn. *Res:* Flow of solids; agglomeration of solids; fluid flow in bulk solids; testing bulk solids properties. *Mailing Add:* 712 Fiero Lane No 37 San Luis Obispo CA 93401

JOHANSON, L(ENNART) N(OBLE), b Salt Lake City, Utah, May 3, 21; m 48; c 3. CHEMICAL ENGINEERING. *Educ:* Univ Utah, BS, 42; Univ Wis, MS, 43, PhD(chem eng), 48. *Prof Exp:* Chem engr, US Bur Mines, Utah, 42; assoc process engr, Richfield Oil Corp, Calif, 44-45, process engr, 48-51; instr chem eng, Univ Wis, 47-48; from asst prof to assoc prof, 51-61, PROF CHEM ENG, UNIV WASH, 61- *Concurrent Pos:* Consult. *Mem:* Am Chem Soc; Am Soc Eng Educ; Am Inst Chem Engrs; Tech Asn Pulp & Paper Indust; Sigma Xi. *Res:* Pulp, paper technology; chemical engineering kinetics; reactor design; fluidization; high temperature technology. *Mailing Add:* Dept Chem Eng BF-10 Univ Wash Seattle WA 98195

JOHANSON, LAMAR, b Kyle, Tex, Oct 31, 35; m 60. PLANT PHYSIOLOGY. *Educ:* Southwest Tex State Col, BS, 57, MA, 58; Tex A&M Univ, PhD(plant physiol), 67. *Prof Exp:* Asst biol, Southwest Tex State Col, 56-58; instr, Tarleton State Univ, 61-63; asst plant physiol, Tex A&M Univ, 63-65; assoc prof, 67-71, PROF BIOL & HEAD DEPT BIOL SCI, TARLETON STATE UNIV, 71- *Mem:* AAAS; Am Soc Plant Physiologists; Scandinavian Soc Plant Physiologists; Am Inst Biol Sci; Am Oil Chemists' Soc; Sigma Xi. *Res:* Nutrition of excised plant tissues and algae, especially calcium and sodium requirements; lateral root formation; biochemistry and physiology of the peanut; mineral nutrition, oil quality and response to irrigation. *Mailing Add:* Dept Biol Sci Box T-9 Tarleton State Univ Stephenville TX 76402

JOHANSON, ROBERT GAIL, b San Francisco, Calif, Aug 26, 36; m 64; c 5. THIN FILM DEPOSITION. *Educ:* Reed Col, AB, 60; Univ Vt, PhD(org chem), 69. *Prof Exp:* Chemist, Aerojet-Gen Corp, 61-66; fel org chem, Case Western Reserve Univ, 69-70; staff mem, Raychem Corp, 70-76; sr mem staff, Signetics Corp, 76-81; mgr head & disk develop, Datapoint Corp, 81-83; consult, Disk Consults, 83-84; mgr appln lab, CPA Inc, 84-88; mgr thin film eng, Akashic Memories Corp, 88-90; DIR MFG ENG, KMI MAGNETICS, INC, 90- *Mem:* Am Chem Soc; Sigma Xi; Inst Elec & Electronics Engrs; Am Vacuum Soc. *Res:* Thin film deposition and analysis. *Mailing Add:* 517 Kenilworth Ct Sunnyvale CA 94087

JOHANSON, WALDEMAR GUSTAVE, JR, b St Paul, Minn, Sept 9, 37; m 60; c 3. INTERNAL MEDICINE, PULMONARY DISEASES. *Educ:* Gustavus Adolphus Col, BS, 59; Univ Minn, Minneapolis, MD, 62. *Prof Exp:* Intern med, Med Ctr, Univ Calif, Los Angeles, 62-63; resident, Minneapolis Vet Admin Hosp & St Paul Ramsey Hosp, 65-67; from instr to assoc prof med, Univ Tex Health Sci Ctr, Dallas, 69-74; assoc prof, 74-78, PROF MED, UNIV TEX HEALTH SCI CTR, SAN ANTONIO, 78-, CHIEF, PULMONARY DIS SECT, 74- *Concurrent Pos:* Nat Inst Arthritis & Infectious Dis fel, Univ Tex Health Sci Ctr, Dallas, 68-71. *Mem:* Am Thoracic Soc; Am Fedn Clin Res. *Res:* Pulmonary disease models; infectious disease of the lungs. *Mailing Add:* Shearn Moody Piz Rm 7135 Galveston TX 77550

JOHANSON, WILLIAM RICHARD, b Oakland, Calif, Aug 8, 48. RARE-EARTH MAGNETISM, FERMI SURFACES. *Educ:* Univ Hawaii, Manoa, BS, 72; Univ Calif, Riverside, MS, 74, PhD(physics), 78. *Prof Exp:* Res fel, Argonne Nat Lab, 78-81 & Los Alamos Nat Lab, 81-83; asst prof physics, Pomona Col, 83-89; res assoc, Univ Calif, Riverside, 83-89; ASSOC PROF PHYSICS & DEPT CHAIR, SANTA CLARA UNIV, 89- *Concurrent Pos:* Instr, Univ NMex, Los Alamos, 83; res assoc, Argonne Nat Lab, 84. *Mem:* Am Asn Physics Teachers; Am Phys Soc; Sigma Xi. *Res:* Low temperature solid state physics; specific heat; Fermi surfaces; magnetic materials; mixed valence; thin films. *Mailing Add:* Dept Physics Santa Clara Univ Santa Clara CA 95053

JOHANSSON, KARL RICHARD, b Bay City, Mich, June 28, 20; m 43; c 3. MICROBIAL ECOLOGY, MEDICAL MICROBIOLOGY. *Educ:* Univ Wis, BS, 42, MS, 46, PhD(bact), 48. *Prof Exp:* Anal chemist, Swift & Co, 42; asst bact, Univ Wis, 42-43, asst vet sci, 46, asst bact, 46-48; instr dairy bact, Univ Calif, Davis, 48-49; from asst prof to assoc prof bact & immunol, Univ Minn, 49-59; exec secy virol & rickettsiol study sect, Div Res Grants, NIH, 59-61; assoc prof environ health eng, Calif Inst Technol, 61-63; chief res grants br, Nat Inst Neurol Dis & Blindness, NIH, 63-65, exec secy virol study sect, Div Res Grants, 65-69; prof microbiol, Univ Tex Med Sch San Antonio, 69-70; dep dir sci affairs, Wistar Inst, 70-73; prof biol sci & chmn dept, NTex State Univ, 73-86; RETIRED. *Concurrent Pos:* Consult, Gen Mills, Inc, 53, Minneapolis-Honeywell, 58-59, Tex Col Osteop Med, 73-86; fac fel, NASA, 83 & 84; vis scholar, Calif Space Inst, Univ Calif, San Diego, 86- *Mem:* AAAS; Am Soc Microbiol; Am Acad Microbiol. *Res:* Pathogenesis, including role of surface proteins, in legionella pneumophila; survival of legionellae in the natural environment; biotransformation and cometabolism of humic compounds; space biology. *Mailing Add:* 825 Santa Regina Solana Beach CA 92075

JOHANSSON, MATS W, b Uppsala, Sweden, Sept 16, 58. CELL ADHESION. *Educ:* Univ Uppsala, Sweden, BSc, 83, PhD(biol), 88. *Prof Exp:* Teaching asst cell biol, Biomed Ctr, Univ Uppsala, 84-85, postdoctoral asst, 88-89; POSTDOCTORAL FEL, LA JOLLA CANCER RES FOUND, 91- *Concurrent Pos:* Res asst, Swed Natural Sci Res Coun, 84-85, grad student fel, 85-88; vis postdoctoral fel, Swed Coun Forestry & Agr Res, La Jolla Cancer Res Found, 89-91. *Mem:* Am Soc Zoologists; Int Soc Develop & Comp Immunol. *Res:* Cell adhesion; invertebrate cellular immunology. *Mailing Add:* La Jolla Cancer Res Found 10901 N Torrey Pines Rd La Jolla CA 92037

JOHANSSON, SUNE, b Falkenberg, Sweden, Apr 8, 28; Can citizen; m 53; c 1. FIRE RETARDANT COATINGS & TREATMENTS, HIGH TEMPERATURE COATINGS. *Educ:* Högre Tekniska Läroverket, BE, 49. *Prof Exp:* Chemist, Acme Paint & Varnish, 51-55, chief chemist, 55-61; chief chemist, Ocean Chemicals, 61-62, tech dir, 62-78; tech dir, Wood-Tech, 78-82; TECH DIR, FLAME CONTROL COATINGS, 82- *Concurrent Pos:* Tech expert, Teltech Resource Network Corp, 89- *Mem:* Am Chem Soc. *Res:* Fire retardant coatings and treatments; high temperature coatings; general chemical coatings. *Mailing Add:* 436 Aberdeen Rd Lewiston NY 14092-1023

JOHANSSON, TAGE SIGVARD KJELL, b Kalstad, Sweden, Aug 8, 19; nat US; m 50. APICULTURE. *Educ:* Beloit Col, BS, 42; Univ Wis, MS, 44, PhD(zool), 47. *Prof Exp:* Asst zool, Univ Wis, 42-46; instr, Grinnell Col, 47-48, Dartmouth Col, 48-50, NY Univ, 50-52; prof, 52-84, chmn dept, 60-63, EMER PROF BIOL, QUEENS COL, NY, 84- *Concurrent Pos:* Entomologist, Bee Cult Lab, Agr Res Serv, USDA, Ariz, 56-57; vis assoc prof, Dept Environ Biol, Univ Guelph, 71-72. *Mem:* Entom Soc Am; Int Bee Res Asn. *Res:* Entomology; apiculture. *Mailing Add:* Rd One Box 256A East Berne NY 12059-9801

JOHAR, J(OGINDAR) S(INGH), b Rawalpindi, West Pakistan, Jan 1, 35; m 60; c 3. ENVIRONMENTAL CHEMISTRY, FLUORINE CHEMISTRY. *Educ:* Panjab Univ, India, BSc, 57, MSc, 59; Univ Fla, PhD(chem), 66. *Prof Exp:* Lectr chem, Govt Col, Ludhiana, India, 59-62; chmn sci & math div, Cleveland State Community Col, 67-68; PROF CHEM, WAYNE STATE COL, 68- *Concurrent Pos:* Fel, Univ Idaho, 66-67. *Mem:* AAAS; Am Chem Soc; Nat Educ Asn. *Res:* Synthesis and study of fluorine compounds containing nitrogen sulfur and phosphorus; volatile products and use of non-aqueous solvents. *Mailing Add:* Head Math-Sci Div Wayne State Col Wayne NE 68787

JOHARI, OM, b Jodhpur, India, Aug 13, 40; m 67; c 2. ELECTRON MICROSCOPY, METALLURGY. *Educ:* Indian Inst Technol, Kharagpur, BTech, 62; Univ Calif, Berkeley, MS, 63, PhD(metall), 65. *Prof Exp:* Res asst metall, Univ Calif, Berkeley, 62-65, res fel & lectr, 65; asst prof, Drexel Inst Technol, 65-66; res metallurgist, IIT Res Inst, 66-68, mgr metal physics, 68-77; pres, Johari Assocs Inc, 78-88; SECY & TREAS, SCANNING MICROS INT, INC, 77- *Concurrent Pos:* Consult, Lockheed-Ga Co, 65-66; ed & managing ed, Scanning Electron Micros & Scanning Micros, 68-, Food Microstruct & Food Struct, 81-, Cell & Mat, 91- *Honors & Awards:* Grossman Award, Am Soc Metals, 66. *Res:* Relationship between structure and properties of materials; applications of scanning and transmission electron microscopy in material sciences and other branches of science and technology; failure analysis of metallic materials. *Mailing Add:* 1034 Alabama Dr Elk Grove Village IL 60666

JOHN, ANDREW, photochemistry, catalysis; deceased, see previous edition for last biography

JOHN, DAVID THOMAS, b Kano, Nigeria, Apr 25, 41; m 63; c 2. PARASITOLOGY. *Educ:* Asbury Col, AB, 63; Univ NC, Chapel Hill, MSPH, 66, PhD(parasitol), 70. *Prof Exp:* NIH malariology training prog grant, Univ Ga, 70-72; asst prof microbiol, 72-78, ASSOC PROF MICROBIOL, MED COL VA, VA COMMONWEALTH UNIV, 78- *Concurrent Pos:* Mem, Raptor Res Found; partic tour, People's Repub China, Am Soc Trop Med & Hyg, 78. *Mem:* Am Soc Parasitol; Wildlife Dis Asn; Soc Protozool; Am Soc Trop Med & Hyg; Sigma Xi. *Res:* Host-parasite relations. *Mailing Add:* Col Osteop Med Okla State Univ 1111 W 17th Tulsa OK 71407-1898

JOHN, E ROY, b Brownsville, Pa, Aug 14, 24; m; c 6. NEUROPHYSIOLOGY, PSYCHOPHYSIOLOGY. *Educ:* Univ Chicago, BS, 48, PhD(physiol psychol), 54. *Prof Exp:* Sr res technician radiochem, Argonne Nat Labs, AEC, 46-51; res asst psychol, Univ Chicago, 51-54; res assoc, Comn Behav Sci, 54-56; assoc res anatomist, Univ Calif, Los Angeles, 56-57, assoc res physiologist, 57-58; assoc prof psychol, Univ Rochester, 59-60, prof psychol & dir, Ctr Brain Res, 60-63; prof psychiat & dir brain res labs, New York Med Col, 63-77, prof physiol, 72-77; PROF PSYCHIAT & DIR BRAIN RES LABS, NEW YORK UNIV MED CTR, 77-; RES SCIENTIST, NATHAN S KLINE INST, 87- *Concurrent Pos:* Res consult chem, C F Pease Co, Chicago, 52-55; City New York Health Res Coun career scientist awards, 64-75; mem, Nat Adv Coun Brain Res; assoc ed, Behav Biol; ed, Brain & Behav Res. *Mem:* Am Physiol Soc; Am Psychopath Soc; Int Brain Res Orgn; Soc Neurosci; Am EEG Soc. *Res:* Mechanisms of learning and memory; automatic computer evaluation of brain activity; assessment of minimal brain dysfunction in children; cognitive deficit in aging. *Mailing Add:* Brain Res Lab NY Univ Med Ctr Belleview 8th Fl 27th St & First Ave New York NY 10016

JOHN, FRITZ, b Berlin, Ger, June 14, 10; nat US; m 33; c 2. MATHEMATICS. *Educ:* Gottingen Univ, PhD(math), 33. *Prof Exp:* Asst, Gottingen Univ, 32; res scholar, Cambridge Univ, 34-35; from asst prof to assoc prof math, Univ Ky, 35-52; mathematician, US War Dept, 43-45; assoc prof math, NY Univ, 46-50; dir res, Inst Numerical Anal, Nat Bur Standards, Los Angeles, 50-51; prof, Courant Inst Math Sci, 51-83, Courant prof, 76-83, EMER PROF MATH, NY UNIV, 83- *Concurrent Pos:* Rockefeller Found fel, NY Univ, 42; Fulbright Lectr, Gottingen Univ, 55; Guggenheim grants, 62-63 & 70; Sherman Fairchild distinguished scholar, 79-80; Humboldt sr US scientist award, 80-81; MacArthur fel, 84. *Honors & Awards:* George David Birkhoff Prize Appl Math, Am Math Soc & Soc Indust & Appl Math, 73; Josiah Willard Gibbs Lectr, Am Math Soc, 75, Leroy P Steel Prize, 82. *Mem:* Nat Acad Sci; Leopoldina Ger Acad Researchers Natural Sci; Am Math Soc; Math Asn Am; fel AAAS. *Res:* Partial differential equations; non-linear elasticity; analysis; geometry. *Mailing Add:* 66 Wellington Ave New Rochelle NY 10804

JOHN, GEORGE, b Nov 24, 21; m; c 2. NUCLEAR ENGINEERING, NUCLEAR RADIATION DETECTION. *Educ:* Ohio State Univ, PhD(nuclear chem), 52. *Prof Exp:* Group leader, Nucleonics Br Mat Lab, 53-56, assoc prof nuclear eng & physics, 56-80, PROF NUCLEAR ENG, AIR FORCE INST TECHNOL, WRIGHT-PATTERSON AFB, 80- *Mem:* Am Chem Soc; Am Nuclear Soc; Am Asn Physics Teachers; Health Physics Soc. *Res:* Nuclear radiation detection; Mossbauer spectrometry. *Mailing Add:* Dept of Eng Physics AFIT/ENP Air Force Inst of Technol Wright-Patterson AFB OH 45433

JOHN, HUGO HERMAN, b Natoma, Kans, Feb 13, 29; m 50; c 3. EDUCATION AND RESEARCH ADMINISTRATION. *Educ:* Univ Minn, BS, 59, MS, 61, PhD(forestry & statist), 64. *Prof Exp:* Instr forestry, Col Forestry, Univ Minn, 62-64, from asst prof to prof res, 65-72, dir & sta statistician, Agr Exp Sta, 67-69; assoc dean & prof admin & res, Col Forestry & Wildlife, Univ Idaho, 73-76; dean & prof admin & res, Sch Natural Resources, Univ Vt, 75-83, dir, Water Resources Res Ctr, 75-77; dean admin, Col Agr & Natural Resources, dir admin, Agr Exp Sta & Coop Exten Serv, 84-87, PROF NATURAL RESOURCES, UNIV CONN, 88- *Concurrent Pos:* Expert, Food & Agr Orgn, UN, Nicaragua, 64-65, & Columbia, 69-71; chmn, nat prog comm, Soc Am Foresters, 79. *Mem:* Soc Am Foresters. *Res:* Development of natural resource and agricultural information and the organizational structures and constraints to their development, management and conservation; land use planning and development. *Mailing Add:* Col Agr & Natural Resources U-87 Univ Conn 1376 Storrs Rd Storrs CT 06268

JOHN, JAMES EDWARD ALBERT, b Montreal, Que, Nov 6, 33; US citizen; m 58; c 4. MECHANICAL ENGINEERING. *Educ:* Princeton Univ, BSE, 55, MSE, 57; Univ Md, PhD(mech eng), 63. *Prof Exp:* Res engr metall div, Air Reduction Co, Inc, NJ, 56-59; from instr to prof mech eng, Univ Md, 59-71; chmn dept, Univ Toledo, 71-77; prof & chmn dept mech eng, Ohio State Univ, 77-83; DEAN, COL ENG, UNIV MASS, AMHERST, 83- *Concurrent Pos:* Consult Goddard Space Flight Ctr, NASA, 63-68; exec dir, Nat Acad Sci comt motor vehicle emissions, 71-72. *Mem:* Am Soc Mech Engrs; Am Soc Eng Educ; Soc Automotive Engr. *Res:* Space simulation; vacuum; cryogenics; automotive emissions; thermal pollution; fluid dynamics. *Mailing Add:* Col Eng Univ Mass Amherst MA 01003

JOHN, JOSEPH, b Madura, India, Mar 14, 38; m 67. NUCLEAR SCIENCE, INSTRUMENTATION. *Educ:* Madras Christian Col, India, BSc, 58; Univ Madras, MA, 60; Fla State Univ, PhD(nuclear physics), 68. *Prof Exp:* Indian AEC fel, Bhabha Atomic Res Ctr, Govt India, 58-59, sci officer nuclear physics, 59-62, jr res officer, 63-64; sr scientist, Gulf Gen Atomic, San Diego, 68-70, staff scientist, 71-72; prog mgr technol appln dept, Gulf Radiation Tech, San Diego, 72-73; mgr NDT technol dept, Intelcom Rad Tech, 73-75, mgr, NDI Systs Dept, 75-76, mgr, NDI Systs Div, 76-78, MGR NUCLEAR SYSTS DIV, IRT CORP, 78-, V PRES, 77- *Concurrent Pos:* Mgr, Californium-252 Demonstration Ctr, San Diego, 72-78. *Mem:* Am Phys Soc; Am Nuclear Soc; Am Soc Nondestructive Test; Am Mgt Asn; Am Soc Test & Mat. *Res:* Applied nuclear physics; applications of nuclear techniques for nondestructive evaluation; neutron radiography, radiation gauging, mineral exploration technology, nuclear materials measurement, automation and computer control of nondestructive inspection systems. *Mailing Add:* 1401 Crestview Dr San Carlos CA 94070

JOHN, KAVANAKUVHIY V, VITAMIN A GLYCOLIPIDS, TUMOR ANTIGENS. *Educ:* Indian Inst Sci, PhD(biochem), 69. *Prof Exp:* DIR RES & DEVELOP, ST JOSEPH'S HOSP, MILWAUKEE, 78- *Res:* DNA probes; flourescent immunoassays; protein electrophoresis; biochemistry of the retinoids. *Mailing Add:* Fransiscan Shared Lab 11020 W Plank Ct Wauwatosa WI 53226

JOHN, MALIYAKAL EAPPEN, GENETIC ENGINEERING. *Educ:* Poona Univ, India, PhD, 75. *Prof Exp:* SCIENTIST, AGRACETUS, 86- *Mem:* Am Soc Biol Chemist. *Res:* Plant gene expression; identification and characterization of agriculturally useful genes; integration and expression in crop plants; plant transformation. *Mailing Add:* Agracetus 8520 University Green Middleton WI 53562

JOHN, PETER WILLIAM MEREDITH, b Porthcawl, Wales, Aug 20, 23; nat US; m 54; c 2. MATHEMATICAL STATISTICS. *Educ:* Oxford Univ, BA, 44, MA, 48, dipl, 49; Univ Okla, PhD(math), 55. *Prof Exp:* Instr math, Univ Okla, 49-52 & 53-55; math master, Casady Sch, Okla, 52-53; asst prof, Univ NMex, 55-57; assoc res statistician, Chevron Res Corp, Stand Oil Calif, 57-58, res statistician, 58-61; from assoc prof to prof, Univ Calif, Davis, 61-67; PROF MATH, UNIV TEX, AUSTIN, 67- *Concurrent Pos:* Vis prof, Univ Calif, Berkeley, 58-61; vis prof, Univ Ky, 70-71. *Mem:* Am Statist Asn; Inst Math Statist; Royal Statist Soc; Int Statist Inst. *Res:* Design of experiments; engineering applications of mathematical statistics; quality assurance. *Mailing Add:* Dept Math Univ Tex Austin Austin TX 78712

JOHN, WALTER, b Okla, Feb 16, 24; m 54; c 4. ENVIRONMENTAL PHYSICS. *Educ:* Univ Calif, PhD, 55. *Prof Exp:* Instr physics, Univ Ill, 55-58; physicist, Lawrence Radiation Lab, Univ Calif, 58-71; prof physics & phys sci & chmn dept, Stanislaus State Col, 71-74; RES SCIENTIST, AIR & INDUST HYG LAB, CALIF DEPT HEALTH, 74- *Concurrent Pos:* Res assoc, Univ Calif, Berkeley, 75- *Mem:* Am Phys Soc; Am Asn Aerosol Res; Am Asn Physics Teachers; Sigma Xi; Am Conf Govt Indust Hygienists. *Res:* Experimental nuclear physics, especially nuclear reactions, fission and bent-crystal gamma ray spectroscopy; photonuclear reactions; x-rays; air pollution; aerosol physics; particulate matter in the atmosphere. *Mailing Add:* Air & Indust Hyg Lab Calif Dept Health Berkeley CA 94704

JOHNK, CARL T(HEODORE) A(DOLF), b Lutterbeck, Ger, Oct 22, 19; US citizen; m 53; c 4. ELECTRICAL ENGINEERING. *Educ:* Shurtleff Col, BS, 41; Mo Sch Mines, BS, 42; Univ Ill, MS, 48, PhD(elec eng), 54. *Prof Exp:* Elec engr, Radio Corp Am, NJ, 42; instr elec eng, Univ Mo, Rolla, 42-44, from instr to asst prof, 45-49; res assoc, Univ Ill, 49-54; assoc prof, 54-65, PROF ELEC ENG, UNIV COLO, BOULDER, 65- *Concurrent Pos:* Consult, Denver Res Inst, 59-62 & Ramo-Wooldridge Corp, 60-61. *Mem:* Inst Elec & Electronics Engrs; Am Soc Eng Educ. *Res:* Antenna and array theory; modeling of antennas above lossy surfaces; modeling of very low frequency propagation in earthionosphere waveguide; electromagnetic fields; passive and active network theory. *Mailing Add:* Dept Elec Eng Univ Colo 425 Boulder CO 80309

JOHNS, DAVID GARRETT, b Prince Rupert, BC, Oct 18, 29; m 62; c 2. PHARMACOLOGY, BIOCHEMISTRY. *Educ:* McGill Univ, BSc, 54, MD, 58, PhD(biochem), 63. *Prof Exp:* Asst prof med, McGill Univ, 62-63; vis fel pharmacol, Sch Med, Yale Univ, 63-65, from asst prof to assoc prof, 65-70; head drug metab sect, Lab Chem Pharmacol, 70-75, actg chief lab med chem & biol, 75-78, CHIEF LAB MED CHEM & BIOL, NAT CANCER INST, 78- *Mem:* Am Soc Clin Invest; Asn Cancer Res; Am Soc Pharmacol & Exp Therapeut; Fedn Clin Res; Can Soc Clin Invest. *Res:* Mode of action and metabolism of cancer chemotherapeutic agents; megaloblastic anemias. *Mailing Add:* Nat Cancer Inst Bldg 37 Rm 5B22 NIH Bethesda MD 20205

JOHNS, HAROLD E, b Chengtu, W China, July 4, 15; m 40; c 3. PHYSICS, BIOPHYSICS. *Educ:* McMaster Univ, BA, 36; Univ Toronto, MA, 37, PhD(physics), 39. *Hon Degrees:* LLD, Univ Sask, 59; DSc, McMaster Univ, 68; Carleton Univ, 76 & Univ Western Ont, 78. *Prof Exp:* Lectr physics, Univ Alta, 39-45; from asst prof to prof, Univ Sask, 45-56; head dept med biophys, Univ Toronto, 62-71; prof physics & med biophys, 58-80; HEAD PHYSICS DIV, ONT CANCER INST, 56- *Concurrent Pos:* Physicist, Sask Cancer Comn, 45-56; mem int comn radiol units, Int Cong Radiol, 52; mem, Nat Cancer Inst Can; Charles Mickle fel, Fac Med, Univ Toronto, 66. *Honors & Awards:* Roentgen Award, Brit Inst Radiol, 53; Medal, Can Asn Physicists, 65; Henry Marshall Tory Medal, Royal Soc Can, 71; Gairdner Int Award, 73; Coolidge Award, Am Asn Physicists in Med, 76; Officer of the Order of Can, 77; Gold Medal, Am Col Radiol, 80; R M Taylor Award, Can Cancer Soc, 82; Medal Honor, Can Med Asn, 83; Radiation Indust Award, Am Nuclear Soc, 83; W B Lewis Award, Can Nuclear Soc, 85. *Mem:* Am Radium Soc; hon fel Am Col Radiol; Can Asn Physicists; Can Asn Med Physicists (pres, 55); Brit Inst Radiol. *Res:* Physical basis of radiotherapy; interaction of radiation with matter; development of cobalt-60 for radiotherapy; physics of radiation therapy and radiology; molecular biology; effects of ultraviolet light on deoxyribonucleic acid and its components; new methods of imaging in diagnostic radiology. *Mailing Add:* Four Boxbury Rd Etobicoke ON M9C 2W2 Can

JOHNS, KENNETH CHARLES, b Montreal, Que, June 26, 44; m 66; c 3. STABILITY TECHNOLOGY TRANSFER & EDUCATORS. *Educ:* McGill Univ, BEng, 66; London Univ, Eng, PhD(civil eng), 70. *Prof Exp:* Consult struct engr, special topics, Destein & Assoc, 70-88; dean faac appl sci, 81-85, DIR INTER COOPERATION, PROF STRUCT ENG, DEPT CIVIL ENG, UNIV SHERBROOKE, 85- *Concurrent Pos:* Sr vis fel, Cranfield Inst Technol, Eng, 73; vis prof civil eng, Univ BC, Vancouver, 80-88; consult, res policy, Quebec Indust Saftey Res Inst, 87-88; educ consult, Can Int Develop Agency, Ottawa, Africa Sect, 84-85. *Mem:* Am Acad Mech. *Res:* Stability and dynamics of civil engineering and aerospace structures; safety of concrete formwork shoring systems; timber structures, fracture and buckling of commercial lumber; Third World technical and engineering education; industrial saftey research policy. *Mailing Add:* Dept Civil Eng Univ Sherbrooke Sherbrooke PQ J1K 2R1 Can

JOHNS, LEWIS E(DWARD), JR, b Pittsburgh, Pa, Dec 13, 35; m 57; c 3. CHEMICAL ENGINEERING. *Educ:* Carnegie Inst Technol, BS, 57, PhD(chem eng), 64. *Prof Exp:* Chem engr, Dow Chem Co, 62-67; asst prof, 67-76, assoc prof, 76-80, PROF CHEM ENG, UNIV FLA, 80- *Concurrent Pos:* Instr, Saginaw Valley Col, 64. *Mem:* Am Inst Chem Engrs; Sigma Xi. *Res:* Fluid mechanics; diffusion. *Mailing Add:* Dept of Chem Eng 3700 NW 12 Ave Gainesville FL 32605

JOHNS, MARTIN WESLEY, b Chengtu, West China, Mar 23, 13; nat Can; m 81; c 4. NUCLEAR PHYSICS. *Educ:* McMaster Univ, BA, 32, MA, 34; Univ Toronto, PhD, 38. *Hon Degrees:* DSc, Brandon Univ, 75. *Prof Exp:* Prof physics, Brandon Col, 37-46; assoc res physicist, Nat Res Coun Can, 46-47; from asst prof to prof physics, McMaster Univ, 47-81, chmn dept, 61-67 & 70-77; RETIRED. *Concurrent Pos:* Nuffield travel grant, Oxford Univ, 59-60; vis scientist, Atomic Energy Can, 67-68. *Mem:* Am Phys Soc; Am Asn Physics Teachers; fel Royal Soc Can; Can Asn Physicists. *Res:* Atomic spectroscopy; neutron physics; nuclear decay schemes; angular correlation of gamma rays; nuclear structure spectroscopy. *Mailing Add:* 115 Dalewood Cres Hamilton ON L8S 4B8 Can

JOHNS, MILTON VERNON, JR, b Berkeley, Calif, Sept 27, 25; m 54; c 2. MATHEMATICAL STATISTICS. *Educ:* Stanford Univ, BA, 49; Columbia Univ, PhD(math, statist), 56. *Prof Exp:* Res assoc, 56-57, from asst prof to assoc prof, 57-66, PROF STATIST, STANFORD UNIV, 66- *Mem:* AAAS; Am Math Soc; Math Asn Am; Inst Math Statist; Am Statist Asn; Sigma Xi. *Res:* Statistical decision theory. *Mailing Add:* Dept Statist Stanford Univ Sequota Hall Stanford CA 94305

JOHNS, PHILIP TIMOTHY, b Bismarck, NDak, July 17, 43; m 73; c 3. ORGANIC CHEMISTRY. *Educ:* Gustavus Adolphus Col, BA, 65; Univ NDak, PhD(biochem), 70. *Prof Exp:* Fla Heart Asn fel biochem, Col Med, Univ Fla, 70-72; asst prof chem, Va Union Univ, 72-76; ASST PROF CHEM, UNIV WIS-WHITEWATER, 76- *Mem:* Am Chem Soc; AAAS. *Res:* Metabolic control; biosynthesis of plasma lipoproteins; enzymology and control of carbohydrate metabolism. *Mailing Add:* Dept Chem Univ Wis Whitewater WI 53190

JOHNS, RICHARD JAMES, b Pendleton, Ore, Aug 19, 25; m 53; c 3. MEDICINE. *Educ:* Univ Ore, BS, 47; Johns Hopkins Univ, MD, 48; Am Bd Internal Med, dipl. *Prof Exp:* Intern med, Johns Hopkins Hosp, 48-49; asst, 51-53, fel, 53-55, from instr to assoc prof, 55-66, asst dean admis, 62-66, dir sub-dept biomed eng, 66-70, PROF MED, JOHNS HOPKINS UNIV, 66-, PRIN PROF STAFF, APPL PHYSICS LAB, 67-, DIR DEPT BIOMED ENG, 70-, MASSEY PROF BIOMED ENG, 80- *Concurrent Pos:* Asst resident physician, Johns Hopkins Hosp, 51-53, resident physician, 55-56, physician, 56- *Honors & Awards:* Centennial Medal, Inst Elec & Electronics Engr, 84. *Mem:* Fel AAAS; Am Soc Clin Invest; Am Clin & Climatol Asn; Biomed Eng Soc; fel Am Col Physicians; Asn Am Physicians; Inst Med. *Res:* Biomedical engineering; chemical sensors. *Mailing Add:* Dept of Biomed Eng Johns Hopkins Univ Sch of Med Baltimore MD 21205-2196

JOHNS, THOMAS RICHARDS, II, neurology, neurophysiology; deceased, see previous edition for last biography

JOHNS, VARNER JAY, JR, b Denver, Colo, Jan 27, 21; m 44; c 3. INTERNAL MEDICINE. *Educ:* La Sierra Col, BS, 44; Col Med Evangelists, MD, 45; Am Bd Internal Med, dipl, 51, cert, 74; Am Bd Cardiovasc Dis, dipl, 66. *Prof Exp:* Intern, White Mem Hosp, 44-45, resident internal med, 45-47; resident path, Loma Linda Sanitarium & Hosp, 47-48; instr internal med, Sch Med, 48-51, asst clin prof & assoc dean, 51-54, from asst prof to to prof med, 54-86, assoc clin prof, 55-56, chief, Med Serv, Univ Hosp, 64-69, assoc dean continuing educ, 75-86, chmn dept med, 56-69 & 80-86, SR PHYSICIAN, SCH MED, LOMA LINDA UNIV, 64- *Concurrent Pos:* Consult, Off Surg Gen, US, 56-67; sr attend physician, Los Angeles County Hosp, 56-64; physician-in-chief internal med, 58-64; vis colleague, Inst Cardiol, London, 62-63; hon vis physician, Nat Heart Hosp, London, 62-63; co-chmn dept med, White Mem Hosp, 78-80. *Mem:* Am Heart Asn; AMA; Am Fedn Clin Res; fel Am Col Physicians; Int Soc Internal Med. *Res:* Cardiology. *Mailing Add:* 11565 Hillcrest Ct Loma Linda CA 92354

JOHNS, WILLIAM DAVIS, b Waynesburg, Pa, Nov 2, 25; m 48; c 4. GEOCHEMISTRY. *Educ:* Col Wooster, AB, 47; Univ Ill, MS, 51, PhD(geol), 52. *Prof Exp:* Spec asst petrol, Eng Exp Sta, Univ Ill, 49-52, asst geol, 52-55; from asst prof to prof, Wash Univ, 55-70, chmn, dept earth sci, 62-69; PROF GEOL, UNIV MO-COLUMBIA, 70- *Concurrent Pos:* Fulbright scholar, Univ Gottingen, 59-60, Univ Heidelberg, 68-69, Univ Vienna, 83-84 & Univ Pittsburgh 90-91. *Honors & Awards:* Alexander von Humboldt US Sr Scientist Award, 77. *Mem:* Fel Geol Soc Am; Am Mineral Soc; Mineral Soc Gt Brit & Ireland; Geochem Soc; Clay Minerals Soc. *Res:* Mineralogy of clays; recent sediments; diagenesis; organic geochemistry; burial diagenesis of pelitic sediments and dispersed organic matter; role played by clay mineral matrix in catalyzing organic reactions involved in transformation of dispersed organic matter in shale source rocks into petroleum hydrocarbons. *Mailing Add:* Dept Geol Univ Mo Columbia MO 65211

JOHNS, WILLIAM E, b Detroit, Mich. COMPUTATIONAL ADHESION SCIENCE. *Educ:* Mich Tech Univ, BS, 66; Univ Mich, MS, 68; Univ Minn, PhD(wood & mat sci), 72. *Prof Exp:* Res technologist, Am Plywood Asn, Tacoma, 72-74; asst wood researcher, Forest Prod Lab, Univ Calif, Berkeley, 74-78; ASSOC PROF MAT SCI, MECH & MAT ENG DEPT, WASH STATE UNIV, 78- *Concurrent Pos:* Vis scientist, Food & Agr Orgn, UN, 83 & Swed Forest Prod Lab, 85-86; Alcoa Res Found researcher, 90. *Mem:* Adhesion Soc; Am Chem Soc. *Res:* Adhesion; wood science. *Mailing Add:* Mech & Mat Eng Dept Wash State Univ Pullman WA 99164-2920

JOHNS, WILLIAM FRANCIS, b Chicago, Ill, Aug 31, 30; m 50; c 3. ORGANIC CHEMISTRY, MEDICINAL CHEMISTRY. *Educ:* Univ Chicago, PhB, 48, MS, 51; Univ Wis, PhD(org chem), 55. *Prof Exp:* Jr res chemist org synthesis, Merck & Co, 51-53; sr res chemist pharmaceut, 55-65, res fel, 65-71, asst dir chem res, 71-73, dir chem res, Searle Labs, G D Searle & Co, 73-82; SR DIR MED CHEM, STERLING WINTHROP RES INST, 82- *Mem:* Am Chem Soc; AAAS; NY Acad Sci. *Res:* Organic synthesis, especially steroids, antialdosterone agents. *Mailing Add:* Nine Dennin Dr Albany NY 12204

JOHNSEN, DENNIS O, b Santa Monica, Calif, Apr 2, 37; m 62; c 3. LABORATORY ANIMAL MEDICINE, VETERINARY MEDICINE. *Educ:* Univ Calif, Davis, BS, 59, DVM, 61; Ohio State Univ, MS, 65. *Prof Exp:* Chief vet med br lab animal med, US Army Res Inst Environ Med, Natick, Mass, 62-63; chief vet med serv, Naval Radiol Defense Lab, San Francisco, 65-68; chief vet med br, SEATO Med Res Lab, Bangkok, 68-72; chief animal resources br, Letterman Army Inst Res, San Francisco, 72-76; exec secy animal resources rev comt, Div Res Resources, NIH, 76-80; sci attache & int health rep, Am Embassy, New Delhi, India, 80-84; chief, vet med & surg sect, Div Res Serv, NIH, 84-85, dir, primate ctrs prog, NIH Nat Ctr Res Resources, 85-88; int health attache, US Mission, Geneva, 88-91; CONSULT, 91- *Concurrent Pos:* Coun mem & consult, Am Asn Accreditation of Lab Animal Care, 74-76; bd dirs & pres, Am Col Lab Animal Med, 77-79, 85-86 & 88-89. *Mem:* Am Vet Med Asn; Am Soc Lab Animal Practitioners; Am Col Lab Animal Med; Am Asn Lab Animal Sci. *Res:* Nonhuman primatology and breeding; spontaneous diseases of laboratory animals; laboratory animal quality assurance; international health; research administration. *Mailing Add:* US Mission Rte de Pregny Case Postale 1292 Chambery Geneva Switzerland

JOHNSEN, EUGENE CARLYLE, b Minneapolis, Minn, Jan 27, 32; m 57. MATHEMATICAL SOCIOLOGY, SOCIAL NETWORKS. *Educ:* Univ Minn, BChem, 54; Ohio State Univ, PhD(math), 61. *Prof Exp:* Instr chem & math, Univ Minn, 56-57; instr math, Ohio State Univ, 62; Nat Acad Sci-Nat Res Coun res assoc, Nat Bur Standards, 62-63; lectr, 63-64, from asst prof to assoc prof, 64-74, PROF MATH, UNIV CALIF, SANTA BARBARA, 74- *Concurrent Pos:* Air Force Off Sci res grants, 64-73, vis lectr, Univ Mich, 68-69; Fulbright Hays res grant, Univ of Tübingen, 69; NSF res grants, 77-78 & 88-90; gen ed, Discovery, Univ Calif, Santa Barbara, J Undergrad Res; guest ed, Social Networks, 83; vis scholar, Harvard Univ, 84-85. *Mem:* AAAS; Am Math Soc; Math Asn Am; Soc Indust & Appl Math; Int Network Soc Network Anal; Am Sociol Asn. *Res:* Mathematical models in the social sciences; social network theory; matrix theory; combinatorial designs and matrices; combinatorial algebraic structures. *Mailing Add:* Dept Math Univ Calif Santa Barbara CA 93106

JOHNSEN, JOHN HERBERT, b Staten Island, NY, Aug 19, 23; m 48; c 3. GEOLOGY. *Educ:* Syracuse Univ, AB, 47, MSc, 48; Lehigh Univ, PhD(geol), 57. *Prof Exp:* Asst geol, Syracuse Univ, 46-48; mining geologist, NJ Zinc Co, Va, 48-49; from instr to assoc prof, 51-67, prof geol, 67-88, chmn dept, 63-66, 69-72, 75-78 & 84-88, EMER PROF GEOL, VASSAR COL, 88- *Concurrent Pos:* Vis prof, Sci Camp, Univ Wyo, 53; del, Int Geol Cong Mex, 56 & Australia, 76; eng geologist, NY State Dept Pub Works, 59; vis prof, St Augustine's Col, 60 & State Univ NY Col New Paltz, 60 & 64-65, geol consult, CPG No 1032, 61-88; assoc dir, Summer Inst Earth Sci, 61, dir, Summer Inst Geol, 62-72; part time geologist-consult, Hudson River Valley Comn, 65, NY State Off of Planning Coord, 66-67, Cent New Region Planning & Develop Bd, 69-70; mem, NSF Conf Geol Lake Superior Region, 63, NSF Conf Geol Southern Can Rockies, 67 & Environ Task Force, 25th Cong Dist, NY, 74-; part time dir, Ecol-Conserv Prog, Vassar Col, 73-80. *Mem:* Emer fel Geol Soc Am; Soc Econ Paleontologists & Mineralogists; Am Geophys Union; Sigma Xi; Nat Asn Geol Teachers; Am Inst Prof Geologists. *Res:* Stratigraphy and petrography of early and middle Paleozoic carbonate rocks of New York; geology of aggregate materials and reclamation and rehabilitation of mined lands. *Mailing Add:* 32 S Tanglewood Spur Sedona AZ 86336

JOHNSEN, PETER BERGHSEY, b Madison, Wis, May 23, 50; m 74; c 2. SENSORY PHYSIOLOGY, FLAVOR CHEMISTRY. *Educ:* Univ Wis-Madison, BS, 74, MS, 76, PhD(zool), 78. *Prof Exp:* Postdoctoral fel, Univ Pa, 78-80, from asst mem to assoc mem, Monell Chem Senses Ctr, 80-86; res physiologist, 86-88, RES LEADER, USDA AGR RES SERV, 88- *Concurrent Pos:* Instr, Col Vet Med, Univ Pa, 80-83 & Dept Biol, 84-86; affil prof, Grad Fac, La State Univ, 88-; Olin fel, Olin Found, Oslo, Norway, 83. *Honors & Awards:* Outstanding Researcher, Catfish Farmers Am. *Mem:* Inst Food Scientists; Sigma Xi. *Res:* Biochemical process for food flavor formation; chemical identification of flavor compounds and relationship to human perception of taste and smell. *Mailing Add:* USDA Southern Regional Res Ctr PO Box 19687 New Orleans LA 70179

JOHNSEN, RAINER, b Kiel, Ger, Jan 23, 40; m 65; c 2. PHYSICS. *Educ:* Univ Kiel, dipl physics, 65, Dr rer nat, 66. *Prof Exp:* Res assoc physics, 66-68, res asst prof, 68-71, RES ASSOC PROF PHYSICS, UNIV PITTSBURGH, 71- *Mem:* Am Phys Soc; AAAS. *Res:* Atomic physics; atomic collisions; physics of upper atmosphere; mass spectroscopy; laser plasma research. *Mailing Add:* Dept Physics & Astron Univ Pittsburgh Pittsburgh PA 15260

JOHNSEN, RICHARD EMANUEL, b Brooklyn, NY, Feb 8, 36; m 57; c 3. INSECTICIDE TOXICOLOGY. *Educ:* St Olaf Col, BA, 57; Iowa State Univ, MS, 59, PhD(entom), 62. *Prof Exp:* Asst entom, Iowa State Univ, 62; asst prof, 65-70, ASSOC PROF ENTOM, COLO STATE UNIV, 70- *Concurrent Pos:* Sabbatical & fel, Dept Pharmacol & Toxicol, Vet Col Norway, Oslo, 76-77. *Mem:* AAAS; Am Chem Soc; Entom Soc Am; NY Acad Sci; Sigma Xi. *Res:* Pesticides and related environmental pollutants, their metabolism, distribution and persistance in plants, soils and the physical environment; microbial degradation; analytical methodology for pollutant studies. *Mailing Add:* Dept of Zool & Entom Colo State Univ Ft Collins CO 80523

JOHNSEN, ROGER CRAIG, b Warren, Pa, Apr 25, 38; m 65. GENETICS. *Educ:* Ohio Wesleyan Univ, BA, 60; Univ Ore, MS, 63; Brown Univ, PhD(genetics), 68. *Prof Exp:* Asst prof, 67-73, ASSOC PROF BIOL, ADELPHI UNIV, 73- *Mem:* AAAS; Soc Study Social Biol; Genetics Soc Am; Am Genetics Asn; Sigma Xi; Amer Soc Human Genetics; NY Acad Sci. *Res:* Chromosome behavior and mechanics; effects of structure and gene action on chromosome recovery during gametogenesis; genetic controls on the competitive behavior of reciprocal gametic types; cancer cytogenetics. *Mailing Add:* Dept of Biol Adelphi Univ Garden City NY 11530

JOHNSEN, RUSSELL HAROLD, b Chicago, Ill, Aug 5, 22; m 48; c 2. RADIATION CHEMISTRY, ACADEMIC ADMINISTRATION. *Educ:* Univ Chicago, BS, 47; Univ Wis, PhD(chem), 51. *Prof Exp:* Res chemist, Ninol Lab, 46-47; from asst prof to assoc prof, 51-61, PROF CHEM, FLA STATE UNIV, 61-, DEAN, GRAD STUDIES, 86- *Concurrent Pos:* Assoc provost, Col Arts & Sci, Fla State Univ, 74-77, assoc dean grad studies, 77-86. *Mem:* Fel AAAS; Am Chem Soc; Am Phys Soc; Radiation Res Soc; Am Soc Mass Spectrometry. *Res:* Kinetics of reactive intermediates, mechanistic studies; free radical reactions in the atmosphere; electron spin resonance studies. *Mailing Add:* 1425 Devil's Dip Tallahassee FL 32308

JOHNSEN, THOMAS NORMAN, JR, b Chicago, Ill, July 3, 29; m 56; c 2. SOIL NUTRIENT CYCLING, PLANT PHYSIOLOGY. *Educ:* Univ Ariz, BS, 50, MS, 54; Duke Univ, PhD(bot), 60. *Prof Exp:* Res scientist range weed control, 56-72, res leader, Agr Res Serv, US Dept Agr, 72-78; RES SCIENTIST RANGE ECOL, TUCSON, ARIZ, 78- *Concurrent Pos:* Adj prof forestry, NAriz Univ, 58-78. *Honors & Awards:* W R Chapling Res Award, Soc Range Mgt, 88. *Mem:* Soil Sci Soc Am; Ecol Soc Am; Weed Sci Soc Am; Soc Range Mgt; Am Soc Agron; Sigma Xi. *Res:* Evaluation and development of methods to revegete semiarid grazing land; fate of herbicides in plant, soils and water; plant life history; plant population changes and trends; plant competition; development of crimson poppy as a crop. *Mailing Add:* 5854 N Wilshire Dr Tucson Tucson AZ 85711

JOHNSGARD, PAUL AUSTIN, b Fargo, NDak, June 28, 31; m 56; c 4. ZOOLOGY. *Educ:* NDak State Univ, BS, 53; Wash State Univ, MS, 55; Cornell Univ, PhD(vert zool), 59. *Prof Exp:* NSF fel zool, Bristol Univ, 59-60, USPHS fel, 60-61; from instr to prof, 61-80, FOUND PROF, UNIV NEBR-LINCOLN, 80- *Concurrent Pos:* NSF res grants, 63-67 & 68-71; Guggenheim Found fel, 71; mem bd dirs, Int Wild Waterfowl Asn, 72-76. *Honors & Awards:* Mari Sandoz Award, 84; Loren Eiseley Award, 87. *Mem:* Am Ornith Union; Wilson Ornith Soc; Cooper Ornith Soc. *Res:* Systematics of birds, especially the family Anatidae; comparative behavior of birds; ecology of vertebrates; speciation and isolating mechanisms; sympatry and hybridization in birds. *Mailing Add:* 7431 Holdrege Lincoln NE 68505

JOHNSON, A(LFRED) BURTRON, JR, b Salt Lake City, Utah, Apr 8, 29; m 54; c 4. CORROSION, NUCLEAR MATERIALS. *Educ:* Univ Utah, BS, 54, PhD(fuel technol), 58. *Prof Exp:* Mem staff, Hanford Labs, Gen Elec Co, 61-65; staff scientist, 65-81, SR STAFF SCIENTIST, PAC NORTHWEST DIV, BATTELLE MEM INST, 81- *Concurrent Pos:* Lectr, Univ Dayton, 60-61, Richland Grad Ctr, 74-; mem staff, Univ Wis, 73; US deleg, Int Atomic Energy Agency Comt, Vienna, Austria; US coord, US/FRG nuclear fuel info exchange. *Mem:* Am Nuclear Soc; Nat Asn Corrosion Engrs; Am Soc Testing & Mat. *Res:* Corrosion in fission and fusion reactors; nuclear plant life extension; corrosion of ancient metals; spent nuclear fuel storage; author or coauthor of over 200 publications and author of one book. *Mailing Add:* Pacific Northwest Div Battelle Mem Inst Richland WA 99352

JOHNSON, A WILLIAM, b Calgary, Alta, Dec 16, 33; US citizen; m 56; c 4. ORGANIC CHEMISTRY. *Educ:* Univ Alta, BSc, 54; Cornell Univ, PhD(chem), 57. *Prof Exp:* Asst chem, Cornell Univ, 55; fel org chem, Mellon Inst, 57-60; from asst prof to assoc prof, Univ NDak, 60-65; assoc prof & chmn dept, Univ Sask, Regina, 65-67; dir res & develop, 67-75, dean grad sch, 67-88, PROF CHEM, UNIV NDAK, 88- *Concurrent Pos:* Dir, NDak Regional Environ Assessment Prog, 75-77. *Mem:* Fel AAAS; Am Chem Soc; Sigma Xi; fel Chem Inst Can. *Res:* Chemistry of ylids; d-orbital interactions; synthetic organic chemistry; environmental assessment; polynuclear aromatic hydrocarbons. *Mailing Add:* PO Box 7185 Grand Forks ND 58202

JOHNSON, ADRIAN EARL, JR, b Port Arthur, Tex, Dec 17, 28; m 49; c 3. CHEMICAL ENGINEERING, MATHEMATICS. *Educ:* La State Univ, BS, 48; Mass Inst Technol, SM, 49; Univ Fla, PhD(chem eng), 58. *Prof Exp:* Process engr, Mobil Oil Co, Tex, 49-53; asst prof chem eng, Lamar State Univ, 53-54; instr, Univ Fla, 54-57; appl scientist, Int Bus Mach Corp, La, 57-60; asst dir comput res ctr-eng res ctr, La State Univ, 60-62; consult & mgr mgt serv dept, Union Carbide Corp, NY, 62-67; staff consult, Real Time Systs, Inc, 67-68; PROF CHEM ENG, LA STATE UNIV, BATON ROUGE, 68- *Mem:* Am Inst Chem Engrs. *Res:* Computer control of petrochemical processes; optimization and control of distillation columns and methanol plants. *Mailing Add:* 14115 Harrell's Ferry Rd Baton Rouge LA 70816

JOHNSON, ALAN ARTHUR, b Beckenham, Eng, Aug 18, 30; m 58, 90; c 5. MATERIALS SCIENCE & ENGINEERING. *Educ:* Univ Reading, BSc, 52; Univ Toronto, MA, 54; Univ of London, Dipl & PhD(metal physics), 60. *Prof Exp:* Demonstr physics, Univ Toronto, 52-54; sci officer, Royal Naval Sci Serv, 54-56; res asst metall, Imp Col, Univ London, 56-57, res fel, 57-60, lectr, 60-62; dir res, Mat Res Corp, NY, 63-65; prof phys metall, Polytech Inst Brooklyn, 65-71, head dept phys & eng metall, 67-71; prof mat sci & chmn dept mat sci & eng, Wash State Univ, 71-75; dean grad sch, 75-76, PROF MAT SCI, UNIV LOUISVILLE, 75- *Concurrent Pos:* Indust consult & expert witness, prod liability personal injury litigations. *Mem:* Fel Inst Physics; Am Soc Metals; Sigma Xi; fel Inst Metals. *Res:* Materials science especially physical metallurgy; failure analysis; accident reconstruction; approximately 100 articles in journals and conference proceedings. *Mailing Add:* Speed Sci Sch Univ of Louisville Louisville KY 40208

JOHNSON, ALAN J, b Washington, DC, Mar 19, 19; m 52; c 3. HEMATOLOGY, BIOCHEMISTRY. *Educ:* Dartmouth Col, BA, 40; Univ Wis, MA, 42; Lond Island Col Med, MD, 45. *Prof Exp:* Fel med, 48-49, asst med, 49-50, from instr med to assoc prof exp med, 50-71, PROF EXP MED, SCH MED, NY UNIV, 71- *Concurrent Pos:* Clin asst vis physician, Bellevue Hosp, New York, 49-52, asst vis physician, 52-; assoc res dir blood prog, Am Nat Red Cross Res Lab, 61-70; assoc attend, Univ Hosp, NY Univ Med Ctr, 61-; mem comt plasma fractionation & related processes, Blood Res Inst, 61-71; mem comt thrombolytic agts, Nat Heart & Lung Inst, 62-68, mem thrombosis adv comt, 68-70, 72- & comt thrombosis, 68-74; consult, Med Serv, Vet Admin Hosp, New York, 64-; consult, WHO, 66-, Community Blood Coun Gtr New York, 68-, Protein Fractionation Ctr, Nat Transfusion Asn Scotland, 69-, Protein Fractionation Unit, Oxford Haemophilia Ctr, Med Res Coun, Eng, 69-, Iranian Nat Transfusion Soc, 71 & Nat Heart & Lung Inst, 72-; mem subcomt factor VIII & IX prep & co-chmn subcomt fibrinolysis, thrombolysis and intravascular coagulation, Int Comt Thrombosis & Haemostasis, 66-71, co-chmn & chmn subcomt standardization, 71-; mem subcomt protocol & standardization & anal biochem data, Streptokinase-Urokinase Pulmonary Embolism Trial, 68-73; mem coun thrombosis, Am Heart Asn; career scientist, NY Health Res Coun, 70-75. *Mem:* Soc Exp Biol & Med; Am Soc Clin Invest; Am Soc Exp Pathologists; Am Fedn Clin Res; Am Physiol Soc. *Res:* Fibrinolysis; blood coagulation; fibrinolytic agents and inhibitors; thromboembolic disease; plasma fractionation with solid-phase reagents; isolation, purification, function and kinetics of proteolytic enzymes and blood coagulation components for clinical and laboratory use; concentration and removal of hepatitis-associated antigen from plasma and plasma fractions; standardization of coagulation and fibrinolytic agents and reagents. *Mailing Add:* Dept Med NY Univ Med Ctr 550 First Ave New York NY 10016

JOHNSON, ALAN KIM, b Altoona, Pa, Aug 15, 42; m 65; c 1. BEHAVIORAL BIOLOGY. *Educ:* Pa State Univ, BS, 64; Temple Univ, MA, 66; Univ Pittsburgh, PhD(psychobiol), 70. *Prof Exp:* Fel psychobiol, Inst Neurol Sci, Univ Pa, 70-73; asst prof, 73-77, ASSOC PROF PSYCHOL, UNIV IOWA, 77- *Concurrent Pos:* NIH fel, 70. *Honors & Awards:* Res Scientist Develop Award, NIMH, 75. *Mem:* Sigma Xi; AAAS; Soc Neurosci. *Res:* Neurobiology and endocrinology of feeding and drinking; physiological bases of motivation and reinforcement. *Mailing Add:* Dept Psychol Univ Iowa Spence Lab Iowa City IA 52242

JOHNSON, ALBERT SYDNEY, III, b Clarkston, Ga, Dec 27, 33; m 59; c 4. WILDLIFE BIOLOGY, WILDLIFE MANAGEMENT. *Educ:* Univ Ga, BS, 59; Auburn Univ, MS, 62, PhD(zool), 69. *Prof Exp:* Fire control aide, US Forest Serv, 58; res asst wildlife biol, Auburn Univ, 59-62, instr wildlife biol & zool, 63-68; wildlife biologist, Ala State Dept Conserv, 62-63; ASSOC DIR, INST NATURAL RESOURCES & PROF FOREST RESOURCES, UNIV GA, 68- *Mem:* AAAS; Wildlife Soc; Am Soc Mammalogists. *Res:* Wildlife habitat biology and management; wildlife foods and habitat relationships; responses to management. *Mailing Add:* Dept Forestry Univ Ga Inst Natural Resources 13 Ecol Bldg Athens GA 30602

JOHNSON, ALBERT W, b Belvidere, Ill, July 29, 26; m 45, 70; c 5. PLANT ECOLOGY. *Educ:* Colo Agr & Mech Col, BS, 49; Univ Colo, MS, 51, PhD(bot), 56. *Prof Exp:* Instr biol, Univ Colo, 54-55; from instr to assoc prof bot, Univ Alaska, 56-62, NSF fac sci fel, 60-61; jr res botanist, Univ Calif, Los Angeles, 62-64; from asst prof to assoc prof, 64-69, dean, Col Sci, 69-77, actg vpres, acad affairs, 77-78, PROF BIOL, SAN DIEGO STATE UNIV, 69-, VPRES, ACAD AFFAIRS, 78- *Mem:* AAAS; Sigma Xi. *Res:* Arctic and alpine plant ecology and taxonomy; cytogenetics. *Mailing Add:* Acad Affairs Dept Biol San Diego State Univ San Diego CA 92182

JOHNSON, ALBERT WAYNE, b Mullins, SC, July 19, 44; m 65; c 3. ENTOMOLOGY. *Educ:* Clemson Univ, BS, 66, MS, 68; Auburn Univ, PhD(entom), 71. *Prof Exp:* Assoc prof tobacco insects, 70-80, assoc prof entom, 74-80, PROF TOBACCO INSECTS & ENTOM, PEE DEE RES & EDUC CTR, CLEMSON UNIV, 80- *Mem:* Entom Soc Am. *Res:* Insecticide screening, economic thresholds, scouting techniques, insect surveys of pests and beneficials, cultural control practices, biological control, host-plant resistance studies and development of insect-resistant varieties; insect control using insecticides and parasites, predators, and pathogens. *Mailing Add:* 3043 Larkspur Rd Florence SC 29501

JOHNSON, ALEXANDER LAWRENCE, b Gisborne, NZ, Oct 13, 31; nat US; m 61; c 3. ORGANIC CHEMISTRY. *Educ:* Victoria Univ, Wellington, BSc, 54, MSc, 55; Univ Rochester, PhD(org chem), 64. *Prof Exp:* Sec sch teacher chem, Rongotai Col, NZ, 55-60; res chemist, cent res dept, 63-81, res assoc, biochem dept, 81-82, res supvr, 82-87, sr res supvr, 87-89, RES MGR, MED PROD DEPT, E I DUPONT DE NEMOURS & CO, INC, 90-, RES MGR, DUPONT MERCK PHARMACEUTICAL CO, 90- *Honors & Awards:* Eastman Kodak Prize, Univ Rochester, 62. *Mem:* Am Chem Soc; NZ Inst Chem. *Res:* Elucidation of the structures of natural products; synthetic organic chemistry relating to these and to heterocyclic systems; application of physical methods to the solution of organic chemical problems; medicinal chemistry. *Mailing Add:* DuPont Merck Pharmaceut Co 353 Bldg Exp Sta Wilmington DE 19880-0353

JOHNSON, ALFRED THEODORE, JR, b Philadelphia, Pa, June 24, 41; m 83. ELECTRICAL ENGINEERING. *Educ:* Drexel Univ, BSEE, 63; Univ Pa, PhD(elec eng), 69. *Prof Exp:* From asst prof to assoc prof, 74-89, PROF ENG, WIDENER UNIV, 90-, CHMN, DEPT ELEC ENG, 90- *Mem:* Inst Elec & Electronics Engrs. *Res:* Approximation problem using analog and digital filters; analog fault analysis; circuit theory. *Mailing Add:* 310 Marlyn Lane Widener Univ Wallingford PA 19086

JOHNSON, ALICE RUFFIN, b Charlottesville, Va, Sept 22, 36; div; c 2. PHARMACOLOGY, IMMUNOLOGY. *Educ:* Univ Va, BS, 58; Emory Univ, MS, 60, PhD(pharmacol), 68. *Prof Exp:* From instr to asst prof pharmacol, Sch Med, Emory Univ, 68-72; vis scientist, Scripps Clin & Res Found, 72-74; ASST PROF PHARMACOL, SOUTHWESTERN MED SCH, UNIV TEX HEALTH SCI CTR, DALLAS, 74- *Concurrent Pos:* Nat Inst Allergy & Infectious Dis res grant, 69-72; NIH spec fel, Scripps Clin & Res Found, 72-74; Nat Heart Lung & Blood Inst res grant, 75- *Mem:* AAAS; Am Soc Pharmacol & Exp Therapeut. *Res:* Release of pharmacologically active substances; mast cells; endothelial cells; peptides; allergy. *Mailing Add:* Dept Biochem Univ Tex Health Sci Ctr PO Box 2003 Tyler TX 75710

JOHNSON, ALLAN ALEXANDER, b Georgetown, Guyana; m; c 2. INTERNATIONAL NUTRITION, FOOD SCIENCE. *Educ:* McGill Univ, BSc, 72; Cornell Univ, MNS, 74, PhD(int nutrit), 78. *Prof Exp:* Res asst, Cornell Univ, 72-78, nutrit sci, 74, biochem, 74-75, ref asst, Albert R Mann Libr, 75-78; asst prof, 78-82, ASSOC PROF HUMAN NUTRIT, HOWARD UNIV, 82- *Concurrent Pos:* Int Health grant, Howard Univ, 78-81. *Mem:* Am Pub Health Asn; Coun Nutrit Anthrop; Nat Coun Int Health; Soc Nutrit Educ. *Res:* Prevalence and etiology of nutritional anemias; programs for control of nutritional anemias; methods for assessment of nutritional status; iron, folacin and zinc status and the immune response in the elderly; nutrition, food choices, lifestyle and the outcome of pregnancy; the impact of a television nutrition education program on the food energy and nutrient intakes; food choices and lifestyle of low income pregnant women. *Mailing Add:* Dept Human Nutrit & Food Howard Univ Washington DC 20059

JOHNSON, ALLEN NEILL, b Colfax, Wash, Dec 29, 44; m 71; c 2. PATHOLOGY. *Educ:* Wash State Univ, DVM, 69; Univ Wis, PhD(vet sci), 77; Am Col Vet Pathologists, dipl, 78. *Prof Exp:* Res asst, Univ Wis-Madison, 71-74; asst prof vet path, Univ Ga Vet Med Col, 74-76; pathologist, Lederle Labs, 76-78; group leader path, Ortho Pharmaceut Corp, 78-83, res mgr, 83-87, asst dir, 87-89; ASST DIR, R W JOHNSON PHARMACEUT RES INST, 89- *Mem:* Am Col Vet Pathologists; Int Acad Path; Am Vet Med Asn; Soc Toxicol Pathologists; Am Soc Vet Clin Pathologists; AAAS. *Res:* Evaluation of tissues from laboratory animals and farm species to determine and resolve pathologic lesions associated with dosing of experimental drugs. *Mailing Add:* 341 Mine Brook Rd Bernardsville NJ 07924

JOHNSON, ALVA WILLIAM, b Tifton, Ga, Nov 8, 36; m 60; c 1. NEMATOLOGY, PLANT PATHOLOGY. *Educ:* Univ Ga, BSA, 63, MS, 64; NC State Univ, PhD(plant path), 67. *Prof Exp:* SUPVRY RES NEMATOLOGIST, COASTAL PLAIN EXP STA, SCI & EDUC ADMIN, US DEPT AGR, 67- *Honors & Awards:* Fel, Soc Nematologists, 90. *Mem:* Soc Nematologists; Am Phytopath Soc; Orgn Trop Am Nematologists. *Res:* Nematode control; population dynamics; nematode-fungus interactions; multiple plant-pest control; nematode resistance in plants; development of integrated pest management systems to manage nematode populations that are effective, economical and environmentally sound. *Mailing Add:* Plant Path Dept US Dept of Agr Coastal Plain Exp Sta Tifton GA 31794

JOHNSON, ANNE BRADSTREET, b Boston, Mass, Mar 5, 27; m 48; c 2. NEUROPATHOLOGY. *Educ:* Cornell Univ, AB, 48, MD, 51. *Prof Exp:* Internal med, self-employed, Cleveland, 55-57; asst prof, 70-77, ASSOC PROF PATH & NEUROSCI, ALBERT EINSTEIN COL MED, 77- *Concurrent Pos:* Prin investr, NIH grants, other grants, 68-; sr vis scientist award, Japan Soc Prom Sci, 90. *Honors & Awards:* Moore Award, Am Asn Neuropathologists; Chem Eng Innovation Award, Am Inst Chem Engrs, 85; Harrison Fac Award, Ohio State Univ, 86. *Mem:* Am Asn Neuropathologists; Soc Neurosci; Histochem Soc; Am Soc Cell Biol; Int Acad Path; AAAS; NY Acad Sci. *Res:* Abnormal nervous systems including; Alzheimer's disease; genetic leukodystrophics; enzyme histochemistry; immunocytochemistry; using tissue and tissue culture, light and electron microscope approaches. *Mailing Add:* Dept Path K 604 Albert Einstein Col Med 1300 Morris Rd UK90 Columbus OH 43220-1180

JOHNSON, ARLO F, b Franklin, Idaho, Dec 2, 15; m 47; c 4. MECHANICAL ENGINEERING. *Educ:* Calif Inst Technol, BS & MS, 42; Stanford Univ, PhD, 52. *Prof Exp:* Aerodynamicist, Douglas Aircraft Corp, 42-45; asst prof aeronaut eng, Univ Ill, 46-47; instr, Univ Utah, 47-48; asst, Stanford Univ, 48-51; from assoc prof to prof mech eng, Univ Utah, 51-63, head dept, 55-57; RETIRED. *Concurrent Pos:* Prof, Bandung Technol Inst, 61-63; aeronaut res engr, Ames Lab, NASA, 57; mem staff, Sandia Corp, 59. *Mem:* Am Soc Eng Educ. *Res:* Boundary layer theory; gas dynamics; applied mechanics. *Mailing Add:* 2070 E 3620 South Salt Lake City UT 84109

JOHNSON, ARMEAD, b Waco, Tex, Dec 16, 42. IMMUNOGENETICS, HISTOCOMPATIBILITY. *Educ:* Univ Tex, BS, 64; Baylor Col Med, MS, 70, PhD(microbiol & immunol), 71. *Prof Exp:* Assoc, Med Ctr, Duke Univ, 74-75, asst prof, 75-80; ASST PROF PEDIAT & MICROBIOL, SCH MED, GEORGETOWN UNIV, 80- DIR, TISSUE TYPING LAB, 85- *Concurrent Pos:* Consult, Blood Bank, Charity Hosp La, New Orleans, 75-; consult, 80-85. *Mem:* Transplantation Soc; Am Asn Immunologists; Am Soc Histocompatibilty & Immunogenetics; Sigma Xi. *Res:* Serological identification, characterization and genetics of antigens within the human major histocompatibility complex and investigation of their role in the immune response. *Mailing Add:* Georgetown Univ Sch Med 3800 Reservoir Rd NW Washington DC 20007

JOHNSON, ARNOLD, VASCULAR INFLAMATION. *Educ:* Albany Med Col, PhD(physiol), 81. *Prof Exp:* ASST PROF RES PULMONARY PHYSIOL, ALBANY MED COL, 84- *Mailing Add:* Albany Med Col 47 New Scottland Ave Albany NY 12208

JOHNSON, ARNOLD I(VAN), b Madison, Nebr, June 3, 19; m 41; c 3. HYDROLOGY, SOIL MECHANICS. *Educ:* Univ Nebr, BS, 49, AB, 50. *Prof Exp:* Supvr mat testing, Omaha Steel Works, Nebr, 40-43; testing engr soils, USN, Construct Batallion, 44-45, Nebr Hwy Testing Lab, 46-48; chief hydrol lab, US Geol Surv, Denver, 48-67, staff hydrologist, Water Resources Div, 67, chief, Water Res Div Training Ctr, 68-70, asst chief, Off Water Data Coord, Washington, DC, 71-79; PRES & CONSULT WATER & SOIL ENG, A IVAN JOHNSON, INC, DENVER, 79- *Concurrent Pos:* Consult,

UNESCO, Turkey, 65, 79 & UN, 79, 83-85, 90; fac affil, Colo State Univ, 69-70; pres, Int Comn Subsurface Water, 72-75 & Int Comn Remote Sensing & Data Transmission, 79-87; dir, Renewable Natural Resources Found, 73-79; AID consult, Oman, 85, AID, Egypt, 87 & 90, Senegal, 87; tech adv & coordr, ASTM/EPA/USGS/USN, Coop Agreement Develop Ground Water Monitoring Standards, 88-91; secy, Hydrol Sect, Am Geophys Union; mem bd dirs, Am Soc Testing & Mat; chmn, Bd Trustees, Inst Standards Res. *Honors & Awards:* Award of Merit, Am Soc Testing & Mat, 82, Frank W Reinhart Award, 83, William T Cavanugh Mem Award, 88; Award of Merit, Dept Interior, 62, Meritorious Serv Award, 77; ICKO Iben Award, Am Water Resources Asn, 86. *Mem:* Am Geophys Union; Int Asn Hydrol Sci (vpres, 75-79, hon pres, 87-); fel Am Soc Civil Engrs; fel Am Soc Testing & Mat; fel Am Water Resources Asn (pres, 72); Nat Soc Prof Engrs. *Res:* Soil moisture; permeability and specific yield of rock and soil materials; land subsidence; waste management; ground water hydrology; artificial recharge. *Mailing Add:* A Ivan Johnson Inc 7474 Upham Ct Arvada CO 80003

JOHNSON, ARNOLD RICHARD, JR, b Allen, Kans, Jan 12, 29; m 53; c 3. ANALYTICAL CHEMISTRY. *Educ:* Fresno State Col, BS, 51; Ore State Univ, PhD(anal chem), 62. *Prof Exp:* Anal chemist, Lab, Socony Mobil Oil Co, Inc, NJ, 54-56; asst prof anal chem, Univ Wyo, 62-65; assoc prof, 65-70, PROF CHEM & HEAD DEPT, MINOT STATE COL, 70- *Mem:* Am Chem Soc; Sigma Xi. *Res:* Differential spectrophotometry; combustion methods of analysis; analytical chemistry of hafnium and zirconium; trace analysis; spot tests. *Mailing Add:* 449 E Brandon Dr Bismarck ND 58501

JOHNSON, ARTHUR ALBIN, b Chicago, Ill, Feb 24, 25; m 51; c 5. PARASITOLOGY. *Educ:* Univ Minn, AB, 50; Univ Ill, MS, 52, PhD(zool), 55. *Prof Exp:* From asst prof to prof biol, 55-81, Harold & Lucy Cabe distinguished prof, 81-90, EMER PROF, HENDRIX COL, 90- *Concurrent Pos:* Vis lectr, Univ Ill, 63 & 64; mem, NSF Radiation Biol Inst, Argonne Nat Lab, 65. *Mem:* Fel AAAS; Am Micros Soc; Am Soc Parasitol; Nat Audubon Soc; Sigma Xi. *Res:* Mermithidae; helminths of birds. *Mailing Add:* 53 Meadowbrook Dr Conway AR 72032

JOHNSON, ARTHUR EDWARD, b Graceville, Minn, July 4, 42; m 65; c 3. BIOPHYSICAL CHEMISTRY, CELL BIOLOGY. *Educ:* Calif Inst Technol, BS, 64; Univ Ore, PhD(biochem), 73. *Prof Exp:* Instr sci, Milton Acad, Mass, 64-69; Helen Hay Whitney res assoc, Chem Dept, Columbia Univ, 74-77; from asst prof to assoc prof, 77-87, PROF, CHEM & BIOCHEM DEPT, UNIV OKLA, 87- *Concurrent Pos:* Prin investr, NIH, Am Heart Asn, Am Chem Soc & Res Corp grants, 79-; vis prof, Dept Biochem & Biophys, Univ Calif, San Francisco, 84-85; NSF Biochem Grant Proposal Adv Panel, 85-87; thrombosis coun, Am Heart Asn; adj asst prof, dept biochem & molecular biol, Univ Okla Health Sci Ctr, 83-87, adj prof, 87- *Mem:* Am Soc Biochem & Molecular Biol; Am Chem Soc; Am Heart Asn; Am Soc Cell Biol. *Res:* Structure, function and regulation of macromolecules and multicomponent complexes (both free and membrane-bound) involved in blood coagulation, protein biosynthesis, protein secretion and membrane protein integration; fluorescence spectroscopy; fluorescence energy transfer; photocrosslinking. *Mailing Add:* Dept Chem & Biochem Univ Okla Norman OK 73019-0370

JOHNSON, ARTHUR FRANKLIN, b Can, Oct 8, 17; US citizen; m 43; c 4. PHYSICS. *Educ:* Univ Alta, BSc, 38; Univ Toronto, MA, 47, PhD(physics), 49. *Prof Exp:* Res physicist tire eng res, US Rubber Co, 49-52 & Honeywell Res Ctr, Minneapolis-Honeywell Regulator Co, 52-55; res supvr, Minn Mining & Mfg Co, 55-64; prof physics, Gustavus Adolphus Col, 64-66; prof physics & chmn dept, Monmouth Col, Ill, 66-78; asst dean, Sch of Eng & Appl Sci, Washington Univ, 78-84; RETIRED. *Concurrent Pos:* Consult, 84- *Mem:* AAAS; Sigma Xi; Am Phys Soc; Am Asn Physics Teachers. *Res:* Magnetism; photoconductivity and electrical properties of solids. *Mailing Add:* HCR 58 Box 197A Spooner WI 54801

JOHNSON, ARTHUR GILBERT, b Eveleth, Minn, Feb 1, 26; m 51; c 4. MICROBIOLOGY. *Educ:* Univ Minn, BA, 50, MSc, 51; Univ Md, PhD(bact), 55. *Prof Exp:* Biochemist, Immunol Div, Walter Reed Army Inst Res, DC, 52-55; from instr to assoc prof, Med Sch, Univ Mich, Ann Arbor, 55-66, prof bact & immunol, 66-78; PROF & HEAD, DEPT MED MICROBIOL & IMMUNOL, SCH MED, UNIV MINN, DULUTH, 78- *Concurrent Pos:* Mem, Nat Inst Dent Res Coun, 72-75; ed, Infection & Immunity, 77-87; mem, Nat Bd Med Examrs, 80-84; mem, bact & mycol study sect, NIH, 83-87, chmn, 85-87. *Mem:* Am Soc Microbiol; Am Asn Immunologists; Soc Exp Biol & Med; Int Soc Immunopharmacol; Reticuloendothelial Soc; Am Acad Microbiol; Infectious Dis Soc; Immunocompromised Hist Soc; Soc Biol Therapy; Int Soc Immunopharmacol. *Res:* Antibody formation; mode of action of bacterial endotoxins; host resistance factors; immunological aspects of aging. *Mailing Add:* Dept Med Microbiol & Immunol Sch Med Univ Minn Duluth MN 55812-2487

JOHNSON, ARTHUR THOMAS, b East Meadow, NY, Feb 21, 41; m 63; c 4. BIOENGINEERING, AGRICULTURAL ENGINEERING. *Educ:* Cornell Univ, BAE, 64, MS, 67, PhD(bioeng), 69. *Prof Exp:* Res bioengr, US Army, Edgewood Arsenal, Md, 71-75; from asst prof to assoc prof agr eng, Univ MD, & assoc prof phys educ, 78-86, PROF AGR ENG & KINESIOLOGY, UNIV MD, 86- *Concurrent Pos:* Grant, Nat Inst Occup Safety & Health, HEW, 78-81; pres, Alliance Eng in Med & Biol, 84-88; chmn, Am Soc Eng Educ, Biol Agr Eng Div, 87-88; consult, Nat Bur Standards Energy Related Devices Prog, 78- & Battelle Mem Inst. *Mem:* Am Soc Agr Engrs; Inst Elec & Electronics Engrs; Am Indust Hygiene Asn; Am Conf Govt Indust Hygienists; Alliance Engrs Med & Biol; Am Soc Eng Educ. *Res:* Instrumentation and control; biological process engineering; respiratory stress and modelling. *Mailing Add:* Dept of Agr Eng Univ of Md College Park MD 20742

JOHNSON, B CONNOR, b Regina, Sask, Apr 28, 11; US citizen. BIOCHEMISTRY, NUTRITION. *Educ:* McMaster Univ, BA, 33, MA, 34; Univ Wis-Madison, PhD(biochem), 40. *Prof Exp:* From asst prof to prof animal nutrit, Univ Ill, 43-65; prof biochem & head dept, Col Med, Univ Okla Sci Ctr, 65-82; res scientist, Dept Pediat, Univ SFla, St Petersburg, 85-87; MEM BIOCHEM, OKLA MED RES FOUND, 65-, DISTINGUISHED CAREER SCIENTIST, 82-85 & 87- *Concurrent Pos:* Guggenheim Found fel, Nat Inst Res Dairying, Reading, Eng, 55; consult mem, President's Second Atom for Peace Mission to SAm, 56; consult, Cent Res Labs, Armour & Co, Chicago, 57-63; US Dept State consult, Orgn Europ Econ Coop, Paris, 58; consult, Merck & Co, NJ, 60, Agr Res Coun of Fedn Rhodesia & Nyasaland, 62, Nutrit Div, US Army Natick Labs, 63-72 & SE Asian Ministers Educ-Univ Indonesia, 74, Nutrit Inst, USDA, 77; NSF sr fel, Inst Chem Natural Substances, Nat Ctr Sci Res, Paris, 61-62; vis prof, Inst Biol Chem, Univ Strasbourg, 72. *Honors & Awards:* Am Feed Mfg Asn-Nutrit Coun US Award, 60; Purkyne Medal, Czech Acad Sci, 69; Osborne-Mendel Award, Am Inst Nutrit, 74. *Mem:* Am Soc Biol Chemists; Am Inst Nutrit; Am Chem Soc; Brit Biochem Soc; Soc Exp Biol & Med; Soc Endocrinol; AAAS; NY Acad Sci; Int Soc Thrombosis & Haemostasis. *Res:* Nutritional biochemistry; metabolic functions of vitamins A, K, B-12 and E; starvation-refeeding; nutrition and enzyme induction; calorie intake restriction and longevity. *Mailing Add:* Okla Med Res Found 825 NE 13th Oklahoma City OK 73104

JOHNSON, B LAMAR, JR, b Minneapolis, Minn, May 31, 30; m 54; c 4. INTERNAL MEDICINE, INFECTIOUS DISEASES. *Educ:* Denison Univ, BA, 51; Univ Calif, Los Angeles, MD, 55. *Prof Exp:* Asst prof, 62-69, asst dean, 63-65, PROF MED, SCH MED, UNIV CALIF, LOS ANGELES, 69- *Concurrent Pos:* Attend med, Wadsworth Vet Admin Hosp, 63- *Res:* Drug induced nephropathy; endocarditis. *Mailing Add:* Dept Infectious Dis Univ Calif Sch Med Los Angeles CA 90024

JOHNSON, B(ENJAMIN) M(ARTINEAU), b Chiralla, South India, Oct 28, 30; nat US; m 54; c 2. CHEMICAL ENGINEERING, MECHANICAL ENGINEERING. *Educ:* Cornell Univ, BChE, 52; Univ Wis, MS, 53, PhD(chem eng), 56. *Prof Exp:* Sr engr, Chem Res & Develop, Hanford Labs, Gen Elec Co, 56-64; mgr, Eng Anal Unit, 65-67, mgr, Sodium Fluid Syst Sect, Fast Flux Text Facil, 67-69, mgr, Fluid & Energy Systs, 69-74, sr engr prog mgr, 74-86, MGR ENERGY SCI DEPT, PAC NORTHWEST LABS, 86- *Concurrent Pos:* Coordr chem eng joint ctr grad study, Univ Wash & Wash State Univ, 65-76, affiliate assoc prof, 65-74, affiliate prof, 74-; mem, Coord Comt, US/USSR Coop Prog in Thermal Power Plant Heat Rejection Systs, 75-80. *Mem:* Fel Am Inst Chem Engrs; Sigma Xi; Am Nuclear Soc. *Res:* Heat and mass transfer; fluid mechanics; economic analysis; nuclear reactor technology; project (development, design, construction) management. *Mailing Add:* 2336 Davison Richland WA 99352

JOHNSON, BARRY LEE, b Sanders, Ky, Oct 24, 38; m 60; c 5. RESEARCH ADMINISTRATION, BIOMEDICAL ENGINEERING. *Educ:* Univ Ky, BS, 60; Iowa State Univ, MS, 62, PhD(elec eng), 67. *Prof Exp:* Biomed engr, USPHS, 62-64, 67-74; biomed engr, Nat Inst Occup Safety & Health, 74-78, res adminr, 78-86; ASST ADMINR, AGENCY TOXIC SUBSTANCES & DIS REGISTRY, 86- *Concurrent Pos:* Lectr, Univ Cincinnati, 68-69; consult ed, Arch Environ Health, Neurotoxicol, J Indust Health & Toxicol, J Clean Technol & Environ Sci; mem, Permanent Comn Occup Health; mem, Am Conf Govt Indust Hygientists. *Mem:* Sigma Xi; Am Pub Health Asn; Am Col Toxicol. *Res:* Behavioral toxicology; sensory evoked potentials; electroencephalography; mathematical modelling of physiological systems; occupational safety and health; neurotoxicology. *Mailing Add:* Agency for Toxic Substances & Dis Registry 1600 Clifton Rd Atlanta GA 30333

JOHNSON, BECKY BEARD, b Denver, Colo, May 4, 42; m 62. PHYSIOLOGY. *Educ:* Okla State Univ, BS, 64; Univ Ill, Urbana, MS, 66, PhD(physiol), 68. *Prof Exp:* asst prof biol, 74-80, ASSOC PROF BOT, OKLA STATE UNIV, 80- *Concurrent Pos:* NIH fel, 69-70. *Mem:* Am Soc Plant Physiologists; Am Tissue Cult Asn; Sigma Xi. *Res:* Plant tissue culture and protoplast fusion for use in plant breeding and genetics. *Mailing Add:* Dept Bot Okla State Univ 318 Life Sci E Stillwater OK 74078

JOHNSON, BEN BUTLER, b Brooklyn, NY, May 23, 20; m 62; c 5. INTERNAL MEDICINE. *Educ:* Harvard Univ, AB, 42; Harvard Med Sch, MD, 44; Am Bd Internal Med, dipl & cert nephrology. *Prof Exp:* Intern path, NY Hosp, 44-45; asst, Med Col, Cornell Univ, 46-47; asst resident med, NY Univ Div, Bellevue Hosp, 47-49; from instr to asst prof med, Stanford Univ, 55-59; from asst prof to assoc prof med, 59-90, head, Div Renal Dis, 59-90, CLIN ASSOC PROF MED, SCH MED, UNIV MISS, 90- *Concurrent Pos:* Res fels, Bassett Hosp, Cooperstown, NY, 49-50 & Sch Med, Stanford Univ, 50-53; head diabetes clin, Univ Hosps, Stanford Univ, 56-59, dir, Grad Training Prog Metab Dis, Univ, 55-59. *Mem:* Int Soc Nephrology; Am Soc Nephrology; fel Am Col Physicians; Endocrine Soc; AMA; Sigma Xi. *Res:* Renal disease; aldosterone and edema; metabolic diseases. *Mailing Add:* Dept of Med Univ of Miss Med Ctr Jackson MS 39216

JOHNSON, BEN FRANCIS, b Sacramento, Calif, Jan 4, 43; m 65; c 2. MICROBIAL GENETICS. *Educ:* Univ Calif, Davis, BA, 65, Berkeley, MA, 67, PhD(bact), 70. *Prof Exp:* Fel microbial genetics, 71-74, ACTG CHIEF MICROBIOL DIV, PALO ALTO MED RES FOUND, 74- *Mem:* Genetics Soc Am; Am Soc Microbiol; Sigma Xi. *Res:* Genetical and biochemical investigation into the nature of the processes of DNA repair, mutagenesis and regulation of cell division. *Mailing Add:* Palo Alto Med Res Found 300 Homer Ave Palo Alto CA 94310

JOHNSON, BEN S(LEMMONS), JR, b Greensburg, Pa, Nov 25, 17; m 43; c 3. NUCLEAR WASTE MANAGEMENT. *Educ:* Univ WVa, BS, 39, BSChE, 40. *Prof Exp:* Asst to metallurgist, Weirton Steel Co, WVa, 40-42; process control engr, Morgantown Ord Works, E I du Pont de Nemours & Co, Inc, 42-44, asst chief chemist, 44-45, instrument engr, Belle Works, WVa, 45-51, process control supvr, Dana Plant, Ind, 51-53, from res supvr to process control supvr, Savannah River Plant, SC, 53-82; RETIRED.

Concurrent Pos: Instr physics, Univ SC-Aiken, 85-86. *Res:* Chemical and isotope separation and purification from atomic reactor fuels and targets; analytical and process control instrumentation; radioactive waste management. *Mailing Add:* 203 Dunbarton Circle Aiken SC 29803

JOHNSON, BERTIL LENNART, b Dawson, Minn, June 11, 09; m 35; c 2. GENETICS. *Educ:* Univ Minn, BS, 38, PhD(bot), 43. *Prof Exp:* Assoc geneticist, USDA, 43-47; asst prof floricult, Univ Calif, Los Angeles, 47-50, assoc prof, 50-56, prof genetics, 56-69; prof genetics, Univ Calif, Riverside, 69-76; RETIRED. *Honors & Awards:* Vaughan Award, Am Soc Hort Sci, 55; Award, Am Carnation Soc, 60. *Mem:* Bot Soc Am; Genetics Soc Am; Am Soc Agron. *Res:* Genetics of Matthiola; cytogenetics of the Gramineae; biochemical-phylogenetic studies in the Triticinae. *Mailing Add:* 1176 Lyndhurst Dr Riverside CA 92507

JOHNSON, BOB DUELL, b Pocahontas, Ark, June 24, 36; c 2. CYTOLOGY, TOXINOLOGY. *Educ:* Ark State Univ, BS, 58; Ariz State Univ, MS, 64, PhD(zool), 67. *Prof Exp:* Teacher, Northeast Independent Sch Dist, Tex, 58-62; partic zool, Acad Year Inst, Ariz State Univ, 62-63, res asst toxinol, 63-66, res assoc, 66-67; asst prof, 67-74, assoc prof, 74-80, PROF ZOOL, ARK STATE UNIV, 80- *Mem:* Int Soc Toxinol; Sigma Xi. *Res:* Effects of toxins on enzyme systems and morphology of cells. *Mailing Add:* Box 720 State University AR 72467

JOHNSON, BOBBY RAY, b Oakwood, Okla, Oct 30, 41; m 62; c 2. LIPID SCIENCE, FLAVOR CHEMISTRY. *Educ:* Okla State Univ, BS, 63, MS, 66, PhD(biochem), 70. *Prof Exp:* Instr chem, Okla Christian Col, 66-67; instr biochem, Okla State Univ, 67-68; USPHS fel, Univ Calif, Davis, 69-70; asst prof food sci, NC State Univ, 70-76; ASST MGR, CAMPBELL INST RES TECHNOL, CAMPBELL SOUP CO, 76- *Mem:* Am Oil Chemists Soc; Inst Food Technol; Am Dairy Sci Asn. *Res:* Fats and oil chemistry; natural antioxidants; dairy science. *Mailing Add:* 129 Kipling Rd Cherry Hill NJ 08003

JOHNSON, BRANN, b Annapolis, Md, Dec 4, 46; m 88; c 2. STRUCTURAL GEOLOGY. *Educ:* Univ Calif, Berkeley, BA, 68; Pa State Univ, MEng, 73, PhD(geol), 75. *Prof Exp:* Geologist, Marine Geol & Hydrol Div, US Geol Surv, 68; asst geol, Dept Geol & Geophysics, Pa State Univ, 68-71; instr geol, Div Geol & Planetary Sci, Calif Tech, 74-75; asst prof geol, 75-80, civil eng, 76-79 & geophysics, 79-80, ASSOC PROF GEOL & GEOPHYSICS, DEPT GEOL & GEOPHYSICS, TEX A&M UNIV, 80- *Concurrent Pos:* Vis staff scientist, Los Alamos Sci Lab, 76-81; subpanel mem, Nat Res Coun, 79-80; prin investr, Cambridge Labs, US Air Force, 75-77, Los Alamos Sci Lab, 76-78 & Div Basic Energy Res, Dept Energy, 79-; res assoc, Ctr Tectonophysics, Tex A&M Univ, 75- *Mem:* Int Glaciol Soc; Am Geophys Union. *Res:* Crustal geologic processes; development of mathematical models; glacial abrasion cracks; landslide mechanics; thermal cracking of rock; fracture permeability; water and rock interaction; mechanics of geologic discontinuites. *Mailing Add:* Ctr Tectonophysics Tex A&M Univ College Station TX 77843

JOHNSON, BRANT MONTGOMERY, b Houston, Tex, Aug 25, 49; m 73; c 2. ATOMIC PHYSICS. *Educ:* Univ Tex, Austin, BS, 71, MA, 74, PhD(physics), 75. *Prof Exp:* Res sci assoc II, Univ Tex, Austin, 71-73, Welch Found Fel, 73-75; res assoc, 75-77, asst physicist, 77-79, assoc physicist, 79-80, PHYSICIST, PHYSICS DEPT, BROOKHAVEN NAT LAB, 81- *Concurrent Pos:* Vis scientist, Lawrence Berkeley Lab & Oak Ridge Nat Lab, 77 & Triumf Lab, BC, 80; lectr, Brookhaven Semester Prog, Brookhaven Nat Lab, 81; German Acad Exchange Serv study visit, German Foreign Exchange Serv, Heidelberg, Fed Repub Germany, 84; assoc ed, Phys Rev Letters, 88-; assoc ed, Phys Rev A, 89-; vis prof, Latin Am Sch Physics, 89; prin investr, Atomic Phys Res, Dept Educ, 90-; NSLS subgroup rep for atom & molecular sci, 91-; chmn, NSLS workshop atom & molecular sci, BNL, 91. *Mem:* Am Phys Soc. *Res:* Ion-atom, ion-electron and ion-photon (synchrotron radiation) collisions and the structure of highly ionized atoms with emphasis on atomic processes relevant to high temperature plasma research. *Mailing Add:* DAS Bldg 815 Brookhaven Nat Lab Upton NY 11973

JOHNSON, BRIAN JOHN, b Reading, Eng, Oct 28, 38; m 68. BIOCHEMISTRY, EDUCATIONAL ADMINISTRATION. *Educ:* Univ Leeds, BSc, 60; Univ London, PhD(org chem) & dipl, Imp Col, 63; Univ London, DSc, 77; Inst Educ Mgt, Harvard Univ, 81. *Prof Exp:* Res assoc org chem, State Univ NY Buffalo, 63-64; res assoc, St John's Univ, 64-65; res assoc, Mass Inst Technol, 65-66; asst prof chem, Tufts Univ, 66-71; ASSOC PROF MICROBIOL, MED SCH, UNIV ALA, BIRMINGHAM, 71-, CO-DIR GRAD PROG, 78- *Mem:* Am Chem Soc. *Res:* Synthesis, structure and biological properties of peptides and proteins; biochemistry of lipid-protein interactions; immunopharmacology; complement; allergy. *Mailing Add:* 3724 Woodvale Rd Birmingham AL 35223

JOHNSON, BRUCE, b Hawarden, Iowa, Sept 4, 32; m 55; c 2. NAVAL ARCHITECTURE, OCEAN ENGINEERING. *Educ:* Iowa State Univ, BSME, 55; Purdue Univ, MSME, 62, PhD(mech eng), 65. *Prof Exp:* Instr marine eng, US Naval Acad, 57-59; mech eng, Purdue Univ, 59-64; assoc prof, 64-70, dir, Hydromech Lab, 76-87, PROF ENG, US NAVAL ACAD, 70- *Concurrent Pos:* Western Elec Fund award eng teaching, 71; chmn, 18th Am Towing Tank Conf, 77; US Rep Info Comt of Int Towing Tank Conf, 75-84, chmn, Symbols & Truminology Group, 85-; Co-auth, Introd to Naval Archit. *Mem:* Am Soc Mech Engrs; Am Soc Eng Educ; Soc Naval Architects & Marine Engrs; Am Soc Naval Engrs. *Res:* Hydrodynamics, ship model testing and brain wave analysis. *Mailing Add:* USAF Hg Af/Le-Rd Pentagon Washington DC 20330

JOHNSON, BRUCE FLETCHER, b Brooklyn, NY, Apr 5, 56; m 84; c 1. ORGANIC CHEMICALS, NEW CHELATORS. *Educ:* Mass Inst Technol, BS, 78; Harvard Univ, MS, 80, PhD(chem), 84. *Prof Exp:* Fel, Dept Chem, Columbia Univ, 84-86; STATE MEM, CORP RES & DEVELOP, GEN ELEC CO, 86- *Res:* Development of efficient synthetic routes leading to organic chemicals; investigation of novel monomers and polymers; synthesis and study of new chelators and binding materials. *Mailing Add:* 36 Irving Rd Schenectady NY 12302

JOHNSON, BRUCE MCDOUGALL, b Ottawa, Ill, Sept 24, 43; m 63; c 3. ANALYTICAL CHEMISTRY. *Educ:* Univ Wis-Madison, BS, 66, MS, 67, PhD(chem), 72. *Prof Exp:* Asst prof clin oncol, Ctr Health Sci, Univ Wis-Madison, 72-75, asst prof human oncol, 75-77; from res scientist to sr res scientist, 77-83, sr res investr, 83-85, SECT HEAD, ANALYSIS RES DEPT, PFIZER INC, 85- *Mem:* AAAS; Am Chem Soc; Am Soc Mass Spectrometry. *Res:* Metabolism of antineoplastic drugs and carcinogens; analysis of pharmaceuticals; application of gas chromatography and mass spectrometry to biomedical and biological problems. *Mailing Add:* Pfizer Eastern Point Rd Analysis Res Bldg No 118 Groton CT 06340

JOHNSON, BRUCE PAUL, b Lewiston, Maine, Aug 8, 38; m 61; c 4. SOLID STATE ELECTRONICS, COMPUTER AIDED DESIGN. *Educ:* Bates Col, BS, 60; Univ NH, MS, 63; Univ Mo-Columbia, PhD(physics), 67. *Prof Exp:* Instr physics, Hobart & William Smith Cols, 62-64; advan physicist, Gen Elec Co, 67-72, supvr, Solid State Lamp Proj, 72-74; assoc prof, 74-78, chmn, 78-83, PROF ELEC ENG, UNIV NEV, RENO, 78-; VPRES, RES & DEVELOP, CADDO ENTERPRISES INC, 90- *Concurrent Pos:* Presidential appointment, US Metric Bd, 78-80 & 80-82; student activ coordr, Region 6, Inst Elec & Electronics Engrs. *Mem:* Inst Elec & Electronics Engrs; Am Soc Eng Educ; Sigma Xi; Int Soc Mini & Microcomputer. *Res:* Biomedical Instrumentation; solid state electronic materials and devices; electronic computer aided design and manufacturing; high frequency electronic design. *Mailing Add:* 3190 W Seventh Reno NV 89503

JOHNSON, BRYAN HUGH, b Hammond, La, Aug 15, 40; m 62; c 2. ENDOCRINE PHYSIOLOGY. *Educ:* Southeastern La Univ, BS, 63; La State Univ, MS, 66; Okla State Univ, PhD(reproduction), 69. *Prof Exp:* NIH res fel, Okla State Univ, 60-71; ASSOC PROF ENDOCRINE PHYSIOL, NC STATE UNIV, 71- *Concurrent Pos:* Biomed res grant, NDak State Univ, 76 & 78. *Mem:* Soc Study Reproduction; Am Soc Animal Sci; Sigma Xi. *Res:* Testicular steroidogenesis; adrenal-testicular interrelationship. *Mailing Add:* Dept Animal Sci NC State Univ 231 Polk Hall Raleigh NC 27650

JOHNSON, BRYCE VINCENT, b Minneapolis, Minn, Oct 24, 49; m 71. INORGANIC CHEMISTRY, ORGANOMETALLIC CHEMISTRY. *Educ:* St Olaf Col, BA, 71; Yale Univ, MS & MPhil, 72, PhD(chem), 75; Univ Chicago, MBA, 86. *Prof Exp:* asst prof chem, Univ Louisville, 75-79; sr chemist, Amoco Res Ctr, Amoco Chem Corp, 79-86; prod develop specialist, 86-89, tech mgr, 89-90, BUS DEVELOP MGR, 3M CORP, 91- *Mem:* Am Chem Soc; Sigma Xi; Soc Plastics Engrs. *Res:* Organometallic synthesis; transition metal isocyanide complexes; homogeneous catalysis; fluxional systems; polyolefin additives; fluoroplastics and elastomers. *Mailing Add:* 3M Ctr 236-6S-04 St Paul MN 55144-1000

JOHNSON, BYRON F, b St Mary's, Pa, July 25, 28; Can citizen; m 52, 67; c 4. CELL BIOLOGY, MICROBIOLOGY. *Educ:* Pa State Univ, BS, 50; Univ Calif, Los Angeles, MA, 58, PhD(zool), 60. *Prof Exp:* Nat Cancer Inst fel zool, Univ Edinburgh, 60-62; res officer, Div Biol Sci, Nat Res Coun Can, 62-90; PROF, DEPT BIOL, CARLETON UNIV, 90- *Concurrent Pos:* Vis scientist, Nat Inst Med Res, London, Eng, 68-69; mem, Int Comn Yeast & Yeast-Like Organisms, 80-; adv bd, CRC Crit Reviews Biotechnol, 81-, Cell & Tissue Kinetics, 84- *Mem:* Am Soc Cell Biol; Am Soc Microbiol; Can Soc Cell Biol; Genetics Soc Can; Int Cell Cycle Soc; Can Soc Microbiologists. *Res:* Cell cycle; cellular growth and division; growth of cell organelles; biosynthesis of wall polysaccharides; regulation of cell size; temperature effects in biological systems; cytoplasmic genetics; chemostat culture; flocculation. *Mailing Add:* Div Biol Carleton Univ Ottawa ON K1S 5B6 Can

JOHNSON, C(HARLES) BRUCE, b Sioux City, Iowa, Aug 5, 35; m 56; c 2. PHYSICS, ELECTRICAL ENGINEERING. *Educ:* Iowa State Univ, BS, 57; Univ Minn, Minneapolis, MSEE, 63, PhD(elec eng), 67. *Prof Exp:* Assoc scientist, Electronics Group, Gen Mills, Inc, 58-61; res asst gaseous electronics, Univ Minn, 61-63, res fel, 63-67; engr, RCA Electronics Components, 67-70; sr staff engr, Bendix Res Labs, 70-74; prin engr, 74-77, TECH DIR, ELECTRO-OPTICAL PRODS DIV, IT&T, FT WAYNE, 78-, SR TECH STAFF ENG, AERO SPACE DIV, 91- *Mem:* AAAS; Am Phys Soc; fel Inst Elec & Electronics Engrs; Optical Soc Am; Soc Photo-Optical Instrumentation; Am Astron Soc. *Res:* Space-charge-effects in vacuum and gases; charged particle optics; electro-optical image transfer characteristics, especially modulation transfer function studies; high resolution image-intensifier/camera-tube development; charged particle transport in gases; high altitude instrumentation; infrared studies; electrical-optical sensor analysis; photon-counting imaging. *Mailing Add:* 6521 Centerton Dr Ft Wayne IN 46815-7830

JOHNSON, C SCOTT, b Sullivan, Mo, Feb 4, 32. PHYSICS, BIOPHYSICS. *Educ:* Univ Mo-Rolla, BS, 54; Wash Univ, PhD(physics), 59. *Prof Exp:* Res assoc physics, Fermi Inst Nuclear Studies, Univ Chicago, 59-63; physicist, Naval Ord Test Sta, 63-67; sr res scientist, 67-69, HEAD MARINE BIOSCI DIV, NAVAL UNDERSEA CTR, 69- *Mem:* Am Phys Soc; Acoust Soc Am; Sigma Xi. *Res:* Marine mammal bioacoustics; shark behavioral studies; nuclear physics. *Mailing Add:* 1876 Sefton Pl San Diego CA 92107

JOHNSON, C WALTER, b Warren, Pa, Apr 19, 10. METALLURGICAL ENGINEERING. *Prof Exp:* Mem staff, Nat Forge, 29-66, dir metall, 60-66; consult, Autoclave Eng, McInnes Steel Co, 66-83; RETIRED. *Mem:* Fel Am Soc Metals. *Mailing Add:* Five E Third Ave Warren PA 16365

JOHNSON, CALVIN KEITH, b Litchfield, Minn, Dec 15, 37; m 60; c 3. THERMOSETTING RESINS, ORGANIC CHEMISTRY. *Educ:* Olivet Nazarene Univ, AB, 59; Mich State Univ, PhD(org chem), 63. *Prof Exp:* NIH fel org chem, Columbia Univ, 63-64; res chemist, Minn Mining & Mfg Co, 64-67; group leader polymer res, CPC Int, 67-69; tech dir res, Acme Resin Corp, 69-77, vpres res & develop, 77-85, SR VPRES & TECH DIR, ACME RESIN CORP, 85- *Mem:* Am Chem Soc; Am Inst Chem; Soc Petrol Eng; Am Foundrymens Soc. *Res:* Organic photochemistry; synthesis and reactions of small ring compounds; mechanisms of polymer decomposition; latent curing resin systems; phenolic, thermosetting and foundry resins; polymers; molding compounds. *Mailing Add:* Acme Resin Corp 1401 Circle Ave Forest Park IL 60130

JOHNSON, CANDACE SUE, b Columbus, Ohio, Apr 10, 49. LEUKEMIA RESEARCH. *Educ:* Ohio State Univ, PhD(microbiol), 77. *Prof Exp:* Sr scientist, AMC Cancer Res Ctr, Lakewood, Colo, 81-89, lab chief exp hemat, 88-89; ASSOC PROF DEPTS OTOLARYNGOL & PATH, SCH MED, UNIV PITTSBURGH, 89- *Res:* T-cells; monoclonal antibodies. *Mailing Add:* Dept Otolaryngol Eye & Ear Inst Pittsburgh Sch Med Univ Pittsburgh 203 Lothrop St Pittsburgh PA 15213

JOHNSON, CARL ARNOLD, b Bend, Ore, Mar 5, 25; m 49; c 2. ORGANIC CHEMISTRY. *Educ:* Reed Col, BA, 50; State Col Wash, MS, 52, PhD(chem), 56. *Prof Exp:* Fel org synthesis, Mellon Inst, 56-59; chief forest prod res, Owens-Ill Co, 59-60, chief org chem res, 60-64, mgr appln res, 64-68, consult coatings & optical mat, 68-69, proj mgr glass fiber reinforcements technol & chem develop, 69-74, sr scientist chem support, 74-75, RES ASSOC CHEM SUPPORT, TEXTILE OPERS, TECH CTR, OWENS CORNING FIBERGLAS CO, 75- *Concurrent Pos:* Dir, Bd of Dirs, Toastmasters Int, 78-80. *Mem:* Am Chem Soc; Sigma Xi; AAAS. *Res:* Development of new glass fiber size systems. *Mailing Add:* 51 Meredith St Port Ludlow WA 98365-9506

JOHNSON, CARL BOONE, b Jacksonville, Fla, Mar 11, 38; m 64; c 1. TOXICOLOGY, ENVIRONMENTAL HEALTH. *Educ:* Fla State Univ, Tallahassee, BS, 59; Am Univ, MS, 67; Georgetown Univ, PhD(biochem), 74. *Prof Exp:* Res chemist, Nat Naval Med Ctr, 63-72; res scientist, Microbiol Assocs, 74-76; sci adminr, 76-81, TOXICOLOGIST, CFSAN, FOOD & DRUG ADMIN, 81- *Mem:* AAAS. *Res:* Solubilized and partially purified a glucagon-binding protein from rat liver plasma membranes; uptake of drugs by rat kidney lysosomes. *Mailing Add:* 12800 Teaberry Rd Silver Spring MD 20906

JOHNSON, CARL EDWARD, b Marshalltown, Iowa, Nov 27, 46; m 73; c 2. SEISMICITY, SEISMIC NETWORKS. *Educ:* Mass Inst Technol, BS & MS, 72; Calif Inst Technol, PhD(geophysics), 79. *Prof Exp:* Geophysicist, Off Earthquake Studies, US Geol Surv, 79-88. *Concurrent Pos:* Vis res assoc, Seismol Lab, Calif Inst Technol, 79- *Mem:* Seismol Soc; Am Geophys Union. *Res:* Seismicity studies related to earthquake prediction research including the development of real-time data acquisition and earthquake data base systems. *Mailing Add:* Dept Geol Univ Hawaii 523 W Lanikaula St Hilo HI 96720

JOHNSON, CARL EDWIN, b Jamestown, Kans, June 24, 17; m 46; c 2. ORGANIC CHEMISTRY. *Educ:* Bethany Col, Kans, BS, 38; Univ Kans, MA, 41, PhD(chem), 43. *Prof Exp:* Chemist, Standard Oil Co Ind, 43-46; dir res, Mich Chem Corp, 46-47; sect leader, Standard Oil Co, Ind, 47-59, dir new chem res, 59-61; dir org chem res, Amoco Chem Corp, 61-65; coordr res & develop, Standard Oil Co Ind, 65-67; gen mgr, Amoco Chem Corp, 67-69, vpres res & develop, 69-80; RETIRED. *Mem:* Am Chem Soc; Am Inst Chem Engrs; Am Inst Chemists; Sigma Xi. *Res:* Chemicals; polymers; hydrocarbon separations; conversions. *Mailing Add:* 1217 Somerset Lane Elk Grove Village IL 60007

JOHNSON, CARL EMIL, JR, b Coleraine, Minn, Dec 11, 21; m 50; c 2. PHYSICAL CHEMISTRY. *Educ:* Univ Calif, BS, 43; Univ Calif, Los Angeles, PhD(chem), 50. *Prof Exp:* Instr chem, Univ Ill, 50-52; from res chemist to supv res chemist, Calif Res Corp, 52-68; sr res assoc, Chevron Oil Field Res Co, La Habra, 68-85; RETIRED. *Concurrent Pos:* Lectr, Univ Southern Calif, 60-; mem city coun, Laguna Beach, Calif, 72-78. *Mem:* Am Chem Soc; Soc Petrol Engrs. *Res:* Kinetics of reactions in solution; chemistry of surfaces and surface active agents; fluid flow through porous media. *Mailing Add:* 616 Mystic View Laguna Beach CA 92651

JOHNSON, CARL ERICK, b Chicago, Ill, Feb 17, 14; m 41; c 2. ORGANIC CHEMISTRY. *Educ:* Univ Chicago, BS, 38, MS, 49. *Prof Exp:* Chemist, Western Shade Cloth Co, Ill, 38-45; chief org chemist, Nat Aluminate Corp, 45-52, dir inorg res, 52-56, sr technol adv, 56-59; sect head, Cent Res, Nalco Chem Co, 59-60, res mgr, Metal Indust Div, 60-68, res assoc, 68-71, res assoc Miss, 71-74, res mgr, Brookhaven Res Lab, 74-76; RETIRED. *Concurrent Pos:* Consult, 76- *Mem:* AAAS; Am Chem Soc; Nat Assn Corrosion Engrs. *Res:* Water and textile treatment; flotation of minerals; synthesis of organic compounds; measurement of the film pressure of insoluble films; organic chemistry of boiler water treatment; aqueous corrosion; ion exchange; industrial lubrication, especially metal rolling and emulsion technology. *Mailing Add:* 401 McNair Ave Brookhaven MS 39601-3744

JOHNSON, CARL J, epidemiology, health administration; deceased, see previous edition for last biography

JOHNSON, CARL LYNN, b Beaumont, Tex, Aug 22, 41. PHARMACOLOGY, BIOCHEMISTRY. *Educ:* Rice Univ, BA, 64; Univ Houston, MS, 68; Baylor Col Med, PhD(pharmacol), 71. *Prof Exp:* Instr, 71-72, assoc, 72-73, asst prof pharmacol, Mt Sinai Sch Med, 73-77; ASSOC PROF PHARMACOL, COL MED, UNIV CINCINNATI, 77- *Res:* Hormone receptors and adenylate cyclase; molecular pharmacology. *Mailing Add:* Dept Pharmacol & Cell Biophys Univ Cincinnati Col Med 231 Bethesda Ave Cincinnati OH 45267

JOHNSON, CARL RANDOLPH, b Charlottesville, Va, Apr 28, 37; m 66; c 1. ORGANIC CHEMISTRY. *Educ:* Med Col Va, BS, 58; Univ Ill, PhD(chem), 62. *Prof Exp:* NSF res fel chem, Harvard Univ, 62; from asst prof to prof, 62-90, DISTINGUISHED PROF CHEM, WAYNE STATE UNIV, 90- *Concurrent Pos:* Alfred P Sloan fel, 65-68; adv bd, J of Org Chem, 76-81; assoc ed, J Am Chem Soc, 84-89; bd dirs, Organic Synthesis, Inc, 81-; Humboldt sr scientist, 91. *Mem:* Am Chem Soc; Royal Soc Chem. *Res:* Organic sulfur chemistry, especially sulfoxides & sulfoximines; exploratory synthetic chemistry; synthesis of compounds of potential medicinal activity; organometallic chemistry; synthesis of natural products; enzymes in synthesis. *Mailing Add:* Dept of Chem Wayne State Univ Detroit MI 48202

JOHNSON, CARL WILLIAM, b Mound Valley, Kans, Feb 11, 42; m 68; c 2. PLANT BREEDING. *Educ:* Kans State Univ, BS, 65; NDak State Univ, MS, 67; Univ Nebr, PhD(agron), 74. *Prof Exp:* PLANT BREEDER, CALIF COOP RICE RES FOUND, 74- *Honors & Awards:* Distinguished Rice Res & Educ Award; First McCaughey Mem Inst Vis Scientists Award, Australia. *Mem:* Crop Sci Soc Am; Am Soc Agronomy; Coun Agr Sci & Technol; Sigma Xi. *Res:* Development of rice varieties for the California rice industry. *Mailing Add:* Rice Exp Sta PO Box 306 Biggs CA 95917

JOHNSON, CARLTON E(GBERT), agricultural engineering; deceased, see previous edition for last biography

JOHNSON, CARLTON ROBERT, b Chicago, Ill, Sept 19, 26; m 51; c 2. PETROLEUM GEOLOGY, GROUNDWATER GEOLOGY. *Educ:* Monmouth Col, Ill, BA, 49; Univ Iowa, MS, 54, PhD, 56. *Prof Exp:* Geologist, US Geol Surv, 50-56; res geologist, Jersey Prod Res Co, 56-65; sr res geologist, Esso Prod Res Co, 65-69, res assoc, Exxon Prod Res Co, 69-72, sr res assoc, 72-79, res adv, 79-91, 3D MODELING PETROL RESERVOIR DESCRIPTION RES, EXXON PROD RES CO, 91- *Mem:* soc Petrol Engrs. *Res:* Computer mapping and modeling programs; geology and performance of oil, gas and water reservoirs; well testing procedures and instrumentation. *Mailing Add:* Exxon Prod Res Co PO Box 2189 Houston TX 77001-2189

JOHNSON, CARROLL KENNETH, b Greeley, Colo, Sept 18, 29; m 51; c 5. CRYSTALLOGRAPHY, BIOPHYSICS. *Educ:* Colo State Univ, BS, 55; Mass Inst Technol, PhD(biophys), 59. *Prof Exp:* Asst biol, Mass Inst Technol, 55-56, asst biophys, 56-59; Am Cancer Soc res fel x-ray crystallog, Inst Cancer Res, Pa, 59-62; RES CHEMIST NEUTRON CRYSTALLOG, CHEM DIV, OAK RIDGE NAT LAB, 62- *Mem:* Am Crystallog Asn. *Res:* Neutron diffraction; stereochemistry of enzyme substrates; diffraction theory of biological polymers, crystallographic computing; automated graphics for illustrating crystal structures; crystallographic thermal-motion analysis; basic crystallographic theory. *Mailing Add:* 344 East Dr Oak Ridge TN 37830

JOHNSON, CECIL GRAY, b Nanafalia, Ala, Feb 26, 22; m 48; c 3. INDUSTRIAL & SYSTEMS ENGINEERING. *Educ:* Ga Inst Technol, BS, 48 & 49, MS, 57. *Prof Exp:* Indust engr, Gen Shoe Corp, 49-50 & Am Art Metals Co, 50-55; PROF INDUST & SYSTS ENG, GA INST TECHNOL, 55- *Concurrent Pos:* Mgt & systs eng consult, 55-; ed-in-chief, J Am Inst Indust Engrs, 55-65; res assoc, Off Naval Res, Univ Calif, Los Angeles, 59; consult, HEW, Univ Ga, 69; partic, Stanford-Ames NASA-ASEE Educ Res Study, 74; prin, Atlanta Assessment Proj, Atlanta Pub Schs, HEW, 76; prin, Prime DOC Commun Syst, Fulton County, Ga, 80; prin, Systs Study Rehab Serv Blind, State Ga, 80-83; mem, 11E Fel Scholar Comt, 85-87; mem, 11E Heritage Task Force, 86-87, chmn, 87-88. *Honors & Awards:* Award, Am Inst Indust Engrs. *Mem:* Am Inst Indust Engrs (vpres, 65-67); Nat Soc Prof Engrs. *Res:* Human performance and organizational theory; educational systems, especially American universities; analysis and design methodology for complex systems; improving productivity from mental and physical activity, especially among university educated individuals. *Mailing Add:* 3211 Argonne Dr NW Atlanta GA 30305

JOHNSON, CHARLES ANDREW, b Chicago, Ill, May 8, 15; m 40; c 2. MATHEMATICS. *Educ:* Northern Ill Univ, BEd, 37; Northwestern Univ, MA, 40; Univ Kans, PhD(math), 50. *Prof Exp:* Teacher & prin pub schs, Ill, 38-40, teacher, 40-43; PROF MATH, UNIV MO-ROLLA, 46- *Concurrent Pos:* Instr, Univ Kans, 48-50; res assoc, Argonne Nat Lab, 62. *Mem:* Am Soc Eng Educ; Am Math Asn; Sigma Xi. *Res:* Mathematical education. *Mailing Add:* Dept of Math Univ of Mo Rolla MO 65401

JOHNSON, CHARLES C, JR, ENVIRONMENTAL ENGINEERING. *Educ:* Purdue Univ, BS, 47, MS, 57; Am Acad Environ Engrs, dipl. *Prof Exp:* Asst surgeon gen, USPHS, 47-71; assoc exec dir, Am Pub Health Asn, 71-72; vpres, Wash Tech Inst, 72-74; vpres, Malcolm Pirnie, Inc, 74-79; PRES, C C JOHNSON & MALHOTRA, PC, 79- *Concurrent Pos:* Comnr, Nat Capital Planning Comn, 71-74; Comn Educ Health Admin, 72-74; mem tech adv group, Munic Wastewater Systs, Environ Protection Agency, 73-75; mem, Task Groups on Eng Career Develop, Surgeon Gen; chmn, Nat Drinking Water Adv Coun, 75-81; adj assoc prof, Sch Environ Med, NY Univ, 76; mem, Adv Comt Water Data Pub Use, US Geol Surv, Dept Interior, 79-80; mem, Safe Drinking Water Act Amendments Tech Adv Workgroup, Am Water Works Asn, 89, Strategic Planning Comt, 90-91; mem, Clean Water Act Reauthorization Comt, Water Pollution Control Fedn, 90; mem, Task Comt Int Relations, Am Acad Environ Engrs, 90-91; mem, Comt Hazardous Waste in Hwy Rights of Way, Transp Res Bd, Nat Res Coun, 90-, Comt Rev Environ Protection Agency's Monitoring & Assessment Prog, 91-; mem bd dirs, Water for People, 91-, Nat Sanit Found, 91- *Honors & Awards:* Walter F Snyder Award, Nat Environ Health Asn & Nat Sanit Found, 77; Award for Except Achievement, US Dept Health & Human Serv, 89; George Warren Fuller Award, Am Water Works Asn, 90. *Mem:* Nat Acad Eng; fel Am Pub Health Asn; hon mem Am Water Works Asn; hon mem Nat Environ Health Asn; Water Pollution Control Fedn. *Res:* Administration of environmental programs; implementation of water supply and waste disposal construction programs. *Mailing Add:* C C Johnson & Malhotra PC 601 Wheaton Plaza S Silver Spring MD 20902

JOHNSON, CHARLES EDWARD, b Pennington Gap, Va, Nov 19, 40; m 70; c 2. ATOMIC PHYSICS. *Educ:* Yale Univ, BS, 62, MS, 65, PhD(physics), 67. *Prof Exp:* Res physicist, Lawrence Radiation Lab, Univ Calif, Berkeley, 67-72; from asst prof to assoc prof, 73-83, PROF PHYSICS, NC STATE UNIV, 83- *Mem:* Am Phys Soc. *Res:* Measurement of the fundamental properties of free atoms and molecules using the techniques of optical pumping and atomic beam magnetic resonance. *Mailing Add:* Dept Physics NC State Univ Raleigh NC 27695

JOHNSON, CHARLES F, b Chicago, Ill, Sept 15, 27; m 61; c 1. MEDICINE, ELECTRON MICROSCOPY. *Educ:* Univ Chicago, PhB, 49, MD, 54; Am Bd Internal Med, dipl, 62. *Prof Exp:* From instr to asst prof, Sch Med, Univ Chicago, 58-67; ASSOC PROF MED, IND UNIV-PURDUE UNIV, INDIANAPOLIS, 67- *Concurrent Pos:* Asst head gastroenterol, Vet Admin Hosp, Indianapolis, 67-71, head, 71-, consult; head sect gastroenterol, St Vincent's Hosp, Indianapolis. *Mem:* AAAS; Am Soc Cell Biologists; Electron Micros Soc Am. *Res:* Electron microscopy of lipid absorption and various human gastrointestinal diseases. *Mailing Add:* 5654 N Pennsylvania St Indianapolis IN 46220

JOHNSON, CHARLES HENRY, b Chicago, Ill, June 12, 25; m 48; c 3. MATHEMATICAL STATISTICS. *Educ:* Bradley Univ, BA, 49, MS, 50; Okla State Univ, PhD(math), 63. *Prof Exp:* Asst math, Univ Pittsburgh, 50-52; sect chief, Continental Casualty Co, 52-55; from asst prof to assoc prof math & astron, DePauw Univ, 55-67; PROF MATH & CHMN DEPT, UNIV WIS-STEVENS POINT, 67- *Res:* Astronomy. *Mailing Add:* 2755 S County Trunk Stevens Point WI 54481

JOHNSON, CHARLES MINOR, b Nashville, Tenn, May 31, 23; m 48; c 1. PHYSICS, SYSTEM ANALYSIS. *Educ:* Vanderbilt Univ, BE, 44; Duke Univ, PhD(physics), 51. *Prof Exp:* Res assoc, Radiation Lab, Johns Hopkins Univ, 51-53, res scientist, 53-56; res mgr, Electronic Commun, Inc, 56-61; res dir, Emerson Res Lab, 60-61; dep safeguard syst mgr, Sci & Technol, Dept Army, 67-73; res mgr, IBM Corp, 61-67, dep dir, World Wide Mil Command & Control Syst Archit Develop, 73-86; prin scientist, Anser, 86-88; PRIN ENGR, MITRE CORP, 88- *Concurrent Pos:* Consult, Sperry-Rand Corp, 55 & Eng Res & Develop Lab, US Army, 59-; sci adv, Joint Strategic Target Planning Staff, 72-81; consult, Develop & Readiness Command, US Army, 76; external adv, Ga Tech Res Inst, 80-85. *Honors & Awards:* Dept Army Medal Exceptional Civilian Serv, 73. *Mem:* Am Phys Soc; Inst Elec & Electronics Engrs. *Res:* Microwave physics, ferrite devices, phased array radars, millimeter wave techniques, microwave spectroscopy; radiation scattering, lasers and optics; semiconductor devices; ballistic missile defense; command and control systems. *Mailing Add:* 11220 Leatherwood Dr Reston VA 22091

JOHNSON, CHARLES NELSON, JR, b Mt Hope, Kans, June 17, 15; m 41; c 3. APPLIED PHYSICS. *Educ:* Friends Univ, AB, 38. *Prof Exp:* Jr instr eng physics, Johns Hopkins Univ, 38-41; physicist, Bur Ord, US Dept Navy, Washington, DC & Naval Operating Base, Norfolk, Va, 41-42, physicist, Norfolk Navy Yard, Va, 42-46, sr physicist, Aviation Ord Dept, 46-51, sr physicist, Ballistic Instrumentation Dept, Naval Proving Ground, 51-55; supvr res physicist, US Army Engr Res & Develop Ctr, 55-67, chief detection br, Intrusion, Detection & Sensor Lab, 67-71, chief phys sci group, Countermine-Counter Intrusion Dept, 71-73; CONSULT PHYSICIST, ENVIRON RES INST MICH, SEARLE CONSORTIUM, 74- *Mem:* AAAS; Am Phys Soc; Sigma Xi. *Res:* Interior and exterior ballistic measurements; weapons systems evaluation and counter-measures; barrier and intrusion detection systems; remote multiband sensor systems; land mines, concealed explosives, letter bombs and booby trap detectors. *Mailing Add:* 3100 N Oxford St Arlington VA 22207

JOHNSON, CHARLES RICHARD, JR, b Macon, Ga, May 27, 50. ADAPTIVE CONTROL, ADAPTIVE SIGNAL PROCESSING. *Educ:* Ga Inst Technol, BEE, 73; Stanford Univ, MS, 75, PhD(elec eng), 77. *Prof Exp:* Asst prof elec eng, Va Polytech Inst & State Univ, 77-81; assoc prof, 81-87, PROF ELEC ENG, CORNELL UNIV, 87-, ASSOC DIR, SCH ELEC ENG, 88- *Concurrent Pos:* Prin investr, NSF grants, 79-; assoc ed, IEEE Transactions Acoustics, Speech, Signal Processing, 81-83, IEEE Transactions Automatic Control, 82-83, Automatica, 86- & ed, Int J Adaptive Control & Signal Processing, 87-; consult, Tellabs Res Lab, 85-88. *Mem:* Inst Elec & Electronics Engrs. *Res:* Development of adaptive parameter estimation theory useful for applications of digital control and system identification to manufacturing processes and for applications of digital signal processing to telecommunications systems. *Mailing Add:* 1491 Elmira Rd Newfield NY 14867

JOHNSON, CHARLES ROBERT, b Ft Collins, Colo, June 8, 41; m 64; c 2. ORNAMENTAL HORTICULTURE. *Educ:* Colo State Univ, BS, 64; Ore State Univ, PhD(ornamental plant physiol), 70. *Prof Exp:* Res floricult, K Stormly Hansen Greenhouses, Copenhagen, Denmark, 64-65; res & teaching, Dept Hort, Clemson Univ, 70-73; res & teaching ornamental hort, Univ Fla, 73-80, assoc prof, 80-84; prof & head dept, Dept Hort, Univ Ga, 85-90; DEPT ARCHIT LANDSCAPING, WASH STATE UNIV, 90- *Concurrent Pos:* Bd dir, Int Plant Propagators Soc, 80-82. *Honors & Awards:* Porter Henegar Res Award, Nurserymen's Asn, 80. *Mem:* Am Soc Hort Sci; Int Plant Propagators Soc. *Res:* Physiological aspects of plant-soil microbial symbiosis, growth and development, stress physiology and urban horticulture. *Mailing Add:* Dept Archit Landscaping Wash State Univ Pullman WA 99164-6414

JOHNSON, CHARLES ROYAL, b Elkhart, Ind, Jan 28, 48; m 72; c 2. ALGEBRA, APPLIED MATHEMATICS. *Educ:* Northwestern Univ, BA, 69; Calif Inst Technol, PhD(math, econ), 72. *Prof Exp:* Res assoc fel math, Appl Math Div, Nat Bur Standards, 72-74; res prof appl math & econ, Inst For Phys Sci & Technol, Univ Md, College Park, 74 -84; prof math sci, Clemson Univ, 85-87; PROF MATH, COL WILLIAM & MARY, 87- *Concurrent Pos:* Consult, Appl Math Div, Nat Bur Standards, 74-82; vis staff mem, Los Alamos Sci Lab, 74-; consult, Icase, 82- *Mem:* Am Math Soc; Soc Indust & Appl Math; Math Asn Am; Int Linear Algebra Soc. *Res:* Matrix analysis and applications; combinatorics; mathematical economics; combinatorial matrix analysis, eigenvalues, inequalities and norms. *Mailing Add:* Dept Math Col William & Mary Williamsburg VA 23185

JOHNSON, CHARLES SIDNEY, JR, b Albany, Ga, Mar 7, 36; m 58; c 2. PHYSICAL CHEMISTRY. *Educ:* Ga Inst Technol, BS, 58; Mass Inst Technol, PhD(phys chem), 61. *Prof Exp:* Nat Acad Sci-Nat Res Coun fel, 61-62; from asst prof to assoc prof phys chem, Yale Univ, 62-67; prof, 67-88,

M A SMITH, PROF CHEM, UNIV NC, CHAPEL HILL, 88- *Concurrent Pos:* Sloan Found res fel, 66-; ed bd, J Magnetic Resonance, 71-; Guggenheim Found fel, 72. *Mem:* AAAS; fel Am Phys Soc; Am Chem Soc. *Res:* Nuclear magnetic resonance; spin relaxation; chemical rate processes; laser light scattering. *Mailing Add:* Dept of Chem Univ of NC Chapel Hill NC 27599-3290

JOHNSON, CHARLES WILLIAM, b Ennis, Tex, Jan 25, 22; m 43; c 3. MICROBIOLOGY. *Educ:* Prairie View State Col, BS, 42; Univ Southern Calif, MS, 47; Meharry Med Col, MD, 53. *Prof Exp:* Instr bact & parasitol, 47-49, from asst prof to assoc prof, 49-59, chmn dept, 59-73, dean div grad studies & res, 66-81, interim dean, Sch Med, 81-82, PROF MICROBIOL, MEHARRY MED COL, 59-, VPRES ACAD AFFAIRS, 81- *Concurrent Pos:* Actg chmn dept microbiol, Meharry Med Col, 53-59; Rockefeller Found fel, 57-59; consult, Hubbard Hosp, 52-54. *Mem:* AAAS; Am Soc Microbiol; Am Acad Allergy; Am Fedn Clin Res; Am Asn Path; Am Asn Clin Immunol & Allergy; Asn Geront Higher Educ. *Res:* Immunology and mycology. *Mailing Add:* VPres Acad Affairs Meharry Med Col 1005 D B Todd Blvd Nashville TN 37208

JOHNSON, CHRIS ALAN, b Roseburg, Ore, Oct 1, 49; m 71. PSYCHOPHYSICS, PHYSIOLOGICAL OPTICS. *Educ:* Univ Ore, BA, 70; Pa State Univ, MSc, 72, PhD(psychol), 74. *Prof Exp:* Res asst psychol, Univ Ore & Pa State Univ, 70-75; res fel ophthal, Univ Fla, 75-77; res fel, 77-78, from asst prof to assoc prof, 78-89, PROF OPHTHAL, UNIV CALIF, DAVIS, 89- *Concurrent Pos:* Nat Eye Inst, NIH fels, 75 & 77, academic investr award, 78; Nat Eye Inst grant, 79-91. *Honors & Awards:* Distinguished Serv Award, 87, Honor Award, Am Acad Ophthal, 88. *Mem:* Asn Res Vision & Ophthal; Optical Soc Am; Int Perimetric Soc; Am Acad Ophthal. *Res:* Visual psychophysics, analysis of the accommodation mechanism, examination of peripheral visual functions and development and adaptation of psychophysical tests to quantitative perimetry and visual field testing; night vision; vision and driving. *Mailing Add:* Dept of Ophthal Univ of Calif Davis CA 95616

JOHNSON, CLARENCE DANIEL, b Exeter, Calif, July 20, 31; m 51; c 4. SYSTEMATIC ENTOMOLOGY. *Educ:* Fresno State Univ, BA, 53; Ariz State Univ, MS, 61; Univ Calif, Berkeley, PhD(entom), 66. *Prof Exp:* High sch teacher, Calif, 56-63; asst prof, 66-70, PROF ZOOL, NORTHERN ARIZ UNIV, 70- *Concurrent Pos:* Fulbright res award, SAm, 84-85. *Mem:* Soc Study Evolution; Ecol Soc Am; Entom Soc Am; Soc Syst Zool; Asn Trop Biol. *Res:* Systematics, ecology and behavior of the beetle family Bruchidae; insect-plant interactions; effects of habitat modification on arthropods. *Mailing Add:* Dept Biol Sci Northern Ariz Univ Box 5640 Flagstaff AZ 86011

JOHNSON, CLARENCE EUGENE, b Elk City, Okla, Nov 1, 41. ENGINEERING, AGRICULTURE. *Educ:* Okla State Univ, BS, 63; Iowa State Univ, MS, 68, PhD(agr eng), 69. *Prof Exp:* Instr agr eng, Iowa State Univ, 64-69; assoc prof, SDak State Univ, 70-77; agr engr, Columbia Plateum Conserv Res Ctr, USDA Agr Res Serv, Ore, 77-79; PROF AGR ENG, AUBURN UNIV, 79- *Mem:* Am Soc Agr Engrs; Am Soc Eng Educ. *Res:* Soil dynamics; tillage and traction; harvesting systems; machinery system simulation; similitude. *Mailing Add:* Dept of Agr Eng Auburn Univ Auburn AL 36849

JOHNSON, CLARENCE L(EONARD), aeronautical engineering; deceased, see previous edition for last biography

JOHNSON, CLARK E, JR, b Minneapolis, Minn, Aug 3, 30; div; c 7. MAGNETIC RECORDING. *Educ:* Univ Minn, Minneapolis, BS, 50, MS, 61. *Prof Exp:* Sr physicist, Cent Res Labs, Minn Mining & Mfg Co, 50-59; pres res & develop, Leyghton-Paige Corp, 59-61; pres, Telostat Corp, 61-63, Micro-Commun Corp, 72-77; vpres, Minnetech Labs, 63-66; vpres eng, Vibrac Corp Div, USM Corp, 67-72; dir, res & develop, Buckeye Int, Inc, 77-80; pres & chmn, Vertmag Systs Corp, 81-85; CONSULT, 85-; PRES, CARD SYSTS TESTING LABS, 90- *Concurrent Pos:* Consult physicist, Graham Magnetics Inc, 68-74; dir & tech adv, Trans Data Syst; finance comt chmn, Magnetics Soc, 75-80, vpres, 81-82, pres, 83-84; dir, Sciencare Corp, 77-84, Magnum Technol, & Megabyte Storage Systs, 87-90; cong sci fel, 88; chmn bd dirs, Appl Info Systs, 89- & Rastech, Inc. *Mem:* AAAS; fel Inst Elec & Electronics Engrs; Am Phys Soc; Instrument Soc Am; NY Acad Sci. *Res:* Magnetic theory; magnetic recording and recording materials; fine particle magnetic theory; electromagnetic transducers and devices; electro-optic transducers and devices; new techniques for recording information using magnetic properties of materials; perpendicular magnetic recording. *Mailing Add:* PO Box 480344 Denver CO 80248

JOHNSON, CLAYTON HENRY, JR, petrology; deceased, see previous edition for last biography

JOHNSON, CLELAND HOWARD, b Pierpont, SDak, Sept 16, 22; m 44; c 3. NUCLEAR PHYSICS. *Educ:* Hastings Col, BA, 44; Univ Wis, PhD(physics), 51. *Prof Exp:* PHYSICIST, OAK RIDGE NAT LAB, 51- *Mem:* Am Phys Soc. *Res:* Experimental nuclear structure physics. *Mailing Add:* Box 310 Rte 1 Ten Mile TN 37880

JOHNSON, CLIFTON W, b Lewisville, Idaho, Sept 23, 24; m 52; c 6. CIVIL & AGRICULTURAL ENGINEERING. *Educ:* Utah State Univ, BS, 56, MS, 57. *Prof Exp:* Water distribution engr, State Engrs Off, Utah, 57-60; HYDRAUL ENGR, AGR RES SERV, 60- *Mem:* Am Soc Civil Engrs; Am Soc Agr Engrs. *Res:* Hydrology, erosion and sediment transport; design, construction and operation of water measuring devices; irrigation water diversion and use; sediment transport, measurement and studies of arid lands hydrology. *Mailing Add:* 3907 Whitehead Boise ID 83703

JOHNSON, CONOR DEANE, b Charlottesville, Va, Apr 20, 43; m 66; c 2. FINITE ELEMENT ANALYSIS, DYNAMIC & DAMPING ANALYSIS. *Educ:* Va Polytech Inst, BS, 65; Clemson Univ, MS, 67, PhD(eng mech), 69. *Prof Exp:* Sr structural analyst, Anamet Labs, Dayton, Ohio, 73-75, prin engr, San Carlos, Calif, 75-81, vpres, 81-82; PRES, CSA ENG INC, 82- *Concurrent Pos:* Fel, NDEA, 67-68; prog mgr, Aerospace Structures Info & Anal Ctr, 75-82. *Honors & Awards:* Struct & Mat Award, Am Soc Mech Engrs, 81. *Mem:* Am Inst Aeronaut & Astronaut; Am Soc Mech Engrs; Sigma Xi. *Res:* Modal strain energy method for damping analysis using finite element techniques; combined system analysis techniques (integration of finite element techniques, damping analysis, component mode synthesis, other engineer disciplines, experimental data). *Mailing Add:* 3425 Lodge Dr Belmont CA 94002

JOHNSON, CORINNE LESSIG, b Wilmington, Del, Oct 29, 38. MICROBIOLOGY, BIOCHEMISTRY. *Educ:* Wellesley Col, AB, 60; Univ Rochester, MS, 64, PhD(biol), 69. *Prof Exp:* Sci Res Coun res asst & fel biochem, Univ Leicester, 69-70; fel, Albert Einstein Col Med, 70-72; assoc res scientist & instr biochem, Dent Ctr, NY Univ, 72-74, asst prof microbiol, 75; vis asst prof biol, Vassar Col, 75-77; asst prof biol, Carleton Col, 77-78; res assoc microbiol, Sch Med, Boston Univ, 78-79; biol ed & gen mgr, Edutech Inc, 81-83; software specialist, Gibco Labs, 84-85; temp proj dir, educ technol database, 85-86, TRAINING COORDR, DEVELOP COMPUT SERVS, HARVARD UNIV, 86- *Concurrent Pos:* Treas, Alliance Independent Scholars, 80-85. *Mem:* AAAS; Am Soc Microbiol; Am Chem Soc; Asn Women Sci. *Mailing Add:* 36 Highland Ave No 48 Cambridge MA 02139

JOHNSON, CORWIN MCGILLIVRAY, b Berthold, NDak, Mar 30, 24; m 46; c 4. AGRONOMY. *Educ:* State Col Wash, BS, 50, MSA, 51; Cornell Univ, PhD, 53. *Prof Exp:* Res asst agron, State Col Wash, 50-51, asst agronomist, Northwestern Wash Exp Sta, 53-56; asst agronomist & asst prof agron, Miss State Univ, 56-61; asst prof crops prod, Calif Polytech State Univ, San Luis Obispo, 62-87, head crop sci dept, 62-87, prof, 74-87; RETIRED. *Mem:* Am Soc Agron; Crop Sci Soc Am. *Res:* Effect of management practices and soil fertility on the yield and quality of crops. *Mailing Add:* 3604 W Lincoln Yakima WA 98902

JOHNSON, CURTIS ALAN, b Johnstown, Pa, Jan 22, 48; m 69; c 2. MATERIAL SCIENCE, CERAMIC SCIENCE. *Educ:* Pa State Univ, BS, 69, PhD(metall), 74. *Prof Exp:* STAFF SCIENTIST CERAMICS, GEN ELEC CORP RES & DEVELOP CTR, 73- *Mem:* Am Ceramics Soc. *Res:* Mechanical and physical properties of metals and ceramics, in particular high temperature structural ceramics; fabrication methods; densification processes and phase transformations of ceramics. *Mailing Add:* Gen Elec Corp Res & Develop Ctr PO Box 8 Schenectady NY 12301

JOHNSON, CURTIS ALLEN, b Mead, Nebr, Apr 3, 17; m 54; c 2. AGRICULTURAL ENGINEERING. *Educ:* Univ Nebr, BSc, 40; Iowa State Univ, MS, 55. *Prof Exp:* Test engr, Tractor Testing Lab, Int Harvester Co, 40-41; asst prof agr eng, Univ Del, 46-48; prin, Friendsville Acad, Tenn, 49-50; instr agr eng, Iowa State Univ, 50-55; agr workshop adv, US State Dept, Int Coop Admin, Pakistan, 55-57; assoc prof, 57-79, EMER PROF AGR ENG, UNIV MASS, AMHERST, 79- *Mem:* Am Soc Agr Engrs. *Res:* Relationship of milking machines to mastitis; liquid handling of agricultural wastes; world-wide water resources planning; farm homes and buildings; design of economical houses for minimal waste of structural materials and fossil fuel energy inputs and storage of summer time heat for use in winter, with concurrent storage of cold (as ice) for use to air condition homes developed in 76-78. *Mailing Add:* 49 Brainerd St No 4 Woolman Common Mt Holly NJ 08060-1809

JOHNSON, D(AVID) LYNN, b Provo, Utah, Apr 2, 34; m 59; c 5. SINTERING, PROCESSING. *Educ:* Univ Utah, BS, 56, PhD(ceramic eng), 62. *Prof Exp:* Mining engr trainee, US Smelting, Ref & Mining Co, 56; from asst prof to assoc prof, 62-71, PROF MAT SCI, NORTHWESTERN UNIV, 71- *Concurrent Pos:* Consult; Walter P Murphy Prof, 87. *Mem:* fel Am Ceramic Soc; Am Inst Mining, Metall & Petrol Engrs; Mat Res Soc; Am Powder Metall Inst. *Res:* Mechanisms of material transport in the sintering of oxides and metals; impurity effects in sintering; grain boundary diffusion in sintering; plasma and microwave processing of ceramics and ceramic composites; processing of high temperature superconductors. *Mailing Add:* Dept Mat Sci Northwestern Univ Evanston IL 60208

JOHNSON, DALE A, b Chicago, Ill, Nov 18, 37; m 60; c 2. PHYSICAL CHEMISTRY, INORGANIC CHEMISTRY. *Educ:* Univ Ill, BS, 59; Northwestern Univ, PhD(chem), 64. *Prof Exp:* From asst prof to assoc prof, 63-73, PROF CHEM, UNIV ARK, FAYETTEVILLE, 73- *Mem:* Am Chem Soc. *Res:* Thermal and photochemical reactions of transition metal complexes; reactions of coordinated molecules; spectroscopy of inorganic compounds. *Mailing Add:* Dept Chem Univ Ark Fayetteville AR 72701

JOHNSON, DALE HOWARD, b Los Angeles, Calif, Feb 23, 45; m 77; c 3. CONSUMER PRODUCT DEVELOPMENT, COSMETIC CHEMISTRY. *Educ:* Univ Redlands, BS, 66; Northwestern Univ, PhD(org chem), 71. *Prof Exp:* Res chemist toiletries, Alberto-Culver Co, 71-73; sect head, Appln Lab, Armak Indust Chem, Div Akzona Inc, 73-77; mgr prod develop, Helene Curtis Indust Inc, 77-81; sect head oral hyg, Vicks Div Res, Richardson-Vicks Inc, 81-84; bus tech mgr, James Rivver Corp, 85-87; sr group leader, Toiletries, Amway Corp, 88-90; MGR PROD DEVEL, HAIR CARE, HELENE CURTIS, INC, 90- *Mem:* Am Chem Soc; Soc Cosmetic Chemists; Sigma Xi. *Res:* Development of personal care products including hair, skin, oral hygiene, cleansers, fine fragrances and treatment products; cosmetic, toiletries & over the counter topical drug type product development. *Mailing Add:* 1505 Lark Lane Naperville IL 60565-1342

JOHNSON, DALE RICHARD, CADMIUM TRANSPORT ACROSS NEONATAL INTESTINE. *Educ:* Marquette Univ, PhD(physiol), 71. *Prof Exp:* ASSOC PROF PHYSIOL, DEPT ENVIRON HEALTH, UNIV CINCINNATI, 73- *Res:* Mercury induced nephrotoxicity; metal transport across epithelial tissue and semicolon heavy metal toxicology. *Mailing Add:* Dept Environ Health Univ Cincinnati Col Med ML 56 3223 Eden Ave Cincinnati OH 45267

JOHNSON, DALE WALDO, b Nelson, Wis, Apr 11, 15; m 45; c 3. FOOD SCIENCE. *Educ:* Univ Minn, BCh, 37, PhD(bact), 41. *Prof Exp:* Res & teaching asst, Univ Minn, 37-41, Hormel fel, 41-42; chief bacteriologist, Res Lab, Pillsbury Mills, 42-45 & Diversey Corp, 45-48; microbiologist & head dept, Soya Div, Glidden Co, 48-53, liaison & head biol, Cent Org Res Labs, 53-55, res consult, Chemurgy Div, 53-58; mgr edible protein prod, Cent Soya Co, 58-63; vpres, Soypro Int, Inc, 63-69; exec vpres, Crest Prod Inc, 63-73; pres, Elk Grove Village, 73-82, PRES, FOOD INGREDIENTS MINN INC, 82-; PRES, GEL TECH, INC, 88- *Concurrent Pos:* Consult, 37-45 & 62-; bacteriologist, State Dept Health, Minn, 40-41; hon fel, dept chem eng, Grad Sch, Univ Minn, 87- *Mem:* Am Chem Soc; Am Asn Cereal Chemists; Inst Food Technologists; Am Oil Chem Soc; Sigma Xi. *Res:* Microbiology of meat and cereals; development of germicides; sewage treatment and analysis; nutrition of yeast, molds and bacteria; fermentation; vitamin B12; antibiotics; fish solubles; bioconversions; micro-biological assays; soy products; nutrition; foods and feeds; enzymes; milk proteins; sesame products; soybean products and processing technology. *Mailing Add:* 2121 Toledo Ave N Golden Valley MN 55422

JOHNSON, DALLAS EUGENE, b Central City, Nebr, Oct 14, 38; c 2. DATA ANALYSIS, LINEAR MODELS. *Educ:* Kearney State Teachers Col, BS, 60; Western Mich Univ, MS, 66; Colo State Univ, PhD(statist), 71. *Prof Exp:* Instr statist, Colo State Univ, 66-68; asst prof, Univ Mo-Rolla, 71-75; assoc prof, 75-81, PROF STATIST, KANS STATE UNIV, 81- *Mem:* Am Statist Asn; Inst Math Statist. *Res:* Linear models; data analysis; statistical design; biased regression methods; messy data. *Mailing Add:* 5239 W 96 Shawnee Mission KS 66207

JOHNSON, DARELL JAMES, b Brooklyn, NY, Dec 22, 49; m 73; c 3. STATISTICAL MECHANICS, FLUIDS. *Educ:* Univ Calif, Riverside, BS, 71, MS, 72, PhD(math), 73; Mass Inst Technol, PhD(physics), 86. *Prof Exp:* Instr math, Mass Inst Technol, 73-75; asst prof, NMex State Univ, 75-79; ASST PROF MATH, TEX STATE UNIV, 86- *Mem:* Am Phys Soc; Am Math Soc; Soc Indust Appl Math; Am Astron Soc; AAAS. *Res:* Theoretical investigation of a strongly interacting many body model system using predominately numerical simulation experimental techniques. *Mailing Add:* 3012 22nd St Lubbock TX 79410

JOHNSON, DAVID, b Newark, NJ, Sept 3, 44. HIGH ENERGY PHYSICS. *Educ:* Univ Calif, Berkeley, AB, 66; Iowa State Univ, PhD(high energy physics), 72. *Prof Exp:* Res assoc & instr physics, Iowa State Univ, Ames Lab, USAEC, 67-72, assoc & instr physics, 72-73; PHYSICIST & UNIV RES ASSOC, FERMI NAT ACCELERATOR LAB, 73- *Mem:* AAAS; Sigma Xi; Am Phys Soc. *Res:* High energy accelarator design and research; high energy experimental research. *Mailing Add:* Acceleration Div PO Box 500 Fermi Nat Accelerator Lab Batavia IL 60510

JOHNSON, DAVID ALFRED, b Muskegon, Mich, Mar 13, 38; m 60; c 3. ELECTROCHEMISTRY, ENVIRONMENTAL CHEMISTRY. *Educ:* Greenville Col, AB, 60; La State Univ, PhD(chem), 66. *Prof Exp:* Chemist, Pet Milk Res Labs, summer 60; asst prof chem, Greenville Col, 62-64; instr, La State Univ, 64-65; PROF CHEM, SPRING ARBOR COL, 66- *Concurrent Pos:* Fel, Dept Chem, La State Univ, 70-71; Am Chem Soc-PFR fel, 85 & 86; NASA fel & Nat Aerospace Serv Asn fel, 80-82. *Mem:* Am Chem Soc. *Res:* Physical chemistry of electrolytes; five coordinate complexes of transition metals; thermodynamics of biological systems. *Mailing Add:* Dept of Chem Spring Arbor Col Spring Arbor MI 49283

JOHNSON, DAVID ASHBY, b Asheville, NC, Sept 6, 43; m 67; c 2. OCEANOGRAPHY. *Educ:* Mass Inst Technol, SB & SM, 66; Univ Calif, San Diego, PhD(oceanog), 71. *Prof Exp:* NSF fel oceanog, 71-72, from asst scientist to assoc scientist, 72-83, VIS INVESTR, WOODS HOLE OCEANOG INST, 83- *Mem:* AAAS; Geol Soc Am; Am Geophys Union. *Res:* Ocean floor processes; biostratigraphy of pelagic sediments; interactions of sea floor topography, abyssal circulation and bottom sediments. *Mailing Add:* Woods Hole Oceanog Inst Woods Hole MA 02543

JOHNSON, DAVID B, b Big Spring, Tex, Jan 11, 40; m 62; c 3. MECHANICS. *Educ:* Univ Tex, Austin, BSME, 63, MSME, 64; Stanford Univ, PhD(eng mech), 68. *Prof Exp:* Assoc prof mech eng, Southern Methodist Univ, 68-73; assoc prof eng sci & mech, Iowa State Univ, 75-; VPRES, PROD DEVEL, JY TAYLOR MFG; PROF SOUTHERN METHODIST UNIV, DALLAS, 91- *Mem:* Am Soc Mech Engrs; Am Soc Eng Educ; Soc Exp Stress Anal; Soc Eng Sci. *Res:* Dynamics; vibrations; space mechanics; phytomechanics. *Mailing Add:* 280 Colleen Rd Garland TX 75043

JOHNSON, DAVID BARTON, b Providence, RI, June 5, 46; m 70; c 3. BIO-ORGANIC CHEMISTRY, BIO-ANALYTICAL CHEMISTRY. *Educ:* Univ RI, BS, 69; Duke Univ, PhD(org chem), 75. *Prof Exp:* NIH fel biochem pharmacol, Med Sch, Duke Univ, 74-76; sr chemist bio-org chem, Midwest Res Inst, 76-80; RES SCIENTIST II, UPJOHN CO, 80- *Mem:* Am Chem Soc; Sigma Xi. *Res:* Bio-organic chemistry dealing in the synthesis, biosynthesis, analysis, and structural elucidation of xenobiotic metabolites; analysis of metabolites in biological samples; in vitro studies of xenobiotic metabolizing enzymes; radiochemical synthesis. *Mailing Add:* 313 Edinburgh Lawrence KS 66044

JOHNSON, DAVID EDSEL, b Chatham, La, Aug 16, 27; m 59; c 4. ELECTRICAL ENGINEERING, APPLIED MATHEMATICS. *Educ:* La Tech Univ, BS & BA, 49; Auburn Univ, MS, 52, PhD(math), 58. *Prof Exp:* Draftsman, La Power & Light Co, 49-50; mathematician, Nat Bur Standards, 52; assoc prof math, 54-62, prof elec eng, LA State Univ, 62-83; PROF MATH, BIRMINGHAM-SOUTHERN COL, 83- *Concurrent Pos:* NSF fac fel, Stanford Univ, 61-62. *Mem:* Sigma Xi. *Res:* Electric circuits and systems. *Mailing Add:* Div Sci & Math Birmingham-Southern Col Birmingham AL 35254

JOHNSON, DAVID FREEMAN, b Nashville, Tenn, Jan 28, 25; m 47; c 2. BIOCHEMISTRY. *Educ:* Allegheny Col, BS, 47; Howard Univ, MS, 49; Georgetown Univ, PhD(biochem), 57. *Hon Degrees:* DSc, Allegheny Col, 72. *Prof Exp:* Instr chem, Howard Univ, 49-50; res chemist, Freedman's Hosp, Washington, DC, 50-52; res chemist, 52-71, CHIEF, LAB ANALYTICAL CHEM, NAT INST ARTHRITIS & METAB DISEASES, 71- *Concurrent Pos:* Instr, USDA Grad Sch, 58-60 & Found Advan Educ in Sci, Inc, NIH, 60- *Mem:* AAAS; Am Chem Soc; Fedn Am Socs Exp Biol; Am Phys Soc. *Res:* Hormones; plant and animal steroids; metabolism; analytical methods; partition chromatography. *Mailing Add:* Rm B2A-17 Bldg 8 Nat Inst Diabetes Dig & Kidney Dis Bethesda MD 20892

JOHNSON, DAVID GREGORY, b Belvidere, Ill, July 11, 40; m 65; c 3. ENDOCRINOLOGY, CLINICAL PHARMACOLOGY. *Educ:* Yale Univ, BA, 62; Dartmouth Med Sch, BMed Sci, 64; Harvard Univ, MD, 67. *Prof Exp:* Resident, Univ Calif, San Francisco, 67-69; res assoc, NIH, 69-71; fel, Univ Wash, 71-73, from asst prof to assoc prof, 73-78; assoc prof, 78-82, PROF, DEPT INTERNAL MED, HEALTH SCI CTR, UNIV ARIZ, 82- *Concurrent Pos:* Assoc ed, Life Sci, 81-82. *Mem:* Am Diabetes Asn; Am Soc Pharmacol & Exp Therapeut; Endocrine Soc; Am Fedn Clin Res. *Res:* Experimental and clinical research regarding diabetes, pancreatic endocrine secretion, gastro intestinal hormones and catecholamine physiology; development and testing of drugs, particularly for the treatment of diabetes. *Mailing Add:* Dept Internal Med Univ Ariz Health Sci Ctr Tucson AZ 85724

JOHNSON, DAVID HARLEY, b Brooklyn, NY, May 31, 41; m 61; c 2. HEAT TRANSFER, FLUID MECHANICS. *Educ:* Purdue Univ, BS, 63, MS, 64; Cornell Univ, PhD(appl physics), 75. *Prof Exp:* Staff mem, Sandia Corp, 64-67; adj instr hydraul, Tompkins-Cortland Community Col, 69-70; teaching asst statist thermodyn, Cornell Univ, 70-72; sr staff physicist, Appl Physics Lab, Johns Hopkins Univ, 73-79, asst group leader, 78-79; PRIN ENGR, SOLAR ENERGY RES INST, 79-, GROUP MGR, 80- *Concurrent Pos:* Consult, Appl Physics Lab, Johns Hopkins Univ, 79-80 & Flow Industs Inc, 81- *Mem:* Am Soc Mech Engrs; Sigma Xi. *Res:* Dynamics of stratified fluids in the ocean and in solar ponds; direct contact heat transfer phenomena important to the design of open-cycle thermal energy conversion power plants and other heat exchangers. *Mailing Add:* 9658 Master Works Dr Vienna VA 22180

JOHNSON, DAVID LEE, b Benson, Minn, Apr 28, 46; m 76; c 1. SOFTWARE SYSTEMS. *Educ:* Univ Minn, BS, 64; Syracuse Univ, MS, 71; Univ Minn, PhD(math), 76. *Prof Exp:* Asst prof math, Univ Southern Calif, 78-80 & Univ Ark, 80-81; sr mem tech staff, El Sequndo, 81-84; asst dept mgr, Radar Systs Group, 84-85; dept mgr, Hughes Aircraft Co, 85-86; dept software eng mgr, Rockville, 86-87; dept mgr, Nat Ctr Systs Directorate, 87-88; eng orgn mgr, 88-90, DEPT PROG MGR, GTE GOVT SYSTS, 90- *Concurrent Pos:* John Wesley Young res instr, Dartmouth Col, 76-78; mem tech staff, Logicon Inc, 80-81. *Mem:* Am Math Soc; Math Asn Am; Inst Elec & Electronic Engrs. *Res:* Functional analysis dealing with distribution theory; continuous group representations on general locally compact groups. *Mailing Add:* 3303 Lowman Lane Linwood MD 21764

JOHNSON, DAVID LEROY, b Truman, Ark, Sept 10, 47. SOLAR ENERGY. *Educ:* Ark State Univ, BS, 69; Univ Ill Urbana-Champaign, MS, 71, PhD(physics), 78. *Prof Exp:* Fel dept physics, Univ Ill, 78-80; PRIN INVESTR, CONSTRUCTION ENG RES LAB, CORPS ENGRS, 80- *Concurrent Pos:* Adj res asst prof physics, Univ Ill, 81- *Mem:* Am Soc Heating, Refrigerating & Airconditioning Engrs; Int Solar Energy Soc. *Res:* Experimental studies of fundamental defect properties in solids; solar energy systems for space heating and cooling of buildings; heating, ventilating and air conditioning controls and energy conservation techniques for heating, ventilating and air conditioning systems. *Mailing Add:* M/A Com Govt Systs 3033 Science Park Rd San Diego CA 92121

JOHNSON, DAVID LINTON, b Chicago, Ill, July 9, 45; div; c 2. THEORETICAL SOLID STATE PHYSICS. *Educ:* Univ Notre Dame, BS, 67; Univ Chicago, MS, 69, PhD(physics), 74. *Prof Exp:* Fel physics, Michelson Lab, Naval Weapons Ctr, 72-74; fel, Ames Lab, Iowa State Univ, 74-76; asst prof physics, Northeastern Univ, 76-79; res physicist, 79-88, SCI ADV, SCHLUMBERGER DOLL RES CTR, 88- *Concurrent Pos:* Consult, GTE Labs, 79. *Mem:* fel Am Phys Soc; Acoust Soc Am; Soc Exp Geol; Am Geophys Union. *Mailing Add:* Schlumberger Doll Res Ctr Old Quarry Rd Ridgefield CT 06877-4108

JOHNSON, DAVID NORSEEN, b Bronx, NY, Sept 28, 38; m 60; c 2. PHARMACOLOGY. *Educ:* N Park Col, Ill, BS, 60; Univ Louisville, Ky, MS, 67; Med Col Va, PhD(pharmacol), 76. *Prof Exp:* Res scientist pharmacol, 66-69, sr res scientist, 69-75, MGR NEUROPHARMACOL, A H ROBINS PHARMACEUT CO, 76- *Mem:* Am Soc Pharmacol & Exp Therapeut; Am Chem Soc; Soc Neurosci. *Res:* Basic mechanisms underlying mental illness and the methodologies for testing new chemical entities for treating disorders. *Mailing Add:* A H Robins Res Labs 1211 Sherwood Ave Richmond VA 23261-6609

JOHNSON, DAVID RUSSELL, b Manaus, Brazil, Oct 23, 45; m 67; c 2. PHYSICAL CHEMISTRY. *Educ:* Austin Col, BA, 67; Tex Christian Univ, PhD(chem), 70. *Prof Exp:* Fel radiation chem, Baylor Univ, 70-72 & Univ Fla, 72-73; res chemist textile fibers, E I du Pont de Nemours & Co Inc, Waynesboro, Va, 73-75; res chemist separations chem, 75-78, staff chemist, 78-79; res supvr anal chem, 79-81; res supvr hydrogen technol, Savannah River Lab, 81-82; chief supvr tritium technol, 82-84; tech supt pretrochemicals, Cape Fear Plant, 84-85; proj liaison leader, Petrochemicals-AED, 85-88, SR RES ASSOC, DU PONT CHEM, FREON RES & DEVELOP, E I DU PONT DE NEMOURS & CO INC, WILMINGTON, DEL, 88- *Mem:* Am Chem Soc. *Res:* Plutonium soil migration studies; environmental dose-to-man modelling methods; uranium fuel fabrication methods; chemical separations processes for nuclear fuel recycle and waste management programs; process development for CFC alternatives. *Mailing Add:* 718 Burnley Rd Wilmington DE 19803

JOHNSON, DAVID SIMONDS, b Porterville, Calif, June 29, 24; wid. METEOROLOGY. *Educ:* Univ Calif, Los Angeles, AB, 48, MA, 49. *Prof Exp:* Meteorol aid, US Weather Bur, 46-47; asst meteorol, Univ Calif, Los Angeles, 48-52; assoc meteorologist, Pineapple Res Inst, Honolulu, Hawaii, 52-56; chief observ test & develop ctr, US Weather Bur, 56-58, asst chief meterol satellite lab, 58-60, chief, 60-62, from dep dir to dir, Nat Weather Satellite Ctr, 62-65, dir, Nat Environ Satellite Ctr, Environ Sci Serv Admin, 65-70; dir, Nat Environ Satellite Serv, 70-80, asst admin satellites, Nat Oceanic & Atmospheric Admin, 80-82; spec asst to pres, Univ Corp Atmospheric Res, 82-83; pres, Damar Int, Inc, 84-86; STUDY DIR, NAT RES COUN, 86- *Concurrent Pos:* Consult to secy gen, World Meteorol Orgn, 82-86. *Honors & Awards:* Gold Medal, Dept Com, 65; Exceptional Service Medal, NASA, 66; Fed Career Serv Award for Sustained Excellence, Nat Civil Serv League, 74; William T Pecora Award, NASA & Dept Interior, 78; US Presidential Meritorious Exec Award, 80; Achievment Award, Am Astronaut Soc, 81; Brooks Award, Am Meteorol Soc, 82; Silver Medal, Dept Com, 85; Group Award, Nat Res Coun, 87. *Mem:* AAAS; fel Am Meteorol Soc (pres, 74); fel Am Geophys Union; Int Acad Astronaut; Sigma Xi; assoc fel, Am Inst Aeronaut & Astronaut; fel Am Astronaut Soc. *Res:* Meteorological instruments and observing techniques; environmental satellites. *Mailing Add:* 1133 Lake Heron Dr Apt 3A Annapolis MD 21403

JOHNSON, DAVID W, JR, b Windber, Pa, Sept 23, 42; m 64; c 2. CERAMIC PROCESSING, ELECTRONIC CERAMICS. *Educ:* Pa State Univ, BS, 64, PhD(ceramic sci), 68. *Prof Exp:* Mem tech staff, Bell Tel Labs, 68-83; supvr, 84-88, DEPT HEAD, AT&T BELL LABS, 88- *Concurrent Pos:* Adj prof mat sci, Stevens Inst Technol, 82- *Honors & Awards:* Ross Coffin Purdy Award, Am Ceramic Soc, 81 & Fulroth Award, 84; Taylor Lectr, Pa State Univ, 87. *Mem:* Am Ceramic Soc (vpres, 90-92); AAAS; Am Soc Metals; Mat Res Soc; Metall Soc. *Res:* Dielectric relaxation in doped strontium titanate; characterization of fine oxide particles; magnetic ceramics; ionic conductors; sol gel glasses; oxide superconductors. *Mailing Add:* AT&T Bell Labs Rm 6D-321 Murray Hill NJ 07974

JOHNSON, DELWIN PHELPS, b Rocky Mountain, NC, Mar 24, 26; m 54; c 2. ANALYTICAL CHEMISTRY. *Educ:* NC State Univ, BS, 48. *Prof Exp:* Chemist, NC Dept Agr, 48-57; res chemist, Chem Div, Union Carbide Corp, 57-63, res chemist, Olefins Div, 63-64; res staff chemist, R J Reynolds Tobacco Co, 64-69, group leader, New Prod Develop Sect, 69-70, head sect, 70-76, mgr new prod develop, 76-87; RETIRED. *Mem:* Am Chem Soc. *Res:* Development of new and novel micro-analytical techniques for trace constituents in complex mixtures; experimental and commercial pesticide residues in animal and plant products, utilizing ultraviolet, infrared and visible spectrophotometry, gas and liquid phase chromatography. *Mailing Add:* 3021 Saint Claire Rd Winston Salem NC 27106

JOHNSON, DENNIS ALLEN, b Pocatello, Idaho, May 10, 49; m 71; c 3. PLANT PATHOLOGY. *Educ:* Brigham Young Univ, BS, 74; Univ Minn, MS, 75, PhD(plant path), 78. *Prof Exp:* Res & teaching asst bot, Brigham Young Univ, 72-74; teaching asst mycol, Univ Minn, 75-76, res asst plant path, 74-78; asst prof plant path, Tex Agr Exp Sta, 78-80; WASHINGTON STATE UNIV, 80- *Concurrent Pos:* Mem, Hard Red Winter Wheat Improv Comt, USDA, 78- *Mem:* Am Phytopath Soc; Mycol Soc Am; Sigma Xi. *Res:* Disease resistance and epidemiology of small grain diseases. *Mailing Add:* 1421 Scenic Dr Prosser WA 99350

JOHNSON, DENNIS DUANE, b Can, Mar 11, 38; c 4. NEUROSCIENCES. *Educ:* Univ Sask, BSP, 60, MSc, 62; Univ Wash, PhD(pharmacol), 65. *Prof Exp:* From asst prof to assoc prof, 66-75, PROF PHARMACOL, UNIV SASK, 75-, ASST DEAN, COL MED, 81-, HEAD, DEPT PHARMACOL, 86- *Concurrent Pos:* Lectr pharmacol, Univ Sask, 65-66; counr, Med Res Coun Can, 87-; mem adv comt, Pharmaceut Mfrs Asn Can, 88-; chmn, bd dirs, Toxicol Res Ctr, Saskatoon, 89- *Mem:* Nat Cancer Inst Can; Am Soc Neurosci; Am Soc Pharmacol & Exp Therapeut; Pharmacol Soc Can. *Res:* Pharamcology and biochemistry of epilepsy; neurochemistry of neuropharmacology. *Mailing Add:* Dept Pharmacol Univ Sask Saskatoon SK S7N 0W0 Can

JOHNSON, DEWAYNE CARL, b Minneapolis, Minn, Sept 15, 35. PHYSICS. *Educ:* Univ Minn, Minneapolis, BS, 57, MS, 60, PhD(elec eng), 63. *Prof Exp:* Mem tech staff physics, Bell Tel Labs, 64-65; asst prof, 65-69, ASSOC PROF PHYSICS, UNIV WIS-MILWAUKEE, 70- *Mem:* Am Phys Soc. *Res:* Low energy electron diffraction. *Mailing Add:* Dept Physics Univ Wis Milwaukee WI 53201

JOHNSON, DEWEY, JR, b Sapulpa, Okla, Sept 23, 26; m 53; c 4. BIOCHEMISTRY, NUTRITION. *Educ:* Colo State Univ, BS, 50; Univ Conn, MS, 55; Rutgers Univ, PhD, 58; Nat Registry Clin Chemists. *Prof Exp:* Asst poultry nutrit, Rutgers Univ, 55-58; assoc animal nutrit, Lime Crest Res Lab, Limestone Prod Corp Am, NJ, 58-62; nutritionist, Food & Drug Res Lab, 62-63; biochemist, Equitable Live Assurance Soc US, New York, 63-68, dir clin lab, 68-79; supv anal chemist, 81-89, UNDERWRITER, METROPOLITAN LIFE INS CO, NEW YORK, 89- *Mem:* Am Soc Animal Sci; Am Dairy Sci Asn; Poultry Sci Asn. *Res:* Metabolism of amino acids; metabolism of drugs; biochemical changes in alcoholism; automated clinical chemistry techniques; folic acid and vitamin B12 metabolism. *Mailing Add:* 12 Barbara Pl Edison NJ 08817

JOHNSON, DONAL DABELL, b Rigby, Idaho, July 20, 22; m 45; c 3. SOILS, MICROBIOLOGY. *Educ:* Brigham Young Univ, BS, 48; Cornell Univ, MS, 50, PhD(soils), 52. *Prof Exp:* Asst agron, Cornell Univ, 48-51; from asst prof to assoc prof, 52-62, coordr, Nigeria Proj, 64-69, PROF AGRON, COLO STATE UNIV, 62-, DEAN COL AGR SCI, 68-, ASSOC & DEP DIR EXP STA, 69- *Concurrent Pos:* Trustee & chmn, Consortium Int Develop, 74-; chmn, Great Plains Agr Coun, 77. *Mem:* Sigma Xi; Fel AAAS; Am Soc Agron; Soil Sci Soc Am. *Res:* Nitrogen transformations in soil. *Mailing Add:* 1812 Orchard Pl Ft Collins CO 80521

JOHNSON, DONALD CHARLES, b Black River Falls, Wis, Jan 30, 27; m 52; c 1. ENDOCRINOLOGY. *Educ:* Univ Wis, BS, 49; Univ Iowa, MS, 50, PhD(zool), 56. *Prof Exp:* Asst zool, Univ Iowa, 53-56, res assoc, 56-58, res asst prof, 59-63; from asst prof to assoc prof, 63-69, PROF OBSTET, GYNEC & PHYSIOL, SCH MED, UNIV KANS, 69-, RES PROF HUMAN REPRODUCTION, 78- *Mem:* AAAS; Endocrine Soc; Am Physiol Soc; Soc Gynec Invest; Soc Study Reprod; Sigma Xi. *Res:* Reproductive physiology and endocrinology; comparative physiology of gonadotrophins; control gonadal steroidogenic enzymes. *Mailing Add:* Dept Obstet & Gynec Univ Kans Med Ctr Kansas City KS 66103

JOHNSON, DONALD CURTIS, b Minneapolis, Minn, Mar 21, 35; m 56; c 3. ORGANIC CHEMISTRY. *Educ:* Hamline Univ, BS, 57; Univ Minn, PhD(org chem), 62. *Prof Exp:* Res aide org chem, 61-67, res assoc & chmn dept chem, Inst Paper Chem, 67-77 prof org chem, 70-77; SCIENTIFIC SPECIALIST FIBER CHEM, WEYERHAEUSER CO, 77- *Concurrent Pos:* Chmn, Gordon Res Conf Chem & Physics of Paper, 72-74. *Mem:* Am Chem Soc; Tech Asn Pulp & Paper Indust. *Res:* Cellulose chemistry, including reactions in solution and mechanisms of chain degradation; lignin chemistry, particularly delignification processes with selective oxidants. *Mailing Add:* 33936 134th Ave SE Auburn WA 98002

JOHNSON, DONALD ELWOOD, b Joliet, Ill, July 23, 35; m 56; c 2. COMPUTER SCIENCE, SOFTWARE SYSTEMS. *Educ:* N Cent Col, BA, 57; Univ Wis-Madison, MS, 59; Ill Inst Technol, PhD(math), 73. *Prof Exp:* Asst mathematician, Argonne Nat Lab, 59-61; from instr to assoc prof, NCent Col, 61-78, chairperson math, 69-73 & 75-78, prof math, 78-82, chairperson, Div Natural Sci & Math, 78-83, prof & chairperson computer sci, 82-88, PROF COMPUTER SCI, NCENT COL, 82- *Mem:* Math Asn Am; Asn Comput Mach; Am Asn Univ Prof. *Mailing Add:* Dept Computer Sci N Cent Col PO Box 3063 Naperville IL 60566-7063

JOHNSON, DONALD EUGENE, b Sykeston, NDak, Nov 17, 38; m 61; c 3. ANIMAL NUTRITION. *Educ:* NDak State Univ, BS, 60, MS, 63; Colo State Univ, PhD(animal nutrit), 66. *Prof Exp:* Res asst animal nutrit, NDak State Univ, 61-63 & Colo State Univ, 63-66; res assoc, Cornell Univ, 66-68; asst prof ruminant nutrit, Univ Ill, Urbana, 68-72; assoc prof, 72-80, PROF ANIMAL NUTRIT & DIR METAB LAB, COLO STATE UNIV, FOOTHILLS CAMPUS, 80-, PROF ANIMAL SCI. *Mem:* AAAS; Am Soc Animal Sci; Am Dairy Sci Asn. *Res:* Animal energy metabolism. *Mailing Add:* 3416 Arapahoe Dr Ft Collins CO 80521

JOHNSON, DONALD GLEN, b Detroit, Mich, Jan 29, 31; m 53; c 2. MATHEMATICS. *Educ:* Albion Col, AB, 53; Mich State Univ, MS, 57; Purdue Univ, PhD(math), 59. *Prof Exp:* From asst prof to assoc prof math, Pa State Univ, 59-65; from assoc prof to prof, 65-88, EMER PROF MATH, NMEX STATE UNIV, 88- *Mem:* Am Math Soc; Math Asn Am; Nat Coun Teachers Math. *Res:* Lattice ordered rings; rings of continuous functions. *Mailing Add:* PO Box 1134 Glen Rock NJ 07452

JOHNSON, DONALD L(EE), b Denver, Colo, Feb 19, 27; m 47; c 4. METALLURGICAL ENGINEERING. *Educ:* Colo Sch Mines, MetE, 50, MS, 56; Univ Nebr, Lincoln, PhD(chem eng), 68. *Prof Exp:* Trainee, Allis Chalmers Mfg Co, 50-51; metall engr, Mine & Smelter Supply Co, 53-56; asst prof metall eng, Wash State Univ, 56-59; sr metallurgist, NAm Rockwell Corp, 60-63; assoc prof metall, 63-75, PROF MECH ENG, METALL PROG, UNIV NEBR, LINCOLN, 75- *Concurrent Pos:* Consult, Brunswick Corp, 69-; Univ Nebr Res Coun-NASA-Ames Res Ctr fel, Univ Nebr, Lincoln, 71-72. *Mem:* Am Soc Metals; Nat Asn Corrosion Engrs. *Res:* Gas-metal equilibria; hydrogen transport in metallic alloys; polarization analysis of corrosion in aqueous systems; leaching kinetics. *Mailing Add:* 2610 S 60th St No 15 Lincoln NE 68506-3528

JOHNSON, DONALD R, b McPherson, Kans, Apr 1, 30; m 53; c 3. METEOROLOGY. *Educ:* Bethany Col, BS, 52; Univ Wash, BS, 53; Univ Wis, MS, 60, PhD(meteorol), 65. *Prof Exp:* From proj asst to proj assoc, 59-64, from asst prof to assoc prof, 64-66, chmn dept, 73-76, PROF METEOROL, UNIV WIS-MADISON, 70-, ASSOC DIR, SPACE SCI & ENG CTR, 77- *Concurrent Pos:* Vis assoc prof, Pa State Univ, 68-69; chief ed, Monthly Weather Review, 77-80. *Mem:* Nat Weather Asn; Am Meteorol Soc; Am Geophys Union. *Res:* Dynamic climatology and meteorology; secondary and general circulation studies. *Mailing Add:* Dept Meteorol & Space Sci Bldg Univ Wis W Dayton St Madison WI 53706

JOHNSON, DONALD RALPH, b Newport, Wash, Aug 18, 31; m 55; c 3. VERTEBRATE ECOLOGY. *Educ:* Univ Idaho, BS, 53, MS, 58; Colo State Univ, PhD(wildlife ecol), 62. *Prof Exp:* From asst prof to assoc prof biol, Ft Lewis Col, 61-65; assoc prof, Minot State Col, 65-68; assoc prof, 68-75, PROF BIOL, UNIV IDAHO, 75- *Mem:* Am Soc Mammal; Sigma Xi. *Res:* Small mammal ecology; effects of 2, 4-D on rodent food habits; energy relations of pikas; diets of sympatric lizards; osprey ecology. *Mailing Add:* Dept Biol Sci Univ Idaho Moscow ID 83843

JOHNSON, DONALD REX, b Tacoma, Wash, July 19, 38; m 59; c 2. MOLECULAR SPECTROSCOPY, RADIO ASTRONOMY. *Educ:* Univ Puget Sound, BS, 60, Univ Idaho, MS, 62; Univ Okla, PhD(physics), 67. *Prof Exp:* Physicist, Nat Bur Standards, 67-76, prog analyst, 76-78, dep dir progs, 78-80, dep dir resources & opers, 80-82, actg dir, Nat Measurements Lab, 82-90; DIR, NAT MEASUREMENT LAB & TECHNOL SERVS, NAT INST STANDARDS & TECHNOL, 90- *Concurrent Pos:* Technol mgt, technol transfer, state and local economic develop. *Mem:* Am Phys Soc; Am Astron Soc; AAAS; Am Soc Testing & Mats; Int Astron Union. *Res:* Molecular radio astronomy; microwave spectroscopy of free radicals and transient chemical species in the gas phase. *Mailing Add:* Dept Com Technol Serv Nat Inst Standards & Technol Gaithersburg MD 20899

JOHNSON, DONALD RICHARD, b Duluth, Minn, Jan 15, 29; m 56; c 2. CLINICAL CHEMISTRY, POLYMER SCIENCE. *Educ:* Univ Minn, BA, 49; Univ Wis, PhD(anal chem), 54. *Prof Exp:* Chemist, Mat Packaging Sect, Forest Prod Lab, USDA, 52-53; res chemist, Polychem Dept, Res & Develop Div, 53-59, res supvr, 59-62, prod mgr, Instrument Prod Div, 62-65, res mgr, 65-71, mgr res & eng, Photo Prod Dept, 71-74; mgr, New Prod Scouting Res Div, 75-80; dir, New Prod Res & Develop, Clin Systs, 80-83; dir, New Technol Res, Diag & Biores Prod Div, Du Pont Biomed Prod Dept, 83-86; PRES, TECHNOL CONVERSION, 86-; CONSULT, NEW PROD RES & DEVELOP, 86- *Concurrent Pos:* Chmn, Gordon Res Conf Anal Chem, 77; Du Pont Corp Planning Group Life Sci Res, 80-81; chmn, indust adv bd, Ctr Biopolymers at Interfaces, Dept Bioeng, Univ Utah. *Honors & Awards:* IR-100 Award, Automatic Clin Anal, 79, Automated Sample Processor, Du Pont, 79. *Mem:* AAAS; Am Chem Soc; Sigma Xi; Am Asn Clin Chemists. *Res:* Chemical instrumentation; thermal analysis; clinical, analytical, physical and polymer chemistry; infrared spectroscopy; immunodiagnostics and biomaterials. *Mailing Add:* 1005 S Hilton Rd Wilmington DE 19803

JOHNSON, DONALD ROSS, b Chicago,Ill, Feb 9, 20; m 47; c 4. URBAN ENTOMOLOGY, MOSQUITO CONTROL. *Educ:* Univ Ill, BSc, 43; Univ Minn, MSc, 50. *Prof Exp:* Lieutenant malaria control, US Navy, 43-46; asst, insecticide res, dept entom, Univ Minn, 46-48; asst state entomologist insect surv & control, Minn Dept Agr, 48-51; malaria advisor, malaria control, US Econ Coop Admin, US Dept State, Indonesia, 51-53; med entomologist, Pub Health Admin, Div Int Health, USPHS, 53-57; dep chief malaria eradication, USAID, 57-64; chief spec servs, Aedes aegypti eradication, Communicable Dis Ctr, USPHS, 64-66, dir sanit, Malaria Eradification Prog, Ctrs Dis Control, 66-73; RETIRED. *Concurrent Pos:* Captain, USPHS, Foreign Quarantine Prog, US Army, Vietnam, 69-70; malaria res advisor, USAID, Indonesia, 74; entomologist & consult, dis, vectors & pest control, 74-; vis consult & lectr, Int Ctr Pub Health Res, Univ SC, 82- *Honors & Awards:* Incentive Award, USPHS, 57; Meritorious Serv Award, Armed Forces Pest Control Bd, 73, Am Mosquito Control Asn, 74 & 82. *Mem:* Am Mosquito Control Asn; Entom Soc Am. *Res:* Mosquitoes, malaria and control of arthropods of public health importance, domestic and international. *Mailing Add:* 1362 N Decatur Road N E Atlanta GA 30306

JOHNSON, DONALD W, b Worthington, Minn, May 4, 29; m 50; c 2. VETERINARY MEDICINE. *Educ:* Univ Minn, BS, 51, DVM, 53, PhD(microbiol), 63. *Prof Exp:* From instr to assoc prof vet med & clins, Col Vet Med, Univ Minn, 55-67; prof vet med, Univ Mo-Columbia, 67-69; PROF VET MED, UNIV MINN, ST PAUL, 69-, DIR GRAD STUDY, 80- *Res:* Viral and bacterial respiratory diseases of cattle and horses; host response to infectious diseases; cell mediated immune response of the bovine. *Mailing Add:* Dept Vet Med 225 Vet Teaching Hosp 1365 Gortner Ave St Paul MN 55108

JOHNSON, DONOVAN EARL, b Holdrege, Nebr, June 26, 42; m 65; c 2. MICROBIOLOGY, BIOCHEMISTRY. *Educ:* Univ Nebr, BS, 64, MS, 66; Univ Wis-Madison, PhD(microbiol), 72. *Prof Exp:* From res asst to res assoc microbiol, 66-74, proj leader microbiol, Northern Regional Res Lab, 74-81, RES MICROBIOLOGIST, US GRAIN MKT RES LAB, AGR RES SERV, USDA, 81- *Concurrent Pos:* Adj prof chem, Bradley Univ, 74. *Mem:* Sigma Xi; Am Soc Microbiol; Soc Invert Path. *Res:* Biological insecticides; microbiology of insect pathogens; physiology of bacterial sporulation. *Mailing Add:* 1515 College Ave Manhattan KS 66502

JOHNSON, DOUGLAS ALLAN, b Montevideo, Minn, Dec 6, 49; m 72; c 3. PLANT PHYSIOLOGY, RANGE ECOLOGY. *Educ:* Augustana Col, SDak, BA, 71; Utah State Univ, MS, 73, PhD(range ecol), 75. *Prof Exp:* Res asst tundra plant water relations, Dept Range Sci, Utah State Univ, 71-75; res assoc, Dept Biol, Augustana Col, 75-76; PLANT PHYSIOLOGIST RANGE PLANT IMPROV, FORAGE & RANGE RES LAB, AGR RES SERV, USDA, 76- *Concurrent Pos:* NSF grant, 75-76, USDA grant, 79-81, US-Spain grant, 84-88; Nat Defense Educ Act fel, 71-73; Commonwealth Sci & Indust Res Orgn Australia vis scientist award, 81-82; USDA-Coop State Res Serv grant, 85-88 & 87-89; grant, Western Regional IPM, 85-88, USDA-OICD, 91. *Mem:* Crop Sci Soc Am; Am Soc Agron; Soc Range Mgt. *Res:* Development of superior forage plants for the Intermountain West; defining physiological basis of range plant resistance to drought stress; nitrogen fixation in range plant species. *Mailing Add:* USDA-ARS Forage & Range Res Lab Utah State Univ Logan UT 84322-6300

JOHNSON, DOUGLAS L, b Minneapolis, Minn, Jan 21, 25; m 48; c 3. CHEMICAL ENGINEERING. *Educ:* Univ Wis-Madison, BS, 47, MS, 48, PhD(chem eng), 56. *Prof Exp:* Engr chem develop, Procter & Gamble Co, Ohio, 48-51; engr res & develop chem, F G Findley Co, Wis, 51-52; instr chem eng, Univ Wis-Madison, 52-56; sr process chem engr, Courtaulds Inc, Ala, 56-58; group leader fiber develop, Fla, 58-59, develop supvr acrylic fibers, 59-63, develop mgr, 63-68, TECH DIR, AM CYANAMID, NJ, 68- *Mem:* Am Inst Chem Engrs. *Res:* Kinetics of gas-liquid reactions in stirred batch reactors; hydrogenation of vegetable oils; adhesion to cold metallic surfaces; cellulose chemistry; extrusion and orientation of cellulose fibers; acrylic polymer chemistry; polymer solutions; acrylic fibers. *Mailing Add:* 701 Rivenwood Dr Franklin Lakes NJ 07417

JOHNSON, DOUGLAS WILLIAM, b Marion, Ind, Sept 15, 53; m 75; c 2. REMOTE SINSING, ATMOSPHERIC SPECTROSCOPY. *Educ:* Rensselaer Polytech Inst, BS, 75; Univ Fla, PhD(astron), 80. *Prof Exp:* Teaching asst physics, dept physics, Univ Fla, 75-80; teaching res assoc,

Battelle Northwest, 80-82, res scientist, 83-84, sr res scientist, 84-87, PRIN RES SCIENTIST, BATTELLE MEM INST, 87- *Mem:* Am Astron Soc; AAAS; Am Geog Union; Sigma Xi; NY Acad Sci; Am Soc Photogammetry & Remote Sensing. *Res:* Integration of remotely sensod imaging data of diverse types using imaging sattuore to aid in interpretation and understanding. *Mailing Add:* 840 Memorial Dr Cambridge MO 02139

JOHNSON, DUDLEY PAUL, b Burbank, Calif, Sept 22, 40; m 64; c 2. MATHEMATICS, MATHEMATICAL STATISTICS. *Educ:* Yale Univ, BA, 62; Mass Inst Technol, PhD(math), 66. *Prof Exp:* asst prof math, Univ Calif, 66-71; from asst prof to assoc prof, 71-90, PROF MATH, UNIV CALGARY, 90- *Res:* Stochastic processes. *Mailing Add:* 16 Varsplain Pl NW Calgary AB T3A 0A8 Can

JOHNSON, E(WELL) CALVIN, b Tampa, Fla, Apr 18, 26; div; c 1. ELECTRICAL ENGINEERING. *Educ:* Ga Inst Technol, BEE, 47; Mass Inst Technol, SM, 49, EE, 50, ScD(elec eng), 51. *Prof Exp:* Res asst, Mass Inst Technol, 47-51; sr engr, Res Labs Div, Bendix Corp, 51-54, proj engr, 54-56, supvry engr, 56-58, head comput dept, 58-62, mgr info & control systs lab, 62-65, asst gen mgr, 65-67, vpres res & dir labs div, 67-69, vpres eng & res, 69-73; vpres res & develop, Gould, Inc, 73-75; consult, 75-80; pres, Vincent Corp, 80-85, consult, 85-87; dir res & develop, Aerosonic Corp, 88-89; CORP STAFF, UBC, INC, 89- *Mem:* Fel Inst Elec & Electronics Engrs. *Res:* Feedback control systems; analog and digital computers; machine-tool control; photogrammetric instruments; aerospace information and control systems; automotive electronics; industrial automation systems. *Mailing Add:* UBC Inc PO Box 18751 Tampa FL 33679

JOHNSON, E(DWARD) O, physical electronics; deceased, see previous edition for last biography

JOHNSON, EARNEST J, b Philipsburg, Pa, Feb 23, 31; m 56; c 8. OPTICS. *Educ:* Pa State Univ, BS, 53; Purdue Univ, MS, 54, PhD(physics), 64. *Prof Exp:* Staff mem, NAm Aviation, Inc, 55-56 & Hughes Aircraft Co, 56-58; mem res staff, Lincoln Lab, Mass Inst Technol, 64-74; mem staff, GTE Lab, Waltham, Mass, 74-; AT MOTOROLA INC. *Mem:* Optical Soc Am; Am Phys Soc. *Res:* Study of band structure of solids by observation of optical absorption and luminescence and effects of doping, magnetic fields and strains; laser materials; quantum electronics; fiber optic subsystems. *Mailing Add:* 1414 N Hibbert Mesa AZ 85201

JOHNSON, EDGAR GUSTAV, b St Cloud, Minn, Sept 16, 22; m 49; c 5. OPTICS. *Educ:* Univ Minn, BS, 47, MS, 49. *Prof Exp:* Res asst physics, Univ Minn, Minneapolis, 47-49; technologist-physicist, Inst Nuclear Studies, Chicago, 49; from physicist to sr physicist, Minn Mining & Mfg Co, 49-58, physics specialist, 58-68, sr physics specialist, 68-81; RETIRED. *Mem:* Optical Soc Am. *Res:* Photometry; geometric optics; retroreflective optics; vacuum vapor deposition. *Mailing Add:* 2676 E 19th Ave North St Paul MN 55109

JOHNSON, EDWARD MICHAEL, b Kenosha, Wis, Apr 9, 45. CHROMOSOME STRUCTURE, GENE REGULATION. *Educ:* Pomona Col, BA, 67; Yale Univ, PhD(pharmacol), 71. *Prof Exp:* Fel, Rockefeller Univ, 71-73; from asst prof to assoc prof cell biol, 75-85; res assoc, Sloan-Kettering Cancer Ctr, 73-75; PROF MOLECULAR BIOL & PATH, MT SINAI SCH MED, 85- *Concurrent Pos:* Adj prof genetics, Cornell Grad Sch Med Sci, 79. *Honors & Awards:* Fac Res Award, Am Cancer Soc, 82-87. *Mem:* Am Soc Cell Biol; Am Soc Pharmacol & Exp Therapeut; NY Acad Sci; Am Soc Biochem & Molecular Biol. *Res:* Regulation of gene expression in higher organisms; structure and chromosomal organization of individual genes, including ways in which hormones and other developmental signals regulate gene activity during development; control of DNA replication. *Mailing Add:* Brookdale Ctr for Molecular Biol Mt Sinai Sch Med One Gustave L Levy Pl New York NY 10029

JOHNSON, EDWIN WALLACE, b New Ulm, Minn, May 2, 23; m 55; c 2. PHYSICAL CHEMISTRY. *Educ:* Harvard Univ, BS, 44, MA, 48, PhD(phys chem), 50. *Prof Exp:* Res phys chemist, 49-55, adv phys chemist, 55-74, ADV ENG, METALL DEPT, WESTINGHOUSE ELEC CORP, 74- *Mem:* Am Chem Soc; Electrochem Soc; Am Inst Mining, Metall & Petrol Eng. *Res:* Molecular state of carboxylic acid vapors; solubility and diffusivity of hydrogen in metals; hydrogen embrittlement of steels; process metallurgy of titanium base alloys; vacuum arc melting; low pressure arc phenomena; properties of thermoelectric materials; measurement of thermal conductivity of liquid semiconductors at high temperatures; coated-electrode welding of austenitic high temperature alloys. *Mailing Add:* 3388 MacArthur Rd Murrysville PA 15668

JOHNSON, EINER WESLEY, JR, b Bemidji, Minn, July 5, 19; m 51; c 4. ORTHOPEDIC SURGERY. *Educ:* Univ Minn, BA, 41, BS, 42, BM, 44, MD, 45, MA, 50. *Prof Exp:* From instr to assoc prof , 52-71, PROF ORTHOP SURG, MAYO GRAD SCH MED, UNIV MINN, 71- *Concurrent Pos:* Consult, Mayo Clin, Rochester Methodist Hosp & Rochester-St Mary's Hosp. *Mem:* Am Acad Orthop Surgeons; Clin Orthop Soc; Am Orthop Asn. *Mailing Add:* Univ Minn 201 First Ave SW No 411 Univ Minn Rochester MN 55902

JOHNSON, ELIJAH, b Eutawville, SC, Jan 1, 48. PHYSICAL CHEMISTRY. *Educ:* Penn State Univ, BS, 69; Univ Ill, PhD(chem), 76. *Prof Exp:* CHEMIST, OAK RIDGE NAT LAB, 76- *Concurrent Pos:* E P Wigner fel, Oak Ridge Nat Lab, 77- *Mem:* Am Chem Soc; Am Phys Soc. *Res:* Theoretical and experimental studies of liquids and amorphous and polymeric materials using statistical mechanics and x-ray and neutron scattering. *Mailing Add:* Div Chem Oak Ridge Nat Lab PO Box 2008 Bldg 5505 MS6375 Oak Ridge TN 37831

JOHNSON, ELIZABETH BRIGGS, b Bowling Green, Ky, 1921; div; c 1. RESEARCH REACTORS, NUCLEAR CRITICALITY SAFETY. *Educ:* Western Ky Univ, BS, 43; Vanderbilt Univ, MS, 52. *Prof Exp:* Res asst, SAM Lab, Manhattan Proj, 44-45; assoc physicist, Oak Ridge Gaseous Diffusion Plant, 48-50, physicist, Oak Ridge Nat Lab, 50-68, physicist, Oak Ridge Y-12 Plant, 68-75, PHYSICIST, OAK RIDGE NAT LAB, 75- *Concurrent Pos:* admin judge, Atomic Safety & Licensing Bd Panel, US Nuclear Regulatory Comn. *Honors & Awards:* Spec Award, Nuclear Criticality Safety, Am Nuclear Soc. *Mem:* Am Phys Soc; fel Am Nuclear Soc; Sigma Xi; NY Acad Sci. *Res:* Experimental nuclear criticality safety and nuclear research reactors. *Mailing Add:* Oak Ridge Nat Lab PO Box 2008 Oak Ridge TN 37831-6010

JOHNSON, ELLIS LANE, b Athens, Ga, July 26, 38; m 62; c 1. OPERATIONS RESEARCH. *Educ:* Ga Inst Technol, BS, 60; Univ Calif, Berkeley, MA, 62, PhD(eng sci), 65. *Prof Exp:* Asst prof admin sci, Yale Univ, 64-68; MEM RES STAFF MATH SCI, THOMAS J WATSON RES CTR, IBM CORP, 68-; COCA-COLA PROF INDUST & SYSTS ENG, GA INST TECHNOL, 89- *Concurrent Pos:* Vis assoc prof, Univ Waterloo, 60-61, adj prof, 72-; vis engr, Sci Develop, IBM, France, 73-74. *Honors & Awards:* Dantzig Prize, Soc Indust & Appl Math; Lanchester Prize ORSA/IMT. *Mem:* Nat Acad Eng; Math Prog Soc. *Res:* Theory and algorithms for integer programming; study of combinatorial polyhedra and mathematical programming; character recognition. *Mailing Add:* Thomas J Watson Res Ctr IBM Corp PO Box 218 Yorktown Heights NY 10598

JOHNSON, ELMER MARSHALL, b Midlothian, Ill, June 16, 30; m 51; c 4. ANATOMY. *Educ:* Agr & Mech Col Tex, BS, 54, MS, 55; Univ Calif, Berkeley, PhD(anat), 59. *Prof Exp:* Asst zool, microtech & bot, Agr & Mech Col Tex, 53-55; asst gross anat & histol, Univ Calif, 55-58; instr anat & physiol, Contra Costa Col, 58-59; assoc prof anat, Univ Fla, 60-68, prof anat sci, 68-71; prof human morphol & chmn dept, Col Med, Univ Calif, Irvine, 71-72; PROF ANAT & CHMN DEPT, JEFFERSON MED COL, THOMAS JEFFERSON UNIV, 72-, DIR, DANIEL BAUGH INST ANAT, 72- *Concurrent Pos:* Asst researcher histochem, Surg Gen, US Army, 54; mem, Environ Protection Agency Sci Adv Bd, Dept of Defense Life Sci Comt, Nat Acad Sci/NASA, Space Sta Toxicol Comt, Training Group for WHO in Reproductive & Develop Toxicol. *Mem:* AAAS; Teratology Soc (pres, 74-75); Am Asn Anatomists; Soc Toxicol; Am Chem Soc. *Res:* Experimental teratology and nutrition; reproductive physiology; molecular biology; electron microscopy; histochemistry. *Mailing Add:* Dept Anat Jefferson Med Col Philadelphia PA 19107-6799

JOHNSON, ELMER ROGER, b Erwin, SDak, Oct 4, 11; m 40; c 4. CHEMISTRY. *Educ:* SDak State Col, BS, 33; Univ Wis, PhD(chem), 40. *Prof Exp:* Asst gen chem, Univ Wis, 37-40; res chemist, Tex Co, NY, 40-46; assoc prof, 46-55, prof, 55-78, EMER PROF CHEM, SDAK STATE UNIV, 78- *Mem:* AAAS; Am Chem Soc; Sigma Xi. *Res:* Fuel composition and antiknock quality; synthesis of fuel components and additives; synthesis of additives for lubricating oils; the Lange gold sol test; factors influencing the preparation of the gold sol and its use in the Lange test. *Mailing Add:* Box 2202 Univ Sta Brookings SD 57007

JOHNSON, ELSIE ERNEST, b Hackensack, NJ; m 64; c 3. ANESTHESIOLOGY. *Educ:* Women's Med Col Pa, MD, 64; Am Bd Anesthesiol, dipl, 75. *Prof Exp:* Intern, Philadelphia Gen Hosp, 64-65; resident anesthesiol, New Eng Deaconess Hosp, 65-67; narcos underlakare, Malmo Almana Stukhusset, Sweden, 67-68; res fel, 71-73, instr, 73-77, ASST PROF ANESTHESIA, HARVARD MED SCH, 77- *Concurrent Pos:* Staff anesthetist, Beth Israel Hosp, Boston, 73- *Mem:* Am Soc Anesthesiologists; Am Physiol Soc. *Res:* Splanchnic effects of mechanical ventilation with regard to circulation; biliary and hepatic function; opiate effects on biliary pressure; splanchnic endogenous opiate release. *Mailing Add:* Dept Anesthesiol Beth Israel Hosp 303 Brookline Ave Boston MA 02215

JOHNSON, ELWIN L PETE, b Hillsboro, Ill, July 20, 35; m 67; c 5. TECHNICAL PROGRAM MANAGEMENT, SEMICONDUCTOR TECHNOLOGY. *Educ:* Univ Ill, BS, 56, MS, 57, PhD(ceramic eng), 60. *Prof Exp:* Proj mgr, Cent Res, Tex Instruments Inc, 60-62, sci mgr, Mat Res Lab, 62-63, mem tech staff, 63-64, develop mgr, 64-65, eng sect mgr, 65-68, br mgr, 68-70, prog mgr, 74-75, lab dir, 75-84, mgr, Assembly Systs Dept, 84-87, MGR, TEST SYSTS DEPT, TEX INSTRUMENTS INC, 87- *Concurrent Pos:* Chmn, JC-11 Eng Comt, Electronic Indust Asn, 70-82; lectr, Gordon Conf, 82; mem lectr, Case Inst, 82. *Honors & Awards:* Distinguished Contrib, Electronics Indust Asn, 85. *Mem:* Corp mem Am Ceramic Soc; Nat Inst Ceramic Engrs; Sigma Xi. *Res:* Solar energy systems, solar cells, fuel cells and hydrogen storage; semiconductor assembly, test, packaging processes; development of automatic assembly and test equipment for sensor and coatings manufacturing; crystal growth, hybrid modules and chemical vapor deposition. *Mailing Add:* Tex Instruments Inc PO Box 655012 Mail Stop 3613 Dallas TX 75265

JOHNSON, EMMETT JOHN, b New Orleans, La, Apr 17, 29; m 55; c 2. MICROBIOLOGY. *Educ:* Loyola Univ of the South, BS, 52; La State Univ, MS, 54, PhD(bact), 57. *Prof Exp:* Nat Res Coun fel, Sch Med, Stanford Univ, 57-58; from asst prof to assoc prof microbiol, Med Sch, Univ Miss, 58-65; res scientist, Exobiol Div, Ames Res Ctr, NASA, 65-66; res assoc, Bruce Lyon Mem Res Inst, Oakland, Calif, 66-67; assoc prof, 67-70, PROF MICROBIOL & IMMUNOL, MED SCH, TULANE UNIV, 70- *Concurrent Pos:* Res assoc microbiol, Stanford Med Sch, 57-58; teaching assoc, 65-66, lectr, 66-67; Lederle Med Fac award, 62-65; res consult, Oak Ridge Nat Lab, 63-64 & 67-; res assoc molecular biol, Pasteur Inst, Paris, France, 74-75. *Mem:* AAAS; fel Am Acad Microbiol; Am Soc Microbiol; Am Soc Biol Chemists; Am Chem Soc. *Res:* Molecular mechanisms of genetic and biochemical regulation; biochemical basis of chemolithotrophic autotrophy; genetic and biochemical characterization of common enterobacterial antigens. *Mailing Add:* Dept Microbiol & Immunol Tulane Univ Med Sch 1430 Tulane Ave New Orleans LA 70112

JOHNSON, EMORY EMANUEL, civil engineering; deceased, see previous edition for last biography

JOHNSON, ERIC G, JR, b Klamath Falls, Ore, June 17, 36; div; c 2. LASERS. *Educ:* Mass Inst Technol, BS, 57; Harvard Univ, MA, 60, PhD(physics), 63. *Prof Exp:* GEN PHYSICIST, BOULDER LABS, NAT BUR STANDARDS, 62- *Mem:* Sigma Xi; Laser Inst Am. *Res:* Measurement theory; unitary matrix field theory; laser properties measurements. *Mailing Add:* 6730 Lake View Dr Boulder CO 80303

JOHNSON, ERIC RICHARD, b Elkhart, Ind, Mar 11, 47; m 68, 90; c 3. PROTEIN CHEMISTRY, ENZYMOLOGY. *Educ:* Rose-Hulman Inst Technol, BS, 69; Univ Minn, PhD(biochem), 74. *Prof Exp:* Asst chemist, Uniroyal, Inc, 68; USPHS fel biochem, Univ Minn, 69-74; res assoc biochem, Duke Univ Med Ctr, 74-76; from asst prof to assoc prof, 76-88, PROF CHEM, BALL STATE UNIV, 88- *Concurrent Pos:* Fel, Nat Inst Environ Health Sci, Duke Univ Med Ctr, 75-76; vis assoc prof biol chem, Univ Calif, Los Angeles, 84-85; res fel, USAF, 88. *Mem:* Am Soc Biol Chem; Sigma Xi. *Res:* Protein and peptide chemistry; high pressure liquid chromatography of protein and peptides; protease enzymology. *Mailing Add:* Dept Chem Ball State Univ Muncie IN 47306

JOHNSON, ERIC ROBERT, b Windom, Minn, Nov 17, 47; m 69. ANALYTICAL CHEMISTRY, CHEMICAL INSTRUMENTATION. *Educ:* Hamline Univ, BS, 69; Fla State Univ, PhD(anal chem), 75. *Prof Exp:* Res assoc anal chem, Mich State Univ, 74-76; mem staff, Mass Spectrometry Div, Varian Assocs, 76-77; MGR LAB COMPUT NETWORK, WAYNE STATE UNIV, 77- *Mem:* Am Chem Soc; Soc Appl Spectros; Sigma Xi. *Res:* Application of minicomputers to laboratory instrumentation; design of special purpose digital, analog and hybrid instrumentation systems; study of atomic absorption, emission and fluorescence spectroscopic methods of trace metal analysis. *Mailing Add:* 213 Bobby Dr Franklin TN 37064

JOHNSON, ERIC VAN, b Medford, Mass, Mar 11, 43; div; c 2. ORNITHOLOGY. *Educ:* Brown Univ, AB, 64; Cornell Univ, PhD(wildlife sci), 69. *Prof Exp:* From asst prof to assoc prof, 69-79, PROF BIOL, CALIF POLYTECH STATE UNIV, SAN LUIS OBISPO, 79- *Mem:* Cooper Ornith Soc; Am Ornith Union; Wilson Ornith Soc. *Res:* Avian taxonomy, behavior and population ecology; endangered species biology. *Mailing Add:* Dept Biol Sci Calif Polytech State Univ San Luis Obispo CA 93407

JOHNSON, ERNEST F(REDERICK), (JR), b Jamestown, NY, Apr 4, 18; m 44; c 4. CHEMICAL ENGINEERING. *Educ:* Lehigh Univ, BS, 40; Univ Pa, PhD(chem eng), 49. *Prof Exp:* From res & develop engr to tech supvr synthetic org chem mfg, Barrett Div, Allied Chem & Dye Corp, 40-46; from asst prof to prof, Princeton Univ, 48-86, assoc dean fac, 62-66, dir grad studies, Dept Chem Eng, 69-74, chmn dept, 77-78, assoc, Plasma Physics Lab, 55-88, clerk fac, 83-86, EMER PROF CHEM ENG, PRINCETON UNIV, 86-, SR ADV PRES, 88- *Concurrent Pos:* Consult chem engr, 50-; trustee, Assoc Univs, Inc, 62-68, chmn, 65-67; dir, Autodynamics, Inc, 67-85. *Honors & Awards:* Jubilee Medal, Am Inst Chem Engrs, 83. *Mem:* Fel AAAS; Am Chem Soc; fel Am Inst Chem Engrs; fel Am Inst Chemists. *Res:* Thermodynamic and transport properties of fluids; automatic process control; industrial wastes management; technological aspects of controlled thermonuclear fusion. *Mailing Add:* Dept Chem Eng Princeton Univ Eng Quadrangle Princeton NJ 08544

JOHNSON, ERNEST WALTER, b Paterson, NJ, Dec 15, 43; m 77; c 4. DIABETES, ENDOCRINOLOGY. *Educ:* Muhlenberg Col, BS, 65; Univ Vt, PhD(physiol, biophys), 70. *Prof Exp:* Res assoc physiol, Med Ctr, Univ Colo, 70-72; asst prof, 72-75; AAAS fel & legis asst health sci, US Senate, 75-76; grants assoc, NIH, 76-77, sect chief diabetes, 77-79, br chief diabetes, Nat Inst Arthritis, Metab & Digestive Dis, 79-84, dir, Div Diabetes, Endocrinol & Metab Dis, Nat Inst Diabetes, Digestive & Kidney Dis, 84-91; DIR, CTR GRANTS & CONTACTS, PA STATE UNIV, 91- *Concurrent Pos:* NIH fel, Med Ctr, Univ Colo, 70-72; prin investr, Nat Inst Neurol & Commun Disorders & Stroke, 73-75. *Mem:* AAAS; Am Diabetes Asn. *Res:* Etiology and pathophysiology of diabetes; hormone synthesis, secretion, action and metabolism; metabolic regulation; insulin delivery systems; islet cell transplantation; neurosecretary processes; receptor activation and inhibition; neurotrophic interactions; synaptic transmission. *Mailing Add:* Pa State Univ PO Box 850 Hershey PA 17033

JOHNSON, EUGENE A, b Crosby, Minn, Feb 24, 25; m 47; c 4. BIOSTATISTICS. *Educ:* Univ Minn, BA, 49, MA, 50, PhD(biostatist), 56. *Prof Exp:* From asst prof to assoc prof biostatist, Univ Minn, Minneapolis, 56-60, assoc prof indust eng, 60-62; prof math & head dept, Gustavus Adolphus Col, 62-64; assoc prof & dir biomed data processing unit, 64-69, PROF BIOMET, COL MED SCI, UNIV MINN, MINNEAPOLIS, 69-, DIR GRAD STUDY, 80- *Mem:* Am Statist Asn; Biomet Soc; Inst Math Statist; Sigma Xi. *Res:* Biomedical computing; computing in biology; mathematics; operations research. *Mailing Add:* Dept Biomet Univ Minn Col Med Sci 197 Mayo 420 Delaware St SE Minneapolis MN 55455

JOHNSON, EUGENE MALCOLM, JR, b Baltimore, Md, Oct 20, 43; m 65; c 2. PHARMACOLOGY. *Educ:* Univ Md, BS, 66, PhD(med chem), 70. *Prof Exp:* Fel pharmacol, Sch Med, Wash Univ, 70-73; asst prof pharmacol, Med Col Pa, 73-76; asst prof, 76-78, ASSOC PROF PHARMACOL, SCH MED, WASH UNIV, 78- *Res:* Autonomic pharmacology; role of sympathetic nervous system in hypertension; effect of drugs on development of the sympathetic nervous system. *Mailing Add:* Dept Pharmacol Wash Univ Sch Med 660 S Euclid Ave St Louis MO 63110

JOHNSON, EUGENE W, b El Paso, Tex, May 25, 39; m 59; c 1. ALGEBRA. *Educ:* Univ Calif, Riverside, BA, 63, MA, 64, PhD(algebra), 66. *Prof Exp:* Asst prof math, Eastern NMex Univ, 66; from asst prof to assoc prof, 66-75, PROF MATH, UNIV IOWA, 75- *Mem:* Am Math Soc; Math Asn Am. *Res:* Noetherian rings and abstract ideal theory. *Mailing Add:* Dept Math Univ Iowa Iowa City IA 52240

JOHNSON, EVERT WILLIAM, b Astoria, NY, Apr 6, 21; m 50; c 3. FORESTRY, PHOTOGRAMMETRY. *Educ:* Univ NH, BS, 43; Yale Univ, MF, 47; Syracuse Univ, PhD, 57. *Prof Exp:* Forester chg aerial surv, Sable Mt Corp, Vt, 47-50; from instr to assoc prof forestry, 50-67, asst, 50-53, asst forester, 53-57, prof, 80-86, EMER PROF FORESTRY, AUBURN UNIV, 86- *Concurrent Pos:* Fel, Soc Am Foresters. *Mem:* Soc Am Foresters; Am Soc Photogram; Sigma Xi. *Res:* Applications of photogrammetry, statistics and computer science to forest measurements. *Mailing Add:* 743 Heard Ave Auburn AL 36830

JOHNSON, F BRENT, b Monroe, Utah, Mar 31, 42; m 65; c 5. VIROLOGY. *Educ:* Brigham Young Univ, BS, 66, MS, 67, PhD(microbiol), 70. *Prof Exp:* Fel virol, NIH, 70-72; from asst prof to assoc prof, 72-80, PROF MICROBIOL, BRIGHAM YOUNG UNIV, 80- *Concurrent Pos:* NIH res grants, 73 & 76; res grants, Air Force Off Sci Res, 77-78 & 79-82, Thrasher Fund, 87-90. *Mem:* Am Soc Microbiol; AAAS; Sigma Xi. *Res:* Viral replication; structure of viruses and biology of virus infections; diagnostic virology. *Mailing Add:* 887 Widtsoe Bldg Brigham Young Univ Provo UT 84602

JOHNSON, F CLIFFORD, b Ft Worth, Tex, Nov 4, 32; m 58; c 3. GENETICS, ECOLOGY. *Educ:* Univ Tex, BA, 55, MA, 60, PhD(zool), 61. *Prof Exp:* Instr zool, Duke Univ, 60-61; asst prof genetics, Va Polytech Inst, 61-62; asst prof biol & chmn dept, NMex Inst Mining & Technol, 62-66, assoc prof, 67-70; PROF ZOOL, UNIV FLA, 70- *Res:* Genetics of polymorphic variation. *Mailing Add:* Dept Zool Univ Fla Gainesville FL 32611

JOHNSON, FATIMA NUNES, b Rizal, Philippines, Jan 1, 39; m 67; c 2. ORGANIC CHEMISTRY. *Educ:* Adamson Univ, Manila, BS, 59; Boston Col, MS, 61, PhD(org chem), 64. *Prof Exp:* Proj leader org med chem, Arthur D Little, Inc, Mass, 64-69; res chemist, Org Chem Labs, Edgewood Arsenal, Md, 69-70; scientist drug standards, US Pharmacopeia, Rockville, Md, 71-90; SR STAFF OFFICER, INST MED, NAT ACAD SCI, WASH, DC, 90- *Mem:* Am Chem Soc; Am Asn Pharmaceut Sci. *Res:* Organo-fluorine compounds; organometallics; molecular rearrangements; nitrogen heterocyclics. *Mailing Add:* 5315 Renaissance Ct Burke VA 22015

JOHNSON, FRANCIS, b Bristol, Eng, Mar 12, 30; m 55; c 3. PHARMACOLOGY. *Educ:* Glasgow Univ, BSc, 51, PhD(org chem), 54. *Prof Exp:* Fel org chem, Boston Univ, 54-57; from res chemist to assoc scientist, Eastern Res Lab, Dow Chem Co, 57-69, res scientist, 69-74; PROF PHARMACOL & CHEM, STATE UNIV NY STONY BROOK, 74- *Concurrent Pos:* Eve lectr, Boston Univ, 56-70; consult, Qm Res Corps, US Army, 56-58 & Dow Chem Co, 74-; vis scientist, Oxford Univ, 66-67; consult res dir, Ganes Chemicals Inc, NJ, 90- *Mem:* Am Chem Soc; Royal Soc Chem; assoc Royal Inst Chem; fel Royal Soc Arts; NY Acad Sci. *Res:* Synthetic organic chemistry, especially natural product and aliphatic areas; medicinal chemistry. *Mailing Add:* Dept Pharmacol Sci State Univ NY Stony Brook NY 11794

JOHNSON, FRANCIS SEVERIN, b Omak, Wash, July 20, 18; m 43; c 1. SPACE PHYSICS, METEOROLOGY. *Educ:* Univ Alta, BSc, 40; Univ Calif, Los Angeles, MA, 42, PhD(meteorol), 58. *Prof Exp:* Physicist, US Naval Res Lab, 46-55; space physicist, Lockheed Missiles & Space Co, 55-62; prof & dir, Earth & Planetary Sci Lab, Southwest Ctr Advan Studies, 62-69; asst dir astron, Atmospheric, Earth & Ocean Sci, NSF, Washington, DC, 79-83; actg pres, Univ Tex, Dallas, 69-71, prof & dir, Ctr Advan Studies, 71-74, Cecil H & Ida M Green Hons Prof Natural Sci, 74-89, EMER PROF, UNIV TEX, DALLAS, 90- *Concurrent Pos:* Consult, NASA, 60-79; mem, Panel Adv Cent Radio Propagation Lab, Nat Bur Standards, 62-65, mem panel on weather & climate modification, 64-70, mem adv comt to Air Force Systs Command panel on re-entry physics, 65-68, mem comt solar-terrestrial res, 66-79, chmn, 71-74, mem comt adv to Environ Sci Serv Admin, 66-71, mem space sci bd, 67-80, mem geophys res bd, 71-75, mem comt adv to Nat Oceanic & Atmospheric Admin, 71-72, mem, Climate Res Bd, Nat Acad Sci, 77-79; mem adv panel atmospheric sci, NSF, 62-66; chmn, US Comn IV, Int Union Radio Sci, 64-67, secy, US Nat Comt, 67-70, vchmn, 70-73, chmn, 73-76; mem, res adv comt, Coord Bd Tex Col & Univ Syst, 66-68; mem, Air Force Sci Adv Bd, 68-79; mem, Nat Adv Comt Oceans & Atmosphere, 71-73; mem, climatic impact comt, Nat Acad Sci, 72-76, mem bd Atmospheric Sci & Climate, 84-87, mem comt solar physics, Nat Acad Sci, 86-88; mem Aerocibo Adv bd, Nat Astron & Ionospher Ctr, 85-88; vpres, Comt Space Res, Int Coun Sci Unions, 75-80. *Honors & Awards:* Space Sci Award, Am Inst Aeronaut & Astronaut, 66; Henryk Arctowski Medal, Nat Acad Sci, 72; Except Sci Achievement Medal, NASA, 73; John A Fleming Award, Am Geophys Union, 77. *Mem:* Am Meteorol Soc; Am Physical Soc; Am Geophys Union; Am Inst Aeronaut & Astronaut; Inst Elec & Electronics Engrs; Sigma Xi. *Res:* Upper atmospheric and magnetospheric physics; space science; planetary atmospheres; upper atmosphere and space physics; planetary science; solar radiation; synoptic and physical meteorology. *Mailing Add:* 13619 Sprucewood Dr Dallas TX 75240

JOHNSON, FRANK BACCHUS, b Washington, DC, Feb 1, 19; m 47; c 2. PATHOLOGY. *Educ:* Univ Mich, BS, 40; Howard Univ, MD, 44. *Prof Exp:* From intern to resident path, Med Ctr, Jersey City, 44-46; dir clin labs, Howard Univ, 46-48; res assoc, Univ Chicago, 50-52; pathologist, Armed Forces Inst Path, 52-60, chief basic sci div, 60-72, chief, Histochem Br, 72-74, CHMN, DEPT CHEM PATH, ARMED FORCES INST PATH, 74- *Concurrent Pos:* AEC fel med sci, Univ Chicago, 48-50. *Honors & Awards:* Citation Admin & Tech Proficiency, Vet Admin, 58, Commendation Outstanding Contributions Histochem, 64. *Mem:* Am Crystallog Soc; Am Chem Soc. *Res:* Histochemistry in pathology. *Mailing Add:* 4222 Mathewson Dr Washington DC 20011

JOHNSON, FRANK HARRIS, b Raleigh, NC, July 31, 08; m 33; c 3. MOLECULAR BIOLOGY. *Educ:* Princeton Univ, AB, 31, PhD(biol), 36; Duke Univ, AM, 32. *Prof Exp:* From instr to prof, 37-69, Edwin Grant Conklin prof, 69-77, EMER EDWIN GRANT CONKLIN PROF BIOL,

PRINCETON UNIV, 77-, SR RES BIOLOGIST, 77- *Concurrent Pos:* Rockefeller fel, Univs Delft & Utrecht, 39; Guggenheim fel, Calif Inst Technol & Marine Biol Lab, Woods Hole, 44-46 & Univ Utah, 50-51; prog dir Develop, Environ & Syst Biol, NSF, 52-53, consult, 53-56. *Honors & Awards:* Sci Prize, AAAS, 41. *Mem:* Fel Explorers Club; Soc Gen Physiol; Am Physiol Soc; Soc Exp Biol & Med; Am Soc Microbiol. *Res:* Physical chemistry of biological reactions; action of temperature, hydrostatic pressure, drugs and other factors in luminescence, respiration, growth, enzyme reactions, cell division and other processes; kinetic basis of molecular biology; theory of rate processes in biology. *Mailing Add:* Edwin Grant Conklin Biol Princeton Univ Princeton NJ 08540

JOHNSON, FRANK JUNIOR, b Rosendale, Mo, Aug 24, 30; m 51; c 3. ANALYTICAL CHEMISTRY. *Educ:* Northwest Mo State Col, BS, 52; Univ Mo, MSc, 61. *Prof Exp:* Instr agr chem, Univ Mo, 55-62; anal chemist, 62-69, head serv, 69-86, CHIEF, PE SERV BR, TENN AUTHORITY, 86. *Mem:* Asn Off Anal Chemists (pres-elect 85-86 & pres 86-87); Am Chem Soc; Fertilizer Soc. *Res:* Fertilizer chemistry; investigation of new or improved analytical methods pertaining to fertilizer and related materials. *Mailing Add:* 205 Westmeade Ct Florence AL 35630

JOHNSON, FRANK WALKER, geology; deceased, see previous edition for last biography

JOHNSON, FRANKLIN M, b Cloquet, Minn, Nov 1, 40; m 83; c 4. GENETICS. *Educ:* Univ Minn, Duluth, BA, 62; Univ Hawaii, MS, 64; Univ Tex, Austin, PhD(zool), 66. *Prof Exp:* NIH fel, 66-67; res scientist, Univ Tex, Austin, 67-68; asst prof genetics, NC State Univ, 68-74, sr geneticist, Res Triangle Inst, 74-77; RES GENETICIST, NAT INST ENVIRON HEALTH SCI, 77- *Mem:* AAAS; Genetics Soc Am; Am Soc Human Genetics. *Res:* Genetic variability in natural populations; genotype-environment relationships; developmental variation in the skeleton; mutagenesis; genetic risk; carcinogenesis. *Mailing Add:* Lab Genetics Nat Inst Environ Health Sci Research Triangle Park NC 27709

JOHNSON, FRED LOWERY, JR, b San Angelo, Tex, Oct 24, 27; m 49; c 3. INDUSTRIAL ORGANIC CHEMISTRY. *Educ:* Univ Tex, BS, 51, PhD(chem), 59. *Prof Exp:* Sr process chemist, Am Cyanamid Co, La, 59-62; res chemist, Jefferson Chem Co, Inc, 62-64, sr res chemist, 64-68, proj chemist, 68-76, SR PROJ CHEMIST, TEXACO CHEM CO, 76- *Mem:* Am Chem Soc; Sigma Xi. *Res:* Catalytic research and process development for petrochemicals. *Mailing Add:* 3002 Yellowpine Terr Austin TX 78757

JOHNSON, FRED TULLOCH, physics, for more information see previous edition

JOHNSON, FREDERIC ALLAN, b Concord, NH, Mar 6, 32; m 56; c 3. INORGANIC CHEMISTRY, PHYSICAL CHEMISTRY. *Educ:* Univ NH, BS, 54, MS, 55; Univ Wis, PhD(chem), 58. *Prof Exp:* Lab instr, Univ NH, 54; chemist, Redstone Arsenal Res Div, Rohm & Haas Co, 58-62, group leader anal chem, 62-70; ASSOC PROF CHEM, AUBURN UNIV, 70- *Mem:* Am Chem Soc. *Res:* Fluorine and metal coordination chemistry; nuclear magnetic resonance; kinetics. *Mailing Add:* Dept Chem Auburn Univ Auburn AL 36849

JOHNSON, FREDERIC DUANE, b Chicago, Ill, Oct 24, 25; m 48; c 5. FOREST ECOLOGY. *Educ:* Ore State Col, BS, 50; Univ Idaho, MS, 52. *Prof Exp:* Radioisotopes technologist, 52-56, from instr to asst prof forest mgt, 56-67, assoc prof forest ecol, 67-72, PROF FOREST ECOL, UNIV IDAHO, 72- *Concurrent Pos:* Adj prof, Inst Agron Vet, Morocco. *Mem:* Ecol Soc Am; fel Soc Am Foresters. *Res:* Forest ecology-temperate and tropical, ecologic accessment; temperate and tropical dendrology. *Mailing Add:* Col Forestry Univ Idaho Moscow ID 83843

JOHNSON, FREDERICK ALLAN, b Winnipeg, Man, Nov 7, 23. NUCLEAR PHYSICS. *Educ:* Univ Man, BSc, 45; McGill Univ, PhD(nuclear physics), 52. *Prof Exp:* Res assoc nuclear physics, Radiation Lab, McGill Univ, 52-53; sr engr, Can Aviation Electronics Co, 53-55; Defence Sci Serv officer nuclear physics, Suffield Exp Sta, 55-59, Defence Sci Serv Officer Chem, Biol & Radiation Labs, 59-71, DEFENCE SCI SERV OFFICER NUCLEAR PHYSICS, DEFENCE RES BD, DEFENCE RES ESTAB, 71- *Mem:* Am Phys Soc; Can Asn Physicists. *Res:* Spectroscopy of nuclear radiations from cyclotron-produced cadmium and silver isotopes; auger transitions in silver; industrial design of radiation detectors; nanosecond pulse electronics; neutron time-of-flight spectroscopy; beam pulsing and deflection; pulse-shape discrimination circuits for neutron identification; neutron activation; radiological protection and health physics. *Mailing Add:* Six Esquimalt Ave Nepean ON K2H 6Z3 Can

JOHNSON, FREDERICK ARTHUR, JR, b Pittsburgh, Pa, Sept 8, 23; m 46; c 1. GEOLOGY. *Educ:* Harvard Univ, BS, 44; Univ Chicago, MS, 49, PhD(geol), 51. *Prof Exp:* From assoc geologist to sr geologist stratig sect, Explor Dept, Humble Oil & Ref Co, 51-60, supvry geologist, 60-66; sect supvr struct geol & basin interpretation, Esso Prod Res Co, 66-67; sr explor geologist, Humble Oil & Refining Co, 67-76; GEOL ADV, EXXON CO, 76- *Mem:* Sigma Xi; AAAS; Geol Soc Am; Soc Econ Paleont & Mineral; Am Asn Petrol Geol. *Res:* Carbonate rock; stratigraphic and structural geology of Permian Basin, west Texas; structural geology of Rocky Mountains; regional geology of Alaska, eastern USSR and western Canada. *Mailing Add:* 2753 W Long Dr Apt F Littleton CO 80120

JOHNSON, FREDERICK CARROLL, b Sheridan, Wyo, Oct 23, 40; m 64; c 1. APPLIED MATHEMATICS, RESOURCE MANAGEMENT. *Educ:* Univ NDak, BS, 62; Univ Wash, MS & PhD(appl math), 66. *Prof Exp:* Res analyst real-time data processing, DBA Systs, Inc, 66-68; res scientist appl math, Boeing Sci Res Labs, 68-73; mathematician, Nat Bur Standards, 73-77, chief, Math Anal Div, 77-82; partner, Nat Res Consult, 82-84; chief, Math Anal Div, 84-87, ASSOC DIR COMPUT, NAT INST STANDARDS &

TECHNOL, 87- *Honors & Awards:* Silver Medal, Dept Commerce, 78. *Mem:* Soc Indust & Appl Math; Inst Elec & Electronics Engrs; Asn Comput Mach; Am Fisheries Soc; Sigma Xi. *Res:* Applications of mathematical modeling to physical systems; high performance computing; scientific visualization for mathematical modeling. *Mailing Add:* Off Assoc Dir Comput Nat Inst Standards & Technol Gaithersburg MD 20899

JOHNSON, FREEMAN KEITH, genetics, plant breeding, for more information see previous edition

JOHNSON, G ALLAN, b Champaigne, Ill, Jan 17, 47; m 69; c 2. MEDICAL IMAGING. *Educ:* St Olaf Col, BA, 69; Duke Univ, PhD(physics), 74. *Prof Exp:* Assoc physics, 78, from asst prof to assoc prof, 78-83, PROF PHYSICS, DEPT RADIOL, DUKE UNIV MED CTR, 88- *Concurrent Pos:* Dir diag physics, Dept Radiol, Duke Univ Med Ctr, 78. *Mem:* Sigma Xi; Am Phys Soc; Am Asn Physicists Med. *Res:* Implementation and enhancement of new imaging technologies in medicine; magnetic resonance microscopy and its extension to basic sciences. *Mailing Add:* Duke Univ Med Ctr Box 3302 Durham NC 27710

JOHNSON, GARLAND A, b Laona, Wis, July 16, 36; m 58; c 4. BIOCHEMISTRY, PHARMACOLOGY. *Educ:* Carroll Col, Wis, BS, 58; Ohio State Univ, MSc, 60, PhD(physiol chem), 63. *Prof Exp:* Res assoc biochem, Res Found, Ohio State Univ, 63; staff fel, Nat Inst Neurol Dis & Blindness, 63-64; RES ASSOC, UPJOHN CO, 64- *Mem:* AAAS; Am Soc Pharmacol Exp Therapeut. *Res:* Metabolism of catecholamines and serotonin; effect of drugs on biogenic amines. *Mailing Add:* Dept Hair Growth Res Cardiovasc Dis Res Upjohn Co Kalamazoo MI 49001

JOHNSON, GARY DEAN, b Sioux City, Iowa, Dec 2, 42; m 65; c 1. GEOLOGY. *Educ:* Iowa State Univ, BS, 64, MS, 67, PhD(geol), 71. *Prof Exp:* Instr geol, Iowa State Univ, 69-71; from asst prof to assoc prof, 71-86, chmn dept, 80-83, PROF GEOL, DARTMOUTH COL, 87- *Concurrent Pos:* Res assoc, Iowa State Univ, 71-72. *Mem:* Geol Soc Am; Soc Econ Paleont & Mineral; Int Asn Sedimentologists. *Res:* Stratigraphy and sedimentology; Cenozoic terrestrial deposits of Asia and Africa; geology of the Himalayas; paleopedology; geochronology. *Mailing Add:* Dept Earth Sci Dartmouth Col Hanover NH 03755

JOHNSON, GARY LEE, b Osage City, Kans, Nov 20, 38; m 60; c 2. ELECTRICAL ENGINEERING. *Educ:* Kans State Univ, BS, 61, MS, 63; Okla State Univ, PhD, 66. *Prof Exp:* From asst prof to assoc prof, 66-84, PROF ELEC ENG, KANS STATE UNIV, 84- *Concurrent Pos:* Consult, Kansas City Power & Light Co, 71-72. *Mem:* Inst Elec & Electronics Engrs; Am Wind Energy Asn; Int Tesla Soc; Nat Soc Prof Eng. *Res:* Power systems; wind electric systems. *Mailing Add:* Dept Elec Eng Kans State Univ Manhattan KS 66506

JOHNSON, GEAROLD ROBERT, b Des Moines, Iowa, Jan 11, 40; m 62; c 2. ENGINEERING DESIGN, ENGINEERING COMPUTER GRAPHICS. *Educ:* Purdue Univ, BS, 62, MS, 68, PhD(mech eng), 72. *Prof Exp:* Engr, aerospace, Boeing Co, 62-66; NATO postdoctoral fluid mech, von Karman Inst, 70-71; from asst prof to prof, 73-84, G T ABELL CHAIR, COLO STATE UNIV, 84- *Concurrent Pos:* Vis prof, Univ Kent, Canterbury, UK, 78-79, Calif Inst Technol, 84; vis scientist, Shape Data Ltd, Cambridge, UK, 85-86. *Mem:* Inst Elec & Electronics Engrs; Inst Elec & Electronics Engrs Computer Soc; Math Asn Am. *Res:* Application of computers to engineering problems such as solar energy, fluid mechanics, solid modeling, etc; uses of computer technology to support engineering education in design. *Mailing Add:* Ctr Computer Assisted Eng Colo State Univ Ft Collins CO 80523

JOHNSON, GEORGE, JR, b Wilmington, NC, Apr 6, 26; m 50; c 4. MEDICINE, SURGERY. *Educ:* Univ NC, BS, 49; Cornell Univ, MD, 52; Am Bd Surg, dipl, 60; Am Bd Thoracic Surg, dipl, 63. *Prof Exp:* Instr surg, Cornell Univ, 58-59; from asst prof to prof, 61-73, ROSCOE B G COWPER PROF SURG, SCH MED, UNIV NC, CHAPEL HILL, 73-, CHIEF, DIV GEN SURG, 69-, VCHMN, DEPT SURG, 77- *Concurrent Pos:* Ed, NC Med J. *Mem:* Am Col Surgeons; Soc Univ Surgeons; Soc Vascular Surgeons; Am Asn Surg of Trauma; Asn Acad Surgeons. *Res:* Vascular and thoracic surgery; hemodynamics associated with cirrhosis of the liver; local and systematic hemodynamics of an arteriovenous fistula; gall bladder surgery. *Mailing Add:* Dept Surg Univ NC Sch Med CB No 7210 Chapel Hill NC 27599

JOHNSON, GEORGE ANDREW, b Brooklyn, NY, June 1914; m 65; c 2. FOOD SCIENCE. *Educ:* Tuskegee Inst, Ala, DVM, 61; Cornell Univ, Ithaca, MS, 72. *Prof Exp:* Exec staff officer, USDA, Wash, DC, 70-72; assoc dean, Sch Agr & Natural Resources, Wash Tech Inst, 72-74; assoc dir, Univ DC, 74-77; mem adv coun, Pub Sch Syst, DC, 77-79; PROF FOOD & ANIMAL SCI & CHAIRPERSON, NC AGR & TECH STATE UNIV, 78- *Concurrent Pos:* Consult, Am Coun Comn Accreditation Serv Experience, 72; dean, Wash Tech Inst, 72-73, prof vet sci & chmn, 73-74. *Mem:* Am Vet Med Asn; Am Meat Sci Asn; Am Pub Health Asn; Nat Inst Food Technologists; Am Lab Animal Soc. *Res:* Detectability of sex flavor in a mildly seasoned comminuted product served cold as affected by concentration of boar meat. *Mailing Add:* Dept Animal Sci NC Agr & Tech State Univ Greensboro NC 27411

JOHNSON, GEORGE FREDERICK, b Harmony, Minn, July 15, 16; m 41. ORGANIC CHEMISTRY. *Educ:* Iowa State Univ, BS, 38; Ohio State Univ, PhD(chem), 43. *Prof Exp:* Proj leader, Process Develop Lab, Carbide & Carbon Chem Co, Union Carbide Corp, 53-55, group leader, Chem Div, 55-71, develop scientist chem & plastics, Res & Develop Dept, 71-76, site adminr, agr prod div, Res & Develop Dept, 76-82; RETIRED. *Mem:* Am Chem Soc. *Mailing Add:* 1336 Morningside Dr Charleston WV 25314

JOHNSON, GEORGE LEONARD, b Englewood, NJ, May 18, 31; m 89; c 3. GEOLOGICAL OCEANOGRAPHY. *Educ:* Williams Col, BA, 53; NY Univ, MS, 65; Univ Copenhagen, PhD(marine geol), 75. *Prof Exp:* Res asst marine geol, Lamont-Doherty Geol Observ, 57-65; oceanogr, US Naval Oceanog Off, Md, 65-75; sci adminr arctic prog, 75-80, dir arctic progs, Phys Sci Admin, 80-85, DIR GEOPHYS SCI, OFF NAVAL RES, 85- *Concurrent Pos:* Consult, Polar Res Bd, Natural Acad Sci, 75-, Panel Polar Eng, Nat Res Coun, 77- & Comn Tectonic Chart of World, 77-; sci consult, Intergovt Oceanog Comn, Int Hydrographic Off, 75-; agency rep, Global Atmospheric Res Prog; chmn, Arctic Geol-Geophys Comt, Instrument Calibration Lab, Arctic Geol Comt, Lithosphere Comn, secy, US-USSR Ocean Bilateral. *Mem:* Am Geophys Union; Arctic Inst NAm; Polar Soc. *Res:* Geophysics with specialization in marine geomorphology and physiography of the world's oceans; arctic and antarctic marine geology; naval arctic research; polar regions. *Mailing Add:* Geophys Sci Off Naval Res 800 N Quincy St Arlington VA 22217-5000

JOHNSON, GEORGE PATRICK, b Pine Bluff, Ark, June 16, 32; m 67; c 3. TECHNOLOGY ASSESSMENT, CIVIL ENGINEERING. *Educ:* Univ Miss, BSCE, 54; Stanford Univ, MS, 67, Engr, 69, PhD(civil eng), 71. *Prof Exp:* Res civil engr int develop, C S McCandless & Co, 65-67; oper analyst housing res, Stanford Res Inst, 67-69; res engr water resources, INTASA, Inc, 69-71; water resource engr, US Army Eng Inst Water Resources, 71-74; prog mgr tech assessment, NSF, 74-84, sr policy analyst, 84-88, head, Off Europe, 88-90, SR POLICY ANALYST, NSF, 90- *Concurrent Pos:* Consult, Rand Corp, 70-71. *Mem:* AAAS; Sigma Xi. *Res:* Technology assessment methods and utilization; policy research and analysis; water resources planning; technological forecasting; futures research; structural modeling; decision analysis for public policy. *Mailing Add:* Div Policy Res & Analysis NSF Washington DC 20550

JOHNSON, GEORGE PHILIP, b Minneapolis, Minn, Nov 13, 26; m 51; c 4. MATHEMATICAL ANALYSIS. *Educ:* Univ Minn, BS, 48, MA, 49, PhD(math), 56. *Prof Exp:* Asst math & statist, Univ Minn, 48-51, instr math, 55-56; mathematician, Nat Security Agency, 51-54; sr mathematician, Standard Oil Co Calif, 56-60; assoc prof math, Wesleyan Univ, 60-64 & Univ of the South, 64-65; chmn dept, 65-70, PROF MATH, OAKLAND UNIV, 65-, DEAN GRAD SCH, 69- *Concurrent Pos:* Off Naval Res assoc, 63-64; consult-evaluator, NCent Asn Cols & Schs, 72- *Mem:* Am Math Soc; Math Asn Am; Sigma Xi. *Res:* Abstract harmonic analysis; numerical analysis and computing. *Mailing Add:* 654 E Buell Rd Lake Orion MI 48035

JOHNSON, GEORGE ROBERT, b Caledonia, NY, Aug 2, 17; m 42; c 4. ANIMAL HUSBANDRY. *Educ:* Cornell Univ, BS, 39; Mich State Univ, MS, 47, PhD, 54. *Prof Exp:* Pub sch teacher, NY, 39-42; asst county agt agr, Canton, NY, 42-43; from instr to assoc prof animal husb, Cornell Univ, 43-55; from assoc prof animal sci to prof animal sci & chmn dept, Ohio State Univ, 55-83; RETIRED. *Mem:* Am Soc Animal Sci. *Res:* Administration in animal science, especially teaching, research and extension; sheep production and management. *Mailing Add:* Dept Animal Sci Ohio State Univ Col Agr Columbus OH 43210

JOHNSON, GEORGE S, b Cokato, Minn, Aug 25, 43. CELL BIOLOGY, MOLECULAR BIOLOGY. *Educ:* Mich State Univ, PhD(biochem), 69. *Prof Exp:* RES CHEMIST, NAT CANCER INST, NIH, 74- *Mem:* Am Soc Biol Chem; Am Soc Microbiologists. *Res:* Oncology. *Mailing Add:* Nat Cancer Inst Executive Plaza N Rm 830 Bethesda MD 20892

JOHNSON, GEORGE THOMAS, botany, bacteriology; deceased, see previous edition for last biography

JOHNSON, GERALD GLENN, JR, b Renovo, Pa, Nov 10, 39; m 63; c 2. MATERIALS SCIENCE. *Educ:* John Carroll Univ, BS, 62; Pa State Univ, PhD(mat sci, physics), 65. *Prof Exp:* Jr physicist, Erie Registor Corp, 60-62; asst prof solid state sci, 65-71, ASSOC PROF COMPUT SCI, PA STATE UNIV, 71- *Concurrent Pos:* Mem, Nat Res Coun. *Mem:* AAAS; Am Phys Soc; Am Crystallog Asn; Am Soc Testing & Mat; Sigma Xi. *Res:* Information retrieval as applied to x-ray powder diffraction identification systems; high resolution powder diffraction techniques using Guinier Cameras and automatic microdensitometers. *Mailing Add:* Dept Comput Sci Col Sci Pa State Univ University Park PA 16802

JOHNSON, GERALD WINFORD, b Minneapolis, Minn, Oct 31, 32; m 58; c 3. CIVIL ENGINEERING. *Educ:* Purdue Univ, BS, 55; Ohio State Univ, MS, 60; Univ Wis-Madison, PhD(civil eng), 69. *Prof Exp:* Field serv engr, Boeing Co, Wash, 60-61; programmer analyst, Syst Develop Corp, Calif, 61-65; ASST PROF CIVIL ENG, UNIV MINN, MINNEAPOLIS, 69-, ASSOC PROF, 80- *Mem:* Am Soc Civil Engrs; Am Cong Surv & Mapping; Am Soc Photogram; Arctic Inst N Am; Am Inst Navig. *Res:* Reliability of atmospheric refraction in polar astronavigation; cartography and map rectification in north Greenland; application of computers to survey net adjustments. *Mailing Add:* Dept Civil Eng Univ Minn 500 Pillsbury Dr SE Minneapolis MN 55455

JOHNSON, GLEN ERIC, b Rochester, NY, May 29, 51; m 75; c 2. OPTIMAL MECHANICAL DESIGN. *Educ:* Worcester Polytech Inst, BS, 73; Ga Inst Technol, MSME, 74; Vanderbilt Univ, PhD(mech eng), 78. *Prof Exp:* Mech eng, Machine Design, Tenn Eastman Co, 74-76; asst prof, Vanderbilt Univ, 78-79 & Univ Va, 79-81; assoc prof mech eng, Vanderbilt Univ, 81-89; ASSOC PROF MECH ENG, UNIV MICH, 89- *Concurrent Pos:* Co-prin investr, US Dept Transp, 80-81; prin investr, NSF, 80-; assoc ed, J Mech Design, Am Soc Mech Engrs, 81-82, J Mech Trans Automation Design, 82-83. *Honors & Awards:* Ralph Teetor Award, Soc Automotive Engrs, 84. *Mem:* Am Soc Mech Engrs; Acoust Soc Am; Math Prog Soc; Sigma Xi; Am Gear Mfrs Asn; Soc Automotive Engrs. *Res:* Development of algorithmic and ad hoc optimization strategies; application of optimization theory to the design of mechanical systems and machines; machine design; system modeling and analysis; noise and vibration control. *Mailing Add:* Dept Mech Eng Univ Mich 2250 G G Brown Ann Arbor MI 48109-2125

JOHNSON, GLENN M, b US citizen. ENGINEERING. *Educ:* Pa State Univ, BS, 64; Northwestern Univ, MS, 65; Am Acad Environ Engrs, dipl. *Prof Exp:* Surveyor, US Forest Serv, 60; designer & draftsman, Chicago Bridge & Iron Co, 61-62; asst proj engr, Nat Forge Co, 62-63; proj engr, 65-68, GROUP MGR, ROY WESTON INC, 72- *Mem:* Am Soc Civil Engrs; Am Water Resources Asn. *Res:* Water resources engineering; resource economics; wastewater management systems design. *Mailing Add:* Roy Weston Inc Weston Way West Chester PA 19380

JOHNSON, GLENN RICHARD, b Geneseo, Ill, Feb 19, 38. PLANT BREEDING. *Educ:* Iowa State Univ, BS, 60, PhD(plant breeding), 65. *Prof Exp:* PLANT BREEDER MAIZE & AREA RES DIR, DEKALB-PFIZER GENETICS, INC, 65- *Mem:* AAAS; Sigma Xi; Am Soc Agron; Am Genetic Asn; NY Acad Sci. *Res:* Plant breeding, including applied statistical techniques in relation to plant breeding problems. *Mailing Add:* DeKalb-Pfizer Genetics Inc Thomasboro IL 61878

JOHNSON, GORDON CARLTON, b Newport, RI, Feb 9, 29; m 56; c 3. PHYSICAL CHEMISTRY, SURFACTANTS. *Educ:* City Col New York, BChE, 52; Pace Univ, MBA, 83. *Prof Exp:* Develop engr, Silicones Div, Union Carbide Corp, 52-62, proj leader silicone prod develop & tech serv, 62-66, group leader, 66-77, technol mgr, 77-84; CONSULT, SURFACTANTS, 84- *Mem:* Am Chem Soc; Am Oil Chemists' Soc; Am Asn Textile Technologists. *Res:* Silicone chemistry; polymer synthesis and characterization; emulsification; resin catalysis and cure; rheology; textile applications; paper release coating; fiber lubricant; surfactants; fiber intermediates; ethylene oxide derivates. *Mailing Add:* 50 Cedar Hollow Rd Wakefield RI 02879-1435

JOHNSON, GORDON E, b Welland, Ont, Sept 21, 34; m 58; c 6. PHARMACOLOGY. *Educ:* Univ Toronto, BScPhm, 57, MA, 59, PhD(pharmacol), 61. *Prof Exp:* Med Res Coun Can fel physiol, Karolinska Inst, Sweden, 62-63; from asst prof to prof pharmacol, Univ Toronto, 63-73; prof & head dept, 73-86, PROF PHARMACOL, UNIV SASK, 86- *Mem:* Am Soc Pharmacol & Exp Therapeut; Pharmacol Soc Can; Am Soc Clin Pharmacol; Can Soc Clin Pharmacol; Can Hypertension Soc. *Res:* Catecholamines; thermoregulation and influence of environmental temperature on drug action; drug metabolism. *Mailing Add:* Dept Pharmacol Univ Sask Saskatoon SK S7N 0W0 Can

JOHNSON, GORDON GUSTAV, b Chicago, Ill, June 23, 36; m 57; c 4. MATHEMATICS. *Educ:* Ill Inst Technol, BS, 58; Univ Tenn, PhD(math), 64. *Prof Exp:* Asst prof math, Univ Ga, 64-69; assoc prof, Va Polytech Inst, 69-71; assoc prof, 71-74, PROF MATH, UNIV HOUSTON, 74- *Concurrent Pos:* Managing ed, Houston J Math, 74-79, 84-89; fel, Oak Ridge Inst Nuclear Studies, 63-64; sr resident res, Nat Res Coun, 78-79; assoc, Johnson Space Ctr, NASA, 78-80 & NASA Hq, 80-81; ed, Houston J Math, 74-; vis prof, Emory Univ, 83-84, IDA, 90-91. *Mem:* Swedish Math Soc; Sigma Xi; Am Math Soc. *Res:* Analysis. *Mailing Add:* 2010 Fairwind Rd Houston TX 77062

JOHNSON, GORDON LEE, b Newark, Ohio, Dec 21, 32; m 58, 66; c 2. GENERAL CHEMISTRY. *Educ:* Ohio Univ, BA, 54; Univ Ill, PhD(inorg chem), 58. *Prof Exp:* PROF CHEM, KENYON COL, 62-, CHMN DEPT, 86- *Concurrent Pos:* Chmn, Chem Dept, Kenyon Col, 68-69 & 70-75; NSF fac fel, Iowa State Univ Sci & Technol, 69-70; NSF proj dir, 85-87; vis prof, Iowa State Univ, 69-70, Ohio State Univ, 83-84; vis scientist, Oak Ridge Nat Lab, 75-76. *Mem:* Sigma Xi; Am Chem Soc. *Res:* Metal-ion hydrolysis of esters-bioinorganic chemistry; titanium in molten salt systems; synthetic heme type compounds bioinorganic chemistry; author of 13 articles and books. *Mailing Add:* Dept Chem Kenyon Col Gambier OH 43022

JOHNSON, GORDON OLIVER, b Portland, Ore, June 2, 44; m 71; c 3. SOLID STATE PHYSICS. *Educ:* Walla Walla Col, BS, 66; Calif Inst Technol, MS, 67, PhD(elec eng), 72. *Prof Exp:* Res assoc elec eng, Purdue Univ, 72-74; asst prof, 74-77, assoc prof, 77-80, PROF PHYSICS, WALLA WALLA COL, 80- *Mem:* Inst Elec & Electronics Engrs; Am Phys Soc. *Res:* Magnetic materials; processes of magnetization; magneto resistance phenomena. *Mailing Add:* Dept Physics Walla Walla Col 204 S College Ave College Place WA 99324

JOHNSON, GORDON V, b Harvey, NDak, Jan 9, 40; m 62; c 2. SOIL FERTILITY. *Educ:* NDak State Univ, BS, 63; Univ Nev, MS, 66; Univ Nebr, PhD(agron), 69. *Prof Exp:* From asst prof to assoc prof agr chem & soils, Univ Ariz, 69-77; assoc prof, 77-83, PROF AGRON, OKLA STATE UNIV, 83-; DIR AGRON SERV & STATE SOIL SPECIALIST, EXTEN, 78- *Mem:* Int Turfgrass Soc; Crop Sci Soc Am; Am Soc Agron; Soil Sci Soc Am. *Res:* Evaluation of micro-nutrient supplying status of soils; evaluation of interferences in the spectrophotometric determination of iron with ethylenediamine Di (o-hydroxyphenylacetic acid); turfgrass management and nutrition; subirrigation of turfgrass; soil-turfgrass systems for tertiary sewage effluent treatment; effects of temperature and nitrogen on turfgrass root decline; soil fertility and soil salinity. *Mailing Add:* Dept Agron Okla State Univ Stillwater OK 74078

JOHNSON, GORDON VERLE, b Long Beach, Calif, Sept 5, 33; m 60; c 4. PLANT PHYSIOLOGY. *Educ:* Univ Calif, Berkeley, BS, 55, MS, 59; Univ Ariz, PhD(agr chem, soils), 65. *Prof Exp:* Res assoc bot, Ore State Univ, 63-64, asst prof, 64-65; asst prof, 65-70, ASSOC PROF BIOL, UNIV NMEX, 70- *Mem:* AAAS; Am Soc Plant Physiol; Sigma Xi. *Res:* Absorption and metabolism of iron by plants; physiological effects of stress on plants; algal nutrition; biological nitrogen fixation; plant tissue and cell culture. *Mailing Add:* Dept Biol Univ NMex Albuquerque NM 87131

JOHNSON, GREGORY CONRAD, b Boston, Mass, June 5, 63; m 90. PHYSICAL OCEANOGRAPHY. *Educ:* Bates Col, BS, 85; Mass Inst Technol, PhD(oceanog), 91. *Prof Exp:* Res asst, Woods Hole Oceanog Inst, 85-90, vis investr, 90; POSTDOCTORAL SCIENTIST, APPL PHYSICS LAB, UNIV WASH, 90- *Mem:* Am Geophys Union; Oceanog Soc. *Res:* Structure and dynamics of the deep circulation of the world ocean and the overflows which supply this circulation. *Mailing Add:* Appl Physics Lab Univ Wash 1013 NE 40th St Seattle WA 98105

JOHNSON, GROVER LEON, b Bunn, Ark, Jan 9, 31; m 62; c 3. PHYSICAL CHEMISTRY. *Educ:* Rice Inst, BA, 53; Univ Tex, PhD(phys chem), 60. *Prof Exp:* Sr res chemist corrosion, Socony Mobil Oil Co, 60-64; ASST PROF CHEM, UNIV TEX, ARLINGTON, 64- *Concurrent Pos:* Consult, Socony Mobil Oil Co, 64- *Mem:* Am Chem Soc. *Res:* Electrochemistry; corrosion. *Mailing Add:* 1716 Cheryl St Arlington TX 76013

JOHNSON, GUY, JR, b Dallas, Tex, Mar 11, 22; m 42; c 3. MATHEMATICAL ANALYSIS. *Educ:* Agr & Mech Col Tex, BS, 43, MS, 52; Harvard Univ, MBA, 47; Rice Inst, PhD, 55. *Prof Exp:* Asst eng, Tex Eng Exp Sta, 48-50; from instr to assoc prof math, Rice Univ, 54-66; assoc prof, 66-69, PROF MATH, SYRACUSE UNIV, 69- *Concurrent Pos:* Vis prof, Syracuse Univ, 64-66. *Mem:* Am Math Soc; Math Asn Am. *Res:* Potential theory. *Mailing Add:* RD 1 E Hill Rd Spencer NY 14883

JOHNSON, HAL G(USTAV), b Saginaw, Mich, Apr 30, 15; m 40; c 3. ORGANIC CHEMISTRY, MARKETING. *Educ:* Beloit Col, BS, 36, MS, 38; Univ Wis, PhD(org chem), 41. *Prof Exp:* Instr chem, Beloit Col, 35-38; asst, Univ Wis, 38-41; org chemist, Com Solvents Corp, Ind, 41-45; asst gen mgr, Dykem Co, St Louis, Mo, 45-46; mgr org intermediate & pharmaceuts, Org Develop Dept, Monsanto Chem Co, 46-49, asst dir, Gen Develop Dept, 49-52, dir res & develop, Western Div, Calif, 52-54, dir develop dept, Res & Eng Div, 54-57; dir chem & rubber div, Bus & Defense Serv Admin, US Dept Com, Washington, DC, 57; vpres, Vick Chem Co, 57-59; chem & mgt consult, 59-62; vpres mkt & sales, Southwest Potash Div, Am Metal Climax, Inc, 62-66; mgt consult, Hal Johnson Assocs & Barnes Res Assocs, 66-69; dir chem develop, Chem Plastics Group, Develop Div, Borg Warner Corp, 69-71; from assoc prof to prof, 71-84, EMER PROF MKT, NORTHERN ILL UNIV, 84- *Concurrent Pos:* Educ & mgt consult; guest prof int & indust mkt, Linköping Univ, Sweden, 77-78; mem, bd dir, Marsh Prod, Batavia, Ill, 84-; chmn, bd dir, Ill Bus Hall of Fame, Macomb. *Mem:* AAAS; Am Chem Soc; Com Develop Asn. *Mailing Add:* 1060 S Adams St Lancaster WI 53813

JOHNSON, HAMILTON MCKEE, geology, geophysics; deceased, see previous edition for last biography

JOHNSON, HARLAN BRUCE, b Indianapolis, Ind, July 3, 22; m 44; c 4. PHYSICAL CHEMISTRY. *Educ:* Purdue Univ, BS, 43; Iowa State Col, MS, 48; Kans State Col, PhD(chem), 52. *Prof Exp:* Org res chemist, Eastman Kodak Co, 43-44; prod supvr, Tenn Eastman Corp, 44-46; asst, Atomic Res Inst, 46-48; from instr to asst prof chem, Ft Hays Kans State Col, 48-52; prof & head dept, Washburn Univ, 52-57, chmn sci div, 56-57; res supvr, Petro-Tex Chem Corp, 57-67, asst dir res, 66-67; dir res, Columbia Nitrogen Corp, 67-70; MEM STAFF, PPG INDUSTS, INC, 70- *Mem:* Am Chem Soc; Am Inst Chem Engrs; Sigma Xi. *Res:* Electrolytic solutions; thermodynamics; petrochemicals; electrochemistry. *Mailing Add:* 2920 Seville Rd Rittman OH 44270

JOHNSON, HARLAN PAUL, b Chicago, Ill, Dec 18, 39; m 72; c 2. OCEANOGRAPHY. *Educ:* Univ Ill, BS, 63; Southern Ill Univ, MS, 66; Univ Wash, PhD(geophysics), 72. *Prof Exp:* RES ASSOC PROF, UNIV WASH, 80- *Concurrent Pos:* Vis prof, Inst Geol, Univ Rennes, France, 81. *Res:* Origin and evolution of oceanic crust; rock magnetism; source of marine magnetic anomalies. *Mailing Add:* 6533 45th Ave NE Seattle WA 98115

JOHNSON, HAROLD DAVID, b Verona, Mo, Feb 28, 24; m 49; c 4. PHYSIOLOGY. *Educ:* Drury Col, BS, 49; Univ Mo, MA, 52, PhD(dairy husb), 56. *Prof Exp:* Asst biol, Drury Col, 48-49; drug rep, Kendall Co, Ind, 49-50; asst zool, 51-52, from asst to prof dairy husb, 52-77, PROF ENVIRON PHYSIOL, UNIV MO-COLUMBIA, 77- *Concurrent Pos:* Mem comt bioclimatol & meteorol, Agr Bd, Nat Acad Sci. *Honors & Awards:* Animal Biometeorol Award, Am Meteorol Soc, 72; Peterson Award, Int Soc Biometeorol, 72; Gamma Sigma Delta Fac Res Award, 75. *Mem:* AAAS; Am Soc Animal Sci; Am Physiol Soc; Am Dairy Sci Asn; Int Soc Biometeorol. *Res:* Environmental physiology; investigations on effects of climate and environment on growth and production; related biochemical and physiological reactions of cattle and smaller laboratory mammals. *Mailing Add:* Dept Dairy Univ Mo 114 Animal Sci Res Ctr Columbia MO 65203

JOHNSON, HAROLD HUNT, b Gary, Ind, Sept 20, 29; m 58; c 1. MATHEMATICS. *Educ:* San Jose State Col, BA, 51; Univ Calif, MA, 56, PhD(math), 57. *Prof Exp:* Instr math, Stanford Univ, 57-58 & Princeton Univ, 58-61; assoc prof, Univ Wash, 61-74; vis assoc prof math, George Washington Univ, 74-76; PROF MATH, TRINITY COL, 77- *Mem:* Am Math Soc; Math Asn Am; Soc Indust & Appl Math. *Res:* Differential geometry; systems of exterior differential forms; infinite pseudo-groups. *Mailing Add:* 590 S Brys Grosse Point MI 48236

JOHNSON, HARRY MCCLURE, b Chicago, Ill, May 15, 25; m 48; c 4. METEOROLOGY, OCEANOGRAPHY. *Educ:* Mass Inst Technol, BS, 46; Cornell Univ, MS & PhD(environ ecol, physics), 54. *Prof Exp:* Officer-in-chg, Navy Weather Unit, Alaska, 46; asst physics & math, Cornell Univ, 47-48, math, 48-50 & zool, 50-52, asst prof meteorol & in chg div meteorol, 54-59; assoc meteorologist & oceanogr, Meteorol Div, Inst Geophys, Univ Hawaii, 60-61; res meteorologist, Nat Environ Satellite Ctr, 61-68 & Nat Hurricane Ctr, Fla, 68-74, RES METEOROLOGIST, NAT OCEANIC & ATMOSPHERIC ADMIN, NAT ENVIRON SCI SERV, MD, 74- *Mem:* AAAS; Ecol Soc Am; Wilderness Soc; Cooper Ornith Soc. *Res:* Satellite meteorology; arctic, subarctic, subtropical and tropical meteorology; oceanography; micrometeorology and ocean-atmosphere interactions; environmental ecology and physiology; wilderness and habitat preservation. *Mailing Add:* 13310 Warburton Dr Ft Washington MD 20744

JOHNSON, HARRY WILLIAM, JR, b Waverly, Fla, Jan 2, 27; m 57; c 3. ORGANIC CHEMISTRY. *Educ:* Mass Inst Technol, SB, 51; Univ Ill, PhD(chem), 54. *Prof Exp:* From instr to assoc prof & chmn dept, Univ Calif, 54-67, prof, 67-88, dean, Grad Div, 74-80, assoc dean, Grad Div, 82-90, EMER PROF CHEM, UNIV CALIF, RIVERSIDE, 88- *Mem:* AAAS; Am Chem Soc; Royal Soc Chem. *Res:* Organic reaction mechanisms; reactions of heterocycles; isocyanate and isocyanide chemistry. *Mailing Add:* Dept Chem Univ Calif Riverside CA 92521

JOHNSON, HENRY STANLEY, JR, b Augusta, Ga, Apr 16, 26; m 54; c 4. EXPLORATION GEOLOGY, ECONOMIC GEOLOGY. *Educ:* Univ SC, BS, 47. *Prof Exp:* Geologist, Minerals Deposits Br, US Geol Surv, 48-49 & 52-57; geologist, Zonolite Co, 49-52; chief geologist, Div Geol, SC State Develop Bd, 57-61; state geologist, 61-69; CONSULT ECON GEOLOGIST & PRES, SANDHILL RESOURCES, INC, 69- *Mem:* Geol Soc Am; Soc Econ Geologists; Am Inst Mining & Metall Engrs; Am Inst Prof Geol Scientists; Am Asn Petrol Geologists. *Res:* Economic geology; metallic and non-metallic mineral deposits and petroleum; petrology; structure; stratigraphy; general geology. *Mailing Add:* Deer Run Dr No R6 Oxford MS 38655

JOHNSON, HERBERT GARDNER, b Wessington, SDak, Mar 22, 33; m 53; c 3. ASTHMA, CHRONIC AIRWAYS DISEASES. *Educ:* Univ Ill, Urbana, BS, 58, MS, 59; Univ Mich, Ann Arbor, PhD(immunol), 69. *Prof Exp:* Res asst biochem, Upjohn Co, 59-66, res assoc immunol, 69-75, res scientist, 75-78, sr res scientist, 78-84, SR SCIENTIST IMMUNOL, UPJOHN CO, KALAMAZOO, MICH, 84- *Concurrent Pos:* Vis scholar pharmacol, Univ Calif, San Francisco, 80-81. *Mem:* Am Asn Immunologists; Am Asn Physiologists; Am Soc Microbiol. *Res:* Role of lipoxygenase metabolites of arachidonic acid in chronic airways diseases; immunopharmacology of lipid mediators and their pharmacologic control as related to airways, smooth muscle and glands. *Mailing Add:* Upjohn Co 301 Henrietta Kalamazoo MI 49001

JOHNSON, HERBERT GORDON, b Granite Falls, Minn, Apr 11, 16; m 41; c 2. PLANT PATHOLOGY. *Educ:* Univ Minn, BS, 39, PhD, 53. *Prof Exp:* Agt barberry eradication, USDA, 39-40; plant pathologist & horticulturist, Yoder Bros, Inc, 40-42 & 45-48; asst plant path, Univ Minn, 48-53; plant pathologist, Green Giant Co, 53-56; assoc prof plant path, Univ Minn, St Paul, 56-64, prof, 64-80, exten plant pathologist, 56-80; RETIRED. *Mem:* Am Phytopath Soc; Sigma Xi. *Res:* Applied plant pathology. *Mailing Add:* 2175 Rosewood Lane S St Paul MN 55113

JOHNSON, HERBERT HARRISON, materials science, engineering; deceased, see previous edition for last biography

JOHNSON, HERBERT WINDAL, b Tenn, July 3, 20; m 48; c 3. GENETICS, PLANT BREEDING. *Educ:* Univ Tenn, BSc, 43; Univ Nebr, MSc, 48, PhD(agron), 50. *Prof Exp:* Instr genetics, Univ Nebr, 47-48; agronomist plant breeding, USDA, NC, 48-53, res agronomist, Crops Res Div, Agr Res Serv, 53-64; PROF AGRON & HEAD DEPT AGRON & PLANT GENETICS, INST AGR, UNIV MINN, ST PAUL, 64- *Mem:* Fel Am Soc Agron. *Res:* Quantitative genetics; plant breeding procedures. *Mailing Add:* 11081 Pleasant Valley Rd Sun City AZ 85351

JOHNSON, HERMAN LEONALL, b Whitehall, Wis, Apr 1, 35; m 76. HUMAN NUTRITIONAL STATUS. *Educ:* N Cent Col, Ill, BA, 59; Va Polytech Inst & State Univ, Blacksburg, MS, 61, PhD(biochem-nutrit), 63. *Prof Exp:* Res biochemist, S R Noble Res Fedn, Ardmore, Okla, 63-65; nutrit chemist human res, US Army Med Res & Nutrit Lab, Denver, 65-74 & US Army Western Inst of Res Ctr, Presidio, San Francisco, 74-80; RES PHYSIOLOGIST HUMAN RES, WESTERN HUMAN NUTRIT RES CTR, USDA, PRESIDIO OF SAN FRANCISCO, 80- *Mem:* Am Inst Nutrit; Am Col Nutrit; Am Col Sports Med; Am Soc Clin Nutrit; AAAS; NY Acad Sci. *Res:* New and improved methods for determining human body composition and energy metabolism-expenditure; effects of nutritional status on body composition and energy metabolism in humans. *Mailing Add:* PO Box 29997 Presidio of San Francisco CA 94129

JOHNSON, HILDING REYNOLD, b Sweden, Feb 14, 20; US citizen; m 47. ANALYTICAL CHEMISTRY. *Educ:* Clarkson Col Technol, BS, 42. *Prof Exp:* Chemist, Heyden Chem Corp, 42-48; group leader anal chem, Heyden Newport Chem Corp, 48-70; supvr, Tenneco Chem, Inc, 70-75, mgr anal serv, 75-81; RETIRED. *Mem:* Am Chem Soc. *Mailing Add:* 19 Lois Ct Packanack Lake Wayne NJ 07470

JOHNSON, HOLLIS RALPH, b Tremonton, Utah, Dec 2, 28; m 54; c 6. ASTRONOMY, SPECTROSCOPY & SPECTROMETRY. *Educ:* Brigham Young Univ, BA, 55, MA, 57; Univ Colo, PhD(astrophys), 60. *Prof Exp:* NSF fel, Paris, France, 60-61; res assoc astron, Yale Univ, 61-63; assoc prof, 63-69, PROF ASTRON, IND UNIV, BLOOMINGTON, 69-, CHMN DEPT, 78-82, 90- *Concurrent Pos:* Vis scientist, High Altitude Observ, Nat Ctr Atmospheric Res, 71-72; sr fel, Nat Res Coun, NASA Ames Res Ctr, Moffett Field, Calif, 82-83; F C Donders vis prof, Univ Utrecht, Netherland, 89; vis prof, Niels Bohr Inst, Univ Copenhagen, 90. *Mem:* Int Astron Union; Sigma Xi; Am Astron Soc; AAAS; Am Asn Univ Profs. *Res:* Theory of spectral line formation; stellar chromosphere; cool giant stars; chemical composition of stars; molecular opacities; radiative transfer. *Mailing Add:* Dept Astron Ind Univ Swain Hall W 319 Bloomington IN 47405

JOHNSON, HOLLISTER, JR, b Watertown, NY, Jan 14, 29; m 51; c 2. CHEMISTRY. *Educ:* Univ Rochester, BS, 59. *Prof Exp:* Res assoc, Eastman Kodak Res Labs, 53-86, consult, 86-89; RETIRED. *Mem:* Am Chem Soc. *Res:* Solution formulation and coating technology. *Mailing Add:* 302 Killarney Dr Rochester NY 14616-2435

JOHNSON, HOMER F(IELDS), JR, chemical engineering; deceased, see previous edition for last biography

JOHNSON, HORACE RICHARD, b Jersey City, NJ, Apr 26, 26; m 50; c 5. PHYSICS, ELECTRICAL ENGINEERING. *Educ:* Cornell Univ, BEE, 46; Mass Inst Technol, PhD(physics), 52. *Prof Exp:* Asst physics, Cornell Univ, 46-47; asst, Mass Inst Technol, 51; head microwave tube dept, Res Lab, Hughes Aircraft Co, 52-57; exec vpres, 58-68, pres, 68-87, VCHMN BD, WATKINS-JOHNSON CO, 88- *Concurrent Pos:* Lectr, Univ Calif, Los Angeles, 56-57; lectr, Stanford Univ, 58-68, assoc, Dept Elec Eng, 68-; mem bd dirs, Nat Asn Mfrs. *Mem:* Nat Acad Eng; Am Phys Soc; Sigma Xi; fel Inst

Elec & Electronics Engrs; Res Soc Am. *Res:* Microwave spectroscopy; electron devices; microwave systems; author of 21 technical publications; awarded three patents. *Mailing Add:* Watkins-Johnson Co 3333 Hillview Ave Palo Alto CA 94304-1204

JOHNSON, HORTON ANTON, b Cheyenne, Wyo, Nov 12, 26. MEDICINE, PATHOLOGY. *Educ:* Colo Col, AB, 49; Columbia Univ, MD, 53; Am Bd Path, dipl, 58. *Prof Exp:* Intern, Univ Mich, 53-54, resident path, 54-57; resident, Pondville Hosp, Walpole, Mass, 57-58; res assoc, Brookhaven Nat Lab, 58-60; asst prof, Univ Utah, 60-63; scientist & attend pathologist, Brookhaven Nat Lab, 63-70; prof path, State Univ NY Stony Brook, 70-72; prof, Sch Med, Ind Univ, Indianapolis, 72-75; prof path & chmn dept, Sch Med, Tulane Univ, 75-84; dir path, St Lukes-Roosevelt Hosp Ctr, 84-91; PROF PATH, COLUMBIA UNIV, 84- *Honors & Awards:* Lederle Med Fac Award. *Mem:* Radiation Res Soc; Am Asn Path; Biophys Soc; Col Am Path; Int Acad Path. *Res:* Radiation pathology; kinetics of cell proliferation; thermal injury; information theory. *Mailing Add:* 30 W 61st St Suite 21B New York NY 10023

JOHNSON, HOWARD (LAURENCE), b San Leandro, Calif, Jan 4, 33; m 56; c 4. MEDICINAL CHEMISTRY, PHARMACOLOGY. *Educ:* Univ Calif, BS, 56, PhD, 63. *Prof Exp:* Pharmaceut educ fel, 59-61; fel chem pharmacol, Nat Heart Inst, 63-65; chemist pharmaceut chem, 65-71, sr pharmacol chemist, 72-78, MGR BIOPHYS CHEMOMETRICS, LIFE SCI RES, STANFORD RES INST, 78- *Concurrent Pos:* Clin lab officer, US Air Force, 57-59; res assoc, Med Ctr, Univ Calif, 72-; Fed Aviation Admin licensed private pilot, 81- *Mem:* AAAS; Am Chem Soc; Am Pharmaceut Asn; Acad Pharmaceut Sci; Am Soc Pharmacol & Exp Therapeut. *Res:* Chemistry, pharmacology of autonomic agents; extrapyramidal central nervous system pharmacology; drug distribution, metabolism and mechanisms of action; structure activity relationships; biochemical pharmacology of biogenic amines; histamine; drug-receptor interaction. *Mailing Add:* Bio-Org Chem Lab Life Scis SRI Int Menlo Park CA 94025

JOHNSON, HOWARD ARTHUR, SR, b Ind, Dec 16, 23; m 47; c 2. MATHEMATICS, OPERATIONS RESEARCH. *Educ:* Franklin Col Ind, AB, 49; Wesleyan Univ, MA, 50. *Prof Exp:* Physicist, Naval Ord Plant, Ind, 50-54; opers analyst, Air Proving Ground Command, 54-58; chief opers anal, Hq, 3rd Air Force, Eng, 58, dept chief opers anal, Hq, US Air Forces Europe, 58-61, dir, Opers Model Eval Group Air Force (OMEGA), 61-63; sr staff scientist & mgr comp effectiveness res div, Spindletop Res, Inc, 63-67; res dir, Vitro Servs Div, Vitro Corp Am, 67-68; sci asst to dir testing, Hq, Armament Develop & Test Ctr, 68-70, sci asst electronics test, 70-73, sr opers res scientist, Hq, USAF Tactical Air Warfare Ctr, Elgin AFB, Fla, 73-84; CHIEF EXEC OFFICER, ASSOC CONSULTS, FT WALTON BEACH, FLA, 74- *Concurrent Pos:* Consult, Supreme Hq, Allied Powers Europ, 59-61, Ministry Defense, WGermany, 61, USAF, 64-65, Univ Ky Med Ctr, 66-67 & Gulf South Res Inst, 68-84; mem, Int Exec Serv Corps, 84- *Mem:* Opers Res Soc Am; Mil Opers Res Soc; Am Statist Asn; Sigma Xi; Int Test & Eval Asn; Armed Forces Commun & Electronics Asn. *Res:* Solution of non-recurring operational problems for command or management decision utilizing the scientific method and a quantitative multi-disciplinary approach. *Mailing Add:* 10409 Huntington Dr Eden Prairie MN 55347

JOHNSON, HOWARD B(EATTIE), b Willits, Calif, Apr 27, 36; m 62; c 4. CERAMICS ENGINEERING, PHYSICAL CHEMISTRY. *Educ:* Univ of the Pac, BS, 58; Univ Minn, MS, 66; Univ Utah, PhD(ceramic eng), 66. *Prof Exp:* Res chemist, PPG Indust, Inc, 60-63; sr ceramist, Pittsburgh Corning Corp, 66-69, dir process develop, 69-77; dir res, Vesuviys Crucible Co, 77-82; tech dir, 82-88, VPRES & TECH DIR, CONSOL CERAM PRODS, INC, 88- *Mem:* Am Ceramic Soc; Am Chem Soc; Sigma Xi. *Res:* Manufacturing inorganic thermal insulation materials and alumina graphite refractories, vacuum formed disposable refractories for steel and aluminum; kinetics and thermodynamics of gas-solid reactions; electrical properties of ceramic materials. *Mailing Add:* 8714 Tanager Woods Dr Cincinnati OH 45249

JOHNSON, HOWARD ERNEST, b Livingston, Mont, Sept 21, 35; m 59; c 3. FRESH WATER ECOLOGY, TOXICOLOGY. *Educ:* Mont State Univ, BS, 59, MS, 61; Univ Wash, PhD(fisheries), 67. *Prof Exp:* From asst prof to assoc prof, 67-75, PROF FISHERIES, MICH STATE UNIV, 75-, COORDR ENVIRON CONTAMINATION, PESTICIDE RES CTR, 78- *Concurrent Pos:* Panel mem comt water qual criteria, Nat Acad Sci, 71-72; coordr toxic substances, Mich Serv & Educ Admin Grant Prog, 78- *Mem:* Am Fisheries Soc; Am Inst Fisheries Res Biologists. *Res:* Toxicity tests with aquatic organisms; production and culture of fish. *Mailing Add:* 1041 University Helena MT 59601

JOHNSON, HOWARD MARCELLUS, LYMPHOCYTE FUNCTION, SOLUBLE MEDIATORS. *Educ:* Ohio State Univ, PhD(immunol), 62. *Prof Exp:* PROF COMP EXP PATH, UNIV FLA, 83- *Mailing Add:* Dept Microbiol & Cell Sci Univ Fla 1059 McCarty Hall Gainesville FL 32611

JOHNSON, HOWARD P, b Odebolt, Iowa, Jan 27, 23; m 52; c 3. HYDROLOGY & WATER RESOURCES. *Educ:* Iowa State Univ, BS, 49, MS, 50, PhD(agr & civil eng), 59; Univ Iowa, MS, 54. *Prof Exp:* Lab asst, Iowa State Univ, 49-50, res assoc, 51-53 & 55-58, from asst prof to prof 59-80, head agr eng, 80-88, ANSON MARSTON DISTINGUISHED PROF, IOWA STATE UNIV, 88- *Concurrent Pos:* Vis prof, Univ Mo, Columbia, 66-67; dir, Am Soc Agr Engrs, 76. *Honors & Awards:* Hancor Soil & Water Eng Award, Am Soc Agr Engrs, 78. *Mem:* Fel AAAS; fel Am Soc Agr Engrs; Soil Conserv Soc Am; Am Soc Eng Educ. *Res:* Hydrology, water quality and soil mechanics problems related to irrigation, drainage, erosion control and small watersheds. *Mailing Add:* Dept Agr Eng Iowa State Univ Ames IA 50010

JOHNSON, HUGH MITCHELL, b Des Moines, Iowa, Mar 4, 23; m 51. ASTRONOMY. *Educ:* Univ Chicago, AB, 48, SB, 49, PhD(astron), 53. *Prof Exp:* Asst astron, Yerkes Observ, Univ Chicago, 50-53; asst prof, Univ Iowa, 54-59; assoc prof & assoc astronr, Univ Ariz, 60-62; assoc scientist, Nat Radio Astron Observ, 62-63; staff scientist & mem res lab, Lockheed Missiles & Space Co, 63-86; RETIRED. *Concurrent Pos:* Res assoc, Yerkes Observ, Univ Chicago, 53-60; vis fel, Australian Nat Univ, 58-59; lectr, Stanford Univ, 71-75 & 80-82. *Mem:* Am Astron Soc; Int Astron Union. *Res:* Nebulae; galaxies; x-ray astronomy. *Mailing Add:* 1017 Newell Rd Palo Alto CA 94303

JOHNSON, I BIRGER, b Brooklyn, NY, Feb 29, 13. ENGINEERING. *Educ:* Brooklyn Polytech, BEE, 37, MEE, 39. *Prof Exp:* Grad prof elec eng, Brooklyn Polytech, 37-39; mgr elec eng, Gen Elec, 39-78; consult, 78-88; RETIRED. *Honors & Awards:* William Martin Havirshaw Award, Inst Elec & Electronics Engrs, 66, Centennial Award, 84 & Lamme Medal, 86. *Mem:* Nat Acad Eng; Inst Elec & Electronics Engrs. *Mailing Add:* 1508 Barclay Pl Schenectady NY 12309

JOHNSON, IRVING, b Chicago, Ill, Oct 23, 18; m 42; c 2. PHYSICAL CHEMISTRY. *Educ:* Cornell Col, BA, 41; Columbia Univ, MA, 43, PhD(phys chem), 47. *Prof Exp:* Asst chem, Columbia Univ, 41-42, lect demonstr, 42-43, lectr chem, 43-44, asst, Div War Res, 44-45; from asst prof to assoc prof chem, Okla Agr & Mech Col, 46-53; prin res engr, Ford Motor Co, 53-57; chemist, 57-79, SR CHEMIST, ARGONNE NAT LAB, 79- *Mem:* Am Chem Soc; Am Inst Chemists. *Res:* Kinetics; light scattering; aerosols; thermodynamics of high temperature systems; electrochemistry; chemistry of nuclear fuels; fuel reprocessing. *Mailing Add:* 276 Woodstock Ave Clarendon Hills IL 60514

JOHNSON, IRVING STANLEY, b Grand Junction, Colo, June 30, 25; m 49; c 4. CANCER, RESEARCH ADMINISTRATION. *Educ:* Washburn Univ, AB, 48; Univ Kans, PhD(zool), 53. *Prof Exp:* Asst instr anat, Washburn Univ, 47-48; asst instr parasitol, embryol & zool, Univ Kans, 48-50, asst zool, 50-53; asst dir biol-pharmacol res div, 53-68, dir biol res div, Lilly Res Lab, 68-72, exec dir, Eli Lilly Res Labs, 72-73, VPRES RES, ELI LILLY RES LABS, 73- *Concurrent Pos:* Ed bd, Chemico-Biol Interactions, 68-73; mem consult panel, Nat Cancer Prog, 71; ed adv bd, Cancer Res, 71-73, assoc ed, 74-; mem develop therapeut comt, Nat Cancer Inst, 78- *Mem:* AAAS; Am Asn Cancer Res; Am Soc Cell Biologists; Environ Mutagen Soc; Soc Exp Biol & Med; Sigma Xi. *Res:* Anti-tumor chemotherapy; antiviral chemotherapy; tissue culture techniques; experimental embryology; oncogenic viruses; maintenance of biological function in tissue culture; recombinant DNA and public policy. *Mailing Add:* Dept MC 700 Bldg 28-1 Lilly Res Labs 307 E McCarty St Indianapolis IN 46285

JOHNSON, IVAN M, b Mansfield, Wash, May 30, 40; m 62; c 2. ZOOLOGY, PHYSIOLOGY. *Educ:* Whitworth Col, Wash, BS, 62; Univ Mont, PhD(zool), 69. *Prof Exp:* Asst zool, Univ Mont, 63-69; Nat Inst Gen Med Sci fel biol, Yale Univ, 69-71; asst prof, 71-78, ASSOC PROF BIOL, CONCORDIA COL, 78- *Mem:* Sigma Xi; Raptor Res Found. *Res:* Osmoregulation of vertebrates. *Mailing Add:* Dept Biol Concordia Col Moorhead MN 56560

JOHNSON, J(OSEPH) ALAN, b W Palm Beach, Fla, Feb 1, 33; m 56; c 2. ENDOCRINOLOGY. *Educ:* Butler Univ, BA, 63; Ind Univ Med Ctr, PhD(physiol), 68. *Prof Exp:* USPHS fel, 69-71, from asst prof to assoc prof, 71-85, PROF PHYSIOL, UNIV MO-COLUMBIA, 85-, RES PHYSIOLOGIST, VET ADMIN HOSP, 74- *Mem:* Am Physiol Soc; Endocrine Soc; Am Soc Nephrol; Soc Exp Biol & Med; Am Heart Asn. *Res:* Cardiovascular and endocrine physiology; mechanisms in the production of hypertension in animal models. *Mailing Add:* Res Serv 151 Truman Mem Vet Admin Hosp Columbia MO 65201

JOHNSON, J DAVID, MUSCLE BIOCHEMISTRY, CALCIUM BINDING PROTEINS. *Educ:* Mich State Univ, PhD(biophysics), 77. *Prof Exp:* PROF MED BIOCHEM, MED CTR, OHIO STATE UNIV, 83- *Mailing Add:* Med Ctr Ohio State Univ 5170 Graves Hall 333 W 10th Ave Columbus OH 43210

JOHNSON, J(AMES) DONALD, b Inglewood, Calif, Aug 1, 35; m 55; c 2. STRUCTURAL GEOLOGY, ANALYTICAL CHEMISTRY. *Educ:* Univ Calif, Los Angeles, BS, 57; Univ NC, PhD(anal chem), 62. *Prof Exp:* From asst prof to assoc prof water chem, 61-72, PROF ENVIRON CHEM, SCH PUB HEALTH, UNIV NC, CHAPEL HILL, 72- *Concurrent Pos:* Vis lectr, NC Wesleyan Col, 63-64; environ fel, Gothenburg Univ, Sweden, 70-71 & Nobel symp, 71. *Honors & Awards:* Tanner Award Water Res, US Environ Protection Agency Sci Adv Bd. *Mem:* Am Asn Univ Professors; Am Chem Soc; Am Water Works Asn; Water Pollution Control Fedn. *Res:* Chemistry of natural aqueous solutions; analysis and kinetics of chlorine and bromine; drinking, cooling, and waste-water disinfection chemistry. *Mailing Add:* 7315 Mount Herman Rock Creek Rd Snow Camp NC 27349

JOHNSON, J(AMES) R(OBERT), b Cincinnati, Ohio, Jan 2, 23; m 45; c 6. CERAMICS ENGINEERING. *Educ:* Ohio State Univ, BCerE, 47, MSc, 48, PhD(ceramic eng), 50. *Prof Exp:* Asst instr ceramic eng, Ohio State Univ, 49-50; asst prof, Univ Tex, 50-51; tech adv, Ceramic Lab, Oak Ridge Nat Lab, 51-56; mgr, nuclear lab, Cent Res Labs, 3M Co, 56-62, dir phys sci res, 62-72, exec scientist & dir adv res progs lab, 72-79; CONSULT, 79- *Concurrent Pos:* Adj prof, Univ Minn & Univ Wis-Stout; 3m William L McKnight distinguished prof, Univ Minn, Duluth, 89-90; Nelva Runnalls res award, Univ Wis, 90. *Honors & Awards:* Pace Award, 59; Texnikoi Award, 62; Engrs Achievement Award, Am Soc Mech Engrs, 80; Greaves-Walker Award, 85; JRJ Award, 85; Prakken Award, Int Test & Eval Asn, 89. *Mem:* Nat Acad Eng; hon mem Am Ceramic Soc (pres, 73); Sigma Xi; Nat Inst Ceramic Engrs; AAAS. *Res:* Ceramics; metallurgy; solid state physics; inorganic chemistry; diffusion; glass. *Mailing Add:* Rte 1 Box 231B River Falls WI 54022

JOHNSON, J(OSEPH) STUART, b Gower, Mo, May 8, 12; m 34; c 3. ELECTRICAL ENGINEERING. *Educ:* Univ Mo, BS, 32, MS, 34; Iowa State Col, PhD(elec eng), 37. *Hon Degrees:* DSc, Lawrence Inst Technol, 63. *Prof Exp:* Asst elec eng, Iowa State Col, 34-36, instr mech, 36-37; elec eng, Mo Sch Mines, 37-39, asst prof, 39-46; assoc res engr, Univ Fla, 46-47, prof elec eng & asst dean, 47-54; prof elec eng & head sch, Purdue Univ, 54-57; prof elec eng & dean col eng, Wayne State Univ, 57-67; prof elec eng & dean sch eng, Univ Mo-Rolla, 67-77; RETIRED. *Concurrent Pos:* Instr eng sci mgt defense & war training, Army spec training prog & Sig Corps, Mo Sch Mines, 40-44. *Mem:* Am Soc Eng Educ (vpres, 63-65); Inst Elec & Electronics Engrs. *Res:* Induction motor design; electromagnetic wave propagation; correlation of electrical and thermal properties of building brick; power system analysis. *Mailing Add:* 5181 Marshfield Lane Sarasota FL 34235

JOHNSON, JACK (LAMAR), b Elkhart, Ind, Mar 30, 30; m 56; c 2. ANALYTICAL CHEMISTRY. *Educ:* Western Mich Univ, BS, 52; Wayne State Univ, MS, 54, PhD(anal chem), 59. *Prof Exp:* Anal chemist, Ethyl Corp, Mich, 54; SR RES CHEMIST, RES LABS, GEN MOTORS CORP, 59- *Mem:* AAAS; Am Chem Soc. *Res:* Chemical microscopy; microchemical techniques of analysis; development of instrumental methods for microanalysis and characterization of micro samples; x-ray diffraction analysis of materials. *Mailing Add:* 26026 Newport Warren MI 48089

JOHNSON, JACK DONALD, b Huntington, Ore, Aug 23, 31; m 58; c 4. RESEARCH ADMINISTRATION, ENVIRONMENTAL SCIENCE. *Educ:* San Diego State Col, BA, 59; Univ Minn, MS, 67, PhD(environ health), 71. *Prof Exp:* Proj engr, Humphrey, Inc, 56-60; sect chief aerospace, Martin-Marietta Corp, 60-63; syst engr, Jet Propulsion Labs, Calif Inst Technol, 63-66; res fel, Univ Minn, 67-70; assoc dean, Col Agr, 81-84, DIR, OFF ARID LANDS STUDIES, UNIV ARIZ, 71-, DIR AID NATURAL RESOURCES PROG, 74- *Concurrent Pos:* Desertification consult, AID, 74-; asst coordr, Interdisciplinary Progs, Univ Ariz, 71-; dir off technol trans, 85-; vpres, Ariz Technol Develop Corp, 87- *Mem:* AAAS; Am Water Resources Asn; Am Geophys Union; Inst Environ Sci; Inst Int Develop; Sigma Xi. *Res:* Desertification; less developed country development; utilization of arid land resources; hydrology; natural resources mangement; biomass and bioenergy development. *Mailing Add:* 7380 E Synder Rd Tucson AZ 85715-6208

JOHNSON, JACK WAYNE, b Cannon Falls, Minn, July 8, 50; m 73; c 3. SYNTHETIC INORGANIC, ORGANOMETALLIC CHEMISTRY. *Educ:* Carleton Col, BA, 72; Univ Wis-Madison, MS, 74, PhD(inorg chem), 76. *Prof Exp:* NSF fel inorg chem, Cornell Univ, 76-77; res chemist, 77-79, sr chemist, 79-81, staff chemist, 81-83, SR STAFF CHEMIST, INORG CHEM, CORP RES LABS, EXXON RES & ENG, 81- *Mem:* Am Chem Soc; Sigma Xi; Clay Minerals Soc. *Res:* Intercalation chemistry and layered solids; solid state chemistry; pillared clays. *Mailing Add:* 12 Sunrise Circle Clinton NJ 08809

JOHNSON, JAMES ALLEN, JR, b Selma, Ala, Nov 13, 54; m 80; c 3. ORGANIZATIONAL BEHAVIOR. *Educ:* Univ SAla, BA, 78, MS, 80; Auburn Univ, MPA, 82; Fla State Univ, PhD(org behav), 87. *Prof Exp:* Resident, Vet Admin Hosp, 82-83; sr assoc pub admin, Fla Ctr Productivity, 83-86; asst prof health admin, Memphis State Univ, 86-89; ASSOC PROF HEALTH ADMIN, MED UNIV SC, 89- *Concurrent Pos:* Instr, Fla State Univ, 83-86 & Tusculum Col, 87-89; bd mem, Alliance for Blind, 87-89; consult, Upjohn Healthcare Serv, 87-90; ed, J Mgt Pract, 88-90; health ed, J Health & Human Resources, 88-; assoc chair, Dept Health Admin, Med Univ SC, 90- *Honors & Awards:* Ram Award for Resin Mgt, Acad Mgt, 88. *Mem:* Acad Mgt; Am Pub Health Asn; Am Col Healthcare Execs; Nat Social Sci Asn; Am Soc Pub Admin. *Res:* Health and human resouce management; applied behavioral science research in quality improvement, risk behavior, and AIDS eduction. *Mailing Add:* Col Health Related Professions Med Univ SC Charleston SC 29425

JOHNSON, JAMES CARL, b Madison, Wis. VIROLOGY. *Educ:* Iowa State Univ, BS, 64; Mich State Univ, MS, 67, PhD(microbiol), 71. *Prof Exp:* Fel biochem, Mich State Univ, 71-72; Jane Coffin Childs fel molecular biol, Albert Einstein Col Med, 72-75; ASST PROF BIOL SCI, OLD DOMINION UNIV, 75- *Mem:* Am Soc Microbiol; Tissue Cult Asn. *Res:* Molecular biology of tumor viruses; ecology of viruses; DNA and RNA synthesis in prokaryotes and eukaryotes. *Mailing Add:* 1223 W 35th St Norfolk VA 23508

JOHNSON, JAMES DANIEL, b Toledo, Ohio, Mar 21, 44; m 66; c 1. THEORETICAL PHYSICS, STATISTICAL MECHANICS. *Educ:* Case Inst Technol, BS, 66; State Univ NY, Stony Brook, MA, 68, PhD(physics), 72. *Prof Exp:* Res assoc physics, Rockefeller Univ, 72-74; fel, Los Alamos Nat Lab, 74-76, staff mem physics, 76-89, proj leader, 85-89, dep group leader, 89-90, ACTG GROUP LEADER, PHYSICS, LOS ALAMOS NAT LAB, 90- *Concurrent Pos:* Prin investr, Los Alamos Nat Lab, 81-; mem US Delegation, Nuclear Testing Talks, 88. *Mem:* Am Phys Soc; AAAS. *Res:* Exact models and rigorous results in statistical mechanics; equation of state studies for materials of interest to energy development programs and to detonation physics. *Mailing Add:* T-1 MS-B221 Los Alamos Nat Lab Los Alamos NM 87545

JOHNSON, JAMES EDWARD, b Warren, Pa, Jan 3, 36; m 59; c 3. RADIATION BIOPHYSICS. *Educ:* Houghton Col, BS, 57; Univ Rochester, MS, 59; Colo State Univ, PhD(radiation biol), 65. *Prof Exp:* Chemist, E I du Pont de Nemours & Co, 57; res asst biophys, Univ Rochester, 59-62; instr radiation physics & radiation safety officer, Colo State Univ, 62-66, asst prof animal sci & radiation biol, 66-67; res assoc biophys, Harvard Med Sch, 67-68; from asst prof to assoc prof, 68-74, PROF ANIMAL SCI & RADIATION BIOL, COLO STATE UNIV, 74- *Concurrent Pos:* Lectr, Oak Ridge Mobile Lab, 65-73. *Mem:* Health Physics Soc. *Res:* Alkali metal metabolism; whole-body counting; environmental radioactivity. *Mailing Add:* Dept Animal Sci 135 BRB Colo State Univ Ft Collins CO 80523

JOHNSON, JAMES EDWARD, b Indianapolis, Ind, May 25, 40; m 62; c 1. FISH BIOLOGY, AQUATIC ECOLOGY. *Educ:* Purdue Univ, BS, 62; Butler Univ, MS, 65; Ariz State Univ, PhD(zool), 69. *Prof Exp:* Teacher jr high sch, 62-65; res assoc zool, Ariz State Univ, 69-70; fac growth & res grants, Univ Mass, Amherst, 71, asst prof fish biol, 70-77; BIOLOGIST ENDANGERED SPECIES, US FISH & WILDLIFE SERV, 77- *Mem:* Am Soc Ichthyologists & Herpetologists; Am Fisheries Soc. *Res:* Biology of Dorosoma petenense; population movement and growth of Catostomus commersoni and Alosa pseudoharengus. *Mailing Add:* 292 Alamosa NW Albuquerque NM 87107

JOHNSON, JAMES EDWIN, b Berwind, WVa, June 5, 17; m 55; c 2. PHYSICAL CHEMISTRY. *Educ:* Emory & Henry Col, BS, 42; Va Polytech Inst, MS, 49, PhD, 52. *Prof Exp:* Instr chem, Emory & Henry Col, 46-48; res chemist, Chemstrand Corp, 52-62; from assoc prof to prof chem, Appalachian State Univ, 61-83; RETIRED. *Mem:* Am Chem Soc. *Res:* Solid State physics; physical chemistry of high polymers. *Mailing Add:* 114 Blanwood Dr Boone NC 28607

JOHNSON, JAMES ELVER, b Montevideo, Minn, Dec 27, 37; m 78; c 4. ORGANIC CHEMISTRY. *Educ:* Univ Minn, BChem, 61, MS, 62; Univ Mo, PhD(chem), 66. *Prof Exp:* Asst prof chem, Sam Houston State Univ, 66-70; from asst prof, to assoc prof, 70-77, PROF CHEM, TEX WOMAN'S UNIV, 77- *Mem:* Am Chem Soc; Royal Soc Chem; Sigma Xi; Int Am Photochem Soc; Nat Asn Adv Health Professions. *Res:* Kinetics and mechanisms of reactions of organic nitrogen compounds; mechanisms of nucleophilic substitution at the carbon- nitrogen double bond; mechanisms of Z-E isomerization at the carbon-nitrogen double bond. *Mailing Add:* Dept Chem Tex Woman's Univ Denton TX 76204

JOHNSON, JAMES HARMON, b Martin, Tenn, March 30, 43; m 67; c 2. CLINICAL CHILD PSYCHOLOGY, CHILDHOOD PSYOPATHOLOGY. *Educ:* Murray State Univ, BS, 66, MS, 68; Northern Ill Univ, PhD(psychol), 76. *Prof Exp:* Instr child psychol, Univ Tex Med Br, 72-75; asst prof psychol, Univ Wash, Seattle, 75-79; assoc prof clin psychol, 79-87, PROF PSYCHOL, UNIV FLA, 87- *Concurrent Pos:* Assoc ed, J Clin Child Psychol, 82-86; mem, Sect Clin Child Psychol, Am Psychol Asn, 82-87, pres, 87; mem, planning comt, Nat Conf Training Clin Child Psychologists, 83-85; conf coordr, Fla Conf Child Health Psychol, 88- *Mem:* Am Psychol Asn; Soc Pediat Psychol. *Res:* Linking stress to problems of both physical health and psychological adjustment; assessment of stress; variables that mediate the impact of stress on individuals; the relationship between stress and fluctuations in health status of those with chronic illness; child psychopathology; effects of stress on children. *Mailing Add:* Dept Clin & Health Psychol Box J 165 Univ Fla Gainesville FL 32610

JOHNSON, JAMES LESLIE, b Kipling, NC, Feb 13, 21; m 45; c 2. CHEMISTRY. *Educ:* Univ NC, BS, 43; Univ Ill, PhD(chem), 49. *Prof Exp:* Chemist, Stamford Res Labs, Am Cyanamid Co, 43-46; chemist, 49-62, from div dir to vpres, Upjohn Co, 62-83; RETIRED. *Mem:* AAAS; Am Chem Soc; Sigma Xi. *Res:* Natural products; spectroscopy; quality control; clinical chemistry. *Mailing Add:* 5400 Glen Harbor Rd Kalamazoo MI 49009

JOHNSON, JAMES M(ELTON), b Pittsboro, NC, Dec 29, 15; m 54. CHEMICAL ENGINEERING. *Educ:* NC State Col, BS, 37. *Prof Exp:* Chem engr, Res & Develop Dept, Socony Vacuum Labs, Socony Mobil Oil Co, Inc, 37-44, chief chemist, Bead Catalyst Plant Lab, 44-49, asst chief chemist, Refinery Labs, 49-62, SR PROCESS ENGR CATALYST & SULFUR MFG, MOBIL OIL CO, INC, 62- *Mem:* Am Chem Soc; Am Inst Chem Engrs. *Res:* Bead catalyst; petroleum refinery control; Claus sulfur plant and tail gas unit design and operation. *Mailing Add:* 645 Washington Ave Haddonfield NJ 08033

JOHNSON, JAMES NORMAN, b Tacoma, Wash, Sept 6, 39; m 59; c 3. SOLID MECHANICS. *Educ:* Univ Puget Sound, BS, 61; Wash State Univ, PhD(physics), 66. *Prof Exp:* Res fel physics, Wash State Univ, 66-67; mem tech staff, Sandia Labs, 67-73; staff consult, Terra Tek, Inc, 73-76; MEM TECH STAFF, LOS ALAMOS NAT LAB, 76- *Concurrent Pos:* NATO sr scientist, Cavendish Lab, Cambridge, UK, 85-86. *Mem:* Am Geophys Union; Sigma Xi; Am Phys Soc. *Res:* Theory of wave propagation and dynamic failure in solids including geophysical materials; constitutive relations for solids; initiation of solid explosives. *Mailing Add:* Los Alamos Nat Lab MS-B214 Los Alamos NM 87545

JOHNSON, JAMES W(INSTON), b Quinton, Okla, May 25, 30; m 53; c 2. CORROSION, ELECTROCHEMISTRY. *Educ:* Univ Mo-Rolla, BS, 57, MS, 58; Univ Mo-Columbia, PhD(chem eng), 61. *Prof Exp:* From instr to assoc prof, 58-67, chmn, Dept Chem Eng, 79-90, PROF CHEM EMG, UNIV MO-ROLLA, 67- *Concurrent Pos:* Fel electrochem lab, Univ Pa, 62-63. *Mem:* Am Inst Chem Engrs; Sigma Xi; Am Soc Eng Educ. *Res:* Electrochemical oxidation and reduction of hydrocarbons; kinetics of metal dissolution and deposition; corrosion. *Mailing Add:* Dept Chem Eng Univ Mo Rolla MO 65401

JOHNSON, JANICE KAY, b Burke, SDak, Apr 12, 46; m 68. PHYSICAL SCIENCE, SCIENCE EDUCATION. *Educ:* Dakota State Col, BS, 68; Southern Ill Univ, MS, 69; Syracuse Univ, PhD(sci educ), 76. *Prof Exp:* Grad intern educ, Southern Ill Univ, 68-69; teacher sci, Pine Grove Middle Sch, NY, 69-74; grad intern sci educ, Syracuse Univ, 74-76; instr phys sci, Mesa Community Col, 77-79; prog developer, Rio Salado Col, 78-79; coordr info systs, Ariz State Univ, 79-82; sci coordr, Glendale Union High Sch Dist, 82-91; SUPT, GUNNISON WATERSHED SCH DIST, 91- *Concurrent Pos:* Consult, Ariz State Dept Educ, Energy Res, 78; NSF grant proposal reviewer, 78-81; mem gov bd, Wash Elementary Sch Dist, 85- *Mem:* Nat Sci Teachers Asn; Nat Asn Res Sci Teaching. *Res:* Cognitive development and its relation to science education; business-industry partnerships in education; mastery learning; strategic planning; grant proposal writing. *Mailing Add:* Gunnison Watershed Sch Dist 216 W Georgia Gunnison CO 81230

JOHNSON, JAY ALLAN, b Two Harbors, Minn, July 15, 41; m 71. WOOD SCIENCE, ENGINEERING MECHANICS. *Educ:* Univ Minn, BS, 64; Col Environ Sci & Forestry, Syracuse Univ, MS, 71; Univ Wash, PhD(wood sci), 73. *Prof Exp:* Asst prof wood physics, Va Polytech Inst & State Univ, 73-77; SCI SPECIALIST WOOD COMPOSITE MAT, WEYERHAEUSER CO, 77- *Mem:* AAAS; Soc Wood Sci & Technol; Forest Prods Res Soc; Am Soc Testing & Mat. *Res:* Development of wood particulate materials; modeling stress development in wood during drying; evaluation of fracture mechanics for testing procedures for wood and wood based materials. *Mailing Add:* Col Forest Resources AR-10 Univ Wash Seattle WA 98195

JOHNSON, JEAN ELAINE, b Wilsey, Kans, Mar 11, 25. BEHAVIORAL MEDICINE. *Educ:* Kans State Univ, BS, 48; Yale Univ, MS, 65; Univ Wis-Madison, MS, 69, PhD(social psychol), 71. *Prof Exp:* Instr nursing, var schs nursing, 48-60; in-serv coordr nursing, Gen Rose Hosp, Denver, Colo, 60-63; res asst nursing res, Sch Nursing, Yale Univ, 65-67; from assoc prof to prof res & nursing & dir ctr health res, Col Nursing, Wayne State Univ, 71-79; PROF NURSING & ASSOC DIR NURSING ONCOL & CLIN NURSING CHIEF ONCOL, STRONG MEM HOSP, UNIV ROCHESTER, 79-*Concurrent Pos:* Mem behav med, study sect, NIH, 82-86; site dir, Robert Wood Johnson Clin Nurse Scholars Prog, 84-91. *Mem:* Inst Med Nat Acad Sci; AAAS; Sigma Xi; Am Nurses' Asn; Oncol Nursing Soc; Acad Behav Med Res; Am Psychol Asn. *Res:* Development of psychological theories about reactions to threatening events, and clinical tests of the effects on patient welfare of care activities deduced from such theories. *Mailing Add:* Cancer Ctr Univ Rochester Med Ctr Rochester NY 14642

JOHNSON, JEAN LOUISE, b Memphis, Tenn, June 17, 47; m 69; c 2. MOLYBDENUM ENZYMES. *Educ:* Cornell Col, BA, 69; Duke Univ, PhD(biochem), 74. *Prof Exp:* res assoc, 74-86, RES ASST PROF BIOCHEM, MED CTR, DUKE UNIV, 86- *Mem:* Am Soc Biol Chemists. *Res:* Structure and role of molybdenum cofactor in molybdoenzymes; molybdenum cofactor biosynthesis; molybdenum cofactor deficiency disease. *Mailing Add:* Dept Biochem Duke Univ Med Ctr Durham NC 27710

JOHNSON, JEFFERY LEE, b Milwaukee, Wis, Mar 6, 41; m 68; c 1. NEUROPHYSIOLOGY. *Educ:* Lakeland Col, BS, 64; Ind Univ, Indianapolis, PhD(physiol), 68. *Prof Exp:* Asst prof, 70-76, ASSOC PROF PHYSIOL & PHARMACOL, SCH MED, UNIV SDAK, 76- *Concurrent Pos:* NIH grants, Inst Psychiat Res, Med Ctr, Ind Univ, Indianapolis, 68-70. *Mem:* AAAS; Soc Neurosci. *Res:* Axoplasmic flow; regeneration; transmitter systems; topographic distribution of amino acids and enzymes in nervous system; electrophysiological analysis of nervous system activity. *Mailing Add:* Dept Physiol & Pharmacol Univ SDak Sch Med Vermillion SD 57069

JOHNSON, JEROME H, b Moscow, Idaho, Nov 22, 18; m 43; c 3. ELECTRICAL ENGINEERING. *Educ:* Univ Idaho, BS, 42; Ore State Univ, MS, 47, PhD(elec eng), 53. *Prof Exp:* Asst prof elec eng, Univ Wyo, 46-47 & Wash State Univ, 47-53; staff mem res, Sandia Corp, 53-58; prof & coordr eng sci, Univ Redlands, 58-77; emer prof eng, 77-; RETIRED. *Mem:* Inst Elec & Electronics Engrs; Sigma Xi. *Res:* High energy shock excited pulse generators; solid state lasers; digital-analog computer elements. *Mailing Add:* 1200 E Cotton Ave Redlands CA 92374

JOHNSON, JERRY WAYNE, b Perry, Ga, July 22, 48; m 68. AGRONOMY. *Educ:* Univ Ga, BSA, 70; Purdue Univ, MS, 72, PhD(agron), 74. *Prof Exp:* Res asst hybrid wheat, Purdue Univ, 70-74; asst prof plant breeding & genetics, Univ Md, 74-77; from asst prof to assoc prof, 77-86, PROF PLANT BREEDING & GENETICS, UNIV GA, 86- *Mem:* Am Soc Agron; Crop Sci Soc Agron. *Res:* Development of barley and wheat varieties that are early and have disease resistance and milling and baking quality; a better feed barley being developed with a higher protein content. *Mailing Add:* Dept Agron Georgia Station Griffin GA 30223

JOHNSON, JOE, BIOPHYSICAL CHEMISTRY. *Educ:* Univ Minn, PhD(biochem), 66. *Prof Exp:* VCHANCELLOR RES & DEVELOP, ATLANTA UNIV CTR, 70- *Res:* Molecular biology. *Mailing Add:* Atlanta Univ Ctr Inc 360 Westview Dr SW Atlanta GA 30310

JOHNSON, JOE W, b July 19, 08; US citizen. HYDRAULIC ENGINEERING. *Educ:* Univ Calif, Berkeley, BSCE, 31, MSCE, 34. *Prof Exp:* Res sediment transport by flowing water, Waterways Exp Sta, Vicksburg, Miss, 34-35 & Soil Conserv Serv, Washington, DC, 42-75; from instr to prof, 42-75, EMER PROF HYDRAUL ENG, UNIV CALIF, BERKELEY, 77- *Concurrent Pos:* Ed, Proc Int Conf Coastal Eng, 50-76 & Shore & Beach, 74-88; hon mem, Soc Engrs, Taiwan, Repub China. *Honors & Awards:* Int Coastal Eng Award, Am Soc Clin Nutrit, 87. *Mem:* Nat Acad Eng; hon mem Am Soc Civil Engrs; Am Shore & Beach Preserv Asn. *Res:* Coastal engineering sediment problems. *Mailing Add:* 2605 Windsor Rd Apt 101 Victoria BC V8S 5H9 Can

JOHNSON, JOHN ALAN, b Gary, Ind, Jan 30, 43; m 65; c 2. ULTRASONICS. *Educ:* Grinnell Col, BA, 65; Carnegie-Mellon Univ, MS, 67, PhD(physics), 70. *Prof Exp:* Asst prof physics, Kenyon Col, 69-76; asst prof physics, Wittenberg Univ, 76-79; SCIENTIST, EG&G IDAHO, 79- *Mem:* Acoust Soc Am. *Res:* Nondestructive evaluation; ultrasonics; microcomputers. *Mailing Add:* Box 1625 EG&G Idaho Idaho Falls ID 83415-2209

JOHNSON, JOHN ALEXANDER, physiology; deceased, see previous edition for last biography

JOHNSON, JOHN ARNOLD, b Cusson, Minn, Dec 6, 24; m 51; c 6. SKIN METABOLISM, PHOTOSENSITIVITY. *Educ:* Univ Minn, BA, 51, MS, 64, PhD(med biochem), 71. *Prof Exp:* Chemist, Bemis Bros Bag Co, 55-60; scientist biomed res, Univ Minn, 60-71; asst prof dermat, 71-73, ASSOC PROF DERMAT & BIOCHEM, MED CTR, UNIV NEBR, OMAHA, 73-; ASSOC PROF DERMAT, SCH MED, CREIGHTON UNIV, 75- *Mem:* AAAS; Am Chem Soc; Soc Invest Dermat. *Res:* In vivo skin glucose metabolism in humans; enzymic determination of glucose, oligoglucosides and glycogen in animal tissues; mechanisms of photoprotection. *Mailing Add:* Sect Dermat Univ Nebr Med Ctr Omaha NE 68105

JOHNSON, JOHN CHRISTOPHER, JR, b Gunnison, Colo, Nov 28, 24; m 48; c 3. ECOLOGY, ORNITHOLOGY. *Educ:* Ohio State Univ, BS, 47; Univ Okla, MS, 52, PhD(zool), 57. *Prof Exp:* Pub sch teacher, Ohio, 48; teacher, Sch Dependents, Ramey AFB, PR, 48-49; instr, Univ Okla, 50-56; from asst prof to assoc prof zool, Pittsburgh State Univ, 56-62, actg chmn dept biol, 60-62, from prof to emer prof, 62-87; RETIRED. *Concurrent Pos:* Actg dir,

Rocky Mountain Biol Lab, Colo, 54, trustee, 64-, dir, 68-77, registr, 72- *Mem:* Am Ornithologists' Union; Soc Syst Zool; Am Inst Biol Sci. *Res:* Vertebrate zoology; bioecology; ornithology. *Mailing Add:* Dept Biol Pittsburg State Univ Pittsburg KS 66762

JOHNSON, JOHN CLARK, b Waterbury, Conn, Aug 17, 19; m 41; c 2. APPLIED PHYSICS. *Educ:* Middlebury Col, AB, 41; Mass Inst Technol, SM, 46, ScD(meteorol), 48. *Prof Exp:* From instr to asst prof meteorol, Mass Inst Technol, 47-53; res assoc & lectr physics, Tufts Univ, 53-54; prof physics, 54-81, EMER PROF PHYSICS, WORCESTER POLYTECH INST, 82- *Concurrent Pos:* Res assoc ed, Harvard Univ, 66-67. *Mem:* Am Phys Soc; Optical Soc Am; Am Meteorol Soc; Am Asn Physics Teachers; Sigma Xi. *Res:* Physical meteorology; scattering theory. *Mailing Add:* Two Duncannon Ave Apt 12 Worcester MA 01604

JOHNSON, JOHN E(DWIN), b Detroit, Mich, Jan 18, 31; m 53, 83; c 6. CIVIL & STRUCTURAL ENGINEERING. *Educ:* Gonzaga Univ, BSCE, 56; Stanford Univ, MSCE, 57; Purdue Univ, PhD, 63. *Prof Exp:* Design engr, Detroit Edison Co, 58-60; instr, Purdue Univ, 60-62; res engr, Dow Chem Co, 62-65; from asst prof to assoc prof civil eng, 65-72, PRES ENG FORENSICS & TESTING, PROF CIVIL & ENVIRON ENG, UNIV WIS-MADISON, 72- *Concurrent Pos:* Consult to over 100 companies, 65-; various fellowships. *Honors & Awards:* Z W Craine Award; NSF Award. *Mem:* Am Soc Civil Engrs; Am Soc Eng Educ; Am Concrete Inst; Nat Soc Prof Engrs; Am Soc Testing & Mat; Sigma Xi; Am Inst St Construct. *Res:* Composite behavior; use of plastics as structural materials; analysis and testing of the physical behavior of engineering materials; large number of papers, articles, design manuals and one textbook. *Mailing Add:* 703 Moygara Rd Madison WI 53716

JOHNSON, JOHN E, JR, b Ft Worth, Tex, Aug 21, 45. ELECTRON MICROSCOPY. *Educ:* Tulane Univ, PhD(neurosci), 73. *Prof Exp:* ASST PROF NEUROSCI, TULANE UNIV, 79- *Mailing Add:* 2340 Mistletoe Ave Ft Worth TX 76110

JOHNSON, JOHN ENOCH, physical organic chemistry, for more information see previous edition

JOHNSON, JOHN HAL, b Benjamin, Utah, July 1, 30; m 58; c 4. ORGANIC & FOOD CHEMISTRY. *Educ:* Brigham Young Univ, BS, 55, MS, 57; Ohio State Univ, PhD(food sci), 63. *Prof Exp:* Lab instr chem, Brigham Young Univ, 59-60; res asst food chem, Agr Expr Sta, Ohio State Univ, 60-63; asst biochemist food sci, Agr Exp Sta, Univ Fla, 63-68; PROF FOOD SCI & NUTRIT, BRIGHAM YOUNG UNIV, 69- *Honors & Awards:* Virginia F Cutler Lectr, 78. *Mem:* Inst Food Technologists; Sigma Xi. *Res:* Chemical reactions occurring in foods during processing and storage; effects on functional qualities of cooker extruded soy enriched cereal flours; development of cereal-based complemented foods. *Mailing Add:* Rte 1 Box 403 Spanish Fork UT 84660

JOHNSON, JOHN HAROLD, b Chicago, Ill. LIQUID CHROMATOGRAPHY, SPECTROSCOPY. *Educ:* Monmouth Col, Ill, BA, 68; Univ Ark, PhD(organ chem), 74. *Prof Exp:* Sr scientist, US Environ Protection Agency, 73-77; res investr, Nalco Environ & Chem Sci Corp, 77-78; supvr chem anal, G D Searle Co, 78-80; group leader, Anal Develop, Dupont Critical Care, 80-88; SR RES INVESTR, ANALYTIC DEVELOP, KRAFT GEN FOODS, 88- *Concurrent Pos:* Lectr, Fac Inst, Argonne Nat Lab, 76-79 & Am Chem Soc Speakers Tour, 77. *Mem:* Acad Pharmaceut Sci; Am Chem Soc; Am Inst Chemists; Asn Off Anal Chemists. *Res:* Separation techniques as applied to food compoents water soluble polymers; basic studies into new chromatographic separation, spectroscopic and laboratory automation techniques as applied to food analyis and structure identification. *Mailing Add:* 320 Juniper Parkway Libertyville IL 60048

JOHNSON, JOHN HARRIS, b Fond du Lac, Wis, Feb 10, 37; m 90; c 2. MECHANICAL ENGINEERING, AIR POLLUTION. *Educ:* Univ Wis-Madison, BS, 59, MS, 60, PhD(mech eng), 64. *Prof Exp:* Res asst mech eng, Univ Wis-Madison, 59-64; chief appl eng res, Int Harvester Co, Ill, 66-70; from asst prof to assoc prof, 70-75, prof, 75-80, DISTINGUISHED PRESIDENTIAL PROF, MICH TECHNOL UNIV, 81-, CHMN DEPT, 86- *Concurrent Pos:* Coordr, Res Coun Air Pollution Res Comt, 68-; mem, Mine Health Res Adv Comt, Dept Health, Educ & Welfare, 79-81; consult, US Environ Protection Agency, 71-77, US Bur Mines, 77-79, NASA, 78-81, Nat Acad Sci, 79-81, Off Tech Assess, US Cong, 79, A D Little, 80-81 & Stanford Res Inst, 81-82; mem bd dir, Soc Automotive Engrs, 82-85; mem, Nat Res Coun, 86-89, Dept Labor, 88. *Mem:* Fel Soc Automotive Engrs; Combustion Inst; Am Soc Mech Engrs; Air Pollution Control Asn; Am Soc Eng Educ. *Res:* Experimental combustion studies; computer calculations of single fuel drop motion and vaporization; computer cycle analysis; hybrid engine research; emissions and air pollution; instantaneous temperature measurements in internal combustion engines; tribology; diesel particulate emissions measurement and control; pollutants in underground mining; cooling system modeling. *Mailing Add:* Dept Mech Eng & Eng Mech Mich Technol Univ Houghton MI 49931

JOHNSON, JOHN IRWIN, JR, b Salt Lake City, Utah, Aug 18, 31. ZOOLOGY. *Educ:* Univ Notre Dame, AB, 52; Purdue Univ, MS, 55, PhD(psychol), 57. *Prof Exp:* Instr psychol, Purdue Univ, 56-57; from instr to asst prof, Marquette Univ, 57-60; USPHS spec res fel lab neurophysiol, Univ Wis, 60-61; Fulbright res scholar physiol, Univ Sydney, 64-65; from assoc prof to prof biophys, psychol & zool, 65-81, PROF ANAT, MICH STATE UNIV, 81- *Concurrent Pos:* NIH career develop award, 65-72. *Mem:* AAAS; Am Soc Zool; Soc Neurosci; Am Asn Anat; Am Soc Mammalogists; hon mem Anat Asn Australia & New Zealand. *Res:* Brain function; neuroanatomy; animal behavior. *Mailing Add:* Dept Anat Mich State Univ East Lansing MI 48824-1316

JOHNSON, JOHN LEROY, b Kanawha, Iowa, Oct 28, 36; m 70; c 2. MICROBIOLOGY. *Educ:* Concordia Col, Moorhead, Minn, BA, 58; Mont State Univ, MS, 60, PhD(bact), 64. *Prof Exp:* Fel microbiol, Univ Wash, 64-68; asst prof bact, 68-71, assoc prof microbiol, 71-77, PROF MICROBIOL, VA POLYTECH INST & STATE UNIV, 77- *Honors & Awards:* Bergey Award, 80. *Mem:* AAAS; Am Soc Microbiol; Sigma Xi. *Res:* Obligate anaerobic bacteria; bacterial taxonomy; nucleic acids. *Mailing Add:* Dept Anaerobic Microbiol Va Polytech Inst & State Univ Blacksburg VA 24061-0305

JOHNSON, JOHN LOWELL, b Butte, Mont, Mar 18, 26; m 51; c 3. PLASMA PHYSICS. *Educ:* Mont State Univ, BS, 49; Yale Univ, MS, 50, PhD(physics), 54. *Prof Exp:* Sr scientist, Atomic Power Dept, Westinghouse Elec Corp, 54-64, fel physicist, Res Labs, 64-68, adv scientist, 68-79, consult scientist, Res & Develop Ctr, 79-85; PRIN RES PHYSICIST, PRINCETON UNIV, 85- *Concurrent Pos:* Vis mem res staff, Plasma Physics Lab, Princeton Univ, 55-70, vis sr res physicist, 70-85. *Mem:* Am Phys Soc. *Res:* Theoretical plasma physics associated with the controlled thermonuclear research program with principal emphasis directed towards investigation of the magnetohydrodynamic properties of toroidal confinement configurations. *Mailing Add:* Princeton Univ Plasma Physics Lab Box 451 Princeton NJ 08543

JOHNSON, JOHN MARSHALL, b McCamey, Tex, Aug 10, 44; m 70; c 2. PHYSIOLOGY. *Educ:* Rice Univ, BA, 66; Univ Tex Southwestern Med Sch, PhD(physiol), 72. *Prof Exp:* Sr fel, Sch Med, Univ Wash, 72-74, res assoc physiol, 74-75; from asst prof to assoc prof, 75-89, PROF PHYSIOL, UNIV TEX HEALTH SCI CTR, SAN ANTONIO, 89- *Mem:* Am Heart Asn; fel Am Physiol Soc; NY Acad Sci; AAAS; Am Col Sports Med. *Res:* Reflex control of the circulatory system; cardiovascular physiology regulation of cutaneous blood flow. *Mailing Add:* Dept Physiol Univ Tex Health Sci Ctr San Antonio TX 78284

JOHNSON, JOHN MORRIS, b Boise, Idaho, Mar 16, 37; m 59; c 2. BOTANY & CYTOLOGY, PLANT TAXONOMY. *Educ:* Col Idaho, BS, 59; Ore State Univ, MS, 61, PhD(bot, tissue cult), 64. *Prof Exp:* From asst prof to assoc prof biol, Cent Col Iowa, 64-69; assoc prof, 69-74, PROF BIOL, WESTERN ORE STATE COL, 74-, CHMN, NATURAL SCI & MATH DIV, 85- *Concurrent Pos:* USPHS fel, Univ Chicago, 65-66. *Mem:* AAAS; Bot Soc Am; Am Soc Cell Biologists. *Res:* Plant tissue culture; behavior and function of nucleus and nucleolar vacuoles; plant taxonomy of Oregon plants; ecology of wet-land species especially Juncus. *Mailing Add:* Div Natural Scis & Math Western Ore State Col Monmouth OR 97361

JOHNSON, JOHN RAYMOND, physiology, pharmacology; deceased, see previous edition for last biography

JOHNSON, JOHN RICHARD, b Edmonton, Alta, July 6, 42; m 67; c 2. BIOPHYSICS, HEALTH PHYSICS. *Educ:* Univ BC, BS, 67, MS, 70, PhD(physics), 73. *Prof Exp:* Res officer, 73-81, head, Biomed Res Br, 81-84, HEAD, DOSIMETRIC RES BR, HEALTH SCI DIV, CHALK RIVER NUCLEAR LAB, 84- *Concurrent Pos:* Dir, Can Radiation Protection Asn, 82-83. *Mem:* Radiation Res Soc; Health Physics Soc; Can Radiation Protection Asn (pres, 84-85). *Res:* Dosimetric and metabolic models for internal dosimetry including radon daughters; improvement of instrumentation for measuring internal radioactive contamination in humans; instrumentation for internal dosimetry and radon exposure monitoring; internal contamination control and risk assessment. *Mailing Add:* Batelle Pacific NW Lab PO Box 999 K3-53 Richland WA 99352

JOHNSON, JOHN WEBSTER, JR, b Pawhuska, Okla, Sept 29, 25. ORGANIC CHEMISTRY. *Educ:* Univ Wichita, AB, 52, MA, 53; Univ Ill, PhD(chem), 56. *Prof Exp:* Asst prof, 56-64, ASSOC PROF CHEM, WICHITA STATE UNIV, 64- *Mem:* Am Chem Soc. *Res:* Nucleophilic displacement reactions on epoxides and lactones; synthetic high molecular weight polymers, especially their preparation, properties and uses. *Mailing Add:* Dept Chem Wichita State Univ Wichita KS 67208

JOHNSON, JOHNNY ALBERT, b El Paso, Tex, Mar 6, 38; m 55; c 2. LATTICES, RINGS. *Educ:* Univ Calif, Riverside, BA, 65, MA, 66, PhD(math), 68. *Prof Exp:* NSF fel, Univ Calif, Riverside, 65-68; from asst prof to assoc prof, 68-78, PROF MATH, UNIV HOUSTON-UNIVERSITY PARK, 78- *Concurrent Pos:* Univ Houston res initiation grant, 69; sr engr, Jet Propulsion Lab, Calif Inst Technol, 69-71; assoc managing ed, Houston J Math, 74-84, ed, 84-; Univ Houston res grant, 78, leave grant, 80. *Mem:* Math Asn Am; Am Math Soc. *Res:* Commutative algebra. *Mailing Add:* Dept Math Univ Houston-University Park Houston TX 77204-3476

JOHNSON, JOHNNY R(AY), b Chatham, La, Dec 19, 29; m 60; c 3. APPLIED MATHEMATICS, ELECTRICAL ENGINEERING. *Educ:* La Polytech Inst, EE, 51. *Prof Exp:* Electronic engr, Pitman-Dunn Lab, Frankford Arsenal, 53-54; asst prof math, La Polytech Inst, 58-62; assoc prof, Appalachian State Teachers Col, 62-63; from assoc prof to prof, Elec Eng, La State Univ, Baton Rouge, 70-83; eng specialist, Gen Dynamics, 83-84; RETIRED. *Mem:* Sr mem Inst Elec & Electronics Engrs; Am Asn Univ Prof. *Res:* Special functions; boundary value problems; analog and digital filters. *Mailing Add:* 222 Medoll Crest Dr Florence AL 35630

JOHNSON, JOSEPH ANDREW, III, b Nashville, Tenn, May 26, 40; m 61; c 4. TURBULENCE, STATISTICAL MECHANICS. *Educ:* Fisk Univ, BA, 60; Yale Univ, MS, 61, PhD(physics), 65. *Prof Exp:* Mem tech staff, Bell Labs, Whippany, NJ, 65-68; vis asst prof eng & appl sci, Yale Univ, 68-69; chmn & prof physics, Southern Univ, Baton Rouge, La, 69-72; assoc prof physics, Rutgers Univ, 73-81; PROF PHYSICS, CITY COL NY, 81- *Concurrent Pos:* Consult, Sikorsky Aircraft Corp, 62-65; Gen Appl Sci Lab, 68-69, Von Karman Gas Dynamic Facil, 69-77, Fermi Nat Lab, 73, Yale Univ, 73-75; Bell Labs, 75-76, Res & Develop Ctr, Gen Elec Corp, 78-80 & Grambling State Univ, 80-83. *Mem:* Assoc fel Am Inst Aeronaut & Astronaut; fel Am Phys Soc. *Res:* Non-strange elementary particle interactions; noise propagation; ionized re-entry trails; plasma-photon absorption processes; new approaches to the physics of turbulence. *Mailing Add:* Dept Physics City Col NY Convent Ave at 138th St New York NY 10031

JOHNSON, JOSEPH EGGLESTON, III, b Elberton, Ga, Sept 17, 30; m 56; c 3. INTERNAL MEDICINE, INFECTIOUS DISEASE. *Educ:* Vanderbilt Univ, BA, 51, MD, 54. *Prof Exp:* Intern, Osler Med Serv, Johns Hopkins Hosp, 54-55, fel med, 58-59, asst resident, 57-58 & 59-60, res physician, 60-61; from instr to asst prof, Sch Med, Johns Hopkins Univ, 61-66, asst dean student affairs, 63-66; from assoc prof to prof, Col Med, Univ Fla, 66-72, chief, Infectious Dis Div, 68-72, assoc dean, 70-72; prof med & chmn dept, Bowman Gray Sch Med, 72-81; PROF INTERNAL MED, UNIV MICH, 81- *Concurrent Pos:* Am Col Physicians Mead Johnson scholar, 60-61; John & Mary R Markle scholar acad med, 62-67; prog dir, USPHS Med Student Res Training Grant, 63-66; prin investr, Off Surgeon Gen, US Dept Army res grant, 66-71; consult, US Army Biol Lab, Ft Detrick, Md, 66-71; dir, Nat Insts Allergy & Infectious Dis training grant & contract Food & Drug Admin, 67-72, sabbatical, London Clin Res Ctr, 70-71, Royal Soc Med traveling fel, 70-71; mem, Federated Coun Internal Med, 78-, vchmn, 81- *Mem:* Am Asn Immunologists; Am Soc Microbiol; Soc Exp Biol & Med; fel Royal Soc Med; Am Clin & Climat Asn. *Res:* Pathogenesis of staphylococcal infection; role of bacterial hypersensitivity and immunity in infection; epidemiology of hospital and laboratory acquired infection; pulmonary host defense mechanisms; adverse drug reactions; epidemiology and mechanisms. *Mailing Add:* Univ Mich Med Ctr 3116 Taubman Ctr Ann Arbor MI 48109-0378

JOHNSON, JOSEPH L, medicine; deceased, see previous edition for last biography

JOHNSON, JOSEPH RICHARD, b Fond du Lac, Wis, June 13, 22; m 46; c 3. MEDICINE. *Educ:* Univ Wis, BS & MD, 46. *Prof Exp:* Asst prof internal med, Univ Mich, 56-57; from clin asst prof to clin assoc prof, Univ Wis, 57-66; ASSOC PROF MED, UNIV MINN, MINNEAPOLIS & MEM SR STAFF, PULMONARY DIS SECT, VET ADMIN HOSP, MINNEAPOLIS, 66- *Concurrent Pos:* Chief sect, Vet Admin Hosp, Ann Arbor, Mich, 54-57 & Madison, Wis, 57-66; dir, Hennepin County Chest Clin & mem staff, Hennepin County Gen Hosp. *Mem:* Am Thoracic Soc; fel Am Col Chest Physicians; fel Am Col Physicians; Am Fedn Clin Res. *Mailing Add:* 9901 Dellridge Rd Minneapolis MN 55425

JOHNSON, JULIAN FRANK, b Baxter, Kans, Aug 20, 23; m 43. POLYMER CHEMISTRY. *Educ:* Col Wooster, BA, 43; Brown Univ, PhD(chem), 51. *Prof Exp:* Supvry res chemist, Chevron Res Corp, 50-68; assoc prof, 68-70, PROF CHEM, UNIV CONN, 70-, ASSOC DIR, INST MAT SCI, 71- *Concurrent Pos:* Lectr, Exten Div, Univ Calif, Berkeley, 60-68. *Honors & Awards:* Am Chem Soc Award in Chromatog, 70. *Mem:* Am Chem Soc; Am Phys Soc; Am Soc Rheol; Brit Soc Rheol. *Res:* Physics of polymers; rheology; chromatography. *Mailing Add:* Inst Mat Sci U-136 Univ Conn Storrs CT 06268

JOHNSON, K JEFFREY, inorganic chemistry, for more information see previous edition

JOHNSON, KAREN ELISE, b Balston Spa, NY, Oct 3, 50. PHYSICS. *Educ:* Grinnell Col, BA, 72; Univ Minn, MS, 76, PhD(hist sci), 86. *Prof Exp:* Asst prof physics, Bates Col, 86-88; vis asst prof, Cath Univ, 87; ASST PROF PHYSICS, ST LAWRENCE UNIV, 88- *Mem:* Am Asn Physics Teachers; Hist Sci Soc; Sigma Xi. *Res:* History of 20th century physics; nuclear and chemical physics; history of women in science. *Mailing Add:* Dept Physics St Lawrence Univ Canton NY 13617

JOHNSON, KAREN LOUISE, b Flint, Mich, Feb 4, 41. BOTANY, PLANT ECOLOGY. *Educ:* Swarthmore Col, BA, 63; Univ Ill, Urbana, MS, 65, PhD(bot), 70. *Prof Exp:* Instr biol, Colby Col, 66-68; fel bot, Univ Man, 69-72; CUR BOT, MANITOBA MUS MAN & NATURE, 72- *Mem:* Ecol Soc Am. *Res:* Alpine plant communities and soils; vegetation mapping and description; establishment of ecological reserves and natural areas; boreal forest plant geography. *Mailing Add:* 190 Groupert Winnipeg MB R3B 0N2 Can

JOHNSON, KEITH EDWARD, b Feltham, Eng, Jan 4, 35; m 60; c 2. ANALYTICAL CHEMISTRY. *Educ:* Univ London, BSc & ARCS, 56; Univ London, DIC & PhD(phys chem), 59, DSc(chem), 74. *Prof Exp:* Res assoc anal chem, Univ Ill, 59-62; asst lectr phys chem, Sir John Cass Col, Eng, 62-63, lectr, 63-66; from asst prof to assoc prof, 66-72, PROF INORG & ANALYTICAL CHEM, UNIV REGINA, 72- *Concurrent Pos:* Vis prof, Univ Calif, Riverside, 72-73, Sask Power Corp, 79-80 & Oak Ridge Nat Lab, Oak Ridge, Tenn, 87-88. *Mem:* Electrochem Soc; fel Chem Inst Can; fel Royal Soc Arts; Royal Soc Chem; Am Chem Soc. *Res:* Molten salt electrochemistry; coordination of transition metal ions in melts; structural studies of inorganic complexes; water and soil trace analysis; electrochemical coal conversion; pyrazolone chemistry; azolium ion chemistry. *Mailing Add:* Dept Chem Univ Regina Regina SK S4S 0A2 Can

JOHNSON, KEITH HUBER, b Reading, Pa, May 1, 36; m 62; c 1. QUANTUM CHEMISTRY. *Educ:* Princeton Univ, AB, 58; Temple Univ, MA, 61, PhD(physics), 65. *Prof Exp:* Res fel quantum theory proj, Univ Fla, 65-67; PROF MAT SCI, MASS INST TECHNOL, 67- *Concurrent Pos:* Consult, Gen Elec Corp Res & Develop, 72- & Exxon Res & Eng Co, 75- *Honors & Awards:* Medal, Int Acad Quantum Molecular Sci, 73. *Mem:* Am Phys Soc. *Res:* Electronic structure of molecules and solids; surface chemistry; catalytic chemistry. *Mailing Add:* Ctr for Mat Sci & Eng Mass Inst Technol Cambridge MA 02139

JOHNSON, KENNETH, b Putnam Co, Ind, Aug 18, 07; m 39; c 2. RESEARCH MANAGEMENT, ORGANIC CHEMISTRY. *Educ:* Ind State Teachers Col, AB, 31; Purdue Univ, PhD(org chem), 37. *Prof Exp:* Asst chem, Purdue Univ, 31-35; res chemist, Com Solvents Corp, 36-40, develop chemist & supt prod scale develop plant, 40-51; area supt in charge polymerization, Nylon Plant, Chemstrand Corp, 51-55, from asst plant mgr to plant mgr, Acrilan Plant, 55-58, dir acrilan mfg, 58-60, dir res admin, Chemstrand Res Ctr, Inc, 60-68; prof, 68-74, EMER PROF RES MGT, GEORGE WASHINGTON UNIV, 74- *Mem:* Am Chem Soc; fel Am Inst Chemists. *Res:* Aminohydroxy compounds from nitrohydroxy compounds; plant development for manufacturing nitroparaffins; polymers; chemical fibers; management of technical information. *Mailing Add:* 25317 Carmel Knolls Dr Carmel CA 93923

JOHNSON, KENNETH ALAN, b Duluth, Minn, Mar 26, 31; m 54; c 1. ELEMENTARY PARTICLE PHYSICS. *Educ:* Ill Inst Technol, BS, 52; Harvard Univ, AM, 54, PhD(physics), 55. *Prof Exp:* Res fel & lectr physics, Harvard Univ, 55-57; NSF fel, Univ Copenhagen, 57-58; from asst prof to assoc prof, 58-65, PROF PHYSICS, MASS INST TECHNOL, 65- *Concurrent Pos:* Guggenheim fel, 71-72. *Mem:* Fel Am Phys Soc; fel Am Acad Arts & Sci; Sigma Xi; fel AAAS. *Res:* Quantum electrodynamics; quantum field theory; elementary particle physics. *Mailing Add:* Dept Physics Mass Inst Technol Cambridge MA 02139

JOHNSON, KENNETH ALLEN, b Davenport, Iowa, Mar 10, 49; m 70; c 2. CELL MOTILITY, ENZYME MECHANISMS. *Educ:* Univ Iowa, BS, 71; Univ Wis, PhD(molecular biol), 75. *Prof Exp:* Fel biophysics, Univ Chicago, 75-79; ASST PROF BIOCHEM, PA STATE UNIV, 79- *Concurrent Pos:* Guest scientist, Brookhaven Nat Lab, 81- *Mem:* Biophys Soc; Am Soc Cell Biol. *Res:* Cell motility, especially structure, mechanism and regulation of the dynein adenosine tryphosphatase in cilia and flagella; microtubule assembly pathway; rapid transient kinetic analysis of enzyme reaction pathways; DNA polymerization mechanism. *Mailing Add:* Molec Cell Biol 301 Althouse Lablthouse Lab Pa State Univ University Park PA 16802

JOHNSON, KENNETH DELFORD, air pollution, industrial hygiene; deceased, see previous edition for last biography

JOHNSON, KENNETH DUANE, b Los Angeles, Calif, Jan 18, 44; m 66; c 2. PLANT PHYSIOLOGY. *Educ:* Univ Calif, Santa Barbara, BA, 66, PhD(biol), 69. *Prof Exp:* From asst prof to assoc prof, 72-80, PROF BIOL, SAN DIEGO STATE UNIV, 80- *Mem:* Am Soc Plant Physiol; Sigma Xi. *Res:* Plant cell biology; biochemistry of growth and development; glycoprotein processing. *Mailing Add:* Dept Biol San Diego State Univ San Diego CA 92182

JOHNSON, KENNETH EARL, b Worcester, NY, July 24, 21; m 43; c 2. CHEMISTRY. *Educ:* NY State Col Teachers, Albany, BA, 42; Stanford Univ, MA, 43. *Prof Exp:* Res chemist natural & synthetic rubber, Am Anode Inc, 43-45, tech salesman, 45-47; res chemist, Fabrics & Finishes Dept, 47-58, group leader, 58-60, tech supvr, 60-63, sales prod mgr, 63-67, venture mgr high temp-sheet struct, 67-70, sales mgr adhesives & coatings, 70-71, prod mgr pkg mat sales, 71-77, res assoc, E I Du Pont De Nemours & Co, Inc, 77-; RETIRED. *Mem:* Am Chem Soc. *Res:* Solubilization and micellar formation of soaps; locus of polymerization of synthetic rubber; rubber and polymer chemistry. *Mailing Add:* 29 Cokesbury VLG Hockessin DE 19707-9805

JOHNSON, KENNETH GEORGE, b Oneonta, NY, Feb 22, 30; m 53; c 3. GEOMORPHOLOGY. *Educ:* Union Col, NY, BS, 52; Mich State Univ, MS, 57; Rensselaer Polytech Inst, PhD(geol), 68. *Prof Exp:* Geologist, Western Hemisphere Explor Div, Gulf Oil Corp, 58-61 & Bolivian Gulf Oil Co, 61-64; from asst prof to assoc prof, 66-78, PROF GEOL, SKIDMORE COL, 78-, CHMN DEPT, 66- *Honors & Awards:* Skidmore Fac Res Lectr, 81. *Mem:* Am Asn Petrol Geologists; Soc Econ Paleontologists & Mineralogists; Nat Asn Geol Teachers; Geol Soc Am; Am Quaternary Asn. *Res:* Applications of geomorphology to military geology and petroleum exploration; photogeology in petroleum exploration; coastal depositional systems and nearshore marine processes. *Mailing Add:* Dept Geol Skidmore Col Saratoga Springs NY 12866-1632

JOHNSON, KENNETH GERALD, b New York, NY, Feb 12, 25; m 50. INTERNAL MEDICINE. *Educ:* Manhattan Col, BS, 44; State Univ NY, MD, 50; Darthmouth Col, MA, 74. *Hon Degrees:* ;. *Prof Exp:* From intern to chief resident internal med, Yale-New Haven Med Ctr, 50-54; from instr to assoc prof, Sch Med, Yale Univ, 54-64; chief of med, Atomic Bomb Casualty Comn, Japan, 64-67; prof community med & dir div epidemiol res, Med Col, Cornell Univ, 67-71; prof community med, chmn dept & assoc dean, Dartmouth Med Sch, 71-74; PROF COMMUNITY MED, MT SINAI SCH MED, 74- *Concurrent Pos:* James Hudson Brown fel med physics, 51-52; Nat Heart trainee, 53-54; consult cardiologist, Yale-New Haven Med Ctr & Hosp of St Raphael, New Haven, Conn, 55-65; vis lectr, Col Med, Hiroshima Univ, 64-67; assoc attend physician, New York Hosp, 67-; sr prog consult, Robert Wood Johnson Found, 75-; chmn, NY State Comn Formulate Plan for Pub Med Schs, 75-76; sr prog consult, Robert Wood Johnson Found, 75-; consult, Am Col Obstet & Gynecol, 76-, Surgeon Army, 77-, Dean, State Univ NY, Binghamton clin campus, 77-, NY State Dept Health, 77-, Greenwall Found, 83- & Commonwealth Fund, 85- *Mem:* Fel Am Col Cariol; Am Soc Aging; fel Am Col Prev Med; Am Pub Health Asn. *Res:* Research and development of health services. *Mailing Add:* Health Serv Res Ctr Benedictine Hosp PO Box 1939 Kingston NY 12401

JOHNSON, KENNETH HARVEY, b Hallock, Minn, Feb 17, 36; m 60; c 3. VETERINARY PATHOLOGY. *Educ:* Univ Minn, BS, 58, DVM, 60, PhD(vet path), 65. *Prof Exp:* NIH training fel, 60-65, from asst prof to assoc prof, 65-73, head sect path, 74-76, act chmn dept vet pathobiol, 76-77, chmn dept vet pathobiol, 77-83, PROF VET PATH, COL VET MED, UNIV MINN, ST PAUL, 73- *Concurrent Pos:* Path consult, Minn Mining & Mfg Co, 66-71 & Medtronic, Inc, 72-80; USPHS biomed sci support grant, 68-; consult, Natural-Y Surg Specialties, Inc, Los Angeles. *Honors & Awards:* Norden Award, 70; Beecham Award for Res Excellence, 89; Ralston Purina Small Animal Res Award, 90. *Mem:* Electron Micros Soc Am; Am Asn Feline Practitioners. *Res:* Amyloidosis; feline diseases; ultrastructural studies; polymer tumorigenesis in mice; diabetes mellitus in cats. *Mailing Add:* Dept Vet Pathobiol Univ Minn Col Vet Med St Paul MN 55108

JOHNSON, KENNETH MAURICE, JR, b Houston, Tex, Dec 7, 44; m 68; c 2. NEUROPHARMACOLOGY. *Educ:* Stephen F Austin State Univ, BS, 67; Univ Houston, PhD(biophys sci), 74. *Prof Exp:* Instr physics, Houston Independent Sch Dist, 67-69; fel pharmacol, Med Col Va, 75-77; from asst prof to assoc prof, 77-87, PROF PHARMACOL, UNIV TEX MED BR, 87- *Concurrent Pos:* Fel, Nat Inst Drug Abuse, 76-77; prin investr, Nat Inst Drug Abuse, 79-; NIDA pharmacol review subcomt, 88-92. *Mem:* AAAS; Am Soc Pharmacol & Exp Therapeut; Soc Neurosci; Sigma Xi. *Res:* Neurochemical and behavioral pharmacology of cannabinoids, opiates, hallucinogens, dissociative anesthetics and psychomotor stimulants; biochemistry of excitatory amino acid receptors, regulation of neurotransmitter synthesis, release and receptor; neuroendocrine effects of psychoactive drugs. *Mailing Add:* Dept Pharmacol Univ Tex Med Br Galveston TX 77550

JOHNSON, KENNETH OLAFUR, US citizen. NEUROPHYSIOLOGY, BIOMEDICAL ENGINEERING. *Educ:* Univ Wash, BS, 61; Syracuse Univ, MS, 65; Johns Hopkins Univ, PhD(biomed eng), 70. *Prof Exp:* Engr, Gen Elec Co, 61-65; asst prof physiol & biomed eng, Sch Med, Johns Hopkins Univ, 71-72; staff mem, Univ Melbourne, 72-80; MEM STAFF, JOHNS HOPKINS UNIV, 81- *Mem:* AAAS; Soc Neurosci. *Res:* Neural mechanisms in sensation and perception. *Mailing Add:* Dept Neurosci Johns Hopkins Univ 725 N Wolfe St Baltimore MD 21205

JOHNSON, KENNETH SUTHERLAND, b Brooklyn, NY, Sept 16, 34; m 59; c 3. GEOLOGY. *Educ:* Univ Okla, BS, 59 & 61, MS, 62; Univ Ill, Urbana, PhD(geol), 67. *Prof Exp:* GEOLOGIST, OKLA GEOL SURV, 62-, ASSOC DIR, 78- *Concurrent Pos:* Teaching asst, Univ Okla, 58-61; teaching asst, Univ Ill, Urbana, 65-67; consult geologist, 68-; vis prof geol & geol eng, Univ Okla, 73-; dir, Okla Mining & Mineral Resources Res Inst, 78-80. *Mem:* AAAS; Geol Soc Am; Am Asn Petrol Geologists; Am Inst Prof Geologists; Am Inst Mining, Metall & Petrol Engrs; Asn Eng Geologists; Int Asn Hydrogeologists. *Res:* Economic geology; stratigraphy; field mapping of geologic structures and mineral resources; photogeology; environmental geology; earth-science education; geology of evaporites and redbeds; disposal of radioactive and industrial wastes; hydrogeology. *Mailing Add:* Okla Geol Surv Norman OK 73019

JOHNSON, KENT J, b Minot, NDak, Nov 4, 46. IMMUNOPATHOLOGY. *Educ:* Univ Conn, MD, 76. *Prof Exp:* ASSOC PROF PATH, SCH MED, UNIV MICH, 83- *Mem:* Am Asn Immunologists; Am Asn Pathologists. *Mailing Add:* Dept Path Univ Mich Sch Med 1301 E Catherine St Box 0602 Ann Arbor MI 48109

JOHNSON, KURT EDWARD, b Needham, Mass, July 6, 43; m 67; c 1. DEVELOPMENTAL BIOLOGY. *Educ:* Johns Hopkins Univ, BS, 65; Yale Univ, MPhil, 69, PhD(develop biol), 70. *Prof Exp:* Fel develop biol, Yale Univ, 70-71; asst prof anat, Med Ctr, Duke Univ, 71-77; ASSOC PROF ANAT, MED CTR, GEORGE WASHINGTON UNIV, 77- *Mem:* AAAS; Sigma Xi; Am Soc Cell Biologists; Soc Develop Biologists. *Res:* Experimental morphogenesis; experimental analysis of amphibian gastrulation. *Mailing Add:* Dept Anat Med Ctr 2300 I St NW Washington DC 20037

JOHNSON, KURT P, b Chicago, Ill, Oct 6, 38; m 61; c 1. MECHANICAL ENGINEERING. *Educ:* Northwestern Univ, BS, 60, PhD(mech eng), 63. *Prof Exp:* Sr staff engr, McDonnell Douglas Astronaut Co, 63-75, dir corp diversification technol, McDonnell Douglas Corp, 76-84, dir eng & opers, 84-89, dir laser commun systs, 89-90; GROUP VPRES ENG, FARREL CORP, 90- *Concurrent Pos:* NSF fel. *Mem:* Soc Mfg Engrs. *Res:* Energy systems; transportation systems technology. *Mailing Add:* 14536 Foxham Ct St Louis MO 63017

JOHNSON, L(AWRENCE) D(AVID), b Tacoma, Wash, Jan 26, 37; m 71; c 2. CIVIL ENGINEERING, MATERIAL SCIENCE. *Educ:* Univ Wash, Seattle, BS, 59; Univ Calif, Berkeley, MS, 61, PhD(eng sci), 63; Miss State Univ, MCE, 73. *Prof Exp:* Sr engr graphite res & develop, Gen Elec Co, Wash, 62-64; res scientist, Pac Northwest Labs, Battelle Mem Inst, 64-65; civil engr, 66-74, RES CIVIL ENGR, WATERWAYS EXP STA, US ARMY CORPS ENGRS, 74- *Honors & Awards:* Standards Develop Award, Am Soc Testing & Mat, 89. *Mem:* Soc Am Mil Engrs; Am Soc Civil Engrs; Am Soc Testing & Mat. *Res:* Mechanical behavior of ceramic materials; soil mechanics and foundation engineering. *Mailing Add:* US Army Eng Waterways Exp Sta 3909 Halls Ferry Rd Vicksburg MS 39180-6199

JOHNSON, L(EE) ENSIGN, b New River, Tenn, May 26, 31; m 55; c 4. ELECTRICAL ENGINEERING, BIOENGINEERING. *Educ:* Vanderbilt Univ, BE, 53, BD, 59; Case Western Reserve Univ, MS, 63, PhD, 64. *Prof Exp:* Prod line mgr, Aladdin Electronics, Div Aladdin Indust, 55-59; from instr to assoc prof, 59-72, assoc provost, 70-75, PROF ELEC ENG, VANDERBILT UNIV, 72- *Mem:* Inst Elec & Electronics Engrs. *Res:* Physiological control systems; iron kinetics in humans; reliability modeling and engineering. *Mailing Add:* Vanderbilt Univ PO Box 1722 Sta B Nashville TN 37235

JOHNSON, LADON JEROME, b Gardner, NDak, Sept 11, 34. ANIMAL HUSBANDRY. *Educ:* NDak State Univ, BS, 56, MS, 57; Ohio State Univ, PhD(animal sci), 65. *Prof Exp:* Res asst animal sci, Ohio State Univ, 56-57; asst county agent com agr, NDak Coop Exten Serv, 59-61; res asst animal sci, Ohio Agr Res & Develop Ctr, 61-64, tech aide, 64-65; from asst exten animal husbandman to exten animal husbandman, 66-74, PROF ANIMAL HUSB, COOP EXTEN SERV, NDAK STATE UNIV, 74- *Mem:* AAAS; Am Soc Animal Sci. *Res:* Physiological differences associated with different gaining ability of beef cattle; effect of stage of maturity on yield and nutritive value of corn silage; improvement of corn silage by chemical additives. *Mailing Add:* Dept Animal Sci NDak State Univ State Univ Sta 185 Hultz Hall Fargo ND 58105

JOHNSON, LARRY, b Lumberton, MC, Jan 26, 49; m 74; c 1. SPERMATOGENESIS, MALE FERTILITY. *Educ:* NC State Univ, BS, 71; Va Polytech Inst & State Univ, MS, 74; Colo State Univ, PhD(reproduction biol), 78. *Prof Exp:* Teaching fel, 78-79, asst instr, 79, prin investr, 83-86, ASST PROF CELL BIOL, UNIV TEX HEALTH SCI CTR, DALLAS, 80- *Concurrent Pos:* Panelist, Off Tech Assessment, US Congress, 84-85. *Mem:* AAAS; Am Soc Andrology; Am Soc Animal Sci; Soc Study Reproduction. *Res:* Human male fertility; estimation of sperm production rates; seasonal and age-related changes in sertoli, Leydig and germ cells; human ejaculate studies. *Mailing Add:* Wichita Falls Family Pract Residency Prog 1301 Third St Wichita Falls TX 76301

JOHNSON, LARRY CLAUD, b Roby, Tex, Aug 24, 36; m 56; c 2. PHYSICS. *Educ:* Tex Christian Univ, BA, 58; Mass Inst Technol, SM, 60; Princeton Univ, PhD(astrophys), 66. *Prof Exp:* Res assoc, 66-69, MEM RES STAFF PLASMA PHYSICS, PLASMA PHYSICS LAB, PRINCETON UNIV, 69- *Mem:* AAAS; Am Phys Soc. *Res:* Plasma physics; plasma spectroscopy and laser scattering; atomic collision cross sections. *Mailing Add:* Plasma Physics Lab PO Box 451 Princeton NJ 08544

JOHNSON, LARRY DON, b Winnfield, La, Nov 20, 40; m 62. PHYSICS. *Educ:* La Polytech Inst, BS, 62; Univ Tenn, MS, 64, PhD(physics), 67. *Prof Exp:* ASSOC PROF PHYSICS, NORTHEAST LA UNIV, 67- *Mem:* Am Asn Physics Teachers; Am Phys Soc; AAAS; Sigma Xi. *Res:* Statistical mechanics and phase transitions; human biomechanics. *Mailing Add:* Dept Physics Northeast La Univ Monroe LA 71203

JOHNSON, LARRY K, b Howard, Kans, Aug 6, 36; m 57; c 5. MATHEMATICS. *Educ:* Kans State Teachers Col, BSEd & AB, 58, MS, 60; Univ Ga, EdD(math educ), 63. *Prof Exp:* Instr math educ, Univ Ga, 61-63; asst prof, 63-67, ASSOC PROF MATH, CENT MO STATE COL, 67- *Mem:* Math Asn Am. *Res:* Mathematics education. *Mailing Add:* Dept Math Central Mo State Univ Warrensburg MO 64093

JOHNSON, LARRY RAY, b Atlanta, Ga, Dec 18, 35; m 58; c 3. INDUSTRIAL ENGINEERING. *Educ:* Ga Inst Technol, BCerE, 58, BIE, 60, MSIE, 62; Okla State Univ, PhD(indust eng), 69. *Prof Exp:* Assoc mfg res engr, Lockheed-Ga Co, 61-63; from asst prof to assoc prof indust eng, 63-76, PROF INDUST ENG, MISS STATE UNIV, 76- *Mem:* Am Inst Indust Engrs. *Res:* Hospital systems; occupational safety and health; energy conservation; work methods. *Mailing Add:* Dept Indust Eng Miss State Univ Mississippi State MS 39762

JOHNSON, LARRY REIDAR, b Seattle, Wash, Jan 5, 45; m 68; c 2. PULMONARY PHYSIOLOGY. *Educ:* Univ Wash, Seattle, BS, 66; State Univ NY, Buffalo, PhD(physiol), 73. *Prof Exp:* SR RES ASSOC, ORE HEALTH SCI UNIV, 82- *Concurrent Pos:* Proj coordr, Ore Health Sci Univ Lung Health Study, 86- *Mem:* Am Physiol Soc; Am Thoracic Soc; AAAS. *Res:* Epidemiology of pulmonary function, quality control of spirometry; pulmonary software maintenance and development. *Mailing Add:* Dept Physiol L334A Ore Health Sci Univ 3181 SW Sam Jackson Park Rd Portland OR 97201

JOHNSON, LAVELL R, b Salt Lake City, Utah, Jan 16, 35; m 58; c 6. BIOCHEMISTRY, ORGANIC CHEMISTRY. *Educ:* Univ Utah, BS, 59; Brigham Young Univ, PhD(biochem), 65. *Prof Exp:* Sr scientist biochem, Ames Co Div, Miles Labs, 64-68; assoc res dir dept med, Latter-Day Saints Hosp, 68-71; PRES, JOHNSON RES, 71- *Mem:* AAAS; Am Chem Soc. *Res:* Mechanism of action of adrenocorticotropic hormone; pregnenolone synthesis by adrenal preparations; analysis of growth hormone, testosterone, metanephrine, insulin and adrenocorticotropic hormone. *Mailing Add:* Johnson Res 3201 Teton Dr Salt Lake City UT 84109

JOHNSON, LAWRENCE ALAN, b Columbus, Ohio, Apr 30, 47; m 69; c 2. PROCESSING OF CROPS. *Educ:* Ohio State Univ, BSc, 69; NC State Univ, MSc, 71; Kans State Univ, PhD(food sci), 78. *Prof Exp:* Res asst food sci, NC State Univ, 69-71; food adv, US Army QM Corp, 71-73; res chemist food prod develop, Dwight P Joyce Res Ctr, Durkee Foods, 73-75; res asst grain sci, Food Sci, Kans State Univ, 75-78; asst res chemist, Food Protein Res & Develop Ctr, Tex A&M Univ, 78-83, assoc res chemist, 83-85; assoc prof, 85-88, PROF FOOD TECHNOL, CTR CROPS UTILIZATION RES, IOWA STATE UNIV, 88- *Honors & Awards:* ADM Award, Am Oil Chemists Soc, 87. *Mem:* Am Asn Cereal Chemists; Inst Food Technologists; Am Oil Chemists Soc; Am Soc Agr Engrs. *Res:* Developing new product or processing technologies to utilize agricultural products; product applications include both food and non-food industrial products; processes include new techniques in crop separations, ingredient conversions, and food refabrication; oil extraction. *Mailing Add:* Dept Food Technol Dairy Indust Bldg Iowa State Univ Ames IA 50011

JOHNSON, LAWRENCE ARTHUR, b Luck, Wis, July 9, 36; m 59; c 3. REPRODUCTIVE PHYSIOLOGY. *Educ:* Univ Wis, River Falls, BS, 61; Univ Minn, St Paul, MS, 63; Univ Md, PhD(animal physiol & biochem), 68. *Prof Exp:* Chemist, 64-66, res chemist, 66-72, RES PHYSIOLOGIST ANIMAL SCI, AGR RES SERV, USDA, 72- *Mem:* Soc Study Reproduction; Am Soc Animal Sci; Soc Anal Cytol. *Res:* Reproductive physiology and biochemistry of mammalian semen and fertilization processes; artificial insemination; frozen semen; sex pre-selection. *Mailing Add:* Reproduction Lab USDA Agr Res Serv Beltsville MD 20705

JOHNSON, LAWRENCE LLOYD, b Bangor, Maine, Dec 30, 41; m 76; c 1. IMMUNOGENETICS. *Educ:* Univ Maine, BA, 64, MA, 73, PhD(zool), 80. *Prof Exp:* Fel, McArdle Lab, Univ Wis-Madison, 80-83, lectr, Lab of Genetics, 83-84; asst mem, 84-89, ASSOC MEM, SARANAC LAKE, NY, 90- *Res:* Genetics of mammalian histocompatibility antigens; developmental immunogenetics; theoretical genetics. *Mailing Add:* Trudeau Inst Inc PO Box 59 Saranac Lake NY 12983

JOHNSON, LAWRENCE ROBERT, b Gyor, Hungary, Feb 14, 31; US citizen. ANALYTICAL CHEMISTRY, PHYSICAL CHEMISTRY. *Educ:* Eotvos Lorand Univ, Budapest, dipl, 53; Columbia Univ, PhD(chem), 61. *Prof Exp:* Res chemist, Lever Bros Res Ctr, NJ, 56-57; AEC res asst, Columbia Univ, 57-59; group leader polymer res radioisotopes, Rohm and Haas Co, Pa, 60-62; asst prof instrumental, anal & phys chem, Lafayette Col, 62-65; assoc prof anal & phys chem, Union Col, Ky, 65-78, actg head dept, 69-73; consult, 78-81; CHEMIST WATER TREAT, CITY UTILITIES CO, CORBIN, KY, 81- *Res:* Kinetics of polymer adsorption, flocculation and deflocculation; radioisotopes; instrumental analysis. *Mailing Add:* PO Box 274 Corbin KY 40701

JOHNSON, LAYNE MARK, b Northfield, Minn, June 4, 53; m 78. INFORMATION SCIENCE & SYSTEMS. *Educ:* Dana Col, BA, 75; Iowa State Univ, MS, 78, PhD(microbiol), 80. *Prof Exp:* Postdoctoral fel, Univ Okla, Norman, 80-82; sr res microbiologist, Cytox Corp, Allentown, Pa, 82-84; microbial ecologist, 84-87; sr info scientist, 87-91, MGR TECH INFO, AM CYANAMID CO, PEARL RIVER, NY, 91- *Mem:* Soc Indust Microbiol; Pharmaceut Mfg Asn. *Res:* Management of published information, including scientific literature and patents pertaining to drug development processes within the pharmaceutical industry; manage state-of-the-art end user search program. *Mailing Add:* 233 Steilen Ave Ridgewood NJ 07450

JOHNSON, LEANDER FLOYD, b Lecompte, La, Aug 3, 26; m 48; c 2. SOIL-BORNE PLANT DISEASES. *Educ:* Southwestern La Inst, BS, 48; La State Univ, MS, 51, PhD(plant path), 53. *Prof Exp:* Instr bot, 53-54, from asst prof to assoc prof plant path, 54-70, PROF PLANT PATH, UNIV TENN, KNOXVILLE, 70- *Mem:* Am Phytopath Soc; Sigma Xi. *Res:* Biological control of plant diseases; methods of approach and basic concepts of soil microbiology. *Mailing Add:* Dept Entom & Plant Path Univ Tenn PO Box 1071 Knoxville TN 37901

JOHNSON, LEE FREDERICK, b Philadelphia, Pa, Jan 10, 46; m 67; c 2. MOLECULAR BIOLOGY. *Educ:* Muhlenberg Col, BS, 67; Yale Univ, MPhil, 69, PhD(molecular biophysics), 72. *Prof Exp:* Fel cell biol, Mass Inst Technol, 71-75; from asst prof to assoc prof, 75-85, PROF BIOCHEM, OHIO STATE UNIV, 85-, PROF MOLECULAR GENETICS, 87-, CHMN MOLECULAR GENETICS, 90- *Concurrent Pos:* fel, 72-74, fac res award, Am Cancer Soc, 80-85; fac res award, Am Cancer Soc, 80-85; mem, molecular, cellular & develop biol progs, Ohio State Univ, 76-, molecular biol panel, NSF, 80-84. *Mem:* Am Soc Cell Biol; Am Soc Biochem & Molecular Biol; Am Soc Microbiol. *Res:* Regulation of growth, RNA metabolism and gene expression in cultured mammalian cells; genetic engineering. *Mailing Add:* Dept Molecular Genetics Ohio State Univ Columbus OH 43210

JOHNSON, LEE H(ARNIE), b Houston, Tex, Jan 4, 09; m 40; c 2. CIVIL ENGINEERING. *Educ:* Rice Inst, BA, 30, MA, 31; Harvard Univ, MS, 32, ScD(civil eng), 35. *Prof Exp:* Asst civil eng, Harvard Univ, 32-35; asst eng aide, US Waterways Exp Sta, Miss, 35-36; jr engr & asst to engr in charge design & specifications, US Eng Off, Ala, 36-37; prof civil eng & dean, Sch Eng, Univ Miss, 37-50; prof civil eng & dean, Sch Eng, 50-72, EMER DEAN & W R IRBY PROF ENG, TULANE UNIV, LA, 72- *Concurrent Pos:* Teacher calculus, Newman Sch, La, 79-87. *Honors & Awards:* Tulane Engr Award. *Mem:* Am Soc Civil Engrs; Am Soc Eng Educ. *Res:* Mathematical simplification of design of statically indeterminate structures; new technique of slide rule operation for duplex-type slide rules; simplified nomography; creative approach to engineering education. *Mailing Add:* 211 Fairway Dr New Orleans LA 70124

JOHNSON, LEE MURPHY, b Lufkin, Tex, Sept 11, 34; m 57; c 1. MATHEMATICS. *Educ:* Univ Tex, Austin, BSChE, 57, MA, 65, PhD(math), 68. *Prof Exp:* Res chem engr, Humble Oil & Refining Co, 57-62; asst prof, 67-71, ASSOC PROF MATH, NORTHERN ARIZ UNIV, 71- *Mem:* Am Math Soc; Math Asn Am. *Res:* General measure theory. *Mailing Add:* 1201 E Ponderosa Pkwy Flagstaff AZ 86011

JOHNSON, LEE W, b Appleton, Minn, Oct 25, 38; m 63. MATHEMATICS. *Educ:* La State Univ, BS, 63, MS, 65; Mich State Univ, PhD(math), 67. *Prof Exp:* Asst prof, 67-74, ASSOC PROF MATH, VA POLYTECH INST & STATE UNIV, 74- *Mem:* Am Math Soc; Soc Indust & Appl Math. *Res:* Numerical analysis and approximation theory. *Mailing Add:* Dept Math Va Polytech Inst & State Univ Blacksburg VA 24061

JOHNSON, LELAND GILBERT, b Roseau, Minn, Oct 16, 37; m 78; c 3. COMPARATIVE PHYSIOLOGY, EMBRYOLOGY. *Educ:* Augustana Col, SDak, BA, 59; Northwestern Univ, MS, 61, PhD(biol sci), 65. *Prof Exp:* From asst prof to assoc prof, 64-73, PROF BIOL, AUGUSTANA COL, SDAK, 73- *Concurrent Pos:* NSF sci fac fel, Queen Mary Col, 70-71; George C Marshall fel, Biol Inst, Odense Univ, Denmark, 77, Fulbright scholar, 83, Australian Inst Marine Sci, 89 & 90. *Mem:* AAAS; Am Soc Zool; Soc Develop Biol; Asn Biol Lab Educ. *Res:* Developmental physiology; effects of temperature on developmental processes; author of two general biology texts and a developmental biology laboratory manual. *Mailing Add:* Dept Biol Augustana Col Sioux Falls SD 57197

JOHNSON, LENNART INGEMAR, b Minneapolis, Minn, Dec 23, 24; m 61; c 1. MATERIALS & PROCESS ENGINEERING, SPECIFICATIONS & STANDARDS. *Educ:* Univ Minn, BS, 48. *Prof Exp:* Sr engr, Ordinance Div, Honeywell, 49-67, prin engr, 67-69, supvr, Eng Plastics Lab, 69-87, staff eng, Defense Systs Div, 87-88; CONSULT, SOC AUTOMOTIVE ENGRS, 89- *Concurrent Pos:* Chmn composites comt, Soc Automotive Engrs, 86-88. *Honors & Awards:* Leadership & Serv Award, Soc Automotive Engrs, Dedication & Distinction Award. *Mem:* Soc Automotive Engrs; Am Inst Chem Engrs; fel Am Inst Chemists. *Res:* Development of casting resins involving urethane and epoxy polymers; development of stain-free injection molding of thermoplastic polymers; development of material specifications. *Mailing Add:* 14109 M Terr Minnetonka MN 55345

JOHNSON, LEO FRANCIS, b White Plains, NY, Nov 6, 28; m 62; c 4. LASERS. *Educ:* Univ Vt, BA, 51; Syracuse Univ, MS, 55, PhD(physics), 59. *Prof Exp:* Tech engr, Gen Elec Co, 51-53; res asst physics, Syracuse Univ, 54-59; mem tech staff physics, Bell Tel Labs, 59-86; consult, Amoco Laser Co, 87-89; CONSULT, AMOCO RES CTR, 89- *Mem:* Fel Am Phys Soc; Sigma Xi. *Res:* Photoconductivity of semiconductors; optical spectroscopy of rare earth and transition metal ions in crystals; investigations of laser phenomena in crystals; interference diffraction gratings; sub-micron surface structures; distributed feedback lasers. *Mailing Add:* 150 Riverwood Ave Bedminster NJ 07921

JOHNSON, LEON JOSEPH, b Detroit, Mich, Jan 17, 29; m 52; c 3. SOIL MINERALOGY. *Educ:* Pa State Univ, BS, 54, MS, 55 PhD(agron), 57. *Prof Exp:* Res geologist, Cities Serv Res & Develop Co, 57-59; asst prof soil technol, 59-67, assoc prof, 67-80, PROF SOIL MINERAL, PA STATE UNIV, 80- *Mem:* Am Soc Agron; Clay Minerals Soc. *Res:* Weathering of soil minerals; formation of soil profiles; clay mineralogy. *Mailing Add:* Dept Agron 119 Tyson Bldg Pa State Univ University Park PA 16802

JOHNSON, LEONARD EVANS, b Ogden, Utah, Nov 13, 40; m 87; c 1. GEOPHYSICS. *Educ:* Mass Inst Technol, BS, 62; Univ Calif, San Diego, MS, 67, PhD(geophys), 71. *Prof Exp:* Res assoc geophys, Boeing Sci Res Labs, 62-65; vis fel, Coop Inst Res Environ Sci, Univ Colo, 71-73; vis prof, Univ Calif, Berkeley, 73-74; assoc prog dir, Continental Lithosphere, NSF, 74-79, prog dir geophys, 79-82, prog dir seismol, 82-84, prog dir, 84-89; PROG DIR, CONTINENTAL DYNAMICS, NSF, 89- *Concurrent Pos:* Prof lectr, George Washington Univ, 77-86; mem comt math geophys, Int Union Geod & Geophys; assoc dir, Off Sci & Technol Centers Develop, NSF, 88. *Mem:* Am Geophys Union; Seismol Soc Am; AAAS. *Res:* Theoretical and observational seismology, inverse problems in geophysics. *Mailing Add:* Div Earth Sci NSF Washington DC 20550

JOHNSON, LEONARD ROY, b Chicago, Ill, Jan 31, 42; m; c 3. PHYSIOLOGY. *Educ:* Wabash Col, BA, 63; Univ Mich, Ann Arbor, PhD(physiol), 67. *Prof Exp:* NIH fel & instr physiol, Sch Med, Univ Calif, Los Angeles, 67-69; from asst prof to assoc prof, Sch Med, Univ Okla, 69-72; PROF PHYSIOL, UNIV TEX MED SCH, HOUSTON, 72- *Concurrent Pos:* Res grant, Univ Okla, 70-73; G A Manahan Trust grant, 70-72; NIH res career develop award, 72-77, grant, 73-; ed, Am J Physiol, 79-85; Vet Admin Merit Rev Bd, 87-91; NIH study sect gastroenterol clin nutrit, 80-82; nat bd med examiners, Physiol Test Comt, 83-91, chmn, 88-91. *Honors & Awards:* Hoffmann-LaRoche Prize, Am Physiol Soc. *Mem:* Am Gastroenterol Asn; Am Physiol Soc; Endocrine Soc; Soc Exp Biol & Med. *Res:* Regulation of gastric and pancreatic secretion action of gastrointestinal hormones; regulation of growth of gastrointestinal mucosa and gastrin receptor binding. *Mailing Add:* Dept Physiol Univ Tenn Med Col 894 Union Ave Memphis TN 38163

JOHNSON, LEROY DENNIS, b Langhorne, Pa, Oct 4, 08; m 40; c 2. ORGANIC CHEMISTRY. *Educ:* Lincoln Univ, Pa, AB, 31; Univ Pa, MS, 34, PhD(chem), 54. *Prof Exp:* Prof chem, Storer Col, 34-55, dean, 40-55; assoc prof, 55-56, actg dean, 56-57, DEAN, LINCOLN UNIV, PA, 57-, PROF CHEM, 56-, REGISTR, 69- *Mem:* Am Chem Soc. *Res:* General chemistry demonstrations; general science demonstrations; analytical organic chemistry; environmental science. *Mailing Add:* Lincoln Univ Box 178 Lincoln University PA 19352

JOHNSON, LEROY FRANKLIN, b Seattle, Wash, Feb 4, 33; m 56; c 2. NUCLEAR MAGNETIC RESONANCE. *Educ:* Ore State Univ, BS, 54, MS, 56. *Prof Exp:* Dept mgr, Varian Assocs, 57-72; vpres, Nicolet Magnetics Corp, 72-83; SR SCIENTIST, GEN ELEC CO, 83- *Concurrent Pos:* Prof-in-charge interpretation nuclear magnetic resonance spectra short course, Am Chem, Soc, 66-; mem subcomt E-13 molecular spectros, Am Soc Testing & Mat, 70-; mem exec comt, Exp Nuclear Magnetic Resonance Conf, 77-79. *Mem:* Am Chem Soc; AAAS; Soc Appl Spectros. *Res:* Applications of nuclear magnetic resonance spectroscopy; development of nuclear magnetic resonance instrumentation; utilization of minicomputers with nuclear magnetic resonance instruments. *Mailing Add:* Gen Elec Co 255 Fourier Ave Fremont CA 94539

JOHNSON, LESLIE KILHAM, b New York, NY, June 9, 45; m 76; c 1. ZOOLOGY. *Educ:* Harvard Univ, BA, 67; Univ Calif, Berkeley, PhD(zool), 74. *Prof Exp:* Grad fel zool, NSF, 71-74; res fel, Alexander von Humboldt, Zool Inst, Wurzburg, Ger, 74-75; asst prof zool, Univ Iowa, 75-80, assoc prof biol, 80-88; biologist, Smithsonian Tropical Res Inst, 87-88; res biologist, 88-89, LECTR, PRINCETON UNIV, 89- *Concurrent Pos:* Mem, bd dir, Orgn Trop Studies, 79-88; res assoc, Smithsonian Trop Res Inst, 83-88. *Mem:* Sigma Xi; Asn Trop Biol; Soc Study Evolution. *Res:* Behavioral ecology; aggressive behavior; learning; foraging patterns of social insects; sexual selection in insects. *Mailing Add:* Dept Ecol & Evolution Biol Princeton Univ Princeton NJ 08544

JOHNSON, LESLYE, ALLERGY RESEARCH. *Prof Exp:* BR CHIEF, ENTERIC DIS BR, NAT INST ALLERGY & INFECTIOUS DIS, NIH, 89- *Mailing Add:* NIH Nat Inst Allergy & Infectious Dis Enteric Dis Br Westwood Bldg Rm 748 5333 Westbard Ave Bethesda MD 20892

JOHNSON, LIONEL, limnology, fisheries, for more information see previous edition

JOHNSON, LITTLETON WALES, b Concord Wharf, Va, Oct 17, 29; m 51; c 3. FOOD SCIENCE. *Educ:* Va Polytech Inst, BS, 56, MS, 58. *Prof Exp:* Asst processing engr, Hercules Powder Co, 55-58; assoc prof food technol, Va Polytech Inst, 58-61; plant mgr, Dulany Foods, Inc, 61-67; opers mgr, Glidden-Durkee Div, SCM Corp, 67-69, dir mfg, Food Serv Group, 69-71, Regional mgr, 71-76; vpres, Mrs Smith's Frozen Food Co, Kellogg Co, 76-80, sr vpres mfg, 80-88; CONSULT, 88- *Mem:* Inst Food Technologists; Am Frozen Food Inst; Sigma Xi; Am Mgt Asn; Nat Food Processors Asn. *Res:* Food processing techniques; statistical quality control; submerged acetic fermentations. *Mailing Add:* 292 Continental Dr Pottstown PA 19464

JOHNSON, LLOYD N(EWHALL), b Eureka, Kans, Nov 16, 21; m 45; c 1. PETROLEUM & CHEMICAL ENGINEERING. *Educ:* Univ Kans, BS, 44; Univ Tex, Austin, PhD(petrol eng), 70. *Prof Exp:* Indust chemist, Hercules Powder Co, 44-45; res engr, Core Labs, Inc, 46-52, supvr reservoir fluids lab, Venezuela, 52-55; res engr, petrol res comt, Univ Tex, 55-63, instr math, 63-64; asst prof petrol & natural gas eng, Tex A&I Univ, 65-77; coordr Oil & Gas Technol, Bee County Col, 79-81. *Concurrent Pos:* Consult engr, 64- *Mem:* Simulation Coun; Soc Petrol Engrs; Geochem Soc; Asn Comput Mach. *Res:* Drilling problems; reserves; completion methods; production methods and controls; mathematical models; scientific data processing; economic development and improved recovery in petroleum reservoirs; engineering methods. *Mailing Add:* PO Box 2254 Station One Kingsville TX 78363

JOHNSON, LOERING M, b Dickinson, NDak, Sept 22, 26; m 52; c 4. CONTROL SYSTEMS DESIGN & ANALYSIS, INSTRUMENT SYSTEMS DESIGN TROUBLE SHOOTING. *Educ:* Univ NDak, BS, 52; Rennselaer Polytech Inst, MS, 61. *Prof Exp:* Engr elec syst design, EI duPont de Nemours, Inc, 52-53; eng, elec syst design, Combustion Eng Inc, 55-58, supvr comput appl, 58-60, mgr, inst control & elect syst, 60-70, mgr standards, 70-80, mgr records control, 80-82, mgr office automation, 82-85; ASSOC PROF, TEACHING, UNIV HARTFORD, 86- *Concurrent Pos:* Mem, Standards Bd, Inst Elec & Electronics Engrs, 78-81; secy, Nuclear Power Eng Comt, 66-78. *Mem:* Nat Soc Prof Engrs; Inst Elec & Electronics Engrs; Sigma Xi. *Res:* Text book on practical control system design; published numerous papers and book section on instrumentation and control, standards, and technical writing. *Mailing Add:* PO Box 372 Tariffville CT 06081-0372

JOHNSON, LOUISE H, b Minneota, Minn, Oct 22, 27. MATHEMATICS EDUCATION. *Educ:* Augsburg Col, BA, 49; Univ Northern Colo, MA, 61, DEduc, 71; Univ Ill, MA, 63. *Prof Exp:* Teacher high schs, Minn, 49-62; assoc dean lib arts & sci, 74-76, PROF MATH, ST CLOUD STATE UNIV, 63-, DEAN LIB ARTS & SCI, 76-, DEAN SCI & TECHNOL, 84- *Mem:* Nat Coun Teachers Math. *Mailing Add:* Hedley Hall St Cloud State Univ MS 145 720 Fourth Ave S St Cloud MN 56301

JOHNSON, LOWELL BOYDEN, b Dwight, Ill, Oct 12, 35; m 56; c 2. PLANT MOLECULAR BIOLOGY, PLANT TISSUE CULTURE. *Educ:* Univ Ill, BS, 57; Purdue Univ, West Lafayette, MS, 62, PhD(plant path), 64. *Prof Exp:* Asst res plant pathologist, Univ Calif, Davis, 64-68; from asst prof to assoc prof, 68-82, PROF PLANT PATH, KANS STATE UNIV, 82- *Concurrent Pos:* Vis scholar, div biol sci, Univ Mich, Ann Arbor, 85. *Mem:* AAAS; Am Phytopath Soc; Am Soc Plant Physiol; Int Soc Plant Molecular Biol. *Res:* Plant disease physiology; plant cell culture and regeneration; alfalfa molecular genetics. *Mailing Add:* Dept Plant Path Kans State Univ Throckmorton Hall Manhattan KS 66506-5502

JOHNSON, LOYD, b Somerville, Ala, Mar 18, 27; m 52; c 3. SOIL & WATER MANAGEMENT, AGRICULTURAL EXPERIMENT STATION MANAGEMENT. *Educ:* Ala Polytech Inst, BS, 50, MS, 55. *Prof Exp:* Asst dist supt farm develop, Tela RR Co, 51-52 & 56; asst agr eng, Ala Agr Exp Sta, Auburn, 53-54; asst engr, Gen Off, United Fruit Co, 56-57, sr proj engr, Cia Agricola Guatemala, 56-60; agr engr, Rockefeller Found, 60-82; exp sta develop, 86-89, EXP STA MGT TRAINING COURSE, WINROCK INT, 90- *Concurrent Pos:* Agr engr, Int Rice Res Inst, 60-68 & NC State Univ, 67-68; agr engr, Int Ctr Trop Agr, Colombia, 68-77; vis scientist, La State Univ, 74-75; rice specialist, Ecuador Nat Inst for Land & Cattle Investigations, 77- & Int Agr Develop Serv, 78-81; vis scientist, Int Fertilizer Develop Ctr, 81-82. *Mem:* Am Soc Agr Engrs; Indian Soc Agr Eng; Nat Soc Prof Engrs; Soc Am Mil Engrs. *Res:* Rice specialist and development of irrigation, fertilizer, drainage, roads, bridges, sanitation, machine and processing systems for agricultural experiment stations and food production in the lowland tropics; machinery management. *Mailing Add:* Rte 3 Box 486 Union Rd Somerville AL 35670

JOHNSON, LUTHER ELMAN, electrical engineering; deceased, see previous edition for last biography

JOHNSON, LYNWOOD ALBERT, b Macon, Ga, Oct 4, 33. INDUSTRIAL ENGINEERING, OPERATIONS RESEARCH. *Educ:* Ga Inst Technol, BIE, 55, MS, 59, PhD(indust eng), 65. *Prof Exp:* Indust engr, E I du Pont de Nemours & Co, Inc, 55-57; from instr to asst prof indust eng, Ga Inst Technol, 58-64; supvr opers res, Kurt Salmon Assocs, Inc, 64-66; assoc prof, 66-68, PROF INDUST ENG, GA INST TECHNOL, 68- *Concurrent Pos:* Vis prof, Thayer Sch Eng, Dartmouth Col, 67, Dept Systs & Indust Eng, Univ Ariz, 81-82, Dept Mech Eng, Univ Wash, 85; assoc ed, J Forecasting, 81-86. *Mem:* Am Inst Indust Engrs; Opers Res Soc Am; Inst Mgt Sci; Am Prod & Inventory Control Soc. *Res:* Production systems analysis; systems modeling and simulation; optimization methods; decision theory; engineering design processes. *Mailing Add:* Sch of Indust & Systs Eng Ga Inst Technol Atlanta GA 30332

JOHNSON, MALCOLM JULIUS, b Portland, Ore, Dec 12, 17; m 42; c 2. AGRONOMY. *Educ:* Ore State Univ, BS, 41, MS, 55; Purdue Univ, PhD(crop physiol, ecol), 61. *Prof Exp:* From assoc prof to prof, 49-80, EMER PROF AGRON & SUPT, CENT ORE EXP STA, ORE STATE UNIV, 49-80. *Mem:* Am Soc Agron; Crop Sci Soc Am; Sigma Xi. *Res:* Crop adaptation and management practices, especially crop fertility on pumice soils. *Mailing Add:* 208 N Canyon Dr Redmond OR 97756

JOHNSON, MALCOLM PRATT, b New Haven, Conn, Aug 9, 41; m 64; c 2. INORGANIC CHEMISTRY. *Educ:* Amherst Col, BA, 63; Northwestern Univ, PhD(inorg chem), 67. *Prof Exp:* Res chemist, Chem Div, Union Carbide Corp, 66-69, res chemist, Linde Div, Tarrytown, NY, 69-71; gen mgr Gulf Coast Div, Humphrey Chem Co, 71-77; mgr com develop, Southwest Specialty Chem Inc, 77-80; MGR MKT, DIXIE CHEM CO, 80- *Mem:* Am Chem Soc; NY Acad Sci. *Res:* Oxygen and nitrogen complexes of transition metals; organometallic chemistry; homogeneous catalysis; Lewis basicity; polyethylenimine chemistry; infrared spectroscopy. *Mailing Add:* Dixie Chem Co PO Box 130410 Houston TX 77219

JOHNSON, MARIE-LOUISE T, b New York, NY, July 26, 27. DERMATOLOGY, MEDICAL EDUCATION. *Educ:* Manhattanville Col, BA, 48; Yale Univ, PhD(microbiol), 54, MD, 56. *Prof Exp:* From instr to asst prof med, Sch Med, Yale Univ, 58-64, actg head, Div Dermat, 61-62; chief dermat, Atomic Bomb Casualty Comn, Hiroshima & Nagasaki, 64-67; assoc prof dermat, Sch Med, NY Univ, 67-69, assoc prof clin dermat, 69-70, from assoc prof to prof dermat, 74-80; assoc prof internal med, Dartmouth Med Sch, 71-74; chief dermat serv, Bellevue Hosp, 74-80; vpres med affairs, 80-82, DIR, MED EDUC, BENEDICTINE HOSP, KINGSTON, NY, 80-; CLIN PROF DERMAT, SCH MED, YALE UNIV, 80- *Concurrent Pos:* Pres, Maternity & Early Childhood Found; chief, Dermat Serv, Vet Admin Hosp, White River Jet, Vt, 71-74, chief, Ambulatory Serv, 73-74; head, Div Educ & Commun, Nat Prog Dermat, 73-75; chmn, Med & Sci Comt, Dermat Found, 74-75 & Coun Educ Affairs, Am Acad Dermat, 80-82; mem, Eval Comt, Am Acad Dermat, 76-82, bd dirs, 77-80, Task Force Manpower, 86-89; vis lectr, 79th All-Japan Dermat Meeting, Hiroshima & Postgrad Course Venereal Dis, Yugoslavia, 80, XVI Int Cong Dermat, Tokyo, 82; deleg, Cong Int Physicians Against Nuclear War, Cambridge, 82, Third Cong, Neth, 83, Seventh Cong, Moscow, 87, Eighth Cong, 88 & Ninth Cong, Japan, 89; mem bd dirs, Am Dermat Asn, 86- *Mem:* Inst Med-Nat Acad Sci; Am Dermat Asn (vpres, 91-); Am Acad Dermat; Soc Invest Dermat; Int Physicians Prev Nuclear War; NY Acad Med; AMA; Soc Trop Dermat. *Res:* Epidemiology studies of late radiation effects in Hiroshima and Nagasaki; population studies as with the Health and Nutrition Examination Survey; prevalence of Hansen's Disease in Pohnpei, Micronesia; author of six scientific publications. *Mailing Add:* Benedictine Hosp 105 Mary's Ave Kingston NY 12401

JOHNSON, MARK EDWARD, b Chicago, Ill, June 27, 52; m 76. STATISTICS, OPERATIONS RESEARCH. *Educ:* Univ Iowa, BA, 73, MS, 74, PhD(indust & mgt eng), 76. *Prof Exp:* STAFF MEM STATIST, LOS ALAMOS NAT LAB, 76- *Mem:* Am Statist Asn; Math Asn Am; Inst Math Statist. *Res:* Applied statistics; random variate generation; Monte Carlo methods; probability distributions. *Mailing Add:* 218 Verdi Dr Gallup NM 87301

JOHNSON, MARK SCOTT, b Oakland, Calif, Apr 16, 51. PROGRAMMING LANGUAGES, COMPILER TECHNOLOGY. *Educ:* Univ Calif, BS, 73, MS, 74; Univ BC, PhD(comput sci), 78. *Prof Exp:* Asst prof, San Francisco State Univ, 78-80; mem tech staff, Hewlett-Packard Labs, 80-86; lang prods, Sun Microsysts, 86-90; MGR EDUC SERV, MICROTEC RES, INC, 90- *Concurrent Pos:* Vchair, Sigplan, Asn Comput Mach, 83-87, chair, 87-89, chair Sig bd, 90-, mem coun, 90-; mem, Spec Interest Group Planning Languages & Spec Interest Group Software Eng. *Mem:* Asn Comput Mach. *Res:* Develop and teach course in software engineering, software development methods, and programming tools and environments, particulary software debugging. *Mailing Add:* Microter Res Inc 2350 Mission College Blvd Santa Clara CA 95054

JOHNSON, MARVIN ELROY, b Red Wing, Minn, Nov 3, 45; m 70. PARTICLE PHYSICS. *Educ:* Univ Minn, BS, 67; Yale Univ, MPhil, 69, PhD(physics), 73. *Prof Exp:* PHYSICIST, FERMI NAT ACCELERATOR LAB, 73- *Mem:* Sigma Xi. *Res:* Strong interactions using hybrid bubble chamber techniques. *Mailing Add:* Nat Accelerator Lab MS 357 Batavia IL 60510

JOHNSON, MARVIN FRANCIS LINTON, b Chicago, Ill, June 6, 20; m 43; c 3. PHYSICAL CHEMISTRY. *Educ:* Loyola Univ, Ill, BS, 40, MS, 42. *Prof Exp:* Res chemist, Res & Develop Dept, Sinclair Refining Co, 41-50, res chemist, Sinclair Res Labs, 50-69, sr res chemist, Harvey Tech Ctr, Atlantic Richfield Co, 69-73, res assoc, 73-79, sr res assoc, 79-84, sr res adv, 84-85; CONSULT, 85- *Concurrent Pos:* Chmn, subcomt Phys Chem Catalysts, Am Soc Testing & Mat, 80-85. *Mem:* Catalysis Soc; Am Chem Soc; Am Soc Testing & Mat. *Res:* Heterogeneous catalysis; adsorption of gases by catalysts; pore structures of catalysts; physical-chemical characterizations of catalysts. *Mailing Add:* 1124 Elder Rd Homewood IL 60430

JOHNSON, MARVIN M, b Salt Lake City, Utah, Mar 21, 28; m 51; c 4. KINETICS, CATALYSIS. *Educ:* Univ Utah, BS, 50, PhD(chem eng), 56. *Prof Exp:* Sr res engr, 56-65, mgr hydrocarbon process, 65-68, res assoc, 68-74, sr res assoc, 74-78, sr scientist catalysis, 78-86, CONSULT, PHILLIPS RES CTR, 86- *Concurrent Pos:* Adj prof chem eng, Univ Kans, 81-82; vis prof, Colo Sch Mines, 82. *Honors & Awards:* Nat Medal Technol, 86. *Mem:* Am Chem Soc; Am Inst Chem Engrs; Sigma Xi; Nat Soc Prof Engrs. *Res:* New catalysts and processes related to production and refining of petroleum and petrochemicals. *Mailing Add:* 354 PL Phillips Res Ctr Bartlesville OK 74004

JOHNSON, MARVIN MELROSE, b Neligh, Nebr, Apr 21, 25; m 51; c 5. INDUSTRIAL ENGINEERING, ENGINEERING STATISTICS. *Educ:* Purdue Univ, BS, 49; Univ Iowa, MS, 66, PhD(indust eng), 68. *Prof Exp:* Supvr qual control, Chicago Bumper Div, Houdaille Hershey, 49-52; sr indust engr, Bell & Howell Co, 52-54; chief indust engr, Pioneer Cent Div, Bendix Corp, 54-57, supvr systs & procedures, 57-59, staff asst to asst gen mgr, 59-64; lectr indust & mgt eng, Univ Iowa, 64, instr, 65-66; assoc prof mech eng, Univ Nebr, Lincoln, 68-70, assoc prof, 70-77, prof indust & mgt systs eng, 77-88; VIS PROF & PROG COORDR INDUST ENG, SDAK SCH MINES & TECHNOL, 89- *Concurrent Pos:* Indust eng consult, Lincoln, Nebr & Davenport, Iowa, 64-; consult, Pioneer Cent Div, Bendix Corp 64-68 & Brunswick Corp, 69-; prof & advisor, USAID-Univ Nebr-Omaha Contract-Kabul Univ, Afghanistan, 74-76; vis prof Indust eng, Univ PR, Mayaguez, 82-83. *Mem:* Fel Am Inst Indust Engrs; Am Soc Mech Engrs; Am Statist Asn; Am Soc Eng Educ; Inst Mgt Sci; Opers Res Soc Am; Sigma Xi. *Res:* Systems; vegetable protein isolate; replaceable energy sources; operations research; applied statistics; simulation; quality control and reliability; production planning and control. *Mailing Add:* Indust Eng SDak Tech 501 E St Joseph Rapid City SD 57701-3995

JOHNSON, MARY FRANCES, b Green Bay, Wis, Nov 17, 40. INORGANIC CHEMISTRY. *Educ:* Marquette Univ, BS, 63, MS, 65; St Louis Univ, PhD(inorg chem), 72. *Prof Exp:* PROF & CHAIRPERSON CHEM DEPT, FONTBONNE COL, 72- *Mem:* Am Chem Soc; Sigma Xi. *Res:* Spectroscopy and synthesis of lanthanide chelates involving nitrogen donor ligands. *Mailing Add:* 11725 Wurnall Rd Kansas City MO 64114

JOHNSON, MARY FRANCES, b Milford, Conn, Nov 21, 51; m 78. CLINICAL TRIALS, SURVIVAL ANALYSIS. *Educ:* Tufts Univ, BS, 73; Yale Univ, MPH, 75, PhD(biostatist), 78. *Prof Exp:* Data analyst, Dept Epidemiol & Public Health, Yale Univ, 73-74, res asst, Conn Cancer Epidemiol Unit, 75-76; consult, Waterford Conserv Comn, Conn, 74-75; MATH STATISTICIAN, DIV BIOMET, BUR DRUGS, FOOD & DRUG ADMIN, 78- *Concurrent Pos:* Student ed, Yale J Biol & Med, Yale Univ 75-78, teaching asst, Div Biostatist, 75-77. *Mem:* Am Statist Asn; Biomet Soc. *Res:* Design and statistical analysis of therapeutic drug trials and epidemiological studies; applications of parametric and non-parametric models for failure time data. *Mailing Add:* 111 Danvers Lane Rockville MD 20857

JOHNSON, MARY IDA, b Harlingen, Tex, Oct 30, 42; m 75; c 3. NEUROBIOLOGY, PEDIATRIC NEUROLOGY. *Educ:* Wash State Univ, BS, 64; Johns Hopkins Univ, MD, 68. *Prof Exp:* Intern & resident, Johns Hopkins Hosp, 68-71; fel neurol, 71-74, res asst prof neurol, 74-84, ASSOC PROF PEDIAT, ANAT & NEUROL, WASH UNIV SCH MED, 84- *Mem:* Soc Neurosci; Child Neurol Soc; Am Acad Neurol; Soc Pediat Res. *Res:* Differentiation of neuronal form, growth cone function, dendritic development; development of neurotransmitter function in the autonomic nervous system. *Mailing Add:* Univ Ariz Health Sci Ctr 1501 N Campbell Ave Tucson AZ 85724

JOHNSON, MARY KNETTLES, b Detroit, Mich, Sept 2, 29; m 55; c 2. BACTERIOLOGY. *Educ:* La State Univ, BS, 54, MS, 55, PhD(bact), 57. *Prof Exp:* Res assoc pharmacol, Stanford Univ, 57-58; asst prof microbiol, Sch Med, Univ Miss, 58-65; assoc prof, 67-80, PROF MICROBIOL, SCH MED, TULANE UNIV, LA, 80- *Concurrent Pos:* Instr, Millsaps Col, 58-61. *Mem:* Fel Am Acad Microbiol; Am Soc Microbiol. *Res:* Bacterial physiology; mechanisms of pathogenicity. *Mailing Add:* Dept Microbiol Tulane Univ Sch Med New Orleans LA 70112

JOHNSON, MARY LYNN MILLER, b Pampa, Tex, Mar 12, 38; m 57; c 2. FUEL SCIENCE, AIR POLLUTION. *Educ:* Univ Tex, El Paso, BS, 58; NMex State Univ, MS, 61; Pa State Univ, PhD(fuel sci), 70. *Prof Exp:* Chemist, El Paso City-County Health Unit, Tex, 59-60 & 61-63 & Tex State Health Dept, 63-64; independent consult air pollution, 64-68; asst prof chem, Univ Tex, Arlington, 68-75; instr chem, Hockaday Sch, Dallas, 75-86; instr chem, Brookhaven Col, Dallas, 80-87 & 91; INST CHEM, HIGHLAND PARK HIGH SCH, DALLAS, 86- *Concurrent Pos:* Fel, Am Inst Chemists. *Mem:* Combustion Inst; Am Chem Soc; Am Inst Chemists. *Res:* Investigation of odor counteractants; combustion reactions, especially in the afterburning region, oxides of carbon and sulfur; analytical methods for measurement of air pollutants; air pollution chemistry; flame chemistry; combustion, new energy sources and air pollution. *Mailing Add:* 3004 Croydon Denton TX 76201

JOHNSON, MAURICE VERNER, JR, research administration, agricultural engineering, for more information see previous edition

JOHNSON, MELVIN ANDREW, b Springfield, Ohio, Sept 4, 29; m 53; c 2. MEDICAL PHYSIOLOGY. *Educ:* Cent State Univ, BS, 50; Miami Univ, MS, 55; Jefferson Med Col, PhD(med physiol), 69. *Prof Exp:* Asst anat, Western Reserve Univ, 51-53; grad asst zool, Miami Univ, 54-55; instr biol, Grambling Col, 55-59; from instr to assoc prof, 61-72, chmn dept, 69-85, PROF BIOL, CENT STATE UNIV, 72-, DEAN, COL ARTS & SCI, 85-; prof, 74-85, ADJ PROF PHYSIOL, SCH MED, WRIGHT STATE UNIV, 85- *Concurrent Pos:* Am Heart Asn res grant, 70-72; prog dir minority biomed support grant, NIH, 72-88, prin investr, 72-77 & 80-88, ad hoc consult, Div Res Resources, 73-80; item writer, Educ Testing Serv, 75-77; prog dir, NASA grant, 77-79; res reviewer, Ohio Affiliate, Am Heart Asn, 85-87; prog dir, NIMH grant, NIH, 90- *Mem:* Nat Inst Sci (pres, 79-81, treas, 84-); AAAS; Am Physiol Soc; Sigma Xi; Am Heart Asn; Fedn Am Socs Exp Biol & Med. *Res:* Hemodynamic and metabolic responses to hemorrhagic stress following surgical alterations in liver and splenic tissue; effect of certain atmospheric pollutants on small mammals; effect of calcium channel blockers on peripheral circulation. *Mailing Add:* Col Arts & Sci Cent State Univ Wilberforce OH 45384

JOHNSON, MELVIN CLARK, b Newark, NJ, Aug 29, 38; m 75; c 2. TOXICOLOGY, PHARMACOLOGY. *Educ:* Rutgers Univ, BS, 62; McGill Univ, MS, 68; Howard Univ, PhD(pharmacol), 72; Am Bd Toxicol, dipl. *Prof Exp:* From assoc scientist to scientist pharmacol, Warner-Lambert Res Inst, 62-70; toxicologist med dept, Hercules, Inc, 72-76; DIR TOXICOL, AGR DIV, AM CYANAMID CO, 77- *Mem:* Am Inst Biol Sci; Am Acad Clin Toxicol; NY Acad Sci; Soc Toxicol. *Res:* Toxicology and pharmacology; safety of food additives, pesticides, animal drugs, food packaging materials and other consumer products; evaluation of potential exposures. *Mailing Add:* Am Cyanamid Co Agr Div PO Box 400 Princeton NJ 08540

JOHNSON, MELVIN WALTER, JR, b Chicago, Ill, May 27, 28; m 54; c 2. AGRONOMY, GENETICS. *Educ:* Univ Ill, BS, 50; Univ Wis, MS, 51, PhD(plant breeding), 54. *Prof Exp:* Asst agron, Univ Wis, 50-54; asst prof & asst agronomist, WVa Univ, 56-60, assoc prof & assoc agronomist, 60-65; ASSOC PROF AGRON, PA STATE UNIV, UNIVERSITY PARK, 65- *Mem:* Am Soc Agron; AAAS. *Res:* Plant breeding; plant genetics; corn breeding; basic and applied corn breeding and genetics research. *Mailing Add:* Dept Agron ASI Bldg Pa State Univ University Park PA 16802

JOHNSON, MICHAEL DAVID, b Chicago, Ill, Jan 6, 45; m 67; c 1. INSECT ECOLOGY. *Educ:* Northern Ill Univ, BS, 67; Northwestern Univ, PhD(biol), 70. *Prof Exp:* Res asst entom, Walter Reed Army Inst Res, 71-73; asst prof, 73-79, assoc prof zool, 79-, PROF BIOSCI, DEPAUW UNIV. *Concurrent Pos:* Model implementation prog res grant, Environ Protection Agency, 78-80. *Res:* Ecology and biology of the solitary bees of central Indiana; benthic research in central Indiana. *Mailing Add:* Dept Biol Sci DePauw Univ Greencastle IN 46135

JOHNSON, MICHAEL EVART, b Cody, Wyo, Sept 4, 45; m; c 1. BIOPHYSICS. *Educ:* Univ Wyo, BS, 68; Northwestern Univ, MS, 70, PhD(biophys), 73. *Prof Exp:* Res assoc & NIH fel biophys, Univ Pittsburgh, 73-75; from asst prof to assoc prof, Med Ctr, 76-84, PROF MED CHEM, UNIV ILL, CHICAGO, 84-, ASSOC DEAN, 86- *Concurrent Pos:* Guest scientist, Argonne Nat Lab, 75-; estab investr, Am Heart Asn, 79-84. *Mem:* Biophys Soc; AAAS; Am Chem Soc; Sigma Xi. *Res:* Sickling mechanism in sickle cell anemia; applications of magnetic resonance and computer aided molecular modeling in molecular structure analysis and design. *Mailing Add:* Dept Med Chem-M/C 781 Univ Ill PO Box 6998 Chicago IL 60680

JOHNSON, MICHAEL L, b Myrtle Point, Ore, Nov 12, 47. PROTEIN CHEMISTRY. *Educ:* Univ Conn, PhD(biophysics), 74. *Prof Exp:* From res asst to asst prof, 80-85, ASSOC PROF PHARMACOL, UNIV VA, 85-, DIR, BIOPHYSCS PROG & DIABETES RES & TRAINING CTR, 85- *Mem:* Biophys Soc; Calorimetry Soc; Am Soc Biol Chemists. *Res:* Computer applications. *Mailing Add:* Univ Va Box 448 Univ Va Charlottesville VA 22908

JOHNSON, MICHAEL PAUL, b Oakland, Calif, Sept 13, 37; m 71; c 3. PLANT ECOLOGY. *Educ:* Univ Calif, Davis, BS, 59; Univ Ore, PhD(biol), 66. *Prof Exp:* Instr bot, San Francisco State Col, 60-61; asst prof ecol, Kent State Univ, 65-68; asst prof field sci, Fla State Univ, 68-72; assoc prof biol, Kans State Univ & assoc dir, Konza Prairie Res Natural Area, 72-80; MEM STAFF, SCI EDUC ADMIN, USDA, 80- *Mem:* Soc Study Evolution; Ecol Soc Am; Brit Ecol Soc; Am Soc Naturalists; Sigma Xi. *Res:* Population biology; ecological genetics; botany; ecology. *Mailing Add:* Dept Computer Sci Ore State Univ Corvallis OR 97331

JOHNSON, MICHAEL ROSS, b Detroit, Mich, Oct 27, 44; m 64; c 2. ORGANIC & STRUCTURAL CHEMISTRY, SYNTHETIC & NATURAL PRODUCTS CHEMISTRY. *Educ:* Univ Calif, Berkeley, BS, 67; Univ Calif, Santa Barbara, PhD(org chem), 70. *Prof Exp:* Res chemist, Pfizer Inc, 71-73, sr res scientist, 73-76, sr res investr & proj leader, 76-80, mgr, Cent Nerv Syst & Metab Dis Res, 81-85, asst dir med chem & dir chem 87-89; VPRES, DIV CHEM, GLAXO, INC, 89- *Concurrent Pos:* NSF undergrad res fel, Calif State Col, Los Angeles, 64; NDEA Title IV fel, Univ Calif, Santa Barbara, 68-70; NIH fel, Univ Calif, Berkeley, 70-71, distinguished res fel, 89- *Mem:* Am Chem Soc; Sigma Xi; NY Acad Sci; AAAS; Am Soc Pharmacol & Exp Therapeut; Pharmaceut Mfrs Asn. *Res:* Mechanism and stereochemistry of carbonium ion, carbanion, organometallic and hydride reduction reactions; synthesis of pharmacalogically active heterocycles and natural products; synthesis of cannabinoid derived therapeutants; rational rug design. *Mailing Add:* Glaxo Inc Five Moore Drive Res Triangle Park NC 27709

JOHNSON, MIKKEL BORLAUG, b Waynesboro, Va, Jan 2, 43; m 65; c 2. THEORETICAL NUCLEAR PHYSICS. *Educ:* Va Polytech Inst, BS, 66; Carnegie-Mellon Univ, MS, 68, PhD(physics), 71. *Prof Exp:* Consult physics, Rand Corp, 67 & 68; res assoc, Cornell Univ, 70-72; STAFF MEM PHYSICS, LOS ALAMOS NAT LAB, UNIV CALIF, 72- *Concurrent Pos:* Assoc ed nuclear physics, North-Holland Publ Co, 75-; vis prof, Dept Physics, State Univ NY, Stony Brook, 81-82; consult, Oak Ridge Nat Lab, 86. *Honors & Awards:* Humboldt Award, 86. *Mem:* Fel Am Phys Soc. *Res:* Effective interactions in nuclear physics; intermediate energy nuclear theory. *Mailing Add:* 118 Piedra Loop Los Alamos NM 87544

JOHNSON, MILES F, b Frederic, Wis, Mar 9, 36; m 63, 81. SYSTEMATIC BOTANY. *Educ:* Wis State Univ, River Falls, BS, 58; Univ Wis-Madison, MS, 62; Univ Minn, Minneapolis, PhD(bot), 68. *Prof Exp:* High sch teacher, Wis, 58-60; teaching asst bot, Univ Wis-Madison, 60-62, instr bot & zool, 62-64; teaching asst bot, Univ Minn, Minneapolis, 64-67, instr, 68; from asst prof to assoc prof, 68-80, PROF BIOL, VA COMMONWEALTH UNIV, 80- *Mem:* Bot Soc Am; Am Soc Plant Taxon; Int Soc Plant Taxon. *Res:* Taxonomy and systematics of Compositae; genus Ageratum; flora of Virginia. *Mailing Add:* Dept Biol Va Commonwealth Univ Acad Ctr Richmond VA 23284

JOHNSON, MILLARD WALLACE, JR, b Racine, Wis, Feb 1, 28; m 53; c 4. MATHEMATICS. *Educ:* Univ Wis, BS, 52, MS, 53; Mass Inst Technol, PhD(math), 57. *Prof Exp:* Instr math, Mass Inst Technol, 53-58; from asst prof to assoc prof eng mech, 59-64, PROF ENG MECH & MATH, UNIV WIS-MADISON, 64- *Concurrent Pos:* Mem staff, Math Res Ctr, Univ Wis, 58-; mem exec comt, Rheol Res Ctr, Univ Wis, 69-; mem adv bd, Int Math & Statist Libr, 71- *Mem:* Soc Rheol; Soc Indust & Appl Math; Am Soc Mech Engrs; Soc Eng Sci. *Res:* Applied mathematics; rheology. *Mailing Add:* Univ Wis 1415 Johnson Dr Madison WI 53706

JOHNSON, MILTON R(AYMOND), JR, b Shreveport, La, Nov 5, 19; m 42; c 3. ELECTRONICS ENGINEERING. *Educ:* La Polytech Inst, BS, 40; Okla State Univ, MS, 51; Tex A&M Univ, PhD, 63. *Prof Exp:* Design engr, Gen Elec Co, 41-47; from asst prof to assoc prof, La Tech Univ, 47-54, prof elec eng, 54-86, head dept, 80-85; RETIRED. *Concurrent Pos:* Consult, Delta Res & Develop Corp, 52-60; NSF sci fac fel, 60-61. *Mem:* Am Soc Eng Educ; Inst Elec & Electronics Engrs. *Res:* Electromechanical energy converters; automatic control systems. *Mailing Add:* Dept Elec Eng La Tech Univ Ruston LA 71270

JOHNSON, MONTGOMERY HUNT, physics; deceased, see previous edition for last biography

JOHNSON, MORRIS ALFRED, b International Falls, Minn, Aug 3, 37; m 61; c 4. PLANT BIOCHEMISTRY. *Educ:* NDak State Univ, BS, 60, MS, 62; Ore State Univ, PhD(biochem), 66. *Prof Exp:* Asst prof & res fel biochem, Inst Paper Chem, 66-73, chmn dept, 70-79, assoc prof, 73-89, res assoc, 74-89; PROF, FOX VALLEY TECH COL, 89- *Mem:* Am Chem Soc; Am Soc Plant Physiol; fel Am Inst Chemists; Sigma Xi; Int Plant Growth Substances Asn; Plant Growth Regulator Soc of Am. *Res:* Intermediary metabolism and oxidative phosphorylation in plants; natural plant growth and development regulators; biochemistry of tree callus and suspension cultures. *Mailing Add:* W 7805 School Rd Appleton WI 54915

JOHNSON, MURRAY LEATHERS, b Tacoma, Wash, Oct 16, 14; m 42; c 4. MEDICINE, MAMMALOGY. *Educ:* Univ Ore, BA, 35, MD, 39; Am Bd Surg, dipl. *Prof Exp:* Res surg, Union Mem Hosp, Baltimore, Md, 39-43; cur mammals, Puget Sound Mus Natural Hist, Univ Puget Sound, 48-83; CUR MAMMALS, BURKE MEM WASH STATE MUS, UNIV WA, 83-; SR AFFIL RES ASSOC, NAT MARINE MAMMAL LAB, NAT MAMMAL FOUND SOC, 85. *Concurrent Pos:* Comnr, US Marine Mammal Comn, 79-87; chmn, sr adv comt, US Marine Mammal Comn, 85- *Honors & Awards:* Hartley H T Jackson Award, Am Soc Mammalogists, 86. *Mem:* AAAS; Am Soc Ichthyol & Herpet; Ecol Soc Am; Am Soc Mammalogists; Am Col Surg. *Res:* Natural history of the mammals of the Pacific Northwest; basic biologic relationships of mammals. *Mailing Add:* 501 N Tacoma Ave Tacoma WA 98403

JOHNSON, MYRLE F, b Jerico Springs, Mo, Dec 12, 18; m 57; c 2. PHYSICAL CHEMISTRY. *Educ:* Southwest Mo State Col, AB, 41; Univ Wis, PhD(phys chem), 50. *Prof Exp:* Assoc prof chem, Southwest Mo State Col, 50-53; from res chemist to sr res chemist, Eastman Kodak Co, 53-71, res assoc, 71-83; RETIRED. *Mem:* Am Chem Soc. *Res:* Rheology and colloid chemistry. *Mailing Add:* 29 Margate Dr Rochester NY 14616-5503

JOHNSON, NED KEITH, b Reno, Nev, Nov 3, 32; m 52; c 4. ORNITHOLOGY. *Educ:* Univ Nev, BS, 54; Univ Calif, PhD(zool), 61. *Prof Exp:* From asst prof to assoc prof, 62-74, asst cur birds, Mus Vert Zool, 62-63, actg dir, 81, PROF ZOOL, UNIV CALIF, BERKELEY, 74-, VCHMN DEPT, 68-, CUR BIRDS, MUS VERT ZOOL, 63- *Concurrent Pos:* NSF res grants, 65- *Mem:* Am Soc Zool; Am Ornith Union; Cooper Ornith Soc; Soc Study Evolution; Soc Syst Zool; Am Soc Naturalists. *Res:* Biosystematics; distribution and ecology of New World birds. *Mailing Add:* Mus Vert Zool Univ Calif Berkeley CA 94720

JOHNSON, NEIL FRANCIS, b Heighington, Co Durham, UK, Mar 15, 48; m 70; c 3. PATHOLOGY, CYTOLOGY. *Educ:* London Univ, BSc, 69; City Univ, London, MSc, 71; Glasgow Univ, PhD(exp path), 76. *Prof Exp:* Res asst ocular path, Tennent Inst Ophthal, Glasgow Univ, 71-77; scientist exp path, MRC Pneumoconiosis Unit, Penarth, UK, 77-84; lectr gen path, Inst Sci & Technol, Univ Wales, 80-84; vis scientist exp path, Los Alamos Nat Lab, 84-86; scientist exp path, MRC Toxicol Unit, UK, 86; GROUP SUPVR EXP PATH, MOLECULAR & CELLULAR TOXICOL, INHALATION TOXICOL RES INST, 86- *Concurrent Pos:* Chmn, DOE/OHER Task Group: Molecular Biol Carcinogenesis, 89-; clin assoc prof, Col Pharm, Univ NMex, 91- *Mem:* Am Soc Testing & Mat; Soc Toxicol; Royal Col Pathologists. *Res:* Determining the cells at risk from carcinogenesis from inhaled materials with particular emphasis on radon progeny and natural and manmade mineral fibers. *Mailing Add:* Inhalation Toxicology Research Inst PO Box 5890 Albuquerque NM 87185

JOHNSON, NOAH R, b Kingsport, Tenn, Oct 15, 28; m 50; c 3. PROPERTIES OF HIGHLY EXCITED NUCLEI. *Educ:* ETenn State Univ, BS, 50; Fla State Univ, PhD, 56. *Prof Exp:* Pub sch teacher, Tenn, 50-52; nuclear chemist, 56-80, NUCLEAR PHYSICIST, OAK RIDGE NAT LAB, 80- *Concurrent Pos:* Fulbright scholar & Guggenheim fel at Niels Bohr Inst, Copenhagen, 62-63; Guggenheim fel award, 62-63; Fulbright Scholar Award, 62-63. *Mem:* Fel Am Phys Soc; Sigma Xi; Am Chem Soc. *Res:* Nuclear spectroscopy and reactions; coulomb excitation; Doppler-shift lifetime measurements; studies of high-angular momentum behavior in nuclei; development of complex gamma-ray detector systems. *Mailing Add:* Oak Ridge Nat Lab PO Box 2008 Oak Ridge TN 37831-6371

JOHNSON, NORMAN ELDEN, b Mesa, Ariz, Apr 26, 33; m 54; c 2. FOREST MANAGEMENT, SILVICULTURE. *Educ:* Ore State Univ, BSF, 55, MS, 57; Univ Calif, PhD, 61. *Prof Exp:* Forestry aid, US Forest Serv, 51-52; forest engr, Southwest Lumber Mills, 54-55; forest entom asst, 55; forestry res mgr, Southern Forestry Res Ctr, Weyerhaeuser Co, Ark, 69-75, mgr, tropical forestry & res, 75-78, vpres, Far E region, Indonesia, 78-80, NC region, New Bern, NC, 80-84; forest entomologist, Forestry Res Ctr,56-66, forest bioprotection leader, 66-69, WEYERHAUSER CO, TACOMA, 84- *Concurrent Pos:* Assoc prof dept entom, Cornell Univ, 67-69; adj prof, Sch Forestry Resources, NC State Univ, 72; assoc ed, J Appl Forestry; mem bd dir, Pacific Sci Ctr, Sci Adv Coun, NC State Univ & Ore State Univ; chmn, Coop Forestry Adv Comt, US Dept Agr, McIntire-Stennis Res Prog; mem, Pres Reagan's Agr 7 Forestry Mission, Honduras, 82-83 & Zaire, 85; mem bd dir & Long Range Res Planning Comt, Wash Technol Ctr, US Nat Comt, Man & Biosphere Prog. *Mem:* Soc Am Foresters. *Res:* Forest plantation management. *Mailing Add:* Weyerhaeuser Co MS WTC 1K42 Tacoma WA 98402

JOHNSON, NORMAN L, b Tillamook, Ore, July 27, 39; m 64; c 3. GEOMETRY. *Educ:* Portland State Univ, BA, 64; Wash State Univ, MA, 66, PhD(math), 68. *Prof Exp:* Asst prof math, Eastern Wash State Col, 68-69; asst prof, 69-78, PROF MATH, UNIV IOWA, 78- *Concurrent Pos:* Researcher, NSF fel, 71-72; res fel, Univ Bergen, 73-74; Sci Res Coun researcher, Great Britain, 78- *Res:* Finite projective planes; classification of semitranslation planes and their construction; translation planes; collineation groups. *Mailing Add:* 130 Westminster Univ Iowa Iowa City IA 52242

JOHNSON, NORMAN LLOYD, b Ilford, Eng, Jan 9, 17; m 64. STATISTICS. *Educ:* Univ Col London, BSc, 36 & 37, MSc, 38, PhD(statist), 48, DSc, 63. *Prof Exp:* Asst lectr statist, Univ Col London, 38-39, 45-46, lectr, 46-56, reader, 56-62; prof, 62-82, chmn dept, 71-76, EMER PROF STATIST, UNIV NC, CHAPEL HILL, 82- *Concurrent Pos:* Vis assoc prof, Univ NC, Chapel Hill, 52-53; vacation consult, Road Res Lab, Eng, 56-59; vis prof, Case Inst Technol, 60-61, Univ NSW, Australia, 69; co-ed in chief, Encycl Statist Sci (9 vols), 82-88. *Honors & Awards:* Shewhart Medal, Am Soc Qual Control, 84. *Mem:* Fel Inst Math Statist; fel Am Statist Asn; Am Soc Qual Control; fel Royal Statist Soc; Biomet Soc; Int Statist Inst. *Res:* Systems of frequency distributions; checks on completeness of samples; reliability. *Mailing Add:* Dept Statistics Univ NC Chapel Hill NC 27599-3260

JOHNSON, OGDEN CARL, b Rockford, Ill, Aug 15, 29; m 55; c 4. FOOD TECHNOLOGY, RESEARCH ADMINISTRATION. *Educ:* Univ Ill, BS, 51, MS, 52, PhD(food technol), 56. *Prof Exp:* Res assoc, Univ Ill, 56-57; sr res chemist, A E Staley Mfg Co, 57-60; asst secy, Coun Foods & Nutrit, AMA, 60-64, assoc secy, 64-66; nutrit sect, NIH, 66-68, nutrit prog, Pub Health Surv, Dept Health Educ & Welfare, 68-70; dir div nutrit, Food & Drug Admin, 70-74; vpres sci affairs, Hershey Foods Corp, 74-80, exec vpres, 80-83, sr vpres, 83-88; RETIRED. *Mem:* Am Chem Soc; Am Oil Chem Soc; Inst Food Technologists. *Res:* Food product development; nutritive value of processed foods; nutrition survey; human nutrition; nutrition education. *Mailing Add:* PO Box 810 Hershey PA 17033

JOHNSON, OLIVER, b Edgetts, Mich, Mar 6, 19; m 46; c 4. PHYSICAL CHEMISTRY. *Educ:* NMich Univ, BSc, 39; Univ Mich, PhD, 42. *Prof Exp:* Res chemist, Emeryville Res Ctr, Shell Develop Co, 46-70; asst prof, Res Inst Catalysis, Houkaido Univ, Sapporo, Japan, 71-72, 83, Inst Physics, Uppsala Univ, Uppsala, Swed, 72-73, Dept Physical Chem, Univ Cambridge, 74-75, Carendish Lab, 80-81, Dept Chem, Univ Ga, Athens, Ga, 75-76, Univ Pittsburgh, 76-80; vis prof, Inst des Recherches sur la Catalyse, Villeursbanue, France, 81-82, Inst Molecular Sci, Japan, 82-83 & Dalian Inst Chem Physics, China, 84. *Mem:* Sigma Xi; Am Chem Soc. *Res:* Development of interstitial electron model for electronic structure of metals and metal alloys; interpretation of heterogeneous catalysis with above model. *Mailing Add:* 1626 Hillcrest San Luis Obispo CA 93401

JOHNSON, OLIVER WILLIAM, b Maud, Okla, Mar 30, 30; m 58; c 1. VERTEBRATE ZOOLOGY, PHYSIOLOGY. *Educ:* Fresno State Col, AB, 55; Ore State Univ, MS, 59, PhD(zool), 65. *Prof Exp:* Instr ecol, Ore State Univ, 59-61; asst prof zool, Ariz State Col, 61-63; res assoc entom, Ore State Univ, 63-64; assoc prof, 64-74, PROF ZOOL, NORTHERN ARIZ UNIV, 74- *Mem:* AAAS; Am Soc Mammalogists; Am Soc Ichthyologists & Herpetologists; Sigma Xi. *Res:* Amphibian and reptilian temperature adaptation; biochemical taxonomy. *Mailing Add:* Dept Biol Sci Northern Ariz Univ PO Box 5640 Flagstaff AZ 86011

JOHNSON, ORLAND EUGENE, b Gary, Ind, July 25, 23; m 46. NUCLEAR PHYSICS. *Educ:* Ind Univ, AB, 49, MS, 51, PhD(physics), 56. *Prof Exp:* Res assoc 56, from asst prof to assoc prof, 56-65, PROF PHYSICS, PURDUE UNIV, 65- *Mem:* Am Phys Soc. *Res:* Beta and gamma spectroscopy; nuclear scattering and reactions. *Mailing Add:* Dept Physics Purdue Univ West Lafayette IN 47901

JOHNSON, OSCAR HUGO, organic chemistry; deceased, see previous edition for last biography

JOHNSON, OSCAR WALTER, b Chicago, Ill, Mar 28, 35; m 55; c 2. ORNITHOLOGY, ECOLOGY. *Educ:* Mich State Univ, BS, 57; Wash State Univ, MS, 59, PhD(zool), 64. *Prof Exp:* Asst prof biol, Western State Col Colo, 63-65; from asst prof to prof biol, Moorhead State Univ, 65-90; ADJ PROF BIOL, MONT STATE UNIV, 90- *Concurrent Pos:* NSF res grants, 65-66, 67-69, Ari State Univ, 71-72, Med Sch, Univ Ariz, 75; Res Corp grant, 73; AEC & Dept Energy res grants, Univ Hawaii, 70, 73, 78, 79, & 80, Nat Geog Soc res grants, 82, 84, 87, 88 & 90; mem, Int Comn Avian Anatomical Nomenclature, 73- *Mem:* AAAS; Am Ornith Union; Cooper Ornith Soc; Wilson Ornith Soc. *Res:* Ecology and behavior in shorebirds, particularly long-distance migrant species of the insular Pacific; microanatomical and physiological studies of the bird kidney. *Mailing Add:* Dept Biol Mont State Univ Bozeman MT 59717

JOHNSON, OWEN W, b Provo, Utah, Mar 31, 31; m 57; c 3. SOLID STATE PHYSICS. *Educ:* Univ Utah, BA, 57, PhD(physics), 62. *Prof Exp:* Asst res prof physics, 62-64, asst prof ceramic eng, 64-65, from asst prof to assoc prof physics, 65-76, PROF PHYSICS, UNIV UTAH, 76-, ADJ ASSOC PROF MAT SCI, 68- *Mem:* Am Phys Soc. *Res:* Electronic and optical properties of oxides and semiconductors; infrared spectroscopy; electronic properties of thin films. *Mailing Add:* Dept Physics Univ Utah 201b Fletcher Bldg Salt Lake City UT 84112

JOHNSON, PATRICIA ANN J, b New York, NY, Oct 10, 43; m 64; c 2. CLINICAL NEUROPSYCHOLOGY, PSYCHOLOGY. *Educ:* Univ Houston, BS, MA, PhD(psychol), 77. *Prof Exp:* Exec dir & clin neuropsychologist, 77-80; Found Land & Learning Opportunities, 77-80; PVT PRACT, 77- *Concurrent Pos:* NIH fel, 74-77; clin asst prof psychol, Univ Houston, 78- *Mem:* AAAS; Int Neuropsychol Soc; Soc Neurosci; Am Psychol Asn; Biofeedback Soc Am. *Res:* Etiology and neuropsychology of learning and language disorders in children. *Mailing Add:* 3722 N Main Baytown TX 77521-3304

JOHNSON, PATRICIA R, b Waco, Tex, Feb 28, 31; div; c 2. CELL CULTURE, GENETIC OBESITY. *Educ:* Baylor Univ, AB, 52, MA, 58; Rutgers Univ, PhD(biochem), 67. *Prof Exp:* Health physicist, Rocky Flats Plant, Dow Chem Corp, 52-53; anal chemist, Va Carolina Chem Corp, Tex, 53-54; high sch teacher, Tex, 56-60; instr biol & chem, Malone Col, 60-61; res asst, Bur Biol Res, Rutgers Univ, 61-64; from instr to assoc prof biol, 64-75, PROF BIOL, VASSAR COL, 75-, CHMN, DEPT BIOL, 75-, WILLIAM R KEENAN CHAIR, 81- *Concurrent Pos:* Adj assoc prof, Rockefeller Univ, 71-75, adj prof, 75-80. *Mem:* AAAS; NY Acad Sci; Am Inst Nutrit; Sigma Xi. *Res:* Adipose tissue growth and development in genetically obese mice and rats: behavior; metabolism; cell culture of fetal hepatocytes and precursor adipocyres from the genetically obese zucker rat. *Mailing Add:* Dept Biol Vassar Col Poughkeepsie NY 12601

JOHNSON, PAUL CHRISTIAN, b Ironwood, Mich, Feb 3, 28; m 55; c 3. PHYSIOLOGY. *Educ:* Univ Mich, BS, 51, MA, 53, PhD(physiol), 55. *Hon Degrees:* DrMed(hon), Univ Limburg, Maastricht, Netherlands. *Prof Exp:* Instr physiol, Univ Mich, 55-56; instr, Western Reserve Univ, 56-58; from asst prof to assoc prof, Sch Med, Ind Univ, 58-67; head dept, 67-87, PROF PHYSIOL, COL MED, UNIV ARIZ, 67- *Concurrent Pos:* NIH fel, 65-66; mem physiol study sect, NIH, 68-72; mem steering comt, circulation sect, Am Physiol Soc, 71-74, chmn, 74, mem coun, 78-82, chmn publs comt, 85-89. *Honors & Awards:* Eugene M Landis Res Award, Microcirc Soc, 76; Carl J Wiggers Award, Am Physiol Soc, 81. *Mem:* AAAS; Am Physiol Soc; Microcirc Soc (pres, 67-68). *Res:* Local regulation of blood flow, microcirculation; capillary filtration and exchange. *Mailing Add:* Dept Physiol Univ Ariz Col Med Tucson AZ 85724

JOHNSON, PAUL H(ILTON), b Nevis, Minn, May 2, 16; div; c 2. CHEMICAL ENGINEERING. *Educ:* Univ Minn, BChE, 38. *Prof Exp:* Process engr, Minn Gas Co, 38-41; res engr, Phillips Petrol Co, 41-54, sect chief, Res & Develop Dept, 54-60, mgr, Petrol Process Br, 60-69, mgr, Carbon Black Br, Res Ctr, 69-81; consult carbon black environ health, process & feed stock, 81-91; RETIRED. *Mem:* Am Chem Soc. *Res:* Process development; petroleum refining; petrochemicals; carbon black environmental health; carbon black feed stock. *Mailing Add:* 1951 Southview Bartlesville OK 74003

JOHNSON, PAUL HICKOK, b Syracuse, NY, Mar 3, 43; m 81; c 4. BIOPHYSICS, GENETICS. *Educ:* State Univ NY Buffalo, BA, 65, PhD(biochem), 70. *Prof Exp:* Am Cancer Soc fel, Calif Inst Technol, 70-74; asst prof biochem & molecular biol, Wayne State Univ, 74-78, assoc prof biochem, 78-81; sr molecular biologist, 81-84, DIR, DEPT MOLECULAR BIOL, SRI INT, 84- *Concurrent Pos:* USPHS grant molecular biol, Wayne State Univ, 74-77; NIH Genetics Study Sect, 78-82. *Mem:* AAAS; Am Chem Soc; Am Asn Microbiol; Sigma Xi; Am Soc Biochem & Molecular Biol. *Res:* Protein and nucleic acid biochemistry; genetic engineering; protein drug development; enzymology. *Mailing Add:* Dept Molecular Biol SRI Int 333 Ravenswood Ave Menlo Park CA 94025

JOHNSON, PAUL L, b Witchita, Kans, Apr 30, 00; US citizen. ENGINEERING. *Educ:* Univ Ill, BS, 23. *Prof Exp:* Asst vpres opers, SCalif & Nev, Pacific Tel, 23-60; RETIRED. *Mem:* Fel Inst Elec & Electronics Engrs. *Mailing Add:* 19191 Harvard Apt 328A Irvine CA 92715

JOHNSON, PAUL LORENTZ, b Hawarden, Iowa, Sept 19, 41; m 71; c 3. COMPUTER SOFTWARE. *Educ:* St Olaf Col, BA, 63; Wash State Univ, PhD(phys chem), 68. *Prof Exp:* Fel, Univ Ill, Urbana-Champaign, 68-69; res assoc, Univ Ariz, 69-71, Mich State Univ, 71-72; Royal Norwegian Coun Sci & Indust res fel, Univ Bergen, Norway, 72-73; res assoc, Univ Ariz, 73-75; res assoc, 75-77, COMPUT SCIENTIST, ARGONNE NAT LAB, 77- *Concurrent Pos:* Instr, Lansing Community Col, 72. *Mem:* Asn Comput Mach; Am Crystallog Asn. *Res:* Neutron and x-ray crystallographic experiments applied to structures of organic, biological and inorganic interest; one-dimensional conducting compounds; portability of computer software; scientific applications of computers. *Mailing Add:* Nat Energy Software Ctr Argonne Nat Lab 9700 S Cass Ave Argonne IL 60439-4832

JOHNSON, PETER DEXTER, b Norwich, Conn, July 1, 21; div; c 3. APPLIED PHYSICS. *Educ:* Harvard Univ, SB, 42; Univ NC, MA, 48, PhD(phys chem), 49. *Prof Exp:* Supvr ballistic testing, Hercules Powder Co, Va, 42-43; RES ASSOC, GEN ELEC CO, 49- *Concurrent Pos:* Vis assoc prof, Cornell Univ, 58-59; patent agent, 81- *Mem:* Fel AAAS; fel Am Inst Chemists; Am Chem Soc; fel Am Phys Soc; Optical Soc Am; Sigma Xi. *Res:* Optical properties of phosphors and semiconductors; luminescence theory; optics and optical instrument design; optical properties of gas discharges. *Mailing Add:* 1100 Merlin Dr Schenectady NY 12309

JOHNSON, PETER GRAHAM, b St Helens, Eng, Aug 28, 45; m 67; c 2. GEOMORPHOLOGY. *Educ:* Univ Leeds, BSc, 66, PhD(geog), 69. *Prof Exp:* From asst prof to assoc prof, 69-85, PROF GEOMORPHOL, UNIV OTTAWA, 85- *Mem:* Geol Asn Can; Asn Am Geog; Arctic Inst NAm; Can Asn Geog. *Res:* Alpine hydrology; rock glacier mechanics and drainage systems; ice cored landform formation and degradation; southwest Yukon Territory; glacier hydrology. *Mailing Add:* Dept Geog Univ Ottawa Ottawa ON K1N 6N5 Can

JOHNSON, PETER WADE, plant pathology, plant nematology, for more information see previous edition

JOHNSON, PHILIP L, b Oneonta, NY, May 26, 31. ECOLOGY. *Educ:* Purdue Univ, BS, 53, MS, 55; Duke Univ, PhD(bot), 61. *Prof Exp:* Instr bot, Univ Wyo, 59-61; res botanist, Range Res, US Forest Serv, Wyo, 61-62; res ecologist, Cold Regions Res & Eng Lab, NH, 62-67; assoc prof forest resources, Univ Ga, 67-70; div dir environ systs & resources, NSF, 70-74; exec dir, Oak Ridge Assoc Univs, 74-81; EXEC DIR, JOHN E GRAY INST, LAMAR UNIV, 81- *Concurrent Pos:* Vis asst prof biol, Dartmouth Col, 63 & 65-; res collabr, Brookhaven Nat Lab, 63-65; mem NH Pesticide Control Bd, 65-67; mem primary productivity comt, Int Biol Prog, 67-68, adv comt tundra biome, 68-70, deciduous forest biome coord comt, 68-70; assoc prog dir, environ biol prog, NSF, 68-69; mem environ biol panel foreign currency prog, Smithsonian Inst, 69-70; vchmn interagency comt ecol res, Fed Coun Sci & Technol-Coun Environ Qual, 72; mem US Comt Man & Biosphere Prog, 73-74; mem fel adv panel environ affairs, Rockefeller Found, 74-; mem exec comt, East Tenn Cancer Res Ctr, Knoxville, 75-78; mem regional comt Southeastern Plant Environ Lab, 75-; mem, US Comn, UNESCO, 78-80,

Gov's Task Force Advan Labor & Mgt Relations, 84- & Houston Dist Export Coun, 85-; mem polar res bd, Nat Acad Sci, 81-85. *Mem:* AAAS; Am Inst Biol Sci; Ecol Soc Am; Brit Ecol Soc; fel Arctic Inst NAm. *Res:* Production and processes in arctic and alpine tundra; aerial sensing of ecological patterns; mineral cycling in ecosystems applications of environmental sciences; interdisciplinary research and training; regional economic development; labor and management relations. *Mailing Add:* 3709-N 36th Rd Arlington VA 22207

JOHNSON, PHILIP M, b Vancouver, Wash, Oct 22, 40; m 64; c 2. PHYSICAL CHEMISTRY, MOLECULAR SPECTROSCOPY. *Educ:* Univ Wash, BS, 62; Cornell Univ, PhD(phys chem), 67. *Prof Exp:* NIH fel, Univ Chicago, 66-68; from asst prof to assoc prof, 68-78, PROF CHEM, STATE UNIV NY STONY BROOK, 78- *Concurrent Pos:* Vis fel, Joint Inst Lab Astrophys, Colo, 75-76; Guggenheim fel, 82-83. *Mem:* Am Phys Soc. *Res:* Ultraviolet and vacuum ultraviolet spectroscopy; evolution of electronic energy in molecules; multiphoton ionization spectroscopy. *Mailing Add:* Dept Chem State Univ NY Stony Brook NY 11794

JOHNSON, PHILLIP EUGENE, b Bostic, NC, Feb 25, 37; m 59; c 1. MATHEMATICS. *Educ:* Appalachian State Teachers Col, BS, 59; George Peabody Col, MA, 63, PhD(math), 68; Am Univ, MA, 66. *Prof Exp:* High sch teacher, Va, 60-63; instr math, Univ Richmond, 63-65; from instr to asst prof, Vanderbilt Univ, 66-71; asst prof, 71-76, ASSOC PROF MATH, UNIV NC, CHARLOTTE, 76- *Concurrent Pos:* Vis asst prof, NC State Univ, 71. *Mem:* Math Asn Am; Nat Coun Teachers Math. *Res:* Mathematics history and education. *Mailing Add:* Dept Math Univ NC Charlotte NC 28223

JOHNSON, PHYLLIS ELAINE, b Grafton, NDak, Feb 19, 49; m 69; c 2. MASS SPECTROMETRY, TRACE METAL NUTRITION. *Educ:* Univ NDak, BS, 71, PhD(phys chem), 76. *Prof Exp:* Lab instr chem & biochem, Mary Col, NDak, 71-72; chemist, Univ NDak, 77-79; CHEMIST, AGR RES SERV, 79-, RES LEADER, NUTRIT BIOCHEM & METAB UNIT, USDA HUMAN NUTRIT RES CTR, 87- *Concurrent Pos:* Fel, Univ NDak, 75-77; clin instr, Univ NDak Sch Med, 81- *Honors & Awards:* Arthur S Flemming Award, 89. *Mem:* Am Chem Soc; Am Inst Nutrit; Am Soc Mass Spectrometry; Soc Exp Biol & Med; Sigma Xi; Am Soc Clin Nutrit. *Res:* Trace metal absorption; biological metal-ligand complexes; lactation and infant nutrition; absorption, metabolism and bioavailability of trace metals, especially iron, zinc, copper and manganese, are investigated in humans using stable and radioactive metal isotopes as tracers. *Mailing Add:* USDA-ARS Human Nutrit Res Ctr Box 7166 Univ Sta Grand Forks ND 58202-7166

JOHNSON, PHYLLIS TRUTH, b Salem, Ore, Aug 8, 26. INVERTEBRATE PATHOLOGY. *Educ:* Univ Calif, PhD(parasitol), 54. *Prof Exp:* Parasitologist med entom, Bur Vector Control, State Dept Health, Calif, 48-50; entomologist, Dept Entom, Walter Reed Army Inst Res, Washington, DC, 50-55; entomologist, Entom Res Br, USDA, 55-58; med entomologist, Gorgas Mem Lab, 59-63; from asst res pathobiologist to assoc res pathobiologist, Univ Calif, Irvine, 64-70; res fel, Calif Inst Technol, 70-71; consult, Off Environ Sci, Smithsonian Inst, 71-72; biologist, Nat Marine Fisheries Serv, 72-90; RETIRED. *Concurrent Pos:* Consult, US Naval Med Res Unit 3, Cairo, Egypt, 57-; res assoc, USDA, 58-63; mem comt animal models & genetic stocks, Nat Res Coun, 71-75. *Honors & Awards:* Bronze Medal, US Dept Com, 81. *Mem:* Sigma Xi; fel AAAS; Soc Invert Path (vpres, 78-80 & pres, 81-82); Am Soc Trop Med & Hyg; Am Soc Parasitol. *Res:* Leishmaniasis; taxonomy of Siphonaptera and Anoplura; pathological processes in invertebrates; viruses in crustaceans; histopathology of crustaceans. *Mailing Add:* 4721 E Harbor Dr Friday Harbor WA 98250

JOHNSON, PORTER W, b Chattanooga, Tenn, Sept 4, 42; m 63; c 2. HIGH ENERGY PHYSICS, MATHEMATICAL PHYSICS. *Educ:* Case Inst Technol, BS, 63; Princeton Univ, MA, 65, PhD(physics), 67. *Prof Exp:* Fel, Case Western Reserve Univ, 67-69; from asst prof to assoc prof, 69-83, PROF PHYSICS, ILL INST TECHNOL, 83-, CHMN PHYSICS DEPT, 84- *Mem:* Am Phys Soc. *Res:* Study of mathematical structure of nonlinear equations involved in applications of general principles to elementary particle scattering data. *Mailing Add:* Dept Physics Ill Inst Technol Chicago IL 60616-3573

JOHNSON, PRESTON BENTON, b Benson, NC, Mar 7, 32; m 54; c 3. ELECTRICAL ENGINEERING. *Educ:* NC State Univ, BSEE, 58, MS, 62; Va Polytech Inst, PhD(elec eng), 66. *Prof Exp:* Instr elec eng, NC State Univ, 61-62; asst prof, Va Polytech Inst, 62-66; assoc prof elec eng, Old Dominion Univ, 66-; PRES, JOHNSON BRADLEY RES CORP. *Concurrent Pos:* Chmn, Dept Elec Eng, Old Dominion Univ, 74-77; vpres, Sigma Consults, Inc, 75- *Mem:* Inst Elec & Electronics Engrs; Am Soc Eng Educ; Instrument Soc Am. *Res:* Negative-resistance electronic devices based on superconductive tunneling between thin films; oceanographic instrumentations. *Mailing Add:* 1005 Briarwood Pt Virginia Beach VA 23452

JOHNSON, QUINTIN C, b Excelsior, Minn, July 24, 35; m 57; c 2. CRYSTALLOGRAPHY. *Educ:* St Olaf Col, BA, 57; Univ Calif, Berkeley, PhD(chem), 61. *Prof Exp:* Chemist, Lawrence Livermore Nat Lab, 60-75, actg dep dept head chem, 75-76, assoc dept head chem, 76-80, div leader, 80-84, SECT LEADER, LAWRENCE LIVERMORE NAT LAB, 84-; PRES, MAT DATA, INC, 84- *Mem:* AAAS; Am Crystallog Asn (vpres, 80, pres, 81); Am Phys Soc. *Res:* Automation of powder diffraction; PC software for materials characterization. *Mailing Add:* Mat Data Inc PO 791 Livermore CA 94550

JOHNSON, R(ICHARD) A(LLAN), b Winnipeg, Man, Mar 21, 32; m 57; c 3. ELECTRICAL ENGINEERING. *Educ:* Univ Man, BSc, 54, MSc, 56. *Prof Exp:* From asst prof to assoc prof, 55-66, actg dir planning, 69-70, head Elec Eng Dept, 73-76, provost, 77-82, assoc vpres planning & anal, 82-87, PROF ELEC ENG, UNIV MAN 66-, ASSOC VPRES 87- *Concurrent Pos:* Pres, APEM, 79; chmn, Comt Accepting Eng Curric, Can Council Prof Eng, 60-62, dir, 80-82; mem Can Accreditation Bd, 64-67. *Mem:* Inst Elec & Electronics Engrs. *Res:* Circuits and systems theory; nonlinear oscillations; chaos and catastrophe theory and applications. *Mailing Add:* 208 Admin Bldg Univ Man Winnipeg MB R3T 2N2 Can

JOHNSON, RALEIGH FRANCIS, JR, b Hazard, Ky, Jan 24, 41; m 63; c 2. NUCLEAR MEDICINE, RADIOLOGICAL PHYSICS. *Educ:* Berea Col, AB, 64; Univ Miami, MS, 65; Purdue Univ, PhD(radiol physics), 69. *Prof Exp:* Assoc radiol & nuclear med & physicist, Duke Univ & Vet Admin Hosp, 69-72; asst prof radiol & nuclear med & physicist, 72-84, asst prof radiol & magnetic resonance imaging & tech dir, 84-91, ASSOC PROF RADIOL & MAGNETIC RESONANCE IMAGING PHYSICS & DIR, UNIV TEX MED BR, GALVESTON, 91- *Concurrent Pos:* Consult, Scientists & Engrs for Appalachia, 71- *Mem:* Health Physics Soc; Nuclear Med Soc; Sigma Xi; Creation Res Soc; Am Asn Physicists in Med; Soc Magnetic Resonance Med; Soc Magnetic Resonance Imaging. *Res:* Oblique imaging techniques in magnetic resonance imaging; quality control of magnetic resonance imaging systems; magnetic resonance imaging using contrast enhancement labeled agents; evaluation of high energy collimators for scintillation gamma cameras; evaluation of microprocession controlled automatic well-type scintillation counting system; evaluation of multipeak scintillation imaging; caordiac magnetic resonance imaging; 3D MRI imaging and 3D video display techniques. *Mailing Add:* Magnetic Resonance Imaging Div Univ Tex Med Br Galveston TX 77550

JOHNSON, RALPH ALTON, b Alton, Ill, Sept 14, 19; m 54; c 1. GENERAL ATMOSPHERIC SCIENCES. *Educ:* Hastings Col, BA, 40; Univ Colo, MS, 42; Univ Minn, PhD(chem), 49. *Prof Exp:* Jr chemist, Manhattan Proj, Hanford Eng Works, E I du Pont de Nemours & Co, 44-45; from instr to asst prof anal chem, Univ Ill, 48-55; sr res chemist, Shell Develop Co, 55-83; ENVIRON ODOR CONSULT, 83- *Mem:* Am Chem Soc; Air Pollution Control Asn; Am Soc Testing & Mat; Sigma Xi. *Res:* Psychophysics, odor measurement; wastewater processing and analysis; precipitation studies; spectrophotometric and electron microscopic investigations; neutron activation analysis. *Mailing Add:* PO Box 79068 Houston TX 77079

JOHNSON, RALPH E, physical chemistry, for more information see previous edition

JOHNSON, RALPH M, JR, b Ririe, Idaho, Apr 19, 18; m 40; c 3. NUTRITION. *Educ:* Utah State Agr Col, BS, 40; Univ Wis, MS, 44, PhD(biochem), 48. *Prof Exp:* Asst prof biochem, Col Med, Wayne State Univ, 48-59; from assoc prof to prof physiol chem, Ohio State Univ, 59-68, dir & res prof, Inst Nutrit & Food Technol, 60-68, dir, 63-68, dean, Col Biol Sci, 66-68, res assoc prof, 59-60, dir labs, 59-63; dean, Col Sci & prof chem, Utah State Univ, 68-84; RETIRED. *Mem:* Am Soc Biochem & Molecular Biol; Am Inst Nutrit. *Res:* Lipid metabolism; metabolism of phosphorous compounds; hormonal and hereditary factors in carcinogenesis; biochemical role of vitamin E. *Mailing Add:* 2044 N 13th E Logan UT 84321

JOHNSON, RALPH STERLING, JR, b Shickshinny, Pa, Apr 2, 26; m 51; c 1. MATERIALS SCIENCE, METALLURGICAL ENGINEERING. *Educ:* Univ Akron, BS, 57, MS, 60; Univ Mich, Ann Arbor, PhD(mat sci & metall eng), 70. *Prof Exp:* Sr res engr mat & mfg res, Res & Develop Dept, Goodyear Aerospace Corp, 49-61; sr staff engr, Seismic Equip Dept, Bendix Aerospace Systems Div, Ann Arbor, 62-72; consult mat corrosion & mfg processes, Res & Eng Dept, Bechtel Nat, Inc, San Francisco, 73-79 & Aramco, Dhahran, Saudi Arabia, 79-81; consult mat corrosion & mfg processes & mem, Corrosion Task Force, Sohio Alaska Petrol Co, Anchorage, 81-84; SR CONSULT, DALLAS TECHNOL CTR, SOHIO PETROL CO, 84- *Concurrent Pos:* Mem water qual task force, Bechtel Power Corp, 75-79; mem, sci adv comn, Alaska Found, Univ Alaska, 83-85; consult, Arctic Res Comn. *Honors & Awards:* Apollo Achievement Award, NASA, 69. *Mem:* NY Acad Sci; Am Inst Mining, Metall & Petrol Engrs; Nat Asn Corrosion Engrs; Sigma Xi; Am Soc Metals. *Res:* Materials performance and corrosion of materials in flue gas desulfurization systems; feedwater and steam generating systems in steam electric plants; oil field production facilities materials of construction and corrosion control. *Mailing Add:* 26 Timberlane Dr Chillicothe AK 45601-1941

JOHNSON, RALPH T, JR, b Salina, Kans, Apr 29, 35; m 58; c 4. SOLID STATE PHYSICS, RESEARCH SUPERVISION. *Educ:* Kans State Univ, BS, 57, MS, 59, PhD(physics), 64. *Prof Exp:* Physicist, Aircraft Nuclear Propulsion Dept, Gen Elec Co, 57-58; asst physics, Kans State Univ, 58-63; proj officer, Air Force Weapons Lab, 63-65; staff mem solid state physics, 65-70, res supvr elec transport & electronic properties mat, 70-85, MGR MEASUREMENT STANDARDS, SANDIA NAT LABS, 85- *Concurrent Pos:* Mem energy conversion panel, NMex Gov Energy Task Force, 74; mem nat res coun bd, Assessment Nat Bur Standards, Panel Basic Standards, 87-90. *Mem:* Am Phys Soc; Sigma Xi. *Res:* X-ray diffraction topography; dislocations and martensitic transformations; rocketborne magnetometers and optical spectrometers; semiconductor radiation defects, ionization effects and neutron detectors; electrical properties of amorphous semiconductors; thermoelectrics; solid electrolytes; electronic properties of dielectric materials. *Mailing Add:* 6601 Arroyo del Oso NE Albuquerque NM 87109

JOHNSON, RANDALL ARTHUR, agriculture, computer software, for more information see previous edition

JOHNSON, RANDOLPH MELLUS, b Los Angeles, Calif, Sept 6, 50; m 80; c 3. BIOCHEMICAL PHARMACOLOGY, NEUROPHARMACOLOGY. *Educ:* Calif State Univ, Long Beach, BS, 74, MA, 78; Univ SC, PhD(pharmacol), 84. *Prof Exp:* Res assoc endocrine pharmacol, Sch Med, Univ Va, 84-87, res asst prof endocrine pharmacol, 87-88; scientist biomolecular pharmacol, Genentech, Inc, 88-91; STAFF RESEARCHER II NEUROPHARMACOL, SYNTEX RES, 91- *Concurrent Pos:* Nat Res Serv award, 86-88; Genentech postdoctoral fel award, 89; consult, Quantex Corp, 89-91. *Mem:* Am Soc Pharmacol & Exp Therapeut; Am Soc Biochem & Molecular Biol; Endocrine Soc; AAAS; Sigma Xi. *Res:* Biomolecular mechanisms of growth factors, neurotransmitters and novel experimental therapeutics as it relates to second messenger formation and protein phosphorylation events in cell activation. *Mailing Add:* Dept Neurosci Inst Pharmacol Syntex Res 3401 Hillview Ave Palo Alto CA 94304

JOHNSON, RANDY ALLAN, b Minneapolis, Minn, Feb 9, 47; m 78. HIGH ENERGY PHYSICS. *Educ:* Princeton Univ, AB, 69; Univ Calif, Berkeley, PhD(physics), 75. *Prof Exp:* Fel physics, Lawrence Berkeley Lab, 75-76; ASSOC PHYSICIST, BROOKHAVEN NAT LAB, 76- *Mem:* Sigma Xi. *Res:* Particle scattering at high energies. *Mailing Add:* 745 Avon Fields Lane Cincinnati OH 45229

JOHNSON, RAY EDWIN, b East View, Ky, Aug 9, 36; m 85; c 1. SOIL FERTILITY. *Educ:* Univ Ky, BS, 57, MS, 59; NC State Univ, PhD(mineral nutrit), 62. *Prof Exp:* Res assoc, Mineral Nutrit Pioneering Res Lab, USDA, 62-63, res plant physiologist, US Regional Soybean Lab, Crops Res Div, Agr Res Serv, Ill, 63-67; from asst prof to assoc prof agron, soil fertil & soil chem, 67-73, PROF AGRON, SOIL FERTIL & SOIL CHEM, WESTERN KY UNIV, 73- *Mem:* Am Soc Agron; Sigma Xi. *Res:* Mineral nutrition and interaction in plants; relationship of fertilizer response to soil test results. *Mailing Add:* Dept Agr Western Ky Univ Bowling Green KY 42101

JOHNSON, RAY LELAND, b LaGrange, Ohio, Nov 7, 39; m 62; c 3. PHYSICAL CHEMISTRY, ENVIRONMENTAL & ANALYTICAL CHEMISTRY. *Educ:* Kent State Univ, BS, 61; Ohio Univ, PhD(phys chem), 66. *Prof Exp:* Sr res chemist, PPG Industs Inc, 66-69; asst prof, 69-77, assoc prof & actg chmn, 77-79, PROF CHEM, DIV NATURAL SCI, HILLSDALE COL, 80- *Concurrent Pos:* Consult, Hillsdale Waste Water Treatment Plant, 70-; W K Kellogg Found res grant water qual studies, 71-73. *Mem:* AAAS; Am Chem Soc. *Res:* Thermodynamics and kinetics; surface and colloid chemistry; interaction of pigments with polymers; solution chemistry; chemical investigations of water quality in lakes and streams; chemical methods of waste water treatment and analysis. *Mailing Add:* Dept Chem Hillsdale Col Hillsdale MI 49242

JOHNSON, RAY O, b Kansas City, Mo, May 25, 55. FOURIER SPECTROSCOPY, ELECTROMAGNETICS. *Educ:* Okla State Univ, BS, 84; Air Force Inst Technol, MS, 87, elec eng, 90- *Prof Exp:* Telecommun engr, Foreign Technol Div, USAF, 84-86; elec engr, Hq Strategic Air Command, 86- *Mem:* Inst Elec & Electronics Engrs. *Res:* Spectroscopy; Fourier optics; interhalogen chemical kinetics. *Mailing Add:* Air Force Inst Technol Box 4215 Wright Patterson AFB OH 45433-5001

JOHNSON, RAYMOND C, JR, b Galveston, Tex, Sept 29, 22; c 9. ELECTRONICS. *Educ:* Tex A&M Univ, BS, 45; Univ Fla, MS, 49. *Prof Exp:* From asst prof to assoc prof, 46-68, PROF ELEC ENG, UNIV FLA, 68-, SECT HEAD, ELECTRONIC RES SECT, 59- *Concurrent Pos:* Dir, Electronic Commun Lab, 76- *Res:* Electronics systems. *Mailing Add:* 204 NW 32nd St Gainesville FL 32607

JOHNSON, RAYMOND EARL, b Peru, Nebr, Oct 26, 14; m 41. ZOOLOGY. *Educ:* Doane Col, BA, 36; Univ Nebr, MA, 38; Univ Mich, PhD(zool), 42. *Prof Exp:* Aquatic biologist, US Fish & Wildlife Serv, Univ Minn, 45-46, fisheries res supvr, 47-51, asst fed aid supvr, Bur Sport Fisheries & Wildlife, 51-56, chief fish div, 56-58, chief br fed aid, 58-59, asst dir, Bur Sport Fisheries & Wildlife, 59-71, chief off environ qual, 71-72; dep div dir, NSF, 72-74; CONSULT, NAT WILDLIFE FEDN, 74- *Mem:* Am Soc Ichtyologists & Herpetologists; Am Fisheries Soc; Am Soc Limnol & Oceanog; Wildlife Soc; Sigma Xi. *Res:* Taxonomy and distribution of freshwater fishes in North America; fisheries management; life history of freshwater fishes. *Mailing Add:* 5209 30th St N Arlington VA 22207

JOHNSON, RAYMOND LEWIS, b Alice, Tex, June 25, 43; m 65; c 1. MATHEMATICS. *Educ:* Univ Tex, Austin, BA, 63; Rice Univ, PhD(math), 69. *Prof Exp:* Assoc chmn grad studies, 87-90, from asst prof to assoc prof, 68-78, PROF MATH, UNIV MD, 80- *Concurrent Pos:* Gen Res Bd grant, 68 & 71; sabbatical leave, Inst Mittag-Leffler, DJursholm, 74-75, Howard Univ, 76-78 & McMaster Univ, Hamilton, Can, 83-84. *Mem:* Am Math Soc. *Res:* Parabolic partial differential equations; representation theorems; spaces of functions defined by difference conditions; harmonic analysis. *Mailing Add:* Dept Math Univ Md College Park MD 20742

JOHNSON, RAYMOND NILS, b New York, NY, July 26, 41; m 65; c 2. ANALYTICAL CHEMISTRY. *Educ:* Franklin & Marshall Col, AB, 63; Middlebury Col, MS, 65; Clarkson Univ, PhD(chem), 69. *Prof Exp:* From res assoc anal chem to group leader, 69-75, sect head anal chem, 75-78, asst dir anal res & develop, 78-83, assoc dir, 83- 84, dir, anal res & serv, 85, asst vpres, 85-87, SR DIR, AYERST LABS, INC, 88- *Mem:* Am Chem Soc; Acad Pharmaceut Sci; Sigma Xi. *Res:* Pharmaceutical analysis using gas chromatography, polarography, mass spectrometry and mass fragmentography; emphasis placed on preparation of novel chemical derivatives and development of analytical methods that are precise, accurate and specific; automation; raw material characterization. *Mailing Add:* Analytical Res & Serv Wyeth-Ayerst Labs Inc 64 Maple St Rouses Point NY 12979

JOHNSON, RAYMOND ROY, b Phoenix, Ariz, June 19, 32; m 76; c 5. SYSTEMATIC BOTANY, VERTEBRATE ZOOLOGY. *Educ:* Ariz State Univ, BS, 55; Univ Ariz, MS, 60; Univ Kans, PhD(bot), 64. *Prof Exp:* Asst prof biol, Western NMex Univ, 64-65; assoc prof biol, Univ Tex, El Paso, 65-68; from assoc prof to prof biol, Prescott Col, 68-74; res scientist, Grand Canyon, 74-75, sr res scientist, 76-79, SR RES SCIENTIST, COOP NAT PARK RESOURCES STUDY UNIT, NAT PARK SERV, UNIV ARIZ, 80. *Concurrent Pos:* Prof, Renewable Nat Res, Univ Ariz, 80- *Mem:* Am Ornith Union; Am Soc Mammal; Am Ornith Union. *Res:* Plant taxonomy, conservation biology; animal distribution; riparian ecology; desertification and arid land ecology. *Mailing Add:* Coop Nat Park Res Studies Unit Univ Ariz 125 Biol Sci E Tucson AZ 85721

JOHNSON, REYNOLD B, b Kingston, Minn, July 16, 06; m; c 3. COMPUTER PERIPHERALS. *Educ:* Univ Minn, BS, 29. *Prof Exp:* Res scientist, 34-71, RES FEL, IBM, 71-; PRES, EDUC ENG ASSOCS, 71- *Honors & Awards:* Nat Medal Technol, 86; Computer Pioneer Award, Inst Elec & Electronics Engrs, 87, Magnetic Soc Award, 88. *Mem:* Nat Acad Eng. *Mailing Add:* Educ Eng Assocs 548 E Crescent Dr Palo Alto CA 94301

JOHNSON, RICHARD ALLEN, b Panama City, Fla, Aug 13, 45; m 68; c 2. PHYSICAL CHEMISTRY. *Educ:* Ill Inst Technol, BS, 67; Mich State Univ, PhD(chem physics), 71. *Prof Exp:* Scientist, Control Anal Res & Develop Unit, 71-73, res scientist, 73-74, sr res scientist, 74-76, MGR PROD CONTROL, UPJOHN CO, 76- *Mem:* Am Chem Soc; Am Phys Soc. *Res:* Molecular spectroscopy of solids; solid state chemistry; physical characterization of pharmaceutical solids; application of computers to online data acquisition from analytical laboratory instrumentation. *Mailing Add:* 597 Aquaview Kalamazoo MI 49009

JOHNSON, RICHARD CLAYTON, b Eveleth, Minn, May 9, 30; div; c 2. APPLIED PHYSICS. *Educ:* Ga Inst Technol, BS, 53, MS, 58, PhD(physics), 61. *Prof Exp:* From asst res physicist to sr res physicist, 56-79, head radar br, 63-68, prin res physicist, 67-79, chief electronics div, 68-72, mgr systs & tech dept, 72-75, assoc dir eng exp sta, 75-79, PRIN RES ENGR, GA INST TECHNOL, 79- *Concurrent Pos:* Distinguished lectr, Inst Elec & Electronics Engrs, Antennas & Propagation Soc, 78-79, pres, 80. *Mem:* Am Phys Soc; fel Inst Elec & Electronics Engrs; Sigma Xi. *Res:* Radar and radiometry systems; antenna research and development; microwave theory and techniques; microwave spectroscopy; electromagnetic compatibility. *Mailing Add:* Micros Inc 7069 Regalview Circle Dallas TX 75248

JOHNSON, RICHARD D, b Zanesville, Ohio, Oct 28, 34; m 57; c 4. CHEMISTRY. *Educ:* Oberlin Col, BA, 56; Carnegie Inst Technol, MS, 61, PhD(chem), 62; Mass Inst Tech, SM, 82. *Prof Exp:* Fel phys org chem, Univ Calif, Los Angeles, 61-62; sr scientist, Jet Propulsion Lab, Calif Inst Technol, 62-63; chief flight exp off, Life Sci, NASA, 75-76, chief, Biosystems Div, 76-85, res scientist, Ames Res Ctr, 63-85; sr technol consult, 85-90, PRIN, SRI INT, 90- *Concurrent Pos:* Lectr, Stanford Univ, 74-86; Sloan fel, 81-82. *Honors & Awards:* Except Serv Medal, NASA, 77. *Mem:* AAAS; Am Inst Aeronaut & Astronaut; Am Chem Soc; Inst Elec & Electronics Engrs. *Res:* Exobiology and the detection of extraterrestrial life; Apollo lunar sample analysis; 1976 Viking Mars life detection experiment; space colonies; 1976 Stanford-Ames study on space settlements; space shuttle experiments; space biomedical experiments; space commercialization; aerospace technology; human factors; technology management. *Mailing Add:* 11564 Arroyo Oaks Los Altos Hills CA 94022

JOHNSON, RICHARD DEAN, b DeKalb, Ill, July 8, 36; m 69; c 4. PRODUCT LICENSING, TECHNOLOGY TRANSFER. *Educ:* Univ Calif, Berkeley, BS, 60, PharmD, 61, MS, 62; Univ Calif, San Francisco, PhD(pharm chem), 65; Rockhurst Col, Kansas City, MBA, 84. *Prof Exp:* Pharmacist, Alta Vista Drug Co, 60-61; teaching asst, Univ Calif, San Francisco, 62-64; res chemist & sect head, Allergan Pharmaceut Co, 65-67; assoc dir med serv, Syntex Labs, Inc, 67-68, dir regulatory affairs, 68-73; dir corp licensing, 73-79, vpres licensing, 80-83, CORP VPRES, MARION LABS, INC, 84- *Concurrent Pos:* Borden Co grad award, Univ Calif, San Francisco, 61-62; fels, Am Found Pharmaceut Educ & Henry S Wellcome Mem, 63-65; mem, Pres Comn Exec Interchange, White House, 70-71; lectr, Bus Sch, Univ SC, 75-77 & Bus Sch, Rockhurst Col, Kansas City, 83; mem bd dir, Tanabe-Marion Labs, 84-, Dey Labs, 85-88. *Honors & Awards:* Marion Labs President's Award, 80 & 81. *Mem:* AAAS; Am Pharmaceut Asn; Am Chem Soc; Acad Pharmaceut Sci; Licensing Exec Soc. *Res:* Thermal titration; thermal electric methods for studying physical and chemical properties of solutions. *Mailing Add:* 222 W Gregory Apt 331 Kansas City MO 64114-0480

JOHNSON, RICHARD EVAN, b Pomona, Calif, Nov 9, 36. ORNITHOLOGY, ZOOGEOGRAPHY. *Educ:* Univ Calif, Berkeley, BS, 58; Univ Mont, MS, 68; Univ Calif, Berkeley, PhD(zool), 72. *Prof Exp:* Asst prof, 72-78, ASSOC PROF ZOOL, WASH STATE UNIV, 78-, DIR, CHARLES R CONNER MUS, 72- *Concurrent Pos:* Ed, The Murrelet, 76-80. *Mem:* Am Ornithologists Union; Cooper Ornith Soc; Wilson Ornith Soc; Soc Study Evolution; Soc Syst Zool; Am Soc Mammalogists. *Res:* Zoogeography, ecology and speciation of birds; evolution of arctic and alpine ecosystems; mammals of the Northwest; biogeography of alpine plants. *Mailing Add:* Dept Zool Wash State Univ Pullman WA 99164-4236

JOHNSON, RICHARD HARLAN, b Portland, Ore, Nov 4, 45; m 65; c 2. METEOROLOGY. *Educ:* Ore State Univ, BS, 67; Univ Chicago, MS, 69; Univ Wash, PhD(atmospheric sci), 75. *Prof Exp:* Res meteorologist, Nat Hurricane Res Lab, 76-77; asst prof atmospheric sci, Univ Wis-Milwaukee, 77-79; from asst prof to assoc prof, 80-86, PROF ATMOSPHERIC SCI, COLO STATE UNIV, 86- *Concurrent Pos:* Co-chief ed, J Atmospheric SCi, 86- *Mem:* Am Meteorol Soc; AAAS; Japan Meteorol Soc; Am Geophys Union. *Res:* Atmospheric convection and the planetary boundary layer; mesoscale meteorology; synoptic meteorology; study of precipitating clouds and their interaction with the atmospheric circulation on various scales. *Mailing Add:* 4216 Breakwater Ct Ft Collins CO 80525

JOHNSON, RICHARD LAWRENCE, b Glendale, WVa, Feb 3, 39; m 60; c 2. ORGANIC CHEMISTRY. *Educ:* Washington & Jefferson Col, BA, 60; Univ Ky, MS, 62; Univ Iowa, PhD(org chem), 66. *Prof Exp:* Chemist, Rayonier, Inc, 65-66; chemist, 66-73, sr res chemist, 73-82, RES ASSOC, E I DU PONT DE NEMOURS & CO INC, 82- *Concurrent Pos:* Lectr, Parkersburg Br, WVa Univ, 70-71 & Parkersburg Community Col, 71-73. *Mem:* Am Chem Soc; Sigma Xi. *Res:* Fluorocarbon polymers and fluorocarbon synthesis; nylon polymerization and extrusion compounding; fluorocarbon dispersion applications; acrylic resins. *Mailing Add:* 565 Blennerhassett Heights Rd Parkersburg WV 26101

JOHNSON, RICHARD LEON, b Enid, Okla, June 12, 38; m 62; c 2. SIGNAL PROCESSING, RADIO DIRECTION FINDING. *Educ:* Univ Tex, Arlington, BSEE, 64; Southern Methodist Univ, MSEE, 66; Okla State Univ, PhD(elec eng), 70. *Prof Exp:* Aerosyst engr electronics, Gen Dynamics Corp, Ft Worth, Tex, 64-66; res asst, Okla State Univ, 66-70; STAFF ENGR ELECTROMAGNETICS, SOUTHWEST RES INST, 70- *Mem:* Int Union Radio Sci; Inst Elec & Electronics Engrs; Nat Soc Prof Engrs. *Res:* Superresolution spectrum estimation; digital signal processing; antennas and radio wave propagation analysis. *Mailing Add:* Dept Radiolocation Sci Southwest Res Inst San Antonio TX 78228

JOHNSON, RICHARD NORING, b Wethersfield, Conn, Apr 12, 34; m 60; c 2. BIOMEDICAL ENGINEERING. *Educ:* Tri-State Col, BSc, 61; Worcester Polytech Inst, MSc, 65; Univ Va, DSc(biomed eng), 69. *Prof Exp:* Instr elec technol, Hartford State Tech Col, 61-65; res assoc neurol, Schs Eng & Med, Univ Va, 69-70, instr, 70-71, asst prof biomed eng & neurol, 72-77, assoc prof, 77-79; PROF BIOMED ENG & NEUROLOGY, SCH MED, UNIV NC, CHAPEL HILL, 79- *Concurrent Pos:* Fel biomed eng, Johns Hopkins Univ, 71-72. *Mem:* AAAS; Am Soc Eng Educ; Soc Neurosci; Biomed Eng Soc; Am Epilepsy Soc. *Res:* Neurophysiological control systems; neural models. *Mailing Add:* Six Timber Line Rd Chapel Hill NC 27514

JOHNSON, RICHARD RAY, b Carrol, Iowa, Nov 18, 47; m 68; c 2. AGRONOMY, SOIL SCIENCE. *Educ:* Iowa State Univ, BS, 69, MS, 70; Univ Minn, PhD(plant physiol), 74. *Prof Exp:* Asst prof crop prod, Univ Ill, Urbana, 74-77, assoc prof, 77-80; staff agronomist, 80-84, PRIN SCIENTIST, DEERE & CO, 84- *Mem:* Fel Crop Sci Soc Am; fel Am Soc Agron; Soil Sci Soc Am; Weed Sci Soc; Am Soc Agron Eng. *Res:* Applying new technology in crop production to the design and marketing of agricultural equipment. *Mailing Add:* Deere & Co Tech Ctr 3300 River Dr Moline IL 61265

JOHNSON, RICHARD T, b Grosse Pointe Farms, Mich, July 16, 31; m 54; c 4. NEUROLOGY, VIROLOGY. *Educ:* Univ Colo, AB, 53, MD, 56. *Prof Exp:* Teaching fel neurol & neuropath, Harvard Med Sch, 59-61; fel microbiol, John Curtin Sch Med, Canberra, Australia, 62-64; from asst prof to assoc prof neurol, Sch Med, Case Western Reserve Univ, 64-69; assoc prof microbiol, 69-74, Dwight D Eisenhower-United Cerebal Palsy prof neurol, 69-89, PROF MICROBIOL, SCH MED, JOHNS HOPKINS UNIV, 74-, PROF & DIR NEUROL, 89-; NEUROLOGIST IN CHIEF, JOHNS HOPKINS HOSP, 89- *Concurrent Pos:* First neurol asst, Univ Newcastle, Eng, 61-62; assoc neurologist, Cleveland Metrop Gen Hosp, Ohio, 64-69; asst neurologist, Highland View Hosp, Cleveland, 64-69; mem comn, Asn Res Nervous & Ment Dis, 64, 69-77; hon prof, Univ Peruana Cayetano Heredia, 80; prof, neurosci, Johns Hopkins Hosp, 89. *Honors & Awards:* Weil Award, Am Asn Neuropath, 67; Sydney Farber Res Award, 74 & 76; Humboldt Prize, 75; Weinstein-Goldson Award, 79; Gordon Wilson Medal, 80; Charcot Award, Int Fed MS Soc, 85; Smadel Medal, 86; MS Medal, Asn British Neurol, 86. *Mem:* Am Fedn Clin Res; Am Soc Clin Invest; Int Brain Res Orgn; Am Asn Neuropath; Am Neurol Asn; Asn Am Physicians. *Res:* Clinical neurology; pathogenesis of viral infections of the nervous system; neurologic complications of HIV infection. *Mailing Add:* Dept Neurol Johns Hopkins Univ Med Sch Baltimore MD 21205

JOHNSON, RICHARD T(ERRELL), b Shreveport, La, July 28, 39; m; c 2. MECHANICAL ENGINEERING. *Educ:* Mo Sch Mines, BSME, 62, MS, 64; Univ Iowa, PhD(mech eng), 67. *Prof Exp:* Instr eng mech, Univ Mo-Rolla, 62-64; instr mech eng, Univ Iowa, 64-66; from asst prof to prof mech eng, Univ Mo-Rolla, 67-89, dir, Inst Flexible Mfg & Indust Automation, 84-88; PROF & CHMN MECH ENG, WICHITA STATE UNIV, 89- *Honors & Awards:* Delos Lab Develop Award, Am Soc Eng Educ. *Mem:* Am Soc Mech Engrs; Soc Automotive Engrs; Sigma Xi; Soc Mfg Engrs; Am Soc Eng Educ; Combustion Inst. *Res:* Mechanical engineering design; control systems and instrumentation; alternate fuels for transportation engines; improved efficiency of combustion engines; manufacturing automation and systems integration; applications of artificial intelligence and expert systems to design and manufacturing. *Mailing Add:* Mech Eng Dept Wichita State Univ Box 35 Wichita KS 67208

JOHNSON, RICHARD WILLIAM, b Denver, Colo, July 11, 50; m 75; c 1. BIO-ORGANIC CHEMISTRY, ELECTRO-ORGANIC CHEMISTRY. *Educ:* Northwestern Univ, BA & MS, 72; Columbia Univ, MPhil, 74, PhD(chem), 76. *Prof Exp:* Asst prof org chem, Harvard Univ, 77-83; SECT MGR, ROHM & HAAS CO, 83- *Mem:* Am Chem Soc. *Res:* New synthetic procedures based on organic electrochemical reactions; haptea-antibody interactions as model systems for enzymes. *Mailing Add:* 119 Sandywood Dr Doylestown PA 18901

JOHNSON, ROBERT A, b Chicago, Ill, Sept 27, 32; US citizen. CIRCUIT THEORY, ACOUSTICS. *Educ:* Univ Calif, Los Angeles, BS, 55, MS, 63. *Prof Exp:* Var positions, 57-84, PRIN ENGR & SALES MGR, ROCKWELL INT, 84- *Mem:* Fel Inst Elec Electronics Engrs. *Res:* Electromechanical filters. *Mailing Add:* Rockwell Int Filter Products 2990 Airway Ave Costa Mesa CA 92626

JOHNSON, ROBERT ALAN, b New York, NY, Jan 2, 33; m 54; c 3. SOLID STATE PHYSICS, MATERIALS SCIENCE. *Educ:* Harvard Univ, AB, 54; Rensselaer Polytech Inst, PhD(physics), 62. *Prof Exp:* Scientist physics, Brookhaven Nat Lab, 62-69; PROF MAT SCI, UNIV VA, 69- *Mem:* Am Phys Soc; AAAS; Am Inst Mining, Metall & Petrol Engrs; Mat Res Soc; Sigma Xi. *Res:* Theoretical study of interatomic forces, defects and radiation damage in metals; use is made of computer simulation techniques and computer solutions of kinetic equations. *Mailing Add:* Dept Mat Sci Thornton Hall Univ Va Charlottesville VA 22901

JOHNSON, ROBERT BRITTEN, b Cortland, NY, Sept 24, 24; m 47; c 3. GEOLOGY. *Educ:* Syracuse Univ, AB, 49, MS, 50; Univ Ill, PhD(geol), 54. *Prof Exp:* Asst, Syracuse Univ, 47-50; asst, State Geol Surv, Ill, 51-53, asst geologist, 53-54; asst prof geol & staff geologist, Syracuse Univ, 54-55; sr geologist & geophysicist, C A Bays & Assocs, 55-56; from asst prof to prof geol, Purdue Univ, 56-66; prof geol & head dept geol & geog, DePauw Univ, 66-67; chmn dept geol, Colo State Univ, 69-73, prof geol prog, 73-77, head earth resources actg dept, 79-80, prof geol, 67-88, EMER PROF GEOL, COLO STATE UNIV, 88- *Concurrent Pos:* Lectr, Univ Ill, 55-56; indust consult, 62-; mem comt A2L01, Transp Res Bd; mem comt A2L05, Transp Res Bd, 75-86, chmn comt A2L01, 76-82; geologist, US Geol Surv, 76-88. *Honors & Awards:* E B Burwell Jr Mem Award, Geol Soc Am, 89; C B Holdredge Award, Asn Eng Geologists, 90. *Mem:* Fel Geol Soc Am; Asn Eng Geologists; Int Asn Eng Geol. *Res:* Engineering geology, especially landslides and geophysical and remote sensing applications. *Mailing Add:* Dept Earth Resources Colo State Univ Ft Collins CO 80523

JOHNSON, ROBERT CHANDLER, b Detroit, Mich, Oct 19, 30; m 55; c 4. ANALYTICAL CHEMISTRY, GENERAL PHYSICS. *Educ:* Univ Mich, BS, 52; State Univ Iowa, MA, 57; Stanford Univ, PhD(physics), 62. *Prof Exp:* Res physicist, E I du Pont de Nemours & Co, Inc, 62-73, res physicist, res & develop planning, 73-75, res physicist, thermal anal, 75-78, RES SUPVR ANALYTICAL SCI, CENT RES & DEVELOP DEPT, E I DU PONT DE NEMOURS & CO INC, 78- *Concurrent Pos:* Vis scientist, Am Inst Physics, 72-75. *Mem:* Fel NAm Thermal Anal Soc (secy, 79-81 & pres, 85-); Am Phys Soc; Am Chem Soc; Mat Res Soc. *Res:* Magnetic field effects on triplet excitons; exciton physics of organic crystals; Kapitza resistance in liquid helium; low temperature physics; thermal analysis; x-ray synchrotron applications at advanced photon source. *Mailing Add:* E I du Pont de Nemours & Co Inc Exp Sta Bldg 228 PO Box 80228 Wilmington DE 19880-0228

JOHNSON, ROBERT ED, b Highland Park, Ill, Nov 14, 42; m 64; c 2. MEDICINAL CHEMISTRY. *Educ:* Univ Wis, BS, 64; Univ Minn, PhD(org chem), 68. *Prof Exp:* RES CHEMIST, GROUP LEADER, SECT HEAD & ASSOC RES DIR MED CHEM, STERLING RES GROUP, 68- *Mem:* Am Chem Soc; NY Acad Sci; AAAS. *Res:* Synthesis of novel heterocyclic and aromatic compounds that may have useful medicinal properties. *Mailing Add:* Sterling Res Group Rensselaer NY 12144

JOHNSON, ROBERT EDWARD, b Chicago, Ill, July 3, 39; m 70; c 2. PLANETARY SCIENCE. *Educ:* Colo Col, BA, 61; Wesleyan Univ, MA, 63; Univ Wis, Madison, PhD(physics), 68. *Prof Exp:* Res fel, Queen's Univ, Belfast, Ireland, 68-69; asst prof physics, Southern Ill Univ, 69-71; PROF ENG PHYSICS, UNIV VA, 71- *Concurrent Pos:* NATO fel, Univ Copenhagen, 76; vis scientist, Ctr Earth & Planetary Physics, Harvard Univ, 77-78; fac fel, Argonne Nat Lab, 82; NSF & NASA grants prin investr, 78-; consult, Dept Physics, Denver Univ, 70 & Bell Tel Lab, 79- *Mem:* Am Phys Soc; Am Geol Phys Union; Am Astron Soc. *Res:* Atomic and molecular physics; problems of interest in the Jovian magnetosphere, and interaction of ionizing radiations with solids and surfaces. *Mailing Add:* Dept Nuclear Eng & Eng Physics Univ Va Charlottesville VA 22901

JOHNSON, ROBERT EUGENE, b Conrad, Mont, Apr 8, 11; m 35; c 2. PHYSIOLOGY, NUTRITION. *Educ:* Univ Wash, BS, 31; Oxford Univ, BA, 34, DPhil(biochem), 35; Harvard Univ, MD, 41. *Prof Exp:* Asst & assoc, Fatigue Lab, Harvard Univ, 35-42, asst prof indust physiol, 42-46; dir, Med Nutrit Lab, US Army, 46-49; prof physiol, Univ Ill, Urbana, 49-73; prof biol, Knox Col, 73-79, coordr, Knox-Rush Med Prog, 73-79; PRES, HORN OF THE MOON ENTERPRISES, MONTPELIER, VT, 79-; VIS PROF PHYSIOL, UNIV VT, 84- *Concurrent Pos:* Head, dept physiol, Univ Ill, 49-60, dir hons prog, 58-67; NSF sr res fel, 57-58; Guggenheim fel, 64-65; consult physician, Presby-St Lukes Hosp, Chicago, 73-84. *Mem:* Am Physiol Soc; Am Soc Clin Invest; Hist Sci Soc; Sigma Xi. *Res:* Physiological responses in man to stresses of work, environment and diet; metabolism of poikilotherms; history of environmental physiology. *Mailing Add:* Five E Terrace South Burlington VT 05403

JOHNSON, ROBERT F, b Crestwood, Ky, Mar 20, 29; div; c 4. TEXTILE ENGINEERING. *Educ:* Univ Ky, BS, 51; Ga Inst Technol, MS, 58; Swiss Fed Inst Technol, PhD(indust & eng chem), 63. *Prof Exp:* Res chemist, Dow Chem Co, 58-65; assoc prof textile eng, Ga Inst Technol, 65-66; res sect mgr, Phillips Petrol Co, 66-68, consult, 69-70; prof textile eng & dir, Chem Processes Lab, Textile Res Ctr, Tex Tech Univ, 68-72; dir, Grad Studies, 81-85, PROF TEXTILES & CLOTHING, UNIV MINN, 72- *Mem:* Am Asn Textile Chem & Colorists; Am Chem Soc; Brit Soc Dyers & Colourists; Am Coun Consumer Interests. *Res:* Physical and chemical properties of textile materials; chemistry of dyes; chariterization of fire hazards of clothing. *Mailing Add:* PO Box 8025 St Paul MN 55108

JOHNSON, ROBERT GLENN, b Green Mountain, Iowa, Dec 12, 22; m 49; c 5. ELECTROPHYSICS. *Educ:* Case Inst Technol, BS, 47; Iowa State Col, PhD(physics), 52. *Prof Exp:* Asst physics, Iowa State Col, 49-52; proj engr, Bendix Aviation Corp, 52-55; sr res physicist, Honeywell Sensors & Systs Develop Ctr, Bloomington, Minn, 55-67, staff scientist, 67-90; RETIRED. *Concurrent Pos:* Adj prof, Dept Geol & Geophys, Univ Minn. *Honors & Awards:* H W Sweatt Award, Honeywell Inc, 68 & 85. *Mem:* Inst Elec & Electronics Engrs; AAAS; Sigma Xi; Geol Soc Am. *Res:* Corona degradation of materials; paleoclimatology; gas discharge phenomena; ultraviolet light sensor technology; silicon microstructures. *Mailing Add:* 12814 March Circle Minnetonka MN 55343

JOHNSON, ROBERT GUDWIN, b Milwaukee, Wis, Nov 23, 27; m 58; c 4. ORGANIC CHEMISTRY. *Educ:* Marquette Univ, BS, 49; Iowa State Col, PhD(chem), 54. *Prof Exp:* Asst chem, Iowa State Col, 49-53; from inst to assoc prof 54-65, chmn dept chem, 66-75 & 84-86, PROF CHEM, XAVIER UNIV, OHIO, 65- *Concurrent Pos:* Vis prof, Purdue Univ, 60 & 63. *Mem:* Am Chem Soc. *Res:* Hunsdiecker-Borodine reaction; oxygen-containing heterocycles; hypolipidemic agents; anti-cancer compounds; aromatic substitution. *Mailing Add:* Dept Chem Xavier Univ 3800 Victory Pkwy Cincinnati OH 45207

JOHNSON, ROBERT H, b Montreal, Que, June 23, 36. DENTISTRY. *Educ:* McGill Univ, BSc, 58, DDS, 62; Ind Univ, MSD, 64; Univ Wash, cert periodontics, 71; Am Bd Oral Med, dipl; FRCD(C). *Prof Exp:* Asst prof dent, McGill Univ, 64-66; asst prof dent & dir hosp dent serv, Med Ctr, Univ Ky, 66-69; from assoc prof to prof dent, Univ Western Ont, 71-80, chmn div periodont, 78-80; PROF PERIODONT & CHMN DEPT PERIODONT, UNIV WASHINGTON SCH DENT, SEATTLE, 80- *Concurrent Pos:* Chief oral diag clin, Montreal Gen Hosp, 64-66. *Mem:* Am Acad Periodont; fel Am Acad Oral Path. *Res:* Effects of tetracyclines on teeth and bone; dentinal hypersensitivity; chemotherapeutic plaque and inflamation control. *Mailing Add:* Dept Periodont SM-44 Univ Wash Sch Dent Seattle WA 98195

JOHNSON, ROBERT JOSEPH, b Toppenish, Wash, Feb 8, 15; m 41; c 3. ANATOMY. *Educ:* Iowa State Teachers Col; Univ Iowa, MD, 43. *Prof Exp:* Asst anat, Col Med, Univ Iowa, 38-41; from instr to assoc prof, Sch Med, Univ Wash, 46-57, prof anat & surg, 57; prof gross & neurol anat & head dept, Sch Med, WVa Univ, 57-63; PROF ANAT & CHMN DEPT, GRAD SCH MED, UNIV PA, 63-, PROF SURG, 66- *Concurrent Pos:* Consult, Madigan Army Hosp, Tacoma, Wash, 49-57 & Vet Admin Hosp, Seattle, Wash, 50-57; lectr, Western State Hosp, Tacoma, 51-57; consult, USPHS Hosp, 52-57; assoc prof surg, Sch Med, Univ Wash, 55-57. *Mem:* Fel Am Acad Forensic Sci; Am Asn Anatomists. *Res:* Human anatomy; peripheral nerves; venous system of lower limb; congenital anomalies. *Mailing Add:* Dept Anat Univ Pa Grad Sch Med Philadelphia PA 19104-6058

JOHNSON, ROBERT KARL, b Worthington, Minn, May 7, 44; m 75. ICHTHYOLOGY. *Educ:* Occidental Col, AB, 66; Univ Calif, San Diego, PhD(marine biol), 72. *Prof Exp:* Res assoc ecol, Chesapeake Biol Lab, Univ Md, 71-72; asst cur fishes, 72-75, assoc cur, 75-81, chmn dept zool, 81-84, cur fishes, Field Mus Natural Hist, 81-86; DIR, GRAD PROG MARINE BIOL, GRICE MARINE BIOL LAB. *Concurrent Pos:* Asst prof earth sci, Univ Notre Dame, 74; adj asst prof biol sci, Northern Ill Univ, 74-79, adj assoc prof, 79-; mem Comt Evolutionary Biol, Univ Chicago, 76-86. *Mem:* Am Soc Ichthyologists & Herpetologists; Soc Syst Zool; AAAS. *Res:* Systematics, ecology and zoogeography of marine fishes. *Mailing Add:* Grice Marine Biol Lab Col Charleston Charleston SC 29412

JOHNSON, ROBERT L(AWRENCE), b Glasgow, Mont, June 18, 19; m 45; c 3. MECHANICAL ENGINEERING. *Educ:* Mont State Univ, BS, 42. *Prof Exp:* Mech engr, Langley Mem Aeronaut Lab, Nat Adv Comt Aeronaut, NASA, 42-43, from mech engr to supvr mat res eng, 43-71, chief lubrication br, Lewis Res Ctr, 63-75; RETIRED. *Concurrent Pos:* Lubrication consult, 62-; chmn, Gordon Res Conf Friction, Lubrication & Wear, 74; US deleg & chmn group experts wear eng mat, Orgn Econ Coop & Develop, 64-73; adj prof mech eng, Rensselaer Polytech Inst, 75-87. *Honors & Awards:* IR 100 Award, 66 & 73; Alfred E Hunt Award, Am Soc Lubrication Engrs, 61 & 65; Nat Award, 71; Tribology Gold Medal, Brit Inst Mech Engrs, 76; Mayo D Hersey Award, Am Soc Mech Engrs, 77. *Mem:* Am Soc Testing & Mat; Soc Automotive Engrs; Soc Tribologists & Lubrication Engrs (pres, 68-69); fel Brit Inst Mech Engrs. *Res:* Lubrication, friction and wear tribology in seals, bearings and other mechanical components and lubricants for extreme environments. *Mailing Add:* 5304 W 62nd St Edina MN 55436

JOHNSON, ROBERT L, b Winslow, Ariz, May 16, 20. ENGINEERING. *Educ:* Univ Calif, Berkeley, BS, 41, MS, 42. *Prof Exp:* From mem staff to vpres, Manned Orbiting Lab, Douglas Aircraft Co, 46-69; asst secy army for res & develop, Dept of Army, 69-73; corp vpres eng & res, McDonnell Douglas Corp, 73-75, pres, McDonnell Douglas Astronaut Co, 75-80, corp vpres aerospace group exec, 80-87; RETIRED. *Concurrent Pos:* Mem eng adv coun, Univ Calif. *Honors & Awards:* James H Wyld Mem Award, Am Rocket Soc. *Mem:* Nat Acad Eng; fel Am Inst Aeronaut and Astronaut. *Mailing Add:* 30881 Greens East Dr Laguna Niguel CA 92677

JOHNSON, ROBERT LEE, b Dallas, Tex, Apr 28, 26; m 52; c 2. PHYSIOLOGY. *Educ:* Southern Methodist Univ, BS, 47; Northwestern Univ, MD, 51. *Prof Exp:* Intern, Cook County Hosp, Chicago, 51-55; res fel internal med, Southwestern Med Sch, Univ Tex, 55-56; res fel physiol, Grad Sch Med, Univ Pa, 56-57; from instr to assoc prof, 57-69, PROF INTERNAL MED, UNIV TEX HEALTH SCI CTR, DALLAS, 69- *Concurrent Pos:* Assoc ed, J Clin Invest, 72-77; prog chmn, Cardiopulmonary Coun, Am Heart Asn, 79-81 & chmn, 85-87, mem, Cardiovascular Develop Res Study Comt, 81-83, mem, Heart Lung & Blood Inst Rev Comt, 85- *Mem:* Am Asn Physicians; Am Fedn Clin Res; Am Thoracic Soc; Am Physiol Soc; Am Soc Clin Invest; Am Heart Asn. *Res:* Exercise physiology; adaptation to high altitude; control of capillary circulation and diffusing surface in the lung. *Mailing Add:* Univ Tex Health Sci Ctr 5323 Harry Hines Blvd Dallas TX 75235

JOHNSON, ROBERT LEROY, b Chicago, Ill, Sept 22, 40; m 63; c 2. MATHEMATICS. *Educ:* Augustana Col, Ill, AB, 62; Univ Kans, MA, 65, PhD(math), 67. *Prof Exp:* Asst prof math, Iowa State Univ, 67-68; asst prof, 68-72, assoc prof, 72-80, PROF MATH, AUGUSTANA COL, ILL, 80- *Mem:* Math Asn Am; Am Math Soc; Sigma Xi. *Res:* Topological rings. *Mailing Add:* Dept Math Augustana Col Rock Island IL 61201

JOHNSON, ROBERT M, b Oklahoma City, Okla, Mar 28, 39; m 63; c 1. METALLURGICAL ENGINEERING. *Educ:* Univ Okla, BS, 62, MMetEng, 65, PhD(eng sci), 67. *Prof Exp:* Asst prof eng mech & mat sci, 67-71, assoc prof mat sci, 71-79, PROF MECH ENG & MAT SCI, UNIV TEX, ARLINGTON, 79-, ASSOC DEAN GRAD SCH, 80- *Concurrent Pos:* Sr scientist, Vought Corp Advanced Technol Ctr, 77-78. *Mem:* Am Soc Metals; Am Soc Eng Educ. *Res:* Basic deformation processes in mechanical metallurgy; dislocation mechanisms; fracture mechanics; corrosion. *Mailing Add:* Mech Eng Dept Univ Tex Arlington TX 76019

JOHNSON, ROBERT MICHAEL, b Brooklyn, NY; c 1. BIOCHEMISTRY. *Educ:* Fordham Col, AB, 61; Columbia Univ, PhD(biochem), 70. *Prof Exp:* NIH fel, Cornell Univ, 71-72; instr, 72-73, asst prof, 73-79, ASSOC PROF BIOCHEM, MED SCH, WAYNE STATE UNIV, 79- *Concurrent Pos:* Vis prof, Int Cell Path, Bicêtre, France, 82. *Mem:* Sigma Xi; Am Heart Assoc; Am Soc Biol Chemists; Soc Rheology; AAAS. *Res:* Biochemistry of biological membranes; protein structure, erythrocyte function. *Mailing Add:* Dept Biochem Wayne State Univ Med Sch Detroit MI 48201

JOHNSON, ROBERT OSCAR, b Detroit, Mich, May 7, 26; m 61; c 3. APPLIED MATHEMATICS. *Educ:* Univ Mich, Ann Arbor, BS(eng) & BS(math), 46, MS, 49; Univ Ill, Urbana, MS, 52; Ohio State Univ, PhD(math), 75. *Prof Exp:* Sr engr, ITT Labs, 50-52; procurement rep aircraft systs, Repub Aviation, Inc, 52-53; admin engr, Teterboro Div, Bendix Corp, 54-58; proposal mgr altitude control systs, Aerospace Div, Walter Kidde &

Co, Inc, 58-62; mgr advan design, Arde, Inc, 62-63; res specialist, Columbus Div, NAm Rockwell, Inc, 64-68; prof math, Franklin Univ, 68-86, div chmn, Eng Technol, 81-85; prof comput sci, Marshall Univ, 86-90; PROF COMPUT SCI, FROSTBURG STATE UNIV, 90- *Concurrent Pos:* Teaching assoc, dept math, Ohio State Univ, 64-69; prof engr, Data Control Ctr, Ohio State Hwy Dept, 69-75. *Mem:* Asn Comput Mach; Inst Elec & Electronics Engrs; Math Asn Am. *Res:* Mathematical modeling. *Mailing Add:* PO Box 30011 Gahanna OH 43230-0011

JOHNSON, ROBERT R(OYCE), b Madison, Wis, June 20, 28; m 53; c 3. ENGINEERING. *Educ:* Univ Wis, BS, 50; Yale Univ, MEng, 51; Calif Inst Technol, PhD, 56. *Prof Exp:* Res physicist, Hughes Aircraft Co, Calif, 51-55; engr, Gen Elec Co, 55-56, mgr, Comput Lab, 56-59, mgr eng, Comput Dept, 59-64; dir eng, 64-68, vpres eng, 68-80, vpres advan technol, Burroughs Corp, Detroit, 80-; ENG DEPT, ENERGY CONVERSION DEVICES, INC. *Mem:* Fel Inst Elec & Electronics Engrs; Sigma Xi. *Res:* Computer technology; logic design; sequential analysis. *Mailing Add:* 3190 Morrill Eng Bldg Univ Utah Salt Lake City UT 84112

JOHNSON, ROBERT REINER, b Chicago, Ill, June 8, 32; m 67; c 1. ORGANIC CHEMISTRY. *Educ:* Brown Univ, ScB, 54; Rice Univ, PhD(chem), 58. *Prof Exp:* Res assoc chem, Johns Hopkins Univ, 58-59; group leader, 59-67, SCIENTIST, BROWN & WILLIAMSON TOBACCO CORP, 67- *Mem:* Am Chem Soc; Phytochem Soc NAm. *Res:* Physical organic chemistry and chemistry of natural products. *Mailing Add:* Brown & Williamson Tobacco Corp Res Dept PO Box 35090 Louisville KY 40232

JOHNSON, ROBERT S, b Pikeville, Ky, Nov 23, 37. MATHEMATICS. *Educ:* Georgetown Col, BS, 59; Univ NC, MA, 62, PhD(ring theory), 66. *Prof Exp:* From instr to assoc prof, 65-75, PROF MATH & HEAD DEPT, WASHINGTON & LEE UNIV, 75- *Mem:* Am Math Soc; Sigma Xi. *Res:* Group theory and ring theory; conditions implying commutativity. *Mailing Add:* Dept Math Washington & Lee Univ Lexington VA 24450

JOHNSON, ROBERT SHEPARD, b Wilkinsburg, Pa, Nov 24, 28; m 59; c 3. NUMERICAL ANALYSIS. *Educ:* Northwestern Univ, BS, 50, MS, 51; Univ Pa, PhD(math), 59. *Prof Exp:* Res assoc, Inst Coop Res, Univ Pa, 53-59; ENGR, RCA CORP, 59- *Mem:* Am Math Soc; Soc Indust & Appl Math. *Res:* Approximation theory; moments. *Mailing Add:* 2102 Brandeis Ave Riverton NJ 08077

JOHNSON, ROBERT W, JR, geology, geophysics; deceased, see previous edition for last biography

JOHNSON, ROBERT WALTER, b New York, NY, Mar 11, 30; m 61; c 2. ECOLOGY. *Educ:* Hofstra Univ, BA, 58, MA, 59; Cornell Univ, PhD(wildlife mgt), 73. *Prof Exp:* Res asst nematol, USDA, 56-57; ASSOC PROF BIOL, HOFSTRA UNIV, 59-61 & 65- *Concurrent Pos:* Marine conserv biologist, Town Oyster Bay, NY, 67-; terrestrial ecologist, Grumman Ecosyst Corp, 74-; pres, R W Johnson & Assoc, Environ Anal Inc, 74- *Mem:* Wildlife Soc; Ecol Soc Am. *Res:* Estuarine ecology; ecology of marsh birds; environmental impact analysis. *Mailing Add:* Dept Biol Hofstra Univ 1000 Fulton Ave Hempstead NY 11550

JOHNSON, ROBERT WARD, b Hampton, Va, Dec 19, 29; m 55; c 4. MARINE SCIENCES, OPERATIONS RESEARCH. *Educ:* Va Polytech Inst & State Univ, BS, 50; Pa State Univ, MS, 54; NC State Univ, PhD(marine sci), 75. *Prof Exp:* Develop engr, Philco Corp, 52-53, E I du Pont de Nemours & Co, Inc, 54-57; design develop res engr, Carrier Corp, 57-63; res & develop engr, Langley Res Ctr, NASA, 63-70, res scientist marine sci, 70-90; RETIRED. *Mem:* Am Soc Photogram. *Res:* Application of remote sensing (aircraft and satellite) techniques to monitor pollution sources and to study processes in marine ecosystems; atmospheric studies research. *Mailing Add:* 348 Wrexham Ct Hampton VA 23669

JOHNSON, ROBERT WELLS, b Hartford, Conn, Apr 21, 38; m 64; c 3. NUMBER THEORY, DIOPHANTINE EQUATIONS. *Educ:* Amherst Col, AB, 59; Mass Inst Technol, MS, 61, PhD(math), 64. *Prof Exp:* From instr to assoc prof, 64-75, prof, 75-90, ISAAC HENRY WING PROF MATH, BOWDOIN COL, 90- *Mem:* Math Asn Am; Am Math Soc. *Res:* Algebra and number theory. *Mailing Add:* Dept Math Bowdoin Col Brunswick ME 04011

JOHNSON, ROBERT WILLIAM, JR, b Jacksonville, Fla, Oct 9, 27; m 57; c 4. ORGANIC CHEMISTRY. *Educ:* Univ Fla, BS, 53, PhD(org chem), 59; Purdue Univ, MS, 56. *Prof Exp:* Res chemist, Ethyl Corp, 59-62; supt compound develop dept, Chem Div, Union Bag-Camp Paper Corp, 62-65, supt prod develop dept, 65-67; mgr chem div, prod develop dept, 67-73, sr chemist, process chem dept, 73-86, TECH ASSOC, UNION CAMP CORP, 86- *Mem:* Am Chem Soc; Am Oil Chemists Soc; NY Acad Sci; fel Am Inst Chemists; Am Inst Chem Engrs. *Res:* Tall oil; fatty acids; organic synthesis; organometallics; separation technology; instrumental methods of analysis; rosin and derivatives; hydrogenation. *Mailing Add:* Chem Div Union Camp Corp PO Box 2668 Savannah GA 31402

JOHNSON, RODNEY L, PEPTIDES, AMINO ACIDS. *Educ:* Univ Kans, PhD(med chem), 75. *Prof Exp:* ASSOC PROF MED CHEM, UNIV MINN, 82- *Mailing Add:* HSUF 8-172 Univ Minn 308 Harvard St SE Minneapolis MN 55455

JOHNSON, ROGER D, JR, b Richmond, Va, May 27, 30; m 55; c 3. MATHEMATICS. *Educ:* Dartmouth Col, AB, 51; Univ Va, MA, 53, PhD(math), 56. *Prof Exp:* Instr math, Univ Va, 55-56; asst prof, 56-61, ASSOC PROF MATH, GA INST TECHNOL, 61- *Mem:* Am Math Soc; Math Asn Am. *Res:* Homology theory and its relationship to certain topics of general topology such as connectedness and dimension. *Mailing Add:* Ga Inst Technol Sch Math Atlanta GA 30332

JOHNSON, ROGER W, b Kalamazoo, Mich, May 4, 29; m 58; c 2. VIROLOGY. *Educ:* Valparaiso Univ, BS, 52; Univ Ky, MS, 58, PhD(microbiol), 63. *Prof Exp:* Microbiologist, US Army Biol Labs, 63-69; prin investr, US Army Biol Defense Res Ctr, 69-72; head dept virus prod, Frederick Cancer Res Ctr, Nat Cancer Inst, 72-81; DIR OPERS, WHITTAKER BIOPRODUCTS, 81- *Mem:* AAAS; Am Soc Microbiol; Sigma Xi. *Res:* Parameters of seed stock development; scale up and production of oncogenic or suspected oncogenic viruses from tissue culture; large scale culture of mammalian cells. *Mailing Add:* 7003 Summerfield Dr Rte 7 Frederick MD 21701

JOHNSON, ROLLAND PAUL, b Stewartville, Minn, Jan 1, 41; m 90; c 2. ACCELERATORS. *Educ:* Univ Calif, Berkeley, AB, 64, PhD(physics), 70. *Prof Exp:* Res asst physics, Lawrence Berkeley Lab, Univ Calif, Berkeley, 67-70, res assoc, 70-74; PHYSICIST, FERMI NAT ACCELERATOR LAB, 74- *Concurrent Pos:* Vis scientist, Inst High Energy Physics, Serpukhov, USSR, 72-73 & Europ Orgn Nuclear Res, Geneva, Switzerland, 80-81; mem, Univ Chicago Rev Comt, High Energy Physics Div, Argonne Nat Lab, 84-86. *Mem:* Am Phys Soc. *Res:* Accelerators; experimental particle physics; experimental high energy physics. *Mailing Add:* Fermilab PO Box 500 Batavia IL 60510

JOHNSON, RONALD CARL, b Milwaukee, Wis, Sept 5, 35; m 60; c 2. INORGANIC CHEMISTRY. *Educ:* Lawrence Col, BS, 57; Northwestern Univ, PhD(chem), 61. *Prof Exp:* From asst prof to assoc prof, 61-73, PROF CHEM, EMORY UNIV, 73-, ASSOC DEAN, 90- *Mem:* AAAS; Am Chem Soc. *Res:* Reactions of compounds of transition metals; mechanisms of reactions. *Mailing Add:* Dept Chem Emory Univ Atlanta GA 30322

JOHNSON, RONALD ERNEST, b Portland, Ore, Oct 14, 39; m 68; c 2. PHYSICAL OCEANOGRAPHY. *Educ:* Ore State Univ, BS, 62, MS, 63, PhD(phys oceanog), 72. *Prof Exp:* Assoc sr engr, Lockheed-Calif Co, 63-64; asst prof, 68-78, ASSOC PROF OCEANOG, OLD DOMINION UNIV, 78-, ASSOC DIR GRAD STUDIES, 85- *Mem:* Sigma Xi; Am Geophys Union. *Res:* Circulation and distribution of intermediate waters of the worlds oceans. *Mailing Add:* Dept Oceanog Old Dominion Univ Norfolk VA 23529-0276

JOHNSON, RONALD GENE, b Detroit, Mich, Nov 14, 41; m 64; c 2. RADIATION BIOPHYSICS, RADIOLOGICAL HEALTH. *Educ:* Eastern Mich Univ, AB, 63; Univ Kans, MS, 68, PhD(radiation biophys), 70. *Prof Exp:* High sch teacher, Mich, 64 & Ohio, 64-65; asst prof, 70-74, assoc prof, 74-78, PROF PHYSICS, MALONE COL, 78-, EXEC VPRES, 81- *Concurrent Pos:* Radiation biologist, Aultman Hosp, 73-; consult, Med Physics Serv, Inc, 73-; vis assoc prof radiation biophys, Univ Kans, 76-77; assoc prof clin radiation biophys radiol, Northeastern Ohio Univs Col Med, 78- *Mem:* Am Asn Physics Teachers; Sigma Xi. *Res:* Effect of glucose on the sensitivity of Escherichia coli to Mitomycin C; radiation repair mechanisms; radiation-induced atrophy of bone; quality control in diagnostic radiology; effects of diagnostic x-rays during first trimester of pregnancy. *Mailing Add:* VPres Off Acad Affairs Malone Col Canton OH 44709

JOHNSON, RONALD ROY, b De Smet, SDak, Dec 8, 28; m 55; c 4. BIOCHEMISTRY, ANIMAL NUTRITION. *Educ:* SDak State Col, BS, 50, MS, 52; Ohio State Univ, PhD(biochem), 54. *Prof Exp:* Asst, Exp Sta, SDak State Col, 50-52; from asst prof to prof animal sci, Ohio Agr Res & Develop Ctr, 55-69; prof animal sci & indust, Okla State Univ, 69-74; prof animal sci & head dept, Univ Tenn, Knoxville, 74-81; assoc dir, Okla Agr Exp Sta, 81-89, PROF ANIMAL SCI, OKLA STATE UNIV, 89- *Concurrent Pos:* Consult, Nutrit Surv Team, Comt Nutrit Nat Defense, Spain, 58 & Chile, 60; USPHS sr fel, Univ Calif, Berkeley, 65-66. *Mem:* Am Soc Animal Sci; Am Dairy Sci Asn; Am Inst Nutrit. *Res:* Nutrition, physiology and biochemistry of Rumen microorganisms and ruminant animals; nutrition of farm livestock. *Mailing Add:* 114B Animal Sci Okla State Univ Stillwater OK 74078

JOHNSON, RONALD SANDERS, b Chicago, Ill, March 9, 52. PHYSICAL BIOCHEMISTRY, INORGANIC BIOCHEMISTRY. *Educ:* Northwestern Univ, BA, 73, PhD(biochem & molecularbiol), 78. *Prof Exp:* Instr biochem & res tech, Northwestern Univ, 78; fel, NIH & Miller Inst Basic Res Sci, Univ Calif, Berkeley, 78-81; asst prof biochem & phys biochem, 81-87, ASSOC PROF BIOCHEM, SCH MED, EAST CAROLINA UNIV, 87- *Mem:* Am Chem Soc; Sigma Xi; Am Soc Biol Chemists. *Res:* Application of biophysical techniques to explore the mechanism of gene regulation in the bacterium E coli, encompassing protein-nucleic acid as well as protein-protein interactions. *Mailing Add:* Dept Biochem Sch Med East Carolina Univ Greenville NC 27858-4354

JOHNSON, ROSE MARY, b Ashland, Ky, July 14, 27. ZOOLOGY. *Educ:* Hood Col, AB, 49; Univ Va, MA, 56, PhD(biol), 62. *Prof Exp:* Res asst microbiol, Sch Med, Univ Va, 51-55; instr biol, Sweet Briar Col, 59-61; asst prof, Old Dom Col, 61-62; from assoc prof to prof biol, Mary Washington Col, 61-89; RETIRED. *Mem:* Fel AAAS; Am Soc Microbiol. *Res:* Taxonomic relationships of crayfishes using serological techniques such as agar diffusion and precipitin reactions; calcium deposition and dissolution during the molting cycle of crayfishes. *Mailing Add:* 4791 Watson Lane Fredericksburg VA 22401-1182

JOHNSON, ROSS BYRON, b Ladd, Ill, June 4, 19; m 42. ENVIRONMENTAL GEOLOGY, FUELS GEOLOGY. *Educ:* Univ NMex, BS, 46, MS, 48. *Prof Exp:* Geologist, US Geol Surv, 48-62, res geologist, 62-74; CONSULT GEOLOGIST, 74- *Mem:* Fel Geol Soc Am; Sigma Xi. *Res:* Formation of sand dunes, rock glaciers, joints, and faults and their effects on the environment and engineering structures; stratigraphic, structural, and igneous geology; geologic mapping and photo-geology; petroleum and coal resources of the Southern Rocky Mountains and adjacent high plains of Colorado and New Mexico. *Mailing Add:* 240 Quay St Lakewood CO 80226

JOHNSON, ROSS GLENN, b McKeesport, Pa, Oct 5, 42; m 64; c 2. CELL BIOLOGY. *Educ:* Augustana Col, Ill, BA, 64; Iowa State Univ, MS, 66, PhD(cell biol), 68. *Prof Exp:* Asst prof cytol & zool, 68-73, assoc prof zool, Univ Minn-Minneapolis, 73-76, assoc prof, 76-80, PROF GENETICS & CELL BIOL, UNIV MINN-ST PAUL, 80- *Concurrent Pos:* NIH Predoctoral Fel, Bush sabbatical fel. *Mem:* AAAS; Am Soc Cell Biol. *Res:* Involvement of cell junctions in cell communication; structure and function of cell organelles. *Mailing Add:* Dept Genetics & Cell Biol Univ Minn St Paul MN 55108-1095

JOHNSON, ROTHER RODENIOUS, bacteriology, immunology; deceased, see previous edition for last biography

JOHNSON, ROY ALLEN, b Bemidji, Minn, July 26, 37; m 63; c 2. ORGANIC CHEMISTRY. *Educ:* Univ Minn, BCh, 59, PhD(org chem), 65; Univ BC, MSc, 61. *Prof Exp:* SR SCIENTIST, UPJOHN CO, 65- *Concurrent Pos:* Vis scientist, Mass Inst Technol, 82-83; chem forum lectr, 90. *Mem:* Am Chem Soc; AAAS; NY Acad Sci. *Res:* Synthetic organic chemistry; prostaglandin chemistry; microbial oxidations; stereochemistry, phospholipid chemistry and superoxide chemistry. *Mailing Add:* Upjohn Labs Upjohn Co Kalamazoo MI 49001-3298

JOHNSON, ROY ANDREW, b Oak Park, Ill, Mar 20, 39; m 67; c 2. MATHEMATICS. *Educ:* St Olaf Col, BA, 60; Univ Iowa, PhD(math), 64. *Prof Exp:* Asst lectr math, Univ Lagos, 64-65; asst prof, Univ Col, Addis Ababa, 65-66; from asst prof to assoc prof, 66-84, PROF MATH, WASH STATE UNIV, 84- *Concurrent Pos:* Vis prof, Univ Lódź, Lódź, Poland, 85-86. *Mem:* Am Math Soc; Math Asn Am. *Res:* Measure theory and integration; real functions. *Mailing Add:* Dept Math Wash State Univ Pullman WA 99164-3113

JOHNSON, ROY MELVIN, microbiology, for more information see previous edition

JOHNSON, ROY RAGNAR, b Chicago, Ill, Jan 23, 32; m 63. PLASMA PHYSICS, SOLID STATE PHYSICS. *Educ:* Univ Minn, BEE, 54, MSEE, 56, PhD(elec eng), 59. *Prof Exp:* Asst solid state physics, Univ Minn, 54-56; res scientist, Boeing Sci Res Lab, 59-70; CONSULT, KMS FUSION INC. *Concurrent Pos:* Vis scientist, Royal Inst Technol, Sweden, 63-64. *Mem:* AAAS; Am Phys Soc; Inst Elec & Electronics Engrs. *Res:* Fluids physics; solids fluctuations. *Mailing Add:* 671 Adrienne Lane Ann Arbor MI 48103

JOHNSON, RULON EDWARD, JR, b Logan, Utah, June 5, 29; m 54; c 6. PHYSICAL CHEMISTRY. *Educ:* Utah State Univ, BS, 51; Stanford Univ, PhD(phys chem), 57. *Prof Exp:* From res chemist to sr res chemist, Jackson Lab, 57-69, RES ASSOC, ORG CHEM DEPT, EXP STA, E I DU PONT DE NEMOURS & CO, INC, 69- *Mem:* AAAS; fel Am Chem Soc; Sigma Xi. *Res:* Thermodynamics, especially application to interfaces; molecular interactions at interfaces. *Mailing Add:* 14 River Way Wilmington DE 19809-2473

JOHNSON, RUSSELL CLARENCE, b Wausau, Wis, Aug 3, 30; m 55; c 3. MICROBIOLOGY. *Educ:* Univ Wis, BS, 57, MS, 58, PhD(microbiol), 60. *Prof Exp:* Res assoc microbiol, Univ Wis, 60; Nat Acad Sci-Nat Res Coun res assoc, 60-61; res microbiologist, Ft Detrick, Md, 61-62; from instr to assoc prof, 62-74, PROF MICROBIOL, MED SCH, UNIV MINN, MINNEAPOLIS, 74- *Concurrent Pos:* USPHS spec fel, 63-65, res grant, 66-; mem subcomt taxon Leptospira & subcomt taxon Spirochaetales, Int Comt Syst Bact, 69-; mem comn viral infections, Armed Forces Epidemiol Bd, 70-73. *Mem:* AAAS; Am Soc Microbiol; Soc Exp Biol & Med; Am Leptospirosis Res Conf; fel Am Acad Microbiol; fel Infectious Dis Soc Am. *Res:* Biology of pathogenic spirochetes. *Mailing Add:* Dept Microbiol Univ Minn Med Sch Minneapolis MN 55455

JOHNSON, RUSSELL DEE, JR, b Granite City, Ill, Dec 10, 28; m 53; c 4. OPERATIONS RESEARCH. *Educ:* Univ Rochester, BS, 50; Univ Calif, PhD(phys chem), 54. *Prof Exp:* Chemist, Dow Chem Co, Mich, 53-56; physicist, 56-62, SR SCIENTIST, OPERS RES, INC, 62- *Mem:* Am Chem Soc; Sigma Xi. *Res:* Weapons systems analysis; applied game theory. *Mailing Add:* 1902 Ventura Ave Silver Spring MD 20902

JOHNSON, SAMUEL BRITTON, b Canyon, Tex, Apr 25, 26; m 49; c 3. OPHTHALMOLOGY. *Educ:* WTex State Col, BS, 46; Tulane Univ, 48, dipl ophthal, 50; Am Bd Ophthal, dipl. *Prof Exp:* PROF & CHMN, OPHTHAL DEPT, UNIV MISS MED CTR, 56- *Concurrent Pos:* Trustee, Miss Sch for Deaf, Miss Sch for Blind & Miss Voc Rehab for Blind; lectr, Law Sci Acad, 59-81; asst prof eye path, Univ Miss. *Mem:* Am Col Surgeons; Asn Res Vision & Ophthal; Sigma Xi. *Res:* Eye pathology. *Mailing Add:* Dept Ophthal McBryde Eye Rehab Bldg Univ Miss 2500 N State St Jackson MS 39216

JOHNSON, SAMUEL EDGAR, II, b San Jose, Calif, Sept 27, 44; m 70. MARINE ECOLOGY. *Educ:* Stanford Univ, BS, 66, PhD(biol), 73. *Prof Exp:* Scholar biophys ecol, Dept Bot, Univ Mich, 72-73; asst prof zool, Clark Univ, 73-81; exec dir, Nat Conservancy, 81-87; dir spec projs, Ore Hist Soc, 87-89; DIR PLANNED GIVING, ORE HEALTH SCI UNIV FOUND, 91- *Concurrent Pos:* Res assoc, New Eng Res Inc, 73- *Honors & Awards:* Arthur C Giese Award, Stanford Univ, 73. *Mem:* Am Soc Zoologists; Ecol Soc Am; Am Meteorol Soc; Sigma Xi; Int Biometeorol Soc. *Res:* Biophysical ecology, microclimatology and biometeorology of the marine rocky intertidal region with emphasis on heat and mass transfer processes as they affect intertidal organisms, particularly amphipods and molluscs; estuarine ecology and coastal zone resource management. *Mailing Add:* 1449 SW Davenport St Portland OR 97201

JOHNSON, SANDRA LEE, b Scottsbluff, Nebr, Dec 27, 52. TOPOLOGICAL GRAPH THEORY, CAD-CAM APPROXIMATION THEORY. *Educ:* Nebr Wesleyan Univ, AB, 74; Clemson Univ, MS, 76; Ohio State Univ, PhD(math), 82. *Prof Exp:* Instr math, Univ Kans, 82-84; MATHEMATICIAN, IBM CORP, 84- *Mem:* Asn Women Math; Soc Indust & Appl Math; Am Math Soc. *Res:* Network topology; linear and spline approximations. *Mailing Add:* PO Box 6154 Kingston NY 12401

JOHNSON, SHIRLEY MAE, b Ironwood, Mich, May 26, 40; m 75. REPRODUCTIVE PHYSIOLOGY. *Educ:* Northern Mich Col, BS, 62; Mich State Univ, MS, 65, PhD(physiol), 70; Univ Mich, MPH, 72. *Prof Exp:* Teacher pub sch, Grand Rapids, Mich, 62-63; lab technician, Endocrine Res Unit, 65-70, asst to vpres, Off Res Develop, 71, UNIV PROF FAMILY MED, COL OSTEOP MED, MICH STATE UNIV, 72- *Concurrent Pos:* Res consult, Mich Cancer Found, 70; educ consult, Tri-County Family Planning Ctr, Lansing, Mich, 75- *Mem:* Am Pub Health Asn; Am Asn Sex Educr, Counr & Therapists; Sigma Xi. *Res:* Influence on health care of knowledge, attitudes, concerns and beliefs patients have toward reproductive physiology and family planning; contraceptive use and advertising. *Mailing Add:* Dept Family Med Mich State Univ East Lansing MI 48824

JOHNSON, STANLEY HARRIS, b Fresno, Calif, Dec 3, 38; m 65; c 1. AUTOMATIC CONTROL SYSTEMS. *Educ:* Univ Calif, Berkeley, BS, 62, MS, 67, PhD(mech eng), 73. *Prof Exp:* Design engr physics res, Lawrence Radiation Lab, 61-65; syst engr comput sales, Int Bus Mach Co, 65-67; sr engr comput control, Mobil Res & Develop Corp, 67-70; from asst prof to assoc prof, 73-79, PROF, DEPT MECH ENG & MECH, LEHIGH UNIV, 79- *Concurrent Pos:* Fac fel, Dryden Flight Res Ctr, 74 & 75; DuPont assoc prof, DuPont Univ Sci & Eng grant, 78-80. *Mem:* Am Soc Mech Engrs; Am Asn Univ Prof. *Res:* Numerical simulation of dynamical systems; development of the methodology of simulation; simulation validity and verification; numerical solution of partial differential equations; application of optimal control theory. *Mailing Add:* Dept Mech Eng & Mech Lab 19 Lehigh Univ Bethlehem PA 18015

JOHNSON, STANLEY O(WEN), b Bismarck, NDak, Dec 28, 30; div; c 2. NUCLEAR ENGINEERING. *Educ:* Univ Colo, BS, 53. *Prof Exp:* Student engr, Westinghouse Elec Corp, 53-54, scientist, Bettis Atomic Power Lab, 54-60, supvr nuclear reactor kinetics, 60-61; group leader reactor safety & dynamics anal, Atomic Energy Div, Phillips Petrol Co, 61-63, sect chief anal & data processing, 63-68, mgr, Spert Proj, 68-69; mgr, Idaho Nuclear Corp, 69-71; mgr, Aerojet Nuclear Co, 71-73; pres, Intermountain Technol, Inc, 73-86; dir, ITI-Japan, Inc, 81-86; VPRES, ROCKWOOD GROWTH FUND INC, 85-; VPRES & DIR, ASPEN SECURITY ADV INC, 85- *Concurrent Pos:* Dir, Ene-Con, Inc, 78-81. *Mem:* Fel Am Nuclear Soc; Nat Soc Prof Engrs. *Res:* Nuclear reactor safety research; nuclear reactor dynamics; computer simulation of nuclear reactors. *Mailing Add:* 1312 Azalea Dr Idaho Falls ID 83404

JOHNSON, STEPHEN ALLEN, b Worcester, Mass, April 26, 48; m 70; c 2. COMBUSTION RESEARCH & DEVELOPMENT. *Educ:* Worcester Polytech Inst, BS, 70. *Prof Exp:* Res & develop, E F Laurence Mfg Co, 70-71; sr res engr, Riley Stoker Corp, 71-76; group supvr, Babcock & Wilcox Co, 76-81; prog mgr, Sci Applns, Inc, 81-83; area mgr, 83-90, VPRES, APPL COMBUSTION TECHNOLS, PHYS SCI INC, 90- *Mem:* Am Inst Chem Engrs; Combustion Inst. *Res:* Developed advanced combustion processes to achieve 80 percent reduction; exploring mineral matter transformations in flames to predict and control ash deposition problems in large furnaces; developing processes to control emissions in coal-fueled diesel and gas turbine engines and waste incinerators; effects of fuels on equipment operation; development of technologies to control emission of nitrogen oxides; sulfur dioxide & airbone toxic. *Mailing Add:* 20 New EngBus Ctr Andover MA 01810

JOHNSON, STEPHEN CURTIS, b Philadelphia, Pa, July 13, 44; m 68; c 2. COMPUTER SCIENCE. *Educ:* Haverford Col, AB, 63; Columbia Univ, MS, 64, PhD(math), 68. *Prof Exp:* MEM TECH STAFF COMPUT SCI, BELL TEL LABS, 68- *Mem:* AAAS; Asn Comput Mach; NY Acad Sci; Sigma Xi. *Res:* Cluster analysis; computerized algebra systems; design and implementation of computer languages and systems; integrated circuit design. *Mailing Add:* Bell Lab Rm 3B-415 AT&T 600 Mountain Ave Murray Hill NJ 07974

JOHNSON, STEPHEN THOMAS, b Washington, DC, May 31, 54; m 83. TOOLING ENGINEERING. *Educ:* Northeastern Univ, BMET, 78. *Prof Exp:* Draftsman, Hollingsworth & Vose, 73-74; tech aide, US Army Natick Res & Develop, 75-78; tool designer, Boeing Aircraft Co. 78-81, propulsion engr, 81; SR TOOL DESIGNER, SIKORSKY AIRCRAFT CO, UNITED TECHNOLOGIES, 81- *Mem:* Am Soc Metals; Am Soc Mech Eng. *Res:* Advanced tooling and fabrication concepts for composite aircraft parts. *Mailing Add:* 260 Ferndale Ave Stratford CT 06497

JOHNSON, STURE ARCHIE MANSFIELD, b Morgan, Ore, Apr 24, 07; m 36. MEDICINE. *Educ:* Univ Ore, BA, 34, MD, 38; Am Bd Dermat & Syphilol, dipl. *Prof Exp:* Fel dermat & syphilol, Univ Ore, 39-41; res assoc, Univ Mich, 41-44, asst prof, 44-46; prof dermat & syphilol & head dept, 46-77, EMER PROF DERMAT & SYPHILOL, UNIV WIS-MADISON, 77- *Concurrent Pos:* Consult, Vet Admin Hosp, Madison. *Mem:* Soc Invest Dermat; Am Dermat Asn; AMA; fel Am Col Physicians; fel Am Acad Dermat. *Res:* Mycology; syphilis; dermatology. *Mailing Add:* 10306 Hutton Dr Sun City AZ 85351

JOHNSON, SUSAN BISSETTE, b Austin, Minn, July 20, 51; m 77. FATTY ACIDS. *Educ:* Mankato State Univ, BS, 73. *Prof Exp:* Jr scientist, Hormel Inst, Univ Minn, 73-77, from asst to assoc, 77-85, scientist, 85-91; RES TECH MAY CLINIC, MINNEAPOLIS, 91- *Concurrent Pos:* Consult, Travenol Lab, 84-85. *Res:* Metabolism of fatty acids in normal and disease conditions. *Mailing Add:* 200 First St SW Rochester MN 55905

JOHNSON, SYLVIA MARIAN, b Sydney, Australia, Aug 29, 54; m 85; c 2. SYNTHESIS & PROCESSING. *Educ:* Univ New South Wales, BSc Hons, 77; Univ Calif, Berkeley, MS, 79, PhD(eng & mat sci), 83. *Prof Exp:* Mat scientist, 82-86, sr mat scientist, 86-88, PROG MGR CERAMICS, SRI INT, 88- *Mem:* Am Ceramic Soc; AAAS; Metall Soc. *Res:* Synthesis of oxide and non-oxide ceramic powders; processing of ceramics, especially silicon nitride; characterization and evaluation of structural ceramics; joining of ceramics. *Mailing Add:* SRI Int 333 Ravenswood Ave Menlo Park CA 94025-3493

JOHNSON, TERRELL KENT, b Inglewood, Calif, Nov 23, 47; m 80; c 1. GENETICS. *Educ:* Univ Calif, San Diego, BA, 70; Calif State Univ, Northridge, MS, 72; Univ Tex, Austin, PhD(zool), 76. *Prof Exp:* Res fel genetics, Calif Inst Technol, 76-77; USPHS trainee, 77-79, RES ASSOC GENETICS, KANS STATE UNIV, 79- *Mem:* Genetics Soc Am; Soc Develop Biol; Sigma Xi. *Mailing Add:* Dept Tissue Cult Sigma Chem Co PO Box 14508 St Louis MO 63178

JOHNSON, TERRY CHARLES, b St Paul, Minn, Aug 8, 36; m 58; c 3. MOLECULAR BIOLOGY, VIROLOGY. *Educ:* Hamline Univ, BS, 58; Univ Minn, Minneapolis, MS, 61, PhD(microbiol), 64. *Prof Exp:* USPHS res asst, Univ Minn, Minneapolis, 58-64; USPHS fel molecular biol, Univ Calif, Irvine, 64-66; from asst prof to assoc prof, Med Sch, Northwestern Univ, Chicago, 66-73, prof virol, 73-77; PROF & DIR, DIV BIOL, KANS STATE UNIV, 77- *Concurrent Pos:* Co-dir, Bioserve Space Technol; consult, Digene, Col Park, Md. *Mem:* AAAS; Teratology Soc; Am Soc Neurochem; Am Soc Microbiol; Sigma Xi; NY Acad Sci. *Res:* Developmental aspects of macromolecular synthesis of mammalian cells and its role in host resistance to viral infection; regulation of protein and nucleic acid metabolism in brain; neural development. *Mailing Add:* 205 S Drake Manhattan KS 66502

JOHNSON, TERRY R(OBERT), b Chicago, Ill, Nov 16, 32; m 56; c 4. NUCLEAR FUEL CYCLE. *Educ:* Rice Univ, BA, 54, BS, 55; Univ Mich, MS, 56, PhD(chem eng), 59. *Prof Exp:* Assoc chem engr, Argonne Nat Lab, 58-74; sr process engr, Aglomet, Inc, 74-75; CHEM ENGR, ARGONNE NAT LAB, 75- *Concurrent Pos:* Vis prof, Iowa State Univ, 70. *Mem:* Am Inst Chem Engrs; Am Nuclear Soc; Sigma Xi. *Res:* Radiation chemistry of aqueous systems; nuclear fuel recovery; chemistry of liquid metals and salts; open-cycle MHD. *Mailing Add:* 1424 S Main Wheaton IL 60187-6482

JOHNSON, TERRY WALTER, JR, b Waukegan, Ill, Jan 13, 23; m 48; c 3. MYCOLOGY. *Educ:* Univ Ill, BS, 48; Univ Mich, MS, 49, PhD(bot), 51. *Prof Exp:* Instr bot, Univ Mich, 50-51; mycologist, Chem Corps Biol Labs, Camp Detrick, 51-53; asst prof biol, Univ Miss, 53-54; from asst prof to prof bot, Duke Univ, 54-85, chmn dept, 63-71. *Concurrent Pos:* Guggenheim fel, 60-61; mem systs panel, NSF, 63-66; ed-in-chief, Mycologia, 81- *Mem:* Bot Soc Am; Mycol Soc Am; Brit Mycol Soc. *Res:* Aquatic phycomycetes; Mycetozoa; marine fungi. *Mailing Add:* 3505 Manford Dr Durham NC 27707

JOHNSON, THEODORE REYNOLD, b Willmar, Minn, Mar 20, 46; m 70; c 3. AGING, CANCER BIOLOGY. *Educ:* Augsburg Col, BA, 68; Univ Ill Med Ctr, MS, 70, PhD(microbiol), 73. *Prof Exp:* Res asst microbiol, Rush-Presby St Lukes Hosp, 68-72; asst prof biol, Mankato State Univ, 72-77; PROF BIOL, ST OLAF COL, 77- *Concurrent Pos:* Consult, St Joseph's Hosp, 74-77 & Donaldson Corp, 80-; vis sci, Trudeau Inst, 83-84. *Mem:* Am Soc Microbiol; AAAS; Sigma Xi. *Res:* Study of the immune response of humans to colonic cancer and screening the excretory products for carcinogens; cancer and immune systems of hibernating animals; aging and cancer immunity. *Mailing Add:* Dept Biol St Olaf Col Northfield MN 55057

JOHNSON, THOMAS, b Halletsville, Tex, Feb 12, 36; m 56; c 3. ECONOMETRICS, RESOURCE ECONOMICS. *Educ:* Univ Tex, Austin, BA, 57; Tex Christian Univ, MA, 62; NC State Univ, MES, 67, PhD(economet & statist), 69. *Prof Exp:* Nuclear engr, Convair, Fortworth, Tex, 57-61; eng specialist, LTV-Vought Aeronaut Div, Dallas, Tex, 61-64; oper analyst, Res Triangle Inst, 64-69; assoc prof, 74-78, PROF ECON & STATIST, NC STATE UNIV, 78- *Concurrent Pos:* Asst prof econ & statist, 69-74, assoc prof econ & statist, Southern Methodist Univ, Dallas, Tex, 74. *Mem:* Am Statist Asn; Am Econ Asn; Economet Soc; Am Econ Asn. *Res:* Statistics and mathematics applications to economic questions; analysis of dynamics of economic and biological systems. *Mailing Add:* Dept Agr & Resource Econ NC State Univ Raleigh NC 27695

JOHNSON, THOMAS CHARLES, b Virginia, Minn, Aug 15, 44; m 66; c 2. LIMNOLOGY. *Educ:* Univ Wash, BS, 67; Univ Calif, San Diego, PhD(oceanog), 75. *Prof Exp:* Officer, Off Res & Develop, US Coast Guard, 69-71; res asst, Scripps Inst Oceanog, 71-74, fel, 75; asst prof geol, Univ Minn, Minneapolis, 75-80, assoc prof, 80-81; assoc prof geol, Univ Minn, Duluth, 81-83; assoc prof, 83-89, DIR, OCEANOG CONSORTIUM LAB, DUKE UNIV NC, 83-, PROF, GEOL DEPT, DUKE UNIV MARINE LAB, 89- *Concurrent Pos:* Prin investr, res grants, NSF, Environ Protection Agency & Nat Oceanic & Atmospheric Admin, 75-, Dept Energy, 78-80, Oil Indust, 85-; vis res fel, Chancellor Col, Univ Malawi, Zomba, E Africa, 85, res fel, Lynde & Harry Bradley Found, 88-91. *Honors & Awards:* George C Taylor Award, Univ Minn, 81. *Mem:* AAAS; Geol Soc Am; Am Geophys Union. *Res:* Sedimentological processes in large lakes, including physical processes in deep offshore regions and chemical processes including seismic reflection and side scan sonar profiling, sediment coring. *Mailing Add:* Duke Univ Marine Lab Beaufort NC 28516

JOHNSON, THOMAS EUGENE, b Denver, Colo, June 19, 48; m; c 2. DEVELOPMENTAL GENETICS. *Educ:* Mass Inst Technol, BS, 70; Univ Wash, PhD(genetics), 75. *Prof Exp:* Fel genetics, Cornell Univ, 75-77; fel molecular, cellular & develop biol, Univ Colo, 77-; AT DEPT MOLECULAR BIOL & BIOCHEM, UNIV CALIF. *Mem:* Genetics Soc Am; Soc Develop Biol; AAAS; fel Geront Soc Am; Am Aging Asn; Am Fedn Aging Res. *Res:* Genetics of aging in the nematode; genetics of alcohol sensitivity in the mouse. *Mailing Add:* Lab Molecular Genetic Inst Behav Genetics Univ Colo Boulder CO 80309-0447

JOHNSON, THOMAS F, b Philadelphia, Pa, Mar 10, 17. SPORTS MEDICINE. *Educ:* Springfield Col, BS, 40; NY Univ, MA, 46; Univ Md, PhD(phys educ), 67. *Prof Exp:* Assoc Dean, Grad Sch, Howard Univ, 74-78; RETIRED. *Mem:* Am Physiol Soc. *Mailing Add:* 130 Ingraham St NW Washington DC 20011

JOHNSON, THOMAS HAWKINS, physics; deceased, see previous edition for last biography

JOHNSON, THOMAS L, biology, for more information see previous edition

JOHNSON, THOMAS LYNN, b Westmount, Que, June 20, 19; nat US; m 56. ORGANIC CHEMISTRY. *Educ:* Amherst Col, BA, 42; Univ Wis, MS, 44, PhD(org chem), 46. *Prof Exp:* Patent agent, Sterling-Winthrop Res Inst, 46-90; RETIRED. *Mem:* Am Chem Soc. *Res:* Medicinal chemistry; steroids; patent law. *Mailing Add:* 18 Stonehenge Lane Albany NY 12208-2233

JOHNSON, THOMAS NICK, b Davenport, Iowa, Aug 20, 23; m 55. NEUROANATOMY. *Educ:* St Ambrose Col, BS, 44; Mich State Col, MS, 49; Univ Mich, PhD(neuroanat), 52. *Prof Exp:* From instr to asst prof physiol, Mich State Col, 46-54; from asst prof to assoc prof, 54-69, PROF ANAT, SCH MED, GEORGE WASHINGTON UNIV, 69- *Concurrent Pos:* Nat Inst Neurol Dis & Blindness spec trainee, Dept Anat, Med Ctr, Univ Calif, Los Angeles, 58-59. *Mem:* AAAS; Am Asn Anatomists; assoc Am Acad Neurologists. *Res:* Comparative neurology of mammalian midbrain and forebrain; experimental neurology of extrapyramidal motor systems; electron microscopy of human primary brain tumors. *Mailing Add:* 2300 First NW Washington DC 20006

JOHNSON, THOMAS RAYMOND, b Washington, DC, July 8, 44; m 73; c 2. CELL BIOLOGY. *Educ:* Harvard Univ, BA, 66; Case Western Reserve Univ, PhD(biol), 71. *Prof Exp:* Instr, Univ Ill, Chicago Med Ctr, 71-73; SR RES ASSOC, CASE WESTERN RESERVE UNIV, 73- *Res:* Expression of insulin gene family; structure and function of messenger ribonucleic proteins. *Mailing Add:* 3062 Huntington Shaker Heights OH 44120

JOHNSON, THYS B(RENTWOOD), b Duluth, Minn, Mar 20, 34; m 58; c 3. MINING ENGINEERING, OPERATIONS RESEARCH. *Educ:* Univ Minn-Minneapolis, BS, 56, MS, 58; Univ Calif, Berkeley, PhD(opers res), 68. *Prof Exp:* Mining engr, Minn Ore Opers, US Steel Corp, 58-61, mathematician, 61-64; mining methods res engr, US Bur Mines, 64-68, mining engr, 68, supvry mining engr, 68-69, supvry opers res analyst, 69-72; head dept, 74-85, PROF MINING ENG, COLO SCH MINES, 72-; DIR APPL RES, NAT RESOURCES RES INST & PROF INDUST ENG, UNIV MINN, DULUTH, 85- *Mem:* Opers Res Soc Am; Am Inst Mining, Metall & Petrol Engrs. *Res:* Research and development of operations research techniques as applied to problems of the mineral industry; developed mathematical and dynamic programming techniques for open pit mine planning and production scheduling. *Mailing Add:* 4354 Turner Rd Duluth MN 55803

JOHNSON, TIMOTHY JOHN ALBERT, MEMBRANE BIOCHEMISTRY. *Educ:* Univ Wis, PhD(biochem), 74. *Prof Exp:* ASST PROF, COLO STATE UNIV, 79- *Res:* Gluteraldehyde fixations; collidal gold labbing of membrane proteins. *Mailing Add:* Dept Anat Colo State Univ Ft Collins CO 80523

JOHNSON, TIMOTHY WALTER, b Newington, Conn, Sept 17, 41; m 66; c 2. PHYSICAL CHEMISTRY & POLYMER CHEMISTRY. *Educ:* Trinity Col, BS, 63, MS, 65; Purdue Univ, PhD, 70. *Prof Exp:* Res assoc, Northwestern Univ, 69-73; res chemist, 73-81, RES ASSOC, PHILLIPS PETROLEUM CO, 81- *Mem:* Sigma Xi; Am Chem Soc; Soc Plastics Engrs; Adhesion Soc. *Res:* Physical chemistry of polymer solutions; polymer rheology; electrical properties of polymers; electrically conductive polymers; polymer morphology; polymer properties; polymer composites; composite interfaces; surface science; thermodynamics and material properties; adhesion. *Mailing Add:* Res & Develop Dept Chem & Polymers Lab Phillips Petroleum Co Bartlesville OK 74004

JOHNSON, TOM MILROY, b Northville, Mich, Jan 16, 35; m 59; c 2. INTERNAL MEDICINE. *Educ:* Col Wooster, BA, 56; Northwestern Univ, Ill, MD, 61. *Prof Exp:* Am Thoracic Soc fel pulmonary dis, Med Ctr, Univ Mich, 67-68; asst prof med, 68-71, assoc prof med, Col Human Med, Mich State Univ, 71-77; PROF INTERNAL MED & DEAN SCH MED, UNIV NDAK, 77- *Concurrent Pos:* Asst dean, Grand Rapids Campus, Univ Mich, 71-77. *Mem:* Am Thoracic Soc; fel Am Col Physicians; Am Col Chest Physicians. *Res:* Relationship of community and university medical education; pulmonary disease. *Mailing Add:* A-108 E Fee Hall East Lansing MI 48824

JOHNSON, TORRENCE VAINO, b Rockville Centre, NY, Dec 1, 44; m 67; c 2. PLANETARY SCIENCES, ASTRONOMY. *Educ:* Washington Univ, BS, 66; Calif Technol, PhD(planetary sci), 70. *Prof Exp:* Mem res staff planetary astron, Planetary Astron Lab, Mass Inst Technol, 69-71; Nat Res Coun resident res assoc planetology, Calif Inst Technol, 71-73, sr scientist, 73-74, group supvr optical astron group, 74-85, res scientist, 80-81, SR RES SCIENTIST, JET PROPULSION LAB, CALIF INST TECHNOL, 81- *Concurrent Pos:* Mem Uranus sci adv comt, NASA, 73-75, mem outer planets probe working group, 74-76; scientist, Proj Galileo, NASA, 77- & mem, Voyager Imaging Sci Team, 78-; vis assoc prof planetary sci, Calif Inst Technol, 81-83; pres, Planetology Sect, Am Geophys Union, 90-92. *Honors & Awards:* Sci Achievement Medal, NASA, 80 & 81; Fel, Am Geophys Union. *Mem:* Sigma Xi; AAAS; Am Astron Soc (secy-treas, 77-); Int Astron Union; Am Geophys Union. *Res:* Telescopic observations of planetary surfaces and atmospheres; laboratory studies of silicates and ices; interpretation of planetary spacecraft data. *Mailing Add:* Jet Propulsion Lab 183-501 4800 Oak Grove Dr Pasadena CA 91109

JOHNSON, VARD HAYES, b Pleasant Grove, Utah, Sept 17, 09; m 46; c 2. GEOLOGY. *Educ:* Brigham Young Univ, BS, 32, MS, 33; Univ Ariz, PhD(econ geol), 41. *Prof Exp:* Student instr mineral & geol, Brigham Young Univ, 31-33; jr engr & geologist, New Park Mines Co, Utah, 36 & Park City Consol Mining, 37-39; recorder, US Geol Surv, Ariz, 40-42, geologist, Washington, DC, 43-53; geologist, Columbia Geneva Steel, 53-64; CONSULT GEOLOGIST, 64- *Mem:* Geol Soc Am; Am Inst Prof Geol; Am Inst Mining, Metall & Petrol Eng; Can Inst Mining & Metall. *Res:* Geology of coal, especially coking coals of western North America; geology and ore deposits of Helvetia mining district, Pima County, Arizona; ferrous and non-ferrous ores of western North America. *Mailing Add:* 2784 Bryant Palo Alto CA 94306

JOHNSON, VERN RAY, b Salt Lake City, Utah, Feb 25, 37; m 59; c 4. EDUCATIONAL ADMINISTRATION, SOCIAL TECHNOLOGY. *Educ:* Univ Utah, BS, 60, PhD(elec eng, physics), 65. *Prof Exp:* Res asst, Microwave Devices Lab, Utah, 60-64; res engr, Microwave Electronics Div, Teledyne, Inc, 64-67; assoc prof elec eng, 67-79, ASSOC DEAN, COL ENG & MINES, UNIV ARIZ, 79- *Mem:* Inst Elec & Electronics Engrs; Am Soc Eng Educ. *Res:* Microwave acoustic amplification; photoelastic interactions in solid materials; surface wave acoustics; engineering manpower system simulation and demand projections; communication; application of engineering techniques to social problems. *Mailing Add:* Col Eng & Mines Univ Ariz Tucson AZ 85721

JOHNSON, VERNER CARL, b Chicago, Ill, Sept 14, 43. EXPLORATION GEOLOGY, EXPLORATION GEOPHYSICS. *Educ:* Southern Ill Univ, BA, 67, MS, 70; Univ Tenn, PhD(geol), 75. *Prof Exp:* Instr, Calif State Univ, 72-74; proj geophysicist, Gulf Res & Develop Corp, 74-76; asst prof, Mesa Col, 76-77; GEOLOGIST, BENDIX FIELD ENG CORP, 77- *Concurrent Pos:* Asst prof, Mesa Col, 77- *Mem:* Am Asn Petrol Geologists; Am Geophys Union; Geol Soc Am; Soc Explor Geophysicists. *Res:* Geologic information; rock formation in field; determination of favorable environment for uranium occurrences. *Mailing Add:* Mesa State Col PO Box 2647 Grand Junction CO 81502

JOHNSON, VINCENT ARNOLD, b York, Nebr, Jan 5, 28; m 53; c 3. ZOOLOGY, PHYSIOLOGY. *Educ:* Univ Nebr, BSc, 52, MSc, 55, PhD(zool, physiol), 64. *Prof Exp:* Spec instr biol, Univ Tex, 57-61; asst prof, Augustana Col, Ill, 64-67; from asst prof to assoc prof, 67-72, PROF BIOL, ST CLOUD STATE UNIV, 72- *Mem:* Am Soc Zool; Soc Protozoologists; AAAS; Sigma Xi. *Res:* Cellular growth and metabolism. *Mailing Add:* Dept Biol Sci St Cloud State Univ St Cloud MN 56301

JOHNSON, VIRGIL ALLEN, b Newman Grove, Nebr, June 28, 21; m 43; c 4. AGRONOMY. *Educ:* Univ Nebr, BSc, 48, PhD(agron), 52. *Prof Exp:* Agent, Agr Res Serv, USDA, 51-52, res agronomist, 54-75, supvr res agronomist, 75-78; from asst prof to assoc prof, 52-63, PROF AGRON, UNIV NEBR-LINCOLN, 62-; LEADER, WHEAT RES, SCI & EDUC ADMIN-APPL RES, USDA, 78- *Concurrent Pos:* Organizer, Int Winter Wheat Res & Eval Network, AID, 68; mem, comt genetic vulnerability of major crops, Nat Acad Sci, 71-72; Nat Wheat Indust resource comt, 72-75 & US Nat Comt Int Union Biol Sci, Nat Res Coun, 79-84; co-chmn, resource comt cereals, protein resources study, Mass Inst Technol, 75; prin organizer, Int Wheat Conf, Turkey, 72, Brazil, 74, Yugoslavia, 75 & Spain, 80; mem, bd dirs, Am Soc Agron, 77-79; mem, Agr Sci Panel, 80- *Honors & Awards:* Crop Sci Award, Crop Sci Soc Am, 75; Int Agron Award, Am Soc Agron, 84; DeKalb-Pfizer Crop Sci Award, 85. *Mem:* NY Acad Sci; fel Am Soc Agron; Am Genetics Asn; fel AAAS; Crop Sci Soc Am (pres, 78); Sigma Xi. *Res:* Wheat breeding and genetics; genetics and physiology of wheat; protein quantity and nutritional quality. *Mailing Add:* 3849 Dudley Lincoln NE 68503

JOHNSON, W(ILLIAM) C, b Jamestown, NY, Aug 20, 27; m 55; c 3. CHEMICAL ENGINEERING. *Educ:* Univ Wis, BS, 51, MS, 53, PhD(chem eng), 60. *Prof Exp:* Instr chem eng, Univ Wis, 53-58; sr engr, 58-63, chem engr res specialist, 63-75, SR ENGR SPECIALIST, MINN MINING & MFG CO, 75- *Concurrent Pos:* Adj prof chem eng, Univ Minn, 77- *Mem:* Am Chem Soc; Am Inst Chem Engrs. *Res:* Organic chemical purification and separations; reaction kinetics and catalysis; ultraviolet light catalyzed polymerization. *Mailing Add:* Minn Mining & Mfg Co 3M Center 518-1 St Paul MN 55144

JOHNSON, W REED, b Chattanooga, Tenn, Sept 3, 31; m 56; c 3. NUCLEAR ENGINEERING. *Educ:* Va Mil Inst, BS, 53; Univ Va, DSc(eng physics), 62. *Prof Exp:* Shielding engr, Elec Boat Div, Gen Dynamics Corp, 54-55; nuclear engr, Alco Prod, Inc, 55-57; proj engr reactor facility, Va, 58-62, proj dir, Univ Va-Philippine Atomic Energy Comn Proj, 62-64; res dir, 64-66, assoc prof, 66-68, asst dir reactor facil, 74-77, PROF NUCLEAR ENG, UNIV VA, 68- *Concurrent Pos:* Proj engr, Div Reactor Licensing, US Atomic Energy Comn, 68-69; mem, Atomic Safety & Licensing Appeal Bd, 74- *Mem:* Fel Am Nuclear Soc; Am Soc Eng Educ. *Res:* Radiation shielding; reactor safety; experimental engineering. *Mailing Add:* Dept Nuclear Eng Univ Va Charlottesville VA 22903

JOHNSON, W THOMAS, b Butte, Montana, 45; m 87. NUTRITIONAL BIOCHEMISTRY. *Educ:* Univ NDak, PhD(biochem), 76; Mont State Univ, BS(physics), 68. *Prof Exp:* Res prof biochem, 85-87, RES CHEMIST, HUMAN NUTRIT RES CTR, USDA, 87- *Mem:* Am Inst Nutrit; Am Soc Biochem & Molecular Biol; AAAS. *Res:* Effects of nutrients on biological membranes; roles of nutrients in transmembrane signalling. *Mailing Add:* USDA Human Nutrit Res Ctr PO Box 7166 University Sta Grand Forks ND 58202

JOHNSON, WALLACE DELMAR, b Idaho Falls, Idaho, June 5, 39; m 58; c 6. ORGANIC CHEMISTRY. *Educ:* Brigham Young Univ, BS, 61; Univ Utah, PhD(org chem), 69. *Prof Exp:* Res chemist, 61-64 & 68-72, patent liaison, 72-84, LICENSING COORD, PHILLIPS PETROL CO, 84- *Mem:* Am Chem Soc; Sigma Xi; AAAS. *Res:* Organophosphorus chemistry; synthesis of rubbers and plastics. *Mailing Add:* 3700 Redbud Lane Bartlesville OK 74006

JOHNSON, WALLACE E, b Chisolm, Minn, Feb 28, 25; m 46; c 5. COMPUTER SCIENCE. *Educ:* Univ Minn, BS, 50. *Prof Exp:* Res engr, High Speed Flight Sta, Nat Adv Comt Aeronaut, 51-53; staff mem, Los Alamos Sci Lab, 53-58; staff mem, Gen Atomic Div, Gen Dynamics Corp, 58-59; staff mem, Los Alamos Sci Lab, 59-60; staff mem, Gen Atomic Div, Gen Dynamics Corp, 60-65; prin res scientist, Honeywell Inc, 65-67, Systs, Sci & Software, 67-71 & Sci Applns Inc, 71-73; CONSULT, COMPUT CODE CONSULTS, 73- *Mem:* Am Phys Soc; Am Sci Affiliation. *Res:* Use of high speed computers for the numerical treatment of radiation flow, neutronics and hydrodynamics; one, two and three dimensional hydrodynamic, strength of materials and radiation codes to solve problems in high energy fluid dynamics. *Mailing Add:* 114 Brompton Rd Garden City NY 11530

JOHNSON, WALLACE W, b LaMoure, NDak, Nov 23, 26; m 51; c 4. PHARMACOLOGY, DENTISTRY. *Educ:* NDak State Col, BS, 50; Univ Iowa, DDS, 57, MS, 58. *Prof Exp:* Asst dent, 57-58, from instr to assoc prof oper dent, 58-65, PROF OPER DENT, COL DENT, UNIV IOWA, 65- *Mem:* Am Dent Asn; Am Col Dentists; Am Asn Dent Res; Am Asn Dent Schs. *Res:* Drugs and their use in dentistry; educational research; dental materials research. *Mailing Add:* 720 Greenwood Dr Iowa City IA 52246

JOHNSON, WALTER C(URTIS), b Weikert, Pa, Jan 6, 13; m 34; c 3. ELECTRONICS. *Educ:* Pa State Col, BSE, 34, EE, 42. *Prof Exp:* Student elec engr, Gen Elec Co, NY, 34-37; from instr to prof, 37-81, chmn dept, 51-65, Arthur LeGrand Prof Eng, 63-81, EMER PROF ELEC ENG, PRINCETON UNIV, 81- *Concurrent Pos:* Resident vis, Bell Labs, 68. *Honors & Awards:* Western Elec Award, Am Soc Eng Educ, 67. *Mem:* Am Soc Eng Educ; Am Phys Soc; fel Inst Elec & Electronics Engrs. *Res:* Semiconductor materials and devices; charge transport and trapping in insulators; insulator reliability. *Mailing Add:* 20 McCosh Circle Princeton NJ 08540

JOHNSON, WALTER CURTIS, JR, b Princeton, NJ, Feb 11, 39; m 60; c 2. BIOPHYSICAL CHEMISTRY. *Educ:* Yale Univ, BA, 61; Univ Wash, PhD(phys chem), 66. *Prof Exp:* NSF fel, Univ Calif, Berkeley, 66-68; asst prof biophys, 68-72, assoc prof biophys, 72-78, PROF BIOPHYS, ORE STATE UNIV, 78- *Concurrent Pos:* NSF grant circular dichroism & conformation of biopolymers, 68-; USPHS grant, protein conformation & function, 74-; mem panel equip, NIH, 79, 83, 84; mem panel biol instrumentation, NSF, 80-82; mem BBCA panel, NIH, 88-; advisory bd Biopolymers. *Honors & Awards:* Milton Harris Award. *Mem:* Biophys Soc. *Res:* Spectroscopic properties of biopolymers, principally their circular dichroism, their conformation and resulting biological function. *Mailing Add:* Dept Biochem & Biophys Ore State Univ Corvallis OR 97331-6503

JOHNSON, WALTER HEINRICK, JR, b Minneapolis, Minn, Sept 20, 28; m 58; c 2. PHYSICS, MASS SPECTROMETRY. *Educ:* Univ Minn, BA, 50, MA, 52, PhD(physics), 56. *Prof Exp:* Res assoc, Univ Minn, 56-57; exp physicist, Knolls Atomic Power Lab, Gen Elec Co, 57-58; from asst prof to assoc prof , 58-68, actg chmn, dept physics, 69-70 & 83, assoc dean, 71-77, actg dean, 77-79, PROF PHYSICS, UNIV MINN, MINNEAPOLIS, 68- *Concurrent Pos:* Mem, Comn on Atomic Masses & Fundamental Constants, Int Union Pure & Appl Physics, 66-72, Comn on Atomic Weights, 71-85 & secy, 72-75. *Mem:* AAAS; fel Am Phys Soc; Am Vacuum Soc; Am Asn Physics Teachers. *Res:* Mass spectroscopy; measurement of atomic masses; nuclear binding energy; neutron cross-section measurements. *Mailing Add:* Sch Physics & Astron 116 Church St SE Minneapolis MN 55455

JOHNSON, WALTER K, b Minneapolis, Minn, Aug 28, 23; m 50; c 3. ENVIRONMENTAL & CIVIL ENGINEERING. *Educ:* Univ Minn, BCE, 48, MSCE, 51, PhD(sanit eng), 63; Am Acad Environ Eng, dipl, 65. *Prof Exp:* Civil engr, Greeley & Hansen, Consult Engrs, 48-49; asst, Univ Minn, 49-51; sanit engr, Infilco, Inc, Ariz, 51-52 & Toltz, King, Duvall & Anderson, Consult Engrs, Minn, 52-55; lectr civil eng, Univ Minn, Minneapolis, 55-63, from asst prof to prof, 63-75; dir planning, Metro Waste Control Comn, St Paul, 75-; RETIRED. *Concurrent Pos:* Environ Protection Agency res fel, Brit Water Pollution Res Lab, Stevenage, Eng, 70. *Honors & Awards:* Radebaugh Award, Cent States Water Pollution Control Asn, 65. *Mem:* Am Soc Civil Engrs; Int Asn Water Pollution Res; Am Water Works Asn. *Res:* Biological treatment of waste waters and the removal of nitrogen and phosphorus from waste waters by biological and chemical means. *Mailing Add:* 5321 29th Ave S Minneapolis MN 55417

JOHNSON, WALTER LEE, b Greensboro, NC, May 23, 18; m 50; c 1. AGRONOMY. *Educ:* Agr & Tech Col NC, BS, 42; Univ Ill, MS, 47, PhD, 53. *Prof Exp:* Agronomist, Southern Univ, 47-50; prof agron, 53-77, head dept, 53-67, head dept earth & plant sci, 62-72, PROF & DIR, DIV AGR SCI, FLA A&M, 73- *Mem:* Am Soc Agron; Soil Sci Soc Am. *Res:* Field crops; soils. *Mailing Add:* 609 Gore Ave Tallahassee FL 32304

JOHNSON, WALTER RICHARD, b Richmond, Va, Feb 25, 29; m 52. PHYSICS. *Educ:* Univ Mich, BSE, 52, MS, 53, PhD(physics), 58. *Prof Exp:* Instr physics, Univ Mich, 57-58; from asst prof to assoc prof, 58-67, PROF PHYSICS, UNIV NOTRE DAME, 67- *Mem:* Am Physics Soc. *Res:* Hydrodynamics; atomic physics; quantum electrodynamics. *Mailing Add:* Dept Physics Univ Notre Dame Notre Dame IN 46556

JOHNSON, WALTER ROLAND, b Boston, Mass, Feb 10, 27; m 62; c 3. JET ENGINE TECHNOLOGIES, ELECTRONICS. *Educ:* Mass Inst Technol, BS, 58. *Prof Exp:* Metallurgist, Missile Systs Div, Raytheon Co, 58-69; METALLURGIST & CONSULT, 69- *Mem:* Am Soc Metals. *Res:* Aerospace; electronics. *Mailing Add:* 35 Norseman Ave Gloucester MA 01930

JOHNSON, WARREN THURSTON, b Charleston, WVa, Apr 22, 25; m 49; c 2. ENTOMOLOGY, PLANT PATHOLOGY. *Educ:* Morris Harvey Col, 47; Ohio State Univ, MS, 52; Univ Md, PhD(entom), 56. *Prof Exp:* Asst entom, USPHS, 48-50; from instr to asst prof, Col Agr, Univ Md, 52-62; from asst prof to assoc prof, 62-71, PROF ENTOM, COL AGR, CORNELL UNIV, 71- *Concurrent Pos:* Vis res prof, Univ Calif, Berkeley, 69-70; vis res scientist, Forest Pest Mgt Inst, Can Forest Serv, Ottawa, 77; vis prof, NC State Univ, 85-86. *Mem:* Entom Soc Am; Entom Soc Can; Sigma Xi; AAAS. *Res:* Biology and control of anthropods affecting woody ornamental plants. *Mailing Add:* 1444 Hanshaw Rd Ithaca NY 14850

JOHNSON, WARREN VICTOR, b Duluth, Minn, Sept 26, 51; m 73; c 3. BIOCHEMISTRY. *Educ:* Univ Minn, Duluth, BA, 73; Univ Wis-Milwaukee, MS, 78; Univ Iowa, PhD(biochem), 84. *Prof Exp:* Teacher sci & math, St Michael's Sch, Duluth, Minn, 74-75 & Strandquist High Sch, Minn, 75-76; teaching asst chem, Univ Wis-Milwaukee, 76-78; res asst biochem, Univ Iowa, 78-84; postdoctoral fel, Revlon Biotech Res Ctr, Rockville, Md, 84-86; postdoctoral assoc biochem, Univ Minn, Duluth, 86-87; ASST PROF CHEM, MOLECULAR BIOL & BIOCHEM, UNIV WIS-GREENBAY, 87- *Mem:* Am Soc Biochem & Molecular Biol; Am Chem Soc; Sigma Xi. *Res:* Structure, function and gene of the developmentally regulated glycoprotein fetuin; role of the carbohydrated moieties of glycoproteins; proteolytic processing; molecular diagnosis of phylogenetic relationships. *Mailing Add:* Univ Wis 2420 Nicolet Dr Green Bay WI 54311-7001

JOHNSON, WARREN W, b Ackerman, Miss, Jan 13, 23; m 52; c 1. PATHOLOGY. *Educ:* Millsaps Col, BS, 50; Univ Miss, MS, 52, MD, 57. *Prof Exp:* Instr anat, Univ Miss, 52-55; from instr to asst prof path, Col Med, Univ Tenn, Memphis, 64-79; PROF PATH, SCH MED, UNIV MISS, 79- *Mem:* AMA; NY Acad Sci; Int Acad Path; Am Asn Cancer Res. *Res:* Ultrastructural diagnosis of neoplasia; immunohistologic study of tumors containing myosin; collagenous tumors; complications in leukemia; pituitary cytology in steroid therapy; pathology of neuromuscular diseases. *Mailing Add:* PO Box 219 Terry MS 39216

JOHNSON, WAYNE DOUGLAS, m 72; c 3. ELECTRON TRANSPORT. *Educ:* Lebanon Valley Col, BS, 73; Univ Del, PhD(physics), 78. *Prof Exp:* Res fel, Univ Pa, 78-79; PROJ LEADER, DOW CHEM CO, 79- *Mem:* Am Phys Soc. *Res:* Electron transport. *Mailing Add:* 618 White Thorn Dr Midland MI 48640

JOHNSON, WAYNE JON, b Elroy, Wis, May 14, 39; m 64; c 3. AUTOMOTIVE ENGINEERING. *Educ:* Univ Wis, BS, 61, MS, 62, PhD(elec eng), 68. *Prof Exp:* Engr, Res Dept, Collins Radio Co, Cedar Rapids, 60 & 61, res engr, 62-64; sr res engr, Lab di Cibernetica, Naples, Italy, 68-69; sr res engr, 69-73, prin res eng assoc, Dept Physics, 73-82, prin staff engr, Electronic Syst Dept, 82-87, MGR, CONTROL SYST DEPT, FORD MOTOR CO, 87- *Mem:* Inst Elec & Electronics Engrs; Soc Automotive Engrs; Eng Soc Detroit. *Res:* Nonlinear wave propagation; superconducting devices; semiconductor device physics; combustion research on internal combustion engines; plasma probing techniques; electromagnetic interference phenomena; networking and distributed computing techniques applied to the automobile; dynamic control systems. *Mailing Add:* Mgr Ford Motor Co Res Lab S-2093 PO Box 2053 Dearborn Heights MI 48121

JOHNSON, WAYNE ORRIN, b Valley City, NDak, May 26, 42; m 65; c 1. AGRICULTURAL CHEMISTRY. *Educ:* Concordia Col, BA, 64; Mich State Univ, MS, 66; Univ Ore, PhD(org chem), 69. *Prof Exp:* Group leader, 69-76, mgr res farms & liaison activ, 76-79, mgr hybrid crops, 79-83, pres, Rohm & Haas Seeds, 83-85, RES MGT BIOCIDES & SPEC POLYMERS, AGR PROD RES, AGR CHEM, ROHM & HAAS CO, 85- *Mem:* Am Chem Soc; Plant Growth Regulator Working Group; Am Seed Trade Asn. *Res:* Synthetic structure-activity chemistry related to biological sciences, especially pesticidal research; microbiology; polymer chemistry. *Mailing Add:* Four Beth Dr Gwynedd Hill Lower Gwynedd PA 19002

JOHNSON, WENDEL J, b Oak Park, Ill, July 13, 41; m 82; c 3. ANIMAL ECOLOGY, ZOOGEOGRAPHY. *Educ:* Mich State Univ, BS, 63, MS, 65; Purdue Univ, PhD(mammalian ecol), 69. *Prof Exp:* From asst prof to assoc prof, 69-86 PROF BIOL, UNIV WIS CTR-MARINETTE, 86- *Concurrent Pos:* Sigma Xi grant-in-aid, 64, 70; Wis Alumni Res Found fel, 70. *Mem:* AAAS; Ecol Soc Am; Am Soc Mammal. *Res:* Zoogeographical analysis of reptiles and amphibians in the Northern Peninsula of Michigan; population dynamics of small mammals in Isle Royale National Park; population regulation in small mammals; environmental problems from human numbers; Green Bay lampreys. *Mailing Add:* Dept Biol Univ Wis Ctr 750 Bay Shore Marinette WI 54143

JOHNSON, WENDELL GILBERT, b Wichita, Kans, Feb 19, 22; m 48; c 4. MATHEMATICS. *Educ:* Phillips Univ, BA, 47; Univ Mich, MA, 48; Syracuse Univ, PhD(math), 55. *Prof Exp:* Asst prof math, Phillips Univ, 48-50; asst, Syracuse Univ, 52-55; asst prof, Southern Ill Univ, 55-57; vpres & dean col, 64-69, PROF MATH, HIRAM COL, 57- *Concurrent Pos:* NSF sci fac fel, 61-62. *Mem:* Am Math Soc; Math Asn Am; Asn Comput Mach. *Res:* Algebra; geometry. *Mailing Add:* 88 Hamilton Rd Chapel Hill NC 27514

JOHNSON, WHITNEY LARSEN, b Brigham City, Utah, July 11, 27; m 54; c 11. STATISTICS, COMPUTER SCIENCE. *Educ:* Utah State Univ, BS, 54; Univ Minn, Minneapolis, MS, 57. *Prof Exp:* Assoc prof statist, Va Polytech Inst & State Univ, 62-68; adminr automated data processing systs, State Coun Higher Educ, Va, 69-72; dir mgt info serv, 72-86, DIR DATA & TEL SYSTS, NORTHERN MICH UNIV, 86- *Concurrent Pos:* Coordr comput ctr, Va Polytech Inst & State Univ, 62-64 & dir, 64-68. *Mem:* Am Statist Asn; Asn Comput Mach; Int Asn Comput Educ. *Res:* Moments of serial correlation coefficients and computing networks on regional and statewide basis; management information for education. *Mailing Add:* Data & Tel Systs Northern Mich Univ Marquette MI 49855

JOHNSON, WILBUR VANCE, b Bellingham, Wash, Jan 14, 31; m 53. CHEMICAL PHYSICS. *Educ:* Univ Wash, Seattle, BS, 53; Ore State Univ, PhD, 60. *Prof Exp:* Instr chem, Univ Mont, 57-59; asst prof physics, Cent Wash State Col, 60-62; physicist, Res Anal Corp, 62-65; assoc prof physics

& chmn dept, Cent Wash State Col, 65-70; exec officer, Am Asn Physics Teachers, Washington, DC, 70-72; assoc prof physics, 72-77, chmn dept geol & physics, 74-78, PROF PHYSICS, CENT WASH UNIV, 77- Mem: Fel AAAS. Res: Crystallography; color centers in alkali halides; quantum chemistry; solid state physics. Mailing Add: Dept Physics Cent Wash Univ Ellensburg WA 98926

JOHNSON, WILEY CARROLL, JR, b Asheville, NC, Jan 1, 30; m 51; c 2. PLANT BREEDING. Educ: Wake Forest Col, BS, 49; NC State Col, BS, 51, MS, 53; Cornell Univ, PhD(plant breeding), 56. Prof Exp: Res agronomist, Cornell Univ, 56-57; assoc prof, 57-69, PROF PLANT BREEDING, AUBURN UNIV, 69- Mem: Am Soc Agron; Crop Sci Soc Am; Am Genetics Asn. Res: Genetics and breeding of clovers. Mailing Add: Coastal Plain Exp Sta PO Box 748 Tifton GA 31793

JOHNSON, WILLARD JESSE, biochemistry, for more information see previous edition

JOHNSON, WILLIAM, b Boston, Mass, Oct 6, 41; m 65; c 3. MICROBIOLOGY. Educ: Marietta Col, BS, 63; Miami Univ, MS, 65; Rutgers Univ, PhD(microbiol), 68. Prof Exp: Nat Acad Sci-Nat Res Coun fel, Army Biol Res Ctr, Ft Detrick, Md, 68-70; asst prof microbiol, 70-74, assoc prof, 74-80, PROF MICROBIOL, COL MED, UNIV IOWA, 80- Mem: AAAS; Am Soc Microbiol; NY Acad Sci; Am Acad Microbiol. Res: Pathogenic microbiology; microbial toxins. Mailing Add: Dept Microbiol Univ Iowa Col Med Iowa City IA 52242

JOHNSON, WILLIAM ALEXANDER, b Ennis, Tex, June 22, 22; m 46; c 2. POULTRY SCIENCE. Educ: La State Univ, BS, 43, MS, 47; Univ Minn, PhD(poultry breeding), 52. Prof Exp: Instr poultry husb, La State Univ, 47-49, asst, Univ Minn, 49-52; from asst prof to assoc prof poultry breeding, 52-65, PROF POULTRY BREEDING, LA STATE UNIV, BATON ROUGE, 65-, HEAD DEPT, 81- Mem: Poultry Sci Asn; World Poultry Sci Asn; Am Genetics Asn; Sigma Xi; Nat Asn Cols & Teachers Agr. Res: Poultry breeding and genetics; environmental physiology; catfish breeding and genetics. Mailing Add: PO Box 2058 St Francisville LA 70775

JOHNSON, WILLIAM BOWIE, b Washington, DC, Sept 25, 54; m 87; c 1. HALL EFFECT, ELECTRONIC BAND STRUCTURE. Educ: George Mason Univ, BS, 76; Univ Md, MS, 78, PhD(physics), 82. Prof Exp: SR PHYSICIST PHYSICS, LAB PHYS SCI, 82- Mem: Am Phys Soc; Inst Elec & Electronics Engrs. Res: Characterization of the electrical properties of materials at low temperatures and high magnetic fields; materials under development include semimagnetic semiconductors, quasicrystals, silicon carbide, gallium arsenide and rare earth semiconductors. Mailing Add: Lab Phys Sci 4928 College Ave College Park MD 20740

JOHNSON, WILLIAM BUHMANN, b Palo Alto, Calif, Dec 5, 44; m 68; c 2. MATHEMATICAL ANALYSIS. Educ: Southern Methodist Univ, BA, 66; Iowa State Univ, PhD(math), 69. Prof Exp: Asst prof math, Univ Houston, 69-71; from asst prof to prof math, Ohio State Univ, 71-84; prof, 84-89, DISTINGUISHED PROF MATH & AG, TEX A&M UNIV, 89-, M E OWN CHAIR, 84- Concurrent Pos: Vis prof math, Univ Tex, Austin, 75 & Tex A&M Univ, College Station, 81; fel Inst Advan Studies, Hebrew Univ, Jerusalem, 77-78; ed, Trans Am Math Soc, 82-86, Ill J Math, 87- Mem: Am Math Soc; Math Asn Am. Res: Functional analysis; isomorphic theory of Banach spaces. Mailing Add: Dept Math Tex A&M Univ College Station TX 77840

JOHNSON, WILLIAM CONE, b Eastland, Tex, Nov 20, 26; m 56; c 3. INTERNAL MEDICINE. Educ: NTex State Univ, BS, 49; Univ Tex, MD, 54; Am Bd Internal Med, dipl, 63; Am Bd Pulmonary Dis, dipl, 68. Prof Exp: From intern to chief resident, John Sealy Hosp, Univ Tex Hosps, 54-58; chief med serv, 1604th USAF Hosp, 58-60, pulmonologist, Wilford Hall Hosp, Aerospace Med Div, Lackland AFB, Tex, 60-61, chief pulmonary & infectious dis serv, 61-63; med dir inhalation ther serv, Scott & White Mem Hosp, Temple, Tex, 63-65, dir pulmonary physiol labs, 63-68; dir respiratory ther serv & pulmonary physiol labs, Hendrick Mem Hosp, Abilene, 68-69; CLIN ASST PROF MED, UNIV TEX HEALTH SCI CTR, DALLAS, 69-; MED DIR RESPIRATORY THER SERV & PULMONARY FUNCTION LABS, WTEX MED CTR HOSP, ABILENE, 70- Concurrent Pos: Consult, Sect Clin Physiol, Scott & White Clin, Temple, 63-68, inhalation ther serv, Scott & White Mem Hosp, 65-68; med examr, Fed Aviation Agency, 66-; consult, Vet Admin Hosps, 65-, WTex Med Ctr Hosp, 68-, Hendrick Mem Hosp, Abilene, 68-69 & 70- & Shannon WTex Mem Hosp, San Angelo, 69; mem bd dirs, WTex Med Ctr Res Found, 69; med dir work eval & rehab unit, Methodist Hosp Dallas, 69-70; med dir respiratory ther serv, Cox Mem Hosp, Abilene, 70-75, Rolling Plains Mem Hosp, Sweetwater, 70-, Root Mem Hosp, Colorado City, Tex, 70, Med Ctr Hosp, Big Spring, 73-74, Shepperd Mem Hosp, Burnet, Tex, 76-80, Morris Mem Hosp, Coleman, Tex, 80-; clin asst prof med, Univ Tex Southwestern Med Sch Dallas, 69-; clin assoc prof med, Tex Tech Univ Sch Med, 74- Mem: AAAS; Am Asn Inhalation Therapists; fel Am Col Chest Physicians; fel Am Col Physicians; Am Fedn Clin Res. Res: Pulmonary physiology. Mailing Add: 6250 Humana Plaza Suite 1030 Abilene TX 79606

JOHNSON, WILLIAM E, JR, b Plano, Tex, July 20, 30; m 53; c 3. MEDICAL ENTOMOLOGY. Educ: Huston-Tillotson Col, BS, 51; Univ Okla, MS, 53, PhD(zool), 61. Prof Exp: Instr biol, Tuskegee Inst, 55-57; mus asst, Univ Okla, 57-60; chmn div sci & math, Albany State Col, 60-69; asst vpres acad affairs, 69-74, dean grad studies, 74-80, prof biol, 69-80; AT DEPT VET SCI, TUSKEGEE INST. Mem: Am Mosquito Control Asn. Res: Mosquito ecology. Mailing Add: Dept Vet Sci Tuskegee Inst Tuskegee AL 36088

JOHNSON, WILLIAM EVERETT, b Wallowa, Ore, Oct 22, 21; m 46; c 4. PHARMACOLOGY. Educ: Wash State Univ, BS, 51, MS, 53, PhD(pharmacol), 58. Prof Exp: From instr to asst prof pharm, Col Pharm, Univ Wyo, 53-58, from assoc prof to prof pharmacol, 58-65; assoc prof, 65-72, PROF PHARMACOL, COL PHARM, WASH STATE UNIV, 72- Res: Pharmacology of the cardiovascular system and mechanism of action of teratogens. Mailing Add: Wash State Univ Col Pharm Pullman WA 99164

JOHNSON, WILLIAM HARDING, physiology, biophysics, for more information see previous edition

JOHNSON, WILLIAM HILTON, b Indianapolis, Ind, Feb 14, 35; m 56; c 3. QUATERNARY GEOLOGY. Educ: Earlham Col, AB, 55; Univ Ill, MS, 61, PhD(geol), 62. Prof Exp: From instr to assoc prof, 68-87, PROF GEOL, UNIV ILL, URBANA, 87- Concurrent Pos: Partic, Ill State Geol Surv. Mem: Geol Soc Am; Sigma Xi; Nat Asn Geol Teachers; Am Quaternary Asn. Res: Sedimentology and stratigraphy of Pleistocene deposits and glacial geology; relict Pleistocene periglacial forms; geomorphology. Mailing Add: Dept Geol Univ Ill 1301 W Green St Urbana IL 61801

JOHNSON, WILLIAM HOWARD, b Sidney, Ohio, Sept 3, 22; m 43; c 3. AGRICULTURAL ENGINEERING. Educ: Ohio State Univ, BS, 48, MS, 53; Mich State Univ, PhD(agr eng), 60. Prof Exp: From instr to prof agr eng, Ohio Agr Res & Develop Ctr, assoc chmn dept, 53-68, actg chmn dept, 68-69; prof agr eng & head dept, Kans State Univ, 70-81, dir, Eng Exp Sta, 81-87; RETIRED. Concurrent Pos: Agr eng consult, 57-; vis scientist & lectr, Tex A&M Univ, 69-70. Mem: Fel Am Soc Agr Engrs (pres, 86-87). Res: Power and machinery area of agricultural engineering; determination of functional requirements and design; efficiency of harvesting and tillage machine components. Mailing Add: Dept Agr Eng Kans State Univ Manhattan KS 66506

JOHNSON, WILLIAM HUGH, b Fayetteville, NC, Sept 14, 32; m 58; c 2. CROP PROCESS ENGINEERING, SYSTEM DESIGN AND ANALYSIS. Educ: NC State Univ, BS, 54, MS, 56, PhD(agr eng), 61. Prof Exp: Res instr, 56-61, from asst prof to prof agr eng, 61-83, ASST DIR NC AGR RES SERV, NC STATE UNIV, 83- Concurrent Pos: Partic & spec reporter, Fourth Int Tobacco Sci Cong, Athens, 66, Fifth Int Tobacco Sci Cong, Hamburg, 70; consult, Indian Inst Technol, Ford Found Proj, Kharagpur, India, 66; ed, Tobacco Sci, 80-84. Honors & Awards: Philip Morris Distinguished Achiev in Tobacco Sci, 73. Mem: Am Soc Agr Engrs; Sigma Xi. Res: Bioengineering of plant materials; energy and mass transfer relations during processing; physical, chemical and enzymatic changes in response to dynamic process variables; health-related modifications of tobacco; systems engineering; biological engineering; solar and heat energy recovery systems engineering. Mailing Add: NC Agr Res Serv Box 7643 NC State Univ Raleigh NC 27695-7643

JOHNSON, WILLIAM JACOB, b Gladwin, Mich, June 23, 14; m 44; c 2. INORGANIC CHEMISTRY. Educ: Univ Calif, Davis, BS, 37; Kans State Univ, MS, 48, PhD(plant biochem), 62. Prof Exp: Sci teacher, Kans High Sch, 44-46; instr, 47-50, bus mgr, 50-52, from asst prof to assoc prof, 52-62, PROF CHEM, TABOR COL, 62-, CHMN DIV NATURAL SCI & MATH, 63- Concurrent Pos: Vis sr lectr, Univ Zambia, 70-71. Mem: Am Chem Soc. Res: Nature of zinc and other micronutrients in plant extracts. Mailing Add: 212 S Madison Hillsboro KS 67063

JOHNSON, WILLIAM K, b Kalamazoo, Mich, Jan 4, 27; m 52; c 1. ORGANIC CHEMISTRY. Educ: Univ Mich, BS, 50, MS, 51, PhD(pharm chem), 54. Prof Exp: Chemist res & eng div, Monsanto Co, 53-60, proj mgr, Org Develop Dept, 60-65, mgr mkt res, Org Div, 65-67, mgr commercial develop plasticizers & gen chem, 67-69, mgr technol gen chem, 69-71, dir res & commercial develop, Process Chem Group, Monsanto Indust Chem Co, 71-76, mgr markets & prod, 76-79, mgr mkt res & planning, Monsanto Intermediates Co, 79-82; mgr mkt res & planning, SRI Int, 83-85; dir petrochemicals, Catalytica Assocs, 85-87; CONSULT, 87- Mem: Am Chem Soc; Commercial Develop Asn; Ger Chem Soc; Chem Mkt Res Asn. Res: Organometallics and organic synthesis; intermediates and fine chemicals. Mailing Add: 7356 Via Laguna San Jose CA 95135

JOHNSON, WILLIAM LAWRENCE, b Keene, NH, Aug 28, 36; c 4. RUMINANT NUTRITION, FORAGE UTILIZATION. Educ: Univ NH, BS, 58; Cornell Univ, MS, 64, PhD(dairy cattle nutrit), 66. Prof Exp: From asst prof to assoc prof, 66-82, PROF ANIMAL SCI, NC STATE UNIV, 82- Concurrent Pos: Dairy husb res specialist, Nat Agrarian Univ, La Molina, Peru, 66-69, co-leader forage & animal nutrit prog, Agr Mission to Peru, 70-73; prin investr small ruminants collab res, Indonesia, Morocco & Brazil, 78-83; res & inst develop adv, Nat Inst for Agr Res, Lima, Peru, 88- Mem: Am Soc Animal Sci; Int Goat Asn. Res: Factors influencing utilization of forages, and roughage by products, including tropical feedstuffs, by cattle, sheep and goats. Mailing Add: Dept Animal Sci NC State Univ Raleigh NC 27695-7621

JOHNSON, WILLIAM LEWIS, b Bryan, Tex, July 6, 40; m 63; c 1. SOLID STATE PHYSICS. Educ: Univ Southern Miss, BA, 62; Naval Postgrad Sch, MS, 66, PhD(physics), 69. Prof Exp: Instr physics, Naval Postgrad Sch, 63-69; res assoc, Univ Ill, 69-71; PROF PHYSICS & CHMN DEPT, WESTMINSTER COL, PA, 71- Mem: Am Phys Soc; Am Asn Physics Teachers; Sigma Xi. Res: Critical point phenomena; microcomputers; laboratory automation. Mailing Add: Dept Physics Westminster Col New Wilmington PA 16142

JOHNSON, WILLIAM LEWIS, b Bowling Green, Ohio, July 26, 48; m 84. SOLID STATE PHYSICS, MATERIALS SCIENCE. Educ: Hamilton Col, BA, 70; Calif Inst Technol, PhD(appl physics), 74. Prof Exp: Fel appl physics, Calif Inst Technol, 74-75; fel, T J Watson Res Ctr, IBM Corp, 75-77; from asst prof to prof mat sci, 77-88, RUBEN & DONNA METTLER PROF ENG & APPL SCI, CALIF INST TECHNOL, 89- Concurrent Pos: Consult,

Jet Propulsion Lab, Calif Inst Technol, Pasadena, 80-, Gen Motors, 83-, Hughes Res Labs, 84-, Lawrence Livermore Lab, 84- *Honors & Awards:* Alexander von Humbolt Sr Scientist Award, 88. *Mem:* Am Phys Soc; AAAS; Am Soc Metals; Mat Res Soc. *Res:* Low temperature physics; superconductivity; amorphous materials; properties of metastable metallic materials. *Mailing Add:* Keck Lab of Eng Calif Inst of Technol Pasadena CA 91125

JOHNSON, WILLIAM RANDOLPH, JR, b Oxford, NC, July 25, 30; m 54; c 3. POLYMER CHEMISTRY, THEORETICAL CHEMISTRY. *Educ:* NC Col Durham, BS, 50; Univ Notre Dame, MS, 52; Univ Pa, PhD(chem), 58. *Prof Exp:* Instr, Prairie View A&M Col, 52-53; prof, Fla A&M Univ, 58-61; chemist, W R Grace & Co, 61-63; chemist, Philip Morris Ops Ctr, 63-75, mgr, Chem Res Div, 75-79, mgr spec affairs, 81-90, SR SCIENTIST, PHILIP MORRIS OPS CTR, 90- *Concurrent Pos:* Adj prof chem, 63-73, exec in residence, Va Union Univ, 79-81. *Res:* Polymer synthesis; smoke chemistry; smoke formation mechanisms; pyrolysis mechanisms. *Mailing Add:* Philip Morris USA PO Box 26603 Richmond VA 23261

JOHNSON, WILLIAM ROBERT, b Buffalo, Okla, Sept 24, 39; m 61; c 2. MATERIALS SCIENCE. *Educ:* San Jose State Col, BS, 64; Stanford Univ, MS, 67, PhD(mat sci), 69. *Prof Exp:* Scientist & prod mgr vacuum metallization, St Clair-Field Inc, Mt View, Calif, 69-70; assoc scientist, General Atomics, 70-71, staff assoc, 71-72, sr engr, 72-73, staff engr, 73-81, mgr mats eval, 81-90; SR STAFF SCIENTIST, SEGA, 90- *Mem:* Am Soc Metals. *Res:* Structure of materials; mechanical behavior of materials (fracture, creep and stress rupture); environmental effects on materials. *Mailing Add:* 12243 Riesling Ct San Diego CA 92131

JOHNSON, WILLIAM S(TANLEY), b Camden, Tenn, Dec 9, 39; m 67; c 2. MECHANICAL & ENVIRONMENTAL ENGINEERING. *Educ:* Univ Tenn, Knoxville, BS, 61; Clemson Univ, MS, 65, PhD(eng), 67. *Prof Exp:* Asst design engr, Pratt & Whitney Aircraft Div, United Aircraft Corp, 61-62; from asst prof to assoc prof, 67-77, PROF MECH ENG, UNIV TENN, 77- *Mem:* Am Soc Mech Engrs; Am Soc Heating Vent & Air Conditioning Engrs; Am Soc Eng Educ. *Res:* Application of pulse-jet flow in low area ratio ejectors; determination of velocity characteristics of two-dimensional fluid jets; boundary layer control on submarine surfaces; energy conservation analysis in buildings; heat pump evaluations. *Mailing Add:* Dept Mech & Aerospace Eng Univ Tenn Knoxville TN 37996-2210

JOHNSON, WILLIAM SUMMER, b New Rochelle, NY, Feb 24, 13; m 40. ORGANIC CHEMISTRY. *Educ:* Amherst Col, AB, 36; Harvard Univ, AM, 38, PhD(org chem), 40. *Hon Degrees:* DSc, Amherst Col, 56 & Long Island Univ, 68. *Prof Exp:* Instr chem, Amherst Col, 36-37; from instr to prof org chem, Univ Wis, 40-54, Homer Adkins prof, 54-60; exec head dept chem, Stanford Univ, 60-69, prof, 60-68, Jackson-Wood prof, 65-68, EMER PROF CHEM, STANFORD UNIV, 68- *Concurrent Pos:* Secy org sect, Int Cong Pure & Appl Chem, 51, invited lectr, Eng, 63; Am-Swiss Found lectr, Switz, 51; mem chem adv panel, NSF, 52-56; vis prof, Harvard Univ, 54-55; mem US-Brazil study group grad teaching & res in chem, US-Brazil Sci Coop Prog, Off Foreign Secy, Nat Acad Sci, 68; mem med chem A study sect, NIH, 70-; Centenary lectr, Univ UK, 73-74. *Honors & Awards:* Coover Lectr, Univ Iowa, 52; Edward Clark Lee Lectr, Univ Chicago, 56; Bachmann Lectr, Univ Mich, 57; Mfg Award Creative Res, 63; Max Tishler Lectr, Harvard Univ, 64; Andrews Lectr, Univ NSW, 65; Phillips Lectr, Haverford Col, 69; Treat B Johnson Lectr, Yale Univ, 71; Falk-Plaut Lectr, Columbia Univ, 71; Van't Hoff Centenary Lectr, Univ Leiden, 74; Nat Medal Sci, 87; Award for Creative Work in Synthetic Org Chem, Am Chem Soc, 58, Synthetic Org Chem Mfrs Award, Creative Res Org Chem, 63; Nichols Medal Award, 68. *Mem:* Nat Acad Sci; Am Chem Soc; Am Acad Arts & Sci. *Res:* Asymmetric synthesis methodology; biomimetic polyene cyclizations. *Mailing Add:* Dept Chem Stanford Univ Stanford CA 94305

JOHNSON, WILLIAM W, b Provo, Utah, July 11, 34; m 69; c 2. GEOPHYSICS. *Educ:* Brigham Young Univ, BS, 56; Univ Utah, MS, 58; Univ Pittsburgh, PhD(geophys), 65. *Prof Exp:* Sr res scientist, Sinclair Oil Corp, 65-69; prin res geophysicist, Atlantic Richfield Co, 69-85; RETIRED. *Concurrent Pos:* Res assoc, Lamont-Doherty Geol Observ, Columbia Univ, 66-67. *Mem:* AAAS; Soc Explor Geophysicists. *Res:* Propagation of elastic waves in anisotropic media; geological interpretation of gravity and magnetic data; seismic wave propagation. *Mailing Add:* 707 Parkview Circle Richardson TX 75080

JOHNSON, WILLIAM WAYNE, b Minneapolis, Minn, Oct 12, 34. GENETICS. *Educ:* Univ Minn, BS, 57, MS, 59, PhD(zool), 63. *Prof Exp:* Interim asst prof biol, Univ Fla, 62-63; asst prof, 63-68, ASSOC PROF BIOL, UNIV NMEX, 68- *Mem:* AAAS; Genetics Soc Am; Am Inst Biol Sci; Sigma Xi. *Res:* Experimental population genetics of Drosophila. *Mailing Add:* Dept Biol Univ NMex Albuquerque NM 87131

JOHNSON, WILLIS HUGH, b Parkersburg, Ind, Dec 21, 02; m 29; c 2. ZOOLOGY. *Educ:* Wabash Col, AB, 25; Univ Chicago, MS, 29, PhD(zool), 32. *Hon Degrees:* DSc, Wabash Col, 74. *Prof Exp:* From instr to assoc prof zool, Wabash Col, 25-35; from asst prof to assoc prof biol, Stanford Univ, 35-41, prof biol & dir army specialized training prog, 41-46; prof zool, 46-65, chmn dept biol & sci div, 46-68; Treves Prof biol, 65-73, EMER PROF BIOL, WABASH COL, 73- *Concurrent Pos:* Instr, US Army Univ, Eng, 44-45; vchmn, Comn Undergrad Educ in Biol Sci, 65-66. *Mem:* Am Soc Zool; Am Soc Naturalists; Soc Protozool; fel NY Acad Sci. *Res:* Populations and sterile culture of protozoa; encystment and nutrition in protozoa; nutrition and regeneration in Planaria. *Mailing Add:* Wabash Col Crawfordsville IN 47933

JOHNSON, WOODROW ELDRED, b Forest Lake, Minn, Oct 22, 17; m 42; c 4. PHYSICS. *Educ:* Hamline Univ, BS, 37; Brown Univ, MS, 39, PhD(physics), 42. *Hon Degrees:* DS, Hamline Univ, 61. *Prof Exp:* Asst, Brown Univ, 37-40; from instr to asst prof physics, Syracuse Univ, 41-44, asst prof, 46-47; sr physicist, Tenn Eastman Corp, 44-46 & Manhattan Proj, Oak Ridge Nat Lab, 47-49; sect mgr, Bettis Atomic Power Lab, Westinghouse Corp, 49-51, mgr tech opers, Prototype Reactor Facil, Idaho, 51-53, asst to dir develop, 53-54, corp tech consult, Matahorn Proj, Princeton Univ, 54, corp mem, Indust Atomic Power Study Group, 54-55, proj mgr, Pa Adv Reactor Proj, Atomic Power Dept, 55-59, dir projs, 59-61, gen mgr, 61-64, gen mgr, Astronuclear Lab, 64-68, corp vpres, 67-85, vpres & gen mgr, Astronuclear-Underseas Div, 68-71, vpres & gen mgr, Transp Div, 71-85; RETIRED. *Mem:* Nat Acad Eng; Inst Elec & Electronics Engrs; Am Nuclear Soc; Nat Asn Mfg; Am Phys Soc. *Res:* Photoelectricity; electron diffraction; physics of thin metallic films; effect of radiation on solids. *Mailing Add:* 114 Brompton Rd Garden City NY 11530

JOHNSON-LUSSENBURG, CHRISTINE MARGARET, b Hawkesbury, Ont, Jan 29, 31; m 53, 72; c 6. VIROLOGY, MOLECULAR BIOLOGY. *Educ:* McGill Univ, BSc, 52, MSc, 53; Univ Ottawa, PhD, 67. *Prof Exp:* Asst prof, 67-78, ASSOC PROF MICROBIOL, UNIV OTTAWA, 78- *Mem:* Can Soc Microbiologists; Am Soc Microbiol. *Res:* Structural and antigenic studies of components involved in virus replication, including myxoviruses, herpesvirus and coronavirus. *Mailing Add:* Dept Microbiol Univ Ottawa Ottawa ON K1H 8M5 Can

JOHNSON-WINEGAR, ANNA, b Frederick, Md, May 27, 45; m 80. MICROBIAL TOXINS. *Educ:* Hood Col, BA, 76; Catholic Univ Am, MS, 79, PhD(microbiol), 81. *Prof Exp:* Med technician res, US Army Med Res Inst Infectious Dis, 66-76, microbiologist, 76-85; microbiologist, 85-90, SCI ADMINR, US ARMY MED MAT DEVELOP ACTIV, 90- *Concurrent Pos:* Reviewer, Appl Environ Microbiol, 83-; mem, Comt on Status of Women Microbiologists & Fed Orgn Prof Women, 83-; guest lectr var insts. *Mem:* Fel Am Soc Microbiologists; AAAS; Int Soc Toxinol; NY Acad Sci; Sigma Xi. *Res:* Purification and biochemical analysis of bacterial toxins; immunology of toxin-derived components; pathogenesis of toxins; fermentation techniques; animal models; genetic control of toxin production. *Mailing Add:* Res Area I US Army Med Res & Develop Command Frederick MD 21702-5012

JOHNSON-WINT, BARBARA PAULE, MORPHOGENESIS, CELL INTERACTIONS. *Educ:* Mich State Univ, PhD(zool), 76. *Prof Exp:* ASST PROF, SCH MED, HARVARD UNIV, 85- *Mailing Add:* Dept Biol Scis Northern Ill Univ DeKalb IL 60115-2861

JOHNSTON, A SIDNEY, b Hinton, WVa, Apr 4, 37. NUCLEAR PHYSICS. *Educ:* Va Polytech Inst, BS, 59; Carnegie-Mellon Univ, MS, 61, PhD(physics), 65; Chicago Kent Col Law, JD, 78. *Prof Exp:* Sr scientist physics, Westinghouse Astronuclear, 65-68; asst prof, Pratt Inst, 68-74; MEM STAFF, DEPT NUCLEAR MED, MICHAEL REESE HOSP, 74- *Concurrent Pos:* Private law pract, 78- *Mem:* AAAS; Am Phys Soc. *Res:* Nuclear engineering; solid state physics; science and society. *Mailing Add:* 51 Quaboag Rd Acton MA 01720

JOHNSTON, ALAN ROBERT, b Long Beach, Calif, June 26, 31; m 56; c 3. OPTICAL PHYSICS. *Educ:* Calif Inst Technol, BS, 52, PhD(physics), 56. *Prof Exp:* Res scientist, 56-62, res group supvr optical physics, 62-71, MEM TECH STAFF, JET PROPULSION LAB, 71- *Mem:* Optical Soc Am; Sigma Xi. *Res:* Fiber optic systems; optoelectronic sensors. *Mailing Add:* 1226 Olive Lane La Canada Flintridge CA 91011

JOHNSTON, ANDREA, b Minneapolis, Minn, Mar 13, 21. MATHEMATICS. *Educ:* St Mary Col, Kans, BA, 48; Cath Univ, MS, 52, PhD(math), 54. *Prof Exp:* CHMN DEPT MATH, ST MARY COL, KANS, 54- *Mem:* Math Asn Am; Nat Coun Teachers Math. *Res:* Mathematics teaching; preparation of elementary and secondary teachers. *Mailing Add:* St Mary Col Leavenworth KS 66048-5082

JOHNSTON, BRUCE (GILBERT), civil engineering; deceased, see previous edition for last biography

JOHNSTON, C EDWARD, b Ont; m 64; c 3. AQUACULTURE. *Educ:* Univ NB, BA, 64, PhD(biol), 68. *Prof Exp:* Asst prof, Prince Wales Col, 68-69; from asst prof to assoc prof biol, 69-84, PROF BIOL, UNIV PEI, 84- *Concurrent Pos:* Vis prof, Biol Sta, NB, 79-80; vis prof, Mem Univ, NLFD, 86-87. *Mem:* Am Fisheries Soc; Can Aqua Soc; Fish & Wildlife Soc. *Res:* Effect of low pH on parr-smolt transformation of Atlantic salmon; effect of temperature regimes on salmonid physiology; rapid and slow acclimation procedures on ionoregulatory mechanisms of rainbow trout; Atlantic salmon reproductive physiology and ichthyoplankton; Alosa aestivalis reproductive biology. *Mailing Add:* Biol Dept Univ PEI Charlottetown PQ C1A 4P3 Can

JOHNSTON, CHARLES LOUIS, JR, clinical pathology hematology, for more information see previous edition

JOHNSTON, CHRISTIAN WILLIAM, b Brooklyn, NY, Jan 9, 11; m 46; c 2. PHYSICAL CHEMISTRY. *Educ:* Columbia Univ, AB, 34. *Prof Exp:* Anal chemist, Martin Dennis Co, NJ, 27-28; res chemist, Darco Corp, NY, 28-40; asst to vpres in chg prod, US Indust Chem Co, 40-50, dir resin res & head tech serv, 50-58; supvr polymer appln res, Food Mach & Chem Corp, 58-63; mgr polymer appln res, Plastics Div, Tenneco Mfg Co, 63-68, sr scientist, Tenneco Chem Inc, 68-76; CONSULT, 76- *Concurrent Pos:* Tech dir, Tyndale Plains Hunter Ltd, 80- *Mem:* Soc Plastics Engrs; Am Soc Qual Control; Am Chem Soc; Soc Rheology; Instrument Soc USA. *Res:* Adsorption form solution; optical method colorimetry; reaction rates; rheology; electronic circuits measurements and control; polymer chemistry; chemical engineering equipment design and plant operating statistical analysis; hydrophyllic polymers. *Mailing Add:* PO Box 477 Neshanic Station NJ 08853-0477

JOHNSTON, CLAIR C, b Beaver Falls, Pa, Jan 13, 99; m 28. CIVIL ENGINEERING. *Educ:* Univ Detroit, BCE, 23, CE, 33; Univ Mich, MCE, 36. *Prof Exp:* From asst prof to prof civil eng, Univ Detroit, 27-42; staff engr, Stevenson, Jordan & Harrison, Mgt Consult, 42-47; training dir, Square D Co, Mich, 48-59; supv prof eng develop, Chrysler Missile Plant, 59-61; prof civil eng, Univ Detroit, 61-62; prof, Detroit Inst Technol, 62-70; civil eng consult, 70-75; RETIRED. *Concurrent Pos:* Chmn dept civil eng, Detroit Inst Technol, 62-69, dean col eng, 68-69. *Mem:* Am Soc Civil Engrs. *Res:* Reinforced concrete and indeterminate structures. *Mailing Add:* 475 Colonial Ct Grosse Point Farms MI 48236

JOHNSTON, COLIN DEANE, b Northern Ireland, Apr 28, 40; m 67; c 2. CIVIL ENGINEERING, MATERIALS SCIENCE. *Educ:* Queen's Univ, Belfast, BSc, 62, PhD(civil eng), 67. *Prof Exp:* Site engr, Govt of Northern Ireland, 66-67; tech mgr concrete prod, Pre-Mix Concrete, Ltd, 67; assoc prof, 67-78, PROF CIVIL ENG, UNIV CALGARY, 78- *Honors & Awards:* Wason Medal, Am Concrete Inst, 77. *Mem:* Am Concrete Inst; Am Soc Testing & Mat; fel Am Concrete Inst, 87; Brit Concrete Soc; Transp Res Bd. *Res:* Concrete, fiber reinforced concrete, fly ash, silica fume, and chemical admixtures in concrete, asphalt concrete; structural and paving applications. *Mailing Add:* Dept Civil Eng Univ Calgary Calgary AB T2N 1N4 Can

JOHNSTON, CYRUS CONRAD, JR, b Statesville, NC, July 16, 29; m 60; c 2. INTERNAL MEDICINE, ENDOCRINOLOGY. *Educ:* Duke Univ, AB, 51, MD, 55; Am Bd Internal Med, dipl. *Prof Exp:* Intern med, Duke Hosp, 55-56; resident, Barnes Hosp, St Louis, 56-57; fel endocrinol & metab, 59-61, USPHS career res develop award, 63-68, from instr to assoc prof med, 61-69, assoc dir gen, 62-67, dir gen, Clin Res Ctr, 67-80, PROF MED, MED CTR, IND UNIV, INDIANAPOLIS, 69-, DIR DIV ENDOCRINOL & METAB, 68- *Concurrent Pos:* Res career develop award, USPHS, 63-68. *Mem:* AAAS; Am Fedn Clin Res; Endocrine Soc; fel Am Col Physicians; Am Diabetes Asn; Am Soc Bone & Mineral Res. *Res:* Metabolism of bone both in human subjects and in the experimental animal; osteoporosis. *Mailing Add:* Dept Med Rm 421 Ind Univ Sch Med 545 Barnhill Dr Indianapolis IN 46223

JOHNSTON, DAVID HERVEY, b Syracuse, NY, Aug 25, 51; m 72; c 1. SEISMOLOGY, ROCK PHYSICS. *Educ:* Mass Inst Technol, SB, 73, PhD(geophysics), 79. *Prof Exp:* SR RES SPECIALIST GEOPHYSICS, EXXON PROD RES CO, 79- *Mem:* Am Geophys Union; Soc Explor Geophysicists (secy-treas, 89-90). *Res:* Reflection seismology; seismic processing; velocity analysis and interpretation; applications of geophysics to oil reservoir development and production; rock physics; relationship of rock microstructure to acoustic, electrical and flow properties; extraction of rock properties from seismic data; structure and evolution of planetary interiors. *Mailing Add:* Exxon Prod Res Co PO Box 2189 Houston TX 77252-2189

JOHNSTON, DAVID OWEN, b Franklin, Tenn, July 27, 30; m 50; c 4. PHYSICAL CHEMISTRY. *Educ:* George Peabody Col, BS, 51; Mid Tenn State Col, MA, 58; Univ Miss, PhD(chem), 63. *Prof Exp:* Teacher pub schs, Tenn, 51-53 & 54-58; instr phys sci, Mid Tenn State Col, 58-60; from asst prof to prof chem, David Lipcomb Col, 63-86, Justin Potter distinguished prof, 86-91; RETIRED. *Concurrent Pos:* Fel res, Vanderbilt Univ, 65, 66, 69. *Mem:* Am Chem Soc. *Res:* Transport properties of rare earth salts in nonaqueous solvents; kinetics of inorganic oxidation-reduction reactions. *Mailing Add:* 1492 Clairmont Pl Nashville TN 37215

JOHNSTON, DAVID WARE, b Miami, Fla, Nov 23, 26; m 48; c 3. AVIAN PHYSIOLOGY, AVIAN ECOLOGY. *Educ:* Univ Ga, BS, 49, MS, 50; Univ Calif, PhD, 54. *Prof Exp:* Assoc prof, Mercer Univ, 54-59 & Wake Forest Col, 59-63; assoc prof biol sci & zool, Univ Fla, 63-74, prof zool, 74-; AT BIOL DEPT, GEORGE MASON UNIV. *Mem:* Ecol Soc Am; Cooper Ornith Soc; Am Ornith Union; Nat Audubon Soc. *Res:* Fat deposition in birds; pesticide levels in birds; ecology of insular avifaunas. *Mailing Add:* 5219 Concordia St Fairfax VA 22032

JOHNSTON, DEAN, South Bend, Ind, Apr 12, 47. TUMOR IMMUNOLOGY. *Educ:* Wayne State Univ, PhD(biochem), 74. *Prof Exp:* ASST PROF DERMATOL, NY UNIV MED CTR, 82- *Mem:* Am Soc Biol Chemists; AAAS. *Mailing Add:* Dept Health Sci Hunter Col 425 E 25th St New York NY 10010

JOHNSTON, DENNIS ADDINGTON, b Oak Ridge, Tenn, Sept 17, 44; m 66; c 2. BIOSTATISTICS. *Educ:* Arlington State Col, BS, 65; Univ Tex, Austin, MS, 66; Tex Tech Univ, PhD(math), 71. *Prof Exp:* Asst biomathematician & asst prof biomath, 72-78, ASSOC BIOMATHEMATICIAN & ASSOC PROF BIOMATH, UNIV TEX M D ANDERSON CANCER CTR, 78- *Concurrent Pos:* adj assoc prof, Dept Statist, Rice Univ, 73-; mem fac, Grad Sch Biomed Sci, Univ Tex, Houston, 73- *Mem:* Am Statist Asn; Inst Elec & Electronics Engrs. *Res:* Biomedical image processing; consultant in mathematical and statistical models; biostatistics; automated chromosome analysis. *Mailing Add:* Dept Biomath-237 Univ Tex M D Anderson Cancer Ctr 1515 Holcombe Blvd Houston TX 77030

JOHNSTON, DON RICHARD, b Union City, Ind, Aug 10, 37; c 2. CHEMICAL PHYSICS. *Educ:* Earlham Col, BA, 57; Brown Univ, PhD(chem), 61. *Prof Exp:* Insulation engr, M&P Lab, 60-67, unit mgr phys chem, 68-70, SUBSECT MGR CHEM & ELEC INSULATION, GEN ELEC CO, 70- *Mem:* Am Chem Soc; Inst Elec & Electronics Engrs; Metals Properties Coun. *Res:* Mechanism of corona degradation of electrical insulation; electrical polarization currents for metallic corrosion. *Mailing Add:* 5140 Bliss Rd Ballston Spa NY 12020

JOHNSTON, E(LWOOD) RUSSELL, JR, b Philadelphia, Pa, Dec 26, 25; m 51; c 2. CIVIL ENGINEERING. *Educ:* Univ Del, BCE, 46; Mass Inst Technol, MS, 47, ScD(civil eng), 49. *Prof Exp:* Asst civil eng, Mass Inst Technol, 46-47; struct designer, Fay, Spofford & Thorndike, 47-49; from asst prof to prof, Lehigh Univ, 49-57; prof, Worcester Polytech Inst, 57-63; head dept, 72-77, PROF CIVIL ENG, UNIV CONN, 63- *Concurrent Pos:* Guest prof, Swiss Fed Inst Technol, Zurich, 70 & 77. *Mem:* Am Soc Civil Engrs; Am Soc Eng Educ; Int Asn Bridge & Struct Engrs; Am Acad Mech. *Res:* Structural engineering; applied mechanics; vibrations. *Mailing Add:* PO Box 525 Storrs CT 06268

JOHNSTON, ERNEST RAYMOND, b Dahinda, Ill, Feb 9, 07; m 39. MATHEMATICS. *Educ:* Ill State Norm Univ, BEd, 38; Univ Ill, MS, 39; Univ Minn, PhD(math), 54. *Prof Exp:* Teacher & prin, Pub Schs, Ill, 26-37; teacher, High Sch, Ill, 39-40; instr math, Austin Jr Col, Minn, 40-42 & Univ Minn, 42-44; mech engr, Naval Ord Lab, Md, 44-47; asst prof math & mech, Univ Minn, 47-51, lectr math, 51-53; prof, Wis State Col, Whitewater, 53-55; from assoc prof to prof math, 55-76, head dept math sci, 63-72, EMER PROF MATH, IND UNIV-PURDUE UNIV, INDIANAPOLIS, 76- *Mem:* Am Math Soc; Math Asn Am. *Mailing Add:* 215 Valley View Dr Kerrville TX 78028

JOHNSTON, FRANCIS E, b Paris, Ky, Oct 9, 31; m 55; c 3. PHYSICAL ANTHROPOLOGY. *Educ:* Univ Ky, BA, 59, MA, 60; Univ Pa, PhD(anthrop), 62. *Prof Exp:* From instr phys anthrop to asst prof anthrop, Univ Pa, 60-66, asst cur phys anthrop, Univ Mus, 63-66; fel, Univ London Inst Child Health, 66-67; assoc prof anthrop, Univ Tex, Austin, 68-71; prof anthrop, Temple Univ, 71-73; PROF ANTHROP, UNIV PA, 73- *Concurrent Pos:* Consult growth & develop, Nat Ctr Health Statist, 63-; fel, Inst Cancer Res, 67-68 & 75-; managing ed, Am J Phys Anthrop, 77- *Mem:* Am Asn Phys Anthrop (secy-treas, 64-68, vpres, 70-72); Am Anthrop Asn; Brit Soc Study Human Biol; Human Biol Coun; fel Royal Soc Med. *Res:* Child growth and development; population biology; human genetics; ecology of nutrition in human populations. *Mailing Add:* Dept Anthrop Univ Pa Philadelphia PA 19104

JOHNSTON, FRANCIS J, b Ferryville, Wis, Sept 20, 24; m 48; c 1. PHYSICAL CHEMISTRY. *Educ:* Univ Wis, BS, 47, PhD(chem), 52. *Prof Exp:* Chemist, E I du Pont de Nemours & Co, 52-54; from asst prof to assoc prof chem, Univ Louisville, 54-60; ASSOC PROF CHEM, UNIV GA, 60- *Mem:* Am Chem Soc; Sigma Xi. *Res:* Radiation and surface chemistry; reaction kinetics. *Mailing Add:* Dept Chem Univ Ga Athens GA 30602

JOHNSTON, G(ORDON) W(ILLIAM), b Toronto, Ont, Dec 10, 26; m 55; c 2. ENGINEERING PHYSICS. *Educ:* Univ Toronto, BSc, 48, MASc, 50, PhD(aerophys), 53. *Prof Exp:* Design engr, A V Roe, Co, Ltd, Can, 49-51; asst, Defense Res Bd Can, 52-53; res supvr sci lab, Res Div, Ford Motor Co, Mich, 53-54, head gas dynamics sect, 54-55; proj engr, De Haviland Aircraft Can, Ltd, 55-63, dir short take off & landing res proj, 60-63, head adv proj group, 63-70; ASSOC PROF AEROACOUST, INST AEROSPACE STUDIES, UNIV TORONTO, 70- *Concurrent Pos:* Mem assoc comt aerodyn noise, Nat Res Coun Can; lectr, Inst Aerophys, Univ Toronto, 57-59; aerodyn consult, Plasma Dynamics Dept, United Aircraft Res Labs, Conn, 67-70. *Mem:* Can Aeronaut Inst. *Res:* Transonic and low-speed aerodynamics; boundary layer control; stability and control of fixed and rotating wing aircraft configurations; slipstream wing aerodynamics. *Mailing Add:* 79 Valecrest Dr Toronto ON M9A 4P5 Can

JOHNSTON, GEORGE I, b Bryn Mawr, Pa, May 29, 29; m 61; c 1. ELECTRICAL ENGINEERING. *Educ:* Johns Hopkins Univ, BS, 55. *Prof Exp:* Electronics technician, Sch Med, Johns Hopkins Univ, 48-55; med electronics engr, NIH, 55-58; dir res, Instrument Serv, Med Sch, 58-76, ASSOC PROF & DIR, INSTRUMENT & SAFETY SERV, UNIV ORE, 76- *Concurrent Pos:* Asst sanit engr, USPHS. *Mem:* Inst Elec & Electronics Engrs; Sigma Xi. *Res:* Biomedical engineering. *Mailing Add:* Instrument & Safety Serv Health Sci Ctr Univ Ore 3181 Sam Jackson Med Portland OR 97201

JOHNSTON, GEORGE LAWRENCE, b Los Angeles, Calif, Nov 11, 32; m 59. PLASMA PHYSICS, THEORETICAL PHYSICS. *Educ:* Calif Inst Technol, BS, 54; Univ Calif, Los Angeles, MS, 62, PhD(physics), 67. *Hon Degrees:* JD, Harvard Univ, 57. *Prof Exp:* Mem tech staff, Space Technol Labs, Inc, 57-60 & Aerospace Corp, 60-64; asst res physicist, Univ Calif, Los Angeles, 67-69; asst prof physics, Sonoma State Univ, 69-74, assoc prof, 74-80; RES SCIENTIST, PLASMA FUSION CTR, MASS INST TECHNOL, 80- *Concurrent Pos:* Res assoc, Mass Inst Technol, 75-77; adv comnr, Calif Energy Comn, 77-78. *Mem:* Am Phys Soc; AAAS; Sigma Xi. *Res:* Plasma kinetic theory; nonlinear plasma theory; mathematical physics; free electron lasers; relativistic electron beams. *Mailing Add:* Plasma Fusion Ctr Mass Inst Technol NW16-236 Cambridge MA 02139

JOHNSTON, GEORGE ROBERT, b Salt Lake City, Utah, July 4, 34; m 59; c 3. CYTOGENETICS. *Educ:* Univ Utah, BS, 59, MS, 61, PhD(genetics), 64. *Prof Exp:* Fel genetics, Univ Calif, Berkeley, 64-65; asst prof zool, Univ Wyo, 65-67; from asst prof to assoc prof, 67-77, PROF BIOL, CALIF STATE UNIV, HAYWARD, 77- *Concurrent Pos:* Consult pediat, Kaiser Hosp, Oakland, Calif, 73-; consult, Biomed Div, Lawrence Livermore Lab, 75- *Mem:* Fel AAAS; Asn Cytogenetics Technologists. *Res:* Human chromosome identification linked to clinical defects and the structure of mammalian chromosomes. *Mailing Add:* Dept Biol Calif State Univ Hayward CA 94542

JOHNSTON, GEORGE TAYLOR, b Princeton, WVa, Apr 18, 42; m 66; c 2. OPTICAL PHYSICS. *Educ:* Mich State Univ, BS, 62, MS, 65, PhD(physics), 67; Univ Dayton, MS, 74. *Prof Exp:* Res assoc physics, Brown Univ, 67-69; asst prof, Univ Dayton, 69-72, res physicist, Univ Dayton, Res Inst, 72-81; sr scientist, Rocketdyne, Div Rockwell Int, 81-82; PROG MGR, OPTICAL COATING LAB, INC, 82- *Mem:* Optical Soc Am; Soc Photo-optical Instrumentation Engrs. *Res:* Optical properties of materials; optical instrumentation and metrology; analysis, test and evalutation of high energy laser optical components and component materials, including optical thin films and laser damage mechanisms in coatings, mirrors and transparent materials. *Mailing Add:* 1829 Sherwood Ct Santa Rosa CA 95405

JOHNSTON, GERALD SAMUEL, b Johnstown, Pa, Aug 4, 30; m 56; c 6. NUCLEAR MEDICINE. *Educ:* Univ Pittsburgh, BS, 52, MD, 56. *Prof Exp:* Intern rotating, Walter Reed Gen Hosp, US Army, 56-57, resident internal med, Brooke Gen Hosp, San Antonio, 57-60, comdr, Mobile Army Surg Hosp, Korea, 61-62, chief nuclear medicine, Walter Reed Gen Hosp, 63-69, nuclear med, Letterman Gen Hosp, San Francisco, 69-71; dir nuclear med,

NIH, 71-82; PROF MED & RADIOL NUCLEAR MED, UNIV MD, BALTIMORE, 76-, ACTG CHMN, DIAG RADIOL, 89- *Concurrent Pos:* Clin assoc prof med, Georgetown Univ, 74-; prof radiol & nuclear med, Uniformed Serv, Univ Health Sci, Bethesda, Md, 79- *Mem:* Soc Nuclear Med; Am Col Physicians; Am Med Asn; Am Col Radiol; Am Col Nuclear Med. *Res:* Renal function; renal transplantation; nuclear medicine applications to renal function and cardiac function. *Mailing Add:* 9423 Holland Ave Bethesda MD 20814

JOHNSTON, GORDON ROBERT, b Portland, Ore, July 13, 28; m 60; c 4. ORGANIC CHEMISTRY. *Educ:* Univ Portland, BS, 50, MS, 52; Univ Ill, PhD(org chem), 56. *Prof Exp:* Res org chemist, Dow Chem Co, 56-58; res assoc org chem, Med Sch, Univ Ore, 58-60; res chemist, Crown Zellerbach Corp, 60-62; res chemist, Aerojet-Gen Corp, 62-63; res fel, Calif Inst Technol, 63-64; asst prof org chem, Col Women, San Diego, 64-66; ASST PROF CHEM, PA STATE UNIV, 66- *Concurrent Pos:* Trainee, Mass Inst Technol, 76-77. *Mem:* Am Chem Soc; Am Inst Chemists; Royal Soc Chem. *Mailing Add:* Pa State Univ Beaver Campus Broohead Rd Monaca PA 15061-1347

JOHNSTON, HARLIN DEE, b Ogden, Utah, Mar 16, 42; m 63; c 4. PETROLEUM CHEMISTRY, PILOT PLANT. *Educ:* Brigham Young Univ, BA, 65, PhD(inorg chem), 68. *Prof Exp:* PROJ MGR, PHILLIPS RES CTR, PHILLIPS PETROL CO, 68- *Mem:* Am Chem Soc; Sigma Xi. *Res:* Pilot plant design, operation, and supervision; hetergenous catalysis of solid-gas and solid-liquid-gas systems; laboratory and pilot plant automation. *Mailing Add:* Phillips Res Ctr Phillips Petrol Co Bartlesville OK 74004

JOHNSTON, HAROLD SLEDGE, b Woodstock, Ga, Oct 11, 20; m 48; c 4. PHYSICAL CHEMISTRY. *Educ:* Emory Univ, AB, 41; Calif Inst Technol, PhD(chem), 48. *Hon Degrees:* DSc, Emory Univ, 65. *Prof Exp:* Asst, Nat Defense Res Comt, Calif Inst Technol, 42-45, assoc prof chem, 56-57; from instr to assoc prof, Stanford Univ, 47-56; dean, Col Chem, 66-70, PROF CHEM, UNIV CALIF, BERKELEY, 57- *Concurrent Pos:* Res grants, Off Naval Res, 50-56, M W Kellogg Co, 51-53, Standard Oil Calif, 55-57, NSF, 59-68 & 75-78, USPHS, 63-70, Mat & Molecular Res Div, Lawrence Berkeley Lab, 66- & others; Guggenheim fel, Belg, 60-61; NATO vis prof, Univ Rome, Italy, 64; nat lectr, Sigma Xi, 73; assoc ed, J Geophys Res, 77-; Acad Senate Fac res lectr, Univ Calif, Berkeley, 88-89. *Honors & Awards:* Bourke Lectr, Faraday Soc, 61; George B Kistiakowsky Lectr, Harvard Univ, 73; Pollution Control Award, Am Chem Soc, 74, Award in Chem Contemporary Technol Probs, 85; G N Lewis Lectr, Univ Calif, 75; Cassett Found Lectr, Temple Univ, 78; Tyler Prize, 83. *Mem:* Nat Acad Sci; fel AAAS; Am Chem Soc; fel Am Phys Soc; Am Acad Arts & Sci; fel Am Geophys Union; Sigma Xi. *Res:* Fast gas phase reactions; kinetic isotope effects; photochemistry; unimolecular reactions; atmospheric chemistry; author of 2 books. *Mailing Add:* Dept Chem Univ Calif Berkeley CA 94720

JOHNSTON, HARRY HENRY, microbiology, biochemistry; deceased, see previous edition for last biography

JOHNSTON, HERBERT NORRIS, b Cleveland, Ohio, Aug 9, 28; m 50; c 2. CHEMISTRY. *Educ:* Ohio Univ, BS, 49. *Prof Exp:* Res chemist coatings res, Glidden Co, 49-52; assoc chief, 52-68, chief polymer & paper technol div, Columbus Lab, 68-72, mgr polymer & paper chem, 72-78, MGR INDUST MKT OFF, COLUMBUS LABS, BATTELLE MEM INST, 78- *Mem:* Am Chem Soc; Sigma Xi; Tech Asn Pulp & Paper Indust; Am Mkt Asn; Am Mgt Asn. *Res:* Coatings, polymers, adhesives and inks for paper, wood and metals; powdered polymers, service life of polymeric materials, processing of plastics. *Mailing Add:* 1883 Andover Rd Columbus OH 43212

JOHNSTON, JAMES BAKER, b Baton Rouge, La, Sept 10, 46; m 70; c 3. MARINE SCIENCES, SCIENCE EDUCATION. *Educ:* La State Univ, Baton Rouge, BS, 70, MEd, 71; Univ Southern Miss, PhD(sci educ, biol), 73. *Prof Exp:* Oceanographer marine biol, Bur Land Mgt, New Orleans, 74-76; ecologist, 76-85, SUPV ECOLOGIST, US FISH & WILDLIFE SERV, US DEPT INTERIOR, 85- *Concurrent Pos:* Math instr & NSF consult, Prentiss Inst & Jr Col, 72-73; marine res asst, Univ Southern Miss, Gulf Univs Res Consortium, 72-73; marine fisheries mgt consult, Miss Marine Res Coun, 73-74; consult, Environ Can, 84-86. *Honors & Awards:* Edward H Hillard Award, Nat Wildlife Fedn, 72. *Mem:* Ecol Soc Am; Explorers Club; Estuarine Res Fedn; Coastal Soc. *Res:* Characterization and geophysical mapping of offshore reefs and banks; marine fisheries management; ecosystem characterization and system analysis of coastal regions; development of marine science education programs; studies on wetlands and coastal barriers. *Mailing Add:* 236 S Military Rd No R14 Slidell LA 70461

JOHNSTON, JAMES BENNETT, b San Diego, Calif, Dec 31, 43; m 69; c 2. BIODEGRADATION, GENETIC ENGINEERING. *Educ:* Univ Md, College Park, BS, 66; Univ Wis-Madison, PhD(biochem), 70. *Prof Exp:* Fel, Inst Pasteur, Paris, 70-71; res fel, Univ Kent, Canterbury, UK, 71-74; vis asst prof, Univ Ill, Urbana, 74-76, asst prop, 76-83; res fel, Smith Kline & French, 83-87; RES FEL, ENZYMATICS INC, 87- *Concurrent Pos:* Consult, Cetus Corp, 76-80, Ill Environ Protection Agency, 76-81, Pan Am Health Orgn, 80-, AgroBiotics Corp, 81-; prin investr, var grants, 77- *Mem:* Am Chem Soc; Am Soc Microbiol; Am Asn Clin Chem. *Res:* Recovery, detection and identification of environmental mutagens, especially in potable waters, and the genetics of hydrocarbon degradation by bacteria; manipulation of bacterial DNA to improve biodegradations for waste treatment or for the production of specialty chemicals; invention and development of instrument-dependent, quantitative diagnostic devices. *Mailing Add:* 1309 Cedar Rd Ambler PA 19002

JOHNSTON, JAMES P(AUL), b Pittsburgh, Pa, May 11, 31; m 57; c 5. MECHANICAL ENGINEERING, FLUID DYNAMICS. *Educ:* Mass Inst Technol, BS & MS, 54, ScD(mech eng), 57. *Prof Exp:* Res engr, Ingersoll-Rand Co. NJ, 58-61; from asst prof to assoc prof, 61-73, PROF MECH ENG, STANFORD UNIV, 73- *Concurrent Pos:* Instr, Night Grad Sch Prog, Lehigh Univ, 59-60; Am Soc Mech Engrs-Freeman fel, 67; vis res scientist, Nat Phys Lab, Teddington, Eng, 67-68. *Honors & Awards:* Robert T Knapp Award, Am Soc Mech Engrs, 75. *Mem:* AAAS; Am Soc Mech Engrs; Am Inst Aeronaut & Astronaut. *Res:* Fluid dynamics of real fluids, particularly two and three-dimensional turbulent boundary layers; effects of coordinate system rotation on the turbulent boundary layer; fluid flow in ducts, diffusers and tubomachinery. *Mailing Add:* 813 Cedro Way Stanford CA 94305

JOHNSTON, JEAN VANCE, b Shippensburg, Pa, Feb 17, 12. ORGANIC CHEMISTRY. *Educ:* Smith Col, AB, 34; Yale Univ, PhD(org chem), 38. *Prof Exp:* Asst chem, Smith Col, 39; instr pvt sch, Conn, 40; asst prof chem, Furman Univ, 40-42; from instr to assoc prof chem, 42-74, EMER ASSOC PROF CHEM, CONN COL, 74- *Concurrent Pos:* Fel, Pa State Univ, 69-70; assoc prof, Shippensburg Univ, 76 & 77. *Mem:* Am Chem Soc; Sigma Xi. *Res:* Synthesis of organic compounds of medicinal interest; amidines. *Mailing Add:* 505 W King St Shippensburg PA 17257

JOHNSTON, JOHN, b Newcastle-on-Tyne, Eng, Nov 8, 25; US citizen; m 47; c 3. RUBBER CHEMISTRY. *Educ:* Hull Col Technol, BSc, 58. *Prof Exp:* Teacher, County Educ Authorities, Hull, Eng, 48-49; chemist, Standard Oil Co, Saltend, 49-50 & T J Smith & Nephew Ltd, Hull, 50-58; res chemist, Arno Adhesive Tape Inc, 59-67, asst dir res pressure sensitive adhesives, 67, dir res, 68-71, dir res & develop, 71-73, vpres res, Develop & Tech Opers, 73-77; dir tech serv, Johnson & Johnson, 77-78; DIR RES, TUCK INDUSTS, 78- *Concurrent Pos:* Lectr, Purdue Univ, North Regional Campus, 65-71; chmn, Pressure Sensitive Tape Coun Tech Comt, 74-76. *Mem:* Am Chem Soc. *Res:* Theory and practice of pressure sensitive adhesives. *Mailing Add:* 29 Westview Terr Midland Park NJ 07432

JOHNSTON, JOHN B(EVERLEY), b Los Angeles, Calif, Aug 11, 29; m 66; c 1. COMPUTER SCIENCE. *Educ:* Calif Inst Technol, BS, 51, PhD(math), 55. *Prof Exp:* Instr math, Cornell Univ, 55-57; asst prof, Univ Kansas City, 57-58; from asst prof to assoc prof, Univ Kans, 58-64; mathematician, Gen Elec Res & Develop Ctr, NY, 64-68; assoc prof comput sci, Ind Univ, Bloomington, 68-69; info scientist, Gen Elec Res & Develop Ctr, NY, 69-71; PROF COMPUT SCI, NMEX STATE UNIV, 71- *Mem:* Asn Comput Mach. *Res:* Structure of computation; computer languages; structure of computer systems. *Mailing Add:* Dept Comput Sci NMex State Univ Las Cruces NM 88003-0001

JOHNSTON, JOHN ERIC, b Detroit, Mich, Feb 5, 48; m. POLYMER CHEMISTRY, ANALYTICAL CHEMISTRY. *Educ:* Univ Notre Dame, BS, 70; Univ Akron, PhD(polymer sci), 75. *Prof Exp:* Fel polymer sci, Ctr Macromolecular Res, 75; sr chemist polymer synthesis & characterization, Union Carbide Corp, 76-80; head, Viscosity Index Modifier Res Group, 81-88, head, Polyalkene Tech, 88-89, LEADER, DISCHARGE ELIMINATION TASK FORCE, 89-, COMPONENT MGR, VISCOSITY MODIFIERS, EXXON CHEM CO, 90- *Mem:* Am Chem Soc; Soc Automotive Engrs; AAAS; NY Acad Sci. *Res:* Novel polymer processes and polymer characterization techniques; polymer structure, process and property correlations; polymer colloid morphology; process relationships; lubricant additives; environmentally sound processes. *Mailing Add:* Exxon Chem Co PO Box 536 Linden NJ 07036

JOHNSTON, JOHN MARSHALL, b North Platte, Nebr, Nov 14, 28; m 53; c 3. BIOCHEMISTRY. *Educ:* Hastings Col, BA, 49; Univ Colo, PhD, 53. *Prof Exp:* Res assoc, Walter Reed Inst Res, 53-55; from instr to assoc prof, 55-66, PROF BIOCHEM, UNIV TEX HEALTH SCI CTR, DALLAS, 66-, PROF OBSTET & GYNEC, 74- *Concurrent Pos:* NSF sr res fel, Univ Lund, 62-63. *Mem:* AAAS; Am Chem Soc; Am Soc Biol Chemists; Sigma Xi. *Res:* Lipid metabolism in absorption; fetal lung maturation, partuition and membranes. *Mailing Add:* Dept Biochem Southwest Med Sch 5323 Harry Hines Dallas TX 75235

JOHNSTON, JOHN O'NEAL, b Baltimore, Md, July 21, 39; m 77; c 2. BIOCHEMICAL ENDOCRINOLOGY, NEUROENDOCRINE PHARMACOLOGY. *Educ:* Univ Md, College Park, BS, 61, MS, 65, PhD(reproductive physiol), 70. *Prof Exp:* Res asst reproductive physiol, Agr Res Serv, USDA, 61-65; res asst, dept animal sci, Univ Md, 65-69; res scientist fertil res, Upjohn Co, 69-71; res endocrinologist, 71-73, sect head endocrinol, 73-81, SR RES ENDOCRINOLOGIST, ENDOCRINOL, MARION MERRELL DOW RES INST, MARION MERRELL DOW INC, 81- *Concurrent Pos:* Biol consult, Life Sci Div, Res Triangle Park, NC, 75-78; consult, Vet Pharmaceut, Jensen-Salsbery Labs, Kansas, Mo, 72-79; sci adv bd, Cincinnati Zoo, 85- *Mem:* Soc Study Reproduction; Am Soc Andrology; NY Acad Sci; AAAS; Endocrine Soc. *Res:* Development of therapeutic agents for control of male and female fertility; regulation of hormonal action via receptor mechanism in target tissues; animal growth stimulants; neuroendocrine pharmacology of animal behavior; development of enzyme inhibitors for regulation of endocrine dependent cancer, endocrine hypertension and reproductive processes. *Mailing Add:* Marion Merrell Dow Res Inst Marion Merrell Dow Inc 2110 Galbraith Cincinnati OH 45215-6300

JOHNSTON, JOHN SPENCER, b Phoenix, Ariz, May 27, 44; m 66; c 1. ENTOMOLOGY. *Educ:* Univ Wash, BS, 66; Univ Ariz, PhD(genetics), 72. *Prof Exp:* NIH fel, Univ Tex, Austin, 72-75; asst prof biol, Baylor Univ, Waco, 75-79; ASSOC PROF GENETICS, TEX A&M UNIV, 79- *Concurrent Pos:* Res grant, Energy Res Develop Asn & Univ Tex, Austin, 73- *Mem:* AAAS; Genetics Soc Am; Evolution Soc Am; Soc Am Naturalists. *Res:* Ecological genetics of Drosophila species. *Mailing Add:* 1303 Augustine Ct College Station TX 77841

JOHNSTON, KATHARINE GENTRY, b Minneapolis, Kans, Jan 19, 21; m 50; c 2. INDUSTRIAL ORGANIC CHEMISTRY. *Educ:* Kans State Univ, BS, 42, MS, 50. *Prof Exp:* Res chemist, Org Div, Monsanto Chem Co, 51-71; staff scientist clin diag, Ames Co, Div Miles Labs, 71-86; RETIRED. *Mem:* Sigma Xi; AAAS; Am Chem Soc. *Res:* Organic and enzymatic reactions in clinical diagnostic systems; complex formation and stabilization; ion temperature polymerization. *Mailing Add:* 1633 Woodfield Court Elkhart IN 46514

JOHNSTON, KENNETH JOHN, b New York, NY, Oct 9, 41; m 66. ASTRONOMY. *Educ:* Manhattan Col, BEE, 64; Georgetown Univ, PhD(astron), 69. *Prof Exp:* Nat Acad Sci-Nat Res Coun res assoc astron, 69-71, ASTRONOMER, NAVAL RES LAB, 71- *Mem:* Am Astron Soc; Int Astron Union. *Res:* Radio astronomy; variable stars. *Mailing Add:* Naval Res Lab Code 7132 Washington DC 20390

JOHNSTON, LA VERNE ALBERT, b Hallettsville, Tex, Mar 7, 30; m 61; c 2. BOTANY. *Educ:* Baylor Univ, AB, 51, MA, 57; Southwestern Baptist Theol Sem, MRE, 54. *Prof Exp:* Teacher, Gonzales Ind Sch Dist, 51-52; instr & asst prof biol, Baylor Univ, 54-60; res asst bot, Univ Tex, Austin, 76-80 & 83-90; RETIRED. *Concurrent Pos:* Tech writer, 80-83. *Mem:* Am Inst Biol Sci; Bot Soc Am. *Res:* Phycology; angiosperm taxonomy. *Mailing Add:* 3905 Ave G Austin TX 78751

JOHNSTON, LAURANCE S, b St Paul, Minn, Aug 4, 50; m 75. HEALTH SCIENCE ADMINISTRATION. *Educ:* Hamline Univ, BS, 72; Northwestern Univ, Evanston, MS, 73, PhD(biochem & molecular biol), 77; George Mason Univ, MBA, 85. *Prof Exp:* Fel, Dept Biochem & Molecular Biol, Northwestern Univ, 76-77; fel, Chicago Med Sch, 77-78; consumer safety officer, Off Compliance, Bur Foods, Food & Drug Admin, Washington, DC, 78-81; health scientist admistr & exec secy, 81-86, DIR, DIV SCI REV, NAT INST CHILD HEALTH & HUMAN DEVELOP, NIH, 86- *Mem:* Am Chem Soc; AAAS. *Res:* Health science administration. *Mailing Add:* Nat Inst Child Health & Human Develop NIH 6130 Exec Blvd Rm 520 North Bethesda MD 20892

JOHNSTON, LAWRENCE HARDING, b Tse-Nan-Fu, China, Feb 11, 18; US citizen; m 42; c 5. OPTICAL PHYSICS. *Educ:* Univ Calif, AB, 40, PhD(physics), 50. *Prof Exp:* Res assoc, Radiation Lab, Mass Inst Technol, 40-43; res assoc, Los Alamos Sci Lab, Univ Calif, 43-45; asst, Lawrence Radiation Lab, Univ Calif, 45-50; from instr to assoc prof physics, Univ Minn, Minneapolis, 50-61; sr scientist, Aerospace Corp, 61-63; sr staff mem, Stanford Linear Accelerator Ctr, 63-67; PROF PHYSICS, UNIV IDAHO, 67- *Mem:* Fel Am Phys Soc; fel Am Sci Affil. *Res:* Far infrared physics; molecular spectroscopy; microwave radar; atom bomb development; proton linear accelerator development; nuclear and high energy particle physics; proton-proton scattering; submillimeter wave laser stark spectroscopy. *Mailing Add:* Dept Physics Univ Idaho Moscow ID 83843

JOHNSTON, MALCOLM CAMPBELL, b Montague, PEI, Feb 13, 31; m 55. TERATOLOGY, DEVELOPMENTAL BIOLOGY. *Educ:* Univ Toronto, DDS, 54, MScD, 56; Univ Rochester, PhD(anat), 65. *Prof Exp:* Res assoc clin res, Cleft Palate Res & Treat Ctr, Hosp Sick Children, Toronto, 56-60; asst & assoc prof hist, Sch Dent & Med, Univ Toronto, 64-69; vis scientist, NIH, Bethesda, Md, 69-76; PROF ORTHODONT & ANAT, SCH DENT & MED, UNIV NC, 76- *Concurrent Pos:* Sect ed, Cleft Palate J, 81- *Mem:* Sigma Xi; Am Cleft Palate Asn; Teratology Soc; Am Asn Anatomists. *Res:* Normal and abnormal embryonic craniofacial development in mice, with limited studies on man. *Mailing Add:* Dent Res Ctr Univ NC Chapel Hill NC 27514

JOHNSTON, MANLEY RODERICK, b Edmonton, Alta, Oct 2, 42; m 67. ORGANIC POLYMER CHEMISTRY. *Educ:* Univ Alta, BSc, 64; Univ Ill, Urbana, MS, 66 & PhD (org chem), 69. *Prof Exp:* Sr chemist, 68-72, res specialist, 72-73, supvr, 73-78, tech mgr, Bldg Serv & Cleaning Prod Div, 78-82, lab mgr, Nonwovens Technol Ctr, 82-83, dir Life Sci Res Lab, 83-86, DIR, DISPOSIBLE PROD DIV, 3M CO, 86- *Mem:* Am Chem Soc; Royal Soc Chem. *Res:* Small ring compounds; organic coatings; metal finishing; adhesion; high temperature polymers; fibers; polymerization catalysts; fluorine chemistry. *Mailing Add:* 7525 Currell Blvd Woodbury MN 55125

JOHNSTON, MARILYN FRANCES MEYERS, b Buffalo, NY, Mar 30, 37. BIOCHEMISTRY, IMMUNOLOGY. *Educ:* Dameon Col (Rosary Hill Col), BS, 66; St Louis Univ, PhD(biochem), 70, MD, 75. *Prof Exp:* NIH fel, Wash Univ, 70-72; instr biochem, Sch Med, St Louis Univ, 72-75; resident path, Sch Med, Wash Univ, 75-77; resident path, St John's Mercy Med Ctr, 77-79; med dir, Mo-Ill Regional Red Cross, 83-88; fel path & med, 79-80, asst prof path, 80-86, ASSOC PROF PATH, MED SCH, ST LOUIS UNIV, 86- *Concurrent Pos:* AMA J Goldberger fel, St Louis Univ, 74; prin investr; vchmn inspection & accreditation, Am Asn Blood Banks, 83-; mem comt transfusion pract, 84-89; med dir blood bank, transfusion serv, Apheris, St Louis Univ Hosp. *Mem:* Am Asn Blood Banks; Col Am Pathologists; Am Asn Immunologists; Int Soc Blood Transfusion; Sigma Xi; Am Soc Clin Path. *Res:* Transfusion medicine; red cell surface antigens. *Mailing Add:* Sch Med St Louis Univ Hosps 3635 Vista at Grand Blvd PO Box 15250 St Louis MO 63110-0250

JOHNSTON, MARSHALL CONRING, b San Antonio, Tex, May 10, 30; m 61; c 2. SYSTEMATIC BOTANY. *Educ:* Univ Tex, BS, 51, MA, 52, PhD(bot), 55. *Prof Exp:* Fel, Rice Inst, 55; asst prof biol, Sul Ross State Col, 58-59; res scientist bot, 59-61, assoc prof, 61-72, PROF BOT, UNIV TEX, AUSTIN, 72- *Concurrent Pos:* Sci asst, Univ Munich, 68-69; dir, Rare Plant Study Ctr, 72- *Mem:* AAAS; Bot Soc Am; Am Soc Plant Taxon; Int Soc Plant Taxon; Am Inst Biol Sci; Sigma Xi. *Res:* Distribution of vegetation types; systematics and historical biogeography of vascular plants of southwestern United States and northern Mexico; flora of Texas. *Mailing Add:* Dept Bot Univ Tex Austin TX 78712

JOHNSTON, MARY HELEN, failure analysis, for more information see previous edition

JOHNSTON, MELVIN ROSCOE, b McAlester, Okla, June 23, 21; m 46; c 2. FOOD TECHNOLOGY. *Educ:* Agr & Mech Col Tex, BS, 48; Ore State Col, MS, 50; Univ Mo, PhD, 56. *Prof Exp:* Instr food technol, Ore State Col, 48-50; food technologist, Libby, McNeil & Libby, 50-52; from instr to asst prof hort, Univ Mo, 52-59; prof food technol & adv dept, Univ Tenn, Knoxville, 59-73; chief, Fruit & Veg Br, Div Food Technol, Bur Foods, Food & Drug Admin, Health Educ & Welfare, 73-80, chief, Plant & Protein Prods

Br, 80-86; RETIRED. *Mem:* Fel Inst Food Technologists; Sigma Xi. *Res:* Implementation and support of regulatory action; food color technology; freezing; freeze-drying and thermal processing of foods; ammonia damage to frozen foods. *Mailing Add:* 23 Country Club Circle New Braunfels TX 78130-5376

JOHNSTON, MICHAEL ADAIR, bioorganic chemistry, medicinal chemistry, for more information see previous edition

JOHNSTON, MILES GREGORY, SHOCK, INFLAMMATION. *Educ:* Univ Toronto, PhD(exp path), 78. *Prof Exp:* ASSOC PROF PATH, UNIV TORONTO, 81- *Mailing Add:* Sunnybrook Hosp S-Wing Univ Toronto Toronto ON M4N 3M5 Can

JOHNSTON, MILTON DWYNELL, JR, b Hillsboro, Ore, Nov 4, 43. PHYSICAL CHEMISTRY, MOLECULAR SPECTROSCOPY. *Educ:* Portland State Univ, BA, 65; Princeton Univ, AM, 68, PhD(chem), 69. *Prof Exp:* Res assoc nuclear magnetic resonance, Univ Ariz, 70-71; res assoc, Tex A&M Univ, 71-73; asst prof, 73-80, ASSOC PROF CHEM, UNIV SFLA, 80- *Mem:* Am Chem Soc; Am Phys Soc; Royal Soc Chem; Sigma Xi; NY Acad Sci. *Res:* Nuclear magnetic resonance solvent effects; theory of nuclear magnetic resonance spectral parameters; theory of liquids and liquid solutions and of intermolecular forces. *Mailing Add:* Dept Chem Univ SFla Tampa FL 33620

JOHNSTON, NORMAN JOSEPH, b Charles Town, WVa, Dec 15, 34; m 57; c 4. ORGANIC POLYMER CHEMISTRY. *Educ:* Shepherd Col, BS, 56; Univ Va, PhD(org chem), 63. *Prof Exp:* Chemist insulating mat dept, Gen Elec Co, 61-63; asst prof chem, Va Polytech Inst & State Univ, 63-66; Nat Acad Sci-Nat Res Coun resident res fel, NASA, 66-67, aerospace technologist & chemist, 67-70, head polymer sect, 70-80, sr scientist, 81-88, chief scientist-mats, 88-90, MGR- COMPOSITES TECHNOL, LANGLEY RES CTR, NASA, 91- *Mem:* Am Chem Soc; Soc Aerospace Mat & Process Engrs. *Res:* Synthesis and characterization of high performance polymers and their evaluation as composite matrices; toughened high performance composites; resin property-composite property relationships. *Mailing Add:* Langley Res Ctr NASA Mail Stop 226 Hampton VA 23665-5225

JOHNSTON, NORMAN PAUL, b Salt Lake City, Utah, Apr 5, 41; m 66; c 5. ANIMAL NUTRITION, REPRODUCTION BIOLOGY. *Educ:* Brigham Young Univ, BA, 66; Ore State Univ, MS, 67, PhD(avian nutrit), 71; Univ Utah, MBA, 69. *Prof Exp:* Animal nutritionist, Brookfield Prod Inc, 69-70; PROF ANIMAL SCI, BRIGHAM YOUNG UNIV, 71- *Mem:* Sigma Xi; Poultry Sci; Am Soc Animal Sci; World Poultry Sci. *Res:* Poultry reproduction, in particular artificial insemination; animal nutrition - rabbits, goats, poultry; international agriculture. *Mailing Add:* 1795 South 340E Orem UT 84051

JOHNSTON, NORMAN WILSON, b Pittsburgh, Pa, June 18, 42; m 65; c 3. POLYMER CHEMISTRY. *Educ:* Clarion State Col, BS, 64; Univ Akron, PhD(polymer sci), 68. *Prof Exp:* Chemist polymer chem, Ethyl Corp, 65; sr chemist, Union Carbide Corp, 68-71, proj scientist, 71-72, group leader adhesives, coatings & moldings, 72-76; assoc dir res & develop, Owens-Corning Fiberglass Corp, 76-77, lab dir, 77-79, res dir, 78-81, mgr bus & tech planning, 81-; VPRES TECHNOL & ENG, LIBBEY-OWENS-FORD CO. *Mem:* Am Chem Soc. *Res:* Polymer structure: property relationships, polymer synthesis, coatings, adhesives, fire retardance, molding and extrusion, composites, polymer blends, degradable plastics, cement, foams, insulation. *Mailing Add:* Dept Technol & Eng Libbe-Ownes-Ford Co Tech Ctr 811 Madison Ave Box 799 Toledo OH 43695

JOHNSTON, PAUL BRUNS, b Chicago. Ill, Apr 2, 27; wid; c 3. MEDICAL MICROBIOLOGY. *Educ:* Northwestern Univ. BS, 49; Loyola Univ, Ill, MS, 51; Univ Chicago, PhD(microbiol), 57. *Prof Exp:* Virologist, US Naval Med Res Unit 2, Taiwan, 57-60; asst prof microbiol, Jefferson Med Col, 60-64; ASSOC PROF MICROBIOL, SCH MED, UNIV LOUISVILLE, 64- *Concurrent Pos:* Instr, Univ Chicago, 57-60. *Mem:* Am Soc Microbiol; Soc Exp Biol & Med; Tissue Cult Asn; Am Asn Immunol. *Res:* Nature of latent virus infections; simian foamy virus immunology; adenoviruses. *Mailing Add:* Dept Microbiol Sch Med Univ Louisville Louisville KY 40292

JOHNSTON, PERRY MAX, b Edgewood, Tex, Feb 6, 21; m 43; c 4. VERTEBRATE EMBRYOLOGY. *Educ:* NTex State Col, BS, 40, MS, 42; Univ Mich, PhD(zool), 49. *Prof Exp:* From asst prof to assoc prof zool, 49-54, PROF ZOOL, UNIV ARK, FAYETTEVILLE, 54-, CHMN DEPT, 66- *Concurrent Pos:* Res partic, Oak Ridge Inst Nuclear Studies, 53-54. *Mem:* Am Soc Zool; Am Micros Soc; Sigma Xi. *Res:* Vertebrate embryology; embryology of centrarchid fishes; utilization of radioisotopes by vertebrate embryos. *Mailing Add:* Dept Zool Univ Ark Fayetteville AR 72703

JOHNSTON, PETER RAMSEY, b 1926; m 49; c 4. FLUID FILTRATION. *Educ:* Miami Univ, Ohio, BA, 48. *Prof Exp:* SR PROJ ENG, AMETEK, INC, 80- *Mem:* Am Chem Soc; Am Inst Chem Engrs; Am Soc Testing & Mat; Am Filtration Soc. *Res:* Liquid filtration; electrochemistry; polymer chemistry. *Mailing Add:* Ametek Inc 502 Indiana Ave Sheboygan WI 53081

JOHNSTON, RAYMOND F, b Fenton, Mo, June 29, 13; m 35; c 1. PHYSIOLOGY, PHARMACOLOGY. *Educ:* Univ Mo, BS, 35; Mich State Univ, MS, 48, DVM, 49; Univ Minn, PhD, 59. *Prof Exp:* Instr voc agr, Univ Mo, 35-45; asst path, 45-47, asst physiol, 47-49, PROF PHYSIOL, MICH STATE UNIV, 49- *Concurrent Pos:* Sr mem team vet to Indonesia, 60-62. *Mem:* Fel Am Vet Med Asn. *Res:* Toxicology; cardiovascular physiology; neurophysiology; biomedical communications. *Mailing Add:* 4583 Sequoia Terr Okemos MI 48864

JOHNSTON, RICHARD BOLES, JR, b Atlanta, Ga, Aug 23, 35; m 60; c 3. PEDIATRICS. *Educ:* Vanderbilt Univ, BA, 57, MD, 61. *Prof Exp:* NIH fel, Harvard Med Sch, 67-68, USPHS training grant, 68-69, NIH spec fel, 69-70; from asst prof to assoc prof pediat & microbiol, Univ Ala, Birmingham, 70-77; dir dept pediat, Nat Jewish Hosp & Res Ctr, Denver, 77-86; prof pediat, Sch Med, Univ Colo, 77-86, vchmn dept, 80-86; WILLIAM BENNETT PROF & CHMN DEPT PEDIAT, SCH MED, UNIV PA, 86-; PHYSICIAN-IN-CHIEF, CHILDREN'S HOSP PHILADELPHIA, 86- *Concurrent Pos:* Macy Found scholar, Rockefeller Univ, NY, 76-77; vis prof, Rockefeller Univ, NY, 83-84; dir trustees, Int Pediat Res Found, 83- *Mem:* Soc Pediat Res (pres, 80-81); Am Asn Immunologists; Asn Am Physicians; Am Soc Clin Invest; Am Pediat Soc; Int Pediat Res Found. *Res:* Mechanisms of resistance to infection; phagocyte function; complement. *Mailing Add:* Dept Pediat Univ Pa Children's Hosp 34th & Civic Center Blvd Philadelphia PA 19104

JOHNSTON, RICHARD FOURNESS, b Oakland, Calif. July 27, 25; m 48; c 3. SYSTEMATICS, ECOLOGY. *Educ:* Univ Calif, BA, 50, MA, 53, PhD, 55. *Prof Exp:* Instr biol, NMex State Univ, 56-57; from asst prof to assoc prof zool, 58-67, assoc cur, 63-67, chmn dept, 79-82, PROF ZOOL, UNIV KANS, 67-, CUR BIRDS, MUS NATURAL HIST, 67- *Concurrent Pos:* Ed, Syst Zool, Soc Syst Zool, 67-70; prog dir syst biol, NSF, 68-69; ed, Annual Rev Ecol & Systematics, 68-, Current Ornith, 81-86. *Honors & Awards:* Coues Award, Am Ornith Union, 75. *Mem:* Fel AAAS; Ecol Soc Am; fel Am Ornith Union; Soc Study Evolution; Soc Conserv Biol; Soc Syst Zool (pres, 77-79). *Res:* Systematics; evolutionary biology; behavior and ecology of birds, especially passerines. *Mailing Add:* Univ Kans 602 Dyche Hall Lawrence KS 66045-2454

JOHNSTON, RICHARD H, b Philadelphia, Pa, Apr 7, 29; m 66; c 1. HYDROGEOLOGY. *Educ:* Pa State Univ, BS, 57; Univ Wyoming, MA, 59. *Prof Exp:* Geologist, 59-69, GROUNDWATER HYDROLOGIST, US GEOL SURV, 69- *Mem:* Geol Soc Am; Int Asn Hydro-Geologists; Asn Groundwater Scientists & Engrs. *Res:* Regional aquifer systems; Karst hydrogeology. *Mailing Add:* 4311 Ninth St E Beach St Simons Island GA 31522

JOHNSTON, ROBERT BENJAMIN, b North Platte, Nebr, Mar 21, 22; m 44; c 2. BIOCHEMISTRY. *Educ:* Hastings Col, AB, 44; Univ Chicago, PhD(biochem), 49. *Prof Exp:* Instr physiol chem, Yale Univ, 53; from asst prof to assoc prof, 53-65, PROF CHEM, UNIV NEBR-LINCOLN, 65- *Concurrent Pos:* USPHS spec fel, Max Planck Inst Cell Chem, Munich, 61-62 & Inst Microbiol Biochem, Erlansen, 78-79. *Mem:* AAAS; Am Chem Soc; Brit Biochem Soc; Am Soc Biol Chemists; NY Acad Sci; Sigma Xi. *Res:* Enzyme mechanisms and biological synthesis of peptide bonds; amino acid racemases; releasing factors; enkephalins; peptide antibiotics. *Mailing Add:* 3900 J St Lincoln NE 68510

JOHNSTON, ROBERT E, b Philadelphia, Pa, Apr 16, 42; m 70; c 2. BEHAVIOR, ETHOLOGY & PHEROMONES. *Educ:* Dartmouth Col, AB, 64; Rockefeller Univ, PhD(behav & life sci), 70. *Prof Exp:* From asst prof to assoc prof, 70-87, PROF PSYCHOL & BIOL, CORNELL UNIV, 87- *Concurrent Pos:* Vis prof, Dept Zool, Univ Tex, 80, USSR Acad Sci, A N Severtson Inst Evolutionary Animal Morphol & Ecol, 91; prin investr, numerous grants. *Mem:* Am Inst Biol Sci; Animal Behav Soc; Sigma Xi; Am Soc Mammalogists; Soc Study Reproduction. *Res:* Mechanisms and evolution of behavior, especially reproductive and aggressive behavior; communication, including olfactory (pheromones), auditory and visual signals; relationships between hormones and behavior; neural mechanisms of olfaction; human ethology; mechanism of puberty social influences on endocrine function. *Mailing Add:* Dept Psychol Uris Hall Cornell Univ Ithaca NY 14853

JOHNSTON, ROBERT EDWARD, b Houston, Tex, Sept 19, 47; m 76. MICROBIOLOGY, VIROLOGY. *Educ:* Rice Univ, BA, 68; Univ Tex, Austin, PhD(microbiol), 73. *Prof Exp:* Med Res Coun fel microbiol, Queens Univ, 73-76; asst prof microbiol, 76-80, ASSOC PROF MICROBIOL, NC STATE UNIV, 80- *Concurrent Pos:* NIH Young investr grant, NC State Univ, 78-81. *Mem:* Am Soc Microbiol. *Res:* Host cell influence on virus replication. *Mailing Add:* Dept Microbiol NC State Univ Raleigh NC 27695

JOHNSTON, ROBERT HOWARD, b Martinsville, Ohio, Feb 7, 24; m 56; c 4. MATHEMATICS. *Educ:* Miami Univ, BSEd, 47, MA, 50. *Prof Exp:* From asst & instr math to asst prof air sci, Miami Univ, 47-54; res assoc math & dir res br, Weapons Employ Br, Armed Forces Spec Weapons Command, Albuquerque, 54-57; from instr to asst prof math, USAF Acad, 57-61, res assoc, Frank J Seker Res Lab, Off Aerospace Res, 63-64; from asst prof to assoc prof, Dept Math, 64-70, dep head dept, 69-70; ASST PROF MATH SCI, VA COMMONWEALTH UNIV, 70-, DIR MATH AUDIO TUTORIAL LAB, 72- *Concurrent Pos:* Reader, Educ Testing Serv, 69-74, consult calculus & anal geom tests, 73- *Mem:* Am Math Soc; Nat Coun Teachers Math. *Res:* Mathematical audio tutorial laboratories as a vehicle to enable greater background development in the basic mathematics areas. *Mailing Add:* Dept Math Sci Va Commonwealth Univ Box 2520 Richmond VA 23284

JOHNSTON, ROBERT R, b Oakland, Calif, June 10, 29; m 50; c 3. THEORETICAL PHYSICS. *Educ:* Univ Calif, Berkeley, AB, 54, MA, 56, PhD(physics), 61. *Prof Exp:* Staff mem theoret physics, Los Alamos Sci Lab, 56-63; res scientist res labs, Lockheed Missiles & Space Co, Calif, 63-64; res consult theoret physics, Systs Res & Develop Ctr, Int Bus Mach Corp, 64-65; res scientist, Lockheed Palo Alto Res Labs, 65-73; DEPT PHYSICS, UNIV BC, 73- *Concurrent Pos:* Consult, Los Alamos Sci Lab, Univ Calif, 64- *Mem:* Am Phys Soc. *Res:* Computer applications in theoretical physics; atomic and nuclear reaction phenomena; neutron physics; Monte Carlo techniques. *Mailing Add:* Dept Physics Univ BC Vancouver BC V6T 1Z1 Can

JOHNSTON, ROBERT WARD, b Buffalo, NY, May 27, 25; m 59; c 2. PHYSICS, ACADEMIC ADMINISTRATION. *Educ:* Cornell Univ, BEE, 46, PhD(physics), 52. *Prof Exp:* Asst physics, Cornell Univ, 46-51, physicist, Aeronaut Lab, 47-48; physicist, Electronics Lab, Gen Elec Co, NY, 51-57; mgr sci & tech rels, Adv Res & Develop Div, Avco Corp, Mass, 57-59; asst prog dir physics, NSF, 59-60, assoc prog dir, 60-61, spec asst to asst dir math, phys & eng sci div, 61, spec asst to assoc dir res, 61-65, exec asst to dir, Washington, DC, 65-69; vchancellor res, Wash Univ, 69-73; assoc exec officer, Nat Acad Sci-Nat Res Coun, 73-82, dir personnel & appts, 82-85, dep dir admin, 85-87; RETIRED. *Mem:* AAAS; Am Phys Soc. *Res:* Soft x-ray spectroscopy; magnetic materials; solid state physics; research administration. *Mailing Add:* 12705 Huntsman Way Potomac MD 20854

JOHNSTON, ROGER GLENN, b Lincoln, Nebr, Feb 15, 54. LIGHT SCATTERING, INTERFEROMETRY. *Educ:* Carleton Col, BA, 77; Univ Colo, Boulder, MS, 83, PhD(physics), 83. *Prof Exp:* Grad student, chem physics, Univ Colo, 77-83; postdoctoral fel flow cytometry, 83-85, staff mem biophys, 85-90, STAFF MEM, CHEM & LASER SCI DIV, LOS ALAMOS NAT LAB, 90- *Concurrent Pos:* Prin investr, Los Alamos Nat Lab, 86-, proj leader, 88-; consult, 90- *Mem:* Am Phys Soc. *Res:* Process development; development of new techniques and instrumentation for electro-optics, interferometry, light scattering biometrics, and flow cytometry. *Mailing Add:* Los Alamos Nat Lab MS J565 Los Alamos NM 87545

JOHNSTON, RONALD HARVEY, b Drumheller, Alta, May 11, 39; m 69. ELECTRONICS ENGINEERING. *Educ:* Univ Alta, BSc, 61; Univ London, DIC & PhD(elec eng), 67. *Prof Exp:* Eng trainee, Can Gen Elec, 61-62; res asst electronics, Queen's Univ, Belfast, 64-67; mem scientific staff res & develop, Northern Elec Co Ltd, 67-69; asst prof, 70-77, ASSOC PROF ELECTRONICS, UNIV CALGARY, 77- *Mem:* Inst Elec & Electronics Engrs. *Res:* Frequency multipliers; transistor amplifiers and multipliers; microwave measurements; semiconductor circuits. *Mailing Add:* 2018 Rossland Rd E Whitby ON L1N 5R5 Can

JOHNSTON, ROY G, b Chicago, Ill, Jan 7, 14. STRUCTURAL ENGINEERING, EARTHQUAKE ENGINEERING. *Educ:* Univ Southern Calif, BS, 35. *Prof Exp:* FOUNDER & EXEC VPRES, BRANDOW & JOHNSTON ASSOCS, 44- *Mem:* Nat Acad Eng; fel Am Soc Civil Engrs; fel Am Concrete Inst. *Mailing Add:* Brandow & Johnston Assocs 1660 W Third St Los Angeles CA 90017

JOHNSTON, RUSSELL SHAYNE, b Ft William, Ont, Nov 4, 48; m 76; c 3. PLASMA PHYSICS, APPLIED MATHEMATICS. *Educ:* McGill Univ, BSc, 70; Princeton Univ, PhD(plasma physics), 75. *Prof Exp:* Res asst plasma physics, Plasma Physics Lab, Princeton Univ, 70-74; res fel plasma physics, Lawrence Berkeley Lab, Univ Calif, 74-76; asst prof appl physics, Columbia Univ, 76-83; ASSOC PROF PHYSICS, JACKSON STATE UNIV, 83- *Mem:* Am Phys Soc. *Res:* Theoretical plasma physics, particularly nonlinear interactions among waves and particles. *Mailing Add:* Dept Physics & Atmospheric Sci Jackson State Univ Jackson MS 39217

JOHNSTON, STEPHEN CHARLES, b Vancouver, Wash, Sept 15, 50. HUMAN PERFORMANCE, EXERCISE IN HEALTH. *Educ:* Univ Utah, BS, 74, PhD(physiol exercise), 85. *Prof Exp:* Vis asst prof exercise & sport sci, 84-87, asst prof, 87-91, ASSOC PROF EXERCISE & SPORT SCI, UNIV UTAH, 91- *Concurrent Pos:* Exercise consult, Holy Cross Hosp, Salt Lake City, Utah, 83- & Neuropsychol Dept, Vet Admin Hosp, Salt Lake City, Utah, 85-; dir physiol, Sports Med Coun, US Ski Team, 87-91, dir, Sport Sci, 91-; adj asst prof, Div Foods & Nutrit, Univ Utah, 87-, dir, Human Performance Res Lab, 87-; chair, Sports Med Comt, SW Alliance for Health, Phys Educ, Recreation & Dance, 89-91. *Mem:* AAAS; Am Asn Univ Professors; Am Alliance Health, Phys Educ, Recreation & Dance; Am Col Sports Med. *Res:* Effects of exercise and environment on the muscular, cardiovascular, respiratory, nervous and thermoregulatory systems of the human body; work with training response and optimization of training in elite athletes. *Mailing Add:* Dept Exercise & Sport Sci Univ Utah Salt Lake City UT 84112

JOHNSTON, STEWART ARCHIBALD, physical chemistry, for more information see previous edition

JOHNSTON, TAYLOR JIMMIE, b Newbern, Tenn, May 11, 40; m 66; c 2. AGRONOMY, PLANT PHYSIOLOGY. *Educ:* Univ Tenn, Martin, BS, 63; Univ Ill, MS, 65, PhD(agron), 68. *Prof Exp:* From asst prof to assoc prof, 68-76, asst dean, 81-83, PROF CROP SCI, MICH STATE UNIV, 76-, ASSOC DEAN, COL AGR & NATURAL RESOURCES, 83- *Mem:* Am Soc Agron; Crop Sci Soc Am; Sigma Xi. *Res:* Photosynthesis of soybeans and general crop physiology and ecology. *Mailing Add:* Col Agr & Natural Resources Mich State Univ 121 Agr Hall East Lansing MI 48823

JOHNSTON, THOMAS M(ATKINS), b Okmulgee, Okla, Dec 8, 21; m 43; c 4. ENGINEERING. *Educ:* US Mil Acad, BS, 43; NY Univ, MS, 49. *Prof Exp:* Instr, Eng Sch, US Army, 47-48, instr math, US Mil Acad, 49-52, commanding officer, 14th Eng Battalion, 53-54, exec officer, Ryukyus Command Eng Serv, Okinawa, 54-55, mem staff, Off Dep Chief of Staff Mil Opers, Pentagon, 56-58; res engr systs eng eval & res, Radio Corp Am, NJ, 58-66; mgr systs anal, 66-67; mgr, Info Systs Dept, Raytheon Co, 67-71; chief, Tech Prog Div, Fed Aviation Admin, 71-79; mgr, Air Traffic Control Systs, Westinghouse Elec Corp, 79-81; PRES, ENREAL ENTERPRISES INC, 81- *Concurrent Pos:* Lectr, Univ Calif, 54-55, Univ Md, 56-57 & Am Univ, 56-58. *Res:* Strategic warning systems; advanced missile systems; military communications systems; satellite systems; air defense systems; command, control and communications systems; ground transportation systems; air traffic control. *Mailing Add:* 5403 38th Ave No 4 Chesapeake Beach MD 20732

JOHNSTON, TUDOR WYATT, b Montreal, Que, Jan 17, 32; m 58; c 3. PLASMA PHYSICS. *Educ:* McGill Univ, BEng, 53; Cambridge Univ, PhD(eng physics), 58. *Prof Exp:* Sr res scientist, Microwave & Plasma Physics Lab, RCA Victor Co, Ltd, 58-67, plasma & space physics lab, 67-69; assoc prof physics, Univ Houston, 69-73; PROF, INST NAT RES SCI ENERGIE, UNIV QUE, 73- *Concurrent Pos:* Vis prof, Tex A&M Univ, 67; consult, RCA, 70, Can Dept Commun, 71, KMS Fusion, 80-81, Laser Lab, Univ Rochester, 81-; vis prof, Univ Rochester, 85. *Mem:* Fel Am Phys Soc; Can Asn Physicists. *Res:* Plasma theory; computer simulation; nonlinear wave-plasma; laser-plasma interaction. *Mailing Add:* INRS-Energie CP 1020 Varennes PQ J0L 2P0 Can

JOHNSTON, WALTER EDWARD, b Clarksville, Ark, Apr 8, 39; m 60; c 3. STATISTICS. *Educ:* Tex A&M Univ, BS, 60, MS, 65, PhD(statist), 70. *Prof Exp:* Teacher, Tex High Sch, 60-61; from asst prof to prof exp statist, Clemson Univ, 67-78; ASSOC PROF, SOUTHWEST TEX STATE UNIV, SAN MARCOS, 80- *Mem:* Am Statist Asn. *Res:* Application of statistical methods in agricultural and biological research. *Mailing Add:* 111 Pine Oak Dr Buda TX 78610

JOHNSTON, WARREN E, b Woodland, Calif, May 27, 33; m 59; c 2. LAND ECONOMICS. *Educ:* Univ Calif, Davis, BS, 59; NC State Col, MS, 63; NC State Univ, PhD(agr econ & statist), 64. *Prof Exp:* PROF AGR ECON, UNIV CALIF, DAVIS, 63- *Concurrent Pos:* Alexander von Humboldt res fel, WGer, 69-70; Fulbright res scholar, NZ, 76-77; Dir, Am Agr Econ Asn, 85-88. *Mem:* Am Agr Econ Asn (pres, 90-91); Int Asn Agr Econ; Am Soc Farm Managers & Rural Appraisers. *Res:* Agricultural, natural resources and environmental economics and public policy; commercial and sustainable agriculture; land economics; land markets; adjustments to policy and economic changes. *Mailing Add:* Dept Agr Econ Univ Calif Davis CA 95616

JOHNSTON, WILBUR DEXTER, JR, b New Haven, Conn, July 6, 40; m 63; c 2. PHYSICS, ELECTRICAL ENGINEERING. *Educ:* Yale Univ, BS, 61; Mass Inst Technol, PhD(physics), 66. *Prof Exp:* Res asst electronics, Mass Inst Technol, 61-66; mem tech staff, 66-79, SUPVR, SOLID STATE MAT, AT&T BELL LABS, 80- *Mem:* AAAS; Inst Elec & Electronics Engrs; Electrochem Soc. *Res:* Optical communications; laser physics; non-linear optics; semiconductor lasers; solar cells; heterojunction and compound semiconductor device physics; vapor phase epitaxial growth of semiconductor materials; materials science. *Mailing Add:* Oak Knoll Rd Mendham NJ 07945

JOHNSTON, WILLIAM CARGILL, b Clarinda, Iowa, Aug 31, 17; m 47; c 4. SOLID STATE PHYSICS. *Educ:* Davidson Col, BA, 39; Univ Va, MS, 42, PhD(physics), 43. *Prof Exp:* Res engr, Westinghouse Elec Corp, 43-68; chmn dept, 68-74, PROF PHYSICS, GEORGE MASON UNIV, 74-, DEAN SUMMER SESSION, 70- *Concurrent Pos:* Instr physics, Carnegie Inst Technol, 45- *Mem:* Sigma Xi. *Res:* Flame velocity measurements; fundamental combustion; solidification of metals. *Mailing Add:* 10927 Stuart Mill Rd Oakton VA 22124

JOHNSTON, WILLIAM DWIGHT, b Bellevue, Pa, Jan 17, 28; m 50; c 3. INORGANIC CHEMISTRY. *Educ:* Univ Pittsburgh, BS, 49, MS, 51, PhD(chem), 53. *Prof Exp:* Res engr, Res Labs, Westinghouse Elec Corp, 53-57, adv chemist, 57-62; res chemist, Pittsburgh Corning Corp, 62-65, asst dir res, 65-69, dir res & develop, 69-74, tech dir int opers, 74-89; RETIRED. *Mem:* Am Chem Soc; Am Ceramic Soc. *Res:* Glass research; solid state chemistry; crystallography; semiconductors; magnetic materials; inorganic preparations; phase diagrams; thermodynamics; metal chelates. *Mailing Add:* 2416 Collins Rd Pittsburgh PA 15235-4905

JOHNSTON, WILLIAM V, b Berkeley, Calif, May 6, 27; m 51; c 4. PHYSICAL CHEMISTRY & METALLURGY. *Educ:* Col Wooster, BA, 50; Univ Pittsburgh, PhD(phys chem), 55. *Prof Exp:* Asst, Univ Pittsburgh, 50-55; sr cryogenic operator, Ohio State Univ, 52; res assoc, Knolls Atomic Power Lab, Gen Elec Co, 55-61; res specialist, Atomics Int Div, NAm Aviation Corp, 61-62, group leader phys metall, Sci Ctr, 62-69, prin scientist, NAm Rockwell Corp, 69, mem tech staff, Rocketdyne Div, 69-72; nuclear engr, US Atomic Energy Comn, 72-74; br chief, Fuel Behav Res Br, US Nuclear Regulatory Comn, 74-80, asst dir div eng, 80-85, chief, Eng Br, 85-86, dep div reactor safety, 86-90; RETIRED. *Concurrent Pos:* Chmn working group on nuclear safety, Orgn Econ Coop & Develop/Int Energy Agency, 75-80; chmn, Halden Reactor Proj, Orgn Econ Coop & Develop, 81. *Mem:* Am Chem Soc; Am Inst Mining, Metall & Petrol Engrs; AAAS; Am Nuclear Soc. *Res:* Metal physics; calorimetry; nuclear fuels; solid electrolytes; solution thermodynamics; nuclear safety; nuclear materials. *Mailing Add:* Two Ruth Lane Downingtown PA 19335

JOHNSTON, WILLIAM WEBB, b Statesville, NC, Aug 26, 33. PATHOLOGY, CYTOLOGY. *Educ:* Davidson Col, BS, 54; Duke Univ, MD, 59; Am Bd Path, dipl. *Prof Exp:* Res training prog grant, Duke Univ, 60-61, res fel path, 61-63, assoc, 63-65, from asst prof to assoc prof, 65- 72, PROF PATH, MED CTR, DUKE UNIV, 72-, DIR CYTOPATH, 66-, FAC CLIN CANCER TRAINING PROG, 66- *Concurrent Pos:* Consult path, Durham Vet Admin Hosp, 66-; mem bd dirs, Am Cancer Soc, Durham County. *Honors & Awards:* Ortho Award, Can Soc Cytol, 72. *Mem:* Am Asn Path; Am Soc Cytol (pres, 81-82); fel Am Soc Clin Path; fel Int Acad Cytol. *Res:* Basic diagnostic methods in cytopathology. *Mailing Add:* Dept Path Duke Univ Med Ctr Box 3322 Durham NC 27710

JOHNSTONE, C(HARLES) WILKIN, b Alamosa, Colo, Aug 22, 16; m 47; c 2. NUCLEAR PHYSICS, INSTRUMENTATION. *Educ:* Colo Col, AB, 38; Dartmouth Col, AM, 40. *Prof Exp:* Asst physics, Colo Col, 37-38, Dartmouth Col, 38-40 & Pa State Col, 40-41; proj engr, Navy Dept Proj, Sperry Gyroscope Co, NY, 41-44, Naval Res Lab, Washington, DC, 44-45, in charge marine radar design & develop, NY, 45-47; mem staff electronics res, Los Alamos Sci Lab, Calif, 47-56; develop proj engr, Schlumberger Well Serv, 56-60, sect head nuclear physics, 60-68, sr develop proj engr, Eng Physics Dept, 66-82; RETIRED. *Mem:* Inst Elec & Electronics Engrs. *Res:* Specialized electronic circuits for IFF, radar and nuclear research, and instrumentation; radioactivity techniques and apparatus for well logging. *Mailing Add:* 2055 Brentwood Dr Houston TX 77019

JOHNSTONE, DONALD BOYES, b Newport, RI, July 25, 19; m 49; c 3. MICROBIOLOGY. *Educ:* RI State Col, BS, 42; Rutgers Univ, MS, 43, PhD(microbiol), 48. *Prof Exp:* Bacteriologist, Woods Hole Oceanog Inst, 42-43 & 46; from asst prof to prof microbiol, Univ Vt, 48-85, microbiologist, Agr Exp Sta, 48-85, chmn dept agr biochem, 59-85, dean grad col, 69-85; RETIRED. *Mem:* AAAS; Am Soc Microbiol; fel Am Acad Microbiol. *Res:* Marine bacteriology; antibiotics from higher plants; isolation of streptomycin producing actinomycetes; vitamin B-12 sources; whey utilization; azotobacter metabolism; classification; fluorescent pigments; extra-cellular polysaccharides; pesticide degradation. *Mailing Add:* Eight Rudgate Rd Colchester VT 05446

JOHNSTONE, DONALD LEE, b Bluefield, WVa, Feb 4, 39; m 62; c 1. BACTERIOLOGY, MICROBIOLOGY. *Educ:* Eastern Wash State Col, BA, 64; Wash State Univ, MS, 66, PhD(aquatic bact), 70. *Prof Exp:* Asst prof civil eng & asst sanit scientist, Res Div, Col Eng, 69-76, ASSOC PROF CIVIL & ENVIRON ENG, WASH STATE UNIV, 76- *Concurrent Pos:* Mem, Int Conf Dis Nature Communicable to Man, 64. *Mem:* AAAS; Am Soc Microbiol; Water Pollution Control Fedn; Am Water Works Soc; Sigma Xi. *Res:* Interaction of bacteria and soil; survival of intestinal bacteria in the aquatic environment; effects of hydrocarbons on indicator bacteria in groundwater; ecology of fresh water bacteria. *Mailing Add:* Environ Eng Sloan Hall 141 Wash State Univ Pullman WA 99164

JOHNSTONE, JAMES G(EORGE), b LaPorte, Ind, July 29, 20; m 46; c 1. GEOLOGY, CIVIL ENGINEERING. *Educ:* Colo Sch Mines, Geol Eng, 48; Purdue Univ, MSE, 52. *Prof Exp:* Asst prof geol & civil eng & res engr, Purdue Univ, 48-55; eng geologist, Geophoto Servs, Colo, 55-57; from asst prof to prof, 57-83, EMER PROF CIVIL ENG, COLO SCH MINES, 83- *Concurrent Pos:* Asst to plant engr, Ford Motor Co, Detroit, 42-45. *Honors & Awards:* Alfred J Ryan Award, 90. *Mem:* Am Soc Civil Engrs; Nat Soc Prof Engrs (vpres, 75-76). *Res:* Soil mechanics; engineering geology; applications of geology to engineering projects; computer science. *Mailing Add:* 13079 W Ohio Ave Lakewood CO 80228

JOHNSTONE, ROSE M, b Lodz, Poland, May 14, 28; Can citizen; m 53; c 2. BIOCHEMISTRY. *Educ:* McGill Univ, BSc, 50, PhD(biochem), 53. *Prof Exp:* Res assoc biochem, McGill Univ & Montreal Gen Hosp Res Inst, 53-65, asst prof, 61-65; from asst prof to assoc prof, 65-76, chmn dept biochem, 80-90, PROF BIOCHEM, McGILL UNIV, 77-, GILMAN-CHNEY PROF BIOCHEM, 85- *Concurrent Pos:* Nat Cancer Inst Can fel, 54-57, res grant, 65-; Med Res Coun grant, 65-, NIH, 88- *Honors & Awards:* Queen's Jubilee Medal, 77. *Mem:* Can Fedn Biol Soc; Can Biochem Soc; Am Soc Biol Chemists; NY Acad Sci; Royal Soc Can. *Res:* Transport of organic substances into mammalian cells; development of transport systems; reconstitution of transport systems; transport in membrane vesicles; membrane remodeling during development of red cells. *Mailing Add:* Dept Biochem McGill Univ Montreal PQ H3G 1Y6 Can

JOHNSTON-FELLER, RUTH M, b Polo, Ill, Mar 31, 23; m 75. COLOR SCIENCE. *Educ:* Univ Ill, AB, 47. *Prof Exp:* Anal chemist, USDA, 44-46, Univ Ill, 46-47 & A E Staley Co, 47; res chemist, Pittsburgh Plate Glass Indust, 47-51, proj leader, 56-67; res chemist, Rohm & Haas Co, Pa, 53-54; asst dir, color ctr, Davidson & Hemmendinger, 67-69; dir, applns serv, Kollmorgen Corp, 69-73; mgr, coatings & colorimetry, Ciba-Geigy Corp, 73-85; RES ASSOC, MELLON INST, 75- *Concurrent Pos:* Lectr color sci, Univ Utah, Lehigh Univ, Clemson Univ & Ciba-Geigy Corp, 73-75; consult, pvt & inst, 75-; mem, bd dir, Paint Res Inst, 79-81. *Honors & Awards:* Bruning Award, Fedn Soc Paint Technol, 70; Macbeth Award, Int-Soc Color Coun, 84, Dorothy Nickerson Award, 88; Mattiello Lect, Fedn Soc Coatings Technol, 85; George Baugh Heckel Award, Fedn Soc Coatings Technol, 89. *Mem:* Am Chem Soc; Optical Soc Am; Int & Am Inst Conservation Historic & Artistic Works; Am Soc Testing & Mat; Fedn Soc Coatings Technol; Int-Soc Color Coun. *Res:* Characterization of pigments and dyes in organic coatings; photochemically-induced fading of such systems; optical behavior of colorants in colored paint, plastic, textile and artists' materials. *Mailing Add:* Mellon Inst Carnegie-Mellon Univ 4400 Fifth Ave Pittsburgh PA 15213

JOINER, JASPER NEWTON, horticulture; deceased, see previous edition for last biography

JOINER, R(EGINALD) GRACEN, b Hawkinsville, Ga, Feb 10, 33; m 52; c 2. MIDDLE ATMOSPHERE RESEARCH, SPACE RADIATION EFFECTS. *Educ:* Univ Ga, BS, 58, MS, 59. *Prof Exp:* Physicist, US Naval Ordnance Lab, 59-64, physicist, Off Naval Res, 64-80, supvry phys sci adminr, 80-82, PROG MGR SPACE PHYSICS, OFF NAVAL RES, 82- *Mem:* Am Geophys Union. *Res:* Extremely low frequency-very low frequency radio propagation; ionosphere; space. *Mailing Add:* 2988 Poplar Trail Annapolis MD 21401

JOINER, ROBERT RUSSELL, food chemistry; deceased, see previous edition for last biography

JOINER, WILLIAM CORNELIUS HENRY, b Camden, NJ, June 8, 36; m 64. SOLID STATE PHYSICS. *Educ:* Rutgers Univ, BA, 57, PhD(physics), 62. *Prof Exp:* Sr physicist, Aerospace Div, Westinghouse Elec Co, 61-65, fel physicist, 65; from asst prof to assoc prof, 65-73, PROF PHYSICS, UNIV CINCINNATI, 73-, HEAD DEPT, 74- *Mem:* Am Phys Soc; Sigma Xi. *Res:* Superconductivity; low temperature physics. *Mailing Add:* 7290 Green Farms Dr Cincinnati OH 45224

JOKELA, JALMER JOHN, b Ely, Minn, Sept 20, 21; m 53; c 3. FOREST GENETICS. *Educ:* Univ Minn, BSF, 47, MS, 51; Univ Ill, PhD(agron), 63. *Prof Exp:* Agr aide, Lake States Forest Exp Sta, 46; asst, Univ Minn, 49-51; asst, Univ Ill, Urbana, 47-49 & 51-59, res assoc, 59-69, from instr to assoc prof forest res, 59-86; RETIRED. *Mem:* Soc Am Foresters; Sigma Xi. *Res:* Genetics and breeding of cottonwoods; silviculture; mensuration. *Mailing Add:* 1661 Saari Rd Ely MN 55731

JOKERST, NAN MARIE, b St Louis, Mo, May 11, 61; m 88. OPTICS, SOLID STATE PHYSICS. *Educ:* Creighton Univ, BS, 82; Univ Southern Calif, MS, 84, PhD(elec eng), 89. *Prof Exp:* ASST PROF ELEC ENG, GA INST TECHNOL, 89- *Concurrent Pos:* Consult, Foster-Miller, Inc, 90-; NSF presidential young investr, 90. *Mem:* Optical Soc Am; Inst Elec & Electronics Engrs; Am Phys Soc; Sigma Xi. *Res:* Monolithic deposition of GaAs and InP onto host substrates such as silicon, glass, lithium niobate, polymers for optoelectronic integrated circuits; solar cells; semiconductor lasers; nonlinear optics in semiconductors. *Mailing Add:* Sch Elec Eng Ga Inst Technol Atlanta GA 30332-0250

JOKINEN, EILEEN HOPE, b Detroit, Mich, July 22, 43. INVERTEBRATE ZOOLOGY, PARASITOLOGY. *Educ:* Wayne State Univ, BS, 65, PhD(zool), 71. *Prof Exp:* Instr introd biol, Wayne County Community Col, Detroit, Mich, 71-72; instr comp anat, Univ Mich, Dearborn, Mich, 72; asst & assoc prof invert zool, ecol, parasitol, comp anat, embryol & introd zool, Suffolk Univ, Boston, Mass, 72-80; vis asst prof gen ecol & invert zool, 80-86, ASST DIR, CONN INST WATER RESOURCES, UNIV CONN, STORRS, 87- *Mem:* Am Malacol Soc; Am Soc Zoologists; Am Inst Biol Sci. *Res:* Freshwater malacology; community ecology of freshwater littoral zone benthos; biogeography of freshwater snails. *Mailing Add:* c/o Rch U-42 Univ Conn Storrs CT 06268

JOKIPII, JACK RANDOLPH, b Ironwood, Mich, Sept 10, 39; m 64; c 3. THEORETICAL PHYSICS, ASTROPHYSICS. *Educ:* Univ Mich, BS, 61; Calif Inst Technol, PhD(physics), 65. *Prof Exp:* Res assoc physics, Enrico Fermi Inst Nuclear Studies, Univ Chicago, 65-67, asst prof, Inst & Univ, 67-69; assoc prof theoret physics, Downs Lab Physics, Calif Inst Technol, 69-73; PROF ASTRON & PLANETARY SCI, UNIV ARIZ, 74- *Concurrent Pos:* Alfred P Sloan Found fel, 69. *Honors & Awards:* Fel, Am Phys Soc. *Mem:* Am Phys Soc; Am Geophys Union; Am Astron Soc; Int Astron Union. *Res:* Theoretical space physics; cosmic ray acceleration and propagation; interpretation of space vehicle observations; solar physics; interstellar physics. *Mailing Add:* Dept Planetary Sci Univ Ariz Tucson AZ 85721

JOKL, ERNST, b Breslau, Ger, Aug 3, 07; nat US; m 33; c 2. PHYSIOLOGY. *Educ:* Breslau Univ, MD, 31; Univ Witwatersrand, MB & BCh, 36. *Prof Exp:* Asst exp med, Breslau Univ, 30-31; sr res fel, Int Inst High Altitude Physics, Switz, 31; dir, Res Inst Med & Sport, Breslau Univ, 31-33; sr med officer, Dept Educ & mem, Nat Adv Coun Phys Educ, Union SAfrica, 38-44, med consult, Dir Gen Med Serv & Aviation Med, Union Defense Force, 40-44; DISTINGUISHED PROF REHAB, PHYSIOL & PHYS EDUC, UNIV KY, 54- *Concurrent Pos:* Pres res comt, Int Coun Sport & Phys Educ, UNESCO; hon prof physiol & med, Univs WBerlin, Frankfurt & Cologne; Nat Libr Med res grant fel, Bethesda. *Honors & Awards:* Buckston Browne Brit Empire Prize, 42; Medal, Brit Harveian Soc, 42; Res Awards, Ger Soc Sports Med & Int Coun Mil Sport. *Mem:* Fel Am Col Cardiol; Brit Med Asn; Sigma Xi. *Res:* Clinical physiology of exercise and rehabilitation. *Mailing Add:* 340 Kingsway Dr Lexington KY 40502-1046

JOKL, MILOSLAV VLADIMIR, b Prague, Czech, Feb 9, 33; m 64; c 2. HYGROTHERMAL MICROCLIMATE, ELECTROIONIC MICROCLIMATE. *Educ:* Tech Univ Prague, MSc, 56, PhD(eng), 63, DSc(environ eng), 75. *Prof Exp:* Assoc prof prev med, Inst Hyg & Epidemiol, 56-77; PROF ENVIRON ENG & DEPT HEAD, TECH UNIV PRAGUE, 77- *Concurrent Pos:* Vis prof, Tech Univ Denmark, 84 & Kans State Univ, 86. *Honors & Awards:* Felber Medal, 83. *Mem:* Int Fedn Housing & Planning; Int Soc Biometeorol; corresp mem Am Soc Heating Refrig & Air Conditioning Engrs. *Res:* Theory and practice of indoor climate; hygrothermal microclimate-thermal climate, its non-uniformity and with electroionic as well as psychological indoor constituent. *Mailing Add:* Na Orechovce 60 Prague 166 29 Czechoslovakia

JOKLIK, G FRANK, b Vienna, Austria, May 30, 28. RESEARCH ADMINISTRATION. *Educ:* Univ Sydney, BSc, 49, PhD(geol), 53. *Prof Exp:* Fulbright scholar, Columbia Univ, 53-54; mgr, Amax, Inc, 63-72, corp vpres, 72-74; sr vpres, Metals Mining Standard Oil Co, Ohio, 82-87; pres & chief exec officer, B P Minerals Am, 87-89; explor geologist, Kennecott Corp, New York, 54-63, vpres, 74-79, pres, Salt Lake City, 80-87, PRES & CHIEF EXEC OFFICER, KENNECOTT CORP, 89- *Concurrent Pos:* Dir, First Security Corp; mem bd, Am Mining Cong, Am Inst Mining, Metall & Petrol Engrs & Australasian Inst Mining & Metall. *Mailing Add:* Kennecott Corp Ten SE Temple PO Box 11248 Salt Lake City UT 84147

JOKLIK, WOLFGANG KARL, b Vienna, Austria, Nov 16, 26; m 55, 77; c 2. MOLECULAR BIOLOGY, VIROLOGY. *Educ:* Univ Sydney, BSc, 48, MSc, 49; Oxford Univ, DPhil(biochem), 52. *Prof Exp:* Fel microbiol, Australian Nat Univ, 52-62; USPHS traveling fel, 59-60; from assoc prof to prof cell biol, Albert Einstein Col Med, 62-68; JAMES B DUKE PROF MICROBIOL & IMMUNOL & CHMN DEPT, MED CTR, DUKE UNIV, 68- *Concurrent Pos:* Pres, Virol Div, Am Soc Microbiol, 68-89 & group counr, Group IV, 81-83; chmn, Virol Study Sect, NIH, 73-75, mem, Rec DNA Adv Comt, 82-87; ed-in-chief, Virol; mem exec comn, Int Comm Taxon Viruses, 78-84; assoc ed, J Biol Chem, 78-89; mem, Coun Res & Clin Awards, Am Cancer Soc, 80-83 & 88-91; ed-in-chief, Microbiol Reviews, 90- *Mem:* Nat Acad Sci; Inst Med of Nat Acad Sci; Am Soc Microbiol; Am Med Sch Microbiol Chmns Asn (pres, 79); Am Soc Virol (pres, 82-83). *Res:* Biochemistry of virus multiplication, including the mechanisms of nucleic acid replication, transcription and translation of genetic information, regulation of genetic expression and the mechanisms of protein synthesis; molecular virology; molecular genetics. *Mailing Add:* Dept Microbiol & Immunol Duke Univ Med Ctr Durham NC 27710

JOLICOEUR, PIERRE, b Montreal, Que, Apr 5, 34; m 69; c 3. BIOMATHEMATICS, BIOMETRICS. *Educ:* Univ Montreal, BA, 53, BSc, 56; Univ BC, MA, 58; Univ Chicago, PhD(paleozool), 63. *Prof Exp:* From asst prof to assoc prof biol, 61-72, chmn dept biol sci, 73-77, PROF BIOL, UNIV MONTREAL, 72- *Concurrent Pos:* Vis assoc prof, Univ Kans, 66; vis assoc scientist, NIH, 67. *Mem:* Fel AAAS; Soc Study Evolution; Biomet Soc. *Res:* Biological applications of mathematics and statistics, multivariate analysis; allometry and nonlinear growth curves; vertebrate zoology; ecology of animal populations. *Mailing Add:* Dept Biol Sci Univ Montreal PO Box 6128 Montreal PQ H3C 3J7 Can

JOLIVETTE, PETER LAUSON, b Madison, Wis, May 27, 41; m 67; c 2. NUCLEAR STRUCTURES. *Educ:* Univ Wis-Madison, BS, 63, PhD(physics), 71; Purdue Univ, MS, 65. *Prof Exp:* Res assoc, Univ Notre Dame, 70-76, vis asst prof physics, 75; asst prof, 76-83, ASSOC PROF PHYSICS, HOPE COL, 83- *Mem:* Am Inst Physics; Am Asn Physics Teachers. *Res:* Low and intermediate energy nuclear physics; isospin and charge symmetry effects. *Mailing Add:* Dept Physics Hope Col Holland MI 49423

JOLLES, MITCHELL IRA, b Bronx, NY, Feb 10, 53. SOLID MECHANICS, FRACTURE MECHANICS. *Educ:* Polytech Inst, Brooklyn, BS & MS, 73; Va Polytech Inst & State Univ, PhD(eng mech), 76. *Prof Exp:* Lectr eng sci & mech, 73-74, instr, Va Polytech Inst & State Univ, 74-76; asst prof aerospace & mech eng, Univ Notre Dame, 77-79; assoc prof mech & aero eng/nuclear eng, Univ Mo, 79-82; HEAD, FRACTURE MECH SECT, NAVAL RES LAB, 82- *Concurrent Pos:* Res assoc, Nat Aeronaut & Space Admin, 73-74, Dept Defense, 73-74, NSF, 73-76, Energy Res & Develop Admin, 75-76, Air Force Flight Dynamics Lab, 75, & Cabot Corp, 77; prin investr, NSF, 78-82, Exxon Educ Found, 78-79, Student Competition Relevant Eng, 78-79 & Argonne Nat Lab, 80-82; lectr, George Wash Univ, Eng Ed, 83-; consult, Mat Eng Assoc, 83-; adj fac, Va Polytech Inst & State Univ, 84- *Honors & Awards:* Ralph R Teetor Award, Soc Automotive Engrs, 79. *Mem:* Soc Exp Mech; Am Soc Testing & Mat; Am Soc Eng Educ; Soc Eng Sci; Sigma Xi. *Res:* Fracture mechanics; experimental mechanics; constitutive theory and material damage models; structural integrity methodology. *Mailing Add:* Dept Eng Widener Univ Chester PA 19013

JOLLEY, DAVID KENT, b Park City, Utah, Jan 25, 44; m 70; c 4. SIGNAL PROCESSING ALGORITHMS, GEOLOCATION TECHNIQUES. *Educ:* Univ Utah, BA, 66, PhD(physics), 73. *Prof Exp:* Sr engr, ESL Inc, 72-82, dept mgr, 84-88; sr engr, Advent Inc, 82-84; VPRES ENG, ASTECH, INC, 88- *Res:* Development of signal processing algorithms and signal processing systems for government agencies primarily in the areas of reconnaissance and surveillance. *Mailing Add:* 20 Windsong Sandy UT 84092

JOLLEY, HOMER RICHARD, b Morgan City, La, May 28, 16; m 72. PUBLIC HEALTH ADMINISTRATION, RESEARCH ADMINISTRATION. *Educ:* Univ Gonzaga, AB, 38, MA, 39; Fordham Univ, MS, 41; St Louis Univ, STL, 46; Princeton Univ, PhD(chem), 51. *Prof Exp:* Instr chem, Spring Hill Col, 39-40 & 41-42; asst instr, Princeton Univ, 48-49; from asst prof to prof chem, Loyola Univ, La, 51-70, chmn dept, 56-64, vpres, Univ, 64-66, pres, 66-70; dir, Off Innovation, Med Serv Admin, HEW, 70-75, spec asst to comnr health serv delivery systs, 75-77; asst acad vpres & vis prof admin med, Med Univ SC, 77-81; CO-FOUNDER & EXEC DIR, CHARLESTON RES INST, 82- *Concurrent Pos:* Res partic, Oak Ridge Inst Nuclear Studies, 47, 59 consult, 61-; co-prin investr, Res Prog, US Off Saline Water; mem, Interagency Comt Ment Retardation, 75- & Interagency Regulatory Group Health Maintenance Orgns, 75- *Mem:* Am Chem Soc; Am Inst Chemists. *Res:* Electrophoresis of polyelectrolytes; assymetric resins; properties of aqueous solutions under high pressure and temperature; cost-effectiveness and quality of care in organized forms of the health care delivery system. *Mailing Add:* 33 Cherokee Hills Tuscaloosa AL 35404

JOLLEY, JOHN ERIC, b Blackpool, Eng, June 26, 29; nat US; m 55; c 3. MATERIALS SCIENCE. *Educ:* Univ Liverpool, BS, 50, PhD(phys chem), 53. *Prof Exp:* Fel, Univ Rochester. 53-55 & Univ Calif, Berkeley, 55-57; res chemist film dept, Res Lab, E I du Pont de Nemours & Co, Inc, 58-60, res chemist cent res dept, 60, res supvr, 60-64, tech mgr develop dept, Exp Sta, 64-67, res fel, photog & electronic prod dept, 67-85; CONSULT, 85- *Concurrent Pos:* Consult, 85- *Mem:* Sigma Xi; Am Ceramic Soc; Int Soc Hybrid Microelectronics; Soc Info Display. *Res:* Kinetics; radical reactions; solubility; polymer chemistry; electronic materials; magnetism; photographic science; glasses; ceramics; rheology. *Mailing Add:* 20 Boulder Brook Dr Wilmington DE 19803

JOLLEY, ROBERT LOUIS, b Little Rock, Ark, July 11, 29; m 50; c 2. ENVIRONMENTAL CHEMISTRY. *Educ:* Friends Univ, BA, 50; Univ Tenn, Knoxville, PhD(ecol), 73. *Prof Exp:* Asst chem, Friends Univ, 49-50 & Univ Chicago, 50-51; chemist, Southwest Grease & Oil Co, 51-55; chemist, 56-73, CHEM ECOLOGIST, OAK RIDGE NAT LAB, 73- *Concurrent Pos:* County comnr, Anderson County, Tenn, 59- *Mem:* Am Chem Soc; AAAS; Sigma Xi. *Res:* Measurement and identification of organic constituents in natural and polluted waters; determination of chlorination effects and analysis of chloro-organics in process effluents and condenser cooling waters for electric power plants; evaluation of treatment technologies for low-level radioactive waste and hazardous waste. *Mailing Add:* 120 N Seneca Rd Oak Ridge TN 37830

JOLLEY, WELDON BOSEN, b Gunnison, Utah, Sept 8, 26; m 54, 83; c 5. PHYSIOLOGY. *Educ:* Brigham Young Univ, AB, 52; Univ Southern Calif, PhD(cell physiol), 59. *Prof Exp:* Res assoc, Univ Southern Calif, 53-59, instr, 58-59, asst prof physiol & co-dir, Surg Res Lab, 59-71, PROF PHYSIOL, BIOPHYS & SURG & ASSOC DIR SURG RES LAB, LOMA LINDA UNIV, 71-; PROF PHYSIOL, JERRY L PETTIS VET ADMIN HOSP, 80- *Concurrent Pos:* Instr, Compton Col, 56; dir, Bio Nuclear Corp; mem bd, Life Resources, Inc & ICN Pharmaceut Inc; pres, Nucleic Acid Res Inst, 85-; mem bd dirs, SPI Pharm, Inc, sr vpres bd dirs, ICN Pharm, Inc. *Mem:* AAAS; AMA; Am Fedn Clin Res; Am Physiol Soc; Transplantation Soc. *Res:* Antiviral Agents; immunologic modulators; cancer immunology; transplantation of skin, pancreas, heart, kidneys; endotoxic and hemorrhagic shock; biological effects of pulsed electromagnetic fields. *Mailing Add:* Nucleic Acid Res Inst ICN Plaza 3300 Hyland Ave Cotsa Mesa CA 92626

JOLLICK, JOSEPH DARRYL, b Denbo, Pa, May 15, 41; m 63; c 2. MICROBIAL GENETICS, MEDICAL MICROBIOLOGY. *Educ:* Calif State Col, BS, 63; Am Univ, MS, 66; WVa Univ, PhD(microbiol), 69. *Prof Exp:* Biologist, Nat Cancer Inst, 63-64; Nat Res Coun grant, Biol Sci Lab, Ft Detrick, 69-70; instr, 70-72, asst prof microbiol, Sch Med, Wayne State Univ, 72-77; asst prof microbiol, Sch Med Ohio Univ, 78-80; ASSOC PROF MICROBIOL, COL OSTEOP MED, OHIO UNIV, 80- *Concurrent Pos:* NIH grant, Wayne State Univ, 74- *Mem:* AAAS; Am Soc Microbiol; Am Asn Univ Prof. *Res:* Genetics of Caulobacter; mechanism and transfer of antibiotic resistance in Serratia and Pseudomonas. *Mailing Add:* Dept Basic Sci Grosvenor Hall Col Osteop Med Ohio Univ Athens OH 45701

JOLLIE, MALCOLM THOMAS, b Lakewood, Ohio, July 11, 19; m 50; c 2. COMPARATIVE ANATOMY, ZOOLOGY. *Educ:* Western Reserve Univ, BS, 41; Univ Colo, MS, 43; Stanford Univ, PhD(comp anat), 54. *Prof Exp:* Asst biol, Univ Colo, 41-43; mus technician birds & asst zool, Univ Calif, 43-45; instr sci, Western NMex Teachers Col, 45-47; asst & assoc prof zool, Univ Idaho, 47-56 & Univ Pittsburg, 56-65; prof biol, Northern Ill Univ, 65-88; RETIRED. *Res:* Comparative anatomy relating to origin and phylogeny of chordates and vertebrates; systematic ornithology and ichthyology. *Mailing Add:* Dept Biol Sci Northern Ill Univ De Kalb IL 60115-2861

JOLLIE, WILLIAM PUCETTE, b Passaic, NJ, June 27, 28; m 50; c 2. ANATOMY, CELL & DEVELOPMENTAL BIOLOGY. *Educ:* Lehigh Univ, BA, 50, MS, 52; Harvard Univ, PhD(biol), 59. *Prof Exp:* Lectr histol & embryol, Queen's Univ, Ont, 59-61; from asst prof to prof anat, Sch Med, Tulane Univ, 61-69; PROF ANAT & CHMN DEPT, MED COL VA, VA COMMONWEALTH UNIV, 69- *Mem:* Am Asn Anatomists; Am Soc Cell Biologists; Teratology Soc. *Res:* Controlling mechanisms for placental transport; visualization of placental transport mechanisms; maternal accommodations to implantation and placental formation; effects of alcohol on acquisition of neonatal immunity. *Mailing Add:* Dept Anat Va Commonwealth Univ PO Box 709 MCV Sta Richmond VA 23298-0709

JOLLIFFE, ALFRED WALTON, geology; deceased, see previous edition for last biography

JOLLS, CLAUDIA LEE, b Detroit, Mich, May 20, 53. PLANT ECOLOGY, PLANT POPULATION BIOLOGY. *Educ:* Univ Mich, Ann Arbor, BS, 75; Univ Colo, Boulder, PhD(biol), 80. *Prof Exp:* Teaching asst ecol, bot & human physiol, Dept Environ Pop & Organismic, Univ Colo, 75-80; fel plant pop biol, Mich State Univ, 80-81; resident terrestrial ecologist, Biol Sta, Univ Mich, 81-84; asst prof, 84-89, ASSOC PROF, BIOL DEPT, ECAROLINA UNIV, 89- *Concurrent Pos:* Gardeners asst, Matthaei Bot Gardens, Univ Mich, 72-73; res asst plant ecol, Dept Environ, Pop & Organismic, Univ Colo at Audubon-Whittel Res Ranch, Ariz, 76; adv, Traineeship Prog, Mt Res Sta, Inst Arctic & Alpine Res, Univ Colo, NSF, 77 & 78; postdoctoral, Kellogg Biostation, Mich State Univ, 81; prog coordr, Naturalist-Ecologist Training Prog, Biol Sta, Univ Mich, 81-84; asst prof biol, ECarolina Univ, 84, res assoc, vis fac, Univ Mich Biostation, 85, 87, 89 & 91, res assoc, Univ Colo, 88. *Mem:* Am Inst Biol Sci; Bot Soc Am; Ecol Soc Am; Sigma Xi. *Res:* Plant ecology and population biology: resource allocation patterns, population dynamics, breeding systems, life histories, wetland community structure, forest succession, alpine ecology. *Mailing Add:* Dept Biol ECarolina Univ Greenville NC 27858-4353

JOLLS, KENNETH ROBERT, b Baltimore, Md, Oct 19, 33; c 1. PHASE BEHAVIOR, COMPUTER GRAPHICS. *Educ:* Duke Univ, AB, 58; NC State Univ, BSChE, 61; Univ Ill, MS, 63, PhD(chem eng), 66. *Prof Exp:* Asst chem eng, Univ Ill, 61-65; from asst prof to assoc prof, Polytech Inst Brooklyn, 65-70; assoc prof chem eng, Iowa State Univ, 70-90, PROF CHEM ENG, 90- *Concurrent Pos:* Vis prof chem eng, Univ Calif, Berkeley, 81-83 & Cornell Univ, 84. *Mem:* Am Inst Chem Engrs; Sigma Xi; Am Chem Soc; Asn Comput Mach. *Res:* Fluid mechanics; thermodynamics; application of electronic instrumentation in chemical engineering; computer graphics; scientific visualization. *Mailing Add:* Dept Chem Eng Sweeney Hall Iowa State Univ Ames IA 50011-2230

JOLLY, ALISON BISHOP, b Ithaca, NY, May 9, 37; m 63; c 4. PRIMATE BEHAVIOR. *Educ:* Cornell Univ, BA, 58; Yale Univ, PhD(zool), 62. *Prof Exp:* Res assoc zool, NY Zool Soc, 62-64; res assoc, Sch Biol, Univ Sussex, 68-81; guest investr, Rockefeller Univ, 82-87; VIS LECTR, PRINCETON UNIV, 87- *Concurrent Pos:* NSF res grant, 62-64. *Mem:* Int Primatol Soc; Wildlife Preservation Trust Int; Animal Behav Asn; Sigma Xi; Am Primatology Asn. *Res:* Conservation of natural ecosystems in Madagascar; primate behavior, particularly that of prosimians; evolution of human behavior. *Mailing Add:* Dept EE Biol Princeton Univ Princeton NJ 08544-1003

JOLLY, CLIFFORD J, b Southend, Eng, Jan 21, 39; m 61; c 2. PHYSICAL ANTHROPOLOGY, PRIMATOLOGY. *Educ:* Univ London, BA, 60, PhD(phys anthrop), 65. *Prof Exp:* Res asst phys anthropology, Univ Col, London, 63-65, asst lectr, 65-67; from asst prof to assoc prof, 67-75, PROF PHYS ANTHROPOLOGY, NY UNIV, 75- *Concurrent Pos:* Vis res fel, Makerere Univ Col, Uganda, 65-66. *Mem:* Soc Study Human Biol; Zool Soc London; Royal Anthrop Inst; Sigma Xi. *Res:* Primate functional anatomy; serology and biology. *Mailing Add:* Dept Anthropology 25 Waverly Pl New York NY 10003

JOLLY, JANICE LAURENE WILLARD, b Bakersfield, Calif, July 23, 31; m 56; c 3. ECONOMIC GEOLOGY, PETROLOGY. *Educ:* Univ Ore, BA, 56, MS, 57. *Prof Exp:* Geologist, US Geol Surv, 58-67; res geologist, RST Tech Serv Ltd, Zambia, 67-72; mineral commodity area specialist, US Bur Mines, 73-80, intern, Exec Managerial Develop Prog, Dept Interior, 80-81, chief, Off Geog Statist, 81-83; SR COMMODITY SPECIALIST, COPPER, US BUR MINES, 81- *Mem:* Fel Geol Soc Am; Am Inst Mining, Metall & Petrol Engrs. *Res:* Geology of the Monument Quadrangle, Oregon; petrography of the crystalline rocks, Potomac River gorge, Maryland-Virginia; ore deposit controls in Mississippi and Appalachian Valleys lead-zinc deposits; eastern United States heavy metal and massive sulfide deposits; geochemistry of copper deposits of Zambia; commodity surveys of copper and iron oxide pigments; international mineral industry studies in Africa and Middle East; supply and demand of strategic and critical minerals; mineral economics. *Mailing Add:* US Bur Mines 2401 E St NW Washington DC 20241

JOLLY, STUART MARTIN, b London, Eng, Aug 29, 46; US citizen. SYSTEMS DESIGN. *Educ:* Haverford Col, BA, 68; Cornell Univ, MS, 71. *Prof Exp:* Mem staff, Robinson Assocs Inc, 72-74, vpres, res & develop, 74-77; mem tech staff, Mitre Corp, 77-79, group leader, 79-85; ASSOC DEPT HEAD, MITRE CORP, 85- *Res:* Analysis and design of communications systems, security systems and speech processing. *Mailing Add:* Rte 101 RFD 3 Horace Greeley Hwy Amherst NH 03031

JOLLY, WAYNE TRAVIS, b Jacksonville, Tex, Aug 15, 40. PETROLOGY, VOLCANOLOGY. *Educ:* Univ Tex, Austin, BFA, 63, MA, 67; State Univ NY Binghamton, PhD(geol), 70. *Prof Exp:* Fel geol, Univ Sask, 70-71; asst prof, 71-75, ASSOC PROF GEOL, BROCK UNIV, 75- *Mem:* Geol Soc Am; Geol Asn Can; Mineral Asn Can. *Res:* Geochemical petrology and metamorphic petrology of volcanic rocks with emphasis on prehnite-pumpellyite facies and origin of Archean volcanics. *Mailing Add:* Dept Geol Sci Brock Univ St Catharines ON L2S 3A1 Can

JOLLY, WILLIAM LEE, b Chicago, Ill, Dec 27, 27; m 50; c 3. INORGANIC CHEMISTRY. *Educ:* Univ Ill, BS, 48, MS, 49; Univ Calif, PhD(chem), 52. *Prof Exp:* Instr, 52-53, chemist, Radiation Lab, 53-55, from asst prof to assoc prof, 55-62, PROF CHEM, UNIV CALIF, BERKELEY, 62- *Mem:* AAAS; Am Chem Soc; Royal Soc Chem. *Res:* Liquid ammonia chemistry; chemistry of the volatile hydrides; studies of the bonding in transition metal complexes; x-ray photoelectron spectroscopy; chemistry of the photographic process. *Mailing Add:* Dept Chem Univ Calif Berkeley CA 94720

JOLY, DANIEL JOSE, b Buenos Aires, Arg, Mar 11, 21; m 45; c 2. EPIDEMIOLOGY, ONCOLOGY. *Educ:* Univ Buenos Aires, MD, 44; NY Univ, Johns Hopkins Univ, MPH, 66, DRPH(epidemiol), 73. *Prof Exp:* Res fel surg, Harvard Med Sch, 51; from asst resident to sr resident oncol surg, Mem Hosp Cancer, New York, 52-56; prof surg & chmn dept, Univ of the Andes, Venezuela, 60-61; chmn dept surg, Inst Oncol, Univ Buenos Aires, 62-64; regional adv cancer control, Pan Am Health Orgn-WHO, rep in Cuba, 75-81; CONSULT EPIDEMIOL & PUB HEALTH, 81- *Concurrent Pos:* Res fel surg physiol, Sloan-Kettering Inst Cancer Res, 53-58; assoc, dept epidemiol, Sch Hyg & Pub Health, Johns Hopkins Univ, 73- *Mem:* Arg Soc Cancerology; Am Pub Health Asn. *Res:* Epidemiology of chronic diseases with special emphasis in cancer and environmental factors, such as cigarette smoking. *Mailing Add:* 525 23rd St NW Washington DC 20037

JOLY, GEORGE W(ILFRED), b Montreal, Que, June 5, 17. CIVIL ENGINEERING. *Educ:* Loyola Col, Can, BA, 38; McGill Univ, BEng, 49, MEng, 50. *Prof Exp:* Asst dean faculty eng, McGill Univ, 57-63; dean, Loyola Col, Montreal, 63-77, prof eng & assoc dean fac eng, Loyola Campus, Concordia Univ, 77-85; RETIRED. *Res:* Structural engineering; steel; mechanics. *Mailing Add:* 3644 Du Musee Ave Montreal PQ H3G 2C9 Can

JOLY, LOUIS PHILIPPE, b Montreal, Que, July 23, 28; m 54; c 3. PHARMACEUTICAL CHEMISTRY. *Educ:* Laval Univ, BA, 49, BSc, 53; Univ Bordeaux, France, PhD, 65. *Prof Exp:* Lectr, 57-58, assoc prof, 58-71, PROF MED CHEM, COL PHARM, UNIV LAVAL, 71- *Concurrent Pos:* Chief pharmacist, Robert Giffard Hosp Ctr, 57-69; consult, Neuropsychopharmacol Res Univ, 69- *Mem:* AAAS; NY Acad Sci; Am Pharmaceut Asn; Can Pharmaceut Asn. *Res:* Synthesis and essay by cell culture methods of new alkylating agents as antineoplastics; biotransforms of long acting psychotrophic drugs. *Mailing Add:* 1324 Rue Marechal Foch Quebec PQ G1S 2C4 Can

JOLY, OLGA G, b Argentina; US citizen; m 45; c 2. EPIDEMIOLOGY OF CHRONIC DISEASES, HEALTH SCIENCE ADMINISTRATION. *Educ:* Women's Col, Buenos Aires, Arg, BS, 38; Col Dent, Univ Buenos Aires, DDS, 43; Sch Hyg & Health, Johns Hopkins Univ, ScD(epidemiol), 70. *Prof Exp:* Adj prof periodont, Univ Los Andes, Merida, Venezuela, 60-62; chief instr, Col Dent, Univ Buenos Aires, Arg, 62-64; proj officer, Nat Caries Prog, Nat Inst Dent Res, NIH, 71-74, prog dir cancer, Nat Career Inst, 74-77; investr, Int Asn Res Cancer, 77-81; PROG DIR CANCER, NAT CANCER INST, NIH, 81- *Mem:* Am Pub Health Asn; Am Asn Cancer Educ; Soc Epidemiol Res; Int Epidemiol Asn; Am Soc Prev Oncol. *Mailing Add:* 8413 Westmont Terr Bethesda MD 20817

JONA, FRANCO PAUL, b Pistoia, Italy, Oct 10, 22; nat US; m 52; c 2. SURFACE PHYSICS, SURFACE CHEMISTRY. *Educ:* Swiss Fed Inst Technol, dipl, 45, PhD(physics), 49. *Prof Exp:* Instr physics, Univ Bern, 45-46 & Swiss Fed Inst Technol, 46-52; res assoc, Pa State Univ, 52-54, asst prof, 54-57; res physicist res labs, Westinghouse Elec Corp, 57-59; staff physicist res lab, Int Bus Mach Corp, NY, 59-69; PROF ENG, STATE UNIV NY STONY BROOK, 69- *Mem:* Fel Am Phys Soc; Swiss Phys Soc. *Res:* Ferroelectricity; crystallography; elasticity; piezoelectricity; crystal growth; surface studies. *Mailing Add:* Col Eng State Univ NY Stony Brook NY 11790

JONAH, CHARLES D, b Lafayette, Ind, Mar 19, 43; m 69. RADIATION CHEMISTRY. *Educ:* Oberlin Col, BA, 65; Columbia Univ, PhD(chem), 70. *Prof Exp:* Fel phys chem, Columbia Univ, 69-71; fel, 71-74, asst scientist, 74-77, CHEMIST RADIATION CHEM, ARGONNE NAT LAB, 77- *Mem:* Am Phys Soc; Am Chem Soc. *Res:* Mechanism of reactions; fast kinetic measurements; instrumentation; radiation chemistry of aqueous systems solvation of electrons. *Mailing Add:* Chem Div Argonne Nat Lab 9700 S Cass Ave Argonne IL 60439

JONAH, MARGARET MARTIN, b Berkeley, Calif, Oct 25, 42; m 69. CELL BIOLOGY. *Educ:* Pomona Col, Calif, BA, 64; Columbia Univ, PhD(chem biol), 71. *Prof Exp:* Res fel biochem, Northwestern Univ, Ill, 71-73; appointee biol, 73-75, res assoc, 75-76, RESIDENT ASSOC, ARGONNE NAT LAB, 76-; ASSOC PROF BIOL, ROSARY COL 84- *Concurrent Pos:* Asst prof biol, Rosary Col, 76-84. *Mem:* AAAS; Am Inst Biol Sci; Am Chem Soc; Am Soc Microbiol; Int Radiation Res Soc; Sigma Xi; Am Asn Dent Res. *Res:* Functions of lipids in membrane formation and surface specificity; metabolism of streptococcus mutans; interactions of biologically active molecules with lipids; cell surface receptors; role of heavy metals in cell metabolism; cadmium metabolism in mammals. *Mailing Add:* Dept Natural Sci Rosary Col 7900 W Division St River Forest IL 60305-1099

JONAS, ALBERT MOSHE, b New Haven, Conn, Oct 3, 31; m 54; c 3. PATHOLOGY, COMPARATIVE MEDICINE. *Educ:* Univ Toronto, DVM, 55; Am Col Vet Path, dipl. *Hon Degrees:* MA, Yale Univ, 74. *Prof Exp:* Pvt pract, 55-61; res asst path, Sch Med, Yale Univ, 61-63, from instr to asst prof, 63-68, assoc prof animal sci & chief lab, 68-73, assoc prof path, 71-73, prof animal sci & path, 73-74, prof comp med, Div Health Sci Res & Path & chief sect comp med, 74-77, prof comp med & path & chmn sect comp med, 77-78, dir animal care, 61-78; first dean, Vet Sch Tufts Univ, 78-81, prof exp path, Med Sch, 78-82, chmn comp med & lab animal sci, Vet Sch, 81-82; sr staff scientist & dir, Lab Animal Med & Comp Path, Jackson Lab, 82-86, chief, Sci Resources, 83-86, PRES, RES ANIMAL CONSULTS, INC, 87- *Concurrent Pos:* Mem coun, Inst Lab Animal Resources, Nat Acad Sci, 65-69; mem, Am Asn Accreditation Lab Animal Care, 68-75; mem animal res adv comt, Div Res Resources, NIH, 72-76; mem adv comt comp path, Armed Forces Inst Path, 73-76; chmn comt, Longterm Holding of Lab Rodents, Inst Lab Animal Res, Nat Acad Sci, 73-76 & mem, Orgn Comt Symp on Lab Animal Housing, 74-75; mem tech rev comt, Bioassay Prog, Nat Cancer Inst, 77-78, eval panel, Primate Res Ctrs Prog, NIH, 77-78, cause & prev sci rev comt, Nat Cancer Inst, 78-81, adv bd, Northeast Regional Primate Ctr, 79-82, comt vet med sci, Nat Acad Sci, 79-80; fel, Japanese Soc Prom Sci, 83; vis prof, Univ Tokyo, 83. *Mem:* Am Asn Lab Animal Sci; Am Vet Med Asn; Int Acad Path; Am Col Vet Path; Soc Pharmacol & Environ Path. *Res:* Naturally occurring diseases in laboratory animals with specific interests in animal model systems; pulmonary pathology including pulmonary hemodynamics and infectious diseases with emphasis in pathogenesis. *Mailing Add:* Res Animal Consult Inc 17 Cumberland St Boston MA 02115

JONAS, ANA, b Rokiskis, Lithuania, Nov 24, 43; US citizen; m 68. BIOCHEMISTRY. *Educ:* Univ Ill, Chicago, BS, 66, Urbana, PhD(biochem), 70. *Prof Exp:* NIH trainee, 70-72, from asst prof to assoc prof, 72-84, PROF BIOCHEM, UNIV ILL, URBANA, 84- *Concurrent Pos:* NATO fel, Max Planck Med Res Inst, Heidelberg, WGer, 73; res grants, Nat Heart Lung & Blood Inst, 73-, consult, 78-; estab investorship, Am Heart Asn, 74-79; res grants, Am Heart Asn, 80-83, consult, 85-89; Max Baer res award, 78-; Coun Arteriosclerosis fel, Am Heart Asn; Fogarty fel, Ctr Molecular Biophys, Nat Ctr Res in Sci, Orleans, France, 81; ed bd, J Lipid Res, 84-; consult, Nat Res Coun, 89- *Mem:* Am Heart Asn; Am Soc Biochem & Molecular Biol; Am Chem Soc. *Res:* Structure and function of high density serum lipoproteins; protein-lipid interactions; transport function of serum proteins; applications of fluorescence spectroscopy. *Mailing Add:* 190 Med Sci Bldg Univ Ill 506 S Mathews Urbana IL 61801

JONAS, EDWARD CHARLES, b San Antonio, Tex, July 24, 24; m 49; c 3. CLAY MINERALOGY. *Educ:* Rice Inst, BS, 44; Univ Ill, MS, 52, PhD(geol), 54. *Prof Exp:* Asst geologist, Ill State Geol Surv, 52-54; from asst prof to prof, 54-, EMER PROF GEOL, UNIV TEX, AUSTIN. *Concurrent Pos:* Fulbright sr res award, NZ, 60-61. *Mem:* Fel AAAS; fel Geol Soc Am; fel Mineral Soc Am; Geochem Soc; Mineral Soc Gt Brit & Ireland. *Res:* Mineralogy of clays and uranium deposits in the Texas Gulf Tertiary. *Mailing Add:* Dept Geol Sci Univ Tex Austin TX 78713

JONAS, HERBERT, b Duesseldorf, Ger, May 23, 15; nat US; m. PLANT PHYSIOLOGY. *Educ:* Univ Calif, BS, 41, PhD(plant physiol), 50. *Prof Exp:* Nat Res Coun & AEC fel plant physiol, Biol Div, Oak Ridge Nat Lab, 51-52; res assoc pharmacol, Sch Med, Univ Va, 52-54, res assoc, Cancer Res Lab, 54-58; assoc prof pharmacog, Col Pharm, 58-69, chmn dept, 58-68, prof plant physiol, Col Biol Sci, 69-85, EMER PROF BIOL, COL BIOL SCI, UNIV MINN, ST PAUL, 85- *Concurrent Pos:* Electronics instr, US Naval Installations, Calif, 42-46. *Mem:* AAAS. *Res:* Physiology, biophysics; medicinal plants; secondary plant metabolites; sensitive plants; teaching; plant-man interrelations; economic botany; fresh water algae. *Mailing Add:* Dept Plant Biol Univ Minn St Paul MN 55108

JONAS, JIRI, b Prague, Czech, Apr 1, 32; US citizen; m 68. PHYSICAL CHEMISTRY, MOLECULAR SPECTROSCOPY. *Educ:* Tech Univ Prague, BS, 56; Czech Acad Sci, PhD(chem), 60. *Prof Exp:* Res assoc chem, Czech Acad Sci, 60-63; vis scientist, 63-65, from asst prof to assoc prof, 66-72, PROF CHEM, UNIV ILL, URBANA, 72-, SR STAFF MEM, MAT RES, 70-, DIR, SCH CHEM SCI, 83- *Concurrent Pos:* Fels, Alfred P Sloan Found, 67-69 & J S Guggenheim Found, 72-73; assoc mem ctr advan study, Univ Ill, 76-77. *Honors & Awards:* Joel Henry Hildebrand Award, Theoretical & Exp Chem of Liquids, Am Chem Soc, 83; US Scientist Award, Alexander von Humboldt Found, WGer. *Mem:* Nat Acad Sci; Am Chem Soc; fel Am Phys Soc; fel AAAS; fel Am Inst Chemists; fel Am Acad Arts & Sci. *Res:* Nuclear magnetic resonance; raman spectroscopy; dynamic structure of liquids; glasses, molecular solids and biopolymers; high pressure research; behavior of materials under extreme conditions of pressure and temperature. *Mailing Add:* 166 Roger Adam Lab 1209 W California Urbana IL 61801

JONAS, JOHN JOSEPH, b Budapest, Hungary, Dec 9, 14; nat US; m 41; c 3. ORGANIC CHEMISTRY. *Educ:* Pazmany Peter Univ, PhD(chem, physics. math), 37. *Prof Exp:* Asst & instr org & pharmaceut chem, Pazmany Peter Univ, 36-39; chemist, Darmol Pharmaceut Co, Budapest, 39-43; chemist, Hungary Viscose Corp, 43-45; asst div leader pharmaceut & nutrit res, Inst Heiligenberg, Ger, 46-51; assoc mgr indust prod, Res & Develop Div, Kraft Inc, 51-77, SCI MGT CONSULT, RES & DEVELOP DIV, KRAFT, INC, 77- *Concurrent Pos:* Consult, Protein Resources Study, NSF, 75-76. *Mem:* Am Chem Soc; Inst Food Technol; Am Inst Chemists. *Res:* Metabolic diseases of dairy cattle; seaweed hydrocolloids; food emulsifying systems; high protein foods; dairy analogues, vegetable proteins, synthetic nutrients. *Mailing Add:* Rte 2 Box 246 Meadows of Dan VA 24120

JONAS, JOHN JOSEPH, b Montreal, Que, Dec 12, 32; m 60; c 4. PHYSICAL METALLURGY, MECHANICAL METALLURGY. *Educ:* McGill Univ, BEng, 54; Cambridge Univ, PhD(mech sci), 60. *Prof Exp:* From asst prof to prof phys metall, 60-85, assoc dean, Fac Grad Studies & Res, 71-75, RES PROF MECH METALL, MCGILL UNIV, 85- *Honors & Awards:* Reaumur Medal, French Soc Metall, 80; Hatchett Medal & Award, UK, 82; DoFasco Mat Eng Award, 82; Can Metal Physics Asn Med, 83. *Mem:* Fel Am Soc Metals; Am Inst Mining, Metall & Petrol Engrs; Brit Inst Metals; Iron & Steel Inst Japan; Can Inst Mining & Metall. *Res:* Mechanical metallurgy; elevated temperature deformation of metals and crystalline materials; microstructural changes and stress-strain rate-temperature relationships during hot working; plastic instability; thermal activation analysis, textures, yield surfaces and formabilty. *Mailing Add:* Dept Metall Eng McGill Univ Montreal PQ H3A 2M5 Can

JONAS, LEONARD ABRAHAM, b New York, NY, Feb 6, 20; m 42; c 2. PHYSICAL CHEMISTRY. *Educ:* Brooklyn Col, AB, 40; Univ Md, MS, 69, PhD(phys chem), 70. *Prof Exp:* Phys chemist, US Army Res & Develop Ctr, 42-80; sr res scientist, Nat Cancer Inst, Frederick Cancer Res Facil, Litton Bionetics, Inc, 80-87; SR RES SCIENTIST, HUGHES ASSOC INC, 87- *Concurrent Pos:* Consult air purification, Sch Hyg & Pub Health, Johns Hopkins Univ, 83- *Mem:* Am Carbon Soc; Am Soc Testing & Mat; AAAS; Am Chem Soc; Sigma Xi. *Res:* Equilibrium gas adsorption; adsorption kinetics; aerosol physics and filtration; heterogeneous catalysis; physical protection against carcinogens. *Mailing Add:* 6612 Baythorne Rd Baltimore MD 21209

JONAS, ROBERT JAMES, b Marinette, Wis, June 8, 26; m 50; c 3. ANIMAL ECOLOGY, WILDLIFE MANAGEMENT. *Educ:* Univ Idaho, BS, 50, MS, 55; Mont State Univ, PhD(wildlife mgt), 64. *Prof Exp:* Instr biol, Lewis-Clark Norm Col, 55-57; asst prof, Whitman Col, 64-65 & Univ Idaho, 65-66; coordr wildlife biol, 77-81, chmn, Dept Biol, 81-82, PROF ZOOL, WASH STATE UNIV, 66- *Concurrent Pos:* Danforth Assoc, 68-; environ specialist, Nat Park Serv, 75-76. *Mem:* AAAS; Sigma Xi (vpres, 74-75); Wildlife Soc. *Res:* Populations of wild animals; human impact on natural ecosystems; wild turkeys. *Mailing Add:* 823 E Fifth St Moscow ID 83843

JONASSEN, HANS BOEGH, b Seelze, Ger, Aug 18, 12; nat US; m 39; c 3. INORGANIC CHEMISTRY. *Educ:* Tulane Univ, BS, 42, MS, 44; Univ Ill, PhD(chem), 46. *Prof Exp:* From instr gen & anal chem to asst prof chem, Tulane Univ, 43-48, assoc prof inorg chem, 48-52, chmn dept chem, 62-68, prof inorg chem, 52-80. *Concurrent Pos:* Sci liaison officer, London Br, Off Naval Res, 58-59; mem adv coun col chem, NSF-Am Chem Soc, 62-67; Reilly centennial lectr, Univ Notre Dame, 65; Francis P Dwyer Mem lectr, Univ New South Wales, 67; Australian-Am Educ Found sr scholar, Univ Sydney, 71. *Mem:* AAAS; Am Chem Soc. *Res:* Inorganic and metalorganic chemistry; complex ions; homogenous and heterogeneous catalysis. *Mailing Add:* 7729 Belfast St New Orleans LA 70125

JONCAS, JEAN HARRY, b Montreal, Que, June 27, 30; m 55; c 5. VIROLOGY, INFECTIOUS DISEASES. *Educ:* Jean de Brebeuf Col, BA, 48; Univ Montreal, MD, 55, PhD(microbiol, immunol), 67; Royal Col Physicians & Surgeons, Can, cert pediat, 59, cert microbiol, 76. *Prof Exp:* Fel, Sch Med, Wayne State Univ, 56; res asst, Univ Montreal, 60-70, res assoc virol, inst microbiol & hyg, 71-76, head infectious dis sect, Ste Justine Hosp, Montreal, 70-76, head dept microbiol, 77-85, PROF MICROBIOL & IMMUNOL, SCH MED, UNIV MONTREAL, 70-, HEAD VIROL LAB, STE JUSTINE HOSP, MONTREAL, 85- *Concurrent Pos:* Clin asst, Montreal Children's Hosp, 59-70, consult, 70-; lectr, Sch Med, Univ Montreal, 64-67, asst prof, 67-70; demonstr, Dept Pediat, McGill Univ, 70-; mem adv comt infection, Immunity & Ther, Defence Res Bd Can, 70-74; Med Res Coun Can grant, 70-86; Nat Defence Res grant, 70-76; Fed Prov pub health res grant, 71-75; Nat Cancer Inst Can grant, 75-81; health & welfare contract Epstein Barr virus ref serv, 85- *Mem:* Can Soc Microbiol; Can Pub Health Asn; Am Soc Microbiol; Can Paediat Soc; fel Am Acad Pediat; fel Infectious Dis Soc Am; Can Infectious Dis Soc. *Res:* Etiology and epidemiology of infectious mononucleosis; the Epstein-Barr herpes virus; cell-virus relationship; Epstein-Barr virus and oncogenesis; diagnosis of viral infections by rapid immunological methods and molecular biology; pediatric infectious diseases, epidemiology, diagnosis and treatment. *Mailing Add:* Dept Microbiol & Immunol St Justine Hosp 3175 Ste-Catherine Rd Montreal PQ H3T 1C5 Can

JONDORF, W ROBERT, b Nuremberg, Ger, Dec 2, 28; Brit citizen; m 63; c 5. DRUG METABOLISM, INTERACTION OF DRUGS. *Educ:* Univ Wales, BSc, 50; Univ London, PhD(biochem), 56. *Prof Exp:* Food res chemist, Unilever Ltd, 50-53; vis scientist, Nat Heart Inst, NIH, Bethesda, Md, 57-60; Chester Beatty spec lectr cancer res, Inst London, 61; spec res fel, Dept Biochem, Univ Glasgow, Scotland, 62-63, Leverhulme res fel, Dept Pharmacol, 71-73; Dept Pharmacol, Med Ctr, George Washington Univ, 63-71; res fel, Dept Biochem, Univ Cambridge, Eng, 73-74; res scientist, Racecourse Security Serv Labs, Newmarket & Suffolk, 74-79; guest fac, Dept Pharmacol, Univ Berne, Switz, 79-81, guest fac mem, Inst Pharmacol, 87; scientist, Endocrinol Sect, Bourn Hall Clin, Bourn, Cambridge, 82-85; spec projs assoc, Dept Pharmacol/Oncol, Mayo Clin, Rochester, Minn, 88; RETIRED. *Mem:* Brit Pharmacol Soc; Am Soc Pharmacol & Exp Therapeut. *Res:* Drug metabolism, toxicology; macromolecular biosyntheses; cancer chemotherapy; evolution, species differences; radioimmunoassay methods; age and hormone dependence in response to drugs and environmental chemicals; mechanism of action of drugs. *Mailing Add:* Three Gough Way Cambridge CB3 9LN England

JONEJA, MADAN GOPAL, b Lyallpur, India, Dec 25, 36; Can citizen; m 65. TERATOLOGY, ULTRASTRUCTURAL CELL BIOLOGY. *Educ:* Panjab Univ, India, MSc, 58; Queen's Univ, Ont, PhD(biol), 65. *Prof Exp:* From lectr to assoc prof, 65-76, chmn grad studies anat, 70-80, PROF ANAT, QUEENS UNIV, ONT, 76-, HEAD DEPT ANAT, 81- *Mem:* Teratology Soc; Am Asn Anatomists; Can Asn Anatomists. *Res:* Cytology; mammalian embryogenesis; cytological mechanisms of teratogenesis; scan electron microscope and transmission electron microscope of in vitro differentiation of the neural tube. *Mailing Add:* Dept Anat Queen's Univ Kingston ON K7L 3N6 Can

JONES(STOVER), BETSY, b Salt Lake City, Utah, May 13, 26; div; c 2. RADIOBIOLOGY, PHARMACOLOGY. *Educ:* Univ Utah, BA, 47; Univ Calif, PhD(chem), 50. *Prof Exp:* Asst, Univ Calif, 47-49, asst, Radiation Lab, 48-50; asst res prof chem, Univ Utah, 50-58, assoc res prof, 58-70; from assoc prof to prof pharmacol, Sch Med, Univ NC, Chapel Hill, 70-88, dir, Grad Training Prog, 74-80, fac mem, Curric Toxicol, 79-88; RETIRED. *Concurrent Pos:* Chemist, Radiobiol Lab, Univ Utah, 50-70, adj assoc res prof anat, Univ NC, 70-75, adj res prof, 75-79; consult, Radiobiol Lab, 70-; mem panel eval NSF Grad Fel Applns, Nat Res Coun, 74-76 & Postdoctoral Fel Applns, 78-80; mem adv comt, Health Physics Div, Oak Ridge Nat Lab, 75-77; referee, Am J Physics, 77-80; adj prof pharmacol, Univ Utah, 79-; mem, Task Group 6, Sci Comt 57, Nat Coun Radiation Protection & Measurements, 80-83. *Mem:* Am Phys Soc; Am Chem Soc; Radiation Res Soc; Am Soc Pharmacol & Exp Therapeut; Am Asn Univ Profs; AAAS. *Res:* Toxicology of radionuclides; rate processes in biology. *Mailing Add:* B243 Villa Carolina Meadows Chapel Hill NC 27510

JONES, A(NDREW) R(OSS), b Mt Sterling, Ill, May 3, 21; m 44; c 2. ELECTRICAL ENGINEERING. *Educ:* Clemson Col, BEE, 47; Univ Pittsburgh, MSEE, 53. *Prof Exp:* Engr, 47-58, mgr preliminary plant eng, Atomic Power Dept, 58-64, mgr advan develop & planning, 64-66, mgr projs syst & technol, Astronuclear Labs, 66-71, mgr spec proj, 71-76, mgr eng, Adv Energy Syst Div, 76-78, MGR EMERGING SYSTS STUDIES, WESTINGHOUSE ELEC CORP, 78- *Mem:* Am Nuclear Soc; Soc Naval Archit; Inst Elec & Electronics Engrs; Am Inst Aeronaut & Astronaut. *Res:* Nuclear, solar and fuel cell power plant design and economics, particularly electric utility and cogeneration applications. *Mailing Add:* 6991 Lemington Ave Pittsburgh PA 15206

JONES, ALAN A, b Jamestown, NY, Nov 15, 44; m 72. POLYMER CHEMISTRY. *Educ:* Colgate Univ, AB, 66; Univ Wis, PhD(phys chem), 72. *Prof Exp:* Res instr polymer chem, Dartmouth Col, 72-74; from asst prof to assoc prof, 74-88, chmn dept, 81-87, actg provost, 87-88, PROF CHEM, CLARK UNIV, 81- *Concurrent Pos:* Vis prof, Univ Wis, 85. *Mem:* Am Chem Soc; Am Phys Soc. *Res:* The dynamic properties of macromolecules in solution and in the bulk are probed by nuclear magnetic resonance spectroscopy or dielectric response and discussed in terms of models relating specific motions to the experimental observations. *Mailing Add:* Dept Chem Clark Univ Worcester MA 01610

JONES, ALAN LEE, b Albion, NY, June 23, 39; m 67; c 2. PLANT PATHOLOGY. *Educ:* Cornell Univ, BS, 61, MS, 63; NC State Univ, PhD(plant path), 68. *Prof Exp:* From asst prof to assoc prof, 68-77, PROF PLANT PATH, MICH STATE UNIV, 77- *Concurrent Pos:* Sabbatical leaves, Plant Protection Inst, Agr Res Ctr, USDA, Beltsville, Md, 74-75, Bayer Agr, Leverkusen, Fed Repub Ger, 82 & Dept Plant Path, NY State Agr Exp Sta, Cornell Univ, Geneva, NY, 89. *Honors & Awards:* Nat Award in Agr, Am Phytopath Soc, 78. *Mem:* AAAS; Can Phytopath Soc; fel Am Phytopath Soc. *Res:* Epidemiology and control of tree fruit diseases; phytobacteriology; fungicide and antibiotic resistance in tree fruit pathogens. *Mailing Add:* Dept Bot & Plant Path Mich State Univ East Lansing MI 48824-1312

JONES, ALAN RICHARD, b Denver, Colo, Dec 25, 39; m 64; c 2. PHYSICAL CHEMISTRY, CHEMICAL ENGINEERING. *Educ:* Univ Colo, BSChE, 62; Lawrence Univ, MS, 64, PhD, 67. *Prof Exp:* Res scientist, 67-69, leader paper prod group, 69-74, leader chem processes group, 74-78, asst tech dir, 78-83, tech dir, 83-85, DIV TECH DIR, UNION CAMP CORP, 85- *Honors & Awards:* Hugh D Camp Award, Union Camp Corp, 73. *Mem:* Am Inst Chem Engr; Tech Asn Pulp & Paper Indust. *Res:* Mechanical and optical properties of paper, characterization of papermaking pulps, development and optimization of pulping, by-product chemical, and papermaking processes. *Mailing Add:* Union Camp Corp PO Box 570 Savannah GA 31402

JONES, ALBERT CLEVELAND, b Coalinga, Calif, Aug 18, 29; m 55; c 3. FISHERIES. *Educ:* Univ Wash, BS, 51; Univ Calif, MA, 54, PhD(zool), 59. *Prof Exp:* Asst, Ore Fish Comn, 49-50, biologist, 51; biologist, Fisheries Res Inst, Wash, 52; asst, Sagehen Creek Wildlife Fisheries Sta, Univ Calif, 53-54, asst zool, Univ, 54-58; asst ichthyol, Calif Acad Sci, 58-59; res asst prof fisheries, Univ Miami, 59-65; asst dir, 65-71, prog mgr, Off-in-Chg, 72-76, asst dir fishery mgt, 76-84, dir Miami Lab, 84-85, dir econ & statist, 85-90, DIR RES MGT DIV, SOUTHEAST FISHERIES CTR, NAT MARINE FISHERIES SERV, 90- *Concurrent Pos:* Ministry of Agr, Fisheries & Food fel, Eng, 64-65; adj assoc prof, Univ Miami, 68-76, adj prof, 76- *Mem:* Am Fisheries Soc; Am Soc Ichthyol & Herpet; Am Inst Fishery Res Biologists. *Res:* Population dynamics; biometrics; ecology. *Mailing Add:* 8950 SW 62 CT Miami FL 33156

JONES, ALFRED, b Richmond, Va, Mar 25, 32; m 62; c 2. PLANT GENETICS. *Educ:* Va Polytech Inst, BS, 53; NC State Col, MS, 57, PhD(plant breeding & path), 61. *Prof Exp:* Res asst cotton breeding, field crops dept, NC State Col, 61-62; RES GENETICIST VEG & ORNAMENTALS RES BR, PLANT SCI RES DIV, US VEG BREEDING LAB, USDA, 62- *Concurrent Pos:* L M Ware Res Award, Am Soc Hort Sci, 79. *Mem:* Am Soc Hort Sci; Am Genetic Asn; Crop Sci Soc Am; Am Soc Agron; Sigma Xi. *Res:* Cytogenetics of sweetpotato, especially nature of ploidy in Ipomoea and its relation to speciation, recombination and breeding systems; quantitative genetic techniques of breeding for disease and insect resistant types. *Mailing Add:* 2875 Savanah Hwy Charleston SC 29414

JONES, ALISTER VALLANCE, b Christchurch, NZ, Feb 4, 24; m 51; c 3. AERONOMY, AURORA. *Educ:* Univ NZ, BSc, 45, MSc, 46; Cambridge Univ, PhD(physics), 50. *Prof Exp:* Nat Res Coun Can fel, 49-51; from asst prof to prof physics, Univ Sask, 53-68; sr res officer, Upper Atmosphere Res Sect, Astrophys Br, 68-76, prin res officer, Planetary Sci Sect, 76-89, GUEST WORKER, SOLAR TERRESTRIAL PHYSICS, HERZBERG INST ASTROPHYS, NAT RES COUN CAN, 89- *Concurrent Pos:* Ed, Physics Can, Can Asn Physicists, 63-66, assoc ed, Aeronomy & Space Physics, Can J Physics, 79-; prin investr, Canopus proj, 80-89; chmn, Div 2, Int Asn Geomag & Aeronomy, 87- *Mem:* Can Asn Physicists; Royal Soc Can. *Res:* Infrared, auroral and airglow spectroscopy. *Mailing Add:* Herzberg Inst Astrophys Nat Res Coun Can Ottawa ON K1A 0R6 Can

JONES, ALLAN W, b Scranton, Pa, June, 3, 37; m 62; c 4. BIOLOGY, PHYSIOLOGY. *Educ:* Princeton Univ, BSE, 59; Univ Pa, PhD(physiol), 65. *Prof Exp:* Trainee & instr physiol, Sch Med, Univ Pa, 65-66; fel, Oxford Univ, 66-68; assoc, Sch Med, Univ Pa, 68-69, asst prof, 69-72, assoc dir, Bockus Res Inst, Grad Hosp, 70-72; assoc prof, 72-78, PROF PHYSIOL, SCH MED, UNIV MO-COLUMBIA, 78- *Concurrent Pos:* Estab investr, Am Heart Asn, 74-79; mem, Coun High Blood Pressure Res. *Mem:* Am Pharmacol Soc; Am Physiol Soc. *Res:* Hypertension; electrolyte metabolism of arteries; cardiovascular research. *Mailing Add:* Dept Physiol Univ Mo MA415 Med Sci Bldg Columbia MO 65212

JONES, ALMUT GITTER, b Oldenburg, WGer, Sept 8, 23; US citizen; wid. PLANT TAXONOMY. *Educ:* Univ Ill, Urbana, BS, 58, MS, 60, PhD(bot), 73. *Prof Exp:* Asst prof bot, 74-75 & 79-88, CUR HERBARIUM, UNIV ILL, URBANA, 73-, ASSOC PROF PLANT BIOL, 89- *Mem:* Am Soc Plant Taxonomists; Am Bot Soc; Int Asn Plant Taxon. *Res:* Taxonomy, phytogeography and biosystematics of Aster, Compositae; flora of Illinois. *Mailing Add:* Dept Plant Biol Univ of Ill-Champaign Urbana IL 61801

JONES, ALUN RICHARD, b Ipoh, Malaya, May 6, 28; Can citizen; m 55; c 2. PHYSICS, RADIATION DOSIMETRY. *Educ:* Univ Bristol, BSc, 52; McGill Univ, MSc, 54. *Prof Exp:* Mem physics div, Electronics Br, Atomic Energy Can Ltd, 54-56; exchange worker, Inst Cancer Res, UK, 62-63; mem biol & health div, Dosimetric Res Br, Atomic Energy Can Ltd, 56-90, sr res officer, 65-90; RETIRED. *Concurrent Pos:* Consult to pres comn, Three Mile-Island, 79. *Mem:* Health Physics Soc; Can Radiation Protection Asn. *Res:* External dosimetry of gamma and beta rays; thermoluminescence dosimetry; monitoring of alpha, beta and gamma contamination; detectors of ionising radiation (Geiger Mueller counters and silicon junction detectors); radiation protection. *Mailing Add:* Box 711 Deep River ON K0J 1P0 Can

JONES, ANITA KATHERINE, b Ft Worth, Tex; m; c 2. COMPUTER SCIENCE. *Educ:* Rice Univ, AB, 64; Univ Tex, MA, 66; Carnegie-Mellon Univ, PhD(comput sci), 73. *Prof Exp:* Programmer, IBM Corp, 66-68; asst prof, 73-78, assoc prof comput sci, Carnegie-Mellon Univ, 78-; DEPT CHAIR COMPUTER SCI, TARTAN LABS, PITTSBURGH. *Concurrent Pos:* Consult, NSF, Defense Advan Res Proj Agency, Nat Res Coun & Indust; trustee, Mitre Corp; dir Sci Appln Int Corp; mem Defense Sci Bd. *Mem:* Asn Comput Mach; Inst Elec & Electronic Engrs. *Res:* Design and implementation of programmed systems on computers, including enforcement of security policies on computers, operating systems and scientific data bases. *Mailing Add:* Dept Computer Sci Univ Va Charlottesville VA 22903

JONES, BARBARA ELLEN, b Philadelphia, Pa, Dec 19, 44; m 72; c 1. NEUROSCIENCE, NEUROANATOMY. *Educ:* Univ Del, BA, 66, PhD(psychol), 71. *Prof Exp:* Fel neurochem, Col France, Paris, 70-72; res assoc psychiat, Univ Chicago, 72-74; vis lectr med physiol, Univ Nairobi, 74-75; res assoc & asst prof psychiat, Univ Chicago, 75-77; from asst prof to assoc prof, 77-89, PROF NEUROL, MCGILL UNIV, 89- *Concurrent Pos:* Scholar, Med Res Coun Can, 78-83; scholar, Ctr Med Res, Quebec, 83-86; vis scientist, human anat, Oxford Univ, UK, 84-85. *Mem:* Soc Neurosci; Am Asn Anat; Asn Psychophysiol Study Sleep. *Res:* Neuroanatomical and neurochemical substrates of mechanisms of the sleep-waking cycle. *Mailing Add:* Dept Neurol & Neurosurg McGill Univ Montreal Neurol Inst Montreal PQ H3A 2B4 Can

JONES, BARCLAY G(EORGE), b Lafleche, Sask, May 6, 31; US citizen; m 59; c 4. NUCLEAR ENGINEERING, MECHANICAL ENGINEERING. *Educ:* Univ Sask, BE, 54; Univ Ill, MS, 60, PhD(nuclear eng), 66. *Prof Exp:* Athlone fel, Eng Elec Co, Rugby, Eng, 54-55; Atomic Energy Res Estab, Harwell, 55-57; engr, Nuclear Div, Canadair Ltd, Montreal, Que, 57-58; res asst, 58-60, instr nuclear eng, 63-66, from asst prof to assoc prof, 66-72, PROF NUCLEAR & MECH ENG, UNIV ILL, URBANA-CHAMPAIGN, 72-, HEAD, 87- *Concurrent Pos:* Consult, WVa Pulp & Paper Co, Va, 68, Arnold Res Orgn, Inc, Tullahoma, Tenn, 74-80, Argonne Nat Lab, 76- & Fauske & Assocs, Burr Ridge, Ill, 81-85; assoc chairperson, nuclear eng, Univ Ill, Urbana, 81-86, actg head, 86-87; Helliburton educ award, 83. *Mem:* Am Nuclear Soc; Can Soc Mech Engrs; Am Soc Mech Engrs; Eng Inst Can; Am Inst Aeronaut & Astronaut; Sigma Xi; Am Soc Eng Educ. *Res:* Experimental fluid mechanics and heat transfer, reactor safety, two-phase flow, turbulence, simulation & training. *Mailing Add:* 310 E Holmes Urbana IL 61801

JONES, BENJAMIN A(NGUS), JR, b Mahomet, Ill, Apr 16, 26; m 49; c 2. AGRICULTURAL ENGINEERING. *Educ:* Univ Ill, BS, 49, MS, 50, PhD(civil eng), 58. *Prof Exp:* Asst agr eng, Univ Ill, 49-50; actg chmn, Dept Agr Eng, Univ Vt, 50-51, asst prof & asst agr engr, Agr Exten Serv, 50-52; from instr to assoc prof, 52-64, prof agr eng & head soil & water div, 64-73, ASSOC DIR, ILL AGR EXP STA, UNIV ILL, URBANA-CHAMPAIGN, 73- *Concurrent Pos:* Consult engr, Ill Drainage Dists. *Honors & Awards:* Hancor Award, Am Soc Agr Engrs, 77. *Mem:* fel, Am Soc Agr Engrs; Am Soc Eng Educ; Soil Conserv Soc Am; Sigma Xi; Coun Agr Sci & Technol. *Res:* Agricultural land drainage and irrigation; agricultural hydrology and the hydraulics of erosion control structures. *Mailing Add:* 211 Mumford Hall Univ Ill 1301 W Gregory Dr Urbana IL 61801

JONES, BENJAMIN FRANKLIN, JR, b Apr 15, 36; US citizen; m 57; c 3. MATHEMATICS. *Educ:* Rice Univ, BA, 58, PhD(math), 61. *Prof Exp:* Temp mem, Courant Inst Math Sci, NY Univ, 61-62; from asst prof to assoc prof math, 62-68, PROF MATH, RICE UNIV, 68- *Concurrent Pos:* Mem, Inst Advan Study, 65-66; vis prof, Univ Minn, 69-70; vis mem, Math Inst, Oxford Univ, 85-86. *Mem:* Am Math Soc; Math Asn Am. *Res:* Partial differential equations; singular integral operators. *Mailing Add:* Dept Math Rice Univ Houston TX 77251

JONES, BENJAMIN LEWIS, b Muncy Valley Twp, Pa, Aug 19, 52; m 80; c 1. BIOCHEMISTRY. *Educ:* Pa State Univ, BS, 74; Univ Tenn, MS, 78, PhD(biochem), 80. *Prof Exp:* Postdoctoral res assoc, Kettering Res Lab, 80-82; res chemist, Lipid Metab Lab, Vet Admin Hosp, 82-84; sr res chemist, 84-85; res scientist, 85-90, SR RES SCIENTIST, CAMPBELL INST RES TECHNOL, 90- *Mem:* Am Soc Biochem & Molecular Biol; Am Chem Soc; Am Soc Microbiol; Inst Food Technologists. *Res:* Developing beef and chicken type reaction flavors using yeast extracts, meats, enzymes and natural chemicals; processing methods and conditions for production of reaction flavors; scale up of reaction flavors. *Mailing Add:* Campbell Soup Co Box 57X Campbell Pl Camden NJ 08103

JONES, BERNE LEE, b Rochester, Ind, May 30, 41; m 63; c 3. BIOCHEMISTRY, PROTEIN CHEMISTRY. *Educ:* Wabash Col, BA, 63; Wash State Univ, PhD(chem), 67. *Prof Exp:* Fel biochem, Univ Colo, 67-69 & Univ Alta, 69-72; asst prof plant sci, Univ Man, 72-77; RES CHEMIST BIOCHEM, USDA, 77- *Mem:* Am Asn Cereal Chem; Am Chem Soc; Am Soc Brewing Chemists. *Res:* Biochemistry of cereal proteins; enzymology of malting; endoproteinase biochemistry. *Mailing Add:* Cereal Corp Res Unit 501 N Walnut St Madison WI 53705

JONES, BERWYN E, b Scottsbluff, Nebr, Mar 11, 37; m 58; c 2. ANALYTICAL CHEMISTRY. *Educ:* Nebr Wesleyan Univ, BA, 58; Kans State Univ, PhD(anal chem), 65. *Prof Exp:* From instr to assoc prof chem, Monmouth Col, Ill, 63-75; prof, Upper Iowa Univ, Fayette, 75-77; assoc prof, Longwood Col, 77-78; res chemist, 78-80, asst lab dir, Nat Water Qual Lab, US Geol Surv, Atlanta, 80-85; QUALITY CONTROL OFF, NAT WATER QUAL LAB, US GEOL SURV, DENVER, 85- *Concurrent Pos:* Vis asst prof, Univ Ill, 69-70; res assoc, Argonne Nat Lab, 70-71. *Mem:* AAAS; Am Chem Soc; Soc Appl Spectros; Sigma Xi. *Res:* Absorption and fluorescence spectroscopy; chromatography; atomic emission spectroscopy; molecular fluorescence spectroscopy; quality assuranceal of chemical analysis. *Mailing Add:* 30926 Shawnee Lane Evergreen CO 80439

JONES, BETTY RUTH, b Hernando, Miss, Mar 20, 51; m 73. CELL BIOLOGY, MEDICAL PARASITOLOGY. *Educ:* Rust Col, BS, 73; Atlanta Univ, MS, 75, PhD(biol), 78. *Prof Exp:* MBS partic res biol, Atlanta Univ, 73-78; asst prof biol, Morehouse Col, 78-90; ASST VPRES ACAD AFFAIRS & PROF BIOL, MORRIS BROWN COL, 90. *Concurrent Pos:* Anal chem lab tech trainee, Polaroid Corp, 70; student nurse, Methodist Hosp Sch Nursing, 71-72; res asst microbiol, Argonne Nat Lab, 73; lab coordr, Atlanta Jr Col, 75; Biomed Res Sci grant, 78-79. *Mem:* Am Soc Parasitol; Am Micros Soc; Electron Micros Soc; Am Soc Microbiol. *Res:* Medical parasitology; tropical medicine; ultrastructure; schistosomiasis and cysticercosis; developmental mechanisms; structural and functional host-parasite interactions. *Mailing Add:* Dept Acad Affairs Morris Brown Col 643 M L King Dr NW Atlanta GA 30314

JONES, BLAIR FRANCIS, b Apr 14, 34; m 55; c 2. GEOLOGY. *Educ:* Beloit Col, BA, 55; Johns Hopkins Univ, PhD, 63. *Prof Exp:* Lab instr geol, Beloit Col, 54-55; geologist, US Geol Surv, 55, Deposits Br, 56-57, res geologist, Water Resources Div, 58-74, res adv geochem, Washington, DC, 74-77, Reston, 81-84, SR SCIENTIST, WATER RESOURCES DIV, US GEOL SURV, NAT CTR, 83- *Concurrent Pos:* Instr, Rockford Col, 54; vis prof, State Univ NY Binghamton, 70; guest investr, Rothamsted Exp Sta, Herpenden, Herts, UK, 72. *Honors & Awards:* Meritorious Serv Award, Dept Interior, 81 & Distinguished Serv Award, 86. *Mem:* Fel Geol Soc Am; Geochem Soc; fel Mineral Soc Am; Am Geophys Union; Clay Minerals Soc; Sigma Xi; Mineral Soc UK. *Res:* Hydrogeochemistry; sedimentary petrology; geochemistry of weathering; brines, lacustrine sediments and evaporites; solutes in natural water. *Mailing Add:* 7905 Glenbrook Rd Bethesda MD 20014

JONES, BRIAN HERBERT, b Chester, Eng, Apr 23, 37; US citizen; m 74. MATERIALS SCIENCE, MECHANICAL ENGINEERING. *Educ:* Univ Liverpool, BEng, 61, PhD(appl mech), 65; Univ Calif, Los Angeles, cert bus admin, 69. *Prof Exp:* Group leader mat technol, Douglas Aircraft Co, 66-69; consult, ARAP, Inc, 69-72; vpres eng, Goldsworthy Eng, Inc, 72-75; PRES & CHIEF OPERATING OFFICER, COMPOSITEK CORP, SUBSID SHELL OIL CO, 75- *Concurrent Pos:* Clayton fel, Inst Mech Engrs, London, 63-65; Busk fel, Royal Aeronaut Soc, London, 63-64. *Mem:* Am Inst Aeronaut & Astronaut; Am Soc Mech Engrs; Soc Plastics Indust; Soc Automotive Engrs. *Res:* Composite materials product design and analysis; lightweight structures; plasticity of metals; process development; machine design. *Mailing Add:* 407 Country Club Dr San Gabriel CA 91775

JONES, BURTON FREDRICK, b Manistique, Mich, Oct 28, 42; c 2. ASTROMETRY. *Educ:* Univ Chicago, BS, 65, MS, 68, PhD(astron), 70. *Prof Exp:* Fel astron, Lick Observ, Univ Calif, 70-71; sr res fel, Royal Greenwich Observ, 72-74; res fel, Univ Tex, 74-75; asst res astronomer, 75-79, asst astronomer & prof, 79-87, ASTRONOMER & PROF, LICK OBSERV, UNIV CALIF, SANTA CRUZ, 87- *Mem:* Int Astron Union; Am Astron Soc. *Res:* Stellar proper motions; cluster membership. *Mailing Add:* Lick Observ Univ Calif Santa Cruz CA 95054

JONES, C ROBERT, b Scranton, Pa, May 8, 33; m 57; c 4. CELL PHYSIOLOGY. *Educ:* Univ Scranton, BS, 54; Fordham Univ, MS, 56, PhD(physiol), 62. *Prof Exp:* Res asst chemother, Sloan-Kettering Inst Cancer Res, 58-59; fel, Fordham Univ, 62-63; from instr to asst prof physiol, 63-68,

assoc prof cell biol, 68-74, PROF CELL BIOL, LEHMAN COL, 74-, CHMN DEPT BIOL SCI, 68- *Concurrent Pos:* Lectr, Bronx Community Col, 62-63; NSF res grant, Div Metab Biol, 64-66. *Mem:* AAAS; Entom Soc Am. *Res:* Activity of respiratory enzymes during the metamorphosis of holometabolous insects; identification of Lysosomes in insect tissues; mammalian physiology. *Mailing Add:* Dept Biol Sci Herbert H Lehman Col Bronx NY 10468

JONES, CARL JOSEPH, b Ithaca, NY, Jan 1, 49; m 82; c 2. BIOCONTROL, IMMUNOPARASITOLOGY. *Educ:* Cornell Univ, BS, 70; Univ Wyo, MS, 79, PhD(entom), 82. *Prof Exp:* Experimentalist, Dept Entom, Cornell Univ, 70-74; res specialist, 75-77; postdoctoral res assoc, Dept Entom, Univ Fla, 82; biol adminr, Off Entom, State Fla, 82-89; ASST PROF VET MED ENTOM, COL VET MED, UNIV ILL, 89- *Concurrent Pos:* Adj fac mem, Gulf Coast Community Col, 89; affil, Ill Natural Hist Surv Ctr Econ Entom, 91- *Mem:* Entom Soc Am; Sigma Xi. *Res:* Biological control of arthropods, physiological interactions of vertebrates and their hematophagous arthropod parasites; behavior and ecosystem dynamics of ticks; population dynamics and physiology of anautogenous muscoid flies; genetics of dispersal in arthropods. *Mailing Add:* Dept Vet Pathobiol Col Vet Med Univ Ill Urbana IL 61801

JONES, CARL TRAINER, b Allentown, Pa, Dec 31, 10; m 38; c 2. CHEMISTRY. *Educ:* Washington Missionary Col, BA, 33; Cath Univ, MS, 39; Ore State Col, PhD, 59. *Prof Exp:* Head dept chem, Atlantic Union Col, 37-38; asst prof chem, Washington Missionary Col, 38-46; head dept sci, Philippine Union Col, 47-52; asst prof chem, Walla Walla Col, 52-76, chmn dept, 60-76; RETIRED. *Concurrent Pos:* Teacher, Takoma Acad, 38-46. *Mem:* Am Chem Soc. *Res:* Analytical chemistry. *Mailing Add:* 207 NE A St College Place WA 99324

JONES, CAROL A, b Kremmling, Colo, Sept 10, 36; m 55; c 7. BIOPHYSICS. *Educ:* Univ Colo, BA, 63, PhD(biophys), 69. *Prof Exp:* Fel biophys, USPHS, 69-71, res assoc biophys & genetics, 71-74, asst prof, 74-80, ASSOC PROF BIOPHYS & GENETICS, UNIV COLO HEALTH SCI CTR, 80-, SR FEL, ELEANOR ROOSEVELT INST CANCER RES, 80- *Mem:* AAAS; Am Soc Human Genetics; Am Asn Immunol. *Res:* Cell biology; somatic cell genetics; cell surface molecules. *Mailing Add:* E Roosevelt Inst Cancer Res Univ Colo Med Ctr 1899 Gaylord St Denver CO 80206

JONES, CHARLES, b New York, NY, Feb 27, 26; m 46; c 2. ROTARY ENGINES, TECHNICAL MANAGEMENT. *Educ:* Columbia Univ, BS, 50, MS, 53. *Prof Exp:* Sect head, Appl Mech, Curtiss-Wright Corp, 55-62, chief design engr, 62-68, chief engr, 68-69, dir res eng, 69-84; CHIEF TECHNOLOGIST, JOHN DEERE TECHNOLOGIES INT, 84- *Concurrent Pos:* Adj lectr, Stevens Inst Technol, 86-87. *Honors & Awards:* Edward N Cole Award, Soc Automotive Engrs, 87, Forest McFarland Award, 91. *Mem:* Fel Soc Automotive Engrs. *Res:* Rotary engines design and development; author of over 25 technical publications; awarded 74 patents. *Mailing Add:* 208 Forest Dr Hillsdale NJ 07642

JONES, CHARLES DINGEE, mechanical engineering, for more information see previous edition

JONES, CHARLES E, b Oklahoma City, Okla, June 10, 28; m 49; c 5. MOLECULAR PHYSICS, ATOMIC PHYSICS. *Educ:* Univ Ark, Fayetteville, BS, 51, MS, 55; Tex A&M Univ, PhD(physics), 65. *Prof Exp:* Jr thermo engr, Convair Aircraft Co, 51-52; aerodyn engr, McDonnell Aircraft Co, 52-53; instr physics, Mo Sch Mines, 55-57 & Tex A&M Univ, 57-61; asst & assoc prof, Univ Ark, 61-70; PROF PHYSICS & HEAD DEPT, ETEX STATE UNIV, 70- *Concurrent Pos:* Consult, NSF-Agency Int Develop Summer Inst, 69. *Mem:* Am Asn Physics Teachers; Am Phys Soc. *Res:* Atoms and molecules in inert matrices at low temperatures by means of absorption and emission spectroscopy. *Mailing Add:* Dept Physics ETex State Univ Commerce TX 75428

JONES, CHARLES E(DWARD), b New York, NY, Apr 20, 20; m 44; c 4. HEAT TRANSFER. *Educ:* City Col New York, BS, 47; Agr & Mech Col Tex, MS, 51; Cornell Univ, PhD, 57. *Prof Exp:* Instr, City Col New York, 47-49; asst prof & asst res engr, Agr & Mech Col Tex, 49-52; res engr, Res Ctr, Babcock & Wilcox Co, Ohio, 54-57, head anal eng sect, 58, supt tech serv, 59-61, mgr anal lab, 61-65, mgr thermodynamics lab, 65-67, asst dir, 67-68, dir res ctr, 68-71; vpres opers, Bailey Meter Co, 71, 71-74; asst to exec vpres technol, Indust Prod Group, Babcock & Wilcox Co, 75-80; RETIRED. *Mem:* Fel Am Soc Mech Engrs (pres, 80-81); Sigma Xi. *Res:* Thermodynamics; fluid mechanics; automatic data acquisition. *Mailing Add:* 9200 Idlewood Dr Mentor OH 44060

JONES, CHARLES MILLER, JR, b Atlanta, Ga, Feb 25, 35; m 57; c 2. PHYSICS. *Educ:* Ga Inst Technol, BS, 57; Rice Univ, MA, 59, PhD(physics), 61. *Prof Exp:* Res assoc, Rice Univ, 61-62; PHYSICIST, OAK RIDGE NAT LAB, 62-, TECH DIR, HOLIFIELD HEAVY ION RES FACIL, 83- *Mem:* AAAS. *Res:* Nuclear structure, especially reactions and scattering in the light nuclei; tests of fundamental symmetries using nuclear reactions; application of superconductivity to particle accelerators; physics and technology of electrostatic particle accelerators. *Mailing Add:* Bldg 6000 MS 6368 Oak Ridge Nat Lab PO Box 2008 Oak Ridge TN 37831-6368

JONES, CHESTER GEORGE, operations research, systems analysis; deceased, see previous edition for last biography

JONES, CLAIBORNE STRIBLING, b Petersburg, Va, Dec 20, 14; m 40; c 3. ZOOLOGY. *Educ:* Hampden-Sydney Col, AB, 35; Univ Va, MA, 40, PhD(biol), 44. *Prof Exp:* From asst prof to assoc prof, Univ NC, Chapel Hill, 44-56, asst to chancellor, 77-80, 66-74, vchancellor bus & finance, 74-77, prof zool, 56-80, exec asst to chancellor; RETIRED. *Mailing Add:* 419 Westwood Dr Chapel Hill NC 27514

JONES, CLARENCE S, b Rigby, Idaho, Aug 21, 26; m 48; c 3. PHYSICS, ENGINEERING. *Educ:* Univ Utah, BA, 50, MA, 52. *Prof Exp:* Mem staff, Los Alamos Sci Lab, 52-55; mem tech staff, Ramo-Wooldridge Corp, Calif, 55-57; chief engr, Res & Develop Labs, Link Div, Gen Precision, Inc, 57-62; mgr equip eng, Sylvania Elec Prod Inc, Ecol 62-64; vpres eng, ESL Inc, 64-70; pres, Anal Develop Assocs Corp, 70-76, CHMN BD, ADAC LABS, 76- *Mem:* AAAS; Sigma Xi. *Res:* Design and execution of physical experiments in nuclear physics and electronics; design of electronic systems and circuits; administration of scientific and engineering activities. *Mailing Add:* 991 S Springfield Rd Los Altos CA 94024

JONES, CLIFFORD KENNETH, b London, Eng, Dec 29, 32; US citizen; m 57; c 2. STRATEGIC PLANNING, PROJECT MANAGEMENT. *Educ:* Univ London, BSc, 57, PhD(physics), 60. *Prof Exp:* Mem staff superconductor develop prog, 62-67, mgr cryogenics, 67-75, mgr spec projs, 75-80, dir strategic planning, 80-88, MGR SPEC PROJS, WESTINGHOUSE R&D CTR, 88- *Concurrent Pos:* Asst prof physics, Univ Calif, Los Angeles. *Mem:* Fel Am Phys Soc; Inst Elec & Electronics Engrs. *Res:* Cryogenics and fusion power technology; co-authored more than 80 publications in physical acoustics, metal physics and superconducting technology. *Mailing Add:* Westinghouse Res & Develop Ctr 1310 Beulah Rd Pittsburgh PA 15235

JONES, CLIVE GARETH, b Cirencester, Eng, March 3, 51; div. PLANT-INSECT INTERACTIONS. *Educ:* Univ Salford, Eng, BSc, 74; Univ York, Eng, DPhil(biol), 78. *Prof Exp:* Fel res, Dept Entom, Univ Ga, 78-80; CHEM ECOLOGIST, INST ECOSYSTEM STUDIES, MILLBROOK, 80- *Concurrent Pos:* Travelling fel, Brit Ecol Soc, 87; Winston Churchill fel, 90. *Mem:* AAAS; Entomol Soc Am; Ecol Soc Am; Int Soc Chem Ecol; Brit Ecol Soc. *Res:* Chemical ecology and plant-insect-microbiol interactions. *Mailing Add:* Inst Ecosystem Studies Box AB Millbrook NY 12545

JONES, CLYDE JOE, b Scottsbluff, Nebr, Mar 3, 35; m 55; c 2. MAMMALIAN ECOLOGY, TAXONOMY. *Educ:* Hastings Col, BA, 57; Univ NMex, MS, 60, PhD(biol). 64. *Prof Exp:* Asst cur biol, Univ NMex, 62-65; asst prof, Tulane Univ, 65-70; res assoc, Delta Regional Primate Res Ctr, 67-70; zoologist, Bur Sport Fisheries & Wildlife, Nat Mus Natural Hist, 70-73, dir, Nat Fish & Wildlife Lab, 73-79; dir, Denver Wildlife Res Ctr, 79-82; PROF BIOL SCI, TEX TECH UNIV, 82- *Concurrent Pos:* Res investr field studies, Rio Muni, WAfrica, 84; biologist, Antarctic Inspection, Oper Deepfreeze, 71-; res assoc, Smithsonian Inst, 71- *Mem:* AAAS; Am Soc Mammal; Soc Syst Zool; Ecol Soc Am. *Res:* Systematics and ecology of mammals, especially rodents and bats of America; bats, primates and rodents of West Africa. *Mailing Add:* 3012 60th St Lubbock TX 79413

JONES, CREIGHTON CLINTON, b Mt Oliver, Pa, Feb 13, 13; m 39; c 2. PHYSICS. *Educ:* Carnegie Inst Technol, BS, 34, MS, 35; Univ NC, PhD(physics), 38. *Prof Exp:* Am Philos Soc res assoc, Univ NC, 38-39; res physicist, Nat Carbon Co Div, Union Carbide Corp, 39-40 & 49-54, physicist, 40-49, tech rep, 54-62, appln mgr brush prod, Carbon Prod Div, 62-78; CONSULT, 78- *Mem:* Am Phys Soc; Inst Elec & Electronics Engrs. *Res:* Nuclear physics; energy loss of electrons in collision with nuclei; processing of carbon products; testing and application of carbon and carbon metal brushes on electrical equipment; sliding contacts. *Mailing Add:* 112 Highland Dr Chapel Hill NC 27514

JONES, DALE ROBERT, b Galesburg, Ill, June 17, 24. PHYSICS. *Educ:* Univ Cincinnati, BSc, 48; Wash Univ, PhD, 53. *Prof Exp:* From asst prof to assoc prof, 53-62, PROF PHYSICS, UNIV CINCINNATI, 62- *Mem:* Am Phys Soc; Sigma Xi. *Res:* Cosmic rays; atomic physics. *Mailing Add:* Dept Physics ML11 Univ Cincinnati Cincinnati OH 45221

JONES, DALLAS WAYNE, b Tiplersville, Miss, Sept 13, 38. PHYSICS. *Educ:* Memphis State Univ, BS, 60; Univ Va, MS, 62, PhD(physics), 66. *Prof Exp:* Teaching asst physics, Univ Va, 60-63, res asst, 63-65; res physicist, US Naval Res Lab, 65-69; ASSOC PROF PHYSICS, MEMPHIS STATE UNIV, 69-, DIR, CTR NUCLEAR STUDIES, 73- *Mem:* Am Phys Soc. *Res:* Quantum physics; nuclear spectroscopy; reactor technology. *Mailing Add:* Dept Physics Memphis State Univ Memphis TN 38152

JONES, DANE ROBERT, b Park City, Utah, Nov 27, 47; m 74. PHYSICAL CHEMISTRY. *Educ:* Univ Utah, BA, 69; Stanford Univ, PhD(phys chem), 74. *Prof Exp:* Res assoc phys chem, Phys Chem Inst, Univ Uppsala, 74-75; res assoc & instr phys chem, Univ Utah, 75-76; from asst prof to assoc prof, 76-85, PROF CHEM, CALIF POLYTECH STATE UNIV, SAN LUIS OBISPO, 85- *Mem:* Am Chem Soc. *Res:* Light scattering; surfaces and molecular complexes. *Mailing Add:* Dept Chem Calif Polytech State Univ San Luis Obispo CA 93407

JONES, DANIEL DAVID, b Olney, Ill, Feb 23, 43; m 65. PLANT PHYSIOLOGY, PHYCOLOGY. *Educ:* Purdue Univ, BS, 65, MS, 67; Mich State Univ, PhD(plant physiol), 70. *Prof Exp:* ASSOC PROF BIOL, UNIV ALA, BIRMINGHAM, 70- *Mem:* Am Soc Plant Physiol; Am Inst Biol Scientists. *Res:* Golgi apparatus mediated polysaccharide secretion by outer root cap cells of Zea mays; isolation, chemical characterization and ultrastructural and conformational changes of gasvacuole membranes from Microcystis aeruginosa Kuetz emend Elenkin. *Mailing Add:* Dept Biol Univ Ala 1919 Seventh Ave S Birmingham AL 35294

JONES, DANIEL ELVEN, b New Orleans, La, Sept 9, 43; m 65. PHYSICAL CHEMISTRY, COMPUTER SCIENCE. *Educ:* La State Univ, Baton Rouge, BS, 65; Univ Calif, Berkeley, PhD(phys chem), 70. *Prof Exp:* Res chemist, Am Cyanamid Co, 70-71, res comput specialist, 71-76; res chemist, 76-78, SR RES CHEMIST, FREEPORT-MCMDRAN, 79- *Mem:* Am Chem Soc; Math Asn Am; Sigma Xi. *Res:* Metal ion-nucleotide binding; carbon-13 Fourier transform nuclear magnetic resonance; application of digital computers and computing techniques for improvement of analytical instrumentation; uranium recovery from phosphoric acid; sulfur purification; inorganic chemistry. *Mailing Add:* 37 Park Timbers Dr New Orleans LA 70131

JONES, DANIEL PATRICK, b Lima, Ohio, Aug 21, 41; m 64; c 1. HISTORY OF SCIENCE. *Educ:* Univ Louisville, BS, 63; Harvard Univ, AM, 65; Univ Wis-Madison, PhD(hist sci), 69. *Prof Exp:* Macy fel hist med & biol sci, Johns Hopkins Univ, 69-70; asst prof hist sci, Ore State Univ, 70-78; vis prof, Ctr Humanistic Studies, Med Ctr, Univ Ill, 78-80, asst prof hist sci, 80-84; PROF OFFICER, DEPT HUMANITIES, SCI & TECHNOL, NAT ENDOWMENT HUMANITIES, WASHINGTON, DC, 84- *Mem:* AAAS; Hist Sci Soc; Am Chem Soc. *Res:* History of biochemistry and organic chemistry, 19th and early 20th century; relationships between science and society; history of public health. *Mailing Add:* Dept Humanities Sci & Technol Nat Endowment Humanities 1100 Pennsylvania Ave NW Washington DC 20506

JONES, DANIEL SILAS, JR, b Charlotte, NC, Nov 16, 43; m 67; c 1. PHYSICAL CHEMISTRY, X-RAY CRYSTALLOGRAPHY. *Educ:* Wake Forest Col, BS, 65; Harvard Univ, AM, 66, PhD, 70. *Prof Exp:* Teaching-res assoc, State Univ NY, Buffalo, 70-71; Nat Acad Sci-Nat Res Coun resident res assoc, Lab Struct Matter, Naval Res Lab, Washington, DC, 71-73; asst prof, 73-78, ASSOC PROF CHEM, UNIV NC, CHARLOTTE, 78- *Mem:* Am Chem Soc; Am Crystallog Asn. *Res:* Crystal and molecular structures by single crystal x-ray diffraction techniques. *Mailing Add:* Dept Chem Univ NC Charlotte NC 28223

JONES, DAVID A, JR, b McCook, Nebr, Feb 9, 37; m 76; c 4. ORGANIC CHEMISTRY. *Educ:* Tex A&M Univ, BS, 58; NMex State Univ, MS, 64; Purdue Univ, PhD(org chem), 68. *Prof Exp:* Mem staff, Dept Med Chem, 68-81, res scientist II, G D Searle & Co, 77-86, FIELD SERV ENG, APPLIED BIOSYSTEMS, INC, FOSTER CITY, CALIF, 86- *Mem:* AAAS; Am Chem Soc. *Res:* Organometallic chemistry of silicon, magnesium, lithium; organic chemistry of phosphorus; amino acid and peptide chemistry. *Mailing Add:* 135 Barton Ave Evanston IL 60202

JONES, DAVID B, b Canton, China, Dec 1, 21; US citizen; m 44; c 3. PATHOLOGY. *Educ:* Syracuse Univ, AB, 43, MD, 45; Am Bd Path, dipl. *Prof Exp:* From asst prof to assoc prof, 50-62, PROF PATH, STATE UNIV NY UPSTATE MED CTR, 62- *Mem:* Am Asn Pathologists & Bacteriologists; Am Soc Cytol; Sigma Xi. *Res:* Electron microscopy. *Mailing Add:* 226 Lockwood Rd Syracuse NY 13214

JONES, DAVID HARTLEY, b Kansas City, Mo, Feb 10, 39; m 65; c 3. BIOCHEMISTRY. *Educ:* Bethany Nazarene Col, BS, 61; Univ Okla, MS, 64; Cornell Univ, PhD(biochem), 68. *Prof Exp:* USPHS fel biochem, Univ Calif, Los Angeles, 67-69; asst prof biochem, Albany Med Col, Union Univ, 69-75, assoc prof, 75-78; assoc prof biochem, Oral Roberts Univ, 78-80. *Mem:* AAAS. *Res:* Oxidative phosphorylation in mitochondria; functional state transitions in the mammary gland; mitochondrial biogenesis during functional state transitions in the mammary gland. *Mailing Add:* Dept Chem Grove City Col Grove City PA 16127

JONES, DAVID LAWRENCE, b Chicago, Ill, Nov 12, 30; m 53; c 4. GEOLOGY, PALEONTOLOGY. *Educ:* Yale Univ, BS, 52; Stanford Univ, MS, 53, PhD, 56. *Prof Exp:* GEOLOGIST, WESTERN REGION, US GEOL SURV, 55-; PROF GEOL, DEPT GEOL, UNIV CALIF-BERKELEY. *Mem:* Geol Soc Am; Paleont Soc. *Res:* Molluscan paleontology; Cretaceous of the Pacific coast region of North America; stratigraphy, structural, biostratigraphy and molluscan paleontology of upper Mesozoic rocks of the Pacific Coast of North America. *Mailing Add:* Dept Geol Univ Calif-Berkeley Berkeley CA 94720

JONES, DAVID LLOYD, b Sapporo, Japan, Apr 11, 19; US citizen; m 42; c 3. ENERGY PLANNING, METEOROLOGY & CLIMATOLOGY. *Educ:* Carleton Col, BA, 41; Pa State Univ, MS, 53, PhD(meteorol), 60. *Prof Exp:* Unit leader meteorol, Statist Dept, US Nat Weather Serv, 43-46; instr math, Fisk Univ, 46; meteorologist, Am Airlines, 46-51; asst meteorol, Pa State Univ, 51-55, instr, 55-56; lectr & assoc res meteorologist, Univ Mich, 56-61; sr res scientist, Travelers Res Ctr, 61-65; assoc prof meteorol, Cent Conn State Col, 65; assoc prof, 65-71, dir Europ earth sci study prog, 74 & 76, PROF METEOROL & ENERGY PLANNING, SOUTHERN ILL UNIV, CARBONDALE, 71- *Concurrent Pos:* Lectr, NSF vis Scientist Prog, Am Meteorol Soc & Mich State Univ, 60-61; lectr, Mus Art, Sci & Indust, Conn, 63-65; consult, Educ Film Proj, Am Meteorol Soc, Boston, 64; consult world weather data syst, US Nat Weather Serv, Washington, DC, 65; partic writing conf, Earth Sci Curric Proj, Am Geol Inst, Colo, 65; chmn sect meteorol & climat, Ill State Acad Sci, 70; dir, US Dept Energy Summer Inst on Energy, 77, staff mem, 78. *Mem:* Am Meteorol Soc; Sigma Xi. *Res:* Large scale atmospheric vertical motion; atmospheric pollution by aeroallergens; atmosphere-hydrosphere interactions; meteorological data-processing system design; laboratory and classroom instructional devices; educational films; interdisciplinary problems in the environmental sciences; science museum design; energy planning and community energy planning; alternate energy resources; oceanography. *Mailing Add:* 21300 Heather Ridge Circle Apt 398 Green Valley AZ 85614-5109

JONES, DAVID ROBERT, b Bristol, Eng, Jan 28, 41; Can citizen; m 62; c 2. PHYSIOLOGY OF DIVING ANIMALS, CARDIOVASCULAR DYNAMICS. *Educ:* Southampton Univ, UK, BSc, 62; Univ E Anglia, UK, PhD(biol), 65. *Prof Exp:* Res fel biol, Univ E Anglia, UK, 65-66; lectr zool, Univ Bristol, UK, 66-69; PROF ZOOL, UNIV BC, 69- *Concurrent Pos:* Sr fel, Killiam Found, Can, 73 & 90; comt mem, Can Soc Zoologists, 85-88; vis prof, Univ Melbourne, Australia, 88; mem, Grant Selection Comt, Nat Sci & Eng Res Coun Can, 90-93. *Mem:* Am Soc Zoologists; Am Physiol Soc; Soc Exp Biol. *Res:* Control of cardiovascular and respiratory responses to diving, attitude and exercise in birds and mammals; cardiovascular dynamics of invertebrates and vertebrates. *Mailing Add:* Zool Dept Univ BC 6270 University Blvd Vancouver BC V6T 1Z4 Can

JONES, DEAN PAUL, b Hazard, Ky, Sept 13, 49; m; c 2. HYPOXIA, TOXICOLOGY. *Educ:* Ore Health Sci Univ, PhD(biochem), 76. *Prof Exp:* Asst prof, 79-85, ASSOC PROF BIOCHEM, EMORY UNIV 85- *Honors & Awards:* Levy Res Award. *Mem:* Am Soc Biochem & Molecular Biol; Am Soc Cell Biol; Am Physiol Soc; Am Chem Soc; AAAS; Soc Toxicol. *Res:* Oxygen metabolism in health and disease; diet and cancer; functions of glutathione in detoxification of carcinogens and other toxic compounds. *Mailing Add:* Sch Med Emory Univ Atlanta GA 30322

JONES, DEREK WILLIAM, b Birmingham, Eng, Dec 9, 33; Can citizen; m 57; c 4. DENTAL MATERIALS. *Educ:* Univ Birmingham, BSc, 65, PhD(dent mat sci), 70; Inst Ceramics, AICeram, 70, FICeram, 78; Brit Royal Soc Chem, CChem, FRSC, 85. *Prof Exp:* Instr dent technol & mat, Univ Birmingham, 65-75; assoc prof dent biomat, 75-77, prof-in-chg dent biomat, 77-79, PROF & ACTG CHMN, DEPT APPL ORAL SCI & HEAD DIV DENT BIOMAT SCI, FAC DENT, DALHOUSIE UNIV, 79-, ASST DEAN RES & PROF, COL PHARM. *Concurrent Pos:* Vis lectr, Mathew Boulton Tech Col, 61-68; examr, City & Guilds London Inst, 66-73; mem comt & consult, Brit Stand Comt Dent Mat, 70-75; Brit expert rep, Int Stand Orgn, 73-75, Can rep, 75-, comt mem coun dent mat & devices, 77-, chmn Can adv comt, tech comt 106; chmn comt dent, Can Stand Asn, 78-; mem, Can Standards Steering Comt Health Care Technol, Can Standards Comt Implant Mat, 79 & Med Res Coun, Can Grants Comt, 81; dent schs rep, Dent Mat Group, Int Asn Dent Res; chmn, Can Standards Asn Tech Comt Dentistry, Can Adv Comt, Int Standards Orgn; mem secretariat, Int Standards Comt, 79-; vis prof at six dental schs, lectr in eleven countries; pres, Dental Mat Group Chap, CADR/IADR, 88-92, vpres-pres, Can Asn Dent Res, 90-95, pres, Int Dent Mat Group, IADR, 90-91, chmn, Dent Mat Group Prog, 88; mem, Nat Adv Panel Adv Indust Mat, Fed Govt Can, 90- *Honors & Awards:* Wilmer Souder Distinguished Scientist Award, Int Asn Dental Res, 88. *Mem:* Can Asn Dent Res; Int Asn Dent Res; Soc Biomat; Inst Ceramics; Royal Soc Chem. *Res:* Development of test methodology; evaluating mechanical-physical properties of materials to optomize clinical and laboratory use; biological factors relative to clinical performance; studies of hard and soft polymers, ceramics and metals; author of over 193 scientific publications. *Mailing Add:* Dalhousie Univ 5981 University Ave Halifax NS B3H 3J5 Can

JONES, DONALD AKERS, b Topeka, Kans, Dec 27, 30; m 56; c 4. ACTUARIAL SCIENCE. *Educ:* Iowa State Univ, BS, 52; Univ Iowa, MS, 56, PhD(math), 59. *Prof Exp:* Asst prof, 59-65, ASSOC PROF MATH, UNIV MICH-ANN ARBOR, 65- *Mem:* Am Statist Asn; Soc Actuaries; Am Acad Actuaries. *Res:* Acturial science. *Mailing Add:* Dept Math Univ Mich Ann Arbor MI 48109

JONES, DONALD EUGENE, b South Bend, Ind, Aug 1, 34; m 55; c 3. ANALYTICAL CHEMISTRY. *Educ:* Manchester Col, AB, 57; Purdue Univ, PhD(anal chem), 63. *Prof Exp:* Chemist, Bendix Corp, Ind, 57; res chemist, E I du Pont de Nemours & Co, 60; instr chem, Wabash Col, 61-63; from asst prof to assoc prof, 63-76, head dept, 76-82, PROF CHEM, WESTERN MD COL, 76- *Concurrent Pos:* Vis assoc prof, Purdue Univ, 71-72; chem consult, Carroll County Gen Hosp, 74-79, USN, 84- *Mem:* Am Chem Soc. *Res:* Fluorescence of materials as applied to analytical procedures; trace analysis of materials; computer applications to chemical analysis; analytical chemistry as applied to clinical situations. *Mailing Add:* Dept Chem Western Md Col Westminster MD 21157

JONES, DONLAN F(RANCIS), b San Francisco, Calif, Feb 5, 30; m 57; c 4. ELECTRICAL ENGINEERING, COMPUTER SCIENCES. *Educ:* Univ Santa Clara, BEE, 52; Univ Calif, Los Angeles, MS, 54; Stanford Univ, Engr, 72. *Prof Exp:* Res engr, Hughes Aircraft Co, 52-56; asst prof elec eng, Univ Santa Clara, 56-63; adv develop engr, Sylvania Electronics Systs Div, Gen Tel & Electronics Corp, 63-65, eng specialist, 65-69; eng mgr, Comput Terminal Prods, Tektronix, Inc, 69-74, eng mgr, 4081 & MEG systs, 74-78, eng mgr, mass storage syst, 78-81, eng mgr, data commun, info display div, 81-84, appln mkt mgr, Graphics Work Sta Div, 85-91; MATH INSTR, CLACKSSMAS COMMUNITY COL, 91- *Concurrent Pos:* Adv develop engr, Sylvania Electronics Systs Div, Gen Tel & Electronics Corp, 57-63; NSF sci fac fel, 61-62; consult, Sonoma State Hosp, Calif, 62-63. *Mem:* Inst Elec & Electronics Engrs; Sigma Xi. *Res:* Application of computers to engineering and non-scientific problems; threshold logical design; graphic computer systems. *Mailing Add:* 427 Laurel St Lake Oswego OR 97034

JONES, DOUGLAS EMRON, b Long Beach, Calif, Aug 19, 30; m 55; c 5. PHYSICS. *Educ:* Brigham Young Univ, BS, 57, MS, 59, PhD(physics), 64. *Prof Exp:* Technician radio repair, Southern Calif Edison Co, 54-55; apprentice engr, Collins Radio Co, 56, group supvr, 57; space scientist, Jet Propulsion Lab, Calif Inst Technol, 59-62; from asst prof to assoc prof, 64-74, PROF PHYSICS, BRIGHAM YOUNG UNIV, 74- *Honors & Awards:* Karl G Maeser Res Award. *Mem:* Am Geophys Union; Am Phys Soc. *Res:* Solar physics; interplanetary magnetic fields; planetary atmospheres; experimental space physics; measurement of microwave emission of planets; magnetic fields of comets and planets and in interplanetary space; space plasma simulations; soft x-ray studies of the sun. *Mailing Add:* Dept Physics & Astron Brigham Young Univ Provo UT 84602

JONES, DOUGLAS EPPS, b Tuscaloosa, Ala, May 28, 30; m 55; c 3. GEOLOGY, PALEONTOLOGY. *Educ:* Univ Ala, BS, 52; La State Univ, PhD(geol), 59. *Prof Exp:* Res geologist, La Geol Surv, 55-58; from asst prof to prof geol, 58-66, head, Dept Geol & Geog, 66-69, dean, Col Arts & Sci, 69-84, actg acad vpres, 88-90, PROF GEOL, UNIV ALA, 66-; DIR, ALA MUS NATURAL HIST, 84- *Mem:* Geol Soc Am; Paleont Soc; Am Asn Petrol Geologists. *Res:* Stratigraphy and paleontology of Gulf Coastal plain region of the United States. *Mailing Add:* PO Box 870340 Tuscaloosa AL 35487-0340

JONES, DOUGLAS L, b Calgary, Alta, Nov 3, 48; m 71; c 3. CARDIOVASCULAR PHYSIOLOGY. *Educ:* Univ Alta, BSc, 72; Univ Alberta, MSc, 74; Univ Calgary, PhD(med physiol), 77. *Prof Exp:* Asst prof, 81-86, ASSOC PROF MED PHYSIOL, UNIV WESTERN ONT, 86- *Concurrent Pos:* Lady Davis scholar, 75; MRC fel, 78, 79, 86-; L L scholar, 80; career scientist, Ont Ministry Health, 86 & 89; assoc scientist, J P Robarts Res Inst, London, 86- *Mem:* Fel Am Col Cardiol; Am Health Asn; Am Phys Soc; Asn Advan Med Instrumentation; AAAS; Soc Neurosci. *Res:* Cardiovascular physiology; regulatory and integrative physiology; electrophysiology neuroscience; electrocardiology. *Mailing Add:* Dept Med & Physiol Univ Western Ont London ON N6A 5C1 Can

JONES, DOUGLAS LINWOOD, b Limeton, Va, Dec 26, 37; m 75. ENGINEERING, SOLID MECHANICS. *Educ:* George Washington Univ, BME, 63, MSE, 65, DSc, 70. *Prof Exp:* Univ fel eng, 66-67, from instr to asst prof eng & appl sci, 67-71, from asst res prof to assoc res prof, 71-77, assoc prof, 77-82, chmn mech engr curric, 81-85, PROF ENG, GEORGE WASHINGTON UNIV, 82- *Concurrent Pos:* Consult, Seal & Co, 70-71, Comsat Labs, 74-76, Eng Servs Co, 76-81, Ensco, Inc, 77-78, Systs Technol Labs, Inc, 80-82, Du Pont Corp, 82-85, Alcoa, 83 & Intelsat Corp, 86-88, NKF Eng, 90, US Dept Transp, 90-91; prin or co prin investr res grants, NASA, Dept Defense, NSF. *Honors & Awards:* George Washington Award, George Washington Univ, 85. *Mem:* Am Acad Mech; Am Soc Testing & Mat; Am Soc Mech Engrs; Am Soc Eng Educ; Soc Exp Mech; Sigma Xi. *Res:* Fatigue, fracture and fracture mechanics of metals and composite materials; computer aided design; fractography and failure analysis; experimental stress analysis; evaluation and development of constitutive relations in continuum mechanics; nondestructive inspection methods; composite materials. *Mailing Add:* Acad Ctr T723 George Washington Univ Washington DC 20052

JONES, DUVALL ALBERT, b Hurlock, Md, Oct 17, 33; m 66; c 2. VERTEBRATE ZOOLOGY. *Educ:* Western Md Col, AB, 55; Univ Md, MS, 61; Univ Fla, PhD(zool), 67. *Prof Exp:* Asst prof biol, James Madison Univ, 60-62; asst prof biol, head dept & chmn div natural sci & math, Ferrum Col, 62-65; asst prof biol & actg head dept, West Liberty State Col, 66-67; asst prof, Carnegie-Mellon Univ, 67-73; PROF BIOL, ST JOSEPH'S COL, 73- *Concurrent Pos:* Scaife grant. *Mem:* AAAS; Am Soc Zool; Am Soc Ichthyol & Herpet; Genetics Soc Am; Nat Sci Teachers Asn. *Res:* Physiological ecology of amphibians; environmental genetics; bile pigments and their effects. *Mailing Add:* Dept Biol St Joseph's Col Rensselaer IN 47978

JONES, E(DWARD) M(CCLUNG) T(HOMPSON), b Topeka, Kans, Aug 19, 24; m 49; c 3. MICROWAVE ELECTRONICS. *Educ:* Swarthmore Col, BS, 44; Stanford Univ, MS, 48, PhD(elec eng), 50. *Prof Exp:* Res assoc elec eng, Stanford Univ, 48-50; sr res engr, Stanford Res Inst, 50-57, head microwave group, 57-61; dir eng, Menlo Park Div, TRG, Inc, Control Data Corp, 61-67; eng mgr, Antennas & Transmission Lines Div, Granger Assocs, 67-68; vpres eng & technol, TCI, 68-71, exec vpres, 71-82, vpres develop, 82-90, CHMN BD, TCI, 90- *Mem:* Inst Elec & Electronics Engrs; Sigma Xi. *Res:* Microwave components and antennas; antennas. *Mailing Add:* 2161 Via Escalera Los Altos CA 94022

JONES, EARLE DOUGLAS, b Birmingham, Ala, Apr 10, 31; m 61. ELECTRONICS. *Educ:* Ga Inst Technol, BS, 56; Stanford Univ, MS, 58. *Prof Exp:* Asst math, Ga Inst Technol, 55-56; exec dir, SRI-ASIA, 86-88; REG MTG DIR, KOREA, 88- *Mem:* Inst Elec & Electronics Engrs; Sigma Xi. *Res:* Space electronics; communication systems research in satellite meteorology; display devices and digital control research; bioengineering. *Mailing Add:* SRI Int 333 Ravenswood Ave Menlo Park CA 94025

JONES, EDWARD DAVID, b Rockland, Wis, May 8, 20; m 47; c 4. PLANT ENTOMOLOGY. *Educ:* Univ Wis, BS, 46, MS, 47, PhD(plant path), 53. *Prof Exp:* Instr plant path, Univ Wis, 48-53; plant pathologist, Red Dot Foods, Inc, 53-58; from asst prof to prof, 58-87, HENRY & MILDRED UIHLEIN PROF PLANT PATH, 87- *Concurrent Pos:* In-chg found & cert seed potato prog NY state, 60-, Uihlein Farm, Cornell Univ, 61-, Henry Uihlein II tissue cult facil at Uihlein Farm, Lake Placid, NY, 77- *Mem:* Am Phytopath Soc; hon Potato Asn Am (vpres, 81-82, pres, 84-85); Sigma Xi. *Res:* Production of disease-free nuclear seed stocks by tissue culture; disease problems relating to the production of seed potatoes. *Mailing Add:* Dept Plant Path Cornell Univ Rm 318 Plant Sci Ithaca NY 14850

JONES, EDWARD GEORGE, b Upper Hutt, NZ, Mar 26, 39; m 63; c 2. NEUROBIOLOGY. *Educ:* Univ Otago, NZ, MB, ChB, 62, MD, 70; Oxford Univ, DPhil(anat), 68. *Prof Exp:* Demonstr anat, Univ Otago, NZ, 64-65, from asst lectr to lectr, 65-70, assoc prof, 71-72; assoc prof anat, Washington Univ, 72-75, prof anat & neurobiol, Sch Med, 75-84, prof neurosci, 81-84; prof & chmn, Univ Calif Irvine, 84- *Concurrent Pos:* Nuffield Dom demonstr, Oxford Univ, 65, 65-68, lectr, Balliol Col, 66-68; NZ Med Res Coun grant, Sch Med, Otago Univ, 69-71; assoc ed, J Comp Neurol, 75-80; Macy Found sr fac scholar, Monash Univ, Australia, 78-79; Green vis prof, Univ Tex Med Br, Galveston, 78, Beale Mem Lectr, 80; assoc ed, J Neurosci, 81-88; dir, James O'Leary Div Exp Neurol & Neurological Surg, Washington Univ, George H & Ethel Ronzon scholar in neurosci & sr scientist, McDonnell Ctr for Study of Higher Brain Function, 81-84, assoc, Neurosciences Res Prog, 84- *Honors & Awards:* Symington Mem Prize, Anat Soc Gt Brit & Ireland, 68; Rolleston Mem Prize, Oxford Univ, 70; Cajal Prize, Am Asn Anatomists, 89. *Mem:* Anat Soc Gt Brit & Ireland; Am Asn Anatomists; Soc Neurosci; Anat Soc Australia & NZ; NZ Med Asn. *Res:* Structure, function, and development of sensory systems particularly in primates and with emphasis on cerebral cortex and somatic sensory system. *Mailing Add:* Dept Anat & Neurobiol Medsurg 2 Col Med Univ Calif Irvine Irvine CA 92717

JONES, EDWARD GRANT, b Toronto, Ont, Feb 16, 42; m 72; c 3. CHEMICAL PHYSICS. *Educ:* Univ Toronto, BSc, 65, MSc, 67, PhD(phys chem), 69. *Prof Exp:* Vis res scientist, Ohio State Univ Res Found, 69-71; res assoc, Purdue Univ, 71-72; consult, Systs Res Labs, 71-72, sr res chemist & proj mgr, 72-75; res asst prof chem, 75-77, RES ASSOC PROF CHEM,

WRIGHT STATE UNIV, 77- *Concurrent Pos:* Consult, Systs Res Labs, 75- *Mem:* Sr mem Am Chem Soc; sr mem Am Soc Mass Spectrometry. *Res:* Gas phase kinetics; ion-neutral collision phenomena; unimolecular decomposition; chemiluminescence; thermal degradation of polymers; kinetics of polymerization. *Mailing Add:* 4850 N Piqua Troy Rd Troy OH 45373

JONES, EDWARD O(SCAR), JR, b Dothan, Ala, June 18, 22; m 47; c 2. MECHANICAL ENGINEERING. *Educ:* Auburn Univ, BS, 43 & 46; Univ Ill, MS, 49. *Prof Exp:* Tooling engr, Consol Vultee Aircraft Corp, 43-45; from instr to assoc prof mech eng, 46-61, asst dean eng, 74-78, PROF MECH ENG, AUBURN UNIV, 61-, ASST HEAD DEPT, 65-, ASSOC DEAN ENG, 78- *Mem:* Soc Automotive Engrs; Soc Exp Stress Anal. *Res:* Experimental stress analysis, especially thin-shell pressure vessels. *Mailing Add:* Sch Eng Ramsay Hall Auburn Univ Auburn AL 36830

JONES, EDWARD RAYMOND, b Steubenville, Ohio, Jan 27, 43; m 64; c 1. AGRONOMY. *Educ:* Ohio State Univ, BS, 65; Pa State Univ, MS, 67, PhD(agron), 69. *Prof Exp:* Assoc prof, 71-77, PROF AGRON, DEL STATE COL, 77- *Mem:* Am Soc Agron; Am Forage & Grassland Coun. *Res:* Forage crop management. *Mailing Add:* Del State Col Dover DE 19901-2275

JONES, EDWARD STEPHEN, b Boston, Mass, Apr 17, 31; m 56; c 2. ORGANIC CHEMISTRY, ORGANOFLUORINE CHEMISTRY. *Educ:* Northeastern Univ, BS, 53; Purdue Univ, MS, 56; Wayne State Univ, PhD(org chem), 61. *Prof Exp:* Res chemist, Gen Chem Div, Allied Chem Corp, NJ, 60-69, sr res chemist, Specialty Chem Div, Buffalo, 69-80; SR RES CHEMIST, HALOCARBON PROD CORP, 80- *Mem:* Am Chem Soc. *Res:* Organic fluorine chemistry; applications, process research and development; basic research; product research and development. *Mailing Add:* Halocarbon Prod Corp PO Box 6369 North Augusta SC 29841

JONES, EDWIN C, JR, b Parkersburg, WVa, June 27, 34; m 60; c 3. ELECTRICAL ENGINEERING, EDUCATION. *Educ:* WVa Univ, BS, 55; Imp Col, Univ London, Dipl, 56; Univ Ill, Urbana, PhD(elec eng), 62. *Prof Exp:* Teaching asst elec eng, Univ Ill, 58-59; from instr to asst prof elec eng, Univ Ill, 60-66; from asst prof to assoc prof, 66-72, PROF ELEC ENG, IOWA STATE UNIV, 72- *Concurrent Pos:* Engr, Gen Elec Co, 55, 62 & Westinghouse Elec Co, 59; proc chmn, Nat Electronics Conf, 65, prog chmn, 68, secy, 69, awards chmn, 69-70, vpres continuing educ, 71; secy, Inst Elec & Electronics Engrs Educ Soc, 70-78, vpres, 73-74, pres, 75-76; ed, Inst Elec & Electronics Engrs Trans Educ, 81-84; mem, Eng Accreditation Comn, 80-84, bd dirs, Accreditation Bd Eng & Technol, 84- 87. *Honors & Awards:* Centennial Medal, Inst Elec & Electronics Engrs, 84, Accreditation Activ Award, 86. *Mem:* AAAS; fel Inst Elec & Electronics Engrs, 88; fel Am Soc Eng Educ; AAAS; Soc Hist Technol; fel Accreditation Bd Eng & Technol, 89. *Res:* Circuit theory; experimental engineering techniques; educational methods; technology and social change. *Mailing Add:* Dept Elec & Comput Eng Iowa State Univ Ames IA 50011

JONES, EDWIN C, b Smithburg, Va, July 20, 03. ELECTRICAL ENGINEERING. *Educ:* Univ West Va, BS, 25; Univ Ill, MS, 29. *Prof Exp:* Prof & chmn, elec eng, Univ WVa, 25-70, dept chmn, 48-68; RETIRED. *Mem:* Fel Inst Elec & Electronics Engrs; Am Soc Eng Educ; Nat Soc Prof Engrs. *Mailing Add:* Dept Electrical Eng WVa Univ 1380 Western Ave Morgantown WV 26505

JONES, EDWIN RUDOLPH, JR, b Lumberton, NC, Aug 3, 38; m 60; c 4. SOLID STATE PHYSICS. *Educ:* Clemson Univ, BS, 60; Univ Wis, MS, 62, PhD(physics), 65. *Prof Exp:* From asst prof to assoc prof, 65-77, PROF PHYSICS, UNIV SC, 77- *Concurrent Pos:* Vis prof, Univ de El Salvador, 78; mem, Comt Undergrad Educ, Am Asn Physics Teachers, 91- *Mem:* Am Phys Soc; Am Asn Physics Teachers; Soc Photo-Optical Instrumentation Engrs. *Res:* Low temperature magnetic properties of solids; three dimensional imaging for video and computer displays; author of one textbook. *Mailing Add:* Dept Physics Univ SC Columbia SC 29208

JONES, ELDON MELTON, b Chenoa, Ill, Feb 1, 14; m 39; c 5. MEDICINAL CHEMISTRY. *Educ:* Univ Ill, BS, 36; Pa State Col, MS, 37, PhD(org chem), 40. *Prof Exp:* Asst, Pa State Col, 40-41; res chemist, Parke, Davis & Co, 41-58, head tech info, 58-69, mgr res libr serv, 69-76, mgr sci info, 76-77; RETIRED. *Mem:* AAAS; Am Chem Soc. *Res:* Steroidal compounds; antimalarials; analgesic drugs; polymerization of olefins; anti-inflammatory agents. *Mailing Add:* 2023 Day St Ann Arbor MI 48104

JONES, ELEANOR GREEN DAWLEY, b Norfolk, Va, Aug 10, 29; m 51, 67; c 2. MATHEMATICS. *Educ:* Howard Univ, BS, 49, MS, 50; Syracuse Univ, PhD(math), 66. *Prof Exp:* Instr, Hampton Inst, 55-62, assoc prof, 66-67; teaching asst, Syracuse Univ, 64-66; PROF MATH, NORFOLK STATE COL, 67-, READER, COL BD ADVAN PLACEMENT EXAM MATH, 89- *Concurrent Pos:* Bd gov, Math Asn Am, 83-86; exec bd, Nat Asn Mathematicians, 88-; bd dir, Asn Women Math, 90- *Mem:* Am Math Soc; Math Asn Am; Nat Asn Math (vpres, 75-80); Asn Women Math; Sigma Xi. *Res:* Abelian groups and their endomorphism rings; direct decompositions and quasi-endomorphisms of torsion free abelian groups. *Mailing Add:* 6301 Bucknell Circle Virginia Beach VA 23464

JONES, ELIZABETH W, b Seattle, Wash, Mar 8, 39. MOLECULAR GENETICS, CELL BIOLOGY. *Educ:* Univ Wash, BS, 60, PhD(genetics), 64. *Prof Exp:* USPHS trainee, Univ Wash, 60-64; res assoc, Mass Inst Technol, 64-67, instr, 67-69; asst prof biol & microbiol, Case Western Reserve Univ, 69-74; assoc prof, 74-82, PROF BIOL SCI, CARNEGIE-MELLON UNIV, 82-; ADJ PROF PSYCHIAT, UNIV PITTSBURGH, 85- *Concurrent Pos:* USPHS res grant, 70-, res career develop award, 71-74, 75-77; NIH Genetics Training Comt, 72-73, Genetics Study Sect, 76- 80 & 84-86, chair, 90; assoc ed, Genetics, 80-, Yeast, 85-, Ann Rev Genetics, 90- *Mem:* Fel AAAS; Genetics Soc Am (vpres, 86, pres, 87); Am Soc Microbiol; Am Soc Cell Biol. *Res:* Organization and expression of genetic material in yeast; protein targeting and organellar assembly. *Mailing Add:* Dept Biol Sci Carnegie-Mellon Univ Pittsburgh PA 15213

JONES, ELMER EVERETT, b Hinsdale, Ill, Sept 2, 26; m 56; c 1. ORGANIC CHEMISTRY. *Educ:* Univ Chicago, PhB, 48, BS, 50; Washington Univ, PhD, 57. *Prof Exp:* Asst, Washington Univ, 50-55; res assoc, Tannhauser Lab, Boston Dispensary, 56-58; asst prof chem, 58-62, Northeastern Univ, 58-62, assoc prof chem, 62-90; RETIRED. *Mem:* AAAS; Am Chem Soc; Sigma Xi. *Mailing Add:* 67 Brook Rd Weston MA 02193-1766

JONES, ERIC DANIEL, b Oakland, Calif, Jan 6, 36; m 57; c 5. SOLID STATE PHYSICS. *Educ:* Ore State Univ, BS, 57; Univ Wash, MS, 59, PhD(physics), 62. *Prof Exp:* Mem tech staff, Bell Tel Labs, NJ, 62-65; mem staff solid state physics res, Sandia Corp, Sandia Nat Labs, 65-68, supvr laser effects res, 68-82, mem staff laser res, 82-85, mem tech staff Sandia Semiconductor Physics, 85-90, DISTINGUISHED MEM TECH STAFF, SANDIA NAT LABS, 90- *Concurrent Pos:* Mem, Adv Comt-Optics Prog, Idaho State Univ, 70-76, Adv Comt Grad Studies, Elec & Computer Eng, Univ NMex, 76-88, bd ed, Rev Sci Instruments, 86-89, Users' Comt for the Francis Bitter Nat Magnet Lab, Mass Inst Technol, 90-93, External Adv Comt, Nat High Magnetic Field Lab, Fla State Univ, 91-94; pres, Albuquerque Chap Laser Inst Am, 72-76; adj prof, Physics Dept, Univ NMex, Albuquerque, NMex, 85-; secy-treas, Instruments & Measurement Sci Topical Group, Am Phys Soc, 88-90; Woodrow Wilson fel, 60. *Mem:* Fel Am Phys Soc; sr mem Inst Elec & Electronics Engrs. *Res:* Study of ferromagnetism, antiferromagnetism, paramagnetism in insulators and metals by the use of nuclear magnetic resonance techniques; high power laser energy deposition in solids; ultrashort laser pulse generation and applications; magneto-optics of semiconductors, and pressure effects in semiconductors; semiconductor physics. *Mailing Add:* Sandia Nat Labs Orgn 1143 PO Box 5800 Albuquerque NM 87185-5800

JONES, ERIC MANNING, b Goldsboro, NC, Mar 25, 44. HYDRODYNAMICS, ASTROPHYSICS. *Educ:* Calif Inst Technol, BS, 66; Univ Wis-Madison, PhD(astron), 69. *Prof Exp:* Staff mem hydrodyn, 69-75, group leader, 76-81, LAB FEL, LOS ALAMOS NAT LAB, 82- *Concurrent Pos:* Nat Res Coun Comt on Nuclear Winter; co-ed with Ben R Finney, Interstellar Migration and the Human Experience, Univ of Calif Press, 85. *Mem:* Am Astron Soc. *Res:* Supernova remnants; interstellar medium; nuclear explosion phenomenology; numerical hydrodynamics; space development. *Mailing Add:* Los Alamos Nat Lab MSF 665 PO Box 1663 Los Alamos NM 87545

JONES, ERIC WYNN, b St Martins, Eng, Sept 24, 24; US citizen; m 48; c 1. VETERINARY SURGERY. *Educ:* MRCVS, 46; Cornell Univ, PhD(vet surg), 50; Am Col Vet Surg, dipl, 71; Am Col Vet Anesthesiol, dipl, 78. *Hon Degrees:* FRCVS, London, 87. *Prof Exp:* Consult, Col Vet Med, Miss State Univ, 74-78; vdean, Col Vet Med, Miss State Univ, 78-83, interim dean, 83-84, from asst to vpres, 84-86, prof 86-87, INTERIM DIR & PROF VET MED, MISS STATE UNIV, 87- *Concurrent Pos:* Dir clin res, 56-77. Concurrent. *Mem:* Am Vet Med Asn; Brit Vet Asn; Am Soc Anesthesiol; Am Soc Vet Anesthesiol; Am Acad Vet Pharmacol & Toxicol; Am Asn Equine Practrs. *Res:* Drug and biologic and model development; program and facilities consulting national and international; spleen function in infectious anemia; enteritis; mechanical ventilators; malignant hyperthermia; drug testing. *Mailing Add:* Col Vet Med Drawer V Miss State Univ Mississippi State MS 39762

JONES, ERNEST ADDISON, b Columbia, Ky, June 5, 18; m 43. PHYSICS. *Educ:* Western Ky State Teachers Col, BS, 42; Vanderbilt Univ, MS, 43; Ohio State Univ, PhD(phys chem), 48. *Prof Exp:* Res physicist, Manhattan Dist, Columbia Univ, 43-45; res chemist, Carbide & Carbon Chem Co, 48-50; from asst prof to prof physics, 50-, EMER PROF PHYSICS, VANDERBILT UNIV. *Mem:* Am Phys Soc; Optical Soc Am. *Res:* Infrared and Raman spectroscopy. *Mailing Add:* Dept Physics Vanderbilt Univ Nashville TN 37240

JONES, ERNEST AUSTIN, JR, b Orange, NJ, Mar 24, 60. PETROLEUM GEOLOGY. *Educ:* Princeton Univ, AB, 83; Harvard Univ, AM, 89, PhD(geol), 89. *Prof Exp:* Geologist, US Geol Surv, 84-87; RES GEOLOGIST, MOBIL CORP, 89- *Mem:* Am Asn Petrol Geologists. *Res:* Development of diagenetic models to predict sandstone reservoir quality; clastic petrology and hydrogeology, Anadarko basin, Denver basin, Llamos basin & NW Shelf, Australia. *Mailing Add:* Mobil Res & Develop Corp 13777 Midway Rd Dallas TX 75244-4390

JONES, ERNEST OLIN, b Atlanta, Ga, Feb 1, 23; m 46; c 2. RADIOLOGICAL PHYSICS. *Educ:* Emory Univ, AB, 48, MS, 49; US Naval Postgrad Sch, MS, 59; NC State Univ, PhD(nuclear eng), 64; Am Bd Radiol, dipl radiol physics, 75. *Prof Exp:* Dep dir nuclear med, Walter Reed Army Inst Res, 64-67, dir div biometrics, 67-68; assoc prof radiol, Col Med, 68-72, PROF RADIOL, COL MED, UNIV NEBR, OMAHA, 72-, ASSOC PROF RADIOL, COL DENT, 69- *Mem:* Soc Nuclear Med; Am Asn Physicists in Med; Asn Mil Surgeons US; Am Col Nuclear Physicians; Am Col Med Physics; fel Am Col Radiol. *Res:* Radiation therapy dosimetry; diagnostic x-ray dosage reduction, medical computer applications. *Mailing Add:* 12823 Jones St Omaha NE 68154

JONES, EUGENE LAVERNE, b Adona, Ark, Sept 20, 28; m 50; c 4. GEOPHYSICS, GEOCHEMISTRY. *Educ:* Univ Ark, BS, 51, MS, 52; Univ Okla, PhD, 61. *Prof Exp:* Petrol geologist, Gulf Oil Corp, 52-54; asst prof geol & head dept, Ark Polytech Col, 54-60; sr res geologist, Field Res Lab, Socony Mobil Oil Co, Inc, 60-64, mgr geol-geochem res & tech serv, Mobil Res & Develop Corp, 64-71, vpres & explor mgr, Mobil North Sea Inc, London, 71-73 & Mobil Explor Norway, Inc, 73-75, adv, 78-79, mgr explor res, Mobil Res & Develop Corp, 75-88, mgr explor, Prod Res Div, 79-88; RETIRED. *Concurrent Pos:* Instr, Oklahoma City Univ, 58; fel, Nat Sci Found, 58-59. *Mem:* AAAS; Am Asn Petrol Geologists; Sigma Xi; Geol Soc London. *Res:* Palynology; sedimentation; petroleum exploration, geology, geochemistry and geophysics; stratigraphy. *Mailing Add:* 915 Green Hills Dallas TX 75137

JONES, EVAN EARL, b Wray, Colo, June 8, 35; m 55. BIOCHEMISTRY. *Educ:* Colo State Univ, BS, 60; Univ Ill, MS, 62, PhD(biochem), 64. *Prof Exp:* Fel biochem, Inst Microbiol, Rutgers Univ, 64-66; asst prof nutrit biochem, 66-69, asst prof animal sci & biochem, 69-71, ASSOC PROF BIOCHEM, 71- & PROF ANIMAL SCI, NC STATE UNIV, 77- *Concurrent Pos:* Vis scholar, Stanford Univ, 75-76. *Mem:* AAAS; Am Chem Soc; Am Soc Biol Chemists; Am Soc Microbiol. *Res:* Amino acid biosynthesis, arginine; metabolic control mechanisms. *Mailing Add:* Dept Animal Sci NC State Univ Box 7621 Raleigh NC 27695

JONES, EVERET CLYDE, b West Plains, Mo, Jan 26, 23; m 70; c 1. MARINE BIOLOGY. *Educ:* Hastings Col, AB, 49; Univ Miami, Fla, MS, 52. *Prof Exp:* Fishery biologist, Nat Marine Fisheries Serv, 55-73; asst prof biol, Northeast Mo State Univ, 74-77, asst prof sci, 79-87; RETIRED. *Mem:* Am Inst Fishery Res Biologists. *Res:* Systematics, ecology and zoogeography of marine copepods; mechanisms controlling distribution of marine plankton and tunas; behavior and systematics of sharks; chemistry of marine algae. *Mailing Add:* Rte 2 Box 29A Norwood MO 65717

JONES, EVERETT, b Albany, NY, Jan 25, 30; m 57; c 3. FLUID MECHANICS, HEAT TRANSFER. *Educ:* Rensselaer Polytech Inst, BAE, 56, MAE, 60; Stanford Univ, PhD(aeronaut & astronaut), 68. *Prof Exp:* Advan study scientist, Lockheed Missiles & Space Co, 56-57; res asst aeronaut eng, Rensselaer Polytech Inst, 57-59; sr thermodynamicist, Lockheed Missiles & Space Co, 59-61, sr engr, 61-64, res specialist, 64-69; engr, US Naval Surface Weapons Ctr, 77-83; PROF AEROSPACE ENG, UNIV MD, COLLEGE PARK, 69- *Concurrent Pos:* Res asst, dept aeronaut & astronaut sci, Stanford Univ, 66-68, res assoc, 68-69; lectr heat transfer, San Jose State Col, 67-68; fac res fel, NASA-ASER fac res prog, 70, US Army fac res eng prof, 83 & David W Taylor res prog, Naval Ship Res Ctr, 85. *Mem:* Am Inst Aeronaut & Astronaut; NY Acad Sci. *Res:* Fluid mechanics, heat transfer and aerodynamics; emphasis on applications for aerospace vehicles, computational fluid mechanics and blood flow. *Mailing Add:* Dept Aerospace Eng Univ Md College Park MD 20742

JONES, EVERETT BRUCE, b Ft Collins, Colo, Sept 23, 33; m 56; c 2. HYDROLOGY, WATER RESOURCES ENGINEERING. *Educ:* Univ Wyo, BS, 55; Pa State Univ, MS, 59; Colo State Univ, PhD(watershed mgt), 64. *Prof Exp:* Chief water develop, Wyo Natural Resources Bd, 59-61; engr-hydrologist, Douglas W Barr, Consult Hydraul Engrs, Minn, 64-65; asst dir inst for res on land & water resources, in-chg of water resources ctr & asst prof meteorol, Pa State Univ, 65-68; coordr water resources, Environ Serv Oper, EG&G, Inc, 69; vpres, M W Bittinger & Assocs, Inc, 70-77, pres, 77; pres, Resource Consults, Inc, 77-87; MGR, WATER RESOURCES DEPT, ENVIRON SCI & ENG INC, 87- *Concurrent Pos:* Asst interstate streams comnr, State of Wyo, 61; vpres, Wyo Well. *Mem:* Am Soc Civil Engrs; Am Meteorol Soc; Am Geophys Union. *Res:* Groundwater hydrology; surface-water hydrology and hydrometeorology, especially water resources management aspects. *Mailing Add:* 6129 S Elm Ct Littleton CO 80121

JONES, FABER BENJAMIN, b Dec 4, 32; US citizen; m 54; c 4. POLYMER CHEMISTRY. *Educ:* Ohio State Univ, BSc, 54. *Prof Exp:* Asst div chief polymer res, Battelle Mem Inst, 53-63; tech dir adhesives res, Evans Adhesives Corp, 63-64; mgr, Chem Appln Br, Phillips Petrol Co, 64-79, mgr, Polymer Appln Br, 79, dir polymer mat res, 80-81, div mgr polymer mat res, 82-86, vpres, planning & budgeting, 86-90, VPRES RES & DEVELOP, PHILLIPS PETROL CO, 90- *Mem:* Am Chem Soc; Adhesion Soc; Soc Plastic Engrs. *Res:* Polymer research and technology, especially on adhesives, coatings and reinforced plastic systems. *Mailing Add:* 2412 Kyles Court Bartlesville OK 74006-6339

JONES, FLOYD BURTON, b Cisco, Tex, Nov 22, 10; m 36; c 4. TOPOLOGY. *Educ:* Univ Tex, BA, 32, PhD(math), 35. *Prof Exp:* Instr pure math, Univ Tex, 32-40, asst prof, 40-43, assoc prof, 43-50; prof math, Univ NC, 50-62; prof math, 62-78, EMER PROF MATH, UNIV CALIF, RIVERSIDE, 78- *Concurrent Pos:* Res assoc, Underwater Sound Lab, Harvard Univ, 42-44; sr fel, NSF, 57-58; mem, Inst Advan Study, 57-58; vis fel, Inst Advan Studies, Australian Nat Univ, 68; Fulbright-Hays Fel, NZ, 75; vis fel, Univ Houston, 77; Mary Moody northern chair, VMI, 79; distinguished vis scientist, Auburn Univ, 82. *Mem:* Am Math Soc; Math Asn Am. *Res:* Pointset theoretic topology. *Mailing Add:* Dept Math Univ Calif Riverside CA 92521-0135

JONES, FRANCIS THOMAS, b Pottsville, Pa, Oct 19, 33; m 81; c 2. PHYSICAL CHEMISTRY. *Educ:* Pa State Univ, BS, 55; Polytech Inst Brooklyn, PhD(phys chem), 60. *Hon Degrees:* MEng, Stevens Inst Technol, 75. *Prof Exp:* Gen Elec Co Ltd fel radiation chem, Univ Leeds, 60-62; chemist, Union Carbide Corp, 62-64; from asst prof to assoc prof chem, 64-71, head dept chem & chem eng, 79-90, PROF CHEM, STEVENS INST TECHNOL, 71- *Concurrent Pos:* Adj assoc prof anesthesiol, New York Med Col, 77-89. *Mem:* Am Chem Soc; Am Inst Chem Engrs. *Res:* Radiation chemistry; photochemistry; catalysis; mass spectrometry; kinetics; instrumentation design. *Mailing Add:* Dept Chem Stevens Inst Technol Hoboken NJ 07030

JONES, FRANCIS TUCKER, b Rocklin, Calif, Jan 17, 05; m 42; c 3. CHEMISTRY. *Educ:* Pac Univ, Ore, AB, 28; Univ Ore, AM, 31; Cornell Univ, PhD(chem micros), 34. *Prof Exp:* Asst chem, Pac Univ, Ore, 26-28; teacher pub sch, Ore, 28-29; asst chem, Univ Ore, 29-31 & Cornell Univ, 32-34; prof, Pac Univ, Ore, 34-42; chemist, Mkt & Nutrit Div, Agr Res Serv, USDA, 42-74; RETIRED. *Mem:* Am Chem Soc; Sigma Xi. *Res:* Physical and analytical chemistry; chemical microscopy applied to determination of optical and crystallographic properties and phase relations; scanning electron microscopy. *Mailing Add:* 244 Trinity Ave Berkeley CA 94708

JONES, FRANK CULVER, b Ft Worth, Tex, July 30, 32; m 55; c 2. COSMIC RAY PHYSICS, THEORETICAL ASTROPHYSICS. *Educ:* Rice Inst, BA, 54; Univ Chicago, MS, 55, PhD(physics), 61. *Prof Exp:* Res assoc physics, Univ Chicago, 60; res assoc, Princeton Univ, 60-61, instr, 61-63; Nat Acad Sci-Nat Res Coun resident res assoc, Theoret Studies Group, 63-65, physicist, 65-77, ASTROPHYSICIST, LAB HIGH ENERGY ASTROPHYS, GODDARD SPACE FLIGHT CTR, NASA, 77- *Concurrent Pos:* Vis scientist, Max Planck Inst Nuclear Physics, 77; secy, Astrophys Div, Am Phys Soc, 88-92. *Mem:* AAAS; fel Am Phys Soc; Am Astron Soc. *Res:* Physics of the origin of cosmic rays and related astrophysical problems; statistical physics of cosmic ray origin and propagation in the galaxy. *Mailing Add:* Code 665 Lab for High Energy Astrophys NASA Goddard Space Flight Ctr Greenbelt MD 20771

JONES, FRANK NORTON, b Columbia, Mo, Dec 27, 36; div; c 1. POLYMER SYNTHESIS, MATERIALS. *Educ:* Oberlin Col, AB, 58; Duke Univ, PhD(org chem), 62. *Prof Exp:* Instr org chem, Duke Univ, 61-62; fel, Mass Inst Technol, 62-63; res chemist, Cent Res Dept, E I Du Pont Co, 63-68, staff chemist, 68-70, res supvr, 70-73; tech mgr, Celanese Polymer Specialties Co, 73-79; res & develop mgr, Cargill, Inc, 79-83; prof & chair Dept Polymers & Coatings, NDak State Univ, 83-90; PROF & DIR, NSF COATINGS RES CTR, 90- *Honors & Awards:* Roon Found Prizes, 86, 87. *Mem:* Am Chem Soc; Fedn Soc Coatings Technol. *Res:* Synthetic polymer chemistry; polymer structure/property relationships; polymeric materials; coatings. *Mailing Add:* Coatings Res Inst Eastern Mich Univ Ypsilanti MI 48197

JONES, FRANKLIN DEL, b Hereford, Tex, Sept 22, 35; m 57; c 4. COMBAT STRESS, PSYCHOPHARMACOLOGY. *Educ:* Baylor Univ, BS, 57; Univ Tex, Dallas, MD, 61. *Prof Exp:* Resident psychiat, Walter Reed Army Med Ctr, 61-65, dir res ward, 68-73; CLIN PROF GEORGETOWN UNIV MED SCH, 73-; CLIN PROF, UNIFORMED SERVS, UNIV HEALTH SCI, 75- *Concurrent Pos:* Intern med & surg, Ireland Army Hosp, Ft Knox, Ky, 61-62; chief psychiat sev, Walter Reed Army Med Ctr, 71-73 & forensic psychiat, 73-77, dir pyschiat educ, 73-77 & 85-88; clin consult psychiat, 73-, psychiat & neurol consult, Surgeon Gen Army, 77-81. *Mem:* World Psychiat Asn (pres, 77-83, secy, 83-89). *Res:* Studies of combat stress in Vietnam, Egypt and Israel; post-traumatic stress disorder in combat veterans and rape victims; psychopharmacology of eating disorders and performance. *Mailing Add:* 6508 Tall Tree Terrace Rockville MD 20852-3733

JONES, FRANKLIN M, b Reidsville, NC, Mar 10, 33; m 63; c 2. SCIENCE EDUCATION. *Educ:* Appalachian State Teachers Col, BS, 55, MA, 60; Univ NC, MEd, 60; Univ Ga, EdD(sci educ), 66. *Prof Exp:* Teacher, High Sch, Va, 55-56 & NC, 56-58; prof chem, Ferrum Jr Col, 60-64; assoc prof, 66-68, PROF PHYS SCI, RADFORD UNIV, 68- *Mailing Add:* Dept Phys Sci Radford Univ Radford VA 24142

JONES, FREDERICK GOODWIN, b Utica, NY, Nov 6, 35; m 59; c 3. PERMANENT MAGNETS, POWDER METALLURGY. *Educ:* Cornell Univ, BMetE, 59; Univ Mich, Ann Arbor, MSE & PhD(metall), 69. *Prof Exp:* Staff engr, Crucible Steel Co Am, 59-65; sr develop engr, magnetism, Gen Elec Co, 69-73; sr develop engr, Hitachi Magnetics Corp, 73-84; PRES, F G JONES ASSOC, LTD, 84- *Concurrent Pos:* Tech consult, NAm, Western Europ, Far East, 84- *Mem:* Am Inst Mining, Metall & Petrol Engrs; Am Foundrymen's Soc; Am Inst Elec & Electronics Engrs; Sigma Xi. *Res:* Permanent magnet materials; rare earth-transitional metal alloys; hydrogen-metal reactions; low alloy steels; high temperature alloys. *Mailing Add:* 820 Wright Ave Alma MI 48801

JONES, GALEN EVERTS, b Milwaukee, Wis, Sept 9, 28; m 54, 86; c 3. MARINE MICROBIOLOGY, MICROBIAL BIOGEOCHEMISTRY. *Educ:* Dartmouth Col, AB, 50; Williams Col, MA, 52; Rutgers Univ, PhD(microbiol), 56. *Prof Exp:* Asst, Williams Col, 50-52; res asst, Tex Gulf Sulfur, Rutgers Univ, 52-55; from jr res microbiologist to asst res microbiologist, Div Marine Biol, Scripps Inst Oceanog, Univ Calif, 55-63, Rockefeller fel, 55-57; assoc prof biol, Boston Univ, 63-66; prof microbiol, Univ NH, 66-91, dir Jackson Estuarine Lab, 83-87; CONSULT, 91- *Concurrent Pos:* Res grants, Nat Inst Allergy & Infectious Dis, 57-59, div water supply & pollution control, USPHS, 59-62, 63-66 & NSF 72-74, 75-76 & 77-78, 81-85 & Sea Grant, 83- 85, NOAA, 87-91; consult, Eli Lilly & Co, Ind, 58-59, Bendix-Pac, Calif, 60, Arthur D Little Co, Mass, 69-70 & 73-74 & Normandeau Assocs, Inc, NH, 70-71; Off Naval Res contract, 63-64 & 66-68; nonresident assoc microbiol, Woods Hole Oceanog Inst, 64-72, lectr, Marine Biol Lab, Woods Hole, 71-72, 74-75 & 76-77; mem, Nat Sea-Grant Univ Comt, 65-67; mem, Santa Barbara Oil Spill Panel, Exec Off of the President, 69-70; mem adv panel biol oceanog, NSF, 71-72 & mem oceanog adv panel, 74-75; vis prof oceanog, Univ Liverpool, 72-73; mem, Inst Ecol Adv Panel to Nat Comn Water Qual, Washington, DC, 74-76; mem exec panel oceanog div, NSF, 80; vis prof, Scripps Inst Oceanog, Univ Calif, 81; dir, Jackson Estuarine Lab, 66-72, chmn dept, 75-80 & Marine Sci Labs, 85-87, Interim Sea Grant, Univ NH, 86-87. *Mem:* Fel AAAS; Am Inst Biol Sci; Sigma Xi; Estuarine Res Fedn; Am Soc Limol & Oceanog; Geochem Soc; fel Am Acad Microbiol; Am Soc Micrbiol. *Res:* Biochemicals and trace elements in sea water; chemosynthesis; fractionation of stable isotopes in microorganisms; biogeochemistry; elemental composition of bacteria. *Mailing Add:* Dept Microbiol Univ NH Durham NH 03824

JONES, GARETH HUBERT STANLEY, earth science, for more information see previous edition

JONES, GARTH, b Victoria, BC, Mar 27, 32. NUCLEAR PHYSICS. *Educ:* Univ BC, BA, 53, MSc, 55, PhD(physics), 59. *Prof Exp:* Jr sci officer electronics, Atomic Energy Can, Ltd, 55-56; Rutherford Mem fel, Clarendon Lab, Oxford Univ, 60, Nat Res Coun Can overseas fel nuclear physics, 60-61; from asst prof to assoc prof, 61-69, PROF PHYSICS, UNIV BC, 69- *Concurrent Pos:* Guggenheim fel, 67-68. *Mem:* Can Asn Physicists; Am Phys Soc. *Res:* Nuclear reactions; positron annihilation; intermediate energy physics. *Mailing Add:* Dept Physics Univ BC Vancouver BC V6T 2A6 Can

JONES, GARTH WICKS, b Aberdare, Wales, Sept 23, 40; m 64; c 3. MICROBIOLOGY. *Educ:* Univ Reading, BSc, 69, PhD(microbiol), 72. *Prof Exp:* Sr sci officer microbiol, Inst Res Animal Dis, Brit Agr Res Coun, 72-75; ASST PROF MICROBIOL, UNIV MICH, ANN ARBOR, 75- *Concurrent Pos:* Scholar, Univ Mich, Ann Arbor, 74-75. *Mem:* Soc Gen Microbiol; Brit Soc Appl Bact; Am Soc Microbiol. *Res:* Nature and function of the adhesive properties of bacteria, particularly enteric pathogens, and the composition of the eukaryotic cell components with which bacterial adhesive substances interact. *Mailing Add:* Dept Microbiol 6643 Med Sci Bldg 2 841 Univ Mich 1301 Catherine Rd Ann Arbor MI 48109

JONES, GARY EDWARD, b Metropolis, Ill, June 2, 40; m 64; c 1. GENETICS, CELL BIOLOGY. *Educ:* Univ Ill, Urbana, BS, 62, MS, 64; Univ Calif, Berkeley, PhD(biophys), 70. *Prof Exp:* NIH fel biophys, Pa State Univ, 70-71; staff geneticist, Hosp for Sick Children, Toronto, Ont, 71-73; asst prof biol, 73-78, asst prof genetics, 78-79, ASSOC PROF GENETICS, UNIV CALIF, RIVERSIDE, 79- *Mem:* Sigma Xi; AAAS; Genetic Soc Am; Am Soc Microbiol. *Res:* Biochemical genetics of amino acid utilization in yeast; somatic cell genetics of cultured animal and plant cells. *Mailing Add:* Dept Bot & Plant Sci Univ Calif Riverside CA 92521

JONES, GEOFFREY MELVILL, b Shelford, Eng, Jan 14, 23; m 53; c 4. NEUROSCIENCES, AEROSPACE MEDICINE. *Educ:* Univ Cambridge, MA, 48, MB, BCh, 49. *Prof Exp:* House surgeon, Middlesex Hosp, London, 49-50; surgeon, Ear, Nose & Throat, Addenbrookes Hosp, Cambridge, 50-51; sci officer, Med Res Coun, Gt Brit, 55-61; dir aviation med, Aerospace Med Res Unit, 61-88; assoc prof, 61-67, PROF PHYSIOL, MCGILL UNIV, 67- *Concurrent Pos:* Vis prof, Stanford Univ, 71-72 & Col de France, Paris, 79; sr res assoc, Nat Acad Sci, 71-72; vis scholar, Univ Tex, Galveston, 82; vis prof, Univ Calgary, 83; assoc mem, Ctr Studies Age & Aging, McGill Univ & Dept Neurol & Neurosurg. *Honors & Awards:* Harry G Armstrong Award res in aerospace med, 69; Arnold D Tuttle Award, 71; Quinquennial Gold Medal, Barany Soc, 88; Wilbur Franks Award, 88; Ashton Graybiel Lectr Award, US Naval Aerospace Med Res Lab, 89; Stewart Mem Lectr Award, Royal Aeronaut Soc London, 89; Buchanan-Barbour Award, 90. *Mem:* Can Physiol Soc; UK Physiol Soc; Am Soc Neurosci; fel Can Aeronaut & Space Inst; fel Royal Soc Can; fel Royal Soc London; fel Royal Aeronaut Soc London; fel Am Aerospace Med Asn. *Res:* Neurophysiology of postural control, vestibular and oculomotor systems; respiration at high altitude; long duration flying fatigue; high altitude bail out; pilot disorientation; adaptive plasticity in brainstem reflexes; cognitive management of subcortical reflexes. *Mailing Add:* 3265 Glencoe Ave Montreal PQ H3R 2C5 Can

JONES, GEORGE HENRY, b Muskogee, Okla, Feb 21, 42; m 65. BIOCHEMISTRY, MOLECULAR BIOLOGY. *Educ:* Harvard Univ, BA, 63; Univ Calif, Berkeley, PhD(biochem), 68. *Prof Exp:* Helen Hay Whitney Found fels, NIH, 68-70 & Univ Geneva, 70-71; asst prof zool, Univ Mich, Ann Arbor, 71-74, assoc prof biol sci & cell & molecular biol, 74-90; ASSOC VPRES RES & GRAD STUDIES & DEAN GRAD SCH ARTS & SCI, EMORY UNIV, 90- *Mem:* AAAS. *Res:* Mammalian protein biosynthesis, specifically initiation mechanisms; immunoglobulin biosynthesis; cellular regulatory mechanisms. *Mailing Add:* Grad Sch Arts & Sci 202 Admin Bldg Emory Univ Atlanta GA 30323

JONES, GEORGE R, b Los Angeles, Calif, Aug 16, 30; m 52; c 6. SOLID STATE PHYSICS. *Educ:* Western Md Col, BS, 51; Cath Univ, MS, 53, PhD(physics), 63. *Prof Exp:* Jr electronics engr, Davies Labs, Inc, 52-54; physicist, Diamond Ord Fuze Labs, 54-63, res physicist, Harry Diamond Labs, 63-66; RES PHYSICIST, INFO & SIGNAL PROCESSING SENSOR, CTR NIGHT VISION & ELECTRO OPTICS, 66- *Concurrent Pos:* Consult, Am Mach & Foundry, 61-62; lectr, Am Univ, 69-71. *Mem:* Am Phys Soc; Inst Elec & Electronics Engrs; AAAS. *Res:* Optical spectra of rare earth doped solids; magnetic properties of solids; electromagnetic theory; electromagnetic instrumentation for solid state measurements; artificial intelligence (visual) and mathematical modeling of human vision processes; applied parallel distributed processors and neural networks. *Mailing Add:* 113 Northway Rd Greenbelt MD 20770

JONES, GERALD MURRAY, b Gouverneur, NY, Apr 17, 41; m 63; c 3. DAIRY SCIENCE. *Educ:* Cornell Univ. BS, 62; Univ Maine, MS, 64; Pa State Univ, PhD(dairy sci), 68. *Prof Exp:* Asst prof animal sci, Macdonald Col, McGill Univ, 68-74; assoc prof, 74-79, PROF DAIRY SCI, VA POLYTECH INST & STATE UNIV, 79-, EXTEN DAIRY SCIENTIST, 74- *Honors & Awards:* West Agron Award Mastitis Res, Am Dairy Sci Asn. *Mem:* Nat Mastitis Coun; Am Dairy Sci Asn. *Res:* Milking management, practices and systems; mastitis; calf nutrition and management; dairy cattle nutrition; dairy herd management. *Mailing Add:* Dept Dairy Sci Va Tech Blacksburg VA 24061-0315

JONES, GERALD WALTER, b Utica, NY, June 25, 42; m. PHOTOGRAPHIC CHEMISTRY, POLYMER APPLICATIONS. *Educ:* Hartwick Col, BA, 64; Syracuse Univ, PhD(org chem), 70. *Prof Exp:* Lab asst chem, Hartwick Col, 64-65; fel, Ohio State Univ, 70-74; res chemist, GAF Corp, 74-81; SR ENGR, IBM CORP, 81- *Mem:* Am Chem Soc. *Res:* Photolysis of alpha, beta-unsaturated ketones and carbene chemistry; photographic science, resilient sheet vinyl flooring and photoresists. *Mailing Add:* 1112 Reynolds Rd Johnson City NY 13790

JONES, GIFFIN DENISON, b Fond du Lac, Wis, Dec 16, 18; m 39; c 4. ORGANIC CHEMISTRY. *Educ:* Univ Wis, BS, 39; Univ Ill, PhD(org chem), 42. *Prof Exp:* Instr org chem, Univ Iowa, 42-44; res chemist, Cent Res Labs, Gen Aniline & Film Corp, Pa, 44-47; res chemist, Phys Res Lab, 47-56, dir. 56-68, dir, E C Britton Lab, 68-70, RES SCIENTIST, PHYS RES LAB, DOW CHEM CO, 70- *Concurrent Pos:* Civilian with Off Sci Res & Develop, 44. *Mem:* Am Chem Soc. *Res:* Polymers; organic reaction mechanism. *Mailing Add:* 4002 Cambridge Midland MI 48640

JONES, GILBERT FRED, b Oakland, Calif, Apr 3, 30; m 51; c 6. MARINE ECOLOGY, HISTOLOGY. *Educ:* Col of the Pac, AB, 52; Univ Wis, MS, 54; Univ Southern Calif, PhD(biol), 67. *Prof Exp:* Biologist, Mainland Shelf Surv, Allan Hancock Found, 57-64, from instr to asst prof, 64-70, ASSOC PROF BIOL, UNIV SOUTHERN CALIF, 70- *Mem:* Am Soc Limnol & Oceanog; Marine Biol Asn UK. *Res:* Benthic marine ecology, particularly population ecology; marine nematodes. *Mailing Add:* Dept Biol Sci 0371 Univ Southern Calif Los Angeles CA 90089

JONES, GILDA LYNN, b Water Valley, Miss, May 16, 27. BACTERIOLOGY. *Educ:* Miss State Univ for Women, BS, 51; Univ Mich, MPH, 67; Univ NC, Chapel Hill, PhD(pub health), 74. *Prof Exp:* Med technologist & chief, Bact Lab, Grady Mem Hosp, Atlanta, Ga, 52-60; SUPVY MICROBIOLOGIST, CENTERS DIS CONTROL, ATLANTA, 61- *Honors & Awards:* Betty King Award, Am Soc Microbiol, 79. *Mem:* Am Soc Microbiol; Am Pub Health Asn; Sigma Xi. *Res:* Development of training materials in bacteriology for use by the clinical laboratory worker. *Mailing Add:* Ctr Dis Control 1600 Clifton Rd Atlanta GA 30333

JONES, GLENN CLARK, b Raleigh, NC, Aug 22, 35; m 65; c 2. ORGANIC & POLYMER CHEMISTRY, TECHNICAL MANAGEMENT. *Educ:* Wake Forest Col, BS, 57; Duke Univ, PhD(chem), 62. *Prof Exp:* Res assoc org chem, Duke Univ, 61-62; from res chemist to sr res chemist, 62-84, RES ASSOC, TENN EASTMAN CO, 84- *Mem:* Am Chem Soc; Sigma Xi. *Res:* Base catalyzed rearrangements; polymer feasibility studies; free radical chemistry; organic electrochemistry; hydroquinone, solvent and powder coatings. *Mailing Add:* 3620 Hemlock Park Dr Kingsport TN 37663

JONES, GORDON ERVIN, b Greenwood, Miss, July 23, 36; m 61; c 1. PHYSICS. *Educ:* Miss State Univ, BS, 58; Duke Univ, PhD(physics), 64. *Prof Exp:* From asst prof to assoc prof physics, 64-72, PROF PHYSICS, MISS STATE UNIV, 72- *Mem:* Am Phys Soc; Am Asn Physics Teachers. *Res:* Microwave spectroscopy. *Mailing Add:* Dept Physics Miss State Univ Box 5167 Mississippi State MS 39762

JONES, GORDON HENRY, b Stockport, Eng, Apr 2, 40; m 63; c 2. ORGANIC CHEMISTRY. *Educ:* Cambridge Univ, BA, 62, PhD, 65, MA, 66. *Prof Exp:* Fel org chem, 65-66, RES CHEMIST, SYNTEX INST ORG CHEM, 66- *Mem:* Am Chem Soc; Royal Soc Chem. *Res:* Application of new reactions in carbohydrate and nucleoside chemistry; peptide chemistry. *Mailing Add:* 2635 Park Blvd Palo Alto CA 94306

JONES, GRAHAM ALFRED, b London, Eng, May 8, 35; m 63; c 4. AGRICULTURAL MICROBIOLOGY. *Educ:* Univ Leeds, BSc, 57; McGill Univ, MSc, 58, PhD(agr bact), 63. *Prof Exp:* Lectr agr bact, McGill Univ, 58-60; asst prof dairy sci, Univ Sask, 63-68, from assoc prof to prof dairy & food sci, 67-82, lectr microbiol, 75-79, PROF APPL MICROBIOL & FOOD SCI, UNIV SASK, 82-, ASSOC, DEPT MICROBIOL, SCH MED, 79- *Concurrent Pos:* Vis prof, Nat Res Coun Can, 74; vis scientist, Agr Res Coun Inst Animal Physiol, Babraham, Eng, 78-79; head dept, Univ Sask, 81- *Honors & Awards:* Queen's Jubilee Medal, 77. *Mem:* Am Soc Microbiol; Can Soc Microbiol (secy-treas, 70-73, second vpres, 80-81, first vpres, 81-82, pres, 82-83); Can Inst Food Sci & Technol; Agr Inst Can; Can Soc Animal Sci. *Res:* Rumen microbiology; agricultural fermentations. *Mailing Add:* Dept Appl Microbiol & Food Sci Univ Sask Saskatoon SK S7N 0W0 Can

JONES, GUILFORD, II, b Jackson, Tenn, Nov 24, 43; m 66; c 2. PHYSICAL ORGANIC CHEMISTRY, PHOTOCHEMISTRY. *Educ:* Southwestern at Memphis, BS, 65; Univ Wis-Madison, PhD(chem), 70. *Prof Exp:* Asst prof, 71-77, assoc prof, 77-82, PROF CHEM, BOSTON UNIV, 82- *Concurrent Pos:* NIH fel, Yale Univ, 69-71. *Mem:* Am Chem Soc; Sigma Xi. *Res:* Photochemical conversion of energy; mechanisms and applications of photochemical and thermal reactions; dye photochemistry; photoeffects for polymer-bound chromophores. *Mailing Add:* Dept Chem Boston Univ Boston MA 02215

JONES, GUY LANGSTON, b Kinston, NC, June 7, 23; m 48; c 2. PLANT BREEDING. *Educ:* NC State Col, BS, 47, MS, 50; Univ Minn, PhD, 52. *Prof Exp:* Supt br sta, NC Agr Exp Sta, 47-49; asst dept agron & plant genetics, Univ Minn, 50-52; asst prof dept agron, 52-58, assoc prof field crops, 58-61, prof crop sci, 61-65, head agron exten, 65-75, prof crop sci exten, 75-85, prof crop sci, 61-85, EMER PROF, NC STATE UNIV, 85-, TOBACCO PROD & AGRON EXTEN PROGS, 71- *Concurrent Pos:* With Ministry Agr, Venezuela, 59, Inst Tobacco, Dominican Repub, 63, Agency Int Develop, Guatemala, 64-65 & Philippines, 64-65 & Food & Agr Orgn, Argentina, 74, Inst Soil Sci, Nanjing, China, 85-91, Guatemala, 87 & 88, Dominican Repub, 87 & 88, Mexico, 87-91; assoc ed, Agron J. *Honors & Awards:* Agron Exten Educ Award, Am Soc Agron. *Mem:* Fel Am Soc Agron; Sigma Xi. Crop Sci Soc Am. *Res:* Tobacco genetics; tobacco variety evaluation; agronomy extension. *Mailing Add:* 3435 Blue Ridge Rd NC State Univ Raleigh NC 27612-8014

JONES, GWILYM STRONG, b Cincinnati, Ohio, May 4, 42; m 67; c 3. MAMMALOGY, VERTEBRATE ECOLOGY. *Educ:* Hanover Col, BA, 64; Purdue Univ, MS, 67; Ind State Univ, PhD(mammal syst), 81. *Prof Exp:* Res investr, Naval Med Res Unit 2, Taiwan, 67-69; mus specialist, Smithsonian Inst, 70-71; PROF BIOL, NORTHEASTERN UNIV, 76- *Concurrent Pos:* Collabr mammal div, Smithsonian Inst, 70; adv, Chinese Asn Conserv Nature & Natural Resources, 69-70; grants, Am Inst Biol Sci, 70, Theodore Roosevelt Mem Fund & Am Mus Natural Hist, 74, US Dept Health & Human Serv, 78 & 81, NH Fish Game Dept, 80-86, Pub Archeol Lab, Brown Univ, 80, Nature Conservancy, 84 & US Fish Wildlife Serv, 83. *Mem:* Am Soc Mammalogists; Soc Syst Zool; Sigma Xi; Soc Marine Mammalogy; Wildlife Soc. *Res:* Mammalian systematics; vertebrate food habits; ectoparasites and demographics. *Mailing Add:* Dept Biol Northeastern Univ Boston MA 02115

JONES, HAROLD LESTER, b Nampa, Idaho, June 19, 43; m 65; c 2. ORGANIC CHEMISTRY. *Educ:* Ore State Univ, BS, 65; Univ Colo, PhD(chem), 69. *Prof Exp:* Res asst, Univ Colo, 68; asst prof, 69-76, assoc prof, 76-86, PROF CHEM, COLO COL, 86-, CHMN, 81-82, 85- *Concurrent Pos:* Res assoc, Univ Colo, 72. *Mem:* Am Chem Soc. *Res:* Nuclear magnetic resonance; small ring chemistry, bicyclic systems and cyclopropanols; free radical reactions in cyclopropanols; photochemistry of bicyclic-spiro-compounds. *Mailing Add:* Dept Chem Colo Col 14 E Cache La Poudre Colorado Springs CO 80903

JONES, HAROLD TRAINER, b Allentown, Pa, Dec 22, 25; m 53; c 2. MATHEMATICAL ANALYSIS. *Educ:* Washington Missionary Col, BA, 46; Lehigh Univ, MA, 49; Brown Univ, PhD(appl math), 58. *Prof Exp:* Instr math, Pac Union Col, 46-48; fel appl, Brown Univ, 49-51, asst, 51-52; from asst prof to assoc prof, 52-64, PROF MATH, ANDREWS UNIV, 64- *Concurrent Pos:* NSF fac fel, 65-66; vis asst prof, Ind Univ, 80-81, vis prof, St Mary's Col, 82- 83. *Mem:* Am Math Soc; Sigma Xi; Math Asn Am. *Res:* Potential theory; geometry. *Mailing Add:* 9193 Woodland Dr Berrien Springs MI 49103

JONES, HELENA SPEISER, b Columbus, Ohio, June 26, 40; m 65; c 2. ANATOMY, MEDICAL SCIENCES. *Educ:* Ohio State Univ, BSc, 62, PhD(anat) 68. *Prof Exp:* Instr anat, Med Ctr, Ind Univ, Indianapolis, 68-69; NIH staff fel, Nat Inst Environ Health Sci 72-75; ASST PROF BIOL, UNIV WIS-EAU CLAIRE, 75- *Concurrent Pos:* Consult, Adv Comt Estab Med Histol Technicians Assoc Degree, 76-; fels biomed sci, Washington, DC, 77 & 78; mem review panels, NSF; Eau Claire Community cancer grant, 80, 81 & 82. *Mem:* Am Asn Anatomists; Sigma Xi; NY Acad Sci. *Res:* Skin cancer; endocrinology; bone. *Mailing Add:* 5729 Elm Rd Rte 3 Eau Claire WI 54701

JONES, HOBART WAYNE, b Logansport, Ind, Apr 15, 21; m 43; c 4. ANIMAL BREEDING. *Educ:* Purdue Univ, BSA, 43; Ohio State Univ, MSA, 46, PhD(animal prod), 60. *Prof Exp:* Assoc prof, 50-61, PROF ANIMAL SCI, PURDUE UNIV, 61- *Mem:* Am Soc Animal Sci. *Res:* Animal production; swine nutrition and environmental studies. *Mailing Add:* Dept Animal Sci Purdue Univ West Lafayette IN 47907

JONES, HOWARD, b Bolton, Eng, Apr 6, 37; m 69; c 2. MEDICINAL CHEMISTRY, ORGANIC CHEMISTRY. *Educ:* Univ Leeds, BSc, 59, PhD(org chem), 62. *Prof Exp:* Fel org chem, Univ Calif, Los Angeles, 62-64; asst dir, Merck Sharp & Dohme Res Labs, 64-78; DIR MED CHEM, USV PHARMACEUT CORP, 78- *Mem:* Royal Soc Chem; Am Chem Soc. *Res:* Inflammation and rheumatoid arthritis; vitamin D and bone metabolism; immunology. *Mailing Add:* Five Emory Pl Holmdel NJ 07733

JONES, HOWARD ST CLAIRE, JR, b Richmond, Va, Aug 18, 21; m 46. ELECTRONICS ENGINEERING, MICROWAVE PHYSICS. *Educ:* Va Union Univ, BS, 43; Howard Univ, cert eng, 44; Bucknell Univ, MSEE, 73. *Hon Degrees:* DSc, Va Union Univ, 71. *Prof Exp:* Electronic scientist microwave electronics, Diamond Ord Fuze Lab, Washington, DC, 53-59, supvry electronic engr, 59-68, chief microwave res & develop, 68-80, tech consult, Harry Diamond Labs, Md, 80-; AT DEPT ELEC ENG, HOWARD UNIV. *Concurrent Pos:* Instr physics & math, Hilltop Radio-Electronics Inst, Washington, DC, 46-53; asst prof electronic eng, Sch Eng, Howard Univ, 58-63; tech consult, Phelps Dodge Electronics, Conn, 68-69; Secy of Army fel, 72. *Mem:* Fel Inst Elec & Electronics Engrs; fel AAAS; Antenna & Propagation Soc; Microwave Theory & Techniques Soc. *Res:* Microwave research and development; directing, planning and coordinating research and development programs which involve theoretical and applied microwave research; management of programs and projects relating to major electronic systems. *Mailing Add:* Tech Consult 3001 Veazey Terr NW Apt 1310 Washington DC 20008

JONES, IRA, b Bartow, Fla, Jan 22, 34; m 57; c 2. ZOOLOGY, PARASITOLOGY. *Educ:* Benedict Col, BS, 55; Atlanta Univ, MS, 57; Wayne State Univ, PhD(biol), 66. *Prof Exp:* Instr biol, Savannah State Col, 57-59; assoc prof, Fla Agr & Mech Univ, 64-66 & Inter-Am Univ PR, 66-69; from asst prof to assoc prof, 69-77, PROF BIOL, CALIF STATE UNIV, LONG BEACH, 77- *Concurrent Pos:* USPHS fel, 61; grant, Caribbean Inst & Study Ctr for Latin Am, 68-69; Sigma Xi res grant, 68-69; PR Nuclear Ctr grant, 69; consult, Nat Commun Dis Ctr, 69; Calif State Univ Long Beach Found grant, 69-71; dir & consult parasitol, Jones Biomed & Lab, Long Beach, Ca, 77- *Mem:* Am Soc Parasitol; Am Inst Biol Sci; Soc Protozool. *Res:* Research on the endosymbionts of Sipunculids, including, zoogeography of parasitism, host specificity, life cycles of parasites and the cytochemistry and ultra-structure of Sipunculids sporozoa. *Mailing Add:* Dept Biol Calif State Univ 6101 E Seventh St Long Beach CA 90840

JONES, IRVING WENDELL, b Washington, DC. STRUCTURAL ENGINEERING & MECHANICS. *Educ:* Howard Univ, BS, 53; Columbia Univ, MS, 57; Polytech Inst Brooklyn, PhD(appl mech), 67. *Prof Exp:* Asst civil eng, Columbia Univ, 56-57; struct engr, Grumman Aerospace Corp, 57-62; asst aerospace & mech, Polytech Inst Brooklyn, 62-63; asst dir & partner, Appl Technol Assocs, Inc, 63-69; assoc prof civil eng, 69-72, PROF CIVIL ENG & CHMN DEPT, HOWARD UNIV, 72- *Concurrent Pos:* Consult, space div, Fairchild-Hiller Corp, 62-64 & Dist Eng Serv, Inc, 77-; mem pressure vessel res coun, Welding Res Found, 64-69; lectr, grad sch, Stevens Inst Technol, 68-69. *Mem:* Am Soc Civil Engrs (pres, 69); Am Soc Mech Engrs (pres, 64); Am Soc Eng Educ (pres, 69). *Res:* Developed methods for computer-aided structural analysis including high temperature effects; helped develop shock-absorbing mounts and foundations for sensitive shipboard equipment; developed analysis methods for effects of high temperature on aerospace structures. *Mailing Add:* Dept Civil Eng Sch Eng Howard Univ Washington DC 20059

JONES, IVAN DUNLAVY, b Holdrege, Nebr, Dec 10, 03; m 30; c 2. FOOD SCIENCE. *Educ:* Nebr Wesleyan Univ, AB, 26; Univ Minn, PhD(agr biochem), 31. *Prof Exp:* Instr agr biochem, Univ Minn, 29-30; assoc horticulturist, Exp Sta, 31-45, prof hort, 45-61 & food sci, 61-70, EMER PROF FOOD SCI, NC STATE UNIV, 70- *Concurrent Pos:* Consult food sci & technol, 70-; vis prof, Middle East Tech Univ, Ankara, Turkey 79-80. *Mem:* Am Chem Soc; fel Inst Food Technol; fel Am Pub Health Asn; fel Am Inst Chem; Sigma Xi. *Res:* Chemical composition of fruits and vegetables and their processing by freezing, canning, dehydration and brining; estimation of chlorophylls and their metal derivatives; influence preservation technique on chlorophyll. *Mailing Add:* 2710 Rosedale St Raleigh NC 27607-7122

JONES, J(AMES) B(EVERLY), b Kansas City, Mo, Aug 21, 23; m 45; c 2. MECHANICAL ENGINEERING. *Educ:* Va Polytech Inst, BS, 44; Purdue Univ, MS, 47, PhD(mech eng), 51. *Prof Exp:* Asst mech engr, Eng Bd, US War Dept, Va, 44-45; asst instr mech eng, Purdue Univ, 45-47, instr, 47-51; serv engr, Babcock & Wilcox Co, 48; develop engr, Gen Elec Co, 51-52; asst prof mech eng, Purdue Univ, 51-54; sr proj engr, Allison Div, Gen Motors Corp, 53; assoc prof mech eng, Purdue Univ, 54-57, prof, 57-64; PROF MECH ENG & HEAD DEPT, VA POLYTECH INST & STATE UNIV, 64- *Concurrent Pos:* NSF faculty fel, Swiss Fed Inst Technol, 61-62. *Mem:* Am Soc Mech Engrs; Am Soc Eng Educ; Am Inst Aeronaut & Astronaut; Sigma Xi. *Res:* Fluid mechanics; thermodynamics. *Mailing Add:* Dept Mech Eng Va Polytech Inst & State Univ Blacksburg VA 24061

JONES, J BENTON, JR, b Tyrone, Pa, Apr 4, 30; m 55; c 3. SOIL FERTILITY, PLANT NUTRITION. *Educ:* Univ Ill, BS, 52; Pa State Univ, MS, 56, PhD(agron), 59. *Hon Degrees:* Dr, Univ Hort, Budapest, Hungary, 87. *Prof Exp:* From assoc prof to prof agron, Ohio Agr Res & Develop Ctr, 59-68; div chmn, Dept Hort, Univ Ga, 74-79, prof agron, agr exten-agron, 68-79, mem, Inst Ecol, 75-89, prof, Dept Hort, 79-89; PRES, BENTON LABS, INC, ATHENS, GA, 69-; VPRES, MICRO-MACRO INT, INC, ATHENS, GA, 90- *Concurrent Pos:* Chmn, Micronutrient Comt, 67-69, Soil Testing & Plant Anal Comt, Soil Sci Soc Am, 67-69, Coun Soil Testing & Plant Anal, 69-72, secy-treas, 72-; consult, St Louis Testing Labs, 67-75; assoc referee, Plant Anal Emission Spectros, Asn Off Anal Chemists, 69-83, Plant Preparation, 69-89; exec ed, Commun Soil Sci & Plant Anal, 69-, J Plant Nutrit, 79-; bd mem, Coun Agr Sci & Technol, 73-78, subcomt Environ Qual, 74-77, Agron Comt, Nat Fertilizer Solutions Asn, 76-80. *Mem:* Fel AAAS; Am Soc Agron; Soil Sci Soc Am; Int Soc Soil Sci; Am Soc Hort Sci; Asn Off Anal Chemists; Hydroponic Soc Am; Sigma Xi. *Res:* Soil and plant chemistry, especially the micronutrients; soil fertility and plant nutrition related to crop production; soil testing and plant analysis; techniques of analysis by emission spectroscopy; techniques of giving plants in soilless media and hydroponically; author of numerous articles, books and book chapters. *Mailing Add:* Micro-Macro Int Inc 183 Paradise Blvd Suite 108 Athens GA 30607

JONES, J KNOX, JR, b Lincoln, Nebr, Mar 16, 29; m 53; c 3. VERTEBRATE ZOOLOGY. *Educ:* Univ Nebr, BS, 51; Univ Kans, MA, 53, PhD, 62. *Prof Exp:* From instr to assoc prof zool, Univ Kans, 59-68, prof systs & ecol, 68-71, from asst cur to assoc cur, Mus Natural Hist, 59-68, assoc dir & cur mammals, 68-71; coord mus studies & dean grad sch, 71-84, vpres res & grad studies, 74-84, PROF BIOL, TEX TECH UNIV, 71-, PAUL WHITFIELD HORN DISTINGUISHED PROF, 86- *Concurrent Pos:* Managing ed, Soc Study Evolution, 65-66 & Am Soc Mammal, 67-73; ed, Tex Acad Sci, 86- *Honors & Awards:* C Hart Merriam Award, Am Soc Mammal, 77, H H T Jackson Award, 83. *Mem:* Am Soc Mammal (vpres, 68-72, pres, 72-74); Soc Syst Zool; Soc Study Evolution. *Res:* Mammalogy, especially systematics and biogeography of North and Middle American mammals. *Mailing Add:* Mus Tex Tech Univ Box 4499 Lubbock TX 79409

JONES, J(OHN) L(LOYD), JR, b Henry, Ill, June 5, 18; m 43; c 3. ELECTRICAL ENGINEERING. *Educ:* Univ Ill, BS, 40, MS, 41; Univ Md, MS, 49, PhD, 63. *Prof Exp:* Physicist, US Naval Ord Lab, 42-63; assoc prof elec eng, Bradley Univ, 63-77; RETIRED. *Mem:* Acoust Soc Am; Am Soc Eng Educ. *Res:* Acoustics; circuit theory; electromagnetic theory; shock and vibration. *Mailing Add:* 1110 Warren St Henry IL 61537

JONES, J P, b Los Angeles, Calif, Sept 9, 41; m 64. MATHEMATICS. *Educ:* Univ Wash, BS, 63, MS, 66, PhD(math), 68. *Prof Exp:* Asst prof, 68-75, assoc prof, 75-, PROF MATH, UNIV CALGARY. *Honors & Awards:* Lester R Ford Award, Math Asn Am, 77. *Mem:* Am Math Soc; Math Asn Am; Asn Symbolic Logic; Can Math Cong. *Res:* Mathematical logic. *Mailing Add:* Dept Math & Statist Univ Calgary Calgary AB T2N 1N4 Can

JONES, JACK EARL, b Middleton, Ga, July 30, 25; m 46; c 1. AGRONOMY. *Educ:* Univ Ga, BS, 48, MS, 50; La State Univ, PhD, 61. *Prof Exp:* From asst prof to assoc prof, 50-68, PROF COTTON BREEDING & GENETICS, LA STATE UNIV, BATON ROUGE, 68- *Mem:* Am Soc Agron; Crop Sci Soc Am; Sigma Xi. *Res:* Cotton breeding for superior fiber properties; resistance to diseases and insects; genetics of quantitative characters of cotton; cotton production practices. *Mailing Add:* Dept Agron La State Univ Baton Rouge LA 70803

JONES, JACK EDENFIELD, b Jacksonville, Fla, Oct 24, 29; m 59; c 3. POULTRY SCIENCE. *Educ:* Univ Fla, BS, 51, MS, 64, PhD(physiol), 66. *Prof Exp:* Supvr farm mgt, Farmers Home Admin, 56-58; sanitarian, St Johns County Health Dept, 58-61; asst dir res, Coop Mills, 66-68; from asst prof to assoc prof poultry, Clemson Univ, 68-76, prof, 76-; RETIRED. *Mem:* Poultry Sci Asn. *Res:* Nutrition; physiological-environmental relationships with turkeys and game birds. *Mailing Add:* 501 Issaqueena Tr Clemson SC 29631

JONES, JAMES DARREN, b Oak Ridge, Tenn, June 7, 59; m 82; c 2. ACOUSTICS, VIBRATIONS. *Educ:* Tenn Technol Univ, BS, 81; Va Polytech Inst & State Univ, MS, 82, PhD(mech eng), 87. *Prof Exp:* Res asst mech eng, Va Polytech Inst & State Univ, 81-82, instr, 83-87; res assoc, Acoust & Noise Reduction Div, NASA Langley Res Ctr, 82-83; ASST PROF

MECH ENG, PURDUE UNIV, 87- *Concurrent Pos:* NSF res grant active vibration control, 88-91, presidential young investr award, 89-90, 90-91; res grants, various agencies, 88-91; consult, Douglas Aircraft Co, McDonnell Douglas Corp, 88, Artesian Indust, 88, 90-, Elgin Sweeper Co, 89. *Mem:* Acoust Soc Am; Am Inst Aeronaut & Astronaut; Am Soc Eng Educ; Am Soc Heating Vent Air-conditioning & Refrig Engrs; Am Soc Mech Engrs; Inst Noise Control Eng. *Res:* Acoustics, noise control, vibrations; active noise and vibration control; intelligent structures, distributed sensors and actuators; machinery noise, shell dynamics, structural/acoustics interactions; biomechanics, bionics, prosthetics; author of numerous publications on acoustics, noise control and vibrations. *Mailing Add:* 1077 Ray W Herrick Labs West Lafayette IN 47907-1077

JONES, JAMES DONALD, b Fond du Lac, Wis, Oct 5, 30; m 56; c 3. BIOCHEMISTRY. *Educ:* Ripon Col, AB, 52; Univ Wis, MS, 56, PhD(biochem), 58. *Prof Exp:* Asst prof animal nutrit, Iowa State Univ, 58-60; asst to staff sect biochem, 60-61, CONSULT, SECT CLIN CHEM, MAYO CLIN, 61- *Concurrent Pos:* Mem, Am Bd Clin Chem, 73; prof lab med & assoc prof biochem, Mayo Med Sch. *Mem:* Am Chem Soc; Am Inst Nutrit; Am Asn Clin Chem; fel Am Inst Chemists; fel Nat Acad Clin Biochem. *Res:* Nitrogen and electrolyte metabolism in animals; biochemistry of the young; metabolism of guanidines; inborn errors of metabolism. *Mailing Add:* Sect Clin Chem Mayo Clin 200 First St SW Rochester MN 55905

JONES, JAMES EDWARD, b Columbus, Ohio, June 5, 24; m 44; c 5. VETERINARY MEDICINE. *Educ:* Ohio State Univ, DVM, 50, MS, 75. *Prof Exp:* Gen pract vet med, 50-68; CLINICIAN, OHIO AGR RES & DEVELOP CTR, 68- *Mem:* Am Vet Med Asn; Sigma Xi. *Res:* Atrophic rhinitis in swine; epizootiology. *Mailing Add:* 17 Storms Rd Kettering OH 45429

JONES, JAMES HENRY, b Phoenix, Ariz, Oct 23, 52; m 90. COMPARATIVE PHYSIOLOGY, RESPIRATORY-EXERCISE PHYSIOLOGY. *Educ:* Univ Ariz, BS & BA, 74, MS, 76; Duke Univ, PhD(zool), 79; Colo State Univ, DVM, 83. *Prof Exp:* Lectr biol, Harvard Univ, 83-86; ASST PROF PHYSIOL, UNIV CALIF, DAVIS, 86- *Concurrent Pos:* Vis prof, Anat Inst, Univ Berne, Switz, 86 & Biosci Inst, Univ Sao Paulo, Brazil, 90; bd dirs, Comp Respiratory Soc, 89- *Honors & Awards:* Scholander Award, Am Physiol Soc, 86. *Mem:* Am Physiol Soc; Am Vet Med Asn; AAAS; Am Soc Zoologists. *Res:* Elucidate mechanisms limiting aerobic and anaerobic exercise performance in animals, especially birds and mammals; understand allometric (body-size) relationships between structure and function. *Mailing Add:* VM Physiological Sciences Univ Calif Davis CA 95616

JONES, JAMES HOLDEN, b Parkersburg, WVa, Feb 2, 28; m 51; c 2. ORGANIC CHEMISTRY. *Educ:* WVa Univ, BS, 50; Duke Univ, PhD(chem), 58. *Prof Exp:* Process chemist, Merck & Co, Inc, 58-60, res chemist, 60-64, res fel, 64-80, SR RES FEL, MERCK SHARP & DOHME RES LABS, 80-, SR INVESTR, 84- *Mem:* Am Chem Soc; Sigma Xi; AAAS. *Res:* Diuretics; central nervous system. *Mailing Add:* 6036 Cannon Hill Rd Ft Washington PA 19034

JONES, JAMES JORDAN, b Palo Alto, Calif, Nov 19, 38; div; c 2. ATMOSPHERIC ELECTRICITY. *Educ:* Stanford Univ, BS, 60; Univ Ariz, MS, 63, PhD(physics), 69. *Prof Exp:* Res assoc cosmic ray particle physics, Univ Chicago, 69-72; proj assoc, Univ Wis, 72-73; res physicist, Univ Ariz, 73-81; RES PHYSICIST ATMOSPHERIC ELEC & SR RES ASSOC PHYSICS, NMEX INST MINING & TECHNOL, 81- *Mem:* Am Phys Soc; Am Geophys Union; fel AAAS. *Res:* Investigation of the electrical structure of convective clouds by means of electric field meters mounted on research aircraft; testing of models of electric charge distribution in study clouds by direct comparison with measured electric fields. *Mailing Add:* Physics Dept NMex Inst Mining & Technol Socorro NM 87801

JONES, JAMES L, electronics, communications; deceased, see previous edition for last biography

JONES, JAMES OGDEN, b Punkin Ctr, Electra, Tex; m; c 2. PALEONTOLOGY. *Educ:* Midwestern State Univ, BS, 62; Baylor Univ, MS, 66; Univ Iowa, PhD(geol), 71. *Prof Exp:* Geologist, Shell Oil Co, 59-60; lab asst geol, Midwestern State Univ, 60-62; teaching asst, Baylor Univ, 62-64 & Univ Iowa, 64-68; geologist, Texaco Inc, 66; asst prof, Univ Southern Miss, 71; from asst prof to assoc prof geol, Southern Ark Univ, 71-74, head dept, 71-77; vis asst prof, 77-78, asst prof & prog coordr, 78-82, ASSOC PROF GEOL, UNIV TEX, SAN ANTONIO, 84- *Concurrent Pos:* US Army Air Defense Artillery; consult, oil, gas, water & environ, 71- *Mem:* Am Asn Petrol Geologists; Soc Sedimentary Geol; Nat Asn Geol Teachers; fel Geol Soc Am; Sigma Xi; Am Geophys Union; Int Asn Sedimentologists; Am Inst Prof Geologists. *Res:* Sedimentology and stratigraphy of Lower Permian shelf deposits of North Texas; cretaceous stratigraphyh and sedimentology of Texas and Mexico; paleontology. *Mailing Add:* Geol Dept Univ Tex San Antonio TX 78249-0663

JONES, JAMES ROBERT, b Quicksand, Ky, Dec 8, 31; m 58; c 3. ANIMAL HUSBANDRY, NUTRITION. *Educ:* Univ Ky, BS, 53, MS, 57; Cornell Univ, PhD(animal husb), 61. *Prof Exp:* Experimentalist animal husb, Cornell Univ, 61-64; EXTEN SPECIALIST, NC STATE UNIV, 64-, HEAD, SWINE EXTEN, 80- *Mem:* Am Soc Animal Sci; Am Registry Prof Animal Scientists. *Res:* Swine nutrition. *Mailing Add:* Dept Animal Sci NC State Univ Raleigh NC 27695-7621

JONES, JANICE LORRAINE, b Takoma Park, Md, Mar 10, 43; m 67; c 2. BIOPHYSICS, CELL PHYSIOLOGY. *Educ:* St Bonaventure Univ, BS, 65; Johns Hopkins Univ, PhD(biophys), 70. *Prof Exp:* Asst prof med technol, Univ Vt, 70-74, res assoc, Dept Med, 78; asst prof, 78-85, ASSOC PROF PHYSIOL, SCH MED, CASE WESTERN RESERVE UNIV, 85- *Concurrent Pos:* Mem comt interdisciplinary grad prog cell biol, Univ Vt, 71-; prin investr, NIH, defibrillator waveshape optimization, 79-86 & defibrillator induced dysfunction, 81-87; consult, Physiocontrol Corp, Redmond, Wash & Intermedics, Inc, Freeport, Tex. *Mem:* Biophys Soc; Int Soc Heart Res; Am Physiol Soc. *Res:* Cardiac physiology; physiology of cardiac cells in tissue culture; electrically induced myocardial damage. *Mailing Add:* Dept Physiol Georgetown Univ 3900 Reservoir Rd NW Washington DC 20007

JONES, JEANETTE, b Ft Valley, Ga, Sept 19, 50. MEDICAL MYCOLOGY, HISTOTECHNIQUES. *Educ:* Ft Valley State Col, BSc, 72; Ohio State Univ, MSc, 73, PhD(bot, med mycol), 76. *Prof Exp:* Instr biol, Ft Valley State Col, 72; univ fel, Ohio State Univ, 72-73, grad teaching assoc, 73-75; from asst prof to assoc prof, 76-85, MEM GRAD FAC, ALA A&M UNIV, 76-, PROF BIOL, 86- *Concurrent Pos:* Res apprenticeship organic chem, Forestry Exp Lab, Macon, Ga, 72; consult, Ft Valley State Col, Ft Valley Ga, 76, Northeast Ala State Jr Col, Riville, Ala, 77, NIH, 78-81 & 83, NSF, 79-80, Nat Adv Coun for Sixteen Insts Health Sci Consortium, NC, 79-81 & Southern Asn Cols & Schs Reaffirmation Comt, 82; prin investr, Grad Traineeships, NSF, 79 & Biomed Res Training Prog, NIH, 80; adj prof, Sch Pharm, Fla A&M Univ, 85. *Honors & Awards:* Honors Award, NASA, 85. *Mem:* Med Mycol Soc Am; Sigma Xi; Int Soc Human & Animal Mycosis; Am Soc Microbiol; Am Soc Allied Health Professionals; Mycol Soc Am; AAAS; Med Mycologists Am. *Res:* Isolation and control of growth of pathogenic fungi; nutrition, growth and morphogenesis of pathogenic fungi. *Mailing Add:* Res Admin Ala A&M Univ PO Box 411 Normal AL 35762

JONES, JENNINGS HINCH, b Petrolia, Pa, Aug 19, 13; m 40; c 2. ORGANIC CHEMISTRY. *Educ:* Pa State Col, BS, 34, MS, 37, PhD(chem), 41. *Prof Exp:* Chemist, Org Labs, Pa Coal Prod Co, 34-36; asst petrol ref, 41-44, from instr to assoc prof chem 44-52, assoc res prof, 53-64, prof chem, 64-69, EMER PROF CHEM ENG, PA STATE UNIV, 69- *Mem:* Am Chem Soc; Am Inst Chem Eng; Combustion Inst; Sigma Xi. *Res:* Petroleum chemistry; identification and separation of the products from the vapor phase oxidation of normal heptane; oxidation of organic chemicals, hydrocarbons and petroleum fractions; chemical behavior of hydroperoxides and perfluoroacids; sulfonates from oxidized paraffins. *Mailing Add:* Dept Chem Eng Pa State Univ University Park PA 16802

JONES, JEROLD W, b Salt Lake City, Utah, July 6, 37; m 61; c 5. THERMAL SYSTEMS, FLUID SYSTEMS. *Educ:* Univ Utah, BSME, 62, PhD(mech eng), 70; Stanford Univ, MS, 65. *Prof Exp:* Res scientist heat transfer, Ames Res Ctr, NASA, 62-66; fel, Ohio State Univ, 69-70, asst prof mech eng, 70-73; asst prof arch eng, 73-76, assoc prof mech eng, 76-83, PROF MECH ENG, UNIV TEX, AUSTIN, 83- *Concurrent Pos:* Asst dir & prin investr, Ctr Energy Studies, Univ Tex, Austin, 75-84; mem, Steering Comt Energy Conserv Bldg, Nat Res Coun, 79-80. *Mem:* Am Soc Heating, Refrig & Air-Conditioning Engrs; Am Soc Mech Engrs. *Res:* Heat transfer and thermodynamics with particular applications in systems modeling; design and analysis for improving energy use efficiency of buildings and heating and air conditioning equipment. *Mailing Add:* Dept Mech Eng Eng Teaching Ctr 5160 Univ Tex Austin TX 78712

JONES, JERRY LATHAM, b St Louis, Mo, Oct 20, 46; m 73; c 2. CHEMICAL PROCESS, PRODUCT DEVELOPMENT. *Educ:* Cornell Univ, BS, 68, ME, 69; Stanford Univ, MS, 76. *Prof Exp:* Pilot plants supvr, Monsanto Biodize Systs, 69-71; eng serv mgr, Monsanto Envirochem, 71-73; environ engr, SRI Int, 73-75, sr chem engr, 75-76, mgr environ control group, 76-78, dir environ & biochem eng, 78-82, dir chem eng lab, 82-89, DIR CHEM ENG, DEVELOP CTR, SRI INT, 90- *Mem:* Am Inst Chem Engrs; Am Chem Soc; Water Pollution Control Fedn; Soc Indust Microbiol; Parenteral Drug Asn. *Res:* Manufacturing process development and evaluation; pharmaceuticals and specialty chemical product development; bioprocesses; pollution control technologies and thermal processes; separations technology. *Mailing Add:* Chem Eng Develop Ctr 333 Ravenswood Ave Menlo Park CA 94025

JONES, JERRY LYNN, b Grandfield, Okla, Mar 28, 33; m 54; c 3. ANALYTICAL CHEMISTRY, EDUCATIONAL ADMINISTRATION. *Educ:* Okla State Univ, AB, 57, MS, 60; Univ Ark, PhD(chem), 63. *Prof Exp:* Res asst chem, Puget Sound Pulp & Timber, 56; res partic, Oak Ridge Nat Lab, 64; asst prof, Tex A&M Univ, 62-68; assoc prof, 68-72, coordr acad grants & contracts, 73-78, interim dean res & grad sch, 76-77, PROF CHEM, CENT WASH UNIV, 72-, SPEC ASST TO PRES, 79- *Concurrent Pos:* Assoc ed sci & technol, USA Today, 75- *Mem:* NY Acad Sci; Am Chem Soc. *Res:* Electroanalytical methods; trace metal analysis; atomic absorption spectroscopy; computers in chemistry; electrode reactions; environmental analysis; research design and evaluation. *Mailing Add:* Off Pres Cent Wash Univ Ellensburg WA 98926

JONES, JESS HAROLD, b Melville, La, Mar 30, 35; m 58; c 4. VIBRATION, STATISTICAL ANALYSIS OF DYNAMIC DATA. *Educ:* La State Univ, BS, 58. *Prof Exp:* Engr, Brown Eng Co, 61-64; sr engr, 64; aerospace engr, 64-66, chief, Acoust Sect, 66-72, TEAM LEADER, UNSTEADY FLOW TEAM, ENVIRON BR, GEORGE C MARSHALL SPACE FLIGHT CTR, NASA, 72- *Mem:* Acoust Soc Am; Am Soc Mech Engrs. *Res:* Theoretical and experimental investigations of the basic noise generation mechanisms of rocket exhaust flows and fluctuating pressure fields associated with space vehicles; analysis of random processes; fluid mechanics; wave propagation; sonic boom analysis; structural dynamics; rotating machinery; turbomachinery analysis; diagnostic analysis of dynamic data; fast fourier transforms analysis; ignition overpressure analysis and testing. *Mailing Add:* 707 Fagan Springs Dr SE Huntsville AL 35801

JONES, JESSE W, b Troup, Tex, Jan 16, 31; m 55; c 5. CHEMISTRY. *Educ:* Tex Col, BS, 54; NMex Highlands Univ, MS, 56; Ariz State Univ, PhD(org chem), 63. *Prof Exp:* Asst chem, NMex Highlands Univ, 54-55; asst biochem, Univ Utah, 55-56; asst prof chem, Tex Col, 56-58; res assoc, Ariz State Univ, 58-63; prof, Tex Col, 63-67, head dept, 63-64, head div natural sci, 64-67; PROF CHEM, BISHOP COL, 67- *Concurrent Pos:* Nat Inst Gen Med Sci & Welch Found grants, 63-65. *Mem:* AAAS; Am Chem Soc. *Res:* Synthesis, mechanism of action and biochemical studies of certain nitrogen heterocycles. *Mailing Add:* Dept Chem Baylor Univ Box 7348 Waco TX 76798

JONES, JIMMY BARTHEL, b Selmer, Tenn, Dec 1, 33; m 54; c 3. INTERNAL MEDICINE, LABORATORY ANIMAL MEDICINE. *Educ:* Univ Tenn, Martin, BS, 56; Univ Ill, BS, 61, DVM, 63. *Prof Exp:* Soil conservationist, USDA, 56-59, dist vet, Animal Health Div, 63-64; pvt practice, 64-67; vet biol div, Oak Ridge Nat Lab, 67-68; vet dir, Col Liberal Arts Animal Facil, Col Vet Med & Head, Dept Environ Pract, Univ Tenn, 68-73, dir animal facil, Mem Res Ctr, 68-76, from assoc prof to prof, Dept Urban Pract, 76-88; AT GRAD STUDIES BLDG, UNIV GA, ATHENS, 88- *Concurrent Pos:* Vet-consult, Biol Div, Oak Ridge Nat Lab, 68-69. *Mem:* Am Vet Med Asn; Int Soc Exp Hematol; Soc Exp Biol & Med. *Res:* Internal diseases of animals; administration and design of animal facilities as they interact with research projects; canine cyclic neutropenia; mechanisms of hematologic changes and management of affected animals. *Mailing Add:* Off VPres Grad Studies Bldg Univ Ga Athens GA 30602

JONES, JOE MAXEY, b Herpel, Ark, Mar 20, 42. IMMUNOPATHOLOGY. *Educ:* Wichita State Univ, BS, 64, MS, 66; Univ NC, Chapel Hill, PhD(immunol), 70. *Prof Exp:* Fel immunopath, Scripps Clin Res Found, 70-73, assoc, 73-77; head immunol & immunochem, Nat Ctr Toxicol Res, 77-80; ASSOC PROF IMMUNOL, UNIV ARK MED SCI, 83- *Mem:* Am Asn Immunologists; AAAS; Am Asn Pathologists. *Res:* Tumor immunology; genetic control of immune responses. *Mailing Add:* Univ Ark Med Sci 4301 W Markham Slot 517 Little Rock AR 72205

JONES, JOHN, JR, b Antioch, Tenn, Nov 19, 17; m 49; c 1. MATHEMATICS. *Educ:* Peabody Col, MA, 46; George Washington Univ, PhD(math), 70. *Prof Exp:* Assoc prof math, Southern Miss Univ, 46-55, head dept, 50-55; instr, Univ Tenn, 55-57; assoc prof, USAF Inst Technol, 57-60; mathematician, Hq, USAF, 60-64; asst prof lectr, George Washington Univ, 64-66; prof lectr, Am Univ, 66-68; assoc prof, 68-70, PROF MATH, AIR FORCE INST TECHNOL, 70- *Mem:* Am Math Soc; Math Asn Am. *Res:* Matrix theory; differential equations; functional analysis. *Mailing Add:* 2101 Matrena Dr Dayton OH 45431

JONES, JOHN A(RTHUR), b Port Chester, NY, Apr 6, 32; c 1. ENVIRONMENTAL SCIENCES, AQUATIC ECOLOGY. *Educ:* Univ Ill, Urbana, BS, 54; Tex A&M Univ, MS, 60; Univ Miami, PhD(marine sci), 68. *Prof Exp:* Res assoc marine biol, Univ Miami, 64-67; acting dir, Lake Erie Environ Studies & adj asst prof biol, State Univ NY Col, Fredonia, 68-70; assoc prof environ technol, 70-80, PROF ENVIRON TECHNOL & NATURAL SCI, MIAMI-DADE COMMUNITY COL, 80- *Concurrent Pos:* Seminar assoc, Columbia Univ, 68-; consult, 71- *Mem:* AAAS; Am Geophys Union; Am Soc Limnol & Oceanog. *Res:* Environmental measurements; graphic and statistical analysis of enviromental data, especially in aquatic enviroments; comparative properties of environmental fluids. *Mailing Add:* Miami-Dade Community Col 11011 SW 104th St Miami FL 33176

JONES, JOHN ACKLAND, b Alexandria, Va, Nov 6, 34; m 60; c 1. ENTOMOLOGY. *Educ:* Univ of the South, BS, 56; Univ Va, MS, 63; Iowa State Univ, PhD(entom), 73. *Prof Exp:* Instr biol, Univ of the South, 57-58; med lab technician histol, US Army Med Serv Corps, 58-60; assoc prof biol, Parsons Col, 63-66; state entomologist regulatory, Nebr Dept Agr, 73-78; ASSOC PROF ENTOM, UNIV NEBR, LINCOLN, 78- *Mem:* Entom Soc Am; Sigma Xi. *Res:* Insect morphology and development; insect pests of shelter belts; horticultural pests. *Mailing Add:* Dept Entom Univ Nebr Lincoln NE 68583-0816

JONES, JOHN BRYAN, b Colwyn Bay, NWales, Dec 11, 34; m 62; c 2. ORGANIC CHEMISTRY, ORGANIC BIOCHEMISTRY. *Educ:* Univ Wales, BSc, 55, PhD(chem), 58; Oxford Univ, DPhil(chem), 60. *Prof Exp:* Fel org chem, Mass Inst Technol, 60-61; NIH res fel, Calif Inst Technol, 61-62; Imp Chem Indust fel, Oxford Univ, 62-63; from asst prof to assoc prof, 63-74, PROF ORG CHEM, UNIV TORONTO, 74- *Mem:* Fel Am Chem Soc; Chem Inst Can. *Res:* Organic chemical applications of enzymes; immobilized enzymes. *Mailing Add:* Dept Chem Univ Toronto 80 Saint George St Toronto ON M5S 1A1 Can

JONES, JOHN DEWI, b Carmarthen, Wales, May 3, 26; m 55; c 4. PLANT BIOCHEMISTRY. *Educ:* Univ Wales, BSc, 46, MSc & PhD, 54. *Prof Exp:* Asst biochem, Univ Col Wales, Aberystwyth, 47 & 50-51, asst agr chem, Bangor, 51-54; sci off & chemist, Plant Path Lab, Ministry Agr, Fisheries & Food, Harpenden, Eng, 54-55; res assoc biol, Queen's Univ, Ont, 55-59; res plant biochemist, Ditton Lab, Agr Res Coun, Maidstone, Eng, 60-65; plant biochemist & head sect storage res fruits & veg, 66-70, res scientist, 70-71, plant biochemist, 71-73, SR RES SCIENTIST, FOOD RES INST, CAN DEPT AGR, 73- *Concurrent Pos:* Chemist, Mauritius-Seychelles Fisheries Res Surv, London, Eng, 48-50; fel, Nat Res Coun Can, 57-59. *Mem:* The Chem Soc; Fel Royal Inst Chem; fel Inst Food Sci & Technol UK. *Res:* Chemical and physical methods of removing toxic substances from cruciferous oil seeds; chemistry of glucosinolates, myrosinases, vegetable protein research and development; food ingredients; nutrition of plant proteins; toxicology of oilseeds; evaluation of novel sources of food proteins. *Mailing Add:* Food Res Inst Cent Exp Farm Can Agr Ottawa ON K1A 0C6 Can

JONES, JOHN EVAN, b Mt Pleasant, Utah, Oct 29, 30; m 54; c 3. INTERNAL MEDICINE, ENDOCRINOLOGY. *Educ:* Univ Utah, BS, 52, MD, 55; Am Bd Internal Med, dipl, cert endocrinol & metab, 73. *Prof Exp:* Dir USPHS trainee endocrinol, Univ Minn Hosps, 59; dir USPHS trainee, 60-61, from instr to asst prof med, 61-63, from asst prof to assoc prof endocrinol, 63-70, chmn div metab-endocrinol, 67-74, PROF MED & ENDOCRINOL, SCH MED, WVA UNIV, 70-, DEAN SCH MED, 74- *Mem:* Fel Am Col Physicians; Endocrine Soc; Am Fedn Clin Res; Am Soc Clin Nutrit. *Res:* Mineral metabolism; thyroid metabolism; adrenal hormone metabolism. *Mailing Add:* Dept Med WVa Univ Sch Med 1157 Basic Sci Bldg Morgantown WV 26506

JONES, JOHN F(REDERICK), b Scranton, Pa, Aug 19, 32; m 62; c 3. CHEMICAL ENGINEERING. *Educ:* Pa State Univ, BS, 54; Univ Del, MS, 56; Univ Colo, PhD(chem eng), 60. *Prof Exp:* Chem engr, Esso Res & Eng Co, 56-58; instr chem eng, Univ Colo, 58-60; res chem engr, FMC Corp, 60-63, sr res chem engr, 63-68, asst mgr, Proj COED, 68-72, mgr, 72-74, dir, coal & coke technol, 74-75, bus venture & tech mgr, Philadelphia, 75-77, dir res & develop, Indust Chem Group, 77-87; CONSULT 87- *Mem:* Am Chem Soc; Am Inst Chem Engrs; Indust Res Inst; AAAS; Sigma Xi. *Res:* Petroleum refining; carbonization; gasification and liquefaction of coal; sewage and water treatment; industrial chemicals. *Mailing Add:* Maple St PO Box 116 Stowe VT 05672

JONES, JOHN PAUL, b Warren, Ohio, Dec 10, 24; m 50; c 3. PLANT PATHOLOGY. *Educ:* Ohio Univ, BS, 50; Univ Nebr, MA, 53, PhD, 56. *Prof Exp:* Plant pathologist, Delta Exp Sta, Agr Res Serv, 55-60, PROF PLANT PATH, UNIV ARK, FAYETTEVILLE, 60- *Concurrent Pos:* Plant pathologist, Arab Repub Egypt, 81- *Mem:* Am Phytopath Soc. *Res:* Phytopathology; diseases of field crops; etiology and control of cereal crops diseases. *Mailing Add:* Dept Plant Path Univ Ark Fayetteville AR 72701

JONES, JOHN PAUL, b Stockdale, Ohio, Feb 24, 32; m 61; c 3. VEGETABLE PLANT PATHOLOGY. *Educ:* Ohio State Univ, BS, 53, MS, 55, PhD(plant path), 58. *Prof Exp:* Plant path asst, Ohio Agr Exp Sta, 54-58; from asst prof to assoc prof, 58-69, PROF PLANT PATH, GULF COAST RES & EDUC CTR, UNIV FLA, 72- *Concurrent Pos:* Vis prof, Int Rice Res Inst, Philippines, 80-81. *Mem:* Am Phytopath Soc; Sigma Xi. *Res:* Nature and control of vegetable diseases; biology of plant pathogens. *Mailing Add:* Gulf Coast Res & Educ Ctr Univ Fla 5007 60th St E Bradenton FL 34203

JONES, JOHN PAUL, b Takoma Park, Md, Nov 17, 40; m 66; c 1. MATHEMATICS. *Educ:* Alderson-Broaddus Col, BS, 62; WVa Univ, MA, 64; Pa State Univ, DEd(math), 71. *Prof Exp:* Instr math, Allegheny Col, 64-67; asst prof, 71-74, ASSOC PROF MATH & HEAD DEPT, FROSTBURG STATE COL, 74- *Mem:* Am Math Soc; Math Asn Am. *Res:* Algebra-groups and rings. *Mailing Add:* Dept Math Frostburg State Col Frostburg MD 21532

JONES, JOHN R, medicine, anesthesiology, for more information see previous edition

JONES, JOHN RICHARD, b Bremerton, Wash, Aug 23, 47; m 69; c 1. LIMNOLOGY. *Educ:* Western Wash State Col, BA, 69; Iowa State Univ, MS, 72, PhD(limnol), 74. *Prof Exp:* Fel, Iowa State Univ, 74-75; asst prof, 75-80, ASSOC PROF LIMNOL, UNIV MO-COLUMBIA, 80- *Mem:* Am Soc Limnol & Oceanog; Ecol Soc Am; Am Fisheries Soc; Sigma Xi. *Res:* Eutrophication process in lakes and reservoirs; attention to phosphorous and algal biomass. *Mailing Add:* Sch Forestry Fish & Wildlife Univ Mo 112 Stephens Hall Columbia MO 65201

JONES, JOHN TAYLOR, b Salt Lake City, Utah, Jan 4, 32; m 53; c 5. CERAMICS ENGINEERING, METALLURGY. *Educ:* Univ Utah, BS, 57, PhD(ceramic eng, metall), 65. *Prof Exp:* Res engr, Coors Porcelain Co, Colo, 57-60, prod supt, 60-61, develop engr, 61-62; asst dir res, Vesuvius Crucible Co, Pa, 65-66; assoc prof ceramic eng, Iowa State Univ, 66-74; process develop mgr, Interspace Corp, Calif, 74; res & develop mgr, Pfaltzgraff Co, Pa, 74-78; res & develop mgr, 78-80, VPRES RES & DEVELOP, LENOX MFG DIV, LENOX, INC, NJ, 80- *Mem:* Nat Inst Ceramic Engrs; fel Am Ceramic Soc; Ceramic Educ Coun; Brit Ceramic Soc; Can Ceramic Soc; Am Soc Testing & Mat; Soc Glass Decorators; Int Precious Metals Inst. *Res:* ceramic whitewares. *Mailing Add:* Lenox Mfg Div Tech Ctr 65 Fire Rd Absecon NJ 08201

JONES, JOHN VERRIER, b Shrewsbury, Eng, Nov 29, 30. RHEUMATOLOGY, IMMUNOLOGY. *Educ:* Univ Oxford, MD, 55. *Prof Exp:* HEAD RHEUMATOLOGY, DALHOUSIE UNIV, 82- *Mailing Add:* Dept Rheumatology Dis Halifax Civic Hosp 5938 Univ Halifax NS B3H 1V9 Can

JONES, JOHNNYE M, b Henderson, Tex, Apr 3, 43. ELECTRON MICROSCOPY, MYCOLOGY & PLANT PHYSIOLOGY. *Educ:* Prarie View A&M Univ, BS, 65; Atlanta Univ, MA, 70, PhD(bot & mycol), 79. *Prof Exp:* Instr biol & math, Carthage Public Schs, 65-67, Chicago Public Schs, 67-69, Morgan State Univ, Baltimore, Md, 70-74 & Mercer Univ, Atlanta, Ga, 75-79; ASSOC PROF BIOL, HAMPTON UNIV, 79- *Concurrent Pos:* Fac fel, Nat Inst Gen Med Sci, NIH, 74; res assoc, Brookhaven Nat Lab, 79-81; dir, Minority Access Res Careers Hons Prog. *Mem:* Bot Soc Am; Mycol Soc Am; Nat Minority Health Affairs Asn; Electron Micros Soc Am; Nat Assoc Minority Med Educr. *Res:* Ultrastructural studies on certain species of fungi Ascomycetes and Oomycetes, especially developmental and physiological aspects. *Mailing Add:* Dept Biol Sci Hampton Inst Hampton VA 23668

JONES, JOIE PIERCE, b Brownwood, Tex, Mar 4, 41; m 65. MEDICAL ULTRASONICS, ACOUSTICAL MICROSCOPY. *Educ:* Univ Tex, Austin, BA, 63, MS, 65; Brown Univ, PhD(physics), 70. *Prof Exp:* Sr scientist, Bolt Beraner & Newman, 70-75; assoc prof med physics, Case, Western Reserve Univ, 75-77; PROF RADIOL SCI, UNIV CALIF, IRVINE, 77- *Concurrent Pos:* Consult, var pvt co & govt agencies, 71-; reviewer, NSF & NIH, 75-; mem, Presendents Sci & Technol adv comt, 76-79; vis prof, Kings Col, London, 82 & 89. *Mem:* Acoust Soc Am; Am Inst Ultrasound Med; Am Asn Physicists Med; Inst Elec & Electronics Engrs. *Res:* Medical ultrasonics; ultrasonic tissue characterization; medical imaging; acoustical microscopy. *Mailing Add:* 2094 San Remo Laguna Beach CA 92651

JONES, JOYCE HOWELL, b Roanoke, Va, May 4, 44; m 68. EMBRYOLOGY, ANIMAL SCIENCE & NUTRITION. *Educ:* Va Polytech Inst & State Univ, BS, 66, MS, 71, PhD(genetics), 74. *Prof Exp:* Jr high sch phys sci teacher, 69-70; asst prof biol, Ferrum Col, 74-77; EXTEN SPECIALIST & ASST PROF POULTRY, VA POLYTECH INST & STATE

UNIV, 77- *Mem:* Poultry Sci Asn; Am Genetic Asn; AAAS; Sigma Xi. *Res:* Genetical, physiological and behavioral relationships in avian and mammalian pre and postnatal development. *Mailing Add:* 111 Cherry St Happy Valley E Abingdon VA 24210

JONES, KAY H, b Spokane, Wash, Jan 13, 35; US citizen; c 6. ENVIRONMENTAL HEALTH, TOXICOLOGY. *Educ:* Univ Washington, BS, 56; Univ Calif, Berkeley, MS, 61, PhD(sanit eng), 68. *Prof Exp:* Mem staff, Nat Air Pollution Control Admin, Dept Health, Educ & Welfare, 67-70; mem, Off Air Prog, Environ Protection Agency, 70-74; consult, WHO, 74-75; mem, Coun Environ Qual, Exec Off Pres, 75-79; prof environ eng, Drexel Univ, Pa, 79-81; vpres, Roy Weston Inc, 81-90; PRES ZEPHRY CONSULT, 90. *Honors & Awards:* State-of-the-Art Civil Eng Award, Am Soc Civil Engrs, 75. *Mem:* Am Soc Civil Engrs; Air Pollution Central Asn. *Res:* Ambient air quality data analysis; air pollution impact analysis; population exposure modeling; environmental epidemiology; environmental toxicology; industrial hygiene; air pollution central engineering; risk assessment. *Mailing Add:* Zephry Consult Suite 18 2600 Fairview Ave Seattle WA 98102

JONES, KEITH WARLOW, b Lincoln, Nebr, Aug 30, 28; m 54; c 3. EXPERIMENTAL ATOMIC PHYSICS, APPLIED PHYSICS. *Educ:* Princeton Univ, AB, 50; Univ Wis, MS, 51, PhD(physics), 55. *Prof Exp:* Asst prof physics, Univ NC, 54-55; res assoc, Columbia Univ, 55-58; from asst prof to assoc prof, Ohio State Univ, 58-63; from assoc physicist to physicist, 63-75, SR PHYSICIST, BROOKHAVEN NAT LAB, 75-, DIV HEAD, DEPT APPL SCI, 84- *Concurrent Pos:* Group leader, Dept Physics, Brookhaven Nat Lab, 76-84. *Mem:* Fel Am Phys Soc. *Res:* Beam foil spectroscopy; heavy ion-atom collisions; trace element and isotope identification techniques; microbeam methods and applications; synchrotron radiation experiments. *Mailing Add:* Dept Appl Sci Bldg 815 Brookhaven Nat Lab Upton NY 11973

JONES, KENNETH CHARLES, b San Pedro, Calif, July 20, 34; m 56; c 2. ALGOLOGY, MOLECULAR GENETICS. *Educ:* Univ Calif, Los Angeles, BA, 57, MA, 62, PhD(plant sci), 65. *Prof Exp:* From asst prof to assoc prof, 64-71, actg dean grad studies & res, 77-78, dept chmn, 79-80, PROF BIOL, CALIF STATE UNIV, NORTHRIDGE, 71- *Mem:* AAAS; Sigma Xi. *Res:* Chemical regulation of plant growth; genetic control mechanisms; physiology of germination of Chara. *Mailing Add:* Dept Biol Calif State Univ Northridge CA 91330

JONES, KENNETH LESTER, b Keweenaw Bay, Mich, Dec 3, 05; m 29; c 2. SOIL MICROBIOLOGY, HISTORY. *Educ:* Syracuse Univ, AB, 28; Univ Mich, PhD(bot), 33. *Prof Exp:* Instr, 29-37, from asst prof to prof, 37-77, chmn dept, 50-63, EMER PROF BIOL SCI, UNIV MICH, ANN ARBOR, 77- *Concurrent Pos:* Researcher, Commercial Solvents Corp, 48-50. *Mem:* AAAS; Am Soc Microbiol; Bot Soc Am; Am Acad Microbiol; Sigma Xi. *Res:* Variation, morphology and distribution of streptomyces. *Mailing Add:* Dept Bot Univ Mich 401 W Oakbrook Dr No 205 Ann Arbor MI 48103

JONES, KENNETH WAYNE, b Decatur, Ill, Dec 22, 46; m 68; c 2. CLINICAL MICROBIOLOGY. *Educ:* Southern Conn State Col, BS, 70; Univ NC, MPH, 74, PhD(public health microbiol), 76. *Prof Exp:* Microbiologist, Conn Health Dept, Greenwich, 71-73; res asst, Centers Dis Control, 75-76; chief microbiologist, RI Dept Health, 76-; AT HEALTH LABS. *Mem:* Sigma Xi; Am Soc Microbiol; Am Public Health Asn. *Res:* Diagnostic procedures in clinical and public health microbiology; microbiological methods for monitoring environmental quality. *Mailing Add:* Health Labs 50 Orms St Providence RI 02904

JONES, KEVIN MCDILL, b Washington, DC, Dec 31, 55; m 83. UNDERGRADUATE TEACHING EXPERIMENTS IN OPTICS. *Educ:* Williams Col, BA, 77; Stanford Univ, PhD(physics), 84. *Prof Exp:* Postdoctoral fel, Hydrogen Maser Lab, 83-84, asst prof, 84-91, ASSOC PROF, DEPT PHYSICS, WILLIAMS COL, 91- *Concurrent Pos:* Consult, Lawrence Livermore Nat Lab, 87-88. *Mem:* Am Phys Soc; Sigma Xi; Optical Soc Am; Coun Undergrad Res. *Res:* Laser spectroscopy of atoms and molecules. *Mailing Add:* Physics Dept Williams Col Williamstown MA 01267

JONES, KEVIN SCOTT, b Gainesville, Fla, Feb 20, 58; m 83; c 2. SEMICONDUCTOR RESEARCH, TRANSMISSION ELECTRON MICROSCOPY STUDIES. *Educ:* Univ Fla, BS, 80; Univ Calif, Berkeley, MS, 85, PhD(mat sci eng), 87. *Prof Exp:* Tech proc engr, E I DuPont & Co, Wash Works Plant, 80-82; consult, TRW, Inc, 85-86; teaching & res asst, Univ Calif, Berkeley, 82-87, postdoctoral researcher, 87; ASST PROF SEMICONDUCTOR PROCESSING & STRUCT & CHARACTERIZATION MAT, DEPT MAT SCI & ENG, UNIV FLA, 87- *Concurrent Pos:* Co-organizer, Compound Semiconductor Growth, Processing & Devices 1990's, Japan/US Topical Conf, 87; organizer, meeting session electronic mat, 89, IX Int Conf Ion Implantation Technol, 92; NSF presidential young investr award, 90-95. *Mem:* Am Soc Metals; Electron Micros Soc Am; Mat Res Soc; Metall Soc; Electrochem Soc. *Res:* Processing and characterization of elemental and compound semiconductors; ion implantation; ion beam induced phase transformations; transmission electron microscopy. *Mailing Add:* Dept Mat Sci & Eng Univ Fla 214A Rhines Hall Gainesville FL 32611-2066

JONES, KIRKLAND LEE, b Amarillo, Tex, Oct 1, 41; m 64; c 2. POPULATION ECOLOGY, HERPETOLOGY. *Educ:* Baylor Univ, BA, 64; Univ NMex, MS, 70, PhD(biol), 74. *Prof Exp:* Asst prof biol, Southern Methodist Univ, 74-81; MEM STAFF, LOS ALAMOS TECH ASSOCS, 81- *Concurrent Pos:* Collabr, Nat Park Serv, 70-73; consult, Environ Protection Agency, 72-73; consult, 76- *Mem:* Sigma Xi; Ecol Soc Am; Am Soc Ichthyologists & Herpetologists. *Res:* Effects of competition on niche dimensionality and morphology; energetics of feeding strategies; endangered species. *Mailing Add:* 727 Viento Circle No A Santa Fe NM 87501

JONES, L(LEWELLYN) E(DWARD), b Montreal, Can, Mar 25, 10; m 38; c 2. HYDRAULIC ENGINEERING, NUMERO-GRAPHICAL METHODS. *Educ:* Univ Man, BScCE, 31, Univ Toronto, MASc, 33, PhD(hydraul), 41. *Prof Exp:* Jr engr, Can Pac Rwy Co, 29-30 & Man Prov Govt, 31-33; instr & lectr, appl physics, 36-44, from asst prof to prof mech eng, 44-75, EMER PROF MECH ENG, UNIV TORONTO, 75-, ENG ARCHIVIST & CUR, 70-, ASSOC, INST ENVIRON STUDIES, 71- *Concurrent Pos:* Hydraul engr, Hydro-Elec Power Comn Ont, 41-57; Ford Found res grant, 63; gen consult, 57- *Honors & Awards:* Sons of Martha Medal, 65; Queen's Silver Jubilee Medal, 77. *Mem:* Am Soc Civil Eng; Am Soc Mech Engrs; Royal Can Inst; fel Brit Inst Mech Engrs; fel Eng Inst Can. *Res:* Applied physics; optics; photography; metrology; fluid mechanics; water resources; applied mathematics; data processing and interpretation; computers and numerical methods; technical publication; engineering history; optimal interpretation of experimental data; memorial authorship and calligraphy. *Mailing Add:* 29 Prince George Dr Islington ON M9A 1X9 Can

JONES, LARRY HUDSON, b Dillon, SC, July 3, 48; m 88; c 1. GENETICS, MOLECULAR BIOLOGY. *Educ:* Wofford Col, BS, 70; Univ NC, Chapel Hill, PhD(bot), 76. *Prof Exp:* Res assoc biochem & microbiol, Cook Col, Rutgers Univ, 75-76; vis asst prof biol, Swarthmore Col, 76-77; from asst prof to assoc prof, 77-90, PROF BIOL, UNIV SOUTH, 90-, DEPT CHAIR, 88- *Concurrent Pos:* vis res assoc, USDA Res Ctr, Florence, SC, 84; vis assoc prof biol, Reed Col, Portland, Ore, 89-90. *Mem:* Am Soc Plant Physiologists; Sigma Xi; Int Soc Plant Molecular Biol; Genetics Soc Am; Am Genetic Asn. *Res:* Tissue culture; effects of methylation of RNA on biological systems; coordination of protein synthesis in chloroplasts and mitochondria; plant hormones; genetics. *Mailing Add:* Dept Biol Univ South Sewanee TN 37375

JONES, LARRY PHILIP, b Hamilton, Mont, Dec 11, 34; m 59; c 2. VETERINARY PATHOLOGY. *Educ:* Wash State Univ, BA, 57, DVM, 58; Am Col Vet Path, dipl, 68. *Prof Exp:* Res assoc path, Agr Res Lab, Univ Tenn, 58-60; asst prof vet path, Inst Trop Vet Med, 65-69; pathologist & head dept path, Tex Vet Med Diag Lab, 69-; AT BIOL DEPT, UNIV TEX, EL PASO. *Mem:* Am Vet Med Asn; Wildlife Dis Asn; Wildlife Soc; Sigma Xi. *Res:* Infectious diseases of domestic and wild ruminants. *Mailing Add:* Biol Sci Dept Univ Tex El Paso TX 79968

JONES, LARRY WARNER, b Huntington Co, Ind, Feb 14, 34; m 57; c 3. PLANT PHYSIOLOGY, ENVIRONMENTAL ENGINEERING. *Educ:* Univ Ariz, BS, 55, MS, 59; Univ Tex, PhD(bot), 64. *Prof Exp:* Res scientist, Res Inst Adv Studies, Div Martin Co, 64-65; from asst prof to assoc prof bot, 65-73, PROF BOT & PLANT PHYSIOL & GENETICS, UNIV TENN, KNOXVILLE, 73-, DIR APPL SCI DIV, HAZARDOUS WASTE RES & EDUC INST, 85- *Concurrent Pos:* Vis prof, Ore State Univ, 74-75; vpres, VeriTec Corp; consult, Waterways Exp Sta, US Army Corps Engrs, 76- *Mem:* Am Soc Plant Physiol; Am Pollution Control Asn; Water Pollution Control Fedn. *Res:* Immobilization and stabilization of hazardous wastes; environmental effects and monitoring; algal physiology; photosynthesis and hydrogen production. *Mailing Add:* Dept Bot Waste Mgt Inst Univ Knoxville TN 37996-1100

JONES, LAWRENCE RYMAN, b Terre Haute, Ind, Jan 8, 21; m 43; c 2. ANALYTICAL CHEMISTRY. *Educ:* Ind State Univ, BS, 46. *Prof Exp:* Res chemist, Com Solvents Corp, 43-75; res scientist, Int Mineral & Chem Corp, 75-86; RETIRED. *Concurrent Pos:* Chemist, St Anthony Hosp, 46-50. *Res:* Analytical method research. *Mailing Add:* 1219 Alamito St Rockport TX 78382

JONES, LAWRENCE WILLIAM, b Evanston, Ill, Nov 16, 25; m 50; c 3. HIGH ENERGY PHYSICS. *Educ:* Northwestern Univ, BS, 48, MS, 49; Univ Calif, PhD(physics), 52. *Prof Exp:* Asst, Univ Calif, 50-52; from instr to assoc prof, 52-63, PROF PHYSICS, UNIV MICH, ANN ARBOR, 63- *Concurrent Pos:* Physicist, Lawrence Radiation Lab, Univ Calif, 50-52; physicist, Midwestern Univs Res Asn, 56-57; consult, Space Tech Labs, Inc, & Thompson-Ramo-Wooldridge, Inc; Ford Found fel, Europ Orgn Nuclear Res, 61-62, Guggenheim Found fel, 65; vis prof, Westfield Col, London, 77 & Tata Inst, Bombay, 79; chmn, dept physics, Univ Mich, 82-87; physicist, SSC Cent Design Group, Lawrence Bereley Lab, 87; assoc, Europ Orgn Nuclear Res, 88- *Mem:* Am Phys Soc; AAAS; Int Asn Hydrogen Energy. *Res:* Strong interactions of elementary particles at high energies; cosmic ray physics at very high energies; hadron production of dilepions and prompt neutrinos; hydrogen energy systems; medical physics instrumentation; hadron production of charm mesons; electron positron interactions at high energies. *Mailing Add:* Dept Physics Univ Mich Ann Arbor MI 48109-1120

JONES, LEE BENNETT, b Memphis, Tenn, Mar 14, 38; m 64; c 2. ORGANIC CHEMISTRY. *Educ:* Wabash Col, BA, 60; Mass Inst Technol, PhD(org chem), 64. *Prof Exp:* NSF fel chem, Calif Inst Technol, 64; from asst prof to assoc prof, 64-72, asst head dept, 71-73, head dept, 73-77, dean grad col, 77-80, prof chem, 72-, PROVOST, GRAD COL & HEALTH SCI, UNIV ARIZ, 80- *Mem:* AAAS; Am Chem Soc; Royal Soc Chem. *Res:* Photochemistry; carbonium ion reactions; nucleophilic substitutions; isotope effects. *Mailing Add:* Dept Chem Univ Nebr 3835 Holdrege 106 Varnes Hall Lincoln NE 68583

JONES, LEEROY G(EORGE), b Flint, Mich, Aug 15, 29; m 53; c 4. ENVIRONMENTAL MEDICINE, INTERNAL MEDICINE. *Educ:* Mich State Col, AB, 52; Wayne State Univ, MD, 56; Johns Hopkins Univ, MPH, 84. *Prof Exp:* Intern med, Holy Cross Hosp, Salt Lake City, 56-57; US Army, 57-, resident, Detroit Receiving Hosp, 58-61, asst chief dept gastroenterol, Walter Reed Army Inst Res, 61-62, asst chief med res br, Hq, US Army Med Res & Develop Command, 62-64, chief outpatient serv & dept med, DeWitt Army Hosp, Ft Belvoir, 64-65; team chief trop med, US Army Med Res Team, Vietnam, 67-68, dir physiol lab, Inst Environ Med, 68-70, cmndg officer & sci-tech dir, US Army Res Inst Environ Med, 71-76, dir med res, 76-77, dep comdr, Med Res & Develop Command, Md, 77-80, comdr, Med Dept Activ, US Army, 80-82; HEALTH OFFICER, DORCHESTER

COUNTY HEALTH DEPT, 86- Concurrent Pos: Asst instr, Wayne State Univ, 58-61; fel, Harvard Med Sch, 65-67, res assoc, 69-70, lectr physiol, 73-; army liaison rep appl physiol study sect, NIH, 71-; panel on self regulation, Advan Res Proj Agency, Dept Defense, 71-; consult environ med, Surgeon Gen, USA, 71-76; resident pub health, Johns Hopkins Univ Sch Hyg & Pub Health, 85. Mem: AAAS; Am Physiol Soc; Asn Mil Surgeons US; Am Fedn Clin Res; fel Am Col Physicians. Res: Cardiovascular physiology; psychophysiology; environmental physiology; exercise physiology. Mailing Add: 206 Cambridge Landing Cambridge MD 21613

JONES, LEONARD CLIVE, engineering, physics, for more information see previous edition

JONES, LEONIDAS JOHN, b Warrenton, NC, May 17, 37; m 63; c 3. SOFTWARE SYSTEMS. Educ: Duke Univ, BS, 58; MS, 60, PhD(elec eng), 66. Prof Exp: Scientist, 66-77, adminr, 77-79, MGR, RES & DEVELOP CTR, GEN ELEC CO, 79- Concurrent Pos: Gen Elec rep, Conf Data Syst Lang, 71-77. Mem: Asn Comput Mach; Sigma Xi. Res: Computer-aided design and database systems for industrial automation. Mailing Add: 534 Devils Lane Ballston Spa NY 12020

JONES, LESLIE F, b Warner Robins, Ga, Aug 16, 65; m 86. CARDIOVASCULAR PHARMACOLOGY, CARDIOVASCULAR PHYSIOLOGY. Educ: Univ Ga, BS, 87, PhD(pharmacol), 90. Prof Exp: POSTDOCTORAL FEL PHARMACOL, UNIV IOWA, 90- Mem: Am Soc Pharmacol & Exp Therapeut; Soc Neurosci. Res: Cardiovascular physiology and pharmacology; central control of the coronary circulation and mechanisms of action of antihypertension agents and cocaine; central control of the cardiovascular system with emphasis on the coronary circulation. Mailing Add: Dept Pharmacol 2-272 BSB Univ Iowa Iowa City IA 52242

JONES, LESTER TYLER, b Des Moines, Iowa, Dec 5, 39; m 62; c 3. PHYSICAL CHEMISTRY, RESEARCH ADMINISTRATION. Educ: Univ Iowa, BS, 61; Wash State Univ, PhD(phys chem), 66. Prof Exp: Sr chemist, Cent Res Labs, 65-72, res specialist, 72-73, supvr, 73-74, mgr, 74-81, mgr, Life Sci Sector, 81-82, MGR, TECHNOL ASSESSMENT & UNIV RELS, CORP RES LABS, 3M CO, 82- Mem: Am Chem Soc; Asn Univ Technol Managers. Res: Corrosion of metals; nuclear quadruple resonance; charge transfer complexes; dye adsorption; controlled release; biomaterials; technology transfer. Mailing Add: 2215 S Shore Blvd St Paul MN 55110-3852

JONES, LEWIS HAMMOND, IV, b Cleveland, Ohio, Feb 26, 41. SEMICONDUCTOR PHYSICS. Educ: Ohio Wesleyan Univ, BA, 63; Univ Ill, MS, 65, PhD(physics), 71. Prof Exp: Vis scientist physics, Ctr Nuclear Energy, Saclay, France, 71-72 & Nat Lab, Frascati, Italy, 72-74; res assoc physics, Univ Md, College Park, 74-77; asst res physicist, Univ Calif, Irvine, 78-79; mem res staff, Fairchild Camera & Instrument Corp, Palo Alto, test eng staff, 85-87; sect head, Nat Semiconductor Corp, Santa Clara, Calif, 88-89; SR YIELD ENHANCEMENT ENGR, ADVAN MICRO DEVICES, SANTA CLARA, CALIF, 89- Mem: Am Phys Soc; Inst Elec & Electronics Engrs. Res: Semiconductor parametric test development; semiconductor characterization and modeling; programmable array logic devices. Mailing Add: 693 Madrone Ave Sunnyvale CA 94086

JONES, LEWIS WILLIAM, b Malad, Idaho, July 6, 06; m 28; c 1. BACTERIOLOGY. Educ: Utah State Univ, BS, 36, MS, 37; Stanford Univ, PhD(bact, physiol), 52. Prof Exp: Pub sch prin, Idaho, 26-34; from instr to prof, 37-75, actg head dept, 60-63, EMER PROF BACT, UTAH STATE UNIV, 75- Concurrent Pos: NSF fac fel, 59-60. Mem: AAAS; Am Soc Microbiol; Am Pub Health Asn. Res: Effects of temperature, alkali salts, insecticides and herbicides upon soil microorganisms; gas production in pasteurized dairy products; anaerobic metabolism; denitrification. Mailing Add: 320 North 100 W Mala ID 83252

JONES, LILY ANN, b Montevideo, Minn, July 6, 38; div; c 2. MICROBIAL GENETICS, MOLECULAR BIOLOGY. Educ: Univ Minn, BA, 60, MS, 63, PhD(microbiol), 64. Prof Exp: Instr, 64-70, asst prof microbial genetics, 70-76, ASST PROF IMMUNOL & MICROBIOL, WAYNE STATE UNIV, 76- Mem: AAAS; Am Soc Microbiol. Res: Genetics of Streptomyces; phylogeny of actinomycetes; bacterial resistance to antibiotics; life-cycle and structure of actinophage; bacteriophage classification & taxonomy. Mailing Add: Dept Immunol & Microbiol Wayne State Univ Med Sch Detroit MI 48202

JONES, LINCOLN D, b Los Angeles, Calif, Dec 4, 23; m 44; c 3. ELECTRICAL ENGINEERING. Educ: Univ Ariz, BS, 51, MS, 56; Stanford Univ, Engr, 64. Prof Exp: Instr elec eng, Univ Ariz, 51-54; asst prof, Calif State Polytech Col, 54-56; from asst prof to assoc prof, 56-65, PROF ELEC ENG, SAN JOSE STATE UNIV, 65- Mem: Inst Elec & Electronics Engrs; Am Soc Eng Educ; Soc Comput Simulation. Res: Finding system models for second order nonlinear systems that exhibit jump resonance. Mailing Add: Dept Elec Eng San Jose State Univ San Jose CA 95192

JONES, LLEWELLYN CLAIBORNE, JR, b Chester, Pa, Nov 4, 19; m 45; c 3. ANALYTICAL CHEMISTRY. Educ: Harvard Univ, BS, 43. Prof Exp: Res chemist, Houston Res Lab, Shell Oil Co, 43-44 & Wood River Res Lab, 44-46, group leader, 46-56, res chemist, Thorton Res Ctr, Shell Res, Ltd, Eng, 56-57, asst chief res physicist, Wood River Res Lab, 57-65, head analytic dept, Emeryville Res Ctr, Shell Develop Co, 65-69, head process develop dept, Res Ctr, Shell Berre, France, 69-70, head, Analytic Dept, Royal Dutch Shell Lab, Netherlands, 70-72, analytic mgr, 72-76, mgr loss Control-Logistics, 76-80; RETIRED. Res: Absorption spectroscopy; infrared and vacuum ultraviolet; ion exchange chromatography; instrumental methods of analysis. Mailing Add: 2754 Fontana Houston TX 77043

JONES, LLOYD GEORGE, b Hobart, La, Aug 6, 19. HORTICULTURE. Educ: La State Univ, BS, 49, MS, 50; Purdue Univ, PhD(hort), 53. Prof Exp: Asst, La State Univ, 46-49; asst, Purdue Univ, 50-53; asst horticulturist, 53-55, assoc prof, 56-61, PROF HORT, AGR EXP STA, LA STATE UNIV, BATON ROUGE, 62- Mem: AAAS; Am Soc Plant Physiol; Am Soc Hort Sci. Res: Plant nutrition; soil fertility. Mailing Add: PO Box 55 Watson LA 70786

JONES, LOIS MARILYN, b Berea, Ohio, Sept 6, 34. GEOCHEMISTRY, GEOLOGY. Educ: Ohio State Univ, BS, 55, MS, 59, PhD(geochem), 69. Prof Exp: Lab asst chem, Ohio State Univ, 53-55, asst, 55-59; res anal chem, Exp Sta, E I du Pont de Nemours & Co, Inc, 59-61; asst, Ohio State Univ, 61; lectr, Mem Univ Nfld, 61-63; res anal chemist, US Geol Surv, 63-66; res asst geochem, Ohio State Univ, 66-67, res assoc isotope geol, 69; asst prof geol, Univ Ga, 69-77; sr res scientist, Conoco, Inc, 77-82, res assoc, Petrol Res & Develop Dept, 82-90; RETIRED. Concurrent Pos: Prin investr, Tenn Copper Co grant, 69; proj leader, NSF grant, Inst Polar Studies, Ohio State Univ, & Univ Ga, 69-70. Mem: Am Asn Petrol Geologists; Am Geophys Union; Geochem Soc; Geol Soc Am; Int Asn Geochem & Cosmochem. Res: Isotope geochemistry in hydrocarbon exploration; rubidium-strontium geochronology; strontium isotopes as natural tracers; geochronology, geochemistry, and glacial history of the ice-free valleys and paleolimnology of the saline lakes, Antarctica. Mailing Add: Dept Geol Kans State Univ Manhattan KS 66506

JONES, LORELLA MARGARET, b Toronto, Ont, Feb 22, 43; US citizen. ELEMENTARY PARTICLE PHYSICS. Educ: Radcliffe Col, BA, 64; Calif Inst Technol, MSc, 66, PhD(physics), 68. Prof Exp: Instr & fel high energy theory, Calif Inst Technol, 67-68; from asst prof to assoc prof, 68-78, PROF PHYSICS, UNIV ILL, URBANA, 78- Concurrent Pos: Ann Horton vis res fel, Newnham Col, Univ Cambridge, 74-75; vis scientist, Ger El Synch & Orgn Europ Res Nuclear labs, 81-82. Mem: AAAS; Sigma Xi; Fel Am Phys Soc. Res: Phenomenological applications of high energy theory to strong interactions. Mailing Add: Dept Physics Univ Ill 1110 W Green St Urbana IL 61801

JONES, LOUISE HINRICHSEN, b Ames, Iowa, Dec 24, 30; m 52. APPLIED MATHEMATICS, COMPUTER SCIENCE. Educ: Radcliffe Col, AB, 52, MA, 53; Univ Del, MA, 68, PhD(appl math), 70. Prof Exp: Physicist, Textile Fibers Dept, E I du Pont de Nemours & Co, Inc, 53-59, res physicist, 59-66; asst prof appl math & comput sci, Univ Del, 69-74; mem staff, E I du Pont de Nemours & co, Inc, 74-76, supvr, 76-85; RETIRED. Mem: Am Math Soc; Soc Indust & Appl Math; Asn Comput Mach. Res: Nonlinear eigenvalue problems; numerical solution of integral equations; automata theory; microprogramming; optimization. Mailing Add: 233 Cheltenham Rd Newark DE 19711

JONES, LYLE VINCENT, b Grandview, Wash, Mar 11, 24; m 49; c 3. PSYCHOMETRICS. Educ: Univ Wash, BS, 47, MS, 48; Stanford Univ, PhD(psych), 50. Prof Exp: Fel, 50-51, asst prof psychol, Univ Chicago, 51-57; from assoc prof to prof, 57-69, ALUMNI DISTINGUISHED PROF PSYCHOL, UNIV NC, CHAPEL HILL, 69-, Dir, THURSTONE PSYCHOMETRIC LAB, 79- Concurrent Pos: Postdoctoral fel, Nat Res Coun, 50-51; vis assoc prof psychol, Univ Tex, 56-57; fel, Ctr Advan Study Behav Sci, 64-65 & 81-82; vchancellor & dean, The Grad Sch, Univ NC, Chapel Hill, 69-79. Mem: Inst Med Nat Acad Sci; fel Am Acad Arts & Sci; fel Am Psychol Asn (pres, 63-64); Psychometric Soc (pres, 62-63); Am Statist Asn; Am Educ Res Asn; fel AAAS. Res: Psychological measurement; monitoring student achievement trends, especially in mathematics and science for minority students. Mailing Add: CB No 3270 Davie Hall Univ NC Chapel Hill NC 27599-3270

JONES, MAITLAND, JR, b New York, NY, Nov 23, 37; m 60; c 3. ORGANIC CHEMISTRY. Educ: Yale Univ, BS, 59, MS, 60, PhD(chem), 63. Prof Exp: Fel chem, Univ Wis, 63-64; from instr to assoc prof, 64-73, PROF CHEM, PRINCETON UNIV, 73- Concurrent Pos: Vis prof, Free Univ, Amsterdam, 73-74 & 78. Mem: Am Chem Soc. Res: Chemistry of reactive intermediates; carborane chemistry. Mailing Add: Dept Chem Princeton Univ Princeton NJ 08544

JONES, MALCOLM DAVID, b Orange, Calif, Feb 16, 23; m 45; c 5. RADIOLOGY. Educ: Univ Calif, AB, 43, MD, 46. Prof Exp: Intern, San Diego Naval Hosp, 46-47; from asst resident to resident radiol, Med Ctr, Univ Calif, San Francisco, 50-53, from asst prof to assoc prof, 54-65, from asst radiologist to assoc radiologist, 54-65, prof radiol & radiologist, 65-74; prof radiol & chmn dept, Univ Tex Health Sci Ctr San Antonio, 74-; prof dept diag & roentgenol, 77-; AT DEPT RADIOL, UNIV CALIF. Mem: AMA; Am Col Radiol. Res: Radiographic assessment of age changes in the primate spine. Mailing Add: 1015 Castlegate Lane Santa Ana CA 92705

JONES, MARGARET ZEE, b Swedesboro, NJ, June 24, 36; m 59; c 3. NEUROPATHOLOGY, PATHOLOGY. Educ: Univ Pa, BA, 57; Med Col Va, MD, 61. Prof Exp: Clin asst, Sch Med, Univ Wash, 62-65; resident neuropath, Med Col Va, 66-67, from instr to asst prof, 67-69, actg div dir neuropath, 68-69; from asst prof to assoc prof path, 70-78, PROF PATH, MICH STATE UNIV, 78- Concurrent Pos: From intern to resident path & neuropath, Univ Wash, 62-65; lectr, Sch Med, Yale Univ, 69-; fel biochem, Nat Inst Neurol Dis & Stroke, Mich State Univ, 70-71, grant, 80-83, NIH grant, 80-87; grant, Nat Multiple Sclerosis Soc, 71-72; vis prof, Muscular Dystrophy Res Labs, Newcastle Gen Hosp, England, 76-77; hon consult, Western Gen Hosp & sr lectr, Univ Edinburgh, Edinburgh, Scotland, 83-84; mem Neurol Prog Comt, Nat Inst Neurol & Commun Disorders & Stroke, NIH, 85-88; mem, Inst Lab Animal Resources, Nat Res Coun, Nat Acad Sci, 85-88, Coun, 89- Mem: Am Fedn Clin Res; Am Asn Neuropath; Soc Neurosci; Am Asn Pathologists. Res: Inherited metabolic diseases; developmental neurobiology; medical education; neuropathology, particularly developmental and neuromuscular disorders. Mailing Add: Dept Path Mich State Univ East Lansing MI 48824

JONES, MARJORIE ANN, b Flint, Mich, Dec 11, 44; m 66; c 1. FETAL-MATERNAL INTERACTIONS, PROSTAGLANDINS. *Educ:* Cent Mich Univ, BS, 70, MS, 72; Univ Ill, MS, 73; Univ Tex, PhD(biochem), 82. *Prof Exp:* Res assoc biochem, Health Sci Ctr, Univ Tex, 82-84, sr res assoc, 84-85; asst prof, 85-89, ASSOC PROF BIOCHEM, ILL STATE UNIV, 89- *Mem:* Soc Study Reproduction; Int Embryo Transfer Soc; Am Fertility Soc; Soc Study Fertility; Am Chem Soc; NY Acad Sci. *Res:* Role of lipids in biological processes, especially the regulation and initiation of events involved in reproduction; interaction between developing embryo and the maternal system, with emphasis on signals exchanged between the two separate systems. *Mailing Add:* Ill State Univ 305 Felmley Hall Normal IL 61761

JONES, MARK MARTIN, b Scranton, Pa, Jan 7, 28; m 51; c 2. INORGANIC CHEMISTRY. *Educ:* Lehigh Univ, BS, 48, MS, 49; Univ Kans, PhD(chem), 52. *Prof Exp:* Fel hydrazine chem, Univ Ill, 52-53, instr chem, 53-55; chemist, Picatinny Arsenal, 55-57, chief develop unit, Explosives Res Sect, 57; from asst prof to assoc prof, 57-64, PROF INORG CHEM, VANDERBILT UNIV, 64- *Mem:* AAAS; Am Chem Soc; Soc Toxicol. *Res:* Therapeutic chelating agents for toxic heavy metals. *Mailing Add:* Dept Chem Vanderbilt Univ PO Box 1583 Nashville TN 37203

JONES, MARK WALLON, physics, mathematics; deceased, see previous edition for last biography

JONES, MARTHA OWNBEY, b Colfax, Wash, Dec 10, 40; m 68; c 1. BIOLOGICAL CHEMISTRY, ORGANIC CHEMISTRY. *Educ:* Grinnell Col, BA, 62; Purdue Univ, PhD(chem), 75. *Prof Exp:* Instr, Purdue Univ, 66-67; from instr to asst prof chem, Drew Univ, 68-78; lectr org chem, Princeton Univ, 78-82; ASST PROF CHEM, UNION COL, 82- *Mem:* Am Chem Soc; AAAS; Sigma Xi; Am Asn Univ Profs. *Res:* Protein chemistry and enzymology; protein folding and the relationship between structure and biological activity of proteolytic enzymes. *Mailing Add:* 123 Mountainside Dr Randolph NJ 07869

JONES, MARVIN RICHARD, b Bristow, Okla, Nov 3, 14; m 35; c 3. DEVELOPING PROCEDURES & EQUIPMENT FOR MAKING & TESTING PRODUCTS. *Prof Exp:* Prod draftsman, Am Iron & Mach Works, 36-37; prod designer, Hughes Tool Co, 37-39; prod develop engr, Cameron Iron Works, Inc, 39-43, dir res & mgr eng serv, 56-79; chief engr, Oil Ctr Tool Co, 46-49; pres, Petrol Mech Develop Co, 49-55; vpres res & develop, Koomey Inc, 81-85; CONSULT ENGR, 85- *Honors & Awards:* Oil Drop Award, Am Soc Mech Engrs, 88, Silver Patent Award, 89. *Mem:* Fel Am Soc Mech Engrs; Soc Petrol Engrs. *Res:* Developing high pressure equipment for drilling and producing oil wells; author of numerous publications; 66 US patents and 81 foreign patents. *Mailing Add:* 414 Flintdale Rd Houston TX 77024

JONES, MARVIN THOMAS, b St Louis, Mo, Apr 20, 36; m 58; c 2. PHYSICAL CHEMISTRY. *Educ:* Wash Univ, St Louis, AB, 58, PhD(phys chem), 61. *Prof Exp:* Res chemist, Exp Sta, Cent Res Dept, E I du Pont de Nemours & Co, Inc, 61-66; assoc prof chem, St Louis Univ, 66-69; assoc prof, 69-71, assoc dean, Col Arts & Sci, 76-86, actg dean, 78-79, interim assoc vice chancellor Acad Affairs, 86-87, spec asst, chancellor Budget Planning & Instnl Res, 87-88, PROF CHEM, UNIV MO, ST LOUIS, 71-, DEP TO CHANCELLOR, 88- *Concurrent Pos:* Res assoc, Univ Groningen, Neth, 75-76 & Sheffield Univ, Eng, 76. *Mem:* AAAS; Am Chem Soc; Am Phys Soc; Sigma Xi. *Res:* Spectroscopic techniques, especially magnetic resonance to study problems of chemical and physical interest; low dimensional synthetic metals. *Mailing Add:* Dept Chem Univ Mo St Louis MO 63121

JONES, MARY ELLEN, b La Grange, Ill, Dec 25, 22; c 2. BIOCHEMISTRY. *Educ:* Univ Chicago, BS, 44; Yale Univ, PhD, 51. *Prof Exp:* Res chemist, Armour & Co, 42-48; AEC fel, Biochem Res Lab, Mass Gen Hosp, 51-53, Am Cancer Soc fel, 53-55; assoc biochemist, Biochem Res Lab, Mass Gen Hosp, 55-57; from asst prof to assoc prof biochem, Brandeis Univ, 57-66; from assoc prof to prof, Univ NC, Chapel Hill, 66-71, partic assoc prof zool, 67-69, prof, 69-71; prof biochem, Univ Southern Calif, 71-78; prof, 78-80, CHMN BIOCHEM, UNIV NC, CHAPEL HILL, 78-, KENAN PROF, 80- *Concurrent Pos:* Am Cancer Soc scholar, 57-62; dir NIH dent training grant, Brandeis Univ, 62-66; assoc ed, Can J Biochem, 69-74; mem grants comt, Am Cancer Soc, 71-73; mem biochem study sect, NIH, 71-75; mem merit rev bd basic sci, Vet Admin, 75-78; mem, Life Sci Comt, NASA, 76-78; mem metabolic biol study sect, NSF, 78-81; mem sci adv bd Nat Heart, Lung & Blood Inst, NIH, 80-84; mem, Nat Adv Gen Med Sci Coun, 88- *Honors & Awards:* Wilbur Lucius Cross Medal, Yale Univ, 82. *Mem:* Nat Acad Sci; Nat Inst Med-Nat Acad Sci; Am Chem Soc; Am Soc Biol Chem (pres, 86); Sigma Xi; fel AAAS. *Res:* Enzymology; biosynthetic and transfer reactions; metabolic regulation of enzymes; multifunctional proteins; pyrimidine and amino acid biosynthesis; carbamyl phosphate; acetyl-coenzyme A. *Mailing Add:* Dept Biochem Univ NC Sch Med Chapel Hill NC 27599-7260

JONES, MAURICE HARRY, b London, Eng, Jan 7, 27; m 52; c 2. PHYSICAL CHEMISTRY, ORGANIC CHEMISTRY. *Educ:* Univ London, BSc, 47, PhD(chem), 50. *Prof Exp:* Fel photochem, Nat Res Coun, Can, 50-52; Bakelite res fel polymerization, Univ Birmingham, 52-53; asst dir dept chem, Ont Res Found, Can, 54-63, dir dept phys chem, 63-72, dir, Dept Res Coord & Planning, 72-77, vpres interdept prog, 77-83, vpres opers, 83-84; consult, 84-88; EXEC DIR, CAN RES MGT ASN, 88- *Mem:* Am Chem Soc; fel Chem Inst Can; Am Electroplaters Soc; Can Res Mgt Asn. *Res:* Polymerization; membranes; electroplating; pollution; research management. *Mailing Add:* 4642 Badminton Dr Mississauga ON L5M 3H8 Can

JONES, MELTON RODNEY, b Richmond, Va, Jan 13, 45; m 68; c 2. GENETICS, BIOLOGY. *Educ:* Am Univ, BS, 66; Howard Univ, MS, 68, PhD(zool), 72. *Prof Exp:* Div chmn, Natural Sci & Math, Shaw Univ, 72-75; asst prof biol, Univ Colo, Boulder, 75-77; asst to assoc dean basic sci, Ohio Univ, 77-78; asst to dean, Community Col, Baltimore, 78-86, prof biol & chmn, Dept Sci, 78-86; DEAN ACAD SERV, JOHN TYLER COMMUNITY COL, 86- *Concurrent Pos:* Asst to dean, Col Osteop med, Ohio Univ, 78-86. *Mem:* AAAS; Am Inst Biol Sci; Sigma Xi. *Res:* Biochemical genetics with respect to enzyme activity and gene dosage. *Mailing Add:* Acad Serv John Tyler Community Col Chester VA 23831-5399

JONES, MELVIN D, b Hardisty, Alta, Nov 16, 43; m 83; c 5. PROSTHODONTICS. *Educ:* Univ Alta, DDS, 66; Ind Univ, Indianapolis, MSD, 69. *Prof Exp:* Asst prof dent, Univ Alta, 66-67; instr, Ind Univ, Indianapolis, 68-69; assoc prof dent, Univ Alta, 69-75; MEM STAFF, UNIV CALGARY, 75- *Concurrent Pos:* Can Fund Dent Educ fel, 67-69; mem, Am Asn Dent Schs, 69-70; supvr audiovisual sect, Fac Dent, Univ Alta, 70-71; mentor, Calgary & Dist Grathological Soc. *Mem:* AAAS; Calgary & Dist Dent Soc; Can Dent Asn; Can Acad Restorative Dent; Am Acad Crown & Bridge Prosthodontics; Asn Prosthodontists Can. *Res:* Phosphate-bonded investments for regular gold castings; three dimensional recordings of mandibular movement; interactive video uses in teaching; audiovisual education; failures of cast restorations long term clinical study; myofacial pain syndrome clinical study; differential diagnosis of temporomandibular joint pain. *Mailing Add:* 1321 Hillside Dr Vestal NY 13850

JONES, MERRELL ROBERT, b Salt Lake City, Utah, June 27, 38; m 59; c 6. COMPUTER SCIENCE, PHYSICS INSTRUCTION. *Educ:* Univ Utah, BS, 60, PhD(physics), 70. *Prof Exp:* From asst prof to prof physics, 66-83, chmn dept eng & phys sci, 73-76, dir, Comput Ctr, 74-82, PROF COMPUT SCI, SOUTHERN UTAH STATE COL, 83- *Mem:* Asn Comput Mach; Digital Equip Corp Users Soc. *Res:* Computer science; computers in undergraduate instruction; methods and curricula in astronomy physics and computer science in elementary and secondary schools; piezoelectricity; point defects in crystals. *Mailing Add:* Dept Math Sci Southern Utah State Col Cedar City UT 84720

JONES, MERRILL C(ALVIN), b Salona, Pa, Jan 4, 25; m 50; c 4. REAL TIME SOFTWARE, NERWORK-SYSTEM MANAGEMENT. *Educ:* Univ NMex, BS, 52, MA, 70. *Prof Exp:* Eng technician, indust apparatus div, Sylvania Elec Prod, Inc, 44-48; electronics technician, Sandia Nat Labs, 48-52, measurement engr, 52-53, sect supvr metrol, 53-55, sect supvr & mem tech staff, 55-65, mem tech staff, 65-77, div supvr, 77-85, mem tech staff, 85-90, SR MEM TECH STAFF, SANDIA NAT LABS, 90- *Mem:* Comput Soc; Inst Elec & Electronic Engrs; Asn Comput Mach. *Res:* Measurement of electrical and physical quantities; application of computers to scientific and administrative disciplines; software design and maintenance; real time data collection and reduction; user-computer interface; computer languages/applications; network and system administration/management. *Mailing Add:* Sandia Nat Lab Div 9215 PO Box 5800 Albuquerque NM 87185

JONES, MICHAEL BAXTER, b Seattle, Wash, Oct 19, 44; m 66; c 2. ECONOMIC GEOLOGY. *Educ:* Univ Wash, BS, 66, MS, 69; Ore State Univ, PhD(geol), 75. *Prof Exp:* Geologist, Kennecott Explor Inc, Kennecott Copper Corp, 73-77; GEOLOGIST, FREEPORT EXPLOR CO, 77- *Mem:* Sigma Xi; Geol Soc Am; Asn Explor Geophysicists. *Res:* Structural, lithological, mineralogical and chemical manifestations of base and precious metal ore deposits. *Mailing Add:* Freeport Explor Co PO Box 1911 Reno NV 89505

JONES, MILLARD LAWRENCE, JR, b Aug 14, 33; m 59; c 2. CHEMICAL ENGINEERING. *Educ:* Univ Utah, BS, 55; Univ Mich, Ann Arbor, MS, 58, PhD(chem eng), 61. *Prof Exp:* Res engr, Dow Chem Corp, 61-66; asst prof chem eng, 66-71, ASSOC PROF CHEM ENG, UNIV TOLEDO, 71- *Concurrent Pos:* Consult, Owens-Ill Corp, 67-87. *Mem:* Am Inst Chem Engrs. *Res:* Process dynamics and controls. *Mailing Add:* Dept Chem Eng Univ Toledo 2801 W Bancroft Toledo OH 43606

JONES, MILTON BENNION, b Cedar City, Utah, Jan 15, 26; m 51; c 6. SOIL FERTILITY, RANGE SCIENCE & IMPROVEMENT. *Educ:* Utah State Univ, BS, 52; Ohio State Univ, PhD(soil fertil), 55. *Prof Exp:* Assoc agronomist, Sta, 55-69, AGRONOMIST, HOPLAND FIELD STA & LECTR, UNIV CALIF, 69- *Concurrent Pos:* Res mineral nutrit of trop legumes, IRI Res Inst, Brazil, 63-65; teacher forage crops & range mgt, Univ Calif, Davis, 67-; res sulfur nutrit forage crops, CSIRO, Canberra, Australia, 74; lectr, Univ Evora, Portugal, 84, Dept Agr & Res, Basque Govt, 87. *Mem:* Fel Am Soc Agron, 87; Am Soc Range Mgt; fel Soil Sci Soc Am. *Res:* Range plant nutrition; range soils; range fertilization. *Mailing Add:* 4070 University Rd Univ Calif Hopland CA 95449

JONES, MORRIS THOMPSON, b St Louis, Mo, Dec 17, 16; m 48; c 2. BACTERIOLOGY. *Educ:* Univ Ill, AB, 40, MS, 41, PhD(bact), 44. *Prof Exp:* Appl res bacteriologist, Swift & Co, 44; res bacteriologist & food technologist, Automatic Canteen Co Am, 46-50; phys sci res adminr, Off Naval Res, 50-56; asst to chief extramural progs, Nat Inst Allergy & Infectious Dis, 56-57, asst chief, Nat Inst Arthritis & Metab Dis, 57-60, dep chief training br, 60-64, asst head spec foreign currency prog, Off Int Res, NIH, 64-68, CHIEF SPEC FOREIGN CURRENCY PROG, FOGARTY INT CTR, NIH, 68- *Mem:* AAAS; Am Soc Microbiol; Am Chem Soc; Am Pub Health Asn; Sigma Xi. *Res:* High frequency cooking of foods; bacteriology of meat food products; solubility of dehydrated cream; incidence and distribution of Clostridium botulinium in soils of Illinois; sanitation and corrosion of stainless steels. *Mailing Add:* 6622 Fernwood Ct Bethesda MD 20817

JONES, MORTON EDWARD, b Alhambra, Calif, Apr 12, 28; m 51; c 4. PHYSICAL CHEMISTRY. *Educ:* Univ Calif, BS, 49; Calif Inst Technol, PhD(chem), 53. *Prof Exp:* Mem tech staff, 53-61, sr scientist, Semiconductor Device Tech Sect, Device Res Dept, 61-65, dir, Mat Sci Res Lab, 65-71, dir, Phys Sci Res Lab, 71-79, DIR, RESOURCE DEVELOP, TEX INSTRUMENTS, INC, 79- *Mem:* Sigma Xi. *Mailing Add:* 619 Northill Richardson TX 75080

JONES, NOEL DUANE, b Omaha, Nebr, Aug 4, 37; m 63; c 2. CRYSTALLOGRAPHY. *Educ:* Rensselaer Polytech Inst, BS, 59; Calif Inst Technol, PhD(chem), 64. *Prof Exp:* NIH fel, Univ Berne, 64-66; SR RES SCIENTIST, PHYS CHEM RES DIV, ELI LILLY & CO, 67- *Concurrent Pos:* Vis res scientist, Yale Univ, 84-85. *Mem:* Am Chem Soc; Am Crystallog Asn. *Res:* Crystal and molecular structure of biologically active compounds; automation of protein crystallization. *Mailing Add:* Lilly Res Lab 0403 Eli Lilly & Co Indianapolis IN 46285

JONES, NOLAN T(HOMAS), b Manhattan, Kans, May 5, 27; m 52, 85; c 3. ELECTRICAL ENGINEERING. *Educ:* Univ Nebr, BSc, 51; Mass Inst Technol, SM, 54. *Prof Exp:* Res engr electronic eng, Lincoln Lab, Mass Inst Technol, 51-58; sub-dept head, 59-73, proj leader, 64-66, TECH STAFF, MITRE CORP, BEDFORD, 59- *Mem:* Inst Elec & Electronics Engrs. *Res:* Applications of digital computers for real-time automatic control. *Mailing Add:* MITRE Corp Burlington Rd Bedford MA 01730

JONES, OLIVER PERRY, anatomy; deceased, see previous edition for last biography

JONES, OLIVER WILLIAM, b Ft Smith, Ark, Feb 7, 32; m 55; c 4. MEDICINE, BIOCHEMISTRY. *Educ:* Northeastern State Col, Okla, BS, 54; Univ Okla, MD, 57. *Prof Exp:* From intern to jr resident med, Med Ctr, Duke Univ, 57-59, sr resident, 60-61; res assoc biochem genetics, Nat Heart Inst, 61-63; asst prof med, Duke Univ, 65-66, asst prof biochem, 66-67, co-dir, Res Training Prog, Med Ctr, 66-68, assoc prof med & biochem, 67-68; assoc prof, 68-73, PROF MED & PEDIAT, SCH MED, UNIV CALIF, SAN DIEGO, 73-, DIR DIV MED GENETICS, 68- *Concurrent Pos:* Arthritis & Rheumatism Asn fel, Duke Univ, 59-60; fel biochem, Stanford Univ, 63-65; NIH career develop award, 63-68; Nat Inst Arthritis & Metab Dis res grant, 65-68; Damon Runyon Found res grant, 68-71; Am Cancer Soc res grant, 69-73; NIH grants, 69-75. *Mem:* AAAS; Am Soc Human Genetics; Am Soc Cell Biol; Am Soc Biol Chemists; Soc Pediat Res. *Res:* Genetic counseling; regulation of pyrimidine biosynthesis; cytogenetics. *Mailing Add:* Dept Med M-013 Univ Calif San Diego Basic Sci Bldg Rm 5042 La Jolla CA 92093

JONES, ORDIE REGINAL, b Memphis, Tex, Aug 7, 37; m 57; c 3. CONSERVATION TILLAGE, WIND & WATER EROSION CONTROL. *Educ:* Tex Tech Univ, Lubbock, BS, 60; WTex State Univ, Canyon, MS, 71. *Prof Exp:* SOIL SCIENTIST, AGR RES SERV, BUSHLAND, TEX, US DEPT AGR, 71- *Concurrent Pos:* Field agronomist. *Mem:* Soil Sci Soc Am; Am Soc Agron; fel Soil & Water Conserv Soc; Int Soc Soil Sci. *Res:* Developing soil and water conservation practices for use in dryland cropping in semi-arid areas; cropping and tillage practices that conserv water, reduce erosion and increase water use efficiency of crops. *Mailing Add:* 2319 Larry St Amarillo TX 79016

JONES, ORVAL ELMER, b Ft Morgan, Colo, Apr 9, 34; m 54; c 3. APPLIED MECHANICS, SHOCKWAVE PHYSICS. *Educ:* Colo State Univ, BS, 56; Calif Inst Technol, MS, 57, PhD(mech eng), 61. *Prof Exp:* Tech staff mem, Res & Develop Labs, Hughes Aircraft Co, Calif, 56-57; res engr, Hydromech Lab, Calif Inst Technol, 60-61; staff mem, Sandia Corp, 61-64, div supvr dynamic stress res, 64-68, mgr phys res dept, 68-71, dir solid state sci res, 71-74, dir nuclear security systs, 74-77, dir nuclear waste and environ progs, 77-78, dir eng sci, 78-82, vpres tech support, 82-83, vpres defense progs, 83-86, EXEC VPRES TECH PROGS, SANDIA NAT LABS, 86- *Concurrent Pos:* NSF fel, 59-60; vis lectr, Univ NMex, 64 & 68. *Mem:* AAAS; fel Soc Mech Eng; Am Phys Soc; Sigma Xi. *Res:* electrical and mechanical response of piezoelectrics, ferroelectrics and semiconductors to shock loading; nuclear security safeguards systems for transportation and storage of nuclear materials and weapons; underground repository and technology development for nuclear waste isolation; engineering applications of structural dynamics, transport phenomena, and rock mechanics. *Mailing Add:* 12321 Eastridge Dr NE Albuquerque NM 87112

JONES, OTHA CLYDE, b Emporia, Kans, May 16, 08; m 31. CHEMICAL ENGINEERING, PHYSICAL CHEMISTRY. *Educ:* Univ Utah, BS, 30, MS, 31. *Prof Exp:* Res engr, Monsanto Co, 31-37, group leader, 38-50, asst dir res, 50-64, mgr process tech monomers, 64-70, sr planning analyst, 70-73; mech engr, Arthur G McKee, Inc, 75-77; process engr, Fishstone, Inc, 78-79; CONSULT ENG CONTRACTOR, 79- *Mem:* Am Chem Soc Am; Inst Chem Engrs. *Res:* Production of phosphorus, inorganic phosphates, fluosilicates, phosphorus sulfides, vinyl chloride, polyethylene, and other polymers. *Mailing Add:* 736 Cedar Field Ct Chesterfield MO 63017

JONES, OWEN LLOYD, b Hackensack, NJ, July 31, 35; c 6. PHYSICAL CHEMISTRY, RADIOCHEMISTRY. *Educ:* Drew Univ, AB, 57; WVa Univ, MS, 60, PhD(phys chem), 67. *Prof Exp:* Instr chem, WVa Univ, 60-65; asst prof, 65-73, ASSOC PROF CHEM, US NAVAL ACAD, 73- *Mem:* Am Chem Soc; Sigma Xi. *Res:* Solution kinetics; exchange rate studies using isotopic tracer techniques. *Mailing Add:* Dept Chem US Naval Acad Annapolis MD 21402

JONES, P(HILIP) H(ARRHY), b Tredegar, Gt Brit, Jan 30, 31; Can citizen; m 54; c 4. SANITARY ENGINEERING. *Educ:* Univ Toronto, BASc, 58, Univ Northwestern, MS, 63, PhD(sanit eng), 65. *Prof Exp:* Asst munic eng, Borough Engr's Off, Surbiton, Eng, 50-51; asst engr to Sir William Halcrow, Volta River Proj, Ghana, 51-54; proj engr, Town Planning Consult, Toronto, Ont, 54-58; prin munic engr, Franklin McArthur Assocs, 58-62; from asst prof to prof civil eng, Univ Toronto, 64-90, sch hyg, 65-66, chmn, Inst Environ Studies, 71-74, dir environ educ, 74-80, dir, Int Prog Off, 81-83; PROF & HEAD, SCH ENVIRON ENG, GRIFFITH UNIV, BRISBANE, AUSTRALIA, 90- *Concurrent Pos:* Chmn, Environ Eng Div, Can Soc Civil Eng. *Honors & Awards:* Berry Medal for Environ Eng, Can Soc Civil Eng, 90; Eng Medal Excellence, Asn Prof Eng Ont, 90. *Mem:* Can Soc Civil Engrs; Am Water Works Asn; Can Asn Water Pollution Res & Control; UK Inst Water Pollution Control; Int Asn Water Pollution Res & Control. *Res:* Microbiology of waste treatment, stream, river and lake pollution; ecology of complex living systems and their relationship to man's environment; management of toxic and hazardous wastes. *Mailing Add:* Griffith Univ Nathen Brisbane Queensland 4111 Australia

JONES, PATRICIA H, b Elizabeth, NJ, Nov 9, 38; div; c 2. REPRODUCTIVE PHYSIOLOGY. *Educ:* NMex State Univ, BS, 62; Purdue Univ, PhD(reprod physiol), 67. *Prof Exp:* Asst prof, 67-74, ASSOC PROF ZOOL, OHIO UNIV, 74- *Concurrent Pos:* Consult, Schering Corp, 67. *Mem:* Soc Study Reproduction. *Res:* Reproductive physiology in domestic pig, dairy cow and laboratory rat. *Mailing Add:* 3930 Marion Johnson Rd Athens OH 45701

JONES, PATRICIA PEARCE, IMMUNOLOGY. *Educ:* Oberlin Col, BA, 69; Johns Hopkins Univ, PhD(biol), 74. *Prof Exp:* From asst prof to assoc prof, 78-90, PROF BIOL SCI, STANFORD UNIV, 90- *Honors & Awards:* Founder's Prize, Tex Instruments Found, 84. *Mem:* Am Asn Immunologists. *Res:* Immunogenetics. *Mailing Add:* Dept Biol Sci Stanford Univ Stanford CA 94305-5020

JONES, PATRICK RAY, b Austin, Tex, Oct 22, 43; m 68; c 2. PHYSICAL CHEMISTRY. *Educ:* Univ Tex, Austin, BA & BS, 66; Stanford Univ, PhD(chem), 71. *Prof Exp:* Res fel chem, Nat Acad Sci, 71-73; res fel chem, Calif Inst Technol, 73-74; MEM FAC CHEM, UNIV PAC, 74- *Concurrent Pos:* Vis scholar chem, Stanford Univ, 79-80. *Mem:* Am Chem Soc; Sigma XI. *Res:* Matrix isolation studies of oxygen, fluorine and chlorine atom reactions; electron-impact excitation of gases; combined liquid chromatography and mass-spectrometry; physical chemistry of organometallics. *Mailing Add:* Chem Dept Univ Pac Stockton CA 95211

JONES, PAUL HASTINGS, b Fostoria, Mich, Aug 31, 18; m 41; c 4. GEOLOGY. *Educ:* Mich State Col, BS, 41; La State Univ, MS, 51, PhD, 68. *Prof Exp:* Geophysicist, Halliburton Oil Well Cementing Co, Tex, 41-42; from geologist to geologist in charge groundwater invests, US Geol Surv, La, 42-52, dist geologist in charge, Pa, 52-55, tech adv, Groundwater Geol Surv, Int Co-op Admin, India, 55-57, dist geologist, Tenn, 58-59, res proj chief, Ground Water Br, Nat Reactor Testing Sta, Idaho, 59-62, chief radiohydrol sect, Water Resources Div, DC, 62-64, res hydrologist, Gulf Coastal Plain, 65-74; mem fac, dept geol, La State Univ, 74-77; CONSULT, 77- *Concurrent Pos:* Consult, World Bank, 66 & 83-87, India, Nepal & Bangladesh, UN, 70; mem Nat Acad Sci adv team, India, 71. *Honors & Awards:* Meritorious Serv Award, US Dept Interior, 75. *Mem:* Soc Econ Geol; fel Geol Soc Am; Am Asn Petrol Geol; Soc Petrol Eng; Am Soc Test & Mat. *Res:* Quantitative interpretation of borehole geophysical logs; hydrogeology of deep sedimentary basins; role of geopressure in the fluid hydrocarbon regime; hydrology of waste disposal; enhanced production of petroleum and natural gas; geothermal resources of Northern Gulf of Mexico Basin. *Mailing Add:* 3256 McConnell Dr Baton Rouge LA 70809

JONES, PAUL KENNETH, b Des Moines, Iowa, Jan 5, 43; m 71. BIOMETRICS, BIOSTATISTICS. *Educ:* Grinnell Col, BA, 64; Univ Iowa, MS, 69, PhD(statist), 72. *Prof Exp:* res asst statist, Am Col Testing, 68-72; sr instr, 72-73, asst prof, 73-80, ASSOC PROF EPIDEMIOL & BIOSTATIST, CASE WESTERN RESERVE UNIV, 80- *Mem:* Am Statist Asn; Biomet Soc; Soc Epidemiol Res. *Res:* Quantitative methods in health care/health services research; statistical methods; regression, logistic regression. *Mailing Add:* Dept Epidemiol & Biostatist Case Western Reserve Univ Cleveland OH 44106

JONES, PAUL RAYMOND, b Chicago, Ill, July 19, 30; m 58; c 3. ORGANIC CHEMISTRY. *Educ:* Albion Col, BA, 52; Univ Ill, PhD, 56. *Prof Exp:* From asst prof to assoc prof, 56-65, PROF CHEM, UNIV NH, 65- *Concurrent Pos:* NSF sci fac fel, Max-Planck Inst, Gottingen, 64-65; Fulbright res fel, Chem Inst, Univ Freiburg, Ger, 73; vis prof, Deutsches Mus, Munich, Ger, 82-83. *Mem:* Am Chem Soc. *Res:* Macrocycles; ring-chain tautomerism; anhrydro dimers; history of chemistry. *Mailing Add:* Dept Chem Univ NH Durham NH 03824

JONES, PAUL RONALD, b York, Pa, Dec 19, 40; m 67; c 2. ORGANOMETALLIC CHEMISTRY. *Educ:* Pa State Univ, BS, 62; Purdue Univ, PhD(chem), 66. *Prof Exp:* Res assoc chem, Univ Wis-Madison, 66-67; asst prof, 68-73, assoc prof, 73-79, PROF CHEM, UNIV NTEX, 79- *Honors & Awards:* W T Doherty Award, Dallas-Ft Worth Sect, Am Chem Soc, 85. *Mem:* Am Chem Soc. *Res:* Organometallic chemistry, especially involving synthesis structure and reactions of group IV compounds; bonding in organometallic compounds; stereochemistry and mechanism of the reactions of subvalent organosilicon intermediates; polysilyl polyacetylenes. *Mailing Add:* Dept Chem Univ NTex Denton TX 76203-5068

JONES, PETER D, b Palmerston North, NZ, Apr 27, 40; m 67; c 3. LIPID METABOLISM, FATS IN NUTRITION. *Educ:* Victoria Univ Wellington, BS, 61, MS, 63; Duke Univ, PhD(biochem), 68. *Prof Exp:* Res assoc, Univ Ariz, 68-69; lectr, Victoria Univ Wellington, 69-72; from asst prof to assoc prof biochem, 72-83, asst dean, Acad Affairs, Col Med, 79-85, PROF BIOCHEM, UNIV TENN CTR HEALTH SCI, 83- *Mem:* AAAS; Sigma Xi; Nutrit Today Soc; Am Soc Biol Chemists. *Res:* Oxidative desaturation of long chain fatty acids; structure and function of the electron transport chains of the endoplasmic reticulum; nutritional role of fatty acids and lipids. *Mailing Add:* Dept Biochem Univ Tenn Memphis TN 38163

JONES, PETER FRANK, b Brooklyn, NY, Mar 15, 37; m 78; c 6. PHYSICAL CHEMISTRY, ANALYTICAL CHEMISTRY. *Educ:* Univ Kans, BA, 60; Univ Chicago, MS, 61; Univ Calif, Los Angeles, PhD(high pressure spectros), 67. *Prof Exp:* Mem tech staff, Aerospace Corp, 66-73, head, Forensic Sci Sect, 73-74, Anal Sci Dept, 74-87, SR ENGR, TECHNOL DEVELOP, AEROSPACE CORP, 87- *Concurrent Pos:* Consult forensic sci, 77. *Mem:* Am Chem Soc; AAAS; Soc Appl Spectros. *Res:* Cryogenic technology; radiation effects on polymers; analytical applications of photoluminescence; nonradiative transitions in molecules. *Mailing Add:* Aerospace Corp PO Box 9045 Albuquerque NM 87119-9045

JONES, PETER HADLEY, b Cleveland, Ohio, Aug 14, 34; m 57; c 3. ORGANIC CHEMISTRY, CARDIOVASCULAR DISEASES. *Educ:* Harvard Univ, AB, 56; Univ Calif, Los Angeles, PhD(org chem), 60. *Prof Exp:* Sr res chemist, Abbott Labs, 60-70, assoc res fel, 70-71, mgr med chem, 71-75, head cardiovasc res, 75-79; vpres & dir res, Interx Corp, 79-80; sect head gastrointestinal dis, Searle Labs, G D Searle & Co, 80-85, dir med chem, 85-86, SR DIR, GASTROINTESTINAL DIS RES, 86- *Concurrent Pos:* Instr, Dept Biochem, Med Sch, Northwestern Univ, 69-73; adj prof med chem, Univ Ill, Chicago, 87- *Mem:* Am Chem Soc; AAAS; Sigma Xi; fel Am Inst Chemists; NY Acad Sci. *Res:* Reactions of diphenylcarbenes; chemistry of macrolide antibiotics; chemistry of hypertensive agents; development of new drugs for hypertension angina; chemical drug delivery systems; gastrointestinal drugs and diseases. *Mailing Add:* Gastrointestinal Dis Res G D Searle 4901 Searle Pkwy Skokie IL 60077

JONES, PHILIP ARTHUR, b Prince George, BC, Mar 1, 24. ENTOMOLOGY. *Educ:* Univ BC, BSA, 49; Univ Wis-Madison, MS, 56, PhD(entom), 63. *Prof Exp:* Asst forest biologist sci serv, Can Dept Agr, 49-52; res asst forest entom, Univ Wis, 52-58; res officer, Forest Biol Div, Can Dept Agr, 58-60; from proj asst to proj assoc biol control, Univ Wis, 60-64; from asst prof to assoc prof entom, SDak State Univ, 65-74; tech dir, Agr Chem Div, FMC of Can Ltd, 74-77; ENVIRON SCIENTIST, ENVIRON CAN, 77- *Concurrent Pos:* Mem tech comt, Can Agr Chem Asn, 74-77. *Mem:* Soc Environ Toxicol & Chem. *Res:* environmental assessment of toxic chemicals. *Mailing Add:* Box 6221 Station J Ottawa ON K2A 1T3 Can

JONES, PHILLIP SANFORD, b Elyria, Ohio, Feb 26, 12; m 35; c 4. MATHEMATICS. *Educ:* Univ Mich, AB, 33, AM, 35, PhD(math), 48. *Hon Degrees:* LHD, Northern Mich Univ, 72. *Prof Exp:* Teacher pub sch, Mich, 34-37; teacher math, Edison Inst Technol, 37-43, Univ Mich, 43-44 & Ohio State Univ, 44-45; from instr to prof, 47-82, EMER PROF MATH, UNIV MICH, ANN ARBOR, 82- *Concurrent Pos:* Vis prof, Ain Shams Univ, Cairo, Egypt, 78. *Mem:* Nat Coun Teachers Math (pres, 60-62); Am Math Soc; Hist Sci Soc; Math Asn Am. *Res:* History and teaching of mathematics; development of the mathematical theory of linear perspective; development of number systems. *Mailing Add:* Dept Math Univ Mich Ann Arbor MI 48109

JONES, PHILLIPS RUSSELL, b Troy, NY, Aug 25, 30; m 52; c 3. PHYSICS. *Educ:* Univ Mass, BS, 51; Univ Conn, MA, 56, PhD(physics), 59. *Prof Exp:* Teacher, Mass Pub Sch, 52; asst physics, Univ Conn, 54-58; from asst prof to assoc prof, 58-68, PROF PHYSICS, UNIV MASS, AMHERST, 68- *Mem:* Am Phys Soc. *Res:* Experimental atomic physics. *Mailing Add:* Dept Physics Univ Mass Amherst MA 01003

JONES, PHYLLIS EDITH, b Barrie, Ont, Sept 16, 24. NURSING. *Educ:* Univ Toronto, BScN, 50, MSc, 69. *Prof Exp:* From asst prof to prof, Univ Toronto, 63-89, dean nursing, 79-88, EMER PROF NURSING, UNIV TORONTO, 89- *Concurrent Pos:* mem bd, Von Metro Toronto; res referee, On Ministry Health, Nat Health Res & Develop Prog; consult, WHO; external examr, Univ Ibadan; hon mem, Finnish Soc Prof Nursing, 87. *Mem:* Can Pub Health Asn; fel Am Pub Health Asn; NAm Nursing Diag Asn. *Res:* Innovations in community health nursing, in collaboration with physician services; nursing diagnoses. *Mailing Add:* 81 Sutherland Dr Toronto ON M4G 1H6 Can

JONES, R E DOUGLAS, b Kansas City, Mo, Nov 28, 33; m 54; c 4. MATHEMATICS. *Educ:* Univ Okla, BA, 55, MA, 57; Iowa State Univ, PhD(math), 62. *Prof Exp:* Aerophys engr, Convair-Ft Worth, 56-58; instr math, Iowa State Univ, 58-62; asst prof, Wichita State Univ, 62-65; assoc prof, Univ Mo-Rolla, 65-71; PROF MATH, McKENDREE COL, 71- *Concurrent Pos:* Consult, Boeing Co, Kans, 62-64. *Mem:* Am Math Soc; Math Asn Am. *Res:* Topologies generated by metric densities; opaque sets. *Mailing Add:* Div Sci & Math McKendree Col Lebanon IL 62254

JONES, R(ICHARD) JAMES, b Detroit, Mich, Sept 6, 21. ELECTRICAL ENGINEERING, MECHANICAL ENGINEERING. *Educ:* Mich Col Mining, BS, 44 & 54, EE, 50, MS, 54. *Prof Exp:* Elec equip tester, Allen-Bradley Co, 44; equip design engr, standardization & correlation eng, Globe-Union, Inc, 44-47; from instr to assoc prof elec eng, 47-76, EMER ASSOC PROF ELEC ENG, MICH TECHNOL UNIV, 76- *Mem:* Inst Elec & Electronics Engrs. *Res:* Automatic controls; servomechanisms; electrical illumination; electrical machinery; heating and venting. *Mailing Add:* 305 W Calverley Ave Houghton MI 49931

JONES, R NORMAN, b Manchester, Eng, Mar 20, 13; nat Can; m 39; c 2. SPECTROCHEMISTRY. *Educ:* Univ Manchester, BSc, 33, MSc, 34, PhD(chem), 36. *Hon Degrees:* DSc, Univ Manchester, 54, Univ Poznan, 72 & Tokyo Inst Technol, 82. *Prof Exp:* Tutor biochem, Harvard Univ, 39-41; lectr chem, Queen's Univ, Can, 42-43, asst prof, 43-46; from assoc res officer to prin res officer, Nat Res Coun Can, 46-77; PVT CONSULT CHEM SPECTROS, 77- *Concurrent Pos:* Chmn,Molecular Spectros comn, Int Union Pure & Appl Chem, 67-71; secy task group comput use, Comt Data Sci & Technol, Int Coun Sci Unions, 67-75, mem bur, 70-74, mem adv comt, UNISIST-UNESCO, 74-77; guest prof, Tokyo Inst Technol, Japan, 79-82, guest researcher, 85-86; guest worker, Nat Res Coun Can, 79-; distinguished visitor, Univ Alta, 82-83; adj prof, Queen's Univ, Kingston, Ont, 84. *Honors & Awards:* Herzberg Award, Spec Soc Can; Fisher Award, Chem Soc Can. *Mem:* Am Chem Soc; Royal Soc Can; Chem Inst Can; Royal Soc Chem; Int Union Pure & Appl Chem (vpres to pres, Phys Chem Div, 71-77, emer pres, 77-). *Res:* Molecular spectroscopy; use of ultraviolet, infrared and Raman spectroscopy for the elucidation of molecular structure, with special reference to steroids and other natural products; use of computers for data logging and as aids in evaluation, storage and retrieval of spectral data; molecular structure determination and analysis by vibrational spectroscopy. *Mailing Add:* Suite 601 71 Somerset St W Ottawa ON K2P 2G2 Can

JONES, RALPH, JR, internal medicine, for more information see previous edition

JONES, RALPH WILLIAM, b Coin, Iowa, Aug 29, 21; m 45; c 3. PHARMACEUTICAL CHEMISTRY. *Educ:* St Louis Col Pharm, BS, 48; Purdue Univ, MS, 51. *Prof Exp:* Res pharmacist, Abbott Labs, 51-57, group leader sterile prod res, 57-61, sect head liquid-ointment prod res, 61-64, secy new prod comt, 63-64, dept mgr allied prod res, 64-66 & radiopharmaceutical prod res & develop, 66-67, dept mgr liquid prods res & develop, 67-83; RETIRED. *Mem:* Am Pharmaceut Asn; Sigma Xi. *Res:* Basic and applied research on new pharmaceutical dosage forms and products; pharmacy, chemistry and allied medical sciences. *Mailing Add:* 716 Fairview Ave Libertyville IL 60048

JONES, RANDALL JEFFERIES, b Gould, Okla, Oct 16, 15; m 37; c 2. SOIL CHEMISTRY. *Educ:* Okla State Univ, BS, 36; Univ Wis, MS, 37, PhD(soils), 39. *Prof Exp:* From asst to assoc prof soils, Exp Sta, Auburn Univ, 39-44; chief, Soils & Fertilizer Res Sect, Agr Rels Div, Tenn Valley Authority, 44-49; assoc dir, Exp Sta, Miss State Col, 49-50; dean instr agr, Okla State Univ, 50-81; RETIRED. *Mem:* Am Soc Agron; Soil Sci Soc Am. *Res:* Methods for determination of phosphorus and potassium in soils and plants; chemistry of potash fixation in soils; utilization of nitrogen under various soil conditions; crop response to phosphate fertilizers. *Mailing Add:* Epworth Villa Apt 117 14901 N Pennsylvania Oklahoma City OK 73134

JONES, REBECCA ANNE, b Menominee, Mich, Jan 23, 51. AQUATIC TOXICOLOGY, WATER QUALITY. *Educ:* Southern Methodist Univ, BS, 73; Univ Tex, Dallas, MS, 75, PhD(environ sci), 78. *Prof Exp:* Res asst prof civil eng, Colo State Univ, 78-81; coordr aquatic biol, Fluor Engrs, Irvine, Calif, 82; res assoc & lectr environ eng, Tex Tech Univ, Lubbock, 82-84; ASSOC PROF ENVIRON ENG, WATER QUAL & AQUATIC TOXICOL, NJ INST TECHNOL, 84- *Concurrent Pos:* Proj assoc, G Fred Leed & Assoc, 78- *Honors & Awards:* Charles B Dudley Award, Am Soc Testing & Mat, 84. *Mem:* Am Soc Testing & Mat; Am Water Works Asn; Am Soc Civil Engrs; Am Chem Soc; Am Pub Health Asn; Soc Environ Toxicol & Chem; Water Pollution Control Fedn. *Res:* Chemical and biological aspects of surface and groundwater supplies and quality; sources, significance and fate of chemical contaminants in the environment; chemical and biological aspects of water pollution control in surface and groundwaters, rivers, lakes, estuaries and the oceans. *Mailing Add:* G Fred Lee & Assocs 27298 E El Macero Dr El Macero CA 95618

JONES, REESE TASKER, b Philadelphia, Pa, June 7, 32; m 56; c 3. PSYCHOPHARMACOLOGY, DRUG DEPENDENCE. *Educ:* Univ Mich, BS, 54, MD, 58. *Prof Exp:* Res psychiat, Langley Porter Neuropsychiat Inst, 62-67; asst prof, 67-73, assoc prof res, Med Ctr, 73-76, PROF PSYCHIAT, SCH MED, UNIV CALIF, SAN FRANCISCO, 76- *Concurrent Pos:* Staff psychiatrist, Langley Porter Neuropsychiat Inst, 67-; NIMH res career develop award, 67- *Mem:* Am Psychiat Asn; Psychiat Res Soc; Am Col Neuropsychopharmacol; Soc Biol Psychiat. *Res:* Objective indices of psychopathology; human neurophysiology; psychopharmacology; drug dependence. *Mailing Add:* Dept Psychiat Univ Calif Sch Med San Francisco CA 94143

JONES, RENA TALLEY, b Chipley, Ga, Aug 3, 37. MICROBIOLOGY. *Educ:* Morris Brown Col, BA, 60; Atlanta Univ, MS, 67; Wayne State Univ, PhD(microbiol), 74. *Prof Exp:* Instr biol, Pub Schs, Ga, 60-66; NSF grant, Atlanta Univ, 66; INSTR BIOL, CHMN DEPT & PROJ DIR HEALTH CAREERS PROG, SPELMAN COL, 73- *Mem:* Am Soc Microbiol; AAAS; Sigma Xi. *Res:* Staphylococcal and slime molds; enzymes; immunochemistry. *Mailing Add:* Dept Biol Spelman Col 350 Spelman Lane SW Atlanta GA 30314

JONES, REX H, veterinary medicine, for more information see previous edition

JONES, RICHARD BRADLEY, b Norristown, Pa, June 21, 47. NUCLEAR SCIENCE, APPLIED MATHEMATICS. *Educ:* Va Polytech Inst & State Univ, BS, 70, MS, 71, PhD(nuclear eng), 74. *Prof Exp:* Lectr nuclear eng, Univ Calif, Santa Barbara, 74-75; fel appl math, Univ Ill, 75-77; mem tech staff, nuclear risk anal, Sandia Lab, 77-78; asst prof comput sci, State Univ NY, Plattsburgh, 78-81; CONSULT, 81- *Concurrent Pos:* Consult, W Alton Jones Cell Sci Ctr, 78, Chem-Nuclear Syst Inc, 78; vis prof appl math, Univ Florence, Italy, 81. *Mem:* Am Nuclear Soc; Am Soc Eng Educ; Soc Indust & Appl Math; AAAS. *Res:* Applied mathematics; mathematical modeling of cell regeneration; computer animation effects in education; systems design. *Mailing Add:* Univ New Haven Inst Computer Studies 300 Orange Ave West Haven CT 06516-1916

JONES, RICHARD CONRAD, b Lebanon, NH, Feb 25, 16; m 41; c 1. BOTANY. *Educ:* Dartmouth Col, AB, 38; State Col Wash, PhD(bot), 44. *Prof Exp:* Asst, Agr Exp Sta, Univ NH, 44-45, instr bot, 45-46, asst, 46-48; prof biol, State Univ NY Col New Paltz, 48-66, actg dean, 66-67, dean col, 67-68, vpres acad affairs, 68; assoc univ dean, State Univ NY Albany, 68; pres, State Univ NY Col Cortland, 68-78; exec dir, Montshire Mus Sci, 79-80; RETIRED. *Concurrent Pos:* With NSF Inst Bot, Cornell Univ, 57 & NSF Genetics, Cold Springs Harbor, 59; res fel, State Univ NY Col New Paltz, 58 & 60. *Mem:* AAAS; Bot Soc Am; Mycol Soc Am. *Res:* Genetics; alteration of generations in biological systems. *Mailing Add:* Doglord Rd Etna NH 03750

JONES, RICHARD DELL, CARDIOVASCULAR PHYSIOLOGY, BIOMATERIALS. *Educ:* Case Western Reserve Univ, PhD(physiol), 62. *Prof Exp:* DIR PHYSIOL, DIV SURG RES, ST LUKE'S HOSP, 75- *Mailing Add:* Div Surg Res St Luke's Hosp 11311 Shaker Blvd Cleveland OH 44104

JONES, RICHARD ELMORE, b Rochester, NY, July 16, 44; m 69; c 1. PHYSICAL PHARMACY. *Educ:* Dartmouth Col, AB, 65; Stanford Univ, PhD(phys chem), 70. *Prof Exp:* Sr progammer & analyst, Syntex Corp, 69-71; staff researcher, 71-77, sr staff researcher, 77-78, dept head, Inst Pharmaceut Sci, Syntex Res, 78-84; dir pharmaceut res & develop, Genentech Inc, 84-88;

vpres develop, Liposome Technol Inc, 88-89; VPRES DEVELOP, PHARMETRIX CORP, 89- *Concurrent Pos:* Adj prof, Univ Pac Sch Pharm, 74-75, 80-84. *Mem:* Am Chem Soc; AAAS; NY Acad Sci; Am Asn Pharmaceut Scientists. *Res:* Percutaneous absorption; experimental design in formulation problems; pharmaceutical aerosols; protein drug delivery. *Mailing Add:* 870 Los Robles Ave Palo Alto CA 94306

JONES, RICHARD EVAN, b Sacramento, Calif, May 13, 40; m 80; c 4. COMPARATIVE ENDOCRINOLOGY, REPRODUCTIVE BIOLOGY. *Educ:* Univ Calif, Berkeley, BA, 61, MA, 64, PhD(zool), 68. *Prof Exp:* Asst prof behav sci, Hershey Med Ctr, Pa State Univ, 68-69; from asst prof to assoc prof, 69-80, PROF BIOL, UNIV COLO, BOULDER, 80- *Concurrent Pos:* Res career develop award, NIH, 74-79. *Res:* Control of ovarian follicular growth; reptilian reproduction; control of uterine contraction. *Mailing Add:* Lab Comp Reproduction Univ Colo Box 334 Boulder CO 80309

JONES, RICHARD EVAN, JR, b Oak Park, Ill, Aug 3, 40; m 64; c 1. PHYSICAL CHEMISTRY, INORGANIC CHEMISTRY. *Educ:* Monmouth Col, BA, 62; Univ Hawaii, PhD(chem), 70. *Prof Exp:* Lab asst qual control, Ralph Wells & Co, 61-62; res scientist, Continental Can Co, 62-64; lab mgr plastics, Gat Ke Corp, 64-66; NDEA Title IV fel, 67; assoc prof, 70-80, PROF CHEM, LEWIS & CLARK COMMUNITY COL, 80- *Mem:* Am Chem Soc; AAAS; Nat Sci Teachers Asn. *Res:* Aqueous-nonaqueous solvent extraction of metals; better methods for presentation of chemistry in lower division courses. *Mailing Add:* Div Health & Life Sci Lewis & Clark Community Col Godfrey IL 62035

JONES, RICHARD HUNN, b Ridley Township, Pa, Oct 31, 34; m 81; c 4. BIOSTATISTICS, COMPUTER SCIENCE. *Educ:* Pa State Univ, BS, 56, MS, 57; Brown Univ, PhD(appl math), 61. *Prof Exp:* NSF fel, Univ Stockholm, 61-62; from asst prof to assoc prof statist, Johns Hopkins Univ, 62-68; prof info & comput sci, Univ Hawaii, 68-75, chmn dept, 70-73; dir, Sci & Comput Ctr, 75-82, PROF BIOMET, UNIV COLO MED CTR, DENVER, 75- *Concurrent Pos:* Consult, Swed Meteorol & Hydrol Inst, 62, RCA Serv Co, Fla, 62-66, Tripler Army Hosp, Hawaii, 69-73, Nat Bur Standards, Boulder, Colo, 79-85 & Nat Ctr Atmospheric Res, Boulder, Colo, 84-86. *Mem:* Biomet Soc; fel Am Statist Asn; Inst Math Statist. *Res:* Time series analysis; stochastic processes; statistical data analysis. *Mailing Add:* Dept Biomet Box B-119 Sch Med Univ Colo Denver CO 80262

JONES, RICHARD LAMAR, b Charleston, Miss, May 31, 39; m 64; c 2. INSECT PHYSIOLOGY, INSECT BEHAVIOR. *Educ:* Miss State Univ, BS, 63, MS, 65; Univ Calif, Riverside, PhD(insect toxicol), 68. *Prof Exp:* Insect physiologist, Southern Grain Insect Res Lab, Agr Res Serv, USDA, 68-77; res assoc, Univ Ga, 69-77; assoc prof insect physiol, 77-84, PROF & HEAD, DEPT ENTOM, UNIV MINN, 84- *Concurrent Pos:* Fulbright Scholar, Univ Leiden, Netherlands, 80. *Mem:* AAAS; Entom Soc Am; Am Chem Soc. *Res:* Investigation of chemicals associated with insect behavior. *Mailing Add:* Dept Entom Univ Minn Hudson Hall 219 St Paul MN 55108

JONES, RICHARD LEE, b Warren, Ohio, Apr 9, 29; m 55; c 2. MATERIALS SCIENCE ENGINEERING, METALLURGICAL ENGINEERING. *Educ:* Wayne State Univ, BS, 52; Univ Mich, MS, 53, PhD(metall), 59. *Prof Exp:* Res asst, eng res inst, Univ Mich, 53-56; staff scientist, Gen Dynamics/Astronaut, 58-61; head metallics res br, 61-62, chief mat res, 62-65, mgr res & technol, 65-70, mgr struct & mat res & technol, 70-75, MGR, MAT & PROCESS RES & TECHNOL, AIRCRAFT DIV, NORTHROP CORP, 75- *Mem:* Am Soc Metals; Am Inst Mining, Metall & Petrol Engrs; Am Inst Aeronaut & Astronaut; Soc Advan Mat & Process Eng. *Res:* Structures and materials research. *Mailing Add:* 12041 Pine St Los Alamitos CA 90720

JONES, RICHARD LEE, b Mendota, Ill, June 27, 44; m 80; c 1. LUNG PHYSIOLOGY. *Educ:* St Thomas Col, BS, 66; Marquette Univ, MS, 69, PhD(physiol), 70. *Prof Exp:* Lectr, 71-73, asst prof, 73-77, assoc prof physiol, 77-86, PROF, UNIV ALTA, 86- *Concurrent Pos:* Dir, Pulmonary Lab, Univ Hosp, 73-; sci assoc, Royal Alexandria Hosp, 77- *Mem:* Sigma Xi; Am Physiol Soc; Am Col Chest Physicians; NY Acad Sci; Can Soc Clin Invest. *Res:* High frequency ventilation; regional lung function; exercise training in patients with lung disease. *Mailing Add:* 2E434 WCM Health Sci Ctr Univ Alberta Edmonton AB T6G 2B7 Can

JONES, RICHARD THEODORE, b Portland, Ore, Nov 9, 29; m 53; c 3. MEDICINE, BIOCHEMISTRY. *Educ:* Univ Ore, BS, 53, MS & MD, 56; Calif Inst Technol, PhD(chem), 61. *Prof Exp:* Intern med, Hosp Univ Pa, 56-57; from asst prof to assoc prof exp med & biochem, 61-66, PROF BIOCHEM & CHMN DEPT, MED SCH, UNIV ORE, 66- *Concurrent Pos:* Former mem biochem comt, Nat Bd Med Examrs; med scientist, Training Comt, Nat Inst Gen Med Serv; mem, Biochem Training Comt, NIH, Med Scientist Training Comt, Sickle Cell Ctr Rev Comt & Blood Res Rev Group; actg pres, Univ Ore Health Sci Ctr, 77-78, spec consult to pres, 78-79. *Mem:* Am Soc Hemat; Int Soc Hemat; Am Fedn Clin Res; Am Soc Biol Chemists. *Res:* Medical and biochemical genetics, structure and function of normal and abnormal hemoglobins and other human proteins. *Mailing Add:* Dept Biochem Univ Ore Med Sch Portland OR 97201

JONES, RICHARD VICTOR, b Oakland, Calif, June 8, 29; c 3. SOLID STATE PHYSICS. *Educ:* Univ Calif, AB, 51, PhD(physics), 56. *Hon Degrees:* MA, Harvard, 61. *Prof Exp:* Sr engr, Shockley Semiconductor Lab, Beckman Instruments, Inc, 55-57; from asst prof to assoc prof appl physics, 57-71, assoc dean div eng & appl physics, 69-71, dean grad sch arts & sci, 71-72, PROF APPL PHYSICS, HARVARD UNIV, 71- *Concurrent Pos:* Guggenheim fel, 60-61; vis MacKay Prof, Univ Calif, Berkeley, 67-68. *Mem:* Inst Elec & Electronics Engrs. *Res:* Optical physics and electromagnetic phenomena; electronic and optical materials; ceramics; theory and application of magnetism and ferroelectricity. *Mailing Add:* Cruft Lab 113 Harvard Univ Cambridge MA 02138

JONES, ROBERT ALLAN, b Guilford, Conn, Feb 25, 38; m 60; c 2. OPTICAL FABRICATION, COMPUTER CONTROLLED MANUFACTURING. *Educ:* Union Col, BS, 59; Syracuse Univ, MS, 64. *Prof Exp:* Physicist, Rome Air Develop Ctr, 60-64; sr staff engr, Perkin-Elmer Corp, 64-84; PRIN ENGR, ITEK OPTICAL SYSTS, 84- *Honors & Awards:* Dennis Gabor Award, Soc Photo-Optical Instrumentation Engrs, 85; Eng Excellence Award, Optical Soc Am, 89; Advan Technol Award, Litton Corp, 90. *Mem:* Fel Optical Soc Am; Sigma Xi; Soc Photo-Optical Instrument Engrs. *Res:* Directing the fabrication of large aspheric optics using advanced equipment and techniques; developing new fabrication processes by conducting analyses, computer simulations and experiments; author of numerous technical publications; awarded two patents. *Mailing Add:* One Melissa Dr Westford MA 01886

JONES, ROBERT CLARK, b Toledo, Ohio, June 30, 16; m 38, 77. OPTICS. *Educ:* Harvard Univ, AB, 38, AM, 39, PhD(physics), 41. *Prof Exp:* Mem tech staff, Bell Tel Labs, 41-44; sr physicist, Polaroid Corp, 44-67, res fel physics, 67-82; RETIRED. *Honors & Awards:* Lomb Medal, Optical Soc Am, 44, Frederic Ives Medal, 72; Thomas Young Medal & Prize, Brit Inst Physics, 77. *Mem:* Fel Optical Soc Am; fel Acoust Soc Am; fel Soc Photog Sci & Eng (vpres, 59-63); fel Am Acad Arts & Sci. *Res:* Theoretical physics; theoretical optics; detectivity and detective quantum efficiency of radiation detectors; theoretical models of photographic films; theory of absorption of light by developed photographic films. *Mailing Add:* 1716 Cambridge St Apt 27 Cambridge MA 02138

JONES, ROBERT EDWARD, b Yonkers, NY, Jan 14, 23; m 48; c 2. PHYSICS. *Educ:* Oberlin Col, BA, 48; Univ Mich, MA, 49; Pa State Univ, PhD(physics), 53. *Prof Exp:* Res asst eng res, Ionosphere Res Lab, Pa State Univ, 52-53, instr physics, 53-54, assoc prof & actg dir, 54-55; assoc prof, Linfield Col, 55-63, from actg head to head dept, 56-74, chmn, Div Natural Sci & Math, 74-79, prof, 63-87, EMER PROF PHYSICS, LINFIELD COL, 87- *Concurrent Pos:* Physicist, Linfield Res Inst, 56-, actg dir, 65-68; physicist, Field Emission Corp, 61-63. *Mem:* AAAS; Am Asn Physics Teachers; Optical Soc Am. *Res:* Physics of the upper atmosphere. *Mailing Add:* Dept Physics Linfield Col McMinnville OR 97128-6894

JONES, ROBERT F, b Dallas, Tex, Dec 30, 26; m 53; c 4. ONCOLOGY. *Educ:* Univ Tex, Dallas, MD, 52; Am Bd Surg, cert, 64. *Prof Exp:* Res fel surg, Univ Tex Southwestern Med Sch, Dallas, 55, USPHS fel, 62-63; from instr to assoc prof, Univ Tex Health Sci Ctr Dallas, 63-74; assoc prof, 74-79, PROF SURG, MED SCH, UNIV WASH, 79- *Concurrent Pos:* Am Cancer Soc advan clin fel, 63-66; fel surg oncol, MD Anderson Hosp, Houston, 61-62. *Mem:* Am Col Surgeons; AmSoc Clin Oncol; Am Asn Cancer Educ; Am Cancer Soc; Soc Surg Oncol. *Res:* Viral and surgical oncology; tumor immunology; clinical cancer. *Mailing Add:* Dept Surg RF-25 Univ Wash Med Sch Seattle WA 98195

JONES, ROBERT JAMES, b Dawson, Ga, June 10, 51; m 70; c 1. CROP PHYSIOLOGY, AGRONOMY. *Educ:* Ft Valley State Col, BS, 73; Univ Ga, MS, 75; Univ Mo, PhD(crop physiol), 78. *Prof Exp:* Soil conservationist, Soil Conserv Serv, 70-73; asst prof corn physiol, 78-80, ASSOC PROF ENTOM, FISHERIES & WILDLIFE, UNIV MINN, 80- *Mem:* Crop Sci Soc Am; Am Soc Agron; Am Soc Plant Physiologists. *Res:* Corn physiology; major interest in the relationship between photosynthesis and dark respiration during plant ontogeny and as affected by nutritional and environmental stress factors. *Mailing Add:* Dept Agron & Plant Genetics Univ Minn 411 Agron Bldg St Paul MN 55108

JONES, ROBERT L, b Wellston, Ohio, Jan 26, 36; m 58; c 3. SOIL MINERALOGY. *Educ:* Ohio State Univ, BSc, 58, MSc, 59; Univ Ill, PhD(soil mineral), 62. *Prof Exp:* Res assoc, 62-64, from asst prof to assoc prof, 64-73, PROF SOIL MINERAL & ECOL, UNIV ILL, URBANA, 73- *Mem:* Am Soc Agron; Mineral Soc Am; Wildlife Soc. *Res:* Soil mineral analysis techniques; applied mineralogy in soil genesis studies; biogeochemistry. *Mailing Add:* Dept Agron Univ Ill Urbana IL 61801

JONES, ROBERT LEROY, b Allentown, Pa, May 28, 39; m 64; c 2. HUMAN BIOLOGY, PHYSIOLOGY & HEALTH PSYCHOPHYSIOLOGY. *Educ:* Millersville State Univ, BS, 61; Univ Okla, MS, 65; Pa State Univ, DEd(biol), 70, MCP, 90. *Prof Exp:* Res scientist physiol, Pillsburg Co, 65-66; instr biol, Pa State Univ, 66-68; chmn math & sci & assoc prof biol, Am Col Switz, 68-69; assoc prof, Plymouth State Col, 69-72; dean acad affairs, prof life sci & chmn div, Harrisburg Area Community Col, 72-78; ASSOC PROF FAMILY & COMMUNITY MED & BASIC MED SCI, COL MED, PA STATE UNIV, 78- *Concurrent Pos:* Mem, Human Biol Coun; dir res & prev health prom prog, Col Med, Pa State Univ, dir undergrad med educ. *Mem:* AAAS; Am Inst Biol Sci; Am Asn Univ Profs; Am Col Sports Med; NY Acad Sci; Soc Teachers Family Med; Human Biol Coun. *Res:* Cardiovascular physiology; health promotion and disease prevention; studies on continuing physical exercise during pregnancy: implications of changes in maternal heat balance, body fat storage, and metabolism on the well-being of mother and unborn child. *Mailing Add:* Dept Family Med Pa State Univ Col Med Hershey PA 17033

JONES, ROBERT MILLARD, b Mattoon, Ill, Aug 8, 39; m 63; c 3. SOLID MECHANICS. *Educ:* Univ Ill, Urbana, BS, 60, MS, 61, PhD(theoret & appl mech), 64. *Prof Exp:* Instr theoret & appl mech, Univ Ill, Urbana, 63-64; mem tech staff, Aerospace Corp, 64-70; assoc prof solid mech, Inst Technol, Southern Methodist Univ, 70-74, prof, 75-81; dir, Composite Mat & Struct Ctr, 85-87, PROF ENG SCI & MECH, VA POLYTECH INST & STATE UNIV, 81- *Concurrent Pos:* Consult, Lockheed Missile & Space Co, 76 & 79, USAF Mat Lab, 71-79, Boeing, 80, Bell Helicopter, 85, Atlantic Res, 86-87, Health Techn, 87. *Mem:* Assoc fel Am Inst Aeronaut & Astronaut; Am Acad Mech; Am Soc Mech Engrs. *Res:* Shell buckling; shell stress analysis; mechanics of composite materials; finite element; stress analysis of axisymmetric solids. *Mailing Add:* Eng Sci & Mech Dept Va Polytech Inst & State Univ Blacksburg VA 24061-0219

JONES, ROBERT SIDNEY, b Gatesville, Tex, Dec 17, 36; c 2. MARINE BIOLOGY, ICHTHYOLOGY. *Educ:* Univ Tex, BA, 59, MA, 63, PhD(zool), 67. *Prof Exp:* Dir & prof biol, Univ Guam Marine Lab, 67-74; fisheries biologist, Harbor Br Found Inc, 74-76; prog mgr admin, Univ Tex, 76; dir, Johnson Sci Lab, Harbor Br Found, Inc, 76-82; DIR, MARINE SCI INST, UNIV TEX, 84- *Concurrent Pos:* Managing dir, Harbor Br Found Inc, 82-84. *Mem:* Am Soc Ichthyologists & Herpetologists; Sigma Xi. *Res:* Ecology and behavior of marine fishes. *Mailing Add:* PO Box 1267 Port Aransas TX 78373

JONES, ROBERT THOMAS, b Macon, Mo, May 28, 1910; m 64; c 6. AERONAUTICAL ENGINEERING. *Hon Degrees:* ScD, Univ Colo, 71. *Prof Exp:* Aeronaut res scientist, NACA, Langley Field, Va, 34-46, res scientist, Ames Res Ctr, NACA-NASA, Moffett Field, Calif, 46-62; sr staff scientist, Avco-Everett Res Lab, Everett, Mass, 62-70; sr staff scientist, 70-81, RES ASSOC, NASA, AMES RES CTR, CALIF, 81- *Concurrent Pos:* Consult prof, Stanford Univ, 81. *Honors & Awards:* Reed Award, Inst Aeronaut Sci, 46; Inventions & Contrib Award, NASA, 75; Langley Medal, Smithsonian Inst, 81; Prandtl Ring Award, German Soc Aviation & Astronaut; President's Award Distinguished Fed Civil Serv; Excalibur Award, US Congress; Fluid Dynamics Prize, Am Phys Soc, 86. *Mem:* Nat Acad Sci; Nat Acad Eng; hon fel Am Inst Aeronaut & Astronaut; fel Am Acad Arts & Sci. *Res:* High speed wing theory. *Mailing Add:* 25005 La Loma Dr Los Altos Hills CA 94022

JONES, ROBERT WILLIAM, b Seattle, Wash, Jan 20, 27; m 53; c 3. GEOLOGY. *Educ:* Univ Wash, Seattle, BS, 50, MS, 57, PhD(geol), 59. *Prof Exp:* Ground water geologist, US Geol Surv, 51-55; from instr to asst prof, 58-66, ASSOC PROF GEOL, UNIV IDAHO, 66- *Mem:* Geol Soc Am; Am Asn Petrol Geol; Nat Asn Geol Teachers; Int Asn Volcanology; Sigma Xi. *Res:* Petrology and structure of igneous and metamorphic rocks. *Mailing Add:* Dept Geol Univ Idaho Moscow ID 83844

JONES, ROBERT WILLIAM, b Wenatchee, Wash, Dec 18, 40. THEORETICAL PHYSICS, ENERGY ANALYSIS. *Educ:* Univ Wash, BS, 64; Univ Colo, PhD(physics), 69. *Prof Exp:* Asst prof, 69-74, ASSOC PROF PHYSICS, UNIV SDAK, 74- *Concurrent Pos:* Res assoc, Univ Ariz, 71-72; assoc prof physics, petroleum & minerals, Univ Dhahran, Saudi Arabia, 74-76; vis staff mem, Los Alamos Sci Lab, 78; consult, Oak Ridge Nat Lab, 78. *Mem:* Int Solar Energy Soc. *Res:* Nuclear many-body theory; statistical mechanics; theory of quantum liquids; solar energy applications to heating and cooling of buildings including passive heating and absorption air-conditioning. *Mailing Add:* Box 674 Martin SD 57551

JONES, ROBERT WILLIAM, b Dyersburg, Tenn, Sept 7, 44; m 69; c 1. HIGH ENERGY LASERS, FREE ELECTRON LASERS. *Educ:* Univ Ala, BA, 67, MS, 72, PhD(physics), 83. *Prof Exp:* Physicist, Teledyne-Brown Eng, 68-70; teaching asst physics, Univ Ala, Huntsville, 70-71; res physicist, US Army Missile Command, 71-83, proj engr, 83-86, chief, Free Electron Laser Div, Ground Based Laser Proj, Off, 86-88, CHIEF, LASER SENSOR DIV, AOA PROJ OFF, US ARMY STRATEGIC DEFENSE COMMAND, 88- *Concurrent Pos:* Consult, Eng Math Co, 79-88; instr, Univ Ala, Huntsville, 85-86; chmn, Systems Eng & Tech Assistance Task Force, US Army Strategic Defense Command, 90- *Mem:* Optical Soc Am; Sigma Xi. *Res:* High energy repetitively pulsed and continuous wave lasers; resonator design and transverse mode formation; free electron laser. *Mailing Add:* 807 Argonne Terr Huntsville AL 35802

JONES, ROBIN L(ESLIE), b Stanley, Eng, May 19, 40; m 65; c 1. PHYSICAL METALLURGY, MATERIALS ENGINEERING. *Educ:* Cambridge Univ, BA, 62, MA, 66, PhD(metall), 66. *Prof Exp:* Sr res metallurgist, Res Labs, Franklin Inst, 66-71, group leader, Metall Lab, 71-72; mgr metall prog, SRI Int, 72-78; proj mgr, 78-80, prog mgr nuclear systs & mat, 80-85, SR PROG MGR, CORROSION CONTROL, ELEC POWER RES INST, 85- *Mem:* Am Soc Mech Engrs; Nat Asn Corrosion Engrs; Am Inst Mining, Metall & Petrol Engrs; Am Soc Testing & Mat. *Res:* Physical and mechanical metallurgy, particularly the relations between mechanical properties and fine microstructure; fracture mechanics; ductile fracture of metallic materials; corrosion; environmentally assisted fracture of metals and ceramics; corrosion cracking damage in nuclear power plant materials. *Mailing Add:* Elec Power Res Inst 3412 Hillview Ave Palo Alto CA 94303

JONES, ROBIN RICHARD, b Little Rock, Ark, Oct 18, 37. PATHOLOGY, BIOCHEMISTRY. *Educ:* Univ Ark, BS, 61, MD, 62, MS, 66, PhD(biochem), 67. *Prof Exp:* Spec instr, 65-69, from asst prof to assoc prof path, 67-88, asst prof med technol, 74-76, PROF PATH, MED SCH, UNIV ARK, 88- *Mem:* Sigma Xi. *Res:* Vitamin E deficiency; muscular dystrophy; interactive videodisc in pathology education. *Mailing Add:* Dept Path Univ Ark Med Sch Little Rock AR 72205

JONES, ROGER, b Kimbolton, Herefordshire, Eng, Mar 19, 40; m 65; c 2. BOTANY. *Educ:* Univ Wales, BSc, 62, PhD(ecol), 67; Kans State Univ, MSc, 64. *Prof Exp:* From asst prof to assoc prof, 67-85, PROF BIOL, TRENT UNIV, 85- *Mem:* Brit Ecol Soc; Soc Int Limnol. *Res:* Plant ecology; paleolimnology. *Mailing Add:* Dept Biol Trent Univ Peterborough ON K9J 7B8 Can

JONES, ROGER ALAN, b York, Pa, Mar 25, 47; div; c 1. NUCLEIC ACID CHEMISTRY. *Educ:* Univ Del, BS, 69; Univ Alta, PhD(chem), 74. *Prof Exp:* NIH fel, Dept Chem, Mass Inst Technol, 75-76; from asst prof to assoc prof, 77-88, PROF CHEM, RUTGERS UNIV, 88- *Honors & Awards:* Fac Res Award, Am Cancer Soc, 86-91. *Mem:* AAAS; Am Chem Soc. *Res:* Synthesis and characterization of modified and/or isotopically labelled oligonucleotides to study DNA polymorphism. *Mailing Add:* Dept Chem Rutgers Univ New Brunswick NJ 08903

JONES, ROGER C(LYDE), b Lake Andes, SDak, Aug 17, 19; m 52; c 2. ELECTRICAL ENGINEERING, PLASMA PHYSICS. *Educ:* Univ Nebr, BS, 49; Univ Md, MS, 53, PhD, 63. *Prof Exp:* Electronic engr, Naval Res Lab, 49-50, electronic scientist, 50-57; sr staff engr, Melpar Inc, Westinghouse Air Brake Co, 57-58, consult proj engr, antenna & radiation systs lab, 58-59, head physics sect, phys sci lab, 59-64, chief scientist for physics, electronics-physics res ctr, 64; acting dir, appl res lab, 64-65, adj prof radiol, 78-87, prof elec eng, 64-89, prof radiation-oncol, 87-89, EMER PROF, UNIV ARIZ, 89- *Concurrent Pos:* Guest prof, Kraeftforskningsinstituttet, Aarhus, Denmark, 82-83. *Mem:* Fel AAAS; Am Phys Soc; sr mem Inst Elec & Electronics Engrs; Optical Soc Am; Bioelectromagnetics Soc. *Res:* General physical electronics; infrared engineering; gas and solid state lasers; hyperthermia; bioelectromagnetics. *Mailing Add:* 5809 E Third St Tucson AZ 85711-1519

JONES, ROGER FRANKLIN, b Philadelphia, Pa, Oct 31, 30; m 53; c 3. REINFORCED THERMOPLASTICS, THERMOPLASTICS ALLOYS. *Educ:* Haverford Col, BS, 52. *Prof Exp:* Process eng, E I DuPont De Nemours Co, Inc, 52-54; lieutenant, US Navy, 55-58; develop engr, Atlantic Refining Co, 58-60; sr staff supvr polymer res, Avisun Corp, 60-67; pres & gen mgr, LNP Corp, 67-81; chmn & pres, Inolex Chem Co, 81-83; prin, Concord Assocs Mgt Consults, 83-84; MANAGING DIR ENG PLASTICS, BASF CORP, 84- *Concurrent Pos:* Group mgr chem, Beatrice Foods Co, 76-81; vchmn, Am Inst Chemists, 80-81. *Mem:* Am Inst Chemists (vchmn 80-81, secy, 81-84); Am Chem Soc; Soc Chem Indust; fel Soc Plastics Engrs; Sigma Xi. *Res:* The development of novel reinforced and modified thermoplastic composites; surfactants and polymer intermediates. *Mailing Add:* Franklin Polymers Inc Four Kenny Circle Broomall PA 19008

JONES, ROGER L, b Holland Patent, NY, June 2, 49; m 74. ERGODIC THEORY, MARTINGALES. *Educ:* State Univ NY, Albany, BS, 71; Rutgers Univ, PhD(math), 74. *Prof Exp:* From asst prof to assoc prof, 74-84, PROF MATH, DEPAUL UNIV, 84- *Mem:* Am Math Soc; Math Asn Am. *Res:* Classical harmonic analysis and its application to problems in ergodic theory and probability; maximal inequalities. *Mailing Add:* Math Dept DePaul Univ 2219 N Kenmore Chicago IL 60614

JONES, ROGER STANLEY, b New York, NY, June 17, 34; m 56; c 2. HIGH ENERGY PHYSICS. *Educ:* City Col New York, BS, 55; Univ Ill, MS, 57, PhD(physics), 61. *Prof Exp:* Res assoc physics, Univ Ill, 61-62; USAF Off Sci Res fel, Nat Comt Nuclear Energy Labs, Frascati, 62-63; from asst physicist to assoc physicist, Brookhaven Nat Lab, 64-67; ASSOC PROF PHYSICS, UNIV MINN, MINNEAPOLIS, 67- *Mem:* Am Phys Soc; Sigma Xi. *Res:* High energy experimental physics and elementary particle physics; epistemology and symbolism of physics. *Mailing Add:* Sch Physics & Astron Univ Minn Minneapolis MN 55455

JONES, RONALD DALE, b Stillwater, Okla, July 18, 32; div; c 2. COMPUTER SCIENCE. *Educ:* Okla State Univ, BA, 55; Univ Kans, MPA, 60; Univ Southern Calif, DPA, 64. *Prof Exp:* Admin aide budget & mgt, City of Wichita, Kans, 59-60; admin intern, City of Beverly Hills, Calif, 60-61; sr res engr, Los Angeles Div, NAm Aviation, Inc, 61-63; engr, Rand Corp, 63-65; PROF COMPUT SCI, SCH ADMIN, UNIV MO-KANSAS CITY, 66- *Concurrent Pos:* Lectr, Kansas City Art Inst, 72- *Mem:* Asn Comput Mach. *Res:* Simulation of social-psychological systems, particularly bureaucracies; security of computer systems. *Mailing Add:* Dept Comput Sci Univ Mo 5100 Rockhill Rd Kansas City MO 64110

JONES, RONALD GOLDIN, b Yorkville, Ga, Nov 29, 33; m 57; c 1. ORGANIC CHEMISTRY. *Educ:* Emory Univ, BA, 55, MS, 57; Ga Inst Technol, PhD(org chem), 61. *Prof Exp:* Chem prod develop & qual control, Southern Latex Corp, Ga, 51-55; from asst prof to assoc prof, 61-67, PROF ORG CHEM, GA STATE UNIV, 67- *Mem:* Am Chem Soc; Sigma Xi. *Res:* Applications of nuclear magnetic resonance spectroscopy in organic chemistry and polysaccharides of microbial origin; organic reaction mechanisms. *Mailing Add:* Ga State Univ University Plaza Atlanta GA 30303

JONES, RONALD MCCLUNG, b Palo Alto, Calif, May 6, 51; m 74; c 1. PHYSIOLOGICAL ECOLOGY, RESPIRATORY PHYSIOLOGY. *Educ:* Swarthmore Col, BA, 72; Univ Calif, Riverside, PhD(biol), 78. *Prof Exp:* FEL PHYSIOL, DARTMOUTH MED SCH, HANOVER, 78- *Mem:* AAAS; Am Inst Biol Sci; Am Soc Zool; Sigma Xi. *Res:* Physiological ecology of vertebrates. *Mailing Add:* 13175 Franklin Ave Mountain View CA 94040

JONES, ROSEMARY, b Caerleon, Wales, UK, Dec 26, 41. PULMONARY DISEASE. *Educ:* Univ London, PhD(exp path), 80. *Prof Exp:* ASST PROF PATH, SCH MED, HARVARD UNIV, 80- *Mailing Add:* Mass Gen Hosp 55 Fruit St Boston MA 02114

JONES, ROY CARL, JR, b New York, NY, Aug 3, 39; m 64; c 4. MATHEMATICS. *Educ:* Case Inst Technol, BS, 62; Western Reserve Univ, MS, 64, PhD(math), 66. *Prof Exp:* Asst prof, Univ Fla, 66-69; ASST PROF MATH, FLA TECHNOL UNIV, 69- *Mem:* Am Math Soc; Math Asn Am. *Res:* Approximation theory, characterizing and finding best uniform or Tchebycheff approximations; numerical methods including developing algorithms or iterative procedures that converge to the best approximations. *Mailing Add:* Dept Math Univ Cent Fla PO Box 25000 Orlando FL 32816

JONES, RUFUS SIDNEY, b Warrenton, NC, May 9, 40; m 68; c 1. MEMBRANE SCIENCE & TECHNOLOGY. *Educ:* Duke Univ, BS, 62; Purdue Univ, PhD(org chem), 68. *Prof Exp:* Res chemist, Celanese Corp, 68-72, sr res chemist, 72-80, res assoc, 81-82, qual adminr, 82-83, proj leader/res supvr, 83-87; prog mgr, 88-89, DIR, TECHNOL & BUS DEVELOP, HOECHST CELANESE CORP, 90- *Mem:* Am Chem Soc; NAm Membrane Soc. *Res:* Synthesis, characterization, and application of novel polymers for use in devices and systems which rely upon a unique functional polymer property rather than a structural property; materials for optical data storage; gas and liquid separations. *Mailing Add:* Hoechst Celanese Corp 13800 S Lakes Dr Charlotte NC 28273

JONES, RUSSEL C(AMERON), b Tarentum, Pa, Oct 18, 35; m 58. CIVIL ENGINEERING. *Educ:* Carnegie Inst Technol, BS, 57, MS, 60, PhD(sci of mat), 64. *Prof Exp:* Struct designer, Hunting, Larsen & Dunnels Engrs, Pa, 57-59, assoc engr, Missiles & Space Systs Div, Douglas Aircraft Co, 60; asst prof civil eng, Mass Inst Technol, 63-66, assoc prof, 66-71; prof & chmn dept, Ohio State Univ, 71-77; dean, Sch Eng, Univ Mass, 77-81; VPRES ACAD AFFAIRS, BOSTON UNIV, 81- *Mem:* AAAS; Am Soc Eng Educ; Nat Soc Prof Engrs; Metall Soc; Am Soc Testing & Mat. *Res:* Science of materials; composite materials; building systems; housing; construction management; engineering education. *Mailing Add:* 205 McDowell Hall Univ Del Newark DE 19716

JONES, RUSSELL HOWARD, b Oakland, Calif, July 7, 44; m 68; c 3. MATERIALS SCIENCE, METALLURGY. *Educ:* Calif State Polytech Col, BS, 67; Univ Calif, Berkeley, MS, 68, PhD(metall), 71. *Prof Exp:* Engr metall, Westinghouse Elec Corp, 71-73; sr res scientist, 73-80, staff scientist metall, 80-90, TECH LEADER & SR STAFF SCIENTIST, PAC NORTHWEST DIV, BATTELLE MEM INST, 90- *Mem:* Am Soc Metals; Am Inst Mining, Metall & Petrol Engrs; Mat Res Soc; Electrochem Soc. *Res:* High temperature alloys; radiation damage and stress corrosion; ceramic and metal matrix composites. *Mailing Add:* Pac Northwest Div PO Box 999 Richland WA 99352

JONES, RUSSELL K, b Port Chester, NY, Aug 17, 22; m 49; c 4. VETERINARY PUBLIC HEALTH. *Educ:* Cornell Univ, DVM, 45; Purdue Univ, PhD(path), 54. *Prof Exp:* From instr to assoc prof, Agr Exp Sta, Sch Vet Sci & Med, Purdue Univ, 48-60, assoc prof path, 60-65, prof vet path, 65-88; RETIRED. *Mem:* Am Vet Med Asn; Conf Res Workers Animal Dis. *Res:* Diagnostics, including microbiology and pathology; zoonoses. *Mailing Add:* 1707 Fernleaf Dr West Lafayette IN 47906

JONES, RUSSELL LEWIS, b Dyserth, Wales, May 10, 41; c 3. PLANT PHYSIOLOGY. *Educ:* Univ Col Wales, BSc, 62, PhD(bot), 65. *Prof Exp:* AEC fel, Mich State Univ-AEC Plant Res Lab, 65-66; from asst prof to assoc prof, 66-74, PROF PLANT BIOL, UNIV CALIF, BERKELEY, 74- *Concurrent Pos:* Peer panel mem, adv, Panel Develop Biol, NSF, 70-73; assoc ed, Annual Rev Plant Physiol, 72-; Guggenheim Mem Found fel, 72-73; Miller res prof, Univ Calif, Berkeley, 75-76, chmn, Col Letters & Sci Comt on Courses, 76-79, actg chmn, Dept Instruction Biol, 77-78, mem exec comt, Miller Inst Basic Res, 77-81, chmn dept bot, 81-86, mem comt res, 81-, chmn, Col Letters & Sci Comt Res, 83-84; actg dir, Univ Calif Herbarium, 82-83; mem, Plant Growth & Develop Panel, USDA CRGO, 84-86, prog mgr, 87-88; Alexander von Humbolt sr scientist, Univ Göttingen, Fed Repub Ger, 86-87; mem, Life sci Peer Rev Panel, NASA, 87-, Life Sci Div Working Group, 88- *Mem:* Am Soc Plant Physiol; Int Plant Growth Substances Asn. *Res:* Biochemistry and physiology of the action of gibberellic acid and calcium. *Mailing Add:* Dept Plant Biol Univ Calif Berkeley CA 94720

JONES, RUSSELL STINE, b Corvallis, Ore, June 5, 14; m 40; c 8. PATHOLOGY. *Educ:* Univ Ore, MD, 40; Am Bd Path, dipl, 48. *Prof Exp:* Intern, Ancker Hosp, Minn, 40-41; resident path, Hosp & Clins, Univ Ore, 41-44; from instr to asst prof, Col Med, Univ Tenn, 47-52; assoc prof, Med Sch, Univ Ore, 52-54; prof, Col Med, Univ Utah, 54-66, dir clin labs, 60-66; prof path, Univ Mo-Kansas City, 66-70; dir labs, Dept Path, Providence Hosp, Ore, 70-72; DIR LABS, DEPT PATH, IMPERIAL POINT MED CTR, FT LAUDERDALE, 72- *Concurrent Pos:* Dir labs, WTenn Tuberc Hosp, Memphis, 48-52; consult, Vet Admin Hosp, Memphis, 48-52. *Mem:* Am Asn Cancer Res; Am Rheumatism Asn; Am Soc Clin Pathologists; Am Asn Blood Banks; Int Acad Path. *Res:* Arthritis and rheumatic diseases; experimental tumors; adrenal steroids; scurvy; mycoplasmal infections; activation analysis; radiobiology. *Mailing Add:* 6401 N Federal Hwy Ft Lauderdale FL 33308

JONES, SAMUEL B, JR, b Roswell, Ga, Dec 18, 33; m 55; c 3. BOTANY, HORTICULTURE. *Educ:* Auburn Univ, BS, 55, MS, 61; Univ Ga, PhD(bot), 64. *Prof Exp:* Teacher, SCobb High Sch, 58-59; instr bot, Auburn Univ, 59-61; asst prof, Univ Southern Miss, 64-67; from asst prof to prof bot, Univ Ga, 67-91, dir, Bot Garden, 81-84; CO-OWNER, PICCADILLY FARM PERENNIAL NURSERY, 91- *Concurrent Pos:* NSF res grants; Calloway Found res grant. *Honors & Awards:* Silver Shield Award, Nat Coun State Garden Clubs. *Mem:* Garden Writers Asn Am; Am Soc Plant Taxonomists; Am Asn Nurserymen. *Res:* Systematics of higher plants; landscaping with native plants; flora of southeastern United States; the genus Hosta (Liliacene); shade gardening; perennials. *Mailing Add:* 1971 Whippoorwill Rd Bishop GA 30621

JONES, SAMUEL O'BRIEN, b Louisburg, NC, Mar 14, 11; m 37; c 2. CHEMISTRY, CHEMICAL ENGINEERING. *Educ:* NC State Col, BS, 32; Johns Hopkins Univ, PhD(org chem), 36. *Prof Exp:* Dir org res, R J Reynolds Tobacco Co, 36-52, dir chem eng, 52-64, mgr tobacco prod develop, 64-70, dir tobacco develop, 70-76; RETIRED. *Mem:* Am Chem Soc; Am Inst Chem Engrs; Sigma Xi. *Res:* Tobacco research and process development; reactions of sulfur, hydrogen sulfide and mercaptans with unsaturated hydrocarbons. *Mailing Add:* 310 Plymouth Ave Winston-Salem NC 27104

JONES, SAMUEL STIMPSON, b Buckingham Co, Va, Apr 9, 23. PHYSICAL CHEMISTRY. *Educ:* Hampden-Sydney Col, BS, 43; Cornell Univ, PhD(phys chem), 50. *Prof Exp:* Chemist, Manhattan Proj, Monsanto Chem Co, 44-46; res assoc phys chem, Knolls Atomic Power Lab, Gen Elec Co, 50-57, radiation chemist, Vallecitos Atomic Lab, 57-61, radiation effects specialist, Defense Syst Dept & Electronics Lab, 61-63 & tech specialist, Hanford Labs, 63-65; res assoc Pac Northwest Lab, Battelle Mem Inst, Wash, 65-70; sr staff res scientist, Ctr Technol, Kaiser Aluminum & Chem Corp, 70-81; mgr, carbon & mat res, 81-83, INDUST CARBON CONSULT, ANACONDA ALUMINUM CO & ARCO METALS CO, 83- *Mem:* AAAS; Am Chem Soc; NY Acad Sci; Am Carbon Soc. *Res:* Radiochemistry; complex ions; radiation chemistry; radiation effects on electronic materials; graphite physics and chemistry; chemistry and physics of carbon. *Mailing Add:* PO Box 43698 Tucson AZ 85733

JONES, SANFORD L, b Bulan, Ky, Sept 22, 25; m 56; c 3. REPRODUCTIVE ENDOCRINOLOGY. *Educ:* Eastern Ky State Col, BS, 50; Univ Ky, MS, 56; Univ Tenn, PhD(physiol, biochem), 60. *Prof Exp:* Secondary teacher, Perry County Schs, Ky, 50-55; res assoc physiol, Univ Tenn, 60-61; from asst prof to assoc prof, 61-65, PROF BIOL, EASTERN KY UNIV, 66-, CHMN DEPT, 79- *Mem:* Am Soc Zool; Sigma Xi. *Res:* Effects of antithyroid compounds on metabolism of thyroxine; absorption of iodinated compounds in amphibians and reptiles; radioimmunoassay of luteinizing hormone. *Mailing Add:* Dept Biol Sci Eastern Ky Univ Richmond KY 40475

JONES, SHARON LYNN, b Pasadena, Calif, Dec 19, 60. NEUROPHARMACOLOGY, NEUROANATOMY. *Educ:* Univ NMex, BS, 83; Univ Iowa, PhD(pharmacol), 87. *Prof Exp:* ASST PROF PHARMACOL, UNIV OKLA, 90- *Mem:* Soc Neurosci; Int Asn Study Pain; Am Pain Soc; Sigma Xi. *Res:* Organization and function of endogenous pain suppression systems that modulate spinal nociceptive transmission by using anatomical, pharmacological and physiological techniques. *Mailing Add:* Dept Pharmacol Univ Okla Oklahoma City OK 73190

JONES, STANLEY B, b July 27, 38. MEDICAL ADMINISTRATION. *Educ:* Dartmouth Col, BA, 60. *Prof Exp:* Staff mem to assoc dir, Div Computer Res & Technol, HEW, NIH, 64-69, chief, Planning Systs Br & dir, Off Mgt Policy, Health Serv & Ment Health Admin, 69-71; prog develop officer, Inst Med-Nat Acad Sci, 77-78; founding partner, Fullerton, Jones & Wolkstein, Health Policy Alternatives, 78-80 & 83-86; vpres, Wash Representation, Blue Cross & Blue Shield Associations, 80-83; pres, Consol Consult Group & vpres, Consol Healthcare, Inc, 86-89; CONSULT HEALTH POLICY, 89- *Concurrent Pos:* Fel, Inst Soc, Ethics & Life Sci, Hastings Ctr, 78-89; mem, Robert Wood Johnson Fel Bd & Bd Ment Health & Behav Med, Inst Med, Nat Acad Sci, 80-86, DC Gen Hosp Comn, 85-87 & Robert Wood Johnson Rev Comt Prog Promote Long-Term Care Ins Elderly, 88; chmn, Ad Hoc Comt Educ Health Professionals & Invitational Workshop Utilization Mgt, Inst Med, Nat Acad Sci, 87 & Panel Long Range Planning Dis Res, 89. *Mem:* Inst Med-Nat Acad Sci. *Res:* Author of various publications on health care administration. *Mailing Add:* 2555 Pennsylvania Ave NW Apt 418 Washington DC 20037

JONES, STANLEY BENNETT, b San Francisco, Calif, Jan 14, 22; m 46; c 2. GEOPHYSICS. *Educ:* Univ Calif, PhD(physics), 50. *Prof Exp:* Physicist, Radiation Lab, Univ Calif, 48-50; res physicist oil field res, Chevron Oil Field Res Co, 50-58, sect supvr, Well Logging & Basic Prod Sect, 58-63, sect supvr, Geophys Sect, 63-68, mgr, Geophys Div, Chevron Oil Field Res Co, 68-76, mgr, Develop & Implementation Div, Chevron Geosci Co, 77-78, geophys res consult, Chevron Oil Field Res Co, 79-80, mgr, Systs & Eng Serv Div, Chevron Oil Field Res Co, 81-85; assoc dir tech servs, Soc Explor Geophysicists, 85-89; RETIRED. *Mem:* Am Phys Soc; Am Geophys Union; Soc Explor Geophys; Europ Asn Explor Geophys; Soc Explor Geophysicists (vpres, 84-85). *Res:* Cosmic rays; meson physics; oil well logging, oil producing and geophysics research. *Mailing Add:* 7823 S California Ave Whittier CA 90602-2708

JONES, STANLEY C(ULVER), b Spokane, Wash, Aug 31, 33; m 57; c 3. CHEMICAL ENGINEERING. *Educ:* Wash State Univ, BSE, 56; Univ Mich, MSE, 59, PhD(chem eng), 62. *Prof Exp:* Engr, Kaiser Aluminum & Chem Corp, 56-58; adv res engr, Denver Res Ctr, Marathon Oil Co, 62-71, sr res scientist, 71-75, res assoc, 75-86; AT RES & DEVELOP DEPT, CORE LABS, 86- *Mem:* Soc Petrol Engrs; Sigma Xi; Am Inst Chem Engrs; Soc Corrosive Analysts. *Res:* Anodizing processes for aluminum alloys; behavior of a pulsed extraction column; movement of water through aquifers in contact with natural gas; secondary and tertiary oil recovery processes; reservoir rock properties; petroleum production. *Mailing Add:* 875 Front Range Rd Littleton CO 80120

JONES, STANLEY E, b Mt Vernon, NY, July 20, 39; m 80; c 2. APPLIED MATHEMATICS, ENGINEERING MECHANICS. *Educ:* Univ Del, BA, 63, MS, 66, PhD(appl sci), 67. *Prof Exp:* Prof eng mech, Univ Ky, 67-87; dept head eng mech, Univ Ala, 87-89; distinguished vis prof, USAF Acad, 89-91; RES PROF, UNIV ALA, 91- *Concurrent Pos:* Vis prof, Univ Iowa, 69 & Ga Inst Tech, 79-80; consult, Marshall Space Flight Ctr, NASA, 70, US Air Force, Eglin AFB, Fla, 81-, NSF, 83 & Nat Water Resources Inst, 84; Siam lectr, 83. *Res:* Fluid transients, plasticity analysis and nonlinear mechanics; co-author one book and author of approximately 100 research papers. *Mailing Add:* Dept Eng Mech Univ Fla Tuscaloosa AL 35487-0278

JONES, STANLEY LESLIE, b Waltham, Mass, Mar 22, 19; m 48; c 5. ANALYTICAL CHEMISTRY. *Educ:* Tufts Col, BS, 41, MS, 47; Harvard Univ, AM, 49, PhD(anal chem), 51. *Prof Exp:* Res chemist drying oils, Bird & Son, Inc, 41-42; jr chemist chem anal, US Navy Yard, Mass, 42-44; res chemist, Merck & Co, Inc, 50-55; appl res chemist, Knolls Atomic Power Lab, Gen Elec Co, 55-60, consult chemist, 60-81; RETIRED. *Mem:* Am Chem Soc; AAAS. *Res:* Analytical research; colloid science. *Mailing Add:* 1314 Fox Hollow Rd Schenectady NY 12309

JONES, STANLEY TANNER, b Palo Alto, Calif, Mar 10, 45; m 77; c 3. PHYSICS. *Educ:* Stanford Univ, BS, 66; Univ Ill, Urbana, MS, 68, PhD(physics), 70. *Prof Exp:* From asst prof to assoc prof, 70-83, PROF PHYSICS, UNIV ALA, 83-, ASST DEAN, COL ARTS & SCI, 89- *Mem:* Am Phys Soc; Am Asn Physics Teachers. *Res:* Elementary particle physics. *Mailing Add:* Dept Physics Univ Ala PO Box 870324 Tuscaloosa AL 35487-0324

JONES, STEPHEN BENDER, b Lansing, Mich, Oct 19, 45; m 70; c 3. CIRCULATORY, MEDICAL EDUCATION. *Educ:* Cent Mich Univ, BS, 67, MS, 69; Univ Mo-Columbia, PhD(physiol), 75. *Prof Exp:* Teaching asst biol, Cent Mich Univ, 67-68, asst instr, 68-69, instr, 69-71; res assoc physiol, 75-76, asst prof, 76-82, ASSOC PROF, STRITCH SCH MED, LOYOLA UNIV, CHICAGO, 82- *Concurrent Pos:* Vis prof, NRC, Ottawa, Can, 71,

Dept Pharmacol, Univ Melbourne, Australia, 87. *Mem:* Sigma Xi; Am Physiol Soc; Shock Soc. *Res:* Neural control of circulation in developing control; control of peripheral neurotransmitter; plasma catacholamines. *Mailing Add:* Dept Physiol Loyola Univ Med Ctr 2160 S First Ave Maywood IL 60153

JONES, STEPHEN THOMAS, b Washington, NC, Feb 12, 42; m 63; c 1. ORGANIC CHEMISTRY. *Educ:* E Carolina Univ, AB, 64; Emory Univ, PhD(org chem), 68. *Prof Exp:* Res chemist, 68-70, supvr prod develop, 70-72, mgr prod develop, 72-77 & opers & planning, 77-83, DIR MKT RES, LORILLARD, INC, 83- *Mem:* Sigma Xi; Am Chem Soc; Am Mkt Asn. *Res:* Steroid synthesis; syntheses of hydroazulenes; syntheses of heterocyclic compounds; tobacco chemistry. *Mailing Add:* Lorillard Inc One Park Ave New York NY 10016

JONES, STEPHEN WALLACE, b Steubenville, Ohio, Nov 7, 53. NEUROSCIENCES. *Educ:* Mich State Univ, BS, 74; Cornell Univ, PhD(neurobiol), 80. *Prof Exp:* Assoc teaching fel muscular dystrophy, dept neurobiol & behav, Cornell Univ, 79-82; teaching fel NIH, dept neurobiol & behav, State Univ NY, Stony Brook, 82-84, res asst prof, 84-86; ASST PROF PHYSIOL, DEPT PHYSIOL, CASE WESTERN RES UNIV, CLEVELAND, OHIO, 86- *Concurrent Pos:* Lectr neurobiol, Cornell Univ, 81. *Mem:* Biophys Soc; Soc Neurosci; AAAS. *Res:* Electrophysiology and pharmacology of vertebrate neurons, primarily in autonomic ganglia; voltage-clamp analysis of voltage-dependent currents and of the actions of neurotransmitters. *Mailing Add:* 3088 Euclid Heights Blvd Cleveland OH 44118

JONES, SUSAN MURIEL, EPITHELIAL TRANSPORT, RENAL HORMONES. *Educ:* State Univ NY, PhD(pharmacol), 78. *Prof Exp:* ASST PROF, MED COL, CORNELL UNIV, 87- *Mem:* Am Physiol Soc; NY Acad Sci. *Mailing Add:* Dept Physiol & Biophysics Med Col Cornell Univ 1300 York Ave New York NY 10021

JONES, T(HOMAS) BENJAMIN, b Madison, Md, Aug 26, 12; m 47; c 2. ELECTRICAL ENGINEERING. *Educ:* Johns Hopkins Univ, BE, 33, Dr Eng, 37. *Prof Exp:* Asst instr elec eng, Johns Hopkins Univ, 36-37; mem tech staff, Bell Tel Lab, NY, 37-41; foreign wire rels engr, C & P Tel Co, Md, 41-44, personnel supvr, 44-45, commercial mgr, 45-46; from asst prof to assoc prof elec eng, Johns Hopkins Univ, 46-56; mem tech staff, Bell Tel Labs, 56-58, tech supvr, 58-67, dielectrics specialist, 67-73; sr assoc, Trident Eng Assocs, 73-88; RETIRED. *Concurrent Pos:* Proj engr bur ships, US Navy, 46-50, res contract dir, Off Naval Res, 48-54; res contract dir, Army Ord Corps, 54-56; consult, E I du Pont de Nemours & Co, Inc, 54-56. *Mem:* Inst Elec & Electronics Engrs. *Res:* Dielectrics and insulation; electrical capacitors; electrical discharges, arcs and welding; electrical measurements; oxidation of impregnated paper insulation. *Mailing Add:* 5309 River Crescent Dr Annapolis MD 21056

JONES, TAPPEY HUGHES, b Norfolk, Va, June 6, 48. CHEMICAL ECOLOGY, MICRO ANALYTICAL ORGANIC CHEMISTRY. *Educ:* Va Mil Inst, BS, 70; Univ NC, PhD(org chem), 74. *Prof Exp:* Postdoctoral fel res, Chem Dept, Cornell Univ, 75-77; asst prof teaching & res, Chem Dept, Furman Univ, 77-79; postdoctoral fel res, Entom Dept, Univ Ga, Athens, 79-81; asst prof teaching & res, Chem Dept, US Naval Acad, 81-85; asst prof teaching & res, Chem Dept, Col William & Mary, 85-87; STAFF FEL RES, LAB BIOPHYS CHEM, NAT HEART, LUNG & BLOOD INST, BETHESDA, MD, 87- *Mem:* Am Chem Soc. *Res:* Over 64 publications, with more than 50 of them in the area of insect natural products; venom chemistry of myrmicine ants; developed a number of structure proof methods and syntheses for micro-scale organic analysis. *Mailing Add:* Lab Biophys Chem Bldg 10 7N 314 Nat Heart Lung & Blood Inst Bethesda MD 20892

JONES, TERRY LEE, b Marysville, Kans, Oct 12, 45. REAL-TIME IMAGE PROCESSORS, PHYSICS. *Educ:* Rutgers Univ, BA, 68; Am Univ, MS, 74. *Prof Exp:* Physicist elec sources, Night Vision Lab, Far Infrared Tech Area, 68-74; res physicist MIS Structures, Night Vision Lab, Image Intensification Tech Area, 74-78; res physicist image process, 78-86, SUPVRY PHYSICIST, CTR NIGHT VISION & ELECTRO-OPTICS, 86- *Concurrent Pos:* Mem, Appl Imagery Pattern Recognition Exec Comt, 81-85, Automatic Target Recognizer Working Group, 82-85, chmn, 86-88, Triservice Signal Processing Panel, 85-90; chmn, Triservice Image Processing & Image Processing Devices Working Group, 82-84; mem, DoD working group ATR, 89-; liaison officer, Off asst secy Army, res develop & acquisitions, 89. *Mem:* Inst Elec & Electronics Engrs Comput Soc. *Res:* Automatic target recognition; image processing; electron sources, photo electric emmision, metal oxide or metal insulated semiconductor photocathodes, field assisted photocathodes, night vision devices, infrared sensors, ultra high vacuum; algorithm investigations for automatic target acquisition. *Mailing Add:* Ctr Night Vision & Electro Optics AMSEL-RD-NV-ISP Ft Belvoir VA 22060

JONES, THEODORE CHARLES, b Pittsburgh, Pa, Nov 9, 39; m 62; c 2. GENETICS, BIOCHEMISTRY. *Educ:* Amherst Col, AB, 61; Univ Wash, PhD(genetics), 67. *Prof Exp:* Fel, Univ Wis, 67-69; asst prof biol, Amherst Col, 69-72; asst prof biol sci, Mt Holyoke Col, 72-79; CONSULT, 79- *Mem:* Genetics Soc Am; Am Soc Microbiol; Sigma Xi. *Res:* Regulation of enzyme synthesis and enzyme localization in microorganisms. *Mailing Add:* 1517 Baylor Ave Rockville MD 20850

JONES, THEODORE HAROLD DOUGLAS, b Belfast, North Ireland, Oct 30, 38; m. BIOCHEMISTRY, MICROBIOLOGY. *Educ:* Univ Edinburgh, BSc, 59; Mass Inst Technol, PhD(biochem), 66. *Prof Exp:* Res asst biochem, Detroit Inst Cancer Res, 59-61; Damon Runyon Fund Cancer Res fel, Harvard Med Sch, 66-68; NIH traineeship aging res, Retina Found, 68-70; from asst prof to assoc prof, 70-82, PROF CHEM, UNIV SAN FRANCISCO, 82- *Concurrent Pos:* Damon Runyon fel. *Mem:* NY Acad Sci; AAAS. *Res:* Biochemistry of differentiation, particularly events occurring during germination of spores of the cellular slime molds; biochemistry of membrane proteins and their changes during differentiation; clinical enzymology; peptide hormones in microorganisms; processing of peptide hormones. *Mailing Add:* Dept Chem Univ San Francisco San Francisco CA 94117-1080

JONES, THEODORE SIDNEY, b East Orange, NJ, Apr 28, 11; m 36; c 2. PETROLEUM GEOLOGY. *Educ:* Rutgers Univ, AB, 32; Univ Mich, PhD(geol), 35. *Prof Exp:* Jr soil surveyor, Soil Conserv Serv, USDA, 35-37; geologist, Humble Oil & Refining Co, 37-48; dist geologist, Sohio Petrol Co, 48-50; staff geologist, 50-53, chief stratigrapher, 53-60, mem staff, 60-69, regional stratigrapher, Union Oil Co Calif, 69-76; INDEPENDENT GEOLOGIST, 76- *Mem:* Fel Geol Soc Am; Am Asn Petrol Geol. *Res:* Paleozoic stratigraphy. *Mailing Add:* 111 Bridgewater Circle Midland TX 79707

JONES, THOMAS CARLYLE, b Boise, Idaho, Sept 29, 12; wid; c 3. VETERINARY PATHOLOGY, COMPARATIVE PATHOLOGY. *Educ:* Wash State Univ, BS & DVM, 35. *Hon Degrees:* DSc, Ohio State Univ, 70. *Prof Exp:* Officer in chg, US Army Vet Res Lab, Front Royal Qm Depot, Va, 39-46, chief vet path sect, Armed Forces Inst Path, Washington, DC, 46-50, chief vet dept, Army Med Field Lab, Heidelberg, Ger, 50-53, chief vet path sect, Armed Forces Inst Path, 53-57; dir dept path, Angell Mem Animal Hosp, Boston, 57-67; clin assoc, Med Sch, 57-63, assoc clin prof path, 63-71, prof comp path, New Eng Regional Primate Res Ctr, Harvard Med Sch, 71-82, EMER PROF COMP PATH, HARVARD UNIV, 82- *Concurrent Pos:* Master res, Grad Coun, George Washington Univ, 47-51; res assoc path, Cancer Res Inst, New Eng Deaconess Hosp, 57-67; consult, Armed Forces Inst Path, 58- & Nat Cancer Inst, 61-62; mem comt path training, Nat Inst Gen Med Sci, 60-63; mem comt animal health, Nat Acad Sci-Nat Res Coun, 62-65; mem consult staff, Peter Bent Brigham Hosp, Boston, 62-78; mem adv comt animal resources, NIH, 65-68; mem vis comt sch vet med, Tufts Univ, 83-87. *Mem:* Am Vet Med Asn; Am Col Vet Pathologists (secy-treas, 48-50 & 53-60, pres, 62-63); Am Asn Path; Int Acad Path (pres, 70-71); Conf Res Workers Animal Dis. *Res:* Genetics and cytogenetics applied to disease in animals. *Mailing Add:* New Eng Regional Primate Res Ctr Harvard Med Sch One Pinehill Dr Southborough MA 01772

JONES, THOMAS EVAN, b Basin, Miss, Dec 26, 44; c 2. ANALYTICAL CHEMISTRY, INORGANIC CHEMISTRY. *Educ:* Col Great Falls, BS, 70; Wash State Univ, PhD(chem), 74. *Prof Exp:* Teaching asst chem, Wash State Univ, 70-74; res assoc, Wayne State Univ, 74-75; asst prof chem, Univ NMex, 75-80; staff chemist, Rockwell Hanford Opers, 80-85; SR RES CHEMIST, PAC NORTHWEST LABS, 85- *Mem:* Am Chem Soc; Sigma Xi. *Res:* Development of an analytical chemistry capabilities to provide characterization of mixed radioactive/hazardous waste samples at the Hanford Nuclear Reservation; characterization of mixed hazardous radioactive waste stored in Hanford site single shell tanks. *Mailing Add:* Pac Northwest Labs PO Box 999 Richland WA 99352

JONES, THOMAS HUBBARD, b Batavia, Ill, June 8, 36; m 58; c 2. PHYSICAL CHEMISTRY, LIGHT-SENSITIVE MATERIALS. *Educ:* Augustana Col, Ill, BA, 58; Univ Minn, Minneapolis, PhD(phys chem), 63. *Prof Exp:* Res chemist, Photo Prod Res Lab, E I Du Pont de Nemours & Co, Inc, 63-69; proj chemist, Richardson Co, Melrose Park, 69-78, res supvr, 78-80, res assoc, 80-82; sr chemist, Turtle Wax Inc, Chicago, 84-86; sr chemist, London Chem Co, Bensenville, Ill, 86-90; PROJ LEADER, ILL TOOL WORKS, ITW TECHNOL CTR, GLENVIEW, 90- *Mem:* Am Chem Soc. *Res:* Photopolymerization; applications of polymers and photopolymers in printed circuit resists. *Mailing Add:* 1032 Douglas Ave Naperville IL 60540

JONES, THOMAS OSWELL, b Oshkosh, Wis, May 13, 08; m 50; c 2. CHEMISTRY. *Educ:* Wis State Col, Oshkosh, BE, 30; Univ Wis, PhM, 34, PhD(inorg chem), 37. *Prof Exp:* From instr to prof chem, Haverford Col, 38-56; actg head, Off Sci Info, NSF, 56-58, head, Off Antarctic Prog, 58-65, div dir environ sci, 65-69, dep asst dir nat & int progs, 69-75, spec dep asst dir int activities, 75-78; RETIRED. *Concurrent Pos:* Asst to sect chief, Metall Lab, Univ Chicago, 44-45, sect chief, Info Div, 45-46. *Mem:* Am Chem Soc; Am Geophys Union. *Res:* Cryoscopy; isotopes; micro-analytical methods; solubilities; preparation, properties and reactions of isotopes of hydrogen and oxygen; radiation chemistry. *Mailing Add:* 7504 Holiday Terr Bethesda MD 20034

JONES, THOMAS S, b Oakland, Md, Nov 11, 29; m 55; c 3. GENERAL EARTH SCIENCES. *Educ:* Univ Ill, BS, 50, MS, 51; Pa State Univ, PhD(metall), 61. *Prof Exp:* Res asst metall, Gen Elec Res Lab, 51-56; res assoc, Pa State Univ, 61-62; sr res metallurgist, Allegheny Ludlum Res Ctr, 62-70; staff metallurgist, 71-74, PHYS SCIENTIST, US BUR MINES, 74- *Mem:* Am Inst Mining Metall & Petrol Engrs. *Res:* Phase equilibria in metal and oxide systems at high temperatures; kinetics of reduction in metallurgical systems; steelmaking. *Mailing Add:* 4212 Braeburn Dr Fairfax VA 22032

JONES, THOMAS V, b Pomona, Calif, July 21, 20. RESEARCH ADMINISTRATION. *Educ:* Stanford Univ, BS, 42. *Hon Degrees:* LLD, George Washington, 67. *Prof Exp:* Tech adv, Brazilian Air Ministry, Rio de Janeiro, 47-51; mem staff, Rand Corp, 51-53; asst chief engr & sr vpres develop planning, Northrop, 53-59, pres, 59-60, chief exec officer, 60-63, chmn bd, 63-90, MEM, BD DIRS, NORTHROP, 90- *Concurrent Pos:* Prof & dept head, Aeronaut Inst Technol Brazil, 51-53; mem bd dirs, MCA, Inc; chmn bd govs, Aerospace Indust Asn, 85. *Honors & Awards:* Reed Aeronaut Award, Am Inst Aeronaut & Astronaut, 85. *Mem:* Hon fel Am Inst Aeronaut & Astronaut; Aerospace Indust Asn. *Mailing Add:* 650 N Sepulveda Blvd Los Angeles CA 90049

JONES, THOMAS WALTER, b Odessa, Tex, June 22, 45; m 68; c 1. THEORETICAL ASTROPHYSICS. *Educ:* Univ Tex, Austin, BS, 67; Univ Minn, MS, 69, PhD(physics), 72. *Prof Exp:* Asst res physicist, Univ Calif, San Diego, 72-75; asst scientist, Nat Radio Astron Observ, 75-77, assoc scientist, 77; from asst prof to assoc prof, 78-84, CHMN, DEPT ASTRON, 81-, PROF

ASTRON, UNIV MINN, 84- *Mem:* Am Astron Soc; Int Astron Union; Sigma Xi; Royal Astron Soc. *Res:* Studies of physical processes in cosmic radio and infrared sources; nuclei of galaxies and quasars; radiation transfer in circumstellar environments; numerical hydrodynamics. *Mailing Add:* Dept Astron Univ Minn Minneapolis MN 55455

JONES, TIMOTHY ARTHUR, NEUROPHYSIOLOGY. *Educ:* Univ Calif, PhD(physiol), 80. *Prof Exp:* asst prof, 82-88, ASSOC PROF PHYSIOL, MED CTR & ASSOC PROF, DEPT SPEC EDUC & COMMUN DIS, UNIV NEBR, LINCOLN, 88- *Concurrent Pos:* NASA res assoc award, Stanford Univ, 80 & 81; fac fel, Army Equip Eng Estab, NASA, 88 & 90; mem, Comn Gravitational Physics, Int Union Psychol Sci. *Res:* Ontogeny of sensory systems; gravitational physiology. *Mailing Add:* Dept Oral Biol Col Dent Univ Nebr Med Ctr Lincoln NE 68583

JONES, TREVOR O, b Maidstone, Eng; m. ENGINEERING. *Prof Exp:* Dir, Delco Electronics Div, Gen Motors, 59-70, Automotive Electronic Control Systs, 70-72, Advan Prod Eng, 72-74, Gen Motors Proving Ground, 74-78; vpres eng, Automotive Worldwide Sector, TRW Inc, 78-79, group vpres & gen mgr, Transp Electronics Group, 79-85, group vpres strategic progs, Automotive Worldwide Sector, 85, group vpres sales & mkt, 85-87; CHMN BD, LIBBEY-OWENS-FORD, 87- *Concurrent Pos:* Mem, Safety Res for a Changing Hwy Environ Comt, Nat Acad Eng, Nat Interests in an Age of Global Technol Comt; pres, Int Develop Corp; mem, Nat Motor Vehicle Safety Adv Coun, 71-, vchmn, 72; mem, Nat Hwy Safety Adv Comt, 75-78, chmn, 78; mem, Transp Res Bd Comt, Nat Res Coun. *Honors & Awards:* Hooper Mem Prize, Brit Inst Elec Engrs, 50; Arch T Colwell Award, Soc Automotive Engrs, 74 & 75, Vincent Bendix Automotive Electronics Eng Award, 76, Buckendale Lectr, 86 & Edward N Cole Automotive Eng Award, 88; Safety Award Eng Excellence, US Dept of Transp, 78. *Mem:* Nat Acad Eng; fel Brit Soc Elec Engrs; fel Am Inst Elec & Electronics Engrs; fel Soc Automotive Engrs. *Res:* Automotive safety and electronics. *Mailing Add:* Libbey-Owens-Ford Co One Cleveland Ctr Suite 2900 1375 E Ninth St Cleveland OH 44114

JONES, ULYSSES SIMPSON, JR, b Portsmouth, Va, Feb 14, 18; m 41; c 1. SOIL FERTILITY, ATMOSPHERIC CHEMISTRY & PHYSICS. *Educ:* Va Polytech Inst, BS, 39; Purdue Univ, MS, 42; Univ Wis, PhD(soils), 47. *Prof Exp:* Chemist, F S Royster, 42; asst, Univ Wis, 46-47; assoc prof, Miss State Univ, State Col Miss, 47-53; agronomist, Olin Mathieson Chem Corp, 53-60; head, dept agron & soils, 60-71, PROF AGRON & SOILS, CLEMSON UNIV, 71- *Concurrent Pos:* Guest prof, Oak Ridge Inst Nuclear Studies, 50 & Univ Ark,52; vis scientist, Int Rice Res Inst, 80; dir, Rural Develop Inst, Cuttington Univ, Liberia, W Africa, 83-85; vis prof, Univ Philippines, 79-80; consult, A G Edwards, 62, J P Stevens, 75, Blount Int, 83, UN Develop Prog, 83 & Gilbert, 88- 91. *Honors & Awards:* Fulbright lectr, Aegean Univ, 76-77 & Univ Zimbabwe, 86-87. *Mem:* Fel Soil Sci Soc Am; fel Am Soc Agron; Am Chem Soc; Entom Soc Am. *Res:* Availability of phosphates in soils; soil acidity and organic matter; limestone availability to crops and reaction in soil; use of radioactive elements for soil and fertilizer studies; insecticide and fertilizer mixtures for leaf feeding and pest control; trends in sulfur supply in air, rainwater and soil; environmental impact and monitoring of acid precipitation. *Mailing Add:* 111 Strawberry Lane Clemson SC 29631

JONES, VERNON DOUGLAS, b Florence, Ala, July 15, 37; m 61; c 1. PHARMACOLOGY. *Educ:* Florence State Univ, BA, 58; Vanderbilt Univ, PhD(pharmacol), 64; Univ NMex, JD, 74. *Prof Exp:* NIH fel pharmacol, Sch Med, Vanderbilt Univ, 64-65; NIH fel, Med Col, Cornell Univ, 65-67; asst prof pharmacol, 67-71, asst prof psychiat, 71-73, CLIN ASSOC, DEPT PSYCHIAT, SCH MED, UNIV NMEX, 73-, PHARMACOLOGIST, MENT HEALTH CTR, 73-; ATTORNEY AT LAW, 75- *Res:* Drugs in criminal and civil law. *Mailing Add:* 1400 Central SE Suite 3100 Albuquerque NM 87106

JONES, VICTOR ALAN, b Fremont, Mich, Feb 24, 30; m 54; c 4. FOOD ENGINEERING, FOOD SCIENCE. *Educ:* Mich State Univ, BS, 52, MS, 59, PhD(agr eng), 62. *Prof Exp:* From asst prof to assoc prof, 62-78, PROF FOOD ENG, NC STATE UNIV, 78- *Concurrent Pos:* Vis prof, Ore State Univ, 71. *Mem:* Am Soc Agr Engrs; Inst Food Technologists; Am Dairy Sci Asn; Sigma Xi. *Res:* Unit operations and control for ultrahigh temperature pasteurization or sterilization of foods; packaging materials and equipment. *Mailing Add:* Dept Food Sci NC State Univ Box 7624 Raleigh NC 27695-7624

JONES, WALTER H(ARRISON), b Griffin, Sask, Sept 21, 22; US citizen; wid. CHEMISTRY, SYSTEMS ANALYSIS. *Educ:* Univ Calif, Los Angeles, BS, 44, PhD(phys org chem), 48. *Prof Exp:* Res assoc, Univ Calif, Los Angeles, 48; chemist, Western Regional Res Lab, USDA, 48-51; chemist, Los Alamos Sci Lab, 51-54; sr res engr, NAm Aviation, Inc, 54-56; mgr, Chem Dept, Aeronutronic Div, Ford Motor Co, 56-60; panel chmn, Inst Defense Anal, 60-63; head, Propulsion Dept, Aerospace Corp, 63-64; sr scientist & head adv tech, Hughes Aircraft Co, 64-68; from assoc prof to prof aeronaut syst, 69-75, dir, Corpus Christi Ctr, 69-75, PROF CHEM, UNIV WFLA, 75- *Concurrent Pos:* Chmn, Thermochem Panel, Joint Army-Navy-Air Force-Adv Res Proj Agency, NASA, 60-62; consult, Fla Energy Comt & Solar Energy Ctr, 74-79, Eng Soc Comn Energy, 82-83 & Naval Surface Weapons Ctr, 82-; vis prof, Univ Toronto, 78-79; consult, Eng Soc Comn Energy, 82-83 & Naval Surface Weapons Ctr, 82- *Mem:* Am Chem Soc; fel Am Inst Chemists; NY Acad Sci; AAAS; Int Solar Energy Soc; World Asn Theoret Org Chemists. *Res:* Chemical kinetics; polymer chemistry; thermodynamics; combustion; propulsion; missile and space systems analysis and engineering; energy systems analyses; quantum chemistry; chemistry at high pressures. *Mailing Add:* Dept Chem Univ WFla Pensacola FL 32514

JONES, WALTER LARUE, entomology, for more information see previous edition

JONES, WESLEY MORRIS, b Raymond, Wash, Apr 29, 19; m 54; c 3. PHYSICAL CHEMISTRY. *Educ:* Univ Calif, AB, 40, PhD(chem), 46. *Prof Exp:* Asst chem, Univ Calif, 40-42; staff mem, Los Alamos Nat Lab, 43-86; GUEST SCIENTIST, 86- *Concurrent Pos:* Res fel, Calif Inst Technol, 57-58. *Mem:* Am Chem Soc; Sigma Xi. *Res:* Low temperature specific heats; thermodynamics; gas kinetics; physical chemistry of tritium and hydrogen isotope effects; diffusion; thermochemical cycles for hydrogen production; plutonium environmental chemistry. *Mailing Add:* 4753 Sandia Dr Los Alamos CA 87544

JONES, WILBER CLARK, b Grove City, Pa, Jan 21, 41; m 62; c 2. INORGANIC CHEMISTRY. *Educ:* Westminster Col, Pa, BS, 62; Univ Tenn, PhD(chem), 66. *Prof Exp:* From asst prof to assoc prof, 66-77, PROF CHEM, CONCORD COL, 77-, CHMN, PHYS SCI DEPT, 74- *Mem:* Am Chem Soc. *Res:* Synthesis and structural studies on coordination compounds. *Mailing Add:* Dept Phys Sci Concord Col Athens WV 24712

JONES, WILBUR DOUGLAS, JR, b Augusta, Ga, July 3, 27; m 52; c 2. MICROBIOLOGY. *Educ:* Emory Univ, AB, 49; WVa Univ, MS, 51; Med Col Ga, PhD(microbiol), 68. *Prof Exp:* Asst, WVa Univ, 51; instr sci, Truett-McConnell Jr Col, 51-53; bacteriologist, Ga State Health Dept, 53-60, chief bacteriologist, Training Lab, 60-62; RES MICROBIOLOGIST, CTR DIS CONTROL, USPHS, 62- *Concurrent Pos:* Instr biol, Ga State Univ, 55-63. *Mem:* Am Soc Microbiol; fel Am Acad Microbiol. *Res:* Genetics and phage typing of the mycobacteria; genetics and molecular biology of the mycobacteriophages. *Mailing Add:* Mycobacteriol Br Ctr for Disease Control Atlanta GA 30333

JONES, WILLIAM B, b Spring Hill, Tenn, Sept 24, 31; m 56; c 5. NUMERICAL ANALYSIS, COMPLEX ANALYSIS. *Educ:* Jacksonville State Col, BA, 53; Vanderbilt Univ, MA, 55, PhD(math), 63. *Prof Exp:* Mathematician, Nat Bur Standards, 58-63; actg asst prof math, Univ Colo, 63-64, asst prof appl math, 64-68, assoc prof math, 68-73, assoc chmn dept, 72-74, chmn dept, 87-90, PROF MATH, UNIV COLO, BOULDER, 73- *Concurrent Pos:* Consult, Nat Bur Standards, 64-65 & Environ Sci Servs Admin, 65-70; consult, Off Telecommun-Inst Telecommun Sci, 70-73, mathematician, 70-71; vis prof, Univ Kent, Canterbury, UK, 77; vis Fulbright prof, Univ Trondeim, Norway, 84-85; Fulbright res-scholar award, 84-85; award grant, Norwegian Marshall Fund, 84-85. *Honors & Awards:* Gold Medal Award, US Dept Com, 65. *Mem:* Am Math Soc; Math Asn Am; Soc Indust & Appl Math. *Res:* Complex analysis; numerical analysis; approximation theory; continued fractions; Pade Approximants. *Mailing Add:* 455 Erie Dr Boulder CO 80303

JONES, WILLIAM B(ENJAMIN), JR, b Fairburn, Ga, Sept 17, 24; m 48; c 3. ELECTRICAL ENGINEERING. *Educ:* Ga Inst Technol, BS, 45, MS, 48, PhD(elec eng), 53. *Prof Exp:* Engr, radar develop, R I Sarbacher & Assoc, 47-48; from instr to assoc prof elec eng & res assoc, Ga Inst Technol, 48-54; res engr, Hughes Aircraft Co, 54-58; prof elec eng, Ga Inst Technol, 58-67; head dept, Tex A&M Univ, 67-84, prof elec eng, 67-90; RETIRED. *Concurrent Pos:* Vis prof, Univ Fla, 84-85. *Mem:* Sr mem Inst Elec & Electronics Engrs; Optical Soc Am; Am Soc Eng Educ. *Res:* Communications theory and systems; optical communication systems. *Mailing Add:* 2612 Melba Circle Bryan TX 77802

JONES, WILLIAM B, b Littlefield, Tex, June 10, 37; m 56; c 6. ASSISTIVE DEVICES FOR HANDICAPPED, POLYMER FRACTURE. *Educ:* Tex Tech Univ, BS, 59, MS, 60; Univ Utah, PhD(mech eng), 70. *Prof Exp:* Teaching asst mech eng, Tex Tech Univ, 59-60, assoc prof, 82-89; mem tech staff, Rocketdyne Div, Rockwell Int, 60-75; mat res engr, Wright AFB Aero Labs, 75-82; CHIEF ENGR, B J ENTERPRISES, 89- *Concurrent Pos:* Res asst mech eng, Univ Utah, 66-69; consult, Lockheed Propulsion Co, 67-69, Textron, Bell Helicopter, 74-75 & Fredrick R Harris, 78-79; lectr, Kent State Univ, 78-82 & Univ Calif, Los Angeles, 79-85. *Mem:* Adhesion Soc; Soc Aerospace Mat & Process Engrs; Am Chem Soc; Am Solar Energy Soc. *Res:* Fracture mechanics of polymers and adhesive joints; materials and processes for adhesives and advanced composite materials. *Mailing Add:* B J Enterprises 1200 W 14th St Box 968 Littlefield TX 79339-0968

JONES, WILLIAM BARCLAY, b San Francisco, Calif, Aug 18, 19; m 39; c 3. PHYSICS. *Educ:* Univ Calif, AB, 47, PhD(physics), 64. *Prof Exp:* Accelerator supvr, Crocker Lab, Univ Calif, 55-62; nuclear physicist, Tech Measurement Corp, 62-67; res assoc physics, Yale Univ, 68-75; physicist, Brookhaven Nat Lab, 75-85; CONSULT 85- *Mem:* Am Phys Soc; AAAS. *Res:* Low and intermediate energy; experimental nuclear physics. *Mailing Add:* Rte 2 Creek Rd Wading River NY 11792

JONES, WILLIAM DAVIDSON, b Folsom, Pa, Oct. 26, 53; m 77; c 3. INORGANIC CHEMISTRY, ORGANIC CHEMISTRY. *Educ:* Mass Inst Technol, BS, 75; Calif Tech, PhD(chem), 79. *Prof Exp:* Nat Sci Found Postdoctoral, chem, Univ Wis, Madison, 79-80; PROF CHEM, UNIV ROCHESTER, 80- *Concurrent Pos:* Ap Sloan fel, Sloan Found, 84; Guggenheim fel, John S Guggenheim Found, 88; Camille & Henry Dreyfus fel, 85-88; Fulbright fel, 88-89; Royal Soc fel, 88-89. *Mem:* Am Chem Soc. *Res:* Inorganic and organometallic chemistry; mechanism and thermodynamics of carbon-hydrogen bond activity by homogeneous transition metal complexes. *Mailing Add:* Dept Chem Univ Rochester Rochester NY 14627

JONES, WILLIAM DENVER, b Jenkinjones, WVa, Apr 14, 35; div; c 2. RELATIVISTIC ELECTRON BEAMS. *Educ:* Berea Col, BA, 58; Vanderbilt Univ, MA, 61, PhD(physics), 63. *Prof Exp:* Res assoc plasma physics, Thermonuclear Div, Oak Ridge Nat Lab, 63-70; assoc prof, 70-72, PROF PHYSICS, UNIV SFLA, 72- *Concurrent Pos:* AEC & Energy Res Develop Admin res contracts, 71-76; Air Force Cambridge Res Labs contract, 72-74; consult, Solarkit of Fla, Tampa, 75-80, Naval Res Lab, 84-86. *Mem:* Fel Am Phys Soc; Am Asn Physics Teachers; Sigma Xi. *Res:* Basic and applied research in plasmas; particularly pulsed power, applied research in alternative energy sources, with emphasis on solar energy. *Mailing Add:* Dept Physics Univ SFla Tampa FL 33620

JONES, WILLIAM ERNEST, b Sackville, NB, Can, Aug 7, 36; m 58; c 4. PHYSICAL CHEMISTRY. *Educ:* Mt Allison Univ, BSc, 58, MSc, 59; McGill Univ, PhD(phys chem), 63. *Prof Exp:* Res assoc, Mt Allison Univ, 59-60; from asst prof to prof phys chem, Dalhousie Univ, 62-89, chmn chem dept, 74-83, chmn Dalhousie Senate, 83-89; PROF CHEM & DEAN SCI, ST MARY'S UNIV, 89- *Mem:* Chem Inst Can; Sigma Xi; Can Asn Physicists; Spectros Soc Can. *Res:* Kinetics; spectroscopy; surface chemistry; catalysis; gas phase kinetics of atoms and free radicals; atomic and molecular spectroscopy. *Mailing Add:* Dean Sci St Mary's Univ Halifax NS B3H 3C3 Can

JONES, WILLIAM F, b Sanford, NC, Sept 5, 27; m 51; c 1. GENETICS, SCIENCE EDUCATION. *Educ:* Davis & Elkins Col, BA, 51; Madison Col, MS, 58; Univ Va, EdD(sci educ), 68. *Prof Exp:* Teacher, Louisa County High Sch, Va, 51-53 & Handley High Sch, Winchester, 53-58; assoc prof, 58-78, PROF BIOL, JAMES MADISON UNIV, 78- *Concurrent Pos:* Dir, NSF Inst. *Res:* General zoology; genetics; population genetics; evolution; biometrics; psychometrics; marine sciences; curriculum and instruction. *Mailing Add:* 138 Fairway Dr Harrisonburg VA 22801-8770

JONES, WILLIAM HENRY, JR, b Waycross, Ga, Mar 3, 04; m 61; c 1. CHEMISTRY. *Educ:* Emory Univ, BS, 24; Princeton Univ, MA, 25, PhD(phys chem), 29. *Prof Exp:* Asst, Princeton Univ, 24-27; instr chem, 27-30, from asst prof to prof, 30-72, EMER PROF CHEM & ADMIN ASST, EMORY UNIV, 72- *Concurrent Pos:* Fel, Calif Inst Technol, 41-42; lab dir, thermal diffusion plant, Manhattan Proj, 44-45; dir progs, NSF, Emory Univ, 59-66; dir progs, Ford Found, 62-67. *Honors & Awards:* Thomas Jefferson Award, 67. *Mem:* AAAS; Am Chem Soc; Am Phys Soc; fel Am Inst Chemists; Sigma Xi. *Res:* Chemical reaction kinetics; x-rays and molecular structure; precise calorimetry of latent heats; thermal diffusion of liquids; yields in proton-induced fission; chemical education. *Mailing Add:* Dept Chem Emory Univ Atlanta GA 30322

JONES, WILLIAM HOWRY, b Lancaster, Pa, Nov 6, 20; m 59; c 4. ORGANIC CHEMISTRY. *Educ:* Juniata Col, BS, 42; Columbia Univ, MA, 44; Mass Inst Technol, PhD(org chem), 47. *Prof Exp:* Lab asst chem, Juniata Col, 39-42; lab asst, Columbia Univ, 42-44, lectr, 43, asst, Manhattan Proj, SAM Labs, 44-45; asst, Anti-Malarial Proj, Mass Inst Technol, 45-46; Du Pont fel nuclear alkylation, Univ Ill, 47-48; instr chem, Univ Ill, 48-49; res chemist, Merck Inc, 49-59, RES ASSOC, MERCK SHARP & DOHME RES LABS, 59- *Mem:* Am Chem Soc; fel NY Acad Sci. *Res:* Synthesis of physiologically active compounds; reaction mechanisms; synthesis of substituted diamines and quinoline derivatives; catalytic hydrogenation; high pressure research. *Mailing Add:* 449 Chelsea Circle NE Northeast Atlanta GA 30307

JONES, WILLIAM J, b New York, NY, Mar 23, 15; m 42; c 3. ENGINEERING PHYSICS. *Educ:* Tufts Univ, BS, 41; Newark Col Eng, MS, 50. *Prof Exp:* Lectr physics, Harvard Univ, 63-72; SR STAFF RES ASSOC, ENERGY LAB, MASS INST TECHNOL, 72- *Res:* High energy physics; energy technologies issues and policies. *Mailing Add:* 92 Bullough Park Newton MA 02160

JONES, WILLIAM JONAS, JR, b Whaleyville, Va, Nov 18, 41; m 80. ORGANIC CHEMISTRY, TEXTILE FIBERS. *Educ:* Col William & Mary, BS, 63; Duke Univ, PhD(org chem), 66. *Prof Exp:* RES ASSOC, DACRON RES LAB, E I DU PONT DE NEMOURS & CO, INC, 66- *Mem:* Am Chem Soc. *Res:* Heterocyclic organic compounds; polymers; chemistry of textile fibers; fiber engineering. *Mailing Add:* Rte 13 Box 565 Greenville NC 27858

JONES, WILLIAM MAURICE, b Campbellsville, Ky, Jan 12, 30; m 56; c 3. SYNTHETIC INORGANIC & ORGANOMETALLIC CHEMISTRY. *Educ:* Union Univ, Tenn, BS, 51; Univ Ga, MS, 53; Univ Southern Calif, PhD(org chem), 55. *Prof Exp:* Instr chem, Univ Southern Calif, 55-56; from asst prof to assoc prof, 56-65, chmn dept, 68-73, prof chem, 65-90, DISTINGUISHED SERV PROF, UNIV FLA, 90- *Concurrent Pos:* Sloan fel, 63-67; NATO sr sci fel, 71; mem ed bd, Chem Rev, 71-74, J Organic Chem, 74-79 & Petrol Res Fund Adv Bd, 86-93, chmn, 90-93. *Mem:* Am Chem Soc. *Res:* Mechanisms of organic reactions; strained allenes and their transition metal complexes; transition metal complexes of conjugated carbocyclic carbenes; transition metal organometallic rearrangements. *Mailing Add:* Dept Chem Univ Fla Gainesville FL 32606

JONES, WILLIAM PHILIP, b Chicago, Ill, Oct 2, 42; m 64; c 2. EXPERIMENTAL NUCLEAR PHYSICS. *Educ:* Univ Notre Dame, BS, 64; Univ Mich, MS, 65, PhD(physics), 69. *Prof Exp:* Res assoc, 69-71, STAFF SCIENTIST PHYSICS, IND UNIV, BLOOMINGTON, 71- *Mem:* Am Phys Soc; Sigma Xi; AAAS. *Res:* Medium energy nuclear physics; charged-particle reactions; properties of nuclear energy levels; cyclotron orbit dynamics; charged particle beam optics. *Mailing Add:* 308 S High St Bloomington IN 47401

JONES, WILLIAM VERNON, b Yellville, Ark, Jan 25, 35; m 55; c 3. COSMIC PHYSICS, ASTROPHYSICS. *Educ:* Univ Tulsa, BS, 63; La State Univ, Baton Rouge, PhD(physics), 67. *Prof Exp:* Res assoc physics, La State Univ, Baton Rouge, 67; guest res assoc, Max Planck Inst Extraterrestrial Physics, 67-68; res instr, 69-70, from asst prof to assoc prof, 70-81, PROF PHYSICS & ASTRON, LA STATE UNIV, BATON ROUGE, 81- *Concurrent Pos:* Alexander von Humboldt res stipend, 67-68; res physicist, Goddard Space Flight Ctr, NASA, 75-76; vis res scientist, Univ Tokyo Inst Cosmic Res, 81; vis sr scientist, Jet Propulsion Lab, 86-87, NASA Hq, 85-88. *Mem:* Am Inst Physics; Am Phys Soc. *Res:* Electromagnetic cascade measurements; properties of high-energy nuclear interactions; Monte Carlo simulations of study nuclear cascade processes; cosmic ray composition and energy spectra. *Mailing Add:* Code ES Space Physics Div NASA Hq Washington DC 20546

JONES, WINSTON WILLIAM, b Eclectic, Ala, Jan 18, 10; m 33; c 2. HORTICULTURE. *Educ:* Ala Polytech Inst, BS, 31; Purdue Univ, MS, 33; Univ Chicago, PhD(plant physiol), 36. *Prof Exp:* Asst, Purdue Univ, 33; asst plant physiologist, Exp Sta, Univ Hawaii, 36-38, asst horticulturist, 38-42; from assoc horticulturist to horticulturist, Univ Ariz, 42-46; assoc horticulturist, 46-51, prof & horticulturist, 51-77, EMER PROF HORT & HORTICULTURIST, UNIV CALIF, RIVERSIDE, 77- *Mem:* AAAS; Am Soc Hort Sci. *Res:* Mineral and organic nutrition of citrus; respiration in etiolated wheat seedlings as influenced by phosphorus nutrition. *Mailing Add:* Dept Bot & Plant Sci Univ Calif Riverside CA 92521

JONES, WINTON D, JR, b Terre Haute, Ind, June 23, 41; m 64; c 2. MEDICINAL CHEMISTRY, BIOCHEMISTRY. *Educ:* Butler Univ, BS, 63, MS, 66; Univ Kans, PhD(med chem), 70. *Prof Exp:* ORG CHEMIST, MERRELL-DOW PHARMACEUT, INC, DOW CHEM CO, 80- *Concurrent Pos:* Cong sci consult. *Mem:* Am Chem Soc. *Res:* Medicinal chemistry, antiallergic agents; synthesis of central nervous system, cardiotonic agents, antihypertensives, antiviral and immunological agents. *Mailing Add:* Marion Merrell Dow Res Inst 2110 Galbraith Rd Cincinnati OH 45215

JONG, ING-CHANG, b Yunlin, Taiwan, Feb 5, 38; US citizen; m 66; c 2. SOLID MECHANICS. *Educ:* Nat Taiwan Univ, BS, 61; SDak Sch Mines & Technol, MS, 63; Northwestern Univ, Evanston, PhD(theoret & appl mech), 65. *Prof Exp:* From asst prof to assoc prof eng sci, Univ Ark, Fayetteville, 65-74, prof, 74-; AT MECH ENG DEPT, UNIV ARK. *Concurrent Pos:* Prin investr, eng res initiation grant, NSF, 67-69, eng mech res grant, 69-71. *Mem:* Am Soc Mech Engrs; Am Soc Eng Educ; Am Acad Mech; Sigma Xi. *Res:* Nonconservative stability of damped structures; vibrations and dynamic stability of structural systems exhibiting yielding and hysteresis; senior author of one book. *Mailing Add:* Dept Mech Eng Univ Ark Fayetteville AR 72701

JONG, SHUNG-CHANG, b Taiwan, Nov 12, 36; US citizen; m 65; c 3. MYCOLOGY. *Educ:* Nat Taiwan Univ, BS, 60; Western Ill Univ, MS, 66; Wash State Univ, PhD(mycol), 69. *Prof Exp:* Asst plant pathologist, Taiwan Agr Res Inst, 61-63; asst instr mycol, Nat Taiwan Univ, 63-65; sr mycologist, 69-71, cur fungi, 71-73, CUR & HEAD MYCOL & BOT DEPT, AM TYPE CULTURE COLLECTION, 74- *Concurrent Pos:* Sci Found grants, 75- *Mem:* Brit Mycological Soc; Chinese Med & Health Asn; Int Soc Human & Animal Mycol; Med Mycol Soc Am; Mycol Soc Am; Int Asn Plant Tissue Culture; Int Mushroom Soc for Tropics; Int Soc Plant Molecular Biol; Japan Antibiotics Res Asn; Sigma Xi; Soc Fermentation Technol; US Fedn Culture Collections. *Res:* Preservation and industrial applications of living fungi; biology of fungi in culture. *Mailing Add:* Dept Mycol Am Type Cult Collection 12301 Parklawn Dr Rockville MD 20852

JONI, SAJ-NICOLE A, US citizen. COMPUTER SCIENCE. *Educ:* Univ Calif, San Diego, BA, 73, MS, 75, PhD(math), 77. *Prof Exp:* Instr appl math, Mass Inst Technol, 77-78; asst prof math, Carnegie-Mellon Univ, 78-80 & Mass Inst Technol, 80-81; ASST PROF COMPUT SCI & MATH, WELLESLEY COL, 81- *Mem:* Am Math Soc; Asn Women Math; Asn Comput Mach. *Res:* Formal languages; combinatorial and algebraic solutions to automata-theortic problems; integration and intersection of feminist theory and mathematics computer science; aritifical intelligence. *Mailing Add:* 76 Fayerweather Cambridge MA 02138

JONNARD, AIMISON, b Sewanee, Tenn, Aug 3, 16; m 61; c 4. CHEMICAL ENGINEERING. *Educ:* Kans State Univ, BS, 38; Columbia Univ, MS, 39; Univ Pittsburgh, PhD(chem eng), 49. *Prof Exp:* Engr, Exp Sta, E I du Pont de Nemours & Co, 39-41; instr chem eng, Kans State Univ, 41-45; sr technologist, Shell Chem Co, 49-54, mgr mkt anal, 54-59; mgr mkt res & develop, US Indust Chem Co Div, Nat Distillers & Chem Corp, 59-61; vpres, Celanese Chem Co, 61-63; sr corp planner, Exxon Chem Co, 63-71; CHIEF, ENERGY & CHEM DIV, US INT TRADE COMN, 71- *Mem:* Am Chem Soc; Am Inst Chem Engrs. *Res:* Chemical economics. *Mailing Add:* 1202 Old Stable Rd McLean VA 22102

JONSEN, ALBERT R, b San Francisco, Calif, Apr 4, 31; m 76. MEDICAL ETHICS. *Educ:* Gonzaga Univ, BA, 55, MA, 56; Univ Santa Clara, STM, 63; Yale Univ, PhD, 67. *Prof Exp:* Assoc prof theol & philos, Univ San Francisco, 67-72; prof med ethics, Sch Med, Univ Calif, San Francisco, 72-88; PROF, DEPT MED HIST & ETHICS, UNIV WASH, SCH MED, 88- *Concurrent Pos:* Mem, Nat Comt Protection Human Subj Biomed Behavioral Res, 74-78, Pres Comt Study Ethical Problems in Med, 79-82; consult, Am Bd Internal Med, 78-; mem, Nat Bd Med Examnrs, 85-87; bd dir, Sierra Found, 86- *Honors & Awards:* McGovern Award, Am Osler Soc. *Mem:* Inst Med-Nat Acad Sci; fel Inst Soc Ethics & Life Sci; Soc Health Human Values (pres, 86); Am Soc Law & Med. *Mailing Add:* Dept Med Hist & Ethics A-205 Health Ctr Univ Wash Sch Med Seattle WA 98195

JONSSON, BJARNI, b Draghals, Iceland, Feb 15, 20; m 50, 70; c 3. MATHEMATICS. *Educ:* Univ Calif, Berkeley, AB, 43, PhD(math), 46. *Prof Exp:* From instr to asst prof, Brown Univ, 46-56; from assoc prof to prof, Univ Minn, 56-66; DISTINGUISHED PROF MATH, VANDERBILT UNIV, 66- *Concurrent Pos:* Vis prof, Univ Iceland, 54-55; vis assoc prof, Univ Calif, Berkeley, 55-56, vis prof & res mathematician, 62-63. *Mem:* Am Math Soc; Asn Symbolic Logic; Am Asn Univ Prof. *Res:* Universal algebra; lattice theory. *Mailing Add:* Dept Math Vanderbilt Univ PO Box 1541 Sta B Nashville TN 37235

JONSSON, HALDOR TURNER, JR, b State College, Pa, Jan 5, 29; m 64; c 2. BIOCHEMISTRY. *Educ:* Tex A&M Univ, BS, 52, MS, 61; Baylor Univ, PhD(biochem), 65. *Prof Exp:* Res asst plastics & resins, Shell Chem Corp, 56-59; res asst biochem, Tex A&M Univ, 59-61 & Col Med, Baylor Univ, 61-65; res assoc, Sch Med, Boston Univ, 65-66; asst prof chem, 66-70, ASSOC PROF BIOCHEM, MED UNIV SC, 70- *Concurrent Pos:* Clin chem consult, Vet Admin Hosp, Charleston, 66- *Mem:* Am Chem Soc; Am Soc Biol Chem; Am Oil Chemists' Soc; NY Acad Sci. *Res:* Gonadotropins and their influence on ovarian function; role of prostoglandins and essential fatty acids in wounds; gas-liquid chromatography; long term effects of pesticides on mammals. *Mailing Add:* Dept Biochem Med Univ SC 171 Ashley Ave Charleston SC 29425

JONSSON, JOHN ERIK, b New York, NY, Sept 6, 01; m 23; c 3. ENGINEERING. *Educ:* Rensselaer Polytech Inst, ME, 22. *Hon Degrees:* DEng, Rensselaer Polytech Inst, 59 & Polytech Inst NY, 84; DSc, Hobart & William Smith Col, 61, Austin Col, 63; LLD, Southern Methodist Univ, 64; DL, Carnegie-Mellon Univ, 72 & Skidmore Col, 72; DCL, Univ Dallas, 68; DHL, Okla Christian Col, 73; DS, Tulane Univ, 81. *Prof Exp:* Mem mgt staff, Aluminum Co Am, 22-23 & 29-30,; mem mgt staff, Dumont Motor Car Co, 27-29; mem mgt staff, Tex Instruments, Inc, 30-42, pres, 51-58, chmn bd, 58-66, hon chmn bd, 67-77, HON DIR, TEX INSTRUMENTS, INC, 77- *Concurrent Pos:* Dir, Repub Tex Corp, 54-80, Dallas Power & Light, 55-64, Neiman-Marcus, 56-65 & Equitable Life Assurance Soc Am, 58-73. *Honors & Awards:* Bene Merenti Medal, 66; Gantt Medal, 68; Hoover Medal, 70; Chauncey Rose Medal, Rose-Hulman Inst, 72; Founders Medal, Nat Acad Eng, 74; John Ericsson Medal, Am Soc Swedish Engrs. *Mem:* Nat Acad Eng; Soc Explor Geophysicists; Am Mgt Asn. *Mailing Add:* NCNB Tower II Suite 3300 325 N St Paul Dallas TX 75201

JONSSON, WILBUR JACOB, b Winnipeg, Man, Sept 18, 36. MATHEMATICS. *Educ:* Univ Man, BSc, 58, MSc, 59; Univ Tubingen, Dr rer nat(math), 63. *Prof Exp:* Lectr math, Univ Man, 59-60, asst prof, 62-65; lectr, Univ Birmingham, 65-66; ASSOC PROF MATH, McGILL UNIV, 66- *Res:* Projective planes; group theory; combinatory mathematics; foundations of geometry. *Mailing Add:* Dept Math McGill Univ Box 6070 Sta West Montreal PQ H3A 2M5 Can

JONTE, JOHN HAWORTH, b Moscow, Idaho, Oct 21, 18; m 42; c 4. GEOCHEMISTRY, INORGANIC CHEMISTRY. *Educ:* Univ of the Pac, AB, 40; Wash State Univ, MS, 42; Univ Ark, PhD(chem), 56. *Prof Exp:* Jr chemist, US Bur Mines, Nev, 42-44 & Shell Develop Co, Calif, 44-46; instr chem, Iowa State Univ, 46-51; instr geol, Univ Ark, 54-55; res chemist, Texaco Inc, Tex, 55-61, group leader geochem, 61-66; assoc prof geochem & anal, 66-69, prof chem & head dept, 69-85, EMER PROF CHEM, SDAK SCH MINES & TECHNOL, 85- *Concurrent Pos:* consult, 85- *Mem:* AAAS; Am Chem Soc; Geochem Soc; Am Ins; Sigma Xi. *Res:* Method development of drugs in biological fluids incorporating analytical instrumentation such as gas and liquid chromatography, fluorescence, techniques specializing in electroanalytical chemistry and computerized data reduction; laboratory information management, system management, system analysis and robotics based on implementation of information and method development. *Mailing Add:* Nine Watchung Rd E Brunswick NJ 08816

JONZON, ANDERS, b Stockholm, Sweden, May 6, 48; m 72; c 3. NEONATOLOGY, PEDIATRIC CARDIOLOGY. *Educ:* Uppsala Univ, MedKand, 70, MedDr(physiol), 72, Läkarexamen, 77. *Prof Exp:* CONSULT & LECTR NEONATOLOGY, DEPT PEDIAT, UNIVERSITY HOSP, UNIV UPPSALA, 84-, CONSULT & LECTR PEDIAT CARDIOL & PEDIAT INTENSIVE CARE, 90- *Concurrent Pos:* Julius Comroe Jr fel, Cardiovasc Res Inst, Univ Calif, San Francisco, 84-86. *Mem:* Europ Soc Pediat Res; Scand Physiol Soc; Am Physiol Soc. *Res:* Positive pressure breathing; control of respiration; lung development. *Mailing Add:* Dept Pediat University Hosp Upsala S-751 85 Sweden

JOOS, BARBARA, b South Amboy, NJ, Nov 2, 57. ANIMAL PHYSIOLOGY. *Educ:* Rutgers Univ, BA, 79; Univ Mich, MS, 82, PhD(biol sci), 86. *Prof Exp:* Lectr physiol, Univ Mich, 85; researcher, Cook Col, Rutgers Univ, 86-88, fel, Dept Biol, 88-89; ASST PROF BIOL, DEPT BIOL, MUHLENBERG COL, 89- *Concurrent Pos:* Vis researcher, Cook Col, Rutgers Univ, 89- *Mem:* AAAS; Am Soc Zoologists. *Res:* Energetics of locomotion in insects and mechanics of terrestrial locomotion in larval insects; temperature effects and size effects on energetics and performance of ectothermic insects. *Mailing Add:* Dept Biol Muhlenberg Col 2400 Chew St Allentown PA 18104

JOOS, BÉLA, b Montreal, Que, Aug 7, 53. SOLID STATE PHYSICS. *Educ:* Loyola Montreal, BSc, 74; McGill Univ, PhD(physics), 79. *Prof Exp:* Res fel, Univ Calif, Berkeley, 79-81; res assoc, Simon Fraser Univ, 81-82, asst prof, 82-84; asst prof, 84-87, ASSOC PROF PHYSICS, UNIV OTTAWA, 87- *Concurrent Pos:* Assoc ed, Can J Physics & Physics in Can, 84- *Mem:* Can Asn Physicists; Am Phys Soc. *Res:* Theoretical solid state physics; structural properties of surfaces, interfaces and monolayers. *Mailing Add:* Dept Physics Univ Ottawa Ottawa ON K1N 6N5 Can

JOOS, HOWARD ARTHUR, b Albany, NY, May 27, 22; m 46; c 3. PEDIATRICS, PEDIATRIC CARDIOLOGY. *Educ:* Univ Rochester, MD, 45. *Prof Exp:* Intern pediat, Harriet Lane Home, Johns Hopkins Univ, 45-46; asst resident, Strong Mem Hosp, Univ Rochester, 48-49, chief resident, 50-51, asst pediatrician, 51-54, from instr to asst prof, 51-55; assoc prof, Sch Med, Univ Southern Calif, 55-61; clin asst prof, Univ Colo, 61-64; prof pediat, State Univ NY Downstate Med Ctr, 64-73; prof pediat & chmn dept, Sch Med, Univ NDak, 73-76; clin prof pediat, Med Col Ohio, 76-79; CLIN PROF PEDIAT, STATE UNIV NY DOWNSTATE MED CTR, 80- *Concurrent Pos:* Vet fel, Sch Med, Univ Rochester, 48-49, 50-51; Masonic res fel, 49-50, Markle Found scholar, 51-55; mem staff, Convalescent Hosp Children, Rochester, NY; consult staff, Genesee Hosp; dir cardiovasc res, physiol res & pvt serv, Children's Hosp, Los Angeles, 55-59; chief pediat, Nat Jewish Hosp, Denver, Colo, 61-64; dir pediat serv, Maimonides Hosp Brooklyn, 64-73; dir med educ & pediat, Toledo Hosp, 76-79; dir pediat, Staten Island Hosp, 79- *Mem:* AAAS; Soc Pediat Res; AMA; Am Acad Pediat. *Res:* Cardiovascular diseases of infancy and childhood. *Mailing Add:* Staten Island Hosp 475 Seaview Ave Staten Island NY 10305

JOOS, RICHARD W, b Cologne, Minn, Sept 22, 34; m 60; c 4. BIOCHEMISTRY. *Educ:* Col St Thomas, BS, 58; Univ Minn, PhD(biochem), 64. *Prof Exp:* Teaching asst biochem, Univ Minn, 58-62, res assoc med, 62-66; biochemist, Vet Admin Hosp, Minneapolis, 66-67; RES SPECIALIST, 3M CTR, MINN MINING & MFG CO, 67- *Concurrent Pos:* Instr, Univ Minn Dent Sch, 71- *Mem:* Int Asn Dent Res. *Res:* Ion binding to macromolecules; humoral factors against bacteria; preventive agents for dental disease; dental materials. *Mailing Add:* 3934 Denmark Ave Eagan MN 55124

JOPLING, ALAN VICTOR, b Sydney, Australia, Oct 3, 24; m 50; c 2. GEOMORPHOLOGY, SEDIMENTOLOGY. *Educ:* Univ Sydney, BSc, 46, BE, 47; Harvard Univ, AM, 58, PhD(geol), 61. *Prof Exp:* Asst engr, Water Conserv Comn, NSW, 42-47; teaching fel geol, Univ Sydney, 47-49; lectr, NSW Univ Tech, 49-53; seismologist, Frontier Geophys Co, Alta, 53-54; geologist, Mobil Oil Can Ltd, 54-56 & water resources, US Geol Surv, 58-61; from asst prof to prof geog, 61-88, Univ Toronto, prof geol, 80-88; RETIRED. *Concurrent Pos:* NSF grant photogeol, Harvard Univ, 61. *Mem:* Geol Soc Am; Soc Econ Paleont & Mineral; Int Asn Sedimentol; Can Asn Geog; Eng Inst Can. *Res:* Fluid mechanics of sedimentation; physics of surface processes; processes of erosion, transportation and deposition of sedimentary materials; hydrology and water resources. *Mailing Add:* 1/73 Lauderdale Ave Fairlight New South Wales 2094 Australia

JOPPA, LEONARD ROBERT, b Billings, Mont, Sept 29, 30; m 59; c 4. GENETICS. *Educ:* Mont State Univ, BS, 57, PhD(genetics), 67; Ore State Univ, MS, 62. *Prof Exp:* Asst agron, Mont Agr Exp Sta, 57-62, asst agronomist, 62-64, res asst agron, 64-67; RES GENETICIST PLANTS, AGR RES SERV, USDA, 67- *Honors & Awards:* Res Award, Sigma Xi, 85. *Mem:* Fel AAAS; Am Soc Agron; Crops Sci Soc Am; Genetics Soc Am; Genetics Soc Can; Sigma Xi. *Res:* Genetics and cytogenetics of wheat and its relatives. *Mailing Add:* Northern Crop Sci Lab Box 5677 Fargo ND 58105

JOPPA, RICHARD M, b Littleton, Colo, Sept 29, 29; m 52; c 2. ELECTRICAL ENGINEERING, ELECTRONICS. *Educ:* Colo State Univ, BS, 51; Univ Ill, MS, 57, PhD(elec eng), 63. *Prof Exp:* USAF, 51-71, instr elec eng, USAF Inst Technol, 57-59, chief space physcics br, res directorate, Air Force Spec Weapons Ctr, NMex, 60-61, asst chief space vehicle div, test directorate, 62-64, from asst prof to assoc prof elec eng, USAF Acad, 64-68, chief anal br, survivability div & dir, vulnerability assessment directorate, Air Force Spec Weapons Ctr, Kirtland AFB, 68-71; elec-electronics engr, Los Alamos Nat Labs, Univ Calif, 71-90; RETIRED. *Concurrent Pos:* NSF fel, Univ Santa Clara, 67. *Mem:* Sr mem Inst Elec & Electronics Engrs; Nat Soc Prof Engrs. *Res:* Telemetry; instrumentation; information theory; circuit theory; experiment design, test and integration; control systems; electromagnetic environment energy conversion; research and development financial management; construction project management. *Mailing Add:* Ten Timber Ridge Los Alamos NM 87544

JOPSON, HARRY GORGAS MICHENER, b Philadelphia, Pa, June 23, 11; m 33; c 2. ZOOLOGY. *Educ:* Haverford Col, BS, 32; Cornell Univ, MA, 33, PhD(vert zool), 36. *Hon Degrees:* ScD, Bridgewater Col, 77. *Prof Exp:* Instr biol, Iowa State Teachers Col, 36; from asst prof to assoc prof, 36-46, prof biol, 46-81, EMER PROF, BRIDGEWATER COL, 81- *Concurrent Pos:* Asst dir, Overseas Oper, United Seamen's Serv, 43-46; trustee, Rockingham County Bd Educ, 57-76, chmn, 74-76; trustee, Nat Parks & Consrv Asn, 65-80. *Mem:* Soc Study Amphibians & Reptiles; Am Soc Ichthyol & Herpet; Am Soc Mammal; Am Ornith Union. *Res:* Salamanders of southeastern United States; vertebrate natural history. *Mailing Add:* PO Box 26 Bridgewater VA 22812

JORCH, HARALD HEINRICH, b WGer, Feb 17, 51; Can citizen; m 75; c 3. PHYSICS, SEMICONDUCTOR SURFACE. *Educ:* Univ Waterloo, BSc, 74, BSc, 75; Univ Guelph, MSc, 77, PhD(physics), 82. *Prof Exp:* Fel, Chalk River Nuclear Lab, Atomic Energy Can Co, 81-83; asst prof physics, Royal Roads Mil Col, 83-86; ASST PROF PHYSICS & COMPUT, WILFRID LAURIER UNIV, 86- *Concurrent Pos:* Tech collabr, Brookhaven Nat Lab, 79-81. *Mem:* Can Asn Physicists; Chem Inst Can; Soc Italiana Fisica; Europ Phys Soc. *Res:* Properties of surfaces and interfaces using particle beams (ions and positrons). *Mailing Add:* Dept Physics & Comput Wilfrid Laurier Univ Waterloo ON N2L 3C5 Can

JORDAN, A(NGEL) G(ONI), b Pamplona, Spain, Sept 19, 30; m 56; c 3. SOLID STATE ELECTRONICS & ADMINISTRATION. *Educ:* Univ Zaragoza, MS, 52; Carnegie Inst Technol, MS & PhD(elec eng), 59. *Hon Degrees:* Dr, Polytechnic Univ, Madrid, Spain, 85. *Prof Exp:* Res assoc electronics, Navy Res Lab, Madrid, Spain, 52-53, res fel electronics, 53-56; instr elec eng, Carnegie Inst Technol, 56-58, res fel semiconductors, Mellon Inst, 58-59, asst prof elec eng, Carnegie Inst Technol, 59-62, assoc prof, 62-66, prof phys electronics, 66-69, prof elec eng & head dept, 69-79, U A & Helen Whitaker prof electronics & elec eng, 72-79, dean, 79-83, provost, 83-91, UNIV PROF ELEC & COMPUTER ENG, CARNEGIE-MELLON UNIV, 90- *Concurrent Pos:* Consult var cos, 60-; dir, Allegheny Singer Res Corp, Allegheny Heart Inst, Pittsburgh High Technol Coun, Keithley Inst, Inc, Calif Micro Devices Corp, Magnascreen Corp & Ben Franklin Tech Ctr; distinguished Fulbright Scholar, 87. *Mem:* Fel Inst Elec & Electronics Engrs; Am Phys Soc; Am Soc Eng Educ; fel AAAS. *Res:* Solid state devices; integrated circuits; thin films; intelligent sensing devices and systems; robotics; applied electronics; semiconductor devices; software engineering. *Mailing Add:* Provost's Off Carnegie-Mellon Univ Pittsburgh PA 15213

JORDAN, ALBERT GUSTAV, b Oak Park, Ill, Jan 7, 41; m 64; c 2. MATERIALS SCIENCE, NUCLEAR METALLURGY. *Educ:* Purdue Univ, BS, MetE, 62; Univ Ill, Urbana, MS, 63, PhD(mat sci), 69. *Prof Exp:* ENGR, NUCLEAR METALL, KNOLLS ATOMIC POWER LAB, GEN ELEC CO, 68- *Mem:* Am Soc Metals; Am Nuclear Soc. *Res:* Behavior of nuclear fuel systems; prediction of performance of fuel systems. *Mailing Add:* 1134 Waverly Pl Schenectady NY 12308

JORDAN, ALBERT RAYMOND, b Alma, Kans, Oct 4, 06; m 32; c 3. PHYSICS. *Educ:* Univ Colo, BA, 29, MA, 33, PhD(physics), 40. *Prof Exp:* Instr physics, Univ Colo, 30-36 & Colo Agr & Mech Col, 36-37; from asst prof to assoc prof, Mont State Col, 37-41; physicist, Naval Ord Lab, Washington, DC, 41-42; from assoc prof to prof physics, Mont State Col, 42-52; sr res physicist, Denver Res Inst, Colo, 52-57; dean, 57-72, EMER DEAN, GRAD SCH, COLO SCH MINES, 72- *Concurrent Pos:* Physicist, Curtiss-Wright Res Lab, 44-45. *Honors & Awards:* Mines Medal. *Mem:* Am Geophys Union; Am Meteorol Soc. *Res:* Barometry and anemometry, atmospheric acoustics; geophysics. *Mailing Add:* 1603 S Uinta Way Denver CO 80231

JORDAN, ALEXANDER WALKER, III, b Richmond, Va, Apr 12, 45; m 72. ENDOCRINOLOGY, REPRODUCTIVE PHYSIOLOGY. *Educ:* Roanoke Col, BS, 67; Univ Richmond, MA, 69; Rutgers Univ, PhD(zool), 75. *Prof Exp:* Fel endocrinol, Dept Physiol & Biophysics, Colo State Univ, 75-78; STAFF FEL ENDOCRINOL, FOOD & DRUG ADMIN, HEW, 78- *Concurrent Pos:* Fel, Rockefeller Found, 75-77. *Mem:* Sigma Xi; Soc Study Reproduction. *Res:* Reproductive endocrinology; investigation into the mechanism of action of peptide hormones and prostaglandins on steroidogenesis. *Mailing Add:* 2755 Ordway St No 506 Washington DC 20008

JORDAN, ANDREW G, b Wrens, Ga, May 18, 39; m; c 3. AGRICULTURAL RESEARCH. *Educ:* Univ Ga, BS, 62; Clemson Univ, MS, 72, PhD(eng), 77. *Prof Exp:* Systs engr, Western Elec Co, 62-65; supvr, Advan Technol Training, Lockheed-Ga Aircraft Corp, 65-70; res scientist & instr, Agr Eng, Clemson Univ, 71-76; mgr mkt & processing technol, 76-81, asst dir tech serv, 82-83, DIR TECH SERV, NAT COTTON COUN, 83- *Concurrent Pos:* Cong task force rural transp; joint cotton breeding policy comt. *Mem:* Agr Res Inst; Am Soc Agr Engrs. *Res:* Agricultural engineering. *Mailing Add:* Nat Cotton Council Am PO Box 12285 Memphis TN 38182

JORDAN, ANDREW STEPHEN, b Mezokovesd, Hungary, May 1, 36; US citizen; m 68; c 2. METALLURGY, PHYSICAL CHEMISTRY. *Educ:* Pa State Univ, BS, 59; Univ Pa, PhD(metall), 65. *Prof Exp:* Engr, Philco Corp, Pa, 59-62; res engr, Westinghouse Res Labs, Pa, 62-63; MEM TECH STAFF, BELL TEL LABS, 65-; SUPVR HETEROSCTRUCTURE MAT GROUP, 84- *Mem:* Electrochem Soc; Am Asn Crystal Growth. *Res:* Crystal growth of compound semiconductors; chemical thermodynamics of optoelectronic materials with special emphasis on phase diagrams; impurity incorporation and defect chemistry; reliability of devices; crystal growth modeling; physical characterization; epotoscial growth. *Mailing Add:* 428 White Oak Ridge Rd Short Hills NJ 07974

JORDAN, ANGEL G, b Pamplona, Spain, Sept 19, 30; nat US. ELECTRICAL ENGINEERING, COMPUTER ENGINEERING. *Educ:* Univ Zaragoza, Spain, MS, 52; Univ Madrid, Spain, PhD(physics), 56; Carnegie Mellon Univ, MS, 59, PhD(elec eng), 59. *Prof Exp:* Instr, Dept Elec Eng, Carnegie Mellon Univ, 56-58, res fel, Mellon Inst Indust Res, 58-59, from asst prof to assoc prof, Dept Elec Eng, 59-66, actg chmn, Biomed Eng Prog, 76-78, head, Dept Elec & Computer Eng, 69-79, dean, Carnegie Inst Technol, 79-83, actg pres, Mellon Inst, 83-85, actg dir, Software Eng Inst, 86, actg dean, Mellon Col Sci, 87-88, PROF ELEC & COMPUTER ENG, CARNEGIE MELLON UNIV, 90- *Concurrent Pos:* Adj asst prof electronics, Naval Ord Sch, Madrid, Spain, 53-56; vis prof, Indian Inst Technol, Kampur, India, 71; vis sr scientist, Health & Safety Exec, Sheffield, Eng, 76. *Honors & Awards:* Sr Scientist Award, NATO. *Mem:* Nat Acad Eng; Inst Elec & Electronics Engrs; AAAS; Am Soc Eng Educ; Am Phys Soc; Sigma Xi. *Res:* Solid state devices; integrated circuits; thin films; gas sensing devices and systems; environmental and biomedical instrumentation; intelligent sensors; robotics; automation; knowledge engineering and software engineering focusing on technological change and technology transfer. *Mailing Add:* Off Provost Carnegie Mellon Univ 5000 Forbes Ave Pittsburgh PA 15213

JORDAN, ARTHUR KENT, b Philadelphia, Pa, Dec 28, 32; m 65; c 3. ELECTROMAGNETIC INVERSE SCATTERING, OPTICAL INTEGRATED CIRCUITS. *Educ:* Pa State Univ, BSc, 57; Univ Pa, MSc, 71, PhD(elec eng), 72. *Prof Exp:* Res engr, Res Div, Philco Corp, 58-61; engr, Astro-Electronics Div, Radio Corp Am, 62-64; physicist, Aerospace Physics Lab, Gen Elec Co, 64-69; res asst, Moore Sch Elec Eng, Univ Pa, 69-73; ELECTRONICS ENGR, CTR ADVAN SPACE SENSING, NAVAL RES LAB, 73-; PROG MGR, OFF NAVAL RES, 86- *Concurrent Pos:* Res fel, Dept Elec Eng, Univ Pa, 71-73; assoc ed, Inst Elec & Electronics Engrs Trans Antennas & Propagation, 81; mem, Advan Res Workshop Electromagnetic Imaging, NATO, WGer, 83; vis scientist, Mass Inst Technol, 89- *Mem:* Fel Inst Elec & Electronics Engrs; Electromagnetics Acad; Sigma Xi; Am Phys Soc; Optical Soc Am; Soc Indust & Appl Math; Inst Elec & Electronics Engrs Antennas & Propagation Soc; Inst Elec & Electronics Engrs Lasers & Electro-Optics Soc; Int Union Radio Sci; AAAS. *Res:* Electromagnetic inverse scattering theory; electromagnetic field theory; quantum electronics; optical waveguides and devices; remote sensing theory; author of numerous publications; holder of one US patent. *Mailing Add:* Naval Research Lab Ctr Advan Space Sensing Code 4210 Washington DC 20375-5000

JORDAN, BERNARD WILLIAM, JR, control engineering, for more information see previous edition

JORDAN, BRIGITTE, b Ger. MEDICAL ANTHROPOLOGY, CROSSCULTURAL OBSTETRICS. *Educ:* Calif State Univ, Sacramento, BA, 69, MA, 71; Univ Calif, Irvine, PhD(soc sci), 75. *Prof Exp:* Asst prof, Dept Anthrop & Community Med, 75-80, ASSOC PROF, DEPT ANTHROP & PEDIAT, MICH STATE UNIV, 80- *Concurrent Pos:* Res assoc, Feminist Women's Health Ctr, Santa Ana, Calif, 72-75; prin investr, res grant, Crosscultural Invest Childbirth Pract, Nat Inst Child Health & Human Develop, NIH, 77-79 & Cult Influences Response Physicians Diag, NSF, 84-86; mem exec bd, Soc Med Anthrop, 85-; consult, WHO, Geneva, Switz. *Honors & Awards:* Margaret Mead Award, Soc Appl Anthrop, 80. *Mem:* Fel Am Anthrop Asn; Soc Appl Anthrop; Soc Med Anthrop; Soc Visual Anthrop. *Res:* Design of culturally appropriate maternal and child health care delivery systems; integration of traditional and western medicine in developing countries; methodology, including videographic methods for documentation and analysis; patient-practitioner relationship; alternate systems of health care delivery; symbolic language of advertising; status of women; Maya Indians of Yucatan, Mexico. *Mailing Add:* Dept Anthrop Mich State Univ East Lansing MI 48824

JORDAN, BYRON DALE, b Akron, Ohio, Jan 24, 47; Can citizen; m 69; c 2. COMPUTER VISION, COLORIMETRY. *Educ:* Hiram Col, BA, 69; McMaster Univ, PhD(physics), 75. *Prof Exp:* Physicist, Welwyn Res Ltd, 75-77; HEAD, OPTICS SECT, PULP & PAPER RES INST CAN, 77- *Concurrent Pos:* Auxiliary prof chem eng, McGill Univ, 85- *Mem:* Am Phys Soc; Optical Soc Am; Tech Asn Pulp & Paper Indust; Inst Elec & Electronics Engrs; Soc Photo-Optical Instrumentation Engrs. *Res:* Paper physics; optical properties of paper; application of image processing to study random textures and fiber morphology; colorimetry and optical methods of quality control. *Mailing Add:* Pulp & Paper Res Inst Can 570 St John's Blvd Pointe Claire PQ H9R 3J9 Can

JORDAN, CARL FREDERICK, b New Brunswick, NJ, Dec 10, 35; m 67; c 2. ECOLOGY. *Educ:* Univ Mich, BS, 58; Rutgers Univ, MS, 64, PhD(ecol), 66. *Prof Exp:* Assoc scientist, P R Nuclear Ctr, AEC, 66-69; from asst ecologist to assoc ecologist, Radiol & Environ Res Div, Argonne Nat Lab, 69-74; RES ASSOC, INST ECOL, UNIV GA, 74-, SR ECOLOGIST, 79- *Concurrent Pos:* Vis scientist, Ecol Ctr, Venezuelan Inst Sci Invest, 74- *Honors & Awards:* Mercer Award, Ecol Soc Am, 73. *Mem:* AAAS; Ecol Soc Am; Soil Sci Soc Am; Sigma Xi. *Res:* Movement of chemical elements in soil; radiation recovery and mineral cycling in the tropical rain forest; application of systems analysis techniques to ecology; shifting agriculture in the Amazon Basin; laungya agriculture in Thailand. *Mailing Add:* Inst Ecol Univ Ga Athens GA 30602

JORDAN, CHARLES EDWIN, b South Charleston, Ohio, Jan 2, 27; m 54; c 3. ANIMAL NUTRITION. *Educ:* Ohio State Univ, BS, 51; Purdue Univ, MS, 55, PhD, 58. *Prof Exp:* Instr animal sci, Purdue Univ, 56-58, asst prof, 58-59; sr sceintist animal nutrit res, 59-61, from asst head to head, 61-71, dir agr res, Lilly Res Ctr Ltd, Eng, 71-74, dir prod info, Elanco Prod Co Div, 74-75, EXECUTIVE DIR, LILLY RES LAB DIV, ELI LILLY CO, 75- *Mem:* AAAS; Am Soc Animal Sci; Animal Nutrit Res Coun; Am Inst Biol Sci; Sigma Xi. *Res:* Swine nutrition and physiology; quality control of rations and feedstuffs; chemical regulation of growth. *Mailing Add:* 2191 E 100 S Greenfield IN 46140

JORDAN, CHARLES LEMUEL, b Ash Grove, Mo, May 28, 22; m 51; c 6. METEOROLOGY. *Educ:* Univ Chicago, PhB, 48, BS, 49, SM, 51, PhD(meteorol), 56. *Prof Exp:* Meteorol aide, US Weather Bur, Philippines & Japan, 46-47, res meteorologist, Nat Hurricane Res Proj, 56-57; asst meteorol, Chicago, 51-54; tech consult, Air Weather Serv, USAF, Washington, DC, 54-55; assoc prof meteorol, 57-63, chmn dept, 63-70, PROF METEOROL, FLA STATE UNIV, 63- *Mem:* AAAS; fel Am Metoerol Soc; Am Geophys Union. *Res:* Tropical meteorology and climatology; synoptic meteorology; hurricanes. *Mailing Add:* Dept Meteorol Fla State Univ Tallahassee FL 32306

JORDAN, CHRIS SULLIVAN, b Yangchow, China, Aug 6, 24; US citizen; m 47; c 3. BIOLOGY. *Educ:* Drake Univ, BA, 48; Univ Iowa, MS, 51, PhD(zool), 55. *Prof Exp:* Clin lab technologist, Vet Admin Hosp, Iowa City, Iowa, 52-55; supvr bact & parasitol, Terrell's Labs, Tex, 55-56; prof biol, Howard Payne Col, 56-63; prof, Houston Baptist Col, 63-67; CHMN DIV SCI & MATH, DALLAS BAPTIST COL, 67- *Concurrent Pos:* Res grant, NIH, 59-62. *Mem:* Am Inst Biol Sci; AAAS; Am Soc Parasitol; Am Soc Microbiol; Sigma Xi. *Res:* Parasitology and medical bacteriology. *Mailing Add:* Dallas Baptist Col PO Box 21206 Dallas TX 75211

JORDAN, CONSTANCE (LOUISE) BRINE, b Newton, Mass, Dec 26, 19; m 57; c 5. NUTRITION. *Educ:* Harvard Univ, MPH, 48; Cornell Univ, PhD(food, nutrit), 54. *Prof Exp:* Chief dietitian, Newton-Wellesley Hosp, 43-45; asst dir sch lunch, Pub Schs, Newton, 45-46; asst nutrit, Harvard Univ, 46-48; assoc prof food & nutrit, Univ RI, 48-56; prof home econ & head dept, 56-73, dean grad studies, 73-78, prof, 78-85, EMER PROF, FOOD & NUTRIT, FRAMINGHAM STATE COL, 85- *Concurrent Pos:* Consult, Arthur D Little, Inc & Mkt Res Corp Am, 54-58. *Mem:* AAAS; Am Dietetic Asn; Am Home Econ Asn. *Res:* Absorption of calcium; institutional dietary studies; nutritional status; nontraditional education at graduate level. *Mailing Add:* Eight Beacon Natick MA 01760

JORDAN, DAVID CARLYLE, b Brampton, Ont, July 11, 26; m 54; c 3. BACTERIOLOGY. *Educ:* Univ Toronto, BSA, 50, MSA, 51; Mich State Univ, PhD, 55, Can Col Microbiol, RM, 79. *Prof Exp:* Asst res, Ont Agr Col, Univ Guelph, 50-52, lectr bact, 52-56, from asst prof to prof microbiol, 56-87, chmn dept, 71-81; RETIRED. *Concurrent Pos:* Nuffield traveling fel, 59. *Mem:* Can Soc Microbiol. *Res:* Bacterial physiology as related to rhizobium species. *Mailing Add:* Eight Young St Guelph ON N1G 1M2 Can

JORDAN, DAVID M, b Ashtabula, Ohio, Aug 19, 37; m 61; c 2. ORGANIC CHEMISTRY. *Educ:* Col Wooster, BA, 59; Ohio State Univ, PhD(chem), 65. *Prof Exp:* Assoc prof, 65-90, PROF CHEM, STATE UNIV NY COL POTSDAM, 90- *Mem:* Am Chem Soc; Sigma Xi. *Res:* Diazoacetophenone decompositions; reaction of ketenes; techniques for thin-layer chromatography on cylindrical surfaces; styryl azide decompositions. *Mailing Add:* Rte 4 Box 202 Potsdam NY 13676

JORDAN, DONALD J, b New York, NY, 16. ENGINEERING. *Educ:* NY Univ, BS, 38. *Prof Exp:* Power plant staff engr, Chance Vought Aircraft, 44-48; mem staff, Pratt & Whitney, 48-71, eng mgr, 71-75; eng mgr, Power Systs Div, United Technologies, 75-78; RETIRED. *Mem:* Nat Acad Sci; Nat Acad Eng. *Mailing Add:* 113 Evergreen Lane Glastonbury CT 06033

JORDAN, DUANE PAUL, b Glendale, Calif, July 17, 35; m; c 2. MECHANICAL ENGINEERING. *Educ:* Stanford Univ, BS, 57, MS, 58, PhD(mech eng), 61. *Prof Exp:* Mech engr, Lawrence Radiation Lab, Univ Calif, 60-63; sr engr, integrated controls dept, Electronics Assocs, Inc, 63-64; asst prof mech eng, 64-67, ASSOC PROF MECH ENG, TEX TECH UNIV, 67- *Concurrent Pos:* Consult, Lawrence Radiation Lab, 63-65, Profit Index Systs, Inc, 66-84, Fanning, Fanning, Agnes Consult Engrs, 76-78 & Tex Indust Comn, 78-79. *Honors & Awards:* Dedicated Serv Award, Am Soc Mech Engrs, 88. *Mem:* Am Soc Mech Engrs; Am Soc Eng Educ. *Res:* Thermal, physical and social economic systems analysis and simulation using digital and analog computer techniques. *Mailing Add:* Dept Mech Eng Tex Tech Univ Lubbock TX 79409

JORDAN, EDWARD C(ONRAD), b Edmonton, Alta, Can, Dec 31, 10; nat US; m 41; c 3. ELECTRICAL ENGINEERING. *Educ:* Univ Alta, BSc, 34, MSc, 36; Ohio State Univ, PhD(elec eng), 40. *Prof Exp:* Operator radio sta, CKUA, 28-36; elec engr, Int Nickel Co, 36-37; instr elec eng, Worcester Polytech Inst, 40-41; from instr to asst prof, Ohio State Univ, 41-45; from assoc prof to prof, 45-54, prof & head dept, 54-79, EMER PROF ELEC ENG, UNIV ILL, URBANA, 79- *Honors & Awards:* Educ Medal, Inst Elec & Electronics Engrs. *Mem:* Nat Acad Eng; Int Union Radio Sci; Am Soc Eng Educ; fel Inst Elec & Electronics Engrs. *Res:* Antennas and radio direction finding. *Mailing Add:* Dept Elec Eng Univ Ill Urbana IL 61801

JORDAN, EDWARD DANIEL, b Bridgeport, Conn, Mar 14, 31; m 57; c 5. RELIABILITY ENGINEERING. *Educ:* Fairfield Univ, BS, 53; NY Univ, MS, 55; Univ Md, PhD(nuclear eng), 65. *Prof Exp:* Reactor physicist nuclear eng, Foster Wheeler Corp, 55-57; US Atomic Energy Comn, 57-59; from assoc prof to prof nuclear eng, 59-68, dir info syst & planning off, 68-83, PROF MECH ENG, CATH UNIV AM, 83- *Mem:* Sigma Xi. *Res:* Computer modeling of complex engineering system reliability. *Mailing Add:* 4010 Shinnecock Dr New Bern NC 28562

JORDAN, ELKE, b Ger, Apr 8, 37. MOLECULAR BIOLOGY, GENETICS. *Educ:* Goucher Col, BA, 57; Johns Hopkins Univ, PhD(biochem), 62. *Prof Exp:* Fel, Harvard Univ, 62-64; fel, Univ Cologne, 64-68; res assoc, Univ Wis-Madison, 68-69; res assoc, Univ Calif, Berkeley, 69-72; grants assoc, Nat Cancer Inst, 72-73, coordr collab res, 73-76, prog admin, Nat Inst Gen Med Sci, 76-78, dep dir, genetics prog, 78-81, assoc dir prog activ, 81-88, dir, Off Human Genome Res, 88-89, DEP DIR, NAT CTR HUMAN GENOME RES, NIH, 89- *Concurrent Pos:* NIH fel, 62-65; fel, Helen Hay Whitney Found, 65-68. *Mem:* Genetics Soc Am; AAAS; Am Soc Microbiol; Am Soc Human Genetics. *Res:* Gene regulation in prokaryotes, genetic recombination. *Mailing Add:* Nat Ctr Human Genome Res NIH Bethesda MD 20892

JORDAN, FRANK, b Budapest, Hungary, Jan 28, 41; US citizen; m 75; c 2. BIO-ORGANIC CHEMISTRY, BIOPHYSICAL CHEMISTRY. *Educ:* Drexel Univ, BS, 64; Univ Pa, PhD(chem), 67. *Prof Exp:* NATO fel quantum chem, Univ Paris, France, 67-68; NIH fel bio-org chem, Chem Dept, Harvard Univ, 68-70; asst prof, 70-75, assoc prof, 75-79, PROF CHEM, RUTGERS UNIV, NEWARK, 79- *Concurrent Pos:* NIH grants, USPHS, 74-82. *Mem:* Am Chem Soc; AAAS; Sigma Xi; Biophys Soc; Am Soc Biol Chem. *Res:* Enzyme mechanism studies on enzymes sythesizing and utilizing thiamine diphosphate, purine nucleoside phosphorylase, glyoxalase I and serine proteases. *Mailing Add:* Dept Chem Rutgers Univ Newark Col Arts Newark NJ 07102

JORDAN, FREDDIE L, b Yazoo City, Miss, Aug 14, 54; m; c 3. BIOCHEMISTRY, NEUROSCIENCES. *Educ:* Jackson State Univ, BS, 78; Meharry Med Col, PhD(biochem), 86. *Prof Exp:* Postdoctoral fel neurobiol, Meharry Med Col, 86-88; res assoc, 88-89, VIS ASST PROF, ORAL BIOL DEPT, OHIO STATE UNIV, 89- *Concurrent Pos:* Vis asst prof, Oral Biol Dept, Ohio State Univ, 89-; vis scientist, Fedn Am Socs for Exp Biol, Soc Cell Biol, 90- *Mem:* Soc Neurosci; Int Asn Dent Res; Am Asn Dent Res; Soc Cell Biol; AAAS. *Res:* Transmembrane signalling in the cerebral cortex. *Mailing Add:* Col Dent Ohio State Univ Columbus OH 43201

JORDAN, GARY BLAKE, b Urbana, Ill, Feb 3, 39; m 68; c 2. ELECTRONICS PROGRAM MANAGEMENT, MARKETING. *Educ:* Ohio Univ, BS, 61; Pac Southern Univ, DEE, 77; Sussex Col Technol, Eng, PhD(elec eng), 77. *Prof Exp:* Sr prog mgt engr, Ford Aerospace, 75-79; prog mgr, ESL Subsid TRW Inc, 79-87, Cubic Corp, 87-89, Sci Atlanta, 88-89; DIR, JORDAN & ASSOCS, 89- *Concurrent Pos:* Exec vpres, EW Orgn, 69-75; dir, Nat Intel Agency, 76-79. *Mem:* Fel Am Biog Inst; sr mem Soc Tech Commun; corp mem Radio Soc Gt Brit; Armed Forces Commun & Electronics Asn; AAAS; assoc mem US Naval Inst; Am Radio Relay League; Inst Elec & Electronics Engrs. *Res:* Electronic warfare as applied to electronics in the battlefield and on battlefield training ranges. *Mailing Add:* 13392 Fallen Leaf Rd Poway CA 92064

JORDAN, GEORGE LYMAN, JR, b Kinston, NC, July 10, 21; m 45; c 4. SURGERY. *Educ:* Univ NC, BS, 42; Univ Pa, MD, 44; Tulane Univ, MS, 49. *Prof Exp:* From instr to assoc prof, 52-64, PROF SURG, BAYLOR COL MED, 64- *Concurrent Pos:* Fel surg, Mayo Found, 49-52; from asst chief to chief surg, Vet Admin Hosp, 52-59, actg chief, 59-60, consult, 60-; attend surgeon, Jefferson Davis Hosp, Methodist Hosp, St Luke's Episcopal Hosp & Tex Children's Hosp; dep chief surg, Ben Taub Gen Hosp, 64-; chief of staff, Harris County Hosp Dist, 68- *Mem:* Soc Univ Surgeons; AMA; Am Col Surgeons; Am Surg Asn; Am Asn Surg of Trauma. *Res:* Postgastrectomy syndromes; pancreatic disease; experimental production of atherosclerosis; aortic homografts and prostheses; homotransplantation. *Mailing Add:* Dept Surg Baylor Col Med One Baylor Plaza Houston TX 77030

JORDAN, GEORGE SAMUEL, b Dallas, Tex, Apr 11, 44; m 66. MATHEMATICS. *Educ:* Southern Methodist Univ, BA, 66; Univ Wis-Madison, MS, 69, PhD(math), 71. *Prof Exp:* From asst prof to assoc prof, 71-84, PROF MATH, UNIV TENN, KNOXVILLE, 84-, ASSOC HEAD DEPT, 80- *Mem:* Am Math Soc; Math Asn Am; Sigma Xi; Soc Ind & Appl Math. *Res:* Integral and differential equations; Tauberian theory; functions of a complex variable. *Mailing Add:* Dept Math Univ Tenn Knoxville TN 37916

JORDAN, GILBERT LEROY, range seeding, range plant germination; deceased, see previous edition for last biography

JORDAN, HAROLD VERNON, b Boston, Mass, Aug 18, 24; m 50; c 3. MICROBIOLOGY. *Educ:* Univ NH, BS, 49; Univ Md, MS, 52, PhD(microbiol), 56. *Prof Exp:* Res scientist oral microbiol, Nat Inst Dent Res, 49-69; RES SCIENTIST ORAL MICROBIOL, FORSYTH DENT CTR, 69- *Concurrent Pos:* Vis scientist, Royal Dent Sch, Malmo, Sweden, 65-66. *Mem:* AAAS; fel Am Col Dent; Am Soc Microbiol; Int Asn Dent Res. *Res:* Lactic acid bacteria, metabolism and taxonomy; microbiology of dental caries and periodontal disease; gnotobiotic techniques in dental research; oral microbiology. *Mailing Add:* 84 Pine Bluff Rd Brewster MA 02631

JORDAN, HARRY FREDERICK, b Tacoma Park, Md, Mar 6, 40; m 62; c 2. COMPUTER SCIENCE, ELECTRICAL ENGINEERING. *Educ:* Rice Univ, BA, 61; Univ Ill, MS, 63, PhD(physics), 68. *Prof Exp:* From asst prof to assoc prof, 66-80, PROF ELEC ENG & COMPUT SCI, UNIV COLO, BOULDER, 80- *Mem:* Asn Comput Mach; Am Phys Soc; Inst Elec & Electronics Engrs. *Res:* Computer systems architecture; parallel algorithm design; optical computing; parallel processor design. *Mailing Add:* 6623 Lefthand Canyon Dr Jamestown CO 80455

JORDAN, HELEN ELAINE, b Bridgewater, Va, July 19, 26. VETERINARY PARASITOLOGY. *Educ:* Bridgewater Col, BA, 46; Va Polytech Inst, MS, 55; Univ Ga, DVM, 55, PhD(parasitol), 62. *Prof Exp:* From asst prof to assoc prof vet parasitol, Univ Ga, 55-69; PROF VET PARASITOL, COL VET MED, OKLA STATE UNIV, 69- *Mem:* Am Soc Parasitol; Am Vet Med Asn; Am Asn Vet Parasitol; World Asn Vet Parasitol. *Res:* Life cycle study of flukes; surveillance and epidemiology parasites in wild and domestic animals; parasite-host interactions and parasite ecology. *Mailing Add:* Dept Vet Parasitol & Pub Health Okla State Univ Col Vet Med Stillwater OK 74078

JORDAN, HOWARD EMERSON, b State College, NMex, May 14, 26; m 49; c 2. ELECTRICAL ENGINEERING. *Educ:* Univ Wis, BS, 46; Case Inst Technol, MS, 58, PhD(elec eng), 62. *Prof Exp:* Appln engr, Ray-O-Vac Co, Wis, 46-52; develop engr, Reliance Elec & Eng Co, Cleveland, 54-63, sr develop engr, 63-66, sr engr advan systs develop, 66-71; MGR CORP RES & DEVELOP ENG, AC MACH, 81-, CHIEF ENG, 83-, CORP R & D MGR, 84- *Mem:* Fel Inst Elec & Electronics Engrs. *Res:* Development and design of electrical rotating machinery and electromechanical devices; development of computer methods for design; technology management. *Mailing Add:* 25300 Chatworth Dr Euclid OH 44117

JORDAN, JAMES A, JR, b Berkeley, Calif, Dec 28, 36; m 61; c 2. APPLIED MATHEMATICS. *Educ:* Ohio State Univ, BSc, 58; Univ Mich, MSc, 59, PhD(physics), 64. *Prof Exp:* Physicist, USAF Aeronaut Res Lab, 60; asst res physicist, Univ Mich, 60-64; asst prof atomic physics, Rice Univ, 64-70; sci staff mem, IBM Houston Sci Ctr, 69-71, mgr appl math, 71-74, MGR POWER SYSTEMS ANALYSIS, IBM PALO ALTO SCI CTR, 74- *Mem:* Inst Elec & Electronics Engrs. *Res:* Optical data processing, computer holography; network analysis; atomic collisions; atomic spectroscopy; scientific computations; simulation and control of power systems; data management. *Mailing Add:* IBM Palo Alto Sci Ctr 1530 Page Mill Rd Palo Alto CA 94304

JORDAN, JAMES HENRY, b Sacramento, Calif, Oct 16, 31; m 58; c 3. MATHEMATICS. *Educ:* Southern Ore Col, BS, 53; Univ Ore, MA, 58; Univ Colo, PhD(math), 62. *Prof Exp:* Teacher elem sch, Ore, 53-56; from asst prof to assoc prof, 62-70, PROF MATH, WASH STATE UNIV, 70-, CHMN, PROG SCI & MATH TEACHING, 77- *Mem:* Math Asn Am; Am Math Soc. *Res:* Number theory in general, specifically Kth power reciprocity, consecutive residues, Gaussian integers and simple continued fractions. *Mailing Add:* Dept Math Wash State Univ Pullman WA 99164-2930

JORDAN, JAMES N, b Whiteville, NC, Jan 30, 25; m 58; c 2. EARTH SCIENCE. *Educ:* Univ NC, BS, 47. *Prof Exp:* Lab instr geol, Univ NC, 47-48; geologist, Creole Petrol Corp, Venezuela, 48-50; self employed, mining, Venezuela, 50-52; geophysicist, US Coast & Geodetic Surv, 53-66, chief spec seismol anal br, Earth Sci Lab, Environ Res Labs, Nat Oceanic & Atmospheric Admin, 66-73; SUPVRY GEOPHYSICIST, OFF EARTHQUAKE STUDIES, US GEOL SURV, 73- *Concurrent Pos:* Mem, NSF Earthquake Eng Res Inst Mission to Chile, 60. *Honors & Awards:* Bronze Award, US Dept Com, 66. *Mem:* Am Geophys Union; Seismol Soc Am; Earthquake Eng Res Inst. *Res:* Geological field studies; field investigations of earthquake damage; earthquake location studies; travel times of seismic waves; magnitude and energy studies of earthquakes; seismic studies relating to nuclear disarmament. *Mailing Add:* US Geol Surv Stop 967 Box 25046 Denver Fed Ctr Denver CO 80225

JORDAN, JOHN PATRICK, b Salt Lake City, Utah, Apr 23, 34; m 54; c 8. BIOCHEMISTRY. *Educ:* Univ Calif, Davis, BS, 55, PhD(comp biochem), 63. *Prof Exp:* From asst prof to assoc prof chem, Okla City Univ, 62-68; assoc prof biochem, Colo State Univ, 68-71, assoc dean, Col Natural Sci & dir biol core curriculum, 68-72, dir, Proj Biocotie, 70-83, prof biochem, 71-83, dir, Univ Exp Sta, 72-83; ADMINR COOP, STATE RES SERV, USDA, 83- *Concurrent Pos:* Grant dir, NASA res grant, 63-; Prin investr, Frontiers Sci Found Okla, Inc res grant, 64-65; Okla Heart Asn res grant, 65-66; NIH res grant, 65-68; consult space med, NASA, 65-70; NIH biomed sci support grant, 69-; Nat Sci Found curriculum res grant, 71-; Boettcher Found res grant, 70-71; gen chmn annual meeting, Am Inst Biol Sci, 71. *Honors & Awards:* Bond Award, Am Oil Chem Soc, 67. *Mem:* Fel AAAS; fel Am Inst Chem; Brit Biochem Soc; Am Physiol Soc; Soc Exp Biol & Med. *Res:* Intermediary metabolism, particularly the effects of artificial atmospheres on metabolism; curricular development, especially in biology and chemistry; research administration. *Mailing Add:* Coop State Res Serv USDA Rm 305A Admin Bldg 14 & Independent SW Washington DC 20250

JORDAN, JOHN WILLIAM, b Pittsburgh, Pa, Apr 25, 12; m 36, 79; c 5. INDUSTRIAL CHEMISTRY. *Educ:* Marietta Col, AB, 34; Columbia Univ, PhD(chem), 38. *Hon Degrees:* ScD, Marietta Col, 59. *Prof Exp:* Asst food anal & colloid chem, Columbia Univ, 35-38; fel asst tech glassware, Mellon Inst, 38-39, sr fel lead, 41-51; chemist, Pittsburgh Corning Corp, Pa, 39-41; mgr res labs, 51-56, tech dir, 57-76, consult, Baroid Div, 76-82, CONSULT CLAY-ORGANIC COMPLEXES, NL INDUSTS, INC, 82- *Concurrent Pos:* Dir, Enenco, Inc, 62-76. *Honors & Awards:* Pioneer Clay Sci, Develop Clay-Organic Complexes, Clay Minerals Soc, 90. *Mem:* AAAS; Am Chem Soc; Clay Minerals Soc (pres, 73-74). *Res:* Chemistry of hydrous ferric oxides; structural glass products; synthetic resins and coatings; oil well drilling fluids; organic complexes of clay minerals for gellants and rheological control agents. *Mailing Add:* 1505 Butlercrest Houston TX 77080-7613

JORDAN, JOSEPH, b Timisoara, Rumania, June 29, 19; nat US; m 52; c 4. ANALYTICAL CHEMISTRY. *Educ:* Hebrew Univ, Israel, MSc, 42, PhD, 45. *Prof Exp:* Asst & instr, Hebrew Univ, Israel, 45-49; res fel, Harvard, Univ, 50; res fel & instr, Univ Minn, 51-54; from asst prof to prof, 54-90, EMER PROF CHEM, PA STATE UNIV, UNIVERSITY PARK, 90- *Concurrent Pos:* Vis prof, Univ Calif, Berkeley, 59, Swiss Fed Inst Technol, Zurich, 60-61; Cornell Univ, 65, Sch Physics & Indust Chem & Pierre & Marie Curie Univ, Paris, 75-76 & 85; res collabr, Brookhaven Nat Lab, 60-63; consult, Am Instrument Co, 60-79, Marcel Dekker, Inc, 65- & Int Paper Co, 79-81; chmn comn electrochem, Int Union Pure & Appl Chem, 67-71, mem phys chem div comt, 69-73, mem, Electroanal Chem Comn, 73-85, chmn, 81-85, US Nat Rep, 85-87, titular mem, Anal Div Comt, 87-; Fulbright lectr, Univ Paris, 68-69, Univ Jodhpur, India, 86-87; I M Kolthoff sr fel anal chem, Hebrew Univ, Israel, 72 & 84-85; mem comt symbols, units & terminology, Numerical Data Adv Bd, Nat Res Coun, Nat Acad Sci, 73-76; mem eval panel phys chem div, Nat Bur Standards, 75-78; res assoc, Nat Ctr Sci Res, Paris, 75-76 & 85; co-ed, Standard Potentials Aqueous Solution, IYPAC, 85. *Honors & Awards:* Frontiers Chem Lectr, Wayne State Univ, 59; Robert Guehm Lectr, Swiss Fed Inst Technol, 69; Bennedetti-Pichler Award, Am Microchem Soc, 78. *Mem:* Fel AAAS; Am Chem Soc; fel Am Inst Chemists; fel Royal Soc Chem. *Res:* Polarography; kinetics and mechanisms of electron transfer; electrode reactions; molten salts; thermochemical titrations; enthalpimetric analysis; thermochemistry of immunological and enzymatic processes; instrumental analysis of sulfur compounds in coal process streams; electrochemical photovoltaic cells; amperometric biosensors. *Mailing Add:* Dept Chem 152 Davey Lab Pa State Univ University Park PA 16802

JORDAN, KENNETH A(LLAN), b Plainfield, NJ, June 30, 30; m 52; c 4. AGRICULTURAL ENGINEERING. *Educ:* Purdue Univ, BS, 52, MS, 54, PhD(agr eng), 59. *Prof Exp:* Res instr, Purdue Univ, 57-58; from asst prof to assoc prof farm struct, NC State Univ, 58-67; prof farm struct, Univ Minn, St Paul, 67-84; PROF, UNIV ARIZ, TUCSON, 86- *Concurrent Pos:* Vis prof, Univ Tokyo, 71. *Mem:* Sigma Xi; Am Soc Agr Engrs. *Res:* Animal shelter and greenhouse simulation; weather pattern frequency; plant and animal modeling; machine vision; sensor technology development. *Mailing Add:* 403 Shantz Bldg 38 Univ Ariz Tucson AZ 85721

JORDAN, KENNETH DAVID, b Norwood, Mass, Feb 25, 48; m 81; c 2. PHYSICAL CHEMISTRY, THEORETICAL CHEMISTRY. *Educ:* Northeastern Univ, BA, 70; Mass Inst Technol, PhD(chem), 74. *Prof Exp:* J W Gibbs instr eng & appl sci, Yale Univ, 74-76, asst prof, 76-78; from asst prof to assoc prof, 78-85, PROF CHEM, UNIV PITTSBURGH, 85- *Concurrent Pos:* Vis asst prof, Univ Utah, 76 & 77; Alfred P Sloan Found fel, 77; Camille & Henry Dreyfus Found teacher scholar, 77; John Simon Guggenheim fel, 81; prog dir of theoretical chem physics, NSF, 84-85; adj prof, Carnegie-Mellon Univ, 88-; chmn, Theoret Chem Subdiv, Phys Chem Div, Am Chem Soc, 90-91. *Mem:* Am Chem Soc; Am Phys Soc; Sigma Xi. *Res:* Theoretical studies of the electronic structure of molecules; electron transmission spectroscopic studies of temporary anions; energy transfer in reactions of excited atoms with molecules; properties of atomic and molecular clusters. *Mailing Add:* Dept Chem Univ Pittsburgh Pittsburgh PA 15260

JORDAN, KENNETH GARY, b Anderson, SC, Nov 18, 35; m 57; c 2. PHYSICAL CHEMISTRY. *Educ:* Clemson Univ, BS, 57, MS, 61, PhD(phys chem), 63. *Prof Exp:* Chemist, 57, res chemist, Dacron Technol Div, NC, 63-68, sr res chemist, 69-70, Dacron Textile Res Lab, Wilmington, 70-75, textile fibers end use mkt specialist, 75-77, develop assoc, 77-78, mkt rep, 78-81, sr mkt rep, 81, account mgr, 82-84, SR ACCOUNT MGR, INDUST FIBERS DIV, E I DU PONT DE NEMOURS & CO, INC, 85- *Mem:* Am Chem Soc. *Res:* Semiconductor properties of polymers; polyester catalysis and kinetics; new polymer technology; synthetic fiber and fabric characterization and evaluation; industrial fiber sales and development. *Mailing Add:* 4502 Lanier Ave Anderson SC 29634

JORDAN, KENNETH L(OUIS) JR, b Portland, Maine, May 10, 33; m 62; c 3. ELECTRICAL ENGINEERING, COMMUNICATIONS. *Educ:* Rensselaer Polytech Inst, BEE; Mass Inst Technol, SM, 56, ScD(elec eng), 61. *Prof Exp:* Staff mem commun, Lincoln Lab, Mass Inst Technol, 60-67, asst group leader, 67-68, group leader, 68-76, prin dep asst secy res & develop, Off Secy Air Force, The Pentagon, 76-79; CHIEF SCIENTIST, SCI APPLN, INC, 79- *Mem:* Fel Inst Elec & Electronics Engrs. *Res:* Random processes; modulation and coding; satellite communications. *Mailing Add:* Sci Appln Inc 1710 Goodridge Dr McLean VA 22102

JORDAN, LAWRENCE M, high energy physics, emergency transport modeling, for more information see previous edition

JORDAN, LOWELL STEPHEN, b Vale, Ore, Apr 23, 30; m 50, 80; c 5. PLANT PHYSIOLOGY. *Educ:* Ore State Univ, BS, 54; Univ Minn, PhD(agron, bot), 57. *Prof Exp:* Asst agron, Univ Minn, 54-57; asst prof plant industs, Southern Ill Univ, 57-59; from asst to assoc plant physiologist, 59-70, PHYSIOLOGIST & PROF HORT SCI, UNIV CALIF, RIVERSIDE, 70- *Concurrent Pos:* Pres Exec Comt, Bd Dirs, Coun Agr Sci & Technol. *Mem:* Fel Weed Sci Soc Am; Am Soc Plant Physiol; Am Soc Hort Sci. *Res:* Weed science; herbicide physiology, mechanism of action and metabolism in plants. *Mailing Add:* Dept Bot & Plant Sci Univ Calif Riverside CA 92521

JORDAN, MARK H(ENRY), b Lawrence, Mass, Apr 10, 15; m 39; c 2. CIVIL ENGINEERING. *Educ:* US Naval Acad, BS, 37; Rensselaer Polytech Inst, BCE, 41, MCE, 42, MS, 65, PhD(mgt), 68. *Prof Exp:* Eng adminr, US Navy, 42-60; officer in charge, Naval Civil Engrs Corps Officers Sch, Pt Hueneme, Calif, 60-63; assoc prof civil eng, construct & mgt, Univ Mo, 66-67; dean continuing studies, Rensselaer Polytech Inst, 67-72, prof civil eng, 68-77; CONSULT ENGR, 76- *Concurrent Pos:* Mem, Rensselaer County Charter Comn, 69-71; arbitrator, Am Arbit Asn, 70- *Mem:* Fel Am Soc Civil Engrs; Am Soc Eng Educ; Nat Soc Prof Engrs; Soc Am Mil Engrs; Am Cons Engrs Coun; Am Arbit Asn. *Res:* Industrial management, application of contemporary management concepts to engineering construction. *Mailing Add:* 256 Broadway Third Floor Troy NY 12180

JORDAN, MARY ANN, b Minneapolis, Minn, July 31, 40; m 84; c 2. MICROTUBULES, VINCA ALKALOIDS. *Educ:* Univ Minn, BA, 62, MS, 64 Univ Rochester, PHD(biol), 69. *Prof Exp:* Temp asst prof, Univ Wyo, 68-69; researcher biol, Univ of Mich, 71-72, Utah State Univ, 74-77; postdoctoral fel, 78-82, asst res biologist, 82-90, ASSOC RES BIOLOGIST, UNIV CALIF, SANTA BARBARA, 91- *Mem:* AAAS; Am Soc for Cell Biol. *Res:* Microtubule structure and function; regulation by drug and physiological compounds; elucidating control of microtubule polymerization in vitro and in vivo, especially by drugs, including the vinca alkaloids. *Mailing Add:* Dept Biol Sci Univ Calif Santa Barbara CA 93106

JORDAN, NEAL F(RANCIS), b Franklinville, NY, July 8, 32; m 55; c 2. ENGINEERING PHYSICS. *Educ:* Cornell Univ, BEngPhys, 55; Purdue Univ, MS, 59, PhD(eng sci), 63. *Prof Exp:* Instr continuum mech, Purdue Univ, 60-63; res assoc geophys, Jersey Prod Res Co, Okla, 63-65, SUBSURFACE IMAGING, EXXON PROD RES CO, 65- *Concurrent Pos:* Consult, Gen Tech Corp, Ind, 59-63. *Mem:* Soc Eng Sci; Soc Explor Geophys. *Res:* Geophysics; nonlinear theories of continuous media; elastic wave propagation. *Mailing Add:* 330 Knipp Houston TX 77024

JORDAN, PAUL H, JR, b Bigelow, Ark, Nov 22, 19; m 44; c 3. GASTROENTEROLOGY, SURGERY. *Educ:* Univ Chicago, BS, 41, MD, 44; Univ Ill, MS, 50. *Prof Exp:* Asst prof surg, Sch Med, Univ Calif, Los Angeles, 55-58; assoc prof, Sch Med, Univ Fla, 59-64; chief staff, 69-70, chief surg, Vet Admin Hosp, Houston, 64-83; PROF SURG, BAYLOR COL MED, 64- *Concurrent Pos:* Fel, NIH, 58-59. *Mem:* Am Surg Asn; Soc Exp Biol & Med; Soc Univ Surg; Am Col Surg; Am Gastroenterol Asn. *Res:* Gastrointestinal physiology. *Mailing Add:* Baylor Col Med One Baylor Plaza Houston TX 77030

JORDAN, PETER ALBION, b Oakland, Calif, Jan 2, 30; div; c 3. ECOLOGY, WILDLIFE MANAGEMENT. *Educ:* Univ Calif, Berkeley, AB, 55, PhD(zool), 67. *Prof Exp:* Asst specialist studies migratory deer, Sch Forestry, Univ Calif, Berkeley, 55-61, teaching asst zool, 61-62, instr, 62-63; res assoc ecol moose & wolves, Purdue Univ, 63-66; asst prof wildlife ecol, Sch Forestry, Yale Univ, 67-74; ASSOC PROF, DEPT FISHERIES & WILDLIFE, UNIV MINN, ST PAUL, 74- *Concurrent Pos:* Bd dirs, Minn Zoo. *Mem:* Wildlife Soc; Am Soc Naturalists; Am Soc Mammal. *Res:* Behavior; population dynamics and food habits of wild ungulates and carnivores; impact of herbivorous mammals upon forest vegetation; sodium acquisition and aquatic feeding by forest herbivores; management of big game; ecosystem processes; integration of wildlife habitat with timber management. *Mailing Add:* Dept Fish & Wildlife Univ Minn St Paul MN 55108

JORDAN, PETER C H, b London, Eng, May 3, 36; US citizen; m 79. THEORETICAL CHEMISTRY. *Educ:* Calif Inst Technol, BS, 57; Yale Univ, PhD(quantum mech), 60. *Prof Exp:* NSF fel, 60-62; asst res chemist, Univ Calif, San Diego, 62-64; asst prof chem, 64-70, assoc prof, 70-81, PROF CHEM, BRANDEIS UNIV, 81- *Concurrent Pos:* Guggenheim fel, 71-72; Marion & Jaspar Whiting fel, 78-79; vis prof, Dept Biol, Konstanz Univ, 78-79; vis scientist, Univ Houston, 86-87. *Mem:* AAAS; Am Phys Soc; Biophys Soc; Am Chem Soc. *Res:* Statistical mechanics; quantum chemistry; irreversible thermodynamics; membrane transport. *Mailing Add:* Dept Chem Brandeis Univ Waltham MA 02254

JORDAN, RICHARD CHARLES, b Minneapolis, Minn, Apr 16, 09; m 35; c 3. MECHANICAL ENGINEERING. *Educ:* Univ Minn, BAeroE, 31, MS, 33, PhD(mech eng), 40. *Prof Exp:* Head, Air Conditioning Div, Minneapolis Br, Am Radiator & Standard Sanit Corp, 33-36; instr petrol eng, Univ Tulsa, 36-37; Eng Exp Sta, Univ Minn, Minneapolis, 37-41, asst dir, 41-44, from asst prof to prof mech eng, 41-76, dir, Indust Labs, 44-45, head, Dept Mech Eng, 50-76 & Sch Mech & Aerospace Eng, 66-76, assoc dean, Inst Technol, 77-80, EMER PROF MECH ENG, UNIV MINN, MINNEAPOLIS, 76- *Concurrent Pos:* Consult, Corps Engrs, US Army, 51-52; adv, Panel Eng Sci, NSF, 54-57, chmn, 56-57; mem, Div Eng & Indust Res, Nat Acad Sci-Nat Res Coun, 54-74, chmn div, 61-65, mem-at-large, 65-74; chmn, ad hoc comt eng develop countries, 59-61 & ad hoc comt eng & soc sci, 61-65; deleg, Int Cong Refrig, Paris, 55, Copenhagen, 59, Munich, 63, Madrid, 67, Wash, 71 & Moscow, 75; US deleg, Int Inst Paris, 57-63, US Nat Comt, Moscow & Prague, 58 & 59-63; tech bd, 63-67; Am Standards Asn deleg, Int Standards Orgn, London, 58; deleg, World Power Conf, Melbourne, 62; mem, CENTO Surv Mission, Iran, Pakistan & Turkey, 63; vchmn eng & accreditation comt, Region VII, Eng Coun Prof Develop, 64-66, chmn eng accreditation comt, Region VIII, 66-69; consult, AID, 64-, Int Power Technol, 79-, Onan Corp, 83- & var indust orgn; mem adv comt int orgns & progs, Off Int Rels, Nat Acad Sci, 65-; mem, Fulbright Prog Long-Range Planning Team, US-Brazil Prog, 67; chmn study group indust res, US-Brazil Sci Coop Prog, Nat Acad Sci, Rio de Janeiro & Washington, DC, 67 & Belo Horizonte, Brazil & Houston, 68; mem, Conf Strategy for Technol Develop Latin-Am, 69; mem bd dirs, Onan Corp Div, McGraw-Edison Corp, 72-83; energy consult, Control Data Corp, 77-84; World Bank consult alt energy resources, Brazil, 76; chmn, comt rev prog, Div Conserv & Community Syst, Nat Res Coun, 80- *Honors & Awards:* Wolverine Award, Am Soc Heat, Refrig & Air-Conditioning Engrs, 49, F Paul Anderson Medal & E K Campbell Award, 66. *Mem:* Nat Acad Eng; fel AAAS; fel Am Soc Mech Engrs; Nat Soc Prof Engrs;

fel Am Soc Heat, Refrig & Air-Conditioning Engrs (treas, 50, vpres, 51-52, pres, 53). *Res:* Air filtration; dust analyses; conditions of comfort; heat transmission; refrigeration; quantitative and qualitative analysis of atmospheric dust; solar energy; engineering. *Mailing Add:* 18418 Horseshoe Circle Rio Verde AZ 85263

JORDAN, ROBERT, JR, b Macon, Ga, Sept 11, 20; m 42; c 3. PEDIATRICS. *Educ:* Vanderbilt Univ, BA, 41, MD, 43. *Prof Exp:* Intern pediat, Duke Univ Hosp, Durham, NC, 44; resident pediat, Henrietta Egleston Mem Hosp, Atlanta, Ga, 44-45; res fel pediat, 48-49, from asst prof to assoc prof, 54-66, PROF PEDIAT, UNIV TENN COL MED, 66-; DIR, CHILD DEVELOP CTR, UNIV TENN, 57-, PROF CHILD DEVELOP, 66- *Concurrent Pos:* Supt Arlington Develop Ctr, Tenn, 74-75; mem, President's Comt Ment Retardation. *Mem:* Am Acad Pediat; fel Am Med Asn; Am Asn Mental Deficiency; Asn Univ Affiliated Facil (past pres). *Res:* Neurologically handicapped children. *Mailing Add:* Univ Tenn Child Develop Ctr 711 Jefferson Ave Memphis TN 38105

JORDAN, ROBERT KENNETH, b Clearfield, Pa, Dec 12, 25; m 48; c 3. ULTRA-FINE INORGANIC PARTICLES. *Educ:* Tufts Col, BS, 54. *Prof Exp:* Chemist, energy res & develop, Olin-Mathieson Chem Co, 54-56; mgr, nuclear div & asst mgr chem, plastics & metals, res div, Curtiss-Wright Corp, 56-60; mgr new prod, Gen Tire Chem-Plastics, 60-65; proj mgr new prod, US Steel Corp, 65-73; consult, energy-indust, Brookhaven Nat Lab, 74-80; DIR ENG RES & DEVELOP, ENG RES INST, GANNON UNIV, 80-, DIR METALLIDING INST, SCIENTIST IN RESIDENCE, 83- *Concurrent Pos:* Expert deleg, NATO, Comm Challenges of Modern Soc Conf, Steel Indust, 75-78; indust consult, 75-82. *Res:* Metallurgical (Metalliding), ceramics, minerals extraction, metals processes, organic and inorganic fluorides, organic nitrogenous and inorganic phosphorous fertilizers research and development; crude oil conversion to petrochemicals, organic intermediates, polymers and plastics-elastomers; metals surface and subsurface alloying. *Mailing Add:* Gannon Univ Erie PA 16541

JORDAN, ROBERT LAWRENCE, b Miami, Fla, May 7, 41; div; c 2. ANATOMY, TERATOLOGY. *Educ:* Fla Southern Col, BS, 65; Univ Fla, 65-66; Univ Cincinnati, PhD(anat), 69. *Prof Exp:* Asst prof, 70-74, ASSOC PROF ANAT, MED COL VA, 74- *Concurrent Pos:* USPHS fel teratology & toxicol, Kettering Lab, Univ Cincinnati, 69-70; vis prof anat, St George's Med Sch, Greneda, West Indies, 79- *Mem:* Teratology Soc; Am Asn Anat; Sigma Xi. *Mailing Add:* Dept Anat Med Col Va Va Commonwealth Richmond VA 23219

JORDAN, ROBERT MANSEAU, b Minneapolis, Minn, Feb 13, 20; m 42; c 3. ANIMAL SCIENCE. *Educ:* Univ Minn, BS, 42; SDak State Col, MS, 49; Kans State Col, PhD(animal nutrit), 53. *Prof Exp:* Instr animal husb, Univ Minn, 42; sales rep, Lyon Chem, Minn, 45; instr, SDak State Col, 48-49, from asst prof to assoc prof, 49-54; from asst prof to assoc prof, 54-64, PROF ANIMAL HUSB, UNIV MINN, ST PAUL, 64- *Concurrent Pos:* Mem comt, Nutrient Req Horses, Nat Res Coun, 78, chmn comt, Nutrient Req Sheep, 80-85. *Honors & Awards:* Animal Mgt Award, Am Soc Animal Sci, 83,. *Mem:* Fel Am Soc Animal Sci; Brit Soc Animal Prod; Equine Nutrit & Physiol Soc; Acad Natural Sci. *Res:* Use of hormones in animal production; digestibility studies; sheep nutrition; protein, calcium and phosphorus energy requirements for growth, reproduction, lactation and work of horses; copper toxicity studies with horses; pasture research involving development of low alkaloid varieties of canary grass and energy intakes by lamb grazing legume or grass pastures; silage studies and non-protein sources of nitrogen for horses; milk replacer studies for the committee for lamb nutrition and management; nutrition requirements for Angora goats. *Mailing Add:* Dept Animal Sci Univ Minn St Paul MN 55108

JORDAN, ROBERT R, b New York, NY, June 5, 37; m 58; c 2. GEOLOGY. *Educ:* Hunter Col, AB, 58; Bryn Mawr Col, MA, 62, PhD(geol), 64. *Prof Exp:* From geologist to asst state geologist, Del Geol Surv, 58-69; from instr to assoc prof geol, Univ Del, 62-88; STATE GEOLOGIST & DIR, DEL GEOL SURV, 69-; PROF GEOL, UNIV DEL, 88- *Concurrent Pos:* Gov rep, Outer Continental Shelf Res Mgt Adv Bd, Dept Interior, 74-77 & Policy Comt, 85-; mem, Comt Offshore Energy Technol, Nat Acad Sci-Nat Res Coun, 78-80; regional coordr, Asn Am Petrol Geologists, 78-84; mem, NAm Comn Stratig Nomenclature, 78, vchmn-secy, 78-82 & 89-90, chmn, 82-83 & 90-91; house of delegates, Am Asn Petrol Geologists, 90-; pres-elect, Del Acad Sci, 90, pres, 91. *Honors & Awards:* Autometric Award, Am Soc Photogrammetry, 76; Cert of Merit, Am Asn Petrol Geologists, 87, Distinguished Serv Award, 88; Presidential Cert of Merit, Am Inst Prof Geologists, 90. *Mem:* Asn Am State Geol (secy-treas, 77-81, vpres, 81-82, pres-elect, 82-83, pres, 83-84); Am Inst Prof Geol; fel Geol Soc Am; Soc Econ Paleont & Mineral; Am Asn Petrol Geologists; Nat Asn Geol Teachers; Sigma Xi. *Res:* Sedimentary petrology; stratigraphy; geology of the Atlantic Coastal Plain; ground water supplies. *Mailing Add:* Del Geol Surv Univ Del Newark DE 19716

JORDAN, RUSSELL THOMAS, b Geneseo, NY; m 46; c 6. VIROLOGY, IMMUNOCHEMISTRY. *Educ:* Univ Ark, BS, 49, MS, 51; Univ Mich, PhD(virol), 53. *Prof Exp:* Rackham res fel, Sch Med, Univ Mich, 53, asst prof bact, 53-54; clin asst prof infectious dis, Sch Med, Univ Calif, Los Angeles, 54-59; chief, Dept Exp Immunol, Nat Jewish Hosp, Denver, 60-63; dir res & labs, Biomed Res Labs Div, Bio-Organic Chem Inc, 63-71; vpres sci & technol, C F Kettering Found, dir, C F Kettering Lab & vpres, Kettering Sci Res, Inc, 71-73; pres & chmn bd, Vipont Chem & Res Ctr, 73-75; res dir, Chemex Corp, 75-77; PRES, MED-X-CONSULT, FT COLLINS, 77- *Concurrent Pos:* Chmn dept microbiol, City of Hope Med Ctr, Calif, 54-60; lectr, Univ Calif, Los Angeles, 56-59; res fel immunochem, Calif Inst Technol, 58-59; asst prof, Sch Med, Univ Colo, Denver, 61-71; chief space biomed res, Aerospace Group, Martin-Marietta Corp, Colo, 66-71. *Mem:* Am Asn Immunol; Soc Exp Biol & Med; Am Asn Cancer Res; Sigma Xi; NY Acad Sci. *Res:* Microbiology; interference phenomenon; infection and resistance; virus induced neoplasms; immunochemistry of cancer; immunochemical properties of tumor specific antigens and antibodies. *Mailing Add:* Med-X-Consult Exchange Inc 1809 Indian Meadows Lane Ft Collins CO 80525

JORDAN, SCOTT WILSON, b Iola, Kans, Aug 22, 34; m 55; c 3. PATHOLOGY. *Educ:* Univ Kans, AB, 56, MD, 59. *Prof Exp:* From intern to resident path, Med Ctr, Univ Kans, 59-63; pathologist, Nat Acad Sci-Nat Res Coun Atomic Bomb Casualty Comn, 63-65; from asst prof to assoc prof, 65-82, PROF PATH, SCH MED, UNIV NMEX, 82- *Concurrent Pos:* USPHS fel, Med Ctr, Univ Kans, 60-63; consult, Midwest Res Inst, Mo, 62-63 & Nat Cancer Inst, 74-77. *Mem:* Am Soc Cytol (past pres); Am Asn Path; Int Acad Path; fel Am Bd Path. *Res:* Pathology of radiation injury; diagnostic cytology; digital image analysis. *Mailing Add:* Dept Path Univ NMex Sch Med Albuquerque NM 87131

JORDAN, STANLEY CLARK, b Elkin, NC, Apr 13, 47; m 76; c 1. PEDIATRICS, PEDIATRIC NEPHROLOGY. *Educ:* Univ NC, Chapel Hill, AB, 69, MD, 73. *Prof Exp:* Asst clin prof pediat, Univ Southern Calif, 79-80; ASST PROF PEDIAT, UNIV CALIF, LOS ANGELES, 80- *Concurrent Pos:* Prin investr, NIH clin investr award, 80-83. *Mem:* Am Soc Pediat Nephrol. *Res:* Renal immmunopathology and transplantation immunology. *Mailing Add:* Cedars-Sinai Med Ctr 8700 Beverly Blvd Los Angeles CA 90048

JORDAN, STEVEN LEE, b Jersey City, NJ, Feb 17, 43; m 66; c 2. COMPUTER GRAPHICS, MATHEMATICS EDUCATION. *Educ:* Princeton Univ, AB, 65; Univ Calif, Berkeley, MA, 67, PhD(math), 70. *Prof Exp:* Asst prof, 70-76, ASSOC PROF MATH, UNIV ILL, CHICAGO CIRCLE, 76-, ASSOC PROF ORAL & MAXILLOFACIAL SURG, 87- *Concurrent Pos:* NSF res grant, 70-72; Carnegie Found res grant, 75-76; HEW res grant, 85-86, Dept Educ grants, 86-; co-dir, UIC/CCC Partnership Prog, 87-; consult to bds educ & biochem labs. *Mem:* Am Math Soc; Math Asn Am; Sigma Xi; Nat Coun Teachers Math. *Res:* Differential geometry and complex manifolds; all levels of mathematics teacher education; history of mathematics; applications of computer graphics, statistics, and modelling to radiology and oral surgery. *Mailing Add:* Dept Math Box 4348 M-C 249 Univ Ill Chicago Circle Chicago IL 60680

JORDAN, STUART DAVIS, b St Louis, Mo, July 25, 36; m 61; c 2. SOLAR PHYSICS. *Educ:* Wash Univ, BS, 58; Univ Colo, Boulder, PhD(physics, astrophys), 68. *Prof Exp:* Res adminr plasma dynamics, Air Force Off Sci Res, 59-63; res scientist, Lab Astron & Solar Physics, 68-84, head, Solar Physics Br, 84-89, SR STAFF SCIENTIST, GODDARD SPACE FLIGHT CTR, NASA, 89- *Concurrent Pos:* Actg chief solar physics, NASA hq, 74. *Mem:* Am Phys Soc; Am Astron Soc; Int Astron Union; Sigma Xi. *Res:* Solar wind research; shock wave heating of solar atmosphere; energy balance and temperature structure of stellar atmospheres. *Mailing Add:* Code 680 Lab Astron & Solar Physics 17 Lakeside Dr Greenbelt MD 20770

JORDAN, THOMAS FREDRICK, b Duluth, Minn, June 4, 36; div. THEORETICAL PHYSICS. *Educ:* Univ Minn, Duluth, BA, 58; Univ Rochester, PhD(physics), 62. *Prof Exp:* Res assoc physics, Univ Rochester, 61-62, instr, 62-63; NSF fel, Berne, 63-64; from asst prof to assoc prof, Univ Pittsburgh, 64-70; PROF PHYSICS, UNIV MINN, DULUTH, 70- *Concurrent Pos:* Sloan Found fel, 65-67. *Res:* Mathematical physics; quantum mechanics; field theory; relativistic particle dynamics; scattering theory; elementary particle interactions; quantum theory of optical coherence; hydrodynamics of Great Lakes circulation; general relativity. *Mailing Add:* Dept Physics Univ Minn Duluth MN 55812

JORDAN, THOMAS HILLMAN, b Coco Solo, CZ, Oct 8, 48; m 73; c 1. GEOPHYSICS. *Educ:* Calif Inst Technol, BS, 69, MS, 70, PhD(geophys), 72. *Prof Exp:* Asst prof geophys, Princeton Univ, 72-75; assoc prof geophys, Scripps Inst Oceanog, Univ Calif, San Diego, 75-88; HEAD DEPT, EARTH ATMOSPHERIC & PLANETARY SCIS, MASS INST TECHNOL, 88- *Concurrent Pos:* Alfred P Sloan fel, physics, 80-82. *Honors & Awards:* James B Macelwane Award, Am Geophys Union, 83. *Mem:* fel Am Geophys Union; Geol Soc Am. *Res:* Structure of the earth's interior; earthquake processes; mantle dynamics; wave propagation; inverse theory. *Mailing Add:* Dept of Earth Atmospheric & Planetary Sci Mass Inst Technol Rm 54-918 Cambridge MA 02139

JORDAN, THOMAS L, b Yazoo City, Miss, Apr 27, 43; m 79; c 1. BACTERIOLOGY. *Educ:* Rockhurst Col, Kansas City, Mo, BA, 64; Univ Wis-Madison, MS, 68, PhD(bact), 72. *Prof Exp:* NIH fel bact, Univ Wash, Seattle, 72-75; fac biol, Alcorn State Univ, Miss, 76-78, Dillard Univ, New Orleans, 78-81; FAC BIOL, NC AGR & TECH STATE UNIV, 81- *Concurrent Pos:* Fel, Marine Biol Lab, 79, Sch Med, Tulane Univ, 80-81; asst res prof, Univ SC, Columbia, 83-84. *Mem:* Am Soc Microbiol. *Res:* Relationship between energy metabolism and starvation induced arrest of the cell cycle in Caulobacter crescentus. *Mailing Add:* Dept Biol NC Agr & Tech State Univ 1601 E Market St Greensboro NC 27411

JORDAN, TRUMAN H, b Wayne, Mich, Nov 18, 37; m 61; c 3. PHYSICAL CHEMISTRY. *Educ:* Albion Col, BA, 59; Harvard Univ, MA, 62, PhD(chem), 64. *Prof Exp:* NASA fel & res assoc crystallog, Univ Pittsburgh, 64-66; from asst prof to assoc prof, 66-77, PROF CHEM, CORNELL COL, 77- *Concurrent Pos:* NIH fel, Nat Bur Standards, 72-73; vis prof, Univ Iowa, 78-79. *Mem:* AAAS; Am Chem Soc; Am Crystallog Asn; Int Asn Dent Res. *Res:* Molecular structure by means of x-ray crystallography; dental chemistry. *Mailing Add:* 6001 St West Mt Vernon IA 52314

JORDAN, WADE H(AMPTON), JR, b Edenton, NC, June 1, 32; m 54; c 5. ELECTROCHEMICAL KINETICS, BATTERIES. *Educ:* ECarolina Univ, AB, 54; Univ Tex, PhD(phys chem), 64. *Prof Exp:* Sr chemist, Tracor, Inc, 58-62; res chemist, E I du Pont de Nemours & Co, Inc, 63-70; assoc prof chem, Col Albemarle, 70-79, head dept, 76-79; sr develop engr, Imperial Clevite Corp, 80-82; staff electrochemist, Eveready Battery Co, 82-90; PROJ MGT ENGR, US ARMY, 90- *Mem:* Electrochem Soc; Am Chem Soc; NY Acad Sci. *Res:* Primary and secondary lithium battery systems; kinetic studies of metal deposition and dissolution. *Mailing Add:* One Mountain Top Rd High Knob Front Royal VA 22630

JORDAN, WAYNE ROBERT, b Kankakee, Ill, Jan 7, 40; m 60; c 6. PLANT PHYSIOLOGY, BIOCHEMISTRY. *Educ:* Univ Ill, Urbana, BS, 61, MS, 62; Univ Calif, Davis, PhD(plant physiol), 68. *Prof Exp:* From asst prof to prof, 68-75, prof crop physiol, Tex Agr Exp Sta, 75-, resident dir, Black Land Res Ctr, 80-; DIR, TEX WATER RESOURCES INST, TEX A&M UNIV. *Concurrent Pos:* Assoc ed, Agron J, 78-81. *Mem:* Am Soc Plant Physiol; fel Crop Sci Soc Am; fel Am Soc Agron; Am Water Resources Asn; Am Water Works Asn. *Res:* Plant water relations; drought resistance; root physiology; hormonal regulation of abscission. *Mailing Add:* Tex Water Resources Inst Tex A&M Univ College Sta TX 77843

JORDAN, WILLARD CLAYTON, b Richmond, Ind, May 13, 22; m 46; c 2. NUCLEAR PHYSICS. *Educ:* Miami Univ, BA, 45; Univ Mich, MS, 48, PhD, 54. *Prof Exp:* Res assoc, Argonne Nat Lab, 51-53, asst physicist, Exp Nuclear Physics, 53-54; physicist, Res Labs, Bendix Corp, 54-64; SCIENTIST, LOCKHEED MISSILES & SPACE RES LABS, PALO ALTO, 64-. *Concurrent Pos:* Asst, Univ Mich, 48-53; vis physicist, Lawrence Radiation Lab, Univ Calif, 58-64. *Mem:* Am Phys Soc; Am Nuclear Soc; Am Asn Physics Teachers; Sigma Xi. *Res:* Radioactive decay; nuclear reactors; thermonuclear research; space physics. *Mailing Add:* 24 Oak St Los Altos CA 94022

JORDAN, WILLIAM D(ITMER), b Selma, Ala, Feb 5, 22; m 47; c 3. MECHANICAL ENGINEERING. *Educ:* Univ Ala, BS, 42, MS, 49; Univ Ill, PhD(theoret & appl mech), 52. *Prof Exp:* Asst prof eng mech, Univ Ala, 46-50, asst theoret & appl mech, Univ Ill, 50-52; from assoc prof eng mech to prof, Univ Ala, 52-86, head Dept Eng Mech, 61-86; RETIRED. *Concurrent Pos:* Distinguished eng fel, Univ Ala, 88. *Mem:* Am Soc Eng Educ; fel Am Soc Mech Engrs; Am Inst Aeronaut & Astronaut; Am Acad Mech; Nat Soc Prof Engrs. *Res:* Strength of materials; structures; stress analysis; thermal stresses; buckling. *Mailing Add:* Dept Eng Mech PO Box 870278 Tuscaloosa AL 35487

JORDAN, WILLIAM KIRBY, food science; deceased, see previous edition for last biography

JORDAN, WILLIAM MALCOLM, b Brooklyn, NY, June 19, 36; m 63; c 2. HISTORY OF GEOLOGY. *Educ:* Columbia Univ, BA, 57, MA, 61; Univ Wis, PhD(geol), 65. *Prof Exp:* Res geologist, Jersey Prod Res Co, Okla, 64-65; res geologist, Esso Prod Res Co, Tex, 65-66; assoc prof, 66-69, chmn, 67-72, PROF EARTH SCI, MILLERSVILLE UNIV, 69- *Concurrent Pos:* Mem, US Nat Comt Hist Geol, 84- *Mem:* Am Asn Petrol Geologists; Geol Soc Am; Hist Sci Soc; Nat Asn Geol Teachers; Soc Econ Paleont & Mineral; Hist Earth Sci Soc. *Res:* Mature clastic sediments; paleocurrents and paleoclimatology; sedimentary facies and petroleum accumulation; history of geology. *Mailing Add:* Dept Earth Sci Millersville Univ Millersville PA 17551

JORDAN, WILLIAM R, III, b Denver, Colo, Apr 30, 44; m 66; c 1. ECOLOGICAL RESTORATION, HISTORY OF IDEAS ABOUT RELATIONSHIP BETWEEN HUMANS & ENVIRONMENT. *Educ:* Marquette Univ, BS, 66; Univ Wis-Madison, PhD(bot), 71, MA, 74. *Prof Exp:* Newswriter, Publ & Outreach Prog, Am Chem Soc, 75-76; MGR, ARBORETUM UNIV WIS-MADISON, 77- *Concurrent Pos:* Consult, Environ Adv Prog, Quest Found, 91- *Mem:* Soc Ecol Restoration. *Res:* Ecological restoration as a technique for research, a teaching technique and a performing art; history of ideas about the relationship between human being and the rest of nature, through intellectual history, anthropology, literature and the arts. *Mailing Add:* Univ Wis Arboretum 1207 Seminole Hwy Madison WI 53711

JORDAN, WILLIAM STONE, JR, b Fayetteville, NC, Sept 28, 17; m 47; c 2. PUBLIC HEALTH & EPIDEMIOLOGY. *Educ:* Univ NC, AB, 38; Harvard Univ, MD, 42. *Prof Exp:* From intern to asst resident, 2nd Med Serv, Boston City Hosp, 42-43, resident, 46-47; from instr to assoc prof prev med, Sch Med, Western Reserve Univ, 48-58, from instr to asst prof med, 48-58; prof prev & internal med & chmn dept prev med, Sch Med, Univ Va, 58-67; prof med & community med, Univ Ky, 67-74, dean Col Med, 67-74; dir, 76-87, EMER DIR & VOL, MICROBIOL & INFECTIOUS DIS PROG, NAT INST ALLERGY & INFECTIOUS DIS, 87- *Concurrent Pos:* Teaching fel prev med & med, Sch Med, Western Reserve Univ, 47-48; dir comn acute respiratory dis, Epidemiol Bd, US Armed Forces; consult, Surgeon Gen; mem, comn epidemiol & vet follow-up studies, Nat Acad Sci, 65-72; mem infectious dis adv comt, Nat Inst Allergy & Infectious Dis, 67-71; mem, Panel on Review of Viral & Rickettsial Vaccines, Food & Drug Admin, 73-76. *Mem:* Soc Exp Biol & Med; Am Epidemiol Soc; Am Soc Clin Invest; Am Pub Health Asn; Am Asn Immunol. *Res:* Etiology and epidemiology of acute respiratory disease. *Mailing Add:* NIH CD Bldg Rm 2075 Bethesda MD 20205

JORDAN, WILLIS POPE, JR, b Rossville, Ga, Oct 7, 18; m 52; c 3. UROLOGY. *Educ:* Emory Univ, BS, 39, MD, 43. *Prof Exp:* Intern, Emory Univ Hosp, 43-44; from asst resident to chief resident, Columbus Med Ctr, Ga, 44-45; preceptorship urol under Dr W P Jordan, Sr, 47-52; asst resident surg, Vet Admin Hosp, Atlanta, 52-53, from asst resident to sr resident urol, New Orleans, 53-55, sect chief, Lake City, Fla, 55-68; assoc prof, 68-77, PROF UROL, COL MED, UNIV TENN, MEMPHIS, 77-; SECT CHIEF, VET ADMIN HOSP, MEMPHIS, 68- *Concurrent Pos:* Asst prof, Col Med, Univ Fla, 65-68. *Mem:* Am Urol Asn; Am Col Surg; AMA; Int Soc Urol; Soc Univ Urol. *Res:* Cancer of the prostate and bladder; hydrodynamics of micturition. *Mailing Add:* Vet Admin Hosp Memphis TN 38104

JORDAN-MOLERO, JAIME E, b Utauado, PR, Jan 5, 41; m 72; c 3. PLANT PHYSIOLOGY, WEED CONTROL. *Educ:* Univ PR, BS, 63, MS, 70; Univ Ill Urbana-Champaign, PhD(agron), 77. *Prof Exp:* Res asst agr, 63-69, asst agronomist agr res, 69-79, asst dir, Agr Exp Sta, 77-79, ASSOC PROF AGR RES, ASST DEAN & DIR, AGR EXP STA, UTUADO REGIONAL COL, UNIV PR, 79- *Concurrent Pos:* Prof crop physiol, Mayaguez Campus, Univ PR, 77- *Mem:* Am Soc Agr Sci. *Res:* Crop nutrition; coffee genetics and selection; plant physiology; weed control physiology and ecology. *Mailing Add:* Agr Exp Sta Univ PR PO Box 1449 Utuado PR 00761

JORDEN, JAMES ROY, b Oklahoma City, Okla, Apr 16, 34; m 56; c 2. FORMATION EVALUATION, WELL LOGGING. *Educ:* Univ Tulsa, BS, 57. *Prof Exp:* Mem eng staff, Shell Oil Co, 60-81, sr tech specialist petrophys eng, Head Off Prod, 81-85, mgr, Petrol Eng Res, Shell Develop Co, 85-88, MGR TECH TRAINING, HEAD OFF PROD, SHELL OIL CO, 88- *Concurrent Pos:* Dir, Soc Petrol Engrs, 79-82; bd chmn, Soc Petrol Engrs Serv Corp, 85, pres, 86 & 89-90. *Honors & Awards:* Distinguished Serv Award, Soc Petrol Engrs, 88, DeGolyer Distinguished Serv Medal, 91. *Mem:* Soc Petrol Engrs (pres, 84); Soc Prof Well Log Analysts. *Res:* Co-author of a four-volume series on formation evaluation and well logging. *Mailing Add:* PO Box 481 Houston TX 77001-0481

JORDEN, ROGER M, b Carthage, Mo, Nov 15, 35. ENVIRONMENTAL ENGINEERING, WATER CHEMISTRY. *Educ:* Univ Tex, Austin, BS, 59; Univ Ariz, MS, 62; Univ Ill, Urbana, PhD(civil eng), 68. *Prof Exp:* Res asst hydrol, Inst Water Utilization, Univ Ariz, 59-62; res assoc, Travelers Res Inst, 62-64; res assoc sanit eng, Univ Ill, Urbana, 64-67, res fel, 67-68; asst prof environ eng, Univ Colo, Boulder, 68-75, assoc prof, 75-76; proj mgr, Elec Power Res Inst, Calif, 76-79; proj mgr water mgt, Colo Ute Elec Asn, Inc, 79-81; CONSULT WATER MGT, 82- *Honors & Awards:* Eddy Award, Water Pollution Control Fedn, 76. *Mem:* AAAS; Am Water Works Asn; Water Pollution Control Fedn. *Res:* Water management of zero discharge water systems in power plants; coagulation-flocculation of dilute colloidal suspensions; physical-chemical removal of trace elements; environmental transport of trace elements; water pollution control in oil shale; side-stream lime softening. *Mailing Add:* Water Systs Eng Box 70545 Steamboat Springs CO 80477

JORDIN, MARCUS WAYNE, b Idaho Falls, Idaho, May 23, 27; m 56; c 2. PHARMACOLOGY. *Educ:* Idaho State Col, BS, 49; Purdue Univ, MS, 52, PhD(pharmacol), 54. *Prof Exp:* Assoc prof, Sch Pharm, Univ Ark, Little Rock, 54-62, prof pharmacol & head dept, 62-90; RETIRED. *Mem:* Am Pharmaceut Asn. *Res:* Central nervous system drugs; tranquilizers and psychic energizers. *Mailing Add:* 309 Brookside Little Rock AR 72205

JORDON, ROBERT EARL, b Buffalo, NY, May 7, 38; m 69; c 1. DERMATOLOGY, IMMUNOLOGY. *Educ:* Hamilton Col, BA, 60; State Univ NY Buffalo, MD, 65; Univ Minn, Minneapolis, MS, 70. *Prof Exp:* Asst prof dermat & immunol, Mayo Med Sch Med, Univ Minn, Rochester, 71-77; prof & chmn sect dermat, Med Col Wis, 77-; AT DERMAT DEPT, UNIV TEX HEALTH SCI CTR. *Concurrent Pos:* Training grant dermat, Mayo Clin, 68-69; Nat Inst Arthritis, Metab & Digestive Dis res fel, Univ Minn, 72-73; vis asst prof & assoc mem grad fac, Univ Minn, Minneapolis, 71- *Mem:* AAAS; Am Asn Immunol; Am Fedn Clin Res; Soc Invest Dermat; Sigma Xi. *Res:* Immunopathology of bullous skin diseases using immunofluorescence, and complement research technics. *Mailing Add:* Dermat Dept Univ Tex Health Sci Ctr 6431 Fannin Suite 1-204 Houston TX 77030

JORDY, GEORGE Y, b Pittsburgh, Pa, May 2, 32. ENERGY. *Educ:* Carnegie Mellon Univ, BS, 54; Univ Pennsylvania, MBA, 55; Univ Maryland, PhD, (math education), 76. *Prof Exp:* MEM STAFF, ENERGY RES DEPT, OFF PROG ANALYSIS, DEPT ENERGY, 80- *Mailing Add:* Off Energy Res ER-30 G-236 Washington DC 20545

JORGENSEN, CLIVE D, b Orem, Utah, July 14, 31; m 55; c 3. ENTOMOLOGY. *Educ:* Brigham Young Univ, BS, 54, MS, 57; Ore State Univ, PhD(entom), 64. *Prof Exp:* Field dir ecol res, 60-63, from instr to assoc prof zool & entom, 63-73, PROF ZOOL, BRIGHAM YOUNG UNIV, 73-, CHMN DEPT, 74- *Concurrent Pos:* US Atomic Energy Comn res grant, 63-66; USDA res grant, 65-67; NSF grant, asst prof zool, Iowa State Univ, 71-72; coordr biol instr, Brigham Young Univ, 72-74. *Mem:* AAAS; Entom Soc Am; Am Soc Mammal. *Res:* Ecological research. *Mailing Add:* Dept Zool Brigham Young Univ Provo UT 84601

JORGENSEN, ERIK, b Denmark, Oct 28, 21; nat Can; m 46; c 2. ENVIRONMENTAL MANAGEMENT, FORESTRY. *Educ:* Royal Vet & Agr Col, Denmark, MF, 46. *Prof Exp:* Asst forest path, Royal Vet & Agr Col, Denmark, 49-53, amanuensis, col & proj leader, Forest Exp Sta, 53-55; res officer in chg plantation dis, Can Dept Agr, 55-59; agr prof forest path, Univ Toronto, 59-63, assoc prof, Shade Tree Res Lab, 63-67, prof forestry, 67-73, in chg lab, 63-72; chief urban forestry prog, Forest Mgt Inst, Can Forestry Serv, Dept Environ, 73-78; dir & prof environ biol, Univ Guelph Arboretum, 78-86; CONSULT PROF FORESTER, 87- *Mem:* Can Phytopath Soc; Arboricult Res & Educ Acad (pres, 76-78). *Res:* Urban forestry management and planning; environmental impact on/and by trees; tree diseases, physiology and breeding; arboriculture. *Mailing Add:* Arbor Consults PO Box 84 Guelph ON N1H 6J6 Can

JORGENSEN, GEORGE NORMAN, b Omaha, Nebr, Feb 22, 36; m 70; c 2. BIOCHEMISTRY, VIROLOGY. *Educ:* Univ Nebr, BS, 57; Univ Ill, MS, 68, PhD(biochem), 72. *Prof Exp:* Lab technician, Shell Chem Co, 57-58; res asst radiation biol, M D Anderson Hosp & Tumor Inst, 59-66; res assoc biochem virol, Baylor Col Med, 72-76; res assoc biol, Rice Univ, 77-78; res assoc, Univ Tex Med Sch Houston, Obstet Gynec, 79-80; RETIRED. *Mem:* Am Chem Soc; AAAS; Sigma Xi. *Res:* Studies of enzymes of snail metabolism such as urease, super oxide desmutase and mannitol oxidase concerned with comparative aspects of nitrogen metabolism and energy sources of snails; isolation and studies of hormone relaxin. *Mailing Add:* 8302 Greenbush Houston TX 77025

JORGENSEN, HELMUTH ERIK MILO, b Odense, Denmark, June 19, 27; nat US; m 53; c 3. PHYSICAL CHEMISTRY. *Educ:* Polytech Inst Brooklyn, BS, 50; Rutgers Univ, PhD(phys chem), 59. *Prof Exp:* Develop chemist, Sterling-Winthrop Res Inst, 50-51; res chemist, Schering Corp, 52-53; develop chemist, Am Cyanamid Co, 53-54; asst, Rutgers Univ, 54-58; res chemist, Distillation Prod Industs, Eastman Kodak Co, 58-60; assoc res chemist, Sterling-Winthrop Res Inst, NY, 60-69; from assoc prof to prof chem, 69-84, adj prof chem, Hudson Valley Community Col, 86- *Mem:* Am Chem Soc. *Res:* Physical chemistry of polymers and colloids; new methods of teaching chemistry. *Mailing Add:* Dept Chem Hudson Valley Community Col 80 Vandenburgh Ave Troy NY 12180

JORGENSEN, JAMES D, b Salina, Utah, Mar 23, 48; m 70. SOLID STATE PHYSICS. *Educ:* Brigham Young Univ, BS, 70, PhD(physics), 75. *Prof Exp:* Fel, Argonne Nat Lab, 75-77, asst physicist, 77-78, physicist, 79-90, SR PHYSICIST, ARGONNE NAT LAB, 90- *Concurrent Pos:* Mem, US Nat Comt Crystallog. *Honors & Awards:* Warren Diffraction Physics Award, 91. *Mem:* Fel Am Phys Soc; Am Crystallog Asn. *Res:* Powder neutron diffraction at pulsed neutron sources; ternary superconductors; neutron diffraction at high pressure; oxide superconductors. *Mailing Add:* Mat Sci Div Argonne Nat Lab Bldg 223 Argonne IL 60439

JORGENSEN, JENS ERIK, b Oslo, Norway, July 2, 36; US citizen; m 62; c 2. MECHANICAL ENGINEERING, SYSTEMS ANALYSIS. *Educ:* Mass Inst Technol, SB, 59, MS, 63, ScD(mech eng), 69. *Prof Exp:* Res engr, Cadilac Gage Co, Calif, 59-61; res engr, MHD Inc, Calif, 62; res asst mech eng, Mass Inst Technol, 63-65, instr, 65-68; asst prof, 68-73, assoc prof, 73-79, PROF MECH ENG, UNIV WASH, 79-, BOEING PROF MFG, 87- *Concurrent Pos:* Nat Insts Health fel bioeng, 69; consult, Wash Iron Works, 69-70, Pac Northwest Forest & Range Exp Sta, US Forest Serv, 71-77, Weyerhaeuser Co, 79-81, Metro, 81-83 & Boeing Com Airlines, 83-85. *Honors & Awards:* Ralph Teetor Award, Soc Automotive Engrs, 71. *Mem:* Am Soc Mech Engrs; Soc Mfg Engrs; Sigma Xi. *Res:* Fluid power systems analysis; design and analysis of fluidic devices; control systems analysis; instrumentation design and performance analysis; design and control of large scale off road equipment for logging in national forests; manufacturing systems analysis and automation of manufacturing processes. *Mailing Add:* Dept Mech Eng FU-10 Univ Wash Seattle WA 98195

JORGENSEN, NEAL A, b Luck, Wis, Feb 3, 35; m 55; c 1. AGRICULTURAL ADMINISTRATION, DAIRY SCIENCE. *Educ:* Univ Wis-River Falls, BS, 60; Univ Wis-Madison, MS, 62, PhD(dairy sci), 64. *Prof Exp:* From asst prof to prof dairy sci, Col Agr & Life Sci, Univ Wis-Madison, 68-84, assoc dean, 84-91, asst dir, 84, assoc dir, 84-89, exec dir, 90-91, ACTG DEAN & DIR, COL AGR & LIFE SCI, UNIV WIS-MADISON, 91- *Concurrent Pos:* Mem, Comt Animal Nutrit, Nat Res Coun, 81-88, Bd Agr, 90-; chair, Animal Systs Subcomt, Exp Sta Comt Orgn & Policy, 89- *Mem:* Am Dairy Sci Asn (pres, 90-91); Am Soc Animal Sci. *Res:* Agricultural administration; dairy cattle nutrition and management; preservation and utilization of forage crops; fiber requirements, amount, source, physical form; vitamin D and calcium metabolism. *Mailing Add:* Col Agr & Life Sci Univ Wis Rm 136 Agr Hall Madison WI 53706

JORGENSEN, PALLE E T, b Copenhagen, Denmark, Oct 8, 47; US citizen; m 75; c 3. OPERATOR ALGEBRAS, MATHEMATICAL PHYSICS. *Educ:* Univ Aarhus, Denmark, AB, 68, MS, 70, PhD(math), 73. *Prof Exp:* Fel math, Univ Wash, 73-74 & Univ Pa, 74-77; asst prof, Stanford Univ, 77-80; assoc prof, Aarhus, 79-82; vis assoc prof, Univ Pa, 82-84; PROF MATH, UNIV IOWA, 83- *Concurrent Pos:* Res fel, Danish Res Coun, 76-77 & NSF, 77-; ed, D Reidel Publ Co, 82-; ed, Am Math Soc, 88-; speaker, US-Japan operator algebra conf, Philadelphia, 88. *Mem:* Danish Acad Sci; Am Math Soc; Danish Math Soc; Soc Indust & Appl Math; fel Royal Swed Acad Sci; NY Acad Sci. *Res:* Modern analysis; operator algebras, spectral theory, harmonic analysis, differential geometry, and mathematical physics; various monographs published. *Mailing Add:* Dept Math MacLean Hall Univ Iowa Iowa City IA 52242

JORGENSEN, PAUL J, b Midway, Utah, Sept 1, 30; m 59; c 6. MATERIALS SCIENCE, CHEMISTRY. *Educ:* Brigham Young Univ, BS, 54; Univ Utah, PhD(mat sci), 60. *Prof Exp:* Ceramist, Gen Elec Res Lab, 60-68; chmn, Ceramics Dept, Stanford Res Inst, 68-74, dir, Mat Res Ctr, 74-76, exec dir phys sci, 76-77, vpres phys & life sci, 77-80, sr vpres sci group, 80-88, EXEC VPRES & CHIEF OPERATING OFFICER, SRI INT, 88- *Concurrent Pos:* Lectr, Univ Calif, Berkeley, 69-70; consult, GTE Sylvania, Inc, 71-; mem Comt High Temp Chem, Nat Res Coun-Nat Acad Sci, 71-74 & Nat Mat Adv Bd, 81-84; chmn, Adv Coun, Col Eng, Univ Utah, 83-84; mem, bd dirs, Mirage Systs, 84-, SRI Int, 86-, Plant Cell Res Inst, 87-, Devco, 87-, David Sarnoff Res Ctr, 88-; mem, adv coun, Konsai Res Inst, 87-; mem, Int Panel Adv Technol, Singapore Inst Standards & Indust Res, 90- *Honors & Awards:* I R 100 Award, 67. *Mem:* Fel Am Ceramic Soc; AAAS; Sigma Xi; Am Electronics Asn; Am Mgt Asn. *Res:* Kinetics of transport processes in ceramics, including sintering, solute segregation, grain growth, diffusion, electrical conductivity, oxidation, corrosion and permeation. *Mailing Add:* 333 Ravenswood Ave Menlo Park CA 94025

JORGENSEN, WILLIAM L, b New York, NY, Oct 5, 49. ORGANIC CHEMISTRY, THEORETICAL CHEMISTRY. *Educ:* Princeton Univ, AB, 70; Harvard Univ, PhD(chem physics), 75. *Prof Exp:* From asst prof to prof org chem, Purdue Univ, 75-90, H C Brown prof chem, 85-90; C P WHITEHEAD PROF CHEM, YALE UNIV, 90- *Concurrent Pos:* Dreyfus teacher-scholar, Camille & Henry Dreyfus Found Inc, 78-83; A P Sloan Found fel, 79-81; A C Cope scholar, 90. *Honors & Awards:* Ann Medal, Int Acad Quantum Molecular Sci, 86. *Mem:* Am Chem Soc; AAAS. *Res:* Theoretical organic chemistry; computer simulations of molecular liquids and solutions; computer assisted synthetic analysis. *Mailing Add:* Dept Chem Yale Univ New Haven CT 06511-8118

JORGENSON, EDSEL CARPENTER, b Kamas, Utah, Mar 11, 26; m 47; c 6. ZOOLOGY, NEMATOLOGY. *Educ:* Univ Utah, BS, 53, MS, 56. *Prof Exp:* Nematologist, USDA, 54-62, res nematologist, Utah State Univ, 62-67, zoologist, Utah Exp Sta, 67-70, zoologist, Calif Exp Sta, 70-87; RETIRED. *Mem:* Soc Nematol; Orgn Trop Am Nematol; Am Phytopath Soc. *Res:* Nematode ecology, control, biology and interactions. *Mailing Add:* 7604 Branding Iron Bakersfield CA 93309

JORGENSON, GORDON VICTOR, b Sunburg, Minn, Jan 3, 33; m 57; c 2. THIN FILMS. *Educ:* St Olaf Col, BA, 54. *Prof Exp:* Physicist, Wright Air Develop Ctr, Wright-Patterson AFB, 54-55; prin lab attendant, Univ Minn, 56-57; assoc scientist, Gen Mills, Inc, 57-61, sr scientist, 61-63; sr scientist, Appl Sci Div, Litton Indust, Inc, 63-66; sr physicist, NStar Res & Develop Inst, 66-75; sr physicist, Midwest Res Inst, 75-78; prin res scientist, Honeywell Inc, 78-79, sr prin res scientist, 80-84, res staff scientist, 84-90, SECT CHIEF, HONEYWELL INC, 91- *Mem:* Soc Photo-Optical Instrumentation Engrs; Am Vacuum Soc. *Res:* Surface physics research utilizing sputtering; effects of solar-wind bombardment of bodies in space; electrohydrodynamics; research in vacuum deposited thin films; optical coating technology. *Mailing Add:* 14609 Summit Oaks Dr Burnsville MN 55337

JORGENSON, JAMES WALLACE, b Kenosha, Wis, Sept 9, 52; m 78. CHROMATOGRAPHY, ELECTROPHORESIS. *Educ:* Northern Ill Univ, BS, 74; Ind Univ, PhD(chem), 79. *Prof Exp:* From asst prof to assoc prof, 79-87, PROF CHEM, UNIV NC, CHAPEL HILL, 87- *Mem:* Am Chem Soc; AAAS. *Res:* Chemical separations: fundamental studies of gas chromatography; liquid chromatography and electrophoresis. *Mailing Add:* Chem Dept Venable Hall Univ NC Chapel Hill NC 27599-3290

JORIZZO, JOSEPH L, b Rochester, NY, Oct 6, 51; c 1. IMMUNODERMATOLOGY, NEUTROPHILS & IMMUNE COMPLEX REACTIONS IN THE SKIN. *Educ:* Boston Univ, AB, 72, MD, 75; Am Bd Dermat, 79. *Prof Exp:* Intern internal med, NC Mem Hosp, 75-76, resident dermat, 76-78, chief resident, 78-79; clin asst prof, dept dermat, Univ Tex Med Br, Galveston, 79-80, from asst prof to assoc prof, 80-86; PROF & CHMN DEPT DERMAT, BOWMAN GRAY SCH MED, WAKE FOREST UNIV, WINSTON-SALEM, NC, 86-, DIR DERMAT RESIDENCY PROG, 87- *Concurrent Pos:* Consult, Vet Admin Clin, Winston-Salem, NC, 86-; fel allergy & clin immunol, Univ Tex Med Br, Galveston, 80-86; teacher family pract, Am Acad Family Physicians, Univ Tex Med Br, 82-86 & Bowman Gray Sch Med, Wake Forest Univ, Winston-Salem, NC, 87-; reviewer, J Am Acad Dermat, 81-, Int J Dermat, 84-, J Invest Dermat, 86-; prin investr, Dermat Found, 84 & 86, Nat Inst Dent Res, 85, Am Cyanamid Co, 87 & Herbert Labs, 87; fel, Dermat Inst, London, Eng. *Mem:* Soc Invest Dermat; Am Acad Dermat; Am Col Cryosurg; fel Am Col Physicians; AMA; Int Soc Trop Dermat; corresp mem Italian Soc Dermat & Venereology; Asn Profs Dermat; Am Fedn Clin Res. *Res:* Histamine induced localized vasculitis in patients with reactive vascular dermatoses; circulating immune complex-mediated aspects of secondary syphilis; dermatologic aspects of circulating immune complexes; dermatologic aspects of rheumatoid arthritis; histamine related mediators of inflammation; immunologic investigations in Behcet's disease and bowel disease; immune enhancers, prostaglandin synthesis blockers; collagen vascular diseases-experimental therapies for cutaneous aspects. *Mailing Add:* Dept Dermat Bowman Gray Sch Med Wake Forest Univ Winston-Salem NC 27103

JORNE, JACOB, b Tel-Aviv, Israel, July 24, 41; US citizen; m 85; c 3. ELECTROCHEMISTRY, MICROELECTRONICS PROCESSING. *Educ:* Technion Israel Inst Technol, BSc, 63, MSc, 67; Univ Calif, Berkeley, PhD(chem eng), 72. *Prof Exp:* Res asst chem eng, Lawrence Berkeley Lab, 67-72; prof, Wayne State Univ, 72-82; PROF CHEM ENG, UNIV ROCHESTER, 82- *Concurrent Pos:* Consult, Gulf & Western, 76-82; adj prof, Wayne State Univ, 84- *Honors & Awards:* Battery Res Award, Electrochem Soc, 89. *Mem:* Electrochem Soc; Am Inst Chem Engrs. *Res:* Theoretical biology. *Mailing Add:* Dept Chem Eng Univ Rochester Rochester NY 14627

JORNS, MARILYN SCHUMAN, b New York, NY, Aug 24, 43; m 71. ENZYMOLOGY. *Educ:* State Univ NY Binghamton, BA, 65; Univ Mich, MS, 67, PhD(biochem), 70. *Prof Exp:* Am Cancer Soc fel chem, Univ Konstanz, 70-72; res assoc biochem, Univ Tex Health Sci Ctr Dallas, 72-75; assoc prof chem, Ohio State Univ, 75-82; assoc prof biochem, 82-87, PROF BIOCHEM, HAHNEMANN UNIV SCH MED, 87- *Honors & Awards:* Linus Pauling Award, 87. *Mem:* Am Chem Soc; Am Soc Biol Chemists; AAAS. *Res:* Mechanism of enzymic repair of DNA damaged by ultra violet light; mechanism of catalysis by flavoproteins; flavin and pterin chemistry. *Mailing Add:* Dept Biologic Chem Hahnemann Univ Sch Med Broad & Vine Philadelphia PA 19102-1192

JORSTAD, JOHN LEONARD, b Richmond, Va, Sept 17, 35; c 3. METALLURGY. *Prof Exp:* Technician, Metall Lab, Reynolds Metals Co, 57-65, res scientist, 65-68, develop engr, Prod Develop Lab, 68-81, dept mgr, 81-85, MGR INGOT & FOUNDRY TECHNOL, REYNOLDS METALS CO, 85- *Concurrent Pos:* Chmn, Cast Metals Coun, Soc Mfg Engrs, 78-81 & Tech Coun, Am Foundrymen's Soc, 90-92; mem, Tech Coun, NAm Die Casters Asn, 91- *Honors & Awards:* Achievement Award, NAm Die Casters Asn, 87; Award of Sci Merit, Am Foundrymen's Soc, 90. *Mem:* Fel Am Soc Metals Int; Am Foundrymen's Soc; NAm Die Casters Asn; Soc Mfg Engrs; Soc Automotive Engrs; Am Soc Testing & Mat. *Res:* Aluminum casting alloys; casting processes and processing parameters; casting applications for automotive use, especially engines. *Mailing Add:* Reynolds Metals Co 6603 W Broad St Richmond VA 23261

JORY, FARNHAM STEWART, b Berkeley, Calif, Dec 6, 26; div; c 2. PHYSICS. *Educ:* Univ Calif, AB, 48; Swiss Fed Polytech, dipl, 51; Univ Chicago, MS, 54, PhD(physics), 55. *Prof Exp:* Asst, Enrico Fermi Inst Nuclear Studies, Ill, 52-55; researcher physics, Univ Md, 56-57; res geophysicist, Inst Geophys, Univ Calif, Los Angeles, 57-58; mem tech staff, Space Tech Labs, Thompson-Ramo-Wooldridge Corp, 58-59; asst prof physics, Long Beach State Col, 59-60; consult physics, 60-85; RETIRED. *Concurrent Pos:* Vis asst prof, Univ Calif, Los Angeles, 57-58; physicist, Lawrence Radiation Lab, 60. *Mem:* Am Phys Soc; Am Geophys Union; AAAS; NY Acad Sci. *Res:* Cosmic-ray and upper-atmosphere physics; geomagnetism and solarterrestrial relationships; experimental spectroscopy. *Mailing Add:* 3550 Pacific Ave Apt 1406 Livermore CA 94550

JORY, HOWARD ROBERTS, b Berkeley, Calif, Dec 8, 31; m 57; c 3. MICROWAVE ELECTRON TUBES, ELECTRON ACCELERATOR SYSTEMS FOR CANCER THERAPY. *Educ:* Univ Calif, Berkeley, BS, 54, MS, 55, PhD(elec eng), 60. *Prof Exp:* Officer, US Army Electronics Res &

Develop Lab, 60-62; res & develop engr, 62-72, eng mgr, 72-76, DEVELOP MGR, VARIAN ASSOCS, 76- *Concurrent Pos:* Consult, High Power Microwave Study Panel, Naval Res Lab, 76; chmn, Subcomt Electron Tubes, Inst Elec & Electronics Engrs, 83. *Mem:* Fel Inst Elec & Electronics Engrs; Sigma Xi. *Res:* Commercial gyrotrons; high power millimeter wave generators used in magnetic fusion laboratories. *Mailing Add:* Varian Assoc B-118, 611 Hansen Way Palo Alto CA 94303

JOSE, JORGE V, b Mexico City, Mex, Sept 13, 49; m; c 3. PHYSICS. *Educ:* Nat Univ Mex, BS, 72, MS, 73, PhD(physics), 76. *Prof Exp:* Res assoc physics, Brown Univ, 74-76, asst prof res, 76-77; James Franck fel physics, Univ Chicago, 77-79; asst prof res, Rutgers Univ, 79-80; asst prof, 80-84, assoc prof, 84-88, PROF PHYSICS, NORTHEASTERN UNIV, 88- *Concurrent Pos:* Guest scholar, Kyoto Univ, Japan, 79; consult, Exxon Res Eng, 82; vis scientist, Schlumberger, Dallas, 84 & Inst Physics, Mex, vis prof, 81 & 85. *Mem:* Am Phys Soc; AAAS; Mex Physics Soc. *Res:* Theoretical condensed matter physics. *Mailing Add:* Physics Dept Northeastern Univ Boston MA 02115

JOSE, PEDRO A, b Dingras, Ilocos Norte, Phillippines, Dec 6, 42. PEDIATRIC NEPHROLOGY. *Educ:* Univ Santo Tomas, MD, 65; Georgetown Univ, PhD(physiol), 76. *Prof Exp:* PROF PEDIAT, GEORGETOWN UNIV, 83- *Mem:* Soc Pediat Res; Am Fedn Clin Res; Am Soc Nephrology; Am Soc Pediat Nephrology. *Mailing Add:* Dept Pediat Georgetown Univ Hosp 3800 Reservoir Rd NW Washington DC 20057

JOSELYN, JO ANN CRAM, b St Francis, Kans, Oct 5, 43. SOLAR PHYSICS, SPACE PHYSICS. *Educ:* Univ Colo, BS, 65, MS, 67, PhD(astrogeophys), 78. *Prof Exp:* Physicist res ionospheric physics, 68-75, physicist res magnetospheric physics, 75-76, physicist res solar wind, 76-78, PHYSICIST SOLAR & GEOMAGNETIC FORECASTING, SPACE ENVIRON SERVS CTR, SPACE ENVIRON LAB, NAT OCEANIC & ATMOSPHERIC ADMIN, 78- *Concurrent Pos:* US deleg, Study Group Six, Consult Comt Ionospheric Radio, 81 & 83; topic reporter, Div 5, Int Asn Geomagnetism & Aeronomy. *Mem:* Am Geophys Union; Union Radio Scientists Int; AAAS; Sigma Xi; Am Inst Aeronaut & Astronaut. *Res:* Astro geophysics, especially solar physics and solar wind physics; also solar-terrestrial relationships, especially geomagnetism. *Mailing Add:* NOAA/ERL/SEL R/432 325 Broadway Boulder CO 80303

JOSENHANS, JAMES GROSS, b Toledo, Ohio, Dec 19, 32; m 61; c 2. SOLID STATE ELECTRONICS. *Educ:* Univ Toledo, BSc, 56; Ohio State Univ, MSc, 58, PhD(elec eng), 62. *Prof Exp:* Engr, Storer Broadcasting Co, Ohio, 51-53; teaching asst physics, Univ Toledo, 53-56; from res asst to res assoc, Electron Device Lab, Ohio State Univ, 57-62, asst prof elec eng, 62-63; MEM TECH STAFF, BELL TEL LAB, 63- *Mem:* Am Phys Soc; Inst Elec & Electronics Engrs; Sigma Xi; Am Acoust Soc; Audio Engr Soc. *Res:* Silicon integrated circuit development strategies to satisfy projected telephone systems needs. *Mailing Add:* 397 Diamond Hill Rd Berkeley Heights NJ 07922

JOSENHANS, WILLIAM T, b Wildbad, Ger, Feb 19, 22; Can citizen; m 44; c 4. RESPIRATORY PHYSIOLOGY, RESPIRATORY DISEASES. *Educ:* Univ Tubingen, MD, 50. *Prof Exp:* Physician student health, Univ Bonn, 51-54, asst prof physiol, 54-58; from asst prof to prof physiol, Dalhouse Univ, 64-86; PVT PRACT, 86- *Concurrent Pos:* Dir, Pulmonary Function Lab, Halifax Infirmary. *Mem:* Can Physiol Soc; Am Physiol Soc; Ger Physiol Soc; fel Am Col Chest Physicians. *Res:* Mechanisms of respiration; ballisto cardiography; physiology of exercise; compliance, prevention of diseases. *Mailing Add:* Nine Hauswiesen Wildbad in Schwarzwald 7547 Germany

JOSEPH, ALFRED S, b Cortland, NY, June 27, 32; m 56; c 2. PHYSICS. *Educ:* Union Col, NY, BS, 56; Case Inst Technol, MS, 60, PhD(physics), 62. *Prof Exp:* Instr, Case Inst Technol, 58-62; sr physicist, Atomics Int Div, Corp Eng, Rockwell Int, 62, mem tech staff, 62-68, group leader, sci ctr, 68-72, dir solid state electronics, 72-76, sr tech adv, Autonetics, 76-77, sr eng exec technol appln, 77-80; RETIRED. *Concurrent Pos:* Consult, Gen Elec Co, Ohio, 59-61. *Mem:* Am Phys Soc. *Res:* Studies of the electronic properties of metals through the de Haas-van Alphen effect; superconducting phenomena; semiconductors and semiconductor devices. *Mailing Add:* 688 Laguna Dr Simi Valley CA 93065

JOSEPH, BERNARD WILLIAM, b Detroit, Mich, June 7, 29; m 51; c 3. VEHICLE EMISSION. *Educ:* Wayne State Univ, BA, 68. *Prof Exp:* Assoc sr res physicist, Res Labs, 51-80, SR DEVELOP ENGR, ENG STAFF, GEN MOTORS CORP, 80- *Honors & Awards:* Arch T Colwell Award, Soc Automotive Engrs, 74. *Mem:* Optical Soc Am; Soc Photog Scientists & Engrs. *Res:* Optical properties of materials, optical design, radiometry, photometry and photochemistry; exhaust emision measurement; optical engineering. *Mailing Add:* 1949 Franklin Berkley MI 48072

JOSEPH, DANIEL D, b Chicago, Ill, Mar 26, 29; m 50; c 3. MECHANICAL ENGINEERING. *Educ:* Univ Chicago, MA, 50; Ill Inst Technol, BS, 59, MS, 60, PhD(mech eng), 63. *Prof Exp:* Asst mech eng, Ill Inst Technol, 59-62,; asst prof, 62-63, from asst prof to assoc prof fluid mech, 63-68, PROF AEROSPACE ENG & MECH, UNIV MINN, 68- *Concurrent Pos:* Guggenheim fel, 69 & 70; assoc ed seven journals. *Honors & Awards:* GI Taylor Medal, Soc Eng Sci, 90. *Mem:* Nat Acad Eng; Nat Acad Sci; Soc Natural Philol; Am Soc Mech Engrs; Am Phys Soc. *Res:* Fluid mechanics, flow through porous media and hydrodynamic stability; applied mathematics; theory of hydrodynamic stability and bifurcation theory; rheology of viscoelastic fluids. *Mailing Add:* Dept Mech & Aerospace Eng Univ Minn Minneapolis MN 55455

JOSEPH, DAVID WINRAM, b Evanston, Ill, June 28, 30; m 60; c 3. ELEMENTARY PARTICLE PHYSICS. *Educ:* Roosevelt Col, BS, 52; Univ Chicago, MS, 57, PhD(elem particle physics), 59. *Prof Exp:* Physicist, Ballistic Res Labs, Aberdeen Proving Ground, Md, 53-55; res assoc physics,

Purdue Univ, 59-61; res assoc, US Naval Res Lab, Washington, DC, 61-63; from asst prof to assoc prof, 63-68, PROF PHYSICS, UNIV NEBR-LINCOLN, 68- *Mem:* Fel Am Phys Soc. *Res:* Group and algebraic methods. *Mailing Add:* Dept Physics & Astron Univ Nebr Lincoln NE 68588

JOSEPH, DONALD J, b Summerfield, Ill, Sept 24, 22; m 45; c 2. OTOLARYNGOLOGY. *Educ:* St Louis Univ, MD, 46; Baylor Univ, MS, 53. *Prof Exp:* Assoc prof, 67-71, PROF SURG, SCH MED, UNIV MO-COLUMBIA, 71-, CHIEF OTOLARYNGOL, 67- *Concurrent Pos:* Consult, US Army Surgeon Gen, 61-64; mem comt hearing & bioacoust, Nat Res Coun, 61-67; communicative sci study sect, NIH, 62-64. *Honors & Awards:* Bronze Star Medal, 57. *Mem:* Fel Am Acad Ophthal & Otolaryngol; fel Am Col Surg; fel Am Laryngol, Rhinol & Otol Soc. *Res:* Communicative sciences; audiology; speech pathology. *Mailing Add:* 1026 Merrill Dr Box 159 Lebanon IL 62254

JOSEPH, EARL CLARK, II, b St Paul, Minn, Mar 2, 56; m 74. STRATEGIC MANAGEMENT. *Educ:* Univ Minn, BS, 78, PhD(strategic mgt), 83. *Prof Exp:* Systs planner, Sperry-Univac, 83-84, financial planner, 84-85; prog mgt, Unisys, 86-88; MKT REQ & RES, CRAY RES, INC, 88- *Concurrent Pos:* Vis lectr, Metro Univ, Minn, 83 & AT&T Mgt Training Ctr, Mankato Univ & St Thomas Col, 85. *Res:* The development and application of strategic management tools for use in exploring the potential future of products proposed for research and currently being developed; future product planning for computer systems of all types. *Mailing Add:* 365 Summit Ave St Paul MN 55102

JOSEPH, EDWARD DAVID, psychiatry, psychoanalysis; deceased, see previous edition for last biography

JOSEPH, J MEHSEN, b Whitesville, WVa, Sept 30, 28; m 51; c 5. MEDICAL MICROBIOLOGY. *Educ:* WVa Univ, BA, 48, MSc, 49; Univ Md, PhD(bact, chem), 54; Univ Toledo, BSc, 55. *Prof Exp:* Asst zool & anat, WVa Univ, 48-49; asst prof microbiol, Univ Toledo, 51-54; dir res, Biol Res Inst, 54-57; asst dir bur labs, 63-76, dir labs, Md State Dept Health & Ment Hyg, 77-87; DIR, COMMUNITY HEALTH SURVEILLANCE & LABS ADMIN, 87-; ASSOC PROF MICROBIOL, UNIV MD, 72- *Concurrent Pos:* Div head bur labs, Md State Dept Health & Ment Hyg, 57-63; assoc epidemiol, Johns Hopkins Univ, 79- *Honors & Awards:* Barnett Cohen Award, Microbiol. *Mem:* Hon Mem Am Soc Microbiol (secy, 74); Brit Soc Microbiol; Tissue Cult Asn; Am Pub Health Asn. *Res:* Disinfectants; antiseptics; microbiology of acid mine waters; lysozyme activity in relation to oral infections; public health microbiology; isolation methods for genus Clostridium; acute and chronic toxicity of chlorinated phenols; disease transmission by anesthetizing apparatus; evaluation of fungicidal compounds; diagnostic virology and tissue culture. *Mailing Add:* Md State Labs Admin 201 W Preston St Baltimore MD 21201

JOSEPH, J WALTER, JR, b Oak Park, Ill, Oct 8, 28; m 53; c 2. MECHANICAL ENGINEERING. *Educ:* NC State Col, BS, 50; Pa State Univ, MS, 54. *Prof Exp:* Asst eng res, Pa State Univ, 50-51, thermal res lab, 53-54; mech engr, Savannah River Lab, 54-65, mech engr, Reactor Tech Dept, 65-71, asst chief supvr, 71-76, chief supvr, 76-78, supt, Traffic & Transp Dept, 78-80, supt, LStartup Proj Team, 80-83, supt, Equip Engr Dept, supt, Site Quality Dept, 83-85, SAVANNAH RIVER PLANT, E I DU PONT DE NEMOURS & CO, INC, 85- *Res:* Participatve leadership nuclear reactor project management; mechanical and welding development supporting nuclear operations; remote equipment, nuclear waste management; isotopic heat sources; shipping containers; resistance welding; high pressure gas technology. *Mailing Add:* 340 Cherbourg Pl Aiken SC 29802

JOSEPH, JAMES, b Los Angeles, Calif, Oct 28, 30; m 58; c 2. MARINE BIOLOGY. *Educ:* Humboldt State Col, BS, 56, MS, 58; Univ Wash, PhD, 66. *Hon Degrees:* Dr, Universite de Bretagne Occidental, France Nautilus Award. *Prof Exp:* Scientist, Inter-Am Trop Tuna Comn, 58-63, prin scientist, 64-69, dir invests, 69, RES ASSOC, INST MARINE RES, SCRIPPS INST OCEANOG, 69-; AFFIL PROF, UNIV WASH, 69- *Concurrent Pos:* Served on various panels, comts, etc concerning marine sci and fisheries & as adv to all levels of govt. *Mem:* Am Inst Fishery Res Biol; Sigma Xi. *Res:* Relationship exploitation by man on the dynamics of the stocks of marine fishes; development of international arrangements for the conservation and management of living marine resources. *Mailing Add:* IAT Tuna Comm Scripps Inst Oceanog La Jolla CA 92037

JOSEPH, JEYMOHAN, b Jodhpur, India, Jan 20, 53; m 87; c 1. NEUROIMMUNOLOGY, ENDOTHELIAL CELL BIOLOGY. *Educ:* Univ Wis-Madison, PhD(immunol), 83. *Prof Exp:* Res assoc oncol, Wis Clin Cancer Ctr, 83-85; res assoc neuroimmunol, 85-87, instr, 87-89, ASST PROF NEUROIMMUNOL, THOMAS JEFFERSON UNIV, 89- *Mem:* Int Soc Neuroimmunol; Am Asn Immunologists; AAAS; NY Acad Sci. *Res:* Role of endothelial cells in immune and inflammatory events following virus infection. *Mailing Add:* Dept Neurol Rm 511 Col Bldg Thomas Jefferson Univ 1025 Walnut St Philadelphia PA 19107

JOSEPH, JOHN MUNDANCHERIL, b Kerala, India, Feb 21, 47; US citizen; m 81; c 3. CHROMATOGRAPHY, SPECTROSCOPY. *Educ:* Kerala Univ, India, BS, 68; Univ Jabalpur, India, MS, 71; Drexel Univ, Philadelphia, MS & PhD(biochem), 80. *Prof Exp:* Lectr chem, Kerala Univ, India, 68-69; Chemist, McDowell Distillery, Kerala, 71-73; Midvale Heppenstal, Philadelphia, 73-75; teaching & res asst biochem, Drexel Univ, Philadelphia, 75-80; GROUP LEADER ANALYTICAL CHEM & DOSAGE FORM ANALYSIS, BRISTOL-MYERS SQUIBB PHARMACEUT RES INST, 81- *Concurrent Pos:* Fel biochem, Drexel Univ, Philadelphia, 75-81. *Mem:* Sigma Xi; Am Asn Pharmaceut Scientists; Am Chem Soc. *Res:* Examination of the stereochemical requirements for the biological activity of cholesterol in terms of its metabolism to other functional sterols and its proper fit into biological membranes, investigated in both enzymatic and functional membranous systems; chemical synthesis of cholesterol analogues; conformational and configurational analysis of sterols by hydrogen and carbon 13-nuclear magnetic resonance; in-vitro dissolution testing; pharmaceutical testing and regulatory affairs. *Mailing Add:* Analytical Develop Bristol-Myers Squibb Pharmaceut Res Inst New Brunswick NJ 08903

JOSEPH, MARJORY L, textile chemistry; deceased, see previous edition for last biography

JOSEPH, PETER D(ANIEL), b Brooklyn, NY, Jan 21, 36; m 57; c 1. ELECTRICAL ENGINEERING. *Educ:* Mass Inst Technol, SB & SM, 58; Purdue Univ, PhD(elec eng), 61. *Prof Exp:* Instr elec eng, Purdue Univ, 58-61; mem tech staff, 61-64, head guidance sect, 64-66, mgr syst anal & software dept, 66-70, mgr sensor design & anal dept, 70-76, LAB MGR, TRW SYSTS & ENERGY GROUP, 76- *Mem:* Inst Elec & Electronics Engrs; Am Inst Aeronaut & Astronaut. *Res:* Guidance of missiles and space vehicles; optimal control theory; optimal filter theory; electro-optical sensors. *Mailing Add:* 2740 233 St Torrance CA 90505

JOSEPH, PETER MARON, b Ridley Park, Pa, Mar 26, 39. MEDICAL PHYSICS. *Educ:* Lafayette Col, BS, 59; Harvard Univ, MA, 61, PhD(physics), 67. *Prof Exp:* Instr physics, Cornell Univ, 67-70; asst prof, Carnegie-Mellon Univ, 70-72; NIH fel, Memorial-Sloan Kettering Cancer Ctr, 72-73; instr radiol, Columbia-Presby Med Ctr, 73-75, asst prof, 75-80; assoc prof diag images physics, Univ Md, Baltimore, 80-89; AT RADIOL DEPT, HOSP, UNIV PA, 89- *Mem:* AAAS; Am Phys Soc; Am Asn Physicists in Med. *Res:* High energy electromagnetic phenomena; experimental tests of quantum electrodynamics; photo production of vector mesons; energy range relations; x-ray spectra and attenuation curves; radiographic image quality; computerized axial tomography; magnetic resonance image. *Mailing Add:* Radiol Dept Hosp Univ Pa 3400 Spruce St Philadelphia PA 19104

JOSEPH, RAMON R, b New York, NY, May 17, 30; m 56; c 3. GASTROENTEROLOGY, ENDOSCOPY. *Educ:* Manhattan Col, BS, 52; Cornell Univ, MD, 56. *Prof Exp:* Intern med, Meadowbrook Hosp, Hempstead, 56-57; resident, 59-62, staff physician, 62-64, asst dir dept med, Wayne County Gen Hosp, 64-73; dir gastroenterol sect, Wayne County Gen Hosp, Westland, 62- 85, dir, Dept Med, 73-85; assoc prof internal med, Univ Mich, Ann Arbor, 68-75, asst dean, Med Sch, 73-84, prof internal med, 75-84; MED DIR, HENRY FORD MED CTR, 87- *Concurrent Pos:* Fel, Wayne County Gen Hosp, 61-62; consult, Annapolis Hosp, Wayne, 62-88, St Mary Hosp, Livonia, 62-, Mich Dept Educ, 69-73; from instr to prof internal med, Univ Mich, Ann Arbor, 62-75; chmn res dept, Wayne County Gen Hosp, 64-85, pres med staff, 71-72; chair, gastroenterol, St Marx Hosp, 85-90. *Mem:* AAAS; fel Am Col Physicians; NY Acad Sci; Asn Am Med Col; AMA; Am Gastroenterol Asn; Am Soc Internal Med; Am Soc Gastrointestinal Endoscopy. *Res:* Origin and nature of human serum lactic dehydrogenase; multiple molecular forms of enzyme amylase; biochemical diagnosis in gastroenterology; endoscopic aspects of gastroenterology. *Mailing Add:* Henry Ford Med Ctr 35605 Warren Rd Westland MI 48185

JOSEPH, RICHARD ISAAC, b Brooklyn, NY, May 25, 36; m 61; c 3. SOLID STATE PHYSICS. *Educ:* City Col New York, BS, 57; Harvard Univ, PhD(physics), 62. *Prof Exp:* Sr res scientist solid state physics, Raytheon Co, 61-66; from asst prof to assoc prof, 66-70, PROF ELEC ENG, JOHNS HOPKINS UNIV, 70- *Mem:* AAAS; Am Phys Soc. *Res:* Statistical mechanics; theory of magnetism and properties of magnetic materials; microwave physics; theory of solid state; exchange interactions in solids; critical phenomena; solitons; non-linear wave equations. *Mailing Add:* 2106 Uffington Rd Baltimore MD 21209

JOSEPH, ROSALINE RESNICK, b New York City, NY, Aug 21, 29; m 54; c 2. MEDICINE, HEMATOLOGY. *Educ:* Cornell Univ, AB, 49; Women's Med Col Pa, MD, 53; Temple Univ, MS, 58. *Prof Exp:* Instr hematol, Med Ctr, Temple Univ, 57-60, assoc med, 60-63, assoc prof med, 63-77, course coordr retinculo-endothelial, Syst Interdisciplinary Course Comt, 68-73; PROF MED & CHIEF HEMATOL & ONCOL DEPT, MED COL PA, 77- *Mem:* Am Fedn Clin Res; Am Soc Hematol; Fel Am Col Physicians; Am Asn Cancer Educ; Am Soc Clin Oncol. *Res:* New modalities in the treatment of cancer and hematologic disorders. *Mailing Add:* Med Col Pa 3300 Henry Ave Philadelphia PA 19129

JOSEPH, ROY D, b Fremont, Ohio, July 26, 37; m 69. APPLIED MATHEMATICS. *Educ:* Fenn Col, BEE, 60; Case Inst Technol, MSEE, 62, PhD(eng), 65. *Prof Exp:* Res assoc, Case Inst Technol, 65-66; asst prof eng, State Univ NY Stony Brook, 66-72; asst prof, 73-80, ASSOC PROF ELEC ENG, SPACE INST, UNIV TENN, 80- *Mem:* Inst Elec & Electronics Engrs; Soc Indust & Appl Math; Am Asn Univ Professors. *Res:* Active network synthesis; optimal control signal processing; digital signal processing. *Mailing Add:* Dept Elec Eng Space Inst Univ Tenn Tullahoma TN 37388-8897

JOSEPH, SAMMY WILLIAM, b Jacksonville, Fla, Oct 10, 34; m 67; c 2. MICROBIOLOGY, MEDICAL BACTERIOLOGY. *Educ:* Univ Fla, BSA, 56; St John's Univ, NY, MS, 64, PhD(microbiol), 70. *Prof Exp:* Asst head serol br, US Naval Med Sch, Nat Naval Med Ctr, 57-58, head bact, serol & mycol br, US Naval Hosp, St Albans, NY, 58-63, head bact & mycol br, Naval Med Sch, Nat Naval Med Ctr, 63-67; Navy contract fel bact, St John's Univ, NY, 67-70; microbiologist & exec officer, Naval Med Res Unit 2, Jakarta Detachment, Indonesia, 70-73, comndg officer bact, Naval Unit, Ft Detrick, 73-75, dep chmn microbiol dept, Naval Med Res Inst, 75-78, prog mgr infectious dis, Naval Med Res & Develop Command, Nat Naval Med Ctr, 78-81; chmn dept, 81-89, PROF, DEPT MICROBIOL, UNIV MD, 81- *Concurrent Pos:* Vis scientist, Off Naval Res, London, 77. *Mem:* Am Soc Microbiol; fel Am Acad Microbiol; AAAS; NY Acad Sci; Am Soc Clin Path; Sigma Xi. *Res:* Studies on bacteria of clinical significance, particularly those causing gastroenteritis; purification and characterization of bacterial toxins; role of antibiotics in treatment; bacterial adherence to surfaces; genetic basis of pathogenic mechanisms; bacterial taxonomy and systematic classification. *Mailing Add:* Dept Microbiol Univ Md College Park MD 20742

JOSEPH, SOLOMON, b Brooklyn, NY, Nov 3, 10; m 35; c 2. INDUSTRIAL CHEMISTRY. *Educ:* Columbia Univ, BS, 35, MA, 37; Polytech Inst Brooklyn, PhD(chem), 44. *Prof Exp:* Lab instr chem, Yeshiva Col, 35-38; chemist, NY Bd Transp, 39-45; dir res, Bri-Test, Inc, 45-50; chief chemist, Camp Chem Co, 51-52; chemist, Res Div, Penetone Co, NJ, 52-60; chief chemist, Chem Div, John Sexton & Co, Mich, 60-63; teacher chem, New York Bd Educ, 63-80; RETIRED. *Concurrent Pos:* Instr, Yeshiva Col, 45-50; lectr & adj prof chem, City Univ New York, 67-70. *Mem:* Sigma Xi; Am Chem Soc. *Res:* Emulsion of waxes, resins and polishes; cryoscopic studies of acids and bases in selenium oxychloride; protective coatings; synthetic detergents; corrosion prevention; sanitation chemicals; chemical specialties. *Mailing Add:* 1044 E Fifth St Brooklyn NY 11230

JOSEPH, STANLEY ROBERT, b Jacobus, Pa, May 21, 30; m 56; c 3. MEDICAL ENTOMOLOGY. *Educ:* Gettysburg Col, BA, 52; Pa State Univ, MS, 54; Univ Md, PhD(entom), 68. *Prof Exp:* From asst entomologist to assoc entomologist, Md State Bd Agr, 56-73; entomologist, pest mgt sect, 73-79, CHIEF, MOSQUITO CONTROL SECT, MD DEPT AGR, 80- *Mem:* Am Mosquito Control Asn. *Res:* Methods and insecticides for use in mosquito control in Maryland; ultra low volume insecticide applications with air and ground equipment; biology of Culiseta melanura in Maryland; insect physiology; toxicology of malathion to vertebrates. *Mailing Add:* 1631 Generals Hwy Annapolis MD 21401

JOSEPH, STEPHEN C, HEALTH SCIENCES. *Educ:* Harvard Col, BA, 59; Yale Univ, MD, 63; Johns Hopkins Univ, MPH, 68; Am Bd Pediat, dipl, 68. *Prof Exp:* Intern pediat, Boston Children's Hosp, 63-64, asst resident, 66-67; fel, Comp Child Care Proj, Dept Pediat, Johns Hopkins Sch Med, 67-68; prof pediat & community health, Univ Ctr Health Sci, Yaounde, Cameroon, 71-73; dir, Med Educ & Planning & consult to pres, Univ Wyo, 73-74; asst med, Children's Hosp Med Ctr, 74-78; dep asst adminr, Human Resources Develop, Bur Develop Support, Agency Int Develop, 78-81; consult & lectr, Int Develop Res Ctr, Can, 81-82; chief pediat, Grenfell Regional Health Serv, St Anthony, Can, 82-83; spec coordr, Child Health & Survival, UN Children's Fund, 83-86; comnr health, New York, 86-90; DEAN, SCH PUB HEALTH, UNIV MINN, 91- *Concurrent Pos:* Dir, Off Int Health Progs, Harvard Sch Pub Health, 74-78, lectr, Dept Maternal & Child Health, 74-78; asst med, Children's Hosp Med Ctr, 74-78; mem, Nat Coun Int Health, 75-78; actg asst adminr, Bur Develop Support, Agency Int Develop, 81; mem bd trustees, US Conf Local Health Officers, 87-89; mem, Nat Adv Comt HIV, Ctrs Dis Control, 88-91; Sol Fleischman vis prof med, Harvard Community Health Plan, 90. *Honors & Awards:* Pub Serv for Med Award, Am Col Physicians, 89. *Mem:* Inst Med-Nat Acad Sci; fel Am Pub Health Asn; fel Am Acad Pediat. *Res:* Author of over 350 publications. *Mailing Add:* Box 197 Mayo 420 Delaware St SE Minneapolis MN 55455-0381

JOSEPHS, JESS J, b New York, NY, Jan 4, 17; m 62; c 2. MUSICAL ACOUSTICS. *Educ:* NY Univ, AB, 38, MSc, 40, PhD(phys chem), 43. *Prof Exp:* Res assoc, Northwestern Univ, 45-46, instr phys chem, 46-47; asst prof phys sci, Univ Chicago, 47-50; asst prof physics, Boston Univ, 50-56; prof, 56-87, EMER PROF PHYSICS, SMITH COL, 87- *Mem:* Am Phys Soc; Am Asn Physics Teachers; Audio Eng Soc; Sigma Xi. *Res:* Solid state physics, psychoacoustics; acoustical study of the violin; distortion in electronically reproduced music. *Mailing Add:* 3300 Darby Rd Haverford PA 19041-1095

JOSEPHS, MELVIN JAY, b New York, NY, Apr 26, 26; m 48; c 2. PLANT PHYSIOLOGY, INFORMATION SCIENCE. *Educ:* Rutgers Univ, BSc, 50, MSc, 52, PhD(plant physiol, bot), 54. *Prof Exp:* Plant physiologist, Dow Chem Co, Mich, 54-60; assoc ed, Chem & Eng News, Am Chem Soc, 60-66, managing ed, Environ Sci & Technol, 66-69, managing ed, Chem & Eng News, 69-73; asst dir, Prod & Prog Mgt, Nat Tech Info Serv, US Dept Com, 73-76 & 78-86; chief, Toxicol Data Bank, Nat Libr Med, HEW, 76-78; EXEC DIR, AM SOC PLANT PHYSIOLOGISTS, 86- *Mem:* Sigma Xi; AAAS. *Res:* Boron nutrition and organic acid content; growth regulators; herbicides; algae; aquatic plants. *Mailing Add:* 9109 Friars Rd Bethesda MD 20817

JOSEPHS, ROBERT, b Philadelphia, Pa, June 29, 37; m 74, 90. BIOPHYSICS, STRUCTURAL BIOLOGY. *Educ:* Univ Ill, BS, 59; Hebrew Univ, MSc, 62; Johns Hopkins Univ, PhD(biol), 66. *Prof Exp:* Res assoc muscle biol, Johns Hopkins Univ, 66-67; fel, MRC Lab Molecular Biol, 68-69; fel struct biol, 70-73, scientist, 73-77, SR SCIENTIST & ASSOC, POLYMER DEPT, WEIZMAN INST, 77-; assoc prof biophys & theoret biol, 77-90, PROF MOLECULAR GENETICS & CELL BIOL, UNIV CHICAGO, 90- *Concurrent Pos:* NIH grant, 78-90. *Honors & Awards:* Res Career Develop Award, NIH. *Mem:* Electron Micros Soc Am; Biophys Soc. *Res:* Structural biology; sickle cell anemia; electron crystallography; cryoelectron microscopy. *Mailing Add:* Dept Molecular Genetics & Cell Biol Univ Chicago 920 E 58th St Chicago IL 60637

JOSEPHS, STEVEN F, b St Marys, Pa, May 3, 50; m 73; c 2. MOLECULAR BIOLOGY, BIOCHEMISTRY. *Educ:* Susquehanna Univ, BA, 72; Am Univ, MS, 77, PhD(chem), 82. *Prof Exp:* Chem info & handling specialist, div med chem, Walter Reed Army Inst Res, Walter Reed Army Med Ctr, 72-74; lab instr, Am Univ, 74-75; CHEMIST, NAT CANCER INST, NIH, 75- *Mem:* NY Acad Sci. *Res:* Structural organization of the human c-sis gene; transformation of c-sis cDNA clone; investigation of the Acquired Immune Deficiency Syndrome virus; human T-lymphotropic virus type III structure and function; human herpesvirus 6. *Mailing Add:* Lab Tumor Cell Biol Nat Cancer Inst NIH Bldg 37 Rm 6C-18 Bethesda MD 20851

JOSEPHSON, ALAN S, b Bronx, NY, Nov 30, 30; m 55; c 3. MEDICINE, IMMUNOLOGY. *Educ:* NY Univ, AB, 52, MD, 56; Am Bd Internal Med, dipl, 66. *Prof Exp:* From asst prof to assoc prof, 63-73, PROF MED, STATE UNIV NY DOWNSTATE MED CTR, 73- *Concurrent Pos:* Nat Inst Allergy & Infectious Dis trainee fel, NY Univ, 58-60. *Mem:* Am Fedn Clin Res; Am Rheumatism Asn; Am Asn Immunologists. *Res:* Immunologic properties of penicillin; proteins of secretions; immunologic properties of air pollutants. *Mailing Add:* Dept Med State Univ NY Downstate Med Ctr Brooklyn NY 11203

JOSEPHSON, BRIAN DAVID, b Cardiff, Eng, Jan 4, 40. PHYSICS. *Educ:* Cambridge Univ, PhD(physics). *Prof Exp:* Dir res, 67-72, PROF PHYSICS, CAMBRIDGE UNIV, 74- *Honors & Awards:* Nobel Prize in Physics, 73; Fitz London Award, 72. *Mem:* Am Acad Arts & Sci; Inst Elec Electronics Engrs. *Mailing Add:* Cavendish Lab Madingley Rd Cambridge CB3 0HE England

JOSEPHSON, EDWARD SAMUEL, b Boston, Mass, Sept 30, 15; m 38; c 3. FOOD SCIENCE, RESEARCH ADMINISTRATION. *Educ:* Harvard Univ, AB, 36; Mass Inst Technol, PhD(biochem), 40; Indust Col Armed Forces, dipl, 62. *Prof Exp:* Williams-Waterman fel, Mass Inst Technol, 40-43; biochemist, NIH, 44-52 & US Army Chem Corp, 52-54; chief, Biol & Chem Br, Qm Res & Eng Command, 54, asst chief, Chem & Plastics Div, 54-56, assoc sci dir develop, 56-61, spec asst food & food irradiation, 61-62; assoc dir food radiation, Food Div & dir, Radiation Lab, 62-72, dep tech dir, Food Serv Systs Prog, US Army Natick Res & Develop Labs, 72-75; sr lectr, Dept Appl Biol Sci, Mass Inst Technol, 76-89; ADJ PROF, DEPT FOOD SCI & NUTRIT, UNIV RI, 86- *Concurrent Pos:* Asst prof, Oglethorpe Univ, 43-44 & Med Sch, Emory Univ, 45; lectr, Am Univ, 50-54; adv on preserving food by ionizing radiation, Int Atomic Energy Agency, UN, 64-, Food & Agr Orgn, UN, 64-, Israel AEC, 65-85, Inter-Am Nuclear Energy Comn, Orgn Am States, 68-69, Ministry Econ, Iran, 69-75, Dept Atomic Energy, India, 67-71, Nuclear Energy Comn, Chile, 82-84 & Ministry Light Indust, China, 84; vchmn, Inter-Dept Comt Radiation Preservation Food, 70-71; consult, Universal Sci & Eng, 76-; lectr, Ctr Lifelong Learning, Harvard Univ, 77-82; sci adv, NH House Reps, 77-78. *Mem:* AAAS; Am Chem Soc; Inst Food Technol; Sigma Xi; Coun Agr Sci & Technol; Am Coun Sci & Health; Res & Develop Assocs Mil Food & Packaging Systs. *Res:* Chemotherapy of malaria and other tropical diseases; enzymes; vitamins; nutrition; microbiology; food preservation by ionizing radiations. *Mailing Add:* Dept Food Sci & Nutrit Univ RI 530 Liberty Lane West Kingston RI 02892

JOSEPHSON, LEONARD MELVIN, b Ashland, Wis, Dec 4, 13; m 40; c 2. PLANT BREEDING, AGRONOMY. *Educ:* Univ Wis, BS, 36, PhD(plant path, agron), 41. *Prof Exp:* Agt, USDA, Wis, 33-36, purchasing seed grain agt, Minn, 36-37; asst, Univ Wis, 37-39; secy & field rep, Malt Res Inst, 39-43; from asst agronomist to agronomist, Exp Sta, Univ Ky, 43-51; maize breeder, Union SAfrica, 51-54; prof agron, 54-71, prof plant & soil sci, 71-79, EMER PROF PLANT & SOIL SCI, UNIV TENN, KNOXVILLE, 79- *Concurrent Pos:* Consult, Univ Tenn-AID, India, SAfrica, 56, 82 & 88, Arg, 82 & 84; res agronomist, USDA, 54-71. *Mem:* AAAS; fel Am Soc Agron; Am Phytopath Soc; Am Genetics Asn; Genetics Soc Am; fel Crop Sci Soc Am. *Res:* Corn breeding, breeding methods and genetics; corn pathology; corn insects. *Mailing Add:* Dept Plant & Soil Sci Univ Tenn Knoxville TN 37916

JOSEPHSON, ROBERT KARL, b Somerville, Mass, July 12, 34; m 56; c 3. COMPARATIVE PHYSIOLOGY. *Educ:* Tufts Univ, BS, 56; Univ Calif, Los Angeles, PhD(zool), 60. *Prof Exp:* NATO fel, Univ Tubingen, 61; from asst prof to assoc prof zool, Univ Minn, 62-65; from assoc prof to prof biol, Case Western Reserve Univ, 65-71; PROF BIOL, UNIV CALIF, IRVINE, 71- *Concurrent Pos:* Guggenheim fel, 77-78; mem, Marine Biol Lab Corp. *Mem:* AAAS; Am Soc Zool; Brit Soc Exp Biol; Soc Gen Physiol; Soc Neurosci; Sigma Xi; Soc Biophys; Am Physiol Soc. *Res:* Mechanical power and efficiency of muscle. *Mailing Add:* Sch Biol Sci Univ Calif Irvine CA 92717

JOSEPHSON, RONALD VICTOR, b Bellefonte, Pa, May 19, 42; m 69; c 2. FOOD CHEMISTRY, BIOCHEMISTRY. *Educ:* Pa State Univ, BS, 64; Univ Minn, St Paul, MS, 66, PhD(food sci), 70. *Prof Exp:* Asst prof dairy technol, Ohio State Univ, 70-71, asst prof food sci & nutrit, 71-75; ASSOC PROF & PROF FAMILY STUDIES & CONSUMER SCI, SAN DIEGO STATE UNIV, 75- *Concurrent Pos:* Prin investr grants, Nat Oceanic & Atmospheric Admin, US Army, Nat Fisheries Inst, 77-; chmn, Basic Symposium Comt, Inst Food Technol, 85-87; pres, San Diego chpt, Sigma Xi, 86-87. *Mem:* Am Dairy Sci Asn; Inst Food Technol; Sigma Xi; AAAS. *Res:* Chemistry, analysis and storage stability of foods; milk and diary foods, fish and seafoods, medical foods, and vegetable proteins. *Mailing Add:* Sch Farm Studies & Consumer Sci San Diego State Univ San Diego CA 92182

JOSHI, ARAVIND KRISHNA, b Poona, India, Aug 5, 29; m 63; c 2. COMPUTER & INFORMATION SCIENCE. *Educ:* Univ Poona, BE, 51; Indian Inst Sci, Bangalore, dipl, 52; Univ Pa, MS, 58, PhD(elec eng), 60. *Prof Exp:* Res asst electronics, Indian Inst Sci, Bangalore, 52-53 & Tata Inst Fundamental Res, Bombay, 53; prof engr, Radio Corp Am, NJ, 54-58; assoc ling anal, 58-61, from asst prof to assoc prof elec eng & ling, 61-72, PROF COMPUT & INFO SCI & CHMN DEPT, UNIV PA, 72- *Concurrent Pos:* Assoc, transformations & discourse anal proj, NSF, 58-; consult info theory & ling, Philco Res Lab, Pa, 62-63; ling consult, Western Reserve, 64-; Guggenheim fel, 71-72; mem, Inst Advan Study, 71-72. *Mem:* Asn Comput Mach; fel Inst Elec & Electronics Engrs; Am Math Soc; Sigma Xi. *Res:* Information theory; structural analysis of natural languages and formal linguistics; natural language processing; artificial intelligence; mathematical linguistics. *Mailing Add:* Rm 555 Moore Sch Univ Pa Philadelphia PA 19104

JOSHI, BHAIRAV DATT, b Dungrakot, Almora, India, Mar 5, 39; US citizen; m 67. PHYSICAL CHEMISTRY, QUANTUM CHEMISTRY. *Educ:* Univ Delhi, BS, 59, MS, 61; Univ Chicago, MS, 63, PhD(chem), 64. *Prof Exp:* Fel chem, Dept Phys, Univ Chicago, 65-66, lectr, Indian Inst Technol, Kanpur, 66-67; reader, Univ Delhi, 67-69; res assoc, State Univ NY Stony Brook, 69-70; from asst prof to assoc prof, 70-84, PROF CHEM, STATE UNIV NY COL GENESEO, 85- *Mem:* Am Chem Soc; Sigma Xi. *Res:* Quantum mechanical studies of the electronic structure of small atoms and molecules; use of computers in undergraduate education. *Mailing Add:* Three Mohawk Ave Geneseo NY 14454-9511

JOSHI, JAYANT GOPAL, b Poona, India, July 22, 32; m 58; c 2. BIOCHEMISTRY. *Educ:* Univ Poona, BSc, 52, MSc, 54, PhD(biochem), 57. *Prof Exp:* Indian Coun Med Res-Rockefeller Found fel biochem, Nutrit Res Labs, Coonoor, India, 56-58; jr sci officer, Cent Food Tech Res Inst, Mysore, 58-59; res fel, Duke Univ, 59-64; sci pool officer, Nat Chem Labs, Poona, India, 64-66; assoc, Duke Univ, 66-68, asst prof, 68-70; assoc prof, 70-79, PROF BIOCHEM, UNIV TENN, KNOXVILLE, 79- *Concurrent Pos:* Vis scientist, Inst Clin Chem, Uppsala, Sweden, 72 & 73; consult, Biol Div, Oak Ridge Nat Labs; rev panel, Environ Protection Agency, 89- *Mem:* Am Soc Biol Chemists; Am Soc Microbiol. *Res:* Comparative biochemistry; mechanism of enzyme action; pyridine nucleotide, metabolism, biosynthesis and regulation, biochemical change induced by ultraviolet light; metal toxicity. *Mailing Add:* Dept Biochem Univ Tenn Knoxville TN 37916

JOSHI, MADHUSUDAN SHANKARRAO, b Jamkandi, India, Oct 21, 28; m 53; c 2. REPRODUCTIVE PHYSIOLOGY, ENDOCRINOLOGY. *Educ:* Karnatak Univ, India, BSc, 49; Univ Bombay, MSc, 53; Weizmann Inst Sci, PhD(biol), 70. *Prof Exp:* Asst res officer physiol reprod, Cancer Res Ctr, Parel, Bombay, 56-66; asst prof anat, State Univ NY Downstate Med Ctr, 70-77; mem fac, 77-80, assoc prof, 80-85, PROF, DEPT ANAT, UNIV NDAK, 85- *Honors & Awards:* Golden Apple Award, Am Med Students Asn, 82 & 89. *Mem:* Soc Study Fertil UK; Soc Study Reproduction; Am Asn Anat; Am Physiol Soc. *Res:* Mechanisms involved in fertilization and implantation; study of hormone dependent enzymes in uterus and in oviduct; proteins in cerebrospinal fluid; sperm maturation. *Mailing Add:* Dept Anat Univ NDak Grand Forks ND 58202

JOSHI, MUKUND SHANKAR, b India, June 11, 47; m 73; c 2. PHARMACEUTICAL CHEMISTRY, ORGANIC CHEMISTRY. *Educ:* VJ Tech Inst, Bombay, India, 69; Univ Md, MS, 73, PhD(chem engr), 76. *Prof Exp:* Vis scientist, Danish Atomic Energy Comn, 75-76; res scientist, 76-87, ASSOC DIR, BIOPROCESS RES & DEVELOP, UPJOHN CO, 87- *Mem:* Am Inst Chem Engrs; Am Chem Soc. *Res:* Chemical process research and development work to commercially manufacture pharmaceutical products; synthesis of steroids and prostaglandins; feasibility studies risk analysis, economic evaluation and supervision of laboratory; bioengineering and biomedical eng. *Mailing Add:* Upjohn Co Portage Rd Kalamazoo MI 49001

JOSHI, RAMESH CHANDRA, b May 6, 32; Can citizen; m 55; c 3. MATERIAL SCIENCE, GEOTECHNICAL ENGINEERING. *Educ:* Rajputana Univ, India, BE, 56; Punjab Univ, India, MSc, 66; Iowa State Univ, MSc, 68, PhD(civil eng), 70. *Prof Exp:* Exec engr, Rajasthan Pub Works Dept, India, 56-66; res assoc, Eng Res Inst, Iowa State Univ, 66-70; sr proj engr, Woodward Clyde Consuts, 70-77; PROF GEOTECH ENG, UNIV CALGARY, 77- *Concurrent Pos:* Adj lectr, Univ Mo, 76-77; vis fel, Japan Soc Promotion Sci, 90-91. *Mem:* Fel Am Soc Civil Eng; Can Geotech Soc; Int Soc Soil mech & Found Eng; Eng Inst Can. *Res:* Coal ash, particularly fly ash, utilization; model testing of piles; leachgate migration control; properties of frozen soil; soft soil fabric and consolidation. *Mailing Add:* Univ of Calgary, 2500 University Dr NW Dept Civil Eng Calgary AB T2N 1N4

JOSHI, SADANAND D, b Panwel, India, Mar 15, 50; US citizen; m 79; c 2. HORIZONTAL DRILLING, RESERVOIR ENGINEERING. *Educ:* WCE Col, India, BE, 72; Indian Inst Technol, MTech, 74; Iowa State Univ, Ames, PhD(mech eng), 78. *Prof Exp:* Res engr, Phillips Petrol Co, 80-88; PRES, JOSHI TECHNOLOGIES, INC, 88- *Concurrent Pos:* Indust teacher, Univ Tulsa, 88-, Am Asn Petrol Geologists & Soc Petrol Engrs, 89- *Mem:* Am Soc Mech Engrs; Soc Petrol Engrs. *Res:* Drilling mechanism and petroleum production research using horizontal drilling; author of one book. *Mailing Add:* Joshi Technologies Inc 5801 E 41st St Suite 603 Tulsa OK 74135

JOSHI, SEWA RAM, b Baluana, Punjab, Oct 15, 33; US citizen; m 54; c 1. VETERINARY MEDICINE. *Educ:* Punjab Univ, BVSc, 54; Cornell Univ, MS, 63, PhD(animal physiol), 65. *Prof Exp:* State vet, Civil Vet Dept, Punjab, 54-55; res assoc animal reprod, Ind Vet Res Inst, Izatnagar, 55-61; res assoc path & toxicol, Cancer Res Found, Harvard Med Sch, 65-71; sr staff fel cancer, Nat Cancer Inst, 71-76; PHYSIOLOGIST PHARM & TOXICOL, CTR DRUGS EVAL & RES, FOOD & DRUG ADMIN, 76- *Concurrent Pos:* Assoc prof, Howard Univ, Washington, DC, 72-75. *Mem:* Soc Study Reproduction; NY Acad Sci; Environ Mutagen Soc; Sigma Xi; Soc Toxicol. *Res:* Reproductive toxicology; chemical carcinogenesis and mutagenesis of chemicals in the environment. *Mailing Add:* Div Anti-infective Drug Prod 5600 Fishers Lane Rm 12B16 Rockville MD 20857

JOSHI, SHARAD GOPAL, b Nagpur, India. REPRODUCTIVE ENDOCRINOLOGY & PHYSIOLOGY. *Educ:* Univ Nagpur, India, BS, 52, MS, 54; Univ Bombay, PhD(biochem), 59. *Prof Exp:* Fel endocrinol, Worcester Fedn & Harvard Med Sch, 59-63; assoc prof reproductive physiol, Inst Med Sci, India, 63-65; staff scientist endocrinol, Syntex Inst Hormone Biol, 65-66 & Southwest Fedn Res & Educ, 66-73; assoc prof obstet, gynec & biochem, 73-82, PROF OBSTET & GYNEC, ALBANY MED COL, 82- *Honors & Awards:* Edward Tyler Award, Int Soc Reproductive Med, 81. *Mem:* Endocrine Soc; Soc Study Reproduction; Int Soc Reproductive Med; NY Acad Sci. *Res:* Hormonal control and role of endometrium in human pregnancy; fertility control in human subjects; effects of toxic agents on human pregnancy; role of human endometrium and placenta in pregnancy; development of in vivo and in vitro models to study human placenta and endocrine-related tumors; monitoring human reproductive functions and tumor growth using biochemical markers. *Mailing Add:* Dept Obstet & Gynec Albany Med Col 47 New Scotland Ave Albany NY 12208

JOSHI, SURESH MEGHASHYAM, b Poona, India; US citizen; c 2. CONTROL SYSTEMS RESEARCH, SPACECRAFT DYNAMICS RESEARCH. *Educ:* Banaras Univ, India, BS, 67; Indian Inst Technol, MS, 69; Rensselaer Polytech Inst, PhD(elec eng), 73. *Prof Exp:* Eng, Stone & Webster Corp, 72-73; post doc res fel, Nat Res Coun, Langley Res Ctr, NASA, 73-75; assoc prof elec & mech eng, Old Dominion Univ Res Found, 75-83; SR RES SCIENTIST, LANGLEY RES CTR, NASA, 83- *Concurrent Pos:* Adj prof, George Wash Univ & Pa State Univ, 86-; chmn, aerospace syst & tech panel, Am Soc Mech Engrs, 87-; mem, tech comt astrodyn, Am Inst Aeronauts & Astronauts, 88-; mem, bd govs, Inst Elec & Electronics Engrs, 89- *Honors & Awards:* DuMont Prize, RPI, 73. *Mem:* Sr mem Inst Elec &

Electronics Engrs; Am Soc Mech Engrs; Assoc fel Am Inst Aeronauts & Astronauts. *Res:* Multivariable control theory and its application to NASA's advanced spacecraft concepts which include very large satellites such as large space antennas and space station; author of approximatly 120 technical articles and one book. *Mailing Add:* NASA Langley Res Ctr Mail Stop 230 Hampton VA 23665-5225

JOSHI, VASUDEV CHHOTALAL, b Borsad, Gujarat, June 26, 38; m 65; c 3. BIOCHEMISTRY. *Educ:* Madras Univ, BPharm, 59; Andhra Univ, MPharm, 61; Indian Inst Sci, PhD(biochem), 65. *Prof Exp:* Res assoc biochem, Med Ctr, Duke Univ, 66-70, assoc in pediat, 70-72; asst prof, 72-78, ASSOC PROF BIOCHEM, BAYLOR COL MED, 78- *Concurrent Pos:* Ford Found fel, Indian Inst Sci, Bangalore, 65-66; Fulbright fel, US Educ Found in India, 66; NIH grant, Med Ctr, Duke Univ, 66-70; vis assoc prof, Mass Inst Technol, 80-81; res career develop award, USPHS, 78-83. *Mem:* Brit Biochem Soc; Am Soc Biol Chemists; Soc Biol Chemists, India. *Res:* Enzymatic mechanism of fatty acid synthesis in bacteria and animal tissues; hormonal regulation of fatty acid synthetase and stearoyl coenzyme A desaturase in liver; lipid metabolism in cultured cells; mechanism of insulin action. *Mailing Add:* 9315 Meaux Dr Houston TX 77031

JOSHUA, HENRY, b Hamburg, Ger, Dec 8, 34; US citizen; m 68; c 3. ORGANIC CHEMISTRY, SEPARATION SCIENCE. *Educ:* Bar-Ilan Univ, Israel, BS, 59; NY Univ, MS, 62, PhD(chem), 64. *Prof Exp:* Res chemist, Res Div, Col Eng, NY Univ, 60-61; res fel chem, Princeton Univ, 64-65; sr res chemist, 65-78; res fel, 78-90, SR RES FEL, MERCK SHARP & DOHME RES LABS, 91- *Mem:* Am Chem Soc. *Res:* Isolation of active compounds from biological sources and purification of organic synthetic reaction products; development of instrumentation for liquid chromatography and laboratory automation. *Mailing Add:* Merck Sharp & Dohme Res Lab PO Box 2000 Rahway NJ 07065

JOSIAS, CONRAD S(EYMOUR), b New York, NY, June 12, 30; m 63; c 3. ELECTRICAL ENGINEERING. *Educ:* NY Univ, BEE, 51; Polytech Inst Brooklyn, MEE, 55. *Prof Exp:* Engr electronic develop, Airborne Instruments Lab, Inc, NY, 51-56; from res engr, missile guid systs to eng group supvr, space electronics group, space sci div, Jet Propulsion Lab, Calif Inst Technol, 56-65; pres, Analog Technol Corp, Pasadena, 65-79; PRES, JOSIAS ASSOCS, INC, 79- *Concurrent Pos:* Instr, Pasadena City Col, 59-61. *Honors & Awards:* Award, NASA Inventions & Contrib Bd, 64. *Mem:* Sr mem Inst Elec & Electronics Engrs. *Res:* Electronic devices and systems; scientific instruments for laboratory and space applications; analytical instruments for industrial laboratories. *Mailing Add:* Josias Assoc Inc 4733 Hillard Ave La Canada CA 91011

JOSIASSEN, RICHARD CARLTON, b Oroville, Calif, Apr 29, 47. PSYCHOPATHOLOGY, NEUROPHYSIOLOGY. *Educ:* Westmont Col, BA, 69; Fuller Theol Sem, MA, 75; Fuller Grad Sch Psychol, PhD(psychol), 79. *Prof Exp:* Consult, Dept Defense, 69-73; from asst prof to assoc prof psychiat, Temple Univ Med Sch, 80-87; ASSOC PROF PSYCHIAT, MED COL PA, 87- *Concurrent Pos:* Fel neurosci, Nat Inst Ment Health, 80-82; vis asst prof, dept neurol, Hahnemann Univ, 81-; assoc secy gen, IV World Congress & Biol Psychiat, Philadelphia, 85. *Mem:* Soc Biol Psychiat; Am Psychopathol Asn; Soc Res in Psychopathol; NY Acad Sci. *Res:* Neurophysiological mechanisms which underlie major mental illness. *Mailing Add:* Med Col Pa Eastern Pa Psychiat Inst 3200 Henry Ave Philadelphia PA 19129

JOSLIN, ROBERT SCOTT, b Indianapolis, Ind, May 28, 29; m 84; c 3. PHARMACEUTICAL CHEMISTRY. *Educ:* Purdue Univ, BS, 51, MS, 55, PhD(phys pharm), 59. *Prof Exp:* Assoc pharmaceut chemist, Eli Lilly & Co, 53-54, sr pharmaceut chemist, 58-65; dir prod improv, William H. Rorer, Inc, 65-68, dir depts pharmaceut sci, Res Div & asst dir res, 68-74; assoc dir pharmaceut & chem develop, G D Searle & Co, 74-78; dir pharmaceut & anal res & develop, Baxter Labs, 78; consult, 78-; AT JOSLIN & ASSOC LTD. *Honors & Awards:* Lunsford Richardson Award, 57. *Mem:* Fel Am Inst Chemists; Am Chem Soc; fel Acad Pharmaceut Sci; Int Pharmaceut Fedn; NY Acad Sci; fel Am Asn Pharmaceut Scientists. *Res:* Pharmaceutical formulation; biopharmaceutics; process and product development. *Mailing Add:* Joslin & Assoc Ltd 291 Deer Trail Ct Ste C Barrington IL 60010

JOSLYN, DENNIS JOSEPH, b Chicago, Ill, Apr 29, 47; m 76; c 1. INSECT CYTOGENETICS, EVOLUTIONARY GENETICS OF INSECTS. *Educ:* St Procopius Col, BS, 69; Univ Ill, Urbana, MS, 73, PhD(zool), 78. *Prof Exp:* Asst res scientist insect genetics, Univ Fla, 76-79; res assoc, Insects Affecting Man & Animals Res Lab, USDA, 76-79; asst prof, 79-85, ASSOC PROF ZOOL, RUTGERS UNIV, 85- *Mem:* Genetics Soc Am; Am Genetic Asn; Am Mosquito Control Asn. *Res:* Evolutionary genetics and insect cytogenetics; genetics of insects of medical, veterinary and agricultural importance; biology of eukaryotic chromosomes. *Mailing Add:* Dept Biol Camden Col Arts & Sci Rutgers Univ Camden NJ 08102

JOSS, PAUL CHRISTOPHER, b Brooklyn, NY, May 7, 45; div; c 1. THEORETICAL ASTROPHYSICS, X-RAY ASTRONOMY. *Educ:* Cornell Univ, BA, 66, PhD(astron, space sci), 71. *Prof Exp:* Mem, Inst Advan Study, 71-73; from asst prof to assoc prof, 73-83, PROF, DEPT PHYSICS, MASS INST TECHNOL, 83- *Concurrent Pos:* Vis scientist, Dept Nuclear Physics, Weizmann Inst Sci, 74-75 & 78, Inst Astron, Univ Cambridge, 77; Alfred P Sloan res fel, 76-80; vis staff mem, Los Alamos Nat Lab, 79; consult, Visidyne, Inc, 79-82 & Los Alamos Nat Lab, 80-; mem, Adv Comt, Inst Geophysics & Planetary Physics, Los Alamos Nat Lab, 87-; mem, High Energy Astrophysics Mgt Opers Working Group, Nat Aeronaut & Space Admin, 88-; mem, Sci Coun Astron & Space Physics, Univ Space Res Asn, 88- *Honors & Awards:* Helen B Warner Prize, Am Astron Soc, 80. *Mem:* Am Astron Soc; Am Phys Soc; Int Astron Union. *Res:* Theoretical and observational studies of compact x-ray sources; theoretical research on the structure and evolution of stars and binary stellar systems, extragalactic astrophysics, the spectra of quasars, and the origin and orbital evolution of comets. *Mailing Add:* Rm 6-203 Mass Inst Technol Cambridge MA 02139

JOSSEM, EDMUND LEONARD, b Camden, NJ, May 19, 19. PHYSICS. *Educ:* City Col New York, BS, 38; Cornell Univ, MS, 39, PhD(physics), 50. *Prof Exp:* Asst physics, Cornell Univ, 40-42, instr, 42-45; mem staff, Los Alamos Sci Lab, 45-46; asst physics, Cornell Univ, 46-50, res assoc, 50-55, actg asst prof, 55-56; from asst prof to assoc prof, 56-64, chmn dept, 67-80, PROF PHYSICS, OHIO STATE UNIV, 64- *Concurrent Pos:* Staff physicist, Comn Col Physics, 63-64, exec secy, 64-65, chmn, 66-71; mem, Nat Adv Coun Educ Professions Develop, 67-70; mem bd dirs, Mich-Ohio Educ Lab, 67-69; adv coun educ & manpower, Am Inst Physics, 69-72; mem physics survey comt, panel on educ, Nat Acad Sci-Nat Res Coun, 70-72; mem comn physic educ, Int Union Pure & Applied Physics, 81-; mem hon bd, Int Conf X-ray & Atomic Inner Shell Physics, 81-82. *Mem:* Fel AAAS; Am Phys Soc; Am Asn Physics Teachers (vpres, 71-72, pres, 73-74); Sigma Xi. *Res:* Solid state physics; x-ray physics. *Mailing Add:* 174 W 18th Ave Columbus OH 43210

JOSSI, JACK WILLIAM, b Portland, Ore, Apr 4, 37; c 3. OCEANOGRAPHY, ECOLOGY. *Educ:* Pac Univ, BS, 59; Univ Wash, Seattle, BS, 62; Univ Miami, MS, 72. *Prof Exp:* Phys oceanogr trop oceanog, Washington Biol Lab, US Dept Interior, 62-65, oceanogr, Trop Atlantic Biol Lab, 65-70, mgr fishery climat prog, Southeast Fisheries Ctr, 71-72, chief, Continuous Plankton Recorder Surv, Marine Resources Monitoring, Assessment & Prediction Field Group, US Dept Commerce, 72-74, asst chief, 74-78, RES OCEANOGR OCEAN CLIMAT, ATLANTIC ENVIRON GROUP, NAT OCEANIC & ATMOSPHERIC ADMIN, US DEPT COMMERCE, 78- *Concurrent Pos:* Fel, Univ Miami, 67-68; consult, Smithsonian Inst, 70; mem, Standing Comt Oceanog, Nat Marine Fisheries Serv, US Dept Commerce, 71- *Mem:* Marine Biol Asn UK. *Res:* Ocean climatology; modeling and forecasting of distribution and abundance of living marine resources. *Mailing Add:* Carpenter Lane Saunderstown RI 02874-0427

JOST, DANA NELSON, b Arlington, Mass, May 11, 25; m 47; c 3. PHYCOLOGY, ENVIRONMENTAL BIOLOGY. *Educ:* Univ Mass, BS, 49; Harvard Univ, PhD(biol), 53. *Prof Exp:* Instr, Framingham State Col, 53-55, from asst prof to assoc prof, 55-59, chmn dept, 64-76, prof biol, 59-90; RETIRED. *Mem:* AAAS; Phycol Soc Am; Bot Soc Am; Int Phycol Soc; Mycol Soc Am; Am Inst Biol Sci. *Res:* Growth and reproduction of chlorophycean algae; evolution of microorganisms; environmental influences on algal growth; distribution and identification of freshwater periphyton. *Mailing Add:* Dept Biol Hemenway Hall Framingham State Col Framingham MA 01701

JOST, DONALD E, b Chicago, Ill, Nov 5, 36; m 57; c 3. CHEMICAL ENGINEERING. *Educ:* Univ Del, BS, 58; Princeton Univ, PhD(chem eng), 64. *Prof Exp:* Instr chem eng, Princeton Univ, 61-62; asst prof, Ore State Univ, 62-64 & Univ Pa, 64-67; res engr, 67-70, mgr, eng res, 70-75, mgr, resource develop, res & develop, Sun Oil Co, 75-79, spec assignment, chem strategic planning, Sun Petrol Prod Co, 79-80, mgr, chem group, res & develop, 80-85, mgr appl res, 85-89, MGR TECHNOL, SUN REFINING & MKT CO, 89- *Concurrent Pos:* Chem res engr, US Bur Mines, 63-64; Ford Found fel, 65-66. *Mem:* Am Inst Chem Engrs; Coun Chem Res; Indust Res Inst. *Mailing Add:* Technol Dept Sun Refining & Mkt Co PO Box 1135 Marcus Hook PA 19061-0835

JOST, ERNEST, b El Ferrol, Spain, Sept 6, 28; US citizen; m 55; c 4. PHYSICAL CHEMISTRY. *Educ:* Univ Berne, license, 55, PhD(phys chem), 58. *Prof Exp:* Mem tech staff, Metals & Controls Div, Tex Instruments Inc, 58-61; mgr develop, Ciba A G, Switz, 61-62; dir res & develop, Mat & Elec Prod Group, Tex Instruments Inc, Attleboro, 62-74, dir prod res dept, 74-80; PRES, CHEMET CORP, 80- *Concurrent Pos:* Consult, Nat Acad Sci, 71. *Mem:* Electrochem Soc; Royal Chem Soc. *Res:* Metallurgy; electrochemistry; diffusion; solid state physics; semiconducting ceramics. *Mailing Add:* Nin Mirimichi Plainville MA 02762

JOST, HANS PETER, b Berlin, Ger, Jan 25, 21; Brit citizen; m 48; c 2. TRIBOLOGY, CENTRALIZED LUBRICATION SYSTEMS. *Educ:* Univ Manchester Inst Sci & Technol, HNC, 43. *Hon Degrees:* DSc, Univ Salford, 70, Univ Bath, 90; DTech, Coun Nat Acad Awards, 87; DrSc, Slovak Univ, Bratislava, 87; DEng, Univ Leeds, 89. *Prof Exp:* Methods engr, K & L Steelfounders & Engr Ltd, 43; chief planning engr, Datim Mach Tool Co Ltd, 46-49; gen mgr & dir, Trier Bros Ltd, 49-55; managing dir, 55-89, CHMN, K S PAUL GROUP, 74-, ASSOC TECHNOL GROUP LTD & ENG & GEN EQUIP LTD, 77- *Concurrent Pos:* Managing dir, Centralube Ltd, 55-77, chmn, 74-77; chmn, Lubrication Educ & Res Working Group, Dept Educ & Sci, 64-65, Comt Tribology, Ministry, Technol, Dept Trade & Indust, 66-74, Peppermill Brass Foundry Ltd, 70-76 & Indust Technol Mgt Bd, Dept Trade & Indust, 72-74; lubrication consult, Richard Thomas & Baldwins Ltd, 69-76; dir, Williams Hudson Ltd, 67-75 & Stothert & Pitt Plc, 71-85; mem, Comt Terotechnol, Dept Trade & Indust, 71-72, Found Sci & Technol, 85 & Parliamentary Group Eng Develop, 86; hon indust prof, Liverpool Polytechnic, 83, hon prof mech eng, Univ Wales, 86. *Honors & Awards:* Comdr of the Order of the Brit Empire, HM the Queen, 69; Georg Vogelpohl Insignia, Ger Tribology Soc, 79; First Nuffield Medal, Inst Prod Engrs, 81; Merit Medal, Hungarian Sci Soc Mech Engrs, 83; Gold Insignia of the Order of Merit, Supreme Coun of State of the Polish Repub, 86. *Mem:* Fel Am Soc Mech Engrs; fel Soc Mfg Engrs; fel Inst Metals; hon mem Inst Plant Engrs; hon mem Chinese Mech Eng Soc. *Res:* Surface finish measurement; oil-free steam cylinder lubrication; solid lubricants and surface treatments; tribology. *Mailing Add:* K S Paul Prods Ltd Angel Lodge Labs & Works Eley Estate London N18 3DB England

JOST, JEAN-PIERRE, b Avenches, Switz, Oct 10, 37; US citizen; m 68; c 2. BIOCHEMISTRY CELL & MOLECULAR BIOLOGY. *Educ:* Swiss Fed Inst Technol, MS, 61, PhD(biol, biochem), 64. *Prof Exp:* Proj assoc, McArdle Mem Lab Cancer Res, Univ Wis-Madison, 64-66, fel, Lab Molecular Biol, 67-68; molecular biologist, Nat Jewish Hosp & Res Ctr, Denver, 68-71; GROUP LEADER, FRIEDRICH MIESCHER INST, SWITZ, 71-

Concurrent Pos: Asst prof, Dept Biophys & Genetics, Med Sch, Univ Colo, Denver, 68-71; res grants, Am Cancer Soc, 69, NIH, Health, Educ & Welfare & NSF, 70. *Res:* Mode of action of steroid hormones; hormonal regulation of the expression of specific genes in eukaryotes, DNA methylation and genes expression. *Mailing Add:* Friedrich Miescher Inst Postfach 2543 CH-4002 Basel Switzerland

JOST, PATRICIA COWAN, b St Louis, Mo. BIOPHYSICAL CHEMISTRY, MOLECULAR BIOLOGY. *Educ:* Memphis State Univ, BS, 52; Univ Ore, PhD(biol), 66. *Prof Exp:* Res assoc molecular genetics, 66-68, SR RES ASSOC MOLECULAR BIOL, INST MOLECULAR BIOL, UNIV ORE, 68- *Concurrent Pos:* NIH fel, Univ Ore, 66-68; co-dir, Biophys Prog, NSF, 84-86. *Mem:* Biophys Soc. *Res:* Membrane structure; reporter groups and magnetic resonance; membrane biology, lipid-lipid and lipid-protein interactions. *Mailing Add:* Div Res Grants Referral & Re Br NIH Bldg WB Rm 236A Bethesda MD 20892

JOSTLEIN, HANS, b Munich, WGer, Dec 27, 40; m 67; c 3. VACUUM ENGINEERING, EXPERIMENTAL PHYSICS. *Educ:* Technische Hochschule München, Dipl Eng, 65, Ludwigs Maximilian Univ, W Germany, PhD(physics), 69. *Prof Exp:* Fel high energy res, Univ Munich, 69-70; res asst, Univ Rochester, 70-73; asst prof, State Univ NY, Stony Brook, 73-79; PHYSICIST RES & ACCELERATION SUPPORT, FERMI NAT ACCELERATOR LAB, 79- *Res:* High energy particle physics, specifically muon; muon inelastic scattering; high mass pair production for leptons and hadrons to test quark theory; drell-yan processes and resonance production; very high lumosity pair production. *Mailing Add:* 432 W Jeferson Ave Naperville IL 60540

JOUBIN, FRANC RENAULT, b San Francisco, Calif, Nov 15, 11; nat Can; m 38; c 1. CHEMISTRY. *Educ:* Univ BC, BA, 36, MA, 41. *Hon Degrees:* DSc, Univ BC, 57. *Prof Exp:* From mine geologist to explor geologist, Pioneer Gold Mines Co, 38-48; pres & managing dir, Algom Uranium Mines, 53-56, dir, Rio Algom Mines, 56-58; pres, Bralorne Pioneer Mines, 58-60; consult to adminr, UN Develop Prog, 68-77; RETIRED. *Concurrent Pos:* Managing dir, Pronto, Pater, Rixahabasca, Rexspar, Lake Nordic, Spanish-Am & Panel Uranium Mines, 53-56; geol consult, Rio Tinto Co, Eng, 56-58; dir, Guaranty Trust of Can, 57-60. *Honors & Awards:* Leonard Gold Medal, Eng Inst Can, 55; Blaylock Gold Medal, Can Inst Mining & Metall, 57. *Mem:* Am Inst Mining, Metall & Petrol Eng; Can Inst Mining & Metall (vpres, 57-58); Geol Asn Can; Royal Can Inst. *Res:* Geology; geophysics. *Mailing Add:* 500 Avenue Rd Toronto ON M4V 2J6 Can

JOULLIE, MADELEINE M, b Paris, France, Mar 29, 27; nat US; m 59. ORGANIC CHEMISTRY. *Educ:* Simmons Col, BS, 49; Univ Pa, MS, 50, PhD, 53. *Prof Exp:* From instr to asst prof, 53-68, assoc prof, 68-77, PROF CHEM, UNIV PA, 77- *Mem:* AAAS; Am Chem Soc; Sigma Xi. *Res:* Mechanisms of organic reactions; heterocyclic chemistry; synthesis of potential antimetabolites. *Mailing Add:* 288 St James Pl Philadelphia PA 19106

JOUNG, JOHN JONGIN, b Korea, June 29, 41; m 68; c 2. CHEMICAL PROCESS ENGINEERING, BIOENGINEERING. *Educ:* Seoul Nat Univ, BS, 63, MS, 67; Univ NMex, PhD(chem eng), 70. *Prof Exp:* Res fel, Ames Lab, 70-71; res supvr, Univ Chicago, 71-76; sr scientist, Colgate-Palmolive Co, 76-78; sci adv, Am Hosp Supply Corp, 78-81; SR RES ENGR, AMOCO CORP, 81. *Concurrent Pos:* Vcmndg officer, Chem Smoke Generator Co, Korea, 63-65; assst investr, Korea Inst Sci & Technol, 67-68. *Mem:* Am Inst Chem Engrs; Am Chem Soc; Soc Plastics Engrs; Am Mgt Asn. *Res:* Process and product research in the areas of chemical, polymer, energy and health care business; cancer and clinical pathology; process development for chemical and biological products, including recombinant DNA products; polymer applications for biomedical and personal care products; reaction engineering. *Mailing Add:* 6095 Millbridge Lane Lisle IL 60532

JOURDIAN, GEORGE WILLIAM, b Northampton, Mass, Apr 21, 29; m 54; c 2. BIOCHEMISTRY, MICROBIOLOGY. *Educ:* Amherst Col, BA, 49; Univ Mass, MS, 53; Purdue Univ, PhD(bact), 58. *Prof Exp:* From instr to assoc prof biol chem & biochem, 61-74, res assoc internal med, 65-74, PROF BIOL CHEM, MED SCH, UNIV MICH, ANN ARBOR, 65- *Concurrent Pos:* Arthritis Found fel, 58-61; Fogarty Sr Int fel, 78-79. *Mem:* Am Soc Biol Chem; Am Chem Soc; Soc Complex Carbohydrates. *Res:* Biochemistry of glycosaminoglycans and glycoproteins. *Mailing Add:* Rackham Arthritis Res Unit Univ Mich Med Ctr Ann Arbor MI 48109

JOURNEAY, GLEN EUGENE, b Orange, Tex, June 14, 25; m 48; c 5. MEDICINE. *Educ:* Rice Univ, Houston, BA, 45, BS, 47; Univ Tex, Austin, PhD(org chem), 52, Galveston, MD, 60. *Prof Exp:* Res chemist, Monsanto Chem, 51-56; physician, Beeler Manske Clin, Toxicity, 61-63; PHYSICIAN, PVT PRACT, 63- *Concurrent Pos:* Vis prof biomed eng, Univ Tex, Austin, 64-66, lectr, 66- *Honors & Awards:* M D Anderson Excellency in Oncol Award, 85. *Mem:* AMA; Am Acad Family Physicians; Am Chem Soc; fel Am Inst Chemists. *Res:* Cyanoethylation; hydrocyanic acid reactions; toxicity of acrylonitrile and acrylamide; neurotoxicity; environmental toxicology. *Mailing Add:* 7101 Woodrow Austin TX 78757

JOVANCICEVIC, VLADIMIR, b Belgrad, Yugoslavia, Dec 14, 47; m 69; c 1. ELECTROCHEMISTRY, CORROSION & PASSIVITY. *Educ:* Univ Belgrad, BSc 72, MSc 76; Univ Paris VI, France, PhD(phy chem), 80. *Prof Exp:* Res asst electrochem, Inst Tech Sci, Belgrad, 75-78; res assoc, Electricity of France, 80-81, Nat Ctr Sci Res, Paris, 81-82; sr scientist mat sci, Sasilor-Sollac, Thionville, France, 82-84; sr res assoc electrochem, Tex A&M Univ, 84-87; RES CHEMIST MAT SCI, WR GRACE & CO COLUMBIA, MD, 87- *Mem:* Serbian Chem Soc (secy, 74-77); Electrochem Soc; Am Chem Soc. *Res:* Investigation of the physico-chemical properties of metal-solution interfaces; spectroscopic characterization of the structure, composition and reactivity of the adsorbed and thin surface layers as related to the corrosion, passivation and electrodeposition. *Mailing Add:* WR Grace & Co Res Div 7379 Rte 32 Columbia MD 21044

JOVANOVIC, M(ILAN) K(OSTA), b Belgrade, Yugoslavia, Oct 29, 13; nat US; m 54; c 1. MECHANICAL ENGINEERING. *Educ:* Univ Belgrade, Dipl Ing, 38, Dipl Phys, 45; Northwestern Univ, MS, 54, PhD(mech eng), 57. *Prof Exp:* Asst thermodyn & physics, Univ Belgrade, 39-46, instr physics, 46-47, asst prof thermodyn, 47-51, assoc prof thermodyn & refig mach & asst dean, Sch Mech Eng, 51-52; instr physics, Univ Ill, 55; assoc prof mech eng, SDak Sch Mines & Technol, 56-58; from assoc prof to prof, Okla State Univ, 58-63; prof, Univ Alaska, 63-65; prof, US Naval Acad, 65-68; prof, 68-84, EMER PROF MECH ENG, WICHITA STATE UNIV, 84- *Res:* Thermodynamics. *Mailing Add:* 1026 N Pinecrest Wichita KS 67208

JOVANOVICH, JOVAN VOJISLAV, b Belgrade, Yugoslavia, July 30, 28; m 55; c 2. HIGH ENERGY PHYSICS, NUCLEAR PHYSICS. *Educ:* Univ Belgrade, BSc, 50; Univ Man, MSc, 56; Washington Univ, St Louis, PhD(physics), 61. *Prof Exp:* Asst physics, Inst Brois Kidric, Belgrade, Yugoslavia, 50-54, Univ Belgrade, 53-55 & Univ Man, 55-57; asst, Wash Univ, 57-61; res assoc, Brookhaven Nat Lab, 61-63, vis asst physicist, 63-64, develop officer, Nuclear Physics Lab, Oxford Univ, 64-65; asst prof, 65-67, assoc prof, 67-79, PROF PHYSICS, UNIV MAN, 79- *Concurrent Pos:* Vis scientist, Orgn Europ Nuclear Res, 71-72, 78-79. *Mem:* Am Phys Soc; Can Asn Physicists. *Res:* Investigation of properties of neutral K mesons; elementary inelastic proton-proton interaction; search for quarks, pion properties. *Mailing Add:* Dept Physics Univ Man Winnipeg MB R3T 2N2 Can

JOWETT, DAVID, b Liverpool, Eng, Oct 14, 34; m 57; c 2. STATISTICS, BOTANY. *Educ:* Univ Wales, BSc, 56, PhD(bot), 59. *Prof Exp:* Demonstr agr bot, Univ Col NWales, 56-59; sr sci officer, Plant Breeding, EAfrican Agr & Forestry Res Orgn, Uganda, 59-65; from asst prof to assoc prof statist, Iowa State Univ, 65-70; assoc prof, 70-72, PROF MATH, UNIV WIS-GREEN BAY, 72- *Concurrent Pos:* Rockefeller Found fel, Iowa State Univ, 62-63; Intern, Acad Admin, Am Coun Educ, 76-77. *Mem:* AAAS; Brit Ecol Soc; Am Statist Asn; Inst Math Statist; Am Soc Nat; fel Royal Statist Soc; Sigma Xi. *Res:* Heavy metal resistance and tolerance of low nutrient levels in plants; improved varieties hybrids of sorghum for Africa; sorghum agronomy crown rust epiphytology; biostatistics; biomathematics; statistical computing. *Mailing Add:* Dept Math Univ Wis Green Bay WI 54301

JOY, DAVID CHARLES, b Colchester, Eng, Nov 15, 43; US citizen; m. ELECTRON MICROSCOPY, MATERIALS SCIENCE. *Educ:* Cambridge Univ, BA, 66; Oxford Univ, DPhil(metall), 69. *Prof Exp:* Fac metall, Oxford Univ, 69-74; mem tech staff electron micros, Bell Tel Labs, 74-87; DISTINGUISHED SCIENTIST & PROF, OAK RIDGE NAT LAB & UNIV TENN, 87- *Concurrent Pos:* Res fel, Imperial Chem Industs, 69-71 & Oxford Univ, 69-72; Warren res fel, Royal Soc London, 72-74; gen ed, J Micros, 81-91. *Honors & Awards:* Burton Medal, Electron Micros Soc Am, 78; Birks Award, Microbeam Anal Soc, 85. *Mem:* Electron Micros Soc Am; Royal Micros Soc London. *Res:* Electron microscopy and electron spectroscopy applied to microstructural and microchemical analysis. *Mailing Add:* EM Facil F241 Walters Life Sci Bldg Univ Tenn-Knoxville Stadium Dr Knoxville TN 37966-8310

JOY, EDWARD BENNETT, b Troy, NY, Nov 15, 41; m 66; c 2. ANTENNA MEASUREMENTS, RADOME ELECTROMAGNETIC DESIGN. *Educ:* Ga Inst Technol, BEE, 63, MS, 67, PhD(elec eng), 70. *Prof Exp:* From asst prof to assoc prof, 70-80, PROF ELEC ENG, GA INST TECHNOL SCH ELEC ENG, 80- *Concurrent Pos:* Prin investr, US Army, USAF, Elec Power Res Inst, NSF & Joint Ser Electronics Prog, 70-; lectr, US, Can, Europe, Mid East & Far East; consult, Martin Marietta Aerospace, Sci Atlanta, Ford Aerospace, Harris Corp, Fed Aviation Admin, Westinghouse Elec, Sperry Corp, Gen Elec Corp & Alcoa, 70-; distinguished scientist lectr, Govt India, 82; tech coordr, Antenna Measurement Techniques Asn, 85, co-host ann meeting, 88. *Mem:* Fel Inst Elec & Electronics Engrs; Antenna Measurement Techniques Asn. *Res:* Development of the theory, technique and application of far- field, anechoic chamber, compact and near-field antenna measurements; radome analysis, design and measurement; earth grounding of power delivery systems. *Mailing Add:* Ga Inst Technol Sch Elec Eng Atlanta GA 30332-0250

JOY, GEORGE CECIL, III, b Lincoln, Nebr, Apr 22, 48; m 70; c 1. INORGANIC CHEMISTRY. *Educ:* Grinnell Col, BA, 70; Northwestern Univ, MS, 71, PhD(inorg chem), 75. *Prof Exp:* res chemist, Allied-Signal Inc, 74-76, groupleader, 76-81, mgr appl catalysis res, 81-86, dir, Tech Dept, Automotive Catalyst Div, 87-90, RES SCIENTIST, RES & TECH DIV, ALLIED-SIGNAL INC, 90-, DIR RES & TECHNOL, EUROPE. *Mem:* Am Chem Soc; Nat Catalysis Soc; Soc Automotive Engrs; Inst Chem Engrs (UK). *Res:* Studies in heterogeneous catalysis; inorganic aspects of preparation and characterization of catalysts. *Mailing Add:* Allied-Signal Inc 480 Ave Louise bte 4 B-1050 Brussels Belgium

JOY, JOSEPH WAYNE, b Iowa City, Apr 12, 30; m 58; c 3. PHYSICAL OCEANOGRAPHY. *Educ:* State Col Wash, BA, 55; Univ Calif, San Diego, MS, 58. *Prof Exp:* Res oceanogr, Marine Phys Lab, Univ Calif, San Diego, 56-61; oceanogr, Marine Adv, Bendix Corp, 61-66; res oceanogr, Meteorol Res, Inc, 66-67; sr scientist, Westinghouse Ocean Res Lab, Calif, 67-70; specialist oceanog, Scripps Inst Oceanog, Univ Calif, San Diego, 70-74; staff oceanogr, Intersea Res Corp, 74-84; comput spec, Nat Marine Fisheries Serv, Nat Oceanic & Atomospheric Admin, US Dept Com, 84-88; COMPUT SPEC, US NAVY PERSONNEL RES & DEVELOP CTR, SAN DIEGO, CALIF, 88- *Mem:* AAAS; Am Geophys Union; Am Meteorol Soc. *Res:* Surface waves and currents; radar oceanography; radar as oceanographic tool. *Mailing Add:* 1210 Agate St San Diego CA 92109

JOY, KENNETH WILFRED, b Sunderland, Eng, May 13, 35; c 2. PLANT PHYSIOLOGY. *Educ:* Bristol Univ, BSc, 56, PhD(plant physiol), 59. *Prof Exp:* Res assoc plant physiol, Imp Col, Univ London, 59-64; vis res assoc agron, Univ Ill, Urbana, 64-65; assoc prof bot, Univ Toronto, 66-68; PROF BIOL, CARLETON UNIV, 68- *Concurrent Pos:* Fulbright travel fel, 64-65.

Mem: Am Soc Plant Physiol; Can Soc Plant Physiol (secy, 68-69). Res: Nitrogen metabolism of plants, especially amino acid metabolism; synthesis and utilization of amides. Mailing Add: Dept Biol Carleton Univ Ottawa ON K1S 5B6 Can

JOY, MICHAEL LAWRENCE GRAHAME, b Toronto, Ont, July 31, 40; m 67; c 2. BIOMEDICAL ENGINEERING. Educ: Univ Toronto, BSc, 63, MASc, 68, PhD(elec eng), 70. Prof Exp: Asst prof elec eng, 70-76, ASSOC PROF BIOMED & ELEC ENG, UNIV TORONTO, 76- Concurrent Pos: Univ & Nat Res Coun Can grants, 70-71; vpres, Facets Inc, 70-74, pres, 74- Mem: Can Med & Biol Eng Soc; Inst Elec & Electronics Engrs. Res: Nuclear medical scintigraphy. Mailing Add: Inst Biomed Eng Univ Toronto Toronto ON M5S 1A4 Can

JOY, ROBERT JOHN THOMAS, b South Kingstown, RI, Apr 5, 29; m 52, 85; c 2. INTERNAL MEDICINE, PHYSIOLOGY. Educ: Univ RI, BS, 50; Yale Univ, MD, 54; Harvard Univ, MA, 65. Prof Exp: Med Corps, US Army, 54-81, intern, Walter Reed Gen Hosp, 54-55, resident, 56-58, chief, Bioastronaut Br, Army Med Res Lab, Ft Knox, Ky, 59-61, comdr & mem res staff, Army Res Inst Environ Med, 61-63, chief med res team, Walter Reed Army Inst Res, Vietnam, 65-66, dep dir, US Army Res Inst Environ Med, Mass, 66-68, chief Med Res Div, US Army Med Res & Develop Command, Washington, DC, 68-69, dep biol & med res, Directorate of Defense Res & Eng, 69-71, dep dir, 71-75, dir & comdr, Walter Reed Army Inst Res, 75-76; prof mil med & hist & chmn dept, 76-81, PROF MED HIST & CHMN DEPT, UNIFORMED SERV UNIV SCH MED, 81- Concurrent Pos: Ed, J Hist Med & Allied Sci, 82-87. Honors & Awards: Osler Medal, Am Asn Hist Med, 54; Hoff Mem Medal Mil Med, 59; J S Billings Award, 66; Clements Award, 80. Mem: Am Asn Hist Med; Hist Sci Soc; Am Physiol Soc; fel Am Col Physicians; Osler Soc; Soc Exp Biol & Med. Res: Environmental physiology and medicine; history of medicine; military medical history; research administration and management. Mailing Add: Uniformed Serv Univ Sch Med 4301 Jones Bridge Rd Bethesda MD 20814

JOY, ROBERT MCKERNON, b Troy, NY, May 9, 41. NEUROPHARMACOLOGY, NEUROTOXICOLOGY. Educ: Ore State Univ, BS, 64; Stanford Univ, PhD(pharmacol), 70; Am Bd Toxicol, dipl. Prof Exp: Res pharmacologist, 69-70, asst prof pharmacol, Sch Vet Med, Univ Calif, Davis, 70-77; assoc prof neuropath, Harvard Med Sch, Harvard Univ, 77-78; assoc prof pharmacol, 78-84, PROF PHARMACOL & TOXICOL, SCH VET MED, UNIV CALIF, DAVIS, 84- Concurrent Pos: Assoc res neurosci, Children's Hosp, Boston, 77-78; prof pharmacol, Northwestern Sch Med, Chicago, 85. Mem: AAAS; Soc Neurosci; Am Soc Pharmacol & Exp Therapeut; Soc Toxicol. Res: Epilepsy, anticonvulsant and convulsant drugs; mechanism of action; insecticide actions on central nervous system. Mailing Add: Dept Vet Med Pharmacol & Toxicol Univ Calif Davis CA 95616

JOY, VINCENT ANTHONY, b New York, NY, Feb 22, 20; m 52; c 7. INTERNAL MEDICINE. Educ: Fordham Univ, BS, 46; Duke Univ, MD, 50. Prof Exp: Staff internist, Vet Admin Hosp, East Orange, NJ, 53-54; staff internist & admitting officer, Vet Admin Hosp, New York, 54-59; med dir clin res, Int Div, E R Squibb & Sons, 59-67; sr med dir basic clin res, Int Div, Merck, Sharp & Dohme, Rahway, NJ, 67-; clin asst prof med, Med Col, Cornell Univ, 69-81, emer prof, 81-; RETIRED. Concurrent Pos: Asst physician, Bellevue Hosp, 60-69; clin asst physician, New York Hosp, 69-; dir, Med Clin, Nasau County Hosp, 84-87. Res: Gastroenterology; influence of the vagus nerve on gastric secretion; various classes of drugs on the parietal cell; action of decarboxylase inhibitor in the treatment of patients with Parkinson's disease, including administration of L-dopa to determine action of inhibitor as a means of reducing requirements; lipid-lowering and uricosuric effects of clofibrate-type drugs; beta-blockers. Mailing Add: 12640 N 89th St Scottsdale AZ 85260

JOYCE, BLAINE R, b Jeannette, Pa, Nov 13, 25; m 55; c 3. PHYSICAL CHEMISTRY. Educ: Univ Pittsburgh, BS, 49; Univ Toledo, MS, 65. Prof Exp: Res assoc activated carbon, Mellon Inst, 49-51; develop engr, Union Carbide Corp, 53-60, group leader activated carbon, 60-65, mgr activated carbon prod eng, 65-68, sr develop engr, Carbon Prod Div, 68-80, tech dir activated carbon, 80-86; CONSULT, ACTIVATED TECH SERV CO, 86- Mem: Am Chem Soc. Res: Production and applications of arc carbons for lighting and metal processing applications; development of activated carbon products and cost studies for business product planning. Mailing Add: 453 Cranston Dr Berea OH 44017

JOYCE, EDWIN A, JR, b Hampton, Va, Feb 23, 37; m 78; c 6. MARINE SCIENCE. Educ: Butler Univ, BA, 59; Univ Fla, Gainesville, MS, 61. Prof Exp: Marine biologist, Fla Bd Conserv, 61-67; sr fisheries biologist, 67-68, supvr marine res lab, 68-72, chief bur marine sci & technol, 72-75, DIR DIV MARINE RESOURCES, FLA DEPT NATURAL RESOURCES, 75- Mem: Am Fisheries Soc; fel Am Inst Fishery Res Biologists; Sigma Xi; Nat Shellfisheries Asn. Res: Research and publications on shellfish issues and administrative activities in fishery management and research supervision. Mailing Add: 14130 N Meridian Rd Tallahassee FL 32312

JOYCE, GLENN RUSSELL, b St Louis, Mo, June 24, 39; m 62; c 1. PLASMA PHYSICS. Educ: Cent Methodist Col, BA, 61; Univ Mo, MS, 63, PhD(physics), 66. Prof Exp: Res assoc, Univ Iowa, 66-68; from assoc prof to prof physics, 68-81; SR RES SCIENTIST, NAVAL RES LAB, 81- Concurrent Pos: Vis assoc prof, Hunter Col, 74; res scientist physics, Max Planck Inst Plasma Physics, 69; vis res scientist, Goddard Space Flight Ctr, NASA, 75-79. Mem: Am Phys Soc. Res: Plasma theory; interaction of test particles with plasmas; particle simulation of plasmas; numerical simulation and kinetic theory of plasmas. Mailing Add: Naval Res Lab Code 4790 Washington DC 20375

JOYCE, JAMES MARTIN, b Bayonne, NJ, Jan 27, 42; m 65; c 2. ATOMIC COLLISIONS, COMPUTER APPLICATIONS. Educ: LaSalle Col, BA, 63; Univ Pa, MS, 64, PhD(physics), 67. Prof Exp: Res assoc nuclear physics, Univ NC, Chapel Hill, 67-70; dir accelerator lab, 70-76, from asst prof to assoc prof, 70-79, PROF PHYSICS, ECAROLINA UNIV, 79-, DIR DIGITAL SYST, 76- Mem: Am Phys Soc; Sigma Xi. Res: Experimental atomic physics and applied physics; computer systems and interface design; physics applied to medicine. Mailing Add: 106 Valley Lane Greenville NC 27858

JOYCE, RICHARD ROSS, b Wilmington, Del, June 28, 44; m 79. ASTRONOMY. Educ: Williams Col, BA, 65; Univ Calif, Berkeley, PhD(physics), 70. Prof Exp: Res asst physics, Lawrence Radiation Lab, Univ Calif, Berkeley, 66-70; fel astron, State Univ NY, Stony Brook, 70-72, lectr, 72-73; SUPPORT SCIENTIST ASTRON, KITT PEAK NAT OBSERV, 73- Mem: Am Phys Soc; Astron Soc Pac. Res: Infrared detector development; telescope optimization for infrared use; study of heavily obscured and/or cool sources in infrared; study of infrared emission line sources. Mailing Add: Kitt Peak Nat Observ PO Box 26732 Tucson AZ 85726-6732

JOYCE, ROBERT MICHAEL, b Lincoln, Nebr, Sept 19, 15; m 41; c 2. ORGANIC CHEMISTRY. Educ: Univ Nebr, BS, 35, MS, 36; Univ Ill, PhD(org chem), 38. Prof Exp: Res chemist, Cent Res Dept, E I DuPont de Nemours & Co, Inc, 38-45, res supvr, 45-50, lab dir, 50-53, asst dir res, 53-62, dir res, 63-65, dir res & develop, Film Dept, 65-72, dir sci affairs pharmaceut div, Biochem Dept, 72-78; RETIRED. Mem: AAAS; Am Chem Soc. Res: Synthetic polymers; plastic films. Mailing Add: 1518 Valley Forge Blvd Sun City Ctr FL 33573

JOYCE, WILLIAM B(AXTER), b Columbus, Ohio, Oct 17, 32; m 58; c 4. THEORETICAL PHYSICS. Educ: Cornell Univ, BEP, 55; Ohio State Univ, PhD(physics), 66. Prof Exp: Mgr advan technol, Accuray Corp, Ohio, 63-66; mem tech staff, Bell Labs, 66-81, distinguished mem tech staff, 81-89, FEL, AT&T BELL LABS, 89- Mem: Am Phys Soc. Res: Applied theoretical physics. Mailing Add: AT&T Bell Labs 2D-350 Murray Hill NJ 07974

JOYNER, CLAUDE REUBEN, b Winston-Salem, NC, Dec 4, 25; m 50; c 2. MEDICINE. Educ: Univ NC, BS, 47; Univ Pa, MD, 49; Am Bd Internal Med, dipl, 57; Am Bd Cardiovasc Dis, dipl, 63. Prof Exp: Resdient med, Bowman Gray Sch Med, 50; Nat Heart Inst trainee & asst instr, Sch Med, Univ Pa, 52-53, from instr to assoc prof, 53-71; prof med, Hahnemann Med Col, 71-72; clin prof med, Univ Pittsburgh, 72-88; DIR DEPT MED, ALLEGHENY GEN HOSP, PITTSBURGH, 72-; PROF MED, MED COL PA, 88- Concurrent Pos: Attend cardiologist, Vet Admin Hosp, Philadelphia, 63-; mem coun arteriosclerosis, Am Heart Asn. Mem: Fel AAAS; Am Heart Asn; fel Am Col Physicians; Am Fedn Clin Res; fel Am Col Cardiol. Res: Academic medicine; cardiology; phonocardiography; ultrasound. Mailing Add: Allegheny Gen Hosp Dept Med 320 E North Ave Pittsburgh PA 15212

JOYNER, HOWARD SAJON, b Ft Worth, Tex, June 6, 39; m 69; c 3. ENGINEERING. Educ: Univ Tex, Austin, BS, 62, MA, 64; Univ Mo-Rolla, MS, 67, PhD, 70. Prof Exp: Nuclear engr, Gen Dynamics-Ft Worth, 64; asst prof mech eng, Wichita State Univ, 69-75; dir planning, res & develop, Univ Kans Sch Med, Wichita, 75-77; PRES, KINETIC CORP, 77- Mem: Am Soc Mech Engrs. Mailing Add: Kinetic Corp PO Box 8161 Wichita KS 67208

JOYNER, JOHN T, III, b Winston-Salem, NC, June 27, 28; m 50; c 3. INTERNAL MEDICINE. Educ: Wake Forest Col, BS, 48, MD, 52; Am Bd Internal Med, 62. Prof Exp: Chief resident med, Hosp, Emory Univ, 57-58; staff physician, 58-61, chief nuclear med, 65-75, CHIEF GEN MED, VET ADMIN HOSP, ASHVILLE, 61-, CHIEF MED SERV, 75-; ASST PROF CLIN MED, DUKE UNIV, 77- Mem: AMA; Am Col Physicians; Am Soc Nuclear Med. Mailing Add: Vet Admin Hosp Asheville NC 28805

JOYNER, POWELL AUSTIN, b Dallas, Tex, July 20, 25; m 52. RESEARCH MANAGEMENT. Educ: Centenary Col, BS, 46; Univ Iowa, PhD(phys chem), 51. Prof Exp: Asst, Univ Iowa, 46-50; res chemist, Minneapolis-Honeywell Regulator Co, 50-52; head anal sect, Callery Chem Co, 52-53; head measurements div, 53-54; sr res scientist, Res Ctr, Minneapolis-Honeywell Regulator Co, 54-56, res supvr, 56-58, head chem sect, 58-60, proj mgr fuel cell chemicals, 60-62, staff scientist, Honeywell Res Ctr, 62-63; asst dir, Res Div, Allis Chalmers Mfg Co, 63-64, gen mgr space & defense sci, 64-67, dir planning & eval, 67-68; dir res, Trane Co, 68-79, vpres res, 79-; AT ELEC POWER RES INST. Mem: Am Chem Soc; AAAS. Res: Raman effect; humidity instrumentation; hydrophyllic films; phase studies; thermochemistry; molten salts; fuel cells; air pollution; air conditioning; combustion. Mailing Add: 1224 Fairbrook Dr Mountain View CA 94040

JOYNER, RALPH DELMER, b Derby, Va, Aug 31, 28; m 50; c 3. INORGANIC CHEMISTRY. Educ: Miami Univ, BS, 50, MS, 51; Case Inst Technol, PhD(inorg chem), 61. Prof Exp: Chemist, Monsanto Co, Ohio, 51-52, res chemist, Mound Lab, 52-58; chemist, Solvay Process Div, Allied Chem Corp, NY, 61; sr res chemist, Chem Div, Pittsburgh Plate Glass Co, Ohio, 61-62, res assoc, 62, res supvr, 62-65; from asst prof to prof inorg chem, Ball State Univ, 65-88, dept chmn, 86-88; RETIRED. Concurrent Pos: Instr night sch, Univ Dayton, 52-54; res assoc & grad student adv, Case Inst Technol, 62-63. Mem: Am Chem Soc; Sigma Xi. Res: Silicon and germanium coordination compounds. Mailing Add: 5065 W River Rd Muncie IN 47304

JOYNER, RONALD WAYNE, b Wake Forest, NC, Mar 21, 47; m 69; c 1. CARDIAC ELECTROPHYSIOLOGY, NEUROPHYSIOLOGY. Educ: Univ NC, BS, 69; Duke Univ, MD, 74, PhD(physiol), 73. Prof Exp: Asst prof physiol, Duke Univ, 76-77; asst prof physiol, Univ Iowa, 77-87; PROF PEDIAT & PHYSIOL, EMORY UNIV, 87- Mem: Biophys Soc. Res: Mechanisms of propagation of cardiac action potentials related to cardiac arrythmias, by electrophysiological and numerical simulation methods; synaptic transmission and motor control in squid. Mailing Add: Dept Pediat & Physiol Emory Univ 2040 Ridgewood Dr Northeast Atlanta GA 30323

JOYNER, WEYLAND THOMAS, JR, b Suffolk, Va, Aug 8, 29; m 55; c 2. NUCLEAR PHYSICS, ELECTRONICS. *Educ:* Hampden-Sydney Col, BS, 51; Duke Univ, MA, 52, PhD(physics), 55. *Prof Exp:* Asst physics, Duke Univ, 51-53, fel, 53-54; physicist, Dept of Defense, 54-57; asst prof, Hampden-Sydney, Col, 57-59, assoc prof, 59-63, prof, 63-66; staff physicist, Univ Mich, Ann Arbor, 66-67; PROF PHYSICS, HAMPDEN-SYDNEY COL, 67-, CHMN DEPT, 69- *Concurrent Pos:* Consult, Oak Ridge Inst Nuclear Studies, 60-66; res partic, Ames Lab, AEC, 64, 65; vis prof, Pomona Col, 65 & Dartmouth Col, 81; dir col physics prog, Am Inst Physics, 67-68; chmn physics comt, Col Entrance Exam Bd, 71-72; NASA sr fel, 84 & 85. *Mem:* Am Phys Soc; Am Asn Physics Teachers; Inst Elec & Electronics Engrs; fel AAAS. *Res:* positron lifetimes; picosecond circuitry; x-ray fluorescence; silicon deep impurity devices; radon. *Mailing Add:* Hampden-Sydney Col Hampden-Sydney VA 23943

JOYNER, WILLIAM B, b Casper, Wyo, Dec 27, 29; m 52. GEOPHYSICS. *Educ:* Harvard Univ, AB, 51, AM, 52, PhD(geophys), 58. *Prof Exp:* Geophysicist, Humble Oil & Refining Co, 58-63; sr scientist, Vela Uniform Prog, Dunlap & Assocs, Inc, 64; GEOPHYSICIST, US GEOL SURV, 64- *Mem:* Am Geophys Union; Soc Explor Geophys; Geol Soc Am; Seismol Soc Am; Earthquake Eng Res Inst. *Res:* Solid earth geophysics; gravity and terrestrial magnetism; terrestrial heat flow and temperatures within the earth; engineering seismology. *Mailing Add:* 472 Virginia Ave San Mateo CA 94402

JOYNER, WILLIAM HENRY, JR, b Washington, DC, Sept 21, 46; m 68; c 2. COMPUTER SCIENCE. *Educ:* Univ Va, BS, 68; Harvard Univ, SM, 69, PhD(appl math), 73. *Prof Exp:* RES STAFF MEM COMPUT SCI, IBM THOMAS J WATSON RES CTR, 73- *Mem:* Asn Comput Mach. *Res:* Computer program verification; automated theorem proving; machine description languages; VLSI design. *Mailing Add:* IBM T J Watson Res Ctr Yorktown Heights NY 10598

JOYNER, WILLIAM LYMAN, b Farmville, NC, June 10, 39; m 82; c 3. PHYSIOLOGY. *Educ:* Davidson Col, BS, 65; Univ NC, MSPH, 67, PhD(physiol), 71. *Prof Exp:* Res technician animal med, Univ NC, 65-67, res assoc physiol, 67-69; trainee, Med Ctr, Duke Univ, 71-73; from asst prof to assoc prof physiol, Col Med, Univ Nebr Med Ctr, 73-83, prof, 83-89,; PROF PHYSIOL & CHMN DEPT, COL MED, ETENN STATE UNIV, 89- *Concurrent Pos:* Pharm travel grant, Microcirculatory Soc, 75; pres, Microcircular Soc Inc, 87; NIH reviewer res grants. *Mem:* Am Physiol Soc; Microcurculatory Soc. *Res:* Cardiovascular physiology particularly the microcirculation, molecular transport, vascular reactivity and controlling mechanisms; hypertension, diabetes and alterations in blood coagulation and hemostasis related to hemodynamic responses of the microcirculation in various tissues. *Mailing Add:* Dept Physiol ETenn Univ Col Med PO Box 14780A Johnson City TN 37614

JOYNSON, REUBEN EDWIN, JR, b Winfield, Kans, Dec 27, 26; m. PHYSICS. *Educ:* Kans State Univ, BS, 49, MS, 50; Mass Inst Technol, PhD(physics), 54. *Prof Exp:* Res physicist, Continental Oil Co, 54-60; consult physicist, Comput Lab, Gen Elec Co, 60-63, physicist, Res & Develop Ctr, 63-87; RETIRED. *Mem:* AAAS; Am Phys Soc; Sigma Xi. *Res:* X-ray diffraction; digital computer design and programming; cryogenics; thin film devices; artificial intelligence; visual image processing. *Mailing Add:* 251 Alplaus Ave PO Box 118 Alplaus NY 12008

JOYNT, ROBERT JAMES, b LeMars, Iowa, Dec 22, 25; m 53; c 6. NEUROLOGY. *Educ:* Westmar Col, BA, 49; Univ Iowa, MD, 52, MS, PhD(anat), 63; Am Bd Psychiat & Neurol, dipl, 59. *Hon Degrees:* DSc, Westmar Col, 64. *Prof Exp:* Assoc neurol, Univ Iowa, 57-58, from asst prof to assoc prof, 58-66; prof neurol & anat & chmn dept neurol, Med Ctr, 66-84, dean, Sch Med & Dent, 85-90, vprovost, 85-89, VPRES, HEALTH AFFAIRS, UNIV ROCHESTER, 89- *Concurrent Pos:* Mem res training grant comt, Nat Inst Neurol Dis & Blindness, 63-67 & neurol study sect, div res grants, NIH, 67-72; bd adv, off biometry & epidemiol, NINCDS, 76-84; ed, Arch Neurol, 82- *Honors & Awards:* Netter Award, Am Acad Neurol, 88. *Mem:* Am Neurol Asn; Am Acad Neurol; Am Electroencephalog Soc; Am Med Asn. *Res:* Investigation of fluid control by central nervous system; correlation of performance tests with lesions in brain damaged patients; Alzheimer's disease. *Mailing Add:* Sch Med & Dent Univ Rochester Rochester NY 14642

JOYS, TERENCE MICHAEL, b Hull, Eng, Jan 19, 35; m 60; c 3. RECOMBINANT DNA, VACCINES. *Educ:* Leeds Univ, Eng, BSc, 57; London Univ, Eng, PhD(microbiol gen), 61. *Prof Exp:* Instr bact, Univ Minn, 64-65; asst prof bact genetics, Univ Ore Med Sch, 65-76; assoc prof med microbiol, Penn State Univ, 76; ASSOC PROF IMMUNOL, TEX TECH HEALTH SCI CTR, 76- *Concurrent Pos:* Career develop award, US Pub Health Dept, 69-74; vis scientist, Max Planck Inst, Munich, Ger, 90. *Mem:* Am Soc Microbiol; Am Acad Microbiol; Southern Asn Agr Scientists. *Res:* Structure of bacterial flagellan filament protein and its use in vaccine development. *Mailing Add:* Dept Microbiol Tex Tech Univ Health Sci Ctr Lubbock TX 79430

JU, FREDERICK D, b Shanghai, China, Sept 21, 29; US citizen; m 56; c 3. THEORETICAL & APPLIED MECHANICS. *Educ:* Univ Houston, BS, 53; Univ Ill, MS, 56, PhD(theoret & appl mech), 58. *Prof Exp:* From asst prof to assoc prof, 58-67, chmn, 73-76, PROF MECH ENG, UNIV NMEX, 67- *Concurrent Pos:* Consult, Los Alamos Nat Lab, 62-, vis staff mem, 72-; nat vis prof, Nat Sci Coun, Repub of China, 71-72; prin investr, Air Force Off Sci Res, 61-72 & 81-& Off Naval Res, 81- *Honors & Awards:* Soc Theoret & Appl Mech Award, 86; Chinese Soc Mat Sci Award, 86. *Mem:* Fel Am Soc Mech Engrs; Soc Eng Sci; Sigma Xi. *Res:* Fracture diagnosis in structures; thermomechanical cracking from friction loading; structural safety of reactor; boundary layer theory in plates. *Mailing Add:* Dept Mech Eng Univ NMex Albuquerque NM 87131

JU, JIN SOON, b Ham-hung, Ham-nam Prov, Oct 20, 21; m 48; c 2. PROTEIN NUTRITION, AGING & NUTRITION. *Educ:* Seoul Nat Univ, MD, 47, PhD(nutrit), 59. *Prof Exp:* Researcher nutrit, Nat Chem Lab, Korea, 47-50; prof nutrit, Nat Fisheries Col, Pusan, Korea, 50-53; prof, 53-87, EMER PROF BIOCHEM & NUTRIT, KOREA UNIV MED COL, SEOUL, KOREA, 87-; DIR & PROF NUTRIT, KOREA NUTRIT INST, HALLYM UNIV, CHUN-CHON, KOREA, 87- *Concurrent Pos:* Dir, Korea Geront Ctr, Hallym Univ, Seoul, 90- *Honors & Awards:* Korea Nat Acad Sci Award, 85. *Mem:* Am Inst Nutrit; Int Union Nutrit Sci; Pac Sci Asn; Korea Nat Acad Sci. *Res:* Protein metabolism in human and animals, especially protein requirement of Korean. *Mailing Add:* 691-8 Jayang-dong Hanyang Villa No 202 Sueng-dong-ku Seoul 133-192 Republic of Korea

JUANG, JER-NAN, b Tou-Liu, Taiwan, July 21, 45; US citizen; m 77; c 2. TEST METHODS FOR SPACE STRUCTURES, LINEAR ALGEBRA. *Educ:* Nat Cheng Kung Univ, Taiwan, BS, 69; Tenn Technol Univ, MS, 71; Va Polytech Inst, PhD(eng mech), 74. *Prof Exp:* Res assoc, Va Polytech Inst, 74-75; tech staff engr, Comput Sci Corp, 75-77; staff engr, Martin Marietta Corp, 77-79, sr staff engr, 81-82; tech staff res, Jet Propulsion Lab, 79-81; SR RES SCIENTIST, LANGLEY RES CTR, NASA, 82- *Concurrent Pos:* Assoc ed, J Astronaut Sci, 86-; mem, tech comt composite mats, Am Astronaut soc, 82-, tech comt composite large space struct, 87-; adj prof, Univ Colo, Boulder, 88- *Honors & Awards:* Merit Awards, Am Soc Mech Engrs; Special Achievement, NASA. *Mem:* Fel Am Astronaut Soc; fel Am Inst Aeronaut & Astronaut; Soc Indust & Appl Math. *Res:* Active and passive control, tracking, modal parameter identification; system realization, and other dynamic problems for large space structures; nonlinear control problems and parameter estimation in distributed-parameter systems. *Mailing Add:* NASA Langley Res Ctr Mail Stop 230 Hampton VA 23665-5225

JUANG, LING LING, b Taipei, Taiwan, Oct 10, 49; m 73; c 1. NUMERICAL ANALYSIS. *Educ:* Nat Taiwan Univ, BS, 72; State Univ NY Stony Brook, MS, 73, PhD(appl math), 76. *Prof Exp:* Syst analyst heat transfer prob, KLD Assoc, Inc, 75-76; asst mathematician energy models, Brookhaven Nat Lab, 76-80; COMPUT PROCESS CONTROL SPECIALIST, GEN ELEC CO, 80- *Mem:* Am Commun Mach; Soc Mfg Eng. *Res:* Economic-energy modeling of depletable resources; computer aided manufacturing system design. *Mailing Add:* 4293 Berryhill Lane Cincinnati OH 45242

JUBB, GERALD LOMBARD, JR, b Ayer, Mass, Jan 3, 43; m 67; c 2. PEST MANAGEMENT. *Educ:* NMex Highlands Univ, BA, 65; Univ Ariz, MS, 67, PhD(entom), 70. *Prof Exp:* NSF trainee entom, Univ Ariz, 67-70; from asst prof to prof entom, Erie County Field Res Lab, Pa State Univ, 70-84; prof entom & ctr head, Western Md Res & Educ Ctr, Univ Md, 84-88; PROF ENTOM, AGR EXP STA, VA POLYTECH INST, 88- *Concurrent Pos:* Pres, Entom Soc Pa, 72; secy & treas, eastern br, Entom Soc Am, 78-81 & gov bd, 85-88. *Mem:* Entom Soc Am; NY Entom Soc; Acarological Soc Am; Am Soc Enol & Viticulture; Sigma Xi. *Res:* Insects and mites attacking grapes; economic injury levels; monitoring techniques; pesticide impact in vineyards; small fruit pest management. *Mailing Add:* Agr Exp Sta Va Polytech Inst Bladesburg VA 24061

JUBERG, RICHARD CALDWELL, human genetics, pediatrics; deceased, see previous edition for last biography

JUBERG, RICHARD KENT, b Cooperstown, NDak, May 14, 29; m 56; c 4. MATHEMATICS. *Educ:* Univ Minn, BS, 52, PhD(math), 60. *Prof Exp:* Temporary mem, Courant Inst Math Sci, NY Univ, 57-58; instr, Univ Minn, 58-60, from asst prof to assoc prof, 60-75, PROF MATH, UNIV CALIF, IRVINE, 75- *Concurrent Pos:* NSF sci faculty fel, Pisa, 65-66; vis prof, Univ Sussex, 72-73, Math Inst Technol, Gothenburg, Sweden, 81. *Mem:* Am Math Soc. *Res:* Problems in analysis and partial differential equations. *Mailing Add:* Dept Math Univ Calif Irvine CA 92717

JUBY, PETER FREDERICK, b Great Yarmouth, Eng, Oct 27, 35; US citizen; m 70; c 2. MEDICINAL CHEMISTRY. *Educ:* Univ Nottingham, BSc, 57, PhD(org chem), 61. *Prof Exp:* Nat Res Coun Can Fel, 60-62; sr res scientist, Bristol Labs, Inc, 62-76, prin investr, 76-82, dir gastrointestinal chem, Bristol-Myers Co, 82-87, DIR CNS THERAPEUTIC AREA OPERATIONS, 87- *Mem:* Am Chem Soc. *Res:* Natural product chemistry; biosynthesis of alkaloids; synthesis of medicinal agents; arthritis; allergies; gastrointestinal diseases. *Mailing Add:* Pharmaceut Res & Develop Bristol-Myers Co Five Res Pkwy PO Box 5100 Wallingford CT 06492-7600

JUCHAU, MONT RAWLINGS, b Virginia, Idaho, Nov 11, 34; m 60; c 4. PHARMACOLOGY. *Educ:* Idaho State Univ, BS, 60; Wash State Univ, MS, 63; Univ Iowa, PhD(pharmacol), 66. *Prof Exp:* Pharmacist, Trolinger Pharm, Tick Klock Drug, 60-63; from instr to asst prof biochem pharmacol, State Univ NY Buffalo, 66-69; asst prof, 69-73; assoc prof, 73-80, PROF PHARMACOL, SCH MED, UNIV WASH, 80- *Honors & Awards:* Delbert-Putnam Award, 59; Rexall Award, 60. *Mem:* AAAS; Am Soc Pharmacol & Exp Therapeut; Int Soc Biochem Pharmacol. *Res:* Investigation of the biotransformation of drugs in the human foetoplacental unit. *Mailing Add:* 8419 NE 177th Bothell WA 98011

JUD, HENRY G, b Rochester, NY, Sept 19, 34; wid; c 4. ELECTRICAL ENGINEERING. *Educ:* Valparaiso Univ, BS, 56; Univ Pittsburgh, MS, 59, PhD(elec eng), 62. *Prof Exp:* Electronics technician, Western Elec Co, 56-57; specialist missile systs, Autonetics Div, NAm Aviation Inc, 62-64; from asst prof to assoc prof elec eng, Valparaiso Univ, 64-68; staff engr, IBM Fed Systs Div, NY, 68-70, advan engr, 70-71; develop engr, 71-76, develop engr, IBM Corp, Md, 76-85; sr staff systs eng, 85-89, DIR SYSTS TECHNOL, UNISYS CORP, 89- *Mem:* Inst Elec & Electronics Engrs. *Res:* Missile system error analysis; automatic control; random input control systems; computer systems design. *Mailing Add:* Unisys Corp 8201 Greenshore Dr McLean VA 22102

JUDAY, GLENN PATRICK, b Elwood, Ind, May 4, 50; m 71; c 4. FOREST ECOLOGY, CONSERVATION. *Educ:* Purdue Univ, BS, 72; Ore State Univ, Phd(plant ecol), 77. *Prof Exp:* Fel ecol & conserv, Ore State Univ, 76-77; coordr ecol reserves, Joint Fed-State Land Use Planning Comn, Alaska, 77-78, Arctic Environ Info & Data Ctr, Univ Alaska, 78 & Inst Northern Forestry, Forest Serv, USDA, 78-81; vis assoc prof, Agr & Forestry Exp Sta, 81-87, ASST PROF FOREST ECOL, UNIV ALASKA, FAIRBANKS, 87- *Concurrent Pos:* Pres, Natural Areas Asn, 85-87, 87-88. *Mem:* AAAS; Ecol Soc Am; Soc Am Foresters; Natural Areas Asn. *Res:* Systematic plan for conserving ecological reserves using vegetation, wildlife, and geologic classifications; old-growth forest structure, including stocking, basal area, age height and spacing; post-glacial primary succession; fire effects research. *Mailing Add:* Sch Agr & Land Resources Mgt Univ Alaska 309 O'Neill Bldg Fairbanks AK 99775-0080

JUDAY, RICHARD EVANS, b Madison, Wis, May 28, 18. ORGANIC CHEMISTRY. *Educ:* Harvard Univ, BA, 39; Univ Wis, PhD(org chem), 43. *Prof Exp:* Asst res chemist, Gen Chem Co, NY, 43-44 & Ortho Pharmaceut Corp, 44-47; from asst prof to prof chem, 47-77, EMER PROF CHEM, UNIV MONT, 77- *Mem:* Fel AAAS; Am Chem Soc; Am Soc Limnol & Oceanog. *Res:* Water quality; chemical limnology. *Mailing Add:* Dept Chem Univ Mont Missoula MT 59812-1006

JUDD, BRIAN RAYMOND, b Chelmsford, Eng, Feb 13, 31. THEORETICAL PHYSICS, ATOMIC PHYSICS. *Educ:* Oxford Univ, BA, 52, MA & DPhil, 55. *Prof Exp:* Fel, Magdalen Col, Oxford Univ, 55-62; instr physics, Univ Chicago, 57-58; chemist, Lawrence Radiation Lab, Univ Calif, Berkeley, 59-62; assoc prof spectros, Univ Paris, 62-64; staff mem nuclear chem, Lawrence Radiation Lab, Univ Calif, Berkeley, 64-66; chmn, physics dept, 79-84, PROF PHYSICS, JOHNS HOPKINS UNIV, 66- *Concurrent Pos:* Consult, Argonne Nat Lab, 77-88; hon fel, Brasenose Col, Oxford, 83- *Honors & Awards:* Frank H Spedding Award, Rare Earth Res, 88. *Mem:* Fel Am Phys Soc. *Res:* Theoretical studies of atoms, molecules and solid state physics, particularly application of group theory. *Mailing Add:* Dept Physics & Astron Johns Hopkins Univ Baltimore MD 21218

JUDD, BURKE HAYCOCK, b Kanab, Utah, Sept 5, 27; m 53; c 3. GENETICS. *Educ:* Univ Utah, BS, 50, MS, 51; Calif Inst Technol, PhD(genetics), 54. *Prof Exp:* Am Can Soc Fel, Univ Tex, Austin, 54-56, from instr to assoc prof zool, 56-69, prof, 69-79; CHIEF, LAB GENETICS, NAT INST ENVIRON HEALTH SCI, RESEARCH TRIANGLE PARK, NC, 79- *Concurrent Pos:* Geneticist, AEC, Washington, DC, 68-69; mem panel genetic biol, NSF, 69-72; dir, Genetics Inst, Univ Tex, Austin, 77-79; Gosney vis prof, Div Biol, Calif Inst Technol, 75-76; adj prof biol, Univ NC, Chapel Hill, 79- & Prog Genetics, Duke Univ, Durham, NC, 80-; assoc ed, Genetics, 73-78; corresp ed, Molecular & Gen Genetics, 86- *Mem:* Fel AAAS; Genetics Soc Am (vpres, 79 & pres, 80); Am Soc Nat (secy, 68-70). *Res:* Chromosome organization; gene function and regulation; recombination mechanism; genetics of Drosophila. *Mailing Add:* Lab Genetics Nat Inst Environ Health Sci PO Box 12233 Research Triangle Park NC 27709

JUDD, DAVID LOCKHART, b Chehalis, Wash, Jan 8, 23; m 45; c 2. THEORETICAL PHYSICS. *Educ:* Whitman Col, AB, 43; Calif Inst Technol, MS, 47, PhD(physics), 50. *Hon Degrees:* DSc, Whitman Col, 74. *Prof Exp:* Staff mem, Physics Div, Los Alamos Sci Lab, 45-46; staff mem theoret physics, Nuclear Energy Div, Rand Corp, 49-51; group leader theoret physics, 51-66, from dept head to head physics div, 63-70, assoc dir, 67-70, lectr, 53-61, dept physics, SR LECTR, DEPT PHYSICS, UNIV CALIF, BERKELEY, 62-, SR RES PHYSICIST, LAWRENCE BERKELEY LAB, 70- *Concurrent Pos:* Consult, Northrop Aircraft Co, 47-49, Radiation Lab, Univ Calif, 50-51, Rand Corp, 51-55 & W M Brobeck & Assocs, 60; mem adv comt, Electronuclear & Physics Div, Oak Ridge Nat Lab, 63-65. *Mem:* Fel Am Phys Soc; Sigma Xi. *Res:* Theoretical and mathematical physics; accelerator theory; ion optics; plasma and particle physics; nonlinear mechanics. *Mailing Add:* Dept Physics Univ Calif 366 Le Conte Hall Berkeley CA 94720

JUDD, FLOYD L, b Janesville, Wis, Jan 25, 34; m 62; c 3. HIGH ENERGY PHYSICS, OPTICS. *Educ:* Carroll Col, BS, 56; Iowa State Univ, MS, 60, PhD, 66. *Prof Exp:* Asst prof, Northwestern State Col, La, 59-62, 64-67; asst prof, 67-70, assoc prof physics, Fresno State Col, 70-71; prof physics & chmn, Calif State Univ, Fresno, 71-77. *Mailing Add:* Dept Physics Calif State Univ Fresno CA 93740

JUDD, FRANK WAYNE, b Wichita Falls, Tex, Aug 23, 39; m 69; c 2. POPULATION ECOLOGY, PHYSIOLOGICAL ECOLOGY. *Educ:* Midwestern State Univ, BS, 65; Tex Tech Univ, MS, 68, PhD(zool), 73. *Prof Exp:* Teaching asst biol, Dept Biol, Tex Tech Univ, 65-68, res asst & instr, 69-71; instr, Biol Dept, Pan Am Univ, 68-69, from asst prof to prof, 72-82, dir, Coastal Studies Lab, 84-91; PROF BIOL & DIR, COASTAL STUDIES LAB, UNIV TEX-PAN AM, 84- *Concurrent Pos:* Adj prof, Dept Biol Sci, Tex Tech Univ, 83-; vis prof, Dept Wildlife & Fisheries Sci, Tex A&M Univ, 89- *Mem:* Am Soc Ichthyologists & Herpetologists; Am Soc Mammalogists; Ecol Soc Am; Herpetologists' League. *Res:* Ecology of the coastal zone of southern Texas and northern Mexico; barrier island ecology; black mangrove distribution; oyster reef distribution; tortoise demography. *Mailing Add:* Coastal Studies Lab Univ Tex-Pan Am PO Box 2591 South Padre Island TX 78597

JUDD, GARY, b Humene, Czech, Sept 24, 42; US citizen; m 64; c 3. PHYSICAL METALLURGY, ELECTRON MICROSCOPY. *Educ:* Rensselaer Polytech Inst, BMetE, 63, PhD(phys metall), 67. *Prof Exp:* Res asst, Rensselaer Polytech Inst, 66-67, from asst prof to prof, 67-76, actg chmn dept mat eng, 74-75, vprovost, Plans & Resources, 75-78, actg provost & vpres, acad affairs, 82-83, 85-86, VPROVOST, ACAD AFFAIRS & DEAN GRAD SCH, RENSSELAER POLYTECH INST, 79- *Concurrent Pos:* Consult, Oak Ridge Nat Lab, 68-70 & Watervliet Arsenal, 68-78; metall eng consult ed, McGraw-Hill Sci & Technol Encycl, 75- *Mem:* Am Inst Mining, Metall & Petrol Engrs; Am Soc Metals; Sigma Xi; Microbeam Anal Soc; AAAS. *Res:* Structure sensitive properties of materials, particularly strengthening mechanisms, precipitation kinetics, defect structures, biomaterials and corrosion; electron probe microanalysis; scanning electron microscopy; forensic science. *Mailing Add:* Pittsburgh Bldg Rensselaer Polytech Inst Troy NY 12181

JUDD, JANE HARTER, b Pittsburgh, Pa, Oct 24, 25; m 72. SPECTROSCOPY, ANALYTICAL CHEMISTRY. *Educ:* Carlow Col, BA, 47. *Prof Exp:* Chemist, Dept Res Med, Univ Pittsburgh, 48-50; SR SCIENTIST, WESTINGHOUSE BETTIS ATOMIC POWER PLANT, 51- *Concurrent Pos:* Exec secy, Pittsburgh Conf Anal Chem & Appl Spectroscopy, 64-, pres, 78. *Mailing Add:* Westinghouse Bettis 16A PO Box 79 West Mifflin PA 15122

JUDD, LEWIS LUND, b Los Angeles, Calif, Feb 10, 30; m 74; c 3. PSYCHIATRY, CHILD PSYCHIATRY. *Educ:* Univ Utah, BS, 54; George Washington Univ, 54-56; Univ Calif, Los Angeles, MD, 58. *Prof Exp:* Intern internal med, Ctr Health Sci, Univ Calif, Los Angeles, 58-59, resident psychiat, 59-60, 62-64; asst prof psychol & psychiat, 65-70; assoc prof, 70-73, actg chmn dept, 74-75, co-chmn dept, 75-77, PROF PSYCHIAT, UNIV CALIF, SAN DIEGO, 73-, CHMN DEPT, 77- *Concurrent Pos:* Fel child psychiat, Ctr Health Sci, Univ Calif, Los Angeles, 62-64, State of Calif fel, 66, NIMH fels, 67-69, Scottish Rites Comt on Res in Schizophrenics, 69; supvr psychiat, Adolescent Outpatient Unit, Marion Davies Pediat Clin, Ctr Health Sci, Univ Calif, Los Angeles, 65-70, supvr psychiat consult serv to dept pediat, 65-70, attend physician, Hosp, 65-70, psychiat consult, Dept Phys Med & Rehab, 65-70, dir educ, Child & Adolescent Psychiat, Dept Psychiat, 65-70; psychiat consult, Calif State Bd Rehab, Sacramento, 65-70; mem adv comt eval of drug abuse progs, County of San Diego, 70; dir, Univ Calif Drug Abuse Progs, 70-73; vchmn & dir clin progs, Dept Psychiat, Univ Calif, San Diego, 70-73, chmn Social & Behavioral Sci Course, Sch Med, 70-; psychiat consult, San Diego County Dept Pub Health, 72-; chief psychiat serv, Vet Admin Hosp, San Diego, 72-77; dir, Nat Inst Mental Health, 88-90. *Mem:* Am Psychiat Asn; Am Orthopsychiat Asn; Soc Res Child Develop; Am Soc Adolescent Psychiat; Psychiat Res Soc. *Res:* Substance abuse in adolescents; developmental psychopathology; epidemiology of deviant populations; clinical psychopharmacology. *Mailing Add:* Dept Psychiat Univ Calif San Diego 9500 Ceilman Dr La Jolla CA 92093-0603

JUDD, O'DEAN P, b Austin, Minn, May 26, 37; m 68; c 3. LASERS & OPTICS, ATOMIC & MOLECULAR PHYSICS. *Educ:* St Johns Univ, BS, 59; Univ Calif, Los Angeles, MS, 61, PhD(physics), 68. *Prof Exp:* Staff physicist, Hughes Res Lab, 59-67 & 69-72; consult, 72-74; fel plasma physics, Univ Calif, Los Angeles, 68-69; assoc group leader, theoret div, Los Alamos Nat Lab, Univ Calif, 72-75, group leader advan laser res, 75-77, mem staff, Appl Photochem Div, 77-82, chief scientist, Defense Res, 82-87; chief scientist, strategic defense iniative, Pentagon, Washington, DC, 87-90; CHIEF SCIENTIST, LOS ALAMOS NAT LAB, 90- *Concurrent Pos:* Hughes masters fel, 59, Hughes doctoral fel, 64; consult, 80-; adj prof physics, Univ NMex, 81- *Mem:* Am Phys Soc; fel Inst Elec & Electronic Engrs; AAAS. *Res:* Non-linear optics; laser physics, atomic and molecular physics; plasma physics; quantum electronics and laser chemistry, theoretical and experimental; 3 US patents. *Mailing Add:* 101 Zuni Los Alamos NM 87544

JUDD, ROSS LEONARD, b London, Ont, June 3, 36; m 62; c 2. HEAT TRANSFER. *Educ:* Western Ont Inst Technol, BESc, 58; McMaster Univ, MEng, 63; Univ Mich, PhD(heat transfer), 68. *Prof Exp:* Develop engr, Civilian Atomic Power Dept, Can Gen Elec, 58-61; lectr, 63-67, asst prof, 67-74, assoc prof, 74-80, PROF HEAT TRANSFER & THERMODYN, MCMASTER UNIV, 80- *Honors & Awards:* R R Teetor Award, Soc Automotive Engrs. *Mem:* Am Soc Mech Engrs; Soc Automotive Engrs. *Res:* Boiling heat transfer; two phase flow; flow induced vibrations. *Mailing Add:* Dept Mech Eng McMaster Univ Hamilton ON L8S 4L8 Can

JUDD, STANLEY H, b Denver, Colo, Feb 12, 28; m 50, 77; c 2. ENVIRONMENTAL HEALTH. *Educ:* Univ Calif, Los Angeles, BS, 49; Univ Calif, Berkeley, MPH, 56; Am Bd Indust Hyg, cert, 66. *Prof Exp:* Chemist, State of Calif Dept Pub Health, 49-50, 53-56 & US Army Environ Health Lab, 50-52; res chemist, Chevron Res Corp, 56-58, indust hygienist, 58-65, sr indust hygienist, 65-69, environ health & pollution engr, 69-78, staff indust hygienist, 78-80, MGR, HEALTH SURVEILLANCE SERV, CHEVRON CORP, CALIF, 80- *Concurrent Pos:* Mem fac, Inst Noise Control Eng & Inst Safety & Syst Mgt, Univ Southern Calif Extension. *Mem:* Am Indust Hyg Asn; Am Acad Indust Hyg; Am Chem Soc. *Res:* Engineering noise control at the source to prevent hearing loss, interference with communications and annoyance; safe handling of pesticide and petrochemical products including facilities designs; occupational health information systems design; biomedical surveillance. *Mailing Add:* Four Boston Ship Plaza San Francisco CA 94111

JUDD, WALTER STEPHEN, b Fairbanks, Alaska, Apr 14, 51; m 72. PLANT SYSTEMATICS. *Educ:* Mich State Univ, BS, 73, MS, 74; Harvard Univ, PhD(biol), 79. *Prof Exp:* ASSOC PROF BOT, DEPT BOT, UNIV FLA, 78- *Mem:* Am Soc Plant Taxonomists; Int Asn Plant Taxon; Am Bryological & Lichenological Soc; Bot Soc Am. *Res:* Systematics and evolution of flowering plants with specific interest in the Ericaceae and Melastomataceae; floras of the West Indies and Florida. *Mailing Add:* Dept Bot Univ Fla Gainesville FL 32611

JUDD, WILLIAM ROBERT, b Denver, Colo, Aug 16, 17; m 42; c 5. ROCK MECHANICS, GEOTECHNICAL ENGINEERING. *Educ:* Univ Colo, AB, 41. *Prof Exp:* Eng geologist, US Bur Reclamation, 38-41, head geol sect I, Off Chief Engr, 44-60; head basing technol group, Rand Corp, Calif, 60-65; prof rock mech, Sch Civil Eng, 66-87, head geotech eng, 76-87, EMER PROF CIVIL ENG, PURDUE UNIV, 88-; CONSULT ENGR, 88- *Concurrent Pos:* Eng geologist, Water Conserv Bd, Colo, 41-42 & Denver & Rio Grande West Rwy, 42-44; instr, Lowry AFB, 46-51; consult geologist & engr, 50-; mem adv bd, mountain & arctic warfare, US Army, 56-62; consult to various US & foreign govt agencies & comts & pvt industs, 58-; founder & chmn, US Nat Comt Rock Mech, 63-69, chmn panel awards, 74-78, sr adv panel res, 77-81; mem panel geophys, USAF Sci Adv Bd, 64-68; geo-sci ed, Am Elsevier Publ Co, Inc, 66-71; reviewer, Appl Mech Reviews, 68-73; ed-in-chief, Int J Eng Geol, 72-; tech dir, Underground Explor & Rock Properties Info Ctr, 72-80; mem, Exec Coun, US Comt Large Dams, 77-83, comt Earthquakes, 76-90, Nat Res Coun Comt Dam Safety, 77-78 & Comt Safety Existing Dams, 82-83; mem Adv Bd Applied Phys Math and Biol Sci, NSF, 79-81. *Honors & Awards:* Spec Award Outstanding Contrib Rock Mech Res, US Nat Comt Rock Mech, 82; Alex du Toit Mem Lectr, SAfrica & Rhodesia, 67. *Mem:* Fel SAfrican Inst Mining & Metall; fel Am Soc Civil Engrs; fel Geol Soc Am; Int Soc Rock Mech (vpres, 67-70); hon mem, India Soc Eng Geol; hon mem Asn Eng Geologists; Sigma Xi. *Res:* Seismic effects on underground openings, dam safety; rock tunnels. *Mailing Add:* 200 Quincy St West Lafayette IN 47906

JUDD, WILLIAM WALLACE, b Windsor, NS, Oct 22, 15; m 46; c 4. ENTOMOLOGY. *Educ:* McMaster Univ, BA, 38; Univ Western Ont, MA, 40; Univ Toronto, PhD(zool), 46. *Prof Exp:* Agr asst, Can Dept Agr, 37-42; asst meteorol, Dept Transport, Ottawa, 42-45; lectr zool, McMaster Univ, 46-48, asst prof, 48-50; asst prof, 50-51, assoc prof, 52-64, prof, 65-81, EMER PROF ZOOL, UNIV WESTERN ONT, 81- *Res:* Aquatic insects; insect morphology. *Mailing Add:* Dept Zool Univ Western Ont London ON N6A 5B7 Can

JUDGE, DARRELL L, b Albion, Ill, Nov 2, 34; m 59; c 3. PHYSICS. *Educ:* Eastern Ill State Col, BS, 56; Univ Southern Calif, MS, 63, PhD(physics), 65. *Prof Exp:* Mem tech staff, Thompson-Ramo-Wooldridge Corp, 58-59; lectr math, 61-63, vis asst prof, 65-66, from asst prof to assoc prof, 66-75, PROF PHYSICS, UNIV SOUTHERN CALIF, 75- *Concurrent Pos:* Consult, Space Physics Dept, Thompson-Ramo-Wooldridge Corp, 60-71, Douglas Aircraft Co, 61 & Planetary Atmospheres Adv Subcomt, Space Sci & Applns Steering Comt, NASA, 69-70. *Honors & Awards:* NASA Except Scientific Achievement Medal & NASA Pub Serv Group Achievement Award to Pioneer 10 Scientific Instrument Team, 74. *Mem:* Am Geophys Union; Am Phys Soc. *Res:* Space physics and spectroscopy. *Mailing Add:* Dept Physics Space Sci Ctr SHS 274 Univ Southern Calif Univ Park Los Angeles CA 90089

JUDGE, FRANK D, NUCLEAR ENGINEERING. *Prof Exp:* GEN MGR NUCLEAR OPERS, GEN ELEC CO, 87- *Mem:* Nat Acad Eng. *Mailing Add:* Gen Elec Co 175 Curtner Ave M/C 802 San Jose CA 95125

JUDGE, JOSEPH MALACHI, b Carbondale, Pa, June 10, 30; m 57; c 5. POLYMER CHEMISTRY, INFORMATION RETRIEVAL. *Educ:* Kings Col, Pa, BS, 52; Univ Notre Dame, PhD(chem), 58. *Prof Exp:* Res chemist, Polychems Dept, E I du Pont de Nemours & Co, 55-58; res chemist, Armstrong Cork Co, 58-80, mgr, Tech Info Serv, 80-86 & Technol Transfer, 86-89, TECH INFO CONSULT, ARMSTRONG WORLD INDUST, INC, 89- *Mem:* AAAS; Am Soc Info Sci; Am Chem Soc. *Res:* Vinyl polymerization; polymeric blends; polyvinyl chloride modifications; structure to dynamic properties relationships; elastomer synthesis; high energy radiation; information retrieval. *Mailing Add:* 436 Manor View Dr Millersville PA 17551

JUDGE, LEO FRANCIS, JR, b Washington, DC, Jan 6, 27; m 49; c 4. MICROBIOLOGY. *Educ:* Univ Md, BS, 53, MS, 55, PhD(bact), 58. *Prof Exp:* Asst bact, Univ Md, 53-57; MGR MICROBIOL SERV, PROCTER & GAMBLE CO, 57- *Mem:* Am Soc Microbiol; Soc Indust Microbiol; Am Soc Testing & Mat; Cosmetic Toiletry & Fragrance Asn. *Res:* Bacterial metabolism and physiology; microbial associations; medical microbiology; antiseptics and disinfectants. *Mailing Add:* 5001 Kellogg Ave No C7 Cincinnati OH 45228

JUDGE, MAX DAVID, b Shirley, Ind, Oct 14, 32; m 53; c 3. ANIMAL SCIENCE, FOOD SCIENCE. *Educ:* Purdue Univ, BS, 54, PhD(animal physiol), 62; Ohio State Univ, MSc, 58. *Prof Exp:* From instr to assoc prof, 58-68, PROF ANIMAL SCI, PURDUE UNIV, 68- *Concurrent Pos:* Res fel, Univ Wis, 64-65. *Mem:* Am Meat Sci Asn; Am Soc Animal Sci; Inst Food Technol. *Res:* Physiological and endocrine control of muscle properties and subsequent utilization of muscle as a food. *Mailing Add:* Dept Animal Sci Purdue Univ Smith Hall West Lafayette IN 47907

JUDGE, ROGER JOHN RICHARD, b London, Eng, Nov 22, 38; US citizen; m 74; c 2. AERONAUTICAL ENGINEERING, ASTRONAUTICAL ENGINEERING. *Educ:* Univ Nottingham, BSc, 60, PhD(physics), 64. *Prof Exp:* Staff scientist, Bell Can Labs, 63-67; fel, Nat Res Coun Can, 67-69; res physicist, Univ Calif, San Diego, 69-75; prin engr, Orincon Corp, 78-79; STAFF SCIENTIST, IRT CORP, 79- *Concurrent Pos:* Vis prof, Univ Ottawa, 67-69 & Univ Calif, San Diego, 75- *Mem:* Am Geophys Union. *Res:* Upper atmosphere and space physics; spacecraft technology; nuclear survivability. *Mailing Add:* 1301 Virginia Way La Jolla CA 92037

JUDIS, JOSEPH, b Toledo, Ohio, Sept 23, 29; m 55; c 2. BIOCHEMICAL PHARMACOLOGY. *Educ:* Univ Toledo, BS, 49; Purdue Univ, MS, 51, PhD(bact), 54. *Prof Exp:* Res assoc microbiol, Sch Med, Western Reserve Univ, 53-55; res bacteriologist, Toledo Hosp Inst Med Res, 55-56; chmn dept biol, 64-66, dean, Col Pharm, 66-76, prof pharm, Col Pharm, Univ Toledo, 62-84, prof biol, Col Arts & Sci, 64-84; RETIRED. *Mem:* AAAS; Am Soc Microbiol; Am Chem Soc; Am Pharmaceut Asn; Brit Soc Gen Microbiol. *Res:* Microbial physiology. *Mailing Add:* 17 Hanasi St Apt 10 Hadera 38423 Israel

JUDISH, JOHN PAUL, b Canonsburg, Pa, May 23, 26; m 58; c 3. CHEMICAL PHYSICS. *Educ:* Univ Pittsburgh, BS, 50; Univ Tenn, PhD(physics), 74. *Prof Exp:* Engr Van de Graaff Accelerator, 51-74, RES STAFF MEM PHYSICS, OAK RIDGE NAT LAB, 75- *Mem:* Am Phys Soc; Inst Elec & Electronics Engrs. *Res:* Atomic and molecular physics; visible and vuv spectroscopy; interaction of optical radiation with gases; lasers; excitation and ionization of gases by high energy ions; superconducting resistance at high frequencies. *Mailing Add:* 107 Wendover Circle Oak Ridge TN 37830

JUDKINS, JOSEPH FAULCON, JR, b Richmond, Va, May 12, 38; m 61; c 3. SANITARY ENGINEERING. *Educ:* Va Polytech Inst, BS, 61, MS, 65, PhD(civil eng), 67. *Prof Exp:* Asst prof civil eng, Auburn Univ, 67-71, Gottlieb assoc prof, 71-77, Gottlieb prof civil eng, 77-81; pvt eng pract, 81-89; DIR WATER RESOURCES RES INST, AUBURN UNIV, 89- *Mem:* Am Soc Civil Engrs; Water Pollution Control Fedn; Am Water Works Asn. *Res:* Industrial and domestic waste treatment; water supply engineering. *Mailing Add:* 211 Cary Dr Auburn AL 36830

JUDKINS, RODDIE REAGAN, b Sunbright, Tenn, Dec 31, 41; div; c 3. COAL CONVERSION SYSTEMS. *Educ:* Tenn Polytech Inst, BS, 63, MS, 65; Ga Inst Technol, PhD(phys chem), 70. *Prof Exp:* Eng assoc, Union Carbide Corp, 64, 65 & 66; instr chem, Ga Inst Technol, 65-70; plant mgr, Nuclear Chem & Metals Corp, 70-73; tech assoc, E R Johnson Assocs, Inc, 73-77; develop engr, Union Carbide Corp, 77-81, task leader, 81-84; task leader, Martin Marietta Energy Systs, 84-86; mgr, Fossil Energy Mat Prog Mgr, 86-88, MGR, FOSSIL ENERGY PROG, OAK RIDGE NAT LAB, 88- *Mem:* Am Soc Metals Int; Sigma Xi; Am Soc Mech Engrs. *Res:* Materials of construction for coal conversion and utilization systems; thorium metal process development and improvement; corrosion mechanisms in coal liquefaction processes; nuclear fuel fabrication technology and economics. *Mailing Add:* Oak Ridge Nat Lab Bldg 4508 PO Box 2008 Oak Ridge TN 37830-6084

JUDSON, BURTON FREDERICK, b Boston, Mass, Aug 7, 28; m 51; c 2. CHEMICAL ENGINEERING. *Educ:* Mass Inst Technol, BS, 48. *Prof Exp:* Engr & mgr processing, Hanford Atomic Prods Oper, 48-65, mgr chem eng & plutonium fuels, Vallecitos Nuclear Ctr, 65-72, plant mgr, Midwest Fuel Recovery Oper, 72-75, mgr advan eng, Nuclear Energy Progs, 75-77, VPRES & MGR ENG, GEN ELEC URANIUM MGT CORP, GEN ELEC CORP, 77- *Mem:* Am Inst Chem Engrs; Am Nuclear Soc; AAAS. *Res:* Reprocessing and refabrication of irradiated nuclear fuels. *Mailing Add:* 159 Vista Delmonte Los Gatos CA 95032

JUDSON, CHARLES LEROY, b Lodi, Calif, Oct 21, 26; m 50; c 3. INSECT PHYSIOLOGY. *Educ:* Univ Calif, BA, 51, PhD(entom), 56. *Prof Exp:* Asst prof, 62-70, assoc prof entom, 70-77, ASSOC, EXP STA, UNIV CALIF, DAVIS, 55-; PROF ENTOM, 77-; VECTOR CONTROL SPECIALIST, CALIF DEPT PUB HEALTH, 55- *Mem:* Entom Soc Am. *Res:* Insect biochemistry; physiology of hatching; mosquito eggs. *Mailing Add:* Dept Entom Univ Calif Davis CA 95616

JUDSON, CHARLES MORRILL, b Washington, DC, July 2, 19; m 44; c 2. ANALYTICAL CHEMISTRY. *Educ:* Swarthmore Col, BA, 40; Univ Pa, MS, 42, PhD(phys chem), 47. *Prof Exp:* Asst instr chem, Univ Pa, 40-42; res chemist, Columbia Univ, 42-44; res chemist, Standard Oil Co (Ind), 44-45; res chemist, Am Cyanamid Co, 47-54, group leader, 54-57, mgr chem physics sect, 57-62; chief anal develop sect, Anal & Control Div, Consol Electrodyn Corp, 62-63; dir eng, 63-70; res scientist, Granville Phillips Co, 70-71; consult mass spectrometry, 71-73; scientist, Analog Technol Corp, 73-76, chief scientist, 76-79; mgr, Microtrace Anal Serv, 75-77; res assoc, Univ Southern Calif, 79-80; dir, Mass Spectrometry Lab, Univ Kans, 80-89; RETIRED. *Mem:* Am Chem Soc; Am Phys Soc; Am Soc Mass Spectrometry; Soc Appl Spectros. *Res:* Mass spectrometry; instruments; radio-tracers; electrolytes; surface agents. *Mailing Add:* 608 Seabrook Pl Lawrence KS 66046

JUDSON, HORACE AUGUSTUS, b Miami, Fla, Aug 7, 41. ORGANIC CHEMISTRY. *Educ:* Lincoln Univ, AB, 63; Cornell Univ, PhD(org chem), 70. *Prof Exp:* Asst prof chem, Bethune-Cookman Col, 69; asst prof, 69-74, PROF CHEM & VPRES ACAD AFFAIRS, MORGAN STATE UNIV, 74- *Mem:* Am Chem Soc; Nat Inst Sci. *Res:* Decomposition mechanisms of organic peroxides and peresters; synthesis of strained ring compounds via peresters. *Mailing Add:* Dept Chem Morgan State Col Box 525 Baltimore MD 21239

JUDSON, SHELDON, JR, b Utica, NY, Oct 18, 18; m 43; c 3. GEOLOGY. *Educ:* Princeton Univ, AB, 40; Harvard Univ, AM, 46, PhD(geol), 48. *Prof Exp:* From instr to assoc prof geol, Univ Wis, 48-55; assoc prof, 55-64, chmn, Dept Geol & Geophys Sci, 70-82, chmn univ res bd, 72-77, Knox Taylor prof, 64-87, EMER PROF GEOL, PRINCETON UNIV, 87- *Concurrent Pos:* Fund Advan Educ fel, 54-55; Guggenheim & Fulbright fels, Italy, 60-61; Guggenheim fel, 66-67. *Mem:* AAAS; Geol Soc Am; assoc Arctic Inst NAm; Sigma Xi. *Res:* Glacial geology; geomorphology. *Mailing Add:* Dept Geol & Geophys Sci Princeton Univ Princeton NJ 08544

JUDSON, WALTER EMERY, b Roxbury, Mass, June 5, 16; m 43; c 3. MEDICINE. *Educ:* Tufts Univ, BS, 38; Johns Hopkins Univ, MD, 42; Am Bd Internal Med, dipl, 50; Am Bd Cardiovasc Dis, cert, 50. *Prof Exp:* Asst med, Sch Med, Boston Univ, 48-49, from instr to asst prof, 50-55; assoc prof, 56-65, PROF MED, SCH MED, UNIV IND, INDIANAPOLIS, 65- *Mem:* Fel Am Col Physicians; AMA; Am Heart Asn; Am Fedn Clin Res; Sigma Xi. *Res:* Cardiovascular research. *Mailing Add:* Riley Res Wing Sch Med Ind Univ 702 Barnhill Dr Indianapolis IN 46223

JUENGE, ERIC CARL, b Weehawken, NJ, Jan 12, 27; m 54, 71; c 3. PHARMACEUTICAL & ORGANOMETALLIC CHEMISTRY. *Educ:* NY Univ, BA, 51, PhD(chem), 57. *Prof Exp:* Jr chemist pharmaceut, Hoffmann-La Roche, Inc, 51-52; asst, NY Univ, 55-56; res chemist organometallic field, Ethyl Corp, La, 56-59; sr asst agr chem, Spencer Chem Co, 59-61; from assoc prof to prof chem, Kans State Col, Pittsburg, 61-74; res assoc, Coop State Res Serv, Ft Valley State Col, 74-75, Asst res prof chem, 75-78; chemist, Div Drug Anal, Food & Drug Admin, 78-89; RETIRED. *Mem:* Am Chem Soc. *Res:* Organic chemistry; agricultural chemistry; chemical nitrogen fixation; methods research in pharmaceutical chemistry. *Mailing Add:* Div Drug Analysis 1114 Market St St Louis MO 63101

JUERGENS, JOHN LOUIS, b Mankato, Minn, Mar 29, 25; m 48; c 4. INTERNAL MEDICINE. *Educ:* Univ Minn, Minneapolis, BS, 46, MS, 56; Harvard Univ, MD, 49. *Prof Exp:* Asst prof med, 62-67, assoc prof clin med, 67-78, PROF MED, MAYO SCH MED, UNIV MINN, 78-; CONSULT CARDIOVASC DIS & INTERNAL MED, MAYO CLIN, 56- *Concurrent Pos:* Fel internal med, Mayo Grad Sch Med, Univ Minn, 53-56. *Mem:* Fel Am Col Physicians. *Res:* Peripheral vascular diseases. *Mailing Add:* 2115 Merrihills Dr SW Rochester MN 55902

JUERGENSMEYER, ELIZABETH B, b Columbia, Mo, May 28, 40; m 63; c 2. CELL BIOLOGY, PROTOZOOLOGY. *Educ:* Ore State Univ, BS, 62; Univ Ill, Urbana, MS, 64, PhD(biol), 67. *Prof Exp:* Asst, Univ Ill, Chicago Circle, 65-68; asst prof, Harper Col, 68-69; assoc prof, 69-80, PROF BIOL & FAC MODERATOR, JUDSON COL, ILL, 80- *Mem:* AAAS; Am Inst Biol Sci; Soc Protozool; Genetics Soc Am; Am Soc Microbiol. *Res:* Comparative genetics of tetrahymena. *Mailing Add:* Dept Biol Judson Col Elgin IL 60123

JUGENHEIMER, ROBERT WILLIAM, agronomy; deceased, see previous edition for last biography

JUHASZ, STEPHEN, b Budapest, Hungary, Dec 26, 13; nat US. MECHANICS. *Educ:* Royal Inst Technol, Budapest, dipl Ing, 36; Royal Inst Technol, Sweden, MSc, 49, Tekn Lic, 51. *Prof Exp:* Mgr & engr, Oeconomia Ltd, Combustion & Salgotarjan Coal Mines, Hungary, 36-46; mem staff, Royal Inst Technol, Stockholm, 49-51 & Univ Toronto, 51-52; res assoc, Fuels Res Lab, Mass Inst Technol, 52-53; exec ed, Appl Mech Rev, 53-60, ed, 60-; dir, Southwest Res Inst, 60-84; CONSULT, 84- *Mem:* Fel Am Soc Mech Engrs; Sigma Xi; Am Inst Aeronaut & Astronaut; fel Am Asn Advan Sci. *Res:* Heat transfer; boiler availability; information retrieval. *Mailing Add:* 6100 NW Loop 410 No 402 San Antonio TX 78238

JUHASZ, STEPHEN EUGENE, b Kecskemet, Hungary, Sept 20, 23; Can citizen; m 65. PSYCHIATRY. *Educ:* Med Univ Budapest, MD, 51; McGill Univ, PhD, 62. *Prof Exp:* Instr bact, Med Univ Budapest, 51-54, lectr, 54-56, asst prof, 56; res assoc, Royal Edward Laurentian Hosp, 57; res asst, McGill Univ, 58-62; guest researcher, Univ Lausanne, 62-63; staff researcher, Res Inst Exp Biol & Med, Borstel, WGer, 63-64; asst prof microbiol, Univ BC, 64-66; assoc prof, 67-70, prof microbiol, Stritch Sch Med, Loyola Univ, 70-72; res microbiologist, Vet Admin Hosp, Hines, 67-72, resident psychiat, 72-75; INSTR PSYCHIAT, STRITCH SCH MED, LOYOLA UNIV, 75-; ATTENDING PSYCHIATRIST, ALEXIAN BROS MED CTR, 75- *Mem:* Am Med Asn; Sigma Xi. *Res:* Bacterial cytology; spheroplasts and L forms of Salmonella; morphogenetics of mycobacteria; phage typing and lysogeny in mycobacteria; transduction and transformation in mycobacteria. *Mailing Add:* Alexian Brothers Med Plaza 850 Biesterfield Rd Suite 3001 Elk Grove Village IL 60007

JUHL, WILLIAM G, b Luverne, Minn, June 30, 24; m 56; c 2. CHEMICAL ENGINEERING. *Educ:* Univ Minn, BChE, 45; Iowa State Col, PhD(chem eng), 53. *Prof Exp:* Res engr, Lion Oil Co, 53-55; group leader res, 55-64, process tech mgr, 65-74, PROCESS TECHNOL DIR, MONSANTO CO, 74- *Mem:* Am Inst Chem Engrs; Am Chem Soc. *Res:* Process development studies of hydrocarbons, conversion and petrochemical production. *Mailing Add:* 203 Millcreek El Dorado AR 71730-2705

JUHOLA, CARL, b Bismarck, NDak, Jan 28, 20; m 44; c 3. ELECTRICAL ENGINEERING. *Educ:* Univ Wash, Seattle, BSEE, 43. *Prof Exp:* Surveyor, US Bur Reclamation, 40-41; plant engr, Boeing Aircraft Co, Seattle, 41-43; eng exec, electronic develop, United Shoe Mach Corp, 46-61, dept mgr, indust & mach develop, 61-70; dir, Comput Control Lab, USM Corp, 70-73, mgr, Electronics Lab, 73-81, & Ctr Technol Innovation, 81-84; RETIRED. *Mem:* Sr mem Inst Elec & Electronics Engrs. *Res:* Management; electronic controls; dielectric heating; fasteners; packaging adhesive systems; development of minicomputer and micro-computer control systems; robotics and computer-aided design; computer-aided manufacturing systems. *Mailing Add:* Box 288 Belmont NH 03220-0288

JUKES, THOMAS HUGHES, b Hastings, Eng, Aug 25, 06; nat US; m 42; c 3. BIOCHEMISTRY, NUTRITION. *Educ:* Univ Toronto, BSA, 30, PhD(biochem), 33. *Hon Degrees:* DSc, Univ Guelph, 72. *Prof Exp:* Nat Res Coun fel biochem, Univ Calif, 33-34, instr poultry husb, 34-39, asst prof, 39-42; dir sect nutrit & physiol res, Lederle Labs Div, Am Cyanamid Co, 42-59, dir chem res, Agr Div, 59-63; PROF IN RESIDENCE MED PHYSICS, NUTRIT SCI & RES BIOCHEMIST, SPACE SCI LAB, UNIV CALIF, BERKELEY, 63-, PROF BIOPHYSICS, 81- *Concurrent Pos:* Consult, Chem Warfare Serv, 43-45 & NASA, 69-70; vis sr res fel biochem, Princeton Univ, 62-63; chmn, Interdisciplinary Sci Comn F on Life Sci, Comt Space Res, Int Coun Sci Unions, 78-84; assoc ed, J Molecular Evolution, 86-, biographical ed, J Nutrit, 83- *Honors & Awards:* Borden Award, Poultry Sci Asn, 47; Kenneth A Spencer Award, Am Chem Soc, 76; Agr & Food Chem Award, Am Chem Soc, 79; Bruce F Cain Mem Award, Am Assoc for Cancer Res, 87; Klaus Schwarz Medal, Int Asn Bioinorganic Sci, 88. *Mem:* Am Soc Biol Chem; Soc Exp Biol & Med; fel Am Soc Animal Sci; fel Poultry Sci Asn; fel Am Inst Nutrit; Sigma Xi. *Res:* Vitamin B complex; choline; pantothenic and folic acid; vitamin B12 and antibiotics in nutrition; folic acid antagonists; proteins; genetic code; biochemical evolution; molecular evolution; trace elements; molecular evolution. *Mailing Add:* Space Sci Lab Univ Calif Berkeley CA 94720

JULES, LEONARD HERBERT, b Cleveland, Ohio, Oct 5, 22; m 53; c 2. ORGANIC CHEMISTRY. *Educ:* Univ Southern Calif, AB, 48, MS, 49. *Prof Exp:* Res chemist, Sahyun Labs, 50-55; res chemist, Purex Corp, 55-58; res chemist, Nat Res & Chem Co, 58-62; res chemist, Philip A Hunt Chem Corp, 62-69, plant mgr, 69-85; RETIRED. *Concurrent Pos:* Consult, 85- *Mem:* Am Chem Soc; Soc Photog Sci & Eng. *Res:* Pharmaceuticals; organic intermediates; organic chlorine bleaches; surfactants; asphalt additives; corrosion inhibitors; photographic chemicals. *Mailing Add:* 6035 Wooster Ave Los Angeles CA 90056

JULESZ, BELA, b Budapest, Hungary, Feb 19, 28; US citizen; m 53. VISION, PHYSIOLOGICAL OPTICS. *Educ:* Budapest Tech Univ, dipl, 50; Hungarian Acad Sci, Dr Ing, 56. *Prof Exp:* Asst prof, dept tel commun, Budapest Tech Univ, 50-51; res engr microwave syst, Inst Telecommun Res Inst, Budapest, 51-56; mem tech staff, AT&T Bell Labs, 56-64, head, Sensory & Perceptual Processes Dept, 64-83 & Visual Perception Res Dept, 83-89; STATE NJ PROF PSYCHOL & DIR, LAB VISION RES, RUTGERS UNIV, BUSCH CAMPUS, 89- *Concurrent Pos:* Vis prof exp psychol, Mass Inst Technol, 69, Fed Tech Univ, Zurich, 75-76, vis prof, biol dept, Calif Tech Inst, Pasadena, 85-; Fairchild distinguished scholar, Calif Inst Technol, 77-79 & 87; neurosci assoc, Neurosci Inst, 82; MacArthur Found fel, 83-88. *Honors & Awards:* Dr H P Heinken Prize, Royal Neth Acad Arts & Sci, 85; Karl Spencer Lashley Award, Am Philos Soc, 89. *Mem:* Nat Acad Sci; Inst Elec & Electronics Engrs; fel Optical Soc Am; Psychonomic Soc; fel Am Acad Arts & Sci; fel AAAS; fel Soc Exp Psychologists; hon mem Hungarian Acad Sci. *Res:* Visual perception; binocular depth perception; pattern recognition; optical data processing; psychophysics and neurophysiology of vision; mathematical models; clinical problems of strabismus. *Mailing Add:* Lab Vision Res Rutgers Univ 41 Gordon Rd-Kilmer Campus New Brunswick NJ 08903

JULIAN, DONALD BENJAMIN, b Pelham, Mass, June 6, 22; m 45; c 3. PHOTOGRAPHIC CHEMISTRY, CHEMICAL MICROSCOPY. *Educ:* Univ Mass, BS, 45. *Prof Exp:* Jr chemist, Eastman Kodak Res Labs, 45-48, from res chemist to sr res chemist, 48-62, res assoc chem, 62-74, sr res assoc, 74-80; RETIRED. *Mem:* Am Chem Soc; Soc Photog Scientists & Engrs. *Res:* Determination of image structure of color photographic films and papers by optical microscope methods; dispersions of oil soluble couplers and other components in aqueous gelatin; chemistry and physics of color photography. *Mailing Add:* 1079 Shoemaker Rd Webster NY 14580

JULIAN, EDWARD A, pharmacy; deceased, see previous edition for last biography

JULIAN, GLENN MARCENIA, b Knoxville, Tenn, Oct 1, 39; m 68; c 3. NUCLEAR PHYSICS. *Educ:* Carnegie-Mellon Univ, BS, 61, MS, 63, PhD(physics), 67. *Prof Exp:* Asst prof, 68-72, ASSOC PROF PHYSICS, MIAMI UNIV, 72- *Mem:* Am Phys Soc; Am Geophys Union; Sigma Xi. *Res:* Studies of nuclear structure by means of gamma and beta radiation. *Mailing Add:* Dept Physics Miami Univ Oxford OH 45056

JULIAN, GORDON RAY, b Wenatchee, Wash, May 29, 28; m 52; c 2. CHEMISTRY. *Educ:* Univ Utah, BS, 50; Univ Ore, MA, 55, PhD(chem), 60. *Prof Exp:* Res fel pharmacol, Harvard Med Sch, 60-62, res assoc, 62-64; from asst prof to prof chem, Mont State Univ, 64-86; RETIRED. *Concurrent Pos:* NIH res grant, 65-68; foreign vis scientist, Szeged, Hungary, 74-75; vis prof, Friedrich Miescher Inst, Basel, Switz, 75-76. *Mem:* AAAS; Am Chem Soc. *Res:* Biochemical processes at elevated temperature; in vitro protein synthesis; early biochemical events in plant development. *Mailing Add:* 3369 Bear Canyon Bozeman MT 59715

JULIAN, MAUREEN M, b New York, NY, July 3, 39; m 68; c 2. PHYSICAL CHEMISTRY, CRYSTALLOGRAPHY. *Educ:* Hunter Col, AB, 61; Cornell Univ, PhD(phys chem), 66. *Prof Exp:* Chemist, Univ Col, Univ London, 66-68; sr chemist, Kirtland AFB, NMex, 69-70, consult chem, 71-73; head dept sci, Wood Lawn Sch, 74-; sr scientist, Environ Testing Systs, Inc, Roanoke, Va, 79-; AT DEPT GEOL SCI, VA POLYTECH & STATE UNIV. *Concurrent Pos:* Vis asst prof, Va Polytech Inst, Blacksburg, 78-; asst prof chem, Hollins Col, Va, 78-81. *Mem:* Am Chem Soc; Am Phys Soc; Am Crystallog Asn; AAAS; Royal Soc Chem. *Res:* Zeolites, optical crystallography of anthracene; solid state reactions; photodimerization of anthracene; history of crystallography; teaching off crystallography for undergraduates; symmetry; thermodynamics; computers; software for environmental chemistry. *Mailing Add:* 3863 Red Fox Dr Roanoke VA 24017

JULIAN, WILLIAM H, b Chicago, Ill, Sept 27, 39. MATHEMATICS. *Educ:* Mass Inst Technol, BS, 61, PhD(math), 65. *Prof Exp:* PROF MATH, NMEX STATE UNIV, 69- *Mem:* Am Math Soc; Sigma Xi. *Res:* Research in mathematics as applied to astrophysics; rotation of Halley's Comet. *Mailing Add:* Math Dept NMex State Univ Las Cruces NM 88003

JULIANO, RUDOLPH LAWRENCE, b New York, NY, July 18, 41; m 63; c 2. CELL BIOLOGY, BIOPHYSICS. *Educ:* Cornell Univ, BS, 63; Univ Rochester, PhD(biophys), 70. *Prof Exp:* Engr, Radio Corp Am, 63-64; sci teacher, US Peace Corps, Philippines, 64-66; cancer res scientist cell biol, Roswell Park Mem Inst, 70-72; investr cell biol, Res Inst, Hosp Sick Children, Toronto, 72-80; ASSOC PROF PHARMACOL, UNIV TEX MED SCH, HOUSTON, 78- *Concurrent Pos:* Asst prof, Dept Med Biophys, Univ Toronto, 73- *Mem:* AAAS; Biophys Soc; Can Biochem Soc. *Res:* Surface proteins of mammalian cells; drug delivery systems. *Mailing Add:* Dept Pharmacol Univ NC Med Sch Chapel Hill NC 27514

JULIEN, HIRAM PAUL, b Syracuse, NY, Oct 21, 29; m 51, 83; c 4. PHYSICAL CHEMISTRY. *Educ:* DePauw Univ, AB, 51; Mass Inst Technol, PhD(phys chem), 55. *Prof Exp:* Asst phys chem, Mass Inst Technol, 51-55; res chemist, Prod Res, Esso Res & Eng Co, 55-58, group head, 58-59; mgr, Adv Studies Dept, Bonded Abrasives Div, Carborundum Co, 59-61, Mgr, Develop Dept, 62-64, mgr, Ceramics & Metall Dept, Res & Develop Div, 64-67; mgr, AdvN Technol & Testing Dept, Jim Walter Res Corp, 67-81. *Mem:* Am Chem Soc. *Res:* Thermodynamics; automotive and jet fuels; bonded abrasives; high temperature materials and composites; building materials; cellular plastics; mineral fibers; inorganic fillers. *Mailing Add:* 700 Starkey Rd No 822 Largo FL 34641-2302

JULIEN, HOWARD L, b Oak Park, Ill, Dec 13, 42; m 65; c 2. HEAT TRANSFER, FLUID MECHANICS. *Educ:* Ill Inst Technol, BS, 64, MS, 67; Stanford Univ, PhD(heat & mass transfer), 69. *Prof Exp:* From assoc sr res engr to sr res engr, Gen Motors Res Labs, 69-77; prin engr, Kaiser Engrs, Inc, 77-88; ASSOC PROF MECH ENG, UNIV NMEX, 89- *Mem:* Am Soc Mech Engrs; Sigma Xi; Soc Automotive Engrs; Am Nuclear Soc. *Res:* Basic experimental/analytical convective heat transfer research; gas turbine heat transfer; experimental fluid mechanics; thermodynamic cycle analysis of alternative automotive power plants; heat transfer in nuclear waste processing and storage facilities; cooling of solid state lasers. *Mailing Add:* Dept Mech Eng Univ NMex Box 30001 Dept 3450 Las Cruces NM 88003

JULIEN, JEAN-PAUL, b Quebec, Que, Jan 15, 18; m 55; c 3. FOOD CHEMISTRY. *Educ:* Univ Montreal, BA, 39; Laval Univ, MSc, 45. *Prof Exp:* Prof dairy sci, Dairy Sch, St Hyacinthe, PQ, 45-57; tech dir, Dalpe & Bros, Vercheres, 57-60; prof biochem, Oka Agr Inst, 60-62; prof, forrd chem, Laval Univ, 62-84, dir, Dept Food Sci, 70-80; PLANING RES & DEVELOP, ROSELL INST INC, MONTREAL, CAN, 84- *Mem:* Inst Food Technol; Can Inst Food Sci & Technol. *Res:* Dairy technology relating to lactic cultures. *Mailing Add:* Dept French Univ Sask Saskatoon SK S7N 0W0 Can

JULIEN, LARRY MARLIN, b Nora Springs, Iowa, Aug 16, 37; m 59; c 2. PHYSICAL CHEMISTRY. *Educ:* Wis State Univ, River Falls, BS, 62; Univ Iowa, MS, 65, PhD(chem), 66. *Prof Exp:* Asst prof, 66-73, ASSOC PROF CHEM, MICH TECHNOL UNIV, 73- *Mem:* Am Chem Soc. *Res:* Thermodynamics; molecular spectroscopy and wood-bark surface properties. *Mailing Add:* Dept Chem & Chem Eng Mich Technol Univ Houghton MI 49931

JULIEN, ROBERT M, b Port Townsend, Wash, Mar 24, 42. PHARMACOLOGY, ANESTHESIOLOGY. *Educ:* Univ Wash, Seattle, PhD(pharmacol), 70; Univ Calif, MD, 77. *Prof Exp:* Anesthesiologist, Ore Health Sci Univ, 80-83; STAFF ANESTHESIOLOGIST, ST VINCENT HOSP & MED CTR, 83- *Mailing Add:* Dept Anesthesiol St Vincent Hosp & Med Ctr 9205 SW Barnes Rd Portland OR 97225

JULIENNE, PAUL SEBASTIAN, b Spartanburg, SC, May 8, 44; m 68; c 2. THEORETICAL SPECTROSCOPY, SCATTERING THEORY. *Educ:* Wofford Col, BS, 65; Univ NC, Chapel Hill, PhD(chem), 69. *Prof Exp:* Nat Acad Sci-Nat Res Coun res assoc, Nat Bur Standards, 69-71, res chemist quantum chem, 71-73; res physicist, Plasma Physics Div, Naval Res Lab, Washington, DC, 73-74; res chemist, Phys Chem Div, 74-77, RES CHEMIST, MOLECULAR SPECTROS DIV, NAT BUR STANDARDS, 77- *Mem:* Am Phys Soc; Am Geophys Union. *Res:* Molecular spectroscopy, photodissociation, line broadening theory; atomic collision theory. *Mailing Add:* B268 Physics Bldg Nat Bur Standards & Technol Gaithersburg MD 20899

JULINAO, PETER C, b New Kensington, Pa, Oct 10, 41; m 65; c 3. POLYMER CHEMISTRY. *Educ:* St Vincent Col, BS, 63; WVa Univ, MS, 65; Univ Akron, PhD(polymer sci), 68. *Prof Exp:* Chemist, 68-71, tech coordr, 71-72, MGR RES & DEVELOP, POLYMER PHYSICS & ENG BR, GEN ELEC CO, 76- *Concurrent Pos:* Chmn, Gordon Conf Elastomers, 81. *Mem:* Acct Control Syst; Soc Plastic Engrs. *Res:* Synthesis and properties of block polymers; use of organometallic and organosiloxane intermediates for polymer forming reactions. *Mailing Add:* Res & Develop Ctr Gen Elec Co Glenville NY 12301

JULIUS, STEVO, b Kovin, Yugoslavia, Apr 15, 29; m; c 2. MEDICINE. *Educ:* Univ Zagreb, MD, 53, DMSc, 64. *Hon Degrees:* MD, Univ Goteborg, Sweden, 79. *Prof Exp:* Intern, Univ Zagreb Hosp, 53-54, resident internal med, 55-60; res asst med, Med Ctr, Univ Mich, Ann Arbor, 61-62; sr instr, Univ Zagreb Hosp, 62-64; from instr to assoc prof, 65-74, PROF INTERNAL MED & DIR HYPERTENSION UNIT, MED CTR, UNIV MICH, ANN ARBOR, 74-, PROF PHYSIOL, 80- *Concurrent Pos:* Mem coun Epidemiol & Med Adv Bd-Coun High Blood Pressure Res, Am Heart Asn. *Honors & Awards:* Astra Award, Int Soc Hypertension. *Mem:* Am Col Cardiol; Am Heart Asn; Am Fedn Clin Res; Int Soc Cardiol; Am Physiol Soc; Int Soc Hypertension. *Res:* Hemodynamics of borderline hypertension; pathophysiology of hypertension. *Mailing Add:* Dept Internal Med Univ Mich Hosp 3918 Taubman Ctr Ann Arbor MI 48109-0356

JULIUSSEN, J EGIL, b Stavanger, Norway, May 4, 43; US citizen; m 66; c 1. COMPUTER SCIENCE, INFORMATION SCIENCE. *Educ:* Purdue Univ, BS, 69, MS, 70, PhD(elec eng), 72. *Prof Exp:* Engr, Norden, Div United Technols, 72-73; sr mem tech staff elec eng, Tex Instruments, Inc, 73-; CHMN, FUTURE COMPUT INC. *Mem:* Inst Elec & Electronics Engrs; Asn Comput Mach. *Res:* Memory technologies, magnetic bubble memories, charge-coupled memories, computer system architecture, minicomputers, microprocessors, microprogramming, computer peripheral controllers. *Mailing Add:* 225 Allen Way Incline Village NV 89451

JULL, ANTHONY JOHN TIMOTHY, b Leeds, Eng, Dec 18, 51; Can citizen. RADIOISOTOPE DATING, ACCELERATOR MASS SPECTROMETRY. *Educ:* Univ BC, BSc, 72; Univ Bristol, PhD(chem), 76. *Prof Exp:* Teaching res asst geochem, Org Geochem Unit, Univ Bristol, 75-76; res fel geochem, dept mineral & petrol, Univ Cambridge, 76-79, NATO fel, 77-79; vis scientist geochem, Max Planck Inst Chem, Mainz, 79-81; res assoc, 81-84, RES SCIENTIST GEOCHEM, DEPT GEOSCI, UNIV ARIZ, 84- *Mem:* Am Geophys Union; Meteoritical Soc; Royal Soc Chem; Geol Soc Am. *Res:* Radiocarbon dating by accelerator mass spectrometry; studies of cosmogenic redionuclides in meteorites and terrestial samples; development of Carbon 14 sample preparation methods for accelerator dating; isotope and light element geochemistry. *Mailing Add:* Dept Physics Univ Arizona Bldg No 81 Tucson AZ 85721

JULL, EDWARD VINCENT, b Calgary, Alta, Aug 8, 34; m 65; c 4. ELECTRICAL ENGINEERING. *Educ:* Queen's Univ Ont, BSc, 56; Univ London, PhD(elec eng), 60. *Hon Degrees:* DSc, Univ London, 79. *Prof Exp:* Jr res officer, Radio & Elec Eng Div, Nat Res Coun Can, 56-57; asst prof elec eng, Univ Alta, 60-61; asst res officer, Radio & Elec Eng Div, Nat Res Coun Can, 61-72, assoc res officer, 67-72; assoc prof, 72-80, PROF ELEC ENG, UNIV BC, 80- *Concurrent Pos:* Guest researcher, lab electromagnetic theory, Tech Univ Denmark, 63-65 & microwave dept, Royal Inst Technol, Sweden, 65; chmn, Can comn VI, Int Union Radio Sci, 73-76, chmn Can Mem Comt, 80-86, assoc ed, Radio Science, 80-83; int dir, Electromagnetics Soc, 81-86; vpres, Int Union Radio Sci, 87-90, pres, 90-93. *Honors & Awards:* J T Bolljahn Award, Inst Elec & Electronic Engrs, 65. *Mem:* Fel Inst Elec & Electronics Engrs; Asn Prof Engrs BC. *Res:* Antennas; antenna far-field/near-field prediction; electromagnetic diffraction theory; geometrical theory of diffraction; electromagnetic diffraction gratings; seismic pulse diffraction. *Mailing Add:* Dept Elec Eng Univ BC Vancouver BC V6T 1W5 Can

JULL, GEORGE W(ALTER), b Calgary, Alta, Can, June 22, 29; m 56; c 1. ELECTRONICS ENGINEERING. *Educ:* Univ Alta, BS, 51; Univ London, DIC & PhD(elec eng), 55. *Prof Exp:* Defense scientist, High Frequency Systs, Defense Res Telecommun Estab, 55-71; prog mgr, Info Directorate, 71-75; SR CONSULT, COMMUN RES DIRECTORATE, 75- *Mem:* Optical Soc Am; Inst Elec & Electronics Engrs. *Res:* Advanced communications systems techniques; coherent optical systems; new voice, video and data systems analysis; telecommunications policy research. *Mailing Add:* 72 Stinson Ave Ottawa ON K2H 6N4 Can

JULLIEN, GRAHAM ARNOLD, b Wolverhampton, Eng, June 16, 43; m 70; c 2. SIGNAL PROCESSING, ELECTRONIC SYSTEMS. *Educ:* Loughborough Univ Technol, BTech, 65; Univ Birmingham, MS, 67; Univ Aston, PhD(elec eng), 69. *Prof Exp:* Engr, English Elec Co, 65-66; res asst elec eng, Univ Aston, 67-69; from asst prof to assoc prof, 69-78, PROF ELEC ENG, UNIV WINDSOR, 78- *Concurrent Pos:* Univ grant, Windsor Univ, 69-70, Nat Res Coun Can grant, 69-72, operating grants, 69-, travel fel, 75; sr res engr, Cent Res Labs, EMI Ltd, 75-76; Nat Sci & Eng Res Coun grant, Can, 80-86; pres, Micrel Ltd, 74- *Mem:* Inst Elec & Electronics Engrs; Am Soc Eng Educ. *Res:* Digital signal processing; image processing; high speed digital hardware; microprocessor systems. *Mailing Add:* Dept Elec Eng Univ Windsor 259 Ste Marie St Donnacona Ave GOA 1T0 Windsor ON N9B 3P4 Can

JULYAN, FREDERICK JOHN, b Cleveland, Ohio, May 21, 27; m 60; c 4. ANATOMY, HISTOLOGY. *Educ:* Western Reserve Univ, AB, 50; Ohio State Univ, PhD(zool), 62. *Prof Exp:* Instr zool, Capital Univ, 60-61; from instr to asst prof anat, Ohio State Univ, 62-66; from asst prof to assoc prof, 67-73, prof anat, Chicago Col Osteop Med, 74-75, chmn dept, 73-75; PROF & CHMN DEPT ANAT, KIRKSVILLE COL OSTEOP MED, KIRKSVILLE, MO, 75- *Mem:* AAAS. *Res:* Teaching techniques. *Mailing Add:* RR 1 Box 215E Kirksville MO 63501

JUMARIE, GUY MICHAEL, b Dakar City, Senegal, Mar 18, 39; Can citizen; m 62; c 1. INFORMATION THEORY & APPLICATIONS, INFORMATION SCIENCE & CONTROL SYSTEMS. *Educ:* Univ Lille, D Univ, 72, DSc (physics), 81. *Prof Exp:* Engr missile guidance, SNIA, 62-66; res engr aeronauts, 66-68; group leader, syst eng comput, MATRA, 68-70; assoc prof appl math, 70-76, PROF APPL MATH & STATIST, UNIV QUE, MONTREAL, 76- *Concurrent Pos:* Lectr, Univ Paris, 63-66, invited prof, Nat Inst Statist, Morocco, 74-79, Fed Univ, Rio de Janeiro, 84, Nat Automic Univ Mex, 83, Inst Theoret Physics, Univ Stuttgart, WGer, 87, invited lectr, Int Ctr Theoret Physics, 78 & 87. *Honors & Awards:* Silver Medal, Soc Encouragement Invention Res, 75. *Mem:* Int Asn Cybernetics; Austrian Soc Cybernetic Study. *Res:* All aspects of systems sciences: control, stability, identification, statistics, games and decision making, engineering computer, robotics, communication, fuzzy systems, pattern recognition, artificial intelligence and signal processing; a unified approach via our new relative information theory of patterns and forms. *Mailing Add:* 365 Rue de Chateauguay Apt 252 Longueuil PQ J4H 3X5 Can

JUMARS, PETER ALFRED, b Dinkelsbuhl, Ger, June 3, 48; US citizen; m; c 1. BIOLOGICAL OCEANOGRAPHY. *Educ:* Univ Del, BA, 69; Univ Calif, San Diego, PhD(oceanog), 74. *Prof Exp:* Res assoc oceanog, Allan Hancock Found, Univ Southern Calif, 74-75; from asst prof to assoc prof, 75-83, PROF OCEANOG, UNIV WASH, 83- *Concurrent Pos:* Sci officer, Environ Sci div, Off Naval Res, 80-82; ed, Limnol & Oceanog, 86. *Mem:* AAAS; Am Soc Limnol & Oceanog; Am Statist Asn; Am Geophys Union. *Res:* Biological-physical interactions, deposit feeding, succession, spatial scales of community processes; biology of polychaetes. *Mailing Add:* Sch Oceanog WB-10 Univ Wash Seattle WA 98195

JUMIKIS, ALFREDS RICHARDS, b Riga, Latvia, Dec 7, 07; US citizen; m 34; c 1. SOIL MECHANICS, CIVIL ENGINEERING. *Educ:* Univ Latvia, Riga, CE, 37; Dr EngSc, 42; Vienna Tech Univ, Dr Techn, 63; Univ Stuttgart, Dr Ing, 68. *Prof Exp:* Asst prof eng, mech, Univ Del, 48-52; assoc prof soil mech & found eng, 52-55, prof, 55-73, distinguished prof, 73-78, EMER PROF CIVIL ENG, RUTGERS UNIV, 78- *Concurrent Pos:* Dir joint hwy res proj, Rutgers Univ & NJ State Hwy Dept, 52-55; NSF res grants, 55-69; mem comt on frost action in soils, Hwy Res Bd, Nat Acad Sci-Nat Res Coun, 55-; consult, Jersey Power & Light Co, 62-63; GAF Corp, 70- & Nat Soil Serv, Inc, 70-; mem, Nat Comt on Rock Mech, 71-; mem, US Comt on Large Dams. *Mem:* Am Soc Civil Engrs; Nat Soc Prof Engrs; Int Soc Rock Mech; Am Soc Eng Educ; Sigma Xi. *Res:* Rupture surfaces in soil; thermal soil mechanics; foundation engineering; author or coauthor of more than 110 publications. *Mailing Add:* 817 Hoes Lane Piscataway NJ 08854

JUMP, J ROBERT, b Kansas City, Mo, Feb 15, 37; m 62; c 2. COMPUTER SCIENCE, ELECTRICAL ENGINEERING. *Educ:* Univ Cincinnati, BS, 60, MS, 62; Univ Mich, MS, 65, PhD(comput sci), 68. *Prof Exp:* Elec engr, Avco Corp, 60-61; elec engr, IBM Corp, 62-64; from asst prof to assoc prof,

68-79, PROF ELEC ENG, RICE UNIV, 80-, MASTER, LOVETT COL, 84- *Concurrent Pos:* Assoc ed, Inst Elec & Electronics Engrs Trans Computs, 79-81. *Mem:* Inst Elec & Electronics Engrs; Asn Comput Mach. *Res:* Digital systems design; parallel computing; simulation of computer systems. *Mailing Add:* Dept Elec Eng Rice Univ PO Box 1892 Houston TX 77251

JUMP, JOHN AUSTIN, b Easton, Md, Dec 28, 13; m 43; c 2. MYCOLOGY, PLANT PATHOLOGY. *Educ:* Swarthmore Col, AB, 34; Univ Pa, PhD(bot), 38. *Prof Exp:* Instr biol, Md State Teachers Col, Frostburg, 38-43; mem res dept, Jos E Seagram & Sons, 43-44; res assoc, Univ Pa, 44-46; from asst prof to assoc prof biol, Univ Notre Dame, 46-58; chmn dept, 58-78, prof biol, 58-78, EMER PROF, ELMHURST COL, 78- *Res:* Plant mimicry; Mesembryanthemacea. *Mailing Add:* 230 Larch Ave Elmhurst IL 60126

JUMP, LORIN KEITH, b Bloomington, Ill, Mar 7, 28; m 51; c 4. GENETICS, PLANT BREEDING. *Educ:* Univ Ill, BS & MS, 51. *Prof Exp:* Corn breeder, 51-56, mgr res cent corn belt, 57-62, asst mgr res opers, 63-67, mgr hybrid corn res, 68-70, assoc dir co res, 70-75, assoc res dir, 75-77, MGR SEED PROD RES & FOUND SEED, FUNK SEEDS INT, 78-, MGR PROD DEVELOP, 83- *Concurrent Pos:* Mem, Agr Res Inst, Nat Acad Sci-Nat Res Coun; mem comt intellectual property rights, Am Seed Trade Asn. *Mem:* AAAS; Am Soc Agron; Sigma Xi. *Res:* Development of commercial corn, sorghum and wheat varieties for the United States and the world. *Mailing Add:* Ciba-Geigy Seed Div Bloomington IL 61702

JUMPER, CHARLES FREDERICK, b Prosperity, SC, Nov 4, 34; m 67; c 1. PHYSICAL CHEMISTRY. *Educ:* Univ SC, BS, 56, MS, 57; Fla State Univ, PhD(phys chem), 61. *Prof Exp:* Instr chem, Univ SC, 60-61; res assoc, Bell Tel Labs NJ, 61-62; from asst prof to assoc prof, 62-69, PROF CHEM, THE CITADEL, 69-, HEAD DEPT, 82- *Mem:* Am Chem Soc; Sigma Xi. *Res:* Hydrogen bonding; nuclear magnetic resonance; ion exchange; kinetics of very fast reactions; structure of liquids; vibrational spectroscopy. *Mailing Add:* Dept Chem The Citadel Charleston SC 29409

JUMPER, ERIC J, b Washington, DC, Aug 18, 46; m 67; c 2. GAS DYNAMICS, LASER PHYSICS. *Educ:* Univ NMex, BS, 68; Univ Wyo, MS, 69; Air Force Inst Technol, PhD(mech eng, laser physics), 75. *Prof Exp:* Lab mech engr, Human Eng Div, 6570 Aerospace Med Res Lab, Wright-Patterson AFB, Ohio, 69-72, aerodynamicist fluid dynamics & laser physics, Technol Div, Air Force Weapons Lab, 74-76, assoc prof, Dept Aeronaut, Us Air Force Acad, 76-81, prof, Dept Aeronaut & Astronaut, Air Force Inst Technol, Wright-Patterson AFB, 81-87; LASER PHYSICIST, AIR FORCE WEAPONS LAB, KIRTLAND AFB, 87- *Concurrent Pos:* Co-coodr, Shroud of Turin Proj, 74-84, bd dir, 78-86; consult laser probs, Air Force Weapons Lab, 76-85; ed, Aeronaut Dig, USAF Acad, 78-81, assoc ed, 81-83; consult heat transfer, Broadcast Prod Div, Harris Corp, 79-82. *Mem:* Assoc fel Am Inst Aeronaut & Astronaut; Sigma Xi; Am Soc Eng Educ. *Res:* Supersonic drag predictions; physical chemistry of heterogeneous reactions; gas dynamics of lasers; laser-target interaction physics; unsteady aerodynamics. *Mailing Add:* Dept Aero & Mech Eng Notre Dame IN 46556

JUMPER, SIDNEY ROBERTS, b Gaston, SC, Dec 10, 30; m 54; c 1. RESOURCES, AGRICULTURAL MARKETING. *Educ:* Univ SC, AB, 51, MS, 53; Univ Tenn, PhD(geog), 60. *Prof Exp:* Instr geog, Univ SC, 52-53; teaching asst, Univ Tenn, 55-56, instr, 56-57; from asst prof to prof, Tenn Technol Univ, 57-67, head dept geog, philos & sociol, 62-67; assoc prof, 67-72, PROF GEOG, UNIV TENN, 72-, HEAD DEPT,77- *Concurrent Pos:* Ed, Southeastern Geographer, Asn Am Geographers, 75-79; Chmn Bd, Tenn Geog Alliance, Inc, 86- *Mem:* Asn Am Geographers; Nat Geog Soc; Sigma Xi. *Res:* Marketing of agricultural products, particularly fresh fruits and vegetables; economic geography; departmental administration. *Mailing Add:* Dept Geog Univ Tenn Knoxville TN 37996-1420

JUNCOSA, MARIO LEON, b Lima, Peru, Sept 18, 21; US citizen; m 46; c 3. APPLIED MATHEMATICS, NUMERICAL ANALYSIS. *Educ:* Hofstra Col, BA, 43; Cornell Univ, MS, 45, PhD(appl math), 48. *Prof Exp:* Asst physics, Cornell Univ, 43-47, asst mech, 45-47, instr, 47-48; math, Johns Hopkins Univ, 48-50; mathematician, Exterior Ballistics Lab & Comput Lab, Ballistics Res Labs, Aberdeen Proving Ground, Md, 49-53; mathematician, 53-70, SR MATHEMATICIAN, RAND CORP, 70- *Concurrent Pos:* Instr, Univ Southern Calif Exten, 54; instr & lectr Univ Calif Exten, Los Angeles, 56-58; ed in chief, Asn Comput Mach J, 59-63; consult math & comput sci adv group, AEC, 63-72; chmn satellite tracking accuracy panel, Nat Acad Sci Adv Comt to Air Force Systs Command, 67-69; mem comput adv panel, Nat Ctr Atmospheric Res, 68-; mem adv coun, Southwest Regional Lab Educ Res & Develop, 68-72; consult, Div Comput Res, NSF, 71- *Mem:* Soc Indust & Appl Math. *Res:* Applied math, probability, ordinary and partial differential equations and applications, general numerical and problem analysis, digital computation, mathematical programing and applications, privacy in computerized data banks, and mathematical applications in medicine. *Mailing Add:* Rand Corp Phys Sci Dept 1700 Main St Santa Monica CA 90401

JUNG, CHAN YONG, b Chungju, Korea, Aug 12, 28; US citizen. TRANSMEMBRANE TRANSPORT, INSULIN ACTIONS. *Educ:* Univ Rochester, NY, PhD(biophys), 64. *Prof Exp:* Res assoc radiation biol, Univ Rochester, 64-65; res assox pharmacol, Univ Louisville, 65-67; res asst prof biophys, 67-74, res assoc prof, 74-77, assoc prof, 77-81, PROF BIOPHYS, STATE UNIV NY, BUFFALO, 81-, RES PROF MED, 83-; CHIEF BIOPHYS RES, VET ADMIN MED CTR, BUFFALO, 75- *Mem:* Am Biophys Soc; Am Soc Biol Chem; Am Diabetes Asn; NY Acad Sci; AAAS; Red Cell Club. *Res:* Molecular elucidation of membrane-related functions; transmembrane solute translocation; information transduction; receptor-ligand interactions; insulin-mediated stimulation of glucose transport function in peripheral tissues. *Mailing Add:* Vet Admin Med Ctr 1108C 3495 Bailey Ave Buffalo NY 14215

JUNG, DENNIS WILLIAM, ION TRANSPORT, MITOCHONDRIA. *Educ:* Univ Ill, PhD(biol), 74. *Prof Exp:* RES SCIENTIST PHYSIOL, DEPT PHYSIOL CHEM, OHIO STATE UNIV, 76- *Res:* Heart biochemistry. *Mailing Add:* Dept Physiol Chem Ohio State Univ Columbus OH 43210

JUNG, FREDERIC THEODORE, PHYSIOLOGY. *Educ:* Univ Chicago, PhD(physiol), 25; Northwestern Univ, MD, 32. *Prof Exp:* At dept ed, J AMA; RETIRED. *Mailing Add:* 1555 Oak Ave Evanston IL 60201

JUNG, GERALD ALVIN, b Milwaukee, Wis, July 16, 30; m 57. AGRONOMY, PLANT PHYSIOLOGY. *Educ:* Univ Wis, BS, 52, MS, 54, PhD, 58. *Prof Exp:* Biologist, Chem Warfare Lab, Ft Detrick, Md, 54-56; asst prof agron, WVa Univ, 58-62, assoc prof, 62-67, prof agron & genetics, 67-70; tech adv for soil fertility & physiol & biochem technol, 77-81, RES AGRONOMIST & RES LEADER, US REGIONAL PASTURE RES LAB, AGR RES SERV, USDA, 70- *Concurrent Pos:* Adj prof agron, Pa State Univ. *Honors & Awards:* Res Award, Northeast Br, Am Soc Agron, 84. *Mem:* Fel Am Soc Agron; Am Forage & Grassland Coun; Soc Cryobiol. *Res:* Forage crop adaptation and production on marginal lands; nutritional value of forage crops as influenced by soils and fertilizers; physiology of cold tolerance of alfalfa. *Mailing Add:* US Reg Pasture Res Lab University Park PA 16802

JUNG, GLENN HAROLD, b Lyons, Kans, Oct 11, 24; m 48; c 5. OCEANOGRAPHY, METEOROLOGY. *Educ:* Mass Inst Technol, SB, 49, SM, 52; Tex A&M Univ, PhD(phys oceanog), 55. *Prof Exp:* Asst meteorol, Mass Inst Technol, 50-51, mem staff, Div Indust Co-op, 52; asst oceanog, Tex A&M Univ, 53-54, assoc oceanog & asst prof phys oceanog & phys meteorol, 55-57; assoc prof oceanog, 58-65, prof oceanog, 65-83, EMER PROF, NAVAL POSTGRAD SCH, 83- *Concurrent Pos:* Consult, Texas Co, La, 58 & US Naval Oceanog Off, 65 & 66; vis fac mem, Inst Phys Oceanog, Univ Copenhagen, 71-72, Naval Postgrad Sch, 87-88 & 90-91. *Mem:* Sigma Xi. *Res:* Energy transfer by sea and air, especially across air-sea boundary; oceanographic analysis and forecasting. *Mailing Add:* 25750 Rio Vista Dr Carmel CA 93923

JUNG, HILDA ZIIFLE, b Gretna, La, Sept 7, 22; m 68. PHYSICAL CHEMISTRY, PLASMA PHYSICS. *Educ:* Tulane Univ, BS, 43. *Prof Exp:* Release clerk, Higgins Aircraft Co, 43-44; res physicist, Southern Regional Res Ctr, USDA, 44-79; RETIRED. *Mem:* Sigma Xi; Am Chem Soc; AAAS. *Res:* Physical and physical chemical reactions and properties of natural polymers; kinetics of reactions; crystalline orientation; elasticity; low temperature plasma reactions; application of statistical techniques. *Mailing Add:* 109 Kennedy Dr Gretna LA 70053

JUNG, JAMES MOSER, b Kannapolis, NC, May 25, 28; m 58; c 5. ORGANIC CHEMISTRY. *Educ:* Davidson Col, BS, 49; Univ NC, PhD(org chem), 62. *Prof Exp:* PROF CHEM & CHMN DEPT, CAMPBELL UNIV, 62- *Mem:* AAAS. *Res:* Preparation, properties and cyclization of azomethines. *Mailing Add:* Dept Chem Campbell Univ Buies Creek NC 27506

JUNG, JOHN ANDREW, JR, b Jersey City, NJ, May 3, 38. ORGANIC CHEMISTRY, CATALYSIS. *Educ:* St Peter's Col, BS, 61; Univ Iowa, PhD(org chem), 66. *Prof Exp:* Res chemist, US Army Ballistic Res Labs, Md, 66-67; res chemist, M W Kellogg Co, 67-71; sr res chemist, Chem Systs Inc, 72-81; STAFF CHEMIST, INTERMEDIATES DIV, EXXON CHEM, 81- *Mem:* Am Chem Soc; Catalysis Soc; Sigma Xi. *Res:* Plasticizer research and applications. *Mailing Add:* 6628 Milstone Dr Baton Rouge LA 70808

JUNG, LAWRENCE KWOK LEUNG, b Canton, China, Aug 6, 50; Can citizen; m 77; c 2. IMMUNOLOGY, MEDICAL SCIENCE. *Educ:* Univ Sask, BSc, 71, MD, 75; FRCP(C), 80; Am Acad Pediat, dipl, 80. *Prof Exp:* Sr resident pediatrician, Hosp Sick Children, Toronto, 75-77; sr resident pediatrician, McMaster Univ Med Ctr, Toronto, 77-78, pediat immunol fel, 78-80; res fel, Sloan-Kettering Inst Cancer Res, 80-82; res fel, dept pediat, Okla Children's Mem Hosp, 82-84; sr clin scientist, 82-86, ASST MEM, OKLA MED RES FOUND IMMUNOL PROG, 86-; ASST PROF & CHIEF, CLIN IMMUNOL SERV, OKLA CHILDREN'S MEM HOSP, 86- *Concurrent Pos:* Prin investr, NIH grants, 84- *Mem:* Am Asn Immunologists; Clin Immunol Soc; NY Acad Sci; AAAS; AMA. *Res:* The role of human T cell differentiation antigens on T cell ontogeny, activation and differentiation in normal and immunodeficiency states. *Mailing Add:* Univ Mass Med Sch 55 Lake Ave N Worcester MA 01655

JUNG, MICHAEL ERNEST, b New Orleans, La, May 14, 47; m 69. SYNTHETIC ORGANIC CHEMISTRY, NATURAL PRODUCTS CHEMISTRY. *Educ:* Rice Univ, BA, 69; Columbia Univ, PhD(chem), 73. *Prof Exp:* NATO fel, Swiss Fed Inst Technol, 73-74; from asst prof to assoc prof, 74-83, PROF CHEM, UNIV CALIF, LOS ANGELES, 83- *Concurrent Pos:* Camille & Henry Dreyfus Teacher-Scholar grant, 78-83; Alfred P Sloan Found Res fel, 79-81; Fulbright-Hays grant, US Res Scholar Award, 80-81; mem, Cancer Ctr, Univ Calif, Los Angeles. *Mem:* Am Chem Soc; Royal Chem Soc; Sigma Xi. *Res:* Organic synthesis, particularly of biologically interesting natural products; development of new synthetic methods; chemistry of organic compounds of group IVb metals, such as silicon and tin, and their use in organic synthesis. *Mailing Add:* Dept Chem Univ Calif 405 Hilgard Ave Los Angeles CA 90024

JUNG, RODNEY CLIFTON, b New Orleans, La, Oct 9, 20; m 86. TROPICAL MEDICINE. *Educ:* Tulane Univ, BS, 41, MD, 45, MS, 50, PhD(parasitol), 53. *Prof Exp:* Asst zool, 39-42, asst parasitol, 48-50, from instr to prof trop med, 50-74, CLIN PROF MED, TULANE UNIV, 74- *Concurrent Pos:* Markle scholar, 53-58; dir health, City of New Orleans, 63-70 & 79-; sr vis physician, Charity Hosp, La; consult, US Quarantine Serv. *Honors & Awards:* Geiger Medal Pub Health. *Mem:* Am Soc Trop Med & Hyg; Am Soc Parasitologists; fel Am Col Physicians; Royal Soc Trop Med & Hyg; hon mem Brazilian Soc Trop Med. *Res:* Clinical aspects of parasitic infections. *Mailing Add:* 3600 Chestnut St New Orleans LA 70115

JUNGA, FRANK ARTHUR, b Patchogue, NY, May 15, 34; m 56. SOLID STATE PHYSICS. *Educ:* Yale Univ, BS, 56; Univ Calif, Berkeley, MA, 59, PhD(physics), 63. *Prof Exp:* Grad prog student, 56-63, RES SCIENTIST SOLID STATE PHYSICS, LOCKHEED RES LAB, 63- *Mem:* Am Phys

Soc. *Res:* Basic mechanisms involved in optical and/or electrical phenomena in semiconductors; photoconductivity; photovoltaic effects; laser phenomena; radiation effects in semiconductors; research on surface phenomena in compound semiconductors. *Mailing Add:* B202 Lockheed Res Lab ORG 97-40 3251 Hanover St Palo Alto CA 94304

JUNGALWALA, FIROZE BAMANSHAW, b India, Aug 28, 36; US citizen. BIOCHEMISTRY, NEUROCHEMISTRY. *Educ:* Gujarat Univ, India, BSc, 56, MSc, 58; Indian Inst Sci, Bangalore, PhD(biochem), 63. *Prof Exp:* BIOCHEMIST, EUNICE KENNEDY SHRIVER CTR MENT RETARDATION, INC, 71-; prin res assoc neurol, Harvard Univ, 75-90, ASSOC PROF NEUROSCI, HARVARD MED SCH, 90- *Concurrent Pos:* Res fel physiol chem, Univ Wis-Madison, 64-66; res fel psychiat, Washington Univ, St Louis, 66-68; Nat Multiple Sclerosis Soc res fel, Inst Animal Physiol, Cambridge, Eng, 68-70; assoc ed, J Lipid Res; mem, Neurol Dis Prog Proj Rev, NIH, 85- *Honors & Awards:* Career Develop Award, NIH. *Mem:* Int Soc Neurochem; Am Soc Neurochem; Fedn Am Socs Exp Biol. *Res:* Lipids; role of lipids in membrane formation and organization; cerebral membranes, systhesis function and breakdown of lipids in health and diseases. *Mailing Add:* Dept Biochem E K Shriver Ctr Ment Retardation 200 Trapelo Rd Waltham MA 02254

JUNGAS, ROBERT LEANDO, b Mt Lake, Minn, Sept 25, 34; m 57. BIOCHEMISTRY, PHYSIOLOGY. *Educ:* St Olaf Col, BA, 56; Harvard Univ, PhD(biochem), 61. *Prof Exp:* Instr biochem, 63-65, assoc 65-68, asst prof biol chem, 68-71, assoc prof biol chem, Harvard Med Sch, 71-74; PROF PHYSIOL, UNIV CONN HEALTH CTR, 74-, ASSOC DEAN PRECLIN EDUC, 85- *Concurrent Pos:* USPHS training grant biochem, Harvard Med Sch, 60-63. *Mem:* Am Soc Biol Chemists; Biochem Soc. *Res:* Action of hormones on adipose tissue metabolism. *Mailing Add:* Dept Physiol Univ Conn Health Ctr Farmington CT 06030

JUNGBAUER, MARY ANN, b Phoenix, Ariz, Aug 10, 34; c 1. INORGANIC CHEMISTRY. *Educ:* Immaculate Heart Col, BA, 57; Univ Notre Dame, MS, 61, PhD(inorg chem), 64. *Prof Exp:* Teacher, Pub Schs, 56-59; teaching & res asst chem, Univ Notre Dame, 59-63; instr chem, Immaculate Heart Col, 63-67; asst prof chem, Drew Univ, 67-69; adj assoc prof, 69-72, ASSOC PROF CHEM, BARRY UNIV, 72-, CHMN, DEPT PHYS SCI, 80- *Mem:* Am Chem Soc; Sigma Xi. *Mailing Add:* 704 Jeronimo Dr Coral Gables FL 33146

JUNGCK, GERALD FREDERICK, b Dubuque, Iowa, Mar 1, 29; m 59; c 2. TOPOLOGY. *Educ:* Wartburg Col, BA, 56; Univ Wis, MA, 59; State Univ NY, Binghamton, PhD(math), 78. *Prof Exp:* Instr, 59-61, from asst prof to assoc prof, 61-82, PROF MATH, BRADLEY UNIV, 82-, DEPT CHAIR, 89- *Concurrent Pos:* NSF fac fel, Rutgers Univ, 68. *Mem:* Math Asn Am. *Res:* Fixed point theorems for commuting mappings on metric spaces; local homeomorphisms. *Mailing Add:* Dept Math Bradley Univ Peoria IL 61625

JUNGCK, JOHN RICHARD, b Moorhead, Minn, Aug 17, 44; m 65; c 1. MOLECULAR EVOLUTION. *Educ:* Univ Minn, BS, 66, MS, 68; Univ Miami, PhD(biol & chem), 73. *Prof Exp:* Asst prof biol, Merrimack Col, 71-75; from asst prof to assoc prof, Clarkson Col, 75-79; FROM ASSOC PROF TO PROF & CHMN DEPT BIOL, BELOIT COL, 79- *Concurrent Pos:* Assoc ed, Bulletin of Math Biol, 82-86; ed, Am Biol Teacher, 84-85; Fulbright Scholar, Chiang Mai Univ, Thailand, 85-86; ed midwest bioscene, Asn Midwestern Col Biol Teachers, 88- *Mem:* Soc Study Evolution; Int Soc Study Origins Life; Soc Math Biol; Nat Asn Biol Teachers; Soc Systematic Zool. *Res:* Origins of and mathematical properties of the genetic code; computer analysis of nucleic acid and protein sequences; genetic language; algorithms for analyzing sequence data; philosophy of biology; computer assisted learning-strategic simulations; combinatorics and finite mathematics. *Mailing Add:* Dept Biol Beloit Col Beloit WI 53511

JUNGCLAUS, GREGORY ALAN, b Yankton, SDak, Dec 16, 47; m 66; c 2. ANALYTICAL CHEMISTRY, ENVIRONMENTAL CHEMISTRY. *Educ:* Univ SDak, BA, 70, MA, 72; Ariz State Univ, PhD(chem), 75. *Prof Exp:* Res assoc chem, Mass Inst Technol, 75-77; sr res scientist, Ford Motor Co, 77-78; res chemist, Battelle Columbus Labs, 78-81; SR CHEMIST, MIDWEST RES INST, 81- *Mem:* Am Chem Soc; Am Soc Mass Spectrom; Meteoritical Soc; Sigma Xi. *Res:* Technical direction of projects concerning analytical method development and application; agent demilitarization and installation restoration; gas chromatographic mass spectrometry. *Mailing Add:* Stauffer Chem Co Peiser Labs 8410 Manchester Blvd Houston TX 77012

JUNGE, DOUGLAS, b Milwaukee, Wis, Jan 16, 38; m 60; c 2. BIOPHYSICS, APPLIED MATHEMATICS. *Educ:* Calif Inst Technol, BS, 59; Univ Calif, Los Angeles, PhD(physiol), 65. *Prof Exp:* Asst res zoologist, Scripps Inst Oceanog, 65-67; from asst prof to assoc prof, 67-79, PROF ORAL BIOL, SCH DENT, UNIV CALIF, LOS ANGELES, 79- *Concurrent Pos:* NIH grant, 68-71 & contract, 69-71; NSF grant, 72-74; ed, J Theoret Neurobiol, 81- *Mem:* Am Physiol Soc; Biophys Soc; Soc Neurosci. *Res:* Electrophysiology of excitable membranes; ionic properties of nerve; computer analysis of BMG signals in painful or tender muscles; neurosciences. *Mailing Add:* Sch Dent 73060 CHS Univ Calif 405 Hilgard Ave Los Angeles CA 90024

JUNGER, MIGUEL C, b Dresden, Ger, Jan 29, 23; nat US; m 60; c 2. ACOUSTICS, APPLIED MECHANICS. *Educ:* Mass Inst Technol, BS, 44, MS, 46; Harvard Univ, ScD(appl mech), 51. *Prof Exp:* res fel, Harvard Univ, 51-55; partner, 55-59, pres, 59-90, CHMN BD & PRIN SCIENTIST, CAMBRIDGE ACOUST ASSOC, 90- *Concurrent Pos:* Sr vis lectr, Mass Inst Technol, 68-78; vis prof, Compiegne Technol Univ, France 75 & 77-81; consult, Off Naval Res, London, 75. *Honors & Awards:* Trent-Crede Medal, Acoust Soc Am; Raleigh lectr, Am Soc Mech Engr, 87. *Mem:* Fel Acoust Soc Am; fel Am Soc Mech Engr. *Res:* Physical acoustics, particularly underwater sound; dynamics of elastic systems in an acoustic medium; noise control; mechanical vibrations and shock. *Mailing Add:* Cambridge Acoust Assocs Inc 80 Sherman St Cambridge MA 02140

JUNGERMAN, JOHN (ALBERT), b Modesto, Calif, Dec 28, 21; m 48; c 4. PHYSICS. *Educ:* Univ Calif, AB, 43, PhD(physics), 49. *Prof Exp:* Asst physics, Univ Calif, 43-44, res physicist, Radiation Lab, 44-45, Los Alamos Sci Lab, 45-46 & Lawrence Berkeley Lab, 46-49; AEC fel, Cornell Univ, 49-50; res physicist, Lawrence Berkeley Lab, Univ Calif, 50-51; res physicist, 51-69, dir, Crocker Nuclear Lab, 69-80, chmn, physics dept, 83-87, PROF PHYSICS, UNIV CALIF, DAVIS, 60- *Concurrent Pos:* Assoc prof, Univ Grenoble, France, 72; consult, Int Atomic Energy Agency, Univ Chile, 82. *Mem:* Fel Am Phys Soc; Sigma Xi. *Res:* Charged particle induced fission; beta ray spectroscopy; sector-focused cyclotrons; particle scattering; medical physics; physics and society. *Mailing Add:* Dept Physics Univ Calif Davis CA 95616

JUNGERMANN, ERIC, b Mainz, Ger, Sept 8, 23; US citizen; m 51; c 1. SURFACE ACTIVE AGENTS, ANTIMICROBIAL AGENTS. *Educ:* City Col New York, BS, 49; Polytech Inst Brooklyn, MS, 53, PhD(org chem), 57. *Prof Exp:* Res chemist, Colgate-Palmolive Co, 46-57; sect head chem synthesis, Armour Indust Chem Co, 57-59, mgr res, Soap Div, 59-65, dir household res & develop, Armour & Co, 65-71, vpres res & develop, Armour Dial Inc, 71-75; Dir corp develop, Helene Curtis Indust, 75-78; PRES, JUNGERMANN ASSOCS, INC, 78- *Concurrent Pos:* Assoc ed, J Am Oil Chemists Soc; mem sci adv bd, Lowes Corp, 79-85; mem bd dirs, Lee Pharmaceut Corp, 79-89. *Honors & Awards:* Award of Merit, Am Oil Chemists Soc, 71; Soc of Cosmetic Chemists Award, 75. *Mem:* Am Oil Chemists' Soc; Soc Cosmetic Chemists; Am Chem Soc. *Res:* Soap, detergent and cosmetic technology, both from the fundamental and practical viewpoint; fat based surfactants; fatty acid derivatives; soaps; hair and skin care products; antimicrobials; organophosphorus compounds. *Mailing Add:* 2323 N Central Ave Suite 1001 Phoenix AZ 85004

JUNGKIND, DONALD LEE, b Washington, DC, Apr 16, 43; m 65; c 4. MEDICAL MICROBIOLOGY, CLINICAL LABORATORY ADMINISTRATION. *Educ:* Lamar State Univ, BS, 65; Univ Houston, MS, 68; Univ Tex Med Br Galveston, PhD(microbiol), 72. *Prof Exp:* Fel clin microbiol, Sch Med, Temple Univ, 72-73; from asst prof to assoc prof, 73-88, PROF PATH & MICROBIOL, SCH MED, THOMAS JEFFERSON UNIV, 88-, DIR, CLIN MICROBIOL LAB, 73- *Concurrent Pos:* Pres, Microbiol Consult, Inc, 78- *Mem:* Am Soc Microbiol; Am Soc Clin Path; Asn Clin Scientists. *Res:* Diagnostic microbiology and immunology with emphasis on automation of procedures; microbial physiology and effects of antimicrobial agents on cells; sexually transmitted diseases. *Mailing Add:* Clin Microbiol Lab Thomas Jefferson Univ Hosp Philadelphia PA 19107-4998

JUNGMANN, RICHARD A, b Volklingen, Switz, Oct 29, 28; nat US; m 61. MOLECULAR BIOLOGY. *Educ:* Univ Saarland, 48-50; Univ Basel, PhD, 58. *Prof Exp:* Res staff mem biochem, Worcester Found Exp Biol, Mass, 58-59; dir org res, McGean Chem Co, Ohio, 59-63; res assoc, 63-68, from asst prof to assoc prof biochem, 68-74, PROF BIOCHEM, MED SCH, NORTHWESTERN UNIV, 74-, PROF MOLECULAR BIOL, 80- *Concurrent Pos:* Lectr, Cleveland State Univ, 60-63; sr res investr biochem, Dept Res, Chicago Wesley Mem Hosp, 63-68; consult, McGean Chem Co, 63-64, dept med, Med Sch, Northwestern Univ, 65- *Mem:* AAAS; Am Chem Soc; fel Am Inst Chem; Swiss Chem Soc; Endocrine Soc. *Res:* Mechanism of action of hormones; regulation of nucleoprotein metabolism and gene expression. *Mailing Add:* Northwestern Univ Med Sch 303 E Chicago Ave Chicago IL 60611

JUNI, ELLIOT, b NY, Aug 6, 21; m 44; c 2. BACTERIAL PHYSIOLOGY. *Educ:* City Col New York, BEE, 44; Western Reserve Univ, PhD(microbiol), 51. *Prof Exp:* Asst prof bact, Univ Ill, 51-56; from assoc prof to prof, Sch Med, Emory Univ, 56-66; PROF MICROBIOL, MED SCH, UNIV MICH, ANN ARBOR, 66- *Mem:* AAAS; Am Soc Microbiol; Am Soc Biol Chem. *Res:* Bacterial metabolism; taxonomy. *Mailing Add:* Dept Microbiol Univ Mich Med Sch Sci Bldg 11 Ann Arbor MI 48109

JUNIPER, KERRISON, JR, medicine, for more information see previous edition

JUNK, WILLIAM A(RTHUR), JR, b Uniontown, Pa, Mar 12, 24; m 56; c 2. CHEMICAL ENGINEERING. *Educ:* Pa State Col, BS, 48; Univ Ill, MS, 50, PhD(chem eng), 52. *Prof Exp:* CHEM ENGR, STANDARD OIL CO, 52- *Mem:* Am Chem Soc. *Res:* Physical properties of hydrocarbons; separation processes. *Mailing Add:* 18344 Aberdeen Ave Homewood IL 60430

JUNKER, BOBBY RAY, b San Antonio, Tex, Aug 29, 43; m; c 3. ATOMIC PHYSICS. *Educ:* Univ Southwestern La, BS, 65; Univ Tex, Austin, MA, 67, PhD(chem), 69. *Prof Exp:* Instr chem, Univ Tex, Austin, 69-70; res assoc physics, Univ Pittsburgh, 70-72; asst prof physics, Univ Ga, 72-76; physicist, 77-83, dir, Physic Div, 83-86, DIR MATH, PHYS SCI DIRECTORATE, OFF NAVAL RES, 86- *Mem:* Am Phys Soc; Sigma Xi; AAAS. *Res:* Theoretical atomic physics, including electron-atom and ion- atom collisions. *Mailing Add:* Off Naval Res Code 111 800 N Quincy St Arlington VA 22217

JUNKHAN, GEORGE H, b Peoria, Ill, Jan 30, 29; m 56; c 2. MECHANICAL ENGINEERING. *Educ:* Iowa State Univ, BS, 55, MS, 59, PhD(mech eng, appl mech), 64. *Prof Exp:* From instr to asst prof, 57-66, ASSOC PROF MECH ENG, IOWA STATE UNIV, 66- *Mem:* Am Soc Mech Engrs; Am Soc Eng Educ; Sigma Xi. *Res:* Heat transfer and fluid mechanics, particularly testing of turbomachinery and experimental heat transfer. *Mailing Add:* 320 Riverside Dr Ames IA 50010

JUNKINS, JERRY R, m; c 2. RESEARCH ADMINISTRATION. *Educ:* Iowa State Univ, BS, 59; Southern Methodist Univ, MS, 68. *Prof Exp:* Mem tech staff, Tex Instruments Inc, 59-75, asst vpres, 75-77, vpres, 77-82, exec vpres, 82-84, mem bd dirs, 84-85, PRES & CHIEF EXEC OFFICER, TEX INSTRUMENTS INC, 85-, CHMN, 88- *Concurrent Pos:* Mem, bd dirs, Proctor & Gamble Co & Caterpillar Inc, Nat Adv Coun Semiconductors. *Mem:* Nat Acad Eng. *Res:* Engineering administration. *Mailing Add:* Tex Instruments Inc PO Box 655474 Mail Stop 236 Dallas TX 75265

JUNKINS, JOHN LEE, b Carters, Ga, May 23, 43; m 65; c 2. AEROSPACE ENGINEERING, APPLIED MATHEMATICS. *Educ:* Auburn Univ, BSAE, 65; Univ Calif, Los Angeles, MS, 67, PhD(eng). 69. *Prof Exp:* Aerospace engr, NASA, 61-65; engr & scientist aerospace eng, McDonnell Douglas Astronaut Co, 65-69; assoc prof, Univ Va, 70-78; PROF ENG MECH, VA POLYTECH INST, 78- *Concurrent Pos:* Consult, US Naval Surface Weapons Labs, 70-73, US Army Topog Maps Command, 71-74, Univ Space Res Asn, 75-76, US Engr Topog Lab, 77-, US Defense Mapping Agency, 77-, Univ Va, 78-; assoc ed, J Astronaut Sci, 77- *Mem:* Am Inst Aeronaut & Astronaut; Am Geophys Union; Am Astronaut Soc; Am Soc Photogram. *Res:* Satellite dynamics and control; mathematical modeling of dynamical systems; geophysics; remote sensing. *Mailing Add:* Dept Aeronaut Tex A&M Univ College Station TX 77843

JUO, PEI-SHOW, b Shantung, China, Feb 6, 30; m 65. BIOCHEMISTRY, MOLECULAR BIOLOGY. *Educ:* Taiwan Prov Chung Hsing Univ, BS, 57; Univ Toronto, MS, 63; Univ NH, PhD(virol), 66. *Prof Exp:* Res biochemist, Kitchawan Res Lab, 66-67 & Gulf South Res Inst, 67-68; assoc prof, 68-74, PROF BIOL, STATE UNIV NY COL, POTSDAM, 74- *Mem:* AAAS; Am Soc Microbiol. *Res:* Purification and characterization of proteins and other macromolecules; immunological and electrophoretic analysis of protein antigens. *Mailing Add:* Dept Biol State Univ NY Potsdam NY 13676

JUOLA, ROBERT C, b Astoria, Ore, Aug 8, 40; m 63; c 2. STATISTICS, MATHEMATICS. *Educ:* Univ Ore, BS, 62; Mich State Univ, MS, 64, PhD(statist), 68. *Prof Exp:* Res engr, Boeing Co, 64-66; asst prof math, Univ Tex, Austin, 68-70; assoc chmn dept, 70-77, PROF MATH, BOISE STATE UNIV, 70- *Mem:* AAAS; Inst Math Statist; Am Statist Asn; Sigma Xi. *Res:* Design of experiments; sequential experimentation. *Mailing Add:* Dept Math Boise State Univ Boise ID 83725

JUORIO, AUGUSTO VICTOR, b Buenos Aires, Arg, July 13, 34; m 61; c 2. NEUROPHARMACOLOGY. *Educ:* Univ Buenos Aires, BSP, 58, BSc, 61, PhD(pharmacol), 67. *Prof Exp:* Demonstr pharmacol, Fac Pharm Biochem, Univ Buenos Aires, 59-63; res fel, Inst Animal Physiol, Babraham, Cambridge, Eng, 63-66; asst prof, Fac Pharm Biochem, Univ Buenos Aires, 66-68; res fel Univ Col, Univ London, 68-73; res pharmacologist, 73-77, head basic studies, 77-83, SR SCIENTIST, NEUROPSYCHIAT RES UNIT, UNIV SASK, SASKATOON, SASK, 83- *Mem:* Am Soc Pharmacol & Exp Therapeut; Brit Pharmacol Soc; Int Soc Neurochem; Can Col Neuropsychopharmacology; Pharmacol Soc Can; Soc Neurosci; Int Brain Res Orgn. *Res:* Brain neurotrasnmitters, synthesis, metabolism and functional role; mechanism of action of drugs used to alleviate abnormal mental conditions. *Mailing Add:* Neuropsychiat Res Unit CMR Bldg Univ Sask Saskatoon SK S7N 0W0 Can

JUPNIK, HELEN, b Kenosha, Wis, Sept 30, 15. OPTICS. *Educ:* Univ Wis, BA & MA, 37; Univ Rochester, PhD(physics), 40. *Prof Exp:* Asst physics, Univ Wis, 37; asst instr, Univ Rochester, 37-40; Huff fel, Bryn Mawr Col, 40-41, Berliner res docentship fel, 41-42; asst, Princeton Univ, 42; physicist, Nat Res Corp, 42-43; res physicist, Am Optical Co, 43-48, sr res physicist, 48-58, supvr thin films, 58-59, chief physicist, 59-65, mgr thin film develop, 65-67, MGR THIN FILM TECH, SCI INSTRUMENT DIV, AM OPTICAL CORP, 67- *Mem:* Am Phys Soc; fel Optical Soc Am; Am Vacuum Soc; NY Acad Sci. *Res:* Physical optics; evaporation and properties of thin films; microscopy. *Mailing Add:* 40 Meadowiew Dr Willamsville NY 14221

JURA, MICHAEL ALAN, b Oakland, Calif, Sept 11, 47; m 74; c 1. ASTROPHYSICS. *Educ:* Univ Calif, Berkeley, BA, 67; Harvard Univ, MA, 69, PhD(astron), 71. *Prof Exp:* Res assoc astron, Goddard Space Flight Ctr, NASA, 71 & Princeton Univ Observ, 73-74; asst prof, 74-77, assoc prof, 77-81, PROF ASTRON, UNIV CALIF, LOS ANGELES, 81- *Concurrent Pos:* Alfred P Sloan Found fel, 77-79. *Mem:* Int Astron Union; Am Astron Soc. *Res:* Physics of the interstellar medium and problems in star formation. *Mailing Add:* Dept Astron Univ Calif 405 Hilgard Ave Los Angeles CA 90024

JURAN, JOSEPH M, b Braila, Rumania, Dec 24, 04; US citizen; m 26; c 4. QUALITY MANAGEMENT. *Educ:* Univ Minn, BS, 24. *Hon Degrees:* JD, Loyola Univ, 35, DEng, Stevens Inst Technol, 88. *Prof Exp:* engr, Western Elec Co, 24-41; asst adminr, Foreign Econ Admin, US Govt, 41-45; prof & chmn indust eng, NY Univ, 45-51; consult, 51-79; chmn, 79-87, EMER CHMN, JURAN INST INC, 87- *Mem:* Nat Acad Eng. *Res:* Author, many books on quality management. *Mailing Add:* 11 River Rd Juran Inst Inc Wilton CT 06897-0811

JURAND, JERRY GEORGE, b Gostyn, Poland, Apr 23, 23; US citizen; m 50; c 3. PERIODONTOLOGY, IMMUNOLOGY. *Educ:* Univ Erlangen, DMD, 56; Univ Tenn, Memphis, DDS, 65. *Prof Exp:* Cancer res scientist, Roswell Park Mem Inst, 58-62; res assoc immunol, St Jude Childrens Res Hosp, Memphis, 62-65; assoc prof, 65-70, PROF PERIODONT, COL DENT, UNIV TENN, MEMPHIS, 70- *Concurrent Pos:* Consult, St Jude Childrens Res Hosp, 65- *Mem:* AAAS; NY Acad Sci; Int Asn Dent Res; Am Dent Asn. *Res:* Immunohistochemical identification and localization of antibodies in humans and animals; tumor immunology; immunological reactions in etiology of periodontal disease; growth factors in oro-facial development; collagen in normal and diseased gingiva. *Mailing Add:* Col Dent Univ Tenn Memphis TN 38163

JURASEK, LUBOMIR, b Uzhorod, Czech, June 2, 31; m 57; c 3. PROTEIN CHEMISTRY, MICROBIOLOGY. *Educ:* Purkyne Univ Brno, MSc, 54, PhD(biol), 63. *Prof Exp:* Res officer mycol, State Forest Prod Res Inst, Bratislava, Czech, 54-64; fel enzymol, Nat Res Coun Can, 64-66; res scientist microbiol, State Forest Prod Res Inst, 66-68; res assoc protein chem, Univ Alta, Edmonton, 68-75; RES SCIENTIST MICROBIOL, PULP & PAPER RES INST CAN, 75- *Concurrent Pos:* Lectr, Dept Biochem, Univ Alta, 73-75. *Mem:* Am Soc Microbiol; Soc Indust Microbiol; Am Chem Soc; Can Biochem Soc; Tech Asn Pulp & Paper Indust. *Res:* Molecular mechanism of biological degradation of cellulose, hemicelluloses and lignin; structure and function of xylanases and their production by microorganisms; pulp and paper biotechnology. *Mailing Add:* Pulp & Paper Res Inst Can 570 St John's Blvd Pointe Claire PQ H9R 3J9 Can

JURASKA, JANICE MARIE, b Berwyn, Ill, Feb 9, 49. BIOPSYCHOLOGY. *Educ:* Lawrence Univ, BA, 71; Univ Ill, Champaign, MA, 75; Univ Colo, PhD(biopsychol), 77. *Prof Exp:* Asst prof biopsychol, Dept Psychol, Ind Univ, 80-85, assoc prof, 85; NIMH fel, 78-79, ASSOC PROF DEPT PSYCHOL, UNIV ILL, 86- *Mem:* Int Soc Develop Psychobiol; Soc Neurosci; Int Acad Sex Res. *Res:* Plasticity and development of the nervous system (anatomy) and of behavior; sex differences in the brain. *Mailing Add:* Dept Psychol 603 E Daniel Champaign IN 61820

JURCH, GEORGE RICHARD, JR, b New Britain, Conn, Feb 1, 34; m 61; c 3. PHYSICAL ORGANIC CHEMISTRY. *Educ:* Univ Fla, BS, 57; Univ Ky, MS, 61; Univ Calif, San Diego, PhD(chem), 65. *Prof Exp:* Res chemist, IBM Corp, Ky, 61; NIH fel chem, Yale Univ, 65-66, res assoc, 66; from asst prof to assoc prof, 66-80, PROF CHEM, UNIV SFLA, 80- *Mem:* Am Chem Soc; Sigma Xi. *Res:* Radical-cation intermediates; free radicals; sulfur chemistry, nonaqueous solvent interactions with biological systems; nuclear magnetic resonance conformation studies. *Mailing Add:* Dept Chem Univ SFla Tampa FL 33620

JURD, LEONARD, b Sydney, Australia, Dec 3, 25; US citizen; m 61; c 2. CHEMISTRY. *Educ:* Univ Sydney, BSc, 47; Univ Nottingham, England, PhD(org chem), 53. *Hon Degrees:* DSc, Univ Sydney, 71. *Prof Exp:* Res chemist, 59-70, RES LEADER NAT PROD CHEM, AGR RES SERV, USDA, 70- *Honors & Awards:* Agr Chem Award, Am Chem Soc, 77. *Mem:* Phytochem Soc NAm (pres, 62); Am Chem Soc. *Res:* Isolation and structure of biologically active plant products. *Mailing Add:* 1054 Park Hills Rd Berkeley CA 94708

JURECIC, ANTON, dental research, for more information see previous edition

JURETSCHKE, HELLMUT JOSEPH, b Berlin, Ger, Aug 9, 24; nat US; m 50; c 2. SOLID STATE PHYSICS. *Educ:* Harvard Univ, BS, 44, MA, 47, PhD(physics), 50. *Prof Exp:* From instr to assoc prof, 50-59, actg head dept, 65-66, head dept, 66-77, PROF PHYSICS, POLYTECH UNIV, 59- *Concurrent Pos:* NSF Fac Fel, Grenoble 64-65; vis fel, Royal Melbourne Inst Technol & Univ Melbourne, 85. *Mem:* Fel Am Phys Soc; Am Asn Physics Teachers; Sigma Xi; Am Crystallog Asn. *Res:* Surface properties of metals; electronic structure of metals; dynamical theory of x-rays. *Mailing Add:* 41 Eastern Pkwy Brooklyn NY 11238

JURF, AMIN N, b Syria, Dec 3, 32; US citizen; m 58; c 3. PHYSIOLOGY. *Educ:* Western Md Col, BA, 59; Univ Md, PhD(physiol), 66. *Prof Exp:* Res scientist, 61-66, from instr to asst prof, 66-74, assoc prof renal physiol, 74-76, ASST PROF PHYSIOL, UNIV MD, BALTIMORE CITY, 76- *Res:* Neuro-renal physiology; salt and water metabolism. *Mailing Add:* Dept Physiol Univ Md Sch Med Baltimore MD 21201

JURGELSKI, WILLIAM, JR, b Englishtown, NJ, May 25, 31; m 54; c 2. EXPERIMENTAL CARCINOGENESIS, TOXICOLOGY. *Educ:* Rutgers Univ, BS, 53, MS, 55, PhD(genetics), 58; Duke Univ, MD, 67. *Prof Exp:* Geneticist, Fed Exp Sta in PR, Agr Res Serv, USDA, 57-60; pharmacologist, Food & Drug Admin, 60-63; res assoc path, Med Ctr, Duke Univ, 63-67, intern, 67-68; med officer, Nat Inst Environ Health Sci, 68-; EMERGENCY RM PHYSICIAN, CRAVEN COUNTY HOSP, NEWBERN, NC, 90- *Concurrent Pos:* Fel neuropath, Duke Univ, 67-68. *Mem:* AAAS; Am Asn Pathologists; Am Asn Cancer Res. *Res:* Carcinogenesis and and pediatric cancer; the marsupial as an experimental animal. *Mailing Add:* 3211 Oak Knob Ct Hillsborough NC 27278

JURGENS, MARSHALL HERMAN, b Minden, Nebr, Dec 26, 41; m 64; c 2. ANIMAL SCIENCE & NUTRITION, ANIMAL HUSBANDRY. *Educ:* Univ Nebr, BS, 64, MS, 66, PhD(animal sci), 69. *Prof Exp:* Res asst, animal sci dept, Univ Nebr, 64-67, asst instr, 66; from asst prof to assoc prof, 68-78, PROF, ANIMAL SCI DEPT, IOWA STATE UNIV, 78- *Mem:* Am Soc Animal Sci; Am Asn Feed Microscopists. *Res:* Nutrient-energy relationships in diet of the growing-finishing pig. *Mailing Add:* Iowa State Univ 119 Kildee Hall Ames IA 50010

JURGENSEN, DELBERT F(REDERICK), JR, b St Paul, Minn, Mar 31, 09; m 28; c 3. CHEMICAL ENGINEERING. *Educ:* Univ Minn, BChE, 31, MS, 32, PhD(chem eng), 34. *Prof Exp:* Chemist, Pure Oil Co, 34-35; res supvr, US Gypsum Co, 35-42; chief engr, Chem Warfare Serv Develop Lab, Mass Inst Technol, 42-45; dir chem res & develop & chief engr, Spec Projs Dept, Am Mach & Foundry Co, NY, 46-52; spec asst to vpres res & develop, 60-64; dir eng & develop, Congoleum Nairn, Inc, 52-56; vpres develop & res, Blaw-Knox Co, Pittsburgh, 56-60; CONSULT MGT ENGR, 64- *Mem:* Am Chem Soc; Asn Res Dirs; Am Inst Chemists; Am Inst Chem Engrs. *Res:* Technical administration; heat transfer; development of equipment and processes for industrial promotion; commercial development and research and development administration; chemical process industries. *Mailing Add:* 2700 Coral Springs Dr Coral Springs FL 33065

JURICA, GERALD MICHAEL, b Detroit, Mich, Sept 24, 41; m 78; c 3. ATMOSPHERIC PHYSICS. *Educ:* Univ Detroit, BEE, 63; Univ Ariz, MS, 66, PhD(atmospheric sci), 70. *Prof Exp:* Res assoc, Univ Ariz, 67-70; asst prof, Purdue Univ, 70-75; ASSOC PROF ATMOSPHERIC SCI, TEX TECH UNIV, 75- *Mem:* Am Meteorol Soc; Sigma Xi. *Res:* Satellite meteorology; cloud physical processes. *Mailing Add:* 3708 Elmwood Lubbock TX 79407

JURICIC, DAVOR, b Split, Yugoslavia, Aug 2, 28; m 53; c 1. MECHANICAL SYSTEMS DESIGN, COMPUTER AIDED DESIGN. *Educ:* Univ Belgrade, BSc, 52, DSc, 64. *Prof Exp:* Analyst aircraft struct, Icarus-Belgrade, Yugoslavia, 53-58; res fel aeroelasticity, Inst Aeronaut, Zarkovo-Belgrade, 58-63; from asst prof to assoc prof aeronaut eng, Univ Belgrade, 63-68; from assoc prof to prof mech eng, SDak State Univ, 68-75; vis prof appl mech, Stanford Univ, 75-78; PROF MECH ENG, UNIV TEX, AUSTIN, 78- *Concurrent Pos:* Consult, Elec Power Res Inst, Palo Alto, Flow Res, Inc, Kent, Wash, 76-, Southern Res Inst, Birmingham, Joy Mfg Co, Los Angeles & Dresser Industs, Houston, 78- *Mem:* Am Soc Eng Educ; Ger Soc Appl Math & Mech; Am Soc Mech Engrs; Sigma Xi. *Res:* Aircraft vibration and flutter; dynamics of railway vehicles; dynamics and stability of bipedal locomotion; modeling and simulation of dynamic systems; mechanical systems design; electrostatic precipitators. *Mailing Add:* Dept Mech Eng Univ Tex Austin TX 78712

JURINAK, JEROME JOSEPH, b Cleveland, Ohio, June 3, 27; m 54; c 4. SOIL CHEMISTRY. *Educ:* Colo State Univ, BS, 51; Utah State Univ, MS, 54, PhD(soil chem), 56. *Prof Exp:* Jr soil chemist, Univ Calif, Davis, 56-58, from asst to assoc soil chemist, 58-67; dept head, 78-83, PROF SOIL CHEM, UTAH STATE UNIV, 67- *Concurrent Pos:* Danforth teaching fel; Consult, USAID, UNESCO, Orgn Am States & private indust; vis prof, Rural Univ Rio de Janeiro, Brazil, 75, 77 & Haryana Univ, India; div chmn, Soil Sci Soc Am, 75. *Mem:* Fel Am Soc Agron; fel Soil Sci Soc Am; Int Soc Soil Sci; fel Am Inst Chem; Coun Agr Sci Technol; AAAS. *Res:* Reclamation of disturbed soils and geologic material; trace element chemistry and transport in soils; salt-affected soils; irrigation and return flow water quality; surface chemistry. *Mailing Add:* Dept Plants Soils & Biometeorol Utah State Univ Logan UT 84322-4840

JURINSKI, NEIL B(ERNARD), b Peekskill, NY, Oct 28, 38; m 62; c 3. INDUSTRIAL HYGIENE, ENVIRONMENTAL CHEMISTRY. *Educ:* State Univ NY, Albany, BS, 60; Univ Miss, PhD(phys chem), 63; Am Bd Indust Hyg, cert, 74. *Prof Exp:* Res assoc chem, Mich State Univ, 63-64; asst prof, Boston Col, 64-69; res scientist, Chem Div, J M Huber Corp, 69-72; environ chemist, US Army Environ Hyg Agency, 72-76; sr indust hygienist, SRI Int, 76-77; indust hygienist, Tracor Jitco, Inc, 77-80; PRES, NUCHEM CO, INC, 75- *Concurrent Pos:* Consult, 70- *Mem:* Am Indust Hyg Asn; Am Acad Indust Hyg; Am Chem Soc; Royal Soc Chem; Int Hazard Control Mgt Asn; NY Acad Sci. *Res:* Industrial hygiene surveys and consultation; hazardous and toxic chemical problems; waste disposal and processing; carcinogen/mutagen safety; environmental sampling and analysis; material handling and processing. *Mailing Add:* 9321 Raintree Rd Burke VA 22015

JURIST, JOHN MICHAEL, b Williamsport, Pa, Oct 13, 43; m 64; c 2. BIOPHYSICS. *Educ:* Univ Calif, Los Angeles, AB, 64, MS, 66, PhD(biophys), 68. *Prof Exp:* Proj assoc med physics, 68-69, asst prof orthop surg & space sci & eng, 69-71, asst prof orthop surg, Univ Wis-Madison, 71-76; res assoc, Mont State Univ, 76-77, sr res engr, 77-78, assoc res biophysicist, 78-80, adj pro med sci, 80-82; PRES, CRM INC, 82- *Concurrent Pos:* NASA fel, Univ Wis-Madison, 68-69; prin investr, NIH grants, 70-78; biophysicist, St Vincent Hosp, Billings, 76-82; consult, Deaconess Hosp, Billings, 84-85, Mont State Bd Health, 87- *Mem:* AAAS; Geront Soc; NY Acad Sci; Orthop Res Soc; Aerospace Med Asn. *Res:* Osteoporosis; nondestructive evaluation of skeletal status; application of computers to medical practice; medical thermography. *Mailing Add:* 2520 17th St W No B-4 Billings MT 59102

JURKAT, MARTIN PETER, b Berlin, Germany, July 23, 35; m 58; c 3. MANAGEMENT SCIENCE, MATHEMATICS. *Educ:* Swarthmore Col, BA, 57; Univ NC, MA, 60; Stevens Inst Technol, PhD(math), 72, MEng, 83. *Prof Exp:* Statistician, Marketer's Res Serv, Inc, 59-60; asst engr, Res Lab, Burroughs Corp, 60-61; sr systs analyst, ITT Info Systs Div, 61-64; staff scientist, Davidson Lab, 64-76, dir, Ctr Munic Studies & Servs, 76-78, prof, 78-80, ALEXANDER CROMBIC HUMPHREYS PROF MGT SCI, STEVENS INST TECHNOL, 80- *Concurrent Pos:* Consult, Tank Automotive Command, US Army. *Mem:* Opers Res Soc Am; Comput Soc Inst Elec & Electronics Engrs; Asn Comput Mach. *Res:* Mathematical modelling; simulation; off-road vehicle design and modelling; computer graphics; transportation systems; operations research; expert systems. *Mailing Add:* Dept Mgt Stevens Inst Tech Castle Pt Sta Hoboken NJ 07030

JURKAT, WOLFGANG BERNHARD, b Gerdauen, Ger, Mar 26, 29. MATHEMATICAL ANALYSIS. *Educ:* Univ Tubingen, PhD(math), 50. *Prof Exp:* Asst math, Univ Tübingen, 50-52, dozent, 52-54; assoc prof, Univ Cincinnati, 54-56 & Ohio State Univ, 56-57; assoc prof, 57-58, JOHN RAYMOND FRENCH PROF MATH, SYRACUSE UNIV, 58- *Mem:* Am Math Soc; Ger Math Soc. *Res:* Analysis, particularly summability, Fourier series, analytic number theory, also combinatories and probability. *Mailing Add:* Dept Math Syracuse Univ Syracuse NY 13210

JURKIEWICZ, MAURICE J, b Claremont, NH, Sept 24, 24; m 51; c 2. SURGERY. *Educ:* Univ Md, DDS, 46; Harvard Univ, MD, 52. *Prof Exp:* Instr surg, Sch Med, Washington Univ, 57-59; from asst prof to prof surg, Col Med, Univ Fla, 59-73, chief plastic surg, 64-73; PROF SURG, SCH MED, EMORY UNIV, 73- *Concurrent Pos:* Clin fel plastic surg, Barnes Hosp, St Louis, Mo, 58-59; mem bd sci counr, Nat Inst Dent Res, 66-71; chief surg serv, Vet Admin Hosp, Gainesville, 68-73; lectr & consult, Naval Hosp, Orlando, 69-70; consult plastic surg, Walter Reed Hosp, Washington, DC, 70- *Mem:* AAAS; Am Col Surg; AMA; Am Soc Plastic & Reconstruct Surg; Plastic Surg Res Coun. *Res:* Wound healing; congenital malformations; general and reconstructive surgery; head and neck surgery. *Mailing Add:* 25 Prescott St NE Atlanta GA 30308

JURKUS, ALGIRDAS PETRAS, b Klaipeda, Lithuania, June 11, 35; Can citizen. ELECTRONICS. *Educ:* Univ Montreal, BASc, 57; Univ Sheffield, PhD(elec eng), 60. *Prof Exp:* RES SCIENTIST ELECTRONICS, NAT RES COUN CAN, 60- *Mem:* Inst Elec & Electronics Engrs. *Res:* Establishment and maintenance of primary national standards and development of precise methods of measurement, at high and microwave frequencies, of various electromagnetic quantities. *Mailing Add:* Inst Nat Measurement Standards Nat Res Coun Ottawa ON K1A 0R6 Can

JURMAIN, ROBERT DOUGLAS, b Worcester, Mass, July 20, 48; m 74. PHYSICAL ANTHROPOLOGY. *Educ:* Univ Calif, Los Angeles, AB, 70; Harvard Univ, PhD(anthrop), 75. *Prof Exp:* Asst prof, 75-79, ASSOC PROF ANTHROP, SAN JOSE STATE UNIV, 79- *Mem:* Am Asn Phys Anthropologists. *Res:* Paleopathology of prehistoric human osteological remains; comparative biomechanical studies of primate limb skeletons; paleoanthropological research of early hominids in East Africa. *Mailing Add:* Dept Anthrop San Jose State Univ San Jose CA 95192

JURS, PETER CHRISTIAN, b Oakland, Calif, Apr 13, 43; m 67; c 3. ANALYTICAL CHEMISTRY. *Educ:* Stanford Univ, BS, 65; Univ Wash, PhD(chem), 69. *Prof Exp:* from asst prof to assoc prof, 69-78, PROF CHEM, PA STATE UNIV, 78- *Concurrent Pos:* Prog dir chem anal, Chem Div, NSF, 83-84. *Honors & Awards:* Computers in Chem Award, Am Chem Soc, 90. *Mem:* Am Chem Soc; AAAS; Asn Comput Machinery. *Res:* Computer methods in analytical chemistry; structure-activity studies; computer applications in chemistry; studies of relations between molecular structure and biological activity (pharmacological effects, carcinogenic potential, odor); chemical applications of pattern recognition; computer-assisted structure elucidation. *Mailing Add:* Dept Chem Pa State Univ University Park PA 16802

JURSINIC, PAUL ANDREW, b Joliet, Ill, June 13, 46. BIOPHYSICS. *Educ:* Univ Ill, BS, 69, MS, 71, PhD(biophys), 77. *Prof Exp:* RES SCIENTIST PLANT PHYSIOL, NORTHERN REGIONAL RES CTR, USDA, 77- *Mem:* Am Soc Photobiol. *Res:* The photochemical reactions which make up the light reactions of green plant photosynthesis. *Mailing Add:* 209 W Crestwood Peoria IL 61604

JURTSHUK, PETER, JR, b New York, NY, July 28, 29; m 71; c 2. MICROBIAL PHYSIOLOGY, BIOCHEMISTRY. *Educ:* NY Univ, AB, 51; Creighton Univ, MS, 53; Univ Md, PhD(microbiol), 57. *Prof Exp:* Asst microbiol, Univ Md, 53-56; asst prof pharmacol, Brooklyn Col Pharm, 57-59; fel enzyme chem, Inst Enzyme Res, Univ Wis, 59-63; from asst prof to assoc prof, Univ Tex, Austin, 63-70; assoc prof, 70-76, undergrad chmn biol, 77-81, PROF BIOL, UNIV HOUSTON, 76 - *Honors & Awards:* Distinguished Serv Award, Am Soc Microbiol, 82. *Mem:* AAAS; fel Am Acad Microbiol; Am Soc Microbiol; Am Chem Soc; Am Soc Biol Chem. *Res:* Isolation and characterization of oxidases and oxygenating enzyme complexes from microorganisms; aspects of the microbiological oxidase reaction relating to cellular bioenergetics; microbial taxonomy specifically, on Bacillus and Azotobacter species. *Mailing Add:* Dept Biol Univ Houston Houston TX 77004

JURY, ELIAHU I(BRAHAM), b Bagdad, Iraq, May 23, 23; nat US; m 49; c 2. ELECTRICAL ENGINEERING. *Educ:* Israel Inst Technol, EE, 47; Harvard Univ, MS, 49; Columbia Univ, EngScD, 53. *Hon Degrees:* DrSc Tech, Swiss Fed Inst Tech, Zörich, Switz, 82. *Prof Exp:* From instr to prof elec eng, 53-81, PROF EMER, UNIV CALIF, BERKELEY, 81-; RES PROF EMER, DEPT ELEC & COMPUT ENG, UNIV MIAMI, 88- *Concurrent Pos:* Res engr, Columbia Univ, 53-54; consult, Bell Tel Labs, 56 & Convair Div, Gen Dynamics Corp, 57; vis lectr, Northwestern Univ, 57, Univ Mich, 58, Univ Paris, 58-59, Imperial Col Sci, London, 64-65. *Honors & Awards:* Rufus Oldenburger Medal, Amer Soc Mech Engrs, 86. *Mem:* Fel mem Inst Elec & Electronics Engrs. *Res:* Sampled-data and discrete systems; automatic control; circuit theory; transform methods; information theory; digital and sampled-data control systems. *Mailing Add:* 6039 Collins Ave 520 Miami Beach FL 33140

JURY, WILLIAM AUSTIN, b Highland Park, Mich, Aug 8, 46; m 72. SOIL PHYSICS, ENVIRONMENTAL PHYSICS. *Educ:* Univ Mich, BS, 68; Univ Wis, MS, 70, PhD(physics), 73. *Prof Exp:* Proj assoc soil physics, Dept Soil Sci, Univ Wis, 73-74; from asst prof to assoc prof, 74-82, PROF SOIL PHYSICS, DEPT SOIL & ENVIRON SCI, UNIV CALIF, RIVERSIDE, 82- *Concurrent Pos:* Mem, Nat Comn, Nat Res Coun/Inst Ecol, 74-75; consult, Rockwell Hanford Oper, 76-79 & County of San Diego, 78-79. *Mem:* Am Soc Agron; Soil Sci Soc Am; Int Soil Sci Soc; Am Geophys Union; AAAS. *Res:* Measurement and modeling of water, heat and chemical transport through and reactions in soil. *Mailing Add:* Dept Soil & Environ Sci Univ Calif Riverside CA 92521

JUSINSKI, LEONARD EDWARD, b Oakland, Calif, Aug 12, 55; m 77; c 4. SOLID STATE PHYSICS, PHYSICAL CHEMISTRY. *Educ:* Univ San Francisco, BS, 77. *Prof Exp:* Physicist, USAF Weapons Lab, 77-81; engr, Varian Assocs, 82; PHYSICIST, SRI INT, 83- *Res:* Laser-induced fluorescence; multi-photon ionization; author/co-author of 45 publications. *Mailing Add:* 3432 Little Ct Fremont CA 94538

JUSKO, WILLIAM JOSEPH, b Salamanca, NY, Oct 26, 42; m 64; c 3. PHARMACY. *Educ:* State Univ NY Buffalo, BS, 65, PhD(pharmaceut), 70. *Hon Degrees:* Dr, Med Acad Crakow, 87. *Prof Exp:* Pharmacologist, Vet Admin Hosp, Boston, 70-72; assoc prof, 72-77, vchmn dept, 84-87, PROF PHARMACEUT, SCH PHARM, STATE UNIV NY BUFFALO, 77- *Concurrent Pos:* Dir, Clin Pharmacokinetics Lab, Millard Fillmore Hosp, Buffalo, 72-81; consult, Wyeth Labs, ICI, Cytogen, 75-; Fulbright fel, 78; NIH fel pharmacol, 61-64. *Honors & Awards:* Rawls Palmer Award, ASCPT, 87; Russell Miller Award, Am Col Clin Pharm 88, Distinguished Serv Award, 89. *Mem:* Fel Am Pharmaceut Asn; fel AAAS; Am Soc Clin Pharmacol & Therapeut; Am Soc Microbiol; fel Am Asn Pharmaceut Sci; Am Soc Pharmacol & Exp Therapeut; fel Am Col Clin Pharmacol. *Res:* Basic and clinical pharmacokinetics and pharmacodynamics; biopharmaceutics; drug analysis and pharmacodynamics; chemotherapy. *Mailing Add:* Sch Pharm Dept Pharmaceut State Univ NY Buffalo Buffalo NY 14260

JUST, GEORGE, b Kobe, Japan, May 17, 29; nat Can; m 55; c 3. ORGANIC CHEMISTRY. *Educ:* Swiss Fed Inst Technol, Ing Chim, 51; Univ Western Ont, PhD(chem), 56. *Prof Exp:* Res chemist, Univ Calif, Los Angeles, 56-57 & Monsanto Can, Ltd, 57-58; from asst prof to assoc prof, 58-70, PROF ORG CHEM, McGILL UNIV, 70- *Mem:* Am Chem Soc; fel Chem Inst Can. *Res:* Natural products synthesis. *Mailing Add:* 801 Sherbrooke W Montreal PQ H3A 2K6 Can

JUST, JOHN JOSEF, b Botschar, Jugoslavia, Nov 17, 38; US citizen; m 65; c 3. DEVELOPMENTAL BIOLOGY, ENDOCRINOLOGY. *Educ:* DePaul Univ, Chicago, BS, 62, MS, 64; Univ Iowa, PhD(zool), 68. *Prof Exp:* Fel biochem, Fla State Univ, 68-70; asst prof biol, 70-74, ASSOC PROF BIOL, UNIV KY, 74- *Concurrent Pos:* Schmitt fel, Dept Zool, Univ Iowa, 63-64; NIH fel, Dept Chem, Fla State Univ, 68-70; vis prof, Univ Bern, 76-77. *Mem:* AAAS; Am Zoologist; Sigma Xi. *Res:* Biochemical and morphological methods are used to study hormonal influence on development with particular emphasis on amphibian metamorphosis; physiological triggers for hatching of embryos. *Mailing Add:* Sch Biol Morgan Bldg Univ Ky Lexington KY 40506

JUST, KURT W, b Oels, Ger, May 19, 27; m 53; c 3. THEORETICAL PHYSICS. *Educ:* Free Univ Berlin, Dr rer nat, 54, Dr habil, 58. *Prof Exp:* Asst physics, Free Univ Berlin, 50-60; assoc prof, 61-66, PROF PHYSICS, UNIV ARIZ, 66- *Concurrent Pos:* Sci counsr, Free Univ Berlin, 60-63; consult, Edgerton, Germeshausen & Grier, Nev, 63-64. *Honors & Awards:* First Prize, Gravity Res Found, 65. *Mem:* Int Soc Gen Relativity & Gravitation. *Res:* Quantum field theory; quantized theory of gravity. *Mailing Add:* Dept Physics Univ Ariz Tucson AZ 85721

JUST, RICHARD, b Tulsa, Okla, Feb 18, 48; m 67, 89; c 2. RESOURCE ECONOMICS. *Educ:* Okla State Univ, BS, 69; Univ Calif, Berkeley, MA, 71, PhD(agr economics), 72. *Prof Exp:* Comput programmer, Okla State Univ, 66-69, assoc prof agr economics, 72-75; prof agr econ, Univ Calif, Berkeley, 75-85; PROF AGR ECON, UNIV MD, COL PARK, 85- *Concurrent Pos:* Consult, World Bank, 76, US Gen Acct Off, 79, Safeway Stores, Inc, 83 & Pillsbury, 88. *Honors & Awards:* Res Award, Am Agr Econ, 74, 77, 80, 81, 83 & 89. *Mem:* Am Econ Asn; Economet Soc; Am Agr Econ Asn; Western Agr Econ Asn. *Res:* Welfare economics, distributional effects of agricultural policy; interaction and macroeconomic policy, land price determination, biotechnology, multi output production modeling, effects of exchange rates on agriculture, futures markets. *Mailing Add:* Dept Agr & Resource Econ Univ Md College Park MD 20904

JUSTEN, LEWIS LEO, b Tampa, Fla, June 24, 51; m 71; c 2. PHYSICAL CHEMISTRY, ANALYTICAL CHEMISTRY. *Educ:* Univ SFla, BA, 72; Northwestern Univ, MS, 73, PhD(chem), 78. *Prof Exp:* Res chemist coal anal, Exxon Res & Eng Co, 77-; AT UOP CORP RES CTR. *Mem:* Am Chem Soc. *Res:* Coal science relating to synthetic fuels; thermodynamics of liquids; optical properties of liquids; gas-liquid chromatography. *Mailing Add:* UOP Corp Res Ctr 50 E Algonquin Des Plaines IL 60017

JUSTER, ALLAN, b Albany, NY, Sept 8, 22; m 52; c 1. MECHANICAL ENGINEERING. *Educ:* Pratt Inst, BME, 47; Univ Del, MME, 51. *Prof Exp:* Design engr, Develop Eng Corp, 47-48; instr eng, Univ Conn, 48-51; supvr res, Diamond Ord Fuze Lab, Washington, DC, 51-53; res engr, Armour Res Found, Ill, 53-55; res engr, Sperry Gyroscope Corp, NY, 55-61; sr staff engr, Repub Aviation Corp, 61-66; prof eng technol & chmn dept, NY Inst Technol, 66-70; PROF ENG TECHNOL & CHMN DEPT, FAIRLEIGH DICKINSON UNIV, 70- *Res:* Cryogenics, educational and ultra high vacuum research. *Mailing Add:* PO Box 195 Shohola PA 18458

JUSTER, NORMAN JOEL, b New York, NY, Feb 19, 24; m 60; c 5. ORGANIC CHEMISTRY. *Educ:* Univ Calif, Los Angeles, BA, 43, MS, 47, PhD(org chem), 56. *Prof Exp:* Mem chem fac, John Muir Col, 47-55; dept phys sci, Pasadena City Col, 56-60; sr chemist & head org sect, Motorola, Inc, 60-61, mgr org-polymer labs, Semiconductors Div, 61; prof chem, Pasadena City Col, 61-84, chmn, dept chem & div phys sci, 78-84; PROF CHEM, UNIV CALIF, LOS ANGELES, 84- *Concurrent Pos:* Res dir & consult, Photo Prod Res Lab, E I du Pont de Nemours & Co, 49-52; consult, Witco Chem Co, 53, Silverton & SIlverton, 68-, McInherry Enterprises, 69- & Oncol Div, Med Res Inst Calif, 74-; vis prof chem, Univ Calif, Los Angeles, 59-65, 67-68, 75-77 & Univ Hawaii, 73-74 & 80; Nat Defense Educ Act lectr, 63; res dir & consult, Energy Conversion Devices, 64; vis prof, Univ Ky, 65-66; res dir & consult, Motorola Semiconductor Prod, Inc, 61-; ed jour, Southern Calif Sect, Am Chem Soc, 85- *Honors & Awards:* Sprenger Medal, Am Asn Consult Chemists, 71; Mfg Chemists Asn Award, 74; J Ray Risner Award, 78. *Mem:* AAAS; Am Chem Soc; Asn Consult Chemists & Chem Eng; Soc Plastics Eng. *Res:* Reactions, particularly polymerization of acetylenic studies; reaction mechanisms; organic semiconduction; solid state organic chemistry; electronic structure of molecules; management of cell growth and reproduction with organic semiconductors. *Mailing Add:* Dept Chem & Biochem 513 N Rexford Dr Beverly Hills CA 90210

JUSTESEN, DON ROBERT, b Salt Lake City, Utah, Mar 8, 30; m 58; c 4. NEUROPSYCHOLOGY. *Educ:* Univ Utah, BA, 55, MA, 57, PhD, 60. *Prof Exp:* Asst prof psychol & head dept, Westminster Col, 59-62; CAREER RES SCIENTIST & DIR NEUROPSYCHOL RES LABS, VET ADMIN HOSP, 62-; PROF PSYCHIAT & NEUROPSYCHOLOGY, MED SCH, UNIV KANS, 71- *Concurrent Pos:* From asst prof to prof psychiat, Med Sch, Univ Kans, 63-71; from lectr to prof, Univ Mo-Kansas City, 63-75; vis prof, Univ Colo, 65; consult, NASA, 70-72; mem, Subcomt C-95-1 Safety Stand for Non-ionizing Radiation, Am Nat Stand Inst, 75-89; assoc ed, J Microwave Power & Electromagnetic Energy, 75-88; mem, comn A, nat Acad Sci Int Union of Radio Sci & chmn, Comt Man & Radiation, Inst Elec Electronic Engrs, 78-79; sci advisor, Elec Power Res Inst, 84-87; mem, USA/USSR Sci Exchange, Non-Ionizing Radiation, 84-89; ed-in-chief, Bioelectromagnetics, 88- *Mem:* Fel Am Psychol Asn; fel AAAS; Soc Neurosci; Bioelectromagnetics Soc (pres, 84-85). *Res:* Neurophysiological correlates of behavior; biothermal correlates of emotion and arousal; biological and psychological response to non-ionizing electromagnetic radiation; philosophy of science. *Mailing Add:* Vet Admin Hosp 4801 Linwood Blvd Kansas City MO 64128

JUSTHAM, STEPHEN ALTON, b New Kensington, Pa, May 22, 37; m 68; c 2. CLIMATOLOGY, METEOROLOGY. *Educ:* Indiana Univ Pa, BS, 64, MA, 70; Univ Ill, PhD(geog), 74. *Prof Exp:* Teacher geog polit sci, Washington Twp Sch Dist, Apollo, Pa, 64-66; teacher civics, New Kensington-Arnold Sch Dist, 66-68; ASSOC PROF GEOG, BALL STATE UNIV, 71- *Mem:* Am Meteorol Soc; Am Geophys Union; Asn Am Geogr; Royal Meteorl Soc. *Res:* Climatology and meteorology of severe weather phenomena, specifically tornado preparedness programs; wind resource inventory and modeling. *Mailing Add:* Provost's Off Univ Kutztown Kutztown PA 19530

JUSTICE, JAMES HORACE, b Big Spring, Tex, Aug 31, 41. EXPLORATION GEOPHYSICS. *Educ:* Univ Tex, BA, 63; Univ Md, MA, 66, PhD(math), 68. *Prof Exp:* From asst prof to assoc prof math, Univ Tulsa, 68-78, prof, 78-82; chair explor geophys, Univ Calgary, 82-87; RES SCIENTIST, DALLAS RES LAB, MOBIL RES & DEVELOP, 87- *Concurrent Pos:* Consult, Amoco Prod Co, 69-80; fel, Amax, Inc, Colo Sch Mines, 69-80; geophys consult, 82- *Mem:* Am Math Soc; Math Asn Am. *Res:* Signal processing in exploration geophysics; multidimensional signal processing; digital signal processing. *Mailing Add:* Dallas Res Lab Mobil Res & Develop PO Box 819047 Dallas TX 75381-9047

JUSTICE, KEITH EVANS, b Arkansas City, Kans, Feb 6, 30; m 57; c 3. ECOLOGY, EVOLUTIONARY BIOLOGY. *Educ:* Univ Ariz, BS, 55, MS, 56, PhD(zool), 60. *Prof Exp:* Res assoc pop genetics, Columbia Univ, 59-60 & Ariz-Sonora Desert Mus, 60-62; proj engr, Melpar Inc, 62-65; asst prof biol sci, Univ Calif, Irvine, 65-67, assoc dean grad div, 68-69, actg dean, 69-71, dean spec progs, 74-77, dean prof studies, 77-81, assoc prof ecol & evolutionary biol, 67-87, EMER PROF, UNIV CALIF, IRVINE, 87- *Concurrent Pos:* Res assoc, Univ Ariz, 60-62; vpres, Rocky Mt Biol Lab, Crested Butte, Colo, 66-71. *Mem:* Am Soc Mammal; Sigma Xi. *Res:* Population dynamics and behavior of desert rodents; computer assisted instruction. *Mailing Add:* 2652 E Kael St Mesa AZ 85213

JUSTICE, RAYMOND, b Logansport, Ind, Oct 14, 24; m 46; c 5. ELECTRICAL ENGINEERING. *Educ:* Purdue Univ, BS, 49; Ohio State Univ, MA, 50, PhD, 56. *Prof Exp:* Res assoc, Ohio State Univ, 50-56; sr group engr, Gen Dynamics/Convair, 56-59; supvry engr, Granger Assocs Elec Res, 59-62; staff mem weapons systs, Eval Div, Inst Defense Anal, 62-64; vpres eng, Andrew Corp, 64-69; pres, Justice Assocs, Inc, 69-76; SR DESIGN SPECIALIST, DEFENSE SYSTS, CUBIC CORP, 76- *Mem:* Inst Elec & Electronics Engrs. *Res:* Electromagnetics; radar; communication antennas. *Mailing Add:* Defense Systs Dept Cubic Corp 9233 Balboa Ave San Diego CA 92123

JUSTIN, JAMES ROBERT, b Scranton, Pa, Oct 14, 33; m 57; c 2. AGRONOMY, FIELD CROPS. *Educ:* Pa State Univ, BS, 55, MS, 57; Tex A&M Univ, PhD(plant breeding), 63. *Prof Exp:* Instr agron, Tex A&M Univ, 60-63; exten agronomist, Univ Minn, 63-67, from instr to asst prof agron, 64-67; assoc specialist, 67-75, SPECIALIST, CROPS & SOILS EXTEN, RUTGERS UNIV, NEW BRUNSWICK, 75- *Mem:* Am Forage & Grassland Coun; Am Soc Agron; Crop Sci Soc Am; Am Soybean Asn; Asn Off Seed Certifying Agencies. *Res:* Improvement of field crop and seed production; variety testing and production research with soybeans and variety testing with forages. *Mailing Add:* Dept Crop Sci Rutgers Univ New Brunswick NJ 08903

JUSTINES, GUSTAVO, virology, immunopathology, for more information see previous edition

JUSTUS, CARL GERALD, b Atlanta, Ga, Oct 2, 39; m 63; c 2. ATMOSPHERIC PHYSICS, ATMOSPHERIC DYNAMICS. *Educ:* Ga Inst Technol, BS, 61, MS, 63, PhD(physics), 66. *Prof Exp:* From assoc prof to prof aerospace eng, 65-78, PROF EARTH & ATMOSPHERIC SCI, GA INST TECHNOL, 78- *Mem:* Am Geophys Union; Am Meteorol Soc; AAAS; Sigma Xi. *Res:* Satellite remote sensing; solar radiation; upper atmosphere meteorology; wind energy; atmospheric turbulence. *Mailing Add:* Sch Earth & Atmospheric Sci Ga Inst Technol Atlanta GA 30332-0340

JUSTUS, DAVID ELDON, b Van Buren, Mo, May 22, 36; m 60; c 5. IMMUNOLOGY, PARASITOLOGY. *Educ:* Southeast Mo State Col, BS, 63; Univ Mo-Columbia, MS, 65; Univ Okla, PhD(med parasitol), 68. *Prof Exp:* ASST PROF MICROBIOL & IMMUNOL, UNIV LOUISVILLE, 70- *Concurrent Pos:* Fel microbiol, Sch Med, Univ Louisville, 69-70. *Mem:* Am Soc Microbiol; Am Soc Trop Med & Hyg; Reticuloendothelial Soc. *Res:* Anaphylactic antibody and mast cell responses in mice infected with Trichinella spiralis. *Mailing Add:* Dept Microbiol Univ Louisville Sch Med Louisville KY 40292

JUSTUS, JERRY T, b Chicago, Ill, Oct 13, 32. DEVELOPMENTAL BIOLOGY. *Educ:* Franklin Col, AB, 57; Ind Univ, MA, 62, PhD(endocrinol, develop biol), 65. *Prof Exp:* Teacher zool, Ind Univ, 60-61, res asst endocrinol, 61-63, res fel develop biol, 64-65; sr cancer res scientist, Springville Labs, Roswell Park Mem Inst, 66-68; from asst prof to assoc prof zool, Ariz State Univ, 68-91; PROF BIOL, GRAND CANYON UNIV, 91- *Concurrent Pos:* Am Cancer Soc Inst grant, 67-68; United Health Found grant, 67-68; Am Heart Asn grant, 67-69 & 76-78; estab investr, Am Heart Asn, 71-76. *Mem:* AAAS; NY Acad Sci; Am Soc Zool; Soc Develop Biol. *Mailing Add:* Dept Zool Ariz State Univ Tempe AZ 85287

JUSTUS, NORMAN EDWARD, b Marionville, Mo, Jan 1, 26; m 46; c 2. AGRONOMY. *Educ:* Univ Ark, BSA, 54, MS, 55; Okla State Univ, PhD(genetics, plant breeding), 58. *Prof Exp:* Instr agron, Okla State Univ, 56-58; res agronomist, Cotton Genetics & Breeding Sect, USDA, 58-65; supt southwest ctr & prof agron, Columbia, AT SOUTHWEST MO RES CTR, UNIV MO, 65- *Mem:* Sigma Xi. *Mailing Add:* SW Mo Res Ctr Univ Mo Mt Vernon MO 65712

JUSTUS, PHILIP STANLEY, b New York, NY, Jan 17, 41; m 62; c 2. GEOLOGY, TECTONICS. *Educ:* City Col New York, BS, 62; Univ NC, Chapel Hill, MS, 66, PhD, 71. *Prof Exp:* Teaching asst geol, Univ NC, 62-67, student dir seismog sta, 64-66; instr astron & phys geog, US Mil Acad, 67-68, asst prof astron & geol, 68-70; fel, Rice Univ, 70-71; from asst prof to assoc prof geol, Fairleigh Dickinson Univ, 71-80; SECT LEADER, HIGH-LEVEL RADIOACTIVE WASTE MGT, US NUCLEAR REGULATORY COMN, 81- *Concurrent Pos:* Assoc ed, Geol Sect, NJ Acad Sci Bull & contrib ed, NJ Sci Teachers Asn Bull, 73-80; geologist, US Nuclear Regulatory Comn, 80-81. *Mem:* AAAS; Geol Soc Am; Am Geophys Union; Sigma Xi. *Res:* Deciphering the sequence of textural development and crystallization history of diabase dikes; determining the structural and metamorphic evolution of the Brevard fault zone and Blue Ridge Mountains in North Carolina; radioactive waste disposal. *Mailing Add:* US Nuclear Regulatory Comm Mailstop 4-H-3 Washington DC 20555

JUTILA, JOHN W, b Mullan, Idaho, May 21, 31; m 53; c 4. IMMUNOLOGY, MEDICAL MICROBIOLOGY. *Educ:* Mont State Univ, BA, 53, MA, 54; Univ Wash, PhD(microbiol), 60. *Prof Exp:* Bacteriologist, Rocky Mountain Lab, Mont, 56-57; from asst prof to prof microbiol & microbiologist, Mont State Univ, 61-89, dean, Col Letters & Sci, 74-78, vpres res, 78-89, COORDR DEVELOP, MONT STATE UNIV, 90- *Concurrent Pos:* Fel, Univ Wash, 60-61; NIH grants, 62-69 & 70-76. *Mem:* Fel AAAS; fel Am Acad Microbiol; Am Soc Microbiol; Soc Ext Biol & Med; Am Asn Immunol. *Res:* Pathogenesis of wasting syndromes in mice; immunobiology of the congenitally athymic mouse and immune cell interactions with tumor cells in vivo and tissue culture systems. *Mailing Add:* Alumni Found Mont State Univ Bozeman MT 59715

JUTRAS, MICHEL WILFRID, b Timmins, Ont, Can, May 11, 36; US citizen; m 61; c 2. AGRONOMY. *Educ:* Univ Mass, BS, 58; Univ Conn, MS, 61; Iowa State Univ, PhD(crop physiol), 64. *Prof Exp:* Instr agron, Iowa State Univ, 62-64; asst prof, 64-70, assoc prof, 70-78, PROF AGRON, CLEMSON UNIV, 78- *Concurrent Pos:* Consult, Agr Serv Labs, Inc, Tex. *Mem:* AAAS; Am Soc Agron; Crop Sci Soc Am; Soil Sci Soc Am; Int Soc Soil Sci. *Res:* Photo-thermoperiodic responses of crop plants; crop drought and heat resistance; microclimatology; crop morphology and development; grazing systems; crop physiology and ecology. *Mailing Add:* Dept Agron 275 Plant & Animal Sci Clemson Univ Clemson SC 29631

JUVET, RICHARD SPALDING, JR, b Los Angeles, Calif, Aug 8, 30; m 55, 84; c 4. PLASMA DESORPTION MASS SPECTROMETRY, CHROMATOGRAPHY. *Educ:* Univ Calif, Los Angeles, BS, 52, PhD(chem), 55. *Prof Exp:* From instr to assoc prof anal chem, Univ Ill, Urbana, 55-70; PROF ANALYTICAL CHEM, ARIZ STATE UNIV, 70- *Concurrent Pos:* Vis prof, Univ Calif, Los Angeles, 60, Cambridge Univ, Eng, 64-65, Nat Taiwan Univ, 68, Ecole Polytechnique, France, 76-77, Univ Vienna, Austria, 88-90; NSF sr fel, Cambridge Univ, Eng, 64-65; mem air pollution chem & physics adv comt, US Dept Health, Educ & Welfare, 69-72; ed advan, J Gas Chromatography, 63-69, J Chromatographic Sci, 69-85, Analytica Chimica Acta, 72-74 & Anal Chem, 74-76; chmn, div anal chem, Am Chem Soc, 72-73; nat counr, 78-89, coun, comt Reagent Chemicals, 85-; Sci Exchange Agreement Award lect & travel Eastern Europe, 77; mem adv panel advan chem alarm technol, Develop & Eng Directorate, Defense Systs Div, Edgewood Arsenal, 75. *Mem:* Am Chem Soc (secy-treas, 69-71); fel Am Inst Chemists; Sigma Xi; Am Radio Relay League; Int Platform Asn. *Res:* New applications of gas chromatography; liquid chromatography detectors; photochemistry; organic structural determinations and functional group analysis; chelate chemistry of polyhydroxy compounds; optical rotation measurements in study of metal chelates; inorganic gas chromatography; computer interfacing; plasma desorption mass spectrometry. *Mailing Add:* Dept Chem Ariz State Univ Tempe AZ 85287-1604

JUVINALL, ROBERT C, b Danville, Ill, Apr 11, 17; m 45; c 2. MECHANICAL ENGINEERING. *Educ:* Case Inst Technol, BS, 39; Chrysler Inst Eng, MAE, 41; Univ Ill, MS, 50. *Prof Exp:* Proj engr, engine develop & supvr eng staff, res design dept, Chrysler Corp, 39-48; asst prof mech eng, Univ Ill, 48-50; assoc prof, Ill Inst Technol, 50-51; asst dir eng, Ransburg Electro-Coating Corp, 51-57; from assoc prof to prof mech eng, Univ Mich, Ann Arbor, 57-86; RETIRED. *Concurrent Pos:* Consult engr. *Mem:* Am Soc Mech Engrs; Am Soc Eng Educ. *Res:* Engineering stress analysis and mechanical design. *Mailing Add:* 5907 Tidewood Ave Sarasota FL 34231

JWO, CHIN-HUNG, b Taiwan, Nov 12, 56; c 2. MECHANICAL SYSTEM ANALYSIS, PRODUCT RESEARCH & DEVELOPMENT. *Educ:* Feng Chia Univ, Taiwan, BS, 79; Univ Fla, ME, 85, PhD(mech eng), 89. *Prof Exp:* ADVAN DEVELOP ENGR, NCR COLOR PRINTER DIV, ITHACA, 89- *Mem:* Assoc mem Am Soc Mech Engrs. *Res:* Impact, thermal transfer and inkjet printers for both industrial and general applications; performed mechanical system analysis, printhead technology development and finite element analysis. *Mailing Add:* Seven Alessandro Dr Ithaca NY 14850

JYUNG, WOON HENG, b Korea, Mar 12, 34; m 61; c 2. PLANT PHYSIOLOGY. *Educ:* Seoul Nat Univ, BS, 57; Mich State Univ, MS, 59, PhD(hort physiol), 63. *Prof Exp:* Res assoc hort physiol, Mich State Univ, 63-64; from asst prof to assoc prof, 64-74, PROF BIOL, UNIV TOLEDO, 74-, CHMN DEPT, 76- *Mem:* AAAS; Am Soc Plant Physiol; Japanese Soc Plant Physiol. *Res:* Mechanisms of ion uptake by plant cells; zinc metabolism in higher plants; biology of aging. *Mailing Add:* Dept Biol Univ Toledo Toledo OH 43606

K

KAAE, JAMES LEWIS, b Bell, Calif, Oct 28, 36; m 65; c 1. MATERIALS SCIENCE. *Educ:* Univ Calif, Los Angeles, BS, 59, MS, 61, PhD(eng), 65. *Prof Exp:* Int res fel, Welding Inst, Eng, 65-67; staff mem, metall, Gulf Energy & Environ Systs, Inc, 67-73; tech adv, Gen Atomic Co, 73-82; SR TECH ADV, GA TECHNOL, 82- *Concurrent Pos:* Mem, Boiler & Pressure Vessel Code Comt, Am Soc Mech Engrs; lectr, Univ Calif, San Diego, 87- *Mem:* Am Carbon Soc; AAAS; Sigma Xi; Am Soc Metals; Am Ceramic Soc. *Res:* Structure, properties and irradiation behavior of pyrolytic carbon; welding metallurgy of steels; behavior of materials under high-temperature cyclic loading; behavior of coated particle nuclear fuels; chemical vapor deposition of carbides and nitrides; mechanical properties of materials. *Mailing Add:* 613 Solana Glen Ct Solana Beach CA 92075

KAARSBERG, ERNEST ANDERSEN, b Denmark, Sept 5, 18; US citizen; m 51; c 3. SCIENCE EDUCATION, GENERAL PHYSICS. *Educ:* Queens Univ, Ont, BSc, 49; Univ Toronto, MA, 50; Univ Chicago, PhD(geol, geophys), 58. *Prof Exp:* Assoc prof geophys, Univ Wash, 61-63; Fulbright lectr, Graz Tech Univ, 63-64; assoc prof, Ga Inst Technol, 64-66; mgr geophys lab, Brown Eng Co, 66-69; prof lectr continuing educ, Am Univ, 69-71; assoc prof earth sci, Staten Island Community Col, 71-76; aide to chmn, NY State Comn Energy Systs, 77-90; civil eng, Dept Transp, New York City, 83-90; CONSULT, PRIN OFFICER, KAARSBERG ASSOCS, 81-83 & 90- *Concurrent Pos:* Geophysicist, Imp Oil Co, Calgary, Alta, 50-51; geophys supvr, Brit Am Oil Co, 51-53 & Standard Oil Calif, Calgary, 53-55; sr res geophysicist & geochemist, Geophys Serv Inc, Dallas, Tex, 58-60; eng physicist, Shannon & Wilson Inc, Soil Mech & Found Engrs, Seattle, 60-61. *Mem:* Am Geophys Union; Soc Explor Geophysicists; NY Acad Sci; Explorors Club. *Res:* Ultrasonic, vibrational, X-ray, electronmicroscopic and spectrographic investigations of soils and rocks; development of a varity of methods for determing elastic and other physical parameters of soils, rocks and other materials; seismic scale dynamic modelling of earth and moon and their surface and subsurface structural features; using laboratory facilities now available developing innovative methods for teaching scientific principals to non-science major students; journalistic publications on scientific and technological matters of current public interest and concern. *Mailing Add:* 1317 Portage Rd Niagara Falls NY 14301

KAAS, JON HOWARD, b Fargo, NDak, Sept 13, 37; m 63; c 2. PSYCHOPHYSIOLOGY. *Educ:* Northland Col, BA, 59; Duke Univ, PhD(psychol), 65. *Prof Exp:* Trainee neurophysiol, Univ Wis-Madison, 65-68, asst prof, 68-73; assoc prof, 73-79, PROF PSYCHOL, VANDERBILT UNIV, 79- *Mem:* Soc Neurosci. *Res:* Visual and sensory systems; brain functions and evolution. *Mailing Add:* Dept Psychol Vanderbilt Univ 132 Wesley Hall Nashville TN 37240

KAATZ, MARTIN RICHARD, b Cleveland, Ohio, Apr 16, 24; m 47; c 3. PHYSICAL GEOGRAPHY, GEOMORPHOLOGY. *Educ:* Univ Mich, BA, 48, MA, 49, PhD(geog), 52. *Prof Exp:* From asst prof to assoc prof, 52-64, chmn dept, 62-76, prof, 65-82, EMER PROF GEOG, CENT WASH UNIV, 83- *Concurrent Pos:* Fulbright prof, Trinity Col, Univ Dublin, 65-66. *Mem:* Asn Am Geogr; Am Quaternary Asn; AAAS. *Res:* Significance of mass wasting and periglacial landforms to the modern environment; irrigation and drought in Central Washington. *Mailing Add:* Dept Geog & Land Studies Cent Wash Univ Ellensburg WA 98926

KABACK, DAVID BRIAN, b New York, NY, May 4, 50. RECOMBINANT DNA, DEVELOPMENTAL BIOLOGY. *Educ:* State Univ NY Stony Brook, BS, 71; Brandeis Univ, PhD(biol), 76. *Prof Exp:* Res fel chem, Calif Inst Technol, 76-78; asst prof, 78-85, ASSOC PROF MICROBIOL, NJ MED SCH, UNIV MED & DENT NJ, 85- *Mem:* Am Soc Microbiol; Harvey Soc. *Res:* Molecular genetics of yeast meiosis and sporulation; organization of eucaryotic genes and chromosomes. *Mailing Add:* Dept Microbiol & Molecular Genetics NJ Med Sch Univ Med & Dent NJ 185 S Orange Ave Newark NJ 07103

KABACK, HOWARD RONALD, b Philadelphia, Pa, June 5, 36; m 57; c 3. BIOCHEMISTRY. *Educ:* Haverford Col, BA, 58; Albert Einstein Col Med, MD, 62. *Prof Exp:* Intern pediat, Bronx Munic Hosp Ctr, 62-63; sr res investr, Lab Biochem, Nat Heart Inst, 66-69; assoc mem, Dept Biochem, Roche Inst Molecular Biol, 70-72, mem, 72-89, head, Lab Membrane Biochem, 77-89, head, Dept Biochem, 83-89; PROF, DEPTS PHYSIOL, MICROBIOL & MOLECULAR GENETICS, UNIV CALIF, LOS ANGELES, 89-, INVESTR, HOWARD HUGHES MED INST, 89- *Concurrent Pos:* Edward John Noble Found fel, 62-64; fel physiol, Albert Einstein Col Med, 63-64; res fel biochem, Nat Heart Inst, 64-66; adj assoc prof, Columbia Univ, NY, 73-85, adj prof, 85-89, adj prof, Grad Sch & Univ Ctr, City Univ NY, 76-89; Lady Davis vis prof, Hebrew Univ, Israel, 80, Albert Alberman vis prof, Technion-Israel Inst Technol, 81, Wellcome vis prof, Univ Idaho, 87-88; nat lectr, Am Soc Microbiol, 81; adj prof microbiol, NJ Med Sch, 86-89; mem, Bd Sci Counselors, Nat Inst Diabetes, Digestive & Kidney Dis, 87-91. *Honors & Awards:* S Waksman Award, 73; Lewis Rosenstiel Award, 74; Harvey Lectr, 88; Nathan Kaplan Mem Lectr, Univ Calif, 88; Kenneth Cole Award, Am Biophys Soc, 88; Philips Lectr, Haverford Col, 89; George A Feigen Mem Lectr, 89. *Mem:* Nat Acad Sci; Fedn Am Socs Exp Biol; NY Acad Sci; Am Soc Microbiol; Biophys Soc; fel Am Acad Arts & Sci; AAAS; Am Soc Biol Chemists; Am Chem Soc; Soc Gen Physiologists. *Res:* Transport; membranes; genetics. *Mailing Add:* Howard Hughes Med Inst Univ Calif MBI 655 405 Hilgard Ave Los Angeles CA 90024-1570

KABACK, STUART MARK, b Elizabeth, NJ, June 12, 34; m 55; c 2. ORGANIC CHEMISTRY. *Educ:* Columbia Univ, AB, 55, AM, 56, PhD(org chem), 60. *Prof Exp:* Chemist, Tech Info Div, Esso Res & Eng Co, 60-63, Chem Res Div, 63 & Tech Info Div, 63-68, sr res chemist, Chem Corp Serv, 68-70, sr res chemist, Res Corp Serv, 70-76, res assoc, 76-85, mem staff, Anal

& Info Div, 78-83, sr res assoc, 85-90, MEM STAFF, INFO & OFF SYSTS DIV, EXXON RES & ENG CO, 84-, SCI ADV, 90- *Mem:* Am Chem Soc (asst secy div chem lit, 69-71, actg treas, 71-72); Chem Struct Asn. *Res:* Information retrieval and analysis; patent information on petrochemicals, polymer chemistry; petroleum technology. *Mailing Add:* 222 Denman Rd Cranford NJ 07016

KABADI, BALACHANDRA N, b Gadag, Mysore, India, July 15, 33; m 50; c 4. PHYSICAL PHARMACY, PHARMACEUTICAL CHEMISTRY. *Educ:* Karnatak Univ, India, BSc, 58; Univ Bombay, BSc, 60; Univ Wash, MS, 62, PhD(pharm), 65. *Prof Exp:* Fel chem, Univ SC, 65-66 & Sch Pharm, Univ Mich, 66-67; asst prof pharm, Col Pharm, Fla A&M Univ, 67-68; SR RES SCIENTIST ANALYSIS CONTROL, E R SQUIBB & SONS, 68- *Mem:* Am Pharmaceut Asn; Am Chem Soc. *Res:* Isolation, identification of compounds from natural plant products; pharmaceutical complexation reactions of phenols and water soluble hydrophilic polymers of nonionic nature; analytical chemistry-analysis of pharmaceutical products involving spectrophotometric methods; reference standards. *Mailing Add:* 34 Colin Dr South River NJ 02406

KABAK, IRWIN WILLIAM, b New York, NY, May 21, 36; m 57; c 2. OPERATIONS RESEARCH, INDUSTRIAL ENGINEERING. *Educ:* NY Univ, BIndE, 56, MIndE, 58, PhD(opers res), 64. *Prof Exp:* Mfg trainee & cost control analyst, Mergenthaler Linotype Co, 56-57; opers res engr, Esso Res & Eng Co, 56-58; mem tech staff & specialist traffic studies, Bell Tel Labs, 58-64; from asst prof to prof opers res, NY Univ, 65-89; pres, Modelmetrics, Inc, 67-89; MEM BD DIRS, NJ AUTOMOBILE FULL INS UNDERWRITING ASN, 89- *Concurrent Pos:* Consult various indust & govt; instr & adj asst prof, NY Univ, 63-65; lectr, City Univ New York, 64-65, Am Mgt Asn, 66-69, Purchasing Agents Asn NY, 67 & Diebold Group, 67; corp dir, Taxtronics Inc, 69-; consult & corp dir, Hardboard Fabricators Inc, 70; vis prof, Polytech Inst NY & CW Post Col, 71-72, Rutgers Univ, 78; exec mgt consult, Stat-A-Matrix, 85- *Mem:* Am Inst Indust Engrs; Inst Mgt Sci; Opers Res Soc Am; Nat Soc Prof Engrs. *Res:* Applications of applied probability in queueing, inventory, reliability and simulation; operations management and finance; model building, testing and evaluating quality and production. *Mailing Add:* NY Univ Grad Sch Bus Admin 100 Trinity Pl New York NY 10006

KABALKA, GEORGE WALTER, b Wyandotte, Mich, Feb 1, 43; m 68; c 2. ORGANIC CHEMISTRY, ORGANOMETALLIC CHEMISTRY. *Educ:* Univ Mich, BS, 65; Purdue Univ, PhD(chem), 70. *Prof Exp:* Res assoc, Purdue Univ, 69-70; PROF CHEM, UNIV TENN, KNOXVILLE, 70-, PROF RADIOL, 84- *Concurrent Pos:* Consult, Oak Ridge Nat Lab, 76-, Oak Ridge Assoc Univ, 77-, Brookhaven Nat Lab, 81-, CTI Corp, Knoxville, Tenn, 85-; dir basic sci res, Inst Biomed Imaging, Univ Tenn Hosp, 88-; consult, Squibb Inst, 87-89; bd dir, Int Isotope Soc, 86-; bd dir, Radiopharmaceut Coun Soc Nuclear Med, 87-89; chancellor's res award, Univ Tenn, 84 & sci alliance res award, 86-91. *Mem:* Am Chem Soc; Soc Nuclear Med; Int Isotope Soc; Soc Magnetic Resonance Med; Magnetic Resonance & Imaging Soc; Radiol Soc NAm. *Res:* Organic synthesis; synthesis of radiopharmaceuticals containing short-lived radionuclides; organometallic reaction mechanisms; magnetic pharmaceuticals; pharmaceutical chemistry. *Mailing Add:* Dept Chem Univ Tenn Knoxville TN 37916

KABARA, JON JOSEPH, b Chicago, Ill, Nov 26, 26; m 70; c 6. PHARMACOLOGY, CLINICAL BIOCHEMISTRY. *Educ:* St Mary's Col, Minn, BS, 48; Univ Miami, MS, 50; Univ Chicago, PhD(pharmacol), 59. *Prof Exp:* Asst biochem, Univ Ill, 48; asst chem, Univ Miami, 48-49, microbiol, 50-53; asst med, Univ Chicago, 53-57; from asst prof to prof chem, Univ Detroit, 57-68; prof pharmacol & assoc dean, 69-70, PROF MED, COL OSTEOP MED, MICH STATE UNIV, 70- *Mem:* AAAS; Am Chem Soc; NY Acad Sci; Am Soc Clin Path; Am Asn Clin Chem. *Res:* Cancer and virus chemotherapy; sterol biogenesis and metabolism; biochemistry of the central nervous system; radiobiology and clinical chemistry; biochemistry; venom research; pharmacology in dental research of food preservation and cosmetic. *Mailing Add:* Dept Biomech Mich State Univ East Lansing MI 48824

KABAT, DAVID, b Minneapolis, Minn, Oct 15, 40; m 62; c 2. BIOCHEMISTRY, GENETICS. *Educ:* Brown Univ, ScB, 62; Calif Inst Technol, PhD(biochem), 67. *Prof Exp:* NIH res assoc biophys, Mass Inst Technol, 67-69; asst prof, 69-72, ASSOC PROF BIOCHEM, MED SCH, UNIV ORE, 72- *Mem:* Am Soc Biol Chemists. *Res:* Biochemical genetics of growth and differentiation. *Mailing Add:* Dept Biochem Ore Health Sci Univ Portland OR 97201

KABAT, ELVIN ABRAHAM, b New York, NY, Sept 1, 14; m 42; c 3. BIOCHEMISTRY. *Educ:* City Col New York, BS, 32; Columbia Univ, AM, 34, PhD(biochem), 37. *Hon Degrees:* DL, Univ Glasgow, 76; PhD, Univ Orleans, France, 82; PhD, Weizman Inst Sci, Rehovot, Israel, 82. *Prof Exp:* Lab asst immunochem, Presby Hosp, 33-37; instr path, Med Col, Cornell Univ, 38-41; res assoc biochem, 41-46, from asst prof to assoc prof bact, 46-52, microbiologist, Columbia Presbyterian Hosp, 56-85, prof, 52-85, Higgins prof, 83-85, EMER PROF & HIGGINS EMER PROF MICROBIOL, COL PHYSICIANS & SURGEONS, COLUMBIA UNIV, 85- *Concurrent Pos:* Rockefeller Found fel, Inst Phys Chem, Univ Uppsala, 37-38; Fogarty scholar, NIH, 74-75; mem subcomt shock, Nat Res Coun, 51-53, panel on plasma, 53-59, comt plasma & plasma substitutes, 59-71; biochem adv panel, Am Inst Biol Sci, Off Naval Res, 57-62; Philips lectr, Haverford Col, 60, 74; prof, Col France, 64; mem expert adv panel on immunol, WHO, 65-82; expert, Nat Cancer Inst, NIH, 75-81; res award, City of Hope, 74. *Honors & Awards:* Nat Medal of Sci, 91; Lilly Award, Am Soc Microbiologists, 49; Award, Nat Multiple Sclerosis Soc, 62; Karl Landsteiner Mem Award, 66; L G Horwitz Prize, Harvey Soc, 77; R E Dyer Lectr Award, NIH, 79; Philip Levine Award, Soc Clin Path, 82; Nat Medal Sci, 91. *Mem:* Nat Acad Sci; AAAS; Am Chem Soc; Am Soc Microbiol; Harvey Soc (vpres, 75-76, pres, 76-77). *Res:* Immunochemistry; organic reactions in qualitative analysis; mechanisms of immune reactions; antibody purification; physical chemistry of antibodies; serum and spinal fluid proteins; blood group substances; allergy; multiple sclerosis; dextrans; structure and immunological specificity; secondary structure of proteins; lectin; nature of antibody combining sites; cloning and sequencing of antibody variable regions. *Mailing Add:* Col Physicians & Surgeons Columbia Univ New York NY 10032

KABAT, HUGH F, b Manitowoc, Wis, Oct 3, 32; m 56, 80; c 3. PHARMACY ADMINISTRATION. *Educ:* Univ Mich, BS, 54, MS, 56, PhD(pharm admin), 61. *Prof Exp:* Chief pharm serv, Alaska Native Hosp, USPHS, 56-58; from asst prof to assoc prof pharm technol, Univ Minn, Minneapolis, 61-69, head dept clin pharm, 69-74, prof clin pharm, 69-80, asst dean admin, 74-80, assoc dean, Acad Affairs, 80-84, prof, Col Pharm, 80-86; PROF COL PHARM, UNIV NMEX, ALBUQUERQUE, 84- *Concurrent Pos:* Consult, Vet Admin Hosp, Minneapolis, Hennepin County Med Ctr, St Paul Ramsey Med Ctr, Data Dynamics & Mkt Measurements, Vet Affairs Med Ctr, Univ NMex Hosp, Alpha Data Servs, NIH; contrib ed, Geriatric Nursing, 65-68, Drug Intel, 67-74 & Int Pharmaceut Abstr, 64-81, Topics in Hosp Pharm Mgt, 86-, Ediciones Mayo, 87-, Hosp Pharm, 90-, Pharm Bus, 90- *Honors & Awards:* Mead-Johnson Award, Am Soc Hosp Pharmacists, 69; Hallie Bruce Mem Lectr, 69; Dorothy Dillon Mem Lectr, 90. *Mem:* Am Pharmaceut Asn; Am Soc Hosp Pharmacists; Am Asn Cols Pharm; Am Pub Health Asn. *Res:* Drug utilization review; patient compliance; clinical pharmacy; drugs and the aging. *Mailing Add:* Col Pharm Univ NMex Albuquerque NM 87131

KABAYAMA, MICHIOMI ABRAHAM, b Kanazawa, Japan, Apr 18, 26; m 51; c 5. POLYMER CHEMISTRY. *Educ:* Sir George Williams Col, BSc, 52; Univ Montreal, MSc, 56, DSc, 58. *Prof Exp:* Control chemist, Monsanto of Can, 51-53; demonstr, Univ Montreal, 53-56; res chemist, Dupont of Can, 58 & E I du Pont de Nemours & Co, 58-65; res chemist, Ethicon, Inc, 65-67; res chemist, Res & Develop Labs, Northern Elec Co, Ltd, 67-69, mgr transducer & polymer mat develop, 69-71, Bell-Northern Res Labs, 71-74, mgr plastics eng, Mfg Res Centre, Northern Elec Co, Ltd, 74-76; tech dir, Soc Plastics Indust Can, 76-85, Vinyl Coun Can, 85-88; mgr recycling, Twinpak Inc, 88-89; MGR ENVIRON AFFAIRS, TETRA PAK, INC, 90- *Mem:* Soc Plastics Engrs; fel Chem Inst Can. *Res:* Physical properties and structure of polymers; thermodynamics of solutions of polymers; calorimetry of polymer solutions; plastic material and processing technology; new methods and applications; combustibility and toxicity of combustion products; occupational health and safety; recycling. *Mailing Add:* 434 Winona Dr Toronto ON M6C 3T7

KABE, DATTATRAYA G, b Belgaum, India, Dec 30, 26; m 54; c 3. MATHEMATICAL STATISTICS, OPERATIONS RESEARCH. *Educ:* Univ Bombay, BSc, 48, MSc, 52; Univ Karnatak, India, MSc, 55; Wayne State Univ, PhD(statist), 64. *Prof Exp:* Lectr statist, Vijay Col, India, 52-53 & Karnatak Univ, India, 53-61; assoc math, Wayne State Univ, 61-64; assoc prof math & statist, Northern Mich Univ, 65-66 & Dalhousie Univ, 66-68; assoc prof, 68-75, PROF MATH & STATIST, ST MARY'S UNIV, NS, 75- *Concurrent Pos:* Nat Res Coun Can grants & Can Math Cong res scholar; prof statist, NMex State Univ, 80-81, Bowling Green State Univ, 87-88. *Mem:* Can Statist Asn. *Res:* Distribution theory; design of experiments; multivariate analysis; Pascal computer programming language; sampling techniques; math programming. *Mailing Add:* Dept Math St Mary's Univ Halifax NS B3H 3C3 Can

KABEL, RICHARD HARVEY, b Detroit, Mich, Dec 18, 32; m 60; c 2. TRIBOLOGY, MECHANICAL ENGINEERING. *Educ:* Gen Motors Inst, BSME, 61. *Prof Exp:* Res engr eng oils, Gen Motors Res Labs, 61-67, sr res engr, 67-87, group leader, 74-78, staff res engr eng oils, 78-81, sr staff res engr, eng oils, Gen Motors Res Labs, 81-87; PRES, RICH-LO-CONSULT CORP, 87- *Concurrent Pos:* Gen Motors rep, Lub Rev Comt, US Army & Soc Automotive Engrs; SAE fel, bd dr. *Honors & Awards:* Coop Eng Colwell Medal, Soc Automotive Engrs. *Mem:* Soc Automotive Engrs; Am Soc Testing & Mat. *Res:* Engine oils, formulation of engine oils, field performance, and methods to evaluate them. *Mailing Add:* 11051 Jonathan Lane Romeo MI 48065-9060

KABEL, ROBERT L(YNN), b Champaign, Ill, Apr 3, 32; m 58; c 2. CHEMICAL REACTION ENGINEERING. *Educ:* Univ Ill, BS, 55; Univ Wash, PhD(chem eng), 61. *Prof Exp:* From asst prof to assoc prof, 63-74, PROF CHEM ENG, PA STATE UNIV, 74- *Concurrent Pos:* Royal Norweg Coun Sci & Indust Res fel, Tech Univ Norway, 71-72; consult, Exxon Res & Eng, 76-; vis lectr, Pahlavi Univ, Iran, 78; invitational prof, Ariz State Univ, 84-85. *Mem:* Am Chem Soc; fel Am Inst Chem Engrs; Am Asn Univ Professors; Am Soc Eng Educ. *Res:* Reaction kinetics; adsorption; thermodynamic equilibria; heterogeneous catalysis; chemical reactor dynamics; aerospace life support systems; thermal conductivity; mathematical modeling of natural processes; mass transfer at the earth's surface; air pollution meteorology; scaleup of chemical processes. *Mailing Add:* 164 Fenske Lab Pa State Univ University Park PA 16802

KABIR, PRABAHAN KEMAL, b Calcutta, India, June 30, 33. SYMMETRY, HIGH ENERGY PHYSICS. *Educ:* Univ Delhi, BSc, 51, MSc, 53; Cornell Univ, PhD(theoret physics), 57. *Prof Exp:* Mem Inst Advan Study, Princeton Univ, 56-57; res fel physics, Univ Birmingham, 57-58; sur reader nuclear physics, Univ Calcutta, 58-60; asst prof physics, Carnegie Inst Technol, 60-63; vis scientist, Europ Org Nuclear Res, Geneva, 63-65; prin sci officer, Sci Res Coun, UK, 65-71; PROF PHYSICS, UNIV VA, 71- *Concurrent Pos:* Mem, Ctr Advan Study, Univ VA, 70-73; sr res fel, Univ Sussex, 74; ed, Physics Letters B, 78-81; vis prof, Harvard Univ, 83. *Mem:* Fel Am Phys Soc. *Res:* Application of concepts of symmetry to the classification of elementary particles and investigation of the broken mirror-symmetries of physical laws. *Mailing Add:* Physics Dept J W Beams Lab Univ Va Charlottesville VA 22901

KABISCH, WILLIAM THOMAS, b Bureau, Ill, Nov 10, 19; div; c 3. ANATOMY. *Educ:* Augustana Col, AB, 48; Univ Chicago, SM, 51, PhD(anat), 54. *Prof Exp:* Asst, Univ Chicago, 49-53, from instr to asst prof anat, 54-62; asst to exec officer, Am Asn Advan Sci, 62-67, asst exec officer, 67-70; assoc prof anat, 70-71, from asst dean to assoc dean, 70-72, PROF ANAT & ASSOC DEAN FOR RES, SCH MED, SOUTHERN ILL UNIV, 73- *Concurrent Pos:* Lederle med fac award, 56-59; dir, Eastern Tech Off & sr staff mem, Enviro-Med Calif, Inc, 72-73. *Mem:* AAAS; Am Asn Anat. *Res:* Gross anatomy; phagocytosis; bone regeneration; wound healing. *Mailing Add:* PO Box 5143 Springfield IL 62705

KABLAOUI, MAHMOUD SHAFIQ, b Tarshiha, Palestine, Apr 15, 38; US citizen; m 65; c 3. CHEMISTRY. *Educ:* Am Univ Beirut, BSc, 60; Univ SC, PhD(org chem), 67. *Prof Exp:* From chemist to sr res chemist, 67-80, group leader, 80-87, MGR CHARGE BIOTECHNOL, COAL GASIFICATION & MEMBRANE SEPARATION, BEACON RES LAB, TEXACO INC, 87- *Mem:* Am Chem Soc; Sigma Xi; Am Soc Microbiol. *Res:* Organic synthesis, petrochemicals; aromatization reactions, organic nitrogen compounds and fuels and lubricants technology; biotechnology area involving fermentation, enzyme catalysis and photobioconversion. *Mailing Add:* Amherst Lane Wappingers Falls NY 12590

KABLER, J D, b Wichita, Kans, Dec 29, 26; m 50; c 5. INTERNAL MEDICINE. *Educ:* Univ Kans, AB, 47, MD, 50; Am Bd Internal Med, dipl, 58, recertified, 74. *Prof Exp:* Intern Univ Hosp, 50-51, resident med, 51-52, 54-56, from instr to assoc prof, Med Sch, 57-70, PROF MED, MED SCH, UNIV WIS-MADISON, 70-, DIR HEALTH SERV, 68- *Concurrent Pos:* Fel, Univ Wis-Madison, 56-57. *Mem:* Am Psychosom Soc; AMA; fel Am Col Physicians; Am Venereal Disease Asn; Asn Mil Surgeons. *Res:* Clinical psychophysiology. *Mailing Add:* Univ Wis 1552 University Ave Madison WI 53705

KABLER, MILTON NORRIS, b Roanoke, Va, Apr 30, 32; m 57; c 2. SURFACE SCIENCE, OPTICAL PROPERTIES. *Educ:* Va Polytech Inst, BS, 55; Univ NC, Chapel Hill, PhD(physics), 59. *Prof Exp:* Res asst prof physics, Univ Ill, Urbana, 59-62; physicist, 62-69, assoc supt mat sci div, 75-77, head, Optical Mat Br, 69-79, head, optical probes br, 79-85, HEAD SYNCHROTRON RAD RES GROUP, NAVAL RES LAB, 86- *Concurrent Pos:* Sabbatical leave, Clarendon Lab, Oxford Univ, 73-74. *Honors & Awards:* Pure Sci Award, Sigma Xi-Naval Res Lab, 73. *Mem:* Fel Am Phys Soc; Sigma Xi; AAAS; Optical Soc Am. *Res:* Electronic and optical properties of materials; defects, excitons, semiconductors, radiation effects, surfaces, and insulators; optical technologies; synchrotron radiation; research management. *Mailing Add:* Code 4686 Naval Res Lab Washington DC 20375-5000

KABRA, POKAR MAL, b India, Nov 17, 42; m 66; c 2. CLINICAL CHEMISTRY, ANALYTICAL CHEMISTRY. *Educ:* Univ Bombay, BS, 66; Univ Kans, PhD(med chem), 72. *Prof Exp:* Trainee neuro-surg, Univ Calif, San Francisco, 73-74, trainee lab med, 74-75, assoc specialist, 75-77, asst prof, 77-81, ASSOC PROF LAB MED, UNIV CALIF, SAN FRANCISCO, 81- *Mem:* Am Chem Soc; AAAS; Am Asn Clin Chemists. *Res:* Drug metabolism; application of liquid chromatography in clinical sciences. *Mailing Add:* 109 Birch Ave Corte Madera CA 94925-1020

KACEW, SAM, b Poland, 46. DRUG-INDUCED CHANGES IN NEWBORNS. *Educ:* McGill Univ, Montreal, BS, 67; Univ Ottawa, MS, 70, PhD(pharmacol), 73. *Prof Exp:* PROF PHARMACOL, DEPT PHARMACOL, UNIV OTTAWA, 75- *Res:* Toxicology of newborns. *Mailing Add:* 48 Parkglen Dr Nepean ON K2G 3G8 Can

KACHANOV, MARK L, b Moscow, USSR, Aug 6, 46; US citizen. FRACTURE & DAMAGE MECHANICS, MICROMECHANICS OF MATERIALS. *Educ:* Leningrad Univ, BS & MS, 64; Leningrad Polytech Inst, Cand Sci, 74; Brown Univ, PhD, 81. *Prof Exp:* Asst prof mech & mat sci, Rutgers Univ, 80-82; assoc prof, 82-88, PROF MECH ENG, TUFTS UNIV, 88- *Concurrent Pos:* Vis scientist, Stanford Res Inst, Nat Inst Standards & Technol, Shell Res Lab, Holland & Gen Elec Labs; consult, SRI Int, Gen Tel Labs & Shell Labs; prin investr, Dept Energy, Dept Transp, US Army & Alcoa Found. *Mem:* Am Acad Mech. *Res:* Mechanics of solids with multiple cracks and other defects; micromechanics of brittle materials; fractures accompanied by damage and microcracking; mechanics of damage. *Mailing Add:* Dept Mech Eng Tufts Univ Medford MA 02155

KACHHAL, SWATANTRA KUMAR, b India, July 7, 47; m 77; c 3. INDUSTRIAL ENGINEERING, OPERATIONS RESEARCH. *Educ:* Univ Roorkee, India, BS, 68; Univ Minn, MS, 71, PhD(indust eng & opers res), 74. *Prof Exp:* from instr to assoc prof, 73-87, assoc prof, 78-87, CHMN DEPT, 85-, PROF INDUST & SYST ENG, UNIV MICH, DEARBORN, 87- *Concurrent Pos:* Consult, Corning Glass Works, 78-80, Henry Ford Hosp, 80- *Mem:* Am Inst Indust Eng. *Res:* Facilities planning; warehousing; automated and mechanized storage systems; applications of operations research in healthcare. *Mailing Add:* Dept Indust & Syst Eng Univ Mich Dearborn MI 48128

KACHIKIAN, ROUBEN, b Ardabil, Iran, Dec 19, 26; nat US; m 51; c 2. MICROBIOLOGY, ANALYTICAL CHEMISTRY. *Educ:* Syracuse Univ, BS, 52; Univ Tenn, MS, 54; Univ Mass, PhD(food technol), 57. *Prof Exp:* Instr food technol, Univ Mass, 54-57; develop chemist, Chas Pfizer & Co, 57-61; sr develop engr, Air Prod & Chem Inc, 61-64; prod develop assoc, Merck & Co, Inc, 64-68; group leader, Pepsico Inc, 68-69; mgr res planning & admin, Beech-Nut-Life Savers, Inc, 69-72, mgr tech serv, 72-76; ASSOC DIR TECH SERV, LIFE SAVERS, INC, 84-; GROUP DIR MEASUREMENT SCI TECHNOL, NABISCO BRANDS, INC, 84- *Mem:* Am Soc Microbiol; Am Inst Chemists; Inst Food Technologists. *Res:* Food preservation; food additives; nutritional supplementation and development of high protein products; beverages; cryogenics in foods; confections; chewing gum and candy; baby foods; sensory testing; product and sanitary standards. *Mailing Add:* 48 Birchwood Rd Westwood NJ 07675

KACHINSKY, ROBERT JOSEPH, b Boston, Mass, May 3, 37; m 63; c 4. WATER SUPPLY & WASTEWATER DISPOSAL ENGINEERING. *Educ:* Northeastern Univ, BS, 63, MS, 75. *Prof Exp:* Staff engr, 65-75, proj dir, 75-77, VPRES, CAMP DRESSER & MCKEE INC, 77- *Mem:* Fel Am Soc Civil Engrs; Am Acad Environ Engrs; Water Pollution Control Fedn; Am Waterworks Asn. *Res:* Wastewater treatment & disposal engineering. *Mailing Add:* Camp Dresser & McKee Inc One Cambridge Rd Cambridge MA 02142

KACHMAR, JOHN FREDERICK, b Akron, Ohio, Jan 10, 16; m 46; c 1. PATHOLOGY. *Educ:* Univ Akron, BS, 36; Univ Minn, MS, 47, PhD(biochem), 51. *Prof Exp:* Asst chemist rubber, Victor Gasket Co, 36-37; jr chemist sewage & stream pollution, USPHS, 37-41, asst chemist, 41-42 & 46; asst vitamins & enzymes, Dept Agr Biochem, Univ Minn, 47-50, assoc biochemist Rh serol, Dept Obstet & Gynec, 50-52; chief, clin chem sect, Dept Biochem, Rush-Presby-St Luke's Med Ctr, 58-67, dir lab training, 67-69, asst dir, clin lab, 69-72, CONSULT, CLIN CHEM LAB, 72-83; assoc prof, biochem, Rush Univ, 71-83; assoc prof biochem, Rush Univ, 71-83; RETIRED. *Concurrent Pos:* Asst prof biochem, Univ Ill Med Sch, 65-71. *Honors & Awards:* Natelson Award, 80. *Mem:* AAAS; Am Chem Soc; Am Asn Clin Chemists. *Res:* Activation of enzymes and enzyme kinetics; blood coagulation; clinical chemistry. *Mailing Add:* 12714 Greenwood Ave N Apt 303 Seattle WA 98133

KACKER, RAGHU N, b India, June 24, 51; US citizen; m 79; c 2. STATISTICAL QUALITY ENGINEERING, INDUSTRIAL EXPERIMENTATION. *Educ:* Univ Delhi, India, BS, 71; Agra Univ, India, MStatist, 73; Univ Guelph, Can, MS, 75; Iowa State Univ, PhD(statist), 79. *Prof Exp:* Instr statist, Iowa State Univ, Ames, Iowa, 75-79; asst prof statist, Va Polytech Inst, Blacksburg, 79-80; tech staff qual assurance, AT&T Bell Labs, Holmdel, NJ, 80-85, distinguished tech staff qual assurance, 85-88; MATH STATISTICIAN STATIST, NAT INST STANDARDS & TECHNOL, GAITHERSBURG, MD, 88- *Mem:* Am Statist Asn; Am Soc Qual Control; Am Soc Testing & Mat; Am Ceramic Soc. *Res:* Accelerate advanced materials formulation and processing through statistical quality engineering; engineering designs of instruments for measurement and for processing respond linearly to changes in input signals and make such designs robust to unavoidable noise factors. *Mailing Add:* Nat Inst Standards & Technol Bldg 101 Rm A337 Gaithersburg MD 20899

KACSER, CLAUDE, b 1934. THEORETICAL PHYSICS. *Educ:* Oxford Univ, BA, 55, MA & PhD(physics), 59. *Prof Exp:* Res fel physics, Magdalen Col, Oxford Univ, 58-59; instr, Princeton Univ, 59-61; res fel, Magdalen Col, Oxford Univ, 61-62; asst prof, Columbia Univ, 62-64; asst prof, 64-67, ASSOC PROF PHYSICS, UNIV MD, COLLEGE PARK, 67- *Mem:* Am Asn Physics Teachers. *Res:* Pedagogy; special relativy. *Mailing Add:* Dept Physics Univ Md College Park MD 20742

KACZMARCZYK, ALEXANDER, b Krynica, Poland, Apr 1, 32; US citizen; c 2. INORGANIC CHEMISTRY. *Educ:* Am Univ, Beirut, BA, 54; Wash Univ, PhD(chem), 60. *Prof Exp:* Res asst, Purdue Univ, 60; res fel, Harvard, 60-62, tutor chem, 61-62; asst prof, Dartmouth Col, 62-68; PROF CHEM, TUFTS UNIV, 68- *Concurrent Pos:* Grants, Air Force Off Sci Res & USPHS, 63; vis res fel, Res Lab Archaeol, Oxford Univ, 75-76; Fulbright res fel, 83. *Mem:* AAAS; Am Chem Soc; NY Acad Sci; Am Inst Archeol; Am Asn Univ Professors. *Res:* Applications of chemistry to art and archaeology. *Mailing Add:* Dept Chem Tufts Univ Medford MA 02155

KACZMARCZYK, WALTER J, b New Britain, Conn, Jan 3, 39; m 63; c 3. BIOCHEMICAL GENETICS. *Educ:* Fairfield Univ, BS, 61; St John's Univ, NY, MS, 63; Hahnemann Med Col, PhD(biochem genetics), 67. *Prof Exp:* Nat Acad Sci fel biochem, Plum Island Animal Dis Lab, USDA, 66-69; asst prof, 69-72, assoc prof genetics, 72-76, PROF GENETICS & BIOCHEM, WVA UNIV, 76-, BIOCHEM GENETICIST, 69- *Mem:* AAAS; Genetics Soc Am; Am Chem Soc. *Res:* Molecular biology of endothia parasitica; biochemistry of Heterosis; fungal viruses. *Mailing Add:* Dept Genetics Rm 1104 WVa Univ Evansdale Campus Morgantown WV 26505

KACZOROWSKI, GREGORY JOHN, b South Bend, Ind, Nov 20, 49; m 82. MEMBRANE BIOCHEMISTRY, MEMBRANE BIOPHYSICS. *Educ:* Univ Notre Dame, BS, 72; Mass Inst Technol, PhD(biochem), 77. *Prof Exp:* Helen Hay Whitney fel, Roche Inst Molecular biol, 77-80; sr res biochemist, 80-84, res fel, 84-86, assoc dir, 86-87, DIR, DEPT MEMBRANE BIOCHEM & BIOPHYSICS, MERCK INST THERAPEUTIC RES, 88- *Mem:* Am Soc Biological Chemists; Am Chem Soc; Biophys Soc; AAAS; NY Acad Sci. *Res:* Membrane biochemistry study of ion channels, especially calcium and potassium channels, as well as ion transporting systems in electrically excitable membranes by a combination of biochemical and biolphysical technices; therapeutic drug development with ion channels as targets. *Mailing Add:* Dept Membrane Biochem & Biophys Rm 80N-31C PO Box 2000 Merck Inst Rahway NJ 07065

KADABA, PANKAJA KOOVELI, b Perumbavoor, India, May 15, 28; m 54; c 1. NEUROCHEMISTRY, MEDICINAL CHEMISTRY. *Educ:* Travancore Univ, India, BSc, 47, MSc, 49; Univ Delhi, PhD(org chem), 54. *Prof Exp:* Lectr chem, Am Mission Med Col, Vellore, India, 49-50 & Univ Delhi, 50-53; guest scholar, Univ Ky, 54-55; res assoc, Brown Univ, 57-60; instr biochem, Univ Ky, 64-65; assoc prof chem, Morehead State Univ, 65-66; asst prof, Christian Bros Col, 66-68; assoc res prof med chem, 68-90, RES PROF MED CHEM & PHARMACEUT, COL PHARM, UNIV KY, 90- *Concurrent Pos:* Fulbright-Smith-Mundt fel, 53-54; vis assoc prof chem, Univ Ljubljana, Yugoslavia, 73-74; prin investr, res proj grant, Nat Inst Neurol Commun Dis & Stroke, NIH, 82-91; vis scientist, Mat Res Lab, Wright Patterson AFB, Dayton, Ohio, 82; chmn, 9th Int Cong Heterocyclic Chem, Tokyo, Japan, 83; mem, Res Bd Adv, Am Biog Inst, 86-; assoc, Sanders-Brown Ctr Aging, Univ Ky, 90-93. *Mem:* Am Chem Soc; Int Soc Heterocyclic Chem; India Chemists & Chem Engrs Club. *Res:* Chemistry of heterocyclic compounds, their synthesis, reaction mechanisms and biological activity; 1,2, 3-triazoles, tetrazoles, aziridinesand 1,2,3-triazolines; role of protic and dipolar aprotic solvents in heterocyclic synthesis via 1,3-cycloaddition reactions; borohydride reductions of heterocyclic compounds; direct esterification of acids with alcohols using borontrifluoride-etherate catalyst; rational design of anticonvulsants, their structure-activity relationships, metabolism and pharmacology, mechanism of action and anti-epileptic drug development; nmda antagonists; author of over 80 published papers and abstracts; holder of six patents. *Mailing Add:* Div Med Chem Univ Ky Col Pharm Lexington KY 40536-0082

KADABA, PRASAD KRISHNA, b Bangalore, India, Feb 14, 24; m 54; c 1. PHYSICS, ELECTRONICS. *Educ:* Univ Mysore, BS, 43, MS, 44; Calif Inst Technol, MS, 46; Univ Calif, Los Angeles, PhD(physics), 49. *Prof Exp:* Sci officer electronics, Nat Phys Lab, India, 50-52, asst supt, Tech Develop Estab, 52-53; asst prof elec eng, Univ Ky, 54-57 & Newark Col Eng, 57-59; assoc prof, 59-62, PROF ELEC ENG & DIR RES, UNIV KY, 62- *Concurrent Pos:* Alumni fel physics, Mich State Univ, 53-54; consult, Fed Pac Elec Co, NJ, 58; scholar, Univ Calif, Los Angeles, 58-59; AEC traveling fel, India, 63-64; Ky Res Found spec fel, 63-64; mem conf elec insulation & dielec phenomena, Nat Res Coun; consult, IBM, Ky; resident assoc, Argonne Nat Lab; prin investr microwave proj, Ky Tobacco Res Inst, 71-72; res fel to Yugoslavia, Int Res & Exchange Bd, 73; sr Fulbright-Hays Award, Yugoslav-Am Bi-Nat Comn, 74; vis scientist, Johnson Space Ctr, Houston, 75; prin investr proj, Off Water Resources Res Inst, 76-; Oak Ridge Assoc Univs fel, 76; prin investr microwave spectros proj, Ky Tobacco Res Inst, 77-78; vis prof, intergovernment personnel act prog, Air Force Off Sci Res, Wright Patterson Air Force Base Mat Lab, 80-81, Avionics Lab, 81-82. *Mem:* Am Soc Eng Educ; Brit Inst Elec Eng; sr mem Inst Elec & Electronics Engrs. *Res:* Microwave absorption of organic liquids; non resonant absorption of compressed gases and molecular nature of materials; nuclear quadrupole resonance and nuclear magnetic resonance studies; biological effects of microwaves; pollution studies. *Mailing Add:* Dept Elec Eng Univ Ky Lexington KY 40506

KADABA, PRASANNA V, b Gundlupet, India, July 4, 31; US citizen; m 66; c 1. MECHANICAL ENGINEERING. *Educ:* Univ Mysore, BE, 52 & 54, Univ Ky, MS, 56, Ill Inst Technol, PhD(mech eng), 64. *Prof Exp:* Asst mech eng, Univ Ky, 54-56; asst mech eng, Ill Inst Technol, 56-60, instr, 60-63; sr res engr, Roy C Ingersol Res Ctr, Borg-Warner Corp, 63-67; sr res scientist, Res & Develop Ctr, Westinghouse Elec Corp, Pa, 67-69; ASSOC PROF MECH ENG, GA INST TECHNOL, 69- *Concurrent Pos:* Vis prof, Univ Caraboba, Valencia Venezuela, 73 & 75; adv, Vol Int Tech Assistance, 77-; guest worker, Nat Bur Standards, Gaithersburg, MD, 78; consult, Lawrence Berkeley Lab, Berkeley, CA, 79-80, Copeland Corp, Emerson Elec, Sidney, OH, 84, IPA fel, Lewis Res Ctr, NASA, Cleveland, OH, 86; fac fel, Marshall Space Flight Ctr, NASA, Huntsville, 82, Lewis Res Ctr, Cleveland, OH, 87; vis prof, Univ Cincinnati, 87-90; mem, tech comt, TC6-3 & TC 9.6, Am Soc Heating Refrigerating & Air Conditioning Engrs; mem, Metric Comt, Am Soc Testing Mats. *Mem:* Am Soc Mech Engrs; Am Soc Heating, Refrig & Air-Conditioning Engrs; Am Soc Eng Educ; Soc Am Mil Engrs; India & Am Cult Assoc; Sigma Xi; Am Inst Chem Engrs; Am Soc Testing Mats. *Res:* Solar energy; refrigeration; air conditioning; heat transfer; thermodynamics; mechanical systems for buildings; thermal systems design; energy conservation; productivity and efficiency; photovoltaic total energy system; heat exchangers; optimization; space power coupled photovoltraic heat engine design; thermal sciences laboratory development. *Mailing Add:* George A Woodruff Sch Mech Eng Ga Inst Technol Atlanta GA 30332-0405

KADAN, RANJIT SINGH, b Karnal, Haryana State, India, Jan 1, 35; m 66; c 1. FOOD BIOCHEMISTRY, MICROBIOLOGY. *Educ:* Punjab Univ, DVM, 58; Kans State Univ, MS, 62; Rutgers Univ, New Brunswick, PhD(food sci), 67. *Prof Exp:* Vet surgeon, Punjab State, India, 58-60; res asst food, Food Sci Dept, Rutgers Univ, 62-67, sr group leader, 67-73; group leader food res, Quaker Oats Co, Barrington, Ill, 66-69, sr group leader, 70-73; mgr, Food Products Res, Lubin Maselli Lab, Chicago, 73-75; SR FOOD SCIENTIST, SOUTHERN REGIONAL RES CTR, AGR RES SERV, USDA, NEW ORLEANS, 75- *Concurrent Pos:* Consult, Volunteers Tech Assistance, 66- *Mem:* Inst Food Technologists; Am Oil Chemists Soc; Am Asn Cereal Chemists; Am Dairy Sci Asn; NY Acad Sci. *Res:* Diversified food research activities in the areas of dairy, cereals, oil seeds, snacks, long shelf life food products, beverages, toxic constituents of foods including microbiol toxins, and nutritional attributes of foods. *Mailing Add:* 8554 Fordham Ct New Orleans LA 70127

KADAN, SAVITRI SINGH, b Sonepat, Haryana, India, Aug 12, 34; m 66; c 1. EMERGENCY MEDICINE, OBSTETRICS & GYNECOLOGY. *Educ:* Bihar Univ, India, MD, 58; Dipl, 64; Patna Med Col, Patna Univ, India, MS, 71. *Prof Exp:* Asst prof obstet & gynec, Patna Med Col, 64-67; intern, St Fransic Hosp, Evanston, Ill, 68-69; resident, Lutheran Gen Hosp, Park Ridge, Ill, 69-70; emergency rm physician, Northwest Community Hosp, 73; dir, Cook County Venereal Dis Clin, Maywood, Ill, 74-75; resident, Clarity Hosp, La State Univ, New Orleans, 75-77; EMERGENCY RM PHYSICIAN, ALEXIAN BROTHERS HOSP, ELKGROVE, 74- *Concurrent Pos:* Fel Family pract, Am Acad Family Pract, 78. *Mem:* Am Acad Family Pract. *Res:* Obstetrics; gynecology. *Mailing Add:* 700 W Judge Perez Dr PO Box 43 Chalmelte LA 70043

KADANE, JOSEPH BORN, b Washington, DC, Jan 10, 41. APPLIED STATISTICS, MATHEMATICAL STATISTICS. *Educ:* Harvard Univ, BA, 62; Stanford Univ, PhD(statist), 66. *Prof Exp:* Asst prof statist, Yale Univ, 66-68, actg dir grad studies, 67-68; assoc prof, 71-72, PROF STATIST & SOCIAL SCI, DEPT STATIST, CARNEGIE-MELLON UNIV, 72-, LEONARD J SAVAGE PROF, 85- *Concurrent Pos:* Res staff mem, Cowles Found Res Econ, Yale Univ, 66-68, mem staff, Ctr Naval Anal, 68-71; consult, Long Island Lighting Co, 68, Nat Develop Corp, Govt Tanzania, 68, Ctr Naval Anal, 71-, Bur Labor Statist, US Dept Labor, 72-73 & World Bank Off Mgt & Budget; mem, Comn Behav & Social Sci & Educ, Nat Res Coun, 86-; mem, Nat Sci Found Adv Comt Ethics & Value Studies, 84-87. *Mem:* Fel Am Statist Asn; fel Inst Math Statist; Opers Res Soc; fel Royal Statist Soc; fel AAAS; Biometric Soc; Econometric Soc. *Res:* Theory and use of statistics in economics, political science, sociology, demography and law. *Mailing Add:* Dept Statist Carnegie-Mellon Univ Pittsburgh PA 15213

KADANKA, ZDENEK KAREL, b Rajhrad, Czech, May 24, 33; m 60; c 2. CYTOGENETICS, BIOCHEMISTRY. *Educ:* Purkyne Univ, Brno, dipl chem & RNDr, 57, dipl med & MUDr, 63; Inst Postgrad Studies for Physicians & Pharmacists, Prague, dipl, 66. *Prof Exp:* Lectr histol & embryol, Med Fac, Purkyne Univ, Brno, 59-63; physician allergy & diabetes, Sanatorium, Luhacovice, Czech, 63-66, intern, 66-69; sr res asst, 69-71, RES ASSOC KARYOLOGY, CONNAUGHT MED RES LABS, UNIV TORONTO, 71- *Concurrent Pos:* Gertrude I'Anson fel, Connaught Med Res Labs, Univ Toronto, 69-70; NIH grants, 71-72. *Res:* Karyologic data; cell membrane changes; transformation of human and animal cells cultured in vitro. *Mailing Add:* 64 Givendolen Circle Willowdale ON M2N 2L7 Can

KADANOFF, LEO P, b New York, NY, Jan 14, 37; m 58; c 3. THEORETICAL PHYSICS. *Educ:* Harvard Univ, AB, 57, AM, 58, PhD(physics), 60. *Prof Exp:* Res fel, Bohr Inst Theoret Studies, Copenhagen, 60-62; from asst prof to prof physics, Univ Ill, Urbana, 62-69; univ prof physics, Brown Univ, 69-78, prof eng, 71-78; PROF PHYSICS, UNIV CHICAGO, 78- *Concurrent Pos:* A P Sloan Found fel, 62-67; vis prof, Cambridge Univ, Eng, 65; mem adv comt, Inst Theoret Physics, Santa Barbara, 78-81, Schlumberger Doll Res Lab, 81-86, Univ Minn, 86-; dir mat res lab, Univ Chicago, 81-84; mem bd physics & astron, Nat Res Coun, 83-; vchmn sci & tech adv comt, Argonne Nat Lab, 83-84. *Honors & Awards:* Buckley Prize, Am Phys Soc, 77; Wolf Found Award, 80. *Mem:* Nat Acad Sci; fel Am Phys Soc; fel Am Acad Arts & Sci; fel AAAS. *Res:* Solid state and many particle theory; development of urban growth models; phenomena near phase transitions; behavior of dynamical systems; author of over 130 articles and books. *Mailing Add:* Dept Physics James Franck Inst Univ Chicago 5640 S Ellis Ave Chicago IL 60637

KADAR, DEZSO, b Zazar, Transylvania, July 21, 33; Can citizen; m 57; c 2. PHARMACOLOGY. *Educ:* Univ Toronto, BS, 59, MS, 66, PhD(pharmacol), 68. *Prof Exp:* Res asst pharmacol & toxicol, Connaught Med Res Lab, 60-65, demonstr, 65-68, lectr, 68-70, asst prof, 70-76, ASSOC PROF PHARMACOL, UNIV TORONTO, 76- *Concurrent Pos:* Mem, Drug Adv Comt, Ont Col Pharmacists, 72-, Comt Drugs & Therapeut, St Joseph's Hosp, 74- & Coun Fac Pharm, Univ Toronto, 73- *Mem:* Pharmacol Soc Can. *Res:* Drug metabolism and disposition in man and animals; microsomal drug oxidation in vitro. *Mailing Add:* Dept Pharmacol Univ Toronto Toronto ON M5S 1A8 Can

KADE, CHARLES FREDERICK, JR, b Sheboygan, Wis, Apr 4, 14; m 46; c 6. MEDICAL & HEALTH SCIENCES. *Educ:* Carleton Col, BA, 36; NDak State Univ, MS, 38; Univ Ill, PhD(biochem), 41. *Prof Exp:* Asst chem, Carleton Col, 35-36 & NDak State Univ, 36-38; asst, Univ Ill, 38, 39-41, fel, 41-43; dir biochem res, Frederick Stearns & Co, 43-47; res chemist, Sterling-Winthrop Res Inst, 47-49; dir div med sci, McNeil Labs, Inc, 49-60, vpres & dir res, 60-66; vpres, Johnson & Johnson Int, 66-79; vpres, Janssen Res & Develop, Inc, 72-79; consult, 79-89; RETIRED. *Mem:* AAAS; Am Chem Soc; Asn Res Dirs; Am Pharmaceut Asn; Soc Indust Chem; Am Inst Chemists. *Res:* Intermediary metabolism of amino acids; protein and amino acid requirements of the dog; preparation of hydrolysates for intravenous use. *Mailing Add:* 983 Butler Pike Blue Bell PA 19422

KADEKARO, MASSAKO, b Brazil, Jan 18, 39; m 79. DRINKING BEHAVIOR, SUBFORNICAL ORGAN. *Educ:* Univ Sao Paulo, Brazil, PhD(neural regulation of gastric secretion), 70. *Prof Exp:* ASSOC PROF & DIR NEUROSURG RES LAB, DIV NEUROSURG, MED BR, UNIV TEX, GALVESTON, 84- *Mem:* Soc Neurosci; Am Physiol Soc; NY Acad Sci; Soc Cerebral Blood Flow & Metabolism. *Res:* Neural regulation of water balance. *Mailing Add:* Div Neurosurg Med Br Univ Tex Rte E 17 Galveston TX 77550-2778

KADER, ADEL ABDEL, b Cairo, Egypt, Mar 1, 41; US citizen; m 63; c 2. PLANT PHYSIOLOGY, HORTICULTURE. *Educ:* Univ Ain Shams, Cairo, BSc, 59; Univ Calif, Davis, MSc, 62, PhD(plant physiol), 66. *Prof Exp:* Lectr hort, Univ Ain Shams, Cairo, 66-71; consult, Agr Inst, Kuwait, 71-72; asst res plant physiol, 72-77, from asst prof to assoc prof, 78-82, PROF POMOL, UNIV CALIF, DAVIS, 82- *Honors & Awards:* Asgrow Award, Am Soc Hort Sci, 78; Nat Food Processors Award, Am Soc Hort Sci, 80. *Mem:* Fel Am Soc Hort Sci; Am Soc Plant Physiologists; Inst Food Technologists; Int Soc Hort Sci; Coun Agr Sci & Technol. *Res:* Postharvest biology and biotechnology of horticultural crops; quality evaluation and maintenance of harvested fruits and vegetables. *Mailing Add:* Dept Pomol Univ Calif Davis CA 95616

KADESCH, ROBERT R, b Cedar Falls, Iowa, May 14, 22; m 43, 83; c 3. PHYSICS. *Educ:* Iowa State Teachers Col, BS, 43; Univ Rochester, MS, 49; Univ Wis, PhD(physics), 55. *Prof Exp:* From asst prof to assoc prof physics, 56-65, assoc dean col lett & sci, 66-68, PROF PHYSICS, UNIV UTAH, 65- *Concurrent Pos:* Staff assoc, NSF, Washington, DC, 68-69; vis res physicist, Lawrence Hall Sci, Univ Calif, Berkeley, 73-74. *Res:* Personalized computer-assisted video disc instruction; learning theory; formal thinking. *Mailing Add:* Dept Physics Univ Utah Salt Lake City UT 84112

KADEY, FREDERIC L, JR, b Toronto, Ont, June 21, 18; US citizen; m 50; c 2. ECONOMIC GEOLOGY, EXPLORATION GEOLOGY. *Educ:* Rutgers Univ, BSc, 41; Harvard Univ, MA, 47. *Prof Exp:* Res petrogr, Res Lab, US Steel Corp, 47-51; mineralogist & petrogr, Johns-Manville Res & Eng Ctr, NJ, 51-66, chief fillers sect, 66-71, res assoc geol, Res Ctr, 71-72; explor mgr, Manville Corp, Colo, 72-83; CONSULT INDUST MINERALS, 83- *Concurrent Pos:* Teaching fel mineralogy, Harvard Univ, 46-47; Nat Defense exec reservist, Emergency Minerals Admin, US Dept Interior, 73-90. *Honors & Awards:* Hardinge Award, Soc Mining Engrs, 86. *Mem:* Fel AAAS; distinguished mem Soc Mining Engrs (pres, 84); Mineral Soc Am; Sigma Xi; Am Inst Prof Geologists. *Res:* Microscopy, particle size analysis; technology of mineral fillers and hydro-thermal silicate reactions; economic evaluation of industrial minerals; exploration of diatomite, perlite, talc & kaolin deposits. *Mailing Add:* 14127 Aster Ave Wellington FL 33414-8506

KADIN, HAROLD, b New York, NY, Jan 12, 22; m 53; c 2. ANALYTICAL CHEMISTRY, BIOCHEMISTRY. *Educ:* City Col New York, BS, 48; NY Univ, MS, 53. *Prof Exp:* Chemist, NY Univ-Bellevue Med Ctr, 48-51; chief clin chemist, Meadowbrook Hosp, NY, 52-53; anal chemist, Hoffmann-La Roche Inc, NJ, 53-58; assoc technologist, Gen Foods Res Ctr, NY, 58-60; SR RES INVESTR PHARMACEUT, BRISTOL-MYERS SQUIBB PHARMACEUT RES INST, 60- *Mem:* Assoc Acad Pharmaceut Sci; Am Chem Soc. *Res:* Methods of analysis involving spectrophotometry, spectrofluorometry and chromatography; pharmaceutical analysis involving testing purity and stability of drugs; residues analysis; high performance liquid chromatography; radioimmunoassay; gas chromatography. *Mailing Add:* Analytical Res & Develop Dept Bristol-Myers Squibb Pharmaceut Res Inst New Brunswick NJ 08903

KADIS, BARNEY MORRIS, b Omaha, Nebr, Dec 26, 27. BIOCHEMISTRY. *Educ:* Univ Nebr, BA, 52; Iowa State Univ, PhD(chem), 57. *Prof Exp:* Asst chem, Iowa State Univ, 52-57; asst prof, Dubuque Univ, 57-58; res chemist, Col Med, Univ Nebr, 58-60; assoc prof chem, State Univ NY Albany, 60-61; res asst prof obstet & gynec, Col Med, Univ Nebr, Omaha, 61-66, asst prof biochem, 62-66; assoc prof biol sci, Southern Ill Univ, 69-70, assoc prof, 70-74, chmn dept dent med, 70-73, prof biochem, 74-82; PROF BIOCHEM, SCH MED, MERCER UNIV, 82- *Concurrent Pos:* Fel, Inst Hormone Biol, Syntex Res, 66-67; fel anat; Sch Med, Stanford Univ, 67-69; vis prof, Univ Calif Med Ctr, 75-76. *Mem:* AAAS; Am Chem Soc; Endocrine Soc; Am Soc Biochem & Molecular Biol; Sigma Xi; Am Col Sports Med; Am Soc Bone & Mineral Res; Am Soc Cell Biol. *Res:* Metabolism of bone cells in culture. *Mailing Add:* Div Basic Med Sci Sch Med Mercer Univ Macon GA 31207

KADIS, SOLOMON, b Baltimore, Md, May 17, 23; m 58; c 2. MICROBIOLOGY. *Educ:* St John's Col, Md, BA, 50; Univ Va, MA, 51; Vanderbilt Univ, PhD(cellular physiol), 57. *Prof Exp:* Asst, Vanderbilt Univ, 55-57; res assoc cellular physiol & microbiol, US Vitamin & Pharmaceut Corp, NY, 57-60 & Geront Res Inst, 60-61; assoc prof microbiol & immunol, Sch Med, Temple Univ, 71-72; PROF MED MICROBIOL, COL VET MED, UNIV GA, 72- *Concurrent Pos:* Asst mem, Res Labs, Albert Einstein Med Ctr, 61-63, assoc mem, 63-73. *Mem:* AAAS; Am Soc Microbiol; Am Acad Microbiol; Sigma Xi; Conf Res Workers in Animal Dis. *Res:* Bacterial physiology and toxin production and action; role of bacterial tox is in pathogenesis of respiratory disease; role of dietary iron in susceptibility of animals to bacterial infectious diseases. *Mailing Add:* Dept Med Microbiol Univ Ga Col Vet Med Athens GA 30602

KADIS, VINCENT WILLIAM, b Seinai, Lithuania, Sept 25, 22; Can citizen; m 58. MICROBIOLOGY, BIOCHEMISTRY. *Educ:* Univ Sask, BA, 55; Purdue Univ, MSc, 57, PhD(microbiol, biochem), 60. *Prof Exp:* Microbiologist, Alta Dept Agr, Can, 57-61, dir food lab serv, 61-90; PVT CONSULT, 90- *Concurrent Pos:* Consult, anal fields of foods & other agr commodities. *Mem:* Fel Am Pub Health Asn; Am Inst Food Technologists; Sigma Xi; Can Inst Food Sci & Technol (pres, 77-78); Am Soc Microbiol. *Res:* Bacteriophage of lactic cultures; Q-fever infection in humans and animals; detection and persistence of chlorinated insecticides in human and animal blood; insecticide residues in food; food sanitation, quality and safety; laboratory planning and design. *Mailing Add:* No 53 903-109 St Edmonton AB T6J 6R1 Can

KADISH, KARL MITCHELL, b Detroit, Mich, Feb 4, 45; c 2. ANALYTICAL CHEMISTRY. *Educ:* Univ Mich, BS, 67; Pa State Univ, PhD(chem), 70. *Prof Exp:* Vis asst prof chem, Univ New Orleans, 70-71; res asst, Nat Ctr Sci Res, France, 71-72; asst prof chem, Calif State Univ, Fullerton, 72-76; from asst prof to assoc prof, 76-81, assoc chmn, 79-84, PROF CHEM, UNIV HOUSTON, 81- *Concurrent Pos:* Pres, Intersci Consults, USA, 75-; vis prof, Univ Louis Pasteur, Strasbourg, France, 80-81, Univ Rome La Torgavata, Italy, 80, Ecole Superievre Chem, Lyon, France, 84 & Univ Dijon, France, 85, 87, 88 & 91; titular mem & secy, Int Union Pure & Appl Chem Comn V.5 Electroanal Chem, 80- *Honors & Awards:* Sigma Xi Res Award, 88. *Mem:* Am Chem Soc; Electrochem Soc; Sigma Xi; fel Royal Soc Chem. *Res:* Over 250 publications; analytical chemistry; electro-and bioanalytical chemistry; rates and mechanisms of electron transferin; biologically important compounds; reactions of porphyrin metal complexes; redox reactions of transition metal complexes and dinuclears metal-metal bonded complexes; spectroelectrochemistry. *Mailing Add:* Dept Chem Univ Houston Houston TX 77204-5641

KADISON, RICHARD VINCENT, b New York, NY, July 25, 25; m 56; c 1. MATHEMATICS. *Educ:* Univ Chicago, MS, 47, PhD, 50. *Hon Degrees:* Dr, Univ d'Aix-Marseille, 85, Univ Copenhagen, 87. *Prof Exp:* Nat Res Coun fel math, Inst Advan Study, 50-51, mem, Off Naval Res Contract, 51-52; from asst prof to prof, Columbia Univ, 52-64; KUEMMERLE PROF MATH, UNIV PA, 64- *Concurrent Pos:* Fulbright res grant, Denmark, 54-55; Sloan fel, 58-62; Guggenheim fel, 69-70. *Mem:* Am Math Soc; foreign mem Royal Danish Acad Sci & Lett; Sigma Xi; foreign mem Norwegian Acad Sci & Lett. *Res:* Spectral theory; group representations; topological algebra; non-commutative analysis. *Mailing Add:* Dept Math Univ Pa Philadelphia PA 19104-6395

KADKADE, PRAKASH GOPAL, b Goa, India, Sept 10, 41; US citizen; m 70; c 1. PLANT PHYSIOLOGY, PLANT BIOCHEMISTRY. *Educ:* Bombay Univ, BSc, 62, MSc, 64; St Louis Univ, PhD(biol), 70. *Prof Exp:* Fel plant biochem, St Louis Univ, 70-71; vis scientist natural prod, Cent Am Res Inst, 71-73; sr res chemist cereal chem, Anheuser Busch, Inc, 73-74; MEM TECH STAFF PLANT PHYSIOL, GEN TEL & ELECTRONICS LAB, 74- *Concurrent Pos:* Vis scientist plant biochem, Cent Am Res Inst, 74; vis prof molecular biol, Cath Univ, PR, 74. *Mem:* AAAS; Am Inst Biol Sci; Am Inst Plant Physiologists; Int Soc Plant Cell & Tissue Cult; Sigma Xi. *Res:* Understanding of mechanisms of light actions on certain plant biological and chemical processes. *Mailing Add:* GTE Labs 45 Lambert Circle Marlboro MA 01752

KADLEC, JOHN A, b Racine, Wis, Sept 22, 31; m 54; c 4. WILDLIFE MANAGEMENT, ECOLOGY. *Educ:* Univ Mich, BSF, 52, MS, 56, PhD(wildlife mgt), 60. *Prof Exp:* Res biologist, Mich Dept Conserv, 58-63 & US Bur Sport Fisheries & Wildlife, 63-67; res assoc & asst prof wildlife mgt, Univ Mich, Ann Arbor & prog coordr anal ecosyst, Int Biol Prog, 68-71, from assoc prof to prof resource ecol, 71-74; head dept, 74-80, PROF WILDLIFE SCI, COL NATURAL RESOURCES, UTAH STATE UNIV, 74- *Mem:* AAAS; Wildlife Soc; Ecol Soc Am. *Res:* Applications of population ecology and systems ecology to resource management, especially wildlife; animal habitat studies; wetland ecology. *Mailing Add:* 1166 Cliffside Dr Logan UT 84321

KADLEC, ROBERT HENRY, b Racine, Wis, June 11, 38; m 79; c 5. CHEMICAL ENGINEERING. *Educ:* Univ Wis, BS, 58; Univ Mich, MS, 59, PhD(chem eng), 62. *Prof Exp:* From asst prof to assoc prof, 61-70, PROF CHEM ENG, UNIV MICH, ANN ARBOR, 70- *Concurrent Pos:* Ed, Am Inst Chem Engrs J, 76-85. *Mem:* Am Inst Chem Engrs; Water Pollution Control Fedn; Soc Wetland Sci; Nat Soc Prof Engrs; Am Water Res Asn. *Res:* Chemical reactors; water quality; mathematical modelling; simulation; wetlands; wastewater; automobile emission control. *Mailing Add:* Dept Chem Eng Univ Mich Main Campus Ann Arbor MI 48109-2136

KADLUBAR, FRED F, b Dallas, Tex, Mar 1, 46; m 68; c 2. TOXICOLOGY, ONCOLOGY. *Educ:* Univ Dallas, BA, 68; Univ Tex, Austin, PhD(chem), 73. *Prof Exp:* Fel, McArdle Lab Cancer Res, Univ Wis, Madison, 73-76; chemist, Div Molecular Biol, 76-79, dir, Div Biochem Toxicol, 79-89, ASSOC DIR RES, NAT CTR TOXICOL RES, 89- *Concurrent Pos:* Adj prof, Dept Biochem, Dept Pharmacol & Toxicol, Univ Ark, Little Rock, 77-; mem, Working Cadre Nat Bladder Cancer Proj, 80-84. *Mem:* Am Asn Cancer Res; Sigma Xi; AAAS; Am Chem Soc; Am Soc Biol Chemists. *Res:* Biochemical mechanisms of chemical carcinogenesis with emphasis on aromatic amines and nitroaromatics and liver, bladder, and colon carcinogenisis; detoxification by glutathione and structure properties of carcinogen DNA adducts. *Mailing Add:* Off Res HFT-100 Nat Ctr Toxicol Res Jefferson AR 72079

KADNER, CARL GEORGE, b Oakland, Calif, May 23, 11; m 39; c 3. INSECT PHYSIOLOGY. *Educ:* Univ San Francisco, BS, 33; Univ Calif, Berkeley, MS, 36, PhD(med entom), 41. *Prof Exp:* Instr biol, 36-41, prof & chmn dept, 41-78, EMER PROF BIOL, LOYOLA MARYMOUNT UNIV, 78- *Concurrent Pos:* Parasitologist, US Army, 43-46. *Mem:* Entom Soc Am; Sigma Xi. *Res:* Nutritional requirements of Dipteran larvae. *Mailing Add:* 8100 Loyola Blvd Los Angeles CA 90045-2639

KADNER, ROBERT JOSEPH, b Los Angeles, Calif, Mar 19, 42; m 67; c 2. BIOCHEMICAL GENETICS. *Educ:* Loyola Univ, Los Angeles, BS, 63; Univ Calif, Los Angeles, PhD(biol chem), 67. *Prof Exp:* Nat Cancer Inst fel microbiol, Med Sch, NY Univ, 67-69; from asst prof to assoc prof, 69-80, PROF MICROBIOL, SCH MED, UNIV VA, 80- *Mem:* Am Soc Microbiol; Genetics Soc Am; Am Soc Biol Chemists. *Res:* Genetics and biochemistry of transport in Escherichia coli; bacterial genetics and regulation. *Mailing Add:* Dept Microbiol Sch Med Univ Va Charlottesville VA 22908

KADO, CLARENCE ISAO, b Santa Rosa, Calif, June 10, 36; m 63; c 2. MOLECULAR BIOLOGY, PLANT PATHOLOGY. *Educ:* Univ Calif, Berkeley, BS, 59, PhD, 64. *Prof Exp:* Res fel virus lab, Univ Calif, Berkeley, 64-67, asst res biochemist, 67-68; from asst prof to assoc prof, 66-76, PROF PLANT PATH, UNIV CALIF, DAVIS, 76- *Concurrent Pos:* NATO sr fel, Ctr Study Nuclear Energy, Mol, Belg, 74-75; sabbatical leave, Dept Molecular, Cellular & Develop Biol, Univ Colo, Boulder, 75. *Mem:* AAAS; Am Soc Microbiol; Am Phytopath Soc; NY Acad Sci; Sigma Xi. *Res:* Molecular biology of host-pathogen interactions; molecular mechanism of tumorigenesis and abnormal growth in higher cells; plant bacteriology. *Mailing Add:* Dept Plant Path Univ Calif Davis CA 95616

KADOR, PETER FRITZ, b Regensburg, Ger, Oct 3, 49; US citizen; m 76; c 2. MEDICINAL CHEMISTRY. *Educ:* Capital Univ, BA, 72; Ohio State Univ, PhD(med chem), 76. *Prof Exp:* Staff fel cataract res, 76-79, res chemist, 79-85, CHIEF MOLECULAR PHARMACOL, NAT EYE INST, NIH, 85- *Concurrent Pos:* Rhoto Cataract Res Award, 81. *Honors & Awards:* Alcon Res Found Award, 86. *Mem:* Am Chem Soc; Asn Res Vision & Ophthal; Am Diabetes Asn. *Res:* Cataract development; drug effects on the lens; aldose reductase inhibitors; diabetic complications. *Mailing Add:* Nat Eye Inst NIH Rm 10B04 Bldg 10 Bethesda MD 20892

KADOTA, T THEODORE, b Ehime-ken, Japan, Nov 14, 30; US citizen; m 56; c 3. MATHEMATICS, COMMUNICATIONS. *Educ:* Yokohama Nat Univ, BS, 53; Univ Calif, Berkeley, MS, 56, PhD(elec eng), 60. *Prof Exp:* Teaching asst, Univ Calif, Berkeley, 55-56, res asst, 56-60; MEM STAFF MATH, AT&T BELL LABS, INC, 60- *Mem:* Fel Inst Elec & Electronics Engrs. *Res:* Mathematical research in communication and information theory, specifically, application of probability theory and stochastic processes to detection, estimation, information theory; model making; theorem proving. *Mailing Add:* Math Sci Res Ctr AT&T Bell Labs Inc 600 Mountain Ave Murray Hill NJ 07974

KADOUM, AHMED MOHAMED, b Oct 28, 37; m 65; c 2. ENTOMOLOGY, TOXICOLOGY. *Educ:* Univ Alexandria, BSc, 58; Univ Nebr, MSc, 63, PhD(entom), 66. *Prof Exp:* Instr chem, Univ Alexandria, 58-60; res asst entom, Univ Nebr, 62-65, instr toxicol, 65-66; ASST PROF ENTOM, KANS STATE UNIV, 66- *Res:* Pesticidal chemistry and toxicology. *Mailing Add:* Dept Entom Kansas State Univ Manhattan KS 66506

KADYK, JOHN AMOS, b Springfield, Ill, Nov 10, 29; m 57; c 2. PHYSICS. *Educ:* Williams Col, AB, 52; Mass Inst Technol, BS, 52; Calif Inst Technol, PhD(physics), 57. *Prof Exp:* Instr physics, Univ Mich, 57-59; EXP PHYSICIST, LAWRENCE BERKELEY LAB, UNIV CALIF, 59- *Mem:* Am Phys Soc. *Res:* High energy physics; colliding beams. *Mailing Add:* Lawrence Berkeley Lab Univ Calif Berkeley CA 94720

KAEDING, WARREN WILLIAM, b Milwaukee, Wis, Apr 24, 21; m 48; c 4. ORGANIC CHEMISTRY. *Educ:* Univ Wis-Oshkosh, BS, 42; Univ Calif, Los Angeles, MS, 49, Univ Calif, PhD(chem), 51. *Prof Exp:* Res chemist, Leffingwell Chem Co, 49 & Dow Chem Co, 52-65; res chemist, 65-68, mgr org chem res, 68-72, sr res assoc, 72-78, sr scientist, 78-86, CONSULT, MOBIL CHEM CO, 86- *Mem:* Am Chem Soc. *Res:* Organic synthesis; catalytic oxidation; oxidation mechanisms, carbamate insecticides; hetero-catalysis with zeolites. *Mailing Add:* Six Roseberry Ct Lawrenceville NJ 08648-1058

KAELBER, WILLIAM WALBRIDGE, b Rochester, NY, Aug 6, 23; m 49; c 4. NEUROLOGY. *Educ:* NY Med Col, MD, 48. *Prof Exp:* Instr neurol, 55-56, assoc, 56-58, res asst prof, 58-61, assoc prof 61-68, PROF ANAT & NEUROL, COL MED, UNIV IOWA, 68- *Concurrent Pos:* USPHS spec clin trainee, Univ Iowa, 56-58. *Mem:* Asn Res Nerv & Ment Dis; Am Acad Neurol; Am Asn Anatomists; Soc Neurosci; Sigma Xi. *Res:* Experimental neuroanatomy; neurophysiology; nociceptive and analgesic aspects of nervous system. *Mailing Add:* Dept Anat Univ Iowa Col Med Iowa City IA 52240

KAELBLE, DAVID HARDIE, b Pine City, Minn, June 2, 28; m 51; c 5. PHYSICAL CHEMISTRY, POLYMER PHYSICS. *Educ:* Univ Minn, Minneapolis, BSc, 51. *Prof Exp:* Res chemist, Cent Res Labs, 3M Co, 51-56, sr res chemist, 56-61, res specialist, 61-69; mem tech staff, Sci Ctr, Rockwell Int Corp, 69-75, group leader polymer & Composites Group, 75-80; DIR, ARROYO COMPUT CTR, 80- *Concurrent Pos:* Mem comt damping nomenclature, Am Standards Asn, 63-65 & comt adhesion, Nat Res Coun, 71; chmn, Gordon Conf Adhesion, 71; chmn, Gordon Conf Thermosetts, 81. *Honors & Awards:* Adhesion Award, Am Soc Test & Mat, 63. *Mem:* Am Chem Soc; Soc Rheol. *Res:* Adhesion phenomena including surface chemistry, rheology, and fracture mechanics; polymer physical chemistry and mechanical properties; biophysics and composite material properties; cognitive science; computer modeling. *Mailing Add:* Arroyo Comput Ctr 730 Blue Oak Ave Thousand Oaks CA 91320

KAELBLE, EMMETT FRANK, b St Louis, Mo, July 31, 31; m 55; c 3. ANALYTICAL CHEMISTRY, SPECTROSCOPY. *Educ:* DePauw Univ, BA, 53; Univ Ill, MS, 55, PhD(anal chem), 57. *Prof Exp:* From res chemist to sr res chemist, Monsanto Indust Chem Co, 57-64, res group leader, Res Dept, Inorg Chem Div, Monsanto Co, 64-78, sr res group leader, appl tech, 78-86; ASSOC, CHELAN ASSOCS, 87- *Mem:* Am Chem Soc; Soc Appl Spectros. *Res:* X-ray spectroscopy; chromatography; other instrumental and chemical analytical techniques. *Mailing Add:* 641 Windrush Dr Kirkwood MO 63122-3054

KAELBLING, MARGOT, b Ulm, W Ger, Mar 25, 36; US citizen. MOLECULAR CYTOGENETICS. *Educ:* Ohio State Univ, PhD(genetics), 80. *Hon Degrees:* BSc, Ohio State Univ. *Prof Exp:* ASSOC GENETICS, ALBERT EINSTEIN COL MED, 82- *Concurrent Pos:* Vis adj instr biol, Kenyon Col, 80; instr, Wittenberg Univ, 81; res assoc, Col Physicians & Surgeons, Columbia Univ, 81-82; ed, Cytogenetics & Cell Genetics, 84-89, managing ed, 89- *Mem:* AAAS; Am Soc Human Genetics. *Res:* Molecular, cytogenetic and somatic cell genetic studies of tumor suppressor genes; localization of genes involved in chemically induced T-cell lymphoma of the mouse. *Mailing Add:* Dept Molecular Genetics Albert Einstein Col Med 1300 Morris Park Ave Bronx NY 10461

KAELLIS, JOSEPH, b Philadelphia, Pa, July 6, 25. HEAT TRANSFER. *Educ:* City Col NY, BS, 49; Univ Mo, MS, 50; Ill Inst of Technol, PhD(chem eng), 70. *Prof Exp:* Asst engr, Griscom Russel Co, 50-55; assoc engr, Argonne Nat Lab, 55-77, Advan Reactors Div, Westinghouse Elec Corp, 72-74, C F Braun & Co, 74-76, TRW Inc, 77-80; chief engr, Basic Technol Inc, 80-81; CONSULT, PVT PRACT, 81- *Mem:* Am Inst Chem Engrs; Am Nuclear Soc; Sigma Xi. *Res:* Computer technique used to determine the simultaneous transient developing mass momentum boundary layer resulting from the flow of a fluid through a channel having walls which dissolve. *Mailing Add:* 20939 Anza Ave Apt 369 Torrance CA 90503-4215

KAEMPFFER, FREDERICK AUGUSTUS, b Gorlitz, Germany, Nov 29, 20; nat Can; m 44; c 2. THEORETICAL PHYSICS. *Educ:* Univ Gottingen, dipl physics, 43, Dr rer nat(physics), 48. *Prof Exp:* Lectr, 48, from asst prof to prof, 49-85, EMER PROF PHYSICS, UNIV BC, 86- *Mem:* Am Phys Soc. *Res:* Theory of fields. *Mailing Add:* 2054 Western Pkwy Vancouver BC V6T 1V5 Can

KAESBERG, PAUL JOSEPH, b Engers, Ger, Sept 26, 23; nat US; m 53; c 3. VIROLOGY. *Educ:* Univ Wis, BS, 45, PhD(physics), 49. *Hon Degrees:* DSc, Univ Leiden, Netherlands, 75. *Prof Exp:* From instr to asst prof biomet & physics, Univ Wis-Madison, 49-54, from asst prof to prof biochem, 54-60, prof biophys & biochem, 64-87, chmn, Biophys Lab, 70-87, William W Beeman prof, 82-90, chmn, Molecular Virol Inst, 87-88, prof molecular virol & biochem, 87-90, EMER PROF MOLECULAR VIROL & BIOCHEM, UNIV WIS-MADISON, 90- *Concurrent Pos:* Career prof, NIH Res, 65- *Mem:* Nat Acad Sci; Biophys Soc; Sigma Xi; Am Soc Microbiol; Am Soc Virol (pres, 87-88); Am Soc Biol Chem. *Res:* Structure and synthesis of viruses and macromolecules. *Mailing Add:* Molecular Virol Inst Univ of Wis Madison WI 53706

KAESLER, ROGER LEROY, b Ponca City, Okla, June 22, 37; div; c 3. MICROPALEONTOLOGY. *Educ:* Colo Sch Mines, GeolE, 59; Univ Kans, MS, 62, PhD(micropaleol), 65. *Prof Exp:* From asst prof to assoc prof, 65-73, PROF GEOL & DIR MUS INVERT PALEONT, UNIV KANS, 73- *Mem:* Paleont Soc; Soc Syst Zool; Geol Soc Am; Am Soc Naturalists. *Res:* Paleoecology of Ostracoda; quantitative methods in paleontology and applied aquatic biology. *Mailing Add:* Dept Geol Univ Kans Lawrence KS 66045

KAESZ, HERBERT DAVID, b Alexandria, Egypt, Jan 4, 33; nat US; m 58; c 3. ORGANOMETALLIC CHEMISTRY. *Educ:* NY Univ, BA, 54; Harvard Univ, MA, 56, PhD, 59. *Prof Exp:* Fel inorg chem & adv prog high sch teachers, Harvard Univ, 58-60; from asst prof to assoc prof, 60-68, PROF INORG CHEM, UNIV CALIF, LOS ANGELES, 68- *Concurrent Pos:* Assoc ed, Inorg Chem, Am Chem Soc, 68- *Honors & Awards:* US Scientist Award, Alexander von Humboldt Sr, Fed Repub Ger, 88. *Mem:* Am Chem Soc; Royal Soc Chem; fel AAAS; fel Japan Soc Prom Sci. *Res:* Chemistry of transition metals, especially organometallic complexes, polynuclear metal carbonyl cluster complexes and hydrides; pathways of homogeneous catalysis; organometallic chemical deposition; coal liquefaction. *Mailing Add:* Dept Chem & Biochem Univ Calif Los Angeles CA 90024-1569

KAETZEL, MARCIA ALDYTH, CALMODULIN, CALMODULIN BINDINGS PROTEINS. *Educ:* Baylor Col Med, PhD(cell biol), 85. *Prof Exp:* FEL CELL BIOL, SCH MED, UNIV TEX, 85- *Mailing Add:* Dept Physiol Univ Cincinnati Sch Med ML 576 Cincinnati OH 45267

KAFADAR, KAREN, b Evergreen Park, Ill, July 6, 53. DATA ANALYSIS, ROBUST METHODS. *Educ:* Stanford Univ, BS & MS, 75; Princeton Univ, PhD(statist), 79. *Prof Exp:* Asst prof statist, Ore State Univ, 79-80; math statistician, statist eng div, Nat Bur Standards, 80-83; mem tech staff, Hewlett-Packard, 83-90; CANCER CONTROL FEL, UNIV COLO HEALTH SCI CTR, 90- *Concurrent Pos:* Consult, OEA, Inc, 77- *Mem:* Am Statist Asn; Inst Math Statist. *Res:* New methodology in data analysis particularly robust methods and treatment of outliers; experimental design; spectrum analysis; statistical engineering design. *Mailing Add:* Biomet Dept Univ Colo Health Sci Ctr Denver CO 80265

KAFALAS, PETER, b Newburyport, Mass, Dec 6, 25; m 57; c 2. CHEMICAL PHYSICS. *Educ:* Harvard Univ, AB, 50; Mass Inst Technol, PhD(inorg chem), 54. *Prof Exp:* Assoc chemist, Argonne Nat Lab, 54-59; staff mem, Mitre Corp, Mass, 59-61; sr scientist, Tech Opers, Inc, 61-65; staff mem, Lincoln Lab, Mass Inst Technol, 65-86; RETIRED. *Concurrent Pos:* Consult, Laser Technol. *Mem:* Am Chem Soc; Optical Soc Am. *Res:* Nuclear chemistry, deuteron reactions and neutron reactions; spectroscopy, high-speed spectrography of plasmas; laser technology, laser Q-switching with saturable dyes; laser propagation studies; laser vaporization of fog droplets; laser beam diagnostics. *Mailing Add:* 24 Hickory Rd Sudbury MA 01776

KAFATOS, FOTIS C, b Crete, Greece, Apr 16, 40; nat US; m 67; c 2. DEVELOPMENTAL BIOLOGY. *Educ:* Cornell Univ, BA, 61; Harvard Univ, MA, 62, PhD(biol), 65. *Prof Exp:* Tutor, 62-63, from instr to asst prof, 65-69, chmn cellular & develop biol, 78-81, PROF BIOL, HARVARD UNIV, 69-; PROF BIOL & DIR, INST MOLECULAR BIOL & BIOTECHNOL, RES CTR CRETE, 82- *Concurrent Pos:* Mem, Develop Biol Panel, NSF, 70-72; distinguished lectr, Univ Tex, 77; mem, Cell Biol & Nucleic Acids & Protein Synthesis Adv Comt, Am Cancer Soc, 83-86; mem, Europ Molecular Biol Orgn Coun, 88-, Nat Sci Adv Bd, Greece, 88- *Honors & Awards:* Walter Bauer Mem Lectr, Helen Hay Whitney Found, 78; Rosenberger Distinguished Lectr, Univ Rochester, 81. *Mem:* Nat Acad Sci; fel AAAS; Am Soc Cell Biol; Soc Develop Biol (pres, 81-82); fel Am Acad Arts & Sci; Europ Molecular Biol Orgn; Genetics Soc Am; Am Soc Zoologists; Int Soc Develop Biologists. *Res:* Molecular and cellular aspects of development, cell differentiation during insect metamorphosis; molecular evolution. *Mailing Add:* Biol Labs Harvard Univ 16 Divinity Ave Cambridge MA 02138

KAFATOS, MINAS, b Crete, Greece, Mar 25, 45; m 71; c 3. ASTROPHYSICS. *Educ:* Cornell Univ, BA, 67; Mass Inst Technol, PhD(physics), 72. *Prof Exp:* Res assoc astrophysics, Joint Inst for Lab Astrophys, Univ Colo, 72-73 & Nat Res Coun, Nat Acad Sci, 73-75; from asst prof to assoc prof, 75-84, PROF PHYSICS, GEORGE MASON UNIV, 84- *Concurrent Pos:* Res scientist astrophys, Goddard Space Flight Ctr, NASA, 75- *Mem:* Am Astron Soc; Am Phys Soc; Int Astron Union; Royal Astron Soc. *Res:* Black holes; quasars; active galaxies; interstellar medium; mass loss and long period variables; forbidden line calculations; symbiotic stars; cosmic rays; supernovae; quantum physics. *Mailing Add:* Dept Physics George Mason Univ 4400 University Dr Fairfax VA 22030

KAFER, ENID ROSEMARY, b Sydney, Australia, May 27, 37. ANESTHESIOLOGY. *Educ:* Univ Sydney, BS, 59, MB & BS, 62, MD, 70, FRCS. *Prof Exp:* Med resident, Royal Prince Alfred Hosp, Sydney, 62-63; anesthetic registr, 64; res fel, Dept Med, Univ Sydney, 65-66, lectr, 67, Life Ins Med Res fel, 68-69; sr registr, Dept Anesthetics, Royal Postgrad Med Sch, London, 69-71; fel physiol, Univ Calif, San Francisco, 71-72, asst prof, Dept Anesthesia, 72-73; ASSOC PROF PHYSIOL & ANESTHESIOL, SCH MED, UNIV NC, CHAPEL HILL, 73- *Concurrent Pos:* Mem, Res Rev Comt, NC Heart Asn, 76-79. *Honors & Awards:* Peter Baneroff Award, 70. *Mem:* Fel Royal Australian Col Physicians; Am Physiol Soc; Am Soc Anesthesiologists; Am Thoracic Soc; Sigma Xi. *Res:* Load adjustment mechanisms of the respiratory system, including examination of neural and muscle factors, and the effects of changing chemical stimuli, chemoreceptor denervation and the effects of general anesthesia. *Mailing Add:* Dept Anesthesiol Sch Med Univ NC CB-7010 Chapel Hill NC 27599-7010

KAFER, ETTA (MRS E R BOOTHROYD), b Zurich, Switz, July 31, 25; m 57; c 3. MITOTIC RECOMBINATION, ANEUPLOIDY. *Educ:* Univ Zurich, dipl, 48, PhD(genetics), 52. *Prof Exp:* Res asst microbial genetics, Glasgow Univ, 53-55, res fel, 55-56; res fel, Carnegie Inst, 56, res assoc, 58-63, from lectr to assoc prof, 59-76 PROF MOLECULAR GENETICS, MCGILL UNIV, 76- *Honors & Awards:* Award of Excellence, Genetics Soc Can, 87. *Mem:* Genetics Soc Am; Genetics Soc Can; Swiss Genetics. *Res:* Microbial genetics; mitotic and meiotic recombination; molecular genetics of nucleases; DNA repair mutants and gene cloning in fungi; assays for environmentally induced nondisjunction. *Mailing Add:* Dept Biol McGill Univ 1205 Dr Penfield Ave Montreal PQ H3A 1B1 Can

KAFESJIAN, R(ALPH), b Chicago, Ill, Mar 28, 34; m 55; c 3. CHEMICAL ENGINEERING, PHYSICAL CHEMISTRY. *Educ:* Purdue Univ, BS, 55; Univ Louisville, MChE, 57, PhD(chem eng), 61. *Prof Exp:* Instr chem physics & chem eng, Univ Louisville, 57-60; res chem engr, Monsanto Res Corp, 61-63, sr res chem engr, 63-67, res group leader, 67-69; biomed res lab, 69-71, task leader corp technol, 71-77, sr scientist, 78-80, PRIN SCIENTIST, AM HOSP SUPPLY CORP, 81- *Mem:* AAAS; Am Chem Soc; assoc Am Inst Chem Engrs; Nat Asn Corrosion Engrs; Biomat Soc; Sigma Xi. *Res:* High temperature materials, reactions, processes and technology; electrochemical energy conversion methods; corrosion; electrochemistry; biomedical materials and devices. *Mailing Add:* 18922 Racine Dr Irvine CA 92715

KAFFEZAKIS, JOHN GEORGE, b Athens, Greece, June 15, 29; US citizen; m 55; c 2. FOOD SCIENCE, MICROBIOLOGY. *Educ:* Col Agr, Athens, BS, 50; Iowa State Univ, MS, 52; Univ Md, PhD(food sci), 67. *Prof Exp:* Lab supt, Borden Co, Ltd, Can, 52-55; food technologist, Agr & Fisheries Br, Ottawa, Ont, 55-64; asst prof food technol, Auburn Univ, 64-66; res scientist, Joseph E Seagrams & Sons, Inc, 67-69; dir tech serv, Overseas Div, Nat Can Corp, 69-85; INDEPENDENT CONSULT, FOOD INDUSTRY, 85- *Mem:* AAAS; Inst Food Technologists; Agr Inst Can. *Res:* Processing, microbiology, packaging, quality evaluation, development of new food products and marketing. *Mailing Add:* 29 Selefkou St Thrakomakedones Attiki 136-71 Greece

KAFKA, MARIAN STERN, b Richmond, Va, Mar 30, 27; m 52; c 3. NEUROSCIENCE, PHYSIOLOGY. *Educ:* Conn Col, BA, 48; Univ Chicago, PhD(physiol), 52. *Prof Exp:* Teaching asst zool, Conn Col, 47-48; asst physiol, Sch Med, Univ Chicago, 48-52; res asst, Sch Med, Emory Univ, 52-53, Ill Neuropsychiat Inst, Univ Ill, Chicago, 53-54 & Sch Med, Yale Univ, 54-57; USPHS fel, Endocrinol Br, Nat Heart & Lung Inst, 65-68, physiologist, Hypertension-Endocrine Br, 68-74, sect biochem & pharmacol, Biol Psychiat Br, 74-82, physiologist, Clin Neurosci Br, 82-85, exec secy, Cellular Neurobiol & Psychopharmacol, 85-90, CHIEF CLIN, EPIDEMIOL & SERV REV BR, DIV EXTRAMURAL ACTIV, NIMH, USPHS, 90- *Concurrent Pos:* Marie J Mergler fel physiol, Univ Chicago, Ill, 50; mem, pub info comt, Fedn Am Socs Exp Biol, 77-82; mem, coun NIMH-Nat Inst Neurol & Commun Dis Assembly Scientists, 81-82, pres, 82-83 & mem, clin res rev comt, NIMH, 82-84; mem, Comt Intramural Prog Dir's Conf, NIH, 82-85; reviewer, Med Res Coun Can & NSF & numerous sci publns. *Mem:* Am Physiol Soc; Endocrine Soc; Biophys Soc; AAAS; Sigma Xi; Soc Neurosci; Int Soc Chronobiol. *Res:* Interaction between neurotransmitters, hormones and receptors on neurons and blood cells; central nervous system control of circadian rhythms. *Mailing Add:* NIMH 9C-02 Parklawn Bldg 5600 Fishers Lane Rockville MD 20857

KAFKA, ROBERT W(ILLIAM), b Chicago, Ill, Oct 30, 37; m 66; c 2. DEFENSE ELECTRONICS SYSTEMS, COMMAND & CONTROL SYSTEMS. *Educ:* Univ Ill, BSEE, 58, MSEE, 59, PhD(elec eng), 63. *Prof Exp:* Instr elec eng, Univ Ill, 60-63; mem tech staff, Guidance & Control Dept, Aerospace Corp, 63-66; MGR ADVAN SYSTS PROGS OFF, SYSTS DIV, HUGHES AIRCRAFT CO, 66- *Mem:* Inst Elec & Electronics Engrs; Sigma Xi. *Mailing Add:* 2749 Puente St Fullerton CA 92635

KAFKA, TOMAS, b Praha, Czech, Oct 15, 36; US citizen. ELEMENTARY PARTICLE PHYSICS. *Educ:* Univ Karlova, Praha, Promovany Fyzik, 60; State Univ NY, Stony Brook, PhD(physics), 74. *Prof Exp:* Res assoc, State Univ NY, Stony Brook, 74-82; from res asst prof to res assoc prof, 82-91, RES PROF, TUFTS UNIV, 91- *Mem:* Am Phys Soc. *Res:* Hadron-hadron, lepton-hadron, and photon-hadron interactions at high energies both by optical and electronic detectors; nucleon decay; cosmic rays; particle astronomy. *Mailing Add:* Dept Physics & Astron Tufts Univ Medford MA 02155

KAFRAWY, ADEL, b Cairo, Egypt, Oct 15, 43; US citizen; m 75; c 2. POLYMER & ORGANIC CHEMISTRY, CELLULOSE TECHNOLOGY. *Educ:* Cairo Univ, BSc, 64; Univ Rochester, MS, 71; Univ Mo-Columbia, PhD(org chem), 74; Syracuse Univ, MBA, 83. *Prof Exp:* Demonstr chem, Cairo Univ, 64-68; assoc phys org chem, Syracuse Univ, 74-75; assoc org chem, State Univ NY Col Environ Sci & Forestry, Syracuse, 75-77; res chemist, Cellulose Res, ITT Rayonier, Inc, 77-81; SR SCIENTIST, JOHNSON & JOHNSON, 81- *Res:* Biomedical application of polymers; controlled drug release systems. *Mailing Add:* J&J Orthop PO Box 350 350 Paramount Dr Raynham MA 02767-0350

KAFRI, ODED, b Tel-Aviv, Israel, Dec 13, 44; m 66; c 2. MOIRE EFFECT, AUTHENTICATION. *Educ:* Technion, BSc, 69, DSc, 73. *Prof Exp:* Asst, Dept Chem, Techinon, 69-71, instr, 71-73; instr appl physics, Hebrew Univ, Jerusalem, 73-74; res scientist assoc, Dept Chem, Univ Wis-Madison, 74-76; res scientist, Nuclear Res Ctr-Negev, Beer Sheva, 76-82, group leader, Optical-Non-Destructive Testing, 78-86; pres, Rotlex Optics Ltd, Arava, Israel, 86-89; PRES, CONSTANZA 33O LTD, 90- *Concurrent Pos:* Adv energy, Nat Res Coun, Israel, 78-82; sr res scientist, Nuclear Res Ctr-Negev, Beer Sheva, 82-89; sr physicist, Corp Technol, Allied Corp, Mt Bethel, NJ, 83-84. *Mem:* Optical Soc Am; Int Soc Optical Eng. *Res:* Optical metrology: invention of moire deflectometry, a method used in several industries like ophthalmological industry, wind tunnels, lasers industries and research labs; theoretical aspects of metrology and information theory; author or co-author of one book, over 100 scientific publications and holder of over 20 patents. *Mailing Add:* Ehud 3 Beer Sheva 84234 Israel

KAGAN, BENJAMIN, b New York, NY, Mar 9, 21; m 43; c 2. ORGANIC CHEMISTRY. *Educ:* DePaul Univ, BS, 47; Pa State Univ, MS, 50. *Prof Exp:* Org res chemist, Army Chem Ctr, Md, 50-62; res anal chemist, Div Food Chem, Bur Drugs, Food & Drug Admin, 62-66, chemist, Div Oncol & Radiopharmaceut Drug Prod, 66-73, supvry chemist, 73-77; consult regulatory affairs, 77-83; RETIRED. *Concurrent Pos:* Lectr regulatory affairs, 79-85; counr, Fla Sect, Am Inst Chemists, 88- *Mem:* AAAS; Am Chem Soc; fel Am Inst Chemists. *Res:* Organo-phosphorous and sulfur compounds; nitrogen heterocycles; radiopharmaceuticals. *Mailing Add:* PO Box 352948 Palm Coast FL 32135-2948

KAGAN, BENJAMIN M, b Washington, Pa, July 18, 13; m 40; c 2. PEDIATRICS. *Educ:* Washington & Jefferson Col, AB, 33; Johns Hopkins Univ, MD, 37; Am Bd Nutrit & Am Bd Pediat, cert & recert. *Prof Exp:* Instr, St Phillip Sch Pub Health Nursing, 40-42, instr, Marine Biol Lab, Woods Hole, 34; intern, Sinai Hosp, Baltimore, 37-38; res contagion, Willard Parker Hosp, 38; resident pediat, Babies Hosp, New York, 38-40; instr pediat, Med Col Va, 40-42; asst, Columbia Univ, 46; from clin asst prof to clin prof, Univ Ill, 47-54; prof, Northwestern Univ, 54-55; prof-in-residence, 55-90, vchmn dept, 74-90, EMER PROF PEDIAT, SCH MED, UNIV CALIF, LOS ANGELES, 90- *Concurrent Pos:* Mem staff, Michael Reese Hosp, 46-55, attend pediatrician, & dir, Pediat Res Dept, Inst Med Res, 46-55, chmn dept pediat, 51-55; dir & chmn pediat, Cedars Sinai Med Ctr, 55-; official examr, mem exec comt & chmn written exam comt, Am Bd Pediat; dir, Cystic Fibrosis Ctr, 60- *Honors & Awards:* Distinguished Serv Award, US Army, 45; Citation, Am Acad of Pediat, 78. *Mem:* AAAS; Am Pediat Soc; Soc Pediat Res; Soc Exp Biol & Med; AMA; fel Am Acad Pediat; fel Am Col Physicians; fel Am Col Chest Physicians; fel Infectious Dis Soc Am; fel Pediat Infectious Dis. *Res:* Nutrition and infectious disease. *Mailing Add:* Dept Pediat Cedars Sinai Med Ctr Univ Calif 5005 Finley Ave Los Angeles CA 90048-0750

KAGAN, FRED, b Chicago, Ill, Dec 24, 20; m 45; c 6. MEDICINAL CHEMISTRY, ORGANIC CHEMISTRY. *Educ:* Univ Ill, BS, 42; Mass Inst Technol, PhD(org chem), 49. *Prof Exp:* Res org chemist, Stand Oil Co Ind, 49-52; res org chemist, Upjohn Co, 52-68, mgr cent nerv syst res, 68-78, group mgr, 78-81, dir exp sci & therapeut, 81-82, vpres therapeut, clin res & biotechnol, 82-86; RETIRED. *Mem:* Am Chem Soc; Royal Soc Chem; NY Acad Sci. *Res:* Synthesis; psychopharmacology, drug development from test tube thru obtaining an NDA and obtaining international registrations. *Mailing Add:* 2225 Chevy Chase Kalamazoo MI 49008

KAGAN, HARVEY ALEXANDER, b New York, NY, Sept 25, 37; m 68; c 2. CONSTRUCTION MANAGEMENT, FORENSIC ENGINEERING. *Educ:* Columbia Univ, BS, 58; Univ Ill, MS, 59; NY Univ, EngScD(civil eng), 65; Purdue Univ, MS, 80. *Prof Exp:* Struct engr, Repub Aviation Corp, 59-61; asst civil engr, New York City Bd Educ, 61; struct test engr, Martin Co Div, Martin-Marietta Corp, 61-62, eng specialist, 65-66; test engr, Vertol Div, Boeing Co, 62; struct mech engr, Grumman Aircraft Corp, 66; from asst prof to assoc prof civil eng, Rutgers Univ, 66-74; sr civil engr, C F Braun & Co, 74-75; assoc prof & prog chmn, Univ Evansville, 75-77; sr consult, Wagner-Hohns-Inglis, 78-81; sr consult, Hill Int, Inc, 81-82; pres, Construct Adv Group, Inc, 82-84; prof civil eng, Rutgers Univ, 84-87; ASSOC, S T HUDSON INT, INC, 87- *Concurrent Pos:* NSF res grant, 67-68. *Mem:* Am Soc Civil Engrs; Am Concrete Inst; Sigma Xi. *Mailing Add:* 18 Tanager Way RD 8 Medford NJ 08055

KAGAN, HERBERT MARCUS, b Boston, Mass, Aug 18, 32; m 62; c 2. BIOCHEMISTRY. *Educ:* Univ Mass, Amherst, BS, 54, MS, 56; Tufts Univ, PhD(biochem), 66. *Prof Exp:* Instr microbiol, Purdue Univ, 56-58; res assoc pharmacol, Sch Med, Boston Univ, 59-60; biologist, Arthur D Little, Inc, 60-61; asst prof, 69-72, assoc prof, 72-80, PROF BIOCHEM, SCH MED, BOSTON UNIV, 80- *Concurrent Pos:* Am Cancer Soc res fel biochem, Harvard Med Sch, 66-69; fel, Arteriosclerosis Coun, Am Heart Asn. *Mem:* AAAS; Am Soc Biol Chemists; Am Heart Asn; Sigma Xi. *Res:* Enzymology; protein chemistry; structure-function relationships of enzymes; connective tissue proteins; stereospecifity in catalysis. *Mailing Add:* Dept Biochem Boston Univ Sch Med 80 E Concord St Boston MA 02118

KAGAN, IRVING GEORGE, b New York, NY, June 1, 19; m 40; c 2. PARASITOLOGY, IMMUNOLOGY. *Educ:* Brooklyn Col, AB, 40; Univ Mich, MA, 47, PhD(zool), 50. *Prof Exp:* Asst prof zool, Univ Pa, 52-57; chief helminth unit, 57-62, chief parasitol unit, 62-66, DIR PARASITOL DIV, CTRS DIS CONTROL, 67- *Concurrent Pos:* Nat Res Coun fel, Univ Chicago, 50-52; mem, Scientific Working Group, Epidemology, WHO, 78. *Honors & Awards:* Henry Baldwin Ward Medal, Am Soc Parasitol, 65; Behring-Bilharz Medal, 81. *Mem:* Am Soc Trop Med & Hyg; Am Soc Parasitol; Am Asn Immunol; Am Micros Soc; Sci Res Soc Am. *Res:* Immunodiagnosis of parasitic infections and the immunology of the host parasite interaction. *Mailing Add:* 1074 Oakdale Rd NE Atlanta GA 30307

KAGAN, JACQUES, b Paris, France, Nov 11, 33; US citizen. ORGANIC CHEMISTRY, BIOLOGICAL CHEMISTRY. *Educ:* Sorbonne, BS, 56; Rice Univ, PhD(chem), 60. *Prof Exp:* Res assoc, Mass Inst Technol, 60-61; res chemist, Amoco Chem Co, Ind, 61-62; res scientist & Welch Found fel, Univ Tex, 62-65; from asst prof to assoc prof, 65-73, PROF CHEM & BIOL, UNIV ILL, CHICAGO, 73- *Concurrent Pos:* Vis prof, Univ Geneva, 71-72 & Univ Haute-Alsace, 80; Fulbright grant, 80, 87; Museum Nat d'Histoire Naturelle, Paris, 87. *Honors & Awards:* Cooley Award, Am Soc Plant Taxon, 64. *Mem:* Am Chem Soc; Am Soc Photobiol. *Res:* Organic synthesis and reaction mechanisms; photochemistry and photobiology; environmental chemistry. *Mailing Add:* 1620 Washington Ave Wilmette IL 60091-2419

KAGAN, JEROME, b Newark, NJ, Feb 25, 29; m 51; c 1. DEVELOPMENTAL PSYCHOLOGY. *Educ:* Rutgers Univ, BS, 50; Yale Univ, PhD(psychol), 54; Harvard Univ, MA, 64. *Prof Exp:* Chmn, dept psychol, Fels Res Inst, 57-64; PROF, DEPT PSYCHOL, HARVARD UNIV, 64- *Honors & Awards:* Distinguished Sci Award, Am Psychol Asn, 87. *Mem:* Nat Acad Sci; Am Psychol Asn; Soc Res Child Develop; Am Asn Advan Sci. *Res:* Cognitive and emotional development of children; role of temperament in personality development. *Mailing Add:* Dept Psychol Harvard Univ 33 Kirkland St Cambridge MA 02138

KAGAN, JOEL (DAVID), b New York, NY, Aug 18, 43; div; c 2. MATHEMATICS. *Educ:* Rutgers Col, BA, 66; Stevens Inst Technol, MS, 67 & 84, PhD(math), 70. *Prof Exp:* Instr math, Stevens Inst Technol, 68-70; asst prof, 70-84, ASSOC PROF MATH, UNIV HARTFORD, 84- *Mem:* Soc Symbolic Logic; Math Asn Am. *Res:* Algebraic logic; algebraic structures arising from theories in the sentential and predicate calculus with identity connective. *Mailing Add:* Dept Math Univ Hartford 200 Bloomfield Ave West Hartford CT 06117

KAGANN, ROBERT HOWARD, b New York, NY, June 26, 46. CHEMICAL PHYSICS. *Educ:* NY Univ, BA, 69; Univ Colo, PhD(chem physics), 75. *Prof Exp:* Res fel physics, Univ BC, 75-77; nat res coun assoc physics, molecular spectros div, Nat Bur Standards, 77-81; sci syst & appln, NASA Goddard, 81-85; AT RCA ASTRO ELEC, 85- *Mem:* Optical Soc Am. *Res:* Molecular spectroscopy in the visible, infrared, and microwave regions; laser spectroscopy and lidar. *Mailing Add:* G E Astrospace Div Advan Missions MS 111 PO Box 800 Princeton NJ 08543-0800

KAGANOV, ALAN LAWRENCE, b Brooklyn, NY, Dec 7, 38. BIOENGINEERING, MECHANICAL ENGINEERING. *Educ:* Duke Univ, BS, 60; New York Univ, MBA, 67; Columbia Univ, MS, 72, ScD, 74. *Prof Exp:* Packaging engr, Amstar Co, 60-64; prod develop engr, Johnson & Johnson, 64-69; career fel, NIH, 69-74; DIR MED PROD RES & DEVELOP, DAVIS & GECK, AM CYANAMID CO, 74- *Mem:* Sigma Xi; Am Chem Soc; Am Soc Testing & Mat. *Res:* Development of new implantable materials and medical devices. *Mailing Add:* Baxter Health Care Inc One Baxter Pkwy Deerfield IL 60015

KAGARISE, RONALD EUGENE, b East Freedom, Pa, July 17, 26; m 47; c 3. PHYSICS. *Educ:* Duke Univ, BA, 48; Pa State Col, MS, 49, PhD(physics), 51. *Prof Exp:* Res assoc physics, Pa State Col, 51-52; head chem spectros sect, Naval Res Lab, 52-66; prog dir phys chem, NSF, 66-68; supt chem div, Naval Res Lab, 68-76; DIR DIV MAT RES, NSF, WASHINGTON, DC, 76- *Mem:* Am Phys Soc; Sigma Xi; Am Chem Soc; Coblentz Soc. *Res:* Rotational isomerism; molecular constants; structure. *Mailing Add:* 602 Pine Rd Ft Washington MD 20744

KAGAWA, YUKIO, b Yamagata, Japan, May 8, 35. COMPUTATIONAL ACOUSTICS, NUMERICAL ANALYSIS. *Educ:* Tohoku Univ, Sendai, Japan, BEng, 58, MEng, 60, DrEng, 63. *Prof Exp:* Postdoctoral fel mech eng, Polytechnic Inst, Brooklyn, NY, 64-65; postdoctoral fel acoust eng, Tech Univ Norway, 66-68; res fel acoust eng, Inst Sound & Vibration Res, Southampton Univ, UK, 68-70; prof, 70-90, EMER PROF ELEC ENG, TOYAMA UNIV, JAPAN, 90-; PROF ELEC & ELECTRONIC ENG, OKAYAMA UNIV, 90- *Concurrent Pos:* Leverhulme fel, Univ New South Wales, Australia, 78; vis prof, Indian Inst Technol, Delhi, India, 82, Inst Acoust, Academia Sineca, China, 84; mem bd dirs, Japan Soc Simulation Technol, 86- *Honors & Awards:* Ishikawa Prize, Union Japanese Scientists & Engrs, 89. *Mem:* Fel Inst Elec & Electronics Engrs; fel Inst Acoust; Inst Electronic Info & Commun Engrs. *Res:* Numerical modelling and simulation of electrical acoustical and vibration systems by means of finite element and boundary element method; electrical impedance and ultrasonic computed tomography as an inverse or optimization problem. *Mailing Add:* Dept Elec & Electronic Eng Okayama Univ Okayama 700 Japan

KAGEL, RONALD OLIVER, b Milwaukee, Wis, Jan 16, 36; m 59; c 3. ANALYTICAL CHEMISTRY. *Educ:* Univ Wis, BS, 58; Univ Minn, PhD(phys chem), 64. *Prof Exp:* Res chemist, Dow Chem Co, 64-67, proj leader, Chem Res Lab, 67-69, group leader, 69-71, sr res chemist, 71-72, sr anal specialist, 72-74, supvr spec anal group, 74-76, supt environ control, 76-77, environ systs group leader surface anal, 77-78, res mgr surface anal, 78-79, mgr, Regulatory Affairs-Water, 79-81, mgr, states environ activities, 81, dir environ qual, 81-85, environ dir, Eng Res & Comput Serv, 85-88, ENVIRON CONSULT, DOW CHEM CO, 88- *Concurrent Pos:* Mem fac, Saginaw Valley Col, 64-67; mem adv bd, Raman Newslett, 68-75; secy, Ad-hoc Subpanel Laser Excited Raman Spectra, Nat Res Coun, 68, mem, Numerical Data Adv Bd, Joint Comt Atomic & Molecular Structure-Subcomt Laser Raman Spectros, 70-80; mem, Ad-hoc Panel Micro-Raman Spectros, Nat Bur Stand, 74; assoc ed, Appl Spectros, 75-80; secy, Int Fourier Trans Conf, 77, CMA leader, Task Group Environ Monitoring, 76-81; mem adv comt, Critical Mat Register, State Mich Dept Nat Res, 76-79, CMA work group leader, Environ Audits, 82-84; liason coordr, Am Soc Testing & Mat, 79-81; chmn, Natural Resources Comt, State Govt Affairs Coun, 81, Groundwater Subcomt, 84-85 & Bd dirs, Coalition Responsible Waste Incineration, 87 & 90-; mem, Environ Qual Comn, Synthetic Org Chem Mfg Asn, 82-85; Dupont fel, Univ Wis; res fel, Univ Minn. *Honors & Awards:* Anachem Award, Am Inst Chem, 88. *Mem:* fel Am Inst Chemists; Sigma Xi; Am Chem Soc; AAAS; NY Acad Sci; Coblentz Soc. *Res:* Molecular structure elucidation, surface catalysis and bulk mechanistic studies; remote detection of ambient air emissions by infrared Fourier transform and Raman spectroscopy; infrared chemiluminescence applications; environmental measurements of trace compounds; combustion chemistry and incineration. *Mailing Add:* Environ Qual 2030 Bldg Dow Chem Co Midland MI 48674

KAGEN, HERBERT PAUL, b Worcester, Mass, May 6, 29; m 56; c 4. ORGANIC CHEMISTRY. *Educ:* Mass Inst Technol, SB, 52; Univ RI, MS, 54; Wayne State Univ, PhD, 60. *Prof Exp:* From asst prof to assoc prof chem, Detroit Inst Technol, 57-67; prof org chem, WVa State Col, 67-90, chmn, Dept Chem, 68-71, dir chem technol, 75-90; FIELD SERV PROF ENVIRON HEALTH & ASSOC DIR & SUPVR ENVIRON ANALYSIS CHEM LAB INST ENVIRON HEALTH, UNIV CINCINNATI MED CTR, 87- *Concurrent Pos:* Consult, Chem Serv Corp, 57-, Union Carbide Corp, WVa State Bd Educ & Kanawha County Bd Educ; Mich Heart Asn fel, 60; adj prof, WVa Univ, 70-71; dir, NSF-Coop Col Sch Sci Prog for high sch chem teachers, 70-71; dir, WVa RESA III Prog Gifted High Sch Sr, 75; fac adv, NSF-SOS Prog, Kanawha Valley, 73; adj prof, Grad Studies, WVa Col, 75-77; vis prof, Inst Environ Health Med Ctr, Univ Cincinnati, 87-88; assoc dir & supvr, Anal Chem Lab, 88-89. *Mem:* Am Chem Soc; Nat Sci Teachers Asn; Am Asn Univ Prof; Sigma Xi. *Res:* Reaction of amines with lactides; organic nitrogen chemistry; preparation of lactides; chemical education; environmental. *Mailing Add:* 7017 Mayfield Ave Cincinnati OH 45243

KAGETSU, T(ADASHI) J(ACK), b Vancouver, BC, Apr 22, 31; m 57; c 2. CHEMICAL ENGINEERING. *Educ:* Univ Toronto, BASc, 54, MASc, 55, PhD(appl chem), 57. *Prof Exp:* Asst res chem engr, metals div, Union Carbide Corp, 57-60, res chem engr, 60-61, assoc engr, nuclear div, 61-64, proj engr, mining & metals div, 64-67, staff engr, 67-75, mgr, process eng, 75-77, mgr design eng, 77-78, asst dir tech, Metals Div, 78-86; CONSULT, 87- *Res:* Kinetics of metal dissolution in aqueous media; fused salt electrolysis; gas-solid mass transfer and heat transfer; hydrometallurgy and pyrometallurgy; process simulation by computers. *Mailing Add:* Consult 435 Dutton Dr Lewiston NY 14092

KAGEY-SOBOTKA, ANNE, b Atlanta, Ga, June 11, 40. ALLERGIES. *Educ:* Vanderbilt Univ, BA, 62; Johns Hopkins Univ, PhD(microbiol), 77. *Prof Exp:* From instr to asst prof, 77-85, ASSOC PROF MED, SCH MED, JOHNS HOPKINS UNIV, 85- *Mem:* Am Asn Immunol; fel Am Acad Allergy; Int Col Allergists. *Mailing Add:* Dept Med J H Asthma & Allergy Ctr 301 Bayview Blvd Baltimore MD 21224

KAGHAN, WALTER S(EIDEL), b New York, NY, Apr 16, 19; m 47; c 3. CHEMICAL ENGINEERING. *Educ:* City Col New York, BChE, 40; NY Univ, MChE, 48; Purdue Univ, PhD(chem eng), 52. *Prof Exp:* Res chem engr, St Regis Paper Co, 40-41; assoc chem engr, E I du Pont de Nemours & Co, 42-43; process engr, Kellex Corp, 44-45; dir, Resinous Res Assoc, 45-48; asst prof chem eng, Rose Polytech Inst, group leader & asst sect chief, Develop Sect, Olin Corp, 51-59, dir develop sect, Film Res & Develop Dept, 59-63, dir res & develop, Film Div, 63-75; consult chem engr, 75-86; RETIRED. *Mem:* Am Soc Plastics Engrs; Am Inst Chem Engrs; NY Acad Sci; Am Chem Soc. *Res:* Cellophane; polymers; plastics extrusion; rheology; polyolefins; packaging. *Mailing Add:* 4315 Brandywine Dr Sarasota FL 34241

KAGIWADA, HARRIET HATSUNE, b Honolulu, Hawaii, Sept 2, 37; m 61; c 2. APPLIED MATHEMATICS, SYSTEMS SCIENCE. *Educ:* Univ Hawaii, BA, 59, MSc, 60; Univ Kyoto, PhD(astrophys), 65. *Prof Exp:* Mathematician, Rand Corp, 61-71, consult, 71-77; from res assoc math to adj assoc prof, Univ Southern Calif 71-77; pres, HFS Assocs, 77-87; sr scientist, Hughes Aircraft Co, 82-87; CHIEF ENGR, INFOTEC DEVELOP, INC, 87- *Concurrent Pos:* Assoc ed, Appl Math & Comput. *Honors & Awards:* Hughes Invention Award. *Mem:* Sr mem Inst Elec & Electronics Engrs; Soc Indust & Appl Math; Asn Old Crows; Sigma Xi. *Res:* Optimization; control theory; team decision theory; command and control; operations research; atmospheric temperature estimation; mathematical models; system identification, mathematical and computational methods; dynamic programming, quasilinearization and invariant imbedding; integral equations; boundary value problems; initial value problems; expert and avionics systems; numerical derivatives; nonlinear analysis; author of over 140 technical papers published in journals in the USA, USSR, UK, France and author and co-author of several textbooks on applied mathematics and systems science. *Mailing Add:* Sch Eng & Comput Sci Calif State Univ Fullerton Fullerton CA 92834

KAGIWADA, REYNOLD SHIGERU, b Los Angeles, Calif, July 8, 38; m 61; c 2. ELECTRONICS ENGINEERING, SOLID STATE PHYSICS. *Educ:* Univ Calif, Los Angeles, BS, 60, MS, 63, PhD(physics), 66. *Prof Exp:* Asst prof physics, Univ Calif, Los Angeles, 67-69 & Univ Southern Calif, 69-72; mem prof staff, TRW Systs Group, 72-75, scientist, 75-76, sect head, 76-77, sr scientist, 77-80, mgr, Microwave Prod Dept, 80-83, mgr, Advan Microelectronics Lab, 83-87, asst proj mgr/dep proj mgr, 87-88, prog mgr, 88-89, mimic chief scientist, 89-90, ADVAN TECHNOL MGR, MICROWAVE TECHNOL DEPT OPER, TRW SYSTS GROUP, 90- *Honors & Awards:* Gold Medal Recipient, Ramo Technol Transfer Award, 85. *Mem:* Inst Elec & Electronics Engrs; Asn Old Crows. *Res:* Gallium-arsenic devices; integrated circuits; millimeter-wave devices; microwave acoustic devices; low temperature physics; superconductivity; liquid helium; ultrasonics; acoustics; microwave physics. *Mailing Add:* TRW Electronic & Defense Sect & Electronics Technol Div One Space Park Redondo Beach CA 90278

KAHAN, ARCHIE M, b Denver, Colo, Jan 18, 17; m 44; c 2. METEOROLOGY. *Educ:* Univ Denver, BA, 36, MA, 40; Calif Inst Technol, MS, 42; Agr & Mech Col Tex, PhD(meteorol oceanog), 59. *Prof Exp:* Jr engr, Denison Dist, Corps Engrs, 38-41; hydrologist, US Weather Bur, 45-46; supvr hydrologist in-chg, Mo River Forecast Ctr, 46-51; assoc dir, Am Inst Aerological Res, 51-53; asst prof meteorol, Agr & Mech Col Tex, 53-54; exec dir, Tex A&M Res Found, 54-63, dir, Tex Eng Exp Sta, 62-63; dir, Univ Okla Res Inst, 63-65; gen phys scientist, US Bur Reclamation, 65-70, chief off atmospheric resources mgt, 70-79, assoc chief, div res, 78-79; CONSULT METEOROLOGIST, 79-; SR SCIENTIST, OPHIR CORP, 87. *Concurrent Pos:* Consult, President's Adv Comt Weather Control, 54 & NSF Panel Weather Modification, 65; mem, Adv Comt Weather Modification, State of Colo, 72-83. *Honors & Awards:* Award, Am Meteorol Soc, 62. *Mem:* Am Soc Civil Engrs; AAAS; Am Meteorol Soc; Am Geophys Union. *Res:* Hydrology; hydrometeorology; oceanography; research administration; development of cloud seeding technology for water resource enhancement. *Mailing Add:* 610 S Eldridge Lakewood CO 80228

KAHAN, BARRY D, b Cleveland, Ohio, July 25, 39; m 62. IMMUNOLOGY, SURGERY. *Educ:* Univ Chicago, BS, 60, PhD(physiol), 64, MD, 65. *Prof Exp:* Intern, Mass Gen Hosp, 65-66; staff assoc, NIH, 66-68; surg residency, 68-72; asst prof, 72-74, assoc prof surg, Northwestern Univ, 74-77; PROF, DEPT SURG, UNIV TEX MED SCH, 77- *Mem:* AAAS; Am Asn Immunol; Soc Univ Surg; fel Am Col Surg; Transplantation Soc. *Res:* Electron microscopy; protein biochemistry; transplantation antigens; delayed-typed hypersensitivity; transplantation and tumor-specific antyius. *Mailing Add:* Dept Surg Univ Tex Med Sch PO Box 20036 Houston TX 77225

KAHAN, I HOWARD, b New York, NY, Jan 7, 23; m 54; c 3. AVIAN PATHOLOGY, BACTERIOLOGY. *Educ:* Queens Col, NY, BS, 43; Univ Pa, VMD, 49. *Prof Exp:* Dir vet med, Glenside Animal Hosp, 51-68; assoc prof, Del Valley Col, 68-76, prof poultry path, Animal Dis & Bact & Dir Poultry Diag Lab, 76-; RETIRED. *Concurrent Pos:* Instr, Pa State Univ, Ogontz Campus, 60-63 & Roxborough Mem Hosp, 61-62; inspector poultry, USDA, 63-68. *Mem:* AAAS; Am Asn Avian Path; US Animal Health Asn; Am Vet Med Asn. *Res:* All phases of avian diseases and pathology using disciplined microbiology, virology, serology and histopathology. *Mailing Add:* Dade City Diag Lab Fla Dept Agr 1414 Hwy 52 W Dade City FL 33525

KAHAN, LAWRENCE, b Los Angeles, Calif, May 16, 44; m 69; c 2. BIOCHEMISTRY, IMMUNOLOGY. *Educ:* Univ Calif, Berkeley, BA, 65; Brandeis Univ, PhD(biochem), 71. *Prof Exp:* Fel biochem, 70-73, from asst prof to assoc prof, 73-83, PROF PHYSIOL CHEM, UNIV WIS-MADISON, 83-, DIR HYBRIDOMA FAC & VCHMN, DEPT PHYSIOL CHEM, 85- *Concurrent Pos:* Fel, Am Cancer Soc, 70-72; res grant, NIH, 75-; consult, NSF, 77-80, 82-85 & 87-; NSF res grant, 80-83. *Mem:* Am Soc Biochem & Molecular Biol; Am Chem Soc. *Res:* Structure and function of bacterial and eukaryote ribosomes; cancer associated enzymes; molecular biology. *Mailing Add:* 4106 Hiawatha Dr Madison WI 53711-3040

KAHAN, LINDA BERYL, b San Francisco, Calif, Sept 28, 41. NEUROPHYSIOLOGY. *Educ:* Univ Calif, Berkeley, BA, 63; Stanford Univ, MA, 65, PhD(biol), 67. *Prof Exp:* Fel neurophysiol, Univ Miami, 67-68; asst prof, Antioch Col, 68-71; MEM FAC BIOL, EVERGREEN STATE COL, 71-; AT FRIDAY HARBOR LABS, FRIDAY HARBOR. *Mem:* AAAS; Sigma Xi. *Res:* Physiology and anatomy of invertebrate nervous systems; neural control of behavior. *Mailing Add:* Dept Biol Evergreen St Col 3095 Verba Buena Rd Olympia WA 98505

KAHAN, SIDNEY, b New York, NY, Oct 14, 15; m 44; c 3. DEVELOPMENT NEW FOOD PRODUCT. *Educ:* Columbia Univ, BS, 35; Polytech Inst Brooklyn, MS, 49. *Prof Exp:* Jr chemist ores & metals, Am Smelting & Refinng Co, 36-38; jr chemist food chemist, US Food Drug Admn, 38-55; chief food technologist, Food Chem Prod Develop, B Manischewitz Co, 55-60; chief chemist food chem, Fitelson Lab Inc, 60-82; CONSULT FOOD CHEM, KAHANSULTANTS INC, 82- *Mem:* Fel asn anal chemists; Am Chem Soc; Inst Food Technologists. *Res:* Adulteration of various food products. *Mailing Add:* 66 Peachtree Lane Roslyn Heights NY 11577-2416

KAHAN, WILLIAM M, b Toronto, Ont, June 5, 33; m 54; c 2. MATHEMATICS, COMPUTER SCIENCE. *Educ:* Univ Toronto, BA, 54, MA, 56, PhD(numerical anal), 58. *Prof Exp:* Nat Res Coun Can fel, Cambridge Univ, 58-60; from asst prof to assoc prof math & comput sci, Univ Toronto, 60-67, prof, 68; prof comput sci, 69-72, PROF MATH, ELEC ENG & COMPUT SCI, UNIV CALIF, BERKELEY, 72- *Concurrent Pos:* Nat Res Coun Can res grant, 64-69; vis assoc prof, Stanford Univ, 66; consult, IBM, NY, 67, 72-73, 84-, Hewlett-Packard, Corvallis, Or, 74-86 & Intel, Santa Clara, Calif, 77-, Apple, 87- *Honors & Awards:* A M Turing Award, Asn Comput Mach, 89. *Mem:* Am Math Soc; Asn Comput Mach; Soc Indust & Appl Math. *Res:* Large matrix calculations; trajectory problems; error analysis; design computer arithmetic units; execution-time diagnostic systems for scientific computer systems; general purpose programs to solve standard problems in numerical analysis with highest reliability on electronic computer; IEEE standards 754 and 854 for floating-point arithmetic. *Mailing Add:* Dept Math Evans Hall Univ Calif Berkeley CA 94720

KAHANA, SIDNEY H, b Winnipeg, Man, July 23, 33; m 57; c 2. THEORETICAL PHYSICS. *Educ:* Univ Man, BSc, 54, MSc, 55; Univ Edinburgh, PhD(physics), 57. *Prof Exp:* From asst prof to assoc prof physics, McGill Univ, 57-67; vis scientist, Niels Bohr Inst, 59-61; vis scientist, 65-66, SCIENTIST PHYSICS, BROOKHAVEN NAT LAB, 67- *Concurrent Pos:* Vis scientist, Atomic Energy Can, 57-62, 63 & 64; Guggenheim fel, John Simon Mem Found, 74-75. *Honors & Awards:* Sr Award, Alexander Von Humboldt Stiftone. *Mem:* Fel Am Phys Soc. *Res:* Nuclear theory; structure and reactions; intermediate energy; baryon-anti-baryon systems; many body problems. *Mailing Add:* Dept Physics Brookhaven Nat Lab Upton NY 11973

KAHANER, DAVID KENNETH, b New York, NY, Sept 12, 41; m 64; c 2. APPLIED MATHEMATICS, COMPUTER METHODS. *Educ:* City Col New York, BS, 62; Stevens Inst Technol, MS, 64, PhD(math), 68. *Prof Exp:* Mem staff numerical anal, Los Alamos Nat Lab, 68-80; mem staff, 79-80, GROUP LEADER, NAT BUR STANDARDS, 81- *Concurrent Pos:* Vis prof, Univ Mich, 72-73, Univ Torino, Italy, 77, Swiss Fed Inst Technol, 77-78 & Vienna Tech Univ, 78. *Mem:* Am Math Soc; Asn Comput Mach. *Res:* Numerical analysis; computing methods; mathematical software. *Mailing Add:* Tech Bldg Rm A151 Nat Inst Standards & Technol Gaithersburg MD 20899

KAHLE, ANNE BETTINE, b Auburn, Wash, Mar 30, 34; m 57; c 4. GEOPHYSICS. *Educ:* Univ Alaska, BS, 55, MS, 57; Univ Calif, Los Angeles, PhD(meteorol), 75. *Prof Exp:* Res asst, Rand Corp, Calif, 61-63, from asst phys scientist to assoc phys scientist, 63-67, phys scientist, 67-74; mem tech staff, 74-75, SUPVR GEOLOGY GROUP, JET PROPULSION LAB, 75- *Concurrent Pos:* Consult, Rand Corp, 74-; Xerox Corp, 75- *Honors & Awards:* Gen James Gordon Steese Prize. *Mem:* Sigma Xi; Am Geophys Union; AAAS; Geol Soc Am. *Res:* Remote sensing of geology; volcanology geomagnetic field; solar-terrestrial relationships; atmospheric physics; gravity; atmospheric radiation. *Mailing Add:* 19767 Grandview Dr Topanga CA 90290

KAHLE, CHARLES F, b Toledo, Ohio, June 2, 30; m 57; c 6. GEOLOGY. *Educ:* St Joseph's Col, Ind, BS, 53; Miami Univ, Ohio, MS, 57; Univ Kans, PhD(geol), 62. *Prof Exp:* Geologist, Mobil Petrol Co, Okla, 57-58; asst prof geol, Okla State Univ 62-63 & Univ Toledo, 63-65; from asst prof to assoc prof, 65-74, PROF GEOL, BOWLING GREEN STATE UNIV, 74- *Mem:* Geol Soc Am; Soc Econ Paleont & Mineral. *Res:* Carbonate geology; stratigraphy; scanning electron microscopy; sedimentation. *Mailing Add:* Dept Geol Bowling Green State Univ Main Campus Bowling Green OH 43403

KAHLER, ALBERT COMSTOCK, III, b Bay Shore, NY, June 10, 51; m 72; c 1. NUCLEAR PHYSICS. *Educ:* Gettysburg Col, BA, 73; Univ Tenn, PhD(physics), 78. *Prof Exp:* Res assoc physics, Cyclotron Inst, Tex A&M Univ, 78-81; SR SCIENTIST EXP PHYSICS, WESTINGHOUSE ELEC CORP, BETTIS ATOMIC POWER LAB, 81- *Mem:* Am Phys Soc. *Res:* Nuclear structure studies using gamma-ray spectroscopy techniques. *Mailing Add:* Bettis Atomic Power Lab Westinghouse Elec Corp PO Box 79 West Mifflin PA 15122

KAHLER, ALEX L, b Scottsbluff, Nebr, July 4, 39; m 63; c 2. GENETICS, PLANT BREEDING. *Educ:* Univ Calif, Davis, BS, 65, MS, 67, PhD, 73. *Prof Exp:* Staff res assoc genetics, Univ Calif, Davis, 65-80; prof plant sci, SDak State Univ, Brookings, 80-; plant geneticist, Northern Grain Insects Res Lab, Agr Res Serv, USDA, 80-86 & 86-88; PRES & FOUNDER BIOGENETIC SERV, INC, 88- *Concurrent Pos:* Mgr biotechnol & sr res geneticist, Garst, Slater, Iowa, 86-88. *Mem:* AAAS; Am Genetics Asn; Soc Study Evolution; Genetics Soc Am; Crop Sci Soc Am; Sigma Xi. *Res:* Determining the extent and distribution of genetic variability within and between populations and with measuring the forces which are responsible for the observed variability; improving corn populations for resistance to insects and diseases. *Mailing Add:* Biogenetic Serv Inc 2308 Sixth St E PO Box 710 Brookings SD 57006

KAHLER, RICHARD LEE, b Milltown, NJ, Jan 2, 33; m 58; c 3. CARDIOVASCULAR PHYSIOLOGY. *Educ:* Yale Univ, MD, 57. *Prof Exp:* From asst prof to assoc prof med, Yale Univ, 65-68; ASSOC PROF MED, SCH MED, UNIV CALIF, SAN DIEGO, 68-; HEAD CARDIOVASCULAR DIV, SCRIPPS CLIN & RES FOUND, 68- *Concurrent Pos:* NIH res career develop award, 65-68. *Mem:* AAAS; Am Physiol Soc; Am Heart Asn; Am Col Physicians; Am Col Cardiol; Sigma Xi. *Res:* Cardiovascular pharmacology. *Mailing Add:* PO Box 104 Rancho Santa Fe CA 92067

KAHLES, JOHN FRANK, b Chicago, Ill, Sept 11, 14; m 40; c 7. METALLURGY. *Educ:* Ill Inst Technol, BS, 36; Univ Cincinnati, PhD(metall), 46. *Prof Exp:* From instr to assoc prof metall, Col Eng & Grad Sch, Univ Cincinnati, 39-51, res assoc, Grad Sch, 51-64; vpres, Metcut Res Assocs, Inc, 58-80, sr vpres, 80-90; RETIRED. *Concurrent Pos:* Consult metallurgy, 39-; dir, Info Technol, 67-; adj prof metall, Col Eng & Grad Sch, Univ Cincinnati, 64- *Honors & Awards:* Joseph Whitworth Prize, Brit Inst Mech Engrs, 68. *Mem:* Nat Acad Eng; Am Soc Testing & Mat; Am Inst Mining, Metall & Petrol Engrs; Am Welding Soc; Am Soc Nondestructive Testing; fel Am Soc Metals Int. *Res:* Heat treatment; metal processing; isothermal transformation studies; failure analysis; metallography; machinability; information science; author of more than 90 technical publications. *Mailing Add:* Metcut Res Assocs Inc 3980 Rosslyn Dr Cincinnati OH 45209-1196

KAHLON, PREM SINGH, b Lyallpur, India, June 16, 36; m 67; c 2. BIOLOGY, PLANT GENETICS. *Educ:* Punjab Univ, India, BS, 56; La State Univ, MS, 62, PhD(plant breeding), 64. *Prof Exp:* Asst prof biol, Talladega Col, 64-65; prof, Alcorn Agr & Mech Col, 65-66; PROF BIOL, TENN STATE UNIV, 66- *Mem:* Genetics Soc Am; Environ Mutagen Soc; Indian Soc Genetics & Plant Breeding. *Res:* Inheritance studies in rice, especially cooking quality; mutation genetics; chemical mutagenesis. *Mailing Add:* Dept Biol Tenn State Univ Nashville TN 37203

KAHN, A CLARK, b Pittsburgh, Pa, Dec 16, 37; m 90; c 2. BIOCHEMISTRY. *Educ:* Univ NH, BA, 61, MS, 63; Pa State Univ, PhD(biochem), 66. *Prof Exp:* Lab officer clin chem & biochemist, USPHS Hosp, 66-68; supvr clin path, ICI Am, 68-76; dir res & develop, Precision Systs, 76-77; dir labs, New Eng Med Labs, 77-78; dir clin path, Int Res & Develop Corp, 78-81; sr chemist, Palasades Nuclear Power Plant, 81-84; dir lab, 85-86, PRES & LAB DIR, ENVIRON EVALUATIONS & LAB SERV, ANA-QUAL LABS. *Mem:* AAAS; Am Chem Soc; Am Asn Clin Chem; Am Soc Vet Clin Pathologists; Nat Registry Clin Chemists. *Res:* Clinical laboratory medicine, particularly electrolyte chemistry and intestinal absorption, malabsorption syndrome; laboratory animal clinical pathology; toxicology of hazardous materials. *Mailing Add:* 506 River Rd Paw Paw MI 49079

KAHN, ALAN RICHARD, b Chicago, Ill, Mar 1, 32; m 70; c 2. BIOENGINEERING, PHYSIOLOGY. *Educ:* Univ Ill, MD, 59. *Prof Exp:* Design engr, Offner Electronics, 58-59, dir physiol res, Offner Div, Beckman Instruments, Inc, 62-63, med dir, 63-66; dir biophys res, Hoffman LaRoche Inc, 66-68; vpres res & develop, Health Technol Corp, 68-70; sr vpres cardiovasc res, Medtronic Inc, 70-77; pres, Andersen Med Systs, 77-78; dir biomed res, Nicolet Instrument Corp, 78-80; consult, Appl Electronic Consults, Inc, 80-; AT DEPT ELEC & COMP ENG, UNIV CINCINNATI. *Concurrent Pos:* Clin asst prof, Univ Minn; lectr, Univ Ill; chmn comt med devices, Am Nat Standards Inst. *Mem:* Neuroelec Soc (vpres); AMA; Biomed Eng Soc; Inst Elec & Electronics Engrs; Asn Advan Med Instrumentation. *Res:* Cardiovascular neurological and rehabilitation instrumentation; aerospace medicine; biological electrode techniques, medical device standards, management for research and development programs. *Mailing Add:* Dept Elec Comput Eng 898 Rhodes Hall Univ Cincinnati Cincinnati OH 45221

KAHN, ALBERT, b Wuerburg, Ger, May 21, 31; nat US. DEVELOPMENTAL BIOLOGY. *Educ:* Cornell Univ, BS, 53; Univ Calif, Los Angeles, PhD(bot), 58. *Prof Exp:* Jr res botanist, Univ Calif, Los Angeles, 58-59; fels, NSF, Stockholm, 59-61 & USPHS, 61-62; res fel biol, Calif Inst Technol, 62-64; assoc prof, Purdue Univ, 64-70; ASSOC PROF BIOL, GENETICS INST, COPENHAGEN UNIV, 70 - *Mem:* AAAS; Am Soc Cell Biol; Am Soc Plant Physiol. *Res:* Chloroplast pigments; genetic control of chloroplast development; mutagenesis and somatic recombinogenesis. *Mailing Add:* Genetics Inst Copenhagen Univ Farimagsgade 2A Oster Copenhagen K DK 1353 Denmark

KAHN, ARNOLD HERBERT, b New York, NY, Nov 5, 28; m 58. PHYSICS. *Educ:* Rensselaer Polytech Inst, BS, 50; Univ Calif, MA, 52, PhD(physics), 55. *Prof Exp:* Asst, Univ Calif, 50-52; res assoc, Univ Ill, 54-56; PHYSICIST, NAT BUR STANDARDS, 56- *Concurrent Pos:* Instr, Univ Md, 56-61. *Mem:* Am Phys Soc. *Res:* Theory of solids; electrical, magnetic, and optical properties of solids; electromagnetic methods in nondestructive evaluation. *Mailing Add:* Nat Bur Standards Washington DC 20234

KAHN, ARTHUR B, b New York, NY, Feb 11, 31; m 56; c 3. MANAGEMENT INFO SYSTEMS. *Educ:* City Col New York, BS, 51; Johns Hopkins Univ, MSE, 54, PhD(dynamic meteorol), 59. *Prof Exp:* Res asst dynamic meteorol, Johns Hopkins Univ, 51-54, staff asst, 54-58; sr engr, Air Arm Div, Westinghouse Elec Corp, 58-63, fel engr, Systs Div 63-68, fel engr, Info Processing Dept, 68-70, adv engr, 71-78; ASSOC PROF, INFO QUANT STUDIES, UNIV BALTIMORE, 78- *Concurrent Pos:* Founder & chmn prog eval & rev tech proj, SHARE, 62-64; vis assoc prof, Univ Wis-Madison, 68-71. *Mem:* Asn Comput Mach; Sigma Xi; Opers Res Soc Am. *Res:* Determination of what the nonspecialist should know about computers and development of ways and means to educate him. *Mailing Add:* 4120 Balmoral Circle Baltimore MD 21208

KAHN, ARTHUR JOLE, b New York, NY, May 29, 21; m 50; c 4. PHYSIOLOGY. *Educ:* NY Univ, AB, 47, MS, 49, PhD(biophys), 52. *Prof Exp:* Res assoc, NY Univ, 48-51, instr, 51; asst prof physiol, Col Med & Dent, Georgetown Univ, 51-56; from asst prof to assoc prof, 56-69, PROF PHYSIOL, COL MED & DENT NJ, 69- *Mem:* Am Physiol Soc. *Res:* Elasticity and contractility of cardiac muscle. *Mailing Add:* Dept Physiol Col Med & Dent NJ Med Sch Newark NJ 07103

KAHN, BERND, b Pforzheim, Germany, Aug 16, 28; nat US; m 61; c 2. RADIOCHEMISTRY. *Educ:* Newark Col Eng, BS, 50; Vanderbilt Univ, MS, 52; Mass Inst Technol, PhD(chem), 60. *Prof Exp:* Assoc chemist radiochem, Oak Ridge Nat Lab, 51-54; radiochemist, USPHS, 54-69; radiochemist, Radiochem & Nuclear Eng Br, Environ Protection Agency, Nat Environ Res Ctr, 69-74; DIR, ENVIRON RESOURCES CTR & PROF NUCLEAR ENG & HEALTH PHYSICS, GA INST TECHNOL, 74- *Concurrent Pos:* Mem, Nat Coun Radiation Protection & Measurements. *Mem:* Health Physics Soc; Am Chem Soc; Am Phys Soc. *Res:* Analytical radiochemical methods; behavior of radionuclides in the environment; radioactive effluents from nuclear power stations. *Mailing Add:* Environ Resources Ctr Ga Inst Technol Atlanta GA 30332

KAHN, CARL RONALD, b Louisville, Ky, Jan 14, 44; m 66; c 2. ENDOCRINOLOGY. *Educ:* Univ Louisville, BA, 64, MD, 68. *Prof Exp:* Sr investr, 73-78, chief cellular & molecular physiol, Diabetes Br, Nat Inst Arthritis & Metab Dis, NIH, 78-81; DIR RES, JOSLIN DIABETES CTR, 81-; PROF MED, HARVARD MED SCH, 84- *Concurrent Pos:* Mem, Med Sci Adv Bd, Juv Diabetes Found, 78-; chief, Div Diabetes & Metabolism, Brigham & Women's Hosp, 81. *Honors & Awards:* Davis Rumbough Mem Award, Juv Diabetes Found, 77; Pfizer Biomed Res Award, 86; Edwin B Astwood Lectr, Endocrine Soc, 87. *Mem:* Am Soc Clin Invest; Am Fedn Clin; Asn Am Physicians; Nat Coun Am Soc Clin Invest (pres-elect, 87). *Res:* Endocrine Soc; Am Diabetes Asn. Res: Insulin action and alterations in insulin action in disease states; hypoglycemia and insulin-like peptides in blood. *Mailing Add:* Joslin Diabetes Ctr One Joslin Pl Boston MA 02115

KAHN, CHARLES HOWARD, b Birmingham, Ala, Feb 10, 26; m 56; c 3. ARCHITECTURE, STRUCTURAL ENGINEERING. *Educ:* Univ NC, AB, 46; NC State Col, BCE, 48, BArch, 56; Mass Inst Technol, MS, 49. *Prof Exp:* Assoc prof struct design, sch design, NC State Univ, 52-59; from assoc prof to prof archit, 59-68, dean, Sch Archit & Urban Design, 68-80, PROF ARCHIT, SCH ARCHIT & URBAN DESIGN, UNIV KANS, 80- *Concurrent Pos:* Consult, 56-66; Fulbright fel, Italy, 57-58; pres, Charles H Kahn & Assoc & Cybertechnol, Ltd, 66-68; mem, Kans State Bldg Comn, 78-80. *Honors & Awards:* Annual res award, Nat Inst Archit Educ, 77. *Mem:* Am Soc Eng Educ; Am Soc Civil Engrs; Am Inst Archit; Asn Col Schs Archit. *Res:* Thin shells, membranal structures; architecture; history of the architectural structuralist movement, 1948 to 1965. *Mailing Add:* Sch Archit Univ Kans Lawrence KS 66045

KAHN, DANIEL STEPHEN, b Brooklyn, NY, Nov 20, 35. MATHEMATICS. *Educ:* Princeton Univ, AB, 57; Mass Inst Technol, PhD(math), 64. *Prof Exp:* Res instr math, Univ Chicago, 62-64; asst prof, 64-69, assoc prof, 69-78, PROF MATH, NORTHWESTERN UNIV, ILL, 78- *Mem:* Am Math Soc. *Res:* Algebraic topology, especially field of stable homotopy theory. *Mailing Add:* Dept Math Northwestern Univ Evanston IL 60201

KAHN, DAVID, b Peoria, Ill, Feb 4, 26; m 57; c 3. SOLID STATE PHYSICS, ELECTROCHEMISTRY. *Educ:* Univ Ill, BS, 45; Univ Chicago, MS, 50, PhD(physics), 53. *Prof Exp:* Solid state physicist, Lewis Lab, Nat Adv Comt Aeronaut, 53-56; staff scientist, Res Inst Advan Study, 56-59, sr scientist, 59-70; res assoc, 70-87, SR MEM TECH STAFF, AMP, INC, 87- *Concurrent Pos:* Am Cancer Soc res fel, Norsk Hydro's Inst Cancer Res, Norway, 54-55. *Mem:* Am Phys Soc; Mat Res Soc; Am Soc Metals; NY Acad Sci. *Res:* X-ray diffraction; magnetic susceptibility; electrometallurgy; conductive polymers; stress relaxation in metals; electro deposition modeling. *Mailing Add:* Res Div AMP Inc Mail Stop 21-01 PO Box 3608 Harrisburg PA 17105-3608

KAHN, DAVID, b New York, NY, Apr 27, 33; m 52; c 3. PHYSICS. *Educ:* Brooklyn Col, BS, 57; Yale Univ, MS, 59, PhD(physics), 62. *Prof Exp:* Sr scientist, Raytheon Co, 62-66 & electronics res ctr, NASA, 66-71; group head, Modelling & Anal Group, US Dept Transp, 71-74, sect chief technol, Transp Systs Ctr, 74-86; CONSULT, 86- *Mem:* Am Phys Soc; sr mem Inst Elec & Electronics Engrs; Sigma Xi. *Res:* Kinetic theory and plasma physics; wave propagation in highly rarefied gases; plasma density discontinuity wave coupling; traffic flow theory. *Mailing Add:* 238 Biddle Dr Exton PA 19341

KAHN, DONALD JAY, b Baltimore, Md, Aug 10, 30; m 60; c 3. ORGANIC CHEMISTRY. *Educ:* Princeton Univ, BA, 52; Univ Chicago, PhD(org chem), 57. *Prof Exp:* Asst org chem, Univ Chicago, 52-55; chemist, Esso Res & Eng Co, 57-60, proj leader & sr chemist, 60-63, sect head, 63-67, mgr aviation tech serv, Exxon Int Inc, 68-71, environ conserv sr adv, Exxon Corp, 71-77, solar energy gen mgr, Exxon Enterprises Inc, 78-81, sr technol adv, 81-84, strategic planning, 84-86; RETIRED. *Concurrent Pos:* Consult, Technol Mgt. *Mem:* Am Chem Soc; Sigma Xi. *Res:* Reaction mechanisms; free radical organic chemistry and polymerization; chemical additives for lubricants; industrial lubricants, greases, wax products and asphalt products; solar photovoltics. *Mailing Add:* 62 Spring St Metuchen NJ 08840

KAHN, DONALD R, b Birmingham, Ala, May 21, 29; c 4. CARDIOVASCULAR SURGERY, THORACIC SURGERY. *Educ:* Birmingham-Southern Col, BA, 50; Univ Ala, BA, 57, MD, 54. *Prof Exp:* Intern, St Louis City Hosp, 54-55; resident, Med Ctr, Univ Mich, Ann Arbor, 55-59, instr thoracic surg, Med Sch, 63-64, from asst prof to assoc prof surg, 64-71; head div thoracic & cardiovasc surg, Med Ctr, Univ Wis-Madison, 71-80, prof surg & chmn cardiovasc med, Univ Wis Hosp, 71-80. *Concurrent Pos:* Am Thoracic Soc fel, 61-64; asst, Sch Med, Washington Univ, 54-55; dir clin invest lab, directorate med res, Chem Warfare Lab, 56-58. *Mem:* Fel Am Col Surg; Am Heart Asn; Am Thoracic Soc; Am Asn Thoracic Surg; Soc Thoracic Surg. *Mailing Add:* 817 Princeton Ave SW Suite 300 Birmingham AL 35211

KAHN, DONALD W, b New York, NY, Nov 21, 35; m 56; c 2. MATHEMATICS. *Educ:* Cornell Univ, BA, 57; Yale Univ, PhD(math), 61. *Prof Exp:* Ritt instr math, Columbia Univ, 61-64; from asst prof to assoc prof, 64-84, PROF MATH, UNIV MINN, MINNEAPOLIS, 84- *Concurrent Pos:* Vis Fulbright prof, Univ Heidelberg, 65. *Mem:* Am Math Soc. *Res:* Algebraic topology. *Mailing Add:* Univ Minn Sch Math Minneapolis MN 55455

KAHN, ELLIOTT H, b Brooklyn, NY, Feb 27, 26; m 52; c 3. ECONOMIC STUDIES, APPLIED MECHANICS. *Educ:* City Col New York, BCE, 45; Polytech Inst Brooklyn, MCE, 48. *Prof Exp:* Res engr, Repub Aviation Corp, NY, 45-47; staff engr, D B Steinman, 47-51; sr engr, W L Maxson Corp, 51-57, mgr reliability anal sect, 57-61; sr tech staff scientist, Kollsman Instrument Corp, Syosset, 61-76; MGR, COOPERS & LYBRAND, 76- *Concurrent Pos:* Lectr & consult engr, various times. *Honors & Awards:* Robert Ridgeway Award, Am Soc Civil Engrs, 45. *Mem:* AAAS; NY Acad Sci; Nat Soc Prof Engrs; Am Soc Civil Engrs; Inst Elec & Electronics Engrs. *Res:* Applied mechanics and systems engineering in the fields of structures, reliability, traffic, instruments, laser safety and electro-optical systems. *Mailing Add:* Coopers & Lybrand 1301 Avenue of the Americas New York NY 10019-6013

KAHN, FREDERIC JAY, b Brooklyn, NY, Sept 1, 41; m 67; c 2. ELECTRO-OPTICAL DEVICES & SYSTEMS, LIQUID CRYSTALS. *Educ:* Rensselaer Polytech Inst, BEE, 62; Harvard Univ, AM, 63, PhD(solid state physics), 68. *Prof Exp:* Res asst magneto-optics garnets & orthoferrites, Gordon McKay Lab, Harvard Univ, 65-68; spec researcher liquid crystal displays, Quantum Device Res Lab, Cent Res Lab, Nippon Elec Co, Kawasaki, Japan, 68-69; mem tech staff liquid crystal mat, displays & related technol, Optical Control Devices Dept, Solid State Lab, Bell Labs, 70-73; lab proj mgr, Hewlett Packard Lab, 74-82, dept mgr, optical mat & polymers, Mat Res Lab, 82-83, dept mgr Storage Physics Dept, Mass Memory Lab, 83-84; VPRES, TECHNOL, GREYHAWK SYSTS INC, 84- *Concurrent Pos:* Prin investr, Joint Serv VHSIC, 81-83; ed, Proj Displays, 87, 89. *Mem:* Am Phys Soc; Inst Elec & Electronics Engrs; fel Soc Info Display; Int Soc Optical Eng. *Res:* Liquid crystal display materials, devices and systems; electron-beam and x-ray resists for high resolution lithography; optical memory materials and systems; optical fiber properties and devices; optical disc memories and erasable media; research and development on very high information content (resolution) imaging systems for display, hard copy generation and optical processing based on laser and optical beam addressed liquid crystal projection light valves. *Mailing Add:* Greyhawk Syst Inc 1557 Centre Point Dr Milpitas CA 95035

KAHN, HAROLD A, b New York, NY, Jan 4, 20; m 40; c 3. EPIDEMIOLOGY. *Educ:* City Col New York, BS, 39; Am Univ, MA, 49. *Prof Exp:* Statistician heart dis res, Nat Heart & Lung Inst, 50-51, med care needs, USPHS,51-57, res adminr, NIH, 57-60, heart dis res, Nat Heart & Lung Inst, 60-71, chief off biometry & epidemiol, Nat Eye Inst, 71-75; prof epidemiol, Sch Hyg & Pub Health, Johns Hopkins Univ, 75-78; CONSULT EPIDEMIOL, 78- *Concurrent Pos:* Fel Coun Epidemiol, Am Heart Asn, 64-; USPHS award, 57; Lady Davis vis prof epidemiol, Hebrew Univ, Jerusalem, 80; vis prof epidemiol, Loma Linda Univ, 82, Johns Hopkins Univ, 85- *Mem:* Fel Am Statist Asn; Soc Epidemiol Res. *Res:* Nutrition and other risk factors in relation to chronic disease; problems of data collection in field surveys. *Mailing Add:* 3405 Pendleton Dr Silver Spring MD 20902

KAHN, JACK HENRY, b Bolivar, Tenn, Nov 1, 23; m 52; c 3. ENGINEERING PHYSICS. *Educ:* Univ Tenn, BS, 47, MS, 49, PhD(physics), 51. *Prof Exp:* Classification analyst, Declassification Br, US Atomic Energy Comn, 51-55, asst chief, 68-73, staff asst, 73-75, asst chief, Weapons Prog Br, Div Classification, 75-79, physicist, US Dept Energy, 79-88; CONSULT, HIST ASSOCS, INC, 88- *Mem:* Am Phys Soc; Sigma Xi. *Res:* Nuclear physics. *Mailing Add:* 12118 Whippoorwill Lane Rockville MD 20852-4446

KAHN, JOSEPH STEPHAN, b Ger, Aug 12, 29; m 70; c 2. PLANT BIOCHEMISTRY. *Educ:* Univ Calif, BS, 55; Univ Ill, PhD, 58. *Prof Exp:* Res assoc physiol, Univ Ill, 58-59; res assoc biol, Johns Hopkins Univ, 59-61; from asst prof to assoc prof, 61-71, PROF BOT & BIOCHEM, NC STATE UNIV, 71- *Concurrent Pos:* Fulbright fel, India, 78. *Mem:* Am Soc Plant Physiol; AAAS; Biophys Soc; Am Soc Biol Chem. *Res:* Electron transport and pathway of adenosine triphosphate formation in chloroplasts; localization of enzyme systems in chloroplasts; modification of protozoal membranes by drugs. *Mailing Add:* Dept Biochem NC State Univ Raleigh NC 27695-7622

KAHN, LAWRENCE F, b Oakland, Calif, Jan 26, 45; m 71; c 2. STRUCTURAL ENGINEERING, CIVIL ENGINEERING. *Educ:* Stanford Univ, BS, 66; Univ Ill, Champaign-Urbana, MS, 67; Univ Mich, Ann Arbor, PhD(civil eng), 76. *Prof Exp:* Struct engr undersea struct, US Naval Civil Eng Lab, 67-71; struct engr power plant, Bechtel Power Corp, 76; asst prof, 76-80, ASSOC PROF STRUCT ENG, GA INST TECHNOL, 80-

Concurrent Pos: Prin investr, NSF grant, 77-85; consult, Wiss-Janney-Elstner Assocs. *Honors & Awards:* Raymond Reese Res Prize, Am Soc Civil Eng, 80. *Mem:* Fel Am Soc Civil Engrs; fel Am Concrete Inst; Earthquake Eng Res Inst; Masonry Soc. *Res:* Earthquake engineering; computer aided engineering/computer aided design/computer aided mechanics; reinforced concrete structures; masonry structures; experimental analysis. *Mailing Add:* Sch Civil Eng Ga Inst Technol Atlanta GA 30332-0355

KAHN, LEO DAVID, b Everett, Mass. BIOPHYSICAL CHEMISTRY. *Educ:* Mass Inst Technol, SB, 54; Yale Univ, PhD(chem), 59. *Prof Exp:* RES CHEMIST, EASTERN REGIONAL RES & DEVELOP LAB, USDA, 58- *Mem:* AAAS; Am Chem Soc; Biophys Soc; NY Acad Sci; Sigma Xi. *Res:* Physical chemistry of proteins; electronic instrumentation for use in chemical investigations. *Mailing Add:* 2125 B N John Russell Cir Elkins Park PA 19117

KAHN, LEONARD B, b Johannesburg, SAfrica, July 20, 37; m 63; c 3. SURGICAL PATHOLOGY. *Educ:* Witwaterstrand Univ, Johannesburg, MB, BCh, 60; Univ Cape Town, M Med Path, 65; MRCPath, 77, FRCPath, 86. *Prof Exp:* Intern med & surg, Johannesburg Gen Hosp, 61-62; resident path, Univ Cape Town, 62-66; resident & fel clin asst, Sch Med, Washington Univ, St Louis, 67-69; assoc prof, Dept Path, Sch Med, Univ Cape Town, 74-77; prof & dir surg & path, Sch Med, Univ NC, Chapel Hill, 77-80; CHMN DEPT LABS, LONG ISLAND JEWISH MED CTR, 80-; PROF PATH, STATE UNIV NY STONY BROOK, 80- *Concurrent Pos:* Cecil John Adams Mem travelling fel, Univ Cape Town, 67. *Mem:* Int Acad Path; Gastrointestinal Path Club; Int Skeletal Soc; Arthur Purdy Stout Soc Surg Pathologists; Col Am Pathologists. *Res:* Clinical pathologic, including immunologic and ultrastructural studies of a variety of human neoplasma, especially those involving lymphoreticular tissues, bone, soft tissue and gastrointestinal tract including salivary glands. *Mailing Add:* Long Island Jewish Med Ctr 270-05 76th Ave New Hyde Park NY 11042-1433

KAHN, MANFRED, b Frankfurt am Main, Ger, Feb 21, 26; nat US; m 60; c 3. CERAMIC SCIENCE, ELECTRICAL ENGINEERING. *Educ:* Univ Wis, BSEE, 54; Rensselaer Polytech Inst, MSEE, 59; Pa State Univ, PhD(ceramic sci), 69. *Prof Exp:* Mem tech staff, Ceramic Dept, Sprague Elec Co, 54-74; sr scientist, AVX Ceramics Inc, 74-82; dept head, Hurry County Tech Col, 82-83; SECT HEAD, NAVAL RES LAB, WASHINGTON, DC, 83- *Mem:* Inst Elec & Electronic Engrs; fel Am Ceramic Soc. *Res:* Surface layer capacitors; ohmic contacts and hybrid microcircuits; multilayer capacitors; pilot plant production; base metal electrodes; process development; pilot plant and production; circuits and applications engineering; piezo electric sensors, programed transmission control devices. *Mailing Add:* 3412 Austin Ct Alexandria VA 22310

KAHN, MARVIN WILLIAM, b Cleveland, Ohio, Feb 1, 26; m 82. OTHER MEDICAL & HEALTH SCIENCES. *Educ:* Pa State Univ, BS, 48, MS, 49, PhD (psychol), 52. *Prof Exp:* From instr to asst prof psychol, Yale Univ, 52-54; from asst prof psychol to assoc prof psychiat, Univ Colo Sch Med, 54-64; prof psychol, Ohio Univ, 64-69; PROF PSYCHOL, UNIV ARIZ, 69- *Mem:* Am Psychol Asn. *Mailing Add:* Psychol Dept Univ Ariz Tucson AZ 85721

KAHN, MILTON, b Philadelphia, Pa, Nov 21, 18; m 40; c 1. CHEMISTRY. *Educ:* Univ Calif, BS, 41; Wash Univ, PhD(chem), 50. *Prof Exp:* Anal chemist, Paraffine Co, Inc, 42-43; res chemist, Los Alamos Sci Lab, 43-46; from asst prof to assoc prof chem, Univ N Mex, 48-57, prof, 57-81; RETIRED. *Mem:* AAAS; Am Chem Soc. *Res:* Isotopic exchange reactions; hot atom chemistry; chemical behavior of substances at low concentrations; radiochemistry. *Mailing Add:* 4201 Hannett NE Albuquerque NM 87110

KAHN, NORMAN, b New York, NY, Dec 28, 32; m 58. NEUROPHARMACOLOGY, DENTAL & MEDICAL EDUCATION. *Educ:* Columbia Univ, AB, 54, DDS, 58, PhD(pharmacol), 64. *Prof Exp:* Dent intern, Montefiore Hosp, Bronx, NY, 58-59; from instr to asst prof, Columbia Univ, 62-72, assoc prof pharmacol, 72-80, assoc prof dent, 74-81, PROF PHARMACOL, COL PHYSICIANS & SURGEONS & DENT, SCH DENT & ORAL SURG, COLUMBIA UNIV, 81-, ASSOC DEAN ACAD AFFAIRS, 89- *Concurrent Pos:* NIH trainee neuropharmacol, Columbia Univ, 59-62; Nat Inst Neurol Dis & Blindness spec fels, Columbia Univ, 62-64 & Pisa, 65-66; NIH career develop award, 67-71; vis assoc prof anesthesiol, Univ Calif, Los Angeles, 78; chmn, Inst Rev Bd, Columbia Presbyterain Med Ctr, New York; hon res fel, Univ Col London, 86. *Mem:* Am Dent Asn; Am Physiol Soc; Int Asn Dent Res. *Res:* Physiology and pharmacology of autonomic nervous system; medical and dental education. *Mailing Add:* Col Physicians & Surgeons Columbia Univ New York NY 10032

KAHN, PETER B, b New York, NY, Mar 18, 35; m 56, 88; c 3. THEORETICAL PHYSICS. *Educ:* Union Col, NY, BS, 56; Northwestern Univ, PhD(physics), 60. *Prof Exp:* Res assoc physics, Univ Iowa, 60-61; from asst prof to assoc prof, 61-71, chmn dept, 74-86, PROF PHYSICS, STATE UNIV NY STONY BROOK, 71- *Concurrent Pos:* Sr Weizmann fel, 67-68. *Mem:* Fel Am Phys Soc. *Res:* Mathematical biology; statistical theory of energy level distributions; innovation in physics curricula. *Mailing Add:* Dept Physics State Univ NY Stony Brook NY 11794

KAHN, PETER JACK, b Santiago, Chile, Dec 1, 39; US citizen; m 63. MATHEMATICS. *Educ:* Oberlin Col, BA, 60; Princeton Univ, PhD(math), 64. *Prof Exp:* Actg instr math, Univ Calif, Berkeley, 63, instr, 64-65; asst prof, 65-70, assoc prof, 70-75, PROF MATH, CORNELL UNIV, 75- *Concurrent Pos:* Mem, Inst Advan Study, 69-7; Humboldt sr scientist, Univ Heidelberg, 74-75. *Mem:* Am Math Soc. *Res:* Algebraic and differential topology. *Mailing Add:* White Hall Cornell Univ Ithaca NY 14853

KAHN, RAYMOND HENRY, b New York, NY, Aug 29, 26; m 50; c 3. ENDOCRINOLOGY, HISTOLOGY. *Educ:* Univ Calif, Los Angeles, BA, 48; Univ Calif, Berkeley, MS, 49, PhD(zool), 53. *Prof Exp:* From instr to assoc prof, 54-67, PROF ANAT, UNIV MICH, ANN ARBOR, 67-, DIR RES, HENRY FORD HOSP, 77- *Concurrent Pos:* Am Cancer Soc fel, Strangeways Res Lab, 53-54; USPHS spec fel, Inst Endocrinol, Gunma Univ, Japan, 67-68. *Mem:* Endocrine Soc; Soc Exp Biol & Med; Tissue Cult Asn (secy, 68-72, pres, 74-76); Am Asn Anat; Nat Coun Univ Res Admin. *Res:* Pseudo intimal linings of artificial prostheses; histochemistry; organ culture of prostate; tissue culture. *Mailing Add:* Scripps Res Inst 10666 N Torrey Pines Rd La Jolla CA 92037

KAHN, ROBERT ELLIOT, b Brooklyn, NY, Dec 23, 38; m 80. COMPUTER SCIENCES, ELECTRICAL ENGINEERING. *Educ:* City Col NY, BEE, 60; Princeton Univ, MA, 62, PhD(elec eng), 64. *Prof Exp:* Mem tech staff, Bell Telephone Labs, 60-62; asst prof elec eng, Mass Inst Technol, 64-66; sr scientist, Bolt Beranek & Newman, 66-72; dir dept, div & prog mgr, Defense Advan Res Proj Agency, Info Processing Tech Off, 72-85; PRES, CORP NAT RES INITIATIVES, 86- *Concurrent Pos:* Mem, Air Force Sci Adv Bd, 87- & Comput Sci & Technol Bd, Nat Acad Sci. *Honors & Awards:* Harry Goode Mem Award, Am Fedn Info Processing Soc, 86. *Mem:* Nat Acad Eng; fel Inst Elec & Electronics Engrs; fel Am Asn Artificial Intel. *Res:* Research and development for the national information infrastructive including networking. *Mailing Add:* Corp Nat Res Initiatives 1895 Preston White Dr Suite 100 Reston VA 22091

KAHN, ROBERT PHILLIP, b Chicago, Ill, Apr 20, 24; m 47; c 4. PLANT PATHOLOGY. *Educ:* Univ Ill, BA, 48, PhD(plant path), 51. *Prof Exp:* Asst bot, Univ Ill, 48-52; from plant pathologist to supvry plant pathologist, Chem Warfare Labs, Ft Detrick, 52-57; plant pathologist & sr staff, Animal & Plant Health Inspection Serv, USDA, 57-85; RETIRED. *Concurrent Pos:* Officer-in-chg, EAfrican Plant Quarantine Sta, USDA, Kenya, 70-72; consult, plant path, protection & quarantine. *Mem:* Am Phytopath Soc; Int Soc Plant Pathol; Am Inst Biol Sci; AAAS. *Res:* Virology; plant quarantine pathology; plant tissue culture; agriculture; tropical plant pathology; international exchange of plant germplasm; plant protection and quarantine. *Mailing Add:* 14104 Flint Rock Terr Rockville MD 20853

KAHN, SAMUEL GEORGE, b Belleville, NJ, May 20, 29; m 60; c 2. NUTRITION SCIENCE ADMINISTRATION, NUTRITION POLICY. *Educ:* Ill Wesleyan Univ, BS, 51; Univ Ill, MS, 53 & 54, PhD(animal nutrit), 55. *Prof Exp:* Asst animal nutrit, Univ Ill, 52-55; res assoc, Radio Carbon Lab, 55-56; asst biochem, Squibb Inst Med Res, 56-58, sect head nutrit res, Div Agr Sci, 58-59, sr res scientist, 59-67, res supvr & head nutrit res, 67-69; assoc prof nutrit & food & chmn, Drexel Inst Technol, 69-71; nutrit adv, Res Off & Univ Rels, Tech Assistance Bur, 71-74, SR NUTRIT ADV, OFF NUTRIT SCI & TECH BUR, AID, 74- *Concurrent Pos:* Hon prof nutrit, Rutgers Univ, 67-70; adj prof, Va Polytech Inst & State Univ, 74; secy, Int Vitamin A Consult Group, 74-77, Int Nutrit Anemia Consult Group, 76-; ed, Am Inst Nutrit-Nutrit Notes, 79-; mem, Joint Res Comt, Pres Bd Int Food & Agr Develop, 80-82; mem, Joint Subcomt Human Nutrit Res, Exec Off President US, 80-83 & mem, Nat Comt Human Nutrit Res, 83- *Mem:* Am Inst Nutrit; fel NY Acad Sci; Am Physiol Soc; Soc Exp Biol & Med; Am Soc Clin Nutrit; Am Chem Soc. *Res:* Vitamin and lipid metabolism; atherosclerosis; animal nutrition; nutrition, health and population; international nutrition. *Mailing Add:* 11827 Goya Dr Potomac MD 20854

KAHN, WALTER K(URT), b Mannheim, Ger, Mar 24, 29; US citizen; m 62; c 2. ELECTRICAL ENGINEERING, ELECTROPHYSICS. *Educ:* Cooper Union, BEE, 51; Polytech Inst Brooklyn, MEE, 54, DEE, 60. *Prof Exp:* Engr radar, Wheeler Labs, NY, 51-54; res assoc, microwave res inst, Polytech Inst Brooklyn, 54-60, asst prof elec eng, 60-62, from assoc prof to prof electrophysics, 62-69; DIR, INST INFO SCI & TECHNOL, 83- *Concurrent Pos:* Liaison scientist, Off Naval Res, London, 67-68; mem comn 6, Int Union Radio Sci; ed, trans,Antennas & Propagation, Inst Elec & Electronics Engrs, 77-80; dir, Anro Eng, Inc. *Mem:* AAAS; fel Inst Elec & Electronics Engrs; Optical Soc Am; Soc Photo Optical Instruments Engrs. *Res:* Optical resonators and fiberoptics; lasers; microwave antennas and antenna arrays; waveguide junctions and directional couplers; microwave measurements; monopulse, radar systems; fiber optics. *Mailing Add:* 7709 Hamilton Spring Rd Bethesda MD 20817

KAHNE, STEPHEN JAMES, b New York, NY, Apr 5, 37; m 70; c 2. CONTROL THEORY, SYSTEMS ENGINEERING. *Educ:* Cornell Univ, BEE, 60; Univ Ill, MS, 61, PhD(elec eng), 63. *Prof Exp:* Engr, Defense Syst Dept, Gen Elec Co, 60; res asst control systs, Coord Sci Lab, Univ Ill, 61-62, res assoc & instr elec eng, 62-63; sr engr control, Electronics Div, Westinghouse Elec Corp, 63; lectr math, Northeastern Univ, 63-65; asst prof elec eng, Univ Minn, Minneapolis, 65-69, assoc prof & dir, Hybrid Comput Lab, 69-76; chmn dept systs eng, Case Western Reserve Univ, 76-80, prof, 76-83; prof & dean engr, Polytechnic Inst, NY, 83-85; pres, 85-86, prof, Ore Grad Ctr, 85-89; prof, Ore Grad Ctr, 85-89; CHIEF SCIENTIST, MITRE CORP, 89- *Concurrent Pos:* Consult, NASA Electronics Res Ctr, 64-65 & 20 various corps; ed, Trans Automatic Control, Inst Elec & Electronics Engrs, 75-78; dir, Div Elec, Comput & Systs Eng, NSF, 80-81. *Honors & Awards:* Curtis Award, Am Soc Environ Educ, 75; Centennial Medalist, Inst Elec & Electronics Engrs, 84. *Mem:* Fel Inst Elec & Electronics Engrs; fel AAAS; distinguished mem, Inst Elec & Electronics Engrs Control Systems Soc (pres 81). *Res:* Control theory, systems engineering, military command and control systems. *Mailing Add:* Mitre Corp 7525 Colshire Dr MS2521 McLean VA 22102

KAHNG, DAWON, b Seoul, Korea, May 4, 31; US citizen; m 56; c 5. PHYSICS, SEMICONDUCTORS. *Educ:* Univ Seoul, BSc, 55; Ohio State Univ, MSc, 56, PhD(elec eng), 59. *Prof Exp:* Res assoc elec eng, Ohio State Univ, 57-59, instr, 58-59; mem tech staff, Bell Labs,59-64, supvr physics, 64-83, fel, 83-87; PRES, NEC RES INST, 88- *Honors & Awards:* Stuart Ballantine Medal, Franklin Inst, 75. *Mem:* Fel Inst Elec & Electronics Engrs.

Res: Semiconductor device physics and technology in the area of impurity diffusion; surface field effect transistor; hot electron devices; Schottky barriers; thin film electroluminescence; charge coupled devices; nonvolatile semiconductor memories. *Mailing Add:* NEC Res Inst Inc Four Independence Way Princeton NJ 08540

KAHNG, SEUN KWON, b Seoul, Korea, Jan 16, 36; US citizen; m 70; c 2. ELECTRICAL ENGINEERING, SOLID STATE ELECTRONICS. *Educ:* Seoul Nat Univ, BSEE, 58; Univ Va, MEE, 63, PhD(elec eng), 67. *Prof Exp:* Res fel transducers, Langley Res Ctr, NASA, 67-68; from asst prof to assoc prof, 68-78, PROF ELEC ENG, UNIV OKLA, 78-, DIR ELEC ENG, 80- *Concurrent Pos:* Res fel, Langley Res Ctr, NASA, 75-76. *Mem:* Inst Elec & Electronics Engrs; Optical Soc Am. *Res:* Solid state electronic devices; piezoresistive silicon sensors; silicon-on-sapphire sensors; piezoelectric sensors. *Mailing Add:* Sch Elec Eng & Comput Sci Univ Okla Norman OK 73019

KAHRIZI, MOJTABA, b Arak, Iran, Jan 26, 49; Can citizen; m 73; c 2. COMPUTER PROGRAMMING, VERY LARGE SCALE INTEGRATION TECHNOLOGY. *Educ:* BSc, Tehran Univ, 73, MSc, 75; MSc, Concordia Univ, 80, PhD(physics), 85. *Prof Exp:* Lectr, Arak Col Sci, Iran, 75-78; Lab instr physics, Concordia Univ, 78-83, fac mem, 83-85; postdoctoral fel physics, St Francis Xavier Univ, 85-87, asst prof, 87-90; RES ASSOC PHYSICS, CONCORDIA UNIV, 90- *Mem:* Can Assoc Physicists; Can Inst Neutron Scattering. *Res:* Investigating magnetic resonance properties of electron nuclear spin coupled systems in condensed matters physics; magnetic and structural phase transitions in solid state materials, particularly high-Tc superconductors and rare-earth metals, using magnetic and dilatometric measurements. *Mailing Add:* Physics Dept Concordia Univ 1455 de Maisonneure W Montreal PQ H3G 2P7 Can

KAHRS, ROBERT F, b Lynbrook, NY, June 28, 30; m 53; c 4. VETERINARY MEDICINE, VETERINARY EPIDEMIOLOGY. *Educ:* Cornell Univ, DVM, 54, MS, 63, PhD(virol biomet), 65. *Prof Exp:* Asst vet, private practice, Interlaken, NY, 54-55; vet, Attica, NY, 55-61; res asst vet virol & biomet, 61-65, res assoc vet epidemiol & microbiol, 65-66, asst prof vet epidemiol, 66-70, assoc prof vet epidemiol, NY State Vet Col, Cornell Univ, 70-78; prof vet epidemiol & chmn dept vet prev med, Univ Fla, 78-; DEAN, COL VET MED, UNIV MO. *Honors & Awards:* Nat Academias Pract. *Mem:* Am Vet Med Asn; US Animal Health Asn; Am Vet Epidemiol Soc. *Res:* Epidemiology of virus diseases of livestock. *Mailing Add:* Col Vet Med Univ Mo Columbia MO 65211

KAIGHN, MORRIS EDWARD, b Camden, NJ, Aug 6, 22; m 79. EMBRYOLOGY, VIROLOGY. *Educ:* Brooklyn Col, BS, 56; Mass Inst Technol, PhD(biol), 62. *Prof Exp:* Asst investr embryol, Carnegie Inst, 64-67; res assoc, New York Blood Ctr, 67-70, assoc investr embryol, 70-72; sr scientist cell ctr, 72-75; SR RES INVESTR, PASADENA FOUND MED RES, 75-; AT FREDERICK CANCER RES FACIL, NAT CANCER INST. *Concurrent Pos:* Fel virol, Univ Toronto, 62-64. *Mem:* Am Asn Cancer Res; Am Soc Cell Biol; Tissue Cult Asn. *Res:* Isolation and characterization of epithelial cells; immortalization of epithelial cells by oncogenes; clonal culture of differentiated human liver prostate and cancer cells; growth control by steroids and peptide growth factors; cell biology. *Mailing Add:* Biol Res Fac & Facil 10075-20 Tyler Place Ijamsville MD 21754

KAILATH, THOMAS, b Poona, India, June 7, 35; m 62; c 3. ELECTRICAL ENGINEERING, APPLIED MATHEMATICS. *Educ:* Univ Poona, BE, 56; Mass Inst Technol, SM, 59, ScD(elec eng), 61. *Hon Degrees:* DEng, Link06ping Univ, Sweden, 90. *Prof Exp:* Res specialist, Jet Propulsion Lab, Calif Inst Technol, 61-62; assoc prof elec eng, 63-68, dir, Info Systs Lab, 71-81, assoc chmn, Dept Elec Eng, 81-87, PROF ELEC ENG, STANFORD UNIV, 68-, HITACHI AM PROF ENG, 88- *Concurrent Pos:* Ed, Prentice-Hall Series on Info & Systs Sci, 63-; vis prof, India Inst Sci, Bangalore, 69-70, Katholieke Univ Leuven, 77 & Tech Univ, Delft, 81; Guggenheim fel, Indian Inst Sci, 69-70; consult, Govt India, 70-71; Churchill fel, Statist Lab, Churchill, Eng, 77; chmn bd dirs, Integrated Systs, Inc, 80-89, vchmn, 89-; Erna & Jacob Michael vis chair theoret math, Weizmann Inst, 84; Royal Soc guest res prof, Imp Col, London, 89. *Honors & Awards:* Eng Achievement Award, Nat Fedn Asian Indian Orgns, 86; Tech Achievement & Soc Award, Signal Processing Soc, Inst Elec & Electronics Engrs, 89 & 91. *Mem:* Nat Acad Eng; fel Inst Math Statist; Sigma Xi; fel Inst Elec & Electronics Engrs; hon fel Inst Electronics & Telecommun Engrs India; Am Math Soc; Math Asn Am; Soc Indust & Appl Math; Inst Math Statist; Soc Explor Geophysicists. *Res:* Information theory; communications; computation; control; linear systems; statistical signal processing; very large scale integration systems; stochastic processes; linear algebra; operator theory; author or co-author of over 300 research papers. *Mailing Add:* Durand 117 Stanford CA 94305

KAIMAL, JAGADISH CHANDRAN, b Kuala Lumpur, Malaysia, Nov 18, 30; US citizen; m 57; c 3. METEOROLOGY. *Educ:* Benares Hindu Univ, BSc, 53; Univ Wash, MS, 59, PhD(meteorol), 61. *Prof Exp:* Res physicist, Air Force Cambridge Res Labs, 61-76; CHIEF, ATMOSPHERIC STUDIES PROG, WAVE PROPAGATION LAB, ENERGY RES LAB, NAT OCEANIC & ATMOSPHERIC ADMIN, 76- *Mem:* Am Meteorol Soc. *Res:* Experimental investigations of turbulent fluctuations in the atmospheric boundary layer and the study of the fluxes of momentum and heat within this layer. *Mailing Add:* 3815 Cloverleaf Dr Boulder CO 80302

KAIN, RICHARD YERKES, b Chicago, Ill, Jan 20, 36; m 61, 81; c 3. ELECTRICAL ENGINEERING. *Educ:* Mass Inst Technol, SB, 57, SM, 59, ScD(elec eng), 62. *Prof Exp:* Asst elec eng, Mass Inst Technol, 57-60, from instr to asst prof, 60-66; assoc prof, 66-77, PROF ELEC ENG, UNIV MINN, MINNEAPOLIS, 77- *Concurrent Pos:* Ford fel eng, 62-64; consult, Honeywell Corp, 75-89, Secure Computing Technol Corp, 89- *Mem:* AAAS; Asn Comput Mach; Inst Elec & Electronics Engrs; Sigma Xi; Computer Prof Social Responsibility. *Res:* Computer systems; computer architecture; secure computer systems. *Mailing Add:* Dept Elec Eng Univ Minn 200 Union St SE Minneapolis MN 55455

KAINE, BRIAN PAUL, EVOLUTION OF THERMOPHILIC ORGANISMS. *Educ:* Northwestern Univ, PhD(biol sci), 81. *Prof Exp:* RES ASSOC, UNIV ILL, 81- *Res:* T RNA gene structure in archaebacteria. *Mailing Add:* 131 Burrill Hall Univ Ill 407 S Goodwin Ave Urbana IL 61801

KAINSKI, MERCEDES H, b Kewaunee Co, Wis, Jan 26, 23; m 64. FOOD SCIENCE, NUTRITION. *Educ:* Univ Wis, BS, 44, MS, 55, PhD(foods), 57. *Prof Exp:* Control chemist, Wilson Res Lab, Ill, 44; with res lab, Carnation Co, Wis, 45-47; teacher pub schs, Wis, 48-53; assoc prof foods & nutrit, Kans State Univ, 57-65; assoc prof home econ, Bowling Green State Univ, 65-67; prof food & nutrit, Univ Wis-Stout, 67-85; RETIRED. *Mem:* Am Home Econ Asn; Inst Food Technol; Am Dietetic Asn. *Res:* Minerals, iron and copper in pork; magnesium and fat metabolism; iron, phenols and organic acid in potatoes. *Mailing Add:* 1712 Fifth St W No 108 Menomonie WI 54751

KAISEL, S(TANLEY) F(RANCIS), b St Louis, Mo, Aug 2, 22; m 58; c 2. ELECTRONICS. *Educ:* Wash Univ, BS, 43; Stanford Univ, MA, 46, PhD(elec eng), 49. *Prof Exp:* Instr math, Wash Univ, 42-43; spec res assoc, Radio Res Lab, Harvard Univ, 43-45; res assoc, Stanford Univ, 46-49; res engr, Radio Corp Am Labs, 49-51; res assoc & group leader, Univ Stanford, 51-55; lectr, 53-55; mgr eng, Electron Tube Div, Litton Industs, 55-58; electronics consult engr, 58-59; founder & pres, Microwave Electronics Corp, 59-69; CONSULT, 69- *Mem:* Fel Inst Elec & Electronics Engrs. *Res:* Radio and radar countermeasures; radar; microwave tubes; traveling wave tubes; vacuum tube technology. *Mailing Add:* Albion Assoc PO Box 620153 Woodside CA 94062

KAISER, ARMIN DALE, b Piqua, Ohio, Nov 10, 27; m 53; c 2. BIOCHEMISTRY, GENETICS. *Educ:* Purdue Univ, BSc, 50; Calif Inst Technol, PhD(biol), 54. *Prof Exp:* From instr to asst prof microbiol, Wash Univ, 56-59; from asst prof to assoc prof, 59-66, PROF BIOCHEM, SCH MED, STANFORD UNIV, 66-, PROF DEVELOP BIOL, 89- *Concurrent Pos:* Am Cancer Soc fel, 54-56; NSF sr fel, 64-65; mem genetics study sect, NIH, 63-68; mem genetic biol panel, NSF, 69-72. *Honors & Awards:* Award in Molecular Biol, US Steel Found, 70; Lasker Award, 80; Waterford Biomed Sci Award, 81. *Mem:* Nat Acad Sci; Genetics Soc Am; Am Soc Biol Chemists; Am Acad Arts & Sci. *Res:* Bacteriophage genetics; nucleic acid biochemistry; biochemistry of morphogenesis; author and co-author of over 100 publications. *Mailing Add:* Dept Biochem Stanford Univ Sch Med Stanford CA 94305-5427

KAISER, C WILLIAM, b Troy, NY, Dec 7, 39; m 66; c 3. SURGERY. *Educ:* Colgate Univ, AB, 61; Tufts Univ, MD, 65. *Prof Exp:* Surg resident, Boston City Hosp, 68-72; assoc dir surg, Tufts Surg Serv, 72-76; assoc chief surg, Vet Admin Med Ctr, Northport, NY, 76-78; chief surg, Pondville Hosp, 78-81; CHIEF SURG, VET ADMIN MED CTR, MANCHESTER, NH, 81-; ASST PROF SURG, HARVARD MED SCH, 81- *Mem:* Soc Surg Alimentary Tract; Asn Acad Surg; Asn Vet Admin Surgeons. *Res:* Surgical oncology. *Mailing Add:* Vet Admin Med Ctr 718 Smyth Rd Manchester NH 03104

KAISER, CARL, b Baltimore, Md, Feb 8, 29; m 53; c 3. MEDICINAL CHEMISTRY. *Educ:* Univ Md, BS, 51, MS, 53, PhD(pharmaceut chem), 55. *Prof Exp:* Lab asst pharmaceut chem, Univ Md, 51-53; Smith Kline & French fel, Univ Va, 55-57; sr med chemist, Smith Kline & French Labs, 57-65, med chem group leader, 65-68, sr investr, 68-72, asst dir chem, 72-79, assoc sci dir, 79-81, sr fel, 81-86; DIR, MED CHEM, NOVA PHARM CORP, 86- *Mem:* Am Chem Soc; Am Pharmaceut Asn. *Res:* Design and synthesis of potential drug products, especially substances affecting the central and autonomic nervous systems, enzyme inhibitors, antimetabolites, drug metabolism and small ring compounds. *Mailing Add:* 8470 Woodland Rd Millersville MD 21108

KAISER, CHARLES FREDERICK, b Dec 30, 42. DEVELOPMENTAL PSYCHOLOGY, HEALTH PSYCHOLOGY. *Educ:* City Col of City Univ of NY, BS, 64, MA, 67; Univ Houston, PhD(psychol), 73. *Prof Exp:* Fel, Univ Houston, 66-72; asst prof, 72-77, ASSOC PROF PSYCHOL, COL CHARLESTON, 77- *Concurrent Pos:* Fel, dept psychiat, Baylor Col Med, 66-70; adj asst prof, Dept Phys Med & Rehab, Med Univ SC, 81-; mem comt, Asn Appl Psychophysiol & Biofeedback, 83- *Mem:* Asn Appl Psychophysiol & Biofeedback. *Res:* Research in personality and cognitive abilities of gifted adolescents; depression in college students, gifted adolescents, medical patients and children; effects of biofeedback and relaxation training on pain in medical patients; published numerous articles in various journals. *Mailing Add:* Dept Psychol Col Charleston 66 George St Charleston SC 29424

KAISER, CHRISTOPHER B, b Greenwich, Conn, Oct 16, 41; m 70; c 3. HISTORY & PHILOSOPHY OF SCIENCE, PHYSICS. *Educ:* Harvard Univ, BA, 63; Univ Colo, PhD(astrogeophys), 68; Edinburgh Univ, PhD(theology), 74. *Prof Exp:* Lectr physics, Gordon Col, 68-71 & Edinburgh Univ, 73-74; with Systs Develop, QEI Inc, Bedford, Mass, 75-76; from asst prof to assoc prof, 77-88, PROF, WESTERN THEOL SEM, HOLLAND, 88- *Concurrent Pos:* Mem, Gravity Res Found, 68-; resident mem, Ctr Theol Inquiry, Princeton, NJ, 84, 87. *Mem:* AAAS; Hist Sci Soc; Soc Hist Technol. *Res:* History of science as influenced by religious belief and practice; history of interaction between theological beliefs and scientific progess; systems theory. *Mailing Add:* Western Theol Seminary 86 E 12th St Holland MI 49423

KAISER, DAVID GILBERT, b Detroit, Mich, Aug 25, 28; m 61; c 2. PHARMACEUTICAL CHEMISTRY, DRUG METABOLISM. *Educ:* Detroit Inst Technol, BS, 52; Purdue Univ, MS, 54, PhD(pharmaceut chem), 59. *Prof Exp:* Fel radiochem, Univ Mich, 59; chemist, Upjohn Co, 59-69, res head drug metab, 69-79, res mgr drug metab, 79-85, dir drug metab res, 85-86, SR RES CONSULT, UPJOHN CO, 86- *Mem:* AAAS; Am Chem Soc; Am Pharmaceut Asn; Sigma Xi; Am Soc Mass Spectrometry; NY Acad Sci. *Res:* Drug metabolism and analytical chemistry. *Mailing Add:* 6605 Robinswood Kalamazoo MI 49002

KAISER, EDWARD WILLIAM, JR, b Minneapolis, Minn, May 10, 42; m 68; c 1. PHYSICAL CHEMISTRY. *Educ:* Northwestern Univ, BA, 64; Harvard Univ, MA, 66, PhD(chem), 70. *Prof Exp:* NATO fel, Southampton Univ, 69-70; temp mem tech staff, Bell Labs, 70-72; assoc scientist, Xerox Corp, 72-74; sr res scientist, 74-80, prin res assoc, 80-86, STAFF SCIENTIST, FORD MOTOR CO, 87- *Honors & Awards:* Donald Julius Groen Prize, Inst Mech Engrs, 86. *Mem:* Am Phys Soc; Am Chem Soc; Combustion Inst. *Res:* Chemical kinetics; combustion research; molecular spectroscopy; ion-molecule reactions. *Mailing Add:* Seven Windham Lane Dearborn MI 48120

KAISER, EDWIN MICHAEL, b Youngstown, Ohio, Oct 15, 38; m 60; c 5. ORGANIC CHEMISTRY. *Educ:* Youngstown Univ, BS, 60; Purdue Univ, PhD(org chem), 64. *Prof Exp:* Res assoc org chem, Duke Univ, 64-66; from asst prof to assoc prof, 70-74, PROF ORG CHEM, UNIV MO-COLUMBIA, 74- *Concurrent Pos:* Dir, Hon Col, 84- *Mem:* Am Chem Soc; Sigma Xi; AAAS; Am Inst Chemists. *Res:* Organometallic derivatives of methylated heterocycles; condensations and cyclizations in nonaqueous media. *Mailing Add:* 123 Chem Bldg Univ Mo Columbia MO 65211

KAISER, GEORGE C, b Bronx, NY, July 30, 28; m 53; c 3. THORACIC SURGERY. *Educ:* Lehigh Univ, AB, 49; Johns Hopkins Univ, MD, 53. *Prof Exp:* Intern surg, Johns Hopkins Hosp, Baltimore, Md, 53-54; resident, Vet Admin Hosp, Ft Howard, Md, 54; clin assoc, clin of surg, Nat Heart Inst, 54-56; resident surg, Med Ctr, Ind Univ, 56-61, from instr to asst prof, 61-63; from asst prof to assoc prof, 63-70, PROF SURG, SCH MED, ST LOUIS UNIV, 70- *Concurrent Pos:* Staff surgeon, Vet Admin Hosp, Indianapolis, Ind, 61; dir St Louis Univ surg serv, Vet Admin Hosp, 63-65. *Mem:* Soc Thoracic Surg; AMA; Am Col Surg; Int Cardiovasc Soc; Am Asn Thoracic Surg; Sigma Xi. *Res:* General and thoracic surgical problems, including research in cardiac physiology. *Mailing Add:* St Louis Univ Hosp 3635 Vista at Grand PO Box 15250 St Louis MO 63110-0250

KAISER, GERARD ALAN, b Brooklyn, NY, Dec 9, 32; m 55; c 3. THORACIC SURGERY, CARDIOVASCULAR SURGERY. *Educ:* Princeton Univ, AB, 54; Columbia Univ, MD, 58. *Prof Exp:* Intern, Presby Hosp, New York, 58-59, asst resident gen surg, 59-62, resident, 64-65; resident thoracic surg, Vet Admin Hosp & Bellevue Hosp Ctr, New York, 66 & Presby Hosp, 67; instr surg, Columbia Univ, 67-68; asst prof, Mt Sinai Sch Med, 68-69; assoc prof, Columbia Univ, 69-71; PROF SURG & CHIEF, DIV THORACIC & CARDIOVASCULAR SURG, SCH MED, UNIV MIAMI, 71- *Concurrent Pos:* Fel, NY Tuberc & Health Asn, 66-67; Glorney-Raisbeck fel, NY Acad Med, 68-69; asst surg, Columbia Univ, 65-67; asst vis prof, Delafield Hosp, 68; asst attend surg, Columbia-Presby Med Ctr, 68, assoc attend surg, 69-71; vis asst surg, Elmhurst Hosp, 68-69 & Harlem Hosp Ctr, 69-71; asst attend surg, Mt Sinai Hosp, 68-69; consult, Vet Admin Hosp, 68-71; active attend & chief div thoracic & cardiovasc surg, Jackson Mem Hosp; Otto G Storm estab investr, Am Heart Asn, 70. *Mem:* Soc Univ Surgeons; Soc Thoracic Surgeons; Asn Acad Surg; Am Fedn Clin Res; Int Cardiovasc Soc; Sigma Xi. *Res:* Cardiovascular physiology and pharmacology, especially electrophysiology. *Mailing Add:* Univ Miami Sch Med PO Box 016960 Miami FL 33101

KAISER, IRWIN HERBERT, b New York, NY, Jan 27, 18; m 38; c 6. OBSTETRICS & GYNECOLOGY. *Educ:* Columbia Univ, AB, 38; Johns Hopkins Univ, MD, 42; Univ Minn, PhD, 53. *Prof Exp:* Assoc prof obstet & gynec, Med Sch, Univ Minn, 54-59; prof & head dept, Col Med, Univ Utah, 59-68; PROF OBSTET & GYNEC, ALBERT EINSTEIN COL MED, 68- *Mem:* Soc Gynec Invest; Am Gynec & Obstet Soc; Am Col Obstet & Gynec. *Res:* Physiology of mammalian reproduction; water, electrolyte and gas equilibria between fetal and maternal circulations during pregnancy; contractile activity of pregnant human uterus prior to labor. *Mailing Add:* Dept Obstet & Gynec Belfer Bldg Rm 501 Einstein Col Med Bronx NY 10461

KAISER, IVAN IRVIN, b Stuart, Nebr, Nov 21, 38; m 66; c 2. BIOCHEMISTRY. *Educ:* Wayne State Col, BA, 62; Iowa State Univ, PhD(biochem), 67. *Prof Exp:* From asst prof to assoc prof biochem, 67-75, prof biochem & chem, 75-78, chmn biochem, 79-84, PROF MOLECULAR BIOL & CHEM, UNIV WYO, 85- *Mem:* Am Chem Soc; Am Soc Biol Chemists; Sigma Xi; AAAS; Int Soc Toxinology; Protein Soc. *Res:* Structure and function of ribonucleic acids; selenium biochemistry; natural toxins. *Mailing Add:* Dept Molecular Biol Univ Wyo Laramie WY 82071

KAISER, JACK ALLEN CHARLES, b Chicago, Ill, Nov 15, 35; m 63; c 2. AIR-SEA INTERACTION, INSTRUMENTATION. *Educ:* Ill Inst Technol, BS, 57; Univ Chicago, PhD(geophysics), 69. *Prof Exp:* Weather forecaster, USAF, 58-60; res asst, Univ Chicago, 62-69, res assoc, 69-72; RES PHYSICIST, NAVAL RES LAB, 72- *Concurrent Pos:* Prin investr, Naval Res Lab, 74- *Mem:* Am Meterol Soc; Am Geophys Union; Sigma Xi. *Res:* Experiments on air-sea interaction and upper ocean dynamics, structure and irradiance distribution. *Mailing Add:* 12503 Woostock Dr E Upper Marlboro MD 20772

KAISER, JAMES F(REDERICK), b Piqua, Ohio, Dec 10, 29; m 54; c 4. ELECTRICAL ENGINEERING. *Educ:* Univ Cincinnati, EE, 52; Mass Inst Technol, SM, 54, ScD, 59. *Prof Exp:* Asst, Servomech Lab, Mass Inst Technol, 52-55, from instr to asst prof elec eng, 55-60; mem tech staff, 59-84, DISTINGUISHED MEM TECH STAFF, BELL TEL LABS INC, DIGITAL SYSTS RES DEPT, BELL COKE, 84- *Honors & Awards:* Centennial Medal, Inst Elec & Electronics Engrs, 84, Technical Achievement Award, Acoustics, Speech & Signal Processing, 78, Soc Award, 81. *Mem:* Fel Inst Elec & Electronics Engrs; Asn Comput Mach; AAAS; Soc Indust & Appl Math; Acoust Soc Am. *Res:* Theory of control and signal processing systems; system optimization; application of digital computations to continuous systems; digital signal processing; continuous system modeling; vocal tract modeling; speech technology research. *Mailing Add:* Bell Commun Res 2A-281 445 South St Morristown NJ 07962-1910

KAISER, JOSEPH ANTHONY, b Baltimore, Md, Mar 22, 26; m 51; c 3. PHARMACOLOGY. *Educ:* Univ Md, BS, 50, MS, 52, PhD(pharmacol), 55. *Prof Exp:* Sr res pharmacologist, Pfizer Therapeut Inst, NJ, 55-58; exp therapeut res sect, Lederle Labs, Am Cyanamid Co, 58-63; pharmacologist, Drug Rev Br, Div Toxicol Eval, Bur Sci Standards & Eval, Food & Drug Admin, 63-64, res pharmacologist, Div Pharmacol, Bur Sci Res, 64-66; exec secy pharmacol & endocrinol fels rev sect, 66-69, from asst chief to dep chief, Career Develop Rev Br, 69-73, exec secy spec progs, 73-74, EXEC SECY PHARMACOL STUDY SECT, DIV RES GRANTS, NIH, 74- *Mem:* Am Soc Pharmacol & Exp Therapeut. *Res:* Pharmacology-toxicology; antibiotics; anticholinergics, antihistamines, antiparasiticides and anti-tubercular agents; health science administration; toxicological evaluation. *Mailing Add:* NIH Div Res Grants 5333 Westbard Ave Bethesda MD 20892

KAISER, KLAUS L(EO) E(DUARD), b Kempten, Ger, June 17, 41; Can citizen; c 3. ORGANIC CHEMISTRY, ANALYTICAL CHEMISTRY. *Educ:* Tech Univ, Munich, cand chem, 64, dipl chem, 66, Dr rer nat(chem), 68. *Prof Exp:* Fel organometallic chem, Fonds Ger Chem Indust, 68-69 & Nat Res Coun, McMaster Univ, Ont, 69-71; RES SCIENTIST, CAN CTR INLAND WATERS, ENVIRON CAN, 72-, HEAD, ORG PROP SECT, 80- *Concurrent Pos:* From alt mem to mem, Water Qual Objectives Subcomt, Int Joint Comn, 74-78; mem, Task Force Polychlorinated Biphenyls, Environ Can & NHW Can, 75-76 & Task Force Mirex, 76-77; liaison mem, Task Force Ecol Effects Non-Phosphate Detergent Builders, Int Joint Comn, 78-80; assoc ed, J Great Lakes Res, 80-; ed, Quant Struct Activity Relationship in Environ Toxicol, 84; co-ed, Acta Hydrochimica et Hydrobiologica, 85-; quant struct activity relationships, Environ Toxicol-II, 87; pres, Int Assoc Great Lakes Res, 87-88; Chief, Nearshore-Offshore Interactions Proj, 87-; co-chair, Fate of Toxics Comt, Lake Ont-Niagara River Mgt Plan, 89- *Mem:* Chem Inst Can; Int Asn Great Lakes Res; Ger Chem Soc; Soc Environ Toxicol Chemists; fel Chem Inst Can. *Res:* Chemistry of contaminants in the biosphere, including their analysis, bioaccumulation, metabolic and photochemical transformation and their toxicity; quantitative structure-activity correlation (QSAR) of contaminants; organometallic and environmental chemistry. *Mailing Add:* Nat Water Res Inst PO Box 5050 Burlington ON L7R 4A6 Can

KAISER, MARY AGNES, b Pittston, Pa, June 11, 48; m 79; c 1. ANALYTICAL CHEMISTRY. *Educ:* Wilkes Univ, BS, 70; St Joseph's Univ, Pa, MS, 72; Villanova Univ, PhD(chem), 76. *Prof Exp:* Assoc chem, Univ Ga, 76-77; res chemist, 77-80, res supvr, 80-84, sr supvr, 84-89, TEST MGR, E I DU PONT DE NEMOURS & CO, 89- *Mem:* Am Chem Soc; Sigma Xi. *Res:* Analytical chemistry of separations; spectroscopy; environmental chemistry. *Mailing Add:* Eng Test Ctr E I du Pont de Nemours & Co Inc PO Box 6094 Newark DE 19714-6094

KAISER, MICHAEL LEROY, b Keokuk, Iowa, Dec 28, 41; m 68; c 2. RADIO ASTRONOMY. *Educ:* Univ Iowa, BA, 64; Univ Md, College Park, MS, 73. *Prof Exp:* Comput programmer astron, Nat Radio Astron Observ, 64-65; sci analyst astron celestial mech, Wolf Res & Develop Corp, 65-69; RADIO ASTRONR, GODDARD SPACE FLIGHT CTR, NASA, 69- *Mem:* Am Astron Soc; Am Geophys Union; Int Union Radio Scientists; Inst Elec & Electronics Engrs. *Res:* Planetary radio physics; magnetospheric physics. *Mailing Add:* NASA/GSFC Code 695 Greenbelt MD 20771

KAISER, PETER, b Aschaffenburg, W Ger, 1938; m 66; c 2. ELECTRICAL ENGINEERING. *Educ:* Munich Tech Univ, Diplom Ing, 63; Univ Calif, Berkeley, MS, 65, PhD(elec eng), 66. *Prof Exp:* NATO fel, 63-64; mem staff, Guided Waves Res Lab, 66-79, SUPVR, LIGHTWAVE TECH GROUP, BELL LABS, 79- *Mem:* Inst Elec & Electronics Engrs; Optical Soc Am. *Res:* Frequency independent antennas; optical communication; guided wave transmission. *Mailing Add:* Navesing Eng & Res Rm NV3Z 231 331 Newman Springs Rd Box 7040 Red Bank NJ 07701

KAISER, QUENTIN C, b Ridgewood, NY, Sept 12, 21; m 45; c 3. SOLID STATE PHYSICS. *Educ:* Hofstra Col, BA, 49; Okla Agr & Mech Col, MS, 50. *Prof Exp:* Physicist, Res Lab, Harry Diamond Labs, 50-53, supvry electronics scientist, Develop Lab, 53-59, physicist, Microminiaturization Br, 59-61, supvr res & develop, 61-63, br chief & supvry physicist, 63-80; RETIRED. *Mem:* Am Phys Soc; Inst Elec & Electronics Engrs. *Res:* Dielectric measurements; proximity fuze design; solid state devices. *Mailing Add:* 4114 Byrd Ct Kensington MD 20895

KAISER, REINHOLD, b Duisburg, Ger, Nov 19, 27. MAGNETIC RESONANCE. *Educ:* Univ Gottingen, dipl physics, 53, Dr rer nat, 54. *Prof Exp:* Ger Res Coun fel, Imp Col, Univ London, 55; Can Res Coun fel, Dalhousie Univ, 56; from asst prof to assoc prof, 57-66, PROF PHYSICS, UNIV NB, 66- *Concurrent Pos:* Res fel, Harvard Univ, 64 & Shell Develop Co, Calif, 65; guest prof, Swiss Fed Inst Technol, 71-72. *Res:* Acoustics; magnetic resonance. *Mailing Add:* Dept Physics Univ NB Fredericton NB E3B 5A3 Can

KAISER, RICHARD EDWARD, b Chicago, Ill, Dec 20, 36; m 63; c 3. NUCLEAR ENGINEERING. *Educ:* Northwestern Univ, Evanston, BS, 59; Kans State Univ, MS, 62, PhD(nuclear eng), 68; Univ Idaho, MBA, 78. *Prof Exp:* Mem staff reactor shielding, Atomics Int, 61-64; asst nuclear engr, 67-72, NUCLEAR ENGR, ARGONNE NAT LAB, 72-, ZPPR REACTOR MGR, 78- *Concurrent Pos:* Instr, Aerojet Nuclear Co, 68-; affil prof, Univ Idaho. *Mem:* Sigma Xi; Am Nuclear Soc. *Res:* Doppler effect and sodium void coefficients in fast reactors. *Mailing Add:* Argonne Nat Lab PO Box 2528 Idaho Falls ID 83401

KAISER, ROBERT, b Strasbourg, France, June 22, 34; US citizen; m 70; c 2. CHEMICAL ENGINEERING, APPLIED CHEMISTRY. *Educ:* Mass Inst Technol, SB, 56, MS, 57, ScD(chem eng), 62. *Prof Exp:* Res engr, M W Kellogg Co, 61, res chemist, Res & Develop Ctr, Pullman, Inc, NJ, 62-65, res engr, 65-66; sr staff scientist, Res & Tech Labs, Space Systs Div, Avco Corp, Lowell, 66-71, group leader, Advan Processes Dept, Systs Div, 71-74; CONSULT ENGR, 74-; PRES & FOUNDER, AGORS ASSOCS, INC,

WINCHESTER, MASS, 77- *Mem:* Am Chem Soc; Am Inst Chem Engrs; Am Ceramic Soc; Soc Automotive Engrs. *Res:* Oil/water separation; magnetic liquids; applied surface chemistry; fine powder technology; process development; technology assessment and forecasting; industrial market research. *Mailing Add:* PO Box 397 Winchester MA 01890-0597

KAISER, ROBERT L, b Erie, Pa, Feb 9, 31; m 59; c 3. TROPICAL MEDICINE, EPIDEMIOLOGY. *Educ:* Brown Univ, AB, 53; Yale Univ, MD, 57; Univ London, DTM&H, 63. *Prof Exp:* Chief parasitic dis unit, Commun Dis Ctr, 63-67, dir malaria prog, Ctr Dis Control, USPHS, 67-73; dir Bur Trop Dis, 73-80, DIR DIV PARASITIC DIS, CTR DIS CONTROL, 80- *Honors & Awards:* Gorgas Medal. *Mem:* Am Soc Trop Med & Hyg; Royal Soc Trop Med & Hyg. *Res:* Epidemiology of parasitic diseases including malaria and schistosomiasis. *Mailing Add:* 846 Barton Woods Rd NE Atlanta GA 30307

KAISER, THOMAS BURTON, b St Louis, Mo, May 11, 40; m 67; c 1. PLASMA PHYSICS. *Educ:* St Edward's Univ, BS, 62; Univ Md, College Park, MS, 71, PhD(physics), 73. *Prof Exp:* Sr analyst programming, LTV Aerospace Corp, Mass, 66-68; res assoc space physics, Goddard Space Flight Ctr, Md, 73-75; PHYSICIST PLASMA PHYSICS, LAWRENCE LIVERMORE LAB, 76- *Concurrent Pos:* Resident res assoc, Nat Acad Sci-Nat Res Coun, 73-75. *Mem:* AAAS; Am Phys Soc; Sigma Xi. *Res:* Theoretical plasma physics; computational physics. *Mailing Add:* Lawrence Livermore Lab L-630 PO Box 5511 Livermore CA 94550

KAISER, WILLIAM RICHARD, b Racine, Wis, Aug 15, 37; m 70; c 2. COAL HYDROGEOLOGY, AQUEOUS GEOCHEMISTRY. *Educ:* Univ Wis-Madison, BA, 59, MS, 62; Johns Hopkins Univ, PhD(geol), 72. *Prof Exp:* Geologist micropaleont, Exxon Co, USA, 62-63, geologist petrol geol, 65-68; geologist igneous & metamorphic petrog, Ghana Geol Surv, Accra, Ghana, 63-65; RES SCIENTIST, BUR ECON GEOL, UNIV TEX, 72- *Concurrent Pos:* Lectr, Dept Geol Sci, Univ Tex, Austin, 78-80; mem, Lignite Subcomt & Fossil Energy Adv Comt, Dept Energy, 78; chmn exec comt, Tex Univ Coal Res Consortium, 83-85; mem, Steering Comt Coal Reserves Assesment, Dept Energy, 90- *Mem:* Geol Soc Am. *Res:* Depositional systems; geology of Gulf Coast (Texas) lignite; hydrogeology of coal; underground coal gasification; low-temperature aqueous geochemistry; brine equilibria in the predication of reservoir quality; retardation of radionuclides; coalbed methane. *Mailing Add:* Bur Econ Geol Univ Tex Univ Sta Box X Austin TX 78713-7508

KAISER, WOLFGANG A, b Schoental, Ger, Feb 22, 23; m 51; c 2. TELECOMMUNICATIONS. *Educ:* Univ Stuttgart, dipl ing, 51, Dr ing, 55. *Hon Degrees:* Dr ing Eh, Univ Munich, 85. *Prof Exp:* Res engr telecommun, Standard Elektrik Lorenz, 54-57, lab head, 57-63, res & develop dir, 63-67; PROF TELECOMMUN, UNIV STUTTGART, 67- *Concurrent Pos:* Chmn, Res Coun, Muenchner Kreis, Munich, 77-; mem, Acad Sci, Heidelberg, 82. *Mem:* Fel Inst Elec & Electronics Engrs. *Res:* Evolution of telecommunications; optical transmission systems; wideband networks for speech, text, data, pictures; television; digital audio; data communication in local and metropolitan area networks. *Mailing Add:* Breitscheidstr 2 Stuttgart D7000 Germany

KAISER-KUPFER, MURIEL I, b New York, NY, May 25, 36; m. OPHTHALMIC GENETICS RESEARCH. *Educ:* Wellesley Col, BA, 57; Hopkins Med Sch, MD, 61; Am Bd Pediat, cert, 67; Am Bd Ophthal, cert, 74. *Prof Exp:* Residency pediat, Johns Hopkins Hosp, Baltimore, 63-65, fel child psychiat, 65-66, asst dir & instr, Comprehensive Care Clin, Dept Pediat, 66-68; residency, Ophthal & Consult Congenital Defects Clin, Sch Med, Univ Wash, 68-70; consult eye care delivery facill, Comprehensive Health Care Ctr, 70-71; asst prof, Dept Obstet & Gynec, Sch Med, George Washington Univ, 71-72; sr staff fel, Nat Inst Child Health & Human Develop, NIH, 72-74, med officer ophthal & pediat, Clin Br, Nat Eye Inst, 74-81, chief, Sect Ophthalmic Genetics & Pediat Ophthal, Clin Br, 81-89, BR CHIEF, OPHTHALMIC GENETICS & CLIN SERV BR, NAT EYE INST, NIH, BETHESDA, 89-, DEP CLIN DIR, 91- *Concurrent Pos:* Vis prof, Pan Am Ophthal Asn, 76; comt mem, Pharm & Therapeut Comt Clin Ctr, NIH, 77-89. *Mem:* Am Acad Ophthal; Nat Soc Prev Blindness; Asn Res Vision & Ophthal; Am Ophthal Soc. *Res:* Child psychiatry; ophthalmic genetics research. *Mailing Add:* NIH Nat Eye Inst Ophthal Genetics & Clin Serv Br Bldg 10 Rm 10N228 Bethesda MD 20892

KAISERMAN, HOWARD BRUCE, b Philadelphia, Pa, Oct 10, 57; m 91. ENZYME STABILIZATION, SURFACTANT & PROTEIN INTERACTIONS. *Educ:* Skidmore Col, BA, 80; Emory Univ, PhD(chem), 84. *Prof Exp:* Postdoctoral fel biochem, Dept Biol, Johns Hopkins Univ & NIH, 84-88; RES SCIENTIST BIOCHEM, UNILEVER RES, 88- *Mem:* Am Chem Soc; Am Soc Biochem & Molecular Biol. *Res:* Influence of chemical agents on protein denaturation with the ultimate goal of protecting proteins from denaturants; storage stability of proteins in aggressive environments. *Mailing Add:* Unilever Res 45 River Rd Edgewater NJ 07020

KAISERMAN-ABRAMOF, ITA REBECA, b Belo Horizonte, Brazil, Sept 11, 33; div; c 1. NEUROBIOLOGY, NEUROCYTOLOGY. *Educ:* Univ Minas Gerais, BS, 55, MS, 56, PhD(biol sci), 62. *Prof Exp:* Actg dept chmn biol, Univ Minas Gerais, 62-63, assoc prof cytol, histol & embryol, 65-66; teaching asst histol, Sch Med, Harvard Univ, 66-67; asst prof anat, Sch Med, Boston Univ, 67-71; ASSOC PROF ANAT, SCH MED, CASE WESTERN RESERVE UNIV, 71- *Mem:* Am Inst Biol Sci; Am Soc Cell Biol; Am Asn Anatomists. *Res:* Cytological investigations of the mammalian brain, including visual and motor cerebral cortex and cerebellum; use of electron microscopy with experimental and quantitative analysis of connectivity; anophthalmic mutant mice and mechanisms involved in epilepsy. *Mailing Add:* Anat Dept Case Western Med Sch 2119 Abington Rd Cleveland OH 44106

KAISTHA, KRISHAN K, b Kangra, Himachal, India, Apr 6, 26; US citizen; m 48; c 3. TOXICOLOGY, CLINICAL CHEMISTRY. *Educ:* Punjab Univ, India, BS, 47, BS, 52, MS, 54; Univ Fla, PhD(pharmaceut chem & anal control methods), 62; Am Bd Forensic Sci, dipl. *Prof Exp:* Drug analyst, Punjab State Drug Control Lab, 52-57; chief pharmacist & dir, Pharm Dept, Dept Pub Health, Punjab State, 57-59; res fel, Univ Buffalo, 63-64; head, Pharmaceut Dept, Punjab State Health Dept, 64-65; res scientist anal control methods, Food & Drug Admin, Ont, Can, 65-67; dir, Toxicol Lab, Dept Alcoholism & Substance Abuse, State of Ill, 67-86; DIR, TOXICOL & MED LABS, TOXICOL TECH INC & K K BIOSCI, 86- *Mem:* Fel Acad Clin Biochem; fel NY Acad Sci; fel Am Acad Forensic Sci; Am Asn Clin Chem. *Res:* Published over 20 articles on drug abuse detection. *Mailing Add:* Ten W 35th St Chicago IL 60616

KAITA, ROBERT, b Tokyo, Japan, Sept 2, 52; US citizen; m 80; c 1. PLASMA PHYSICS. *Educ:* State Univ NY, Stonybrook, BSc, 73; Rutgers Univ, PhD(physics), 78. *Prof Exp:* Res assoc, Plasma Physics Lab, Princeton Univ, 78-80, res staff, 80-84, res physicist, 84-90, PRIN RES PHYSICIST, PLASMA PHYSICS LAB, PRINCETON UNIV, 90- *Mem:* Am Physical Soc; Sigma Xi; AAAS. *Res:* Tokamak heating with neutral particle beams and radiofrequency waves; probe beams and particle detectors as plasma diagnostics; computer simulations of thermonuclear plasmas. *Mailing Add:* Princeton Univ Plasma Phys Lab Box 451 Princeton NJ 08543

KAIZER, HERBERT, b Boston, Mass, Sept 30, 30; m 54; c 3. ONCOLOGY, BONE MARROW TRANSPLANTATION. *Educ:* Boston Univ, AB, 51, PhD(exp psychol), 56; Stanford Univ, MD, 65. *Prof Exp:* Assoc psychologist, Int Bus Mach, Inc, 56-58; mem tech staff, Thompson, Ramo, Woolridge, Inc, 58-59; intern & asst resident, Johns Hopkins Hosp, 65-67; fel microbiol, Johns Hopkins Univ, 67-69; sr fel pediat, Univ Tex M D Anderson Hosp & Tumor Inst, 69-70; asst prof pediat & oncol, Sch Med, Johns Hopkins Univ, 70-88; COLEMAN-FANNIE MAY CANDIES FOUND PROF PEDIAT, MED & IMMUNOL, RUSH UNIV & DIR, BONE MARROW TRANSPLANT CTR, RUSH PRESBY, ST LUKE'S, 88- *Mem:* AAAS; Am Soc Microbiol. *Res:* Autologous bone marrow transplantation in cancer. *Mailing Add:* Rush Presby St Luke's 1653 W Cong Pkwy Chicago IL 60612

KAJFEZ, DARKO, b Delnice, Yugoslavia, July 8, 28; m 54; c 2. ELECTRICAL ENGINEERING. *Educ:* Univ Ljubljana, EE, 53; Univ Calif, Berkeley, PhD(eng), 67. *Prof Exp:* Assoc prof elec eng, 67-70; PROF ELEC ENG & RES ENGR, UNIV MISS, 70- *Concurrent Pos:* Vis prof elec eng, Univ Ljubljana, 76-77; consult, Harris-Farinan, San Carlos, Calif, 80-83. *Mem:* Inst Elec & Electronics Engrs; Int Union Radio Sci; Inst Elec Engrs Brit. *Res:* Microwave circuits and antennas. *Mailing Add:* Dept Elec Eng Univ Miss University MS 38677

KAJI, AKIRA, b Tokyo, Japan, Jan 13, 30; m 58; c 4. BIOCHEMISTRY. *Educ:* Univ Tokyo, BS, 53; Johns Hopkins Univ, PhD(biochem), 58. *Prof Exp:* Res assoc microbiol, Sch Med, Vanderbilt Univ, 60-61; assoc, 63, from asst prof to assoc prof, 64-72, PROF MICROBIOL, SCH MED, UNIV PA, 72- *Concurrent Pos:* Res fel ophthal, Sch Med, Johns Hopkins Univ, 58-59, res fel, McCollum Pratt Inst, 59-60; Helen Hay Whitney Found fel, 61-63; vis investr, Rockefeller Inst, 59; vis scientist, Oak Ridge Nat Lab, 62; Helen Hay Whitney estab investrship, 64-69; John Simmon Guggenheim Scholar, Imperial Cancer Res Fund Lab, London & prof, Tokyo Univ, 69- *Mem:* Am Soc Biol Chemists; Am Soc Microbiol. *Res:* Sulfur metabolism; neurochemistry; mechanism of enzyme action; tumorgenesis; protein biosynthesis; nucleic acids. *Mailing Add:* Dept Microbiol Univ Pa Sch Med Philadelphia PA 19104

KAJI, HIDEKO (KATAYAMA), b Tokyo, Japan, Jan 1, 32; m 58; c 4. BIOCHEMISTRY, PHARMACOLOGY. *Educ:* Univ Nebr, MS, 56; Purdue Univ, PhD(pharmacol), 58. *Prof Exp:* Eli Lilly fel, Sch Med, Johns Hopkins Univ, 58-59; from instr to asst prof, Sch Med, Vanderbilt Univ, 60-62; vis scientist, Oak Ridge Nat Lab, 62-63; assoc, Sch Med, Univ Pa, 63-64; res assoc, 65-66, asst mem biochem, Inst Cancer Res, 66-76, assoc prof, 76-83, PROF, JEFFERSON MED COL, THOMAS JEFFERSON UNIV, 83- *Concurrent Pos:* Bd mem, sci counr, NIH. *Mem:* Am Soc Biol Chemists; Am Soc Pharmacol & Exp Therapeut. *Res:* Mechanism of macromolecular synthesis; transport mechanism; genetic regulatory mechanisms of oncogenesis. *Mailing Add:* Dept Pharmacol Jefferson Med Col 1020 Locust St Philadelphia PA 19107

KAK, AVINASH CARL, b Srinagar, Kashmir, Oct 22, 44; m 76; c 2. COMPUTER ENGINEERING. *Educ:* Indian Inst Technol, PhD(elec eng), 70. *Prof Exp:* From asst prof to assoc prof, 70-77, PROF ELEC ENG, ROBOT VISION LAB, PURDUE UNIV, 77- *Mem:* Am Asn Artificial Intel. *Res:* Sensory aspects of robotic intelligence; computer vision; spatial reasoning. *Mailing Add:* Sch Elec Eng Purdue Univ West Lafayette IN 47907

KAK, SUBHASH CHANDRA, b Srinagar, India, Mar 26, 47; m 79; c 2. NEURAL NETWORKS, ARTIFICIAL INTELLIGENCE. *Educ:* Kashmir Univ, BS, 67; Indian Inst Technol, Delhi, PhD(elec eng), 70. *Prof Exp:* Lectr elec eng, Indian Inst Technol, Delhi, 71-74, asst prof, 74-79; assoc prof elec eng, 79-83, PROF ELEC & COMPUTER ENG, LA STATE UNIV, 83- *Concurrent Pos:* Acad visitor, Imp Col, Univ London, 75-76; guest researcher, Bell Labs, Murray Hill, 76 & Tata Inst Fundamental Res, Bombay, 77-78; guest ed, Inst Elec & Electronics Engrs Computer, 83; vis prof, Indian Inst Technol, Delhi, 85-86; consult, UN Develop Prog, 86 & 89-90. *Honors & Awards:* Sci Acad Medal, Indian Nat Sci Acad, 77; Kothari Award, Kothari Sci & Res Inst, 77. *Res:* Information theory; quantum physics; cognitive science; artificial intelligence; neural computing; cryptology and study of ancient scripts; history and philosophy of science. *Mailing Add:* Dept Elec & Computer Eng La State Univ Baton Rouge LA 70803-5901

KAKAR, ANAND SWAROOP, b India, Oct 14, 37; US citizen; m 69; c 1. CONDUCTIVE COATINGS, SURFACE CHEMISTRY. *Educ:* Banaras Hindu Univ, BSc, 60; Indian Inst Technol, MTech, 64; Wayne State Univ, MS, 71, PhD(phys chem), 78. *Prof Exp:* Lectr chem eng, Indian Inst Technol, Delhi, 65-68; technician, Can Gen Elec, 68-69 & Mercury Paint Co, 71-72; mfg develop engr, Ford Motor Co, Mt Clemons, Mich, 72-74; res asst, Wayne State Univ, 74-78; staff chemist, Acheson Colloids, Mich, 78-81; DIR RES, GRAFO COLLOIDS, 81- *Mem:* Electrochem Soc; Am Inst Chem Engrs. *Res:* Heat transfer and hold-up fluidized beds; zone refining and single crystal growth; optical and electrical properties of semiconductor; photovoltaic cells; electroless deposition; size reduction; colloidal dispersion; conductive coatings; solid film lubricants; surface preparation and analysis. *Mailing Add:* PO Box 505 Emlenton PA 16373

KAKAR, RAJESH KUMAR, b New Delhi, India, Oct 2, 50; m 77. STATISTICS, DATA PROCESSING. *Educ:* Univ Delhi, BSc, 70, MS, 72; Tex Tech Univ, DBA(bus statist), 78. *Prof Exp:* Instr bus statist, Tex Tech Univ, 72-78; ASST PROF BUS STATIST, ARIZ STATE UNIV, 78- *Mem:* Am Statist Asn; Inst Mgt Sci; Am Inst Decision Sci. *Res:* Empirical bayesian estimation; assessment of subjective probabilities; forecasting; manpower models; auditing software. *Mailing Add:* 4891 E Butler Dr Paradise Valley AZ 85253

KAKEFUDA, TSUSYOSHI, b Jan 20, 29; m; c 2. INTERNATIONAL COOPERATION ON CANCER RESEARCH & TREATMENT. *Educ:* Tokyo Univ, MD, 52, PhD(path), 58. *Prof Exp:* MED OFFICER, NAT CANCER INST, 67- *Concurrent Pos:* City of Hope Med Ctr, 60-67. *Mem:* Am Asn Cancer Res. *Res:* Cancer etiology. *Mailing Add:* Lab Molecular Carcinogenesis Bldg 37 Rm 3D21 Nat Cancer Inst Bethesda MD 20892

KAKIS, FREDERIC JACOB, b Drama, Greece, Nov 1, 30; US citizen; m 52; c 4. PHYSICAL ORGANIC CHEMISTRY, FOOD SCIENCE. *Educ:* City Col New York, 60; Stanford Univ, PhD(org chem), 64. *Prof Exp:* Chmn dept, 63-68, assoc prof, 66-71, chmn, Div Natural Sci, 78-80, PROF CHEM, CHAPMAN COL, 71-, ASSOC VPRES, 83- *Concurrent Pos:* Grants, NSF, 65, Petrol Res Fund, 65, 66, 76 & 77, Res Corp, 66, 67, 70 & Union Oil Found, 74, 75, 76 & 77; res fel, Oak Ridge Nat Lab, 66; assoc prof, Calif State Col, Long Beach, 66-67 & Calif State Univ, Fullerton, 66-70; res fel, NASA-Ames Res Ctr & Stanford Univ, 69; environmentalist, Defense Contract Admin Serv, 70; vis prof, Lab Org Synthesis, Polytech Sch, Paris, 70-71; NSF res fel, Univ Calif, Riverside; vis prof, Univ Calif, Los Angeles & Univ Calif, Riverside; Fulbright award, 80. *Honors & Awards:* Prof Develop Award, NSF, 77. *Mem:* AAAS; Am Chem Soc; fel Am Inst Chemists; Royal Soc Chem; NY Acad Sci; Inst Food Technologists; soc cosmetic Chemists. *Res:* Study of reaction mechanisms by isotopic labelling; air pollution research; synthetic and mechanistic organic chemistry; heterogeneous catalysis and adsorption; food dehydration. *Mailing Add:* Assoc Vpres Chapman Col Orange CA 92666-1099

KAKO, KYOHEI JOE, b Tokyo, Japan, May 29, 28; Can citizen; m 62; c 2. PHYSIOLOGY. *Educ:* Tokyo Jikei Univ, MD, 53; FRCP. *Prof Exp:* Resident internal med, Tokyo Jikei-Kai Tokyo Hosp, 54-56; res asst med, Sch Med, Washington Univ, 56-57; res assoc, Wayne State Univ, 59-61; from asst prof to assoc prof, 64-75, PROF PHYSIOL, FAC MED, UNIV OTTAWA, 75- *Concurrent Pos:* Mo Heart Asn fel, 57-59; fel, Kanton Hosp, Univ Zurich, 61-63; Alexander von Humboldt fel, I Med Clin, Univ Munich, 63-64; med res assoc, 68-88, residency, Mt Sinai Hosp, 81-82. *Mem:* Am Physiol Soc; Can Cardiovasc Soc; fel Am Col Cardiol; fel Am Col Physicians; fel Am Col Chest Physicians; Royal Col Physicians Can. *Res:* Heart muscle biochemistry; lipid and carbohydrate metabolism; cardiomyopathies; cellular & subcellular function, membrane, calcium fluxes. *Mailing Add:* Dept Physiol Univ Ottawa Fac Med Ottawa ON K1H 8M5 Can

KAKU, MICHIO, b San Jose, Calif, Jan 24, 47. THEORETICAL HIGH ENERGY PHYSICS, NUCLEAR PHYSICS. *Educ:* Harvard Univ, BA, 68; Univ Calif, Berkeley, PhD(physics), 72. *Prof Exp:* Lectr physics, Princeton Univ, 72-73; from asst prof to assoc prof, 73-82, PROF PHYSICS, CITY COL NEW YORK, 82- *Concurrent Pos:* Vis prof, NY Univ, 88, Inst Advan Study, Princeton, 90. *Mem:* Fel Am Phys Soc. *Res:* High energy and nuclear physics; unified field theories; quantum gravity and supergravity; kinetics and neutron transport theory; reactor physics; gauge field theory of superstrings, which will include general covariance and SU(3)X SU(2)X U(1) as subsets, making it a candidate for a unified field theory of all known interactions. *Mailing Add:* Dept Physics City Col New York New York NY 10031

KAKUTANI, SHIZUO, b Osaka, Japan, Aug 28, 11; m 52; c 1. MATHEMATICS. *Educ:* Tohoku Univ, Japan, MA, 34; Osaka Univ, PhD(math), 41. *Hon Degrees:* MA, Yale Univ, 53. *Prof Exp:* Res mem, Inst Adv Study, 40-42; asst prof, Osaka Univ, 42-48; res mem, Inst Advan Study, 48-49; from asst prof to assoc prof, Yale Univ, 49-53, prof math, 53-; RETIRED. *Mem:* Am Math Soc; Math Soc Japan. *Res:* Functional analysis; probability and stochastic processes. *Mailing Add:* 32 Round Hill Rd North Haven CT 06473

KALAB, BRUNO MARIE, b Vienna, Austria, Sept 19, 29; US citizen. PHYSICS, ELECTRONICS ENGINEERING. *Educ:* Univ Vienna, PhD(physics), 64. *Prof Exp:* Supvry physicist nuclear physics & nuclear instrumentation, Naval Radiol Defense Lab, San Francisco, Calif, 64-66, res physicist electronics & nuclear weapons effects, 66-68, head, Electromagnetic Effects Prog, 68-69; res electronics engr nuclear weapons effects, USA Mobility Equip Res & Develop Ctr, Ft Belvoir, Va, 69-71; PHYSICIST NUCLEAR WEAPONS EFFECTS, DEPT ARMY, HARRY DIAMOND LABS, 71- *Mem:* Am Phys Soc; Inst Elec & Electronics Engrs. *Res:* Radiation effects on electronics; nuclear instrumentation. *Mailing Add:* 4712 Exeter St Annandale VA 22003

KALAB, MILOSLAV, b Urcice, Czech, June 12, 29; m 56; c 2. BIOCHEMISTRY. *Educ:* Brno Tech Univ, BSc, 50; Slovak Tech Univ, Bratislava, MSc, 52; Slovak Acad Sci, PhD(chem), 57. *Prof Exp:* Res scientist, Chem Inst, Slovak Acad Sci, 57-58; from asst prof to assoc prof, Sch Med, Palacky Univ, Czech, 58-65, assoc prof, Dept Natural Sci, 65-66; Nat Res Coun Can fel, 66-68; RES SCIENTIST, FOOD RES INST, AGR CAN, 68- *Concurrent Pos:* Ed-in-chief, Food Structure. *Honors & Awards:* Pfizer Inc Award, Am Dairy Sci Asn, 82. *Mem:* Am Dairy Sci Asn; Can Inst Food Sci & Technol; Microscopical Soc Can; Electron Microscopy Soc Am; Inst Food Technologists. *Res:* Food proteins; milk protein gelation, composition, texture; microstructure of dairy products using electron microscopy. *Mailing Add:* Food Res Centre Agr Can Ottawa ON K1A 0C6 Can

KALAFUS, RUDOLPH M, b Jackson, Mich, Dec 17, 37; m 65; c 2. ELECTRICAL ENGINEERING. *Educ:* Univ Mich, BS(elec eng) & BS(eng math), 60, MS, 63, PhD(elec eng), 66. *Prof Exp:* Asst res engr, Radiation Lab, Univ Mich, 64-67 & Electronics Res Ctr, NASA, 67-70; electronics engr, 70-76, SR ELECTRONICS ENGR, TRANSP SYSTS CTR, US DEPT TRANSP, 77- *Mem:* Inst Elec & Electronics Engrs. *Res:* Electrodynamics of moving media; fundamental limitations of antennas; mutual coupling between antennas; cylindrical phased arrays; microwave landing system design; wind sheer detection with CO2 lasers; wave vortex detection; satellite navigation. *Mailing Add:* Trimble Navig PO Box 3642 Sunnyvale CA 94088-3642

KALAI, EHUD, b Tel-Aviv, Israel, Dec 7, 42; US citizen; m 67; c 2. GAME THEORY, MATHEMATICAL ECONOMICS. *Educ:* Univ Calif Berkeley, AB, 67; Cornell Univ, MS, 71 & PhD(math), 72. *Prof Exp:* Asst prof decision theory & oper res, dept statist, Tel-Aviv Univ, 72-76; asst & assoc prof, 76-78, prof, 78-82, MORRISON CHAIR PROF DECISION SCI, KELLOG SCH MGT, NORTHWESTERN UNIV, 82- *Concurrent Pos:* Consult, Div Common, Israeli Army, 74-75; prin investr, NSF grants, 79-; mem bd dirs, First Savings Am & Fed Savings & Loan Asn, 86- *Mem:* Am Math Soc; Econometrics Soc; Pub Choice Soc. *Res:* Author of over 40 articles. *Mailing Add:* Kellog Sch Mgt Northwestern Univ Evanston IL 60208

KALANT, HAROLD, b Toronto, Ont, Nov 15, 23; m 48. PHARMACOLOGY, CELL PHYSIOLOGY. *Educ:* Univ Toronto, MD, 45, BSc, 48, PhD(path chem), 55. *Prof Exp:* Sect head, Defense Res Med Labs, Can, 56-59; assoc prof, 59-64, PROF PHARMACOL, UNIV TORONTO, 64- *Concurrent Pos:* Nat Res Coun Can fel biochem, Cambridge Univ, 55-56; asst res dir, Ont Alcoholism Res Found, 59-62, assoc res dir, 62-89, emer res dir biol sci, 89-; mem res comt NAm Asn of Alcoholism Progs, 62-67; mem alcoholism Study sect, NIMH, Washington, DC, 70-74; res comt non-med use drugs, Dept Nat Health & Welfare, Can, 70-72; mem sci adv bd, Int Coun Alcoholism & Addictions, Lausanne, 72-; mem expert adv panel on drugs of dependence, WHO, 74-84; mem, Comn Prob Drug Dependence, US, 78-; chmn, Bd Sci Counr, Nat Inst Alcohol Abuse & Alcoholism, 83-88; assoc ed, Can J Physiol Pharmacol, 75-81, pharmacol field ed, J Stud Alcohol, 85- *Honors & Awards:* Jellinek Mem Award Res on Alcoholism, 72; Int Gold Medal Res Award, Raleigh Hills Found, 81; Ann Res Award, Res Soc Alcoholism (USA), 83; Upjohn Award, Pharmacol Soc Can, 85; Nathan B Eddy Mem Medal Award, 86. *Mem:* Pharmacol Soc Can; Int Soc Biomed Res Alcoholism (pres, 90-94); fel Royal Soc Can; AAAS. *Res:* Pharmacology of ethanol and other addictive drugs; cell membrane chemistry and physiology; drug-behavior interactions in drug tolerance and dependence. *Mailing Add:* Dept Pharmacol Univ Toronto Toronto ON M5S 1A1 Can

KALANT, NORMAN, b Toronto, Ont; m 48; c 3. ENDOCRINOLOGY, METABOLISM. *Educ:* Univ Toronto, MD, 47, BSc, 49; McGill Univ, PhD(exp med), 54. *Prof Exp:* From asst dir to assoc dir res, Lady Davis Inst Med Res, Jewish Gen Hosp, 55-67, dir inst, 67-90, dir res, 84-90; CONSULT, 90- *Concurrent Pos:* From asst prof to assoc prof, McGill Univ, 58-81, prof, 81- *Mem:* Am Diabetes Asn; Can Soc Clin Invest; Can Biochem Soc; Can Diabetes Asn; Can Atherosclerosis Soc. *Res:* Diabetes; atherosclerosis. *Mailing Add:* 220 Kindersley Ave Montreal PQ H3T 1E2 Can

KALANTAR, ALFRED HUSAYN, b Chicago, Ill, Dec 13, 34; m 61; c 4. SPECTROSCOPY, DATA ANALYSIS. *Educ:* Rutgers Univ, BSc, 56; Cornell Univ, PhD(chem), 63. *Prof Exp:* NSF res fel chem, Calif Inst Technol, 63-64; asst prof, 64-69, ASSOC PROF CHEM, UNIV ALTA, 69- *Concurrent Pos:* Adj assoc prof, State Univ NY Binghamton, 70-71; vis scientist, Nat Res Coun, Ottawa, 85. *Res:* Emission spectroscopy of luminescent organic molecules; analysis of errors in parameters extracted from data; effects of weighting on efficiency of data analysis. *Mailing Add:* Dept Chem Univ Alta Edmonton AB T6G 2G2 Can

KALASINSKY, VICTOR FRANK, b Columbus, Ohio, Dec 30, 49; m 74; c 2. PHYSICAL CHEMISTRY, SPECTROSCOPY. *Educ:* Mass Inst Technol, SB, 72; Univ SC, PhD(phys chem), 75. *Prof Exp:* Asst prof chem, Furman Univ, 76-77; from asst prof to assoc prof, 77-85, PROF CHEM, MISS STATE UNIV, 85- *Concurrent Pos:* Vis scientist, NIH, 87-88. *Honors & Awards:* Res Award,. *Mem:* Am Chem Soc; Am Phys Soc; Soc Appl Spectros; Coblentz Soc; Sigma Xi. *Res:* Raman, infrared and microwave spectroscopy; chemical structure and conformation; intramolecular and intermolecular interactions; applications of the laboratory computer; GC/FTIR and HPLC/FTIR. *Mailing Add:* Dept Chem Miss State Univ Mississippi State MS 39762

KALATHIL, JAMES SAKARIA, b Shertallai, India, Dec 4, 35; US citizen; c 2. ATMOSPHERIC PHYSICS, PHYSICS. *Educ:* Univ Madras, BS, 56; Southern Ill Univ, MS, 63; Univ Nev, PhD(atmospheric physics), 77. *Prof Exp:* Instr physics, Frostburg State Col, Md, 63-65; ASSOC PROF PHYSICS, CALIF POLYTECH STATE UNIV, SAN LUIS OBISPO, 65- *Concurrent Pos:* vis scholar, Univ Calif, Berkeley. *Mem:* Am Meteorol Soc; Am Asn Physics Teachers; Am Geophys Union. *Res:* Cumulus cloud models; history of meteorology; effects of solar activity on weather and climate; climatology. *Mailing Add:* Dept Physics Calif Polytech State Univ San Luis Obispo CA 93407

KALB, G WILLIAM, b Akron, Ohio, Dec 10, 43; m 65; c 2. MINERALOGY, ANALYTICAL CHEMISTRY. *Educ:* Col Wooster, BA, 65; Ohio State Univ, MS, 67, PhD(mineral), 69. *Prof Exp:* Lab mgr mineral, Geol Surv, 69-70; PRES, TRADET INC, 70- *Honors & Awards:* Bituminous Coal Res Award, Am Chem Soc, 72. *Mem:* Am Chem Soc; Am Soc Testing & Mat; Geol Soc Am. *Res:* Determination of volatile trace metals in coal; development of analytical methods for the collection and measurement of volatilized mercury in high 502 concentration gas streams. *Mailing Add:* TraDet Inc PO Box 2019 Wheeling WV 26003

KALB, JOHN W, b Columbus, Ohio, June 6, 18. HIGH VOLTAGE POWER EQUIPMENT. *Educ:* Swarthmore Col, BS, 40. *Prof Exp:* Sr develop engr, Ohio Brass Co, 40-63, dir res, 63-81; RETIRED. *Mem:* Nat Acad Eng; fel Inst Elec & Electronics Engrs. *Mailing Add:* 101 Pier Pont Condos 100 Floyd St St Simons Island GA 31522

KALBACH, CONSTANCE, b Chicago, Ill, Jan 12, 44; m 75. REACTION PHENOMENOLOGY, RADIOACTIVE WASTE DISPOSAL. *Educ:* Univ Rochester, BS, 65, PhD(nuclear chem), 70. *Prof Exp:* Vis res assoc, Nuclear Structure Res Lab, Univ Rochester, & lectr chem, Nazareth Col Rochester, 70-71, fel chem, Univ Rochester, 71-73; guest researcher, Dept Physics, Tech Univ Munich, 72; sr res collabr physics, French AEC, 73-74; asst prof physics, Univ Tenn & res consult physics div, Oak Ridge Nat Lab, Union Carbide, 74-75; guest researcher, Triangle Univs Nuclear Lab, 75-81; vis scholar, 81-85, SR RES SCIENTIST, DEPT PHYSICS, DUKE UNIV, 85- *Concurrent Pos:* Vis asst prof, Dept Chem, NC State Univ, 77; consult, 78-; mem, NC Low Level Radioactive Waste Mgt Authority, 87-, vchmn, 89- *Mem:* Am Chem Soc; Sigma Xi. *Res:* Statistical models of nuclear reactions especially preequilibrium particle emission; nuclear level densities. *Mailing Add:* Physics Dept Duke Univ Durham NC 27706

KALBACH, JOHN FREDERICK, b Seattle, Wash, Jan 2, 14. POWER GROUNDING. *Educ:* Univ Wash, BS, 37. *Prof Exp:* Machine designer, Gen Elec Co, 37-47; staff scientist, Univ Calif, 47-51; mgr eng, William Miller Corp, 51-55; corporate staff consult, Burroughs Corp, 55-79; CONSULT, 79- *Mem:* Fel Inst Elec & Electronics Engrs; fel Inst Advancement Eng; Electrostatic Soc Am. *Res:* Quality of power grounding for computers; analog & digital computers; circuit development & devices. *Mailing Add:* Kalbach Eng 920 Alta Pine Dr Altadena CA 91001

KALBERER, JOHN THEODORE, JR, b New York, NY, Mar 15, 36. PHYSIOLOGY, BIOLOGY. *Educ:* Adelphi Univ, AB, 56; Creighton Univ, MS, 57; NY Univ, PhD(biol, physiol), 66; Dartmouth Col, Inst Grad, 81. *Prof Exp:* Res assoc path, Beth Israel Med Ctr, NY, 54-66; grants assoc, Div Res Grants, 66-67, spec asst to assoc dir extramural activities, 67-73, assoc dir prog planning, Nat Cancer Inst, 74-78, asst dir, Off Med Appln Res, 79-83, coordr dis prev & health prom, off dir, 83-86, DEP DIR, DIV OF DIS PREV, OFF DIR, NIH, 87- *Mem:* Am Soc Zool; Am Asn Anat; Aerospace Med Asn; Am Acad Polit & Soc Sci; NY Acad Sci; Soc Epidemiol Res; Sigma Xi. *Res:* Decompression sickness, especially as it relates to fat embolization to the lung; role of vasoactive substances as they relate to stress conditions; science and society. *Mailing Add:* NIH Rm 258 Bldg 1 Bethesda MD 20892

KALBFLEISCH, GEORGE RANDOLPH, b Long Beach, Calif, Mar 14, 31; m 54; c 4. PARTICLE PHYSICS, HIGH ENERGY PHYSICS. *Educ:* Loyola Univ, Calif, BS, 52; Univ Calif, Berkeley, PhD(physics), 61. *Prof Exp:* Qual control supvr, United Can & Glass, Hunt Foods, Inc, Calif, 52-56; anal chemist, Hales Testing Labs, 57; technician, Lawrence Radiation Lab, Univ Calif, 57-59, asst, 59-61, physicist, 61-64; assoc physicist, 64-67, physicist, Brookhaven Nat Lab, 67-76; physicist, Fermi Nat Accelerator Lab, 76-79; PROF PHYSICS, UNIV OKLA, NORMAN, 79- *Concurrent Pos:* Consult, Anamet Testing Labs, 62-64; mem bd overseers, URA Fermilab, 89-91. *Mem:* Fel Am Phys Soc. *Res:* Neutrino interactions; muon and pion physics; photon physics; superconducting magnets; beauty and charm physics. *Mailing Add:* Dept Physics & Astron Univ Okla 440 W Brooks 131 Norman OK 73019

KALBFLEISCH, JAMES G, b Galt, Ont, Sept 12, 40; m 63; c 3. MATHEMATICAL STATISTICS. *Educ:* Univ Toronto, BSc, 63; Univ Waterloo, MA, 64, PhD(math), 66. *Prof Exp:* Lectr math, 64-66, from asst prof to assoc prof statist, 66-71, chmn dept, 75-79, PROF STATIST, UNIV WATERLOO, 71- *Concurrent Pos:* Adj prof, York Univ, 67; Dept Univ Affairs res grant, 67-70; Nat Res Coun Can res grant, 67-; vis prof, Univ Essex, 68-69; C D Howe fel, 68-69; prof statist, Univ Man, 70-71. *Mem:* Biomet Soc; fel Int Statist Inst; Royal Statist Soc; fel Am Statist Asn. *Res:* Statistical inference; combinatorial mathematics. *Mailing Add:* Dept Statist Univ Waterloo Waterloo ON N2L 3G1 Can

KALCKAR, HERMAN MORITZ, biochemistry cell biology; deceased, see previous edition for last biography

KALDJIAN, MOVSES J(EREMY), b Beirut, Lebanon, Dec 26, 25; US citizen; m 58; c 3. STRUCTURAL MECHANICS, CIVIL ENGINEERING. *Educ:* Am Univ, Beirut, BA, 48, BSc, 49; Univ Man, MSc, 52; Univ Mich, PhD(civil eng), 60. *Prof Exp:* Off engr, Trans-Arabian Pipeline Co, Lebanon, 49-50; civil engr, Dom Bridge Co Ltd, Can, 52-53; lectr struct, Queen's Univ, Ont, 53-54; from instr to assoc prof solid mech, 57-76, ASSOC PROF CIVIL ENG, UNIV MICH, 76- *Concurrent Pos:* Partic, Ford Found comput proj, 61 & Ford Fac Develop adv comt grant, 62; vis prof, Univ Mich-US AID Prog & Indian Inst Technol, Kanpur, 62-64; consult, G C Optronics & Palmer-Shile. *Mem:* Am Soc Civil Engrs. *Res:* Numerical techniques in structural mechanics including finite element methods; response of buildings and dams to earthquake forces; ship structures in ice fields and some experimentation with holography. *Mailing Add:* Dept Civil Eng Univ Mich Main Campus Ann Arbor MI 48109-2125

KALDOR, ANDREW, b Budapest, Hungary, Oct 11, 44; US citizen; m 67; c 2. CLUSTER SCIENCE, CATALYSIS LASER PHYSICS. *Educ:* Univ Calif, Berkeley, BS, 66; Cornell Univ, PhD(chem), 70. *Prof Exp:* Mem staff laser chem, Nat Bur Standards, 70-74; sr res chemist appl physics, 74-77, head chem physics group, 77-81, DIR RESOURCE CHEM LAB, CORP RES LAB, EXXON RES & ENG CO, 81- *Concurrent Pos:* Nat Acad Sci-Nat Res Coun fel, Nat Bur Standards, 70-72; chmn bd trustee, Gordon Res Conf, 89-90. *Honors & Awards:* Silver Medal, Dept Com, 73; Frontiers Chem Lectr, Case Western Reserve, 79; Edwin G Baetjer Lectr, Princeton, 83. *Mem:* AAAS; Am Chem Soc; Am Vacuum Soc; Am Phys Soc. *Res:* Laser chemistry; laser isotope separation; chemical physics; reaction dynamics; molecular spectroscopy; surface chemistry; chister science; materials science. *Mailing Add:* Corp Res Sci Lab Exxon Res & Eng Co Clinton Twp Rte 22 E Annandale NJ 08801

KALDOR, GEORGE, b Budapest, Hungary, Feb 10, 26; nat US; m 63; c 3. PHYSIOLOGY. *Educ:* Med Univ Budapest, MD, 50; Am Bd Clin Chem, dipl, 64; Am Bd Path, dipl clin path, 65, dipl chem path, 78. *Prof Exp:* Asst prof clin biochem, Med Univ Budapest, 54-56; res assoc biochem & head phys chem, Isaac Albert Res Inst, Jewish Chronic Dis Hosp, 59-65; assoc prof physiol, 65-69, prof physiol & biophys, Med Col Pa, 69-75, clin path, 70-75, CHIEF, CLIN LAB SERV, VETERANS ADMIN HOSP, ALLEN PARK, MICH, 75-; PROF PATH, WAYNE STATE UNIV, DETROIT, 75- *Concurrent Pos:* Res fel biochem, Mass Gen Hosp, 57-58 & McArdle Mem Lab, Wis, 58-59. *Mem:* Am Soc Biol Chem; Am Physiol Soc; Am Soc Exp Path; fel Am Soc Clin Pathologists; fel Royal Soc Health. *Res:* Biochemistry of muscular contraction and relaxation; computer assisted medical decision making. *Mailing Add:* Vet Admin Hosp Allen Park MI 48101

KALE, HERBERT WILLIAM, II, b Trenton, NJ, Dec 24, 31; div; c 3. ORNITHOLOGY, CONSERVATION. *Educ:* Rutgers Univ, BSc, 54; Univ Ga, MSc, 61, PhD(zool), 64. *Prof Exp:* Ornithologist, Encephalitis Res Ctr, Fla Div Health, 64-66, vert ecologist, Fla Med Entom Lab, 66-74; VPRES ORNITH RES, FLA AUDUBON SOC, 75- *Concurrent Pos:* Adj asst prof biol sci, Univ Cent Fla, 76-; res assoc, Fla State Mus, 80-; ed, Colonial Waterbirds, 80-85; dir, Fla Breeding Bird Atlas Proj, 85-91. *Mem:* Fel Am Ornith Union; Ecol Soc Am; Cooper Ornith Soc; Wilson Ornith Soc; Soc Wetland Scientists; Colonial Waterbird Soc (pres, 90-91); Wildlife Soc. *Res:* Ecology, animal ecology, breeding biology, density and status of avian populations, bird distribution; conservation of animal populations and habitats, especially colonial nesting birds, rare and endangered species. *Mailing Add:* 1101 Audubon Way Maitland FL 32751

KALELKAR, MOHAN SATISH, b Bombay, India, Apr 24, 48. PHYSICS. *Educ:* Harvard Col, BA, 68; Columbia Univ, MA, 70, PhD(physics), 75. *Prof Exp:* Res assoc physics, Columbia Univ, 75-77, asst prof, 77-78; asst prof, 78-83, assoc chmn dept, 85-89, ASSOC PROF PHYSICS, RUTGERS UNIV, PISCATAWAY, 83- *Mem:* Am Phys Soc; Am Inst Physics. *Res:* Experimental work in elementary particle physics; neutrino interactions and electron-positron collisions. *Mailing Add:* Rutgers Univ Physics Dept PO Box 849 Piscataway NJ 08855-0849

KALENDA, NORMAN WAYNE, b Grand Rapids, Mich, Nov 27, 28; m 57. ORGANIC CHEMISTRY. *Educ:* Univ Mich, BS, 51; Univ Ill, PhD(chem), 55. *Prof Exp:* Res chemist, Mellon Inst, 54-55; res chemist, Eastman Kodak Co, 57-90; RETIRED. *Mem:* Am Chem Soc. *Res:* Organic chemistry; photographic chemistry. *Mailing Add:* 66 Parkmere Rd Rochester NY 14617

KALENSHER, BERNARD EARL, b Beaumont, Tex, May 4, 27. POTENTIAL THEORY, FLUID MECHANICS. *Educ:* Univ Tex, PhD(physics), 54. *Prof Exp:* Sr res engr, Jet Propulsion Lab, Calif Inst Technol, 54-60; sr physicist, Electro-Optical Systs, Xerox Corp, 61-76; SR ANALYSIS PHYSICIST, PHRASOR SCI, INC, 77- *Concurrent Pos:* Auth, J Appl Phys, Vol 53 p 2904, 82; vol 56, p 1347, 84. *Mem:* Am Phys Soc. *Res:* Theory modification of blackbody radiation law as applied to the thermal radiation from micron size, spherical, liquid metal droplets; mathematical analysis of pressure-time history of gas flow between chambers of a dual-chamber thruster and the vacuum of outer space; statistical analyses of charged droplet distributions. *Mailing Add:* 551 B Linwood Ave Monrovia CA 91016-2659

KALER, ERIC WILLIAM, b Burlington, Vt, Sept 23, 56; m 79; c 2. COLLOIDS, SURFACTANTS. *Educ:* Calif Inst Technol, BS, 78; Univ Minn, PhD(chem eng), 82. *Prof Exp:* Res assoc, Chevron Oil Field Res Co, 78; intern, Oak Ridge Nat Lab, 79; res & teaching asst chem eng, Univ Minn, 78-82; from asst prof to assoc prof chem eng, Univ Wash, 82-89; assoc prof, 89-91, PROF CHEM ENG, UNIV DEL, 91- *Concurrent Pos:* Presidential young investr, 84-89. *Mem:* Am Inst Chem Engrs; Am Chem Soc; Am Crystallog Asn; AAAS. *Res:* Colloid and surfactant science; complex fluid thermodynamics; materials synthesis; small-angle scattering. *Mailing Add:* Dept Chem Eng Univ Del Newark DE 19716

KALER, JAMES BAILEY, b Albany, NY, Dec 29, 38; m 60; c 4. ASTRONOMY. *Educ:* Univ Mich, Ann Arbor, AB, 60; Univ Calif, Los Angeles, PhD(astron), 64. *Prof Exp:* From asst prof to assoc prof, 64-76, PROF ASTRON, UNIV ILL, URBANA, 76- *Concurrent Pos:* Guggenheim fel, 72-73. *Mem:* Am Astron Soc; Int Astron Union; Astron Soc Pac. *Res:* Planetary nebulae; nebular spectrophotometry; interstellar medium; chemical abundances. *Mailing Add:* 103 Astron Bldg Univ Ill 1002 W Green Urbana IL 61801

KALEY, GABOR, b Budapest, Hungary, Nov 16, 26; m 53; c 2. PHYSIOLOGY, EXPERIMENTAL PATHOLOGY. *Educ:* Columbia Univ, BS, 50; NY Univ, MS, 57, PhD(exp path), 60. *Prof Exp:* Resident asst surg, Bellevue Hosp, New York, 55-56; res asst path, NY Univ Med Ctr, 56-60; from instr to asst prof, 61-62; assoc prof, 64-70, PROF PHYSIOL, NY MED COL, 70-, CHMN DEPT, 72- *Concurrent Pos:* USPHS fel, NY Univ Med Ctr, 60-62; prin investr, NIH Grants, 74- *Mem:* Am Physiol Soc. *Res:*

Cardiovascular physiology; hypertension; juxtaglomerular cells; renal-adrenal relationships; renen-angiotens in system; erythropoietin; inflammation; microcirculation and prostaglandins; endotoxins; endothelial cells; nature and mechanisms of action of a variety of biochemical and hormonal factors that regulate the function of small blood vessels and local blood flow. *Mailing Add:* Dept Physiol NY Med Col Valhalla NY 10595

KALEY, ROBERT GEORGE, II, b Litchfield, Ill, Nov 8, 45; m 81; c 3. GAS CHROMATOGRAPHY, MASS SPECTROMETRY. *Educ:* Purdue Univ, BS, 68; Univ Ill, MS, 71, PhD(anal chem), 74. *Prof Exp:* Sr res chemist, Indust Chem Co, 73-78, res group leader, 78-81, sr res specialist, Corp Res & Develop Staff, 81-85, prod & environ safety mgr, 85-86, MGR, ENVIRON TECH SUPPORT, CORP ENVIRON POLICY STAFF, MONSANTO CO, 86- *Mem:* Am Chem Soc; Am Soc Mass Spectrometry; AAAS. *Res:* Spectrochemical analysis; gas chromatography-mass spectrometry; environmental analysis. *Mailing Add:* Monsanto A2NE 800 N Lindbergh Blvd St Louis MO 63167-0001

KALF, GEORGE FREDERICK, b New Britain, Conn, Dec 22, 30; m 53; c 2. BIOCHEMISTRY. *Educ:* Upsala Col, BS, 52; Pa State Univ, MS, 54; Yale Univ, PhD(biochem), 57. *Prof Exp:* Enzymologist, Chem & Physics Sect, Animal Dis & Parasite Res Div, USDA, 59; from asst prof to assoc prof biochem, NJ Col Med & Dent, 60-66; prof biochem, 66-86, prof path, 79-86, PROF BIOCHEM & MOLECULAR BIOL, JEFFERSON MED COL, THOMAS JEFFERSON UNIV, 86- *Concurrent Pos:* Nat Found fel, Yale Univ, 57-59; investr, Am Heart Asn, 63-68; adj prof pharm & toxicol, Rutgers Univ, 82. *Mem:* Am Soc Biol Chem; Am Asn Cancer Res; Brit Biochem Soc; Soc Toxicol; Int Soc Study Xenobiotics. *Res:* Carcinogenesis; biochemical oncology; benzene toxicity. *Mailing Add:* Jefferson Med Col Thomas Jefferson Univ Philadelphia PA 19107

KALFAYAN, BERNARD, ANATOMICAL PATHOLOGY. *Educ:* Am Univ, Beirut, MD, 39. *Prof Exp:* Pathologist, Gunderson Clinic, Ltd; RETIRED. *Mailing Add:* Gundersen Clin Ltd 1836 South Ave La Crosse WI 54601

KALFAYAN, LAURA JEAN, oogenesis, developmental gene expression; deceased, see previous edition for last biography

KALFAYAN, SARKIS HAGOP, b Turkey, July 2, 16; US citizen; m 53; c 3. POLYMER CHEMISTRY. *Educ:* Am Univ, Beirut, BA, 40, MA, 42; Case Inst Technol, PhD(chem), 50. *Prof Exp:* Instr chem, Am Univ, Beirut, 42-47; from asst prof to assoc prof, Mt St Mary's Col, Calif, 51-55; fel, Univ Southern Calif, 55-56; group supvr, Prod Res Co, 56-58, chief chemist, 58-60; lab mgr, Chem-Seal Corp, 60-62; sr scientist polymer chem, 62-67, TECH GROUP SUPVR POLYMER CHEM, JET PROPULSION LAB, CALIF INST TECHNOL, 67- *Mem:* AAAS; Am Chem Soc; fel Am Inst Chem; Sigma Xi. *Res:* Polymer synthesis and electrical properties of polymers; spacecraft materials; physical testing of polymers; chemorheology of polymers. *Mailing Add:* 4834 Matley Rd La Canada Flintridge CA 91011

KALFF, JACOB, b Velsen, Neth, Dec 20, 35; Can citizen; m 59; c 2. HYDROBIOLOGY. *Educ:* Univ Toronto, BSA, 59, MSA, 61; Ind Univ, PhD(limnol), 65. *Prof Exp:* From asst prof to assoc prof, 65-76, PROF BIOL, MCGILL UNIV, MONTREAL, 77-, DIR, LIMNOL RES CTR, DEPT BIOL, 82- *Concurrent Pos:* Vis scientist hydrobiol, Inst Nat Rech Agron, France, 72-73; consult, Ecol Adv Comt, Baie James Energy Corp, 77-; vis prof, Dept Bot, Univ Nairobi, Kenya, 79-80; vis scientist, CNRS, Toulouse, France, 86-87; regional ed, Hydrobiol. *Mem:* AAAS; Am Soc Limnol & Oceanog; Int Asn Theoret & Appl Limnol. *Res:* Ecology of algae, bacteria and aquatic higher plants with an emphasis on their productivity and their role in the nutrient and contaminant cycling in lakes. *Mailing Add:* Dept Biol McGill Univ 1205 Dr Penfield Montreal PQ H3A 1B1 Can

KALFOGLOU, GEORGE, b Istanbul, Turkey, June 12, 39; m 68; c 2. SURFACE CHEMISTRY, POLYMER CHEMISTRY. *Educ:* Robert Col, Istanbul, BS, 63; NC State Univ, PhD(chem), 68. *Prof Exp:* RES CHEMIST & RES ASSOC, RES LAB, TEXACO INC, 68- *Mem:* AAAS; Am Chem Soc; Soc Petrol Engrs; Sigma Xi. *Res:* Solution thermodynamics; colloidal chemistry; chemical treatment of water-sensitive minerals; physical and interfacial properties of surfactants; design enhanced petroleum recovery processes in hard brines and elevated temperatures by utilizing proper surfactant systems; tertiary oil recovery by micellar/polymer systems, caustic/polymer floods, thermally stable polymers, design of polymer gels for profile modification and water shut-off treatments. *Mailing Add:* 523 Greenpark Dr Houston TX 77079

KALIA, MADHU P, b Kashmir, India, Sept 11, 40. NEUROBIOLOGY. *Educ:* Univ Delhi, MD, 64, PhD(neurophysiol), 68. *Prof Exp:* Assoc prof, 74-78, PROF PHARMACOL & NEUROSURG, THOMAS JEFFERSON UNIV, 78- *Mem:* Soc Neurosci; Am Physiol Soc; Am Asn Anatomists; German Physiol Soc. *Res:* Respiratory control; cardiovascular control. *Mailing Add:* Dept Pharmacol & Neurosurg Rm 329 Thomas Jefferson Univ 1020 Locust St Philadelphia PA 19107

KALIAKIN, VICTOR NICHOLAS, b Los Angeles, Calif, Nov 1, 56; m 90. GEOMECHANICS, STRUCTURAL MECHANICS. *Educ:* Univ Calif, Davis, BS, 78, PhD(eng mech), 85; Univ Calif, Berkeley, MS, 79. *Prof Exp:* Postdoctoral civil eng, Univ Calif, Davis, 85-86; vis asst prof civil eng, Univ Ariz, 86-87; mem tech staff, Sandia Nat Labs, 87-90; ASST PROF CIVIL ENG, UNIV DEL, 90- *Concurrent Pos:* Mem, Comt Inelastic Behav, Am Soc Civil Engrs, 91- *Mem:* Assoc mem Am Soc Civil Eng; Am Acad Mech; Am Soc Eng Educ. *Res:* Computational mechanics applied to problems in: Geotechnical engineering, composite material and structural engineering. *Mailing Add:* Dept Civil Eng Univ Del Newark DE 19716

KALIKSTEIN, KALMAN, physics; deceased, see previous edition for last biography

KALIL, FORD, b Akron, Ohio, Jan 5, 25; m 50; c 3. ENGINEERING PHYSICS. *Educ:* Univ Akron, BEE, 50; Vanderbilt Univ, MS, 51, PhD(physics), 58. *Prof Exp:* Physicist, USPHS, 51-53; staff mem, Los Alamos Sci Lab, 53-55; supvr lab, Martin Co, 58-59, gen supvr test eval, 59, mgr, 59-60, design engr, 60-63; aerospace technologist, Goddard Space Flight Ctr, 63-68, sr tech asst to br head, Manned Flight Planning & Anal Div, 68-74, sr tech consult, network procedures & eval div, networks directorate, 74-78, systs mgr cosmic background explorer proj, Eng Directorate, 78-80, TECH ENGR & OPERS MGR & PRES, GODDARD SPACE FLIGHT CTR, NASA, 80- *Concurrent Pos:* Adj prof, Drexel Inst, 58-; mem, Apollo Exten Syst, Commun & Navig Traffic Control Panel, 65 & Apollo Exten Syst Working Group, NASA, 65- *Mem:* Am Inst Aeronaut & Astronaut; Am Phys Soc; Sigma Xi. *Res:* Aerospace technology; mission and systems analysis; orbital mechanics; electron and radiation physics; management. *Mailing Add:* 9108 Bridgewater St College Park MD 20741

KALIMI, MOHAMMED YAHYA, b Surat, India; US citizen; m 85; c 1. ENDOCRINOLOGY. *Educ:* Bombay Univ, BS, 61, MS, 64, PhD(biochem), 70. *Prof Exp:* Res fel, Inst Cancer Res, Columbia Univ, 72-74; res assoc, dept cell biol, Baylor Col Med, Houston, 74-75; res asst prof, dept biochem, Albert Einstein Col Med, Bronx, NY, 75-79; from asst prof to assoc prof, 79-89, PROF, DEPT PHYSIOL, MED COL VA, RICHMOND, 89- *Concurrent Pos:* Vis prof, City of Hope Med Ctr, Los Angeles, 83-84. *Honors & Awards:* Res & Career Develop Award, NIH, 80. *Mem:* Am Physiol Soc; Endocrine Soc; AAAS. *Res:* Mechanism of steroid hormone action; isolation, characterization and purification of glucocorticoid receptors; interaction of the steroid-receptor complex with genomic components; developmental and aging related changes in the steroid receptors. *Mailing Add:* Box 551 MCV Sta Richmond VA 23298

KALIN, ROBERT, b Everett, Mass, Dec 11, 21; wid; c 4. MATHEMATICS EDUCATION. *Educ:* Univ Chicago, BS, 47; Harvard Univ, MAT, 48; Fla State Univ, PhD(math educ), 61. *Prof Exp:* Teacher high sch, Mass, 48-49 & pub sch, Mo, 49-52; statistician, Naval Air Tech Training Ctr, Okla, 52-53; test specialist math, Educ Testing Serv, NJ, 53-55, res assoc math educ, 55; exec asst, Comn Math Col Bd, 55-56; from instr to prof math educ, 56-89, chmn math educ prog, 74-78, EMER PROF MATH EDUC, FLA STATE UNIV, 89- *Concurrent Pos:* Mem, NSF Acad Year Insts, 66-71. *Honors & Awards:* Serv Award, Math Asn Am, 91. *Mem:* Nat Coun Teachers Math; Math Asn Am; Int Group Psychol Math Educ. *Res:* Comparison of mathematical performance of elementary school children in Federal Republic of Germany, United Kingdom and USA since 1978; television-text instructional system for elementary school teachers; analytic geometry; elementary mathematics texts for grades 1-8; secondary geometry text. *Mailing Add:* 721 Key Corner St Brownsville TN 38012-2424

KALINA, ROBERT E, b New Prague, Minn, Nov 13, 36; m 59; c 2. OPHTHALMOLOGY. *Educ:* Univ Minn, BA, 57, BS, 60, MD, 60. *Prof Exp:* Intern med, Univ Ore, 60-61, resident ophthal, 61-66; fel retina, Children's Hosp, San Francisco, 66-67 & Harvard Med Sch, 67; chief, Harborview Med Ctr, 68-69; actg chmn, 70-71, PROF & CHMN OPHTHAL, UNIV WASH, SEATTLE, 70- *Concurrent Pos:* Consult, Vet Admin Hosp, 69-, Pub Health Hosp, 69- & Madigan Gen Hosp, US Army, 69-; med dir, Lions Eye Bank, 69-; assoc head, Ophthal Div, Children's Orthopedic Hosp, 75-; dir, Am Bd Ophthal, 82- *Mem:* Am Acad Ophthal; Am Bd Ophthal; Asn Univ Professors Ophthal; Nat Soc Prevent Blindness. *Res:* Diseases and surgery of the retina. *Mailing Add:* Dept Ophthal Univ Wash RJ-10 Seattle WA 98195

KALINOWSKI, MATHEW LAWRENCE, b Chicago, Ill, Jan 1, 15; m 46; c 2. PETROLEUM CHEMISTRY. *Educ:* Univ Chicago, BS, 37; Univ Ill, MS, 40. *Prof Exp:* Chemist, Ill State Geol Surv, 37-41 & Sherwin-Williams Co, 41-48; group leader petrol refining, Stand Oil Co, Ind, 48-66, SR PATENT ADV, AMOCO RES CTR, 66- *Mem:* Am Chem Soc. *Res:* Petroleum refining and products. *Mailing Add:* 734 S Sleight Naperville IL 60540-6629

KALINSKE, A(NTON) A(DAM), environmental engineering, fluid mechanics; deceased, see previous edition for last biography

KALINSKY, ROBERT GEORGE, b Cleveland, Ohio, Sept 18, 45; m 71. PHYCOLOGY, ENVIRONMENTAL BIOLOGY. *Educ:* Univ Dayton, BS, 67; Ohio State Univ, MSc, 69, PhD(bot), 73. *Prof Exp:* Lectr bot, Ohio State Univ, 73-74; ASST PROF BIOL, LA STATE UNIV, SHREVEPORT, 74- *Concurrent Pos:* Biol consult, Dames & Moore Engrs, Cincinnati, 73-74. *Mem:* Sigma Xi; Phycol Soc Am; Brit Phycol Soc; Bot Soc Am. *Res:* Systematic revision of the diatom genus Nitzschia; ecological studies of the major waterways of Northwestern Louisiana. *Mailing Add:* Dept Biol Sci La State Univ 8518 Youree Dr Shreveport LA 71105

KALISCH, GERHARD KARL, b Breslau, Ger, Dec 21, 14; nat US; m 42; c 2. MATHEMATICS. *Educ:* Univ Iowa, BA, 38, MS, 39; Univ Chicago, PhD(math), 42. *Prof Exp:* Asst, Inst Adv Study, 41-42; instr math, Univ Kans, 42-44 & Cornell Univ, 44-46; from asst prof to prof, Univ Minn, 46-65; PROF MATH, UNIV CALIF, IRVINE, 65- *Mem:* AAAS; Am Math Soc; Math Asn Am; Math Soc France. *Res:* Functional analysis; topological algebra. *Mailing Add:* Dept Math Univ Calif Irvine CA 92717

KALISH, DAVID, b New York, NY, Aug 15, 39; c 1. LIGHTGUIDE SYSTEMS ENGINEERING, GLASS TECHNOLOGY. *Educ:* Mass Inst Technol, SB, 60, SM, 64, ScD, 66. *Prof Exp:* Engr, ManLabs, Inc, 60-63; instr metall, Mass Inst Technol, 63-65; staff scientist, ManLabs, Inc, 65-68; scientist & team leader phys metall, Lockheed-Ga Co, 68-71; mem tech staff mat eng & chem, 71-73, supvr metall eng, 73-79, supvr lightguide glass technol, 80-85, SUPVR LIGHTGUIDE SYST ENG, BELL TEL LABS, INC, 86- *Concurrent Pos:* Vis assoc prof, Ga Inst Technol, 70, lectr mech eng, 73-81. *Mem:* Fel Am Soc Metals. *Res:* Optical fibers for communications and systems engineering, optical fibers measurements and fabrication; strength and static fatigue of glass. *Mailing Add:* Bell Tel Labs Inc 2000 NE Expressway Norcross GA 30071

KALISH, HERBERT S(AUL), b New York, NY, Aug 11, 22; m 50; c 2. METALLURGICAL ENGINEERING. *Educ:* Univ Mo, BS, 43, MetE, 53; Univ Pa, MS, 48. *Prof Exp:* Metall observer open hearth steel, Carnegie-Ill Steel Corp, 43; res engr, Alloy Develop, Battelle Mem Inst, 43-44 & 46; asst, Thermodyn Res Lab, Univ Pa, 46-48; res metallurgist lead alloys, Elec Storage Battery Co, 48; sr engr spec prod, Sylvania Elec Prod, Inc, 48-50, engr-in-charge, Zirconium Sect, 50-52, sect head appl metall, 52-55, eng mgr metal fabrication & assembly, 55-57; eng mgr, Sylvania-Corning Nuclear Corp, 57; sect chief mat, Nuclear Fuel Res Labs, Olin Mathieson Chem Corp, 57-60, chief nuclear metall, Nuclear Fuel Res, 60-62; mgr, Com Fuel Dept, United Nuclear Corp, Conn, 62-65; asst to pres, 65-71, VPRES, HERTEL CUTTING TECHNOLOGIES INC, 71- *Concurrent Pos:* Trustee, Am Soc Metals, 85-88. *Mem:* Fel Am Soc Metals; Am Welding Soc; Sigma Xi; Am Soc Test & Mat; Am Inst Mining, Metall & Petrol Engrs. *Res:* Nuclear materials; high temperature materials; powder metallurgy, particularly alloy development; physical metallurgy; fabrication research; cemented carbide. *Mailing Add:* 65 Falmouth St Short Hills NJ 07078

KALISKI, MARTIN EDWARD, b New York, NY, Oct 22, 45; m 69. MICROPROCESSOR-BASED CONTROL, SYSTEMS THEORY. *Educ:* Mass Inst Technol, BS, 66 & 68, SM, 68, PhD(elec eng), 71. *Prof Exp:* Asst prof comput sci, City Col New York, 71-73; asst prof, 73-76, ASSOC PROF ELEC ENG, NORTHEASTERN UNIV, 76- *Concurrent Pos:* Fulbright scholar, France, 80-81. *Mem:* Inst Elec & Electronics Engrs; Sigma Xi. *Res:* Systems theory; industrial automation & robotics; microprocessor-based control; software engineering. *Mailing Add:* 888 Vista Brisa San Luis Obispo CA 93405

KALISS, NATHAN, b New York, NY, Aug 1, 07; m 28; c 3. IMMUNOLOGY. *Educ:* City Col New York, BS, 29; Columbia Univ, MA, 31, PhD(zool), 38. *Prof Exp:* Asst zool, Columbia Univ, 31-38 & Univ Pa, 39-40; asst path, Med Col, Cornell Univ, 41-43; instr zool, George Washington Univ, 46-47; Am Cancer Soc sr fel, 47-50, staff scientist cancer res, 50-59, asst dir res, 58-62 & 75-76, sr staff scientist, 59-72, EMER SR STAFF SCIENTIST, 72-, GUEST INVESTR, JACKSON LAB, 76- *Concurrent Pos:* Med statistician, US Vet Admin, 46-47; vis fel, Sloan-Kettering Inst, 48; ed, Transplantation Bull, 53-63 & Transplantation, 63-75; Guggenheim fel, 56-57; res assoc surg, Harvard Med Sch, 66-73. *Honors & Awards:* Silver Medal, Am Med Assoc, Sci Res Exhibit, 55. *Mem:* Soc Exp Biol & Med; Am Asn Cancer Res; Transplantation Soc; Reticuloendothelial Soc. *Res:* Homograft immunity; etiology of carcinogenesis; biology of cancer; immunology of tissue grafting. *Mailing Add:* Seely Rd Bar Harbor ME 04609-1506

KALIVODA, FRANK E, JR, b Cicero, Ill, Mar 17, 30; m 57; c 3. MECHANICAL ENGINEERING. *Educ:* Ill Inst Technol, BSME, 54, MSME, 58. *Prof Exp:* Coop trainee prod, Link-Belt Corp, 49-52; asst res engr, Heat-Power Res Dept, Armour Res Found, 54-58; sr res engr, Roy C Ingersoll Res Ctr, Borg-Warner Corp, 58-65; mgr component develop, Res Div, Cummins Engine Co, 65, mgr adv fuel systs, Eng Div, 65-66, mgr thermosci, Res Div, 66-68; mgr res & develop, South Wind Div, Stewart Warner Corp, 68-70, chief engr, 70-73; mgr prod eval, 73-81, MGR ENG, COPELAND CORP, 81- *Concurrent Pos:* Instr quality control & mfg processes, Edison State Univ, 75-78. *Mem:* Am Soc Mech Engrs; Am Soc Heating, Refrig & Air-Conditioning Engrs; Sigma Xi; Am Soc Qual Control. *Res:* Development of air conditioners, heat exchangers, compressors, combustion heaters, hydraulic equipment, engine fuel injectors and pumps; research on engine heat transfer, hydrodynamic bearings; manufacturing processes; reliability engineering. *Mailing Add:* Copeland Corp 1675 W Campbell Rd Sidney OH 45365

KALKA, MORRIS, b Landsberg, Ger, May 4, 49; US citizen; m 71. MATHEMATICAL ANALYSIS. *Educ:* Yeshiva Univ, BA, 71; NY Univ, MS, 73, PhD(math), 75. *Prof Exp:* Instr math, Univ Utah, 75-77; asst prof math, Johns Hopkins Univ, 77-80; mem fac math, Tulane Univ, 80- *Mem:* Am Math Soc. *Res:* Study of holomorphic functions of several complex variables using methods of differential geometry and partial differential equations. *Mailing Add:* Dept Math Tulane Univ La New Orleans LA 70118

KALKHOFF, RONALD KENNETH, endocrinology, metabolism; deceased, see previous edition for last biography

KALKOFEN, WOLFGANG, b Mainz, Ger, Nov 15, 31; US citizen; m 60; c 2. ASTROPHYSICS. *Educ:* Univ Frankfurt, BS, 56; Harvard Univ, MA, 61, PhD(physics), 63. *Prof Exp:* Engr, Raytheon Mfg Co, 57-59; PHYSICIST, SMITHSONIAN ASTROPHYS OBSERV, SMITHSONIAN INST, 63-; PHYSICIST, CTR FOR ASTROPHYS, HARVARD COL OBSERV, 74- *Concurrent Pos:* Res fel, Harvard Col Observ, 64-66; vis lectr astron, Yale Univ, 65; lectr, Harvard Univ, 65-; vis fel, Univ Heidelberg, 79-80; sr US scientist award, Alexander von Humboldt Found, 79. *Mem:* Am Astron Soc; AAAS; Int Astron Union. *Res:* Theoretical astrophysics; radiative transfer; gas dynamics. *Mailing Add:* Astrophys Observ Harvard Univ Cambridge MA 02138

KALKSTEIN, LAURENCE SAUL, b Brooklyn, NY, Jan 29, 48; m 71; c 1. APPLIED CLIMATOLOGY, BIOCLIMATOLOGY. *Educ:* Rutgers State Univ, BA, 69; La State Univ, MA, 72, PhD(geog, climat), 74. *Prof Exp:* Instr geog, La State Univ, 73; asst prof geog & ecosystems anal, Univ Calif, Los Angeles, 73-75; assoc prof Geog & Climat, 75-88, PROF GEOG & CLIMAT, UNIV DEL, 88- *Concurrent Pos:* Prin investr, climate & socio-econ assessment, Nat Oceanic & Atmospheric Admin, 80-86; vis scientist, Nat Oceanic & Atmospheric Admin, 82-83, Environ Protection Agency Climate Change Div, 89-90; coordr, global warming/human health prog, Environ Protection Agency, synoptic climatology-pollution relationships, climate-visibility relationships, Salt River Proj & Southern pine beetle- climate anal, US Forest Serv, 85-; mem, comt applied climatol, Am Meteorol Soc, comn climatol, World Metrol Orgn. *Mem:* Asn Am Geographers; Am Meteorol Soc. *Res:* Climate's impact on human health and well-being; impact of climate on organism population fluctuations; development of applied climatological indices; socio-economic impacts of a global warming. *Mailing Add:* Dept Geog Univ Del Newark DE 19716

KALKWARF, DONALD RILEY, b Portland, Ore, Aug 17, 24; m 49; c 4. BIOPHYSICAL CHEMISTRY. *Educ:* Reed Col, BA, 47; Northwestern Univ, PhD(chem), 51. *Prof Exp:* Chemist, Gen Elec Co, 51-53, sr scientist, 54-62, res specialist, 62-65; res assoc chem dept, 65-66, mgr radiation chem unit, Environ & Radio Sci Dept, 66-68, res assoc radiol sci dept, 68-78, STAFF SCIENTIST BIOL & CHEM DEPT, BATTELLE-NORTHWEST, 78- *Mem:* Am Chem Soc; Royal Soc Chem; Sigma Xi. *Res:* Radiation biochemistry; synthetic biomembranes; electron spin resonance spectroscopy; controlled release of pharmaceuticals; kinetics of pollutant transformation. *Mailing Add:* 324 Bldg Battelle-Northwest Richland WA 99352

KALKWARF, KENNETH LEE, b Lincoln, Nebr, Apr 12, 46; m 74; c 2. PERIODONTOLOGY. *Educ:* Univ Nebr, DDS, 70, MS, 73. *Prof Exp:* Asst prof periodont, Col Dent, Univ Nebr, 73-78; assoc prof, Col Dent, Univ Okla, 78-81; prof periodont & dir grad periodont, Col Dent, Univ Nebr, 81-87, dir grad & postgrad studies, 83-87; assoc dean adv educ, 87-88, DEAN DENT SCH, UNIV TEX HEALTH SCI CTR, SAN ANTONIO, 88- *Concurrent Pos:* Consult, Nebr Vet Admin Hosp, Lincoln, 73-, Grand Island, 81-87, Omaha, 86-, Cent Regional Dent Testing Serv, 80-87, Periodont Case Reports, 82- Cent Community Col, Hastings, Nebr, 83-87, J Periodontol, 84-; vis prof, Independent Univ Guadalajara, Mex, 80-82; adv bd, Southeast Nebt Community Col, 81-87; mem, Nebr Bd Dent Examr, 85-87. *Mem:* Int Asn Dent Res; Am Acad Periodont; Am Asn Dent Schs; Am Dent Asn; fel, Am Col Dentists. *Res:* Longitudinal evaluation of periodontal therapy; histologic evaluation of oral wound healing. *Mailing Add:* Dent Sch Univ Tex Health Sci Ctr 7703 Floyd Carl Dr San Antonio TX 78284-7906

KALLAHER, MICHAEL JOSEPH, b Cincinnati, Ohio, Sept 4, 40; m 63; c 5. MATHEMATICS. *Educ:* Xavier Univ Ohio, BS, 61; Syracuse Univ, MS, 63, PhD(math), 67. *Prof Exp:* Instr math, Syracuse Univ, 66-67; fel Univ Man, 67-68, asst prof, 68-69; assoc dean sci, 79-84, from asst prof to assoc prof, 69-76, PROF MATH, WASH STATE UNIV, 76-, CHMN DEPT, 84- *Concurrent Pos:* Fulbright-Hays fel, Fulbright Comn, 75-76. *Mem:* AAAS; Am Math Soc; Math Asn Am; Sigma Xi; NY Acad Sci; Soc Indust & Appl Math. *Res:* Non-associative algebras; finite geometries, particularly finite projective planes and finite affine planes. *Mailing Add:* Dept Math Wash State Univ Pullman WA 99164-2930

KALLAI-SANFACON, MARY-ANN, b Montreal, Que, Apr 16, 49; m 77; c 2. BIOCHEMISTRY, ENDOCRINOLOGY. *Educ:* McGill Univ, Montreal, BSc, 70; Laval Univ, Que, MSc, 73; Univ Toronto, PhD(physiol), 77. *Prof Exp:* Res fel biochem, Erindale Col, Univ Toronto, 77-78; res fel, Ayerst Res Lab Montreal, 78-80, sr scientist, 80-82; dir, Lab Med Anal, BioEndo Labs, 82-83; dir, Clin Chem Lab, Louis-H Lafontaine Hosp, 83-85; SECT HEAD PHARMACOL & TOXICOL, CLIN CHEM LAB, SACRE-COEUR HOSP, MONTREAL, 85- *Concurrent Pos:* Indust fel, Nat Res Coun Can, 78-80; grant, Med Res Coun, 85; mem, prof affairs comt, Asn Biochem Que, 85-91, ad hoc comt on drugs in the workplace, Order of Chemists of Que, 91; lectr, Continuous Educ Prog Med Technologists, Pharmacol & Toxicol, 89; tutor in pharmacol & toxicol for postdoctoral trainees in clin chem, 89-90; counc, Can Soc Clin Chemists, 91- *Mem:* Can Biochem Soc; NY Acad Sci. *Res:* Lipid metabolism; the development of hypolipidemic agents and the elucidation of their mode of action; carbohydrate metabolism; development of oral hypoglycaemic agents; hypertension; development of antihypertensive agents particularly those associated with angiotensin converting enzyme inhibition. *Mailing Add:* Hosp Sacre-Coeur de Montreal 5400 boul Gouin W Montreal PQ H4J 1C5 Can

KALLAL, R(OBERT) J(OHN), b Chesterfield, Ill, Feb 1, 21; m 46; c 5. CHEMICAL ENGINEERING. *Educ:* Univ Ill, BS, 43, MS, 46; Mass Inst Technol, ScD(chem eng), 49. *Prof Exp:* Asst munitions develop, Univ Ill, 43-45; asst chem eng, Mass Inst Technol, 46-47, instr, 47; chem engr plastics develop, 49-55, asst tech supt, 55-58, sr supvr res & develop, 58-70, asst plants tech mgr, 70-74, res mgr, 74-75, PLANNING MGR, E I DU PONT DE NEMOURS & CO, INC, 75- *Mem:* AAAS; Am Chem Soc; Am Inst Chem Engrs. *Res:* Organic and inorganic chemical process development. *Mailing Add:* 518 Kerfoot Farm Rd Wilmington DE 19803-2444

KALLAND, GENE ARNOLD, b Ashland, Wis, Dec 15, 36; m; c 2. REPRODUCTIVE ENDOCRINOLOGY. *Educ:* Calif State Univ, Northridge, BA, 62; Ind Univ, PhD(zool), 66. *Prof Exp:* Assoc engr, Rocketdyne Div, NAm Aviation, Inc, 57-62; from asst prof to assoc prof, 66-76, chairperson dept, 72-75, PROF BIOL, CALIF STATE UNIV, DOMINGUEZ HILLS, 71- *Concurrent Pos:* Vis asst prof, Univ Southern Calif, 71; res assoc endocrinol, Harbor-Univ Calif, Los Angeles Med Ctr, 75-80. *Mem:* AAAS; Sigma Xi. *Res:* computer use in science instruction; biology of childhood and adolescence; amphibian hatching mechanisms. *Mailing Add:* Dept Biol Calif State Univ Dominguez Hills Carson CA 90747

KALLANDER, JOHN WILLIAM, b Bessemer, Mich, June 20, 27; m 69. COMPUTER SCIENCE. *Educ:* Mich Technol Univ, BS, 48; Univ Cincinnati, MS, 50, PhD(appl sci), 52. *Prof Exp:* Develop engr analog computs, Gen Elec Co, 52; res physicist magnetic amplifiers, 53-56, US Naval Res Lab, high-energy radiation effects, 56-61, proj leader, prog sect, 61-63, consult prog lang & physics to head res comput ctr, 63-64, head prog systs sect, 64-68, supvry mathematician, 68-82; RETIRED. *Mem:* Sigma Xi; Inst Elec & Electronics Engrs; Asn Comput Mach. *Res:* Digital computer systems analysis and programming; data base management analysis and programming. *Mailing Add:* 1138 Kensington Ave Flint MI 48504

KALLELIS, THEODORE S, b Peabody, Mass, Oct 12, 12; m 47. PHARMACOGNOSY. *Educ:* Tufts Univ, BS, 35; Mass Col Pharm, BS, 52; Temple Univ, MS, 52; Univ Md, PhD, 56. *Prof Exp:* Assoc prof pharm & chmn dept, Fordham Univ, 56-65; assoc prof, 65-71, chmn dept pharmacog & toxicol, 65-69, prof, 71-80, EMER PROF PHARMACOG, TEMPLE UNIV, 80- *Mem:* Am Chem Soc; Sigma Xi; Am Pharmaceut Asn; Am Soc Pharmacog; NY Acad Sci. *Res:* Isolation of plant constituents. *Mailing Add:* 921 Beverly Rd Rydal PA 19046

KALLEN, FRANK CLEMENTS, b Colonie, NY, May 27, 28; m 58; c 2. VERTEBRATE ANATOMY. *Educ:* Cornell Univ, BA, 49, PhD(zool), 61. *Prof Exp:* Instr anat, Sch Med & Dent, Univ Rochester, 60-64; from asst prof to assoc prof, 64-75, PROF ANAT, STATE UNIV NY BUFFALO, 75- *Mem:* Am Asn Anat; Am Soc Mammal. *Res:* Blood volume, fluid balance, hibernation, hypothermia and functional morphology in bats and other vertebrates. *Mailing Add:* Dept Anat Farber Hall State Univ NY 303 Sherman Buffalo NY 14214

KALLEN, ROLAND GILBERT, b Glasgow, Scotland, July 3, 35; US citizen; m 63; c 2. MEDICINE, BIOCHEMISTRY. *Educ:* Amherst Col, AB, 56; Columbia Univ, MD, 60; Brandeis Univ, PhD(biochem), 65. *Hon Degrees:* MS, Univ Pa. *Prof Exp:* Intern, NY Univ-Bellevue Med Ctr, 60-61; from asst prof to assoc prof biochem, 65-77, PROF BIOCHEM & BIOPHYS, SCH MED, UNIV PA, 78- *Concurrent Pos:* Fogerty Int fel, 81. *Mem:* Am Chem Soc; Am Soc Biochem & Molecular Biol. *Res:* Biochemistry of cancer; mechanisms and regulation of enzymatic activity; mechanisms of hormone action; mechanisms and regulation of gene expression. *Mailing Add:* Dept Biochem and Biophys Univ Pa Sch Med Philadelphia PA 19104-6059

KALLEN, THOMAS WILLIAM, b Hammond, Ind, Oct 26, 38; m 58; c 2. INORGANIC CHEMISTRY. *Educ:* Beloit Col, BS, 65; Wash State Univ, PhD(chem), 68. *Prof Exp:* Res assoc chem, Georgetown Univ, 68-70; from asst prof to assoc prof, 70-80, PROF CHEM, STATE UNIV NY COL BROCKPORT, 80- *Mem:* Sigma Xi; Am Chem Soc. *Res:* Kinetics and mechanisms of substitution reactions and oxidation-reduction reactions of transition-metal complex ions; catalysis by transition-metal ions; ion-exchange chromatography. *Mailing Add:* 707 Ellis Dr Brockport NY 14420

KALLENBACH, ERNST ADOLF THEODOR, b Minden, Ger, Feb 21, 26; m 57; c 3. HISTOLOGY, ELECTRON MICROSCOPY. *Educ:* George Williams Col, BSc, 58; McGill Univ, MSc, 60, PhD(anat), 63. *Prof Exp:* Res assoc histol, McGill Univ, 60-63; instr anat, 63-64, asst prof, 64-67, assoc prof path, 67-77, assoc prof anat, 77-81, PROF ANAT, COL MED, UNIV FLA, 81- *Concurrent Pos:* Assoc ed, Anat Rec. *Mem:* Am Asn Anatomists; Electron Micros Soc Am. *Res:* Lymphocyte production in thymus gland; presence and arrangement of cytoplasmic fibrils within epithelial cells; formation of enamel; fine structure of enamel organ. *Mailing Add:* Dept Anat Univ Fla Col Med Gainesville FL 32610

KALLENBACH, NEVILLE R, b Johannesburg, SAfrica, Jan 30, 38; US citizen; m 59; c 1. BIOPHYSICAL CHEMISTRY. *Educ:* Rutgers Univ, BS, 58; Yale Univ, PhD(chem), 61. *Prof Exp:* NSF fel biophys chem, Univ Calif, San Diego, 61-62, NIH fel, 62-64; from asst prof to assoc prof, 64-71, PROF BIOL, UNIV PA, 71- *Concurrent Pos:* Guggenheim Mem Found fel, 72-73. *Mem:* Biophys Soc; Am Soc Biol Chemists. *Res:* Structure and function of nucleic acids. *Mailing Add:* Dept Biol 204 KW G6 Univ Pa Philadelphia PA 19104

KALLER, CECIL LOUIS, b Humboldt, Sask, Mar 26, 30; m 62; c 4. MATHEMATICS, EDUCATION. *Educ:* Univ Sask, BA & BEd, 54, MA, 56; Purdue Univ, PhD(math statist), 60. *Prof Exp:* Teacher elem & high schs, Sask, 48-52; from asst prof to assoc prof math, Univ Sask, 60-65, from assoc prof to prof, Univ Regina, 65-70, chmn dept, 65-70; prof math & pres, Notre Dame Univ of Nelson, BC, 70-76; MATHEMATICIAN, OKANAGAN COL, BC, 76- *Concurrent Pos:* Asst res statistician, Educ Div, Dominion Bur Statist, Ottawa, 55; res statistician, UpJohn Co, Kalamazoo, 58, 59 & 60. *Mem:* Am Math Soc; Math Asn Am; Am Statist Asn; Biomet Soc; Sigma Xi; Can Math Soc. *Res:* Mathematical statistics and probability theory; statistical models in biological sciences. *Mailing Add:* 2056 Pandosy St Kelowna BC V1Y 1S3 Can

KALLEY, GORDON S, b Windham, Conn, Dec 4, 53; m 78; c 1. APPLIED AI, OFFICE AUTOMATION. *Educ:* Mitchell Col, AS, 75; Univ New Haven, BS, 77; NJ Inst Technol, BS, 81, MS, 85. *Prof Exp:* Spec lectr indust eng, 81-90, ASST PROF ENG TECHNOL, NJ INST TECHNOL, 90- *Concurrent Pos:* Staff consult, Modern Human Resources, 81-87; pres, Cruise Productivity Eng Inc, 87- *Mem:* Soc Mech Engrs; Inst Indust Engrs; Am Soc Elec Engrs. *Res:* Office automation in both the manufacturing and service sector; system integration; applications of AI; major publications in computer graphics and productivity science; application of control theory to ethics and productivity. *Mailing Add:* Two Beekman Rd Bridgewater NJ 08807

KALLFELZ, FRANCIS A, b Syracuse, NY, July 17, 38; m 65; c 3. NUTRITION. *Educ:* Cornell Univ, DVM, 62, PhD(phys biol), 66. *Prof Exp:* From asst prof to assoc prof, dept large animal med obstet & surg, 66-80, PROF CLIN NUTRIT, DEPT CLIN SCI, NY STATE COL VET MED, CORNELL UNIV, 80- *Mem:* Am Vet Med Asn; Am Inst Nutrit; Soc Nuclear Med; Am Acad Vet Nutritionists; Am Col Vet Med. *Res:* Alkaline earth metabolism, the role of vitamin D in calcium metabolism; applications of radioisotopes in clinical veterinary medicine; mineral metabolism in domestic animals; metabolic diseases. *Mailing Add:* Dept Clin Sci Vet Col Cornell Ithaca NY 14853-6401

KALLFELZ, JOHN MICHAEL, b Atlanta, Ga, Nov 21, 34; m 59; c 3. NUCLEAR ENGINEERING. *Educ:* US Mil Acad, BS, 56; Calif Inst Technol, MS, 61; Univ Karlsruhe, Dr Ing(nuclear eng), 66. *Prof Exp:* Res engr, Atomics Int Div, NAm Aviation, Inc, 61-62; res in reactor physics, Inst Appl Nuclear Physics, Nuclear Res Ctr, Univ Karlsruhe, 63-65, scientist, 66; asst prof nuclear eng, Ga Inst Technol, 67; asst nuclear engr, Reactor Physics Div, Argonne Nat Lab, 67-69; asst prof, 69-71, assoc prof, 71-80, PROF NUCLEAR ENG, GA INST TECHNOL, 80- *Mem:* Am Nuclear Soc; Sigma Xi. *Res:* Neutron thermalization and transport theory; fast reactor physics. *Mailing Add:* Dept Nuclear Eng Ga Inst Technol Atlanta GA 30332

KALLIANOS, ANDREW GEORGE, b Piraeus, Greece, Sept 14, 30; nat US; m 56; c 4. ORGANIC CHEMISTRY. *Educ:* Hendrix Col, 51; Univ Okla, MS, 56, PhD(chem), 58. *Prof Exp:* Asst chem, Univ Okla, 54-58; res org chemist & supvr res, Liggett & Myers Inc, 58-80; ASSOC PRIN SCIENTIST, PHILIP MORRIS USA, 80- *Mem:* Am Chem Soc; Am Col Toxicol. *Res:* Separation and identification of naturally occurring substances; synthesis and evaluation of odorous materials and determination of their pyrolysis products. *Mailing Add:* 13439 Glendower Rd Midlothian VA 23113-3808

KALLIANPUR, GOPINATH, b Mangalore, India, Apr 16, 25; m 53; c 1. MATHEMATICS. *Educ:* Univ Madras, BA, 45, MA, 46; Univ NC, PhD(math statist), 51. *Prof Exp:* Lectr statist, Univ Calif, 51-52; mem, Inst Advan Study, 52-53; reader statist, Indian Statist Inst, Calcutta, 53-56; vis assoc prof, Mich State Univ, 56-59; assoc prof math, Ind Univ, 59-61; prof statist, Mich State Univ, 61-63; prof math & statist, Univ Minn, Minneapolis, 63-; AT DEPT STATIST, UNIV NC, CHAPEL HILL. *Mem:* AAAS; fel Inst Math Statist. *Res:* Probability theory; stochastic processes; statistics. *Mailing Add:* Dept Statist Univ NC 318 Phillips 039a Chapel Hill NC 27514

KALLIO, REINO EMIL, b Worcester, Mass, July 6, 19; m 42; c 3. MICROBIAL PHYSIOLOGY. *Educ:* Univ Ala, BS, 41; Univ Iowa, MS, 48, PhD(bact), 50. *Prof Exp:* Sr supvr, Trojan Powder Co, 41-44; asst, Univ Iowa, 46-48, instr microbiol, 48-50, from asst prof to prof, 50-65; DIR SCH LIFE SCI, UNIV ILL, URBANA, 65- *Concurrent Pos:* Consult, Am Petrol Inst, 70- *Mem:* Am Chem Soc; Am Soc Microbiol; Brit Soc Gen Microbiol; Sigma Xi. *Res:* Microbial hydrocarbon metabolism; fatty acids; lipids. *Mailing Add:* Microbiol 131 Burrill H Univ Ill Urbana IL 61801

KALLIOKOSKI, JORMA OSMO KALERVO, b Harma, Finland, Nov 23, 23; nat US; m 49; c 3. ECONOMIC GEOLOGY. *Educ:* Univ Western Ont, BSc, 47; Princeton Univ, PhD(geol), 51. *Prof Exp:* Geologist, Geol Surv Can, 49-53 & Newmont Explor, Ltd, 53-56; from asst prof to assoc prof geol, Princeton Univ, 56-68; head, Geol Eng Dept, 68-81, PROF GEOL, MICH TECHNOL UNIV, 68- *Concurrent Pos:* Bus ed, Econ Geol Publ Co, 71-77. *Mem:* Geol Soc Am; Soc Econ Geol (secy, 65-67, pres elect, 79, pres, 80); Can Inst Mining & Metall; Geol Soc Finland; Am Inst Min Metall. *Res:* Relationship between structure and mineral deposits; Precambrian geology in Canada, the United States and Venezuela; uranium geology. *Mailing Add:* 1010 Seventh Ave Houghton MI 49931

KALLMAN, BURTON JAY, b New York, NY, Nov 1, 27; m 58; c 2. BIOCHEMISTRY, NUTRITION. *Educ:* Bethany Col, WVa, BS, 47; Univ Southern Calif, MS, 51, PhD(biochem), 58. *Prof Exp:* Instr chem, Sch Dent, Univ Southern Calif, 52-53, res assoc, Sch Med, 53-58; fel, Attend Staff Asn, Los Angeles County Gen Hosp, 58-59; biochemist, Fish-pesticide Res Lab, US Fish & Wildlife Serv, 59-63 & Vet Admin Ctr, 63-67; biochemist, TRW Inc, 67-76; sr scientist, Sci Appln Inc, 76-80; prin, Interdisciplinary Sci Assoc, Inc, 80-82; dir, Appl Biol Sci Labs, 82-85; DIR, SCI & TECHNOL, NAT NUTRIT FOODS ASSOC, 85- *Concurrent Pos:* Consult, Childrens Asthma Res Inst, 62-63, Behav Health Serv, 74-76, US State Dept, 78, IWG Corp, 80-, Sci Appln Inc, 80- *Mem:* AAAS; Am Chem Soc; Inst Food Technologists; Am Oil Chemists Asn. *Res:* Adrenocorticotropic hormone release; thyroid hormone metabolism and physiology; pesticide biochemistry and pharmacology; immunochemistry; bioconversion of energy; fate and effects of petroleum on marine ecosystems; potential health effects of shale oil industry. *Mailing Add:* 23214 Robert Rd Torrance CA 90505

KALLMAN, KLAUS D, b Berlin, Ger, July 20, 28; US citizen; m 65. GENETICS, ICHTHYOLOGY. *Educ:* Queens Col, NY, BS, 52; NY Univ, MS, 55, PhD(genetics), 59. *Prof Exp:* Res assoc, 60-62, GENETICIST, OSBORN LABS MARINE SCI, NY AQUARIUM, 63-; RES ASSOC ICHTHYOL, AM MUS NATURAL HIST, NEW YORK, 65- *Concurrent Pos:* USPHS fel, 59-61; lectr, City Col New York, 60-66. *Mem:* Genetics Soc Am; Am Soc Zool; Am Soc Ichthyol & Herpet. *Res:* Tissue transplantation; sex determination; evolution; pigment cell biology. *Mailing Add:* Osborn Labs Marine Sci New York Aquarium Brooklyn NY 11224

KALLMAN, MARY JEANNE, b Alexandria, Va, May 27, 48; m 69. PSYCHOBIOLOGY, PSYCHOPHARMACOLOGY. *Educ:* Lynchburg Col, BS, 70; Univ Ga, MS, 74, PhD(biopsychol), 76. *Prof Exp:* Res asst psychiat, Med Ctr, Univ Miss, 73-74; fel, Med Col Va, 76-79, res assoc, 79-80, asst prof pharmacol, 80-83; ASSOC PROF PSYCHOL & PHARMACOL, UNIV MISS, 83- *Concurrent Pos:* Mem adj fac, Dept Psychol, Va Commonwealth Univ, 75-76. *Mem:* AAAS; Soc Neurosci; Soc Stimulus Properties Drugs; Am Psychol Asn; Am Asn Pharmacol & Exp Therapeut; Behav Pharmacol Soc. *Res:* State dependency of drugs and stimulus control; central nervous system function and sites of drug action; comparative and developmental central nervous system differences; central sensory processing; psychophysiology; behavioral toxicology and teratology. *Mailing Add:* Dept Psychol Univ Miss University MS 38677

KALLMAN, RALPH ARTHUR, b Holdrege, Nebr, Sept 17, 34. MATHEMATICS. *Educ:* Univ Minn, BA, 56, MA, 61, PhD(math), 65. *Prof Exp:* Asst prof math, Univ Minn, Duluth, 65-67; asst prof, 67-73, ASSOC PROF MATH, BALL STATE UNIV, 73- *Mem:* Am Math Soc; Math Asn Am; Asn Comput Mach. *Res:* Real and functional analysis; probability and statistics; computational combinatorics; computational probability and statistics. *Mailing Add:* Dept Math Sci Ball State Univ Muncie IN 47306

KALLMAN, ROBERT FRIEND, b New York, NY, May 21, 22; m 48, 69; c 3. RADIOBIOLOGY. *Educ:* Hofstra Col, AB, 43; NY Univ, MS, 49, PhD, 52. *Prof Exp:* Instr biol, Brooklyn Col, 47-48; asst res physiologist, Univ Calif, 52-56; res assoc, 56-60, from asst prof to assoc prof, 60-72, dir radiobiol res div, 59-77, PROF RADIOL, STANFORD UNIV, 72- *Mem:* AAAS; Radiation Res Soc (pres, 76-77); Am Asn Cancer Res; Am Asn Lab Animal Sci; fel NY Acad Sci; Sigma Xi; Am Soc Therapeut Radiol & Oncol; Soc Anal Cytol. *Res:* Radiation effects in mammals; mechanisms of biological action of radiation; experimental cancer therapy; carcinogenesis. *Mailing Add:* Dept Radiation Oncol Stanford Univ Stanford CA 94305

KALLMAN, WILLIAM MICHAEL, b New York, NY, Mar 18, 47; m 69. PSYCHOPHYSIOLOGY. *Educ:* Lynchburg Col, BA, 69; Univ Ga, MS, 73, PhD(psychol), 75. *Prof Exp:* Asst instr psychol, Lynchburg Col, 69-70; asst prof, Va Commonwealth Univ, 75-80, dir, Psychol Serv Ctr, 78-80, assoc prof psychol, 80-83; CONSULT, 83- *Concurrent Pos:* Prog consult, Crossroads Drug Treatment Ctr, 73-74; consult, Va Dept Corrections, 75-79, Va Hosp, 79-, Va Dept Ment Health, 79-80; fel, Gestalt Inst Richmond, NY Acad Sci; adj assoc prof, Univ Miss, 83-, vis assoc prof, 84-85 & 87- *Mem:* Am Psychol Asn; Sigma Xi. *Res:* Interrelationships of the autonomic nervous system and the central nervous system functioning in psychopathology; biofeedback. *Mailing Add:* 404 S 11th St Oxford MS 38655

KALLMANN, SILVE, b Ger, Feb 13, 15; nat US; m 41; c 1. ANALYTICAL CHEMISTRY. *Educ:* Real Gymnasium Bitburg, Ger, 35; Univ Cologne, 37. *Prof Exp:* Anal chemist, Dr Oppenheimer Lab, Dusseldorf, Ger, 35-37; analyst, Walker & Whyte, Inc, NY, 37-41; analyst, 41-51, res dir, 51-78, vpres, 61-78, ASN CONSULT, CHEM & CHEM ENERGY, LEDOUX & CO, INC, 78- *Mem:* Am Chem Soc; Am Soc Test & Mat; Asn Consult Chemists & Chem Eng; Inst Mining, Metall & Petrol Eng. *Res:* Analysis of metallurgical products by chemical and instrumental methods. *Mailing Add:* 552 Beverly Rd Teaneck NJ 07666

KALLO, ROBERT MAX, b San Francisco, Calif, Oct 6, 23; m 47; c 2. PHYSICAL CHEMISTRY. *Educ:* Univ Calif, Berkeley, BS, 45, PhD, 50. *Prof Exp:* From instr to assoc prof, Calif State Univ, Fresno, 50-60, prof chem, 60-; RETIRED. *Res:* Thermodynamics. *Mailing Add:* 5236 N Poplar Fresno CA 93740

KALLOK, MICHAEL JOHN, b Gary, Ind, April 9, 48. CARDIOVASCULAR PHYSIOLOGY, PULMONARY MECHANICS. *Educ:* Univ Colo, BS, 70; Purdue Univ, MS, 74; Univ Minn, PhD(biomed eng), 78. *Prof Exp:* Design engr, Pratt & Whitney Aircraft, 70-71; instr math, Andrean High Sch, 71-72; staff eng, Chicago Metallic Corp, 72-74; res asst, Univ Minn, 74-77; res fel, Mayo Clin, 77-79; sr engr, 79-81, sr scientist, Medtronic Inc, 81-85, tachycordia res scientist, 85-87, SR RES SCIENTIST, MEDTRONIC INC, 87- *Concurrent Pos:* Eng consult, Ind Health Eng Assocs, Inc, 76; instr physiol, Mayo Med Sch, 79; assoc fel, Medtronic Inc, 83-88, fel, 88-; NIH, 79. *Mem:* Am Physiol Soc; Am Soc Mech Engrs; Biomed Eng Soc; NY Acad Sci; Am Heart Asn; Am Col Cardiol. *Res:* Pulmonary mechanics; cardiac tachyarrhythmia mechanisms; detection and therapy of ventricular tachycardia and ventricular fibrillation; applications of engineering for medicine and physiology. *Mailing Add:* Medtronic Inc 7000 Central Ave NE Minneapolis MN 55432

KALLOS, GEORGE J, b Greece, May 21, 36; US citizen; m 63; c 2. ANALYTICAL CHEMISTRY. *Educ:* Cent Mich Univ, BS, 60; Univ Detroit, MS, 62. *Prof Exp:* Res chemist, Dow Chem Co, 63-71, sr res chemist, 71-75, sr res specialist, 75-78, ASSOC SCIENTIST CHEM, DOW CHEM USA, 78- *Concurrent Pos:* Chmn, Carcinogen Safety Monograph Rev Panel, Nat Cancer Inst, 78. *Mem:* Am Chem Soc; Sigma Xi. *Res:* Development of new analytical technology for trace analysis; application of mass spectrometry to the elucidation of organic structure; monitoring of environmental pollutants. *Mailing Add:* 1897 Bldg Dow Chem Co Midland MI 48667

KALLSEN, HENRY ALVIN, b Jasper, Minn, Mar 25, 26; m 50; c 4. ENGINEERING ECONOMY, COMPUTER SCIENCE. *Educ:* Iowa State Univ, BS, 48; Univ Wis, MS, 52, PhD, 56. *Prof Exp:* Asst engr, Wabash RR Co, 48-49; from instr to asst prof civil eng, Univ Wis, 49-59; from assoc prof to prof, La Polytech Inst, 59-64, actg head dept, 62-63; asst exec secy, Am Soc Eng Educ, 64-65; prof civil eng & asst dean eng, 65-72, actg head dept, 85-86, PROF INDUST ENG, UNIV ALA, 72- *Concurrent Pos:* Mem adv res coun, La Hwy Dept, 62-64; consult bd eng educ, Comn Higher Educ, State of Tenn, 69; co-dir sch transp proj Ala Dept Econ & Community Affairs, 87- *Mem:* Sigma Xi; Am Soc Eng Educ; Am Soc Civil Engrs; Int Indust Engrs. *Res:* Engineering economy; geometronics; safety; computer science. *Mailing Add:* Indust Eng Box 6316 Univ Ala Tuscaloosa AL 35487-6316

KALLUS, FRANK THEODORE, b La Grange, Tex, Aug 19, 36; wid; c 2. PHYSIOLOGY, ANESTHESIOLOGY. *Educ:* Tex A&M Univ, BS, 66; Univ Tex Southwestern Med Sch, MD, 61, PhD(physiol), 70. *Prof Exp:* Intern, Methodist Hosp Dallas, 61-62; resident, Parkland Mem Hosp, 64-66; instr physiol, Univ Tex Southwestern Med Sch, 69-70; asst prof physiol, Sch Med, La State Univ, Shreveport, 70-72; ASST PROF ANESTHESIOL & DIR ANESTHESIOL LABS, UNIV TEX SOUTHWESTERN MED SCH, DALLAS, 72- *Concurrent Pos:* Res fel, Parkland Mem Hosp, Tex, 65-66; USPHS spec fel, 66-70. *Mem:* Am Physiol Soc. *Res:* Physiology of anesthesia; potassium and sodium metabolism and fluid and electrolyte balance. *Mailing Add:* Southwestern Med Sch 4020 Marquette Dallas TX 75235

KALM, MAX JOHN, b Munich, Germany, Nov 27, 28; US citizen; m 69; c 2. ORGANIC CHEMISTRY, PHARMACEUTICAL CHEMISTRY. *Educ:* Univ Calif, Berkeley, BS, 52, PhD(chem), 54. *Prof Exp:* Res assoc chem, Univ Mich, 54-55; sr investr, G D Searle & Co, 55-65; dir sci liaison, Cutter Labs, Inc, 65-71, dir qual control, 71-74, dir qual assurance, 74-77, vpres qual assurance, 77-82; VPRES QUAL ASSURANCE, SCHERING CORP, 82- *Concurrent Pos:* Mem adv comt, Dept Health, Calif, 76-78. *Mem:* AAAS; Am Chem Soc; Acad Pharmaceut Sci; fel Am Inst Chem; Pharmaceut Mfrs Asn; Parenteral Drug Asn; Am Pharm Asn; NY Acad Sci. *Res:* Photochemistry of benzene; psycho-stimulants and anorexics; synthesis of steroids with hormonal activity. *Mailing Add:* Schering Corp 1011 Morris Ave Union NJ 07083

KALMA, ARNE HAERTER, b Long Branch, NJ, May 26, 41; m 65; c 1. NUCLEAR RADIATION EFFECTS, OPTICAL SENSORS. *Educ:* Rensselaer Polytech Inst, BS, 63, MS, 65, PhD(nuclear sci eng), 68. *Prof Exp:* Res scientist physics, Univ Paris, 68-69; res scientist group leader physics, IRT Corp, 69-79; MEM STAFF, NORTHROP RES & TECH CTR, 79- *Mem:* Am Phys Soc; Optical Soc Am; Inst Elec & Electronics Engrs; Sigma Xi. *Res:* Radiation effects in materials, particularly infrared detectors, optical materials and semiconductors; design and use of radiation hard fiber optics systems; radiation testing; nuclear radiation effects; optical sensors. *Mailing Add:* c/o Mission Res Corp 4935 N 30th St Colorado Springs CO 80919

KALMAN, CALVIN SHEA, b Montreal, Que, Oct 29, 44; m 66; c 2. HIGH ENERGY PHYSICS. *Educ:* McGill Univ, BSc, 65; Univ Rochester, MA, 67, PhD(physics), 71. *Prof Exp:* Asst prof, Loyola Col Montreal, 68-75; chmn dept, 83-89, PROF PHYSICS, SIR GEORGE CAMPUS, CONCORDIA UNIV, 75- *Concurrent Pos:* Vis assoc prof, Ind Univ, 76-77. *Mem:* Am Phys Soc; Am Asn Physics Teachers; Inst Particle Physics. *Res:* Supersymmetric Gauge Field Theory, subquark structure. *Mailing Add:* Dept Physics Concordia Univ Montreal PQ H3G 1M8 Can

KALMAN, GABOR J, b Budapest, Hungary, Dec 12, 29; US citizen; c 2. PLASMA PHYSICS, THEORETICAL PHYSICS. *Educ:* Polytech Univ, Budapest, dipl, 52; Israel Inst Technol, DSc, 61. *Prof Exp:* From jr res scientist to res scientist, Cent Res Inst Physics, Budapest, Hungary, 52-56; res assoc plasma physics, Israel Inst Technol, 57-58, lectr physics, 58-61; prof, Univ Paris, 61-66, dir res, 66-68; vis prof, Brandeis Univ, 66-70; RES PROF PHYSICS, BOSTON COL, 70- *Concurrent Pos:* Dir, Orsay Summer Inst Plasma Physics, France, 62; vis fel, Joint Lab Astrophys, 65-66; expert, Air Force Cambridge Res Lab, 67-68; vis scientist, Paris Observ, Meudon, France, 73-74 & Ionosphere Res Group, Nat Ctr Sci Res, Orleans, France, 74; sr vis fel, Univ Oxford, 75; exchange prof, Univ Paris, 76; assoc, Ctr Astrophys, Harvard Univ, 73-77; dir NATO Advan Study Inst Strongly Coupled Plasmas, Orleans, France, 77; res leader, Int Ctr Theoret Physics, 81 & 84; co-dir, NSF-CNRS Workshop Spectral Diag Turbulence Solar Flares, Boston, 82; prin investr grants & contracts, NSF, Dept Energy, Air Force Off Sci Res, Air Force Geophys Lab & Israel-US Brit Nuclear Forum, NATO. *Mem:* AAAS; Am Astron Soc; European Phys Soc; fel Am Phys Soc; fel Fr Phys Soc; fel NY Acad Sci. *Res:* Strongly coupled plasmas; plasma physics; plasma response functions; plasma astrophysics; models for high density; astrophysical many body systems; solid state plasmas. *Mailing Add:* Dept Physics Boston Col Chestnut Hill MA 02167

KALMAN, RUDOLF EMIL, b Budapest, Hungary, May 19, 30; nat US; m 59; c 2. MATHEMATICS, ENGINEERING. *Educ:* Mass Inst Technol, SB, 53, SM, 54; Columbia Univ, DSci, 57. *Prof Exp:* Asst Servomechanisms Lab, Mass Inst Technol, 53-54; res engr process control res, E I du Pont de Nemours & Co, Del, 54-55; instr, Columbia Univ, 55-57; staff engr, Int Bus Mach Corp, NY, 57-58; staff mathematician, Res Inst Adv Studies, 58-62, head ctr control theory, 62-64; prof eng mech & elec eng, Stanford Univ, 64-67, prof math syst theory & opers res, 67-71; GRAD RES PROF & DIR CTR MATH SYST THEORY, UNIV FLA, 71- *Concurrent Pos:* Prof math system theory, Swiss Fed Inst Technol, Zurich, 73- *Honors & Awards:* Rufus Oldenburger Medal, Am Soc Mech Eng, 76. *Mem:* Nat Acad Eng; Am Math Soc; Inst Elec & Electronics Engrs; Hungarian Acad Sci. *Res:* Automatic control; network and information theory; mathematical statistics; automata; nonlinear dynamic systems; calculus of variations; stochastic processes; engineering science; algebraic system theory. *Mailing Add:* Ctr Math Theory Univ Fla Gainesville FL 32611

KALMAN, SUMNER MYRON, b Boston, Mass, Nov 14, 18; m 52; c 1. BIOLOGICAL CHEMISTRY. *Educ:* Harvard Univ, AB, 40; Stanford Univ, MD, 51. *Prof Exp:* Intern, Mt Zion Hosp, Calif, 50-51; fel pharmacol, Stanford Univ, 51-53; NIH, Carlsberg Lab, Denmark, 53-54; physiol, Univ Copenhagen, 54; Graham fel pharmacol, Stanford Univ Sch Med, 54-59; from instr to assoc prof pharmacol, Sch Med, Stanford Univ, 54-67, prof, 67-88; RETIRED. *Concurrent Pos:* Res Career Develop Award, Div Gen Med Sci, US Pub Health Serv, 59-69; consult, US Food & Drug Admin, 72-75; dir drug assay lab, Stanford Univ, 73-85. *Mem:* AAAS; Am Soc Pharmacol & Exp Therapeut; Am Soc Biol Chem; Am Fedn Clin Res; Soc Gen Physiol. *Res:* Human pharmacology; measurement of drug dosage responses in man and basis for individual differences in sensitivity; metabolic fate and elimination, especially digitalis as a model for detecting such differences. *Mailing Add:* 2299 Tasso Palo Alto CA 94301

KALMAN, THOMAS IVAN, b Budapest, Hungary, Jan 20, 36; nat US; m 63; c 2. MOLECULAR PHARMACOLOGY, BIO-ORGANIC CHEMISTRY. *Educ:* Dipl ChE, Tech Univ Budapest, 59; State Univ NY Buffalo, PhD(biochem, pharmacol), 68. *Prof Exp:* Res chemist, Hungary, 59-62; res asst, Res Found, State Univ NY, 63-66; fel, NIH, 67-68; adj asst prof & sr res assoc health sci, 68-70, asst prof biochem pharmacol, 70-75, ASSOC PROF MED CHEM, STATE UNIV NY BUFFALO, 76-, ASSOC PROF BIOCHEM PHARMACOL, 80- *Concurrent Pos:* NIH res career develop award, 71-76; vis assoc prof, Dept Pharmacol, Sch Med, Yale Univ, 75-76. *Mem:* AAAS; Am Chem Soc; NY Acad Sci; fel Am Inst Chem. *Res:* Mechanisms of enzyme and drug action; design and synthesis of selective enzyme inhibitors; biosynthesis of purine and pyrimidine nucleotides; folic acid metabolism; drug design; experimental chemotherapy; cancer research; antiviral agents. *Mailing Add:* 955 Pinetree Ct East Amherst NY 14051

KALMANSON, KENNETH, b Brooklyn, NY, Mar 26, 43; m 66; c 1. MATHEMATICS. *Educ:* Brooklyn Col, BS, 64; City Col New York, PhD(math), 70. *Prof Exp:* Teacher high schs, NY, 65-66; ASST PROF MATH, MONTCLAIR STATE COL, 70- *Mem:* Am Math Soc; Math Asn Am. *Res:* Combinatorial geometry, especially extreme Hamiltonian lines with respect to metric spaces. *Mailing Add:* 102 Squire Hill Rd Upper Montclair NJ 07043

KALMBACH, SYDNEY HOBART, b Fond du Lac, Wis, June 8, 13; m 40; c 1. PHYSICS. *Educ:* Marquette Univ, BS, 34, MS, 39. *Prof Exp:* Asst prof physics, Elmhurst Col, 40-43 & US Naval Acad, 47-52; from assoc prof to prof, 52-84, EMER PROF PHYSICS, NAVAL POSTGRAD SCH, 84- *Mem:* Am Phys Soc; Am Asn Physics Teachers. *Res:* Infrared spectroscopy; atmospheric optics. *Mailing Add:* Dept Physics Naval Postgrad Sch Monterey CA 93940

KALME, JOHN S, b Riga, Latvia, June 20, 38; US citizen. MATHEMATICS. *Educ:* Univ Pa, BA, 61, MA, 64, PhD(math), 66. *Prof Exp:* Teaching fel math, Univ Pa, 61-64; instr, Drexel Inst, US Naval Acad, Annapolis, 65-66, from asst prof to assoc prof math, 69-79; consult & comput prog, David Taylor, US Naval Ship Res & Develop Ctr, Annapolis, 79-82; CONSULT, 82- *Res:* Mathematical analysis, chiefly probability theory and integral operators; stochastic processes, time series analysis; use of time series analysis in source and path identification of structure-borne noise on ships; vibration analysis. *Mailing Add:* Dept Math US Naval Acad Annapolis MD 21402

KALMUS, GERHARD WOLFGANG, b Berlin, Germany, Dec 19, 42; m 67. DEVELOPMENTAL BIOLOGY. *Educ:* Univ Calif, Berkeley, BA, 67; Rutgers Univ, Camden, MS, 74; Rutgers Univ, New Brunswick, PhD(zool), 77. *Prof Exp:* Res asst physiol, Univ Pa, 67-68; asst proj mgr transl, Info Intersci Inc, 69-70; teacher biol, Quakertown Community High Sch, 72; res asst embryol, Temple Univ, 73; teaching asst zool, Rutgers Univ, 73-77; asst prof develop, 77-83, ASSOC PROF BIOL, ECAROLINA UNIV, 83- *Mem:* AAAS; Am Soc Zoologists; Am Inst Biol Sci; Tissue Cult Asn; Soc Develop Biol; Sigma Xi. *Res:* Transplantation immunology; chick primordial germ cells; fetal alcohol syndrome; differentiation in cell culture. *Mailing Add:* Dept Biol ECarolina Univ Greenville NC 27858-4353

KALMUS, HENRY P(AUL), b Vienna, Austria, Jan 9, 06; nat US; m 54; c 3. PHYSICS, ELECTRICAL ENGINEERING. *Educ:* Vienna Tech Univ, Dipl Eng, 30, Dr Tech, 60. *Prof Exp:* Researcher, Orion Radio Corp, Budapest, 30-38; develop engr, Emerson Radio Corp, NY, 38-41; res physicist, Zenith Radio Corp, Ill, 41-48; physicist, Nat Bur Standards, US Dept Commerce, 48-53; CHIEF SCIENTIST & ASSOC DIR, HARRY DIAMOND LABS, US ARMY MATERIEL COMMAND, 53- *Concurrent Pos:* Consult, W M Welch Mfg Co, Ill, 48- *Honors & Awards:* Gravity Res Found Award, 64; Prod Eng Master Designer Award, 67; Sperry Award, Instrument Soc Am, 70. *Mem:* Fel Inst Elec & Electronics Engrs. *Res:* Electronics, measurement techniques; mathematics. *Mailing Add:* 3000 University Terr NW Washington DC 20016

KALNAJS, AGRIS JANIS, astronomy, for more information see previous edition

KALNAY, EUGENIA, b Buenos Aires, Argentina, Oct 10, 42; US citizen; m 81; c 1. NUMERICAL WEATHER PREDICTION, GENERAL CIRCULATION MODELING. *Educ:* Univ Buenos Aires, Lic, 65; Mass Inst Technol, PhD(meteorol), 71. *Prof Exp:* Asst prof meteorol, Univ Montevideo, Uruguay, 71-73; res assoc, Mass Inst Technol, 73-74, from asst prof to assoc prof, 75-78; sr res meteorologist, NASA Goddard Lab Atmosphere, 79-83, head, Global Modeling & Simulation Br, 83-86; DEVELOP DIV CHIEF, NAT METEROL CTR, NAT OCEANIC & ATMOSPHERIC ADMIN, 87- *Concurrent Pos:* Prin investr, NASA res projs, 73-; adj prof meteorol, Univ Md, 80-83; mem, First Global Atmospheric Res Prog Global Exp Panel, Nat Acad Sci, 83-; assoc ed, J Atmospheric Sci, 84-90; mem, Bd Atmospheric Sci & Climate, Nat Acad Sci. *Honors & Awards:* Except Sci Achievement Medal, NASA, 81; Silver Medal, Dept Com, 90. *Mem:* Am Meteorol Soc. *Mailing Add:* Nat Meterol Ctr Nat Oceanic & Atmospheric Admin Washington DC 20233

KALNIN, ILMAR L, b Riga, Latvia, Jan 23, 26; US citizen; m 54; c 3. MATERIALS SCIENCE. *Educ:* Westminster Col, Pa, BA, 52; Ill Inst Technol, PhD(chem), 57. *Prof Exp:* Mem tech staff, Bell Tel Labs, 57-62; res scientist, Am-Standard Co, 62-66; sr res scientist, Celanese Res Co, 66-67, res assoc, 67-86, RES ASSOC, HOECHST-CELANESE CO, 86-91. *Mem:* Am Chem Soc; Am Ceramic Soc; Am Soc Testing & Mat. *Res:* Mat Res Soc; Functional polymers; structural composites; fiber reinforced plastics; advanced ceramics. *Mailing Add:* 135 Haas Rd Basking Ridge NJ 07920

KALNINS, ARTURS, b Riga, Latvia, Feb 13, 31; US citizen; m 56. MECHANICS. *Educ:* Univ Mich, BS, 55, MS, 56, PhD(eng mech), 60. *Prof Exp:* Res asst eng mech, Univ Mich, 56-58; res engr appl mech, Univ Calif, Berkeley, 58-60; asst prof eng & appl sci, Yale Univ, 60-65; assoc prof mech, 65-67, PROF MECH, LEHIGH UNIV, 67- *Concurrent Pos:* Assoc ed, J Acoust Soc Am, 70-; Fulbright-Hayes fel, Univ Innsbruck, 77. *Mem:* Am Soc Mech Engrs; Acoust Soc Am. *Res:* Stress analysis, pressure vessel design. *Mailing Add:* Dept Mech Eng & Mech Lehigh Univ Bethlehem PA 18015

KALNITSKY, GEORGE, b Brooklyn, NY, Oct 22, 17; m 40; c 3. BIOCHEMISTRY. *Educ:* Brooklyn Col, BA, 39; Iowa State Col, PhD(physiol bact), 43. *Prof Exp:* Asst, Iowa State Col, 42-43, res assoc, 43; res assoc & instr biochem, Univ Chicago, 43-45; from instr to assoc prof, 46-57, PROF BIOCHEM, UNIV IOWA, 57- *Concurrent Pos:* Univ col med travelling fel, Oxford Univ, 56-57; Guggenheim fel, Weizmann Inst Sci, Israel, 65-66; Welcome res travel award, 80. *Mem:* Fel AAAS; Am Chem Soc; Soc Exp Biol & Med; Am Soc Biol Chemists; Sigma Xi. *Res:* Bacterial metabolism; intermediary carbohydrate metabolism; mechanism of action of enzymes; proteolytic enzymes; intracellular proteases, lung, and protease inhibitors. *Mailing Add:* Dept Biochem Univ Iowa Col Med Iowa City IA 52242

KALOGEROPOULOS, THEODORE E, b Megalopolis, Greece, Jan 20, 31; m 54. HIGH ENERGY PHYSICS. *Educ:* Dipl physics, Nat Univ Athens, 54; dipl electronics, Radio-Eng Sch, Athens, 51; Univ Calif, Berkeley, PhD(physics), 59. *Prof Exp:* Res assoc physics, Lawrence Radiation Lab, Univ Calif, 59; instr, Columbia Univ, 59-62; from asst prof to assoc prof, 62-69, PROF PHYSICS, SYRACUSE UNIV, 69- *Concurrent Pos:* Vis physicist, Argonne Nat Lab, 65, Brookhaven Nat Lab, 60-85; vis prof, Nuclear Res Ctr Democritos, 69-70. *Mem:* Am Phys Soc; AAAS. *Res:* Elementary particle physics; investigations on antinucleon-nucleon interactions, using emulsions, bubble chambers, spark chambers, counters and drift chambers; medical applications of antiprotons; optimization algorithms and applications. *Mailing Add:* Dept Physics Syracuse Univ Syracuse NY 13244

KALONJI, GRETCHEN, US citizen. ATOMISTIC COMPUTER SIMULATIONS. *Educ:* Mass Inst Technol, BS, 80, PhD(mat sci), 82. *Prof Exp:* Mat engr, Nat Bur Standards, 79-81; from asst prof to assoc prof, mat sci, Mass Inst Technol, 82-90, KYOCERA PROF, UNIV WASH, 90- *Honors & Awards:* Presidential Young Investr Award, NSF, 84. *Mem:* Am Ceramics Soc; Mat Res Soc; Am Phys Soc; AAAS. *Res:* Theory of defects in crystalline solids; computer simulation techniques in materials science; rapid solidification of ceramics. *Mailing Add:* Dept Mat Engr Mass Inst Technol Cambridge MA 02139

KALOOSTIAN, GEORGE H, b Kaisarea, Turkey, Jan 12, 12; nat US; m 35; c 2. ENTOMOLOGY. *Educ:* Fresno State Col, BA, 35; Ore State Col, MS, 39. *Prof Exp:* Res leader, Calif, Fruit & Veg Insect Res, Agr Res Serv, USDA, 38-40, Wash, 40-48, Utah, 48-57, Ga, 57-61, res leader, Calif, 61-79; RETIRED. *Mem:* Entom Soc Am; Am Phytopath Soc; Sigma Xi. *Res:* Life history and control of dried fruit insects; plant and insect survey methods; insects in relation to fruit tree diseases caused by virus and mycoplasmalike organisms. *Mailing Add:* 4066 Mt Vernon Ave Riverside CA 92507

KALOS, MALVIN HOWARD, b New York, NY, Aug 5, 28; m 49; c 2. THEORETICAL PHYSICS, COMPUTER SCIENCE. *Educ:* Queens Col, NY, BS, 48; Univ Ill, MS, 49, PhD(physics), 52. *Prof Exp:* Res assoc physics, Univ Ill, 52-53 & Cornell Univ, 53-55; adv scientist, Nuclear Develop Corp, 55-64; res prof, Courant Inst Math, 64-89; DIR, CORNELL THEORY CTR & PROF DEPT PHYSICS, CORNELL UNIV, 89- *Concurrent Pos:* Lectr, Univ Paris, 70-77. *Honors & Awards:* Eugene Feenberg Award, 89. *Mem:* Am Phys Soc; fel NY Acad Sci; AAAS; fel Am Nuclear Soc. *Res:* Nuclear physics; statistical physics; neutron interactions; parallel computers; application of computers to physics, especially Monte Carlo methods. *Mailing Add:* Cornell Theory Ctr Cornell Univ Ithaca NY 14853-3801

KALOTA, DENNIS JEROME, b North Tonawanda, NY, Nov 15, 45; m 67; c 3. ORGANIC CHEMISTRY, FLUORINE CHEMISTRY. *Educ:* Niagara Univ, BS, 68; Univ Detroit, MS, 71, PhD(chem), 74. *Prof Exp:* SR RES SPECIALIST, MONSANTO CHEM CO, 74- *Mem:* Am Chem Soc; Sigma Xi. *Res:* Process development, new products and heterogeneous catalysis research in the area of chlorinated aromatic hydrocarbons and their derivatives; process development, exploratory and applications research in the area of fluorinated organic compounds; new products research in the area of amino acid derivatives; process development, heterogeneous catalysis and new products research in the area of chlorinated aromatic hydrocarbons. *Mailing Add:* 1306 Green Mist Dr Fenton MO 63026

KALOW, WERNER, b Cottbus, Ger, Feb 15, 17; nat Can; m 46; c 2. PHARMACOLOGY, PHARMACO GENETICS. *Educ:* Univ Koenigsberg, MD, 41. *Prof Exp:* Sci asst, Univ Berlin, 47-48; sci asst, Free Univ Berlin, 49; instr, Univ Pa, 51; from lectr to assoc prof, 52-62, chmn dept, 66-77, PROF PHARMACOL, UNIV TORONTO, 62- *Concurrent Pos:* Res fel pharmacol, Univ Pa, 50; dir biol res, C H Boehringer Sohn, Ingelheim, WGer, 65-66; fel Royal Soc Can, 76- *Honors & Awards:* Upjohn Award, Pharmacol Soc Can, 81. *Mem:* Am Soc Pharmacol & Exp Therapeut; NY Acad Sci; Can Physiol Soc; Pharmacol Soc Can (secy-treas, 56-58, pres, 63-64); Ger Pharmacol Soc. *Res:* Bile secretion; serum cholinesterase; curare; local anesthetics; genetics and drug response. *Mailing Add:* Dept Pharmacol Univ Toronto Toronto ON M5S 1A8 Can

KALPAKJIAN, SEROPE, b Istanbul, Turkey, May 6, 28; US citizen; m 62; c 2. MANUFACTURING ENGINEERING. *Educ:* Robert Col, Istanbul, BSc, 49; Harvard Univ, SM, 51; Mass Inst Technol, SM, 53. *Prof Exp:* Res engr, Mass Inst Technol, 53-54; res supvr metal forming, Cincinnati Milacron, Inc, 57-63; PROF MECH & AEROSPACE ENG, ILL INST TECHNOL, 63- *Concurrent Pos:* Consult, Ill Inst Technol Res Inst, 63-80, Continental Can Co, 69-76, Xerox Corp, 79 & A Finkl & Sons, 84-86; co-ed, J Appl Metalworking, 78-86; assoc ed, J Tribology, 86-89, Mfg Rev, 90- *Honors & Awards:* Centennial Medallion, Am Soc Mech Engrs. *Mem:* Fel Am Soc Mech Engrs; fel Am Soc Metals; Soc Mfg Engrs; Int Inst Prod Eng Res. *Res:* Grinding and lubrication; machining and forming; manufacturing processes. *Mailing Add:* Dept Mech & Aerospace Eng Ill Inst Technol Ten W 32nd St Chicago IL 60616-3793

KALRA, JAWAHAR, b Aligarh, India, Apr 2, 49; Can citizen; m 86; c 1. CLINICAL PATHOLOGY, MEDICAL BIOCHEMISTRY. *Educ:* Aligarh Univ, India, BSc, 67, MSc, 69; Mem Univ Nfld, 72, PhD(biochem), 76; Can Soc Clin Chemists, cert, 86; Royal Col Physicians & Surgeons Can, cert, 86; FACB, 87; FRCPC, 87; FICA, 88; FACA, 89; FCACB, 89. *Prof Exp:* From asst prof to assoc prof, 85-91, actg head, 90-91, PROF & HEAD, DEPT PATH COL MED, UNIV SASK, 91- *Concurrent Pos:* Lab instr, 3rd & 4th yr Biochem Lab courses, Mem Univ, Nfld, 72-74, res asst, Mem Univ Res Unit, Fac med, 74-76, res assoc, 76-77; Burrough-Wellcome scholar, 79; jr resident, Dept Lab Med, Ottawa Civic Hosp, 82-83 & 84, Dept Med, 83, actg chief resident internal med, endocrinol & metab, Hosp, 84, sr resident, Dept Med, 84-85 & Dept Lab Med, 85; active med staff, Royal Univ Hosp, 85-, mem med adv comt, 90-; Schering travelling award, Can Soc Clin Invest, 88; dir, residency training progs gen & anat path, Univ Sask, 90-91, postgrad training prog, Dept Path, 90- & mem Adv Comt Acad Enrichment Progs, Col Med, 90- *Mem:* Int Soc Free Radical Res-Oxygen Soc; Am Soc Clin Pathologists; Int Soc Heart Research; NY Acad Sci; Nat Acad Clin Biochem; Am Asn Clin Chem; Can Med Asn. *Res:* Role of oxygen free radicals in various clinical diseases especially earth failure, atherosclecrsis and Parkinson's disease; earlier work on cardiac glycoside (digoxin) biotransformation and thyroid function testing. *Mailing Add:* Dept Path Royal Univ Hosp Saskatoon SK S7N 0X0 Can

KALRA, S(URINDRA) N(ATH), b Lahore, India, May 12, 27; nat Can; m 59; c 2. COMMUNICATIONS. *Educ:* Panjab Univ, India, BS, 46; Univ Ill, MS, 47, PhD, 50. *Prof Exp:* Reader in electronics, Phys Res Lab, India, 50-52; fel, Nat Res Coun Can, 52-53, head high frequency physics lab, Div Appl Physics, 53-62; assoc prof elec eng, Univ Windsor, 62-67, dir interdisciplinary studies in commun, 65-67; assoc prof, 67-69, PROF ELEC ENG, UNIV WATERLOO, 69- *Mem:* Sr mem Inst Elec & Electronics Engrs; Brit Inst Elec Engrs. *Res:* Communication sciences; electronics; computers. *Mailing Add:* Dept Elec & Computer Eng Univ Waterloo Waterloo ON N2L 3G1 Can

KALRA, SATYA PAUL, b Mari Indus, WPakistan, Jan 1, 39; m 69; c 1. NEUROENDOCRINOLOGY & ENDOCRINOLOGY, BEHAVIOR. *Educ:* Univ Delhi, BSc, 60, MSc, 62, PhD(physiol), 66. *Prof Exp:* Res asst physiol reproduction, Univ Delhi, 66-68; from asst prof to to assoc prof, 71-82, PROF REPRODUCTIVE BIOL, COL MED, UNIV FLA, 82- *Concurrent Pos:* Ford Found fel anat, Col Med, Univ Calif, Los Angeles, 68-69 & fel physiol, Southwestern Med Sch, Univ Tex Health Sci Ctr, Dallas, 69-71; NIH grant obstet & gynec, Univ Fla, 71- *Mem:* Endocrine Soc; Am Physiol Soc; Soc Gynec Invest; Int Soc Neuroendocrinol; Int Soc Neurosci; Am Andrology Soc; Soc Neurosci. *Res:* Neuroendocrinology of reproduction; regulation of pituitary gonadotropin functions by gonadal steroids; monoamines and hypothalamic releasing hormones; neuroendocrine and feeding and sexual behavior. *Mailing Add:* Dept Obstet & Gynec Univ Fla Gainesville FL 32610

KALRA, VIJAY KUMAR, b Multan, WPakistan, Aug 26, 42; m 71; c 2. BIOCHEMISTRY. *Educ:* Univ Delhi, BSc, 61, MSc, 63, PhD(chem), 67. *Prof Exp:* From asst prof to assoc prof, 71-84, PROF BIOCHEM & SEN, FAC SENATE, SCH MED, UNIV SOUTHERN CALIF, 84- *Concurrent Pos:* USDA res fel, Ctr Advan Studies Chem of Natural Prod, Univ Delhi, 67; USPHS res fel, Sch Med, Univ Southern Calif, 67-70. *Mem:* Am Chem Soc; Am Soc Biol Chem. *Res:* Structure and function of membranes; oxidative phosphorylation, mechanism of transport of amino acids in bacterial and mammalian cells; sterol metabolism in animal and human cells and relationship to atherosclerosis; structure and function of membranes of red blood cells and sickle cells; structure and function of endothelial cells. *Mailing Add:* Dept Biochem Univ Southern Calif Sch Med 2025 Zonal Ave Los Angeles CA 90033

KALSBECK, JOHN EDWARD, b Grand Rapids, Mich, May 20, 27; m 59; c 3. NEUROSURGERY. *Educ:* Calvin Col, AB, 49; Univ Mich, MD, 53; Am Bd Neurol Surg, dipl, 68. *Prof Exp:* From instr to assoc prof, 62-80, PROF SURG NEUROSURG, IND UNIV, INDIANAPOLIS, 80- *Concurrent Pos:* NIH fel, Nat Hosp, Queen Sq, London, 60-61; consult, New Castle State Hosp, 63- *Mem:* Am Asn Neurol Surg; Cong Neurol Surg. *Mailing Add:* Ind Univ Sch Med & Neurol Surg Riley Childrens Hosp Rm 213 702 Barnhill Dr Indianapolis IN 46223

KALSER, MARTIN, b Pittsburgh, Pa, Jan 7, 23; m 53; c 3. GASTROENTEROLOGY. *Educ:* Univ Pittsburgh, BS, 42, MD, 46; Univ Ill, MS, 51, PhD(physiol), 53; Am Bd Internal Med, dipl, 55; Am Bd Gastroenterol, dipl, 57. *Prof Exp:* From instr to assoc prof gastroenterol, Grad Sch Med, Univ Pa, 54-56; assoc prof, 59-63, PROF GASTROENTEROL & PHYSIOL, SCH MED, UNIV MIAMI, 63-, CHIEF DIV GASTROENTEROL, 61- *Concurrent Pos:* Res fel clin sci, Univ Ill, 50-53; res fel gastroenterol, Grad Sch Med, Univ Pa, 54-55; consult, Montefiore Hosp, Pittsburgh, 47-50, Vet Admin Clin, Dayton, Ohio & dean's comt, Col Med, Univ Cincinnati, 50-52, Cook County Hosp, Chicago, 52-53, Grad Hosp, Philadelphia, 54-55; mem comn enteric infections, Armed Forces Epidemiol Bd, 63-67. *Mem:* AMA; Am Gastroenterol Asn; Am Col Physicians. *Res:* Internal medicine; physiology. *Mailing Add:* 12145 SW 69th Pl Miami FL 33156

KALSER, SARAH CHINN, b Connellsville, Pa, June 11, 29; m 52. PHARMACOLOGY. *Educ:* Pa State Univ, BS, 51; Northwestern Univ, MS, 53; Univ Pittsburgh, PhD(pharmacol), 61. *Prof Exp:* Biochemist, Med Labs, US Army Chem Ctr, 53-58; res assoc pharmacol, Sch Med, Univ Pittsburgh, 58-60, from instr to asst prof, 61-68; LIVER DIS PROG DIR, NAT INST ARTHRITIS, DIABETES & DIGESTIVE & KIDNEY DIS, NIH, 68- *Mem:* AAAS; Am Gastroenterol Asn; Am Asn Study Liver Dis; Am Soc Exp Pharmacol & Therapeut. *Res:* Glutathione synthesis in trauma; atropine metabolism; drug metabolism in hypothermia and in cold acclimatization. *Mailing Add:* Westwood Bldg Rm 3A17 Nat Inst Diabetes Digestive & Kidney NIH Bethesda MD 20892

KALSNER, STANLEY, b New York, NY, Aug 21, 36; m 63; c 3. PHARMACOLOGY. *Educ:* NY Univ, AB, 58; Univ Man, PhD(pharmacol), 66. *Prof Exp:* Asst pharmacologist, Schering Corp, 62; from asst prof to prof pharmacol, Univ Ottawa, 67-85; PROF & CHMN DEPT PHYSIOL, CITY UNIV NEW YORK MED SCH, 85- *Concurrent Pos:* Fel pharmacol, Cambridge Univ, 66-67; Med Res Coun Can grant, 67-; Ont Heart Found grant, 70- *Mem:* Pharmacol Soc Can; Am Soc Pharmacol & Exp Therapeut. *Res:* Autonomic and cardiovascular pharmacology and physiology; biogenic amines; supersensitivity of autonomic effectors to drugs and denervation; receptor mechanisms; coronary artery disease; vascular smooth muscle; hypertension. *Mailing Add:* Dept Physiol City Univ New York Med Sch 138th St & Convent Ave New York NY 10031

KALSOW, CAROLYN MARIE, b Elgin, Ill, July 9, 43; m 81; c 5. MICROBIOLOGY, IMMUNOLOGY. *Educ:* Iowa State Univ, BS, 65; Univ Tex Med Br, MA, 67; Univ Louisville, PhD(microbiol), 70. *Prof Exp:* Instr microbiol, Univ Louisville, 70-71, lectr biol, 70-72, res assoc ophthal, 71-72, instr, 72-73, asst prof, 73-79, assoc prof, 79-81; adj assoc prof, Hope Col, 81-85; DIR CLIN OPHTHAL MICROBIOL, UNIV ROCHESTER, 86-; AT MISSION RES CORP, COLORADO SPRINGS & SR SCIENTISTS, UNIV ROCHESTER, 86- *Concurrent Pos:* Vis asst prof med microbiol, SFla Med Sch, 72. *Mem:* Am Ureitis Soc; Asn Res Vision & Ophthal; Sigma Xi; AAAS; Am Asn Immunol; Int Soc Eye Res. *Res:* Ocular immunology and microbiology. *Mailing Add:* Ophthal Dept Univ Rochester Box 314 Rochester NY 14642

KALT, MARVIN ROBERT, b Elizabeth, NJ, Aug 25, 45; m 67; c 1. RESEARCH GRANTS & CONTRACTS POLICY. *Educ:* Lafayette Col, AB, 67; Case Western Reserve Univ, PhD(anat), 71. *Prof Exp:* USPHS fel cell & molecular biol, dept biol, Yale Univ, 71-73; asst prof anat, Univ Conn Health Ctr, 73-80; exec secy, grants & contracts rev, Nat Inst Aging, NIH, 80- 82, chief, sci rev, 82-90, DEP DIR, DIV EXTRAMURAL ACTIV, NAT CANCER INST, NIH, 90-; MEM, SR EXEC SERV, US AM, 90- *Concurrent Pos:* Prin investr grant awards, USPHS, Nat Inst Child Health Human Develop, 73-77, NSF, 77-80; mem rev policy comt, Off of the Dir, NIH, 82-90, referral & training officer, Nat Inst Aging, 80-83. *Mem:* Am Soc Cell Biol; Sigma Xi; AAAS; Soc Res Adminrs; Soc Study Reproduction; Geront Soc Am. *Res:* Vertebrate germ cell development; vertebrate morphogenesis; cell biology; gerontology; federal grants and contract review and administration; human subjects and animal welfare policy; national scientific research policy. *Mailing Add:* Nat Cancer Inst NIH Bldg 31 RM 10A03 Bethesda MD 20892-4500

KALTENBACH, CARL COLIN, b Buffalo, Wyo, Mar 22, 39; m 64; c 2. REPRODUCTIVE PHYSIOLOGY. *Educ:* Univ Wyo, BSc, 61; Univ Nebr, MSc, 63; Univ Ill, PhD(animal physiol), 67. *Prof Exp:* Australian Wool Bd fel, Univ Melbourne, 67-69; actg head, Div Animal Sci, Univ Wyo, 78, assoc dir res, Col Agr, 80, assoc dean & dir, Wyo Agr Exp Sta, 84, prof animal physiol, 69-89; VICE DEAN & DIR, AGR EXP STA, UNIV ARIZ, 89- *Mem:* AAAS; Am Soc Animal Sci; Soc Study Fertil; Soc Study Reprod; Sigma Xi. *Res:* Luteotrophic and steroidogenotrophic properties of pituitary hormones; corpus luteum function; radioimmunoassay for protein and steroid hormones; experimental surgery; fetal growth and development. *Mailing Add:* Forbes Bldg Rm 314 Univ Ariz Tucson AZ 85721

KALTENBACH, JOHN PAUL, b Rockford, Ill, Feb 28, 20; m 53; c 1. BIOCHEMISTRY. *Educ:* Beloit Col, BS, 44; Univ Iowa, MS, 48, PhD, 50. *Prof Exp:* Chief cell metab lab, Vet Admin Res Hosp, Chicago, Ill, 54-56; asst prof, 56-60, Northwestern Univ, assoc prof path & biochem, 60-80, dir, Dent Sch Interview Prog, 77-79, actg dir admin, 78-79, prof path, 80-82, EMER PROF PATH, MED SCH, NORTHWESTERN UNIV, 82- *Concurrent Pos:* Brittingham fel cancer res, McArdle Mem Lab, Wis, 50-52; USPHS fel, Karolinska Inst, Stockholm, Sweden, 53-54. *Mem:* AAAS; Am Asn Cancer Res; Am Soc Exp Path; Am Soc Cell Biol; Asn Clin Sci. *Res:* Metabolism of whole cells, normal and neoplastic; metabolism of ischemic myocardium. *Mailing Add:* 8CR 58 171A Spooner WI 54801

KALTENBORN, HOWARD SCHOLL, b Pittsburgh, Pa, Jan 21, 07; m 37; c 2. MATHEMATICS. *Educ:* Carnegie Inst Technol, BS, 28; Univ Mich, MS, 31, PhD(math), 34. *Prof Exp:* Instr math, Carnegie Inst Technol, 29-32 & Univ Mich, 34-37; instr appl math, Univ Tex, 38-39; assoc prof math, La Polytech Inst, 39-43 & Univ Idaho, 45-46; prof & chmn dept, 46-72, EMER PROF MATH, MEMPHIS STATE UNIV, 72- *Res:* Mathematical analysis; mathematics education. *Mailing Add:* 169 S Mendenhall Rd Memphis TN 38117

KALTENBRONN, JAMES S, b New Baden, Ill, Nov 21, 34. ORGANIC CHEMISTRY. *Educ:* Univ Ill, BS, 56; Mass Inst Technol, PhD(org chem), 60. *Prof Exp:* RES CHEMIST, PARKE, DAVIS & CO, 60- *Mem:* Am Chem Soc. *Res:* Medicinal chemistry; natural products; stereochemistry. *Mailing Add:* Res Labs Parke Davis & Co Ann Arbor MI 48106-1047

KALTER, HAROLD, b New York, NY, Feb 26, 24; m 45; c 3. GENETICS, TERATOLOGY. *Educ:* Sir George Williams Col, BA, 49; McGill Univ, MSc, 51, PhD(genetics), 53. *Prof Exp:* From asst prof to assoc prof, 58-70, PROF RES PEDIAT, COL MED, UNIV CINCINNATI, 70-; RES ASSOC, CHILDREN'S HOSP RES FOUND, 55- *Concurrent Pos:* Nat Cancer Inst fel, McGill Univ, 53-55; mem human embryol & develop study sect, NIH, 66-70; ed, Teratol, 67-76; mem Secy's Comn Pesticides, 69; adv comt, 2, 4, 5-T, Environ Protection Agency, 71; consult, comt biol effects atmospheric pollutants, Nat Res Coun Panel Vapor Phase Organic Air Pollutants from Hydrocarbon, 71; mem adv comt, Dept Safety Assessment, Merck Inst Therapeut Res, 75-80; mem panel qual criteria water reuse, Nat Acad Sci, 79-81; ed, Issues & Rev in Teratology, 82- *Mem:* Teratology Soc. *Res:* Experimental mammalian teratology. *Mailing Add:* Children's Hosp Res Found Elland Ave & Bethesda Cincinnati OH 45229

KALTER, SEYMOUR SANFORD, b New York, NY, Mar 19, 18; m 46; c 3. VIROLOGY. *Educ:* St Joseph's Col, Pa, BS, 40; Univ Kans, MA, 43; Syracuse Univ, PhD(med bact), 47; Am Bd Microbiol, dipl. *Prof Exp:* Asst instr bact, Univ Kans, 41-43; asst med bact, Sch Med, Univ Pa, 43-45; from instr to assoc prof microbiol, State Univ NY Upstate Med Ctr, 45-56; chief, Virus Diag Methodology Unit, Commun Dis Ctr, USPHS, Ga, 56-61; chief, Virol Sect, US Air Force Sch Aerospace Med, Brooks AFB, Tex, 61-63; chmn, Dept Microbiol, 63- 66, dir, Div Microbiol & Infectious Dis, 66-88, DIR VIR REF LAB, 88- *Concurrent Pos:* Bacteriologist, Syracuse Dept Health, NY, 45-56; asst prof prev med, Sch Med, Emory Univ, 56-60; lectr, Med Sch, Baylor Univ, 61-63; adj prof, Trinity Univ & Univ Tex Health Sci Ctr, San Antonio, adj prof pediat & microbiol; adj prof, Dent Sci Inst, Univ Tex Health Sci Ctr, Houston; consult, Pan Am Sanit Bur; consult, Neurol Inst, Univ Cologne & Off Pesticide Progs, Environ Protection Agency; consult simian & pox viruses, WHO, chmn, comt simian viruses; consult virol, Univ Tex Syst Cancer Ctr, Houston; mem bd dir, Cancer Ther & Res Found, San Antonio; mem, Fedn US Culture Collections. *Mem:* Fel AAAS; Am Acad Microbiol; Soc Exp Biol & Med; Am Asn Immunol; fel Am Pub Health Asn. *Res:* Enteroviruses; respiratory viruses; oncogenic viruses; virus diagnosis; simian virology; comparative primate virology; oncogenic viruses; latent viruses. *Mailing Add:* Virus Ref Lab Inc 7540 Louis Pasteur San Antonio TX 78229

KALTHOFF, KLAUS OTTO, b WGer, Feb 5, 41. DEVELOPMENTAL BIOLOGY. *Educ:* Univ Hamburg, BA, 64; Univ Freiburg, MA, 67, PhD(zool), 71. *Prof Exp:* Asst prof zool, Univ Freiburg, 71-76, assoc prof, 76-77; assoc prof, 78-80, PROF ZOOL, UNIV TEX, AUSTIN, 80- *Mem:* Europ Develop Biologist Orgn; Soc Develop Biol; Am Soc Photobiol; AAAS. *Res:* Role of localized cytoplasmic determinants in embryogenesis using insect embryos and ultra violet irradiation, microinjection and molecular techniques. *Mailing Add:* Dept Zool Univ Tex Austin TX 78712

KALTOFEN, ERICH L, b Linz, Austria, Dec 21, 55; m 81. COMPUTER ALGEBRA, ALGORITHM DESIGN & ANALYSIS. *Educ:* Rensselaer Polytech Inst, MS, 79, PhD(comput sci), 82. *Prof Exp:* Lectr comput sci, Univ Del, 81-82; asst prof, Univ Toronto, 82-83; asst prof, 84-87, ASSOC PROF COMPUT SCI, RENNSELAER POLYTECH INST, 88- *Concurrent Pos:* Res assoc, Kent State Univ, 82; vis scientist, Tektronix Inc, 85; res fel, Math Sci Res Inst, 85; fac develop award, IBM, 85-87. *Mem:* Asn Comput Mkt (SIGSAM secy, 86-87, vchmn, 87-88); Soc Indust & Appl Math. *Res:* Computational algebra and number theory; design and analysis of sequential and parallel algorithms; symbolic manipulation systems and languages. *Mailing Add:* Dept Comp Sci Rensselaer Polytech Inst Troy NY 12180-3590

KALTON, ROBERT RANKIN, b Minneapolis, Minn, Oct 28, 20; m 42; c 5. CROP BREEDING. *Educ:* Univ Minn, BS, 41, Iowa State Col, MS, 45, PhD(crop breeding), 47. *Prof Exp:* Asst crop prod, Iowa State Col, 41-42, res assoc, 46-47; jr supvr grain inspection, Grain Mkt Br, USDA, 42-44, agent, Regional Soybean Lab, Ames, Iowa, 44-46; assoc agronomist, Tex Res Found, 47-50; from assoc prof to prof, farm crops, Iowa State Univ, 50-60; res dir, Rudy-Patrick Seed Div, W R Grace & Co, 60-70; dir agron res, Land O' Lakes Inc, 70-85; RES DIR, RES SEEDS INC, 85- *Concurrent Pos:* Mem, Nat Coun Com Plant Breeders, pres, 78-79; pres, Nat Alfalfa Improv Conf, 82-84; res agronomist, Vista, 87-; pres, Seed Asn, 79-80; vchmn, Nat Plant Genetics Resource Bd, 82-87. *Mem:* Fel AAAS; fel Am Soc Agron; Am Genetic Asn; Crop Sci Soc Am; Nat Plant Genetic Resource Bd (vchmn, 82-87). *Res:* Crop breeding, genetics and seed production, especially forage grasses, sorghums and soybeans. *Mailing Add:* 814 N Terrace Dr Webster City IA 50595

KALTREIDER, D FRANK, obstetrics; deceased, see previous edition for last biography

KALTSIKES, PANTOUSES JOHN, b Nigrita, Greece, Feb 19, 38; Can citizen; m 70; c 1. CYTOGENETICS, PLANT BREEDING. *Educ:* Univ Thessaloniki, BSc, 61; MSc, Univ Man, MSc, 66, PhD(plant sci), 68. *Prof Exp:* Agronomist, Hellenic Sugar Indust, 64-65; asst prof, Univ Man, 68-73, assoc prof cytogenetics, 73-77, assoc prof plant sci, 77-80; PROF PLANT BREEDING & BIOMETRY & HEAD DEPT, ATHENS AGR UNIV, GREECE, 80-, DEAN FAC AGR PROD, 85- *Concurrent Pos:* Nat Res Coun res grant, Univ Man, 68-74; vis prof, Univ Athens, Greece, 74-75; ed, Bull Genetics Soc Can, 74-; consult seed co, 80-85. *Mem:* Can Soc Agron; Genetics Soc Can; Crop Sci Soc Am. *Res:* Cytogenetics of synthetic amphiploids; biometrical genetics. *Mailing Add:* Agr Univ Athens IERA ODOS 75 Athens 11855 Greece

KALU, DIKE NDUKWE, b Nigeria, Jan 3, 38; m 67; c 3. PHYSIOLOGY. *Educ:* Univ London, BS, 67, PhD(biochem), 71. *Prof Exp:* Sci officer, Royal Postgrad Med Sch, Univ London, 67-71; fel Sch Med, Johns Hopkins Univ, 72-75; asst prof, 75-80, ASSOC PROF PHYSIOL, UNIV TEX HEALTH SCI CTR, SAN ANTONIO, 80- *Concurrent Pos:* Fel, Inst Med Lab Sci, UK. *Mem:* Inst Med Lab Sci UK; AAAS; The Endocrine Soc; Fed Am Soc Exp Biol; NY Acad Sci. *Res:* Hormonal control of calcium and skeletal metabolism and the effects of aging. *Mailing Add:* Univ Tex Health Sci Ctr 7703 Floyd Curl Dr San Antonio TX 78284

KALUZIENSKI, LOUIS JOSEPH, b Union Beach, NJ, July 28, 48. X-RAY ASTRONOMY. *Educ:* Rutgers Univ, BA, 70; Univ Md, MS, 74, PhD(physics), 77. *Prof Exp:* Grad res asst x-ray astron, Goddard Space Flight Ctr, Univ Md, 74-77, res assoc, 77-78; STAFF SCIENTIST HIGH ENERGY ASTROPHYS, NASA HQ, 78- *Mem:* Am Astron Soc. *Res:* Transient x-ray sources; x-ray binaries. *Mailing Add:* Code SZ NASA Hq Washington DC 20546

KALVINSKAS, JOHN J(OSEPH), b Philadelphia, Pa, Jan 14, 27; m 55; c 1. CHEMICAL ENGINEERING, COAL DESULFURIZATION. *Educ:* Mass Inst Technol, BS, 51, MS, 52; Calif Inst Technol, PhD(chem eng), 59. *Prof Exp:* Res engr, Eastern Lab, E I du Pont de Nemours & Co, NJ, 52-55, 59-60; asst chem eng, Calif Inst Technol, 55-59; res specialist, Rocketdyne Div, Rockwell Int Corp, 60-61, supvr basic studies, 61-64, group scientist propellant eng, 64-67, dir environ health systs, Life Sci Opers, 67-70; pres, Resource Dynamics Corp, 70-74; proj mgr, Holmes & Narver Inc, 74; TASK MGR & GROUP SUPVR, JET PROPULSION LAB, CALIF INST TECHNOL, 74- *Concurrent Pos:* Corp res dir, Monogram Indust, Inc, 72; Tech Contribs (8), NASA, 77-85. *Mem:* NY Acad Sci; Am Chem Soc; Am Inst Chem Engrs; Sigma Xi. *Res:* Chemical reaction kinetics; machine computation; chemical engineering process design; heat transfer; rocket propulsion; environmental engineering; coal beneficiation; bioconversion and bioenergy; hazardous material disposal; environmental monitoring. *Mailing Add:* 316 Pasadena Ave South Pasadena CA 91030

KALYAN-RAMAN, KRISHNA, b Madras, India, June 2, 35; m 63; c 2. NEUROLOGY. *Educ:* Univ Madras, MBBS, 58, DM, 71; Univ Delhi, MD, 62. *Prof Exp:* Asst prof med, Thanjavur Med Col, Madras Univ, 62-65; asst prof neurol & hon asst, Inst Neurol, Govt Gen Hosp, Madras, 69-71; clin asst prof, 71-72, from asst prof to assoc prof neurol, Sch Med, State Univ NY Buffalo, 72-76; assoc prof neurol, 76-83, PROF CLIN NEUROL, COL MED, UNIV ILL, PEORIA, 83- *Concurrent Pos:* Neurologist, E J Meyer Mem Hosp, Buffalo, 71-; neurologist, Out Patient Serv, Vet Admin Hosp, Buffalo, 72-; consult neurologist, West Seneca Develop Ctr, 73-; vis lectr, Inst Neurol, Govt Gen Hosp, 75-, mem staff, St Francis Hosp Med Ctr, Methodist Med Ctr Ill & Proctor Community Hosp, Peoria, 76-; dir, Muscular Dystrophy Asn Neuromuscular Clin, Col Med, Univ Ill, Peoria, 79- *Mem:* Neurol Soc India; Am Acad Neurol; fel Am Col Physicians; assoc Am Asn Electromyog & Electrodiag. *Res:* Nerve and muscle involvement in systemic disorders and effects of upper motor neurone lesions on histochemical pattern of muscle. *Mailing Add:* 5522 N Briarcrest Peoria IL 61614

KAM, GAR LAI, b Canton, China, Aug, 36; US citizen; m 69; c 1. DIGITAL SIGNAL, SPEECH ANALYSIS. *Educ:* Nat Taiwan Univ, BSEE, 61; Univ Tenn, MSEE, 66. *Prof Exp:* Sr electronics engr, Lockheed Co, Ga, 66-69; mem tech staff, Hughes Aircraft Co, 69-71; engr III, Jet Propulsion Lab, Calif Inst Technol, 71; sr engr, Martin-Marietta Aerospace Corp, 71-74; eng specialist, Singer-Kearfott Co, 74-80; prin engr, Xybion Corp, 80-; AT AT&T BELL LABS, WHIPPANY, NJ. *Mem:* Inst Elec & Electronics Engrs; sr mem Am Inst Aeronaut & Astronautics. *Res:* Developing models and algorithms of complex aerospace engineering applications and implementing these models and algorithms in comprehensive computer simulations for system performance analyses. *Mailing Add:* 16 Cold Hill Rd Morris Plains NJ 07950

KAM, JAMES TING-KONG, b Hong Kong, July 29, 45; m 74; c 2. COMPUTER MODELING, OPERATIONS RESEARCH. *Educ:* Univ Manitoba, Can, BSc, 70; Univ Calif, Berkeley, PhD(soil physics & hydrol), 74. *Prof Exp:* Res fel, Water Eng, Univ Calif, Davis, 74-75; hydrologist, Geol Eng, Morrison Knudsen, 75-79; sr engr hydrol, Sci Applns Inc, 79-81; staff specialist, consult, hydrol & geol, Davy McKee-Davy Inc, 81-85; PRIN HYDROLOGIST, MORRISON KNUDSEN, 85- *Concurrent Pos:* Geotech consult, 84-85. *Mem:* Am Soc Civil Engrs; Soc Petrol Engrs; Am Water Resources Asn; Sigma Xi. *Res:* Computer modeling and analyses of hydrologic and geological engineering problems; general civil (hydrologic) engineering design. *Mailing Add:* 2430 35th Ave San Francisco CA 94116

KAM, MOSHE, b Tel Aviv, Israel, Oct 3, 55. NEURAL NETWORKS, DECISION THEORY. *Educ:* Tel Aviv Univ, BSc, 77; Drexel Univ, MS, 85, PhD(elec eng), 87. *Prof Exp:* Res & develop engr, Israeli Defense Forces, 76-83; asst prof, 87-90, ASSOC PROF, ELEC & COMPUTER ENG DEPT, DREXEL UNIV, 90- *Concurrent Pos:* Vpres, Reshet Inc, 87-; asst dept head develop, Elec & Computer Eng, Drexel Univ, 90-; NSF presidential young investr, 90. *Mem:* Inst Elec & Electronics Engrs. *Res:* Synthesis of neural networks and sensor fusion architectures with objective functions that include computational and hardware complexities explicitly; modulation recognition over fading channels in digital communications. *Mailing Add:* Dept Elec & Computer Eng Drexel Univ Philadelphia PA 19104

KAMACK, H(ARRY) J(OSEPH), b Conn, Dec 5, 18. CHEMICAL ENGINEERING. *Educ:* Ga Inst Technol, BS, 41; Univ Del, MS, 56. *Prof Exp:* Chem engr, Gen Chem Co, 41 & Ord Dept, US Army, 42; chem engr, E I Du Pont De Nemours & Co, Inc, 42-46, res engr, 46-54, process design engr, 54-69, sr design consult, 69-73, prin design consult, 73-78; RETIRED. *Mem:* Am Inst Chem Engrs. *Res:* Chemical plant design; atomic energy design; particle size reduction and measurement. *Mailing Add:* 490 Stamford Dr Apt 304 Newark DE 19711-2774

KAMAL, ABDUL NAIM, b Dhaka, Bangladesh, Oct 28, 35; m 62; c 3. PARTICLE PHYSICS. *Educ:* Univ Dhaka, BSc, 55, MSc, 56; Univ Liverpool, PhD(theoret physics), 62. *Prof Exp:* Lectr physics, Univ Dhaka, 62-63; fel theoret physics, Univ Liverpool, 63 & Theoret Physics Inst, Edmonton, Alta, 63-64; from asst prof to assoc prof physics, 64-73, dir, Theoret Physics Inst, 79-80, chmn physics, 80-84, actg chmn, 88-89, McCalla prof, 90-91, PROF PHYSICS, UNIV ALTA, 73- *Concurrent Pos:* Sr sci officer, Rutherford Lab, UK, 68-69, vis scientist, 71 & 78; vis scientist, Int Ctr Theoret Physics, Trieste, Italy, 72; vis prof, Stanford Linear Accelerator Ctr, 79, vis scientist, 84-85. *Mem:* Am Phys Soc; Can Asn Physicists. *Res:* Theoretical particle physics; weak and strong interactions. *Mailing Add:* Dept Physics Univ Alta Edmonton AB T6G 2J1 Can

KAMAL, MOUNIR MARK, b Beirut, Lebanon, Feb 13, 36; US citizen; m 62; c 3. ENGINEERING MECHANICS. *Educ:* Robert Col, Istanbul, BS, 56; Univ Mich, Ann Arbor, MS, 58, MS, 62, PhD(eng), 65. *Prof Exp:* Assoc sr res engr, 65-67, sr res engr, 67-68, supv res engr, 68-71, asst head dept, 71-77, head dept eng mech, 77-82, tech dir mech & elec eng, 82-87, EXEC DIR ENG SCI, GEN MOTORS RES LABS, 88- *Concurrent Pos:* Mem, Univ Mich Indust Comt, 70-78. *Mem:* Sigma Xi; Soc Automotive Engrs; Am Soc Mech Engrs; Am Acad Mech; Soc Mfg Engrs. *Res:* Mechanical engineering; internal combustion engines; fluid mechanics and vehicle structural mechanics; vehicle crash dynamics. *Mailing Add:* 1615 Dutton Rd Rochester MI 48064

KAMAL, MUSA RASIM, b Tulkarm, Jordan, Dec 8, 34; m 61; c 2. POLYMER ENGINEERING, MATERIALS SCIENCE ENGINEERING. *Educ:* Univ Ill, BS, 58; Carnegie Inst Technol, MS, 59, PhD(chem eng), 62. *Prof Exp:* Teacher elem sch, Kuwait, 52-54; res chem engr, Stamford Res Labs, Am Cyanamid Co, Conn, 61-65, group leader, Wallingford Develop Lab, 65-67; assoc prof, 67-73, PROF CHEM ENG, MCGILL UNIV, 73-, CHMN CHEM ENG, 83- *Concurrent Pos:* Vis prof, Am Univ of Beirut, 74-75; Dir Microeconomics & Sectoral Sect, Morocco Indust Develop Plan, Dar Al-Handasah Consults, 79; pres, Talkarm Enterprises Ltd, 78-; dir, Brace Res Inst, 86- *Honors & Awards:* Int Educ Award, Soc Plastics Engr, 84; Kuwait Prize, Appl Sci & Technol, 83; CANPLAST Award, Soc Plastics Indust, Can, 85. *Mem:* Am Inst Chem Engrs; Am Chem Soc; Soc Plastics Engrs; Soc Rheol; NY Acad Sci; Sigma Xi; fel Chem Inst Can; Am Acad Mech; AAAS. *Res:* Polymer engineering; plastics processing; injection molding; rheology; heat transfer; non-Newtonian flow; thermoset and thermoplastic processing; weatherability of plastics systems; properties of polymers; microstructure development and control in plastics processing; blends; composites; computer simulation; project evaluation and planning. *Mailing Add:* Dept Chem Eng McGill Univ 3480 University St Montreal PQ H3A 2A7 Can

KAMAN, CHARLES HENRY, b Brookline, Mass, 43. COMPUTER SCIENCE. *Educ:* Harvard Col, AB, 65; Polytech Inst Brooklyn, MS, 67, PhD(syst sci), 74. *Prof Exp:* Consult engr comput archit, Digital Equip Corp, Tewksbury, 69-81. *Mem:* Asn Comput Mach; Inst Elec & Electronics Engrs; Soc Indust & Appl Math; Am Math Soc; Math Asn Am. *Res:* Computer architecture; computer implementation; programming languages and semantics; computer algorithms; application of theoretical computer science to practical computer design and implementation. *Mailing Add:* 274 Dedham St Newton Highlands MA 02161

KAMAN, CHARLES HURON, b Washington, DC, June 15, 19; m 45; c 3. AERONAUTICAL ENGINEERING. *Educ:* Catholic Univ, BAeroE, 40. *Hon Degrees:* DSc, Univ Colo, 84 & Univ Hartford, 85; LLD, Univ Conn, 85. *Prof Exp:* Aerodyn, Hamilton Standard Div, United Aircraft Corp, 40-45; pres, 45-90, CHMN & CHIEF EXEC OFFICER, KAMAN CORP, 90- *Concurrent Pos:* Chmn, Helicopter Coun, Aerospace Indust Asn, 54 & Vertical Lift Aircraft Coun, 64; bd regents, Catholic Univ Am, bd gove; bd dirs, Conn Bus & Indust Asn; dir, Emhart Corp, Hartford Nat Corp, Conn Nat Bank, Inst Living & Security Conn Ins Co; adv bd, World Affairs Ctr Honors; founder, Univ Hartford; indust comt mem, Greater Hartford YMCA; pres & dir, Fidelco Guide Dog Found Coun, 64; corporator, Inst Living, Hartford Hosp. *Honors & Awards:* Asn Award, Navy Helicopter Asn, 75; Dr Alexander Klemin Award, Am Helicopter Soc, 81; Fleet Adm Chester W Nimitz Award, Navy League US, 86. *Mem:* Nat Acad Eng; hon mem Navy Helicopter Asn; hon fel Am Helicopter Soc; fel Am Inst Aeronauts & Astronauts. *Mailing Add:* Kaman Corp Blue Hills Ave PO Box 1 Bloomfield CT 06002

KAMAN, ROBERT LAWRENCE, b New York, NY, June 26, 41; m 67; c 2. BIOCHEMISTRY. *Educ:* Univ Pa, AB, 63; Va Polytech Inst, MS, 67, PhD(biochem), 69. *Prof Exp:* NIH fel biochem, Sch Med, Univ Mich, 69-72, res assoc, Sch Pub Health, 72-73; asst prof biochem, 73-77, ASST PROF BASIC HEALTH SCI, TEX COL OSTEOP MED, NTEX STATE UNIV, 77- *Mem:* Am Osteop Acad Sports Med; Am Col Sports Med; Am Asn Fitness Dirs. *Res:* Chemotherapy for atherosclerosis; effect of exercise on health and fitness; diet and exercise; exercise programs for firemen; exercise and alcohol rehabilitation. *Mailing Add:* Dept Physiol Tex Col Osteop Med 3500 Camp Bowie Blvd Ft Worth TX 76107-2690

KAMAT, PRASHANT V, b Binaga, Karnataka, India, Jul 6, 53; m 83; c 2. COLLOID & SURFACE SCIENCE. *Prof Exp:* Res asst, catalysis, Hindustan Lever Res Ctr, Bombay, 77-79; res assoc phys chem, chem dept, Boston Univ, 79-81, Univ Tex, Austin, 81-83; asst prof specialist, 83-88, PRIN INVEST, NOTRE DAME RADIATION LAB, 83-, ASSOC PROF SPECIALIST PHYS CHEM, 88- *Concurrent Pos:* Reviewer, Am Chem Soc Pubs, 83- *Mem:* Sigma Xi; Am Chem Soc; Soc Electroanal Chem; Electrochem Soc; Interamerican Photochemical Soc. *Res:* Investigating dynamics of interfacial processes in semiconductor particulate systems; surface photochemistry and heterogeneous photocatalysis; conducting polymers; polymer modified electrodes; excited state behavior of polymers and dyes; electrochemistry and interfacial processes; solar energy conversion. *Mailing Add:* Radiation Lab Univ Notre Dame Notre Dame IN 46556

KAMATH, KRISHNA, b Shertallai, India, Aug 24, 20; m 69. PETROLEUM ENGINEERING, PHYSICAL CHEMISTRY. *Educ:* Univ Travancore, India, BSc, 41; Banaras Univ, MSc, 44; Pa State Univ, MS, 57, PhD(petrol eng), 60. *Prof Exp:* Anal chemist, Govt India, 44-46; res chemist, Alembic Chem Works, India, 47-49; prof chem, Petlad Col, India, 49-52; sr sci asst phys chem, Nat Chem Lab India, Poona, 52-54; res engr, Gulf Res & Develop Co, 59-61; asst prof petrol technol & chmn dept, Indian Sch Mines & Appl Geol, Dhanbad, 61-63; vis assoc prof petrol eng, Stanford Univ, 63-66; res engr, Continental Oil Co, Okla, 66-68 & IIT Res Inst, 69-72; environ protection engr, Ill Environ Protection Agency, 74-76; PETROL ENGR, MORGANTOWN ENERGY TECHNOL CTR, US DEPT ENERGY, 76- *Concurrent Pos:* Adj assoc prof petrol eng, WVa Univ, Morgantown, 77- *Mem:* AAAS; Am Chem Soc; Soc Petrol Engrs; Am Water Works Asn. *Res:* Preparative electrochemistry; education; surface and colloid chemistry relating to petroleum recovery and multiphase fluid flow through porous media; enhanced petroleum recovery; surface and colloid chemistry. *Mailing Add:* 1519 W Polk St Chicago IL 60607-3120

KAMATH, SAVITRI KRISHNA, b Kanhangad, India, Sept 22, 30; US citizen; m 69. BIOCHEMISTRY, CLINICAL NUTRITION. *Educ:* Bombay Univ, BSc, 50; Univ Baroda, MSc, 63; Iowa State Univ, PhD(nutrit), 67. *Prof Exp:* Lectr chem, St Agnes Col, India, 51-58; reader nutrit, Univ Baroda, 67-69; res biochemist, Hektoen Med Res Inst, Chicago, 71-72; from asst prof to assoc prof, 72-81, PROF NUTRIT, UNIV ILL MED CTR, 81-, DEPT HEAD, 79- *Mem:* Am Dietetic Asn; Soc Nutrit Educ; NY Acad Sci; fel Am Col Nutrit; Am Inst Nutrit. *Res:* Ascorbic acid, lipid metabolism; nutrition and cancer; nutritional assessment of population groups; dietetic education and practice. *Mailing Add:* 1519 W Polk St Chicago IL 60607

KAMATH, VASANTH RATHNAKAR, b Mangalore, India, July 16, 44; m 74; c 2. POLYMER CHEMISTRY. *Educ:* Univ Bombay, BS, 66; Univ Akron, MS, 68, PhD(polymers), 73. *Prof Exp:* Group leader, Lucidol Dir, Pennwalt Corp, 73-77; MGR APPLNS, ORG PEROXIDE DIV, ATOCHEM NAM, 77- *Res:* Cellular plastics, rubber reinforced polymer systems, novel free radical initiators including polymeric peroxides, emulsion polymerization characteristics and polymerization processes in general including polymer-crosslinking, acrylic coatings and stabilizers. *Mailing Add:* Org Peroxide Div Atochem NAm 1740 Mil Rd Buffalo NY 14240-1048

KAMATH, YASHAVANTH KATAPADY, b Katapady, India, Apr 15, 38; m 72; c 1. PHYSICAL CHEMISTRY, POLYMER CHEMISTRY. *Educ:* Univ Bombay, BSc, 59 & 61, MSc, 64; Univ Conn, PhD(phys chem), 73. *Prof Exp:* Assoc lectr plastic technol, Dept Chem Technol, Univ Bombay, 63-66; fel, 72-74, staff scientist, 74-76, sr scientist, 76-87, PRIN SCIENTIST, TEXTILE FINISHING, TEXTILE RES INST, 87- *Honors & Awards:* Lit Award, Soc Cosmetic Chemists. *Mem:* Am Chem Soc; Fiber Soc. *Res:* Surface chemical properties of human hair; effect of polymers and surfactants on the surface wettability of human hair; fractography of human hair; compressibility of fiber bundles; environmental fading of dyes in nylon; microspectrophotocetry of dyes in monofilaments; mechanisms of formaldehyde release in durable press fabric; finish distribution in fibers and textiles. *Mailing Add:* Textile Res Inst 601 Prospect Ave Princeton NJ 08540

KAMB, WALTER BARCLAY, b San Jose, Calif, Dec 17, 31; m 57; c 4. MINERALOGY, GLACIOLOGY. *Educ:* Calif Inst Technol, BS, 52, PhD(geol), 56. *Prof Exp:* From asst prof to assoc prof geol, 56-62, chmn div geol & planetary sci, 72-83, vpres & provost, 87-89, PROF GEOL & GEOPHYS, CALIF INST TECHNOL, 62- *Concurrent Pos:* Guggenheim Mem Found fel, 60; Sloan fel, 63. *Honors & Awards:* MSA Award, Mineral Soc Am, 68; Seligman Award, Int Glaciol Soc, 77. *Mem:* Fel Nat Acad Sci; Geol Soc Am; Am Geophys Union; Mineral Soc Am; Am Asn Petrol Geologists; AAAS; fel Am Acad Arts & Sci. *Res:* Crystallography; tectonophysics; structural geology and petrology; glaciology; mineralogy and x-ray crystallography; crystal optics. *Mailing Add:* Div Geol & Planetary Sci Calif Inst Technol 1201 E Cal Blvd Pasadena CA 91109

KAMBARA, GEORGE KIYOSHI, b Sacramento, Calif, Feb 23, 16; m 41; c 4. OPHTHALMOLOGY. *Educ:* Stanford Univ, AB, 37, MD, 41; Am Bd Ophthal, dipl, 47. *Prof Exp:* Teaching asst surg, Sch Med, Stanford Univ, 41-42; instr ophthal, Med Sch, Univ Wis, 46-48; asst prof & instr ophthal, Sch Med, Loma Linda Univ, 49-56, from assoc clin prof to clin prof, 56-59; chief ophthal serv, Rancho Los Amigos Hosp, 65-87; clin prof ophthal, Univ Calif, Irvine, Calif Col Med, 66-69; clin prof ophthal, 69-88, EMER PROF OPHTHAL, SCH MED, UNIV SOUTHERN CALIF, 88-; EMER PROF OPHTHAL, SCH MED, LOMA LINDA UNIV, 87- *Concurrent Pos:* Mem attend staff, Los Angeles County Gen Hosp, 50-67, Olive View Gen Hosp, Rancho Los Amigos Hosp; chief ophthal serv, White Mem Med Ctr, 58-65, chmn opthal dept, 65-83. *Mem:* AMA; fel Am Col Surgeons; Am Ophthal Soc. *Mailing Add:* 2232 Silver Lake Blvd Los Angeles CA 90039

KAMBAYASHI, TATSUJI, b Kyoto, Japan, Nov 8, 33; m 63. ALGEBRAIC GEOMETRY. *Educ:* Univ Tokyo, ScB, 57; Northwestern Univ, PhD(math), 62. *Prof Exp:* Instr math, Brown Univ, 61-63; lectr, Ind Univ, 63-64, asst prof, 64-67; assoc prof, 67-74, PROF MATH, NORTHERN ILL UNIV, 74- *Concurrent Pos:* Mem res staff, Res Ctr Physics & Math, Pisa, Italy, 65-67; res mem, Res Inst Math Sci, Kyoto Univ, Japan, 72-73; res grant, NSF, 73- *Mem:* Am Math Soc; Math Soc Japan. *Res:* Algebraic groups. *Mailing Add:* Math Sci Dept Tokyo Denki Univ Hiki-gun Saitama 350-03 Japan

KAMBOUR, ROGER PEABODY, b Wilmington, Mass, Apr 1, 32; m; c 3. PHYSICAL CHEMISTRY, POLYMER PHYSICS. *Educ:* Amherst Col, BA, 54; Univ NH, PhD(chem), 60. *Prof Exp:* Res assoc, 60-70, mgr polymer studies unit, 70-74, RES ASSOC, GENERAL ELECTRIC RES & DEVELOP CTR, 75- *Honors & Awards:* Union Carbide Chem Award, Am Chem Soc, 68; Ford High Polymer Physics Prize, Am Phys Soc, 85. *Mem:* Fel Am Phys Soc; Am Chem Soc. *Res:* Diffusion of gases and vapors in polymers; polymer crazing and fracture; properties of block polymers; crystallization; polymer flame retardance; polymer blend thermodynamics; mobility and toughness in plasticized resins. *Mailing Add:* Polymer Physics & Eng Lab Gen Elec Co PO Box 8 Schenectady NY 12301

KAMBYSELLIS, MICHAEL PANAGIOTIS, b Antissa, Greece, Mar 1, 35. DEVELOPMENTAL GENETICS. *Educ:* Nat Univ Athens, BSc, 60; Yale Univ, MS, 65; Univ Tex, PhD(zool), 67. *Prof Exp:* Res asst genetics, Univ Tex, 65-67, res assoc, 67-68; res fel insect physiol, Harvard Univ, 68-70, lectr biol, 71; asst prof, 71-73, assoc prof develop biol, 73-80, PROF BIOL, NY UNIV, 80- *Concurrent Pos:* Vis prof, Athens Univ, Greece, 74-75. *Mem:* AAAS; Genetics Soc Am; Am Inst Biol Sci; NY Acad Sci; Soc Develop Biol. *Res:* Physiological genetics; Drosophila genetics and evolution; insect tissue transplantations; Drosophila ovarian development; insect tissue cultures; hormonal control of insect reproduction. *Mailing Add:* Dept Biol NY Univ 952 Brown Blvd New York NY 10003

KAMDAR, MADHUSUDAN H, b Bombay, India, Mar 28, 30; m 63; c 1. MATERIALS SCIENCE & ENGINEERING. *Educ:* Univ Bombay, BSc, 53; Univ Wash, MS, 57; Mass Inst Technol, DSc(metall), 61. *Prof Exp:* Scientist, Res Inst Advan Study, 61-74; SR RES SCIENTIST, BENET WEAPONS LAB, WATERVLIET ARSENAL, 74- *Mem:* Am Soc Metals; Am Inst Mining, Metall & Petrol Engrs; Am Soc Testing & Mat. *Res:* Environmental sensitive fracture behavior of materials. *Mailing Add:* 41 Longview Dr Elmira NY 12189

KAMEGAI, MINAO, b Koshu, Korea, July 7, 32; US citizen; m 58; c 2. COMPUTATIONAL PHYSICS. *Educ:* Univ Hawaii, BA, 57; Univ Chicago, MS, 60, PhD(physics), 63. *Prof Exp:* Res assoc nuclear physics, Enrico Fermi Inst, Univ Chicago, 63; physicist, Knolls Atomic Power Lab, Gen Elec Co, 63-66; SR PHYSICIST, LAWRENCE LIVERMORE LAB, 66- *Concurrent Pos:* Vis scientist, Nat Chem Lab, Tsukuba Res Ctr, Tsukuba City, Ibaraki, Japan. *Mem:* Am Phys Soc; Sigma Xi. *Res:* Theoretical and computational physics; materials science, hydrodynamics and laser optics; computer modeling in shock hydrodynamics. *Mailing Add:* Lawrence Livermore Lab PO Box 808 Livermore CA 94550

KAMEGO, ALBERT AMIL, b Detroit, Mich, June 11, 41; m 73. PHYSICAL ORGANIC CHEMISTRY. *Educ:* Wayne State Univ, BS, 64; Calif State Univ, Long Beach, MS, 68; Univ Calif, Santa Barbara, PhD(chem), 74. *Prof Exp:* Chemist paints, Ford Motor Co, 65-67; teaching asst chem, Calif State Univ, Long Beach, 67-68; from teaching asst chem to staff res assoc, Univ Calif, Santa Barbara, 68-73; fel chem, State Univ NY Buffalo, 73-75; lectr chem, Univ Mont, 75-76; instr chem, Univ Tex, Arlington, 76-77; sr res chemist, 77-85, LTV MISSILES & ELECTRONICS GROUP, CORE LABS, INC, 85- *Mem:* Sigma Xi; Soc Petrol Engrs; AAAS; Am Chem Soc; Am Ceram Soc; Soc Advan Mat Process Eng. *Mailing Add:* 832 Spring Brook Dr Bedford TX 76021

KAMEL, HYMAN, b Philadelphia, Pa, Dec 2, 19; m 41, 68; c 2. MATHEMATICS. *Educ:* Univ Pa, AB, 41, NY Univ, MS, 44; Univ Pa, PhD(math), 52. *Prof Exp:* Instr math, Univ Pa, 50-52 & Cornell Univ, 52-54; asst prof, Rensselaer Polytech Inst, 54-61; assoc prof, Howard Univ, 61-67; PROF MATH, WIDENER COL, 67- *Mem:* Am Math Soc; Math Asn Am; Sigma Xi. *Res:* Functional analysis. *Mailing Add:* 2316 Waverly St Philadelphia PA 19146

KAMEMOTO, FRED ISAMU, b Honolulu, Hawaii, Mar 8, 28; m 63; c 3. ZOOLOGY, COMPARATIVE ENDOCRINOLOGY. *Educ:* George Washington Univ, AB, 50, MS, 51; Purdue Univ, PhD(zool), 54. *Prof Exp:* Res assoc zoophysiol, Wash State Col, 57-59; asst prof zool, Univ Mo, 59-62; from asst prof to assoc prof, 62-69, PROF ZOOL, UNIV HAWAII, 69- *Concurrent Pos:* Vis res scholar, Ocean Res Inst, Univ Tokyo, 68-69; vis scholar, Dept Biol, Wesleyan Univ, Middletown, Conn, 75-76; vis sr scientist, Dept Fisheries, Nihon Univ, Tokyo, 86. *Mem:* AAAS; Am Soc Zoologists; Sigma Xi; Zool Soc Japan; Western Soc Naturalists; World Aquaculture Soc. *Res:* Neurosecretion and osmoregulation; crustacean biology. *Mailing Add:* Dept Zool Univ Hawaii 2538 The Mall Honolulu HI 96822

KAMEMOTO, HARUYUKI, b Honolulu, Hawaii, Jan 18, 22; m 52; c 3. HORTICULTURE. *Educ:* Univ Hawaii, BS, 44, MS, 47; Cornell Univ, PhD, 50. *Prof Exp:* From asst horticulturist to horticulturist, 50-86, prof hort & chmn dept, 69-75, EMER PROF HORT, UNIV HAWAII, 86- *Concurrent Pos:* Fulbright award, 56-57; consult, Food & Agr Orgn, UN, 71, 80. *Honors & Awards:* Gold Medal, Malayan Orchid Soc, 63; Norman Jay Colman Award, 77; Orchid Soc Thailand Medal of Honor, 78; Alex Laurie Award, 82; Childers Award, 84; Gold Medal, Am Orchid Soc, 89. *Mem:* Fel AAAS; fel Am Soc Hort Sci; fel Am Orchid Soc; Int Aroid Soc; Am Genetic Asn; Bot Soc Am. *Res:* Cytogenetics and breeding of tropical ornamentals. *Mailing Add:* Dept Hort Univ Hawaii 3190 Maile Way Honolulu HI 96822

KAMEN, EDWARD WALTER, b Mansfield, Ohio, Oct 2, 45. ENGINEERING, MATHEMATICS. *Educ:* Ga Inst Technol, BEE, 67; Stanford Univ, MS, 69, PhD(elec eng), 71. *Prof Exp:* Engr, Argo Systs Inc, Calif, 70-71; asst prof, 71-76, assoc prof elec eng, Ga Inst Technol, 76-; AT DEPT ELEC ENG, UNIV FLA, GAINESVILLE. *Concurrent Pos:* NSF res initiation grant, Ga Inst Technol, 72-74; res specialist, Inst Res Info & Automatic Control, France, 72- *Mem:* AAAS; Inst Elec & Electronics Engrs. *Res:* Mathematical system theory; network theory; receiver systems. *Mailing Add:* Elec Eng Dept Univ Pittsburgh Pittsburgh PA 15261

KAMEN, GARY P, MOTOR CONTROL, NEUROMUSCULAR PHYSIOLOGY. *Educ:* Univ Mass, PhD(exercise sci), 79. *Prof Exp:* ASSOC RES PROF, BOSTON UNIV, 88- *Mem:* AAAS; fel Am Col Sports Med; Int Soc Biomech; Soc Neurosci; Sigma Xi; Am Physiol Soc. *Res:* Neuromuscular physiology and motor control with applications in gerontology, sports medicine and rehabilitation medicine; consulting in sports medicine; exercise science. *Mailing Add:* Boston Univ 635 Commonwealth Ave Boston MA 02215

KAMEN, MARTIN DAVID, b Toronto, Ont, Aug 27, 13; nat US; wid; c 1. PHYSICAL BIOCHEMISTRY. *Educ:* Univ Chicago, BS, 33, PhD(phys chem), 36. *Hon Degrees:* ScD, Univ Chicago, 69; hon Dr, Univ Paris, 69; ScD, Wash Univ, 77, Univ Ill, Chicago Circle, 78, Univ Freiburg, Ger 79, Brandeis Univ, 88. *Prof Exp:* Fel nuclear chem, Radiation Lab, Univ Calif, 37-39, res assoc, 39-41; marine test engr, Kaiser Cargo, Calif, 44-45; assoc prof biochem, Wash Univ, 45-46, assoc prof chem & chemist, Mallinckrodt Inst, 45-57; prof biochem, Brandeis Univ, 57-61; prof chem, Univ Calif, San Diego, 61-74, chmn dept, 71-73; prof 74-78, EMER PROF BIOL SCIV SOUTHERN CALIF, 78-; EMER PROF CHEM, UNIV CALIF, SAN DIEGO, 78- *Concurrent Pos:* NSF sr fel, 56; Guggenheim fel, 56 & 72; Fogarty Scholar, 88-89. *Honors & Awards:* Am Chem Soc Award, 63; C F Kettering Award, Am Soc Plant Physiol, 69; Merck Award, Am Soc Biol Chemists, 82. *Mem:* Nat Acad Sci; Am Chem Soc; Am Soc Biol Chemists; Am Acad Arts & Sci; fel Am Inst Chem; Am Philos Soc. *Res:* Application of biophysical chemical methods, including isotopic tracer methodology, to study bacterial metabolism; energy storage, especially in photosynthesis; comparative biochemical iron proteins. *Mailing Add:* Dept Chem B-017 Univ Calif-San Diego La Jolla CA 92093

KAMENETZ, HERMAN LEO, b Kovno, Russia, Sept 1, 07; US citizen; m 47. PHYSICAL MEDICINE & REHABILITATION. *Educ:* Univ Paris, BA, 44, MS, 45, MD, 52. *Prof Exp:* Intern med & surg, St Anthony's Hosp, Rockford, Ill, 53-54; resident phys med & rehab, State of Conn Vet Hosp, Rocky Hill, 54-56, Yale Univ Med Ctr, 56-57 & Ga Warm Springs Found, 58; psychiatrist, Woodruff Rehab Ctr, New Haven, Conn, 58-59; chief of staff, State of Conn Vet Home & Hosp, 60-66, chief phys med & rehab, 59-75; chief, 75-83, CONSULT, REHAB MED SERV, US VET ADMIN HOSP, WASHINGTON, DC, 83- *Concurrent Pos:* Physicians Recognition Awards, AMA, 70-73, 73-76, 76-79 & 79-82; asst attend physician, Med Ctr, Yale Univ, 57-60, physician, Outpatient Dept, 60-75, clin instr med, Sch Med, 58-64, asst clin prof phys med, 64-68; consult phys med, St Francis Hosp, Hartford, 64-75, Waterbury Hosp, 68-75 & Gaylord Rehab Hosp, Wallingford, 70-75; clin prof med, George Washington Univ Sch Med & Health Sci, Washington, DC, 75-, prof lectr phys med & rehab, 76- *Honors & Awards:* Silver Award Exhibit, Am Cong Rehab Med, Montreal, 68. *Mem:* Int Soc Hist Med; Int Rehab Med Asn; Am Med Writers Asn; emer mem Am Cong Rehab Med; hon mem Int Balneological Soc. *Res:* Lexicography in English and French in physical medicine and rehabilitation; translation and editing. *Mailing Add:* 4501 Arlington Blvd No 824 4501 Arlington Blvd Arlington VA 22203

KAMEN-KAYE, MAURICE, b London, Eng, Aug, 17, 05. GLOBAL PALEOGEOGRAPHY. *Educ:* Royal Col Sci, London, BSc, ARES, 26; Royal Sch Mines, London, ARSM(petrol geol), 29. *Prof Exp:* Explor geologist, Caracas Petrol Corp, Venezuela, 30-40, chief geologist, 40-50; consult geologist, US & Canada, 50-53; chief geologist, sedimentary sect, Province Sask, Can, 54; explor res geologist, Conorada Petrol Corp, NY, 65-67; explor res geologist, Amerada Petrol Corp, Tulsa, 65-67; CONSULT GEOLOGIST, 68- *Mem:* Fel Geol Soc Am; Am Asn Petrol Geologists; Soc Explor Geophys. *Res:* Architecture & petroleum productivity of sedimentary basins; global paleogeography; petroleum geology of African borderlands of the Indian Ocean. *Mailing Add:* 8401 Waterhous St No 5 Cambridge MA 02138

KAMENTSKY, LOUIS A, b Newark, NJ, July 28, 30; m 55; c 3. ENGINEERING PHYSICS, BIOPHYSICS. *Educ:* Newark Col Eng, BS, 52; Cornell Univ, PhD(eng physics), 56. *Prof Exp:* Res asst physics, Brookhaven Nat Lab, 51; res asst eng physics, Cornell Univ, 52-54; mem staff, Electronics Res Lab, Columbia Univ, 54-55; mem staff, Bell Tel Labs, Inc, 56-60 & Watson Lab, IBM Corp, 60-68; pres, Bio/Physics Systs, Inc, 68-76; vpres res & develop, Ortho Instruments, 76-80, vpres res, Ortho Diagnostics Systs, 80-88; dir, Cambridge Res Lab, 83-88; PRES, COMPUCYTE CORP, 88- *Concurrent Pos:* Physicist, Res Lab, US Steel Corp, 53; vis scientist, Karolinska Inst, 66; adj assoc prof, dept path, Med Ctr, NY Univ, 69-73; consult, dept path, Mem Hosp for Cancer, NY, 73-; sr res scientist, Res Lab Electronics, Mass Inst Technol, 80-88. *Mem:* AAAS; Inst Elec & Electronics Engrs; NY Acad Sci. *Res:* Information and computer theories; solid state physics; optics; pattern recognition; medical instrumentation; research administration. *Mailing Add:* 180 Beacon St Boston MA 02116

KAMIEN, C ZELMAN, b Bialystock, Poland, Feb 24, 27; US citizen; m 55; c 2. MECHANICAL ENGINEERING. *Educ:* Purdue Univ, BS, 55, MSEng, 56, PhD(eng), 60. *Prof Exp:* Instr mech eng, Purdue Univ, 58-60; sr scientist, Res & Adv Develop Div, Avco Corp, 60-63; proj mgr thermionics, Thermo Electron Eng Corp, 63-66; assoc prof mech eng, Lowell Technol Inst, 66-81, ASSOC PROF & CHMN, MECH ENG DEPT, UNIV LOWELL, 81- *Mem:* AAAS; Am Soc Mech Engrs; Am Soc Eng Educ; Sigma Xi. *Res:* Effect of pressure and temperature on viscosity of refrigerants; effect of rain intensity on prevention of icing on transmission cables; solar thermionic power systems development. *Mailing Add:* 29 Arbutus Ave Chelmosford MA 01824

KAMIEN, ETHEL N, b New York, NY, July 1, 30; m 55; c 2. PLANT PHYSIOLOGY, HUMAN SEXUALITY. *Educ:* Brooklyn Col, BA, 50; Univ Wis, MS, 52, PhD(bot), 54. *Prof Exp:* Asst bot, Univ Wis, 50-54; instr hort, Purdue Univ, 54-58; from instr to assoc prof biol, 60-66, chmn dept biol & phys sci, 65-75, PROF BIOL, UNIV LOWELL-NORTH CAMPUS, 66- *Concurrent Pos:* Coop teacher, Ed Serv, Inc, 65-; mem, US Sex Info & Educ Coun. *Mem:* Am Soc Plant Physiol; Am Inst Biol Sci; Am Asn Sex Educrs, Counrs & Therapists; Sigma Xi; AAAS. *Res:* Chemical control of plant growth; plant tissue culture. *Mailing Add:* 29 Arbutus Ave Chelmsford MA 01824

KAMIL, ALAN CURTIS, b Bronx, NY, Nov 20, 41; m 63; c 2. ANIMAL BEHAVIOR, BEHAVIORAL ECOLOGY. *Educ:* Hofstra Univ, BA, 63; Univ Wis-Madison, MS, 66, PhD(psychol), 67. *Prof Exp:* From asst prof to assoc prof, 67-79, PROF PSYCHOL & ZOOL, UNIV MASS, AMHERST, 79- *Concurrent Pos:* Vis assoc prof psychol, Univ Calif, Berkeley, 76-77; prin investr grants, NSF, 71- & NIH, 78-79 & 88- *Mem:* Animal Behav Soc; Ecol Soc Am; Am Ornithologists Union; fel Am Psychol Asn. *Res:* Mechanisms of foraging behavior. *Mailing Add:* Dept Psychol Univ Mass Amherst MA 01003

KAMILLI, DIANA CHAPMAN, b New York, NY, Sept 5, 41; m 69; c 2. ARCHAEOMETRY, PETROLOGY. *Educ:* Vassar Col, BA, 63; Rutgers Univ, MS, 66, PhD(igneous petrol, metamorphic petrol), 68. *Prof Exp:* Instr geol, Vassar Col, 68; asst prof mineral, City Col New York, 68-69; asst prof geol, Wellesley Col, 69-75, chmn dept geol, 69-74; res assoc archeol mat anal, Mass Inst Technol & res fel archeol mat anal, Harvard Univ Peabody Mus, 75-77; res assoc archaeol mat anal, Univ Colo Mus, 77-83; INDEPENDENT CONSULT GEOL, ARCHEOL GEOL & MAT ANAL, 83- *Concurrent Pos:* NSF grant, Colo Plateau, 71; Fisher fel fund & fac grants, Wellesley Col, 71-72; consult, Sardis Expedition, Harvard Univ, 74-76; NSF res grant, 75; consult, Amax Molybdenum Co, 76-77 & 79-80. *Mem:* Geol Soc Am; Mineral Soc Am; Soc Am Archaeol. *Res:* Petrology and geochemistry of granitic and metagranitic rocks, New Jersey, Ontario and Colorado; mineralogic and chemical analysis of ancient mesopotamian, North and Central American ceramics; geochemistry and correlation of Ubaid, Samarran and Halaf ceramics, Mesopotamia. *Mailing Add:* 5050 N Siesta Dr Tucson AZ 85715

KAMILLI, ROBERT JOSEPH, b Philadelphia, Pa, June 14, 47; m 69; c 2. ECONOMIC GEOLOGY, GEOCHEMISTRY. *Educ:* Rutgers Univ, BA, 69; Harvard Univ, AM, 71, PhD(geol), 76. *Prof Exp:* geologist, 76-79, asst resident geologist, 79-80, proj geologist, AMAX, Inc, 80-83; adj prof, Univ Colo, Boulder, 81-83; geologist, US Geol Surv Mission, Jeddah, Saudi Arabia, 83-87, mission chief & geologist, 87-89, RES GEOLOGIST, US GEOL SURV, UNIV ARIZ, TUCSON, 89- *Concurrent Pos:* Consult geologist, Huampar Mines, 73-76 & Buenaventura Mines, 75-76. *Mem:* Geol Soc Am; Mineral Soc Am; Am Inst Mining, Metall & Petrol Engrs; AAAS; Soc Econ Geologists. *Res:* Investigation of the origin of vein and intrusion-related metal deposits, especially molybdenum, tin, tungsten, gold and silver; emphasis on the geochemistry of such ore deposits. *Mailing Add:* US Geol Surv Tucson Field Off Gould-Simpson Bldg 77 Univ Ariz Tucson AZ 85721

KAMIN, HENRY, biochemistry; deceased, see previous edition for last biography

KAMINER, BENJAMIN, b Slonim, Poland, May 1, 24; m 48; c 2. PHYSIOLOGY. *Educ:* Univ Witwatersrand, MB, BCh, 46; Royal Col Physicians & Surgeons, dipl child health, 50. *Prof Exp:* Intern med, surg & pediat, Johannesburg Hosp, SAfrica, 47-48; house physician & registr pediat, Edgeware Hosp, London, Eng, 49-50; res asst endocrinol, Postgrad Med Sch, Univ London, 50-51; from lectr to sr lectr physiol, Med Sch, Univ Witwatersrand, 51-59; investr muscle, Inst Muscle Res, 59-69; lectr anat, Harvard Med Sch, 69-70; PROF PHYSIOL & CHMN DEPT, SCH MED, BOSTON UNIV, 70- *Concurrent Pos:* Rockefeller Found fel, 59-60. *Mem:* Soc Gen Physiol; Biophys Soc; Am Soc Cell Biol; Am Physiol Soc; Corp Marine Biol Lab. *Res:* Physiology and biochemistry of muscle; endoplasmic reticulum and intracellular calcium regulation. *Mailing Add:* Dept Physiol 80 E Concord St Boston MA 02118

KAMINETZKY, HAROLD ALEXANDER, b Chicago, Ill, Sept 6, 23; m 57; c 2. OBSTETRICS & GYNECOLOGY. *Educ:* Univ Ill, BS, 48, MD, 50; Am Bd Obstet & Gynec, dipl. *Prof Exp:* From instr to prof, Col Med, Univ Ill, 54-68; prof obstet & gynec, chmn Dept Col Med & Dent NJ, Newark, 68-85, from actg dean to dean, 72-74; DIR, PRACT ACTIV, AM COL OBSTET & GYNEC, 85- *Concurrent Pos:* Mem cancer adv comt, Chicago Bd Health; ed, Int J Gynaecology & Obstet, 78; prog dir, Greater Newark Family Planning Prog, 70-81; pres, Am Col Obstet & Gynecologists, 78; pres, 10th World Cong of Obstet & Gynecol, 82. *Mem:* AMA; Am Col Obstet & Gynec; Am Col Surg; Sigma Xi; Am Gynecologists & Obstetricians Soc. *Res:* Experimental dysplasia and carcinogenesis of the uterine cervix; maternal nutrition; vitamin profiles of mothers and newborns at partuition; nutrition during pregnancy. *Mailing Add:* ACOG 409 12th St SW Washington DC 20024

KAMINKER, JEROME ALVIN, b Chicago, Ill, May 10, 41; m 67; c 2. TOPOLOGY, MATHEMATICAL ANALYSIS. *Educ:* Univ Calif, Berkeley, BA, 63; Univ Calif, Los Angeles, MA, 65, PhD(math), 68. *Prof Exp:* Asst prof math, Ind Univ, Bloomington, 68-73; from asst prof to assoc prof, 73-79, PROF MATH, IND UNIV-PURDUE UNIV, 79- *Concurrent Pos:* Vis prof, Univ Calif, Los Angeles, 83; mem, Math Sci Res Inst, 84-85. *Mem:* Am Math Soc; Sigma Xi. *Res:* Development of relations between algebraic topology and functional analysis; application of K-theory to the theory of linear operators on Hilbert space. *Mailing Add:* Dept Math Ind Univ-Purdue Univ Indianapolis IN 46202

KAMINOW, IVAN PAUL, b Union City, NJ, Mar 3, 30; m 52; c 3. FIBER OPTICS, SEMICONDUCTOR LASERS. *Educ:* Union Univ NY, BS, 52; Univ Calif, Los Angeles, MS, 54; Harvard Univ, AM, 57, PhD(appl physics), 60. *Prof Exp:* Mem tech staff, Hughes Aircraft Co, 52-54; mem tech staff, Bell Labs, Inc, 54-84, DEPT HEAD, AT&T BELL LABS, INC, 84- *Concurrent Pos:* Vis lectr, Princeton Univ, 68, & Univ Calif, Berkeley, 77; mem Nat Acad Sci-Nat Bur Stand eval panel, Optical Physics Div, 72-75, Ctr for Electronics & Elec Eng, 84-87; assoc ed, J Quantum Electronics, 77-83; adj prof, Columbia Univ, 86; visi prof, Univ Tokyo, 90. *Honors & Awards:* Quantum Electronics Award, Inst Elec & Electronics Engrs, 83. *Mem:* Nat Acad Eng; fel Am Phys Soc; fel Optical Soc Am; fel Inst Elec & Electronics Engrs; Am Bd Laser Surg. *Res:* Microwave antennas; ferrites; high pressure physics; ferroelectrics; optical lasers and communication techniques; light modulation; Raman scattering; photopolymers; integrated optics; optical fibers; semiconductor lasers; photonic networks. *Mailing Add:* Bell Labs Inc Box 400 Holmdel NJ 07733

KAMINS, THEODORE I, b San Francisco, Calif, Nov 11, 41. SOLID-STATE ELECTRONICS, SILICON INTEGRATED CIRCUITS. *Educ:* Univ Calif, Berkeley, BS, 63, MS, 64, PhD(elec eng), 68. *Prof Exp:* Act asst prof elec eng, Univ Calif, Berkeley, 68-69; mem res staff, Fairchild Semiconductor, 69-74; mem tech staff, 74-81, PROJ LEADER ELEC ENG, HEWLETT-PACKARD CO, 81- *Concurrent Pos:* Vis lectr, dept elec eng, Stanford Univ, 74; consult, Stanford Electronic Lab, Stanford Univ, 75-82; vis scholar, Ctr Integrated Systs, Stanford Univ, 86-87; consult prof, Dept Elec Eng, Stanford Univ, 90- *Honors & Awards:* Electronics Div Award, Electrochem Soc, 89. *Mem:* Fel Inst Elec & Electronic Engrs; fel Electrochem Soc. *Res:* Research and development of materials and devices for silicon integrated circuits; especially, polycrystalline silicon, silicon-on-insulator, photodiode arrays, epitaxial techniques, silicon-germanium devices; author of books on device electronics and polycrystalline silicon. *Mailing Add:* Hewlett-Packard Co PO Box 10350 Palo Alto CA 94303-0867

KAMINSKAS, EDVARDAS, b Kaunas, Lithuania, Sept 12, 35; US citizen; m 65. PHARMOCOLOGY, MOLECULAR BIOLOGY. *Educ:* Seton Hall Univ, NJ, AB, 55; Sch Med, Yale Univ, MD, 59. *Prof Exp:* Intern & fel hemat, Michael Reese Hosp, Chicago, Ill, 61-63; resident med, Vet Admin Res Hosp, Chicago, Ill, 61-63; chief med, US Air Force Hosp, 63-65; res fel, Mass Inst Technol, 65-68; from instr to asst prof med, Harvard Med Sch, Beth Israel Hosp, 68-74; from assoc prof to prof, Univ Wis, Mt Sinai Med Ctr, 74-81; ASSOC PROF MED, HARVARD MED SCH, BETH ISRAEL HOSP, 81-; PHYSICIAN-IN-CHIEF, HEBREW REHAB CTR AGED, 81- *Mem:* Am Soc Biol Chemists; Am Asn Cancer Res; Am Fedn Clin Res; Geront Soc Am; Am Geriat Soc; AAAS. *Res:* Regulation of tumor cell growth; effects of anti-neoplastic agents on tumor cells; molecular biology of Alzheimer's disease cells. *Mailing Add:* 1200 Centre St Boston MA 02131

KAMINSKI, DONALD LEON, b Elba, Nebr, Nov 9, 40; m 65; c 4. SURGERY. *Educ:* Creighton Univ, BS, 62, MD, 66. *Prof Exp:* Asst prof, 71-75, assoc prof, 75-80, PROF SURG, ST LOUIS UNIV, 80- *Mem:* Am Physiol Soc; Am Gastroenterol Soc; Soc Univ Surgeons; Asn Acad Surg; Am Col Surgeons. *Res:* Gastrointestinal physiology, studying the hormonal control of hepatic bile flow. *Mailing Add:* St Louis Univ Med Ctr 3636 Vista at Grand Blvd St Louis MO 63110

KAMINSKI, EDWARD JOZEF, b Torun, Poland, Mar 24, 26; US citizen; m 51; c 2. CHEMISTRY. *Educ:* Northwestern Univ, PhB, 60, PhD(chem), 64, Am Bd Toxicol, Dipl. *Prof Exp:* Res technologist, Royal Cancer Hosp, London, Eng, 51-53; res asst path, Mt Sinai Hosp, Toronto, Can, 53-56; res technologist, Dent Sch, 56-60, res assoc path, 64-67, asst prof, 67-71, assoc prof, 71-79, PROF PATH, DENT & MED SCH, NORTHWESTERN UNIV, 79- *Concurrent Pos:* Consult toxicology, 64- *Mem:* Soc Toxicol; Am Chem Soc. *Res:* Toxicology of materials used in the human body; study on the mechanism of absorption of substances from the environment. *Mailing Add:* Dept Path Northwestern Univ Dent & Med Sch Chicago IL 60611-3010

KAMINSKI, JAMES JOSEPH, b Buffalo, NY, June 5, 47; m 68. MEDICINAL & THEORETICAL CHEMISTRY. *Educ:* State Univ NY Col Fredonia, BS, 69; Univ NH, PhD(org chem), 72. *Prof Exp:* Sr res chemist, Interx Res Corp, 72-78; res chemist, Schering Corp, 78-80, SECT LEADER, SCHERING-PLOUGH CORP, 80- *Concurrent Pos:* Adj assoc prof med chem, Univ Fla, Gainesville, Fla; chair-elect, Med & Natural Prod Chem, Am Asn Pharmaceut Scientists, 91. *Mem:* Am Chem Soc; Am Asn Pharmaceut

Scientists. *Res:* Physical-chemical approach to pharmaceutical problems; chemical modification of drug to improve drug delivery and development of soft medicinal agents; computer-assisted drug design. *Mailing Add:* Schering-Plough Corp 60 Orange St Bloomfield NJ 07003

KAMINSKI, JOAN MARY, b Darby, Pa, May 3, 47. ORGANIC CHEMISTRY. *Educ:* West Chester State Col, BA, 69; Drexel Univ, PhD(org chem), 75. *Prof Exp:* Asst res chemist, Univ Ill, Urbana, 67 & 69; teaching asst chem, Drexel Univ, 70-74; Nat Res Coun-Agr Res Serv res assoc org chem, 74-75, res scientist org chem, fats & proteins res found, Eastern Regional Res Ctr, Agr Res Serv, USDA, 75-77; res chemist, Mobil Res & Develop Corp, 77-80. *Mem:* Am Chem Soc; Sigma Xi; Am Oil Chemists' Soc. *Res:* Lubricants, petroleum chemistry, organic surfactants; amphoteric lime soap dispersing agents; fatty acid derivatives; lipid chemistry; gel chromatography; chemistry of the sulfur-nitrogen bond. *Mailing Add:* RR2-Box 398 G Mullica Hill NJ 08062

KAMINSKI, ZIGMUND CHARLES, b Hartford, Conn, Jan 15, 29. MEDICAL MICROBIOLOGY, INFECTIOUS DISEASES. *Educ:* Univ Conn, BA, 52; Hahnemann Med Col, MS, 54, PhD(microbiol), 57; Am Bd Med Microbiol, dipl pub health & med microbiol, 72. *Prof Exp:* From instr to asst prof microbiol, Col Med, Seton Hall Univ, 57-65; asst prof, 68-71, ASSOC PROF MICROBIOL & PATH, MED SCH, UNIV MED & DENT, NJ, 71-; DIR CLIN MICROBIOL, UNIV HOSP, 68- *Mem:* Am Soc Microbiol; Acad Clin Lab Physicians & Scientists; Am Venereal Dis Asn; NY Acad Sci; Infectious Dis Soc Am. *Res:* Non-specific urethritis; candidalbicans; morphogenesis. *Mailing Add:* Dept Path Med Sch Univ Med & Dent NJ Newark NJ 07103

KAMINSKY, LAURENCE SAMUEL, b Cape Town, SAfrica, Dec 25, 40; m 66; c 2. BIOLOGICAL SCIENCES, TOXICOLOGY. *Educ:* Univ Cape Town, BS, 62, Hons, 63, PhD(chem), 66. *Prof Exp:* Res assoc biochem, Sch Med, Yale Univ, 67-68; assoc prof med biochem, Med Sch, Univ Cape Town, 68-75; prin res scientist biochem, 75-77, dir biochem toxicol, 78-83, CHIEF BIOCHEM GENETIC TOXICOL, NY STATE DEPT HEALTH, 83-; PROF, STATE UNIV NY, 85- *Concurrent Pos:* Vis res prof biochem, State Univ NY Albany, 74; adj assoc prof, Albany Med Col, 76- *Honors & Awards:* Frank Blood Award, Soc Toxicol. *Mem:* Brit Biochem Soc; Am Soc Biochem & Molecular Biol; Soc Toxicol; Int Soc Study Xenobiotics. *Res:* Investigations into the role of of the heme proteins hepatic microsomal cytochrome P-450 in the metabolism of drugs; toxifying and detoxifying properties of cytochromes P-450. *Mailing Add:* Wadsworth Ctr for Labs & Res NY State Dept of Health Albany NY 12201-0509

KAMINSKY, MANFRED STEPHAN, b Koenigsberg, Ger, June 4, 29; m 57; c 2. EXPERIMENTAL PHYSICS, SURFACE PHYSICS. *Educ:* Univ Rostock, Dipl, 51, Univ Marburg, Ger, PhD(physics), 57. *Prof Exp:* Asst physics, Univ Rostock, 50-52, lectr, Med Tech Sch, 52; res asst, Phys Inst, Univ Marburg, 53-57, sr asst, 57-58; res assoc, Argonne Nat Lab, 58-59, asst physicist, 59-62, assoc physicist, 62-70, dir, Surface Sci Ctr, 74-80, sr physicist, 70-76; PROP, SURFACE TREATMENT SCI INT, 86- *Concurrent Pos:* Invited prof, Inst Energy, Univ Quebec, Montreal-Varennes, 76-83; Coop US scientist, Div Int Prog, NSF, 78-83; mem, Task Group Plasma-Wall Interactions, Off Fusion Energy, Dept Energy, 79-80; chmn, Steering Comt Fusion Technol, Int Union Vacuum Sci, Technol & Appln, 81-83; res fel, Japanese Soc Prom Sci, 82; proj mgr, US Dept Energy E Cut Tribology Proj & mgr tribology proface, Argonne Nat Lab, 84-86. *Honors & Awards:* E W Mueller Lectr, Univ Wis-Milwaukee, 78. *Mem:* AAAS; fel Am Phys Soc; Am Chem Soc; Sigma Xi; sr & hon mem Am Vacuum Soc. *Res:* Atomic and ionic impact phenomena on solids; channeling phenomena; nuclear polarization; surface science in thermonuclear research; mass spectrometry; ultrahigh vacuum technology; ionic processes in electrolytic solutions; radiation effects on solid surfaces; surface effects in controlled fusion devices; author of numerous books on surface physics. *Mailing Add:* 906 South Park Hinsdale IL 60521

KAMIYAMA, MIKIO, b Kyoto, Japan, Mar 25, 36; m 71; c 3. IMMUNOCHEMISTRY, IMMUNOHEMATOLOGY. *Educ:* Kyoto Prefectural Univ, BS, 62; Univ Tokyo, PhD(biochem) & DMSc, 67. *Prof Exp:* Fel biochem, Princeton Univ, 67-68; res assoc microbiol, Albert Einstein Med Ctr, 68; res assoc biochem, State Univ NY, Buffalo, 69-70; sr researcher molecular biol, Inst Molecular Path, Univ Paris, 71-74; res assoc hemat, St Luke's Roosevelt Hosp Ctr & Columbia Univ, 74-77, attend staff immunol, 77-88; DIR RES, BLOOD RES INST, ST MICHAEL'S MED CTR, 88-; PROF, SCH GRAD MED EDUC, SETON HALL, 88- *Concurrent Pos:* Vis lectr biochem, Inst Physiol Chem, Univ Marburg, 72-73. *Mem:* Am Asn Immunologists; sr mem Am Fedn Clin Res; Asn Med Lab Immunologists; NY Acad Sci; Harvey Soc; Am Heart Asn Coun Thrombosis; Int Soc Thrombosis & Haemostasis. *Res:* Immunological and biochemical characterization of platelet glycoproteins; cell membrane studies; monoclonal antibody preparations directed to human platelet glycoproteins; non-radioactive immune assays. *Mailing Add:* Blood Res Inst Saint Michael's Med Ctr Seton Hall Univ 268 Dr Martin Luther King Jr Blvd Newark NJ 07102

KAMLET, MORTIMER JACOB, physical organic chemistry, for more information see previous edition

KAMM, DONALD E, b Rochester, NY, July 10, 29; m 56; c 4. NEPHROLOGY, PHYSIOLOGY. *Educ:* Cortland State Col, BS, 51; Albany Med Col, MD, 60. *Prof Exp:* Intern & resident med, Beth Israel Hosp, Boston, 60-62; from instr to asst prof, 66-71, ASSOC PROF MED, SCH MED & DENT, UNIV ROCHESTER, 71- *Concurrent Pos:* NIH res fel renal physiol, Sch Med, Boston Univ, 62-64; NIH fel metab, Harvard Med Sch, 64-66. *Mem:* Am Fedn Clin Res; Am Physiol Soc Nephrology; Int Soc Nephrology. *Res:* Effects of potassium balance on the regulation of urea production and renal ammonia production. *Mailing Add:* Dept Med Univ Rochester Med Ctr Rochester Gen Hosp 1425 Portland Ave Rochester NY 14621

KAMM, GILBERT G(EORGE), b Emington, Ill, Aug 6, 25; m 54; c 1. CHEMICAL ENGINEERING. *Educ:* Univ Ill, BS, 49. *Prof Exp:* Group leader metals, Res Div, Am Nat Can Co, Ill, 58-66, mgr metals sect, Res & Develop Ctr, 66-68, asst to assoc dir, Mat Sci Sect, 68-71, assoc dir, Princeton Res Ctr, 71-78, assoc dir metall, 79-81, dir, metal mat technol, 82-84, dir mat technol, Barrington Tech Ctr, 84-88; CONSULT, 88- *Mem:* Am Chem Soc; Electrochem Soc; Am Soc Metals. *Res:* Metal cleaning and plating; corrosion, especially electrolytic tin plate and other metals used in containers; metallurgical properties of container materials; plastic-metal composites; energy related research in pulp and paper processes; organic coatings and sealing compounds. *Mailing Add:* 21032 N Crestview Dr Barrington IL 60010-2924

KAMM, JAMES A, b Ft Collins, Colo. ECONOMIC ENTOMOLOGY. *Educ:* Univ Wyo, BS, 61, MS, 63, PhD(entom), 67. *Prof Exp:* RES ENTOMOLOGIST, AGR RES SERV, USDA, 67- *Concurrent Pos:* Assoc prof, Ore State Univ, 70- *Mem:* Entom Soc Am; AAAS. *Res:* Grassland insects. *Mailing Add:* 440 NW Witham Dr Corvallis OR 97333

KAMM, JEROME J, b New York, NY, June 4, 33; m 56. BIOCHEMICAL PHARMACOLOGY. *Educ:* Brooklyn Col, BS, 54; Georgetown Univ, MS, 59, PhD(biochem), 64. *Prof Exp:* Chemist, Nat Heart Inst, 55-64; sr biochemist, Smith, Kline & French Labs, 64-67; SECT HEAD, HOFFMANN-LA ROCHE, INC, 67- *Mem:* AAAS; Am Chem Soc; Acad Pharmaceut Sci; NY Acad Sci; Am Soc Pharmacol & Exp Therapeut; Sigma Xi. *Res:* Drug metabolism and mechanisms of drug metabolism; toxicology. *Mailing Add:* Dept Toxicol Hoffmann-La Roche Inc Nutley NJ 07110

KAMM, ROGER DALE, b Ashland, Wis, Oct 10, 50; m 74; c 1. BIOMEDICAL ENGINEERING, FLUID MECHANICS. *Educ:* Northwestern Univ, BS, 72; Mass Inst Technol, SM, 73, PhD(mech eng), 77. *Prof Exp:* Instr, 77, lectr & res assoc, 77-78, asst prof, 78-81, assoc prof, 81-88, PROF MECH ENG, MASS INST TECHNOL, 88- *Mem:* Am Soc Mech Engrs; Asn Res Vision & Ophthal; Am Physiol Soc; Biomed Eng Soc. *Res:* Biomedical fluid mechanics, specifically physiology and pathophysiology of venous circulation, respiratory tract and eye. *Mailing Add:* 31 Nonesuch Rd Weston MA 02193

KAMMANN, KARL PHILIP, JR, b St Louis, Mo, Mar 30, 36; m 60; c 4. CHEMISTRY. *Educ:* Washington Univ, AB, 57; La State Univ, MS, 60, PhD(chem), 62. *Prof Exp:* Sr res chemist, Cities Serv Co, 62-64; group leader, Emery Indust, Inc, 64-74; PROD DEVELOP MGR, KEIL CHEM DIV, FERRO CORP, 75- *Mem:* Am Chem Soc; Am Oil Chemists Soc; Am Soc Lubrication Engrs. *Res:* Sulfurization and chlorosulfurization; fats and oils, and derivatives. *Mailing Add:* Keil Chem Div Ferro Corp 3000 Sheffield Ave Hammond IN 46320

KAMMASH, TERRY, b Salt, Jordan, Jan 27, 27; nat US; m 56. NUCLEAR ENGINEERING & ENGINEERING MECHANICS, APPLIED PHYSICS. *Educ:* Pa State Univ, BS, 52, MS, 54; Univ Mich, PhD(nuclear eng), 58. *Prof Exp:* Asst aerodyn, Pa State Univ, 52-53, instr eng mech, 53-54; asst aircraft propulsion lab, Eng Res Inst, 54-55, from instr to assoc prof nuclear eng & eng mech, 55-67, prof nuclear eng, 67-77, STEPHEN S ATWOOD PROF NUCLEAR ENG & ACTG CHMN DEPT, UNIV MICH, ANN ARBOR, 77- *Concurrent Pos:* Vis scientist, Lawrence Radiation Lab, 62-63, 63-67. *Honors & Awards:* Arthur Holly Compton Award & Am Nuclear Soc, 77. *Mem:* Fel Am Phys Soc; Am Nuclear Soc; Am Soc Eng Educ; Soc Eng Sci. *Res:* Magnetohydrodynamics; plasma physics; plasticity and physics of ionized gases; controlled fusion; space application of fusion energy. *Mailing Add:* Dept Nuclear Eng Univ Mich Ann Arbor MI 48109

KAMMEN, HAROLD OSCAR, b New York, NY, July 28, 27; m 59; c 2. BIOCHEMISTRY. *Educ:* Bethany Col, WVa, BS, 46; Stanford Univ, PhD(chem), 60. *Prof Exp:* Res asst biochem, M D Anderson Hosp, Tex, 56-59; fel pharmacol, Sch Med, Yale Univ, 59, res asst, 59; asst res biochemist, 64-69, ASSOC RES BIOCHEMIST, UNIV CALIF, BERKELEY, 69- *Concurrent Pos:* Lectr, Univ Calif, Berkeley, 71, 75- *Mem:* AAAS; Am Chem Soc; Am Soc Biol Chem; fel Am Inst Chem. *Res:* Enzymes and regulation of nucleic acid metabolism. *Mailing Add:* Bldg 155 Richmond Field Sta Univ Calif 1301 S 46th St Richmond CA 94804

KAMMER, ANN EMMA, b Auburn, NY, Dec 26, 35. NEUROBIOLOGY. *Educ:* State Univ NY Teachers Albany, BS, 56; Univ NH, MS, 58; Univ Calif, Berkeley, PhD(zool), 66. *Prof Exp:* Lectr zool, Univ Calif, Davis, 65; from actg asst prof to asst prof, Kans State Univ, 65-72, from assoc prof to prof biol, 72-85; PROF ZOOL & CHMN DEPT, ARIZONA STATE UNIV, 86- *Concurrent Pos:* Vis res fel, Univ Saarland, 70 & Univ Cologne, 71; vis assoc prof, Univ Calif, Los Angeles, 78-79 & vis lect, Univ, Irvine, 82; mem steering comt, Sect G, AAAS, 83-85. *Honors & Awards:* Masua Hon lectr, 84-85. *Mem:* Am Soc Zoologists (secy, 85-); Soc Exp Biol; Soc Neurosci; Am Physiol Soc; fel AAAS; Sigma Xi. *Res:* Neural control of muscles; development of central nervous system and neuromuscular junctions in insects. *Mailing Add:* Dept Biol Kans State Univ Manhattan KS 66506

KAMMER, PAUL A, b Auburn, NY. WELDING METALLURGY, THERMAL SPRAYING. *Educ:* Rensselaer Polytech Inst, BEng Met, 59; Ohio State Univ, MS, 66, MBA, 67. *Prof Exp:* Asst mgr metall welding, Battelle Mem Inst, 61-67; dir, res & develop, Teledyne McKay, 67-76, metall welding, Gorham Int, 76-78; dir, 78-80, vpres, 80-84, SR VPRES, RES & DEVELOP, EUTECTIC CORP, 84- *Concurrent Pos:* Vpres & gen mgr, Wear Control Technol, Inc, 82-85; mem, Welding Res Coun. *Mem:* Am Welding Soc; Am Soc Metals; Metall Soc; Am Powder Metall Inst. *Res:* Manage and conduct research and development in fields of metallurgy, welding and thermal spraying. *Mailing Add:* 40-40 172nd St Flushing NY 11358

KAMMERAAD, ADRIAN, b Holland, Mich, Feb 25, 12; wid; c 3. EXPERIMENTAL BIOLOGY, RESEARCH ADMINISTRATION. *Educ:* Hope Col, AB, 33; Yale Univ, PhD(exp embryol), 40. *Prof Exp:* Asst zool, Yale Univ, 33-36; instr, Dartmouth Col, 36-38; asst anat, Sch Med, La State Univ, 38-40, instr, 40-42; sci dir, Van Patten Pharmaceut Co, 46-50; dir res & prod control, Kremers-Urban Co, 50-56; in chg pharmaceut prod develop, Dow Chem Co, 56-60, dir pharm lab, 59-60, in chg res admin, 60-62, mgr drug regulatory sect, Res Ctr, 62-77; RETIRED. *Concurrent Pos:* Instr, Sch Med, Northwestern Univ, 47-50. *Res:* Administration and supervision of drug regulatory activities. *Mailing Add:* 9115 Washington Blvd Indianapolis IN 46240

KAMMERER, WILLIAM JOHN, b Rochester, NY, Oct 8, 31; m 57; c 2. MATHEMATICS. *Educ:* Univ Rochester, BA, 54; Univ Wis, Madison, MS, 55, PhD(math), 59. *Prof Exp:* Asst prof oper res, Case Western Reserve Univ, 59-60; from asst prof to prof math, Ga Inst Technol, 60-90; CONSULT, 90- *Concurrent Pos:* Soc Indust & Appl Math lectr, 71-72. *Mem:* Am Math Soc; Soc Indust & Appl Math; Asn Comput Mach. *Res:* Numerical analysis; approximation theory and optimization. *Mailing Add:* 3430 Pin Oak Circle Atlanta GA 30340

KAMMERMEIER, MARTIN A, b Cold Spring, Minn, Oct 23, 31; m 59; c 3. SPEECH PATHOLOGY. *Educ:* St Cloud State Col, BS, 58, MS, 63; Univ Minn, Minneapolis, PhD(speech path), 69. *Prof Exp:* Speech therapist, Pub Sch, 58-62; from instr to asst prof, 62-73, PROF SPEECH PATH, ST CLOUD STATE UNIV, 73-, CHMN DEPT SPEECH SCI, PATH & AUDIOL, 70- *Mem:* Am Speech & Hearing Asn. *Res:* Acoustic analysis of the voices of speakers with a variety of central nervous system disorders. *Mailing Add:* Dept Commun Disorders St Cloud State Univ St Cloud MN 56301

KAMMERMEYER, KARL, b Nurnberg, Ger, June 15, 04; nat US; m 30; c 1. CHEMICAL ENGINEERING. *Educ:* Univ Mich, BSChE & BS(math), 29, MSE, 31, DSc(chem eng), 32. *Prof Exp:* Res assoc eng, Univ Mich, 30-32; develop engr, Standard Oil Co, Inc, 33-36; refinery chief chemist & chem engr, Pure Oil Co, Ohio, 36-39; asst prof chem eng, Drexel Inst, 39-42; dir res, Publicker Industs Inc, 42-47; mgr res & develop, Chem Div, Glenn L Martin Co, 47-49; prof chem eng & head dept, 49-73, EMER PROF CHEM ENG, UNIV IOWA, 73- *Concurrent Pos:* Sci consult, US Dept Com, Ger, 46; consult, Vet Admin Hosp, Iowa City; chmn, Gordon Res Conf Separation & Purification, 57. *Mem:* AAAS; Am Chem Soc; fel Am Inst Chem Engrs; Am Soc Eng Educ; Sigma Xi. *Res:* Separation processes; membrane separations and distillation; air purification in submarines and space capsules; genetic engineering for technologists. *Mailing Add:* Dept Chem Eng Univ Iowa Iowa City IA 52242

KAMMERUD, RONALD CLAIRE, b Monroe, Wis, July 10, 42; m 63; c 2. SOLAR ENERGY. *Educ:* Drexel Univ, BS, 66; Ind Univ, Bloomington, MS, 69, PhD(physics), 70. *Prof Exp:* Res assoc exp high energy physics, Ohio State Univ, 70; res assoc exp high energy physics, Fermi Nat Accelerator Lab, 73-76; SOLAR ENERGY RES PHYSICIST, LAWRENCE BERKELEY LAB, 76- *Mem:* Am Phys Soc; Int Solar Energy Soc. *Res:* Heat transfer analysis as applied to passive solar building energy analysis. *Mailing Add:* 4032 Waterhouse Rd Oakland CA 94602

KAMMLER, DAVID W, b Belleville, Ill, Oct 29, 40; m 65; c 2. NUMERICAL ANALYSIS, APPROXIMATION THEORY. *Educ:* Southern Ill Univ, Carbondale, BA, 62; Southern Methodist Univ, MS, 69; Univ Mich, Ann Arbor, PhD(math), 71. *Prof Exp:* Mem tech staff, Tex Instruments Inc, 65-68; from asst prof to assoc prof, 71-78, PROF MATH, SOUTHERN ILL UNIV, CARBONDALE, 78- *Concurrent Pos:* Fel, Rome Air Develop Ctr, 80-81. *Mem:* Am Math Soc; Soc Indust & Appl Math; Math Asn Am; Sigma Xi. *Res:* Approximation with sums of exponentials; transient analysis; numerical analysis; discrete fourier analysis. *Mailing Add:* Dept Math Southern Ill Univ Carbondale IL 62901

KAMMULA, RAJU G, b May 1, 35; m 66; c 3. TOXICOLOGY, PHYSIOLOGY. *Educ:* Madras Univ, DVM, 57; Univ Minn, PhD(physiol), 64. *Prof Exp:* Res asst, Univ Minn, 59-64; prof physiol, Col Vet Med, Tuskegee Univ, 64-76; vet practr, Calif, 76-78; CHIEF OBSTET & GYNEC MED DEVICES BR, CTR DEVICES & RADIOL HEALTH, FOOD & DRUG ADMIN, 79- *Mem:* Am Physiol Soc; Am Vet Med Asn; Am Col Toxicol; DC Vet Med Asn. *Res:* Pharmacology. *Mailing Add:* Food & Drug Admin 1390 Piccard Dr Rockland MD 20850

KAMON, ELIEZER, b Jerusalem; US citizen; m 67; c 3. ERGONOMICS, OCCUPATIONAL HEALTH. *Educ:* Israel State Col, T dipl, 52; Hebrew Univ, Jerusalem, MSc, 59, PhD(zool), 64. *Prof Exp:* Res asst occup health, Univ Pittsburgh, 66-67, res assoc, 68-73; PROF PHYSIOL, PA STATE UNIV, 74- *Concurrent Pos:* Fel ergonomics, Loughboro Univ Technolog, 65. *Mem:* Am Physiol Soc; Am Indust Hyg Asn; fel Am Col Sports Med; Ergonomics Soc; Human Factors Soc. *Res:* Human thermal physiology and man's adaptability to his working conditions as they relate to muscular strength, cardiovascular capacity and respiratory functions. *Mailing Add:* 453 Park Lane State College PA 16803

KAMOUN, MALEK, LABORATORY MEDICINE. *Educ:* Pierre & Marie Curie Univ, Paris, France, MD, 75. *Prof Exp:* ASST PROF PATH & LAB MED, DEPT IMMUNOL & HEMAT, UNIV PA, 83- *Res:* Lymphocyte growth and differentiation. *Mailing Add:* Dept Path Univ Pa 3400 Spruce St Seven Founders Philadelphia PA 19104

KAMOWITZ, HERBERT M, b Brooklyn, NY, Dec 31, 31; m 55; c 3. MATHEMATICS. *Educ:* City Col New York, BS, 52; Brown Univ, ScM, 54, PhD(math), 60. *Prof Exp:* Assoc scientist math, Res & Advan Develop Div, Avco Corp, 57-60, sr scientist, 60-61, from staff scientist to sr staff scientist, 61-66; assoc prof, 66-70, PROF MATH, UNIV MASS, BOSTON, 70- *Concurrent Pos:* Vis scientist, Pure Math Dept, Weizmann Inst Sci, Rehovot, Israel, 73 & 80. *Mem:* Am Math Soc. *Res:* Functional analysis. *Mailing Add:* Dept Math Univ Mass Dorchester MA 02125

KAMP, DAVID ALLEN, b St Louis, Mo, Sept 26, 47; m 73. ORGANIC CHEMISTRY, POLYMER CHEMISTRY. *Educ:* Univ Calif, Los Angeles, BSc, 70; Univ Ore, PhD(chem), 76, Univ Calif, Santa Cruz, cert hazardous mat mgt, 91. *Prof Exp:* Assoc chem, Cornell Univ, 76-77; staff org chemist, Res & Develop Ctr, Gen Elec Co, 77-80; staff scientist, 80-88, CONSULT, ORG & POLYMER CHEM, RAYCHEM CORP, 87- *Mem:* Am Chem Soc; Asn Consult Chemists & Chem Engrs; Prof & Tech Consults Asn; Int Soc Optical Eng. *Res:* Organic synthesis; organic and polymer chemistry; materials science. *Mailing Add:* 886 Ticonderoga Dr Sunnyvale CA 94087

KAMPAS, FRANK JAMES, b Buffalo, NY, Apr 24, 46. PHYSICAL CHEMISTRY, SOLID STATE PHYSICS. *Educ:* Univ Pa, BA & MS, 68; Stanford Univ, PhD(physics), 74. *Prof Exp:* Res assoc chem, Univ Wash, 74-77; res assoc, Molecular Sci Div, Brookhaven Nat Lab, 77-78, from asst scientist to scientist, Dept Energy & Environ, Mat Sci Div, 78-85; sr scientist, Chronar Co, 85-86, mgr, Process Develop Group, 86-87, mgr, process develop & eng, 88-89; DIR ENCAPSULATION, AM PHYS SOC, 90- *Mem:* Am Phys Soc; Mat Res Soc; Inst Elec & Electronic Engrs. *Res:* Physics of organic molecules in the solid state and in solution; plasma chemistry as applied to thin film deposition; optical spectroscopy; solar cell device physics. *Mailing Add:* Am Phys Soc PO Box 7093 Princeton NJ 08543

KAMPE, DENNIS JAMES, b Brooklyn, NY, July 16, 45; div; c 2. ELECTRON MICROSCOPY. *Educ:* State Univ NY Stony Brook, BESc, 67; Univ Va, MMSc, 69, PhD(mat sci), 72. *Prof Exp:* Staff scientist, 72-73, dept head micros & phys testing, 73-75, res scientist, 75-82, SR STAFF SCIENTIST, CARBON PROD DIV, UNION CARBIDE CORP, 83- *Mem:* Electron Micros Soc Am; Am Soc Testing & Mat; Sigma Xi. *Res:* Optical and electron microscopy and physical properties determinations of carbons, graphites and fibers; structure and performance capabilities of porous air-oxygen cathodes in chlor-alkali cells and battery devices. *Mailing Add:* UCAR Carbon PO Box 6116 Cleveland OH 44101

KAMPER, ROBERT ANDREW, b Surbiton, Eng, Mar 14, 33; nat US; m 55; c 3. PHYSICS. *Educ:* Oxford Univ, BA, 54, MA & DPhil(physics), 57. *Prof Exp:* Imp Chem Industs res fel, Oxford Univ, 57-61; physicist, Cent Elec Generating Bd, Eng, 61-63; physicist, Cryogenics Div, Nat Bur Standards, 63-74, assoc chief, Electromagnetics Div, 74-78, CHIEF ELECTROMAGNETIC TECHNOL DIV, NAT INST STANDARDS & TECHNOL, 78-, DIR, BOULDER LABS, 82- *Concurrent Pos:* Fulbright travel grant, Univ Calif, Berkeley, 58-59. *Honors & Awards:* Arnold O Beckman Award, Instrument Soc Am, 74; Gold Medal, US Dept Com, 75. *Mem:* Fel Inst Elec & Electronics Engrs. *Res:* Cryoelectronics; superconductivity; electrical measurement technique; electron spin resonance. *Mailing Add:* Nat Inst Standards & Technol 325 Broadway Boulder CO 80303

KAMPHOEFNER, FRED J(OHN), b San Francisco, Calif, Mar 23, 21; m 60. ELECTRONICS. *Educ:* Univ Calif, BS, 43; Stanford Univ, MA, 47, PhD, 49. *Prof Exp:* Res assoc, Radio Res Lab, Harvard Univ, 43-45 & Stanford Univ, 46-49; mgr, control systs lab, SRI Int, 49-71, dir eng sci lab, 71-86, dep dir, advan technol div, 86-90; RETIRED. *Concurrent Pos:* Res analyst, Opers Res Off, Johns Hopkins Univ, 52. *Mem:* Sigma Xi; sr mem Inst Elec & Electronics Engrs. *Res:* Instrumentation; data systems; bioengineering; mechanized data entry and non-impact printing. *Mailing Add:* 175 Ravenswood Ave Atherton CA 94027

KAMPHUIS, J(OHN) WILLIAM, b Vollenhove, Netherlands, Sept 9, 38; Can citizen; m 60; c 2. COASTAL & OCEANOGRAPHIC ENGINEERING. *Educ:* Queen's Univ, Ont, BSc, 61, MSc, 63, PhD(civil eng), 66; Delft Technol Univ, dipl hydraul eng, 64. *Prof Exp:* Asst res officer, Nat Res Coun Can, 65-68; from asst prof to assoc prof, 68-72, PROF CIVIL ENG, QUEEN'S UNIV, ONT, 74- *Concurrent Pos:* Lectr, Carleton Univ, 65-68; specialist consult coastal eng, 69- *Mem:* Am Soc Civil Engrs; Int Asn Hydraul Res; Can Soc Civil Engrs. *Res:* Wave mechanics; interaction of waves and coasts; coastal sediment transport by waves and tides; model analysis; tidal propagation and numerical analysis; marina design. *Mailing Add:* Dept Civil Eng Ellis Hall Queen's Univ Kingston ON K7L 3N6 Can

KAMPMEIER, JACK A, b Cedar Rapids, Iowa, June 11, 35; m 58; c 3. ORGANIC CHEMISTRY. *Educ:* Amherst Col, BA, 57; Univ Ill, PhD(org chem), 60. *Prof Exp:* From instr to assoc prof, 60-71, chmn dept, 75-79, PROF CHEM, UNIV ROCHESTER, 71-, assoc dean grad studies, Col Arts & Sci, 82-86, DEAN, COL ARTS & SCI, 88- *Concurrent Pos:* NSF sci fac fel, Univ Calif, Berkeley, 71-72; NSF fel, 58-60; Fulbright-Hays, sr res fel, Univ Freiburg, 79-80; sr scientist, NATO, 79-80. *Mem:* Am Chem Soc; Sigma Xi. *Res:* Mechanistic organic chemistry; free radical reactions; organometallic reactions; photochemistry. *Mailing Add:* Dept Chem Univ Rochester River Rochester NY 14627-1001

KAMPRATH, EUGENE JOHN, b Seward, Nebr, Jan 9, 26; m 56; c 2. SOIL FERTILITY. *Educ:* Univ Nebr, BS, 50, MS, 52; NC State Col, PhD(soils), 55. *Hon Degrees:* DSc Univ Nebr, 87. *Prof Exp:* From asst prof to prof, 55-81, WILLIAM NEAL REYNOLDS PROF SOIL SCI, NC STATE UNIV, 81-, HEAD DEPT, 90- *Concurrent Pos:* Dir soil testing div, State Dept Agr, NC, 57-62; ed-in-chief, Soil Science Soc Am, 69-74. *Honors & Awards:* Soil Sci Appl Res Award, Soil Sci Soc of Am, 86. *Mem:* Fel Am Soc Agron; fel Soil Sci Am; Sigma Xi. *Res:* Soil fertility; soil chemistry relationships in soils as they affect availability of nutrients to plants. *Mailing Add:* Dept Soil Sci Box 7619 NC State Univ Raleigh NC 27695-7619

KAMPSCHMIDT, RALPH FRED, b Gerald, Mo, May 6, 23; m 54; c 4. BIOCHEMISTRY. *Educ:* Univ Mo, BS, 47, MS, 49, PhD(agr chem), 51. *Prof Exp:* Instr animal husb, Univ Mo, 47-51; res chemist biochem, Armour & Co, 51-55; sect head, Samuel Roberts Noble Found Inc, 55-85; RETIRED. *Mem:* Fel AAAS; Am Asn Cancer Res; Soc Exp Biol & Med; hon mem Reticuloendothelial Soc; NY Acad Sci. *Res:* Cancer research; iron metabolism; reticuloendothelial system; endotoxin; monokines. *Mailing Add:* 1400 Oakridge Ardmore OK 73401

KAMRA, OM PERKASH, b Lahore, India, Mar 18, 35; m 59; c 2. RADIATION GENETICS, CYTOGENETICS. *Educ:* Univ Delhi, BSc, 54; NC State Univ, MS, 56; Wash State Univ, PhD(genetics), 59; Univ Lund, dipl, 59. *Prof Exp:* Fel Swedish Agr Res Coun, 59; secy, Food & Agr Orgn-Swedish Int Training Ctr Genetics, Univ Lund, 59-60; res officer radiation genetics, Atomic Energy Estab, Trombay, India, 60-61; res assoc plant genetics, Univ Man, 61-63; assoc prof, 63-72, PROF RADIATION BIOL, DALHOUSIE UNIV, 72- *Concurrent Pos:* Fel, Swedish Agr Res Coun, 59-60; chmn biol subcomt, Atlantic Prov Inter-Univ Comt Sci, 67-68; consult, Int Atomic Energy Agency, Vienna, 69-70 & UN Develop Prog, Indonesia, 72; mem comt int exchange, Nat Res Coun, 73-76; vis prof, Belgium Nuclear Res Estab, 77, 78 & 81. *Honors & Awards:* Travel Award, Can Genetics Soc, 63. *Mem:* Sigma Xi. *Res:* Cytology; genetics; mutations; food additives; effect of laser beams on biological systems. *Mailing Add:* 1588 Cambridge St Halifax NS B3H 4A6 Can

KAMRAN, MERVYN ARTHUR, b Sialkot, Pakistan, Nov 8, 38. ENTOMOLOGY, ECOLOGY. *Educ:* Punjab Univ, Pakistan, BS, 57, MS, 59; Univ Hawaii, PhD(entom), 65. *Prof Exp:* Lectr biol, Pakistan Ed Serv, Lahore, 59-61; Pakistan AEC scholar, Cent Treaty Orgn Inst Nuclear Sci, Tehran, Iran, 61; res fel entom, Int Rice Res Inst, Philippines, 65-67; asst prof entom, Pa State Univ, 68; asst prof biol, 68-72, assoc prof, 72-79, PROF BIOL, DOWLING COL, 79- *Mem:* AAAS; Am Entom Soc; Entom Soc Am. *Res:* Ecology and biological control of insect pests; taxonomy of tachinid flies. *Mailing Add:* Dept Biol Dowling Col Idle Hour Blvd Oakdale NY 11769

KAMRASS, MURRAY, operations research, systems analysis; deceased, see previous edition for last biography

KAMRIN, MICHAEL ARNOLD, b Brooklyn, NY, Aug 5, 40; m 64; c 2. PUBLIC EDUCATION IN SCIENCE, SCIENCE POLICY. *Educ:* Cornell Univ, BA, 60; Yale Univ, MS, 62, PhD(chem), 65. *Prof Exp:* Res assoc biol, Biol Div, Oak Ridge Nat Lab, 63-64, consult, 65-66; NIH trainee, Hopkins Marine Sta, Stanford Univ, 66-67; from asst prof to prof natural sci, 67-89, PROF, INST ENVIRON TOXICOL, MICH STATE UNIV, 82-, PROF, RES DEVELOP, 90- *Concurrent Pos:* Vis scientist, Mich Legis Off Sci Adv, 80-81. *Honors & Awards:* Mem Medal, Univ Turku, Finland. *Mem:* AAAS; Am Chem Soc; Sigma Xi; Soc Environ Toxicol & Chem; Soc Toxicol; Soc Risk Anal. *Res:* Risk assessment, risk management, risk communication; science education for non-scientists; science policy. *Mailing Add:* Inst Environ Toxicol C-231 Holden Hall Mich State Univ East Lansing MI 48824-1206

KAMYKOWSKI, DANIEL, b Chicago, Ill, Nov 23, 45; m; c 2. BIOLOGICAL OCEANOGRAPHY. *Educ:* Loyola Univ, Chicago, BS, 67; Univ Calif, San Diego, PhD(oceanog), 73. *Prof Exp:* Killam res assoc oceanog, Dalhousie Univ, 73-75; asst prof bot & marine sci, Univ Tex, Austin, 75-79; assoc prof 79-86, PROF MARINE EARTH & ATMOSPHERIC SCI, NC STATE UNIV, 86- *Mem:* Am Soc Limnol & Oceanog; Am Geophys Union; Phycol Soc Am; Oceanog Soc. *Res:* Physiology and behavior of marine dinoflagellates in response to physical processes; global patterns in hydrographic factors, plant nutrients and phytoplankton species composition; phytoplankton physiology in the upper mixed layer of the ocean. *Mailing Add:* Dept Marine Earth & Atmospheric Sci NC State Univ Box 8208 Raleigh NC 27695-8208

KAN, JOSEPH RUCE, b Shanghai, China, Feb 10, 38; US citizen; c 3. SPACE PLASMA PHYSICS. *Educ:* Nat Cheng-Kung Univ, Taiwan, BS, 61; Wash State Univ, MS, 66; Univ Calif, San Diego, PhD(appl physics), 69. *Prof Exp:* Fel space physics, Dartmouth Col, 69-72; asst prof, 72-76, assoc prof, 76-81, head, Space Physics & Atmospheric Sci Prog, 77-80, PROF GEOPHYSICS, GEOPHYSICS INST, UNIV ALASKA, 81- *Concurrent Pos:* Consult, Aerospace Corp, 80-81; vis prof, Inst Geophys & Planetary Physics, Univ Calif, Los Angeles, 80-81; assoc ed, J Geophys Res, 84- *Mem:* AAAS; Sigma Xi; Am Phys Soc; Am Geophys Union. *Res:* Space physics; plasma physics. *Mailing Add:* Dept Physics Univ Alaska Fairbanks AK 99701

KAN, LOU SING, b Honan, China, Feb 28, 43; m 70. PHYSICAL CHEMISTRY, BIOPHYSICS. *Educ:* Nat Taiwan Univ, BS, 64; Duquesne Univ, PhD(phys chem), 70. *Prof Exp:* Asst chem, Nat Taiwan Univ, 65; asst phys chem, Duquesne Univ, 66-70, fel, 70-71; fel chem, 71-74, res assoc, 74-77, asst prof, 77-81, ASSOC PROF, DIV BIOPHYS, SCH HYG & PUB HEALTH, JOHNS HOPKINS UNIV, 81- *Concurrent Pos:* Vis fel, Mellon Inst, Carnegie-Mellon Univ, 70-71. *Mem:* Am Chem Soc; Biophys Soc; Sigma Xi. *Res:* Studies of structure and backbone conformations of oligoribonucleotides and deoxyribonucleotides; conformation of nucleic acids; modified nucleic acids. *Mailing Add:* Dept Biochem Johns Hopkins Univ 615 N Wolfe St Baltimore MD 21205

KAN, PETER TAI YUEN, b Canton, China, Apr 12, 27; US citizen; m 51; c 2. ORGANIC CHEMISTRY. *Educ:* Gannon Col, BS, 49; Univ Mich, MS, 51; Wayne State Univ, PhD(org chem), 58. *Prof Exp:* Sr anal chemist, R P Scherer Corp, Mich, 51-55; from res chemist to sr res chemist, Wyandotte Chem Corp, 57-68, res assoc, 68-76, res supvr, 76-85, RES MGR, BASF WYANDOTTE CORP, 86- *Mem:* AAAS; Am Chem Soc; fel Am Inst Chem. *Res:* Isocyanates; isocyanurates; polyurethanes; organo-metallics; general organic synthesis. *Mailing Add:* Polymer Res & Develop BASF Corp 1419 Biddle Ave Wyandotte MI 48192

KAN, YUET WAI, b Hong Kong, China, June 11, 36; m 64; c 2. GENETICS, HEMATOLOGY. *Educ:* Univ Hong Kong, MB & BS, 58, DSc, 80. *Hon Degrees:* MD, Univ Cagliari, Italy, 81; DSc, Chinese Univ Hong Kong, 81 & Univ Hong Kong, 87. *Prof Exp:* Asst prof pediat, Harvard Med Sch, 70-72; chief, Hemat Sect, San Francisco Gen Hosp, 72-79; PROF MED, UNIV CALIF, SAN FRANCISCO, 77- *Concurrent Pos:* Assoc prof med, Depts Med & Lab Med, Univ Calif, San Francisco, 72-77; investr, Howard Hughes Med Inst Lab, 76-; lectr, Harvey Soc, 80. *Honors & Awards:* Damashek Award, Am Soc Hemat, 79; Stratton lectr, Int Soc Hemat, 80; George Thorn

Award, Howard Hughes Med Inst, 80; Allan Award, Am Soc of Human Genetics, 84; Gairdner Found Int Award, 84; Lita Annenberg Hazen Award, 84; Waterford Award, 87; Am Col of Phys Award, 88; Warren Alpert Award, 89; San Remos Int Award, 89. *Mem:* Nat Acad Sci; Am Fedn Clin Res; Am Soc Clin Invest; Asn Am Physicians; fel Royal Soc London; Third World Acad Sci. *Res:* Control of globin synthesis and genetic defects in homoglobinopathies and thalassemia. *Mailing Add:* 20 Yerba Buena Ave San Francisco CA 94127

KANA, ALFRED JAN, applied statistics, analytical statistics, for more information see previous edition

KANA, DANIEL D(AVID), b Cuero, Tex, Sept 22, 34; m 58; c 3. ENGINEERING MECHANICS, MECHANICAL ENGINEERING. *Educ:* Univ Tex, BS, 58, PhD(eng mech), 67; Univ NMex, MS, 61. *Prof Exp:* From res engr to sr res engr, 61-70, group leader, 70-71, mgr struct dynamics & acoust, 71-83, INST ENGR, SOUTHWEST RES INST, 83- *Mem:* Fel Am Soc Mech Engrs; fel Am Inst Aeronaut & Astronaut. *Res:* Liquid and structure dynamic interaction; linear and non-linear vibrations; dynamic response and stability of structures; general structural dynamics and acoustics; environmental testing; earthquake engineering. *Mailing Add:* 10410 Mt Hope St San Antonio TX 78284

KANA'AN, ADLI SADEQ, physical chemistry; deceased, see previous edition for last biography

KANABROCKI, EUGENE LADISLAUS, b Chicago, Ill, Apr 18, 22; m 50; c 2. BIOCHEMISTRY. *Educ:* DePaul Univ, BS, 47; Loyola Univ Chicago, MS, 69; Jagiellonian Univ, Poland, DSc, 83; Nat Registry Clin Chem, cert. *Prof Exp:* Asst chief chemist, Clin Lab, Hines Vet Admin Hosp, 46-48, asst chief biochemist, 48-56, biochemist nuclear med, 56-73, chief clin chemist, Four Hosp Complex, 73-83, RES CHEMIST, NUCLEAR MED SERV, HINES VET ADMIN HOSP, 83- *Concurrent Pos:* Consult, 56-; res assoc, Loyola Univ Chicago; chemist, US Customs Lab, Chicago; WHO investr, Int Atomic Energy Agency. *Mem:* Am Chem Soc; Int Soc Chronobiol; Sigma Xi; Health Physics Soc. *Res:* Etiology of arteriosclerosis; chronobiology; trace elements. *Mailing Add:* 151 Braddock Dr Melrose Park IL 60160

KANAKKANATT, ANTONY, b Cochin, India, Mar 6, 35; m 65; c 3. POLYMER CHEMISTRY. *Educ:* Univ Kerala, India, BSc, 56; Marquette Univ, MS, 60; Univ Akron, PhD, 63. *Prof Exp:* Sr res chemist, Monsanto Co, 63-65; res asst, 65-67, SR ASSOC INDEXER, CHEM ABSTRACTS SERV, 67- *Mem:* Am Chem Soc. *Res:* Catalytic isomerization; sequence distribution in polypeptides; reinforcement in elastomers. *Mailing Add:* Chem Abstracts Serv Dept 64 PO Box 3012 Columbus OH 43210-0012

KANAKKANATT, SEBASTIAN VARGHESE, b Kerala, India, Jan 20, 29; US citizen; m 58; c 1. POLYMER PHYSICS. *Educ:* Univ Madras, BSc, 50; Univ Akron, MS, 65, PhD(polymer physics), 69. *Prof Exp:* Instr chem, Univ Madras, 50-52 & Univ Addis Ababa, 52-64; from asst prof to assoc prof, 69-80, asst dir environ mgt lab, 76-80, PROF GEN TECHNOL, UNIV AKRON, 80-; PRES, UNIQUE TECHNOL, INC, 83- *Concurrent Pos:* Bd chmn, Unique Technol, Inc, 82-83. *Mem:* Am Chem Soc; Rheol Soc; Sigma Xi. *Res:* Diffusion in polymeric matrices used in the controlled release of pesticides, herbicides, fertilizers and photochromic dyes; cancer therapeutic agents. *Mailing Add:* 2459 Audubon Rd Akron OH 44320

KANAL, LAVEEN NANIK, b Dhond, India, Sept 29, 31; m 60; c 3. INTELLIGENT SYSTEMS. *Educ:* Univ Wash, BS, 51, MS, 53; Univ Pa, PhD(elec eng), 60. *Prof Exp:* Develop engr, Can Gen Elec Co, Ont, 53-55; instr elec eng, Moore Sch Elec Eng, Univ Pa, 55-60; mgr mach intel lab, Gen Dynamics/Electronics, NY, 60-62; res mgr info sci, Philco Appl Res Lab, Philco-Ford Corp, 62-65, mgr advan eng & res activ, Commun & Electronics Div, Philco-Ford Corp, 65-69; pres, 69-70, MANAGING DIR, L N K CORP, RIVERDALE, 70-; PROF COMPUT SCI, UNIV MD, COLLEGE PARK, 70- *Concurrent Pos:* Lectr, Wharton Grad Sch, Univ Pa, 63-64, vis assoc prof, 64-66, vis prof, 66-74; adj prof, Lehigh Univ, 65-70; admin comt, Systs & Cybernet, Inst Elec & Electronics Engrs, 72-74,77-81 bd govs, info theory group, 73-79. *Mem:* Fel AAAS; Asn Comput Mach; Pattern Recognition Soc; Soc Mfg Engr; fel Inst Elec & Electronics Engrs; Am Asn Artificial Intel; Soc Photooptical Engrs. *Res:* Information science; machine recognition of patterns; stochastic learning models; statistical classification theory and applications of artificial intelligence; pattern recognition and image processing in remote sennsing and automated digital cartography. *Mailing Add:* Comput Sci Dept Univ Md College Park MD 20742

KANAMORI, HIROO, b Tokyo, Japan, Oct 17, 36; m 64; c 2. SEISMOLOGY, GEOPHYSICS. *Educ:* Univ Tokyo, BS, 59, MS, 61, PhD(geophys), 64. *Prof Exp:* Res fel geophys, Calif Inst Technol, 65-66; from assoc prof to prof, Univ Tokyo, 66-72; PROF GEOPHYS, CALIF INST TECHNOL, 72- *Concurrent Pos:* Vis assoc prof geophys, Mass Inst Technol, 69-70. *Mem:* Fel Am Geophys Union; Seismol Soc Am; Seismol Soc Japan. *Res:* Mechanism of earthquakes; earthquake prediction; application of seismology to earthquake engineering. *Mailing Add:* Seismol Lab PO Bin 2 Calif Inst Technol Pasadena CA 91109

KANAMUELLER, JOSEPH M, b Chicago, Ill, July 4, 38. INORGANIC CHEMISTRY. *Educ:* St Joseph's Col Ind, BS, 60; Univ Minn, PhD(inorg chem), 65. *Prof Exp:* Res assoc inorg chloramine chem, Univ Fla, 65-66; asst prof chem, 66-72, assoc prof, 72-80, PROF CHEM, WESTERN MICH UNIV, 80- *Concurrent Pos:* Vis prof inorg chem, Vienna Tech Univ, Austria, 74. *Mem:* AAAS; Am Chem Soc. *Res:* Synthetic inorganic chemistry, especially sulfur-nitrogen and phosphorus-nitrogen compounds. *Mailing Add:* Dept Chem Western Mich Univ Kalamazoo MI 49008

KANAREK, ROBIN BETH, b Apr 8, 46; m 86; c 2. REGULATION OF FOOD INTAKE, NUTRITION & BEHAVIOR. *Educ:* Antioch Col, BA, Rutgers Univ, MS, PhD(psychol), 74. *Prof Exp:* Assoc prof, 83-89, PROF PSYCHOL & ADJ PROF NUTRIT, TUFTS UNIV, 89- *Concurrent Pos:* Prin investr, NIH Res Grants, 78-; ed-in-chief, Nutrit & Behav, 80-87. *Mem:* Am Inst Nutrit; Behav Pharmacol Soc; Am Col Nutrit; NY Acad Sci; N Am Soc Study Obesity; Soc Study Ingestive Behav. *Res:* Physiological psychology; author. *Mailing Add:* Dept Psychol Tufts Univ 490 Boston Ave Medford MA 02155

KANARIK, ROSELLA, b Hungary, Feb 7, 09; US citizen; m 36; c 2. MATHEMATICS. *Educ:* Univ Pittsburgh, BA, 30, MA, 31, PhD(math), 34. *Prof Exp:* Asst math, Univ Pittsburgh, 31-33; teacher 46-53; instr, 53-56, counr, 56-62, from assoc prof to prof, 62-74, EMER PROF MATH, LOS ANGELES CITY COL, 74- *Mem:* Am Math Soc; Math Assn Am. *Res:* Differential equations; group theory. *Mailing Add:* 238 S Mansfield Ave Los Angeles CA 90036

KANAROWSKI, S(TANLEY) M(ARTIN), b Beausejour, Man, Dec 12, 12; nat US; m 36; c 3. MATERIALS ENGINEERING, POLYMER CHEMISTRY. *Educ:* Univ Toledo, BS, 34. *Prof Exp:* From chemist to chief chemist, Dept Liquor Control, Ohio, 36-42; consult & sr chemist, Ord Plant, Firestone Tire & Rubber Co, Nebr, 42-43, asst dir corp gen lab, Ohio, 43, chief factory prod chem engr, 43-46, res & develop compounding chem engr, 46-49; lab dir & asst res & develop mgr, Fremont Rubber Co, 49-52; res & develop chem engr, Glass Fibers Inc, 52-53; chief res & develop chemist-engr & qual control mgr, Dairypak Butler Inc, 53-60; chief chemist, Northern Region, Enforcement Div, Ohio, 60-62; res & develop chem engr, Consol Paper Co, 62-63; sewage & indust wastes chemist, City of Toledo, 63-64; proj engr & head chemist, Invests Sect, Ohio River Div Labs, Chem & Thermal Effects Br, US Army Eng Div Corps Eng, 64-69, proj leader & prin investr, Eng Mat Div, US Army Construct Eng Res Lab, 69-86; RETIRED. *Mem:* Am Chem Soc; Am Inst Chem Engrs; Am Defense Preparedness Asn. *Res:* Construction materials application research and development; sealants application; waterproofing materials; reflective solar control films for windows; maple gymnasium floor finishes; rubber and synthetic rubber products; elastomers; polymers; coatings; paint-test kit for paint quality evaluation; plastics; glass fibers; resins; paperboard; compounding; quality control; laboratory and factory operations. *Mailing Add:* 1329 Excaliber Ln Sandy Spring MD 20860-1117

KANASEWICH, ERNEST RAYMOND, b Eatonia, Sask, Mar 4, 31; m 69; c 2. GEOPHYSICS. *Educ:* Univ Alta, BSc, 52, MSc, 60; Univ BC, PhD(geophys), 62. *Prof Exp:* Seismologist, Tex Instruments-Geophys Serv, Inc, 53-58; fel, Univ BC, 62-63; from asst prof to assoc prof, 63-71, from asst chmn to actg chmn dept, 69-74, PROF PHYSICS, UNIV ALTA, 71- *Concurrent Pos:* Mem cubcomt glaciol, Nat Res Coun, 63-, subcomt seismol, 67-; mem adv comt explor tech, Northern Alta Inst Technol, 64- & subcomt phys methods appl to geol probs, Nat Adv Comt Res Geol Sci, 65-; vis assoc prof, Dept Earth & Planetary Sci, Calif Inst Technol, 70-71. *Mem:* Am Geophys Union; Soc Explor Geophys; Seismol Soc Am; Can Asn Physicists. *Res:* Gravity; seismology; isotope dating techniques; glaciology. *Mailing Add:* Dept Physics Univ Alta Edmonton AB T6G 2M7 Can

KANATZAR, CHARLES LEPLIE, b St Elmo, Ill, Apr 12, 14; m 40; c 2. ZOOLOGY. *Educ:* Eastern Ill State Teachers Col, BEduc, 35; Univ Ill, MS, 36, PhD(protozool), 40. *Hon Degrees:* DH, Mac Murray Col, 86. *Prof Exp:* Asst zool, Univ Ill, 35-38; asst prof biol, 46-48, prof & head dept, 48-61, dean fac, 61-67, dean col, 67-74, EMER DEAN, MACMURRAY COL, 74- *Concurrent Pos:* Pres, Ill State Acad Sci, 59-60; mem bd adv, Ill State Mus, 62-74, chmn, 70-74; mem bd dirs, Ill State Mus Soc, 85- *Res:* Free-living protozoa; general zoology; general education in science. *Mailing Add:* 1841 Mound Rd Jacksonville IL 62650

KANATZIDIS, MERCOURI G, b Thessaloniki, Greece, Aug 6, 57; m 87. SOLID STATE CHEMISTRY. *Educ:* Univ Thessaloniki, BS, 79; Univ Iowa, PhD(chem), 84. *Prof Exp:* Res assoc chem, Univ Mich, 84-85 & Northwestern Univ, 85-87; ASST PROF CHEM, MICH STATE UNIV, 87- *Concurrent Pos:* NSF presidential young investr, 89; Alfred P Sloan Found fel, 91. *Mem:* Am Chem Soc; Am Crystallog Soc; Mat Res Soc. *Res:* Inorganic chemistry systhesis of novel molecular and solid state compounds of sulfur, selenium and tellurium; intercalation chemistry; conductive polymers; solid state chemistry; crystallography. *Mailing Add:* Dept Chem Mich State Univ East Lansing MI 48824

KANCIRUK, PAUL, b New York, NY, Oct 10, 47; m 77. TECHNICAL MANAGEMENT. *Educ:* City Col New York, BS, 69; Fla State Univ, PhD, 76. *Prof Exp:* Marine biol, Nova Ocean Sci Ctr, 76-77; RES ASSOC AQUATIC ECOL, OAK RIDGE NAT LAB, 78- *Mem:* Ecol Soc Am. *Res:* Management of very large environmental research databases. *Mailing Add:* X-10 Area Bldg 1000 Mail Stop 6335 Oak Ridge Nat Lab Oak Ridge TN 37831

KANCZAK, NORBERT M, b Buffalo, NY, Feb 12, 31; m 58; c 4. ANATOMY. *Educ:* State Univ NY Buffalo, BA, 58, PhD(anat), 64. *Prof Exp:* Sr instr anat, Sch Med, Tufts Univ, 65-66, asst prof, 66-70; ASSOC PROF ANAT, SCH DENT MED, UNIV PITTSBURGH, 70- *Concurrent Pos:* Nat Inst Arthritis & Metab Dis fel, Ohio State Univ, 63-64, Nat Cancer Inst fel, 64-65. *Res:* Morphology and physiology of transitional epithelium; oncology of transitional cell carcinoma; electron microscopic histochemistry. *Mailing Add:* Dept Anat & Histol Univ Pittsburgh 4200 Fifth Ave Pittsburgh PA 15260

KANDA, MOTOHISA, b Kanagawa, Japan, Sept 10, 43; US citizen; m 71; c 4. ELECTROMAGNETIC THEORY, ANTENNA THEORY. *Educ:* Keio Univ, BS, 66; Univ Colo, MS, 68, PhD(elec eng), 71. *Prof Exp:* Technician, Keio Univ, Tokyo, 65-66; tech asst, Electrotech Lab, Tokyo, 66; res asst, Univ Colo, Boulder, 67-71; SECT CHIEF ELECTRONICS ENG, NAT INST

STANDARDS & TECHNOL, BOULDER, COLO, 71- *Concurrent Pos:* Assoc adj prof, Univ Colo, 74-78, adj prof, 78-; chmn, Int Union Radio Sci, 87-, comt A, 90- *Mem:* Fel Inst Elec & Electronics Engrs; Int Union Radio Sci; Sigma Xi. *Res:* Electromagnetic field strength calibrations procedures and their associated accuracy statements; directs calibration services for near field parameters: field strength, antenna factor, pattern, gain; technical design and analysis of international intercomparisons of electromagnetic field standard parameters. *Mailing Add:* Electromagnetic Fields Div Nat Inst Standards & Technol Boulder CO 80303-3328

KANDEL, ABRAHAM, b Tel-Aviv, Israel, Oct 6, 41; US citizen; m 66; c 3. COMPUTER SCIENCES. *Educ:* Technion-Israel Inst Technol, BSc, 66; Univ Calif, Santa Barbara, MSc, 68; Univ NMex, PhD(elec eng, comput sci), 77. *Prof Exp:* From instr to assoc prof comput sci, NMex Inst Mining & Technol, 70-78; assoc prof, 78-80, dir comput sci, dept math & comput sci, 78-84, DIR & PROF COMPUT SCI, FLA STATE UNIV, 80-, CHMN DEPT, 84- *Concurrent Pos:* Vis sr lectr, Ben Gurion Univ Negev & Tel-Aviv Univ, Israel, 76-77; consult, Sandia Labs, Albuquerque, NMex, 76, TASC Corp, Mass & EL-AL, Tadiran; distinguished vis, Inst Elec & Electronics Engrs-Comput Soc, 81-85. *Mem:* Sr mem Inst Elec & Electronics Engrs; Asn Comput Mach; Pattern Recognition Soc; Am Soc Eng Educ. *Res:* Fuzzy sets and systems; computer architecture; performance evaluation; pattern recognition; fault-tolerant systems; switching, microprocessors and logic design; expert systems; applied artificial intelligence. *Mailing Add:* Dept Comput Sci Fla State Univ Tallahassee FL 32306

KANDEL, ERIC RICHARD, b Vienna, Austria, Nov 7, 29; nat US; m 56; c 2. NEUROBIOLOGY, PSYCHIATRY. *Educ:* Harvard Univ, AB, 52; NY Univ, MD, 56. *Hon Degrees:* DSc, Hahnemann Univ & State Univ NY, 86; DHL, Johns Hopkins Univ, 86. *Prof Exp:* Intern, Montefiore Hosp, NY, 56-57; res assoc, Lab Neurophysiol, NIH, 57-60; dir, Mass Ment Health Ctr, 60-65, res psychiat, 60-62, 63-64; from assoc prof physiol & psychiat to prof, Sch Med, NY Univ, 65-74; prof physiol & psychiat, Col Physicians & Surgeons, 74-83, UNIV PROF, COLUMBIA UNIV, 83-; SR INVESTR, HOWARD HUGHES MED INST, 84- *Concurrent Pos:* Numerous lectrs, US & foreign, 59-90; teaching fel psychiat, Harvard Med Sch, 60-61 Milton res fel, 61-62; USPHS spec fel, Lab Gen Neurophysiol, Col France, 62-63; res assoc psychiat, Med Sch, Harvard Univ, 63-64, instr, 64-65; chief, Dept Neurobiol & Behav, Pub Health Res Inst, New York, 68-74; mem, Comt Life Sci, Nat Acad Sci-Nat Res Coun, 68-69; mem, Neuropsychol Res Rev Comt, NIMH, 69-72; assoc ed, J Neurophysiol, 77-80, J Neurosci, 81-83. *Honors & Awards:* Moses Award, 59; Lester N Hofheimer Prize, Am Psychiat Asn, 77, Spec Presidential Commendation, 86; Karl Spencer Lashley Prize in Neurobiol, Am Philos Soc, 81; NY Acad Sci Award in Biol & Med Sci, 82; Albert Lasker Basic Med Res Award, 83; Howard Crosby Warren Medal, Soc Exp Psychologists, 84; J Murray Luck Award for Sci Reviewing, Nat Acad Sci, 88; Nat Medal Sci, 88; Distinguished Serv Award, Am Psychiat Asn, 89. *Mem:* Nat Acad Sci; Inst Med; Am Acad Arts & Sci; fel AAAS; Am Philos Soc; hon mem Am Neurol Asn; Int Brain Res Orgn; Am Psychiat Asn; Soc Neurosci (pres, 80-81). *Res:* Electrophysiology of central neurons; neurosecretion; cellular mechanisms of behavior; neuronal plasticity; author or co-author of seven books and over 220 publications. *Mailing Add:* Ctr Neurobiol & Behav-Col Physicians & Surgeons Columbia Univ 722 W 168th St Res Annex New York NY 10032

KANDEL, RICHARD JOSHUA, b New York, NY, Apr 30, 24; m 48; c 4. PHYSICAL CHEMISTRY. *Educ:* NY Univ, BS, 46, PhD(chem), 50. *Prof Exp:* Instr gen chem, NY Univ, 49-50; mem staff phys chem, Los Alamos Sci Lab, 50-67; mem staff chem prog br, Div Res, 67-69, chief radiation, Isotope & Phys Chem Br, 69-75, chief chem & atomic physics br, div phys res, US Energy Res & Develop Admin, 75-77; chief, Fundamental Interactions Br, Div Chem Sci, US Dept Energy, 77-86; RETIRED. *Concurrent Pos:* Consult. *Mem:* Am Chem Soc. *Res:* Mass spectrometry; kinetics; radiation chemistry. *Mailing Add:* 818 Fordham St Rockville MD 20850

KANDHAL, PRITHVI SINGH, b Bikaner, India, May 6, 35; US citizen; m 58; c 2. PAVEMENT MATERIALS, ASPHALT TECHNOLOGY. *Educ:* Univ Rajasthan, India, Bachelor Engineering, 57; Iowa State Univ, MS, 69. *Prof Exp:* Asst dist engr, Rajasthan Pub Works Dept, India, 57-65, dist engr, 65-68; hwy engr, Berger Assocs, Camp Hill, Pa, 69-70; chief asphalt engr, Pa Dept Transp, Harrisburg, 70-88; ASST DIR, NAT CTR ASPHALT TECHNOL, AUBURN UNIV, 88- *Concurrent Pos:* Chmn, Transp Res Bd Comt A2DO2 Asphalt Mixtures, 82-88, Subcomt Joint & Crack Sealers, Am Soc Testing & Mat, 88-, Subcomt Bituminous Mat, Am Soc Civil Engrs, 89- *Honors & Awards:* W J Emmons Award, Asn Asphalt Paving Technologists, 89. *Mem:* Asn Asphalt Paving Technologists; Am Soc Testing & Mat; Transp Res Bd; Am Soc Civil Engrs. *Res:* Author of over 70 publications; highway pavement materials; mix design; construction; maintenance. *Mailing Add:* 635 Woody Dr Auburn AL 36830

KANDINER, HAROLD J(ACK), b Detroit, Mich, Mar 23, 17; m 41; c 2. CHEMICAL ENGINEERING. *Educ:* Cooper Union, BChE, 38; Univ Mich, MS, 39. *Prof Exp:* Res chemist, Ansbacher Siegle, NY, 39; chemist in charge anal sect, US Treas Dept, Washington, DC, 39-42; chem engr, Navy Dept, 42-43, US Bur Mines, Md, 43-45 & Pa, 45-51; chem engr, Barrett Div, Allied Chem Corp, 51-58, dir qual control, 58-66, asst tech dir, 66-67, asst tech dir, Fabricated Prod Div, 67, sr tech assoc Corp Res, 67-82; bus mgr, Selexo Syst, Norton Corp, 82-83; RETIRED. *Concurrent Pos:* Spec lectr, Univ Pittsburgh, 47-49. *Mem:* Am Chem Soc; Am Inst Chem Engrs; Air Pollution Control Asn. *Res:* Coal hydrogenation and high pressure equipment design; metallurgy of alumina; hydrocarbon phase equilibria; industrial wastes; quality control; coal conversion technology. *Mailing Add:* Old Coach Rd Summit NJ 07901

KANDOIAN, A(RMIG) G(HEVONT), b Van, Armenia, Nov 28, 11; nat US; m 45; c 3. ENGINEERING. *Educ:* Harvard Univ, BS, 34, MS, 35. *Hon Degrees:* DEng, Newark Col Eng, 67. *Prof Exp:* From jr engr to head dept radio commun equip, Int Tel & Tel Corp, 35-46, head radio commun lab, Fed Telecommun Labs, Inc, 46-58, vpres commun systs, Int Tel & Tel Labs, 58-59, vpres & gen mgr, 60-64, vpres eng, 64-65; vpres & gen mgr, Commun Systs Inc, Comput Sci Corp, 65-66, pres, 66-67, exec staff vpres, 67-68; vpres & dir, Scanwell Labs, Inc, 68-70; consult telecommun, US Dept Commerce, 70, dir off telecommun, 70-73; INDEPENDENT CONSULT, TELECOMMUN ENG, 73- *Concurrent Pos:* Exec secy, Cable Television Tech Adv Comt, Fed Commun Comn; mem, Panel on Telecommun Res in US, Nat Acad Eng. *Honors & Awards:* Award, Inst Elec & Electronics Engrs, 65. *Mem:* Fel Inst Elec & Electronics Engrs. *Res:* Radio aids to air navigation; instrument landing systems; radio antenna systems, particularly in the range; design of radio transmitters and receivers for the very high frequency regions; radio ranges for point to point flight of aircraft; radar components; radio communication systems; satellite communication. *Mailing Add:* 194 Orchard Pl Ridgewood NJ 17450

KANDUTSCH, ANDREW AUGUST, b Kennan, Wis, Oct 10, 26; m 52; c 2. BIOCHEMISTRY. *Educ:* Ripon Col, BA, 50; Univ Wis, MS, 52, PhD(biochem), 54. *Prof Exp:* Res fel biochem, Roscoe B Jackson Mem Lab, 54-55, res assoc, 55-57, staff scientist, 57-64, asst dir res, 65-66, dep dir, 81-85, SR STAFF SCIENTIST, JACKSON LAB, 64- *Mem:* Am Soc Biol Chemists. *Res:* Animal sterols and their metabolism; sterol biosynthesis and its regulation; membrane structure and function; relationships between sterols, cancer and atherosclerosis; regulation of dolichol biosynthesis; regulation of the cell replication cycle. *Mailing Add:* Jackson Lab Bar Harbor ME 04609

KANE, AGNES BREZAK, b Danbury, Conn, Nov 3, 46; m. CELL INJURY, EXPERIMENTAL CARCINOGENESIS. *Educ:* Temple Univ Sch Med, MD, 74, PhD(exp path), 76. *Prof Exp:* Postdoctoral fel, Karolinska Inst, Stockholm, Swed, 76-77; staff pathologist, Temple Univ Hosp, 79-82; asst prof, 82-87, ASSOC PROF PATH, BROWN UNIV, 87- *Concurrent Pos:* Lucretia Mott grad fel, 69-71; Sci consult, RI Comn Safety & Occup Health, 86-; consult, Identify Occup Dis which pose a major health threat to workers in RI, 87-90; mem, Environ Health Sci Rev Comt, Nat Inst Environ Health Sci, NIH, 88-92. *Mem:* Sigma Xi; Am Asn Pathologists; Int Acad Path; AAAS; Am Med Women's Asn; Am Thoracic Soc. *Res:* Experimental pathology. *Mailing Add:* Dept Path & Lab Med Brown Univ Providence RI 02912

KANE, BERNARD JAMES, b New York, NY, Sept 22, 32; m 55; c 6. ORGANIC CHEMISTRY. *Educ:* Iona Col, BS, 54; Adelphi Col, MS, 56. *Prof Exp:* Chemist, 57-62, mgr develop terpene chem, 62-71, DIR RES, GLIDDEN-DURKEE, DIV SCM CORP, 71- *Honors & Awards:* D P Joyce Award, Glidden-Durkee, Div SCM Corp, 70. *Mem:* Am Chem Soc. *Res:* Terpene chemical and organic research and development. *Mailing Add:* 333 Ocean Blvd Atlantic Beach FL 32233-5279

KANE, DANIEL E(DWIN), b Iowa Park, Tex, Aug 12, 23; m 53; c 2. CHEMICAL ENGINEERING, STERILIZATION ENGINEERING. *Educ:* Iowa State Univ, BS, 47; Lawrence Col, MS, 50, PhD(chem eng), 53. *Prof Exp:* Chem engr, Phillips Petrol Co, 47-48; sr chem engr, Fibreboard Paper Prod Corp, 53-59; res assoc res & develop, Nat Vulcanized Fibre Co, 59-66, tech mgr, NVF Co, 66-70; supvr paper/coatings, Bus Equip Div, SCM Corp, 70-73, mgr pilot prod, 73-76; group leader mfg technol, Miles Inc, 76-80, mgr process develop, 80-82, mgr sterilization eng, 82-89; RETIRED. *Concurrent Pos:* pharmaceutical process validation, GMPs. *Mem:* Tech Asn Pulp & Paper Indust; Am Chem Soc. *Res:* Kraft chemical recovery; vulcanized fibre; saturating and specialty papers; water and air pollution abatement; reprographics; coatings; hospital supplies; intravenous solutions; steam sterilization of intravenous solutions and process equipment; ETO sterilization; sterilize in place; clean in place. *Mailing Add:* 18680 Quailridge Rd Cottonwood CA 96022

KANE, E(NEAS) D(ILLON), b San Francisco, Calif, Jan 8, 17; m 44; c 8. ENGINEERING. *Educ:* Univ Calif, BS, 38, PhD(mech eng), 49; Kans State Col, MS, 39. *Prof Exp:* Student & design engr, Westinghouse Mfg Co, Philadelphia, 39-40; mech engr, Radiation Lab, Univ Calif, 42-43; group engr, Clinton Eng Works, Tenn Eastman Corp, Oak Ridge, 43-45; asst prof eng design, Univ Calif, 45-47, lectr, 47-48, assoc prof, 50-51, assoc prof radiation lab, 51-52; supvr process eval, Calif Res & Develop Co, 52-53; res, Calif Res Corp, 54-63; mgr prod res, Cheron Res Co, 63-64, vpres prod res, 65-67, pres, 67-70, vpres res, 70-16; asst secy exec comt Standard Oil Co, Calif, 64-65, vpres technol, 77-75; RETIRED. *Mem:* Nat Acad Eng; Am Inst Chem Engrs; Am Inst Mining, Metall & Petrol Engrs; Am Soc Mech Engrs. *Res:* Flow of gases at low pressures; process design and evaluation in nuclear energy and petroleum refining; oil field research. *Mailing Add:* 781 Balra Dr El Cerrito CA 94530

KANE, EDWARD R, b Schenectady, NY, Sept 13, 18; m 48; c 2. TECHNICAL MANAGEMENT. *Educ:* Union Col, BS, 40; Mass Inst Technol, PhD(phys chem), 43. *Prof Exp:* Pres, E I du Pont de Nemours & Co, 73-79, dir, 69-89; RETIRED. *Concurrent Pos:* Dir, Inco Ltd, 81-89, Tex Instruments, 80-89, Mead Corp, 80-88; mem corp, Mass Inst Technol, 79-89; coun mem, Nat Acad Eng, 86-; mem gov bd, Nat Res Coun, 90- *Honors & Awards:* Int Paladium Medal, Soc Indust Chem, 79. *Mem:* Nat Acad Eng (treas, 86-); Soc Chem Indust (pres, 79-80). *Mailing Add:* E I du Pont de Nemours & Co 1007 Market St Wilmington DE 19898

KANE, EVAN O, solid state physics, for more information see previous edition

KANE, FRANCIS JOSEPH, JR, b New York, NY, Mar 29, 29; m 55; c 7. PSYCHIATRY. *Educ:* Iona Col, BS, 49; NY Med Col, MD, 53. *Prof Exp:* Intern med, Mercy Hosp, Wilkes Barre, Pa, 53-54; resident path, NY Med Col, 54-55; resident, psychiat, Inst Living, 57-60, staff psychiatrist, 60-61; from instr to prof psychiat, Sch Med, Univ NC, Chapel Hill, 61-71; clin prof, Sch Med, Tulane Univ, 71-74; PROF PSYCHIAT, BAYLOR COL MED, 74-; DEP CHIEF PSYCHIAT, METHODIST HOSP, TEX MED CTR, 74-

Concurrent Pos: NIH career teacher award, 63-65; med dir, DePaul Hosp, 71-74. Mem: AAAS; AMA; Am Psychosom Soc; Am Psychiat Asn; Acad Psychoanal. Res: Relationship between gonadal hormones and behavior in the normal and mentally ill. Mailing Add: 6565 Fannin MS 706 Houston TX 77030

KANE, GEORGE E, industrial engineering; deceased, see previous edition for last biography

KANE, GORDON LEON, b St Paul, Minn, Jan 19, 37; m 58. HIGH ENERGY PHYSICS. Educ: Univ Minn, BA, 58; Univ Ill, MS, 61, PhD(physics), 63. Prof Exp: Res assoc physics, Johns Hopkins Univ, 63-65; asst prof, 65-75, PROF PHYSICS, UNIV MICH, ANN ARBOR, 75- Concurrent Pos: Guggenheim Mem Found fel, 71-72. Mem: Fel Am Phys Soc. Res: Theoretical high energy physics. Mailing Add: Dept Physics Univ Mich Ann Arbor MI 48109

KANE, GORDON PHILO, b New York, NY, Feb 21, 26; m 52; c 3. PESTICIDE FORMULATION. Educ: Adelphi Univ, BA, 49; Polytech Inst Brooklyn, MS, 53. Prof Exp: Chemist, Warner-Lambert Pharmaceut Corp, NY, 51-54; indust chemist, Barrett Div, Allied Chem Corp, 56; res dir, Valchem Div, United Merchants & Mfrs, SC, 56-62, res scientist, Res Ctr, 62-64; sr org chemist, Columbia Nitrogen Corp, 64-68; sr chemist, Ciba-Geigy Corp, 68-73, group leader, 73-78, sr staff chemist, 78-87; CONSULT, 87- Mem: Am Chem Soc; fel Am Inst Chemists. Res: Pesticide formulation and process development. Mailing Add: 38127 Monticello Dr Prairieville LA 70769

KANE, HARRISON, b Brooklyn, NY, Jan 2, 25; m 52; c 3. GEOTECHNICAL ENGINEERING. Educ: City Col New York, BCE, 47; Columbia Univ, MS, 48; Univ Ill, PhD(soil mech), 61. Prof Exp: Design engr, Parsons, Brinckerhoff Hall & Macdonald, 48-51 & 53-56; planning engr, US Dept Army, Ger, 51-53; lectr struct, City Col New York, 56; asst prof civil eng, Pa State Univ, 56-61; asst prof, Univ Ill, 61-64; from assoc prof to prof, 64-70, chmn, civil eng dept, 71-83, PROF CIVIL & ENVIRON ENG, UNIV IOWA, 70- Concurrent Pos: NSF res grant, 67-72. Mem: Fel Am Soc Civil Engrs; Am Soc Eng Educ. Res: Engineering properties of loess; earth pressures on braced excavations. Mailing Add: Dept Civil Eng Univ Iowa Iowa City IA 52242

KANE, HENRY EDWARD, b New Orleans, La, Dec 18, 17; m 53; c 2. GEOLOGY. Educ: La State Univ, BS, 45, MS, 48; Univ Calif, Los Angeles, PhD, 65. Prof Exp: Geol scout, Tex Co, 44; field geologist, Miss River Comn, 45; asst, La State Univ, 46-48; subsurface geologist, Stanolind Oil & Gas Co, Standard Oil Co Ind, 48; asst, Univ Calif, Los Angeles, 49-51; geologist, Lloyd Corp Ltd, 52-56; asst prof geol, Lamar State Col Technol, 56-60 & Ft Hays Kans State Col, 60-61; from asst prof to prof, 61-83, EMER PROF GEOL, BALL STATE UNIV, 83- Concurrent Pos: Res grants, Shell Res & Develop Co, Shell Oil Co, 56-59, Univ Calif, Los Angeles, 64-65, Ball State Univ, 65-67, 68-69 & 70-71, Ind Acad Sci, 66 & Non-Western Studies, 67; Univ Sci Improv Prog grant, 70-71; Partic, NSF teachers cong, Univ Ore, 59, Am Univ, 61, field geol inst, Ind Univ, 60, res adv undergrad res prog, 59 & NSF geol of Gulf Coast, Rice Univ, 67. Mem: Fel Geol Soc Am. Res: Recent sedimentation, microfaunas, quaternary geomorphology and geology of the Gulf Coast and the Southern Rockies; fluviatile geomorphology; Kentucky River Basin; geology, Eastern Indiana. Mailing Add: 4109 N Redding Rd Muncie IN 47304

KANE, HOWARD L, b Pittsburgh, Pa, Dec 11, 11; m 32; c 2. ORGANIC CHEMISTRY. Educ: Univ Pittsburgh, BS, 32, PhD(org chem), 36. Prof Exp: Chief res chemist, Nat Starch Prods, Inc, 36-42, chief chemist, Synthetics Dept, 42-46; vpres & dir res, Polymer Indust, Inc, 46-64; chmn dept sci, 64-67, chmn div basic sci, 67-75, prof chem, 75-83, EMER PROF CHEM, EDISON COMMUNITY COL, 83- Honors & Awards: Phillips Medal, 32. Mem: AAAS; Am Chem Soc. Res: Adhesives; modifications of starch; adhesive compositions of matter. Mailing Add: 1495 Whiskey Creek Dr Ft Myers FL 33919

KANE, JAMES F, b Philadelphia, Pa, Nov 22, 42; m 66; c 3. MICROBIAL BIOCHEMISTRY & GENETICS. Educ: State Univ NY, Buffalo, PhD(biol), 69. Prof Exp: Asst prof, 70-75, assoc prof microbiol & immunol, Ctr Health Sci, Univ Tenn, 75-81; sr scientist, Bethesda Res Lab, 81-82; sr res specialist, 82-84, sr res group leader, 84-88, FEL, MONSANTO CO, 88- Mem: Am Soc Microbiol; Am Soc Biol Chemists; Sigma Xi. Res: Development of microbial systems to express recombinant gene products and scaling up of these systems to operate in large volume fermentes. Mailing Add: 1845 Walnut Way Dr St Louis MO 63146

KANE, JAMES JOSEPH, b New York, NY, Mar 4, 29; m 67. ORGANIC CHEMISTRY. Educ: Upsala Col, BS, 54; Ohio State Univ, PhD(org chem), 60. Prof Exp: Res chemist, E I du Pont de Nemours & Co, 60-64; asst prof, 64-65, coordr, 65-66, asst prof, 66-70, ASSOC PROF CHEM, WRIGHT STATE UNIV, 70- Mem: Am Chem Soc; Sigma Xi. Res: Chemistry of small and medium carbocyclic and heterocyclic systems; polymers with high thermal stability. Mailing Add: 925 Talus Dr Yellow Spring OH 45387

KANE, JOHN JOSEPH, b Key West, Fla, Jan 13, 15; m 39; c 1. RADIOLOGY. Educ: Col Charleston, BS, 35; Med Col SC, MD, 38; Am Bd Radiol, dipl, 57. Prof Exp: From instr to assoc prof, 57-66, prof radiol, 66-77, EMER PROF RADIOL, MED UNIV SC, 77-; radiologist, Med Univ Hosp, 57-77. Concurrent Pos: Consult, Mat Air Transport Serv, US AFB, Charleston, 62-85. Mem: Am Col Radiol; Radiol Soc NAm. Res: Cardiac radiology. Mailing Add: 1875 Capri Dr Charleston SC 29407

KANE, JOHN POWER, b West Point, NY, July 15, 32; m 66; c 3. MEDICINE, BIOCHEMISTRY. Educ: Ore State Univ, BS, 55; Univ Ore, MS & MD, 57; Univ Calif, San Francisco, PhD(biochem), 71. Prof Exp: Intern med, Santa Clara County Hosp, San Jose, 57-58; asst resident internal

med, Hosps, Stanford Univ, 58-59; asst resident, Hosps, 59-60, asst prof, 71-76, ASSOC PROF MED, CARDIOVASC RES INST, UNIV CALIF, SAN FRANCISCO, 76-; 415-476-1517. Concurrent Pos: Am Heart Asn estab investr, Cardiovasc Res Inst, Univ Calif, San Francisco, 71-76; mem coun arteriosclerosis, Am Heart Asn. Mem: AAAS; Am Chem Soc; Biophys Soc; Am Soc Clin Invest; Am Fedn Clin Res; Am Asn Physicians. Res: Structure and function of serum lipoproteins; lipid and carbohydrate metabolism; arteriosclerosis. Mailing Add: Med Ctr 1327M Univ Calif San Francisco CA 94143

KANE, JOHN ROBERT, b Washington, DC, May 16, 36; m. NUCLEAR PHYSICS. Educ: Loyola Col, Md, BS, 59; Carnegie-Mellon Univ, MS, 62, PhD, 64. Prof Exp: Res assoc, 64-68, asst prof, 68-71, assoc prof nuclear physics, 71-78, PROF, COL WILLIAM & MARY, 79- Mem: Am Phys Soc. Res: Muonic and hadronic atom x-ray studies; muonium in vacuum measurements; rare kaon decays. Mailing Add: Dept Physics Col William & Mary Williamsburg VA 23185

KANE, JOHN VINCENT, JR, b Philadelphia, Pa, Feb 13, 28; m 57; c 2. THREE PARTICLE REACTIONS, NUCLEAR ELECTRONICS. Educ: Villanova Univ, BS, 50; Univ Pa, MS, 52, PhD(physics), 57. Prof Exp: Scientist nuclear physics, Brookhaven Nat Lab, 57-61 & Bell Telephone Labs, 61-66; asst prof nuclear physics, Mich State Univ, 66-68; vis prof nuclear physics, Univ Sask, 68-69; guest prof nuclear physics, Univ Munich, 69-70; researcher ion implanters, Extrion Inc, 73-74; prof nuclear physics, Mass Inst Technol, 74-75; PRES NUCLEAR PHYSICS, J V KANE & CO, 75. Honors & Awards: Two Diamonds with Gold, Soaring Soc Am, 70. Mem: Am Phys Soc; Soaring Soc Am. Res: Nuclear physics; low energy and high energy; electronics; computers; aviation theory and practice; radio communication; particle accelerators. Mailing Add: Two Cricket Ave Apt No 7 Ardmore PA 19003

KANE, MARTIN FRANCIS, b Portland, Maine, Sept 9, 28; m 57; c 5. GEOPHYSICS. Educ: St Francis Xavier Univ, BSc, 51; St Louis Univ, PhD, 70. Prof Exp: Geophysicist, Regional Geophys Br, US Geol Surv, 52-63 & Astrogeol Br, 64-67, supvry geophysicist, Regional Geophys Br, 68-70, geologist-in-chg, Marine Geol Br, 70-71, chief, Regional Geophys Br, 72-78, res scientist, 78-83, chief geophysicist, US Geol Surv Mission Jeddah Saudi Arabia, 84-86, CONSULT, US GEOL SURV, 86- Mem: Geol Soc Am; Am Geophys Union; Soc Explor Geophys; Europ Asn Explor Geophys. Res: Regional geophysics; marine geophysics; planetary geophysics. Mailing Add: 1347 S Routtway Lakewood CO 80232

KANE, ROBERT B, b Oak Park, Ill, July 27, 28; m 50; c 5. MATHEMATICS. Educ: Univ Ill, BS, 50, MS, 58, PhD(math educ), 60. Prof Exp: Teacher pub schs, Ill, 50-51; mgr trainee, Stand Oil Co, 53-55, indust engr, 55-57; res asst bur educ res, Univ Ill, 59-60; from asst prof to assoc prof, 60-69, PROF MATH & EDUC, PURDUE UNIV, 69-, DIR TEACHER EDUC, 75- Concurrent Pos: Consult sch dists, 61-; vis res prof, Univ Canterbury, 69. Mem: Nat Coun Teachers Math; Am Educ Res Asn. Res: Linguistic factors in learning and teaching mathematics; cognitive development and mathematics learning. Mailing Add: Purdue Univ Main Campus Purdue Univ West Lafayette IN 47907

KANE, ROBERT EDWARD, b Erie, Pa, Mar 22, 31; m 53, 70. EMBRYOLOGY. Educ: Mass Inst Technol, SB, 53; Johns Hopkins Univ, PhD(develop biol), 57. Prof Exp: Asst prof biochem, Brandeis Univ, 58-61; asst prof cytol, Dartmouth Med Sch, 61-66; assoc prof, 66-69, PROF BIOL, UNIV HAWAII, 69-, ASSOC DIR PAC BIOMED RES CTR, KEWALO LAB, 66- Mem: Am Soc Cell Biol; Int Soc Develop Biologists; fel AAAS. Res: Cell division; cell motility; electrophysiology of fertilization; biochemistry of development. Mailing Add: Kewalo Lab Pac Biomed Res Ctr Univ Hawaii Honolulu HI 96822

KANE, RONALD S(TEVEN), b New York, NY, Feb 11, 44; m 68; c 2. THERMAL FLUID ANALYSIS, ENERGY SYSTEMS. Educ: City Col New York, BME, 65, MME, 69; City Univ New York, PhD(eng), 73. Prof Exp: Mech engr heat transfer, Pratt & Whitney Aircraft, 65-66; proj engr mech, Esso Res & Eng, 66-70; grad asst fluid mech, City Col New York, 70-73; consult thermal sci, Polytech Design Co, 73-74; from asst prof to prof mech eng, Manhattan Col, 74-85, chmn dept & grad dir reactor admin, 81-85; dean grad studies, res & continuing prod educ, prof mech eng, Stevens Inst Technol, 85-90, ASST VPRES ACAD AFFAIRS, NJ INST TECHNOL, 90- Concurrent Pos: Res asst, City Col New York Res Found, 70-73; consult, Polytech Design Co, Foster Wheeler Energy Corp, 74-; reviewer, McGraw-Hill Publ Co, 76-; consult, Burns & Roe, 80-; consult, Transnuclear Corp, Gen Elec, 85- Mem: Am Soc Mech Engrs; Am Inst Chem Engrs; Sigma Xi; Am Soc Eng Educ. Res: Drag reduction in particulate suspensions; coal gasification; liquid metal heat transfer and fluid mechanics; ocean thermal energy conversion; heat pipe development; new energy resource development; advanced reactor safety. Mailing Add: 98 Algonquin Trail Oakland NJ 07436

KANE, STEPHEN SHIMMON, b Chicago, Ill, Nov 5, 17; m 46; c 2. CHEMISTRY. Educ: Univ Chicago, BS, 37, PhD(org chem), 41. Prof Exp: Res chemist, Phillips Petrol Co, Okla, 40-42 & Apex Smelting Co, Chicago, 42-44; instr chem, Jersey City Jr Col, 46-48, San Bernardino Valley Col, 48-50 & East Los Angeles Col, 50-59; res chemist, Zolatone Process, Inc, 59-63; instr chem, East Los Angeles Col, 63-68; asst dean instr, West Los Angeles Col, 68-75, dean student serv, 75-80; RETIRED. Mem: AAAS; Am Chem Soc; Soc Coating Technol. Res: Organic coatings; petroleum; aluminum magnesium metallurgy; organic synthesis. Mailing Add: 4333 Redwood Ave No 5 Marina Del Rey CA 90292

KANE, THOMAS R(EIF), b Vienna, Austria, Mar 23, 24; nat US; m 51; c 2. MECHANICS, AEROSPACE ENGINEERING. Educ: Columbia Univ, BS, 49 & 50, MS, 52, PhD(appl mech), 53. Hon Degrees: Dr Tech Sci, Tech Univ Vienna, Austria, 90. Prof Exp: Res assoc, Columbia Univ, 52-53; from asst prof to assoc prof mech eng, Univ Pa, 53-60; PROF ENG MECH, STANFORD UNIV, 61- Concurrent Pos: Fulbright lectr, Victoria Univ Manchester, 58-59; vis prof, Fed Univ Rio de Janeiro, 71-72. Honors & Awards: Dirk Brouwer Award, Am Astron Soc; Alexander von Humboldt Prize. Mem: Am Soc Mech Engrs; Am Astronaut Soc. Res: Dynamics; human motion; mechanics of continua. Mailing Add: Dept Mech Eng Stanford Univ Stanford CA 94305

KANE, WALTER REILLY, b Ithaca, NY, Nov 3, 26; m 53; c 1. NUCLEAR PHYSICS. Educ: Stanford Univ, BS, 49; Univ Wash, MS, 51; Harvard Univ, PhD(physics), 59. Prof Exp: Physicist, Nat Bur Stand, 51-52; mem staff, Los Alamos Sci Lab, 52-54; physicist, Avco Mfg Corp, 56-57; res assoc, 58-60, from asst physicist to assoc physicist, 60-66, PHYSICIST, BROOKHAVEN NAT LAB, 66- Mem: Fel Am Phys Soc. Res: Nuclear spectroscopy; neutron physics. Mailing Add: Dept Nuclear Energy Brookhaven Nat Lab Upton NY 11973

KANE, WILLIAM J, b Brooklyn, NY, Feb 22, 33; m 60; c 5. ORTHOPEDIC SURGERY, PHYSIOLOGY. Educ: Col of the Holy Cross, AB, 54; Columbia Univ, MD, 58; Univ Minn, Minneapolis, PhD(orthop surg), 65. Prof Exp: From instr to assoc prof orthop surg, Univ Minn, Minneapolis, 64-71; Ryerson prof orthop surg & chmn dept, 71-78, PROF ORTHOP SURG, MED SCH, NORTHWESTERN UNIV, 78- Honors & Awards: Kappa Delta Award, Am Acad Orthop Surg, 66. Mem: Am Orthop Asn; Am Acad Orthop Surg; Scoliosis Res Soc (pres, 79-80); Int Soc Study Lumbar Spine; Am Col Surgeons. Res: Bone blood flow; degenerative diseases of the spine; pelvic fractures; scoliosis. Mailing Add: 701 Park Ave Minneapolis MN 55415

KANE, WILLIAM THEODORE, b Jamaica, NY, Sept 8, 32; m 54; c 4. X-RAY DIFFRACTION, SPECTROSCOPY. Educ: Univ Kans, BA, 60; Univ Mo, PhD(x-ray crystallog), 66. Prof Exp: Crystallographer, 66-67, res crystallographer, 67-73, supvr, X-ray Anal, 73-74, RES SUPVR, CORNING GLASS WORKS, 74- Concurrent Pos: Treas, Nat Conf Electron Probe Anal, 69-71. Mem: Mineral Soc Am; Am Crystallog Asn; Microbeam Anal Soc. Res: Instrumental analysis of materials including electron probe, x-ray diffraction, scanning electron micorscopy, electron transmission microscopy; laboratory automation and computerization. Mailing Add: MP-RO-2-2 Corning Glass Works Corning NY 14830

KANEDA, TOSHI, b Utsunomiya-shi, Japan, May 21, 25; m 59; c 2. BIOCHEMISTRY, MICROBIOLOGY. Educ: Tokyo Inst Technol, BEng, 50; Univ Tokyo, DSc(biochem), 62. Prof Exp: Fel microbiol, Prairie Regional Lab, Nat Res Coun Can, 56-58; fel biochem, Sch Med, Western Reserve Univ, 58-60; RES MICROBIOLOGIST, ALTA RES COUN, 60-, HEAD BIOL, 81-, RES FEL, 84- Concurrent Pos: Med Res Coun Can grant, 64-; hon prof med bact, Univ Alta, 75-, Heilongjiang Acad Sci, China, 88- Mem: Am Chem Soc; Can Biochem Soc. Res: Microbiology of fossil fuels; low temperature microbiology; biosynthesis and functions of iso and anteiso series of fatty acids in bacteria. Mailing Add: 250 Karl Clark Rd Edmonton AB T6N 1E4 Can

KANEHIRO, YOSHINORI, b Puuloa, Hawaii, Oct 6, 19; m 47; c 2. SOIL SCIENCE. Educ: Univ Hawaii, BS, 42, MS, 48, PhD(soil sci), 64. Prof Exp: Asst soils & agr chem, 42-48, jr soil scientist, 48-57, from asst soil scientist & asst prof to assoc soil scientist & assoc prof, 57-70, SOIL SCIENTIST & PROF SOIL SCI, UNIV HAWAII, 70- Concurrent Pos: Tech consult, Olin Mathieson Chem Corp, 57; res soil scientist, Agr Res Serv, USDA, Ft Collins, Colo, 68-69. Mem: Am Soc Agron; Soil Sci Soc Am; Sigma Xi. Res: Nitrogen transformation in soils; minor elements, especially zinc in soils; clay mineralogy of soils. Mailing Add: Dept Agron & Soil Sci Univ Hawaii 2500 Campus Rd Honolulu HI 96822

KANEKO, JIRO JERRY, b Stockton, Calif, Nov 20, 24; m 50; c 3. PHYSIOLOGY, CLINICAL BIOCHEMISTRY. Educ: Univ Calif, AB, 52, DVM, 56, PhD(comp path), 59. Hon Degrees: DVSc, Belg, 80. Prof Exp: Asst specialist, Exp Sta, 56-59, from asst prof to assoc prof, 59-69, PROF CLIN PATH & CHMN DEPT, SCH VET MED, UNIV CALIF, DAVIS, 69- Concurrent Pos: Lectr, 57-59. Mem: Soc Exp Biol & Med; Am Physiol Soc; Am Asn Clin Chem; Am Col Vet Pathologists; Sigma Xi; Am Chem Soc. Res: Biochemistry of erythrocyte and hemoglobin of animals; blood dyscrasias of animals; metabolic diseases; kinetics of erythropoiesis and granulopoiesis; organ functions. Mailing Add: UM-Clin Path Univ Calif Davis CA 95616

KANEKO, THOMAS MOTOMI, b Tokyo, Japan, Aug 15, 14; US citizen; m 57. ORGANIC CHEMISTRY, POLYMER CHEMISTRY. Educ: Univ Utah, BSChE, 36, PhD(metall), 56. Prof Exp: Assayer, Nev Mines Div, Kennecott Copper Corp, 36-39; res chemist, Cent Res Labs, Mitsubishi Chem Industs Ltd, Japan, 39-41; res engr, Res & Planning Dept, Mitsubishi Rayon Co, Ltd, 50-52; res metallurgist, Union Carbide Nuclear Div, 56-57; res chemist, Nat Distillers & Chem Corp, 57-59; res metallurgist, Basf Wyandotte Corp, 59-65, sr res chemist, 65-78, res assoc, 78-84; RETIRED. Concurrent Pos: Task force chmn, Am Soc Test & Mat, 77- Mem: Fel AAAS; Am Chem Soc; fel Am Inst Chem; NY Acad Sci; Weed Sci Soc Am. Res: Extractive metallurgy; reaction kinetics; colloid and surfactant chemistry; surfactant applications research in the formulating and evaluating of detergents, pesticides and metal processing compounds. Mailing Add: 1224 Bracebridge Ct Campbell CA 95008-6427

KANE-MAGUIRE, NOEL ANDREW PATRICK, b Brisbane, Queensland, Australia, May 4, 42; m 69; c 1. INORGANIC CHEMISTRY. Educ: Univ Queensland, BSc, 63, Hons, 64, PhD(chem), 69. Prof Exp: Assoc inorg chem, Boston Univ, 68-69 & Wayne State Univ, 69-70; assoc, Carleton Univ, 70-73; asst prof, 73-76, ASSOC PROF CHEM, FURMAN UNIV, 77- Mem: Am Chem Soc. Res: Reaction mechanisms of inorganic compounds; transition metal photochemistry. Mailing Add: Dept Chem Furman Univ Poinsett Hwy Greenville SC 29613

KANEMASU, EDWARD TSUKASA, b Hood River, Ore, Nov 16, 40; m; c 3. AGRICULTURE. Educ: Mont State Univ, BS, 62, MS, 64; Univ Wis-Madison, PhD(soil physics), 69. Prof Exp: From asst prof to assoc prof, Kans State Univ, 69-78, prof agron, Evapo-transportation Lab, 78-89; PROF & HEAD, AGRON DEPT, UNIV GA, 89- Honors & Awards: Agron Res Award, Am Soc Agron; Fel, AAAS. Mem: AAAS; fel Am Soc Agron; Am Meteorol Soc. Res: Stomatal diffusion resistance as influenced by leaf water potential and light; evapo-transpiration and water use efficiency of agronomic crops. Mailing Add: Agron Dept Univ Ga Griffin GA 30223

KANES, WILLIAM H, b New York, NY, Oct 15, 34; m 59, 84; c 6. TECTONICS. Educ: City Col New York, BS, 56; Univ WVa, MS, 58, PhD(geol), 65. Prof Exp: Sr res geologist, Exxon Prod Res Co, 60-66, area geologist, Exxon Stan Libya, 67-69; asst prof geol, WVa Univ, 69-70; assoc prof, 71-75, dir, Earth Sci & Resource Inst, 75, PROF GEOL, UNIV SC, 75-, DISTINGUISHED PROF EARTH RESOURCES, 85- Concurrent Pos: NSF resident res prof, Acad Sci Res & Technol, Cairo, Egypt, 76-77; vis prof fel, Univ Col Swansea, Univ Wales, 77-80, hon prof fel, Univ Col Aberystwyth, 79-83 & Univ Col Swansea, 80-86, co-dir, Earth Resources Inst, Univ Col Swansea, 80-86; co-dir, ESRI-UK, Univ Bristol, 86-88, hon prof fel, 86-89; vis prof & adv, Postgrad Res Inst Sedimentology & exec dir, ESRI-UK, Univ Reading, 89- Mem: Assoc Am Asn Petrol Geol; Soc Econ Paleontologists & Mineralogists; fel Geol Soc Am; fel AAAS; Am Geophys Union. Res: Stratigraphy, sedimentation and structural geology in the Appalachian Region; African and circum Mediterranean regional geology and tectonics; South American tectonics and petroleum geology; tectonics and petroleum geology of Eastern Europe and the USSR. Mailing Add: Earth Sci & Resources Inst Univ SC 901 Sumpter St Columbia SC 29208

KANESHIGE, HARRY MASATO, b Aiea, Oahu, Hawaii, July 11, 29; m 63, 81; c 2. CIVIL ENGINEERING. Educ: Univ Wis, BS, 51, MS, 52, PhD(civil eng), 59. Prof Exp: Instr civil eng, Univ Wis, 54-56, 57-58, proj assoc, 56-57; from asst prof to assoc prof, 58-74, PROF CIVIL ENG, OHIO UNIV, 74- Mem: AAAS; Am Soc Eng Educ; Am Soc Civil Engrs; Am Water Works Asn; Water Pollution Control Fedn; Am Water Resources Asn. Res: Environmental health and sanitation; surveying and mapping; water and wastewater treatment. Mailing Add: Dept Civil Eng Ohio Univ Athens OH 45701

KANESHIRO, EDNA SAYOMI, b Hilo, Hawaii, Dec 20, 37. CELL BIOLOGY, BIOCHEMISTRY. Educ: Syracuse Univ, BS, 57, MS, 62, PhD(zool), 68. Prof Exp: USPHS fel cell biol, Univ Chicago, 68-70; NSF fel biochem, Bryn Mawr Col, 70-72; asst prof, 72-78, ASSOC PROF BIOL, UNIV CINCINNATI, 78- Concurrent Pos: Mem corp, Marine Biol Lab, Woods Hole, 73-; sr res microbiologist, Nat Inst Allergy & Infectious Dis, 80-81. Mem: Am Soc Cell Biol; Soc Protozoologists; AAAS; NY Acad Sci; Sigma Xi. Res: Structure and function of protozoans; membrane structure, function and biochemistry; osmoregulation in marine organisms. Mailing Add: Dept Biol Sci Univ Cincinnati Cincinnati OH 45221

KANESHIRO, KENNETH YOSHIMITSU, b Honolulu, Hawaii, Dec 15, 43; m 67; c 2. EVOLUTIONARY BIOLOGY. Educ: Univ Hawaii, BA, 65, MS, 68, PhD(entom), 74. Prof Exp: COORDR, HAWAIIAN DROSOPHILA PROJ, DEPT ENTOM, UNIV HAWAII, 70-, DIR, HAWAIIAN EVOLUTIONARY BIOL PROG, 85- Mem: Entom Soc Am; Soc Study Evolution; Am Soc Naturalists. Res: Basic mechanisms of speciation processes in Hawaiian Drosophila; tools and techniques for the formulation of a biosystematic classification of the endemic Hawaiian Drosophilidae; sexual behavior and inferences of directions of evolution. Mailing Add: Hawaiian Evolutionary Biol Prog 3050 Maile Way Honolulu HI 96822

KANESHIRO, TSUNEO, microbiology; deceased, see previous edition for last biography

KANEY, ANTHONY ROLLAND, b Centralia, Ill, Mar 8, 40. GENETICS. Educ: Wabash Col, AB, 61; Univ Ill, Urbana, PhD(microbiol), 66. Prof Exp: Res assoc microbiol, Univ Ill, Urbana, 66-67; asst prof biol, Univ PR, San Juan, 67-69; from asst prof to assoc prof, 69-80, PROF BIOL, BRYN MAWR COL, 80- Concurrent Pos: NATO Fel, Brit Inst Carlsberg Found, Copenhagen, 76-77. Mem: Genetics Soc Am; Soc Protozool. Res: Developmental genetics of Tetrahymena thermophila. Mailing Add: Dept Biol Bryn Mawr Col Bryn Mawr PA 19010

KANFER, JULIAN NORMAN, b Brooklyn, NY, May 23, 30; m 87; c 2. BIOCHEMISTRY. Educ: Brooklyn Col, BS, 54; George Washington Univ, MS, 58, PhD(biochem), 61. Prof Exp: Lab asst biol, Brooklyn Col, 54-55; chemist, Nat Inst Neurol Dis & Blindness, 55-61, biochemist, 62-69; assoc biochemist, Mass Gen Hosp, Boston, 69-71; biochemist, 71-75; dir biochem, Eunice Kennedy Shriver Ctr, 71-75; dept head, 75-86, PROF BIOCHEM, FAC MED UNIV MAN, 75- Concurrent Pos: NIH res fel, Harvard Med Sch, 61-62, NSF res fel, 62-63; assoc prof, Med Sch, Duke Univ, 68-69 & Harvard Med Sch; adj assoc prof Brandeis Univ; mem med adv bd, Nat Tay-Sachs Found. Honors & Awards: Vis Scientist Award, Med Res Coun Can, 83, 84. Mem: AAAS; Am Chem Soc; Am Soc Biol Chem; Am Soc Neurochem; Int Soc Neurochem. Res: Sphingolipid metabolism in relationship to the sphingolipidosis; membrane phosopholipids of brain; alzheimers disease, multiple sclerosis. Mailing Add: Dept Biochem Univ Man Winnipeg MB R3T 2N2 Can

KANG, C YONG, b Hadong, Korea, Nov 28, 40; Can citizen; m 66; c 3. VIROLOGY, MOLECULAR BIOLOGY. Educ: Malling Agr Col, Denmark, DiplVSci, 63; Kon-Kuk Univ, Seoul, Korea, BSA, 65; McMaster Univ, Hamilton, Can, PhD(virol), 71. Prof Exp: Asst viral oncol, McArdle Lab, Univ Wis, 71-74; from asst prof to assoc prof virol, Southwestern Med Sch, Univ Tex, 74-82; PROF MICROBIOL & CHMN DEPT, SCH MED, UNIV OTTAWA, 82- Concurrent Pos: Dir, Univ Ottawa Biotechnol Res Inst, 87- Mem: Am Soc Virol; Am Soc Microbiol; AAAS; Can Soc Microbiologists;

Genetic Soc Can; NY Acad Sci. *Res:* Studies of molecular mechanisms of viral interference mediated by defective interfering virus particles; investigation of molecular genetics of Hantaviruses; studies on cellular transformation by reticuloendotheliosis virus transforming gene rel. *Mailing Add:* Dept Microbiol & Immunol Univ Ottawa 451 Smyth Rd Ottawa ON K1H 8M5 Can

KANG, CHANG-YUIL, b Korea, Nov 28, 54; m 84; c 2. IDIOTYPE, VIRAL IMMUNOLOGY. *Educ:* Seoul Nat Univ, BS, 77, MS, 81; State Univ NY, Buffalo, PhD(microbiol, immunol), 86. *Prof Exp:* Teaching asst microbiol, Seoul Nat Univ Sch Pharm, 79-81, instr, 81-82; res affil, Roswell Park Mem Inst, 83-87; SCIENTIST, IDEC PHARMACEUT CORP, 87- *Mem:* Am Asn Immunologists. *Res:* Analysis of humoral immune response in HIV-1 infected individuals; characterization of different neutralizing antibodies with various epitope specificity and idiotype characteristics to define their roles for host defense mechanism. *Mailing Add:* 11099 N Torrey Pines Rd No 160 La Jolla CA 92037

KANG, CHIA-CHEN CHU, b China, Apr 14, 23; m 51; c 2. FUEL SCIENCE, PETROLEUM SCIENCE. *Educ:* Shanghai Univ, BS, 44; Univ Ill, MS, 49, PhD(anal chem), 51. *Prof Exp:* Res chemist, M W Kellogg Co, 51-52, supvr, 53-57, sect head, 57-67, mgr sci serv, 67-70; consult chemist, 70-74; asst to sr vpres res & develop, Hydrocarbon Res, Inc, 74-78; vpres, Catalysis Res Corp, 78-81; PRES, KANG ASSOCS, INC, 81- *Concurrent Pos:* Consult catalysis, United Catalysts, Inc, 71- *Mem:* Am Chem Soc. *Res:* Catalysis; coal liquefaction process development; petroleum and petrochemical process development; trouble shooting for ammonia unit; chemical and instrumental analytical method development. *Mailing Add:* 301 Gallup Rd Princeton NJ 08540

KANG, DAVID SOOSANG, b Yiryong, Korea, Nov 7, 31; US citizen; m 58; c 4. PEDIATRICS, GENETICS. *Educ:* Seoul Nat Univ, MD, 53, PhD(pharmacol), 63. *Prof Exp:* Asst prof genetics, Ill Inst Technol, 71-75; from asst prof to assoc prof, 77-88, PROF PEDIAT & GENETICS, COL MED, RUSH UNIV, 88- *Mem:* Am Soc Human Genetics. *Res:* Genetic and biochemical studies of genetic disease and common diseases: interrelations of basic amino acids and their metabolites in urea cycle disorder, and role of protein-bound homocystine in common diseases. *Mailing Add:* Dept Pediat Rush Univ 600 S Paulina St Chicago IL 60612

KANG, IK-JU, b Korea, Nov 13, 28; m 55; c 3. ATOMIC PHYSICS. *Educ:* Yonsei Univ, Korea, BS, 55, MS, 57; Northwestern Univ, PhD(physics), 62. *Prof Exp:* Instr physics, Yonsei Univ, 55-59; res assoc, Brandeis Univ, 62-63; asst prof, Univ Mass, Amherst, 63-67; assoc prof, Carbondale, 67-69, assoc prof, Edwardsville, 69-70, PROF PHYSICS, SOUTHERN ILL UNIV, EDWARDSVILLE, 70-, CHMN, 80- *Mem:* Am Phys Soc; Am Asn Physics Teachers. *Res:* Theoretical atomic physics and scattering theory. *Mailing Add:* Dept Physics Southern Ill Univ Edwardsville IL 62026

KANG, JOOHEE, b Seoul, Korea. SUPERCONDUCTIVITY, THIN FILM TECHNOLOGY. *Educ:* Seoul Nat Univ, BA, 77; Korea Advan Inst Sci, MS, 79; Univ Minn, PhD(physics), 87. *Prof Exp:* Asst prof physics, Ulsan Inst Technol, 79-82; postdoctoral superconductivity, Argonne Nat Lab, 87-89; SR SCIENTIST, WESTINGHOUSE SCI & TECHNOL CTR, 89- *Mem:* Am Phys Soc. *Res:* Electromagnetic properties of exotic superconducting materials; superconducting thin films fabrication; superconducting electronic circuits design and fabrication. *Mailing Add:* Westinghouse Sci & Technol Ctr 401-3X9B 1310 Beulah Rd Pittsburg PA 15235

KANG, JUNG WONG, b Tokyo, Japan, July 25, 33; nat US; m 55; c 3. POLYMER ORGANIC CHEMISTRY. *Educ:* Kinki Univ, Japan, BSc, 56; Osaka Univ, MSc, 59, PhD(org chem), 62. *Prof Exp:* Instr chem, Nara Med Col, Japan, 62-63; fel org chem, Harvard Univ, 63-64; fel organometallic chem, Univ NC, 64-66; res assoc, McMaster Univ, 66-70; res scientist chem, 70-75, sr res scientist, 75-78, assoc scientist, 78-82, RES ASSOC, FIRSTONE TIRE & RUBBER CO, 83- *Mem:* Am Chem Soc; Japanese Chem Soc; Korean Chem & Chem Eng NAm. *Res:* Polymer chemistry; organometallic chemistry. *Mailing Add:* Firestone Fire & Rubber Co 1200 Firestone Pkwy Akron OH 44317

KANG, KENNETH S, b Seoul, Korea, Nov 20, 33; US citizen; m 59; c 3. MICROBIAL PHYSIOLOGY. *Educ:* Yonsei Univ, Korea, BS, 57; Univ Del, MS, 63, PhD(microbiol), 65. *Prof Exp:* Sr res chemist, 66-68, proj leader, 68-69, sect head biochem develop, 69-78, mgr biochem res, Kelco Div, 78-79, assoc dir, 79-86, DIR, MSR, MERCK & CO, 86- *Concurrent Pos:* Fel, Pa State Univ, 75-76. *Mem:* AAAS; Am Chem Soc; Sigma Xi; Am Soc Microbiol. *Res:* Development, production and characterization of microbial polysaccharides for industrial applications; authored 28 scientific publications and 19 US patents. *Mailing Add:* Kelco Civ Merck & Co 8225 Aero Dr San Diego CA 92123-1716

KANG, KEWON, b Andong, Korea, July 3, 34; US citizen; m 60; c 2. GENETICS. *Educ:* Seoul Univ, Korea, BS, 57, MS, 60; NC State Univ, PhD(genetics), 66. *Prof Exp:* Res assoc forest genetics, Inst Genetics, Korea, 57-60; res asst genetics, NC State Univ, 61-66; fel, Ind Univ, Indianapolis, 66-69, from asst prof to assoc prof med genetics, 69-86; AT KOREA INST TECH. *Concurrent Pos:* Vis assoc prof, Seoul Univ, 77; vis researcher, Univ Hawaii, 77. *Mem:* Am Soc Human Genetics; Am Soc Oil Chemists; Am Soc Genetics; Korea Soc Scientists & Engrs. *Res:* Human quantitative genetics and computer application in medical genetics. *Mailing Add:* Korea Inst Technol 400 Kusong/Dong Chung/Gu Taejon/Shi Chung Chong Nam-Do 300-31 Republic of Korea

KANG, KYUNGSIK, b Jochiwon, Korea, July 12, 36; m 63; c 3. ELEMENTARY PARTICLE PHYSICS, THEORETICAL PHYSICS. *Educ:* Seoul Nat Univ, BS, 59; Ind Univ, PhD(theoret physics), 64. *Prof Exp:* Res assoc, 64-66, from asst prof to assoc prof, 66-73, PROF PHYSICS, BROWN UNIV, 73- *Concurrent Pos:* Vis prof, Univ Paris, 72-73 & 86-87, Seoul Nat Univ, 78, Korea Advan Inst Sci & Technol, 80, 83, 85 & 88; vis scientist, Europ Orgn Nuclear Res, 73 & 87; vis physicist, Fermi Nat Accelerator Lab, 74, 77 & 81, Argonne Nat Lab, 68 & 84, Brookhaven Nat Lab, 68 & 79 Neils Bohr Inst, 71 & 87, Los Alamos Nat Lab, 77 & 86; hon prof, Yanbian Univ, Yanji, China, 90. *Honors & Awards:* Camellia Medal, Legion of Honor, Korea, 85. *Mem:* Korean Phys Soc; Korean Scientists & Engrs Am (pres, 82-83); fel Am Phys Soc. *Res:* Phenomenological descriptions of high energy elementary particle physics; grand unification theories; flavor dynamics; composite models; electroweak gauge theories; lepton-induced reactions; supersymmetric string theories; cosmology. *Mailing Add:* Dept Physics Brown Univ Providence RI 02912

KANG, SUNG-MO STEVE, b Seoul, Korea, Feb 25, 45; US citizen; m 72; c 2. COMPUTER-AIDED DESIGN, MODELING & SIMULATION. *Educ:* Fairleigh Dickinson Univ, BS, 70; State Univ NY, Buffalo, MS, 72; Univ Calif, Berkeley, PhD(elec eng), 75. *Prof Exp:* Asst prof elec eng, Rutgers Univ, 75-77; mem tech staff, AT&T Bell Labs, Murray Hill, 77-81, supvr, 82-85; PROF ELEC & COMPUTER ENG, UNIV ILL, URBANA-CHAMPAIGN, 85-, ASSOC DIR MICROELECTRONICS, 88-, PROF COMPUTER SCI, 90- *Concurrent Pos:* Vis prof, Swiss Fed Inst Technol, Lausanne, 89; consult, Teltech, Inc, 89-, MCC, Austin, 89, Motorola, Inc & AT&T Bell Labs, 90. *Honors & Awards:* Meritorious Serv Award, Inst Elec & Electronics Engrs Computer Soc, 90. *Mem:* Fel Inst Elec & Electronics Engrs; Inst Elec & Electronics Engrs Circuits & Systs Soc (pres, 91); Inst Elec & Electronics Engrs Computer Soc; Inst Elec & Electronics Engrs Lasers & Electrooptical Soc. *Res:* Computer-aided design of very-large scale integrated circuits and systems; modeling and simulation of optoelectronic and novel devices and circuits; analog and digital microelectronics. *Mailing Add:* 1909 Trout Valley Rd Champaign IL 61821

KANG, SUNGZONG, b Puyo, Korea, Mar 1, 37; m 65. BIOPHYSICS, BIOCHEMISTRY. *Educ:* Univ Tubingen, PhD(chem), 64. *Prof Exp:* Res assoc chem, Chem Inst, Univ Tubingen, 64-66, Univ Notre Dame, 66-67 & NY Univ, 67-68; from instr to asst prof, 68-72, ASSOC PROF, MT SINAI SCH MED, CITY UNIV NEW YORK, 72- & BRONX VET ADMIN MED CTR, 80- *Concurrent Pos:* Vis prof, Max Plank Inst Biophys Chem, Gottingen, 76-77, Max Plank Inst Physiol, Dortmund, 77-78, Seoul Nat Univ, 78-79 & AID, 78-79; Fogarty sr int scholar, 76-77; Alexander von Humboldt US sr scientist fel, 77-78. *Mem:* AAAS; Am Chem Soc; NY Acad Sci; Am Soc Pharmacol & Exp Therapeut. *Res:* Stability, structure and function of biological macromolecules, membranes, proteins and nucleic acids; brain research; molecular pharmacology; quantum biochemistry. *Mailing Add:* Hanhyo Inst CPO Box 1751 Seoul Republic of Korea

KANG, TAE WHA, b Chejoodo, Korea; m 73; c 2. INDUSTRIAL TOXICOLOGY, MEDICAL TECHNOLOGY. *Educ:* Yonsei Univ, BS, 70; Ill Inst Technol, PhD(biol), 76. *Prof Exp:* Fel molecular biol, Univ Edinburgh, 76-77; chmn biochem, Bio-Technics Labs, Inc, 77-78; DIR LAB, BIO-SCI RES INST, 78- *Concurrent Pos:* Adj prof, Am Int Univ, 79-80 & Pac Western Univ, 82- *Mem:* Sigma Xi; Am Soc Microbiol; AAAS. *Res:* Testing of food, drug and cosmetics: product label validation and shelf-life stability studies (Rx drugs and OTC products); pre-clinical studies of new drugs and toxicological assessment of environment chemicals (acute, subchronic & chronic animal studies); mutagenicity and carcinogenicity studies of food additives and color additives; safety tests and potency assay of human leukocyte interferon; water and wastewater tests. *Mailing Add:* Bio-Sci Res Inst Inc 4813 Cheyenne Way Chino CA 91710-5510

KANG, UAN GEN, b Chonpuk, Korea, March 2, 38; US citizen; m 68; c 2. POLYMER CHEMISTRY. *Educ:* Chanpuk Nat Univ, BS, 60; Mich State Univ, MS, 69, PhD(org chem), 72; Johns Hopkins Univ, MAS, 82. *Prof Exp:* Vis res assoc biochem, Ohio State Univ, 73-75 & org chem, 75-77; res chem, Wacker Silicone Co, 77-81, sr res chemist polymer, W R Grace Washington Res Ctr, 81-90, SR RES CHEMIST, WACKER SILICONE CO, 90- *Mem:* Am Chem Soc. *Res:* Organic reaction mechanism; mechanism of enzyme action; carbine chemistry; polymer chemistry. *Mailing Add:* 3077 Signature Blvd No K Ann Arbor MI 48103-6433

KANG, YUAN-HSU, CELL BIOLOGY. *Educ:* Brigham Young Univ, PhD(zoology), 72. *Prof Exp:* CHIEF ELECTRON MICROS SECT, NAVAL MED RES INST, 84- *Res:* Natural killer cells; reticuendothelial cells in septic shock. *Mailing Add:* 2305 Deckman Lane Silver Spring MD 20906

KANGAS, DONALD ARNE, b Detroit, Mich, Feb 12, 29; m 54; c 5. CHEMISTRY. *Educ:* Mich Technol Univ, BS, 50; Mich State Univ, MS, 58. *Prof Exp:* Anal chemist, R P Scherer Corp, 50-51; develop chemist, US Army-Chem Corp, 51-53; sr res specialist, 53-80, RES ASSOC POLYMER, DOW CHEM USA, 80- *Res:* Hydrophobic colloids formed from vinyl monomers; kinetics of polymerization; characterization of dispersion; morphology of particles; properties of films and composites; hydrophillic polyelectrolytes from ionizable vinyl monomers; kinetics characterization and properties. *Mailing Add:* 5112 Nurmi Dr Midland MI 48640

KANICK, VIRGINIA, b Coaldale, Pa, Nov 10, 25. RADIOLOGY. *Educ:* Barnard Col, AB, 47, Columbia Univ, MD, 51; Am Bd Radiol, dipl, 55. *Prof Exp:* Attend physician, 55-63, DEP DIR, DEPT RADIOL, ST LUKE'S ROOSEVELT HOSP, NY, 64-; PROF CLIN RADIOL, COLUMBIA UNIV, 75- *Concurrent Pos:* assoc prof, Columbia Univ, 74-75. *Mem:* Fel Am Col Radiol; Radiol Soc NAm; Am Roentgen Soc. *Res:* Clinical research in angiology; use of contrast media in radiology. *Mailing Add:* 560 Riverside Dr New York NY 10027

KANIECKI, THADDEUS JOHN, b Brooklyn, NY, Mar 24, 31; m 55; c 3. ORGANIC CHEMISTRY. *Educ:* NY Univ, BA, 53, MS, 55, PhD(org chem), 60. *Prof Exp:* Res assoc, Res & Develop Div, Lever Bros Co, NJ, 60-67; res supvr, Armour-Dial Inc, Chicago, 67-69, res mgr, 69-72; sect mgr, Am Cyanamid Co, 72-75; sr chemist res & develop, Stauffer Chem Co, 75-81;

sr tech serv rep, Brent Chem Corp, 81-82, dir res & develop, 82-84; dir res & develop, Clenesco Div, Chemed Corp, 84-87, DIR OPERS, CLENESCO PROD CORP, 87- *Mem:* Am Chem Soc; Soc Cosmetic Chem. *Res:* Synthesis and applications of surface active molecules; detergents, toiletries and consumer products, both basic and applied research; preparation and uses of disinfectants, sanitizers and biocides. *Mailing Add:* Two Van Alen Pl Pompton Plains NJ 07444

KANIG, JOSEPH LOUIS, b New York, NY, July 11, 21; m 47; c 3. PHARMACEUTICS. *Educ:* LI Univ, BS, 42, Columbia Univ, MS, 49; NY Univ, PhD, 60; Am Bd Pharm, dipl. *Prof Exp:* Assoc pharm, 49-51, from asst prof to assoc prof, 51-62, dir aerosol res lab, 58-67, PROF PHARM, COLUMBIA UNIV, 62-, DEAN COL, 65-, DIR INDUST PHARM LAB, 59- *Concurrent Pos:* Consult, UNESCO, 62-65 & pharmaceut, chem & cosmetic industs; mem, NY State Coun Hosp Pharmacists. *Mem:* Fel AAAS; fel Am Col Apothecaries; fel Am Inst Chem; Am Pharmaceut Asn; Soc Cosmetic Chem. *Res:* Pharmaceutical sciences; industrial pharmaceutical processing; product development and research in dosage design and evaluation; process machinery and evaluative instruments; biopharmaceutics. *Mailing Add:* 79 Beaver Brook Rd Ridgefield CT 06877

KANIZAY, STEPHEN PETER, b Cleveland, Ohio, Feb 3, 24; m 48; c 3. GEOLOGY. *Educ:* Miami Univ Ohio, AB, 49, MS, 50; Colo Sch Mines, DSc, 56. *Prof Exp:* From instr to asst prof geol, Colo Sch Mines, 52-58; geologist, Eng Geol Br, US Geol Surv, 58-83; RETIRED. *Concurrent Pos:* Consult, 54-58. *Mem:* Am Geophys Union; Asn Eng Geol. *Res:* Engineering and structural geology; rock and soil mechanics. *Mailing Add:* 625 S Parfet Lakewood CO 80226

KANKEL, DOUGLAS RAY, b Waterbury, Conn, Jan 22, 44. DEVELOPMENTAL BIOLOGY, NEUROBIOLOGY. *Educ:* Brown Univ, PhD(biol), 70. *Prof Exp:* Res fel neurobiol, Calif Inst Technol, 70-74; MEM FAC BIOL DEPT, YALE UNIV, 74- *Mailing Add:* Dept Biol 1018 Kbt Yale Univ New Haven CT 06520

KAN-MITCHELL, JUNE, b Hong Kong, June 3, 49; US citizen; m 67; c 1. CELLULAR IMMUNITY, MELANOMA. *Educ:* Yale Univ, PhD(pharmacol), 77. *Prof Exp:* ASST PROF MICROBIOL, UNIV SOUTHERN CALIF SCH MED, 85-, ASSOC PROF PATH, 90- *Concurrent Pos:* Assoc ed, J Lab Clin Invests, Human Antibodies & Hybridomas & Vaccine Res; prin investr, Nat Eye Inst. *Mem:* Am Asn Cancer Res; Asn Res Vision & Ophthal; Fedn Am Soc Exp Biol; Am Asn Immunol. *Res:* Melanocyte transformation; immune response to melanoma in man. *Mailing Add:* Norris 710 Univ Southern Calif Sch Med 2025 Zonal Ave Los Angeles CA 90033

KANNAPPAN, PALANIAPPAN, b Nattarasan Kottai, India, June 28, 34; m 52; c 5. MATHEMATICS. *Educ:* Annamalai Univ, Madras, BSc(Hons), 55, MA, 57; Univ Wash, Seattle, MS & PhD(math), 64. *Prof Exp:* Lectr math, Annamalai Univ, Madras, 55-61, reader, 64-67; asst, Univ Wash, 61-64; assoc prof, 67-77, PROF MATH, UNIV WATERLOO, 77- *Concurrent Pos:* Consult, Dept Univ Affairs, Can, 68-69 & Nat Res Coun Can, 69- *Mem:* Am Math Soc; Indian Math Soc. *Res:* Functional analysis and functional equations; linear algebra and quasigroups; information theory. *Mailing Add:* Dept Math Univ Waterloo Waterloo ON N2L 3G1 Can

KANNEL, WILLIAM B, b Brooklyn, NY, Dec 13, 23; m 42; c 4. CARDIOVASCULAR DISEASES, INTERNAL MEDICINE. *Educ:* Ga Med Col, MD, 49; Harvard Univ, MPH, 59. *Hon Degrees:* MD, Gothenberg Univ, 85. *Prof Exp:* Intern & resident, USPHS Hosp, Staten Island, NY, 49-50, 53-56; clin investr heart dis epidemiol study, NIH, Framingham, Mass, 50-51; med officer, Newton Heart Prog, Mass, 51-52; assoc dir, Framingham Unit, NIH, 55-65, dir, 65-79; chief sect epidemiol & prev med, 79-90, PROF MED & PUB HEALTH, BOSTON UNIV, 79- *Concurrent Pos:* Fel, Harvard Med Sch, 56-59; asst med, Peter Bent Brigham Hosp, 56-62; instr, Harvard Med Sch, 59-60, assoc prev med, 60-70, lectr, 70-; consult, Cushing State Hosp, 56-73 & Framingham Union Hosp, 64- *Honors & Awards:* Dana Award, 72 & 86; Einthoven Award, 73; Gaidner Int Award, 76; Paul Dudley White Award, 77; Copernicus Award, 77; Ciba Award, 81; J D Bruce Award, Am Col Physicians, 82; Distinguished Serv Award, Am Col Cardiol, 88. *Mem:* Fel Am Col Prev Med; fel Am Heart Asn; fel Am Col Physicians; fel Am Col Cardiol; fel Am Col Epidemiol. *Res:* Cardiovascular epidemiology; investigation of factors of risk and natural history of coronary heart disease, hypertension, stroke and peripheral vascular disease; preventive medicine. *Mailing Add:* Boston Univ Med Ctr 720 Harrison Ave Boston MA 02218

KANNENBERG, LLOYD C, b Sarasota, Fla, Mar 23, 39; m 63; c 1. PHYSICS, OPTICS. *Educ:* Mass Inst Technol, SB, 61; Univ Fla, MS, 63; Northeastern Univ, PhD(physics), 67. *Prof Exp:* Instr physics, Lowell Technol Inst, 66-67 & Northeastern Univ, 67-68; from instr to assoc prof, Lowell Technol Inst, 68-77, PROF PHYSICS, UNIV LOWELL, 77- *Concurrent Pos:* Vis res assoc, Northeastern Univ, 72-80. *Mem:* Am Phys Soc; Int Soc Gen Relativity & Gravitation; Sigma Xi. *Res:* General relativity, field theory. *Mailing Add:* Dept Physics Univ Lowell Lowell MA 01854

KANNENBERG, LYNDON WILLIAM, b Chicago, Ill, Oct 15, 31; m 54; c 5. PLANT BREEDING. *Educ:* Mich State Univ, BSc, 57, MS, 59; Univ Calif, Davis, PhD(genetics), 64. *Prof Exp:* NIH fel, Univ Calif, Davis, 64-65; from asst prof to assoc prof, 65-77, PROF CORN BREEDING, UNIV GUELPH, 77- *Concurrent Pos:* Nat Res Coun Can Grants Comt, 71-73; mem, Expert Comt on Plant Gene Resources, 83-; maize consult, Can Int Develop Agency, Bangladesh, 85; vis lectr, China, 88. *Mem:* Am Soc Agron; Genetics Soc Can; Sigma Xi; Can Soc Agron. *Res:* Development of short season corn breeding populations. *Mailing Add:* Dept Crop Sci Univ Guelph Guelph ON N1G 2W1 Can

KANNEWURF, CARL RAESIDE, b Waukegan, Ill, Mar 24, 31; m 83. SOLID STATE ELECTRONICS. *Educ:* Lake Forest Col, BA, 53; Univ Ill, MS, 54; Northwestern Univ, PhD(physics), 60. *Prof Exp:* Res assoc physics, 60-62, from asst prof to assoc prof elec eng, 63-71, PROF ELEC ENG & COMP SCI, NORTHWESTERN UNIV, EVANSTON, 71- *Mem:* Am Phys Soc; Sigma Xi; sr mem Inst Elec & Electronics Engrs; Mat Res Soc. *Res:* Study of various electrical and optical phenomena in semiconductors, metals, molecular metals and conducting polymers; transport phenomena and superconductivity; development of optical materials and devices. *Mailing Add:* Dept Elec Eng Tech Inst Northwestern Univ Evanston IL 60208-3118

KANNINEN, MELVIN FRED, b Ely, Minn, Jan 31, 35; m 57; c 2. ENGINEERING MECHANICS, MATERIAL SCIENCE ENGINEERING. *Educ:* Univ Minn, BS, 57, MS, 59, Stanford Univ PhD(Eng Mech), 66. *Prof Exp:* Res, Battelle's Columbus Lab, 66-75, sr res, 75-79, res leader, 79-83; inst scientist, 83-91, PROG DIR ENG MECH, SOUTHWEST RES INST, 91- *Concurrent Pos:* Mem, Nat Mat Adv Bd, Nat Res Coun & numerous prof soc & govt adv comts; lectr, numerous US & foreign univs. *Mem:* Nat Acad Eng; Am Soc Testing Mat; Int Asn Struct Mech; Am Welding Soc; fel Am Soc Mech Engrs; Soc Eng Sci. *Res:* Fracture mechanics for fast fracture arrest; elastic-plastic fracture mechanics; lifetime predictions for fiber reinforced composites; residual stress and cracking of welds; applications to nuclear pressure vessels, cryogenic storage tanks, railroad equipment, gas transmission pipelines, plastic pipe and joints; author of more than 150 technical publications. *Mailing Add:* Southwest Res Inst PO Drawer 28510 San Antonio TX 78228-0510

KANNOWSKI, PAUL BRUNO, b Grand Forks, NDak, Aug 11, 27; m 53; c 2. ZOOLOGY. *Educ:* Univ NDak, BS, 49, MS, 52; Univ Mich, PhD(zool), 57. *Prof Exp:* Asst biol, Univ NDak, 50-52; instr, Bowling Green State Univ, 56-57; from asst prof to prof biol, Univ NDak, 57-90, chmn dept, 63-70 & 82-88, dir, Inst Ecol Studies, 65-81, EMER PROF BIOL, UNIV NDAK, 91- *Concurrent Pos:* Res assoc, Harvard Univ, 66-67; NSF sr fel, 66-67; vis scientist, Smithsonian Trop Res Inst, 67, 68; entom consult, Lystads Pest Control, Inc, 68-75; ed, Prairie Naturalist, 68-; natural resources consult, US Congressman Mark Andrews, NDak, 69-70. *Honors & Awards:* Prof Award, Wildlife Soc, 89. *Mem:* Am Inst Biol Sci; Am Soc Zoologists; AAAS; Entom Soc Am; Ecol Soc Am. *Res:* Ecology; biogeography; animal behavior; myrmecology; chemical communication. *Mailing Add:* Dept Biol Univ NDak Grand Forks ND 58202-8238

KANO, ADELINE KYOKO, b Mitchell, Nebr, Nov 22, 27. CHEMISTRY, ACADEMIC ADMINISTRATION. *Educ:* Univ Nebr, BA, 48. *Prof Exp:* Instr chem & jr chemist, 55-60, asst chemist, 60-66, asst prof, 60-73, ADMIN ASST & FAC AFFIL, COLO STATE UNIV, 73- *Mem:* Am Chem Soc; Sigma Xi. *Res:* Factors imposed on chicks and rats, their alleviation and relation to blood and tissue content of amino acids. *Mailing Add:* Dept Biochem Colo State Univ Ft Collins CO 80523

KANOFSKY, ALVIN SHELDON, b Philadelphia, Pa, July 5, 39; m 64; c 2. ELEMENTARY PARTICLE PHYSICS. *Educ:* Univ Pa, BA, 61, MS, 62, PhD(physics), 66. *Prof Exp:* From asst prof to assoc prof, 67-76, PROF PHYSICS, LEHIGH UNIV, 76- *Concurrent Pos:* Res collabr, Brookhaven Nat Lab, Fermilab; pres, Res & Develop Co. *Mem:* Fel Am Phys Soc; Sigma Xi; AAAS; Rotary Int Chambers Com. *Res:* Research in eta decay; proton-proton scattering; mu magnetic moment, Glauber calculations in particles, high energy particle channeling in crystals, particle-nuclei interactions, hypernuclei, jet production, cosmic rays and instrumentation; accelerator research; intermediate energy physics. *Mailing Add:* Dept Physics Lehigh Univ Bldg 16 Bethlehem PA 18015

KANOJIA, RAMESH MAGANLAL, b Mangrol, India, Feb 15, 33; m 67. MEDICINAL CHEMISTRY. *Educ:* Bombay Univ, BSc, 54, BScTech, 56, MScTech, 61; Univ Wis, PhD(pharmaceut chem), 66. *Prof Exp:* Instr pharmaceut chem, Bombay Univ, 58-59, hon lectr tech pharmaceut & fine chem, 59-60; res asst pharmaceut chem, Sch Pharm, Univ Wis, 61-66; assoc scientist, 66-70, scientist, 70-76, sr scientist, 77-79, RES FEL, ORTHO PHARMACEUT CORP, 80- *Honors & Awards:* Philip B Hoffman Award, Johnson & Johnson Co, 77. *Mem:* Am Chem Soc. *Res:* Isolation, characterization and synthesis of natural and synthetic organic medicinal compounds; antifertility compounds of natural and synthetic origin; synthesis of cardiovascular-active drugs. *Mailing Add:* R W Johnson Pharmaceut Res Inst Raritan NJ 08869

KANOPOULOS, NICK, b Drama, Greece, Aug 11, 56; US citizen; m 84; c 1. VLSI DESIGN, INTERGRATED CIRCUIT TESTING. *Educ:* Univ Patras, Greece, BS, 79; Duke Univ, MS, 80, PhD(elect eng), 84. *Prof Exp:* Res asst, Duke Univ, 79-81; engr, Bendix Corp, 81-82; from engr to sr engr, 82-85, coordr, 85-87, MGR, RES TRIANGLE INST, 87- *Concurrent Pos:* Adj asst prof, Duke Univ, 85- *Mem:* AAAS; Inst Elec & Electronics Engrs. *Res:* Design of very large scale, application specific integrated circuits; design of parallel signal processor architectures for high-performance, real-time applications; developed design for testability and built in self-test techniques and structures. *Mailing Add:* Eight Ludwell Pl Durham NC 27705

KANO-SUEOKA, TAMIKO, b Kyoto, Japan, June 26, 32; m 56; c 1. MOLECULAR BIOLOGY. *Educ:* Kyoto Univ, Japan, BA, 56; Radcliffe Col, MA, 60; Univ Ill, Urbana, PhD(molecular biol), 63. *Prof Exp:* Res asst biol, Calif Inst Technol, 56-58; res assoc biochem, Princeton Univ, 63-67, res staff, 68-72; from asst prof to assoc prof, 73- 85, PROF CELLULAR & MOLECULAR BIOL, UNIV COLO, 85-, SR RES ASSOC, 85- *Mem:* Am Soc Biochem & Molecular Biol; Am Asn Cancer Res; Am Tissue Cult Asn. *Res:* Regulation of growth of normal and neoplastic mammary cells, in particular the involvement of membrane phospholipids in the control of cell proliferation. *Mailing Add:* Dept Molecular Cellular & Develop Biol Univ Colo Boulder CO 80309-0347

KANT, FRED H(UGO), b Vienna, Austria, Jan 11, 30; nat US; m 52; c 2. CHEMICAL ENGINEERING. *Educ:* Columbia Univ, BS, 51, MS, 53, DEngSc, 57. *Prof Exp:* Sr staff adv, New Areas Staff, Esso Res & Eng Co, 54-66, dir new invests res lab, Linden, 66-69, proj mgr, Corp Res Staff, 69-72, sr staff adv, 72-75, planning mgr, Govt Res, 75-78, sr staff adv, Corp Res, Exxon Res & Eng Co, 78-80, sr tech adv, Exxon Corp, 80-84; RETIRED. *Concurrent Pos:* Consult, 84- *Mem:* Am Chem Soc; AAAS; NY Acad Sci. *Res:* Fuels process; staff research coordination, planning and project evaluation in new areas; research administration. *Mailing Add:* 31 Rutgers Rd Cranford NJ 07016

KANT, KENNETH JAMES, b Elyria, Ohio, July 14, 35; m 58; c 4. PHYSIOLOGY, VETERINARY MEDICINE. *Educ:* Ohio State Univ, BS, 58; Univ Ill, Urbana, MS, 64, PhD(physiol), 67. *Prof Exp:* From instr to asst prof physiol, State Univ NY Buffalo, 67-74; ASSOC PROF PHYSIOL, UNIV TENN, KNOXVILLE, 74- *Mem:* Am Vet Med Asn; Am Physiol Soc; Soc Neurosci. *Res:* Neurophysiology and behavior, especially limbic structures. *Mailing Add:* Sam Houston Schooll Rd Maryville TN 37801

KANTACK, BENJAMIN H, b Greenleaf, Kans, Sept 26, 27; m 53; c 7. ENTOMOLOGY, AGRONOMY. *Educ:* Kans State Univ, BS, 51; Okla State Univ, MS, 54; Univ Nebr, PhD(entom), 63. *Prof Exp:* Trainee agron, Libby, McNeill & Libby, Hawaii, 51-52; asst entom, Okla State Univ, 52-54; instr & entomologist, Univ RI, 54-55; entomologist, USDA, Ga, 55-58; instr entom, Univ Nebr, 58-62; from asst prof to prof entom, 62-91, exten entomologist, 63-91, EMER PROF ENTOM, SDAK STATE UNIV, 91- *Honors & Awards:* F O Butler Award. *Mem:* Entom Soc Am. *Res:* Stored grain, vegetable and field crop insects; livestock ecto parasites. *Mailing Add:* Dept Plant Sci & Entomol SDak State Univ Brookings SD 57006

KANTAK, KATHLEEN MARY, b Syracuse, NY, Nov 11, 51; m 75; c 2. PSYCHOBIOLOGY. *Educ:* State Univ NY Potsdam, BA, 73; Syracuse Univ, PhD (biopsychol), 77. *Prof Exp:* Res asst prof biopsychol, Syracuse Univ, 78; res assoc behav neurochem, Univ Wis-Madison, 78-81; res assoc psychol, Tufts Univ, 81-; AT DEPT PSYCHOL, BOSTON UNIV. *Concurrent Pos:* Pvt invest, Nat Inst Drug Abuse grant. *Mem:* Soc Neurosci; Sigma Xi; Int Soc Res Aggression; Behav Pharmacol Soc. *Res:* Neurochemical correlates of aggression in terms of how these measures are affected by nutritional factors; nutritional aspects of drug abuse; animal models of tardiue dyskinesia. *Mailing Add:* Dept Psychol Boston Univ 64 Cummington St Boston MA 02215

KANTER, GERALD SIDNEY, b New York, NY, Dec 7, 25; m 56; c 2. PHYSIOLOGY. *Educ:* Long Island Univ, BS, 47; Univ Rochester, PhD(physiol), 52. *Prof Exp:* Jr instr physiol, Sch Med, Univ Rochester, 52; from instr to assoc prof physiol, 52-63, lectr biochem, 53-55, asst to dean, 66-67, asst dean, 67-69, PROF PHYSIOL, ALBANY MED COL, 63-, ASSOC DEAN, 69- *Concurrent Pos:* Chief physiol br, US Army Inst Environ Med, 63-64. *Mem:* AAAS; Am Physiol Soc; Sigma Xi. *Res:* Body fluid and electrolyte regulation; kidney function; thirst; temperature regulation and environmental physiology. *Mailing Add:* 81 Meadowland St Delmar NY 12054

KANTER, HELMUT, b Hamburg, Ger, Jan 19, 28; US citizen. ELECTRON PHYSICS. *Educ:* Univ Marburg, MS, 53, PhD(physics), 56. *Prof Exp:* Res physicist, Res Labs, Westinghouse Elec Corp, 57-64; mem tech staff, Lab Div, 64-74, MEM STAFF, ELECTRONIC RES LAB, AEROSPACE CORP, 74- *Mem:* Am Phys Soc; Ger Phys Soc. *Res:* Electron scattering; photo and secondary electron emission; electron transport; photoconductivity; imaging tubes. *Mailing Add:* Aerospace Corp M2244 PO Box 92957 Los Angeles CA 90009

KANTER, IRA E, b Chicago, Ill, Oct, 31, 31; m 68; c 2. ENGINEERING PHYSICS. *Educ:* Ill Inst Technol, BS, 53, MS, 54. *Prof Exp:* Asst chem engr, US Army Chem Corps, 54-56; engr high temperature, Vanguard X405 Rocket, Flight Propulsion Lab, Gen Elec, 57-59; res staff high temperature chem, Univ Wis, 60-61; sr engr NERVA nuclear reactors, Astronuclear Labs, Westinghouse, 63-69, sr engr chem eng, 69-77, sr engr chem physics, Res & Develop Labs, 77-87, SR ENGR ENVIRON CONTROL OFFICER, SOLID OXIDE FUEL CELLS, SCI & TECH CTR, WESTINGHOUSE, 87- *Concurrent Pos:* Environ sci. *Mem:* Am Nuclear Soc; Inst Elec & Electronics Engrs; Air & Waste Mgt Asn; Int Technol Inst. *Res:* Radio-chemical reactions for synthesis; chemical thermodynamics; inertially and magnetically confined fusion reactor blanket systems; radio-gas waste treatment and storage; glow discharge chemistry; physical chemistry; pollution control systems. *Mailing Add:* Westinghouse Sci & Technol Ctr-Solid Oxide Fuel Cells 2000 Tech Ctr Dr Monroeville PA 15146

KANTER, IRVING, b New York, NY, Oct 30, 24; m 51; c 5. RADAR, ELECTRONIC COUNTER MEASURES & COUNTER COUNTER MEASURES. *Educ:* Brooklyn Col, AB, 44; Brown Univ, PhD(appl math), 53. *Prof Exp:* Physicist, Kellex Corp, 44-45, Union Carbide & US Army, 45-47; systs specialist, Lockheed Aircraft Corp, 51-54; systs engr, Radio Corp Am, 54-66; CONSULT ENGR, RAYTHEON CO, 66- *Concurrent Pos:* Adj lectr, Univ Calif, Los Angeles, Temple Univ, Univ Pa & Northeastern Univ. *Mem:* Fel Inst Elec & Electronics Engrs; Inst Elec & Electronics Engrs Aerospace & Electronic Systs; Inst Elec & Electronics Engrs Info Theory Soc. *Res:* Detection and estimation; monopulse radar. *Mailing Add:* Missille Syst Div T3-TA3 Raytheon Co Apple Hill Dr Tewksbury MA 01876

KANTER, MANUEL ALLEN, b Boston, Mass, Jan 18, 24; m 90; c 4. INTERNATIONAL TRAINING. *Educ:* Northeastern Univ, BS, 44; Ill Inst Technol, MS, 49, PhD, 55. *Prof Exp:* Assoc chemist, Argonne Nat Lab, 46-68, training coordr, Argonne Ctr Educ Affairs, 69-75, dir, Int Atomic Energy Agency Nuclear Power Training, 75-86, CONSULT, EDUC, TRAINING, SCI & TECHNOL, ARGONNE NAT LAB, 86- *Mem:* AAAS; Am Nuclear Soc; Inst Nuclear Mat Mgt; Am Phys Soc; Sigma Xi. *Res:* High temperature chemistry; galvanomagnetic effects; actinide compounds; diffusion; nuclear material safeguards. *Mailing Add:* 5733 Sheridan Rd No 17C Chicago IL 60660

KANTHA, LAKSHMI, US citizen; m 74; c 1. OCEANIC CIRCULATION. *Educ:* Bangalore Univ, BE, 67; Indian Inst Sci, ME, 69; Mass Inst Technol, PhD(aerospace & astron), 73. *Prof Exp:* Fel, Johns Hopkins Univ, 74-75, assoc res scientist, 75-79, res scientist, 79-80; RES SCIENTIST, DYNALYSIS PRINCETON, 80- *Mem:* Am Meteorol Soc; Am Geophys Union. *Res:* Turbulence and wave motions in the atmosphere and the oceans; ocean circulation in the coastal regions; oceanic mixing and influence of the ice cover on polar oceans; numerical and experimental modeling of oceanic and atmospheric processes. *Mailing Add:* APAS Campus Box 391 Univ Colo Boulder CO 80309

KANTOR, FRED STUART, b New York, NY, July 2, 31; m 58; c 3. IMMUNOLOGY. *Educ:* Union Col NY, BS, 52; NY Univ, MD, 56; Am Bd Internal Med, dipl, 64; Am Bd Allergy, dipl, 66, Yale Univ, MA, 73. *Prof Exp:* Intern, Ward Med Serv, Barnes Hosp, St Louis, Mo, 56-57; res assoc, Nat Inst Allergy & Infectious Dis, 57-59; asst res med, Grace New Haven Hosp, Sch Med, Yale Univ, 59-60; from instr to prof, 62-83, PAUL B BEESON PROF MED, SCH MED, YALE UNIV, 83- *Concurrent Pos:* Whitney fel, Sch Med, Yale Univ, 60-61 & NY Univ, 61-62; USPHS career develop awardee, 62-; vis scientist with Dr Gustave Nossal, Walter & Eliza Hall Inst, Melbourne, Australia, 68-69; mem coun, Am Heart Asn. *Mem:* Asn Am Physicians; Am Soc Clin Invest; Am Acad Allergy; Am Asn Immunol. *Res:* Immune response in man, including both delayed and immediate types of immunity. *Mailing Add:* Yale Univ Sch Med 333 Cedar St New Haven CT 06510

KANTOR, GEORGE JOSEPH, b Titusville, Pa, Jan 24, 37; m 66; c 3. BIOPHYSICS, MOLECULAR BIOLOGY. *Educ:* Slippery Rock State Col, BS, 58; NMex Highlands Univ, MS, 62; Pa State Univ, PhD(biophys), 67. *Prof Exp:* NIH fel biophys, Pa State Univ, 67-68; fel, Biomed Res Group, Los Alamos Sci Lab, 68-70; from asst prof to assoc prof, 70-80, PROF BIOL SCI, WRIGHT STATE UNIV, 80- *Mem:* Biophys Soc; AAAS; Sigma Xi; Tissue Culture Asn; Am Soc Photobiol. *Res:* Effects of radiation on biological systems with emphasis on human cells cultured in vitro; DNA repair in human cells. *Mailing Add:* Dept Biol Sci Wright State Univ Dayton OH 45435

KANTOR, GIDEON, b Vienna, Austria, Mar 30, 25; US citizen; m 67; c 2. NEUROMUSCULAR ELECTRICAL STIMULATION. *Educ:* NY Univ, BEE, 48; Polytech Univ NY, MEE, 50; Cornell Univ, PhD(elec eng), 63. *Prof Exp:* Res assoc, Microwave Res Inst, Brooklyn, 50-55; res asst, Cornell Univ, 55-59; physicist, Air Force Cambridge Lab, 59-65; staff mem, Avco, Lowell, 65-68 & Mitre, Bedford, 68-72; PHYSICIST, CTR DEVICES & RADIOL HEALTH, FOOD & DRUG ADMIN, ROCKVILLE, 72- *Concurrent Pos:* Engr, Gen Elec, Ithaca, 55-56. *Honors & Awards:* Bicentennial Medal, Inst Elec & Electronics Engrs, 84. *Mem:* Fel Inst Elec & Electronics Engrs; Sigma Xi. *Res:* Neuromuscular electrical stimulation with emphasis on the role of electrical parameters such as current and phase charge on safety and effectiveness. *Mailing Add:* Ctr Devices & Radiol Health 12721 Twinbrook Pkwy Rockville MD 20857

KANTOR, HARVEY SHERWIN, b New York, NY, Apr 30, 38; div. INFECTIOUS DISEASES, MICROBIOLOGY. *Educ:* Wash Univ, MD, 62; Am Bd Internal Med, dipl, 68. *Prof Exp:* Asst med, Sch Med, Wash Univ, 62-63; res fel, New Eng Med Ctr Hosp, Tufts Univ, 66-69, asst, Sch Med, 66-69; res educ assoc, Vet Admin, 70-71; asst prof med & microbiol, Univ Ill Med Ctr, 71-75; assoc prof med & dir div infectious dis, 75-85, ASSOC PROF PATH, CHICAGO MED SCH, 78-, DIR, DIV MED MICROBIOL, 85-; CHIEF, MED MICROBIOL LAB, VET ADMIN MED CTR, NORTH CHICAGO, 85- *Concurrent Pos:* Actg dir div infectious dis, Cook County Hosp, 72-74; consult, Highland Park Hosp, Highland Park & US Naval Hosp, Great Lakes, Ill, 75-; chief infectious dis sect, Vet Admin Med Ctr, North Chicago, 75-85. *Mem:* Fel Am Col Physicians; Sigma Xi; Am Fedn Clin Res; Am Soc Microbiol; NY Acad Sci; fel Infectious Dis Soc Am; Assoc Hosp Epidemiologists Am. *Res:* Bacterial toxins and their mechanism of action; their influence on cyclic nucleotides and prostaglandin interactions; hospital infection control. *Mailing Add:* Univ Health Sci Chicago Med Sch Dept Med North Chicago IL 60064

KANTOR, PAUL B, b Washington, DC, Nov 27, 38; m 62; c 2. DETECTION AND DECISION SYSTEMS. *Educ:* Columbia Univ, AB, 59; Princeton Univ, PhD(physics), 63. *Prof Exp:* Res assoc physics, Brookhaven Nat Lab, 63-65; vis asst prof, State Univ NY, Stony Brook, 65-67; asst prof, Case Western Reserve Univ, 67-69, assoc prof physics, 69-74, assoc prof oper res, Libr & Info Sci, 74-77, prog dir, Complex Systs Inst, 73-74, sr res assoc, systs eng, 77-81; PRES, TANTALUS INC, 77- *Concurrent Pos:* Guest physicist, Brookhaven Nat Lab, 65-; adj assoc prof libr & info sci, Kent State Univ, 78-81; sr lectr, Weatherhead Sch Mgt, Case Western Reserve Univ, 81-; distinguished vis scholar, Online Computer Libr Ctr, Ohio, 87; vis prof, Info Systs, Rutgers Univ, 90. *Mem:* Am Phys Soc; Am Statist Asn; Am Soc Info Sci; NY Acad Sci; Am Libr Asn. *Res:* Stability of complex systems; economics of information; information retrieval, large databases; distributed detection and decision systems; industrial learning phenomena. *Mailing Add:* Tantalus, Inc 3257 Ormond Rd Cleveland OH 44118

KANTOR, SIDNEY, b New York, NY, Feb 1, 24; m 49, 73; c 5. PARASITOLOGY, PROTOZOOLOGY. *Educ:* George Washington Univ, BA, 47, MA, 49; Univ Ill, PhD(zool), 56. *Prof Exp:* Invert zoologist, Acad Natural Sci, Pa, 51-53; asst parasitol, Col Vet Med, Univ Ill, 53-55; coop agent, USDA, Ill, 55-56; from parasitologist to group leader, Am Cynamid Co, 56-73, sr res biologist protozool chemother, Agr Res Div, 70-73, prin res biologist, 77-89; RETIRED. *Mem:* Am Soc Parasitol; Soc Protozool; Sigma Xi. *Res:* Parasitic chemotherapy; veterinary entomology. *Mailing Add:* 4A Van Buren Dr Cranbury NJ 08512

KANTOR, SIMON WILLIAM, b Brussels, Belg, Mar 23, 25; nat US; m 70; c 2. POLYMER CHEMISTRY, ORGANIC CHEMISTRY. *Educ:* City Col NY, BS, 45; Duke Univ, PhD(org chem), 49. *Prof Exp:* Fel, Duke Univ, 49-51; res assoc, Gen Elec Co, Schenectady, NY, 51-60, sect mgr, 60-65, br mgr, 65-72; vpres res & develop, GAF Corp, Wayne, 72-82; RES PROF, UNIV MASS, AMHERST, 82- *Honors & Awards:* Gold Patent Medallion, Gen Elec Co, 66. *Mem:* AAAS; Am Chem Soc. *Res:* Organic reactions of carbanions; organosilicon polymers; synthesis of aromatic condensation polymers; liquid crystal polymers. *Mailing Add:* Polymer Sci & Eng Dept Univ Mass Amherst MA 01003

KANTOROVITZ, SHMUEL, b Casablanca, Morocco, Sept 17, 35; m 60; c 4. OPERATOR THEORY. *Educ:* Hebrew Univ Israel, MSc, 56; Univ Minn, Minneapolis, PhD(math), 62. *Prof Exp:* Instr math, Princeton Univ, 62-63; mem, Inst Advan Study, 63-64; asst prof, Yale Univ, 64-67; from assoc prof to prof math, Univ Ill, Chicago Circle, 70-78; PROF MATH, BAR ILAN UNIV, ISRAEL, 72- *Concurrent Pos:* Dept chmn, Bar Ilan Univ, 77-79 & 85-87. *Mem:* Am Math Soc. *Res:* Functional analysis. *Mailing Add:* Bar Ilan Univ Ramat Gan Israel

KANTOWSKI, RONALD, b Shreveport, La, Dec 18, 39; m 61; c 3. PHYSICS. *Educ:* Univ Tex, Austin, BS, 62, PhD(physics), 66. *Prof Exp:* Res scientist med, Univ Tex Med Br, Galveston, 62-63; teaching asst physics, Univ Tex, Austin, 63-66, asst prof, 66-67; res assoc, Southwest Ctr Advan Studies, 67-68; from asst prof to assoc prof, 68-81, PROF PHYSICS, UNIV OKLA, 81- *Mem:* Am Phys Soc. *Res:* Quantum field theory; gravity theories. *Mailing Add:* Dept Physics & Astron 440 W Brooks Norman OK 73019

KANTROWITZ, ADRIAN, b New York, NY, Oct 4, 18; m 48; c 3. SURGERY. *Educ:* NY Univ, AB, 40; Long Island Col Med, MD, 43; Am Bd Surg, dipl. *Prof Exp:* Intern, Jewish Hosp, Brooklyn, 44; surg resident, Mt Sinai Hosp, 47; resident, Montefiore Hosp, 48-50; from asst prof to prof surg, State Univ NY Col Med, 55-70; PROF SURG, COL MED, WAYNE STATE UNIV, 70- *Concurrent Pos:* USPHS fel cardiovasc res & teaching fel physiol, Dept Physiol, Western Reserve Univ, 51-52; dir cardiovasc surg, Maimonides Med Ctr, Brooklyn, 55-64, dir surg, 64-70; chmn dept surg, Sinai Hosp, Detroit, 70- *Honors & Awards:* H L Moses Prize, Montefiore Alumnus, 49; Exhibit Prize, NY State Med Soc, 52; Theodore & Susan B Cummings Award, Am Col Cardiol, 67; Gold Plate Award, Am Acad Achievement, 66; Max Berg Award, 69. *Mem:* Fel NY Acad Sci; Am Chem Soc; Int Soc Angiol; Am Soc Artificial Internal Organs (pres, 68-69); Harvey Soc. *Res:* Cardiac pacemakers; heart transplants; human balloon pump; partial human mechanical heart. *Mailing Add:* 70 Gallogly Rd Pontiac MI 48055

KANTROWITZ, ARTHUR (ROBERT), b Bronx, NY, Oct 20, 13; m 43, 80; c 3. GAS DYNAMICS, LASER PROPULSION. *Educ:* Columbia Univ, BS, 34, MA, 36, PhD(physics), 47. *Hon Degrees:* DE, Mont Col Mineral Sci & Technol, 75; DSc, NJ Inst Technol, 81. *Prof Exp:* Physicist, Nat Adv Comt Aeronaut, 36-46; from assoc prof to prof aeronaut eng & eng physics, Cornell Univ, 46-56; founder, dir, chief exec officer & chmn, Avco Everett Res Lab, 55-78, sr vpres, bd dir, Avco Corp, 56-78; PROF ENG, THAYER SCH ENG, DARTMOUTH COL, 78- *Concurrent Pos:* Vis lectr, Harvard Univ, 52; Fulbright scholar & Guggenheim fel, Cambridge Univ & Univ Manchester, 54; vis inst prof & fel, Sch Advan Study, Mas Inst Technol, 57; hon trustee, Univ Rochester, 71; mem adv coun, Dept Aeronaut Eng, Princeton Univ, 59-77; mem bd overseers, Thayer Sch Eng, Dartmouth Col, 75-82; presidential adv group, Anticipated Advances in Sci & Technol, Sci Court Task Force chmn, 75-76; eng adv bd mem, Stanford Univ, 66-82 & Rensselaer Polytech Inst, 81-86; hon prof, Huazhang Inst Technol, Wuhan, China, 80. *Honors & Awards:* Theodore Roosevelt Asn Medal of Honor, Distinguished Service Sci, 67; Carl F Kayan Medal, Columbia Univ, 73; Messenger lectr, Cornell Univ, 78; Fluid & Plasma Dynamics Award & Medal, Am Inst Aeronaut & Astronaut, 81, Aerospace Contrib to Soc Award & Medal, 90; MHD Faraday Mem Medal, UNESCO, 83. *Mem:* Nat Acad Sci; Nat Acad Eng; fel Am Phys Soc; fel Am Inst Aeronaut & Astronaut; fel Am Acad Arts & Sci; Int Acad Astronaut. *Res:* Physical gas dynamics; magneto-hydrodynamics power; high power lasers; cardiac assist devices; strategic technology; social control of technology. *Mailing Add:* Four Downing Rd Hanover NH 03755-1902

KANTROWITZ, IRWIN H, b Brooklyn, NY, Oct 12, 37. GEOLOGY, HYDROLOGY. *Educ:* Brooklyn Col, BS, 58; Ohio State Univ, MS, 59. *Prof Exp:* From geologist to chief hydrologist, 59-80, DISTRICT CHIEF, US GEOL SURV, FLA, 80- *Concurrent Pos:* Mem, US Geol Surv Water Resources Adv Bd. *Mem:* Fel Geol Soc Am; Am Geophys Union; Asn Groundwater Scientists & Engrs. *Mailing Add:* 227 N Bronough St No 3015 Tallahassee FL 32301

KANTZ, PAUL THOMAS, JR, b Jacksonville, Tex, Jan 21, 41; m 62; c 3. PHYCOLOGY. *Educ:* Univ Tex, BA, 63, MA, 65, PhD(phycol), 67. *Prof Exp:* From asst prof to assoc prof, 67-81, dept chair, 88-90, PROF BIOL, CALIF STATE UNIV, SACRAMENTO, 81- *Mem:* Phycol Soc Am; Bot Soc Am; NY Acad Sci; Sigma Xi. *Res:* Taxonomy and morphology of blue green algae. *Mailing Add:* 525 42nd St Sacramento CA 95819

KANTZES, JAMES (GEORGE), b Bertha, Pa, Mar 29, 24; m 54; c 3. PLANT PATHOLOGY. *Educ:* Univ Md, BS, 51, MS, 54, PhD(plant path), 57. *Prof Exp:* From instr to assoc prof, 52-69, PROF PLANT PATH, COL AGR, UNIV MD, 69- *Mem:* Am Phytopath Soc; Sigma Xi. *Res:* Agriculture; control of vegetable diseases. *Mailing Add:* 751 Richwill Dr Salisbury MD 21801-5627

KANWAL, RAM PRAKASH, b India, July 4, 24; m 54; c 2. GENERALIZED FUNCTIONS. *Educ:* Punjab Univ, India, BA, 45, MA, 48; Ind Univ, PhD, 57. *Prof Exp:* Asst, Ministry of Agr, Govt of India, 48-50; lectr math, Daynand Anglo Vernacular Col, India, 50-51; asst prof, Birla Col, Pilani, 51-52; asst lectr, Indian Inst Technol, Kharagpur, 52-54; res assoc appl math, Ind Univ, 54-57; asst prof, Math Res Ctr, Univ Wis, 57-59; sr scientist, Oak Ridge Nat Lab, 59; assoc prof 59-62, PROF MATH, PA STATE UNIV, 62- *Concurrent Pos:* Vis prof, Tech Univ Denmark, 65-66 & Royal Inst Technol, Stockholm, 66. *Mem:* Soc Indust & Appl Math; Allahabed Math Soc. *Res:* Hydrodynamics; aerodynamics; magnetohydrodynamics; elasticity; diffraction; integral and differential equations. *Mailing Add:* Dept Math Pa State Univ University Park PA 16802

KANZELMEYER, JAMES HERBERT, b Manila, Philippines, Aug 9, 26; m 49; c 9. ANALYTICAL CHEMISTRY. *Educ:* Univ Calif, AB, 47; Ore State Col, PhD(anal chem), 55. *Prof Exp:* Chemist, Beacon res labs, Tex Co, 47-49; instr chem, Ore State Col, 53-54; asst prof, NMex Highlands Univ, 54-57; anal res chemist, St Joe Zinc Co, 57-63, chief chemist smelting div, 63-80, chief chemist, Corp Anal Serv, St Joe Minerals Corp, 80-87; CONSULT, 87- *Concurrent Pos:* Mem, Nat Res Coun Eval Panel, Anal Chem Div, Nat Bur Standards, 75-78. *Mem:* Am Soc Testing & Mat; Sigma Xi. *Res:* Analytical chemistry of zinc-containing materials, including chemical, optical-emission and x-ray spectrographic methods. *Mailing Add:* 5219 Webb St Aliquippa PA 15001

KANZLER, WALTER WILHELM, b Jersey City, NJ, Sept 17, 38. ANIMAL BEHAVIOR, BIOETHICS. *Educ:* Montclair State Col, BA, 60, MA, 63; Marshall Univ, MA, 64; Univ Cincinnati, PhD(ecol, behav), 72. *Prof Exp:* Instr biol, Union City High Schs, NJ, 60-65; asst prof, Trenton State Col, 65-66; from instr to asst prof, 66-76, assoc prof biol, 76-84, PROF BIOL & CHMN, WAGNER COL, 84- *Concurrent Pos:* NASA fel, 69-70; NSF grant, Nat Primate Ctr, Univ Calif, Davis, 71; sr res assoc, Nat Ctr Bioethics, Drew Univ, 76; consult, Scientists Ctr Animal Welfare, 80- *Mem:* Animal Behav Soc; Sigma Xi; AAAS; Nat Wildlife Fed. *Res:* Insect, gerbil and primate behavior, zoo animal behavior; history of biology and medicine; social issues in biology and medicine. *Mailing Add:* Dept Biol Wagner Col Staten Island NY 10301-4495

KAO, CHARLES K, b Shanghai, China, Nov 4, 33; US citizen; m 59; c 2. FIBER OPTICS. *Educ:* Univ London, BSc, 57, PhD(elec eng), 65. *Hon Degrees:* DSc, Chinese Univ Hong Kong, 85, Univ Sussex, 90. *Prof Exp:* Engr, Standard Tel & Cables Ltd, UK, 57-60; res engr, Standard Telecommun Labs Ltd & ITT Cent Europ Lab, 61-70; chmn, Dept Electronics, Chinese Univ Hong Kong, 70-74; chief scientist & dir eng, Electro-Optical Prod Div, Itt, Va, 74-82, exec scientist & corp dir res, Advan Technol Ctr, Conn, 82-87; VCHANCELLOR & PRES, CHINESE UNIV HONG KONG, 87- *Concurrent Pos:* adj prof & fel, Trumbull Col, Yale Univ, 85. *Honors & Awards:* Morey Award, Am Ceramic Soc, 76; Morris H Liebmann Mem Award, Inst Elec & Electronics Engrs, 78, Alexander Graham Bell Medal, 85; L M Ericsson Int Prize, Sweden, 79; Gold Medal, Armed Forces Commun & Electronics Asn, 80; Int Prize New Mat, Am Phys Soc, 89. *Mem:* Nat Acad Eng; Inst Elec & Electronics Engrs; Royal Swed Acad Sci; Royal Soc Arts. *Res:* Fiber optics. *Mailing Add:* The Chinese Univ Hong Kong Shatin New Territories Hong Kong Hong Kong

KAO, CHIEN YUAN, b Shanghai, China, Dec 20, 27; nat US; m 57; c 2. NEUROTOXINS, SMOOTH MUSCLES. *Educ:* Univ Southern Calif, BA, 48; State Univ NY, MD, 52. *Prof Exp:* Intern & asst path, NY Hosp-Cornell Med Ctr, 52-53, res assoc neurol, Col Physicians & Surgeons, Columbia Univ, 53-54; instr, State Univ NY Downstate Med Ctr, 55-56; asst, Rockefeller Inst, 56-57; asst to assoc prof pharmacol, 57-69, PROF PHARMACOL, STATE UNIV NY DOWNSTATE MED CTR, 69-, PROF NEUROSCI, 74- *Concurrent Pos:* Fel physiol, State Univ NY Downstate Med Ctr, 54-55. *Mem:* Am Physiol Soc; Am Soc Pharmacol & Exp Therapeut; Soc Gen Physiol; Biophys Soc. *Res:* Tetrodotoxin, saxitoxin and related toxins on sodium channel; ionic-channel functions of mammalian smooth muscles, and actions of drugs thereon. *Mailing Add:* State Univ NY Downstate Med Ctr 450 Clarkson Ave Box 29 Brooklyn NY 11203-9967

KAO, DAVID TEH-YU, b Shanghai, China, May 1, 36; m 68; c 2. CIVIL ENGINEERING. *Educ:* Cheng Kung Univ, Taiwan, BS, 59; Duke Univ, MS, 65, PhD(civil eng), 67. *Prof Exp:* From instr to asst prof, 66-76, PROF CIVIL ENG, UNIV KY, 77- *Mem:* Am Soc Civil Engrs. *Res:* Effect of transient flow conditions on solids moving through viscous fluid; dynamic drainage of fluid under moving bodies; mechanics of gully erosion of soil due to rain fall and runoff. *Mailing Add:* Dept Civil Eng Univ Ky Lexington KY 40506

KAO, FA-TEN, b Hankow, China, Apr 20, 34; nat US; m 60; c 1. GENETICS. *Educ:* Nat Taiwan Univ, BS, 55; Univ Minn, St Paul, PhD(genetics), 64. *Prof Exp:* Asst prof, Univ Col Med Ctr, 67-70, assoc prof biophys, 70-81, PROF BIOCHEM BIOPHYS & GENETICS, UNIV COLO HEALTH SCI CTR, DENVER, 81- *Concurrent Pos:* Nat Cancer Inst fel, Univ Colo Med Ctr, Denver, 65-67; sr fel, Eleanor Roosevelt Inst Cancer Res, Denver, 65-; Int Union Against Cancer-Eleanor Roosevelt Int Cancer fel, Univ Oxford, 73-74; res scientist, Europ Molecular Biol Lab, Heidelberg, 85; hon consult prof med molecular genetics, Harbin Med Univ, People's Rep China, 87; vis prof human molecular genetics, Tonji Med Univ, People's Repub China, 88. *Mem:* Genetics Soc Am; Am Soc Cell Biol; Am Soc Human Genetics; Am Asn Cancer Res; Tissue Cult Asn. *Res:* In vitro genetic studies of somatic mammalian cells; somatic cell and molecular genetic analysis of the human genome; mapping of human genes; use of recombinant DNA technology in human genetic studies; molecular analysis of genetic diseases. *Mailing Add:* Eleanor Roosevelt Inst Cancer Res 1899 Gaylord St Denver CO 80206

KAO, FREDERICK, b Peking, China, Jan 29, 19; nat US; m 49; c 1. PHYSIOLOGY. *Educ:* Yenching Univ, China, BS, 43; WChina Union Univ, MD, 47; Northwestern Univ, MS, 50, PhD(physiol), 52. *Prof Exp:* Instr physiol, Northwestern Univ, 51; instr med physics, Yale Univ, 52; from instr to assoc prof, 52-65, PROF PHYSIOL, STATE UNIV NY DOWNSTATE MED CTR, 65- *Concurrent Pos:* Vis prof, Med Col, Nat Taiwan Univ, 56-57, Inst Clin Physiol, Univ Goettingen, 65, Hacettape Univ, Turkey, 68 & Shangai First Med Col, 79; vis scientist, Lab Physiol, Oxford Univ, 59; career scientist, Health Res Coun NY, 61-71; vis lectr, Univ Hong Kong, 64; spec consult, Vet Admin Hosp, Brooklyn, NY, 65; vis scientist, Nobel Inst Neurophysiol, Karolinska Stockholm, Sweden, 76; expert panel, WHO, 77-; hon prof, Shanghai Med Univ, People's Repub China, 80-, Univ Hong Kong, 85. *Honors & Awards:* Morse Medal, WChina Union Univ, 47. *Mem:* AAAS; Am Physiol Soc; Soc Exp Biol & Med; fel NY Acad Sci; Harvey Soc; Int Brain Res Orgn; hon mem Brit Brain Res Asn; hon mem Europ Brain & Behav Soc; Sigma Xi. *Res:* Pneumo-hemodynamics during muscular activity; cardiac output during metabolic adjustments; central and peripheral control of respiration; cardiovascular and respiratory functions during environmental stress; sensitivity of respiratory centers to analeptics; brain neurotransmitter and respiration. *Mailing Add:* SUNY Health Sci Ctr Dist Prof Hofstra Univ Hempstead NY 11550

KAO, JOHN Y, b Hong Kong, Dec 12, 48; m; c 2. ANIMAL & EXPLORATORY DRUG METABOLISM. *Educ:* Univ Surrey, Eng, BSc, 73, PhD(biochem), 77. *Prof Exp:* NIH fel, Lab Reprod & Develop Toxicol, Nat Inst Environ Health Sci, 77-80; staff scientist, Biol Div, Oak Ridge Nat Lab, 80-86; sr investr, Dept Drug Metab, SmithKline Beecham Pharmaceut, 86-91; SR RES FEL, DEPT ANIMAL & EXPLOR DRUG METAB, MERCK SHARP & DOHME RES LABS, 91- *Concurrent Pos:* Fac mem, Traveling Lect Prog, Oak Ridge Assoc Univs, 84-86; mem, Arthropod Repellent Subcomt & Bd Environ Studies & Toxicol, Nat Res Coun, 86-87. *Honors & Awards:* Frank Blood Award, Soc Toxicol, 87. *Mem:* Soc Toxicol; Int Soc Study Xenobiotics; Am Asn Pharmaceut Scientists; Am Soc Pharmacol & Exp Therapeut. *Res:* Absorption, metabolism and toxicokinetics of xenobiotics; mechanisms of chemical toxicity and safety evaluation; dermatotoxicology, percutaneous absorption and transdermal delivery of drugs; drug development, drug metabolism and pharmacokinetics; author of numerous technical publications. *Mailing Add:* Dept Animal & Explor Drug Metab Merck Sharp & Dohme Res Labs PO Box 2000 Rahway NJ 07065

KAO, KUNG-YING TANG, b Nanking, China, May 8, 17; nat US; m 43; c 4. BIOCHEMISTRY. *Educ:* Nat Kiangsu Med Col, China, MD, 40; Purdue Univ, MS, 50; Univ Md, PhD(chem), 53. *Prof Exp:* Res assoc, Johns Hopkins Univ, 52-53; instr, Univ Md, 53; dir protein lab, Miami Heart Inst, Fla, 53-56; investr, Howard Hughes Med Inst, 56-57; chief res chemist, Geriat Res Lab, Vet Admin Ctr, Martinsburg, WVa, 57-78; RETIRED. *Mem:* Am Chem Soc; Geront Soc; Soc Exp Biol & Med; Am Inst Nutrit; Sigma Xi. *Res:* Biochemical studies on aging processes, especially of the connective tissue. *Mailing Add:* Gum Spring Hollow Brunswick MD 21716

KAO, KWAN CHI, b Chungshan, China, Oct 11, 26; Can citizen; m 57; c 4. SEMICONDUCTORS & DIELECTRICS. *Educ:* Univ Nanking, BSc, 48; Univ Mich, MSc, 50; Univ Birmingham, Eng, PhD (elec eng), 57,. *Hon Degrees:* DSc, Univ Birmingham, Eng, 84. *Prof Exp:* Res fel mat sci, Univ Col Swansea, UK, 57-58; res engr non-linear control systs, Nelson Res Lab, Eng Elec Co Ltd, Eng, 58-60; group leader & sr res engr dielec mats, Brush Elec Eng Co, Ltd, Eng, 60-61; sr lectr elec eng, Univ Salford, Eng, 62-65; assoc prof, 66-68, PROF ELEC ENG, UNIV MAN, 69- *Concurrent Pos:* Vis prof, Nat Defense Acad Japan, 80-81, Nat Univ Singapore & Xian Jiaotong Univ, China, 81; external referee, Res Grants Comts Can Coun & Natural Sci & Eng Res Coun Can, Ottawa, 76- *Mem:* Fel Inst Elec Engrs UK; fel Inst Physics UK; sr mem Inst Elec & Electronics Engrs; Can Asn Physicists; Sigma Xi; Asn Prof Engrs Can. *Res:* Electronic and optical properties of semiconductors and insulators (crystalline, poly-crystalline, micro-crystalline, and non-crystalline, organic and inorganic) in bulk and film forms with and without doping, and their applications for devices; high-field conduction and breakdown phenomena in dielectrics; photo-electric properties of polymers and ceramics incorporated with various impurities. *Mailing Add:* Dept Elec Eng Univ Man Winnipeg MB R3T 2N2 Can

KAO, MING-HSIUNG, b Taipei, Taiwan, Jan 10, 44; Can citizen; c 2. ANIMAL PHYSIOLOGY. *Educ:* Nat Taiwan Univ, BSc, 64; Mem Univ Nfld, MSc, 70, PhD(biol), 79. *Prof Exp:* Asst prof, 81-88, ASSOC PROF BIOL, MEM UNIV NFLD, 88- *Res:* Protein & glycoprotein antifreeze activities in marine teleosts. *Mailing Add:* 35 Oakridge Dr St Johns NF A1A 4R9 Can

KAO, RACE LI-CHAN, b Chungking, China, Dec 1, 43; m 69; c 2. MYOCARDIAL METABOLISM, CARDIOVASCULAR DISEASE. *Educ:* Nat Taiwan Univ, BS, 65; Univ Ill, MS, 71, PhD(biochem & physiol), 72. *Prof Exp:* Res assoc animal sci, Univ Ill, 72; res assoc physiol, M S Hershey Med Ctr, 72-75, asst prof, 75-77; asst prof surg & physiol, Univ Tex Med Br, 77-82; assoc prof surg, Wash Univ, 82-83; DIR SURG RES, ALLEGHENY-SINGER RES INST, 83-; PROF SURG, MED COL PA, 88- *Concurrent Pos:* Mem, Coun Circulation, Am Heart Asn. *Mem:* Am Heart Asn; Am Physiol Soc; AAAS; Am Soc Artificial Internal Organs; Int Soc Heart Res; NY Acad Sci. *Res:* Utilizing isolated heart muscle cells, isolated perfused organs, and several animal models to study regulation of myocardial metabolism, gene expression, hypertrophy, function, and the recovery of the failing heart. *Mailing Add:* Allegheny-Singer Res Inst 320 E North Ave Pittsburgh PA 15212

KAO, SAMUEL CHUNG-SIUNG, b Kaohsiung, Taiwan, June 12, 41; c 2. MATHEMATICAL STATISTICS, MATHEMATICS. *Educ:* Nat Taiwan Univ, BS, 64; Nat Tsing Hua Univ, MS, 66; Columbia Univ, PhD(statist), 72. *Prof Exp:* Lectr math, Nat Tsing Hua Univ, 66-67; statist assoc, Biomet Res, NY State Psychiat Inst, 71-73; asst prof statist, Univ Mass, Amherst, 73-74; STATISTICIAN, BROOKHAVEN NAT LAB, 74- *Mem:* Am Statist Asn; Inst Math Statist; Biomet Soc; Sigma Xi; NY Acad Sci. *Res:* Sequential experimentation; robust statistical procedures; applied probability. *Mailing Add:* 26 Cornwallis Rd East Setauket NY 11733

KAO, TAI-WU, b China, Jan 5, 35; m 59; c 5. COMMUNICATION. *Educ:* Nat Taiwan Univ, BS, 58; Chiao Tung Univ, MS, 61; Univ Utah, PhD(elec eng), 65. *Prof Exp:* From asst prof to assoc prof, 65-77, PROF ELEC ENG, LOYOLA MARYMOUNT UNIV, 77- *Concurrent Pos:* Consult, Teledyne Systs Control Syst, 65-68, DWP, Los Angeles, 71-72, Dept Navy, 73-74 & TRW, 75- *Mem:* Inst Elec & Electronics Engrs. *Res:* Electromagnetics and semiconductors; communication systems. *Mailing Add:* 8221 Zitola Terr Playa Del Rey CA 90293

KAO, TIMOTHY WU, b Shanghai, China, July 20, 37; US citizen; m 65; c 2. FLUID MECHANICS, CIVIL ENGINEERING. *Educ:* Univ Hong Kong, BSc, 59; Univ Mich, MSE, 60, PhD(eng mech), 63. *Prof Exp:* Asst prof space sci, 64-66, assoc prof atmospheric sci, 66-70, PROF CIVIL ENG, CATH UNIV AM, 70-, CHMN, DEPT CIVIL ENG, 81- *Concurrent Pos:* Prin investr, NSF grants, 65- & Off Naval Res Contracts, 74- *Mem:* Am Meteorol Soc; fel Am Soc Civil Engrs. *Res:* Physical oceanography; air-sea interaction; mountain waves; environmental fluid mechanics. *Mailing Add:* Dept Civil Eng Cath Univ Am 620 Michigan Ave Washington DC 20064

KAO, WEN-HONG, b Taipei, Taiwan, Mar 15, 54; m 79; c 3. ELECTROCHEMISTRY, ENERGY STORAGE & TRANSFER SYSTEMS. *Educ:* Nat Tsing Hua Univ, Hsinchu, Taiwan, BS, 76; Ohio State Univ, Columbus, PhD(anal chem), 84. *Prof Exp:* LEAD SCIENTIST RES & DEVELOP, RAYOVAC CORP, 84- *Concurrent Pos:* Res fel, Ohio State Univ, 84. *Mem:* Am Chem Soc; Electrochem Soc. *Res:* Chemical and electrochemical analysis of battery electrode materials and systems; models of chemical and electrochemical reactions; production-related battery problems. *Mailing Add:* 5356 W Silverleaf Lane Milwaukee WI 53223-1651

KAO, YI-HAN, b Foochow, China, Jan 27, 31; m 57; c 2. PHYSICS. *Educ:* Nat Taiwan Univ, BS, 55; Okla State Univ, MS, 58; Columbia Univ, PhD(physics), 62. *Prof Exp:* Res assoc physics, Thomas J Watson Lab, Int Bus Mach Corp, 62-63; from asst prof to assoc prof, 63-71, PROF PHYSICS, STATE UNIV NY, STONY BROOK, 71- *Mem:* Am Phys Soc. *Res:* Low temperature solid state physics; superconductivity; physics of thin films; transport phenomena. *Mailing Add:* Dept Physics State Univ NY Stony Brook NY 11794

KAO, YUEN-KOH, b Liaoning, China, Apr 3, 41; US citizen; m 66; c 2. CHEMICAL ENGINEERING, CONTROL ENGINEERING. *Educ:* Nat Taiwan Univ, BS, 64; Northwestern Univ, MS, 68, PhD(chem eng), 73. *Prof Exp:* Assoc chem eng, Rensselaer Polytech Inst, 73-75; asst prof, 75-81, ASSOC PROF CHEM ENG, UNIV CINCINNATI, 81- *Concurrent Pos:* Consult, Mound Lab, Monsanto Res Corp, 77, Columbia Gas, 76, R Katzen Asn, 80. *Mem:* Am Inst Chem Engrs; Sigma Xi; Electrochem Soc. *Res:* Process simulation and control; electrochemical engineering; boiling heat transfer; reaction engineering. *Mailing Add:* 2930 Scioto St Apt 1401 Cincinnati OH 45219

KAPADIA, ABHAYSINGH J, b Bombay, India, Mar 14, 29; m 55; c 2. PHARMACY. *Educ:* L M Col Pharm, Ahmedabad, India, 52; Univ Mich, Ann Arbor, MS, 58; Univ Tex, Austin, PhD(pharm), 63. *Prof Exp:* Res chemist, Univ Mich, 57-58; sect head anal res, Alcon Labs, 63-65; res assoc, 65-74, MGR STABILITY TESTING & PROD EVAL, A H ROBINS CO INC, 74- *Mem:* Am Pharmaceut Asn. *Res:* Analytical methods development; stability testing of pharmaceuticals; preformulation studies of pharmaceuticals; drug plastic interactions. *Mailing Add:* 2221 Walhala Dr Richmond VA 23236

KAPANIA, RAKESH KUMAR, b Nakodar, Punjab, Aug 3, 56; m 85; c 1. STRUCTURAL MECHANICS, PLATES & SHELLS. *Educ:* Punjab Univ, India, BS, 77; Indian Inst Sci, MS, 79; Purdue Univ, PhD(aerospace), 85. *Prof Exp:* Grad asst aerospace struct, Purdue Univ, 79-85; asst prof, 85-90, ASSOC PROF AEROSPACE STRUCT, VA POLYTECH INST & STATE UNIV, 90- *Concurrent Pos:* NASA-ASEE fel, NASA-Langley Res Ctr, 85. *Mem:* Am Inst Aeronaut & Astronaut; Am Soc Civil Engrs; Sigma Xi; Soc Indust & Appl Math. *Res:* Application and development of state-of-the-art computational methods to solve problems in aeroelasticity, wave propagation in composites, impact response of laminated structures and plates and shells with emphasis on finite element method. *Mailing Add:* Aerospace & Ocean Eng Va Polytech Inst & State Univ 215 Randolph Hall Blacksburg VA 24061

KAPANY, NARINDER SINGH, b Moga, India, Oct 31, 27; nat US; m 54; c 2. PHYSICS, OPTICS. *Educ:* DAV Col, Dehra Dun, BS, 48; Imp Col, London, dipl, 52; Univ London, PhD(optics), 54. *Prof Exp:* Supvr, Ord Factory, India, 49-51; lens designer, Barr & Stroud Optical Co, Scotland, 52; res assoc physics, Imp Col, London, 54-55; res assoc, Inst Optics, Rochester, 55-57; mgr optics sect, Ill Inst Technol Res Inst, 57-61; pres & dir res, Optics Technol, Inc, 61-73; CHMN BD & PRES, KAPTRON INC, 73- *Concurrent Pos:* Consult, Bausch & Lomb Optical Co, 55-57, Argus Camera Co, 55-57 & Johns Hopkins Hosp, 56-57; res assoc, Palo Alto Med Res Found, Calif, 62-; vis scholar, Stanford Univ, 73-74; regents prof, Univ Calif, Santa Cruz, 76-77, dir, Ctr Innovation & Entrepreneurial Develop, 78. *Mem:* AAAS; fel Am Phys Soc; fel Optical Soc Am; sr mem Inst Elec & Electronics Engrs; fel Brit Inst Phys; Sigma Xi. *Res:* Geometrical and physical optics; fiber optics with applications in medicine, photoelectronics, photography, high-speed photography; infrared fiber optics communications, local area networks; laser and its applications; image evaluation and optical information processing; photoelectronics; aspherics; interference microscopy; refractometry; solar energy. *Mailing Add:* Kaptron Inc 2525 E Bayshore Rd Palo Alto CA 94303

KAPECKI, JON ALFRED, b Chicago, Ill, June 8, 42; m 82. PHYSICAL ORGANIC CHEMISTRY, PHOTOGRAPHIC SCIENCE. *Educ:* Col St Thomas, BS, 64; Univ Vienna, Dipl, 64; Univ Ill, Urbana, PhD(org chem), 69. *Prof Exp:* NIH fel chem, Cornell Univ, 68-71, fel, 71-72; sr chemist, Eastman Kodak Co, 72-80, res assoc, 80-85, LAB HEAD, EASTMAN KODAK CO, 85- *Concurrent Pos:* Mem staff, X-ray Clinic, State Univ NY, Albany, 75-; lectr, Univ Rochester, 72-80, sr lectr, 80-84, vis assoc prof, 84-86. *Mem:* Am Chem Soc; Am Crystallog Asn. *Res:* Organic cycloaddition mechanisms; solid state reactions; molecular orbital theory; x-ray crystallography; computer applications to organic chemistry; models for reactive intermediates; structure-pharmacological activity relationships; reaction mechanisms. *Mailing Add:* 161 Crosman Terr Rochester NY 14620

KAPER, HANS G, b Alkmaar, Neth, June 10, 36; US citizen; m 62; c 2. APPLIED MATHEMATICS, MATHEMATICAL ANALYSIS. *Educ:* State Univ Groningen, Neth, MSc, 60, PhD(math), 65. *Prof Exp:* Asst prof appl math, State Univ Groningen, Neth, 65-66, assoc prof, 67-69; res assoc, Stanford Univ, 66-67; SR MATHEMATICIAN, ARGONNE NAT LAB, 69-, DIV DIR, 87- *Concurrent Pos:* Vis prof, Univ van Amsterdam, Neth, 76-77, Univ Vienna, Austria, 77 & Northwestern Univ, 78-80 & 84-85; adj prof,

Northern Ill Univ, 83- *Mem:* Am Math Soc; Soc Indust & Appl Math; Math Soc Neth; corresp mem Royal Neth Acad Sci. *Res:* Applied analysis. *Mailing Add:* Math & Comput Sci Div Argonne Nat Lab 9700 S Case Ave Argonne IL 60439

KAPER, JACOBUS M, b Madjalenka, Indonesia, Dec 9, 31; US citizen; m 55; c 1. BIOCHEMISTRY, MOLECULAR BIOLOGY. *Educ:* Univ Leiden, BS, 51, Drs, 54, PhD(biochem), 57. *Prof Exp:* Asst biochem, Univ Leiden, 54-57, sr res biochemist, 59-62; res fel, Virus Lab, Univ Calif, 57-59; biochemist, Plant Sci Res Div, 62-69, RES CHEMIST, PLANT PROTECTION INST, AGR RES SERV, USDA, 69- *Concurrent Pos:* Fel, Neth Orgn Pure Res, 57-58; USPHS trainee, 58; assoc res prof, George Washington Univ, 62-69; prin investr, NIH res grants, 62-69 & USDA res grant, 78-81, 82-85 & 85-87. *Mem:* Am Soc Virol; Am Chem Soc; Am Soc Biol Chemists; Am Soc Microbiol; corresp mem Royal Neth Acad Sci; Int Soc Plant Molecular Biol. *Res:* Molecular organization and stabilizing interactions of viruses; structural biochemistry of proteins nucleic acids; divided genome viruses; mechanisms of viral disease regulation. *Mailing Add:* 115 Hedgewood Dr Greenbelt MD 20070-1610

KAPER, JAMES BENNETT, b Havre de Grace, Md, Aug 25, 52. VACCINE DEVELOPMENT, INFECTIOUS DISEASES. *Educ:* Univ Md College Park, BS, 73, PhD(microbiol), 79. *Prof Exp:* Lab technician, England Labs, Beltsville, Md, 74-75; res asst, micros dept, Univ Md, College Park, 75-79; res fel, Dept Microbiol & Immunol, Univ Wash, 79-81; asst prof med & microbiol, Univ Md, 81-84, asst prof biol chem, 83-85, assoc prof med, microbiol & biol chem, 84-90, CHIEF, BACT GENETICS SECT, CTR VACCINE DEVELOP, UNIV MD, 83-, PROF MED, MICROBIOL & BIOCHEM, 90- *Concurrent Pos:* Adj prof micros, Univ RI, 82-83; consult, NIH, 82-83, 88- & WHO, 84. *Mem:* Am Soc Microbiol; AAAS. *Res:* Development of vaccines and improved diagnostic tests for infectious diseases using recombinant DNA techniques. *Mailing Add:* Ctr Vaccine Develop Sch Med Univ Md Ten S Pine Baltimore MD 21201

KAPETANAKOS, CHRISTOS ANASTASIOS, b Sparta, Greece, Jan 2, 36; US citizen; c 2. PLASMA PHYSICS. *Educ:* Nat Univ Greece, Bachelor, 60; Mass Inst Technol, Master, 64; Univ Md, PhD(physics), 70. *Prof Exp:* Res physicist plasma physics, Univ Tex, 70-71; supvry res physicist plasma physics, Naval Res Lab, 71-80, head, Beam Dynamics Prog, 80-85, head, Adv Beam Technol Br, 85-89, SECT HEAD, NAVAL RES LAB, 89- *Concurrent Pos:* Energy Res & Develop Admin grant, 75-77, Dept Energy, 77-80, Off Naval Res, 80-84, SDID, 84, Defense Adv Res Proj Agency, 84 & Spawar, 85- *Honors & Awards:* Outstanding Performance Award, Naval Res Lab, 72, Res Publ Awards, 73, 76, 79, 86 & 90. *Mem:* Fel Am Phys Soc. *Res:* Supervise and contact research on intense, relativistic electron and ion beams and fusion reactors based on reversed magnetic field configurations; free electron lasers and ultra-high current accelerators. *Mailing Add:* Code 4795 Naval Res Lab Washington DC 20375

KAPETANOVIC, IZET MICHAEL, DRUG METABOLISM, EPILEPSY. *Educ:* Northwestern Univ, Chicago, PhD(pharmacol), 78. *Prof Exp:* PHARMACIST, PRECLIN EPILEPSY BR, NIH, 78- *Mailing Add:* 4204 S End Rd Rockville MD 20853-2047

KAPIKIAN, ALBERT ZAVEN, b New York, NY, May 9, 30; m 60; c 3. EPIDEMIOLOGY, VIROLOGY. *Educ:* Queens Col, NY, BS, 52; Cornell Univ, MD, 56. *Prof Exp:* Intern, Meadowbrook Hosp, Hempstead, NY, 56-57; actg head epidemiol, 64-67, RES EPIDEMIOLOGIST, LAB INFECTIOUS DIS, NAT INST ALLERGY & INFECTIOUS DIS, 57-, HEAD EPIDEMIOL SECT & ASST CHIEF, 67- *Concurrent Pos:* Guest worker virol, Royal Postgrad Med Sch, Univ London, 70; res prof Child Health & Develop, Sch Med & Health Serv, George Washington Univ, 77- *Honors & Awards:* Kabakjian Award, Armenian Students' Asn Am, 74; Stitt Award, Asn Mil Surgeons US, 74; Behring Diag Award, Am Soc Microbiol, 87. *Mem:* Am Pub Health Asn; Am Soc Microbiol; Am Epidemiol Soc; Infectious Dis Soc Am; Soc Epidemiol Res; AAAS. *Res:* Epidemiologic investigations of infectious diseases; viral gastroenteritis. *Mailing Add:* 11201 Marcliff Rd Rockville MD 20852

KAPILA, ASHWANI KUMAR, b Ludhiana, India, Aug 26, 46; US citizen; m 71; c 2. ASYMPTOTICS. *Educ:* Punjabi Univ, India, BS, 68; Univ Sask, Can, MS, 70; Cornell Univ, PhD(theoret & appl mech), 75. *Prof Exp:* Instr & res assoc mech, Cornell Univ, 74-76; from asst prof to assoc prof, 76-88, PROF MATH, RENSSELAER POLYTECH INST, 88- *Concurrent Pos:* Vis asst prof, Northwestern Univ, 78, Math Res Ctr, Univ Wis-Madison, 79-80; vis assoc prop, Inst Math Appln, Univ Minn, 86-87. *Mem:* Combustion Inst; Soc Indust & Appl Math. *Res:* Applied mathematics, especially asymptotics, perturbation theory and numerics; application to problems in mechanics, chemically reactive flows and combustion theory. *Mailing Add:* Dept Math Sci Rensselaer Polytech Inst Troy NY 12180-3590

KAPLAN, ABNER, b New York, NY, June 21, 23; m 50; c 3. AERONAUTICS. *Educ:* Calif Inst Technol, BS, 48, MS, 49, PhD(aeronaut), 54. *Prof Exp:* Res engr, Struct Res Group, Northrop Aircraft Corp, 52-55; mem tech staff & mgr struct dept, 55-70, STAFF ENGR, TRW SYSTS GROUP, 70- *Mem:* Am Soc Mech Engrs; Am Inst Aeronaut & Astronaut. *Res:* Nonlinear buckling; thin shells; pressure vessels. *Mailing Add:* TRRW 0556 Bldg R4 Rm 2142 1 Space Park Redondo Beach CA 90278

KAPLAN, ALAN MARC, b Brooklyn, NY, Dec 10, 40; m 72; c 1. IMMUNOLOGY. *Educ:* Tufts Univ, BS, 63; Purdue Univ, PhD(immunol), 69. *Prof Exp:* Res asst tumor immunol, Sloan-Kettering Inst, 63-65; asst prof, Med Col Va, 72-75, assoc prof surg & microbiol, 75-79, coordr, Tumor Immunol Sect, 74-82, prof surg & microbiol, 79-82, dep chmn, Dept Microbiol, 81-82, assoc dir res, Va Commonwealth Univ Cancer Ctr, 80-82; PROF & CHMN, DEPT MICROBIOL & IMMUNOL, SCH MED, UNIV KY, 82- *Concurrent Pos:* Can Med Res Coun fel immunol, Univ Toronto, 69-72. *Mem:* Am Soc Microbiol; NY Acad Sci; Can Soc Immunol; Am Asn Immunol; Am Asn Cancer Res; Reticuloendothelial Soc. *Res:* Tumor and cellular immunology; autoimmunity; immunoadjuvants; immunogenetics; macrophage differentiation. *Mailing Add:* Dept Microbiol & Immunol Col Med MS411 Univ Ky Lexington KY 40536

KAPLAN, ALBERT SYDNEY, virology; deceased, see previous edition for last biography

KAPLAN, ALEX, b New York, NY, May 22, 10; m 40; c 3. PHYSIOLOGY, CHEMISTRY. *Educ:* Univ Calif, Los Angeles, AB, 32; Univ Calif, PhD(physiol), 36. *Prof Exp:* Res assoc, Mt Zion Hosp, San Francisco, 37-39; asst physiol, Univ Calif, 39-40; asst dir, Harold Brunn Inst Cardiovasc Res, San Francisco, 40-42; chief lab, Vio-Bin Corp, 46-50; asst dir dept biochem, Michael Reese Hosp, Ill, 50-57; chief chemist, Children's Hosp, San Francisco, 57-60; assoc prof biochem, 60-69, prof biochem & lab med & dir, Chem Div, 69-80, dir, Hosp Chem Labs, 60-80, EMER PROF LAB MED, SCH MED, UNIV WASH, 80- *Mem:* AAAS; Soc Exp Biol & Med; Am Physiol Soc; Am Asn Clin Chem (pres, 71). *Res:* Lipid metabolism; clinical chemistry. *Mailing Add:* BB245 Univ Hosp SB-10 Sch Med Univ Wash Seattle WA 98195

KAPLAN, ALEXANDER E, b Kiev, USSR, June 9, 38; US citizen. OPTICS, THEORETICAL PHYSICS. *Educ:* Moscow Phys-Tech Inst, MS, 61; USSR Acad Sci, Moscow & Gorky State Univ, PhD(physics & math), 67. *Prof Exp:* Mem res staff, USSR Acad Sci, Moscow, 63-79; mem res staff, Francis Bitter Nat Magnet Lab, Mass Inst Technol, 79-82; prof elec eng, Purdue Univ, 82-87; PROF DEPT ELEC & COMPUT ENG, JOHN HOPKINS UNIV, 87- *Concurrent Pos:* Vis scientist, Max Planck Inst, Garching, WGer; consult, Bell Labs, 80-81, Los Alamos Nat Lab, 81 & Honeywell, 82; prin investr res proj, Off Sci Res, USAF, 80-, Mat Res Lab, NSF, 82-87 & CST Ind, 86-87. *Mem:* Am Phys Soc; fel Optical Soc Am; Lasers & Electroptical Soc. *Res:* Quantum electronics and nonlinear optics, nonlinear and quantum optics of a single electron; theory of two-level systems in a strong field; self-focusing and self-bending effects; theory of solitons; interaction of light with nonlinear interfaces and theory of cavityless optical bistability; light induced nonreciprocity, sagnac effect in nonlinear ring resonators and optical gyroscopes; nonlinear optical effects in superlattices; x-ray radiation by fast electron beams in periodical structrees (in particular in superlattices); four-wave mixing instabilities and multistability; the switching and steering of laser beams. *Mailing Add:* Dept Elec & Comput Eng John Hopkins Univ Baltimore MD 21218

KAPLAN, ALLEN P, b Jersey City, NJ, Oct 27, 40; m; c 2. MEDICINE. *Educ:* Columbia Univ, BA, 61; Downstate Med Sch, MD, 65; Am Bd Internal Med, cert, 72, Rheumat, cert, 72, Allergy & Clin Immunol, cert, 74, Diag Lab Immunol, cert, 86. *Prof Exp:* Intern med, Strong Mem Hosp, Rochester, NY, 65-66, asst resident med, 66-67; clin assoc, Nat Inst Arthritis & Metab Dis, NIH, Bethesda, Md, 67-69, head, Allergic Dis Sect, Lab Clin Invest, 72-78; res fel med, Peter Bent & Robert B Brigham Hosps, Harvard Med Sch, 69-72; prof med & head, Div Allergy, Rheumat & Clin Immunol, State Univ NY, Stony Brook, 78-87; CHMN, DEPT MED, STATE UNIV NY, STONY BROOK HEALTH SCI CTR, 87- *Concurrent Pos:* Spec fel, NIH, 70-72; prof lectr, Dept Biochem, Georgetown Univ, Washington, DC, 76-78; prof med & head, Div Allergy, Rheumat & Clin Immunol, Northport Vet Admin Hosp, NY, 78-87; grants, NIH, 78-; mem study sect, Nat Heart, Lung & Blood Inst, NIH, 81-85, Am Rheumatism Asn, 81-84, Am Heart Asn, 83-87; chmn, Res Coun, Am Acad Allergy & Immunol, 81-84; dir, Am Bd Allergy & Clin Immunol, 85-88; mem, Allergy, Immunol & Transplantation Res Comt, Nat Inst Allergy & Infectious Dis, 88-, chmn, 90- *Mem:* AAAS; Am Col Rheumat; Am Asn Immunologists; Am Thoracic Soc; Am Soc Exp Path; Am Soc Pharmacol & Exp Therapeut; Am Soc Hemat; Am Soc Clin Res; fel Am Acad Allergy; fel Am Col Physicians. *Res:* Coagulation, fibrinolysis and inflammation; immunochemistry and immunopathology; biochemical mechanisms of allergic reactions; cytokines, mast cells and rheumatic disease; author or co-author of over 180 publications. *Mailing Add:* Dept Med State Univ NY Health Sci Ctr Stony Brook NY 11794

KAPLAN, ANN ESTHER, b New York, NY, Dec 28, 26. BIOCHEMISTRY, BIOPHYSICS. *Educ:* Hunter Col, BA, 47; Mt Holyoke Col, MA, 49; Univ Pa, PhD(biochem), 59. *Prof Exp:* Instr neurol, Albert Einstein Col Med, 62-63; res assoc, Rockefeller Inst, 63-65; asst prof doctoral fac biochem, City Univ New York, 65-67; sr res assoc, Salk Inst Biol Studies, 67-72; biochemist, 72-77, RES BIOCHEMIST, NAT CANCER INST, 77- *Concurrent Pos:* Dazian Found fel microbiol, Sch Med, NY Univ, 59-60; NIH sr fel physiol, Albert Einstein Col Med, 60-62; chmn & comr rep, sci manpower comn, Am Soc Exp Biol, 78-81. *Mem:* AAAS; Am Chem Soc; NY Acad Sci; Am Soc Biol Chem; Biophys Soc; Am Soc Physiol; Soc Gen Physiologists. *Res:* Biosynthesis of lipids in biological membranes; serum lipid factors in cell growth; oxygenase pathway in heart mitochondria; lactate dehydrogenace in central nervous system and serum in experimental allergic encephalomyelitis; modification of lactate dehydrogenace in hepatocyte lines with neoplastic transformation; aerobic glycolysis in neoplastic cell lines. *Mailing Add:* 4242 East-West Hwy Apt 514 Chevy Chase MD 20815

KAPLAN, ARNOLD, b New York, NY, Dec 20, 39; m 63; c 2. LYSOSOMOLOGY, ORGANELLE BIOGENESIS. *Educ:* City Col New York, BS, 61; George Washington Univ, MS, 63, PhD(biochem), 66; Univ Calif, PhD(biochem), 68. *Prof Exp:* from asst prof to assoc prof, 68-84, PROF MICROBIOL, MED SCH, ST LOUIS UNIV, 84- *Concurrent Pos:* NIH fel, 66-68 & spec res fel, 70-72; fel, Nat Cancer Soc, 66-68; vis scientist biophys, Weissman Inst, 72; vis assoc prof pediat, Wash Univ Med Sch, 76; consult, Weissman Inst, Calbiochem, Monsanto & Childrens Inst; prog dir, Cell Biol Prog, NSF, 86. *Mem:* Am Soc Biol Chemists; Am Soc Cell Biol. *Mailing Add:* Dept Microbiol Sch Med St Louis Univ 1402 S Grand Blvd St Louis MO 63104

KAPLAN, ARTHUR LEWIS, b Boston, Mass, Mar 13, 33; m 57; c 2. HEALTH PHYSICS, NUCLEAR ENGINEERING. *Educ:* Mass Inst Technol, BS, 54, MS, 55. *Prof Exp:* Proj engr, Aircraft Nuclear Propulsion Prog, Wright Air Develop Ctr, Ohio, 55-57; physicist, Tech Opers, Inc, 57-60; physicist & unit mgr, Gen Elec Co, 60-64; physicist & proj leader, Tech Opers Res, 64-69; tech dir, Systs Sci & Eng, Inc, 69-72; consult engr, Gen Elec Co Nuclear Fuel Dept, 72-78, mgr licensing & compliance, Mfg Dept, 78-81, mgr environ control, Lighting Bus Group, 81-90, SR ENVIRON ENG, GEN ELEC LIGHTING, GEN ELEC CO, CLEVELAND, 91- *Mem:* Am Soc Photobiology. *Res:* Theoretical and experimental radiation shielding; biological effects of ionizing radiation; long range environmental effects of radioactive fallout; radiation effects in electronics and materials; health physics and radiation safety; health physics, licensing and compliance in uranium fabrication plants; environmental safety engineering. *Mailing Add:* 25422 Bryden Rd Beachwood OH 44122

KAPLAN, BARRY HUBERT, b Brooklyn, NY, Nov 16, 38; m 62; c 2. ONCOLOGY, BIOCHEMISTRY. *Educ:* NY Univ, BA, 58; Johns Hopkins Univ, MD, 62, PhD(physiol chem), 67. *Prof Exp:* Intern med, Johns Hopkins Hosp, Baltimore, 62-63; res assoc biochem, Nat Heart Inst, 64-66; resident med, Bronx Munic Hosp Ctr, NY, 66-67; assoc, 67-70, asst prof med, 71-75, actg, Div Oncol, 74-80, asst prof biochem, 73-83, assoc prof med, 75-83, dir, Div Oncil, 81-83, CLIN ASSOC PROF MED, ALBERT EINSTEIN COL MED, 83-; PHYSICIAN-IN-CHARGE, MED ONCOL BOOTH MEM MED CTR, FLUSHING, 86- *Concurrent Pos:* USPHS fel physiol chem, Sch Med, Johns Hopkins Univ, 63-64; USPHS training grant, Albert Einstein Col Med, 67-69; attend physician, Bronx Munic Hosp Ctr, NY, 67-; Am Heart Asn estab investr, Albert Einstein Col Med, 69-74. *Mem:* AAAS; Harvey Soc; Am Asn Cancer Res; Am Soc Clin Oncol. *Mailing Add:* 199 Weaver St Scarsdale NY 10583

KAPLAN, BERNARD, b New York, NY, Feb 16, 21; m 54; c 3. MATHEMATICAL PHYSICS. *Educ:* City Col New York, BS, 47; Ohio State Univ, PhD(physics), 53. *Prof Exp:* Prin engr math eng anal, Aircraft Nuclear Propulsion Dept, Gen Elec Co, 53-61; from assoc prof to prof physics, USAF Inst Technol, 61-89; ADJ PROF, WRIGHT STATE UNIV, 90- *Mem:* Am Asn Physics Teachers. *Res:* Numerical solution with digital computers of boundary value and initial value problems of engineering and applied physics, especially in heat transfer and nuclear engineering. *Mailing Add:* 416 Rendale Pl Trotwood OH 45246

KAPLAN, BERTON HARRIS, b Winchester, Va, June, 30; c 2. EPIDEMIOLOGY. *Educ:* Va Polytechnic Inst, BS, 51; Univ NC, Chapel Hill, MS, 52, PhD(sociol), 62. *Prof Exp:* PROF SOCIAL EPIDEMIOL, UNIV NC, CHAPEL HILL, 60- *Concurrent Pos:* Fel, Social Res Coun, 65-66. *Mem:* AAAS; Soc Behav Med; Am Anthrop Asn; Am Sociol Asn; Soc Epidemiol Res. *Res:* Behavioral factors in coronary disease. *Mailing Add:* Dept Epidemiol CB 7400 Sch Pub Health Univ NC Chapel Hill NC 27514

KAPLAN, DANIEL ELIOT, b San Mateo, Calif, Aug 17, 32; m 59; c 4. SOLID STATE PHYSICS, PLASMA PHYSICS. *Educ:* Univ Calif, AB, 53, MA, 55, PhD(physics), 58. *Prof Exp:* Res scientist, 58-67, SR STAFF SCIENTIST, LOCKHEED PALO ALTO RES LAB, 67- *Mem:* Am Phys Soc; Sigma Xi. *Res:* Paramagnetic and ferrimagnetic resonance; plasma resonance phenomena. *Mailing Add:* 27000 Appaloosa Way Los Altos Hills CA 94022

KAPLAN, DANIEL MOSHE, b Philadelphia, Pa, May 21, 53; m 86; c 1. ELEMENTARY PARTICLE PHYSICS. *Educ:* Haverford Col, BA, 74; State Univ NY, Stony Brook, PhD(physics), 79. *Prof Exp:* Res asst, State Univ NY, Stony Brook, 74-78; res assoc, Columbia Univ, 79-82; assoc scientist, Fermilab, 82-84; comput res specialist, Fla State Univ, 84-86; asst prof, 87-90, ASSOC PROF PHYSICS, NORTHERN ILL UNIV, 90- *Mem:* Am Phys Soc; Sigma Xi. *Res:* Elementary particle physics carrying out a series of fixed-target experiments studying the interactions of quarks at large momentum transfer. *Mailing Add:* Physics Dept Northern Ill Univ Faraday Hall DeKalb IL 60115

KAPLAN, DAVID GILBERT, b Chicago, Ill, Nov 13, 44; m 71; c 2. PHYSICAL CHEMISTRY. *Educ:* Univ Ill, Urbana, BS, 65; Univ Southern Calif, PhD(phys chem), 72; JD, Loyola Sch Law, 80. *Prof Exp:* Fel biochem, Sch Med, Univ Calif, Los Angeles, 72-74; res coordr radiopharm, Sch Pharm, Univ Southern Calif, 75-76; unit head gelatin res, Banner Gelatin Prod Corp, 76-81; patent chemist, Union Oil Calif, 81-83; sr scientist, Cilgo Inc, Pomona, Calif, 83-87; TAX LAW SPECIALIST, INTERNAL REVENUE SERV, LOS ANGELES, CALIF, 87- *Mem:* Am Chem Soc. *Res:* Studies on the viscoelastic behavior of connective tissue and lipid metabolism of biological systems. *Mailing Add:* 4215 Los Springs Dr Calavasas Hills CA 91301-5328

KAPLAN, DAVID JEREMY, b Honolulu, Hawaii, Oct 8, 34; c 3. OPERATIONS RESEARCH, SYSTEMS ANALYSIS. *Educ:* State Univ Iowa, BA, 56; Univ Calif, Berkeley, MA, 58. *Prof Exp:* Mathematician, Ames Res Ctr, NASA, 58-60; mathematician syst anal, Stanford Res Inst, 60-69; OPERS RES ANALYST, NAVAL RES LAB, 69- *Mem:* Fel AAAS; Sigma Xi; Opers Res Soc Am; Math Asn Am. *Res:* Developing the symbolic framework that underlies command and control systems; data-driven language representations that form the interface between signal coordinating systems and their environment. *Mailing Add:* Code 5155 Naval Res Lab Washington DC 20375-5000

KAPLAN, DAVID L, b New York, NY, Feb 14, 18; m 51; c 2. STATISTICS, DEMOGRAPHY. *Educ:* NY Univ, BA, 39. *Prof Exp:* Statistician pop div, US Bur Census, 40-56, census planner, 57-62, asst chief, 62-68, coordr 1970 census pop & housing, 66-73, chief demog census staff, 71-78, asst dir, 74-79, CONSULT STATISTICIAN, US BUR CENSUS, 79- *Honors & Awards:* Silver Medal, Dept Com, 52, Gold Medal, 71. *Mem:* Fel AAAS; fel Am Statist Asn; Pop Asn Am; Int Asn Surv Statisticians; Int Union Sci Study Pop. *Res:* Techniques for collecting, processing and disseminating demographic data from censuses and surveys. *Mailing Add:* 15100 Interlachen Dr No 1022 Silver Spring MD 20906

KAPLAN, DAVID LEE, b Glen Cove, NY, Mar 18, 53. BIODEGRADATION-BIOTRANSFORMATIONS, BIOPOLYMERS. *Educ:* Univ Albany, BS, 75; State Univ NY, PhD(biochem), 78. *Prof Exp:* Teaching asst biol, Col Environ Sci & Forestry, State Univ NY, 76, res asst, 76-78, res assoc fel, Res Found, 78-79; RES MICROBIOLOGIST & CHEMIST, NATICK RES & DEVELOP CTR, US ARMY, 79- *Concurrent Pos:* Reviewer, NSF grants, 80-, Environ Sci & Technol J, US Army Res Off grants & US Army Toxic & Hazardous Mat Agency proposals, 83- & US Environ Protection Agency grants, 84- *Mem:* Am Soc Microbiol; Am Soc Indust Microbiol; AAAS; Water Pollution Control Fedn; Sigma Xi. *Res:* Applied and environmental microbiology and biochemistry; prevention of deterioration and contamination of materials; biotransformations and biochemical reactions of natural and synthetic organic compounds in soils and waters; analytical chemistry techniques for separations and indentification of organic compounds; sewage sludge stabilization; enzymatic reactions with aromatic compounds; pollution abatement; installation restoration approaches for hazardous wastes. *Mailing Add:* 25 Hallock Point Rd Stow MA 01775

KAPLAN, DONALD ROBERT, b Chicago, Ill, Jan 17, 38; m 64; c 2. PLANT MORPHOLOGY. *Educ:* Northwestern Univ, BA, 60; Univ Calif, Berkeley, PhD(bot), 65. *Prof Exp:* Asst prof biol sci, Univ Calif, Irvine, 65-68; from asst prof to assoc prof, 68-77, PROF BOT, UNIV CALIF, BERKELEY, 77- *Concurrent Pos:* NSF fel, Royal Bot Garden, Eng, 65; Guggenheim fel, 87-88; Alexander von Humboldt sr US scientist award, Heidelburg, Ger, 88-89. *Mem:* AAAS; Bot Soc Am; Int Soc Plant Morphol; Am Soc Cell Biol; fel Linnean Soc London; Sigma Xi. *Res:* Comparative and developmental morphology of plants. *Mailing Add:* Dept Plant Biol Univ Calif Berkeley CA 94720

KAPLAN, EDWARD LYNN, b Philadelphia, Pa, May 11, 20; div. MATHEMATICAL PROGRAMMING. *Educ:* Carnegie Inst Technol, BS, 41; Princeton Univ, PhD(math), 51. *Prof Exp:* Mathematician, US Naval Ord Lab, 41-48; asst, Princeton Univ, 48-50; mem tech staff, Bell Tel Labs, 50-57; mathematician, Lawrence Radiation Lab, Univ Calif, 57-61; from assoc prof to prof, 61-81, EMER PROF MATH, ORE STATE UNIV, 81- *Mem:* Am Math Soc. *Res:* Random sequences; elliptic-integral tables; probability; statistics; Monte Carlo methods; computation; mathematical programming; optimization; musical symbology. *Mailing Add:* 727 NW 11th St Corvallis OR 97330

KAPLAN, EHUD, b Jerusalem, Israel, Dec 29, 42; m 66; c 2. NEUROPHYSIOLOGY, SENSORY PROCESS. *Educ:* Hebrew Univ, Jeruselem, Israel, BA, 67; Syracuse Univ, PhD(neurophysiol), 73. *Prof Exp:* Res asst vision, Hadassah Hosp, Jerusalem Israel, 63-65 & Syracuse Univ, 68-73; fel, 73-76, asst prof, 77-83, ASSOC PROF, ROCKEFELLER UNIV, 83- *Concurrent Pos:* Instr, Marine Biol Lab, Woods Hole, Mass, 77- *Mem:* NY Acad Sci; Asn Res Vision & Opthal; Sigma Xi. *Res:* Information processing by the brain especially in the visual system; the way photoreceptors transduce light into electrical energy. *Mailing Add:* Dept Biophys Rockefeller Univ 1230 York Ave New York NY 10021

KAPLAN, EMANUEL, b Clearfield, Pa, Mar 12, 10; m 34; c 2. BIOCHEMISTRY. *Educ:* Johns Hopkins Univ, AB, 31, ScD(biochem), 34. *Prof Exp:* Spec asst biochem, Sch Hyg & Pub Health, Johns Hopkins Univ, 30-32, asst, 31-34; chief div chem, Bur Labs, Baltimore City Health Dept, 34-57, asst dir, 57-65; chief, Div Biochem, Bur Labs, Md State Dept Health, 65-76; RETIRED. *Concurrent Pos:* Instr, Sch Nursing, Sinai Hosp, 39-44. *Mem:* Am Chem Soc; fel Am Pub Health Asn; Am Asn Clin Chem. *Res:* Environmental chemistry; clinical chemistry; completely edible dentifrice. *Mailing Add:* Three Stonehenge Circle Apt 1 Baltimore MD 21208

KAPLAN, EPHRAIM HENRY, b New York, NY, Nov 16, 18; m 52; c 3. ANALYTICAL CHEMISTRY. *Educ:* City Col New York, BS, 38; Univ Iowa, MS, 40; Univ Pittsburgh, PhD(org chem, biochem), 45; Northwestern Univ, MM, 74. *Prof Exp:* Asst sci aide, Eastern Regional Res Lab, Bur Agr & Indust Chem, USDA, 41-42; org chemist, Bur Mines, Pittsburgh, 42-45; res chemist, Sinclair Refining Co, Ind, 45-47; res assoc, Polytech Inst Brooklyn, 47-49; Res Corp fel enzyme chem, Inst Enzyme Res, Univ Wis, 49-50; res assoc, Med Sch, Northwestern Univ, 50-53; res chemist, Vico Prod Co, 53-56; res assoc, Inst Tuberc Res, Univ Ill, 56-57; tech dir, Hodag Chem Corp, 57-60; res chemist, Velsicol Chem Corp, Chicago, 60-69; toxicologist & supvr Biochem Sect, Chicago Bd Health, 70-89; RETIRED. *Mem:* Am Chem Soc. *Res:* Surface active chemicals; enzymes; proteins; intermediary metabolism; sorption of vapors by and permeation through polymers; organic synthesis; analysis of hydrocarbon mixtures; pesticides; instrumental, drug and clinical analysis. *Mailing Add:* 9526 Kostner Ave Skokie IL 60076

KAPLAN, ERVIN, b Independence, Iowa, June 19, 18; m 45; c 2. INTERNAL MEDICINE, NUCLEAR MEDICINE. *Educ:* Univ Ill, BS, 47, MS & MD, 49; Am Bd Internal Med, dipl; Am Bd Nuclear Med, dipl. *Prof Exp:* Intern, Mt Sinai Hosp, Chicago, 49-50; clin asst, Univ Ill Col Med,52-53, clin instr, 53-57, from clin asst prof to prof med & physiol, 59-75; resident internal med, 50-52, actg assoc dir radioisotope serv, 52-59, chief, 59-70, sr physician, 71-75, CHIEF NUCLEAR MED SERV, VET ADMIN HOSP, 59- *Concurrent Pos:* Physician-in-charge, Radioisotope Clin & assoc attend physician, Mt Sinai Hosp, Chicago, 53-59, consult, 67-84; physician-in-charge, Radioisotope Lab, Michael Reese Hosp, Chicago, 56-59; assoc attend physician, Cook County Hosp, 59-64, attend physician, 64-; bd trustees & chmn, standing comt Technol Nuclear Med, Soc Nuclear Med, 60-67; consult nuclear med, Louis Weiss Mem Hosp, 70-78; lectr, Chicago Med Sch, 71-78 & Stritch Sch Med, Loyola Univ; mem ,Post World Cong Nuclear Med, Vet Gen Hosp, Taiwan, 74-78; dist dir, Nat Asn Vet Admin Physicians, 76-80; mem, liaison comt between Soc Nuclear Med & World Fedn Nuclear Med & Biol, 78; vis prof, Albert Einstein Col Med, 82; sci corresp, Rev Biol & Nuclear Med, Uruguay; dep dir nuclear med US Vet Admin Hosp Region 4. *Honors & Awards:* First Prize Award, Gema Czerniak; Award Nuclear Med & Radiopharmacol, Ahavat Zion Found, Israel, 74. *Mem:* AAAS; Soc Exp Biol & Med; Soc Nuclear Med; Am Col Nuclear Physicians. *Res:* Application of radioisotopes in medicine and biological research. *Mailing Add:* 2600 Wilmette Ave Wilmette IL 60091

KAPLAN, EUGENE HERBERT, b Brooklyn, NY, June 26, 32; m 58; c 2. PARASITOLOGY, SCIENCE EDUCATION. *Educ:* Brooklyn Col, BS, 54; Hofstra Col, MS, 56; Ny Univ, PhD(sci educ), 63. *Prof Exp:* Teacher high sch, NY, 56-58; from lectr to assoc prof, 58-74, PROF BIOL, HOFSTRA UNIV, 75- *Concurrent Pos:* NSF sci fac fel, 63-64; UNESCO expert elem sci, Israel, 71-72. *Honors & Awards:* Marine Educ Award, Nat Marine Educr Asn, 89. *Mem:* Am Soc Parasitol; Nat Asn Res Sci Teaching. *Res:* Marine ecology, especially coral reef invertebrates; effects of dredging on benthos; introductory science courses for non-science majors; measuring aspects of scientific thinking. *Mailing Add:* Dept Biol Hofstra Univ Hempstead NY 11550

KAPLAN, FRED, b Brooklyn, NY, Sept 2, 34; m 73; c 4. ORGANIC CHEMISTRY. *Educ:* NY Univ, BA, 55; Yale Univ, PhD(chem), 60. *Prof Exp:* USPHS res fel chem, Swiss Fed Inst Technol, 59-60; univ fel, Calif Inst Technol, 60-61; from instr to assoc prof, 61-68, PROF CHEM, UNIV CINCINNATI, 68- *Mem:* AAAS; Am Chem Soc; Am Asn Univ Professors. *Res:* Applications of ion cyclotron resonance spectroscopy; gas phase ion-molecule reactions; gas phase properties of organic species; electron deficient species. *Mailing Add:* 835 Cathcart Rd Blue Bell PA 19422

KAPLAN, GEORGE HARRY, b Hagerstown, Md, Apr 24, 48; m 72. ASTROMETRY, RADIO & OPTICAL INTERFEROMETRY. *Educ:* Univ Md, BS, 69, MS, 76, PhD, 85. *Prof Exp:* ASTRONR, US NAVAL OBSERV, 71- *Mem:* AAAS; Am Astron Soc; Sigma Xi; Int Astron Union. *Res:* Radio and optical astrometry; radio and optical interferometry; earth rotation; solar system dynamics and ephemerides; reference frames; ephemerides. *Mailing Add:* US Naval Observ 34th St & Massachusetts Ave NW Washington DC 20392-5100

KAPLAN, GERALD, b Brooklyn, NY, Dec 21, 39; m 67; c 4. ANALYTICAL CHEMISTRY. *Educ:* Columbia Univ, BS, 61, MS, 63; Rutgers Univ, PhD(pharmaceut chem), 68. *Prof Exp:* Res scientist, Johnson & Johnson Res Ctr, 67-71, sr res scientist anal chem, 71-72, group leader methods develop, 72-75, asst mgr, 75-79, mgr anal labs, 79-88, dir, 89, MGR, JOHNSON & JOHNSON CONSUMER PRODS, 90- *Mem:* Am Chem Soc; Sigma Xi. *Res:* Separations sciences. *Mailing Add:* Johnson & Johnson Consumer Prod 199 Grandview Rd Skillman NJ 08558-9418

KAPLAN, HAROLD IRWIN, b Brooklyn, NY, Oct 1, 27; m 81; c 3. PSYCHIATRY, PSYCHOANALYSIS. *Educ:* NY Med Col, MD, 49; Am Bd Psychiat & Neurol, dipl, 57; NY Univ, BA, 88. *Prof Exp:* Intern med, Brooklyn Jewish Hosp, 49-50; resident psychiat, Bronx Vet Admin Hosp, 50-53; resident psychiat, Mt Sinai Hosp, New York, 52-53; from instr to prof psychiat, New York Med Col, 54-80; PROF PSYCHIAT, SCH MED, NY UNIV, 80- *Concurrent Pos:* Attend psychiatrist, Metrop Hosp, 54-80, Flower & Fifth Hosp, 54-80 & Bird S Coler Hosp, 54-80; attend psychiatrist, Univ Hosp & Bellevue Hosp, New York Univ Med Ctr, NY. *Honors & Awards:* Distinguished Serv Award, Asn Psychiat Outpatient Centers Am, 82. *Mem:* Am Psychiat Asn; Am Acad Psychoanal; Am Col Physicians; NY Acad Med; Am Med Writers Asn. *Res:* Education of women physicians; psychiatric education research; psychosomatic medicine. *Mailing Add:* 50 E 78th St New York NY 10021

KAPLAN, HAROLD M, b Boston, Mass, Sept 4, 08; m 34; c 3. PHYSIOLOGY. *Educ:* Dartmouth Col, AB, 30; Harvard Univ, AM, 31, PhD(physiol), 33. *Prof Exp:* Asst instr zool, Harvard Univ, 33-34; instr physiol, Middlesex Univ, 34-37, prof, Med Sch, 37-45 & Vet Sch, 45-47; prof, Brandeis Univ, 47; assoc prof, Univ Mass, 47-49; chmn dept, Southern Ill Univ, Carbondale, 49-71, prof physiol, 49-77; RETIRED. *Concurrent Pos:* Pvt res with Dr E V Enzmann, Biol Labs, Harvard Univ, 35-37; writer, Wash Inst Med, 46-49; vis prof, Sch Med, Southern Ill Univ, 77- *Mem:* AAAS; Am Physiol Soc; Am Soc Zool; Electron Micros Soc Am. *Res:* Laboratory animal medicine. *Mailing Add:* Sch Med Southern Ill Univ Carbondale IL 62901

KAPLAN, HARRY ARTHUR, b Duluth, Minn, Oct 25, 11. NEUROSURGERY. *Educ:* Univ Minn, BS, 35, MD, 38. *Prof Exp:* From instr to assoc prof, Col Med, State Univ NY Downstate Med Ctr, 51-60; prof, 63-82, EMER PROF NEUROSURG, UNIV MED & DENT NJ, NEWARK, 82- *Concurrent Pos:* Fel neurosurg, Long Island Col Hosp, 49-51. *Mem:* Am EEG Soc; Am Asn Neurol Surg; AMA; Asn Res Nerv & Ment Dis; Am Asn Neuropath. *Res:* Neurological sciences; neurovascular and trauma of the nervous system. *Mailing Add:* Div Neurosurg Col Med & Dent NJ Med Sch 185 S Orange Ave Newark NJ 07103-2757

KAPLAN, HARVEY, b New York, NY, Nov 29, 24; m 47; c 3. THEORETICAL PHYSICS. *Educ:* City Col New York, BS, 48; Univ Calif, PhD(physics), 52. *Prof Exp:* Res assoc, Mass Inst Technol, 52-54; asst prof physics, Univ Buffalo, 54-59; from asst prof to assoc prof, 59-65, PROF PHYSICS, SYRACUSE UNIV, 65- *Mem:* Am Phys Soc. *Res:* Theory of solid state; dynamical systems. *Mailing Add:* Dept Physics Syracuse Univ Syracuse NY 13244

KAPLAN, HARVEY, b New York, NY, May 24, 40; Can citizen; m 66; c 2. BIOCHEMISTRY. *Educ:* Queen's Univ, Ont, BSc, 62; Univ Ottawa, PhD(kinetics), 66. *Prof Exp:* Nat Res Coun Can fel, 66-67; fel, Lab Molecular Biol, Cambridge Univ, 67-68; asst res officer biochem, Nat Res Coun Can, 68-71; ASST PROF BIOCHEM, UNIV OTTAWA, 71- *Mem:* Can Biochem Soc. *Res:* Kinetics and mechanism of enzyme action; structure and function of serine proteases; ionization constants and reactivity of functional groups in proteins. *Mailing Add:* Dept Biochem Univ Ottawa Ottawa ON K1N 6N5 Can

KAPLAN, HARVEY ROBERT, b New Brunswick, NJ, Aug 21, 41; m 64; c 2. PHARMACOLOGY. *Educ:* Philadelphia Col Pharm, BSc, 63; Univ Conn, MSc, 65, PhD(pharmacol), 66. *Prof Exp:* Assoc dir, Dept Pharmacol, 67-80, DIR CARDIOVASC SECT, WARNER-LAMBERT RES INST, 80- *Concurrent Pos:* NIH fel, Univ Pittsburgh, 66-67. *Mem:* AAAS; NY Acad Sci; Am Soc Pharmacol & Exp Therapeut. *Res:* Cardiovascular and autonomic pharmacology; cardiac arrhythmias and antiarrhythmic drugs; central cardiovascular mechanisms; evaluation and assay of synthetics as well as natural products isolated form both plant and animals. *Mailing Add:* Dept Pharmacol Parke-Davis Pharmaceut Res Div Warner Lambert Co 2800 Plymouth Rd Ann Arbor MI 48105

KAPLAN, HELEN SINGER, b Vienna, Austria, Feb 6, 29; US citizen; div; c 3. PSYCHIATRY, SEXUAL DISORDERS. *Educ:* Syracuse Univ, BFA, 49; Columbia Univ, MA, 51, PhD(psychol), 55; NY Med Col, MD, 59; Am Bd Psychiat & Neurol, dipl. *Prof Exp:* Instr dept pharmacol & physiol, NY Med Col, 59-61; fel psychiat, Bellevue Hosp, 61-62, NY Med Col, Metropolitan Hosp Ctr, 62-64; Nat Inst Mental Health career teacher psychiat, NY Med Col, Metrop Hosp Ctr, 64-66; from asst prof to assoc prof psychiat, NY Med Col, 66-70, chmn behav sci topic teaching block, 69-70; coordr undergrad teaching psychiat, 70-76, CLIN PROF PSYCHIAT, CORNELL UNIV MED COL, 70-; ATTEND PSYCHIATRIST, PAYNE WHITNEY CLIN NY HOSP, 70- *Concurrent Pos:* Assoc attend psychiatrist, Metrop, Flower Fifth & B S Coler Hosp, 66-70; dir, Human Sexuality Prog, Payne Whitney Clin, 70-, Helen S Kaplan Assocs, 70- *Honors & Awards:* Outstanding Prof Contrib Field Sexuality, Am Asn Sex Educrs & Counrs, 83. *Mem:* Am Psychol Asn; fel Am Psychiat Asn; Psychosom Soc; Acad Psychoanal; Sigma Xi; AMA; Am Group Psychother Asn; Asn Advan Behav Ther; NY Acad Sci; Am Asn Sex Educrs & Counrs; NY Acad Med. *Res:* Psychopharmacology; psychosomatic medicine; treatment of sexual disorders; psychiatric education. *Mailing Add:* 960 Fifth Ave New York NY 10021

KAPLAN, HENRY J, b New York, NY, Dec 29, 42; m 66; c 3. OPHTHALMOLOGY. *Educ:* Columbia Univ, AB, 64; Cornell Univ, MD, 68; Am Bd Ophthal, cert, 79. *Prof Exp:* Intern med, Lakeside Hosp, Univ Hosps, Cleveland, Case Western Reserve Univ, 68-69; surg resident, Bellevue Hosp, NY Univ Med Ctr, New York, NY, 69-70; res fel immunol, NIH, Dept Cell Biol, Univ Tex Med Sch, Dallas, 72-74, asst prof, Dept Cell Biol, 74-75; assoc prof, Dept Ophthal, Sch Med, Emory Univ, Atlanta, Ga, 79-84, prof, 84-88; PROF & CHMN, DEPT OPHTHAL & VISUAL SCI, SCH MED, WASH UNIV, ST LOUIS, MO, 88- *Concurrent Pos:* Affil scientist path & immunol, Yerkes Regional Primate Res Ctr, Atlanta, Ga, 81-; dir res, Dept Ophthal, Sch Med, Emory Univ, Atlanta, Ga, 84-88, assoc prof, Dept Microbiol, 85-88; adj prof, Dept Small Animal Med, Univ Ga, Athens, 85-; mem, Visual Dis Study Sect A-1, Nat Eye Inst, NIH, 85-89, chmn, 87-89; ophthalmologist-in-chief, Barnes Hosp, 88-; active staff ophthalmologist, St Louis Children's Hosp, 88-; assoc attend ophthalmologist, Jewish Hosp, St Louis, 88-; mem bd dirs, Med Alliance Corp, Wash Univ, 89-, Partners Health Plan, 89-; assoc examr, Am Bd Ophthal, 90- *Honors & Awards:* Honor Award, Am Acad Ophthal, 84. *Mem:* Asn Res Vision & Ophthal; fel Am Acad Ophthal; AAAS; AMA; NY Acad Sci; Am Asn Immunologists; Am Uveitis Soc; fel Am Col Surgeons; Am Soc Contemp Ophthal. *Res:* Retina-vitreous; immunology-uveitis; author or co-author of over 80 publications. *Mailing Add:* Dept Ophthal Sch Med Wash Univ 660 S Euclid Ave Box 8096 St Louis MO 63110

KAPLAN, HERMAN, b New York, NY, Aug 17, 28; m 56; c 3. PSYCHIATRY, DENTISTRY. *Educ:* NY Univ, AB, 48, DDS, 53; State Univ NY Upstate Med Ctr, MD, 63. *Prof Exp:* From asst prof to assoc prof oral surg, 64-75, ASSOC PROF MED, DEPT ORAL DIAG, SCH DENT, UNIV OF THE PAC, 75-, ADJ PROF DIAG SCI & MED, 81- *Concurrent Pos:* Consult, Cowell Hosp & Mt Zion Hosp; resident psychiat, Presby Hosp-Pac Med Ctr, 72-75; mem, Div Behav Sci, Sch Dent, Univ of the Pac & Facial Pain Res Ctr, 82-; consult, WHO, 80. *Mem:* Fel Acad Dent Int; AAAS. *Res:* Rapid stick method to detect diabetes mellitus; vital dye staining technique for oral cancer; use of a thermal sensing device in assessing and modifying inflammation in tissues; use of intravenous agents in ambulatory anesthesia. *Mailing Add:* 38 Arroyo Dr Moraga CA 94556

KAPLAN, HESH J, botany, engineering, for more information see previous edition

KAPLAN, IRVING, b New York, NY, Dec 1, 12; m 45; c 3. THEORETICAL PHYSICS. *Educ:* Columbia Univ, AB, 33, AM, 34, PhD(chem), 37. *Prof Exp:* Res chemist, Michael Reese Hosp, Chicago, 37-41; theoret physicist, Div War Res, Manhattan Proj, 41-44; sr scientist & head, Reactor Physics Div, Nuclear Eng Dept, Brookhaven Nat Lab, 45-57; vis prof, Mass Inst Technol, 57-58, prof nuclear eng, 58-78, secy fac, Dept Nuclear Eng, 74-75, EMER PROF, MASS INST TECHNOL, 78- *Concurrent Pos:* Theoret physicist, SAM Labs, Columbia Univ & Carbide & Carbon Chem Corp, 44-46; vis lectr, Harvard Univ, 56. *Mem:* Am Phys Soc; Am Nuclear Soc; Am Acad Arts & Sci. *Res:* Nuclear physics; nuclear reactor physics; history of physics. *Mailing Add:* 77 Massachussetts Ave Location 24-207 A Cambridge MA 02139

KAPLAN, ISSAC R, b Baranowicze, Poland, July 10, 29; m 55; c 2. GEOCHEMISTRY. *Educ:* Univ Canterbury, BSc, 52, MSc, 53; Univ Southern Calif, PhD(biogeochem), 61. *Prof Exp:* Res officer oceanog, Commonwealth Sci & Indust Res Orgn, Australia, 53-57; res fel geochem, Calif Inst Technol, 61-62; guest lectr & Ziskind scholar microbiol & geochem, Hebrew Univ, Israel, 63-65; assoc prof geol & geophys, 65-69, PROF GEOL & GEOCHEM, UNIV CALIF, LOS ANGELES, 69-; PRES, GLOBAL GEOCHEM CORP, 77- *Concurrent Pos:* Assoc ed, Geochem Soc, 66-70 & Chem Geol, 66-67; mem planetary biol subcomt, Space Sci & Appln Steering Comt, NASA, 67-; Guggenheim Mem Found res fel, Mineral Res Labs, Commonwealth Sci & Indust Res Orgn, Japan, New Caledonia, NZ & Australia, 70-71; assoc ed, Marine Chem, 72- & Geochem J, 76-; mem exobiol panel, Space Sci Bd, Nat Acad Sci. prin investr lunar return mat, Apollo 11, 12, 14 & 15; chmn, Org Geochem Div, Geochem Soc, 76-77. *Mem:* Geochem Soc; fel Am Inst Chemists; Am Asn Petrol Geologists; fel AAAS; Am Chem Soc; fel Geol Soc Am. *Res:* Biogeochemistry of recent sediments; factors controlling the distribution of elements in the ocean; isotope geochemistry and organic geochemistry of terrestrial rocks and meteorites; biological fractionation of stable isotopes; atmospheric chemistry and atmospheric pollution. *Mailing Add:* 5853 Slichter Hall Univ Calif Los Angeles CA 90024

KAPLAN, JACOB GORDIN, cell biology, molecular biology; deceased, see previous edition for last biography

KAPLAN, JAMES, mathematics, for more information see previous edition

KAPLAN, JEROME I, b New York, NY, July 28, 26; m 65; c 1. SOLID STATE PHYSICS. *Educ:* Univ Mich, BS, 50; Univ Calif, Berkeley, PhD(physics), 54. *Prof Exp:* Res scientist, Naval Res Lab, 54-59; res assoc physics, Brandeis Univ, 59-62, asst prof, 62-63; Fulbright lectr, Univ Col, Rhodesia & Nyasaland, 63-64; assoc res prof, Brown Univ, 64-67; res fel, Battelle-Columbus, 67-74; res assoc, Krannert Inst Cardiol, 74-80; PROF PHYSICS, IND UNIV-PURDUE UNIV INDIANAPOLIS, 74-; SR FEL, INDIANAPOLIS CTR ADVAN RES, 80- *Concurrent Pos:* Louis Lipsky fel physics, Weizmann Inst, 56-57; consult, Lincoln Lab, Mass Inst Technol, 61; vis scientist, Naval Res Lab, 81; consult, Hercules Chem Co. *Mem:* Am Phys Soc. *Res:* Magnetic properties of solids; nuclear and ferromagnetic wave resonance; electron, nuclear and ferromagnetic spin resonance phenomena; magnetic resonance in liquids and liquid crystals; nuclear magnetic resonance in heart muscle; solar heating design; nuclear magnetic resonance theory. *Mailing Add:* 4417 N Pennsylvania Indianapolis IN 46205

KAPLAN, JERRY, CELL BIOLOGY. *Educ:* Purdue Univ, PhD(biol sci), 71. *Prof Exp:* PROF PATH, MED CTR, UNIV UTAH, 80- *Res:* Membrane dynamics; biochemistry. *Mailing Add:* Med Ctr Rm SC239 Univ Utah Col Med Salt Lake City UT 84132

KAPLAN, JOEL HOWARD, b New York, NY, Apr 6, 41; m 63; c 3. IMMUNOLOGY, CELL BIOLOGY. *Educ:* City Col New York, BS, 62; Johns Hopkins Univ, PhD(biochem), 67. *Prof Exp:* Nat Cancer Inst fel cancer biochem, McArdle Lab Cancer Res, 67-69; STAFF SCIENTIST MED SCI, GEN ELEC RES & DEVELOP CTR, 69- *Mem:* Am Asn Immunol; AAAS. *Res:* In vitro methods in cell-mediated immunity; electrokinetic properties of lymphocyte subpopulations; cancer-immunodiagnosis. *Mailing Add:* NY State Health Dept Two University Pl Albany NY 12203

KAPLAN, JOEL HOWARD, b Paterson, NJ, Sept 8, 38; c 2. CHEMICAL ENGINEERING. *Educ:* Newark Col Eng, BS, 61, MS, 62, DSc(chem eng), 66. *Prof Exp:* Res chem engr, 66-69, group leader, Process Anal Sect, 69-71, mgr systs anal dept, 71-80, proj leader process develop, Org Chem Div, Am Cynamid Co, Bound Brook, 80-83; MGR MAT & RES PLANNING, LONZA INC, FAIRLAWN, 83- *Concurrent Pos:* Adj prof chem eng, NJ Inst Technol, 78- *Mem:* Am Inst Chem Engrs; Am Chem Soc; NY Acad Sci. *Res:* Kinetics and reactor design of industrial processes; process development; application of computer control to industrial processes; research management concerned with plant and laboratory automation and development of research strategy models; inventory and production planning systems, MRP II. *Mailing Add:* 26 Bianculli Dr South Plainfield NJ 07080

KAPLAN, JOHN ERVIN, b Chicago, Ill, Dec 4, 50; m 73; c 2. PHAGOCYTOSIS, THROMBOSIS. *Educ:* Univ Ill, BS, 72; Albany Med Col, PhD(physiol), 76. *Prof Exp:* Fel, 75-77, instr, 76-77, asst prof, 77-80, assoc prof physiol, Albany Med Col, 80-87, ASSOC PROF, GRAD SCH PUB HEALTH, 85-; PROF PHYSIOL & CELL BIOL, ALBANY MED COL, 87-, VCHMN, 90- *Concurrent Pos:* Prin investr res grants, New York Heart Asn, 77-78 & 90-, NIH, 78- & Shared Instrumentation Prog, 79-82; Sinsheimer Fund Scholar, 80-83. *Mem:* Sigma Xi; Reticuloendothelial Soc; Am Heart Asn; Am Physiol Soc; Am Soc Cell Biol. *Res:* Mechanisms by which macrophages and adhesive proterms act as physiological anti-thrombotic mechanisms, and the role of these mechanisms in sepsis, trauma, intravascular coagulation and vascular injury; interaction of platelets and endothelial cells; fibronectin receptors. *Mailing Add:* Dept Physiol & Cell Biol A-134 Albany Med Col Albany NY 12208

KAPLAN, JOSEPH, physics, geophysics; deceased, see previous edition for last biography

KAPLAN, JOSEPH, b Boston, Mass, March 7, 41; m 86; c 2. PEDIATRICS, HEMOTOLOGY & ONCOLOGY. *Educ:* NY Univ, BA, 62; Johns Hopkins Univ, MD, 66. *Prof Exp:* From asst prof to assoc prof pediat, 72-87, PROF PEDIAT, IMMUNOL & MICROBIOL, WAYNE STATE UNIV, 87-, PROF MED, 88- *Mem:* Am Asn Immunologists; Am Pediat Soc; Am Soc Hematol. *Res:* Clinical immunology; role of natural killer cells in health and disease such as in prevention of graft-version hour disease and bone marrow graft rejection. *Mailing Add:* Childrens Hosp Mich 3901 Beaubien Detroit MI 48201

KAPLAN, LAWRENCE, b Chicago, Ill, Apr 14, 26; m 46; c 1. BOTANY. *Educ:* Univ Iowa, BA, 49, MS, 51; Univ Chicago, PhD(bot), 56. *Prof Exp:* Assoc cur, Mus Useful Plants, Mo Bot Gardens, 55; instr biol, Wright Jr Col, 56; asst prof, Roosevelt Univ, 57-65; assoc prof, 65-68, PROF BIOL, UNIV MASS, BOSTON, 68- *Concurrent Pos:* Ed, Econ Bot, 91- *Mem:* Fel AAAS; Bot Soc Am; Sigma Xi; Soc Econ Bot (pres, 87); Soc Am Archeol. *Res:* Ethnobotany; systematics. *Mailing Add:* Dept Biol Univ Mass Boston MA 02125

KAPLAN, LAWRENCE JAY, b Newark, NJ, Mar 20, 43; m 65; c 2. BIOCHEMISTRY, FORENSIC SCIENCE. *Educ:* Univ Pittsburgh, BS, 64; Purdue Univ, PhD(chem), 70. *Prof Exp:* Fel, Univ Mass, Amherst, 70-71; from asst prof to assoc prof, 71-84, PROF CHEM, WILLIAMS COL, 84-, CHMN, CHEM DEPT & CO-CHMN, PROG BIOCHEM & MOLECULAR BIOL, 88- *Concurrent Pos:* Res scientist, Weizmann Inst, Israel, 76-77; vis assoc prof, Biochem Dept, Brandeis Univ, 80-81. *Mem:* AAAS; Am Chem Soc; Am Soc Biochem & Molecular Biol. *Res:* Physical biochemistry of proteins; conformational transitions of macromolecules; structure of chromatin. *Mailing Add:* Dept Chem Williams Col Williamstown MA 01267

KAPLAN, LEONARD, b Brooklyn, NY, July 18, 39. ORGANIC CHEMISTRY, CATALYSIS. *Educ:* Cooper Union, BChE, 60; Univ Ill, PhD(org chem), 64. *Prof Exp:* Res assoc org chem, Columbia Univ, 64-65; instr, Univ Chicago, 65-67, asst prof, 67-74; proj scientist, 74-75, res scientist,

75-79, SR RES SCIENTIST, UNION CARBIDE CORP, 79- *Concurrent Pos:* NSF fel, 64-65; Alfred P Sloan Found fel. *Mem:* Am Chem Soc; The Chem Soc. *Res:* Homogeneous catalysis; organometallic chemistry; mechanistic and physical organic chemistry; exploratory synthesis; free radical chemistry. *Mailing Add:* 227 Walker Dr Dunbar WV 25064-1913

KAPLAN, LEONARD LOUIS, b New York, NY, Oct 10, 28; m 68; c 2. PHARMACEUTICS. *Educ:* NY Univ, BA, 48, PhD(statist, mgt sci), 68; Ohio State Univ, BScPharm, 52; City Col New York, MBA, 63. *Prof Exp:* Res assoc pharm, Sterling-Winthrop Res Inst, 55-59; dir develop, Walker Labs Div, Richardson-Merrell, 59-63; group mgr, Vick Div Res, 63-69; dir res & develop, Health Care Div, Ortho Pharm Corp, Johnson & Johnson-Domestic Oper Co, 69-78, dir res & develop, Advan Care Prods, 78-83, vpres, 84-89; vpres res & develop, Over The Counter Drugs, Sterling Drug, 89-90; SR VPRES BUS & TECH DEVELOP, D M GRAHAM LABS, 91- *Concurrent Pos:* Adj assoc prof, Brooklyn Col Pharm, 66-69 & Col Pharm Rutgers Univ, 79- *Mem:* Acad Pharmaceut Sci; Soc Cosmetic Chemists; Am Asn Clin Chemists. *Res:* Pharmaceutical research specializing in areas of analgesics, oral hygiene, sports medicine, dermatology and deodorancy; contraceptives; diagnostics. *Mailing Add:* One Minuteman Ct East Brunswick NJ 08816

KAPLAN, LEWIS DAVID, b Brooklyn, NY, June 21, 17; m 42; c 1. ATMOSPHERIC SENSING FOR NUMERICAL WEATHER PREDICTION. *Educ:* Brooklyn Col, BA, 39; Univ Chicago, SM, 47, PhD(meteorol), 51. *Prof Exp:* Observer, US Weather Bur, 40-45, meteorologist, 47-54; mem, Inst Advan Study, 54-56; guest, Imperial Col, London, 56-57; mem res staff, Mass Inst Technol, 57-61; chief planetary studies & staff scientist, Jet Propulsion Lab, Calif Inst Technol, 62-70; prof, 70-79, EMER PROF METEOROL, UNIV CHICAGO, 80-; PRIN SCIENTIST, ATMOSPHERIC & ENVIRON RES INC, 81- *Concurrent Pos:* Res assoc & instr, Univ NMex, 48-49; instr, Univ Chicago, 50; prof, Univ Nev, 61-64; prin experimenter, Mariner II Infrared Exp, 61-63; vis prof, Oxford Univ, 64-65, Univ Oslo, 65, Mass Inst Technol, 68 & Univ Paris, 69-70 & Univ Sci & Technol, China, 82; sr staff scientist, Goddard Space Flight Ctr, NASA, 78-81; distinguished vis scientist, Calif Inst Technol, 82- *Honors & Awards:* Except Sci Achievement Medal, NASA, 68; Second Half-Century Award, Am Meteorol Soc, 70; Silver Medal, US Dept Com, 85. *Mem:* fel Am Meteorol Soc; fel Royal Meteorol Soc (UK). *Res:* Radiative heat transfer in planetary atmospheres; analysis of infrared spectra; composition and structure of planetary atmospheres; atmospheric dynamics and energy exchange; numerical weather prediction. *Mailing Add:* Atmospheric & Environ Res Inc Cambridge MA 02139

KAPLAN, MANUEL E, b New York, NY, Nov 6, 28; m 55; c 3. INTERNAL MEDICINE, HEMATOLOGY. *Educ:* Univ Ariz, BS, 50; Harvard Med Sch, MD, 54. *Prof Exp:* Intern med, Boston City Hosp, 54-55, from asst resident to sr resident, 55-59; res assoc, Mt Sinai Hosp, NY, 62-63, asst dir hemat, 63-65; asst prof med, Sch Med, Wash Univ, 65-69; assoc prof, 69-73, PROF MED, MED SCH, UNIV MINN, MINNEAPOLIS, 73-; CHIEF HEMAT & ONCOL, VET ADMIN MED CTR, 69- *Concurrent Pos:* Fel hemat, Thorndike Mem Lab, Boston City Hosp, 59-62; USPHS res grants, 63-64 & 66-, career develop award, 67-69; res fel microbiol, Col Physicians & Surgeons, Columbia Univ, 63-65; chief hemat, Jewish Hosp of St Louis, 65-69. *Mem:* AAAS; Am Fedn Clin; Am Soc Clin Invest; Am Soc Hemat; Am Asn Immunol. *Res:* Immunohematology; lymphocyte structure and function; hematopoiesis. *Mailing Add:* Vet Admin Med Ctr One Vet Dr Minneapolis MN 55417

KAPLAN, MARK STEVEN, b New York, NY, Feb 25, 47; m 88; c 2. PHOTOGRAPHIC CHEMISTRY. *Educ:* Bucknell Univ, BS & MS, 67; Univ Ore, PhD(org chem), 71. *Prof Exp:* SR RES SCIENTIST ADVAN CHEM TECHNOLOGIES, EASTMAN KODAK CO, 71- *Mem:* Am Chem Soc; Am Defense Preparedness Asn. *Res:* Novel lithographic systems; use of lasers in graphic arts; photoresists; chemical defense research; non silver imaging systems. *Mailing Add:* Eastman Kodak Co Res Labs B-82 Kodak Park Rochester NY 14650-2156

KAPLAN, MARSHALL HARVEY, b Detroit, Mich, Nov 5, 39; m 61; c 2. AERONAUTICS, ASTRONAUTICS. *Educ:* Wayne State Univ, BS, 61; Mass Inst Technol, SM, 62; Stanford Univ, PhD(aeronaut, astronaut), 68. *Prof Exp:* Mem tech staff, Hughes Res Lab, Calif, 62-64; mem tech staff, Space Systs Div, Hughes Aircraft Co, 64-65; sr engr, Western Develop Labs, Philco Corp, 65-66; from asst prof to assoc prof aerospace eng, 68-78, PROF AEROSPACE ENG, PA STATE UNIV, 78-; CONSULT, SPACETECH INC. *Honors & Awards:* Outstanding Res Award, Pa State Univ, 78. *Mem:* Am Inst Aeronaut & Astronaut; Am Astronaut Soc; Am Soc Eng Educ. *Res:* Space systems synthesis and engineering; astrodynamics; propulsion; satellite dynamics and control. *Mailing Add:* 233 Hammond Bldg Pa State Univ University Park PA 16802

KAPLAN, MARTIN CHARLES, b 1953. COMPUTER IMAGE PROCESSING, PHOTOGRAPHIC SCIENCE. *Educ:* Mass Inst Tech, BSc, 75 & PhD(physics), 80. *Prof Exp:* NSF grad fel physics, Mass Inst Technol, 75-78, IBM grad fel, 78-79; asst physicist, Brookhaven Nat Lab, 80-82; RES SCIENTIST PHYSICS & IMAGE PROCESSING, KODAK RES LABS, 82- *Mem:* Am Phys Soc; Am Asn Advan Sci; Sigma Xi. *Res:* Computer image processing and photographic science related to amateur and professional photography. *Mailing Add:* Kodak Res 65-1407 Rochester NY 14650-1816

KAPLAN, MARTIN L, b New York, NY, Apr 7, 23; m 48; c 2. COMPARATIVE PHYSIOLOGY, INSECT PATHOLOGY. *Educ:* Brooklyn Col, AB, 49; NY Univ, MS, 54, PhD(exp zool), 58. *Prof Exp:* Lectr biol, Brooklyn Col, 53-56, tutor, 56-57; instr biol, Fairleigh Dickinson Univ, 57-58, asst prof anat, Sch Dent, 58-59; assoc path, St Vincent's Hosp, New York, 59-62; from asst prof to assoc prof, 62-71, asst dean sch gen studies, 70-75, PROF BIOL, QUEENS COL, NY, 71- *Concurrent Pos:* Lectr, Sch Gen Studies, Brooklyn Col, 59-62. *Mem:* AAAS; Am Soc Zool; Sigma Xi. *Res:* Histogenesis and biochemistry of melanotic tumors in Drosophila. *Mailing Add:* Dept of Biol Queens Col Flushing NY 11367

KAPLAN, MARTIN L, b New York, NY, Dec 27, 35; m 64; c 1. PHYSICAL ORGANIC CHEMISTRY. *Educ:* City Col New York, BS, 56; Fla State Univ, MS, 60; Seton Hall Univ, JD, 70. *Prof Exp:* Res technician microbiol, Columbia Univ, 56-57 & Sloan-Kettering Inst Cancer Res, 57-58; res assoc phys chem, Fla State Univ, 60; chemist, Richfield Oil Corp, 60-62; vol sci teaching, US Peace Corps, 62-64; assoc mem staff, 64-77, MEM STAFF, BELL LABS, 77- *Concurrent Pos:* Atty at law, NJ, 70- *Mem:* Am Chem Soc. *Res:* Mechanisms of organic reactions; rates of conformational isomerization of organic molecules by nuclear magnetic resonance; epoxy resin reactions; reactions by singlet oxygen with polymers; electrical conductivity of organic molecules and polymers. *Mailing Add:* AT&T Bell Labs IC-358 Murray Hill NJ 07974

KAPLAN, MARTIN MARK, b Philadelphia, Pa, June 23, 15; m 44; c 3. PUBLIC HEALTH. *Educ:* Univ Pa, VMD, 40, AB, 41, MPH, 42. *Hon Degrees:* DrMedVet, Hannover Vet Sch, 63, DSc Univ Wisc, 87. *Prof Exp:* Assoc prof vet prev med & pub health, Sch Vet Med, Middlesex Univ, 42-44; chief livestock sect, UN Relief & Rehab Admin, Greece, 45-47; vet consult, UN Food & Agr Orgn, 47-49; chief vet pub health sect, WHO, Geneva, Switz, 49-71, spec adv res develop, 62-69, dir off sci & technol, 69-71, dir off res prom & develop, Off Dir-Gen, 71-76, Sec-Gen Pugwash Confs Sci & World Affairs, London & Geneva, 76-88; RETIRED. *Concurrent Pos:* NIH fel, 59-60. *Honors & Awards:* K F Meyer Award, 67; Schofield Mem Medal, 74; Medal of French Comt World Vet Asn, 74; Candau Medal, WHO, 83; Gregor Mendel Medal, 88. *Mem:* AAAS; hon assoc Royal Col Vet Surgeons. *Res:* Animal diseases transmissible to man; food hygiene; comparative medicine; virology, influenza, rabies, AIDS. *Mailing Add:* Pugwash Confs 11 A Ave De La Paix Geneva 1202 Switzerland

KAPLAN, MAURICE, b Gorzd, Lithuania, Oct 10, 07; nat US. PSYCHIATRY. *Educ:* Univ Ill, BS, 32, MD, 35, MS, 39; Am Bd Psychiat & Neurol, dipl psychiat, 48, cert, child psychiat, 59. *Prof Exp:* Asst instr psychiat, Col Med, Univ Ill, 36-39; staff psychiatrist, Inst Juv Res, Ill, 39-42; asst med dir, Am Joint Dist Comn, France, 46-47; asst dir, Children's Div, Langley Porter Clin & lectr psychiat, Clin & Sch Med, Univ Calif, 48-50, from clin instr to assoc clin prof, 50-62; assoc clin prof psychiat, Univ Ill Col Med, 62-67; LECTR, NORTHWESTERN UNIV, 69- *Concurrent Pos:* Fel, Neuropsychiat Inst, Univ Ill, 36-39; instr, Col Med, Univ Ill, 40-42; dir, Child Guid Clin, Children's Hosp, San Francisco, Calif, 48-60; pvt pract, 49-59; consult, San Francisco Unified Sch Dist, 50-59; consult dept psychiat, US Army Letterman Gen Hosp, San Francisco, 58-59; Fulbright lectr, India, 59; dir, South Coast Child Guid Clin, 67-69. *Mem:* Am Med Asn; fel Am Psychiat Asn; fel Am Orthopsychiat Asn; Am Psychoanal Asn; Am Acad Child Psychiat. *Res:* Child psychiatry; psychoanalysis; psychotherapy. *Mailing Add:* 572 Cherokee Highland Park IL 60035

KAPLAN, MELVIN, b Brooklyn, NY, Nov 11, 27; m 53; c 4. ORGANIC CHEMISTRY. *Educ:* Brooklyn Col, BS, 50; Ohio State Univ, PhD(org chem), 54. *Prof Exp:* From proj leader to res supvr urethane applns, Indust Chem Div, Allied Chem Corp, Buffalo, 54-70, mgr tech serv & develop urethanes, Specialty Chem Div, 70-81, mgr res serv, 81-89; CONSULT, URETHANES & DIISOCYANATES, M KAPLAN ASSOCS, BUFFALO, 89- *Mem:* Am Chem Soc; Am Soc Testing & Mat; Int Isocyanate Inst; Soc Plastics Indust. *Res:* Isocyanate and urethane polymer chemistry; plastics; organic synthesis; kinetics of chemical reactions; blowing agents; plastic foams. *Mailing Add:* 292 Culpepper Rd Williamsville NY 14221

KAPLAN, MELVIN HYMAN, b Malden, Mass, Dec 23, 20. RHEUMATOLOGY, IMMUNOLOGY. *Educ:* Harvard Univ, AB, 42, MD, 52. *Prof Exp:* Intern med, Boston City Hosp, 52-53; asst bact & immunol, Harvard Med Sch, 53-54; from asst prof to prof, Sch Med, Case Western Reserve Univ, 58-74; dir div Rheumatology & Immunol, 74-81, PROF MED, MED SCH, UNIV MASS, 74- *Concurrent Pos:* Resident, Med Boston City Hosp, Harvard, 52-53, res fel, House Good Samaritan, Boston, 53-54; USPHS res career award, 64-74; res assoc, House Good Samaritan, Boston, 54-57; instr, Harvard Med Sch, 54-56, assoc, 57-58; estab investr, Am Heart Asn, 54-64; assoc mem comn streptococcal dis, Armed Forces Epidemiol Bd, US Dept Defense, 56-72; temp adv, WHO, 65-66; assoc ed, J Lab & Clin Med, 63-69, J Clin & Exp Immunol, Exp Path. *Mem:* Am Soc Clin Invest; Am Asn Immunol; Am Col Rheumatol; Infectious Dis Soc Am. *Res:* Microbiology; pathogenesis of rheumatic diseases, rheumatic fever, lupus, rheumatoid arthritis, particularly in the role of immunologic mechanisms; clinical immunology and rheumatology. *Mailing Add:* Dept Med Div Rheumatol Univ Mass Med Sch 55 Lake Ave N Worcester MA 01605

KAPLAN, MICHAEL, b New York, NY, Nov 7, 37; m 68; c 2. RADIATION CHEMISTRY. *Educ:* Rensselaer Polytech Inst, BS, 59; Columbia Univ, MA, 61, PhD(electron spin resonance), 65. *Prof Exp:* Mem tech staff, RCA Labs, 65-87; EDUC TESTING SERV, 87- *Mem:* Am Chem Soc; Am Phys Soc; NY Acad Sci; fel Am Inst Chemists. *Res:* Electron spin resonance of organic materials; interaction of charged particles with thin films; electron-beam lithography; x-ray lithography. *Mailing Add:* 645 Copper Mine Rd Princeton NJ 08540

KAPLAN, MILTON TEMKIN, medical microbiology, for more information see previous edition

KAPLAN, MORTON, b Chicago, Ill, Nov 21, 33; m 57; c 2. NUCLEAR REACTIONS, HYPERFINE INTERACTIONS. *Educ:* Univ Chicago, AB, 54, SM, 56; Mass Inst Technol, PhD(phys chem), 60. *Prof Exp:* Res assoc chem, Mass Inst Technol, 60; res staff chem, Lawrence Radiation Lab, 60-62; from asst prof to assoc prof, Yale Univ, 62-70; assoc prof, 70, PROF CHEM, CARNEGIE MELLON UNIV, 71- *Concurrent Pos:* Alfred P Sloan res fel, 65-69; vis scientist, Univ Oxford, 71-72. *Mem:* Am Phys Soc; Am Chem Soc; AAAS. *Res:* Nuclear reactions induced by heavy ions; Mossbauer effect; perturbed angular correlations of gamma rays; magnetic properties and chemical bonding at low temperatures; nuclear spectroscopy; low temperature nuclear orientation. *Mailing Add:* Dept Chem Carnegie Mellon Univ Pittsburgh PA 15213-3890

KAPLAN, MURRAY LEE, b Jan 9, 41; m 65; c 2. NUTRITION, METABOLISM. *Educ:* Alfred Univ, NY, BA, 62; City Univ New York, PhD(biol), 72. *Prof Exp:* Lectr biol, Brooklyn Col, 66-71; res assoc nutrit, Dept Food Sci & Human Nutrit, Mich State Univ, 71-74; asst prof nutrit, Rutgers Univ, 74-80; from assoc prof to prof food & nutrit, 81-90, PROF FOOD SCI & HUMAN NUTRIT, IOWA STATE UNIV, 90- *Concurrent Pos:* NIH res fel, Dept Food Sci & Human Nutrit, Mich State Univ, 72-74. *Mem:* AAAS; Am Inst Nutrit; NY Acad Sci; NAm Asn Obesity; Am Diabetes Asn. *Res:* Role of early nutritional experiences on the development of regulation of carbohydrate, lipid metabolism and obesity; Adipocyte metabolism. *Mailing Add:* Dept Food Sci & Human Nutrit Iowa State Univ Ames IA 50011

KAPLAN, NORMAN M, b Dallas, Tex, Jan 2, 31; m 50; c 6. INTERNAL MEDICINE. *Educ:* Univ Tex, BS, 50, MD, 54; Am Bd Internal Med, dipl & cert endocrinol & metab. *Prof Exp:* Res physician, Parkland Mem Hosp, Dallas, 55-58; from instr to assoc prof, 61-70, PROF MED, HEALTH SCI CTR, UNIV TEX, DALLAS, 70-, HEAD HYPERTENSION SECT, 78- *Concurrent Pos:* USPHS res fel, Clin Endocrinol Br, Nat Heart Inst, 60-61; USPHS grants, 62-70; dep vpres res progs, Am Heart Asn, 75-76; NIH acad award, 79-84. *Mem:* Am Fedn Clin Res; Endocrine Soc; Am Col Physicians; Am Soc Clin Invest; Coun High Blood Pressure Res. *Res:* Mechanisms controlling biosyntheses of adrenal cortical hormones particularly aldosterone; relationship of renin-angiotension system to hypertension; sodium restriction and other non-drug modalities in treatment of hypertension. *Mailing Add:* Univ Tex Health Sci Ctr 5323 Harry Hines Blvd Dallas TX 75235-8899

KAPLAN, PAUL, b New York, NY, Dec 6, 29; m 56; c 2. FLUID DYNAMICS, APPLIED MATHEMATICS. *Educ:* City Col New York, BS, 50; Stevens Inst Technol, MS, 51, DSc(appl mech), 55. *Prof Exp:* Physicist, Stevens Inst Technol, 50-56, res asst prof math & mech eng, 56-59, staff scientist, 56-57, head, Math Studies Div, 57-58 & Fluid Dynamics Div, 58-59; chief hydrodynamicist, TRG, Inc, 59-61; pres, Oceanics, Inc, Plainview, 61-79; PRES, HYDROMECH, INC, PLAINVIEW, 79- *Concurrent Pos:* Lectr, City Col New York, 54-56; adj prof, Webb Inst Naval Archit, 63-66; consult, Edo Corp, NY, 58, Vitro Lab, Md, Westinghouse Elec Corp, Calif, Elec Boat Div, Gen Dynamics Corp, Conn, 59 & Marine Adv, Inc, 61-62. *Mem:* Soc Naval Archit & Marine Engrs; Am Inst Aeronaut & Astronaut; Marine Technol Soc; Royal Inst Naval Architects. *Res:* Hydrodynamics; wave motion; stability and control; hydroelasticity; dynamics; random vibration and motion analysis; acoustics; analog, digital and hybrid computation; oceanography; automatic control systems; structural dynamics; offshore structures. *Mailing Add:* Hydromech Inc PO Box H Plainview NY 11803

KAPLAN, PHYLLIS DEEN, b Everett, Wash, Feb 9, 31; c 2. CHEMISTRY, BIOCHEMISTRY. *Educ:* Univ Wash, BA, 53; Brandeis Univ, MA, 56; Univ Cincinnati, PhD(chem), 66. *Prof Exp:* Spectroscopist chem, Syntex Corp, 67-68; res assoc, Med Ctr, Univ Cincinnati, 68-71, asst prof environ health, 71-77; sr res toxicologist, Am Cyanamid Co, 77-82; consult engr, Am Standards Testing Prog, 82-83; adminr preclin int res & develop, Allergan, 83-87, dir int res & develop Europe & MIddle East, 87-90; MGT CONSULT, 90- *Concurrent Pos:* Am Chem Soc Petroleum Res Fund grant, Univ Cincinnati, 65-66; Nat Inst Occup Safety & Health res grant, 71-76; lectr, Col Arts & Sci, Univ Cincinnati, 72-73 & Col Nursing, 73-74; adj prof, NY Univ, 82-83; consult, Baker Chem Co. *Mem:* AAAS; Am Chem Soc; Sigma Xi. *Res:* Metabolism, binding and structural identity of transition metal compounds in the body, with a special interest in elucidating the mechanisms determining toxicity and essentiality of metals within living systems; pharmaceutical product development; eye and skin care; inhalation toxicology. *Mailing Add:* Five Rana Irvine CA 92715

KAPLAN, RALPH BENJAMIN, b New York, NY, Feb 20, 20; m 53; c 2. ORGANIC CHEMISTRY. *Educ:* Brooklyn Col, BA, 42; Ohio State Univ, PhD(chem), 50. *Prof Exp:* Res assoc, Res Found, Ohio State Univ, 50-51; res chemist, Org Chem Dept, Jackson Lab, 51-77, patent assoc, Chem & Pigments Dept, E I Du Pont De Nemours & Co, 78-88; CONSULT, PATTENT MATTERS, 88- *Honors & Awards:* US Dept Navy Award, 62. *Mem:* Am Chem Soc. *Res:* Synthesis and chemistry of nitro and polynitro aliphatic compounds and anthraquinon vat dyes; properties of synthetic elastomers; petroleum chemicals; organometallic compounds; electrochemistry; patent law. *Mailing Add:* 1409 Silverside Rd Wilmington DE 19810

KAPLAN, RAPHAEL, b New York, NY, Mar 26, 36. SOLID STATE PHYSICS. *Educ:* Syracuse Univ, AB, 57; Brown Univ, PhD(physics), 63. *Prof Exp:* PHYSICIST, SEMICONDUCTORS BR, SOLID STATE DIV, US NAVAL RES LAB, 63- *Mem:* Am Phys Soc. *Res:* Spin resonance of color centers in irradiated crystals; far infrared and millimeter wave spectroscopy in semiconductors and other materials. *Mailing Add:* US Naval Res Lab Code 6834 Washington DC 20375

KAPLAN, RAYMOND, b New York, NY, Jan 26, 29; m 65; c 2. SOLID STATE PHYSICS. *Educ:* City Col New York, BS, 50; Columbia Univ, MA, 52, PhD(physics), 59. *Prof Exp:* Jr res physicist, Univ Calif, 58-59; res physicist, Airborne Instruments Lab, 60-62; asst prof physics, Adelphi Univ, 62-64; res physicist, US Rubber Co, NJ, 64-68; asst prof, Cooper Union, 68-71, assoc prof physics, 71-77; adj assoc prof physics, York Col, 77-79; asst prof physics, Maritime Col, State Univ NY, 79-80; ASST PROF PHYSICS, FORDHAM UNIV, 81- *Concurrent Pos:* Consult, Info Div, Am Inst Physics & Electronic Semiconductor Co. *Mem:* Am Phys Soc; Am Asn Physics Teachers; NY Acad Sci. *Res:* Superconductivity; cryogenics. *Mailing Add:* 2408 Hawthorne Dr Yorktown Heights NY 10598

KAPLAN, RICHARD E, b Philadelphia, Pa, July 4, 38; m 60; c 2. AEROSPACE ENGINEERING, FLUID MECHANICS. *Educ:* Mass Inst Technol, BS & MS, 61, ScD(aerospace eng), 64. *Prof Exp:* From asst prof to assoc prof, 64-73, dir systs simulation lab, 69-71, dir Eng Comput Lab, 82-83, dept chmn, 83-86, assoc dean, 84-86, VPROVOST ACAD COMPUT, 87-, PROF AEROSPACE ENG, UNIV SOUTHERN CALIF, 73- *Concurrent Pos:* Fulbright lectr & Guggenheim fel, 71-72; Fulbright lectr, 75-76. *Mem:* Am Inst Aeronaut & Astronaut; Am Phys Soc; Am Soc Eng. *Res:* Fluid dynamic stability theory and turbulence experimentation; numerical methods in fluid mechanics; digital techniques in turbulence experimentation; aerosonics and jet noise. *Mailing Add:* Dept Aerospace Eng Univ Southern Calif Los Angeles CA 90007

KAPLAN, RICHARD STEPHEN, b Pittsburgh, Pa, Aug 24, 45; m 70. MEDICAL ONCOLOGY. *Educ:* Univ Pittsburgh, BA, 66; Univ Miami, MD, 70; Am Bd Internal Med, dipl, 74; Am Bd Med Oncol, dipl, 75. *Prof Exp:* Clin assoc oncol, Nat Cancer Inst, 71-73; fel, Univ Miami, 74-75, asst prof oncol, 75-79; ASSOC PROF ONCOL & MED, UNIV MD, 79- *Concurrent Pos:* Surgeon, USPHS, 71-73; consult oncologist, Miami Vet Admin Hosp & sr staff mem, Comprehensive Cancer Ctr, Fla, 75-79; sr investr, Nat Cancer Inst, 79-81. *Mem:* Am Asn Cancer Res; Am Soc Clin Oncol; NY Acad Sci; AAAS; Am Soc Hemat; Am Fed Clin Res. *Res:* Clinical and laboratory research in clinical oncology: neuro-oncology, malignant lymphomas and gastrointestinal malignancy. *Mailing Add:* Univ Md Cancer Ctr 22 S Greene St Baltimore MD 21201

KAPLAN, ROBERT JOEL, b New York, NY, Sept 13, 47. DERMATOLOGY. *Educ:* Franklin & Marshall Col, BA, 69; Univ Tenn, Memphis, MD, 73. *Prof Exp:* Tech asst, Englewood Hosp, 68-69; internship, Geisinger Med Ctr, Danville, Pa, 73-74; residency, Univ Tenn, 74-77, asst prof dermat, 77-79; PVT PRACT, MEMPHIS, TENN, 79- *Mem:* Am Acad Dermat; AMA. *Res:* Clinical studies involving cutaneous levels of cyclic adenosine monophate in atopic dermatitis. *Mailing Add:* 910 Madison Suite 922 Memphis TN 38103

KAPLAN, ROBERT LEWIS, b Long Branch, NJ, Oct 5, 28; c 1. OPERATIONS RESEARCH, SYSTEMS ENGINEERING. *Educ:* US Mil Acad, BS, 53; Mass Inst Technol, MS, 60. *Prof Exp:* vpres opers res, Actuarial Res Corp, 76-80; PRES, RUMSON CORP, 80- *Concurrent Pos:* dep dir mat plans & prog & dep chief staff res, develop & aquisition, Hq, Dept Army, 63- *Mem:* Am Inst Aeronaut & Astronaut; Am Helicopter Soc; Sigma Xi. *Res:* Low speed aeronautical research c/w v-stol, helicopters, and aircushion vehicles; command and control operations research; quantitative measurement of subjective judgements. *Mailing Add:* PO Box 1943 Middleburg VA 22117

KAPLAN, ROBERT S, b New York, NY, July 13, 40; m 67; c 2. EXTRACTIVE METALLURGY, CHEMICAL METALLURGY. *Educ:* Univ Mich, BSE, 62, MSE, 64; Carnegie-Mellon Univ, PhD(metall, mat sci), 68. *Prof Exp:* Res metallurgist, Battelle Columbus Labs, 68-71; staff metallurgist, Bur Mines, US Dept Interior, 71-76; proj mgr res recovery, Off Tech Assess, US Cong, 76; res supvr metall, Resource Recovery, 76-78, staff minimum policy rev, 78-79, mgr extractive nonfuel minerals processes, 80-82, MGR RECYCLING TECHNOL, BUR MINES, US DEPT INTERIOR, 82- *Mem:* Am Inst Mining, Metall & Petrol Engrs. *Res:* Iron-making slags; decarburization of steels; steel refining; inclusions in steels; recovery of metals from nonferrous metal scrap and wastes. *Mailing Add:* Bur of Mines 2401 E St NW Mail Stop 6204 Washington DC 20241

KAPLAN, RONALD M, b Los Angeles, Calif, July 15, 46; m 70; c 2. COMPUTER SCIENCE, COMPUTATIONAL LINGUISTICS. *Educ:* Univ Calif, Berkeley, BA, 68; Harvard Univ, MA, 70, PhD(social psychol), 75. *Prof Exp:* Consult, Rand Corp, 68-72 & Info Sci Inst, Univ Southern Calif, 72-73; consult, 73-74, res scientist psycholing, Palo Alto Res Ctr, 74-86, RES FEL, XEROX CORP, 86- *Concurrent Pos:* Res assoc, Harvard Univ, 73-74; vis scholar cognitive sci, Mass Inst Technol, 78; consult prof ling, Stanford Univ, 88-; mem, adv comt sci & technol ctrs, NSF, 88- *Mem:* Asn Comput Mach; Asn Comput Ling; Linguistics Soc Am; Am Asn Artificial Intel; Cognitive Sci Soc. *Res:* Computational models of human language comprehension; linguistics; psycholinguistics. *Mailing Add:* Xerox Palo Alto Res Ctr 3333 Coyote Hill Rd Palo Alto CA 94304

KAPLAN, RONALD S, b New York, NY, July 12, 57; m 87; c 1. BIOENERGETICS. *Educ:* NY Univ, BA, 73, MS, 75, PhD(biol), 81. *Prof Exp:* Fel, Johns Hopkins Sch Med, 80-86; ASST PROF PHARMACOL, COL MED, UNIV SOUTH ALA, 86- *Mem:* AAAS; Am Chem Soc; Biophys Soc; Am Soc Biochem & Molecular Biol; Am Diabetes Asn. *Res:* Elucidate the structure, function and mechanisms of regulation of mitochondrial transport proteins in normal and diseased States. *Mailing Add:* Col Med Univ SAla MSB Rm 3130 Mobile AL 36688

KAPLAN, SAM H, b Chicago, Ill, Jan 16, 15; m 41; c 2. ELECTRONICS, PHYSICAL CHEMISTRY. *Educ:* Armour Inst Technol, BS, 37. *Prof Exp:* Proj engr monochrome & color tubes, Rauland Div, 52-57, mgr advan color tube develop, 57-62, mgr color tube res, 62-80, SR TECHNOL CONSULT DISPLAY DEVICE, RES & DEVELOP, ZENITH RADIO CORP, 80- *Honors & Awards:* Vladimir K Zworykin award, Inst Elec & Electronics Engrs, 78; IR-100 Award, 70; Eugene McDonald award, Zenith Radio Corp, 70. *Mem:* Fel Inst Elec & Electronics Engrs; Electrochem Soc; Soc Info Display; Soc Motion Picture & TV Engrs. *Res:* Improving performance and reducing cost of color television display tube; improvement of brightness, contrast, and color under high ambient illumination; development of simplified means to make present very complicated tube. *Mailing Add:* 4601 W Touhy Ave Lincolnwood IL 60646

KAPLAN, SAMUEL, b Detroit, Mich, Sept 13, 16; m 53; c 2. MATHEMATICS. *Educ:* Univ Mich, BS, 37, MS, 38, PhD(math), 42. *Prof Exp:* Instr math, Univ Mich, 46; Rackham fel, Princeton Univ, 46-47; researcher, Inst Adv Study, 47-48, 56-57; from asst prof to prof, Wayne State Univ, 48-61; PROF MATH, PURDUE UNIV, 61- *Mem:* Am Math Soc. *Res:* Homology theory; topological groups; topological spaces; duality; functional analysis. *Mailing Add:* Dept Math Purdue Univ West Lafayette IN 47907

KAPLAN, SAMUEL, b Johannesburg, SAfrica, Mar 28, 22; nat US; m 52. CARDIOLOGY. *Educ:* Univ Witwatersrand, MD & MB, BCh, 44. *Prof Exp:* Lectr physiol, Univ Witwatersrand, 46-47, lectr internal med, 47-49; registr cardiol, Postgrad Med Sch, Univ London, 49-50; sr res assoc pediat, 51-54, asst prof, 54-61, from asst prof to assoc prof internal med, 54-82, assoc prof pediat, 61-66, PROF PEDIAT, UNIV CINCINNATI, 66-, PROF INTERNAL MED, 82-; DIR DIV CARDIOL, CHILDREN'S HOSP, 53- *Concurrent Pos:* Consult, NIH. *Mem:* Soc Pediat Res; Am Pediat Soc; Am Fedn Clin Res. *Res:* Hemodynamics and extracorporeal circulation. *Mailing Add:* 6969 Glenmeadow Lane Cincinnati OH 45237

KAPLAN, SANDRA SOLON, b Chicago, Ill, Sept 9, 34; m 57; c 2. CLINICAL PATHOLOGY, HEMATOLOGY. *Educ:* Roosevelt Univ, BS, 55; Boston Univ, MD, 59. *Prof Exp:* Res assoc, Yale Univ, 65-69; asst res prof path & med, 70-78, from asst prof to assoc prof, 78-90, PROF PATH, SCH MED, UNIV PITTSBURGH, 91- *Concurrent Pos:* USPHS res fel hemat, Children's Hosp, San Francisco, 61-62; USPHS res training grant, Med Sch, Yale Univ, 63-64; res worker, Sir William Dunn Sch Path, Oxford Univ, 71-72; dir hemat, Magee Womens Hosp, Pittsburgh, 78- *Mem:* Am Soc Hemat; Am Soc Clin Path. *Res:* Mechanisms of leukocyte activation associated with phagocytosis; mechanisms of chemotaxis and bacterial killing by leukocytes. *Mailing Add:* 326 Orchard Dr Mt Lebanon PA 15228

KAPLAN, SANFORD SANDY, b New York, NY, Oct 2, 50; m; c 5. STRATIGRAPHY, SEDIMENTATION. *Educ:* Lafayette Col, AB, 71; Lehigh Univ, MS, 76; Univ Pittsburgh, PhD(geol), 81; Salve Regina Col, MA, 87. *Prof Exp:* teaching asst, Lehigh Univ, 75-76; lectr gen geol, Univ Nebr, Lincoln, 77-78; vis lectr coal geol, Univ Pittsburgh, 80; geologist, Coal Prep Div, Pittsburgh Mining Technol Ctr, US Dept Energy, 79-80; geologist, Penzoil Explor & Prod Co, 80-86; PRES, EARTHSOURCE CONSULT, INC, 87-; RES ASSOC PROF, UNIV NEBR, LINCOLN, 89- *Concurrent Pos:* Surface Warfare Officer, US Navy, 69-; vis lectr geol, Northampton Co Area Community Col, 74-75. *Mem:* Am Asn Petrol Geologists; Geol Soc Am; Soc Econ Paleontologists & Mineralogists; Sigma Xi; AAAS; Am Econ Asn; Int Asn Sedimentologists. *Res:* Interpreting ancient environments of deposition of sedimentary sequences especially those containing coal, oil and gas and deducing their tetonic setting from such evidence; environmental geology; site assessments; hydrology. *Mailing Add:* 3720 Stockwell Circle Lincoln NE 68506

KAPLAN, SELIG N(EIL), b Chicago, Ill, June 30, 32; m 54; c 2. NUCLEAR PHYSICS & ENGINEERING. *Educ:* Univ Ariz, BS, 52; Univ Calif, MA, 54, PhD(physics), 57. *Prof Exp:* Physicist, Lab, 57-68, from asst prof to assoc prof nuclear eng, 65-75, PROF NUCLEAR ENG, UNIV CALIF, BERKELEY, 75-, SR PHYSICIST, LAWRENCE BERKELEY LAB, 68- *Concurrent Pos:* Lectr nuclear eng, Univ Calif, Berkeley, 62-65. *Mem:* Am Phys Soc; Am Nuclear Soc; Inst Elec & Electronics Engrs. *Res:* Nuclear instrumentation; neutronics; interaction of muons with nuclei. *Mailing Add:* Dept Nuclear Eng Univ Calif Berkeley CA 94720

KAPLAN, SELNA L, b Brooklyn, NY, Apr 8, 27. PEDIATRICS, ENDOCRINOLOGY. *Educ:* Brooklyn Col, BA, 48; Wash Univ, MA, 50, PhD(anat), 53, MD, 55. *Prof Exp:* Asst anat, Sch Med, Wash Univ, 51-52; instr pediat, Col Physicians & Surgeons, Columbia Univ, 61-63, assoc, 63-65, asst prof, 65-66; from asst prof to assoc prof, 66-74, PROF PEDIAT, SCH MED, UNIV CALIF, SAN FRANCISCO, 74- *Concurrent Pos:* NIH fel, 58-61, career develop award, 62-71. *Honors & Awards:* Ayerst Award, Endocrinol Soc, 87. *Mem:* Endocrine Soc; Soc Pediat Res; NY Acad Sci; Am Pediat Soc. *Res:* Growth disorders in children; immunochemistry of pituitary human growth hormone; ontogenesis of human fetal hormones; pubertal development. *Mailing Add:* Dept Pediat Univ Calif San Francisco CA 94143

KAPLAN, SOLOMON ALEXANDER, b SAfrica, Feb 5, 24; nat US; m 57. MEDICINE. *Educ:* Univ Witwatersrand, MB & BCh, 46. *Prof Exp:* Instr pediat, Univ Cincinnati, 51-53; from asst prof to assoc prof, State Univ NY, 53-59; from assoc prof to prof, Sch Med, Univ Southern Calif, 59-68; PROF PEDIAT, MED CTR, UNIV CALIF, LOS ANGELES, 68- *Concurrent Pos:* Res fel pediat, Univ Cincinnati, 49-51. *Mem:* AAAS; Am Physiol Soc; Soc Pediat Res; Am Pediat Soc; Brit Soc Endocrinol. *Res:* Pediatrics; endocrinology; biochemistry. *Mailing Add:* Dept Pediat Univ Calif Med Ctr Los Angeles CA 90024

KAPLAN, STANLEY, b Canton, Ohio, Apr 28, 36; c 3. TERATOLOGY, HUMAN DEVELOPMENT. *Educ:* Univ Miami, BS & BEd, 62, PhD(teratology), 67. *Prof Exp:* Instr & postdoctoral fel, Col Med, Univ Fla, 66-67, asst prof anat sci, 67-69; from asst prof to assoc prof, Med Col Wis, 69-82, vchmn, 72-82, actg chmn, 82-84, PROF ANAT, MED COL WIS, 82-, ASSOC CHMN, 84- *Concurrent Pos:* Vis prof, Univ Man, Can, 74 & Hebrew Univ, Israel, 88; prin investr, NIH, 85-86. *Mem:* Am Asn Anat; Am Soc Zool; Am Inst Biol Sci; Europ Teratology Soc; Teratology Soc; Toxicol Soc. *Res:* Mechanisms by which chemical and physical environmental agents produce congenital malformations. *Mailing Add:* Dept Anat & Cellular Biol Med Col Wis 8701 Watertown Plank Rd Milwaukee WI 53226

KAPLAN, STANLEY A, b New York, NY, Sept 28, 38; m 60; c 3. PHARMACEUTICAL RESEARCH & DEVELOPMENT, ADMINISTRATION. *Educ:* Columbia Univ, BS, 59, MS, 61; Univ Calif, San Francisco, PhD(pharmaceut chem), 65. *Prof Exp:* Postdoctoral fel, NIH, 65; several positions, dept dir & assoc div dir exp therapeut, Hoffmann-La Roche Inc, Nutley, NJ, 66-84; exec dir develop, Med Res Div, Lederle Labs, Am

Cyanamid, 84-87; sr vpres res & develop, Liposome Technol, Inc, 87-89; PRES & CHIEF OPER OFFICER, PHARMETRIX CORP, 89- *Mem:* Am Asn Pharmaceut Scientists; Am Soc Clin Pharmacol & Therapeut; Am Col Clin Pharmacol; Am Soc Pharmacol & Exp Therapeut; AAAS; NY Acad Sci; Controlled Release Soc. *Res:* Pharmacokinetics; biopharmaceutics; drug metabolism; drug development; novel drug delivery systems; analytical methodology; design and implement programs to develop new drugs and drug products; over 90 publications, presentations or chapters in books. *Mailing Add:* Pharmetrix Corp 1330 O'Brien Dr Menlo Park CA 94025

KAPLAN, STANLEY BARUCH, b Memphis, Tenn, Jan 6, 31; m. MEDICINE, RHEUMATOLOGY. *Educ:* Univ Tenn, MD, 54. *Prof Exp:* Intern med, Jefferson Med Col, 55; from asst resident to chief resident, 58-62, from instr to assoc prof, 61-73, PROF MED & RHEUMATOL, SCH MED, UNIV TENN, MEMPHIS, 73- *Concurrent Pos:* Fel rheumatol, Sch Med, Univ Tenn, 60-62; attend physician, Vet Admin Hosp, 67- *Mem:* AMA; Am Col Rheumatol. *Res:* Clinical investigation in rheumatic diseases. *Mailing Add:* Dept Med Health Sci Ctr Univ Tenn 800 Madison Ave Memphis TN 38163

KAPLAN, STANLEY MEISEL, b Cincinnati, Ohio, May 10, 22; m 50; c 3. PSYCHIATRY. *Educ:* Univ Cincinnati, BS, 43, MD, 46. *Prof Exp:* Intern med, Cincinnati Jewish Hosp, 46-47, resident, 47-48, resident psychiat, Cincinnati Gen Hosp, 49-51; from instr to assoc prof, 52-69, actg dir, 75-77, PROF PSYCHIAT, COL MED, UNIV CINCINNATI, 69- *Concurrent Pos:* Res fel, May Inst, 48-49; fel psychosom, Cincinnati Gen Hosp, 51-52; NIMH spec res fel, 54-56 & Inst Psychoanal, 61-67. *Mem:* AAAS; Am Psychosom Soc; Am Med Asn; fel Am Psychiat Asn. *Res:* Psychosomatic medicine; psychiatry; psychoanalysis. *Mailing Add:* Dept Psychiat Univ Cincinnati Col Med Cincinnati OH 45267

KAPLAN, STEPHEN ROBERT, b Brooklyn, NY, May 18, 37; m 61; c 3. RHEUMATOLOGY, MEDICAL EDUCATION. *Educ:* Wesleyan Univ, BA, 59; New York Univ, MD, 63; Brown Univ, MA, 77. *Prof Exp:* Instr med, 69-70, from asst prof to assoc prof rheumatology, 70-84, PROF MED, PROG MED, BROWN UNIV, 84-, ASSOC DEAN MED, 85- *Concurrent Pos:* Adj prof pharmacol, Prog Med, Brown Univ, 78-, adj dean med, 82-85; mem, Nat Arthritis Info Clearing House, 79- & Pharm Panel Anti-Rheumatic Drugs, 78- *Mem:* Am Fedn Clin Res; Am Rheumatism Asn; Arthritis Health Prof Asn; AAAS; Am Col Physicians. *Res:* Mechanism and use of anti-rheumatic and immunomanipulative drugs; medical education and the teaching of rheumatology. *Mailing Add:* Roger Williams Gen Hosp 825 Chalkstone Ave Providence RI 02908

KAPLAN, THOMAS ABRAHAM, b Philadelphia, Pa, Feb 24, 26; m 56; c 3. SOLID STATE PHYSICS. *Educ:* Univ Pa, BS, 48, PhD(physics), 54. *Prof Exp:* Res asst physics, Willow Run Res Ctr, Univ Mich, 54-55, res assoc, Eng Res Inst, 55-56; res assoc, Pa State Univ at Brookhaven Nat Lab, 56-59; staff mem, Lincoln Lab, Mass Inst Technol, 59-70; PROF PHYSICS, MICH STATE UNIV, 70- *Concurrent Pos:* Alexander von Humboldt sr scientist award, 81-82; Consult, Naval Res Lab, Washington, DC, 79 & 80; vis scientist, Max-Planck- Institut für Festkörperforschung, Stuttgart, Ger, 81-82, 83-84, 88-89, Institut für Festkörperforschung der KFA Jülich, Ger, 82; distinguished vis prof, Univ Tsukuba, Japan, 89. *Mem:* Fel Am Phys Soc; Sigma Xi. *Res:* Quantum theory of solids; magnetism; numerical solution of models of highly-correlated-electron systems, e.g. Heisenberg spin models, Hubbard models; possible relations to high-Tc superconductivity are being explored. *Mailing Add:* Dept Physics & Astron Mich State Univ East Lansing MI 48824

KAPLAN, WILFRED, b Boston, Mass, Nov 28, 15; m 38; c 2. MATHEMATICS. *Educ:* Harvard Univ, AB, 36, AM, 36, PhD(math), 39. *Prof Exp:* Instr math, Col William & Mary, 39-40; from instr to assoc prof, 40-57, prof, 57-86, EMER PROF MATH, UNIV MICH, ANN ARBOR, 88- *Concurrent Pos:* Res assoc, Brown Univ, 44-45; Guggenheim Found fel, 49-50. *Mem:* AAAS; Am Phys Soc; Am Math Soc; Math Asn Am; Math Soc France. *Res:* Non-linear differential equations; dynamics; kiemann surfaces; statistical mechanics. *Mailing Add:* 317 W Eng Bldg Univ Mich Ann Arbor MI 48109

KAPLAN, WILLIAM, b New York, NY, Apr 27, 22; m 53. MEDICAL MYCOLOGY. *Educ:* Cornell Univ, BS, 43, DVM, 46; Univ Minn, Minneapolis, MPH, 51. *Prof Exp:* Vet, UN Relief & Rehab Admin, 46-47; vet, USDA, 47-50; COMMISSIONED OFFICER, MYCOL DIV, CTR DIS CONTROL, USPHS, 51- *Concurrent Pos:* Adj field prof, Sch Pub Health, Univ NC, 69-; adj assoc prof, Ga State Univ, 71-; lectr, Sch Vet Med, Univ Pa, 71- *Mem:* Am Soc Microbiol; Am Pub Health Asn; Int Soc Human & Animal Mycol (vpres, 67-71); Med Mycol Soc Am; Am Vet Med Asn. *Res:* Selected zoonosis, with emphasis on epidemiology; diagnostic procedures for mycotic diseases, with emphasis on immunofluorescence. *Mailing Add:* 1222 Briar Hill Dr Atlanta GA 30329

KAPLAN, WILLIAM DAVID, b New York, NY, Aug 24, 14; m 52; c 2. GENETICS. *Educ:* Brooklyn Col, BA, 36; Harvard Univ, MA, 37; Univ Calif, PhD(zool), 51. *Prof Exp:* Chief genetics sect, City Hope Res Inst, 53-58, chmn dept genetics, 58-62, asst chmn & sr res geneticist, dept biol, 62-83; RETIRED. *Concurrent Pos:* Agr Res Coun sr res fel, Inst Animal Genetics, Univ Edinburgh, 51-53; Fulbright res scholar, Norsk Hydro's Inst Cancer Res, Oslo, Norway, 63-64; lectr, Univ Calif, 51; vis prof, Univ Calif, Los Angeles, 62-63; vis prof, Max-Planck Inst Biol Cybernet, Tubingen, Ger, 74-75; Consult, 83- *Mem:* Genetics Soc Am; AAAS; Behav Genetics Asn; Soc Neurosci; Sigma Xi; Am Soc Human Genetics. *Res:* Mutation and biochemical genetics; Drosophila melanogaster; mammalian cytology; behavioral genetics. *Mailing Add:* Dept Biol City Hope Res Inst Duarte CA 91010

KAPLANSKY, IRVING, b Toronto, Ont, Mar 22, 17; m 51; c 3. MATHEMATICS. *Educ:* Univ Toronto, BA, 38, MA, 39; Harvard Univ, PhD(math), 41. *Hon Degrees:* DMath, Univ Waterloo, 68; DSc, Queens Univ, Ont, 69. *Prof Exp:* Instr math, Harvard Univ, 41-44; res mathematician, Appl Math Group, Nat Defense Res Comt, Columbia Univ, 44-45; from instr to prof math, Univ Chicago, 45-69, chmn dept, 62-67, George Herbert Mead Distinguished Serv prof, 69-84; DIR MATH, MATH SCI RES INST, BERKELEY, 84- *Concurrent Pos:* Guggenheim Found fel, 48-49. *Mem:* Nat Acad Sci; Am Math Soc; Math Asn Am. *Res:* Algebra. *Mailing Add:* Math Sci Res Inst 1000 Centennial Dr Berkeley CA 94720

KAPLER, JOSPEH EDWARD, b Cresco, Iowa, Mar 13, 24; m 59; c 4. BIOLOGY. *Educ:* Loras Col, BS, 48; Marquette Univ, MS, 53; Univ Wis, PhD(zool), 58. *Prof Exp:* Instr biol, Loras Col, 48-51; asst zool, Marquette Univ, 51-53, instr, 53-54; instr biol, Loras Col, 54-55; asst entom, Univ Wis, 55-57; from asst prof to prof biol, 57-89, EMER PROF BIOL, LORAS CO, 89- *Mem:* Entom Soc Am. *Res:* Biology and ecology of forest insects. *Mailing Add:* Dept Biol Loras Col 1450 Alta Vista Dubuque IA 52001

KAPLON, MORTON FISCHEL, b Philadelphia, Pa, Feb 11, 21; m 46; c 3. PHYSICS. *Educ:* Lehigh Univ, BS, 41, MS, 47; Univ Rochester, PhD(physics), 51. *Prof Exp:* Res assoc physics, Univ Rochester, 51-52, from asst prof to prof, 52-71, assoc dean col arts & sci, 63-65, chmn dept physics & astron, 64-69; assoc provost, City Col New York, 71-75, vpres admin affairs, 75-86; RETIRED. *Concurrent Pos:* NSF sr fel, 59-60. *Mem:* AAAS; Am Phys Soc; Am Geophys Union; Ital Phys Soc; Am Astron Soc. *Res:* Cosmic ray physics; fundamental particle physics; high energy nuclear physics. *Mailing Add:* 1047 Johnston Dr New York NY 18017

KAPLOW, LEONARD SAMUEL, b New York, NY, Feb 11, 20; m 55; c 2. FLOW CYTOMETRY. *Educ:* Rutgers Univ, BS, 41; Univ Vt, MS, 55, MD, 59. *Hon Degrees:* MS, Yale Univ, 75. *Prof Exp:* Asst prof path, Med Col Va, 63-64; assoc clin fac, Quinnipiac Col, 68-78; assoc prof path & lab med, 70-75, PROF PATH & LAB MED, SCH MED, YALE UNIV, 75-; DIR, MED TECH PROG, HOUSATONIC COMMUNITY COL, 77- *Concurrent Pos:* Pathologist, Vet Admin Med Ctr, 64-66, chief clin path, 66-74, actg assoc chief staff, 74-77, chief lab serv, 74-; chmn hemat, Nat Comt Clin Lab Standards, 73-76; pres, Asn Vet Admin Chiefs Lab Servm 78-80. *Mem:* Fel Am Soc Clin Pathologists; fel Col Am Pathologists; Histochem Soc (pres, 84); Int Acad Path; NY Acad Sci; Royal Micros Soc. *Res:* Development and clinical application of cytochemical assays for leukocyte enzymes; discoverer of dialysis induced leukopenia. *Mailing Add:* Dept Path & Lab Med Yale Univ Sch Med 333 Cedar St New Haven CT 06510-8023

KAPNER, ROBERT S(IDNEY), b New York, NY, Dec 23, 27; m 55; c 2. CHEMICAL ENGINEERING, PHYSICAL CHEMISTRY. *Educ:* Polytech Inst Brooklyn, BChE, 50; Univ Cincinnati, MSc, 52; Johns Hopkins Univ, DEng, 59. *Prof Exp:* Develop engr, Cent Res Labs, Gen Foods Corp, 52-53; res & develop engr, Gen Aniline & Film Corp, 53-55; assoc prof chem eng, Rennselaer Polytech Inst, 59-68; prof chem eng & head dept, Cooper Union, 68-; RETIRED. *Concurrent Pos:* Instr, McCoy Col, Johns Hopkins Univ, 56-57; consult, Chem & Metall Div, Gen Elec Co, 60- *Mem:* AAAS; Am Inst Chem Engrs; Am Chem Soc; Am Soc Eng Educ; Electrochem Soc. *Res:* Chemical engineering and kinetics; catalysis; reactor design; properties of materials. *Mailing Add:* Seven Woodmont Rd Upper Montclair NJ 07043

KAPOOR, AMRIT LAL, b Amritsar, India, Oct 15, 31; m 59; c 2. MEDICINAL CHEMISTRY, PHARMACEUTICAL CHEMISTRY. *Educ:* Punjab Univ, India, BS, 52, MS, 54; Swiss Fed Inst Technol, ScD(pharmaceut chem), 56. *Prof Exp:* Teaching fel, Sorbonne, 56-57; fel, Wayne State Univ, 57-58; sci officer, Nat Chem Labs, India, 58-59; chief chemist, Merck, Sharpe & Dohme Int, NY, 59-63; res fel chem, 63-66, PROF PHARMACEUT CHEM, COL PHARM & ALLIED HEALTH PROFESSIONS, ST JOHN'S UNIV, NY, 66- *Mem:* AAAS; Am Chem Soc; Am Pharmaceut Asn. *Res:* Natural products; synthesis of biologically active peptides and polypeptides. *Mailing Add:* 16 Richlaurne Lane Melville NY 11747

KAPOOR, BRIJ M, b Chawli, India, Mar 3, 36; m 63; c 3. PLANT CYTOLOGY. *Educ:* Univ Delhi, BSc, 57, MSc, 59, PhD(cytol), 63. *Prof Exp:* Nat Res Coun Can res assoc plant biosyst, Univ Montreal, 63-65; NSF res assoc cytogenetics, Univ Colo, 65-66, vis prof biol, 66-67; asst prof, 67-68; from asst prof to assoc prof, 68-80, PROF BIOL, ST MARY'S UNIV, NS, 80-, CHMN DEPT, 89- *Mem:* Bot Soc Am; Genetics Soc Can; Am Inst Biol Sci; Soc Econ Bot. *Res:* Cytomorphological development of the endosperm of angiosperms; cytomorphological studies of the genus Solidago; cytogenetics of Eastern North American plants with special emphasis on compositae. *Mailing Add:* Dept Biol St Mary's Univ Halifax NS B3H 3C3 Can

KAPOOR, CHIRANJIV L, signal transduction, messenger protein phosphorylation, for more information see previous edition

KAPOOR, INDER PRAKASH, b Multan, India, Sept 9, 37; m 70; c 2. METABOLISM, INSECT TOXICOLOGY. *Educ:* Univ Delhi, BSc, 57; Univ Ill, Urbana, PhD(entom), 70. *Prof Exp:* Tech asst entom, Ministry Food & Agr, India, 57-66; res asst, Univ Calif, Riverside, 66-68; from res asst to res assoc entom metab, Univ Ill, Urbana, 68-72; res chemist, 73-75, group leader, 76-80, mgr, 80-85, DIR, PLANT INDUST DISCOVERY, AM CYANAMID CO, 85- *Mem:* Am Chem Soc. *Res:* Metabolism of pesticides in the environment and its elements; biological screening and development of new pesticides, plant growth regulants and biotechnology research. *Mailing Add:* Am Cyanamid Co PO Box 400 Princeton NJ 08540

KAPOOR, NARINDER N, b Calcutta, India, Sept 4, 37; Can citizen; m 69; c 3. ZOOLOGY, PHYSIOLOGY. *Educ:* Panjab Univ, India, BSc, 60, MSc, 61; McMaster Univ, PhD(animal behav & physiol), 68. *Prof Exp:* Lectr zool, Govt Col, Panjab, India, 61-62; demonstr physiol, McMaster Univ, 66-68; lectr, Univ Waterloo, 68-69, asst prof, 69-73, asst prof, 73-76, ASSOC PROF

BIOL, CONCORDIA UNIV, 76- *Mem:* Micros Soc Can. *Res:* Respiratory physiology and behavior of stream animals; morphology, osmoregulation, scanning and transmission electron microscopy; Plecoptera; sense organs and feeding behavior. *Mailing Add:* Dept Biol Concordia Univ 1455 de Maisonneuve Blvd Montreal PQ H3G 1M8 Can

KAPOOR, S F, b Bombay, India, Sept 7, 34. MATHEMATICS. *Educ:* Univ Bombay, BSc, 55, MSc, 57, LLB, 63; Mich State Univ, PhD(math), 67. *Prof Exp:* Staff asst, State Bank India, 58-61; lectr math, Kirti Col, Univ Bombay, 61-63; asst, Mich State Univ, 63-67; from asst prof to assoc prof, 67-81, PROF MATH, WESTERN MICH UNIV, 81- *Mem:* Math Asn Am. *Res:* Topology; graph theory. *Mailing Add:* Dept Math Western Mich Univ Kalamazoo MI 49008

KAPOS, ERVIN, b Brashov, Rumania, June 21, 31; US citizen; m 52; c 1. OPERATIONS RESEARCH, SYSTEMS ANALYSIS. *Educ:* Ind Univ, AB, 54. *Prof Exp:* Assoc math, Ind Univ, 53-58; analyst, Opers Eval Group, Mass Inst Technol, 58-59, rep to oper test & eval force, US Pac Fleet, 59-60, rep to comdr 1st Fleet, 60-61, head command & control sect, Ctr Naval Anal, 62-66, rep to comdr-in-chief, US Pac Fleet, 66-67, dir, Southeast Asia Combat Anal Div, Opers Eval Group, 67-68 & Marine Corps Anal Group, 68-69, dir Opers Eval Group, 69-72; exec vpres & dir, Washington Opers, Ketron, Inc, 72-80, pres, 80-83; PRES, KAPOS ASSOCS INC, 84- *Concurrent Pos:* Assoc mem, Defense Sci Bd, 74-76, 82-87; panel mem marine bd, Nat Acad Sci, 78-80; mem, adv bd, Nat Security Agency, 79-82; mem, panels sci & tech policy & crisis mgt, Ctr Strategic & Int Studies. *Mem:* Am Math Soc; Opers Res Soc Am; fel Mil Opers Res Soc. *Res:* Military operations research, particularly in command, control and communications; surveillance, intelligence and electronic warfare; human information processing and problem-solving; gaming simulation. *Mailing Add:* 908 Turkey Run Rd McLean VA 22101

KAPP, ROBERT WESLEY, JR, b Point Pleasant, NJ; m 67; c 3. TOXICOLOGY, GENETICS. *Educ:* Syracuse Univ, AB, 67; George Washington Univ, MS, 74, PhD(genetic toxicol), 79. *Prof Exp:* Head cytogenetics, Nat Naval Med Ctr, Md, 69-72; med technician med, Group Health Asn, Washington, DC, 69-73; staff scientist toxicol, Hazleton Lab Am, Va, 73-78, sr toxicologist, 78-79; assoc dir, Toxicol Lab, Exxon Corp, 79-82, dir, 82-89; PRES, ROBERT KAPP ASSOCS, 89- *Concurrent Pos:* Vis fac, Cancer Ctr, Med Br, Univ Tex, 78-, adj prof genetic toxicol, 85-; consult, Genetic Toxicol Ctr, 79-; mem, Dominant Lethal Comt & Sperm Anal Comt, Genetic Toxicol Prog, Environ Protection Agency, 79-; chmn, Med Toxicol Comt, Am Soc Testing & Mat, 79-; reviewer, Nat March of Dimes, 80-; consult, Dept Health & Human Servs, 80-82, bd sci counrs, Nat Toxicol Prog, Nat Inst Occup Safety & Health, 82-; rev panel, Energy Health Sci, Environ Protection Agency, 81-; adj prof genetic toxicol, Med Br, Univ Tex, 85- *Mem:* NY Acad Sci; Soc Toxicol; Am Soc Testing & Mat; Am Col Toxicol; Sigma Xi; Environ Mutagen Soc. *Res:* Development of clinical and nonclinical methodology to determine occupational carcinogenesis and mutagenesis; evaluation of general and genetic toxicological procedures for safety assessment; laboratory and research management. *Mailing Add:* 52 Hoagland Dr Belle Mead NJ 08502

KAPPAGODA, C TISSA, CARDIOLOGY, PHYSIOLOGY. *Educ:* Univ Leeds, Eng, PhD(cardiovasc physiol), 72. *Hon Degrees:* FRCP(Lond), 88, FRCP(C), 80. *Prof Exp:* Prof med cardiol, Univ Alta, 78-90; PROF MED, UNIV CALIF, DAVIS, 90- *Mem:* Physiol Soc Am; Physiol Soc Gt Brit; fel Am Col Cardiol. *Res:* reflex regulation of the circulatiai and exercin physiology. *Mailing Add:* Internal Med Div Cardiovasc Med TB 172 Bioletti Way Univ Calif Davis CA 95616

KAPPAS, ATTALLAH, b Union City, NJ, Nov 4, 26; m 63; c 3. METABOLISM, PHARMACOLOGY. *Educ:* Columbia Univ, AB, 47; Univ Chicago, MD, 50; Am Bd Internal Med, dipl, 58. *Hon Degrees:* DSc, NY Med Col, 78. *Prof Exp:* Intern med, Univ Serv, Kings County Hosp, New York, 50-51; med resident, Peter Bent Brigham Hosp, 54-56; assoc, Sloan-Kettering Inst, 56-57; from asst prof to assoc prof, Sch Med, Univ Chicago, 57-67; assoc prof & physician, 67-71, sr physician, 71-74, PROF ROCKELLER UNIV, 71-, PHYSICIAN-IN-CHIEF, 74-, SHERMAN FAIRCHILD PROF, 81-, VPRES, 83- *Concurrent Pos:* Res fel, Sloan-Kettering Inst, 51-54; Commonwealth Fund fel, Courtauld Inst, Middlesex Hosp Med Sch, London, Eng, 61-62; John Simon Guggenheim Found fel & guest investr, Rockefeller Univ, 66-67; vis prof clin pharmacol, Med Sch, Johns Hopkins Univ, 75; Vincent Astor prof clin sci, Mem Sloan-Kettering Cancer Ctr, Cornell Univ Med Col, 79-81; Nicholson exchange fel, Karolinska Inst, 85. *Honors & Awards:* Spec Award Clin Pharmacol, Burroughs Wellcome Fund, 73; Sir Henry Hallet Dale Mem Lectr, Med Sch, Johns Hopkins Univ, 75; Pfizer Lectr, Peter Bent Brigham Hosp, Harvard Med Sch, 77; Res Award, Am Soc Pharmacol & Exp Therapeut, 78; Pfizer Lectr, Hershey Med Ctr, Pa State Univ, 80; Glaxo Lectr, Cornell Univ Med Col, 84. *Mem:* Am Soc Clin Invest; Endocrine Soc; Harvey Soc; Asn Am Physicians; Am Clin & Climat Asn; Am Soc Pharmacol & Exp Therapeut; fel Am Col Physicians; Endocrine Soc. *Res:* Metabolic-genetic diseases; hormone biology; drug metabolism toxicology and disorders of porphyrin-heme metabolism. *Mailing Add:* Lab Metab & Pharm Rockfeller Univ Hosp 1230 York Ave New York NY 10021

KAPPE, DAVID SYME, b Philadelphia, Pa, Sept 28, 35; c 3. PHYSICAL CHEMISTRY, RADIOCHEMISTRY. *Educ:* Univ Md, College Park, BS, 59; Pa State Univ, PhD(phys chem), 65. *Prof Exp:* Phys sci aid, Metall Div, Nat Bur Standards, 58-59; chief radiation appln sect, Hittman Assocs, Inc, 66-67; res dir, 67-85, CHMN & CEO, KAPPE ASSOCS, INC, 86- *Concurrent Pos:* Consult, Am Acad Environ Eng-Environ Protection Agency Manpower Training Proj, 71; tech rev res proposals, USEPA, 75, monitoring progs hazardous waste incinerators. *Mem:* Am Chem Soc; Am Inst Physics; Am Water Works Asn; Water Pollution Control Fedn. *Res:* Reclamation of spent nuclear reactor fuels; measurement of thermal neutron cross sections; development and application of radionuclide-phosphor self-luminescent light sources; treatment of domestic, industrial and agricultural waste-waters; sludge composting. *Mailing Add:* Kappe Assoc Inc 100 Wormans Mill Ct Frederick MD 21701

KAPPEL, ELLEN SUE, b Brooklyn, NY, Oct 22, 59. MARINE GEOLOGY, GEOPHYSICS, SONAR SWATH MAPPING. *Educ:* Cornell Univ, AB, 80; Columbia Univ, MA, 82, MPhil & PhD(geol), 85. *Prof Exp:* Postdoctoral assoc res scientist, Lamont-Doharty Geol Observ, Columbia Univ, 85-86; STAFF SCI ASSOC, JOINT OCEANOG INSTITS INC, 86-, PROG MGT, 501-US SCI SUPPORT PROG, 88- *Concurrent Pos:* Co-chief scientist, Hawaii-EEF Cruise, US Geol Surv, 88. *Mem:* Am Geophys Union; Sigma Xi. *Res:* Relationship between midocean ridge tectonics and hydrothermal mineralization and MOR tectonics and petrology. *Mailing Add:* 5610 Gloster Rd Bethesda MD 20816

KAPPENMAN, RUSSELL FRANCIS, b Lennox, SDak, Sept 2, 38; m 64; c 4. STATISTICS. *Educ:* Univ SDak, BA, 60; Univ Iowa, MS, 62; State Univ NY Buffalo, PhD(statist), 69. *Prof Exp:* asst prof statist, Pa State Univ, 69-76; MATH STATISTICIAN, NORTHWEST & ALASKA FISHERIES CTR, SEATTLE, 76- *Mem:* Am Statist Asn. *Res:* Statistical inference. *Mailing Add:* Northwest & Alaska Fisheries Ctr 2725 Montlake Blvd East Seattle WA 98112

KAPPERS, LAWRENCE ALLEN, b Hingham, Wis, May 27, 41; m 63. SOLID STATE PHYSICS. *Educ:* Cent Col, Iowa, BA, 63; Univ Mo-Columbia, MS, 66, PhD(physics), 70. *Prof Exp:* Air Force Off Sci res fel physics, Univ Minn, Minneapolis, 70-72; NSF res assoc, Okla State Univ, 72-73; from asst prof to assoc prof, 73-80, PROF PHYSICS, UNIV CONN, 80- *Mem:* Am Phys Soc. *Res:* Electronic structure of defects in ionic crystals; optical absorption; luminescence and electron paramagnetic resonance; production and decay mechanisms of color centers, additive coloration and radiation damage; lasers; high pressure diamond anvil studies. *Mailing Add:* 84 Village Hill Rd Willington CT 06279

KAPPLER, JOHN W, MEDICINE. *Prof Exp:* PROF, HEALTH SCI CTR, UNIV COLO, 85-; INVESTR, HOWARD HUGHES MED INST, 85- *Mem:* Nat Acad Sci. *Mailing Add:* Dept Med & Basic Immunol Nat Jewish Ctr 1400 Jackson St Denver CO 80206

KAPP-PIERCE, JUDITH A, PATHOLOGY. *Prof Exp:* Assoc path-immunol, Harvard Med Sch, 73-76; from asst prof to assoc prof, 76-84, PROF, DEPTS PATH, MICROBIOL & IMMUNOL, SCH MED, WASHINGTON UNIV, 84-; ASSOC STAFF, JEWISH HOSP, ST LOUIS, MO, 76- *Concurrent Pos:* Assoc ed, J Immunol, 79-83; res career develop award, NIH, 79-84; mem, Transplantation Biol & Immunol Comt, Nat Inst Allergy & Infectious Dis, NIH, 80-84, Cancer Preclin Prog Proj Rev Comt, Nat Cancer Inst, 86-88; mem, immunobiol study sect, Nat Inst Allergy & Infectious Dis, NIH, 88-; mem comt fundamental res, Nat Multiple Sclerosis Soc, 88- *Mem:* Sigma Xi; Am Asn Immunologists; Am Asn Pathologists. *Res:* Immunology; microbiology; author or co-author of over 140 publications. *Mailing Add:* Dept Path Jewish Hosp 216 S Kingshighway St Louis MO 63110

KAPPUS, KARL DANIEL, b Cleveland, Ohio, July 2, 38. MEDICAL ENTOMOLOGY, EPIDEMIOLOGY. *Educ:* Ohio State Univ, BSc, 60, MSc, 62, PhD(entom), 64. *Prof Exp:* Res asst mosquito biol, Res Found, Ohio State Univ, 61-64; Nat Res Coun fel arborvirus infection, US Army Biol Labs, 65-66; res assoc mosquito biol, Res Found, Ohio State Univ, 66-67; res entomologist, Nat Commun Dis Ctr, 67-69, biologist, 69-76, EPIDEMIOLOGIST, CTR DIS CONTROL, 76- *Mem:* Am Soc Trop Med & Hyg; Entom Soc Am; Sigma Xi. *Res:* Animal photoperiodism; viral infection in arthropods; mosquito behavior; viral zoonoses; epidemiology of viral infections; human intestinal protozoa; disease eradication. *Mailing Add:* 216 Glendale Decatur GA 30030

KAPRAL, FRANK ALBERT, b Philadelphia, Pa, Mar 12, 28; m 51; c 3. MEDICAL MICROBIOLOGY, IMMUNOLOGY. *Educ:* Philadelphia Col Pharm & Sci, BS, 52; Univ Pa, PhD(med microbiol), 56. *Prof Exp:* Asst instr microbiol, Univ Pa, 52-56; from instr to assoc prof, 56-69, actg chmn dept, 73-78, PROF MED MICROBIOL, OHIO STATE UNIV, 69- *Concurrent Pos:* NIH grants, Ohio State Univ, 67-, NIH training grant, 68-71; assoc microbiol, Philadelphia Gen Hosp, 62-64, chief microbiol res, 64-66, chief microbiol, 65-66; asst chief microbiol res, Vet Admin Hosp, Philadelphia, 62-66. *Mem:* AAAS; Am Soc Microbiol; Am Asn Immunol; Infectious Dis Soc; Soc Exp Biol & Med; Am Acad Microbiol. *Res:* Pathogenesis of staphylococcal infections; bacterial host-parasite interactions; bacterial toxins; lipids as immune mechanisms. *Mailing Add:* Dept Med Microbiol & Immunol 5065 Graves Hall Ohio State Univ Columbus OH 43210-1239

KAPRAL, RAYMOND EDWARD, b Swoyersville, Pa, Mar 21, 42; m 64; c 1. PHYSICAL CHEMISTRY. *Educ:* King's Col, Pa, BS, 64; Princeton Univ PhD(chem), 67. *Prof Exp:* Res assoc chem, Princeton Univ, 67 & Mass Inst Technol, 68-69; asst prof, 69-74, assoc prof, 74-80, PROF CHEM, UNIV TORONTO, 80- *Honors & Awards:* Noranda Award, Chem Inst Can, 81. *Mem:* Am Phys Soc; Sigma Xi. *Res:* Statistical mechanics; quantum mechanics; chemical kinetics. *Mailing Add:* Dept Chem Univ Toronto 80 St George St Toronto ON M5S 1A1 Can

KAPRAUN, DONALD FREDERICK, b Spring Valley, Ill, Sept 13, 45; m 75. PHYCOLOGY. *Educ:* Eastern Ill Univ, BS, 66; Univ Tex, PhD(bot), 69. *Prof Exp:* Asst prof bot, Univ Southwestern La, 69-71; assoc prof bot & phycol, 71-77, ASSOC PROF BIOL, UNIV NC, WILMINGTON, 77- *Mem:* Phycol Soc Am; Int Phycol Soc; Brit Phycol Soc. *Res:* Ecology and reproductive periodicity of benthic marine algae in North Carolina. *Mailing Add:* Dept Biol Univ NC 601 S College Rd Wilmington NC 28403

KAPRELIAN, EDWARD KARNIG, b Union Hill, NJ, June 20, 13; m 36; c 3. OPTICS, PHOTOGRAPHY. *Educ:* Stevens Inst Technol, ME, 34. *Prof Exp:* Patent exam, US Patent Off, 36-42; physicist, Bd Econ Warfare, 42-45; patent adv, Off Chief Signal Officer, 45-46; chief photog res br, US Army Signal Eng Lab, 46-52; dir res & eng, Kalart Co, 52-55; dir res, Kaprelian Res & Develop Co, 55-57; dept dir res, US Army Signal Res & Develop Lab, 57-

62, tech dir, US Army Limited War Lab, 62-67; vpres & tech dir, Keuffel & Esser Co, Morristown, 68-73; PRES, KAPRELIAN RES & DEVELOP CO, 73- *Concurrent Pos:* Mem, Nat Acad Sci-Nat Res Coun, 52-58. *Mem:* Fel Soc Photog Sci & Eng (pres, 48-52); Optical Soc Am; Am Soc Mech Eng; Soc Motion Picture & TV Eng; sr mem Inst Elec & Electronics Engrs. *Res:* Applied physics; photographic processes and apparatus; electronics; patent law; optical instruments. *Mailing Add:* 15 Lowery Lane Mendham NJ 07945-9514

KAPRON, FELIX PAUL, b St Catharines, Ont, Nov 29, 40; US citizen; m 71. FIBER OPTICS. *Educ:* Univ Toronto, BASc, 62; Univ Waterloo, MSc, 63, PhD(physics), 67. *Prof Exp:* Physicist, Corning Glass Works, 67-72; sr scientist & mgr, Bell-Northern Res, 73-82; STAFF SCIENTIST, ELECTRO-OPTICAL PROD DIV, INT TEL & TEL, 82- *Concurrent Pos:* Lectr, Carleton Univ, 76-82. *Mem:* Optical Soc Am; Soc Photo-Optical Instrumentation Engrs; Can Asn Physicists; sr mem Inst Elec & Electronics Engrs. *Res:* Optical communications and sensors; fiber-optical waveguides; optical devices properties of solids, particularly emitters, modulators, detectors and couplers. *Mailing Add:* Electro-Optical Prod Div Int Tel & Tel 7635 Plantation Rd Roanoke VA 24019

KAPSALIS, JOHN GEORGE, b Mytilene, Greece, Jan 27, 27; US citizen; m 54; c 2. FOOD SCIENCE, BIOCHEMISTRY. *Educ:* Athens Col Agr, BS, MS, 54; Univ Fla, MAgr, 55; Tex A&M Univ, PhD(food sci), 59. *Prof Exp:* Asst prof & fel dairy tech, Ohio State Univ, 59-60; food technologist, Armed Forces Food & Container Inst, US Army Natick Res & Develop Labs, 60-62, res chemist, 62-63, chief, food biochem lab, 63-74, chief, Biochem Br, Sci & Advan Technol Lab, 74-87; CONSULT, 87- *Concurrent Pos:* Secy Army res & study fel, 65-66. *Honors & Awards:* Scientific Dir Silver Key Award Res, US Army Natick Lab, 69. *Mem:* Am Chem Soc; Sigma Xi; fel Am Inst Chemists; NY Acad Sci. *Res:* Quality parameters of dehydrated foods; effect of water vapor equilibrium on chemical and rheological properties of foods; nondestructive methods of measurement in foods; chemical and rheological properties of lipids. *Mailing Add:* 5776 Deauville Circle C-308 Naples FL 33962

KAPUR, BHUSHAN M, b Amritsar, India, Feb 23, 38; m 68; c 2. TOXICOLOGY, CLINICAL BIOCHEMISTRY. *Educ:* Bombay Univ, BSc, 59; Univ Basel, PhD(org chem), 67; ARIC, 72, MRIC, 76, FRSC, 79, FACB, 83, FCACB, 89. *Prof Exp:* Res assoc, Univ Basel, 67; fac pharm, Univ Toronto, 68-71; sr chemist, 71-72, DIR LABS, CLIN INST, ADDICTION RES FOUND, 72- *Concurrent Pos:* Lectr, dept clin biochem, Fac Med, Univ Toronto, 74 & 76-78, asst prof, 78; instr, Toronto Inst Med Technol, 74-79. *Mem:* Soc Toxicol; The Chem Soc; Can Soc Clin Chem; Am Asn Clin Chem; Can Soc Sci. *Res:* Natural product chemistry; clinical biochemistry; toxicology; biochemical changes due to alcohol use. *Mailing Add:* Clin Inst Addiction Res Found 33 Russell St Toronto ON M5S 2S1 Can

KAPUR, KAILASH C, b India, Aug 17, 41; US citizen; c 2. QUALITY & PRODUCTIVITY IMPROVEMENT, DESIGN OF EXPERIMENTS. *Educ:* Delhi Univ, India, BS, 63; India Inst Technol, MTech, 65; Univ Calif, Berkeley, MS, 67, PhD(oper res & indust eng), 69. *Prof Exp:* Sr res engr res, Gen Motors Res Labs, 69-70; from asst prof to assoc prof teaching & res, Wayne State Univ, 70-80, assoc chmn, 75-76, prof, 80-89; sr reliability engr res, Tank Automotive Command, 78; DIR & PROF TEACHING & RES, UNIV OKLA, 89- *Concurrent Pos:* Vis scholar, Ford Motor Co, 73; vis assoc prof, Univ Waterloo, 77-78. *Honors & Awards:* Allan Chop Tech Advan Award, Am Soc Qual Control, 87; Craig Award, Am Soc Qual Control, 89. *Mem:* Inst Indust Engrs; Opers Res Soc Am; Inst Mgt Sci; fel Am Soc Qual Control. *Res:* Quality engineering; product and process design optimization; reliability engineering; design of experiments; author of various publications. *Mailing Add:* Univ Okla 202 W Boyd Suite 124 Norman OK 73019

KAPUR, KRISHAN KISHORE, b Jullundur, India, Mar 14, 30; US citizen; m 59; c 3. PHYSIOLOGY, PROSTHODONTICS. *Educ:* Panjab Univ, India, BSc, 48; Univ Bombay, BDS, 54; Tufts Univ, MS, 56, DMD, 58; Am Bd Prosthodont, dipl, 71. *Prof Exp:* Res assoc prosthetics, Sch Dent Med, Tufts Univ, 56-58; from instr to assoc prof, 58-62; assoc prof dent, Univ Detroit, 64-65, dir dent res, 64-67, prof oral biol & chmn dept, 65-67; assoc clin prof prosthetics, Harvard Sch Dent Med, 67-75, asst dean, Vet Admin Progs, 69-75; chief dent serv, Vet Admin Hosp, West Roxbury Mass, 71-75; chief dent serv, Vet Admin Hosp, Sepulveda, Calif, 75-88; prof in residence prosthodont, Los Angeles, Calif, 75-88, chief dent serv, West Los Angeles, Calif, 88-90, PROF PROSTHODONT, SCH DENT, UNIV CALIF, LOS ANGELES, 90- *Concurrent Pos:* Teaching fel, Sch Dent Med, Tufts Univ, 56-58; pvt pract, Boston, 58-63; consult, Warner-Lambert Pharmaceut Co, 65-67 & Vet Admin Oral Dis Res Prog Eval Comt, Washington, DC, 67-72; prog dir dent res, Vet Admin Outpatient Clin, Boston, 67-75, chief dent serv, 67-71 & 74-75; consult, Comn Dent Accreditation, Am Dent Asn, Coun Dent Mat, Instruments & Equip; consult, Food & Drug Admin. *Honors & Awards:* Prosthodont Sci Award, Int Asn for Dent Res, 74; Carl A Schlack Award, Asn Mil Surg US, 86. *Mem:* Am Dent Asn; Int Asn Dent Res; Acad Dent Prosthetics; Fedn Int Col Dent; Am Asn Hosp Dentists; Geront Soc Am. *Res:* Oral physiology; oral sensations; prosthodontics. *Mailing Add:* 3935 Bon Homme Rd Woodland Hills CA 91364

KAPUR, SHAKTI PRAKASH, b Ludhiana, Panjab, India, Aug 20, 32; m 66; c 2. HUMAN ANATOMY, HISTOLOGY. *Educ:* Panjab Univ, India, BSc, 53, MSc, 54; McGill Univ, PhD(zool), 64. *Prof Exp:* Lectr biol, Govt Col, Panjab, 55-61; sr teaching asst zool, McGill Univ, 61-64; exp biologist, Ayerst Drug Res Labs, Can, 64-66; asst prof zool, Panjab Univ, India, 66-71; res assoc, 71-72, asst prof, 72-78, ASSOC PROF ANAT, GEORGETOWN UNIV, 78- *Mem:* Sigma Xi; Am Asn Anatomists; Soc Exp Biol & Med; AAAS. *Res:* Electron microscopy; histochemistry of thyroid-parathyroid; endocrine mechanisms controlling calcium homeostasis; calcification in biological systems; zinc homeostasis; immunocytochemistry; radioimmunoassay. *Mailing Add:* Dept Anat Georgetown Univ Washington DC 20007

KAPUSCINSKI, JAN, b Warsaw, Poland, June 2, 36; US citizen; m 63. FLUORESCENCE SPECTROSCOPY. *Educ:* Polytech Univ Lodz, Poland, BS, 61, MS, 61, PhD(chem), 65. *Prof Exp:* Asst prof org chem, Polytech Univ Lodz, Poland, 61-65, assoc prof, 65-68; head lab res & develop, Pharmaceut & Cosmetic Indust, 69-78; fel, NY Univ Med Ctr, 78-89; res assoc, 89-90, ASST MEM, SLOAN-KETTERING CANCER CTR, 90-; ASSOC PROF MED, CANCER RES INST, NY MED COL, 90- *Concurrent Pos:* Adj scientist phys biochem, Nencki Inst Exp Biol, Warsaw, 75-78. *Mem:* Polish Chem Soc; Soc Anal Cytol; Am Chem Soc; fel Am Inst Chemists. *Res:* Fluorescence methods of nucleic acids assay; the mechanisms of interactions between nucleic acids and dyes and antitumor agents; organic synthesis; physical biochemistry and medicinal chemistry. *Mailing Add:* Cancer Res Inst NY Med Col 100 Grasslands Rd Elmsford NY 10523

KAPUSTA, GEORGE, b Max, NDak, Nov 20, 32; m 58; c 4. AGRONOMY, WEED CONTROL. *Educ:* NDak State Univ, BS, 54; Univ Minn, MS, 57; Southern Ill Univ, PhD(bot), 75. *Prof Exp:* Agronomist, NDak State Univ, 58-64; assoc prof, 64-80, PROF AGRON, SOUTHERN ILL UNIV, 80- *Honors & Awards:* Outstanding Res & Exten Award, Land of Lincoln Soybean Asn, 78. *Mem:* Agron Soc Am; Soil Sci Soc Am; Weed Sci Soc Am; Sigma Xi. *Res:* Weed control in field and forage crops; minimum and zero-tillage; culture, especially plant density and geometry, cultivars and growth regulators; nitrification inhibition; symbiotic nitrogen fixation. *Mailing Add:* Plant Indust Dept Southern Ill Univ Carbondale IL 62901

KAPUSTKA, LAWRENCE A, b Elyria, Nebr, Apr 17, 48; m 83; c 2. PLANT-MICROBE INTERACTIONS, BIOTECHNOLOGY. *Educ:* Univ Nebr, Lincoln, BS, 70, MS, 72; Univ Okla, PhD(bot), 75. *Prof Exp:* Acad staff biol res, Univ Wis-Superior, 75-78; from asst prof to assoc prof, 78-86, PROF BOT, MIAMI UNIV, 86- *Concurrent Pos:* Pres, Modular Genes of Miami, Inc, Oxford, Ohio, 84-88; res ecologist & team leader, Corvallis Environ Res Lab, US Environ Protection Agency, 88- *Mem:* AAAS; Am Soc Plant Pathologists; Bot Soc Am; Brit Ecol Soc; Ecol Soc Am; Int Soc Plant Molecular Biol; Am Soc Microbiol; Nature Conservancy; Sigma Xi. *Res:* Physiological ecology of non-pathogenic plant-microbe interactions, especially in native grasslands and in agricultural ecosystems; dinitrogen fixation; mycorrhizal fungi; biotechnology; production of synthetic proteins for enriching foods and animal feeds. *Mailing Add:* 2040 Abbott Rd Brookville IN 47012

KAR, NIKHILES, solid state physics, for more information see previous edition

KARABATSOS, GERASIMOS J, b Chomatada, Greece, May 17, 32; US citizen; m 56; c 4. ORGANIC CHEMISTRY. *Educ:* Adelphi Col, BA, 54; Harvard Univ, PhD(org chem), 59. *Prof Exp:* From asst prof to assoc prof, 59-66, chmn dept, 75-86, PROF CHEM, MICH STATE UNIV, 66- *Concurrent Pos:* Sloan Found res fel, 63-66; NSF sr fel, 65-66; sci dir, Greek Atomic Energy Comn, 74-75. *Honors & Awards:* Petrol Chem Award, Am Chem Soc, 71. *Mem:* Am Chem Soc; corresp mem Acad Athens; The Chem Soc. *Res:* Carbonium ions; nuclear magnetic resonance spectroscopy; isotope effects; stereochemistry of enzymatic reactions. *Mailing Add:* Dept Chem Mich State Univ East Lansing MI 48823

KARACAN, ISMET, b Istanbul, Turkey, July 23, 27; m 62; c 5. SLEEP DISORDERS. *Educ:* Univ Istanbul, BS, 48, MD, 53; State Univ NY Downstate Med Ctr, DSc(med), 65; Turkish Bd Neuropsychiat, 60; Am Bd Psychiat & Neurol cert psychiat, 63. *Prof Exp:* From assoc prof to prof psychiat & dir, Sleep Labs, Univ Fla, Gainville, 66-73; PROF PSYCHIAT & DIR, SLEEP DIS & RES CTR, BAYLOR COL MED, TEX MED CTR, HOUSTON, 73-; ASSOC CHIEF STAFF, RES & DEVELOP & DIR, SLEEP RES LAB, VET ADMIN MED CTR, HOUSTON, 73- *Honors & Awards:* Nathaniel Kleitman Prize, Asn Sleep Dis Ctrs, 81. *Mem:* Fel Am Psychiat Asn; AMA; AAAS; fel Am Col Physicians; Sleep Res Soc (pres, 76-79); NY Acad Sci; Am Col Neuropsychopharmacol; Brit Asn Psychopharmacol. *Res:* Psychological and physiological mechanisms of male impotence; neurophysiological and biochemical mechanisms responsible for male erectile failure; pharmacology of human sleep. *Mailing Add:* Baylor Col Med Psych Houston TX 77030

KARADBIL, LEON NATHAN, b New York, NY, Apr 2, 20; m 41; c 3. RESOURCE MANAGEMENT. *Educ:* City Col New York, BS, 40. *Prof Exp:* Asst sect chief, Census Bur, US Dept Com, 40-42, economist & statistician, War Prod Bd, 42-45, economist, Civilian Prod Admin, 45-46, statistician, War Assets Admin, 46-48, economist, Econ Coop Admin, 48-51, indust specialist, Nat Prod Auth, Defense Prod Admin, 51-53 & Off Defense Mobilization, 53-57; consult & analyst, Opers Res Off, Johns Hopkins Univ, 57-61; opers analyst & study chmn opers res, Res Anal Corp, Va, 61-72; study chmn, Gen Res Corp, Va, 72-76; dir, Emergency Preparedness Div, Int Trade Admin, US Dept Com, 76-85; RETIRED. *Concurrent Pos:* Consult, Opers Res Off, Johns Hopkins Univ, 57-58; sr fel, Nat Defense, Univ Ft McNair, Wash. *Mem:* Opers Res Soc Am. *Res:* Military operations research in logistics and costing; analysis of industrial resources. *Mailing Add:* 6909 Winterberry Lane Bethesda MD 20817

KARADI, GABOR, b Budapest, Hungary, Sept 12, 24; m 51; c 2. CIVIL ENGINEERING. *Educ:* Tech Univ Budapest, BSc, 50, MSc, 54, PhD(civil eng), 60; Hungarian Acad Sci, DSc(hydraul), 64. *Prof Exp:* Asst hydraul engr, Hungarian Dept Hydraul Eng, 50-51; sr engr, Inst Water Resources Eng, Budapest, 54-58; sr engr, Inst Hwy Eng, 58-59; sr engr, Water Resources Co, 59-60, chief develop engr, 60-63; lectr civil eng, Univ Khartoum, 63-64, sr lectr, 65-66; vis assoc prof, Northwestern Univ, Evanston, 66-67; assoc prof, 67-69, prof eng mech, 69-77, PROF CIVIL ENG & CHMN DEPT, UNIV WIS-MILWAUKEE, 77- *Concurrent Pos:* Consult, Agr Res Inst, 60-63, Inst Chem Eng, Budapest, Hungary, 61-63 & Northwestern Univ, Evanston, 67-68; mem US comn, Int Comn Irrig & Drainage, 68. *Mem:* Am Soc Civil Engrs; Am Water Resources Asn. *Res:* Hydrodynamics of groundwater flow; watershed hydrology; urban hydrology. *Mailing Add:* Dept Civil Eng Univ Wis Milwaukee WI 53201

KARADY, GEORGE GYORGY, b Budapest, Hungary, Aug 17, 30; US citizen. POWER ELECTRONICS, TRANSMISSION & DISTRIBUTION. *Educ:* Tech Univ Budapest, Dipl Eng, 52, Dr Eng, 60. *Prof Exp:* Assoc prof elec eng, Tech Univ Budapest, 52-68; lectr, Univ Salford, 68-69; prog mgr res, Hydro Que Inst Res, Montreal, Can, 69-77; dir energy conversion & sr consult engr, 77-78, chief elec consult engr, 79-82, chief engr comput technol, 82-84, mgr elec systs, Ebasco Serv Inc, 84-86, SRP CHAIR PROF, ARIZ STATE UNIV, 86- *Concurrent Pos:* Consult high voltage res, Inst Elec Energy Res, Budapest, 58-63; consult power syst anal, Elec Bd, Budapest, 63-66; vis prof, Univ Iraq, Baghdad, 66-68; adj prof, Univ Montreal, 71-77 & McGill Univ, 72-77; adj prof, Polytech Inst Brooklyn, 78-; secy-treas, US Nat Comt, CIGRE, 85- *Honors & Awards:* T&D Comt Outstanding Working Group Chmn Award, 88. *Mem:* Fel Inst Elec & Electronics Engrs; Can Elec Asn; Conf Int Grandes Reseaux Electriques. *Res:* High voltage technic, insulation pollution and special insulators; high voltage thyristor valves and high voltage direct current technology; rectifier and inverter systems; pulsed power supplies. *Mailing Add:* Elec Eng Dept Ariz State Univ Tempe AZ 85287

KARADY, SANDOR, b Budapest, Hungary, Aug 18, 33; US citizen; m 63; c 2. ORGANIC CHEMISTRY. *Educ:* Eotvos Lorand Univ, Lorand, Budapest, BSc, 56; Mass Inst Technol, PhD(org chem), 63. *Prof Exp:* Chemist, Pharmaceut Res Labs, Budapest, Hungary, 55-56; chemist, Merck & Co, 57-59, res chemist, 66-75, sr chemist, 75-78, res fel, 78-83, SR INVESTR, RES LABS, MERCK & CO, 66-, SR RES FEL, 83- *Concurrent Pos:* NIH fel, Mass Inst Technol, 63-64, Inst Org Chem, Gif sur Yvette, France, 64-65. *Honors & Awards:* Thomas Alva Edison Patent Award, 85. *Mem:* Am Chem Soc. *Res:* Synthetic organic chemistry; natural products; pharmaceuticals; cephalosporin chemistry; heterocycles; synthetic electrochemistry; thienamycins. *Mailing Add:* 348 Longview Dr Mountainside NJ 07092

KARAFIN, LESTER, b Philadelphia, Pa, Sept 26, 26; m 50; c 3. UROLOGY. *Educ:* Temple Univ, MD, 49, MSc, 56. *Prof Exp:* PROF UROL, MED COL PA, 64- *Concurrent Pos:* Consult, Vet Admin Hosp, Philadelphia, 64-; prof, Med Ctr, Temple Univ. *Mem:* AMA; Am Urol Asn. *Res:* General urology. *Mailing Add:* Med Col Pa 3300 Henry Ave Philadelphia PA 19129

KARAGIANES, MANUEL TOM, b Boise, Idaho, Sept 22, 32; m 57; c 3. TOXICOLOGY EXPERIMENTAL SURGERY, BIOMATERIAL. *Educ:* Wash State Univ, BS, 61, DVM, 63. *Prof Exp:* Pvt pract, Sunset Animal Clin, Idaho, 63-67; res scientist, 67-70, res assoc, 70-78, mgr, Inhalation Technol & Toxicol, 78-85, MGR SPECIAL OPERATIONS FACILITY, PAC NW LABS, BATTLE MEM INST, 85- *Honors & Awards:* I-R 100 Award, 71. *Res:* Development of intravascular implant operations for bioengineering-biomaterials research; use of porous metals as bone substitutes for orthopedic and dental prostheses. *Mailing Add:* 519 Holly St Richland WA 99352

KARAKASH, JOHN J, b Istanbul, Turkey, June 14, 14; nat US; m 45; c 1. ELECTRICAL ENGINEERING. *Educ:* Duke Univ, BS, 37; Univ Pa, MS, 38. *Hon Degrees:* DEng, Lehigh Univ, 71. *Prof Exp:* Instr, Univ Pa, 38-40; with Am TV Inc, Ill, 40-42; ed dir, 6th Serv Comn, Signal Corps Radar Sch, 42-44; instr & proj engr, Moore Sch Elec Eng, Univ Pa, 44-46; from asst prof to prof elec eng, Lehigh Univ, 46-62, head dept, 55-68, distinguished prof elec eng, 62-81, dean eng, 66-81, EMER DISTINGUISHED PROF ELEC & COMPUT ENG, LEHIGH UNIV, 81-, EMER DEAN 81- *Concurrent Pos:* Consult, Bell Tel Labs, NY, 50-55; proj engr, Signal Corps, 50-54, proj dir, 54-61; mem hon adv bd, Pergamon Inst; mem, Nat Accreditation Coun Eng Cols; consult, Dept Educ, Commonwealth of PR, 72-76 & Gen State Authority, Commonwealth of Pa, 74-77; resident consult, Int Bus Mach, 80- *Honors & Awards:* Noble Robinson Award, Lehigh Univ, 48, Hillman Award, 63 & 80; Distinguished Engr Award, Nat Soc Prof Engrs, 65; Centennial Award, Inst Elec & Electronics Engrs, 84. *Mem:* Fel Inst Elec & Electronics Engrs. *Res:* Electrical networks; microwaves; transmission line theory; filter networks. *Mailing Add:* Packard Lab Bldg 19 Col Eng & Phys Sci Lehigh Univ Bethlehem PA 18015

KARAKASHIAN, ARAM SIMON, b Philadelphia, Pa, Nov 16, 39; m 75; c 2. SOLID STATE PHYSICS, OPTICAL DEVICES. *Educ:* Temple Univ, BA, 61, MA, 63; Univ Md, PhD(physics), 70. *Prof Exp:* From asst prof to assoc prof, 70-82, PROF PHYSICS, UNIV LOWELL, 82-, CHMN, DEPT PHYSICS & APPL PHYSICS, 87- *Mem:* Am Phys Soc; Sigma Xi. *Res:* Optical properties of metals and semiconductors; surface plasma oscillations; photovoltaic, electro-optic and acousto-optic devices. *Mailing Add:* Dept Physics & Appl Physics Univ Lowell Lowell MA 01854

KARAKAWA, WALTER WATARU, CAPSULAR POLYSACCHARIDES. *Educ:* State Univ Iowa, PhD(immunol), 60. *Prof Exp:* ASSOC PROF IMMUNOL & BIOCHEM, PA STATE UNIV, 73- *Res:* Specificity and protection of monoclonal antibodies against capsules. *Mailing Add:* 206 S Frear Bldg Pa State Univ University Park PA 16802

KARAL, FRANK CHARLES, JR, b Philadelphia, Pa, Aug 3, 26. APPLIED MATHEMATICS. *Educ:* Univ Colo, BS, 46; Univ Tex, Austin, PhD(physics), 50. *Prof Exp:* Res physicist, Defense Res Lab, 49-50; res engr, Hughes Aircraft Co, 50-51; res physicist, Defense Res Lab, 51-52; sr res technologist, Mobil Oil Corp, 52-57; post doctoral fel (math), 57-59, assoc res scientist, Courant Inst Math Sci, 59-61, from asst prof to assoc prof, 61-70, PROF MATH, NY UNIV, 71-, UNDERGRAD CHMN DEPT, 85- *Concurrent Pos:* Geophys consult. *Mem:* Am Math Soc; Am Phys Soc; Am Geophys Union. *Res:* Geophysics; electromagnetic theory; computer assisted instruction. *Mailing Add:* NY Univ Courant Inst 251 Mercer St New York NY 10012

KARAM, JIM DANIEL, b Kumasi, Ghana. MOLECULAR GENETICS, BIOCHEMISTRY. *Educ:* Am Univ Beirut, BS, 58; Univ NC, PhD(biochem), 65. *Prof Exp:* Res asst biochem, Am Univ Beirut, 59-60; res asst prof genetics & cell biol sect, Univ Conn, 67-68; res assoc, Sloan-Kettering Inst Cancer Res, 68-71; assoc prof, 71-80, PROF BIOCHEM, MED UNIV SC, 80- *Concurrent Pos:* USPHS fel, Cold Spring Harbor Lab Quant Biol, 65-67; USPHS res career develop award, 74- *Mem:* Genetics Soc Am; Am Soc Biol Chemists; Am Soc Microbiol. *Res:* Genetic control of DNA replication of phage T4. *Mailing Add:* Dept Biochem Med Univ SC 171 Ashley Ave Charleston SC 29425

KARAM, JOHN HARVEY, b Shreveport, La, July 20, 29; m 55; c 2. ENDOCRINOLOGY. *Educ:* St Louis Univ, BS, 49; Tulane Univ, MD, 53. *Prof Exp:* Resident physician internal med, Bronx Vet Admin Hosp, Bronx, NY, 54-56; res fel endocrinol, Hammersmith Hosp, London, Eng, 59-60; res fel, 60-63, asst prof, 63-74, ASSOC PROF INTERNAL MED, UNIV CALIF, SAN FRANCISCO, 74- *Concurrent Pos:* Fulbright vis prof, Univ Baghdad, 65-67. *Mem:* Endocrine Soc; Am Fedn Clin Res; Am Diabetes Asn. *Res:* Diabetes mellitus, particularly relating to disorders of insulin secretion; obesity and factors relating to insulin insensitivity; hypoglycemia and its management. *Mailing Add:* 1141 HSW Univ Calif San Francisco CA 94143

KARAM, RATIB A(BRAHAM), b Miniara, Lebanon, Mar 8, 34; US citizen; m 60; c 1. NUCLEAR ENGINEERING, MATHEMATICS. *Educ:* Univ Fla, BChE, 58, MSE, 60, PhD(nuclear eng), 63. *Prof Exp:* Res asst nuclear field, Fla, 58-63; asst nuclear engr, Argonne Nat Lab, 63-67, assoc nuclear engr, 67-72; PROF NUCLEAR ENG, GA INST TECHNOL, 72- *Mem:* Am Nuclear Soc; Am Phys Soc. *Res:* Fast reactor physics; neutron transport; alternate fuel cycles; new breeder concepts and heterogeneity effects. *Mailing Add:* 3134 Embry Hills Dr Chamblee GA 30341

KARAMCHETI, K(RISHNAMURTY), b India, Feb 8, 23; m 45; c 3. AERONAUTICAL ENGINEERING. *Educ:* Benares Hindu Univ, BS, 44; Indian Inst Sci, Bangalore, dipl, 46; Calif Inst Technol, MS, 52, PhD(aeronaut, physics), 56. *Prof Exp:* Lectr & asst aeronaut & appl mech, Indian Inst Sci, Bangalore, 47-48, sci officer aeronaut, 50-51; asst prof mech eng & in chg lab, Birla Eng Col, India, 48-49; res asst aeronaut, Calif Inst Technol, 51-55; from asst prof to prof, 58-90, EMER PROF AERONAUT & ASTRONAUT, STANFORD UNIV, 90- *Mem:* Assoc fel Am Inst Aeronaut & Astronaut. *Res:* Mechanics of ideal fluids; compressible fluid flows; physical gas dynamics; mechanics of high temperature and rarefied gases. *Mailing Add:* Dept Aeronaut & Astronaut Stanford Univ Stanford CA 94305

KARARA, H(OUSSAM) M, b Cairo, Egypt, Sept 5, 28; m 55; c 2. PHOTOGRAMMETRY. *Educ:* Univ Cairo, BSc, 49; Swiss Fed Inst Technol, MSc, 53, DSc(geod sci), 56. *Prof Exp:* Field civil engr, Dept Reservoirs, Pub Works Ministry, Egypt, 49-51 & La Grande Dixence, Sion, Switz, 55; sci collabr & asst photogram, Swiss Fed Inst Technol, 55-57; sci collabr, Wild-Heerbrugg Survey Instruments Co, Switz, 57; from asst prof to prof civil eng, 57-89, EMER PROF CIVIL ENG, UNIV ILL, URBANA, 89- *Concurrent Pos:* Assoc mem, Ctr Adv Study, Univ Ill, 65-66. *Honors & Awards:* Talbert Abrams Award, Am Soc Photogram, 59 & 61; Res Prize, Am Soc Civil Engrs, 64. *Mem:* Am Soc Photogram; Am Soc Civil Engrs; Am Cong Surv & Mapping; Can Inst Surv. *Res:* Photogrammetry, especially close range; surveying. *Mailing Add:* Dept Civil Eng Univ Ill 210 Woodshop Bldg Urbana IL 61801

KARAS, JAMES GLYNN, b Chicago, Ill, Feb 24, 33. HORTICULTURE, BIOLOGY. *Educ:* Univ Ill, BS, 56; Mich State Univ, MS, 58, PhD(hort, bot), 62. *Prof Exp:* Teaching asst hort, Mich State Univ, 56-61, res assoc biochem, hort & bot, 62-64, asst prof natural sci, 64-67; asst prof hort, NMex State Univ, 67-69; ASSOC PROF BIOL SCI, YOUNGSTOWN STATE UNIV, 69- *Mem:* Am Soc Hort Sci; Am Inst Biol Sci. *Res:* Plant physiology; grauperceptions in plants; natural products; electron microscopy; plant nutrition; seed germination. *Mailing Add:* Dept Biol Sci Youngstown State Univ 410 Wick Ave Youngstown OH 44555

KARAS, JOHN ATHAN, b Lebanon, Pa, Apr 7, 22; m 58; c 2. SCIENCE COMMUNICATIONS. *Educ:* Lehigh Univ, BS, 43, MS, 47; Lowell Tech Inst, DSc, 60. *Prof Exp:* Instr physics, Lehigh Univ, 43-44, 45-50; asst prof, Univ NH, 51-57; PRES, JONATHAN KARAS & ASSOCS, 60-; PRES, SCI HOUSE, 65- *Concurrent Pos:* Sci dir, WBZ-TV, Boston, Mass; asst, USAF Cambridge Res Ctr, 43-44; res proj dir, Univ NH; consult commun of sci & technol; pres, Neutral Territory, 83- *Honors & Awards:* Bausch & Lomb Sci Medal. *Mem:* AAAS; Am Phys Soc; Am Asn Physics Teachers. *Res:* Accident reconstruction; science and information films and television programs; science museum design and concepts. *Mailing Add:* Sci House Manchester MA 01944

KARASAKI, SHUICHI, b Kure, Japan, Nov 27, 31; m 61; c 1. DEVELOPMENTAL BIOLOGY, CANCER. *Educ:* Nagoya Univ, Japan, BSc, 54, MSc, 56, PhD(biol), 59. *Prof Exp:* Asst prof chem, Col Gen Educ, Nagoya Univ, 59-61; vis investr, Biol Div, Oak Ridge Nat Lab, Tenn, 61-65; staff mem, Putnam Mem Hosp Inst Med Res, Bennington, Vt, 65; from assoc prof to prof anat, Univ Montreal, 68-79; mem staff, Montreal Cancer Inst, Notre Dame Hosp, 65-79 & Dept Pathol, Chiba Cancer Ctr, Res Inst, Japan, 79-84; MEM STAFF, KORIYAMA INST MED IMMUNOL, JAPAN, 84. *Concurrent Pos:* Res assoc, Nat Cancer Inst Can, 70-79. *Mem:* Soc Develop Biol; Am Soc Cell Biol; Am Asn Cancer Res; Int Soc Develop Biologists; NY Acad Sci; AAAS; Histochem Soc; Am Soc Microbiol. *Res:* surface replica electron microscopy of human tissue cells in culture. *Mailing Add:* Koriyama Inst Med Immunol z-11-1 Zukei Koriyama 963 Japan

KARASEK, FRANCIS WARREN, b Council Bluffs, Iowa, Dec 11, 19; m 42; c 7. ANALYTICAL CHEMISTRY. *Educ:* Elmhurst Col, BS, 42; Ore State Col, PhD(chem), 52. *Prof Exp:* Sr chemist, Res & Develop Labs, Pure Oil Co, 42-48; mgr instrument develop, Phillips Petrol Co, 51-68; PROF CHEM, UNIV WATERLOO, 68- *Mem:* Am Chem Soc; Am Soc Mass Spectrometry. *Res:* Mass spectroscopy; chromatography; analytical instrumentation; ion mobility spectrometry; environmental sciences. *Mailing Add:* Dept Chem Univ Waterloo Waterloo ON N2L 3G1 Can

KARASEK, MARVIN A, b Chicago, Ill, Mar 8, 31. BIOCHEMISTRY. *Educ:* Purdue Univ, BS, 53; Univ Calif, PhD(biochem), 56. *Prof Exp:* Asst prof biochem, Tufts Univ, 57-60; asst prof, 60-68, assoc prof biochem & res dermat, 68-85, PROF BIOCHEM IN DERMAT, SCH MED, STANFORD UNIV, 86- *Concurrent Pos:* Boston Med Found fel, 58-61. *Mem:* AAAS; Soc Invest Dermat; Soc Cell Biol. *Res:* Protein synthesis; nucleotide metabolism; biochemistry of virus infections; blood vessel metabolism; sebaceous gland metabolism. *Mailing Add:* Dept Dermat Stanford Univ Sch Med Palo Alto CA 94305

KARASZ, FRANK ERWIN, b Vienna, Austria, July 23, 33; m 58; c 2. POLYMER SCIENCE, BIOPHYSICAL CHEMISTRY. *Educ:* Univ London, BSc, 54, DSc(chem), 72; Univ Wash, PhD(phys chem), 58. *Prof Exp:* Fel, Univ Ore, 58-59; sr res fel, Basic Physics Div, Nat Phys Lab, Eng, 59-61; res chemist, Gen Elec Co, 61-67; assoc prof, 67-71, co-dir Mat Res Lab, 73-85, prof polymer sci & eng, 71-86, DISTINGUISHED UNIV PROF, UNIV MASS, AMHERST, 86- *Concurrent Pos:* Mem comt fire safety of polymeric mat, Nat Acad Sci, 74-82. *Honors & Awards:* Mettler Award, NAm Thermal Analysis Soc, 75; High Polymer Physics Prize, Am Phys Soc, 84. *Mem:* Nat Acad Eng; Am Phys Soc; Am Chem Soc. *Res:* Physical chemistry of polymers; thermodynamics and statistical thermodynamics of liquids; biological macromolecules. *Mailing Add:* Grad Res Ctr Rm 701 Univ Mass Amherst MA 01003

KARATZAS, IOANNIS, b Kallithea, Greece, May 29, 51; m 75. STOCHASTIC PROCESSES & CONTROL. *Educ:* Nat Tech Univ Athens, dipl, 75; Columbia Univ, MSc, 76, PhD(math statist), 80. *Prof Exp:* Vis asst prof appl math, Brown Univ, 79-80; asst prof, 80-83, ASSOC PROF MATH STATIST, COLUMBIA UNIV, 83- *Concurrent Pos:* Vis scientist, MIT, 84-85. *Mem:* Sigma Xi; Inst Math Statist; Inst Elec & Electronics Engrs. *Mailing Add:* Dept Math Statist Columbia Univ New York NY 10027

KARAVOLAS, HARRY J, b Peabody, Mass, Feb 21, 36; m 62. BIOCHEMISTRY. *Educ:* Mass Col Pharm, BS, 57, MS, 59; St Louis Univ, PhD(biochem), 63. *Prof Exp:* Res fel biol chem, Harvard Med Sch, 63-66, res assoc & tutor biochem sci, Harvard Univ, 66-68; from asst prof to assoc prof, 68-75, PROF PHYSIOL CHEM & CHMN DEPT, SCH MED, UNIV WIS-MADISON, 75- *Concurrent Pos:* Nat Inst Child Health & Human Develop res career develop award, 72-75. *Mem:* AAAS; Am Chem Soc; Am Soc Biol Chem; Soc Neurosci; Endocrine Soc. *Res:* Mechanism of action of steroid hormones; steroid metabolic patterns in neural and pituitary tissues; enzymology. *Mailing Add:* Dept Physiol Chem Sch Med Univ Wis Madison WI 53706

KARAYANNIS, NICHOLAS M, b Athems, Greece, May 30, 31; m 55; c 2. INORGANIC CHEMISTRY. *Educ:* Nat Tech Univ Athens, BS, 55; Univ London, PhD(chem), 60. *Prof Exp:* Sci collabr, Hellenic Nat Defense Gen Staff, 61-62 & Greek Ministry of Coord, 62-65; NIH res fel anal chem, Johns Hopkins Univ, 65-67; US Army Edgewood Arsenal res fel inorg chem, Drexel Univ, 67-70; res chemist, Amoco Chem Co, 70-72, sr res chemist, 72-76, res assoc, 76-88, SR RES ASSOC, AMOCO CHEM CO, 88- *Mem:* AAAS; Am Chem Soc; NY Acad Sci; Greek Tech Chamber. *Res:* Coordination chemistry; metal complexes of neutral and acidic phosphoryl ligands and aromatic amine oxides; catalysis; catalysts for olefin polymerization; analytical chemistry, high pressure gas chromatography; organic chemistry, redox systems; bioinorganic chemistry. *Mailing Add:* Amoco Chem Co Res & Develop PO Box 3011 Naperville IL 60566-7011

KARCHMER, JEAN HERSCHEL, b Dallas, Tex, Dec 28, 14; m; m 39; c 2. ANALYTICAL CHEMISTRY. *Educ:* Southern Methodist Univ, BS, 36. *Prof Exp:* Jr engr, Dept Agr, Tex, 38; chief chemist, Nat Chemsearch Co, 39-42; asst chemist, Tenn Valley Authority, Ala, 42-44; sr analyst & res chemist, Humble Oil & Refining Co, 44-50, sr res chemist, 50-55, res specialist, 55-63, res assoc, 63-77, consult anal chem, Exxon Res & Eng Co, 78-84; RETIRED. *Concurrent Pos:* Consult chemist, 84- *Mem:* Am Chem Soc. *Res:* Analytical chemistry of sulfur compounds; analysis of petroleum, coal, polymers; elemental analysis; polarography. *Mailing Add:* 3018 Castlewood Houston TX 77025

KARCZMAR, ALEXANDER GEORGE, b Warsaw, Poland, May 9, 18; nat US; m 46; c 2. PHARMACOLOGY, PHYSIOLOGY. *Educ:* Warsaw & Free Polish Univ, Med Sci, 39; Columbia Univ, MA, 41, PhD(biophysics, embryol), 46. *Prof Exp:* Res assoc, Amherst Col, 44-45; asst, Columbia Univ, 46; from instr to assoc prof pharmacol, Sch Med, Georgetown Univ, 47-54; assoc mem, Sterling-Winthrop Res Inst, 54-56; sr co-dir Inst Mind, Drugs & Behav, 65-68; prof pharmacol & therapeut & chmn dept, Loyola Univ, 56-85, sr co-dir, Inst Mind, Drugs & Behav, 65-85, assoc dean res & assoc dean grad training, Stritch Sch Med, 81-86, EMER PROF, LOYOLA UNIV, 86- *Concurrent Pos:* Am Philos Soc grant, NY Univ, 42-44; NIH grants, 47-85; mem, pharmacol A study sect, NIH, 68-72, mult sclerosis study sect, 73- & Alzheimer dis study sect, 84; Guggenheim fel, 69-70; vis prof, Sorbonne, Paris, 69 & 75 & Polish Acad Sci, 82; hon prof, Kurume Univ, 80-; mem nat toxicol panel, Nat Res Coun, 80-; consult, US Army Res Develop Co, 80-, Off US Surgeon Gen, 80-, consult res serv, Hines Vet Admin Hosp. *Mem:* AAAS; Am Soc Pharmacol & Exp Therapeut; Soc Exp Biol & Med; fel Am Col Neuropsychopharmacol; Int Brain Res Orgn; Soc Neurosci. *Res:* Physiology and pharmacology of synaptic transmission; cholinesterases and anticholinesterase drugs; cholinergic system and behavior; neuropsychopharmacology. *Mailing Add:* Dept Pharmacol Stritch Med Sch Maywood IL 60153

KARDONSKY, STANLEY, b Brooklyn, NY, Nov 21, 41; c 3. NUCLEAR CHEMISTRY, ANALYTICAL CHEMISTRY. *Educ:* Long Island Univ, BS, 63; Univ Fla, MS, 64; City Univ New York, PhD(nuclear chem), 70. *Prof Exp:* From instr to assoc prof, Long Island Univ, 64-73, prof chem & dean admin, 73-84; VPRES ADMIN & PROF CHEM, SAN FRANCISCO STATE UNIV, 84- *Mem:* AAAS; Am Chem Soc. *Res:* Neutron activation analysis; nuclear reactions; radiation effects on biologically active materials; mechanisms of genetic damage caused by ultraviolet radiation. *Mailing Add:* San Francisco State Univ 1600 Holloway Ave San Francisco CA 94132

KARDOS, GEZA, b Tolna, Hungary, Mar 2, 26; Can citizen; m 49; c 3. MECHANICAL ENGINEERING. *Educ:* Univ Sask, BSc, 48; McGill Univ, ME, 57, PhD(mech eng), 65. *Prof Exp:* Jr res officer fire hazards, Nat Res Coun Can, 48-50; proj engr, Tamper Ltd, 50-54; proj engr, Aviation Elec Ltd, 54-56, eng supvr, 56-62, staff engr, 65-66; site mgr, HARP, McGill Univ, 62-63; assoc prof design, McMaster Univ, 66-71; PROF ENG, CARLETON UNIV, 71- *Concurrent Pos:* Assoc dir res, Ctr Appl Res & Eng Design, McMaster Univ, 67-69; vis prof, Royal Col Arts; consult, Vitro-Tech, Monterey, Mex, 80-82; vis scholar, Univ Stellenbosch, SA; consult, Nat Res Coun, Can Energy Group, 83-84, BBC Eng Res Ltd, 86, Elma Eng Serv, 87-; vis assoc prof, Stanford Univ, 71. *Honors & Awards:* Fred Merryfield Design Award, Am Soc Eng Educ, 83; Fel, Am Soc Mech Engrs, 88. *Mem:* Am Soc Mech Engrs; Am Soc Eng Educ. *Res:* Mechanical pressure elements; high strain rates; design and computer aided design; case method of engineering teaching; metal physics; fracture mechanics; creative problem solving; systematic design. *Mailing Add:* Eng Fac Carleton Univ Carleton Univ Ottawa ON K1S 5B6 Can

KARDOS, JOHN LOUIS, b Colfax, Wash, Apr 19, 39; m 66; c 3. CHEMICAL ENGINEERING, POLYMER SCIENCE. *Educ:* Pa State Univ, BS, 61; Univ Ill, MS, 62; Case Inst Technol, PhD(polymer physics), 65. *Prof Exp:* From asst prof to assoc prof, 65-74, actg chmn, 77-78, PROF CHEM ENG, WASH UNIV, 74-, DIR, MAT RES LAB, 70-, CHMN, MAT SCI & ENG PROG, 71- *Concurrent Pos:* Chmn, Gordon Conf on Composite Mat, 83. *Honors & Awards:* Mat Eng & Sci Div Award, Am Inst Chem Engrs, 81. *Mem:* Am Chem Soc; Soc Rheology; Am Phys Soc; Am Inst Chem Engrs; Soc Plastics Engrs; Soc Advan Mat Process Eng. *Res:* Chemistry, physics and processing science of composite materials; structure-property relations in reinforced plastics; process modelling for composite materials; materials characterization techniques. *Mailing Add:* Dept Chem Eng Wash Univ St Louis MO 63130

KARDOS, OTTO, b Vienna, Austria, Feb 7, 07; nat US; m 30; c 1. CHEMISTRY. *Educ:* Univ Vienna, PhD(chem), 32. *Prof Exp:* Electrochemist, Galvapol, Vienna, 35-38; electroplating consult, France, 38-42; res electrochemist, Conmar Prod, NJ, 43-44; res electrochemist, Hanson-Van Winkle-Munning Co, 44-52, chief res electrochemist, 53-58, sr scientist, 58-64; res assoc, 64-69, sr res assoc, 69-72, CONSULT, M&T CHEM INC, 72- *Honors & Awards:* Heussner Award, Am Electroplaters Soc, 56-57. *Mem:* Am Chem Soc; Electrochem Soc; Am Electroplaters Soc. *Res:* Electrodeposition of metals. *Mailing Add:* 10004 Vernon Ave Huntington Woods MI 48070

KARE, MORLEY RICHARD, physiology, nutrition; deceased, see previous edition for last biography

KAREEM, AHSAN, b Lahore, Pakistan, Sept 29, 47; m; c 2. PROBABILISTIC DYNAMICS, STRUCTURAL ENGINEERING. *Educ:* WPakistan Univ Eng & Technol, BSc, 68; Univ Hawaii, Honolulu, MSc, 75; Colo State Univ, PhD(civil eng), 78. *Prof Exp:* Design engr, Harza Eng Co Int, Pakistan, 68-71; res assoc, Colo State Univ, 77-78; from asst prof to prof civil eng, Univ Houston, 78-90; PROF CIVIL ENG, UNIV NOTRE DAME, 90- *Concurrent Pos:* Gen consult, Aerovironment Inc, 79-; pres young investr, White House Off Sci & Technol, NSF, 83. *Mem:* Am Soc Civil Engrs; Sigma Xi; Am Inst Aeronaut & Astronaut. *Res:* Analysis and design of civil engineering and ocean engineering structures subjected to stochastic excitation due to wind, waves and earthquakes; reliability based design and digital simulation of civil engineering systems; design of vibration mitigation devices; wind energy. *Mailing Add:* Dept Civil Eng Univ Notre Dame Notre Dame IN 46556-0767

KAREIVA, PETER MICHAEL, b Utica, NY, Sept 20, 51. INSECT POPULATION BIOLOGY, AGRICULTURAL ECOLOGY. *Educ:* Duke Univ, BS, 73; Univ Calif, Irvine, MS, 76; Cornell Univ, PhD(ecol & evolution), 81. *Prof Exp:* Lectr environ biol, Calif State Univ, Los Angeles, 76; asst prof theoret ecol & math modelling, Brown Univ, 81-; PROF, DEPT ZOOL, UNIV WASH, SEATTLE. *Mem:* Ecol Soc Am; Entom Soc Am. *Res:* Population biology of herbivorus insects; mathematical models of insect dispersal; the influence of vegetation texture on herbivore dynamics. *Mailing Add:* Zool Dept Univ Wash Seattle WA 98195

KAREL, KARIN JOHNSON, b Portland, Ore, Aug 9, 50; m 72; c 3. ORGANOMETALLIC CHEMISTRY. *Educ:* Univ Chicago, BS, 72; Princeton Univ, MA, 74, PhD(chem), 78. *Prof Exp:* NSF fel, Univ Ill, 78-79; CHEMIST, CENT RES & DEVELOP, E I DU PONT DE NEMOURS & CO, 80- *Mem:* Am Chem Soc. *Res:* Organometallic reagents for organic synthesis; preparation and characterization of novel organometallic species. *Mailing Add:* 104 Country Club Dr Wilmington DE 19803-2918

KAREL, LEONARD, b Baltimore, Md, Jan 23, 12; m 42; c 3. PHARMACOLOGY. *Educ:* Johns Hopkins Univ, AB, 32; Univ Md, PhD(pharmacol), 41. *Prof Exp:* Toxicologist & actg chief toxicol sect, Med Div, Chem Corps, US Army, 46-47; exec secy div res grants, NIH, 47-51, chief extramural progs, Nat Inst Allergy & Infectious Dis, 51-61; spec asst to assoc dir res, NSF, 61, assoc head sci resources planning off, 61-63, actg head, 63-64; chief bibliog serv div, Nat Libr Med, 64-66, asst toxicol, Info Prog, 68-69, spec asst to assoc dir libr opers, 66-74; RETIRED. *Concurrent Pos:* Consult chem warfare, US War Dept, 47; lectr, Sch Med, Univ Md, 47-51; mem fels speciality bd, NIH, 51-55, mem fels cent qualifications, 51-59, mem exec comt extramural res & training activ, 51-61; mem civil defense lab resources comt, Commun Dis Ctr, USPHS, 56; consult, Jewish Nat Home Asthmatic Children, 59; mem bd dirs, Common Cold Found, 56-60; consult, Pan-Am Health Orgn, WHO & interim dir regional med libr for SAm, 67-68; asst to chmn comt sci & technol info, Off Sci & Technol, Exec Off President, 68, exec secy panel info sci technol, 69-70. *Mem:* Fel AAAS; Am Soc Pharmacol & Exp Therapeut; Am Pub Health Asn. *Res:* Toxicology; chemotherapy; science resources planning; educational manpower statistics; information science; automatic data processing publications; science administration. *Mailing Add:* 11405 Commonwealth Dr Apt T-3 Rockville MD 20852

KAREL, MARCUS, b Lwow, Poland, May 17, 28; US citizen; m 58; c 4. FOOD SCIENCE. *Educ:* Boston Univ, AB, 55; Mass Inst Technol, PhD, 60. *Prof Exp:* Res assoc food tech, Mass Inst Technol, 57-61, from asst prof to prof food eng, 61-88, head, Dept Nutrit Food Sci, 74-79, prof chem & food eng, 88-89, EMER PROF CHEM ENG, MASS INST TECHNOL, 89-; PROF FOOD SCI, STATE NJ, RUTGERS UNIV, 89- *Concurrent Pos:* Consult var food & chem co, 60-; distinguished vis prof, Rutgers Univ, 86- *Honors & Awards:* William V Cruess Award, Inst Food Technol, 70; Food Eng Award, Am Soc Agr Engrs & Dairy & Food Industs Supply Asn, 78; Nicholas Appert Medal, Inst Food Technol, 86- *Mem:* Nat Acad Sci Arg; fel Inst Food Technol; fel Brit Inst Food Sci & Technol; NY Acad Sci; Am Inst Chem Eng; hon fel Int Asn Eng & Food; Am Chem Soc. *Res:* Food engineering; autoxidation of lipids; diffusion of gases and vapors through polymeric membranes; physicochemical properties of foods; heat and mass transfer aspects of food processing; controlled drug release. *Mailing Add:* 349 Linwood Ave Newton MA 02160

KAREL, MARTIN LEWIS, b Baltimore, Md, Mar 15, 44; m 72; c 3. NUMBER THEORY. *Educ:* Johns Hopkins Univ, BA, 66; Univ Chicago, MA, 67, PhD(math), 72. *Prof Exp:* Asst math, Inst Advan Study, Princeton Univ, 72-73, mem, 73-74; asst prof math, Univ NC, Chapel Hill, 74-80; asst prof, 80-83, ASSOC PROF, RUTGERS UNIV, CAMDEN COL ARTS & SCI, 83- *Concurrent Pos:* NSF fel, 75-80 & 81-86; res assoc, Univ Ill, Urbana-Champaign, 79; mem, Inst Advan Study, Princeton, 83. *Mem:* Am Math Soc. *Res:* Arithmetical theory of automorphic forms. *Mailing Add:* Camden Col Arts & Sci Dept Math Rutgers Univ Camden NJ 08102

KARFAKIS, MARIO GEORGE, b Iskenderun, Turkey, Sept 12, 50; m 83; c 2. ROCK MASS CHARACTERIZATION, MINE SUBSIDENCE. *Educ:* Univ Grenoble, France, BS, 75; Univ Wis-Madison, MS, 78, PhD(mining eng), 83. *Prof Exp:* Res asst rock mech, Univ Wis, 76, & 78-79, res assoc, 77 & teaching asst mining eng, 80-83; ASST PROF MINING ENG, UNIV WYO, 83- *Concurrent Pos:* Consult, Chrome-Alloy Eng Instrumentation Wastewater Treatment plant, 78, site eval, Wash, Rockwell Int, 80 & DEQ Subsidence Eval, Wyo, 85, 87, 88. *Mem:* Instrument Soc Am; Int Soc Rock Mech; Sigma Xi; assoc mem Am Inst Mining, Metall & Petrol Engrs. *Res:* Application of acoustic emission techniques for stability monitoring of rock structures, in the mechanics of chimney subsidence, in the in situ stress determination, and in drilling mechanism of layered rocks. *Mailing Add:* Dept Petrol & Mining Eng Va Polytech Inst & State Univ Blacksburg VA 24061

KARG, GERHART, b New York, NY, Jan 21, 36; m 66; c 5. PHYSICAL CHEMISTRY, COSMETIC CHEMISTRY. *Educ:* Manhattan Col, BS, 57; Polytech Inst Brooklyn, PhD(phys chem), 63. *Prof Exp:* Res chemist, M W Kellogg Co, 62-64; res chemist, Ultra Chem Co Div, Witco Chem Co, 64-69; develop chemist, Avon Prod Inc, 69-70, sr phys chemist, 71-78, sr develop chemist, 78-88; sr prod develop chemist, 88-90, MGR COSMETIC & TREAT PROD, COTY RES & DEVELOP, PFIZER, INC, 90- *Mem:* Soc Cosmetic Chemists. *Res:* Photochemistry; surface chemistry; hair properties; skin care. *Mailing Add:* Pfizer Inc 100 Jefferson Rd Parsippany NJ 07054

KARGER, BARRY LLOYD, b Boston, Mass, Apr 2, 39; m 61; c 2. ANALYTICAL CHEMISTRY. *Educ:* Mass Inst Technol, BS, 60; Cornell Univ, PhD(anal chem), 63. *Prof Exp:* From asst prof to assoc prof, 63-72, prof chem 72-, DIR, BARNETT INST CHEM ANALYSIS & MAT SCI, NORTHEASTERN UNIV, 73-, JAMES L WATER PROF ANALYTICAL CHEM, NORTHEASTERN UNIV, 85- *Concurrent Pos:* NIH res grant, 69-; NSF res grant, 66-85; Fed Water Pollution Control Admin res grant, 67-70; Off Naval Res grants, 69-74; sci adv, Food & Drug Admin, 73-76; consult, Technicon Instruments Corp, 77-82, Cambridge Anal, 85-87, Beckman 88-, Genentech 87- *Honors & Awards:* Gulf Res Award, 71; Steven Dal Nogare Mem Award, Delaware Valley Chromatog Form, 75; Chromatography Award, Am Chem Soc, 82; Tswett Mem Medal, USSR, 80; Tswett Medal, USA, 87; Fisher Award, Am Chem Soc, 89; Martin Medal, 90. *Mem:* Fel AAAS; Am Chem Soc; NY Acad Sci; Sigma Xi. *Res:* High performance liquid chromatography; biochemical applications of high performance liquid chromatography; fundamentals of biopolymer separations; capillary electrophoresis, sequencing, electrophoresis and mass spectrometry; separation science. *Mailing Add:* Barnett Inst Northeastern Univ Boston MA 02115

KARGL, THOMAS E, b Des Plaines, Ill, Feb 25, 32; m 57; c 8. BIOCHEMISTRY. *Educ:* St Ambrose Col, BA, 54; Purdue Univ, MS, 56, PhD(biochem), 59. *Prof Exp:* From asst prof to assoc prof, 59-70, head dept, 64-68, PROF CHEM, BELLARMINE COL, 70- *Mem:* Sigma Xi. *Res:* Carotenoid pigments of tomatoes; Lewis acid catalyzed reactions of methyl vinyl ketone; structure determination of complex polyenes; slow release fertilizers. *Mailing Add:* 9214 Old Six Mile Lane Jeffersontown KY 40299

KARICKHOFF, SAMUEL WOODFORD, b Buckhannon, WVa, Oct 22, 43; m 64; c 2. PHYSICAL CHEMISTRY. *Educ:* WVa Wesleyan Col, BS, 65; Fla State Univ, PhD(phys chem), 71. *Prof Exp:* RES CHEMIST, ENVIRON RES LAB, ENVIRON PROTECTION AGENCY, 71- *Mem:* Am Chem Soc; Soc Environ Toxicol & Chem. *Res:* Fate and transport of pollutants in the environment; computer modeling of chemical reactions. *Mailing Add:* Environ Res Lab Environ Protection Agency Athens GA 30613

KARIG, DANIEL EDMUND, b Irvington, NJ, July 20, 37; m 71; c 1. GEOLOGY. *Educ:* Colo Sch Mines, GeolE, 59, MSc, 64; Scripps Inst Oceanog, PhD(earth sci), 70. *Prof Exp:* NSF grant & asst res geologist, Scripps Inst Oceanog, 70-71; asst prof geol sci, Univ Calif, Santa Barbara, 71-74; asst prof, 74-75, assoc prof, 75-80, PROF GEOL SCI, CORNELL UNIV, 80- *Mem:* Geol Soc Am; Am Geophys Union; Sigma Xi; Geol Soc Malaysia. *Res:* Marine geology and geophysics of marginal basins and island arc systems; genesis of rift zones; environmental problems in streams and small lagoons; structure and evolution of island arcs and young mountain belts. *Mailing Add:* Dept Geol Sci Snee Hall Cornell Univ Ithaca NY 14853

KARIM, AZIZ, b Dar es Salaam, Tanzania, Aug 20, 39; m 64; c 2. PHARMACEUTICAL CHEMISTRY. *Educ:* Univ London, BPharm, 64, PhD(pharmaceut chem), 67. *Prof Exp:* NIH fel, Univ Wis, 67-69; sr res investr drug metab, Dept Biochem Res, 69-72, group leader, 72-74, res fel, dept drug metab, 74-79, DIR CLIN BIOAVAILABILTY & PHARMACOKINETICS, G D SEARLE & CO, 79- *Mem:* Am Chem Soc; Pharmaceut Soc Gt Brit; Am Soc Clin Pharmacol & Therapeut; Am Soc Exp Pharmacol & Therapeut; Acad Am Pharm Sci. *Res:* Drug metabolism; study of biotransformation and pharmacokinetics of drugs; isolation and structural elucidation of natural products possessing biological activities. *Mailing Add:* Dept Drug Metab G D Searle & Co 4901 Searle Pkwy Chicago IL 60077

KARIM, GHAZI A, b Baghdad, Iraq, Jan 2, 34; m 57; c 3. ENGINEERING, COMBUSTION. *Educ:* Univ Durham, BSc, 56; Univ London, DIC & PhD(mech eng), 60. *Hon Degrees:* DSc, Univ London, 72. *Prof Exp:* Trainee prime movers, Eng Elec Co, 56-57; res asst mech eng, Imp Col, Univ London, 57-60; consult engr, Ministry of Indust Repub Iraq, 60-61; UN tech fels, 61-62; lectr mech eng, Imp Col, Univ London, 62-68, chmn combustion res group, 64-68; assoc prof, 68-69, PROF MECH ENG, UNIV CALGARY, 69- *Concurrent Pos:* Vis prof, Univ Cambridge, 72-73; consult, Can Petrol Asn, 75-76, Chevron Can Ltd, 77-78, C N G Ltd, 80-, Nova Corp, 83, Canterra Energy Ltd, 84 & PetroCanada, 85. *Honors & Awards:* Unwin Award in Mech Eng, Eng, 60. *Mem:* Soc Automotive Engrs; Combustion Inst. *Res:* Utilization of natural gas and other gaseous fuels for power in internal combustion engines; chemical kinetics of common gaseous fuels; air pollution from combustion processes; fire and explosion research; engineering education; liquid natural gas utilization; thermodynamics; coal, oil sands and heavy oil. *Mailing Add:* Dept Mech Eng Univ Calgary Calgary AB T2N 1N4 Can

KARIM, KHONDKAR REZAUL, b Bangladesh, Feb 8, 50; c 1. ATOMIC & MOLECULAR PHYSICS, ASTROPHYSICS. *Educ:* Dhaka Univ, BS, 72, MS, 74; Univ Ore, MS, 80, PhD(physics), 83. *Prof Exp:* Res assoc, Univ Ore, 83-85; RES ASSOC, KANS STATE UNIV, 85- *Mem:* Am Phys Soc. *Res:* Dielectronic recombination; plasma diagnostics; x-ray and Auger transition rates; resonant transfer and excitation; electron and position scattering in rare gases. *Mailing Add:* Physics Dept Ill State Univ Normal IL 61761

KARIMAN, KHALIL, b Mashad, Iran, Feb 1, 44. PULMONARY IMMUNOLOGY, OXYGEN TRANSPORT. *Educ:* Mashad Med Sch, Iran, MD, 68. *Prof Exp:* ASST PROF MED, MED CTR, DUKE UNIV, 80- *Mem:* Am Fedn Clin Res; Am Thoracic Soc; fel Am Col Chest Physicians; NY Acad Sci. *Mailing Add:* Durham Med Ctr 1901 Hillandale Rd Durham NC 27705

KARIN, SIDNEY, b Baltimore, Md, July 8, 43. COMPUTER SCIENCE. *Educ:* City Col NY, BE, 66; Univ Mich, MSE, 67, PhD(nuclear eng), 73. *Prof Exp:* Computer programmer/nuclear engr, ESZ Assocs, Inc, 68-72; sr engr & sect leader, Gen Atomics, 73-75, mgr, Fusion Div Computer Ctr, 75-82, dir, Info Systs Div, 82-85, DIR, SAN DIEGO SUPERCOMPUTER CTR, GEN ATOMICS, 85-, VPRES ADVAN COMPUT, 87- *Concurrent Pos:* Mem comput rev panel, Fusion Energy Div, Oak Ridge Nat Lab, 80 & Plasma Physics Lab, Princeton Univ, 83-85; mem, Tech Adv Group Supercomputer Ctrs, NSF, 84-85, ann rev panel, Computer Ctr, Lawrence Livermore Nat Lab, 86 & Indust Liaison Coun, Dept Nuclear Eng & Eng Physics, Univ Wis, 87-89; mem, Sci Computer Systs Tech Adv Panel, 84-89; mem, Nat Res Coun Panel, Nat Bur Standards Comput, 86, Computer Sci Res Prog Rev Panel, Nat Res Coun-NASA, 87 & Comput Sci & Technol Bd, Nat Res Coun, 88-; chmn, Nat Res Coun Panel, Nat Bur Standards Comput, 87-88. *Mem:* AAAS; Asn Comput Mach; Inst Elec & Electronics Engrs; Am Nuclear Soc. *Res:* Scientific computing; computer systems; computer hardware and software; networking and communications; distributed computing. *Mailing Add:* San Diego Supercomputer Ctr Gen Atomics PO Box 85608 San Diego CA 92186-9784

KARINATTU, JOSEPH J, b Kerala, India, Aug 6, 38; US citizen; m 63; c 2. CLINICAL BIOCHEMISTRY, PATHOLOGY. *Educ:* Univ Kerala, BSc, 61; Univ Delhi, MSc, 63; St Thomas Inst, MS, 65, PhD(biochem), 67; Univ Autonoma, MD, 80. *Prof Exp:* Biochemist, Jewish Hosp, Cincinnati, 67-69; biochemist, St Therese Hosp, 69-78; PHYSICIAN-RESEARCHER, EAST CENT ILL EDUC FOUND & UNIV ILL, CHAMPAIGN, 80- *Concurrent Pos:* Biochemist consult, Our Lady of Mercy Hosp, Cincinnati, Ohio, 66-69; vis prof clin chem, Col of Lake County, Grayslake, Ill, 73- *Mem:* AMA; Am Chem Soc; fel Am Asn Clin Scientists; Am Asn Clin Chemists; NY Acad Sci. *Res:* Diagnostic methods in laboratory medicine; trace metals. *Mailing Add:* 52 Maywood Dr Danville IL 61832

KARINEN, ARTHUR ELI, b Comptche, Calif, Feb 25, 19; m 46; c 4. PHYSICAL GEOGRAPHY, CARTOGRAPHY. *Educ:* Univ Calif, Berkeley, AB, 44, MA, 48; Univ Md, College Park, PhD(geog), 58. *Prof Exp:* Cartogr, US Govt, 42-43; instr geog, Ohio State Univ, 46-47; asst prof, Univ Md, 48-59; from asst prof to prof, 59-86, chmn dept, 67-72, EMER PROF GEOG, CALIF STATE UNIV, CHICO, 86- *Concurrent Pos:* Fulbright lectr, Univ Oulu, Finland, 70, lectr, 80; vis prof, Helsinki Sch Econ, 79-80. *Mem:* Corresp mem Finnish Geog Soc; Asn Am Geogrs; Am Cong Surv & Mapping; AAAS. *Res:* Geography of Europe and California; world food production; Finnish settlement in the US. *Mailing Add:* 834 Arbutus Ave Chico CA 95926

KARIPIDES, ANASTAS, b Canton, Ohio, July 4, 37. INORGANIC CHEMISTRY, PHYSICAL CHEMISTRY. *Educ:* Oberlin Col, BA, 59; Univ Ill, MS, 61, PhD(chem), 64. *Prof Exp:* Res chemist, David Sarnoff Res Ctr, RCA Corp, NJ, 64-66; fel, Cornell Univ, 66-68; asst prof, 68-73, ASSOC PROF CHEM, MIAMI UNIV, 73- *Res:* Co-ordination chemistry; spectroscopy; crystallography. *Mailing Add:* Dept Chem Miami Univ Oxford OH 45056-1618

KARIV-MILLER, ESSIE, b Sofia, Bulgaria. ORGANIC ELECTROCHEMISTRY. *Educ:* Hebrew Univ, Jerusalem, MSc, 63; Weizman Inst Sci, PhD(chem), 69. *Prof Exp:* Sr lectr chem, Tel Aviv Univ, 69-77; PROF CHEM, UNIV MINN, 81- *Honors & Awards:* Int Exchange Award, Nat Acad Sci, 87; Career Adv Award, NSF, 87. *Res:* Studies of the electrochemical behavior of organic compounds; synthesis by means of electrochemistry; conducting organic solids; chemistry of organic radicals. *Mailing Add:* Dept Chem Univ Minn Minneapolis MN 55455

KARIYA, TAKASHI, b Belmont, Calif, June 23, 25; m 51; c 2. BIOCHEMISTRY. *Educ:* Drake Univ, BS, 52, MA, 54. *Prof Exp:* Res asst atherosclerosis res, Merrell Dow Res Inst, Dow Chem Co, 55-61, biochemist, 61-65, sect head, Lipid Metab Sect, Biochem Dept, Merrell-Nat Labs, Div Richardson-Merrell, Inc, 65-80, sr pharmacologist, Pharmacol Dept, 80-86; RETIRED. *Mem:* NY Acad Sci. *Res:* Pharmacological control of metabolism of cholesterol and other lipids in relation to the treatment of atherosclerosis; biochemical approaches to the regulation of cardiovascular function by pharmaceutical agents. *Mailing Add:* 5809 Bluespruce Lane College Hill OH 45224

KARK, ROBERT ADRIAAN PIETER, b Boston, Mass, Dec 3, 40; c 5. NEUROLOGY, NEUROCHEMISTRY. *Educ:* Oxford Univ, BA, 62, MA, 67; Harvard Univ, MD, 65. *Prof Exp:* asst prof neurol, Reed Neurol Res Ctr, Neuropsychiat Inst, Sch Med, 72-80, assoc prof Neuro-Psychiat Inst Hosp, Univ Calif, Los Angeles & dir, Ataxia Ctr, 80-83; assoc prof neurol, Sch Med, La State Univ, Shreveport, 83-90; chief, Neurol Serv, Vet Admin Med Ctr, Shreveport, 83-90; CLIN DIR, PERFORMING ARTS ASN, CENT NY, 90- *Concurrent Pos:* Clin assoc & guest scientist neurol & neurochem, Med Neurol Br, Nat Inst Neurol Dis & Stroke, NIH, 68-71; investr, Neurobiochem Group, Ment Retardation Prog, Neuropsychiat Inst, Univ Calif, Los Angeles, 72-83, assoc co-dir, Clin Neuromuscular Dis, 74-76, dir, Friedreichs Ataxia Clin, 75-83; consult, Wadsworth Vet Admin Hosp, 73- & Friedreichs Ataxia Group Am, 75-; mem, Med Adv Bd, Nat Ataxia Found, 75-; mem, Med Adv Bd, Nat Ataxia Found, Western Regional Chap, 77-82 & Joseph's Dis Found, 78- *Mem:* Am Acad Neurol; Am Fedn Clin Res; Am Soc Neurochem; Int Soc Neurochem; Soc Neurosci; fel Am Col Physicians. *Res:* Enzymatic defects, metabolic changes, pathophysiology and treatment of inherited forms of ataxia, mental retardation and neuromuscular disease; neurochemistry of mercurial poisoning; biochemical aspects of neuromuscular trophic effects. *Mailing Add:* Neurol Consults PO Box 505 Dewitt NY 13214

KARKALITS, OLIN CARROLL, JR, b Pauls Valley, Okla, May 31, 16; m 61; c 2. CHEMICAL ENGINEERING. *Educ:* Rice Inst, BS, 38; Univ Mich, MS, 41, PhD(chem eng), 50. *Prof Exp:* Jr res chemist, Shell Oil Co, 37-42; instr chem eng, Univ Mich, 45-47; group leader process develop, Am Cyanamid Co, 48-56; supvr res, Petro-Tex Chem Corp, 56-63, mgr, 63-66, asst dir technol, 66-72; DEAN, COL ENG & TECHNOL, McNEESE STATE UNIV, 72- *Mem:* AAAS; fel Am Inst Chem Engrs; Am Soc Eng Educ; Am Asn Cost Engrs; Nat Soc Prof Engrs. *Res:* Catalysis; geothermal energy. *Mailing Add:* Dean Engr & Tech McNeese State Univ 4100 Ryan St Lake Charles LA 70609

KARKHECK, JOHN PETER, b New York, NY, Apr 26, 45; m 69; c 3. PHYSICS. *Educ:* Le Moyne Col, BS, 66; State Univ NY Buffalo, MS, 72; State Univ NY Stony Brook, PhD(physics), 78. *Prof Exp:* Physics assoc, Brookhaven Nat Lab, 75-79; asst prof, Gen Motors Inst, 81-84, ASSOC PROF PHYSICS, GMI ENG & MGT INST, 84-, PROF & DIR PHYSICS, 88-, DEPT HEAD SCI & MATH, 89- *Concurrent Pos:* Fel, State Univ NY, Stony Brook, 78, res assoc, 79-81; consult, Brookhaven Nat Lab, 79-85, STS, 83, BID Ctr, 85, 87; guest scientist, RWTH Aachen, 83-85, Rijksuniversiteit Utrecht, 86; acad assoc, 88 & 90, Mich State Univ, vis scholar, 89. *Mem:* Am Phys Soc; AAAS. *Res:* Transport theory; energy modeling; kinetic theory; optical properties of composites; optoelectronics; physics education. *Mailing Add:* Sci & Math Dept GMI Eng & Mgr Inst Flint MI 48504-4898

KARKLINS, OLGERTS LONGINS, b Tukums, Latvia, Oct 3, 24; US citizen; m 56; c 1. GEOLOGY, BIOLOGY. *Educ:* Columbia Univ, BS, 57, Univ Minn, MS, 61, PhD, 66. *Prof Exp:* GEOLOGIST BIOSTRATIG PALEONT, US GEOL SURV, 63- *Concurrent Pos:* Asst prof lectr, Col Gen Studies, George Washington Univ, 69-72. *Mem:* Paleont Soc; Am Geol Inst; Geol Soc Am; AAAS; Int Bryozool Asn; Sigma Xi. *Res:* Invertebrate paleontology; biostratigraphy; use of paleobiology, stratigraphy and paleogeography of Paleozoic Ectoprocta in regional correlations. *Mailing Add:* 11301 Hawhill End Potomac MD 20854

KARL, CURTIS LEE, b Milwaukee, Wis, Apr 27, 40; m 63. ORGANIC CHEMISTRY, POLYMER CHEMISTRY. *Educ:* St Olaf Col, BA, 62; Mich State Univ, PhD(org chem), 67. *Prof Exp:* Res assoc, Polymer Develop Dept, Gen Mills Chem Inc, 67-80; res assoc, Polymer Develop Dept, Henkel Corp, 80-; MEM STAFF, UNION CARBIDE CORP. *Mem:* Am Chem Soc. *Res:* Preparation and chemistry of synthetic water soluble polymers; polysaccharide chemistry; commercial applications of water soluble polymers; rheology of aqueous polymer solutions. *Mailing Add:* Union Carbide Corp Box 670 Bound Brook NJ 08805-0670

KARL, DAVID JOSEPH, physical chemistry, materials engineering; deceased, see previous edition for last biography

KARL, GABRIEL, b Cluj, Romania, Apr 30, 37; Can citizen; m 65; c 1. THEORETICAL PHYSICS. *Educ:* Univ Cluj, BA, 58; Univ Toronto, PhD(chem), 64. *Prof Exp:* Fel molecular physics, Univ Toronto, 64-66; fel high energy physics, Oxford Univ, 66-69; from asst prof to assoc prof physics, 69-75, PROF PHYSICS, UNIV GUELPH, 75- *Concurrent Pos:* Vis scientist, Europ Orgn Nuclear Res, Geneva, 74, 83. *Mem:* Am Phys Soc; Can Asn Physicists; fel Royal Soc Can. *Res:* High energy physics; atomic physics. *Mailing Add:* Dept Physics Univ Guelph Guelph ON N1G 2W1 Can

KARL, HERMAN ADOLF, b New York, NY, Mar 24, 47; m 70. MARINE GEOLOGY, SEDIMENTOLOGY. *Educ:* Colgate Univ, BS, 69; Univ Nebr, MS, 71; Univ Southern Calif, PhD(geol sci), 77. *Prof Exp:* Explor geologist petrol explor, Humble Oil & Refining Co, 71; res geologist, Esso Prod Res Co, 72; Nat Res Coun res assoc marine geol, 77, MARINE GEOLOGIST, PAC-ARCTIC BR MARINE GEOL, US GEOL SURV, 77- *Mem:* AAAS; Geol Soc Am; Soc Econ Paleontologists & Mineralogists; Int Asn Sedimentologists; Am Geophys Union; Sigma Xi. *Res:* Dynamics of depositional processes and sediment transport on continental margins. *Mailing Add:* US Geol Surv Marine Geol-MS99 345 Middlefield Rd Menlo Park CA 94025

KARL, MICHAEL M, b Milwaukee, Wis, Jan 30, 15; m; c 2. CLINICAL MEDICINE. *Educ:* Univ Wis, BS, 36; Univ Louisville, MD, 38; Am Bd Internal Med, cert, 46. *Prof Exp:* Intern, St Louis City Hosp, Sch Med, Wash Univ, 38-42, resident internal med, 40-42; pract internal med, Md Med Group, St Louis, Mo, 42-87; PROF CLIN MED, SCH MED, WASH UNIV, 72-, DIR CLIN AFFAIRS, DEPT MED, 87- *Concurrent Pos:* Dir, Third Yr Med Clerkship, St Louis City Hosp, 42-44 & Dept Med, Jewish Hosp St Louis, 63-64; med dir, Red Cross Mobile Blood Unit, 42-44; mem, Munic Nursing Bd, City St Louis, 60-62; consult internal med, USAF, 62-64; counr, Soc Internal Med, 67; co-organizer, Jeff-Vander-Lou Med Clin, 67-72; pres, Fac Ctr, Wash Univ, 69, mem exec fac, Sch Med, 75-76 & 85-86; chmn, Comt Serv to Elderly, Nat Coun Jewish Fedns, 76-81; mem, White House Conf Families, 78-80; Am Col Physicians rep, Coun Med Specialty Socs, 86-; mem, Accreditation Coun Continuing Med Educ, 87-, chmn, 91; mem prog comt, Inst Med-Nat Acad Sci, 88-90; Irene & Michael Karl prof endocrinol. *Honors & Awards:* Laureate Award, Am Col Physicians, 88; Ralph O Claypoole Sr Mem Award, Am Col Physicians, 90; Irene & Michael Karl Lectr. *Mem:* Inst Med-Nat Acad Sci; fel & master Am Col Physicians; AMA; Cent Soc Clin Res; Am Asn Study Liver Dis; Am Soc Internal Med. *Mailing Add:* Dept Med Wash Univ Sch Med 660 S Euclid Ave Box 8121 St Louis MO 63110

KARL, RICHARD C, b Albany, NY, Feb 16, 20; m 44; c 3. SURGERY. *Educ:* Columbia Univ, AB, 42; Cornell Univ, MD, 44; Am Bd Surg, dipl, 52. *Prof Exp:* Instr anat, Med Col, Cornell Univ, 46-47, asst surg, 48-51, from instr to assoc prof, 52-70; DIR SURG, DARTMOUTH-HITCHCOCK AFFIL HOSPS, 70-; PROF SURG & CHMN DEPT, DARTMOUTH MED SCH, 70- *Concurrent Pos:* From asst to assoc attend surgeon, NY Hosp, 54-70; dir second surg div, Bellevue Hosp, NY, 63-67; dir surg, North Shore Hosp, 67-70; consult, NY Vet Admin Hosp, 63-70, USPHS Hosp, Staten Island, 64-70 & Vet Admin Hosp, White River Junction, Vt, 70- *Mem:* Fel Am Col Surg. *Res:* Academic educational surgery. *Mailing Add:* Dartmouth Med Sch Hanover NH 03755

KARL, ROBERT RAYMOND, JR, b Sewickley, Pa, June 15, 45; c 2. ATMOSPHERIC CHEMISTRY, PHYSICS. *Educ:* Pa State Univ, BS, 67; Cornell Univ, PhD(phys chem), 74. *Prof Exp:* Res assoc surface adsorption, Chem Dept, Pa State Univ, 66-67; res asst molecular struct, Chem Dept, Cornell Univ, 68-73, res asst chem laser, 73-74; postdoctoral, spectros, Isotope Sepn, State Univ NY, Binghamton, 74-76; STAFF SCIENTIST, LOS ALAMOS NAT LAB, 76- *Mem:* Am Inst Physics; Am Chem Soc; Am Phys Soc. *Res:* Spectroscopy, photochemistry; remote atmospheric sensing; remote lidar sensing; fluorescence spectroscopy; remote beamdiagnostics; remote exoarmospheric diagnostics of weapons tests and ionospheric plasmas. *Mailing Add:* Los Alamos Nat Lab Los Alamos NM 87545

KARL, SUSAN MARGARET, b Pittsburg, Pa, Oct 7, 51; m 85; c 1. REGIONAL GEOLOGY OF ALASKA, MARINE GEOLOGY & GEOCHEMISTRY. *Educ:* Middlebury Col, BA 73; Stanford Univ, PhD(geol), 82. *Prof Exp:* GEOLOGIST, US GEOL SURV, 77- *Concurrent Pos:* Mem, proj 187, Int Geol Correlation Prog, ODP Leg 129. *Honors & Awards:* Harold Stearns Award, Geol Soc Am, 78. *Mem:* Geol Asn Can; Am Geophys Union; Geol Soc Am. *Res:* Geochemistry, sedimentology and environmental interpretation of siliceous rocks; sedimentologic analysis of turbidite deposits; paleoenvironmental analysis of sedimentary basins; sedimentary and tectonic processes in accretionary complexes; textural and mineralogic evolution of migmatites. *Mailing Add:* US Geol Surv 4200 University Dr Anchorage AK 99508

KARL, THOMAS RICHARD, b Evergreen Park, Ill, Nov 22, 51; m 73; c 2. CLIMATOLOGICAL TIME SERIES, SECULAR CLIMATE CHANGE. *Educ:* Northern Ill Univ, BS, 73; Univ Wis-Madison, MS, 74. *Prof Exp:* Meteorologist, air qual res, Environ Sci Res lab, 75-79, weather forecasting & anal, Nat Weather Serv, 79-80; METEOROLOGIST CLIMATE RES & APPLN, NAT CLIMATE DATA CTR, 80- *Concurrent Pos:* Rapporteur, Climat Time Series, World Meteorol Orgn, 83- *Honors & Awards:* Bronze Medal, Dept Com, 87, Gold Medal, 91. *Mem:* Am Metrol Soc; Am Geophys Union. *Res:* The analysis and reconstruction of the 20th century climate record for identifying climate change for basic climate research; design and management strategies of various environmentally sensitive systems. *Mailing Add:* Global Climate Lab Fed Bldg Nat Climatic Data Ctr Asheville NC 28801

KARLANDER, EDWARD P, b Manchester, Vt, Nov 30, 31; wid; c 6. BOTANY. *Educ:* Univ Vt, BS, 60; Univ Md, MS, 62, PhD, 64. *Prof Exp:* From res asst to res assoc, 60-65, asst prof, 66-69, ASSOC PROF ALGAL PHYSIOL, UNIV MD, COLLEGE PARK, 69-, ASST CHMN DEPT, 82- *Concurrent Pos:* Prog officer NSF, 79-80; actg dir, Md Water Resources Res Ctr, 80-81. *Mem:* Phycol Soc Am; Am Inst Biol Sci. *Res:* Ecological biophysics; algal physiology; responses of organisms to light; cell growth. *Mailing Add:* 107 Lakeside Dr Greenbelt MD 20770

KARLE, HARRY P, b Sanger, Calif, Jan 4, 27; m 56; c 4. PLANT PATHOLOGY, VITICULTURE. *Educ:* Fresno State Col, BS, 50; Univ Calif, Davis, MS, 59, PhD(plant path), 65. *Prof Exp:* Instr, High Sch, Calif, 50-51; foreman viticulture, Fresno State Col, 51-53; lab helper plant path, Univ Calif, Davis, 54-55, lab asst, 55-58, res asst, 58-59, lab technician, 59-62; from asst prof to assoc prof, 62-69, prof plant path & chmn dept plant sci, 69-86, ASSOC DEAN AGR OPERS, CALIF STATE UNIV, FRESNO, 86- *Concurrent Pos:* Consult res & study comt, Calif Raisin Adv Bd, 65-84. *Mem:* AAAS; Am Phytopath Soc; Am Soc Hort Sci; Am Inst Biol Sci; Am Soc Agron; Sigma Xi. *Res:* Grape diseases; non-cultivation studies. *Mailing Add:* Sch Agr Sci & Technol Calif State Univ Sch Agr Fresno CA 93740-0085

KARLE, ISABELLA LUGOSKI, b Detroit, Mich, Dec 2, 21; m 42; c 3. MOLECULAR BIOLOGY. *Educ:* Univ Mich, BS, 41, MS, 42, PhD(phys chem), 44. *Hon Degrees:* DSc, Univ Mich, 76, Wayne State Univ, 79 & Univ Md, 86; LHD, Georgetown Univ, 84. *Prof Exp:* Assoc chemist, Univ Chicago, 44; instr, Univ Mich, 44-46; physicist, 46-59, HEAD, X-RAY ANALYSIS SECT, US NAVAL RES LAB, 59- *Concurrent Pos:* Mem, Nat Comt Crystallog, Nat Acad Sci-Nat Res Coun, 74-77; mem, Exec Comt, Am Peptide Symposium, 76-81; mem bd, Int Orgn & Progs, Nat Acad Sci, 80-83; mem adv bd, Off Chem & Chem Tech, Nat Res Coun, 78-81, Corp Vis Comt, Mass Inst Technol, 82-90. *Honors & Awards:* Sci Res Soc Am Award, 67; Hillebrand Award, Am Chem Soc, 69, Garvan Award, 76; Fed Woman's Award, US Govt, 73; Dexter Conrad Award, Off Naval Res, 80; Pioneer Award, Am Inst Chemists, 84; Gregori Aminoff Prize, Royal Swedish Acad Sci, 88; Rear Admiral William S Parsons Award, Navy League of US, 88. *Mem:* Nat Acad Sci; Am Phys Soc; Biophys Soc; Am Crystallog Asn (vpres, 75, pres, 76); Am Chem Soc. *Res:* Application of electron and x-ray diffraction to structure problems; phase determination in crystallography; elucidation of molecular formulae; peptides; configurations and conformations of natural products and biologically active materials. *Mailing Add:* Lab Struct Matter Code 6030 US Naval Res Lab Washington DC 20375-5000

KARLE, JEAN MARIANNE, b Washington, DC, Nov 14, 50. X-RAY CRYSTALLOGRAPHY. *Educ:* Univ Mich, BS, 71; Duke Univ, PhD(chem), 76. *Prof Exp:* Pub health serv fel, Nat Inst Arthritis, Diabetes, Digestive & Kidney Dis, 76-78, staff fel, Nat Cancer Inst, NIH, 78-83; AT DEPT PHARMACOL, WALTER REED ARMY INST RES, WASHINGTON, 83- *Mem:* Am Chem Soc; Am Asn Cancer Res; Int Soc Study Xenobiotics; Am Crystallographic Asn; Am Soc Trop Med Hyg. *Res:* Three-dimensional structure of biologically active small molecules; regulation of the glucocorticoid receptor; chiral chromatographic methods development; drug development of antimalarials. *Mailing Add:* Dept Pharmacol Div Exp Therapeut Walter Reed Army Inst Res Washington DC 20307

KARLE, JEROME, b New York, NY, June 18, 18; m 42; c 3. CRYSTALLOGRAPHY. *Educ:* City Col New York, BS, 37; Harvard Univ, AM, 38; Univ Mich, MS, 42, PhD(phys chem), 44. *Hon Degrees:* LHD, Georgetown Univ, 84; DHC, Univ Md, City Univ NY, 86, Univ Mich, 89. *Prof Exp:* Lab asst, State Dept Health, NY, 39-40; res assoc, Manhattan Proj, Chicago, 43-44 & US Navy Proj, Mich, 44-46; head electron diffraction sect, 46-58, head diffraction br, 58-67, CHIEF SCIENTIST, LAB FOR STRUCT OF MATTER, US NAVAL RES LAB, 67- *Concurrent Pos:* Mem, Nat Res Coun, 54-56 & 67-87; prof, Univ Md, 51-70; chmn, USA Nat Comt Crystallog, Nat Acad Sci-Nat Res Coun, 73-75; mem exec comt, Int Union Crystallog, 78-87, pres, 81-84; chmn chem sect, Nat Acad Sci, 89-91. *Honors & Awards:* Nobel Prize in Chem,85; Sigma XI, Award, 59; Chair of Sci Award, 68; Hillebrand Award, Am Chem Soc, 69; Robert Dexter Conrad Award, 86; Patterson Award, Am Crystallog Asn, 86; Albert A Michelson Award, 86; Rear Admiral William S Persons Award, 86; Townsend Harris Award, 86; Nat Libr Med Medal, 86. *Mem:* Nat Acad Sci; Am Chem Soc; fel Am Phys Soc; Am Crystallog Asn (treas, 50-52, vpres, 71, pres, 72); Am Math Soc; British Crystallog Asn. *Res:* Structure of atoms, molecules, glasses, crystals and solid surfaces. *Mailing Add:* Lab Struct Matter Code 6030 US Naval Res Lab Washington DC 20375-5000

KARLEKAR, BHALCHANDRA VASUDEO, b Baroda, India, Jan 19, 39; m 64; c 2. MECHANICAL ENGINEERING. *Educ:* Univ Baroda, BE, 58; Univ Ill, Urbana, MS, 59, PhD(mech eng), 62. *Prof Exp:* Lectr mech eng, Indian Inst Technol, 62-63; consult, Ibcon Pvt Ltd, Bombay, 63-66; PROF MECH ENG, ROCHESTER INST TECHNOL, 66- *Concurrent Pos:* Consult, Eastman Kodak Co, 66-70, A Burgart Inc, 70-71, Xerox, 74-77 & Chapin Co, 78; actg chmn, Chapin Co, 76-77; chmn energy task force, Rochester Inst Technol, prof & head mech engr dept. *Mem:* Am Soc Mech Engrs; Am Soc Eng Educ; Sigma Xi. *Res:* Heat transfer; energy conservation. *Mailing Add:* 30 Kitty Hawk Dr Pittsford NY 14534

KARLEN, DOUGLAS LAWRENCE, b Monroe, Wis, Aug 28, 51; m 73; c 3. SOIL MANAGEMENT, CROP MANAGEMENT. *Educ:* Univ Wis-Madison, BS, 73; Mich State Univ, MS, 75; Kans State Univ, PhD(agron), 78. *Prof Exp:* Res soil scientist, Coastal Plains Soil & Water Conserv Res Ctr, 78-87, Nat Soil Tilth Lab, USDA Agr Res Serv, 87- *Concurrent Pos:* Adj prof, Agron Dept, Clemson Univ, 83-; collab & prof, Agron Dept, Iowa State Univ, 87-; assoc ed, Crop Sci Soc Am, 87- *Honors & Awards:* Scarseth Mem Award, Scarseth Mem Found, 77. *Mem:* Am Soc Agron; Soil Sci Soc Am; Crop Sci Soc Am; Coun Agr Sci & Technol; Soil & Water Conserv Soc Am; Coun Soil Testing & Plant Anal. *Res:* Evaluation of the interactions among soil, crop, water and nutrient management practices as they affect nutrient and pesticide losses from the soil and assessing the effects of conservation tillage and other management practices on soil tilth. *Mailing Add:* Nat Soil Tilth Lab USDA-ARS 2150 Pammel Dr Ames IA 50011

KARLER, RALPH, b Mishawaka, Ind, Nov 11, 28; m 53. PHARMACOLOGY. *Educ:* Univ Chicago, AB, 47; Ind Univ, BA, 50; Univ Calif, MS, 53, PhD(physiol), 59. *Prof Exp:* Res instr, 59-63, from asst prof to assoc prof, 63-76, PROF PHARMACOL, COL MED, UNIV UTAH, 76- *Concurrent Pos:* USPHS spec res fel, 61-62, USPHS res career develop award, 62-72. *Mem:* Am Soc Pharmacol & Exp Therapeut; assoc Am Physiol Soc; Int Soc Biochem Pharmacol. *Res:* Pharmacology of drugs affecting the nervous system and muscle; role of calcium in contraction; drug metabolism. *Mailing Add:* Univ Utah Sch Med Rm 2C219 Salt Lake City UT 84132

KARLIN, ALVAN A, b Newark, NJ, May 3, 50; m; c 2. EVOLUTION, SYSTEMATICS. *Educ:* Rutgers Univ, AB, 72; Ind State Univ, MA, 75; Miami Univ, PhD(zool), 78. *Prof Exp:* STAFF BIOLOGIST GENETICS, TALL TIMBERS RES STA, 78-; AT DEPT BIOL, UNIV ARK, LITTLE ROCK. *Concurrent Pos:* Adj asst prof, Fla State Univ, 79- *Mem:* Soc Study Evolution; Soc Syst Zoologists; Am Soc Ichthyologists & Herpetologists; Soc Study Amphibians & Reptiles; Sigma Xi. *Res:* Evolutionary biology, population genetics and ecological genetics; vertebrate biology and sociobiology. *Mailing Add:* Dept Biol Univ Ark Little Rock AR 72204

KARLIN, ARTHUR, b Philadelphia, Pa, Jan 14, 36; m 58, 77; c 6. NEUROBIOLOGY, RECEPTORS. *Educ:* Swarthmore Col, BA, 57; Rockefeller Univ, PhD(biol), 62. *Prof Exp:* From res asst to res assoc neurol, 62-64, from asst prof to assoc prof physiol, 65-74, assoc prof neurochem, 74-78, PROF BIOCHEM & NEUROL, COLUMBIA UNIV, 78-, DIR CTR MOLECULAR RECOGNITION & HIGGINS PROF BIOCHEM & MOLECULAR BIOPHYS, 89- *Concurrent Pos:* New York City Health Res Coun career scientist award, Columbia Univ, 70-72; mem bd rev, Fedn Proc, 74, JBC, 79-84, 87-92, Proteins, 86-; chmn, Gordon conf molecular pharmacol, 75; Grass travelling scientist, 79; Quastel vis prof, McGill Univ, 84; Krantz lectr pharmacol & exp therapeut, Univ Maryland, 85; dir, MBL neurobiol course, 85-89. *Honors & Awards:* Lucy Moses Prize Basic Neurol, 75; Louis & Bert Freedman Found Award Res Biochem, 85. *Mem:* Am Soc Biol Chem; Am Soc Pharmacol & Exp Therapeut; Soc Neurosci; Harvey Soc; fel AAAS; NY Acad Sci; Soc Gen Physiologists. *Res:* Molecular mechanisms of acetylcholine receptors and dopamine receptors. *Mailing Add:* Ctr Molecular Recognition Col Phys & Surg Columbia Univ 630 W 168th St New York NY 10032

KARLIN, KENNETH DANIEL, b Pasadena, Calif, Oct 30, 48; c 2. INORGANIC CHEMISTRY. *Educ:* Stanford Univ, BS, 70; Columbia Univ, PhD(inorg chem), 75. *Prof Exp:* Res assoc & NATO fel organometallic chem, Cambridge Univ, Eng, 75-77; prof inorg chem, State Univ NY Albany, 77-90; PROF INORG CHEM, JOHNS HOPKINS UNIV, BALTIMORE, 90- *Concurrent Pos:* Hon US Ramsey fel, 76-77. *Honors & Awards:* Buck-Whitney Award, 91. *Mem:* Am Chem Soc; The Chem Soc. *Res:* Bioinorganic chemistry; chemistry of copper I; binuclear copper complexes; activation of molecular oxygen; models for copper metalloproteins; organometallic chemistry. *Mailing Add:* Dept Chem Johns Hopkins Univ Baltimore MD 21218

KARLIN, SAMUEL, b Yonava, Poland, June 8, 24; nat US; m 47; c 3. MATHEMATICAL STATISTICS, STATISTICS. *Educ:* Ill Inst Technol, BS, 44; Princeton Univ, PhD(math), 47. *Hon Degrees:* DSc, Technion-Israel Inst Technol, Haifa, Israel, 85. *Prof Exp:* Asst prof math, Calif Inst Technol, 49-50 & 51-54, assoc prof, 54-56; vis asst prof, Princeton Univ, 50-51; dean, fac math & chmn, Dept Math, Weizmann Inst Sci, Rehovot, Israel, 70-76; prof math & statist, 56-74, prof math, 74-78, ROBERT GRIMMITT PROF MATH, STANFORD UNIV, 78- *Concurrent Pos:* Consult, Rand Corp, Calif, 48-; Andrew D White Prof-at-large, Cornell Univ, 75-81. *Honors & Awards:* Wilkes Lectr, Princeton Univ, 77, Seymour Sherman Mem Lectr, 78, Gibbs Lectr, Am Math Soc 1st Mahalanobis Mem Lectr, Indian Statist Inst & 11th Fisher Mem Lectr, London, 83, Britton Lectr, McMasters Univ, Ont, Can; John Von Neumann Theory Prize, Opers Res Soc Am, 87; Nat Medal Sci, 89. *Mem:* Nat Acad Sci; Am Math Soc; Inst Math Statist (pres-elect, 77, pres, 78-79); Am Statist Asn; Am Soc Human Genetics; Genetic Soc Am; Am Naturalist Soc. *Res:* Problems in mathematics, statistics, genetics and biology. *Mailing Add:* Dept Math Stanford Univ Stanford CA 94305

KARLINER, JERROLD, b Stanislawow, Poland, Mar 5, 40; US citizen; m 63; c 2. STRUCTURE ELUCIDATION, APPLIED SPECTROSCOPY. *Educ:* City Col New York, BS, 62; Stanford Univ, PhD(org mass spectrometry), 66. *Prof Exp:* Res assoc mass spectrometry, Lederle Labs Div, Am Cyanamid Co, 66-68; group leader spectros, 68-78, DEPT HEAD, ANALYTICAL RES DEPT, CIBA-GEIGY CORP, 78- *Mem:* Am Chem Soc; Am Soc Mass Spectrometry. *Res:* Structure elucidation of organic compounds by physical methods; analysis of organic compounds by mass spectrometry, nuclear magnetic resonance, optical rotatory dispersion and circular dichroism; analytical methods development. *Mailing Add:* Ciba Geigy Corp 444 Sawmill River Rd Ardsley NY 10502-2690

KARLL, ROBERT E, b Davenport, Iowa, Apr 29, 24; m 46; c 3. ORGANIC CHEMISTRY. *Educ:* St Ambrose Col, BS, 45; Univ Iowa, MS, 47, PhD(org chem), 49. *Prof Exp:* Res chemist, Standard Oil Co (Ind), 49-54, group leader, 54-65, SECT LEADER, AMOCO CHEM CORP, 65- *Mem:* Am Chem Soc. *Res:* Surfactants; motor oil additives; tertiary oil chemicals. *Mailing Add:* 1171 Lexington Lane Batavia IL 60510-3358

KARLOF, JOHN KNOX, b Rochester, NY, Nov 9, 46; m 69; c 2. MATHEMATICS. *Educ:* State Univ NY Col Oswego, BA, 68; Univ Colo, MA, 70, PhD(math), 73. *Prof Exp:* Asst prof math, Univ Nebr, Omaha, 74-77, assoc prof, 77-80; assoc prof math & comput sci, State Univ NY, Stony Brook, 80-87; PROF MATH SCI, UNIV NC, 87- *Mem:* Am Math Soc; Math Asn Am. *Res:* Gaussian channel coding theory; algebraic coding theory; group theory. *Mailing Add:* Math Sci Univ NC 601 S College Rd Wilmington NC 28403

KARLOVITZ, BELA, b Papa, Hungary, Nov 9, 04; nat US; m 29; c 3. MECHANICAL & ELECTRICAL ENGINEERING. *Educ:* Budapest Tech Univ, ME, 26; Swiss Fed Inst Technol, EE, 28. *Prof Exp:* Sect engr, Elec Power Co, Hungary, 29-38; res engr, Westinghouse Elec Corp, 38-47; sect chief, US Bur Mines, 47-53; PARTNER, COMBUSTION & EXPLOSIVES RES, INC, 53- *Honors & Awards:* Gold Medal, Combustion Inst, 70; Int MHD Faraday Mem Medal, 86. *Mem:* Am Phys Soc; Combustion Inst. *Res:* Magnetohydrodynamic power generation; combustion; turbulent flames; propulsion systems; electrically augmented flames; high power dispersed electrical discharge; plasma phenomena. *Mailing Add:* 1510 Scenery Ridge Dr Pittsburgh PA 15241

KARLOW, EDWIN ANTHONY, b Glendale, Calif, May 13, 42; m 64; c 2. PHYSICS. *Educ:* Walla Walla Col, BS, 66; Wash State Univ, MS, 68, PhD(physics), 71. *Prof Exp:* Chmn dept math & physics, Columbia Union Col, 72-78; chmn dept physics, Loma Linda Univ, 78-90; CONSULT, 90- *Mem:* Am Asn Physics Teachers; Am Phys Soc; Optical Soc Am; Am Sci Affil. *Res:* Analog and digital processing of signals. *Mailing Add:* Dept Physics Loma Linda Univ Riverside CA 92515-8247

KARLSON, ALFRED GUSTAV, medical microbiology, pathology; deceased, see previous edition for last biography

KARLSON, ESKIL LEANNART, b Johnkeping, Sweden; Jan 5, 20; US citizen; m 42; c 3. BIOPHYSICS, ZOOLOGY. *Educ:* Univ Pittsburgh, BS, 46, MS, 48; Occidental Univ, St Louis, DSc(physics, zool), 70. *Prof Exp:* Lab leader develop radiation instrumentation, Savana River Plant, AEC, 50-55; group leader, Reactor Inst, Greenwich Plant, AMF Inc, 55-57; chief appl physics atomic bomb tests, Las Vegas Labs, EG&G, 57-61; pres gas analyzers, Precision Res, 61-67; vpres res ozone systs, Pollution Control Industs, 67-71; PRES ION EXCHANGE, LIFE SUPPORT INC, 71- *Concurrent Pos:* Consult to reactor control, 74-77; res & develop adv ozone, Iconex, Inc, Stamford, Conn, 75-78; consult ion exchange, Facet Enterprises, Tulsa, Okla, 77-78. *Mem:* Optical Soc Am; Inst Soc Am; Health Physics Soc; Am Nuclear Soc. *Res:* Developed first digital pressure transducer, first eight gas analyzer, first automatic inbedable heat pump; developed first continuous separation system for oil, blood or water employing the chromatographic phenomena; developed the first sterilizer employing ozone as the sterilizing agent. *Mailing Add:* Life Support Inc 2926 State St Erie PA 16508

KARLSON, KARL EUGENE, b Worcester, Mass, July 30, 20; m 47; c 6. SURGERY. *Educ:* Univ Minn, BS, 42, MB, 44, MD, 45; Am Bd Surg, dipl, 53; Am Bd Thoracic Surg, dipl, 56. *Prof Exp:* Intern surg, Univ Hosps, Univ Minn, 44-45, resident, 47-51; from instr to prof, Col Med, State Univ NY Downstate Med Ctr, 51-71; prof, 71-90, EMER PROF MED SCI & SURG, BROWN UNIV, 90-; CONSULT, 85- *Concurrent Pos:* From asst vis surgeon to vis surgeon, Kings County Hosp, Brooklyn, 51-71; mem cent adv comt, Coun Cardiovasc Surg, Am Heart Asn, 62-; surgeon & surgeon-in-charge thoracic & cardiovasc surg, RI Hosp, Providence, 71-85. *Mem:* Am Asn Thoracic Surg; Int Cardiovasc Soc; Am Surg Asn; Soc Vascular Surg; Soc Univ Surg. *Res:* Heart surgery and cardiovascular physiology. *Mailing Add:* Dept Surg RI Hosp Providence RI 02903

KARLSON, RONALD HENRY, b Coalinga, Calif, Oct 13, 47; m 77; c 2. MARINE ECOLOGY, BENTHIC ECOLOGY. *Educ:* Pomona Col, BA, 69; Duke Univ, MA, 72, PhD(zool), 75. *Prof Exp:* Fel, Johns Hopkins Univ, 76-78; asst prof, 78-84, ASSOC PROF INVERT ECOL, UNIV DEL, 84- *Concurrent Pos:* Vis fac, coral reef ecol, Discovery Bay Marine Lab, Univ WI, Jamaica, 84; vis scientist, Dept Zool, Univ Adelaide, 85, Victorian Inst Marine Sci, 86, Sch Biol Sci, Univ Sydney, 86, Australian Inst Marine Sci, 86, Mountain Lake Biol Sta, Univ Va, 90; ed adv, Marine Ecol Progress Ser, 90- *Mem:* Am Soc Zoologists; AAAS; Am Soc Naturalists; NAm Benthological Soc. *Res:* Ecological and evolutionary questions involving biological interactions; physical disturbance, and life history strategies of sessile colonial invertebrates. *Mailing Add:* Ecol Prog Sch Life Health & Sci Univ Del Newark DE 19716

KARLSSON, STURE KARL FREDRIK, b Sodra Vi, Sweden, Oct 11, 25; US citizen; m 49; c 2. FLUID MECHANICS. *Educ:* Johns Hopkins Univ, PhD(aeronaut), 58. *Prof Exp:* Fel aeronaut, Johns Hopkins Univ, 58-59; NATO fel, Royal Inst Technol Sweden, 59-60; from asst prof to assoc prof eng, 60-71, PROF ENG, BROWN UNIV, 71- *Mem:* Am Phys Soc. *Res:* Turbulent flows; laminar stability. *Mailing Add:* Div Eng Brown Univ Providence RI 02912

KARLSSON, ULF LENNART, b Uppsala, Sweden, Sept 11, 35; m 60, 80; c 4. ANATOMY, NEUROBIOLOGY. *Educ:* Karolinska Inst, Sweden, MK, 58, Doc, 66, ML, 68, Radiol Oncol Bd, cert, 86. *Hon Degrees:* DrMed, Royal Univ Umea, Sweden, 69. *Prof Exp:* Res zoologist, Univ Calif, Los Angeles, 61-65; teacher anat, Univ Umea, Sweden, 65-69; from assoc prof to prof anat & dent, Col Med & Dent, Univ Iosa, 69-84; ASST PROF RADIOL ONCOL, HAHNEMANN UNIV, PHILADELPHIA, 84- *Mem:* Sigma Xi; Scand Radiation Ther Soc, 82-; Am Soc Therapeut Radiol & Oncol. *Res:* Brain tumor treatment. *Mailing Add:* 2217-6 Green St Philadelphia PA 19130

KARLSTROM, ERNEST LEONARD, b Seattle, Wash, May 18, 28; m 50; c 3. HERPETOLOGY, ECOLOGY. *Educ:* Augustana Col, AB, 49; Univ Wash, Seattle, MS, 52; Univ Calif, Berkeley, PhD(zool), 56. *Prof Exp:* Assoc zool, Univ Calif, Berkeley, 55-56; from asst prof to assoc prof biol, Augustana Col, 56-61; assoc prof, 61-64, PROF BIOL, UNIV PUGET SOUND, 64- *Concurrent Pos:* Arctic Inst NAm res grant, 59-61; NSF basic res grants, 62-64. *Mem:* Am Soc Ichthyol & Herpet; Sigma Xi; Western Soc Naturalists. *Res:* Comparative anatomy of reptiles; ecology and systematics of amphibians; basic marine ecology; radioactive tracer methods; ecological recovery Mount Saint Helens, Washington. *Mailing Add:* Dept Biol Univ Puget Sound Tacoma WA 98416

KARLSTROM, THOR NELS VINCENT, b Seattle, Wash, Mar 10, 20; m 48; c 4. GEOLOGY. *Educ:* Augustana Col, AB, 43; Univ Chicago, PhD(geol), 53. *Prof Exp:* Assoc prof geol, Upsala Col, 46-49, dean men, 48-49; geologist, Alaska Terrain & Permafrost Sect, Mil Br, US Geol Surv, 49-65; geologist, Astrogeol Br, 65-88; RETIRED. *Concurrent Pos:* Adj prof, Dept Geol, Northern Ariz Univ, 69- *Mem:* Fel Geog Soc Am. *Res:* Structural and quaternary geology; paleoclimatology; photogrammetry; astrogeology. *Mailing Add:* 561 Klamath Dr La Conner WA 95257

KARMALI, RASHIDA A, NUTRITION, ENDOCRINE PATHOPHYSIOLOGY. *Educ:* Univ Newcastle-upon-Tyne, Eng, PhD(biochem), 76. *Prof Exp:* Vis ASSOC RES PROF, COOK COL, RUTGERS UNIV, 84- *Concurrent Pos:* Adj assoc prof, Sloan-Kettering Cancer Ctr, 80-90, consult, 90-; student legal specialist, City Law Dept Tort Div, NY, 90- *Mailing Add:* Sloan-Kettering Cancer Ctr 1275 York Ave New York NY 10021

KARMAS, ENDEL, food science; deceased, see previous edition for last biography

KARMAS, GEORGE, b Rochester, NY, Dec 18, 20; m 51. ORGANIC CHEMISTRY. *Educ:* Univ Rochester, BS, 42; NY Univ, MS, 45; Polytech Inst Brooklyn, PhD(org chem), 55. *Prof Exp:* Res chemist, Manhattan Dist Proj, Iowa State Col, 42-44; asst, 44-64, RES FEL, ORTHO PHARMACEUT CORP, RARITAN, 64- *Mem:* Am Chem Soc. *Res:* Synthetic medicinal chemistry, especially antimicrobials; heterocyclic and steroid chemistry. *Mailing Add:* 757 Cedarcrest Dr Bound Brook NJ 08805-1103

KARMAZYN, MORRIS, b Wloclawek, Poland, Apr 5, 50; Can citizen; m 88. HEART RESEARCH, EICOSANOIDS. *Educ:* Loyola Col, BSc, 74; McGill Univ, MSc, 76, PhD(physiol), 79. *Prof Exp:* Fel physiol, Univ Man, 78-81; from asst prof to assoc prof pharmacol, Dalhousie Univ, 81-89; assoc prof, 89-90, PROF PHARMACOL & TOXICOL & CAREER INVESTR, HEART & STROKE FOUND ONT, UNIV WESTERN ONT, 90- *Concurrent Pos:* Vis scientist, Weis Ctr Res, Geisinger Clin, Danville, Pa, 87-88. *Honors & Awards:* Merck Frosst Award, 90. *Mem:* Int Soc Heart Res; Am Soc Pharmacol & Exp Ther; Can Pharmacol Soc; Am Heart Asn; AAAS. *Res:* Study of the role of eicosanoids and Na/H exchange in cardiac injury associated with ischemia and reperfusion. *Mailing Add:* Dept Pharmacol & Toxicol Univ Western Ont London ON N6A 5C1 Can

KARMEN, ARTHUR, b New York, NY, Feb 25, 30; m 55; c 3. MEDICINE, CLINICAL PATHOLOGY. *Educ:* NY Univ, AB, 50, MD, 54. *Prof Exp:* Resident & intern med, Bellevue Hosp, NY, 54-56; res investr, Nat Heart Inst, 56-63; assoc prof radiol, radiol sci & med, Johns Hopkins Univ, 63-68; prof path & med, Sch Med & dir clin labs, Univ Hosp, NY Univ & Bellevue Hosp, 68-71; PROF & CHMN DEPT LAB MED, ALBERT EINSTEIN COL MED, 71- *Concurrent Pos:* Dir clin labs, Bronx Munic Hosp Ctr & Hosp Albert Einstein Col Med, 71- *Honors & Awards:* Sloan Award Cancer Res, 57; Van Slyke Award, Am Asn Clin Chemists, 79; Tswett Medal Chromatography, 82. *Res:* Analytical biochemistry, clinical pathology and chemistry, lipid metabolism and clinical enzymology; nuclear medicine; biochemistry. *Mailing Add:* Albert Einstein Col Med Yeshiva Univ Bronx NY 10461

KARMIS, MICHAEL E, b Athens, Greece, June 9, 48; m 72; c 3. MINING ENGINEERING. *Educ:* Univ Strathclyde, BSc, 71, PhD(rock mech), 74. *Prof Exp:* Royal Soc Brit fel rock mech, Dept Mining Eng, Univ Strathclyde, 74-75; asst prof mining eng, Nat Tech Univ Athens, Greece, 75-78; asst prof, 78-81, ASSOC PROF MINING ENG, VA POLYTECH INST & STATE UNIV, 81- *Mem:* Am Inst Mining, Metall & Petrol Engrs; Inst Mining, Metall & Petrol Engrs; Int Soc Rock Mech. *Res:* Stress analysis around mining excavations using theoretical and experimental methods; design of instrumentation for monitoring underground stresses and strains; in-situ investigations; mining subsidence; geotechnical techniques; mine design. *Mailing Add:* Dept Mining & Minerals Eng Va Polytech Inst & State Univ Blacksburg VA 24061

KARN, JAMES FREDERICK, b Columbus, Ohio, Jan 28, 39; m 60; c 2. RANGE RUMINANT NUTRITION, FORAGE EVALUATION. *Educ:* Ohio State Univ, BS, 62, MS, 64; Univ Nebr, PhD(ruminant nutrit), 76. *Prof Exp:* Res technician, North Platte Sta, Univ Nebr, 67-76; RES ANIMAL SCIENTIST BEEF CATTLE NUTRIT, NORTHERN GREAT PLAINS RES LAB, AGR RES SERV, USDA, 76- *Mem:* Am Soc Animal Sci; Soc Range Mgt; Am Regist Prof Animal Scientists. *Res:* Improving the efficiency of producing beef cattle on rangelands; forage evaluation; clarifying the nutrient requirements of range cattle. *Mailing Add:* Northern Great Plains Res Lab PO Box 459 Mandan ND 58554

KARN, ROBERT CAMERON, b Berwyn, Ill, Mar 12, 45; m 66; c 2. MAMMALIAN GENETICS, BIOCHEMICAL GENETICS. *Educ:* Ind Univ, BA, 67, MA, 70, PhD(zool), 72. *Prof Exp:* From instr to assoc prof med genetics, Sch Med, Ind Univ, 74-86, dir Genotyping Labs, Dept Med Genetics, 75-81, grad adv, 81-86; PROF & HEAD BIOL SCI, BUTLER UNIV, 86- *Concurrent Pos:* NIH fel, Sch Med, Ind Univ, 74-75, career develop award, 77-82. *Mem:* Sigma Xi; Genetics Soc Am; Am Inst Biol Sci; Am Soc Biol Chem & Molecular Biol. *Res:* Molecular genetics of salivary and prostate proteins; evolution by gene duplication. *Mailing Add:* Biol Sci Dept Butler Univ 4600 Sunset Ave Indianapolis IN 46208

KARNAKY, KARL JOHN, JR, b Houston, Tex, Sept 2, 43. EPITHELIAL TRANSPORT. *Educ:* Rice Univ, PhD(biol), 72. *Prof Exp:* Asst prof anat & cell biol, Sch Med, Temple Univ, 76-80; asst prof physiol, Sch Med, Univ Tex, Houston, 80-86; ASSOC PROF ANAT & CELL BIOL, MED UNIV SC, 86- *Mem:* Am Soc Biol Chemists. *Mailing Add:* Dept Physiol & Cell Biol Univ Tex PO Box 20036 Houston TX 77225

KARNAUGH, MAURICE, b New York, NY, Oct 4, 24; m 70; c 2. HEURISTIC SEARCH, KNOWLEDGE REPRESENTATION. *Educ:* City Col NY, BS, 48; Yale Univ, MS, 50, PhD(physics), 52. *Prof Exp:* Res staff, Bell Tel Labs, 52-66; res & develop mgr, Fed Systs Div, 66-70, RES STAFF, IBM, YORKTOWN HEIGHTS, NY, 70- *Concurrent Pos:* Distinguished adj prof comput sci, Polytech Inst NY, 81- *Mem:* Fel Inst Elec & Electronics Engrs; Am Asn Artificial Intel; Sigma Xi. *Res:* Techniques for implementing knowledge based systems in computers; knowledge representations and search methods. *Mailing Add:* IBM Watson Res Ctr PO Box 218 Yorktown Heights NY 10598

KARNER, FRANK RICHARD, b Elmhurst, Ill, Aug 14, 34; m 58; c 5. GEOLOGY. *Educ:* Wheaton Col, BS, 57; Univ Ill, PhD(geol), 63. *Prof Exp:* From asst prof to assoc prof, 62-69, PROF GEOL, UNIV NDAK, 69- *Mem:* AAAS; Geol Soc Am; Sigma Xi. *Res:* Mineralogy and petrology of igneous, sedimentary and metamorphic rocks. *Mailing Add:* Geol Dept Univ NDak Box 8068 Grand Forks ND 58201

KARNEY, CHARLES FIELDING FINCH, b Eng, Nov 7, 51; m. RADIO-FREQUENCY HEATING. *Educ:* Cambridge Univ, Eng, BA, 72; Mass Inst Technol, SM, 74, PhD(elec eng & comp sci), 77. *Prof Exp:* Res assoc, Dept Elec Eng & Comp Sci, Mass Inst Technol, 77; res assoc, 77-79, res staff, 79-88, PRIN RES PHYSICIST, PLASMA PHYSICS LAB & LECTR/PROF, DEPT ASTROPHYS, PRINCETON UNIV, 88- *Mem:* Am Phys Soc. *Res:* Plasma physics, especially radio-frequency heating; intrinsic stochasticity with application to plasma physics. *Mailing Add:* Plasma Physics Lab Princeton Univ PO Box 451 Princeton NJ 08543-0451

KARNI, SHLOMO, b June 23, 32; US citizen; m 61; c 2. ELECTRICAL ENGINEERING. *Educ:* Israel Inst Technol, BS, 56; Yale Univ, MEng, 57; Univ Ill, PhD(elec eng), 60. *Prof Exp:* Testing engr, Palestine Power Co, 55-56; asst elec eng, Yale Univ, 56-57; from instr to asst prof, Univ Ill, 57-61; from asst prof to assoc prof, 61-69, dir grad studies, 71-87, PROF ELEC ENG, UNIV NMEX, 69-, DIR UNDERGRAD STUDIES, 87- *Concurrent Pos:* Mem circuits group, Univ Ill, 60-61; consult, Los Alamos Nat Lab, Dept Energy, Westinghouse, var publ houses & Kirtland AFB; vis prof, Univ Hawaii, 68-69, Tel Aviv Univ, 70-71 & Technion, 77- *Mem:* AAAS; fel Inst Elec & Electronics Engrs; Am Soc Eng Educ. *Res:* Network theory; system theory; power and energy modelling; engineering education. *Mailing Add:* Dept Elec Eng & Comput Eng Univ NMex Albuquerque NM 87131

KARNOPP, BRUCE HARVEY, b Milwaukee, Wis, June 13, 38; m 63; c 3. ENGINEERING MECHANICS, APPLIED MATHEMATICS. *Educ:* Mass Inst Technol, SB, 60; Brown Univ, ScM, 63; Univ Wis, PhD(eng mech), 65. *Prof Exp:* Engr, AC Spark Plug, Gen Motors Corp, Wis, 60-61; engr, Sanders Assocs, NH, 61; instr eng mech, Univ Wis, 62-65; asst prof mech eng, Univ Toronto, 65-68; asst prof eng mech, 68-77, ASSOC PROF ENG MECH, UNIV MICH, ANN ARBOR, 77- *Mem:* Acoust Soc Am; Tensor Soc. *Res:* Variational methods in mechanics, vibrations and dynamics. *Mailing Add:* Dept Mech Eng Univ Mich Main Campus Ann Arbor MI 48109-2125

KARNOPP, DEAN CHARLES, b Milwaukee, Wis, June 12, 34; m 58; c 2. MECHANICAL ENGINEERING. *Educ:* Mass Inst Technol, BS & MS, 57, PhD(mech eng), 61. *Prof Exp:* Asst appl mech, Mass Inst Technol, 57-59, instr, 59-61, asst prof & Ford fel, 61-63; develop engr, Siemens Schuckert Res Ctr, Ger, 63-64; from asst prof to assoc prof syst dynamics & control, Mass Inst Technol, 64-69, prof syst dynamics & control, 69-80, PROF MECH ENG, UNIV CALIF, DAVIS, 80- *Concurrent Pos:* Vis prof, Univ Stuttgart, Ger, 75-76. *Honors & Awards:* Levy Medal, Franklin Inst, 69; Sr US Scientist Award, Humbolt Found, 75. *Mem:* Am Soc Mech Engrs. *Res:* Dynamic systems; random vibrations; search and optimization theory; control; computation; bond graph modeling of engineering systems. *Mailing Add:* 1521 41st St Sacramento CA 95819

KARNOSKY, DAVID FRANK, b Rhinelander, Wis, Oct 12, 49; m 70; c 2. FOREST GENETICS. *Educ:* Univ Wis-Madison, BS, 71, MS, 72, PhD(forest genetics), 75. *Prof Exp:* Forest geneticist, Cary Arboretum, NY Bot Garden, 75-83; AT DEPT FORESTRY, MICH TECH UNIV, HOUGHTON, 83- *Mem:* Int Soc Arboriculture; Tissue Culture Asn; Soc Am Foresters; Int Tissue Cult Asn; Sigma Xi; Int Plant Propagators Asn. *Res:* Variation in air pollution tolerance of trees; cytogenetic and tissue culture studies of elms; developing urban hardy trees; interspecific hybridization of Ulmus and Larix species. *Mailing Add:* Dept Forestry Mich Tech Univ Houghton MI 49931

KARNOVSKY, MANFRED L, b Johannesburg, SAfrica, Dec 14, 18; nat US; m 52; c 1. BIOCHEMISTRY. *Educ:* Univ Witwatersrand, BSc, 51, hons, 42, MSc, 43; Univ Capetown, PhD(org chem), 47. *Prof Exp:* Jr lectr chem, Univ Witwatersrand, 41-42; chief chemist & inspector, Brit Ministry Aircraft Prod, SAfrica, 42-43; asst, Univ Capetown, 44-47; from res assoc to assoc, 50-51, from asst prof to prof, 52-65, chmn dept, 69-73, Harold T White prof, 65-89, EMER HAROLD T WHITE PROF BIOL CHEM, HARVARD MED SCH, 89- *Concurrent Pos:* Res fel, Univ Wis, 47-48; Lederle med fac award, 55-58. *Honors & Awards:* Glycerine Producers Asn Second Award, 53; Gold Medal, Reticuloendothelial Soc, 66. *Mem:* Am Soc Biol Chemists; fel Am Acad Arts & Sci; Histochem Soc (secy, 82-85); Am Chem Soc; Am Soc Cell Biol. *Res:* Biochemistry of phagocytosis, pinocytosis and other transport phenomena; biochemistry of sleep. *Mailing Add:* Dept Biol Chem Harvard Med Sch 25 Shattuck St Boston MA 02115

KARNOVSKY, MORRIS JOHN, b Johannesburg, SAfrica, June 28, 26; nat US; m 51; c 2. PATHOLOGY. *Educ:* Univ Witwatersrand, BSc, 46, MB, BCh, 50, DSc, 84; Univ London, dipl clin path, 54. *Hon Degrees:* MA, Harvard Univ, 65. *Prof Exp:* House officer med & surg, Johannesburg Gen Hosp, 51; asst resident path, Beth Israel Hosp, 55-56; assoc, 61-63, from asst prof to assoc prof, 63-68, PROF PATH, HARVARD MED SCH, 68- *Concurrent Pos:* Res fel, Harvard Med Sch, 56-60; Lederle med fac award, 63-66; assoc, Peter Bent Brigham Hosp, 58-60; sci collabr, Sch Med, Univ Geneva, 61-63; mem study group path, USPHS, 65-69. *Honors & Awards:* Distinguished Scientist Award, Electron Micros Soc Am, 88; E B Wilson Award, Am Soc Cell Biol, 90. *Mem:* Nat Acad Sci-Inst Med; fel Am Acad Arts & Sci; Am Soc Exp Path; Histochem Soc; Int Acad Path; Am Soc Cell Biol. *Res:* Histochemistry; electron microscopy; ultrastructural cytochemistry; cell surface topography and modulation; cell junctions; metabolism and structure of kidney; structure and function of capillaries; growth regulation in blood vessels. *Mailing Add:* Dept Path Harvard Med Sch 25 Shattuck St Boston MA 02115

KARNS, CHARLES W(ESLEY), b Waynesboro, Pa, July 15, 20; m 46; c 3. OPERATIONS ANALYSIS. *Educ:* Dickinson Col, BA, 41; Northwestern Univ, MA, 48. *Prof Exp:* Asst math, Northwestern Univ, 46-51; mem staff, Opers Eval Group, Div Sponsored Res, Mass Inst Technol, 51-53, 54-62, Opers Res Group, 53-54; mem staff opers eval group, Ctr Naval Anal, Franklin Inst, 62-63, naval warfare anal group, 63-64, opers eval group, 64-67; mem staff, Opers Eval Group, Ctr Naval Anal, Univ Rochester, 67-71; staff asst off dep dir test & eval, Off Dir Defense Res & Eng, Off Secy Defense, 71-78; staff specialist prog & financial matters, Off Dir Test & Eval, Off Undersecy Defense Res & Eng, 78-85; RETIRED. *Mem:* Opers Res Soc Am; Math Asn Am. *Res:* Military operations research. *Mailing Add:* 8629 Redwood Dr Vienna VA 22180

KARO, ARNOLD MITCHELL, b Wayne, Nebr, May 14, 28; m 66; c 2. THEORETICAL CHEMISTRY, SOLID STATE PHYSICS. *Educ:* Stanford Univ, BS, 49; Mass Inst Technol, PhD(chem phys), 53. *Prof Exp:* Mem staff, Lincoln Lab, Mass Inst Technol, 55-58; mem chem staff, 58-70, group leader theoret chem, 70-76, SR RES SCIENTIST, LAWRENCE LIVERMORE LAB, UNIV CALIF, 76- *Concurrent Pos:* Vis res scientist, Europ Ctr Atomic & Molecular Theory, Univ Paris, Orsay, 75; lectr, Univ Calif, Davis/Livermore, 88. *Mem:* fel NY Acad Sci; fel Am Phys Soc; fel Am Inst Chem; Sigma Xi. *Res:* Atomic and molecular physics; theoretical solid state physics; lattice and molecular dynamics; quantum chemistry; computer characterization of the elementary excitations and the optical and defect properties of solid materials; application of molecular dynamics to shock and detonation phenomena in condensed matter and to plasma-surface interactions. *Mailing Add:* Dept Chem & Mat Sci L-325 Lawrence Livermore Nat Lab Livermore CA 94550

KARO, DOUGLAS PAUL, b Seattle, Wash, Aug 24, 47; m 71; c 1. STATISTICAL MECHANICS, OPTICS. *Educ:* Stanford Univ, BS, 69; Mass Inst Technol, PhD(physics), 73, MS, 80. *Prof Exp:* Physicist, Harry Diamond Lab, US Army, 71; sr staff scientist physics, Avco Everett Res Lab Inc, 73-78; staff mgt consult, Texton Defense Syst, 80-87; TECH CONSULT NAT AFFAIRS, 87-; TECH STAFF, DRAPER LAB, 89- *Mem:* Am Phys Soc; AAAS; Sigma Xi; Int Inst Strategic Studies; Am Inst Aeronaut & Astronaut. *Res:* Defense science and technology; systems analysis; management of research and development. *Mailing Add:* 37 Warren St Medford MA 02155

KARO, WOLF, b Altona-Hamburg, Ger, Apr 2, 24; nat US; m 55; c 1. INDUSTRIAL ORGANIC CHEMISTRY. *Educ:* Cornell Univ, AB, 45, PhD(org chem), 49. *Prof Exp:* Aeronaut res scientist jet fuel, Nat Adv Comt Aeronaut, 49-53; aeronaut res scientist fuel synthesis, Monomer-Polymer, Inc, 53-55; group leader contract res, synthesis & polymerization sects, Borden Chem Co, 55-61, develop mgr, Monomer-Polymer & Dajac Labs, 61-68; sr sci specialist, Scott Paper Co, 68-69; new prod mgr, Sartomer Resins, Inc, 70-71; supvr qual control, Lactona Corp Div, Warner-Lambert Pharmaceut Co, 72-75; res supvr, Haven Chem Co, 75-76; MGR RES & DEVELOP, POLYSCI INC, 76- *Mem:* AAAS; Am Chem Soc; Sigma Xi. *Res:* Reaction kinetics and mechanisms in organic chemistry; organic functional group synthesis; emulsion and anaerobic polymerization; adhesives; coatings; product and process development; materials for radiation-induced polymerization; anionic polymerization; organic polymer chemistry; monodispersed polymer latices for biotechnology; magnetizable latex and application. *Mailing Add:* 328 Rockledge Ave Huntingdon Valley PA 19006

KAROL, FREDERICK J, b Norton, Mass, Feb 28, 33; m 58; c 3. POLYMER CHEMISTRY. *Educ:* Boston Univ, BS, 54; Mass Inst Technol, PhD(org chem), 62. *Prof Exp:* Chemist, chem & plastics group, 56-59 & 62-65, proj scientist, 65-67, res scientist, 67-69, group leader chem & plastics, 69-78, res assoc & group supvr, 78-81, corp fel, 81-84, SR CORP FEL, UNION CARBIDE CORP, 84- *Honors & Awards:* Chem Pioneer Award, Am Inst Chemists, 88; Perkin Medal, Soc Chem Indust, 89; Int Gold Medal, Soc Plastics Engrs, 90; Award for Creative Invention, Am Chem Soc, 91. *Mem:* Am Chem Soc; Am Inst Chemists; Sigma Xi. *Res:* Heterogeneous and polyolefin catalyses; mechanism of polymerization; production of high density polyethylene and low density polyethylene; polypropylene; new polymers. *Mailing Add:* Box 131 Rd 1 Hiland Dr Belle Mead NJ 08502

KAROL, MARK J, b Jersey City, NJ, Feb 28, 59; m 87; c 1. COMMUNICATIONS SYSTEMS RESEARCH, NETWORK SYSTEMS RESEARCH. *Educ:* Case Inst Technol, BS & BSEE, 81; Princeton Univ, MS, 82, MA, 83, PhD(elec eng), 86. *Prof Exp:* MEM TECH STAFF, AT&T BELL LABS, 85- *Concurrent Pos:* Assoc ed, Inst Elec & Electronics Engrs J Lightwave Technol, 91; secy, Tech Comt Computer Commun, Inst Elec & Electronics Engrs, 89-91, vchmn, 91-93. *Mem:* Inst Elec & Electronics Engrs; Math Asn Am. *Mailing Add:* AT&T Bell Labs Rm 4F-529 Crawfords Corner Rd Holmdel NJ 07733-3030

KAROL, MERYL HELENE, b New York, NY; m 63; c 3. IMMUNOCHEMISTRY, TOXICOLOGY. *Educ:* Cornell Univ, BS, 61; Columbia Univ, PhD(microbiol), 67. *Prof Exp:* Fel biochem, State Univ NY, Stony Brook, 67-68; res assoc epidemiol, 74-76, res asst prof toxicol, 76-78, assoc prof, 79-85, PROF IMMUNOTOXICOL, UNIV PITTSBURGH, 85- *Concurrent Pos:* Sci adv govt & indust, NIH Study Sect. *Honors & Awards:* Frank R Blood, 81. *Mem:* Am Chem Soc; Am Thoracic Soc; AAAS; Soc Toxicol; NY Acad Sci; Am Asn Immunol. *Res:* Chemical and industrial allergens; environmental lung disease; occupational disease; diagnostic radioimmunoassays. *Mailing Add:* Dept Environ & Occup Health Univ Pittsburgh Pittsburgh PA 15261

KAROL, PAUL J(ASON), b New York, NY, Mar 18, 41; m 63; c 3. PHYSICAL CHEMISTRY, ANALYTICAL CHEMISTRY. *Educ:* Johns Hopkins Univ, BA, 61; Columbia Univ, MS, 62, PhD(chem), 67. *Prof Exp:* Res assoc nuclear chem, Brookhaven Nat Lab, 67-69; asst prof chem, 69-74, assoc dean sci, 81-86, ASSOC PROF CHEM, CARNEGIE-MELLON UNIV, 74- *Concurrent Pos:* Res collabr, Brookhaven Nat Lab, 69-72; consult, Westinghouse Elec Corp, 72-85; mem Comt Nuclear Radiochem, Nat Res Coun, 81-87, assoc mem, Comn on Radiochem, Int Union Pure & Appl Chem, 85-93. *Mem:* AAAS; Am Chem Soc; Am Phys Soc; fel Am Inst Chemists. *Res:* Mechanisms of high energy nuclear reactions; rapid radiochemical separations; high performance column chromatography. *Mailing Add:* Dept Chem 4400 Fifth Ave Pittsburgh PA 15213

KAROL, ROBIN A, b Bronx, NY, Sept 29, 51. MONOCLONAL ANTIBODY PRODUCTION. *Prof Exp:* Res immunologist, 82-85, SR RES IMMUNOLOGIST & GROUP LEADER, DEPT BIOMED PROD, E I DU PONT DE NEMOURS & CO INC, WILMINGTON, DEL, 85- *Mem:* AAAS; Am Asn Immunologists. *Mailing Add:* Med Prod Dept Virol Res Glasgow Res Lab Box 713 E I du Pont de Nemours & Co Inc Wilmington DE 19898

KAROLY, GABRIEL, b Budapest, Hungary, May 19, 30; US citizen. CHEMICAL ENGINEERING, POLYMER CHEMISTRY. *Educ:* Budapest Tech Univ, MS, 52. *Prof Exp:* Sr engr, Esso Res & Eng Co, 56-66; res chemist, Union Carbide Corp, 66-69; RES SUPVR, M&T CHEM INC, 69- *Mem:* Am Chem Soc. *Res:* Block and graft copolymers; anionic polymerization; telechelic polymers; coordination polymerization; synthetic fibers; coatings. *Mailing Add:* 255 Baltusrol Way Springfield NJ 07081

KARON, JOHN MARSHALL, b Milwaukee, Wis, Nov 6, 41. BIOSTATISTICS. *Educ:* Carleton Col, BA, 63; Stanford Univ, MS, 65, PhD(math), 68. *Prof Exp:* Asst prof math, Syracuse Univ, 68-70; res assoc, Stanford Univ, 70-71; asst prof math, Colo Col, 71-77; fel biostatist, Univ NC, 77-80, res assoc prof, 80-84; STATISTICIAN, CTR DIS CONTROL, ATLANTA, 84- *Concurrent Pos:* Vis lectr, Tel Aviv Univ, 72-73. *Mem:* AAAS; Soc Indust & Appl Math; Am Statist Asn. *Res:* Evaluation of statistical methods; statistical epidemiology. *Mailing Add:* Univ NC 423 Blanton Rd Atlanta GA 30342

KAROW, ARMAND MONFORT, JR, b New Orleans, La, Nov 11, 41; m 64; c 2. PHARMACOLOGY, CRYOBIOLOGY. *Educ:* Duke Univ, BA, 62; Univ Miss, PhD(pharmacol), 68. *Prof Exp:* Instr nursing & pharmacol, Med Ctr, Univ Miss, 65-68; res instr, Dept Surg, 68-71, res asst prof, 71-77, asst prof, Dept Pharmacol, 68-70, assoc prof, 70-75, dir grad studies, 73-80, RES ASSOC PROF, DEPT SURG, MED COL GA, 77-, PROF, DEPT PHARMACOL, 75- *Concurrent Pos:* Ed, Organ Preserv for Transplantation, 74, 81, Biophys of Organ Cryopreservation, 87; officer, Xytex Corp, Augusta, Ga, 75-; res grants, NIH & USPHS; mem, Nat Endowment Humanities, 80; Fogarty sr int fel award, NIH, 81; consult, Cryolife Inc, Marietta, Ga, 85-; pres, Tissue Technol Asn, 89- *Mem:* Fel AAAS; Am Chem Soc; Soc Cryobiol (secy, 77-80); Am Soc Pharmacol & Exp Therapeut. *Res:* Freezing organs and tissues. *Mailing Add:* Dept Pharmacol Med Col Ga Augusta GA 30912

KARP, ABRAHAM E, b New York, NY, Mar 11, 15; m 40; c 2. RESEARCH ADMINISTRATION, MATHEMATICS. *Educ:* City Col New York, BS, 36, MS, 37. *Prof Exp:* Mathematician, Aberdeen Proving Ground, Dept Army, 40-50, chief anal sect math statist, 50-55, dir anal lab, 55-62, chief gaming div, Strategy & Tactics Anal Group, 62-66; tech dir, opers res & systs anal progs, Nat Bur Standards, 66-69; dir, Ctr Criminal Justice Opers & Mgt, Law Enforcement Assistance Admin, 69-70; dir, Tech Anal Div, Nat Bur Standards, 70-71; pvt consult, Systs Anal & Opers Res, 71-80; RETIRED. *Concurrent Pos:* Mem, US Civil Serv Bd Exam, 63-71 & Army Math Steering Comt, 63-66; mem comt govt & bus exec policy & progs, Brookings Inst, 69-71; consult, prof bus orgn on prog develop & tech mgt, 72-85. *Mem:* Opers Res Soc Am. *Res:* Operations research and systems analysis in the areas of transportation, other public systems and military defense systems. *Mailing Add:* 10308 Green Trail Dr N Boynton Beach FL 33436

KARP, ALAN H, b Syracuse, NY, Aug 6, 46; m 70; c 1. ASTROPHYSICS. *Educ:* Rensselaer Polytech Inst, BS, 68; Univ Md, College Park, PhD(astron), 74. *Prof Exp:* Fel astron, IBM Res, Yorktown Heights, NY, 74-76; asst prof physics, Dartmouth Col, 76-77; MEM STAFF PHYSICS, IBM SCI CTR, 77- *Concurrent Pos:* Consult, IBM Res, Yorktown Heights, NY, 76-77. *Mem:* Am Astron Soc; Asn Computer Mach; Astron Soc Pac; Inst Elec & Electronic Engrs Computer Soc. *Res:* Algorithms for parallel processors; radiative transfer in moving stellar atmospheres; radiative transfer in planetary atmospheres containing dust. *Mailing Add:* IBM Sci Ctr 1530 Page Mill Rd Palo Alto CA 94304

KARP, ARTHUR, b New York, NY, Apr 26, 28. HIGH FREQUENCY PHYSICS, MICROWAVE ELECTRONICS. *Educ:* City Col New York, BEE, 48; Mass Inst Technol, SM, 50; Cambridge Univ, PhD(elec eng), 62. *Prof Exp:* Jr engr, A Alford Consult Engrs, Mass, 48; res asst electronics, Mass Inst Technol, 48-50; res asst cent lab, Int Tel & Tel, Paris, France, 50-51, mem tech staff, Bell Tel Labs, Inc, 51-56; engr lab, Cambridge Univ, 56-59; res engr, W W Hansen Labs, Stanford, 60-64; sr res engr, SRI Int, 64-77; SR ENGR, VARIAN ASSOCS INC, 77- *Concurrent Pos:* Consult, Sylvania Elec Prod Inc, Calif, 60-62, Varian Assocs, 62-63 & Goodyear Aerospace Corp, Ariz, 63-64. *Honors & Awards:* B J Thompson Mem Prize, Inst Radio Engrs, 58. *Mem:* AAAS; Inst Elec & Electronics Engrs. *Res:* Electron devices; ultrahigh frequency, microwave and millimeter-wave techniques, components, circuits, electron tubes, bio-effects; color perception and display techniques including color encryption. *Mailing Add:* 1470 Sand Mill Rd Apt 301 Palo Alto CA 94304

KARP, BENNETT C, b Brooklyn, NY, May 15, 54. NUCLEAR PHYSICS. *Educ:* State Univ NY, Binghampton, Ba, 76; Univ Pittsburgh, MS, 78, PhD(physics), 82. *Prof Exp:* postdoc res assoc, Univ NC, 82-84; DISTINGUISHED MEM TECH STAFF, AT&T BELL LABS, 84- *Mem:* Am Phys Soc. *Mailing Add:* 812 Wellington Pl Aberdeen NJ 07747

KARP, HERBERT RUBIN, b Atlanta, Ga, Apr 13, 21; m 48; c 3. NEUROLOGY. *Educ:* Emory Univ, AB, 43; MD, 51; Am Bd Psychiat & Neurol, dipl, 60. *Prof Exp:* Intern & jr asst resident internal med, Grady Mem Hosp, 51-53; fel metab dis, Sch Med, Emory Univ, 53-54; resident neurol, Univ Hosp, Duke Univ, 54-56; clin & res fel, Harvard Med Sch, 56-57, res fel neuropath, 57-58; asst prof med, 58-63, PROF NEUROL, SCH MED, EMORY UNIV, 63- *Concurrent Pos:* Nat Inst Neurol Dis & Blindness spec trainee, 56-58; consult, Vet Admin Hosp, Atlanta, Ga. *Mem:* AAAS; Am Neurol Asn; fel Am Acad Neurol. *Res:* Cerebrovascular disease from the standpoint of further understanding of underlying pathophysiology as well as evaluation of current methods of therapy; age-dependent degenerative diseases of the nervous system. *Mailing Add:* Div Geriat Med Emory Univ Wesley Woods Ctr Atlanta GA 30329

KARP, HOWARD, b Pittsburgh, Pa, Sept 26, 26; m 52; c 4. ANALYTICAL CHEMISTRY. *Educ:* Univ Pittsburgh, BS, 49. *Prof Exp:* assoc res consult chem, US Steel Corp, 49-85; RETIRED. *Concurrent Pos:* Mem, comt chem anal metals, Am Soc Testing & Mat. *Res:* Analytical chemistry as it pertains to steel chemistry. *Mailing Add:* 151 Kelvington Dr Monroeville PA 15146

KARP, LAURENCE EDWARD, b Paterson, NJ, Apr 26, 39; m 62; c 2. OBSTETRICS & GYNECOLOGY, MEDICAL GENETICS. *Educ:* NY Univ, MD, 63. *Prof Exp:* Instr obstet & gynec, Sch Med, Univ Tex, San Antonio, 69-70; sr fel reprod genetics, Sch Med, Univ Wash, 70-72, asst prof obstet & gynec, 72-76; assoc prof, Harbor Gen Hosp, Univ Calif, Los Angeles, 76-77; ASSOC PROF OBSTET & GYNEC, SCH MED, UNIV WASH, 77-; DIR EDUC OBSTET & GYNEC, SWED HOSP MED CTR, SEATTLE, 77- *Mem:* Fel Am Col Obstet & Gynec; AAAS; Am Soc Human Genetics. *Res:* Investigation of chromosomal anomalies in gametes and preimplantation embryos; also, advancement of procedures and techniques for prenatal diagnosis. *Mailing Add:* 2557 Perkins Lane W Seattle WA 98199

KARP, RICHARD DALE, b Minneapolis, Minn, June 19, 43; m 68; c 3. IMMUNOLOGY. *Educ:* Univ Minn, BA, 65, MS, 68, PhD(microbiol), 72. *Prof Exp:* Res assoc microbiol, Univ Minn, 66-72; NIH & C D Rogers fels, Univ Calif, Los Angeles, 73-75; from asst prof to assoc prof, 75-86, PROF BIOL SCI, UNIV CINCINNATI, 86- *Mem:* Am Soc Microbiol; Am Soc Zoologists; Am Asn Immunologists; AAAS; NY Acad Sci; Int Soc Develop Comp Immunologists; Entom Soc Am. *Res:* Evolution of humoral and cell-mediated immunity. *Mailing Add:* Dept Biol Sci Univ Cincinnati Cincinnati OH 45221

KARP, RICHARD M, b Boston, Mass, Jan 3, 35; m 79; c 1. COMPUTER THEORY, ALGORITHMS & COMPUTATIONAL COMPLEXITY. *Educ:* Harvard Univ, AB, 55, SM, 56, PhD(appl math), 59. *Prof Exp:* Mem res staff, Watson Res Ctr, Int Bus Mach Corp, 59-68; adj assoc prof indust & mgt eng, Columbia Univ, 67-68; assoc chmn dept, 73-75, Miller res prof, 80-81, fac res lectr, 81-82, PROF COMPUT SCI & OPER RES, UNIV CALIF, BERKELEY, 68-, PROF MATH, 80-; RES SCIENTIST, INT COMPUTER SCI INST, 88- *Concurrent Pos:* Vis assoc prof elec eng, Univ Mich, 64-65; vis assoc prof, Polytechnic Inst Brooklyn, 65-68, vis prof, 68. *Honors & Awards:* Lanchester Prize, 77; Fulkerson Prize, 79; Turing Award, Asn Comput Mach, 85; von Neumann lectr, Soc Indust & Appl Math, 87; von Neumann Theory Prize, Opers Res Soc Am, 90. *Mem:* Nat Acad Sci; NY Acad Sci; Am Acad Arts & Sci; Am Math Soc; Asn Comput Mach; Comput Prof Social Responsibility; Soc Indust & Appl Math. *Res:* Construction of computational algorithms and the determination of the inherent computational complexity of problems with particular emphasis on combinatorial problems. *Mailing Add:* Comput Sci Dept Univ Calif Berkeley CA 94720

KARP, SAMUEL NOAH, b Brooklyn, NY, Feb 13, 24; m 46; c 2. APPLIED MATHEMATICS. *Educ:* Brown Univ, MSc, 45, PhD, 48. *Prof Exp:* Res assoc compressible fluids & flutter, Brown Univ, 46-48; sr res scientist, Div Electromagnetic Res, 48-55, from res asst prof to res assoc prof math, 55-61, instr, Wash Square Col, 48-55, PROF MATH, COURANT INST MATH SCI, NY UNIV, WASHINGTON SQUARE, 61- *Concurrent Pos:* Indust consult. *Mem:* Am Math Soc. *Res:* Electromagnetic theory; diffraction; boundary value problems; ship resistance and motions; surface waves; far field expansions of radiated fields; multiple impedance; higher order eigen functions of integral equations; inverse scattering. *Mailing Add:* Dept Math NY Univ New York NY 10003

KARP, STEWART, b New York, NY, Mar 17, 32; m 57; c 2. ANALYTICAL CHEMISTRY. *Educ:* Queens Col, NY, BS, 53; Polytech Inst Brooklyn, MS, 60, PhD(chem), 67. *Prof Exp:* Chemist, Sperry Gyroscope Co, 57-60; anal chemist, Colgate-Palmolive Co, 60-62; sr chemist, Am Cyanamid Co, 67-68; asst prof, 68-71, assoc prof, 71-82, chmn, Dept Chem, 81-88, PROF CHEM, C W POST COL, LONG ISLAND UNIV, 82- *Mem:* Am Chem Soc; Sigma Xi; AAAS. *Res:* Electroanalytical chemistry; analytical methods. *Mailing Add:* Dept Chem Long Island Univ C W Post Ctr Greenvale NY 11548

KARP, WARREN B, b Brooklyn, NY, Feb 12, 44; m 76; c 2. BIOCHEMISTRY. *Educ:* Pace Univ, BS, 65; Ohio State Univ, PhD(physiol chem), 70; Med Col Ga, DMD, 77. *Prof Exp:* Teaching asst physiol chem, Ohio State Univ, 66-68, res assoc, 68-70, res assoc pediat, 70-71; pediat res instr, Sch Med, Med Col Ga, 71-73, instr cell & molecular biol, 72-73, asst res prof pediat & asst prof cell & molecular biol, Sch Med & asst prof, Sch Grad Studies, 73-79, asst prof biochem, Sch Dent, 74-79, assoc res prof. pediat, Sch Med, 79-88, assoc prof oral biol biochem & oral med, Sch Dent, 79-88, PROF PEDIAT, ORAL BIOL, ORAL DIAGNOSIS/PATIENT SERV, CELL & MOLECULAR BIOL, SCH MED, DENT & GRAD STUDIES, MED COL GA, 88- *Concurrent Pos:* Environ Protection Agency grant, Med Col Ga, 71-74, dir pediat res, 77-, dir clin perinatal lab, 78- *Mem:* AAAS; Sigma Xi; Am Chem Soc; NY Acad Sci; Int Dent Res Soc; Am Inst Nutrit; Am Soc Clin Nutrit. *Res:* Environmental effects on human placental enzymology; human placental amino acid metabolism; the effect of bilirubin on brain metabolism; human placental lipid metabolism. *Mailing Add:* Med Col Ga Augusta GA 30912

KARPATI, GEORGE, Can citizen; c 2. HISTOCHEMISTRY, MUSCLE BIOLOGY. *Educ:* Dalhousie Univ, MD, 60. *Prof Exp:* KILLAM CHAIR NEUROL, MONTREAL NEUROL INST, MCGILL UNIV, 85-, ASSOC DIR RES, 85-, PROF PEDIAT, 90- *Concurrent Pos:* Coordr, Neuromuscular Res Group, Montreal Neurol Inst, 85-; chmn, Bd Examiners, Royal Col Physicians & Surgeons Can, 86-89. *Mem:* Hon mem French Neurol Soc; Am Acad Neurol; Am Neurol Asn; Can Cong Neurol Sci; Histochem Soc; Royal Col Physicians & Surgeons Can; World Fedn Neurol. *Res:* Neuromuscular system using histochemical, cytochemical, immunological and physiological techniques; cell therapy of inherited muscle diseases; gene therapy. *Mailing Add:* 3801 University St Montreal PQ H3A 2B4 Can

KARPATKIN, SIMON, b Brooklyn, NY, Sept 6, 33; m 65. BIOCHEMISTRY, PHYSIOLOGY. *Educ:* Brooklyn Col, BS, 54; NY Univ, MD, 58. *Prof Exp:* Intern med, Bellevue Hosp, NY Univ, 58-59, resident, 59-60; resident, Einstein Med Ctr, Bronx, 60-61; from instr to assoc prof, 64-74, PROF MED, SCH MED, NY UNIV, 74- *Concurrent Pos:* Fel hemat, Sch Med, Wash Univ, 61-62; fel biochem, 62-64; USPHS trainee, 61-62; Am Cancer Soc fel, 62-64; res grants, Health Res Coun City of New York, 66, Muscular Dystrophy Asn Am, 66-68, NY Heart Asn, 67-70, NIH, 70-84 & NSF, 78-82; career scientist, Health Res Coun City of New York, 66-71. *Mem:* Am Soc Hemat; Am Fedn Clin Res; Am Soc Physiol; Am Soc Clin Invest; Am Soc Biol Chem. *Res:* Regulation and organization of glycolytic enzymes in platelets; platelet biochemical interactions during hemostasis; biochemical and physiological aspects of human platelet senescence; regulation of platelet production; autoimmune platelet disorders; role of platelets in cancer. *Mailing Add:* NY Univ Sch Med 550 First Ave UH 411 New York NY 10016

KARPEL, RICHARD LESLIE, b New York, NY, May 31, 44; m 68; c 1. MOLECULAR BIOLOGY, BIOPHYSICS. *Educ:* Queens Col, NY, BA, 65; Brandeis Univ, PhD(chem), 70. *Prof Exp:* Res assoc, Princeton Univ, 70-71, NIH res fel, 71-72, res assoc, 72-74, NIH res fel biochem sci, 74-76; asst prof, 76-81, ASSOC PROF CHEM, UNIV MD, BALTIMORE COUNTY, 81- *Concurrent Pos:* Sr fel, Nat Res Coun, Nat Cancer Inst, Frederick Cancer Res Facil, NIH, 82-83. *Mem:* AAAS; Am Chem Soc; Sigma Xi; Am Soc Biochem Molecular Biol. *Res:* Protein-nucleic acid interactions; nucleic acid-interactive enzymes; structure-function studies on nucleic acid helix-destabilizing proteins; retroviral nucleic acid binding proteins; metal-nucleic acid interactions. *Mailing Add:* Dept Chem & Biochem Univ Md 5401 Wilkens Ave Baltimore MD 21228

KARPETSKY, TIMOTHY PAUL, CHEMICAL WARFARE. *Educ:* Johns Hopkins Univ, PhD(org chem), 70. *Prof Exp:* SUPVRY PHYS SCIENTIST, US ARMY CHEM RES & ENG CTR, 82- *Res:* Convention compliance monitoring. *Mailing Add:* Technol GP Inc 1400 Taylor Ave Baltimore MD 21209

KARPIAK, STEPHEN EDWARD, b Hartford, Conn, Aug 13, 47. PSYCHIATRY. *Educ:* Col of the Holy Cross, BA, 69; Fordham Univ, MA, 71, PhD(exp psychol), 72. *Prof Exp:* Fel neuroimmunol, Parkinson Dis Found, Dept Neurol, Col Physicians & Surgeons & vis fel, Dept Psychiat, Columbia Univ, 72-74; SR RES SCIENTIST, DIV NEUROSCI, NY STATE PSYCHIAT INST, 74-; ASSOC PROF PSYCHIAT, COLUMBIA UNIV, 78- *Concurrent Pos:* Asst prof psychol, Manhattan Col, 72-78. *Mem:* Soc Neurosci; Am Psychol Asn. *Res:* Development of immunological tools for the study of brain function and pathology, specifically the use of brain antibodies to study behavior and electrophysiology; effects of exogenous administration of gangliosides on central nervous system pathology. *Mailing Add:* Dept Psychiat Box 64 Columbia Univ 722 W 168th St New York NY 10032

KARPLUS, MARTIN, b Vienna, Austria, Mar 15, 30; nat US; m 81; c 3. PHYSICAL CHEMISTRY. *Educ:* Harvard Univ, BA, 51; Calif Inst Technol, PhD(chem), 53. *Prof Exp:* NSF fel chem, Oxford Univ, 53-55; from instr to assoc prof phys chem, Univ Ill, 55-60; from assoc prof to prof, Columbia Univ, 60-66; prof chem, 66-79, THEODORE WILLIAM RICHARDS PROF CHEM, HARVARD UNIV, 79- *Concurrent Pos:* NSF sr fel, 65-66; vis prof, Univ Paris, 72-73 & 80-81, prof, 74-75. *Honors & Awards:* Fresenius Award, 65; Harrison Howe Award, 66. *Mem:* Nat Acad Sci; Am Acad Arts & Sci; Int Acad Quantum Molecular Sci. *Res:* Theory of molecular structure and spectra with emphasis on biologically important molecules. *Mailing Add:* Dept Chem Harvard Univ Cambridge MA 02138

KARPLUS, ROBERT, educational physics; deceased, see previous edition for last biography

KARPLUS, WALTER J, b Vienna, Austria, Apr 23, 27; nat US; m 69; c 2. COMPUTER SCIENCE. *Educ:* Cornell Univ, BEE, 49; Univ Calif, MS, 51; Univ Calif, Los Angeles, PhD(elec eng), 55. *Prof Exp:* Field party chief, Sun Oil Co, 49-50; res engr, Int Geophys Inc, 51-52; chmn, Comput Sci Dept, 71-79, PROF COMPUT, ELEC CIRCUITS & ELECTRONICS, UNIV CALIF, LOS ANGELES, 52- *Concurrent Pos:* Fulbright res fel, 61; Guggenheim fel, 68. *Honors & Awards:* Sr Sci Simulation Award, Soc Comput Simulation. *Mailing Add:* Dept Comput Sci Univ Calif 3732 Boelter Hall Los Angeles CA 90024

KARR, ALAN FRANCIS, b Bryn Mawr, Pa, July 12, 47. INFERENCE FOR STOCHASTER PROCESSES. *Educ:* Northwestern Univ, BS, 69, MS, 70 & PhD(appl math), 73. *Prof Exp:* PROF MATH SCI, JOHN HOPKINS UNIV, 73-, ASSOC DEAN ENG, 86- *Mem:* Fel Inst Math Statist; Am Statist Asn. *Res:* Statistical inference for stochastic processes; image analysis and processing. *Mailing Add:* QWC Whiting Sch Eng John Hopkins Univ Baltimore MD 21218

KARR, CLARENCE, JR, b St Louis, Mo, May 12, 23; m 47; c 4. CHEMISTRY. *Educ:* St Louis Univ, BS, 44; Johns Hopkins Univ, PhD(chem), 50. *Prof Exp:* Fel petrol chem, Mellon Inst, 50-55; supvry res chemist low temperature tar, US Bur Mines, 55-66, coal chemistry, 66-75; supvry res chemist coal liquefaction, Energy Res & Develop Admin, US Dept Energy, 75-77, res chemist synthetic fuels, 77-80, proj mgr advan gasification, Morgantown Energy Technol Ctr, 80-83; RETIRED. *Concurrent Pos:* Prin investr, Apollo 11, 12, 14 & 15 Lunar Sample Prog, 69-72. *Honors & Awards:* Award, US Dept Interior, 65 & 66. *Mem:* Fel Am Inst Chem; Am Chem Soc. *Res:* Composition of low temperature coal tar, petroleum; organic synthesis; chromatography; infrared ultraviolet spectroscopy; air pollution; coal minerals; synthetic fuels from coal; lunar minerals; liquid fuels from coal; coal gasification. *Mailing Add:* 624 Vista Pl Morgantown WV 26505

KARR, JAMES PRESBY, b Nashua, NH, July 24, 41; m 62; c 2. REPRODUCTIVE PHYSIOLOGY, STEROID BIOCHEMISTRY. *Educ:* Univ Vt, BA, 64, MS, 66; Pa State Univ, PhD(reprod physiol), 70. *Prof Exp:* Res assoc reprod physiol, Pa State Univ, 70-71; asst prof animal breeding, Haille Selassie I Univ, 71-73; cancer res scientist reprod physiol, Roswell Park Mem Inst, 73-74 & 76-77; asst prof animal breeding & reprod physiol, Am Univ, Beirut, 74-75; dep dir sci affairs cancer res, Nat Prostatic Cancer Proj, 78-84, assoc dir, Organ Systs Coord Ctr, 84-85, DIR ORGAN SYSTS COORD CTR, ROSWELL PARK MEM INST, 85-; ASST PROF PHYSIOL, UNIV BUFFALO & NIAGARA UNIV, 82- *Mem:* AAAS; Am Asn Cancer Res; NY Acad Sci. *Res:* Reproductive endocrinology, steroid biochemistry, plasma steroid binding proteins and hormone receptors in the normal physiology and disease states of the human prostate. *Mailing Add:* Roswell Park Mem Inst 666 Elm St Buffalo NY 14263

KARR, JAMES RICHARD, b Shelby, Ohio, Dec 26, 43; m 63, 84; c 2. TROPICAL BIOLOGY, CONSERVATION BIOLOGY. *Educ:* Iowa State Univ, BSc, 65; Univ Ill, Urbana-Champaign, MSc, 67, PhD(zool), 70. *Prof Exp:* Fel, Princeton Univ, 70-71 & Smithsonian Tropical Res Inst, 71-72; asst prof ecol, Purdue Univ, 72-75; from assoc prof to prof ecol, Univ Ill, Urbana-Champaign, 75-84; dep dir, Smithsonian Trop Res Inst, Balboa, Panama, 84-87, actg dir, 87-88; Harold H Bailey prof biol, Va Polytech Inst & State Univ, Blacksburg, Va, 88-91; DIR, INST ENVIRON STUDIES, UNIV WASH, 91- *Concurrent Pos:* Mem, Eval Panel, Instrnl Sci Equip Prog, NSF, 75, Undergrad Res Participation, 76, conserv biol, 90; consult, Orgn Am States, 80; prin investr grants, Nat Sci Found, Environ Protection Agency, Nat Geog Soc, Am Philos Soc, US Fish & Wildlife Serv, US Forest Serv & Off Water Resources Technol, Tenn Valley Authority, 73-; affil, Ill Natural Hist Surv, 81-; ed, Trop Ecol, 77-81, Ecol, 81-84 & Biosci, 85- *Mem:* Ecol Soc Am; fel Am Ornithologists Union; Wilson Ornith Soc; Int Soc Trop Ecol; fel AAAS. *Res:* Community ecology from both basic and applied perspectives with emphasis on studies of tropical forest birds and stream fishes, including a wide range of land use and water resource problems; improving knowledge of biological communities and to apply that knowledge to solution of selected environmental and natural resource problems. *Mailing Add:* Dir Inst Environ Studies Eng Annex FM-12 Univ Wash Seattle WA 98195

KARR, REYNOLD MICHAEL, JR, b New York, NY, June 24, 42; m 76; c 2. ALLERGY, RHEUMATOLOGY. *Educ:* Johns Hopkins Univ, BA, 64; Univ Md, MD, 69. *Prof Exp:* Assoc prof med, Clin Immunol Sect, Sch Med, Tulane Univ, 76-80; ASSOC PROF MED, UNIV WASH, 80- *Concurrent Pos:* Vis consult, Vet Admin Hosp, New Orleans, 76-; vis physician, Charity Hosp, New Orleans, 76- *Mem:* Fel Am Col Physicians; Am Acad Allergy; Am Thoracic Soc; Am Rheumatism Asn. *Res:* Arthritis; occupational lung disease; bronchoprovocation. *Mailing Add:* Med Dept Univ Wash 9706 171st Ave SE Snohomish WA 98290

KARRAKER, ROBERT HARRELD, b Carbondale, Ill, May 6, 31; m 53; c 2. INORGANIC CHEMISTRY. *Educ:* Southern Ill Univ, BA, 53; Iowa State Univ, PhD(inorg chem), 61. *Prof Exp:* Chemist, Olin-Mathieson Chem Corp, NY, 53-55; asst prof chem, Memphis State Univ, 61-67; assoc prof chem, 67-80, PROF CHEM, EASTERN ILL UNIV, 80- *Mem:* Am Chem Soc; Sigma Xi. *Res:* Chemistry of rare earth elements. *Mailing Add:* Dept Chem Eastern Ill Univ Charleston IL 61920

KARRAS, THOMAS WILLIAM, b Chicago, Ill, Jan 4, 36; m 60; c 2. LASERS, ELECTRO-OPTICS. *Educ:* Univ Chicago, BS, 57; Ill Inst Technol, MS, 61; Univ Calif, Los Angeles, PhD(physics), 64. *Prof Exp:* Physicist elec propulsion, Rocketdyne Div, NAm Aviation, 59-61; physicist, 64-72, mgr laser & plasma physics, 72-79, laser res, 79-83, MGR ELECTRO-OPTICS ANALYSIS, ASTRO-SPACE DIV, GEN ELEC CO, KING OF PRUSSIA, PA, 85- *Mem:* Am Phys Soc; Am Inst Aeronaut & Astronaut; Sigma Xi; Optical Soc Am. *Res:* Metal vapor lasers; nanosecond discharges. *Mailing Add:* 231 Wooded Way Berwyn PA 19312

KARREMAN, GEORGE, b Rotterdam, Holland, Nov 4, 20; US citizen; m 53; c 3. MATHEMATICAL BIOLOGY. *Educ:* Univ Leiden, BS, 39, MS, 41; Univ Chicago, PhD(math biol), 51. *Prof Exp:* Instr math & physics, Col Tech Sci, Rotterdam, Holland, 46-48; res assoc math biol, Univ Chicago, 51-53, asst prof, 53-54; res assoc theoret biol, Inst Muscle Res, Mass, 54-57; med res scientist, Eastern Pa Psychiat Inst, 57-62; from assoc prof to prof physiol, 62-83, EMER PROF MATH BIOL, SCH MED, UNIV PA, 83- *Mem:* Soc Math Biol (pres, 73-81). *Res:* Physiological irritability; biological energy transfer; quantum biology; system analysis of cardiovascular, central nervous renal and endocrine systems; cooperative phenomena; threshold mechanisms; bioelectric phenomena; adsorption mechanism. *Mailing Add:* 435 S Woodbine Ave Penn Valley PA 19072

KARREMAN, HERMAN FELIX, b Rotterdam, Neth, June 21, 13; US citizen; m 38, 74; c 2. APPLIED MATHEMATICS. *Educ:* Neth Sch Econ, Drs, 49. *Prof Exp:* Staff employee, Royal Packet Navig Co, Dutch E Indies, 37-47; sr officer, Cent Planning Bur, Neth, 49-54; res assoc, Nat Bur Econ Res, NY, 54-56; res assoc, Economet Res Prog, Princeton Univ, 56-63; mem, Math Res Ctr, Univ Wis-Madison, 63-67, prof, Sch Bus, 65-83, prof, Col Eng, 68-83, EMER PROF, UNIV WIS-MADISON, 83- *Honors & Awards:* Lanchester Prize, Opers Res Soc Am, 60. *Mem:* Opers Res Soc Am; Math Asn Am; Am Math Soc. *Mailing Add:* 3412 Blackhawk Dr Madison WI 53705

KARREN, KENNETH W, b Vernal, Utah, May 20, 32; m 53; c 7. STRUCTURAL DESIGN. *Educ:* Univ Utah, BS, 53, MS, 61; Cornell Univ, PhD(civil eng), 65. *Prof Exp:* Proj develop engr, Pipeline Div, Phillips Petrol, 56-57; chief engr, Otto Buehner Co, 57-61; asst prof civil eng, Brigham Young Univ, 61-62, prof, 65-70; grad student civil eng, Cornell Univ, 62-65; sr engr, Hercules Inc, 70-71; PRES, STRUCT ENGRS, KARREN & ASSOCS, 78- *Concurrent Pos:* Consult, Hercules Inc, 66-75. *Mem:* Am Soc Civil Engrs. *Res:* Cold-forming of sheet steel led to provisions included in Am Iron and Steel Institute specifications. *Mailing Add:* 424 E 4750 N Provo UT 84604

KARRER, KATHLEEN MARIE, b Grosse Pointe Farms, Mich, June 16, 49. MOLECULAR BIOLOGY, DEVELOPMENTAL BIOLOGY. *Educ:* Marquette Univ, BS, 71; Yale Univ, PhD(biol), 76. *Prof Exp:* Fel biol, Ind Univ, 76-80; asst prof biol, Brandeis Univ, 80-89; CLARE BOOTH LUCE PROF, MARQUETTE UNIV, 89- *Concurrent Pos:* Jane Coffin Childs Mem Fund Med Res fel, 76-78; NIH fel, 78-79. *Honors & Awards:* John Spangler Nicholas Prize Exp Zool, Yale Univ, 77. *Mem:* AAAS; Am Soc Cell Biol; Soc Develop Biol; Soc Protozoologists. *Res:* Eukaryotic chromosome structure and function; DNA rearrangement; molecular biology of ciliates; DNA methylation. *Mailing Add:* Dept Biol Marquette Univ Milwaukee WI 53233

KARROW, PAUL FREDERICK, b St Thomas, Ont, Sept 14, 30; m 62; c 4. GEOLOGY. *Educ:* Queen's Univ, Ont, BSc, 54; Univ Ill, PhD(geol), 57. *Prof Exp:* Geologist, Ont Dept Mines, 57-63; prof civil eng, 63-65, assoc prof earth sci, 65-69, chmn dept, 65-70, PROF EARTH SCI, UNIV WATERLOO, 69- *Concurrent Pos:* Geologist, Ont Dept Mines, 64 & 73; vis scientist, Scripps Inst Oceanog, La Jolla, Calif, 70 & 76; vis prof, Univ SFla, Tampa, 84. *Mem:* Geol Soc Am; Soc Econ Paleont & Mineral; Geol Asn Can; Int Asn Gt Lakes; Am Asn Quaternary Environ. *Res:* Quaternary geology; glacial geology; geomorphology; urban geology; paleoecology. *Mailing Add:* Dept Earth Sci Univ Waterloo Waterloo ON N2L 3G1 Can

KARSCH, FRED JOSEPH, b New York, NY, Aug 8, 42; m 67; c 2. REPRODUCTIVE ENDOCRINOLOGY, NEUROENDOCRINOLOGY. *Educ:* Juniata Col, BS, 64; Univ Maine, MS, 66; Univ Ill, PhD(animal sci, biochem & physiol), 70. *Prof Exp:* From asst prof to assoc prof, 72-82, PROF PHYSIOL, UNIV MICH, ANN ARBOR, 82- *Concurrent Pos:* Ford Found fel, Med Sch, Univ Pittsburgh, 70-71, NIH fel, 71-72. *Mem:* Endocrine Soc; Soc Study Reprod; Soc Study Fertil. *Res:* Neuroendocrine control of gonadotropin secretion; seasonal reproduction; developmental endocrinology. *Mailing Add:* Dept Path Univ Mich Ann Arbor MI 48109

KARSON, JEFFREY ALAN, b Akron, Ohio, Nov 3, 49; m 78; c 1. STRUCTURAL GEOLOGY. *Educ:* Case Inst Technol, BS, 72; State Univ NY, Albany, MS, 75, PhD(geol), 77. *Prof Exp:* Asst instr, State Univ NY, Albany, 72-75, res asst, 75-77; fel, Erindale Col & Univ Toronto, 77-79; scholar, 79-80, ASST SCIENTIST, GEOL, WOODS HOLE OCEANOG INST, 80- *Concurrent Pos:* Vis lectr, Bridgewater State Col, Mass, 81- *Mem:* Geol Soc Am; Am Geophys Union. *Res:* Internal structure of the oceanic lithosphere via direct observation of the sea floor and structural analysis of ophiolites. *Mailing Add:* Dept Geol Duke Univ Durham NC 27706

KARSTAD, LARS, veterinary science, for more information see previous edition

KARSTEN, KENNETH STEPHEN, b Holland, Mich, July 24, 13; m 39; c 4. PLANT PHYSIOLOGY, CHEMISTRY. *Educ:* Hope Col, AB, 35; Univ Nev, MS, 37; Univ Wis, PhD(plant physiol), 39. *Prof Exp:* Chemist-analyst, Sullivan Mining Co, Idaho, 37; asst, Univ Wis, 37-39; tutor biol, Brooklyn Col, 39-41; dir org res, Niagara Sprayer & Chem Co, 41-45; insecticide chemist, Rohm and Haas Co, Pa, 45-47; dept mgr, R T Vanderbilt Co, Inc, 48-72, dir res & develop, 72-78, vpres res & develop, 78-81; CONSULT, INDUST MINERALS & CHEMICALS, 81- *Mem:* Am Chem Soc. *Res:* Organic syntheses; insecticide and fungicide formulation and development; plant hormones; plant physiology; root activity and oxygen in relation to soil fertility; fungicides; bactericides; sap stain control chemicals; bacteriostats for soap. *Mailing Add:* 5397 Keysville Ave Spring Hill FL 33526

KARSTENS, ANDRES INGVER, b Pendleton, Ore, Dec 17, 11; m 42, 83. AEROSPACE MEDICINE. *Educ:* Univ Ore, BA, 38, MD & MS, 43; Am Bd Prev Med, dipl aerospace med, 56. *Prof Exp:* Res aviation med, high altitude physiol & environ physiol, Aero Med Lab, Wright-Patterson AFB, USAF, 46-50, commanding officer, Arctic Aero Med Lab, Ladd AFB, Alaska, 50-55, dir res, USAF Sch Aviation Med, Randolph AFB, Tex, 56-58, from asst chief to chief, Aerospace Med Res Lab, Wright-Patterson AFB, 58-64, dir Bioastronaut, Manned Orbiting Lab, Space & Missiles Syst Orgn, Los Angeles, 64-69, dir res & develop, Aerospace Med Div, Air Force Systs Command, Brooks AFB, Tex, 69-71; staff physician & chief flight surgeon, Med Support Serv, NASA Johnson Space Ctr, Houston, 76; RETIRED. *Honors & Awards:* Hubertus Stughold Award Aerospace Med, 73. *Mem:* Fel AAAS; AMA; fel Aerospace Med Asn; fel Am Col Prev Med. *Res:* Autonomic control of cardiac function and intestinal motility; high altitude and environment physiology; environmental protection and life support; arctic physiology and ecology; life support in aeronautical and space flight. *Mailing Add:* 887 Mocking Bird College Place WA 99324-1828

KARTEN, HARVEY J, b New York, NY, July 13, 35; m 64; c 3. NEUROANATOMY. *Educ:* Yeshiva Col, BA, 55, Albert Einstein Col Med, MD, 59. *Prof Exp:* Intern med, Univ Utah, 59-60; resident psychiat, Univ Colo, 60-61; res assoc neurophysiol, Walter Reed Army Inst Res, 61-65; res assoc neuroanat, Mass Inst Technol, 65-73; sr res assoc, 73-74; prof psychiat & anat sci, State Univ NY, Stony Brook, 74-86, prof neurobiol, 79-86; PROF NEUROSCI & PSYCHIAT, UNIV CALIF, SAN DIEGO, 86. *Concurrent Pos:* USPHS fel, Univ Colo, 60-61; NIMH career develop award, 61-65; Nat Inst Child Health & Human Develop career develop award, 65-74; res assoc, Lab Neuropsychol, Wash Sch Psychiat, 63-65; adj prof, Salk Inst & Univ Utah, 86. *Honors & Awards:* Herrick Award, Am Asn Anat, 68. *Mem:* Am Soc Zool; Am Asn Anat; Soc Neurosci. *Res:* Neuroanatomy. *Mailing Add:* Dept Neurosci 0608 Univ Calif San Diego La Jolla CA 92093

KARTEN, MARVIN J, b New York, NY, Apr 26, 31; m 56; c 2. MEDICINAL CHEMISTRY. *Educ:* Brooklyn Col, BS, 54; Univ Pittsburgh, PhD(chem), 58. *Prof Exp:* Res chemist, Monsanto Chem Co, Ohio, 58-59, Mass, 59-60; sr res chemist, USV Pharmaceut Corp, 60-67, group leader, 67-70; HEALTH SCI ADMINR, NAT INST CHILD HEALTH & HUMAN DEVELOP, NIH, 71- *Mem:* Am Chem Soc. *Res:* Medicinal chemistry; organic synthesis; synthesis and biological evaluation of new contraceptive agents. *Mailing Add:* Nat Inst Child Health & Human Develop Nat Inst Health Bethesda MD 20892

KARTHA, KUTTY KRISHNAN, b Shertallai, India, Aug 9, 41; Can citizen; m 72; c 2. PLANT BIOTECHNOLOGY, PLANT CELL & TISSUE CULTURE. *Educ:* Saugar Univ, India, BSc, 62; Jawaharal Nehru Agr Univ, India, MSc, 65; India Agr Res Inst, PhD(plant path), 69. *Prof Exp:* Postdoctoral fel, Nat Inst Agr Res, France, 70-72; vis scientist, Prairie Regional Lab, Nat Res Coun, Saskatoon, 73-74, asst res officer, Plant Biotechnol Inst, 74-76, assoc res officer, 76-81, SR RES OFFICER, PLANT BIOTECHNOL INST, NAT RES COUN, 81-, HEAD, CELL TECHNOL SECT, 85- *Concurrent Pos:* Nat corresp, Int Asn Plant Tissue Cult, 82-86; ed, J Plant Physiol, 87; adj prof, Univ Sask, Saskatoon, 87-; mem, Can Agr Res Coun, 90-93. *Honors & Awards:* George M Darrow Award, Am Soc Hort Sci, 81. *Mem:* Int Asn Plant Tissue Cult; Can Soc Plant Physiologists. *Res:* Plant biotechnology especially the genetic engineering of crops such as cereals and strawberry; cryopreservation of plant cells and organs; plant tissue culture. *Mailing Add:* Plant Biotechnol Inst Nat Res Coun 110 Gymnasium Pl Saskatoon SK S7N 0W9

KARTHA, MUKUND K, b Pattanakad, Kerala, India, July 31, 36; US citizen; m 63; c 2. RADIOLOGY, BIOPHYSICS. *Educ:* Univ Kerala, BSc, 58; Univ Sagar, India, MSc, 61; Univ Western Ont, PhD(radiol physics), 69. *Prof Exp:* Sci officer radiol physics, India Atomic Energy Comn, 61-63; cancer res fel, Ont Cancer Found, 63-68; asst prof, 68-73, ASSOC PROF RADIOL, OHIO STATE UNIV, 73-, ASSOC PROF ALLIED MED PROF, 76- *Concurrent Pos:* Am Cancer Soc fel, Ohio State Univ, 70-71, Nat Cancer Inst fel, 74-77; co-dir, Radiation Ther Consult Prog, Cancer Res Ctr, Ohio State Univ, 73- *Mem:* Radiol Soc NAm; Radiation Res Soc; Am Asn Physicists in Med; Am Col Radiol; Am Soc Therapeut Radiol & Oncol; Am Onocol Assoc Inc (pres). *Res:* Experimental and clinical research in radiation therapy; investigation of cancer treatment using radiation; radiation therapy. *Mailing Add:* 5003 11th Ave Vienna WV 26105-3152

KARTHA, SREEHARAN, b Kerala, India, 1948. GROWTH DIFFERENTIATION, GROWTH REGULATION. *Educ:* Nehru Univ, India, PhD(cell biol), 78. *Prof Exp:* ASST PROF GROWTH REGULATION, UNIV CHICAGO, 83- *Concurrent Pos:* Res assoc, Johns Hopkins Med Sch, Baltimore, 80-83. *Res:* Growth factors; oncogenes. *Mailing Add:* Dept Med Box 453 Univ Chicago 5841 S Maryland Ave Chicago IL 60637

KARTZMARK, ELINOR MARY, b Selkirk, Man, May 16, 26; c 1. PHYSICAL CHEMISTRY. *Educ:* Univ Man, BSc, 49, MSc, 50, PhD(chem), 52. *Prof Exp:* Mem staff, Univ Man, 47-50, from lectr to prof phys chem, 51-88; RETIRED. *Mem:* Fel Chem Inst Can. *Res:* Electrolytic conductance; heterogeneous equilibria. *Mailing Add:* Box 2 Grp 30 RR 1 Lockport MB R0C 1W0 Can

KARUKSTIS, KERRY KATHLEEN, b Buffalo, NY, 55. PHOTOSYNTHESIS, FLUORESCENCE SPECTROSCOPY. *Educ:* Duke Univ, BS, 77, PhD(chem), 81. *Prof Exp:* Research fel chem, Lab Chem Biodynamics, Univ Calif, Berkeley, 81-84; asst prof, 84-89, ASSOC PROF CHEM, HARVEY MUDD COL, 89- *Mem:* Am Chem Soc; Biophys Soc; Am Soc Photobiol; Sigma Xi. *Res:* Use of steady-state and time-resolved fluorescence and absorbance measurements to monitor the organization of chloroplast photosynthetic membranes and the processes of excitation transfer and electron transport in photosynthesis. *Mailing Add:* Dept Chem Harvey Mudd Col Claremont CA 91711-5990

KARULKAR, PRAMOD C, b Maharashtra State, India, 50; US citizen. PHYSICS. *Educ:* Univ Poona, BSc, 69; Indian Inst Technol, MSc, 71; Portland State Univ, MS, 75; Univ Wis, PhD(mat sci), 79. *Prof Exp:* Mem staff, Rockwell Int Corp, Anaheim, Calif, 80-84; mem staff, Hughes Aircraft Co, Newport Beach, Calif, 84-85; MEM STAFF, LINCOLN LAB, MASS INST TECHNOL, 85- *Concurrent Pos:* Res assoc, Saha Inst Nuclear Physics, India, 71-72; res scholar low temp physics, Indian Inst Technol, Bombay, 72-73; lectr elec eng, Calif State Polytech Univ, Pomona, 81-85. *Mem:* Sigma Xi; Am Phys Soc; Am Vacuum Soc; Inst Elec & Electronics Engrs. *Res:* Fabrication and analysis of electronic materials; fabrication of solid state devices; fabrication of VLSI circuits; plasma processing; thin film technology. *Mailing Add:* Lincoln Lab Mass Inst Technol Box 18 MS B-167 Lexington MA 02173

KARUNASIRI, GAMANI, b Colombo, Sri Lanka, Apr 14, 56; m 84; c 2. SOLID STATE ELECTRONICS, QUANTUM MECHANICS. *Educ:* Univ Colombo, Sri Lanka, BS, 79; Univ Pittsburgh, MS, 81, PhD(physics), 84. *Prof Exp:* Asst lectr physics, Univ Colombo, 79-80; res assoc, Microtonics Assocs, 85-87; RES ENG, UNIV CALIF, LOS ANGELES, 87- *Mem:* Am Phys Soc; Inst Elec & Electronics Engrs. *Res:* Physics and device application of semiconductor; quantum wells and superlattices; characterization of the devices using photo current and fourier transform infrared spectroscopies. *Mailing Add:* Elec Eng Dept Univ Calif Los Angeles 7619 Boelter Hall Los Angeles CA 90024-1594

KARUSH, FRED, b Chicago, Ill, July 12, 14; m 36; c 3. MONOCLONAL ANTIBODIES, GENE EXPRESSION. *Educ:* Univ Chicago, BS, 35, PhD(chem), 38. *Prof Exp:* Res physicist, E I du Pont de Nemours & Co, Inc, NJ, 41-46; from asst prof to assoc prof immunol, Dept Pediat, 50-57, prof immunochem, 57-85, prof microbiol, 60-85, EMER PROF, UNIV PA, 85- *Concurrent Pos:* Rockefeller fel enzyme kinetics, Mass Inst Technol, 39-40; Harrison fel biophysics, Univ Pa, 40-41; Am Cancer Soc sr fel, Col Med, NY Univ, 47-48, Sloan-Kettering Inst Cancer Res, 49 & Col Physicians & Surgeons, Columbia Univ, 50; res career award, NIH, 62-85. *Mem:* Am Chem Soc; Am Soc Biol Chem; Am Asn Immunol. *Res:* Photoelectric polarimetry; enzyme kinetics; physics of pigments; protein interactions; bacterial synthesis of proteins; molecular immunology; affinity analysis of monoclonal antibodies. *Mailing Add:* Dept Microbiol Univ Pa Sch Med Philadelphia PA 19104

KARUSH, WILLIAM, b Chicago, Ill, Mar 1, 17; m 39; c 2. MATHEMATICS. *Educ:* Univ Chicago, BS, 38, MS, 39, PhD(math), 42. *Prof Exp:* Mathematician, Geophys Lab, Carnegie Inst, 42-43; physicist, Metall Lab, Univ Chicago, 43-45, from instr to assoc prof math, 45-56; mem sr staff, Ramo-Wooldridge Corp, 56-57; sr opers res scientist, Syst Develop Corp, 58-62, prin scientist, 62-67; prof math, 67-87, EMER PROF MATH, CALIF STATE UNIV, NORTHRIDGE, 87- *Concurrent Pos:* Mathematician, Inst Numerical Anal, Nat Bur Standards, Univ Calif, Los Angeles, 49-52; mem tech staff, Res & Develop Labs, Hughes Aircraft Co, 53 & Ramo-Wooldridge Corp, 54-55; Ford fac fel, Univ Calif, Los Angeles, 55-56. *Mem:* Am Math Soc; Opers Res Soc Am. *Res:* Operations research; calculus of variations; applied mathematics. *Mailing Add:* Dept Math Calif State Univ 18111 Nordoff St Northridge CA 91330

KARUZA, SARUNAS KAZYS, b Kaunas, Lithuania, Jan 19, 40; m 77; c 2. ATOMIC FREQUENCY STANDARDS, NAVIGATION SYSTEMS. *Educ:* Univ Southern Calif, BSEE, 63, MSEE, 66, PhD(elec eng), 72. *Prof Exp:* Mem tech staff, Commun Div, Hughes Aircraft Co, 63-65, Ground Systs Div, 65-66, Aeronaut Systs Div, 66-67; dir & prof staff assoc, Electronics Lab, Rancho Los Amigos Hosp, 72-80; MEM TECH STAFF, ELECTRONIC RES LABS, AEROSPACE CORP, 80- *Concurrent Pos:* Consult, Fullerton Internal Med Clin, 78; adj asst prof med & adj assoc prof biomed eng, Univ Southern Calif, 75- *Mem:* Sigma Xi; Inst Elec & Electronics Engrs. *Res:* Precision atomic frequency standards (cesium-rubidium) which are used in the Navstar Global Positioning System (GPS) satellites for world wide navigation; stability properties of these standards as they are influenced by their electronics and environmental factors. *Mailing Add:* Aerospace Corp PO Box 92957 Los Angeles CA 90009

KARVE, MOHAN DATTATREYA, b Kupwad, India, Aug 14, 39; m 67; c 2. INDUSTRIAL MICROBIOLOGY, MYCOLOGY. *Educ:* Univ Poona, BSc, 59, Hons, 60, MSc, 61; Ohio State Univ, PhD(mycol), 65. *Prof Exp:* Asia-Pac area rep, 65-71, gen mgr, Northern Asia-Pac Area, 71-86, VPRES, JAPAN/KOREA, BUCKMAN LABS, INC, 86- *Mem:* Am Chem Soc; Tech Asn Pulp & Paper Indust. *Res:* Microbial physiology; fungal proteins and amino acids; microbial deterioration. *Mailing Add:* Buckman Labs Inc PO Box 8305 Memphis TN 38108-0305

KARWAN, MARK HENRY, b Cleveland, Ohio, Nov 16, 51; m 73; c 4. INTEGER PROGRAMMING. *Educ:* Johns Hopkins Univ, BES, 74, MSE, 75; Ga Inst Technol, PhD(opers res), 76. *Prof Exp:* From asst to assoc prof, 76-86, PROF, OPERS RES, DEPT INDUST ENG, STATE UNIV NY BUFFALO, 86-, DEPT CHMN, 88- *Concurrent Pos:* Prin investr, NSF, 78-82 & Off Naval Res, 85-87. *Mem:* Opers Res Soc Am; Inst Mgt Sci; Inst Ind Eng. *Res:* Discrete optimization, routing and scheduling; multicriteria decision making; multilevel decentralized planning; redundancy in mathematical programming; industrial inspection. *Mailing Add:* 342 Bell Hall Dept Indust Eng State Univ NY Buffalo NY 14260

KARWEIK, DALE HERBERT, b Milwaukee, Wis, May 27, 48; m 70; c 2. ANALYTICAL CHEMISTRY. *Educ:* Univ Wis-Milwaukee, BS, 70; Purdue Univ, PhD(anal chem), 75. *Prof Exp:* Asst prof, Wayne State Univ, 75-80; MEM FAC, DEPT CHEM, OHIO STATE UNIV, 80- *Mem:* Am Chem Soc. *Res:* Measurement of homogeneous electron transfer rates; electrochemistry of porphyrins and related compounds with mechanistic studies. *Mailing Add:* Dept Chem Ohio State Univ 120 W 18th Ave Columbus OH 43210-1106

KARY, CHRISTINA DOLORES, biochemical pharmacology, inhalation toxicology, for more information see previous edition

KARZON, DAVID T, b New York, NY, July 8, 20; m 50; c 2. VIROLOGY, PEDIATRICS. *Educ:* Ohio State Univ, BS, 40, MS, 41; Johns Hopkins Univ, MD, 44; Am Bd Pediat, dipl; Am Bd Microbiol, dipl, 64. *Prof Exp:* Instr contagious dis, Johns Hopkins Univ Hosp, 45, 48, instr virol, Sch Med, 49-50; from asst prof to prof pediat, Sch Med, State Univ NY Buffalo, 52-68, from asst prof to prof virol, Dept Bact & Immunol, 54-68, dir virol lab, 52-68; PROF PEDIAT & CHMN DEPT, SCH MED, VANDERBILT UNIV, 68-, MED DIR, CHILDREN'S HOSP, UNIV, 71- *Concurrent Pos:* Lowell Palmer fel, 52-54; res career develop award, NIH, 62-68; Markle scholar, 56-61; spec consult, Nat Commun Dis Ctr, USPHS, Atlanta, Ga, 59-62, mem surgeon-gen spec adv comt immunization pract, 64-70; consult res reagents comt, Nat Inst Allergy & Infectious Dis, 63-67, chmn, 66-67, mem virol & rickettsiol study sect, 67-69; prog consult growth & develop sect, Nat Inst Child Health & Human Develop, 64-68; assoc ed, Am J Epidemiol, 66-78; mem biol rev steering comt, Food & Drug Admin, 72- *Mem:* Soc Pediat Res; Soc Exp Biol & Med; Fedn Am Soc Exp Biol; Am Epidemiol Soc; Infectious Dis Soc; Am Asn Immunologists; Am Acad Microbiol; Am Soc Virol; Am Soc Microbiol. *Res:* Animal virology; tissue culture. *Mailing Add:* Dept Pediat Vanderbilt Univ Sch Med T3313 Med Ctr N Nashville TN 37232

KAS, ARNOLD, b Washington, DC, July 18, 40. MATHEMATICS. *Educ:* Johns Hopkins Univ, BA, 62; Stanford Univ, PhD(math), 66. *Prof Exp:* Instr math, Stanford Univ, 66-67; Air Force Off Sci Res fel, Math Inst, State Univ Leiden, 67-69; asst prof, Univ Calif, Berkeley, 69-73; assoc prof math, Ore State Univ, 73-80, prof, 80-; AT BOEING HELICOPTERS. *Res:* Complex manifolds; algebraic geometry. *Mailing Add:* 171 McGraw St Seattle WA 98109

KASAHARA, AKIRA, b Tokyo, Japan, Oct 11, 26; US citizen; m 52; c 2. METEOROLOGY. *Educ:* Univ Tokyo, BS, 48, MS, 50, DSc, 54. *Prof Exp:* Asst geophys inst, Univ Tokyo, 48-53, res assoc, 53-54; res assoc oceanog & meteorol, Agr & Mech Col Tex, 54-56; res assoc meteorol, Univ Chicago, 56-62; res scientist, Courant Inst Math Sci, NY Univ, 62-63; prog scientist, 63-73, SR SCIENTIST, NAT CTR ATMOSPHERIC RES, 73- *Concurrent Pos:* Affil prof dept meteorol, Tex A&M Univ, 67-70; assoc ed, J Appl Meteorol, 67-72; vis lectr, Inst Meteorol, Univ Stockholm, Sweden, 71-72; adj prof, Dept Meteorol, Univ Utah, 79-; external examr, Dept Meteorol, Univ Nairobi, Kenya, 77-79. *Honors & Awards:* Award, Meteorol Soc Japan, 61. *Mem:* Am Geophys Union; Meteorol Soc Japan; fel Am Meteorol Soc; AAAS. *Res:* Dynamic meteorology; development of weather prediction methods with the numerical integration of thermo-hydro-dynamical equations. *Mailing Add:* Nat Ctr Atmospheric Res PO Box 3000 Boulder CO 80307-3000

KASAI, PAUL HARUO, b Osaka, Japan, Jan 30, 32, nat US; m 59; c 2. PHYSICAL CHEMISTRY. *Educ:* Univ Denver, BS, 55; Univ Calif, Berkeley, PhD(chem), 59. *Prof Exp:* Mem res staff, Hitachi Cent Res Lab, Japan, 59-62; res inst, Union Carbide Corp, 62-66; assoc prof chem, Univ Calif, Santa Cruz, 66-67; mem res staff, Res Inst, Union Carbide Corp, 67-75, group leader, 75-77, sr scientist, Tarrytown Tech Ctr, 77-79; mgr tech support, IBM Instruments Inc, Danbury, 79-85; Thomas J Watson res ctr, IBM Corp, Yorktown Heights, NY, 85-86, RES STAFF MEM, IBM ALMADEN RES CTR, 86- *Mem:* Am Chem Soc. *Res:* Magnetic resonance studies of polymer synthesis and degradation, organometallic complexes, free radicals and surface states; zeolites. *Mailing Add:* IBM Corp Almaden Res Ctr San Jose CA 95120

KASAMATSU, HARUMI, MOLECULAR BIOLOGY. *Educ:* Osaka Univ, Japan, PhD(molecular biol), 69. *Prof Exp:* PROF MOLECULAR BIOL, UNIV CALIF, LOS ANGELES, 84- *Mailing Add:* Dept Molecular Biol Univ Calif 405 Hilgard Ave Los Angeles CA 90024

KASAMEYER, PAUL WILLIAM, b Detroit, Mich, Sept 9, 43; m 65; c 3. GEOPHYSICS. *Educ:* Mass Inst Technol, BS, 65, PhD(geophys), 74; Yale Univ, MS, 66. *Prof Exp:* GEOPHYSICIST, LAWRENCE LIVERMORE NAT LAB, 74- *Mem:* Soc Explor Geophysicists; Sigma Xi; Am Geophys Union. *Res:* Collection and interpretation of geophysical data; thermal modeling; magnetotellurics; experimental studies of gravity. *Mailing Add:* Lawrence Livermore Lab PO Box 808-Mail Stop L-208 Livermore CA 94550

KASAPLIGIL, BAKI, b Istanbul, Turkey, Nov 13, 18; US citizen; m 55; c 2. BOTANY. *Educ:* Istanbul Univ, BSc, 41; Univ Calif, Berkeley, PhD(bot), 50; Univ Ankara, cert, 53. *Prof Exp:* Engr, Forest Serv Turkey, 41-42; asst botanist, Agr Col Ankara, 44-46; cur & botanist, Univ Ankara, 50-53, asst prof bot, 53-54; forest ecologist, Food & Agr Orgn, UN, 54-56; from asst prof to assoc prof, 56-66, PROF BOT, MILLS COL, 66- *Mem:* Bot Soc Am. *Res:* Floristics of the Middle East; ontogenetic studies in Lauraceae; histotaxonomy of Corylus; forest ecology of the Mediterranean region; tertiary flora of Asia Minor; systematics of Corylus and Quercus. *Mailing Add:* 1304 Albina Ave Berkeley CA 94706

KASARDA, DONALD DAVID, b Kingston, Pa, Oct 12, 33; m 64; c 1. PROTEIN CHEMISTRY. *Educ:* King's Col, Pa, BS, 55; Boston Col, MS, 57; Princeton Univ, MA, 59, PhD(phys chem), 61. *Prof Exp:* Mem Tech Staff, Bell Tel Labs, NJ, 61-63; Cardiovasc Res Inst fel, Sch Med, Univ Calif, San Francisco, 63-64; res chemist, 64-72, res leader, 72-85, RES CHEMIST, FOOD QUAL RES UNIT, WESTERN REGIONAL RES CTR, AGR RES SERV, USDA, 85- *Concurrent Pos:* Assoc Exp Sta, Dept Agron & Range Sci, Univ Calif, Davis, 74- *Mem:* AAAS; Am Chem Soc; Am Asn Cereal Chem. *Res:* Protein chemistry; wheat genetics. *Mailing Add:* Western Regional Res Ctr USDA Albany CA 94710

KASARSKIS, EDWARD JOSEPH, b Chicago, Ill, Oct 9, 46; m 69; c 4. NEUROLOGY. *Educ:* Col St Thomas, BA, 68; Univ Wis-Madison, MD, 74, PhD(biochem), 75. *Prof Exp:* Resident internal med, Univ Wis-Madison Hosp, 74-76; resident neurol, Univ Va Hosp, 76-79; asst prof neurol, Sch Med, La State Univ, 79-80; asst prof neurol, 80-85, ASSOC PROF NEUROL & TOXICOL, UNIV KY, 85- *Concurrent Pos:* Staff neurologist, Vet Admin Hosp, Lexington, Ky, 80- *Mem:* Sigma Xi; Am Acad Neurol; Soc Neurosci; Am Soc Neurochem; Int Soc Neurochem; Int Soc Bioinorg Sci. *Res:* Role of trace metals in the function of the brain; investigative factors that modify zinc metabolism; role of trace metals in Amyotrophic lateral sclerosis. *Mailing Add:* Dept Neurol Univ Ky Col Med 800 Rose St Lexington KY 40536-0084

KASBEKAR, DINKAR KASHINATH, b Bombay, India, Apr 3, 32; m 61; c 1. PHYSIOLOGY, BIOCHEMISTRY. *Educ:* Univ Bombay, BSc, 52 & 54, MSc, 57; Univ Calif, PhD(biochem), 61. *Prof Exp:* Jr res biochemist, Univ Calif, San Francisco, 61-63, asst res biochemist, Cardiovasc Res Inst, 63-65; asst res physiologist, Univ Calif, Berkeley, 66-68; from asst prof to assoc prof physiol & biophys, 69-84, PROF PHYSIOL & BIOPHYS, SCH MED & DENT, GEORGETOWN UNIV, 84- *Concurrent Pos:* San Francisco Heart Asn fel, 64-65; prin investr, Washington Heart Asn, 70-71, NSF, 70-; vis sr fel, Nat Inst Arthritis, Metab & Digestive Dis, NIH, 81. *Mem:* Biophys Soc; Am Physiol Soc; NY Acad Sci. *Res:* Ion transport, zymogen secretion, specifically in the area of gastric secretion. *Mailing Add:* Dept Physiol 253 Basic Sci Georgetown Univ 37th & O Sts NW Washington DC 20057

KASCIC, MICHAEL JOSEPH, JR, mathematics; deceased, see previous edition for last biography

KASCSAK, RICHARD JOHN, b Whitestone, NY, Sept 20, 47; m 72; c 2. SLOW VIRUS & PERSISTENT INFECTIONS. *Educ:* St Francis Col, NY, BS, 69; Adelphi Univ, MS, 71; Cornell Univ Med Col, PhD(virol), 76. *Prof Exp:* Teaching asst, Adelphi Univ, NY, 69-71; training fel, Cornell Univ, 71-75; RES SCIENTIST, NY STATE INST BASIC RES DEVELOP DISABILITIES, 75- *Mem:* Am Soc Microbiol; Am Soc Virol. *Res:* Slow viral infections of the central nervous system with emphasis on the creation of model systems relevant to human disease; unconventional slow virus diseases; scropie. *Mailing Add:* The NY State Inst Basic Res Develop Disabilities 1050 Forest Hill Rd Staten Island NY 10314

KASE, KENNETH RAYMOND, b Oak Park, Ill, July 13, 38; m 62; c 2. MEDICAL PHYSICS, HEALTH PHYSICS. *Educ:* Ga Inst Technol, BS, 61; Univ Calif, Berkeley, MS, 63; Stanford Univ, PhD(biophys), 75; Am Bd Health Physics, cert, 69; Am Bd Radiol, cert, 81. *Prof Exp:* Scientist reactors environ, Lockheed Missiles & Space Co, 61-62; health physicist radiation safety, Lawrence Livermore Lab, Univ Calif, 63-67, chief radiation safety, 67-69; health physicist, Stanford Linear Accelerator, Stanford Univ, 69-73; asst prof & chief dosimetry & radiation safety radiol physics, Harvard Med Sch, 75-84; PROF & DIR PHYSICS, DEPT RADIATION & ONCOL, MED CTR, UNIV MASS, 85- *Concurrent Pos:* Ed, Health Physics J, Health Physics Soc, 77-82; adj prof, Lowell Univ, 78-; mem, sci comt, Nat Coun Radiation, Protection & Measurement, 46 & 83-; treas, Am Asn Physicists in Med, 86-; mem, Nat Coun Radiation Protection & Measurement, 87- *Honors & Awards:* Elda E Anderson Award, Health Physics Soc, 78. *Mem:* Health Physics Soc; Am Asn Physicists Med (treas, 86-); Radiation Res Soc; fel Am Col Radiol; Am Col Med Physics; Am Acad Health Physics. *Res:* Radiation measurement and dosimetry; biological effects of radiation; application of new treatment modalities to cancer therapy. *Mailing Add:* Dept Radiation Oncol Med Ctr Univ Mass 55 Lake Ave N Worcester MA 01655

KASEL, JULIUS ALBERT, b Homestead, Pa, Dec 7, 23; m 50; c 3. MICROBIOLOGY, VIROLOGY. *Educ:* Univ Pittsburgh, BS, 49; Georgetown Univ, MS, 58, PhD(microbiol, virol), 60. *Prof Exp:* Head med virol sect, Nat Inst Allergy & Infectious Dis, 50-72; PROF MICROBIOL & IMMUNOL, BAYLOR COL MED, 72- *Concurrent Pos:* Assoc dir, Influenza Res Ctr. *Mem:* Infectious Dis Soc Am; Am Asn Immunologists; Am Soc Microbiol; Soc Exp Biol & Med; Soc Gen Microbiol; Sigma Xi; fel Am Acad Microbiol. *Res:* Virological research related to respiratory viral infections in man. *Mailing Add:* Dept Microbiol Immunol Baylor Col Med Houston TX 77030

KASER, J(OHN) D(ONALD), b Oak Park, Ill, Nov 21, 29; div; c 2. CHEMICAL ENGINEERING. *Educ:* Augustana Col, BA, 56; Univ Iowa, BS, 58, MS, 60, PhD(chem eng), 63. *Prof Exp:* Sr develop engr, Battelle-Northwest, 63-76; staff engr, 76-80, PRIN ENGR, ROCKWELL HANFORD OPER, 80- *Concurrent Pos:* Mem fac, Joint Ctr for Grad Study, 68-; mem steering comn on shallow land burial of radioactive waste, Dept Energy, 77-78. *Mem:* Am Inst Chem Engrs; Am Nuclear Soc. *Res:* Solidification and disposal of radioactive waste from nuclear fuel reprocessing; decontamination; solvent extraction; radioactive waste management; heat transfer. *Mailing Add:* Pac Northwest Labs PO Box 999 P7-41 Richland WA 99352

KASETA, FRANCIS WILLIAM, b Norwood, Mass, June 6, 33; m 60; c 4. SOLID STATE PHYSICS, AMATEUR RADIO. *Educ:* Boston Col, BS, 55; Mass Inst Technol, PhD(solid state physics), 62. *Prof Exp:* Asst prof elec eng, Mass Inst Technol, 62-64; asst prof physics, 64-67, chmn dept, 77-83, ASSOC PROF PHYSICS, COL HOLY CROSS, 67- *Concurrent Pos:* Ford Found fel, 62-64; consult, Mass Inst Technol, 64-65. *Mem:* Am Phys Soc. *Res:* Dielectric breakdown; conduction processes in semiconductors and dielectrics; electrooptics. *Mailing Add:* Dept Physics Col Holy Cross College St Worcester MA 01610

KASH, KATHLEEN, b Corona, Calif, Nov 28, 53; m 78; c 2. SYSTEMS OF REDUCED DIMENSIONALITY, ULTRAFAST PHENOMENA. *Educ:* Middlebury Col, BA, 75; Mass Inst Technol, PhD(physics), 82. *Prof Exp:* Postdoctoral, AT&T Bell Labs, 82-84; MEM TECH STAFF, BELLCORE, 84- *Mem:* Am Phys Soc. *Res:* Confinement of excitons to quantum wires and dots; carrier relaxation-diffusion in semiconductors; optical properties of quantum wells; strain confinement of excitons in semiconductors. *Mailing Add:* Bellcore 3G113 331 Newman Springs Rd Red Bank NJ 07701-7040

KASHA, HENRY, b Warsaw, Poland. HIGH ENERGY PHYSICS, COSMIC RAY PHYSICS. *Educ:* Hebrew Univ, Jerusalem, MSc, 54; Israel Inst Technol, DSc(physics), 60. *Prof Exp:* Lectr physics, Israel Inst Technol, 60-63; asst physicist, Brookhaven Nat Lab, 64-66, assoc physicist, 66-70; sr res assoc, 70-73, SR RES PHYSICIST, YALE UNIV, 73- *Mem:* AAAS; Am Phys Soc. *Mailing Add:* Dept Physics Yale Univ PO Box 6666 New Haven CT 06511

KASHA, KENNETH JOHN, b Lacombe, Alta, May 6, 33; m 58; c 2. PLANT CYTOGENETICS, PLANT CELL CULTURE. *Educ:* Univ Alta, BSc, 57, MSc, 58; Univ Minn, PhD(plant genetics), 62. *Hon Degrees:* LLD, Univ Calgary, 86. *Prof Exp:* Teaching asst, Univ Minn, 60-61; res officer 2, Ottawa Res Sta, Can Dept Agr, 62-64, res officer 3, 64-66, res scientist 1, 66; dir, Plant Biotech Ctr Guelph, Waterloo Biotech, 84-87; asst prof crop sci & crop cytogeneticist, 66-69, assoc prof crop cytogenetics, 69-74, PROF CROP CYTOGENETICS, UNIV GUELPH, 74- *Concurrent Pos:* Orgn chmn, Int Symposium Haploids in Higher Plants, Guelph, 74; vis scientist, Plant Indust, Commonwealth Sci & Indust Orgn, Canberra, Australia, 85-86; prog chmn, XVI Int Cong of Genetics, Toronto, 88; nat corresp, Int Asn Plant Cell & Tissue Cult, 90- *Honors & Awards:* Grindley Medal, Agr Inst Can, 77; E C Manning Award, 83; Nilsson-Ehle Lectr, Sweden, 87. *Mem:* Genetics Soc Am; Genetics Soc Can (secy, 66-69, dir, 70-72, vpres, 75, pres, 76); Am Soc Agron; fel Royal Soc Can; Int Asn Plant Cell & Tissue Cult; Can Soc Plant Molecular Biol. *Res:* Crop plant cytogenetics; haploidy in cereals; molecular cytology; interspecific hybridization and chromosome pairing in Hordeum, Triticum and Secale; linkage and RFLP mapping of barley; plant cell culture in cereals; male sterility and self-incompatibility; molecular biology. *Mailing Add:* Dept Crop Sci Univ Guelph Guelph ON N1G 2W1 Can

KASHA, MICHAEL, b Elizabeth, NJ, Dec 6, 20; m 47; c 1. CHEMICAL PHYSICS, SPECTROSCOPY & MOLECULAR ELECTRONIC PHENOMENA. *Educ:* Univ Mich, BS, 43; Univ Calif, PhD(phys chem), 45. *Hon Degrees:* DSc, Gonzaga Univ, 88. *Prof Exp:* Lab asst, Res Lab, Merck & Co, Inc, 38-41; res chemist, Plutonium Proj, Univ Calif, 44-46, univ fel & instr, 46, res assoc, 46-49; AEC fel, Univ Chicago, 49-50; Guggenheim fel & spec lectr, Univ Manchester, 50-51; chmn, Dept Chem, 59-62, dir, Inst Molecular Biophys, 60-80, PROF CHEM, FLA STATE UNIV, 51-, ROBERT O LAWTON DISTINGUISHED PROF, 62- *Concurrent Pos:* Vis prof, Harvard Univ, 59-60; Charles F Ketteing Res Award, Gen Motors Corp, 63-69; vis prof chem, Univ Mich, 69; Nat Sci bd, Pres Carter, 79-84, France, 80-; exec comt, Inst La Vie, Paris, France, 80-; foreign coun, Inst Molecular Sci, Okazaki, Japan, 82-85; mem, Sci & Tech Adv Comt, Argonne Nat Lab, 83-8; sci adv to Gov Bob Graham, Fla, 83-87. *Honors & Awards:* Phillips Lectr, Haverford Col, 59; Reilly Lectr, Univ Note Dame, 59; S C Lind Lectr, Oak Ridge Nat Lab, 61; Charles F Kettering Res Award, 63-69; George Porter Medal, 90; Robert Mulliken Lectr, Univ Ga, 90. *Mem:* Nat Acad Sci; fel Am Acad Arts & Sci. *Res:* Molecular biophysics and electronic spectroscopy; triplet states of molecules; emission spectroscopy of molecules; classification of electronic transitions; spin-intercombinations; n-pi transitions; radiationless transitions; theoretical photochemistry; molecular excitons and energy transfer; biological molecular interactions. *Mailing Add:* Inst Molecular Biophys Fla State Univ Tallahassee FL 32306-3014

KASHAR, LAWRENCE JOSEPH, b Brooklyn, NY, June 1, 33; m 81; c 3. METALLURGY, MATERIALS SCIENCE. *Educ:* Rensselaer Polytech Inst, BMetE, 55; Stevens Inst Technol, MS, 59; Carnegie Inst Technol, MS, 61; Carnegie-Mellon Univ, PhD(metall, mat sci), 70. *Prof Exp:* Assoc metallurgist, AMAX Res & Develop Co, Inc, 55-59; res assoc, Carnegie Inst Technol, 60-64; sr res metallurgist, US Steel Appl Res Lab, 64-70; mem tech staff, B-1 Div, Rockwell Int, 71-72; staff engr, Orlando Div, Martin Marietta Corp, 73; dir metall serv, Scanning Electron Anal Labs, Inc, 73-79, dir tech serv, 79-83, vpres technol, 73-91; PRES, KASHAR TECH SERV, 91- *Concurrent Pos:* Adj lectr, Univ Southern Calif, 75-81; vpres & secy, ATFA, Inc, 77-80; vpres, Litigation Consults Int, 78-81; chmn, struct anal prog, Int Symp Testing & Failure Anal, 80-87; chmn, Westec Conf, Am Soc Metals, 83. *Mem:* Am Soc Metals; Am Soc Testing & Mat; Inst Elec & Electronics Engrs; Sigma Xi; Int Soc Testing & Failure Anal; Am Chem Soc; Electron Micros Soc Am. *Res:* Causes and prevention of failures of metal structures; development of microanalytical and surface analysis techniques for practical materials problem solving. *Mailing Add:* Kashar Tech Serv Inc 250 N Nash St El Segundo CA 90245

KASHATUS, WILLIAM C, b Nanticoke, Pa, Apr 23, 29; m 54; c 2. PATHOLOGY, HEMATOLOGY. *Educ:* Wilkes Col, BS, 51; Bucknell Univ, MS, 53; Hahnemann Med Col, MD, 59. *Prof Exp:* Instr chem, Bucknell Univ, 51-52; from instr to assoc prof path, Hahnemann Med Col, 63-73, dir, Sch Med Technol, 64-71, vchmn dept, 73-79, PROF PATH, HAHNEMANN MED COL, 72- *Concurrent Pos:* Mary Bailey Heart Found fel, 56-58; Am Cancer Soc fel, 62-64; dir labs, Hahnemann Hosp, 64-70; mem tech adv bd, Southeast Pa Div, Am Red Cross, 68-72 & West Co, 69-71; med dir, SBCL, Philadelphia, 71- *Mem:* AMA; fel Col Am Path; Am Soc Clin Path; Acad Clin Lab Physicians & Scientists. *Res:* Hematology, especially cancer chemotherapy; blood banking, especially immunochematology; tissue typing. *Mailing Add:* 400 Egypt Rd Norristown PA 19403

KASHDAN, DAVID STUART, b New York, NY, Oct 21, 50; m 83; c 1. PHARMACEUTICAL APPLICATIONS OF POLYMERS, DRUG DELIVERY METHODS. *Educ:* Stevens Inst Technol, BS, 72; Univ Vt, PhD(org chem), 77. *Prof Exp:* Vis instr org chem, Univ Vt, 76-77; res fel, Univ Calif, Berkeley, 77-79; res chemist, 79-82, sr res chemist, 82-88, RES ASSOC, EASTMAN CHEM DIV, EASTMAN KODAK CO, 88- *Mem:* Am Chem Soc. *Res:* Organic synthesis; synthetic methods; organolithium reagents and halogenation of aromatics; synthesis of morphinans and isoquinolines; development of new polymers for use in pharmaceutical applications. *Mailing Add:* 408 Harding Rd Kingsport TN 37763

KASHEF, A(BDEL-AZIZ) I(SMAIL), b Cairo, Egypt, Feb 10, 19; m 48. GEOTECHNICAL ENGINEERING, GROUNDWATER SCIENCES. *Educ:* Univ Cairo, BS, 40, MS, 48; Purdue Univ, PhD(soil mech), 51. *Prof Exp:* Irrig engr, Egyptian Govt, 40-45 & 48-51; instr struct, Univ Cairo, 45-48; sr lectr, Ein Shams Univ, 51-54 & Univ Cairo, 54-56; prof, Am Univ Beirut, 56-60; vis prof soil mech & ground water, 62-67, prof, 67-80, EMER PROF CIVIL ENG, NC STATE UNIV, 80- *Concurrent Pos:* Consult soil engr, 52-; mem water-well comt, Nat Prod Coun, Govt Egypt, 53, mem, Nat Hydraul Comt, 56, mem tech comt, River Harbors Comt, 60-62; soil consult, High Aswan Dam Auth, Egypt, 54-56; dir, Consult Eng Off, Saudi Arabia, 59-60; ed, Water Resources Bull, 70-73. *Mem:* Fel Am Water Resources Asn; fel Am Soc Civil Engrs; Am Geophys Union. *Res:* Water resources research, especially in ground-water field and geotechnical engineering. *Mailing Add:* 5504 N Hills Dr Raleigh NC 27612

KASHGARIAN, MICHAEL, b New York, NY, Sept 20, 33; m 60; c 2. PATHOLOGY. *Educ:* NY Univ, BA, 54; Yale Univ, MD, 58. *Prof Exp:* Asst med, Sch Med, Wash Univ, 58-59; asst resident path, 59-61, from instr to assoc prof, 62-74, vchmn dept, 76-89, PROF PATH, YALE UNIV, 74- *Concurrent Pos:* Life Ins Med Res fel physiol, Univ Gottingen, 61-62; USPHS spec fel, 63-65 & res career award, 65-75; assoc pathologist, Yale New Haven Hosp, 63-66, asst attend pathologist, 66-69, attend pathologist, 69-, assoc chief pathologist, 76-86. *Mem:* Am Asn Path; Am Physiol Soc; Am Soc Clin Path; Am Soc Nephrology; Int Acad Path. *Res:* Pathology and physiology of the kidney. *Mailing Add:* Dept Path Sch Med Yale Univ PO Box 3333 New Haven CT 06510-8023

KASHIN, PHILIP, b New York, NY, Oct 27, 30; m 58; c 3. NEUROPHYSIOLOGY, PHARMACOLOGY. *Educ:* Brooklyn Col, BA, 53; Columbia Univ, MA, 58; NY Univ, MS, 61; Ill Inst Technol, PhD(physiol), 70. *Prof Exp:* Res asst immunochem, Hosp for Spec Surg, New York, 61-62; res asst biochem, State Univ NY Downstate Med Ctr, 62-63; asst biochemist, IIT Res Inst, 63-64, assoc biochemist, 64-69, res biochemist, 69-70; Nat Inst Neurol Dis & Stroke spec res fel neurophysiol, Univ Ore, 70-71; asst prof biol, Queens Col, NY, 71-76; sr asst dir clin res, USV Pharmaceut Corp, 76-82; clin monitor, Abbott Lab, 83-84; asst dir, Pfizer Int Corp, 84-86; ASSOC DIR CLIN AFFAIRS, PFIZER HOSP PROD GROUP, 86- *Concurrent Pos:* Consult, 82-83. *Honors & Awards:* I R 100 Award, 67. *Mem:* AAAS; Soc Neurosci; NY Acad Sci; Am Physiol Soc. *Res:* Methods to assay mosquito repellents; cardiovascular and anti-infective clinical research; mechanism of action of carbon dioxide with neurotransmitters in the central nervous system; clinical research. *Mailing Add:* 47 Glen Cove Dr Glen Head NY 11545

KASHIWA, BRYAN ANDREW, b Oswego, NY, Feb 20, 52; m 69; c 3. NUMERICAL FLUID DYNAMICS. *Educ:* Worcester Polytech Inst, BS, 73; Univ Wash, MS, 78, PhD, 87. *Prof Exp:* Engr, K2 Corp, 73-79; MEM STAFF, LOS ALAMOS NAT LAB, 79- *Mem:* Am Soc Mech Engrs. *Res:* Application and development of methods in numerical fluid dynamics with emphasis on multifield flows. *Mailing Add:* Gr T-3 MS-216 Los Alamos Nat Lab Los Alamos NM 87545

KASHIWA, HERBERT KORO, b Waialua, Hawaii, Nov 12, 28; m 47; c 3. ANATOMY, HISTOCHEMISTRY. *Educ:* Univ Hawaii, BS, 50; George Washington Univ, MS, 54, PhD(anat), 60. *Prof Exp:* From instr to asst prof anat, Sch Med, Univ Louisville, 59-66; asst prof, 66-71, ASSOC PROF ANAT, SCH MED, UNIV WASH, 71- *Mem:* Am Asn Anat; Histochem Soc; Biol Stain Comn; Sigma Xi. *Res:* Mineral metabolism; hormonal control of magnesium; development of the glyoxal bis method for labile calcium; role of cells as depositors of calcium during ossification, calcification and odontogenesis. *Mailing Add:* Biol Structure SM-20 Univ Wash Seattle WA 98195

KASHKARI, CHAMAN NATH, b Srinagar, India, Aug 27, 33; m 63; c 2. ELECTRICAL ENGINEERING. *Educ:* Univ Jammu & Kashmir, India, BA, 52; Univ Rajasthan, BS, 57; Univ Detroit, MS, 65; Univ Mich, Ann Arbor, PhD(elec eng), 69. *Prof Exp:* Asst engr, Gen Elec Co, India, 57-58, grad trainee, Eng, 58-60, plant engr, India, 61-64; teaching fel, Univ Detroit, 64-65 & Univ Mich, Ann Arbor, 66-69; asst prof elec eng, 69-75, ASSOC PROF ELEC ENG, UNIV AKRON, 75- *Concurrent Pos:* Energy consult, Govt Nepal, NSF, 75. *Mem:* Inst Elec & Electronics Engrs; Am Soc Eng Educ; Sigma Xi. *Res:* Energy planning in developing countries; solar energy; biogas plants; mini power plants; energy conservation; electric power systems engineering. *Mailing Add:* Dept Elec Eng Univ Akron Akron OH 44304

KASHKET, EVA RUTH, b Zagreb, Yugoslavia, Mar 1, 36; US citizen; m 57; c 2. BACTERIOLOGY, BIOCHEMISTRY. *Educ:* McGill Univ, BSc, 56, MSc, 57; Harvard Univ, PhD(med sci), 63. *Prof Exp:* Res assoc, Dept Physiol, Harvard Med Sch, 57-74; assoc prof, 74-82, PROF MICROBIOL, SCH MED, BOSTON UNIV, 82- *Concurrent Pos:* Fel biochem pharmacol, Sch Med Tufts Univ, 62-65. *Mem:* Am Soc Microbiol; Am Soc Biochem & Molecular Biol. *Res:* Bioenergetics, fermentations; clostridia, lactobacilli. *Mailing Add:* Dept Microbiol Boston Univ Sch Med Boston MA 02118

KASHKET, SHELBY, b Montreal, Que, Feb 1, 31; m 57; c 2. BIOCHEMISTRY. *Educ:* McGill Univ, BSc, 52, MSc, 53, PhD(biochem), 56. *Prof Exp:* Res fel biochem, McGill Univ, 56-57; res fel bact, Harvard Med Sch, 57-59; asst biochemist, Mass Gen Hosp, 63-67; biochemist, USPHS, 67-70; ASST RES PROF, SCH MED, BOSTON UNIV, 67-, ASST RES PROF, SCH GRAD DENT, 71-; ASSOC MEM STAFF, FORSYTH DENT CTR, 80- *Concurrent Pos:* Res fel med, Harvard Med Sch, 59-60, res assoc, 60-77, lectr, 79- *Mem:* AAAS; Int Asn Dent Res; Am Chem Soc; Am Soc Biochem & Molecular Biol; Sigma Xi. *Res:* Metabolic effects of fluoride and food products; biological adhesion; oral biology; intermediary metabolism; biochemical and clinical methods; foods and dental disease. *Mailing Add:* Forsyth Dent Ctr 140 The Fenway Boston MA 02115

KASHKOUSH, ISMAIL I, b Egypt, Nov 10, 58; m 90. UNDERWATER ACOUSTICS, MICROCONTAMINATION CONTROL. *Educ:* Cairo Univ, Egypt, BS, 82, MS, 86. *Prof Exp:* Teaching asst, Dept Mech Design, Cairo Univ, 82-86; instr solid mech, Acad Defense, Cairo, 87-88; RES ASST MICROCONTAMINATION CONTROL, CLARKSON UNIV, 88- *Concurrent Pos:* Assoc lectr, Dept Mech Design, Cairo Univ, 86-88; instr thermodyn & fluid mech, Clarkson Univ, 90 & 91. *Mem:* Am Inst Aeronaut & Astronaut; Am Soc Mech Engrs; Am Soc Prof Engrs; Inst Environ Sci. *Res:* Microcontamination control in the clean room environment using various removal techniques; detection of particulates on silicon wafers; research in different aspects of contamination; establishing numerical simulation of the surface cleaning using sonic methods; developing new models that accurately describe the cleaning mechanism; author of several publications. *Mailing Add:* MAE Dept Clarkson Univ Potsdam NY 13699-5725

KASHNOW, RICHARD ALLEN, b Worcester, Mass, Mar 26, 42; m 63; c 2. PHYSICS. *Educ:* Worcester Polytech Inst, BS, 63; Tufts Univ, PhD(physics), 68. *Prof Exp:* US Army, 68-70. *Mem:* Am Phys Soc; Inst Elec & Electronics Engrs. *Res:* Liquid crystals; quantum electronics; lattice dynamics. *Mailing Add:* 5116 S Kenton Way Englewood CO 80111

KASHY, EDWIN, b Beirut, Lebanon, July 8, 34; US citizen; m 57; c 2. EXPERIMENTAL NUCLEAR PHYSICS. *Educ:* Rice Univ, BA, 56, MA, 57, PhD(physics), 59. *Prof Exp:* NSF fel physics, Mass Inst Technol, 59-60, instr, 60-62; asst prof, Princeton Univ, 62-64; assoc prof, 64-67, PROF PHYSICS, MICH STATE UNIV, 67- *Concurrent Pos:* Guggenheim fel, Niels Bohr Inst, Copenhagen, 70-71. *Mem:* Am Phys Soc. *Res:* Experimental investigations of nuclear spectroscopy and nuclear reaction mechanisms by means of charged particle and gamma ray studies. *Mailing Add:* Cyclotron Lab Dept Physics Mich State Univ East Lansing MI 48823

KASHYAP, MOTI LAL, b Singapore, Feb 19, 39; m 70; c 3. GERONTOLOGY, LIPIDOLOGY. *Educ:* Univ Singapore, MB, BS, 64; McGill Univ, MS, 67; FRCP(C), 69, FACP, 78. *Prof Exp:* Intern med & surg, Teaching Hosps, Univ Singapore, 64-65; resident internal med & fel endocrinol & metab, Royal Victoria Hosp, McGill Univ, 65-69, lectr, Fac Med, 69-70; sr fel, Cardiovasc Res Inst, Moffit Hosp, Sch Med, Univ Calif, San Francisco, 70-71; sr lectr med physiol, Fac Med, Univ Singapore, 71-74; from asst prof to assoc prof, 74-81, prof med, Col Med, Univ Cincinnati, 81-86; PROF MED, CALIF COL MED, UNIV CALIF, IRVINE, 86- *Concurrent Pos:* Sr res fel, Am Heart Asn, 70-71, fel, Arteriosclerosis Coun; assoc dir-Lipid Res Clin, Cincinnati, 74-; dir, Apoliprotein Res Labs, Cincinnati, 74-86; dir, Coronary Primary Prevention Trial, NIH, 74-78; chief of geriatric res, 86-88, chief gerontol, Vet Admin Med Ctr, Long Beach, Calif, 88-; res awards, Nat Inst Health & Am Heart Asn; Irvine Page Young Investr Award, Am Heart Asn, 76. *Mem:* Am Fedn Clin Res; Int Soc Cardiol; Can Soc Endocrinol & Metab; Cent Soc Clin Res. *Res:* Atherosclerosis; preventive cardiology; risk factors; geriatrics; apolipoproteins; immuno-assays; lipoproteins; cholesterol; metabolism; high density lipoproteins; reverse cholesterol transport. *Mailing Add:* Vet Admin Med Ctr 111GE 5901 E Seventh St Long Beach CA 90822

KASHYAP, RANGASAMI LAKSMINARAYANA, b Mysore, India, Mar 28, 38. ELECTRICAL ENGINEERING. *Educ:* Univ Mysore, BSc, 58; Indian Inst Sci, Bangalore, Dipl, 61, MEng, 63; Harvard Univ, PhD(eng), 66. *Prof Exp:* Res asst control systs, Harvard Univ, 63-65, teaching fel, 64, res fel eng, 65-66; from asst prof to assoc prof elec eng, 66-74, PROF ELEC ENG, PURDUE UNIV, 74- *Concurrent Pos:* Gordon McKay fel, 62-63; consult, Gen Elec Co, Ind, 67-68. *Honors & Awards:* King Sun Fu Res Award, Int Asn Pattern Recognition, 90. *Mem:* Fel Inst Elec & Electronics Engrs; Asn Comput Mach; Sigma Xi. *Res:* Systems science; pattern recognition; learning systems; statistical inference; image processing. *Mailing Add:* Sch Elec Eng Purdue Univ West Lafayette IN 47907

KASHYAP, TAPESHWAR S, b Kapurthala, India, Oct 15, 29; US citizen; m 59; c 2. POPULATION GENETICS, POULTRY BREEDING. *Educ:* Punjab Agr Col, India, BSc, 50; Univ Minn, St Paul, MSc, 56, PhD(animal husb), 58. *Prof Exp:* Asst animal husb, Univ Minn, St Paul, 55-58; geneticist & head data processing dept, 59-73, DIR GENETICS DEVELOP, KIMBER FARMS INC, 73- *Concurrent Pos:* FAO consult, Poultry Proj POL/71/515, Poznan, Poland, 76-78; consult geneticist poultry res proj, Animal Sci Dept, Univ Nebr, 78-; systs consult, Bank Am, 81- *Mem:* Genetics Soc Am; Am Genetics Asn; AAAS. *Res:* Animal breeding; improving livestock performance with the aid of principles of genetics; statistical analysis of data, using modern computers, to evaluate and seek answers to various problems in poultry breeding. *Mailing Add:* 41532 Pasco Padre Pkwy Fremont CA 94538

KASI, LEELA PESHKAR, b Bombay, India, July 15, 39; US citizen; m 71. RADIOPHARMACEUTICAL CHEMISTRY, NUCLEAR MEDICINE. *Educ:* Univ Bombay, India, BS, 58; Univ Marburg, W Ger, PhD(pharmaceut chem), 68. *Prof Exp:* Sr chemist, pharmaceut quality control, Boehringer-Knoll Ltd, Bombay, India, 69-71; dir quality control, pharmaceut chem, Health Care Indust, Mich City, Ind, 72-77; asst chemist radiopharmaceut chem, 79-90, ASST PROF NUCLEAR MED, M D ANDERSON CANCER CTR, UNIV TEX, 82-, ASSOC CHEMIST, 90- *Concurrent Pos:* Asst prof clin radiol, Univ Tex Med Sch, Houston, 82-, mem grad fac, Grad Sch Biomed Sci, 84-89; asst ed, J Nuclear Med, 84-89. *Mem:* Soc Nuclear Med; AAAS; Sigma Xi; NY Acad Sci. *Res:* Development and evaluation (in vitro and in vivo) of new radiolabeled substances for use in diagnostic imaging, radioimmunotherapy or biokinetic studies in cancer patients; radioimmunoimaging, pharmacology and nuclear magnetic resonance; radiopharmaceutic chemistry. *Mailing Add:* M D Anderson Cancer Ctr Univ Tex 1515 Holcombe Blvd Box 57 Houston TX 77030

KASIK, JOHN EDWARD, b Chicago, Ill, Aug 9, 27; m 45; c 6. MEDICINE, PHARMACOLOGY. *Educ:* Roosevelt Univ, BS, 49; Univ Chicago, MS, 53, MD, 54, PhD(pharmacol), 62; Am Bd Internal Med, dipl. *Prof Exp:* Intern, Clins Univ Chicago, 54-55, from jr asst to asst resident med, 55-56, from asst prof to assoc prof, Grad Sch Med, 59-70; assoc prof, 70-73, PROF MED, COL MED, UNIV IOWA, 73-, ASSOC DEAN, 80-; CHIEF OF STAFF, VET ADMIN MED CTR, IOWA CITY, 80- *Concurrent Pos:* Miller fel, Univ Chicago, 57-59; Fulbright scholar, Dunn Sch, Oxford Univ, 66-67; med dir, Kirchwood Col-Vet Admin Hosp Iowa City Sch Respiratory Ther. *Honors & Awards:* Walter L Bierring Award, Am Thoracic Soc, Iowa, 76. *Mem:* Am Thoracic Soc; Am Fedn Clin Res; fel Am Col Physicians; fel Am Soc Clin Pharmacol & Therapeut; Am Acad Clin Toxicol. *Res:* Pharmacology of immunosuppressent drugs; pharmacology of antibiotics. *Mailing Add:* Dept Internal Med Univ Iowa Iowa City IA 52240

KASINSKY, HAROLD EDWARD, b New York, NY, Jan 20, 41; m 67, 84; c 3. BIOCHEMISTRY, ZOOLOGY. *Educ:* Columbia Univ, BA, 61; Univ Calif, Berkeley, PhD(biochem), 67. *Prof Exp:* NIH fel, Dept Embryol, Carnegie Inst, 67-69; asst prof, 69-81, ASSOC PROF ZOOL, UNIV BC, 81- *Concurrent Pos:* Vis prof, Univ Calif, Berkeley, Univ Calgary, Univ Amsterdam & Polytech Univ Barcelona. *Mem:* AAAS; Am Soc Zool; Soc Develop Biol; Can Soc Cell Biol; Am Soc Cell Biol; Can Soc Zool. *Res:* Comparative aspects of sperm protein diversity in vertebrates and invertebrates; characterization of sperm basic protein and their genes in reptiles, amphibians, fish, tunicates, mollusks, worms and plants. *Mailing Add:* Dept Zool Univ BC Vancouver BC V6T 2A9 Can

KASK, JOHN LAURENCE, b Red Deer, Alta, Mar 21, 06; nat US; m 35; c 2. FISHERIES. *Educ:* Univ BC, BA, 28; Univ Wash, Seattle, PhD(zool), 36. *Prof Exp:* Asst, Biol Bd Can, 28; asst scientist, Int Fisheries Comn, 29-38; assoc scientist & asst dir, Int Salmon Comn, 39-43; cur aquatic biol, Calif Acad Sci, 45-48; chief biologist, Fisheries Div, Food & Agr Orgn, UN, 48-50; chief invests & asst dir, Pac Oceanic Fish Invest, Honolulu, 51; chief officer of foreign activ, US Fish & Wildlife Serv, & asst dir fish, Washington, DC, 51-53; chmn & sci adminstr, Fisheries Res Bd Can, 53-63; dir invests, Inter-Am Trop Tuna Comn, 63-71; consult, fisheries & biol oceanog, Food & Agr Orgn, UN, 71-81; RETIRED. *Concurrent Pos:* Lectr, Univ Hawaii, 51; lectr sch fisheries, Univ Wash, 35-43; assoc, Scripps Inst Oceanog, Univ Calif, San Diego; consult, Govt of Costa Rica, 47 & US Dept State, 47-48. *Mem:* AAAS; Am Fisheries Soc; Am Soc Ichthyol & Herpet; Am Soc Limnol & Oceanog. *Res:* Population dynamics; genetics; scientific administration. *Mailing Add:* 5877 Honors Dr San Diego CA 92122

KASK, UNO, b Sadala, Estonia, Sept 16, 22; US citizen. INORGANIC CHEMISTRY. *Educ:* Univ Ga, BS, 50; Univ Minn, MA, 56; Univ Tex, PhD(inorg chem), 63. *Prof Exp:* Instr chem, Armstrong Col, 52-55 & Eureka Col, 55-56; asst prof, Valdosta State Col, 56-57, Am Int Col, 57-60, Ind State Col, 61-62 & Univ Ga, 63-66; assoc prof, 66-69, chmn dept, 70-72, PROF

CHEM, TOWSON STATE UNIV, 69- *Mem:* Am Chem Soc. *Res:* Nonaqueous reactions and reaction mechanisms of transition metal compounds with liquid sulfur; science education; science writing. *Mailing Add:* Dept Chem Towson State Univ Towson MD 21204

KASKA, HAROLD VICTOR, b Brooklyn, NY, Jan 11, 26; m 50; c 3. GEOLOGY. *Educ:* NY Univ, BA, 50; Univ Ind, MA, 52. *Prof Exp:* Paleont asst, Univ Ind, 50-52; micropaleontologist, Dominion Oil Ltd, 52-53, chief paleontologist, 53-56; explor paleontologist, Calif Explor Co, 56-57; stratig paleontologist, Compania Guatemala Calif de Petroleo, 57-62; palynologist, Calif Explor Co, 62-65; paleontologist, Chevron Explor Co, 65-67; sr paleontologist, Standard Oil Co, Calif, 68-71; staff paleontologist, Chevron Overseas Petrol Inc, 71-85; RETIRED. *Mem:* Am Asn Petrol Geol; Am Inst Prof Geol; Paleont Soc; Am Asn Stratig Palynologist; Brit Micropaleont Soc; Swiss Geol Soc. *Res:* palynology, paleozoic, mezoic and tertiary spores, pollen dinoflagellates and acritards; micropaleontology, Mesozoic and Tertiary foraminifera, planktonics, tintinnids, calcareous nannoplankton; application of micropaleontology and palynology to petroleum exploration, dating, correlation and facies studies. *Mailing Add:* 50 Nottinghsm Circle Cayton CA 94517

KASKA, WILLIAM CHARLES, b Ancon, CZ, May 13, 35; m 64; c 4. CHEMISTRY. *Educ:* Loyola Univ, Calif, BS, 57; Univ Mich, PhD(chem), 63. *Prof Exp:* Res assoc chem, Pa State Univ, 63-64; asst prof, 65-74, assoc prof, 74-79, PROF CHEM, UNIV CALIF, SANTA BARBARA, 79- *Mem:* Am Chem Soc; Sigma Xi. *Res:* Synthesis and chemistry of organometallic compounds of the transition elements. *Mailing Add:* Dept Chem Univ Calif Santa Barbara CA 93106

KASKAS, JAMES, b Detroit, Mich, Jan 30, 39. THEORETICAL PHYSICS. *Educ:* Wayne State Univ, BS, 60, MS, 61, PhD(physics), 64. *Prof Exp:* from instr to assoc prof physics, Detroit Inst Technol, 63-89; RETIRED. *Mem:* Am Phys Soc; Am Asn Physics Teachers. *Res:* Elementary particle theory; quantum field theory. *Mailing Add:* 14255 Mark Twain St Detroit MI 48227

KASLER, FRANZ JOHANN, b Vienna, Austria, Jan 1, 30; m 64; c 2. ANALYTICAL CHEMISTRY. *Educ:* Univ Vienna, PhD(org microanal), 59. *Prof Exp:* Asst prof, 59-65, ASSOC PROF CHEM, UNIV MD, COLLEGE PARK, 65- *Mem:* AAAS; Am Chem Soc; Am Microchem Soc; Austrian Chem Soc; Sigma Xi. *Res:* Quantitative nuclear magnetic resonance; organic elemental analysis of classic and instrumental methods. *Mailing Add:* Dept Chem Univ Md College Park MD 20742

KASLICK, RALPH SIDNEY, b Brooklyn, NY, Oct 17, 35. PERIODONTICS, ORAL MEDICINE. *Educ:* Columbia Univ, AB, 56, DDS, 59. *Prof Exp:* From instr to prof periodont & oral med, Sch Dent, Fairleigh Dickinson Univ, 65-88, from asst dean to dean acad affairs, 72-88; DIR DENT, GOLDWATER MEM HOSP, MED CTR, NY UNIV, 88-, CLIN PROF PERIODONT, SCH DENT, 88- *Concurrent Pos:* Res grants from various indust firms & founds, 65-; res consult for various indust firms, 69- *Honors & Awards:* Jour Award, Int Col Dent, 72; Medallion, Japan Stomatol Soc, 78. *Mem:* Fel Am Col Dent; Am Acad Periodont; Int Asn Dent Res; Sigma Xi; Am Asn Dent Schs. *Res:* Genetic studies of periodontal diseases in young adults; quantitative analysis of gingival fluid; clinical testing of therapeutic dentifrices, ointments and hygiene aids. *Mailing Add:* Goldwater Mem Hosp NY Univ Med Ctr Roosevelt Island New York NY 10044

KASLOW, ARTHUR L, multiple sclerosis, rheumatoid arthritis, for more information see previous edition

KASLOW, CHRISTIAN EDWARD, b Mora, Minn, Mar 20, 03; m 26; c 2. ORGANIC CHEMISTRY. *Educ:* Hamline Univ, BS, 24; Univ Iowa, MS, 29; Univ Minn, PhD(org chem), 43. *Prof Exp:* Teacher pub sch, Minn, 24-30; instr chem, NDak Col, 30-41; from instr to prof, Ind Univ, Bloomington, 41-71, dir labs, 63-68, emer prof chem, 71; RETIRED. *Mem:* Fel AAAS; Am Chem Soc; Sigma Xi. *Res:* Quinoline chemistry. *Mailing Add:* 2455 Tamarack Trail No 202 Bloomington IN 47408

KASLOW, DAVID EDWARD, b Bloomington, Ind, Sept 27, 42. AEROSPACE SYSTEM ENGINEERING. *Educ:* Ind Univ, Bloomington, AB, 64, MS, 66; Univ Mich, Ann Arbor, PhD(physics), 71. *Prof Exp:* Res assoc physics, Lehigh Univ, 71-73; ANALYST & MGR, VALLEY FORGE SPACE CTR, GEN ELEC CO, 73- *Mailing Add:* Valley Forge Space Ctr Gen Elec Co PO Box 8555 Philadelphia PA 19101

KASLOW, RICHARD ALAN, b Omaha, Nebr, Mar 1, 43. MEDICINE. *Educ:* Yale Col, BA, 65; Harvard Med Sch, MD, 69; Harvard Sch Pub Health, MPH, 76. *Prof Exp:* Epidemiologist, Sidney Farber Cancer Ctr, 75-76; chief, Arthritis & Immunol Dis Activ, Ctr Dis Control, 76-79; clin asst prof med, Emory Univ Sch Med, 76-79; sr surgeon, US Pub Health Serv, 76-85; CHIEF, EPIDEMIOL & BIOMET BR, NAT INST ALLERGY & INFECTIONS DIS, 80-; CAPT, USPHS, 85- *Concurrent Pos:* Diabetes mellitus coord comt, 79-81, epidemiol comt, NIH, 79-; Reye Syndrome Task Force, 82-; Prev Activ, Hosp Infections Prog, Ctr Dis Control, 76-78; chmn adv comt, Study AIDS Natural Hist, NIH, 83-85; task force AIDS epidemiol, USPHS, 84-86; workshop on epidemiol & dis burden AIDS, Inst Med, Nat Acad Sci, 86, Human Immunodeficiency Virus Infection, Nat Inst Allergy & Infectious Dis, India, 86-; sr attend phys, Nat Inst Allergy & Infectious Dis, 81-; adj prof, Uniformed Serv Univ Health Sci, 88-, George Washington Univ, 88- *Honors & Awards:* A Conger Goodyear Prize, 65. *Mem:* Am Col Physicians; Infectious Dis Soc; Am Epidemiol Soc; Am Col Epidemiol; Am Fed Clin Res; AAAS. *Res:* Epidemiologic research on infectious and immune disease including AIDS and other genital infections, Lyme disease, autoimmune diseases and asthma. *Mailing Add:* 3204 Rolling Rd Chevy Chase MD 20815

KASNER, FRED E, b New York, NY, July 8, 26; m 53; c 1. PHYSICAL CHEMISTRY. *Educ:* City Col New York, BS, 48; Univ Chicago, MS, 49, PhD(phys chem), 61. *Prof Exp:* From instr to asst prof natural sci, Univ Chicago, 54-59; from instr to prof chem, Fenger Jr Col, 61-70; PROF CHEM, OLIVE-HARVEY COL, 71- *Concurrent Pos:* Consult, USAF, 59; vis scholar, Northwestern Univ, 71; vis scientist, Argonne Nat Lab, 79. *Mem:* Am Chem Soc; Am Soc Testing & Mat. *Res:* Laser Raman spectroscopy of aqueous solutions; thermal conductivity and its temperature coefficient; heats of dilution of aqueous solutions of strong electrolytes. *Mailing Add:* 320 17th St Wilmette IL 60091-3224

KASNER, WILLIAM HENRY, b Killbuck, Ohio, Jan 27, 29; m 51; c 1. PHYSICS. *Educ:* Case Western Reserve Univ, BS, 51; Univ Pittsburgh, PhD(physics), 58. *Prof Exp:* Res assoc physics, Univ Md, 58-59, asst res prof, 59-61; FEL SCIENTIST, GAS LASERS RES & DEVELOP, WESTINGHOUSE RES & DEVELOP CTR, 61- *Mem:* Am Phys Soc. *Res:* Atomic physics, especially atomic and electronic collision phenomena; ultraviolet spectroscopy; gas discharges; optics; laser development and application. *Mailing Add:* Westinghouse Res & Develop Ctr 1310 Beulah Rd Pittsburgh PA 15235

KASOWSKI, ROBERT V, b Bremond, Tex, Feb 14, 44; m 69; c 1. SOLID STATE PHYSICS. *Educ:* Tex A&M Univ, BS, 66; Univ Chicago, PhD(physics), 69. *Prof Exp:* Physicist, E I du Pont de Nemours & Co, Inc, 69-80. *Mem:* Am Phys Soc. *Res:* Calculating the electronic properties of molecules adsorbed onto metal or semiconductor surfaces using linear combination of muffin tin orbitals method. *Mailing Add:* 2153 Brinton's Bridge Rd West Chester PA 19382

KASPAREK, STANLEY VACLAV, b Prague, Czech, June 11, 29. ORGANIC CHEMISTRY, MEDICINAL CHEMISTRY. *Educ:* Charles Univ, Prague, Dr rer nat, 65. *Prof Exp:* Chemist, Res Inst Pharm & Biochem, Prague, 50-59 & Czech Acad Sci, 62-65; fel & chemist, Nat Res Coun Can, 65-67; abstractor, Chem Abstracts Serv, Ohio, 68; info scientist, Hoffmann-La Roche Inc, 68-81, tech fel, 82-85; PRES, CHEMINFO, 85- *Mem:* Am Chem Soc. *Res:* Scientific information. *Mailing Add:* Three Rockledge Pl Cedar Grove NJ 07009

KASPER, ANDREW E, JR, b Bridgeport, Conn, Oct 29, 42; m 67; c 2. PALEOBOTANY. *Educ:* Duquesne Univ, BA, 65; Univ Conn, MS, 68, PhD(bot), 70. *Prof Exp:* Asst prof, 70-75, ASSOC PROF BOT, RUTGERS UNIV, NEWARK, 75- *Mem:* Bot Soc Am; Sigma Xi; Paleont Soc. *Res:* Description and classification of Devonian age plant fossils. *Mailing Add:* Dept Biol Sci Rutgers Univ 101 Warren St Newark NJ 07102

KASPER, CHARLES BOYER, b Joliet, Ill, Apr 27, 35; m 57; c 4. BIOCHEMISTRY. *Educ:* Univ Ill, BS, 58; Univ Wis, PhD(physiol chem), 62. *Prof Exp:* Asst prof biol chem, Univ Calif, Los Angeles, 64-65; from asst prof to assoc prof, 65-82, PROF ONCOL, McARDLE LAB, UNIV WIS-MADISON, 82- *Concurrent Pos:* NIH fels, Univ Utah, 62-63 & Univ Calif, Los Angeles, 63-64. *Mem:* Am Soc Biol Chemists; Am Asn Cancer Res. *Res:* Regulation of enzymes responsible for the metabolic activation of chemical carcinogens; molecular basis of enzyme induction. *Mailing Add:* Dept Oncol 421a McArdle Res Univ Wis Med Sch Madison WI 53706

KASPER, DENNIS LEE, BACTERIAL DISEASES, IMMUNOCHEMISTRY. *Educ:* Univ Ill, Chicago, MD, 67. *Prof Exp:* PROF MED, HARVARD MED SCH, 85-, ASSOC DIR, CHANNING LAB, 82-; CHIEF INFECTIOUS DIS, BETH ISRAEL HOSP, BOSTON, MASS, 81- *Res:* Infectious diseases. *Mailing Add:* Dept Med Channing Lab Harvard Med Sch 180 Longwood Ave Boston MA 02115

KASPER, GERHARD, b Salzburg, Austria, June 6, 49; m; c 2. AEROSOL SCIENCE & TECHNOLOGY, CONTAMINATION CONTROL. *Educ:* Univ Vienna, Austria, PhD(physics), 77. *Prof Exp:* asst prof physics, Inst Exp Physics, Univ Vienna, Austria, 77-78; from res asst prof to asst prof elec engr, State Univ NY, Buffalo, 78-83; sr sci, 83-85, DIR RES & DEVELOP, CHICAGO RES CTR, AM AIR LIQUIDE INC, COUNTRYSIDE, ILL, 85- *Concurrent Pos:* Adj prof, Univ Vienna, Austria, 83-; chmn, particulate standards subcomt Semicond Equip & Mat Inst, 84-; ed-in-chief, J Aerosol Sci, 85- *Mem:* Europ Asn Aerosol Res; Am Asn Aerosol Res; Inst Environ Sci; Am Soc Testing & Mat. *Res:* Aerosol science and particle technology including instrumentation and measurement techniques; filtration; contamination control; surface-particle interactions; aerosol generation and sampling; dynamics of irregular particles. *Mailing Add:* Chicago Res Ctr Am Air Liquide Inc 5230 S East Ave Countryside IL 60525

KASPER, JOHN SIMON, b Newark, NJ, May 27, 15. PHYSICAL CHEMISTRY. *Educ:* Johns Hopkins Univ, AB, 37, PhD(chem), 41. *Prof Exp:* Jr instr chem, Johns Hopkins Univ, 37-41; instr, St Louis Univ, 41-42; res chemist, Nat Defense Res Comt contract, Johns Hopkins Univ, 42-43 & Manhattan Dist Proj, SAM Labs, Columbia Univ, 43-45; res assoc, Res Lab, Gen Elec Co, 45-58, phys chemist, 58-80; RETIRED. *Mem:* AAAS; Am Chem Soc; Am Phys Soc; Am Crystallog Asn. *Res:* X-ray diffraction; structures of crystals; structural chemistry; fluorocarbon chemistry; gas adsorption; physical properties; aqueous solutions; neutron diffraction. *Mailing Add:* 18 Cumberland Pl Scotia NY 12302

KASPER, JOSEPH EMIL, b Cedar Rapids, Iowa, May 2, 20; m 57; c 5. SCIENCE EDUCATION. *Educ:* Coe Col, BA, 51; Univ Iowa, MS, 54, PhD(physics), 58. *Prof Exp:* Asst physics, Univ Iowa, 51-57, instr, 58; asst prof, 59-63, PROF PHYSICS, COE COL, 63- *Mem:* Am Phys Soc; Am Asn Physics Teachers. *Res:* Geomagnetic theory; cosmic radiation; soft radiations in upper atmosphere; rocketry. *Mailing Add:* 3344 Carlisle St NE Cedar Rapids IA 52402

KASPER, JOSEPH F, JR, b Baltimore, Md, Dec 9, 43; m 66; c 2. NAVIGATION, APPLIED PHYSICS. *Educ:* Mass Inst Technol, BS, 64, MS, 66, DSc(instrumentation), 68. *Prof Exp:* Mgr spec projs, Anal Sci Corp, 75-78, mgr, Navig Systs, 78-79, dir, Stratig Syst Div, 79-80, dir sea-launched ballistic missile, 80-90; PRES & CHIEF EXEC OFFICER, FOUND INFORMED MED DECISION MAKING, 90- *Mem:* Am Geophys Union; Inst Navig; Inst Elec & Electronics Engrs; Am Inst Aeronaut & Astronaut. *Res:* Mathematical modeling of very low frequency propagation anomalies; statistical description of geodetic phenomena; analysis of radio and inertial navigation error behavior; applied Kalman filtering. *Mailing Add:* Found Informed Med Decision Making PO Box C17 Hanover NH 03755

KASPERBAUER, MICHAEL J, b Manning, Iowa, Oct 8, 29; m 62; c 4. PLANT PHYSIOLOGY, PHOTOBIOLOGY. *Educ:* Iowa State Univ, BS, 54, MS, 57, PhD(plant physiol), 61. *Prof Exp:* NSF fel, Univ Md, 61-62; res plant physiologist, Pioneering Res Lab Plant Physiol, USDA, Beltsville, Md, 62-63, plant physiologist, Agr Res Serv, Lexington, Ky, 63-83, PLANT PHYSIOLOGIST, AGR RES SERV, USDA, FLORENCE, SC, 83-; PROF PLANT PHYSIOL, CLEMSON UNIV, 83- *Concurrent Pos:* Adj prof agron, Univ Ky, 65-83; assoc ed, Agron J, 75-83; adj prof plant physiol, Clemson Univ, 83- *Honors & Awards:* L M Ware Res Award, 90; Crop Sci Res Award, 90. *Mem:* Am Soc Plant Physiol; fel Crop Sci Soc Am; fel Am Soc Agron; Am Soc Photobiol; Sigma Xi. *Res:* Interaction of light and temperature on plant growth, development and composition; phytochrome control of plant physiological processes; haploid and doubled haploid utilization in crop improvement. *Mailing Add:* 1717 Williamsburg Ct Lexington KY 40504

KASPEREK, GEORGE JAMES, b Albert Lea, Minn, June 1, 44; m 66; c 2. BIOCHEMISTRY, EXERCISE. *Educ:* Mankato State Col, BA, 66; Ore State Univ, PhD(org chem), 69. *Prof Exp:* Fel bioorg, Univ Calif, Santa Barbara, 69-72; asst prof biochem, Conn Col, 72-77, assoc prof, 77-78; assoc prof, 78-88, PROF, BIOCHEM, SCH MED, ECAROLINA UNIV, 88- *Concurrent Pos:* NIH fel, 69-71; vis prof, Sch Med, ECarolina Univ, 78-79. *Mem:* Am Col Sports Med. *Res:* Regulation of metabolism during exercise; amino acid metabolism; protein metabolism. *Mailing Add:* Dept Biochem Sch Med ECarolina Univ Greenville NC 27858-4354

KASPROW, BARBARA ANN, b Hartford, Conn, Apr 23, 36. MICRO ANATOMY, REPRODUCTIVE BIOLOGY. *Educ:* Albertus Magnus Col, BA, 58; Loyola Univ, Ill, PhD(anat), 69. *Prof Exp:* USPHS training scholar, Yale Univ, 59-60, res asst anat & reproductive biol, Sch Med, 61; res assoc, NY Med Col, 61-62; from res assoc to sr res & admin assoc, Inst Study Human Reproduction, Ohio, 62-67; sr res assoc, Stritch Sch Med, Loyola Univ, Chicago, 67-69, asst prof anat, 69-76; WRITER, OPERS SPECIALIST, 76- *Concurrent Pos:* Co-ed, Biol of Reproduction, 73. *Mem:* AAAS; Am Asn Anatomists; Am Soc Zoologists; NY Acad Sci; Sigma Xi. *Res:* Reproductive phenomena in the mammalian female; growth mechanisms, pathologic variants in reproductive organs and endocrinologic interrelationships; cytophysiology; research administration; science education. *Mailing Add:* 1021 S Hwy 83 Elmhurst IL 60126

KASPRZAK, LUCIAN A, b Scranton, Pa, July 22, 43; m 67; c 2. DESIGN FOR RELIABILITY. *Educ:* Stevens Inst Technol, BS, 65, PhD(mat), 72; Syracuse Univ, MS, 70. *Prof Exp:* Engr, Phys & Elec Failure Anal, 65-70; resident fel, IBM, 70-72; engr, Reliability Eng, 72-77, mgr, Mosfet Memory, 77-82; tech asst, Tech Personnel Develop, 82-83; mgr, Memory Technol, 83-84; prog mgr, Syst Technol Support, 84-85; PROG MGR, TECH PROF RELS, IBM CORP, 85- *Concurrent Pos:* Resident fel, IBM, 70-72. *Mem:* Fel Inst Elec & Electronics Engrs; Am Phys Soc. *Res:* Technology management, reliability of very-large-scale integration (mosfet and bipolar); measurement, qualification and building in high reliability by systematic study during development and manufacture; joint discovery of the hot electron effect in short channel mosfets; materials, devices and processes. *Mailing Add:* IBM Corp 500 Columbus Ave Thornwood NY 10594

KASRIEL, ROBERT H, b Tampa, Fla, Oct 18, 18; m 46; c 2. MATHEMATICS. *Educ:* Univ Tampa, BS, 40; Univ Va, MA, 49, PhD(math), 53. *Prof Exp:* Coordr war-training courses, Univ Tampa, 40-42; aeronaut res scientist, Nat Adv Comt Aeronaut, 52-54; from asst prof to prof, 54-84, EMER PROF MATH, GA INST TECHNOL, 85- *Honors & Awards:* Ferst Res Award, Sigma Xi, 62. *Mem:* Am Math Soc; Math Asn Am; Sigma Xi. *Res:* Analytic topology; fixed point theorems; mapping theorems. *Mailing Add:* Math Dept Ga Tech Atlanta GA 30332

KASS, EDWARD HAROLD, infectious diseases; deceased, see previous edition for last biography

KASS, GUSS SIGMUND, b Chicago, Ill, Oct 19, 15; m 38; c 2. COSMETIC CHEMISTRY, TOPICAL DRUGS. *Educ:* Univ Chicago, BS, 38. *Prof Exp:* Res chemist, Munic Tuberc Sanitarium, 38-39; chemist, Prod Corp Am, 39-40; res chemist, Acme Cosmetic Corp, 40-41; chief chemist, Duart Mfg Co, Ltd, Calif, 41-42 & 46-48; asst res dir, Helene Curtis Industs, Inc, 48-54; res dir & vpres, Lanolin Plus, Inc, 54-60; from tech dir to vpres & dir corp res & develop, Alberto-Culver Co, 60-74; PRES, G S KASS & ASSOCS, LTD, 74- *Concurrent Pos:* Lectr, Univ Chicago, 58; lectr, Sch Med, Univ Ill, 65-86. *Mem:* Am Acad Dermat; Am Chem Soc; fel Soc Cosmetic Chem; fel Am Inst Chem. *Res:* Cosmetics; toiletries; proprietary pharmaceuticals. *Mailing Add:* 8938 N Keeler Ave Skokie IL 60076

KASS, IRVING, medicine, for more information see previous edition

KASS, LEON RICHARD, b Chicago, Ill, Feb 12, 39; m 61; c 2. MEDICAL ETHICS. *Educ:* Univ Chicago, BS, 58, MD, 62; Harvard Univ, PhD(biochem, molecular biol), 67. *Prof Exp:* Intern med, Beth Israel Hosp, Boston, Mass, 62-63; staff assoc molecular biol, Nat Inst Arthritis & Metab Dis, 67-69, sr staff fel, 69-70; exec secy comt life sci & social policy, Nat Acad Sci, 70-72; tutor, St John's Col, Md, 72-76; Joseph P Kennedy, Sr res prof bioethics, Kennedy Inst & assoc prof neurol & philos, Georgetown Univ, 74-76; Henry

R Luce prof lib arts of human biol, 76-84, prof, 84-90, ADDIE CLARK HARDING, PROF, COL & COMT SOCIAL THOUGHT, UNIV CHICAGO, 90- *Concurrent Pos:* Guggenheim fel, 72-73; fel & mem bd dir, Inst Soc, Ethics & Life Sci, 70-; mem bd gov, US-Israel Binat Sci Found, 82-88; mem, Nat Humanities Coun, 84-; fel, Nat Humanities Ctr, 84-85. *Mem:* AAAS. *Res:* Ethical and social implications of advances in biomedical science and technology; philosophy of biology and medicine; philosophical anthropology. *Mailing Add:* Univ Chicago 1116 E 59th St Chicago IL 60637

KASS, ROBERT S, b New York, NY, June 13, 46; m 82; c 2. MEMBRANE BIOPHYSICS. *Educ:* Univ Ill, BSc, 68; Univ Mich, MSc, 69, PhD(physics), 72. *Prof Exp:* Fel physiol, Univ Mich, 72-74; fel membrane biophysics, Marine Biol Labs, Mass, 73; fel physiol, Yale Univ, 74-77; from asst prof to assoc prof physiol, 77-90, PROF PHYSIOL & PEDIAT, UNIV ROCHESTER, 90- *Mem:* Biophys Soc; Am Heart Asn; Soc Gen Physiologists; NY Acad Sci. *Res:* Physiology and biophysics of excitable membranes with a particular interest in the membranes of heart muscle cells; regulation of ion channels by hormones and drug molecules. *Mailing Add:* Dept Physiol Univ Rochester Med Ctr Rochester NY 14642

KASS, SEYMOUR, b New York, NY, Apr 13, 26; m 55; c 2. ALGEBRA. *Educ:* Brooklyn Col, BA, 48; Stanford Univ, MS, 57; Univ Chicago, SM, 65; Ill Inst Technol, PhD(math), 66. *Prof Exp:* Mathematician, Curtiss Wright Corp, 52-55 & Stanford Res Inst, 55-57; from instr to asst prof math, Ill Inst Technol, 60-71; from assoc prof to prof, Boston State Col, 77-82, chmn dept, 72-75; dir eng prog, 84-85; PROF MATH, UNIV MASS, BOSTON, 82- *Concurrent Pos:* NSF res grant, 69 & 70. *Mem:* Am Math Soc; Math Asn Am; Sigma Xi. *Res:* Algebra; geometry; statistics. *Mailing Add:* 118 York Terr Brookline MA 02146

KASSAKHIAN, GARABET HAROUTIOUN, b Jerusalem, Palestine, Aug 15, 44; Can citizen; m 69; c 3. ENVIRONMENTAL CHEMISTRY, ENVIRONMENTAL REMEDIATION. *Educ:* Yerevan State Univ, Armenia, MSc, 67; Harvard Univ, AM, 70, PhD(anal chem), 75. *Prof Exp:* Asst prof chem, Univ Mass, Boston, 75-78; environ chemist, Eldorado Nuclear Ltd, Ottawa, 78-80; mgr, Amerada Hess Corp, Woodbridge, NJ, 80-85; assoc, Lockman & Assocs, Monterey Park, Calif, 85-87; SR ENVIRON SCIENTIST, LAW ENVIRON INC, BURBANK, CALIF, 88- *Concurrent Pos:* Asst prof environ chem, Univ Laverne, Calif, 85-87. *Honors & Awards:* Willem Rudolfs Medal, Water Pollution Control Fedn, 82. *Mem:* Am Chem Soc; Water Pollution Control Fedn; Air & Waster Mgt Asn. *Res:* Uranium mine and mill tailings; methane gas mitigation; air and water pollution; radionuclides in environment; water purification; wastewater treatment; contaminated soil remediation; environmental permitting; drinking water standards. *Mailing Add:* 221 N Cedar St No 39 Glendale CA 91206

KASSAKIAN, JOHN GABRIEL, b Mar 27, 43; US citizen; c 2. POWER ELECTRONICS, ELECTRICAL ENGINEERING. *Educ:* Mass Inst Technol, SB, 65, SM, 67, EE, 67, ScD(elec eng), 73. *Prof Exp:* Tech rep to Univac naval data syst, USN, 69-71; from asst prof to assoc prof elec eng, 73-84, assoc dir, Elec Power Syst Eng Lab, 79-83, PROF ELEC ENG, MASS INST TECHNOL, 84-, ACTG DIR, LAB ELECTROMAGNETIC & ELECTRONIC SYST, 84- *Concurrent Pos:* Consult, Gould Labs, Gould Inc, 75-, Lutron Electronics, 78-; mgr spec proj, Amerada Hess Corp, 80-85; dir, Ault Inc, 84-; Sheldahl Inc, 85-; assoc, Lockman & Assoc, 85-87. *Honors & Awards:* Centennial Medal, Inst Elec & Electronics Engrs, 84. *Mem:* Fel Inst Elec & Electronics Engrs; Sigma Xi. *Res:* Simulation, analysis, synthesis of electronic energy conversion systems; pulsed power supplies for fusion research; power semiconductor devices. *Mailing Add:* Mass Inst Technol Bldg 10-098 77 Massachusetts Ave Cambridge MA 02139-4309

KASSAL, ROBERT JAMES, b Berwick, Pa, Oct 23, 36; m 58; c 4. ORGANIC CHEMISTRY. *Educ:* Hofstra Univ, BA, 58; Univ Fla, PhD(org chem), 64. *Prof Exp:* Chemist, Am Cyanamid Co, 58-60; res asst fluorine chem, Univ Fla, 60-63; res chemist, 63-68, sr res chemist, Plastics Dept, 68-69, res supvr, 69-75, mem staff, Elastomers Dept, 75-78, res assoc, Polymer Prod Dept, 75-85, SR res assoc & group leader, 85-90, RES FEL, POLYMER PROD DEPT, E I DU PONT DE NEMOURS & CO INC, 90- *Mem:* Soc Plastics Indust. *Res:* Fluorine chemistry; polymer preparation; fluorocarbon heterocyclic polymers; intermediates and monomer exploratory research; new product development; polymer synthesis; polymer modification; high performance composite development; novel elastomer systems; polymer toughening; solvent resistant polymers; engineering polymers; polymer compounding and processing; failure analysis of polymer systems. *Mailing Add:* Exp Sta Lab Bldg 323/227 E I du Pont de Nemours & Co Inc Wilmington DE 19880-0323

KASSAM, SALEEM ABDULALI, b Dar es Salaam, Tanzania, June 16, 49; US citizen; m 78; c 3. SIGNAL PROCESSING, COMMUNICATION THEORY. *Educ:* Swarthmore Col, BS, 72; Princeton Univ, MSE & MA, 74, PhD(elec eng), 75. *Hon Degrees:* MA, Univ Pa, 80. *Prof Exp:* From asst prof to assoc prof, 75-86, PROF, MOORE SCH ELEC ENG, UNIV PA, 86- *Concurrent Pos:* Prin investr, Air Force Off Sci Res res grants, Univ Pa, 76-, NSF grant, 77-79; Naval Res Lab grants, 78-80 & Off Naval Res grant, 80-; consult, RCA, 80-81, Interspec, 82-88; Naval Air Dev Cen, 87- & Bio Rad Inc, 89-90; vis assoc prof, Univ BC, 83. *Mem:* Sr mem Inst Elec & Electronics Engrs; Sigma Xi. *Res:* Signal processing and communication theory; nonparametric detection; quantization, robust signal processing; image processing; microwave and ultrasonic imaging; spectrum estimation; author of numerous technical papers and a book. *Mailing Add:* Moore Sch Elec Eng Univ Pa Philadelphia PA 19104

KASSANDER, ARNO RICHARD, JR, b Carbondale, Pa, Sept 10, 20; m 43; c 1. GEOLOGY. *Educ:* Amherst Col, BA, 41; Univ Okla, MS, 43; Iowa State Col, PhD(physics), 50. *Hon Degrees:* DSc, Amherst Col, 71. *Prof Exp:* Asst geologist, Tex Co, 41; asst geophysics, Magnolia Petrol Co, 43; asst prof physics, Iowa State Col, 50-54; assoc dir, Inst Atmospheric Physics, 54-57, dir, 57-73, head dept atmospheric sci, 58-73, VPRES RES, UNIV ARIZ, 72-, PROF ATMOSPHERIC SCI, 76- *Concurrent Pos:* Dir, Water Resources Res Ctr, Univ Ariz, 64-72; mem panel environ, President's Sci Adv Comt; trustee, Ariz Sonora Desert Mus; chmn, Univ Corp Atmospheric Res, 58-68; dir, Burr Brown Res Corp, 74- *Mem:* Fel AAAS; Am Phys Soc; fel Am Meteorol Soc. *Res:* General geophysical instrumentation; recording and automatic analysis of statistical data. *Mailing Add:* 3341 E Fourth St Tucson AZ 85716

KASSCHAU, MARGARET RAMSEY, b Cambridge, Mass, Sept 9, 42; div; c 2. COMPARATIVE PHYSIOLOGY, CELL PHYSIOLOGY. *Educ:* Univ Rochester, AB, 64; Univ SC, MS, 70, PhD(biol), 73. *Prof Exp:* Res asst biol, Oak Ridge Nat Lab, 64-67; guest worker parasitol, NIH, 73-74; fel res physics, M D Anderson Hosp & Tumor Inst, 74-75; from asst prof to assoc prof, 75-88, PROF BIOL, UNIV HOUSTON, CLEAR LAKE CITY, 89- *Concurrent Pos:* Prin investr, Sea Grant Col Prog, 78-80; vis res assoc, Med Sch, Stanford Univ, 80-81; prin investr, NSF Grant, RUI Prog, 87-91; NSF vis prof women, Harvard Med Sch, 90-91. *Mem:* Am Soc Zoologists; Am Soc Parasitologists; AAAS; Soc Environ Toxicol & Chem; Biophys Soc; Soc Toxicol. *Res:* Hemolysis by schistosoma parasites; cell to cell adhesion; aquatic toxicology of marine animals; cellular toxicology. *Mailing Add:* Univ Houston Clearlake PO Box 40 2700 Bay Area Blvd Houston TX 77058

KASSIRER, JEROME PAUL, b Buffalo, NY, Dec 19, 32; c 6. NEPHROLOGY, INTERNAL MEDICINE. *Educ:* Univ Buffalo, BA, 53, MD, 57. *Prof Exp:* Asst physician, New Eng Med Ctr Hosp, Tufts Univ, 62-65, physician, 69-74, assoc physician-in-chief, 71-76, actg physician-in-chief, 76-77, instr med & nephrology, Sch Med, 62-65, from asst prof to assoc prof med, 65-74, assoc chmn, 71-76, actg chmn, 74-75 & 76-77, PROF MED, SCH MED, TUFTS UNIV, 74-, ASSOC CHMN, 77-, ASSOC PHYSICIAN-IN-CHIEF, NEW ENG MED CTR, 77- *Concurrent Pos:* Mem med sci panel, Am Inst Biol Sci, 77-80; co-ed, Kidney Int, Nephrology Forum, 78-; bd sci counr, Nat Libr Med, 86-; Sara Murray Jordan prof, 87- *Mem:* Inst Med-Nat Acad Sci; Am Col Informatics. *Res:* Renal, electrolytes and acid-base physiology; clinical nephrology; decision analysis and clinical cognition. *Mailing Add:* Dept Med New Eng Med Ctr 750 Washington St Boston MA 02111

KASSNER, JAMES LYLE, JR, b Tuscaloosa, Ala, May 1, 31; m 56; c 5. CLOUD PHYSICS. *Educ:* Univ Ala, BS, 52, MS, 53, PhD(physics), 57. *Prof Exp:* Asst physics, Univ Ala, 53-54, res assoc, 54-56; asst prof, Mo Sch Mines, 56-59; assoc prof, 59-66, dir, Grad Ctr Cloud Physics Res, 68-84, PROF PHYSICS, UNIV MO-ROLLA, 66-; MFG MGT, FIBREFORM CONTAINERS INC, 90- *Concurrent Pos:* Mem subcomt nucleation, Int Asn Meteorol & Atmospheric Physics; mem subcomn IV, Ions, Aerosols & Radioactivity, Int Comn Atmospheric Elec. *Mem:* Fel Am Phys Soc; Am Meteorol Soc; Am Geophys Union; Sigma Xi. *Res:* Atmospheric condensation; homogeneous and heterogeneous nucleation from the vapor; mobility of cluster ions; laboratory simulation of cloud formation; measurements on atmospheric particulates; nucleation of ice. *Mailing Add:* 11947 Graceland Acres Northport AL 35476

KASSNER, RICHARD J, b Chicago, Ill, July 1, 39; m 62; c 3. BIOCHEMISTRY. *Educ:* Purdue Univ, BS, 61; Yale Univ, MS, 63, PhD(biophys chem), 66. *Prof Exp:* NIH fel chem, Univ Calif, San Diego, 66-68, instr, 68-69; asst prof, 69-74, ASSOC PROF CHEM, UNIV ILL, CHICAGO CIRCLE, 74- *Res:* Structural basis for the properties of heme and iron-sulfur proteins; model systems for the active sites of hemeproteins; heme and chlorophyll biosynthesis. *Mailing Add:* Dept Chem Box 4348 Univ Ill Chicago Circle Box 4348 Chicago IL 60680-4348

KASSOY, DAVID R, b Brooklyn, NY, Jan 29, 38; m 64; c 2. FLUID MECHANICS, COMBUSTION. *Educ:* Polytech Inst Brooklyn, BAE, 59; Univ Mich, MSAE, 61, PhD(aerospace eng), 65. *Prof Exp:* Asst res engr, Univ Calif, San Diego, 65-67, asst prof aerospace & mech eng sci, 68-69; from asst prof to assoc prof, 69-78, PROF MECH ENG, UNIV COLO, BOULDER, 78-, ASST VCHANCELLOR RES, 88- *Concurrent Pos:* Guggenheim fel, 73, Fulbright Res fel, Tech Univ Delft, 83; vis fel, Sci & Eng Res Coun, Univ EAnglia, Eng, 82-83 & Japan Soc Prom Sci, Nagoya Univ, Japan, 85; vchmn exec comn, Div Fluid Dynamics, Am Phys Soc, 89-90, mem exec comn, 89- *Mem:* fel Am Phys Soc; Soc Indust & Appl Math; Combustion Inst; Nat Coun Univ Res Admin. *Res:* Combustion theory; ignition and explosion; perturbation methods; theoretical fluid mechanics. *Mailing Add:* Grad Sch B-26 Univ Colo Boulder CO 80309

KASTELLA, KENNETH GEORGE, b Kalispell, Mont, May 27, 33; div; c 1. PHYSIOLOGY. *Educ:* Univ Wash, BS, 59, MS, 65, PhD(physiol), 69. *Prof Exp:* NIH training grant neurophysiol, Univ Wash, 69-70, res assoc, 70; asst prof physiol, Univ NMex, 70-76; assoc prof, Univ Alaska, 76-87; TEACHING ASSOC, DEPT BIOL STRUCT, SCH MED, UNIV WASH, 87- *Mem:* AAAS; Inst Elec & Electronics Engrs; Biophys Soc. *Res:* Neurophysiology, especially central control of blood pressure; temperature regulation. *Mailing Add:* Dept Biol Struct Univ Wash Sch Med Seattle WA 98195

KASTEN, FREDERICK H, b New York, NY, Mar 7, 27; m 49; c 4. CYTOLOGY. *Educ:* Univ Houston, BA, 50; Univ Tex, MS, 51, PhD(zool), 54. *Prof Exp:* Scientist cancer res, Roswell Park Mem Inst, NY, 54-56; asst prof zool, Agr & Mech Col, Tex, 56-61; res coordr & dir ultrastruct cytochem dept, Pasadena Found Med Res, 63-70; PROF ANAT, MED CTR, LA STATE UNIV, 70- *Concurrent Pos:* NSF sr fel, Giessen, Ger, 61-62; NSF sr fel, Inst Cancer Res, Villejuif, France, 62-63, NIH spec res fel, 62-63; from adj asst prof to adj assoc prof, Univ Southern Calif, 63-70; from asst clin prof to assoc clin prof, Loma Linda Univ, 65-70; partic, W Alton Jones Cell Sci Ctr, 71; consult, Nat Heart & Lung Inst, 71- & Nat Cancer Inst, 73-; assoc coordr, La Cancer Ctr, 73-78; rev ed, In Vitro, 75-78; vis prof anat, ETenn State Univ Med Sch, 79-80, Ain-Shams Univ Fac Med, Cairo Egypt, 87 & 90; pres, Biol Stain Comn, 86-, trustee, 73-; vis prof histol, Alex Univ Med Sch, Alexandria, Egypt, 86 & 87; adv bd, TCA Tech Manual, 74-80; res grants comt, Cancer Asn GNO, 75-; pres, Am Asn Dent Res, NO Sect, 83-84; vis prof zool, Jagiellonian Univ, Krakow, Poland, 89. *Mem:* Biol Stain Comn; Am Soc Cell Biol; AAAS; Ger Histochem Soc; Tissue Cult Asn; Sigma Xi. *Res:* Quantitative cytochemistry of nucleic acids; absorption curve analyses of stained cells; development of new staining techniques; electron microscopy; cytochemistry of viral infections; dye impurities; fluorescence microscopy; cancer; tissue culture; history of medicine. *Mailing Add:* Dept Anat La State Univ Med Ctr 1100 Florida Ave New Orleans LA 70119

KASTEN, PAUL R(UDOLPH), b Jackson, Mo, Dec 10, 23; m 47; c 3. REACTOR EVALUATION, REACTOR TECHNOLOGY. *Educ:* Univ Mo Sch Mines, BS, 44, MS, 47; Univ Minn, PhD(chem eng), 50. *Prof Exp:* Staff mem, Oak Ridge Nat Lab, 50-55, sect chief, 55-61, assoc dept head, 61-63, assoc dir molten salt reactor prog, 65-70, dir, gas cooled reactor & thorium utilization progs & mgr, Alternate Fuel Cycle Eval, 77-79, dir gas cooled reactor progs, 79-85, tech dir, 85-88; guest dir, Inst Reactor Develop, WGer, 63-65; PROF, COL ENG, UNIV TENN, KNOXVILLE, 88-; CONSULT, 88- *Concurrent Pos:* Lectr, Univ Tenn, Knoxville, 53-60, prof, 65- *Honors & Awards:* Outstanding Serv Award, Oak Ridge Chap, Tenn Soc Prof Engrs, 88. *Mem:* Fel Am Nuclear Soc; fel AAAS; Nat Soc Prof Engrs; Sigma Xi. *Res:* Nuclear engineering; very high temperature reactors. *Mailing Add:* 341 Louisiana Ave Oak Ridge TN 37830

KASTENBAUM, MARVIN AARON, b New York, NY, Jan 16, 26; m 55; c 2. BIOMETRICS-BIOSTATISTICS. *Educ:* City Col New York, BS, 48; NC State Col, MS, 50, PhD(statist), 56. *Prof Exp:* Asst statistician, US Bur Census, 48 & 50; chief statistician mkt res, Dun & Bradstreet, Inc, 52; biostatistician, Atomic Bomb Casualty Comm, 53-54; sr res statistician, Oak Ridge Nat Lab, Tenn, 56-70; dir statist, Tobacco Inst, 70-87; CONSULT, 87- *Concurrent Pos:* Mem math res ctr, Univ Wis, 65-66; NSF vis lectr, 66-71; mem statist dept, Stanford Univ, 69; med statist, biomet & epidemiol consult. *Mem:* Fel AAAS; Biomet Soc (secy-treas, Eastern NAm region, 59-60, gen treas, 60-63); fel Am Statist Asn; Inst Math Statist; fel Royal Statist Soc; fel NY Acad Sci. *Res:* Medical statistics; biometry; epidemiology. *Mailing Add:* 16933 Timberlakes Dr SW Ft Myers FL 33908-4339

KASTENBERG, WILLIAM EDWARD, b New York, NY, June 25, 39; m 63; c 3. NUCLEAR ENGINEERING. *Educ:* Univ Calif, Los Angeles, BS, 62, MS, 63; Univ Calif, Berkeley, PhD(eng), 66. *Prof Exp:* From asst prof eng to assoc prof eng & appl sci, Univ Calif, Los Angeles, 66-75, vchmn dept chem, nuclear & thermal eng, 77-78, asst dean grad studies, 81-85, chmn, 85-88, PROF MECH, AEROSPACE & NUCLEAR ENG, UNIV CALIF, LOS ANGELES, 75- *Mem:* Fel Am Nuclear Soc; fel AAAS. *Res:* Nuclear reactor safety; fusion technology; risk assessment; toxic waste control; environmental risk assessment. *Mailing Add:* Sch Eng & Appl Sci Univ Calif 405 Hilgard Ave Los Angeles CA 90024

KASTENHOLZ, CLAUDE E(DWARD), b Milwaukee, Wis, Nov 27, 36; m 59; c 5. ELECTRICAL ENGINEERING. *Educ:* Marquette Univ, BEE, 58; Univ Southern Calif, MS, 60; Univ Wis, PhD(elec eng), 63; Pepperdine Univ, MBA, 70. *Prof Exp:* Mem tech staff, Hughes Aircraft Co, Calif, 58-60; res engr, Autonetics Div, NAm Aviation, Inc, 60-61; res asst circuit design, Univ Wis, 61-62; res specialist, Autonetics Div, NAm Rockwell Corp, 63-67; sr staff engr, 67-75, sr scientist, 75-81, PROJ MGR, HUGHES AIRCRAFT CO, 82- *Mem:* Inst Elec & Electronics Engrs. *Res:* Sonar systems engineering; fire control system design; system testing. *Mailing Add:* Hughes Aircraft Co Inc PO Box 3310 Fullerton CA 90045

KASTENS, KIM ANNE, b Menlo Park, Calif, May 19, 54. MARINE GEOLOGY, MARINE GEOPHYSICS. *Educ:* Yale Univ, BS, 75; Scripps Inst Oceanog, PhD(oceanog), 81. *Prof Exp:* RES ASSOC, LAMONT-DOHERTY GEOL OBSERV, COLUMBIA UNIV, 81- *Mem:* Am Geophys Union; Geol Soc Am; Sigma Xi. *Res:* Tectonic and sedimentological processes in the deep sea. *Mailing Add:* Lamont-Doherty Geol Observ 116 Oceanog Palisades NY 10964

KASTIN, ABBA J, b Cleveland, Ohio, Dec 24, 34. ENDOCRINOLOGY, NEUROSCIENCES. *Educ:* Harvard Col, AB, 56; Harvard Med Sch, MD, 60. *Hon Degrees:* Dr, Univ Nat Federico Villarreal, Lima, Peru, 80; DSc, Univ New Orleans, 80. *Prof Exp:* Clin assoc endocrinol, NIH, 62-64; NIH spec fel, Sch Med, Tulane Univ, 64-65; clin investr med, Vet Admin Hosp, New Orleans, 65-68; assoc prof, 71-74, PROF MED, SCH MED, TULANE UNIV, 74-; CHIEF ENDOCRINOL, VET ADMIN HOSP, 68- *Concurrent Pos:* Vis physician, Charity Hosp New Orleans, 69-78, sr vis physician, 79-89; mem med adv bd, Nat Pituitary Agency, 74-77; assoc mem grad fac, Univ New Orleans, 76-, consult prof, Dept Psychol, 86-; mem res adv comt, Nat Asn Retarded Citizens, 78-79; consult, Food & Drug Admin, 79-80; ed-in-chief, Peptides, 80-; Wellcome vis prof, Fedn Am Soc Exp Biol, ETex State Univ, 89-90. *Honors & Awards:* Edward T Tyler Fertil Award, Int Fertil Soc, 75; Copernicus Medal, Poland, 79; William S Middleton Award, Vet Admin, 82; Talmage Lectr, Aspen Allergy Conf, 86. *Mem:* Endocrine Soc; Am Physiol Soc; Soc Exp Biol & Med; Soc Neurosci; Int Soc Psychoneuroendocrinol; hon mem Endocrine Socs of Chile, Philippines, Peru, Poland & Hungary; Int Soc Neuroendocrinol; Int Pigment Cell Soc; Am Peptide Soc; Am Geriat Soc. *Res:* Brain peptides; neuroendocrinology; hypothalamic hormones. *Mailing Add:* Vet Affairs Med Ctr 1601 Perdido St New Orleans LA 70146

KASTING, JAMES FRASER, b Schenectady, NY, Jan 2, 53; m 80; c 2. ATMOSPHERIC EVOLUTION, RADIATIVE TRANSFER. *Educ:* Harvard Univ, AB, 75; Univ Mich, MS(phys) & MS(atmospheric sci), 78, PhD(atmospheric sci), 79. *Prof Exp:* Res fel, Nat Ctr Atmospheric Res, 79-81; res fel, 81-83, RES SCIENTIST, AMES RES CTR, NASA, 83-; PROF GEOSCI, PA STATE UNIV. *Mem:* Am Geophys Union; Int Soc Study Origin Life; AAAS. *Res:* Evolution of planetary atmospheres; history of the earth and why it is different from that of Mars and Venus. *Mailing Add:* 1072 Crabapple Dr State College PA 16801

KASTL, PETER ROBERT, b Alexandria, La, July 25, 49; m 74; c 2. OPHTHALMOLOGY, BIOCHEMISTRY. *Educ:* Centenary Col La, BS, 71; Tulane Univ, MD, 74, PhD(biochem), 78. *Prof Exp:* Instr biochem, 75-81, instr ophthal, 77-81, asst prof, 81-85, assoc prof biochem & ophthal, 85-88, PROF BIOCHEM & OPHTHAL, TULANE UNIV, 89- *Concurrent Pos:* Fel, Nat Inst Gen Med Sci, 75-76; NIH res grant, 81-83. *Mem:* Sigma Xi; AMA; Southern Med Asn; Am Acad Ophthal; Asn Res Vision & Ophthal. *Res:* Microsomal treatment of ingested toxins; pharmacologic prevention of cataracts; design of new types of ophthalmologic prosthetic devices; tear analysis; contact lenses. *Mailing Add:* Dept Ophthal Tulane Med Ctr New Orleans LA 70112

KASTNER, CURTIS LYNN, b Altus, Okla, Sept 21, 44; m 66; c 2. MEAT SCIENCE, MUSCLE BIOLOGY. *Educ:* Okla State Univ, BS, 67, MS, 69, PhD(food sci), 72. *Prof Exp:* Asst prof food sci, Wash State Univ, 72-75; PROF & RES COORDR, ANIMAL & FOOD SCI, KANS STATE UNIV, 75- *Concurrent Pos:* Sect chmn, Inst Food Technologists, 82, mem exec bd, Muscle & Food Div, 91; chmn, Am Meat Sci Asn Ann Conf, 88, chair, 89, exec bd, 88-89. *Honors & Awards:* US Key Res Scientist, AAAS, 84. *Mem:* Am Meat Sci Asn; Am Soc Animal Sci; Inst Food Technologists. *Res:* Technology, development, processing and preservation of meat and meat products, including hot processing, tenderization, microbial sampling, shelf life extension, packaging sanitation, microbiology, chemical residues, meat safety and low fat technology. *Mailing Add:* Dept Animal Sci & Indust Kans State Univ Weber Hall Manhattan KS 66506-0201

KASTNER, MARC AARON, b Toronto, Ont, Nov 20, 45; m 67; c 2. SEMICONDUCTORS, HIGH-TEMPERATURE SUPERCONDUCTIVITY. *Educ:* Univ Chicago, BS, 67, MS, 69 & PhD(physics), 73. *Prof Exp:* Res fel, div eng & appl physics, Harvard Univ, 72-73; asst prof physics, 73-77, assoc prof, 77-83, prof, 83-89, DONNER PROF PHYSICS, MASS INST TECHNOL, 89- *Concurrent Pos:* Head div atomic, condensed matter & plasma physics, dept physics, Mass Inst Technol, 83-87; counr, Am Phys Soc, 90-94; assoc dir, Consortium Superconducting Electronics. *Mem:* Fel Am Phys Soc; AAAS. *Res:* Electronic and optical studies of amorphous semiconductors led to the Valence Alteration Model; measurements of conductivity of nanometer-size semiconductor devices; magnetic optical and transport studies of high-temperature superconductors. *Mailing Add:* Rm 13-2142 Mass Inst Technol Cambridge MA 02139

KASTNER, MIRIAM, b Bratislava, Czech; US citizen. GEOLOGY. *Educ:* Hebrew Univ Jerusalem, Israel, BSc & MSc, 64; Harvard Univ, PhD(geol), 70. *Hon Degrees:* Dr, Univ Paris XI, 84. *Prof Exp:* Fel geol, Harvard Univ, 70-71, Univ Chicago, 71-72; from asst prof to assoc prof, 72-82, PROF GEOL, SCRIPPS INST OCEANOG, 82-, CHMN, GEOL RES DIV, 89- *Concurrent Pos:* Assoc ed, J Sedimentary Petrol, 80-88; distinguished lectr, Am Asn Petrol Geologists, 83 & 84; vis prof, Hebrew Univ, Jerusalem, Israel, 86; planning comt mem, Ocean Drilling Proj, 84-; mem, NSF Adv Comt, Earth Sci, 86-88, steering comt, 2nd Conf Sci Ocean Drilling, 86-87; mem-at-large, Sect Geol & Geog, AAAS, 87-; chmn, Gordon Res Conf Chem Oceanog, 89. *Honors & Awards:* Newcomb Cleveland Prize, AAAS. *Mem:* AAAS; Geochem Soc; Am Geophys Union; Int Asn Geochem & Cosmochem; Sigma Xi; Int Asn Sedimentologists. *Res:* Origin, mineralogy and geochemistry of silicates and carbonates in marine and non-marine environments; stable isotopes for diagenesis; processes that cause metal enrichment in oceanic sediments; surface chemistry in diagenesis; hydrothermal deposits in the submarine environment; fluids in convergent plate margins. *Mailing Add:* Geol Res Div A-012 Scripps Inst Oceanog La Jolla CA 92093-0212

KASTNER, SIDNEY OSCAR, b Winnipeg, Man, Apr 20, 26; m 51; c 3. PHYSICS. *Educ:* McGill Univ, BSc, 50; Syracuse Univ, MS, 55, PhD(physics), 60. *Prof Exp:* Jr res officer, Nat Res Coun Can, 50-52; physicist, Gen Elec Res Lab, 55-57; physicist, Goddard Space Flight Ctr, NASA, 59-82; CONSULT SCIENTIST, 82- *Mem:* AAAS; Am Phys Soc; NY Acad Sci; AAAS. *Res:* Atomic physics and spectroscopy; solar physics and astrophysics; theoretical calculations and applications of atomics physics to observe visible, ultraviolet and infrared spectra of astronomical sources; primarily fluorescent spectra of nebulae, cataclysmic, variables, symbiotic stars and novae. *Mailing Add:* 1-A Ridge Rd Greenbelt MD 20770

KASUBA, ROMUALDAS, US citizen. MECHANICAL ENGINEERING, APPLIED MECHANICS. *Educ:* Univ Ill, Urbana, BS, 54, MS, 57, PhD(mech eng), 62. *Prof Exp:* Res asst dynamics, mech eng dept, Univ Ill, Urbana, 58-62; head stress & dynamics group power systs div, TRW Inc, Ohio, 62-68; from assoc prof to prof mech eng, 68-78, CHAIRPERSON DEPT MECH ENG, CLEVELAND STATE UNIV, 78-, PROF MECH ENG, 81- *Concurrent Pos:* Indust consult; lectr, US & Can; consult, Indust Fasteners Inst, 70-75 & Am Nat Standards Inst, 71-74. *Mem:* Am Soc Mech Engrs. *Res:* Vibration and noise studies in industrial machines; dynamic loads in geared systems; development of optimum threaded fastener system; dynamic simulation of geared systems; dynamic simulation of machine tool structures and drives; application of finite element techniques. *Mailing Add:* Mech Eng Dept Eng Bldg 120 Northern Ill Univ De Kalb IL 60115-2854

KASUBE, HERBERT EMIL, b Chicago, Ill, Mar 23, 49; m 71; c 1. MATHEMATICS EDUCATION. *Educ:* MacMurray Col, BA, 71; Univ Ill, MA, 73; Univ Mont, PhD(math), 79. *Prof Exp:* Teaching asst math, Univ Ill, 71-75 & Univ Mont, 75-78; ASSOC PROF MATH, BRADLEY UNIV, PEORIA, ILL, 78- *Mem:* Sigma Xi; Math Asn Am. *Res:* Discrete mathematics; number theory; college mathematics education. *Mailing Add:* Dept Math Bradley Univ 1501 W Bradley Ave Peoria IL 61625

KASUPSKI, GEORGE JOSEPH, b Boston, Mass, July 26, 46; m 79; c 1. DIAGNOSTIC VIROLOGY, MOLECULAR VIROLOGY. *Educ:* McGill Univ, BSc, 69, PhD(molecular biol), 75. *Prof Exp:* Lectr genetics, Dept Biol, McGill Univ & fel virol, Dept Microbiol, Royal Victoria Hosp, 75-78; STAFF MICROBIOLOGIST VIROL, WELLESLEY HOSP, UNIV TORONTO, 78- *Concurrent Pos:* Lectr, Dept Med Microbiol, Univ Toronto, 78-81, asst prof, 81; adv, Toronto Inst Med Technol, 81, Wellesley Hosp Res Inst, 81; dir, Lakeshore Labs Ltd, 84. *Mem:* Am Soc Microbiol; Pan Am Group Rapid Viral Diag; NY Acad Sci. *Res:* Pathogenesis and diagnosis of viral infections of the lower respiratory tract in immunocompromised adults. *Mailing Add:* Dept Microbiol Wellesley Hosp 160 Wellesley St E Toronto ON M4Y 1J3 Can

KASVINSKY, PETER JOHN, b Bridgeport, Conn, Dec 7, 42; m 74; c 2. ENZYME REGULATION, CALCIUM CONTROL. *Educ:* Bucknell Univ, BSc, 64; Univ Vt, PhD(biochem), 70. *Prof Exp:* Biochemist, US Army Aeromed Res Lab, 69-72; instr biochem, Sch Med, Wayne State Univ, 72-74; sr res assoc biochem, Univ Alta, 74-79, instr, Dept Biochem, 77-79; asst prof, 79-82, dir res develop, 86-88, ASSOC PROF BIOCHEM, MARSHALL UNIV 82-, DIR RES DEVELOP & GRAD STUDIES, SCH MED, 88- *Concurrent Pos:* Radiol control officer, US Army Aeromed Res Lab, 70-72; adj asst prof, 80-82, adj assoc prof biomed sci, WVa Univ, 82-; prin investr, NIH grants, 81. *Mem:* Am Chem Soc; AAAS; Can Biochem Soc; Am Soc Biol Chemists; Sigma Xi; Nat Coun Univ Res Admin. *Res:* Enzymology of covalent modification of proteins, enzyme regulation, structure function and allosteric control, especially as applied to the regulation of enzymes of glycogen metabolism. *Mailing Add:* Dept Grad Studies & Res 109 Waller Admin Bldg Bloomsburg Univ Bloomsburg PA 17815

KATARIA, YASH P, b 1936. PULMONARY DISEASES, INTERNAL MEDICINE. *Educ:* Glancy Med Col, India, MD, 59; Liverpool Sch Med, Eng, DTM & H, 63; Welsh Nat Sch Med, Univ Wales, DTCD, 65; FRCP, 79. *Prof Exp:* Registr, Welsh Nat Sch Med, 67-69; assoc med, Chicago Med Sch, 70-71; from instr to asst prof med & pulmonary dis, Col Med, Ohio State Univ, 72-78; assoc prof, 78-82, PROF INTERNAL MED, ECAROLINA UNIV SCH MED, 82-, VCHMN DEPT MED, 87- *Concurrent Pos:* Actg chmn dept med, E Carolina Univ Sch Med, 86-87, sect head pulmonary div, 78-; med dir, Spec Servs & Respiratory Ther, 88- *Mem:* Fel Am Col Chest Physicians; Am Lung Asn; NY Acad Sci; Am Fedn Clin Res; AMA; Sigma Xi. *Res:* Clinical and immunologic aspects of sarcoidosis exploring its pathogenesis and etiology; immunologic work involves studies of peripheral blood and bronchoalveolar lavage (BAL), T and B-cell quantitation; examination of sarcoidal granuloma for its cellular components and production of lymphokines, etc; production of Kreim antigen from auto logous BAL cells. *Mailing Add:* Dept Med Pulmonary Dis ECarolina Univ Sch Med Greenville NC 27858-4354

KATAYAMA, DANIEL HIDEO, b Honolulu, Hawaii, Sept 26, 39; m 63. MOLECULAR SPECTROSCOPY. *Educ:* Univ Hawaii, BS, 62, MS, 64; Tufts Univ, PhD(physics), 70. *Prof Exp:* Physicist aeronomy, Air Force Cambridge Res Lab, 63-66; physicist solid state physics, Gillette Co, 70-71; PHYSICIST, AIR FORCE GEOPHYS LAB, 71- *Concurrent Pos:* Vis scientist, Bell Lab, Murray Hill, NJ, 78-79 & Mass Inst Tech, Cambridge, 80-82. *Mem:* AAAS; Am Phys Soc; Optical Soc Am. *Res:* Laser induced fluorescence of molecules and ions; absorption-photoionization cross sections and spectroscopy of atmospheric gases in vacuum ultraviolet; phonon scattering in solids; elastic constants of materials. *Mailing Add:* Two Fox-Run Rd Bedford MA 01730

KATCHEN, BERNARD, b New York, NY, May 20, 28; m 51; c 2. BIOCHEMISTRY. *Educ:* City Col New York BS, 49; Ohio State Univ, MSc, 51; NY Univ, PhD(biochem), 56. *Prof Exp:* Chemist, Clairol Inc, 54-56; sr chemist, Nat Cash Register, 56-61; PRIN SCIENTIST, SCHERING CORP, 62- *Mem:* AAAS; Am Chem Soc; NY Acad Sci. *Res:* Pharmacokinetics; drug metabolism; biopharmaceutics. *Mailing Add:* 18 Chestnut St Livingston NJ 07039-4402

KATCHER, DAVID ABRAHAM, b New York, NY, Apr 28, 15; m 47, 84; c 2. WRITING & EDITING. *Educ:* Univ Wis, BA, 36. *Prof Exp:* Tech ed, Naval Ord Lab, 41-43; founding ed, Physics Today, Am Inst Physics, 47-51; assoc ed, Opers Res Off, 51-56; assoc ed, Jour Opers Res Soc Am, 52; ed, Weapons Systs Eval Group, Inst Defense Anal, 56-60, exec secy, Jason Div, 60-66; mem sr staff, Opers Res Sect, Arthur D Little, Inc, Mass, 66-72; sr staff mem, Secretariat of Nat Adv Comt Oceans & Atmospheres, Nat Oceanic & Atmospheric Admin, 72-76; sr policy analyst, Off Sci & Technol Policy, Exec Off of the Pres, 76-77; spec asst to undersecy state for security assistance, Sci & Technol, Dept State, 77-80; CONSULT, 80- *Mem:* AAAS. *Mailing Add:* 5608 Warwick Pl Chevy Chase MD 20815

KATCHMAN, ARTHUR, b New York, NY, Oct 4, 24; m 60; c 2. ORGANIC CHEMISTRY. *Educ:* NY Univ, BA, 49; Polytech Inst Brooklyn, PhD(chem), 55. *Prof Exp:* Asst instr biochem, NY Med Col, 49-50, instr, 50-52, assoc, 52-54; res chemist, Hooker Chem Co, 55-56; res assoc, Polytech Inst Brooklyn, 57-58; res assoc, Gen Elec Res Lab, Gen Elec Co, 59-62, mgr mat physics & chem, Capacitor Dept, 63-66, mgr polymer chem, Chem Develop Oper, 66-68, mgr advan res, Plastics Dept, 68-70, mgr prod develop, Plastics Dept, 70-75, mgr chem develop, 76-79, mgr opers control & planning, 80-81, mgr, Dept Qual Assurance, 81-90; RETIRED. *Concurrent Pos:* Hooker fel, 57-58. *Mem:* Am Chem Soc. *Res:* Mechanism and kinetics of polymerization; polymer structure and properties; stereospecific polymerization. *Mailing Add:* 6904 Country Lakes Circle Sarasota FL 34243-3803

KATCOFF, SEYMOUR, b Chicago, Ill, Aug 19, 18; m 51; c 2. NUCLEAR CHEMISTRY, SCIENCE EDUCATION. *Educ:* Univ Chicago, BS, 40, PhD(phys chem), 44. *Prof Exp:* Assoc chemist, Metall Lab, Univ Chicago, 43-45; staff mem, Los Alamos Sci Lab, 45-48; SR CHEMIST, BROOKHAVEN NAT LAB, 48- *Concurrent Pos:* Fel, Weizmann Inst Sci, 58-59 & Guggenheim fel, 67-68; vis scientist, City Col NY, 84-85. *Mem:* AAAS; Am Chem Soc; Am Phys Soc; Sigma Xi. *Res:* High-energy nuclear reactions; nuclear track detectors; cross section measurements; neutron-rich isotope studies; nuclear spectroscopy; heavy ion reactions. *Mailing Add:* Brookhaven Nat Lab Upton NY 11973

KATEKARU, JAMES, b Kauai, Hawaii, June 10, 35; m 64; c 2. ANALYTICAL CHEMISTRY, NUCLEAR CHEMISTRY. *Educ:* Univ Ore, BS, 56; Univ Ariz, MS, 61; Univ Cincinnati, PhD(chem), 65. *Prof Exp:* Anal chemist, Food & Drug Admin, 62-63; res chemist, Rocketdyne Div, NAm Rockwell, 65-66; index ed, Chem Abstr Serv, 66-67; res mgr, Naval Radiol Defense Lab, 67-69; from asst prof to assoc prof, 69-80, PROF CHEM, CALIF POLYTECH STATE UNIV, 80- *Concurrent Pos:* Res consult, Trapelo West Div, Lab Electronics, 70-71. *Mem:* Am Chem Soc; Am Inst Physics. *Res:* Solvent extraction; polarography; rocket exhaust product analysis; catalysis of non-hypergollic propellant combinations; nuclear fallout phenomenology; detection and diagnosis of nuclear weapons. *Mailing Add:* Dept Chem Calif Polytech State Univ San Luis Obispo CA 93407

KATELL, SIDNEY, chemical engineering; deceased, see previous edition for last biography

KATEN, PAUL C, b Lawrence, Mass, Feb 7, 43; m 68. MICROMETEOROLOGY, AIR POLLUTION. *Educ:* Lowell Technol Inst, BS, 64; Trinity Col, MS, 68; Colo State Univ, PhD(atmospheric sci), 77. *Prof Exp:* Test engr, Pratt & Whitney Aircraft, 64-69; res assoc, Dept Atmospheric Sci, Ore State Univ, 77-84; exp scientist, Div Environ Mech, Commonwealth Sci & Indust Res Orgn, Canberra, Australia, 84-85; CONSULT, 85- *Mem:* Am Meteorol Soc; Am Geophys Union; Air Pollution Control Asn. *Res:* Instrumentation development with recent applications to the dry-deposition of gases and particles; micrometeorology; turbulence; indoor air pollution; industrial hygiene; wind power resource assessment. *Mailing Add:* Met One 481 California Ave Grants Pass OR 97256

KATER, STANLEY B, b Cleveland, Ohio, June 12, 43; div; c 1. NEUROSCIENCES. *Educ:* Case Western Reserve Univ, BA, 65; Univ Va, PhD(biol), 68. *Prof Exp:* NIH fel biol, Univ Ore, 68-69; from asst prof to prof zool, Univ Iowa, 79-90; PROF ANAT & NEUROBIOL, COLO STATE UNIV, 90- *Concurrent Pos:* Javits neurosci investr award, NIH sponsored res. *Honors & Awards:* Alexander Von Humboldt Res Scientist Award. *Mem:* AAAS; Soc Neurosci. *Res:* Developmental neurobiology; control of neuronal growth cores. *Mailing Add:* Dept Anat & Neurobiol Colo State Univ Ft Collins CO 80523

KATES, JOSEF, b Vienna, Austria, May 5, 21; nat Can; m 44; c 4. SCIENCE POLICY, SYSTEMS SCIENCE. *Educ:* Univ Toronto, BA, 48, MA, 49, PhD(physics), 51. *Hon Degrees:* LLD, Concordia Univ, Can, 81. *Prof Exp:* Supvr, Imp Optical Co, 42-44; proj engr, Rogers Electronic Tubes, 44-48; res engr, Univ Toronto, 48-54; pres, KCS Ltd & Traffic Res Corp, 54-66; dep managing partner, Kates, Peat, Marwick & Co, 67-68, assoc, 69-73; PRES, JOSEF KATES ASSOCS, INC, 74- *Concurrent Pos:* Pres, Setak Comput Servs Co, 67-; chmn, Teleride Sage Corp, 78-, mem, Sci Coun Can, 68-74, chmn, 75-78; consult, US AEC; chancellor, Univ Waterloo, 79-85. *Mem:* Inst Mgt Sci; Can Oper Res Soc; fel Eng Inst Can; Sci, Eng & Technol Community Can; fel Inst Mgt Consult. *Res:* Application of scientific methods, especially computers and operations research to industrial, scientific and engineering problems. *Mailing Add:* 265 Upper Highland Cr Willowdale ON M2P 1V4 Can

KATES, MORRIS, b Galati, Roumania, Sept 30, 23; Can citizen; m 57; c 3. LIPID CHEMISTRY. *Educ:* Univ Toronto, BA, 45, MA, 46, PhD(biochem), 48. *Prof Exp:* Asst, Banting & Best Med Res, 48-49; Nat Res Labs fel, Nat Res Coun Can, 49-51, asst res officer, Div Appl Biol, 51-55, assoc res officer, 55-61, sr res officer, 61-68; prof biochem, Univ Ottawa, 68-69, vdean res, 78-82, chmn dept, 82-85, prof, 69-89, EMER PROF BIOCHEM, UNIV OTTAWA, 89- *Concurrent Pos:* Co-ed, Can J Biochem, 74-84; staff res lectr, Univ Ottawa, 81. *Honors & Awards:* Supelco Award, lipid res, Am Oil Chemists' Soc, 84. *Mem:* Am Chem Soc; Am Soc Biol Chem; Can Biochem Soc; Brit Biochem Soc; Royal Soc Can; Am Oil Chemists' Soc. *Res:* Synthesis of lecithins and related compounds; structure of the alkaloid, gelsemine; plant lecithinases and plant phospholipids; glycerides; lipases; bacterial lipids; phospholipid desaturases; biosynthesis of phospholipids; diphytanyl glycerol ether lipids. *Mailing Add:* Dept Biochem Univ Ottawa Ottawa ON K1N 6N5 Can

KATES, ROBERT, b Brooklyn, NY, Jan 31, 29; m 48; c 3. LONG TERM POPULATION DYNAMICS. *Educ:* Univ Chicago, MA, 58, PhD(geog), 62. *Prof Exp:* From asst prof to prof, Grad Sch Geog, Clark Univ, 62-87; UNIV PROF, & DIR, ALAN SHAWN FEINSTEIN WORLD HUNGER PROG, BROWN UNIV, 86- *Concurrent Pos:* Dir, Bur Resource Assessment & Land Use Planning, Univ Col, Dar Es Salaam, Tanzania; univ prof, Clark Univ, 74-80; chmn, Comn Human Rights, NAS, 76-79, mem, 79-85; fel, Woodrow Wilson Int Ctr Scholars, 79; Mac Arthur Prize fel, 81-85; fel, distinguished scholar exchange, comt scholarly common people's Rep China, 85; res prof, Ctr Technol, Environ & Develop, Clark Univ, 81-87; mem, bd dirs, Comt Prob & Policy, Social Sci Res Coun & Comt Int Appl Syst Anal, Am Acad Arts & Sci, 82; mem, nat coun, Fedn Am Scientists & Bd Sci & Technol Int Develop, Nat Res Coun, 86-89, comt Global Change, 89- *Honors & Awards:* Nat Medal of Sci, 91; Honors Award, Asn Am Geographers. *Mem:* Nat Acad Sci; Asn Am Geographers; fel AAAS; Fedn Am Sci; Am Acad Arts & Sci. *Res:* The prevalance and persistence of hunger; long term population dynamics; sustainability of the biosphere, climate impact assessment; theory of the human environment; author of many books on hunger, environment and technology. *Mailing Add:* Alan Shawn Feinstein World Hunger Prog Brown Univ Box 1831 Providence RI 02912

KATH, WILLIAM LAWRENCE, b Pasadena, Calif, June 23, 57; m 84. WAVE PROPAGATION, ASYMPTOTIC METHODS. *Educ:* Mass Inst Technol, SB, 78; Calif Inst Technol, PhD(appl math), 81. *Prof Exp:* NSF res fel appl math, Calif Inst Technol, 81-82, Von Karman instr, 82-84; ASST PROF ENG SCI & APPL MATH, NORTHWESTERN UNIV, 84- *Concurrent Pos:* Prin investr, NSF Presidential young investr award, 85- *Mem:* Soc Indust & Appl Math; Sigma Xi. *Res:* Application of asymptotic and singular perturbation methods to linear and nonlinear wave propagation (and related phenomena) in fluids, fiber optics and quantum optical devices. *Mailing Add:* ES-AM Technol Inst Northwestern Univ Evanston IL 60201

KATHAN, RALPH HERMAN, b Chicago, Ill, Feb 1, 29; div; c 2. CLINICAL BIOCHEMISTRY. *Educ:* Univ Chicago, SB, 49; Univ Ill, MS, 59, PhD(biochem), 61. *Prof Exp:* Res asst biol chem, Univ Chicago, 48-49; biochemist res labs, Kraft Foods Div, Nat Dairy Prod Corp, 49-51; tech serv, Am Can Co, 56-57; res asst, 57-61, res assoc, 61-62, asst prof, 62-68, ASSOC PROF BIOL CHEM, COL MED, UNIV ILL, 68-; CHMN DIV BIOCHEM, COOK COUNTY HOSP, 71- *Concurrent Pos:* Consult comn influenza, Armed Forces Epidemiol Bd, 64-68. *Mem:* Nat Acad Clin Biochemists; Soc Complex Carbohydrates; Am Soc Biol Chem; Am Asn Clin Chem. *Res:* Protein structure; mechanisms of viral infection; bacterial metabolism; carbohydrate absorption; plasma expanders; diagnostic biochemistry. *Mailing Add:* Div Biochem Cook County Hosp 627 S Wood St Chicago IL 60612

KATHMAN, R DEEDEE, b Stamford, NY, Apr 16, 48; m 85. TARDIGRADOLOGY, FRESHWATER POLLUTION ASSESSMENT. *Educ:* State Univ NY, Oswego, BA, 70; Tenn Technol Univ, MS, 81; Univ Victoria, BC, PhD, 89. *Prof Exp:* Technician, Weather Observ, Lake Ont Environ Lab, 72-73; biologist environ conserv, LMS Engrs Inc, 73-; proj mgr environ conserv, AWARE Corp, 78-81; regional dir environ conserv, EVS Consult, Ltd, 81-84; UNIV VICTORIA, BC, 84- *Mem:* Int Asn Meiobenthologists; Am Micros Soc; NAm Benthological Soc; Sigma Xi. *Res:* Taxonomy, ecology, evolution and distribution of tardigrades; taxonomy and ecology of oligochaetes and chironomid larvae; freshwater benthic taxonomy and ecology; freshwater pollution assessment. *Mailing Add:* 4256 Warren Rd Franklin TN 37064

KATHOLI, CHARLES ROBINSON, b Charleston, WVa, Jan 2, 41; m 80; c 1. BIOMATHEMATICS. *Educ:* Lehigh Univ, BA, 63; Adelphi Univ, MS, 65, PhD(math, appl anal), 70. *Prof Exp:* Asst, Adelphi Univ, 64-66; instr math, Suffolk County Community Col, 66-67; instr, Adelphi Univ, 67-70; asst prof biomath, 70-77, asst prof info sci, 73-76, ASSOC PROF BIOMATH, UNIV ALA, BIRMINGHAM, 77- *Mem:* AAAS; Soc Indust & Appl Math; NY Acad Sci; Asn Comput Mach; Acoust Soc Am; Sigma Xi. *Res:* Mathematical modelling and computer simulations in the field of cardiovascular research; computational methods for special functions. *Mailing Add:* 315 Poinciana Dr Birmingham AL 35209

KATHREN, RONALD LAURENCE, b Windsor, Ont, June 6, 37; US citizen; m 64; c 2. HEALTH PHYSICS, HISTORY OF SCIENCE. *Educ:* Univ Calif, Los Angeles, BS, 57; Univ Pittsburgh, MS, 62; Am Bd Health Physics, dipl, 66; Am Acad Environ Eng, dipl, 78; Soc Radiol Protection, cert appl health physics, 85; Am Bd Med Physics, dipl, 89. *Prof Exp:* Supvr health physicist, Mare Island Naval Shipyard, USN, 59-61; health physicist, Lawrence Radiation Lab, Univ Calif, 62-67; sect mgr & sr res scientist radiation dosimetry, Pac Northwest Div, Battelle Mem Inst, 67-72; corp health physicist, Portland Gen Elec Co, 72-78; staff scientist, PAC Northwest Div, Battelle Mem Inst, 78-87; dir, Health Physics, 87-89, DIR, US TRANSURANIUM & URANIUM REGISTRIES, HANFORD ENVIRON HEALTH FOUND, 89-, ACTG MGR RES, 90- *Concurrent Pos:* Abstractor, Chem Abstr, 62-78; lectr, Tri-Cities Univ Ctr, Univ Wash, 71-72, affil assoc prof, 78-, coordr radiol sci, 80-82, 86-87; adj prof, Ore State Div Continuing Educ, 72-77; health physicist, Reed Col, 73-78; mem traineeship adv comt, US AEC, 73-74; mem, Nat Adv Comt Nuclear Technicians, Tech Educ Res Ctr, 75-81; ed, Health Physics J, Health Physics Soc, 76-80; 055302460Radiation Adv Comt, 77-78; consult, Int Atomic Energy Agency tech expert, 77, US Nuclear Regulatory Comn, 78, US Adv Comt Reactor Safeguards, 78-, US Adv Comt Nuclear Wastes, 88- & US Transuranium & Uranium Registries, 84-87; mem panel examr, Am Bd Health Physics, 78-80; mem, Am Bd Health Physics, 82-84, secy-treas, 84; mem, Nat Coun Radiation Protection & Measurements Task Force on Alarm & Access Control Systs, 83-, chmn SC 1-3 comt collective dose, 90-; bd dir, Am Acad Health Physics, 85-86; lectr, Wash State Univ, 87-; mem comt film badge dosimetry, Nat Res Coun, 88-89; int adv bd, J Radiation Protection, 88-; ed, Radiation Prof Dosimetry, 91- *Honors & Awards:* Elda E Anderson Award, Health Physics Soc, 77, Founders Award, 85; Arthur Humm Award, Nat Registry Radiation Protection Technologists, 88. *Mem:* Fel Health Physics Soc; Am Asn Physicists Med; AAAS; fel Soc Radiol Protection; Am Acad Health Physics; Sigma Xi; Health Physics Soc (bd dir, 73-76, pres-elect, 88-89 & pres, 89-90). *Res:* Biokinetics and dosimetry of actinides; applied health physics; radiological dosimetry; environmental radioactivity; history of radiation protection and physics. *Mailing Add:* Hanford Environ Health Found PO Box 100 Richland WA 99352

KATNER, ALLEN SAMUEL, organic chemistry, for more information see previous edition

KATO, IKUNOSHIN, PROTEIN CHEMISTRY. *Educ:* Univ Osaka, Japan, PhD(biochem), 69. *Prof Exp:* PROTEIN CHEMIST, CENTOCOR, INC, 83- *Res:* Sequencing of protein; monoclonal antibody. *Mailing Add:* Takara Shuzo Co Ltd Ctrl Res SETA 3-4-1 Otsu Shiga 520-21 Japan

KATO, TOSIO, b Kanuma, Japan, Aug 25, 17; m 44. MATHEMATICS. *Educ:* Univ Tokyo, BS, 41, DSc(math physics), 51. *Prof Exp:* From asst to prof physics, Univ Tokyo, 43-62; PROF MATH, UNIV CALIF, BERKELEY, 62- *Honors & Awards:* Asahi Award, 60; Norbert Wiemer Prize, 80. *Mem:* Am Math Soc; Math Soc Japan. *Res:* Functional analysis and applications; mathematical physics. *Mailing Add:* Univ Calif Berkeley CA 94720

KATO, WALTER YONEO, b Chicago, Ill, Aug 19, 24; m 53; c 3. REACTOR SAFETY. *Educ:* Haverford Col, BS, 46; Univ Ill, MS, 49; Pa State Col, PhD(physics), 54. *Prof Exp:* Res assoc hydrodyn, Ord Res Lab, Sch Eng, Pa State Col, 49-52; jr res assoc neutron physics, Brookhaven Nat Lab, 52-53; asst physicist nuclear & reactor physics, Argonne Nat Lab, 53; assoc physicist, Reactor Eng Div, 53-63, assoc physicist, Reactor Physics Div, 63-68, sr physicist, 68-75, sr physicist, Appl Physics Div, 69-75, head fast reactor exps sect, 63-70; assoc chmn & sr nuclear engr, 75-80, dept chmn, 80-88, CHMN, DEPT NUCLEAR ENERGY, BROOKHAVEN NAT LAB, 88- *Concurrent Pos:* Fulbright res scholar, Japan Atomic Energy Res Inst & Univ Tokyo,

58-59; vis prof nuclear eng, Univ Mich, 74-75; consult, Off Nuclear Regulatory Res, Nuclear Regulatory Comn, 74-85; Fulbright Fel, 58. *Mem:* AAAS; Am Phys Soc; fel Am Nuclear Soc. *Res:* Hydrodynamics; cavitation studies; neutron resonance phenomenon; neutron total cross section measurements; neutron inelastic scattering studies; reactor physics; critical assembly experiments; fast reactor physics and safety; reactor safety research. *Mailing Add:* Brookhaven Nat Lab Upton NY 11973

KATOCS, ANDREW STEPHEN, JR, b Passaic, NJ, Oct 7, 44; m 67; c 2. HYPERTENSION, ULTRASOUND. *Educ:* Rutgers Univ, BPh & BS, 67; Marquette Univ, PhD(pharmacol), 72. *Prof Exp:* Nat Heart & Lung Inst fel & instr pharmacol, Sch Med, Ind Univ, 72-73; sr res pharmacologist atherosclerosis, 73-84, sr res pharmacologist osteo-arthritis, 86-89, SR RES PHARMACOLOGIST HYPERTENSION, LEDERLE LABS, AM CYANAMID CO, 89- *Mem:* Am Heart Asn; Am Soc Pharmacol & Exp Therapeut. *Res:* Development of animal models of atherosclerosis; evaluation of compounds for anti-atherosclerotic activity; real-time ultrasonic imaging of arterial vasculature; development of animal models for the evaluation of compounds for anti osteo-arthritic activity; tissue culture; receptor binding assay. *Mailing Add:* Lederle Labs Med Res Div Middletown Rd Pearl River NY 10965

KATOH, ARTHUR, b Honolulu, Hawaii, Aug 24, 33; m 63; c 1. DEVELOPMENTAL BIOLOGY. *Educ:* Syracuse Univ, AB, 54; Univ Ill, MS, 56, PhD(zool), 60; Univ Pittsburgh, MPH, 86. *Prof Exp:* NSF fel, 60-61; res assoc zool, Univ Ill, 61-62; asst prof biol, Univ Toledo, 62-63; res assoc, Argonne Nat Lab, 63-66; dir oncol lab, Dept Radiother, 66-73, DIR, DIV NUCLEAR PATH & ONCOL, MERCY HOSP, 73- *Mem:* Soc Develop Biol; NY Acad Sci; Am Soc Cell Biol. *Res:* Developmental biology; cellular differentiation in amphibian and chick embryos; cancer metastasis. *Mailing Add:* Div Nuclear Path & Oncol Mercy Hosp Pittsburgh PA 15219

KATON, JOHN EDWARD, b Toledo, Ohio, Jan 5, 29; m 55; c 3. PHYSICAL CHEMISTRY. *Educ:* Bowling Green State Univ, BA, 51; Kans State Univ, MS, 55; Univ Md, PhD(chem), 58. *Prof Exp:* Sr res chemist, Monsanto Chem Co, 58-61, res group leader, Monsanto Res Corp, Ohio, 61-68; assoc prof, 68-72, PROF CHEM, MIAMI UNIV, 72-, DIR MOLECULAR MICROS LAB, 84- *Concurrent Pos:* Consult, US Air Force, 60-61 & 68-69 & NIH, 70; chmn, Fedn Anal Chem & Spectros Soc, 81. *Honors & Awards:* Chemist Award, Am Chem Soc, 79. *Mem:* Am Chem Soc; Microbeam Analysis Soc; Soc Appl Spectros; Coblentz Soc; Soc Appl Spectros (pres, 76). *Res:* Molecular spectroscopy and structure; microspectroscopy; spectroscopic identification of materials. *Mailing Add:* Dept Chem Miami Univ Oxford OH 45056

KATONA, PETER GEZA, b Budapest, Hungary, June 25, 37; US citizen; m 66; c 2. BIOMEDICAL CONTROL SYSTEMS. *Educ:* Univ Mich, BS, 60; Mass Inst Technol, SM, 62, ScD(elec eng), 65. *Prof Exp:* From instr to asst prof elec eng, Mass Inst Technol, 63-69; assoc prof biomed eng, 69-78, chmn dept biomed eng, 80-87, PROF BIOMED ENG, CASE WESTERN RESERVE UNIV, 78- *Concurrent Pos:* Ford res fel, 65-67; consult, Biosysts, Inc, Mass, 65-67 & Mass Gen Hosp, Boston, 66-69; Fogarty sr int fel, 78-79; vis scientist, Univ Heidelberg, WGer, 87-88; prog dir bioeng, NSF, 89-91; fel Cardiovasc Sect, Am Physiol Soc. *Honors & Awards:* Alexander von Humboldt Sr US Scientist Award, 87. *Mem:* AAAS; Inst Elec & Electronics Engrs; Biomed Eng Soc (pres, 84-85); Am Physiol Soc; Am Soc Eng Educ. *Res:* Neural control of the cardiovascular system; interaction of cardiovascular and respiratory control mechanisms; automated control of drug infusion; noninvasive diagnostic techniques. *Mailing Add:* Dept Biomed Eng Case Western Reserve Univ Cleveland OH 44106

KATOVIC, VLADIMIR, b Bihac, Yugoslavia, Dec 19, 35; US citizen; m 71; c 1. ELECTROCHEMISTRY. *Educ:* Univ Zagreb, BS, 61, PhD(chem), 65. *Prof Exp:* Res assoc inorg chem, Ohio State Univ, 68-71; assoc prof anal chem, Univ Zagreb, 71-76; res assoc inorg chem, Iowa State Univ, 76-78; assoc prof, 78-91, PROF INORG CHEM, WRIGHT STATE UNIV, 91- *Mem:* Am Chem Soc; Croatian Chem Soc. *Res:* Synthetic and structural studies of coordination compounds of biological or catalytic interest; metal-metal bonding and metal-metal compounds. *Mailing Add:* Chem Dept Wright State Univ Dayton OH 45435

KATOVICH, MICHAEL J, b San Jose, Calif, July 16, 48; div; c 2. PHARMACODYNAMICS. *Educ:* Univ Calif, Davis, BS, 70, MS, 73, PhD(physiol), 76. *Prof Exp:* Teaching asst, Dept Physiol, Univ Calif, 72-73, assoc, 73-74, res assoc, 74-75, res physiologist, 75-76; assoc physiol, Col Med, Univ Fla, 76-77, Am Heart fel, 77-79, asst prof, Dept Pharmaceut Biol, 79-84, assoc prof, 84-88, PROF PHARMACODYNAMICS, COL PHARM, UNIV FLA, 89-, CHMN DEPT, 88- *Concurrent Pos:* Prin investr, Am Heart Asn, 79-81 & 84-92, Nat Inst Child Health & Human Develop, 83-91 & 85-88, NIH, 89-92; mem, Coun Complications, Am Diabetes Asn, 90. *Honors & Awards:* Irving I Hertzendorf Mem Award in Physiol, 76. *Mem:* Sigma Xi; Aerospace Med Asn; AAAS; Am Physiol Soc; Endocrine Soc; Am Diabetes Asn; Am Asn Cols Pharm. *Res:* Hypertension; metabolic phenomena; diabetes; environmental physiology with emphasis on temperature regulation; endocrinology; thirst control mechanisms; morphine dependency and withdrawal; acceleration biology; author of more than 150 technical publications. *Mailing Add:* Col Pharm Univ Fla Box J-487 JHMHC Gainesville FL 32610

KATRITZKY, ALAN R, b London, Eng, Aug 18, 28; US citizen; m 52; c 4. CHEMISTRY. *Educ:* Univ Oxford, BSc, 52, MA, 54, DPhil, 54; Univ Cambridge, PhD, 58, ScD, 63. *Hon Degrees:* Dr, Nat Univ, Madrid, Spain, 86 & Univ Poznan, Poland, 90. *Prof Exp:* Lectr, Pembroke Col, 56-58; univ demonstr, Univ Cambridge, 58-62, lectr, 62-63; prof chem, Univ E Anglia, 63-80, dean Sch Chem Sci, 63-70 & 76-80; KENAN PROF CHEM, UNIV FLA, 80- *Concurrent Pos:* Fel, Churchill Col, 60-63. *Honors & Awards:* Tilden Medal, 75; Royal Soc Chem Award in Heterocyclic Chem, 82. *Mem:* Hon fel Italian Chem Soc; fel Royal Soc; foreign fel Royal Australian Inst Chem; hon fel Polish Chem Soc. *Res:* Heteroaromatic tautomerism & aromaticity; heteroaromatic rearrangements; electrophilic substitution; conformational analysis of heterocycles; intermolecular interactions; infrared intensites; cycloadditions to heterocyclic betaines; pyrylium & pyridinium chemistry; mechanism of nucleophilic substitution reactions. *Mailing Add:* Dept Chem Univ Fla Gainesville FL 32611

KATSAMPES, CHRIS PETER, b Rochester, NY, Aug 23, 10; m 45; c 3. PEDIATRICS. *Educ:* Cornell Univ, BS, 31; Univ Rochester, MD, 36; Am Bd Pediat, dipl, 44. *Prof Exp:* From instr to asst prof pediat, Univ Rochester, 39-60; asst dir clin res, Warner-Lambert Res Inst, 60-72; from asst clin prof to assoc clin prof, Col Physicians & Surgeons, Columbia Univ, 61-75; dir clin invest, Warner-Chilcott Med Dept, 72-75, consult, 75-83; RETIRED. *Mem:* Soc Pediat Res; Am Pediat Soc; Am Acad Pediat; Sigma Xi. *Res:* Bacteriology and serology in infectious diseases; giardiasis; rheumatic fever; vitamin A in infections; antibacterial agents; new drugs; bronchodilators. *Mailing Add:* 48 Holly Dr Short Hills NJ 07078-1318

KATSANIS, D(AVID) J(OHN), b Philadelphia, Pa, Sept 28, 26; m 48; c 3. PHYSICS, PRODUCTION OPERATIONS MANAGEMENT. *Educ:* Temple Univ, BA, 52, MA, 54, PhD, 62, George Washington Univ, MEng, 80. *Prof Exp:* Physicist fluid dynamics, Naval Air Mat Ctr, 52-54; physicist ballistics, Frankford Arsenal, 54-57, chief, gas mech sect, 57-58, theoret ballistics sect, 58-59, systs ballistics sect, 59-63, LASH Proj, 63-65, advan concepts br, 65-66 & spec prods lab, 66-68, chief physicist, laser safety team, 68-69; chief physicist, US Army Small Arms Systs Agency, 69-73, chief, suppressive shielding, Edgewood Arsenal, 73-77, chief mech process, Chem Systs Lab, US Army Small Arms Systs Agency, Aberdeen Proving Ground, 77-81; physicist & chief producibility Eng Br, T&E Int, Inc, Bel-air, 81-86, sr physicist mech, 86-91; SR PHYSICIST MECH, SHIELDING TECHNOLOGIES, INC, 91- *Mem:* Am Phys Soc; Int Asn Bomb Technicians & Investigators; Asn US Army. *Res:* Weapon systems analysis; fluid dynamics; mechanics; thermodynamics; design of experiments; production technology; venteted suppressive shielding. *Mailing Add:* 4047 Heaps Sch Rd Pylesville MD 21132

KATSANIS, ELEFTHERIOS P, b Mytilene, Greece, Sept 28, 44; US citizen; div; c 1. COLLOID CHEMISTRY, SURFACE CHEMISTRY. *Educ:* Lehigh Univ, BS, 67, MS, 70; Clarkson Col Tech, PhD, 81. *Prof Exp:* SR CHEMIST, PHILADELPHIA QUARTZ CO, 75- *Mem:* Am Chem Soc; Sigma Xi; Soc Petrol Engrs. *Res:* Preparation of hydrous metal oxide sols via precipitation techniques; zeolites; stabilization of colloidal dispersions and their application to practical systems; enhanced oil recovery; water chemistry. *Mailing Add:* 121 Green Hill Rd King of Prussia PA 19406

KATSAROS, KRISTINA B, b Gothenburg, Sweden, July 24, 38; m 59; c 2. ATMOSPHERIC PHYSICS. *Educ:* Univ Wash, BS, 60, PhD(atmospheric sci), 69. *Prof Exp:* Res asst atmospheric sci, Univ Wash, 60 & 67-68, res assoc, 69-74, from res asst prof to res assoc prof, 74-83, assoc prof, 83-89, PROF ATMOSPHERIC SCI, UNIV WASH, 89- *Concurrent Pos:* Vis prof women, NSF, 83; vis scientist, Risø Nat Lab, Denmark, 80, Royal Dutch Meteorol Soc, 84 & 85, Univ Paris, 87 & 90. *Honors & Awards:* NDEA Fel, 63-65; NSF Vis Prof Women, 83-85. *Mem:* Am Geophys Union; fel Am Meteorol Soc; AAAS; Swed Geophys Soc; Europ Geophys Soc; Oceanog Soc. *Res:* Turbulent fluxes; free convection; remote sensing of atmosphere. *Mailing Add:* Dept Atmospheric Sci AK-40 Univ Wash Seattle WA 98195

KATSH, SEYMOUR, b New York, NY, Jan 13, 18; m 46; c 3. BIOLOGY. *Educ:* NY Univ, BA, 44, MS, 48, PhD, 50. *Prof Exp:* Instr zool & physiol, Univ Mass, 49-50; vis assoc biologist, Brookhaven Nat Lab, 50-51; instr zool & physiol, Univ Mass, 51-52; res fel physiol & biochem, Calif Inst Technol, 52-55; researcher, Carnegie Inst, 55-58; from asst prof to prof pharmacol, Med Ctr, Univ Colo, Denver, 58-86, actg chmn dept, 64-67, assoc dean grad & res affairs, 69-83; RETIRED. *Mem:* AAAS; Endocrine Soc; Am Soc Zool; Soc Exp Biol & Med; Am Physiol Soc. *Res:* Transplantation; endocrinology; pharmacology; immunology; reproduction. *Mailing Add:* Dept Pharmacol & Pathol Univ Colo Sch Med 4200 E Ninth Ave Denver CO 80262

KATSOYANNIS, PANAYOTIS G, b Greece, Jan 7, 24; nat US; m 55; c 2. BIOCHEMISTRY. *Educ:* Nat Univ Athens, MS, 48, PhD(chem), 52. *Prof Exp:* Res assoc, Med Col, Cornell Univ, 52-56, asst prof biochem, 56-58; assoc res prof, Sch Med, Univ Pittsburgh, 58-64; head div biochem, Med Res Ctr, Brookhaven Nat Lab, 64-68; PROF BIOCHEM & CHMN DEPT, MT SINAI SCH MED, 68- *Concurrent Pos:* Corresp mem, Nat Acad Greece. *Honors & Awards:* 50th Anniversary Award, Am Diabetes Asn. *Mem:* Am Chem Soc; Am Soc Biol Chem; NY Acad Sci; Royal Soc Chem; Brit Biochem Soc. *Res:* Biologically active polypeptides; isolation, characterization and synthesis; insulin synthesis. *Mailing Add:* Dept Biochem Mt Sinai Sch Med One Gustave L Levy Pl New York NY 10029

KATSUMOTO, KIYOSHI, b Oakland, Calif, May 4, 36; m 68; c 2. CHEMISTRY. *Educ:* San Jose State Col, BS, 64; Univ Calif, Berkeley, PhD(chem), 68. *Prof Exp:* From res chemist to sr res chemist, 67-75, SR RES ASSOC, CHEVRON RES CO, 75- *Mem:* Am Chem Soc. *Res:* Reaction mechanisms of cyclopropane ring openings; heterogeneous catalytic oxidations of hydrocarbons. *Mailing Add:* 2615 Brooks Ave El Cerrito CA 94530-1416

KATTAKUZHY, GEORGE CHACKO, b Kottayam, SIndia, Oct 1, 44; US citizen; m 74; c 3. RATES & PROPORTIONS, MODEL FITTING & PREDICTION. *Educ:* Kerala Univ, Kerala, India, BSc, 66, MSc, 68; Temple Univ, MA, 73, PhD(math statist), 75. *Prof Exp:* Instr math, Temple Univ, Philadelphia, Pa, 72-75, lectr statist, 73-75; asst prof statist, Pahlavi Univ, Shiraz, Iran, 75-76; asst prof math, Philadelphia Col Pharm & Sci, 76-78; STATISTICIAN STATIST, HEALTH CARE FINANCING ADMIN, 78- *Mem:* Am Statist Asn (treas, 89-90). *Mailing Add:* OPAI HSQB Health Care Financing Admin 6325 Security Blvd Baltimore MD 21207

KATTAMIS, THEODOULOS ZENON, b Kythrea, Cyprus, May 7, 35; US citizen; div; c 2. PHYSICS, METALLURGY. *Educ:* Univ Liege, Mining Engr, 60, Geol Engr, 61, Metall Engr, 62; Mass Inst Technol, MS, 63, ScD(metall), 65. *Prof Exp:* Res assoc metall, Mass Inst Technol, 65-69; from asst prof to assoc prof, 69-75, PROF METALL, UNIV CONN, 75- *Concurrent Pos:* Grants, NASA, 72-74, NSF, 73-76 & Air Force Off Sci Res, 77-81; contract, Continental Can Co, 82-85. *Honors & Awards:* Cert for Innovation in Metal Casting, NASA. *Mem:* Am Soc Metals; Metall Soc; Am Foundrymen's Soc. *Res:* Solidification and properties of materials, eutectics, composite materials, joining, powder metallurgy, vapor deposition, single crystal growth and materials processing; microstructure-property relationships. *Mailing Add:* Dept Metall U-136 Univ Conn Storrs CT 06269-3136

KATTAN, AHMED A, b Cairo, Egypt, Mar 21, 25; nat US; m 51; c 3. HORTICULTURE, FOOD SCIENCE. *Educ:* Cairo Univ, BSc, 45; Univ Md, MS, 50, PhD(hort), 52. *Prof Exp:* Asst, Cairo Univ, 46-48, lectr, 53-54; asst veg crops, Univ Md, 51-52, res assoc, 52-53, asst prof, 54-55; from asst prof to prof hort food sci, 55-, head dept, 68-, EMER PROF, UNIV ARK, FAYETTEVILLE. *Honors & Awards:* Woodbury Award, Am Soc Hort Sci, 59; Gourley Award, 79. *Mem:* Fel Am Soc Hort Sci; Inst Food Technol. *Res:* Pre- and post-harvest physiology of horticultural crops; methods of quality evaluation of raw and processed fruits and vegetables; methods of handling and mechanical harvesting of fruits and vegetables. *Mailing Add:* 1625 Halsell Rd Fayetteville AR 72701

KATTAWAR, GEORGE W, b Beaumont, Tex, Aug 10, 37; m 61; c 3. OPTICAL PHYSICS. *Educ:* Lamar State Col, BS, 59; Tex A&M Univ, MS, 61, PhD(physics), 64. *Prof Exp:* Theoretical physicist, Los Alamos Sci Lab, 63-64; sr res physicist, Esso Prod Res, 64-66; asst prof physics, NTex State Univ, 66-68; assoc prof, 68-73, PROF PHYSICS, TEX A&M UNIV, 73- *Concurrent Pos:* Consult, Navy & Jet Propulsion Lab. *Mem:* Fel Optical Soc Am; Sigma Xi. *Res:* Electromagnetic scattering theory; hydrologic optics. *Mailing Add:* Dept Physics Tex A&M Univ College Station TX 77843

KATTERMAN, FRANK REINALD HUGH, b Paia, Hawaii, June 28, 29; m 56; c 5. PLANT PHYSIOLOGY. *Educ:* Univ Hawaii, BA, 54; Tex A&M Univ, PhD(plant physiol), 60. *Prof Exp:* Plant physiologist, Agr Res Serv, USDA, 59-67; assoc prof plant breeding, 67-70, PROF AGRON & PLANT GENETICS, UNIV ARIZ, 70-, PLANT BREEDER, AGR EXP STA, 74- *Concurrent Pos:* Mem, Nat Cotton Coun Am. *Mem:* AAAS. *Res:* Composition and biochemistry of the nucleic acids in higher plants. *Mailing Add:* Dept Agron & Plant Genetics Univ Ariz Tucson AZ 85721

KATTI, SHRINIWAS KESHAV, b Bijapur, India, June 20, 36; US citizen; m 60; c 2. ANALYTICAL STATISTICS, APPLIED STATISTICS. *Educ:* Univ Delhi, BA, 56; Iowa State Univ, MA, 58, PhD(statist), 60. *Prof Exp:* From asst prof to assoc prof statist, Fla State Univ, 60-69; PROF STATIST, UNIV MO-COLUMBIA, 69- *Concurrent Pos:* US Air Force fel, Fla State Univ, 60-62, USPHS fel, 64-66 & USDA fel, 67-69; consult, Underwriter's Nat Assurance Co, 62-64 & Scot Res Lab, Perkesie, 67-70; assoc ed, Biomet Soc, 67-72; vis prof, Univ New South Wales, 71. *Mem:* Biomet Soc; fel Am Statist Asn; Am Inst Biol Sci; Am Math Soc; Inst Math Statist; Sigma Xi. *Res:* Inference; methods of tested priors; adaptive estimators. *Mailing Add:* Dept Statist 314 Math Sci Bldg Univ Mo Columbia MO 65211-0001

KATTUS, J ROBERT, b Cincinnati, Ohio, Aug 25, 22; m 46; c 5. FAILURE ANALYSIS, METALLOGRAPHY. *Educ:* Purdue Univ, BS, 44. *Prof Exp:* Metallurgist, US Naval Res Lab, 44-46 & Aluminum Industs Inc, 46-48; chief metallurgist, Anderson Elec Corp, 48-52; dir metall res, Southern Res Inst, 52-66; gen mgr, Bethea Castings Co, 66-68; consult metallurgist, 68-80; CONSULT METALLURGIST & VPRES, ASSOC METALL CONSULTS INC, 80- *Concurrent Pos:* Chmn, Test Methods Panel, Joint Comt Effects Temperature Properties Metals, Am Soc Testing & Mat, 63-68, chmn, Comt A-4 Iron Castings, 65-69; nat trustee, Am Soc Metals Int; mem, Eng Manpower Comn, 65-68. *Honors & Awards:* Award of Merit, Am Soc Testing & Mat, 71; Allen Ray Putnam Award, Am Soc Metals Int, 90. *Mem:* Fel Am Soc Testing & Mat; fel Am Soc Metals Int. *Res:* Elevated-temperature properties of metals under conditions of rapid and rapid loading simulating aerospace conditions; effects of composition and foundry practice on the quality of ferrous and non-ferrous castings. *Mailing Add:* Assoc Metall Consults Inc 810 Fifth Ave N Birmingham AL 35203

KATZ, ADRIAN I, b Bucharest, Romania, Aug 3, 32; m 65; c 2. INTERNAL MEDICINE, PHYSIOLOGY. *Educ:* Hebrew Univ Jerusalem, MD, 62. *Prof Exp:* House officer internal med, Belinson Med Ctr, Sch Med, Tel-Aviv Univ, 62-65; res fel med, Sch Med, Yale Univ, 65-67; res fel, Harvard Med Sch, 67-68, asst prof & attend physician, 68-71, assoc prof, 71-74, PROF MED, SCH MED, UNIV CHICAGO, 75- *Concurrent Pos:* Asst med, Peter Bent Brigham Hosp, Boston, 67-68; head sect nephrology, Univ Chicago, 73-82. *Mem:* Am Soc Clin Invest; Am Fedn Clin Res; Am Soc Nephrology; NY Acad Sci; fel Am Col Physicians; Asn Am Physicians. *Res:* Renal physiology, especially biochemical mechanisms of renal tubular sodium transport; renal handling of polypeptide hormones; kidney function in pregnancy; clinical nephrology; biochemistry. *Mailing Add:* Dept Med Box 453 Univ Chicago Chicago IL 60637

KATZ, ALAN JEFFREY, b Columbus, Ohio, Oct 2, 47; m 68; c 3. GENETIC TOXICOLOGY, BIOSTATISTICS. *Educ:* Ohio State Univ, BS, 69, MS, 70, PhD(genetics), 74. *Prof Exp:* NIH fel pop genetics, Dept Genetics & Cell Biol, Univ Minn, 74-75; from asst prof to assoc prof, 75-85, PROF GENETICS, DEPT BIOL SCI, ILL STATE UNIV, 85- *Concurrent Pos:* vis res fel, Swiss Fed Inst Toxicol, Schwerzenbach, 82, res grants, March Dimes, 82-86, NIH, 87-89. *Mem:* Genetics Soc Am; Biomet Soc; Environ Mutagens Soc; AAAS; Sigma Xi. *Res:* Identification and study of chemical mutagens and antimutagens in the somatic tissue of drosophila. *Mailing Add:* Dept Biol Sci Ill State Univ Normal IL 61761

KATZ, ARNOLD MARTIN, b Chicago, Ill, July 30, 32; m 59; c 4. MEDICINE, PHYSIOLOGY. *Educ:* Univ Chicago, BA, 52; Harvard Univ, MD, 56. *Prof Exp:* Intern med, Mass Gen Hosp, 56-57; res assoc, Nat Heart Inst, 57-59; asst resident, Mass Gen Hosp, 59-60; hon registr, Inst Cardiol, London, 60-61; res fel med, Med Sch, Univ Calif, Los Angeles, 61-64; asst prof physiol, Col Physicians & Surgeons, Columbia Univ, 63-67; assoc prof med & physiol, Univ Chicago, 67-69; Philip J & Harriet L Goodhart prof med-cardiol, Mt Sinai Sch Med, 69-77; PROF MED & HEAD DIV CARDIOL, HEALTH CTR, UNIV CONN, 77- *Concurrent Pos:* Mosely traveling fel from Harvard Univ, 60-61; Am Heart Asn res fel, 61-63; estab investr, Am Heart Asn, 63-68, mem exec coun, Coun Basic Sci, 68-71; asst physician, Med Serv, Presby Hosp, 63-67; mem, Comt Myocardial Infarction, Nat Heart Inst, 66, Ad Hoc Comt Rev Proposals Myocardial Infarction Study Ctrs, 67, Heart Prog Proj B Comt, 67-69 & prog proj comt A, 80-; session chmn, Gordon Res Conf Cellular Control Cardiac Contraction, 68 & Gordon Res Conf Cardiac Muscle, 70; consult, Vet Admin, 70-; vchmn task group cardiac failure, Nat Heart, Blood, Lung & Blood Vessel Prog, NIH. *Mem:* Am Physiol Soc; Am Soc Pharmacol & Exp Therapeut; Harvey Soc; Cardiac Muscle Soc (pres, 69-71); Am Soc Biol Chemists; Sigma Xi. *Res:* Cardiology; cardiovascular physiology; muscle biochemistry. *Mailing Add:* Dept Med Univ Conn Health Ctr Farmington CT 06032

KATZ, BERNARD, b Leipzig, Ger, Mar 26, 11. BIOPHYSICAL RESEARCH. *Prof Exp:* Dir res biophys & Henry Head res fel, Royal Soc, 46-50; reader physiol, Univ Col London, 50-51, prof biophys & head dept, 52-78; RETIRED. *Concurrent Pos:* Royal Australian Air Force, 42-45; Agr Res Coun, 67-77. *Honors & Awards:* Nobel Prize in Med, 70; Baly Medal, Royal Col Physicians, 46-50; Herter Lectr, Johns Hopkins Univ, 58; Dumham Lectr, Harvard Univ, 61; Croonian Lectr, Royal Soc, 61; Copley Medal, Royal Soc, 67. *Mem:* Royal Soc. *Mailing Add:* Dept Physiol Univ Col Gower St London England

KATZ, DAVID HARVEY, b Richmond, Va, Feb 17, 43; m 63; c 2. MEDICINE, IMMUNOLOGY. *Educ:* Univ Va, AB, 63, Duke Univ, MD, 68. *Prof Exp:* Med house officer, Johns Hopkins Hosp, 68-69; staff assoc immunol, NIH, 69-71; from instr to assoc prof immunol & path, Harvard Med Sch, 71-76; chmn & mem immunol staff, Scripps Clin & Res Found, 76-81; CHIEF EXEC OFF & PRES, QUIDEL, 81-; PRES & DIR, MED BIOL INST, LA JOLLA, 81- *Concurrent Pos:* Mem adv comt cancer ctrs, Nat Cancer Inst, 72-74; mem allergy & immunol study sect, NIH, 77; mem human cell biol adv panel, NSF, 77-78. *Mem:* Am Asn Immunologists; Am Soc Clin Invest; AAAS; Am Asn Pathologists; Am Fedn Clin Res. *Res:* Basic immunology; allergy; tumor immunology; developmental biology. *Mailing Add:* Dept Immunol Med Biol Inst 11077 N Torrey Pines Rd La Jolla CA 92037

KATZ, DONALD L(AVERNE), chemical engineering; deceased, see previous edition for last biography

KATZ, EDWARD, b New York, NY, Aug 10, 23; m 51; c 1. MICROBIOLOGY, BIOCHEMISTRY. *Educ:* NY Univ, BA, 47; Rutgers Univ, PhD(microbiol), 51. *Prof Exp:* Asst supvr antibiotics, Heyden Chem Corp, 48; asst prof microbiol, Univ NH, 51-54 & Rutgers Univ, 54-60; Nat Heart Inst sr res fel biochem, 60-62; assoc prof, 62-69, PROF MICROBIOL, SCHS MED & DENT, GEORGETOWN UNIV, 69- *Concurrent Pos:* Am Med Asn Educ & Res Found grant, 63; Nat Cancer Inst res grant, 63-; consult, Nat Heart Inst, 63-65, Civil Div, US Justice Dept, Schering-Plough Corp & Schwarz BioRes Corp; vchmn & chmn div appl & environ microbiol, Am Soc Microbiol. *Mem:* Am Soc Microbiol; Am Soc Biol Chemists. *Res:* Antibiotics; biosynthesis of peptide antibiotics; tryptophan metabolism; imino acid biosynthesis. *Mailing Add:* 1101 Wyoming St St Louis MO 63118

KATZ, ELI JOEL, b Brooklyn, NY, Jan 12, 37; m 57; c 3. PHYSICAL OCEANOGRAPHY. *Educ:* Polytech Inst Brooklyn, BSME, 57; Pa State Univ, MS, 59; Johns Hopkins Univ, PhD(fluid mech), 62. *Prof Exp:* Res assoc mech, Johns Hopkins Univ, 62-63; vis lectr meteorol, Hebrew Univ Jersualem, 63-65; res specialist acoustics, Gen Dynamics Corp, 65-66; asst scientist, Woods Hole Oceanog Inst, 66-69, assoc scientist phys oceanog, 70-78; sr lectr mech, Tel Aviv Univ, 69-70; sr res assoc, 79-83, SR SCIENTIST, LAMONT-DOHERTY GEOL OBSERV, 84- *Concurrent Pos:* Co-ed, J Phys Oceanog, 86- *Res:* Ocean dynamics and ocean role in world climate. *Mailing Add:* Lamont-Doherty Geol Observ Columbia Univ Palisades NY 10964

KATZ, ERNST, b Maehr-Ostrau, Austria, July 23, 13; Netherlands citizen; m 39; c 1. SOLID STATE PHYSICS. *Educ:* Univ Utrecht, BS, 33, MS, 37, PhD(physics), 41. *Prof Exp:* Asst physics, Univ Utrecht, 38-47, Rockefeller Biophys Res Group, 37-41; with Neth Instrument & Elec Apparatus Co, 41-45, dir res, 45-47; from asst prof to prof, 47-80, EMER PROF PHYSICS, UNIV MICH, ANN ARBOR, 80- *Mem:* Fel Am Phys Soc. *Res:* Ionic solids; crystallization; magneto-resistance; general solid state problems. *Mailing Add:* 33 Ridgeway Ann Arbor MI 48104

KATZ, EUGENE RICHARD, b Brooklyn, NY, Apr 10, 42; m 69; c 2. GENETICS. *Educ:* Univ Wis, Madison, BS, 62; Univ Cambridge, Eng, PhD(molecular genetics), 69. *Prof Exp:* From asst prof to assoc prof, Dept Biol, 70-80, assoc prof, 80-85, PROF GENETICS, DEPT MICROBIOL, SU, NY, STONYBROOK, 85-, DEAN, DIV BIOL SCI, 88- *Concurrent Pos:* Vis scientist, Mass Inst Technol, 70; dir, Grad Prog Cellular & Develop Biol, SU NY at Stony Brook, 75-80, Grad Prog Genetics, 80-88; vis prof, Univ Nijmegen, Netherlands, 77-78. *Res:* Genetic control of development using the cellular slime mold; Dictyostelium discoideum as a model system; formal genetics and biochemical analysis of mutants affecting development. *Mailing Add:* Dept Microbiol Grad Biol Bldg Sch Med SU New York Stony Brook NY 11794-5200

KATZ, FRANCES R, b LeRoy, Il, Aug 16, 37; m 83; c 1. CARBOHYDRATE CHEMISTRY, PATENT AFFAIRS. *Educ:* Ind Cent Univ, BS, 61; Univ Chicago, MBA-XP, 82. *Prof Exp:* Food technologist, Durkee Foods, SCM Corp, 61-65, Continental Coffee Co, 65-69; assoc ed, Putnam Publ Co, 69-74; ed dir, Gorman Publ Co, 74-78; VPRES RES, AM MAIZE PROD CO, 78- *Concurrent Pos:* Exec ed, Am Asn Cereal Chemists, 79-82, columnist, 82- *Mem:* Am Chem Soc; Inst Food Technologists; Corn Refiner's Asn; Am Asn Cereal Chemists. *Res:* Carbohydrate research; applications of carbohydrates; formulation of research policy. *Mailing Add:* Am Maize Prod Co 1110 Indianapolis Blvd Hammond IN 46320-1094

KATZ, FRANK FRED, b Philadelphia, Pa, July 19, 27; m 55; c 2. PARASITOLOGY. *Educ:* Philadelphia Col Pharm, BS, 51; Tulane Univ, MS, 53; Univ Pa, PhD(parasitol), 56. *Prof Exp:* Asst zool, Philadelphia Col Pharm, 51; asst instr parasitol, Univ Pa, 54-55; jr res assoc biol, Brookhaven Nat Lab, 55-56; sr parasitologist, Eaton Labs, Norwich Pharmacal Co, 56-57; asst prof microbiol, Jefferson Med Col, 57-62; from asst prof to assoc prof, 62-70, actg chmn dept, 71-72, chmn dept, 72-83, PROF BIOL, SETON HALL UNIV, 70-, ASSOC DEAN, COL ARTS & SCI, 85- *Mem:* AAAS; Am Soc Parasitol; Am Soc Trop Med & Hyg; Micros Soc Am. *Res:* Helminthology; protozoology; experimental parasitology; biology of Strongyloides, Trichinella, Plasmodium and trypanosomes. *Mailing Add:* Dept Biol Seton Hall Univ South Orange NJ 07079

KATZ, FRED H, b Essen, Ger, Apr 7, 30; US citizen; m 60; c 3. INTERNAL MEDICINE, ENDOCRINOLOGY. *Educ:* Columbia Univ, AB, 52, MD, 56; Am Bd Internal Med, dipl, 64. *Prof Exp:* Asst prof med, Sch Med, Univ Chicago, 63-66; assoc prof med & chief endocrinol, Stritch Sch Med, Loyola Univ, 66-69; assoc prof, 69-75, head div endocrinol, 72-76, PROF MED, UNIV COLO, 75-, CLIN PROF MED, 76- *Concurrent Pos:* Nat Found fel, Presby Hosp, New York, 59-60, Nat Inst Arthritis & Metab Dis trainee, 60-61; chief endocrinol, Vet Admin Hosp, Denver, 69-76; mem med adv bd, Coun High Blood Pressure, Cent Soc Clin Res. *Mem:* Endocrine Soc; Soc Exp Biol & Med; fel Am Col Physicians. *Res:* Steroid hormone metabolism, physiology and pharmacology. *Mailing Add:* 3535 Cherry Creek N Dr No 307 Denver CO 80209

KATZ, GARY VICTOR, b New York, NY, July 12, 43; m 84; c 2. INHALATION TOXICOLOGY, INDUSTRIAL HYGIENE. *Educ:* City Col New York, BS, 65; NY Univ, MS, 68, PhD(biol & environ health sci), 75. *Prof Exp:* From asst res to assoc res scientist inhalation toxicol & chem carcinogenesis, Dept Environ Med, NY Univ Med Ctr, 69-77; toxicologist, 77-81, MGR, INHALATION TOXICOL, EASTMAN KODAK CO, 81-, DIR CHEM & REGULATORY INFO, 90- *Concurrent Pos:* Adj asst prof environ med, Dept Environ Med, NY Univ Med Ctr, 77-; tech assoc, Eastman Kodak Co, 81-, asst to the dir & div vpres, 88-; chmn, Chem Manufacturers Asn (CMA) Toxicol Task Group, 84-; mem, CMA Integrated Risk Info Syst Task Group, 88-; Am Indust Health Coun (ATHC) Air Toxics Work Group. *Mem:* Am Indust Hygiene Asn; Soc Risk Anal. *Res:* Chemical carcinogenesis; neurotoxicology; quantitative risk assessment. *Mailing Add:* Toxicol Sci Lab Health & Environ Labs Eastman Kodak Co Rochester NY 14652-3615

KATZ, GEORGE MAXIM, b Mar 26, 22; US citizen; m 49; c 2. ENGINEERING, NEUROPHYSIOLOGY. *Educ:* City Col New York, BEE, 42; Polytech Inst Brooklyn, MEE, 57; Columbia Univ, EE, 61, PhD(physiol), 67. *Prof Exp:* Engr, Gen Elec Co, NY, 42-47; lectr elec eng, City Col New York, 47-50; sr engr, Advan Develop Lab, Allen B Dumont Co, NJ, 50-51; res assoc surg, Col Physicians & Surgeons, Columbia Univ, 52-61, sr res assoc & asst prof neurol, 62-; RETIRED. *Concurrent Pos:* Consult, St Vincent's Hosp & Med Ctr, New York, 68- *Mem:* Biophys Soc. *Res:* Methodology and instrumentation for conducting research; mathematical analysis of data. *Mailing Add:* 325 Allaire Ave Leonia NJ 07605

KATZ, GERALD, b Brooklyn, NY; m 58; c 3. SOLID STATE SCIENCE, X-RAY CRYSTALLOGRAPHY. *Educ:* Brooklyn Col, BA, 44; Pa State Univ, PhD(solid state sci), 65. *Prof Exp:* Physicist weapons res, US Army Chem Warfare Serv, 48-49; physicist solid state mat res, US Signal Corps Res & Develop Labs, US Electronics Command, 49-58; res fel struct of metals & alloys, Israel Inst Technol, 58-59; staff mem x-ray crystallog res uranium & intermetallic compounds, Dept Metall, Israel AEC, 59-61; from res asst to res assoc solid state mat res, Mat Res Lab, Pa State Univ, 61-66; staff scientist crystal films & epitaxy, IBM Watson Res Ctr, 66-68; dir x-ray lab, Inst Res & Develop, Israel Mining Industs Ltd, 68-73; res assoc solid state mat res, dept mat eng, Israel Inst Technol, 73-79; sr res lectr, dept inorg chem, Hebrew Univ, Jerusalem, 79-84; CONSULT, 85- *Mem:* Am Crystallog Asn; Israel Soc Crystal Growth & Thin Films; Am Asn Crystal Growth. *Res:* Synthesis and/or crystal growth of inorganic materials, metals, intermetallic compounds including thin films; their structural characterization by x-ray crystallography and their associated magnetic, electrical and optical properties; solid state reaction mechanisms; topotaxy; optical spectra of rare earth doped glasses and crystals. *Mailing Add:* 21 Raanan St Haifa 34384 Israel

KATZ, HAROLD W(ILLIAM), electrical engineering; deceased, see previous edition for last biography

KATZ, HERBERT M(ARVIN), b Brooklyn, NY, Apr 4, 26; m 54; c 2. CHEMICAL ENGINEERING. *Educ:* City Col New York, BChE, 49; Univ Cincinnati, MS, 50, PhD(chem eng), 54. *Prof Exp:* Asst chem engr, Argonne Nat Lab, 54-56; staff engr, Eng Ctr, Univ Columbia, 56-57; chem engr, Brookhaven Nat Lab, 57-67; chem engr, Res Div, W R Grace & Co, 67-68; chmn dept, Howard Univ, 68-73, prof chem eng, 73-86; RETIRED. *Concurrent Pos:* Consult, Nuclear Safety Asn & Brookhaven Nat Lab. *Mem:* Sigma Xi; Am Inst Chem Engrs. *Res:* Chemical reprocessing of nuclear reactor fuels; treatment of radioactive wastes; fluidized bed technology. *Mailing Add:* 1602 Sherwood Rd Silver Spring MD 20902

KATZ, IRA, b New York, NY, Nov 10, 33; m 55; c 3. FOOD SCIENCE. *Educ:* Univ Ga, BSA, 57; Univ Md, MS, 59, PhD, 62. *Prof Exp:* Res asst lipid chem, Univ Md, 61-62, res assoc, 62-65, asst prof, 65-67; proj leader, 67-71, groupleader, 71-73, dir, 73-80, VPRES & DIR, RES & DEVELOP, INT FLAVORS & FRAGRANCES, INC, 80- *Mem:* AAAS; Am Dairy Sci Asn; Am Oil Chem Soc; Am Chem Soc; Inst Food Technologists; Sigma Xi. *Res:* Flavor of food and fragrance systems. *Mailing Add:* Two Lawley Ct West Long Branch NJ 07764

KATZ, IRVING, b Brooklyn, NY, Oct 25, 33; m 57; c 3. MATHEMATICS. *Educ:* Brooklyn Col, BS, 56; Ohio State Univ, MA, 58; Univ Md, PhD(math), 64. *Prof Exp:* Mathematician, Nat Security Agency, 58-59 & Opers Res Inc, 59-60; from instr to assoc prof, Am Univ, 61-66; from asst prof to assoc prof, 66-77, PROF MATH, GEORGE WASHINGTON UNIV, 77- *Mem:* Am Math Soc; Math Asn Am. *Res:* Matrix theory. *Mailing Add:* George Washington Univ Washington DC 20052

KATZ, IRWIN ALAN, b Malden, Mass, Feb 13, 40; m 64; c 4. PHARMACEUTICS, PHARMACY. *Educ:* Boston Univ, AB, 61; Mass Col Pharm, BS, 68, MS, 69, PhD(pharm), 71. *Prof Exp:* Res investr formulation design, E R Squibb & Sons Inc, 71-73; hosp pharmacist, JFK Med Ctr, 74-75; res investr liquid/parenteral formulation develop, 73-76, process develop mgr, 76-77, liquid, sterile, semi-solid develop supvr, 77-88, tech doc off supvr, 78-82, tech doc sect head, 82-85, TECH DOC SECT MGR, PHARM TECHNOL DEPT, E R SQUIBB & SONS INC MFG, 85- *Concurrent Pos:* Retail pharmacist, Wash Park Pharm, Newtonville, Mass, 68-71; hosp pharmacist, Boston Lying In Hosp, 70-71 & Raritan Valley Hosp, 75- *Mem:* Am Pharmaceut Asn; Acad Pharmceut Sci. *Res:* Parenteral research and development; lyophilization; parenteral particulate matter monitoring; pharmaceutical formulation design; physical pharmacy and biopharmaceutics. *Mailing Add:* Three Noel Lane East Brunswick NJ 08816

KATZ, ISRAEL, b New York, NY, Nov 30, 17; m 42; c 3. MECHANICAL ENGINEERING, ENGINEERING MANAGEMENT. *Educ:* Northeastern Univ, BSME, 41; Mass Inst Technol, cert naval archit, 42; Cornell Univ, MME, 44. *Prof Exp:* Test engr, Gen Elec Co, 38-42; engr-in-charge, submarine propulsion machinery, US Naval Diesel Eng Lab, Cornell Univ, 42-46, asst prof grad sch aeronaut eng, 46-48, assoc prof mech eng & prof-in-charge aircraft powerplants lab, 48-57; mgr-liason & consult engr, Gen Elec Advan Electronics Ctr, Cornell Univ, 57-63; dir, Benwill Pub Corp, 64-82; dean, Ctr Continuing Educ, 67-74, prof & dir advan eng progs, 74-88, PROF EMER ENG TECHNOL, NORTHEASTERN UNIV, 88- *Concurrent Pos:* Consult eng designer, Pratt & Whitney Aircraft Co, 47-52; chmn subcomt res & educ comt Mat Sci Appln & Coord, Nat Res Count, 72-73; ed-in-chief, 80; consult engr, Nat Acad Sci, 71- *Honors & Awards:* Pioneer Award, Am Soc Eng Educ, 78. *Mem:* Am Soc Eng Educ; Inst Elec & Electronics Engrs; AAAS; Sigma Xi. *Res:* Engineering thermodynamics; electromechanical systems; aerospace technology; continuing education for scientists and engineers; author of numerous publications. *Mailing Add:* 40 Auburn St Brookline MA 02146

KATZ, ISRAEL NORMAN, b New York, NY, Apr 14, 32; m 57; c 2. MATHEMATICS, STATISTICS. *Educ:* Yeshiva Univ, BA & MS, 52; Mass Inst Technol, PhD(math), 59. *Prof Exp:* Asst math, Yeshiva Univ, 52-54; asst, Mass Inst Technol, 55-58, res asst, 58-59; sr staff scientist, Res & Adv Develop Div, Avco Corp, 59-63, chief math anal sect, 63-65, mgr math dept, 66-67; assoc prof appl math & comput sci, 67-74, PROF APPL MATH & SYSTS SCI, WASH UNIV, 74- *Concurrent Pos:* Lectr, Math Asn Am; vis consult, Soc Indust & Appl Math; consult, McDonnell Aircraft Co. *Mem:* Opers Res Soc Am; Am Math Soc; Math Asn Am; Soc Indust & Appl Math. *Res:* Applied math; numerical analysis; facility location; finite elements; biomathematics; ordinary and partial differential equations; algorithms for parallel computation. *Mailing Add:* Dept Systs Sci & Math Wash Univ Box 1040 St Louis MO 63130

KATZ, J LAWRENCE, b Brooklyn, NY, Dec 18, 27; m 50; c 3. BONE BIOMECHANICS, BONE BIOMATERIALS. *Educ:* Polytech Inst Brooklyn, BS, 50, MS, 51, PhD(physics), 57. *Prof Exp:* Instr math, Polytech Inst Brooklyn, 52-56; from asst prof to prof physics, Rensselaer Polytech Inst, 61-73, prof biophys & biomed eng, 73-89, dir, Ctr Biomed Eng, 74-84, chmn, Dept Biomed Eng, 83-85; DEAN ENG & PROF BIOMED ENG, CASE INST TECHNOL, CASE WESTERN RESERVE UNIV, 89- *Concurrent Pos:* Consult, Ernest F Fullam, Inc, NY, 58-83, Bio-Anal Labs, Inc, 61-83, Orthop Panel, Food & Drug Admin, 76-78 & Orthop Div, Johnson & Johnson Co, 79-80; NSF sci fac fel & hon res asst crystallog, Univ Col, London, 59-60; mem, eng in biol & med training comt, NIH, 68-71; vis prof, Univ Miami, 69-70, Sao Carlos Inst Physics & Chem, Univ Sao Paulo, Brasil, 78, Univ London, Eng, 85-86 & Fac Med Lariboisiere, Paris, 86; mem equip & mat for med radiation appln & chmn subcomt diag radiol, Am Nat Standards Inst, 69-74; consult & site vis, Nat Inst Dent Res & Nat Inst Gen Med Sci; partic vis sci prog physics, Am Asn Physics Teachers-Am Inst Physics, 70-71; prof surg, Albany Med Col, 75-89; Jerome Fischbach travel grant, Rensselaer Polytech Inst, 76; Guggenheim fel, Harvard Univ, 78, vis lectr orthop, Sch Med, 78; vis biophysicist, Orthopaedics Res Lab, Children's Hosp, Boston, 78; E Leon Watkins vis prof, Wichita State Univ, Kans, 78; mem, Coun Alliance Eng Med & Biol, 78-81; mem, Sci Rev & Eval Bd for Rehab Eng Res & Develop, Vet Admin, 81-83, 90-; assoc ed, Biomat, Biomech & Rehab Eng, Annals of Biomed Eng, 84-89; NIH sr int fel, 85-86; vis prof, Dept Mat, Queen Mary Col, Univ London, Eng, 85-86 & Lab Orthop Res, Fac Med Lariboisiere-Saint-Louis, Paris, France, 86. *Honors & Awards:* 3rd Annual Award for Outstanding Contributions to Tech Lit Biomat, Soc Biomat & Clemson Univ, 75; Outstanding Biomed Eng Educator Award, Am Soc Eng Educ, 88; George Winter Award for Outstanding Res, Europ Soc Biomat, 89. *Mem:* Am Crystallog Asn; Am Phys Soc; Int Asn Dent Res; Sigma Xi; Soc Biomat (pres, 78-79); Biomed Eng Soc (pres, 83-84). *Res:* Bone biomechanics and biomaterials, especially the correlation between structure and properties of the various calcified tissues and of synthetic materials used as implant biomaterials; biomechanics of calcified and connective tissues;

electromechanical properties of bone and bone remodeling; rehabilitation engineering; scanning electron microscopy; X-ray diffraction and ultrasonic studies of bone and teeth; biomedical materials; rehabilitation engineering. *Mailing Add:* Sch Eng Case Inst Technol Case Western Reserve Univ Cleveland OH 44106

KATZ, JACK, b New York, NY, Mar 25, 34; m 56; c 2. AUDIOLOGY, CENTRAL AUDITORY PROCESSING. *Educ:* Brooklyn Col, BA, 56; Syracuse Univ, 57; Univ Pittsburgh, PhD(audiol), 61. *Prof Exp:* Therapist speech & hearing, Bd Educ, Cayuga County, NY, 57-58; res audiologist, Univ Pittsburgh, 60-61; asst prof audiol, Northern Ill Univ, 61-62; asst prof speech path & audiol, Tulane Univ, 62-65; dir audiol lab, Menorah Med Ctr, Kansas City, 65-74; clin prof, 74-76, chmn commun dis & sci, 82-87, PROF COMMUN DIS & SCI, SU NY, BUFFALO, 76- *Concurrent Pos:* Consult audiol, Univ Pittsburgh, 61-62, Menorah Med Ctr, 74-75, Roswell Park Mem Inst, 79-, Veteran Admin Med Ctr, Buffalo, NY, 84-; assoc clin prof, Univ Mo-Kansas City, 70-74 & Univ Kans, 71-74, vis prof Univ Kans Med Ctr, 87-88; Fulbright lectr, Ankara, Turkey, 72-73; mem spec med staff, Chedoke-McMaster Hosps, Hamilton, Ont, 81-84; mem bd dir Orton Soc Western NY, 87-90 & Buffalo Hearing Speech Ctr; vpres univ & labs, NY State Speech Lang-Hearing Asn, 81-85. *Honors & Awards:* Fulbright-Hays Sr Lectr, Ankara, Turkey, 72-73. *Mem:* Fel Am Speech-Lang-Hearing Asn; NY Acad Sci. *Res:* Evaluation of central auditory integrity, binaural hearing, low level adaptation, auditory perception, learning disabilities, listening problems in incarcerated populations and influence of conductive hearing loss. *Mailing Add:* Dept Commun Disorders & Sci State Univ NY Buffalo NY 14260

KATZ, JAY, b Zwickau, Ger, Oct 20, 22; nat US; m 52; c 3. PSYCHIATRY. *Educ:* Univ Vt, BA, 44; Harvard Univ, MD, 49; Am Bd Psychiat & Neurol, dipl. *Prof Exp:* Intern, Mt Sinai Hosp, New York, 49-50; asst resident psychiat, SU NY & Northport Vet Admin Hosp, Long Island, 50-51; asst resident psychiat, Sch Med, Yale Univ, 53-54, chief resident outpatient clin, 54-55, from instr to asst prof psychiat, 55-58, asst prof psychiat & law, Law Sch, 58-60, assoc prof law & assoc clin prof psychiat, 60-67, adj prof law & psychiat, 67-84, John A Gorver prof law & psychoanalyst, 84-90, ELIZABETH DOLLARD PROF LAW, MED PSYCHIAT, LAW SCH, YALE UNIV, 90-; TRAINING & SUPV PSYCHOANALYST, WESTERN NEW ENG INST PSYCHOANAL, 72- *Concurrent Pos:* Asst investr, USPHS res grant hypnotic dreams, 53-56; attend psychiatrist, Yale-New Haven Med Ctr, 57-; chmn adv comt ment health, Woodbridge Bd Educ, Conn, 64-68; staff psychoanalyst, Psychoanal Clin, Western New Eng Inst Psychoanal, 66-69, trustee, 68-71; fel, Ctr Advan Psychoanal Studies, 67-; fel, Morse Col, Yale Univ, 68- *Honors & Awards:* Isaac Ray Award, Am Psychiat Asn, 75; William C Menninger Award, Am Col Physicians, 83; Am Soc Law & Med Award, 87. *Mem:* Inst Med-Nat Acad Sci; fel Am Psychiat Asn; Am Orthopsychiat Asn; Am Col Psychiat; Am Psychoanal Asn. *Mailing Add:* Yale Law Scht Law Sch New Haven CT 06520

KATZ, JONATHAN ISAAC, b New York, NY, Jan 5, 51; m 82; c 3. ASTROPHYSICS. *Educ:* Cornell Univ, AB, 70, MA, 71, PhD(astron), 73. *Prof Exp:* Mem staff astrophysics, Inst Advan Study, 73-76; assoc prof astron & geophysics, Univ Calif, Los Angeles, 76-81; ASSOC PROF PHYSICS, WASH UNIV, MO, 81- *Concurrent Pos:* Consult, Lawrence Livermore Lab, 73-, SRI Int, 74-82 & adv coun, NASA, 83-86; Sloan Found fel, 77-79; MITRE, 82- *Mem:* Am Phys Soc; Am Astron Soc. *Res:* Theoretical high energy astrophysics. *Mailing Add:* Dept Physics Wash Univ St Louis MO 63130

KATZ, JOSEPH, b Vilno, Lithuania, Jan 15; US citizen; c 5. BIOCHEMISTRY. *Educ:* Univ Calif, Berkeley, BS, 43, PhD, 49. *Prof Exp:* Asst res physiologist, Univ Calif, Berkeley, 53-55; from res assoc to sr res assoc, Inst Med Res, Cedars of Lebanon Hosp, 55-70, SR RES SCIENTIST, MED RES INST, CEDARS-SINAI MED CTR, 70- *Concurrent Pos:* Res fel biochem, Univ Calif, Berkeley, 49-51, fel physiol, 51-53; advan res fel, Cedars-Sinai Med Ctr, 59-61; estab investr, Am Heart Asn, 61-66; adj prof, Univ Southern Calif, 69-88, Univ Calif, Los Angeles, 88- *Mem:* Am Chem Soc; Am Soc Biol Chemists; Am Nutrit Soc; Am Physiol Soc; Brit Biochem Soc. *Res:* Carbohydrate metabolism determination of pathways of glucose utilization; the interrelationship between lipogenesis and glucose utilization; plasma protein metabolism. *Mailing Add:* Harbor Univ Calif Los Angeles Med Ctr REI A-17 1000 W Carson St Torrance CA 90509

KATZ, JOSEPH J, b Detroit, Mich, Apr 19, 12; m 44; c 4. PHYSICAL CHEMISTRY. *Educ:* Wayne State Univ, BS, 32; Univ Chicago, PhD(chem), 42. *Prof Exp:* Chemist, Univ Chicago, 42-43, chemist metall lab, 43-46; sr chemist, 46-82, EMER DISTINGUISHED SR SCIENTIST, ARGONNE NAT LAB, 82- *Concurrent Pos:* Guggenheim fel, 57-58; Am ed, J Inorg & Nuclear Chem; ed-in-chief, Inorg & Nuclear Chem Letters, 55-81. *Honors & Awards:* Nuclear Appln Award, Am Chem Soc, 61, Midwest Award, 69. *Mem:* Nat Acad Sci; AAAS; Am Chem Soc. *Res:* Chemistry of uranium and transuranium elements; deterium isotope studies; chlorophyll chemistry; photosynthesis; solar energy. *Mailing Add:* Argonne Nat Lab 9700 S Cass Ave Argonne IL 60439

KATZ, JOSEPH L, b Colon, Panama, Aug 4, 38; US citizen; m 65; c 2. NUCLEATION, CHEMICAL ENGINEERING. *Educ:* Univ Chicago, BS, 60, PhD(phys chem), 63. *Prof Exp:* Asst prof phys chem, Univ Copenhagen, 63-64; mem tech staff, N Am Rockwell Sci Ctr, 64-70; prof chem eng, Clarkson Col Technol, 70-79; Dir, Energy Res Inst, 81- 83, dept chmn, 81-84, PROF CHEM ENG, JOHNS HOPKINS UNIV, 79- *Concurrent Pos:* Guggenheim fel, 76-77. *Honors & Awards:* John W Graham Prize, 75; Maryland Chemist Award, 82. *Mem:* Fel AAAS; fel Am Phys Soc; Am Chem Soc; Am Inst Chem Engrs; Sigma Xi. *Res:* Nucleation; equations of state; thermal conductivity and flame generation of ceramic powders. *Mailing Add:* Dept Chem Eng Johns Hopkins Univ Baltimore MD 21218

KATZ, LAURENCE BARRY, b Syracuse, NY, Oct 3, 54; m 81; c 2. GASTROENTEROLOGY, CARDIOVASCULAR PHARMACOLOGY. *Educ:* Univ Pa, BA, 76; Philadelphia Col Pharm & Sci, MS, 79 & PhD(pharmacol), 82. *Prof Exp:* Res assoc toxicol, Univ Wis, 81-83; res scientist pharmacol, Ortho Pharmaceut Corp, 83-84, sr res scientist, 84-89; prin scientist, 89-90, PROJ MGR, R W JOHNSON PHARMACEUT RES INST, 90- *Mem:* Am Soc Pharmacol & Exp Therapeut; Am Asn Advan Sci; Gastrointestinal Res Group; Sigma Xi. *Res:* Drugs useful in peptic ulcer disease and hypertension including prostaglandins, histamine H2-receptor antagonist and vasodilators. *Mailing Add:* R W Johnson Pharmaceut Res Inst PO Box 300 Raritan NJ 08869-0602

KATZ, LEON, b Poland, Aug 9, 09; Can citizen; m 41; c 4. SCIENCE POLICY. *Educ:* Queen's Univ, Ont, BSc, 34, MSc, 37; Calif Inst Technol, PhD(physics), 42. *Hon Degrees:* DSc, Univ Sask, 90. *Prof Exp:* Res lead acid batteries, Monarch Battery Co, Ont, 31-33, plant foreman, 34-36; res engr, Westinghouse Elec Corp, 42-46; from assoc prof to prof physics, Univ Sask, 46-75, head dept, 65-75, dir accelerator lab, 61-75; dir, Sci Policy Secretariat, Govt Sask, 75-80; RETIRED. *Concurrent Pos:* Mem, Sci Coun Can, 66-72; mem coun trustees, Inst Res Pub Policy, 74-89. *Mem:* Fel Am Phys Soc; Can Asn Physicist (past pres); fel Royal Soc Can. *Res:* Thermodynamics; radar; ratio of specific heats of gases; nuclear physics. *Mailing Add:* 203 Ball Crs Saskatoon SK S7K 6E1 Can

KATZ, LEON, b Springfield, Mass, Aug 27, 21; m 47; c 3. ORGANIC CHEMISTRY. *Educ:* Trinity Col, Conn, BS, 44; Univ Ill, PhD(org chem), 47. *Prof Exp:* Chemist, Am Cyanamid Co, 47-49; mgr org chem, Schenley Labs, 49-53; res chemist, Gen Aniline & Film Corp, NY, 53-55, sect mgr dyes & pigments, 55-58, prod mgr pigments, 58-59, tech dir, 59-62, dir res, Dyestuff & Chem Div, 62-65, corp dir res, 65-66, vpres, 66-69; exec vpres, Rockwood Industs, Conn, 69-71; vpres corp develop, Polychrome Corp, Yonkers, NY, 71-73; vpres res & develop & packaging, Am Can Co, Greenwich, Conn, 73-82; sr vpres corp res & develop, 82-86, vpres corp technol, James River Corp, 87-88; CONSULT, 89- *Mem:* AAAS; Indust Res Inst; Am Chem Soc; NY Acad Sci; Am Inst Chem; Sigma Xi. *Res:* Alkaloids; pharmaceuticals; dyestuffs; pigments; reprographics; photography; specialty chemicals packaging; pulp paper. *Mailing Add:* 195 Dogwood Ct Stamford CT 06903

KATZ, LEWIS, b Fond du Lac, Wis, Mar 19, 23; m 48; c 2. X-RAY CRYSTALLOGRAPHY. *Educ:* Univ Minn, BChem, 46, PhD(phys chem), 51. *Prof Exp:* Asst chem, Univ Minn, 46-50; res fel, Calif Inst Technol, 51-52; from instr to prof phys chem, Univ Conn, 52-88, actg vpres grad educ & res, 81-83, assoc provost, 85-88, EMER PROF PHYS & CHEM, UNIV CONN, 88- *Concurrent Pos:* NSF sci fac fel, Cambridge Univ, 61-62; guest scientist, Weizmann Inst, Univ Leyden & Univ Stockholm, 69. *Mem:* Am Chem Soc; Am Crystallog Asn. *Res:* X-ray diffraction by crystals. *Mailing Add:* Dept Chem Univ Conn Storrs CT 06269

KATZ, LEWIS E, b Philadelphia, Pa, July 9, 40; m 69; c 1. PHYSICAL METALLURGY, MATERIALS SCIENCE. *Educ:* Drexel Inst, BS, 63; Univ Pa, MS, 64, PhD(mat sci), 67. *Prof Exp:* Fel diffusion, Lawrence Radiation Lab, Univ Calif, 67-69; mem tech staff, 69-80, SUPVR, BELL LABS, 80- *Mem:* Electrochem Soc. *Res:* Radiation damage; phase transformations; diffusion; electron microscopy; x-ray analysis; crystal growth; semiconductor development. *Mailing Add:* 1530 Hampton Rd Allentown PA 18104

KATZ, LOUIS, b New York, NY, Aug 3, 32; m 68; c 1. MOLECULAR BIOLOGY, COMPUTER SCIENCE. *Educ:* City Col New York, BS, 53; Univ Wis, MS, 55, PhD(physics), 59. *Prof Exp:* Physicist, Union Carbide Metals Co, 58-60, res physicist, Visking Co Div, Union Carbide Corp, 60-61; res assoc biol, Mass Inst Technol, 61-64; res assoc biochem, Albert Einstein Col Med, 64-65 & Mass Inst Technol, 65-68; sr res assoc & dir comput graphics facil, Dept Biol Sci, Col Physicians & Surgeons, Columbia Univ, 68-; SOFTWARE ENG MGR, RGB SPECTRUM, BERKELEY, CALIF. *Mem:* Biophys Soc; Asn Comput Mach. *Res:* Small angle x-ray scattering; molecular biophysics; interactive computer graphics; macromolecular structure. *Mailing Add:* RGB Spectrum 2550 Ninth St Berkeley CA 94710

KATZ, MANFRED, b Ger, Feb 16, 29; nat US; m 53; c 4. POLYMER CHEMISTRY, TEXTILE CHEMISTRY. *Educ:* Okla State Univ, BS, 50, MS, 51; Univ Del, PhD, 61. *Prof Exp:* Res supvr, 61-67, tech supvr, 67-71, res assoc, 71-81, sr res assoc, 81-86, RES FEL, E I DU PONT DE NEMOURS & CO, INC, 86- *Mem:* AAAS; Am Chem Soc; Sci Res Soc Am; NY Acad Sci. *Res:* Condensation polymers; synthetic fibers; non-woven fabrics; composites; carbon fibers. *Mailing Add:* 310 Brockton Rd Sharpley Wilmington DE 19803

KATZ, MARTIN, b New York, NY, May 16, 27; m 47; c 2. PHARMACEUTICS. *Educ:* St John's Univ, NY, BS, 47; Columbia Univ, MS, 48, MA, 52; Philadelphia Col Pharm, DSc, 54. *Prof Exp:* Instr pharm, Columbia Univ, 48-52; group leader, Pharmaceut Res & Develop, Chas Pfizer & Co, Inc, 53-55; group leader cosmetics, Revlon, Inc, 55-57, Thayer Labs Div, 57-60; Middle Atlantic Dir, R A Gosselin & Co, Boston, 50-60; dir pharmaceut res & develop, Syntex Labs, 60-64, vpres, Syntex Res & dir, Inst Pharmaceut Sci, 64-77, dir, Prof Prod Group & sr vpres, Syntex Res, 77-81, sr vpres res & develop & int mkt, Syntex Beauty Care Inc, 81-85; SR VPRES, ADVAN POLYMER SYSTS INC, REDWOOD CITY, 85- *Honors & Awards:* Ebert Award, Am Pharmaceut Asn, 54; IFF Award, Soc Cosmetic Chem, 72. *Mem:* Am Chem Soc; Am Pharmaceut Asn; Acad Pharmaceut Sci; Soc Cosmetic Chem. *Res:* Pharmaceutical and cosmetic development; sublingual absorption; percutaneous absorption. *Mailing Add:* Advan Polymer Systs Inc 3696-C Haven Ave Redwood City CA 94063-4604

KATZ, MARVIN L(AVERNE), b Tulsa, Okla, Dec 12, 35; m 55; c 3. CHEMICAL ENGINEERING. *Educ:* Univ Mich, BS, 56, MS, 58, PhD(chem eng), 60. *Prof Exp:* Res engr, Sinclair Res Inc, 60-64, adminr sci comput, 64-69; mgr admin dept, N Am Producing Div, Atlantic Richfield Co, 69-72, mgr res & develop dept, 72-78, vpres res & develop dept, Arco Oil-Gas

Co Div, 79-82, vpres planning & eval, 82-84, vpres eng, Arco Oil Gas Co Div, 84-86; VPRES ENG, CHIEF PETROL CO, 86- *Concurrent Pos:* Chmn, technol task group, Nat Petrol Coun Comt on enhanced recovery techniques for oil & gas in the US, 76. *Mem:* Soc Petrol Engrs (pres, 80); Am Inst Chem Engrs. *Res:* Fluid flow through porous media; heat transfer; computer science. *Mailing Add:* 6924 Leameadow Dallas TX 75248

KATZ, MAURICE JOSEPH, b Brooklyn, NY, May 17, 37; m 87. ENERGY, RADIATION EFFECTS. *Educ:* Columbia Univ, AB, 58, MA, 61, PhD(physics), 65. *Prof Exp:* Res physicist, IBM Res Lab, Zurich, Switz, 61-63; fel, appointee physics div, Los Alamos Sci Lab, Univ Calif, 65-67, staff mem, 67-70, prog mgr, 70-72, group leader, 72-73, asst div leader, Energy & Controlled Thermonuclear Res Divisions, 73-74; tech asst to dir, Controlled Thermonuclear Res Div, US Atomic Energy Agency & US Energy Res & Develop Admin, US Dept Energy, 75-76, spec asst to asst adminr, Solar, Geothermal & Advan Energy Syst, 77-79, dep dir, Distributed Technol, 79-80, off dir, Solar Power Appln, 80, off dir, Solar Elec Technol, 81-82, dep dir, Military Appln, 82-85, actg dep asst secy, 84; COUNR NUCLEAR TECHNOL, US MISSION UN SYSTS ORGN, VIENNA, 85- *Mem:* Am Phys Soc; Sigma Xi; Am Nuclear Soc. *Res:* Physics of the solid and liquid states; neutron scattering; structure of liquids; x-ray effects in materials and structures, controlled thermonuclear research, advanced energy systems; energy storage systems; photovoltaic, wind, and ocean thermal energy systems; nuclear energy; international scientific cooperation. *Mailing Add:* US Embassy Vienna APO New York NY 09108

KATZ, MAX, b Seattle, Wash, Mar 27, 19; m 46; c 4. FISHERIES. *Educ:* Univ Wash, Seattle, BS, 39, MS, 42, PhD(fisheries biol), 49. *Prof Exp:* Fisheries biologist, State Dept Fisheries, Wash, 40-42; asst, Inst Paper Chem, Wis, 46-47; from fisheries res biologist to pollution biologist, USPHS, Cincinnati, 49-53, Corvallis, 53-60; actg assoc prof fisheries, Univ Wash, 60-66, res prof, 66-73, dir water resources info ctr, 71-73; res dir, Seattle Marine Labs, Inc, res dir, Parametrix, Inc, 74-76; PRES, ENVIRON INFO SERV, INC, 76-; AFFIL PROF FISHERIES, UNIV WASH, 73- *Concurrent Pos:* Hon assoc prof, Dept Fish & Game Mgt, Ore State Col, 54-60; consult, Calif Water Pollution Control Bd, 58, Rayonier, Inc, Wash, 59, Northwest Pulp & Paper Asn, 61, off resource develop, USPHS, 62-64, Simpson Timber Co, Wash, 63, Libby, McNeil & Libby, Ill, 63-65, Health Plating Co, Wash, 64-65 & various other companies. *Mem:* Am Fisheries Soc; Am Soc Ichthyol & Herpet; Am Inst Fishery Res Biol; Water Pollution Control Fedn; Marine Biol Asn UK; Sigma Xi. *Res:* Water quality requirements of fish; fish toxicology; biological effects of water pollution; blood parasites of fish; hematology of fish; fish diseases. *Mailing Add:* 730 A Heritage Village Southbury CT 06488

KATZ, MICHAEL, b Lwow, Poland, Feb 13, 28; nat US; m 86; c 1. PEDIATRICS, VIROLOGY. *Educ:* Univ Pa, AB, 49; State Univ NY, MD, 56; Columbia Univ, MS, 63. *Prof Exp:* Instr biol, Queen's Col, NY, 51-52; intern, Med Ctr, Univ Calif, Los Angeles, 56-57; resident pediat, Babies Hosp, New York, 60-62; instr, Col Physicians & Surgeons, Columbia Univ, 64-65; assoc, Sch Med, Univ Pa, 65-66, asst prof, 66-70; prof pediat, 71-77, PROF TROP MED, COL PHYSICIANS & SURGEONS, COLUMBIA UNIV, 70-, REUBEN S CARPENTER PROF PEDIAT & CHMN DEPT, 77- *Concurrent Pos:* Hon lectr, Makerere Univ, Uganda & hon pediat specialist, Mulago Hosp, Kampala, Uganda, 63-64; consults, Peace Corps vols, Princeton Univ, 64, WHO, UNICEF & USAID; assoc vis pediatrician, Harlem Hosp Ctr, New York, & asst pediatrician, Babies Hosp, 64-65, assoc mem, Wistar Inst, Philadelphia, 65-70; assoc physician, Children's Hosp Philadelphia, 66-70; attend pediatrician, Presby Hosp, 70-77, dir pediat serv, 77-; mem, Subcomt Interactions Nutrit & Infections, Nat Acad Sci, 71-74, chmn, 74-75, consult, 75-80; vis prof, Inst Virol, Univ Wierzborg, Ger, 88; mem, comt exam adverse effects of vaccines, 90-91. *Honors & Awards:* Med Award, Turzykowski Found, 84; Sr Scientist Award, Humboldt Found, 87. *Mem:* Inst Med-Nat Acad Sci; Am Soc Trop Med & Hyg; NY Acad Sci; Am Pediat Soc; Soc Pediat Res; Infectious Dis Soc Am; fel AAAS. *Res:* Relationship of malnutrition to infection; antibody production and other host defense responses in protein deficiency; etiology of diarrhea; rubella and vaccine production; nature of slow virus infections. *Mailing Add:* Dept Pediat Columbia Univ New York NY 10032

KATZ, MORRIS HOWARD, b Milwaukee, Wis, Jan 12, 20; m 54; c 2. FOOD SCIENCE, BIOCHEMISTRY. *Educ:* Univ Wis, BS, 43. *Prof Exp:* Chemist, Nat Syrup Prod Co, 46-47; chief chemist, Martin Food Prod, Inc, 47-49; asst res dir, Orange Crush Co, 49-52; mfg dir, B A Railton Co, 52-53; flavor res chemist, Fries & Fries, Inc, 53-58; sr chemist, Pillsbury Co, 58-59, head flavor sect, 59-61, sr scientist, 61-68, res assoc res labs, 68-80; PRES, M H KATZ CONSULT INC, 80- *Concurrent Pos:* Guest lectr, Ill Inst Technol, 66 & 70; lectr flavor technol, Ctr Prof Advan, 74 & 76, Univ Minn, 76, Int Microwave Power Inst, 77 & Am Asn Cereal Chemists, 79. *Mem:* Am Chem Soc; Inst Food Technol; Soc Flavor Chemists. *Res:* Flavor chemistry; food texture, ingredient systems and processes; flavor and food products development. *Mailing Add:* 2700 S Yosemite Ave Minneapolis MN 55416-1856

KATZ, MORTON, b New York, NY, Apr 25, 34; m 61; c 3. ORGANIC CHEMISTRY, POLYMER CHEMISTRY. *Educ:* State Univ NY Albany, BS, 56, MS, 61; Wayne State Univ, PhD(org chem), 68. *Prof Exp:* Teacher high sch, 56-58; technician, Gen Elec Res Lab, 59-61; res chemist, Buffalo, 67-70, res chemist, Plastic Prod Dept, 70-85, SR RES CHEMIST, E I DU PONT DE NEMOURS & CO, INC, 85- *Mem:* Am Chem Soc. *Res:* Chemistry of bicyclic and tricyclic molecules; high temperature polymers; market research. *Mailing Add:* Electronics Dept E I du Pont de Nemours & Co Inc Circleville OH 43113

KATZ, MURRAY ALAN, b Albuquerque, NMex, June 15, 41; m 64; c 2. NEPHROLOGY, MICROCIRCULATORY PATHOPHYSIOLOGY. *Educ:* Johns Hopkins Univ, BA, 63, MD, 66. *Prof Exp:* Intern med, Osler Ward Serv, Johns Hopkins Univ Hosp, 66-67, resident, 67-68; fel nephrol, Univ Tex Southwestern Med Sch, 68-70; asst prof, Sch Med, Temple Univ, 71-74, actg chief nephrol, 73-74; from asst prof to assoc prof, 74-81, PROF NEPHROL, SCH MED, UNIV ARIZ, 81-; ASSOC CHIEF STAFF RES, TUCSON VET AFFAIRS MED CTR, 82-; PROF PHYSIOL, SCH MED, UNIV ARIZ, 87- *Concurrent Pos:* Res career develop award, Pub Health Serv, 71-76; clin investr award, Vet Admin, 76; staff physician med & nephrology, Vet Admin Hosp, Tucson, 74-; dir, B W Zweifach Microcirculation Labs, 86- *Mem:* Am Soc Nephrol; Am Fedn Clin Res; AAAS; Microcirculatory Soc; Am Physiol Soc; Int Soc Lymphology; W Soc Clin Invest (pres, 87); W Asn Physicians. *Res:* General microcirculatory physiology and pathophysiology; control of microcirculatory dynamics, hypertension, capillaropathies, vasculitis; diabetes. *Mailing Add:* Res Serv (151) Dept Vet Affairs Med Ctr Tucson AZ 85723

KATZ, NORMAN L, b Boston, Mass. PHARMACOLOGY. *Educ:* Mass Col Pharm, BS, 63; Albany Med Col, Union Univ, PhD(pharmacol), 69. *Prof Exp:* Fel neurophysiol, State Univ NY Albany, 69-72; asst prof, 72-81, ASSOC PROF PHARMACOL, UNIV ILL MED CTR, 81- *Mem:* Soc Neurosci; Am Soc Pharmacol Exp Therapeut. *Res:* Behavioral Pharmacology; regulation of feeding behavior. *Mailing Add:* Dept Pharmacodynamics M/C 865 Univ Ill Col Pharm 833 S Wood St Chicago IL 60612

KATZ, OWEN M, b Baltimore, Md, Dec 21, 32. FAILURE ANALYSIS, ELECTRON MICROSCOPY. *Educ:* Carnegie Mellon Inst, BS, 54, MS, 58; Univ Pittsburgh, PhD(metall eng), 63. *Prof Exp:* MEM STAFF, BETTIS ATOMIC POWER LAB. *Concurrent Pos:* Adj prof metall eng, Univ Pittsburgh, 70-75. *Mem:* Fel Am Soc Metals; Electron Micros Soc Am; Int Metallog Soc. *Mailing Add:* Bettis Atomic Power Lab PO Box 79 West Mifflin PA 15122

KATZ, PAUL K, INTERNAL MEDICINE, IMMUNOLOGY. *Educ:* Georgetown Univ, MD, 73. *Prof Exp:* ASST PROF MED, & CHIEF, ALLERGY DIV, GEORGETOWN UNIV HOSP, 84- *Mailing Add:* Dept Med Georgetown Univ 3800 Reservoir Rd NW Washington DC 20057

KATZ, RALPH VERNE, b Jersey City, NJ, Mar 20, 44; m 68; c 1. EPIDEMIOLOGY, GERONTOLOGY. *Educ:* Trinity Col, Conn, BS, 65; Tufts Univ, DMD, 69; Univ Minn, MPH, 71, PhD(epidemiol), 76. *Prof Exp:* Chief, Div Prev Dent, Inst Dent Res, US Army, Washington, DC, 74-76; dir dent serv, Phys Med & Rehab Unit, Univ Minn, 72-74, assoc prof, dept health ecol, Sch Dent, 76-82; assoc prof & head, Dept Restorative Dent, 82-89, ASSOC PROF, DEPT BEHAV SCI & COMMUNITY HEALTH, UNIV CONN, 89- *Concurrent Pos:* NIH & Nat Res Serv Awards fel, 70-74; assoc prof, Prog Dent Pub Health, Sch Pub Health, Univ Minn, 76-82, dept epidemiol, Grad Sch, 78-82, dir & prin investr, Cardiol Training Prog, 80-82, dir grad studies, Oral Health Serv Older Adults, 81-82; co-dir, Oral Epidemiol Training Grant, Nat Inst Dent Res, 85- *Mem:* Am Public Health Asn; Int Asn Dent Res; Soc Epidemiol Res; Am Asn Dent Sch; Sigma Xi; fel Am Col Epidemiol. *Res:* Epidimiology of root caries and coronal caries; clinical trials of preventive agents for dental caries; oral health of older adults. *Mailing Add:* Sch Dent Med Univ Conn Farmington CT 06032

KATZ, RICHARD WHITMORE, b Williamsburg, Va, Sept 12, 48. STATISTICS, ATMOSPHERIC SCIENCE. *Educ:* Univ Va, BA, 70; Pa State Univ, PhD(statist), 74. *Prof Exp:* Statistician climatic & environ assessment, Environ Data Serv, Nat Oceanic & Atmospheric Admin, 74-75; scientist & statistician environ & societal impacts, Nat Ctr Atmospheric Res, 75-79; ASST PROF, DEPT ATMOSPHERIC SCI, ORE STATE UNIV, 79- *Concurrent Pos:* Fel, Nat Ctr Atmospheric Res, 75-76; consult, NASA, 77; adj prof, Dept Econ, Univ Colo, 77-79; prin investr, NSF grant, 80-; consult, Lawrence Livermore Nat Lab, 81. *Honors & Awards:* Spec Achievement Award, Environ Data Serv, 75. *Mem:* Am Statist Asn; Inst Math Statist; AAAS; Am Meteorol Soc. *Res:* Meteorological statistics; probabilistic models for hydrological variables; applied probability theory; climatic impacts. *Mailing Add:* Dep Math & Comput Sci Calif State Univ 5151 State Univ Dr Los Angeles CA 90032

KATZ, ROBERT, b New York, NY, July 17, 17; m 43; c 2. PHYSICS. *Educ:* Brooklyn Col, AB, 37; Columbia Univ, AM, 38; Univ Ill, PhD(physics), 49. *Prof Exp:* Radiologist, US Army Air Force, Wright Field, 39-43, physicist, 43-46; from asst prof to prof, Kans State Univ, 49-66; vchmn dept, 68-73, EMER PROF PHYSICS, UNIV NEBR, LINCOLN, 66- *Mem:* fel Am Phys Soc; Radiation Res Soc. *Res:* Radiography; precipitation static radio interference; cereal technology; nuclear physics; structure of particle tracks; theory of relative biological effectiveness. *Mailing Add:* 5850 Sunrise Rd Lincoln NE 68510-4049

KATZ, ROBERT, b Walsenburg, Colo, Jan 14, 28; c 2. MECHANICAL & NUCLEAR ENGINEERING. *Educ:* Univ Colo, BS, 48; Calif Inst Technol, MS, 54, PhD(mech eng), 58. *Prof Exp:* Design engr, Colo Fuel & Iron Corp, 49-50; develop engr, Jet Propulsion Lab, Calif Inst Technol, 54, res asst turbomach, 55-58; staff mem, Gen Atomic Co, 58-64, dep div mgr eng anal, 64-68, br mgr steam generator res, 68-71, sr tech adv res & develop, 71-82; SR STAFF ENGR, GA TECHNOL INC, 82- *Mem:* Am Soc Mech Engrs; Am Inst Aeronaut & Astronaut. *Res:* Nuclear reactor heat transfer and fluid mechanics; nuclear reactor safety analysis; fluid mechanics of axial turbomachinery; heat transfer in solar and fusion power plants; thermodynamics of energy conversion systems. *Mailing Add:* 6177 Stresemann St San Diego CA 92122

KATZ, RONALD LEWIS, b New York, NY, Apr 22, 32; m 54; c 3. ANESTHESIOLOGY. *Educ:* Univ Wis, BA, 48; Boston Univ, MD, 52; Am Bd Anesthesiol, dipl, 62; FRCPS, 81. *Prof Exp:* Intern, Staten Island Pub Health Serv, NY, 56-57; resident anesthesiol, Columbia-Presby Med Ctr Hosp, 57-59; instr, Col Physicians & Surgeons, Columbia Univ, 60-61, assoc, 61-62, from asst prof to prof, 62-73; chmn dept anesthesiol, 73-90, PROF ANESTHESIOL, MED SCH, UNIV CALIF, LOS ANGELES, 73- *Concurrent Pos:* Fel pharmacol, Col Physicians & Surgeons, Columbia Univ, 59-60; Guggenheim fel, 68-69; consult, Coun Drugs, AMA, 62-; consult anesthesiol res grant comt, NIH, 65-, mem anesthesiol res training grant comt, 70-; vis prof, Royal Postgrad Med Sch, Univ London, 68-69. *Mem:* Am Soc Anesthesiol; fel Am Col Anesthesiol; Am Soc Pharmacol & Exp Therapeut; Am Physiol Soc; Sigma Xi. *Res:* Physiology; pharmacology; respiratory neurophysiology and neuropharmacology; cardiovascular physiology and pharmacology; neuromuscular transmission. *Mailing Add:* Dept Anesthesiol Univ Calif Sch Med Los Angeles CA 90024

KATZ, SAMUEL, b Berlin, Ger, Feb 13, 23; nat US; m 53; c 3. GEOPHYSICS. *Educ:* Univ Mich, BS, 43; Columbia Univ, AM, 47, PhD, 55. *Prof Exp:* Mem staff, Radiation Lab, Mass Inst Technol, 43-46; asst, Lamont Geol Observ, Columbia Univ, 48-53; physicist, Stanford Res Inst, 53-56, sr physicist, 56-57; from assoc prof to prof, 57-85, chmn dept, 63-68, EMER PROF GEOPHYS, RENSSELAER POLYTECH INST, 86- *Honors & Awards:* Kunz Prize, NY Acad Sci, 53. *Mem:* AAAS; Sigma Xi; Am Geophys Union. *Res:* Marine sciences; seismology; underwater sound propagation; high pressure; exploration geophysics. *Mailing Add:* 908 Karenwald Lane Schenectady NY 12309-6416

KATZ, SAMUEL LAWRENCE, b Manchester, NH, May 29, 27; div; c 9. VIROLOGY, PEDIATRICS. *Educ:* Dartmouth Col AB, 48; Harvard Univ, MD, 52. *Prof Exp:* Intern, Med Serv, Beth Israel Hosp, Boston, Mass, 52-53; jr asst resident, Children's Hosp Med Ctr, 53-54, resident, 55; asst resident, Children's Med Serv, Mass Gen Hosp, 54-55; exchange registr from Children's Hosp Med Ctr to pediat unit, St Mary's Hosp Med Sch, London, 56; instr pediat, Harvard Med Sch, 58-59, assoc, 59-63, tutor med sci, 61-63, asst prof pediat, 63-68; prof pediat & chmn dept, 68-90, WILBURT C DAVISON PROF, SCH MED, DUKE UNIV, 72- *Concurrent Pos:* Nat Found Infantile Paralysis res fel pediat, Res Div Infectious Dis, Children's Hosp Med Ctr - Harvard Med Sch, 56-58; pediatrician-in-chief, Beth Israel Hosp, 58-61; assoc physician, Children's Med Ctr, 58-63, res assoc, Res Div Infectious Dis, 58-68, chief newborn div, 61-68, sr assoc med, 65-68; consult coun drugs, AMA, 63-65; Nat Inst Allergy & Infectious Dis career develop award, 65-68; mem, Vaccine Develop Bd, Nat Inst Allergy & Infectious Dis, 67-71, Armed Forces Epidemiol Bd, Comn Immunization, 69-73, Gen Clin Res Ctr Comt, NIH, 71-74 & Nat Adv Coun Child Health & Human Develop, 74-77; chmn & pres, Asn Med Sch Pediat Dept, 77-79; chmn, adv comt Immunization Pract, US Pub Health Serv, 85-89. *Honors & Awards:* Grulee Medal, Am Acad Pediat; Jacobi Award, Am Med Asn & Am Acad Pediat; Saint Geme Award, Am Pediat Soc & Soc Pediat Res; Bristol Award, Infectious Dis Soc Am. *Mem:* Inst Med-Nat Acad Sci; Infectious Dis Soc Am; Am Epidemiol Soc; Am Pediat Soc (vpres, 85-86, pres, 86-87); Am Soc Clin Invest; Am Soc Virol; Am Acad Pediat. *Res:* Tissue culture studies of measles virus variants; development of live attenuated measles virus vaccine; central nervous system viral infections; AIDS & human immuno-deficiency virus infections. *Mailing Add:* Dept Pediat Duke Univ Sch Med Box 2925 Durham NC 27710

KATZ, SHELDON LANE, b Philadelphia, Pa, Oct 6, 48; m 73; c 3. PHYSICS. *Educ:* Temple Univ, BA, 69, MA, 73, PhD(physics), 77. *Prof Exp:* Lab instr physics, Temple Univ, Philadelphia, 69-74, res asst, 74-77; vis lectr, Lafayette Col, 77-78, asst prof physics, 78-83; asst prof physics, Villanova Univ, 83-84; SR MEM ENG STAFF, GE-GOVT ELECTRONIC SYSTS DIV, GE AEROSPACE, MOORESTOWN, 84- *Mem:* Am Phys Soc; Inst Elec & Electronics Engrs. *Res:* Dynamics of first and second order phase transitions; renormalization group; Monte Carlo simulations; radar performance analysis. *Mailing Add:* GE-Govt Electronic Systs Div MS 108-130 Moorestown NJ 08057

KATZ, SIDNEY, b Winnipeg, Man, Aug 17, 09; US citizen; m 37; c 2. PHYSICAL CHEMISTRY. *Educ:* Univ Man, BSc, 34, MSc, 35; McGill Univ, PhD(phys chem), 37. *Prof Exp:* Royal Soc Can traveling fel, London, 37-38; res chemist, British Thompson-Houston Co, Eng, 38-39, Pfanstiehl Chem Co, Ill, 40-41 & Goldsmith Bros, Chicago, 41-44; res assoc, Manhattan proj, 44-45; assoc prof chem, Inst Gas Technol, 45-52; res phys chemist, Ill Inst Technol Res Inst, 52-58, sci adv, 58-69, sr sci adv, 69-87, TECH CONSULT, IIT RES INST, 87- *Concurrent Pos:* Lectr eng, Sci & Mgt War Training, Ill Inst Technol, 42-45. *Mem:* Am Chem Soc; Sigma Xi. *Res:* Light scattering; kinetics of gaseous reactions; kinetics and thermodynamics; spectroscopy; aerosol technology. *Mailing Add:* 5532 SS Shore Dr Apt 19F Chicago IL 60637

KATZ, SIDNEY, chemistry; deceased, see previous edition for last biography

KATZ, SIDNEY, b Brooklyn, NY, Dec 23, 30; m 57. PHYSIOLOGY, NEUROPHYSIOLOGY. *Educ:* NY Univ, BA, 57, MS, 59, PhD(physiol), 63. *Prof Exp:* Teaching asst neuroanat, Sch Med, NY Univ, 61-62, instr physiol, Col Dent, 62-63; from asst prof to assoc prof, 65-77, PROF PHYSIOL, MED UNIV SC, 77- *Concurrent Pos:* Nat Heart Inst fel, 63-65. *Mem:* Am Physiol Soc; Soc Neurosci. *Res:* Central control of respiration and circulation; modulation of medullary neuron discharge patterns; ionic permeabiltiy of muscle studies with electrophysiological methods; electrophysiology of spinal cord injury. *Mailing Add:* Dept Physiol Med Univ SC 171 Ashley Ave Charleston SC 29425

KATZ, SIDNEY, b Cleveland, Ohio, Feb 4, 24; m 46; c 4. MEDICINE. *Educ:* Case Western Reserve Univ, MD, 48; Brown Univ, MA, 84; Am Col Epidemiol, cert, 82. *Prof Exp:* Intern & resident internal med, Case Western Reserve Univ, Univ Hosp Cleveland, 48-50, Am Cancer Soc fel path, 50-51, from instr to prof, dept prev med & med, Sch Med, 52-71; prof med & dir, Off Health Serv Educ & Res, Col Human Med, Mich State Univ, 71-77, prof community health & chmn dept, 78-82; assoc dean med & prof community health & med, Brown Univ, 82-87; prof architectonic & med, Case Western Reserve Univ, 87-89; EMER PROF GERIAT, COLUMBIA UNIV, 89- *Concurrent Pos:* Assoc dir, Dept Community Health, 69-71, dir, Univ Health Serv, Sch Med, Case Western Reserve Univ, 66-71; prof, Dept Med, 71-82, dir, Ctr Policy Anal in Aging & Long Term Care, Col Med, Mich State Univ, 80-82; dir, Long Term Care Geront Ctr, 82-87, assoc, Pop Studies & Training Ctr, Brown Univ, 84-87; spec adv, White House Conf Aging, 80-81, sr adv, US Prev Serv Task Force, 85-; consult, Health Care Financing Admin, Nat Ctr Health Statist, Nat Ctr Health Serv Res, Rand Corp, Dykewood Corp, Appl Mgt Sci, Inc, Morgan Mgt Systs, Inc, Geomet, Inc, Urban Inst, Herman Miller Res Corp. *Honors & Awards:* Sidney Katz lectr geriat/geront, Brown Univ; Robert Weiss Award. *Mem:* Inst Med-Nat Acad Sci; fel Geront Soc; Am Geriatrics Soc; Int Epidemiol Asn; Soc Epidemiol Res; AMA. *Res:* Author of four books and over 60 journal articles; gerontology; clinical epidemiology. *Mailing Add:* Ctr Geriat Columbia Univ New York NY 10032

KATZ, SIDNEY A, b Camden, NJ, June 4, 35; m 57; c 2. RADIOCHEMISTRY, ANALYTICAL CHEMISTRY. *Educ:* Rutgers Univ, AB, 58; Univ Pa, PhD(chem), 62. *Prof Exp:* Chemist, R H Hollingshead Corp, 53-58; asst instr chem, Univ Pa, 58-60; instr, 60-62, from asst prof to assoc prof, 62-71, PROF CHEM, RUTGERS UNIV, 71- *Concurrent Pos:* Res chemist, E I du Pont de Nemours & Co, Inc, 66; res assoc, Univ Pa Hosp, 60-70; prof, Temple Univ, 66-70; vis prof, The Univ, Reading, Berkshire, UK, 73; NATO sr fel sci, NSF, 73; consult, ACCU Test & Consult Lab, 74-75, John G Reutter & Assocs, 74-75, Rossnagel & Assocs, 77-78 & Jack McCormick & Assocs, 78; vis prof, Trace Anal Res Ctr, Dalhousie Univ, 77 & ATOMKI, Hungarian Acad Sci, 84. *Mem:* AAAS; Am Chem Soc; Am Nuclear Soc. *Res:* Environmental and biochemical effects of trace elements. *Mailing Add:* Dept Chem Rutgers Univ Camden NJ 08102

KATZ, SOL, b New York, NY, Mar 29, 13; m 46; c 3. MEDICINE. *Educ:* City Col New York, BS, 35; Georgetown Univ, MD, 39; Am Bd Internal Med, dipl, 48. *Hon Degrees:* DSc, Sch Med, Georgetown Univ, 79. *Prof Exp:* Adj clin prof, Georgetown Univ, 45-58, assoc prof, 58-65, dir pulmonary dis, 70-78, PROF LECTR, Georgetown Univ, 58-, PROF MED & KOBER LECTR, 65-, PROF PULMONARY MED, 78- *Concurrent Pos:* Chief med serv, Vet Admin Hosp, DC, 59-70; clin prof med, Sch Med, Howard Univ, 67-; consult, NIH, 58-, Children's Hosp, Walter Reed Army Hosp & Bethesda Naval Hosp; vis consult, Cardiothoracic Inst, Brompton Hosp, London, 74-75. *Mem:* Am Thoracic Soc; Am Col Physicians; Am Col Chest Physicians; Am Fedn Clin Res; Brit Thoracic Asn. *Res:* Pulmonary diseases. *Mailing Add:* Pulmonary Div Georgetown Univ Hosp Washington DC 20007

KATZ, SOLOMON H, b Beverly, Mass, July 27, 39; m 64; c 2. BIOLOGICAL ANTHROPOLOGY. *Educ:* Northeastern Univ, BA, 63; Univ Pa, MA, 66, PhD(phys anthrop), 67. *Prof Exp:* DIR, W M KROGMAN CTR RES CHILD GROWTH & DEVELOP, 71-; PROF BIOL ANTHROP, DEPT ORTHODONTIC & PEDODONTICS, SCH DENT MED & ANTHROP, UNIV PA, 76-, CUR PHYS ANTHROP, UNIV MUS, 76-; PROF EPIDEMIOL, EASTERN PA PSYCHIAT INST, MED COL PA, 81- *Concurrent Pos:* Sr med res scientist, Eastern Pa Psychiat Inst, 67-80; prin investr, NIH grants, 71-74; pres, Corp Inst Continuous Study Man, 74-; assoc ed, Zygon, 77- *Mem:* AAAS; Human Biol Asn; Am Anthrop Asn. *Res:* Evolutionary and population analyses of physiological and psychosocial studies of human growth, development and aging and their consequences for long term pathology including hypertension and mental disorders; biocultural adaptation, evolution and nutritional significance of human food patterns. *Mailing Add:* Dept Anthrop Univ Mus Univ Pa Philadelphia PA 19104

KATZ, STEPHEN I, DERMATOLOGY. *Prof Exp:* CHIEF DERMATOLOGIST, NAT CANCER INST, NIH, 74- *Mailing Add:* NIH Bldg 10 Rm 12N238 9000 Rockville Pike Bethesda MD 20892

KATZ, THOMAS JOSEPH, b Prague, Czech, Mar 21, 36; US citizen; m 63; c 1. ORGANIC CHEMISTRY, ORGANOMETALLIC CHEMISTRY. *Educ:* Univ Wis, BA, 56; Harvard Univ, MA, 57, PhD(chem), 59. *Prof Exp:* Instr, 59-61, from asst prof to assoc prof, 61-68, PROF CHEM, COLUMBIA UNIV, 68- *Concurrent Pos:* Sloan fel, 62-66; Guggenheim fel, 67-68. *Mem:* Am Chem Soc; Royal Soc Chem (UK). *Res:* Non-benzenoid aromatic compounds; organometallic compounds; organic synthesis; catalysis by metals. *Mailing Add:* Dept Chem Columbia Univ New York NY 10027

KATZ, VICTOR JOSEPH, b Philadelphia, Pa, Dec 31, 42; m 69; c 3. HISTORY OF MATHEMATICS. *Educ:* Princeton Univ, AB, 63; Brandeis Univ, MS, 65, PhD, 68. *Prof Exp:* Asst prof, Fed City Col, 68-73; assoc prof, 73-80, PROF MATH UNIV DC, 80- *Concurrent Pos:* Vis res assoc, Int Hist & Philos Sci, Univ Toronto, 78-79; vis prof, Math Dept, Boston Univ, 85-86; mem coun, Can Soc Hist & Philos Math, 87-89, 90- *Mem:* Am Math Soc; Math Asn Am; Hist Sci Soc. *Res:* History of mathematics. *Mailing Add:* 841 Bromley St Silver Spring MD 20902-3019

KATZ, WILLIAM, b Dayton, Ohio, Dec 10, 53; m 79. SURFACE ANALYSIS, ION BEAM METHODS. *Educ:* Earlham Col, Richmond, Ind, BA, 75; Univ Ill, Urbana, MS, 77, PhD(anal chem), 79; State Univ NY, Albany, MBA, 86. *Prof Exp:* Res chemist, Exxon Res & Develop Lab, 78-80; mat scientist, Gen Elec, 80-84, mgr anal chem, 84-86; dir labs, Perkin Elmer Corp, 86-90; OWNER, EVANS CENT, 90- *Concurrent Pos:* Adj prof physics, SU NY, Albany, 81-; chmn, corp affil comt, Mat Res Soc. *Mem:* Am Inst Physics; Am Vacuum Soc; Am Chem Soc; Microbeam Anal Soc. *Res:* Application of ion beams for the characterization of electronic materials; ion-solid interactions; sputtering; secondary ionization mechanisms. *Mailing Add:* Evans Cent 5909 Baker Rd Suite 580 Minnetonka MN 55345

KATZ, WILLIAM J(ACOB), b Chicago, Ill, Jan 19, 25; m 48; c 3. SANITARY & CHEMICAL ENGINEERING. *Educ:* Univ Ill, BS, 48; Univ Wis, MS, 49, PhD(chem eng), 53. *Prof Exp:* Proj assoc indust waste & treatment, Univ Wis, 49-52; dir sanit & indust wastes res & consult, Envirex Inc Div, Rexnord Inc, 53-63, tech dir & mgr water treatment & water pollution control res, 63-70, mgr, Ecol Div, Rex Chainbelt Inc Div, 70-75, vpres res & develop, Envirex Inc, 76-77; dir tech serv, Milwaukee Metrop Sewerage Dist, 77-81; PRES, ENVIRON PLANNING & SCI DIV, CAMP DRESSER & MCKEE INC, 81-; PRES, WJK ASSOCS, LTD, 81- *Concurrent Pos:* Vis prof civil eng, Univ Wis, 65-67; adj prof civil eng, Marquette Univ. *Honors & Awards:* Eddy Medal, Water Pollution Control

Fedn, 55, Gascoigne Medal. *Mem:* Water Pollution Control Fedn; Nat Soc Prof Engrs; Am Soc Civil Engrs; Am Inst Chem Engrs; Am Water Works Asn. *Res:* Water and industrial waste treatment; packing plant; mechanism of activated sludge; foundries; refineries. *Mailing Add:* 220 W Cherokee Circle Milwaukee WI 53217

KATZ, YALE H, b Milwaukee, Wis, Mar 15, 20; m 45; c 2. METEOROLOGY. *Educ:* Univ Wis, BS, 47, MS, 48; Pa State Col, MS, 51. *Prof Exp:* Asst, Univ Wis, 47-48; meteorologist, US Air Force Air Weather Serv, 51, asst to chief, Data Integration Br, 51-52, specialist climatic res, 52-54, tech consult, 54-56; res assoc, Phys Res Labs, Univ Boston, 56-57; sr staff meteorologist, Itek Corp, 58-60; phys scientist, Rand Corp, Calif, 60-67; sr staff mem, TRW Systs Group, Redondo Beach, 67-73, sr systs engr, 73-76; dir, Appl Res Assocs, 76-77; tech dir, Soc Photo-Optical Instrumentation Engrs, 77-81; DIR, INST FOR TECHNOL COMMUNICATION, 82- *Concurrent Pos:* Partner & assoc, Appl Res Assocs, 54-60, 76-77; lectr, Univ Chicago, 54; lectr, Air Force Sr Technol Specialist Sch, Mather AFB, 56; lectr, MIT, 56; lectr, Univ Calif, Los Angeles, 61, 63, 65; lectr, Univ Calif, San Diego, 70; ed, J Soc Photo-Optical Instrumentation Engrs, 70-72, assoc ed, Optical Eng, 72-78. *Mem:* Am Meteorol Soc; Am Geophys Union; fel Soc Photo-Optical Instrument Engrs (vpres, 71-75, gov, 70-71 & 75); Coun Eng Sci Soc Execs; NAm Infrared Thermographic Asn. *Res:* Development and direction of educational and training programs for medical, legal and industrial professionals; technology transfer; applied optical and electro-optical engineering; environmental engineering; utilization of solar energy; atmospheric pollution; statistical meteorology and climatology; utilization of solar energy; environmental engineering. *Mailing Add:* 2800 Woodridge Dr Bellingham WA 98226

KATZBERG, ALLAN ALFRED, b Can, July 6, 13; nat US; m 48; c 4. ANATOMY. *Educ:* Univ Man, BSc, 43; Inst Divi Thomae, MS, 49; Univ Okla, PhD(med sci), 56. *Hon Degrees:* DSc, St Thomas Inst, 86. *Prof Exp:* Instr histol & embryol, Sch Med, Univ Okla, 49-51, from instr to asst prof anat, 51-59; res assoc prof & head cellular biol sect, Aerospace Med Ctr, US Air Force, Brooks AFB, Tex, 59-63, dep chief, Astrobiol Div, 60-63; assoc prof physiol, Med Sch, Univ Sask, 63-64; head anat div, Southwest Found for Res & Educ, 64-65, chmn dept, 65-68; assoc prof biol, Western Ill Univ, 68-69; assoc prof anat, 69-77, prof, 77-83, EMER PROF ANAT, MED CTR, IND UNIV-PURDUE UNIV, INDIANAPOLIS, 83- *Concurrent Pos:* Consult, Arctic Aeromed Lab, US Air Force, 61 & Fed Aviation Admin-US Air Force 6571st Aeromed Res Lab, Holloman AFB, 67; actg chmn dept anat, Ind Univ-Purdue Univ, Indianapolis, 70-71. *Honors & Awards:* Eli Lilly Award, 72. *Mem:* AAAS; Am Asn Anat; hon mem Mex Soc Anat; fel Royal Micros Soc; Int Primatol Soc. *Res:* Aging processes; tissue culture; regeneration; aerospace medicine; primate histology. *Mailing Add:* 944 E Main St Carmel IN 46032

KATZE, JON R, b Portland, Ore, Nov 21, 39; m 82; c 2. NEOPLASIA, Q NUCLEOSIDE. *Educ:* Univ Calif, Berkeley, BS, 61, Univ Calif, Los Angeles, PhD(physiol chem), 66. *Prof Exp:* Post doctoral biochem, Yale Univ, 66-69; Asst prof microbiol, Univ Southern Calif, 69-76; assoc prof microbiol & immunol, 76-83, PROF MICROBIOL & IMMUNOL, UNIV TENN, MEMPHIS, 83- *Mem:* AAAS; Am Soc Microbiol; Am Soc Biochem & Molecular Biol. *Res:* The association of defective Q nucleoside metabolism with neoplasia; the linkage of Q deficiency with tumor promotion; the wide distribution of Q base in the biosphere but absence of synthesis in eucayotes. *Mailing Add:* Univ Tenn Memphis TN 38163

KATZEN, HOWARD M, b Baltimore, Md, May 2, 29; m 60; c 3. BIOCHEMISTRY. *Educ:* Johns Hopkins Univ, BS, 56; George Washington Univ, MS, 58, PhD(enzym), 62. *Prof Exp:* Biochemist, Nat Inst Arthritis & Metab Dis, 54-62; instr pediat, Sch Med, Johns Hopkins Univ & asst dir pediat res, Sinai Hosp Baltimore, Inc, 63-65; sr res biochemist, 65-75, SR RES FEL, MERCK INST THERAPEUT RES, RAHWAY, 75- *Concurrent Pos:* Am Cancer Soc res fel enzym, Mass Inst Technol, 62-63; vis scientist, Nat Inst Arthritis & Metab Dis, 65; guest scientist, Imp Col, Univ London, 69. *Mem:* AAAS; Am Chem Soc; Am Diabetes Asn; Am Soc Biol Chem; Am Inst Chem. *Res:* Glycogen structure and metabolism; isulin metabolism and disulfide biosynthesis; methionine biosynthesis; folate-vitamin B-twelve regulations; isoenzymes; insulin action; coenzymes; enzyme regulation; membrane receptors; hexokinase. *Mailing Add:* Merck Inst Therapeut Res Rahway NJ 07065

KATZEN, RAPHAEL, b Baltimore, Md, July 28, 15; m 38; c 1. CHEMICAL ENGINEERING. *Educ:* Polytech Inst Brooklyn, BChE, 36, MChE, 38, DChE(chem eng), 42. *Prof Exp:* Chem supvr, Chem Prod, North Wood Chem Co, 37-40, dir res, Diamond Alkali Co, 42-44; proj mgr chem plant & mgr design & construct, Eng Div, Vulcan-Cincinnati, 44-53; prin chem process, 53, PRES CONSULT & DESIGN, RAPHAEL KATZEN ASSOCS INT, INC, 53- *Concurrent Pos:* Mem adv bd, Expos Chem Indust, 55-; mem nat panel, Am Arbit Asn, 70- *Honors & Awards:* Chem Eng Pract Award, Am Inst Chem Engrs, 86, Robert L Jacks Mem Award, 90. *Mem:* Fel Am Inst Chem Engrs; fel Am Inst Chemists; Tech Asn Pulp & Paper Indust; Can Pulp & Paper Asn. *Res:* Research and development of new organosolv pulping process; new technology for enzymatic conversion of cellulosic materials to sugar and ethanol; advanced technology in scrubbing and recovery of sulfur dioxide emmissions; biomass conversion to fuels and chemicals technology. *Mailing Add:* Raphael Katzen Assocs Int Inc 7162 Reading Rd Suite 1200 Cincinnatti OH 45237

KATZENELLENBOGEN, BENITA SCHULMAN, b New York, NY, Apr 11, 45; m 67; c 2. REPRODUCTIVE ENDOCRINOLOGY, CANCER BIOLOGY. *Educ:* Brooklyn Col, BA, 65; Harvard Univ, MA, 66, PhD(biol), 70. *Prof Exp:* NIH res fel endocrinol, Dept Physiol & Biophys, 70-71, from assoc prof to prof, 70-82, PROF PHYSIOL, CELL & STRUCT BIOL, COL MED, UNIV ILL, PHYSIOL & BIOPHYS, URBANA, 82- *Concurrent Pos:* Vis prof, Dept Biochem & Biophys, Univ Calif, San Francisco, 77-78; endocrinol study sect, NIH, 79-83, NIH-Nat Inst Diabetes & Digestive & Kidney Dis Bd Sci Counselors, 85-89; publications comt, Endocrine Soc, 81-83, prog comt, 83-87; int org comt, Int Cong on Hormones & Cancer, 87-; Am Asn Cancer Res Task Force Endocrinol, 87-89; co-chmn Gordon Res Conf Hormone Action, 88; mem, Central Comt, Int Soc Endocrinol, 88-; mem adv comt, Biochem & Endocrinol, Am Cancer Soc, 89-; mem coun, Endocrine Soc, 89-; mem, Waterman Award Comt, NSF; Breast Cancer Res Award, Susan A Komen Found, 88-89. *Honors & Awards:* Ernst Oppenheimer Mem Award, Endocrine Soc. *Mem:* Am Physiol Soc; Endocrine Soc; Am Asn Cancer Res; Soc Study Reproduction. *Res:* Regulation of the growth and function of reproductive tissues and tumors by reproductive hormones and antihormones. *Mailing Add:* Dept Physiol & Biophys Univ Ill 524 Burrill Hall 407 S Goodwin Ave Urbana IL 61801

KATZENELLENBOGEN, JOHN ALBERT, b Poughkeepsie, NY, May 10, 44; m 67; c 2. BIO-ORGANIC CHEMISTRY, SYNTHETIC ORGANIC CHEMISTRY. *Educ:* Harvard Univ, BA, 66, MA, 67, PhD(chem), 69. *Prof Exp:* From asst prof to assoc prof, 69-79, PROF CHEM, UNIV ILL, URBANA, 79- *Concurrent Pos:* Sloan Fel, 74-76; Guggenheim Fel, 77-78; Chmn, study sect, NIH. *Honors & Awards:* Teacher Scholar Award, Camille & Henry Dreyfus Found, 74-79. *Mem:* AAAS; Am Chem Soc; NY Acad Sci; Am Soc Biol Chemists; Royal Soc Chem; Soc Nuclear Med. *Res:* New synthetic methods; organometallic chemistry; natural product synthesis; mechanism of hormone action; affinity labeling; tumor localizing agents; radiopharmaceutical development; flourescence; enzyme inhibitors. *Mailing Add:* Sch Chem Sci Univ Ill 1209 W California St Urbana IL 61801

KATZIN, GERALD HOWARD, b Winston-Salem, NC, Aug 2, 32; m 58; c 2. PHYSICS. *Educ:* NC State Univ, BS, 54, MS, 56, PhD(relativity), 63. *Prof Exp:* Assoc nuclear reactor theory, Astra, Inc, Conn, 57-58; from asst prof to assoc prof, 63-76 PROF PHYSICS, NC STATE UNIV, 76- *Concurrent Pos:* NSF grant, 67. *Mem:* Am Phys Soc. *Res:* Study of the relations between symmetries and conservation laws; theoretical mechanics; Riemannian geometry and tensor analysis; differential geometry; general relativity; classical electrodynamics. *Mailing Add:* Dept Physics NC State Univ Raleigh NC 27695-8202

KATZIN, LEONARD ISAAC, b Eau Claire, Wis, Jan 18, 15; m 38; c 4. PHYSICAL INORGANIC CHEMISTRY. *Educ:* Univ Calif, Los Angeles, AB, 35; Univ Calif, PhD(phys chem biol), 38. *Prof Exp:* Asst zool, Univ Calif, 35-37, fel, 38-40; jr biologist, USPHS, 40-41, asst biologist, 41-42; fel radiol, Sch Med & Dent, Univ Rochester, 42-43; res assoc, Metall Lab, Univ Chicago, 43-46; sr chemist, Argonne Nat Lab, 46-78. *Concurrent Pos:* Vis prof, Univ Chicago, 56-57; exchange fel, Atomic Energy Res Estab, Harwell, Eng, 58-59; exten lectr, Univ Ill, 61; vis prof inorg chem, Hebrew Univ Jerusalem, 69-70 & Inst Chem, Tel Aviv Univ, 69-70. *Mem:* AAAS; Am Phys Soc; Am Chem Soc; Sigma Xi. *Res:* Nuclear chemistry; heavy element chemistry; optical rotation; coordination chemistry; nuclear waste management. *Mailing Add:* 428 Hudson Lane Port Hueneme CA 93041-2140

KATZMAN, PHILIP AARON, b Omaha, Nebr, May 18, 06; m 33; c 2. BIOCHEMISTRY. *Educ:* Kalamazoo Col, AB, 27; St Louis Univ, PhD(biochem), 32. *Prof Exp:* Res instr, 32-36, sr instr, 36-41, from asst prof to prof, 41-74, EMER PROF BIOCHEM, MED SCH, ST LOUIS UNIV, 74- *Concurrent Pos:* US Pharmacopoeia Comt, 40-50; lectr, Univ Kans, 52; Merck Sharpe & Dohme vis prof, Univ SDak, 63. *Mem:* AAAS; Am Soc Biol Chem; Am Chem Soc; Endocrine Soc; Soc Study Reproduction. *Res:* Chorionic gonadotropin; reproduction; antihormones; antibiotics; hydrolysis of conjugated steroids; biological properties of estrogens; molecular action of ovarian hormones. *Mailing Add:* Dept Biochem St Louis Univ Med Sch 1402 S Grand Blvd St Louis MO 63104

KATZMAN, ROBERT, b Denver, Colo, Nov 29, 25; m 47; c 2. NEUROLOGY. *Educ:* Univ Chicago, BS, 49, MS, 51; Harvard Med Sch, MD, 53; Am Bd Psychiat & Neurol, dipl & cert neurol, 59. *Prof Exp:* Intern, Harvard Med Serv, Boston City Hosp, 53-54; asst resident neurol, Neurol Inst, Columbia Presby Hosp, 54-56, chief resident neurologist, 56-57; instr neurol, Albert Einstein Col Med, 57-58, assoc, 58-60, from asst prof to prof, 60-84, chmn dept, 64-81, dir, Hosp, 74-84; chair dept, 84-90, PROF NEUROSCI, UNIV CALIF, SAN DIEGO, 84- *Concurrent Pos:* Nat Mult Sclerosis Soc fel, 57-59; USPHS sr fel neurophysiol, 61-62, USPHS career res develop award, 62-66; asst neurol, Columbia Univ, 56-57; guest scholar, Polytech Inst Brooklyn, 60-61; consult, Jewish Bd Guardians, 61-65; assoc attend neurologist, Bronx Munic Hosp Ctr, 62-64, attend neurologist, 64-84, dir neurol serv, 70-84; mem res rev panel, Nat Mult Sclerosis Soc, 64-70; asst examr, Am Bd Psychiat & Neurol, 63-70; chair, Neurochem Sect, Am Acad Neurol, 65-67; consult, Montefiore Hosp & Med Ctr, 66-82, attend neurologist, 82-84; mem, Neurol Prog Proj A Comn, Nat Inst Neurol & Commun Dis & Stroke, 69-72, chmn, Neurol Dis Prog Proj Rev Comt, 72-73, mem, Aging Rev Comt, 76-81, chmn, 80-81; mem, adv coun, Nat Inst Aging, 82-85, Adv Bd Alzheimer's Dis, US Cong, 87 & Adv Coun Alzheimer's Dis, Dept Health & Human Serv, 89-; Florence Riford prof res, Alzheimer's Dis, Sch Med, Univ Calif, San Diego, 84, attend neurologist, San Diego Med Ctr & San Diego Vet Admin Med Ctr, 84- *Honors & Awards:* S Weir Mitchell Award, Am Acad Neurol, 60; Ann Prize, Am Asn Neuropathologists, 62; Humanitarian Award & Allied Achievement Aging Award, Alzheimer's Dis & Related Dis Asn, 85; Henderson Mem Award, Geriat Soc, 86; George W Jacoby Award, Am Neurol Asn, 89; Distinguished Serv Award, Alzheimer's Asn, 89. *Mem:* Inst Med-Nat Acad Sci; fel AAAS; fel Am Acad Neurol; Am Asn Neuropathologists; In Soc Neurochem; Soc Neurosci; Am Neurol Asn (pres, 84-85). *Mailing Add:* Dept Neurosci Univ Calif San Diego La Jolla CA 92093

KATZMANN, FRED L, b Magdeburg, Ger, Apr 10, 29; US citizen; m 55; c 3. ELECTRONICS. *Educ:* City Col New York, BEE, 52; Stevens Inst Technol, MSc, 62. *Prof Exp:* Res eng, Allen B DuMont Labs Div, Fairchild Camera & Instrument Corp, 52-59, eng mgr, Corp, 59-65; dir electronic instruments dept, Electronic Prod & Controls Div, Monsanto Co, 66-69, dir & gen mgr, 69-70; vpres mkt, Electronic Prods Div, Singer Co, 70-71; PRES,

BALLANTINE LABS, INC, 71- *Mem:* Inst Elec & Electronics Engrs; Precision Measurement Asn; Int Standards Asn; Instrument Soc Am. *Res:* Electronic instrumentation, particularly oscilloscopes, voltmeters and digital frequency counters, ultra fast transient recording instrumentation systems and precision alternating current measurements. *Mailing Add:* Instrument Div Ballantine Labs Box 97 Boonton NJ 07005

KATZOFF, SAMUEL, b Baltimore, Md, Aug 3, 09. AERODYNAMICS. *Educ:* Johns Hopkins Univ, BS, 29, PhD(chem), 34. *Prof Exp:* Lab technician, Rockefeller Inst, 29-30; res chemist, Baltimore Paint & Color Works, 34-53; Jones fel biophys, Cold Spring Harbor, 35-36; physicist, Nat Adv Comt Aeronaut, NASA, 36-58, res scientist, 58-60, asst chief, Appl Mat & Physics Div, 60-64, from sr staff scientist to chief scientist, Langley Res Ctr, 64-72; RETIRED. *Concurrent Pos:* Ed, Proc Symposium on Thermal Radiation of Solids, 64 & Remote Measurement of Pollution, 71; instr, Gifted Children, Hampton, Va, & John Hopkins Univ, 72- *Res:* X-ray crystallography; x-ray studies of the molecular arrangements in liquids; colloids; general aerodynamics; stability; electrical analogies; wind-tunnel interference; cascades; helicopters; space sciences; thermal control of spacecraft. *Mailing Add:* 285-G Clemwood Pkwy Hampton VA 23669

KATZUNG, BERTRAM GEORGE, b Floral Park, NY, June 11, 32; m 57; c 2. PHARMACOLOGY. *Educ:* Syracuse Univ, BA, 53; State Univ NY, MD, 57; Univ Calif, PhD, 62. *Prof Exp:* Lectr, 60-62, from asst prof to assoc prof, 62-71, acting chmn, 80-83, PROF PHARMACOL, MED CTR, UNIV CALIF, SAN FRANSICO, 71-, VCHMN DEPT, 67- *Concurrent Pos:* Ed, Basic & Clin Pharmacol; Markle Scholar, 66-71. *Mem:* AAAS; Soc Pharmacol & Exp Therapeut; Biophys Soc; Soc Gen Physiologists. *Res:* Cardiovascular pharmacology; electrophysiology. *Mailing Add:* Dept Pharmacol & Exp Therapeut Univ Calif San Francisco CA 94143-0450

KAUDER, OTTO SAMUEL, b Vienna, Austria, Nov 26, 26; US citizen; m 56; c 2. ORGANIC CHEMISTRY. *Educ:* City Col New York, BS, 46; Polytech Inst Brooklyn, MS, 49; Oxford Univ, DPhil(org chem), 52. *Prof Exp:* Chemist, Polychem Labs, NY, 46-48; chemist plastics additives, 52-59, res group leader, 59-68, patent liaison & toxicol supvr, 62-68, vpres res & develop, 68-79, TECH VPRES, ARGUS CHEM CORP, 79- *Mem:* Am Chem Soc. *Res:* Time-dependent properties of organic compounds and effect of additives and contaminants thereon; synthesis of organic compounds containing phosphorus, cadmium, tin and antimony. *Mailing Add:* 633 Court St Brooklyn NY 11231-2107

KAUER, JAMES CHARLES, b Cleveland, Ohio, Jan 17, 27; m 54; c 5. ORGANIC CHEMISTRY, NEUROCHEMISTRY. *Educ:* Case Western Reserve Univ, BS, 51; Univ Ill, PhD(chem), 55. *Prof Exp:* Res chemist, E I du Pont de Nemours & Co, Inc, 55-87, dir chem, Cent Res Dept, Mex Sta, 87-89; RETIRED. *Mem:* Am Chem Soc; Soc Neurosci; NY Acad Sci. *Res:* Organic synthesis; heterocycles; peptides; antiviral agents; neurotransmitter analogs; Neuropeptides. *Mailing Add:* Saverys Hill Kennett Square PA 19348

KAUER, JOHN STUART, b New York, NY, Dec 26, 43; m; c 2. SENSORY PHYSIOLOGY. *Educ:* Clark Univ, BA, 67, MA, 69; Univ Pa, PhD(anat), 73. *Prof Exp:* Fel neurophysiol, 73-75, res assoc & dir lab studies, 76-78, asst prof, sect neurosurg, Sch Med, Yale Univ, 78-80; PROF NEUROSCI, ANAT & CELL BIOL, NEW ENG MED CTR, TUFTS UNIV, 80- *Mem:* Soc Neurosci; Am Asn Anatomists. *Res:* Central synaptic organization of the olfactory and other sensory systems; mathematical models of neurophysiology. *Mailing Add:* Dept Neurosurg Med Sch NE Med Ctr Tufts Univ 750 Washington St Boston MA 02111

KAUFERT, FRANK HENRY, forest products; deceased, see previous edition for last biography

KAUFERT, JOSEPH MOSSMAN, b Minneapolis, Minn, Feb 10, 43; m 70; c 1. MEDICAL ANTHROPOLOGY. *Educ:* Univ Minn, BA, 66; Northwestern Univ, MA, 68, PhD(polit sci, anthrop), 73. *Prof Exp:* Asst prof health admin, Baylor Univ, 71-72; asst prof med sociol & social psychiat, Med Sch, Univ Tex, San Antonio, 72-74; head soc sci sect, St Thomas Hosp Med Sch, Univ London, 74-76; PROF COMM MED, COMMUNITY HEALTH SCIS, FAC MED, UNIV MAN, 76- *Concurrent Pos:* Leverhulme fel, Univ Birmingham, 73-74; consult, Welsh Off, Brit Health & Social Serv, UK, 74-76, Nat Haemophilia Soc UK, 74-76 & Ment Health Man, 77-78; adj prof, Dept Anthrop, Univ Man, 77-, prof, Dept Social & Prev Med, 85; vis health scientist, Univ Toronto, 85; vis prof Can Studies, Leeds Univ. *Honors & Awards:* Keith L Were Award, Nat Media Develop Trust, 72. *Mem:* Soc Social Med; fel Soc Appl Anthrop; Brit Sociol Asn; Soc Med Anthrop; Can Asn Med Anthrop (pres). *Res:* Social epidemiology; medical sociology; illness behavior; the sociology of disability; social gerontology. *Mailing Add:* Dept Community Health Sci Fac Med Univ of Man Winnipeg MB R3T 2N2 Can

KAUFFELD, NORBERT M, b Trivandrum, Kerala, India, Jan 30, 23; US citizen; m 45; c 5. ENTOMOLOGY, APICULTURE. *Educ:* Kans State Univ, BS & MS, 49, PhD(entom), 67. *Prof Exp:* Rancher, 49-58; instr high sch, Kans, 58-62; instr apicult, Kans State Univ, 62-66; res entomologist, Univ Wis, 66-67, invest leader, Bee Breeding Invest, La State Univ, Baton Rouge, 67-74, RES ENTOMOLOGIST, BEE RES LAB, AGR RES SERV, TUCSON, ARIZ, 74- *Concurrent Pos:* Apiarist entom div, Kans State Bd Agr, 61-63. *Mem:* AAAS; Entom Soc Am; Bee Res Asn. *Res:* Pollination of crops, honey bee behavior, nutrition, and the interrelationship between them. *Mailing Add:* 1471 W Rancho Feliz Pl Tucson AZ 85704-2355

KAUFFMAN, CAROL A, b Columbia, Pa, Oct 9, 43. INFECTIOUS DISEASES, HOST-DEFENSE MECHANISMS. *Educ:* Penn State Univ, BS, 65; Univ Mich, MD, 69. *Prof Exp:* CHIEF, DIV INFECTIOUS DIS, VET ADMIN MED CTR, ANN ARBOR, MICH, 77-; PROF MED, UNIV MICH, 81,asst dean student affairs, 86- *Mem:* Am Asn Immunologists; Infectious Dis Soc Am; Am Soc Microbiol; Am Fedn Clin Res; Am Col Physicians; Cent Soc Clin Res. *Res:* Infectious in the elderly; effects of aging and malnutrition in fabrile response to infection. *Mailing Add:* Vet Admin Med Ctr 2215 Fuller Rd Ann Arbor MI 48105

KAUFFMAN, ELLWOOD, b Philadelphia, Pa, Mar 18, 28; m 50; c 4. COMPUTER APPLICATIONS, MATHEMATICS. *Educ:* Temple Univ, AB, 52. *Prof Exp:* Sr analyst, Remington Rand, Inc, NY, 52-55; comput applns officer, Chesapeake & Ohio Rwy Co, Va, 55-57; sr programmer digital comput, Elec Assocs, Inc, 57-58; comput consult, 58-59; pres, Appl Data Res, Inc, 59-63 & Comput Mgt Corp, 63-65; tech dir, Mgt Info Systs, Inc, 65-69; exec vpres, Mainstem, Inc, 65-78; exec vpres, K-Squared Systs, Inc, 78-84; PRES, MAINTENANCE DATABASE SYSTS, INC, 85-, DIR, PARHAM GROUP, 88- *Mem:* Am Pub Work Asn. *Res:* Utility concept of computer problem solving; application of digital computing systems to the solutions of commercial and scientific problems; automatic programming procedures for digital computers; application of mainframe, mini- & personal computers to maintenance management applications. *Mailing Add:* 148 Library Pl Princeton NJ 08540

KAUFFMAN, ERLE GALEN, b Washington, DC, Feb 9, 33; m 56; c 3. PALEOBIOLOGY, PALEOECOLOGY. *Educ:* Univ Mich, BS, 55, MS, 56, PhD(geol, paleont, stratig), 61. *Hon Degrees:* MS, Oxford Univ, 70. *Prof Exp:* Asst cur, US Nat Mus, Smithsonian Inst 60-61, assoc cur, 61-67, cur dept paleobiol, 67-; AT GEOL SCI STOP, UNIV COLO, BOULDER. *Concurrent Pos:* Lectr, George Washington Univ, 63-64, adj prof, 65-; NSF res grant, 63-71; Am Geol Inst vis lectr, 65-66; Smithsonian Res Found grants, 65-; Paleont Soc rep, Comt Earth Sci, Nat Res Coun-Nat Acad Sci, 65-71; vis prof, Oxford Univ, 70-71, Univ Tübingen, 74 & Univ Colo, 76-78; res assoc, Mus Paleontol, Univ Mich, 75-; adj prof, Univ Colo, 76- *Mem:* AAAS; Brit Palaeont Asn; Malacol Soc London; Int Palaeont Union; Paleont Soc. *Res:* Systematics, evolution and paleoecology of Mesozoic-Cenozoic Mollusca; ecology of Recent Mollusca; Mesozoic-Cenozoic stratigraphy, biostratigraphy and sedimentation. *Mailing Add:* Geol Box 250 Univ Colo Boulder Boulder CO 80309-0250

KAUFFMAN, FREDERICK C, b Chicago, Ill, July 9, 36; m 61; c 2. BIOCHEMISTRY, PHARMACOLOGY. *Educ:* Knox Col, Ill, BA, 58; Univ Ill, Chicago, PhD(pharmacol), 65. *Prof Exp:* Asst prof & assoc pharmacol, SU NY, Buffalo, 67-74; assoc prof, 74-78, prof pharmacol, Sch Med, Univ MD, Baltimore, 78-88; DISTINGUISHED PROF, PHARMACOL TOXICOL, RUTGERS UNIV, NEW BRUNSWICK, NJ. *Concurrent Pos:* USPHS fel pharmacol, Wash Univ, 65-67. *Mem:* AAAS; Am Chem Soc; Soc Neurosci; Am Soc Pharmacol & Exp Therapeut; Am Soc Neurochem. *Res:* Biochemical pharmacology; neurochemistry. *Mailing Add:* Dept Pharmacol Univ Md Sch Med Baltimore MD 21201

KAUFFMAN, GEORGE BERNARD, b Philadelphia, Pa, Sept 4, 30; m 52, 69; c 5. INORGANIC CHEMISTRY, HISTORY OF SCIENCE. *Educ:* Univ Pa, BA, 51; Univ Fla, PhD(chem), 56. *Prof Exp:* Asst chem, Univ Fla, 51-55; instr, Univ Tex, 55-56; from asst prof to assoc prof, 56-66, PROF CHEM, CALIF STATE UNIV, FRESNO, 66- *Concurrent Pos:* Res corp grant, 55, 57, 59 & 69; NSF res grants, 60 & 67-69, Zurich, 63 & Berkeley, 76-77; NSF undergrad res partic dir, 72; Am Chem Soc Petrol Res Fund grant, 62 & 65; Am Philos Soc grant, 63 & 69; tour speaker, Am Chem Soc, 71; fel, John Simon Guggenheim Mem Found, 72, grant, 75; contributing ed, J Col Sci Teaching, 73-; co-ed, Topics Hist Chem, Lectures on Tape Series, Am Chem Soc, 75-78, ed, 78-81; vis scholar, Univ Calif, Berkeley, 76 & Univ Puget Sound, 78; contrib ed, The Hexagon, 80-, Polyhedron, 82-85, Indust Chemist, 85, J Chem Educ, 87- & Today's Chemist, 89-; Nat Endowment Humanities grant, 82-83; Strindberg Fel, Svenska Inst, 82-83. *Honors & Awards:* Lev Aleksandrovich Chugaev Jubilee Dipl & Bronze Medal, USSR Acad Sci, 76; Dexter Award in the Hist of Chem, 78. *Mem:* AAAS; Am Chem Soc (chmn div hist chem, 69); Hist Sci Soc; Soc Study Alchemy & Chem; Mensa. *Res:* Inorganic synthesis; stereochemistry; coordination compounds; chromatography separations; ion exchange; platinum metals; lanthanides; chemical education; unusual oxidation states; history of chemistry; biographies of chemists; translations of classics of chemistry. *Mailing Add:* Dept Chem Calif State Univ Fresno CA 93740-0070

KAUFFMAN, GLENN MONROE, b Goshen, Ind, Apr 8, 38. PHYSICAL ORGANIC CHEMISTRY. *Educ:* Goshen Col, BA, 61; Univ Pa, PhD(org chem), 66. *Prof Exp:* PROF CHEM, EASTERN MENNONITE COL, 65-, CHMN DEPT, 66- *Concurrent Pos:* Acad exten grant, Univ Fla & Eastern Mennonite Col, 68-70; Res Corp res grant, Eastern Mennonite Col, 68-69; res fel, Univ Fla, 75-76. *Mem:* AAAS; Am Chem Soc. *Res:* Conformational analysis of cyclopentane compounds; mechanisms of epoxidation reactions and ring-opening reaction of epoxides; hydrogen-deuterium exchange of pyridine-N-oxides. *Mailing Add:* Dept Chem Eastern Mennonite Col Harrisonburg VA 22801

KAUFFMAN, HAROLD, b West Liberty, Ohio, Apr 23, 39; m 61; c 3. PHYTOPATHOLOGY, AGRONOMY. *Educ:* Goshen Col, BS, 61; Mich State Univ, PhD(plant path), 67. *Prof Exp:* plant pathologist, Int Rice Res Inst, 67-81, joint coordr, Int Rice Testing Prog, 75-81; DIR, INT SOYBEAN PROG, UNIV ILL, URBANA, 81- *Mem:* Am Phytopath Soc. *Res:* International agriculture; bacterial diseases of rice plants; crop improvement. *Mailing Add:* Int Soybean Prog 113 Mumford Hall Univ Ill Urbana IL 61801

KAUFFMAN, JAMES FRANK, b St Joseph, Mo, Jan 29, 37; m 64; c 2. ELECTRICAL ENGINEERING. *Educ:* Univ Mo-Rolla, BS, 60; Univ Ill, Urbana, MS, 64; NC State Univ, PhD(elec eng), 70. *Prof Exp:* Engr airborne radar, Westinghouse Elec Corp, 60-62; res asst lens antennas, Antenna lab, Univ Ill, Urbana, 64-65; sr engr, Electronics Res Lab, Corning Glass Works, 67-70; from asst prof to assoc prof, 70-89, PROF ELEC ENG, NC STATE UNIV, 89- *Mem:* Inst Elec & Electronics Engrs; Sigma Xi. *Res:* Electromagnetics, especially antennas and microwave transmission. *Mailing Add:* 6932 Valley Lake Dr Raleigh NC 27612

KAUFFMAN, JOEL MERVIN, b Philadelphia, Pa, Jan 3, 37; m 66, 81; c 2. LASER DYES, FLUORESCENT TAGS. *Educ:* Philadelphia Col Pharm, BS, 58; Mass Inst Technol, PhD(org chem), 63. *Prof Exp:* Chemist, Reaction Motors Div, Thiokol Chem Corp, 63-64; USPHS fel antiradiation drugs, Mass

Col Pharm, 64-66; res chemist, ICI Am Inc, Mass, 66-69; res & develop dir, Pilot Chem Div, New Eng Nuclear Corp, Watertown, 69-76; res assoc, Mass Col Pharm, 77-79; asst prof, 79-85, ASSOC PROF CHEM, PHILADELPHIA COL PHARM & SCI, 85- Concurrent Pos: Consult, Mass Div, Am Automobile Asn, 64-; consult liquid scintillation counting, 76- & laser dyes, 82- Mem: Am Chem Soc; Am Asn Univ Professors. Res: Synthesis of fluors, fluorescent dyes, laser dyes, scintillators, antiallergenic drugs, antimalarials, anticancer drugs and other medicinal chemicals; laser dyes; scintillators; blocked amino acids; heterocyclics; vinyl monomers; photochemical reactions; terpenes; peptides; boron cage compounds; antiradiation-anticancer drugs; radiation sensitizers; plasticizers; fatty acid derivatives; formulation of liquid scintillators, solubilizers, decontaminants. Mailing Add: Philadelphia Col Pharm & Sci 43 St & Woodland Ave Philadelphia PA 19104-4495

KAUFFMAN, JOHN W, b Washington, DC, Mar 28, 25; m 59; c 1. BIOPHYSICS. Educ: George Washington Univ, BS, 47; Univ Md, MS, 49; Univ Ill, PhD, 55. Prof Exp: Res physicist, US Naval Res Lab, 48-50; asst, Univ Ill, 50-55; prof biomed eng, Northwestern Univ, 55-; RETIRED. Res: Biomaterials. Mailing Add: PO Box 218 Green Valley AZ 85622

KAUFFMAN, LEON A, b Philadelphia, Pa, July 26, 34; m 69; c 2. PULMONARY DISEASES, INTERNAL MEDICINE. Educ: Temple Univ, AB, 57, MD, 61; Am Bd Internal Med, dipl, 73, cert pulmonary med, 78. Prof Exp: Resident path, SDiv, Einstein Med Ctr, Philadelphia, 62-63; resident internal med, Hahnemann Med Col & Hosp, 63-65, fel pulmonary med & pulmonary physiol, 65-66, instr med, 66-68, sr instr & dir pulmonary function lab, 68-70, asst dir pulmonary dis div & dir respiratory intensive care unit, 69-73, asst prof med, 70-77; med dir sect respiratory ther, St Agnes Hosp, 73-78; chmn, Div Pulmonary Med, Metrop Hosp, Philadelphia, 73-83; ASSOC PROF MED, HAHNEMANN UNIV, 77- Concurrent Pos: Pa Thoracic Soc fel pulmonary physiol & clin chest dis, 65-66; clin asst pulmonary med, Hahnemann Div, Philadelphia Gen Hosp, Pa, 66-78; mem ad hoc comt to evaluate med care in state tuberc hosp syst, Pa, 67; mem fel & res comt, Pa Thoracic Soc, 68-74; pulmonary consult, Shock & Trauma Unit, Hahnemann Med Col & Hosp, 70-82; attend pulmonary med, St Agnes Hosp, 73- Mem: AMA; Am Thoracic Soc; fel Am Col Physicians; Am Soc Internal Med. Res: Respiratory intensive care and respiratory failure in man; design of systems for delivery of care and treatment of repiratory failure; respiratory therapy; clinical pulmonary physiology. Mailing Add: 1930 Pine St Philadelphia PA 19103

KAUFFMAN, MARVIN EARL, b Lancaster, Pa, Aug 31, 33; m 53; c 7. GEOLOGY. Educ: Franklin & Marshall Col, BS, 55; Northwestern Univ, MS, 57; Princeton Univ, PhD, 60. Prof Exp: Asst geologist, Alaskan Br, US Geol Surv, 53; asst geologist, Bethlehem Steel Co, 53-55; prof geol, Franklin & Marshall Col, 59-88; EXEC DIR, LEARNING CTR APPL ENVIRON TECHNOL, 85- Concurrent Pos: NSF sci fel, State Univ Utrecht, 65-66; consult geologist various co; pres, Yellowstone-Bighorn Res Asn. Mem: Nat Asn Geol Teachers (past-pres); Int Asn Sedimentologists; Geol Soc Am; Am Inst Prof Geol; Am Asn Petrol Geol; AAAS. Res: Structure and stratigraphy of the Garnet Range and Marine Jurassic of western Montana; cambrian stratigraphy of southeastern Pennsylvania. Mailing Add: RE Wright Assoc Inc 3240 Schoolhouse Rd Middletown PA 17057-3595

KAUFFMAN, RALPH EZRA, DRUG METABOLISM, PHARMACOKINETICS. Educ: Univ Kans, MD, 65. Prof Exp: PROF PEDIAT & PHARMACOL, SCH MED, WAYNE STATE UNIV, 82- Mailing Add: Dept Pediat & Pharmacol Wayne State Univ Childrens Hosp Mich Detroit MI 48201

KAUFFMAN, RAYMOND F, b Dayton, Ohio, Aug 20, 52; m 74; c 2. CARDIOVASCULAR PHARMACOLOGY, ATHEROSCLEROSIS. Educ: Univ Dayton, BS, 73; Univ Wis-Madison, PhD(biochem), 78. Prof Exp: Postdoctoral fel biochem, Enzyme Inst, Univ Wis, 79; postdoctoral fel biochem, Hormel Inst, Univ Minn, 79-81; sr pharmacologist cardiovasc res, 81-86, RES SCIENTIST CARDIOVASC RES, LILLY RES LABS, ELI LILLY & CO, 87- Mem: Am Chem Soc; Am Soc Pharmacol & Exp Therapeut; Sigma Xi; Am Heart Asn Arteriosclerosis Coun. Res: Vascular occlusive disorders, including atherosclerosis and chronic restenosis following balloon angioplasty; discovery of new mechanisms/drugs for prevention of vascular proliferative disorders and for modulating serum lipoproteins. Mailing Add: Cardiovasc Res Lilly Corp Ctr Eli Lilly & Co Indianapolis IN 46285

KAUFFMAN, ROBERT GILLER, b St Joseph, Mo, Dec 29, 32; m 55; c 2. MEAT SCIENCE. Educ: Iowa State Univ, BS, 54; Univ Wis, MS, 58, PhD(animal sci), 61. Prof Exp: Asst prof animal sci, Univ Ill, 61-66; prof animal sci, 66-81, PROF MEAT & ANIMAL SCI, UNIV WIS-MADISON, 81- Mem: Am Soc Animal Sci; Am Meat Sci Asn; Inst Food Technol. Res: Lipid transport in striated muscle; composition of meat animals and carcasses. Mailing Add: Animal Sci/270 Meat Sci Lab Univ Wis Madison WI 53706

KAUFFMAN, SHIRLEY LOUISE, b Grand Junction, Colo, Sept 10, 24. PATHOLOGY. Educ: Univ Chicago, BS, 46, MS, 48; Univ Kans, MD, 55. Prof Exp: Asst path, Med Col, Cornell Univ, 55-57; Nat Cancer Inst trainee path, Francis Delafield Hosp, 57-59; instr, Albert Einstein Col Med, 59-60; fel, 60, from asst prof to assoc prof, 61-70, PROF PATH, SU NY DOWNSTATE MED CTR, 70- Concurrent Pos: Provisional asst pathologist, NY Hosp, 55-57; vis prof, Dept Anat, Univ Berne, 69; Path Inst Rikshospitalet, Oslo, Norway, 76- & Inst Cancer Res Royal Marsden, Sutton Surrey, UK, 77. Mem: Am Soc Exp Path; NY Acad Sci; Int Soc Stereology; Am Asn Path & Bact. Res: Mammalian embryogenesis; lung morphometry; cell differentiation and proliferation; neoplasia. Mailing Add: Dept Path SU NY Health Sci Ctr 450 Clarkson Ave Brooklyn NY 11203

KAUFFMAN, STUART ALAN, b Sacramento, Calif, Sept 28, 39; m 67; c 1. MEDICINE, THEORETICAL BIOLOGY. Educ: Dartmouth Col, BA, 61; Oxford Univ, BA, 63; Univ Calif, San Francisco, MD, 68. Prof Exp: Vis scientist, Mass Inst Technol, 67-68; intern, Cincinnati Gen Hosp, 68-69; asst prof theoret biol, Univ Chicago, 69-73, asst prof med, 70-73; res assoc, Lab Theoret Biol, Nat Cancer Inst, 73-75; assoc prof biochem-biol, 75-81, ASSOC PROF BIOCHEM & BIOPHYS, UNIV PA, 81- Concurrent Pos: Fel genetics, Univ Cincinnati, 68-69. Honors & Awards: Norbert Wiener Gold Medal, Am Soc Cybernet, 70. Mem: Philos Sci Asn. Res: Theory of organization of eukaryotic gene regulation networks; control of DNA synthesis. Mailing Add: Dept Biochem 411 Anat Chem G3 Univ Pa Philadelphia PA 19104

KAUFFMANN, STEVEN KENNETH, nuclear & particle physics, for more information see previous edition

KAUFMAN, ALBERT IRVING, b New York, NY, July 22, 38; m 62; c 2. MEDICAL PHYSIOLOGY, EPITHELIAL TRANSPORT. Educ: Cooper Union, BEE, 61; Drexel Inst Technol, MS, 62; Temple Univ, PhD(physiol), 68. Prof Exp: Instr, 66-69, asst prof, 69-80, ASSOC PROF PHYSIOL, SUNY DOWNSTATE MED CTR, 80-, ASST DEAN, 90- Concurrent Pos: Vis prof, Tokyo Metrop Inst Geront, 75-76. Mem: Sigma Xi; NY Acad Sci; AAAS; Am Physiolog Soc. Res: Transepithelial water and electrolyte movement. Mailing Add: SUNY Health Sci Ctr 450 Clarkson Ave Box 31 Brooklyn NY 11203

KAUFMAN, ALLAN N, b Chicago, Ill, July 21, 27; m 57; c 2. PLASMA PHYSICS. Educ: Univ Chicago, PhD(physics), 53. Prof Exp: Assoc, 65-67, PROF PHYSICS, UNIV CALIF, BERKELEY, 67-, STAFF PHYSICIST, LAWRENCE BERKELEY LAB, 65- Concurrent Pos: Chmn, Div Plasma Physics, Am Phys Soc, 81-82; sr fel, Goddard Space Flight Ctr, 67. Mem: Fel Am Phys Soc. Res: Basic plasma physics theory. Mailing Add: Dept Physics Univ Calif Berkeley CA 94720

KAUFMAN, ALVIN B(ERYL), b Jacksonville, Fla, Oct 9, 17; m 48; c 2. ELECTRICAL ENGINEERING. Educ: Los Angeles City Col, AA, 38. Prof Exp: Res analyst, Douglas Aircraft Co, Calif, 39-52; group engr, Northrop Aircraft Corp, 52-55; chief develop engr, Arnoux Corp, 55-58; head, Mat & Devices Res Sect, Res & Analysis Dept, Litton Industs, 58-65, mem tech staff, Litton Systs Div, 65-70; sr mil systs engr, Lockheed Aircraft Corp, 70-74; proj engr, Inet Div, Teledyne Corp, 74-78; mem tech staff, Hughes Aircraft Co, 78-81; WRITING, CONSULT & LECTR, 81- Res: Development of instrumentation equipment and systems; development and test of night vision systems, radar and uninterruptible power supply systems; technical writing; nuclear effects on guidance and control systems. Mailing Add: 22420 Philiprimm St Woodland Hills CA 91367

KAUFMAN, ARNOLD, b Nuremberg, Ger, Dec 23, 28; US citizen; m 52; c 3. RESEARCH ADMINISTRATION, BIOCHEMICAL ENGINEERING. Educ: Columbia Univ, BS, 50, MA, 60. Prof Exp: Chem engr, Merck Sharp & Dohme Res Lab, 50-58, group leader chem eng res & develop, 58-62, res assoc, 62-64, sect mgr, 64-66, sr res fel, 66-69, mgr eng develop & process automation, 69-71, assoc dir chem eng res & develop, 71-77, dir, chem eng res & develop, 77-81, SR DIR RES PLANNING & ANALYSIS, MERCK SHARP & DOHME RES LABS, MERCK & CO, INC, RAHWAY, 81- Mem: Am Inst Chem Engrs; Am Chem Soc. Res: Processes to manufacture organic chemical and naturally derived products for pharmaceutical use; biological development; heterogeneous catalysis; separations techniques; solids processing; process control using digital computers; fermentation technology; production of vaccines; planning and analysis of research and development activities; chemical engineering. Mailing Add: 2269 Stocker Lane Scotch Plains NJ 07076-1299

KAUFMAN, BERNARD, b Chicago, Ill, Aug 17, 32; m 56; c 2. BIOCHEMISTRY. Educ: Univ Ill, BSc, 54, MSc, 56; Ind Univ, PhD(microbiol), 61. Prof Exp: Asst microbiol, Univ Ill, 54-56; asst, Ind Univ, 56-61; Arthritis & Rheumatism Found fel, Univ Mich, Ann Arbor, 61-64; lectr bot, 64-66; asst prof biochem, Johns Hopkins Univ, 66-68; ASSOC PROF BIOCHEM, MED CTR, DUKE UNIV, 68- Mem: AAAS; Am Soc Microbiol. Res: Elucidation of the reactions concerned in the synthesis of brain gangliosides; relationship of axonal gangliosides to myelination; chemistry and biosynthesis of cell surface glycolipids and glycoproteins. Mailing Add: Dept Biochem Box 3711 Duke Univ Med Ctr Durham NC 27710

KAUFMAN, BERNARD TOBIAS, b Richmond, Va, Nov 30, 27; m 57; c 3. BIOCHEMISTRY. Educ: Univ Va, BS & MS; Univ Calif, Los Angeles, PhD(biochem), 57. Prof Exp: Res asst biochem, Univ Va, 48-50; Am Cancer Soc fel, Brandeis Univ, 57-60; res scientist biochem, 60-67, chief sect vitamin metab, 67-74, CHIEF NUTRIT BIOCHEM, LAB CELLULAR & DEVELOP BIOL, NAT INST DIABETES, DIGESTIVE & KIDNEY DIS, NIH, 74- Mem: Am Soc Biol Chem; AAAS; Am Chem Soc; Sigma Xi. Res: Purification and properties of folic acid enzymes; characterization of naturally occurring folic acid derivatives; metabolism of folic acid in disease; anti-fol drugs. Mailing Add: Bldg 6 Rm B1-05 NIH Bethesda MD 20892

KAUFMAN, BORIS, b New York, NY, Jan 10, 26; m 49; c 1. MECHANICAL ENGINEERING. Educ: Univ Cincinnati, BS, 50, MechEng, 60; Ill Inst Technol, MS, 52. Prof Exp: Asst engr, Armour Res Found, Ill Inst Technol, 50-53, design engr, US Air Conditioning Corp, Ohio, 53-57, chief engr, 57-59; asst prof mech eng, Univ Idaho, 59-61; from asst prof to assoc prof, 61-69, PROF MECH ENG, CALIF STATE UNIV, SACRAMENTO, 69- Concurrent Pos: Co-dir, Bio-Eng Sect, Cardio-Pulmonary Div, Sutter Hosps Med Res Found, Sacramento, 62- Mem: Am Soc Heating, Refrig & Air-Conditioning Engrs. Res: Air conditioning theory and design; bio-engineering aspects of prosthetic heart devices. Mailing Add: Dept Mech Eng Calif State Univ 6000 J St Sacramento CA 95819

KAUFMAN, C(HARLES) W(ESLEY), b Thomas, WVa, Nov 26, 11; m 35; c 1. CHEMICAL ENGINEERING. *Educ:* Wash & Lee Univ, BS, 33; Univ Ariz, MS, 63, PhD, 67. *Prof Exp:* Chemist, Nat Fruit Prod Corp, 33-34; chem engr, Nat Canners Asn, 35-39; lab mgr food technol, Gen Foods Corp, 39-43, dir res, 44-48, vpres, 48-50, dir, 49-50; dir res, Kraft Foods Corp, 50-51, vpres res & develop, 51-57; dir, Nat Dairy Prod Corp, 58-61; vpres, Foremost Dairies, 61-62; assoc prof dairy sci, Univ Ariz, 64-67; vpres res & develop, Mars, Inc, 67-70; CONSULT, CJC CONSULT, 70- *Concurrent Pos:* Consult, Off Qm Gen, US Army, 40-44; lectr NY Univ, 47; vpres, Res & Develop Assoc Food & Container Inst, 50-52; mem bd trustees, Shimer Col, 58-63; vpres admin, Pima Community Col, Tucson, 70-71, assoc fac math & statist, 78-90. *Mem:* Inst Food Technol; hon mem Packaging Inst (pres, 59-60); fel Am Inst Chemists. *Res:* Food technology and engineering; colloid chemistry; research administration. *Mailing Add:* 2601 Camino Valle Verde Tucson AZ 85715

KAUFMAN, CHARLES, b Brooklyn, NY, June 4, 37. THEORETICAL PHYSICS. *Educ:* Univ Wis, BS, 56; Pa State Univ, MS, 59, PhD(physics), 63. *Prof Exp:* Instr physics, Pa State Univ, 63-64; from asst prof to assoc prof, 64-81, PROF PHYSICS, UNIV RI, 81- *Concurrent Pos:* Physicist, US Naval Underwater Systs Ctr, 69-71; guest lectr, Univ Vienna, 71-72; consult, Raytheon Corp, 78-81; fel, Off Naval Res-Am Soc Eng Educ, Naval Underwater Systs Ctr, New London, 83-84; sr visitor, DAMTP, Univ Cambridge, 85-86. *Mem:* NY Acad Sci; Acoust Soc Am; Am Phys Soc; Am Asn Univ Professors. *Res:* Electrodynamics; quantum field theory; atomic and elementary particle physics; turbulence theory; quantum chaos. *Mailing Add:* Dept Physics Univ RI Kingston RI 02881-0817

KAUFMAN, CLEMENS MARCUS, b Moundridge, Kans, Aug 27, 09; m 41; c 3. SILVICULTURE, FOREST ECOLOGY. *Educ:* Bethel Col, Kans, AB, 36; Univ Minn, MS, 38, PhD(forestry), 43. *Prof Exp:* Asst, Div Forestry, Univ Minn, 39, asst forester, Agr Exten Serv, 40-42, asst, Cloquet Forest Exp Sta, 42-43; from asst res prof to assoc res prof forestry, NC State Col, 43-48, prof forest mgt, 48-51; dir sch forestry, 51-62, prof, 62-76, EMER PROF FORESTRY, UNIV FLA, 76- *Concurrent Pos:* Res prog develop, Col Forestry, Univ NY, 79-81. *Mem:* Soc Am Foresters; Ecol Soc Am. *Res:* Physiological ecology; growth of slash pine. *Mailing Add:* 13026 NW 49th Ave Gainesville FL 32606

KAUFMAN, DANIEL, b Washington, DC, Mar 8, 20; m 43; c 5. ORGANIC CHEMISTRY. *Educ:* Univ Md, MS, 41. *Prof Exp:* Asst, Univ Md, 41-42; chemist, US Bur Mines, Utah, 42-44, NC, 46-47; chemist, Manhattan Proj, Los Alamos, NMex, 44-46; res chemist, Nat Lead Co, 47-59; supvr inorg res, Res Div, Wyandotte Chem Corp, 59-68; res assoc, Kerr Mfg Co Div, Ritter Pfaudler Corp, 68-69, dir res & develop, Kerr Mfg Co Div, Sybron Corp, 69-71, vpres res & develop, 71-75; RES DIR, KAUFMAN DEVELOP CO, 75- *Concurrent Pos:* Mem nat adv bd biomat res, Clemson Univ; US mem, Fedn Dentaire Int Comn, 75-77. *Mem:* Am Chem Soc; Int Asn Dent Res; AAAS. *Res:* Hydrogenation and hydrogenolysis of furfural; production of elemental boron; hydrometallurgy of manganese ores; titanium chemistry; cyclopentadienyl metal compounds; catalysis and olefin polymerization; metallurgical chemistry; dental materials. *Mailing Add:* 2242 Newquist Ct Camarillo CA 93010

KAUFMAN, DAVID GORDON, b Jersey City, NJ, May 28, 43; m 66; c 2. PATHOLOGY, BIOCHEMISTRY. *Educ:* Reed Col, BA, 66; Wash Univ, MD, 68, PhD(exp path), 73. *Prof Exp:* Intern path, Barnes Hosp, Wash Univ, 68-69, resident, 69-70; res assoc carcinogenesis, Nat Cancer Inst, 70-73, res scientist, 73-75; assoc prof, 75-80, PROF PATH, UNIV NC, 80- *Concurrent Pos:* Mem, Path B Study Sect, NIH, 77-79, Chem Path Study Sect, 79-83; mem prototype explicit anal pesticides comt, Nat Acad Sci, 78-80; res career develop award, Nat Cancer Inst, 78-83, mem Cancer Ctr support rev comt, 84- *Mem:* Am Asn Cancer Res; Am Asn Pathologists; Am Soc Cell Biol; Am Col Toxicol; NY Acad Sci; Soc Toxicol. *Res:* Chemical carcinogenesis; eukaryotic DNA replication and repair; cell biology of respiratory tract and female genital tract tissues; toxicology. *Mailing Add:* Dept Path Sch Med Univ NC Chapel Hill NC 27514

KAUFMAN, DON ALLEN, b Wahoo, Nebr, Aug 4, 40; m 63; c 1. ORGANIC CHEMISTRY. *Educ:* Univ Nebr, BS, 61; Univ Colo, MBS, 65; Colo State Univ, PhD(chem), 69. *Prof Exp:* Instr gen sci, Omaha Pub Schs, Nebr, 61-63; instr chem, Chandler High Sch, Ariz, 63-64; PROF ORG CHEM, KEARNEY STATE COL, 69- *Concurrent Pos:* NSF fac prof develop grant, Univ Nebr-Lincoln, 77-78. *Mem:* Am Chem Soc. *Res:* Use of crown ethers in organic synthesis; stability of vinyl cations; pH of precipitation; synthesis and reactions of Bunte salts. *Mailing Add:* Dept Chem Kearney State Col Kearney NE 68849

KAUFMAN, DONALD BARRY, b Los Angeles, Calif, Aug 5, 37. IMMUNOLOGY, NEPHROLOGY. *Educ:* Univ Calif, MD, 63. *Prof Exp:* PROF MED, COL HUMAN MED, MICH STATE UNIV, 81- *Mailing Add:* Dept Pediat & Human Develop Mich State Univ East Lansing MI 48824

KAUFMAN, DONALD DEVERE, b Wooster, Ohio, Dec 2, 33; m 57. SOIL MICROBIOLOGY, AGRICULTURAL CHEMISTRY. *Educ:* Kent State Univ, BA, 55, MA, 58; Ohio State Univ, PhD(plant path), 62. *Prof Exp:* Res technician, Plant Path, Ohio Agr Exp Sta, 56-57, res asst, 58-62; soil microbiologist pesticides, Plant Indust Sta, 62-73, soil microbiologist, Agr Environ Qual Inst, Beltsville Agr Res Ctr W, Sci & Educ Admin-Agr Res, USDA, 73-84, RES LEADER & SOIL MICROBIOLOGIST, SOIL-MICROBIAL SYSTS LAB, AGR ENVIRON QUAL INST, AGR RES SERV, USDA, 84- *Concurrent Pos:* Fulbright lectr soil microbiol, Khonkaen Univ, Thailand, 67-68. *Mem:* AAAS; Am Phytopath Soc; Am Soc Microbiol; Am Chem Soc; Weed Sci Soc Am. *Res:* Microbial decomposition of pesticides; effects of pesticides on soil microorganisms; soil microbiology of root diseases. *Mailing Add:* Agr Environ Qual Inst USDA Agric Res Serv Barc-East Beltsville MD 20705

KAUFMAN, DONALD WAYNE, b Abilene, Tex, June 7, 43; m 67; c 1. MAMMALIAN ECOLOGY, EVOLUTIONARY BIOLOGY. *Educ:* Ft Hays Kans State Col, BS, 65, MS, 67; Univ Ga, PhD(zool), 72. *Prof Exp:* Fel genetics, Univ Tex, 71-73; vis scientist, Savannah River Ecology Lab, Aiken, SC, 73-74; asst prof zool, Univ Ark, 74-75; asst prof biol, SUNY, Binghamton, 75-77; dir pop biol & physiol ecol prog, NSF, 77-80; asst prof, 80-84, ASSOC PROF BIOL, KANS STATE UNIV, 84- *Concurrent Pos:* Mem rev panel, Environ Protection Agency, 81-85; assoc dir, Konza Prairie Res Natural Area, 81-85, actg dir, 85-86, dir, 90; res grant, NSF, 83-; mem, External Oversight Comt Pop Biol & Ecol Progs, NSF, 84; Kans Nongame Wildlife Adv Coun, 85-88; Kans Natural & Sci Areas Adv Bd, 85-88; proj dir, Konza Prairie Long-term Ecol Res Prog, 85-90; mem, bd dirs, Am Soc Mammalogists, 89- *Honors & Awards:* Am Soc Mammalogists Award, 72. *Mem:* Am Soc Mammalogists; Ecol Soc Am; AAAS; Soc Study Evolution; Am Inst Biol Scientists; Wildlife Soc. *Res:* Ecology of rodents, evolutionary ecology, ecological genetics; grassland ecology; ecological effects of disturbances in grasslands; physical impacts of mammals on grassland ecosystems; mammalian herbivory; plant-mammal interactions. *Mailing Add:* Div Biol Ackert Hall Kans State Univ Manhattan KS 66506

KAUFMAN, EDWARD GODFREY, b New York, NY, June 8, 19; m 38; c 3. DENTISTRY. *Educ:* UCLA 39, NY Univ, DDs, 43. *Prof Exp:* From instr to assoc prof prosthodont, 46-66, from asst dean to assoc dean,69-84, PROF PROSTHODONT, COL DENT, NY UNIV, 67-, CHMN DEPT, 74-, DEAN, 84- *Concurrent Pos:* Pvt practice, 46-; res assoc, Mat Res Lab, Murry & Leone Guggenheim Inst Dent Res, 56-; Williams Ref Corp grant, 64-; pvt pract, 46-; Gordon Res Found grant, 57-60; consult dent asst training prog, NIH, 60-, co-prin investr, NIH Grant, 66-71; consult outpatient clin USPHS, 66- *Mem:* AAAS; Am Dent Asn; Am Soc Metals; Am Ceramic Soc; Sigma Xi. *Res:* Clinical and laboratory research in fields of dental materials, stress analysis and applied technology. *Mailing Add:* 98 Cutter Mill Rd Great Neck NY 11021

KAUFMAN, ELAINE ELKINS, b Cincinnati, Ohio, June 23, 23; m 48; c 3. BIOSYNTHESIS. *Educ:* Wellesley Col, BA, 45; Duke Univ, PhD(biochem), 49. *Prof Exp:* Fel, USPHS, 47-48; res chemist, NIMH, 65-67 & NIH, 67-69; RES CHEMIST, NIMH, 70- *Mem:* Sigma Xi; Am Soc Neurochem; Int Soc Neurochem; Am Soc Biochem & Molecular Biol. *Res:* Neuronal-astroglial interactions and interdependence; biosynthesis and degradation of the neuromodulator-hydroxybutyrate; biochemical basis for the physiological effects of -hydroxybutyrate. *Mailing Add:* NIH Bldg 36 Rm 1A21 9000 Rockville Pike Bethesda MD 20892

KAUFMAN, ERNEST D, b Cologne, Germany, Sept 21, 31; US citizen; m 59; c 3. PHYSICAL CHEMISTRY. *Educ:* Ill Inst Technol, BS, 53; Loyola Univ, Ill, MS, 58, PhD(phys chem), 62. *Prof Exp:* Res chemist, Dearborn Chem Co, Ill, 53-57; sr engr, Cook Elec Co, 57-61; lectr phys sci, Roosevelt Univ, 61-62; from asst prof to assoc prof chem, St Mary's Col, Minn, 62-69; res chemist, Am Cyanamid Co, 69-74, proj leader, 74-77, proj mgr, 77-78, dir, Bradford Tech Ctr, Cyanamid Int, UK, 78-80, tech dir, Cyanamid BV, Neth, 80-82, dir, Chem Technol Assessment, 82-85, dir, New Prod Develop, 85-89, TECH DIR, INT CHEM DIV, CYANAMID, 90- *Concurrent Pos:* Lectr phys chem, Mundelein Col, 61-62; Fulbright lectr, Univ Ceylon, 66-67. *Mem:* Am Chem Soc; Indust Mkt Res Asn; Indian Chem Soc; Soc Chem Indust; Royal Dutch Chem Soc; Licensing Executives Soc. *Res:* Polymerization kinetics; process development; sodium atom reactions; protein conformation; rotation of collagen model compounds; polymer-cellulose interactions. *Mailing Add:* Cyanamid Int Chem Div One Cyanamid Plaza Wayne NJ 07470

KAUFMAN, FRANK B, b June 23, 43; US citizen. PHYSICAL CHEMISTRY, ORGANIC CHEMISTRY. *Educ:* Univ Rochester, BS, 65; Johns Hopkins Univ, PhD(chem), 71. *Prof Exp:* NIH fel phys chem, Royal Inst Great Brit, 70-72; vis prof, Univ Ill, 72-73; MEM RES STAFF PHYS CHEM, T J WATSON RES CTR, IBM CORP, 73- *Mem:* Am Chem Soc; NY Acad Sci. *Res:* Design, synthesis and properties of new monomeric and polymeric materials with novel electronic properties. *Mailing Add:* T J Watson Res Ctr IBM Corp Yorktown Heights NY 10598

KAUFMAN, GLENNIS ANN, b Deshler, Nebr, Nov 13, 47; m 67; c 1. MATING SYSTEMS, SOCIAL ORGANIZATION. *Educ:* Kans State Univ, BS, 84, PhD(biol), 90. *Prof Exp:* Res asst, Savannah River Ecol Lab, 68-71 & 73-74; res asst zool, Univ Tex, 71-73; res asst biol, 81-84, grad asst, 84-90, INSTR GEN BIOL, KANS STATE UNIV, 91-, ASST SCIENTIST, 91- *Honors & Awards:* A Brazier Howell Award, Am Soc Mammalogists, 89. *Mem:* Am Soc Mammalogists; Am Behav Soc; Ecol Soc Am; Soc Study Evolution; Sigma Xi. *Res:* Mammalian behavior and population biology; community organization of small mammals in grasslands; biostatistics. *Mailing Add:* Div Biol Kans State Univ Ackert Hall Manhattan KS 66506

KAUFMAN, HAROLD ALEXANDER, b Brooklyn, NY, Jan 27, 33; m 56; c 2. ORGANIC CHEMISTRY, AGRICULTURAL. *Educ:* Brooklyn Col, BS, 55; Univ Pittsburgh, PhD(chem), 61. *Prof Exp:* Instr chem, Brooklyn Col, 55; res chemist, AMP, Inc, 55-56; mgr pesticides synthesis, screening & develop, Mobil Chem Co, 61-76; V PRES RES & DEVELOP, J T BAKER CHEM CO, 76- *Mem:* Am Chem Soc; The Chem Soc; AAAS. *Mailing Add:* 142 Fountain Ave Piscataway NJ 08854

KAUFMAN, HAROLD RICHARD, b Audubon, Iowa, Nov 24, 26; m 48; c 4. PLASMA PHYSICS, MECHANICAL ENGINEERING. *Educ:* Northwestern Univ, BS, 51; Colo State Univ, PhD(mech eng), 71. *Prof Exp:* Res engr, Nat Adv Comt Aeronaut, 51-57, sect head aircraft propulsion res, 57-58, res engr, NASA, 58-60, sect head, Space Propulsion Res, 60-64, br chief, 64-67, asst div chief, 67-74; prof mech eng & physics, 74-81, prof physics & chmn dept, 81-84, EMER PROF, COLO STATE UNIV, 84-; PRES, FRONT RANGE RES, 84- *Honors & Awards:* James H Wyld Award, Am Inst Aeronaut & Astronaut, 69; Medal Except Sci Achievement, NASA, 71. *Mem:* Assoc fel Am Inst Aeronaut & Astronaut; Am Phys Soc; Am Vacuum Soc. *Res:* Electric space propulsion; ion beams and ion sources. *Mailing Add:* 5920 Obenchain La Porte CO 80505

KAUFMAN, HERBERT EDWARD, b New York, NY, Sept 28, 31; m 77; c 3. OPHTHALMOLOGY. *Educ:* Princeton Univ, AB, 52; Harvard Med Sch, MD, 56; Am Bd Ophthal, dipl, 63. *Prof Exp:* Head & lectr ophthal, Uveitis Lab, Mass Eye & Ear Infirmary, 59-62; prof ophthal & pharmacol & chmn Ophthal, Col Med, Univ Fla, 62-77; BOYD PROF OPHTHAL & PHARMACOL & HEAD OPHTHAL, LA STATE UNIV MED CTR, NEW ORLEANS, 78- *Concurrent Pos:* Dir outpatient clins, Univ Fla; bd dir, Eye Bank Asn Am; ed, Invest Ophthal; ed, Am J Ophthal & Chemotherapy, Metabolic Ophthal & Ann Ophthal; med dir, Eye & Ear Inst La, 79-84. *Honors & Awards:* Knapp Award, AMA, 63; Albion O Bernstein Award, NY State Med Soc, 63; Lions Int Humanitarian Award, 68; Conrad Berens Award, 75; Proctor Award, 78; Jackson Mem Lectr, 79; Pocklington Lectr, 79; Proctor Lectr, 81; Twentieth Annual Edwin B Dunphy Lectr, 83; First Annual Wohl Lectr Ophthal, 83; R Townley Paton Award, 83; G Victor Simpson Lectr, 84; Peter Kronfeld Mem Lectr, 84. *Mem:* AAAS; AMA; Asn Res Vision & Ophthal (secy-treas, 64-73, pres, 75); Am Asn Immunol; Am Fedn Clin Res; Am Acad Ophthal; Asn Univ Prof Ophthal; Am Soc Contemporary Ophthal; Int Soc Refractory Surg. *Mailing Add:* LSU Eye Ctr 2020 Gravier St Suite B New Orleans LA 70112

KAUFMAN, HERBERT S, b Salina, Kans, Apr 30, 35; m 63; c 3. MEDICINE, ALLERGY. *Educ:* Univ Kans, BA, 57; Baylor Univ, MD, 61; Am Bd Pediat, dipl & cert pediat allergy, 66. *Prof Exp:* Intern pediat, St Louis Childrens Hosp, Mo, 61-62, resident, 62-63; resident, Tex Childrens Hosp, 63-64; NIH fel allergy & immunol, 64-66, CLIN INSTR DERMAT, MED CTR, UNIV CALIF, SAN FRANCISCO, 66- *Concurrent Pos:* Res grant, Univ Calif, San Francisco, 66-; comt mem & course chmn, Continuing Educ Dept, Pac Presby Med Ctr, 66-, dir pediat allergy clin, 66-68; consult, Dept Rehab, State of Calif & Letterman Gen Hosp, 66-; chief allergy & immunol clin, Dept Pediat, Children's Hosp San Francisco, 68-71. *Mem:* AMA; Am Acad Allergy; Am Col Allergists; Brit Soc Allergy; Brit Soc Immunol. *Res:* Immunoglobin defects in allergic individuals; complement levels in allergic disease; organic components of mental illness. *Mailing Add:* 2352 Post San Francisco CA 94115

KAUFMAN, HERMAN S, polymer science, science education; deceased, see previous edition for last biography

KAUFMAN, HOWARD NORMAN, b Boston, Mass, Jan 2, 26; m 47; c 3. MECHANICAL ENGINEERING, TRIBOLOGY. *Educ:* Northeastern Univ, BS, 45; Carnegie-Mellon Univ, MS, 52. *Prof Exp:* Res fel engr mech eng & tribology, Westinghouse Sci & Technol Ctr, Westinghouse Elec Corp, 47-86; PVT CONSULT, 87- *Concurrent Pos:* Instr mech eng, Carnegie-Mellon Univ, 52-56 & Allegheny County Community Col, 66-69; lectr, Am Soc Lubrication Engrs, 82- *Honors & Awards:* Walter D Hodson Award, Am Soc Lubrication Engrs, 55. *Mem:* Am Soc Lubrication Engrs. *Res:* Research and development in the field of friction, wear, and lubrication mechanics encompassing the theory, test, and application of journal and thrust bearings, rolling contact bearings and seals. *Mailing Add:* 1233 Northwestern Dr Monroeville PA 15146

KAUFMAN, HYMAN, b Lachine, Que, Feb 2, 20; m 59. MATHEMATICS. *Educ:* McGill Univ, BSc, 41, MSc, 45, PhD(physics), 48. *Prof Exp:* Lectr math, McGill Univ, 41-48; fel & asst instr elec eng, Yale Univ, 48-49; geophysicist, Continental Oil Co, 49-51; engr, Lab, Fox Electronics, Inc, 51-52; prof math, McGill Univ, 52-80; RETIRED. *Mem:* Soc Indust & Appl Math; Soc Explor Geophys; Asn Comput Mach; Math Asn Am; Inst Elec & Electronics Engrs. *Res:* Applied mathematics in engineering. *Mailing Add:* 400 Stewart Apt 2211 Ottawa ON K1N 6L2 Can

KAUFMAN, IRVING, b Geinsheim, Ger, Jan 11, 25; US citizen; m 50; c 3. ELECTRICAL ENGINEERING. *Educ:* Vanderbilt Univ, BE, 45; Univ Ill, MS, 49, PhD(elec eng), 57. *Prof Exp:* Engr, RCA Victor, 45-48; asst elec eng, Univ Ill, 48-49, instr, 49-53, res assoc, 53-56; mem tech staff, Ramo-Wooldridge & Space Tech Labs, TRW, Inc, 57-64, head microwave res, 61-64; dir, Solid State Res Lab, 68-78, PROF ENG, ARIZ STATE UNIV, 65- *Concurrent Pos:* Fulbright sr res fel, Italy, 64-65 & 73-74; collaborating scientist, Consiglio Nazionale delle Ricerche, Florence, Italy, 73-74; vis prof, Univ Auckland, New Zealand, 74; liaison scientist, Off Naval Res, London, 78-80; distinguished res award, Grad Col, Ariz State Univ, 86-87. *Honors & Awards:* Ann Achievement Award Outstanding Contrib Elec Eng, Inst Elec & Electronics Engrs, 68. *Mem:* Fel Inst Elec & Electronics Engrs; Am Inst Physics; Sigma Xi. *Res:* Microwave electronics; electronic and optical device research; displays; non-destructive evaluation. *Mailing Add:* Dept Elec & Comput Eng Ariz State Univ Tempe AZ 85287-5706

KAUFMAN, JANICE NORTON, b Denver, Colo, June 22, 23. PSYCHIATRY. *Educ:* Univ Utah, BA, 48, MD, 51; Am Bd Psychiat & Neurol, dipl, 58. *Prof Exp:* Intern med, Strong Mem Hosp, Rochester, NY, 51-52, resident psychiat, 52-55; from instr to assoc prof, Univ Colo Med Ctr, Denver, 55-72; dir, 69-72, MEM FAC, DENVER INST PSYCHOANALYSIS, 72-; PROF PSYCHIAT, UNIV COLO MED CTR, DENVER, 72- *Concurrent Pos:* USPHS career teacher fel, 56-58; mem fac, Chicago Inst Psychoanalysis, 63- *Mem:* Fel Am Psychiat Asn; Am Psychoanal Asn. *Res:* Practice and teaching of psychoanalysis and psychiatry. *Mailing Add:* 1112 Santa Rufing Ct Solana Beach CA 92075

KAUFMAN, JOHN GILBERT, JR, b Baltimore, Md, Oct 14, 31; m 53; c 3. COMPUTERIZED MATERIAL SCIENCE DATA, FRACTURE MECHANICS. *Educ:* Carnegie Inst Technol, BSCE, 53, MS, 54; Carnegie Mellon Univ, MS, 75. *Prof Exp:* Mgr, eng properties, Aluminum Co Am, 54-83, technol develop, 84-89 & fabrication technol, 89-90; PRES, NAT MAT PROPERTIES DATA NETWORK, 85- & MAT PROPERTY DATABASES, 87- *Concurrent Pos:* Dir res & develop, Anaconda Aluminum Co, 80-83 & Arco Metals, Atlantic Ritchfield, 83-85; mem, Struct Comt, Nat Res Coun, 80-85 & Nat Mat Adv Bd Comt Appln Computers Mat Sci, 91-; chmn, Comt E24, Am Soc Testing & Mat, 80-84 & Comt E49, 86-88; mgr, Numeric & Mat Properties Serv, Am Soc Chem, 91- *Mem:* Fel Am Soc Testing & Mat; fel Am Soc Metals; Sigma Xi; Am Soc Mech Engrs; Comt Data Sci & Technol; Asn Comput Mach. *Res:* Materials development, notably aluminum alloys; fracture mechanics and applications; networking of materials databases. *Mailing Add:* 3662 Pevensey Dr Columbus OH 43220

KAUFMAN, JOSEPH J, b New Haven, Conn, Feb 10, 21; m 42; c 2. UROLOGY, SURGERY. *Educ:* Univ Calif, Los Angeles, BA, 42; Univ Calif, San Francisco, MD, 45; Univ Guadalajara, Mex, MD Hons, 80. *Prof Exp:* From asst prof to prof surg & urol, 66-76, CHIEF, DIV UROL, SCH MED, UNIV CALIF, LOS ANGELES, 70- *Concurrent Pos:* Hon mem fac, Sch Med, Univ Chile, 66. *Mem:* Am Urol Asn; Int Soc Urol; Soc Clin Urol; Soc Univ Urol; Am Asn Genito-Urinary Surg. *Res:* Renovascular hypertension; kidney transplantation; urological oncology. *Mailing Add:* Ctr Health Sci Univ Calif Los Angeles CA 90024

KAUFMAN, JOYCE J, b New York, NY, June 21, 29; m 48; c 1. QUANTUM CHEMISTRY, PSYCHOPHARMACOLOGY. *Educ:* Johns Hopkins Univ, BS, 49, MA, 59, PhD(chem, chem physics), 60; Sorbonne Univ, DES (theoret physics), 63. *Prof Exp:* Chemist, US Army Chem Ctr, Md, 49-52; res asst, Johns Hopkins Univ, 52-60; staff scientist, Res Inst Advan Studies, 60-62, head quantum chem group, 62-69; ASSOC PROF ANESTHESIOL, SCH MED & PRIN RES SCIENTIST CHEM, JOHNS HOPKINS UNIV, 69- *Concurrent Pos:* Vis staff mem, Ctr Appl Quantum Mech, France, 62; Soroptimist fel int study, 62; mem heavy ion sources comt, Nat Acad Sci, 73-75; US deleg, Int Atomic Energy Symp, Vienna, Austria, 64; mem corresp, Acad Europ Sci, Arts & Letters. *Honors & Awards:* Gold Medal, Martin Co, 64, 65 & 66; Dame Chevalier, Nat Ctr Sci Res, France, 69; Garvan Medal, Am Chem Soc, 74; Lucy Pickett Award, 75; Md Chemist Award, 74. *Mem:* Am Chem Soc; fel Am Phys Soc; fel Am Inst Chem; Int Soc Quantum Biol; Am Soc Pharmacol Exp Therapeut. *Res:* Physicochemistry and theory of drugs which affect the central nervous system; computer systems; experimental chemical physics; chemical effects of nuclear transformation; isotopic exchange reactions of boron hydrides; quantum chemistry. *Mailing Add:* Dept Chem Johns Hopkins Univ Baltimore MD 21218

KAUFMAN, KARL LINCOLN, b Attica, Ohio, Aug 18, 11; m 36; c 3. PHARMACY, MEDICINAL CHEMISTRY. *Educ:* Ohio State Univ, BSc, 33; Purdue Univ, PhD(pharm), 36. *Prof Exp:* Spec investr revision comt, US Pharmacopoeia, 36; from instr to asst prof pharm, Wash State Univ, 36-40; prof pharm & pharm chem & head dept, Med Col, Va, 40-49; exec officer & prof, Butler Univ, 49-52, dean col pharm, 52-76; state pharm dir, Ind Dept Ment Health, 76-85; EXEC DIR & CO-FOUNDER, IND SCI EDUC FUND, 85- *Concurrent Pos:* Indust consult, 41-; consult, Vet Admin, 60-; mem bd, Coun Aging, 84-; founder, exec dir, Sci Educ Found, Ind. *Mem:* Assoc Am Chem Soc; Am Pharmaceut Asn; Am Heart Asn (past pres); Int Asn Torch Clubs (past int pres); Sigma Xi. *Res:* Drug deterioration; drug analysis; professional ethics; product development and quality control; drug abuse and control. *Mailing Add:* 8616 W 10th St No 302 Indianapolis IN 46234

KAUFMAN, LARRY, b Brooklyn, NY, June 6, 31; m 55; c 3. THERMODYNAMICS, MATERIALS SCIENCE. *Educ:* Polytech Inst Brooklyn, BMetE, 52; Mass Inst Technol, ScD(metall), 55. *Prof Exp:* Mem res staff, Lincoln Lab, Mass Inst Technol, 55-58; sr metallurgist, Mfg Labs, Inc, 58-63, dir res, Manlabs Inc, 63-76, vpres, 76-84, pres, 84-91, PRIN SCIENTIST, MANLABS DIV, ALCAN ALUMINUM CORP, 91- *Concurrent Pos:* Ed-in-chief, Calphad, 77- *Honors & Awards:* Rossiter Raymond Award, Am Inst Mining, Metall & Petrol Engrs, 64. *Mem:* Am Soc Metals; Am Inst Mining, Metall & Petrol Engrs; Sigma Xi. *Res:* Kinetics; phase equilibria; high pressure and temperature; transformations; computer calculation of phase diagrams. *Mailing Add:* Alcan Aluminum Corp 21 Erie St Cambridge MA 02139

KAUFMAN, LEO, b New York, NY, Jan 20, 30; m 52; c 3. MEDICAL MYCOLOGY. *Educ:* Brooklyn Col, BS, 52; Univ Ky, MS, 55, PhD(bact), 59. *Prof Exp:* Instr bact, Univ Ky, 58-59; microbiologist med res, USPHS, 59-62, in-chg, Fungus Serol Lab, Mycol Unit, 63-67, chief fungus immunol br, Mycol Div, Ctr Dis Control, 67-90, ASST CHIEF MYCOTIC DIS BR, DIV BACT & MYCOTIC DIS, USPHS, 90- *Concurrent Pos:* Dir, Nat Ctr Fungal Serol; mem fac, Univ NC, Sch Pub Health, Ga State Univ & Emory Univ. *Honors & Awards:* Kimble Methodology Res Award, 74; Meridian Award, Med Mycol Soc Am, 84; Int Soc Human Animal Mycol Award, 85. *Mem:* Am Asn Immunol; Am Thoracic Soc; Am Soc Microbiol; Int Soc Human & Animal Mycol; Sigma Xi. *Res:* Immunological procedures for diagnosis of systemic fungus infections and for identification of fungal pathogens. *Mailing Add:* Ctr Dis Control G-11 Div Bact & Mycotic Dis Bldg 5 B-13 Atlanta GA 30333

KAUFMAN, LINDA, b Fall River, Mass, Mar 20, 47; m 81. NUMERICAL ANALYSIS, COMPUTER SCIENCE. *Educ:* Brown Univ, ScB, 69; Stanford Univ, MS, 71, PhD(comput sci), 73. *Prof Exp:* Asst prof comput sci, Univ Colo, Boulder, 73-76; MEM STAFF, BELL LABS, 76- *Concurrent Pos:* Vis lectr comput sci, Univ Aarhus, 73-74; prin investr, NSF Grant, 76- *Mem:* Asn Comput Mach; Soc Indust & Appl Math. *Res:* Development of algorithms in numerical linear algebra and function minimization. *Mailing Add:* AT&T Bell Labs 600 Mountain Ave Rm 2C461 Murray Hill NJ 07974

KAUFMAN, MARC P, PHYSIOLOGY & PSYCHOLOGY. *Educ:* Univ Miami, PhD(physiol psychol), 77. *Prof Exp:* Asst prof physiol, Health Sci Ctr, Univ Tex, Dallas, 80-87; assoc prof, 87-91, PROF INTERNAL HUMAN PHYSIOL, UNIV CALIF, DAVIS, 91- *Res:* neurol control of cardiovascular and respiratory system. *Mailing Add:* Div Cardiovasc Med Univ Calif TB-172 Davis CA 95616

KAUFMAN, MARTIN, b Boston, Mass, Dec 6, 40; m 68; c 2. HISTORY OF AMERICAN MEDICINE & PUBLIC HEALTH. *Educ:* Boston Univ, AB, 62; Univ Pittsburgh, MA, 64; Tulane Univ, PhD(hist), 69. *Prof Exp:* Instr hist, Worcester State Col, 68-69; from asst prof to assoc prof hist, 69-76, chairperson dept, 82-90, PROF HIST, WESTFIELD STATE COL, 76-

Concurrent Pos: Vis prof, Univ Vt Col Med, 73-74; dir, Inst Mass Studies, 80- *Mem:* Am Asn Hist Med; Hist Sci Soc; Orgn Am Historians; Nat Educ Asn. *Res:* History of American medicine and public health; unorthodox medicine in the 19th century; history of medical education; biographies of physicians and nurses throughout history. *Mailing Add:* Dept Hist Westfield State Col Westfield MA 01086

KAUFMAN, MAVIS ANDERSON, b Yonkers, NY, July 14, 19. NEUROPATHOLOGY. *Educ:* Radcliffe Col, AB, 41; NY Med Col, MD, 44. *Prof Exp:* Intern med & surg, Flower & Fifth Ave Hosp, 44-45; resident path, NY Postgrad Hosp, 45-46, resident med & surg, Northern Westchester Hosp, 46-47; resident internal med, Aultman Hosp, Canton, Ohio, 47-48; resident psychiat, Kings County Hosp, NY, 48-50; resident path, Vet Admin Hosp, 50-53; from instr to asst prof, 53-69, assoc, 55-56, ASSOC PROF NEUROPATH, COL PHYSICIANS & SURGEONS, COLUMBIA UNIV, 69- *Concurrent Pos:* Assoc res scientist, NY State Psychiat Inst, 56- *Mem:* Am Asn Neuropath; Am Acad Neurol. *Res:* Surgical and autopsy specimens of central nervous system; demyelinating and degenerative diseases; effects of psychopharmacologic agents; effects of aging. *Mailing Add:* Dept Neuropath NY State Psychiat Inst 722 W 168 St New York NY 10032

KAUFMAN, MIRON, b Oct 31, 50; Israeli citizen; m 75; c 2. POLYMER PHYSICS, SUPERCONDUCTORS. *Educ:* Tel Aviv Univ, Israel, BSc, 73, MSc, 77; Carnegie Mellon Univ, PhD(physics), 81. *Prof Exp:* Res physicist statist mech, Carnegie-Mellon Univ, 81-82; Bantrell fel surface physics, Mass Inst Tech, 83-85; asst prof, 85-89, ASSOC PROF, CLEVELAND STATE UNIV, 89- *Concurrent Pos:* Vis asst prof, Boston Univ, 85. *Mem:* Am Phys Soc. *Res:* Research activity on condensed matter and statistical physics focused on following topics: random magnets, liquid mixtures, spin (ISING-POTTS) models, percolation, superconductivity, polymers and fractals; research methods: scaling theory of critical phenomena, renormalization-group technique. *Mailing Add:* Physics Dept Cleveland OH 44115

KAUFMAN, MYRON JAY, b New York, NY, Mar 24, 37; m 67. PHYSICAL CHEMISTRY. *Educ:* Rensselaer Polytech Inst, BS, 58; Harvard Univ, MS, 63, PhD(chem physics), 65. *Prof Exp:* Trainee, Gen Elec Co, 58-59; res fel chem, Harvard Univ, 64-66; asst prof, Princeton Univ, 66-72; assoc prof, 72-78, PROF CHEM, EMORY UNIV, 78- *Mem:* Am Chem Soc. *Res:* Chemical kinetics; molecular beams; atmospheric chemistry; combustion; coal chemistry. *Mailing Add:* Dept Chem Emory Univ Atlanta GA 30322

KAUFMAN, NATHAN, b Lachine, Que, Aug 3, 15; m 46; c 5. PATHOLOGY. *Educ:* McGill Univ, BSc, 37, MD & CM, 41; Am Bd Path, dipl, 50. *Prof Exp:* Intern, Royal Victoria Hosp, Montreal, 41-42; resident path, Jewish Gen Hosp, 46-47; asst resident, Cleveland City Hosp, 47-48; from instr to assoc prof, Med Sch, Western Reserve Univ, 48-60; prof, Sch Med, Duke Univ, 60-67; prof & head dept, 67-79, EMER PROF PATH, QUEEN'S UNIV, ONT, 81-; SECY-TREAS, US-CAN DIV, INT ACAD PATH, 79- *Concurrent Pos:* Asst pathologist, Cleveland Metrop Gen Hosp, 48-52, pathologist in chg, 52-60; pathologist-in-chief, Kingston Gen Hosp, 67-79; mem grants comt path & morphol, Med Res Coun, 68-74, chmn, 71-74, mem coun, 71-77, exec, 71-74; consult, Lennox & Addington County Gen Hosp, Napanee, Ont & Hotel Dieu Hosp, 69-79 & Ont Cancer Treatment & Res Found, Kingston Clin, 71-79; mem grants panel, Nat Cancer Inst Can, 70-74; ed, Lab Invest, 72-75. *Mem:* Int Acad Path (vpres, 72-74, pres-elect, 74-76, pres, 76-78); Can Asn Path; Am Asn Path; Soc Exp Biol & Med; US & Can Acad Path (pres, 73-74, secy-treas, 79-). *Mailing Add:* US-Can Acad Path 3643 Walton Way Ext Augusta GA 30909

KAUFMAN, PAUL LEON, b New York, NY, Sept 16, 43; m 70; c 1. OPHTHALMOLOGY, GLAUCOMA. *Educ:* NY Univ, MD, 67. *Prof Exp:* Staff assoc, Nat Cancer Inst, 68-70; resident ophthal, Wash Univ, 70-73; res fel, Univ Uppsala, Sweden, 73-75; from asst prof to assoc prof, 75-83, PROF OPHTHAL, UNIV WIS-MADISON, 83- *Concurrent Pos:* Mem, Glaucoma Prog Planning Panel, Nat Adv Eye Coun, Nat Eye Inst, NIH, 80-82, Special Study Sects & Site Visit Teams, 80-82, Vis Sci A Study Sect, 82-85; prin investr, Nat Eye Inst Res Grants, 79-; consult, Retinal & Choroidal Dis Prog Planning Panel, Nat Adv Eye Coun, NIH, 77. *Honors & Awards:* Alcon Res Inst Award for Outstanding Contrib to Vision Res, 85. *Mem:* Asn Res Vision & Ophthal; Am Acad Ophtal. *Res:* Anatomy, physiology, and pharmacology of aqueous humor formation and drainage, as related to pathophysiology and treatment of glaucoma. *Mailing Add:* Clin Sci Ctr Univ Wis 600 Highland Ave Madison WI 53792

KAUFMAN, PETER BISHOP, b San Francisco, Calif, Feb 25, 28; m 58; c 2. PLANT PHYSIOLOGY. *Educ:* Cornell Univ, BS, 49; Univ Calif, PhD(bot), 54. *Prof Exp:* Res technician hort, Cornell Univ, 45-49; res technician, Shell Develop Co, 49; res technician & asst bot, Univ Calif, 49-54; Muellhaupt scholar, Ohio State Univ, 54; res assoc, 56-57, from instr to assoc prof bot, 56-73, cur, Bot Gardens, 57-73, PROF BOT, UNIV MICH, ANN ARBOR, 73-, PROF CELL & MOL BIOL, 83- *Concurrent Pos:* NSF grants, 59-61, 75 & 80-83; grants, Inst Plant Physiol, Univ Lund, 64-66, Am Cancer Soc, 68-71, Inst Environ Qual, Univ Mich, 71-72 & NASA, 79-88; vis prof cell, Molecular & Develop Biol, Univ Colo, Boulder, 74 & Nagoya Univ, Japan, 81; vis scientist, Int Rice Res Inst, Los Banos, Philippines, 81 & US Dept Agr, Beltsville, Md, 81, Univ Calgary, 85. *Mem:* Am Soc Plant Physiol; Soc Develop Biol; Bot Soc Am; fel Am Asn Advan Sci; Am Asn Cereal Chemists; Tissue Culture Asn; Am Soc Gravitational & Space Biol; Int Plant Growth Substances Asn. *Res:* Scanning electron microscopy, electron microprobe analysis and neutron activation analysis as related to silicification mechanisms in rice, oats, sugarcane and other grasses; studies on hormonal interactions and primary mode of action of gibberellin hormone regulation of stem elongation in grasses; mechanism of negative gravitropic response in grasses under NASA Space Biology programs; molecular biology of rice seed proteins, heat-shock proteins in rice; use of plant cell cultures for production of useful secondary compounds; development of life support systems for NASA space station and moon-mars bases. *Mailing Add:* Dept Biol Univ Mich Ann Arbor MI 48109

KAUFMAN, PRISCILLA C, b Shanghai, China, Oct 5, 30; US citizen; m 60; c 2. ORGANIC CHEMISTRY, RADIATION CHEMISTRY. *Educ:* MacMurray Col, BA, 53; Univ Chicago, PhD(chem), 58. *Prof Exp:* Res assoc chem, Brookhaven Nat Lab, 58-60, assoc chemist, 60-62; res chemist, Atlas Chem Indust, NJ, 63-66; vis asst prof chem, Elmhurst Col, 67. lectr, Rosary Col, 67-71; lectr, 71-77, INSTR CHEM, COL DUPAGE, 77- *Mem:* Am Chem Soc. *Res:* Radiolysis of organic compounds; radiation induced polymerizations. *Mailing Add:* Dept Chem Col Dupage Glen Ellyn IL 60137

KAUFMAN, RAYMOND, b Aug 30, 17; US citizen; m 42; c 3. EXPERIMENTAL PHYSICS. *Educ:* City Col NY, BS, 42, MS, 46; NY Univ, PhD(physics), 50. *Prof Exp:* Asst physicist labs, US Signal Corps, 42-43; tutor physics, City Col NY, 43-44, 47-49; asst physicist, Farrand Optical Co, Inc, 44-47, physicist, 49-58; dir res, Del Electronics Corp, Mt Vernon, 58-76, pres & chief exec officer, 76-82, chmn bd, 82-85; RETIRED. *Concurrent Pos:* Assoc prof, City Col NY. *Mem:* Am Phys Soc; Am Asn Physics Teachers. *Res:* Ionization potentials of molecules; electron optics; mass spectroscopy; ultrahigh vacuum techniques; infrared; high voltage phenomena. *Mailing Add:* 3755 Henry Hudson Pkwy Riverdale NY 10463

KAUFMAN, RAYMOND H, b Brooklyn, NY, Nov 24, 25; m 46; c 4. OBSTETRICS & GYNECOLOGY. *Educ:* Univ Md, MD, 48; Am Bd Obstet & Gynec, dipl. *Prof Exp:* Resident obstet & gynec, Beth Israel Hosp, New York, 48-53; from asst prof to assoc prof Obstet, Gynec & Path, 58-73, actg chmn dept obstet & gynec, 68-72, PROF PATH, OBSTET & GYNEC, & ERNST W BERTNER CHMN OBSTET & GYNEC, BAYLOR MED COL, 73- *Concurrent Pos:* Fel path, Methodist Hosp, Houston, 55-58. *Mem:* Am Col Obstet & Gynec; fel Am Col Surg. *Res:* Gynecologic pathology; cytopathology. *Mailing Add:* Baylor Col Med One Baylor Plaza Houston TX 77030

KAUFMAN, SAMUEL, b Toledo, Ohio, Jan 29, 13; m 41. PHYSICAL & ANALYTICAL CHEMISTRY, CHEMICAL EDUCATION. *Educ:* Univ Toledo, BEd, 37, BSc, 40, MSc, 47. *Prof Exp:* Control & prod supvr ceramics, Save Elec Corp, 37-38; control analyst, US Gypsum Co, 41; chemist, Engr Corps, US Army, 41-45; chemist, Nat Bur Standards, 45-48; chemist, 48-75, consult, US Naval Res Lab, 76-82; CONSULT, UNIV MD, 82- *Concurrent Pos:* Instr, Montgomery Jr Col, 46-47. *Mem:* Am Chem Soc; Sigma Xi. *Res:* Isopycnic ultracentrifugation; lubricant additives; water pollution abatement; nonaqueous micelle formation and solubilization; reactions of amines in nonaqueous media; analysis of fluorine-bearing silicates; concrete curing agents; nonaqueous titrations; carbon fiber composites. *Mailing Add:* 919 Hyde Rd Silver Spring MD 20902

KAUFMAN, SEYMOUR, b NY, Mar 13, 24; m 48; c 3. BIOCHEMISTRY. *Educ:* Univ Ill, BS, 45, MS, 46; Duke Univ, PhD(biochem), 49. *Prof Exp:* Res fel, Sch Med, NY Univ, 49, from instr to asst prof, 50-53; chief sect cellular regulatory mechanisms, 54-68, CHIEF LAB NEUROCHEM, NIMH, 68- *Concurrent Pos:* NSF travel award, Int Cong Biochem, Paris, 52. *Mem:* Nat Acad Sci; Am Chem Soc; Harvey Soc; Am Soc Neurochem; Int Soc Neurochem; Am Soc Biol Chemists; Am Acad Arts & Sci. *Res:* Mechanism of action of enzymes; intermediary metabolism of amino acids, phenylketonuria, neurotransmitter biosynthesis, tetrahydrobiopterin. *Mailing Add:* NIMH 36/3D-30 Bethesda MD 20205

KAUFMAN, SHELDON BERNARD, b Los Angeles, Calif, June 7, 29; m 60; c 2. NUCLEAR CHEMISTRY. *Educ:* Univ Chicago, MS, 51, PhD, 53. *Prof Exp:* Res assoc chem, Columbia Univ, 55-57; from instr to asst prof, Princeton Univ, 57-66; from assoc chemist to chemist, 66-86, SR PHYSICIST, ARGONNE NAT LAB, 86- *Mem:* Am Phys Soc; AAAS; Am Chem Soc. *Res:* Radiochemical studies of low and high energy nuclear reactions; hot-atom chemistry; tracer applications to inorganic chemistry; high-energy nuclear reactions; nuclear fission; pi-meson reactions; reactions of complex nuclei with energetic protons, pi-mesons, and heavy ions; nuclear fission. *Mailing Add:* 910 W Elm St Wheaton IL 60187

KAUFMAN, SIDNEY, b Passaic, NJ, Aug 10, 08; m 36; c 2. CRUSTAL STUDIES, GEOTHERMAL STUDIES. *Educ:* Cornell Univ, AB, 30, PhD(physics, math), 35. *Prof Exp:* Asst physics, Cornell Univ, 30-33, Coffin Found fel, 35; geophysicist, Shell Oil Co, 36-41, sr physicist, Shell Develop Co, 46-58, head geophys instrumentation dept, 58-61, sr res assoc, 61-65, asst to vpres, 65-74; PROF GEOPHYSICS, CORNELL UNIV, 74- *Concurrent Pos:* Prin physicist, Naval Res Labs, 46; consult, Adv Res Projs Agency, US Dept Defense, 61-73; mem geophys adv panel, Air Force Off Sci Res, 61-74, chmn, 64-66; mem comt seismol, Nat Acad Sci-Nat Res Coun, 66-71, 74-77; consult, Energy Res & Develop Admin, 75-; dir, Geothermal Resources Coun, 73-77. *Honors & Awards:* Gold Medal Award, Soc Explor Geophys, 83. *Mem:* Hon mem Soc Explor Geophys; Seismol Soc Am; Am Geophys Union; Sigma Xi; Europ Asn Explor Geophysicists. *Res:* Geophysical exploration; deep crustal seismic profiling; geothermal resource assessment. *Mailing Add:* Kimball Hall-Geol Cornell Univ Main Campus Ithaca NY 14853-1504

KAUFMAN, SOL, b New York, NY, Mar 2, 28; m 54; c 5. SIMULATION MODELING, EXPERT SYSTEMS. *Educ:* Wash Univ, AB, 51; Cornell Univ, PhD(math), 65. *Prof Exp:* Physicist, Nat Bur Standards, 51-52; systs analyst & asst head opers res dept, Cornell Aeronaut Lab/Calspan Corp, 53-62 & 65-73; coordr res & eval, Niagara Falls Community Ment Health Ctr, 73-74; cancer control network coordr, SUNY Buffalo, 74-79; mem staff, Falcon Res & Develop Co, 79-84; systs analyst, XMCO, Inc, 84-86; SYSTS ANALYST, ANALYSIS SIMULATION, INC, 86- *Concurrent Pos:* Lectr indust eng & social & preventive med, State Univ NY Buffalo, 70-82, res asst prof otolaryngol, 80-; independent consult, 90- *Mem:* Math Asn Am; Asn Comput Mach. *Res:* Expert system application to simulation modeling; discrete event simulation, statistical analysis of complex survey data; cancer epidemiology and outcome analysis; public systems research; aerosol transport and diffusion models. *Mailing Add:* 1201 Stolle Rd Elma NY 14059

KAUFMAN, STANLEY, b New York, NY, Oct 30, 41; m 64; c 2. MATERIALS SCIENCE. *Educ:* City Col NY, BS, 63; Brown Univ, PhD(chem), 70. *Prof Exp:* Res scientist, Uniroyal Res Ctr, 68-70; mem tech staff, 70-77, SUPVR CHEM, CHEM GROUP, AT&T BELL LABS, 77- *Honors & Awards:* Akzo Chemie Award, UK, 80. *Mem:* Am Phys Soc; Am Soc Testing & Mat; Nat Fire Protection Asn. *Res:* Materials for communications use. *Mailing Add:* AT&T Bell Labs 2000 Northeast Expressway Norcross GA 30071

KAUFMAN, STEPHEN J, b New York, NY, Jan 3, 43; div; c 2. CELL BIOLOGY, DEVELOPMENTAL BIOLOGY. *Educ:* State Univ NY, Binghamton, BA, 64, MA, 66; Univ Colo, PhD(microbiol), 71. *Prof Exp:* Fel molecular biol, Mass Inst Technol, 71-74; asst prof microbiol, 74-81, asst prof cell biol, 77-81, ASSOC PROF MICROBIOL & CELL BIOL, UNIV ILL, 81- *Concurrent Pos:* Jane Coffin Childs Found fel, 71-73; Muscular Dystrophy Asn fel, 73-74; prin investr, 75-78; prin investr, Basil O'Connor Grant, 75-78; consult, Nat Birth Defect Found, 78-81; prin investr, NIH grant, 79-; sr fel, Fogarty Int Ctr, 84-85; vis scientist, Max Planck Inst for Biophys Chem, Goettingen, FRG, 84-85; ed, Exp Cell Res, 85-91. *Mem:* AAAS; Soc Develop Biol; Am Soc Cell Biol; Am Soc Microbiol; Am Soc Biol Chem. *Res:* Muscle differentiation; development of specialized cells; cell fusion; transformation; effect of viruses on development. *Mailing Add:* Dept Cell & Struct Biol 506 Morrill Hall Univ Ill Urbana IL 61801

KAUFMAN, THOMAS CHARLES, b Chicago, Ill, July 28, 44; m 67; c 1. GENETICS. *Educ:* San Fernando Valley State Col, BA, 67; Univ Tex, Austin, MA, 69, PhD(genetics), 71. *Prof Exp:* Nat Res Coun Can res assoc zool, Univ BC, 71-73; lectr, 73-74; asst prof zool, 75-81, PROF GENETICS, IND UNIV, BLOOMINGTON, 81-; INVESTR, HOWARD HUGHES MED INST, 90- *Concurrent Pos:* Adj prof med genetics, Med Sch, Ind Univ. *Mem:* AAAS; Genetics Soc Am. *Res:* Mutagenesis; genetic fine structure in eucaryotic organisms; cytology of dipterin polytene salivary gland chromosomes; position effect variegation and developmental genetics of drosophila; genetics and control of redundant genes. *Mailing Add:* Howard Hughes Med Inst/Dept Biol Ind Univ Bloomington IL 47405

KAUFMAN, VICTOR, b New York, NY, Sept 27, 25; m 49; c 4. ATOMIC PHYSICS. *Educ:* Kans State Univ, BS, 49, MS, 50; Purdue Univ, PhD(physics), 59. *Prof Exp:* Instr physics, Univ NDak, 50-55; res assoc, Purdue Univ, 59-60; PHYSICIST, NAT BUR STANDARDS, 60- *Mem:* Fel Optical Soc Am. *Res:* Atomic emission spectroscopy; interferometry; wavelength standards by precision measurement and by calculation from atomic energy levels. *Mailing Add:* Aromic & Plasma Physics Div Nat Inst Standards & Technol Gaithersburg MD 20899

KAUFMAN, WILLIAM, b Pittsburgh, Pa, Dec 31, 31; m 53; c 3. ELECTRICAL ENGINEERING. *Educ:* Carnegie Inst Technol, BS & MS, 53, PhD(elec eng), 55. *Prof Exp:* Instr, Carnegie Inst Technol, 53-54, res engr, 54-55; engr, Westinghouse Elec Corp, 55-57, res mathematician, 57-59, supvry engr, 59-62; dir res, Gen Instrument Corp, 62-65; consult engr, Gen Elec Co, 65-66; mgr, Med Eng Dept, Hittman Assocs, Inc, 66-71; vpres eng, Ensco, Inc, 71-84; VPRES APPL RES & DIR, CARNEGIE MELLON RES INST, 85- *Mem:* Inst Elec & Electronics Engrs. *Res:* The application of solid state materials to electronic and electromechanical systems; artificial organs; medical instrumentation; data acquisition and processing; transportation safety research; railroad track geometry automated inspection. *Mailing Add:* Carnegie Mellon Res Inst 4400 Fifth Ave Pittsburgh PA 15213

KAUFMAN, WILLIAM, b New York, NY, Dec 30, 10; m 40. FOOD ALLERGY, PSYCHOSOMATIC MEDICINE. *Educ:* Univ Pa, BA, 31; Univ Mich, MA, 32, MD, 38, PhD(physiol), 37. *Prof Exp:* Intern med, Barnes Hosp, St Louis, 38-39; asst resident & resident, Mt Sinai Hosp, NY, 39-40; Dazian Found fel physiol, Med Sch, Yale Univ, 40-42; pvt med pract, Bridgeport, Conn, 40-64; assoc med dir, L W Frohlich & Co-Intercon Int Inc, 64-65, med dir, 65-67, dir med affairs, 67-68; assoc med dir, Klemtner Casey, Inc, 69-70, dir med affairs, 70-71, vpres & dir med affairs, Klemtner Advert, Inc, 71, sr vpres & dir sci & med affairs, 71-81. *Concurrent Pos:* Am ed-in-chief, Int Arch Allergy Appl & Immunol, 55-67; film consult, Family Film Ctr Conn Inc, 67-74. *Honors & Awards:* Tom Spies Award Nutrit & Mem Lectr, Int Acad Prevent Med, 78; Merit Award, Am Col Allergy & Immunol, 81. *Mem:* Fel AAAS; fel Am Col Physicians; fel Am Col Allergy & Immunol; fel Am Col Nutrit; fel Gerontol Soc Am; Nat Asn Sci Writers. *Res:* Reflex physiology; tissue conductivity; electrocardiography; psychosomatic medicine; food allergy; human nutrition including niacinamide therapy of osteoarthritis. *Mailing Add:* 3180 Grady Street Winston-Salem NC 27104

KAUFMAN, WILLIAM CARL, JR, b Appleton, Minn, Jan 21, 23; m 46; c 2. HUMAN PHYSIOLOGY, BIOPHYSICS. *Educ:* Univ Minn, BA, 48; Univ Ill, MS, 53; Univ Wash, PhD(physiol), 61. *Prof Exp:* Instr aviation physiol, Wright-Patterson AFB, 50-51, proj officer altitude suits, Aeromed Lab, 53-56, res biologist thermal environ, Aerospace Med Res Labs, 58-66, chief, Byodynamics Br, Aeromed Res Lab, Holloman AFB, 66-68; Nat Inst Med Res spec res fel, Hampstead Labs, London, Eng, 68-69; prof human adaptability & chmn dept, 69-78, chmn res coun, 78-81, prof, 78-86, EMER PROF HUMAN BIOL, UNIV WIS, GREEN BAY, 86- *Concurrent Pos:* Asst prof prev med, Ohio State Univ, 62-67; mem nuclear weapons effects res comt, Defense Atomic Support Agency, 65-68; consult to pvt indust; NIH special res fel, London, Eng, 68-69. *Mem:* AAAS; Aerospace Med Asn; Am Physiol Soc. *Res:* Temperature regulation and peripheral circulation; thermal and space environments; respiration; evaluation and development of cold weather protective equipment; evaluation and development of protective and recreational clothing. *Mailing Add:* 19228 NE 202nd St Woodinville WA 98072

KAUFMAN, WILLIAM MORRIS, b Pittsburgh, Pa, Dec 31, 31; m 53; c 3. INSTRUMENTATION & TESTING FOR TRANSPORTATION SAFETY. *Educ:* Carnegie Inst Technol, BS, 53, MS, 53, PhD(elec eng), 53. *Prof Exp:* Supvr, Westinghouse Elec Corp, 62-65; dir res, Gen Instrument Corp, 62-65; consult engr, Gen Elec Co, 65-66; mgr, Med Eng Dept, Hittman Assocs, Inc, 66-71; vpres eng, Ensco, 71-83; vpres, Ocean Data Systs, Inc, 84-85; VPRES APPL RES & DIR, CARNEGIE-MELLON RES INST, CARNEGIE-MELLON UNIV, 85- *Mem:* Inst Elec & Electronics Engrs; Am Soc Metals. *Res:* Computer applications; materials; biotechnology; information management. *Mailing Add:* 4400 Fifth Ave Pittsburgh PA 15213

KAUFMANN, ALVERN WALTER, b Cleveland, Ohio, Feb 21, 24; m 46; c 3. MATHEMATICS. *Educ:* Greenville Col, BA, 47; Ohio State Univ, MA, 48, PhD, 60. *Prof Exp:* Instr math, Aurora Col, 48-50; instr math & physics, Cent Col, Kans, 50-52; teacher, Pub Sch, Ohio, 52-54; asst instr math, Ohio State Univ, 54-57; assoc prof math & physics, Roberts Wesleyan Col, 57-65, prof math, 65-81, acad dean, 74-81; prof math, Mt Vernon Nazarene Col, Ohio, 81-86; RETIRED. *Mem:* Math Asn Am; Am Sci Affil; Nat Coun Teachers Math. *Res:* Meaning and definition in mathematics. *Mailing Add:* 224 Sychar Rd Mt Vernon OH 43050

KAUFMANN, ANTHONY J, b Millen, Ga, Aug 19, 36; m 66. MICROBIOLOGY, BIOCHEMISTRY. *Educ:* Univ Ga, BS, 59, MS, 61; La State Univ, PhD(microbiol), 67. *Prof Exp:* Med microbiologist, Nat Communicable Dis Ctr, 62-63; fel microbiol, La State Univ, 63-67; assoc prof health sci, Etenn State Univ, 67-69; assoc prof, 69-74, PROF BIOL, ST MARY'S UNIV, SAN ANTONIO, 74- *Concurrent Pos:* Res assoc, La State Univ, 67; consult, Southwest Res Found & Inst, San Antonio, 71- *Mem:* AAAS; Am Soc Microbiol; Am Inst Biol Sci. *Res:* Microorganisms capable of degrading certain solid waste products such as cellulose, paper products and certain plastics. *Mailing Add:* St Mary's Univ One Camino Santa Maria San Antonio TX 78284

KAUFMANN, ARNOLD FRANCIS, b Dubuque, Iowa, Feb 24, 36; div; c 3. EPIDEMIOLOGY. *Educ:* Iowa State Univ, DVM, 60; Univ Minn, MS, 68; Am Col Vet Path, dipl. *Prof Exp:* Vet, 62-63; vet epidemiologist, Ctr Dis Control, 63-67, vet pathologist, 68-70, chief bact zoonoses act, 71-90, CHIEF, MYCOTIC DIS BR, CTR DIS CONTROL, 90- *Mem:* Am Vet Med Asn; Am Asn Lab Animal Sci; Am Col Vet Pathologists. *Res:* Pathology and epidemiology of infectious diseases; molecular biology of leptospires. *Mailing Add:* Div Bact & Mycotic Dis Ctr Dis Control Atlanta GA 30333

KAUFMANN, ELTON NEIL, b Cleveland, Ohio, Mar 18, 43. PHYSICS, MATERIALS SCIENCE. *Educ:* Rensselaer Polytech Inst, BS, 64; Calif Inst Technol, PhD(physics), 69. *Prof Exp:* Mem tech staff, Bell Tel Labs, 68-81; physicist, Lawrence Livermore Nat Lab, 81-89; ASSOC DIR, STRATEGIC PLANNING GROUP, OFF DIR, ARGONNE NAT LAB, 89- *Concurrent Pos:* Ed, Hyperfine Interactions, 80- *Mem:* Am Phys Soc; Metall Soc; Mat Res Soc (pres, 85). *Res:* Hyperfine interactions using nuclear spectroscopic methods; particle-solid interactions including ion-beam channeling and ion-implantation; directed energy beam materials modification. *Mailing Add:* Off Dir Argonne Nat Lab 9700 S Cass Ave Argonne IL 60439-4832

KAUFMANN, ESTEBAN, organic chemistry, for more information see previous edition

KAUFMANN, GERALD WAYNE, b Dubuque, Iowa, Sept 18, 40; m 66; c 3. ANIMAL BEHAVIOR, ECOLOGY. *Educ:* Loras Col, BS, 62; Iowa State Univ, MS, 64; Univ Minn, Minneapolis, PhD(biol), 71. *Prof Exp:* From instr to assoc prof, 64-79, PROF BIOL, LORAS COL, 79- *Mem:* Am Ornith Union; Wilson Ornith Soc. *Res:* Marsh ecology; behavior of soras and Virginia rails; behavior of Weddell seals; marsh and river ecology. *Mailing Add:* Dept Biol Loras Col 1450 Alta Vista Dubuque IA 52001

KAUFMANN, JOHN HENRY, b Baltimore, Md, Jan 7, 34; div; c 3. VERTEBRATE ZOOLOGY, BEHAVIORAL ECOLOGY. *Educ:* Cornell Univ, BS, 56; Univ Calif, Berkeley, PhD(vert zool), 61. *Prof Exp:* Biologist animal ecol, Nat Inst Neurol Dis & Blindness, 61-63; from asst prof to assoc prof zool, 63-74, PROF ZOOL, UNIV FLA, 74- *Mem:* Am Soc Mammal; Ecol Soc Am; Animal Behav Soc. *Res:* Social behavior and ecology of Emydidae, Procyonidae, Mustelidae, Primates and Macropodidae, including home range, movements and dominance behavior including territoriality, food habits and activity cycles. *Mailing Add:* Dept Zool Univ Fla Gainesville FL 32611

KAUFMANN, JOHN SIMPSON, b Raleigh, NC, Apr 18, 31; m 59; c 3. CLINICAL PHARMACOLOGY, INTERNAL MEDICINE. *Educ:* Wake Forest Univ, BS, 53, MD, 56, PhD(pharmacol), 68; Am Bd Internal Med, dipl, 64. *Prof Exp:* Instr med, Wake Forest Univ, 62-64; instr pharmacol & assoc med, 64-70, from asst prof to assoc prof med & pharmacol, Bowman Gray Sch Med, 75-87; CONSULT, 87- *Concurrent Pos:* USPHS spec fel & vis asst prof, Vanderbilt Univ, 68-70. *Mem:* AAAS. *Res:* Interaction of drugs in man; mechanisms of action of antihypertensive agents; platelet amine uptake and aggregation; actions of hematologic and oncolytic agents; neuronal amine uptake and psychoactive drugs. *Mailing Add:* 4210 Briarcliffe Rd Winston-Salem NC 27106

KAUFMANN, KENNETH JAMES, b New York, NY, May 2, 47. INSTRUMENTATION. *Educ:* City Col NY, BS, 68; Mass Inst Technol, PhD(chem), 73. *Prof Exp:* Fel chem, Calif Inst Technol, 73-74; Bell Tel Labs, 74-75; asst prof chem, Univ Ill, Urbana, 76-80; MEM STAFF, WORTHINGTON GROUP, MCGRAW EDISON CO, 80- *Mem:* Am Phys Soc; Am Chem Soc; Sigma Xi. *Res:* Picosecond kinetics of biological and chemical reaction. *Mailing Add:* 360 Foothill Rd No 6910 Bridgewater NJ 08807-0910

KAUFMANN, MAURICE JOHN, b Hopedale, Ill, Nov 11, 29. PLANT PATHOLOGY. *Educ:* Bluffton Col, BS, 52; Univ Ill, MS, 55, PhD(plant path, bot), 57. *Prof Exp:* Plant pathologist, Agr Res Serv, USDA, Wis, 57-63; mem fac biol, 63-70, assoc prof, 70-71, PROF BIOL, BLUFFTON COL, 71- *Mem:* Am Phytopath Soc. *Res:* Diseases of soybeans and forage grasses. *Mailing Add:* Dept Biol Bluffton Col Bluffton OH 45817

KAUFMANN, MERRILL R, b Paxton, Ill, June 17, 41; m 62, 88; c 2. PHYSIOLOGICAL ECOLOGY, FOREST HYDROLOGY. *Educ:* Univ Ill, BS, 63; Duke Univ, MF, 65, PhD(forestry), 67. *Prof Exp:* Asst prof plant physiol & asst plant physiologist, Univ Calif, Riverside, 67-73, assoc prof plant physiol & assoc plant physiologist, 73-77; PRIN PLANT PHYSIOLOGIST, ROCKY MOUNTAIN FOREST & RANGE EXP STA, US FOREST SERV, USDA, 77- *Concurrent Pos:* Chair, Whole Plant Physiol Working Party, Int Union Forestry Res Orgn, 91- *Mem:* Int Union Forestry Res Orgn; Am Soc Plant Physiologists. *Res:* Plant water relations; plant-environment interaction; carbon balance and allocation in relation to growth efficiency; forest effects on hydrologic cycle in subalpine forest watersheds. *Mailing Add:* USDA Forest Serv Rocky Mountain Forest & Range 240 W Prospect St Ft Collins CO 80526

KAUFMANN, PETER JOHN, b Amsterdam, Holland, Oct 30, 35; US citizen; m 63; c 2. COSMETIC CHEMISTRY. *Educ:* Univ Ill, Urbana, BS, 59. *Prof Exp:* Chief chemist, Dr P Fahrney & Sons, Chicago, 63-65; res chemist cosmetics, Alberto-Culver Co, Ill, 65-67; lab dir, Marcelle Cosmetics Div, Borden, Inc, 67-70; dir prod develop, Max Factor & Co, 70-; AT ALMAY INC. *Mem:* Am Chem Soc; Soc Cosmetic Chemists. *Res:* Emulsion technology; formulation, development and manufacture of makeup and skin care products; efficacy and safety of cosmetics. *Mailing Add:* Almay Inc 1501 Williamsboro Rd PO Box 611 Oxford NC 27565-3461

KAUFMANN, RICHARD L, b Honolulu, Hawaii, June 11, 35; m 63; c 2. PHYSICS. *Educ:* Calif Inst Technol, BS, 57; Yale Univ, MS, 58, PhD(chem), 60. *Prof Exp:* From asst prof to assoc prof, 63-73, PROF PHYSICS, UNIV NH, 73- *Mem:* Am Phys Soc. *Res:* Space physics. *Mailing Add:* Dept Physics Univ NH Durham NH 03824

KAUFMANN, ROBERT FRANK, b Valley Stream, NY, June 23, 40; m 64; c 3. GEOLOGY, HYDROGEOLOGY. *Educ:* Villanova Univ, BS, 62; Ohio State Univ, MS, 64; Univ Wis-Madison, MS, 69, PhD(geol), 70. *Prof Exp:* Assoc res prof hydrol, Desert Res Inst, Univ Nev Syst, 70-74; hydrogeologist, Off Radiation Progs, US Environ Protection Agency, 74-80; prin geologist, Converse Consult Inc, 80-84; PRIN, THE MARK GROUP ENGRS & GEOLOGISTS, 84- *Mem:* Am Geophys Union; Am Water Resources Asn; Am Inst Hydrol; Nat Water Well Asn; Am Inst Prof Geologists. *Res:* Hydrogeology of solid and liquid waste disposal; ground water management; ground water quality. *Mailing Add:* 7265 W Coley Ave Las Vegas NV 89117

KAUFMANN, THOMAS G(ERALD), b Szombathely; Hungary, Jan 10, 38; US citizen; m 60; c 2. CHEMICAL ENGINEERING. *Educ:* Columbia Univ, BS, 60, MS, 62, PhD(chem eng), 65. *Prof Exp:* Teaching asst thermodyn, Columbia Univ, 60-62; engr, Esso Res & Eng Co, Florham Park, 64-74, dir process develop, Exxon Res & Develop Labs, 75-77, mgr process eng, 77-79, mgr, Corp Bus Serv, 79-83, asst gen mgr, 83-86, MGR PROD RES, EXXON RES & ENG CO, 86- *Concurrent Pos:* USPHS res asst, 64-65. *Mem:* Am Inst Chem Engrs; Am Chem Soc. *Res:* Thermodynamics; irreversible thermodynamics; membrane transport; gaseous diffusion; fractionation; tower internals; computer systems; catalytic cracking; fluid/solid systems; heavy crude upgrading; flexicoking; hydrodesulfurization; fluid coking; hydrofining; environmental control; energy conservation; fuels and lubes product quality; automotive emissions control; additives technology. *Mailing Add:* 11 E Cheryl Rd Pine Brook NJ 07058

KAUFMANN, WILLIAM B, b San Francisco, Calif, Nov 11, 36; m 68; c 2. ELEMENTARY PARTICLE PHYSICS. *Educ:* Univ Calif, Berkeley, PhD(physics), 68. *Prof Exp:* From asst prof to assoc prof, 69-87, PROF PHYSICS, ARIZ STATE UNIV, 87- *Mem:* Am Phys Soc; AAAS. *Res:* Theoretical medium-energy nuclear physics. *Mailing Add:* Dept Physics Tempe AZ 85287

KAUFMANN, WILLIAM KARL, b Richland, Wash, Aug 13, 51. MOLECULAR BIOLOGY. *Educ:* Yale Univ, BS, 73; Univ NC, PhD(path), 79. *Prof Exp:* Fel biol, Lab Radiol & Environ Health, Univ Calif, San Francisco, 82-87; ASST PROF, DEPT PATH, UNIV NC, CHAPEL HILL, 88- *Mem:* AAAS; Sigma Xi. *Res:* Mechanisms of DNA replication and repair and their importance in carcinogenesis. *Mailing Add:* 4206 Trotter Ridge Rd Durham NC 27707-5532

KAUGERTS, JURIS E, b Riga, Latvia, Sept 24, 40; US citizen; m 69; c 3. LOW TEMPERATURE PHYSICS. *Educ:* Stevens Inst Technol, BS, 62, MS, 64, PhD(physics), 72. *Prof Exp:* Presidential intern superconductivity, Lawrence Berkeley Lab, Univ Calif, 72-73; res assoc, Plasma Physics Lab, Princeton Univ, 73-75; asst physicist superconductivity, Brookhaven Nat Lab, 75-77, assoc physicist superconductivity, 77-79, physicist, 80-82; sr scientist, Oxford Superconducting Technol, 84-87; SUPERCOLLIDER CENT DESIGN GROUP, 87- *Mem:* Am Phys Soc. *Res:* Superconducting accelerator magnet research, design and development. *Mailing Add:* 34 Frederick Dr West Milford NJ 07480

KAUKER, MICHAEL LAJOS, b Szerecseny, Hungary, Jan 24, 35; US citizen; m 61; c 4. PHARMACOLOGY. *Educ:* Univ Ala, Birmingham, PhD(pharmacol), 67. *Prof Exp:* From asst prof to assoc prof pharmacol, Ctr Health Sci, Univ Tenn, Memphis, 69-83; PROF DEPT PHYSIOL & PHARMACOL, UNIV SDAK, VERMILLION, 83- *Concurrent Pos:* NIH fel, Univ NC, Chapel Hill, 67-69. *Mem:* Am Soc Pharmacol & Exp Therapeut; Soc Exp Biol & Med; Am Soc Nephrology; Int Soc Nephrology. *Res:* Electrolyte and water metabolism; renal micropuncture; mechanism of action of antidiuretic hormone; renal effects of diuretic drugs; regulation of body fluid compartments. *Mailing Add:* Dept Physiol & Pharmacol Sch Med Univ SDak Vermillion SD 57069

KAUL, MAHARAJ KRISHEN, b India, Nov 11, 40; m 69; c 1. ENGINEERING MECHANICS, APPLIED MATHEMATICS. *Educ:* Punjab Univ, India, BS, 62; SUNY, Stony Brook, MS, 67; Univ Calif, Berkeley, PhD(civil eng), 72. *Prof Exp:* Sr engr, EDS Nuclear, Inc, 72-77;

consult engr, Quadrex Corp, 77-80; vpres, Enconi, Inc, 80; PRES, ENG MECH RES, INC, 81- *Concurrent Pos:* Consult, Nuclear Energy Div, Gen Elec, 87-88. *Res:* Vibrations; earthquake engineering; finite elements; probabilistic and stochastic methods; numerical analysis; applied mathematics. *Mailing Add:* 43670 Vista Del Mar Fremont CA 94539

KAUL, PUSHKAR N, BIOLOGICAL SCIENCE. *Educ:* Banaras Univ, India, BS, 54, MS, 55; Univ Calif, San Francisco, PhD(pharmaceut chem), 60. *Prof Exp:* Res & teaching asst, Med Ctr, Univ Calif, San Francisco, 57- 60; lectr & res assoc pharmacol, Med Sch, Univ Melbourne, Australia, 60-61; chief, Pharmacol & Stability Studies Div, H A Ltd, India, 61-65; group leader pharmacol, Med Res Div, Farbwerke Hoechst AG, Frankfurt, Ger, 65-68; prof pharmocodynamics & toxicol, Univ Okla Health Sci Ctr, Oklahoma City, 68-81; chmn & prof, Dept Pharmacol, spec asst to pres & dir, Off Res Admin, Morehouse Sch Med, Atlanta, Ga, 81-84; assoc vpres acad affairs & res admin & develop & prof, 84-87, TENURED PROF BIOL SCI, ATLANTA UNIV, GA, 84- *Concurrent Pos:* Adj prof pediat & res med, Univ Okla Health Sci Ctr, Oklahoma City, 68-81; numerous vis prof, US & foreign, 70-90; career develop award, NIH, 72, nat res serv award, Nat Inst Ment Health, 75; mem Biol Resources Comt, Nat Task Force Marine Biomed Res, US Marine Technol Soc, 74-78; vis scientist, Nat Inst Ment Health, US Dept Health & Human Serv, Rockville, Md, 80-81; assoc vpres res, dir, Off Res Admin & prof health sci, Wichita State Univ, Kans, 87-89. *Mem:* Asn Study Higher Educ; Int Soc Biochem Pharmacol; Int Soc Toxicol; Am Soc Pharmacol & Exp Therapeut; Am Acad Pharmaceut Sci; NY Acad Sci; AAAS; Am Asn Lab Animal Sci. *Res:* Drug metabolism; biochemical pharmacology and toxicology; preclinical pharmacology; clinical dose-response relationships; marine pharmacology; author or co-author of over 70 publications. *Mailing Add:* Dept Biol Clark-Atlanta Univ 223 J P Brawley Dr SW Atlanta GA 30314

KAUL, PUSHKAR NATH, b Srinagar, India, June 29, 33; m 61; c 4. PHARMACOLOGY, CLINICAL PHARMACOLOGY. *Educ:* Banaras Hindu Univ, BPharm, 54, MPharm, 55; Univ Calif, San Francisco, PhD(pharmacol, pharmaceut chem), 60. *Prof Exp:* Asst prof pharmaceut, Birla Inst Technol, India, 55-57; asst pharmaceut chem, Med Ctr, Univ Calif, San Francisco, 57-58, asst pharmacol, 58-60; res assoc pharmacol & vis scientist, Med Sch, Univ Melbourne, 60-61; chief res pharmacol, Antibiotics Res Ctr, India, 61-65; group leader, Farbwerke Hoechst, Ger, 65-68; assoc prof pharmacol, res med & res pediat, Univ Okla, 68-75, prof pharmacol, 75-77, prof pharmacodyn & toxicol, 77-81; prof & chmn pharmacol & asst to pres, Res & Spec progs, Sch Med, Morehouse Univ, 81-; off res admin, Atlanta Univ; DEPT CHEM, FITCHBURG STATE COL. *Concurrent Pos:* Lectr, Univ Poona, 62-63; dir marine pharmacol & adj prof pediat & res med, Univ Okla; dir drug metab, Cent State Hosp, Norman; chmn, Nat Task Force Marine Biomed. *Honors & Awards:* Aruna & Malaviya Prizes, 54; Lunsford Richardson Pharm Award, 60; Ebert Prize Cert, 62; Univ Okla Alumni Res Award, 69 & 70. *Mem:* Assoc fel Royal Australian Chem Inst; Am Soc Pharmacol & Exp Therapeut; Int Soc Biochem Pharmacol; Acad Pharmaceut Sci. *Res:* Biotransformation of drugs; mechanism of drug action; screening of pharmacologically active substances from the sea; antibiotics; psychotropic drugs. *Mailing Add:* Dept Chem Fitchburg State Col 160 Pearl St Fitchburg MA 01420

KAUL, ROBERT BRUCE, b Faribault, Minn, Jan 28, 35; m 76. BOTANY, ECOLOGY. *Educ:* Univ Minn, 57, PhD(bot), 64. *Prof Exp:* Asst bot, Univ Minn, 57-60, instr, 61-62; from asst prof to assoc prof, 64-72, PROF BOT, UNIV NEBR, LINCOLN, 72-; VICE DIR, BIOL SCI, 89- *Concurrent Pos:* Ed, Trans Nebr Acad Sci, 88- *Mem:* Bot Soc Am; Am Soc Plant Taxon; Am Inst Biol Sci. *Res:* Morphology and life history of angiosperms, especially trees and aquatic plants; floristics of the Great Plains. *Mailing Add:* Sch Bio Sci Univ Nebr Lincoln NE 68588-0118

KAUL, S K, b Lucknow, India, Dec 25, 36; Can citizen; m 63; c 2. PURE MATHEMATICS. *Educ:* Univ Lucknow, BSc, 54, MSc, 55; Univ Delhi, PhD(math), 59. *Prof Exp:* Instr math, Hampton Inst, 58-59; instr math, Univ Rochester, 59-60, univ fel, 60-61; instr math, Univ Utah, 62; from asst prof to assoc prof, 63-71, PROF MATH, UNIV REGINA, 71- *Mem:* Am Math Soc; Can Math Cong; Math Asn Am. *Res:* Studying topological structures associated with differential equations, like flows, semi-flows and generalized dynamical systems and various stability notions using a flow associated with a semiflow and a semiflow associated with a generalized dynamical system. *Mailing Add:* Dept Math Univ Regina Regina SK S4S 0A2 Can

KAUL, SANJIV, b Kashmir, India, Aug 18, 51; m 81; c 1. CARDIOLOGY. *Educ:* Univ Delhi, MBBS, 75. *Prof Exp:* Intern med, Chicago Med Sch, 77-78; resident med, Univ Vt, 78-80; fel cardiol, Univ Calif Sch Med, Los Angeles, 80-82 & Harvard Med Sch, 82-84; asst prof, 84-88, ASSOC PROF MED, UNIV VA, 88- *Concurrent Pos:* Mem, Coun Clin Cardiol, Am Heart Asn, 85. *Mem:* Fel Am Col Physicians; fel Am Col Cardiol; fel Am Col Chest Physicians; Physicians for Social Responsibility; Am Fedn Clin Res; fel Am Heart Asn; Am Soc Echocardiography. *Res:* Assessment of regional myocardial flow-function relationships using non-invasive techniques. *Mailing Add:* 722 Broomley Rd Charlottesville VA 22901

KAULA, WILLIAM MASON, b Sydney, Australia, May 19, 26; US citizen; m 49, 78; c 4. GEOPHYSICS. *Educ:* US Mil Acad, BS, 48; Ohio State Univ, MS, 53. *Hon Degrees:* DSc, Ohio State Univ, 75. *Prof Exp:* Geodesist, Army Map Serv, Washington DC, 57-58, chief geod res & anal div, 58-60; geophysicist geod, Celestial Mech & Planetary Interiors, Goddard Space Flight Ctr, NASA, Md, 60-63; PROF GEOPHYS, DEPT EARTH & SPACE SCI, INST GEOPHYS & PLANETARY PHYSICS, UNIV CALIF, LOS ANGELES, 63- *Concurrent Pos:* Chief, Nat Geodetic Survey, NOAA, Rockville, Md, 84-87. *Honors & Awards:* Whitten Medal, Am Geophys Union; Brouwer Medal, Am Astron Soc. *Mem:* Nat Acad Sci; Fel Am Geophys Union; Am Astron Soc; Geol Soc Am. *Res:* Gravitational fields of the earth and moon; origin and evolution of the earth, moon and planets; mantle convection; dynamics of the solar system. *Mailing Add:* Dept Earth & Space Sci Univ Calif Los Angeles CA 90024-1567

KAUNE, WILLIAM TYLER, b Everett, Wash, Aug 31, 40; m 72; c 2. BIOENGINEERING. *Educ:* Univ Wash, BS, 66; Stanford Univ, PhD(physics), 73. *Prof Exp:* Res asst high energy physics, Stanford Linear Accelerator Ctr, 68-72; res assoc, Univ Wash, 72-73; asst prof physics, Loyola Marymount Univ, 73-75; sr res engr, 75-80, staff engr bioeng, Pac Northwest Div, Battelle Mem Inst, 80-87; physicist, Nat Bur Standards, Boulder, Colo, 87-88; VPRES, ENERTECH CONSULTS, CAMPBELL, CALIF, 88- *Mem:* Bioelectromagnetics Soc; Inst Elec & Electronics Engrs. *Res:* Biological effects of electromagnetic radiation; exposure systems and dosimetry. *Mailing Add:* Enertech Consults 300 Orchard City Dr Suite 132 Campbell CA 95008

KAUNITZ, HANS, b Vienna, Austria, Oct 20, 05; US citizen; m 43. NUTRITION. *Educ:* Vienna Univ, MD, 30. *Prof Exp:* Attending physician & head clin lab, Dept Med, Sch Med, Univ Vienna, 35-38; assoc prof med & head clin lab, Univ Philippines, 38-40; clin prof path, Columbia Univ, 41-73; RETIRED. *Concurrent Pos:* NIH grants, 55-73; consult several food firms, 53-78; consult geront, Rutgers Univ, 78. *Honors & Awards:* Achievement award, Am Oil Chem Soc, 70; Presidential Merit Medal Philippines, 73; Big Sign Honor, Repub Austria, 73; Alton E Bailey Award, Am Oil Chem Soc, 81. *Mem:* Sigma Xi; Am Soc Exp Path; Harvey Soc; Am Oil Chem Soc; Am Inst Nutrit. *Res:* Biological effect of edible fats, especially medium chain triglycerides; function of cholesterol in arteriosclerosis; biological effects of sodium chloride; philosophy of science. *Mailing Add:* 152 E 94th St Apt 7A New York NY 10128

KAUP, DAVID JAMES, b Marionville, Mo, Apr 8, 39; m 82; c 3. INTEGRABLE SYSTEMS, NONLINEAR STUDIES. *Educ:* Univ Okla, BS, 60, MS, 62; Univ Md, PhD(physics), 67. *Prof Exp:* From res asst prof to res assoc prof, 74-76, FROM ASST PROF TO PROF PHYSICS, CLARKSON UNIV, 67. *Concurrent Pos:* vis res geophysicist, Univ Calif Los Angeles, 81, vis prof math, Lab Phys Math, Univ Sci & Tech Langs, Montpellier, France, 87; consult, Varian Assocs, 83; Res scientist, Dynamics Technol, 80-81. *Mem:* Am Phys Soc; Sigma Xi; Am Math Soc; Soc Indust Appl Math. *Res:* Soliton theory; inverse scattering; nonlinear optics; plasma physics; mathematical physics. *Mailing Add:* Dept Physics Clarkson Univ Potsdam NY 13676

KAUP, EDGAR GEORGE, b Irvington, NJ, Oct 5, 27; m 53; c 2. CHEMICAL ENGINEERING. *Educ:* Lehigh Univ, BS, 50; Neward Col Eng, BS, 58, MS, 63. *Prof Exp:* Phys chemist, Hoffmann-La Roche, NJ, 52-54; spectroscopist, Air Reduction Lab, 54-55, res chem engr, 55-62; develop engr, Celanese Plastic Co, 62-65; sr chem engr, 65-69, resident eng mgr, 69-76, SR CHEM ENGR, BURNS & ROE, INC, BURNS & ROE CONSTRUCT CORP DIV, CONTRACTOR TO OFF SALINE WATER, US DEPT INTERIOR, 76- *Mem:* Am Chem Soc; Am Inst Chem Engrs; Am Soc Test & Mat. *Res:* Desalting and water pollution abatement; reverse osmosis evaluations and applications; water and waste treatment by ion exchange and evaporative methods. *Mailing Add:* Eight Essex Rd Essex Fells NJ 07021-1104

KAUPP, VERNE H, b Denver, Colo, Apr 15, 40; m 66; c 2. ELECTRICAL ENGINEERING. *Educ:* Univ Md, BS, 71; Univ Kans, DEng, 79. *Prof Exp:* Engr microwave sensor, Martin Marietta Corp, 71-75; eng consult microwave sensor, Earth Resources Technol, 75; sr res engr microwave remote sensing, Ctr for Res, Inc, 75-80; PROF ELEC ENG, UNIV ARK, 80- *Concurrent Pos:* Consult, Systs Technol/Appl Res Corp, 77-80, Ark Res Consults, Inc, 80- *Mem:* Inst Elec & Electronics Engrs; Sigma Xi; Am Soc Photogram & Remote Sensing; Am Soc Eng Educ. *Res:* Microwave remote sensing; electromagnetics; digital signal processing. *Mailing Add:* Univ Ark Dept Elec Eng Fayetteville AR 72701

KAUPPILA, RAYMOND WILLIAM, b Iron Mountain, Mich, Feb 17, 29; m 52; c 4. ENGINEERING MECHANICS. *Educ:* Univ Mich, BS(mech eng) & BS(eng math), 51; Mich Col Mining & Technol, MS, 61; Univ Mich, PhD, 68. *Prof Exp:* Maintenance, develop & inspection engr, Standard Oil Div, Am Oil Co, Ind, 51-55; plant engr, Cliffs Dow Chem Co, Mich, 55-57; from asst prof to prof, 57-89, EMER PROF MECH ENG, MICH TECHNOL UNIV, 89- *Concurrent Pos:* Design consult; expert witness, failure analysis. *Res:* Machine design; dynamics and vibrations of machinery; stress analysis; thermal stresses; plasticity in forming operations. *Mailing Add:* 424 W Ridge St Marquette MI 49855

KAUPPILA, WALTER ERIC, b Hancock, Mich, Sept 11, 42; m 66; c 2. POSITRON & ELECTRON SCATTERING EXPERIMENTS. *Educ:* Mich Technol Univ, BS, 64; Univ Pittsburgh, PhD(physics), 69. *Prof Exp:* Res assoc, Joint Inst Lab Astrophys, Univ Colo, 69-71; asst prof physics, Univ Mo, Rolla, 71-72; from asst prof to assoc prof, 72-83, PROF PHYSICS, WAYNE STATE UNIV, 83- *Concurrent Pos:* Co-prin investr, NSF supported res grants, 75- *Honors & Awards:* Fel, Sigma Xi. *Mem:* fel Am Phys Soc. *Res:* Experimental studies of elastic, inelastic and total scattering for positrons and electrons colliding with atoms and molecules. *Mailing Add:* Dept Physics & Astron Wayne State Univ Detroit MI 48202

KAUS, PETER EDWARD, b Vienna, Austria, Oct 9, 24; nat US; m 50; c 3. THEORETICAL PHYSICS. *Educ:* Univ Calif, Los Angeles, BS, 47, MA, 52, PhD, 55. *Prof Exp:* Asst physics, Univ Calif, Los Angeles, 51-53; res physicist, Labs, Radio Corp Am, NJ, 54-58; asst prof physics, Univ Southern Calif, 58-62; assoc prof, 62-67, PROF PHYSICS, UNIV CALIF, RIVERSIDE, 67- *Concurrent Pos:* Consult, Hughes Aircraft Co, 58-59, Jet Propulsion Lab, Pasadena, 59-61 & Los Alamos Nat Lab, 80-; trustee, Aspen Ctr Physics, 64-, vpres, 70-82, pres, 81-83; Fulbright res scholar, Denmark, 65-66. *Honors & Awards:* Sarnof Achievement Medal, 57. *Mem:* Fel Am Phys Soc. *Res:* Field theory; elementary particle theory; biophysics; biologic rhythms. *Mailing Add:* Dept Physics Univ Calif Riverside CA 92521

KAUSHIK, AZAD KUMAR, b Dhauj, India, Sept 8, 55; m 82; c 1. AUTOIMMUNITY-AUTOIMMUNE DISEASES, VETERINARY CLINICAL IMMUNOLOGY. *Educ:* Pasteur Inst, DSc, 87; Haryana Agril Univ, MVSc, 78, BVSc, 76. *Prof Exp:* Asst prof immunol, Haryana Agril Univ, Hisar, India, 79-83; res assoc immunol, Pasteur Inst, Paris, France, 83-87; res scientist immunol, Mt Sinai Sch Med, NY, 87-90; asst prof immunol, Med Sch, Univ Geneva, 90-91; ASST PROF IMMUNOL, UNIV GUELPH, CAN, 91- *Concurrent Pos:* Prin investr, SLE Found, NY, 89-90. *Mem:* Am Asn Immunologists. *Res:* Natural autoimmunity and autoimmune disorders; immunoglobulin molecular genetics; idiotypy; protective immunity and clinical veterinary immunology. *Mailing Add:* Dept Vet Microbiol & Immunol Univ Guelph Guelph ON N1G 2W1 Can

KAUSHIK, NARINDER KUMAR, Can citizen; m; c 2. ECOLOGY, HYDROBIOLOGY. *Educ:* Univ Delhi, India, BS, 54, MS, 56; Univ Waterloo, MS, 66, PhD(biol), 69. *Prof Exp:* asst prof, 73-77, PROF ENVIRON BIOL, UNIV GUELPH, 77-; res & sr res asst sanit biol, Cent Pub Health Eng Res Inst, 61-64; fel ecol, Univ Toronto, 69-71; asst prof biol, Univ Waterloo, 71-72; asst prof, 73-77, ASSOC PROF ENVIRON BIOL, UNIV GUELPH, 77- *Mem:* Can Water Resource Asn; NAm Benthol Soc; Can Soc Zoologists; Int Soc Theoret & Appl Limnol. *Res:* Role of autumn shed leaves in secondary production in streams; nitrogen transport and transformations in streams; use of limnocorrals for pesticide impact assessment; ecotoxicology. *Mailing Add:* Dept Environ Biol Univ Guelph Guelph ON N1G 2W1 Can

KAUTZ, FREDERICK ALTON, II, b Knoxville, Tenn, Aug 27, 50; m 77; c 2. LOW DENSITY GAS DYNAMICS, AEROTHERMODYNAMICS. *Educ:* Univ Tenn, BSc, 72; Mass Inst Technol, SM & NucE, 83. *Prof Exp:* Staff mem, Oak Ridge Nat Lab, 72-73; staff scientist, Off Sci & Weapons Res, Cent Intel Agency, 77-86; res asst, 74-77, STAFF MEM, LINCOLN LAB, MASS INST TECHNOL, 86- *Concurrent Pos:* Mem, Themophys Tech Comt, Am Inst Aeronaut & Astronaut, 87-90, secy, 87; mem, NASA-Langley ad hoc comt reentry plasmas, 88-; reviewer, J Spacecraft & Rockets, Am Inst Aeronaut & Astronaut, 89- *Mem:* NY Acad Sci; Am Inst Aeronaut & Astronaut; Am Phys Soc; Am Geophys Union; Inst Elec & Electronics Engrs; Soc Indust & Appl Math; AAAS; Sigma Xi. *Res:* Low density gas dynamics; physics and chemistry of reentry and planetary entry plasmas; computational aerothermodynamics; spacecraft-environment interactions. *Mailing Add:* Lincoln Lab Mass Inst Technol 244 Wood St Rm D-382 Lexington MA 02173-0073

KAUZLARICH, JAMES J(OSEPH), b Des Moines, Iowa, Sept, 27, 27; m 52; c 4. MECHANICAL ENGINEERING, TRIBOLOGY. *Educ:* Univ Iowa, BS, 50; Columbia Univ, MS, 52; Northwestern Univ, PhD(mech eng), 58. *Prof Exp:* Lab asst, Columbia Univ, 50-52; develop engr, Gen Elec Co, NY, 52-54; instr mech eng, Northwestern Univ, 54-57; from asst prof to assoc prof, Worcester Polytech Inst, 58-61; assoc prof, Univ Wash, Seattle, 61-63; chmn dept, 63-75, PROF MECH ENG, UNIV VA, 63- *Concurrent Pos:* Engr, Boeing Corp, 62 & 63; vis res, Cambridge Univ, 70-71 & Swansea Univ, 84-85 & 88-89. *Mem:* Am Soc Eng Educ; fel Am Soc Mech Engrs; Am Soc Lubrication Engrs; Sigma Xi. *Res:* Fluid mechanics; heat transfer; rehabilitation engineering. *Mailing Add:* Dept Mech Eng Univ Va Charlottesville VA 22903-2442

KAUZMANN, WALTER (JOSEPH), b Mt Vernon, NY, Aug 18, 16; m 51; c 3. PHYSICAL CHEMISTRY, PROTEIN CHEMISTRY. *Educ:* Cornell Univ, BA, 37; Princeton Univ, PhD(phys chem), 40. *Prof Exp:* Fel, Westinghouse Elec & Mfg Co, 40-42; chemist, Nat Defense Res Comt Contract, Pa, 42-43; engr, Manhattan Dist Proj, Los Alamos, 44-46; from asst to assoc prof chem, Princeton Univ, 46-63, chmn dept, 64-68, David B Jones prof chem, 63-82, chmn, Dept Biochem Sci, 80-82; RETIRED. *Concurrent Pos:* Guggenheim fel, 57 & 74-75; vis prof, Univ Ibadan, 75; vis scientist, Nat Res Coun Can, Halifax, 83. *Honors & Awards:* Linderstrom-Lang Medal. *Mem:* Nat Acad Sci; Am Geophys Union; Am Acad Arts & Sci; Am Chem Soc; Am Phys Soc; Hist Sci Soc. *Res:* Physical chemistry of proteins; theory of water; properties of matter at high pressures; geochemistry. *Mailing Add:* 302 N Harriston St Suite 152 Princeton NJ 08540

KAVALER, FREDERIC, b New York, NY, Feb 2, 26; m 55; c 2. PHYSIOLOGY. *Educ:* Columbia Univ, AB, 47; Johns Hopkins Univ, MD, 51. *Prof Exp:* Intern med, Maimonides Hosp, Brooklyn, 51-52; resident, Goldwater Mem Hosp, NY, 52-54; NY Heart Asn fel, 54-55; fel physiol, Col Med, Cornell Univ, 55-56, instr physiol, 56-57; from instr to assoc prof, 58-65, PROF PHYSIOL, STATE UNIV NY DOWNSTATE MED CTR, 65- *Mem:* Am Physiol Soc. *Res:* Electrophysiology of the heart; cardiac physiology. *Mailing Add:* SUNY Downstate Med Ctr Brooklyn NY 11203

KAVALJIAN, LEE GREGORY, b Chicago, Ill, Feb 6, 26. PLANT MORPHOLOGY. *Educ:* Univ Chicago, PhB, 47, BS, 48, PhD(bot), 51. *Prof Exp:* Res assoc bot, Brooklyn Bot Garden, NY, 51-52; vis res assoc, Brookhaven Nat Lab, NY, 52; asst to chief chemist, Modern Agr Crop Serv, Calif, 53; Ford Found teaching intern natural sci, Univ Chicago, 53-54, instr, 54; from instr to assoc prof biol sci, 54-64, PROF BIOL SCI, CALIF STATE UNIV, SACRAMENTO, 64- *Mem:* Bot Soc Am; Soc Econ Bot. *Res:* Plant tissue cultures; floral morphology; cytochemistry; ethnobotany. *Mailing Add:* Dept Biol Calif State Univ 6000 Jay St Sacramento CA 95819

KAVANAGH, RALPH WILLIAM, b Seattle, Wash, July 15, 24; m 48; c 5. NUCLEAR PHYSICS. *Educ:* Reed Col, BA, 50; Univ Ore, MA, 52; Calif Inst Technol, PhD, 56. *Prof Exp:* From res fel to sr res fel, 56-60, from asst prof to assoc prof, 60-70, PROF PHYSICS, KELLOGG LAB, CALIF INST TECHNOL, 70- *Mem:* Am Phys Soc. *Res:* Spectroscopy of light nuclei using electrostatic accelerators. *Mailing Add:* Kellogg Lab 1201 E California Pasadena CA 91125

KAVANAGH, ROBERT JOHN, b Whitchurch, Hants, Eng, Oct 7, 31; m 56; c 2. RESEARCH MANPOWER. *Educ:* Univ NB, BSc, 53; Univ Toronto, MASc, 54, PhD(elec eng), 57; Imperial Col, London, DIC, 60. *Prof Exp:* Lectr elec eng, Univ Toronto, 57-59, asst prof, 60-62; from assoc prof to prof elec eng, Univ NB, 62-84, assoc dean, 69-71, actg vpres acad, 78-80, dean grad studies & res, 71-84; DIR-GEN, SCHOLARSHIPS & INTERNAT

PROG, NATURAL SCI & ENG, RES COUN, 84- *Concurrent Pos:* NATO fel, Imp Col, Univ London, 59-60; guest worker, Control Eng Div, Warren Spring Lab, Eng, 68-69; vis scientist, Natural Sci & Eng Res Coun, 82-83. *Mem:* Sr mem Inst Elec & Electronics Engrs; fel NY Acad Sci. *Mailing Add:* Scholar & Int Prog Directorate Natural Sci & Eng Res Coun 200 Kent St Ottawa ON K1A 1H5 Can

KAVANAU, JULIAN LEE, b Detroit, Mich, Jan 21, 22; c 3. ETHOLOGY. *Educ:* Univ Mich, BS, 43; Univ Calif, MS & PhD(zool), 52. *Prof Exp:* Asst physics, Univ Mich, 41-43; physicist, Univ Calif, 43; mem res staff physics, Calif Inst Technol, 43-45; asst math, Univ Calif, Los Angeles, 46-47; asst zool, Univ Calif, 49-51; USPHS fel, Wenner-Gren Inst, Stockholm, Sweden, 52-54; res assoc develop Rockefeller Inst, 55-57; from asst prof to assoc prof biol, 57-67, PROF BIOL, UNIV CALIF, LOS ANGELES, 67- *Mem:* AAAS; Animal Behav Soc; Am Ornithol Union. *Res:* Instrumentation for behavior research; influences of environmental variables on mammalian activity; symmetry of curves and figures; behavior and evolution of psittaciforms. *Mailing Add:* Dept Biol Univ Calif Los Angeles CA 90024-1606

KAVANAUGH, DAVID HENRY, b San Francisco, Calif, Apr 7, 45; m 65; c 5. SYSTEMATIC ENTOMOLOGY, BIOGEOGRAPHY. *Educ:* San Jose State Univ, BA, 67; Univ Colo, Denver, MA, 70; Univ Alta, PhD(entom), 78. *Prof Exp:* From asst cur to assoc cur, 74-84, chmn dept, 79-83, dir res, 86-88 CUR ENTOM, CALIF ACAD SCI, 84- *Concurrent Pos:* Fel, Nat Res Con Can 72-74. *Mem:* Soc Syst Zool; Entom Soc Am; Coleopterists Soc. *Res:* Classification, phylogeny, zoogeography and natural history of ground beetles; biogeography and evolution of high altitude biota, especially the coleoptera faunas of western North America; theory and practice of systematic zoology. *Mailing Add:* Dept Entom Calif Acad Sci San Francisco CA 94118

KAVARNOS, GEORGE JAMES, b New London, Conn. CLINICAL CHEMISTRY. *Educ:* Clark Univ, BA, 64; Univ RI, PhD(org chem), 68; Dipl, Am Bd Clin Chem. *Prof Exp:* NIH fel, Columbia Univ, 68-71; chief chemist, New London, Cyto-Roche, Div Hoffmann La-Roche, 71-74; assoc dir & clin chemist, Cyto Med Lab Inc, Norwich, 74-89; RES CHEMIST, NAVAL UNDERWATER SYSTS CHIEF NAVAL TRAINING, 89- *Concurrent Pos:* Vpres, Bio-Anal Labs, 73-; adj prof chem, Univ RI, 78-; lectr, St Joseph Col, 85- *Mem:* Am Chem Soc; Am Asn Clin Chemists. *Res:* Photochemistry; clinical chemistry; photoinduced electron transfer material. *Mailing Add:* 121 Riverview Ave New London CT 06320

KAVASSALIS, TOM A, b Toronto, Ont, Feb 3, 58; m 83; c 2. PHYSICAL CHEMISTRY. *Educ:* Univ Toronto, BSc, 80; Mass Inst Technol, PhD(phys chem), 85. *Prof Exp:* Chemist, Ont Hydro Res, 85-87; MEM RES, XEROX RES CTR, CAN, 87- *Concurrent Pos:* Adj prof, Dept Chem Eng, Univ Waterloo, 91- *Mem:* Am Chem Soc; Am Phys Soc. *Res:* Theoretical and computational methods for material science applications; simulation of surfactants; theories of polymer structure, morphology and dynamics; mechanical properties of polymers; theory of transport in fluids. *Mailing Add:* Xerox Res Ctr Can 2660 Speakman Dr Mississauga ON L5K 2L1

KAVATHAS, PAULA, b Evanston, Ill, May 30, 50; m; c 2. IMMUNOLOGY, GENETICS. *Educ:* Univ Wis-Madison, PhD(genetics), 80. *Prof Exp:* ASST PROF, LAB MED, IMMUNOBIOL & GENETICS, SCH MED, YALE UNIV, 86- *Res:* Immunobiology of T lymphocyte co-receptor molecule CD8. *Mailing Add:* Dept Lab Med Yale Univ Sch Med New Haven CT 06510

KAVEH, MOSTAFA, b Karadj, Iran, Apr 18, 47. STATISTICAL SIGNAL PROCESSING, IMAGE PROCESSING. *Educ:* Purdue Univ, BS, 69; PhD(elec eng), 74; Univ Calif, Berkeley, MS, 70. *Prof Exp:* Postdoc res assoc, Purdue Univ, 75; various ranks, 75-85, PROF ELEC ENG, UNIV MINN, 85-, HEAD ELEC ENG, 90- *Concurrent Pos:* Consult. *Honors & Awards:* Sr Award, Inst Elec & Electronics Engrs Signal Processing Soc, 86, Meritorious Serv Award, 88. *Mem:* Fel Inst Elec & Electronics Engrs. *Res:* Sensor array signal processing. *Mailing Add:* Dept Elec Eng Univ Minn Minneapolis MN 55455

KAVENOFF, RUTH, b New York, NY, Aug 11, 44. BIOPHYSICAL CHEMISTRY, VIROLOGY. *Educ:* Reed Col, BA, 67; Albert Einstein Col Med, PhD(cell biol), 71. *Prof Exp:* UN Int Agency Res Cancer-WHO fel virol, Univ Auckland, NZ, 72-73; fel phys chem, 71-72 & 73-79, res assoc virol, 79-80, RES ASSOC BIOL, UNIV CALIF, SAN DIEGO, 80- *Concurrent Pos:* Anna Fuller Found fel, 74-75; NIH fel, 75-78. *Res:* Chromosome structure; nucleic acids. *Mailing Add:* Designer Genes Posters Ltd PO Box 100 Del Mar CA 92014

KAVESH, SHELDON, b New York, NY, Jan 15, 33; m 57; c 2. CHEMICAL ENGINEERING. *Educ:* Mass Inst Technol, BSChE, 57; Polytech Inst Brooklyn, MChE, 65; Univ Del, PhD(chem eng), 68. *Prof Exp:* Res engr, Celanese Corp Am, 57-60, Foster Grant Co, Inc, 60-62 & Avisun Corp, 62-65; proj leader polymers, Films Packaging Div, Union Carbide Corp, 68-70; res assoc, 70-80, SR RES ASSOC, ALLIED CHEM CORP, MORRISTOWN, 80- *Mem:* Am Inst Chem Engrs; Am Phys Soc; Am Chem Soc. *Res:* Polymer physics; transport phenomena; materials science. *Mailing Add:* 16 N Pond Rd Whippany NJ 07981

KAWAHARA, FRED KATSUMI, b Penngrove, Calif, Feb 26, 21; m 52; c 3. ENVIRONMENTAL POLLUTION. *Educ:* Univ Tex, BS, 44; Univ Wis, MS, 46, PhD(chem), 48. *Prof Exp:* Assoc chemist, USDA, 48-51; fel org chem, Univ Chicago, 51-53; sr res scientist, Standard Oil Co, Ind, 53-65; org chemist, Anal Qual Control Lab, 68-71, spec consult, Method Develop & Qual Assurance Lab, 72-74, SPEC CONSULT OIL IDENTIFICATION, ANAL QUAL CONTROL LAB, ENVIRON PROTECTION AGENCY, 71- *Concurrent Pos:* Expert witness, petroleum fuels; dep gov, Am Biographic Inst Res Asn. *Honors & Awards:* Group Superior Serv Award, Bur Agr & Indust Chem, USDA, 52; First Five Hundred Gold medal, IBC, Cambridge, Eng. *Mem:* Fel Am Inst Chem; Am Chem Soc; Am Biographic Inst Res Asn;

fel Int Biographic Asn. *Res:* Synthetic fuels, coal liquefaction; Lubricants; phosphorus; fluorocarbons; gasoline additives; waxes; carcinogens; chromatography, infrared, ultraviolet, synthesis, identification; insecticides; greases; phenols; mercaptans; oil pollution; soy bean oil flavor reversion; aromatic amines; peroxides; methods development; auto-oxidation; laser-fiber optics. *Mailing Add:* Environ Protection Agency Environ Monitor & Systs Lab 26 W Martin Luther King Cincinnati OH 45268

KAWAI, MASATAKA, b Gifu, Japan, June 13, 43; m 69; c 2. ELECTRICAL ENGINEERING, COMPUTER CONTROLLED EXPERIMENTS. *Educ:* Tokyo Univ, BSc, 66; Princeton Univ, PhD(biol), 71. *Prof Exp:* Res assoc, Columbia Univ, 71-78, asst prof muscle physiol, Dept Neurol, 78-83, asst prof anat & cell biol, 83-87; ASSOC PROF ANAT, UNIV IOWA, 87- *Concurrent Pos:* Prin investr cross-bridge kinetics res, Dept Neurol, Columbia Univ, 76- *Mem:* Biophys Soc; Gen Physiol Soc. *Res:* Cross-bridge kinetics in chemically skinned muscle fibers by use of sinnsoidal analysis which changes the length and detects concomitant amplitude and phase shift in tension. *Mailing Add:* Dept Anat Univ Iowa Iowa City IA 52242

KAWALEK, JOSEPH CASIMIR, JR, b Stockton, Calif, Dec 21, 45; m 72; c 2. BIOCHEMISTRY, BIOCHEMICAL PHARMACOLOGY. *Educ:* St Francis Col, BS, 67; Univ Pittsburgh, PhD(biochem), 74. *Prof Exp:* Res asst biochem, Univ Pittsburgh, 70-73; res assoc, Hoffmann-La Roche, Inc, 74-75; staff scientist chem carcinogen, Frederick Cancer Res Ctr, Litton Bionetics, Inc, 76-80; RES CHEMIST, DIV VET MED RES, CTR VET MED, FOOD & DRUG ADMIN, 80- *Mem:* Am Chem Soc; AAAS; Am Inst Biol Sci; Sigma Xi; NY Acad Sci; Soc Toxicol; Am Col Toxicol; Am Acad Vet Comp Toxicol. *Res:* Factors affecting drug metabolism in food producing and companion animals. *Mailing Add:* Agr Res Ctr Barc E Bldg 328A Center Rd Beltsville MD 20705

KAWAMURA, HIROSHI, b Antong, China, Jan 26, 27; Japanese citizen; m 71; c 2. BRAIN MECHANISMS OF BEHAVIOR, CIRCADIAN RHYTHMS. *Educ:* Univ Tokyo, MD, 54, DMed Sc(neurophysiol), 59. *Prof Exp:* Instr neurophysiol, Brain Res Inst, Univ Tokyo, 59-61; assoc prof, Yokohama Univ Sch Med, 61-63; asst res anatomist, Dept Anat, Univ Calif, Los Angeles, 63-65; UNESCO fel neurophysiol, Inst Physiol, Univ Pisa, 65-66; res assoc, Dept Pharmacol, Univ Mich, 66-71; chief, Neurophysiol Lab, 72-81, dir, Dept Neurosci, 81-90, DISTINGUISHED SCIENTIST, MITSUBISHI KASEI INST LIFE SCI, 90- *Concurrent Pos:* Chief, Neurophysiol Sect, Lafayette Clin, 67-71; vis scientist, Dept Res Anesthesia, McGill Univ, 71-72. *Honors & Awards:* Mainischi Award, Mainischi Shinbunsha, Tokyo. *Mem:* Am Physiol Soc; Soc Neurosci; NY Acad Sci. *Res:* Hypothalmic mechanisms of sleep-wakefulness and circadian rhythm generation of the suprachiasmatic nucleus; brain tissue transplantation and brainstem transection. *Mailing Add:* Mitsubishi Kasei Life Sci Inst 11 Minamiooya, Machida-shi Tokyo 194 Japan

KAWAMURA, KAZUHIKO, b Nagoya, Japan, Feb 4, 39; m 71. INTELLIGENT ROBOTICS, INTELLIGENT TRAINING SYSTEMS. *Educ:* Waseda Univ, Japan, BEng, 63; Univ Calif, Berkeley, MS, 66; Univ Mich, Ann Arbor, PhD(elec eng), 72. *Prof Exp:* Lectr elec eng, Univ Mich-Dearborn, 72-73; res specialist exp vehicles, Ford Motor Co, 73; prin researcher tech assessment, Columbus Div, Battelle Mem Inst, 73-81; assoc prof elec eng & mgt technol, 80-88, ASSOC DIR, CTR INTEL SYSTS, VANDERBILT UNIV, 85-, PROF ELEC ENG & DIR GRAD STUDIES, 88-, PROF MGT TECHNOL, DEPT ELEC ENG, 88- *Concurrent Pos:* Sr res fel, Japan Soc Prom Sci, 80; vis prof, Kyoto Univ, Japan, 80-81; consult, Saudi Arabian Nat Ctr Sci & Technol, 81-83; orgn coordr, Int Asn Impact Assessment, 81-83; mem, AAAS Comt Sci, Eng, & Pub Policy, 83- *Mem:* Inst Elec & Electronics Engrs; AAAS; Sigma Xi; Am Asn Artificial Intel. *Res:* Expert systems; intelligent robotics; intelligent tutoring systems; risk analysis; computer vision. *Mailing Add:* 5908 Robert E Lee Dr Nashville TN 37215-5224

KAWANISHI, HIDENORI, CELLULAR IMMUNOLOGY, IMMUNOCHEMISTRY. *Educ:* Kyoto Med Sch, Japan, MD & PhD(exp path), 60. *Prof Exp:* Assoc prof med, Health Sci Ctr, SUNY, Stony Brook, 82-; PROF MED, RUTGERS UNIV. *Mailing Add:* 4061 Bay Berry Ct Monmouth Junction NJ 08852

KAWASAKI, EDWIN POPE, b Sikeston, Mo, Jan 25, 26; m 48; c 2. CHEMICAL ENGINEERING, PHYSICAL CHEMISTRY. *Educ:* Case Inst Technol, BS, 54, MS, 58, PhD(chem eng), 60. *Prof Exp:* Res engr, res ctr, 54-58, supvr chem processing, 58-63, div head surface chem, 63-73, div head processing, 73-75, asst dir res, Repub Steel Res Ctr, 75-84. *Mem:* Nat Asn Corrosion Engrs. *Res:* Corrosion of ferrous metals; chemical processing; environmental control; iron and steel making. *Mailing Add:* 4250 Meadow/Gateway Brecksville OH 44141

KAWASE, MAKOTO, plant physiology, horticulture; deceased, see previous edition for last biography

KAWATA, KAZUYOSHI, b Portland, Ore, Jan 2, 24; m 49; c 3. SANITARY ENGINEERING, ENVIRONMENTAL HEALTH. *Educ:* Ore State Col, BS, 49; Univ Minn, MS, 50; Univ Calif, Berkeley, MPH, 58; Johns Hopkins Univ, DrPH(sanit eng), 65. *Prof Exp:* Civil-sanit engr, Bd Missions, Methodist Church, 50-66; from asst prof to assoc prof, 66-80, PROF ENVIRON HEALTH ENG & PROF INT HEALTH, JOHNS HOPKINS UNIV, 80- *Concurrent Pos:* Consult, WHO, Bangladesh, 73 & Philippines, 75, consult & lectr, Egypt, 76; expert health sci, AID, 76-78, consult, 79-81. *Mem:* Am Soc Civil Engrs; Am Pub Health Asn; Am Water Works Asn; Water Pollution Control Fedn; Am Acad Environ Engrs. *Res:* Water and waste-water treatment processes; disinfection kinetics; tropical environmental health. *Mailing Add:* 1215 Brixton Rd Baltimore MD 21239

KAWATERS, WOODY H, b Hoboken, NJ, Jan 14, 51; c 5. AIR POLLUTION IMPACTS, WASTE MANAGEMENT IMPACTS. *Educ:* Univ Md, BA, 74. *Prof Exp:* Sr anal, Chi-Comput Horizons, 74-78; prin consult, 78-82, vpres environment sci div dir, 82-87, VPRES & SR OPERATOR OFFICER, TRC ENVIRON CONSULT. *Concurrent Pos:* Consult, Govs Infostructure Task Force, 85-87. *Mem:* Air pollution Control Asn; Nat Asbestos Coun; Nat Asn Manufacturers. *Res:* Research physical and engineering science aspects of air and waste pollution; environmental policy analysis; data tracking system design for pollution information; research designs; cost of complying with regulations; management of technical personnel and projects. *Mailing Add:* TRC Environmental Consults Inc 800 Connecticut Blvd East Hartford CT 06108

KAWATRA, MAHENDRA P, b Wazirabad, India, June 22, 35; US citizen; m 62; c 3. PHYSICS. *Educ:* Univ Delhi, BSc, 55, MSc, 57, PhD(physics), 62. *Prof Exp:* Lectr physics, Univ Delhi, 57-63; Smith-Mundt scholar & Fulbright grant, Mass Inst Technol, 63-64; assoc res scientist, Courant Inst Math Sci, NY Univ, 64-66; asst prof, Fordham Univ, 66-71; dir educ & comput technol, 80-83, PROF PHYSICS, MEDGAR EVERS COL, CITY UNIV NEW YORK, 71- *Concurrent Pos:* Res assoc, Univ Ill, 64; Smith-Mundt scholar. *Mem:* Am Phys Soc; NY Acad Sci. *Res:* Quantum-statistical mechanics; many-body problem; liquid helium; thin film and theory of superconductivity; low-temperature physics. *Mailing Add:* Dept Phys Medgar Evers Col 1150 Carroll St Brooklyn NY 11225

KAWOOYA, JOHN KASAJJA, b Ft Portal, Uganda, May 16, 52. PROTEIN FOLDING, RECONSTITUTION OF PROTEINS WITH LIPIDS. *Educ:* Makerere Univ Kampala, BSc, 73; Univ Nairobi, MSc, 79; Univ Ill, Urbana-Champagne, PhD(entomol), 82. *Prof Exp:* Res assoc, Univ Ariz, 83-84 & Univ Chicago, 84-86; res asst prof, Univ Ariz, 86-88; LEADING RES SCIENTIST, UPJOHN CO, 89- *Concurrent Pos:* Vis scientist, Univ Utrecht, Holland, 87-88, Univ Rio de Jeneiro, Brazil, 88-89. *Mem:* Am Soc Biochem & Molecular Biol; AAAS. *Res:* Isolation and refolding of recounbinant proteins-these proteins represent a new generation of biopharmaceutical therapeutics designed to combat various ailments. *Mailing Add:* Upjohn Co 1410-89-1 Kalamazoo MI 49001

KAY, ALVIN JOHN, b Luling, Tex, June 10, 38; m 67; c 4. MATHEMATICAL ANALYSIS. *Educ:* Southwest Tex State Univ, BS, 61, MA, 65; Univ Houston, PhD(math), 75. *Prof Exp:* Teacher math, Woodsboro High Sch, 61-64; asst, Southwest Tex State Univ, 64-65; instr, San Jacinto Col, 65-69; asst, Univ Houston, 69-70; from instr to asst prof, 70-78, ASSOC PROF MATH, TEX A&I UNIV, 78- *Mem:* Am Math Soc. *Res:* Integral equations and product integral. *Mailing Add:* Dept Math Tex A&M Univ Kingsville TX 78363

KAY, BONNIE JEAN, b Chicago, Ill, Nov 13, 41. HEALTH SYSTEMS PLANNING, POLICY ANALYSIS. *Educ:* Oberlin Col, AB, 63; Univ Chicago, MS, 66; Northwestern Univ, PhD(urban systs eng), 75. *Prof Exp:* Instr chem, US Peace Corps, 63-65, Ministry Educ, Ghana, 67-68 & Chicago City Col, 68-70; chemist, US Customs Lab, 70-72; asst prof health systs, GA Inst Technol, 75-80; ASST PROF, SCH PUB HEALTH, UNIV MICH, 80- *Mem:* Am Pub Health Asn; Am Health Planning Asn; Sigma Xi; AAAS. *Res:* Evaluation research of programs and public policy related to population; women's health and poverty. *Mailing Add:* Dept Pub Health Policy & Admin Sch Pub Health Univ Mich Ann Arbor MI 48109

KAY, CYRIL MAX, b Calgary, Alta, Oct 3, 31; m 53; c 2. BIOCHEMISTRY. *Educ:* McGill Univ, BSc, 52; Harvard Univ, PhD(biochem), 56. *Prof Exp:* Fel, Life Ins Med Res Fund, Cambridge Univ, 56-57; res phys biochemist, Eli Lilly & Co, 57-58; from asst prof to assoc prof, 58-67, PROF BIOCHEM, UNIV ALTA, 67- *Concurrent Pos:* Med Res Coun Can vis prof, Weizmann Inst Sci, Rehovot, Israel, 69-70; co-dir, Med Res Coun Group Protein Structure & Function Function, Univ Alta, 74-; mem, Protein Eng Network Ctr Excellence, 90. *Honors & Awards:* Ayerst Award Biochem, 70. *Mem:* Fel NY Acad Sci; fel Royal Soc Can; Brit Biochem Soc; Am Soc Biol Chem; Can Biochem Soc. *Res:* Protein physical chemistry; hydrodynamic and optical properties of macromolecules; correlation of physico-chemical properties with biological function for muscle proteins. *Mailing Add:* Dept Biochem Univ Alta Edmonton AB T6G 2E8 Can

KAY, DAVID CLIFFORD, b Oklahoma City, Okla, July 26, 33; m 55, 78; c 3. COMBINATORICS, FINITE MATHEMATICS. *Educ:* Otterbein Col, BS, 55; Univ Pittsburgh, MS, 59; Mich State Univ, PhD(math), 63. *Prof Exp:* Asst prof math, Univ Wyo, 63-66; AT DEPT MATH, UNIV NC. *Concurrent Pos:* Res Coun award, Univ Wyo, 65; dir, Reg NSF Conf, Convexity, Nat Sci Found, 71. *Mem:* Sigma Xi; Am Math Soc; Math Asn Am. *Res:* Problems regarding curve-curvature in metric spaces; axiomatic convexity, matroids and geometric problems in topological linear spaces. *Mailing Add:* Dept Math Univ NC Ashville NC 28804

KAY, DAVID CYRIL, b Sault Ste Marie, Mich, Sept 5, 32; m 61; c 4. PSYCHIATRY, PSYCHOPHARMACOLOGY. *Educ:* Wheaton Col, Ill, BS, 54; Univ Ill, Chicago, MD, 58; Am Bd Psychiat & Neurol, dipl, 69. *Prof Exp:* Intern, Presby-St Luke's Hosp, Chicago, 58-59; staff physician, USPHS Hosp, Ft Worth, Tex, 59-61; res psychiatrist, Ill State Psychiat Inst, 61-64; chief exp psychiat unit, Addiction Res Ctr, Nat Inst Drug Abuse, 66-69, chief exp psychiat sect, 69-80; DIR DRUG ABUSE PROG, HOUSTON VET ADMIN HOSP, 80-; ASSOC PROF PSYCHIAT & PHARMACOL, BAYLOR COL MED, 80- *Concurrent Pos:* USPHS Ment health career develop fel, 61-66; fel, Addiction Res Ctr, Nat Inst Ment Health, 64-66; vis lectr, Asbury Theol Sem, 64-66; clin assoc prof, Med Ctr, Univ Ky, 64-80. *Mem:* Am Soc Pharmacol & Exp Therapeut; Int Brain Res Orgn; Sigma Xi; Am Psychiat Asn; Am Soc Clin Pharmacol & Therapeut; Soc Neurosci. *Res:* Behavioral and physiological investigation of psychoactive drugs and individuals who abuse them; interaction of sleep with drugs and sexual function. *Mailing Add:* 1313 Campbell Rd Bldg C Houston TX 77055

KAY, EDWARD LEO, b Cleveland, Ohio, Sept 23, 24; m 55; c 4. ORGANIC CHEMISTRY. *Educ:* Case Western Reserve Univ, BS, 47, MS, 53, PhD(org chem), 55. *Prof Exp:* Res scientist org chem, Texaco Inc, 55-60; sr res assoc org chem, Firestone Tire & Rubber Co, 60-85; RETIRED. *Mem:* Am Chem Soc. *Res:* Organic chemicals synthesis; organic chemical process development; oxidation studies; vulcanization accelerators; adhesion studies; vapor phase oxidation of hydrocarbons; scrap rubber disposal processes; fire and smoke suppressants; guayule natural rubber processing; reinforcement of polyurethanes. *Mailing Add:* 79 S Tamarack Akron OH 44319

KAY, ELIZABETH ALISON, b Kauai, Hawaii, Sept 27, 28. BIOLOGY. *Educ:* Mills Col, BA, 50; Cambridge Univ, BA, 52, MA, 56; Univ Hawaii, PhD, 57. *Prof Exp:* From asst prof to assoc prof sci, 57-66, assoc dean, Grad Div, 75-79, actg vchancellor, 84-85, PROF ZOOL, UNIV HAWAII, MANOA, 70-,; HON ASSOC MALACOL, B P BISHOP MUS, 58- *Concurrent Pos:* Fulbright scholar, 50-52. *Mem:* Fel AAAS; Soc Syst Zool; Marine Biol Asn UK; Challenger Soc; Malacol Soc Australia; fel Linmean Soc. *Res:* Functional morphology of marine gastropods; molluscan ecology and systematics; biogeography. *Mailing Add:* Dept Zool Univ Hawaii Manoa Honolulu HI 96822

KAY, ERIC, b Heidelberg, Ger, Nov 23, 26; nat US; m 53; c 3. INORGANIC CHEMISTRY, PHYSICAL CHEMISTRY. *Educ:* Univ Calif, BS, 53; Univ Wash, Seattle, PhD(chem), 58. *Prof Exp:* Res chemist, Best Co, Oakland, Calif, 48-52 & Lawrence Radiation Lab, Univ Calif, 52-54; asst phys chem, Univ Wash, Seattle, 54-55; staff res chemist, 58-65, HEAD MAT SCI DEPT, RES LAB, IBM CORP, 65- *Concurrent Pos:* Vis prof, Univ Calif, Berkeley, 68-69; mem, tech rev panel, Nat Bur Standards, 76-79 & rev comt, Argonne Univ Asn Math Sci & Technol Div, Argonne Nat Lab, 81-86 & tech rev bd, Nat Submicron Facil, Cornell Univ, 85- *Mem:* Am Vacuum Soc; Am Phys Soc; Sigma Xi. *Res:* Ion impact phenomena on condensed phases; plasma chemistry; surface phenomena; chemistry and physics of thin films; surface magnetism; cluster science. *Mailing Add:* 20280 Via Santa Teresa San Jose CA 95120

KAY, ERNEST ROBERT MACKENZIE, biochemistry, for more information see previous edition

KAY, FENTON RAY, b Pacoima, Calif, Oct 10, 42; m; c 2. PHYSIOLOGICAL ECOLOGY, VERTEBRATE ZOOLOGY. *Educ:* Nev Southern Univ, BS, 67; Univ Nev, Las Vegas, MS, 69; NMex State Univ, PhD(biol), 75. *Prof Exp:* Res asst, Desert Res Inst, Univ Nev, Las Vegas, 70, US Int Biol Prog, Desert Biome, Dept Biol, NMex State Univ, 70-74; NIH trainee, Dept Physiol, Col Med, Univ Fla, 74-76; asst prof biol, Calif State Univ, Los Angeles, 76-78; Habitat staff biol, Nev Dept Wildlife, 84-90; INDEPENDENT CONSULT, BIOL & ENVIRON, 90- *Concurrent Pos:* Comput programmer & opers mgr, OAO Corp (White Sands Missile Range), 80-83; independent comput consult, 83-84; instr, Truckee Meadows Community Col, Reno, 84- *Mem:* Ecol Soc Am; Am Soc Mammalogists; Herpetologists League; Sigma Xi; Soc Conserv Biol. *Res:* Respiratory and metabolic responses of rodents to carbon dioxide; thermal biology of desert animals; desert animal community composition; economic value of wildlife. *Mailing Add:* 20 Smithridge Park Reno NV 89502

KAY, H DAVID, b Glendale, Ohio, Sept 6, 43. CELLULAR IMMUNOLOGY, LEUKEMIA. *Educ:* Iowa State Univ, PhD(immunobiol), 72. *Prof Exp:* Asst tumor immunol, M D Anderson Hosp & Tumor Inst, Univ Tex, 72-73; assoc prof med & dir, Exp Immunol Lab, Med Ctr, Univ Nebr, 83-90, chmn, Immunol Coun, 84-90; LEGAL CONSULT, 90- *Mem:* Am Asn Immunologists; Am Rheumatism Asn; Am Asn Cancer Res; Am Fedn Clin Res; Int League Against Rheumatism. *Res:* Regulation of the immune response to cancer. *Mailing Add:* 12829 O St Omaha NE 68137-1834

KAY, IRVIN (WILLIAM), b Savannah, Ga, Apr 19, 24; m 54; c 2. APPLIED MATHEMATICS. *Educ:* NY Univ, BA, 48, MS, 49, PhD(math), 53. *Prof Exp:* Res assoc math, NY Univ, 52-58, from asst prof to assoc prof, 59-62; sr res mathematician, Conductron Corp, 62-64, dept head advan systs, 64-68, dir independent res & develop, 68-71; prof elec eng, Wayne State Univ, 71-73; STAFF MEM, INST DEFENSE ANALYSIS, 73- *Mem:* Am Math Soc; Am Phys Soc. *Res:* Electromagnetic theory; systems analysis; optics. *Mailing Add:* 6111 Wooten Dr Falls Church VA 22044

KAY, JACK GARVIN, b Scott City, Kans, July 11, 30; m 52; c 2. PHYSICAL & INORGANIC CHEMISTRY, NUCLEAR & ATMOSPHERIC CHEMISTRY. *Educ:* Univ Kans, AB, 52, PhD(phys chem), 60. *Prof Exp:* From instr to asst prof inorg chem, Univ Ill, Urbana, 59-66; prof chem, Univ Toledo, 66-69, chmn dept, 66-68; head dept, 69-85, PROF CHEM, DREXEL UNIV, 69- *Concurrent Pos:* Consult, Chemotronics, Inc, Avco, Inc, Charlestown Twp, Chester County, Pa, Alex C Fergusson Co & Dwight & Wilson Co; prin investr, Atomic Energy Comn, 60-69 & NSF, 84-; rep Coun Chem Res, 80-; rep Pa Asn Cols & Univs on Task Force to Develop Pa Right-to-Know legis, 83-84. *Mem:* AAAS; Am Chem Soc; Am Phys Soc; Faraday Soc; fel Am Inst Chem; Am Geophys Union. *Res:* Electronic spectroscopy of gaseous diatomic molecules; matrix-isolation spectroscopy; flash heating and kinetic spectroscopy; flash photolysis; high temperature chemistry; solar furnaces; radiation chemistry; hot atom chemistry in inorganic crystals; nuclear and radiochemistry; radon and decay products in the atmosphere and oceans. *Mailing Add:* 118 Reveille Rd Wayne PA 19087

KAY, KENNETH GEORGE, b New York, NY, Oct 13, 43; m 68; c 3. INTRAMOLECULAR ENERGY TRANSFER. *Educ:* Polytech Inst, Brooklyn, BS, & MS, 65; Johns Hopkins Univ, PhD(chem), 70. *Prof Exp:* Res assoc chem, Univ Chicago, 70-71; from asst prof to prof chem, Kans State Univ, 71-81; PROF CHEM, BAY-ILAN UNIV, 87- *Concurrent Pos:* Vis assoc prof, Tel-Aviv Univ, 79-80; vis prof, Univ Toronto, 82-83. *Mem:* Am Phys Soc; Am Chem Soc; Am Asn Univ Profs; Sigma Xi. *Res:* Theory of molecular reaction dynamics; unimolecular dissociation; intramolecular vibrational energy transfer; quantum ergodic theory; theoretical models for photodecomposition reactions. *Mailing Add:* Chem Dept Bar Ilan Univ Ramat Gan Israel

KAY, MARGUERITE M B, b Washington, DC, May 13, 47. GERIATRICS & GERONTOLOGY, MEMBRANE BIOLOGY. *Educ:* Univ Calif, Berkeley, BA, 70; Univ Calif, San Francisco, MD, 74; NIA Nat Inst Health, cert, 76. *Prof Exp:* Staff fel, Gerontology Res Ctr, NICHD, NIH Baltimore City Hosp, 74, US Pub Health Serv Off & Chief High Resolution Membrane Lab, 75-77; chief, Lab Molecular & Clinical Immunology, VA Wadsworth Med Ctr, Los Angeles Calif; prof med & prof med biochem & genetics, Tex A&M Univ, Col Med, 81-90, prof micro biol & immunol, 81-91; REGENTS PROF, MICROBIOL & IMMUNOL, COL MED, UNIV ARIZ, 90- *Concurrent Pos:* Consult, immuno-electron microscopy, Dept Basic & Clin Immunol & Microbiol, Med Univ SC, 74-81, biol consult, Electron Microscopy Lab, Enrico Fermi Inst, Univ Chicago, 74-75, reviewer, Nat Sci Found & US-Israel Binational Sci Found, 77-, reviewer, NIH. *Mem:* Am Soc Clin Investigation; Am Soc Biol Chem; Am Asn Immunologists; Am Geriatrics Soc; Am Soc Cell Biol; Am Soc Hemat. *Res:* Molecular and cell biology of aging. *Mailing Add:* Dept Microbiol & Immunol Rm 644 LSN Univ Tex Col Med 1501 N Campbell Ave Tucson AZ 85724

KAY, MICHAEL AARON, b San Francisco, Calif, May 7, 43; m 76; c 2. RADIOANALYTICAL CHEMISTRY, HEALTH PHYSICS. *Educ:* Univ Calif, Berkeley, BS, 65. *Prof Exp:* Radiochemist, US Naval Radiol Defense Lab, Calif, 65; sr chemist, Res Reactor Facility, Univ Mo, Columbia, 70-75, sr res scientist, 75-78; mem fac, Univ Mo-Columbia, 75-78; sr scientist, Rockwell Hanford Opers, Wash, 78-80; assoc prof chem & dir,Reed Reactor Facil, Reed Col, 80-86; mgr chem res & develop, Hannah Car Wash Int, 86-88; PRES, AMBRY, INC, 88- *Concurrent Pos:* Consult forensic sci & health physics, 81-; radiol instr, Level III, Fed Emergency Mgt Agency, 85- *Mem:* AAAS; Am Nuclear Soc; Am Chem Soc; Health Physics Soc. *Res:* Neutron activation analysis; trace elements in the environment; environmental radiation monitoring; radioisotope production; radiation protection. *Mailing Add:* PO Box 22266 Milwaukie OR 97222-0266

KAY, MORTIMER ISAIA, b Bronx, NY, Aug 27, 30; m 64. CRYSTALLOGRAPHY. *Educ:* Brooklyn Col, BA, 52; Purdue Univ, MS, 53; Univ Conn, PhD(phys chem), 58. *Prof Exp:* Asst chem, Purdue Univ, 52-53; asst, Univ Conn, 53-57; res assoc, Pa State Univ, 57-59; fel, Royal Norweg Coun Sci & Indust Res, 59-60; res scientist, NASA, 60-61; res assoc prof, Ga Inst Technol, 62-64; sr scientist & head neutron diffraction prog, PR Nuclear Ctr, 64-77, head, Mats Sci Div, PR Ctr Energy & Environ Res, 76-79; head sea water-surfactant project, ocean thermal energy conversion, 79-80, PHYS SCIENTIST, US DEPT ENERGY, 81- *Concurrent Pos:* Vis assoc physicist, Brookhaven Nat Lab, 57-59. *Mem:* AAAS; Am Chem Soc; Am Phys Soc; Am Crystallog Asn; Sigma Xi. *Res:* Molecular and crystal structure; diffraction studies of ferroelectric transitions; ocean thermal energy; pyroelectric energy conversion; surface chemistry; atomic energy. *Mailing Add:* 70 Oak Shade Rd Gaithersburg MD 20878

KAY, PETER STEVEN, b Milwaukee, Wis, Sept 24, 37. CLINICAL CHEMISTRY, INSTRUMENTATION. *Educ:* Cornell Univ, AB, 59; Purdue Univ, PhD(org chem), 66. *Prof Exp:* Tech asst coal chem, US Steel Res Ctr, 59-60; teaching asst chem, Purdue Univ, 60-62; res chemist, Textile Fibers Dept, 66-69, col rels rep, Employee Rels Dept, 69-70; res chemist, Textile Fibers Dept, E I Du Pont de Nemours & Co, Inc, 70-72, mkt rep, 72-73, dist mgr, 73-76, nat sales mgr, Sci Instruments, Inst Prod, 76-77, mgr thermal anal, 77-79, mgr liquid chromatography, 79-80, nat sales mgr, Electronic Div, 80-82; dir mkt Harshaw Chem Co (Gulf Oil), 82-84; gen mgr x-ray prod, Picker Int, 85-86; PRIN, STRATEGIC MGR ADV GROUP, 86- *Mem:* AAAS; Am Chem Soc; Sigma Xi. *Res:* Kinetics and product distributions in solvolyses of allylic halides and esters; polyamide fibers; biocomponent fibers; polyamide textile fibers; robotics; digital x-ray techniques; clinical laboratory automated analyzers and medical imaging devices. *Mailing Add:* Six Mariners Cove Cincinnati OH 45249-1791

KAY, ROBERT EUGENE, b Missoula, Mont, Dec 23, 25; m 51; c 3. EPITAXIAL CRYSTAL GROWTH, INFRARED DETECTORS. *Educ:* Univ Calif, Los Angeles, BS, 48, PhD(bot sci), 52. *Prof Exp:* Asst bot, Univ Calif, Los Angeles, 51-52; supvry chemist, US Naval Radiol Defense Lab, 54-61; mgr, Biosci Dept, 61-80, TECH CONSULT, SEMICONDUCTOR TECHNOL DEPT AERONUTRONIC DIV, FORD AEROSPACE CORP, 61- *Res:* Radiobiology; lipid metabolism; isolated perfused organs; adaptive enzyme systems; olfactory transduction in insects; model membrane systems; interactions of dyes with biological macromolecules; immobilized enzymes; organic semiconductors; infrared detectors; missile-vehicle integration; missile systs mgt; high temperature batteries; hetero epitaxial crystal growth. *Mailing Add:* 1515 Warwick Ln Newport Beach CA 92660

KAY, ROBERT LEO, b Hamilton, Ont, Dec 13, 24; m 52; c 3. PHYSICAL CHEMISTRY. *Educ:* Univ Toronto, MA, 50, PhD(phys chem), 52. *Prof Exp:* Merck fel, Rockefeller Inst, 52, res asst, 53-56; asst prof chem, Brown Univ, 56-63; sr fel, Mellon Inst, 63-67, actg dir, Ctr Spec Studies, Univ, 73-74, head dept, 74-83, PROF CHEM, CARNEGIE-MELLOW UNIV, 67- *Concurrent Pos:* Ed, Jour Solution Chem, 71- *Mem:* AAAS; Am Chem Soc; Biophys Soc. *Res:* Transport properties of electrolyte solutions; structure of liquids; electrophoresis; solutions at high pressure and temperature. *Mailing Add:* Carnegie Mellon Univ 4400 Fifth Ave Pittsburgh PA 15213-3876

KAY, ROBERT WOODBURY, b New York, NY, Jan 21, 43; m 75; c 2. GEOCHEMISTRY. *Educ:* Brown Univ, AB, 64; Columbia Univ, PhD(geol), 70. *Prof Exp:* Asst prof geol, Columbia Univ, 70-75; asst res geophysicist, Univ Calif, Los Angeles, 75-76; from asst prof to assoc prof, 76-86, PROF GEOL, CORNELL UNIV, 86- *Mem:* Fel Geol Soc Am; Am Geophys Union; Geochem Soc. *Res:* Geochemistry of rare earth elements in volcanic rocks; regional geology of the Aleutian Islands, Alaska; chemistry of the lower crust. *Mailing Add:* Dept Geol Sci Snee Hall Cornell Univ Ithaca NY 14853

KAY, RUTH MCPHERSON, nutritional biochemistry, for more information see previous edition

KAY, SAUL, b New York, NY, Feb 13, 14; m 40; c 1. SURGICAL PATHOLOGY. *Educ:* NY Univ, BA, 36, MD, 39; Am Bd Path, dipl & cert path anat & clin path. *Prof Exp:* Resident path, Fordham Hosp, 41-42 & NY Postgrad Hosp, 46-48; resident surg path, Presby Med Ctr, Columbia Univ, 48-50; assoc prof, 50-52, PROF SURG PATH, MED COL VA, 52- *Concurrent Pos:* Ed staff, Am J Surg Path; consult, VA Hosp, Richmond Eye & Ear Hosp. *Mem:* Am Asn Path & Bact; fel Am Col Path; fel Am Soc Clin Path; Am Soc Cytol; Int Acad Path; AMA. *Res:* Granulomas and oncology; electron microscopy. *Mailing Add:* 322 Charmain Rd Richmond VA 23226

KAY, SUZANNE MAHLBURG, b Rockford, Ill, May 30, 47; m 75; c 2. PETROLOGY, MINERALOGY. *Educ:* Univ Ill, Urbana, BS, 69, MS, 72; Brown Univ, PhD(geol), 75. *Prof Exp:* Fel geol, Univ Calif, Los Angeles, 75-76; res assoc geol, 76-82, SR RES ASSOC, INST STUDY CONTINENTS, CORNELL UNIV, ITHACA, NY, 83- *Concurrent Pos:* Vis assoc petrol, Calif Inst Technol, 82; vis prof, Univ Buenos Aires, Arg, 89; Fulbright fel, 89-90. *Mem:* Fel Geol Soc Am; fel Mineral Soc Am; Am Geophys Union; Sigma Xi. *Res:* Study of natural and experimentally produced intergrowths in ternary and plagioclase feldspars, genesis of magmutic rocks in the Aleutian Islands, Alaska; study of lower crustal xenoliths; volcanism and tectonism in the Andes of Argentina and Chile. *Mailing Add:* Dept Geol Sci Cornell Univ Ithaca NY 14853

KAY, WEBSTER BICE, b Hammond, Ind, Dec 8, 00; m 39; c 2. CHEMICAL ENGINEERING, PHYSICAL CHEMISTRY. *Educ:* Ohio State Univ, BChE, 22; Univ Chicago, PhD(phys chem), 26. *Prof Exp:* Res engr phys properties, Standard Oil Co, Ind, 26-47; prof thermodyn, 47-71, EMER PROF PHYS PROPERTIES, DEPT CHEM ENG, OHIO STATE UNIV, 71- *Concurrent Pos:* Dir & prin investr grants, Am Chem Soc Res Fund, 54-58, NSF, 57-71 & Am Petrol Inst, 65-71. *Mem:* AAAS; Am Chem Soc; Am Inst Chem Engrs. *Res:* Phase behavior of mixtures at high temperature and pressure; thermodynamic properties of pure compounds and mixtures; critical properties of vapor-liquid mixtures. *Mailing Add:* Dept Chem Eng Ohio State Univ Columbus OH 43210

KAYA, AZMI, b Acik, Turkey, Feb 1, 33; m 64; c 1. SYSTEMS ENGINEERING. *Educ:* Tech Col Men, Ankara, Dipl, 52; Univ Wis, Madison, MS, 62; Univ Minn, Minneapolis, MS & PhD, 70. *Prof Exp:* Control engr, Honeywell, Inc, 62-68; teaching assoc, Univ Minn, Minneapolis, 70; asst prof mech eng, 70-75, ASSOC PROF MECH ENG, UNIV AKRON, 75- *Concurrent Pos:* NSF res grant, Univ Akron, 71-72; NATO vis expert, 74. *Mem:* Am Soc Mech Engrs; Sigma Xi; Inst Elec & Electronics Engrs; Am Soc Eng Educ; Tech Asn Pulp & Paper Indust. *Res:* Modeling, control and optimization of large scale systems; control theory applications; energy management systems. *Mailing Add:* Dept Mech Eng Univ Akron Akron OH 44325

KAYA, HARRY KAZUYOSHI, b Honolulu, Hawaii, Nov 20, 40; m 64; c 2. INSECT PATHOLOGY. *Educ:* Univ Hawaii, BS, 62, MS, 66; Univ Calif, Berkeley, PhD(insect path), 70. *Prof Exp:* Asst entomologist, Conn Agr Exp Sta, 71-76; from asst prof to assoc prof, 76-84, PROF, DEPT NEMATOL, UNIV CALIF, DAVIS, 84- *Mem:* Soc Invert Path; Entom Soc Am; Int Orgn Biol Control; Soc Nemat. *Res:* Biological control of insects; epizootiology in insect populations; use of microorganisms to control insects; insect nematology. *Mailing Add:* Dept Nemat Univ Calif Davis CA 95616

KAYANI, JOSEPH THOMAS, b Kuravilangad, India, Mar 8, 45; US citizen; m 69; c 2. APPLIED MECHANICS, STRESS ANALYSIS. *Educ:* Univ Kerala, India, BSc, 67; Polytech Inst Brooklyn, MS, 71; Polytech Inst NY, PhD(mech eng), 75. *Prof Exp:* Engr supvr construct, Telecommun Dept, Govt India, 67-69; mech engr, Acoustics & Vibrations Lab, Souncoat Co, Inc, Brooklyn, 72-74, stress analysis, Nuclear Power Servs, Inc, New York, 74-75, Burns & Roe, Inc, 75-78; PRIN ENGR APPL MECH, EBASCO SERV, INC, 78- *Concurrent Pos:* Res asst, Polytech Inst Brooklyn, 73-75; res fel, Polytech Inst New York, 75- *Mem:* Am Soc Mech Engrs; Am Acad Mech. *Res:* Stress and vibration analysis of pressure vessels and piping; nonlinear random vibrations; acoustics and noise control of machines and structural components. *Mailing Add:* 300 Ellen Pl Jericho NY 11753

KAYAR, SUSAN RENNIE, b Highland, Ill, May 17, 53. RESPIRATORY PHYSIOLOGY, MICROCIRCULATION. *Educ:* Univ Miami, BS, 74, PhD(biol), 78. *Prof Exp:* Res asst, Everglades Nat Park, 78-79; res assoc, Univ Colo, Boulder, 79-81; postdoctoral fel, Med Sch, Univ Colo, Denver, 81-84; res asst prof, Univ Bern, Switz, 84-89; instr, Univ Med Dent, NJ, 89-90; RES PHYSIOLOGIST, NAVAL MED RES INST, NAT NAVAL MED CTR, BETHESDA, MD, 90- *Mem:* Am Physiol Soc; Am Soc Zoologists; Int Soc Oxygen Transport Tissues; Microcirculatory Soc; Sigma Xi; AAAS. *Mailing Add:* Diving Biomed Technol Naval Med Res Inst Nat Naval Med Ctr Bethesda MD 20814-5055

KAYDEN, HERBERT J, b New York, NY, Jan 30, 20; m 51; c 2. MEDICINE. *Educ:* Columbia Col, AB, 40; NY Univ, MD, 43. *Prof Exp:* Asst, 49-51, clin instr, 51-54, from asst prof to assoc prof, 54-70, PROF MED, SCH MED, NY UNIV, 70- *Concurrent Pos:* From asst vis physician to assoc vis physician, Goldwater Mem Hosp, 50-56, vis physician, 56-, assoc dir, 58-62; from asst attend physician to assoc attend physician, NY Univ Hosp, 51-71, attend physician, 71-; from asst vis physician to assoc vis physician, Bellevue Hosp, 52-71, vis physician, 71-; attend physician, Manhattan Vet Admin Hosp, 56-; mem coun arteriosclerosis, Am Heart Asn. *Mem:* Am Soc Pharmacol & Exp Therapeut; Harvey Soc; Am Fedn Clin Res; fel Am Col Physicians; fel NY Acad Med. *Res:* Cardiovascular diseases, especially disorders of cardiac rhythm; pharmacology of antiarrhythmic drugs; lipid metabolism in humans, including studies of serum and tissue lipoproteins; vitamin E metabolism in humans. *Mailing Add:* Dept Med NY Univ Med Ctr 550 First Ave New York NY 10016

KAYE, ALBERT L(OUIS), b New York, NY, Mar 16, 09; m 34; c 3. ELECTROCHEMICAL ENGINEERING. *Educ:* Mass Inst Technol, SB, 31, MS, 32, ScD(electrochem), 34. *Prof Exp:* Asst chem, Calif Inst Technol, 32-33; res assoc, Div Indust Coop, Mass Inst Technol, 34-37, secy comt corrosion metals, 35-37; metallurgist, Carnegie-Ill Steel Corp, Chicago, 37-41, mgr alloy bur, Metall Div, Chicago Dist, 41-44, metall engr, Pittsburgh, 44-45; vpres & gen mgr, Beckman Supply Co, 45-67; from assoc prof to prof, 67-74, spec asst to chancellor, 76-81, EMER PROF METALL ENG TECHNOL, PURDUE UNIV, 74- *Concurrent Pos:* Gerard Swope fel physics, Mass Inst Technol, 31-32; indust metall consult, 82-88; Paul Harris fel, Rotary Int, 86; mem, City of Hammond Redevelopment Comn, 66-70. *Mem:* AAAS; Am Soc Testing & Mat; Am Soc Metals; Am Inst Mining, Metall & Petrol Engrs; Royal Photog Soc of Gt Brit. *Res:* Development of alloy steels for automotive, aircraft and high temperature uses; electrochemistry of the alkaline-earth metals. *Mailing Add:* 6618 Forest Ave Hammond IN 46324-1003

KAYE, ALVIN MAURICE, b New York, NY, Sept 18, 30; m 58. REPRODUCTIVE ENDROCRINOLOGY, DEVELOPMENTAL BIOLOGY. *Educ:* Columbia Univ, AB, 51, AM, 55; Univ Pa, PhD, 56. *Prof Exp:* Asst cytol, Columbia Univ, 51-52; asst cell physiol, Univ Pa, 53-55; mem res staff biochem cancer, 56-68, sr scientist biodynamics, 68-77, assoc prof, hormone res, 77-86, JOSEPH MOSS PROF, MOLECULAR ENDOCRINOL, HORMONE RES, WEIZMANN INST SCI, 86- *Concurrent Pos:* corresp ed, J Steroid Biochem, steroids & J Endocrine Invest; mem, Int Orgn Comt, Hormones & Cancer Congresses. *Honors & Awards:* Bernhard Zondek Mem Plenary Lectr, VI Int Cong Hormonal Steroids, 82. *Mem:* Israel Chem Soc; Biochem Soc Israel; Am Soc Cell Biol; Endocrine Soc; Sigma Xi; Biochem Soc. *Res:* Enzymic modification of nucleic acids and proteins; hormonal induction of protein synthesis; enzyme catabolism; mechanism of carcinogenesis by ethyl carbamate; regulation of creatine kinase amd ornithine decarboxylase genes; hormones & osteoporosis. *Mailing Add:* Dept Hormone Res Weizmann Inst Sci Rehovot 76100 Israel

KAYE, BRIAN H, b Hull, Eng, July 8, 32; m 57; c 4. PHYSICS. *Educ:* Univ Hull, BSc, 53, MSc, 55; Univ London, PhD(physics), 62. *Prof Exp:* Sci officer, Brit Atomic Weapons Res Estab, 55-59; lectr physics, Univ Nottingham, 59-62; res officer, Welwyn Res Asn, 62-63; sr physicist, IIT Res Inst, 63-68; PROF PHYSICS & DIR INST FINE PARTICLES RES, LAURENTIAN UNIV, 68- *Concurrent Pos:* Consult, Brit Atomic Energy Authority, 61-63; managing dir, Brian Kaye Assocs Ltd; educ & res consult. *Mem:* Am Soc Testing & Mat. *Res:* Particle size analysis of powders; physical and chemical properties of powder systems and aerosols; fractal geometry powder mixing; author of two books. *Mailing Add:* Dept Physics Laurentian Univ Sudbury ON P3E 2C6 Can

KAYE, CHRISTOPHER J, b Wilkes Barre, Pa, June 18, 57. MATERIALS SCIENCE ENGINEERING. *Educ:* Univ Dayton, BS, 79, MS, 88. *Prof Exp:* Qual control engr, Monsanto Res Corp, 79-81; prod engr, 81-87, sr prod engr, 87-90, SR MFG ENGR, EG&G MOUND APPL TECHNOL, 90- *Honors & Awards:* Weapons Recognition Excellence, Dept Energy, 88 & Qual Improv Award, 89. *Mem:* Soc Plastics Engrs; Soc Advan Mat & Process Eng; Soc Mfg Engrs. *Res:* New product applications utilizing alternate materials for improved manufacturability and reduced cost special emphasis on the characterization and use of engineering thermoplastics and thermosets to replace parts traditionally made of metals. *Mailing Add:* EG&G Mound Appl Tech PO Box 3000 Mail Stop E206A Miamisburg OH 45343

KAYE, DONALD, b New York, NY, Aug 12, 31; m 55; c 4. MEDICINE. *Educ:* Yale Univ, AB, 53; NY Univ, MD, 57; Am Bd Internal Med, dipl, 64, cert infectious dis, 74. *Prof Exp:* From asst prof to assoc prof med, Cornell Univ, 63-69; PROF MED & CHMN DEPT, MED COL PA & CHIEF MED, HOSP, 69- *Concurrent Pos:* NIH fel, Cornell Univ Med Col, 60-62, spec fel, 62-63, NY Health Res Coun career scientist award, 66-69; from asst attend physician to assoc attend physician, NY Hosp, 63-69; hon prof, Fed Univ Bahia, Salvadore, Brazil. *Honors & Awards:* Lindback Award. *Mem:* Asn Am Physicians; Asn Profs Med; Infectious Dis Soc Am; master Am Col Physicians; Am Soc Clin Invest. *Res:* Research in infectious diseases with special interest in pathogenesis of bacterial infections and host defense mechanisms against bacterial infection. *Mailing Add:* 1535 Sweet Briar Rd Gladwyne PA 19035

KAYE, GEORGE THOMAS, b Lorain, Ohio, Dec 11, 44; m 67; c 2. SYSTEMS DESIGN. *Educ:* US Naval Acad, BS, 66; Univ Mich, MS, 72, PhD(oceanog), 74. *Prof Exp:* Asst res oceanogr, Sea Grant Prog, Univ Mich, 71-74; asst res oceanogr, Marine Phys Lab, Scripps Inst Oceanog, Univ Calif, San Diego, 74-78; br head, 79-84, PROG MGR, NAVAL OCEAN SYSTS CTR, SAN DIEGO, 85- *Mem:* Acoust Soc Am. *Res:* High-frequency sound scattering from biota and water density structures; theoretical acoustics; upper-ocean measurements with drifting arrays; acoustic noise generation by storms; modal decomposition of internal waves; information processing and data fusion for systems application; optical propagation in the upper ocean. *Mailing Add:* 6414 Corsica Way San Diego CA 92111

KAYE, GORDON I, b New York, NY, Aug 13, 35; m 56; c 2. ANATOMY, PATHOLOGY. *Educ:* Columbia Col, AB, 55; Columbia Univ, AM, 57, PhD(anat), 61. *Prof Exp:* Res assoc anat, Columbia Univ, 61-63, assoc surg path, 63-66, from asst prof to assoc prof, 66-76; prof & chmn, Dept Anat, 76-87, PROF PATH, ALBANY MED COL, 81-, ALDEN MARCH PROF ANAT, CELL BIOL & NEUROBIOL, 87- *Concurrent Pos:* Career scientist, Health Res Coun New York, 63-72; dir, F H Cabot Lab Electron Micros, Columbia Univ, 63-76; consult, NY Vet Admin Hosp, 65-; metab & digestive dis res career award, Nat Inst Arthritis, 72-76; consult surg, 76-78; affil attend surg, Albany Med Ctr Hosp, 78-; prof, Sch Pub Health, State Univ NY, Albany, 85- *Honors & Awards:* Charles Huebschman Prize, Columbia Univ, 54; Tousimis Prize, 81; Raymond C Truex Distinguished lectr, Hahneman Med Col, 87. *Mem:* Am Soc Cell Biol; Am Asn Anat; Asn Anat Chmn (pres, 80-81); Harvey Soc; Sigma Xi; Electron Microscopy Soc Am. *Res:* Electron microscopy; fluid transport; epithelial-mesenchymal interactions in gastrointestinal tissue differentiation; collagen-glycoconjugate interaction in cornea; cell biology of soft tissue tumors. *Mailing Add:* Dept Anat Albany Med Col Albany NY 12208

KAYE, HOWARD, b New York, NY, Dec 9, 38; m 66; c 2. POLYMER CHEMISTRY, INDUSTRIAL CHEMISTRY. *Educ:* Polytech Inst Brooklyn, BS, 60, PhD(polymer chem), 65. *Prof Exp:* NIH fel, Cambridge Univ, 65-67; asst prof chem, Tex A&M Univ, 67-73; pres, Howard Kaye & Assoc, 73-80; DIR, POLYHEDRON LABS, 80- *Concurrent Pos:* Consult chem indust. *Mem:* Fel Am Inst Chemists; Am Chem Soc; Soc Plastic Engrs; Royal Soc Chem; Am Soc Testing Mat; Royal Micros Soc. *Res:* Synthesis and properties of macromolecules; new syntheses for the manufacture of industrially important materials and chemicals; process and product improvement research; characterization of high polymers; automatic chemical analysis; chemical, physical and thermal testing of plastics. *Mailing Add:* PO Box 11669 Houston TX 77293

KAYE, IRVING ALLAN, organic chemistry, for more information see previous edition

KAYE, JACK ALAN, b Brooklyn, NY, Nov 3, 54; m 84; c 1. STRATOSPHERIC MODELING, SATELLITE DATA ANALYSIS. *Educ:* Adelphi Univ, BA, 76; Calif Inst Technol, PhD(chem), 82. *Prof Exp:* Res assoc, Plasma Physics Div, Naval Res Lab, Washington, DC, 82-83; SPACE SCIENTIST, LAB ATMOSPHERES, NASA GODDARD SPACE FLIGHT CTR, 83- *Concurrent Pos:* Co-investr, Satellite Data Anal Proposal, 85-, Strasopheric Modeling Proposal, 86- & Upper Atmosphere Res Satellite, NASA, 88-, Earth Observing Syst Interdisciplinary Sci; prin investr, Diag Anal Trace Constituent & Temperature Variability in Atmosphere. *Mem:* Am Chem Soc; Am Geophys Union; Am Phys Soc. *Res:* The computational modeling and analysis of satellite data of the earth's stratosphere focusing on the interaction of chemical and dynamical processes. *Mailing Add:* NASA Goddard Space Flight Ctr Code 916 Greenbelt MD 20771

KAYE, JAMES HERBERT, b Seattle, Wash, Sept 3, 37; m 65; c 3. RADIOCHEMISTRY. *Educ:* Univ Wash, BS, 58; Carnegie Inst Technol, MS, 61, PhD(nuclear chem), 63. *Prof Exp:* RES SCIENTIST RADIOCHEM, PAC NORTHWEST LABS, BATTELLE MEM INST, 63- *Mem:* Am Chem Soc; Am Nuclear Soc; AAAS. *Res:* Development of highly sensitive instrumentation and techniques for measurement of trace substances in the environment. *Mailing Add:* 2119 Newcomer Ave Richland WA 99352-1830

KAYE, JEROME SIDNEY, b Hartford, Conn, June 15, 30; m 55. CELL BIOLOGY. *Educ:* Columbia Univ, AB, 52, MA, 54, PhD, 57. *Prof Exp:* Instr zool, Univ Calif, Los Angeles, 57-59; from asst prof to assoc prof biol, 59-75, PROF BIOL, UNIV ROCHESTER, 75- *Concurrent Pos:* Lalor Found fel, 58. *Res:* Transacting factors controlling gene transeription. *Mailing Add:* Dept Biol Univ Rochester Rochester NY 14627

KAYE, JOHN, engineering, management science, for more information see previous edition

KAYE, MICHAEL PETER, b Chicago, Ill, Feb 10, 35; m 60; c 5. PHYSIOLOGY, SURGERY. *Educ:* Loyola Univ, Chicago, MS & MD, 59. *Prof Exp:* Clin assoc physiol, Stritch Sch Med, Loyola Univ, 61-62, from asst prof to assoc prof surg & physiol, Med Ctr, 67-71; sci dir artificial heart prog, Res Inst, Ill Inst Technol, 71-74; from assoc prof to prof surg, Mayo Med Sch, Univ Minn, 74-85, dir cardiovasc res, Mayo Clin, 74-85, prof surg & dir res, Heart & Lung Inst, 85-89; PROF SURG, MED CTR, UNIV CALIF, SAN DIEGO, 89- *Concurrent Pos:* Assoc thoracic & cardiovasc, Cook County Hosp, Chicago, 67-74; adj assoc prof physiol, Loyola Univ, 71-74. *Mem:* Am Asn Thoracic Surg; Am Col Surg; Am Col Cardiol; Int Soc Heart Transplantation; Am Physiol Soc; Soc Thoracic Surg. *Res:* Cardiovascular surgery and physiology; neural control of the heart. *Mailing Add:* Div Cardiothoracic Surg Univ Calif Med Ctr 225 Center St San Diego CA 92103

KAYE, NANCY WEBER, b Englewood, NJ, Sept 14, 29; m 56; c 2. EMBRYOLOGY, ANIMAL PHYSIOLOGY. *Educ:* Swarthmore Col, BA, 51; Hunter Col, MEd, 54; Columbia Univ, MA, 58, PhD(zool), 60. *Prof Exp:* Res worker, Col Physicians & Surgeons, Columbia Univ, 62-76; res assoc, Albany Med Col, 76-82, asst prof, Dept Anat, 82-89, res asst prof, Dept Med, 82-89, CONSULT, DEPT MED, ALBANY MED COL, 89- *Concurrent Pos:* Fel neuroanat, Columbia Univ, 60-61; vis lectr, Dept Biol, Rensselaer Polytech Inst, 84. *Mem:* Sigma Xi; AAAS; Am Soc Cell Biol; Asn Res Vision & Ophthal. *Res:* spleen and liver pathobiology and fine structure; gene mapping connective tissue genes; fine structure of connective tissue; corneal fine structure and physiology. *Mailing Add:* 212 Pinewood Ave Troy NY 12180

KAYE, NORMAN JOSEPH, b Milwaukee, Wis, Apr 24, 23; m 47; c 8. APPLIED BUSINESS STATISTICS, MATH APPLIED TO BUSINESS. *Educ:* Marquette Univ, BS, 48; Univ Mich, MBA, 51; Univ Wis, PhD(com), 56. *Prof Exp:* Asst instr math, Marquette Univ, 49-50, asst prof math & statist, 51-55, from assoc prof to prof grad statist, 56-87, EMER PROF GRAD STATIST, MARQUETTE UNIV, 88- *Concurrent Pos:* Commun & Statist consult, 51-88; dept chair, Mgt Dept, Marquette Univ, 56-57 & 69-71. *Mem:* Am Statist Soc. *Res:* Communications and applied statistics. *Mailing Add:* 3137 S 30th St Milwaukee WI 53215

KAYE, ROBERT, b New York, NY, July 17, 17; m 42; c 3. PEDIATRICS. *Educ:* Johns Hopkins Univ, AB, 39, MD, 43. *Prof Exp:* From intern to chief resident pediat, Johns Hopkins Hosp, 43-45; instr, Med Sch, Johns Hopkins Univ, 45; assoc physiol, Sch Pub Health, Harvard Univ, 46-47; instr pediat, Sch Med, Univ Pa, 48-50, assoc, 50-51, from asst prof to prof, 51-73; PROF PEDIAT & CHMN DEPT, HAHNEMANN MED COL & HOSP, 73- *Concurrent Pos:* Asst, Harvard Med Sch, 46-47; asst physician, Children's

Hosp, 48-51, sr physician, 51-, dir clins & clin teaching, 52-57, dep physician-in-chief, 64; asst chmn dept, Sch Med, Univ Pa, 64-73; chmn, Nat Med Adv Bd, Juvenile Diabetes Found, 73-76. *Mem:* AAAS; Am Pediat Soc; Soc Pediat Res; AMA; Am Diabetes Asn. *Res:* Nutrition and metabolism. *Mailing Add:* 34th & Civic Center Blvd Philadelphia PA 19104

KAYE, SAMUEL, b Canton, Ohio, Dec 18, 17; m 41; c 4. ORGANIC CHEMISTRY. *Educ:* Mt Union Col, BS, 40; Ohio State Univ, MS, 41, PhD(chem), 48. *Prof Exp:* Anal chemist, Repub Steel Corp, 41; inspector powder & explosives, Ind Ord Works, 42; aeronaut res scientist chem, Nat Adv Comt Aeronaut, 48-56; tech specialist, Aerojet Gen Corp, 56-57; STAFF SCIENTIST, SPACE SCI DEPT, GEN DYNAMICS/CONVAIR, 57- *Mem:* AAAS; Am Chem Soc; Am Inst Aeronaut & Astronaut; fel Am Inst Chem; Combustion Inst. *Res:* Pollution detection; materials sciences; space manufacturing. *Mailing Add:* 5626 Albalone Pl La Jolla CA 92037-7501

KAYE, SAUL, b Montreal, Que, May 23, 20; m 41; c 2. STERILIZATION, ASEPSIS. *Educ:* Brooklyn Col, BA, 41; Univ Chicago, MS, 69. *Prof Exp:* Chemist, Biol Labs, Chem Corps, US Army, 43-48; res assoc med, Univ Chicago, 48-50; chief decontamination br, Chem Corps, US Army, Ft Detrick, 50-56; res dir, Sterilants Ben Venue Labs, 56-58; res dir, US Movidyn Co, 58-61; PRES, KAYE RES INC, 61- *Honors & Awards:* Kilmer Award. *Mem:* AAAS; Am Chem Soc; Am Soc Microbiol; Sigma Xi; Soc Indust Microbiol; Parenteral Drug Asn. *Res:* Disinfection and sterilization, basic principles; methods of application; development of new methods and devices; aseptic processes. *Mailing Add:* Kaye Res Inc 838 Mich Ave Evanston IL 60202

KAYE, SIDNEY, b Brooklyn, NY, Mar 10, 12; m 51; c 2. TOXICOLOGY. *Educ:* NY Univ, BS, 35, MSc, 39; Med Col Va, PhD(pharmacol), 56; Am Bd Clin Chem, dipl, 52; Nat Registry Clin Chem, cert, 68. *Prof Exp:* Lab asst chem, NY Univ, 31-35, teaching fel, 35-38; res asst toxicol lab, City Off Chief Med Exam, New York, 38-41; instr path, Sch Med, Wash Univ, 46-47; toxicologist & dir toxicol labs, Off Chief Med Exam, Va, 47-62, dir, Richmond Poison Control Ctr, 58-62; prof toxicol, Pharmacol and Legal med & assoc dir, Inst Legal Med, Univ PR, San Juan, 62-82, emer prof pharmacol, Toxicol & Path, 82; RETIRED. *Concurrent Pos:* Toxicologist & assoc dir, Sci Crime Detection Lab, St Louis Police Dept, 46-47; from asst prof to assoc prof, Med Col Va, 47-62; mem subcomt alcohol & drugs, Nat Safety Coun, 52-; lectr, Armed Forces Inst Path, 58-62 & 71; coordr poison control ctrs, Community PR, 64-84; consult toxicologist, Vet Admin Hosp, Richmond, US Army Hosp, San Juan, 64-70, Vet Admin Hosp, San Juan, 69-, USAF Hosp, 71-75 & USN Hosp, 71-82; emer consult toxicol, Dept of US Army, 70-; exec res liaison officer, Defense Civil Prep Agency, Fed Emergency Mgt Admin, Dept Defense, 73- *Honors & Awards:* Award of Merit, Am Acad Forensic Soc, 73, Gettler Outstanding Achievement Award, 85; Milton Helpern Award, Nat Asn Med Examnrs, 89. *Mem:* Assoc fel Am Soc Clin Path; Asn Mil Surg US; Soc Toxicol; Pan Am Med Asn; Sigma Xi. *Res:* Analytical method for detection of lead poisoning; identification of seminal stains; diagnosis of poisoning; alcohol and its effects on man. *Mailing Add:* Condominium Mundo Feliz Isla Verde PR 00913

KAYE, STEPHEN VINCENT, b Rahway, NJ, Sept 17, 35; m 59; c 3. HEALTH PHYSICS, RADIOECOLOGY. *Educ:* Rutgers Univ, BS, 57; NC State Univ, MS, 59; Univ Rochester, PhD(radiation biol), 66. *Prof Exp:* Res staff health physics & radioecol, Ecol Div & Health Physics Div, Oak Ridge Nat Labs, 60-72, proj supvr environ impacts, Environ Sci Div, 73-75, sect head radiol assessments, 75-77, div dir, biol div, 87-88, DIV DIR, HEALTH & SAFETY RES, OAK RIDGE NAT LAB, 77- *Concurrent Pos:* Adv health physics fel, Univ Rochester, 63-66; mem nuclear fuel subgroup, Comt Nuclear & Alternative Energy Systs, Nat Res Coun, 76-78; chmn radiol data working group, US Dept Energy Reactor Safety Data Coord Group, 76-79; mem support group to develop proc guide for probabilistic risk assesment, Inst Elec & Electronics Engrs, Am Nuclear Soc, Nuclear Regulatory Comn, & nuclear indust, 81-82; mem, Environ Protection Agency High Level Radioactive Waste Comt, Sci Adv Bd, 82-84; mem, Tech Adv, Fla Phosphate Res Inst, 86-87; consult, US Vet Admin, 90- *Mem:* Health Physics Soc; Soc Risk Analysis; Am Nuclear Soc; Sigma Xi; AAAS. *Res:* Transport of radionuclides in the environment and estimation of dose to man from ingestion or external exposure; assessments and comparisons of health and environmental issues related to all energy technologies. *Mailing Add:* Health & Safety Res Div Oak Ridge Nat Lab PO Box 2008 Oak Ridge TN 37831-0124

KAYE, WILBUR (IRVING), b Pelham Manor, NY, Jan 28, 22; m 44; c 2. CHEMISTRY, INSTRUMENTATION. *Educ:* Stetson Univ, BS, 42; Univ Ill, PhD(chem), 45. *Prof Exp:* Asst chem, Univ Ill, 42-44; sr res chemist, Tenn Eastman Corp, 45-55; dir res, Sci Instruments Div, Fullerton, 56-68, dir sci res, Corp Res Activity, 68-73, sr scientist, 73-80, prin staff scientist, Beckman Instruments Inc, Irvine, 80- 87; RETIRED. *Concurrent Pos:* Beckman fel. *Mem:* Am Chem Soc; Optical Soc Am; Soc Appl Spectros. *Res:* Infrared and ultraviolet spectroscopy; chromatography; instrument development. *Mailing Add:* 3607 Surfview Lane Corona del Mar CA 92625

KAYES, STEPHEN GEOFFREY, b Madison, Wis, May 1, 46. IMMUNOPARASITOLOGY. *Educ:* Univ Wis-Madison, BS, 71; Tulane Univ, MS, 73; Univ Iowa, PhD(anat), 77. *Prof Exp:* Fel immunol, Sch Med, Vanderbilt Univ, 77-81; asst prof neuroanat, Dept Anat & asst prof parasitol, 81-86, ASSOC PROF ANAT & CELL BIOL, 86-, ASSOC PROF PARASITOL, DEPT MICROBIOL, UNIV SOUTH ALA, 86- *Concurrent Pos:* Res assoc, Vet Admin Med Ctr, 77-81. *Mem:* Am Soc Trop Med & Hyg; Am Soc Parasitologists; AAAS; Am Assoc Anat. *Res:* Immunologic basis of the host-parasite relationship by correlation of the host's immune status with the pathology elicited by the parasite. *Mailing Add:* Dept Anat Univ SAla Mobile AL 36688

KAYHART, MARION, b Butler, NJ, Sept 14, 26. GENETICS. *Educ:* Drew Univ, BA, 47; Univ Pa, MA, 49, PhD(zool), 54. *Prof Exp:* From instr to asst prof biol, Roanoke Col, 49-52; from asst prof to assoc prof, 54-57, PROF BIOL & CHMN DEPT, CEDAR CREST COL, 57- *Mem:* AAAS; Genetics Soc Am; Nat Asn Biol Teachers; Sigma Xi. *Res:* Radiation genetics. *Mailing Add:* Dept Biol Cedar Crest Col Allentown PA 18104

KAYLL, ALBERT JAMES, b Vancouver, BC, Jan 21, 35; m 62; c 2. FORESTRY. *Educ:* Univ BC, BSF, 59; Duke Univ, MF, 60; Aberdeen Univ, PhD(fire ecol), 64. *Prof Exp:* Res scientist fire ecol, Can Dept Forestry, 60-68; asst prof fire ecol, Univ NB, 68-71, actg chmn, Dept Forest Resources, 75-76, assoc prof fire ecol, 71-77, co-dir, Fire Sci Ctr, 70-78, prof fire ecol & chmn, Dept Forest Resources, 77-80; prof & dir, Sch Forestry, 81-87, PROF, LAKEHEAD UNIV, 87- *Concurrent Pos:* Mem fire mgt working group, NAm Forestry Comn, Food & Agr Orgn, UN, 73-75; Can Forestry Accreditation Bd, 89- *Honors & Awards:* H R MacMillan Prize, 59. *Mem:* Soc Am Foresters; Can Inst Forestry (pres, 87-88); Asn Univ Forestry Schs Can. *Res:* Ecological and physiological effects of fire on forest vegetation. *Mailing Add:* Sch Forestry Lakehead Univ Thunder Bay ON P7B 5E1 Can

KAYLOR, HOYT MCCOY, b Alexander City, Ala, Aug 17, 23; m 57; c 2. OPTICAL PHYSICS. *Educ:* Birmingham-Southern Col, BS, 43; Univ Tenn, MS, 49, PhD(physics), 53. *Prof Exp:* Assoc prof, 52-58, prof physics, 58-81, PROF PHYSICS & MATH, BIRMINGHAM-SOUTHERN COL, 81- *Mem:* Fel AAAS; Am Phys Soc; Optical Soc Am; Am Asn Physics Teachers. *Res:* High dispersion infrared spectroscopy; physical properties of optical materials. *Mailing Add:* Birmingham-Southern Col PO Box A36 Birmingham AL 35254

KAYNE, FREDRICK JAY, b Washington, DC, Jan 19, 41; m 65; c 1. BIOCHEMISTRY. *Educ:* Ill Inst Technol, BS, 62; Mich State Univ, PhD(biochem), 66. *Prof Exp:* NATO fel, Max Planck Inst Phys Chem, 67, res assoc phys chem, 67-69; asst prof phys biochem, Johnson Res Found, 69-74, assoc prof biochem & biophys, Univ Pa, 74-77; ASSOC PROF PATH & LAB MED, HAHNEMANN MED COL, 77- *Concurrent Pos:* Fulbright res fel, Max Planck Inst Biophys, 85-86. *Mem:* Am Asn Clin Chemists; Am Soc Biochem & Molecular Biol. *Res:* Enzyme mechanisms; chemical relaxation; clinical chemistry; control of cell metabolism. *Mailing Add:* Dept Path Broad & Vine Sts Philadelphia PA 19102

KAYNE, HERBERT LAWRENCE, b Chicago, Ill, Sept 22, 34; m 62; c 2. PHYSIOLOGY. *Educ:* Univ Ill, BS, 55, MS, 58, PhD(physiol), 62. *Prof Exp:* From instr to asst prof, 62-69, ASSOC PROF PHYSIOL, SCH MED, BOSTON UNIV, 69- *Mem:* Am Physiol Soc; Biomet Soc. *Res:* Biostatistics. *Mailing Add:* Dept Physiol Boston Univ Sch Med 80 E Conard St Boston MA 02118

KAYNE, MARLENE STEINMETZ, b Bronx, NY, July 6, 41; m 65. BIOCHEMISTRY, MOLECULAR BIOLOGY. *Educ:* St John's Univ, BS, 62; Mich State Univ, PhD(biochem), 66. *Prof Exp:* Fel immunol, Max Planck Inst Exp Med, 67-69; res assoc enzym, Dept Biophysics, Univ Pa, 70-74 & Dept Biol, 74-77; ASST PROF MOLECULAR BIOL & CHMN DEPT BIOL, TRENTON STATE COL, 77- *Concurrent Pos:* Ger Res Asn fel, 67-69; NSF res grants, 76-77 & 78- *Mem:* Sigma Xi. *Res:* Purification and characterization of procaryotic enzymes required in protein biosynthesis. *Mailing Add:* Four St Ames Pl Yardley PA 19067

KAYS, M ALLAN, b Princeton, Ind, May 13, 34; m 55; c 3. GEOLOGY, PETROLOGY. *Educ:* Southern Ill Univ, BA, 56; Univ Wash, St Louis, MA, 58, PhD(geol), 61. *Prof Exp:* From asst prof to assoc prof, 61-80, PROF GEOL, UNIV ORE, 80- *Concurrent Pos:* Vis geologist, Precambrian Geol Div, Dept Mineral Resources, Prov Sask, Can, 70-71; part-time geologist, US Geol Surv, 79, 80. *Mem:* Am Geophys Union; Geol Soc Am. *Res:* Petrology of xenoliths and their fused products in margins of basic intrusions; petrology and structural relations of Archaean supracrustal metamorphic and plutonic rocks, East Greenland; petrology of migmatized gneisses of Canada, Finland and East Greenland; metamorphism and structure of convergent plate marginal sequences in cordillera of western North America. *Mailing Add:* Dept Geol Univ Ore Eugene OR 97403-1272

KAYS, STANLEY J, b Stillwater, Okla, Feb 3, 45. HORTICULTURE, VEGETABLE CROP PHYSIOLOGY. *Educ:* Okla State Univ, BS, 68; Mich State Univ, MS, 69, PhD(hort), 71. *Prof Exp:* Researcher plant biol, Dept Biol, Tex A&M Univ, 71; researcher, Sch Plant Biol, Univ Col Northern Wales, UK, 71-72; asst prof veg crops, Dept Hort, Univ Ga, Tifton, 73-75; assoc prof, Dept Hort Food Sci, Univ Ark, 76-77; ASSOC PROF VEG CROPS & POST-HARVEST, DEPT HORT, UNIV GA, 77- *Concurrent Pos:* Grants, Nat Pecan Shellers, 76, Gilroy Foods Inc, 78, Woolfolk Chem Works Inc, 78, Sci Educ Adm, USDA, 81, AID 86; vis scientist, Dept Appl Biol, Cambridge Univ, Cambridge, Eng; vis scholar, Wolfson Col, Cambridge. *Mem:* Am Soc Hort Sci; Int Hort Soc; AAAS; Int Trop Root Crops Soc; Sigma Xi. *Res:* Developmental and post-harvest physiology of vegetable crops. *Mailing Add:* Dept Hort-Plant Sci Univ Ga Athens GA 30602

KAYS, WILLIAM MORROW, b Norfolk, Va, July 29, 20; m 47, 83; c 4. MECHANICAL ENGINEERING. *Educ:* Stanford Univ, AB, 42, MS, 47, PhD(mech eng), 51. *Prof Exp:* Res assoc, 47-51, from asst prof to assoc prof, 51-57, head dept, 61-72, dean eng, 72-84, PROF MECH ENG, STANFORD UNIV, 57- *Concurrent Pos:* Fulbright lectr, Imp Col London, 59-60; NSF sr fel, 66-67. *Honors & Awards:* Am Soc Mech Engrs Mem Award, 65. *Mem:* Nat Acad Eng; Fel Am Soc Mech Engrs; fel Am Soc Eng Educ. *Res:* Heat transfer to fluids, especially turbulent boundary layers. *Mailing Add:* Sch Eng Stanford Univ Stanford CA 94305

KAYSER, BORIS JULES, b New York, NY, June 2, 38; m 60. THEORETICAL ELEMENTARY PARTICLE PHYSICS. *Educ:* Princeton Univ, AB, 60; Calif Inst Technol, PhD(physics), 64. *Prof Exp:* Res assoc physics, Univ Calif, Berkeley, 64-66; asst prof, State Univ NY Stony Brook, 66-69; asst prof, Northwestern Univ, Evanston, 69-74; assoc prog dir theoret physics, 72-75, PROG DIR THEORET PHYSICS, NSF, 75- *Mem:* Fel Am Phys Soc. *Res:* Weak interactions. *Mailing Add:* Div Physics Nat Sci Found Washington DC 20550

KAYSER, FRANCIS X, b Toledo, Ohio, Feb 10, 27; m 52; c 5. METALLURGICAL ENGINEERING. *Educ:* Univ Notre Dame, BS, 48; Mass Inst Technol, MS, 50, DSc(metall), 63. *Prof Exp:* Metallurgist, Unitcast Corp, Ohio, 48-49; res metallurgist, Res Labs Div, Gen Motors Corp, Mich, 50-55; res metallurgist, sci labs, Ford Motor Co, 55-58; asst prof phys metall, Inst Atomic Res, Ames Lab, 63-70; assoc prof, 70-75, PROF METALL, IOWA STATE UNIV, 75- *Concurrent Pos:* Assoc metallurgist, Inst Atomic Res, Ames Lab, 63- *Honors & Awards:* Presidents Award, Am Soc Metals Int, 88. *Mem:* Am Soc Metals; Sigma Xi. *Res:* Nature of solid solutions; diffraction; phase transformations in solids; elastic and plastic behavior of metallic and non-metallic materials. *Mailing Add:* 2818 Torrey Pines Rd Ames IA 50010

KAYSER, RICHARD FRANCIS, b Toledo, Ohio, Feb 24, 25; m 50; c 9. TECHNICAL MANAGEMENT. *Educ:* Univ Cincinnati, PhD(chem eng), 52. *Prof Exp:* Engr, Linde Div, 52-65, prod mgr, Silicones Div, 65-72, opers mgr, 72-74, dir, Res & Develop, Chem & Plastics Div, 74-80, vpres technol, Ethylene Oxide/Glycol Div, Union Carbide Corp, 80-; ACTG DEP DIR CTR CHEM TECHNOL, NAT BUR STANDARDS TECHNOL, GAITHERSBURG. *Mem:* Sigma Xi. *Res:* Silicones processes; low pressure oxo process and oxo alcohols; new ethylene oxide catalyst developments. *Mailing Add:* A105 Phys Bldg Nat Inst Standards Technol Gaithersburg MD 20899

KAYSER, ROBERT HELMUT, b Orange, NJ, Aug 21, 48; m 72; c 1. BIO-ORGANIC CHEMISTRY. *Educ:* Stevens Inst Technol, BS, 70; Georgetown Univ, PhD(org chem), 75. *Prof Exp:* Res asst org chem, Georgetown Univ, 74; res assoc bio-org chem, Univ Md, Baltimore County, 74-80; MEM STAFF US ENVIRON PROTECTION AGENCY, WASHINGTON, DC, 80- *Mem:* Am Chem Soc. *Res:* Designing model systems in an attempt to mimic various enzymatic processes and elucidate enzymatic mechanisms. *Mailing Add:* 5355 Iron Pen Pl Columbia MD 21044-3310

KAYTON, MYRON, b New York, NY, Apr 26, 34; m 54; c 2. VEHICLE SYSTEM DESIGN, VEHICLE ELECTRONICS. *Educ:* Cooper Union, BS, 55; Harvard Univ, SM, 56; Mass Inst Technol, PhD(aeronaut & astronaut), 60. *Prof Exp:* Design engr, Res & Develop Div, Avco, 56-58; res asst navig systs, Draper Lab, Mass Inst Technol, 58-60; sect head, Guid & Control Div, Litton, 60-65; mgr, NASA Johnson Space Ctr, 65-69; sr staff mem, TRW Defense & Space Sector, 69-81; PRES, KAYTON ENG CO, 81- *Concurrent Pos:* Lectr, Univ Calif, Los Angeles, 69-88; mem, bd gov, Inst Elec & Electronics Engrs, 83- & Space Serv Working Group, NASA, 90- *Honors & Awards:* M B Carlton Award, Inst Elec & Electronics Engrs, 88. *Mem:* Inst Elec & Electronics Engrs; Am Soc Mech Engrs; Inst Navig. *Res:* Electronic system design and testing for high-value vehicles and plants, emphasizing communications, navigation, control, fault-tolerance; author of two books. *Mailing Add:* Kayton Eng Co PO Box 802 Santa Monica CA 90406

KAZAHAYA, MASAHIRO MATT, b Yonago, Japan, Jan 24, 32; US citizen; m 60; c 3. MANAGING EXPERIENCE OF RESEARCH & DEVELOPMENT ENGINEERS, COMPUTER-TO-COMPUTER COMMUNICATION. *Educ:* Okayama Univ-Japan, BS, 54, PhD(technol), 89; Univ Pa, MBA, 77. *Prof Exp:* Dir res, 85-91, VPRES TECHNOL MKT & STRATEGY, FISCHER & PORTER CO, WARMINSTER, PA, 91- *Concurrent Pos:* Lectr corp planning, Wharton Grad Sch, Univ Pa, 78-89; voting mem, SP50 Fieldbu Standard Making Comt, Instrument Soc Am, 85-; voting mem & US Rep, Int Electro-Tech Comn, SC65 WG6 Comt, 86- *Mem:* Sr mem Instrument Soc Am; sr mem Inst Elec & Electronics Engrs; Soc Mfg Engrs. *Res:* Develop new measuring instruments and control systems; corporate planning and business modeling. *Mailing Add:* 1275 Woods Rd Southampton PA 18966

KAZAKIA, JACOB YAKOVOS, b Istanbul, Turkey, Feb 27, 45; m 72; c 2. APPLIED MECHANICS. *Educ:* Istanbul Tech Univ, MS, 68; Lehigh Univ, PhD(appl mech), 72. *Prof Exp:* Res assoc, 72-74, asst prof, 74-79, ASSOC PROF, CTR APPLN MATH, LEHIGH UNIV, 79- *Mem:* Am Acad Mech; Am Soc Mech Engrs. *Res:* Nonlinear wave propagation in fluids; viscoelastic fluid flows; stability of liquid filled shells; run-up and spin-up problems. *Mailing Add:* Dept Math Lehigh Univ Bethlehem PA 18015

KAZAKS, PETER ALEXANDER, b Riga, Latvia, Feb 22, 40; US citizen; m 68; c 6. THEORETICAL NUCLEAR PHYSICS. *Educ:* McGill Univ, BSc, 62; Yale Univ, MS, 63; Univ Calif, Davis, PhD(physics), 68. *Prof Exp:* Res assoc, Ohio Univ, 68-70; asst prof physics, St Lawrence Univ, 70-73; asst prof, 73-75, PROF PHYSICS, UNIV SFLA, 80- *Concurrent Pos:* NSF res partic, Univ Fla, 71; NSF grant, St Lawrence Univ, 71-72; Res Corp grant, 75; vis prof/scholar, Univ Penn, 85; chmn, Div Natural Sci, New Col, Univ SFla, 80-85. *Mem:* Am Phys Soc. *Res:* Three-body models of nuclear reactions; electron-atom collision collisions; pion-nucleus scattering; proton-proton scattering; spin physics. *Mailing Add:* Div Nat Sci New Col Univ SFla Sarasota FL 34243

KAZAL, LOUIS ANTHONY, b Newark, NJ, July 2, 12; m 42; c 4. BIOCHEMISTRY, HEMATOLOGY. *Educ:* Seton Hall Col, BS, 35; Rutgers Univ, PhD(biochem, physiol), 40. *Prof Exp:* Asst physiol & biochem, Rutgers Univ, 37-40; res biochemist, Merck Sharp & Dohme, Inc, 40-50, dir biol develop, 50-54, mgr tech info, 54-55, tech asst to med dir, 55-56; head sect blood plasma fractionation, Cardeza Found, Jefferson Med Col, 56-60, from asst prof to prof physiol, 57-78, assoc prof med & assoc dir Cardeza Found Hemat Res, 60-78, prof physiol, Col Grad Studies, 70-78, HON PROF PHYSIOL & HON ASSOC PROF MED, JEFFERSON MED COL, THOMAS JEFFERSON UNIV, 78- *Concurrent Pos:* Chmn blood coagulation sessions, Fed Am Soc Exp Biol, 63-68. *Honors & Awards:* Co-recipient Rorer Award, Am J Gastroenterol, 66. *Mem:* Am Chem Soc; Soc Exp Biol & Med; Am Soc Biol Chem; fel NY Acad Sci; Int Soc Thrombosis & Homeostasis; fel AAAS; Sigma Xi (pres, 67); Asn Clin Scientists. *Res:* Blood coagulation; proteins; blood group specific substances; ion-exchange resins; erythropoietin inhibitors; lipids; trypsin inhibitor; isolation and crystalization of pancreatic secretory trypsin inhibitor; gastric juice and saliva thromboplastin; fibrinogen-glycine; human typing serum; author of numerous publications and recipient of one US patent. *Mailing Add:* 18215 Organ Pipe Dr Sun City AZ 85373

KAZAN, BENJAMIN, b New York, NY, May 8, 17; div; c 1. PHYSICS. *Educ:* Calif Inst Technol, BS, 38; Columbia Univ, MA, 40; Munich Tech Univ, Dr rer nat, 61. *Prof Exp:* Radio engr, Signal Corps Eng Labs, 40-44, chief spec purpose tube sect, 44-50; physicist, RCA Labs, 50-58; head solid-state display sect, Res Labs, Hughes Aircraft Co, 58-61; chief scientist aerospace electronics div, Electro-Optical Systs, Inc, 61-68; mgr explor display dept, Thomas J Watson Res Ctr, IBM Corp, 68-74; head, Display Group, Xerox Corp, 74-84; CONSULT, 84- *Concurrent Pos:* Ed, Advan in Image Pickup & Display, 72-84; assoc ed, Inst Elec & Electronics Engrs, Trans on Electron Devices, 78-84 & Advan Electronics & Electron Physics, 85-; consult adv group electron devices, Defense Dept, 73-82; adj prof, Univ RI, 70-74. *Honors & Awards:* Silver Medal, Am Roentgen Ray Soc, 57; Coolidge Award, Gen Elec Co, 58. *Mem:* Sigma Xi; Am Phys Soc; Inst Elec & Electronics Engrs; Soc Info Display. *Res:* Electronic image storage; image pickup and display devices; display devices; solid-state image devices. *Mailing Add:* 557 Tyndall St Los Altos CA 94022

KAZANJIAN, ARMEN ROUPEN, b New Haven, Conn, Feb 13, 28; m 62; c 3. PHYSICAL CHEMISTRY. *Educ:* Northeastern Univ, BS, 51; Univ Calif, Los Angeles, PhD(phys chem), 65. *Prof Exp:* Chemist, Raw Mat Develop Lab, AEC, 51-56; res chemist, Rocketdyne Div, NAm Aviation, Inc, 56-60; res chemist, Rocket Power Inc, 66; res chemist, Rocky Flats Div, Dow Chem Co, 66-75; res chemist, Rocky Flats Div, Rockwell Int, 75-90; RETIRED. *Mem:* AAAS; Radiation Res Soc; Sigma Xi; Nuclear Soc Am. *Res:* Chemical effects of nuclear transformations; radiation chemistry; plutonium chemistry. *Mailing Add:* 1596 Snee-Oosh Rd La Conner WA 98257

KAZARIAN, LEON EDWARD, b Norwalk, Conn. BIOMECHANICS. *Educ:* Northrop Inst Technol, BSAAE, 66; Karolinska Inst, Sweden, Dr Ing(orthop biomech), 72. *Prof Exp:* Res scientist biomech, Aerospace Med Res Labs, Wright Patterson AFB, 67-91; RETIRED. *Mem:* Orthop Res Soc; Aerospace Med Asn. *Res:* Hard and soft tissue mechanics as related to dynamic environments. *Mailing Add:* 1062 Geneva Rd Beaver Creek OH 45385

KAZARINOFF, MICHAEL N, b Ann Arbor, Mich, Mar 24, 49; m 70; c 3. BIOCHEMISTRY. *Educ:* Yale Univ, BS, 70; Cornell Univ, PhD(biochem), 75. *Prof Exp:* Fel biochem, Univ Calif, Berkeley, 75-76; fel microbiol, Univ Tex, Austin, 76-78; asst prof, 78-84, ASSOC PROF NUTRIT BIOCHEM, CORNELL UNIV, 85- *Concurrent Pos:* Actg Dir, Div Nutrit Sci, Cornell Univ, 87-88; assoc dir, Pew Nat Nutrit Prog, 87- *Mem:* Am Chem Soc; AAAS; Am Inst Nutrit; Am Soc Biol Chem. *Res:* Enzymology; protein-coenzyme interactions; protein turnover; coenzyme mechanisms; purification and properties of enzymes of vitamin metabolism; nutrition and cancer. *Mailing Add:* Div Nutrit Sci Cornell Univ Ithaca NY 14853

KAZARINOFF, NICHOLAS D, b Ann Arbor, Mich, Aug 12, 29; m 48; c 6. MATHEMATICS. *Educ:* Univ Mich, BS, 50, MS, 51; Univ Wis, PhD, 54. *Prof Exp:* Asst math, Univ Wis, 51-53; from instr to asst prof, Purdue Univ, 53-56; from asst prof to prof, Univ Mich, 56-71, consult elec eng, Radiation Lab, 58-65; Martin prof, 72-77, chmn dept, 71-75, PROF MATH, STATE UNIV NY BUFFALO, 71- *Concurrent Pos:* Res assoc, Univ Wis, 54; exchange prof, Steklov Math Inst, Moscow, 60-61 & 65, vis Ctr Nat Res Prof, Trento, Italy, 78, 80. *Mem:* Soc Indust & Appl Math; Am Math Soc; Math Asn Am; Am Acad Mech; fel AAAS. *Res:* Ordinary and partial differential equations and applications; bifurcation theory; geometry; inequalities; scattering and diffraction problems; iteration of complex maps. *Mailing Add:* 106 Dienfendorf Hall State Univ NY Buffalo NY 14214-3093

KAZDA, LOUIS F(RANK), b Dayton, Ohio, Sept 21, 16; m 40; c 3. ELECTRICAL ENGINEERING. *Educ:* Univ Cincinnati, EE, 40, MSE, 43; Syracuse Univ, PhD(elec eng), 62. *Prof Exp:* Res & develop engr elec eng, Bendix Aviation Corp, 43-46; from instr to assoc prof, 47-60, prof elec eng, 60-81, PROF ELEC & COMPUT ENG, UNIV MICH, ANN ARBOR, 81- *Concurrent Pos:* Consult, Cook Res Labs, 51-53, Willow Run Labs, Mich, 54-, USAF, 57-59, Maxitrol Corp, 58-68, Clark Equipment Co, 60-62, Ford Motor Co, 62 & Conduction Corp, 63. *Mem:* AAAS; fel Inst Elec & Electronics Engrs. *Res:* Feedback control systems of linear, nonlinear or adaptive type; inertial navigation systems; application of system engineering techniques to societal problems. *Mailing Add:* 3013 Ronna Dr Las Cruces NM 88001

KAZDAN, JERRY LAWRENCE, b Detroit, Mich, Oct 31, 37. GEOMETRY, MATHEMATICAL ANALYSIS. *Educ:* Rensselaer Polytech Inst, BS, 59; NY Univ, MS, 61, PhD(math), 63. *Prof Exp:* Instr math, NY Univ, 63; Benjamin Peirce instr, Harvard Univ, 63-66; from asst prof to assoc prof, 66-74, PROF MATH, UNIV PA, 74-, CHAIR, 89- *Concurrent Pos:* Vis assoc prof, Harvard Univ, 71-72; vis prof, Univ Calif, Berkeley, 74-76, Univ Paris, 81, Kyoto Univ, 85-86. *Mem:* Am Math Soc. *Res:* Partial differential equations; differential geometry. *Mailing Add:* Dept Math Univ Pa Philadelphia PA 19104-6395

KAZEM, SAYYED M, b Kabul, Afghanistan, July 28, 38; US & Afghan citizen; m 72; c 4. MATERIALS TESTING, REFRIGERATION. *Educ:* Tulsa Univ, BS, 63, MS, 64; Purdue Univ, MS, 71. *Prof Exp:* Asst prof chem eng, Fac Eng, Kabul Univ, Afghanistan, 67-77; asst prof, 80-87, ASSOC PROF HEAT POWER & MAT, MET DEPT, SCH TECHNOL, PURDUE UNIV, 87- *Concurrent Pos:* Mem, Nat Res Coun, Kabul Univ, 71-73 & 75-77, head appl eng res & consult, Fac Eng, 75-77. *Mem:* Am Soc Mech Engrs; Am Soc Eng Educ; Am Soc Heating Refrig & Air-Conditioning Engrs. *Res:* Purification of vegetable oils; refrigeration, pressure drops and heat transfer; steam power plants; heat power; materials. *Mailing Add:* MET Dept Knoy Hall West Lafayette IN 47906

KAZEMI, HOMAYOUN, b Teheran, Iran, Sept 28, 34; US citizen; m 58; c 2. PULMONARY MEDICINE. *Educ:* Univ London , MB, 53; Lafayette Col, BA, 54; Columbia Univ, MD, 58. *Hon Degrees:* MSc, Harvard Univ, 90. *Prof Exp:* Intern, 58-59, asst res med, Bassett Hosp, 59-6, res fel, 60- 61, Am Heart Asn res fel, 61-62; resident med, 63, Am Heart Asn res fel, 64, chief pulmonary unit, 67-88, CHIEF, PULMONARY & CRITICAL CARE UNIT, MASS GEN HOSP, 88-; PROF MED, HARVARD MED SCH, HARVARD-MASS INST TECHNOL DIV HEALTH SCI & TECHNOL, 80- *Concurrent Pos:* Vis fel, Hammersmith Hosp, Royal Postgrad Med Sch, London, 65; Consult, Brigham & Women's Hosp, Boston, 65-82, Nat Heart Lung & Blood Inst, 73-, Fed Aviation Agency, 87- *Honors & Awards:* P D Agarwal Orator, Calcutta, India, 87. *Mem:* Am Thoracic Soc; Am Physiol Soc; Am Heart Asn; Am Soc Clin Invest. *Res:* Central chemical control of ventilation as it relates to biochemistry of the respiratory centers; role of brain metabolism & amino acid neurotransmitters in determining the central ventilatory drive. *Mailing Add:* Mass Gen Hosp Pulmonary Unit 32 Fruit St Boston MA 02114

KAZEMI, HOSSEIN, b Iran, Mar 11, 38; m 64; c 3. PETROLEUM ENGINEERING. *Educ:* Univ Tex, BS, 61, PhD(petrol eng), 63. *Prof Exp:* Sr res scientist, reservoir eng, Tulas Res Ctr, Sinclair Oil & Gas Co, 63-69, res scientist, Atlantic Richfield Co, 69; adv res scientist, sr res scientist & res assoc, Marathon Oil Co, 69-80, mgr eng dept, 81-86, mgr, Reservoir Mgt Dept, Explor & Prod Technol Ctr, 86-88, ASSOC DIR, PROD TECHNOL, PROD TECHNOL CTR, MARATHON OIL CO, 88- *Concurrent Pos:* Eng fel, Univ Tex, 61; lectr math, Univ Tulsa, 67-69; adj prof petrol eng, Colo Sch Mines, 81- *Honors & Awards:* Henry Matlson Technical Award, Soc Petrol Engrs, 80; John Franklin Carll Award, 87. *Mem:* Soc Petrol Engrs. *Res:* Solution mining; pressure transient testing of oil and gas wells; reservoir simulation; enhanced oil recovery, naturally fractured reservoirs. *Mailing Add:* Petrol Technol Ctr Marathon Oil Co PO Box 269 Littleton CO 80160

KAZEROUNIAN, KAZEM, b Shiraz, Iran, Nov 3, 56; US citizen; m 76; c 2. OPTIMIZATION, REDUNDANCY RESOLUTION. *Educ:* Univ Ill, BS, 80, MS, 81, PhD(mech design). *Prof Exp:* Res asst, Univ Ill, 80-84; asst prof, 84-89, ASSOC PROF DESIGN & ROBOTICS, UNIV CONN, 89- *Concurrent Pos:* Consult, specialized mechanics, Gen Elec, Bran Rex & Rogers, Inc, 84-91; reviewer, Inst Elec & Electronics Engrs, Int J Robotics Res, Robotics & Automation, Am Soc Mech Engrs, 84-91. *Res:* Analysis and design optimization of robotic systems and mechanisms, theoretical and applied kinematics. *Mailing Add:* 165 Davis Rd Storrs CT 06268

KAZES, EMIL, b Istanbul, Turkey, June 13, 26; nat US; m 54; c 3. THEORETICAL PHYSICS. *Educ:* Univ Wis, BS, 49, MS, 50; Univ Chicago, PhD(physics), 56. *Prof Exp:* Proj assoc physics, Univ Wis, 57-59; from asst prof to assoc prof, 59-66, PROF PHYSICS, PA STATE UNIV, 66- *Mem:* Fel Am Phys Soc. *Res:* Electrodynamics; general relativity; soluble field theories; elementary particle theory; pion nucleon interaction, current algebra. *Mailing Add:* 337 Davey Lab Penn State Univ University Park PA 16802

KAZHDAN, DAVID, MATHEMATICS. *Prof Exp:* PROF MATH, HARVARD UNIV. *Mem:* Nat Acad Sci. *Mailing Add:* Math Dept Sci Ctr 325 Harvard Univ Cambridge MA 02138

KAZI, ABDUL HALIM, b Kreuzlingen, Switz, Jan 12, 35; US citizen; m 59; c 2. RADIATION EFFECTS, NEUTRON SOURCES. *Educ:* Am Univ, Cairo, Egypt, BSc, 54; Rensselaer Polytech Inst, MS, 56; Mass Inst Technol, SM, 59, PhD(nuclear eng), 61. *Prof Exp:* Staff mem, Gen Atomics, La Jolla, Calif, 61-63; sect chief, United Nuclear, White Plains, NY, 63-66; prin investr, Army Pulse Radiation Facil, 66-87, DIR, NUCLEAR EFFECTS DIRECTORATE, US ARMY COMBAT SYSTS TEST ACTIV, ABERDEEN PROVING GROUND, 87- *Concurrent Pos:* Mem, NATO Panel VII & VIII Res Study Groups, 76-, Radiation Dosimetry Standards Comt, Am Soc Testing & Mat, 84- & Multi Serv Test & Res Investment Comt Nuclear Effects, 89- *Mem:* Am Nuclear Soc. *Res:* Design, operation and utilization of nuclear weapon radiation simulators; neutron sources; flash gamma accelerators; radiation dosimetry; radiation effects. *Mailing Add:* 2813 Rocks Rd Jarrettsville MD 21084

KAZIMI, MUJID S, b Jerusalem, Palestine, Nov 20, 47; US citizen; m 73; c 3. THERMAL ENGINEERING, SAFETY ENGINEERING. *Educ:* Univ Alexandria, Egypt, BEng, 69; Mass Inst Technol, MS, 71, PhD(nuclear eng), 73. *Prof Exp:* Sr engr, Westinghouse Elec Corp, 73-74; assoc scientist, Brookhaven Nat Lab, 74-76; from asst prof to assoc prof, 76-86, PROF NUCLEAR ENG, MASS INST TECHNOL, 86-, DEPT HEAD NUCLEAR ENG, 89- *Concurrent Pos:* Consult, Brookhaven Nat Lab, 76-, Elec Power Res Inst, 89- & Argonne Nat Lab, 89-; pres, Asn Arab Am Univ Graduates, Inc, 80 & 87; chmn, High Level Waste Tech Adv Panel, Dept Energy, 90- *Mem:* Am Nuclear Soc; Am Soc Mech Engrs; Am Inst Chem Engrs; Am Soc Eng Educ. *Res:* Thermal design and safety of nuclear facilities, including nuclear power reactors, nuclear waste storage facilities and nuclear fusion research facilities. *Mailing Add:* Mass Inst Technol 77 Massachusetts Ave Cambridge MA 02139

KAZMAIER, HAROLD EUGENE, b Bowling Green, Ohio, Feb 17, 24; m 49; c 3. ENVIRONMENTAL SCIENCES, ENVIRONMENTAL MANAGEMENT. *Educ:* Ohio State Univ, BS, 49, MS, 51, PhD(bot), 60. *Prof Exp:* Asst plant path, Agr Exp Sta, Ohio State Univ, 50-52; sr res plant pathologist, Battelle-Columbus, 52-72; CHIEF TECH ASSISTANCE, PESTICIDE BR, US ENVIRON PROTECTION AGENCY, 72- *Mem:* Am Phytopath Soc; Soc Nematol; Sigma Xi. *Res:* Pesticides; plant pest control; registration support data. *Mailing Add:* Four Evans Dr Wilmington MA 01887

KAZMAIER, PETER MICHAEL, b Neustadt, Ger, Apr 19, 51; Can citizen; m; c 3. ELECTRONIC MATERIALS, MOLECULAR MODELING. *Educ:* Univ Calgary, BSc Hons, 73; Queen's Univ, PhD(chem), 78. *Prof Exp:* Killam postdoctoral fel chem, Univ BC, 78-79; MEM RES STAFF CHEM, XEROX RES CTR CAN, XEROX CORP, 79- *Concurrent Pos:* Mem comt, Can Soc Chem Publ, 91- *Mem:* Chem Inst Can; Am Chem Soc. *Res:* Electronic materials; novel photogenerator materials; structure-property relationships in photogeneration; hole and electron transport; use of molecular orbital calculations in the prediction of solution and solid state properties of pigments and the solid state and solution nuclear magnetic resonance investigation of electronic materials. *Mailing Add:* Xerox Res Ctr Can 2660 Speakman Dr Mississauga ON L5K 2L1 Can

KAZMANN, RAPHAEL GABRIEL, b Brooklyn, NY, Oct 16, 16; m 42; c 3. HYDROLOGY, HYDROLOGIC ENGINEERING. *Educ:* Carnegie-Mellon Univ, BS, 39. *Prof Exp:* Hydraul engr ground water, US Geol Surv, Washington, DC, 40-45; chief hydraul engr ground water explor, Ranney Method Water Supplies, Inc, Columbus, Ohio, 46-50; consult engr ground water, 50-63; prof, 63-82, EMER PROF CIVIL ENG, LA STATE UNIV, 82- *Concurrent Pos:* Consult, 82- *Mem:* Am Soc Civil Engrs; Soc Mining Engrs; Nat Water Well Asn; Am Water Works Asn; Am Geophys Union. *Res:* Cyclic storage of fresh water in saline aquifers; storage and retrieval of heated and superheated water in saline aquifers; miscible displacement processes and their application to solution mining and deepwell disposal of wastes; monitoring of leachates from landfills; geomorphology and national water policy; book: Modern Hydrology 3rd ed pub 5/88. *Mailing Add:* 231 Duplantier Blvd Baton Rouge LA 70808

KAZMERSKI, LAWRENCE L, b Chicago, Ill, June 9, 45; m 68; c 2. PHOTOVOLTAICS, SURFACE SCIENCE. *Educ:* Univ Notre Dame, BSEE, 67, MSEE, 68, PhD(elec eng), 70. *Prof Exp:* Res fel, Am Eng Coun, Notre Dame Radiation Lab, 71; asst prof teaching res, Univ Maine, Orono, 71-74, assoc prof, 74-77; sr scientist, 77-79, PRIN SCIENITIST, SOLAR ENERGY RES INST, 79-, BR MGR, 80- *Concurrent Pos:* Adj prof, Univ Colo, 79-, Colo Sch Mines, 80-; ed, J Solar Cells, 79-, Polycrystalline & Amorphous Thin Films & Devices, 80; chmn, Nat Am Vacuum Soc Symposium, 82, IEEE PVSC, 87. *Honors & Awards:* Peter Mark Mem Award, Am Vacuum Soc, 80; Res Develop IR-100 Award, 85; Res Develop R&D 100 Award, 89. *Mem:* Am Vacuum Soc (pres, 91); fel Inst Elec & Electronics Engrs; fel Am Phys Soc; Sigma Xi. *Res:* Photovoltaic devices and solid-state physics, with emphasis on the correlation of compositional/chemical properties and electrical characteristics of interfaces in solar cells and other semiconductor devices; scanning tunneling microscopy and surface analysis. *Mailing Add:* Solar Energy Res Inst 1617 Cole Blvd Golden CO 80401

KAZNOFF, ALEXIS I(VAN), b Harbin, China, Oct 22, 33; US citizen; m 80. METALLURGICAL ENGINEERING, CHEMICAL ENGINEERING. *Educ:* Univ Calif, Berkeley, BS, 55, PhD(phys metall), 61; Calif Inst Technol, MS, 56. *Prof Exp:* Scientist mat sci, Gen Elec Co, 60-64, mgr ceramics & electronic mat, 64-66, mat sci & develop, Nucleonics Lab, 66-69, metall & ceramics lab, Nuclear Technol & Appln Oper, 69-73, consult engr, prod & qual assurance oper, nuclear energy bus group, 73-75, mgr, 75-82; DIR MATS ENG, NAVAL SEA SYSTS COMMAND, 82- *Mem:* Am Soc Metals; Am Ceramic Soc; Am Chem Soc; Am Welding Soc; Am Mgt Asn. *Res:* Nuclear fuel technology; structural materials for nuclear plants and ships; nuclear materials; welding and materials processing; marine corrosion; corrosion control; fuels and lubricants. *Mailing Add:* Naval Sea Syst Command Code 514 Washington DC 20362-5101

KAZURA, JAMES, b Cleveland, Ohio, 1946. ONCOLOGY, TROPICAL HEALTH. *Educ:* Ohio State Univ, MD, 72. *Prof Exp:* ASSOC PROF MED, UNIV HOSP, CASE WESTERN RESERVE UNIV, 83- *Mem:* Am Soc Trop Med & Hyg; Am Fedn Clin Res; Infectious Dis Soc Am. *Mailing Add:* Univ Hosp Cleveland Cleveland OH 44106

KE, PAUL JENN, b Ahwei Prov, China, Jan 16, 34; Can citizen; m 61; c 2. ANALYTICAL BIOCHEMISTRY, FOOD TECHNOLOGY. *Educ:* Nat Cheng-Kung Univ, Taiwan, BEng, 59; Nat Taiwan Univ, MSc, 63; Mem Univ Nfld, MSc, 66; Univ Windsor, PhD(anal biochem), 72. *Prof Exp:* Res & develop chem engr, Taiwan Sugar Res Inst, Taiwan, 59-61; instr, Nat Taiwan Univ, 62-64; anal chemist, Fish Res Bd Can, 66-69; res scientist, Halifax Lab, Fisheries & Oceans Can, 72-83, sr scientist & head tech studies dept, 83-87; PROF BIOCHEM, MEM UNIV, 88- *Concurrent Pos:* Prof, Tech Col NS, 81- *Mem:* Chem Inst Can; Can Soc Chem Engrs; Inst Food Technologists; Can Inst Food Sci & Technol; Am Oil Chemists Soc; fel Can Sci Coun. *Res:* Biochemical study on kinetics of lipid oxidation and various rancidity reactions; methodological studies for determination of biochemical parameters and contaminates in various fishery products and waters; quality science studies for sea foods; preservation biochemistry investigation; fish engineering sciences. *Mailing Add:* Mem Univ PO Box 4200 AIC 557 St Johns NF B3M 2P6 Can

KEAGY, PAMELA M, FOLIC ACID & NUTRIENT BIOAVAILABILITY. *Educ:* Univ Calif, Berkeley, PhD, 81. *Prof Exp:* PROJ LEADER FOOD QUAL RES, AGR RES SERV-USDA, 78- *Mem:* Am Inst Nutrit; Inst Food Technologists; Am Asn Cereal Chemists. *Mailing Add:* Food Quality Res Western Region Res Ctr ARS USDA 800 Buchanan St Albany CA 94710

KEAHEY, KENNETH KARL, b Covington, Okla, Sept 17, 23; m 56; c 3. VETERINARY PATHOLOGY. *Educ:* Okla State Univ, BS, 48, DVM, 54; Mich State Univ, PhD(vet path), 63. *Prof Exp:* Adv vet med, Imp Ethiopian Col Agr & Mech Arts, 54-56, head dept animal sci, 56-57, dean, 57-58, actg pres, 58-60; NIH fel, Mich State Univ, 60-63, from asst prof to prof vet path, 63-90, dir, Anal Health Diag Lab, 77-90; RETIRED. *Mem:* AAAS; Am Vet Med Asn. *Res:* Infectious diseases and nutritional deficiencies in swine. *Mailing Add:* 1817 Cahill Dr East Lansing MI 48823

KEAIRNS, DALE LEE, b Vincennes, Ind, Nov 20, 40; m 67; c 1. CHEMICAL ENGINEERING, RESEARCH ADMINISTRATION. *Educ:* Okla State Univ, BS, 62; Carnegie Inst Technol, MS, 64, PhD(chem eng), 67. *Prof Exp:* Assoc develop engr, Oak Ridge Nat Lab, 62 & Gaseous Diffusion Plant, Tenn,

63; sr engr, Res & Develop Ctr, 67-73, mgr, Fluidized Bed Eng, 73-78, mgr, fossil fuel & fluidized bed processing, 78-83, MGR, CHEM & PROCESS ENG, RES & DEVELOP CTR, WESTINGHOUSE ELEC CORP, 83- *Concurrent Pos:* Chmn, First Int Fluidization Conf, Eng Found, 75, co-chmn, 78; mem, fossil fuel adv comt, Oak Ridge Nat Lab, 79-83, chem & process eng adv comt,NSF, 82-83; mem, adv panel, chem eng dept, Univ Pittsburgh, 84-; mem, tech comt, Particulate Solids Res Inst; mem, adv comt, Carnegie Inst & Carnegie Libr, 84-85, coun, Carnegie Mus Nat Hist, 84- *Mem:* Am Inst Chem Engrs; Am Chem Soc; AAAS; Soc Hist Technol. *Res:* Hydrodynamic, heat transfer and reaction rate studies on fluidized bed systems; pilot plant engineering and design; gasification and fluidized bed combustion systems development; gas cleaning. *Mailing Add:* 5419 N Umberland St Pittsburgh PA 15217-1128

KEAMMERER, WARREN ROY, b Gary, Ind, Nov 25, 46; m 70. PLANT ECOLOGY. *Educ:* Capital Univ, BS, 68; NDak State Univ, PhD(bot), 72. *Prof Exp:* Lectr biol, Capital Univ, 71-72; fel ecol, Univ Colo, 72-73; ECOL CONSULT, STOECKER-KEAMMERER & ASSOCS, 73- *Concurrent Pos:* Consult with var projs. *Mem:* Ecol Soc Am; Brit Ecol Soc; Sigma Xi; Wilderness Soc; Soc Range Mgt. *Res:* Preparation of baseline plant ecological reports designed to provide necessary data for impact analysis and permit applications; study areas are located in eastern Wyoming, western Colorado, Utah, New Mexico and North Dakota; monitoring revegetation success on reclaimed lands using comprehensive microcomputer program. *Mailing Add:* 5858 Woodbourne Hollow Rd Boulder CO 80301

KEAN, BENJAMIN HARRISON, b Chicago, Ill, Dec 2, 12; m 48. TROPICAL MEDICINE. *Educ:* Univ Calif, AB 33; Columbia Univ, MD, 37; Am Bd Path, dipl, 45; Am Bd Microbiol, dipl, 63. *Prof Exp:* From intern to resident, Gorgas Hosp, CZ, 37-39, sr pathologist, 40-45; asst prof path, Postgrad Med Sch, NY Univ, 46-52; from asst prof to assoc prof trop med, 52-65, prof trop med, Med Col, 65-78, prof pub health, 72-78, EMER CLIN PROF TROP MED & PUB HEALTH, CORNELL UNIV, 78- *Concurrent Pos:* Attend pathologist, Col Med, NY Univ-Bellevue Med Ctr, 46-52; attend physician & parasitologist, New York Hosp, 52-, dir parasitol lab, 52-78; attend physician, Doctors Hosp, 52-; asst prof parasitol pub health & prev med, Med Col, Cornell Univ, 54-65. *Mem:* Am Soc Trop Med & Hyg; Am Soc Clin Path; fel AMA. *Res:* Medical parasitology. *Mailing Add:* Cornell Univ Med Col 1300 York Ave New York NY 10021

KEAN, CHESTER EUGENE, b Chicago, Ill, Oct 16, 25; m 49, 72; c 6. FOOD CHEMISTRY. *Educ:* Univ Ill, BS, 48; Ore State Col, MS, 50; Univ Calif, PhD(agr chem), 54. *Prof Exp:* Asst chem, Ore State Col, 48-50; food technologist, Univ Calif, 50-53; assoc technologist, Calif & Hawaiian Sugar Refining Corp, 53-58, technologist, 58-61, new prod technologist, 61-66, sr technologist, 66-78, chief chemist prod develop, C&H Sugar Co, Crockett, 78-86; RETIRED. *Mem:* Inst Food Technologists. *Res:* Copper clouding in wines; fungal amylases in butanol acetone fermentation; organic acids in wine; method for determining the sub-sieve particle size distribution of pulverized sugar; carbohydrate chemistry; product development based on sugar properties. *Mailing Add:* 667 Byrdee Way Lafayette CA 94549

KEAN, EDWARD LOUIS, b Philadelphia, Pa, Oct 19, 25; m 62; c 4. BIOCHEMISTRY. *Educ:* Univ Pa, BA, 49, PhD(biochem), 61; Drexel Univ, MS, 56. *Prof Exp:* Chemist, Sharp & Dohme Inc, 49-52 & Smith Kline & French Labs, 52-56; asst instr biochem, Univ Pa, 56-57; res assoc, Univ Mich, 61-64; sr cancer res scientist, Roswell Park Mem Inst, NY, 64-65; from sr instr to assoc prof, 65-79, PROF OPHTHAL & BIOCHEM, SCH MED, CASE WESTERN RESERVE UNIV, 79- *Concurrent Pos:* Arthritis & Rheumatism Found fel, 61-64; Nat Inst Neurol Dis & Stroke & Nat Eye Inst res grants, 68-; exchange scientist, Japan, 81; exchange scientist fel, Japan Soc Prom Sci, 81; Fogarty sr int res fel, 86-87; Erna & Jakob Michael, Vis Professorship Award, Weizman Inst Sci, Israel, 86-87. *Mem:* Am Chem Soc; Am Soc Biol Chem; Asn Res Vision & Ophthal; Int Soc Eye Res; Soc Complex Carbohydrates. *Res:* Biosynthesis, subcellular location and degradation of cytosine monophosphate-sialic acid; glycolipid sulfation and vitamin A deficiency; biosynthesis and enzymatic degradation of rhodopsin; multiple forms of bovine and bacterio-rhodopsins; dolichol pathway and rhodopsin glycosylation; regulation of the dolichol pathway; role of carbohydrates in phagocytosis. *Mailing Add:* Dept Ophthal Case Western Reserve Univ Cleveland OH 44106

KEANA, JOHN F W, b St Joseph, Mich, Sept 14, 39; m 66; c 2. ORGANIC CHEMISTRY. *Educ:* Kalamazoo Col, BA, 61; Stanford Univ, PhD(chem), 65. *Prof Exp:* NSF fel, Columbia Univ, 64-65; from asst prof to assoc prof, 65-77, PROF CHEM, UNIV ORE, 77- *Concurrent Pos:* Guggenheim fel; A P Sloan fel; res career award, NIH. *Mem:* Am Chem Soc; Soc Magnetic Res Med; Int Soc Heterocyclic Chem. *Res:* Biological membranes; new synthetic reactions; preparation and properties of unusual organic molecules; chemistry and biophysics of nitroxide free radical spin-labels; neurochemistry; magnetic research imaging; photoresists and electron beam lithography. *Mailing Add:* Dept Chem Univ Ore Eugene OR 97403

KEANE, J R, b Washington, DC, Mar 12, 37. NEURO-OPHTHALMOLOGY, BRAIN STEM NEUROLOGY. *Educ:* Univ Utah, BS, 58; Harvard Med Sch, MD, 61. *Prof Exp:* Intern, Bellevue Hosp, Univ Calif Med Ctr, 61-62 & 64-65; resident neurol, NY Neurol Inst, Columbia-Presby, 65-67; fel res, Mt Sinai Hosp, NY, 67-68; fel neuro-ophthal, Univ Calif Med Ctr, San Francisco, 68-69; from instr to assoc prof, 70-82, PROF NEUROL, UNIV SOUTHERN CALIF SCH MED, LOS ANGELES, 82- *Concurrent Pos:* Bd examr, Am Bd Neurol & Psychiat, 74- *Mem:* Am Acad Neurol; Am Neurol Asn. *Res:* Clinical-anatomic correlations, with emphasis on eye movement neuropathology. *Mailing Add:* LAC-USC Med Ctr 1200 N State St Rm 5641 Los Angeles CA 90033

KEANE, JOHN FRANCIS, JR, b Milford, Mass, Feb 3, 22; m 48; c 2. BIOPHYSICS, PHYSIOLOGY. *Educ:* Boston Col, BS, 43; Fordham Univ, MS, 49; Univ St Louis, PhD(biol chem), 54. *Prof Exp:* Asst biol, Fordham Univ, 48-49 & Cytochem Sect, Sloan-Kettering Inst, 49-50; asst biol, Biophys Inst, Univ St Louis, 50-54, res assoc, 54-56; from instr to assoc prof physics, St Louis Col Pharm, 55-80; RETIRED. *Mem:* AAAS; NY Acad Sci. *Res:* Physical properties and chemical constitution of crystalline inclusions in giant Amoebae; ultraviolet microspectrography of normal and malignant, desquammated and cultured cells; protective and other action of chemical agents of biological and physical systems subjected to subfreezing temperatures; studies of the action of amides. *Mailing Add:* 1105 Missouri Ave Kirkwood MO 63122

KEANE, KENNETH WILLIAM, b Newark, NJ, Aug 27, 21; wid; c 3. NUTRITION, BIOCHEMISTRY. *Educ:* Murray State Col, BS, 44; Univ Ill, MS, 51, PhD(nutrit biochem), 53. *Prof Exp:* Chemist, Nutrit Res, Raritan Labs, 46-48 & Lederle Labs, Am Cyanamid Co, 48-50; sr res radiochemist, Int Minerals & Chem Corp, 53-56; head, Div Nutrit Res, Campbell Soup Co, 56-71; dir nutrit res, Champion Valley Farms, Moorestown, 71-; AT DEPT NUTRIT, CAMDEN COUNTY COL, BLACKWOOD, NJ. *Concurrent Pos:* Emer prof, Camden County Col, 74- *Mem:* AAAS; Am Chem Soc; Asn Vitamin Chem; Animal Nutrit Res Coun; Am Inst Nutrit. *Res:* New growth factors and intermediary metabolism; nutritional biochemistry. *Mailing Add:* Camden Co Col PO Box 200 Blackwood NJ 08012

KEANE, ROBERT W, NEUROIMMUNOLOGY, DEVELOPMENTAL NEUROBIOLOGY. *Educ:* Univ Calif, Davis, PhD, 76. *Prof Exp:* ASSOC PROF PHYSIOL, SCH MED, UNIV MIAMI, 82- *Mem:* Soc Neurosci; Soc Develop Biol; Soc Cell Biol. *Mailing Add:* Dept Physiol & Biophys Univ Miami Sch Med PO Box 016430 Miami FL 33101

KEAR, BERNARD HENRY, b Port Talbot, SWales, July 5, 31; US citizen; m 59; c 4. MATERIALS & TECHNOLOGY. *Educ:* Birmingham Univ, BSc, 54, PhD, 57, DSc, 70. *Prof Exp:* Res metallurgist, Tube Investments Ltd, UK, 57-59; fel, Franklin Inst, Philadelphia, 59-63; mem staff, Com Prod Div, Pratt & Whitney Aircraft, 63-76; sr consult scientist, United Technologies Res Ctr, 76-81; sci adv, Exxon Res & Eng Co, 81-87; PROF MAT SCI & TECHNOL, CHMN DEPT MECH & MAT SCI & DIR ADVAN TECHNOL, CTR SURFACE ENG MAT, RUTGERS UNIV, 87-, DIR, CTR MAT SYNTHESIS, 89- *Concurrent Pos:* Chmn, Gordon Res Conf Phys Metall, 74; fel, Am Soc Metals, 76; chmn nat mat adv bd, Nat Res Coun, 86- *Honors & Awards:* Mathewson Gold Medal, Am Inst Mining, Metall & Petrol Engrs, 71; Howe Medal, Am Soc Metals, 70; John Dorn Mem Lectr, 80; Henry Krumb Mem Lectr, 83. *Mem:* Nat Acad Eng; fel Am Soc Metals; Am Inst Mining, Metall & Petrol Engrs; Mat Res Soc. *Res:* New phenomena associated with chemical vapor deposition and solidification of materials surfaces; laser processing of materials; structure and property relationships in nickel base superalloys; author of 150 technical publications; awarded 27 patents. *Mailing Add:* Dept Mat Sci & Eng Col Eng Rutgers Univ Po Box 909 Piscataway NJ 08855-0909

KEAR, EDWARD B, JR, b Yonkers, NY, Mar 23, 32; m 54; c 3. MECHANICAL ENGINEERING, SYSTEMS ANALYSIS. *Educ:* Clarkson Tech Univ, BME, 54; Cornell Univ, MS, 56, PhD, 69. *Prof Exp:* From asst prof to assoc prof control systs anal, 58-76, ASSOC PROF MECH & INDUST ENG, CLARKSON COL TECHNOL, 76-, EXEC OFFICER, MECH ENG DEPT, 71- *Mem:* Am Soc Eng Educ. *Res:* Control systems analysis; variation of hand-eye coordination with age. *Mailing Add:* Dean/Dir Summer Sch Clarkson Univ Potsdam NY 13676

KEARL, WILLIS GORDON, b Laketown, Utah, May 11, 27. FARM AND RANCH MANAGEMENT, MARKETING AND PRICES. *Educ:* Utah State Univ, BS, 49, MS, 51; Univ Calif, Berkeley, PhD(agr econ), 68. *Prof Exp:* Agr economist, res water develop, Bur Agr Econ USDA, Utah State Univ, 50-51, res fluoride damage, 54-55, digest staff, Doan Agr Serv, 51, agr econ res, Econ Res Serv, USDA, Calif & Wyo, 58-62; first lieutenant radar maintenance, USAF, 51-53; prof ranch mgt, Range Econ & Livestock Mkt, Univ Wyo, 62-90; RETIRED. *Mem:* Am Agr Econ Asn; Am Soc Range Mgt; Am Soc Animal Sci; Am Agr Law Asn. *Res:* Ranch management and economics of range improvements. *Mailing Add:* Box 3983 Univ Sta Laramie WY 82071

KEARLEY, FRANCIS JOSEPH, JR, b Mobile, Ala, July 7, 21; m 54. ORGANIC CHEMISTRY. *Educ:* Spring Hill Col, BS, 42; Vanderbilt Univ, MS, 44, PhD(org chem), 50. *Prof Exp:* Assoc prof, 53-66, PROF CHEM & CHMN DEPT, SPRING HILL COL, 66- *Mailing Add:* 4121 Ursuline Dr Mobile AL 36608-2494

KEARNEY, JOHN F, b Orroroo, SAustralia, Mar 30, 45. IMMUNOBIOLOGY, MICROBIOLOGY. *Educ:* Univ Adelaide, SAustralia, BDS Hons, 69; Univ Melbourne, PhD, 73. *Prof Exp:* Res assoc, Dept Pediat, Univ Ala, Birmingham, 74-76; asst prof, Dept Microbiol, 76-80, assoc scientist, Comprehensive Cancer Ctr, 76-83, ASSOC PROF, CELLULAR IMMUNOBIOL UNIT, SR SCIENTIST, COMPREHENSIVE CANCER CTR & PROF, DIV DEVELOP & CLIN IMMUNOL, DEPT MICROBIOL, UNIV ALA, BIRMINGHAM, 83- *Concurrent Pos:* Vis foreign dent scientist, Dept Pediat, Univ Ala, Birmingham, 73-74; Europ Molecular Biol Orgn vis sr fel, Dept Genetics, Univ Cologne, WGer, 78; assoc scientist, Multipurpose Arthritis Ctr, Univ Ala, Birmingham, 79; consult, Becton-Dickinson, 79-84 & Idec Inc, Calif, 86-; mem adv bd, Am Type Cult Asn, 84-, Basel Inst Immunol, 85-86 & Allergy & Immunol Study Sect, 84-; Am Cancer Soc & Eleanor Roosevelt int cancer fels, 85-86. *Mem:* Am Asn Univ Prof; Am Asn Immunologists; Am Asn Pathologists; AAAS. *Mailing Add:* Div Develop & Clin Immunobiol Univ Ala 263 Tumor Inst Birmingham AL 35294

KEARNEY, JOSEPH W(ILLIAM), b Denver, Colo, Apr 6, 22. SYSTEM ENGINEERING, MICROWAVE ENGINEERING. *Educ:* Univ Colo, BS, 43. *Prof Exp:* Spec res assoc, Radio Res Lab, Harvard Univ, 43-45; engr, Airborne Instruments Lab, Div Cutler-Hammer, Inc, 46-53, asst supvr engr, 53-55, sect head, 55-58, dept head, 58-62, dept div dir, 62-64, dir new progs, 64-65, div dir, 65-67, vpres, 67-79; dir bus develop, Instruments & Systs Oper, Eaton Corp, 79-86; RETIRED. *Concurrent Pos:* Chmn reconnaissance panel, electronic warfare adv group, US Air Force, 59-60. *Mem:* AAAS; fel Inst Elec & Electronics Engrs; Am Inst Aeronaut & Astronaut; Am Chem Soc. *Res:* Electronic warfare systems for the military services and electronic systems for space exploration. *Mailing Add:* 432 Calle Mayor Redondo Beach CA 90277

KEARNEY, MICHAEL SEAN, b Chicago, Ill, May 12, 47; m 80; c 1. COASTAL GEOMORPHOLOGY, PALYNOLOGY. *Educ:* Univ Ill, Urbana, AB, 73; Western Ill Univ, MA, 76; Univ Western Ont, PhD(geog), 81. *Prof Exp:* Lectr, 80-81; asst prof, 81-87, ASSOC PROF GEOMORPHOL, UNIV MD, COLLEGE PARK, 87- *Concurrent Pos:* Consult, Environ Can, 77-78; US Fish & Wildlife Serv, 84-85 & Cult Triangle Proj, UNESCO, 84-; prin investr, US Environ Protection Agency, 83-84 & Off Water Policy, US Dept Interior, 83-85 US Fish & Wildlife Serv, 87- *Mem:* AAAS; Am Asn Geogr; Am Quaternary Asn; Am Asn Stratig Palynologists. *Res:* Coastal and quaternary geomorphology and paleoecology, with emphasis on the Holocene; coastal marshes; estuaries; sea-level rise. *Mailing Add:* Dept Geog Lab Coastal Res Univ Md College Park MD 20742

KEARNEY, PHILIP C, b Baltimore, Md, Dec 31, 32; m 55; c 2. BIOCHEMISTRY, AGRICULTURE. *Educ:* Univ Md, BS, 55, MS, 57; Cornell Univ, PhD(agr), 60. *Prof Exp:* NSF fel biochem, 60-62; chief pesticide degradation lab, 72-88, BIOCHEMIST PESTICIDES, AGR RES CTR-WEST, USDA, 62-; DEP AREA DIR, NAT RESOURCES INST, 88- *Concurrent Pos:* Unit leader, Pesticide Degradation Lab, 65-72; adj prof chem & biochem, Univ Md, 83. *Honors & Awards:* Int Award Res Pesticide Chem, Am Chem Soc, 81. *Mem:* Am Chem Soc; Int Union Pure & Appl Chem; AAAS; Weed Sci Soc Am; Asn Off Anal Chemists. *Res:* Pesticides; metabolism of organic pesticides by soil microorganisms; enzymology of pesticides. *Mailing Add:* Bldg 001 BARC-West USDA Beltsville MD 20705

KEARNEY, PHILIP DANIEL, b Detroit, Mich, Nov 21, 33; m 58; c 4. PHYSICS. *Educ:* Univ Mich, BS, 58, MS, 60, PhD(physics), 64. *Prof Exp:* Asst prof, 64-74, ASSOC PROF PHYSICS, COLO STATE UNIV, 74-; CONSULT, ARGONNE NAT LAB, 81-, CHEM-NUCLEAR CORP, 84-, CORE LAB, 87- *Concurrent Pos:* Sabbatical leave, Solar Particle Physics, Los Alamos Sci Lab, 71-72, Environ Radiation Measurements, Argonne Nat Lab, 80-81; vpres, Environ Radiation Serv, Inc. *Mem:* Health Physics Soc. *Res:* Environmental radiation measurements; radon, radon flux density, soil radium measurements. *Mailing Add:* Dept Physics Colo State Univ Ft Collins CO 80523

KEARNEY, ROBERT EDWARD, b Montreal, Que, Jan 19, 47. BIOMEDICAL ENGINEERING. *Educ:* McGill Univ, BEng, 68, MEng, 71, PhD(biomed eng), 76. *Prof Exp:* Computer systs engr, Div Neurol, Montreal Gen Hosp, 74-77; res asst, Biomed Eng Unit, McGill Univ, 76-77, postdoctoral fel, Aviation Med Res Unit, 77-78, lectr, Biomed Eng Unit, 78, fac lectr, Dept Physiol, 78-79, asst prof Biomed Eng, 78-83 & Dept Physiol, 79-83, assoc prof, Biomed Eng & Dept Physiol, 83-90, PROF DEPT BIOMED ENG & PHYSIOL, MCGILL UNIV, 90-, CHMN BIOMED ENG, 90- *Concurrent Pos:* Assoc mem, Sch Phys & Occup Therapy, McGill Univ, 81-, dir, Biomed eng Unit, 85-89, assoc mem, Dept Mech Eng, 86-, Dept Elec Eng, 89, actg chmn, Dept Biomed Eng, 89-90; assoc ed, Inst Elec & Electronics Engrs Trans in Biomed Eng; mem Med Res Coun Grants Comt Biomed Eng, 89- *Honors & Awards:* Geddes Prize in Biomed Eng, 72. *Mem:* Inst Elec & Electronics Engrs; Soc Neurosci; Biomed Eng Soc. *Res:* Biomedical engineering; medical imaging; human joint dynamics; motor control system; medical and biological engineering and computing; numerous technical publications. *Mailing Add:* Biomed Eng Dept McGill Univ 3775 University St Montreal PQ H3A 2B4

KEARNEY, ROBERT JAMES, b Manchester, NH, Oct 5, 35; m 61; c 4. SOLID STATE PHYSICS. *Educ:* Univ NH, BS, 57, MS, 59; Iowa State Univ, PhD(physics), 64. *Prof Exp:* Asst physics, Ames Lab, AEC, 60-64; from asst prof to assoc prof, 64-73, PROF PHYSICS, UNIV IDAHO, 73- *Concurrent Pos:* Vis prof, Univ Milan, Italy, 72-73. *Mem:* AAAS; Am Asn Physics Teachers; Am Phys Soc. *Res:* Electronic structure of metals and semi-conductors; optical spectroscopy of molecules. *Mailing Add:* Dept Physics Univ Idaho Moscow ID 83843

KEARNS, DAVID R, b Urbana, Ill, Mar 20, 35; m 58; c 2. BIOPHYSICAL CHEMISTRY. *Educ:* Univ Ill, BS, 56; Univ Calif, Berkeley, PhD(phys chem), 60. *Prof Exp:* Fel theoret chem, Univ Chicago, 60-61; fel, Mass Inst Technol, 61-62; from asst prof to prof phys chem, Univ Calif, Riverside, 62-75; chmn, Chem Dept, 88-89, PROF CHEM, UNIV CALIF, SAN DIEGO, 75- *Concurrent Pos:* A P Sloan fel, 65-67; lectr comt biophys, Harvard Med Sch, 65; Guggenheim fel, 69-70; assoc ed, photochem & photobiol, 71-75, molecular photochem, 72-78, chem rev, Anal Biochem, 77-82; adv bd, Biopolymers, 74. *Mem:* Am Chem Soc; Am Photobiol Soc; Biophys Soc. *Res:* Physical biochemistry; spectroscopy. *Mailing Add:* Dept Chem Univ Calif San Diego La Jolla CA 92093-0342

KEARNS, DONALD ALLEN, b New Bedford, Mass, Sept 10, 23; m 47; c 7. MATHEMATICS. *Educ:* Boston Univ, AB, 47, PhD(math), 55; Brown Univ, MA, 50. *Prof Exp:* From instr to asst prof math, Merrimack Col, 48-53; asst prof math, Univ Maine, 53-58; PROF MATH, MERRIMACK COL, 58- *Mem:* Am Math Soc; Math Asn Am. *Res:* Differential equations. *Mailing Add:* Merrimack Col North Andover MA 01845

KEARNS, LANCE EDWARD, b Greensburg, Pa, May 22, 49; m 71; c 2. GEOLOGY, MINERALOGY. *Educ:* Waynesburg Col, BS, 71; Univ Del, MS, 73, PhD(mineral), 77. *Prof Exp:* ASSOC PROF GEOL, JAMES MADISON UNIV, 76- *Mem:* Am Mineral Soc. *Res:* Mineral chemistry, especially fluorine effects in high temperature metacarbonates. *Mailing Add:* Dept Geol & Geog James Madison Univ Harrisonburg VA 22807

KEARNS, ROBERT J, b Feb 9, 46; m; c 3. CELLULAR IMMUNOLOGY. *Educ:* Wash State Univ, PhD, 78. *Prof Exp:* asst prof, 84-90, ASSOC PROF BIOL, UNIV DAYTON, 90- *Mem:* Am Soc Microbiologists; Am Asn Immunologists; Reticuloendothial Soc. *Res:* Cellular immunology. *Mailing Add:* Dept Biol Univ Dayton Dayton OH 45469-0001

KEARNS, ROBERT WILLIAM, b Gary, Ind, Mar 10, 27; m 53; c 6. MECHANICAL ENGINEERING. *Educ:* Univ Detroit, BSME, 52; Wayne State Univ, MSEM, 57; Case Inst Technol, PhD(eng), 64. *Prof Exp:* Jr engr digital computers, Burroughs Corp Res Labs, 52-53; engr servo-mechanisms, Bendix Corp Res Labs, 53-57; from asst prof to assoc prof eng mech, Dept Eng Mech, Wayne State Univ, 57-67; comnr inspection bldg, Dept Bldg & Safety Eng, City Detroit, 67-71; prin investr, Bur Standards, US Dept Com, 71-76; INVENTOR INTERMITTENT WINDSHIELD WIPER CONTROLS, KEARNS ENGRS, 62- *Concurrent Pos:* Mfr & partner, Kearns & Law Engrs, 57-61; prof engr, Kearns Engrs, 63-76; mfr, Computer Cent, 65-76; supvr, Wayne County, Mich Bd Supvr, 67-69; comnr, Comn Housing Law Rev, State Mich, 68-71. *Mem:* Sigma Xi; fel NSF. *Res:* Intermittent windshield wiper systems including those whose pause time varies automatically with the degree-of-dryness of the windshield-yet, no moisture sensor is utilized. *Mailing Add:* 2350 Bering Dr Apt No 124 Houston TX 77057

KEARNS, THOMAS J, b Evanston, Ill, June 1, 40; m 63; c 3. ALGEBRA. *Educ:* Univ Santa Clara, BS, 62; Univ Ill, MS, 64, PhD(math), 68. *Prof Exp:* From instr to asst prof math, Univ Del, 67-75; asst prof, 75-77, ASSOC PROF MATH, NORTHERN KY UNIV, 77-, CHAIRPERSON DEPT, 81- *Mem:* Am Math Soc; Math Asn Am. *Res:* Representation theory for Lie algebras of classical type. *Mailing Add:* Dept Math & Comput Sci Northern Ky Univ Highland Heights KY 41076

KEARNS, THOMAS P, b Louisville, Ky, Apr 2, 22; m 44; c 2. OPHTHALMOLOGY. *Educ:* Univ Louisville, AB, 44, MD, 46; Univ Minn, MS, 52. *Prof Exp:* From asst prof to assoc prof, 53-77, MEM FAC OPHTHAL, MAYO MED SCH UNIV MINN, 77- *Concurrent Pos:* Consult, Mayo Clin, 53- *Res:* Diseases of the brain and eye. *Mailing Add:* Mayo Clin Rochester MN 55905

KEARSLEY, ELLIOT ARMSTRONG, b Springfield, Mass, Jan 15, 27; m 57, 64; c 2. RHEOLOGY, CONTINUUM MECHANICS. *Educ:* Harvard Univ, AB, 49, MA, 50; Brown Univ, PhD(physics), 55. *Prof Exp:* Physicist, Res Lab, Bendix Aviation Corp, Mich, 53-55; physicist, Nat Acad Sci-Nat Res Coun, 55-76; sr liaison scientist, Off Naval Res, Tokyo, 76-78; PHYSICIST POLYMERS DIV, NAT BUR STANDARDS, 78- *Mem:* Fel Am Phys Soc; Soc Rheol. *Mailing Add:* 1653 Park Rd NW Washington DC 20010

KEASLING, HUGH HILARY, pharmacology; deceased, see previous edition for last biography

KEAST, DAVID N(ORRIS), b Pittsburgh, Pa, Jan 8, 31; m 55; c 4. ENVIRONMENTAL NOISE, EMERGENCY WARNING. *Educ:* Amherst Col, BA, 52; Mass Inst Technol, BS & MS, 54. *Prof Exp:* Engr acoust, Bolt Beranek & Newman, Inc, Los Angeles, 54-57, sr engr, 57-60, consult acoust & instrumentation, Calif, 60-64, supvr consult, 64-66, vpres Data Equip Div, 66-71; vpres develop, MFE Corp, Wilmington, 71-73; mgr, environ acoust, Bolt Beranek & Newman, 73-83; vpres, HMM Assocs, Inc, 83-88; CONSULT, 88- *Mem:* Fel Acoust Soc Am; sr mem Inst Elec & Electronics Engrs; Inst Noise Control Engrs. *Res:* Acoustic-meteorological interactions and processing techniques for high-frequency dynamic data; effects of sound on the human environment; community noise and emergency warning. *Mailing Add:* 657 Westford Rd Carlisle MA 01741

KEASTER, ARMON JOSEPH, b Lilbourn, Mo, Mar 12, 33; m 56; c 2. ENTOMOLOGY. *Educ:* Univ Mo, BS, 59, MS, 61, PhD(entom), 65. *Prof Exp:* From instr to assoc prof, 70-76, PROF ENTOM, UNIV MO-COLUMBIA, 76- *Mem:* Sigma Xi; Entom Soc Am. *Res:* Biology and management of soil and foliar pests attacking corn and other field crops; dispersal migration of Noctuidae, biodegradation of soil- applied pesticides. *Mailing Add:* Dept Entom Univ Mo Columbia MO 65211

KEAT, PAUL POWELL, b Elizabeth, NJ, Nov 29, 23; m 52; c 3. CERAMICS, PHYSICAL CHEMISTRY. *Educ:* Rutgers Univ, BSc, 47, MSc, 50, PhD(ceramics), 56. *Prof Exp:* Res engr, 53-58, sr res engr, 58-63, res assoc, 63-86, SR RES ASSOC, NORTON CO, 86- *Res:* Design of ultrahigh pressure equipment; synthesis at ultrahigh pressure; hydrothermal synthesis; abrasive bond development. *Mailing Add:* Res & Develop Norton Co New Bond St Worcester MA 01606

KEATING, BARBARA HELEN, b Brooksville, Fla, Dec 25, 50. PALEOMAGNETISM, ARCHAEOLOGY. *Educ:* Fla State Univ, BA, 71; Univ Tex, Dallas, MS, 75, PhD(geosci), 76. *Prof Exp:* RESEARCHER GEOPHYSICS, UNIV HAWAII, 76-, PROF OCEANOG, 81- *Mem:* Geol Soc Am; Am Geophys Union; Int Asn Geomagnetism & Aeronomy; Soc Econ Paleontologist & Mineralogist. *Res:* Paleomagnetism and marine geology of the Pacific Ocean basin. *Mailing Add:* Hawaii Inst Geophysics Univ Hawaii 2525 Correa Rd Honolulu HI 96822

KEATING, EUGENE KNEELAND, b Liberal, Kans, Feb 15, 28; m 51; c 2. AGRICULTURAL BIOCHEMISTRY, NUTRITION. *Educ:* Kans State Univ, BS, 53, MS, 54; Univ Ariz, PhD(ruminant nutrit), 64. *Prof Exp:* Instr animal sci, Midwestern Univ, 57-60, asst farm mgr, 57-59, farm mgr, 59-60; from asst prof to assoc prof, 64-71, chmn dept animal sci, 71-78, PROF RUMINANT NUTRIT, CALIF STATE POLYTECH UNIV, POMONA, 78- *Concurrent Pos:* mem coun, Agr Sci & Technol, Am Inst Chem. *Mem:* Am Soc Animal Sci; Brit Soc Animal Prod; fel Am Inst Chem. *Res:* Ruminant nutrition, particularly in cattle. *Mailing Add:* Dept Animal Sci Calif State Polytech Univ Pomona CA 91768

KEATING, JAMES T, b Oak Park, Ill, Jan 21, 41; m 70; c 3. ORGANIC CHEMISTRY, POLYMER CHEMISTRY. *Educ:* St Mary's Col, Minn, BA, 62; Pa State Univ, PhD(chem), 68. *Prof Exp:* Res chemist, Plastic Prod & Resins Dept, Exp Sta, Wilmington, Del, 68-76, sr chemist, Seneca, 76-81, sr chemist, 81-83, RES ASSOC, E I DU PONT DE NEMOURS & CO, INC, WILMINGTON, DEL, 83- *Concurrent Pos:* NSF fel, 62-66. *Mem:* Am Chem Soc. *Res:* Carbene chemistry; aliphatic carbonium ion reactions; fluorinated free radicals; electrochemistry; Friedel-Crafts-type polymerizations; organic and inorganic coatings; emulsion polymerization; occupational safety and health. *Mailing Add:* Du Pont Sta Exp 323 Wilmington DE 19880

KEATING, JOHN JOSEPH, b Montrose, SDak, Jan 17, 38; m 61; c 4. NUCLEAR REACTOR FUELS. *Educ:* SDak State Col, BS, 60; Iowa State Univ, MS, 66, PhD(nuclear eng), 68. *Prof Exp:* Nuclear engr, Idaho Opers Off, US Atomic Energy Comn, 68-73, reactor fuels engr, Div Reactor Develop & Technol, 73-74, asst dir engr technol & fuels, Fast Flux Test Fac Proj Off, US Atomic Energy Comn, US Energy Res & Develop Agency, 74-78; dir, Reactor Technol Div, Fast Flux Test Fac Proj Off, 78-81, DIR, FUELS SUPPLY DIV, RICHLAND OPERS OFF, US DEPT ENERGY, 81- *Mem:* Am Nuclear Soc. *Res:* Development and production of core components for liquid metal fast breeder reactors; core components include fuel, blanket, absorber and reflector assemblies. *Mailing Add:* PO Box 550 MS A552 Richland WA 99352

KEATING, KATHLEEN IRWIN, b NJ, Mar 7, 38; m 62; c 1. PLANKTON CULTURE, TRACE ELEMENT NUTRITION. *Educ:* Cornell Univ, BA, 60; William Patterson Col, MS, 70; Yale Univ, MPh, 72, PhD(limnol), 75. *Prof Exp:* Teacher sci & math, Dumont Pub Sch Syst, 62-68; PROF LIMNOL & ENVIRON SCI, DEPT ENVIRON SCI, RUTGERS UNIV, 74- *Concurrent Pos:* Prin investr, NJ Agr Exp Sta, 76-, NSF Ecol Prog, 79-82; consult, Dow Chem Co, 82- 89. *Mem:* AAAS; Am Soc Limnol & Oceanog; Crustacean Soc; Ecol Soc Am; Soc Environ Toxicol & Chem; Int Soc Chem Ecol. *Res:* Roles of trace element nutrition and allelochemistry in plankton (phytoplankton and zooplankton) community structure; use of highly controlled, defined cultures to isolate critical factors significant to in situ community structure; in vivo trace element interaction. *Mailing Add:* Dept Environ Sci Cook Rutgers Univ New Brunswick NJ 08903

KEATING, KENNETH L(EE), b Chicago, Ill, May 19, 23; m 47; c 2. MATERIALS SCIENCE ENGINEERING. *Educ:* Mass Inst Technol, SB, 47; Univ Mo, MS, 50; Stanford Univ, PhD(metall), 54. *Prof Exp:* Metallurgist, Titanium Div, Nat Lead Co, 47-48; asst metall, Univ Mo, 48-49, instr, 49-50; instr, Stanford Univ, 51-54; metallurgist, Bell Tel Labs, Inc, 54-55; metallurgist, Semiconductor Prod Div, Motorola, Inc, Ariz, 55-61; consult, Cabot Corp, 78-86; assoc prof, 61-67, prof, 67-90, EMER PROF METALL ENG, UNIV ARIZ, 90- *Mem:* Am Soc Metals; Electrochem Soc; Am Inst Mining, Metall & Petrol Engrs; Am Ceramic Soc; Nat Asn Corrosion Engrs. *Res:* Corrosion of metals; phase relations between materials; solid state metallurgy. *Mailing Add:* 5256 Camino de la Cumbre Tucson AZ 85715-1506

KEATING, PATRICK NORMAN, b Newcastle, UK, Feb 18, 39; div; c 2. APPLIED PHYSICS, ELECTRONICS ENGINEERING. *Educ:* Univ Nottingham, BSc, 59, MSc, 61; Univ Mich, PhD(physics), 69. *Prof Exp:* Physicist, Assoc Elec Industs, UK, 60-63; physicist, Tyco Labs, Inc, Waltham, Mass, 63-65; proj physicist, 65-70, head laser optics & acoustics dept, 70-74, dir, Appl Physics Dept, Bendix Res Lab, 74-79, assoc dir res Bendix Advan Technol Ctr, 80-83, DIR & GEN MGR, ALLIED-SIGNAL AEROSPACE TECHNOL CTR, 83- *Mem:* Am Mgt Asn; Inst Elec & Electronics Engrs. *Res:* Computer science, acoustics, underwater acoustics, and acoustic signal processing; sensors, optics; lattice dynamics; solid state physics. *Mailing Add:* Allied-Signal Aerospace Technol Ctr 9140 Old Annapolis Rd Columbia MD 21045

KEATING, RICHARD CLARK, b St Paul, Minn, Aug 6, 37; m 61; c 2. SYSTEMATIC BOTANY. *Educ:* Colgate Univ, AB, 59; Univ Cincinnati, MS, 62, PhD(bot), 65. *Prof Exp:* Asst prof biol, Wis State Univ, Platteville, 64-65; vis asst prof bot, Univ Cincinnati, 65-66; from asst prof to assoc prof, 66-75, PROF BIOL, SOUTHERN ILL UNIV, EDWARDSVILLE, 75-, ACTG COORDR, DEAN'S COL, 90- *Concurrent Pos:* Res assoc, Mo Bot Garden, St Louis, 69-; dir, Trop Biol Prog, Assoc Univs Int Educ, 70; res assoc, Marie Selby Bot Garden, Sarasota, 74-80; consult, Syst Panel, NSF, 82-84; elected fel, Ill State Acad Sci, 84. *Mem:* AAAS; Bot Soc Am; Int Asn Wood Anatomists; Int Asn Plant Taxon; Int Aroid Soc; Soc Conserv Biol; Am Soc Plant Taxon. *Res:* Anatomical investigations on the evolution and classification of vascular plants; Ranales, Solanaceae and Araceae. *Mailing Add:* Dept Biol Southern Ill Univ Edwardsville IL 62026

KEATING, ROBERT JOSEPH, b Forest Hills, NY, Oct 24, 44. ENDOCRINOLOGY, GENITOURINARY MALIGNANCIES. *Educ:* Univ St Thomas, BA, 69; Univ Houston, MS, 72, PhD(reprod physiol), 75. *Prof Exp:* Supvr & coordr bacterial monitoring prog respiratory ther, Methodist Hosp, Houston, 67-73; res asst reprod physiol, Univ Houston, 73-75; fel reprod endocrin, Univ Tex Med Sch, 75-78; sr subj specialist oncol, 78-84, ASST BIOLOGIST, UROL DEPT, M D ANDERSON HOSP & TUMOR INST, UNIV TEX SYST CANCER CTR, 84- *Concurrent Pos:* Teaching asst biol, Univ Houston, 69-72; mem, Study Group Prostatic Cancer, M D Anderson Hosp & Tumor Inst, Animal Resources & Utilization Comt, Urol Oncol Res Group,; mem, Reproductive Biol Training Group, Univ Tex Med Sch. *Mem:* AAAS; Soc Study Reprod; Andrology Soc; Endocrine Soc; Am Soc Clin Oncol; Tissue Culture Soc. *Res:* In vivo dynamics of circulatory hormones; total parenteral nutrition; oncology (clinical); computerized data storage and retrieval; hormonal regulation of prostatic growth and metabolism. *Mailing Add:* 10634 Ella Lee Lane Houston TX 77042

KEATON, CLARK M, b LaGrande, Ore, feb 24, 10; m 38; c 2. PHYSICAL CHEMISTRY. *Educ:* Univ Kans, BS, 31; Univ Idaho, MS, 36; State Col Wash, PhD(soil chem), 38. *Prof Exp:* Asst, State Col Wash, 36-40; res chemist, Grange Powder Co, 40-42; res chemist, Rayonier, Inc, 42-43; res chemist, Am Marietta Col, 43-62; res dir, Atwood Adhesives Inc, 62-75; RETIRED. *Res:* Phenol-formaldehyde resins; sulfite waste liquor; oxidation-reduction potentials and acid determination of soils; liquid industrial adhesives. *Mailing Add:* 2637 30th W Seattle WA 98199

KEATON, MICHAEL JOHN, b Alton, Ill, April 30, 45; m 81. CHEMICAL ENGINEERING. *Educ:* Univ Calif, Berkeley, BS, 66, MS, 67; Princeton Univ, PhD(chem eng), 70. *Prof Exp:* Vpres, Energy & Environ Eng, 70-72, exec vpres, Teknekron Industs, 72-74; PRES & CHMN BD, TERA CORP, 74-, CHIEF EXEC OFFICER, 82- *Res:* Integrated line of proprietary computer systems including hardware and software, automated document storage and retrieval systems, and related computer-aided engineering services, principally for the energy industry; fuel exploration, production, processing and transportation. *Mailing Add:* 1950 Mancanita Dr Oakland CA 94611

KEATON, PAUL W, JR, b Roanoke, Va, Oct 1, 35; m 57; c 3. NUCLEAR PHYSICS. *Educ:* Emory & Henry Col, BS, 57; Johns Hopkins Univ, PhD(physics), 63. *Prof Exp:* Res assoc nuclear physics, Johns Hopkins Univ, 63-65; staff mem, Physics Div, 65-73, leader, Electron Div, 73-79, asst to the dir, 79-80, STRATEGIC PLANNING & POLICY ANALYSIS, LOS ALAMOS NAT LAB, 80- *Concurrent Pos:* Vis scientist, Ctr Europ Nuclear Res, Geneva, Switz, 72-73. *Mem:* Fel Am Phys Soc; sr mem Inst Elec & Electronics Engrs. *Res:* Experimental research with charged particle polarization, Mossbauer effect, direct reactions, electronics and fast neutron cross sections. *Mailing Add:* 137 Piedra Loop Los Alamos NM 87544

KEATS, ARTHUR STANLEY, b New Brunswick, NJ, May 31, 23; m 46; c 4. PHARMACOLOGY. *Educ:* Rutgers Univ, BS, 43; Univ Pa, MD, 46. *Prof Exp:* Asst instr, Sch Med, Univ Pa, 46; asst anesthetist, Mass Gen Hosp, 51; anesthesiologist, House Sisters Red Cross, Switz, 52-53; anesthesiologist, Mary Imogene Bassett Hosp, 53-55; prof anesthesiol & chmn dept, Baylor Col Med, 55-74, clin prof, 74-75; CLIN PROF ANESTHESIOL, UNIV TEX HEALTH SCI CTR HOUSTON, 78-; CHIEF ANESTHESIA, DIV CARDIOVASC ANESTHESIA, TEX HEART INST, 74- *Concurrent Pos:* Assoc anesthesiol, Col Physicians & Surgeons, Columbia Univ, 53-55; dir anesthesiol, Ben Taub Gen Hosp, 55- *Mem:* AAAS; Am Soc Pharmacol & Exp Therapeut; Am Soc Anesthesiol. *Res:* Opiates; analgesics. *Mailing Add:* Div Cardiovasc Anesthesia Tex Heart Inst Houston TX 77025

KEATS, JOHN BERT, b New York, NY, Sept 14, 36; m 68; c 2. INDUSTRIAL ENGINEERING, STATISTICS. *Educ:* Lehigh Univ, BS, 59; Fla State Univ, MS, 64, PhD(educ res), 70. *Prof Exp:* Indust engr, US Steel Corp, Ill, 59-61; asst prof indust eng, La Tech Univ, 64-66; assoc, Advan Proj Dept, Syst Develop Corp, Calif, 68; assoc prof, La Tech Univ, 69-80; mem fac, Sch Indust Eng & Mgt, Okla State Univ, 80-83; AT ARIZ STATE UNIV, 83- *Concurrent Pos:* Consult, Southern Regional Off, Col Entrance Exam Bd, Ga, 70- *Mem:* Sr mem Am Inst Indust Engrs; Am Statist Asn; Am Educ Res Asn. *Res:* Educational and operations research; computer assisted instruction. *Mailing Add:* Comput Integrated Systs Mfg Res Ctr Ariz State Univ Tempe AZ 85287-5106

KEATS, THEODORE ELIOT, b New Brunswick, NJ, June 26, 24; m 74; c 2. RADIOLOGY. *Educ:* Rutgers Univ, BS, 45; Univ Pa, MD, 47; Am Bd Radiol, dipl. *Prof Exp:* Intern, Hosp Univ Pa, 47-48; resident radiol, Univ Mich Hosp, 48-51; from instr to asst prof, Sch Med, Univ Calif, 53-56; from assoc prof to prof, Sch Med, Univ Mo, 56-63; vis prof, Karolinska Inst, Sweden, 63-64; PROF RADIOL & CHMN DEPT, UNIV HOSP, UNIV VA, 64- *Concurrent Pos:* Trustee, Am Bd Radiol, 73-85. *Mem:* Radiol Soc NAm; Roentgen Ray Soc; AMA; fel Am Col Radiol; Asn Univ Radiol; Int Skeletal Soc. *Res:* Pediatric and skeletal radiology. *Mailing Add:* Dept Radiol Univ Va Hosp Charlottesville VA 22901

KEAVENEY, WILLIAM PATRICK, b New York, Dec 25, 36; m 61; c 5. ORGANIC CHEMISTRY. *Educ:* Manhattan Col, BS, 58; Fordham Univ, PhD(org chem), 64. *Prof Exp:* res assoc, Inmont Corp, Clifton, 62-90; SR RESEARCHER, SUN CHEM CORP, CARLSTADT, 91- *Mem:* Am Chem Soc. *Res:* Pyrodoxine determination; synthesis of dichloro-diphenyl-trichlorethane analogs; norbornylene polymerization; ozonolysis; radiation curing; polymer chemistry. *Mailing Add:* Sun Chem Corp Tech Ctr 631 Central Ave Carlstadt NJ 07072

KEAY, LEONARD, b Crayford, Eng, Nov 26, 32; nat US; m 54; c 3. BIOCHEMISTRY, MICROBIOLOGY. *Educ:* Univ London, BSc, 53, PhD(chem), 55, MSc, 56. *Prof Exp:* Hon asst biochem, Univ Col, Univ London, 56-58; hon fel, Sch Advan Study & res fel enzymol, Mass Inst Technol, 58-60; res biochemist, Monsanto Co, 60-71; res biochemist, Sch Med, Washington Univ, 71-74, dir, Basic Cancer Res Ctr, 71-80; mem staff, McDonnell Douglas Astronaut Co, 80-89; PROG MGR, US DEPT ENERGY, 88- *Concurrent Pos:* Salters Inst Indust Chem res fel, Univ Col, Univ London, 56-58. *Mem:* Am Chem Soc; Tissue Cult Asn. *Res:* Organo phosphorus chemistry and biochemistry; enzymology; applied biochemistry and microbiology; animal cell culture; tissue culture; fermentation technology. *Mailing Add:* 2221 Brandon St Idaho Falls ID 83402

KEBABIAN, JOHN WILLIS, b New York, NY, Sept 20, 46; m 75; c 2. ENDOCRINOLOGY, DOPAMINE RECEPTORS. *Educ:* Yale Univ, BS, 68, MPhil, 70, PhD(pharmacol), 73. *Prof Exp:* Res assoc, NIH, 74-76, sr staff fel, 76-78, pharmacologist, 78-81, sect chief, 81-86; proj leader, 86-88, SR PROJ LEADER NEUROSCIENCES, ABBOTT LABS, 88- *Concurrent Pos:* Bd dirs, Yale Alumni Fund, 83- *Mem:* Am Soc Biol Chemists; Am Soc Pharm & Exp Therapeut; Brit Pharmacol Soc; Endocrine Soc; AAAS. *Res:* Receptors for neurotransmitters; dopamine receptors; pituitary gland; author of 3 publications. *Mailing Add:* Abbott Labs Abbott Park IL 60064

KEBARLE, PAUL, b Sofia, Bulgaria, Sept 21, 26; m 55; c 1. PHYSICAL CHEMISTRY. *Educ:* Swiss Fed Inst Technol, Dipl Ing Chem, 52; Univ BC, PhD, 55. *Prof Exp:* Nat Res Coun Can fel, 55-58; from asst prof to assoc prof, 58-68, PROF CHEM, UNIV ALTA, 68- *Mem:* Fel Royal Soc Can. *Res:* Application of mass spectrometry to reaction kinetics in the gas phase; ion-molecule interactions at high pressure; ionic solvation and ionic reactivity in the gas phase; ion-molecule equilibria. *Mailing Add:* Dept Chem Univ Alberta Edmonton AB T6G 2E2 Can

KEBLAWI, FEISAL SAID, b Acre, Palestine, July 11, 35; US citizen; m 73; c 4. COMMUNICATIONS SYSTEMS ENGINEERING. *Educ:* Am Univ Beirut, BS, 57; NC State Univ, MS, 62, PhD(elec eng), 65. *Prof Exp:* Mem tech staff satellite systs eng, RCA Corp, 65-68; SATELLITE COMMUN SYSTS ENGR, MITRE CORP, 68- *Concurrent Pos:* US deleg, US/USSR Working Group on Air Traffic Control, Moscow, 78; cong fel, Inst Elec & Electronics Engrs, 81-82; staff asst defense, Senator Thurmond, 81-; US deleg, US/Ger group on air defense. *Mem:* Sr mem Inst Elec & Electronics Engrs; Planetary Soc. *Res:* Satellite communications deep space and tactical systems; tactical communications systems; air traffic control; forward air defense; control systems engineering; stabilization of heat transfer process in nuclear reactors; legislation in civil defense; researcher of major foreign policy and arms sales issues. *Mailing Add:* Mitre Corp C3 Div 1820 Dolley Madison Blvd McLean VA 22101

KEBLER, RICHARD WILLIAM, b Owosso, Mich, Nov 25, 20; m 50; c 4. APPLIED PHYSICS. *Educ:* Univ Mich, BSE, 42, MS, 47, PhD(physics), 54. *Prof Exp:* Res assoc, Eng Res Inst, Univ Mich, 42; res physicist, Linde Co Div, Union Carbide Corp, 53-55, develop supvr, 56-59; res physicist, Gen Motors Res Lab, 59; group leader, Union Carbide Res Inst, 60-66, mgr mat res & develop, Space Sci & Eng Lab, 66-68, sr res assoc, Linde Co Div, 68-70, PROG MGR COMPOSITES, LINDE CO DIV, UNION CARBIDE CORP, 70- *Mem:* Am Phys Soc; Optical Soc Am; Am Ceramic Soc; Metall Soc; Sigma Xi. *Res:* Spectro-chemical analysis; extreme ultraviolet spectroscopy; crystal growth; refractory and composite materials; turbine engine compressors. *Mailing Add:* 9785 E Skyview Dr Tucson AZ 85748

KECECIOGLU, D(IMITRI) B(ASIL), b Istanbul, Turkey, Dec 26, 22; nat US; m 51; c 2. ENGINEERING MECHANICS, RELIABILITY ENGINEERING. *Educ:* Robert Col, Istanbul, BS, 42; Purdue Univ, MS, 48, PhD(eng mech), 53. *Prof Exp:* Instr mech, Purdue Univ, 47, asst metal cutting, 49-52, asst instr eng drawing & descriptive geom, 50-52, asst instr mach tool lab, 51; eng scientist-in-chg mech lab, Res Labs, Allis-Chalmers Mfg Co, 52-57, asst to dir mech eng industs group, 57-63, dir reliability & corp consult, 60-63; PROF AEROSPACE & MECH ENG, UNIV ARIZ, 63-, PROF, RELIABILITY ENG PROG, 69- *Concurrent Pos:* Fulbright scholar, Greece, 71-72; dir, Reliability Eng & Mgt Insts; consult reliability & maintainability eng, indust & govt; hon prof, Phi Kappa Phi, Shangai Univ Technol. *Honors & Awards:* Ralph Teetor Award, Soc Automotive Engrs, 77; Allen Chop Award, Am Soc Qual Control, 81; Excellence Award, Soc Reliability Eng; Reliability Educ Advan Award, Am Soc Qual Control; Anderson Prize, Univ Ariz Col Eng & Mines. *Mem:* AAAS; Am Soc Mech Engrs; Am Soc Eng Educ; Soc Exp Stress Anal; Inst Elec & Electronics Engrs; Sigma Xi; Soc Reliability Engrs; Inst Environ Sci; Soc Automotive Engrs. *Res:* System effectiveness, reliability, maintainability; quality control; statistics; probability; design; production engineering; design by reliability; tooling engineering; applied mathematics. *Mailing Add:* Dept Aerospace & Mech Eng Univ Ariz Tucson AZ 85721-0663

KECK, DONALD BRUCE, b Lansing, Mich, Jan 2, 41; m 65; c 2. PHYSICS. *Educ:* Mich State Univ, BS, 62, MS, 64, PhD(physics), 67. *Prof Exp:* Res physicist, 68-74, res assoc physics, 74-76, mgr appl physics dept, 76-86, DIR APPL PHYSICS RES, CORNING GLASS WORKS, 86- *Honors & Awards:* Technol Achievement Award, Soc Photo-optical Instrumentation Engrs, 81; IR-100 Award, 81; Eng Achievement Award, Am Soc Metall, 83. *Mem:* Fel Optical Soc Am; Inst Elec & Electronics Engrs; Soc Photo-optical Instrumentation Engrs. *Res:* Near infrared molecular spectroscopy; magnetic rotation spectroscopy; fiber optics; propagation in fiber optic waveguides; gradient index imaging; guided wave optics. *Mailing Add:* FR 17 Corning Glass Works Sullivan Park Corning NY 14830

KECK, JAMES COLLYER, b New York, NY, June 11, 24; m 47; c 2. PHYSICS. *Educ:* Cornell Univ, BA, 47, PhD, 51. *Prof Exp:* Res assoc physics, Cornell Univ, 51-52; res fel, Calif Inst Technol, 52-55; prin scientist, Avco-Everett Res Lab, Mass, 55-65, dep dir, 60-64; Ford prof mech eng, 65-89, PROF EMER & SR LECTR, MASS INST TECHNOL, 89- *Mem:* Am Phys Soc; AAAS; Sigma Xi; fel Am Acad Arts & Sci; Combustion Inst. *Res:* Atomic and molecular kinetics; high temperature gas dynamics; combustion; nonequilibrium thermodynamics; high energy nuclear physics. *Mailing Add:* 52 Harold Parker Rd Andover MA 01810

KECK, KONRAD, b Vienna, Austria, Mar 13, 28; nat US; m 57; c 2. BIOLOGY. *Educ:* Univ Vienna, PhD(biol, chem), 52. *Prof Exp:* Instr physiol chem, Med Sch, Univ Vienna, 53-54; Int Co-op Admin fel, Univ Wis, 54-56; res assoc, Max Planck Inst Marine Biol, Ger, 56-57; res assoc, New Eng Med Ctr, 57-58; res assoc chem, Ind Univ, 58-59; asst prof biol, Johns Hopkins Univ, 59-64; assoc prof zool, 64-68, PROF MOLECULAR CELL BIOL, UNIV ARIZ, 68- *Res:* Nucleocytoplasmic interactions in microorganisms; structure and function of nucleic acids in viral reproduction. *Mailing Add:* Dept Cell Develop Biol Univ Ariz Tucson AZ 85721

KECK, MAX HANS, b Konstanz, Ger, May 7, 19; US citizen; m 49; c 2. POLYMER CHEMISTRY. *Educ:* Col Wooster, BA, 41; Univ Akron, MSc, 45. *Prof Exp:* Sr res chemist polyester chem, Goodyear Tire & Rubber Co, 42-67, res scientist, Fiber Tech Ctr, 67-83; RETIRED. *Mem:* Am Chem Soc. *Res:* Linear polyester research, preparation of new linear polyesters and new monomers; catalysis studies; dyeable and specialty polyester fibers; cross-linkable polyesters for coatings. *Mailing Add:* 3117 Mayfield Rd Silver Lake Cuyahoga Falls OH 44224

KECK, MAX JOHANN, b Feb 22, 39; US citizen. VISUAL PSYCHOPHYSICS. *Educ:* Mass Inst Technol, BS, 61; Purdue Univ, MS, 64, PhD(physics), 68. *Prof Exp:* From asst prof to assoc prof, 68-78, PROF PHYSICS, JOHN CARROLL UNIV, 78-; DEAN OF STUDENT DEVELOP, JOHN CARROLL UNIV. *Concurrent Pos:* Adj staff mem ophthalmol, Cleveland Clin Found, Ohio, 78-; prin investr res grant, Nat Eye Inst, NIH, 79- *Mem:* Am Phys Soc; Asn Res Vision & Ophthal. *Res:* Binocular vision; spatial vision; amblyopia and strabismus. *Mailing Add:* Physics Dept John Carroll Univ Cleveland OH 44118

KECK, ROBERT WILLIAM, b Manchester, Iowa, Jan 2, 41; m 64; c 2. SCIENCE ADMINISTRATION, RESEARCH ADMINISTRATION. *Educ:* Univ Iowa, BS, 62, MS, 64; Ohio State Univ, PhD(plant physiol), 68. *Prof Exp:* Researcher photosynthesis, Charles F Kettering Res Lab, 68-70; researcher hort, Univ Ill, 70-71, researcher bot, 71-72, lectr bot, 72; from asst to prof biol, Ind Univ-Purdue Univ, 72-82, asst dean, 82-84, assoc dean, 84-88, 87-90, actg dean, 88-89, PROF BIOL, IND UNIV-PURDUE UNIV, 90- *Mem:* Am Soc Plant Physiologists; Crop Sci Soc Am. *Res:* Photosynthesis, membrane physiology. *Mailing Add:* Dept Biol Ind Univ Purdue Univ Indianapolis IN 46202-5167

KECK, WINFIELD, b Clifton Heights, Pa, Sept 15, 17; m 44; c 4. PHYSICS. *Educ:* Amherst Col, AB, 37; Univ Pa, MA, 38; Brown Univ, PhD(physics), 49. *Prof Exp:* Instr math, Franklin & Marshall Col, 39-40; instr physics, Muhlenberg Col, 41-46; instr, Brown Univ, 46-48; from asst prof to assoc prof, Lafayette Col, 49-61, prof physics, 61-83, chmn dept, 60-82; RETIRED. *Concurrent Pos:* Vis assoc prof, Brown Univ, 58-59. *Mem:* Am Asn Physics Teachers; Sigma Xi. *Res:* Acoustic wave propagation. *Mailing Add:* RD #2 Box 124 Boyertown PA 19512-9417

KEDDY, JAMES RICHARD, b Boston, Mass, Oct 18, 36; m 61; c 3. COMPUTER SCIENCES. *Educ:* Colby Col, BA, 58. *Prof Exp:* Mem staff air defense, Syst Develop Corp, 60-62; sect mgr satellite control, 63-67; sect mgr oper systs, Sci Data Systs, 68-71; mem advan design staf, Xerox Data Systs, 72-73; prin engr off systs, Xerox Corp, 74-80; proj leader, 80-84, DIR ENG, TERADATA CORP, 85- *Res:* Word processing systems; information storage and retrieval; operating systems; database management systems; communications. *Mailing Add:* 100 N Sepulveda Blvd El Segundo CA 90245

KEDDY, PAUL ANTHONY, b London, Ont, May 29, 53; c 2. BOTANY, ENVIRONMENTAL SCIENCES. *Educ:* York Univ, Toronto, BSc, 74; Dalhousie Univ, Halifax, PhD(ecol), 78. *Prof Exp:* Asst prof ecol, Dept Bot & Genetics, Univ Guelph, 78-82; from asst prof to assoc prof, 82-89, PROF ECOL, DEPT BIOL, UNIV OTTAWA, 89- *Concurrent Pos:* Vis lectr, Dept Bot & Microbiol, Univ London, 85-86 & comp plant ecol unit, Univ Sheffield, 86; mem, subcomt plants, Comt Status Endangered Wildlife Can, 85-91; chmn; sci comt, Can Coun Ecol Areas, 86-90; mem, Sci Adv Comt, World Wildlife Fund, Can, 86-90; grant comt, Population Biol Grant Selection Comt, Nat Sci & Eng Res Coun, Can, 86-89; coordr, Inst Res Environ & Econ, Univ Ottawa, 89-90. *Mem:* Ecol Soc Am; Brit Ecol Soc; Can Bot Asn; Int Asn Veg Sci. *Res:* Plant community ecology; competition and plant traits; assembly rules; wetland ecology; conservation of endangered wetland plants and habitats. *Mailing Add:* Dept Biol Univ Ottawa Ottawa ON K1N 6N5 Can

KEDER, WILBERT EUGENE, b Columbus, Nebr, July 29, 28; m 51; c 4. GENERAL ENVIRONMENTAL SCIENCES. *Educ:* Doane Col, BA, 50; Univ Pittsburgh, PhD(chem), 56. *Prof Exp:* Asst, Univ Pittsburgh, 50-56; res chemist, Hanford Labs, Gen Elec Co, 56-64; sr res scientist, Pac Northwest Labs, Battelle-Northwest, 65-68; adj assoc prof, Wash State Univ, 68-69; res assoc, Battelle-Northwest, 69-71; from asst prof to prof chem, 71-86, dir Petrol Technol Prog, 75-86, EMER PROF CHEM, UNIV PITTSBURGH, BRADFORD, 87- *Mem:* Am Chem Soc. *Res:* Solution chemistry; solvent extraction; chemistry of the actinide elements; petroleum technology; environmental effect of energy utilization. *Mailing Add:* 971 Terrace Circle Colo Springs CO 80904-2841

KEDES, LAURENCE H, b Hartford, Conn, July 19, 37; m 58; c 3. MOLECULAR GENETIC RESEARCH, GENE EXPRESSION. *Educ:* Stanford Univ, BS, 61, MD, 62. *Prof Exp:* Res assoc, Lab Biochem, Nat Cancer Inst, 64-66; postdoctoral biol, Mass Inst Technol, 67-69; from asst prof to prof med, Stanford Univ, 70-89; WILLIAM KECK PROF & CHMN, DEPT BIOCHEM & DIR, INST GENETIC MED, UNIV SOUTHERN CALIF, 88- *Concurrent Pos:* Leukemia Soc Am scholar, 69-74; staff physician, Vet Admin Med Ctr, Palo Alto, 70-89; investr, Howard Hughes Med Inst, 74-82; vis scientist, Imp Cancer Res Fund, London, 76-77; assoc ed, J Biol Chem, Molecular & Cellular Biol, J Molecular Evolution & Oxford Surv Eukaryotic Genes; fel, John Simon Guggenheim Found, 76-77. *Mem:* Am Soc Microbiol; Am Soc Biochem & Molecular Biol; Int Soc Develop Biol; Am Soc Clin Invest; Asn Am Physicians; Am Soc Hemat. *Res:* Biotechnology; gene expression in animal cell differentiation. *Mailing Add:* Dept Biochem HMR No 413 Univ Southern Calif 2011 Zonal Ave Los Angeles CA 90033

KEDZIE, DONALD P, ENG ADMINISTRATION. *Prof Exp:* PROF MECH ENG, ARK STATE UNIV, 84- *Mailing Add:* Off Dean Eng Ark State Univ PO Box 1990 State University AR 72467

KEDZIE, ROBERT WALTER, b Milwaukee, Wis, Sept 16, 32; m 56; c 4. SOLID STATE PHYSICS. *Educ:* Marquette Univ, BS, 54; Univ Calif, MA, 56, PhD(physics), 59. *Prof Exp:* Asst neutron diffraction, Oak Ridge Nat Lab, 54; instr physics, Univ Calif, Berkeley, 54-56; res physicist, Exp Sta, E I du Pont de Nemours & Co, 59-61; staff physicist, Sperry Rand Res Ctr, 61-68; mgr spectrometer eng, Magnion Div, Ventron Instrument Corp, 68-69; assoc prof physics, Univ Detroit, 69-79, chmn dept, 75-79; sr staff physicist, 79-84, asst proj mgr, 84-85, proj mgr, 85-88, ACTG DEPT MGR, HUGHES AIRCRAFT CO, 88- *Concurrent Pos:* Asst, Los Alamos Sci Lab, 55. *Mem:* Am Phys Soc. *Res:* Electron and nuclear magnetic resonance; electric and magnetic phenomena in polymers, defect soild state, magneto-acoustic and photochromic phenomena. *Mailing Add:* 551 S Colt St Anaheim CA 92806

KEE, DAVID THOMAS, b Escanaba, Mich, July 25, 29; m 62; c 2. ORNITHOLOGY. *Educ:* Northern Mich Univ, AB, 58; Univ Ark, MS, 60; Mich State Univ, PhD(ornith), 64. *Prof Exp:* Asst prof, 64-67, ASSOC PROF ZOOL, NORTHEAST LA STATE UNIV, 67- *Mem:* Cooper Ornith Soc; Am Ornith Union; Wildlife Soc; Am Soc Mammal. *Res:* Taxonomy and distribution of Louisiana birds and mammals; pesticide studies, birds and mammals. *Mailing Add:* Dept Biol Northeast La State Univ Monroe LA 71209

KEEDY, CURTIS RUSSELL, b Selma, Calif, Sept 14, 38; m 76; c 3. PHYSICAL CHEMISTRY, RADIOCHEMISTRY. *Educ:* Occidental Col, BA, 60; Univ Wis, PhD(phys chem), 65. *Prof Exp:* Resident res assoc nuclear chem, Chem Eng Div, Argonne Nat Lab, 64-66; asst prof chem, Reed Col, 66-70, reactor supvr, 68-72; asst prof, 72-75, assoc prof, 75-85, PROF CHEM, LEWIS & CLARK COL, 85-. *Concurrent Pos:* Vis lectr, Lewis & Clark Col, 71-72. *Mem:* Am Chem Soc; Sigma Xi; Am Asn Univ Prof. *Res:* Nuclear chemistry; neutron activation analysis as applied to geochemical systems and environmental areas. *Mailing Add:* 0666 SW Palatine Hill Rd Portland OR 97219-7831

KEEDY, HUGH F(ORREST), b Berkeley Springs, WVa, Sept 22, 26; m 48; c 2. ENGINEERING MECHANICS. *Educ:* George Peabody Col, BS, 51, MA, 52; Univ Mich, MSE, 62, PhD, 67. *Prof Exp:* Asst prof appl math, 51-68, assoc prof eng sci, 68-74, assoc dean instr, 69-71, PROF ENG SCI, VANDERBILT UNIV, 74- *Concurrent Pos:* Consult various industs, 54-; asst, Univ Mich, 62-63, instr, 63-65; tech ed & writer, Lawrence Livermore Nat Lab, 80-81; pres, Southeastern Sect, Am Soc Eng Educ, Zone II chmn, 88-90; author. *Mem:* Am Soc Eng Educ; Soc Tech Commun. *Res:* Fluid mechanics; engineering education; technical communication. *Mailing Add:* Dept Eng Sci Vanderbilt Univ Nashville TN 37240

KEEDY, MERVIN LAVERNE, b Bushnell, Nebr, Aug 2, 20; m 41; c 2. MATHEMATICS. *Educ:* Univ Chicago, BS, 46; Univ Nebr, MA, 50, PhD, 57. *Prof Exp:* Teacher, Pub Sch, Idaho, 47-49; asst math, Univ Nebr, 49-50; supvr lab sch, Nebr State Teachers Col, Peru, 50; asst prof physics, NDak State Col, 51-53; instr math, Univ Nebr, 53-55, instr physics, 55-56, counr sci tech improv prog, 56-57; assoc dir math prog, Univ Md, 57-60; supvr math & sci, Jr High Sch, 60-61; emer prof math, Purdue Univ, West Lafayette, 61-87; EXEC DIR & FOUNDER, TEXTBOOK AUTH ASN, 87- *Concurrent Pos:* Ground & flight instr, Rodman Aircraft Co, 48-49. *Mem:* Am Math Soc; Math Asn Am. *Res:* Mathematics education. *Mailing Add:* Box 535 Orange Spring FL 32182

KEEFE, DEBORAH LYNN, b Oklahoma City, Okla, Nov 23, 50; m 71; c 3. CARDIOVASCULAR PHARMACOLOGY, NONINVASIVE CARDIOLOGY. *Educ:* Rice Univ, BA, 73; NY Med Col, MD, 76; Columbia Univ, MPH, 90. *Prof Exp:* Resident internal med, St Vincents Hosp, New York, 76-79; fel cardiol, Stanford Univ, 79-81; asst prof med, Albert Einstein Col Med, 81-87; assoc dir clin invest, Am Cynamid Med Res, 87-88; ASSOC ATTEND, MEM SLOAN KETTERING, 88-; ASSOC PROF MED, MED COL, CORNELL UNIV, 88- *Concurrent Pos:* Dir coronary care, Bronx Municipal Hosp Ctr, 81-; atten physician, Hosp Albert Einstein Col Med, 81-87; assoc ed, J Clin Pharamacol, 85-; asst clin prof, Albert Einstein Col Med, 87-88; assoc mem, Mem Sloan Kettering, 88- *Mem:* Fel Am Col Cardiol; fel Am Col Chest Physicians; fel Am Heart Asn; fel Am Col Angiolog; fel Am Col Clin Pharmacol; fel Am Soc Clin Pharmacol & Therapeut. *Res:* Clinical investigation of cardiovascular therapeutic agents including clinical trials; pharmacokinetics and pharmacodynamics; epidemiology and the prevention of heart disease. *Mailing Add:* 276 Trenor Dr New Rochelle NY 10804-3814

KEEFE, DENIS, high energy physics, accelerator physics; deceased, see previous edition for last biography

KEEFE, JOHN RICHARD, b Sandusky, Ohio, Jan 31, 35; m 59; c 2. DEVELOPMENTAL BIOLOGY, CELL BIOLOGY. *Educ:* John Carroll Univ, BS, 60, MS, 65; Case Western Reserve Univ, PhD(anat), 69. *Prof Exp:* Res assoc, Basic Res Div, Cleveland Psychiat Inst, Ohio, 61-65; from instr to asst prof anat, Med Sch, Univ Va, 69-75; assoc prof anat, Med Sch, Univ Louisville, 75-77; assoc prof, 77-81, ASST PROF ANAT, DEVELOP BIOL CTR, CASE WESTERN RESERVE UNIV, 81- *Concurrent Pos:* NIH grant, Univ Va, 69-71, Nat Eye Inst grant, 71-75 & 75-79; NASA contractor, 74-81. *Mem:* Am Soc Cell Biol; Soc Develop Biol; Asn Res Vision & Ophthal; NY Acad Sci; Am Asn Anat. *Res:* Cytological studies of retinal and vestibular development and regeneration in amphibian, avian and mammalion species utilizing autoradiography, electron microscopy and cytochemistry. *Mailing Add:* Dept Anat Sch Med Case Western Reserve Univ Cleveland OH 44106

KEEFE, THOMAS J, b Algona, Iowa, Dec 4, 37; m 65; c 4. VETERINARY MEDICINE. *Educ:* Univ Mo, BS & DVM, 63. *Prof Exp:* Pvt vet pract, 66-67; livestock consult, Livestock Servs, Ralston Purina, 67-69; mgr clin res, Bristol Labs, 69-74; DIR VET MED, BEECHAM LABS, 74- *Mem:* Am Vet Med Asn; Am Asn Swine Practitioners (pres, 71); Am Asn Bovine Practitioners; fel Am Col Pharmacol & Therapeut; Indust Vet Asn; Sigma Xi. *Res:* Pharmacology; pathology; diagnostic medicine. *Mailing Add:* 2431 Newport Ct Ft Collins CO 80526

KEEFE, THOMAS LEEVEN, b Columbia, SC, Jan 22, 37; m 64. BOTANY, BIOLOGY. *Educ:* Univ SC, BS, 59, MS, 61; Univ Ga, PhD(bot), 67. *Prof Exp:* Asst prof biol, Newberry Col, 62; ASST PROF BIOL, EASTERN KY UNIV, 66- *Mem:* Am Forestry Asn; Am Inst Biol Sci. *Res:* Shoot development in forest trees; radiation inducted mutations in insects. *Mailing Add:* 235 Moore Bldg Eastern Ky Univ Richmond KY 40475

KEEFE, WILLIAM EDWARD, b Norfolk, Va, Feb 23, 23; m 46; c 2. BIOPHYSICS, CRYSTALLOGRAPHY. *Educ:* Va Polytech Inst, BS, 59, MS, 64; Med Col Va, PhD(biophys), 67. *Prof Exp:* Asst prof biophys, 66-76, asst prof, 76-81, ASSOC PROF MICROBIOL, MED COL VA, VA

COMMONWEALTH UNIV, 81- *Mem:* Am Crystallog Asn; Am Inst Physics; Am Phys Soc; Sigma Xi; Int Solar Energy Soc. *Res:* Formation of kidney stone nuclei; interaction of fast neutrons with biological materials; determination of molecular structure of biologically important compounds; computer programming to solve crystal structures; model building of proteins. *Mailing Add:* Dept Biosta Med Col Va Box 32-MCV Sta Richmond VA 23298

KEEFER, CAROL LYNDON, b Columbia, SC, Jan 20, 53. IN VITRO FERTILIZATION, PREIMPLANTATION DEVELOPMENT. *Educ:* Univ SC, BS, 74; Univ Del, PhD(biol sci), 81. *Prof Exp:* Postdoctoral fel reproductive physiol, Sch Hyg & Pub Health, Johns Hopkins, 81-82 & Sch Vet Med, Univ Pa, 82-83; reproductive biologist, Reproductive Biol Assocs, 83-86; asst physiologist reproductive physiol, Col Vet Med, Univ Ga, 84-85, asst prof, 85-89; RES SCIENTIST, AM BREEDERS SERV SPECIALTY GENETICS, 89- *Mem:* Soc Develop Biol; Int Embryo Transfer Soc; Soc Study Reproduction; Am Soc Cell Biol. *Res:* Bovine oocyte maturation; in vitro fertilization and embryo culture; bovine embryo nuclear transfer; assisted fertilization. *Mailing Add:* Am Breeders Serv Specialty Genetics PO Box 459 6908 River Rd De Forest WI 53532-0459

KEEFER, DENNIS RALPH, b Winter Haven, Fla, Sept 22, 38; m 57; c 3. AEROSPACE ENGINEERING, LASER PROPULSION. *Educ:* Univ Fla, BES, 62, MSE, 63, PhD(aerospace eng), 67. *Prof Exp:* Asst prof, 67-76, assoc prof aerospace eng, Univ Fla, 76-78; PROF ENG SCI & MECH, UNIV TENN, 78- *Mem:* Am Phys Soc; Sigma Xi. *Res:* Electrodeless arcs and discharges; gas lasers; plasma spectroscopy. *Mailing Add:* Univ Tenn Space Inst Tullahoma TN 37388

KEEFER, DONALD WALKER, b Idaho Falls, Idaho, Nov 7, 31; m 54; c 2. METALLURGY. *Educ:* Univ Idaho, BS, 54; Univ Ill, MS, 57, PhD(metall), 61. *Prof Exp:* Mem tech staff, Atomics Int, 61-77; mgr, Off Res Mgt, Fuels & Mat Div, 77-81, MGR MAT SCI BR, MAT TECH DIV, EG&G IDAHO INC, 81- *Mem:* Am Asn Advan Sci; Am Soc Metals. *Res:* Studies of point defects in metals and alloys by means of anelastic techniques; studies of void formation in irradiated reactor cladding materials; environmental effects on materials. *Mailing Add:* 6731 E Lincoln Rd Idaho Falls ID 83401

KEEFER, LARRY KAY, b Akron, Ohio, Oct 28, 39; m 62; c 2. NITROSAMINES, NITRIC OXIDE. *Educ:* Oberlin Col, BA, 61; Univ NH, PhD(org chem), 65. *Prof Exp:* Asst prof oncol, Inst Med Res, Chicago Med Sch, 65-68; asst prof biochem, Col Med, Univ Nebr, 68-71; head, Anal Chem Sect, Lab Carcinogen Metab, 71-83, HEAD, CHEM SECT, LAB COMP CARCINOGENESIS, NAT CANCER INST, 83- *Mem:* Am Chem Soc; AAAS; Am Asn Cancer Res. *Res:* Chemistry of nitrosamines and related carcinogens including mechanisms of formation, their chemical reactivity and the pathways responsible for their biological action; chemistry and pharmacology of nitric oxide and its progenitors. *Mailing Add:* 7016 River Rd Bethesda MD 20817

KEEFER, RAYMOND MARSH, b Twin Falls, Idaho, Apr 29, 13; m 43; c 3. PHYSICAL ORGANIC CHEMISTRY, MOLECULAR COMPLEXES. *Educ:* Univ Calif, BS, 34, PhD(chem), 40. *Prof Exp:* From asst to assoc prof, 36-41, from instr to assoc prof, 41-56, chmn dept, 62-74, prof, 56-81, EMER PROF CHEM, UNIV CALIF, DAVIS, 81- *Mem:* Am Chem Soc; Sigma Xi. *Res:* Molecular complexes; electrophilic aromatic halogenation; participation by ortho substituents in reactions at aromatic side chains; medium effects on nucleophilic solvolytic displacement reactions. *Mailing Add:* Dept Chem Univ Calif Davis CA 95616

KEEFER, ROBERT FARIS, b Wheeling, WVa, May 27, 30; c 6. SOIL SCIENCE. *Educ:* Cornell Univ, BS, 52; Ohio State Univ, MS, 61, PhD(agron), 63. *Prof Exp:* Res agronomist, Hercules Powder Co, 63-65; from asst prof to assoc prof soil sci, 65-74, assoc prof & assoc agronomist, 74-76, PROF AGRON & AGRONOMIST, WVA UNIV, 76- *Concurrent Pos:* Dept Health, Educ & Welfare grant, 66-70. *Mem:* Am Soc Agron; Soil Sci Soc Am; Int Soil Sci Soc; Int Humic Sub Soc. *Res:* Soil organic matter; soil fertility, particularly micronutrient nutrition; sewage sludge, fly ash and strip mine reclamation. *Mailing Add:* 3473 University Ave Morgantown WV 26505

KEEFER, WILLIAM RICHARD, b Fayette, Ohio, June 7, 24; m 45; c 2. GEOLOGY. *Educ:* Univ Wyo, BA, 48, MA, 52, PhD, 57. *Prof Exp:* geologist, US Geol Surv, Denver, 48-81; explor adv, Mitchell Energy Corp, 81-; RETIRED. *Mem:* Geol Soc Am; Am Asn Petrol Geologists. *Res:* Regional stratigraphy and structure, especially in sedimentary rocks. *Mailing Add:* 5693 Xeno Way Arvada CO 80002

KEEFFE, JAMES RICHARD, b Visalia, Calif, Nov 13, 37; m 61; c 1. ORGANIC CHEMISTRY. *Educ:* Univ Calif, Santa Barbara, BA, 59; Univ Wash, Seattle, PhD(chem), 64. *Prof Exp:* NIH res fel, 64-65; from asst prof to assoc prof, 65-74, PROF CHEM, SAN FRANCISCO STATE UNIV, 74- *Concurrent Pos:* Res grants, Petrol Res Fund, 65-66 & Res Corp, 66-68. *Res:* Kinetic hydrogen isotope effects; acid-base catalysis; organic reaction mechanisms. *Mailing Add:* Dept Chem San Francisco State Univ San Francisco CA 94132-2999

KEEGAN, ACHSAH D, RESEARCH. *Prof Exp:* GUEST RESEARCHER, NAT INST ALLERGY & INFECTIOUS DIS, NIH, 89- *Mailing Add:* Lab Immunol Nat Inst Allergy & Infectious Dis NIH 9000 Rockville Pike Bldg 10 Rm 11N311 Bethesda MD 20892

KEEGSTRA, KENNETH G, b Grand Rapids, Mich, Aug 10, 45; m 65; c 3. BIOMEMBRANES, CHLOROPLAST BIOGENESIS. *Educ:* Hope Col, BA, 67; Univ Colo, PhD(biochem), 71. *Prof Exp:* Fel biochem, Mass Inst Technol, 71-73; asst prof microbiol, State Univ NY, 73-77; from asst prof to assoc prof, 77-83, PROF PLANT PHYSIOL, UNIV WIS-MADISON, 84- *Honors & Awards:* George Olmsted Award, Am Paper Inst, 73. *Mem:* Am Soc Plant Physiologists; Int Soc Plant Molecular Biol. *Res:* Structure, function and biogenesis of plastid envelope membranes; import of cytoplasmically synthesized proteins into chloroplasts. *Mailing Add:* Dept Bot Birge Hall Univ Wis Madison WI 53706

KEEHN, PHILIP MOSES, b Brooklyn, NY, Mar 22, 43. ORGANIC CHEMISTRY, PHYSICAL-ORGANIC CHEMISTRY. *Educ:* Yeshiva Col, BA, 64; Yale Univ, MA, 67, PhD(chem), 69. *Prof Exp:* NIH res fel chem, Harvard Univ, 69-71; asst prof, 71-78, Wolfson Professorship, 79-80, assoc prof chem, 78-86, PROF CHEM, BRANDEIS UNIV, 86- *Concurrent Pos:* Consult, Am Optical Corp, 76-, US Army, 79-80, Olive Corp, 82-85; Dreyfus teacher-scholar, 79-84; Nat Acad Sci E Europ Exchange fel, Yugoslavia, 85; Acad Sinica Lecturship, 87; Fulbright scholar, 88. *Honors & Awards:* Alfred Bader Award, 80. *Mem:* Am Chem Soc; Sigma Xi. *Res:* Synthesis of strained rings and theoretically interesting molecules; synthetic methods; application of nuclear magnetic resonance spectroscopy to organic systems; photooxidation; thermal chemistry; pure and applied laser chemistry of organic systems; host-guest chemistry. *Mailing Add:* 121 Gibbs St Brandeis Univ Newton MA 02159-1927

KEELE, DOMAN KENT, pediatrics, medical genetics; deceased, see previous edition for last biography

KEELER, CLYDE EDGAR, b Marion, Ohio, Apr 11, 00; m 39; c 1. ZOOLOGY. *Educ:* Denison Univ, BS, 23, MS, 25; Harvard Univ, MA, 25, ScD(genetics), 26. *Hon Degrees:* DSc, Denison Univ, 73. *Prof Exp:* Asst zool, Harvard Univ, 24-25, instr ophthal res, Howe Lab, Harvard Med Sch, 27-39; fel, Wistar Inst, 39-42, curator animal colony, 40-42; mem staff, Edgewood Sch, 42-43; instr biol, Woman's Col, Univ NC, 43-44; prof & head dept, Wesleyan Col, 44-45; prof, 45-61, EMER PROF, GA STATE COL WOMEN, 82-; dir res dept, 61-75, head genetics lab, 61-76, MEM RES COMT, CENT STATE HOSP, 61-, CONSULT GENETICS, 80- *Concurrent Pos:* Asst, Radcliffe Col, 28-30; Sheldon fel, Paris & Berlin, 26-27, Turkey, 30; Milton fel, China & Japan, 36; Bache Fund, 37-38; Guggenheim Mem Found fel, Europe, 38; Rockefeller Found grant-in-aid, 47-54; Southern Fund fel, 55-56; recipient Numerous grants & fellowships; Guggenheim fel; Paul Harris fel, Rotary Int. *Mem:* Zool Soc France; Am Genetic Asn; fel AAAS. *Res:* Medical genetics; mammalian heredity; vertebrate eye defects; effect of coat color genes on morphology and behavior in mammals; rodless retina, an ophthalmic mutation in the house mouse; Cuna Indian culture and moonchildren; albinism in man; pre-Columbus cultures in America; traces traditional cuna religion. *Mailing Add:* 130 N Tattnall St Milledgeville GA 31061

KEELER, EMMETT BROWN, b West Point, NY, Sept 28, 41; c 3. SYSTEMS ANALYSIS. *Educ:* Oberlin Col, BA, 62; Harvard Univ, MA, 67, PhD(math), 69. *Prof Exp:* MEM STAFF, RAND CORP, 68-; FAC RAND GRAD SCH, 77- *Concurrent Pos:* Vis assoc prof econ, Univ Chicago, 73; vis res assoc, Sch Pub Health, Harvard Univ, 74-75 & 82. *Mem:* Asn Pub Policy Mgt; Math Asn Am. *Res:* Utility theory; mathematical statistics; health economics; operations research; medical decision-making. *Mailing Add:* Rand Corp 1700 Main St Santa Monica CA 90406

KEELER, JOHN S(COTT), b Toronto, Ont, Can, Aug 12, 29; m 51; c 3. ELECTRICAL ENGINEERING. *Educ:* Univ Toronto, BASc, 51, MASc, 63. *Prof Exp:* Res officer, Nat Res Coun Can, 51-55; chief engr, Hallman Organs, Ont, 55-59; from lectr to asst prof, 60-64, ASSOC PROF ELEC ENG, UNIV WATERLOO, 64- *Concurrent Pos:* Acoust consult, 59- *Mem:* Sr mem Inst Elec & Electronics Engrs; Audio Eng Soc. *Res:* Numerical analysis and synthesis of sound particularly noise and music; effects of noise on man; acoustical instrumentation; environmental noise. *Mailing Add:* Dept Elec Eng Univ Waterloo Waterloo ON N2L 3G1 Can

KEELER, MARTIN HARVEY, b New York, NY, June 16, 27; m 53; c 3. MEDICINE. *Educ:* NY Univ, BA, 49; NY Med Col, MD, 53; Am Bd Psychiat & Neurol, dipl, 59. *Prof Exp:* Intern, State Univ NY Upstate Med Ctr, 53-54; from asst resident to resident psychiat, Sch Med, Univ NC, Chapel Hill, 54-57, from instr to assoc prof, 57-69; prof, NY Med Col, 69-70; prof psychiat, Med Univ, SC, 70-77; DIR ALCOHOLIC TREAT PROG HOUSTON, VA HOSP, 77-; PROF PSYCHIAT, BAYLOR COL MED, 77- *Concurrent Pos:* Res grants, 61 & 62- *Mem:* AMA; Am Psychoanal Asn; Am Med Soc Alcoholism; Am Psychiat Asn. *Res:* Defining of the psychological abnormalities in schizophrenia as specific to the individual or to the disease process and the pharmacological manipulation of these differences in schizophrenic and normal populations. *Mailing Add:* Dept Psychiat Vet Admin Hosp Houston TX 77211

KEELER, RALPH, b Norwich, Eng, Jan 11, 30; m 56; c 2. PHYSIOLOGY. *Educ:* Univ Birmingham, BSc, 53, PhD(physiol), 56. *Prof Exp:* Asst lectr physiol, Univ Birmingham, 55-56; lectr, Univ Ibadan, 56-59; lectr, Univ Newcastle, 59-66; from asst prof to assoc prof, 66-74, PROF PHYSIOL, UNIV BC, 74- *Mem:* Can Physiol Soc; Am Physiol Soc. *Res:* Pathophysiology of renal function; control of sodium excretion. *Mailing Add:* Dept Physiol Univ BC 2146 Health Sci Mall Vancouver BC V6T 1W5 Can

KEELER, RICHARD FAIRBANKS, b Provo, Utah, Jan 24, 30; m 52; c 5. NATURAL PRODUCTS CHEMISTRY. *Educ:* Brigham Young Univ, BS, 54; Ohio State Univ, MS, 55, PhD(biochem), 57. *Prof Exp:* Asst biochemist, Mont State Col, 57-61; res chemist biochem, Nat Animal Dis Lab, 61-65; RES CHEMIST BIOCHEM, USDA, UTAH STATE UNIV, 65- *Mem:* Am Chem Soc; Soc Exp Biol & Med; AAAS; Teratology Soc. *Res:* Molybdenum-tungsten metabolism; silicon-mucoprotein interaction in urolithiasis; muscular dystrophy; cytochemistry of Listeria and Vibrio; products of Nocardia; steroidal, quinolizidine and piperidine alkaloid chemistry and metabolic effects; chemistry of poisonous and teratogenic plants. *Mailing Add:* 125 Quarter Circle Logan UT 84321

KEELER, ROBERT ADOLPH, b New York, NY, Feb 4, 20; m 49; c 2. PHYSICAL INORGANIC CHEMISTRY. *Educ:* Queens Col, NY, BS, 42. *Prof Exp:* Chief chemist, NY Testing Labs, 42-45 & Pub Serv Testing Labs, 45; supvr anal chem, Allied Chem & Dye Corp, 45-50; group leader nuclear chem, Vitro Labs Div, Vitro Corp, 50-63; proj supvr propellant chem, Reaction Motors Div, Thiokol Chem Corp, 63-68; sr res chemist, Radiation Safety Off, Syracuse Res Lab, Allied Signal Corp, 68-86; RETIRED. *Concurrent Pos:* Mem, Air Pollution Control Bd, Newark, NJ, 59-69. *Mem:* AAAS; Am Chem Soc. *Res:* Abatement of industrial atmospheric pollutants and recovery as useful materials. *Mailing Add:* 301 Oakridge Dr Camillus NY 13031

KEELER, ROGER NORRIS, b Houston, Tex, Aug 12, 30; m 57, 87; c 4. TECHNICAL MARKETING. *Educ:* Rice Univ, Houston, BA, BS, 47-51; Univ Colo, Boulder, MS, 57-58; Univ Calif, PhD(chem eng), 63. *Prof Exp:* From staff mem to head, Physics Dept, Lawrence Livermore Lab, Livermore, Calif, 63-75; staff of dir, Lawrence Livermore Lab, Univ Calif, 78-80; prin sci adv, 87-88 dir, Tech Mkt, KAMAN DIVERSIFIED TECHNOL CORP, WASH, DC & BLOOMFIELD, CONN, 87-88. *Concurrent Pos:* Consult, Nat Tech & various orgns, 60-; lectr, Dept Applied Sci, Univ Calif, 67-75, Dept Chem Eng, 60-; lectr, Int Sch Physics, Enrico Fermi, Varenna, Italy, 68, 70, 79; adj prof, Physics & Chem, US Naval postgrad Sch, Calif, 79- 81; lectr, Nat Strategy Info Ctr, Wash, DC, 87. *Honors & Awards:* Ford Found Prof Chem Eng, Univ Mex City, Mex City, DF, 68. *Mem:* Sigma Xi; Res Soc Am; Am Inst Chemists; Am Phys Soc. *Res:* Turbulence, chemical kinetics, catalysis, cryogenics and cryogenic engineering; thermodynamics of Phase Equilibria; high pressure equation of state; optical and electronic properties of condensed media at high pressure; high pressure geophysics; high pressure fabrication of materials, anti-submarine warfare; advanced sensor technology. *Mailing Add:* 6652 Hampton Park Ct McLean VA 22101

KEELER, STUART P, b Wausau, Wis, Sept 1, 34. MECHANICAL ENGINEERING. *Educ:* Ripon Col, BA, 56; Mass Inst Technol, BS, 57, DSc(mech metal), 61. *Prof Exp:* mgr automotive res, Nat Steel Corp, 63-67; MGR METAL TECHNOL, BUDD CO TECH CTR, 87- *Concurrent Pos:* Instr, Sheet Metal Formability, Univ Wis, 84- *Mem:* Fel, Am Soc Metal; fel Soc Automotive Eng; Amer Inst Mining, Metal & Petrol Engrs. *Mailing Add:* Budd Co 1515 Alantic Blvd Auburn Hills MI 48326

KEELEY, DEAN FRANCIS, b Chicago, Ill, Nov 16, 26; m 51; c 1. ANALYTICAL CHEMISTRY, HEADSPACE ANALYSIS. *Educ:* Univ Ill, BS, 52; Fla State Univ, PhD(chem), 57. *Prof Exp:* From asst prof to assoc prof, 57-77, PROF CHEM, UNIV SOUTHWESTERN LA, 77- *Mem:* Am Chem Soc; Am Asn Univ Prof. *Res:* Physical properties by headspace analysis. *Mailing Add:* Drawer 44250 Univ Southwestern La Lafayette LA 70504

KEELEY, FRED W, b Winnipeg, Man, Mar 21, 44; m 66; c 3. BIOCHEMISTRY. *Educ:* Univ Man, BSc, 65, PhD(pharmacol), 70. *Prof Exp:* Med Res Coun Can fels, Agr Res Coun, Langford, Eng, 70-72 & Res Inst, Hosp Sick Children, Toronto, 72-73; assoc prof, 83-90, PROF, RES INST, HOSP SICK CHILDREN, TORONTO, 90- *Concurrent Pos:* Med Res Coun Can scholarship, Res Inst, Hosp Sick Children, 73-; assoc prof 83-90, prof biochem, Univ Toronto, 90-, assoc prof, 87-90, prof clin biochem, 90- *Mem:* Can Biochem Soc; NY Acad Sci. *Res:* Biosynthesis of elastin; calcification of aortic tissue in atherosclerosis; effects of hypertension on vascular connective tissue. *Mailing Add:* 92 Brentcliffe Rd Toronto ON M4G 3Y8 Can

KEELEY, JOHN L, b Streator, Ill, Apr 12, 04; m 37; c 3. SURGERY. *Educ:* Loyola Univ, Ill, BS, 27, MD, 29. *Prof Exp:* Instr surg, Med Sch, Univ Wis & Wis Gen Hosp, 35-36; Arthur Tracy Cabot fel, Harvard Med Sch-Peter Bent Brigham Hosp, 36-38, Harvey Cushing & Univ res fels, 37-38; from instr to asst prof, Sch Med, La State Univ, 39-41; from asst clin prof to assoc clin prof, 41-54, asst, Dept Chem, 54-58, chmn dept, 58-69, PROF SURG, STRITCH SCH MED, LOYOLA UNIV CHICAGO, 54- *Mem:* Int Soc Surg; Am Surg Asn; fel Am Col Surgeons; Am Asn Thoracic Surg; Soc Vascular Surg; Am Col Chest Physicians; Am Heart Asn. *Mailing Add:* Loyola Univ Med 2160 S First Ave Maywood IL 60153

KEELEY, JON E, b Chula Vista, Calif, Aug 11, 49; m 73, 90. POPULATION ECOLOGY. *Educ:* San Diego State Univ, BS, 71, MS, 73; Univ Ga, PhD(bot), 77. *Prof Exp:* Lectr bot, Univ Ga, 76-77; asst prof, 77-83, assoc prof, 83-88, PROF BIOL, OCCIDENTAL COL, 88- *Concurrent Pos:* Res grants, NSF, 79-88; consult, EIRs Publ; Guggenheim fel, 85. *Mem:* Ecol Soc Am; Bot Soc Am; Am Soc Plant Physiologists; Am Soc Naturalists. *Res:* Aquatic plant photosynthesis; reproductive biology and demography of plants; fire ecology of mediterranean vegetation. *Mailing Add:* Dept Biol Occidental Col Los Angeles CA 90041

KEELEY, LARRY LEE, b South Bend, Ind, Jan 3, 39; m 59; c 5. INSECT PHYSIOLOGY. *Educ:* Univ Notre Dame, BS, 62; Purdue Univ, PhD(entom), 66. *Prof Exp:* From asst prof to prof entom, 66-85, PROF ENTOM & BIOCHEM, TEX A&M UNIV, 85- *Concurrent Pos:* NSF grants, 74-78, 81-83, 85-88, 91-94; NIH grant, 78-81 & 84-87; Sea grant, 85-87, 87-89, 89-91; head, Lab Invertebrate Neuroendocrine Res, Dept Entom, Texas A&M Univ. *Mem:* Fel AAAS; Entom Soc Am; Am Soc Zool; Soc Neurosci; Am Soc Biochem & Molecular Biol. *Res:* Hormonal regulation of metabolism; neuroendocrine regulation of mitochondrial development and functions; identification and action of neurohormones on physiological functions of insects and other invertebrates; neurohormone isolation; isolation, characterization and regulation of insect and shrimp reproduction; molecular biology of neurohormone genes. *Mailing Add:* Dept Entom Tex A&M Univ College Station TX 77843

KEELEY, STERLING CARTER, b San Francisco, Calif, Oct 23, 48. SYSTEMATIC BOTANY, ECOLOGY. *Educ:* Stanford Univ, AB, 70; San Diego State Univ, MS, 73; Univ Ga, PhD(bot), 77. *Prof Exp:* Res asst ecol, Int Biol Prog Struct Ecosysts, 70-73; lectr bot, Calif State Univ, Long Beach, 78-79; assoc prof biol, 79-90, PROF BIOL, WHITTIER COL, 90- *Concurrent Pos:* NSF dissertation improv grant, 74-77; consult flora, Southern Calif Ocean Studies Consortium of Calif State Univ & Cols, 78-; res assoc, Los Angeles County Mus Natural Hist, 78-; consult salt marsh veg, Port of Los Angeles, 79-81; NSF grants, Systs Neotropical Vernonia, 79-81, 82-86; chloroplast DNA, Vernonia, NSF grant, 88-89; vis prof biol, Univ Conn,

89-90. *Mem:* Soc Study Evolution; Am Soc Plant Taxonomists; Am Bot Soc; AAAS; Sigma Xi; Ecol Soc Am. *Res:* Systematics and biogeography of neotropical species of the genus Vernonia, Compositae; ecology and reproductive biology of mediterranean climate plants in relation to fire. *Mailing Add:* Dept Biol Whittier Col Whittier CA 90608

KEELING, BOBBIE LEE, b Durant, Okla, Apr 22, 31; div; c 2. PLANT PATHOLOGY. *Educ:* Southeastern State Col, BS, 56; Okla State Univ, MS, 59; Univ Minn, Minneapolis, PhD(plant path), 66. *Prof Exp:* RES PLANT PATHOLOGIST, DELTA BR EXP STA, USDA, 66- *Mem:* Am Phytopath Soc; Sigma Xi. *Res:* Diseases of soybeans with emphasis on pathogenic variation, nature of host resistance and host-parasite interaction. *Mailing Add:* USDA Delta Br Exp Sta Stoneville MS 38776

KEELING, CHARLES DAVID, b Scranton, Pa, Apr 20, 28; m 54; c 5. PHYSICAL CHEMISTRY, MARINE GEOCHEMISTRY. *Educ:* Univ Ill, BA, 48; Northwestern Univ, PhD(chem), 53. *Prof Exp:* Fel geochem, Calif Inst Technol, 53-56; asst res chemist, 56-60, from assoc res chemist to assoc prof, 60-68, PROF OCEANOG, SCRIPPS INST OCEANOG, UNIV CALIF, SAN DIEGO, 68- *Concurrent Pos:* Guest prof oceanog, Univ Heidelberg, 69-70; mem, Comn Atmospheric Chem & Global Pollution, Int Asn Meteorol & Atmospheric Physics, 67-85; mem, Panel on Energy & Climate of US, Nat Acad Sci, 74-77; mem, Interim Sci Directorate Carbon Dioxide Res Prog, US Dept Energy, 77-80; guest prof, Phys Inst, Univ Bern, Switz, 79-80; dir, Cent Carbon Dioxide Lab, World Meteorol Orgn, 75-; mem, CO2 Panel, Ocean Sci Bd, Nat Acad Sci, 87-; convenor carbon dioxide activ, Int Global Atmospheric Comt, 90- *Honors & Awards:* Second Half Century Award, Am Meteorol Soc, 80. *Mem:* Fel AAAS; Am Geophys Union; fel Am Acad Arts & Sci. *Res:* Marine chemistry; geochemistry of carbon and oxygen; atmospheric chemistry; influence of atmospheric carbon dioxide on carbon cycle and on world climate. *Mailing Add:* Scripps Inst Oceanog Mail Code A020 La Jolla CA 92037

KEELING, RICHARD PAIRE, b Crawfordsville, Ind, Sept 17, 31; m 52; c 2. MYCOLOGY. *Educ:* Wabash Col, AB, 57; Purdue Univ, MS, 60, PhD(bot), 63. *Prof Exp:* Asst prof microbiol, 63-70, assoc prof biol, 70-74, PROF BIOL, EMPORIA KANS STATE COL, 74- *Mem:* AAAS; Mycol Soc Am; Am Soc Microbiol; Soc Indust Microbiol; Japanese Mycol Soc; Sigma Xi. *Res:* Fungus physiology; metabolism. *Mailing Add:* 1219 Frontier Way Emporia KS 66801

KEELING, ROLLAND OTIS, JR, b Hillsboro, Ind, Aug 13, 25; m 46; c 2. PHYSICS. *Educ:* Wabash Col, AB, 50; Pa State Univ, MS, 52, PhD(physics), 54. *Prof Exp:* Instr physics, Pa State Univ, 53-54; physicist, Gulf Res & Develop Co, Pa, 54-61; dept head, 80-84, prof physics, 61-84, EMER PROF PHYSICS, MICH TECHNOL UNIV, 84- *Mem:* AAAS; Am Phys Soc; Am Asn Physics Teachers; Am Crystallog Asn. *Res:* X-ray diffraction; crystallography; x-ray spectroscopy; magnetic structures. *Mailing Add:* RR 1, Box 61 Royalewood Houghton MI 49931

KEELUNG, HONG, MEMBRANE FUSION. *Educ:* Univ Calif, Berkeley, PhD(chem), 75. *Prof Exp:* RES BIOCHEMIST, CANCER RES INST, UNIV CALIF, SAN FRANCISCO, 81- *Mem:* Biophys Soc; Am Chem Soc. *Mailing Add:* Cancer Res Inst Univ Calif San Francisco CA 94143

KEELY, WILLIAM MARTIN, b Louisville, Ky, Nov 10, 24; m 58; c 5. PHYSICAL CHEMISTRY. *Educ:* Univ Louisville, BA, 45; Ind Univ, MA, 46; Univ Ky, PhD(phys chem), 49. *Prof Exp:* Asst chem, Ind Univ, 45-46; chemist, Girdler Chem, Inc, 49-51, group leader, 51-55, supvr phys measurements, 55-78; MGR ANALYSIS DEPT, UNITED CATALYSTS, INC, 78- *Concurrent Pos:* Instr, Univ Louisville, 61-75, Ind Univ Southwest, 76-80; dir catalyst selection & evaluation, Ctr Prof Advan, 72- *Mem:* Am Chem Soc. *Res:* Catalysts; solid state reactions; x-ray diffraction; surface area and pore volume; thermogravimetric analysis; differential thermal analysis; magnetic inductance investigations and microscopic studies; fats and oils; selective hydrogenations; chemisorption and physical adsorption investigations; reformer; water gas shift; dehydrogenation, hydrogenation; zeolites. *Mailing Add:* 819 Keswick Blvd Louisville KY 40217

KEEM, JOHN EDWARD, b Buffalo, NY, May 31, 48; m 80. SOLID STATE PHYSICS. *Educ:* Syracuse Univ, BS, 70; Purdue Univ, PhD(physics), 76. *Prof Exp:* Fel Dept Physics, Purdue Univ, 76-77; Devices Phys Dept, Gen Motors Res, 77-80, mgr, Superconductivity Res, Energy Conversion, 80-83; DIR RES & DEVELOP, OVONIC SYNTHETIC MAT CO, 84- *Concurrent Pos:* Mat res adv bd mat sub-micron struct, Nat Acad Sci. *Mem:* Am Phys Soc; Am Soc Metals; Mat Res Soc. *Res:* Synthesis of multilayer structures by sputtering and ion beam deposition; x-ray scattering from multilayer structures; melt spinning of Nd2 Fel4B permanent magnets; magnetic interactions between grains in Wd2 Fel4B materials; solid lubricating materials. *Mailing Add:* 1788 Northwood Dr Troy MI 48084

KEEN, CARL L, DEVELOPMENTAL NUTRITION, MINERALS. *Educ:* Univ Calif, Davis, PhD, 79. *Prof Exp:* ASSOC PROF NUTRIT, UNIV CALIF, DAVIS, 84- *Mem:* Am Inst Nutrit; Teratology Soc; Soc Exp Biol & Med. *Mailing Add:* Dept Nutrit Univ Calif Davis CA 95616

KEEN, CHARLOTTE ELIZABETH, b Halifax, NS, June 22, 43; m 63. MARINE GEOPHYSICS. *Educ:* Dalhousie Univ, BSc, 64, MSc, 66; Cambridge Univ, PhD(geophys), 70. *Prof Exp:* RES SCIENTIST MARINE GEOPHYS, ATLANTIC GEOSCI CENTRE, BEDFORD INST, 70- *Concurrent Pos:* Mem working group 8 of inter-union comn on geodynamics, Int Union Geod & Geophys & Int Union Geol Sci, 72-; chmn study group NW Atlantic Continental Margin, Inter-Union Comn Geodynamics; assoc ed, Can J of Earth Sci; chmn, Can Nat Lithosphere Comn, 81- *Mem:* Geol Asn Can; Royal Soc Can; Am Geophys Union. *Res:* Surface wave propagation in Canadian shield and along mid-ocean ridges; plate tectonics of Baffin Bay region; continental-oceanic transition in the North West Atlantic; application of Backus-Gilbert inversion to upper mantle properties at ocean-continent transition; subsidence and thermal history of continental margins. *Mailing Add:* Nine Wenlock Grove Dartmouth NS B3P 1P6 Can

KEEN, DOROTHY JEAN, b Lancaster, Pa, June 19, 22. PHYSICAL OCEANOGRAPHY. *Educ:* Swarthmore Col, BA, 44. *Prof Exp:* Res asst chem & phys oceanog, Woods Hole Oceanog Inst, 44-53; phys oceanogr mil appln, Hydrographic Off, US Naval Oceanog Off, 53-58, oceanogr ocean prediction, 58-66, head systs anal group, Ocean Prediction, 66-71, sci staff asst, Ocean Sci Dept, Plans & Requirements Off, 71-78; phys sci adminr, Naval Ocean Res & Develop Activ, 78-82; RETIRED. *Mem:* AAAS; Marine Tech Soc; Am Geophys Union. *Res:* Military oceanography, plans and analysis; ocean prediction; fleet environmental support programs; effects on acoustic systems and tactics; ocean and estuarine dynamics; chemical analyses of sea water. *Mailing Add:* 7214 Ahi Ct Bay St Louis MS 39520

KEEN, JAMES H, b New York, NY, Feb 14, 48; m 73; c 2. CELL BIOLOGY. *Educ:* Cornell Univ, PhD, 76. *Prof Exp:* ASSOC PROF FELS RES INST, SCH MED, TEMPLE UNIV, 85- *Concurrent Pos:* Chmn grad studies, Prog Molecular Biol & Genetics, Sch Med, Temple Univ. *Mem:* AAAS; NY Acad Sci; Am Soc Cell Biol; Am Soc Biol Chem. *Res:* Structure and function of clathrin coated membranes; role in receptor-mediated endocytosis and membrane dynamics. *Mailing Add:* Fels Res Inst Temple Univ Sch Med 3420 N Broad St Philadelphia PA 19140

KEEN, LINDA, b New York, NY, Aug 9, 40; c 2. MATHEMATICS. *Educ:* City Col NY, BS, 60; NY Univ, MS, 62, PhD(math), 64. *Prof Exp:* NSF fel math, Inst Advan Study, 64-65; asst prof, Hunter Col, 65-67; from asst prof to assoc prof, 67-74, PROF MATH, CUNY, LEHMAN COL, GRAD CTR, 74- *Concurrent Pos:* Vis prof, 80-81, NSF partial res grant, 82-88, vis prof, Boston Univ, 87, vis scientist, IBM, 88, vis mem Nat Sci Res Inst, 86, Max Planck Inst, 88. *Mem:* Am Math Soc; Asn Women Math (pres 85-86). *Res:* Complex analysis; Riemann surfaces; discontinuous groups; Teichmüller spaces; dynamical systems. *Mailing Add:* 180 Ames Ave Leonia NJ 07605

KEEN, MICHAEL J, geophysics, oceanography; deceased, see previous edition for last biography

KEEN, NOEL THOMAS, b Marshalltown, Iowa, Aug 13, 40; m 86. PLANT PATHOLOGY. *Educ:* Iowa State Univ, BS, 63, MS, 65; Univ Wis, Madison, PhD(plant path), 68. *Prof Exp:* PROF PLANT PATH, UNIV CALIF, RIVERSIDE, 68- *Mem:* AAAS; Am Phytopath Soc; Am Chem Soc; Am Soc Plant Physiologists; Int Soc Plant Path. *Res:* Mechanisms of pathogenesis by plant parasitic microorganisms; mechanisms of disease resistance. *Mailing Add:* Dept Plant Path Univ Calif Riverside CA 92521

KEEN, RAY ALBERT, b Valley Falls, Kans, Oct 9, 15; m 43; c 4. ORNAMENTAL HORTICULTURE. *Educ:* Kans State Col, BS, 42; Ohio State Univ, MSc, 47, PhD, 56. *Prof Exp:* From asst prof to assoc prof, 47-62, prof, 62-81, EMER PROF HORT, KANS STATE UNIV, 81-, RES HORTICULTURIST, AGR EXP STA, 72- *Mem:* Am Soc Hort Sci; Am Soc Agron; Sigma Xi. *Res:* Turf grass genetics; shade trees; woody ornamentals; propagation; soils and mineral nutrition. *Mailing Add:* 1916 Blue Hills Rd Manhattan KS 66502

KEEN, ROBERT ERIC, b Oakland, Calif, May 29, 44; m 72; c 2. LIMNOLOGY, POPULATION ECOLOGY. *Educ:* Kans State Univ, BS, 65; Mich State Univ, MS, 67, PhD(zool), 71. *Prof Exp:* Asst prof zool, Univ Vt, 71-76; vis asst prof biol, Kans State Univ, 76-77; vis asst prof zool, Ind Univ, 77; asst prof, 77-81, ASSOC PROF BIOL, MICH TECHNOL UNIV, 81- *Concurrent Pos:* Fel, Philadelphia Acad Natural Sci, 70-71; partic, Advan Inst Statist Ecol, Pa State Univ, 72; staff consult, Nat Comn Water Qual, 75; consult, Vt Inst Water Resources Res, 76-77. *Mem:* Am Soc Limnol & Oceanog; Ecol Soc Am; Int Soc Limnol; Soc Pop Ecol. *Res:* Population ecology of zooplankton; limnology of Lake Superior; toxicity tests with Ceriodaphnia. *Mailing Add:* Dept Biol Sci Mich Technol Univ Houghton MI 49931

KEEN, VERYL F, b Stilwell, Okla, Jan 14, 23; m 59; c 2. BIOLOGY. *Educ:* Northeastern State Col, BS, 50; Okla State Univ, MS, 54; Univ Colo, MA, 62, PhD(zool), 65. *Prof Exp:* Teacher, Maramec Sch, Okla, 50-52, supt schs, 52-57; instr biol, Univ Colo, 59-60, vis lectr, 65-66; assoc prof, Northeastern Mo State Col, 63-65; asst prof & coordr, 66-71, PROF BIOL & CHMN DEPT, ADAMS STATE COL, 71- *Mem:* AAAS; Am Soc Mammal; Nat Asn Biol Teachers. *Res:* Small mammal population ecology; rodent botflies of Colorado; ecology of small mammals of San Luis Valley in Colorado. *Mailing Add:* 12551 Chamisa Trail Alamosa CO 81101

KEEN, WILLIAM HUBERT, b Jewell Ridge, Va, Sept 2, 44; m 68; c 1. ECOLOGY, ZOOLOGY. *Educ:* Pikeville Col, BA, 67; Eastern Ky Univ, MS, 71; Kent State Univ, PhD(ecol), 75. *Prof Exp:* Teacher biol & sci, Buchanan County Pub Sch, Grundy, Va, 67-68; coordr interdisciplinary field studies, Pikeville Col, Ky, 68-69; teacher biol & sci, Jefferson County Pub Sch, Louisville, Ky, 69-70; instr, Kent State Univ, Ohio, 75-76; from asst prof to assoc prof, State Univ NY Cortland, 76-86, chmn dept, 83-88, PROF BIOL, STATE UNIV NY CORTLAND, 86-, DEAN ARTS & SCI, 88- *Concurrent Pos:* Lectr, Cuyahoga Community Col, Cleveland, Ohio, 75-76. *Mem:* Sigma Xi; AAAS; Ecol Soc Am; Soc Study Amphibians & Reptiles; Am Soc Ichthyologists & Herpetologists; Animal Behav Soc. *Res:* Population ecology and behavior of lower vertebrates; thermoregulation in amphibians; functions of fish schooling; interspecific interactions. *Mailing Add:* Dept Biol Sci State Univ NY Cortland NY 13045

KEENAN, CHARLES WILLIAM, b Ft Worth, Tex, Apr 10, 22; m 45; c 2. INORGANIC CHEMISTRY. *Educ:* Centenary Col, BS, 43; Univ Tex, PhD(chem), 49. *Prof Exp:* Chemist, La Ord Plant, 42-43; asst, Univ Tex, 43-44, instr, 46-49; chemist, Naval Ord Plant, 44-46; from asst prof to assoc prof, 49-58, assoc dean liberal arts, 73-78, PROF CHEM, UNIV TENN, KNOXVILLE, 58- *Concurrent Pos:* NSF sci fac fels, Cambridge Univ, 57-58, 64-65. *Mem:* AAAS; Am Chem Soc; Hist Sci Soc; Am Asn Univ Profs; Sigma Xi. *Res:* Reactions in non-aqueous solvents. *Mailing Add:* 4501 Appleby Rd Knoxville TN 37920

KEENAN, EDWARD JAMES, b Shelton, Wash, Sept 6, 48. PHARMACOLOGY, ENDOCRINOLOGY. *Educ:* Creighton Univ, BS, 70, MS, 72; WVa Univ, PhD(pharmacol), 75. *Prof Exp:* Res assoc, 75-76, ASST PROF SURG, UNIV ORE HEALTH SCI CTR, 76-, ASST PROF PHARMACOL, 78-, DEPT SURG. *Concurrent Pos:* Dir, Clin Res Ctr Lab & Hormone Res Lab, Univ Ore Health Sci Ctr, 76-, instr pharmacol, 77-78. *Mem:* Am Soc Andrology; AAAS; Sigma Xi. *Res:* Significance of steroid hormones in cancer of the breast and prostate gland; mechanism of steroid hormone action; role of prolactin in male accessory sex organ function. *Mailing Add:* Ore Health Sci Univ 3181 SW Sam Jackson Park Rd Portland OR 97201

KEENAN, EDWARD MILTON, mathematics, for more information see previous edition

KEENAN, JOHN DOUGLAS, b Sarnia, Ont, Mar 16, 44; US citizen; m 62; c 4. WATER & WASTEWATER TREATMENT, ALTERNATIVE ENERGY SOURCES. *Educ:* State Univ NY, Buffalo, BA, 67; Syracuse Univ, MS, 70, PhD(civil eng), 72. *Hon Degrees:* MA, Univ Pa, 78. *Prof Exp:* Instr water & wastewater treat, Syracuse Univ, 70-72; from asst prof to assoc prof civil eng, 73-86, assoc prof, 86-90, PROF CIVIL ENG SYSTS & ASSOC DEAN UNDERGRAD EDUC, UNIV PA, 90- *Concurrent Pos:* Vis asst prof civil eng, Univ Pa, 72-73. *Res:* Environmental systems engineering; water and wastewater engineering; biological and health effects of pollutants; alternative energy sources. *Mailing Add:* Dept Systs Univ Pa Philadelphia PA 19104-6391

KEENAN, JOSEPH ALOYSIUS, b Washington, DC, Aug 5, 38; m 68; c 2. NUCLEAR CHEMISTRY. *Educ:* Spring Hill Col, BS, 64; Clark Univ, PhD(nuclear chem), 71. *Prof Exp:* mem res staff, 69-80, SR MEM TECH STAFF, MAT SCI LAB, TEX INSTRUMENTS INC, 80- *Mem:* Am Chem Soc; Electro Chem Soc; Am Vacuum Soc. *Res:* Instrumental neutron activation analysis, radiotracer techniques and x-ray fluorescence analysis in materials characterization; design and building of mini computer systems for manufacturing and laboratory automation; auger spectroscopy; surface science ion backseat housing; nuclear reaction analysis. *Mailing Add:* 1214 Cherokee Dr Richardson TX 75080-3906

KEENAN, KATHLEEN MARGARET, b St Paul, Minn, May 24, 34. BIOSTATISTICS. *Educ:* St Catherine Col, BA, 56; Univ Minn, MS, 58, PhD(biostatist), 64. *Prof Exp:* Asst prof, 64-69, ASSOC PROF ORAL SCI, SCH DENT, UNIV MINN, 69- *Mem:* AAAS; Am Statist Asn; Biom Soc; fel Am Pub Health Asn; Am Soc Human Genetics; Sigma Xi. *Res:* Biostatistical applications in dental research. *Mailing Add:* 1768 Field Ave St Paul MN 55116

KEENAN, PHILIP CHILDS, b Bellevue, Pa, Mar 31, 08. ASTRONOMY. *Educ:* Univ Ariz, BS, 29, MS, 30; Univ Chicago, PhD(astrophysics), 32. *Hon Degrees:* Dr, Univ Cordoba, 71. *Prof Exp:* Asst astron, Yerkes Observ, Univ Chicago, 29-35, instr, 36-42; instr Perkins Observ, Ohio State Univ & Ohio Wesleyan Univ, 35-36; physicist, Bur Ord, US Dept Navy, 42-46; from asst prof to prof, 46-76, actg dir, 57-59, EMER PROF ASTRON, PERKINS OBSERV, OHIO STATE UNIV, 76- *Mem:* fel Royal Astron Soc. *Res:* Stellar spectroscopy; spectral classification; history of astronomy. *Mailing Add:* Astron 174 W 18th Ave Ohio State Univ Columbus OH 43210

KEENAN, ROBERT GREGORY, b St Albans, Vt, Dec 19, 15; m 44; c 3. ENVIRONMENTAL CHEMISTRY. *Educ:* Catholic Univ, BS, 37; Univ Md, MS, 52; Am Bd Indust Hyg, 62. *Prof Exp:* Lab helper chem, Div Occup Health, USPHS, 38-40, from jr chemist to assoc chemist, 40-45, sr asst scientist, 45-49, scientist, 49-53, chief phys anal unit, 53-56, asst chief anal serv, 56-60, chief, 60-66, dep chief res & med affairs, 66-67, assoc chief div occup health, 67-69; vpres & dir lab serv, George D Clayton & Assocs, 69-76; RETIRED. *Concurrent Pos:* Guest worker, Anal Chem Div, Radiochem Anal Sect, Nat Bur Standards, 63-64. *Honors & Awards:* Moyer D Thomas Award, Am Soc Testing & Mat, 77. *Mem:* emer mem Am Chem Soc; emer mem Am Indust Hyg Asn; Am Soc Testing & Mat. *Res:* Spectrography; determination of cobalt in dust samples; determination of iron in welding fume samples; quantitative analytical methods in emission spectroscopy; activation analysis; atomic absorption; beryllium in air; biological materials and ores; analytical techniques for industrial hygiene and air pollution; author or coauthor of over 75 scientific publications. *Mailing Add:* 122 Country Club Dr E South Burlington VT 05403-5838

KEENAN, ROBERT KENNETH, b Pueblo, Colo, Nov 29, 38; m 60; c 3. ELECTRONICS ENGINEERING. *Educ:* Univ Calif, Los Angeles, BSc, 62; Calif Inst Technol, MSEE, 63; Monash Univ, Australia, PhD(elec eng), 67. *Prof Exp:* Staff engr, Commun Div, Hughes Aircraft Co, 62-64; lectr elec eng, Monash Univ, 64-67; eng specialist, Electronics Div, Gen Dynamics Corp, 67-68; mgr res & adv develop, Electronic Commun Inc, Nat Cash Register Co, 68-71; mem tech staff, Mitre Corp, 71-75; aerospace corp, 75-77; dir systs sci, BDM Corp, 77-78; sr prof eng, TRW Inc, 78-81; VPRES ENG, KEENAN CORP, 79- *Mem:* Inst Elec & Electronics Engrs; Am Inst Aeronaut & Astronaut; Sigma Xi. *Res:* Communications and communications electronics, particularly signal processing theory and practice. *Mailing Add:* TKC 8609 66th St North Pinellas Park FL 34666

KEENAN, ROY W, METABOLISM. *Educ:* Ohio State Univ, PhD, 60. *Prof Exp:* PROF BIOCHEM, UNIV TEX HEALTH SCI CTR, 82- *Mailing Add:* 9118 Saddle Trail San Antonio TX 78255

KEENAN, THOMAS AQUINAS, b Rochester, NY, Mar 8, 27; m 48; c 3. COMPUTER SCIENCE. *Educ:* Univ Rochester, BA, 47; Purdue Univ, MS, 50, PhD(physics), 55. *Prof Exp:* Instr physics, Purdue Univ, 50-55; dir comput ctr, Rochester Univ, 55-66, chmn prog appl math, 58-66, asst prof physics, 57-62; dir systs planning, Interuniv, Commun Coun, 66-68, exec dir educ info network, 68-69; PROG DIR SOFTWARE SYSTS SCI, NSF, 69- *Concurrent Pos:* Exec dir, comt on uses of comput, Nat Acad Sci-Nat Res Coun, 62-63;

consult, Sch Math Study Group, 65-66. *Mem:* AAAS; Asn Comput Mach; Inst Elec & Electronics Engrs. *Res:* Computation; formal languages; symbol manipulation; information retrieval; data structure; phase transitions; combinatorial mathematics. *Mailing Add:* 12433 Over Ridge Rd Potomac MD 20854-3047

KEENAN, THOMAS K, b Ft Dodge, Iowa, Oct 8, 24; m 52; c 4. INORGANIC CHEMISTRY. *Educ:* SDak Sch Mines & Technol, BS, 48; Univ NMex, MS, 50, PhD(chem), 54. *Prof Exp:* Asst, Univ NMex, 49-53; mem staff, Los Alamos Nat Lab, 54-75, group leader waste mgt, 75-81, asst to dep assoc dir, 81-84, assoc group leader waste mgt, 84-86; RETIRED. *Mem:* Am Nuclear Soc. *Mailing Add:* 289 Venado Los Alamos NM 87545

KEENAN, THOMAS WILLIAM, b Johnstown, Pa, May 12, 42; m 64; c 3. FOOD SCIENCE. *Educ:* Pa State Univ, BS, 64; Ore State Univ, MS, 65, PhD(food sci), 67. *Prof Exp:* Assoc prof food sci, 67-73, PROF ANIMAL SCI, PURDUE UNIV, 73-, ASST DEAN GRAD SCH, 77- *Mem:* Am Chem Soc; Am Dairy Sci Asn; Inst Food Technol. *Res:* Membrane function; microbial biochemistry; lipid metabolism. *Mailing Add:* Dept Biochem & Nutrit VPI and State Univ Blacksburg VA 24061

KEENAN, WILLIAM JEROME, b Rawlins, Wyo, Sept 8, 39; m 65; c 7. PEDIATRICS, NEONATAL-PERINATAL MEDICINE. *Educ:* Loyola-Stritch Sch Med, MD, 64. *Prof Exp:* Fel pediat, Univ Cincinnati, 67-69, from asst prof to prof, 69-80; PROF PEDIAT-OBSTET, ST LOUIS UNIV, 80- *Concurrent Pos:* Dir neonatology, Cardinal Glennon Children's Hosp, 80-; dir pediat, St Mary's Health Ctr, 81-; dir, Southern Ill grant, 84-; consult, Mo Dept Health, 85. *Mem:* Soc Pediat Res; Am Fedn Clin Res; AAAS; Sigma Xi. *Res:* Neonatal hyperbilirubinemia; critical care of neonatal patients; developmental biology. *Mailing Add:* Dept Pediat St Louis Univ Med Sch 1465 S Grand Blvd St Louis MO 63104

KEENE, CLIFFORD H, b Buffalo, NY, Jan 28, 10. INTERNAL MEDICINE. *Educ:* Univ Mich, AB, 32, MD, 34, MSc, 38. *Prof Exp:* Intern, Univ Hosp, Ann Arbor, 34-36, teaching res, surg & path, 36-39; pres, Kaiser Found Hosp & Health Plan, 68-75; MEM STAFF, COMMUNITY HOSP, MONTEREY, CALIF, 75- *Concurrent Pos:* Instr, Surg & Path, Univ Mich, 36-39 & 46-54. *Mem:* Inst Med-Nat Acad Sci; Am Med Asn; Am Col Occup Med; fel Am Col Surgeons. *Mailing Add:* 3978 Ronda Rd PO Box 961 Pebble Beach CA 93953

KEENE, HARRIS J, b Brooklyn, NY, Apr 13, 31; m 56; c 3. ORAL PATHOLOGY. *Educ:* Univ Md, DDS, 55. *Prof Exp:* Res officer dent, US Navy, 60-77, dir, Naval Dent Res Inst, 68-77; MEM STAFF, UNIV TEX DENT BR, DENT SCI INST, 77- *Mem:* Fel AAAS; Am Dent Asn; Am Asn Phys Anthrop; Int Asn Dent Res. *Res:* Epidemiology of oral diseases; dental oncology; dental anthropology; paleopathology; oral physiology. *Mailing Add:* Univ Tex Dental Br Rm 4131 PO Box 20068 Houston TX 77225

KEENE, JACK DONALD, b Jacksonville, Fla, June 21, 47; m 69; c 2. MOLECULAR VIROLOGY, GENE EXPRESSION. *Educ:* Univ Calif, Riverside, AB, 69; Univ Wash, Seattle, PhD(microbiol & immunol), 74. *Prof Exp:* Staff fel molecular virol, Lab Molecular Genetics, Nat Inst Neurol Dis & Stroke, NIH, 74-78; from asst prof to assoc prof virol & molecular genetics, 79-88, ASST PROF RHEUMATOL & IMMUNOL, DEPT MED, DUKE UNIV MED CTR, DURHAM, 85-, PROF VIROL & MOLECULAR GENETICS, DEPT MICROBIOL & IMMUNOL, 88- *Concurrent Pos:* Fac Res Award, Am Can Soc, 81-86; Assoc ed, Virol, 83-; spec reviewer, Virol Study Sect, NIH, 84-, mem, Exp Virol Study Sect, 85-88; Pew Scholar Biomed Sci, 86-90; mem, Molecular Biol Study Sect & Arthritis Found Study Sect, Fel Comn & Res Comn. *Mem:* Am Soc Microbiol; Am Soc Virol; Am Soc Biochem & Molecular Biol. *Res:* RNA metabolism and processing; nature of autoimmunity and genetic regulation; virus-host interactions as models of cellular gene expression. *Mailing Add:* Med Ctr Box 3020 Duke Univ Durham NC 27710

KEENE, JAMES H, b Epps, La, May 8, 30; m 58; c 2. POULTRY NUTRITION. *Educ:* Univ Ark, BS, 57; La State Univ, MS, 59, PhD(poultry nutrit), 62. *Prof Exp:* Nutritionist, George B Matthews & Sons Inc, 62-64; from asst prof to assoc prof, 64-70, PROF POULTRY & DAIRYING, ARK STATE UNIV, 70- *Mem:* Poultry Sci Asn. *Res:* Poultry production and physiology. *Mailing Add:* Ark State Univ Box 202 State University AR 72467

KEENE, OWEN DAVID, b New Eagle, Pa, Apr 28, 34; m 57; c 1. POULTRY NUTRITION. *Educ:* Pa State Univ, BS, 55; Univ Md, MS, 59, PhD(poultry nutrit), 63. *Prof Exp:* Sr biochemist, Abbott Labs, 63-69; asst prof, 69-75, ASSOC PROF POULTRY SCI EXTEN, PA STATE UNIV, 75- *Concurrent Pos:* Agr prog leader, Poultry Exten Prog, USDA/Exten Serv, Washington, DC, 84-85 & Residual Avoidance Prog, 85. *Mem:* Poultry Sci Asn; World Poultry Sci Asn. *Res:* Product development relating to the nutrition of poultry. *Mailing Add:* 224 Henhing Pa State Univ University Park PA 16802

KEENE, WAYNE HARTUNG, b Boothbay Harbor, Maine, Apr 29, 37; m 57; c 2. PHYSICS, LASER RADAR. *Educ:* Univ Maine, BS, 58, MS, 61; Worcester Polytech Inst, PhD(physics), 64. *Prof Exp:* Physicist elec boat div, Gen Dynamics Corp, Conn, 58-59; sr engr, Defense & Space Ctr, Westinghouse Elec Corp, Md, 64-65; sr engr, 65-67, prin engr, 67-69, sect mgr, 69-79, CONSULT SCIENTIST, ELECTROOPTICS LAB, EQUIP DIV, RAYTHEON CO, 79- *Mem:* Optical Soc Am. *Res:* Time-resolved spectroscopy of neodymium glass laser emission; laser dynamics; x-ray crystallography; design and development of laser scanners, imagers, modulators, coherent optics and Doppler laser radar systems. *Mailing Add:* Electro-Optics Lab MS-1K9 Equip Div Raytheon Co Sudbury MA 01776

KEENE, WILLIS RIGGS, b Woodbine, Ga, Jan 30, 32; c 4. INTERNAL MEDICINE, HEMATOLOGY. *Educ:* Emory Univ, BA, 53; Johns Hopkins Univ, MD, 57. *Prof Exp:* From intern to resident, Johns Hopkins Hosp, 57-59; with Nat Cancer Inst, 59-60 & USPHS Hosp, Boston, 60-61; fel med, Harvard

Univ, 61-63, instr, 63-64; staff physician, Dept Internal Med, Lahey Clin, 64-68; prof med, Col Med, Univ Fla, 71-75, assoc chmn dept internal med, 74-75; CLIN PRACTICE, 75- *Mem:* Am Col Physicians; Am Soc Hemat. *Res:* Blood platelet physiology; iron metabolism. *Mailing Add:* 1001 N Third St Folkston GA 31537

KEENER, CARL SAMUEL, b Columbia, Pa, Apr 12, 31; m 55; c 3. BOTANY. *Educ:* Eastern Mennonite Col, AB, 57; Univ Pa, MS, 60; NC State Univ, PhD(bot), 66. *Prof Exp:* Asst prof biol, Eastern Mennonite Col, 60-63; asst prof, 66-71, ASSOC PROF BOT, PA STATE UNIV, 71-, CUR SEED PLANTS HERBARIUM, 78- *Concurrent Pos:* Vis lectr, Univ Va, 66, 68, 72, 76 & 78; NFS grad fel bot, 63-66. *Honors & Awards:* Jesse M Greeman Award, 68; Henry Allan Gleason Award, 84. *Mem:* Am Soc Plant Taxon; Int Asn Plant Taxon; Int Orgn Biosyst; Systs Asn; Sigma Xi. *Res:* Evolutionary patterns in the shale barren endemics of eastern US; floristics of Pennsylvania; Ranunculaceae of North America. *Mailing Add:* 208 Mueller Lab Pa State Univ University Park PA 16802

KEENER, E(VERETT) L(EE), b Grafton, WVa, Jan 30, 22; m 42. ELECTRICAL ENGINEERING. *Educ:* Univ WVa, BSEE, 44; Purdue Univ, MSEE, 49. *Prof Exp:* Test engr, Gen Elec Co, 44; from instr to asst prof elec eng, Univ WVa, 46-55; instr, Purdue Univ, 47-48; sr res engr, Analog & Hybrid Comput, 55-66, US Steel Corp, 55-66, assoc res consult, Res Ctr, 66-82; RETIRED. *Concurrent Pos:* Sales, 87- *Mem:* Sr mem Inst Elec & Electronics Engrs; Soc Construct Superintendents. *Res:* Instrumentation, control and electrical analogs for steel industry processes. *Mailing Add:* 303 McGraw Ave Grafton WV 26354

KEENER, HAROLD MARION, b Ashland, Ohio, July 28, 43; m 65; c 4. AGRICULTURAL ENGINEERING. *Educ:* Ohio State Univ, BS, 67, MS, 68, PhD(agr eng), 73. *Prof Exp:* From inst to asst prof, 68-80, ASSOC PROF AGR ENG, OHIO AGR RES & DEVELOP CTR, 80- *Mem:* Am Soc Agr Engrs. *Res:* Biomass combustion systems; fluidized bed combustion applied to small scale energy systems; grain and solar grain drying; analysis of total energy consumption in crop production systems and livestock enterprises; composting agricultural and/or yard waste. *Mailing Add:* Dept Agr Eng Ohio Agr Res & Develop Ctr Wooster OH 44691

KEENER, HARRY ALLAN, b Greensboro, Pa, Dec 22, 13; m 41; c 2. ANIMAL NUTRITION. *Educ:* Pa State Univ, BS, 36, PhD(animal nutrit, dairy husb), 41; WVa Univ, MS, 38. *Prof Exp:* Asst dairy husb, WVa Univ, 36-38 & Pa State Univ, 38-41; from instr to asst prof animal & dairy husb, 41-45, assoc prof dairy husb, 45-50, prof animal sci, 50-78, dir agr exp sta, 58-78, dean, Col Life Sci & Agr, 61-78, EMER PROF ANIMAL SCI & EMER DEAN, COL LIFE SCI & AGR, UNIV NH, 78- *Mem:* Am Soc Animal Sci; Am Dairy Sci Asn; NY Acad Sci. *Res:* Trace elements; cobalt; vitamin D; nitrogen and energy metabolism. *Mailing Add:* PO Box 165 Durham NH 03824-0165

KEENER, MARVIN STANFORD, b Birmingham, Ala, Oct 25, 43; m 65; c 2. MATHEMATICAL ANALYSIS. *Educ:* Birmingham-Southern Col, BS, 65; Univ Mo-Columbia, MA, 67. PhD(math), 70. *Prof Exp:* From asst prof to assoc prof, 70-79, PROF MATH, OKLA STATE UNIV, 79- *Mem:* Am Math Soc; Sigma Xi; Soc Indust & Appl Math; Math Asn Am. *Res:* Ordinary differential equations. *Mailing Add:* 823 Ranch Dr Stillwater OK 74074

KEENEY, ARTHUR HAIL, b Louisville, Ky, Jan 20, 20; m 42; c 3. OPHTHALMOLOGY. *Educ:* Col William & Mary, BS, 41; Univ Louisville, MD, 44; Univ Pa, MS, 52, DSc(med), 55; Am Bd Ophthal, dipl, 51. *Prof Exp:* Res surgeon, Wills Eye Hosp, Philadelphia, Pa, 49-51; dir res, Sect Ophthal, Sch Med, Univ Louisville, 52-59, from asst prof to assoc prof ophthal, 59-65; prof & chmn dept, Sch Med, Temple Univ, 66-74; dean, 73-80, EMER DEAN & DISTINGUISHED PROF OPHTHAL, SCH MED, UNIV LOUISVILLE, 80- *Concurrent Pos:* Area consult, US Vet Admin, 54-; Alvaro lectr to SAm, 60; consult ed, Am J Ophthal; mem sci adv comt, Nat Coun Combat Blindness; life trustee, J G Brown Found; ophthalmologist-in-chief, Wills Eye Hosp & Res Inst, 65-73; mem Nat Adv Coun to US Secy Transportation, 67-71; pres, Int Cong Ultrasound in Ophthal, 68; mem grants adv coun, The Seeing Eye, 68-71; vchmn, Residency Rev Comt Ophthal, 69-73; chmn Z80 comt on ophthal standards, Am Nat Standards Inst, 70-86; trustee, Med Found Jefferson County Med Soc, 83-, pres, 87-89; med adv opthal, Off of Hearings & Appeals, Social Security Admin, 81-; mem, Nat Res Coun-Comt on Vision, Working Group on Mobility Aids for Visually Impaired, 85-86; mem, Joint Comn on Allied Health Personnel in Opthal, pres, 87. *Honors & Awards:* Zentmayer Award, Col of Physicians, Pa, 51; Second Ann Dora Bornstein Mem Award, Fight for Sight, Pa, 69; 11th Lucien Howe Gold Medal, State Univ Buffalo, 73. *Mem:* Fel Am Ophthal Soc; AMA; fel Am Asn Hist Med; fel Am Acad Ophthal (pres, 84-86); Am Asn Automotive Med (pres, 67). *Res:* Light damage; ocular injuries; ultrasound; diabetic retinopathy; safety lens materials; macular disease; dyslexia; strabismus; ocular tumors. *Mailing Add:* Deans Off H-S-C Univ Louisville Louisville KY 40292

KEENEY, CLIFFORD EMERSON, b Springfield, Mass, June 28, 21; m 50; c 2. PHYSIOLOGY. *Educ:* Springfield Col, BS, 48, MEd, 49; Rutgers Univ, MS, 51; NY UNiv, PhD(phys ed), 59. *Prof Exp:* Teacher high sch, Mass, 49-50; instr biol, Springfield Col, 51-52; biologist, Lederle Labs Div, Am Cyanamid Co, NY, 52-55; asst prof physiol, 55-57, from asst prof to assoc prof biol, 57-65, dir div arts & sci, 62-64, PROF BIOL, SPRINGFIELD COL, 65- *Concurrent Pos:* NSF sci fac fel, 64-65. *Mem:* AAAS; Nat Asn Biol Teachers; NY Acad Sci. *Res:* Cytological changes induced by exercise. *Mailing Add:* 47 Old Coach Rd Hampden MA 01109

KEENEY, DENNIS RAYMOND, b Osceola, Iowa, July 2, 37; m 59; c 2. SOIL FERTILITY, BIOCHEMISTRY. *Educ:* Iowa State Univ, BS, 59, PhD(soil fertil), 65; Univ Wis, MS, 61. *Prof Exp:* Fel soil biochem, Iowa State Univ, 65-66; from asst prof to assoc prof, Univ Wis, 66-74, chmn dept, 79-84, prof soils, 74-88; chmn land resources, Inst Environ Sci, 85-88; DIR

LEOPOLD CTR SUSTAINABLE AGR, IOWA STATE UNIV, 88- *Concurrent Pos:* Romnes grad sch fel, Univ Wis Grad Sch, 75; sr res fel, Dept Sci & Indust Res, Grasslands, Palmerston N, Nz, 76-77. *Mem:* Fel Am Soc Agron; fel Soil Sci Soc Am; Soil Conserv Soc Am; fel AAAS. *Res:* Modeling of N cycle; elucidation of nitrogen transformation in soils and waters; sustainable agriculture; land application of solid and liquid municipal and industrial wastes. *Mailing Add:* Leopold Ctr Sustainable Agr 126 Soil Tilth Bldg Iowa State Univ Ames IA 50011

KEENEY, JOE, energy conversion, cosmology, for more information see previous edition

KEENEY, MARK, b Sharon, Pa, May 18, 21; m 51. DAIRY SCIENCE. *Educ:* Pa State Univ, BS, 42, PhD(dairy husb), 50; Ohio State Univ, MS, 47. *Prof Exp:* Chemist, Borden Co, 43-44; asst, Pa State Univ, 48-50; asst prof dairy mfg, 50-54, from assoc prof to prof dairy sci, 54-74, prof chem & dairy sci, 74-81, PROF BIOCHEM & CHMN DEPT, UNIV MD, COLLEGE PARK, 81- *Mem:* Am Dairy Sci Asn; Am Soc Biol Chemists; NY Acad Sci; Sigma Xi. *Res:* Food chemistry; lipid metabolism; characterization and metabolism of rumen microbial lipids and of oxygenated fatty acids in animals. *Mailing Add:* Dept Chem Univ Md College Park MD 20742

KEENEY, NORWOOD HENRY, JR, b Hartford, Conn, July 10, 24; m 46; c 1. CHEMICAL ENGINEERING, PULP AND PAPER-FOREST PRODUCTS. *Educ:* Trinity Col, Conn, BS, 48; Univ Maine, Orono, MS, 50; Victoria Univ, Manchester, PhD, 62. *Prof Exp:* Paper chemist, Fram Corp, 50-53; from asst prof to assoc prof, Univ Lowell, 53-64, chmn dept, 76-83, prof chem eng, 64-86. *Mem:* Am Inst Chem Engrs; Tech Asn Pulp & Paper Indust; Sigma Xi; Soc Prof Engrs. *Res:* Chemical engineering applications to pulp and paper industry; porous media; filtration of compressibles; zeta potentials; stress-strain properties of fibers and fibrous structures. *Mailing Add:* Skyline Rd Unity NY 03773

KEENEY, PHILIP G, b Caldwell, NJ, Feb 28, 25; m 57; c 1. FOOD SCIENCE. *Educ:* Univ Nebr, BSc, 49; Ohio State Univ, MSc, 53; Pa State Univ, PhD(dairy sci), 55. *Prof Exp:* Asst prof dairy sci, 55-63, assoc prof, 63-68, prof food sci, 68-85, EMER PROF, PA STATE UNIV, 68- *Concurrent Pos:* Consult, Sci & Technol Ice Cream & Chocolate. *Mem:* Fel AAAS; Am Chem Soc; Am Dairy Sci Asn; Inst Food Technol. *Res:* Food technology and chemistry; ice cream; chocolate products. *Mailing Add:* 1449 Curtin St State College PA 16803

KEENEY, RALPH LYONS, b Lewistown, Mont, Jan 29, 44. RISK ANALYSIS, DECISION ANALYSIS. *Educ:* Univ Calif, Los Angeles, BS, 66; Mass Inst Technol, MS, 67, EE, 68, PhD(opers res), 69. *Prof Exp:* Engr, Bell Tel Labs, 66-69; asst prof civil eng & staff mem, Opers Res Ctr, Mass Inst Technol, 69-72, assoc prof mgt & opers res, 72-74; res scholar, Int Inst Appl Systs Anal, Laxenburg, Austria, 74-76; head, Decision Anal, Woodward-Cycle Consults, 76-83, vpres, 80-83; PROF SYSTS SCI, UNIV SOUTHERN CALIF, 83- *Concurrent Pos:* Pvt consult, 69- *Mem:* Opers Res Soc Am; Inst Mgt Sci; Soc Risk Anal. *Res:* Decision analysis, risk analysis; probabilistic models. *Mailing Add:* 101 Lombard St No 7042 San Francisco CA 94111

KEENLEYSIDE, MILES HUGH ALSTON, b Ottawa, Ont, Apr 8, 29; m 51; c 2. ZOOLOGY. *Educ:* Univ BC, BA, 52, MA, 53; Univ Groningen, PhD(zool), 55. *Prof Exp:* Asst scientist fisheries biol, Biol Sta, Fisheries Res Bd Can, 55-57, assoc scientist, 57-61; from asst prof to assoc prof, 61-72, PROF ZOOL, UNIV WESTERN ONT, 72- *Concurrent Pos:* Sr Queen's Fel Marine Sci, Australia, 83. *Mem:* Can Soc Zoologists; fel Animal Behav Soc; Int Soc Behav Ecol. *Res:* Social and reproductive behavior of fishes; parent-young interactions; social organization, mating systems and ecology; mate choice; correlates of reproductive success. *Mailing Add:* Dept Zool Univ Western Ont London ON N6A 5B7 Can

KEENLYNE, KENT DOUGLAS, b Durand, Wis, May 28, 41; m 64; c 2. WILDLIFE ECOLOGY, GEOLOGY. *Educ:* Univ Wis-River Falls, BS, 64; Univ Minn, MS, 68, MAPA, 71, PhD(wildlife ecol), 76. *Prof Exp:* Wildlife biologist river basin studies, US Fish & Wildlife Serv, Minneapolis, Minn, 70-72; coordr interagency coord, Upper Miss River Conserv Comt, 72-74; herpetologist herpetol studies & res, Fla Game & Fresh Water Fish Comn, 75-76; fish & wildlife biologist ecol serv & proj planning, Rock Island, Ill, 76, coal coordr mineral develop, Casper, Wyo, 76-77, area supvr, Pierre, SD, 77-87, MISSOURI RIVER COORDR, US FISH & WILDLIFE SERV, 87- *Concurrent Pos:* Big game biologist, Minn Dept Natural Resources, 78; Interior Coal rep, Interior Task Force Strip Mine Legis Coal Develop, 78; Adj prof wildlife, SDak State univ, Brookings, 87- *Honors & Awards:* Spec Achievement Award, US Fish & Wildlife Serv, 76 & 89; Nat Wildlife Vol Award, 91. *Mem:* Am Fisheries Soc. *Res:* Whitetailed deer reproduction; natural history and populations of turtles; reproduction and life history of rattlesnakes; alligator attacks; physiology of whitetailed deer; telemetry studies on whitetailed deer; large river ecology. *Mailing Add:* US Fish & Wildlife Serv PO Box 986 Pierre SD 57501

KEENMON, KENDALL ANDREWS, b Detroit, Mich, Sept 13, 20; m 42; c 4. PETROLEUM GEOLOGY, REMOTE SENSING. *Educ:* Univ Mich, BS, 47, MS, 48, PhD(geol), 50. *Prof Exp:* Div geologist, Shell Oil Co, 50-60, sr geologist, 60-61, sr res geologist, Shell Develop Co, 61-67, sr geologist, Shell Oil Co, 67-69 & Shell Develop Co, 69-72, sr geologist, Int Region, 72-77, staff geologist, Pecten Int Co, 78-85; OWNER & CONSULT, K-TECHNOL, 85- *Mem:* Geol Soc Am; Am Asn Petrol Geol. *Res:* Structural geology; stratigraphy. *Mailing Add:* 5158 Imogene St Houston TX 77096

KEENS, THOMAS GEORGE, b Altadena, Calif, Nov 22, 46; m 72; c 2. PEDIATRIC PULMONOLOGY, RESPIRATORY PHYSIOLOGY. *Educ:* St John's Col, Santa Fe, NMex, 68; Univ Calif, San Diego, MD, 72. *Prof Exp:* Pediat intern & resident, Childrens Hosp Los Angeles, Calif, 72-75; res fel pediat respiratory physiol, Res Inst, Hosp Sick Children, Toronto, Ont, 75-77; asst prof, 77-83, ASSOC PROF PEDIAT PULMONOL, SCH MED, UNIV

SOUTHERN CALIF & CHILDRENS HOSP LOS ANGELES, 83- *Concurrent Pos:* Pediat pulmonologist & neonatologist, Childrens Hosp Los Angeles, 77-; mem, Southern Adv Coun, Calif Sudden Infant Death Syndrome Info & Coun Proj, 79-, moderator, 84-85; pediat pulmonary consult, Kern Med Ctr, Bakersfield, Calif, 81-85. *Mem:* Am Physiol Soc; Soc Pediat Res; fel Am Acad Pediat; Am Thoracic Soc; Am Col Chest Physicians; Sleep Res Soc. *Res:* Sudden infant death syndrome; abnormal arousal responses in response to a hypoxic challenge in infants at high risk for sudden infant death syndrome; chronic lung disease; ventilatory muscle function; bronchopulmonary dysplasia; pediatric pulmonary function. *Mailing Add:* Childrens Hosp Los Angeles 4650 Sunset Blvd Los Angeles CA 90027

KEENY, SPURGEON MILTON, JR, b New York, NY, Oct 24, 24; m 52; c 3. PHYSICS. *Educ:* Columbia Univ, BA, 44, MA, 46. *Prof Exp:* Asst physics, Columbia Univ, 44-46; intel analyst, Directorate of Intel Hq, US Air Force, 50-52, chief Spec Weapons Sect, 52-55; mem staff, Panel Peaceful Uses Atomic Energy, 55-56; chief atomic energy div, Off Asst Secy Defense Res & Eng, 56-57; mem, Gaither Security Resources Panel, 57; tech asst, President's Sci Adv, 58-69; sr staff mem, Nat Security Coun, 63-69; asst dir sci & technol, US Arms Control & Disarmament Agency, 69-73; dir policy & prog develop, Mitre Corp, 73-77; dep dir, US Arms Control & Disarmament Agency, 77-81; SCHOLAR-IN-RESIDENCE, NAT ACAD SCI, 81- *Concurrent Pos:* Mem US del, Geneva Conf Experts Nuclear Test Detection, 58, Conf Discontinuance Nuclear Weapon Tests, 58-60,; Am Phys Soc Study Group Light-Water Reactor Safety, 74-75; dep chmn, Nat Acad Sci, Comt on Environ Decision Making, 74; chmn, Ford-Mitre Nuclear Energy Policy Study, 75-77; head US deleg, US/Soviet Theater Nuclear Force Talks, 80; mem Comt Int Security & Arms Control, Nat Acad Sci, 81- *Mem:* Fel Am Acad Arts & Sci; Coun Foreign Relations; Am Phys Soc. *Res:* Arms control and disarmament; defense policy; military and civilian applications of atomic energy; energy and environmental policy. *Mailing Add:* 3600 Albemarle St NW Washington DC 20008

KEEPIN, GEORGE ROBERT, JR, b Oak Park, Ill, Dec 5, 23; m 48; c 5. NUCLEAR PHYSICS, INSTRUMENTATION. *Educ:* Univ Chicago, PhB, 45; Mass Inst Technol, BS & MS, 47; Northwestern Univ, PhD(physics), 49. *Prof Exp:* Consult, Argonne Nat Lab, 48-49; AEC fel radiation lab, Univ Calif, 50; res physicist, Los Alamos Sci Lab, 52-62; head physics sect, Int Atomic Energy Agency, Vienna, 63-65; group leader nuclear assay res, 66-75, prog dir nuclear safeguards, 75-79, prog mgr, Nuclear Safeguards Affairs, Los Alamos Sci Lab, 79-82; spec adv dep dir gen, Int Atomic Energy Agency, Vienna, 82-85; SR ADV, LOS ALAMOS SAFEGUARDS & SECURITY PROG, 85- *Concurrent Pos:* Deleg, Atoms for Peace Conf, Geneva, 55, 64 & 71; tech adv, Int Atomic Energy Agency, Geneva, 64; nat prog chmn, Inst Nuclear Mat Mgt, 74-76, nat chmn, 78-80; app fel, Los Alamos Nat Lab, 85- *Honors & Awards:* Am Nuclear Soc Spec Award, 73; Distinguished Serv Award, Inst Nuclear Mat Mgmt, 84. *Mem:* NY Acad Sci; fel Am Phys Soc; fel Am Nuclear Soc; fel Inst Nuclear Mat Mgt. *Res:* Fission physics; reactor dynamics; pulsed neutron research, nuclear safeguards research and development; development of non-destructive assay techniques for domestic and international inspection and safeguards of fissionable materials; development and implementation of nondestructive assay technology for stringent nuclear safeguards and nonproliferation of nuclear weapons. *Mailing Add:* 600 La Bajada Los Alamos NM 87544

KEEPLER, MANUEL, b Atlanta, Ga, Nov 4, 44; m 66; c 1. RANDOM EVOLUTIONS, MARKOV PROCESSES. *Educ:* Morehouse Col, BS, 65; Columbia Univ, MA, 67; Univ NMex, PhD(math), 73; Univ SC, MEd. *Prof Exp:* Asst prof math, Va State Univ, 70-71; assoc prof & chmn, Laugston Univ, 71-73; from assoc prof to prof math & comput sci, SC State Col, 73, chmn dept, 81-89; VIS PROF MATH, CORNELL UNIV, 89- *Concurrent Pos:* Vis prof, Dilliard Univ, 76; fel comput, Lawrence Livermore Lab, 77, comput sci, Langley Res Ctr, 81; SC rep, Nat Tech Asn, 80-89; mem comput adv comt, Comn Higher Educ State SC, 81-86; Comt mem, Am Statist Asn, 87-89; exec bd, Pancomp, USA, 87-89; dir, Sloan Proj, 87-89. *Mem:* Opers Res Soc Am; Am Statist Asn; Inst Math Statist; Math Assoc Am. *Res:* Random evolutions on Markov processes; mathematics; statistics. *Mailing Add:* Dept Math Cornell Univ White Hall Ithaca NY 14853-7901

KEEPORTS, DAVID, b York, Pa, June 15, 51. PHYSICS EDUCATION SOFTWARE. *Educ:* Univ Del, BS, 73; Yale Univ, MS, 74; Univ Wash, PhD(phys chem), 82. *Prof Exp:* Lectr physics/chem/math, Quinnipiac Col, 75-79; asst prof, 82-87, ASSOC PROF PHYSICS/CHEM, MILLS COL, 87- *Concurrent Pos:* Lectr math, Southern Conn State Col, 76-78; lectr physics/ math, Univ New Haven, 76-79. *Mem:* Am Chem Soc; Am Asn Physics Teachers; Sigma Xi. *Res:* Author of numerous publications on the practice and theory of molecular spectroscopy and numerous publications in physics, chemistry and math education journals; development of software for physics education. *Mailing Add:* Dept Chem & Physics Mills Col Oakland CA 94613

KEER, LEON M, b Los Angeles, Calif, Sept 13, 34; m 56; c 4. CIVIL ENGINEERING, ENGINEERING MECHANICS. *Educ:* Calif Inst Technol, BS, 56, MS, 58; Univ Minn, PhD(eng mech), 62. *Prof Exp:* Mem tech staff, Hughes Aircraft Co, 56-59; NATO fel eng mech, Newcastle, 62-63; preceptor, Columbia Univ, 63-64; from asst prof to assoc prof, 64-70, PROF CIVIL ENG, NORTHWESTERN UNIV, 70-, ASSOC DEAN GRAD STUDIES & RES, 85- *Concurrent Pos:* Guggenheim sr vis fel, Dept Math, Univ Glasgow, 72-73; JSPS fel, 86; tech ed, J Appl Mech, ASME, 88-; NATO fel. *Mem:* Fel Am Soc Mech Engrs; Acoust Soc Am; fel Am Soc Civil Engrs; fel Am Acad Mech (secy, 81-85, pres, 88-89). *Res:* Contact stress and fracture problems; contact mechanics; composite materials; wave propagation. *Mailing Add:* Dept Civil Eng Northwestern Univ Evanston IL 60201

KEES, KENNETH LEWIS, b Highland Park, Mich, Jan 17, 50. SYNTHESIS OF NON-INSULIN RELEASING ANTIDIABETIC AGENTS & NOVEL ANTIINFLAMMATORY AGENTS. *Educ:* Wayne State Univ, BS, 73; Univ Calif, Santa Cruz, PhD(chem), 79. *Prof Exp:* Vis lectr chem, Univ Calif, Santa Cruz, 80; postdoctoral chem, Univ Calif, Berkeley, 81; res assoc & adj lectr

chem, Univ Santa Clara, Calif, 82; res scientist immunoinflammatory, 83-85, res scientist & supvr diabetes, 85-88, PRIN SCIENTIST METABOLIC DISORDERS, AM HOME PROD, WYETH-AYERST RES, 88- *Concurrent Pos:* Contrib bd mem, Chemtracts, Org Chem, 89- *Mem:* Am Chem Soc. *Res:* Design and synthesis of novel (patentable) medicinal chemicals using state of the art synthetic and analytical techniques; synthesis of lipophillic, acidic heterocycles-some are new bioisosteres of carboxylic acids-for application to diabetes, inflammation and cardiovascular metabolic derangements. *Mailing Add:* 1015 Aspen Dr Plainsboro NJ 08536

KEESE, CHARLES RICHARD, b Cooperstown, NY, Mar 4, 44; m 67; c 2. BIOCHEMISTRY, BIOPHYSICS. *Educ:* State Univ NY, Albany, BS, 67; Rensselaer Polytech Inst, PhD(biol), 71. *Prof Exp:* From asst prof to assoc prof physics, State Univ NY, Cobleskill, 71-79, prof biol, 71-83; staff scientist, Res & Develop, Gen Elec Corp, 83-; AT RENSSELAER POLYTECHNIC INST. *Concurrent Pos:* NSF fel sci fac prof develop award, Gen Elec Corp Res & Develop, 77-78; assoc investr, Nat Found Cancer Res, 81-82. *Honors & Awards:* IR-100 Award, 84, Sci Digest, Outstanding Investr, 84-85. *Mem:* Sigma Xi; Am Soc Microbiol; AAAS. *Res:* Behavior of cells in culture; properties of proteins on surfaces; biocatalysis-synthesis of special chemicals using biological systems; biosensors. *Mailing Add:* Rensselaer Polytech Inst 110 Eighth St Troy NY 12180

KEESEE, JOHN WILLIAM, mathematics; deceased, see previous edition for last biography

KEESEE, ROBERT GEORGE, b Spokane, Wash, Nov 10, 53. ION CHEMISTRY, AEROSOL SCIENCE. *Educ:* Univ Ariz, BS, 75; Univ Colo, PhD(phys chem), 79. *Prof Exp:* Teaching asst chem, Univ Colo, 75-76, res asst, Coop Inst Res in Environ Sci, 76-79; Nat Res Coun res assoc, Space Sci Div, NASA-Ames Res Ctr, 79-81; res asst prof chem, Pa State Univ, 82-91; assoc prof dir, Atmospheric Chem Prog, NSF, 88-89; ASSOC PROF ATMOSPHERIC SCI, STATE UNIV NY, 91- *Concurrent Pos:* Sr res assoc chem, Univ Colo, 82. *Mem:* Am Chem Soc; Am Geophysical Union; AAAS; Sigma Xi. *Res:* Chemistry of planetary atmospheres; nucleation phenomena; ion solvation; gas-surface interactions; ion-molecule and ion-aerosol interactions; chemical and physical properties of molecular clusters and aerosols. *Mailing Add:* Earth Sci 218 State Univ NY Albany NY 12222

KEESEY, RICHARD E, b York, Pa, Oct 14, 34; c 1. NUTRITION. *Educ:* Dartmouth Col, AB, 56; Brown Univ, ScM, 58, PhD, 60. *Prof Exp:* From asst prof to assoc prof, 62-69, PROF, UNIV WIS-MADISON, 69- *Concurrent Pos:* Vis lectr, Sydney Univ, 74. *Mem:* Am Inst Nutrit; NAm Soc Study Obesity; Am Psychol Soc; Soc Study Ingestive Behav. *Res:* Physiology of body weight regulation; central nervous control of energy expenditure; obesity and other disorders of energy regulation. *Mailing Add:* Dept Psychol & Nutrit Univ Wis 1202 W Johnson St Madison WI 53706

KEESLING, JAMES EDGAR, b Indianapolis, Ind, June 26, 42; m 63; c 4. TOPOLOGY, APPLIED MATHEMATICS. *Educ:* Univ Miami, Fla, BSIE, 64, MS, 66, PhD(math), 68. *Prof Exp:* Teaching asst, Univ Miami, Fla, 64-65; NASA fel, Univ Miami, 65-67; from asst prof to assoc prof, 67-75, PROF MATH, UNIV FLA, 75- *Mem:* AAAS; Am Math Soc; Math Asn Am; Soc Indust & Appl Math. *Res:* Topology; numerical analysis; biomathematics; over 50 professional publications. *Mailing Add:* Dept Math Univ Fla Gainesville FL 32611

KEESOM, PIETER HENDRIK, b Leiden, Netherlands, Feb 10, 17; m 46; c 4. PHYSICS. *Educ:* Univ Leiden, PhD(physics), 48. *Prof Exp:* Asst, Kamerlingh Onnes Lab, Univ Leiden, 39-46; physicist, State Mines of Netherlands, 46-48; vis prof, 48-50, from asst prof to assoc prof, 50-57, PROF PHYSICS, PURDUE UNIV, 57- *Concurrent Pos:* Guggenheim fel, 60-61. *Mem:* Am Phys Soc; Netherlands Phys Soc. *Res:* Low temperatures; specific heat of elements. *Mailing Add:* Dept Physics Purdue Univ Lafayette IN 47907

KEETON, KENT T, ANTIHYPERTENSIVE DRUGS, CARDIOVASCULAR PHARMACOLOGY. *Educ:* Univ Tex, Dallas, PhD(pharmacol), 75. *Prof Exp:* ASSOC PROF PHARMACOL, UNIV TEX HEALTH SCI CTR, 83- *Mem:* Am Fed Clin Res; Am Soc Pharmacol & Exp Therapeut. *Mailing Add:* Dept Pharmacol Univ Tex Health Sci Ctr 7703 Floyd Curl Dr San Antonio TX 78284

KEEVER, CAROLYN ANNE, b Norfolk, Va, March 18, 48. CELLULAR IMMUNOLOGY, IMMUNOGENETICS. *Educ:* Old Dominion Univ, BS, 70; Wake Forest Univ, PhD(immunol), 84. *Prof Exp:* Res fel, 84-87, RES ASSOC IMMUNOL, MEM SLOAN-KETTERING CANCER CTR, 87- *Concurrent Pos:* Prin investr, res grants, 86-; chmn, Am Bd Histocompatability & Immunogenetics. *Mem:* Am Soc Histocompatibility & Immunogenetics; NY Acad. *Res:* Immunology of bone marrow transplantation; the effects of interleukin Z on T-cell and NK-cell activities; the role of NK-cells in graft rejection and leukemia relapse; the kinetics of immune reconstitution following bone marrow transplantation; donor-host and host-donor tolerance. *Mailing Add:* Tara Hill Dr Columbus OH 43221

KEEVIL, NORMAN BELL, chemistry, physics; deceased, see previous edition for last biography

KEEVIL, THOMAS ALAN, b Long Branch, NJ, Feb 11, 47; m 69; c 2. ORGANIC CHEMISTRY, BIOCHEMISTRY. *Educ:* Bucknell Univ, BS, 68; Univ Calif, PhD(chem), 72. *Prof Exp:* Res assoc biochem, Med Sch, Univ Ore, 72-74; asst prof, 74-78, ASSOC PROF CHEM, SOUTHERN ORE STATE COL, 78- *Mem:* Am Chem Soc; AAAS. *Res:* Oxidase mechanisms. *Mailing Add:* Dept Chem Southern Ore State Col Ashland OR 97520-5029

KEFALIDES, NICHOLAS ALEXANDER, b Alexandroupolis, Greece, Jan 17, 27; US citizen; m 49; c 3. BIOLOGICAL CHEMISTRY, INTERNAL MEDICINE. *Educ:* Augustana Col, Ill, AB, 51; Univ Ill, BS, 54, MD & MS, 56, PhD(biochem), 65. *Hon Degrees:* MA, Univ Pa, 71; Dr, Univ Reims, France, 87. *Prof Exp:* Intern med, Res & Educ Hosps, Univ Ill, 56-57; dir res proj in burns, USPHS, Peru, 57-60; resident internal med, Res & Educ Hosps, Univ Ill, 60-63, instr, Col Med, 64-65; from asst prof to assoc prof, La Rabida Inst & Dept Med, Univ Chicago, 65-70; assoc prof, 70-74, PROF MED, UNIV PA, 74- PROF BIOCHEM & BIOPHYS, 75-; DIR CONNECTIVE TISSUE RES INST, 77- *Concurrent Pos:* USPHS fel, 62-64; assoc attend physician, Cook County Hosp, Chicago, Ill, 62-65; chief infectious dis sect, Vet Admin Hosp, Hines, 64-65; chief infectious dis consult serv, Univ Chicago Hosp, 65-70; attend physician, Philadelphia Gen Hosp, 70-77, dir, Gen Clin Res Ctr, 72-76; dir, Connective Tissue Res Inst, Univ Pa, 75-; vis prof, Oxford Univ, Eng, 77-78 & 84-85; Guggenheim fel, 77-78. *Honors & Awards:* Borden Award, 56. *Mem:* Int Soc Nephrology; Am Chem Soc; Am Asn Path; Am Soc Clin Invest; Am Soc Biochem & Molecular Biol; Am Soc Cell Biol. *Res:* Chemistry of glycoproteins and basement membranes; molecular biology of collagen; metabolism of endothelial cells. *Mailing Add:* Univ City Sci Ctr 3624 Market St Philadelphia PA 19104

KEFFER, CHARLES JOSEPH, b Philadelphia, Pa, Aug 7, 41; m 66; c 4. SOLID STATE PHYSICS, CRYSTALLOGRAPHY. *Educ:* Univ Scranton, BS, 63; Harvard Univ, AM, 64, PhD(solid state physics), 69. *Prof Exp:* From instr to asst prof physics, Univ Scranton, 69-73; dean, 73-77, PROVOST, UNIV ST THOMAS, 77- *Mem:* Am Phys Soc; Sigma Xi. *Mailing Add:* Col St Thomas 2115 Summit Ave St Paul MN 55105

KEFFER, FREDERIC, b Anaconda, Mont, May 23, 19; m 49; c 2. SOLID STATE PHYSICS. *Educ:* State Col Wash, BS, 45; Univ Calif, PhD(physics), 52. *Prof Exp:* Res assoc physics, Univ Calif, 52; from asst prof to assoc prof, 52-59, chmn dept, 63-69, PROF PHYSICS, UNIV PITTSBURGH, 59- *Concurrent Pos:* Res physicist, Westinghouse Res Labs, 52-55; vis prof, Univ Calif, 59-60. *Mem:* AAAS; Am Phys Soc. *Res:* Magnetism; theory of solids. *Mailing Add:* 412 Woodside Rd Pittsburgh PA 15221-3642

KEFFER, JAMES F, b Toronto, Ont, Dec 15, 33; m 55; c 2. MECHANICAL ENGINEERING. *Educ:* Univ Toronto, BASc, 56, MASc, 58, PhD(mech eng), 62. *Prof Exp:* Nat Res Coun Can fel physics, Cambridge Univ, 62-64; from asst prof to assoc prof, 64-73, PROF MECH ENG & VPRES RES, UNIV TORONTO, 73- *Concurrent Pos:* Consult, Pulp & Paper Res Co Can, 65-; vis prof, Inst Mechnique Statisique Turbulence, Marseille, France, 73-74. *Mem:* Soc Automotive Engrs. *Res:* Fluid mechanics; heat and mass transfer; turbulent flows; wind erosion, building aerodynamics turbulent combustion. *Mailing Add:* Simcoe Hall Rm 112 27 Kings Col Circle Toronto ON M5S 1A1 Can

KEFFORD, NOEL PRICE, b Melbourne, Victoria, Australia, Feb 5, 27; m 50; c 3. PLANT PHYSIOLOGY. *Educ:* Univ Melbourne, BSc, 48, MSc, 50; Univ London, PhD(bot, plant physiol), 54. *Prof Exp:* Res officer, Australian Paper Mfrs, Ltd, 50-51; res officer div plant indust, Commonwealth Sci & Indust Res Orgn, 54-59, sr res officer, 59-64, prin res officer, 64-65; PROF BOT & CHMN DEPT, UNIV HAWAII, 65- *Concurrent Pos:* Fulbright sr res fel & res assoc, Univ Calif, 62-63; res assoc, Univ Calif, Santa Cruz, 71; actg assoc dir, Hawaii Agr Exp Sta, 76- *Mem:* Am Soc Plant Physiol; Brit Soc Exp Biol; Australian Soc Plant Physiol; Sigma Xi. *Res:* Plant growth and development; hormonal regulation; research management and administration. *Mailing Add:* Inst Trop Agr-Human Res 3050 Maile Way Honolulu HI 96822

KEGEL, GUNTER HEINRICH REINHARD, b Herborn, Ger, June 16, 29; m 57; c 2. NUCLEAR PHYSICS. *Educ:* Rio de Janeiro, BS, 51; Mass Inst Technol, PhD(physics), 61. *Prof Exp:* Engr, Nat Inst Technol, Brazil, 51-56; prof physics, Cath Univ, Rio de Janeiro, 61-64; prof physics, 64-66, prof nuclear eng, 66-71, chmn dept physics & appl physics, 71-81, PROF PHYSICS, UNIV LOWELL, 71- *Concurrent Pos:* Prof, Rio de Janeiro, 52-56 & 61-64; res asst, Lab Nuclear Sci, Mass Inst Technol, 58-61; consult, Millipore Corp, Bedford, Mass, 69-70. *Mem:* Brazilian Acad Sci; Am Phys Soc; Electrochem Soc; Am Nuclear Soc; Inst Elec & Electronics Engrs; Am Vacuum Soc; Mat Res Soc. *Res:* Nuclear spectroscopy; Rutherford backscattering spectroscopy (RBS); proton induced x-ray emission (PIXE); neutron physics; neutron radiation damage. *Mailing Add:* Dept Physics & Appl Physics Univ Lowell Lowell MA 01854

KEGELES, GERSON, b New Haven, Conn, Apr 23, 17; m 44; c 5. BIOPHYSICAL CHEMISTRY. *Educ:* Yale Univ, BS, 37, PhD(phys chem), 40. *Prof Exp:* Fel, Yale Univ, 40-41; fel, Univ Wis, 45-47; phys chemist, Nat Cancer Inst, 47-51; from assoc prof to prof chem, Clark Univ, 51-68, chmn, Chem Dept, 58-61; prof, 68-82, EMER PROF SECT BIOCHEM & BIOPHYS, UNIV CONN, 82- *Concurrent Pos:* Vis prof chem, Yale Univ, 68; mem, Study Sect Biophys Chem, NIH, 67-70. *Mem:* Am Chem Soc; Biophys Soc; Am Acad Arts & Sci. *Res:* Ultracentrifugation; countercurrent distribution; equilibria and kinetics of protein interactions; diffusion and optical methods for its study; pressure effects in protein transport experiments. *Mailing Add:* RFD 1 Box 156 Groveton NH 03582

KEGELES, LAWRENCE STEVEN, b Madison, Wis, Feb 9, 47; m 87; c 1. NUCLEAR MEDICINE, THEORETICAL ASTROPHYSICS. *Educ:* Princeton Univ, AB, 69; Univ Pa, PhD(physics), 74. *Prof Exp:* Res assoc physics, Univ Pa & Naval Res Lab, 74-76 & Univ Alta, 76-78; res assoc physics, Stevens Inst Technol, 78-80; mem tech staff, Bell Labs, 81-87; MT SINAI SCH MED, 87- *Mem:* Sigma Xi; NY Acad Sci; Am Phys Soc. *Res:* Perturbations of spacetimes; pulsar magnetospheres; nonspherical gravitational collapse; equations of motion and radiation damping; tracer kinetics in nuclear medicine. *Mailing Add:* 57-59 E 96th St New York NY 10128

KEGELMAN, MATTHEW ROLAND, b New York, NY, June 24, 28; m 53; c 10. ELECTROCHEMISTRY, POLYMER CHEMISTRY. *Educ:* Fordham Univ, BS, 48, MS, 49, PhD(org chem), 53. *Prof Exp:* From res chemist to sr res chemist, E I du Pont de Nemours & Co, 53-71, res assoc petrochem dept, 71-78, sr res assoc, 71-90; RETIRED. *Concurrent Pos:* Prin investr, Dielectric Gases Proj, Elec Power Res Inst. *Mem:* AAAS; Am Chem Soc; Sigma Xi. *Res:* Heterocyclics; pinacol rearrangement; petroleum chemicals; dielectric fluids; high-energy batteries; electro-organic synthesis; polymer intermediates; polyesters; polyamides. *Mailing Add:* 204 N Pembrey Dr Wilmington DE 19803

KEHEW, ALAN EVERETT, b Pittsburgh, Pa, Sept 17, 47; m 74; c 3. HYDROGEOCHEMISTRY, CONTAMINANT HYDROGEOLOGY. *Educ:* Bucknell Univ, BS, 69; Montana State Univ, MS, 71; Univ Idaho, PhD(geol), 77. *Prof Exp:* Geologist, NDak Geol Surv, 77-80; from asst prof to assoc prof geol, Univ NDak, 80-84; ASSOC PROF GEOL, WESTERN MICH UNIV, 86- *Mem:* Geol Soc Am; Nat Water Well Asn; Am Geophys Union; Am Quaternary Asn. *Res:* Groundwater contamination by waste disposal; chemical evolution of groundwater in glacial terrains. *Mailing Add:* Dept Geol Western Mich Univ Kalamazoo MI 49008

KEHL, THEODORE H, b Racine, Wis, Apr 1, 33; m 54; c 2. COMPUTER SCIENCE, BIOPHYSICS. *Educ:* Univ Wis, BS, 56, MS, 58, PhD(zool), 61. *Prof Exp:* Wis Alumni Res Found res assoc, Univ Wis, 56-61, NIH fel, 61; NIH fel, Sch Med, 61-63, from instr to assoc prof physiol & biophys, 63-73, assoc prof, 73-77, PROF PHYSIOL, BIOPHYS & COMPUTER SCI, SCH MED, UNIV WASH, 77- *Mem:* AAAS; Asn Comput Mach. *Res:* Implementation of computer science to quantitative physiology and biophysics. *Mailing Add:* Dept Comput Sci Univ Wash Seattle WA 98195

KEHL, WILLIAM BRUNNER, b Pittsburgh, Pa, Apr 8, 19; m 44; c 2. COMPUTER SCIENCE. *Educ:* Harvard Univ, SB, 40, AM, 42 & 48. *Prof Exp:* Instr math, Ga Inst Tech, 43-46; instr, Mass Inst Technol, 48-54, head anal group, Instrumentation Lab, 54-56; prof comput sci & dir comput & data processing ctr, Univ Pittsburgh, 56-66; assoc prof elec eng & assoc dir comput ctr, Mass Inst Technol, 66-67; DIR ACADEMIC COMPUTING, UNIV CALIF, LOS ANGELES, 67- *Concurrent Pos:* Consult, USAF, 57-59 & comt use of comput, Nat Acad Sci; consult, NSF & NIH; mem sci adv bd, Regional Indust Develop Corp. *Mem:* Asn Comput Mach. *Res:* Computers; applied mathematics. *Mailing Add:* Dir Comput Univ Calif Los Angeles 405 Hilgard Ave Palisades CA 90272

KEHLENBECK, MANFRED MAX, b Bremen, Germany, Jan 16, 37; US citizen; m 68. STRUCTURAL GEOLOGY, METAMORPHIC GEOLOGY. *Educ:* Hofstra Univ, BA, 59; Syracuse Univ, MS, 64; Queen's Univ, PhD(geol), 71. *Prof Exp:* Vis prof geol, Univ NB, 69-70; from asst prof to assoc prof, 71-86, PROF GEOL, LAKEHEAD UNIV, 86- *Concurrent Pos:* Nat Res Coun Can grant, 72-; Energy, Mines & Resources grant, 86-; Northern Ont Rural Develop Agreement grant, 83-85; dept chmn, Lakehead Univ, 76-80. *Mem:* Am Geol Inst; Geol Asn Can. *Res:* Structural evolution of Archean gneissic terrains; polyphase folding in volcano sedimentary belts in northwestern Ontario; archean subprovince margins and boundaries; transpressional basins. *Mailing Add:* Dept Geol Lakehead Univ Thunder Bay ON P7B 5E1 Can

KEHLER, PHILIP LEROY, b Lyons, NY, June 15, 36; m 65; c 2. STRATIGRAPHY, SEDIMENTOLOGY. *Educ:* Purdue Univ, BS, 59, MS, 61; Southern Methodist Univ, PhD(geol), 70. *Prof Exp:* Asst, Southern Methodist Univ, 65-69; asst prof geol, ETex State Univ, 69-73, fac res grant, 71-72; chairperson, 73-78, ASSOC PROF EARTH SCI, UNIV ARK, LITTLE ROCK, 73-, CHAIRPERSON, 83- *Concurrent Pos:* NSF student originated studies grant, 75; lignite consult, Shell Mining, 79-81; young scholar, NSF, 89-90. *Mem:* Geol Soc Am; Soc Econ Paleontologists & Mineralogists; Nat Asn Geol Teachers. *Res:* Regional studies of Jurassic and Cretaceous rocks in western North America; stratigraphic relationships associated with widespread unconformities within these rocks; statigraphic relationships, Ouachita Mountains of Arkansas and Oklahoma; earthquake mitigation in northeast Arkansas. *Mailing Add:* Dept Earth Sci Univ Ark 2801 S University Ave Little Rock AR 72204

KEHOE, BRANDT, b Cleveland, Ohio, Nov 20, 33; m 61; c 2. PHYSICS. *Educ:* Cornell Univ, BA, 56; Univ Wis, MS, 59, PhD(physics), 63. *Prof Exp:* Res asst physics, Los Alamos Sci Lab, 56-57; from asst prof to assoc prof, Univ Md, 62-72; dean Sch Natural Sci, 72-83, PROF PHYSICS, CALIF STATE UNIV, FRESNO, 72-; pres, Deep Springs Col, 83-87; dept chmn comput sci, 87-89, PROF PHYSICS, CALIF STATE UNIV, FRESNO, 87-, DEPT CHMN PHYSICS, 89- *Mem:* Am Am Physics Teachers; AAAS. *Res:* High energy physics; weak interactions, rare decay modes and low energy K-N and high energy P-P interactions; neural networks; modeling of biological systems. *Mailing Add:* 4793 Sunset Fresno CA 93704

KEHOE, LAWRENCE JOSEPH, b Rock Island, Ill, Nov 17, 40; m 66; c 3. ORGANIC CHEMISTRY. *Educ:* St Ambrose Col, AB, 62; Univ Iowa, MA, 65, PhD(org chem), 67. *Prof Exp:* Sr res chemist, 66-78, prod mgr, 78-82, MKT MGR, ETHYL CORP, 82- *Mem:* Am Chem Soc; Am Soc Enologists. *Res:* Pharmaceutical intermediates. *Mailing Add:* Ethyl Corp Performance Prod Div 451 Florida Blvd Baton Rouge LA 70801

KEHOE, THOMAS J, b Bisbee, Ariz, June 16, 19; m 50; c 6. ANALYTICAL CHEMISTRY, INORGANIC CHEMISTRY. *Educ:* Loyola Univ, BS, 41. *Prof Exp:* Res chemist, Am Potash & Chem Co, 42-48; lab supvr, 48-50; consult waste treatment, Pomeroy & Assocs, 50-54; appln engr, 54-61, MGR APPLN ENG, BECKMAN INSTRUMENTS, INC, 61- *Mem:* Fel Instrument Soc Am (pres, 69-70); Am Inst Chem; Am Chem Soc; Am Inst Chem Eng. *Res:* Process analytical instrumentation; industrial waste treatment; phase rule studies in inorganic chemistry of Searles Lake brine. *Mailing Add:* 1506 Victoria Way Placentia CA 92670-2335

KEHR, CLIFTON LEROY, b Brodbeck, Pa, May 25, 26; m 48; c 3. ORGANIC CHEMISTRY, POLYMER CHEMISTRY. *Educ:* Gettysburg Col, AB, 49; Univ Del, MS, 50, PhD(org chem), 52. *Prof Exp:* Res asst synthetic org chem, Forrestal Res Ctr, Princeton Univ, 52-53; res chemist, Org Chem Dept, E I du Pont de Nemours & Co, 53-57, elastomers chem dept, 57-59; RES CHEMIST, RES & DEVELOP DIV, WASHINGTON RES CTR, W R GRACE & CO, 59-, RES DIR, 69- *Mem:* Am Chem Soc; Sigma Xi. *Res:* Mechanisms of organic reactions; polyurethanes; elastomers; polyolefins; isocyanate chemistry; polymers from chloroprene and related monomers; foam technology; radiation curable polymers; water based coatings; biocompatible polymers. *Mailing Add:* 1216 Ednor Rd Silver Spring MD 20905

KEHRER, JAMES PAUL, b Watertown, Wis, Aug 25, 51; m 77; c 2. PULMONARY TOXICOLOGY, CARDIAC TOXICOLOGY. *Educ:* Purdue Univ, BS, 74; Univ Iowa, PhD(pharm), 78. *Prof Exp:* Investr, Biol Div, Oak Ridge Nat Lab, 78-80; from asst prof to assoc prof, 80-90, PROF PHARMACOL & TOXICOL, COL PHARM, UNIV TEX, AUSTIN, 90- *Concurrent Pos:* Consult, Radian Corp, 82-84; res career develop award, NIH, 84-89; Gustavus Pfeiffer Centennial fel pharmocol, 85-; assoc ed rev, Toxicol Letters, 88- *Honors & Awards:* Achievement Award, Soc Toxicol, 89. *Mem:* Soc Toxicol; AAAS; Am Soc Pharmacol & Theraput. *Res:* Collagen synthesis and degradation during the development of pulmonary fibrosis after acute lung damage; lung toxicity of anti-cancer drugs; oxidative stress in cardiac reperfusion injury. *Mailing Add:* Div Pharmacol & Toxicol Univ Tex Austin TX 78712-1074

KEHRES, PAUL W(ILLIAM), b Milwaukee, Wis, Dec 14, 22; m 50; c 2. ANALYTICAL CHEMISTRY, MATERIALS TESTING. *Educ:* Univ Wis-Milwaukee, BS, 47; Marquette Univ, MS, 48. *Prof Exp:* Res assoc chem, Ames Lab, AEC, 48-51; instr, Iowa State Univ, 50-51; intermediate scientist, Atomic Power Div, Westinghouse Elec Corp, 51-52; asst dir anal res, A O Smith Corp, 52-62, supvr chem processes, 62-66, mgr, 66-68, asst mgr plastics res & develop, 68-76, proj coordr, 76-82; consult, 82-88; RETIRED. *Concurrent Pos:* Chief chemist, Glendale Crime Lab, 66-, dir, 76- *Honors & Awards:* John R Hunt Award, Am Acad Forensic Sci, 85. *Mem:* Am Chem Soc; Soc Appl Spectros; Am Acad Forensic Sci. *Res:* X-ray and optical emission spectroscopy; mass spectrometry; x-ray diffraction; infrared and ultraviolet spectrophotometry and radioisotopes; paints and other coatings; mastic and tape sealers; fiberglass reinforced plastics; drug analysis. *Mailing Add:* 5745 N Witte Lane Glendale WI 53209-4555

KEHRL, HOWARD H, b Detroit, Mich, Feb 2, 23. MECHANICAL ENGINEERING. *Educ:* Ill Inst Technol, BS, 44; Univ Notre Dame, MS, 48; Mass Inst Technol, MS, 60. *Prof Exp:* Vpres & gen mgr, Oldsmobile Div, 72-73, vpres & group exec, 73-74, exec vpres, 74-81, VCHMN, GEN MOTORS CORP, 81- *Mem:* Nat Acad Eng; Soc Automotive Engrs; Motor Vehicle Mfrs Asn US. *Mailing Add:* Gen Motors Corp 3044 Gen Motors Blvd Detroit MI 48202

KEICHER, WILLIAM EUGENE, b Pittsburgh, Pa, Dec 28, 47; m 72; c 3. ELECTRO-OPTICAL SYSTEMS, LASER RADAR SYSTEMS. *Educ:* Carnegie-Mellon Univ, BS, 69, MS, 70, PhD(elec eng), 74. *Prof Exp:* Elec engr, Manned Spacecraft Ctr, NASA, 69; tech asst, Kodak Res Labs, 70; sr elec engr, CBS Labs, 73-75,; staff mem, 75-83, asst group leader, 83-85; GROUP LEADER, LINCOLN LABS, MASS INST TECHNOL, 85- *Mem:* Sr mem Inst Elec & Electronics Engrs; Optical Soc Am. *Res:* Leader of the (laser radar measurements) group; long range, high power laser radar systems; imaging radar; infrared detection systems; electro-optic modulators; atmospheric propagation. *Mailing Add:* Six Winn Valley Dr Burlington MA 01803

KEIDEL, FREDERICK ANDREW, b New Brunswick, NJ, Feb 4, 26. PHYSICAL CHEMISTRY. *Educ:* Rutgers Univ, BS, 46; Cornell Univ, PhD(phys chem), 51. *Prof Exp:* Res engr phys chem, 51-56; res proj engr, E I du Pont de Nemours & Co, Inc, 56-60, sr res phys chemist, 61-80, res assoc, 80-85; CONSULT, 85- *Honors & Awards:* Longstreth Medal, Franklin Inst, 60. *Mem:* Am Chem Soc; Sigma Xi; Mineral Soc Am; Am Inst Chemists. *Res:* Electron diffraction; electrochemical processes; chemical physics of surfaces; mineralogy. *Mailing Add:* 705 Prospect Ave Wilmington DE 19809-2334

KEIDERLING, TIMOTHY ALLEN, b Waterloo, Iowa, June 22, 47; m 76; c 1. SPECTROSCOPY, OPTICAL ACTIVITY. *Educ:* Loras Col, BS, 69; Princeton Univ, MA, 71, PhD(phys chem), 74. *Prof Exp:* Res assoc optical activity, Univ Southern Calif, 73-76; from asst prof to assoc prof, 76-85, PROF CHEM, UNIV ILL, CHICAGO, 85- *Concurrent Pos:* Guest prof, Max Planck Inst Quantenoptik; Fulbright fel, WGer, 84. *Mem:* Am Chem Soc; Am Phys Soc; Biophys Soc. *Res:* Experimental and theoretical studies of vibrational optical activity of small chiral molecules; magnetic vibrational circular dichroism and associated vibronic coupling effects; spectroscopic studies of protein and nucleic acid conformations. *Mailing Add:* Dept Chem Univ Ill Box 4348 Chicago IL 60680

KEIFFER, DAVID GOFORTH, b New Orleans, La, July 24, 31; m 56; c 6. PHYSICS. *Educ:* Loyola Univ, La, BS, 52; Univ Notre Dame, MS, 54, PhD(physics), 56. *Prof Exp:* Asst prof physics, Canisius Col, 56-64; ASSOC PROF PHYSICS, LOYOLA UNIV, LA, 64-, CHMN DEPT, 73- *Mem:* Am Phys Soc. *Res:* Radiation damage in glass. *Mailing Add:* Dept Physics Loyola Univ New Orleans LA 70118

KEIGHER, WILLIAM FRANCIS, b Montclair, NJ, Oct 28, 45; m 68; c 3. ALGEBRA. *Educ:* Montclair State Col, BA, 67; Univ Ill, AM, 69, PhD(math), 73. *Prof Exp:* Lectr math, Southern Ill Univ, Carbondale, 73-74; asst prof mat, Univ Tenn, Knoxville, 74-78; ASST PROF MATH, RUTGERS UNIV, 78- *Mem:* Am Math Soc; Math Asn Am. *Res:* Category theory and its applications to differential algebra. *Mailing Add:* Dept Math & Comput Sci Rutgers Univ Newark NJ 07102

KEIGHIN, CHARLES WILLIAM, b Pontiac, Ill, Aug 29, 32; m 60; c 3. GEOCHEMISTRY, ECONOMIC GEOLOGY. *Educ:* Oberlin Col, BA, 54; Univ Colo, MS, 60, PhD(geol), 66. *Prof Exp:* Res mineralogist, Cerro de Pasco Corp, La Oroya, Peru, SAm, 60-62, geologist, 62-63; asst prof geol, Northern Ill Univ, 66-72; vis asst prof geol & mineral, Ohio State Univ, 72-73; asst prof geosci, Northeastern La Univ, 73-74; GEOLOGIST, US GEOL SURV, DENVER, 74- *Mem:* Am Asn Petrol Geologists; Soc Econ Paleontologists & Mineralogists; Soc Petrol Engrs; Can Soc Petrol Geologists. *Res:* Trace element migration; diagenesis of clastic rocks; inorganic geochemistry; oil shale resource evaluation. *Mailing Add:* 1666 S Holland Court Lakewood CO 80232

KEIGLER, JOHN EDWARD, b Baltimore, Md, July 10, 29; m 55; c 6. AEROSPACE ENGINEERING, ELECTRICAL ENGINEERING. *Educ:* Johns Hopkins Univ, BE, 50, MS, 51; Stanford Univ, PhD(elec eng), 58. *Prof Exp:* Aeronaut res scientist, Ames Res Lab, NACA, 56; res assoc commun theory, Stanford Electronics Labs, 56-58; systs engr space electronics, 58-63, mgr spacecraft systs eng, 63-69, mgr systs eng, 69-73, mgr commun satellite systs, 73-84, CHIEF SCIENTIST, GE ASTRO SPACE DIV, 85- *Concurrent Pos:* Trustee, Nat Asn Search & Rescue; mem, Adv Coun Space Electronics, NASA, study comt, Nat Res Coun. *Honors & Awards:* David Sarnoff Medal Tech Achievement, 76; Aerospace Commun Award, Am Inst Aeronaut & Astronaut, 90. *Mem:* Fel Inst Elec & Electronics Engrs; fel Am Inst Aeronaut & Astronaut; Int Acad Astronaut. *Res:* Communication and information theory; video systems; spacecraft systems design, attitude control and stabilization; satellite telecommunications systems. *Mailing Add:* GE Astro Space Div PO Box 800 Princeton NJ 08543-0800

KEIHN, FREDERICK GEORGE, b Scranton, Pa, Aug 29, 23; m 48; c 4. SOLID STATE CHEMISTRY. *Educ:* Randolph Macon Col, BS, 47; Lehigh Univ, MS, 49; Syracuse Univ, PhD(chem), 53. *Prof Exp:* Asst electrochem, Lehigh Univ, 48-49; asst chem, Syracuse Univ, 49-51; inorg chemist electronics lab, Gen Elec Co, 52-57; res chemist ceramics lab, Corning Glass Works, 57-59; mem staff, Union Carbide Res Inst, NY, 59-65; assoc prof chem, Presby Col, SC, 65-67; PROF CHEM, BRIDGEWATER COL, 67- *Mem:* AAAS; Am Chem Soc; Am Crystallog Asn. *Res:* Double crystal x-ray diffractometry high temperature phase and mechanical properties studies; physical science curriculum development; applied ecology. *Mailing Add:* 1551 Central Ave Harrisonburg VA 22801-2808

KEIL, ALFRED ADOLF HEINRICH, b Kdhradswaldau, Ger, May 1, 13; m; c 2. ENGINEERING. *Educ:* Univ Breslau, GDr, 1939. *Prof Exp:* Chief scientist, underwater explosive res, Norfolk Naval Shipyard, 47-59; tech dir, Structural Mech, 59-63, tech dir basin, David Taylor Model Basin, 63-66; prof & head, Dept Naval Archit & Marine Eng, Mass Inst Technol, 66-71, dean Sch Eng, 71-77, Ford prof eng, 77-78, EMER PROF ENG, MASS INST TECHNOL, 78- *Honors & Awards:* Gold Medal, Am Soc Naval Eng, 64; Gibbs Bros Gold Medal, Naval Archit-Nat Acad Sci, 67. *Mem:* Nat Acad Eng; Am Soc Naval Eng. *Mailing Add:* Dept Eng Mass Inst Technol Cambridge MA 02139

KEIL, DAVID JOHN, b Elmhurst, Ill, Dec 13, 46; m. PLANT TAXONOMY. *Educ:* Ariz State Univ, BS, 68, MS, 70; Ohio State Univ, PhD(bot), 73. *Prof Exp:* Vis asst prof biol, Grand Valley State Col, 73-74; asst prof, Franklin Col, 75; lectr, 76-78, from asst prof to assoc prof, 78-85, PROF BIOL, CALIF POLYTECH STATE UNIV, 85- *Concurrent Pos:* Vis assoc prof bot, Ohio State Univ, 80; ed, Madrono, 88-90. *Mem:* Bot Soc Am; Am Soc Plant Taxon; Int Asn Plant Taxon; Torrey Bot Club; Asn Trop Biol; Sigma Xi. *Res:* Cytology and systematics of compositae; cytology, taxonomy, evolution and biogeography of genus Pectis. *Mailing Add:* Dept Biol Sci Calif Polytech State Univ San Luis Obispo CA 93407

KEIL, JULIAN E, b Charleston, SC, Oct 30, 26; m 48; c 3. EPIDEMIOLOGY. *Educ:* Clemson Univ, BS, 49, MS, 68; Univ NC, Chapel Hill, PhD(epidemiol), 75. *Prof Exp:* Entomologist, W R Grace & Co, 49-55, pesticide dept mgr, 56-67; instr, 67-69, assoc, 70-72, asst prof prev med, 73-77, assoc prof epidemiol, 77-83, PROF EPIDEMIOL, SCH PUB HEALTH, UNIV SC, 84- *Concurrent Pos:* Coun epidemiol, Am Heart Asn. *Mem:* Am Pub Health Asn; AAAS; Entom Soc Am; Asn Teachers Prev Med; Soc Epidemiol Res; Int Epidemiol Asn. *Res:* Cardiovascular epidemiology; environmental epidemiology. *Mailing Add:* 16 Sheridan Rd Charleston SC 29407

KEIL, KLAUS, b Hamburg, Ger, Nov 15, 34; US citizen; m 61, 84; c 2. METEORITICS, PETROLOGY. *Educ:* Univ Jena, MSc, 58; Univ Mainz, PhD(mineral, meteoritics), 61. *Prof Exp:* Res assoc & instr mineral & meteoritics, Mineral Inst, Univ Jena, 58-60; res assoc, Meteoritics & Cosmochem, Max Planck Inst Chem, 61 & Univ Calif, San Diego, 61-63; Nat Acad Sci-Nat Res Coun resident res assoc, Space Sci Div, Ames Res Ctr, NASA, 63-64, staff res scientist, 64-68; lectr, Dept Geol, San Jose State Col, 66-67; prof geol & dir inst meteorites, Univ NMex, 68-90, Presidential prof, 85-90, chmn, Dept Geol, 86-89; HEAD & PROF, UNIV HAWAII, 90- *Concurrent Pos:* Mem, Nat Steering Comt, 66-68; prin investr, Electron Microprobe Study of Returned Lunar Sample, NASA, 67-68, mem planetology adv subcomt, space sci & applns steering comt, 68; US rep, Comt Cosmic Mineral, Int Mineral Asn, 67-70, secy, 70-; rep, Comt Meteorites, Int Union Geol Sci, 68-72; John Wesley Powell invited lectr, Ariz Acad Sci, 71; mem lunar sci review bd, 71-73; mem & chmn, US Nat Comt Geochem, Nat Acad Sci, 71-75; invited speaker, Int Geol Cong, 72; assoc ed, Chem Geol, 73-85 & J Geophys Res, 82-85, & J Earth Chem, 84-86; mem geophys res bd, Nat Acad Sci, 74-75; mem & chmn, Lunar Sample Anal Planning Team, 74-78 & chmn facil subcomt, 75-76; distinguished vis prof, Inst Earth Sci, 74, 76, 77 & 78 & Dept Astonomy & Geophys, Univ Sao Paulo, Brazil, 81; mem, Viking Mars Flight Team, 76-78; vis assoc geochem, Div Geol Planet Sci, Calif Inst Technol, 76-77; honorary res assoc, Dept Mineral Sci, Am Mus Natural Hist, 77-; mem, Lunar & Planet, Sci Coun, 77-79, 87-90, chmn, 80-84; mem, Antarctic Meteorite Working Group, NSF, 78-84; dir, Caswell Silver Found, Univ NMex, 80-90; distinguished vis scientist, Jet Propulsion Lab,

Pasadena, 81; ann res lectr, Univ NMex, 81; mem adv comt, Comparative Planetology, Inst Geol Sci, 81-82, & Inst Geophys, Planet, Physics, Univ Calif, Los Alamos Nat Lab, 84-90; mem, NASA adv comt, minority graduate researchers, 84; chmn rev panel, space sta planetology exp & Mars observer, NASA, 85; mem panel, Lunar & Planetology, Geoscience Review, 87-; assoc ed, Meteoritics. *Honors & Awards:* Apollo Achievement Award, NASA, 70; George P Merrill Award, Nat Acad Sci, 70; NASA Except Sci Achievement Medal, 77; Group Achievement Award, NASA, 84; Leonard Medal, Meteoritical Soc, 88. *Mem:* Fel AAAS; Geol Soc Brazil; Ger Mineral Soc; Microbeam Anal Soc; Int Asn Geochem & Cosmochem (secy, 72-76); Planetary Soc; fel Mineral Soc; Am Geophys Union; Geochemical Soc; fel Meteoritical Soc (pres, 68-70). *Res:* Lunar geology; chemistry, geology and mineralogy of extraterrestrial materials, such as meteorites, cosmic dust and lunar surface; application of electron microprobe, laser microprobe and ion microprobe to study of rocks and minerals; geology of Mars. *Mailing Add:* Planetary Geosci Div Dept Geol & Geophysics SOEST Univ Hawaii Honolulu HI 96844

KEIL, LANNY CHARLES, b Elgin, Nebr, Apr 16, 36; m 66; c 3. PHYSIOLOGY, ENDOCRINOLOGY. *Educ:* Creighton Univ, BS, 63, MS, 66; Univ Calif, Davis, PhD(physiol), 73. *Prof Exp:* Res scientist, Physiol Br, 67-72, RES SCIENTIST ENDOCRINOL, BIOMED RES DIV, AMES RES CTR NASA, 72- *Mem:* Edocrine Soc; AAAS; Am Physiol Soc. *Res:* Hormonal control of water and electrolyte metabolism; gravitational biology; acceleration stress physiology. *Mailing Add:* Life Sci Div MS 239-17 Ames Res Ctr NASA Moffett Field CA 94035

KEIL, ROBERT GERALD, b New Rochelle, NY, May 7, 41; m 64; c 2. PHYSICAL CHEMISTRY. *Educ:* Villanova Univ, BS, 63; Temple Univ, PhD(phys chem), 67. *Prof Exp:* Res chemist, Org Chem Dept, E I du Pont de Nemours & Co, Inc, 67-69; from asst prof to assoc prof, 69-84, PROF CHEM, UNIV DAYTON, 84-, CHMN, 88- *Mem:* Electrochem Soc; Am Chem Soc. *Res:* Anodic oxide films; kinetic study of film growth; polarography of complexes in aqueous and nonaqueous solutions; general physical chemistry problems. *Mailing Add:* Dept Chem Univ Dayton Dayton OH 45469

KEIL, STEPHEN LESLEY, b Billings, Mont, Feb 21, 47; m 71. SOLAR PHYSICS. *Educ:* Univ Calif, Berkeley, AB, 69; Boston Univ, AM, 71, PhD(physics & astron), 75. *Prof Exp:* Nat Acad Sci-Nat Res Coun res fel, Air Force Geophysics Lab, 78-80, scholar physicist, 80-83, RES ASSOC SOLAR PHYSICS, AIR FORCE CAMBRIDGE RES LAB, SACRAMENTO PEAK OBSERV, 75-, CHIEF, GEOPHYSICS DIR, SOLAR RES BR, 83- *Concurrent Pos:* Res fel appl math, Univ Sydney, Australia, 76-77. *Mem:* Am Astron Soc; Int Astron Union. *Res:* Solar atmospheric inhomogenonities; multidimensional stellar atmospheres; high resolution solar observations; mathematical models of solar atmospheric structure. *Mailing Add:* Geophys Directurate Sacramento Peak Observ Sunspot NM 88349

KEIL, THOMAS H, b Philadelphia, Pa, July 24, 39; m 64. SOLID STATE PHYSICS. *Educ:* Calif Inst Technol, BS, 61; Univ Rochester, PhD(optics), 65. *Prof Exp:* Sloan fel solid state physics, Princeton Univ, 65-66, Sloan vis lectr, 66-67; asst prof, 67-72, assoc prof & chmn dept physics, 72-78, PROF PHYSICS, WORCESTER POLYTECH INST, 78- *Mem:* Am Phys Soc; Sigma Xi. *Res:* Solid state theory; optics. *Mailing Add:* Dept Physics Worcester Polytech Inst 100 Institute Rd Worcester MA 01609

KEILIN, BERTRAM, b New York, NY, Oct 18, 22; div; c 3. WATER CHEMISTRY, PSYCHOTHERAPY. *Educ:* NY Univ, BA, 42; Calif Inst Technol, MS, 45, PhD(phys chem), 50; Pepperdine Univ, MA, 82. *Prof Exp:* Res engr, Radar Labs, US Signal Corps, 42-43; res chemist, Aridye Corp, 43-44, Nat Defense Res Coun, 45-47 & Douglas Aircraft Co, 48-50; fel & res assoc, Univ Southern Calif, 50-51; sr res engr, Jet Propulsion Lab, Calif Inst Technol, 51-53; res chemist, Olin Mathieson Chem Corp, 53-56; mgr, Water Resources Dept, Chem & Struct Prod Div, Aerojet-Gen Corp Div, Gen Tire & Rubber Co, 56-66; vpres process develop, Amicon Corp, 66-68; dir, Environ Sci Dept, Tyco Labs, Inc, 68-69; mgr filter div, Bohna Eng & Res, Inc, 69-71; exec dir, Pac Water Qual Asn, 71-80; PRES, ENVIRON MGT SERV CO, 80-; PARTNER, NEWPORT THER GROUP, 82- *Concurrent Pos:* Adj prof, Grad Sch Educ & Psychol, Pepperdine Univ, 86- *Mem:* AAAS; Am Chem Soc; Am Water Works Asn; Water Pollution Control Fedn; Asn Marriage & Family Ther. *Res:* Ion exchange; molecular structure; electron diffraction; polarography; boron hydrides; rocket propulsion; adsorption; hydraulic fluids; membrane technology; water desalting; couples dynamics; intimacy; water softening. *Mailing Add:* 18 Springwater Irvine CA 92714

KEILSON, JULIAN, b Brooklyn, NY, Nov 19, 24; m 54; c 2. MATHEMATICAL STATISTICS, OPERATIONS RESEARCH. *Educ:* Brooklyn Col, BS, 47; Harvard Univ, PhD(physics), 50. *Prof Exp:* Res fel electronics, Harvard Univ, 50-52; mem staff, Lincoln Lab, Mass Inst Technol, 52-56; staff consult & sr eng specialist, Gen Tel & Electronics Labs, Inc, 56-62, sr scientist, 62-66; PROF STATIST, UNIV ROCHESTER, 66- *Concurrent Pos:* Lectr, Boston Univ, 56-; res fel, Univ Birmingham, 63; dir , Ctr Syst Sci; ed, Stochastic Processes and Their Appln, 73-79; fel, Japan Soc Prom Sci, 85. *Mem:* Am Phys Soc; sr mem Inst Elec & Electronics Eng; fel Inst Math Statist; fel Royal Statist Soc; fel Int Statist Inst; Opers Res Soc Am (pres-elect, 84-). *Res:* Semiconductor diffusion; electronic noise; Brownian motion; information theory; stochastic processes; electromagnetic propagation; probability theory; queuing theory; reliability theory. *Mailing Add:* Dept Statistics Univ Rochester Rochester NY 14642

KEILY, HUBERT JOSEPH, b Worcester, Mass, Jan 29, 21; m 45, 71; c 2. DRUG REGULATORY AFFAIRS. *Educ:* Niagra Univ, BS, 49; Union Col, MS, 51; Mass Inst Technol, PhD(anal chem), 56. *Prof Exp:* Lab asst, Res Lab, Linde Co, 40-43; asst, Union Col, 49-51; asst, Mass Inst Technol, 51-55; anal chemist, Gen Elec Co, 56-58; sect chief anal methods, Res Ctr, Lever Bros Co, 58-64; dept head anal chem, Merrell Dow Pharmaceut Inc, 64-82; mgr, Submissions & Develop Qual Control, Adria Lab Inc, 82- 84; REV CHEMIST, FOOD & DRUG ADMIN, 84- *Concurrent Pos:* Adj asst prof pharmaceut chem, Col Pharm, Univ Cincinnati, 82-83. *Mem:* Am Chem Soc; fel Am Inst Chem. *Res:* Evaluations of manufacturing and controls information submitted in support of investigational new drugs and new drug applications for drug products; analytical methods; packaging guidelines; computer assisted reviews. *Mailing Add:* 12907 Crookston Lane Apt B-1 Rockville MD 20851-2005

KEIM, BARBARA HOWELL, b Detroit, Mich, Mar 9, 46; m 75; c 2. ECOLOGICAL GENETICS. *Educ:* Univ NC, Greensboro, BA, 67; Rutgers Univ, MS, 69; Univ Va, PhD(genetics), 76; Bradley Univ, MBA, 88. *Prof Exp:* Asst ed biol, Biol Sci Info Serv, Philadelphia, 69-70; instr, Dept Biol, Wheaton Col, Norton, Mass, 75-76; asst prof, Dept Biol, Bradley Univ, 76-77, asst prof, Dept Nursing, 77-78, adj prof, Dept Biol, 79-80; from asst prof to assoc prof biol, Eureka Col, Ill, 80-87; DIR FINANCIAL DEVELOP, AM RED CROSS, 89- *Res:* Disruptive selection; speciation; polymorphisms. *Mailing Add:* 516 W Stratford Dr Peoria IL 61614

KEIM, CHRISTOPHER PETER, b Tecumseh, Nebr, Apr 6, 06; m 29; c 2. CHEMISTRY. *Educ:* Nebr Wesleyan Univ, AB, 27; Univ Nebr, MSc, 32, PhD(chem), 40. *Hon Degrees:* DSc, Nebr Wesleyan Univ, 59. *Prof Exp:* Head dept phys sci, York Col, 33-37; instr chem, Univ Tulsa, 40-41; res engr, Sylvania Corp, Mass, 41-42; res chemist & fel, Mellon Inst, 42-44; res physicist & adminr, Tenn Eastman Corp, 44-47; dir stable isotope res & prod div, Oak Ridge Nat Lab, 47-57, tech info div, 57-71; consult, Roane State Community Col, 71-81; RETIRED. *Concurrent Pos:* Pres & gen mgr, Mgt Servs, Inc, 73-75; consult, Hiwassee Col, 78-79; dir rowing, Spec Olympics Int, 86-89. *Mem:* Fel AAAS; Am Chem Soc; fel Am Phys Soc; Sigma Xi. *Res:* Isotope separations and properties; monomolecular surface films; electrical discharge in gases; surface chemistry; spreading of organic liquids and mixtures on water in the presence of monomolecular surface films; technical information. *Mailing Add:* 102 Orchard Lane Oak Ridge TN 37830

KEIM, GERALD INMAN, b Mt Berry, Ga, Oct 28, 10; m 43; c 1. PAPER CHEMISTRY. *Educ:* Univ Ga, BS, 32; MS, 35; Polytech Inst Brooklyn, PhD(chem), 44. *Prof Exp:* Instr, Univ Ga, 32-35; anal chemist, Warner-Quinlan Ref Co, NJ, 35-36; anal chemist, Colgate-Palmolive-Peet Co, 36-39, res chemist, 39-43; res chemist, Nat Defense Res Comt contract, 43-44; from res chemist to sr res chemist, Hercules Powder Co, 44-64, res assoc, Res Ctr, Hercules Inc, 64-75; RETIRED. *Mem:* Am Chem Soc; Tech Asn Pulp & Paper Indust. *Res:* C2 C3 elastomers; sterioregular olefin polymers; wet strength resins for paper; special sizes for paper; synthetic detergents; heterocyclic nitrogen compounds; functional and decorative coatings. *Mailing Add:* 41 Harleck Dr Anglesey-Ender Wilmington DE 19807-2507

KEIM, JOHN EUGENE, b Columbus, Ohio, Dec 3, 41; m 65; c 3. ELECTRONICS ENGINEERING. *Educ:* Valparaiso Tech Inst, BS, 69; Pac Western Univ, MS, 84, PhD(appl physics), 85. *Prof Exp:* Proj engr, Gilford Instrument Labs, 67-76; chief engr, Advan Weight Systs, 76-78; res & develop engr, Scott & Fetzer, 78-81; sr develop engr, Westinghouse Corp, 81-89; CORP VPRES ENG, CIRCLE PRIME MFG CO, 90- *Concurrent Pos:* Consult, Inservco, 72, Advan Weight Systs, 75- *Mem:* Inst Elec & Electronics Engrs. *Res:* Medical electronics; power controls; RF, military communications devices; seven patents. *Mailing Add:* 9319 Avon Lake Rd Lodi OH 44254

KEIM, LON WILLIAM, b Washington, DC, June 1, 43. PULMONARY DISEASE. *Educ:* Med Col Va, BS, 66, MD, 70. *Prof Exp:* Intern med, Univ Kans Med Ctr, Kansas City, 70-71; resident, Med Col Va, 71-73; fel, Col Med, Univ Iowa, 73-75, assoc pulmonary dis, 75-76; ASST PROF INTERNAL MED, UNIV NEBR, 76-; MEM MED STAFF, BISHOP CLARKSON MEM HOSP, 76- *Mem:* Fel Am Col Physicians; fel Am Col Chest Physicians; Am Thoracic Soc; Undersea Med Soc; Int Union Aganist Tuberc; Am Med Asn. *Res:* Tuberculosis and atypical mycobacteria; pulmonary diagnostic techniques, fiberoptic bronchoscopy; hyperbaric oxygen therapy. *Mailing Add:* 4242 Farnham St Omaha NE 68131

KEIM, WAYNE FRANKLIN, b Ithaca, NY, May 14, 23; m 47; c 3. CROP BREEDING, PLANT BREEDING AND GENETICS. *Educ:* Univ Nebr, BS, 47; Cornell Univ, MS, 49, PhD(plant breeding), 52. *Prof Exp:* From instr to asst prof bot, Iowa State Col, 52-56; from asst prof to prof agron, Purdue Univ, 56-75; head dept, 75-85, PROF AGRON, COLO STATE UNIV, 85- *Concurrent Pos:* NSF sci fac fel, Inst Genetics, Univ Lund, 62-63. *Honors & Awards:* Agron Educ Award, Am Soc Agron, 71. *Mem:* Fel AAAS; fel Am Soc Agron; Genetics Soc Am; Am Genetic Asn; fel Crop Sci Soc Am (pres, 83-84). *Res:* Legume genetics; breeding. *Mailing Add:* Dept Agron Colo State Univ Ft Collins CO 80523

KEINATH, GERALD E, b Grand Rapids, Mich, May 1, 24; m 56; c 3. MECHANICAL ENGINEERING. *Educ:* Northwestern Univ, BSME, 49. *Prof Exp:* Engr in training, Chicago & Northwestern RR, 46-49; res engr, Battelle Mem Inst, Ohio, 49-52, bus mgr Europ opers, Frankfurt & Geneva, 52-58, asst supvr contract prep, Ohio, 58-63; vpres, NStar Res & Develop Inst, 63-72; PRES, NOVUS INC, 72- *Mem:* Am Soc Mech Engrs; Am Soc Automotive Engrs. *Res:* Product development; glass repair; research management. *Mailing Add:* 5261 Lochloy Dr Edina MN 55436-2023

KEINATH, STEVEN ERNEST, b Saginaw, Mich, Sept 10, 54; m 87. POLYMER SCIENCE. *Educ:* Saginaw Valley State Col, BS, 76, MBA, 81; Univ Mass, MS, 78; Cent Mich Univ, MA, 85. *Prof Exp:* Asst to dir instrumentation, Dept Polymer Sci & Eng, Univ Mass, 78; sr res asst, Mich Molecular Inst, 78-84, asst ed, MMI Press, 81-82, adminr grants & contracts, 83-84, INDEPENDENT RESEARCHER, MICH MOLECULAR INST, 84-, INSTR, 85- *Mem:* Am Chem Soc; Soc Advan Mat & Process Eng; NY Acad Sci; Sigma Xi; Soc Plastics Engrs; Am Soc Composites. *Res:* Thermal analysis; polymer transitions and relaxations; binary and ternary polymer blends; characterization of glass-filled epoxy composites; effects of absorbed moisture in high-performance organic fibers. *Mailing Add:* Mich Molecular Inst 1910 W St Andrews Rd Midland MI 48640-2696

KEINATH, THOMAS M, b Frankenmuth, Mich, Jan 5, 41; m 63; c 1. ENVIRONMENTAL ENGINEERING. *Educ:* Univ Mich, Ann Arbor, BSE, 63, MSE, 64, PhD(water resources eng), 68. *Prof Exp:* Inst Sci & Technol fel, Univ Mich, 68-69; PROF ENVIRON SYSTS ENG & HEAD DEPT, CLEMSON UNIV, 69- *Concurrent Pos:* Consult, Waverly Assocs, 68-69, Westvaco Inc, 70-, Gaston Co, Dyeing Mach Co, 71-, Eng Sci, Inc, 74- & UNESCO, 80; expert sci adv, Environ Protection Agency, 75-76. *Honors & Awards:* Huber Prize, Am Soc Civil Engrs, 86. *Mem:* Am Chem Soc; Am Inst Chem Engrs; Am Water Works Asn; Am Soc Civil Engrs; Asn Environ Eng Prof; Sigma Xi; Water Pollution Control Fedn; Int Asn Water Pollution Res & Control; Am Soc Engr Educ. *Res:* Physiochemical processes of water and waterwaste treatment; automation and control of water and wastewater treatment systems. *Mailing Add:* Dept Environ Systs Eng Clemson Univ Clemson SC 29634-0919

KEIPER, RONALD R, b Allentown, Pa, Sept 21, 41; m 64; c 2. ANIMAL BEHAVIOR. *Educ:* Muhlenberg Col, BS, 63; Univ Mass, MS, 66, PhD(zool), 68. *Prof Exp:* Asst prof zool & biol, 68-73, assoc prof zool, 73-82, PROF ZOOL, PA STATE UNIV, 82-, DIR ACAD AFFAIRS & DISTINGUISHED PROF BIOL, 90- *Concurrent Pos:* Theodore Roosevelt Mem Fund-Am Mus Natural Hist grant, 68-69; Frank M Chapman Mem Fund-Am Mus Natural Hist grant, 68-70; Nat Park Serv study grants; Fulbright fel, 84-85. *Mem:* Animal Behav Soc; Lepidopterists Soc. *Res:* Causes and functions of the abnormal stereotyped behaviors shown by caged birds; effects of early experience on bird behavior; natural behavior of cyptic moths; studying the behavior, ecology and social organization of feral horses and Prezewalski horses. *Mailing Add:* 11570 E Airport Rd Waynesboro PA 17268

KEIRANS, JAMES EDWARD, b Worcester, Mass, Apr 4, 35; m 63; c 2. MEDICAL ENTOMOLOGY, ACAROLOGY. *Educ:* Boston Univ, AB, 60, AM, 63; Univ NH, PhD(zool), 66. *Prof Exp:* Res asst parasitol, Boston Univ, 60-63; res asst entom, Univ NH, 65-66; res entomologist, Commun Dis Ctr, USPHS, 66-69; res entomologist, NIH, 69-90; RES PROF & CUR, US NAT TICK COLLECTION, 90- *Concurrent Pos:* Res entomologist, Brit Mus Nat Hist, 77-78. *Mem:* Am Soc Parasitol; Entom Soc Am; Acarological Soc Am. *Res:* Arthropods of public health significance; Ixodoidea taxonomy. *Mailing Add:* Inst Arthropodology & Parasitol Ga Southern Univ Statesboro GA 30460

KEIRNS, JAMES JEFFERY, b New Haven, Conn, July 1, 47; m 67, 75; c 3. BIOCHEMISTRY. *Educ:* Rice Univ, BA, 68; Yale Univ, MPhil, 70, PhD(molecular biophys & biochem), 72. *Prof Exp:* Jane Coffin Childs Mem Fund Med Res fel biochem, Dept Path, Sch Med, Yale Univ, 72-75; sr res biochemist & proj leader allergy res, Lederle Labs Div, Am Cyanamid Co, 75-79; dir, Dept Biochem, 79-88, DIR, DEPT DRUG METAB & PHARMACOKINETICS, BOEHRINGER INGELHEIM LTD, 88- *Mem:* Am Chem Soc; AAAS; NY Acad Sci; Am Acad Allergy; Health Physics Soc. *Res:* Biochemical aspects of metabolic, immunological and viral diseases; cyclic nucleotides; mechanism of enzyme reactions; inflammation; immediate hypersensitivity; pharmacokinetics and drug metabolism. *Mailing Add:* Boehringer Ingelheim Pharmaceut Inc 90 E Ridge Rd Ridgefield CT 06877

KEIRNS, MARY HULL, b Jacksonville, Fla, Mar 16, 47; m 67; c 3. CHEMISTRY, AUTOMOTIVE ENGINEERING. *Educ:* Yale Univ, MPhil, 71, PhD(chem), 75. *Prof Exp:* RES CHEMIST, EXXON RES & ENG CO, 74- *Mem:* Am Chem Soc; Soc Automotive Engrs. *Res:* Pollution control systems and fuel economy; aviation fuel quality and handling. *Mailing Add:* RD3 Sherwood Hill Rd Brewster NY 10509

KEIRS, RUSSELL JOHN, b Springfield, Ill, Aug 27, 15; m 41; c 1. ANALYTICAL CHEMISTRY. *Educ:* Univ Ill, BS, 37, MS, 38, PhD, 41. *Prof Exp:* Chemist, Continental Can Co, 41-42; assoc prof chem, Fla State Univ, 50-65, assoc dean, Grad Sch & dir res, 62-69, prof, 65-81; RETIRED. *Mem:* Am Chem Soc. *Res:* Molecular phosphorescence analysis at low temperatures; instrumental analysis. *Mailing Add:* 1506 Golf Ter Dr Tallahassee FL 32301

KEISCH, BERNARD, b Brooklyn, NY, Aug 1, 32; m 54; c 3. RADIOCHEMISTRY. *Educ:* Rensselaer Polytech Inst, BS, 53; Wash Univ, St Louis, PhD(chem), 57. *Prof Exp:* Res chemist, Idaho Chem Processing Plant, Phillips Petrol Co, 57-59, mat testing reactor, 59-62; sr scientist, Nuclear Sci & Eng Corp, 62-66; from fel to sr fel, Carnegie-Mellon Univ, 66-74, sr fel, Carnegie-Mellon Inst Res, 74-78; CHEMIST, BROOKHAVEN NAT LAB, 78- *Mem:* Sigma Xi; AAAS; Am Chem Soc. *Res:* Nuclear applications in art and archaeology; activation analysis; isotope mass spectrometry; carbon-14 dating; Mossbauer effect; nuclear safeguards. *Mailing Add:* Dept NUC Bldg 197 Brookhaven Nat Lab Upton NY 11973

KEISER, BERNHARD E(DWARD), b Richmond Heights, Mo, Nov 14, 28; m 55; c 5. TELECOMUNICATIONS ENGINEERING, ELECTRICAL ENGINEERING. *Educ:* Wash Univ, St Louis, BS, 50, MS, 51, DSc, 53. *Prof Exp:* Proj engr, White-Rodgers Elec Co, Mo, 53-56, Petrolite Corp, 56-57 & Mo Res Labs, 57-59; group leader new commun systs, RCA Corp, 59-64, mgr plans & prog sect, Kennedy Space Ctr Commun Proj, RCA Serv Co, Fla, 64-67, adminr advan tech planning, RCA Missile & Surface Radar Div, NJ, 67-69; vpres systs res & eng, Page Commun Engrs, Va, 69-70; dir advan systs electronics & commun, Atlantic Res Corp, Alexandria, 71-72; dir anal, Fairchild Space & Electronics Co, 72-75; PRES, KEISER ENG, INC, 75- *Mem:* Fel Inst Elec & Electronics Engrs. *Res:* Telecommunications; electronic systems; engineering management and consulting. *Mailing Add:* 2046 Carrhill Rd Vienna VA 22181-2917

KEISER, EDMUND DAVIS, JR, b Appalachia, Va, Feb, 18, 34; c 4. VERTEBRATE ZOOLOGY, WETLANDS ECOLOGY. *Educ:* Southern Ill Univ, BA, 56, MS, 61; La State Univ, PhD(vert zool), 67. *Prof Exp:* Teacher high sch, Ill, 56-57, pub schs, 57; sci instr & dist sci coordr, Dist 70, Freeburg, Ill, 58-62; instr zool & anat, La Salle-Peru-Oglesby Jr Col, 62-64; teaching asst zool, La State Univ, 64-66; dir biol, Physics & Chem, NSF Coop Col-Sch Sci Prog biol, 69-70; from asst prof to prof comp anat & syst zool, Univ Southwestern La, 66-76; assoc prof, 76, chmn dept, 76-87, PROF BIOL, UNIV MISS, 76- *Concurrent Pos:* Teaching asst, Southern Ill Univ, 61; sci ed consult, Southwestern La Parish Schs, 66-71; res assoc, Gulf South Res Inst, Baton Rouge, 72-75; consult & proj dir, US Fish & Wildlife Serv, Atchafalaya Basin Surv, 73-76; dir, Lafayette Natural Hist Mus, 73; consult, La Chenier Plain Study, US Fish & Wildlife Serv, 78; comnr, Miss Dept Wildlife Conserv, 78-79 & 80-84, chmn, 83-84, mem, Miss Wildlife Heritage Comt, 80-84; mem, Governor's Select Comt Radioactive Waste & Waste Depository, 79; environ consult, Lockheed Eng, Abort Solid Rocket Motor Proj, NASA, 90-91. *Mem:* Am Soc Ichthyol & Herpet; Inst Caribbean Sci; Sigma Xi; Soc Study Amphibians & Reptiles; Herpetologist League; fel Explorers Club. *Res:* Systematics, ecology and developmental morphology of vertebrates, especially amphibians and reptiles of the United States and the Neotropics; wetlands ecology and management; developmental embryology. *Mailing Add:* Dept Biol Univ Miss University MS 38677

KEISER, GEORGE MCCURRACH, b Plainfield, NJ, July 21, 47. GRAVITATIONAL & ATOMIC PHYSICS. *Educ:* Middlebury Col, AB, 69; Duke Univ, PhD(physics), 76. *Prof Exp:* Res assoc, Joint Inst Lab Astrophys, 76-77; Nat Res Coun fel, Nat Bur Standards, 77-80; MEM FAC, DEPT PHYSICS, STANFORD UNIV, 80- *Concurrent Pos:* Lectr, Univ Colo, 77-78. *Mem:* Am Phys Soc. *Res:* High precision measurements in gravitational and atomic physics. *Mailing Add:* Hansen Labs GP-B Stanford Univ Stanford CA 94305

KEISER, HAROLD D, RHEUMATOLOGY. *Educ:* New York Univ, MD, 64. *Prof Exp:* PROF MED, ALBERT EINSTEIN COL MED, 83- *Mem:* Am Rheumatism Asn; Am Asn Immunologists; Am Asn Clin Res; Soc Complex Carbohydrates; Am Soc Clin Invest. *Mailing Add:* Albert Einstein Col Med 1300 Morris Park Ave Bronx NY 10461

KEISER, HARRY R, b Aug 9, 33; m 65; c 2. CLINICAL PHARMACOLOGY. *Educ:* Northwestern Univ, MD, 58. *Prof Exp:* CLIN DIR, NAT HEART, LUNG & BLOOD INST, NIH, 76- *Mem:* Am Bd Internal Med; Am Fedn Clin Res. *Mailing Add:* Nat Heart Lung & Blood Inst NIH Bldg 10 Rm 8C103 Bethesda MD 20892

KEISER, JEFFREY E, b Kalamazoo, Mich, Feb 25, 41; m 88; c 5. ORGANIC CHEMISTRY. *Educ:* Kalamazoo Col, AB, 62; Wayne State Univ, PhD(org chem), 66. *Prof Exp:* From asst prof to prof chem, Coe Col, 66-88, chmn dept, 76-88; SR RES ASSOC, PENFORD PROD CO, 88- *Mem:* Am Chem Soc; AAAS. *Res:* Organic analytical chemistry; starch chemistry. *Mailing Add:* PO Box 428 Penford Prod Co Cedar Rapids IA 52406

KEISLER, HOWARD JEROME, b Seattle, Wash, Dec 3, 36; m 59; c 3. MATHEMATICAL LOGIC. *Educ:* Calif Inst Technol, BS, 59; Univ Calif, Berkeley, PhD(math), 61. *Prof Exp:* Mathematician, Commun Res Div, Inst Defense Anal, 61-62; from asst prof to assoc prof, 62-67, PROF MATH, UNIV WIS, MADISON, 67- *Concurrent Pos:* Vis res assoc, Princeton Univ, 61-62; Alfred P Sloan fel, 66-69; vis prof, Univ Calif, Los Angeles, 67-68; John S Guggenheim fel, 76-77; vis prof, Univ of Colo, 85. *Mem:* Am Math Soc; Asn Symbolic Logic (vpres, 77-80). *Res:* Model theory; set theory; applications of model theory to probability theory and mathematical economics. *Mailing Add:* Math 303 Van Vleck Univ Wis Madison WI 53706

KEISLER, JAMES EDWIN, b Spartanburg, SC, Aug 20, 29; m 50; c 3. MATHEMATICS. *Educ:* Midland Col, BS, 49; Univ Mich, MA, 54, PhD(math), 59. *Prof Exp:* Teacher high sch, Nebr, 49-51; from asst prof to assoc prof, 59-73, PROF MATH, LA STATE UNIV, BATON ROUGE, 73- *Mem:* Am Math Soc; Math Asn Am. *Res:* Point-set topology; fixed point problems and characterizations of spaces. *Mailing Add:* Dept Math La State Univ Baton Rouge LA 70803

KEISTER, DONALD LEE, b Beckley, WVa, Dec 10, 33; m 62; c 3. MICROBIAL BIOCHEMISTRY. *Educ:* WVa Wesleyan Col, BS, 54; Univ Md, MS, 56, PhD, 59. *Prof Exp:* Fel, McCollum-Pratt Inst, Johns Hopkins Univ, 58-61; fel, Res Inst Adv Study, Md, 61-62; assoc prof biochem, Antioch Col, 62-80; sr investr, Charles F Kettering Res Lab, 62-84; RES LEADER, SOYBEAN & ALFALFA RES LAB, USDA AGR RES SERV, 84- *Concurrent Pos:* Nat Found fel, 58-60; chmn, Gordon Res Conf Photosynthesis, 69, Beltsville Symp Agr, 89; grad fac, Wright State Univ, 80-84. *Mem:* Am Soc Biol Chemists; Am Soc Plant Physiol; Am Soc Microbiol. *Res:* Mechanisms of pyridine nucleotide reduction in photosynthetic organisms; structure and function in photosynthetic organelles; control mechanisms in nitrogen fixation; symbiotic nitrogen fixation in legumes; author or co-author of over 100 publications. *Mailing Add:* Soybean & Alfalfa Res Lab USDA Agr Res Serv Bldg 011 HH19 BARC-W Beltsville MD 20705

KEISTER, JAMES E, b Coburg, Iowa, July 11, 14; m 35; c 3. ELECTRICAL ENGINEERING. *Educ:* Cornell Univ, BS, 35. *Prof Exp:* RETIRED. *Concurrent Pos:* Semiconductor Standards Comt; US Radio Tech Planning Bd. *Honors & Awards:* Apollo Achievement Award, NASA, 69. *Mem:* Fel Inst Elec & Electronics Engrs; Sigma Xi. *Res:* Development of television transmitters; radar counter measures, semiconductors; ground support equipment for Apollo & Skylab Prog. *Mailing Add:* 5566 Dry Ridge Rd Cincinnati OH 45252

KEISTER, JAMIESON CHARLES, b Schenectady, NY, Feb 28, 38; m 60; c 4. DIFFERENTIAL EQUATIONS. *Educ:* Cornell Univ, BS, 60; Georgetown Univ, MS, 67, PhD(physics), 70. *Prof Exp:* Engr, Nuclear Prop Div, Buships, 60-64; jr tech assoc, NUS Corp, 64-66; sr field res physicist, Melpar Div, West Airbrake Co, 66-67; prof physics & math, Covenant Col, 70-84; PRIN SCIENTIST, BASIC RES, ALCON LAB, 84- *Concurrent Pos:* Math consult, Miami Valley Labs, Procter & Gamble, Inc, 81- *Res:* Tunneling in super conductors; mathematical modeling and experiments for diffusion problems; complex variables; conformal mapping. *Mailing Add:* 7690 Floyd Hampton Rd Crowley TX 76036

KEISTER, JEROME BAIRD, b Baton Rouge, La, Mar 28, 53. INORGANIC CHEMISTRY, ORGANOMETALLIC CHEMISTRY. *Educ:* La State Univ, Baton Rouge, BS, 73; Univ Ill, Urbana-Champaign, PhD(chem), 78. *Prof Exp:* res chemist organometallic catalysis, Corp Pioneering Res, Exxon Res & Eng Co, 77-80; asst prof, 80-86, ASSOC PROF INORG CHEM, STATE UNIV NY, BUFFALO, 86- *Mem:* Am Chem Soc. *Res:* Homogeneous catalysis, organometallic chemistry; metal cluster chemistry. *Mailing Add:* Dept Chem State Univ NY Acheson Hall Buffalo NY 14214

KEITEL, GLENN H(OWARD), b Chicago, Ill, Feb 16, 30; m 53; c 2. ELECTRICAL ENGINEERING, OFFICE AUTOMATION. *Educ:* Wash Univ, BS, 52, MS, 54; Stanford Univ, PhD(elec eng), 55. *Prof Exp:* Fulbright fel, Cavendish Lab, Cambridge Univ, 55-56; engr advan studies, Microwave Lab, Gen Elec Co, 56-59; eng dept mgr, Western Develop Lab, Philco Corp, Calif, 59-62; assoc prof elec eng, San Jose State Col, 62-66, prof & chmn dept, 66-69; prof elec eng & chmn elec eng curric, Drexel Univ, 69-71; prof elec eng & dean eng, Bucknell Univ, 71-79; dir eng & technol planning, CPT Corp, 80-81; PRES, OFF AUTOMATION CONSULT, INC, 81- *Concurrent Pos:* Consult, Stanford Res Inst, 62-63 & 64-69 & Western Develop Labs, Philco Corp, 62-63; electronics liaison scientist, Off Naval Res, Br Off, London, 63-64; adv scientist, Lockheed Missiles & Space Co, 65-69. *Res:* Functional and engineering design, selection and implementation of office automation systems, with training and applications development. *Mailing Add:* 220 Birch Bluff Rd Excelsior MN 55331

KEITER, ELLEN ANN, INORGANIC CHEMISTRY, PHYSICAL CHEMISTRY. *Educ:* Augsburg Col, BA, 64; Univ Md, MS, 68; Univ Ill, PhD(inorg chem), 86. *Prof Exp:* From instr to asst prof, 78-87, ASSOC PROF CHEM, EASTERN ILL UNIV, 87- *Mem:* Am Chem Soc; AAAS. *Res:* Nuclear quadrupole double resonance spectroscopy; hydrogen bonds; metal hydrides; metal-protein complexes (model compounds); metal-nucleic acid complexes (model compounds). *Mailing Add:* Dept Chem Eastern Ill Univ Charleston IL 61920

KEITER, RICHARD LEE, b Winchester, Va, Jan 10, 39; m 66; c 2. INORGANIC CHEMISTRY. *Educ:* Shepherd Col, BS, 61; WVa Univ, MS, 64; Univ Md, PhD(inorg chem), 67. *Prof Exp:* Assoc inorg chem, Iowa State Univ, 67-69; assoc prof, 69-79, PROF INORG CHEM, EASTERN ILL UNIV, 79- *Concurrent Pos:* Vis prof, Univ Wis, 72 & 77, Univ Exeter, Eng, 75, Univ Ill, 80 & Colo State Univ, 90. *Mem:* Am Chem Soc; AAAS. *Res:* Coordination chemistry of trivalent phosphorous ligands; transition metal carbonyls; polydentate phosphorus ligand control; synthetic inorganic and organometallic chemistry; phosphido-bridged complexes. *Mailing Add:* Dept Chem Eastern Ill Univ Charleston IL 61920

KEITH, CHARLES HERBERT, physical chemistry; deceased, see previous edition for last biography

KEITH, DAVID ALEXANDER, b Chelmsford, Essex, Eng, Aug 28, 44; m 76; c 2. ORAL & MAXILLOFACIAL SURGERY. *Educ:* Univ London, BDS, 66; FDSRCS(Eng), 70; DMD, Harvard Univ, 83. *Prof Exp:* Lectr oral surg, Hosp Dent Sch, Kings Col, London, 71-73; res fel, Mass Gen Hosp, Boston, 73-74; res fel, 75-77, asst prof, 78-84, ASSOC PROF ORAL SURG, HARVARD SCH DENT MED, 85- *Concurrent Pos:* Res assoc orthop surg, Children's Hosp Med Ctr, 77; clin assoc oral surg, Mass Gen Hosp, 78-84, asst surgeon oral & maxillofacial surg, 84- *Honors & Awards:* Malleson Prize, 66; Brit Asn Oral Surgeons Award, 73. *Mem:* Brit Dent Asn; Brit Asn Oral Surgeons; Int Asn Dent Res; Am Dent Asn; Int Asn Oral Surgeons. *Res:* Biochemistry of craniofacial development. *Mailing Add:* Ambulatory Care Ctr #230 Mass Gen Hosp Fruit St Boston MA 02114

KEITH, DAVID LEE, b Mankato, Minn, Dec 7, 40; m 61; c 4. ENTOMOLOGY. *Educ:* Gustavus Adolphus Col, BSc, 62; Univ Minn, MSc, 65; Univ Nebr, Lincoln, PhD(entom), 71. *Prof Exp:* EXTEN ENTOMOLOGIST, UNIV NEBR, LINCOLN, 67- *Mem:* Sigma Xi; Entom Soc Am. *Res:* Biology, ecology and control of cutworms; development of integrated pest management projects on Nebraska field crops. *Mailing Add:* 225 Keim Hall IANR/UNL Lincoln NE 68583-0917

KEITH, DENNIS DALTON, b Hartford, Conn, July 11, 43; c 2. ORGANIC SYNTHESIS, ANTIBIOTICS. *Educ:* Bates Col, BS, 65; Yale Univ, MS, 67, MPh, 69, PhD(org chem), 69. *Prof Exp:* NIH fel, Harvard Univ, 69-71; sr res chemist, 71-76, res fel, 76-81, RES GROUP CHIEF, HOFFMANN-LA ROCHE INC, 81- *Mem:* Am Chem Soc; Am Soc Microbiol; Sigma Xi. *Res:* Synthesis of natural products; heterocyclic chemistry; synthetic methods. *Mailing Add:* Eight Mendl Terr Montclair NJ 07042

KEITH, DONALD EDWARDS, b Ft Worth, Tex, Oct 7, 38; m 59. INVERTEBRATE ECOLOGY. *Educ:* Tex Christian Univ, BA, 62, MS, 64; Univ Southern Calif, PhD(biol), 68. *Prof Exp:* NSF res grant, summer 61; asst prof biol, Tex Christian Univ, 68-75, dir environ sci prog, 69-71; asst prof to assoc prof, 75-85, PROF ECOL TARLETON STATE UNIV, 85- *Concurrent Pos:* Consult, US Army CEngrs, Lake Proctor, 76-; res grant, Tarleton State Univ, 75-88; Tex Acad Sci fel, 77. *Honors & Awards:* Tex Acad Sci fel, 77. *Mem:* AAAS; Sigma Xi. *Res:* Benthic ecology; substrate selection, feeding and functional digestive tract morphology of Caprellid amphipods; amphipod phylogeny; effects of industrial effluents on benthic invertebrate communities; corals of the Swan Islands, Honduras; brachyuran crabs of Roatan and The Swan Islands, Honduras; octocorals of Roatan. *Mailing Add:* Dept Biol Sci Tarleton State Univ Stephenville TX 76402

KEITH, ERNEST ALEXANDER, b Fayetteville, Tenn, Dec 19, 51; m 72. RUMINANT NUTRITION. *Educ:* Univ Ark, BS, 73, MS, 74; Purdue Univ, PhD(ruminant nutrit), 78. *Prof Exp:* ASST PROF DAIRY NUTRIT, DEPT DAIRY SCI, LA STATE UNIV, 78- *Mem:* Sigma Xi; Am Dairy Sci Asn; Am Soc Animal Sci; Am Forage & Grassland Coun. *Res:* Forage nutrition of dairy cattle. *Mailing Add:* 945 E University Springfield MO 65807

KEITH, FREDERICK W(ALTER), JR, b Chicago, Ill, Jan 20, 21; m 43; c 1. CHEMICAL ENGINEERING. *Educ:* Yale Univ, BS, 42; Univ Pa, PhD(chem eng), 51. *Prof Exp:* Chem engr process develop, E I du Pont de Nemours & Co, 42-44; chem engr res & develop, Sharples Res Lab, 44-48; asst instr, Univ Pa, 49; chem engr process develop, Pennwalt Chem Equip Div, Sharples Corp, 50-71, dir environ technol, Sharples Div, Pennwalt Corp, Warminster, 71-79; consult, 79-88; RETIRED. *Mem:* Am Chem Soc; Am Inst Chem Engrs; Sigma Xi. *Res:* Development and evaluation of centrifuges; waste and sewage process development; separations in synfuel processing. *Mailing Add:* 454 Conshohocken State Rd Gladwyne PA 19035

KEITH, H(ARVEY) DOUGLAS, b Belfast, Northern Ireland, Mar 10, 27; m 53; c 2. CRYSTALLINE MORPHOLOGY. *Educ:* Queen's Univ, Belfast, BSc, 48; Univ Bristol, Eng, PhD(physics), 51. *Prof Exp:* Lectr physics, Univ Bristol, 51-56; res physicist, Am Visase Corp, 57-60; mem tech staff, Bell Tel Labs, 60-88; RES PROF MAT SCI, UNIV CONN, 88- *Concurrent Pos:* Lectr physics, St Joseph's Col, Philadelphia, Pa, 58-60; div counr, Am Phys Soc, 77-85; consult, AT&T Bell Labs, 88- *Honors & Awards:* High-Polymer Physics Prize, Am Phys Soc, 73. *Mem:* Fel Am Phys Soc; fel AAAS. *Res:* Optical and electron microscopy; x-ray and electron diffraction of structure and morphology of crystalline polymers and relationships to properties. *Mailing Add:* 51 Gail Lane South Windsor CT 06074

KEITH, HARVEY DOUGLAS, b Belfast, Ireland, Mar 10, 27; US citizen; m 53, 84; c 2. EXPERIMENTAL PHYSICS. *Educ:* Queen's Univ, Belfast, BS, 48; Bristol Univ, PhD(physics), 51. *Prof Exp:* Lectr physics, Bristol Univ, 52-57; res physicist, Am Viscose Corp, Pa, 57-59, leader phys sect, 59-60; head, Org Mat Res Dept, 69-82, MEM TECH STAFF, BELL LABS, MURRAY HILL, 60- *Concurrent Pos:* Lectr, St Joseph's Col, Pa, 58-60. *Honors & Awards:* High Polymer Physics Prize, Am Phys Soc, 73. *Mem:* Fel Am Phys Soc; fel AAAS. *Res:* Solid state physics; crystallography; optics; high polymers. *Mailing Add:* Rm 1A324 Bell Labs PO Box 261 Murray Hill NJ 07974-2070

KEITH, JAMES OLIVER, b Pasadena, Calif, Mar 20, 32; m 50; c 5. POLLUTION ECOLOGY. *Educ:* Univ Calif, Berkeley, AB, 53; Univ Ariz, MS, 56; Ohio State Univ, PhD(ecol), 78. *Prof Exp:* Wildlife res biologist, Rocky Mt Forest & Range Exp Sta, US Forest Serv, 56-61; wildlife res biologist, Denver Wildlife Res Ctr, US Fish & Wildlife Serv, 61-76, chief, 69-73, Wildlife res biologist environ contaminants, Patuxent Wildlife Res Ctr, 76-81, wildlife res biologist, Int Prog, Denver Wildlife Res Ctr, 81-90; CONSULT, 91- *Concurrent Pos:* Res assoc, Agr Exp Sta, Univ Calif, 61-65; consult, World Wildlife Fund, Galapagos Islands, Food & Agr Orgn, Sudan, Kenya & Argentina, US Aid, Haiti, Senegal & Morocco, Nat Geog, Chile, govt Bahamas. *Mem:* Am Soc Mammalogists; Soc Conserv Biol; Wildlife Soc; Am Ornith Union; Soc Ecosyst Restoration & Mgt. *Res:* Ecological effects of land management practices; influence of logging, grazing, agriculture and pesticides on wildlife and their habitats; restoring altered ecosystems; control of introduced predators. *Mailing Add:* Wildlife Res Ctr Bldg 16 Denver Fed Ctr Denver CO 80225

KEITH, JENNIE, b Carmel, Calif, Nov 15, 42; m 68, 80; c 2. GERONTOLOGY, ANTHROPOLOGY. *Educ:* Pomona Col, BA, 64; Northwestern Univ, MA, 66, PhD(anthrop), 68. *Prof Exp:* From asst prof to prof, 70-90, CENTENNIAL PROF ANTHROP, SWARTHMORE COL, 90- *Concurrent Pos:* Mem, Res Rev Comt, NIMH, 79-82; task group leader, Nat Res Plan Aging, 81; co-dir, Proj AGE, Nat Inst Aging, 82-90; mem, Aging & Human Develop Rev Panel, NIH, 85-90; exec comt Behav & Social Sci sect, Geront Soc Am, 85-87, prog chair, 89, chair, 90; assoc ed, J Gerontol, 86-; sr adv coun, Brookdale Found, 89- *Mem:* Geront Soc Am; Am Anthrop Asn; Asn Anthrop & Geront. *Res:* Cross-cultural comparative research on social and cultural influences on aging and old age. *Mailing Add:* 612 Ogden Ave Swarthmore PA 19081

KEITH, JERRY M, b Salt Lake City, Utah, Oct 22, 40. ENZYMOLOGY, VIROLOGY. *Educ:* Univ Calif, Berkeley, BA, 73, PhD(comparative biochem), 76. *Prof Exp:* Staff fel, Lab Biol Viruses, Nat Inst Allergy & Infectious Dis, NIH, 76-78; asst prof biochem, Col Dent, NY Univ, 78-; AT DEPT PATH, ROCKY MOUNTAIN LABS, NAT INST ALLERGY & INFECTIOUS DIS, NIH, HAMILTON, MONT. *Concurrent Pos:* Adj asst prof biol doctoral fac, City Univ New York, 81-; prin investr, gen med-biochem, NIH, 81-84. *Mem:* Sigma Xi; AAAS; Am Soc Microbiol; Am Soc Virol. *Res:* Structure and function of biologically active nucleic acids and proteins, with a particular interest in the isolation and characerization of the enzymes related to the synthesis, processing and post-transcriptional modification of MRNA's. *Mailing Add:* Intramural Res Prog Bldg 30 Rm 316 Nat Inst Dent Res Bethesda MD 20892

KEITH, LAWRENCE H, b Morris, Ill, Apr 5, 38; m 69. ENVIRONMENTAL CHEMISTRY. *Educ:* Stetson Univ, BS, 60; Clemson Univ, MS, 63; Univ Ga, PhD(natural prod chem), 66. *Prof Exp:* Res chemist, Environ Protection Agency, 66-77; head, Org Chem Dept, 77-78, mgr, Anal Chem Div, 79-81, chem develop coordr, 81-84, SR PROG MGR, RADIAN CORP, 85- *Concurrent Pos:* Pres, KCP, 73-83; vchmn, Gordon Res Conf Environ Sci & Water, 73-; mem, Comt Mil Environ Res & Subcomt Indust Hyg, Nat Res Coun, 81; Chmn, Am Chem Soc Div Environ Chem, 79, Am Chem Soc Subcomt on Environ Monitoring & Anal, 81; adv bd, ES&T, 82-85, Environ Lab, 89-, Environ Protection, 90- *Honors & Awards:* Chemist of the Year, Am Chem Soc, 75; Distinguished Serv Award, Am Chem Soc Div Environ Chem, 86. *Mem:* Am Chem Soc; Sigma Xi; Am Soc Testing & Mat. *Res:* Chemical changes produced by pollution treatment; nuclear magnetic resonance of pesticides; mass spectrometry; identification of organic chemical pollutants; computerized GC-MS analysis of pollutants; industrial pollutants; electronic book publishing; artificial intelligence. *Mailing Add:* Radian Corp PO Box 201088 Austin TX 78720-1088

KEITH, LLOYD BURROWS, b Victoria, BC, Nov 29, 31; m 54; c 4. WILDLIFE MANAGEMENT. *Educ:* Univ Alta, BSc, 53, MSc, 55; Univ Wis, PhD(wildlife mgt), 59. *Prof Exp:* Asst forestry & wildlife mgt, 55-59, fel, 59-60, from instr to assoc prof, 60-70, PROF WILDLIFE ECOL, UNIV WIS-MADISON, 70- *Mem:* Wildlife Soc; Am Soc Mammal; Ecol Soc Am. *Res:* Natural regulation of animal populations; ten-year cycle of northern fur-bearers and grouse. *Mailing Add:* c/o Wildlife Ecol Univ Wis Madison WI 53706

KEITH, MACKENZIE LAWRENCE, b Edmonton, Alta, Oct 12, 12; nat US; m 40; c 5. GEOCHEMISTRY, GEOPHYSICS. *Educ:* Univ Alta, BSc, 34; Queen's Univ, Can, MSc, 36; Mass Inst Technol, PhD(geol), 39. *Prof Exp:* Field geologist, Ventures, Ltd, 37, Geol Surv Can, 38 & US Smelting, Ref & Mining Co, 39-40; asst prof geol, Queen's Univ, Can, 40-47; petrologist, Geophys Lab, Carnegie Inst, 47-50; prof geochem, 50-, EMER PROF GEOCHEM, PA STATE UNIV. *Concurrent Pos:* Field geologist, McIntyre Mines, Ont, 41, Aluminum Co Can, Montreal, 42-43 & Ont Dept Mines, Toronto, 45-47. *Mem:* Fel Geol Soc Am; Mineral Soc Am; Geochem Soc; Geol Asn Can. *Res:* Petrology of alkaline rocks; staining methods for silicate minerals; mineral deposits; silicate chemistry, including system $MgO-Cr_2O_3-SiO_2$; element distribution; geochemical prospecting; isotope ratios in limestone and fossils; geochemistry of sedimentary rocks; trace element and isotopic criteria for differentiating marine and fresh water sediments; global tectonics; evidence against plate tectonics. *Mailing Add:* Dept Geosci Pa State Univ 309 Deike Bldg University Park PA 16802

KEITH, PAULA MYERS, b Wheeling, WVa, Jan 13, 50; m 73. MICROBIOLOGY, BACTERIAL PHYSIOLOGY. *Educ:* W Liberty State Col, AB, 71; WVa Univ, MS, 73; Va Polytech Inst, PhD(microbiol), 78. *Prof Exp:* Clin microbiologist, Pub Health Sect, WVa State Ref Lab, 73-75; res asst bact, Va Polytech Inst, 75-76, res assoc, 75-77; res microbiologist, 79-82, MGR, MICROBIOL RES & DEVELOP, PITMAN MOORE, INC, 82- *Concurrent Pos:* Consult, WHO, 79; sr ed, J Ind Microbiol. *Mem:* Am Soc Microbiol; Sigma Xi; Soc Indust Microbiol (secy, 85-88). *Res:* Microbial control agents of insects; fermentation microbiology; bacterial physiology; industrial microbiology; animal growth promotants and vaccines; recombinant DNA; veterinary products; immunomodifiers; animal infectious disease; antibiotics and anti-microbiology products. *Mailing Add:* Microbiol Res & Develop Pitman Moore Inc PO Box 207 Terre Haute IN 47808

KEITH, ROBERT ALLEN, b Brea, Calif, Mar 16, 24; m 49; c 2. PSYCHOLOGY, MEDICAL REHABILITATION. *Educ:* Univ Calif, Los Angeles, BA, 48, MA, 51, PhD(psychol), 53, Dipl(clin psychol) Am Bd Prof Psychol. *Prof Exp:* From asst prof to assoc prof, 53-66; psychol consult, 55-67, DIR CTR REHAB RES & PLANNING, CASA COLINA HOSP, 68-; PROF, 66-89, PROF EMER PHYCHOL, CLAREMONT GRAD SCH, 89- *Concurrent Pos:* Res fel Dept Nutrit, Sch Pub Health, Harvard Univ, 60-61; vis scholar, Dept Child Develop, Univ London, 67-68; fel, Div Rehab Psychol, Am Psychol Asn, 84, Int Exchange Experts & Info in Rehab, World Rehab Fund, 87. *Mem:* Am Psychol Asn; Am Asn Univ Profs; Am Congr Rehab Med; Asn Health Servs Res; Am Pub Health Asn. *Res:* Treatment effectiveness for brain injury, strokes, spinal cord injury; organizational analysis of operations of the rehabilitation hospital; market research and strategic planning. *Mailing Add:* Ctr Rehab Res & Planning Casa Colina Hosp 2850 N Garey Ave Pomona CA 91767

KEITH, TERRY EUGENE CLARK, b Redlands, Calif, Jan 28, 40; m 66; c 2. HYDROTHERMAL ALTERATION, MINERALOGY. *Educ:* Univ Ariz, BS, 62; Univ Ore, MS, 64. *Prof Exp:* Assoc chief, BR Igneous & Geothermal Processes, 87-90, GEOLOGIST, US GEOL SURV, 64- *Mem:* Am Geophys Union; Geothermal Res Coun; Clay Mineral Soc. *Res:* Hydrothermal and fumarolic alteration mineralogy, primarily in Yellowstone National Park, the Pacific Northwest Cascade Range, and Alaskan volcanoes; field distribution and petrography of ultramafic rocks in the Yukon-Tanana Upland, Alaska. *Mailing Add:* Br Igneous & Geothermal Processes 345 Middlefield Rd MS910 Menlo Park CA 94025

KEITH, THEO GORDON, JR, b Cleveland, Ohio, July 2, 39; m 60; c 2. THERMAL SCIENCES, NUMERICAL ANALYSIS. *Educ:* Fen Col, BME, 64; Univ Md, MSME, 68, PhD(mech eng). *Prof Exp:* Mech engr, Naval Ship Res & Develop Ctr, Annapolis, Md, 64-71; PROF & CHMN MECH ENG, UNIV TOLEDO, 71- *Concurrent Pos:* Prin investr pumping ring seal grant, Lewis Res Ctr, NASA, 77-81 & wind energy grant, 79-, co-prin investr devicing grant, 80- *Mem:* Am Soc Mech Engrs; Am Soc Eng Educ; Am Inst Aeronaut & Astronaut; Soc Automotive Engrs; Sigma Xi. *Mailing Add:* 3866 Laplante Rd Monclova OH 43542

KEITH, WARREN GRAY, b Anamosa, Iowa, Sept 16, 08; m 37; c 2. CIVIL ENGINEERING, ENGINEERING EDUCATION. *Educ:* Iowa State Col, BS, 34; Univ Mo, MS, 48. *Prof Exp:* Asst county hwy engr, Lyon Co, Minn, 27-34; area engr admin & construct, Fed Emergency Relief Admin & Works Progress Admin, 34-37; county hwy engr, Chisago Co, 37-41; asst prof drawing, surv & structures, Univ Ala, 41-44; sr stress analyst aircraft structures, Goodyear Aircraft Corp, Ohio, 44-45; assoc prof structures, 45-51, prof civil eng, 51-66, head dept, 66-70, prof civil eng & structures, 70-72, dir eng technol progs, 72-74, EMER PROF CIVIL ENG, UNIV ALA, TUSCALOOSA, 74- *Mem:* Am Soc Civil Engrs; Nat Soc Prof Engrs. *Mailing Add:* 1611 27th Ave East Tuscaloosa AL 35404

KEITHLY, JANET SUE, b Jefferson City, Mo, Nov 29, 41; m 73; c 2. MICROBIOLOGY. *Educ:* Cent Mo State Univ, BSc, 63; Iowa State Univ, PhD(zool), 68. *Prof Exp:* Fel parasitol, Rutgers Univ, 68-70, Rockefeller Univ, 70-72; from asst prof to assoc prof biol, Herbert H Lehman Col, City Univ NY, 72-78; vis assoc prof microbiol, 79, res assoc med, 79-80, ASST PROF MICROBIOL MED, MED COL, CORNELL UNIV, 81- *Concurrent Pos:* Adj asst prof biochem cytol, Rockefeller Univ, 78-80; vis asst prof, Seattle Biomed Res Inst, 86-87. *Mem:* Am Soc Microbiol. *Res:* Chemotherapeutic strategies in treatment of leishmaniasis; drug mode of action against the human blood protozoa; factors influencing virulence of leishmania species; metabolic pathways of Leishmania as unique targets for chemotherapy; cloning the genes for and studying the expression of the rate-controlling enzymes in these pathways eg polyamine and trypanothione metabolism; ornithine decarboxylase and trypanothione reductase genes. *Mailing Add:* 48 Elm St Albany NY 12202

KEITT, GEORGE WANNAMAKER, JR, b Madison, Wis, Sept 11, 28; m 57; c 4. PLANT PHYSIOLOGY, FORENSIC SCIENCE. *Educ:* Harvard Univ, AB, 50; Univ Wis, MS, 52, PhD(bot), 57. *Prof Exp:* Res assoc, Ford Agr Plant Nutrit Proj, Mich, 57-59; asst prof bot, Fla State Univ, 59-67; sr fel, Mackinac Col, 67-70; res dept, Brooklyn Botanic Garden, 70-75, chmn, 70-74; PLANT PHYSIOLOGIST BIOL & ECON ANALYTICAL DIV, PESTICIDE PROG, ENVIRON PROTECTION AGENCY, 75- *Concurrent Pos:* Vis investr, Princeton Univ, 70. *Mem:* NY Acad Sci; Sigma Xi; Bot Soc Am; Am Inst Biol Sci; Scand Soc Plant Physiol; Am Soc Plant Physiol. *Res:* Chemical control of plant growth and differentiation. *Mailing Add:* Biol Analytical Br Bead H7503W EPA 401 M St SW Washington DC 20460

KEIZER, CLIFFORD RICHARD, b Hudsonville, Mich, Mar 19, 18; m 43; c 2. PHYSICAL CHEMISTRY. *Educ:* Hope Col, AB, 39; Univ Ill, MS, 41, PhD(phys chem), 43. *Prof Exp:* Jr res physicist, Monsanto Chem Co, 43-44; instr chem, Univ Ill, 44-46; from instr to asst prof chem, Western Reserve Univ, 46-48; prof chem & chmn Div Nat Sci, Cent Col Iowa, 48-57; prof chem, Ky Contract Team to Univ Indonesia, 57-62, from actg chief to chief, 58-62; prof chem, Lindenwood Col, 62-64; head dept, 64-70, actg dean col, 66-67 & 74-75, actg vpres acad affairs, 76-77, PROF CHEM, NMEX INST MINING & TECHNOL, 64- *Mem:* Am Chem Soc; Sigma Xi. *Res:* Electrochem. *Mailing Add:* 405 Col Ave Socorro NM 87801

KEIZER, EUGENE O(RVILLE), b LeMars, Iowa, Sept 13, 18; m 41; c 2. VIDEO SYSTEMS, COLOR TELEVISION. *Educ:* Iowa State Col, BS, 40. *Prof Exp:* Asst, Exp Sta, Iowa State Col, 37-40; res engr, RCA Labs, 40-64, head TV res group, Systs Res Lab, 64-67, head video systs res group, Consumer Electronics Res Lab, 67-77, head microtopographics res, 77-78, head video recording res, Commun Res Labs, 78-79, staff scientist RCA selecta-vision video disc oper, 79-80, STAFF SCIENTIST, VIDEO DISC SYSTS RES LAB, RCA CORP, 80- *Concurrent Pos:* Instr war training prog, Rutgers Univ, 41-45. *Honors & Awards:* David Sarnoff Award, RCA Corp, 77 & 81; Edvard Rhein Award, 80. *Mem:* Inst Elec & Electronics Engrs; Sigma Xi. *Res:* Television; radar; microwave; radio frequency receivers and radio frequency circuits; color television; information storage and retrieval; video disc systems. *Mailing Add:* 732 Kingston Rd Princeton NJ 08540

KEIZER, JOEL EDWARD, b North Bend, Ore, Aug 31, 42; m 64; c 2. PHYSICAL CHEMISTRY, STATISTICAL PHYSICS. *Educ:* Reed Col, BA, 64; Univ Ore, PhD(chem physics), 69. *Prof Exp:* Actg instr, Univ Calif, Santa Cruz, 69-71; from asst prof to assoc prof, 71-78, PROF CHEM, UNIV CALIF, DAVIS, 78-, DIR, INST THEORET DYNAMICS, 85- *Concurrent Pos:* Vis scientist, NIH, 78-79 & 86-87; assoc ed, Accounts Chem Res, 78-; J S Guggenheim mem fel, 86-87; Battelle Mem Inst fel, 69-71. *Mem:* AAAS; Biophys Soc. *Res:* Molecular origins and nature of macroscopic dynamic phenomena in chemical, physical and biological systems; nonlinear, nonequilibrium thermodynamics; fluctuations and stochastic processes. *Mailing Add:* Inst Theoret Dynamics Univ Calif Davis CA 95616

KELBER, JEFFRY ALAN, b Philadelphia, Pa, Dec 17, 52. SURFACE CHEMISTRY. *Educ:* Calif Inst Technol, BS, 75; Univ Ill, Urbana-Champaign, PHD(inorg chem), 79. *Prof Exp:* MEM TECH STAFF, SANDIA NAT LABS, 79- *Mem:* Am Vacuum Soc. *Res:* Photoelectron spectroscopy of organic systems; auger lineshape analysis; electron and photon stimulated desorption. *Mailing Add:* 9687 Asbury Lane Albuquerque NM 87114

KELCH, WALTER L, b Dayton, Ohio, Oct 27, 48; m 70; c 1. ENGINEERING, ASTROPHYSICS. *Educ:* Miami Univ, AB, 70; Ind Univ, MA, 73, PhD(astrophys), 75. *Prof Exp:* Instr astron, Kean Col, NJ, 75-76; res assoc, Joint Inst Lab Astrophys, Univ Colo, 76-78; ENGR & ANALYST, CENT INTEL AGENCY, LANGLEY, VA, 78- *Mem:* Am Astron Soc. *Res:* Spectral line formation in stellar atmospheres; solar and stellar atmosphere models; radiative transport. *Mailing Add:* 2103 Sugarloaf Ct Herndon VA 22070

KELCHNER, BURTON L(EWIS), b Bethlehem, Pa, Nov 15, 21; m 44; c 3. CHEMICAL ENGINEERING. *Educ:* Moravian Col, BS, 43; Va Polytech Inst, BS, 44. *Prof Exp:* Sect leader, Los Alamos Sci Lab, 46-51; supt dept, Rocky Flats Div, Dow Chem Co, 52-65; mfg tech mgr, 65-68, sr res engr, facilities eng, 68-70, proj mgr, 70-75; mgr nuclear waste processing, Rockwell Int Corp, 75-79, proj mgr, Long-Range Rocky Flats Utilization Study, 79-83; RETIRED. *Concurrent Pos:* Consult, plutonium processing. *Res:* Nuclear waste processing; uranium and plutonium processing. *Mailing Add:* 5357 S Cody St Littleton CO 80123

KELE, ROGER ALAN, b Waterbury, Conn, Jan 24, 43; m 72; c 2. INDUSTRIAL MICROBIOLOGY. *Educ:* Clark Univ, BA, 64; Harvard Univ, MA, 66; Univ Wis, PhD(bact), 70. *Prof Exp:* RES MICROBIOLOGIST, LEDERLE LABS DIV, AM CYANAMID CO, 70- *Honors & Awards:* Am Cyanamid Sci Achievement Award, 80. *Mem:* Am Soc Microbiol; Soc Indust Microbiol. *Res:* Strain improvement work on the tetracycline antibiotics. *Mailing Add:* Lederle Labs Pearl River NY 10965

KELEHER, J J, b Winnipeg, Man, Feb 9, 26; m 53; c 3. ENVIRONMENTAL SCIENCES. *Educ:* Univ Man, BA, 48; Univ Toronto, MA, 50. *Prof Exp:* Biologist, Fisheries Res Bd Can, 50-68; chief fisheries biologist, Man Dept Mines & Natural Resources, 68-69, chief, Fisheries Opers, 70-71, spec asst, 72-82, environ officer, Man Dept Consumer, Corp Affairs & Environ, 83-85; RETIRED. *Concurrent Pos:* Exec secy, Man Environ Coun, 73-83. *Mem:* Am Fisheries Soc; Am Inst Fishery Res Biol. *Res:* Environmental management. *Mailing Add:* Ten Baldry Bay Winnipeg MB R3T 3C4 Can

KELEMEN, CHARLES F, b Mt Vernon, NY, Jan 7, 43; m 75; c 2. MATHEMATICS. *Educ:* Valparaiso Univ, BA, 64; Pa State Univ, MA, 66, PhD(math), 69. *Prof Exp:* From asst prof to assoc prof math, Ithaca Col, 69-80; assoc prof, 80-82, prof comput sci, Lemoyne Col, 82-84; PROF MATH & COMPUT SCI DIR COMPUT SCI PROG PROG, SWARTHMORE COL, 84- *Concurrent Pos:* Res assoc, Dept Comput Sci, Cornell Univ, 75-76, vis assoc prof, 77-81; NSF grant, 77-81. *Mem:* Inst Elec & Electronics Engrs; Math Asn Am; Am Math Soc; Asn Comput Mach; Soc Indust & Appl Math; Sigma Xi. *Res:* Computational complexity; analysis of algorithms; computer science education. *Mailing Add:* 2105 N Providence Rd Media PA 19063

KELEMEN, DENIS GEORGE, b Budapest, Hungary, June 18, 25; nat US; m 51; c 1. PHYSICAL CHEMISTRY. *Educ:* Princeton Univ, PhD(chem), 51. *Prof Exp:* Ed asst tables of chem kinetics, Nat Res Coun, 48-50; res chemist, 50-57, res supvr, 57-68, res mgr electronic prod div, Electrochem Dept, 68-70, planning mgr, Photoprod Dept, 70-72, prod mgr, 72-78, develop mgr, Electronic Prod Div, Photoprod Dept, 78-80, prin consult, 80-84, sr consult, electronic mat div, E I du Pont de Nemours & Co, Inc, 84-87; INDEPENDENT CONSULT, TECHNOL APPRAISALS, 88- *Mem:* Am Chem Soc; Int Soc Hybrid Microelectronics; Inst Elec & Electronics Engrs. *Res:* Physical chemistry of solids. *Mailing Add:* PO Box 283 Lyme NH 03768

KELIHER, THOMAS FRANCIS, medicine; deceased, see previous edition for last biography

KELISKY, RICHARD PAUL, b St Louis, Mo, Nov 27, 29; m 60; c 1. MATHEMATICS, DATA PROCESSING. *Educ:* Tex Tech Col, BS, 51; Univ Tex, MA, 53, PhD(math), 57. *Prof Exp:* Asst appl math, Univ Tex, 52-55, lectr, 55-57, asst prof, 57-58; dir comput systs dept, 71-82, dir lab opers, 82-86, THOMAS J WATSON RES CTR, IBM CORP, 58-, RES MATHEMATICIAN DIR PLAN & QUAL, 86- *Concurrent Pos:* Adj prof, Grad Div, City Univ New York, 65-72. *Mem:* Am Math Soc; Math Asn Am; Asn Comput Mach. *Res:* Theory of numbers; numerical analysis; computing center management. *Mailing Add:* T J Watson Res Ctr IBM PO Box 218 Yorktown Heights NY 10598

KELKER, DOUGLAS, b Logan, Utah, Mar 23, 40; m 75; c 2. STATISTICS. *Educ:* Hiram Col, BA, 61; Univ Ore, MA, 63, PhD(math), 68. *Prof Exp:* Asst prof probability & statist, Mich State Univ, 68; asst prof math, Wash State Univ, 68-73; vis asst prof, 73-76, asst prof math, 76-81, ASSOC PROF STATIST, UNIV ALTA, 81- *Mem:* Inst Math Statist; Am Statist Asn; Can Statist Soc. *Res:* Characterization theorems; infinite divisibility; distributions on the unit sphere applied to geological data. *Mailing Add:* Dept Statist Appl Prob Univ Alta Edmonton AB T6G 2G1 Can

KELL, ROBERT M, b Piqua, Ohio, Nov 27, 22; m 49; c 3. EMULSION POLYMERIZATION, EXPERIMENTAL DESIGN. *Educ:* Ohio State Univ, BChE, 47, MSc, 48. *Prof Exp:* Jr chem engr, Olin Corp, 48-52; res chemist, Battelle Mem Inst, 52-62, sr res chemist, 62-68; res chemist, 68-79, SR RES ASSOC, FRANKLIN INT, 79- *Mem:* Am Chem Soc; Tech Asn Pulp & Paper Indust. *Res:* Adhesives; physical chemistry of polymers; plastics applications; vinyl polymerization. *Mailing Add:* Franklin Int 2020 Bruck St Columbus OH 43207-2329

KELLAND, DAVID ROSS, b East Orange, NJ, July 29, 35; m 56; c 3. PHYSICS. *Educ:* Montclair State Col, BA, 57, MA, 60; Salford Univ, PhD, 89. *Prof Exp:* Teacher high sch, NJ, 57-60; instr physics, Simmons Col, 61-63; asst prof, Emmanuel Col, Mass, 63-67; staff mem, 67-77, asst group leader, 77-78, co-group leader, 78-80, GROUP LEADER, FRANCIS BITTER NAT MAGNET LAB, MASS INST TECHNOL 80- *Concurrent Pos:* Prog mgr, NSF, 87- *Mem:* Am Phys Soc; Inst Elec & Electronics Engrs. *Res:* Applied magnetism and low temperature physics. *Mailing Add:* Mass Inst Technol NW14-3101 Cambridge MA 02139

KELLAR, KENNETH JON, b Baltimore, Md, Feb 13, 45; m 72; c 2. NEUROPHARMACOLOGY, MOLECULAR PHARMACOLOGY. *Educ:* Johns Hopkins Univ, BS, 66; Ohio State Univ, PhD(pharmacol), 74. *Prof Exp:* asst prof, 76-81, ASSOC PROF PHARMACOL, SCH MED, GEORGETOWN UNIV, 81- *Mem:* Soc Neurosci; Am Soc Pharmacol Exp Therapeut. *Res:* Regulation of neurotransmission. *Mailing Add:* Dept Pharmacol Georgetown Univ Sch Med 3900 Reservoir Rd NW Washington DC 20007

KELLAWAY, PETER, b Johannesburg, SAfrica, Oct 20, 20; nat US; m 58; c 5. NEUROPHYSIOLOGY. *Educ:* Occidental Col, BA, 41, MA, 42; McGill Univ, PhD(physiol), 47, Am Bd Clin Neurophysiol, 52. *Hon Degrees:* MD, Gothenburg Univ, Sweden, 77. *Prof Exp:* Demonstr physiol, McGill Univ, 44-46, lectr, 46-47, asst prof, 47-48; from assoc prof to prof physiol, 48-77, PROF NEUROL, BAYLOR COL MED, 77-, CHIEF, SECT NEUROPHYSIOL, 48-, PROF DIV NEUROSCI, 89- *Concurrent Pos:* Dir, Blue Bird Children's Epilepsy Clin, 49-60; dir dept electroencephalog, Methodist Hosp, 49-71, chief & sr attend, Neurophysiol Serv, 71-; consult, US Vet Admin Hosp, 49-75, Hermann Hosp, 55-73 & St Luke's Hosp, 71-; ed, EEG & Clin Neurophysiol, 58-; consult ed, 72-75; dir, EEG Lab, Ben Taub Gen Hosp, 65-; chief Neurophysiol Serv, Dept Med, Tex Children's Hosp, 72- & St Luke's Hosp, 73-; dir, Epilepsy Res Ctr, Baylor Col Med & Methodist Hosp, 75- *Honors & Awards:* Sir William Osler Medal, Am Asn Hist Med, 46; Lennox Lectr, Am Epilepsy Soc, 84; Merrit-Putnam Lectr, 85; First Distinguished Clin Investr Award, Am Epilepsy Soc, 89. *Mem:* Am Physiol Soc; Am Electroencephalog Soc (treas, 58-68, pres elect, 62-63, pres, 63-64); Soc Neurosci; Am Neurol Asn; Int League Against Epilepsy (secy-treas, 55-58); Am Epilepsy Soc (pres, 60); hon fel Am Acad Pediat; Am Acad Neurol; Child Neurol Soc. *Res:* Genesis and ontogenesis of electrical activity of the brain and of the epileptic process; epilepsy. *Mailing Add:* Dept Neurol Sect Neurophysiol Baylor Col Med Houston TX 77030

KELLEHER, DENNIS L, ENVIRONMENTAL PHYSIOLOGY. *Educ:* Univ Fla, PhD(physiol), 78. *Prof Exp:* ASST PROF, DEPT PHYSIOL, UNIFORMED SERV UNIV, BETHESDA, 85- *Mem:* Am Physiol Soc; Aerospace Med Asn; Sigma Xi. *Mailing Add:* Naval Health Res Ctr PO Box 85122 San Diego CA 92138-9174

KELLEHER, JAMES JOSEPH, b Hudson, Mass, Sept, 12, 38; m 63; c 4. MOLECULAR BIOLOGY, IMMUNOLOGY. *Educ:* Boston Col, BS, 60, MS, 63; Rutgers Univ, PhD(microbiol), 68. *Prof Exp:* Instr microbiol, Rutgers Univ, 66-67; res asst, Woods Hole Oceanog Inst, 67-68; from asst prof to assoc prof microbiol, 68-80, PROF MICROBIOL/IMMUNOL, SCH MED, UNIV NDAK, 80-, CHMN, 89- *Concurrent Pos:* Fel, Woods Hole Oceanog Inst, 67-68; consult diag virol, 72-, environ virol, 74-78. *Mem:* AAAS; Am Soc Microbiol; Sigma Xi; NY Acad Sci; Am Heart Asn. *Res:* Nutrition, viral infection and immune response; clinical diagnosis of viral infections; herpes virus latency in cell culture and animal model systems; virus transmission by the water route; virology; nucleic acid probes; immunological diagnosis. *Mailing Add:* Dept Microbiol & Immunol Univ NDak Grand Forks ND 58201

KELLEHER, MATTHEW D(ENNIS), b Flushing, NY, Feb 1, 39; m 69; c 2. MECHANICAL ENGINEERING, HEAT TRANSFER. *Educ:* Univ Notre Dame, BS, 61, MS, 63, PhD(mech eng), 66. *Prof Exp:* Asst prof mech eng, Univ Notre Dame, 65-66; Ford Found fel eng, Dartmouth Col, 66-67; from asst prof to assoc prof, 67-82, PROF MECH ENG, NAVAL POSTGRAD SCH, 82- *Concurrent Pos:* Consult, Apple Comput, 84 & Kaiser Engrs, 85; coord comt, Nat Heat Transfer Conf, 84-88; vis prof, Univ Notre Dame, 87; sr acad visitor, Oxford Univ, 88-89. *Mem:* Fel Am Soc Mech Engrs; Sigma Xi. *Res:* Heat transfer and fluid mechanics, specifically convection and radiation; heat pipes; electronics cooling. *Mailing Add:* Dept Mech Eng Naval Postgrad Sch Code ME Kk Monterey CA 93943

KELLEHER, PHILIP CONBOY, b New Rochelle, NY, July 23, 28; m 55; c 3. TUMOR MARKERS, GLYCOCONJUGATES. *Educ:* Georgetown Univ, BS, 50, MD, 54. *Prof Exp:* Resident physician, State Univ NY Upstate Med Ctr, 55-58; res fel biochem, Harvard Med Sch, 60-63; from instr to asst prof, 63-68, ASSOC PROF MED, COL MED, UNIV VT, 69- *Concurrent Pos:* Tutor, Harvard Med Sch, 62-63; clin fel med, Mass Gen Hosp, 61-63. *Mem:* AAAS; Am Fedn Clin Res; Int Soc NCo-develop Mental Biol & Med; Am Asn Cancer Res. *Res:* Serum protein metabolism; glycoconjugates; specific fetal serum proteins; carcinoembryonic antigens; collagen metabolism. *Mailing Add:* Dept Med Col Med Univ Vt Burlington VT 05405

KELLEHER, RAYMOND JOSEPH, JR, b Fall River, Mass, Sept 27, 39; m 64; c 3. GENETICS, BIOCHEMISTRY. *Educ:* Col Holy Cross, AB, 61; Boston Col, MS, 64; Univ NC, Chapel Hill, PhD(genetics), 69. *Prof Exp:* NIH fel genetics & biochem, 69-73, res assoc, 73-75, sr res assoc, Salk Inst Biol Studies, 75-76; res fel, Univ Calif, San Diego, 76-77; ASST PROF, STATE UNIV NY, BUFFALO, 77- *Mem:* AAAS; Genetics Soc Am. *Res:* Somatic cell genetics; molecular endocrinology; eukaryotic gene regulation. *Mailing Add:* 97 Jeanmoor Dr Amherst NY 14150

KELLEHER, ROGER THOMSON, b New Haven, Conn, Dec 28, 26; m 52; c 3. PHARMACOLOGY. *Educ:* Univ Conn, BA, 50; NY Univ, MA, 53, PhD(exp psychol), 55. *Prof Exp:* Asst psychol, NY Univ, 52-55; asst exp psychol, Yerkes Labs Primate Biol, 55-56, res assoc, 56-57; sr pharmacologist, Smith Kline & French Labs, NJ, 57-61; from asst prof to assoc prof pharmacol, 61-72, PROF PSYCHOBIOL, HARVARD MED SCH, 72- *Concurrent Pos:* Specific field ed, J Am Soc Pharmacol & Exp Therapeut. *Mem:* AAAS; Am Soc Pharmacol & Exp Therapeut; Sigma Xi. *Res:* Behavioral pharmacology; behavioral physiology; effects of drugs on cardiovascular regulation. *Mailing Add:* Lab Psychobiol Harvard Med Sch 25 Shattuck St Boston MA 02115

KELLEHER, WILLIAM JOSEPH, b Hartford, Conn, July 18, 29. BIOCHEMISTRY, PHARMACOGNOSY. *Educ:* Univ Conn, BS, 51, MS, 53; Univ Wis, PhD(biochem), 60. *Prof Exp:* Asst pharm, Univ Conn, 51-53; asst biochem, Univ Wis, 56-60; from asst prof to assoc prof, Sch Pharm, Univ 60-70, chmn, Med Chem & Pharmacog Sect, 71-76, asst dean, 76-81, prof pharmacog, 70-88; CONSULT, VICKS RES CTR, 85- *Concurrent Pos:* Mem, Nat Formulary Adv Panel Pharmacog, 64-71; guest prof, Univ Freiburg, 70-71 & 77-78; assoc ed, Lloydia, 71-76; vis scientist, Vicks Res Ctr, 84-85. *Mem:* Am Chem Soc; Am Soc Pharmacog; Brit Biochem Soc. *Res:* Microbial chemistry and the production and biosynthesis of alkaloids and other medicinal products by fermentation processes; formulation and dosage form development. *Mailing Add:* PO Box 205 Storrs CT 06268

KELLEMS, RODNEY E, MOLECULAR GENETICS. *Educ:* Princeton Univ, PhD(biochem), 75. *Prof Exp:* ASSOC PROF BIOCHEM, BAYLOR COL MED, 84- *Mem:* Am Soc Biol Chem; Am Chem Soc; Am Soc Cell Biol; Am Soc Human Genetics. *Mailing Add:* Dept Biochem Baylor Col Med Houston TX 77030-3498

KELLER, ARTHUR CHARLES, b New York, Aug 18, 01; m 28; c 1. ELECTRICAL ENGINEERING. *Educ:* Cooper Union, BS, 23, EE, 26; Yale Univ, MS, 25. *Prof Exp:* Lab asst, Bell Tel Labs, 17-23, mem tech staff, 23-42, spec apparatus engr, 42-46, switching apparatus engr, 46-49, dir switching apparatus develop, 49-55, 56-58, dir components, 55-56, dir switching systs develop, 58-61, dir switching apparatus lab, 61-66; RES & DEVELOP CONSULT, A C KELLER CO, 66- *Concurrent Pos:* Consult, Munitions Bd, 51-53, consult res & develop, Dept Defense, 55-61; mem bd dirs, Waukesha Motor Co, 63-; dir, Fifth Dimension, Inc, NJ, 67- *Honors & Awards:* US Navy Bur Ships & Bur Ord Awards; Emile Berliner Award, Audio Eng Soc, 62. *Mem:* Fel Inst Elec & Electronics Engrs; fel Acoust Soc; Am Soc Motion Picture & TV Engrs; Am Phys Soc. *Res:* Design, development and preparation for manufacture of electromechanical apparatus; sound recording; sonar; electrical measurements; telephone apparatus, radio frequency heating, telephone signaling and switching systems. *Mailing Add:* Res & Develop Consult A C Keller Consult 125 White Plains Rd Bronxville NY 10708

KELLER, BARRY LEE, b Chicago, Ill, Nov 15, 37; m 62; c 2. WILDLIFE ECOLOGY. *Educ:* Western Mich Univ, BA, 61, MA, 62; Ind Univ, PhD(ecol), 68. *Prof Exp:* Fel ecol, Ind Univ, 68-69; assoc prof biol, Keen State Col, 69-70; from asst prof to assoc prof, 70-80, PROF POP ECOL, IDAHO STATE UNIV, 80- *Concurrent Pos:* prin investr, 65 grants, 70-; cur mammals, Idaho Mus Natural Hist, 79-; ed, Tebiwa, The J of Idaho Mus Nat Hist, 84-; vis scholar, Utah Mus Nat Hist, 87-88. *Mem:* Am Soc Mammalogists; Brit Ecol Soc; Ecol Soc Am; Soc Pop Ecol; Wildlife Soc; Sigma Xi. *Res:* Wildlife ecology, with emphasis on population ecology of non-game species; ecology of desert mammals, with emphasis on small mammals residing on radioactive waste disposal sites; powerline corridor analyses. *Mailing Add:* Dept Biol Sci Idaho State Univ Pocatello ID 83201

KELLER, BERNARD GERARD, JR, b New Orleans, La, Dec 18, 36. PHARMACY. *Educ:* Loyola Univ, BS, 59; Univ Miss, MS, 64, PhD(pharm admin), 66. *Prof Exp:* Asst prof pharm & pharm admin, Southern Col Pharm, 65-67, assoc prof pharm admin, 67-69; asst dean clin progs, 72-81, PROF PHARMACEUT & CHMN DEPT, SCH PHARM, SOUTHWESTERN OKLA STATE UNIV, 69-, CHMN, DIV PHARMACEUT & PHARM ADMIN, 70-, DEAN, 81- *Mem:* Am Pharmaceut Asn; Am Soc Hosp Pharmacists; Am Col Apothecaries; Nat Asn Retail Druggists. *Res:* Pharmacy administration; motivation research; the pharmacist's relationship to the terminal patient; medical ethics. *Mailing Add:* 715 E Eureka Ave Weatherford OK 73096

KELLER, CHARLES A(LBERT), b Columbus, Ohio, June 30, 19; m 41; c 2. CHEMICAL ENGINEERING. *Educ:* Ohio State Univ, BChE, 41. *Prof Exp:* Analyst, E I du Pont de Nemours & Co, Va, 41; asst declassification officer, US AEC, 46-47, declassification officer & actg asst chief declassification br, 47-48, chief, 48-50, chem engr, Oak Ridge Prod Div, 50-52, asst chief opers div, Portsmouth Area, 52-55, dep dir, Prod Div, 55-57, dir, 57-75, asst mgr opers, 75-79, asst mgr mfg & support, 79-81, tech adv to mgr, Oak Ridge Opers Off, 81-83, consult energy & mgt, US Dept Energy, 84-88; consult, Seehuus & Hart Assoc, 88-90; SR ENG, RUSS ELMOS TECH ASSOCS, 90- *Mem:* Am Inst Chem Engrs; fel Am Inst Chem. *Res:* Absorption of volatile hydrocarbons from air mixtures; absorption and desorption of a volatile nontoxic, noninflammable organic solvent from an air-gas mixture. *Mailing Add:* 106 Norwood Lane Oak Ridge TN 37830

KELLER, D STEVEN, b Syracuse, NY, July 15, 58; m 81; c 1. FINE PARTICLE & COLLOID SCIENCE, SUSPENSION RHEOLOGY. *Educ:* Syracuse Univ, BS, 80. *Prof Exp:* Asst chemist, Champion Chem, 80-81, regional tech serv coordr, 81-82; assoc res chemist, Otisca Industs, Ltd, 82-90; RES SPECIALIST PAPER CHEM, COL ENVIRON SCI & FORESTRY, STATE UNIV NY, 90- *Concurrent Pos:* Prin investr, SBIR Res Proj, Dept Energy, 89-90. *Mem:* Am Chem Soc; Soc Rheology. *Res:* Investigation of surface properties of micron size precipitated and naturally occuring calcium carbonate compounds using inverse gas chromatography; kinetics of rheological instabilities in shear-thickening, non-Newtonian concentrated fine particle suspensions. *Mailing Add:* Col Environ Sci & Forestry State Univ NY Syracuse NY 13210

KELLER, DOLORES ELAINE, b New York, NY, Oct 29, 26; m 46; c 3. REPRODUCTIVE PHYSIOLOGY, MICROBIOLOGY. *Educ:* Long Island Univ, BS, 45; NY Univ, MA, 47, PhD(sex educ), 56; Univ Hawaii, cert, 64; Univ Calif, Berkeley, cert, 66. *Prof Exp:* Teacher biol & chmn dept, NY Pub Sch, 49-52; instr biol, French & lang & asst dean women, Long Island Univ, 52-56; from instr to assoc prof & chmn, Dept Sci, Fairleigh Dickinson Univ, 56-65; chmn dept biol, 65-68, PROF BIOL & DIR ALLIED HEALTH PROGS, PACE UNIV, 65-, DIR, NSF INSERV INST CELL PHYSIOL & GENETICS, WESTCHESTER CAMPUS, 66- *Concurrent Pos:* US deleg, Int Oceanog Conf, 59; NSF grants, 63-; res assoc, Haskins Labs, Carnegie Found, 63-; consult, Rennselaer Polytech Prog Intgerdisciplinary Col Sci, Charles Kettering Found, 64-; curriculum chmn, Bergen County Community Col, 64-; NSF-AEC grant marine & radiation biol, Univ Hawaii, 64-; res assoc, Lamont Geol Lab, Columbia Univ, 67-; NSF partic, Conf Primate Behav, Univ Calif, Davis, 71; spec consult, UN Comt Human Environ, 71-72; clin asst prof biol in psychiat, Dept Psychiat, Med Col, Cornell Univ, 74-; sr therapist, Payne Whitney Sexual Disorder Clin & pvt pract marriage counr sexual dysfunction, NJ, 74- *Mem:* Fel AAAS; Nat Sci Teachers Asn; Soc Protozool; Int Soc Clin & Exp Hypnosis; Am Asn Sex Educ Counrs. *Res:* Protozoology; fresh water and marine microbiology; science curriculum and education; sex education; human sexuality. *Mailing Add:* Dept Biol Pace Univ Pleasantville NY 10570

KELLER, DONALD V, b Centralia, Wash, Aug 17, 30; m 59; c 2. EXPERIMENTAL PHYSICS. *Educ:* Harvard Univ, AB, 52; Univ Calif, Berkeley, PhD(physics), 57. *Prof Exp:* Chief shock dynamics, Boeing Co, 57-62; chief tech exp physics, Northrop Corp, Calif, 62-66; mem tech staff, Defense Res Corp, 66-69; pres, Effects Technol, Inc, 69-71; PRES, KTECH CORP, 71- *Mem:* Am Phys Soc. *Res:* High energy nuclear physics; shock hydrodynamics; laser physics; dynamic mechanic and thermal properties of materials. *Mailing Add:* Ktech Corp 901 Pennsylvania NE Albuquerque NM 87110

KELLER, DOUGLAS VERN, JR, b Syracuse, NY, Feb 8, 28; m 53; c 4. PHYSICAL CHEMISTRY. *Educ:* Univ Buffalo, BA, 55; Syracuse Univ, PhD(chem), 58. *Prof Exp:* Asst prof metall, Mont Sch Mines, 58-59; from asst prof to assoc prof metall eng, 59-69, prof mat sci, Syracuse Univ, 69-78; VPRES TECHNOL, OTISCA INDUSTS LTD, 78- *Concurrent Pos:* Chmn bd dirs, Otisca Industs Ltd, NY, 73-; adj prof mat sci, Syracuse Univ, 78- *Mem:* Am Chem Soc; Am Phys Soc; Am Soc Metals. *Res:* Physical chemistry of surfaces; coal physical chemistry and fuels benification. *Mailing Add:* Otisca Industs Ltd 501 Butternut St Syracuse NY 13208

KELLER, EDWARD ANTHONY, b Los Angeles, Calif, June 6, 42; m 66; c 2. GEOMORPHOLOGY. *Educ:* Calif State Univ, Fresno, BS, 65, BA, 68; Univ Calif, Davis, MS, 69; Purdue Univ, PhD(geol), 73. *Prof Exp:* Asst prof geol, Calif State Univ, Fresno, 69-70; instr, Purdue Univ, 70-73, res asst,

71-73; asst prof, Univ NC, Charlotte, 73-76; PROF ENVIRON STUDIES & GEOL SCI, UNIV CALIF, SANTA BARBARA, 76- *Concurrent Pos:* Hartley vis prof, Southampton Univ, UK. *Mem:* Geol Soc Am; Sigma Xi. *Res:* Fluvial processes in geomorphology; environmental geology; tectonic geomorphology. *Mailing Add:* Dept Environ Studies & Geol Sci Univ Calif Santa Barbara CA 93106

KELLER, EDWARD CLARENCE, JR, b Freehold, NJ, Oct 8, 32; m 50; c 2. ECOLOGY, BIOSTATISTICS. *Educ:* Pa State Univ, BSc, 56, MSc, 59, PhD(genetics), 61. *Hon Degrees:* ScD, Salem Col, 78. *Prof Exp:* Asst genetics, Pa State Univ, 56-61; NIH trainee, Med Sch, Univ NC, 61-62, res assoc, 62, NIH fel, 62-64; asst prof zool, Univ Md, Col Park, 64-67; mgr biostatist, NUS Corp, 66-68; chmn dept, 69-74, PROF BIOL, WVA UNIV, 68- *Concurrent Pos:* Staff biologist, Comn Undergrad Educ Biol Sci, 65-66; pres, WVa Acad Sci, 75-76; vpres, Ecometrics Corp, 73-79; pres Found Sci & the Handicapped, 77; chmn, spec ed adv comt, Nat Sci Teachers Asn; expert, disabled affairs, EHR Directorate, NSF, 91-92. *Mem:* AAAS; Nat Sci Teachers Asn; Ecol Soc Am; Am Statist Asn; Am Inst Biol Sci; Found Sci & the Handicapped. *Res:* Aquatic ecology; quantitative inheritance of biochemical traits in Drosophila; vibration stress in organisms; ecosystem analysis and simulation; environmental influences on human health; handicapped in science; genetics. *Mailing Add:* Dept Biol WVa Univ Morgantown WV 26506-6057

KELLER, EDWARD LEE, b Glade Springs, Va, Nov 23, 41; m 68; c 2. APPLIED MATHEMATICS. *Educ:* Duke Univ, BS, 64; Univ Mich, Ann Arbor, MA, 66, PhD(math), 69. *Prof Exp:* Res asst, Univ Mich, Ann Arbor, 64-68; from asst prof to assoc prof, 69-80, PROF MATH, CALIF STATE UNIV, HAYWARD, 80-, DEPT CHAIR, 89- *Mem:* Am Math Soc; Math Asn Am; Soc Indust & Appl Math. *Res:* Mathematical programming, particularly quadratic programming; matrix theory; mathematics of population. *Mailing Add:* Dept Math & Computer Sci Calif State Univ Hayward CA 94542

KELLER, EDWARD LOWELL, b Rapid City, SDak, Mar 6, 39; m 65; c 3. BIOMEDICAL ENGINEERING, NEUROBIOLOGY. *Educ:* US Naval Acad, BS, 61; Johns Hopkins Univ, PhD(biomed eng), 71. *Prof Exp:* Asst prof, 71-76, assoc prof, 76-79, PROF ELEC ENG, UNIV CALIF, BERKELEY, 79- *Concurrent Pos:* Dir, Smith-Kettlewell Ctr for Vision Res & sr scientist, Smith-Kettlewell Inst Visual Sci, 80. *Mem:* AAAS; Asn Res Vision & Ophthal; Inst Elec & Electronic Engrs; Soc Neurosci. *Res:* Neurophysiological studies of the central organization of the primate oculomotor system; mathematical modelling of neuromuscular control systems. *Mailing Add:* Smith-Kettlewell Inst Visual Sci 2232 Webster St San Franciso CA 94115

KELLER, ELDON LEWIS, b Tiffin, Ohio, Dec 25, 34; m 61; c 1. NUCLEAR PHYSICS, RESEARCH ADMINISTRATION. *Educ:* Heidelberg Col, BS, 56; Univ Pittsburgh, MS, 60. *Prof Exp:* From assoc scientist to sr scientist, 60-69, res prog adminr, 69-74, ASST TO RES DIR, WESTINGHOUSE RES & DEVELOP CTR, 74- *Mem:* Am Phys Soc; Am Nuclear Soc. *Res:* Low-temperature radiation effects in superconductors; gamma-ray imaging using image intensifiers; semiconductor gamma-ray monitor; thickness gauging; gamma-ray spectrometry. *Mailing Add:* Westinghouse Res & Develop Ctr 1310 Beulah Rd Pittsburgh PA 15235

KELLER, ELIZABETH BEACH, b Diongloh, China, Dec 28, 17; US citizen; m 41. BIOCHEMISTRY. *Educ:* Univ Chicago, BS, 40; George Washington Univ, MS, 45; Cornell Univ, PhD(biochem), 48. *Prof Exp:* Asst, Med Col, Cornell Univ, 46-48; Atomic Energy Comn fel, Col Med, Ohio State Univ, 48-49; Huntington Mem Lab, Mass Gen Hosp, 49-50, res fel, Harvard Univ, 50-52, res assoc, 52-58, USPHS spec fel & res fel, 58-60; res assoc, Mass Inst Technol, 60-62; res specialist, 62-65, asst prof, 65-71, ASSOC PROF BIOCHEM, CORNELL UNIV, 71- *Mem:* Am Soc Biol Chem. *Res:* Mechanism of the biosynthesis and functions of nucleic acids. *Mailing Add:* Dept Biochem Molecular & Cell Biol Biotechnol Bldg Cornell Univ Ithaca NY 14853

KELLER, EVELYN FOX, mathematical biology, for more information see previous edition

KELLER, FREDERICK ALBERT, JR, b New York, NY; m 66. BIOCHEMISTRY, BIOCHEMICAL ENGINEERING. *Educ:* Stevens Inst Technol, BE, 61; Rutgers Univ, New Brunswick, MS, 67, PhD(microbial biochem & eng), 68. *Prof Exp:* Chemist polymerization develop lab, Hercules Inc, 61-62; USPHS fel, 68-70; biochem engr, Biol & Med Sci Lab, Gen Elec Co, 70-75; sr biochem engr, Union Carbide Corp, 75-79; sect leader, CPC Int, 79-87, Cambridge Biosci Corp, 87-90; sr consult, M G Pappas Co, 90-91; TASK LEADER & SR BIOCHEM ENGR, SOLAR ENERGY RES INST, 91- *Mem:* Am Inst Chem Engrs; Am Chem Soc; Am Soc Microbiol; Soc Indust Microbiol. *Res:* Biosynthesis, bioregulation and biodegradation of structural and storage macromolecules, including cellulo-lignins, starch, hemicellulose, chitin, glycogen; biochemical engineering processing; separation processes; regenerable raw materials; SCP; commercialization of chemicals by fermentation processes; biotechnology process development. *Mailing Add:* Solar Energy Res Inst BTRF 1617 Cole Blvd Golden CO 80401

KELLER, FREDERICK JACOB, b Huntington, WVa, May 10, 34; m 54; c 4. EXPERIMENTAL SOLID STATE PHYSICS. *Educ:* Marshall Univ, BS, 60; Univ Tenn, MS, 62, PhD(physics), 66. *Prof Exp:* Teaching asst physics, Univ Tenn, 63-64; physicist, Oak Ridge Nat Lab, 66; from asst prof to assoc prof, 66-77, PROF PHYSICS, CLEMSON UNIV, 77- *Mem:* Am Phys Soc. *Res:* Color centers in alkali halides. *Mailing Add:* Dept Physics & Astron Clemson Univ Main Campus Clemson SC 29634

KELLER, GEOFFREY, b New York, NY, June 12, 18; m 50; c 2. ASTRONOMY. *Educ:* Swarthmore Col, BS, 38; Columbia Univ, PhD(astron), 48. *Prof Exp:* Asst physics, Columbia Univ, 38-41; assoc physicist, Bur Ord, USN, 41-45; from instr to prof physics & astron, Ohio State Univ, 48-59, dir Perkins Observ, 53-59; prog dir astron, NSF, 59-61, div dir math & phys sci, 61-66, dep planning dir, 66-68; dean col math & phys sci, Ohio State Univ, 68-71, prof astron, 72-89. *Concurrent Pos:* Instr, Ohio Wesleyan Univ, 48-49. *Mem:* Am Astron Soc. *Res:* Internal constitution of stars; fluid turbulence. *Mailing Add:* 410 Higate Ave Columbus OH 43085

KELLER, GEORGE E, II, b Charleston, WVa, June 4, 33. ENGINEERING ADMINISTRATION, CHEMICAL ENGINEERING. *Educ:* Va Polytech Inst, BS, 55; Penn State Univ, MS, 58, PhD(chem eng), 64. *Prof Exp:* Res assoc, Res & Develop, 76-81, corp res fel, 81-87, SR CORP RES FEL, UNION CARBIDE CORP, 87- *Concurrent Pos:* Adj assoc prof, reaction eng, WVa Col Grad Studies, 63-; lectr, Am Chem Inst, 85; chair, Gordon Res Conf Separation & Purification, 85, Eng Found Int Conf Separation Technol, 91. *Mem:* Nat Acad Eng; Am Inst Chem Engrs; Am Chem Soc. *Res:* Advanced technique hydrocarbon cracking; separation, process research and development; development of medical oxygen technology. *Mailing Add:* Union Carbide Corp PO Box 8361 Charleston WV 25303

KELLER, GEORGE EARL, b Baton Rouge, La, Nov 6, 40; m 64. NUCLEAR PHYSICS. *Educ:* La State Univ, BS, 62, PhD(physics), 69. *Prof Exp:* ASST PROF PHYSICS, W GA COL, 69- *Mem:* Am Phys Soc; Am Asn Physics Teachers; Am Inst Physics; Sigma Xi. *Res:* Gamma ray spectroscopy; determination of the properties of the excited states of the doubly even deformed nuclei. *Mailing Add:* 3510 Garrett Ct Aberdeen MD 21001

KELLER, GEORGE H, b Hartford, Conn, Sept 9, 31; m 55; c 2. MARINE GEOLOGY. *Educ:* Univ Conn, AB, 54; Univ Utah, MS, 56; Univ Ill, PhD(marine geol), 66. *Prof Exp:* Geologist, Stand Oil Co Tex, 57-59; geol oceanogr, US Naval Oceanog Off, DC, 59-67; res oceanogr, Inst Oceanog, Md, 67-69, res oceanogr, Atlantic & Meteorol Oceanog Labs, Nat Oceanic & Atmospheric Admin, Fla, 69-75; assoc dean, sch oceanog, 75-81, dean res, 81-85, VPRES RES & GRAD STUDIES, ORE STATE UNIV, 85- *Concurrent Pos:* Mem Mid-Atlantic Ridge explor, Nat Oceanic & Atmospheric Admin & others, 74-77. *Mem:* Geol Soc Am; Int Asn Sedimentol; Am Geophys Union. *Res:* Marine geology and oceanography of the Malacca Strait, Malaysia; marine geotechnique, study of the mass physical and engineering properties of deep sea sediments and bottom material stability. *Mailing Add:* Res Off Ore State Univ Corvallis OR 97331

KELLER, GEORGE RANDY, JR, b Muskogee, Okla, Apr 17, 46; m 67; c 1. SEISMOLOGY. *Educ:* Tex Tech Univ, BS, 68, MS, 69, PhD(geophys), 73. *Prof Exp:* Instr geophys, Tex Tech Univ, 70-71; res asst prof, Univ Utah, 72-73; asst prof, Univ Ky, 73-76; from asst prof to assoc prof, 76-82, PROF GEOPHYS, UNIV TEX, EL PASO, 82-, CHMN, 81- *Mem:* Am Geophys Union; Am Asn Petrol Geologists; Geol Soc Am; Seismol Soc Am; Soc Explor Geophysicist; Royal Astron Soc. *Res:* Solid earth geophysics (seismology, gravity and geomagnetism); specifically the crustal structure, tectonics of North America and extensional terranes. *Mailing Add:* Dept Geol Sci Univ Tex El Paso TX 79968

KELLER, HAROLD WILLARD, b Newton, Kans, Dec 10, 37; m 65; c 2. MYCOLOGY. *Educ:* Kans Wesleyan Col, BA, 60; Univ Kans, MA, 63; Univ Iowa, PhD(bot), 71. *Prof Exp:* Fel bot, Grad Sch, Univ Fla, 71-72; asst prof biol, Wright State Univ, Ohio, 72-78, from asst dir to assoc dir, Univ Res Serv, 78-82, adj assoc prof, Dept Microbiol & Immunol, 80-82; dir res & assoc prof biol, Univ NC, Wilmington, 82-83; DIR OFF SPONSORED PROJS & ASSOC PROF BIOL, UNIV TEX, ARLINGTON, 83- *Concurrent Pos:* Conf partic, Comn Undergrad Educ Biol Sci, 70; NSF fel, Summer Inst Systematics V, 71; Ohio Biol Surv, 74-75; NSF grants, 75-78 & 83-84; panel reviewer, Comprehensive Assistance Undergrad Sci Educ, NSF, 81 & Instrnl Sci Equip Prog, 80; gov's appointment, State NC Marine Res Ctr Admin Bd, 82-83; chmn planning team, Southeastern NC Regional Forum Sci & Technol, 83; mem, NC Bd Sci & Technol, 83; field reader & reviewer G-pop proposals, Dept Educ, 85. *Mem:* Asn Southeastern Biologists; Mycol Soc Am; Sigma Xi; NAm Mycol Soc; Bot Soc Am; Nat Coun Univ Res Adminr; Soc Res Admin. *Res:* Taxonomic, monographic and floristic studies of the corticolous myxomycetes; myxomycetes of Central America and Mexico. *Mailing Add:* Off Sponsored Projs Univ Tex Arlington PO Box 19145 Arlington TX 76019

KELLER, HERBERT BISHOP, b Paterson, NJ, June 19, 25; m 53; c 2. APPLIED MATHEMATICS, NUMERICAL ANALYSIS. *Educ:* Ga Inst Technol, BEE, 45; NY Univ, MA, 48, PhD(math), 54. *Prof Exp:* Instr physics & math, Ga Inst Technol, 46-47; res scientist, Div Electromagnetic Res, Inst Math Sci, NY Univ, 48-53, lectr math, Wash Sq Col, 57-59, assoc prof, Univ, 59-61, prof appl math, Courant Inst Math Sci, 61-67, assoc dir, AEC Comput & Appl Math Ctr, 64-67; PROF APPL MATH, CALIF INST TECHNOL, 67- *Concurrent Pos:* Head dept math, Sarah Lawrence Col, 51-53; assoc ed, J Appl Math, Soc Indust & Appl Math, 61-66, ed, J Numerical Anal, 64-71; ed, Monogr Ser, Asn Comput Mach, 63-65, Numerical Math, 81-; vis prof, Calif Inst Technol, 65-66; mem math div, Nat Res Coun, 69-72; mem coun, Conf Bd Math Sci, 71-73; assoc ed, J Comput & Systs Sci, 71-74, ed, 74-; consult, var indust & govt concerns; Guggenheim fel, 79-80; assoc ed, Japan J Appl Math, 84- *Mem:* Soc Indust & Appl Math (pres, 75-76); Asn Comput Mach; Am Math Soc; Math Asn Am; fel Am Acad Arts & Sci. *Res:* Numerical analysis; fluid mechanics; nuclear and chemical reactors; applied mechanics; computing machinery; bifurcation theory. *Mailing Add:* Dept Appl Math Calif Inst Technol Pasadena CA 91125

KELLER, JACK, b Roanoke, Va, Jan 5, 28; m 54; c 3. AGRICULTURAL & IRRIGATION ENGINEERING. *Educ:* Univ Colo, BS, 53; Colo State Univ, MS, 55; Utah State Univ, PhD(irrigation eng), 67. *Prof Exp:* PROF AGR & IRRIGATION ENG, UTAH STATE UNIV, 60-; CHIEF EXEC OFFICER, KELLER-BLIESNER ENG, LOGAN, UTAH, 62- *Concurrent Pos:* Sales engr, South Irrigation Co, Miss, 55-56; asst irrigation engr, indust sales mgr & eng coordr, WR Ames Co, Denver, Colo & San Jose, Calif, 56-60; chmn sprinkler irrigation comt, Soil & Water Br, Am Asn Agr Eng, 66-72, mem, 72-, mem comt consult eng, 76-; co-dir water mgt synthesis proj, Utah State Univ, 78-, head dept agr & irrigation eng, 79-85; mem tech control bd, Jordan Valley Irrigation Proj, Jordan, 78-80; mem comt water resources res, Water Sci Technol Bd, Nat Res Coun, 84-86; assoc ed, Irrigation & Drainage Systs J, Hague, Neth, 84-; vis prof, fac agr sci & fac appl sic, KU Leuven, Belg, 86, 90 & 91; mem Int Comn Irrigation & Drainage. *Mem:* Nat Acad Eng; Am Soc Agr Engrs; Am Soc Civil Engrs; Am Soc Eng Educ; AAAS; Irrigation Asn; Sigma Xi. *Res:* Sprinkle and trickle irrigation; socio-technical assistance for transferring irrigation technologies worldwide; improving irrigated agriculture in developing countries; author of 81 technical papers, 15 articles and 6 books. *Mailing Add:* Agr & Irrigation Eng Dept Utah State Univ Logan UT 84322-4105

KELLER, JAIME, b Mexico, DF, Mex, Nov 10, 36; m 67; c 3. MATERIALS SCIENCE, PHYSICS & CHEMISTRY. *Educ:* Nat Autonomous Univ Mex, Chem Eng, 59; Univ Bristol, UK, PhD(physics), 72. *Prof Exp:* Res engr chem physics, Indust Chem Pennsalt, 59-63; tech dir chem process, Der Macrochem, 63-69; res fel physics, Univ Bristol, UK, 70-71; PROF THEORET PHYSICS, NAT AUTONOMOUS UNIV MEX, 72-, HEAD DEPT THEORET CHEM, 74- *Concurrent Pos:* Fac, IBM Res Lab, San Jose, Calif, 72; res visitor, Fed Polytech Inst, Zurich, 73- & Univ Geneve; mem vd, Mex Nat Res Syst & fel, 84-; dir, Fac Study, Superior Cuauthlan, Nat Automous Univ Mex. *Honors & Awards:* Nat Prize Chem, Chem Soc Mex, 80; Nat Award Chem Sci, Mex Fed Govt, 82; Jose Gomez-Ibanez lectr, Wesleyan Univ, 82. *Mem:* Am Phys Soc; Chem Soc Mex; Mex Soc Physics; Ital Soc Physics; Mex Acad Sci Invest; Europ Acad Art & Sci. *Res:* Chemistry and physics of condensed matter especially of metals in the liquid, amorphous and crystalline state; fundamental theory behind chemistry and physics; foundations of quantum and elementary particles theory. *Mailing Add:* Div Postgrad Studies Chem Fac Nat Autonomous Univ Mex PO Box 70-528 Mexico Federal District 04510 Mexico

KELLER, JAMES LLOYD, b Kittanning, Pa, June 21, 18; m 41, 60; c 4. ALCOHOL FUELS, ALTERNATIVE FUELS. *Educ:* Pa State Col, BS, 39; Emory Univ, MS, 40; Univ Calif, Los Angeles, PhD(phys org chem), 48. *Prof Exp:* Chemist, Magnolia Petrol Co, 40 & Koppers Co, 41-42; lab dept head synthetic rubber, US Rubber Co, 42-45; sr res assoc, 48-80, STAFF CONSULT, SCI & TECHNOL DIV, UNION OIL CO, 80- *Mem:* Am Chem Soc; Soc Automotive Engrs; Am Petrol Inst. *Res:* Petroleum product development and application; alternative fuels evaluation. *Mailing Add:* 1425 Longview Dr Fullerton CA 92631-1001

KELLER, JEFFREY THOMAS, b Cincinnati, Ohio, Oct 17, 46; m 76; c 2. NEUROANATOMY, NEUROSURGICAL RESEARCH. *Educ:* Univ Cincinnati, BA, 69, MS, 72, PhD(anat), 75. *Prof Exp:* Asst biol, Univ Cincinnati, 69-71, asst anat, Col Med, 71-75; DIR NEUROANAT RES, MAYFIELD NEUROL INST, 75-; dir, Div Neurosurg Res, 84-89, DIR, DIV NEUROSURG EDUC, DEPT NEUROSURG, COL MED, UNIV CINCINNATI, 89- *Concurrent Pos:* Adj asst prof anat, Col Med, Univ Cincinnati, 75-79, post doctoral fel, 78, 79-80, adj assoc prof, 79-, res assoc prof neurosurg, 84-89, res prof neurosurg, 90- *Mem:* Sigma Xi; fel Am Heart Asn; Am Asn Anatomists; Soc Neurosci. *Res:* Post-operative cicatrix and the spinal dura; spinal dura repair; basal ganglia; trigeminal system; cephalgias; facial neuralgias; cranial nerves and their brainstem circuitry; applying state of the art neuroanatomical tract tracing techniques and immunocytochemistry to examine the neuronal circuitry of cranial nerves involved with cerebral vasculature including the dura mater and cephalgias. *Mailing Add:* James N Gamble Inst Med Res 2141 Auburn Ave Cincinnati OH 45219

KELLER, JOHN RANDALL, b Ogdensburg, NY, Dec 14, 25; m 60. MICROBIOLOGY. *Educ:* Cornell Univ, BS, 47, PhD, 52. *Prof Exp:* Asst prof plant path, Univ Md, 52-54; teacher pub schs, 55-59; res assoc phys chem, Clarkson Col Technol, 59-60; from asst prof to assoc prof microbiol, 60-64, ASSOC PROF BIOL, SETON HALL UNIV, 64- *Concurrent Pos:* Res assoc phys chem, Clarkson Tech, 59- *Honors & Awards:* Vaughn Award, Am Soc Hort Sci, 52. *Mem:* Mycol Soc; Am Soc Microbiol. *Res:* Microbiology; plasmids; cyclic adenosine monophosphate. *Mailing Add:* Dept Biol Seton Hall Univ South Orange NJ 07079

KELLER, JOSEPH BISHOP, b Paterson, NJ, July 31, 23; m 63; c 2. MATHEMATICS. *Educ:* NY Univ, BA, 43, MS, 46, PhD(math), 48. *Hon Degrees:* Dr Tech, Tech Univ, Copenhagen. *Prof Exp:* Instr physics, Princeton Univ, 43-44; asst, Div War Res, Columbia Univ, 44-45; mathematician, Inst Math & Mech, 45-52; asst, Wash Sq Col, 46-47; asst prof, NY Univ, 48-52, assoc res prof, 52-56, prof, 56-79, chmn dept, 67-73, dir, Div Electromagnetic Res, Courant Inst Math Sci, 66-79; PROF MATH, STANFORD UNIV, 79- *Concurrent Pos:* Lectr, Grad Sch, Stevens Inst Technol, 48; head, Math Br, Off Naval Res, 53-54; vis prof, Stanford Univ, 69-70 & 76-78; res assoc, Woods Hole Oceanog Inst, 69-; vis scholar, Calif Inst Technol, 73-74; consult, var indust & govt concerns. *Honors & Awards:* Gibbs Lectr, Am Math Soc; Von Karman Prize, Soc Ind Appl Math; Timoshenko Medal, Am Soc Mech Eng; Hedric lectr, Math Asn Am; Von Neuman lectr, Soc Ind Appl Math. *Mem:* Nat Acad Sci; Am Phys Soc; Am Math Soc; Soc Indust & Appl Math (vpres, 78-79); Am Acad Arts & Sci; Royal Soc London. *Res:* Applied mathematics; acoustics; electromagnetic theory; fluid dynamics; geometrical optics. *Mailing Add:* 820 Sonoma Terr Stanford CA 94305

KELLER, JOSEPH EDWARD, JR, b La Crosse, Wis, Mar 31, 36. APPLIED MECHANICS, MECHANICAL ENGINEERING. *Educ:* Swarthmore Col, BS, 58; Univ Kans, MS, 60, PhD(eng mech), 64. *Prof Exp:* Engr, 64-74, dep leader, Weapons Prog, 74, dep div leader, 74-78, DIV LEADER, LAWRENCE LIVERMORE NAT LAB, 78- *Mem:* Am Soc Mech Engrs; Sigma Xi; AAAS. *Res:* Development and implementation of numerical techniques. *Mailing Add:* 786 Mirador Ct Pleasanton CA 94566

KELLER, JOSEPH HERBERT, b Bristol, Va, Sept 25, 46; m 69; c 2. PHYSICAL CHEMISTRY. *Educ:* King Col, BS, 68; Univ Ill, Urbana, MS, 70, PhD(phys chem), 74. *Prof Exp:* Res assoc & NSF fel chem, Univ Tenn, Knoxville, 73-75; res assoc catalysis, Oxy-Catalyst, Inc, 75-77, sr res assoc, 77-80; mgr, catalyst dept, 80-85, MGR PLANT & DEPT CATALYSTS, MET-PRO CORP, 85- *Mem:* Am Chem Soc; Sigma Xi; Catalysis Club; Org Reactions Catalysis Soc. *Res:* Heterogeneous catalysis; kinetic isotope effects of hydrogen and carbon; vapor pressure isotope effects; surface and media effect on reaction rates; wastewater analysis and treatment; catalyst manufacturing; nitrogen-oxygen catalyst; woodstove catalyst; volatile organic compound and polycyclic organic matter catalyst abatement, testing and manufacture. *Mailing Add:* 522 N Maryland Ave West Chester PA 19380

KELLER, KENNETH F, b Louisville, Ky, July 3, 21; m 46; c 7. MICROBIOLOGY. *Educ:* Univ Louisville, BA, 43, MS, 57, PhD(microbiol), 65. *Prof Exp:* Inst, Univ Louisville, 61-66, from asst prof to assoc prof microbiol, Sch Med, 66-88, assoc prof path, 74-88. *Concurrent Pos:* Consult, Gen Elec Co, 60-63 & Stand Oil Co Ky, 63. *Mem:* Am Soc Microbiol. *Res:* Biological and antigenic properties of the inclusion conjunctivitis agent; adrenergic receptors of mouse adipose tissue; use of lectins in diagnostic microbiology. *Mailing Add:* 12812 W Port Rd Anchorage KY 40222

KELLER, KENNETH H(ARRISON), b New York, NY, Oct 19, 34; m 57, 81; c 4. SCIENCE POLICY. *Educ:* Columbia Univ, BA, 56, BS, 57; Johns Hopkins Univ, MSE, 63, PhD(chem eng), 64. *Prof Exp:* Engr, Div Reactor Develop, AEC, Washington, DC, 57-61; from asst prof to assoc prof chem eng, Univ Minn, 64-71, actg dean, Grad Sch, 74-75, head, Dept Chem Eng & Mat Sci, 78-80, vpres acad affairs, 80-84, pres, 85-88, PROF CHEM ENG, UNIV MINN, MINNEAPOLIS, 71- *Concurrent Pos:* NIH spec fel, 72-73; mem surg & bioeng study sect, NIH, 76-80; Sigma Xi nat lectr, 78-79; vis fel, Woodrow Wilson Sch Pub & Int Affairs, Princeton Univ, 88-90; Philip D Reed sr fel sci & technol, Coun Foreign Rels, NY, 90- *Mem:* Am Inst Chem Engrs; Am Soc Artificial Internal Organs. *Res:* Transport phenomena in biological systems; artificial internal organ development. *Mailing Add:* 24 Roper Rd Princeton NJ 08540

KELLER, LAURA R, CELL BIOLOGY, DEVELOPMENTAL BIOLOGY. *Educ:* Univ Va, PhD, 80. *Prof Exp:* ASST PROF BIOL, FLA STATE UNIV, 86- *Mem:* Am Soc Cell Biol; Soc Develop Biol. *Mailing Add:* Dept Biol Sci Fla State Univ Tallahassee FL 32306

KELLER, LELAND EDWARD, b Carnegie, Okla, Jan 21, 23; m 49; c 3. HISTORY OF MEDICINE, QUACKERY. *Educ:* Univ Wichita, BA, 50; Univ Kans, PhD(anat), 58. *Prof Exp:* From asst prof to prof anat & physiol, 57-74, prof, 74-87, EMER PROF BIOL, PITTSBURG STATE UNIV, 87- *Concurrent Pos:* Presents illustrated lectr-demonstrations, early med quack devices. *Mem:* Am Asn Hist Med. *Res:* Study of early medical quack and electrotherapeutic devices in private collection; histology; physiology; history and use of old medical quack devices. *Mailing Add:* 1205 Imperial Dr Pittsburg KS 66762

KELLER, MARGARET ANNE, b Boston, Mass, May 29, 47; m 71; c 2. INFECTIOUS DISEASES. *Educ:* Mass Inst Technol, SB, 68; Albert Einstein Col Med, MD, 72. *Prof Exp:* From intern to resident pediat, Med Ctr, Univ Calif, San Diego, 72-75, chief resident, 75, fel infectious dis, Dept Pediat, 75-76; fel immunol & infectious dis, Dept Pediat, 76-78, asst prof, 78-85, ASSOC PROF PEDIAT, HARBOR-UNIV CALIF LOS ANGELES MED CTR, 85- *Mem:* Fel Am Acad Pediat; Am Fedn Clin Res; Soc Pediat Res; Am Phys Soc; Am Soc Microbiol; fel Infectious Dis Soc Am; Am Asn Immunol. *Res:* Immunology of human breast milk; neonatal immunity; idiotypic networks. *Mailing Add:* 7018 Crest Rd Rancho Palos Verdes CA 90274

KELLER, MARION WILES, b Dayton, Ohio, June 21, 05; m 27, 54, 83; c 1. MATHEMATICS. *Educ:* Ohio Wesleyan Univ, AB, 26; Ind Univ, MA, 29, PhD(anal), 32. *Prof Exp:* Teacher, Ohio Pub Schs, 26-27; head dept math, Kans Wesleyan Univ, 27-28; asst, Ind Univ, 28-29, instr, 29-30, asst, 30-31; teacher, Ohio Pub Schs, 33-36; from instr to prof, 36-71, asst head dept, 60-71, EMER PROF MATH, PURDUE UNIV, 71- *Mem:* Nat Coun Teachers Math; Am Soc Eng Educ; Math Asn Am; Am Math Soc. *Res:* Student errors; placement; testing; angular velocity between the foci in Keplerian elliptic motion. *Mailing Add:* 2741 N Salisbury Apt 3416 Lafayette IN 47907

KELLER, MARTIN DAVID, b New York, NY, Apr 7, 23; m 53; c 3. EPIDEMIOLOGY. *Educ:* Yeshiva Univ, AB, 44; NY Univ, MS, 46, PhD(biol), 53; Cornell Univ, MD, 52; Columbia Univ, MPH, 58. *Prof Exp:* Intern pediat, New York Hosp-Cornell Med Ctr, 52-53; med resident internal med, Vet Admin Hosp, New York, 55-56; resident med serv, Columbia Univ, 57; actg dir chronic dis div, Ohio Dept Health, 57-58, dir res training, 58-60; dir clin serv, Beth Israel Hosp, Boston, Mass, 60-62; assoc prof prev med, 62-66, head div epidemiol & biomet, 66-67, ASST PROF MED, OHIO STATE UNIV, 62-, PROF PREV MED, 66-, HEAD DIV COMMUNITY HEALTH, 67- *Concurrent Pos:* Lectr, Harvard Med Sch, 60-62; consult, Ohio Dept Health, 64- *Mem:* Am Pub Health Asn; Am Col Prev Med; NY Acad Sci. *Res:* Environmental and host factors affecting distribution of human disease entities. *Mailing Add:* Dept Prev Med Ohio State Univ Med Ctr 410 W Tenth Ave Columbus OH 43210

KELLER, OSWALD LEWIN, b New York, NY, May 24, 30; m 53; c 4. PHYSICAL CHEMISTRY, RESEARCH ADMINISTRATION. *Educ:* Univ of the South, BS, 51; Mass Inst Technol, PhD(phys chem), 59. *Prof Exp:* USPHS res fel, 59-60; chemist, Oak Ridge Nat Lab, 60-67, dir, Transuranium Res Lab, 67-74 & 84-89 & Chem Div, 74-84; RETIRED. *Concurrent Pos:* Mem nuclear physics panel, Physics Surv Comt, Nat Acad Sci, 69-72. *Mem:* Fel AAAS; Am Chem Soc; Am Phys Soc; Sigma Xi. *Res:* Physical chemistry of proteins; chemistry of transuranium elements; molecular spectroscopy; preparation and characterization of compounds; heavy ion reactions; administration of nuclear research. *Mailing Add:* 1550 Terrell Mill Rd Apt 12I Marietta GA 30067

KELLER, PATRICIA J, b Detroit, Mich, Nov 16, 23. BIOCHEMISTRY. *Educ:* Univ Detroit, BS, 45; Wash Univ, PhD(biochem), 53. *Prof Exp:* Res assoc biochem, Sch Med, 55-56, instr, 56-57, res asst prof, 57-62, assoc prof, Sch Dent, 62-67, assoc dean, Grad Sch, 74-77, PROF ORAL BIOL, SCH DENT, UNIV WASH, 67-, CHMN ORAL BIOL, 79- *Concurrent Pos:* USPHS fels, Wash Univ, 53-54 & Univ Wash, 54-55; vis fel, Inst Marine Biochem, Aberdeen, Scotland, 78-79. *Mem:* AAAS; Am Soc Biol Chem; Am Soc Cell Biol; Am Chem Soc; Int Asn Dent Res. *Res:* Structure, function and biosynthesis of enzyme proteins. *Mailing Add:* Dept Oral Biol Univ Wash Sch Dent SB22 Seattle WA 98195

KELLER, PHILIP CHARLES, b San Francisco, Calif, Mar 10, 39; m 65. INORGANIC CHEMISTRY. *Educ:* Univ Calif, Berkeley, BA, 61; Ind Univ, PhD(boron chem), 66. *Prof Exp:* From asst prof to assoc prof, 66-75, PROF CHEM, UNIV ARIZ, 75- *Mem:* Am Chem Soc; Royal Soc Chem. *Res:* Boron hydride chemistry; chemistry of Group III elements. *Mailing Add:* Dept Chem Univ Ariz Tucson AZ 85721-0002

KELLER, PHILIP JOSEPH, b New Brunswick, NJ, Sept 21, 41; m 63; c 2. PHYSICAL CHEMISTRY. *Educ:* Temple Univ, AB, 64, PhD(phys chem), 70. *Prof Exp:* Sr anal chemist, Merck Sharp & Dohme Res Labs, Pa, 64-66; res chemist, E I Du Pont de Nemours & Co, 69-72, develop rep, 71-72, prod technologist, 73-74, mkt res specialist, 75, planning specialist, Indust Chems Dept, 75-76, purchasing agent, 76-79, sr purchasing agent, 79-82, regional mgr, 82-84, mgr imaging med & electronics purchasing, Mat & Logistics Dept, 84-89, MGR ENG PROCUREMENT, MAT LOGISTICS & SERV, E I DU PONT DE NEMOURS & CO. *Concurrent Pos:* Adj asst prof, Temple Univ, 69-70. *Mem:* Am Chem Soc. *Res:* Waste water chemistry; fused salts; electrochemistry; polyelectrolytes; surface chemistry. *Mailing Add:* E I du Pont de Nemours & Co Mat Logistics & Serv Louviers Bldg Rm 1160 Newark DE 19714-6090

KELLER, RAYMOND E, b Cape Girardeau, Mo, May 25, 45; c 2. DEVELOPMENTAL BIOLOGY, CELL BIOLOGY. *Educ:* Southeast Mo State Univ, BS, 67; Univ Ill, Urbana, MS, 69, PhD(develop), 75. *Prof Exp:* Assoc, Lab of Prof J P Trinkaus, Dept Biol, Yale Univ, 75-76, Am Cancer Soc fel, 76-77; vis scientist, Ind Univ, 77-80; ASSOC PROF, DEPT ZOOL, UNIV CALIF, BERKELEY, 80- *Mem:* Soc Cell Biol; Soc Zool. *Res:* Analysis of the mechanisms of metazoan morphogenetic cell movements. *Mailing Add:* Dept Zool Univ Calif Berkeley CA 94720

KELLER, REED THEODORE, b Aberdeen, SDak, May 26, 38; m 59; c 3. GASTROENTEROLOGY, INTERNAL MEDICINE. *Educ:* Univ NDak, BA, 60, BS, 61; Harvard Med Sch, MD, 63; cert, Am Bd Internal Med, 66 & 77; cert, Am Bd Gastroenterol, 75. *Prof Exp:* From intern to resident med, Univ Hosp, Cleveland, 63-66, chief resident, 67-68; asst prof, Med Sch, Case Western Reserve Univ, 70-73; PROF MED & CHMN DEPT, SCH MED, UNIV NDAK, 73-, CHIEF MED, MED CTR REHAB HOSP, 74- *Concurrent Pos:* Fel gastroenterol, Vet Admin Hosp, Cleveland, 66-67, Vet Admin grant, 72-73; grant, Univ NDak, 73-; vis prof med, Univ Guadalajara, 74; chief med, Fargo Vet Admin Hosp, 74-76, consult gastroenterol, 74-75 & 76-; mem, Nat Bd Med Examrs, 84-88; Mem, NDak State Bd of Med Examrs, 87- *Honors & Awards:* Physicians Recognition Award, AMA, 88; Laureate Award, Am Col Physicians. *Mem:* Am Soc Gastrointestinal Endoscopy; Soc Exp Biol & Med; AMA; fel Am Col Gastroenterol; fel Am Col Physicians. *Res:* Use of acrylic polymers to control gastrointestinal hemorrhage. *Mailing Add:* Dept Med Univ NDak Sch Med Grand Forks ND 58201

KELLER, RICHARD ALAN, b Pittsburgh, Pa, Nov 28, 34; m 56; c 3. ANALYTICAL CHEMISTRY. *Educ:* Allegheny Col, BS, 56; Univ Calif, Berkeley, PhD(phys chem), 61. *Prof Exp:* Asst prof chem, Univ Ore, 59-63; staff mem, Div Phys Chem, Nat Bur Standards, 63-76; STAFF MEM CHEM, LOS ALAMOS NAT LABS, 76- *Concurrent Pos:* fel, Los Alamos Nat Lab, 83- *Mem:* Am Chem Soc. *Res:* Laser induced chemistry; laser induced isotope enrichment; laser based analytical techniques. *Mailing Add:* CLS-2 MS G738 Los Alamos Nat Lab Los Alamos NM 87545

KELLER, ROBERT B, b Wichita, Kans, Nov 8, 24; m 52; c 2. MECHANICAL ENGINEERING. *Educ:* Univ Wichita, BS, 48; Univ Mich, MS, 51, PhD(mech eng), 62. *Prof Exp:* Res engr, N Am Aviation, Inc, 51-54; supvr turbine engines, Allison Div, Gen Motors Corp, 54-59; ASSOC PROF MECH ENG, UNIV MICH, ANN ARBOR, 62- *Concurrent Pos:* Consult, Ford Motor Co, Babcock & Wilcox, Clark Equip Co & Navistar Corp. *Mem:* AAAS; Am Soc Mech Engrs; Am Inst Aeronaut & Astronaut; Soc Automotive Engrs. *Res:* Rocket propulsion systems; gas turbine technology; fluid control systems; micro-processor real-time control. *Mailing Add:* Dept Mech Eng Univ Mich Main Campus Ann Arbor MI 48109-2125

KELLER, ROBERT ELLIS, b Marshalltown, Iowa, Jan 10, 23; m 46; c 2. ANALYTICAL INSTRUMENTATION, SPECTROSCOPY. *Educ:* Univ Iowa, BA, 47, MS, 49, PhD(anal chem), 51. *Prof Exp:* Res chemist anal chem, Smith, Kline & French Labs, 50-52; res chemist, Monsanto Co, 52-54, proj leader, 54-55, from group leader to sr res group leader, 55-67, sect mgr, 67-69, mgr appl sci, 69-82, mgr appl technol, 82-86; RETIRED. *Mem:* Am Chem Soc; Soc Appl Spectros. *Res:* Analytical and physical chemistry-separation; characterization and measurement of chemical species by chromatography, spectroscopy, general instrumental and chemical techniques; on line process; process monitoring and control, computerized data management. *Mailing Add:* 10142 Glenfield Terr St Louis MO 63126

KELLER, ROBERT H, HEMATOLOGY, IMMUNOLOGY. *Educ:* Temple Univ, MD, 70. *Prof Exp:* ASSOC PROF MED & BIOPHYS, MED COL WIS, VAMC, 82- *Mem:* Fel Am Col Physicians; Int Soc Hemat; Am Soc Hemat; Am Soc Immunol. *Mailing Add:* 5000 W National Ave Milwaukee WI 53293

KELLER, ROBERT MARION, computer science, electrical engineering, for more information see previous edition

KELLER, ROY ALAN, b Davenport, Iowa, Feb 5, 28; m 52; c 4. ANALYTICAL CHEMISTRY. *Educ:* Univ Ariz, BSc, 50, MS, 52; Univ Utah, PhD(chem), 57. *Prof Exp:* Instr chem, Univ Ariz, 51-53, from asst prof to assoc prof, 56-68; chmn dept, 68-74, PROF CHEM, STATE UNIV NY COL FREDONIA, 68-, CHMN DEPT, 87- *Concurrent Pos:* Petrol Res Fund Int Award, Free Univ Brussels, 62-63; ed, J Chromatog Sci. *Mem:* Am Chem Soc. *Res:* Chromatography. *Mailing Add:* Dept Chem State Univ NY Col Fredonia NY 14063

KELLER, ROY FRED, b Cape Girardeau, Mo, Apr 3, 27; m 49; c 2. MATHEMATICS, COMPUTER SCIENCE. *Educ:* Southeast Mo State Univ, BS, 50; Univ Mo, AM, 58, PhD(math), 62. *Prof Exp:* Instr math, Univ Mo-Columbia, 56-57, actg dir comput res ctr, 59-62, asst prof math & dir comput ctr, 62-67; assoc prof math & comput sci, Iowa State Univ, 67-71, prof, 71-81; PROF COMPUT SCI & CHMN DEPT, UNIV NEBR, 81- *Mem:* Asn Comput Mach; Inst Elec & Electronics Engrs. *Res:* Iterative methods for solving systems of equations; programming and programming languages. *Mailing Add:* Dept Comput Sci Univ Nebr Lincoln NE 68588-0115

KELLER, RUDOLF, b Winterthur, Switz, Dec 27, 33; US citizn. ELECTROMETALLURGY, ELECTROCHEMISTRY. *Educ:* Kantonsschule Winterthur, Matura, 52; Swiss Fed Inst Technol, Zürich, dipl, 56, DSc nat, 60. *Prof Exp:* Scientist, Stanford Res Inst, 61-63; mem tech staff, Rocketdyne Div, Rockwell Int, 63-70; group leader, Swiss Aluminium Ltd, 70-77, Argonne Nat Lab, 77-79; staff scientist, Alcoa Labs, 79-83; SOLE PROPRIETOR, EMEC CONSULT, 84- *Concurrent Pos:* Res assoc, Carnegie Mellon Univ, 88- *Mem:* Int Soc Electrochem (treas, 76-79); Electrochem Soc; Minerals, Metals & Mat Soc; Am Chem Soc; Space Studies Inst. *Res:* Research and consulting in electrometallurgy, electrochemistry and related areas; principal investigator on government funded research, and development efforts on topics such as electrochemical processes to utilize lunar resources, neodymium oxide electrolysis; cyanide formation in Hall-Heroult cells; heat exchange in microgravity. *Mailing Add:* RD 3 Roundtop Rd Export PA 15632

KELLER, SEYMOUR PAUL, b New York, NY, July 5, 22; m 49; c 4. SOLID STATE PHYSICS. *Educ:* Univ Chicago, BS, 47, MS, 48, PhD(chem, physics), 51. *Prof Exp:* Du Pont fel chem, Univ Wis, 51-52; res assoc, Columbia Univ, 52-53; staff mem, 53-63, mgr solid state physics & chem, 62-64, dir tech planning res, 64-66, dir phys sci dept, 66-72, CONSULT TO DIR RES, THOMAS J WATSON RES CTR, IBM CORP, 72- *Mem:* Fel Am Phys Soc. *Res:* Optical and electrical properties of dielectric and semiconducting solids; luminescent materials; paramagnetic resonance of solids; wave function calculations. *Mailing Add:* 29 Garey Dr Chappaqua NY 10514

KELLER, STANLEY E, b Medford, Mass, Sept 9, 21; m 45; c 4. DENTISTRY. *Educ:* Tufts Univ, DMD, 44; Univ Ala, MS, 62. *Prof Exp:* Instr dent, Sch Dent Med, Tufts Univ, 47-48; assoc prof dent, Sch Dent, 57-61, chmn div restorative & prosthetic dent, 61-62, chmn, Dept Oral Diag, 62-74, prof, 61-82, dir clins, 62-82, EMER PROF, SCH DENT, UNIV ALA, 82- *Concurrent Pos:* Consult, Vet Admin Hosp, Birmingham, 67- *Mem:* Am Asn Dent Schs; Int Asn Dent Res; Sigma Xi; Am Dent Asn; fel Am Col Dentists. *Res:* Studies of the affect and removal of dental plaque. *Mailing Add:* 4266 Hoffman Rd Mobile AL 36619

KELLER, STEPHEN JAY, b Philadelphia, Pa, July 30, 40; m 62; c 2. MOLECULAR BIOLOGY, BIOCHEMISTRY. *Educ:* Univ Pa, AB, 63; State Univ NY Stony Brook, PhD(biol), 70. *Prof Exp:* NSF fel, 69-71, asst prof, 69-75, ASSOC PROF BIOL, UNIV CINCINNATI, 75- *Concurrent Pos:* Am Cancer Soc grant, 69-71; NSF grants, 70-78; NIH grant, 75-77, EPA grant, 80-84; vis assoc prof microbiol, State Univ NY, Stony Brook, 78-79; consult biotechnol, Protatek Inc, Sperti Drug, MDH Labs, Arel Pharmaceuticals & Hy-Gene Inc, 84- *Mem:* Sigma Xi; Am Soc Virol; AAAS. *Res:* Molecular biology of development and differentiation of animal cell cultures and viruses. *Mailing Add:* Dept Biol Univ Cincinnati Cincinnati OH 45221

KELLER, TEDDY MONROE, b Parrottsville, Tenn, Nov 20, 44. POLYMER CHEMISTRY, ORGANIC CHEMISTRY. *Educ:* E Tenn State Univ, BS, 66; Univ SC, PhD(org chem), 72. *Prof Exp:* Nat Defense Educ Act fel, Univ SC, 66-69; fel, Univ Fla, 72-74; chief chemist leather, A C Lawrence Leather Co, 74-75; RES CHEMIST POLYMER CHEM, NAVAL RES LAB, 77. *Mem:* Am Chem Soc. *Res:* Monomer synthesis, polymerization, and unusual polymer properties such as exceptional thermal and oxidative stability of phthalonitrile resins and polyamides; electrical conductivity of infinite network, fully conjugated polymers; fluoropolymers. *Mailing Add:* Dept Navy Naval Res Lab Code 6124 Washington DC 20375

KELLER, THOMAS C S, b June 20, 50; m; c 1. CELL BIOLOGY. *Educ:* Univ Va, PhD, 81. *Prof Exp:* Asst prof biol, Wesleyan Univ, Conn, 84-86; ASST PROF BIOL SCI, FLA STATE UNIV, 86- *Mem:* NY Acad Sci; AAAS; Am Soc Cell Biol. *Res:* Cell and molecular biology of the cytoskeleton. *Mailing Add:* Fla State Univ B-157 Tallahassee FL 32306-3050

KELLER, THOMAS W, b 1949; m; c 3. COMPUTER SCIENCE. *Educ:* Univ Tex Austin, BS, 71, MA, 72, PhD(computer sci), 76. *Prof Exp:* Staff mem, Los Alamos Sci Lab Computer Div, Los Alamos, NMex, 76-79; vpres, Info Res Assoc, Austin, Tex, 79-84; res assoc, Computation Ctr, Univ Tex, Austin, 79-81, assoc dir appln support, 81-84; sr mem tech staff, Advan Computer Archit Prog, Microelectronics & Computer Technol Corp, Austin, 84-89; SR SCIENTIST, IBM, AIX PERFORMANCE, AUSTIN, 89- *Concurrent Pos:* Secy, Spec Interest Group Measurement & Eval, Asn Comput Mech, 81-85, vchair, 85-89, chair, 89-91. *Mem:* Asn Comput Mech; Inst Elec & Electronics Engrs. *Res:* Computer performance measurement, analysis and modeling; computer systems design; numerous technical publications. *Mailing Add:* IBM Advan Worksta Div 11400 Burnett Rd Austin TX 78758

KELLER, WALDO FRANK, b Hicksville, Ohio, Apr 13, 29; m 58; c 2. VETERINARY SURGERY. *Educ:* Ohio State Univ, DVM, 53; Mich State Univ, MS, 61; Am Col Vet Ophthal, dipl. *Prof Exp:* Instr, Mich State Univ, 53-55 & 57-61, univ clin res grant, 61-65, from asst prof to assoc prof vet surg & med, 61-70, assoc dean, 79-83 & 84-88, actg dean, 83-84, PROF VET SURG & MED, MICH STATE UNIV, 70-, CHMN DEPT SMALL ANIMAL SURG & MED, 68-, PROF OPHTHAL, 88- *Concurrent Pos:* Trainee, Div Ophthal, Sch Med, Stanford Univ, 65-66. *Mem:* Am Asn Vet Clinicians; Am Soc Vet Ophthal; Am Col Vet Ophthalmologists. *Res:* Veterinary ophthalmology; growth of cornea in tissue culture and pathology of eye tissues in evaluating surgical techniques. *Mailing Add:* Vet Clin Ctr Mich State Univ East Lansing MI 48824

KELLER, WALTER DAVID, b North Kansas City, Mo, Mar 13, 00; m 36; c 2. GEOLOGY, CERAMICS. *Educ:* Univ Mo, AB, 25, AM, 26, BS, 30, PhD(geol), 33; Harvard Univ, AM, 32. *Hon Degrees:* DEngr, Univ Mo, Rolla, 88. *Prof Exp:* Instr, Univ Mo, 26-29; ceramic technologist, A P Green Fire Brick Co, 29-31; from asst prof to prof, 32-70, chmn dept, 42-45, EMER PROF GEOL, UNIV MO, COLUMBIA, 70- *Concurrent Pos:* Chmn clay minerals comt, Nat Res Coun-Nat Acad Sci, 57-60; distinguished prof geol, Univ Mo; vis prof, Univ SFla, 70-73. *Honors & Awards:* Neil A Miner Award, Nat Asn Geol Teachers, 67; Hardinge Award, Am Inst Mining, Metall & Petrol Engrs, 79; William H Twenhofel Award, Soc Econ Paleontologists & Mineralogists. *Mem:* Fel AAAS; Am Ceramic Soc; fel Geol Soc Am; fel Mineral Soc Am; Soc Econ Paleont & Mineral. *Res:* Clay mineralogy; fire clay; sedimentary petrology; optical mineralogy; mineral and rock soil amendments; lunar sample research. *Mailing Add:* 305 Geol Bldg Univ of Mo Columbia MO 65211

KELLER, WAYNE HICKS, physical chemistry, for more information see previous edition

KELLER, WILLIAM EDWARD, b Cleveland, Ohio, Mar 11, 25; m 47; c 4. QUANTUM PHYSICS. *Educ:* Harvard Univ, AB, 45, AM, 47, PhD(chem), 48. *Prof Exp:* Assoc supvr mil sponsored res, Res Found, Ohio State Univ, 48-50; mem staff, 50-70, group leader cryogenics, 70-83, asst div leader physics, Los Alamos Sci Lab, 83-87; RETIRED. *Concurrent Pos:* Vis scientist, Univ Sussex, England, 67-68; author of one publication, 68; Nat Acad Sci/Nat Res Coun Panel to evaluate Nat Bur Standards, 70-76; chair, Appl Superconductivity Conf, 79-80; US/USSR comt superconducting power transmission, Pres Nixon's Sci Exchange Prog, 72-78. *Mem:* Am Chem Soc; fel Am Phys Soc; fel Am Inst Chemists; Sigma Xi. *Res:* Infrared spectroscopy; low temperature physics; liquid helium hydrodynamics; applications of superconductivity to electric power systems. *Mailing Add:* 1090 Old Taos Hwy Santa Fe NM 87501

KELLER, WILLIAM JOHN, b Meridian, Miss, Sept 26, 20; m 46; c 2. ORGANIC CHEMISTRY. *Educ:* Miss State Col, BS, 43; Mass Inst Technol, PhD(org chem), 51. *Prof Exp:* Chemist org res, E I du Pont de Nemours & Co, Inc, 51-79; RETIRED. *Mem:* Fel AAAS; Am Chem Soc. *Res:* Synthetic rubber; noble metal catalysis. *Mailing Add:* Rte 4 Box 2176 Branson MO 65616

KELLERHALS, GLEN E, b Vinton, Iowa, Sept 29, 45; m 73; c 2. PHYSICAL CHEMISTRY. *Educ:* Upper Iowa Col, BS, 67; Okla State Univ, PhD(chem), 74; Tulsa Univ, MBA, 80. *Prof Exp:* Res chemist, 74-77, group leader, 77-80, strategic planner, 80-81, SPECIAL PROJECTS ENGR, CITIES SERV CO, 81- *Mem:* Am Chem Soc; Soc Petrol Engrs; Am Petrol Inst. *Res:* Enhanced oil recovery. *Mailing Add:* 4304 Valley Dr Midland TX 79707

KELLERMAN, KARL F(REDERIC), b Washington, DC, May 11, 08; m 34; c 1. ENGINEERING. *Educ:* Cornell Univ, EE, 29. *Prof Exp:* Commun engr, NY Tel Co, 29-36; head electronics coord br, Bur Aeronaut, USN, Washington, DC, 42-46; engr, Aircraft Radio Corp, 46-47; exec dir comt guided missiles; Res & Develop Bd, 47-49; engr, Brush Develop Co, 49-53; pres, Educ Labs, Kellerman & Co, 53-55; asst vpres eng, Bendix Corp, 55-62; pres, Microwave Devices, Inc, 62-64; sci adv, USAF Systs Command, 64-73; pres, Low Country Guild, Inc, 73-78; PRES, KELLERMAN & ASSOCS, 78- *Mem:* AAAS; Inst Elec & Electronics Engrs; Sigma Xi. *Res:* Communication and navigation equipment; guided missiles; space systems; electronic measuring equipment. *Mailing Add:* 300 Woodhaven Dr Apt 1301 Hilton Head Island SC 29928

KELLERMAN, MARTIN, b New York, NY, Feb 11, 32; m 63; c 2. PHYSICAL CHEMISTRY. *Educ:* Polytech Inst Brooklyn, BS, 53; Univ Wash, PhD(chem), 66. *Prof Exp:* Anal chemist, Continental Baking Co, 58-61; NIH res traineeship, Univ Calif, San Diego, 66-68; ASSOC PROF CHEM, CALIF POLYTECH STATE UNIV, SAN LUIS OBISPO, 68- *Mem:* AAAS. *Res:* X-ray crystal structure analysis; structure of metal chelate compounds; circular dichroism studies on structure of molecules of biological interest. *Mailing Add:* Chem Dept Calif Polytech State Univ San Luis Obispo CA 93407

KELLERMANN, KENNETH IRWIN, b New York, NY, July 1, 37; wid; c 1. RADIO ASTRONOMY. *Educ:* Mass Inst Technol, SB, 59; Calif Inst Technol, PhD(physics & astron), 63. *Prof Exp:* Res scientist, Radiophys Lab, Commonwealth Sci & Indust Res Orgn, 63-65; from asst scientist to assoc scientist, 65-69, scientist, 69-77, asst dir, 77, SR SCIENTIST, NAT RADIO ASTRON OBSERV, 78- *Concurrent Pos:* NSF fel, 65-66; lectr, Leiden Univ, 67; res assoc, Calif Inst Technol, 69; adj prof, Univ Ariz, 70-73; dir, Max Planck Inst Radio Astron, 77-79; vpres, Comn 40, Int Radio Sci Union, 79-82, pres, 82-85; chmn, Comn J, US Nat Comt. *Honors & Awards:* Calif Inst Technol-Eastman Kodak Corp Eastman Kodak Prize, 63; Rumford Prize, Am Acad Arts & Sci, 70; Helen B Warner Prize, Am Astron Soc, 71; B A Gould Prize, Nat Acad Sci, 73. *Mem:* Nat Acad Sci; Am Astron Soc; Am Acad Arts & Sci; Int Astron Union; Int Radio Sci Union. *Res:* Extragalactic radio sources; galaxies; quasars; cosmology; instrumentation. *Mailing Add:* Nat Radio Astron Observ Edgemont Rd Charlottesville VA 22901

KELLERS, CHARLES FREDERICK, b Montclair, NJ, Sept 12, 30; m 58; c 2. PHYSICS. *Educ:* Swarthmore Col, BA, 53; Duke Univ, PhD(physics), 60. *Prof Exp:* Engr, Gen Elec Co, 53-55; res assoc, Duke Univ, 60; asst prof physics, Wells Col, 61-65; sr res assoc, Cornell Univ, 65-68; assoc prof, 68-70, chmn dept, 71-80, PROF PHYSICS, CALIF STATE UNIV, SAN BERNARDINO, 70- *Mem:* Sigma Xi. *Mailing Add:* 694 E 39th St San Bernardino CA 92404

KELLERSTRASS, ERNST JUNIOR, b Peoria, Ill, Jan 9, 33; m 54; c 5. CIVIL ENGINEERING, GEOPHYSICS. *Educ:* Bradley Univ, BSCE, 54; St Louis Univ, MS, 62; George Washington Univ, MS, 67. *Prof Exp:* Engr, US Army Corps Engrs, 54, US Air Force, 54-, aeronaut meteorologist, 55-57, analyst & forecaster, Weather Cent Japan, 57-60, asst staff meteorologist for environ eng, 62-64, geophysicist, Electronic Syst Div, 64-66, geophysicist VELA prog, 67-71, chief planning remote piloted vehicle syst prog off, Aeronaut Systs Div, 71-74, staff scientist, Advan Res Br, Foreign Technol Div, Wright-Patterson AFB, 74-79; Instr Univ Dayton, 79-80; civil engr, Sanitary Landfills, Bowse Morner Testing Lab, 80-81, CIVIL ENGR ENVIRON, SYSTECH CORP, 81- *Mem:* Am Meteorol Soc; Am Geophys Union; Am Soc Civil Eng; Sigma Xi. *Res:* Meteorology, forecasting and environmental engineering; management of research in meteorological sensors, environmental effects, electric systems survivability-vulnerability, seismological instrumentation-field experiments and aeronautical systems; research and development management; geotechnical engineering; hydrology; environmental engineering. *Mailing Add:* 2547 Sugarloaf Ct Xenia OH 45385

KELLETT, CLAUD MARVIN, b Memphis, Tenn, Sept 5, 28; m 48; c 4. RESEARCH ADMINISTRATION. *Educ:* Ga Inst Technol, BEE, 50; Purdue Univ, MS, 57. *Prof Exp:* Res test engr, Allison Div, Gen Motors Corp, 50-54; sr engr, Tex Instruments, Inc, 56-62; develop engr, Semiconductor Div, Raytheon Co, 62-63; prod eng supvr, Sperry Semiconductor Div, Sperry Rand Corp, 63-64; prod supvr transistor mfg, Crystalonics, Inc, 64-65; sr res physicist, Ion Physics Corp, 65-66; physicist, Electronics Res Ctr, NASA, 66-70; sr res scientist, Tyco Corp Technol Ctr, 71; prog mgt officer, NSF, 72-88; RETIRED. *Mem:* Inst Elec & Electronics Engrs. *Res:* Electrical measurements, especially Hall and photoelectromagnetic effects of semiconductor materials; modification of electrical properties of semiconductor materials by ion implantation. *Mailing Add:* 5203 Faraday Ct Fairfax VA 22032-2708

KELLEY, ALBERT J(OSEPH), b Boston, Mass, July 27, 24; wid; c 3. SYSTEMS ENGINEERING, TECHNICAL MANAGEMENT. *Educ:* USN Acad, BS, 45; Mass Inst Technol, BS, 48, ScD(aeronaut & electronics eng), 56. *Prof Exp:* Exp test pilot & proj dir, USN Air Test Ctr, Patuxent River, Md, 51-53; asst head, Air-to-Air Missile Br, Navy Bur Aeronaut, Washington, DC, 56-57 & Guided Missile Guidance Br, 57-58, proj mgr, Eagle Missile Syst, 58-60; proj mgr, Navy Bur Weapons, 60; prog mgr, Agena Launch Vehicle, NASA, Washington, DC, 60-61, dir electronics & control, 61-64, dep dir, Electronics Res Ctr, Cambridge, 64-67, consult, 67-70,; dean, Sch Mgt, Boston Col, 67-77; pres, Arthur D Little Prog Systs Mgt Co, 77-85, chmn, 85-88, sr group vpres, Arthur D Little Inc, 85-88; sr vpres strategic planning, United Technol Corp, 88-90; DEP UNDERSECY DEFENSE, INT PROGS, US DEPT DEFENSE, PENTAGON, 90- *Concurrent Pos:* Chmn bd econ adv, Mass, 70-75; consult, Dept Transp, 71-77; mem bd vis, US Defense Systs Mgt Col, 74-78; dir, State St Bank & Trust Co, State St Boston Corp, 75-; mem, C S Draper Lab Corp, 75-; mem, Space Appln Bd, Nat Acad Eng, 77-83. *Honors & Awards:* Except Serv Medal, NASA, 67. *Mem:* Assoc fel Am Inst Aeronaut & Astronaut; fel Inst Elec & Electronics Engrs; Int Acad Astronaut; Sigma Xi. *Res:* Guided missiles and space vehicles; aircraft flight test; control systems engineering; electronics; strategic planning and management; project management. *Mailing Add:* Dep Undersecy Defense/Int Progs The Pentagon Rm 3E1082 Washington DC 20301-3070

KELLEY, ALEC ERVIN, b Pharr, Tex, Oct 28, 23; m 70. ORGANIC CHEMISTRY. *Educ:* Univ Tex, BS, 44; Purdue Univ, MS, 49, PhD, 56. *Prof Exp:* Asst chem, Univ Tex, 43-44; asst, Metall Lab, Univ Chicago, 44-45, jr chemist, 45-46; asst chem, Purdue Univ, 47-49, res fel, 49-52; from instr to prof, 52-83, asst dept head, 68-83, EMER PROF CHEM, UNIV ARIZ, 84 - *Mem:* Am Chem Soc; Royal Soc Chem. *Res:* Organic fluorine compounds; diazonium salts; kinetics and mechanisms of organic reactions. *Mailing Add:* 3738 E Garden Tucson AZ 85713-2454

KELLEY, ALLEN FREDERICK, JR, b Franklin, NH, July 1, 33. MATHEMATICS, FORESTRY. *Educ:* Mont State Univ, BS, 55; Univ Calif, Berkeley, PhD(math), 63. *Prof Exp:* Instr math, Univ Calif, Berkeley, 63-64; partic, Exchange Prog, US Nat Acad Sci-USSR Acad Sci, 64-65; mem fac, Inst Advan Study, 65-66; asst prof, 66-69, ASSOC PROF MATH, UNIV CALIF, SANTA CRUZ, 69- *Mem:* Soc Am Foresters; Am Math Soc. *Res:* Differential equations and celestial mechanics; wood technology and engineering. *Mailing Add:* Dept Math Univ Calif Santa Cruz CA 95064

KELLEY, ARNOLD E, chemical engineering, for more information see previous edition

KELLEY, CHARLES JOSEPH, b Akron, Ohio, Feb 2, 43; c 2. ORGANIC CHEMISTRY. *Educ:* St Joseph's Col, Ind, BA, 64; Ind Univ, Bloomington, PhD(org chem), 70. *Prof Exp:* Res assoc org synthesis, Ind Univ, 70-75; asst prof chem, Ball State Univ, 75-76; res assoc natural prod, Northeastern Univ, 76-77; ASST PROF CHEM, MASS COL PHARM, 77- *Mem:* Am Chem Soc; Am Soc Pharmacog. *Res:* Isolation from plant sources of potential pharmaceutical agents; laser dye and scintillator synthesis. *Mailing Add:* Mass Col Pharm 179 Longwood Ave Boston MA 02115

KELLEY, CHARLES THOMAS, JR, b Boston, Mass, Feb 9, 40; m 64; c 3. SYSTEMS ANALYSIS, PHYSICS. *Educ:* Univ Notre Dame, BS, 61; Univ Mass, MS, 63; Ind Univ, PhD(nuclear physics), 67. *Prof Exp:* Res asst nuclear physics, Cyclotron Lab, Ind Univ, 65-67; physicist, Anal Serv Inc, Falls Church, Va, 67-71; analyst, Washington Defense Res Div, Washington, DC, 71-73, phys scientist, 73-77, dir, Ground Warfare Prog, 77-79, SR PHYS SCIENTIST, RAND CORP, 79- *Mem:* Am Phys Soc; Opers Res Soc Am; Sigma Xi. *Res:* Weapon systems analysis; operations research; nuclear physics. *Mailing Add:* 909 Glenhaven Dr Pacific Palisades CA 90272

KELLEY, DARSHAN SINGH, b Ludhiana, Punjab, India, Feb 5, 47; US citizen; m 80; c 2. DIET & IMMUNO-COMPETENCE, NUTRIENT REQUIREMENTS. *Educ:* Punjab Agr Univ Ludhiana, India, BSc, 67, MSc, 69 ; Okla Univ, PhD(biochem), 74. *Prof Exp:* Consult, res, Okla Med Res Found, 74-75; postdoctoral res assoc, McArdle Lab, Univ Wis, 75-80; asst prof, biochem, WVa Univ Med Ctr, 80-83; RES CHEMIST, WESTERN HUMAN NUTRIT RES CTR, AGR RES SERV, USDA, 83- *Mem:* Am Instit Nutrit; Am Soc Biochem & Molecular Biol; AAAS; Sigma Xi. *Res:* Nutritional regulation of immune-status in humans and animals; nutritional and hormonal regulation of hepatic gene expression. *Mailing Add:* Western Human Nutrit Res Ctr ARS/USDA PO Box 29997 Presidio San Francisco CA 94129

KELLEY, DONALD CLIFFORD, b St John, Kans, June 23, 13; m 36; c 2. VETERINARY MEDICINE. *Educ:* Kans State Univ, DVM, 35, MS, 52; Am Bd Vet Pub Health, dipl. *Prof Exp:* Assoc prof vet pub health, Col Vet Med, Kans State Univ 58-69, prof infectious dis, 69-78, emer prof, 78; RETIRED. *Concurrent Pos:* Secy-treas, Am Bd Vet Pub Health, 63-66, pres, 66-; fels, 63- & 66- *Mem:* Am Vet Med Asn; Conf Pub Health Vets; Asn Mil Surgeons US; Am Col Vet Toxicologists; Med Mycol Soc Am. *Res:* Mycology and epidemiology as related to zoonotic diseases. *Mailing Add:* 2027 Thackrey St Manhattan KS 66502

KELLEY, FENTON CROSLAND, b Chicago, Ill, Aug 24, 26. MAMMALIAN PHYSIOLOGY, ENVIRONMENTAL PHYSIOLOGY. *Educ:* Univ NMex, BSc, 51, MSc, 54; Univ Calif, Berkeley, PhD(physiol), 67. *Prof Exp:* Fisheries biologist, Calif State Dept Fish & Game, 54-57; res assoc physiol, Inst Environ Stress, Univ Calif, Santa Barbara, 57-58, lectr, Dept Phys Educ & Ergonomics, 58-59; asst prof, 76-88, ASSOC PROF ZOOL, BOISE STATE UNIV, 88-; CONSULT, US CENGRS, 73- *Concurrent Pos:* consult fisheries, Stearns, Rogers, Inc, Denver, 75-, City of Boise, 80, 81 & 82 & M K Eng Co-Water Qual, Boise, 85- *Res:* Fresh water fisheries biology; aquatic ecology; various aspects of adaptation to environmental extremes in mammals; anatomy; physiological effects of selective pesticides; aquaculture and invertebrates. *Mailing Add:* 2260 Berkeley St Boise ID 83705

KELLEY, FRANK NICHOLAS, b Akron, Ohio, Jan 19, 35; m 60; c 3. ADVANCED NON-METALLIC MATERIALS, SOLID ROCKET PROPELLANTS. *Educ:* Univ Akron, BS, 58, MS, 59, PhD(polymer chem), 61. *Prof Exp:* Res chemist, Air Force Rocket Propulsion Lab, 64-66, br chief solid rockets, 66-70, chief advan plans, 70-71, chief scientist, 71-73; chief scientist, Dept Defense, Wright-Patterson AFB, 73-76, dir, Air Force Mat Lab, 76-78; DIR & PROF, INST POLYMER SCI, UNIV AKRON, 78- *Concurrent Pos:* Chmn, Interagency Working Group Mech Behav, 63-65; consult, NSF, 79- & Dept Energy Progs, Jet Propulsion Lab, Midwest Res Inst, Solar Energy Res Inst, 79- *Honors & Awards:* Rubber Age Award; Outstanding Tech Contrib Award, Am Ins Aeronaut & Astronaut. *Mem:* Assoc fel Am Inst Aeronaut & Astronaut; Am Chem Soc. *Res:* Polymer physics; structure-property relationships of elastomers and thermosetting resins; mechanical properties of solid propellants. *Mailing Add:* Univ Akron Univ Akron Akron OH 44325-0002

KELLEY, GEORGE G, b Pottstown, Pa, Feb 23, 20. PLASMA PHYSICS, ELECTRONIC INSTRUMENTATION. *Educ:* Calif Inst Technol, BS, 47. *Prof Exp:* Prin scientist, Oak Ridge Nat Labs, 74-79; partner, 84-87, CHMN BD, SCI ENDEAVORS CORP, 87- *Mem:* Am Physsoc. *Mailing Add:* Rte 4 Box 79 Kingston TN 37763

KELLEY, GEORGE GREENE, b Philadelphia, Pa, Nov 6, 18; m 47; c 4. BIOLOGICAL CHEMISTRY. *Educ:* Fla State Univ, BS, 50, MS, 51, PhD(biochem, food, nutrit), 56. *Prof Exp:* Tech res asst chem, Fla State Univ, 52-54; asst prof, Univ Mo, 56-57; instr pharmacol, Howard Col, 58-61; assoc prof chem, 63-67, chmn div sci & math, 64-73, PROF CHEM, JACKSONVILLE UNIV, 67- *Concurrent Pos:* Sr biochemist, Southern Res Inst, 57-62; instr, Exten Ctr, Univ Ala, 59-62, guest lectr, Med & Dent Schs, 61. *Mem:* AAAS; Am Chem Soc; Am Soc Limnol & Oceanog; NY Acad Sci; fel Am Inst Chemists. *Res:* Study of the life cycle, the propagation and large scale cultivation of three species of the large fresh water shrimp, genus Macrobrachium. *Mailing Add:* 5332 Setton Ave Jacksonville FL 32211

KELLEY, GEORGE W, JR, b Winfield, Kans, Dec 5, 21; m 42; c 3. ZOOLOGY. *Educ:* Univ Nebr, BS, 48, PhD(zool), 53; Univ Ky, MS, 50. *Prof Exp:* Asst parasitologist, Univ Ky, 48-50; parasitologist, State Dept Health, Nebr, 50-53; from asst prof to assoc prof parasitol, Univ Nebr, 53-64; tech sales assoc, Eli Lilly Int Corp, 64-67; chmn dept, 67-74, prof, 67-82, EMER PROF BIOL, YOUNGSTOWN STATE UNIV, 82- *Concurrent Pos:* Consult-evaluator, NCen Asn Sch, 72-82. *Mem:* AAAS; Am Soc Parasitol; Am Inst Biol Sci; Nat Asn Teachers. *Res:* Epidemiology of parasites of domestic animals. *Mailing Add:* Rte 1 Box 195 Sunman IN 47041

KELLEY, GREGORY M, b Boston, Mass, Nov 18, 33; m 74. DESIGN ENGINEERING, VACUUM SYSTEM DESIGN. *Educ:* Wentworth Inst Technol, Boston, AEng, 58 & 65; Col Santa Fe, BCS, 85. *Prof Exp:* Sr technician, Avco Everett Res Lab, 61-66; sr technologist, 66-89, SPEC PROD TECHNOLOGIST, LOS ALAMOS NAT LAB, 89-; INSTR VACUUM TECHNOL, UNIV NMEX, LOS ALAMOS, 84- *Concurrent Pos:* Group criticality safety officer, Los Alamos Nat Lab, 91- *Honors & Awards:* Award of Excellence, US Dept Energy, 89. *Mem:* Soc Mfg Engrs. *Res:* Automated measurement of radioactive samples by gamma and neutron counting and calorimetry. *Mailing Add:* 1710-37th St Los Alamos NM 87544-2152

KELLEY, HENRY J, aerospace engineering, mathematics; deceased, see previous edition for last biography

KELLEY, JAMES CHARLES, b Los Angeles, Calif, Oct 5, 40; m 63; c 2. OCEANOGRAPHY. *Educ:* Pomona Col, BA, 63; Univ Wyo, PhD(geol), 66. *Prof Exp:* From asst prof to assoc prof oceanog, Biomath & Geol Sci, Univ Wash, 66-75; DEAN SCI, SAN FRANCISCO STATE UNIV, 75-*Concurrent Pos:* Fulbright prof, Univ Athens, 71; pres, Calif Acad Sci, 86-*Mem:* AAAS; Am Geophys Union; Pac Sci Cong. *Res:* Coastal upwelling; structural petrology; statistics; computer science. *Mailing Add:* Sch Sci San Francisco State Univ San Francisco CA 94132

KELLEY, JAMES DURRETT, b Louisville, Ky, June 10, 29. ORNAMENTAL HORTICULTURE. *Educ:* Univ Ky, BS, 52; Iowa State Col, MS, 54; Mich State Univ, PhD, 57. *Prof Exp:* Res asst, Iowa State Col, 52-54 & Mich State Univ, 54-57; from asst prof to assoc prof hort, Univ Ky, 57-67; ASSOC PROF HORT, IOWA STATE UNIV, 67- *Mem:* Am Soc Hort Sci; Am Asn Bot Gardens & Arboretums; Int Plant Propagators' Soc. *Res:* Plant nutrition and marketing of nursery crops. *Mailing Add:* Dept Hort 139 Hort Iowa State Univ Ames IA 50011

KELLEY, JAMES LEROY, b San Diego, Calif, Nov 12, 43; m 67; c 3. MEDICINAL CHEMISTRY. *Educ:* Fresno State Col, BS, 67; Univ Calif, Santa Barbara, PhD(chem), 70. *Prof Exp:* ASST DIV DIR, BURROUGHS WELLCOME & CO, 70- *Mem:* Am Chem Soc; AAAS; NY Acad Sci. *Res:* Chemistry on purine, pyrimidine and imidazole heterocycles and antiviral agents especially acyclic nucleosides; design and synthesis of enzyme inhibitors and novel CNS active agents. *Mailing Add:* Burroughs Wellcome Co 3030 Cornwallis Rd Research Triangle Park NC 27709

KELLEY, JASON, b Buffalo, NY, Sept 8, 43; m 65; c 3. CELL BIOLOGY. *Educ:* Harvard Univ, AB, 66; Univ Tex, Dallas, MD, 72. *Prof Exp:* From asst prof to prof cell biol & med, 83-88, PROF, PULMONARY DIS SECT, UNIV VT, 88- *Mem:* NY Acad Sci; Biochem Soc; Am Soc Cell Biol; Reticuloendothelial Soc; Am Thoracic Soc; AAAS. *Res:* Cellular and molecular mechanisms of tissue remodelling in growth and disease. *Mailing Add:* Pulmonary Sect (111N) Univ Vt Given C-321 Burlington VT 05405

KELLEY, JAY HILARY, b Greensburg, Pa, Mar 9, 20; m 49; c 9. MINING ENGINEERING, COMPUTER SCIENCE. *Educ:* Pa State Univ, BS, 42, MS, 47, PhD, 52. *Prof Exp:* Res asst, Pa State Univ, 46-49, instr, sr res engr, Joy Mfg Co, Pa, 52-57; sr engr, Westinghouse Elec Corp, 57-62; staff scientist, Off Sci & Technol, Exec Off President, Washington, DC, 62-65; prof info sci & assoc dir, Bur Info Sci Res, Rutgers Univ, 65-66; mgr comput & asst instr, Philco-Ford Corp, Pa, 66-69; pres, Urbdata Inc, Pa, 69-80; pres, Kelastic Mine Beam Co, Greensburg, 69-86; dean, 70-78, distinguishd prof, 78-87, DISTINGUISHED EMER PROF, COL MINERAL & ENERGY RESOURCES, WVA UNIV, 87- *Concurrent Pos:* Vpres, Mammoth Coal & Coke Co, 46-54; dir, Leonard Express, Inc, 54-; exec secy panel sci info, President's Sci Adv Comt, 62-63, panel drug info, 64-66; comt sci & technol info, Fed Coun Sci & Technol, 62-65; trustee, Engrs Index, Inc, 67-80, dir, 69-78, pres, 76-78; ptnr, Mining Eng Consult, Greensburg, Pa; chmn, Coal Mining Sect, Nat Safety Coun, Chicago, 79-80; chmn proj comt, Eng Found, New York. *Mem:* AAAS; Inst Elec & Electronics Engrs; Opers Res Soc Am; Am Soc Info Sci; Am Inst Mining, Metall & Petrol Engrs; Am Mining Congress; asn Comput Mach; Cosmos Club. *Res:* Mine design; mineral resource economics; entropic systems; computer science and applications; mine roof control; machinery design and development; bulk handling; spontaneous combustion. *Mailing Add:* Maplewood Terr Greensburg PA 15601

KELLEY, JIM LEE, b Ada, Okla, Oct 20, 47; m 68; c 3. BIOCHEMISTRY, CHEMISTRY. *Educ:* Bethany Nazarene Col, BS, 69; Univ Okla, PhD(biochem), 73. *Prof Exp:* Fel, 73-76, staff scientist med res, Okla Med Res Found, 76-79; ASST PROF, UNIV TEX, SAN ANTONIO, 80-*Concurrent Pos:* Grants, Am Heart Asn, Okla Affil, Inc, 76-77 & 77-78 & HEW Pub Health Serv, 78- *Mem:* AAAS; Am Soc Cell Biol; Am Heart Asn; Am Asn Pathobiologists. *Res:* Metabolism of plasma lipoproteins by macrophages; regulation of cellular lipoprotein receptors. *Mailing Add:* Dept Path Univ Texas Health Sci Ctr 7703 Floyd Curl Drive San Antonio TX 78284

KELLEY, JOHN DANIEL, b Chicago, Ill, July 30, 37; m 60; c 3. CHEMICAL PHYSICS. *Educ:* St Louis Univ, BS, 59; Georgetown Univ, PhD(phys chem), 64. *Prof Exp:* Res assoc theoret chem, Georgetown Univ, 63-64; res fel, Brookhaven Nat Lab, 64-67; res scientist, 67-70, assoc scientist, 70-73, sr scientist, 73-81, PRIN SCIENTIST, MCDONNELL DOUGLAS RES LABS, 81- *Concurrent Pos:* Adj prof, Dept Chem & Physics, Univ Mo, St Louis, 77- *Mem:* AAAS; Am Chem Soc. *Res:* Theoretical and experimental reaction kinetics; molecular quantum mechanics; inter-molecular energy transfer processes. *Mailing Add:* McDonnell Douglas Bldg 110 PO Box 516 St Louis MO 63166-0516

KELLEY, JOHN ERNEST, b Milwaukee, Wis, June 27, 19; m 50; c 4. MATHEMATICS. *Educ:* Univ Wis, BS, 41; Marquette Univ, MS, 48; Univ Mich, PhD(math), 60. *Prof Exp:* Instr math, Univ Miami, 49-51; analyst, US Dept Defense, 53-54; from instr to asst prof math & chmn dept, Marquette Univ, 54-64; chmn dept, 66-69, ASSOC PROF MATH, UNIV SFLA, 64-*Mem:* Math Asn Am; Nat Coun Teachers of Math. *Res:* Mathematical logic. *Mailing Add:* 1120 Michigan Blvd Dunedin FL 34698

KELLEY, JOHN FRANCIS, JR, b Boston, Mass, July 10, 20; m 52; c 2. BIOCHEMISTRY, ACADEMIC ADMINISTRATION. *Educ:* Boston Col, BS, 46, MS, 47; Georgetown Univ, PhD(chem), 52. *Prof Exp:* Asst chem, Boston Col, 46-47; from instr to assoc prof, 47-58, PROF CHEM, 58-, asst dean, 68-86, ASSOC DEAN ACADEMIC AFFAIRS, US NAVAL ACAD, 86- *Res:* Organic iodine compounds; identification of organic compounds. *Mailing Add:* Off Acad Dean US Naval Acad Annapolis MD 21402-5000

KELLEY, JOHN FREDRIC, b Gay, WVa, Sept 17, 31; m 60; c 3. PSYCHIATRY. *Educ:* Marietta Col, AB, 54; McGill Univ, MD, 58; Am Bd Psychiat & Neurol, cert psychiat, 66, cert child psychiat, 71. *Prof Exp:* Resident psychiat, Health Ctr, Ohio State Univ, 59-62; staff psychiatrist, Patuxent Inst, Md, 64-66; fel child psychiat, Worcester Youth Guid Ctr, Mass, 66-68; assoc prof, 68-74, dir child psychiat prog, Med Ctr, 68-83 & 87-90, PROF BEHAV MED, PSYCHIAT & PEDIAT, MED SCH, WVA UNIV, 74- *Mem:* Am Psychiat Asn; AMA; Am Acad Child Psychiat. *Res:* Child psychiatry. *Mailing Add:* Dept Behav Med & Psychiat WVa Univ Health Scis Morgantown WV 26506

KELLEY, JOHN JOSEPH, II, b Philadelphia, Pa, Jan 4, 33; m 70. OCEANOGRAPHY, METEOROLOGY. *Educ:* Pa State Univ, BS, 58; Univ Nagoya, Japan, PhD(oceanog), 71. *Prof Exp:* Sr scientist meteorol, Univ Wash, 60-68; oceanogr, 68-73, asst prof oceanog, Univ Alaska, 73-77, dir, Naval Arctic Res Lab, 77-80, ASSOC PROF MARINE SCI, UNIV ALASKA, 80- *Concurrent Pos:* Prog mgr, NSF Off Polar Prog, 74-76; chmn, Sci Adv Comt, Northslope Borough/Alaska Eskimo Whaling Comn, 80-*Mem:* Am Geophys Union; Am Soc Limnol & Oceanog; Am Polar Soc; Arctic Inst NAm; Sigma Xi. *Res:* Exchange processes; polar ecosystems; coastal upwelling phenomena; air-sea exchange processes. *Mailing Add:* Inst Marine Sci Univ Alaska Fairbanks AK 99775-1080

KELLEY, JOHN LE ROY, b Kans, Dec 6, 16; m 38, 63; c 5. MATHEMATICS. *Educ:* Univ Calif, Los Angeles, AB, 36, MA, 37; Univ Va, PhD(math), 40. *Prof Exp:* Asst prof math, Univ Notre Dame, 40-42; mathematician, Ballistic Res Lab, Aberdeen Proving Grounds, 42-45; asst prof math, Univ Chicago, 45-47; assoc prof, Univ Calif, 47-50; vis assoc prof, Tulane Univ, 50-52 & Univ Kans, 52-53; chmn dept, 57-60, 75-78, PROF, 53-, EMER PROF MATH, UNIV CALIF, BERKELEY, 85- *Concurrent Pos:* Fel, Inst Advan Study, 45-46; NSF fel, 53-54; Fulbright res prof, Cambridge Univ, 57-58; Am-Kanpur Prog lectr, Indian Inst Technol, Kanpur, 64-65; nat teacher, Continental Classroom, NBC, 60. *Mem:* Fel AAAS; Nat Coun Teachers Math; Math Asn Am. *Res:* Topology; functional analysis. *Mailing Add:* Dept Math Univ Calif Berkeley CA 94720

KELLEY, JOHN MICHAEL, b Lynchburg, Va, Feb 28, 48; m 70; c 1. MICROBIOLOGY, TOXICOLOGY. *Educ:* Va Polytech Inst & State Univ, BS, 70; Univ Va, MS, 72, PhD(microbiol), 75. *Prof Exp:* Fel virol, Duke Univ, 74-76; prin investr virol, Meloy Labs Inc, 76-79; SR SCIENTIST & PROJ MGR, JRB ASSOC, 79- *Concurrent Pos:* NIH fel, NIH grant neurosci, 75-76. *Mem:* Am Soc Microbiologists. *Res:* Chief fields of interest are type and group specific radioimmunoassays for type-D retroviruses, and indentification and localization of tumor specific or viral antigens within tumor tissue by immunoperoxidase staining. *Mailing Add:* 6239 Winward Berk VA 22015

KELLEY, JOSEPH MATTHEW, b Baltimore, Md, Dec 10, 29; m 55; c 3. POLYMER CHEMISTRY. *Educ:* Loyola Col, Md, BS, 50; Fordham Univ, MS, 52, PhD(chem), 56; NY Univ, BSChE, 66. *Prof Exp:* Res chemist, Chem Res Div, Esso Res & Eng Co, NJ, 55-57, proj leader polyolefins, 57-61; supvr polyolefins res, Rexall Chem Co, Paramus, 61-63; mgr develop res, Dart Industs Chem Group, 63-65; mgr polymer res, 65-66; asst dir styrenic polymer develop, 66-67, dir, ABS res & develop, 67-69, dir mkt admin rexene polymers, 69-70, vpres res & develop, 70-79; vpres res & develop, El Paso Polyolefins Co, 79-84; VPRES, MOJAVE RES & DEVELOP CO, 84- *Mem:* Am Chem Soc; Sigma Xi. *Res:* Polymerization of olefins; stabilization of polymers; organic synthesis; enzyme chemistry; heterogeneous catalysis; styrene type polymers; chlorine oxide chemistry. *Mailing Add:* 1321 E Broad St Westfield NJ 07090

KELLEY, KEITH WAYNE, b Bloomington, Ill, Nov 5, 47; m 78; c 1. IMMUNOPHYSIOLOGY. *Educ:* Ill State Univ, BS, 69; Univ Ill, MS, 73, PhD(animal physiol), 76. *Prof Exp:* From asst prof to assoc prof, Wash State Univ, 76-84; PROF, UNIV ILL, 84- *Concurrent Pos:* Invited res scientist, Nat Inst Agron Res, Paris, France, 82-83; Nat Inst Med Res, Bordeaux, France, 87. *Honors & Awards:* Animal Mgt Award, Am Soc Animal Sci, 87. *Mem:* Am Asn Immunologists; Am Soc Animal Sci; Soc Exp Biol & Med; AAAS; Am Asn Vet Immunologists. *Res:* Neuroimmunomodulation; influence of hormones and neurotransmitters on regulation of T and B cell function in young and aged mammals. *Mailing Add:* Dept Animal Sci Univ Ill 162 ASL 1207 W Gregory Dr Urbana IL 61801

KELLEY, LEON A, b Madison, Wis, June 28, 23; m 45; c 4. BIOCHEMISTRY. *Educ:* Univ Wis, BS, 48, MS, 49, PhD(agr chem), 51. *Prof Exp:* Asst prof agr chem, Univ Calif, Davis, 53-54; from asst prof to assoc prof, 54-63, PROF CHEM, SAN JOSE STATE UNIV, 63- *Mem:* Am Chem Soc; Am Dairy Sci Asn. *Res:* Dairy chemistry; factors affecting rennet coagulation; rancidity in milk; development of instant milk products. *Mailing Add:* Dept Chem San Jose State Univ San Jose CA 95192

KELLEY, MAURICE JOSEPH, b Danielson, Conn, Aug 6, 16; m 45. SURFACE ACTIVE CHEMICALS, PHOTOGRAPHIC AND GRAPHICS ARTS CHEMICALS. *Educ:* La Salle Univ, AB, 36; Fordham Univ, MA, 40; Univ Pa, PhD(org chem), 42; NY Univ, MBA, 65. *Prof Exp:* Res chemist, Nopco Chem Co, 36-44, chief chemist chg sales develop lab, 44-48, dir indust develop lab, 48-53, dir indust specialties lab, 53-58, dir proj coord dept, 58-61; dir res labs, Diversey Corp, Ill, 61-65; dir electrostatics res, Philip A Hunt Chem Corp, 66-69, dir res, 69-75, asst vpres & dir res, 75-82; RETIRED. *Mem:* AAAS; Am Chem Soc; Chem Mgt & Resources Asn; Soc Imaging Sci & Technol; fel Am Inst Chemists; fel Royal Soc Chem. *Res:* Management of research and development; surface active agents; nitrogen compounds oil and fat derivatives; emulsion polymers; plastics; metallic soaps; synthetic detergents; bactericides; metal processing chemicals; photographic and graphic arts chemicals. *Mailing Add:* 4709 N 76th Pl Scottsdale AZ 85251-1565

KELLEY, MAURICE LESLIE, JR, b Indianapolis, Ind, June 29, 24. MEDICINE. *Educ:* Univ Rochester, MD, 49. *Prof Exp:* From instr to assoc prof med, Univ Rochester, 55-67; assoc prof, 67-74, PROF CLIN MED, DARTMOUTH MED SCH, 74-; STAFF MEM, MARY HITCHCOCK MEM HOSP CLIN, 67- *Concurrent Pos:* Fel, Mayo Clin, 57-59; from asst physician to assoc physician, Strong Mem Hosp, 55-59, sr assoc physician, 63-; consult, Vet Admin Hosp, Canandaigua, NY, Genesee Hosp & Rochester Gen Hosp. *Mem:* AMA; Am Gastroenterol Asn; Am Fedn Clin Res; Am Col Physicians; Am Physiol Soc. *Res:* Gastroenterology; notility of the esophagus. *Mailing Add:* 15 Ledge Rd Hanover NH 03755

KELLEY, MICHAEL C, b Toledo, Ohio, Dec 21, 43; m 66; c 3. GEOPHYSICS. *Educ:* Kent State Univ, Ohio, 64; Univ Calif, Berkeley, PhD(physics), 70. *Prof Exp:* From asst prof to assoc prof, 75-82, PROF, SCH ELEC ENG, CORNELL UNIV, ITHACA, 82- *Concurrent Pos:* Alexander von Humboldt fel, 74-75; mem, Nat Res Coun Comt on Jicamara Radar Observ, 76-79; assoc ed, J Geophys Res, 79-83; Nat Acad Sci Comt on Solar Space Plasmas, 80-83; NSF Atmospheric Sic Adv Comt, 82-84; proj scientist, NASA-Peru Rocket campaign, 81-83, NASA-Greenland I Rocket campaign, 84-85, Greenland II, 86-87; investr, NASA-CRRES satellite working group, 86- *Honors & Awards:* James MacElware Award, Am Geophys Union, 79. *Mem:* Fel Am Geophys Union. *Res:* AC-DC electrical field experiments in space; supplied electron and/or analyzing results for sixty rocket flights, four satellite missions and numerous balloon flights; author of two textbooks and over 100 articles. *Mailing Add:* Sch Elec Engr Cornell Univ 318 E & TC Ithaca NY 14853

KELLEY, MICHAEL J, CHEMISTRY. *Prof Exp:* POSTDOCTORAL RESEARCHER, DEPT CHEM, UNIV CALIF, SAN DIEGO, 89- *Mailing Add:* Dept Chem Univ Calif San Diego La Jolla CA 92093

KELLEY, MYRON TRUMAN, b Allerton, Iowa, Mar 9, 12; m 37. ANALYTICAL CHEMISTRY. *Educ:* Univ Nebr, BSc, 32, MSc, 33; Iowa State Univ, PhD(phys chem), 37. *Prof Exp:* Asst chief anal chemist, Queeny Plant Monsanto Chem Co, Mo, 37-41, chief anal chemist, 41-45; asst sect chief chem process develop sect, Clinton Lab, 45-48; dir anal chem div, 48-72, CONSULT, OAK RIDGE NAT LAB, 73-; CONSULT, HARSHAW CHEMICAL CO, 81- *Concurrent Pos:* Consult, Tennecomp Systs, Inc, 73-79. *Honors & Awards:* Chem Instrumentation Award, 73. *Mem:* AAAS; Am Nuclear Soc; Am Chem Soc; Sigma Xi. *Res:* Analytical instrumentation; instrumental methods of analysis, especially applications of small computers; analysis of highly radioactive materials. *Mailing Add:* 1814 Village Lane Naples FL 33963

KELLEY, NEIL DAVIS, b Clayton, Mo, Jan 8, 42. METEOROLOGICAL MEASUREMENTS, ACOUSTICS. *Educ:* St Louis Univ, BS, 63; Pa State Univ, MS, 68. *Prof Exp:* Meteorologist, Meteorol Res Inc, 63-66; instr meteorol, Pa State Univ, 66-71; group chief airborne measurements, Nat Ctr Atmospheric Res, 72-77; br chief measurements, 77-80, PRIN SCIENTIST WIND ENERGY, SOLAR ENERGY RES INST, 80- *Concurrent Pos:* Prog supvr, ESSO (Exxon) Res & Eng Co, 67-68. *Mem:* Am Meteorol Soc; Instrument Soc Am; Am Inst Aeronaut & Astronaut; AAAS; Sigma Xi; Inst Environ Sci. *Res:* Developing a physical understanding of the role of atmospheric turbulence on the energy conversion efficiency and structural component lifetime of wind energy conversion systems. *Mailing Add:* 605 S 42nd St Boulder CO 80303

KELLEY, PATRICIA HAGELIN, b Cleveland, Ohio, Dec 8, 53; m 77; c 2. INVERTEBRATE PALEONTOLOGY, EVOLUTIONARY PALEONTOLOGY. *Educ:* Col Wooster, BA, 75; Harvard Univ, AM, 77, PhD(geol), 79. *Prof Exp:* Instr, New England Col, 79; asst prof, 79-85, ASSOC PROF GEOL, UNIV MISS, 85- *Concurrent Pos:* Prin investr, NSF Grant. *Mem:* Paleont Soc; Geol Soc Am; AAAS; Sigma Xi. *Res:* Evolutionary patterns, including modes and rates of evolution; origin of macroevolutionary trends, particularly as exhibited by Miocene molluscs; gastropod predation and coevolution; sexual dimorphism; biometric analysis; coastal plain biostratigraphy; carboniferous biogeography. *Mailing Add:* Dept Geol Univ Miss Main Campus University MS 38677

KELLEY, PAUL LEON, b Philadelphia, Pa, Dec 8, 34; m 58; c 2. LASERS, NONLINEAR OPTICS. *Educ:* Rutgers Univ, BA, 56; Cornell Univ, MS, 59; Mass Inst Technol, PhD(physics), 62. *Prof Exp:* Teaching asst physics, Cornell Univ, 56-58; staff assoc, 58-62; staff mem, Lincoln Lab, 62-69, asst group leader, 69-71, ASSOC GROUP LEADER, LINCOLN LAB, MASS INST TECHNOL, 71- *Concurrent Pos:* Lectr, Northeastern Univ, 63-64, Mass Inst Technol, 66-67; vis lectr, Univ Calif, Berkeley, 68-69; ed, Optics Letters, Optical Soc Am, 84-89, chair bd ed, 90-; vis indust prof, Tufts Univ, 85- *Mem:* Am Phys Soc; Optical Soc Am; Sigma Xi; AAAS. *Res:* Laser, nonlinear optics; ultrafast electro-optical devices. *Mailing Add:* Lincoln Lab Mass Inst Technol PO Box 73 Lexington MA 02173

KELLEY, RALPH EDWARD, b Greenville, SC, Mar 6, 30; m 68; c 3. ATOMIC PHYSICS, MOLECULAR PHYSICS. *Educ:* Furman Univ, BA, 51, BS, 55; Univ Va, MS, 57, PhD(physics), 60. *Prof Exp:* Sr scientist theoret anal, Res Labs Eng Sci, Univ Va, 60-66; PROG MGR PHYSICS DIRECTORATE, OFF SCI RES, USAF, 66- *Mem:* Am Phys Soc; Sigma Xi. *Res:* Annihilation radiation of positrons in crystals; polarization effects in scattering. *Mailing Add:* 7551 Marshall Dr Annandale VA 22003

KELLEY, RAYMOND H, b Roscoe, Mont, July 10, 22; m 51; c 1. NUCLEAR PHYSICS. *Educ:* Mont State Col, BS, 50; Ohio State Univ, MS, 55, PhD(nuclear physics), 63. *Prof Exp:* Instr electronics, Ellington AFB, Tex, USAF, 51-54; nuclear res officer, Modern Physics Br, Wright-Patterson AFB, Ohio, 55 & Aeronaut Res Lab, 57-60, staff scientist, Brookhaven Nat Lab, 55-57, from instr to assoc prof physics, USAF Acad, 62-69; part-time instr math, 69-71, from assoc prof to prof, 71-83, EMER PROF PHYSICS & MATH, SOUTHWESTERN ORE COMMUNITY COL, 83- *Mem:* Am Phys Soc; Am Asn Physics Teachers. *Res:* Helium filled scintillation detectors; radiation damage to semiconductor materials; particle accelerators; gamma ray spectroscopy. *Mailing Add:* PO Box 335 Bandon OR 97411

KELLEY, ROBERT LEE, b East St Louis, Ill, Mar 20, 37. MATHEMATICAL PHYSICS. *Educ:* Univ Ill, Urbana, BS, 58; Univ Miami, MS, 60; Univ Mich, PhD(math), 66. *Prof Exp:* From instr to asst prof, 64-72, ASSOC PROF MATH, UNIV MIAMI, 72- *Mem:* Am Math Soc; Math Asn Am; Sigma Xi; Soc Indust & Appl Math; Asn Comput Mach. *Res:* Mathematical physics; functional analysis; mathematical biology; theory of algorithms. *Mailing Add:* Dept Math Univ Miami PO Box 249085 Coral Gables FL 33124

KELLEY, ROBERT OTIS, b Santa Monica, Calif, Apr 30, 44; m 65; c 2. DEVELOPMENTAL BIOLOGY, CELL BIOLOGY. *Educ:* Abilene Christian Col, BS, 65; Univ Calif, Berkeley, MA, 66, PhD(zool), 69. *Prof Exp:* Assoc zool, Univ Calif, Berkeley, 67-68, actg asst prof, 69; from instr to assoc prof, 69-79, PROF ANAT, SCH MED, UNIV NMEX, 79-, CHMN DEPT, 81- *Concurrent Pos:* NIH grant, 70- & res career develop award, 72; res fel, Hubrecht Lab, Utrecht, Neth, 72-73; distinguished res prof, Nat Inst Basic Biol, Okazaki Nat Res Inst, Japan, 85. *Mem:* Soc Develop biol; Am Soc Cell Biologists; Am Asn Anat; Electron Micros Soc Am; Biophys Soc. *Res:* Fine structural associations between interacting cell layers during early amphibian development; ultrastructure and cell biology of vertebrate limb mesenchyme and associated limb morphogenesis; biology of the aging cell surface; organization of the cytoskeleton, cell imaging. *Mailing Add:* Dept Anat Univ NMex Sch Med Albuquerque NM 87131

KELLEY, RUSSELL VICTOR, b Norfolk, Va, Dec 21, 34; m 56; c 3. BIOLOGY, SCIENCE EDUCATION. *Educ:* Va State Col, BS, 57; NY Univ, MA, 64; Purdue Univ, PhD(biol sci), 72. *Prof Exp:* Teacher chem & biol, Baltimore Pub Schs, 60-62; teacher biol, Plainview, Long Island Pub Schs, 62-66; ASSOC PROF BIOL SCI, MORGAN STATE UNIV, 66- *Concurrent Pos:* Consult, NASA, 67 & 68; lectr contemp biol, Towson State Univ, 72 & 75; lectr zool & biol, Community Col Baltimore, 74-; consult sci, Md State Dept Educ Bicentennial Comt, 76 & Sci Curric Adv Comt, Div Instr & Curric, Baltimore City Pub Sch Syst, 78-; chmn bd adv, Math Eng Sci Achievement, 77- *Mem:* Nat Sci Teachers Asn; AAAS. *Res:* Instructional strategies in science teaching and population genetics. *Mailing Add:* 3400 Olympia Ave Baltimore MD 21215

KELLEY, THOMAS F, b Melrose, Mass, Mar 23, 32; m 56; c 3. BIOMEDICAL ENGINEERING, CLINICAL CHEMISTRY. *Educ:* Boston Univ, AB, 54, MA, 55; Brown Univ, PhD(biol), 59. *Prof Exp:* Sr res assoc, Bio-Res Inst, Inc, 58-68; prog mgr, 68-80, DIR APPL RES, INSTRUMENTATION LAB, INC, 80- *Mem:* Am Asn Clin Chemists; Am Soc Clin Path; Sigma Xi. *Res:* Development of hospital, medical and laboratory instrumentation. *Mailing Add:* 460 N Tenth St Albemarle NC 28001

KELLEY, THOMAS NEIL, b Toledo, Ohio, Feb 1, 29; m 55; c 3. PROCESS METALLURGY, ENGINEERING STATISTICS. *Educ:* Purdue Univ, BS, 51; Univ Pittsburgh, MS, 61. *Prof Exp:* Res metallurgist, NAm Aircraft Corp, Columbus, Ohio, 54-55; process develop supvr, Universal-Cyclops Div, Cyclops Steel Corp, 55-62; wrought alloy mgr, Austenal Div, Howmet Corp, 62-67; vpres tech, Guterl Specialty Steel Corp, 67-83; chief metallurgist, Stellite Div, Cabot Corp, 67-79, plant mgr, Reading Plant, 83-86; TECH DIR, FT WAYNE SPECIALTY STEEL DIV, SLATER STEEL CORP, 87- *Mem:* Am Inst Mining, Metall & Petrol Engrs, Fel Am Soc Metals. *Res:* Wrought special purpose steel and nickel or cobalt base alloy mill process development; statistical process and product control. *Mailing Add:* 2400 Taylor St W Ft Wayne IN 46801

KELLEY, VICKI E, AUTOIMMUNITY. *Educ:* Univ Pittsburgh, PhD, 77. *Prof Exp:* ASSOC PROF MED, SCH MED, HARVARD UNIV, 81- *Mem:* Am Asn Immunologists; Am Asn Pathologists. *Mailing Add:* Dept Med Lab Immunogene & Trasplantation Sch Med Harvard Univ Brigham & Women's Hosp Boston MA 02115

KELLEY, VINCENT CHARLES, b Tyler, Minn, Jan 23, 16; m 42; c 7. PEDIATRICS. *Educ:* Univ NDak, BA, 34, MS, 35; Univ Minn, BS, 36, PhD(biochem), 42, BS, 44, MB, 45, MD, 46. *Prof Exp:* Asst chem, Univ NDak, 34-35; asst biochem, Univ Minn, 40-41; asst prof org chem, Col St Thomas, 42-43; chief dept biophys, USAF Sch Aerospace Med, 46-47, res med, 47-48; instr pediat, Univ Minn, 49-50; from asst prof to assoc prof, Univ Utah, 50-58; prof, 58-86, EMER PROF PEDIAT, SCH MED, UNIV WASH, 86- *Concurrent Pos:* Swift fel pediat, Univ Minn, 48-50. *Mem:* AAAS; Am Chem Soc; Am Pediat Soc; Soc Pediat Res; Soc Exp Biol & Med. *Res:* Starch chemistry; renal and liver function; deceleration injuries; aviation medicine; protein chemistry; physiochemical studies of electrodialyzed starches; pituitary-adrenal function; endocrinology. *Mailing Add:* 8611 45th Ave NE Seattle WA 98115

KELLEY, VINCENT COOPER, economic geology; deceased, see previous edition for last biography

KELLEY, WILLIAM NIMMONS, b Atlanta, Ga, June 23, 39; m 59; c 4. INTERNAL MEDICINE, RHEUMATOLOGY. *Educ:* Emory Univ, MD, 63; Am Bd Internal Med, dipl, 69. *Prof Exp:* Intern med, Parkland Mem Hosp, 63-64, resident, 64-65; clin assoc, Nat Inst Arthritis & Metab Dis, 65-67; sr resident med, Mass Gen Hosp, 67-68; from asst prof to prof med, Med Ctr, Duke Univ, 68-75; from asst prof to assoc prof biochem, 69-75, chief div rheumatic & genetic dis, 70-75; John G Searle prof & chmn, Dept Internal Med & prof biol chem, Med Sch, Univ Mich, 75-89; EXEC VPRES & DEAN, SCH MED, CHIEF EXEC OFFICER, MED CTR & ROBERT G DUNLOP PROF MED, BIOCHEM & BIOPHYS, UNIV PA, 89- *Concurrent Pos:* Mosby scholar award, 63-; Am Col Physicians Mead-Johnson scholar, 67-68; teaching fel med, Harvard Med Sch, 67-68; Am Rheumatism Asn clin scholar, 69-72; res career develop award, 72-75; Macy Fac scholar, Oxford Univ, 74-75; mem, numerous sci comts & councils, NIH, 74-; vis prof & lectr, numerous univs, 79-90; consult med, numerous univs & indust, 82-88; chmn, Am Bd Int Med, 85-86 & Sect 4, Inst Med Nat Acad Sci, 87-89; pres, Cent

Soc Clin Res, 86-87; master, Am Col Physicians, 88. *Honors & Awards:* John D Lane Award, USPHS, 69; Geigy Int Prize Rheumatology, 69; Heinz Karger Prize, 73; Numerous Named Lectr, US & Foreign Univs, 73-90; John Phillips Mem Award & Medal, Am Col Physicians, 90. *Mem:* Inst Med-Nat Acad Sci; Am Fedn Clin Res (pres, 79-80); Am Rheumatism Asn (pres, 86-87); Am Soc Biol Chemists; Am Soc Clin Invest (pres, 83-84); AAAS; Asn Am Physicians; Asn Prof Med (secy-treas, 87-). *Res:* Human biochemical genetics; rheumatology; author of numerous articles and books. *Mailing Add:* Univ Pa 21 Penn Tower 34th St & Civic Center Blvd Philadelphia PA 19104-6055

KELLEY, WILLIAM RUSSELL, b Universal, Pa, Feb 3, 14; m 42; c 5. BOTANY. *Educ:* Ind Univ Pa, BS, 39; Univ Pittsburgh, MLitt, 48; Cornell Univ, MS, 49, PhD(bot), 51. *Prof Exp:* Pub sch teacher, Pa, 39-41 & 45-48; asst bot, Cornell Univ, 49-51; from asst prof to assoc prof, Univ SC, 51-59; from assoc to prof bot, Shippensburg Univ, Pa, 61-77, prof biol, 77; RETIRED. *Mailing Add:* 9554 Forest Ridge Rd Shippensburg PA 17257

KELLEY, WILLIAM S, b Washington, Pa, Nov 30, 41; m 68; c 1. MICROBIAL GENETICS. *Educ:* Haverford Col, BS, 63; Mass Inst Technol, MS; Tufts Univ, PhD(microbiol), 68. *Prof Exp:* From asst prof to assoc prof biol, Dept Biol Sci, Carnegie-Mellon Univ, Pittsburgh, Pa, 71-80; sr scientist, 81-82, asst dir res for admin & molecular biol, 82-83, VPRES, PROD & PROCESS DEVELOP, BIOGEN RES CORP, CAMBRIDGE, MASS, 84- *Concurrent Pos:* US Pub Health Serv, postdoc fel, Dept Molecular Biol, Edinburgh Univ, Scotland, 68-70; dept fel, grad dept biochem, Brandeis Univ, Mass, 70-71. *Mem:* Am Chem Soc; Am Soc Microbiol; AAAS. *Res:* Responsible for transfer of projects from research into pharmaceutical development, overseeing the groups who devise the manufacturing processes and implement them at pilot scale sufficiently large to support clinical trials; head technology transfer teams for interaction with other companies. *Mailing Add:* Biogen Res Corp 14 Cambridge Ctr Cambridge MA 02142-1401

KELLGREN, JOHN, b New York, NY, Dec 26, 40; m 71; c 2. ORGANIC CHEMISTRY, RUBBER CHEMISTRY. *Educ:* Rutgers Univ, BS, 62; Columbia Univ, PhD(org chem), 66; Univ New Haven, MBA, 77 & MS, 84. *Prof Exp:* Res chemist, Uniroyal Res Ctr, 67-75, tech supt, Uniroyal Inc, 76-79, res scientist indust prod, 79-84; AT WYROUGH-LASER INC, 84- *Mem:* Am Inst Chemists; The Chem Soc; Sigma Xi. *Res:* Free radical reactions; polyurethane chemistry; oxidation of organic compounds; aging; vulcanization. *Mailing Add:* Wyrough & Loser 1008 Whitehead Rd Ext Trenton NJ 08638

KELLIHER, GERALD JAMES, b Taunton, Mass, May 31, 42; m 65; c 2. PHARMACOLOGY. *Educ:* Univ RI, BS, 65; Duquesne Univ, MS, 67; Univ Pittsburgh, PhD(pharmacol), 69. *Prof Exp:* From asst prof to prof pharmacol, 70-78, ASSOC PROF MED, MED COL PA, 75-, ASSOC DEAD, MED EDUC. *Concurrent Pos:* Fel pharmacol, Sch Med, Univ Pittsburgh, 69-70; Southeast Pa Heart Asn, Del Heart Asn, Heart & Lung Found, Ayerst Co & Shering Co grants, Med Col Pa, 71-81; Whitehall Found grants, 71-77; Nat Heart & Lung Inst grants, 71-75 & 73-77, 76-80; Nat Inst Age grant, 76-80; Nat Inst Child Health & Human Develop grant, 72-76; consult, Vet Admin, 78-80; educ consult, Smith Kline Corp. *Mem:* Am Soc Pharmacol & Exp Therapeut; assoc fel Am Col Cardiol; Geront Soc; fel Am Col Clin Pharmacol; Am Fedn Clin Res. *Res:* Cardiovascular and autonomic pharmacology with emphasis on the mechanisms and treatment of cardiac arrhythmias and hypertension. *Mailing Add:* Assoc Dean Med Educ Med Col Pa 3300 Henry Ave Philadelphia PA 19129

KELLING, CLAYTON LYNN, b Killdeer, NDak, Mar 26, 46; m 74. VETERINARY VIROLOGY. *Educ:* NDak State Univ, BS, 68, MS, 71, PhD(pharm chem), 75. *Prof Exp:* Technician vet virol, Dept Vet Sci, NDak State Univ, 68-; AT DEPT VET SCI, UNIV NEBR, LINCOLN. *Mem:* Am Soc Microbiol; Sigma Xi. *Res:* Veterinary microbiology concerned with respiratory and reproductive diseases of animals; antiviral agents; virological diagnostic techniques. *Mailing Add:* RSD No 8 Lincoln NE 68583

KELLISON, ROBERT CLAY, b Marlinton, WVa, Nov 20, 31; m 65; c 2. FOREST GENETICS. *Educ:* WVa Univ, BSF, 59; NC State Univ, MS, 66, PhD(forest genetics), 70. *Prof Exp:* Forest supt, WVa Univ, 59-61; liaison geneticist, 63-67, assoc dir coop prog, 66-77, DIR HARDWOOD COOP, NC STATE UNIV, 77- *Concurrent Pos:* Fel, Am-Scand Found, 65; panel expert, Food & Agr Orgn-Int Breeding Prog for Preserv Forest Gene Resources, 68; scientist, NZ Forest Serv, 73-74; mem Panel Forest Tree Breeding, Peoples Repub China, 81, 83; adv forestry, Venezuela, 75, 77, Brazil, 80, Portugal, 83, 85, Taiwan, 85. *Res:* Selection and breeding of forest trees for improved volume yields, quality, adaptability and resistance to frost, drought and environmental pollution; preservation of forest gene resources. *Mailing Add:* Box 8002 NC State Univ Sch Forest Resources Raleigh NC 27695

KELLMAN, RAYMOND, b Staten Island, NY, Feb 27, 42; m 78; c 2. POLYMER CHEMISTRY, ORGANIC CHEMISTRY. *Educ:* St Peter's Col, NJ, BS, 63; Univ Colo, Boulder, PhD(org chem), 68. *Prof Exp:* Res assoc chem, Univ Wis-Madison, 67-69; res chemist, Uniroyal Inc, 69-72; res assoc, 72-75, lectr chem, Univ Ariz, 75-77; asst prof polymer chem, Univ Tex, 77-82; PROF CHEM, SAN JOSE STATE UNIV, 82- *Concurrent Pos:* Fulbright Res Scholar, Univ Queensland, Australia, 88-89. *Mem:* AAAS; Am Chem Soc. *Res:* Synthesis of new monomers; new methods of condensation polymerization; synthesis of thermally stable electroactive and biocompatable polymers; physical organic chemistry. *Mailing Add:* Dept Chem San Jose State Univ San Jose CA 95192

KELLMAN, SIMON, b Brooklyn, NY, July 26, 34; m 59; c 2. REACTOR PHYSICS. *Educ:* Carnegie Inst Technol, BS, 55, MS, 58, PhD(physics), 61. *Prof Exp:* Sr physicist, Lawrence Radiation Lab, Univ Calif, Livermore, 61-64 & United Nuclear Corp, 64-69; fel scientist, 69-72, mgr math & programming, 72-76, acting mgr methods develop, 76-78, MGR, SAFEGUARDS RELIABILITY & APPLN, PWR SYSTS DIV, WESTINGHOUSE ELEC CORP, MONROEVILLE NUCLEAR CTR, 78- *Mem:* Am Nuclear Soc. *Res:* Depletion calculations; nuclear safety analysis; Monte Carlo techniques in neutron transport. *Mailing Add:* 5441 Fair Oaks St Pittsburgh PA 15217

KELLN, ELMER, b Sask, Can, Nov 6, 26; m 51; c 3. ORAL PATHOLOGY, CANCER. *Educ:* Univ Nebr, BSc & DDS, 49; Univ Minn, MSD(path), 60. *Prof Exp:* Assoc prof path, Sch Med, WVa Univ, 60-66; prof oral med, Univ, 66-71, PROF ORAL MED, GRAD SCH & ASSOC DEAN SCH DENT, LOMA LINDA UNIV, 71- *Concurrent Pos:* Grants wound healing, 60-63 & age studies, 61-; cancer coordr, Sch Dent, WVa Univ & mem tumor bd, WVa Univ Hosp, 63-66. *Mem:* Fel Am Acad Oral Path; Am Dent Asn; Int Asn Dent Res. *Res:* Cancer behavior; disease processes of oral diseases, particularly wound healing, cancer treatment and behavior, and vascular degeneration. *Mailing Add:* 25246 Lawton Ave Loma Linda CA 92354

KELLNER, AARON, b New York, NY, Sept 24, 14; m 42; c 3. PATHOLOGY. *Educ:* Yeshiva Univ, BA, 34; Columbia Univ, MS, 36; Univ Chicago, MD, 39. *Prof Exp:* Res assoc exp cardiovasc path, 47-50, from asst prof to assoc prof path, 50-68, PROF PATH, MED COL, CORNELL UNIV, 68-; DIR NEW YORK BLOOD CTR, 64- *Concurrent Pos:* Life Inst Res Fund fel, 47-48; dir cent labs, NY Hosp, 48-64; mem coun arteriosclerosis, Am Heart Asn. *Mem:* Soc Study Blood (pres, 53); Am Soc Exp Path; Am Asn Path & Bact; Am Asn Blood Banks (pres, 54); Am Heart Asn (pres, 60). *Res:* Experimental cardiovascular pathology and arteriosclerosis. *Mailing Add:* 310 E 67th St New York NY 10021

KELLNER, HENRY L(OUIS), b Philadelphia, Pa, Sept 18, 05; m 35; c 2. CHEMICAL ENGINEERING. *Educ:* Pa State Col, BS, 26; Yale Univ, PhD(chem eng), 30. *Prof Exp:* Res chemist, Scovill Mfg Co, 27-29; chem engr, Doherty Res Co, 30-33 & Eastern Eng Co, 33-35; from chem engr to secy & tech dir, Lea Mfg Co, 35-56, vpres, 56-68, tech dir, 56-70, exec vpres, 68-76, pres, 76-78, mgr foreign opers, 70-80, vchmn, 78-82. *Concurrent Pos:* Pres & chmn, Lea Ronal Inc, 53-70. *Mem:* Am Chem Soc; Am Electroplaters Soc. *Res:* Buffing and polishing compositions; chemical and electroplating processes. *Mailing Add:* 10 Fairgrounds Rd Woodbridge CT 06525

KELLNER, JORDAN DAVID, b New York, NY, Aug 25, 38; m 60; c 2. PHYSICAL CHEMISTRY. *Educ:* City Col New York, BS, 58; NY Univ, MS, 62, PhD(phys chem), 64; Rensselaer Polytech Inst, MS, 72. *Prof Exp:* Jr chemist, Kings County Hosp, 58-59; res asst biochem res, St Catherine's Hosp, 59-61; res asst phys chem, NY Univ, 61-64; sr chemist, Atomics Int Div, NAm Aviation, 64-68; res scientist, Hamilton Standard Div, United Technologies Corp, 68-71, sr res scientist, 71-78, supvr chem processes, Res Ctr, 78-80, supvr, 80-81; res assoc, 81-82, sr res assoc, 83-91, SR SCIENTIST, KENDALL RES LAB, 91- *Concurrent Pos:* Adj prof chem, Univ Hartford, 80-81; trustee, Boston sect, Nat Asn Corrosion Engrs, 87-90, career develop chmn, Northeast Region, 90- *Honors & Awards:* SAm Tour Award, Am Soc Testing & Mat. *Mem:* Sigma Xi; Am Chem Soc; Nat Asn Corrosion Engrs; Am Soc Testing & Mat. *Res:* Transport processes in fused salts and metal-metal salt mixtures; Soret effect and viscosity; electrodeposition of semi-metals and their compounds from fused fluorides; electrochemical techniques of corrosion measurement including electrochemical impedance spectroscopy. *Mailing Add:* 38 Grove St Wayland MA 01778

KELLNER, STEPHAN MARIA EDUARD, b Friedberg, Ger, Feb 1, 33; US citizen; m 60; c 12. PHYSICAL CHEMISTRY. *Educ:* Univ Rochester, BS, 55, PhD(phys chem), 60. *Prof Exp:* From asst prof to assoc prof, 59-69, chmn dept, 71-75, PROF CHEM, ST MICHAEL'S COL, VT, 69- *Mem:* Am Chem Soc; Sigma Xi. *Res:* Rates and mechanisms of homogeneous gas phase reactions. *Mailing Add:* Dept Chem St Michael's Col Winsooki Park Colchester VT 05439

KELLOGG, CHARLES NATHANIEL, b Albuquerque, NMex, June 29, 38; m 57; c 3. MATHEMATICS. *Educ:* NMex Inst Mining & Technol, BS, 60; La State Univ, PhD(math), 64. *Prof Exp:* Teaching asst math, La State Univ, 63-64; asst prof, Univ Ky, 64-70; ASSOC PROF MATH, TEX TECH UNIV, 70- *Mem:* Am Math Soc; Math Asn Am. *Res:* Harmonic analysis; theory of multiplier operators; Banach algebras. *Mailing Add:* Dept Math Tex Tech Univ Lubbock TX 79409

KELLOGG, CRAIG KENT, b Westfield, Mass, Dec 3, 37; m 60; c 3. ORGANIC CHEMISTRY. *Educ:* Ga Inst Technol, BS, 59, PhD(org chem), 63. *Prof Exp:* Res chemist, E I du Pont de Nemours & Co, Inc, 63-66; asst prof, 66-70, ASSOC PROF CHEM, GA SOUTHERN COL, 70- *Mem:* Am Chem Soc. *Res:* Natural products; dioxetanes. *Mailing Add:* 113 Herty Dr Statesboro GA 30458-5434

KELLOGG, DAVID WAYNE, b Seymour, Mo, Aug 19, 41; m 64; c 4. NUTRITION, ANIMAL PHYSIOLOGY. *Educ:* Univ Mo-Columbia, BS, 63, MS, 64; Univ Nebr-Lincoln, PhD(nutrit), 68. *Prof Exp:* From asst prof to prof dairy sci, NMex State Univ, 67-81; head, Dept Animal Sci, 81-86, PROF, DEPT ANIMAL SCI, UNIV ARK, FAYETTEVILLE, 86- *Mem:* Am Dairy Sci Asn; Am Soc Animal Sci; Am Forage & Grassland Coun; Am Regist Prof Animal Scientists. *Res:* Nutrition of dairy calves; nutritive value of alfalfa varieties; mineral nutrition; improvement of forage digestion by ruminants. *Mailing Add:* Dept Animal Sci Univ Ark Fayetteville AR 72701

KELLOGG, DOUGLAS SHELDON, JR, b Washington, DC, May 22, 26; m 54; c 3. MEDICAL MICROBIOLOGY. *Educ:* Univ Ill, BS, 50; Univ Tex, MS, 54, PhD(bact), 57. *Prof Exp:* Res scientist biochem, Univ Tex, 56-57, res assoc cytol, M D Anderson Hosp & Tumor Inst, 57-60; res supvry microbiologist, Venereal Disease Res Br, Ctr for Disease Control, USPHS, 60-86; RETIRED. *Mem:* Am Soc Microbiol; Sigma Xi; Am Venereal Disease Asn. *Res:* Immunology; biochemical genetics; host-parasite interrelationships. *Mailing Add:* 2581 Mercedes Dr Atlanta GA 30345

KELLOGG, EDWIN M, b New York, NY, Feb 3, 39; m 60, 74, 81; c 6. ELECTRON & ION BEAM SYSTEMS. *Educ:* Rensselaer Polytech Inst, BS, 60; Univ Pa, MS, 63, PhD(physics), 66. *Prof Exp:* Physicist, Radiation Dynamics Inc, 61-62; sr scientist, Am Sci & Eng, Mass, 65-69, sr staff scientist, 69-73, mem, Inst Advan Study, 73; proj mgr, Micro-Bit Div, Control Data Corp, Mass, 80, software mgr, 81-82; staff scientist, Ion Beam Technol

Inc, 82-83, vpres eng, 83-86; astrophysicist, 73-79, ASTROPHYSICIST, SMITHSONIAN ASTROPHYS OBSERV, 88- Concurrent Pos: Lectr astron, Harvard Univ, 73-79. Honors & Awards: Newton Lacey Pierce Prize, Am Astron Soc, 74. Mem: Fel Am Phys Soc; Inst Elec & Electronics Engrs; Am Vacuum Soc; Sigma Xi; Am Astron Soc. Res: Ion beam assisted surface phenomena; liquid metal ion sources; electron beam lithography; x-ray astronomy. Mailing Add: Ctr Astrophysics 60 Garden St MS-3 Cambridge MA 02138

KELLOGG, HERBERT H(UMPHREY), b New York, NY, Feb 24, 20; c 4. EXTRACTIVE METALLURGY. Educ: Columbia Univ, BS, 41, MS, 43. Prof Exp: Jr engr, Dorr Co, Conn, 41; asst mineral dressing, Columbia Univ, 41-42; instr, Pa State Col, 42-44, asst prof mineral preparation, 44-46; assoc prof extractive metall, Columbia Univ, 46-56, prof, 56-68, Stanley-Thompson prof chem metall, 68-90, EMER STANLEY-THOMPSON PROF, COLUMBIA UNIV, 90- Concurrent Pos: Chmn titanium adv comt, Off Defense Mobilization, 54-58; consult, Int Nickel Co & Am Smelting & Refining Co, 69- Honors & Awards: James Douglas Gold Medal, Am Inst Mining, Metall & Petrol Engrs, 73. Mem: Nat Acad Eng; Am Inst Mining, Metall & Petrol Engrs; fel Inst Mining & Metall, London. Res: Thermodynamics and kinetics of metallurgical reactions; high-temperature chemistry; equilibria in the systems Cu-S-O, Ni-Fe-S; slag chemistry; computer modeling of metallurgical processes. Mailing Add: Henry Krumb Sch Mines Columbia Univ New York NY 10027

KELLOGG, LILLIAN MARIE, b Detroit, Mich, Mar 6, 39. SOLID STATE CHEMISTRY. Educ: Ariz State Univ, BS, 61; Wayne State Univ, PhD(phys chem), 67. Prof Exp: Assoc scientist chem, Aeroneutronic Div, Ford Motor Co, 61-62; RES ASSOC RES LABS, EASTMAN KODAK CO, 68- Concurrent Pos: Adj fac, Rochester Inst Technol, 70-75. Mem: AAAS; Am Chem Soc; Am Phys Soc; Soc Photog Scientists & Engrs; NY Acad Sci. Res: Solid state chemistry; light interactions in solids; photoconductivity, photochemical and photographic studies. Mailing Add: 1786 Lake Rd Webster NY 14580

KELLOGG, PAUL JESSE, b Tacoma, Wash, Nov 6, 27; m 69; c 4. PLASMA PHYSICS. Educ: Mass Inst Technol, BS, 50; Cornell Univ, PhD(theoret physics), 55. Prof Exp: Nat Res Coun fel, Naval Res Lab, 55-56; res assoc, 56-57, from asst prof to assoc prof, 57-64, PROF PHYSICS, UNIV MINN, MINNEAPOLIS, 64- Concurrent Pos: Guggenheim Mem Found fel, 62-63; NATO fel; fel, Minna-James-Heineman Stiftung, 73; res fel, Australian Nat Univ, 84 & 87. Mem: Fel Am Phys Soc; fel Am Geophys Union. Res: Plasma physics as applied to the earth's upper atmosphere and the interplanetary medium; beam-plasma interaction, waves in plasma; antennas in flowing plasma. Mailing Add: Sch Physics & Astron Univ Minn Minneapolis MN 55455

KELLOGG, RALPH HENDERSON, b New London, Conn, June 7, 20. PHYSIOLOGY. Educ: Univ Rochester, BA, 40, MD, 43; Harvard Med Sch, PhD(physiol), 53. Prof Exp: Intern med, Cleveland Univ Hosps, 44; investr physiol, Naval Med Res Inst, 46; instr, Harvard Med Sch, 47-53; actg comn dept physiol, 66-70, from asst prof to prof, Sch Med, 53-90, EMER PROF PHYSIOL, UNIV CALIF, SAN FRANCISCO, 90- Concurrent Pos: Sr res fel, Sch Pub Health, Harvard Univ, 66-70; mem physiol study sect, NIH, 66-70, physiol test comt, Nat Bd Med Examrs, 66-73 & chmn, 69-73; vis fel, Corpus Christi Col, Oxford Univ, 70-71; vis scientist, Lab Physiol Respiratory, Cent Nat Res Sci, Strasbourg, France, 77; adj lectr, Hist Health Sci Dept, Univ Calif, 78- & actg chmn, 84-85. Mem: AAAS; Am Physiol Soc; Am Asn Hist Med; History Sci Soc; Sigma Xi. Res: Isotonic and osmotic diuresis in rats; respiration at altitude; history of physiology. Mailing Add: Dept Physiol Box 0444 Univ Calif San Francisco CA 94143-0444

KELLOGG, RICHARD MORRISON, b Los Angeles, Calif, Dec 24, 39; m 67; c 2. ORGANIC CHEMISTRY. Educ: Kans State Teachers Col, AB, 61; Univ Kans, PhD(org chem), 65. Prof Exp: Res fel chem, Univ Kans, 65; res fel, 65-70, assoc prof, 70-75, PROF CHEM, STATE UNIV GRONINGEN, 75- Mem: Am Chem Soc; Royal Dutch Chem Soc. Res: Synthetic organic chemistry; photochemistry; bio-organic chemistry; synthesis of unusual organic molecules and models for mechanisms of enzymic reactions. Mailing Add: Dept Org Chem Nyenborgh 16 Groningen 9747 AG Netherlands

KELLOGG, ROYAL BRUCE, b Chicago, Ill, Dec 28, 30; m 56; c 3. APPLIED MATHEMATICS. Educ: Mass Inst Technol, BS, 52; Univ Chicago, MS, 53, PhD(math), 59. Prof Exp: Mathematician, Combustion Eng, Inc, 58-61; mathematician, Westinghouse Elec Corp, 61-66; from assoc prof to prof math, 66-74, res prof, Inst Fluid Dynamics & Applied Math, 74-80, RES PROF MATH & INST PHYS SCI & TECHNOL, UNIV MD, COLLEGE PARK, 80- Mem: Am Math Soc; Soc Indust & Appl Math. Res: Numerical analysis. Mailing Add: 7519 Maple Ave Silver Spring MD 20912

KELLOGG, SPENCER, II, b Buffalo, NY, Dec 9, 13; m 38; c 5. AERONAUTICAL ENGINEERING. Educ: Cornell Univ, ME, 37; Polytech Inst Brooklyn, MSEE, 67. Prof Exp: Field serv engr, Sperry Gyroscope Co, 37-39, flight test engr, 39-40, from gyropilot engr to dept head, 40-50, dept head flight instrument eng, 50-59, asst chief engr, Aeronaut Equip Div, 59-67; INDEPENDENT AVIATION CONSULT, 67- Honors & Awards: Pioneer Award, Inst Elec & Electronics Engrs Aerospace & Elec Systs Soc, 76. Res: Gyropilot and gyroscopic flight instruments; altitude control for aircraft; turn error control of gyroscopes; erection mechanism for gyroscopes; flight directors. Mailing Add: 90 Arrival Ave Ronkonkoma NY 11779

KELLOGG, THOMAS B, b New York, NY, Apr 30, 42; m 67; c 3. PALEO-OCEANOGRAPHY, PALEO-CLIMATOLOGY. Educ: Columbia Univ, BA, 68, PhD(geol), 73. Prof Exp: Res assoc, geol sci, Brown Univ, 73-75; res assoc, 75-78, from asst prof to assoc prof, 78-89, PROF GEOL SCI & QUARTERNARY STUDIES, UNIV MAINE, 89- Concurrent Pos: mem, Cushman Found Foraminiferal Res, Climap Long Range Invest, Mapping &

Prediction, 71-80. Honors & Awards: Antarctic Serv Medal, US Congress, 79. Mem: Fel Geol Soc Am. Res: High latitude marine sediment & how it is used to determine the past extent of ice sheets, ice shelves & icebergs. Mailing Add: Geol Dept Univ Maine Orono ME 04469-0110

KELLOGG, THOMAS FLOYD, b Aurora, Ill, Apr 7, 34; m 82; c 3. BIOCHEMISTRY. Educ: Iowa State Univ, BS, 59, MS, 60; Univ Wis-Madison, PhD(biochem), 64. Prof Exp: Res asst biochem, Univ Wis, 60-64; fel microbiol, Lobund Lab, Univ Notre Dame, 64-65, res scientist, 65-68, asst prof, 68-70, assoc prof, 70-78, PROF BIOCHEM, MISS STATE UNIV, 78- Concurrent Pos: NIH Spec Res fel, 67-68. Mem: Am Chem Soc; Asn Gnotobiotics (vpres & pres-elect, 81-82, pres, 82-83). Res: Cholesterol and bile acid metabolism; liquid scintillation counting. Mailing Add: Dept Biochem Miss State Univ PO Drawer BB Mississippi State MS 39762

KELLOGG, WILLIAM WELCH, b New York Mills, NY, Feb 14, 17; m 42; c 5. METEOROLOGY. Educ: Yale Univ, AB, 39; Univ Calif, Los Angeles, MS, 42, PhD(meteorol), 49. Prof Exp: Teacher prep sch, Mass, 39-40; asst optics lab, Univ Calif, Berkeley, 40-41; instr meteorol, Univ Calif, Los Angeles, 42-43; res asst, 47-48, res assoc, Inst Geophys, 48-49; asst prof, 49-52; phys scientist & dept head, Rand Corp, Calif, 47-64; assoc dir, Nat Ctr Atmospheric Res, 64-73; sr scientist, 73-87; RETIRED. Concurrent Pos: Mem, Upper Atmosphere Comt, Nat Adv Comt Aeronaut, 53-55; mem, Comt Meteorol Aspects Effects Atomic Radiation, Nat Acad Sci, 56-64, Space Sci Bd, 59-66, Comt Atmospheric Sci, 63-67, Spec Comt Int Years Quiet Sun, 63-66 & Polar Res Bd, 75-78; Rocket & Satellite Res Panel, 57-62; chmn int comn meteorol upper atmosphere, Int Union Geod & Geophys, 60-75; mem tech panel, Earth Satellite Prog, Int Geophys Year, 57-58; consult & mem sci adv bd, US Air Force, 57-65, mem sci adv group, Off Aerospace Res, 65-70; chmn meteorol satellite comt, Adv Res Projs Agency, 58-59; mem planetary atmospheres subcomt, NASA, 61-65; chmn working group upper atmosphere, World Meteorol Orgn, 61-65; mem consult group potentially harmful effects of space exp, Comt Space Res, 62-68; mem tech adv bd, US Dept Com, 62-64; mem adv group supporting tech oper meteorol satellites, NASA-US Weather Bur, 64-75; mem panel on environ, President's Sci Adv Comt, 68-70; chmn meteorol adv comt, Environ Protection Agency, 70-74; chmn adv comt, Div Polar Progs, NSF, 82-85. Honors & Awards: Special Award, Am Meteorol Soc, 61; Decoration Except Civilian Serv, Dept Air Force, 66; Commemorative Medal, Soviet Geophys Comt, 85; Spec Citation, Garden Club Am, 88. Mem: Fel AAAS; fel Am Geophys Union; Sigma Xi; fel Am Meteorol Soc (pres, 73-74); Int Acad Astronaut. Res: Physics of the atmosphere; turbulence and structure of the upper atmosphere; scientific uses of rockets, satellites and space probes; atmospheres of Mars and Venus; causes of climate change. Mailing Add: 445 College Ave Boulder CO 80302

KELLS, LYMAN FRANCIS, b Seattle, Wash, May 19, 17; div; c 2. VARIABLE STARS, ASTROMETRY. Educ: Univ Wash, BS, 38, PhD(phys chem), 44. Prof Exp: Teaching fel, Univ Wash, 38-44; res scientist, Manhattan Proj, Kellex Corp, Carbide & Carbon Corp & Columbia Univ, 44-46; res chemist, Standard Oil Develop Co, 46-48; mem fac, Hunter Col, 48-49; asst prof, Iona Col, 49-51; res chemist, Gen Chem Div, Allied Chem Corp, 51-61; spec lectr, Newark Col Eng, 61; assoc prof chem, East Tenn State Univ, 62-64; prof chem, Westmar Col, 64-74; RES, 74- Mem: Am Chem Soc; Am Astron Soc; Astron Soc Pac; AAAS. Res: General astronomy and physics; variable stars, light, relativity, gravity and astrometry; philosophy of science; atomic and molecular structure, reaction kinetics and mechanisms, catalysis and non-ideality in solutions and gases. Mailing Add: 13716 12th Ave SW Apt 47 Seattle WA 98166-1143

KELLS, MILTON CARLISLE, b Seattle, Wash, May 7, 20; m 49; c 3. PHYSICAL CHEMISTRY. Educ: Univ Wash, BS, 42; Mass Inst Technol, PhD(chem), 48. Prof Exp: Res scientist, Gaseous Diffusion Studies, Manhattan Proj, Kellex Corp, 44-46, Atomic Energy Process Develop, 48-51; contract adminr, Res Div, AEC, 51-54; atomic energy process develop, Sylvania Elec Co, 54-56; head detonations sect, Stanford Res Inst, 56-61; scientist, Ames Res Ctr, NASA, 61-66; prof phys sci, Calif Univ, Pa, 66-91. Mem: AAAS; Am Chem Soc; Sigma Xi. Res: Vacuum ultraviolet spectroscopy. Mailing Add: 11840 26th Ave S NO 211 Seattle WA 98168

KELLY, ALAN, DEVELOPMENTAL BIOLOGY. Educ: Univ Pa, PhD(path), 68. Prof Exp: PROF PATH, SCH VET MED, UNIV PA, 80- Mem: Am Soc Cell Biol; Biophys Soc Am. Mailing Add: Path Lab Univ Pa Sch Vet Med Philadelphia PA 19174

KELLY, AMY SCHICK, b Rochester, NY, Nov 11, 40; m 71; c 2. NEUROBIOLOGY, NEUROPHYSIOLOGY. Educ: Mt Holyoke Col, AB, 62; Brown Univ, MSc, 64, PhD(psychol), 67. Prof Exp: Fel psychol, Northeastern Univ, 67-68, asst prof, 68-71; fel neurobiol, Univ Calif, Berkeley, 72-74; fel neurobiol, Med Ctr, Stanford Univ, 74-78; ASST PROF PHYSIOL, UNIV CALIF, SAN FRANCISCO, 78- Concurrent Pos: Fel, Northeastern Univ, 67-68; spec res fel, Med Sch, Stanford Univ, 74-77. Mem: Soc Neurosci; Asn Res Vision & Ophthal; AAAS. Res: Organization of the mammalian central visual system; development of the central visual pathways and visual centers; plasticity of central connections in the mammalian visual system. Mailing Add: 1315 Fourth Ave San Francisco CA 94122

KELLY, B(ERNARD) WAYNE, b Corning, NY, Oct 7, 18; m 45; c 3. AGRICULTURAL ECONOMICS. Educ: Pa State Univ, BS, 49, MS, 50. Prof Exp: Instr & asst county agt, Agr Exten, Univ Md, 50-53, asst prof & county agt, 54-56; from asst prof to prof, 56-83, EMER PROF FARM MGT EXTEN, PA STATE UNIV, UNIV PARK, 83- Concurrent Pos: Consult, 83- Mem: Am Agr Econ Asn; Am Soc Farm Mgrs & Rural Appraisers. Res: Cost of production; fruits and vegetable crops; taxation, insurance, investments and credits. Mailing Add: Dept Farm Mgt Exten 1427 S Pugh St State College PA 16801

KELLY, CLARK ANDREW, b Rocky Ford, Colo, Sept 14, 25; m 54; c 1. ANALYTICAL CHEMISTRY, PHARMACEUTICAL CHEMISTRY. *Educ:* Univ Colo, BS, 46; Temple Univ, MS, 51; Univ Minn, PhD(anal pharm chem), 58. *Prof Exp:* Res asst, Sterling-Winthrop Res Inst, Rensselaer, 46-48 & summers, 51-54, res assoc, 56-63, res chemist, 63-68, sr res chemist & group leader, 68-88, sr res chemist & group leader, Sterling Res Group, 88-91; RES ASSOC, CHEM DEPT, NMEX STATE UNIV, LAS CRUCES, 91- *Mem:* Am Chem Soc; Am Pharmaceut Asn; fel Acad Pharmaceut Sci; NY Acad Sci; Sigma Xi. *Res:* Polarography of organic compounds; ion exchange separations of organic compounds; colorimetric and spectrophotometric studies. *Mailing Add:* 4024 Shadow Run Las Cruces NM 88001-7696

KELLY, CONRAD MICHAEL, b Bradford, Pa, Nov 26, 44; m 66; c 1. CHEMICAL ENGINEERING. *Educ:* Mich State Univ, BS, 66, MS, 67, PhD(chem eng), 70. *Prof Exp:* Asst prof, 69-75, assoc prof, 75-80, PROF CHEM ENG, VILLANOVA UNIV, 80-; ASSOC PROF, AIR PROD & CHEM INC, 80- *Mem:* Am Inst Chem Engrs; Am Soc Eng Educ; Sigma Xi. *Res:* Molecular diffusion; air and water pollution abatement; mathematical modeling. *Mailing Add:* Chem Eng Dept Villanova Univ Villanova PA 19085

KELLY, DENNIS D, neuropsychology; deceased, see previous edition for last biography

KELLY, DONALD C, b Poland, Ohio, Aug 18, 33; m 55; c 4. THEORETICAL PHYSICS. *Educ:* Miami Univ, Ohio, AB, 55, MA, 56; Yale Univ, PhD(physics), 59. *Prof Exp:* Res assoc physics, Yale Lab Marine Physics, 59-60; from asst prof to assoc prof, 60-69, PROF PHYSICS, MIAMI UNIV, OHIO, 69- *Concurrent Pos:* Nat Acad Sci sr res assoc, Inst Space Studies, 70-71. *Mem:* Am Phys Soc; Am Asn Physics Teachers. *Res:* Classical and quantum kinetic theory; scattering theory; theoretical plasma physics. *Mailing Add:* Dept Physics Miami Univ Oxford OH 45056

KELLY, DONALD HORTON, b Erie, Pa, May 6, 23; m 50; c 1. VISION. *Educ:* Univ Rochester, BS, 44; Univ Calif, Los Angeles, PhD(eng), 60. *Prof Exp:* Engr, Mitchell Camera Corp, 44; photog res engr, Technicolor Corp, 46-52, sr staff mem res, 53-61; sr staff mem optics res div, Itek Corp, Mass, 61-63, mgr info systs dept, Vidya Div, Calif, 63-66; STAFF SCIENTIST, VISUAL SCI PROG, SRI INT, 66- *Concurrent Pos:* Mem comt vision, Armed Forces-Nat Res Coun, 62-64; vis prof & NIH spec fel, Ctr Visual Sci, Univ Rochester, 71-72; mem visual sci B study sect, NIH, 73-77. *Honors & Awards:* Edgar B Tillyer Award, Optical Soc Am, 86. *Mem:* AAAS; fel Optical Soc Am; Asn Res Vision & Ophthal. *Res:* Vision research; visual instruments; spatio-temporal interactions in the visual process; stabilized retinal images and automated psychophysical techniques. *Mailing Add:* Visual Sci Prog SRI Int Menlo Park CA 94025

KELLY, DOROTHY HELEN, b Fitchburg, Mass, July 29, 44. PEDIATRICS, PULMONOLOGY. *Educ:* Fitchburg State Col, BSN, 66; Wayne State Univ, BS, 68, MD, 72. *Prof Exp:* Intern, Dept Pediat, Mass Gen Hosp, 72-73, resident, 73-75; from instr to assoc prof pediat, Harvard Med Sch, 75-89; ASSOC PEDIATRICIAN, MASS GEN HOSP, 85- *Concurrent Pos:* Fel pediat pulmonary med, Mass Gen Hosp, 76-79, asst pediat, 75-79, co-dir, Pediat Pulmonary Lab, 77-86, asst pediatrician 79- 84, assoc dir, Pediat Pulmonary Unit, 88; consult, Sudden Infant Death Syndrome Proj, Bur Commun Health Serv, Dept Health, Educ & Welfare, 79-80, prof, Orgn Soc Pediat Res, 85-, FDA, Health Devices, 85, ECRI, Apnea Monitoring Standards, 87-88, FDA, Health Devices, 88; chmn, Apnea Adv Comt, Nat Sudden Infant Death Syndrome Found, 79-81, mem, Sci Rev Comt, 81. *Mem:* Fel Am Acad Pediat; Asn Psychophysiol Study; Int Pediat Soc; Am Med Women's Asn; Am Thoracic Soc; Soc Pediat Res. *Res:* Control of ventilation; Sudden Infant Death Syndrome; sleep apnea. *Mailing Add:* Mass Gen Hosp Pediat Pulmonary Unit Boston MA 02114

KELLY, DOUGLAS ELLIOTT, b Cheyenne, Wyo, Nov 13, 32; m 54; c 5. DEVELOPMENTAL ANATOMY, MICROSCOPIC ANATOMY. *Educ:* Colo State Univ, BS, 54; Stanford Univ, PhD(biol sci), 58. *Prof Exp:* From instr to asst prof biol, Univ Colo, 58-63; from asst prof to assoc prof biol struct, Sch Med, Univ Wash, 63-70; prof & chmn dept, Sch Med, Univ Miami, 70-74; PROF ANAT & CELL BIOL & CHMN DEPT, SCH MED, UNIV SOUTHERN CALIF, 74- *Concurrent Pos:* USPHS res fel, Zool Lab, State Univ Utrecht, 59-60; NSF & NIH res grants, Univ Colo, 60-63, Univ Wash, 63-70, Univ Miami, 70-74, & Univ Southern Calif, 77-; Univ Colo fac res fel, Univ Wash, 62-63; mem anat comt, Nat Bd Med Exam, 70-74; NIH Human Embryol & Develop Study Sect, 78-82 & chmn, 83-85; pres, Asn Anat Chmn, 79-80 & Am Asn Anat, 86-87; mem admin bd, coun acad sci, Asn Am Med Col, 82-84 & 85-88, chmn, 87-88. *Honors & Awards:* Medal, Japan Asn Anat, 84. *Mem:* Am Asn Anat; Soc Develop Biol; Am Soc Zool; Am Soc Cell Biol. *Res:* Electron microscopy; development and ultrastructure of junctional complexes; ultrastructure of muscle and eye. *Mailing Add:* Am Med Col One DuPont Circle NW Suite 200 Washington DC 90033

KELLY, EDGAR PRESTON, JR, b Beaumont, Tex, Aug 5, 33; m 54; c 3. MATHEMATICS. *Educ:* Stephen F Austin State Col, BS, 55; Fla State Univ, MS, 56; Okla State Univ, PhD(math), 60. *Prof Exp:* Mathematician comput ctr, Socony Mobil Oil Co, 56-57; asst prof math, Stephen F Austin State Col, 60-62; prof & dir comput ctr & dean basic col, Univ Southern Miss, 62-64, chmn dept math, 64-67; prof math, 67-80, PROF MATH & STATIST, LA TECH UNIV, 80- *Mem:* Am Math Soc; Math Asn Am. *Res:* Infinite series and summability methods. *Mailing Add:* Dept Math La Tech Univ Ruston LA 71272

KELLY, EDWARD JOSEPH, b Baltimore, Md, Mar 4, 34; m 67; c 2. PHYSICAL CHEMISTRY. *Educ:* Johns Hopkins Univ, BES, 56, MAT, 62, MS, 68; Purdue Univ, MS, 67, PhD(chem, physics), 72. *Prof Exp:* Engr, Bendix Radio Corp, 60-61; teacher, Mt St Joseph High Sch, Md, 62-65; asst prof math & physics, Mt Marty Col, 72-75; asst prof, 75-80, ASSOC PROF CHEM, MARIAN COL, 80- *Mem:* AAAS; Am Chem Soc; Am Asn Physics Teachers; Sigma Xi. *Res:* Exploring alternatives in science teaching; quantum mechanics of small molecules. *Mailing Add:* 4440 Manning Rd Indianapolis IN 46208

KELLY, ERNEST L, b DuBois, Pa, Jan 6, 50; m 69; c 3. PHYSICAL PHARMACY, ANALYTICAL METHODS DEVELOPMENT. *Educ:* Millersville State Col, BA, 71; Villanova Univ, MS, 74, PhD(phys chem), 77. *Prof Exp:* Res asst, McNeil Labs, 72-74; sr anal chem, Merck Sharp & Dohme Res Labs, 74-79; sect head, Wm H Rorer, Inc, 79-81; DIR, ROGER CENT RES, 81- *Mem:* Am Chem Soc; Am Pharmaceut Asn; Acad Pharmaceut Sci; Am Asn Pharmaceut Soc. *Res:* Development of analytical and microscopic methods for the analysis of pharmaceutical drug substances and raw materials; evaluation of physical chemical properties of pharmaceutical drug substances in relationship to the formulation and stability of the drug. *Mailing Add:* 159 Pine Lane Yardley PA 19067

KELLY, FLOYD W, JR, b Greeley, Colo, Dec 30, 41; m 65; c 2. ORGANIC CHEMISTRY. *Educ:* Colo State Univ, BS, 63; Univ Ore, MS, 65; Univ Idaho, PhD(org chem), 68. *Prof Exp:* Fel chem, Utah State Univ, 68-69; INSTR CHEM, CASPER COL, 69- *Mem:* Am Chem Soc. *Res:* Organic synthesis; organic photochemistry; gas phase homolyses. *Mailing Add:* Dept Chem Casper Col Casper WY 82601

KELLY, FRANCIS JOSEPH, b Baltimore, Md, Oct 12, 40; m 64; c 3. LONG WAVE PROPAGATION, MAGNETO SPHERIC PROPAGATION. *Educ:* Cath Univ Am, Washington, DC, BA, 62, PhD(physics), 66. *Prof Exp:* Physicist, Nat Bur Standards, Washington, DC, 62-65 & Naval Ord Lab, White Oak, Silver Spring, Md, 65-68; PHYSICIST, NAVAL RES LAB, WASHINGTON, DC, 68- *Concurrent Pos:* Lectr, Va Polytech Inst, 76-84. *Mem:* Am Phys Soc; Inst Elec & Electronics Engrs; Union Radio Sci Int. *Res:* Propagation of long electromagnetic waves and systems for transmitting and receiving them; constructed models of atmospheric noise and studied the propagation of such waves from a satellite to the earth; nuclear structure effects on electron and neutrino reactions; author of various publications. *Mailing Add:* Space Sci Div Code 4101 Naval Res Lab Washington DC 20375-5000

KELLY, GEORGE EUGENE, b Brooklyn, NY, Mar 28, 44; m 70; c 3. MECHANICAL ENGINEERING, STATISTICAL MECHANICS. *Educ:* State Univ NY, Stony Brook, BES, 65; Northwestern Univ, PhD(mech eng), 70. *Prof Exp:* Res assoc, Nat Res Coun, Nat Bur Standards 70-72; MECH ENGR, NAT INST STANDARDS & TECHNOL, 72- *Honors & Awards:* Silver Medal, Dept Commerce, 78. *Mem:* Am Phys Soc; Am Soc Heating, Refrig & Air-Conditioning Engrs. *Res:* Theoretical, laboratory and field research on the performance of heating and cooling equipment and systems, controls, and energy management systems in buildings and residences; thermodynamics, fluid mechanics, heat transfer and methods of numerical and analytical analysis. *Mailing Add:* Mech Systs Group Rm B114 Bldg 226 Nat Inst Standards & Technol Washington DC 20899

KELLY, GREGORY, b McKeesport, Pa, Dec 7, 54; m 76; c 2. CELL-CYCLE REGULATION, CARCINOGENESIS. *Educ:* Univ Pittsburgh, BS, 77; Purdue Univ, PhD(biochem), 83. *Prof Exp:* Postdoctoral fel, Dept Biochem, Sch Med, Univ Iowa, 83-86; STAFF SCIENTIST, INHALATION TOXICOL RES INST, 86- *Concurrent Pos:* Adj asst prof, Dept Vet Path, Purdue Univ, 89-; clin asst prof, Toxicol Prog, Sch Pharm, Univ NMex, 90- *Mem:* Am Asn Cancer Res; Am Soc Biochem & Molecular Biol; AAAS. *Res:* Pulmonary carcinogenesis; development of the tracheo-bronchial epithelium; role of cell-cycle controlling genes in the development of neoplasia. *Mailing Add:* Inhalation Toxicol Res Inst PO Box 5890 Albuquerque NM 87185

KELLY, GREGORY M, PHARMACOLOGY. *Prof Exp:* SR FEL, DEPT PHARMACOL, UNIV WASH, 89- *Mailing Add:* Dept Pharmacol Univ Wash Med Ctr SJ-30 Seattle WA 98195

KELLY, HENRY CHARLES, b Boston, Mass, July 10, 45; m 69; c 2. TECHNOLOGY & ECONOMIC GROWTH, ENERGY EFFICIENCY. *Educ:* Cornell Univ, BA, 67; Harvard Univ, PhD(physics), 72. *Prof Exp:* Physicist, US Arms Control & Disarmament Agency, 71-74; tech adv to dir, Off Technol Assessment, 75-78, dir technol & int rels, 78-79; assoc dir, Solar Energy Res Inst, 79-81; SR ASSOC, OFF TECHNOL ASSESSMENT, 81- *Concurrent Pos:* AAAS Cong Sci fel, 74-75; chmn, Lawrence Berkely Lab Appl Sci Div Rev Comt, 89 & 90; mem sci adv bd, Risk Reduction Subcomt, Environ Protection Agency. *Mem:* Fel AAAS; fel Am Phys Soc; Fedn Am Specialists. *Res:* Theory and application of light scattering techniques; photovoltaic and other solar energy equipment; energy conservation technologies; international relations; policy, technology and structural economic change including publications on federal statistics, textiles and apparel, information technolgy, technology and education, energy efficiency technology, construction, and renewable energy; strategic arms control, nuclear effects, and quantumelectrodynamics. *Mailing Add:* Off Technol Assessment US Cong Washington DC 20510

KELLY, HENRY CURTIS, b Providence, RI, May 17, 30; m 56; c 3. INORGANIC CHEMISTRY. *Educ:* Bates Col, BS, 51; Brown Univ, PhD(chem), 62. *Prof Exp:* Anal chemist, Metal Hydrides Inc, 51-52, res chemist, 52-58, sr res chemist, 62-64; from asst to instr chem, Brown Univ, 58-62; from asst prof to assoc prof, 64-74, PROF CHEM, TEX CHRISTIAN UNIV, 74-, CHMN, DEPT CHEM, 89- *Concurrent Pos:* Dir, honors prog, Tex Christian Univ, 81-87. *Mem:* AAAS; Am Chem Soc; The Chem Soc; Sigma Xi; NY Acad Sci; Royal Soc Chem. *Res:* Chemistry of boron and silicon hydrides; boron-nitrogen compounds; kinetics and mechanisms of hydride reactions in solution; amineborane solvolysis and oxidation; kinetics of peroxidatic activity of metal-porphyrins and enzymes; cyclodextrin inclusion compound function. *Mailing Add:* Dept Chem Tex Christian Univ Ft Worth TX 76129

KELLY, HUGH P, b Boston, Mass, Sept 3, 31; m 55; c 9. PHYSICS. *Educ:* Harvard Univ, AB, 53; Univ Calif, Los Angeles, MS, 54; Univ Calif, Berkeley, PhD(physics), 63. *Prof Exp:* Physicist, Univ Calif, Berkeley, 62-63; res physicist, Univ Calif, San Diego, 63-64, res asst prof, 64-65; from asst prof to assoc prof & assoc dean, 65-70, prof, 70-77, chmn dept, 74, Commonwealth prof physics, 77-89, dean fac arts & sci, 85-89, ALUMNI COUN THOMAS JEFFERSON PROF, UNIV VA, 89-, PROVOST, 89- Concurrent Pos: Lectr, Univ Calif, San Diego, 65. Honors & Awards: J W Beams Award, Am Physics Soc, 85. Mem: Fel Am Phys Soc. Res: Theoretical physics, particularly many-body and atomic physics. Mailing Add: Dept Physics Univ Va Charlottesville VA 22903

KELLY, JAMES L(ESLIE), b New Orleans, La, Dec 20, 32; m 56; c 4. CHEMICAL ENGINEERING, MATERIALS SCIENCE ENGINEERING. Educ: Tulane Univ, BS, 54; La State Univ, MS, 60, PhD(chem eng), 62. Prof Exp: Process engr, Kaiser Aluminum & Chem Corp, 56-57 & Ormet Corp, 57-59; chem tech div, Oak Ridge Nat Lab, 62-64; assoc prof, 64-72, PROF NUCLEAR ENG, SCH ENG & APPL SCI, UNIV VA, 72- Mem: Am Inst Chem Engrs; Am Nuclear Soc; Nat Asn Corrosion Engrs. Res: Radiation processing; reactor materials; nuclear chemical engineering; radioactive waste disposal. Mailing Add: Rte 2 Box 670 Keswick VA 22947

KELLY, JEFFREY JOHN, b Portland, Ore, Nov 2, 42; m 66; c 3. BIOCHEMISTRY, CHEMISTRY. Educ: Harvey Mudd Col, BS, 64; Univ Calif, Berkeley, PhD(chem), 68. Prof Exp: Asst prof chem, Reed Col, 68-72; FAC MEM, DEPT CHEM, EVERGREEN STATE COL, 72- Concurrent Pos: Vis prof chem, Harvey Mudd Col, 80-81; vis prof chem, Ctr Process Anal Chem, Univ Wash, 87-88. Mem: AAAS; Sigma Xi. Res: Physical and chemical processes of photosynthesis; biomedical spectroscopy; analytical near infrared spectroscopy. Mailing Add: 5735 Cedar Flats Rd SW Olympia WA 98502

KELLY, JOHN BECKWITH, b New York, NY, Aug 30, 21. MATHEMATICS. Educ: Columbia Univ, AB, 42; Mass Inst Technol, PhD, 48. Prof Exp: Instr math, Univ Wis, 48-50; mem, Inst Advan Study, 50-51; from instr to assoc prof, Mich State Univ, 51-62; assoc prof, 62-66, PROF MATH, ARIZ STATE UNIV, TEMPE, 66- Mem: Sigma Xi. Res: Number theory; graph theory; combinatorial analysis. Mailing Add: 6707 E McDonald Dr Scottsdale AZ 85253

KELLY, JOHN FRANCIS, b Chicago, Ill, Nov 28, 31; m 59; c 9. HORTICULTURE, OLERICULTURE. Educ: Mich State Univ, BS, 53, MS, 57; Univ Wis, PhD(hort, plant physiol), 60. Prof Exp: Agr res asst, Campbell Soup Co, 52-53; asst, Mich State Univ, 56 & Univ Wis, 57-59; asst prof veg crops & soils, Southern Ill Univ, 59-62; soils technologist, Campbell Soup Co, 62-64; dir pioneer plant res, 65-66, vpres pioneer res, Campbell Inst Agr Res, 66-72; prof veg crops & chmn dept, Univ Fla, 72-78; chmn dept, 78-90, PROF HORT, MICH STATE UNIV, 78- Mem: Fel AAAS; fel Am Soc Hort Sci (pres, 85-86); Am Hort Soc. Res: Culture, physiology, nutrition and chemical composition of vegetable crops; quality of food crops. Mailing Add: Dept Hort Mich State Univ East Lansing MI 48824-1112

KELLY, JOHN HENRY, b Tonawanda, NY, Sept 26, 52. DIFFRACTION, NON-LINEAR PROPAGATION. Educ: Univ Buffalo, BS, 74; Univ Rochester, MS, 76, PhD(optics), 80. Prof Exp: RES ASSOC, LAB LASER ENERGETICS, 80- Mem: Inst Elec & Electronics Engrs; Optical Soc Am; Sigma Xi. Res: Diffraction and the propagation of light in large laser systems; resonant energy transfer in both crystalline and amorphous materials. Mailing Add: 85 Derick Dr Fishkill NY 12524

KELLY, JOHN RUSSELL, b Nashua, NH, Jan 25, 52. OCEANOGRAPHY. Educ: Univ NH, BA, 74; Univ RI, PhD(oceanog), 82. Prof Exp: Res asst, Grad Sch Oceanog, Univ RI, 75-81; RES ASSOC, ECOSYSTS RES CTR, CORNELL UNIV, 81- Mem: Am Soc Limnol & Oceanog; Sigma Xi. Res: Elemental cycling in marine, aquatic and terrestrial systems. Mailing Add: 237 Corson Hall Cornell Univ Ithaca NY 14853

KELLY, JOHN V, b London, Ont, Aug 21, 26; nat US. OBSTETRICS, GYNECOLOGY. Educ: Wayne State Univ, BS, 48, MD, 51; Am Bd Obstet & Gynec, dipl, 61. Prof Exp: Intern, Metrop Hosp, NY Med Col, 51-52; resident obstet & gynec, 52-55; from instr to asst prof, Sch Med, Univ Calif, Los Angeles, 57-64; med missionary, St Luke's Hosp, Anua, ENigeria, 64-66; prof obstet & gynec, Sch Med, Univ Pa, 67-75; CHMN DEPT OBSTET & GYNEC, MARICOPA COUNTY HOSP, PHOENIX, 75- Concurrent Pos: Res fel, Harvard Med Sch, 55; Graves fel, Free Hosp Women, Brookline, Mass, 55; Fulbright fel, Stockholm, Sweden, 56; adj prof obstet & gynec, Sch Med, Univ Ariz, 75- Mem: Am Fertil Soc; Am Med Asn; Am Fedn Clin Res. Res: Dynamics of uterine muscle contraction. Mailing Add: Maricopa County Hosp Box 5099 Phoenix AZ 85010

KELLY, KENNETH C, b New York, NY, March 6, 28; div; c 2. ARRAY ANTENNAS FOR MICROWAVES, MICROWAVE FILTERS. Educ: Polytech Inst Brooklyn, BSEE, 53; Univ Calif, Los Angeles, MS, 63. Prof Exp: Engr, Polytech Res & Develop, 51-53; mem tech staff, Hughes Aircraft Co, 53-58, section head, 58-62; mem tech staff, Rantec Div, Emerson Elec Co, 62-70, dept head, 70-73; consult, 73-86; SR SCIENTIST, HUGHES AIRCRAFT CO, 86- Honors & Awards: Community Serv Award, Inst Elec & Electronics Engrs, 75. Mem: Res Engrs Soc Am; Inst Elec & Electronics Engrs. Res: Microwave optics, microwave antennas and various passive microwave components and devices. Mailing Add: Hughes Aircraft Co Bldg RO1 MS-10G24 PO Box 92426 Los Angeles CA 90009-2426

KELLY, KENNETH WILLIAM, b New York, NY. ORGANIC CHEMISTRY. Educ: St John's Univ, NY, BS, 61, MS, 63; Rutgers Univ, PhD(chem), 69. Prof Exp: Chemist synthesis, Merck & Co, Rahway, NJ, 63-69; DIR RES & DEVELOP, KAY-FRIES, INC, 69- Mem: Am Chem Soc. Res: Organic synthesis; organic analysis. Mailing Add: PO Box 246 Tomkins Cove NY 10986

KELLY, KEVIN ANTHONY, b United Kingdom, March 29, 45; US citizen; m 81. SCIENCE & PUBLIC POLICY. Educ: Notre Dame Univ, BS, 67; Yale Univ, MS, 68; Ohio State Univ, PhD(nuclear eng), 76. Prof Exp: Sr res assoc, Mech Eng Dept, Ohio State Univ, 76-77; ASSOC DIR RES, NAT

REGULATORY RES INST, 77- Concurrent Pos: Adj asst prof nuclear eng, Ohio State Univ, 77. Mem: Am Nuclear Soc; AAAS. Res: Regulation of electric and gas utilities, the economic, technological and sociological impacts of regulation. Mailing Add: 5594 Longbow Lane Columbus OH 43235

KELLY, LEROY MILTON, b Leominster, Mass, May 8, 14; m 44; c 2. MATHEMATICS. Educ: Northeastern Univ, BSCE, 38; Boston Univ, MA, 40; Univ Mo, PhD(math), 48. Prof Exp: Instr math, Univ Mo, 45-48; from asst prof to assoc prof, 48-58, PROF MATH, MICH STATE UNIV, 58- Mem: Am Math Soc; Math Asn Am. Res: Metric geometry; topology; lattice theory. Mailing Add: Mich State Univ East Lansing MI 48823

KELLY, MAHLON GEORGE, JR, b Plymouth, NH, Mar 24, 39; m 70. AQUATIC ECOLOGY, LIMNOLOGY. Educ: Harvard Univ, AB, 60, PhD(biol), 68; Univ NH, MS, 62. Prof Exp: Sci staff, R/V Anton Bruun, 62-63; res asst biol, Woods Hole Oceanog Inst, 63 & Harvard Univ, 64-67; staff oceanogr, Mass Inst Technol, 68; vis asst prof environ biol, Univ Miami, 68-69; asst prof biol, NY Univ, 69-70; asst prof, 70-75, ASSOC PROF ENVIRON SCI, UNIV VA, 75- Concurrent Pos: Vis scientist, Scottish Marine Biol Lab, 76, Danish Fresh Water Lab, 78 & 80-84, Freshwater Biol Asn, UK, 84, 85 & 88-90. Mem: AAAS; Am Soc Limnol & Oceanog; Ecol Soc Am; Sigma Xi; Freshwater Biol Asn UK; Scottish Marine Biol Asn. Res: Photosynthetic behavior of algae and aquatic macrophytes; marine bioluminescence; remote sensing of benthic communities; aquatic nutrient cycling. Mailing Add: Dept Environ Sci Clark Hall Univ Va Charlottesville VA 22903

KELLY, MARTIN JOSEPH, b New York, NY, Sept 27, 24. PHYSICS. Educ: St John's Univ, NY, BS, 49; NY Univ, PhD, 58. Prof Exp: Physicist, Naval Mat Lab, 51-53; assoc, Nucleonics, Inc, 54-59; assoc, Tech Res Group, Inc, 59; mem fac, Manhattan Col, 59-64; chmn dept, 64-74, PROF PHYSICS, C W POST COL, LONG ISLAND UNIV, 64- Mem: Am Phys Soc. Res: Neutron physics; reactors; shielding. Mailing Add: Dept Physics C W Post Col Greenvale NY 11548

KELLY, MICHAEL DAVID, b Wis, Jan 26, 38; m 63; c 3. INTELLIGENT DATABASES, KNOWLEDGE-BASED SYSTEMS. Educ: Ga Tech, BS, 64; Stanford Univ, MS, 67, PhD(computer sci), 70. Prof Exp: Asst prof computer sci, Ga Tech, 70-75; CHIEF SCIENTIST, BDM INT, INC, 75- Concurrent Pos: Consult, IBM Corp, 68 & UNESCO, Caracas, Venezuela, 75; vis scientist, Carnegie-Mellon Univ, 73 & Centre Mondial, Paris, France, 83. Mem: Am Asn Artificial Intel; Asn Comput Mach; Inst Elec & Electronics Engrs. Mailing Add: 1670 Moorings Dr Reston VA 22090

KELLY, MICHAEL THOMAS, b Indianapolis, Ind, Mar 8, 43; m 65; c 4. CLINICAL MICROBIOLOGY, IMMUNOLOGY. Educ: Purdue Univ, BS, 65; Ind Univ, PhD(microbiol), 69, MD, 73. Prof Exp: Fel, Sch Med, Ind Univ, 69-71, res assoc infectious dis, 71-73; intern path, Scg Med, Univ Minn, 73-74; comn officer res, Rocky Mountain Lab, NIH, USPHS, 74-76; asst prof path, Sch Med, Univ Utah, 76-78; assoc prof path, Univ Tex Med Br, Galveston, 78-; HEAD, MICROBIOL DEPT, METRO MCNAIR CLIN LAB. Mem: AAAS; Reticuloendothelial Soc; Am Asn Immunologists; Am Soc Microbiol; Am Fedn Clin Res. Res: Host-parasite relationships; modulation of macrophage function by microbial agents; immunopotentiation by microbial agents; mechanism of macrophage activation; clinical microbiology; antimicrobial susceptibility testing; marine microbiology. Mailing Add: Metro McNair Clin Lab 660 W Seventh Ave Vancouver BC V5Z 1B5 Can

KELLY, MINTON J, b Liberty, Mo, Feb 14, 21; m 49; c 3. HIGH TEMPERATURE CHEMISTRY. Educ: Tex A&M Univ, BS, 47, MS, 50, PhD(phys chem), 56. Prof Exp: Field party chief oceanog res found, Tex A&M Univ, 47-48, consult instrumentation, 50-54; teaching fel chem univ, 54-55; develop engr instrumentation & controls div, Oak Ridge Nat Lab, 55-59, group leader reactor chem div, 59-62; group supvr instrumentation, Aerospace Div, Boeing Co, 62-63; chemist, Reactor Chem Div, 63-74, RES ASSOC, CHEM TECHNOL DIV, OAK RIDGE NAT LAB, 74- Concurrent Pos: Engr, Arabian-Am Oil Co, 48-49. Mem: AAAS; Sigma Xi; fel Am Inst Chemists. Res: Instrumental measurements under nuclear conditions. Mailing Add: 114 Lewis Lane Oak Ridge TN 37830

KELLY, NELSON ALLEN, b Lakewood, Ohio, Aug 6, 51; m 82; c 3. PHOTOCHEMISTRY, GAS-PHASE KINETICS. Educ: Miami Univ, Ohio, BS, 73; Pa State Univ, PhD(phys chem), 77. Prof Exp: Sr res scientist, 77-82, STAFF RES SCIENTIST, GEN MOTORS RES LABS, 82- Concurrent Pos: Mem, Chem Comt, Air Pollution Control Asn, 82- Honors & Awards: Joseph P Culler Prize. Mem: Sigma Xi; Am Chem Soc; InterAm Photochem Soc; Air Pollution Control Asn. Res: The chemistry of ozone formation and transport in the atmosphere; the role of automobiles in the formation of photochemical smog. Mailing Add: 13748 Imperial Ct Sterling Hts MI 48312

KELLY, PATRICK CLARKE, analytical chemistry, for more information see previous edition

KELLY, PATRICK JOSEPH, b Minneapolis, Minn, Feb 12, 26; m 50; c 8. ORTHOPEDIC SURGERY. Educ: St Lawrence Univ, BS, 45; St Louis Univ, MD, 49; Univ Minn, MS, 58. Prof Exp: Prof orthop surg, Mayo Grad Sch, 69-73; prof orthop surg, Mayo Med Sch, 73-90; CONSULT, MAYO CLIN, 57- Concurrent Pos: Am Orthop Asn Traveling Fel; pres, Bd Trustees, Orthopaedic Res & Educ Found; mem, Am Bd Orthop Surg; mem, Am Inst Biol Sci Adv Panel, NASA. Mem: Am Acad Orthop Surg; Am Orthop Asn; Orthop Res Soc (past pres); Am Physiol Soc. Res: Circulation and physiology of bone; bone metabolism. Mailing Add: Mayo Clin 200 First St SW Rochester MN 55905

KELLY, PAUL ALAN, b Washington, DC, June 3, 43; m 69; c 1. MEDICAL RESEARCH. *Educ:* Western Mich Univ, BS, 66, MS, 68; Univ Wis, PhD(endocrinol & reprod physiol), 72. *Prof Exp:* Fel endocrinol, McGill Univ, 72-74; fel, 74-75, asst prof, 75-80, assoc prof physiol, Laval Univ, 80-82; prof, Dept Med & Physiol, McGill Univ, 83-91; dir, Lab Molecular Endocrinol, Royal Victoria Hosp, Montreal, 83-91; DIR PROF, HOPITAL NECKER ENFANTS MALADES, PARIS, 91- *Concurrent Pos:* Sr mem, Med Res Coun Group Molecular Endocrinol, 75-; Med Res Coun Can scholar, 75-80; dir Lab Molecular Endocrinol, 83-91, dir L'Unite 344 D'Endocrinol Moleculaire. *Mem:* Endocrine Soc; Am Physiol Soc; Can Soc Clin Invest; Int Soc Neuroendocrinol; Can Soc Endocrinol & Metab. *Res:* Endocrine control of mammary carcinoma; neuroendocrine regulation of pituitary hormone secretion; control of polypeptide and steroid hormone receptors in target tissues. *Mailing Add:* Dept Med Royal Victoria Hosp 687 Pine Ave W Montreal PQ H3A 1A1 Can

KELLY, PAUL J, b Riverside, Calif, June 26, 16; m 46; c 2. MATHEMATICS. *Educ:* Univ Calif, Los Angeles, AB, 37, MA, 39; Univ Wis, PhD(math), 42. *Prof Exp:* Instr math, Univ Southern Calif, 46-49; from asst prof to assoc prof, 49-59, PROF MATH, UNIV CALIF, SANTA BARBARA, 59- *Res:* Metric geometry. *Mailing Add:* 30 Winchester Canyon Rd Apt 68 Goleta CA 93117

KELLY, PAUL JAMES, b Montreal, Que, July 19, 34; m 60; c 5. PHYSICS. *Educ:* Sir George Williams Univ, BSc, 60; Carleton Univ, MSc, 62, PhD(physics), 65. *Prof Exp:* From asst res officer to assoc res officer, 65-76, SR RES OFFICER PHYSICS, NAT RES COUN, 76- *Concurrent Pos:* Asst invest officer, Energy Res & Develop Admin consult grant, Wash State Univ, 75-77; Air Force consult grant, Wash State Univ, 78- *Mem:* Am Phys Soc. *Res:* Thermally stimulated processes; interaction of high-intensity laser pulses with solids. *Mailing Add:* 310 Smyth Rd Ottawa ON K1H 5A3 Can

KELLY, PAUL SHERWOOD, b Erie, Pa, Dec 22, 27; m 56; c 3. ATOMIC PHYSICS, QUANTUM MECHANICS. *Educ:* Haverford Col, AB, 49; Yale Univ, MS, 50; Univ Calif, Los Angeles, PhD(physics), 61. *Prof Exp:* Physicist, US Naval Ord Lab, Md, 50-51; electronic scientist, Nat Bur Standards, Calif, 51-53; res scientist, Lockheed Missiles & Space Co, Palo Alto, 60-68; PROF PHYSICS, HUMBOLDT STATE UNIV, 68- *Mem:* Am Phys Soc; Sigma Xi. *Res:* Calculation of atomic wave functions and related atomic parameters; nuclear structure calculations. *Mailing Add:* Dept Physics Humboldt State Univ Arcata CA 95521

KELLY, PETER MICHAEL, b New York, NY, July 6, 22; m 46; c 3. PHYSICS, ELECTRICAL ENGINEERING. *Educ:* Union Col, NY, BS, 50; Calif Inst Technol, MS, 52, PhD(physics) & PhD(elec eng), 60. *Prof Exp:* Design engr, jet propulsion lab, Calif Inst Technol, 51-52; proj engr, electronics div, Century Metalcraft Co, 53-54; mem tech staff, Hughes Aircraft Co, 54-56; from design engr to prin scientist, aeronutronic div, Ford Motor Co, 56-61; mgr elec dept, Astropower, Douglas Aircraft Co, 61-62; from assoc dir res to chief engr, Systs Tech Ctr, Philco-Ford, 62-69; pres & chmn bd, Kelly Sci Corp, 69-80; PROF ELEC ENG & DIR, TELECOMMUN CTR, GEORGE WASHINGTON UNIV, WASHINGTON, DC, 80- *Concurrent Pos:* Consult, NSF, 60-61; consult, President's Crime Comn, 66 & President's Commun Task Force, 68. *Mem:* AAAS; Asn Comput Mach. *Res:* Radar; network synthesis; data processing. *Mailing Add:* 3431 Emerson St Arlington VA 22207

KELLY, RAYMOND CRAIN, b Portland, Ore, Sept 4, 45; m 68; c 1. TOXICOLOGY. *Educ:* Wash State Univ, BS, 67; Univ Ore, PhD(chem), 75. *Prof Exp:* Develop chemist, Sacred Heart Gen Hosp, 69-71; assoc toxicologist, Cuyahoga County Coroners Off, Ohio, 75-77; head toxicol & statist, Lab Procedures, Upjohn Co, 77-78; asst dir, Dept Toxicol & Emergency Serv, Bio-Sci Lab, Van Nuys, Calif, 78-83; LAB & SCI DIR, MEDTOX LAB INC, 89- *Concurrent Pos:* Nat Res Serv fel, Nat Inst Drug Abuse, NIH, 76; lab inspector, Nat Inst Drug Abuse, 91-; mem, Clin Lab Tech Adv Comt, Calif Dept Health Serv. *Mem:* Am Acad Forensic Sci; Am Asn Clin Chem; Soc Forensic Toxicologists; Col Am Pathologists. *Res:* Devising of novel methods for the analysis of drugs in biological fluids, characterization of drug metabolites, mechanisms of drug toxicity and monitoring of therapeutic drug concentrations in man; pharmacology of drug abuse in man. *Mailing Add:* Medtox Lab Inc 9176 Independence Ave Chatsworth CA 91311

KELLY, RAYMOND LEROY, b Rockford, Ill, Feb 2, 21; m 43; c 2. ATOMIC SPECTROSCOPY. *Educ:* Univ Wis, PhD(physics), 51. *Prof Exp:* Asst, Univ Wis, 47-51; res physicist, Stanford Res Inst, 51-60; assoc prof, 60-68, PROF PHYSICS, NAVAL POSTGRAD SCH, 68- *Concurrent Pos:* Mem comt line spectra elements, Nat Res Coun, 71-78. *Mem:* Am Phys Soc; fel Optical Soc Am. *Res:* Infrared; spectroscopy of the ultraviolet. *Mailing Add:* Dept Physics Naval Postgrad Sch Monterey CA 93943

KELLY, REGIS BAKER, b Edinburgh, Scotland, May 26, 40; div; c 3. CELL BIOLOGY. *Educ:* Univ Edinburgh, BSc, 61, dipl, 62; Calif Inst Technol, PhD(biophys), 67. *Prof Exp:* Instr neurobiol, Harvard Med Sch, 69-71; asst prof biochem & biophys, 71-74, assoc prof, 74-78, PROF BIOCHEM & BIOPHYS, UNIV CALIF, SAN FRANCISCO, 78- *Concurrent Pos:* Helen Hay Whitney Found fel, Sch Med, Stanford Univ, 67-69 & Harvard Med Sch, 69-70; Multiple Sclerosis fel, 70-71; mem adv panel, NEI; mem, Study Sect, NIH; rev, Am Cancer Soc; vis prof, MIT, 86. *Mem:* Soc Neurosci; Am Soc Biol Chem; Am Soc Cell Biol. *Res:* Membrane traffice in cells; protein sorting; molecular events in nerve terminals; synaptic vesicles; development of the neuron. *Mailing Add:* Dept Biochem & Biophys Univ Calif Med Ctr San Francisco CA 94143

KELLY, RICHARD DELMER, b Kingston, NY, Aug 24, 35; m 54; c 4. BIOLOGY, SCIENCE EDUCATION. *Educ:* State Univ NY Albany, BS, 55, MS, 56; Syracuse Univ, EdD(biol, sci educ), 65. *Prof Exp:* High sch teacher, NY, 56-63; PROF BIOL, STATE UNIV NY ALBANY, 63- *Concurrent Pos:* Consult, NY State Educ Dept, 60- & NSF Summer Progs, 60-65; vis fel, Col Educ, Kingston Upon Hull, Eng, 73-74 & Rosentiel Inst, Univ Miami, 82. *Mem:* AAAS; Am Inst Biol Sci. *Res:* Instructional technology; television, audio-tutorial; cetaceans and whaling history. *Mailing Add:* Dept Biol State Univ NY Albany 1400 Washington Ave Albany NY 12222

KELLY, RICHARD W(ALTER), b Iowa City, Iowa, Sept 6, 35; m 64. ELECTRICAL ENGINEERING. *Educ:* Univ Iowa, BSEE, 58, MS, 62, PhD(elec eng), 65. *Prof Exp:* From instr to asst prof elec eng, Univ Iowa, 58-65; assoc prof, 65-70, PROF ELEC ENG, ARIZ STATE UNIV, 70-, ASST DEAN, 80- *Concurrent Pos:* Sr Fulbright-Hays lectureship, Trinity Col, Dublin, 72-73. *Mem:* Inst Elec & Electronics Engrs; Am Soc Eng Educ. *Res:* Application of modern signal theory; detection and estimation theory. *Mailing Add:* Dept Elec Eng Ariz State Univ Tempe AZ 85281

KELLY, ROBERT CHARLES, b St Joseph, Mich, Nov 28, 39; m 60; c 2. ORGANIC CHEMISTRY. *Educ:* Kalamazoo Col, BA, 61; Harvard Univ, MA, 63, PhD(chem), 66. *Prof Exp:* RES ASSOC ORG CHEM, UPJOHN CO, 65- *Mem:* Am Chem Soc. *Res:* Organic synthesis and structure determination, particularly of cyclic hydrocarbons; terpenes and oxygen heterocycles; natural products chemistry; prostaglandins. *Mailing Add:* Dept Chem 7252-209-6 Upjohn Co 301 Henrietta St Kalamazoo MI 49001

KELLY, ROBERT EDWARD, b Abington, Pa, Oct 20, 34; m 64; c 2. FLUID MECHANICS, HEAT TRANSFER. *Educ:* Franklin & Marshall Col, BA, 57; Rensselaer Polytech Inst, BS, 57; Mass Inst Technol, AE, 59, ScD(aeronaut eng), 64. *Prof Exp:* Guest scientist, Nat Phys Lab, UK, 60-61, UK Civil Serv sr res fel fluid mech, 64-66; res asst aeronaut eng, Mass Inst Technol, 61-64; asst res geophysicist, Inst Geophys & Planetary Physics, Univ Calif, San Diego, 66-67; from asst prof to assoc prof, 67-75, PROF ENG, UNIV CALIF, LOS ANGELES, 75- *Concurrent Pos:* Sci Res Coun sr vis fel, Dept Math, Imp Col, London, 73-74; consult, Hughes Aircraft Co, 76-83; assoc ed, Physics Fluids, 81-83. *Mem:* Fel Am Phys Soc; Am Inst Aeronaut & Aeronaut; Am Soc Mech Engrs; Sigma Xi. *Res:* Viscous flow; flow instabilities; fluid wave motion; stratified and rotating flow phenomena; thermal convection. *Mailing Add:* Dept Mech Aero & Nuclear Engr Engr IV Rm 46-147B Univ Calif Los Angeles CA 90024-1597

KELLY, ROBERT EMMETT, b Cape Girardeau, Mo, Nov 26, 29; m 62; c 3. PHYSICS. *Educ:* Southeast Mo State Univ, BS, 50; Univ Mo-Rolla, MS, 52; Univ Conn, PhD(physics), 59. *Prof Exp:* Prof physics, Univ Miss, 59-88; PHYSICIST, LOS ALAMOS NAT LAB, 88- *Concurrent Pos:* Consult, Boeing Co, 54, E I du Pont de Nemours & Co, Inc, 57 & Am Optical Co, 59; Richland fac fel, Hanford Lab, 65; vis investr oceanog, Woods Hole Oceanog Inst, 67; prof, NMex Highlands Univ, 68; physicist, Gen Elec Co, 52 & Marshall Space Flight Ctr, NASA, 70 & 71; consult, Los Alamos Sci Lab, 75-82 & Lawrence Livermore Lab, 75-79; vis scientist, Ctr d'Etudes Bruyeres-le-Chatel, Serv Physique Nucleaire, France, 81 & 82. *Mem:* Am Geophys Union; Acoust Soc Am; Int Soc Optical Eng. *Res:* Electromagnetic theory; physical optics; atmospheric and mathematical physics; mathematical approach to transient radiation damage in optical fibers; energy deposition and profiles of particle beams, plus topics in musical acoustics. *Mailing Add:* 75 Tesuque Los Alamos NM 87544

KELLY, ROBERT FRANK, b Fond du Lac, Wis, May 21, 19; m 44; c 6. BIOCHEMISTRY, ANIMAL HUSBANDRY. *Educ:* Univ Wis, BS, 48, MS, 53, PhD(biochem, animal husb), 55. *Prof Exp:* Pub sch instr, Wis, 48-51; asst, Univ Wis, 51-55; assoc prof, 55-58, PROF FOOD SCI & TECHNOL, VA POLYTECH INST & STATE UNIV, 58- *Concurrent Pos:* Williams-Waterman Scientist, Haiti, 63; AED Prof, Sri Lanka, 86. *Honors & Awards:* Signal Serv Award, Am Meat Sci Asn, 84. *Mem:* Fel AAAS; Am Meat Sci Asn; Am Soc Animal Sci; Am Inst Food Technologists; NY Acad Sci; Am Coun Sci & Health. *Res:* Food science and nutrition. *Mailing Add:* Dept Food Sci & Technol Rm 109 Va Polytech Inst & State Univ Blacksburg VA 24061-0324

KELLY, ROBERT JAMES, b New York, NY, Dec 2, 23; m 52; c 7. ORGANIC CHEMISTRY. *Educ:* Trinity Col, BS, 43; NY Univ, MS, 47, PhD(chem), 52. *Prof Exp:* Asst, NY Univ, 46-51; res chemist, 51-62, sr res scientist, 62-65, mgr new fiber res & develop, Uniroyal Fiber & Textile Div, 65-69, mgr tire cord res & develop, 69-76, TECH DIR UNIROYAL FIBER & TEXTILE DIV, 76- *Mem:* Am Chem Soc. *Res:* Synthetic rubber and fibers. *Mailing Add:* 4018 Sandwood Dr Columbia SC 29206-2222

KELLY, ROBERT LINCOLN, elementary particle physics, for more information see previous edition

KELLY, ROBERT P, b Dover, NJ, Mar 17, 38; m 63. CELL PHYSIOLOGY. *Educ:* Fairleigh Dickinson Univ, BS, 62; Fordham Univ, MS, 65, PhD(biol), 66. *Prof Exp:* From instr to asst prof, 64-70, assoc prof, 70-80, PROF BIOL & CHMN DEPT & COORDR PROG BIOL CHEM, ST PETER'S COL, NJ, 80- *Concurrent Pos:* Fac res grant, St Peter's Col, NJ, 64-65 & 67. *Res:* Insect and cell physiology with emphasis on nutrition and enzyme chemistry. *Mailing Add:* Dept Biol St Peter's Col 2641 Kennedy Blvd Jersey City NJ 07306

KELLY, ROBERT WITHERS, b Stanford, Ky, Oct 20, 26; m 48; c 2. ZOOLOGY, ECOLOGY. *Educ:* Centre Col, BA, 49; Univ Ore, MS, 50; Univ Mo, PhD(zool), 56. *Prof Exp:* Head sci dept, Campbellsville Jr Col, 51-53; assoc prof biol, Southeastern La Col, 56-63 & Ariz State Col, 63-64; prof biol, Furman Univ, 64-88, chmn dept, 74-85; RETIRED. *Mem:* Am Soc Zool. *Res:* Invertebrate ecology, especially freshwater forms. *Mailing Add:* 85 Elizabeth Dr Travelers Rest SC 29690

KELLY, RONALD BURGER, b Fairvale, NB, May 26, 20; m 45. ORGANIC CHEMISTRY. *Educ:* Univ NB, MSc, 51, PhD(chem), 53. *Prof Exp:* Beaverbrook overseas scholar, Univ London, 53-54; Nat Res Coun Can fel, Queen's Univ, Ont, 54-55; sr res chemist, Merck & Co, Ltd, Can, 55-58; res assoc chem, Upjohn Co, 58-67; prof, 67-85, chmn div sci & math, 73-79,

EMER PROF CHEM, UNIV NB, 85- *Mem:* AAAS; Am Chem Soc; Royal Soc Chem; fel Chem Inst Can; NY Acad Sci. *Res:* Structure determination of organic molecules; synthesis of natural products. *Mailing Add:* Dept Chem Univ NB St John NB E2L 4L5 Can

KELLY, SALLY MARIE, b Bridgeport, Conn. CLINICAL PATHOLOGY, BIOCHEMICAL MEDICAL GENETICS. *Educ:* Conn Col, AB, 43; Univ Wis, MA, 44, PhD(bot), 46; NY Univ, MD, 63; Am Bd Path, dipl, 71. *Prof Exp:* Instr, Simmons Col, 47-48; asst prof plant sci, Vassar Col, 48-51; sr res scientist, 51-64, assoc res scientist, 64-67, RES PHYSICIAN, WADSWORTH CTR LABS & RES, NY STATE DEPT HEALTH, 67- *Concurrent Pos:* Fel, Brooklyn Bot Garden, 45-47; fel, Harvard Univ, 47-48; Brown-Hazen Fund fel, 58-59 & 60-63; res assoc prof pediat, Albany Med Col, 68- *Mem:* Col Am Pathologist; fel AAAS. *Res:* Cell physiology; enteroviruses; biochemical medical genetics; clincal pathology. *Mailing Add:* Wadsworth Ctr Labs & Res NY State Dept Health Albany NY 12201

KELLY, SIDNEY J(OHN), energy conversion and conservation, for more information see previous edition

KELLY, SUSAN JEAN, b Cincinnati, Ohio, Oct 2, 47; div; c 2. ENZYMOLOGY, BIOCHEMICAL ENGINEERING. *Educ:* Col Mt St Joseph, AB, 69; Purdue Univ, PhD(biochem), 74. *Prof Exp:* Res assoc enzyme eng, Dept Biochem, Purdue Univ, 74-80; res assoc, Univ NC, Chapel Hill, 80-84, workshop coordr, Carolina Workshop, 84-90; BIOL, DURHAM NC, 90- *Concurrent Pos:* Consult, 80-; lectr & lab mgr, Duke Univ, Howard Hughes Lab Molecular. *Mem:* Sigma Xi. *Res:* Enzyme-catalyzed synthesis of sucrose and other economically important physiological compounds; enzymic mechanism of phosphatases; phosphonate analogs of phosphatase substrates; relationship of phosphatases to developmental changes and to cancer. *Mailing Add:* 8104 Lair Ct Chapel Hill NC 27516

KELLY, THADDEUS ELLIOTT, b New York, NY, Oct 7, 37; m 60; c 3. MEDICAL GENETICS. *Educ:* Davidson Col, BS, 59; Med Col SC, MD, 63; Johns Hopkins Univ, PhD(genetics), 75. *Prof Exp:* Asst prof med & pediat, Sch Med, Johns Hopkins Univ, 73-75; assoc prof, 75-80, PROF PEDIAT, UNIV VA, 80- *Concurrent Pos:* Dir, Div Med Genetics, Univ Va, 75-, assoc dir, Clin Res Ctr, 76-81; chmn, Genetics Adv Comt, State Va, 79- *Mem:* Soc Pediat Res; Am Pediat Soc; Am Soc Human Genetics; Am Fedn Clin Res; Clin Genetics & Birth Defects Soc. *Res:* Biochemical genetic analysis of genetic heterogenity; genetic disorders in large family studies; molecular biology of X chromosome. *Mailing Add:* Dept Pediat Sch Med Univ Va Hosp Box 386 2488A Charlottesville VA 22903

KELLY, THOMAS J, b Birmingham Al, Nov 21, 41; m 69; c 2. MOLECULAR BIOLOGY. *Educ:* Johns Hopkins Univ, BA, 62, PhD(biophys), 68, MD, 69. *Prof Exp:* Staff assoc, NIH, 70-72; from asst prof to assoc prof microbiol, 72-79, PROF MOLECULAR BIOL & GENETICS, JOHNS HOPKINS UNIV, SCH MED, 81-, CHMN DEPT, 82- *Concurrent Pos:* Mem, Virol Study Sect, NIH, 80-84, chmn, 88-90; bd dirs, Passano Found, 87-; NIH career develop award, 72-77; mem, Awards Assembly, Gen Motors Cancer Prize, 86-89, Bd Sci Counr, Nat Ctr Biotechnol Info; Harvey Soc lectr, 90. *Mem:* Am Soc Microbiol; Am Soc Biol Chemists; Am Soc Virol; fel Am Acad Arts & Sci. *Res:* Molecular genetics of animal cells & viruses. *Mailing Add:* Dept Molecular Biol & Genetics Johns Hopkins Univ Sch Med Baltimore MD 21205

KELLY, THOMAS JOSEPH, b Brooklyn, NY, June 14, 29; m 51; c 6. AEROSPACE ENGINEERING, INFORMATION RESOURCE MANAGEMENT. *Educ:* Cornell Univ, BME, 51; Columbia Univ, MSME, 56; Mass Inst Technol, MS in IM, 70. *Hon Degrees:* DSc, State Univ NY, 83. *Prof Exp:* Propulsion engr, Rigel Missile prog, Grumman Aerospace Corp, 51-53, group leader jet air induction, 53-56; performance engr, Wright Patterson AFB, 56-58; group leader rocket propulsion, Lockheed Aircraft Corp, 58-59; asst chief propulsion, Grumman Aerospace Corp, 59-60, eng proj leader Apollo & Lunar Module studies & proposals, 60-62, proj engr, eng mgr & dep prog mgr, Lunar Module Prog, 62-70, dep dir space shuttle prog, 70-72, dir space progs, 72-76, vpres eng, 76-81, vpres tech opers, 81-86, VPRES IRM, DATA SYSTS DIV, GRUMMAN AEROSPACE CORP, 86- *Concurrent Pos:* Mem NASA panel on space vehicles, Res & Technol Adv Coun, 75-77; mem aeronaut & space eng bd ad hoc comt on technol large space syst, Nat Res Coun, 78. *Honors & Awards:* Cert Appreciation, NASA, 69, Distinguished Pub Serv Medal, 83; Spacecraft Design Award, 73, fel, Am Inst Aeronaut & Astronaut. *Mem:* Nat Acad Eng; Am Soc Mech Engrs; Am Inst Aeronaut & Astronaut. *Res:* Development of manned spacecraft; engineering effort of Project Apollo Lunar Module; development and production engineering of a variety of military aircraft; information systems planning. *Mailing Add:* Grumman Corp 111 Stewart Ave Bethpage NY 11714-3584

KELLY, THOMAS MICHAEL, b Watertown, NY, May 16, 41; m 62; c 3. PHYSICS, SOLID STATE SCIENCE. *Educ:* Le Moyne Col, NY, BS, 62; Wayne State Univ, PhD(physics), 66. *Prof Exp:* AEC res assoc positron annihilation, New Eng Inst Med Res, 66-68; sr res physicist, Eastman Kodak Res Labs, 68-74, res assoc, 74-88, lab head, 82-84, asst div dir, 84-88, VPRES & DIR RES & DEVELOP, EASTMAN KODAK (JAPAN) LTD, 88- *Mem:* Inst Elec & Electronics Engrs. *Res:* Physics of solid state imaging; design and fabrication. *Mailing Add:* Eastman Kodak Co 343 State St NJ160 Rochester NY 14650

KELLY, THOMAS ROSS, b New York, NY, Apr 26, 42; m 66; c 2. ORGANIC CHEMISTRY. *Educ:* Col of the Holy Cross, BS, 64; Univ Calif, Berkeley, PhD(org chem), 68. *Prof Exp:* Asst prof, 69-74, assoc prof, 74-80, PROF CHEM, BOSTON COL, 80- *Res:* Organic synthesis; natural products. *Mailing Add:* Dept Chem Boston Col Chestnut Hill MA 02167-3800

KELLY, WALTER JAMES, b Cleveland, Ohio, Feb 25, 41; m 71. ORGANIC CHEMISTRY, POLYMER CHEMISTRY. *Educ:* Case Inst Technol, BS, 63, Case Western Reserve Univ, PhD(phys org chem), 70. *Prof Exp:* Sr res chemist, Polymer Res, Goodyear Tire & Rubber Co, 69-78; staff scientist, Polymer Technol, 78-80, proj engr, 80-81, PROJ MGR, FOIL DV, GOULD INC, 81- *Mem:* Am Chem Soc; Electrochemical Soc. *Res:* Dynamic properties of elastomers; structure-property correlations; polymer rheology and processing; crosslinking mechanisms; post polymerization reactions; adhesion, polymer modification electrodeposition. *Mailing Add:* 1545 Forest Lane Marion IN 46952-9810

KELLY, WILLIAM ALBERT, b Cincinnati, Ohio, July 16, 27; m 52; c 2. NEUROSURGERY. *Educ:* Ohio Wesleyan Univ, BA, 50; Univ Cincinnati, MD, 54. *Prof Exp:* Res fel neurosurg, Univ Chicago Clins, 56-57; resident, 57-59, chief resident & clin asst, 60-61, from instr to assoc prof, 61-77, PROF NEUROSURG, UNIV WASH, 77- *Concurrent Pos:* Res fel, Univ Wash, 59-60. *Mem:* Am Asn Neurol Surg. *Res:* Pituitary tumors; medical education on student and resident level. *Mailing Add:* Dept Neurol Surg Univ Wash Sch Med R1-20 Seattle WA 98195

KELLY, WILLIAM ALVA, b Cullman, Ala, Feb 24, 37; div; c 5. VETERINARY PATHOLOGY. *Educ:* Auburn Univ, DVM, 62; Purdue Univ, PhD(vet path), 71. *Prof Exp:* Vet, pvt pract, 62-66; instr vet path, Purdue Univ, 66-70; vet pathologist, Mead Johnson & Co, 70-81; SR VET PATHOLOGIST, BRISTOL MYERS SQUIBB, 87- *Mem:* Am Col Vet Pathologists; Int Acad Path. *Res:* Experimental toxicologic pathology; pathology of laboratory animals; nutritionally-induced pathology; chemical carcinogenesis. *Mailing Add:* 3249 Lower New Harmony Rd Mt Vernon IN 47620

KELLY, WILLIAM CARY, b Memphis, Tenn, June 14, 19; m 42; c 4. VEGETABLE CROPS. *Educ:* Univ Tenn, BS, 40; Ohio State Univ, MS, 41; Cornell Univ, PhD(veg crops), 45. *Prof Exp:* Asst, Ohio State Univ, 40-41 & Cornell Univ, 42-45; assoc agronomist, Plant Soil & Nutrit Lab, USDA, 45-46; horticulturist, 46-52; assoc prof, 52-55, prof, Dept Veg Crops, 55-84, EMER PROF, CORNELL UNIV, 84- *Concurrent Pos:* Vis prof, Univ Philippines, 59-60. *Mem:* Fel AAAS; Am Soc Hort Sci; fel Am Soc Plant Physiol; Sigma Xi. *Res:* Plant growth and development; yield and composition of vegetables as influenced by environment and mineral nutrition. *Mailing Add:* 159A Plant Sci Cornell Univ Ithaca NY 14850

KELLY, WILLIAM CLARK, b Braddock, Pa, Mar 18, 22; m 47; c 2. PHYSICS. *Educ:* Univ Pittsburgh, BS, 43, MS, 46, PhD(physics), 51. *Prof Exp:* From asst to assoc prof physics, Univ Pittsburgh, 46-58; dir dept educ & manpower, Am Inst Physics, 58-65; fel officer, Nat Acad Sci-Nat Res Coun, 65-67, dir off sci personnel, 67-74; exec dir comn human resources, Nat Res Coun, 74-83; spec asst, Am Assoc Physics Teachers, 84-89; RETIRED. *Concurrent Pos:* Ford fac fel, 54-55; mem subcomt prof sci & technol manpower, Dept Labor, 71-72; secy comn physics educ, Int Union Pure & Appl Physics, 66-72, chmn, 72-75; mem coun on teaching sci, Int Coun Sci Unions, 75-78. *Mem:* AAAS; Am Phys Soc; Am Asn Physics Teachers. *Res:* Measurement of spectral emissivities of metals; beta and gamma ray spectroscopy; improvements in the teaching of science; manpower studies; human-resource supply and demand, especially in science and engineering. *Mailing Add:* 9320 Renshaw Dr Bethesda MD 20817

KELLY, WILLIAM CROWLEY, b Philadelphia, Pa, May 10, 29; m 59. ECONOMIC GEOLOGY. *Educ:* Columbia Univ, AB, 51, MA, 53, PhD(geol), 54. *Prof Exp:* Asst econ geol, Columbia Univ, 51-53; instr geol, Hunter Col, 54; opers analyst, Opers Res Off, Johns Hopkins Univ, 54-56; from instr to assoc prof geol, 56-67, prof geol & mineral, 67-80, prof & chmn geol sci, 80-90, VPRES RES, UNIV MICH, ANN ARBOR, 90- *Concurrent Pos:* Ed, Geochem News, 61-63. *Mem:* Geol Soc Am; Geochem Soc; Mineral Soc Am; Soc Econ Geol; Geol Soc France. *Res:* Chemical weathering; telluride ore deposits; oxidation of lead-zinc ores; mineralogy of iron oxides; ore microscopy. *Mailing Add:* Fleming Admin Bldg Rm 4080 Univ Mich Ann Arbor MI 48109

KELLY, WILLIAM DANIEL, b St Paul, Minn, Oct 28, 22; m 51; c 6. SURGERY. *Educ:* Univ Minn, BS, 43, MB, 45, MD, 46, PhD(surg), 55; Am Bd Surg, dipl, 55; Am Bd Thoracic Surg, dipl, 59. *Prof Exp:* From instr to assoc prof, sr surg, 61-80, mem surg staff, 42-80, CLIN PROF SURG, UNIV HOSPS, 80- *Concurrent Pos:* Dir exp surg lab, Vet Admin Hosp, Minneapolis, 59-60, chief surg, 60-62. *Mem:* AAAS; Soc Exp Biol & Med; Soc Univ Surgeons; AMA; NY Acad Sci. *Res:* Homotransplantation; cardiovascular physiology and surgery. *Mailing Add:* 3838 Zenith Ave S Minneapolis MN 55410

KELLY, WILLIAM H, b Rich Hill, Mo, July 2, 26; m 50; c 3. EXPERIMENTAL NUCLEAR PHYSICS, PHYSICS PEDAGOGY. *Educ:* Graceland Col, AA, 48; Univ Mich, BSE, 50, MS, 51, PhD(physics), 55. *Prof Exp:* Asst physics, Eng Res Inst, Univ Mich, 51-54; from asst prof to prof physics, Mich State Univ, 55-79, from assoc chmn to chmn dept, 68-79; prof physics, Mont State Univ, 79-83, dean col letters & sci, 79-83; dean, Col Sci & Humanities, 83-89, PROF PHYSICS, IOWA STATE UNIV, AMES, 83- *Concurrent Pos:* Physicist, Naval Res Lab, 56, Lawrence Radiation Lab, Univ Calif, 61-62, 67-68 & Oak Ridge Nat Lab, 64; mem bd trustees, Graceland Col, 78-90; mem, Spec Adv Comt on medium energy electron accelerator fac, Argonne Univ Asn, 81-83; mem bd trustees, Univ Res Asn; guest scientist, Lawrence Berkeley Lab, 89-90. *Mem:* AAAS; fel Am Phys Soc; Am Asn Physics Teachers (vpres, 79, pres elect, 80, pres, 81); Am Soc Eng Educ. *Res:* Nuclear spectroscopy; nuclear structure; gamma ray spectroscopy; physics pedagogy. *Mailing Add:* Dept Physics & Astron Physics Bldg Iowa State Univ Ames IA 50011

KELLY, WILLIAM ROBERT, b Norfolk, Va, July 24, 44; m 69; c 1. GEOCHEMISTRY, ANALYTICAL CHEMISTRY. *Prof Exp:* vis assoc geochem, Calif Inst Technol, 75-77, res fel, 77-79; RES CHEMIST ANALYTICAL CHEM, NAT INST STANDARDS & TECHNOL, 79- *Honors & Awards:* Nininger Meteorite Award, 74; IR100 Award, 84. *Mem:* Am Chem Soc; Meteoritical Soc; Geochem Soc; Sigma Xi. *Res:* Cosmochemistry; thermal ionization mass spectrometry; environmental chemistry; analytical chemistry. *Mailing Add:* Ctr Analytical Chem Nat Inst Standards & Technol Gaithersburg MD 20899

KELLY-FRY, ELIZABETH, wid; c 2. BIOACOUSTICS, MEDICAL ULTRASOUND. *Educ:* Howard Univ, ScM, 53; Sarasota Univ, EdD(sci educ), 75. *Prof Exp:* Biophys res lab, Univ Ill, 54-64; assoc dir res, Intersci Res Inst, 64-67, vpres, 68-71; assoc prof surg, 72-80, assoc prof radiol, Sch Med, Ind Univ, Indianapolis, 80-; RES SCIENTIST, IND CTR ADVAN RES, 72-, WISHARD HOSP, IND UNIV MED CTR, 80- *Concurrent Pos:* Consult, Bur Radiol Health, 79-81, ultrasound corp, 80-; mem, NIH Diag Res Adv Comt, 80-81; assoc ed, J Clin Ultrasound, 75-79. *Honors & Awards:* Japan Soc US Med Award, 76; Presidential Award, Am Inst Ultrasound Med, 80; World Fedn US Med Award, 88. *Mem:* Am Phys Soc; Acoust Soc Am; Biophys Soc; fel Am Inst Ultrasound Med; NY Acad Sci. *Res:* Ultrasound breast examination; design of ultrasound instrumentation. *Mailing Add:* Dept Radiol Wishard Hosp Ind Univ Med Ctr Indianapolis IN 46202

KELMAN, ARTHUR, b Providence, RI, Dec 11, 18; m 49; c 1. PHYTOBACTERIOLOGY. *Educ:* Univ RI, BS, 41; NC State Univ, MS, 46, PhD(plant path), 49. *Hon Degrees:* DSc, Univ RI, 77. *Prof Exp:* From instr to prof plant path, NC State Univ, 48-62, Reynolds Distinguished prof, 62-65; prof & chmn dept, 65-75, L R Jones distinguished prof, 75-85, Wis Alumni Res Found sr distinguished prof plant path, Univ Wis-Madison, 85-89, prof bact, 78-89; UNIV DISTINGUISHED SCHOLAR, NC STATE UNIV, 90- *Concurrent Pos:* Vis investr, Rockefeller Inst, 53-54; vis lectr, Am Inst Biol Sci, 58-60; NSF sr fel, Cambridge Univ, 71-72, vis prof, Dept Biochem, 71-72; mem US nat comt, Int Union Biol Sci; chmn, Div Biol Sci, Assembly Life Sci, Nat Res Coun, 81-84; chmn, Sect Appl Biol & Agr Sci, Comn Life Sci, 81-83, Bd Basic Biol, 84-85; coun, Nat Acad Sci, 86-89; chmn, Class VI, Appl Biol & Agr Sci, Nat Acad Sci, 88-91. *Honors & Awards:* Fel Award, Am Phytopath Soc, 69, Award of Distinction, 83; Stakman Award, 87. *Mem:* Nat Acad Sci (coun, 86-89); Am Acad Arts & Sci; Am Inst Biol Sci; Soc Gen Microbiol; hon mem Int Soc Plant Path (vpres, 68-73, pres, 73-78); Am Phytopath Soc (vpres, 66, pres, 67); fel AAAS; Sigma Xi. *Res:* Physiology of parasitism; bacterial diseases of plants; nature of resistance to bacterial soft rot of potatoes. *Mailing Add:* Dept Plant Path NC State Univ Box 7616 Raleigh NC 27695-7616

KELMAN, BRUCE JERRY, b Chicago, Ill, July 1, 47; m 72. TERATOLOGY, COMPARATIVE TOXICOLOGY. *Educ:* Univ Ill, BS, 69, MS, 71, PhD(vet med sci), 75; Am Bd Toxicol, cert, 80, 85. *Prof Exp:* Res asst physiol, Univ Ill, 69-74; res assoc toxicol, Comp Animal Res Lab, Oak Ridge, Tenn, 74-76, asst prof prenatal toxicol, 76-79; sr res scientist develop toxicol, Pac Northwest Labs, Battelle Mem Inst, 79-80, assoc mgr, 80-81 & 83-84, mgr, 81-84, mgr biol & chem dept, 85-90; DEPT TOXICOL, FAILURE ANALYSIS ASSOC, 90- *Mem:* Soc Toxicol; Soc Exp Biol & Med; Teratology Soc; Am Soc Pharmacol & Exp Therapeut; Am Physiol Soc; Am Acad Vet & Comp Toxicol. *Res:* Toxicology of chemicals (including chemical mixtures and metals) and radiation; teratology and other developmental effects of toxic materials including radionuclides; transplacental movements of materials; toxicology of electrical and magnetic fields. *Mailing Add:* Failure Anal Assoc 149 Commonwealth Dr Menlo Park CA 94025

KELMAN, L(EROY) R, b Minneapolis, Minn, Aug 16, 19; m 42; c 3. ENGINEERING. *Educ:* Univ Minn, BS, 42. *Prof Exp:* Metallurgist, Caterpillar Tractor Co, 42-44; group leader, Metall Div, Argonne Nat Lab, 47-66, mgr fuels & mat sect, Liquid Metal Fast Breeder Reactor Prog Off, 66-70, prog planner, Mat Sci Div, 70-73, proj leader & prog coordr, Safety of Light Water Reactor Fuels, 75-78 SR METALLURGIST, ARGONNE NAT LAB, 59- *Mem:* Am Nuclear Soc; Am Soc Metals; Am Inst Mining, Metall & Petrol Engrs; Sigma Xi. *Res:* Metallurgy of materials for nuclear reactors; fuel and structural materials and liquid metal coolants; behavior of nuclear fuels under transient and hypothetical accident conditions; fuels and structural materials for nuclear reactors; materials behavior in liquid metal coolants. *Mailing Add:* 1030 E Prairie Ave Naperville IL 60540

KELMAN, ROBERT BERNARD, b Ansonia, Conn, Aug 12, 30; m 57; c 2. MATHEMATICS, COMPUTER SCIENCE. *Educ:* Univ Calif, Berkeley, AB, 53, MA, 55, PhD(math), 58. *Prof Exp:* Comput engr, NAm Aviation, Inc, 55-56; instr, Univ Ill, 57-58; mathematician, Int Bus Mach Corp, 58-61; mgr biomath res, Univac Div, Sperry Rand Corp, 61-63; res asst prof math, Univ Md, 63-66; assoc prof math, Colo State Univ, 66-68, prof comput sci, 72-88, chmn, 81-88; assoc prof prev med, Univ Colo Med Ctr, 67-72, prof, 72-80; DIR, KLMN CONSULT, 88- *Concurrent Pos:* Lectr, Howard Univ, 61-65; consult, Exec Off President Eisenhower, 60-61. *Mem:* Fel AAAS; Am Math Soc; Math Asn Am; Soc Indust & Appl Math; Am Soc Nephrology; Am Fed Clin Res. *Res:* Differential equations; theoretical renal physiology; computer modeling. *Mailing Add:* 1312 Robertson St Ft Collins CO 80524

KELMERS, ANDREW DONALD, b New York, NY, Mar 24, 29; m 52; c 4. INORGANIC CHEMISTRY. *Educ:* Antioch Col, BS, 52; Ohio State Univ, MS, 54. *Prof Exp:* Chemist, Oak Ridge Nat Lab, 54-57 & Union Carbide Nuclear Co, NY, 57-62; res scientist, Astropower Inc Div, Douglas Aircraft Co, Inc, Calif, 62-63; group leader, 63-78, PROG MGR PROCESS CHEM, OAK RIDGE NAT LAB, 79- *Mem:* AAAS; Am Chem Soc; Am Soc Biol Chemists. *Res:* Program planning and coordination; nuclear fuel cycle process research and development; solvent extraction, transition element and nitrogen-compound chemistry as well as uranium and plutonium behavior; resource recovery and extrusive metallurgy. *Mailing Add:* 130 Cumberland View Dr Oak Ridge TN 37830

KELNER, ALBERT, b Philadelphia, Pa, Sept 7, 12; m 46; c 3. BIOLOGY. *Educ:* Univ Pa, BA, 40, PhD(biol), 43; NC State Col, MSc, 42. *Prof Exp:* William Pepper Lab Clin Med fel, Sch Med, Univ Pa, 43-46; bacteriologist, Biol Lab Cold Spring Harbor, NY, 46-49; USPHS fel, Harvard Univ, 49-51; from asst prof to assoc prof, 51-61, prof, 61-81, ABRAHAM S & GERTURDE BURG EMER PROF BIOL, BRANDEIS UNIV, 81- *Mem:* AAAS; Am Soc Microbiol; Am Soc Photobiol. *Res:* Microbiology; microbiological genetics; photobiology; genetics; DNA repair, especially photoreactivation, genetic and evolutionary aspects; science education for non-scientists, especially humanistic biology. *Mailing Add:* 303 Florence Rd Waltham MA 02154

KELNHOFER, WILLIAM JOSEPH, b Manitowoc, Wis, Nov 24, 30. MECHANICAL ENGINEERING, FLUID MECHANICS. *Educ:* Marquette Univ, BME, 56; Catholic Univ, MME, 60, DEng, 66. *Prof Exp:* Proj engr, US Navy Bur Ships, 56-59; from instr to assoc prof, 60-73, ORD PROF MECH ENG, CATH UNIV AM, 73-, DEPT CHMN, 83- *Concurrent Pos:* Prin investr, US Navy contract, 62-; assoc, US Army contract, 63-; prin investr, Off Naval Res contract, 66 & Nat Bur Standards Contracts, 74-76; res prof, Max Planck Inst, Goettingen, 66-67 & Munich Tech Univ, 69-70. *Mem:* Am Soc Mech Engrs; Nat Soc Prof Engrs; Am Soc Heating, Refrig & Air Conditioning Engrs; Sigma Xi. *Res:* Heat transfer; boundary layer theory; thermal systems; energy conservation. *Mailing Add:* Dept Mech Eng Cath Univ Washington DC 20017

KELSAY, JUNE LAVELLE, b Jacksboro, Tex, June 29, 25. NUTRITION. *Educ:* NTex State Univ, BS, 46, MS, 47; Univ Wis, PhD(foods & nutrit), 67. *Prof Exp:* Instr nutrit, NTex State Univ, 47-50; technician nutrit res, Tex Agr Exp Sta, 51-52; nutrit specialist, USDA, 54-62, res nutritionist, Agr Res Serv, 67-87; RETIRED. *Honors & Awards:* Borden Award, Am Home Econ Asn, 82. *Mem:* Am Inst Nutrit; Am Soc Clin Nutrit; Am Home Econ Asn. *Res:* Preadolescent children; folic and pantothenic acid; vitamin B-6 deficiency in man; effect of protein level; forms of vitamin B-6; excretion of niacin metabolites; nutritional status; carbohydrate response in human subjects; effects of fiber in human subjects; mineral balances; oxalic acid and mineral bioavailability; fiber and nutrient intakes. *Mailing Add:* 10401 Grosvenor Pl #1315 Rockville MD 20852

KELSEY, CHARLES ANDREW, b Norfolk, Nebr, July 9, 35; m 60; c 4. MEDICAL PHYSICS. *Educ:* St Edward's Col, BS, 57; Univ Notre Dame, PhD(physics), 62; Am Bd Radiol, dipl. *Prof Exp:* Res assoc physics, Univ Notre Dame, 62; res assoc, Univ Wis-Madison, 62-63, from instr to asst prof physics, 63-65, from asst prof to prof radiol, 65-75; chief biomed physics, 75-80, PROF RADIOL, UNIV NMEX, 75- *Mem:* AAAS; Am Phys Soc; Am Acad Phys Med & Rehab; Am Inst Ultrasonics in Med; Radiol Soc NAm; fel Am Col Radiol. *Res:* Application of physics technology to medical problems. *Mailing Add:* Dept Radiol Univ NMex Albuquerque NM 87131

KELSEY, EDWARD JOSEPH, b Washington, DC, Dec 10, 48. SOFTWARE SYSTEMS, ATOMIC & MOLECULAR PHYSICS. *Educ:* Wesleyan Univ, BA, 70; Univ Md, MS, 72, PhD(physics), 74. *Prof Exp:* Res assoc physics, Univ Nebr, Lincoln, 74-76; assoc res scientist, NY Univ, 76-78; MEM TECH STAFF COMPUT SCI, AT&T BELL LABS, 79- *Mem:* Am Phys Soc; Asn Comput Mach. *Res:* Quantum electrodynamic theory applied to problems concerning one and two electron systems; operation systems which maintain and test the telephone network. *Mailing Add:* 26 Malibu Dr Eatontown NJ 07724

KELSEY, EUGENE LLOYD, b Ponca City, Okla, May 10, 32; m 73; c 2. ELECTRICAL ENGINEERING, AERO-SPACE ENGINEERING. *Educ:* Okla State Univ, BSEE, 58; Va Polytech Inst, MSEE, 66. *Prof Exp:* Jr engr guidance, Autonetics-NAm Aviation, 56-57; test engr B-58 radar guidance, Gen Dynamics, Fort Worth, 58-62; aerospace technologist, 62-72, ENG SUPVR SYSTS DEVELOP-ELEC FLIGHT SYSTS, NASA, 72- *Concurrent Pos:* Adj instr math, Christopher Newport Col, 76-79. *Mem:* Soc Automotive Engrs-Aerospace. *Res:* Design, development and analysis of aerospace stabilization control and pointing systems for aircraft, satellite and research projects; unique requirements-unique solutions. *Mailing Add:* Langley Res Ctr NASA MS-432 Hampton VA 23665

KELSEY, FRANCES OLDHAM, b Cobble Hill, BC, Can, July 24, 14; nat US; m 43; c 2. PHARMACOLOGY. *Educ:* McGill Univ, BSc, 34, MSc, 35; Univ Chicago, PhD(pharmacol), 38, MD, 50. *Hon Degrees:* DSc, Hood Col, 62, Univ NB, 64, Western Col Women, 64, Middlebury Col, 66, Wilson Col, 67, St Mary's Col, 69, Drexel Univ, 73, McGill Univ, 84 & Univ SDak, 82. *Prof Exp:* Asst prof pharmacol, Univ Chicago, 46; assoc prof med, Sch Med, Univ SDak, 54-57; pvt pract, 57-60; DIR DIV SCI INVEST, OFF SCI EVAL, BUR DRUGS, FOOD & DRUG ADMIN, DEPT HEALTH, EDUC & WELFARE, 60- *Concurrent Pos:* Lederle award, 54-57. *Honors & Awards:* President's Award Distinguished Fed Civilian Serv, 62. *Mem:* Am Soc Pharmacol & Exp Therapeut; Soc Exp Biol & Med; Teratology Soc; Am Med Women's Asn; Am Med Writers' Asn; Sigma Xi. *Res:* Posterior pituitary; chemotherapy of malaria; radioisotopes. *Mailing Add:* 5811 Brookside Dr Chevy Chase MD 20815

KELSEY, JENNIFER LOUISE, b Montclair, NJ, Aug 27, 42. EPIDEMIOLOGY. *Educ:* Smith Col, BA, 64; Yale Univ, MPH, 66, MPhil, 68, PhD(chronic dis, epidemiol), 69. *Prof Exp:* Asst prof epidemiol, Sch Med, Yale Univ, 69-75, assoc prof, 75-; AT DEPT EPIDEMIOL, COLUMBIA UNIV, NY. *Mem:* Int Epidemiol Asn; Soc Epidemiol Res; Am Epidemiol Soc; Am Col Epidemiol. *Res:* Epidemiology of chronic diseases, especially those of the musculo-skeletal system; herniated lumbar intervertebral discs; epidemiology of cancers of the female reproductive system. *Mailing Add:* Dept Epidemiol Columbia Univ 360 W 168 St New York NY 10032

KELSEY, JOHN EDWARD, b Beloit, Wis, Oct 28, 42; m 65; c 3. SCIENTIFIC INFORMATION, ENVIRONMENTAL HEALTH. *Educ:* Univ Wis, BS, 65, PhD(pharm chem), 69. *Prof Exp:* Nat Cancer Inst overseas fel, 68-70; res chemist, 70-74, sr res chemist, 75-78, dir occup health, 78-80, dir tech serv div, 81-84, DIR TECH & ADMIN OPERS, BURROUGHS WELLCOME, CO, INC, 84- *Mem:* Am Chem Soc; Am Indust Health Coun. *Res:* Management of science information, automation, administration, radiation, and unit facilities services functions in the pharmaceutical research, development and manufacturing industry. *Mailing Add:* Burroughs-Wellcome Co Burroughs-Wellcome Co Research Triangle Park NC 27709-4498

KELSEY, LEWIS PRESTON, entomology; deceased, see previous edition for last biography

KELSEY, MORRIS IRWIN, b Easton, Pa, Aug 14, 39; m 64; c 2. IMMUNOLOGY. *Educ:* Lehigh Univ, BA, 61; Univ Mass, MS, 64; Univ Pittsburgh, PhD(biochem), 69. *Prof Exp:* Chemist starch chem, Nat Starch & Chem Corp, 63-65; asst prof biochem, Mo Inst Psychiat, 71-73; sr scientist, Frederick Cancer Res Ctr, 73-75; sect head chem carcinogenesis prog, 75-80, exec secy, Exp Therapeut Study Sect, Div Res Grants, 85-89, ASST COORDR ENVIRON CANCER, NAT CANCER INST, NIH, 80-, PROG DIR, BIOL RESPONSE MODIFIERS PROG, 89- *Concurrent Pos:* Fel biochem, St Louis Univ, 69-71; adj asst prof, Univ Mo, 72-73; adj assoc prof agr biochem, WVa Univ, 77-79. *Mem:* Am Chem Soc; AAAS; Sigma Xi; Am Soc Biol Chemists. *Res:* Biotransformation of neutral sterols and bile acids by enterohepatic enzyme systems; effects of metabolism of endogenous steroid metabolites on the metabolic activation of chemical carcinogens; mechanism of action of chemotherapeutic agents. *Mailing Add:* Div Res Grants NIH Westwood Bldg Rm 221 Bethesda MD 20892

KELSEY, RICK GUY, b Libby, Mont, Aug 14, 48; m 82; c 4. PLANT CHEMISTRY, BIOMASS UTILIZATION. *Educ:* Univ Mont, BS, 70, PhD(forestry), 74. *Prof Exp:* res assoc plant chem, Wood Chem Lab, 74-80; from res asst prof to res assoc prof, Dept Chem, Univ Mont, 81-86; res assoc prof, Entomol Dept, Ore State Univ, 86-89; RES SCIENTIST, PNW RES STA, USDA FOREST SERV, 89- *Concurrent Pos:* Prin investr, McKnight Found Individual Award in plant biol, 83-86. *Mem:* NAm Phytochem Soc. *Res:* Isolation and identification of plant allelochemicals, their physiological and ecological function in the plant and their potential use to man. *Mailing Add:* PNW Res Sta USDA Forest Serv 3200 Jefferson Way Corvallis OR 97331

KELSEY, RONALD A(LBERT), b Oakville, Conn, Mar 29, 23; m 47; c 3. CIVIL ENGINEERING. *Educ:* Polytech Inst Brooklyn, BS, 49; Carnegie Inst Technol, MS, 52. *Prof Exp:* Res engr, Alcoa Res Labs, Aluminum Co Am, 49-55; nuclear engr, Gen Dynamics Corp, 49-55, sr res engr, 60-70, eng assoc, 70-74, sect head, 74-81, sr tech specialist, 81-83; RETIRED. *Concurrent Pos:* Chmn, Aluminum Alloys Comn, Welding Res Coun, 68-84; mem, Joint USA/USSR Comn on Properties of Welds for Low Temp Appln, 75-81; chmn, Int Comt Fatigue Data Exchange & Eval, 80-84; mem, Tech Adv Comt Metals Prop Coun, 80-83; pres, Seniors Helping Seniors, Inc, 90- *Mem:* Soc Exp Stress Anal; fel Am Soc Metals; Am Welding Soc. *Res:* Deformation and fracture mechanics of materials and structures; development of metal deformation process; armor development. *Mailing Add:* PO Box 608 West Falmouth MA 02574

KELSEY, RUBEN CLIFFORD, b Park Falls, Wis, May 26, 23; m 60; c 2. COMPARATIVE ENDOCRINOLOGY. *Educ:* Univ Wis, PhB, 49, MS, 50, PhD(zool), 59. *Prof Exp:* Sr res scientist biochem, Smith, Kline & French Labs, 59-63; asst prof biol sci, Drexel Inst Technol, 63-68; head dept, E Stroudsburg Univ, 68-74 & 80-83, prof biol, 68-88; RETIRED. *Mem:* AAAS; Am Soc Zool; Am Inst Biol Sci. *Res:* Physiology of mammalian reproduction; function. *Mailing Add:* 51 Club Ct Stroudsburg PA 18360

KELSEY, STEPHEN JORGENSEN, b Salt Lake City, Utah, Jan 15, 40; m 65; c 5. DESIGN & DEVELOPMENT, CONSTRUCTION OF CHEMICAL PLANTS. *Educ:* Univ Utah, BSChE, 65, MES, 69, PhD(chem eng), 71. *Prof Exp:* Process engr, Celanese Chem Co, Pampa, Tex, 65-66; GS-4, GS-9, US Bur Mines, Salt Lake City, Utah, 66-67; asst res prof, Div Mat Sci Mech Eng, Col Eng, Univ Utah, 71-76, asst res prof, Div Artificial Surg, Organs, Col Med, 71-76, assoc res prof, Computer Sci, Dept Elec Eng, Col Eng, 71-76; plant supt, Wasatch Chem Co, Salt Lake City, Utah, 76-78; MGR ENG, THATCHER CO, SALT LAKE CITY, UTAH, 78- *Mem:* Am Inst Chem Engrs. *Res:* Use of computer graphics with numerical analysis in solving flow problems applied to artificial heart work; chlorine and sulfur dioxide handling and facility design. *Mailing Add:* 5368 Cottonwood Club Dr Salt Lake City UT 84117

KELSH, DENNIS J, b Valley City, NDak, Dec 24, 36; m 61; c 3. PHYSICAL CHEMISTRY. *Educ:* St John's Univ, Minn, BA & BS, 58; Iowa State Univ, PhD(phys chem), 62. *Prof Exp:* From instr to assoc prof, 62-72, chmn dept, 68-74, 81-84, 85-86, 89-90, PROF CHEM, GONZAGA UNIV, 72- *Concurrent Pos:* Res chemist, Spokane Mining Res Ctr, US Bur Mines, 65-; assoc res scientist, NY Univ, 66-67; Am Coun Educ fel acad admin & spec asst to dean grad sch, Wash State Univ, 74-75; coun mem, Am Chem Soc, 74-; vis scientist, Univ Wash, 85. *Mem:* Am Chem Soc. *Res:* Electrical properties of surfaces; adsorption from solution; solid-liquid separations by electrokinetics. *Mailing Add:* Dept Chem Gonzaga Univ Spokane WA 99258

KELSO, ALBERT FREDERICK, b Ft Wayne, Ind, Nov 19, 17; m 43; c 3. PHYSIOLOGY. *Educ:* George Williams Col, BA, 43, MS, 46; Loyola Univ, PhD, 59. *Hon Degrees:* DSc, Kirksville Col Osteop & Surg, 70. *Prof Exp:* Instr physiol, George Williams Col, 46-47; from instr to prof physiol, Chicago Col Osteop Med, 46-90, actg chmn dept, 54-59, chmn dept, 58-90, dir res affairs, 76-90; RETIRED. *Concurrent Pos:* Consult, Nat Bd Osteopath Exam, 65-78; res consult, Am Acad Osteop; educ consult, Am Osteopath Asn. *Honors & Awards:* Louisa Burns Mem Lectr, 81; Guttensohn-Denslow Prize, 84; Phillips Medal Honor. *Mem:* AAAS; Am Physiol Soc; Soc Exp Biol & Med; Am Heart Asn; Inst Elec & Electronics Eng; Sigma Xi. *Res:* Sensorimotor performance; circulation; tissue respiration; family medicine theory and practice. *Mailing Add:* 15443 University Dolton IL 60419

KELSO, ALEC JOHN (JACK), b Chicago, Ill, Dec 5, 30; m 51; c 2. PHYSICAL ANTHROPOLOGY. *Educ:* Northern Ill Univ, BS, 52; Univ Mich, MA, 54, PhD, 58. *Prof Exp:* From instr to assoc prof anthrop, 58-75, chmn dept, 63-68 & 71-81, dir, Semester at Sea Prog, 78-79, PROF ANTHROP, UNIV COLO, BOULDER, 75- *Concurrent Pos:* Consult, Coun Grad Schs US, 64-78; mem training comt, Nat Inst Child Health & Human Develop, 64-66; NIH spec fel, Univ Hawaii, 65-66; distinguished vis prof, Ore State Univ, 71; vchancellor acad affairs, Univ Colo, Colorado Springs, 75-77; dir, Farrand Hall Residential Acad Prog, 83-88; chmn biol unit, Am Anthrop Asn, 85-87; dir, Young Scholars Summer Session, 85-89, Honors Prog, 88-; Presidents Teaching Scholar, 90. *Mem:* Am Anthrop Asn; Am Asn Phys Anthropologists (vpres, 72-74). *Res:* Selection and blood groups; human sexuality; healthy people. *Mailing Add:* Dept Anthrop Univ Colo Boulder CO 80303

KELSO, DONALD PRESTON, b Pulaski, Va, Aug 12, 40; m 63; c 2. MARINE ECOLOGY. *Educ:* Univ Tenn, Knoxville, BS, 62; Univ Fla, MS, 65; Univ Hawaii, PhD(zool), 70. *Prof Exp:* Asst prof, 70-77, ASSOC PROF BIOL, GEORGE MASON UNIV, 77- *Mem:* Am Inst Biol Sci; Am Soc Zool; Ecol Soc Am. *Res:* Inshore marine ecology; evolution of echinoderms; reproductive cycles of tropical animals. *Mailing Add:* Dept Biol George Mason Univ 4400 University Dr Fairfax VA 22030

KELSO, EDWARD ALBERT, b Galveston, Tex, Sept 17, 14; m 44; c 2. PETROLEUM CHEMISTRY. *Educ:* Univ Tex, BA, 36, MA, 38, PhD(phys chem), 41. *Prof Exp:* Tutor chem, Univ Tex, 36-38, instr, 38-39, res assoc res inst, 40-41; chem engr, Tech Serv Div, Humble Oil & Refining Co, Exxon Chem Co, 41-45, res chemist, Res & Develop Div, 45-49, sr res chemist, 49-52, res specialist, 52-59, sr res specialist, 59-64, res assoc, Esso Res & Eng Co, 64-76, res assoc, 76-79; CONSULT, 79- *Mem:* Am Chem Soc. *Res:* High pressure thermodynamics; hydrocarbon isomerization and conversions; relationships of some isomeric hexanes. *Mailing Add:* 1310 E James No 15 Baytown TX 77520

KELSO, JOHN MORRIS, b Punxsutawney, Pa, Mar 12, 22; m 45; c 1. RADIO PHYSICS. *Educ:* Gettysburg Col, AB, 43; Pa State Univ, MS, 45, PhD(physics), 49. *Prof Exp:* Instr physics, Pa State Univ, 43-45, asst, 45-48, eng res assoc, 48-49, from asst prof to assoc prof eng res, 49-54; eval specialist, Martin Co, 54-55; mem tech staff, Ramo-Wooldridge, Inc, 55-58 & Space Tech Labs, 58-62; dir res, Electro-Physics Labs, ACF Indust, Inc, 62-66; dir res, ITT Electro-Physics Lab, Inc, 66-68, vpres & dir res, 68-75; consult, Off Telecommun Policy, Exec Off of the President, 76-78; chief scientist, Signal Anal Ctr, Honeywell Inc, Annapolis, MD, 78-87; RETIRED. *Concurrent Pos:* Vis observer, Chalmers Univ Technol, Sweden, 51-52; mem, Arecibo Eval Panel, 67-69; mem comn III & US del numerous Gen Assemblies, Int Union Radio Sci, chmn Comn G, 73-75; mem, US Nat Comt, 73-78. *Mem:* Am Geophys Union; Am Phys Soc; fel Inst Elec & Electronics Eng. *Res:* Radio wave propagation; ionospheric physics; space vehicle instrumentation; systems engineering; operational evaluation of weapon systems; space physics; electricity and magnetism. *Mailing Add:* 7801 Sylvan Dr Hudson FL 34667

KELSO, JOHN RICHARD MURRAY, b Kingston, Ont, Feb 9, 45; m 66; c 2. ENVIRONMENTAL SCIENCE. *Educ:* Univ Guelph, BSc, 67, MSc, 69; Univ Man, PhD(zool), 71. *Prof Exp:* Dir, Nanticoke Proj, Ont Ministry Natural Resources, 71-73; RES SCIENTIST, CAN DEPT FISHERIES & OCEANS, 73- *Concurrent Pos:* Assoc ed, Can J Fishery Aquatic Sci, 88- *Honors & Awards:* Chandler-Misener Award, Int Asn Great Lakes Res, 82. *Mem:* Can Soc Zoologists; Am Fisheries Soc; Int Asn Great Lakes Res. *Res:* Effects of environmental perturbations and natural factors on the community structure, biomass and production of freshwater fish communities. *Mailing Add:* Great Lakes Lab Fisheries & Aquatic Sci Sea Lamprey Control Ctr Can Dept Fisheries & Oceans Canal Dr Ship Canal PO Sault St Marie ON P6A 1P0

KELSO, RICHARD MILES, b Knoxville, Tenn, Jan 20, 37; m 60; c 3. HEATING VENTILATION & AIR CONDITIONING, BUILDING SYSTEMS. *Educ:* Univ Tenn, BS, 60, MS, 61. *Prof Exp:* Sales engr, Trane Corp, 60-68; vpres, George S Campbell & Assoc, 68-71; mech engr, Facil Planning, 71-76, PROF ENVIRON CONTROLS, SCH ARCHIT, UNIV TENN, 76- *Concurrent Pos:* Pres, Richard Kelso & Assoc, Consult Engrs, 74-89 & Kelso-Regen Assoc, Consult Engrs, 89-; regional vchmn, Am Soc Heating Refrig & Air Conditioning Engrs, 87-90. *Mem:* Am Soc Heating Refrig & Air Conditioning Engrs; Am Soc Plumbing Engrs. *Res:* Thermal and moisture transfer in buildings; indoor air quality; energy consumption; numerical modeling of building systems. *Mailing Add:* Sch Archit Univ Tenn Knoxville TN 37996-2400

KELTIE, RICHARD FRANCIS, b Alexandria, Va, Aug 1, 51; m 73; c 1. STRUCTURAL DYNAMICS. *Educ:* NC State Univ, BS, 73, MS, 75, PhD(mech eng), 78. *Prof Exp:* Engr, Appl Physics Lab, Johns Hopkins Univ, 78-81; from asst prof to assoc prof, 81-90, PROF MECH ENG, NC STATE UNIV, 90- *Mem:* Acoust Soc Am; Am Soc Mech Engrs. *Res:* Mechanical design; structural dynamics; forced acoustics radiation from large structures; structural acoustics; structural dynamics, acoustic radiation and acoustic emission. *Mailing Add:* NC State Univ Campus Box 7910 Raleigh NC 27695

KELTING, RALPH WALTER, b Bridgeport, Nebr, Sept 27, 18; c 2. PLANT ECOLOGY. *Educ:* Southwestern Inst Technol, BA, 41; Univ Okla, MS, 48, PhD(plant sci), 51. *Prof Exp:* Instr plant sci, Univ Okla, 51-53; from asst prof to assoc prof bot, Univ Tulsa, 53-61; chief, Pine Hills Field Sta, Southern Ill Univ, 61-62; prof, 62-89, EMER PROF BOT, PITTSBURG STATE UNIV, KANS, 89- *Mem:* Ecol Soc Am (treas, 63-66). *Res:* Grassland ecology; plant taxonomy. *Mailing Add:* Dept Biol Pittsburg State Univ Pittsburg KS 66762

KELTON, DIANE ELIZABETH, b Holden, Mass, Dec 4, 24. GENETICS, CANCER. *Educ:* Univ Mass, BS, 45, PhD(zool), 61. *Prof Exp:* Res asst genetics, Jackson Lab, 47-50, sr res asst, 50-53; res asst cancer res, Univ Mass, Amherst, 53-56, histol, 56-58 & genetics, 58-61, res assoc genetics & neuropath, 61-74; staff scientist, Mason Res Inst, 74-83; RETIRED. *Mem:* AAAS; Am Inst Biol Sci; Am Genetic Asn; Genetics Soc Am; Environ Mutagen Soc; Sigma Xi. *Res:* Genetics; tumor biology; cancer chemotherapy. *Mailing Add:* 15 Heatherstone Rd RFD No4 Amherst MA 01002

KELTON, FRANK CALEB, b Berkeley, Calif, Sept 1, 15; m 45, 59; c 4. PETROLEUM ENGINEERING. *Educ:* Univ Ariz, BS, 36; Johns Hopkins Univ, PhD(phys chem), 41. *Prof Exp:* Res engr, 40-45, mgr eng & res, 45-55, spec probs consult, 55-59, SR PROJ ENGR, CORE LABS, INC, 59- *Mem:* Am Inst Mining, Metall & Petrol Engrs. *Res:* Catalysis; oil well core analysis; multiphase fluid flow through porous media; electrical model studies of petroleum reservoirs; petroleum reservoir engineering; secondary recovery; pressure maintenance; electronic data processing. *Mailing Add:* 4161 Wilada Dallas TX 75220

KELTS, LARRY JIM, b Westfield, Pa, Aug 13, 37; m 67; c 2. MARINE BIOLOGY, ENTOMOLOGY. *Educ:* Cornell Univ, BS, 59; Southeastern Mass Univ, MS, 71; Univ NH, PhD(zool), 77. *Prof Exp:* Res asst marine biol, Marine Lab, Duke Univ, 60-61; res asst plant path, Agr Exp Sta, Cornell Univ, 64-69; ASST PROF MARINE BIOL & ECOL, DEPT BIOL, MERRIMACK COL, 77- *Mem:* Am Inst Biol Sci; Ecol Soc Am; Nat Wildlife Fedn. *Res:* Faunal and floral community structure and composition in stressed aquatic environments, such as salt-marsh pannes, supratidal rock pools, mixohaline and oligohaline lotic systems, bogs, temporary woodland pools and creek beds; salt-marsh dragonfly ecology. *Mailing Add:* Dept Biol Merrimack Col North Andover MA 01845

KELTY, MIRIAM C, b New York, NY, Nov 4, 38. PSYCHOLOGY & PSYCHOBIOLOGY, AGING. *Educ:* City Col NY, BA, 60, MA, 62; Rutgers Univ, PhD(psychol & psychobiol), 65. *Prof Exp:* ASSOC DIR, NAT INST AGING, NIH, 86- *Concurrent Pos:* Ed, Am Psychol J. *Mem:* Fel Am Psychol Asn; fel AAAS; World Future Soc; Geront Asn Am. *Res:* Aging, hormones and behavior; health research; ethics of research; science policy. *Mailing Add:* Off Extramural Affairs Nat Inst Aging NIH Bldg 31 Rm 5C02 Bethesda MD 20892

KEMELHOR, ROBERT ELIAS, b New York, NY, May 19, 19; m 47; c 3. MECHANICAL ENGINEERING. *Educ:* George Washington Univ, BSME, 49. *Prof Exp:* Design engr, Dept Navy, 39-52; chief engr, McLean Develop Lab, 52-57; dir res & develop, 57-58, Pesco Div, Borg Warner Corp, sect supvr, Polaris Prog, 58-62, proj engr, Landing Force Support Weapon, 62-66, prog mgr, Pershing Weapon Syst, 66-76, prog mgr, Ocean Data Acquisition Prog, 76-82, br supvr, Design & Fabrication Br, 82-85; chief engr, Appl Physics Lab, Tech Serv Dept, Johns Hopkins Univ, 86-91; CONSULT, ENG MECH, 91- *Concurrent Pos:* Consult, Thompson Ramo Wooldridge, Inc, 61-62, Cleveland Pneumatic, Inc, 62-63; US del, Int Standards Orgn Tech Comt; mem, Comput Automation Systs Asn of Soc of Mfg Engrs; chmn, DC Chap Soc Man Eng, Wash. *Mem:* Assoc fel Am Inst Aeronaut & Astronaut; sr mem Soc Mfg Engrs; sr mem Am Astronaut Soc. *Res:* Magnetic fluids and mechanisms for use in shock and vibration absorbing devices; demonstrate the feasibility of electrically controlling spring rates and damping constants by electronically reactive fluids. *Mailing Add:* 6211 Aedwing Ct Bethesda MD 20817

KEMENY, GABOR, b Budapest, Hungary, Feb 6, 33; US citizen; m 58; c 2. THEORETICAL PHYSICS. *Educ:* Eotvos Lorand Univ, Budapest, dipl, 56; NY Univ, PhD(physics), 62. *Prof Exp:* Assoc scientist, Cent Res Inst Physics, Hungarian Acad Sci, 55-56; assoc scientist, Res Dept, Lamp Div, Westinghouse Elec Corp, 57-61; res scientist, Am-Standard, 61-62; res scientist, Ledgemont Lab, Kennecott Copper Corp, 63-68; assoc prof elec eng, 68-70, assoc prof Metall, Mech & Mat Sci, 68-74, assoc prof, 70-74, PROF BIOPHYS, MICH STATE UNIV, 74- *Concurrent Pos:* Sr vis, Cavendish Lab, Cambridge Univ, 66. *Mem:* AAAS; Am Phys Soc. *Res:* Quantum mechanics and electronics; many-body problem; solid state and mathematical physics; electrical conductivity in biomacromolecules; protein denaturation; microwave interactions with biological systems. *Mailing Add:* Dept Physics Mich State Univ East Lansing MI 48824

KEMENY, LORANT, b Abony, Hungary, May 28, 13; US citizen; m 50; c 1. VETERINARY MICROBIOLOGY. *Educ:* Royal Hungarian Vet Col, dipl, 36, DVM, 39. *Prof Exp:* Head antisera prod dept, Phylaxia State Serum Inst, Budapest, 39-56; res fel virol, Rockefeller Found, 57-58; asst dir vet biol, Colo Serum Co, 58-63; res vet, Nat Animal Dis Ctr, USDA, 63-87; RETIRED. *Mem:* Am Vet Med Asn. *Res:* Virology and immunology; production, control testing and research of new veterinary biologicals; isolation and adaptation of new animal viruses to tissue culture systems; characterization of viruses. *Mailing Add:* 2017 Northwestern Ave Ames IA 50010

KEMENY, NANCY E, b Elizabeth, NJ, Jan 18, 45; m 77; c 3. COLORECTAL CARCINOMA. *Educ:* Univ Pa, BA, 67; NJ Col Med, MD, 71. *Prof Exp:* Assoc prof clin med, Cornell Univ, 82; ASSOC ATTENDING, MEM SLOAN KETTERING CANCER CTR, 83- *Concurrent Pos:* chmn mem comt, Am Soc Clin Oncol, 87- *Mem:* Am Soc Clin Oncol; Am Asn Cancer Res. *Res:* Chemotherapeutic treatments for metastatic colorectal carcinoma. *Mailing Add:* 1275 York Ave New York NY 10021

KEMIC, STEPHEN BRUCE, b Boston, Mass, Dec 31, 46. ASTROPHYSICS. *Educ:* Univ NC, BS, 68; Univ Colo, MS, 70, PhD(astrophys), 73. *Prof Exp:* STAFF MEM PHYSICS, LOS ALAMOS NAT LAB, 74- *Mem:* Am Astron Soc; Am Asn Physics Teachers. *Res:* Spectroscopy of magnetic white dwarfs; laser fusion. *Mailing Add:* 1213 San Ildefonso Rd Los Alamos NM 87544

KEMMERER, ARTHUR RUSSELL, nutrition, biochemistry; deceased, see previous edition for last biography

KEMMERLY, JACK E(LLSWORTH), b Marion, Ohio, Aug 19, 24; m 45; c 5. SYSTEMS ENGINEERING. *Educ:* Cath Univ Am, BEE, 50; Univ Denver, MS, 52; Purdue Univ, PhD(elec eng), 58. *Prof Exp:* Asst res engr electronics, Denver Res Inst, Colo, 51-53; instr elec eng, Purdue Univ, 53-58; sr proj engr electronics, AC Spark Plug Div, Gen Motors Corp, 58-59; asst prof elec eng, Purdue Univ, 59-61; prin engr, Aeronutronic Div, Philco Corp, 61-68; sr staff engr, Ground Systs Group, Hughes Aircraft Co, 68; assoc prof elec eng, 68-70, prof eng, 70-79 & 81-85, chmn elec eng fac, 72-77, chmn div eng, 77-79, EMER PROF ENG, CALIF STATE UNIV, FULLERTON, 85- *Concurrent Pos:* Consult, AC Spark Plug Div, Gen Motors Corp; lectr, Univ Calif, Los Angeles; vis prof eng, Fort Lewis Col, Durango Colo, 80-81. *Mem:* Inst Elec & Electronics Engrs; Am Soc Eng Educ. *Res:* Systems analysis; applied probability; noise and circuit theory. *Mailing Add:* 323 Snowshoe Lane Durango CO 81301

KEMNITZ, JOSEPH WILLIAM, b Baltimore, Md, Mar 15, 47; m 90. PHYSIOLOGICAL PSYCHOLOGY. *Educ:* Univ Wis, BA, 69, MS, 74, PhD(physiol psychol), 76. *Prof Exp:* Proj specialist psychol, Univ Wis, 69-71, teaching & res asst, 71-76; res assoc, 77-79, asst scientist, 79-84, ASSOC SCIENTIST, WIS REGIONAL PRIMATE RES CTR, UNIV WIS, 84-, AFFIL SCIENTIST, INST AGING & ADULT LIFE, 89- *Concurrent Pos:* Mem task force animal models in diabetes res, NIH, 80-81; vis scientist, Div Endocrinol, Med Ctr, Univ Calif, Los Angeles, 81; consult, Div Diabetes & Clin Nutrit, Univ Southern Calif Med Ctr, 84-, Div Comp Med, Caribbean Primate Res Ctr, 85-; mem, NIH Spec Study Sect, 87-; assoc ed, Hormones & Behav, 87- *Mem:* Soc Neurosci; Am Soc Primatologists; Int Primatological Asn; AAAS; Gerontol Soc Am; Am Inst Nutrit; NY Acad Sci; Am Physiol Soc. *Res:* Regulation of energy balance, emphasizing obesity, diabetes, caloric restriction and aging, particularly in Rhesus monkeys. *Mailing Add:* Wis Regional Primate Res Ctr Univ Wis 1223 Capitol Ct Madison WI 53715-1299

KEMP, ALBERT RAYMOND, food science, for more information see previous edition

KEMP, ARNE K, b Kajaani, Finland, Mar 5, 18; US citizen; m 43; c 2. FOREST PRODUCTS. *Educ:* Univ Georgia, BSF, 48; Duke Univ, MF, 49; Univ Minn, PhD(wood tech), 57. *Prof Exp:* Instr wood tech, Sch Forestry, Univ Minn, 49-53; assoc prof, La State Univ, 53-55; head dept forestry, Stephen F Austin State Col, 55-63; chief div forest prod utilization & mkt & eng res, Lake State Forest Exp Sta, 63-65, asst dir forest prod utilization & mkt, Eng & Genetics Res, 65-73, ASST DIR RES, N CENT FOREST EXP STA, US FOREST SERV, 73- *Concurrent Pos:* NSF grant, 62. *Mem:* Forest Prod Res Soc; Soc Wood Sci & Technol; Int Union Forest Res Orgns. *Res:* Wood seasoning, preservation and anatomy; wood liquid relationships; forest products marketing. *Mailing Add:* Six Captains Crossing Savannah GA 31411

KEMP, DANIEL SCHAEFFER, b Portland, Ore, Oct 20, 36. ORGANIC CHEMISTRY. *Educ:* Reed Col, BA, 58; Harvard Univ, PhD(org chem), 64. *Prof Exp:* From asst prof to assoc prof, 64-72, PROF CHEM, MASS INST TECHNOL, 72- *Concurrent Pos:* A P Sloan Found fel, 68-70; vis asst prof, Univ Calif, San Diego, 69; Camile & Henry Dreyfus fel, 70. *Mem:* Am Chem Soc. *Res:* Peptide chemistry. *Mailing Add:* Dept Chem 18-584 Mass Inst Technol 77 Mass Ave Cambridge MA 02139

KEMP, EMORY LELAND, b Chicago, Ill, Oct 1, 31; m 58; c 3. STRUCTURAL ENGINEERING. *Educ:* Univ Ill, BSc, 52, PhD(theoret & appl mech), 62; Univ London, DIC, 55, MSc, 58. *Prof Exp:* Asst engr, Ill State Water Surv, 52; struct engr, consult firms, London, 56-59; instr theoret & appl mech, Univ Ill, 59-62; assoc prof, 62-65, chmn dept 67-74, PROF CIVIL ENG, WVA UNIV, 65-, PROF HIST SCi & TECHNOL, 77- *Concurrent Pos:* Fel, Am Coun Learned Socs, 75-76; regents' fel, Smithsonian Inst, 83-84. *Mem:* fel Am Soc Civil Engrs; Fel Brit Inst Civil Engrs; Fel Am Concrete Inst; Brit Inst Struct Engrs. *Res:* History of technology; industrial archeology; structural engineering. *Mailing Add:* 614 Woodburn Hall WVa Univ Morgantown WV 26506

KEMP, GORDON ARTHUR, b Newark, NJ, Dec 12, 32; m 58; c 3. MICROBIOLOGY, RESEARCH ADMINISTRATION. *Educ:* Lehigh Univ, AB, 54; Rutgers Univ, PhD(microbiol), 61. *Prof Exp:* Res scientist microbiol, 61-64, group leader chemother, 64-70, mgr chemother res, 70-73, mgr animal indust res, 73-76, dir, animal indust res & develop, Am Cyanamid Co, 76-82; PFIZER CENT RES, GROTON, 83- *Mem:* Fel AAAS; Am Soc Microbiol; Sigma Xi. *Res:* Pathogenesis of disease; prophylaxis and therapy of experimental infections; veterinary microbiology and immunology; protozoal and helminth infections of domestic animals; non-medical uses of antibiotics. *Mailing Add:* Pfizer Cent Res Eastern Pt Rd Groton CT 06340

KEMP, GRAHAM ELMORE, b Alta, Can, Jan 8, 27; US citizen; m 48. VETERINARY PUBLIC HEALTH. *Educ:* Univ Toronto, DVM, 51; Univ Calif, Berkeley, MPH, 58. *Prof Exp:* Mem staff, Div Livestock Indust, Univ Ill, 51-52; pvt pract, Ill, 52-57; mem staff, Epidemiol Bur Commun Dis, Div Prev Med Serv, Calif State Dept Pub Health, 58-64; staff mem, Rockefeller Found, Virus Res Lab, Fac Med, Univ Ibadan, 64-72; chief virol unit, San Juan Trop Dis Labs, PR, 73-75; dir, Bur Labs, Vector-Borne Dis Div, Commun Dis Ctr, USPHS, 75-80; RETIRED. *Concurrent Pos:* Lectr, Sch Pub Health, Univ Calif, Berkeley & Sch Vet Med, Univ Calif, Davis, 57-64; consult zoonoses, State of Calif, 58-64; hon sr scientist, PR Nuclear Ctr, 73-74; mem animal res comt, San Juan Vet Admin Hosp, 73-74. *Mem:* Am Vet Med Asn; Am Pub Health Asn; Conf Pub Health Vets; Am Soc Trop Med & Hyg. *Res:* Arbovirus; food-borne disease and zoonoses. *Mailing Add:* 808 Inverness St Ft Collins CO 80524

KEMP, HAROLD STEEN, b Ishpeming, Mich, May 19, 17; m 44; c 4. CHEMICAL ENGINEERING. *Educ:* Univ Minn, BChE, 39; Univ Mich, MS, 40, PhD(chem eng), 44. *Prof Exp:* Asst, Univ Mich, 40-41; res chem engr, E I du Pont de Nemours & Co, Inc, 43-53; res proj supvr, 53-55, res supvr, 55-60, consult supvr, 60-62, consult mgr, 62-84; RETIRED. *Concurrent Pos:* Chmn, Admin Comn Design Inst Emergency Relief Systs,

76-87; adj prof chem eng, Univ Del, 78-84; dir Am Inst Chem Eng, 81-83, chmn admin & Tech Comt, Ctr Chem Process Safety, 85-86, bd dir Found, 89- *Honors & Awards:* Founders Award, Am Inst Chem Engrs, 77, F J & Dorothy Van Antwerpen Award, 88. *Mem:* Fel Am Inst Chem Engrs (vpres, 85, pres, 86, past pres, 87); Sigma Xi; Am Asn Eng Soc (secy-treas, 88). *Res:* Thermal properties of hydrocarbon systems; liquid flow across distillation column plates; factors influencing distillation column plate efficiency; heat transfer; fluid flow. *Mailing Add:* 20 Crestfield Rd Crestfield Wilmington DE 19810

KEMP, JAMES CHALMERS, astrophysics; deceased, see previous edition for last biography

KEMP, JAMES DILLON, b Pickett, Ky, Feb 6, 23; m 47; c 2. ANIMAL SCIENCE. *Educ:* Univ Ky, BS, 48, MS, 49; Univ Ill, PhD(animal sci), 52. *Prof Exp:* Asst animal sci, Univ Ill, 49-52; from asst prof to assoc prof animal husb, Univ Ky, 52-59, prof animal sci, 59-89, coord, food sci prog, 66-89, EMER PROF ANIMAL SCI, UNIV KY, 89- *Concurrent Pos:* Fulbright res scholar, NZ, 64; consult, Thailand, 74 & Italy, 85. *Honors & Awards:* Res Award, Am Soc Animal Sci, 71; Pollock Award, Am Meat Sci Asn, 88. *Mem:* Fel Am Soc Animal Sci; fel Inst Food Technologists; Sigma Xi; Am Meat Sci Asn (pres, 75-76). *Res:* Meats teaching and research; composition and processing characteristics of red meats. *Mailing Add:* Dept Animal Sci Univ Ky Lexington KY 40546

KEMP, JOHN DANIEL, b Minneapolis, Minn, Jan 20, 40; m 75; c 3. BIOCHEMISTRY. *Educ:* Univ Calif, Los Angeles, BS, 62, PhD(biochem), 65. *Prof Exp:* NIH fel biochem, Univ Wash, 65-67, res assoc, 67-68; asst prof, 68-72, assoc prof, 72-77, prof plant path, Univ Wis-Madison, 77-; res chemist, USDA, 68-; AT AGRIGENETICS CORP, MADISON, WIS. *Mem:* Am Soc Plant Physiologists; Sigma Xi. *Res:* Molecular mechanisms of normal and abnormal plant growth and development; plant genetic engineering by novel approaches. *Mailing Add:* PGEL Box 3GL NMex State Univ Las Cruces NM 88003

KEMP, JOHN WILMER, b Midvale, Utah, July 28, 20; m 52; c 2. PHYSIOLOGICAL CHEMISTRY, PHARMACOLOGY. *Educ:* Westminster Col, AB, 50; Univ Calif, PhD(physiol chem), 57. *Prof Exp:* Res pharmacologic metab of heroin, Sch Med, Univ Calif, 57-59; assoc prof pharmacol, 59-83, ASSOC PROF PHYSIOL, COL MED, UNIV UTAH, 83- *Mem:* Am Soc Pharmacol & Exp Therapeut. *Res:* Biochemistry and pharmacology of the central nervous system; nucleic acids; membrane transport; neuronal excitability; anticonvulsants; physiology. *Mailing Add:* Dept Physiol Col Med Univ Utah Salt Lake City UT 84108

KEMP, KENNETH COURTNEY, b Chicago, Ill, Aug 7, 25. PHYSICAL ORGANIC CHEMISTRY. *Educ:* Northwestern Univ, BS, 50; Ill Inst Technol, PhD(chem), 56. *Prof Exp:* Asst chem, Ill Inst Technol, 50-55; vchmn dept, 76-80, from instr to prof, 55-90, EMER PROF CHEM, UNIV NEV, 90- *Mem:* Am Chem Soc; Chem Soc London; Sigma Xi. *Res:* Organic mechanisms; neighboring group reactions. *Mailing Add:* Dept Chem Univ Nev Reno NV 89557

KEMP, KENNETH E, b Detroit, Mich, Aug 24, 41; m 64, 78; c 2. STATISTICS. *Educ:* Mich State Univ, BS, 63, MS, 65, PhD(animal husb), 67. *Prof Exp:* Sr statist programmer, Biomet Serv, Agr Res Serv, USDA, 67-68; asst prof, 68-71, assoc prof, 71-79, PROF STATIST, KANS STATE UNIV, 79- *Mem:* Biomet Soc; Am Statist Asn. *Res:* Algorithms and techniques for statistical analysis on the digital computer. *Mailing Add:* Statist Dickens Hall Kans State Univ Manhattan KS 66506

KEMP, L(EBBEUS) C(OURTRIGHT), JR, b Houston, Tex, Oct 8, 07; m 36; c 2. CHEMICAL ENGINEERING. *Educ:* Rice Inst, BS, 29. *Prof Exp:* Res chemist, Texaco, Inc, Tex, 29-33, res supvr, 33-38, asst chief chemist, 38-40, asst supt res labs, NY, 40-41, dir res, 41-53, asst to vpres, 53-54, asst to sr vpres, 54-55, gen mgr petrochem, 55-57, vpres petrochem, 57-59, vpres res & technol, 59-68, vpres spec assignments, 68-71; vpres, Texaco, Inc, Houston Area, 71-72; RETIRED. *Concurrent Pos:* Mem indust adv comt, US Bur Mines, 44; trustee, United Eng Trustees, 58-60; consult, 71- *Mem:* AAAS; Am Chem Soc; Am Inst Aeronaut & Astronaut; Soc Chem Indust; fel Am Inst Chem Engrs; Sigma Xi. *Res:* Petroleum refining including work on both product and process development; alkylation of isobutane with olefins; hydrocarbon synthesis from carbon monoxide and hydrogen; petrochemicals; synthetic liquid and gaseous fuels. *Mailing Add:* 12318 Huntingwick Dr Houston TX 77024

KEMP, LOUIS FRANKLIN, JR, b New York, NY, Mar 19, 40; m 64; c 2. MATHEMATICAL STATISTICS. *Educ:* Princeton Univ, BS, 62; Polytech Inst NY, MS, 65, PhD(math), 67. *Prof Exp:* Engr, Grumman Aircraft Corp, 62-63; asst prof math, Polytech Inst NY, 67-69; RES ASSOC, AMOCO PROD CO, 69- *Concurrent Pos:* Adj prof math, Univ Tulsa, 79-84 & 89-91. *Mem:* Soc Indust & Appl Math; Math Asn Am; Sigma Xi; Soc Petrol Engrs; Asn Comput Mach. *Res:* Numerical analysis; engineering statistics; neural networks. *Mailing Add:* 5334 S 74th E Ave Tulsa OK 74145

KEMP, MARWIN K, b Strong, Ark, Nov 23, 42; m 61; c 2. PHYSICAL CHEMISTRY, GEOCHEMISTRY. *Educ:* Univ Ark, BS, 64; Univ Ill, MS, 65, PhD(phys chem), 68. *Prof Exp:* From asst prof to assoc prof phys chem, Univ Tulsa, 68-81; sr res scientist, 81-85, staff res scientist, 85-88, RES SUPVR, AMOCO PROD CO, 88- *Mem:* Am Chem Soc; AAAS; Sigma Xi. *Res:* Geochemistry; natural gas equilibrium in underground reservoirs; thermodynamics; infrared spectroscopy. *Mailing Add:* Amoco Prod Co PO Box 3385 Tulsa OK 74104

KEMP, NELSON HARVEY, fluid mechanics, heat transfer; deceased, see previous edition for last biography

KEMP, NORMAN EVERETT, b Otisfield, Maine, June 20, 16; m 42. DEVELOPMENTAL BIOLOGY. *Educ:* Bates Col, BS, 37; Univ Calif, PhD(zool), 41. *Prof Exp:* Asst zool, Univ Calif, 37-41; instr, Wayne Univ, 46-47; from instr to assoc prof, 47-61, prof, 61-86, EMER PROF ZOOL, UNIV MICH, ANN ARBOR, 86- *Concurrent Pos:* Res assoc, Argonne Nat Lab, 54 & 55; vis investr, Rockefeller Inst, 58; vis colleague, Univ Hawaii, 65, 72; vis investr, Lab, Marine Biol Asn UK, Plymouth, Eng, 79. *Mem:* Am Soc Zool; Am Asn Anat; Electron Micros Soc Am; Am Soc Cell Biol; Soc Develop Biol. *Res:* Electron microscopy of differentiating cells; fertilization of amphibian eggs; differentiation of digestive tract and skin; calcification of calcified cartilage, bones, scales and teeth in aquatic vertebrates; fibrillogenesis of collagen; regeneration of fish fins. *Mailing Add:* Dept Biol Univ Mich Ann Arbor MI 48109-1048

KEMP, PAUL JAMES, b Inglewood, Calif, June 26, 42; m 90; c 4. IDENTIFICATION OF ODOPHORIC COMPOUNDS, DETOXIFICATION REACTIONS. *Educ:* Iowa State Univ, BS, 65; Ore State Univ, MS, 69. *Prof Exp:* Grad teaching asst gen & anal chem, Ore State Univ, 65-69; gen mgr & chief exec officer food packaging prod, Outrite Plastics Inc, 69-74; grad teaching fel anal chem, Univ Hawaii, 75-77; dist mgr food processing & water treatment, Olin Water Serv, Olin Corp, 77-81; regional mgr food processing & water treatment, Assoc Chem & Serv, Inc, 81-89; TECH SALES DEVELOP ENGR INDUST CHEM, BREWER ENVIRON INDUSTS, INC, 90- *Concurrent Pos:* Dir, Lens, Inc. *Mem:* Am Chem Soc; Am Waterworks Asn; Water Pollution Control Fedn; Sigma Xi. *Res:* Industrial odor control chemistry; non-halogen oxidative sterilants for industrial and potable water; industrial wastewater management. *Mailing Add:* 58110 Mamo St Haleiwa HI 96712-9746

KEMP, PAUL RAYMOND, b Denver, Colo, Mar 14, 49; m 77. PLANT ECOLOGY, PLANT WATER RELATIONS. *Educ:* Colo State Univ, BS, 71, MS, 73; Wash State Univ, PhD(bot), 77. *Prof Exp:* Res assoc plant physiol, N Mex State Univ, 77-83; curator bot, NMex Mus Nat Hist, 83-88; RES PLANT ECOLOGIST, SYSTS ECOL RES GROUP, SAN DIEGO STATE UNIV, 88- *Concurrent Pos:* Lectr, NMex State Univ, 80-81. *Mem:* Sigma Xi; Ecol Soc Am; Am Soc Plant Physiologists; Bot Soc Am. *Res:* Physiological plant ecology; vegetation response to climate; photosynthetic adaptations; stress physiology. *Mailing Add:* Systs Ecol Res Group San Diego State Univ San Diego CA 92182-0401

KEMP, ROBERT GRANT, b Massillon, Ohio, Feb 12, 37; m 67, 85; c 2. BIOCHEMISTRY. *Educ:* Col Wooster, BA, 59, Yale Univ, PhD(biochem), 64. *Prof Exp:* Res assoc biochem, Univ Wash, 64-66; from asst prof to prof biochem, Med Col Wis, 66-76, chmn, 76-88, PROF BIOCHEM, UNIV HEALTH SCI, CHICAGO MED SCH, 76- *Concurrent Pos:* Estab investr, Am Heart Asn, 68-73; Fulbright fel, 71. *Mem:* AAAS; Am Chem Soc; Am Soc Biol Chem; Sigma Xi; Protein Soc. *Res:* Control of carbohydrate metabolism; structure-activity relationships of enzymes. *Mailing Add:* Dept Biochem Chicago Med Sch 3333 Green Bay Rd North Chicago IL 60064

KEMP, WALTER MICHAEL, b Big Spring, Tex, Aug 26, 44; m 83; c 4. IMMUNOBIOLOGY, PARASITOLOGY. *Educ:* Abilene Christian Col, BSE, 66; Tulane Univ, PhD(biol), 70. *Prof Exp:* Cell biol trainee biol, Tulane Univ, 68-70; asst prof, Abilene Christian Col, 70-75; from asst prof to assoc prof, 75-82, PROF BIOL, TEX A&M UNIV, 82-, ASSOC DEAN SCI, 89- *Concurrent Pos:* Res Corp res grant, 70-71; res assoc immunol, Southwest Found Res & Educ, 70-73 & Univ Ga, 74; NIH grants, 72-75, 78-83 & 86-89; Clark Found grants, 76-78, 78-82 & 82-85; mem, Study Sect Trop Med Parasitol, NIH, 81-85, chmn, 83-85. *Honors & Awards:* Henry Baldwin Ward Medal, Am Soc Parasitologists, 83. *Mem:* AAAS; Am Soc Parasitologists (pres, 89); Am Soc Trop Med & Hyg; Am Asn Immunologists. *Res:* Immune responses to parasites, particularly schistosomes and trypanosomes; host-parasite antigen sharing and parasite immune escape mechanisms. *Mailing Add:* Dept Biol Tex A&M Univ College Station TX 77843

KEMP, WILLIAM MICHAEL, b Washington, DC, May 16, 47. ECOLOGY. *Educ:* Ga Inst Technol, BA, 69, MA, 71; Univ Fla, PhD(environ sci), 76. *Prof Exp:* Environ engr eval, US Environ Protection Agency, 71-72; SYSTS ECOLOGIST ENVIRON RES, CTR ENVIRON & ESTUARINE STUDIES, UNIV MD, 77- *Mem:* Ecol Soc Am; Am Soc Limnol & Oceanog; AAAS; Sigma Xi. *Res:* Ecosystem modeling; productivity and nutrient dynamics of estuaries; structure of ecological trophic webs; economics and energetics of environment. *Mailing Add:* Horn Pt Environ Labs Univ Md PO Box 775 Cambridge MD 21613

KEMPE, LLOYD L(UTE), b Pueblo, Colo, Nov 26, 11; m; c 1. CHEMICAL ENGINEERING. *Educ:* Univ Minn, BChE, 32, MS, 38, PhD(chem eng), 48. *Prof Exp:* Asst, Div Soils, Univ Minn, 34-35, res assoc, 40-41, asst chem eng, 46-48; asst sanit engr, State Dept Health, Minn, 35-40; instr bact, Univ Mich, 48-49, asst prof, 49-50; asst prof food tech, Univ Ill, 50-52; from asst prof to prof chem eng & bact, 52-60, prof chem eng & sanit eng, 60-63, prof, 63-81, prof microbiol & immunol, Med Sch, 67-81, EMER PROF CHEM ENG, UNIV MICH, ANN ARBOR, 63- *Concurrent Pos:* Consult engr; mem comt microbiol, Adv Bd Qm Res & Develop, Nat Acad Sci-Nat Res Coun; mem comt irradiation preservation of foods, US AEC, Am Inst Biol Sci; mem comt botulism hazards, US Food & Drug Admin. *Mem:* AAAS; Am Chem Soc; Am Soc Microbiol; Am Inst Chem Engrs; Inst Food Technol; Sigma Xi. *Res:* Biochemical engineering; irradiation processing of foods; botulism hazards in foods; industrial waste treatment. *Mailing Add:* 2861 Maplewood Dr SE Grand Rapids MI 49506-5001

KEMPE, LUDWIG GEORGE, b Brandenburg, Ger, Oct 16, 15; US citizen; m 55. NEUROSURGERY, NEUROANATOMY. *Educ:* Univ Berne, MD, 42. *Prof Exp:* Assoc clin prof neurosurg, George Washington Univ, 60-73; PROF NEUROSURG & ANAT, MED UNIV SC, 73- *Concurrent Pos:* Mem adv bd, Coun Neurosurg, 68- *Mem:* Am Asn Neurol Surgeons; Cong Neurol Surgeons; Soc Neurol Surgeons; Am Asn Anatomists; Am Col Surgeons. *Res:* Mesoscopic neuroanatomy. *Mailing Add:* Dept Neurosurg-Anat Med Univ SC 171 Ashley Ave Charleston SC 29425

KEMPEN, RENE RICHARD, b Kankakee, Ill, Mar 24, 28; m 69; c 1. PHARMACOLOGY. *Educ:* St Joseph's Col, Ind, BS, 50; Loyola Univ Chicago, MS, 55, PhD(pharmacol), 62. *Prof Exp:* Chemist, Chicago Biol Res Lab, 54-55; lab instr pharmacol, Stritch Sch Med, Loyola Univ Chicago, 56-59; instr, Col Med, Baylor Univ, 61-63; ASST PROF, SCH MED, UNIV TEX MED BR GALVESTON, 63-, ASSOC DIR TOXICOL LAB, 73-, ASST PROF, SCH ALLIED HEALTH, 76- *Concurrent Pos:* Instr, St Anne's Hosp Sch Nursing, 55-57, Loyola Univ Sch Nursing, 56 & St Elizabeth's Hosp Sch Nursing, 57. *Mem:* AAAS; Am Asn Lab Animal Sci; Am Heart Asn; Sigma Xi. *Res:* Action of drugs on cardiac electrophysiological parameters; muscle contraction; toxicology; effect of drugs on endocrine pancrease. *Mailing Add:* Dept Pharmacol Univ Tex Med Br J31 Galveston TX 77550

KEMPER, GENE ALLEN, b Drake, NDak, Apr 12, 33; m 69; c 1. NUMERICAL ANALYSIS, PERFORMANCE ANALYSIS. *Educ:* Univ NDak, BS, 56, MS, 59; Iowa State Univ, PhD(appl math), 65. *Prof Exp:* Instr math, Univ NDak, 56-59; exten lectr math, Univ Wash, Seattle, 60-61 & 65-66; mathematician, Boeing Co, 60-61, sr res specialist & vis staff mem, Boeing Sci Res Labs, 65-66; instr math, Iowa State Univ, 61-65; assoc prof, 66-72, sr consult, Comput Ctr, 69-79, dir, Inst Comput Use Educ, 74-81, assoc dir, Comput Ctr, 79-81, asst vpres acad affairs, 81-82, PROF MATH, UNIV NDAK, 72-, ASSOC VPRES ACAD AFFAIRS, 82- *Concurrent Pos:* NSF Col Sci Improv Prog grant, Univ NDak, 68-71; vis soc indust & appl math lectr, 75-77, 79-80, 80-82; vis scientist, Atomic Energy Comn Lab, Iowa State Univ, Ames, Iowa. *Mem:* Soc Indust & Appl Math; Asn Comput Mach. *Res:* Numerical solution of functional differential equations and integral equations; mathematical modeling of biological systems. *Mailing Add:* Univ Station Grand Forks ND 58202-8232

KEMPER, JOHN D(USTIN), b Portland, Ore, May 29, 24; m 47; c 1. MECHANICAL ENGINEERING. *Educ:* Univ Calif, Los Angeles, BS, 49, MS, 59; Univ Colo, PhD(struct mech), 69. *Prof Exp:* Engr, Telecomput Corp, Calif, 49-50, proj engr, 50-52, chief mech engr, 52-55; chief mech engr, H A Wagner Co, 55-56; asst to vpres eng, Marchant Calculators, Inc, 56-58, chief engr, Marchant Div, Smith-Corona Marchant Inc, 58-59, vpres eng, 59-62; assoc prof, 62-67, dean, Col Eng, 69-83, PROF ENG, UNIV CALIF, DAVIS, 67- *Concurrent Pos:* Chmn, Panel Grad Educ & Res, Nat Res Coun, 83-85, Task Force Prep Teaching Eng, Am Soc Eng Educ, 84-86; dir, Plantronics Inc, 83-86. *Honors & Awards:* Alex Laurie Award, Am Soc Hort Sci, 74. *Mem:* Fel AAAS; Nat Soc Prof Engrs; fel Am Soc Mech Engrs; Am Soc Eng Educ. *Res:* mechanical design; structural mechanics; writings on engineering profession, introduction to engineering profession, ethics, creativity, graduation education and research, preparation for teaching. *Mailing Add:* PO Box 729 Woodland CA 95695

KEMPER, KATHLEEN ANN, b Peoria, Ill, July 6, 66. SONOCHEMISTRY, THERMAL ANALYSIS. *Educ:* Ill State Univ, BS, 87, MS, 89. *Prof Exp:* Grad teaching asst gen chem, Ill State Univ, 88-89; chemist, Bloomington-Normal Sanit Dist, 88-89; grad teaching asst instrumental anal, 90-91, GRAD RES ASST, UNIV ILL, URBANA-CHAMPAIGN, 91- *Mem:* Am Chem Soc. *Res:* Sonoluminescence from inorganic systems; chemical and physical effects of ultrasound; phase transitions in inorganic solids; rates and mechanisms of solid state reactions. *Mailing Add:* Noyes Lab Box 11-1 505 S Mathews Urbana IL 61801

KEMPER, KIRBY WAYNE, b New York, NY, Apr 13, 40; m 64; c 3. NUCLEAR PHYSICS, ION SOURCE DEVELOPMENT. *Educ:* Va Polytech Inst, BS, 62; Ind Univ, MS, 64, PhD(physics), 68. *Prof Exp:* Res assoc nuclear physics, 68-71, asst prof, 71-75, assoc prof, 75-79, PROF PHYSICS, FLA STATE UNIV, 79- *Mem:* Sigma Xi; Am Phys Soc. *Res:* Selective population of states with heavy ions; polarized ion source development; laser induced atomic polarizations; spin effects in nuclear reactions. *Mailing Add:* Dept Physics Fla State Univ Tallahassee FL 32306

KEMPER, ROBERT SCHOOLEY, JR, b Oakland, Calif, Feb 20, 27; m 49, 66; c 4. MECHANICAL & METALLURGICAL ENGINEERING. *Educ:* Ore State Col, BS, 51, MS, 52. *Prof Exp:* Tech specialist, Hanford Atomic Prods Oper, Gen Elec Co, 52-64; res assoc, 65-66, unit mgr, 66-69, mgr mat develop, 69-85, MGR RES OPERS, MAT SCI & TECHNOL DEPT, PAC NORTHWEST LABS, BATTELLE MEM INST, 85 - *Mem:* Am Soc Metals; Soc Mfg Engrs; Am Defense Preparedness Asn; AAAS; Am Soc Testing & Mat. *Res:* Irradiation damage in fuel and structural materials; metallic fabrication development. *Mailing Add:* 1623 Alder Richland WA 99352

KEMPER, WILLIAM ALEXANDER, b Baltimore, Md, Jan 1, 11; m 56, 73. ENVIRONMENTAL CHEMISTRY, BALLISTICS. *Educ:* Johns Hopkins Univ, PhD(phys chem), 34. *Prof Exp:* Chemist, Res Dept, Baltimore Gas & Elec Co, 34-43; physicist, US Naval Weapons Lab, 46-72; sci adv, Comdr Cruiser Destroyer Forces Atlantic Fleet, 72-73; sr ballistician, Navy Surface Weapons Ctr-Dahlgren Lab, 73-75; asst prof physics, Metrop State Col, 76-77; RETIRED. *Concurrent Pos:* Chmn, USN Aeroballistics Adv Comt, 67 & 68; US Nat Leader, US, Australia, Can & UK Coop Prog in Exterior Ballistics, 67-74. *Mem:* AAAS; Am Phys Soc; Am Defense Prep Asn. *Res:* Ballistics; fire control. *Mailing Add:* 7363 W 26th Place Denver CO 80215

KEMPERMAN, JOHANNES HENRICUS BERNARDUS, b Amsterdam, Netherlands, July 16, 24; m 53; c 5. MATHEMATICS. *Educ:* Univ Amsterdam, BS, 45, MS, 48, PhD(math), 50. *Prof Exp:* Res assoc appl math, Math Ctr, Amsterdam, 48-51; from asst prof to prof math, Purdue Univ, 51-61; prof math, Univ Rochester, 61-85; PROF MATH, RUTGERS UNIV, 86- *Concurrent Pos:* On leave, Univ Amsterdam, 58-59 & 72-73, Univ Wis, 60-61, & Stanford Univ, 66-67 & Univ Tex, Austin, 77-78; corresp mem Royal Neth Acad Sci. *Mem:* Am Statist Asn; fel Inst Math Statist; Dutch Math Soc; Math Asn Am; fel Am Asn Advan Sci. *Res:* Analysis; probability; statistics. *Mailing Add:* Dept Statist Rutgers Univ New Brunswick NJ 08901

KEMPH, JOHN PATTERSON, b Lima, Ohio, Dec 17, 19; m 43; c 4. PSYCHIATRY, PHYSIOLOGY. *Educ:* Ohio Northern Univ, AB, 47; Ohio State Univ, BSc, 47, MSc, 48, MD, 53; Am Bd Psychiat & Neurol, dipl psychiat, 60, dipl child psychiat, 62. *Prof Exp:* Res asst, Res Found, Ohio State Univ, 47-48, res assoc, 51-55; resident psychiat, Med Ctr, Univ Mich, 55-56, jr clin instr, 56-57; dir, Ohio Northwest Guid Ctr, 57-60; from instr to assoc prof psychiat, Univ Mich, 60-68; prof psychiat & dir child & adolescent psychiat, State Univ NY Downstate Med Ctr, 68-72; prof psychiat & chmn dept, Med Col Ohio, 72-74, vpres acad affairs, dean med fac & prof psychiat, 74-86; PROF PSYCHIAT, UNIV FLA, 86- *Concurrent Pos:* Res fel, Ohio State Univ, 48-51; fel child psychiat, Med Ctr, Univ Mich, Ann Arbor, 60-61; intern, Mt Carmel Hosp, Columbus, Ohio, 53-54; resident psychiat, Columbus State Hosp, 54-55; mem active staff, Mem Hosp, Lima, Ohio, 57-60; vice chief staff psychiat, St Rita's Hosp, 58-60; instr child psychiat, Med Ctr, Univ Mich, Ann Arbor, 60-61, infections control officer, 61-68, lectr human growth & behav, Sch Social Work & lectr psychosom med, Univ Hosp, 62-68, mem clin serv comt, 63-68; dir in-patient serv & coord res, Children's Psychiat Hosp, 61-65, clin dir, 65-68; consult, Cent Mich Coun Continuing Psychiat Educ, 64-68; chmn clin serv comt, Children's Psychiat Hosp, 65-68; emer prof & dean, Med Col Ohio, 86- *Mem:* Fel Am Psychiat Asn; AMA; Am Orthopsychiat Asn; Am Asn Ment Deficiency; Am Psychosom Soc; Sigma Xi. *Res:* Child psychiatry; applied cardiovascular and respiratory physiology; study of physiological and psychological correlates of behavior. *Mailing Add:* Dept Psychiat Univ Fla Gainesville FL 32611

KEMPHUES, KENNETH J, b Cincinnati, Ohio, July 3, 50; m 84; c 1. ANIMAL DEVELOPMENT. *Educ:* Univ Va, BA, 76; Ind Univ, PhD(genetics), 81. *Prof Exp:* Teaching fel, Univ Colo, 81-84; ASST PROF DEVELOP BIOL, CORNELL UNIV, 84- *Mem:* Genetics Soc Am; Soc Develop Am; AAAS. *Res:* The problem of determination in animal development; identification and characterization of gene encoding functions necessary for determination; the use of molecular, ultrastructural and biochemical techniques to exploit mutations in the genes. *Mailing Add:* Genetics & Develop Sect Cornell Univ 101 Biotech Bldg Ithaca NY 14853

KEMPLE, MARVIN DAVID, b Indianapolis, Ind, Sept 2, 42; m 64; c 2. MAGNETIC RESONANCE. *Educ:* Purdue Univ, BS, 64; Univ Ill, Urbana-Champaign, MS, 65, PhD(physics), 71. *Prof Exp:* Enrico Fermi fel chem & physics, Enrico Fermi Inst, Univ Chicago, 71-72, res assoc, Dept Chem, 72-76; Nat Res Coun res assoc, Nat Bur Standards, 76-77; ASST PROF PHYSICS, IND UNIV-PURDUE UNIV, INDIANAPOLIS, 77- *Mem:* Am Phys Soc; AAAS; Sigma Xi. *Res:* Application of electron paramagnetic resonance and electron nuclear double resonance to the study of ions and molecules in ionic crystals, organic crystals, protein crystals, and intact, live biological systems. *Mailing Add:* Dept Physics KB033 IUPUI 1125 E 38 St Indianapolis IN 46205-2810

KEMPLER, WALTER, b New York City, NY, Sept 9, 23; c 5. FAMILY COUNSELING. *Educ:* Univ Tex, BS, 46, MD, 47. *Prof Exp:* FOUNDER & DIR FAMILY COUN, KEMPLER INST, 60- *Res:* The structure, dynamics and treatment of the family. *Mailing Add:* Kempler Inst PO Box 1692 Costa Mesa CA 92628

KEMPNER, DAVID H, IMMUNOCHEMISTRY, BIOCHEMISTRY. *Educ:* Tufts Univ, PhD(chem), 75. *Prof Exp:* Prin consult, Bernard Wolnak & Assocs. 83-88; SR ASSOC, STRATEGIC TECH INT, 88- *Mailing Add:* Strategic Tech Int 1204 Allanson Rd Mundelein IL 60060

KEMPNER, ELLIS STANLEY, b New York, NY, Mar 20, 32; m 61; c 3. BIOPHYSICS. *Educ:* Brooklyn Col, BS, 53; Yale Univ, MS, 55, PhD(biophys), 59. *Prof Exp:* Asst scientist bionucleonics, 58-61, physicist, Nat Inst Arthritis, Diabetes & Digestive & Kidney Dis, NIH, 61-86; CHIEF, SECT MACROMOLECULAR BIOPHYSICS, NIAMS, 86- *Concurrent Pos:* Lectr, Univ Calif, Davis, 68-69. *Mem:* Biophys Soc. *Res:* Radiation effects on macromolecules; macromolecular synthesis; growth under extreme conditions; cellular organization. *Mailing Add:* NIAMS NIH Bethesda MD 20892

KEMPNER, JOSEPH, b Brooklyn, NY, Apr 25, 23; m 47; c 2. APPLIED MECHANICS, AEROSPACE ENGINEERING. *Educ:* Polytech Inst Brooklyn, BAeroE, 43, MAeroE, 47, PhD(appl mech), 50. *Prof Exp:* Aeronaut engr struct res, Nat Adv Comt Aeronaut, 43-47; from res asst to res assoc appl mech, 47-50, from instr to assoc prof, 50-57, head dept aerospace & appl mech, 66-76, prof appl mech, 57-90, EMER PROF APPL MECH, POLYTECH INST NY, 90- *Concurrent Pos:* Prin investr, Off Naval Res & Air Force Off Sci Res grants & contracts, 58-77; consult, USN, 70-; mem adv group, Ship Res Comt, Maritime Transp Res Bd, Nat Acad Sci, 73-76; mem comt basic res, Adv Army Res Off, 73-76 & 82-85. *Honors & Awards:* Citation Distinguished Res, Sigma Xi, 73; Laskowitz Gold Medal Res Aerospace Eng, NY Acad Sci, 73. *Mem:* Assoc fel Am Inst Aeronaut & Astronaut; Am Soc Mech Eng; fel Am Acad Mech; fel NY Acad Sci; Am Soc Eng Educ. *Res:* Structural research related to aerospace vehicles, submersible vessels and pressure vessels; statics and dynamics of plates and shells, including large deformation and elevated temperature effects; applied mechanics. *Mailing Add:* 1163 E 13th St Brooklyn NY 11230

KEMPNER, WALTER, b Berlin, Ger, Jan 25, 03; nat US. INTERNAL MEDICINE, CELL PHYSIOLOGY. *Educ:* Univ Heidelberg, MD, 26. *Prof Exp:* Res asst, Kaiser Wilhelm Inst, 27-28; asst physician, Univ Berlin, 28-33; res assoc, Kaiser Wilhelm Inst, 33-34; from assoc med & physiol to prof med, 34-72, EMER PROF MED, MED CTR, DUKE UNIV, 72- *Honors & Awards:* Ciba Award, Am Heart Asn, 75. *Mem:* Am Physiol Soc; AMA; fel Am Col Physician; Am Soc Internal Med. *Res:* Cellular respiration and fermentation; pathological physiology; the rice diet and other dietary regimes for the treatment of heart, kidney disease, hypertensive and arteriosclerotic vascular disease, vascular retinopathy, diabetes mellitus and obesity. *Mailing Add:* 1821 Green St Durham NC 27705

KEMPSON, STEPHEN ALLAN, b Walsall, Eng, July 2, 48; US citizen; m 73; c 2. MEMBRANE & EPITHELIAL TRANSPORT. *Educ:* Lancaster Univ, UK, BA, 70; Warwick Univ, MSc, 71; London Univ, PhD(biochem), 75. *Prof Exp:* Fel biochem, Univ Rochester, 75-77; fel physiol, Mayo Clinic Found, 77-80, asst prof, Mayo Med Sch, 79-80; asst prof med, Univ Pittsburgh, 80-82; asst prof, 82-87, ASSOC PROF PHYSIOL, IND UNIV MED SCH, 87- *Concurrent Pos:* Prin investr, Pittsburgh Health Res & Serv Found grant, 81-82; grant, NIH, 82- *Mem:* Am Physiol Soc; Am Soc Renal Biochem & Metab; Am Soc Nephrol; Am Fedn Clin Res. *Res:* Biochemistry and physiology of the kidney, specifically the cellular control mechanisms which regulate the transport of inorganic phosphate by the kidney. *Mailing Add:* Dept Physiol & Biophys Ind Univ Sch Med 635 Barnhill Dr Indianapolis IN 46202-5120

KEMPTER, CHARLES PRENTISS, b Burlington, Vt, Feb 12, 25; m 77; c 3. TOXICOLOGY. *Educ:* Stanford Univ, BS, 49, MS, 50, PhD(chem), 56. *Prof Exp:* Asst phys sci, Stanford Univ, 49-50; phys chemist, Dow Chem Co, 50-53; staff mem, Los Alamos Sci Lab, Univ Calif, 56-71; sci consult, 71-73; tech dir, Kempter-Rossman Int, 73-75, SCI ADV, 75- *Concurrent Pos:* Vis scientist, Inst Phys Chem, Vienna, 63-64; thesis adv, Los Alamos Grad Ctr, Univ NMex, 59-71; gov's adv, NMex State Crime Lab, 71-73. *Mem:* Fel Inst Chemists; Am Chem Soc; AAAS; Sigma Xi; Int Plansee Soc Powder Metall. *Res:* Biomedical literature research primarily in toxicology. *Mailing Add:* 6202 Agee St San Diego CA 92122

KEMPTHORNE, OSCAR, b Cornwall, Eng, Jan 31, 19; nat US; m 49; c 3. STATISTICS, MATHEMATICAL BIOLOGY. *Educ:* Cambridge Univ, BA, 40, MA, 43, ScD, 60. *Prof Exp:* Statistician, Rothamsted Exp Sta, Eng, 41-46; assoc prof, 47-51, prof statist, 51-89, distinguished prof sci & humanities, 64-89, EMER PROF STATIST & SCI & HUMANITIES, IOWA STATE UNIV, 89. *Mem:* Int Statist Inst; fel Am Statist Asn; fel Inst Math Statist; hon fel Royal Statist Soc; Biomet Soc (past pres); Inst Math Statist (pres, 84-85). *Res:* Design of experiments; statistical inference; biological statistics. *Mailing Add:* Statist Lab Snedecor Hall Iowa State Univ Ames IA 50011

KEMPTON, JOHN P(AUL), b Buffalo, NY, Aug 14, 32; m 54; c 2. GROUNDWATER GEOLOGY. *Educ:* Denison Univ, BS, 54; Ohio State Univ, MA, 56; Univ Ill, PhD, 62. *Prof Exp:* Asst, Ohio State Univ, 54-56; geologist, Ohio Div Water, 55-56; from asst geologist to assoc geologist, Ill Geol Surv, 56-71, geologist, Hydrogeol & Geophys Sect, 71-84, SSC geol task force leader, 84-86, sr geologist & spec proj leader, 86-88, SR GEOLOGIST & HEAD, QUATERNARY FRAMEWORK STUDIES SECT, ILL GEOL SURV, 88- *Concurrent Pos:* Vis prof, Northern Ill Univ, 73. *Mem:* AAAS; Geol Soc Am; Asn Eng Geologists; Sigma Xi; Am Quaternary Asn. *Res:* Quaternary stratigraphy and mapping, environmental and groundwater geology; three-dimensional, lithostratigraphic stackunit geologic maps of quaternary sediments for direct interpretation for ground-water resources development and protection, siting, and other land uses. *Mailing Add:* Natural Resources Bldg Ill Geol Surv 615 E Peabody Champaign IL 61820

KENAGA, CLARE BURTON, b Cadillac, Mich, Jan 9, 27; m 52; c 4. PLANT PATHOLOGY. *Educ:* Western Mich Col, BS, 50; Univ Mich, MS, 52; Mich State Univ, PhD(plant path), 57. *Prof Exp:* Plant pathologist, Morton Chem Co, 57-66; vis assoc prof, 66-67, assoc prof, 67-77, prof plant path, 77-81, EDUC ADMIN, PURDUE UNIV, 81- *Concurrent Pos:* Indust consult; mem, Coun Agr Sci & Technol. *Mem:* Fel Nat Asn Cols & Teachers Agr; Am Phytopath Soc; Soc Nematol. *Res:* Mechanisms of action of fungicides and soil fumigants; teaching. *Mailing Add:* Dept Bot & Plant Path Purdue Univ West Lafayette IN 47907

KENAGA, DUANE LEROY, b Midland, Mich, Mar 9, 20; m 83; c 4. WOOD TECHNOLOGY, PULP & PAPER TECHNOLOGY. *Educ:* Univ Mich, BSChe, 43, MWT, 48. *Prof Exp:* Asst, Univ Mich, 47-48; wood technologist, Wood & Paper Sect, Southern Res Inst, 48-51; wood technologist, biochem res lab, Dow Chem Co, 51-65, sr res wood chemist, 65-69, res specialist, 69-78, res assoc, Designed Prod Dept, 78-85; RETIRED. *Mem:* Forest Prod Res Soc; Soc Wood Sci & Tech; Tech Asn Pulp & Paper Indust. *Res:* Chemical utilization of wood; chemical modification of wood to promote dimensional stability; paper and fiber treatments; wet end additives in paper systems, including bulking aids, retention aids and high filler sheets. *Mailing Add:* 4622 Chatham Court Midland MI 48640

KENAGY, GEORGE JAMES, b Los Angeles, Calif, July 9, 45; m 69; c 2. ECOLOGY, BEHAVIOR PHYSIOLOGY. *Educ:* Pomona Col, BA, 67; Univ Calif, Los Angeles, PhD(zool), 72. *Prof Exp:* Fel, Max Planck Inst Behav Physiol, Ger, 72-73; res biologist, Univ Calif, Los Angeles, 74; res biologist, Scripps Inst Oceanog, Univ Calif, San Diego, 74-76; from asst prof to assoc prof, 76-86, PROF, DEPT ZOOL, UNIV WASH, 86- *Mem:* AAAS; Ecol Soc Am; Animal Behav Soc; Am Soc Mammalogists; Soc Study Evolution. *Res:* Bahavior and population biology of small mammals; daily and seasonal rhythms of animals; reproduction; hibernation; energetics. *Mailing Add:* Dept Zool Univ Wash Seattle WA 98195

KENAHAN, CHARLES BORROMEO, metallurgy, for more information see previous edition

KENAN, RICHARD P, b Waycross, Ga, Dec 25, 31; m 68; c 3. INTEGRATED OPTICS, FIBER OPTICS. *Educ:* Ga Inst Technol, BA, 55; Ohio State Univ, PhD(physics), 62. *Prof Exp:* Res physicist, 62-63, sr physicist, 63-69, fel, 69-75, prin res scientist, 75-81, assoc sect mgr, Battelle Mem Inst, 81-86; PROF, ELEC ENG, GA INST TECHNOL, 86- *Mem:* Am Phys Soc; Am Asn Physics Teachers; Soc Photo-Optical Instrumentation Engrs; Optical Soc Am; sr mem Inst Elec & Electronics Engrs. *Res:* Physics; solid state; theory of magnetism; band structure; integrated optics; optical processing. *Mailing Add:* Sch Elec Eng Ga Inst Technol 777 Atlantic Dr NW Atlanta GA 30332-0250

KENAT, THOMAS ARTHUR, b Cleveland, Ohio, Aug 6, 42; m 64; c 2. CHEMICAL ENGINEERING, POLYMER SCIENCE. *Educ:* Carnegie-Mellon Univ, BS, 64, MS, 65, PhD(chem eng), 68. *Prof Exp:* Res engr, Chemstrand Res Ctr, Inc, NC, 68-69; res engr, 69-74, sr res & develop engr, 74-80, sr eng scientist, 81-83, sr res & develop assoc, B F Goodrich Co, 83-88; sr res & develop assoc, Camet Co, 88-89; SR PROJ MGR, QUANTUM TECHNOL INC, 89- *Mem:* Am Inst Chem Engrs; Am Chem Soc; Nat Asn Corrosion Engrs; Tech Asn Pulp & Paper Indust. *Res:* Dynamics and control of polymerization reactions; design and development of chemical reaction systems; processing of polymer composites; corrosion testing to select materials of construction for chemical process applications; design and development of chemical process concepts; synthetic rubber R & D; paper pulp bleaching chemicals. *Mailing Add:* 745 Falling Oaks Dr Medina OH 44256

KENDALL, BRUCE REGINALD FRANCIS, b Guildford, Western Australia, July 23, 34; m 56, 88; c 3. SPACE PHYSICS. *Educ:* Univ Western Australia, BSc, 54, PhD (physics), 60. *Prof Exp:* Nat Res Coun Can, 59-60, asst res officer, 60-61; sr res scientist, Nuclide Corp, 61-62, dir new prod develop, 62-64; assoc prof, 64-69, PROF PHYSICS, PA STATE UNIV, 69- *Concurrent Pos:* Consult, 64-; chmn, Vacuum Technol Div, Am Vacuum Soc, 88-89. *Mem:* Am Phys Soc; Am Vacuum Soc. *Res:* Electron, vacuum and space physics; mass spectrometry; vacuum and electrical properties of spacecraft materials. *Mailing Add:* Dept Physics Pa State Univ University Park PA 16802

KENDALL, BURTON NATHANIEL, b San Francisco, Calif, Dec 15, 40; m 63, 79; c 3. NETWORKS, DISTRIBUTED DATA BASES. *Educ:* Stanford Univ, BS, 62; Brown Univ, PhD(physics), 69. *Prof Exp:* Res aide microwave design, Stanford Univ, 59-62; res asst physics, Brown Univ, 62-69; lectr, Univ Calif, Santa Barbara, 69-71, asst prof, 71-73; sr res scientist, Systs Control, Inc, 73-78; sr staff scientist, Measurex Corp, 78-80, prin scientist, 80-89; SYST ARCHITECT, OCTEL COMMUN CORP, 89- *Concurrent Pos:* Consult, Libr Automation, 79-; pres, Delta Res Found, 80-; mem, NASA adv panel knowledge based systs verification & validation, 87-89. *Mem:* AAAS; Am Phys Soc; Instrument Soc Am. *Res:* Computer hardware and software design; electronics and microwave design; large scale systems design and modelling; indust process control syst design; voice processing system design. *Mailing Add:* Octel Commun Corp 890 Tasman Dr Milpitas CA 95035

KENDALL, DAVID NELSON, b Gardner, Mass, Oct 20, 16; m 42; c 3. PHYSICAL CHEMISTRY, SPECTROSCOPY. *Educ:* Wesleyan Univ, BA, 38, MA, 39; Johns Hopkins Univ, PhD(chem, physics), 43. *Prof Exp:* Res phys chemist, Titanium Div, Nat Lead Co, NJ, 43-44; res chem physicist, Calco Chem Div, Am Cyanamid Co, 44-46, head infrared spectros labs, 46-53; consult chemist & spectroscopist & infrared specialist, founder, Kendall Infrared Labs, 53-87; CONSULT, DAVE KENDALL ASSOCS, 87- *Concurrent Pos:* Asst ed, Appl Spectros, Soc Appl Spectros, 46-48; ed, Your Consult, Newslett Asn Consult Chemists & Chem Engrs, Inc, 74-83. *Honors & Awards:* Gold Medal, Soc Appl Spectros, 73. *Mem:* Am Chem Soc; Soc Appl Spectros (pres, 55-56); Coblentz Soc; Asn Consult Chemists & Chem Eng (pres, 64-66); fel Am Inst Chem; Sigma Xi. *Res:* Colloid chemistry; x-ray crystallography; pigment particle size determination; colored oil smokes; lightfastness of dyes and pigments; infrared, Raman, visual and ultraviolet spectroscopy; analytical chemistry. *Mailing Add:* Ten Myrtle-Warbler Rd Hilton Head SC 29928

KENDALL, FRANCIS M, image analysis, computer applications; deceased, see previous edition for last biography

KENDALL, H(AROLD) B(ENNE), b Midland, Mich, Apr 27, 23; m 48; c 4. CHEMICAL ENGINEERING. *Educ:* Grove City Col, BS, 48; Case Inst Technol, MS, 50, PhD(chem eng), 56. *Prof Exp:* Instr chem & metall eng, Univ Mich, 50-51; instr chem & chem eng, Case Inst Technol, 51-55, asst prof chem eng, 55-60; chmn dept chem engr, Ohio Univ, 61-67, 71-72 & 82-83, PROF CHEM ENG, OHIO UNIV, 60- *Mem:* Am Chem Soc; Soc Hist Technol; Am Inst Chem Engrs. *Res:* Reaction kinetics in flow reactors; catalytic processing; heterogeneous catalysis; history of technology. *Mailing Add:* 69 Morris Ave Athens OH 45701

KENDALL, HARRY WHITE, b Sopchoppy, Fla, Oct 9, 24; m 50; c 3. PHYSICS. *Educ:* Tusculum Col, BA, 48; Fla State Univ, MS, 50; Univ Fla, PhD(electronics, physics), 61. *Prof Exp:* Instr physics, Chipola Jr Col, 50-51; asst prof, Emory & Henry Col, 51-54, assoc prof & head dept, 57-59; teaching asst, Univ Fla, 54-57, NSF fac fel, 59-60; assoc prof & chmn dept, 60-63, PROF PHYSICS, UNIV S FLA, 63-, ACTG CHMN DEPT, 78- *Mem:* Am Phys Soc; Am Asn Physics Teachers; Sigma Xi. *Res:* Electrical breakdown of gases. *Mailing Add:* RR 1 Greenwood FL 32443

KENDALL, HENRY WAY, b Boston, Mass, Dec 9, 26. PHYSICS. *Educ:* Amherst Col, BA, 50; Mass Inst Technol, PhD(nuclear physics), 55. *Hon Degrees:* DSc, Amherst Col, 75. *Prof Exp:* NSF fel, Mass Inst Technol, 54-56; res assoc, High Energy Lab, Stanford Univ, 56-57, lectr physics, 57-58, asst prof, 58-61; from asst prof to assoc prof, 61-67, PROF PHYSICS, MASS INST TECHNOL, 67- *Concurrent Pos:* Chmn, Union Concerned Scientists, 75- *Honors & Awards:* Nobel Prize Physics, 90; Leo Szilard Award, Am Phys Soc, 81; Bertrand Russell Soc Award, 82; WKH Panofsky Prize, 89. *Mem:* Fel Am Phys Soc; fel Am Acad Arts & Sci; fel AAAS. *Res:* Nucleon structure; high energy electron scattering; meson and neutrino physics. *Mailing Add:* 24-514 Mass Inst Technol Cambridge MA 02139

KENDALL, JAMES TYLDESLEY, b New York, NY, Dec 14, 16; m 50; c 3. PHYSICS, PHYSICAL CHEMISTRY. *Educ:* Cambridge Univ, BA, 39, MA, 42; Univ London, PhD(physics), 53. *Prof Exp:* Scientist, Assoc Elec Industs, Ltd, 39-53; mgr semiconductor res, Plessey Co, Ltd, 53-57; gen mgr, Tex Instruments, Ltd, 57-61; managing dir, Microwave Assocs, Ltd, 61-62; gen mgr, SGS-Fairchild Ltd, 62-64; managing dir, Edwards High Vacuum Int Ltd, 64-65; mgr phys res, Nat Cash Register Co, 66-68, gen mgr integrated circuits,

69-71, asst to gen mgr, microelectronics div, NCR Corp, 71-; RETIRED. *Mem:* Sr mem Inst Elec & Electronics Engrs; fel Royal Soc Edinburgh. *Res:* Semiconductor research, particularly electronic properties of silicon carbide and gray tin; metal-oxide-silicon integrated circuits. *Mailing Add:* 38 Second Ave Frinton-on-Sea Essex Co 13 England

KENDALL, JOHN HUGH, b Mt Pleasant, Tex, Sept 30, 42; m 65; c 2. FOOD SCIENCE, CEREAL CHEMISTRY. *Educ:* La State Univ, BS, 64, MS, 69, PhD(food sci), 73. *Prof Exp:* Qual control rep, Borden Inc, 71-73; SR FOOD TECHNOLOGIST, RIVIANA FOODS INC, 73- *Concurrent Pos:* Adj asst prof, Univ Houston, 75- *Mem:* Inst Food Technologists; Am Asn Cereal Chemists; Int Asn Milk Food & Environ Sanitarians; Am Soc Microbiol. *Res:* Rice processing and by-product utilization. *Mailing Add:* Riviana Foods Inc 1702 Taylor Houston TX 77007

KENDALL, JOHN WALKER, JR, b Bellingham, Wash, Mar 19, 29; m 54; c 3. ENDOCRINOLOGY. *Educ:* Yale Univ, BA, 52; Univ Wash, MD, 56. *Prof Exp:* USPHS trainee, 59-62; from asst prof to assoc prof, 62-71, prof med & head, Div Metab, Med Sch, 71-80, prof med & asst dean res, 80-83, DEAN, SCH MED, ORE HEALTH SCI UNIV, 83- *Concurrent Pos:* Instr med, Sch Med, Vanderbilt Univ, 59-60; assoc chief of staff for res, Vet Admin Hosp, Portland, 71-83; chmn pro tem, dept med, Med Sch, Univ Ore, 75-76; mem, Vet Admin Res Adv Comt, 80-83; pres, Ore Found Med Excellence, 89-91. *Mem:* Asn Am Physicians; Endocrine Soc; Am Soc Clin Invest; Am Fedn Clin Res; Soc Exp Biol & Med. *Res:* Neural control of pituitary function. *Mailing Add:* Dean's Off Sch Med Ore Health Sci Univ Portland OR 97201

KENDALL, MICHAEL WELT, b Glendale, Ariz, Jan 30, 43; m 65; c 1. GROSS ANATOMY, MICROSCOPIC ANATOMY. *Educ:* Univ Northern Iowa, BA, 65; Univ Louisville, MS, 69, PhD(anat), 72. *Prof Exp:* Asst prof anat, Med Ctr, Univ Miss, 72-74; from asst prof to assoc prof anat, Sch Med Sci, Univ Nev, Reno, 74-84, chmn dept, 75-77; CONSULT, 84- *Concurrent Pos:* Pesticide consult, Dept Agr, Univ Nev, 75-76; consult gross anat, Int Cong Col Physicians & Surgeons, 76- *Mem:* AAAS; Am Asn Anatomists; Am Heart Asn. *Res:* Ultrastructural descriptive analysis of carcinogenesis induced by aflatoxin-B, in rat liver; ultrastructural hepatotoxic effects of mirex in rats; scanning electron microscopy of human knee joints. *Mailing Add:* 5106 Arden Rd Amarillo TX 79110

KENDALL, NORMAN, b Philadelphia, Pa, May 7, 12; m 44; c 2. PEDIATRICS. *Educ:* Temple Univ, MD, 36, MS, 41. *Prof Exp:* Intern, Hosp, Temple Univ, 38, resident, 41, instr pediat, Sch Med, 41-47, from asst prof to assoc prof, 47-67, prof pediat, 67-79; consult, NJ Dept Health, 79-82; RETIRED. *Concurrent Pos:* Attend pediatrician, St Christopher's Hosp for Children; chmn dept neonatology, Temple Univ Hosp; consult, 82. *Mem:* Am Acad Pediat. *Res:* Newborn infants. *Mailing Add:* 2401 Pennsylvania Ave Apt 20-B-22 Philadelphia PA 19130

KENDALL, PERRY E(UGENE), b Paoli, Ind, July 27, 21; m 42; c 3. ELECTRICAL ENGINEERING. *Educ:* Purdue Univ, BS, 48, MS, 49, PhD(elec eng), 53. *Prof Exp:* Instr elec eng, Purdue Univ, 47-52; sr engr, res in servomechanisms, Cook Res Lab, 52-54; res proj engr, Capehart-Farnsworth Co Div, Int Tel & Tel Corp, 54-56, sr engr, Farnsworth Electronics Co, 56-58, head systs anal & design labs, 58-60, lab dir guidance & control lab, Astrionics Ctr, ITT Fed Labs, 60-63; mgr res & develop, Fed Systs Div, Indust Nucleonics Corp, 63-66; mgr guid & control, Missile Div, NAm Aviation Corp, 66-70; mgr advan develop, Indust Systs Res & Develop, Indust Nucleonics Corp, 70-74; mem sr tech staff, TRW Defense & Space Systs Group, 74-79, proj engr, 79-81; RETIRED. *Res:* Servomechanisms and electronics; circuits for photoemissive electron tubes and infrared cells; guidance for guided missiles and space vehicles; application of digital computers for process control; high energy laser systems. *Mailing Add:* RR 4 Mitchell IN 47446

KENDALL, PHILIP C, m 74; c 2. CHILD & ADOLESCENT CLINICAL PSYCHOLOGY. *Educ:* Old Dominion Univ, BS, 72; Va Commonwealth Univ, PhD(clin psychol), 77. *Prof Exp:* Prof psychol, Univ Minn, 77-84; PROF PSYCHOL, TEMPLE UNIV, 84-, DIR CHILD & ADOLESCENT ANXIETY DISORDERS CLIN, 84- *Concurrent Pos:* Fel, Ctr Aduan Study Behav Sci, Stanford, Calif, 77; head Div Psychol, Temple Univ, 84- *Mem:* Am Psychol Asn; Asn Advan Behav Ther (pres-elect, 88). *Res:* Child and adolescent clinical psychology with a special focus on the cognitive and behavioral aspects of psychopathology; design and evaluation of psychotherapeutic programs; self control, anxiety and depression. *Mailing Add:* 238 Meeting House Lane Merion Station PA 19066

KENDALL, ROBERT MCCUTCHEON, b Pasadena, Calif, Dec 29, 31; m 57; c 4. CHEMICAL ENGINEERING. *Educ:* Stanford Univ, BS, 52, MS, 53; Mass Inst Technol, ScD, 59. *Prof Exp:* Thermodyn specialist, Calif Adv Propulsion Systs Oper, 56-60; sect mgr, Vidya Inc, 60-65; vpres & div mgr, Aerotherm Corp, 65-76, sr vpres, chief scientist & mgr combustion technol, Acurex Corp, 76-82; PRES, ALZETA CORP, 82- *Concurrent Pos:* Lectr, exten, Univ Calif, 59-62 & Stanford Univ, 63-64; consult prof, Stanford Univ, 78- *Mem:* Am Inst Chem Engrs; Am Inst Aeronaut & Astronaut; Combustion Inst. *Res:* Application of advanced experimental and computational techniques to the study of problems of mass, energy and momentum exchange in combusting or chemically active fluid dynamic systems. *Mailing Add:* 1097 Enderby Way Sunnyvale CA 94087

KENDALL, WILLIAM ANDERSON, b Fitchburg, Mass, Sept 24, 24; m 52; c 3. PLANT PHYSIOLOGY. *Educ:* Univ Maine, BS, 49; Ohio State Univ, PhD(bot), 54. *Prof Exp:* Asst, Ohio State Univ, 49-54; PLANT PHYSIOLOGIST, AGR RES SERV, USDA, 54- Prof Exp: Adj prof, Univ Ky, 54-70 & Pa State Univ, 70- *Mem:* Am Soc Agron; Am Soc Plant Physiol. *Res:* Interactions of genotypes and environments on growth and development of forage crops. *Mailing Add:* 714 Devonshire Dr State College PA 16803

KENDE, ANDREW S, b Budapest, Hungary, July 17, 32; nat US; m 54; c 1. ORGANIC CHEMISTRY, ORGANIC SYNTHESIS. *Educ:* Univ Chicago, BA, 50; Harvard Univ, MA, 54, PhD, 57; Glasgow Univ, postdoc, 58; Univ Munchen, postdoc, 60. *Prof Exp:* Res chemist, Lederle Labs, Am Cyanamid Co, 57-62, res assoc, 63-67, res fel, 67-68; prof chem, 68-81, chmn, 79-83, CHARLES F HOUGHTON PROF CHEM, UNIV ROCHESTER, 81- *Concurrent Pos:* Vis prof, Mich State Univ, 68 & Univ Geneve, 74; consult, Lederle Labs, 68-, Dow Chem Co, 74- & Eastman Kodak, 88-; consult, Med Chem Study Sect, NIH, 72-76 & 85-86, chmn, 74-76; Guggenheim fel, 78-79; chmn, Org Div, Am Chem Soc, 78-79; Japan Soc Promotion Sci Award, 85-86. *Mem:* Am Chem Soc; Chem Soc; Europ Photochem Asn. *Res:* Thermal and photochemical rearrangements, total synthesis of alkaloids and antibiotics, synthetic methods, chemistry of polyhaloaromatic environmental contaminants; chemistry of antitumor compounds. *Mailing Add:* Dept Chem Univ Rochester Rochester NY 14627

KENDE, HANS JANOS, b Szekesfehervar, Hungary, Jan 18, 37; US citizen; m 60; c 3. PLANT PHYSIOLOGY. *Educ:* Univ Zurich, PhD(bot), 60. *Prof Exp:* Res fel plant physiol, Nat Res Coun Can, 60-61; res fel, Div Biol, Calif Inst Technol, 61-63; plant physiologist, Negev Inst Arid Zone Res, Israel, 63-65; dir, 85-88, from assoc prof to prof, 65-90, DISTINGUISHED PROF BOT & PLANT PATH, DEPT ENERGY PLANT RES LAB, MICH STATE UNIV, 90- *Concurrent Pos:* Guggenheim Mem Found fel & vis prof, Swiss Fed Inst Technol, 72-73; mem, Adv Panel Develop Biol, NSF, 74-78; vis prof, Swiss Fed Inst Technol, 79-80. *Mem:* AAAS; Am Soc Plant Physiol; Soc Develop Biol; Ger Acad Sci. *Res:* Function, biosynthesis and action mechanism of plant growth regulators. *Mailing Add:* Dept Energy Plant Res Lab Mich State Univ East Lansing MI 48824-1312

KENDER, DONALD NICHOLAS, b Passaic, NJ, Aug 30, 48; m 71. ANALYTICAL CHEMISTRY. *Educ:* Ohio State Univ, BA, 70; Georgetown Univ, PhD(chem), 75. *Prof Exp:* Chemist & Nat Res Coun assoc, Naval Surface Weapons Ctr, 75-76; MGR, CIBA-GEIGY CORP, 76- *Mem:* Am Chem Soc; Soc Appl Spectros; Sigma Xi. *Res:* Isolation, identification and physical organic chemistry of pharmaceuticals. *Mailing Add:* 556 Morris Ave Summit NJ 07901-1330

KENDER, WALTER JOHN, b Camden, NJ, Dec 20, 35; m 57; c 2. ADMINISTRATION, CITRUS. *Educ:* Del Valley Col, BS, 57; Rutgers Univ, MS, 59, PhD(plant nutrit), 62. *Prof Exp:* From asst prof to assoc prof hort, Univ Maine, Orono, 62-69; assoc prof, 69-75, prof pomol & chmn dept, Cornell Univ, 75-82, head dept pomol & viticult, NY Agr Exp Sta 72-82; PROF & DIR CITRUS RES & EDUC CTR, UNIV FLA, LAKE ALFRED, 82- *Concurrent Pos:* Assoc ed, Am Soc Hort Sci, 72-76; distinguished scientist, ARIC Univ, Wageningen The Netherlands, 74; bd dir Am Soc Hort Sci, 75-80, 82-85; adv comt, Farm Bur Citrus, 85-; Consult, US-AID-Pakistan, 89; prin investor, ARS Coop Agreement Citrus Exotic Dis, 89-94. *Honors & Awards:* Darrow Award, Am Soc Hort Sci, 83. *Mem:* Am Pomol Soc; fel AAAS; fel Am Soc Hort Sci; Sigma Xi. *Res:* Physiology and culture of fruit crops; air pollution effects on agricultural crops; emphasis on impacts of fossil fuel effluents and acid rain on fruit crop productivity and economic assessment; citrus; agricultural administration. *Mailing Add:* Citrus Res & Educ Ctr Univ Fla 700 Exp Sta Rd Lake Alfred FL 33850

KENDIG, EDWIN LAWRENCE, JR, b Victoria, Va, Nov 12, 11; m 41; c 2. PEDIATRICS, RESPIRATORY DISORDERS. *Educ:* Hampden-Sydney Col, BA, 32, BS, 33; Univ Va, MD, 36. *Hon Degrees:* DSc, Hampden-Sydney Col, 71. *Prof Exp:* Assoc prof, 58-61, PROF PEDIAT, MED COL VA, VA COMMONWEALTH UNIV, 61-, DIR CHILD CHEST CLIN, COL HOSP, 58- *Concurrent Pos:* Instr, Johns Hopkins Hosp, 44; mem comt med educ & founding mem sect dis in childhood, Am Thoracic Soc; mem bd visitors, Univ Va; mem Va steering comt, White House Conf, 60; chmn, Richmond City Bd Health. *Res:* Pediatrics; sarcoidosis; unclassified mycobacteria. *Mailing Add:* 5801 Bremo Rd Richmond VA 23226

KENDIG, JOAN JOHNSTON, b Derby, Conn, May 1, 39; m 64; c 2. NEUROBIOLOGY. *Educ:* Smith Col, BA, 60; Stanford Univ, PhD(biol sci), 66. *Prof Exp:* NSF fel neurophysiol, Univ Calif, Berkeley, 65-67; res assoc, 67-71, from asst prof to assoc prof, 71-86, PROF BIOL ANESTHESIA, SCH MED, STANFORD UNIV, 86- *Concurrent Pos:* Melion fac fel, Stanford Univ, 76; vis scientist, Clin Res Ctr, Northwick Park, UK, 84, vis prof, Ben Gurion Univ, Israel, 88; NIH physiol study sect, 81-85; Javits Neurosci Investr Award, 88- *Mem:* Am Soc Pharmacol & Exp Therapeut; Undersea Med Soc; Biophys Soc; Soc Neurosci. *Res:* Neuropharmacology of anesthetic and analgesic drugs; cellular effects of anesthetic agents; effects of hyperbaric pressures on excitable cells. *Mailing Add:* Dept Anesthesia Sch Med Stanford Univ Stanford CA 94305

KENDIG, MARTIN WILLIAM, b Danville, Pa, Oct 20, 45; m 69; c 2. PHYSICAL CHEMISTRY, CORROSION SCIENCE. *Educ:* Franklin & Marshall Col, AB, 67; Brown Univ, PhD(phys chem), 74. *Prof Exp:* Res assoc, Ctr Surface & Coatings Res, Lehigh Univ, 73-76; asst chemist, 76-78, assoc chemist corrosion sci, Brookhaven Nat Lab, 78-80; MEM TECH STAFF, ROCKWELL INT SCI CTR, 80- *Honors & Awards:* Melvin Romanoff Award, Nat Asn Corrosion Engrs. *Mem:* Electrochem Soc; Am Chem Soc; Sigma Xi. *Res:* Electrochemical aspects of surface energy, wetting wear and environmental fracture; the chemistry and physics of localized corrosion and corrosion protection; corrosion monitoring; polymer coatings. *Mailing Add:* Rockwell Int Sci Ctr Thousand Oaks CA 91360

KENDRICK, AARON BAKER, physiological chemistry; deceased, see previous edition for last biography

KENDRICK, BRYCE, b Liverpool, Eng, Dec 3, 33; m 57, 77; c 2. MYCOLOGY. *Educ:* Univ Liverpool, BSc, 55, PhD(mycol), 58. *Hon Degrees:* DSc, Univ Liverpool, 80. *Prof Exp:* Fel taxonomic mycol, Nat Res Coun Can, 58-59; mycologist, Plant Res Inst, Res Br, Can Dept Agr, 59-65; from asst prof to assoc prof, 65-71, PROF BIOL, UNIV WATERLOO, 71-,

ASSOC DEAN GRAD STUDIES, 85- *Concurrent Pos:* Chmn, Plant Biol Grant Selection Comt, Nat Sci & Eng Res Coun, Can, 79; Guggenheim fel, 79-80; secy, Acad Sci, Royal Soc Can, 85-; hon prof, Nanjing Forestry Univ, Nanjing, China, 88. *Mem:* Mycol Soc Am; Can Bot Asn; fel Royal Soc Can. *Res:* Computer simulations; systematics of hyphomycetes; development, ecology, karyology, and toxicology of microfungi; mycorrhizae. *Mailing Add:* Dept Biol Univ Waterloo Waterloo ON N2L 3G1 Can

KENDRICK, FRANCIS JOSEPH, b St Petersburg, Fla, Oct 19, 26; m 53; c 2. PATHOLOGY, DENTISTRY. *Educ:* Northwestern Univ, DDS, 52, PhD(path), 63. *Prof Exp:* Intern dent, USPHS Hosp, Seattle, 52-53; pvt practr, 53-55; instr dent, Northwestern Univ, 55-56; res assoc teratology, Nat Inst Dent Res, 60-63, pathologist, Nat Inst Child Health & Human Develop, 63-69, asst to chief gen res support br, 69-70, spec asst prog planning & eval, 70-74, actg dir gen res support prog, div res resources, 74-75, dir, biomed res support prog, 75-80, asst dir, manpower & resource develop, 81-82, DEPT CHIEF REV, DIV RES RESOURCES, NIH, 83- *Mem:* AAAS; Teratology Soc; fel Am Acad Oral Path; Int Asn Dent; NY Acad Sci. *Res:* Experimental carcinogenesis and teratology; oral and general pathology. *Mailing Add:* 11809 Devilwood Dr Rockville MD 20854

KENDRICK, HUGH, b Ewell, Eng, Jan 25, 40; m 63; c 2. NUCLEAR ENGINEERING, SOLID STATE PHYSICS. *Educ:* Univ London, BSc, 61; Calif Inst Technol, MS, 62; Univ Mich, PhD(nuclear eng), 68. *Prof Exp:* Teaching res asst mech eng, Calif Inst Technol, 61-62; scientist, Vickers Res Ltd, 62-63; sr physicist, Radiation Transport Group, Gulf Radiation Technol, 68-72; dep mgr, Div Environ & Safety, Sci Applns, Inc, 72-75, mgr, Div Safeguards & Nuclear Fuels, 75-77; spec asst, Off Fuel Cycle Eval, US Dept Energy, 77-79, dir, Off Plans & Anal, 79-81; vpres, 81-84, dep chief operating officer, 85-89, CORP VPRES, SCI APPLICATIONS INT CORP, 84-, ASST PRES, 89- *Mem:* Am Phys Soc; Inst Nuclear Mat Mgt; Am Nuclear Soc; Sigma Xi. *Res:* Investigation of magnetic materials through neutron diffraction; pulsed neutron investigation of radiation transport in shields; spectroscopy and unfolding techniques; nuclear materials assay; assessment of proliferation risks of nuclear technology; nuclear safeguards system effectiveness evaluation; environmental economic safety assessment of technology. *Mailing Add:* 13062 Caminito Pt Del Mar Del Mar CA 92014-3853

KENDRICK, JAMES BLAIR, JR, b Lafayette, Ind, Oct 21, 20; m 42; c 2. PLANT PATHOLOGY. *Educ:* Univ Calif, Berkeley, BA, 42; Univ Wis, PhD(plant path), 47. *Prof Exp:* Jr plant pathologist, Univ Calif, Riverside, 47-49, from asst plant pathologist to prof plant pathologist, 49-68; vpres agr sci, 68-77, dir Agr Exp Sta, 73-80, dir Coop Exten, 75-80, vpres agr, 77-82, vpres agr & natural resources, 82-86, EMER PROF PLANT PATH, UNIV CALIF, 86-, EMER VPRES AGR & NATURAL RESOURCES, 86- *Concurrent Pos:* NSF sr fel, 61-62; mem, Calif State Bd Food & Agr, 68- & US Forest Serv Adv Comt, Region V, 70-75; chmn div agr, Nat Asn State Univs & Land Grant Cols, 73, mem exec comt, 74-76, 79; mem, Agr Res Policy Adv Comt, 74-76, Gov Bd, Agr Res Inst, 74-76; chmn adv panel, Off Technol Assessment, US Cong, 81-; chmn sci rev panel, Calif Air Resources Bd, 87- *Mem:* Fel AAAS; Am Phytopath Soc; Sigma Xi. *Res:* Soil fungicides; biology of root decaying organisms; diseases of vegetables. *Mailing Add:* 615 Spruce St Berkeley CA 94707

KENDRICK, JOHN EDSEL, b Scott City, Kans, Dec 23, 28; m 54; c 3. PHYSIOLOGY. *Educ:* Univ Kans, AB, 52, PhD(physiol), 57. *Prof Exp:* From instr to asst prof, 57-67, ASSOC PROF PHYSIOL, SCH MED, UNIV WISMADISON, 67- *Mem:* Am Physiol Soc. *Res:* Cardiovascular physiology. *Mailing Add:* Dept Physiol Univ Wis 1300 University Ave Madison WI 53706

KENDZIORSKI, FRANCIS RICHARD, b Alpena, Mich, Apr 2, 31; m 64; c 2. NUCLEAR PHYSICS. *Educ:* Univ Detroit, BS, 53; Cornell Univ, PhD(physics), 61. *Prof Exp:* Asst prof physics, Univ Detroit, 61-63; asst prof, Univ Dayton, 63-67; assoc prof, 67-77, chmn Dept Physics & Astron, 78-85, PROF PHYSICS, WESTERN CONN STATE COL, 77- *Mem:* AAAS; Am Phys Soc; Am Asn Physics Teachers. *Res:* Low energy studies of nuclear structure in intermediate weight nuclei; extensive cosmic ray air showers; elementary education. *Mailing Add:* 37 Farview Ave Danbury CT 06810

KENEALY, MICHAEL DOUGLAS, b Council Bluffs, Iowa, May 7, 47; m 69; c 2. ANIMAL NUTRITION, PHYSIOLOGY. *Educ:* Iowa State Univ, BS, 69, PhD(animal nutrit & physiol), 74. *Prof Exp:* Nutritionist, Dr Macdonalds Feed Co, 74-75; from asst prof to assoc prof, 75-84, PROF ANIMAL SCI, IOWA STATE UNIV, 84- *Mem:* Sigma Xi; Am Dairy Sci Asn; Am Soc Animal Sci. *Res:* International work in China, Costa Rica, Taiwan, and the Soviet Union. *Mailing Add:* Iowa State Univ 123 Kildee Hall Ames IA 50011

KENEALY, PATRICK FRANCIS, b Chicago, Ill, Aug 4, 39. MATHEMATICS EDUCATION. *Educ:* Loyola Univ, Ill, BS, 61; Univ Notre Dame, PhD(physics), 67. *Prof Exp:* From asst prof to assoc prof physics, Wayne State Univ, 67-88; PROF PHYSICS & SCI EDUC, CALIF STATE UNIV, LONG BEACH, 88- *Concurrent Pos:* Vis assoc prof, Stanford Univ, 76-77; NSF fac fel sci, 76-77; vis scholar, grad group sci & math educ, Univ Calif, Berkeley; sr sci consult, Detroit Sci Ctr, 88. *Mem:* AAAS; Am Phys Soc; Am Asn Physics Teachers; Am Educ Res Asn. *Res:* Research on the role of language in science and mathematics learning; teacher training in science and mathematics; informal science and mathematics instruction in museum settings; use of computers in teaching physics; theoretical and experimental analysis of physics instruction and learning. *Mailing Add:* Dept Physics Calif State Univ Long Beach CA 90840-3901

KENEFICK, ROBERT ARTHUR, b Syracuse, NY, Mar 9, 37; m 60; c 3. TRAPPED IONS. *Educ:* Mass Inst Technol, BS, 59; Fla State Univ, PhD(physics), 62. *Prof Exp:* Res assoc nuclear physics, Univ Colo, 62-63, asst prof, 63-64; asst prof, Univ Mich, 64-65; from asst prof to assoc prof, 65-74, PROF PHYSICS, TEX A&M UNIV, 74- *Mem:* Am Phys Soc; AAAS. *Res:* Atomic collisions; particle detectors; ion sources. *Mailing Add:* Dept Physics Tex A&M Univ College Station TX 77843

KENELLY, JOHN WILLIS, JR, b Bogalusa, La, Nov 22, 35; m 56; c 2. MATHEMATICS. *Educ:* Southeastern La Col, BS, 57; Univ Miss, MS, 57; Univ Fla, PhD(math), 61. *Prof Exp:* Instr math, Univ Fla, 59-61; asst prof, Univ Southwestern La, 61-63; assoc prof, Clemson Univ, 63-68; prof & chmn dept, La State Univ, 68-69; head dept, 69-77, prof, 69-85, ALUMNI PROF MATH SCI, CLEMSON UNIV, 85- *Concurrent Pos:* Vis lectr & curric consult, Math Asn Am, 70-, ed, Placement Test Newslett, 79-85; chief reader, Advan placement Prog Math, Educ Testing Serv, 75-79, dir advan placement reading, 85-; chmn, Advan Placement Math Comt, Col Bd, 79-83, Math Sci Adv Comt, 83- & coun acad affairs, 85-87; prog officer, NSF, 88; interim dir, Col & Advan Placement Prog, 89-90; chmn, Southeastern Sect, Math Asn Am. *Mem:* Am Math Soc; Math Asn Am; Nat Coun Teachers Math; Nat humanities Fac. *Res:* Geometry; convexity; operations research. *Mailing Add:* 327 Woodland Way Clemson SC 29631

KENESHEA, FRANCIS JOSEPH, b Providence, RI, June 25, 21; wid; c 2. PHYSICAL CHEMISTRY. *Educ:* RI State Col, BS, 43, MS, 48; Univ NMex, PhD(chem), 51. *Prof Exp:* Instr chem, RI State Col, 46-47; asst, Cornell Univ, 47-48; sr res engr, NAm Aviation, Inc, 51-55; sr chemist, Stanford Res Inst, 55-71; consult engr, Quadrex Corp, 74-80, sr consult, 80-85; PVT CONSULT. *Concurrent Pos:* Sr res assoc, Ore State Univ, 63-64. *Mem:* Sigma Xi; Am Chem Soc; Am Nuclear Soc. *Res:* Thermodynamics of vaporization; chemical diffusion; chemistry of molten salts and metal-salt solutions; nuclear and radiochemistry; nuclear technology. *Mailing Add:* 20 Bear Paw Portola Valley CA 94028-8014

KENETT, RON, b Zurich, Switz, Oct 20, 50; Israel citizen; m 76; c 4. STATISTICAL METHODS, DESIGN OF EXPERIMENTS. *Educ:* London Univ, BS, 74; Weizmann Inst Sci, PhD(math), 78. *Prof Exp:* Lectr, Dept Statist, Univ Wis-Madison, 78-80; mem tech staff statist, Bell Labs, NJ, 80-81; dir statist methods, indust statist, Tadiran Co, 81-90; PROF QUAL MGT, SCH MGT, STATE UNIV NY, 87- *Concurrent Pos:* Adj prof, Tel Aviv Univ, 82-89, Technion, 86; consult statist methods & qual mgt, 87- *Mem:* Am Statist Asn; Am Soc Qual Control. *Res:* Quality management; industrial statistics; statistical process control; design of experiments; performance appraisal systems; software reliability; multivariate quality control. *Mailing Add:* Sch Mgt State Univ NY Binghamton NY 13901

KENG, PETER C, b Kinagsu, China, Aug 12, 46; m 72; c 3. CELL SEPARATION. *Educ:* Tunghai Univ, BS, 68; Colo State Univ, PhD(radiation biol), 78. *Prof Exp:* Asst prof, dept radiation oncol, 80-85, asst prof, dept radiation biol & biophys, 81-85, ASSOC PROF, DEPT RADIATION ONCOL & DEPT RADIATION BIOL & BIOPHYS, UNIV ROCHESTER, 85-, DIR RES, CELL SEPARATION & FLOW CYTOMETRY FACIL, 81- *Mem:* Radiation Res Soc; Anal Cytometry; Am Soc Cell Biol; AAAS. *Res:* Separation of cell subpopulations from solid tumors, bone marrow and tissue culture cells into various host cells; neoplastic cells and cells at different stages of the cell cycle to study the DNA damage of these cells. *Mailing Add:* Dept Radiation Oncol Univ Rochester Sch Med 601 Elmwood Ave Box 74 Rochester NY 14642

KENIG, MARVIN JERRY, b Philadelphia, Pa, Sept 20, 36; m 59; c 2. APPLIED MECHANICS, MATERIALS SCIENCE. *Educ:* Drexel Univ, BSME, 59, MSME, 63; Princeton Univ, MA, 63, PhD(eng), 65. *Prof Exp:* From assoc prof to prof mech eng, Drexel Univ, 69-82, asst to pres, 74-83; PROF & CHMN, DEPT MECH ENG, 83-, PROF & CHMN, DEPT AIRCRAFT & AUTOMOTIVE ENG, WESTERN MICH UNIV, 87- *Concurrent Pos:* Consult, J P Oat & Sons, Inc, 68- & US Army Frankford Arsenal, 70-75. *Mem:* Am Soc Mech Engrs; Am Acad Mech; Am Soc Eng Educators; Am Defense Preparedness Asn; AAAS; Sigma Xi. *Res:* Effect Portevin-Le Chatalier phenomenon on plastic potential theory of yielding; implications with respect to propagation of small stress increments; creep; fatigue; quantum mechanics modeling of dislocation motion; response of orthotropic plates under lateral pressure pulse; inelastic buckling of nonprismatic columns; bending of prismatic unsymmetric eccentrically loaded columns. *Mailing Add:* Dept Mech Eng Western Mich Univ Kalamazoo MI 49008

KENIMER, JAMES G, GENERAL PHYSIOLOGY. *Prof Exp:* ACTG CHIEF LAB ALLERGY & IMMUNOL CHEM, BIOL EVAL & RES, FOOD & DRUG ADMIN, 91- *Mailing Add:* Biol Eval & Res Food & Drug Admin 8800 Rockville Pike Bethesda MD 20892

KENK, ROMAN, invertebrate zoology; deceased, see previous edition for last biography

KENK, VIDA CARMEN, b San Juan, PR, Dec 24, 39; m 74; c 2. INVERTEBRATE ZOOLOGY. *Educ:* Col William & Mary, BS, 61; Radcliffe Col, AM, 62; Harvard Univ, PhD(biol), 67. *Prof Exp:* Asst prof, 66-70, assoc prof, 70-77, PROF BIOL, SAN JOSE STATE UNIV, 77- *Mem:* AAAS; Am Malacol Union; Western Soc Malacologists (pres, 80). *Res:* Systematics, ecology and functional anatomy of bivalve molluscs. *Mailing Add:* 18596 Paseo Pueblo Dr Saratoga CA 95120

KENKARE, DIVAKER B, b Goa, India, May 25, 36; US citizen; m 66; c 2. BIOLOGICAL CHEMISTRY, PHYSICAL CHEMISTRY. *Educ:* Univ Poona, BSc, 59; Sardar Patel Univ, India, MSc, 61; Ohio State Univ, MSc, 63, PhD(food chem), 66. *Prof Exp:* Res assoc protein chem, Univ Ill, Urbana, 66-68; res chemist, Colgate Palmolive Res Ctr, Piscataway, 68-71, sr res chemist, 71-78, res assoc, 78-85, SR ASSOC, COLGATE PALMOLIVE RES CTR, 85- *Mem:* Am Chem Soc. *Res:* Changes of protein at elevated temperatures; characterization, physical chemical behavior, and modification of proteins. *Mailing Add:* 909 River Rd PO Box 343 Piscataway NJ 08855

KENKEL, JOHN V, b Harlan, Iowa, Apr 20, 48; m 75; c 3. ANALYTICAL CHEMISTRY. *Educ:* Iowa State Univ, BS, 70; Univ Tex, Austin, MA, 72. *Prof Exp:* Sr staff assoc, Sci Ctr, Rockwell Int, 73-77; PROF SUPVR CHEM, SOUTHEAST COMMUNITY COL, 77- *Concurrent Pos:* Mem, Comt Chem

2 Yr Col, Am Chem Soc, 82-, Comt Educ, Div Anal Chem, 88-; Burlington Northern Found fac achievement award, 85; Chem Mfg Asn regional catalyst award, 88. *Honors & Awards:* Gustav Ohans Award, Nat Sci Teachers Asn, 90. *Mem:* Am Chem Soc. *Res:* Analytical chemistry; author of two books and two publications. *Mailing Add:* Southeast Community Col 8800 O St Lincoln NE 68520

KENKNIGHT, GLENN, b Canby, Ore, Nov 26, 10; m 40; c 2. PLANT PATHOLOGY. *Educ:* Carleton Col, BA, 34; Mich State Col, MS, 37, PhD(bot, plant path), 39. *Prof Exp:* Asst plant path, Univ Minn, 35; asst, Exp Sta, Mich State Col, 35-39; plant pathologist, Exp Sta, Agr & Mech Col Tex, 40-42; assoc plant pathologist, Exp Sta, Univ Idaho, 42-45; assoc plant pathologist, Calif Dept Agr, Indio, 45-48; plant pathologist, Hort Field Lab, USDA, 48-62, res plant pathologist, Pecan Field Lab, 62-73; CONSULT, 73- *Mem:* Am Phytopath Soc. *Res:* Soil actinomyces in relation to potato scab; fungicidal action of mercury compounds; breeding vegetable crops for disease resistance; virus diseases of stone fruits; witches' broom disease of trees; pecan diseases. *Mailing Add:* 9517 Palmetto Lane Shreveport LA 71118

KENKRE, VASUDEV MANGESH, b Panjim, India, Sept 21, 46; m 69; c 2. THEORETICAL SOLID STATE PHYSICS. *Educ:* Indian Inst Technol, Bombay, BTech, 68; State Univ NY, Stony Brook, MA, 71, PhD(physics), 71. *Prof Exp:* Instr physics, State Univ NY, Stony Brook, 71-72; res assoc, 72-74, asst prof, 74-79, ASSOC PROF PHYSICS, UNIV ROCHESTER, 79-, FEL, INST FUNDAMENTAL STUDIES, 72- *Res:* Transport and response theories, master equations, random walks; charge, excitation and energy transfer in organic and amorphous solids; interaction of light with matter; polaron and exciton motion; size quantization effect; statistical mechanics. *Mailing Add:* Dept Physics & Astron Univ NMex Albuquerque NM 87131

KENLEY, RICHARD ALAN, b Chicago, Ill, Jan 17, 47; m 73. ORGANIC CHEMISTRY. *Educ:* Univ Ill, Champaign, BS, 69; Univ Calif, San Diego, PhD(org chem), 73. *Prof Exp:* Phys org chem, SRI Int, 75-; assoc dir parenteral res & develop, Baxter Health Care Corp, 85-90; DIR, PHARMACEUT DEVELOP, GENETICS INST, 90- *Mem:* Am Chem Soc. *Res:* Free radical reactions in gas and solution phase. *Mailing Add:* Genetics Inst One Burtt Rd Andover MA 01810

KENNA, BERNARD THOMAS, b Hays, Kans, Jan 4, 35; m 60; c 2. NUCLEAR CHEMISTRY, GEOCHEMISTRY. *Educ:* Ariz State Col, BS, 56; Univ Miss, MS, 58; Univ Ark, PhD(nuclear chem), 61. *Prof Exp:* Res asst radiochem, Univ Ark, 59-61; nuclear chemist in-charge, Radiochem & Activiation Anal Lab, 61-81, proj mgr, Environ Res, 81-85, PROJ MGR, EXPLOSIVES DETECTION, SANDIA NAT LAB, 85-; TECH CONSULT, FED AVIATION ADMIN, 88- *Concurrent Pos:* Assoc prof, Univ NMex, 67-79. *Mem:* AAAS; Am Chem Soc; Am Nuclear Soc. *Res:* Theoretical and practical application of nuclear and radiochemistry to analytical chemistry; ion exchange; inorganic coordination chemistry; nuclear geochemistry; environmental research (atmosphere and water); detection of explosives. *Mailing Add:* Div 5248 Sandia Nat Lab Albuquerque NM 87185

KENNAMER, JAMES EARL, b Fairfield, Ala, Aug 6, 42; m 67; c 2. WILDLIFE ECOLOGY. *Educ:* Auburn Univ, BS, 64; Miss State Univ, MS, 67, PhD(wildlife mgt), 70. *Prof Exp:* Instr wildlife mgt, Miss State Univ, 69-70; from asst prof to assoc prof wild life ecol, Auburn Univ, 70-80; DIR RES & MGT, NAT WILD TURKEY FEDN, 80- *Mem:* Wildlife Soc; Sigma Xi. *Res:* Wild turkey ecology and physiology; Canada goose, white-tailed deer and fallow deer ecology and physiology. *Mailing Add:* 429 Stonehenge Circle Edgefield SC 29824

KENNARD, KENNETH CLAYTON, b Battle Creek, Mich, Dec 18, 26; m 49; c 3. ORGANIC CHEMISTY, BIOCHEMISTY. *Educ:* Univ Notre Dame, BS, 49; Univ Nebr, MS, 52, PhD(org chem), 54; Mass Inst Technol, SM, 64. *Prof Exp:* Res chemist, Eastmam Kodak Co, 54-65, asst div head, Emulsion Res Div, 65-69, staff asst to dir res, 69-75, dir biosci div, Kodak Res Labs, 75-84, gen mgr & vpres, Bio-Prod Div, 8-87; RETIRED. *Mem:* Am Chem Soc; Am Asn Clin Chemists; AAAS. *Res:* Organic chemistry of phosphorous and sulfur compounds; preparation and properties of light sensitive materials; biotechnology. *Mailing Add:* 19 Veldor Park No A Rochester NY 14612

KENNARD, WILLIAM CRAWFORD, b Centreville, Md, Nov 29, 21; m 43; c 3. PLANT PHYSIOLOGY, HORTICULTURE. *Educ:* Univ Del, BS, 43; Pa State Univ, MS, 48, PhD(plant physiol, soils), 56; Oak Ridge Inst Nuclear Studies, cert, 60. *Prof Exp:* Res fel pomol, Pa State Univ, 46-48, instr, 48-52; horticulturist, Mayaguez Inst Trop Agr, Mayaguez, PR, 52-57; prin horticulturist & res adminr, US Off Exp Sta, Washington, DC, 57-62; prof hort & assoc dir res admin, Univ Conn, 62-74, dir inst water resources, 65-74, prof plant physiol, 62-91. *Concurrent Pos:* Vis prof, Univ PR, 56; actg dir, Inst Water Resources, Univ Conn, 64-65; assoc seminars, Columbia Univ, 69- *Mem:* Fel AAAS; Am Inst Biol Sci; Am Soc Hort Sci. *Res:* Physiology and culture of temperate zone fruit crops; physiology of flowering; growth and development of tropical plants, including fruits, drug crops, insecticidal crops and bamboo; remote sensing of the environment. *Mailing Add:* 70 Lynnwood Rd Storrs Mansfield CT 06268

KENNEDY, ALBERT JOSEPH, b Spring Valley, Ill, July 2, 43; m 67; c 4. RADIOCHEMISTRY, CORROSION. *Educ:* Univ Ill, Champaign, BS, 66; Purdue Univ, PhD(nuclear chem), 72. *Prof Exp:* Fel nuclear chem, Lawrence Berkeley Lab, 72-73; sr res chemist, Babcock & Wilcox Co, 73-78; prin chemist, 78-80, supvr chem, 80-89, CHEM ASSESSMENT ADMINR, COMMONWEALTH EDISON, 89- *Mem:* Am Chem Soc; Am Nuclear Soc. *Res:* Corrosion chemistry; corrosion product deposition; activation analysis, radiochemistry, quality control and nuclear fuel evaluation. *Mailing Add:* Commonwealth Edison Co PO Box 767 Chicago IL 60690-0767

KENNEDY, ANDREW JOHN, b Budapest, Hungary, May 16, 35; US citizen; m 58; c 2. SOLID STATE PHYSICS, PHYSICAL ELECTRONICS. *Educ:* Wash State Univ, BS, 61; Univ Wash, MS, 64. *Prof Exp:* Assoc res engr A, Boeing Co, 61-64; RES PHYSICIST, CTR NIGHT VISION & ELECTRO-OPTICS, 64- *Mem:* Inst Elec & Electronics Engrs; Am Inst Physics. *Res:* Solid state infrared detector physics and technology, intensified charge coupled devices; imaging focal plane technology. *Mailing Add:* Ctr Night Vision & Electro-Optics Amsel-Rd-NV-IT Ft Belvoir VA 22060-5677

KENNEDY, ANN RANDTKE, b Rochester, NY, Dec 24, 46; m 73; c 2. CARCINOGENESIS. *Educ:* Vassar Col, AB, 69; Harvard Univ, SM, 71, SD, 73. *Prof Exp:* Res assoc, 73-75, asst prof, 76-80, ASSOC PROF RADIOBIOL, HARVARD UNIV, 80- *Concurrent Pos:* Comt mem, Nat Coun Radiation Protection Public Educ, 80-; mem, chem pathol study sect, consult, workshops & prin investr grants, NIH, 81-88. *Honors & Awards:* Outstanding Res Award, Radiation Res Soc, 84. *Mem:* Am Asn Cancer Res & Radiation Res; Sigma Xi; Free Radical Res Soc. *Res:* Radiobiology; mechanism of carcinogenesis with the ultimate aim of preventing cancer in human populations. *Mailing Add:* 407 Huntington Ave Boston MA 02115

KENNEDY, ANTHONY JOHN, b Brooklyn, NY, Dec 1, 32; m 60; c 6. DIGITAL SIGNAL PROCESSING, ADOPTIVE CONROL SYSTEMS. *Educ:* Univ Notre Dame, BS, 54; Carnegie Inst Technol, MS, 56, PhD(physics), 62. *Prof Exp:* Scientist, Nuclear Div, Martin Marietta Corp, 60-65 & Space Div, Chrysler Corp, 65-70; consult, Boland & Boyce, Inc, 70-74; scientist, Space Sci Lab, Gen Elec Co, 74-77; PRIN SCIENTIST, XYBION CORP, 77- *Res:* Design of signal processing systems for the detection of signals in ocean noise; computer systems for processing oceanographic information. *Mailing Add:* Xybion Corp 240 Cedar Knolls Rd Cedar Knolls NJ 07927

KENNEDY, BARBARA MAE, nutrition; deceased, see previous edition for last biography

KENNEDY, BILL WADE, b Dallas, Tex, Mar 21, 29; m 51; c 4. PLANT PATHOLOGY. *Educ:* Southeastern State Col, BS, 51; Okla State Univ, MS, 55; Univ Minn, PhD(plant path), 61. *Prof Exp:* Asst plant path, Okla State Univ, 51-52 & 54-55; sr technician, Univ Calif, 55-58; res asst, 58-59, res fel, 59-60, res assoc, 61-63, from asst prof to assoc prof, 63-72, PROF PLANT PATH, UNIV MINN, 72- *Concurrent Pos:* Res grants, Grad Sch, 64-66; Coop, US Regional Soybean Lab, Ill, 64-; USDA grant, 67; leaves for advan study, Univ Calif, Berkeley, 67, Eng, 71 & Italy, 78; sr ed, Phytopath, 73-76. *Mem:* Am Phytopath Soc; Am Inst Biol Sci; Am Soybean Asn. *Res:* Chemical control of cotton seedling blight; root-rot studies; physiology of reproduction in Phytophthora, identity and epidemiology of bacterial blight on strawberry; seed pathology; ecology of bacteria associated with soybean. *Mailing Add:* 2020 Eldridge Ave St Paul MN 55113

KENNEDY, BURTON MACK, b St Louis, Mo, Apr 15, 49; m 74; c 1. ISOTOPE GEOLOGY, GEOCHRONOLOGY. *Educ:* Washington Univ, BA, 74, PhD(earth & planetary sci), 81. *Prof Exp:* Res assoc, 81-85, RES GEOPHYSICIST, DEPT PHYSICS, UNIV CALIF, BERKELEY, 85- *Mem:* Meteoritical Soc; Am Geophys Union. *Res:* High sensitivity rare gas mass spectrometry as applied to meteoritic, lunar and terrestrial samples to investigate the origin and early history of the earth and solar system. *Mailing Add:* Dept Physics Univ Calif Berkeley CA 94720

KENNEDY, BYRL JAMES, b Plainview, Minn, June 24, 21; m 50; c 4. INTERNAL MEDICINE, ONCOLOGY. *Educ:* Univ Minn, BA & BS, 43, BM, 45, MD, 46; McGill Univ, MSc, 51; Am Bd Internal Med, dipl, Am Bd Med Oncol, cert, 79. *Prof Exp:* Intern & asst resident med, Mass Gen Hosp, 45-46; resident, Mass Gen Hosp & Harvard Univ, 51-52; from asst prof to assoc prof, Med Ctr, 52-67, prof med, Health Sci Ctr, Dept Med, 70-88, masonic prof oncol, 70-91, REGENTS PROF MED, SCH MED, UNIV MINN, MINNEAPOLIS, 88- *Concurrent Pos:* Fel, Mass Gen Hosp & Harvard Univ, 47-49; fel, Med Sch, McGill Univ, 49-50; fel, Med Sch, Cornell Univ, 50-51. *Honors & Awards:* Nat Div Award, Am Cancer Soc, 75; Margaret H Edwards Achievement Medal, Am Asn Cancer Educ, 90. *Mem:* AAAS; Am Col Phys; Am Asn Cancer Res; AMA; Am Asn Cancer Educ; Am Soc Clin Oncol. *Res:* Medical oncology. *Mailing Add:* Univ Hosps Box 286 Univ Minn Minneapolis MN 55455

KENNEDY, CHARLES, b Buffalo, NY, Aug 27, 20; m 46; c 3. PEDIATRIC NEUROLOGY. *Educ:* Princeton Univ, AB, 42; Univ Rochester, MD, 45. *Prof Exp:* Instr path, Sch Med, Yale Univ, 45-46; resident pediat, Children's Hosp, Buffalo, 48-51; resident neurol, Hosp Univ Pa, 53-54; asst neurologist, Children's Hosp, Philadelphia, 56-58; neurologist, 58-67; from asst prof to assoc prof neurol pediat, Sch Med, Univ Pa, 58-70; PROF PEDIAT, SCH MED, GEORGETOWN UNIV, 71- *Concurrent Pos:* Life Insurance Med Res Fund Fel, 51-52; fel physiol, Grad Sch Med, Univ Pa, 51-53; vis fel, Neurol Inst, Columbia-Presby Med Ctr, 57-58; guest worker, Lab Cerebral Metab, NIMH, 68-; sr res scientist, NIMH, 79- *Mem:* AAAS; Soc Pediat Res; Am Neurol Asn; Am Pediat Soc; Soc Neurosci. *Res:* Cerebral circulation; developmental neurology; energy metabolism of developing brain. *Mailing Add:* Georgetown Univ Hosp 3800 Reservoir Rd Washington DC 20007

KENNEDY, DAVID P, b Boston, Mass, Nov 15, 23. ELECTRICAL ENGINEERING. *Educ:* Mass Inst Technol, MA, 74. *Prof Exp:* Prof elec eng, Fla Univ; owner, D P Kennedy & Assoc In, 82-85; RETIRED. *Mem:* Fel Inst Elec & Electronics Engrs. *Mailing Add:* D P Kennedy & Assoc Inc 2227 NW 16 Ave Gainesville FL 32605

KENNEDY, DONALD, b New York, NY, Aug 18, 31; m 53; c 2. BIOLOGY. *Educ:* Harvard Univ, AB, 52, AM, 54, PhD(biol sci), 56. *Hon Degrees:* DSc, Columbia Univ, 79, Williams Col, 80, Univ Mich, 82, Univ Rochester, 84, Univ Ariz, 85; LLD, Reed Col, 86. *Prof Exp:* From asst prof to assoc prof zool, Syracuse Univ, 56-60; from asst prof to prof zool, Stanford Univ, 60-77, chmn dept, 65-72; comnr, Food & Drug Admin, 77-79; vpres & provost, 79-80, PRES, STANFORD UNIV, 80- *Concurrent Pos:* Ed bd, J Exp Zool, 65-70, J Comp Physiol, 66-77, J Neurophysiol, 70-76, Science, 73-78; Nat lectr, Sigma Xi, 69-70. *Honors & Awards:* Bowditch Lectr, Am Physiol Soc, 70. *Mem:* Nat Acad Sci; Inst Med-Nat Acad Sci; Am Soc Zool; Soc Gen Physiol; fel Am Acad Arts & Sci; Soc Exp Biol UK; Am Physiol Soc; Am Inst Biol Sci; fel AAAS; Marine Biol Asn UK. *Res:* Comparative physiology of sense organs, especially visual systems; central nervous system of Crustacea; over sixty articles and publications. *Mailing Add:* Off Pres Stanford Univ Bldg 10 Stanford CA 94305-2060

KENNEDY, DONALD ALEXANDER, medical anthropology, for more information see previous edition

KENNEDY, DUNCAN TILLY, b Brooklyn, NY, May 13, 30; m 55; c 3. NEUROPHYSIOLOGY, NEUROANATOMY. *Educ:* Columbia Univ, BSc, 55; Stanford Univ, AM, 64; Wayne State Univ, PhD(anat), 66. *Prof Exp:* Phys therapist, King's Daughters Hosp, Ashland, Ky, 55-57; phys therapist, Marmet Hosp, WVa, 58-60; instr phys ther, Stanford Univ, 62; from instr to asst prof anat, Wayne State Univ, 66-75, instr neuroanat & neurophys, Div Phys & Occup Ther, 64-75; asst prof, 75-78, asst dir, Ctr Med Educ, 80, ASSOC PROF PHYSIOL & HEALTH SCI, BALL STATE UNIV, 78-, ASSOC PROF MED EDUC & PHYSIOL & HEALTH SCI, 84- *Concurrent Pos:* NIH spec fel, Lab Neurophysiol, Univ Wis, 70-72; fac res awards, Wayne State Univ, 74 & Ball State Univ, 75; instr, Sch Nursing, Moorhead State Col, 55-57; lectr, Mich Phys Ther Continuing Educ Prog, 74-75. *Mem:* AAAS; Soc Neurosci; Sigma Xi; Am Asn Anatomists. *Res:* Neurophysiology and neuroanatomy of sensory motor systems; electrophysiology of the central auditory systems; biomechanics and kinesiology. *Mailing Add:* 2610 Ethel Ave Muncie IN 47303

KENNEDY, EDWARD EARL, b Evansville, Ind, Jan 7, 25; m 50; c 2. ANALYTICAL CHEMISTRY. *Educ:* Purdue Univ, BS, 45; Ind Univ, MA, 48. *Prof Exp:* Anal chemist, Eli Lilly & Co, 45-46; instr, Ind Univ, 47-48; anal chemist, Eli Lilly & Co, 48-50, head, Dept Assay Methods Develop, 50-52, head anal res & develop, 52-56, head anal develop & spec servs, 56-62, asst dir anal res & develop, 62-66, dir corp qual assurance, 66-69, dir, Park Fletcher Plant, 70, dir biochem mfg, 70-80, dir biosynthetic oper, 80-83, dir qual assurance, 83-85; RETIRED. *Mem:* AAAS; Am Pharmaceut Asn; Am Soc Qual Control; Am Chem Soc. *Res:* Instrumentation of analytical chemistry, particularly field of spectrophotometry. *Mailing Add:* 8305 Reef Ct Rd Indianapolis IN 46236-9539

KENNEDY, EDWARD FRANCIS, b Chicago, Ill, Jan 2, 32; m 56; c 6. NUCLEAR PHYSICS. *Educ:* Loyola Univ, Ill, BS, 54; Univ Notre Dame, PhD(nuclear physics), 60. *Prof Exp:* Technician, Argonne Nat Lab, 52-54; asst physics, Univ Notre Dame, 54-58, res assoc, 58-60; from asst prof to assoc prof physics, 60-70, actg chmn dept, 63-64, chmn dept, 64-76, PROF PHYSICS, COL OF THE HOLY CROSS, 70- *Concurrent Pos:* Consult, Air Force Cambridge Res Labs, 62-71; vis scientist, Cavendish Lab, Cambridge, 68-69, Fraunhofer-Inst, Munich, Ger, 82-83, Univ Aarhus, Denmark, 90; vis assoc, Calif Inst Technol, 75-76, 77 & 78; vis prof, Cornell Univ, 83, 84, 85; vis res physicist, Univ Calif, San Diego, 87. *Mem:* Am Phys Soc; Am Asn Physics Teachers; Sigma Xi. *Res:* Ion channeling in crystals; surface physics; nuclear fluorescence; radiation damage. *Mailing Add:* Physics Dept Holy Cross Col Worcester MA 01610

KENNEDY, EDWIN RUSSELL, b Los Angeles, Calif, Nov 4, 11; wid; c 5. ENVIRONMENTAL CHEMISTRY. *Educ:* Calif Inst Technol, BS, 33, MS, 34, PhD(chem), 36. *Prof Exp:* Technologist, Shell Oil Co, 36-42; group leader res, Shell Chem Co, 46-48, res coordr, 48-56, sr technologist, 56-62; CONSULT, 77- *Mem:* Am Chem Soc; Am Inst Aeronaut & Astronaut; Am Ord Asn; Nat Asn Corrosion Eng. *Res:* Air pollution; corrosion; protective coatings; petrochemicals; rocket propellants; metallurgy; petroleum technology. *Mailing Add:* 240 Forest Ave Rye NY 10580-4124

KENNEDY, ELDREDGE JOHNSON, b Fayetteville, Tenn, Sept 19, 35; m 61; c 3. SOLID STATE ELECTRONICS, ELECTRICAL ENGINEERING. *Educ:* Univ Tenn, BS, 58, MS, 59, PhD(eng sci), 67. *Prof Exp:* Coop student, Arnold Eng Develop Ctr, ARO Inc, Tenn, 53-57; asst elec eng, Univ Tenn, 58-59, instr, 59-63, res engr, Exp Sta, 60-63; design engr, Instrumentation & Controls Div, Oak Ridge Nat Lab, 63-70; assoc prof, 69-75, PROF ELEC ENG, UNIV TENN, KNOXVILLE, 75- *Concurrent Pos:* Ford Found assoc prof, 68-69; consult, Oak Ridge Nat Lab, 70- *Mem:* Inst Elec & Electronics Engrs; Sigma Xi; Int Soc Hybrid Microelectronics. *Res:* Electronic solid state circuit design; low-current meaurements; hybrid thick-film integrated circuits; high-speed pulse amplifiers, low-noise electronics; radiation effects in integrated circuits and devices. *Mailing Add:* Dept Elec Eng Univ Tenn Knoxville TN 37996

KENNEDY, ELHART JAMES, b Lincoln, Nebr, Feb 15, 23; m 48; c 2. AGRICULTURAL MICROBIOLOGY, BOTANY. *Educ:* Colo Agr & Mech Col, BS, 50; Cornell Univ, PhD(veg crops), 53. *Prof Exp:* Dir res agr, Spud Chips, Inc, Colo, 53-59; chmn div sci & math, 68-81, dir continuing educ, 81-83, prof biol, N Park Col, 59-88; RETIRED. *Concurrent Pos:* Mem prod & tech div, Nat Potato Chip Inst, chmn potato div, 55-57; agr consult, Envirodyne, Inc, 74-88. *Mem:* AAAS; Am Sci Affiliation; Am Soc Microbiol; Am Inst Biol Sci. *Res:* Physiology of the potato, including tuberization, pathology and irradiation effects of clostridium botulinum; microbiology of surface waters. *Mailing Add:* 324 Kempton St No 656 Spring Valley CA 92077

KENNEDY, EUGENE P, b Chicago, Ill, Sept 4, 19; c 3. BIOCHEMISTRY. *Educ:* Univ Chicago, PhD(biochem), 49. *Hon Degrees:* MA, Harvard Univ, 60; DSc, Univ Chicago, 77. *Prof Exp:* From asst prof to prof, Dept Biochem & Ben May Lab, Univ Chicago, 51-60; HAMILTON KUHN PROF BIOL, HARVARD MED SCH, 60- *Concurrent Pos:* Am Cancer Soc fel, Univ Calif, 49-50; Am Chem Soc res award, 55; NSF fel, 59-60. *Honors & Awards:* Paul Lewis Award, Am Chem Soc, 58; Lipid Res Award, Am Oil Chem Soc, 70; Gairdner Found Award, 76; Ledlie Prize, 76; Passano Award, 86; Wieland Prize, 86. *Mem:* Nat Acad Sci; Am Acad Arts & Sci; Am Chem Soc; Am Soc Biol Chemists (pres, 70-71). *Res:* Metabolism and function of lipids; membrane function. *Mailing Add:* Dept Biol Chem Harvard Med Sch Boston MA 02115

KENNEDY, EUGENE RICHARD, b Scranton, Pa, July 3, 19; m 45; c 3. BACTERIOLOGY. *Educ:* Univ Scranton, BS, 41; Cath Univ Am, MS, 43; Brown Univ, PhD, 49; Am Bd Med Microbiol, dipl, 64. *Prof Exp:* Asst bact, Cath Univ Am, 41-43; instr, Brown Univ, 46-48; from instr to prof bact & immunol, 49-85, dean Sch Arts & Sci, 73-85, EMER PROF BACT & IMMUNOL, CATH UNIV AM, 85- *Concurrent Pos:* Serologist, US Army Med Ctr, DC, 42; instr, RI Hosp, 46-48; bacteriologist, US Food & Drug Admin, 49; consult bacteriologist, Providence Hosp, DC, 54-58, staff microbiologist, 58-77, consult microbiologist, 81- *Mem:* AAAS; Am Soc Microbiol; Sigma Xi. *Res:* Vi antigen; quantitative dye adsorption; quantitative gram reaction; staphylococcus autogenous vaccine; in vivo and in vitro staphylococci. *Mailing Add:* Sch Arts & Sci Cath Univ Am Washington DC 20064

KENNEDY, FLYNT, b Chillicothe, Tex, May 25, 31; m 57; c 1. ORGANIC CHEMISTRY. *Educ:* Tex Christian Univ, BA, 52; Rice Univ, PhD(org chem), 56. *Prof Exp:* Res Corp fel, Calif Inst Technol, 56-57; res chemist, 57-60, sr res chemist, 60-61, res group leader, 61-64, supv res scientist, 64-69, mgr, Chem Res Div, Conoco Inc, 69-82; gen mgr, coal & chem res develop, 84-87, VPRES RES & DEVELOP, CONSOL COAL CO, 87- *Mem:* Am Chem Soc. *Res:* Investigation of reactions of organometallic compounds; synthesis of three and four membered compounds; upgrading of hydrocarbons; chemicals from coal, polyvinyl chloride and polyolefins; coal seam degasification; improved coal mining technology; coal processing and combustion; Sulfurdioxide. *Mailing Add:* 4000 Brownsville Rd Library PA 15129-9545

KENNEDY, FRANK METLER, b Woodstock, Ont, Dec 6, 16; nat US; m 42; c 4. CHEMISTRY. *Educ:* Univ Toronto, BA, 41; Columbia Univ, dipl, 57. *Prof Exp:* Chemist, Res Lab, Mining & Smelting Div, Int Nickel Co Can, Ltd, 41-42, chemist platinum metals, Nickel Refining Div, 42-45; res engr, Zinc Smelting Div, St Joseph Lead Co, Pa, 46, asst dir res, 46-60, mgt engr, 60-65; economist, Tenn Valley Authority, 67-75, chem engr, 75-82; RETIRED. *Mem:* Am Chem Soc; Am Inst Chem Engrs; NY Acad Sci. *Res:* Relationships among fertilizer use, agricultural production, food and population; sulfur dioxide recovery. *Mailing Add:* 2410 Chickasaw Dr Florence AL 35630-1654

KENNEDY, FRANK SCOTT, b Washington, DC, Oct 16, 44; m 80; c 2. METALLOENZYMES, TRACE METALS. *Educ:* Wash & Lee Univ, BS, 66; Univ Ill, Urbana, PhD(biochem), 70. *Prof Exp:* Res assoc biochem, Harvard Med Sch, 74-76; asst prof, 76-78, ASSOC PROF BIOCHEM, SCH MED, LA STATE UNIV, SHREVEPORT, 78-,. *Concurrent Pos:* Coordr student admis, Sch Med, La State Univ, 87-88, asst dean, 88- *Mem:* Am Chem Soc; AAAS; Sigma Xi. *Res:* Intermediary metabolism in cardiac tissue; role of copper in normal iron metabolism. *Mailing Add:* PO Box 33932-Biochem Shreveport LA 71130-3932

KENNEDY, FREDERICK JAMES, b Lowell, Mass, Mar 20, 37; m 67; c 3. THEORETICAL PHYSICS. *Educ:* Lowell Tech Inst, BS, 60; Univ Del, MS, 65, PhD(physics), 67. *Prof Exp:* Asst prof physics, Univ Bridgeport, 67-68; fel, Theoret Physics Inst, Univ Alta, 68-73; SCI LIBRN, KILLAM LIBR-SIC, DALHOUSIE UNIV, 73- *Concurrent Pos:* Lectr, Dept Math, Statist & Comput Sci, Dalhousie Univ, 82-86. *Res:* Classical mechanics and electrodynamics. *Mailing Add:* Killam Libr Sci Dalhousie Univ Halifax NS B3H 8J5 Can

KENNEDY, GEORGE ARLIE, b Chicago, Ill, Jan 11, 40; m 72. VETERINARY PATHOLOGY. *Educ:* Univ NMex, BS, 62; Wast State Univ, DVM, 67; Kans State Univ, PhD(path), 75. *Prof Exp:* Res pathologist, US Army Med Res & Nutrit Lab, 67-70; instr vet path, Dept Path, Col Vet Med, 70-75, ASST PROF, VET DIAG LAB, KANS STATE UNIV, 72- *Concurrent Pos:* Clinician, Kans State Univ Vet Teaching Hosp, 70-72. *Mem:* Am Vet Med Asn; Am Col Vet Path; Sigma Xi. *Res:* Transmission and scanning electron microscipic study of swine enteric diseases, particularly swine dysentery and diseases of the large intestine. *Mailing Add:* 1029 Bertrand Manhattan KS 66502

KENNEDY, GEORGE GRADY, b Amityville, NY, Mar 23, 48; m 73; c 2. ECONOMIC ENTOMOLOGY. *Educ:* Ore State Univ, BS, 70; Cornell Univ, PhD(entom), 74. *Prof Exp:* Asst prof entom, Univ Calif, Riverside, 74-75; from asst prof to assoc prof, 76-84, PROF ENTOM, NC STATE UNIV, 84- *Concurrent Pos:* Prog mgr, USDA Competitive Grants Prog Entom/Nematol, 84-86; L M Ware res award hort, 89. *Mem:* Entom Soc Am; Am Inst Biol Sci. *Res:* Pest management; insect/plant interactions. *Mailing Add:* Dept Entom NC State Univ Raleigh NC 27695-7630

KENNEDY, GEORGE HUNT, b Seattle, Wash, Apr 24, 36; m 61; c 2. SURFACE CHEMISTRY. *Educ:* Univ Ore, BS, 59; Ore State Univ, MS, 62, PhD(phys chem), 66. *Prof Exp:* Res chemist, Chevron Res Corp Div, Chevron Oil Co, 61-62; from asst prof to assoc prof, 65-76, head dept chem & geochem, 76-88, PROF CHEM, COLO SCH MINES, 77- *Mem:* Am Chem Soc. *Res:* Physical adsorption of gases on solid adsorbents; gas chromatography; sorption of vapors on liquid coated adsorbents. *Mailing Add:* Dept Chem Colo Sch Mines Golden CO 80401

KENNEDY, HARVEY EDWARD, b Goldsboro, NC, Oct 2, 28; m 51; c 2. MICROBIOLOGY, INFORMATION SCIENCE. *Educ:* Atlantic Christian Col, BA, 48; NC State Univ, MS, 52, PhD(bact), 54. *Prof Exp:* Assoc res bacteriologist, NC Sanatorium Syst, Med Ctr, Univ NC, 54-57, sr res scientist, 57-59; asst prof & res assoc, Ohio State Univ, 59-61; dir dairy prod res, Johnson & Johnson, 61-65, dir prod develop, Vetco Div, 65-67; asst dir-

dir sci affairs, 67-75, exec dir, 75-79, PRES, BIOSCI INFO SERV, 80- *Concurrent Pos:* USPHS res grants, 54-58 & 59-61; pres, Nat Fedn Abstracting & Info Serv, 74-75. *Honors & Awards:* Miles Conrad Award, Nat Fedn Abstracting & Info Servs; Meritorious Serv Award, Coun Biol Ed. *Mem:* AAAS; Am Soc Microbiol; Am Inst Biol Sci (pres, 86-87); Am Soc Info Sci; Coun Biol Ed (secy, 73-74, pres, 78-79); Int Coun Sci & Tech Info (pres, 89-). *Res:* Information science and communications applied to biological and biomedical research literature; bacterial nutrition and metabolism; virulence of pathogens; antimicrobial agents; pharmaceutical and agricultural product development. *Mailing Add:* 205 Haverford Ave Swarthmore PA 19081

KENNEDY, IAN MANNING, b Brisbane, Australia, Sept 11, 52; m 81; c 2. COMBUSTION. *Educ:* Sydney Univ, BEng, 75, PhD(mech eng), 80. *Prof Exp:* Mem res staff, Princeton Univ, 80-83; res scientist, Aeronaut Res Labs, 83-86; asst prof, 86-88, ASSOC PROF ENG, UNIV CALIF, DAVIS, 88- *Concurrent Pos:* NSF presidential young investr, 88. *Mem:* Am Inst Aeronaut & Astronaut; Combustion Inst. *Res:* Fundamental combustion phenomena; turbulent reacting flows; formation of pollutants such as soot in flames; application of laser and optics to measurements in flames; dynamics of aerosol systems. *Mailing Add:* Mech Aero & Mat Univ Calif Davis CA 95616

KENNEDY, J(OHN) R(OBERT), b Frederick, Md, Mar 25, 25; m 45; c 4. INDUSTRIAL & MANUFACTURING ENGINEERING. *Educ:* Purdue Univ, BS, 49. *Prof Exp:* Aeronaut engr, Chem Corps, Ft Detrick, Md, 49-50, physicist, 50-51, biol test engr, 51-55, mech engr, 55-58; gen engr, Nat Animal Disease Ctr, USDA, 58-60, chief, Eng & Plant Mgt, 58-79, supvry gen engr, 60-79; CONSULT ENGR, 79- *Res:* Maintenance engineering; design of laboratory facilities. *Mailing Add:* 510 Nicholas St Vincennes IN 47591-1057

KENNEDY, JAMES A, b Rochester, Minn, July 3, 35; m 65. ENZYMOLOGY. *Educ:* Univ Notre Dame, BS, 57; St Louis Univ, MD, 61. *Prof Exp:* NIH fel med, Med Sch, Univ Kans, 63-65; assoc internal med, Col Physicians & Surgeons, Columbia Univ, 68-71, asst prof, 71-73; assoc prof, 73-81, PROF MED, UNIV KANS MED CTR, 81- *Concurrent Pos:* Assoc Ed, J Lab Clin Med, 82-; mem, VA Res Adv Group A, 83-, chrmn, 85- *Mem:* Am Soc Biol Chemists. *Res:* Urea cycle; superoxide; regulation of pyrimidine biosynthesis in mammals; electron transport. *Mailing Add:* Dept Med Univ Kans Med Ctr 39th & Rainbow Blvd Kansas City KS 66103

KENNEDY, JAMES CECIL, b Toronto, Ont, Mar 14, 35; m 66; c 7. EXPERIMENTAL CANCER THERAPIES. *Educ:* Univ Toronto, BA, 57, MD, 61, PhD(biophys), 66. *Prof Exp:* Intern, Wellesley Hosp, Toronto, 61; asst prof, 69-73, ASSOC PROF PATH, QUEENS UNIV, ONT, 73-, ASSOC PROF RADIATION ONCOL, 77- *Concurrent Pos:* Res fel, Nat Cancer Inst Can, 66-68, res scholar, 69-71, res assoc, 72-77; res assoc, Ont Cancer Treat & Res Found, 77-83, career scientist, 84- *Mem:* Am Soc Photobiol; Europ Soc Photobiol. *Res:* Photoradiation therapy for cancer; fluorescence detection of cancer; chemistry and pharmacology of photosensitizing agents. *Mailing Add:* Dept Oncol Queen's Univ Kingston ON K7L 3N6 Can

KENNEDY, JAMES M, b Ottawa, Ont, Apr 25, 28; m 50. COMPUTER SCIENCE, ADMINISTRATION. *Educ:* Univ Toronto, BA, 49, MA, 50; Princeton Univ, PhD(physics), 53. *Prof Exp:* Res officer, Theoret Physics Br, Atomic Energy Can Ltd, 52-66, supvr, Comput Ctr, 56-66; dir, Comput Ctr, 66-80, vpres, 80-84, PROF COMPUTER SCI, UNIV BC, 68- *Mem:* Can Math Soc; hon mem Can Info Processing Soc (pres, 71-72); Can Asn Physicists. *Res:* Numerical and non-numerical computer methods. *Mailing Add:* Dept Comput Sci Univ BC 2075 Wesbrook Pl Vancouver BC V6T 1W5 Can

KENNEDY, JAMES VERN, b Jessup, Pa, May 4, 34; m 62; c 2. CHEMISTRY, RESEARCH ADMINISTRATION. *Educ:* Pa State Univ, BS, 55; Univ Pittsburgh, PhD(chem), 72. *Prof Exp:* Res assoc phys chem, Mellon Inst, 55-63; technologist, Baroid Div, Nat Lead Co, 63-69, sect leader catalysis res, 70, supvr catalysis labs, 70-71, tech mgr mineral synthesis dept, NL Industs, Inc, 71-73, catalyst prod mgr, Baroid Div, 72-73; group leader petrol prod res, Engelhard Minerals & Chem Corp, 73-74, mgr prod res, Minerals & Chem Div, 74-78, dir res-existing bus, 78, dir res-new bus, 79-80; dir catalysis res, Chemicals & Minerals Div, Gulf Sci & Technol Co, 80-85; SR RES ASSOC, CHEVRON RES CO, CHEVRON OIL CORP, 85- *Mem:* Am Chem Soc; fel Am Inst Chem; Catalysis Soc; Clay Minerals Soc; NY Acad Sci. *Res:* Fluidized cracking catalyst research and development for petroleum redefining; catalysis by layer-lattice silicates; alteration and synthesis of clay minerals; infrared characterization of synthetic clays; applications of minerals; new product development in catalyst, ceramic, industrial and paper products. *Mailing Add:* Chevron Res Co PO Box 1627 Richmond CA 94802-0627

KENNEDY, JERRY DEAN, b Oklahoma City, Okla, June 23, 34; m 57; c 2. PHYSICS. *Educ:* Univ Okla, BS, 56; Univ Calif, Berkeley, MA, 59; Lehigh Univ, PhD(physics), 63. *Prof Exp:* Adv study scientist, Lockheed Missile & Space Co, 56-59; engr, Autonetics Div, NAm Aviation, Inc, 59; mem tech staff physics, 63-69, supvr test exp div, 69-71, supvr exp planning div, 71-73, MGR ENG SCI DEPT, SANDIA LABS, 73- *Mem:* Am Phys Soc. *Res:* Dynamic high pressure solid state physics in semiconductors; shock wave phenomena in solids. *Mailing Add:* Orgn 7130 Sandia Labs Albuquerque NM 87115

KENNEDY, JOHN B, b Baghdad, Iraq, Jan 7, 32; m 57; c 3. ENGINEERING MECHANICS, STRUCTURAL ENGINEERING. *Educ:* Univ Wales, BSc, 55; Univ Toronto, PhD(civil eng), 61. *Hon Degrees:* DSc, Univ Wales, 84. *Prof Exp:* Asst engr, Develop Bd Iraq, 55-57; res asst skewed bridges, Univ Toronto, 57-61; asst prof civil eng, Univ Sask, 61-63; assoc prof, 63-66, head dept, 66-76, PROF CIVIL ENG, UNIV WINDSOR, 66- *Concurrent Pos:* Consult engr, Ministry Transp & Commun, Ont. *Honors & Awards:* T Y Lin Award, Am Soc Civil Engrs, 83; Duggan Medal, Eng Inst Can, 78. *Mem:* Am Soc Civil Engrs; Am Concrete Inst; Eng Inst Can. *Res:* Structural mechanics; skewed slab structures; waffle-slab bridges; cold-bending of HSS beams; reinforced-earth supporting soil-steel arch structures. *Mailing Add:* Dept Civil Eng Univ Windsor Windsor ON N9B 3P4 Can

KENNEDY, JOHN EDWARD, b Kemptville, Ont, Sept 12, 16; m 41; c 3. PHYSICS, ASTRONOMY. *Educ:* Queen's Univ, Ont, BA, 37; McGill Univ, MSc, 42. *Prof Exp:* Jr res physicist, Physics Div, Nat Res Coun Can, 41-45; from asst prof to prof physics, Univ NB, 45-56; sci serv officer, Defence Res Med Labs, Defence Res Bd, 56-65, head physics group, 61-65; from assoc prof to prof physics, Univ Sask, 65-84, asst head dept, 66-67, asst dean, Col Arts & Sci, 67-81, emer prof physics, 84-; RETIRED. *Concurrent Pos:* Consult physicist, NB Dept Health, 50-52; mem, Comn 41, Hist of Astron, Comn 46, Teaching Astron, Int Astron Union, 70-; Can Coun leave fel, 73-74. *Honors & Awards:* Can Silver Jubilee Medal, 78; Serv Award, Royal Astron Soc Can. *Mem:* Royal Astron Soc Can (nat secy, 58-64, 2nd vpres, 64-66, 1st vpres, 66-68, pres, 68-70); Can Astron Soc; Fel Royal Astron Soc London. *Res:* Spectroscopy; stellar physics; physics of clothing and footwear; history of early Canadian astronomy; history of early interest in solar-terrestrial interactions; history of the boundary survey Maine-New Brunswick of the 1840's. *Mailing Add:* 323 Lake Crescent Saskatoon SK S7H 3A1 Can

KENNEDY, JOHN ELMO, JR, b Louisville, Ky, June 21, 32; div; c 4. TECHNICAL MANAGEMENT, INDUSTRIAL & MANUFACTURING TECHNOLOGY. *Educ:* Univ Louisville, BS, 59, PhD(org chem), 63. *Prof Exp:* Lab technician anal chem, Schenley Distillers, Inc, 55-56; chemist, Ky Color & Chem Co, 56-59; chemist, Dept Exp Med, Sch Med, Univ Louisville, 59-61; res chemist, 63-64, group leader org chem, 64-67, sr group leader biol chem, 67-70, res area supvr, Brown & Williamson Tobacco Corp, 70-76; instr org chem, Univ Louisville, 76-77; PRIVATE CONSULT, 77- *Concurrent Pos:* Expert witness, Liability & Environ Litigation, Gen, Forensic & Environ Toxicol. *Mem:* AAAS; fel Am Inst Chem; NY Acad Sci; Am Chem Soc; Phytochem Soc NAm; Sigma Xi. *Res:* Biological chemistry; pharmacology; natural products; synthesis; steroids; alkaloids; alicyclics; biosynthetic routes; reaction mechanisms; psychopharmacology; information science; science writing; toxicology. *Mailing Add:* 3501 Pimlico Pkwy No 26 Lexington KY 40517

KENNEDY, JOHN FISHER, b Farmington, NMex, Dec 17, 33; m 59; c 4. CIVIL ENGINEERING, FLUID MECHANICS. *Educ:* Univ Notre Dame, BS, 55; Calif Inst Technol, S, 56, PhD(hydraul, fluid mech), 60. *Hon Degrees:* Hon Doctorate, Univ Notre Dame, 89; Hon Prof, E China Tech Univ, 85. *Prof Exp:* Res fel civil eng, Calif Inst Technol, 60-61; asst prof hydraul, Mass Inst Technol, 61-64, assoc prof, civil eng, 64-66; chmn dept, 74-76, prof, 66-81, DIR INST HYDRAUL RES, 66-, Carver distinguished prof, 81-87, HUNTER ROUSE PROF HYDRAUL, UNIV IOWA, 87- *Concurrent Pos:* Consult engr, 61-; mem, CEngrs, Hydral Consult Bd, Int Comn Experts Rev Leningrad Stormsurge Barrier, 90-; James Hardie eminent speaker, Aust Inst Engrs, 90. *Honors & Awards:* J C Stevens Award, Am Soc Civil Engrs, 62; Hilgard Prize, 74 & 78; Hunter Rouse Hydrol Engr Lect Award, Am Soc Civil Engrs, 81; W L Huber Award, 65. *Mem:* Nat Acad Eng; Am Soc Mech Engrs; Am Soc Eng Educ; Am Soc Civil Engrs; hon mem Int Asn Hydraul Res (pres, 80-84); hon mem Hungarian Hydrol Soc; hon fel Inst Water Conservancy & Hydroelec Power China; Chinese Hydraul Eng Soc. *Res:* Flow in alluvial channels, including sediment transport, channel roughness and the mechanics of ripples and dunes; turbulence, especially turbulent wakes; waterhammer in centrifugal pump systems; ice processes in rivers and oceans; thermal pollution of rivers; cooling towers. *Mailing Add:* Inst Hydraul Res Univ Iowa Iowa City IA 52242

KENNEDY, JOHN HARVEY, b Oak Park, Ill, Apr 24, 33; m 56, 70; c 5. ELECTROCHEMISTRY, ANALYTICAL CHEMISTRY. *Educ:* Univ Calif, Los Angeles, BS, 54; Harvard Univ, PhD(anal chem), 57. *Prof Exp:* Res chemist, E I du Pont de Nemours & Co, Del, 57-61; asst prof, Univ Calif, Santa Barbara, 61-63; assoc prof, Boston Col, 63-64; head inorg chem, Gen Motors Defense Res Labs, 64-67; asst prof, 67-71, assoc prof, 71-76, PROF CHEM, UNIV CALIF, SANTA BARBARA, 76- *Concurrent Pos:* Tech adv, Bissett-Berman Corp, 67-71, chmn dept, 82-85. *Mem:* Am Chem Soc; Electrochem Soc. *Res:* Solid electrolytes; fused salts; electrochemistry; instrumental methods of analysis; photoelectrochemistry. *Mailing Add:* Dept Chem Univ Calif Santa Barbara CA 93106

KENNEDY, JOHN HINES, b Washington, DC, Nov 1, 25; m 47, 72; c 4. THORACIC SURGERY, CARDIOVASCULAR SURGERY. *Educ:* Harvard Med Sch, MD, 49; Am Bd Surg, dipl, 57; Am Bd Surg, 60; Imperial Col, London, MPhil, 90. *Prof Exp:* Intern, Mass Gen Hosp, 49-50, asst resident, 50-51 & 53-54, resident, 54-55; sr registr thoracic unit, Frenchay Hosp, Bristol, Eng, 59-60; asst prof thoracic surg, Sch Med, Case Western Reserve Univ, 62-69; prof surg, Baylor Col Med, 69-76, dir Taub Labs Mech Circulatory Support, 70-76, mem, Admis Comt, 71-76; adj prof, Biomed Eng, Rice Univ, 75-76; fac mem, Dept Macro Molecular Sci, Case Western Reserve Univ, 76-77; consult surgeon, Middlesex Hosp Med Sch, Wembley & Central Middlesex Hosps, London, 78-82; physiol flow studies unit, Imp Col, Univ London, 82-85; vis scientist, INSERM 141, Paris, 86-87; VIS SCIENTIST, DEPT MOLECULAR PHYSIOL, BABRAHAM HALL, CAMB, 87- *Concurrent Pos:* USPHS grant, Baylor Col Med, 69-72; clin asst, Bristol Royal Infirm, Bristol Univ, 59-60; dir, Div Thoracic Surg, Cleveland Metrop Gen Hosp, 62-69; res assoc, Eng Design Ctr, Case Western Reserve Univ, 66-69; dir, Circulatory Assistance Proj Group, Artificial Heart-Myocardial Infarction Prog, NIH contract, Case Western Reserve Univ, 67-69 & Baylor Med Col, 69-71; mem, Tech Adv Group, Artificial Heart-Myocardial Infarction Prog, Nat Heart & Lung Inst, 68, consult site visitor prog proj grants, 70-, prin investr grant, 70-; mem, Coun Cardiovasc Surg & ed, surg supplement Circulation, Am Heart Asn, 68-70; adj prof, Rice Univ, 69-75; vis prof, Dept Macromolecular Sci, Case Western Reserve Univ, 75-76; consult, President's Panel Biomed Res, 75-; med dir, Moat House Hosp, Gt Easton, Dunmow, Essex, 80-86. *Mem:* Am Asn Thoracic Surg; Soc Thoracic Surgeons; Western Surg Asn; fel Am Col Surg; Royal Soc Med. *Res:* Physiology; interstitium; physiology. *Mailing Add:* Dept Molecular Physiol Agr & Food Res Coun Inst Babraham Cambridge CB2 4AT England

KENNEDY, JOHN ROBERT, b Cleveland, Ohio, July 17, 37; m 60, 89; c 2. CYTOLOGY. *Educ:* Univ Mich, BS, 59, MS, 61; Univ Iowa, PhD(zool), 64. *Prof Exp:* From instr to asst prof anat, Bowman Gray Sch Med, 64-69; assoc prof, 69-77, PROF ZOOL, UNIV TENN, KNOXVILLE, 77- *Mem:* Am Soc Cell Biol; Electron Micros Soc Am. *Res:* Effect of physiological factors on tracheal cell fine structure and function; ciliary cell physiology; consulting in electron microscopy and toxicology. *Mailing Add:* Dept Zool Univ Tenn Knoxville TN 37916

KENNEDY, JOSEPH PATRICK, b Houston, Tex, Mar 9, 32. ANATOMY, ECOLOGY. *Educ:* Univ St Thomas, Tex, BA, 54; Univ Tex, MA, 55, PhD(zool), 58. *Prof Exp:* Chmn dept biol, Univ St Thomas, Tex, 58-60; from asst prof to assoc prof, 60-68, prof animal ecol & chmn dept, Grad Sch Biomed Sci, 69-77, prof anat, Univ Tex Health Sci Ctr-Houston Med Sch, 76-77, PROF ANAT, UNIV TEX DENT BR, HOUSTON, 68- *Concurrent Pos:* Vis prof, Mt Lake Biol Sta, Univ Va, 62, Terra Alta Biol Sta, WVa Univ, 68; lectr, Univ Houston, 63; ed, J Herpet, 68-79; chmn, Adv Comt, Univ Tex Environ Sci Park, 69-71, dir, 71-76; prof ecol, M D Anderson Hosp & Tumor Inst, 69-76; mem exec comt & bd trustees, Armand Bayou Nature Ctr, Inc, 74-77; mem, Coun Biol Eds; prof anat, Tex Health Sci Ctr Dent Br. *Mem:* Am Soc Ichthyol & Herpet; Soc Study Amphibians & Reptiles; Coun Biol Ed; Sigma Xi. *Res:* Ecology, evolution and behavior; herpetology; literary criticism. *Mailing Add:* Dept Anat Sci Dent Br Health Sci Ctr Univ Tex PO Box 20068 Houston TX 77225

KENNEDY, JOSEPH PAUL, b Budapest, Hungary, May 18, 28; US citizen; m 57; c 3. POLYMER CHEMISTRY. *Educ:* Univ Vienna, PhD(chem), 55; Rutgers Univ, MBA, 61. *Prof Exp:* Fel biochem, Sorbonne, 55-56; res assoc, McGill Univ, 56-57; res chemist, Celanese Corp Am, 57-59; res chemist, Easso Res & Eng Co, 59-62, sr res chemist, 62-65, res assoc, 65-69, sr res assoc, 69-70; DISTINGUISHED PROF POLYMER SCI & CHEM, UNIV AKRON & RES ASSOC INST POLYMER SCI, 70- *Honors & Awards:* Morley Medal & Award, Cleveland Sect, Am Chem Soc, 83; Polymer Chem Award, Am Chem Soc, 85. *Mem:* Am Chem Soc. *Res:* Cationic polymerizations; carbenium ion chemistry; polymer synthesis; polymerization mechanisms; elastomer chemistry, particularly butyl rubber and polyisobutylene; blocks and grafts; terminally functional liquids; derivatization of polymers; living polymerizations; biomaterials. *Mailing Add:* Inst Polymer Sci Univ Akron Akron OH 44325-3909

KENNEDY, KATHERINE ASH, b Bryn Mawr, Pa, Mar 24, 50; m 87. TOXICOLOGY, CANCER CHEMOTHERAPY. *Educ:* Vanderbilt Univ, BA, 73; Univ Iowa, PhD(pharmacol), 77. *Prof Exp:* Fel, Sch Med, Yale Univ, 78-81; asst prof pharmacol, 81-87, ASSOC PROF, SCH MED, GEORGE WASHINGTON UNIV, 87- *Mem:* AAAS; Am Asn Cancer Res; NY Acad Sci; Radiation Res Soc; Am Soc Pharmacol & Exp Ther. *Res:* Role of biotransformation for drug activity and toxicity; mechanisms for antitumor agents in normally aerated and hypozic cells; mechanisms of drug induced cytotoxicity; tumor microenvironmental effects on drug induced toxcicity. *Mailing Add:* Dept Pharmacol Sch Med George Washington Univ 2300 Eye St NW Washington DC 20037

KENNEDY, KEN, b Aug 12, 45. RESEARCH ADMINISTRATION. *Educ:* Rice Univ, BA, 67; NY Univ, MS, 69, PhD(computer sci), 71. *Prof Exp:* From asst prof to prof, Dept Math Sci, 71-84, chmn, Dept Computer Sci, 84-88, DIR, COMPUTER & INFO TECH INST, RICE UNIV, 89-, CHAIR, DEPT COMPUTER SCI, 90- *Concurrent Pos:* Prin investr, numerous grants, 73-87; vpres, RM Thrall & Assoc, Inc, 74-81, pres, 81-; vis scientist, Space Shuttle Prog Lead Off, NASA, 75; vis staff mem, Computer Div, Los Alamos Sci Lab, 77; mem, Panel Computer Sci & Eng Res, Div Computer Res, NSF, 75-77 & Adv Comt Computer Res, NSF, 84- *Mem:* Nat Acad Eng; Asn Comput Mach; Soc Indust & Appl Math; Inst Elec & Electronics Engrs; Sigma Xi; AAAS. *Res:* Numerous publications; computer science. *Mailing Add:* Rice Univ PO Box 1892 Houston TX 77251

KENNEDY, KENNETH ADRIAN RAINE, b Oakland, Calif, June 26, 30; m 61, 69. PHYSICAL & FORENSIC ANTHROPOLOGY, ARCHAEOLOGY. *Educ:* Univ Calif, Berkeley, BA, 53, MA, 54, PhD (anthrop), 62; Am Acad Forensic Sci, dipl, 78. *Prof Exp:* Actg instr phys anthrop, Univ Calif, Berkeley, 62-63; vis prof, Deccan Col Post-Grad & Res Inst, 63-64; from asst prof to assoc prof anthrop, 64-81, assoc prof div biol sci, 69-81, PROF ECOL DIV BIOL SCI, ANTHROP & ASIAN STUDIES, CORNELL UNIV, 81- *Concurrent Pos:* NSF fels, Deccan Col Post-Grad & Res Inst, Univ Poona, India, 63-64 & 71, 80-81, Univ Calif, Berkeley, 68-69 & Cornell Univ, 72; Cornell Univ fac res grant, Brit Mus Natural Hist, London, 65-66; vis prof, Univ Ariz, 79 & 85; vis prof anthrop, Univ Ariz, Tuscon, 79 & 85; vis fel, Kings Col, Cambridge Univ, 81; fel, Am Inst Indian Studies, 66, 71-72, & 80-88; actg ed-in-chief, Am J Phys Anthrop, 85; elected chmn, Biol Anthrop Unit, Am Anthrop Asn, 86-88. *Honors & Awards:* T Dale Stewart Award Forensic Anthrop, Am Acad Forensic Sci, 87. *Mem:* AAAS; fel Am Anthrop Asn; Am Asn Phys Anthrop; fel Royal Anthrop Inst Gt Brit; Int Asn Human Biol. *Res:* Human evolution in South Asia, particularly the hominid osteological fossil record; history of biological sciences, especially human evolution and physical anthropology; palaeodemography of South Asia; forensic anthropology. *Mailing Add:* Ecol & Systs E-231 Corson Hall Cornell Univ Ithaca NY 14853

KENNEDY, LAWRENCE A, b Detroit, Mich, May 31, 37; m 57; c 6. FLUID MECHANICS, COMBUSTION. *Educ:* Univ Detroit, BS, 60; Northwestern Univ, MS, 62, PhD(mech eng), 64. *Prof Exp:* Res engr, Mech Res & Develop Div, Gen Am Transp Corp, Ill, 63-64; dir aerospace eng, State Univ NY, Buffalo, 69-71, prof mech eng, 64-83; DEPT MECH ENG, OHIO STATE UNIV, COLUMBUS, 83- *Concurrent Pos:* NSF sci fac fel, 68-69, NATO sr fel sci, 71-72; consult, Cornell Aero Labs, 66-71, Adv Group Aerospace Res & Develop, NATO, 71, MGB Res Corp, 76- & Air Preheater Div, Combustion Eng Corp, 78-; ed, J Exp Methods in Thermal & Fluid Sci. *Honors & Awards:* AT&T Found Award, Am Soc Eng Educ; Ralph R Teetor Award, Soc Auto Engrs. *Mem:* Combustion Inst; assoc fel Am Inst Aeronaut & Astronaut; Am Phys Soc; fel Am Soc Mech Engrs; Soc Automotive Engrs; Am Soc Engr Educ. *Res:* High temperature gas dynamics; chemical reacting flow; magnetohydrodynamics and combustion; radiative transfer; combustion generated pollutants; manufacturing processes; extensive work in the areas of reacting flows and optical dragnistics; specific studies turbulent combustion, radiation transfer with application to coal combustion and catalytic combustion. *Mailing Add:* Dept Mech Eng Ohio State Univ Robinson Columbus OH 43210

KENNEDY, M(ALDON) KEITH, b Little Rock, Ark, July 26, 47. ENTOMOLOGY. *Educ:* Hendrix Col, BA, 69; Cornell Univ, MS, 71, PhD(insect ecol), 76. *Prof Exp:* asst prof entom, Mich State Univ, 75-81; SECT MGR, ENTOM RES CTR, SC JOHNSON WAX CO, 84- *Mem:* Entom Soc Am; Ecol Soc Am; Acarological Soc Am. *Res:* Biology of the Sciaridae Diptera, especially those of economic importance. *Mailing Add:* SC Johnson Wax Co Entom Res Ctr 1525 Howe St Racine WI 53402

KENNEDY, MARGARET WIENER, b Arlington, Mass, Dec 16, 29; m 64. TOXICOLOGY, ENDOCRINOLOGY. *Educ:* Jackson Col, BS, 50; Boston Univ, MA, 54; Albany Med Col, PhD(toxicol), 72. *Prof Exp:* Res asst biochem, Schering Corp, 54-57; res asst endocrinol, Children's Mem Hosp, Chicago, 57-58; res assoc, Dept Obstet & Gyncol, Univ Chicago Clin, 59-63; res asst prof, Dept Obstet & Gyncol, Albany Med Col, 63-68, res asst prof toxicol, Inst Exp Path & Toxicol, 68-76; res scientist toxicol, Health Res Inc, NY State Dept Health, 77-80; FORENSIC TOXICOLOGIST, NY STATE POLICE CRIME LAB, 85- *Concurrent Pos:* Res assoc, NIH res grant, 64-67 & 72-76, prin investr, 72-75; prin investr, USAEC contract, 67-70. *Res:* Invitro metabolism of individual polychlorinated biphenyls; control of placental hormone synthesis. *Mailing Add:* Seven N Lyons Ave Menands NY 12204

KENNEDY, MARY BERNADETTE, b Pontiac, Mich, July 4, 47. BRAIN BIOCHEMISTRY, LEARNING MECHANISMS. *Educ:* St Mary's Col, BS, 69; Johns Hopkins Univ, PhD(biochem), 75. *Prof Exp:* Postdoctoral fel, Harvard Med Sch, 75-78, Yale Med Sch, 78-80; asst prof, 81-84, ASSOC PROF NEUROBIOL, CALIF INST TECHNOL, 84- *Concurrent Pos:* Mem sci adv bd, Hereditary Dis Found, 84-87, chmn, 86-87; assoc ed, J Neurosci, 86-89; mem & consult, ss1 study sect, NIH Neurol Sci, 90-; vchmn, Gordon Conf Neural Plasticity, 91. *Mem:* Soc Neurosci; Am Soc Biochem & Molecular Biol; Am Soc Cell Biol; fel AAAS. *Res:* Biochemical mechanisms by which calcium regulates neuronal function in the central nervous system; regulation of type II calcium/calmodulin-dependent protein kinase and the molecular structure of the postsynaptic density. *Mailing Add:* Div Biol 216-76 Calif Inst Technol Pasadena CA 91125

KENNEDY, MAURICE VENSON, b Pontotoc, Miss, Nov 23, 25; m 48; c 2. BIOCHEMISTRY. *Educ:* Miss State Univ, BS, 49, MS, 54, PhD(biochem), 67. *Prof Exp:* Dir microbiol & chem, Miss Dept Agr Lab, 49-62; instr microbiol, 62-66, assoc prof biochem, 66-83, prof biochem, Miss State Univ, 84-86; RETIRED. *Concurrent Pos:* Consult, NATO Sponsored Symp Pesticides, Lethbridge, Can, 70. *Mem:* AAAS; Am Chem Soc; Am Soc Microbiol; Sigma Xi. *Res:* Biochemical mechanisms of toxic substances, metabolic pathways in food poisoning microorganisms, production of useful substances from animal waste, and degradation and disposal of waste pesticides. *Mailing Add:* 410 S Washington St Starkville MS 39759

KENNEDY, MICHAEL CRAIG, b Buffalo, NY, Dec 5, 46; c 3. NEUROBIOLOGY. *Educ:* Rice Univ BA, 68; Univ Rochester, MS, 71, PhD(biol, neurobiol), 74. *Prof Exp:* Fel comp neuroanat, NY Univ Med Ctr, 74-76; asst prof biol, NY Univ, 76-81; ASSOC PROF ANAT, HAHNEMANN UNIV, 81- *Mem:* AAAS; Am Asn Anatomists; Soc Neurosci. *Res:* Investigations of the anatomy and physiology of neural pathways in nonmammalian vertebrates; neural substrates of auditory communication in the Tokay Gecko; developmental neurobiology of the reptilian auditory system; the visual system in the sea lamprey, Petromyzon marinus; influence of norepinephrine on blood-forming cells in bone marrow. *Mailing Add:* Dept Anat MS 408 Hahnemann Univ Broad & Vine Philadelphia PA 19102-1192

KENNEDY, MICHAEL JOHN, b London, Eng, Jan 21, 40; m 66; c 3. STRUCTURAL GEOLOGY, TECTONICS. *Educ:* Univ Dublin, BS, 63, MA & PhD(struct geol), 66. *Prof Exp:* Nat Res Coun Can fel, Geol Surv Can, 66-67; from asst prof to assoc prof, 67-74, prof geol, Mem Univ Newf, 74-76; prof & chmn, Dept Geol Sci, Brock Univ, 76-80; PROF & HEAD, DEPT GEOL, UNIV COL, DUBLIN, 80- *Concurrent Pos:* Co-chmn, Int Geodynamics Working Group 9, Appalachian-Caledonian Group, 72-80; chmn, invest comt IGCP Proj 233. *Honors & Awards:* Young Award, Atlantic Provinces Inter-Univ Comt on the Sciences, 73. *Mem:* Fel Geol Asn Can; fel Geol Soc London; fel Geol Soc Am. *Res:* Structural geology of metamorphic rocks, particularly Caledonian and Appalachian systems; petrofabrics and the relationship of deformation with metamorphism; metamorphic complexes of Appalachians and Caledonides; structural development of south east Ireland; terwanes in Caledonian and Appalachian orogens. *Mailing Add:* Dept Geol Univ Col Belfield Dublin 4 Ireland

KENNEDY, MICHAEL LYNN, b Scotts Hill, Tenn, Jan 31, 42. VERTEBRATE ZOOLOGY. *Educ:* Memphis State Univ, BS, 66, MS, 68; Univ Okla, PhD(vert zool), 75. *Prof Exp:* Asst vert zool, Univ Okla, 69-74; asst prof, 74-80, ASSOC PROF BIOL, MEMPHIS STATE UNIV, 80- *Mem:* Am Soc Mammalogists; Soc Syst Zool; Am Ornithologists Union. *Res:* Mammalian systematics; geographic variation studies with small mammals. *Mailing Add:* 3075 Charles Bryan Rd Memphis TN 38134

KENNEDY, PATRICK JAMES, b Louisville, Ky, Sept 8, 45. METEOROLOGY, SCIENCE EDUCATION. *Educ:* Univ Notre Dame, BS, 67; Univ Colo, MS, 69. *Prof Exp:* RES SCIENTIST METEOROL, NAT CTR ATMOSPHERIC RES, 70- *Concurrent Pos:* Sci lectr, Boulder Valley Sch Dist, 70-, Miacatlan, Morelos, Mexico, 83- *Mem:* Am Geophys Union.

Res: Mesoscale meteorology, including jet stream structure, cyclogenesis, fronts, and windstorms; aircraft observations of weather systems; global modelling of surface physical processes and hydrology; pre-college science education. *Mailing Add:* Nat Ctr Atmospheric Res Boulder CO 80307-3000

KENNEDY, PETER CARLETON, b Berkeley, Calif, June 19, 23; m 46; c 4. VETERINARY PATHOLOGY. *Educ:* Kans State Univ, DVM, 49; Cornell Univ, PhD(vet path), 54. *Prof Exp:* Intern, Angell Mem Animal Hosp, Mass, 50; asst large animal surg, Cornell Univ, 51, asst path, 52-53; from lectr to assoc prof, 54-65, PROF VET PATH, UNIV CALIF, DAVIS, 65-, LECTR PATH, MED SCH, 57-, PATHOLOGIST, EXP STA, 70- *Mem:* Am Col Vet Path. *Res:* Pathology of infectious diseases and endocrinopathies of domestic animals. *Mailing Add:* 2535 Loyola Dr Davis CA 95616

KENNEDY, ROBERT A, b Benson, Minn, Sept 29, 46; m 84; c 4. PLANT STRESS PHYSIOLOGY. *Educ:* Univ Minn, BS, 68; Univ Calif, Berkeley, PhD(bot), 74. *Prof Exp:* Asst prof bot, Univ Iowa, 74-78; from assoc prof to prof plant physiol, Wash State Univ, 79-85, asst dir res, 84-85; prof hort & chair, Ohio State Univ, 85-87; prog dir, NSF, 87-89; ASSOC VCHAIR & DIR, AGR RES, UNIV MD, 89- *Concurrent Pos:* Res assoc, US Army Med Res & Nutrit Lab, 69-71; mem fac, Biotechnol Ctr, Ohio State Univ, 86-88; consult, NSF, 89-90. *Honors & Awards:* Kenneth Post Award, Am Soc Hort Sci, 83. *Mem:* Am Soc Plant Physiologists; Am Soc Hort Sci; AAAS; Bot Soc Am; Sigma Xi; Am Inst Biol Sci. *Res:* Plant biochemistry and physiology, especially anaerobic or flooding metabolism; metabolic adaptation to flooding; induction and coordination of metabolic pathways during anoxia; energy relations and regulations of protein synthesis without oxygen. *Mailing Add:* Agr Exp Sta Univ Md Symons Hall Rm 1118 College Park MD 20742

KENNEDY, ROBERT ALAN, b Benson, Minn, Sept 29, 46; m 68. PLANT PHYSIOLOGY. *Educ:* Univ Minn, BS, 68; Univ Calif, PhD(bot), 74. *Prof Exp:* Asst prof bot, Univ Iowa, 74-80; mem fac, hort dept, Washington State Univ, 80-; AT DEPT HORT, OHIO STATE UNIV, COLUMBUS. *Concurrent Pos:* Sigma Xi res grant, 74; NSF grant, 75. *Mem:* AAAS; Am Soc Plant Physiologists; Bot Soc Am; Sigma Xi. *Res:* Physiology of plants with the C4 pathway of photosynthesis, particularly the effects of anatomy, age and waterstress on operation of the C4 pathway. *Mailing Add:* Dept Hort Ohio State Univ 2001 Fyffe Ct Columbus OH 43210

KENNEDY, ROBERT E, b Santa Monica, Calif, June 5, 39; m 61; c 5. THEORETICAL PHYSICS, THERMAL PHYSICS. *Educ:* Loyola Univ, BS, 61; Univ Notre Dame, PhD(physics), 66. *Prof Exp:* Res asst, Univ Notre Dame, 63-66; asst prof, 66-72, chmn dept, 73-81, ASSOC PROF PHYSICS, CREIGHTON UNIV, 73- *Mem:* Am Inst Physics; Am Phys Soc; Am Asn Physics Teachers. *Res:* Non-equilibrium thermodynamics; impact of Albert Einstein on modern physics. *Mailing Add:* Dept Physics Creighton Univ Omaha NE 68131

KENNEDY, ROBERT MICHAEL, organic chemistry, for more information see previous edition

KENNEDY, ROBERT P, b Glendale, Calif, Apr 2, 39. SEISMIC LOADING. *Educ:* Stamford Univ, BA, 60, MA, 61, PhD(struct eng), 67. *Prof Exp:* Res engr, Northrop Corp, 61-64; dir eng mech, Holmes & Narber, 66-76; vpres eng div, Anal Corp, 76-80; pres, Struct Mech Assoc, 80-86; PRES, R P K STRUCT MECH CONSULT, 86- *Mem:* Nat Acad Eng; Inst Elec & Electronics Engrs; Am Concrete Inst; Earthquake Eng Res Inst; Am Soc Civil Engrs. *Mailing Add:* R P K Struct Mech Consult 18971 Villa Terr Yorba Linda CA 92686

KENNEDY, ROBERT SPAYDE, b Augusta, Kans, Dec 9, 33; m 55; c 3. ELECTRICAL ENGINEERING, COMMUNICATIONS. *Educ:* Univ Kans, BS, 55; Mass Inst Technol, SM, 59, ScD(info theory), 63. *Prof Exp:* Nuclear engr, Naval Reactors Br, US AEC, 55-57; Ford Found grad fel, 59-60; staff mem, Lincoln Lab, 63-64; from asst prof to assoc prof, 64-76, PROF ELEC ENG, MASS INST TECHNOL, 76- *Concurrent Pos:* Ford fel, Mass Inst Technol, 64-65; dir, Commun Form, Mass Inst Technol, 84-88. *Mem:* Fel Inst Elec & Electronics Engrs; Optical Soc Am. *Res:* Extraordinary wide band fiber optic networks. *Mailing Add:* Dept Elect Eng Mass Inst Technol Cambridge MA 02139

KENNEDY, ROBERT WILLIAM, b Syracuse, NY, Sept 13, 31; m 56; c 3. WOOD SCIENCE, TECHNOLOGY. *Educ:* State Univ NY, BS, 53; Univ BC, MF, 55; Yale Univ, PhD(wood tech), 62. *Prof Exp:* Instr wood tech, Univ BC, 55-56 & 57-61; from asst prof to assoc prof, Univ Toronto, 62-66; head wood biol sect, Western Forest Prod Lab, 66-69, prog mgr, Protection & Prod Div, 69-71, from assoc dir to dir, 71-79; prof wood sci & indust, 79-83, dean, Fac Forestry, 83-90, PROF WOOD SCI, UNIV BC, 90- *Concurrent Pos:* Consult, Forestry & Forest Prod Div, Food & Agr Orgn, UN, 64, 82. *Mem:* Forest Prod Res Soc; fel Int Acad Wood Sci (pres 87-); Can Inst Forestry; fel Inst Wood Sci. *Res:* Physiology of wood formation; wood structure and properties at micro level; wood utilization. *Mailing Add:* Fac Forestry 270-2357 Main Mall Univ BC Vancouver BC V6T 1Z4 Can

KENNEDY, ROBERT WILSON, b Tampa, Fla, Sept 9, 27; m 49; c 1. ORGANIC CHEMISTRY. *Educ:* Emory Univ, AB, 53, MS, 54, PhD(chem), 56. *Prof Exp:* Develop res chemist, 56-58, sr chemist, 59-72, dept supt, Intermediates Dept, 72-73, asst div supt, Polymers Div, 73-74, asst to the works mgr, 74, proj mgr, New Prod Div, 74-79, PROJ MGR, ORG CHEM DIV, TENN EASTMAN CO, 79- *Mem:* Am Chem Soc; Sigma Xi. *Res:* Developmental research in industrial organic chemistry; mechanisms of organic reactions; naturally occurring organic compounds, particularly pine resin acids; new products marketing aspects; new products development. *Mailing Add:* 3357 Ridgeview St Kingsport TN 37764-3467

KENNEDY, RUSSELL JORDAN, b Dunrobin, Ont, Nov 23, 17; m 46; c 4. CIVIL ENGINEERING, HYDRAULICS. *Educ:* Queen's Univ, Ont, BSc, 41; Univ Iowa, MS, 49. *Prof Exp:* Lectr, 46-48, from asst prof to prof civil eng, 49-84, assoc dean sch grad studies, 68-70, vprin admin, 70-76, exec dir, Alumni Asn, 81-86, EMER PROF, QUEEN'S UNIV, ONT, 83- *Concurrent Pos:* Consult, Ont Paper Co, 48-51; Pulp & Paper Res Inst Can, 51-64, Dept Energy, Mines & Resources, 68-71; Irving Pulp & Paper Ltd, St John, NB, 79-80 & North West Hydraul Consults, Vancouver, BC, 79-80. *Honors & Awards:* Angus Medal, Eng Inst Can, 58. *Mem:* Eng Inst Can; Int Asn Hydraul Res. *Res:* Ice control; improvement of design criteria for air bubbler systems. *Mailing Add:* Ellis Hall Queen's Univ Kingston ON K7L 3N6 Can

KENNEDY, THELMA TEMY, b Chicago, Ill, Oct 18, 25. NEUROPHYSIOLOGY. *Educ:* Univ Chicago, PhB & BS, 47, MS, 49, PhD(biopsychol), 55. *Prof Exp:* Asst neurosurg, Univ Chicago, 51-56; from instr to prof, 58-88, assoc dean grad sch, 69-72, EMER PROF PHYSIOL, UNIV WASH, 88- *Concurrent Pos:* USPHS fel neurophysiol, Univ Wash, 56-58. *Mem:* AAAS; Psychonomic Soc; Soc Neurosci; Am Physiol Soc. *Res:* Cerebral cortex organization; unit activity; motor systems; sensory physiology. *Mailing Add:* Dep Physiol & Biophys SJ-40 Univ Wash Seattle WA 98195

KENNEDY, THOMAS JAMES, JR, b Washington, DC, June 24, 20; m 50; c 5. PHYSIOLOGY, NEPHROLOGY. *Educ:* Cath Univ, BS, 40; Johns Hopkins Univ, MD, 43; Am Bd Internal Med, dipl, 56. *Prof Exp:* Asst med, Col Med, NY Univ, 45-47; asst med, Col Physicians & Surgeons, Columbia Univ, 47-50; investr, Lab Kidney & Electrolyte Physiol, Nat Heart Inst, 50-60, mem staff, Off of Dir, NIH, 60-65, chief, Div Res Facil & Resources, 65-68, assoc dir prog planning & eval, 68-74; exec dir assembly life sci, Nat Acad Sci-Nat Res Coun, 74-76; dir, Dept Planning & Policy Develop, Asn Am Med Col, 76-86, assoc vpres, 86-90; RETIRED. *Concurrent Pos:* Res assoc, Sch Med, George Washington Univ, 51-65. *Mem:* Am Physiol Soc; Am Fedn Clin Res. *Res:* Renal physiology, especially mechanisms for excretion of electrolytes; electrolyte physiology; clinical disorders of renal and electrolyte physiology; administration of research. *Mailing Add:* 10703 Weymouth St Box 427 Garrett Park MD 20896-0427

KENNEDY, THOMAS WILLIAM, b Danville, Ill, Jan 7, 38; div; c 2. CIVIL ENGINEERING. *Educ:* Univ Ill, BS, 60, MS, 62, PhD(civil eng), 65. *Prof Exp:* From asst to instr civil eng, Univ Ill, 62-65; from asst prof to assoc prof, 65-74, dir, Coun Advan Transp Studies, 75-78, asst vpres res, 78-79, PROF CIVIL ENG, UNIV TEX, AUSTIN, 74-, ASSOC DEAN ENG RES PLANNING, 79-, ENG FOUND PROF, 85- *Concurrent Pos:* Hwy Res Bd, Nat Acad Sci-Nat Res Coun, 65-; mem, Transp Res Bd. *Mem:* Am Soc Civil Engrs; Am Concrete Inst; Am Soc Testing & Mat; Asn Asphalt Paving Technologists. *Res:* Materials; pavements; transportation; civil engineering. *Mailing Add:* Dept Civil Eng Univ Tex Austin TX 78712-1080

KENNEDY, VANCE CLIFFORD, b Big Run, Pa, May 18, 23; m 48; c 4. GEOCHEMISTRY, HYDROLOGY. *Educ:* Pa State Univ, BS, 48, MS, 49; Univ Colo, PhD(geol), 61. *Prof Exp:* Geologist geochem, 49-52 & 55-60, RES GEOLOGIST GEOCHEM & HYDROL, US GEOL SURV, 60- *Mem:* AAAS; Am Geophys Union; Geochem Soc. *Res:* Geochemical prospecting; uranium geology and geochemistry; transport of stream sediment; effects of stream sediment on the chemistry of water; chemistry of rainfall-runoff. *Mailing Add:* US Geol Surv 345 Middlefield Rd 345 Middlefield Rd MS 495 CA 94025

KENNEDY, W KEITH, JR, b Phoenix, Ariz, Sept 19, 43; m 65; c 2. ELECTRICAL ENGINEERING, SOLID STATE PHYSICS. *Educ:* Cornell Univ, BEE & MS, 65, PhD(elec eng), 68. *Prof Exp:* Mem tech staff, 68-69, mgr, Solid State Res & Develop Dept, 71-74, Solid State Div, 74-78, vpres, 77, vpres Devices Group, 78-86, Shareowner Relations & Planning Coord, 86-88, PRES & CHIEF EXEC OFFICER, WATKINS-JOHNSON CO, 88- *Mem:* Inst Elec & Electronics Engrs. *Res:* Microwave power generation and amplication with semiconductor devices; microwave integrated circuits; microwave systems. *Mailing Add:* Watkins-Johnson Co 3333 Hillview Ave Palo Alto CA 94304

KENNEDY, WILBERT KEITH, b Vancouver, Wash, Jan 4, 19; m 41; c 2. AGRONOMY. *Educ:* State Col Wash, BS, 40; Cornell Univ, MSA, 41, PhD(agron), 47. *Prof Exp:* Asst, Cornell Univ, 40-42 & 46-47; asst prof & asst agronomist to assoc prof & assoc agronomist, exp sta, State Col, Wash, 47-49; assoc dir, Res & Agr Exp Sta, Col Agr, 59, dir, 59-65, assoc dean col, 65-67, vprovost univ, 67-72, dean col, 72-78, provost univ, 78-84, PROF AGRON, NY STATE COL AGR & LIFE SCI, CORNELL UNIV, 49-, EMER PROVOST, 84- *Concurrent Pos:* Fulbright res scholar & Guggenheim fel, 56-57. *Honors & Awards:* NY Farmers Award, 57. *Mem:* Fel AAAS; fel Am Soc Agron; Sigma Xi. *Res:* Chemistry; botany; factors influencing yield and nutritive value of farm crops; grazing management practices and their relationship to the behavior and grazing habits of cattle; measuring, harvesting and storage losses in hay and silage; accumulation of nitrates in forage plants; nitrate toxicity. *Mailing Add:* 3 Sandra Pl Ithaca NY 14850

KENNEDY, WILLIAM ALEXANDER, b Merlin, Ont, Can, Sept 28, 15; m 42, 89; c 5. FISH BIOLOGY. *Educ:* Univ Toronto, BA, 37, PhD(ichthyol), 41. *Prof Exp:* Asst, Ont Fish Res Labs, 36-41; from asst scientist to dir, Cent Fish Res Sta, 45-56, dir biol sta, 57-66, res scientist, Fisheries Res Bd Can, Nanaimo Biol Sta, 66-76; consult, 77-85; CO DIR, HAGENSBORG RESOURCES, 86- *Concurrent Pos:* War res, Nat Res Coun Can, 41-45. *Mem:* Aquaculture Asn Can. *Res:* Fish farming and related fields. *Mailing Add:* 999 Beach Dr Nanaimo BC V9S 2Y4 Can

KENNEDY, WILLIAM ROBERT, b Chicago, Ill, Nov 2, 27; m 57; c 5. NEUROLOGY. *Educ:* Univ Ill, BS, 51; Univ Wis, MS, 52; Marquette Univ, MD, 58. *Prof Exp:* From asst prof to assoc prof, 64-71, PROF NEUROL, MED CTR, UNIV MINN, MINNEAPOLIS, 71- *Concurrent Pos:* Fel internal med, Mayo Clin, 59-60, fel neurol, 60-64. *Mem:* Am Acad Neurol;

Am Neurol Asn; Am Electroencephalog Soc; Am Asn Electromyog & Electrodiag (past pres). *Res:* Clinical-pathological-physiological research on neuromuscular disorders. *Mailing Add:* Box 187 UMHC 420 SE Delaware St Minneapolis MN 55455

KENNEL, CHARLES FREDERICK, b Cambridge, Mass, Aug 20, 39; m 64; c 2. PLASMA PHYSICS, SPACE PHYSICS. *Educ:* Harvard Univ, AB, 59; Princeton Univ, PhD(astrophys sci), 64. *Prof Exp:* Asst res scientist, Avco-Everett Res Lab, 60-61, staff mem, 64-65, prin res scientist, 66-67; assoc prof, 67-71, chmn, 83-86, PROF PHYSICS, UNIV CALIF, LOS ANGELES, 71-, MEM, INST GEOPHYS & PLANETARY PHYSICS, 72- *Concurrent Pos:* NSF fel, 65-66; vis scientist, Int Ctr Theoret Physics, Trieste, 65-66; Alfred P Sloan Found fac fel, 68-70; vis prof, Ctr Phys Theory, Polytech Sch, Paris, 74-75; mem sci adv group, NASA, 71-72; mem physics res eval group, Air Force Off Sci Res, 70-78; mem, Space Sci Bd, Nat Acad Sci-Nat Res Coun, 77-80, chmn, Comt Space Physics, 77-80, space & earth sci adv comt, NASA, 87-89; Fairchild Prof, Calif Inst Tech, 87; Guggenheim fel, 88; consult, TRW Systs Group, Calif, 67-; vis scientist, Space Res Ctr, Moscow, 88-; vis scholar, Univ Alaska, 88, 89 & 90; mem, Fusion Policy Adv Comt, Dept Energy, 90. *Honors & Awards:* Fulbright Sr lectr, Brazil, 85. *Mem:* Nat Acad Sci; fel Am Phys Soc; Int Union Radio Sci; Am Astron Soc; fel Am Geophys Union. *Res:* Plasma turbulence theory; solar system and astrophysical plasma physics. *Mailing Add:* Dept Physics Univ Calif Los Angeles CA 90024

KENNEL, JOHN MAURICE, b Sioux City, Iowa, Oct 7, 27; m 52; c 4. AEROSPACE NAVIGATION & GUIDANCE, STAR TRACKERS & INERTIAL INSTRUMENTS. *Educ:* Miami Univ, AB, 48; Univ Tex, PhD(physics), 55. *Prof Exp:* Physicist, US Naval Ord Lab, Md, 49-51; res engr, NAm Aviation, Inc, 51-52, res specialist inertial navig, Autonetics Div, 55-58, supvr phys res, 58-60, group leader eng proposals, 60-62, sr scientist, 62-67, res prog mgr, NAm Rockwell Microelectronics Co, 67-70, mgr microelectronic process develop, Products Div, 70-75, mem tech staff, Autonetics Div, Rockwell Int Corp, 75-84, mgr Astroinertial Navig Design, Northrop Electronics Div, 84-89, RES ENGR, NORTHROP ELECTRONICS DIV, 89- *Mem:* AAAS; Am Phys Soc; Am Inst Aeronaut & Astronautics; Astron Soc Pac. *Res:* Star trackers; scientific instruments and measurements; microelectronic processes; liquid crystal displays; inertial navigation and guidance. *Mailing Add:* 11591 Surburnas Way Santa Ana CA 92705

KENNEL, STEPHEN JOHN, b Peoria, Ill, Jan 15, 45; m 66; c 3. TUMOR IMMUNOLOGY. *Educ:* Univ Ill, BS, 67; Univ Calif, San Diego, MS, 68, PhD(chem), 71. *Prof Exp:* USPHS Grad fel trainee fel, 67-71; res fel, Dept Exp Path, Scripps Clin & Res Found, 71-73, res asst, 73-74, res assoc, Dept Immunopath, 74-76; staff mem, 76-81, SR STAFF SCIENTIST, BIOL DIV, OAK RIDGE NAT LAB, 81- *Mem:* Am Asn Cancer Res. *Res:* Antibody directed specific chemotherapy and specific immunotherapy of malignancies; analysis of cell surface proteins; leukemia virus proteins and radioimmunoassay. *Mailing Add:* Y-12 Area Bldg 9220 Mail Stop 8077 Oak Ridge Nat Lab Box 2009 Oak Ridge TN 37831-8077

KENNEL, WILLIAM E(LMER), b St Louis, Mo, Aug 11, 17; m 39; c 2. CHEMICAL ENGINEERING. *Educ:* Univ Ill, BS, 40; Mass Inst Technol, MS, 47, DSc(chem eng), 49. *Prof Exp:* Chem engr, A E Staley Mfg Co, 39-41; chem engr, Standard Oil Co, Ind, 48-51, res group leader, 51-52, res sect leader, 52-57; mgr tech develop, 57-60, dir chem res, 60-61, vpres res & develop & dir, 61-67, vpres & dir, 67-68, vpres plastics, 68-70, group vpres & dir, 70-72, vpres mkt & dir, 72-75, EXEC VPRES & DIR, AMOCO CHEM CORP, 75- *Mem:* Am Chem Soc; Am Inst Chem Engrs. *Res:* Technical development of petrochemicals. *Mailing Add:* 208 Inverness Lane Schererville IN 46375

KENNELL, DAVID EPPERSON, b Syracuse, NY, May 23, 32; c 3. MOLECULAR BIOLOGY. *Educ:* Univ Calif, Berkeley, AB, 54, PhD(biophysics), 59. *Prof Exp:* Res engr mineral tech, Univ Calif, Berkeley, 56-57; res fel bact & immunol, Harvard Med Sch, 59-60; res assoc, Mass Inst Technol, 60-61; from instr to assoc prof, 61-73, PROF MICROBIOL, SCH MED, WASH UNIV, 73- *Concurrent Pos:* Nat Cancer Inst fel, 57-61; NIH res career develop award, 69-74; mem, Microbial Physiol Study Sect, 81-85. *Mem:* Am Soc Microbiol; Am Soc Biol Chem; Biophys Soc; Am Soc Cell Biol; Am Asn Univ Professors. *Res:* Ribonucleic acid metabolism in bacteria and regulation of energy metabolism in cultured mammalian cells. *Mailing Add:* Dept Molecular Microbiol Box 8230 Wash Univ Sch Med St Louis MO 63110

KENNELL, JOHN HAWKS, b Reading, Pa, Jan 9, 22; m 49; c 3. PEDIATRICS, BEHAVIORAL SCIENCE. *Educ:* Univ Rochester, BS, 44, MD, 46. *Prof Exp:* Intern pediat, Children's Hosp, Boston, 46-47; asst resident, Children's Hosp Med Ctr, 49-50, chief resident med out-patient dept, 50, dir dept, 52, chief med resident, 51; dir, Family Clin, 52-60, dir, Pediat Clin, 60-70, sr instr, 52-55, from asst prof to assoc prof, 55-73, PROF PEDIAT, CASE WESTERN RESERVE UNIV, 73-, ASSOC PEDIATRICIAN, 56- *Concurrent Pos:* Nat Inst Child Health & Human Develop spec res fel, Univ London, 66-67; consult, Headstart, 68-; dir, Neonatal Nurseries, Univ Hosp, Cleveland, 52-67. *Honors & Awards:* George Armstrong Award, Ambulatory Pediat Asn. *Mem:* Am Acad Pediat; Am Pediat Soc; Asn Child Care Hosps (vpres, 73-75); Soc Res Child Develop; Ambulatory Pediat Asn (pres, 70). *Res:* Child development; medical education; social and psychological factors in medicine; effects of mother-infant separation on maternal attachment; effects of perinatal death on parents; effect of supportive companion during labor and delivery; medical and ambulatory care. *Mailing Add:* 2074 Abington Rd Cleveland OH 44106

KENNELLEY, JAMES A, b Rochester, NY, Aug 23, 28; m 55; c 3. METALLURGY, EXECUTIVE EDUCATION & DEVELOPMENT. *Educ:* Col Wooster, BA, 45; Mich State Univ, PhD(chem), 55. *Prof Exp:* Res chemist uranium, Mallinckrodt, Inc, 55-57, group leader res, 57-59, mgr res, 59-62, tech dir div, 62-65; asst to pres, Que Iron & Titanium Corp, 65-73,

vpres, 73-75, group vpres, 75-78; pres, Direct Reduction Corp, 78-83; ASSOC DIR EXEC EDUC, COLUMBIA BUS SCH, 83- *Concurrent Pos:* Consult. *Mem:* Am Chem Soc; Am Soc Metals; Am Inst Mining, Metall & Petrol Engrs. *Res:* Rare earths; uranium chemistry and metallurgy; titanium; raw materials; ilmenite smelting; titanium dioxide pigments; direct reduction of iron ore. *Mailing Add:* 17 Huguenot Dr Larchmont NY 10538

KENNELLEY, KEVIN JAMES, b St Louis, Mo, Aug 6, 58; m 82; c 3. NONMETALLICS, ELASTOMERICS. *Educ:* Univ Okla, BS, 80, MS, 85, PhD(metall eng & mat sci), 86. *Prof Exp:* Sr field engr, Duncan Dist, Schlumberger Well Serv, 80-82; res specialist, Mat Sect, Exxon Prod Res Co, 86-90; SR RES ENGR, MAT SECT, ARCO OIL & GAS CO, 90- *Concurrent Pos:* Prog chmn ann conf, Nat Asn Corrosion Engrs, 88-89 & 89-90. *Mem:* Sigma Xi; Metall Soc; Am Soc Metals; Nat Asn Corrosion Engrs. *Res:* Controlling corrosion of materials in oil and gas production that are exposed to high temperature, high pressure sour hydrocarbon fluids; cathodic protection; nonmetallics. *Mailing Add:* Arco Oil & Gas Co 2300 W Plano Pkwy Plano TX 75075

KENNELLY, MARY MARINA, b Chicago, Ill, Nov 12, 19. INORGANIC CHEMISTRY, ORGANIC CHEMISTRY. *Educ:* Mundelein Col, BS, 42; Univ Notre Dame, MS, 50, PhD(chem), 59. *Prof Exp:* Asst prof, 50-57, chmn dept, 59-69, PROF CHEM, MUNDELEIN COL, 59- *Concurrent Pos:* NSF sci fac summer fels, London, 60, Fla State, 61 & Seattle, 62. *Mem:* AAAS; Am Chem Soc; Nat Sci Teachers Asn; Sigma Xi. *Res:* Coordination chemistry; infrared studies of metal complexes of amino acids. *Mailing Add:* Dept Chem 6363 Sheridan Rd Chicago IL 60626

KENNELLY, WILLIAM J, b Cleveland, Ohio, Aug 22, 48; m 77; c 2. CHEMISTRY. *Educ:* Mass Inst Technol, BS, 70; Northwestern Univ, PhD(chem), 75; Temple Univ, MBA, 86. *Prof Exp:* Fel, Univ NDak, Grand Forks, 75-76, Mass Inst Technol, 76-77; res scientist, Rohm & Haas Co, 77-84; sr res chemist, 84-87, Amax Mat Res Ctr, DIR, TECHNOL & OPERS, AMAX POLYMER ADDITIVES GROUP, ANN ARBOR, MICH, 87- *Mem:* Am Chem Soc; Soc Plastics Eng; Am Soc Testing & Mat. *Res:* Polymer additives especially flame retardants and smoke suppressants. *Mailing Add:* 447 Marlpool Dr Saline MI 48176

KENNER, CHARLES THOMAS, b Waxahachie, Tex, Oct 20, 10; m 42; c 1. ANALYTICAL CHEMISTRY. *Educ:* Trinity Univ, BS, 32; Univ Tenn, MS, 35; Univ Tex, PhD(phys chem), 39. *Prof Exp:* Instr gen chem, Univ Tex, 35-38; asst prof chem, The Citadel, 38-42; chief chemist & metallurgist, Thor Corp, 46; dir res & control labs, Cent Testing Co, Chicago, 47; from asst prof to prof, 48-76, EMER PROF CHEM, SOUTHERN METHODIST UNIV, 85- *Concurrent Pos:* Consult, Food & Drug Admin, 67-76 & RSR Corp, 70-; vis prof, NTex State Univ, 78 & 81; adj prof chem, Southern Methodist Univ, 85. *Mem:* Am Chem Soc; Soc Appl Spectros. *Res:* Ion exchange separations; chelometry; spectrophotometric methods; trace metal analysis; pharmaceutical analysis. *Mailing Add:* Dept Chem Southern Methodist Univ Dallas TX 75275

KENNER, MORTON ROY, b Rochester, NY, June 10, 25; m 54; c 2. MATHEMATICS. *Educ:* Univ Rochester, AB, 49; Univ Minn, MA, 51; PhD(math, math ed), 58. *Prof Exp:* From asst prof to assoc prof math, Southern Ill Univ, 52-67; prof math & chmn dept, Stephens Col, 67-70; PROF MATH & CHMN DEPT, NORTHWEST MO STATE UNIV, 70-, CHMN DIV MATH & COMPUT SCI, 78- *Concurrent Pos:* Dir develop proj sec math, Southern Ill Univ, 58-67; consult, Opers Res Group, Ohio State Univ, 60-; dir Nairobi Math Ctr, Kenya, 64 & 65. *Mem:* Hist Sci Soc; Math Asn Am; Sigma Xi. *Res:* Foundations of mathematics; systems models; mathematical education; history of mathematics. *Mailing Add:* Math Sci Dept Northwest Mo State Univ Maryville MO 64468

KENNERLY, GEORGE WARREN, b Boston, Mass, Mar 11, 22; m 49; c 2. INDUSTRIAL CHEMISTRY. *Educ:* Harvard Univ, BS, 44, MA, 47, PhD, 49. *Prof Exp:* Res chemist, 49-54, group leader, 54-59, mgr, 59-68, dir, Am Cyanamid Co, 68-84; RETIRED. *Mem:* AAAS; Am Chem Soc. *Res:* Auto-oxidation; peroxide chemistry; photochemistry; electrochemistry; luminescence; reaction kinetics. *Mailing Add:* 42 Holly Lane Darien CT 06820-3306

KENNET, HAIM, b Jerusalem, Israel, Apr 20, 35; US citizen. AERONAUTICS, ASTRONAUTICS. *Educ:* Mass Inst Technol, SB & SM, 57, ScD(aeronaut & astronaut), 61. *Prof Exp:* Res engr gas dynamics, Aero-Space Div, 61-63, sr group engr aerothermodyn, 63-65, sr supvr Mars Explor, 65-67, systs eng mgr, Space Div, 67-69, PROJ MGR SPACE EXPLOR SYSTS, AEROSPACE GROUP, BOEING CO, 69- *Mem:* Am Inst Aeronaut & Astronaut. *Res:* Flight mechanics, flight control, propulsion and thermal control aspects of unmanned space probes. *Mailing Add:* 2040 43rd No 408 Seattle WA 98112

KENNETT, JAMES PETER, b Wellington, NZ, Sept 3, 40; m 64; c 2. MICROPALEONTOLOGY, PALEOECOLOGY. *Educ:* New Zealand Univ, BSc, 62; Victoria Univ, Hons, 63, PhD(geol), 65, DSc, 76. *Prof Exp:* Sci officer, NZ Oceanog Inst, 65-66; NSF fel micropaleont, Allan Hancock Found, Univ Southern Calif, 66-68; asst prof, Fla State Univ, 68-70; assoc prof, 70-74, PROF, GRAD SCH OCEANOG, UNIV RI, 74- *Concurrent Pos:* Mem adv comt, Antarctic Deep-Sea Drilling. *Honors & Awards:* McKay Hammer Award, Geol Soc NZ, 88. *Mem:* AAAS; Geol Soc Am; Am Asn Petrol Geol; Soc Econ Paleont & Mineral; Int Quaternary Asn; Sigma Xi. *Res:* Marine geology; foraminiferal ecology and paleoecology; stratigraphic paleontology and stratigraphy of the Cenozoic; geology of the Antarctic continent. *Mailing Add:* Dir Marine Sci Inst Univ Calif Santa Barbara CA 93106

KENNETT, ROGER H, b Lakewood, NJ, Dec 27, 40; m 66; c 3. GENETICS. *Educ:* Eastern Col, AB, 64; Princeton Univ, PhD(biochem sci), 70; Univ Pa, MSEd, 86. *Prof Exp:* Demonstr, Genetics Lab, Oxford Univ, 72-73, res officer, 73-76; asst prof, 76-80, ASSOC PROF HUMAN GENETICS, SCH MED, UNIV PA, 80-, DIR, HUMAN GENETICS CELL CTR, 73- *Mem:* Am Soc Human Genetics; AAAS; Am Asn Immunologists; Am Soc Cell Biol. *Res:* Use of combination of immunological, biochemical and molecular genetic techniques to study molecular changes related to oncogenesis, and to the growth and differentiation of human neuroblastoma cells. *Mailing Add:* Dept Human Genetics Sch Med Univ Pa Philadelphia PA 19104

KENNETT, TERENCE JAMES, b Toronto, Ont, Aug 8, 27; m 49; c 2. NUCLEAR PHYSICS. *Educ:* McMaster Univ, BSc, 53, MSc, 54, PhD(physics), 56. *Prof Exp:* Fel physics, McMaster Univ, 56-57; assoc physicist, Argonne Nat Lab, 57-59; from asst prof to assoc prof, 59-66, PROF ENG PHYSICS & PHYSICS, MCMASTER UNIV, 66- *Mem:* Am Phys Soc. *Res:* Neutron physics; decay scheme studies; neutron capture gamma rays; instrumentation and detector development. *Mailing Add:* McMaster Univ Hamilton ON L8S 4M1 Can

KENNEY, DONALD J, b Chicago, Ill, Aug 26, 25; m 48; c 7. PHYSICAL CHEMISTRY. *Educ:* Loyola Univ, Ill, BS, 49; Iowa State Univ, PhD(phys chem), 53. *Prof Exp:* Res engr, Steel Div, Ford Motor Co, 53-54; PROF CHEM, UNIV DETROIT, 54- *Concurrent Pos:* Dir govt projs, 54-65. *Mem:* Am Chem Soc. *Res:* Iron complexes; metallurgy. *Mailing Add:* Dept Chem Univ Detroit Detroit MI 48221

KENNEY, FRANCIS T, b Springfield, Mass, Mar 16, 28; m 51; c 2. BIOCHEMISTRY. *Educ:* St Michael's Col, BS, 50; Univ Notre Dame, MS, 53; Johns Hopkins Univ, PhD(biochem), 57. *Prof Exp:* Instr biol, St Michael's Col, 50-51; res assoc pediat, Med Col, Cornell Univ, 57-59; from biochemist to sr biochemist, 59-69, sci dir, Carcinogenesis Prog, 69-75, SR STAFF SCIENTIST, OAK RIDGE NAT LAB, 75- *Mem:* AAAS; Am Soc Biol Chemists; Am Chem Soc; Sigma Xi; Am Soc Cell Biol. *Res:* Mammalian biochemistry; regulation. *Mailing Add:* Biol Div Oak Ridge Nat Lab Oak Ridge TN 37831-8077

KENNEY, GARY DALE, industrial hygiene, loss prevention, for more information see previous edition

KENNEY, GERALD, b Seattle, Wash, Dec 14, 34; m; c 4. SURGERY, MEDICINE. *Educ:* Notre Dame Univ, BS, 56; Northwestern Univ, MD, 60. *Prof Exp:* Intern, Cook County Hosp, 60-61, resident gen surg, 61-62 & urol, 64-67; asst prof, State Univ NY, 68-71; CLIN INSTR UROL, UNIV WASH, 71- *Concurrent Pos:* Gen med officer, Ft Sam Houston, 62 & Little Rock AFB, 62 & 64; fel, Inst Urol, Univ London, 67-68; priv pract, OA Nelson Med Group, 71-72, Belleuve Urol, Inc, 72-85, Overlake Urol, Inc, 85-; chief transplantation serv, Swedish Hosp Med Ctr, 71-; secy-treas, NW Urol Soc, 80-83, pres, 84; Urol Adv Coun, Nat Kidney Found, 82-85; med review bd, Nat Transplant Prog, 84-; bd dirs, NW Kidney Ctr, 85- & Lake Wash Kidney Ctr, 84-; treas, Seattle Surg Soc, 85-86. *Mem:* Am Med Asn; Am Urol Asn; Am Col Surgeons; Am Soc Transplant Surgens; Soc Univ Urol; Soc Acad Surg; Am Soc Clin Urologists; Int Urol Soc. *Res:* Evaluation of diagnostic modalities for urological cancer which includes transrectal prostatic ultrasound and serum blood studies; methods for early detection of organ transplantation rejection. *Mailing Add:* Pac NW Res Found 720 Broadway Seattle WA 98022

KENNEY, JAMES FRANCIS, b Buffalo, NY, Sept 3, 26; m 57; c 3. GEOPHYSICS. *Educ:* Union Col, NY, BS, 51; Univ NMex, MS, 53, PhD(physics), 57. *Prof Exp:* Res assoc physics, Univ NMex, 57-58, instr, 59-60; res fel, Univ NMex & Lab Cosmic Physics, La Paz, 58-59; staff mem geoastrophys res, Boeing Sci Res Lab, Boeing Aerospace Co, 60-69, head, Geoastrophys Dept, 69-70, head, Environ Sci Dept, 70-73, mgr, Laser & Environ Sci Lab, 73-75, resources develop mgr, 75-76, mgr, radiation physics, 76-79, chief scientist, 79-90, CHIEF ENG MISSILES SYSTS, BOEING AEROSPACE CO, 90- *Concurrent Pos:* BSD technol mgr, Boeing Aerospace Co. *Mem:* Am Inst Aeronaut & Astronaut; AAAS; Am Geophys Union. *Res:* Cosmic rays; ionospheric properties and radio propagation; nuclear, space and solar physics; magnetic fields and micropulsations; environmental science, urban studies; remote sensing, laser physics; military sciences; radiation physics; countermeasures. *Mailing Add:* Boeing Aerospace Co Box 3999 MS-3J74 Seattle WA 98124-2499

KENNEY, JAMES FRANKLIN, b Richmond, Va, Aug 4, 34; m 58; c 3. POLYMER CHEMISTRY. *Educ:* Howard Univ, BS, 56, MS, 58; Univ Akron, PhD(polymer chem), 64. *Prof Exp:* Res chemist, US Air Force Mat Lab, 58-61; res fel, Inst Rubber Res, Akron, 61-64; res chemist, Chemstrand Res Ctr, Inc, 64-68; res assoc, Plastics Div, Allied Chem Corp, 68-71 & M & T Chem, Inc, 71-78; sr proj leader, 78-79, mgr polymers & mat sci, 79-91, MGR POLYMER & ANALYTICAL CHEM, VISTAKON DIV, JOHNSON & JOHNSON VISION PRODS, INC, 91- *Mem:* Am Chem Soc. *Res:* Synthesis, structure, property and performance of polymers; adhesion to skin; adhesive tapes; emulsion, condensation and addition polymerization; graft and block copolymers; polyblends; impact modification; processing characteristics of polymers; economic and technical evaluation of research; developing catalysts for polyesters and polyurethanes; emulsion and hot melt adhesives. *Mailing Add:* Vistakon PO Box 10157 Jacksonville FL 32247

KENNEY, JOHN T, physical chemistry, for more information see previous edition

KENNEY, MALCOLM EDWARD, b Berkeley, Calif, Oct 7, 28; m 51. ORGANOSILICON CHEMISTRY. *Educ:* Univ Redlands, BS, 50; Cornell Univ, PhD(chem), 54. *Prof Exp:* Asst, Cornell Univ, 50-52, fel, 54; from instr to assoc prof, 56-66, PROF CHEM, CASE WESTERN RESERVE UNIV, 66- *Concurrent Pos:* John Teagle prof fel, 64-66; Hurlbut prof chem, 91. *Mem:* Am Chem Soc. *Res:* Metal complexes and inorganic polymers. *Mailing Add:* Dept Chem Case Western Reserve Univ 2040 Adelbert Rd Cleveland OH 44106-7078

KENNEY, MARGARET JUNE, b Boston, Mass, June 7, 35. MATHEMATICS. *Educ:* Boston Col, Chestnut Hill, BS, 57, MA, 59; Boston Univ, PhD(math), 77. *Prof Exp:* Instr, 59-63, lectr, 63-70, asst prof, 70-79, ASSOC PROF MATH, BOSTON COL, 79- *Concurrent Pos:* Asst dir, Math Inst, Boston Col, 57- *Mem:* Nat Coun Teachers Math; Math Asn Am; Am Math Soc; Nat Coun Supvrs Math; Asn Women Math. *Res:* Mathematics education at the pre-college level; problem solving; number theoretic applications of mathematics to art. *Mailing Add:* Math Inst Boston Col Chestnut Hill MA 02167

KENNEY, MARY ALICE, b Lubbock, Tex, May 16, 38. NUTRITIONAL STATUS, NUTRITION & IMMUNOCOMPETENCE. *Educ:* Tex Tech Univ, BS, 58; Iowa State Univ, MS, 60, PhD(nutrit), 63. *Prof Exp:* Instr nutrit, Purdue Univ, 60-61; from asst prof to assoc prof, Iowa State Univ, 63-73; prof food & nutrit, Tex Tech Univ, 73-78; prof food, nutrit & inst admin, Okla State Univ, 78-84; PROF HOME ECON, UNIV ARK, 85- *Mem:* AAAS; Am Inst Nutrit; NY Acad Sci; Am Dietetic Asn; Am Pub Health Asn; Am Col Nutrit. *Res:* Immunoglobulin levels; assessment of nutritional status; magnesium nutrition. *Mailing Add:* Dept Home Econ Univ Ark 118 HOEC Bldg Fayetteville AR 72701

KENNEY, NANCY JANE, b Wilkes-Barre, Pa. PSYCHOBIOLOGY, BEHAVIORAL ENDOCRINOLOGY. *Educ:* Wilkes Col, BA, 70; Univ Va, MA, 72, PhD(psychol), 74. *Prof Exp:* Fel neurol sci, Univ Pa, 74-76; asst prof, 76-82, ASSOC PROF PSYCHOL & WOMEN STUDIES, UNIV WASH, 83- *Concurrent Pos:* NIH res fel, Nat Inst Child Health & Human Develop, 74-76; prin investr, Nat Inst Arthritis, Metab & Digestive Dis, 78-84. *Mem:* Soc Neurosci; Am Psychol Asn; Nat Women Studies Asn. *Res:* Neural, endocrine and behavioral control of food intake and body weight. *Mailing Add:* Dept Psychol NI-25 Univ Wash Seattle WA 98195

KENNEY, RICHARD ALEC, b Coventry, Eng, Oct 4, 24; m 59; c 1. MEDICAL PHYSIOLOGY. *Educ:* Univ Birmingham, BSc, 45, PhD(physiol), 47. *Prof Exp:* Lectr physiol, Univ Leeds, 47-51; sr sci officer, Colonial Res Serv, Nigeria, 51-54; prof physiol, Univ Rangoon, 55-58, Univ NSumatra, 58-60 & Univ Singapore, 60-65; reader, Univ Melbourne, 65-68; prof, 68-90, EMER PROF PHYSIOL, MED CTR, GEORGE WASHINGTON UNIV, 90- *Mem:* Brit Physiol Soc; Brit Renal Asn; Am Physiol Soc; Int Soc Nephrology. *Res:* Renal, climatic and cardio-vascular physiology. *Mailing Add:* Dept Physiol George Washington Univ Med Ctr Washington DC 20037

KENNEY, ROBERT WARNER, b Portland, Ore, Nov 9, 22; m 50; c 2. HIGH ENERGY PHYSICS, PARTICLE PHYSICS. *Educ:* Univ Calif, Los Angeles, BA, 44; Calif Inst Technol, BS, 47; Univ Calif, PhD(physics), 52. *Prof Exp:* Mem staff, Los Alamos Sci Lab, 52-53; SR STAFF PHYSICIST, LAWRENCE BERKELEY LAB, UNIV CALIF, BERKELEY, 53- *Concurrent Pos:* Lectr, Univ Calif, Berkeley; consult, Marquardt Corp; NASA. *Mem:* Am Phys Soc; AAAS; Sigma Xi. *Res:* High energy particle physics; electromagnetic interactions at high energies; particle accelerators; pion nuclear interaction; weak interaction and symmetry principles. *Mailing Add:* 122 Scenic Dr Orinda CA 94563

KENNEY, T CAMERON, b Montreal, Que, Mar 26, 31; m 60; c 4. CIVIL & GEOTECHNICAL ENGINEERING. *Educ:* McGill Univ, BEng, 53; Univ London, DIC, 54, MSc, 56, PhD(civil eng), 67. *Prof Exp:* Geotech engr, Acres Ltd, Niagara Falls, Ont, 56-61; res engr, Norweg Geotech Inst, Oslo, 61-67; assoc prof civil eng, 67-68, chmn dept, 68-74, PROF CIVIL ENG, UNIV TORONTO, 68- *Concurrent Pos:* Eng consult, 68- *Honors & Awards:* Walter L Huber Prize, Am Soc Civil Engrs, 67; First Bjerrum Lectr, Norweg Geotech Soc, 75; Silver Jubilee Medal, Govt Can, 77; Keefer Medal, Can Soc Civil Eng, 83; Can Geotech Prize, Can Geotech Soc, 85. *Mem:* Fel Eng Inst Can; Can Geotech Soc (pres 74-76). *Res:* Engineering properties of natural soils; landslides; engineering geology; dams; tailings dams. *Mailing Add:* Dept Civil Eng Univ Toronto Toronto ON M5S 1A4 Can

KENNEY, VINCENT PAUL, b New York, NY, Sept 15, 27; m 54; c 4. PHYSICS. *Educ:* Iona Col, AB, 48; Fordham Univ, MS, 50, PhD(physics), 56. *Prof Exp:* Predoctoral res assoc, Brookhaven Nat Lab, 53-55; asst prof physics, Univ Ky, 55-60, assoc prof, 60-63; assoc prof, 63-66, PROF PHYSICS, UNIV NOTRE DAME, 66- *Concurrent Pos:* Vis physicist, Brookhaven Nat Lab, 57-; Oak Ridge Inst Nuclear Studies, 58-64; European Organization for Nuclear Res, 61-62 & 82-85; Argonne Nat Lab, 65-82; Fermi Nat Accelerator Lab, 71-; fel, Max Planck Inst Physics & Astrophys, Munich, 61-62 & 72; sr vis fel, Cavendish Lab, Cambridge Univ, 82-83, life mem, Clare Hall, Cambridge Univ, 85-; sr physicist, Dept Energy, 86-88. *Mem:* Fel Am Phys Soc; AAAS; Sigma Xi. *Res:* High energy particle physics; energy studies. *Mailing Add:* Dept Physics Univ Notre Dame Notre Dame IN 46556

KENNEY, WILLIAM CLARK, b Grand Forks, NDak, Feb 25, 40. BIOCHEMISTRY. *Educ:* Carleton Col, BA, 62; Univ Calif, Berkeley, PhD(biochem), 67. *Prof Exp:* Teaching & res asst, Univ Calif, Berkeley, 63-67; asst & assoc res biochemist, Univ Calif, San Francisco, 70-79, res chemist, Vet Admin Med Ctr, 79-84; RES SCIENTIST, AMGEN, THOUSAND OAKS, CALIF, 84-, LAB HEAD, PROTEIN CHEM, 89- *Concurrent Pos:* adj assoc prof biochem med, Univ Calif, San Francisco, 79-84; fel, Am Cancer Soc, 68-69. *Honors & Awards:* Alcoholism Res Award, Vet Admin, 79. *Mem:* Am Soc Biochem & Molecular Biol; AAAS; Am Chem Soc; NY Acad Sci; Res Soc Alcoholism; Sigma Xi; Am Cancer Soc; Protein Soc. *Res:* Enzymology; protein biochemistry and biophysics; molecular biology; structure and function of enzymes; biological oxidations. *Mailing Add:* Amgen 1840 Dehavilland Dr Thousand Oaks CA 91320

KENNEY-WALLACE, GERALDINE ANNE, b London, Eng, Mar 29, 43. CHEMICAL PHYSICS, CHEMICAL DYNAMICS OPTICS. *Educ:* Royal Inst Chem, ARIC, 65; Univ BC, MS, 68, PhD(chem), 70. *Hon Degrees:* DSc, DLitt, Univ Toronto, 88. *Prof Exp:* Res assoc biophys, Oxford Univ, 64-66;

fel chem, Univ BC, 70-71; assoc, Radiation Lab, Univ Notre Dame, 71-72; from instr to asst prof, Yale Univ, 72-74; from asst to assoc prof chem, 74-80, PROF CHEM & PHYSICS, UNIV TORONTO, 80- *Concurrent Pos:* Vis scientist chem, Argonne Nat Lab, 73-; Alfred P Sloan fel, 77-79; Killam Res fel, 79-81; vis scientist, École Polytech Paris, 81; Guggenheim fel, 83; vis prof, Stanford Univ, 85-86; chmn, res bd, Univ Toronto, 85-87; chmn, Sci Coun Can, 87-92. *Honors & Awards:* Corday-Morgan Medal, UK, 79; Noranda Award, 84; E W R Steacie Award, 84. *Mem:* Royal Soc Chem; Am Chem Soc; Am Phys Soc; Sigma Xi; Optical Soc Am; InterAm Photochem Soc; Nat Adv Bd Sci & Technol. *Res:* Molecular photophysics, energy transfer and molecular dynamics studied via picosecond, femtosecond laser spectroscopy; electronic and molecular structure of electrons in fluids; laser-induced electron transfer; holography, picosecond non linear optics. *Mailing Add:* Chem Dept Univ Toronto 80 St George St Toronto ON M5S 1A1 Can

KENNICK, WALTER HERBERT, b Hampton, Va, Aug 23, 20; m 43; c 3. MEAT SCIENCE. *Educ:* Clemson Col, BS, 48; Ore State Col, MS, 58, PhD(animal husb, meats), 59. *Prof Exp:* Res fel, 56-59, asst prof, 59-70, ASSOC PROF ANIMAL HUSB, ORE STATE UNIV, 70- *Concurrent Pos:* Res scientist, Agr Inst, 73; Fulbright-Hays, Animal Prod Res Ctr, Dublin, Ireland, 74. *Mem:* AAAS; Am Soc Animal Sci; Am Meat Sci Asn; Inst Food Technol. *Res:* Quantitative yield of lean tissue from meat animals, methods of evaluating quality and quantity of such tissue and factors contributing to their variation; post slaughter treatments to improve eating quality. *Mailing Add:* 3290 SW Chintimini Ave Corvallis OR 97333

KENNICOTT, PHILIP RAY, physical chemistry, for more information see previous edition

KENNINGTON, GARTH STANFORD, b Afton, Wyo, Apr 19, 15; m 43. ANIMAL ECOLOGY, ANIMAL PHYSIOLOGY. *Educ:* Univ Wyo, BS, 40; Univ Chicago, MS, 48, PhD(zool), 52. *Prof Exp:* Instr biol, George Williams Col, 47-48; instr, Roosevelt Col, 48-49; from asst prof to assoc prof zool, Lawrence Col, 52-60; assoc prof zool, Univ Wyo, 60-62, prof, 62-81, prof physiol, 74-81; RETIRED. *Concurrent Pos:* Consult, Kimberly-Clark Corp, 58-; NSF fel, Donner Lab, Univ Calif, 59-60; Fulbright lectr, Aligarh Muslim Univ, India, 63-64; Assoc Rocky Mt Univs fac grant, Nat Reactor Testing Sta, Idaho Falls, Idaho, 65-67; Oak Ridge Assoc Univs res grant, Environ Sci Div, Oak Ridge Nat Lab, 72-73. *Mem:* Fel AAAS; Am Soc Zool; Ecol Soc Am; Sigma Xi. *Res:* Ecology and physiology of the high arid plains; assessment of effects of mining and milling on biological communities in Wyoming. *Mailing Add:* 1404 Bridger Laramie WY 82070

KENNINGTON, MACK HUMPHERYS, b Kamas, Utah, Apr 6, 23. ANIMAL SCIENCE. *Educ:* Univ Idaho, BS, 46; Purdue Univ, MS, 56, PhD(animal nutrit), 58. *Prof Exp:* County exten agent, Exten Serv, Univ Idaho, 48-54; from asst prof to assoc prof, 58-68, PROF ANIMAL SCI, CALIF STATE POLYTECH UNIV, POMONA, 69- *Concurrent Pos:* Vis prof animal sci, Cornell Univ, 66-67. *Mem:* Am Soc Animal Sci; Can Soc Animal Prod; Am Dairy Sci Asn; Brit Soc Animal Prod. *Res:* Animal production and nutrition. *Mailing Add:* Dept Animal Sci Calif State Polytech Univ Pomona CA 91768

KENNISH, JOHN M, b Vineland, NJ, Oct 6, 45; m 67; c 1. ANALYTICAL CHEMISTRY. *Educ:* Rutgers Univ, AB, 67; Shippensburg Univ, MS, 73; Portland State Univ, PhD(environ chem), 78. *Prof Exp:* Res assoc res, Ore Health Sci Ctr, 77-79; from asst prof to assoc prof, 79-88, PROF TEACHING & RES, UNIV ALASKA ANCHORAGE, 88- *Concurrent Pos:* Vis fel res, Univ Colo, Boulder, 82-83; vis prof res, Ore State Univ, 88-89. *Mem:* Am Chem Soc; AAAS; Sigma Xi; Nat Asn Environ Prof. *Res:* Fish biochemistry related to evaluating the response of major hepatic enzymes to toxic agents; the biochemistry of post-mortem changes in fish tissue especially lipids and the application of analytical methods to environmental chemistry. *Mailing Add:* Dept Chem & Physics Univ Alaska Anchorage 3211 Providence Dr Anchorage AK 99508

KENNISON, JOHN FREDERICK, b New York, NY, Oct 7, 38; m 64; c 2. MATHEMATICS. *Educ:* Queens Col, NY, BS, 59; Harvard Univ, AM, 60, PhD(topology), 63. *Prof Exp:* PROF MATH, CLARK UNIV, 63- *Mem:* Am Math Soc. *Res:* Category theory. *Mailing Add:* Clark Univ Worcester MA 01610

KENNY, ALEXANDER DONOVAN, b London, Eng, Mar 4, 25; nat US; m 50; c 4. PHARMACOLOGY, ENDOCRINOLOGY. *Educ:* Imp Col, Univ London, BSc, 45, DSc, 82; Athenaeum of Ohio, MS, 49, PhD(biochem), 50. *Prof Exp:* Sr chemist, Metab Labs, Univ Col Hosp, London, 50-51; chief chem lab, Mass Gen Hosp, 52; asst dent sci, Sch Dent Med, Harvard Univ, 52-53, instr, 53-55; assoc pharmacol, Harvard Med Sch, 55-59; from assoc prof to prof, Med Ctr, WVa Univ, 59-67; prof, Sch Med, & investr, Space Sci Res Ctr, Univ Mo, Columbia, 67-71, prof biochem, 71-74, investr, Dalton Res Ctr, 67-74; prof pharmacol med br, Univ Tex, 74-75; PROF & CHMN PHARMACOL, SCH MED, HEALTH SCI CTR, TEX TECH UNIV, 76- *Concurrent Pos:* USPHS Spec fel, Royal Postgrad Med Sch, London, 67-68; chmn, NIH Special Study Section, 75; actg dir, Tarbox Parkinson's Dis Inst, 76-78, dir, 78-87. *Mem:* Endocrine Soc; Biophys Soc; Am Soc Pharmacol & Exp Therapeut; Am Chem Soc; Brit Biochem Soc. *Res:* Calcium and bone metabolism; vitamin D; parathyroid hormone; calcitonin; hormone assay. *Mailing Add:* Dept Pharmacol Tex Tech Univ Health Sci Ctr Lubbock TX 79430-0001

KENNY, ANDREW AUGUSTINE, b Chicago, Ill, July 21, 34; m 66; c 2. AUTOMOTIVE CONTROL DESIGN & DEVELOPMENT, FACILLITATE QUALITY FUNCTION DEPLOYMENT. *Educ:* Univ Ill, BS, 61. *Prof Exp:* Proj engr, Bastian-Blessing, 63-70; proj engr, Controls Div, Eaton, 70-86, eng supvr, 81-86, chief engr, 86-89, QUAL ASSURANCE MGR, CONTROLS DIV, EATON, 89-91. *Mem:* Soc Automotive Engrs; Am Soc Qual Control. *Res:* Seventeen patents and a variety of automotive controls; published many articles on engineering, management and quality control. *Mailing Add:* 451 S Park St Roselle IL 60172

KENNY, DAVID HERMAN, b Lake Linden, Mich, Oct 6, 27. ORGANIC CHEMISTRY. *Educ:* Cornell Univ, AB, 49; Univ Mich, MS, 55, PhD, 59. *Prof Exp:* Asst prof chem, Eastern Mich Univ, 58-60; Smith-Mundt lectr, Baghdad, 60-62; chmn, Org Chem Dept, 74-81, ASSOC PROF CHEM, MICH TECHNOL UNIV, 62- *Mem:* AAAS; Am Chem Soc. *Res:* Organic nitrogen chemistry. *Mailing Add:* 1220 Military Rd Houghton MI 49931

KENNY, GEORGE EDWARD, b Dickinson, NDak, Sept 23, 30; m 58; c 6. MICROBIOLOGY. *Educ:* Fordham Univ, BS, 52; Univ NDak, MS, 57; Univ Minn, Minneapolis, PhD(microbiol), 61. *Prof Exp:* Res instr prev med, 61-63, from asst prof to assoc prof, 63-71, PROF & CHMN DEPT PATHOBIOL, UNIV WASH, 71- *Honors & Awards:* Kimble Methodology Award, Am Pub Health Asn, 71. *Mem:* Am Soc Microbiol; Soc Exp Biol & Med; Am Asn Immunol; Infectious Dis Soc Am; Int Orgn Mycoplasmology; Sigma Xi; fel Am Acad Microbiol. *Res:* Antigenic analysis of microorganisms; host-parasite relationships of animal cells and microorganisms; biology of the mycoplasmatales. *Mailing Add:* 1504 37th Ave Seattle WA 98122

KENNY, JAMES JOSEPH, IMMUNE DEFICIENCY, B-LYMPHOCYTES. *Educ:* Univ Calif, Los Angeles, PhD(immunol), 77. *Prof Exp:* ASST PROF MICROBIOL & IMMUNOL, UNIFORMED SERV UNIV HEALTH SCI, 79- *Mailing Add:* NCI-FCRDC Prog Res Inc PO Box B Bldg 567 Rm 227 Frederick MD 21701

KENNY, MICHAEL THOMAS, b San Francisco, Calif, Oct 3, 38; m 63; c 1. IMMUNOMODULATION, INFLAMMATORY DISEASE. *Educ:* Univ San Francisco, BS, 60; Univ Del, PhD(microbiol), 64. *Prof Exp:* Sr virologist, Biohazards Dept, Pitman-Moore Div, Marion Merrell Dow Pharmaceut Inc, 66-67, res virologist, Dow Human Health Res & Develop Labs, 67-71, sr res virologist, Dept Infectious Dis, 71-74, sr res immunologist, Dow Diag Res & Develop, 74-78, clin res assoc, Med Dept, 78-81, ASSOC SCIENTIST, PHARMACOL DEPT, MARION MERRELL DOW PHARMACEUT, 81- *Mem:* Am Soc Microbiol; AAAS; Am Asn Immunologists; AAAS; Int Soc Antiviral Res; Reticuloendothelial Soc. *Res:* Invertebrate microbiology and invertebrate tissue culture; diagnostic virology; radioimmunoassay; laboratory safety; viral immunology; virus vaccine development; development of antibiotics and anti-viral compounds; immunomodulation and inflammatory disease. *Mailing Add:* Marion Merrell Dow Pharmaceut Inc PO Box 68470 Indianapolis IN 46268-0470

KENSEK, RONALD P, b Buffalo, NY, July 9, 58. NUCLEAR ENGINEERING, COMPUTER SIMULATION. *Educ:* State Univ, NY, Buffalo, BA, 80, BS, 80; Univ Mich, PhD(nuclear eng), 86. *Prof Exp:* MEM, TECH STAFF, SANDIA NAT LAB, 86- *Mem:* AAAS; Am Phys Soc. *Res:* Pursuing the development of inertial confinement fusion and developing new methods of radiation transport computer simulation. *Mailing Add:* Sandia Nat Lab Org 1231 Albuquerque NM 87185

KENSHALO, DANIEL RALPH, b West Frankfort, Ill, July 27, 22; m 70; c 4. PSYCHOPHYSIOLOGY. *Educ:* Wash Univ, BA, 47, PhD(exp psychol), 53. *Prof Exp:* Instr psychol, Wash Univ, St Louis, 48-49; actg asst prof, 50-53, from asst prof to assoc prof, 53-59, PROF PSYCHOL, FLA STATE UNIV, 59- *Concurrent Pos:* Vis prof physiol, Univ Marburg, Ger, 69, Univ Claude Bernard, France, 73 & Peking Univ, People's Repub China. *Mem:* Soc Neurosci; Fel Am Psychol Asn; Fel AAAS; Fel NY Acad Sci; Am Physiol Soc. *Res:* Psychophysical and electrophysiological investigation of the skin senses. *Mailing Add:* Dept Psychol Fla State Univ Tallahassee FL 32306

KENSLER, CHARLES JOSEPH, b New York, NY, Jan 21, 15; m 44; c 2. PHARMACOLOGY, BIOCHEMISTRY. *Educ:* Columbia Univ, AB, 37, MA, 38; Cornell Univ, PhD(pharmacol), 48. *Prof Exp:* Chem asst, Rockefeller Inst, 38-39; res assoc biochem, Mem Hosp, NY, 39-43, researcher, Off Sci Res & Develop, 42-43; from instr to assoc prof, Med Col, Cornell Univ, 43-53; head biol labs, Arthur D Little, Inc, 54-57; prof & chief exec officer, Arthur D Little Int, Inc, 73-85; prof pharmacol & exp therapeut & chmn dept, 57-60, PROF PHARMACOL, SCH MED, BOSTON UNIV, 60- *Concurrent Pos:* Traveling fel, Oxford Univ, 49-50; Sloan scholar, 51-54; lectr, Harvard Med Sch, 54-57; mem drug evaluation panel & chmn pharmacol comt, Cancer Chemother Nat Serv Ctr, 57-61; mem sub-comt carcinogenesis, Nat Acad Sci-Nat Res Coun, 57-; consult, Food & Drug Admin, 57, 60 & 71- & Nat Cancer Inst, 62-70; sr vpres prof oper, Arthur D Little, Inc, 60-85; vis prof, Mass Inst Technol, 72-84; trustee, Gordon Res Conf Coun, 77-83, chmn, 79-80; pres, Mass Health Res Inst, 77-80, dir, 60-85. *Mem:* AAAS; Am Asn Cancer Res; Am Soc Pharmacol & Exp Therapeut. *Res:* Nutrition and cancer; tissue metabolism; mode of action of carcinogenic agents; activity and mode of action of cancer chemotherapeutic agents; biochemical aspects of pharmacology; industrial toxicology; research and development management. *Mailing Add:* Woodland Rd North Hampton NH 03862

KENSON, ROBERT EARL, b Stoneham, Mass, Apr 15, 39; m 68. PHYSICAL CHEMISTRY. *Educ:* Boston Univ, AB, 61; Purdue Univ, PhD(phys chem), 65. *Prof Exp:* Sr res chemist, Olin Mathieson Chem Corp, 65-69 & Engelhard Minerals & Chem Corp, Newark, 69-74; sr proj scientist, TRC, The Res Corp New Eng, 74-78; mgr engineered systs div, Oxy-Catalyst Inc, 78-80; DEVELOP DIR, MET-PRO CORP, 80- *Mem:* Am Chem Soc; Am Inst Chem Eng; Catalysis Soc; Air Pollution Control Asn; Am Mgt Asn. *Res:* Catalysis; kinetics of gas and solution reactions; petroleum chemistry and petrochemicals; energy systems; environmental control. *Mailing Add:* 1125 E Cardinal Dr West Chester PA 19380

KENT, ALLEN, b New York, NY, Oct 24, 21; m 43; c 4. INFORMATION SCIENCE. *Educ:* City Col New York, BS, 42. *Prof Exp:* Res assoc info sci, Mass Inst Technol, 51-53; prin doc engr, Battelle Mem Inst, 53-55; assoc dir & prof ctr for doc & commun res, Western Reserve Univ, 55-63; dir off commun progs, 63-80, chmn, interdisciplinary dept info sci, 68-80, interim dean, 85-86, DISTINGUISHED SERV PROF, UNIV PITTSBURGH, 76- *Concurrent Pos:* Consult, Diebold, Inc, 62-72; consult info sci ment retardation, Spec Asst to Presidents Kennedy & Johnson, 63-64; coun mem, Ctr Res Libr, 66-86; chmn nat adv comt, Nat Inst Neurol Dis & Stroke, 67-70; mem bd dirs, Marcel Dekker Inc, 79- *Honors & Awards:* Info Technol Award, Eastman Kodak Co, 68; Award of Merit, Am Soc Info Sci, 76, Pioneer of Info Sci Award, 84. *Mem:* Am Soc Info Sci; fel AAAS; Asn Comput Mach; Am Libr Asn. *Res:* Quantitative studies of information transfer; modelling and simulation of library networks. *Mailing Add:* Rm 801 LIS Bldg Univ Pittsburgh Pittsburgh PA 15260

KENT, BARBARA, b Decatur, Ill, July 29, 40. PHYSIOLOGY, GENERAL MEDICAL SCIENCES. *Educ:* Emory Univ, BA, 62, MS, 64, PhD(physiol), 70. *Prof Exp:* res physiologist, Bronx Vet Admin Hosp, 70-83, dir, surg res, 72-83; DEPT PHYSIOL, MT SINAI SCH MED, NEW YORK, ASSOC PROF GERIAT & ADV DEVELOP, 83- *Concurrent Pos:* Investr, Mt Desert Island Biol Lab, 68-; res assoc prof surg, Mt Sinai Sch Med, NY, 71-, assoc prof physiol, 78; vis scientist, Jackson Lab, 82. *Mem:* Am Physiol Soc; NY Acad Sci. *Res:* Cardiovascular control systems; patho physiology of respiration; comparative cardiovascular physiology; aging. *Mailing Add:* Dept Physiol & Biophys Mt Sinai Sch Med One Gustave L Levy Pl New York NY 10029

KENT, BION H, b Utica, NY, Oct 19, 25. STRATIGRAPHY, COAL GEOLOGY. *Educ:* Cornell Univ, BA, 49; Stanford Univ, MS, 52. *Prof Exp:* Geologist, US Geol Surv, 49-86; RETIRED. *Mem:* Fel Geol Soc Am; Am Asn Petrol Geologists. *Mailing Add:* 13511 W Alaska Place Denver CO 80228

KENT, CLAUDIA, b South Bend, Ind, Oct 6, 45; m 81; c 2. BIOCHEMISTRY. *Educ:* St Mary's Col, Ind, BS, 67; Johns Hopkins Univ, PhD(biochem), 72. *Prof Exp:* Am Cancer Soc fel biochem, Dept Biol Chem, Sch Med, Wash Univ, 72-75; NIH fel, 74-75; from asst prof to assoc prof biochem, Purdue Univ, 75-85, prof biochem, 85-91; PROF BIOCHEM, UNIV MICH, 91- *Concurrent Pos:* SAC Biochem & Carcinogenesis, Am Cancer Soc; Physiol Chem Study Sect, NIH. *Mem:* Am Soc Biochem Molecular Biol; Am Soc Cell Biol. *Res:* Regulation of phospholipid metabolism. *Mailing Add:* Dept Biol Chem Univ Mich Sch Med Ann Arbor MI 48109-0606

KENT, CLEMENT F, b Charleston, SC, Mar 15, 27; m 48; c 3. THEORETICAL COMPUTER SCIENCE, LOGIC. *Educ:* Ga Inst Technol, BS, 48, MS, 50; Mass Inst Technol, PhD(math), 60. *Prof Exp:* Instr physics, Ga Inst Technol, 48-50; sci staff mem, Opers Eval Group, Mass Inst Technol, 51-62, assoc prof math, Case Western Reserve Univ, 62-68; chmn dept math, 68-72, 76-79 & 85-87, PROF MATH SCI, LAKEHEAD UNIV, 68 - *Concurrent Pos:* Vis prof, Univ Bristol, 72-73, Univ Laval, 79-80. *Mem:* Math Asn Am; Am Math Soc; Asn Symbolic Logic; Can Math Cong; Asn Comput Mach. *Res:* Mathematical logic; theoretical computer science; proof theory; recursive functions. *Mailing Add:* Dept Math Lakehead Univ Thunder Bay ON P7B 5E1 Can

KENT, CLIFFORD EUGENE, b Butler Co, Kans, Oct 11, 20; m 42; c 3. CHEMICAL ENGINEERING, ELECTROCHEMISTRY. *Educ:* Purdue Univ, BSChE, 42, BA(Theol), 80. *Prof Exp:* Pilot plant supvr, Merck & Co, NJ, 42-45; develop engr, Western Prod, Inc, 45-46; develop engr, Gen Elec Co, 46-49, design engr, 49-52, prog planning supvr, 52-56, mgr proj eng, 56-61, prog mgr advan fuel cell technol, 61-65, prog mgr electrochem eng, 65-67, mgr chem systs design, Nuclear Energy Div, 67-81; INTERIM RECTOR, ST PATRICK'S, KENWOOD, CALIF, 89- *Concurrent Pos:* Res placement counsellor, 81-82; assoc, St Andrew's, Saratoga, Calif, 83-86. *Mem:* Am Inst Chem Engrs; Am Chem Soc; Am Nuclear Soc; Electrochem Soc. *Res:* Fuel cells; radiochemical plants; equipment design; chemical and gas processes; nuclear plant effluent control; water treatment. *Mailing Add:* 5555 Montgomery Dr, No 64 Santa Rosa CA 95409

KENT, DENNIS V, b Prague, Czech, Nov 4, 46; m 71; c 1. PALEOMAGNETISM. *Educ:* City Col NY, BS, 68; Columbia Univ, PhD(geophys), 74. *Prof Exp:* Res assoc, 74-79, sr res scientist, 79-84, DOHERTY SR SCIENTIST, LAMONT-DOHERTY GEOL OBSERV, COLUMBIA UNIV, 84- *Concurrent Pos:* Adj assoc prof paleomagnetism, dept geol sci, Columbia Univ, 81-86, adj prof, 87-, assoc dir, 87-89, actg & interim dir, Lamont-Doherty Geol Observ, 88-90; guest prof, Inst Geophys, ETH, Zurich, 82. *Mem:* Fel Am Geophys Union; fel Geol Soc Am; AAAS. *Res:* Paleomagnetism and rock magnetism and their application to geomagnetic and geologic problems. *Mailing Add:* Lamont Doherty Geol Observ Columbia Univ Palisades NY 10964

KENT, DONALD MARTIN, b Medicine Hat, Alta, Can, Jan 25, 33; m 87; c 8. SCANNING ELECTRON MICROSCOPE, CARBONATE ROCKS. *Educ:* Univ Sask, BSc, 57, MSc, 59; Univ Alta, PhD(geol), 68. *Prof Exp:* Res geologist, Sask Dept Mineral Resources, 58-68, sr res geologist, 68-71; assoc prof, 71-77, PROF GEOL, UNIV REGINA, 77-, HEAD DEPT, 82- *Concurrent Pos:* Consult, Can Occidental Petrol, Union Oil Can, 76-81; pres, D M Kent Consult Geologist Ltd, 81-; assoc ed, Bull Can Petrol Geol, 86-88 & Geol Atlas of Western Can Sedimentary Basin, 89- *Mem:* Am Asn Petrol Geologists; Int Asn Sedimentologists; Can Soc Petrol Geologists; Geol Asn Can; Sask Geol Soc (pres, 65, 69 & 78). *Res:* Microfacies, diagenesis and nature of pore systems in carbonate hydrocarbon reservoir rocks; depositional settings of mixed carbonate-siliciclastic and carbonate-evaporite sequences; application of scanning electron microscope to diagenetic studies of carbonate rocks; paleotectonic controls on sedimentation; application of cathodo luminescence to carbonate diagenesis. *Mailing Add:* Dept Geol Univ Regina Regina SK S4S 0A2 Can

KENT, DONALD WETHERALD, JR, b Philadelphia, Pa, June 26, 26; m 58; c 3. PHYSICS. *Educ:* Yale Univ, BSc, 49; Temple Univ, PhD(physics), 60. *Prof Exp:* Researcher cosmic radiation, H H Wills Lab, Eng, 49-52; RES PHYSICIST, BARTOL RES FOUND, FRANKLIN INST, 52- *Mem:* Am Phys Soc; Am Geophys Union; Sigma Xi. *Res:* Cosmic ray physics. *Mailing Add:* Bartol Res Found Univ Del Newark DE 19711

KENT, DOUGLAS CHARLES, b Hastings, Nebr, Sept 26, 39; m 62; c 2. GEOLOGY. *Educ:* Univ Nebr, BSc, 61, MSc, 63; Iowa State Univ, PhD(water resources, geol), 69. *Prof Exp:* Prod geologist, Gulf Oil Corp, 64-66; instr, Iowa State Univ, 67-68, res assoc, 68-69; asst prof geol, 69-72, ASSOC PROF GEOL, OKLA STATE UNIV, 72- *Concurrent Pos:* Groundwater consult, 71- *Mem:* Geol Soc Am; Am Geophys Union; Am Water Resources Asn; Nat Water Well Asn; Sigma Xi. *Res:* Application of remote sensing to groundwater exploration and water resources; geochemistry of aquifers; application of mathematical modeling to groundwater management; stratigraphy; groundwater geology; applied geophysics and water resources. *Mailing Add:* Dept Geol Main Campus Okla State Univ Stillwater OK 74078

KENT, EARLE LEWIS, b Adrian, Tex, May 22, 10; m 35; c 2. ACOUSTICS, DATA PROCESSING. *Educ:* Kans State Univ, BS, 35, MS, 36; Univ Mich, Ann Arbor, PhD(elec eng), 52. *Prof Exp:* Instr elec eng, Armour Inst Technol, 36-40; res engr acoust, C G Conn, Ltd, Elkhart, Ind, 40-41, dir res, develop & design, 41-70; syst analyst & comput programmer data processing, Oaklawn Psychiat Ctr, 70-75; CONSULT ENGR, 75- *Mem:* Fel Acoust Soc Am; fel Audio Eng Soc; sr mem Inst Elec & Electronics Engrs; Catgut Acoust Soc. *Res:* Acoustics of wind musical instruments; acoustics of pianos; electronic organ technology. *Mailing Add:* 2510 Riverview Place Elkhart IN 46516

KENT, GEOFFREY, b Amsterdam, Holland, Jan 30, 14; nat US; m 44; c 4. PATHOLOGY. *Educ:* Univ Amsterdam, MD, 39; Univ Manchester, MSc, 44; Northwestern Univ, PhD(path), 57. *Prof Exp:* Chief asst hemat, Res Dept, Manchester Royal Infirmary, 40-43, asst dir, 43-44; pathologist, London Hosp, 47-50; sr pathologist, Cook County Hosp, Chicago, 53-56, assoc dir path, 56-57; from asst prof to prof path, Northwestern Univ, 60-84, chmn dept, 72-77, EMER PROF PATH, MED SCH, NORTHWESTERN UNIV, 84- *Concurrent Pos:* Chief pathologist, WSuburban Hosp, 58-69; chmn dept path, Chicago Wesley Mem Hosp, 69-72. *Mem:* AAAS; Am Soc Exp Path; Am Asn Pathologists; Col Am Path; Am Soc Cell Biol. *Res:* Iron metabolism; liver disease. *Mailing Add:* Northwestern Univ 303 Chicago Ave Chicago IL 60611

KENT, GEORGE CANTINE, JR, b Kingston, NY, July 25, 14; m 37; c 1. VERTEBRATE MORPHOLOGY. *Educ:* Maryville Col, AB, 37; Vanderbilt Univ, MA, 38, PhD(zool), 42. *Prof Exp:* From instr to prof comp anat, 42-67, chmn dept zool, 60-63, Alumni Prof, 67-79, EMER ALUMNI PROF ZOOL, LA STATE UNIV, 79- *Concurrent Pos:* Consult, Consult Bur, Comn Undergrad Educ in Biol Sci, Southern Asn Cols & Schs, 70-79; vchmn, Ctr Res in Col Instruct in Sci & Math, Tallahassee, 66-67; ed, Proc La Acad Sci, 47-55. *Mem:* Am Soc Zool; Soc Exp Biol & Med; Endocrine Soc; Sigma Xi. *Mailing Add:* 482 Stanford Ave Baton Rouge LA 70808

KENT, GORDON, b Pittsfield, Mass, Oct 1, 20; m 57; c 1. ELECTRICAL ENGINEERING. *Educ:* Univ Wis, BS, 47; Stanford Univ, MS, 49, PhD(elec eng), 52. *Prof Exp:* Res engr comput design, Inst Adv Study, 51-53; res fel electronics, Gordon McKay Lab, Harvard Univ, 53-57; assoc prof elec eng, Syracuse Univ, 57-63, prof, 63-85; CONSULT, 83- *Mem:* Inst Elec & Electronics Engrs. *Mailing Add:* 1995 Stanley Rd Cazenovia NY 13035

KENT, HARRY CHRISTISON, b Los Angeles, Calif, May 20, 30; m 56; c 2. GEOLOGY. *Educ:* Colo Sch Mines, Geol E, 52; Stanford Univ, MS, 53; Univ Colo, PhD(geol), 65. *Prof Exp:* Geologist, Calif Co, Fla & La, 53-56; from instr to assoc prof, 56-69, head dept geol, 69-75, prof, 69-89, EMER PROF GEOL, COLO SCH MINES, 89-, DIR, INST ENERGY RESOURCE STUDIES, POTENTIAL GAS AGENCY, 76- *Mem:* Am Asn Petrol Geol; Geol Soc Am; Soc Petrol Engrs. *Res:* Gas and oil resources; micropaleontology; marine geology. *Mailing Add:* 5131 Jellison Ct Arvada CO 80002-3257

KENT, HENRY PETER, radiology; deceased, see previous edition for last biography

KENT, JAMES RONALD FRASER, b Halifax, NS, Feb 29, 12; nat US. MATHEMATICAL ANALYSIS. *Educ:* Queen's Univ, Ont, BA, 33, MA, 34; Univ Ill, PhD(math), 47. *Prof Exp:* Asst math, Syracuse Univ, 34-35 & Univ Ill, 35-39; instr, Univ Ark, 39-42; asst prof math, Univ BC, 46-48; assoc prof in charge dept math, Triple Cities Col, Syracuse, 48-50; from assoc prof to prof, 50-79, chmn dept, 50-64, EMER PROF MATH, STATE UNIV NY, BINGHAMTON, 79- *Mem:* Am Math Soc; Math Asn Am; Can Math Soc; Sigma Xi. *Res:* Differential equations. *Mailing Add:* Two Martha Rd Vestal NY 13850

KENT, JOHN FRANKLIN, b Franklin, Ind, Apr 30, 21; m 42; c 3. ZOOLOGY. *Educ:* Franklin Col, AB, 41; Cornell Univ, PhD(zool), 49. *Prof Exp:* Asst zool, Cornell Univ, 41-42 & 45-49; from instr to asst prof anat, Univ Mich, 49-57; LUCRETIA L ALLYN PROF ZOOL, CONN COL, 57- *Mem:* AAAS; Am Asn Anat. *Res:* Ultrastructure; irradiation biology; hematology. *Mailing Add:* Four N Ridge Rd New London CT 06320-4115

KENT, JOSEPH C(HAN), b Victoria, BC, Jan 16, 22; nat US; m 52; c 3. CIVIL ENGINEERING. *Educ:* Univ BC, BS, 45; Stanford Univ, MS, 48; Univ Calif, PhD(fluid mech), 52. *Prof Exp:* Hydrographic surveyor, Dept Mines & Resources, Can, 45-47; asst civil eng, Univ Calif, 49-52; from instr to asst prof, 52-61, ASSOC PROF CIVIL ENG, UNIV WASH, 61- *Mem:* Am Soc Eng Educ. *Res:* Fluid mechanics specializing in waves; drag of submerged bodies and fluid flow. *Mailing Add:* Dept Civil Eng Univ Wash Seattle WA 98195

KENT, JOSEPH FRANCIS, b Richmond, Va, Feb 13, 44; m 66. TOPOLOGY. *Educ:* Univ Va, BA, 66, MA, 67, PhD(math), 70. *Prof Exp:* Asst prof math, Univ Fla, 70-73; asst prof, 73-80, ASSOC PROF MATH, UNIV RICHMOND, 80- *Mem:* Am Math Soc; Math Asn Am; Sigma Xi. *Res:* Topological dynamics and the study of ergodic flows; differentiability of norms on Banach spaces. *Mailing Add:* 15 Quail Run Dr Manakin-Sabot VA 23103

KENT, LOIS SCHOONOVER, b Marietta, Ohio, Dec 1, 12; m 43; c 1. GEOLOGY, PALEONTOLOGY. Educ: Oberlin Col, AB, 34; Cornell Univ, AM, 36; Bryn Mawr Col, PhD, 40. Prof Exp: Demonstr geol, Bryn Mawr Col, 36-40; jr geologist, Sect Metalliferous Deposits, US Geol Surv, 41-43, asst geologist, 43-45; ed asst, Geol Soc Am, 46; instr geol, Univ Ill, 54-55; from asst geologist to assoc geologist, Ill State Geol Surv, 56-85, cur, 56-85; RETIRED. Mem: Paleont Res Inst; Paleont Soc; Geol Soc Am; Soc Econ Paleont & Mineral. Res: Miocene mollusks of Maryland; Pennsylvanian fossils of Illinois. Mailing Add: 1003 Lincolnshire Dr Champaign IL 61821

KENT, RAYMOND D, b Red Lodge, Mont, Dec 21, 42. COMMUNICATIVE DISORDERS, SPEECH DEVELOPMENT. Educ: Univ Mont, BA, 65; Univ Iowa, MA, 69, PhD(speech path), 70. Prof Exp: Fel, Res Lab Electronics, Mass Inst Technol, 70-71; prof commun disorders, Univ Wis-Madison, 71-79; sr res assoc, Boys Town Inst Commun Disorders in Children, 79-; DEPT COMMUN DISORDERS, UNIV WIS-MADISON. Concurrent Pos: Ed, J Speech & Hearing Res, 77-81; prin investr, NIH grants, 73-; mem, Ad Hoc Adv Comt Commun Disorders Prog, Nat Inst Neurol & Commun Disorders & Stroke, 81- Mem: Am Speech-Language-Hearing Asn; Am Asn Phonetic Sci; Acoust Soc Am; Sigma Xi; NY Acad Sci. Res: Production and perception of speech, especially to speech development in children, neurologic speech and language disorders, and theories of speech production. Mailing Add: One Kingsbury Ct Madison WI 53711

KENT, RONALD ALLAN, b New York, NY, Feb 23, 35; m 60; c 2. CATALYSIS, INORGANIC CHEMISTRY. Educ: Cornell Univ, AB, 54. Prof Exp: Develop chemist catalytic chlorination, Gen Chem Res Lab, Allied Chem, 55-57; res assoc inorg sulfur & nitrogen, Dept Chem, Univ Pa, 57-68; res chemist platinum metal catalysis, Matthey Bishop Inc, 68-71; prin chemist heterogeneous catalysis, Dart Industs, 71-75; sr res assoc catalysis, 75-78, dir systs technol res & appl catalytic chem, 78-81, dir tech & develop, 81-83; dir com develop, Phillips Petrol, 83-85; tech mgr, Process Develop, PQ Corp, 86-87; PRES, TECHNOL NETWORK, 87- Concurrent Pos: Instr chem, Spring Garden Inst Technol, 60-61; chief chemist, Chem Info & Doc Serv, Univ Pa, 65-67. Mem: Am Chem Soc (secy treas, inorg sect, 70); NY Acad Sci; AAAS; Com Develop Asn; Catalysis Soc NAm. Res: Hydrogenation and isomerization catalysis; polymerization catalysis; oxide catalyst processes; ozone preparation and utilization; sulfur dioxide control; noble metal utilization and recovery; secondary copper recovery processes; polyolefins; silica products; composites for invivo use. Mailing Add: 314 SW 297th Federal Way WA 98063

KENT, STEPHEN BRIAN HENRY, b Wellington, NZ, Dec 12, 45; c 3. BIOLOGICAL CHEMISTRY, PROTEIN CHEMISTRY. Educ: Victoria Univ, Wellington, BSc, 68; Massey Univ, NZ, MSc, 70; Univ Calif, Berkeley, PhD(org chem), 75. Prof Exp: Res assoc biochem, Rockefeller Univ, 74-77, asst prof, 77-81; dir, protein chem, Molecular Genetics, Inc, 82; sr res assoc biol div, Calif Inst Tech, 83-89; MEM STAFF, BOND UNIV, AUSTRALIA, 89- Mem: NY Acad Sci; Am Chem Soc; Harvey Soc; Protein Soc; AAAS. Res: Development of methodology in protein chemistry and its application to biological problems involving the relationship of protein structure to function. Mailing Add: Bond Univ Grad Sch Sci & Technol Gold Coast 4229 Queensland Australia

KENT, STEPHEN MATTHEW, b West Orange, NJ, Dec 2, 52. EXTRAGALACTIC ASTRONOMY. Educ: Mass Inst Technol, BS, 74; Calif Tech, PhD(astron), 80. Prof Exp: Res fel astron, Ctr Astrophys, 79-81; teaching fel astron, Mass Inst Technol, 81-83; ASST PROF ASTRON, HARVARD UNIV, 83- Mem: Am Astron Soc; Int Astron Union. Res: Extragalactic astronomy; internal dynamics of galaxies and clusters of galaxies. Mailing Add: Seven Bacon Arlington MA 02174

KENT, THOMAS HUGH, b Iowa City, Iowa, Aug 17, 34; m 57; c 3. PATHOLOGY. Educ: Univ Iowa, BA, 56, MD, 59; Am Bd Path, dipl, 65. Prof Exp: Intern, Methodist Hosp, Indianapolis, 59-60; resident path, Univ Iowa, 60-64; assoc pathologist, Walter Reed Army Inst Res, 64-66; from asst prof to assoc prof, 66-72, PROF PATH, COL MED, UNIV IOWA, 72- Mem: AMA; Col Am Path; Am Asn Pathologists & Bacteriologists; Am Soc Clin Path. Res: Gastroenterology. Mailing Add: Univ Iowa Col Med Iowa City IA 52240

KENTFIELD, JOHN ALAN CHARLES, b Hitchin, Eng, Mar 4, 30; m 66. MECHANICAL ENGINEERING. Educ: Univ Southampton, BSc, 59; Univ London, DIC & PhD(mech eng), 63. Prof Exp: Trainee, C V A Kearney & Trecker Ltd, Eng, 50-52; asst tester, Ricardo & Co Ltd, 52-56; asst lectr res & educ, Imp Col, Univ London, 62-63; proj engr, Curtiss-Wright Corp, 63-66; lectr res & educ, Imp Col, Univ London, 66-70; assoc prof, 70-79, PROF ENG, UNIV CALGARY, 81- Concurrent Pos: Nat Res Coun grant, Univ Calgary, 71-72, operating grant, 77 & 80 & Nat Sci & Eng Rec Coun Can grant, 83, 86 & 89; Killam Resident fel, Univ Calgary, 80; mem Tech Adv Comt Wind Energy, 86-88, Adv Comt Propulsion, Nat Res Coun, 84-90. Mem: Am Inst Aeronaut & Astronaut; Am Soc Mech Engrs; Brit Inst Mech Engrs; Am Wind Energy Asn. Res: Non-steady flow of compressible fluid and the application of non-steady flow phenomena in engineering equipment, such as pressure exchangers and pulsating combustors; wind-turbines and wind-energy systems; author of one hundred and twenty research publications. Mailing Add: Dept Mech Eng Fac Eng Univ Calgary Calgary AB T2N 1N4 Can

KENTZER, CZESLAW P(AWEL), b Poland, June 29, 25; nat US; m 58; c 3. GAS DYNAMICS, COMPUTATIONAL FLUID MECHANICS. Educ: San Diego State Col, BS, 52; Purdue Univ, MS, 54, PhD(aerospace eng), 58. Prof Exp: Asst prof, 58-60, ASSOC PROF AERODYN, PURDUE UNIV, 60- Concurrent Pos: Consult, Missiles & Space Systs Div, Douglas Aircraft Corp. Mem: Am Math Soc; Am Acad Mech; Am Inst Aeronaut & Astronaut. Res: Fluid mechanics; transonic aerodynamics; acoustics; nonlinear waves; turbulence theories; computational fluid mechanics; geophysical fluid mechanics. Mailing Add: Dept Aeronaut Purdue Univ Main Campus West Lafayette IN 47907

KENWORTHY, ALVIN LAWRENCE, b Shiatook, Okla, Sept 6, 15; m 37; c 2. HORTICULTURE. Educ: Okla State Univ, BS, 37; Kansas State Univ, MS, 39; Wash State Univ, PhD(hort), 48. Prof Exp: Asst prof hort, Univ Fla, 43-44; assoc prof hort, Univ Del, 45-47; assoc prof, Mich State Univ, 48-52, prof, 52-80; CONSULT, 80- Mem: Am Soc Hort Sci. Res: Nutrient element requirements of fruit trees; nutrient element interrelationships and balance; spectrographic plant analysis; soil management; irrigation and growth chemicals. Mailing Add: 4647 Meridian Williamston MI 48895

KENYON, ALAN J, b Whitehall, Wis, Sept 10, 29; m 54; c 3. BIOCHEMISTRY, IMMUNOLOGY. Educ: Univ Minn, BS, 54, DVM, 57, PhD(bact & biochem), 61. Prof Exp: From assoc prof to prof biochem, Univ Conn, 61-73; prof biol, Med Sch, Cornell Univ, 73-80; MEM, WALKER LAB, SLOAN-KETTERING INST CANCER RES, 73- Concurrent Pos: NIH res grants, 63- & career develop award, 65-75; consult, Manned Spacecraft Ctr, Apollo Prog, NASA, 68-; Path Dept, Hartford Hosp, 70- & Res Inst, Ill Inst Technol, 71-; sabbatical leave pediat & path, Sch Med, Univ Minn, Minneapolis, 71-72. Mem: AAAS; NY Acad Sci; Am Soc Exp Path; Reticuloendothelial Soc; Am Asn Lab Animal Sci. Res: Immunochemistry as applied to lymphoproliferative diseases and neoplasias of man and animals; comparative biochemistry of Mustelidae; diseases of marine mammals; immunological deficiency. Mailing Add: 822 The Parkway Mamaroneck NY 10543

KENYON, ALLEN STEWART, b Constance, Ky, Mar 6, 16; m 42; c 4. POLYMER SCIENCE. Educ: Univ Ky, BS, 38, MS, 39; Columbia Univ, PhD(phys chem), 47. Prof Exp: Asst phys chem, Univ Ky, 38-39, instr chem, 39-40; asst, Columbia Univ, 40-42; res chemist, 47-58, group leader, 58-69, sr res specialist, 69-80, fel, Monsanto Co, 80-85; instr physics, St Louis Univ, 85-88; CONSULT, PHARMACEUT ANALYSIS, FOOD & DRUG ADMIN, 89- Concurrent Pos: Asst, Nat Defense Res Comt Proj, Columbia Univ, 41-42. Mem: Am Chem Soc; Sigma Xi. Res: Mechanical properties; light scattering on high polymers; nondispersed sulfur sols; higher order Tyndall spectra; physical chemistry, structure and property relations of polymers; polymer characterization; chemistry of interfaces in composite materials; separating of polymers; instrumental analysis; pharmaceutical analysis. Mailing Add: 1318 Lindgate Kirkwood MO 63122

KENYON, GEORGE LOMMEL, b Wilmington, Del, Aug 29, 39; m 81; c 1. BIOCHEMISTRY. Educ: Bucknell Univ, BS, 61; Harvard Univ, MA, 63, PhD(org chem), 65. Prof Exp: Fel biochem, Mass Inst Technol, 65-66; asst prof chem, Univ Calif, Berkeley, 66-72; asst prof pharm chem, 72-74, assoc prof, 74-77, PROF PHARM CHEM, UNIV CALIF, SAN FRANCISCO, 77- Honors & Awards: Merit Award, NIH, 86. Mem: Fel AAAS; Am Chem Soc; Am Soc Biol Chemists; fel NY Acad Sci. Res: Enzyme mechanisms; organophosphorus chemistry; bio-organic chemistry of nucleotides; application of nuclear magnetic resonance spectroscopy to biological problems; design of reagents for protein modification; design of specific enzyme inhibitors. Mailing Add: Dept Pharmaceut Chem Univ Calif San Francisco CA 94143

KENYON, HEWITT, b Marysville, Calif, Aug 31, 20; m 47, 61; c 5. MATHEMATICS. Educ: Univ Calif, BS, 42, PhD(math), 54. Prof Exp: From instr to asst prof math, Univ Rochester, 52-61; from asst prof to assoc prof, 61-67, chmn dept, 67-71, PROF MATH, GEORGE WASHINGTON UNIV, 67- Concurrent Pos: Vis assoc prof, Univ Calif, Berkeley, 66-67. Mem: Am Math Soc; Math Asn Am; Sigma Xi. Res: Convergence in topology; differentiation of set functions. Mailing Add: 1611 Kennedy Pl NW Washington DC 20011

KENYON, KERN ELLSWORTH, b Kansas City, Mo, May 24, 38; m 66; c 2. PHYSICAL OCEANOGRAPHY. Educ: Mass Inst Technol, BS & MS, 61; Scripps Inst Oceanog, PhD(oceanog), 66. Prof Exp: Asst prof, Grad Sch Oceanog, Univ RI, 67-73; ASST RES OCEANOGR, SCRIPPS INST OCEANOG, 73- Mem: Am Geophys Union; AAAS; Oceanog Soc Japan; Sigma Xi. Res: Large-scale ocean circulation and air sea interaction; wave-wave and wave-current interactions. Mailing Add: 4632 North Lane Del Mar CA 92014

KENYON, RICHARD A(LBERT), b Syracuse, NY, Apr 8, 33; m 54; c 3. MECHANICAL ENGINEERING, THERMODYNAMICS. Educ: Clarkson Col Technol, BME, 54; Cornell Univ, MS, 56; Syracuse Univ, PhD(mech eng), 65. Prof Exp: Instr mech eng, Cornell Univ, 54-56; asst prof, Clarkson Col Technol, 56-62, assoc prof, 62-70, assoc dean grad sch & assoc dir, Div Res, 66-68, chmn dept & exec officer, 68-70; head dept mech eng, 70-72, PROF MECH ENG, ROCHESTER INST TECHNOL, 70-, DEAN COL ENG, 71- Mem: Fel Am Soc Mech Engrs (vpres, 78-84); Am Soc Eng Educ; Nat Soc Prof Engrs; Sigma Xi. Res: Technology transfer from the academic and private sectors to local government; solid waste engineering and resource recovery. Mailing Add: Dept Eng Rochester Inst Technol Rochester NY 14623

KENYON, RICHARD H, b Blakely, Pa, Nov 22, 42; m 68. VIROLOGY, IMMUNOLOGY. Educ: Bucknell Univ, BS, 64; Pa State Univ, MS, 66, PhD(microbiol), 68. Prof Exp: Virologist-immunologist, US Army Biol Labs, Ft Detrick, 69-71; Virol Lab, US Army Med Res Inst Infectious Dis, 71, rickettsiologist-immunologist, Virol Lab, 71-77, ASST CHIEF, RICKETTSIOLOGY DIV, US ARMY MED RES INST INFECTIOUS DIS, 78-, VIROL DIV, 80- & DIS ASSESSMENT DIV, 88- Res: Vaccine development; Rocky Mountain spotted fever; Rickettsiae; Junn virus; Argentine hemorrliagic fever. Mailing Add: US Army Med Res Int Infect Dis Ft Detrick Frederick MD 21701

KENYON, RICHARD R(EID), b Middletown, Ohio, Oct 6, 28. ELECTRICAL ENGINEERING, COMPUTER SCIENCE. Educ: Purdue Univ, BS, 50, MS, 51, PhD(elec eng), 61. Prof Exp: Mem tech staff, Bell Tel Labs, NJ, 51-55; res asst comput sci, Purdue Univ, 55-58, instr, 58-61, asst prof, 61-65; consult comput sci, McDonnell Douglas Astronaut Co, 80-86;

PVT CONSULT, 86- *Mem:* Asn Comput Mach; Inst Elec & Electronics Engrs; Sigma Xi. *Res:* Circuit theory; numerical methods of approximation; computer architecture, programming languages. *Mailing Add:* 17781 Crestmoor Ln Huntington Beach CA 92649

KENYON, STEPHEN C, b Coronado, Calif, May 9, 48; m 70; c 1. DIGITAL SIGNAL PROCESSING, NEURAL NETWORKS. *Educ:* Va Polytech Inst, BS, 71. *Prof Exp:* Elec engr, US Army, 71-75; chief eng, Ensco Inc, 75-83; vpres eng, Digital Signal Corp, 83-90; PRES, CREATIVE ENG CONCEPTS, INC, 90- *Mem:* Inst Elec & Electronics Engrs Systs Man & Cybernetics Soc; Inst Elec & Electronics Engrs Control Systs Soc; Inst Elec & Electronics Engrs Acoust Speech & Signal Processing Soc; Inst Elec & Electronics Engrs Comput Soc; Inst Elec & Electronics Engrs Commun Soc; Int Neural Network Soc. *Res:* Research and Development in electronic, optical and software signal processing; intelligent sensors, pattern recognition and machine intelligence; adaptive control for autonomous systems; three United States patents. *Mailing Add:* 230 Tacketts Mill Rd Stafford VA 22554

KEOGH, FRANK RICHARD, mathematics, for more information see previous edition

KEOGH, MICHAEL JOHN, b Bronx, NY, May 26, 37; m 75; c 3. ORGANIC CHEMISTRY, POLYMER CHEMISTRY. *Educ:* Manhattan Col, BS, 59; Purdue Univ, PhD(org chem), 63. *Prof Exp:* Chemist, 63-67, proj scientist, 67-70, res scientist, 70-78, sr res scientist, 78-82, res assoc, Polyolefins Div, 82-88, CORP RES FEL, UNION CARBIDE CORP, 88- *Mem:* Am Chem Soc; Fire Retardant Chem Asn. *Res:* Synthetic, organic and polymer chemistry; fluorocarbons; condensation monomer synthesis; organometallic and anionic polymerization systems; epoxide and other thermosetting polymerization systems; polymers engineered for pollution control; wire and cable technology; flame retardant technology. *Mailing Add:* Res & Develop Dept Union Carbide Corp One Riverview Dr Somerset NJ 08873

KEOGH, RICHARD NEIL, b Nashua, NH, Apr 21, 40. CELL BIOLOGY. *Educ:* Tufts Univ, BS, 62; Brown Univ, PhD(biol), 67. *Prof Exp:* From asst prof to assoc prof, 67-77, PROF BIOL, RI COL, 77- *Mem:* AAAS; Am Asn Biol Teachers; Am Inst Biol Sci; Sigma Xi. *Res:* Mammalian pigment cell biology; teaching of biology via television; multimedia methods of instruction. *Mailing Add:* Off Res & Grants Admin RI Col Providence RI 02908

KEON, WILBERT JOSEPH, b Sheenboro, Que, May 17, 35; m 60; c 3. CARDIOVASCULAR SURGERY, DISEASE COSTING. *Educ:* St Pat's Col, Ottawa, BSc, 57; Univ Ottawa, MD, 61; McGill Univ, MSc, 63; FRCPS(C), 66. *Prof Exp:* Assoc prof surg, Univ Ottawa, 69-76, dir, Cardiac Unit, 69-83, prof surg & chmn dept, 76-91, CHMN, DIV CARDIOVASC & THORACIC SURG, UNIV OTTAWA, 69-, DIR-GEN, HEART INST, 83-; CHIEF, DIV CARDIOTHORACIC SURG, OTTAWA CIVIC HOSP, 69- *Concurrent Pos:* Sr fel, Ont Heart Found, 70-76; surg fel, James IV Asn Surgeons, 79; vchmn, med adv bd, Can Heart Found, 79-80, chmn, 80-83, med vpres, 83-85; assoc ed, Can J Cardiol, 84-; mem bd dirs, Transplant Int Inc; surgeon-in-chief, Ottawa Civic Hosp, 77-83; vpres, Med Res Coun Can, 85-90; sen, govt Can, 90. *Honors & Awards:* Officer, Order of Can, Can Govt, 85; Hippocrates Award, Am Hellenic Educ Progressive Asn, 85. *Mem:* Am Surg Soc; Am Asn Thoracic Surg; Int Asn Cardiac Biol Implants; Int Cardiovasc Soc; Can Cardiovasc Soc (pres, 88-89); Soc Cardiothoracic Surgeons. *Res:* Surgical treatment in the presence of acute myocardial infarction; cardiac muscle mechanics; lasers for treatment of coronary artery disease; artificial heart as a bridge to transplantation. *Mailing Add:* Dept Surg Rm 111 CPC Ottawa Civic Hosp 1053 Carling Ave Ottawa ON K1Y 4E9 Can

KEONJIAN, EDWARD, b Tiflis, Russia, Aug 14, 09; US citizen; m 36; c 1. ELECTRONICS, SOLID STATE PHYSICS. *Educ:* Leningrad Inst, BS & MS, 32, PhD, 33. *Prof Exp:* Res engr electronics, Leningrad Cent Radio Lab, 32-36; scientist, Leningrad Res Inst Electronics, 36-40; asst prof elec commun, Leningrad Inst Elec Eng, 40-42; lectr elec commun, City Col New York, 49-51; electron develop engr, Gen Elec Co, Syracuse, NY, 51-57; solid circuit develop engr, Am Bosch Arma Corp, Garden City, NY, 57-63; staff scientist, Grumman Aircraft Corp, Bethpage, NY, 63-64, eng consult, 64-73; chief, Microelectronics Failure Anal, UN Develop Proj, India, 73-74, eng expert & chief tech adv, 75-76; ENG CONSULT, 77- *Concurrent Pos:* US mem & dir lectr series, Adv Group Aeronaut Res & Develop, NATO, 62-74; NSF vis prof, Carro Univ, 77; chmn various tech comts & distinguished colleague, Aerospace Indust Asn Am. *Mem:* Fel Inst Elec & Electronics Engrs; Explorers Club; Circumnavigators Club; Aerospace Inst Asn. *Res:* Research and development of solid state technology. *Mailing Add:* 40 Stoner Ave Great Neck NY 11021

KEOUGH, ALLEN HENRY, b Chelsea, Mass, Apr 24, 29; m 52; c 6. ORGANIC CHEMISTRY. *Educ:* Univ Mass, BS, 50; Univ NH, MS, 52; Mass Inst Technol, PhD(org chem), 56. *Prof Exp:* Res chemist, Johnson & Johnson, 55-58; sr res engr, Explor Res Div, Norton Co, Worcester, 58-62, res assoc, 62, asst dir res & develop, 62-63; res assoc, Nat Res Corp, 63-66, asst dir res, 66-68; pres, Chem-Tech Assocs, Inc, Mass, 68-71; sect head, Res Div, 71-74, sect head advan develop div, Dennison Mfg Co, 74-78; pres, Design Cote Co, 78-81; tech dir, Metall Prods Div, Household Mfg, Inc, 81-86; TECH DIR, METALLIZED PRODS INC, 86- *Concurrent Pos:* Chmn, Radiation Curing Div, Asn Finishing Processes, Soc Mfg Engrs, 85-86; bd dirs, Rad Tech Int, 86, 87-88, pres, 86; bd dir, Vitronics Corp, 89- *Mem:* AAAS; Am Chem Soc; Asn Finishing Processes Soc Mfg Engrs. *Res:* Synthetic organic chemistry; organometallic compounds; heterogeneous catalysis surface active agents; polymer chemistry; radiation curing printing inks and coatings. *Mailing Add:* 53 Patricia Rd Sudbury MA 01776

KEOUGH, KEVIN MICHAEL WILLIAM, b St George's, Nfld, Aug 2, 43; m 67; c 2. BIOCHEMISTRY. *Educ:* Univ Toronto, BSc, 65, MSc, 67, PhD(biochem), 71. *Prof Exp:* Muscular Dystrophy Asn Can fel phys biochem, Univ Sheffield, 71-72; from asst prof to assoc prof biochem, 72-82, assoc prof pediat, 80-82, PROF BIOCHEM & PEDIAT, MEM UNIV NFLD, 82-, HEAD, DEPT BIOCHEM, 86- *Concurrent Pos:* Pres, Can Fed Biol Soc, 90-91. *Mem:* Can Biochem Soc (pres 88-89); Can Lung Asn; Biochem Soc; Am Soc Biol Chemists; Am Chem Soc; Biophys Soc; Sigma Xi; Biophys Soc Can; Am Oil Chem Soc. *Res:* Molecular organization in membranes and lung surfactant; synthetic lung surfactant; liposomal biotechnology. *Mailing Add:* Dept Biochem Mem Univ Nfld St John's NF A1B 3X9 Can

KEOWN, ERNEST RAY, b Thurber, Tex, Mar 17, 21; m 43; c 2. APPLIED MATHEMATICS. *Educ:* Univ Tex, BS, 46; Mass Inst Technol, PhD(math), 50; Univ Ark, JD, 81. *Prof Exp:* Sr aerophysics engr, Consol Vultee Aircraft Co, 51-52; asst prof math, Tex A&M, 52-57; comput specialist, Douglas Aircraft Co, 57-59; tech specialist, Aerojet Gen Corp Div, Gen Tire & Rubber Co, 59-60; prof math, Tex A&M Univ, 60-67; PROF MATH, UNIV ARK, FAYETTEVILLE, 67- *Concurrent Pos:* Consult, AEC, Standard Oil Co, Tex & Magnolia Petrol Co; mem solid state & molecular theory group, Mass Inst Technol, 63-64. *Mem:* Am Math Soc. *Res:* Group representation theory; applications of group representation theory in physics; numerical analysis; computers and law; gauge theory. *Mailing Add:* Dept Math SE 301 Univ Ark Fayetteville AR 72701

KEOWN, KENNETH K, anesthesiology; deceased, see previous edition for last biography

KEOWN, ROBERT WILLIAM, b Louisville, Ky, Apr 23, 29; m 49; c 3. PACKAGING CHEMISTRY, FOOD CHEMISTRY. *Educ:* Univ Louisville, BS, 51, MS, 52, PhD(chem), 54. *Prof Exp:* Res chemist, E I Du Pont de Nemours & Co, Inc, 54-71, tech assoc, 71-74, supvr, Adhesives & Fluids Div, 74-81, res assoc, 81-82, sr res assoc, 82-85; prof, 85-87, PROF & ACTG CHAIR, DEPT FOOD SCI, UNIV DEL, 87- *Mem:* Am Chem Soc; AAAS; Adhesion Soc; Inst Food Technologists; Soc of Pack. *Res:* Polymer chemistry; food interactions with polymeric packaging materials, specifically flavor loss and adhesion effects. *Mailing Add:* 724 Foulkslone Wilmington DE 19803-2226

KEPECS, JOSEPH GOODMAN, b Philadelphia, Pa, Oct 8, 12; m 44; c 2. PSYCHIATRY, PSYCHOANALYSIS. *Educ:* Univ Chicago, BS, 35, MD, 37; Chicago Inst Psychoanal, cert, 49. *Prof Exp:* Pvt practr, 46-65; PROF PSYCHIAT, MED SCH, UNIV WIS-MADISON, 65- *Concurrent Pos:* Vis lectr, Univ Cincinnati, 56; lectr, Chicago Inst Psychoanal, 57-60; prof lectr, Univ Chicago, 60-65; consult, Univ Wis, 60-65. *Mem:* Am Psychiat Asn; Am Psychoanal Asn; Am Psychosom Soc. *Res:* Applications of psychiatry to medicine; sociological studies of changes in therapists and patients; psychiatry in developing countries. *Mailing Add:* 600 Highland Ave Madison WI 53792

KEPES, JOHN J, b Budapest, Hungary, Mar 31, 28; US citizen; m 50; c 1. PATHOLOGY, NEUROPATHOLOGY. *Educ:* Univ Budapest, MD, 52. *Prof Exp:* Pathologist-in-chief, Nat Inst Neurosurg, Hungary, 54-56; from asst prof to assoc prof, 60-68, PROF PATH, UNIV KANS MED CTR, KANSAS CITY, 68- *Concurrent Pos:* Spec fel neuropath, Mayo Found, Univ Minn, 57-58; consult, Vet Admin Hosp, Kansas City, 60-; vis prof, Neurol Inst, Univ Vienna, 68-69. *Mem:* Am Acad Neurol; Am Asn Neuropathol (vpres, 78-79, pres, 84-85); Am Asn Neurol Surgeons. *Res:* Histological differential diagnosis of brain tumors; electron microscopic studies of meningiomas; spinal cord circulation; primary malignant lymphomas of central nervous system; histiocytosis and xanthosarcomas of central nervous system; pathogenesis of central pontine myelinolysis; etiology and pathogenesis of the Arnold-Chiari malformation. *Mailing Add:* Univ Kans Med Ctr 39th St & Rainbow Blvd Kansas City KS 66103

KEPES, JOSEPH JOHN, b Cleveland, Ohio, Jan 25, 31; m 54; c 6. NUCLEAR & REACTOR PHYSICS. *Educ:* Case Inst Technol, BS, 53; Univ Notre Dame, PhD(nuclear physics), 58. *Prof Exp:* Sr scientist, Bettis Atomic Power Lab, 57-62; assoc prof, 62-71, chmn dept, 62-75, PROF PHYSICS, UNIV DAYTON, 71- *Mem:* Am Phys Soc; Am Asn Physics Teachers. *Res:* Electron-electron scattering at low energies; resonance escape probabilities in natural uranium plates. *Mailing Add:* 4741 Ackerman Blvd Dayton OH 45429

KEPHART, ROBERT DAVID, b Phillipsburg, Pa, Nov 27, 49; m 71; c 2. HIGH ENERGY PHYSICS, CRYOGENICS. *Educ:* Va Polytech Inst & State Univ, BS, 71; State Univ NY, Stony Brook, MS, 73, PhD(physics), 75. *Prof Exp:* Res asst physics, Dept Physics, Va Polytech Inst & State Univ, 69-71; res asst, Dept Physics, State Univ NY, Stony Brook, 71-75, fel, 75-77; staff physicist & group leader, 77-88, CDF DEPT HEAD, FERMI NAT ACCELERATOR LAB, 88- *Concurrent Pos:* Proj mgr, Chicago Cyclotron Super Conducting Magnet Proj & leader, Fermilab Superconducting Analysis Magnetic Group; proj physicist, Collider Defector Facil, 3M0X5M Superconditioning Solenoid Proj. *Mem:* Am Phys Soc. *Res:* Detector development; liquid argon calorimetry; super conducting magnet design and research and development; high purity gas systems; dimuon high energy physics; engineering physics; p collider physics; super conducting solenoid. *Mailing Add:* CDF Dept MS 318 PO Box 500 Batavia IL 60510

KEPLER, CAROL R, b Berea, Ohio, Oct 21, 37; m 59; c 2. LIPID CHEMISTRY. *Educ:* Oberlin Col, AB, 59; Univ NC, PhD(zool), 65. *Prof Exp:* Asst prof nutrit, NC State Univ, 65-66, asst prof biochem, 66-81; asst prof, Meredith Col, 81-84; ASST PURCHASING ADMINR, STATE NC, 84- *Concurrent Pos:* NIH fel, 65-66. *Mem:* Sigma Xi; Am Soc Zoologists. *Res:* Rumen bacteria; hydrogenation and isomerization of unsaturated fatty acids; specificity of triglyceride synthesis. *Mailing Add:* Dept Administration State NC 116 W Jones St Raleigh NC 27603-8002

KEPLER, HAROLD B(ENTON), b Dayton, Ohio, Jan 3, 22; m 46; c 2. MECHANICAL ENGINEERING. *Educ:* Sinclair Col, AEA, 53, BBA, 56; Xavier Univ, Ohio, MBA, 58. *Prof Exp:* Eng draftsman, Eng Div, Wright Field, Ohio, 40-42, design checker, Aircraft Lab, Wright-Patterson AFB,

46-47; prod designer, Ohmer Corp, 47-48; from instr to assoc prof mech eng, 48-77, EMER PROF, USAF INST TECHNOL, 77- *Concurrent Pos:* Lectr, Sinclair Col, Ohio, 60- & Univ Dayton, 84- *Mem:* Am Soc Eng Educ. *Res:* Engineering graphics; mechanisms and reliability engineering. *Mailing Add:* 90 Sheldon Dr Centerville OH 45459

KEPLER, RAYMOND GLEN, b Long Beach, Calif, Sept 10, 28; m 53; c 4. EXPERIMENTAL SOLID STATE PHYSICS. *Educ:* Stanford Univ, BS, 50; Univ Calif, MS, 55, PhD(physics), 57. *Prof Exp:* Res physicist, Cent Res Dept, E I du Pont de Nemours & Co, 57-64; div supvr, 64-69, DEPT MGR, SANDIA NAT LABS, 69- *Concurrent Pos:* Mem, Solid State Sci Panel, Nat Acad Sci, 77-82; mem, comt educ, Am Phys Soc, 78-80, chmn, 79-80, applns physics comt, 79-81; mem, Eval Panel Mat Sci, Nat Bur Standards, 82-88; vchmn, Panel 2 Comt on Mat Sci & Eng, Nat Acad Sci-Nat Res Coun. *Mem:* AAAS; fel Am Phys Soc. *Res:* Photoconductivity; conductivity; excitons and other solid state properties, primarily of organic solids; piezoelectricity, pyroelectricity and ferroelectricity in polymers. *Mailing Add:* 404 Turner Dr NE Albuquerque NM 87123

KEPLINGER, MORENO LAVON, b Ulysses, Kans, May 25, 29; m 50; c 3. TOXICOLOGY. *Educ:* Univ Kans, BS, 51, MS, 52; Northwestern Univ, PhD(pharmacol), 56; Am Bd Toxicol, dipl, 80. *Prof Exp:* Res instr pharmacol, Univ Miami, 56-59, asst prof, 59-60; toxicologist, Hercules Powder Co, 60-64; pharmacol, Univ Miami, 64-68; asst dir, Indust Bio-Test Labs, Inc, 68-70, mgr toxicol, 70-77; CONSULT TOXICOL, 78- *Mem:* Europ Soc Toxicol; Soc Toxicol; Am Soc Pharmacol & Exp Therapeut; Am Indust Hyg Asn. *Res:* Experimental and industrial toxicology; pharmacology. *Mailing Add:* Keplinger PO Box 1299 Hilltop Lakes TX 77871-1299

KEPLINGER, ORIN CLAWSON, b Carlinville, Ill, Oct 7, 18; m 39; c 3. POLYMER CHEMISTRY, EXPLOSIVES. *Educ:* Southern Ill Univ, BEd, 40; Univ Ill, MS, 48, PhD(chem), 49. *Prof Exp:* From jr chemist to chemist explosives, Western Cartridge Co, 40-44; spec asst, Univ Ill, 46-49; sr res chemist, Polymers, Gen Tire & Rubber Co, 49-59; chem supvr, Sherwin-Williams Co, 59-64, dir varnish resin lab, 64-66, dir prod develop, 66-75, dir resin technol, 75-78; tech dir, Chem Coatings Div, Valspar, 79-81; RETIRED. *Concurrent Pos:* Pres, Paint Res Inst, 75-81. *Mem:* Am Chem Soc; Fedn Soc Coatings Technol; Soc Chem & Indust London. *Res:* Explosives process development; synthetic rubber and plastics preparation; resins and polymers for coatings; paint formulation. *Mailing Add:* Laurel Lake Dr No 366 Hudson OH 44236

KEPNER, RICHARD EDWIN, b Los Angeles, Calif, July 27, 16; m 46; c 2. ORGANIC CHEMISTRY. *Educ:* Univ Calif, BS, 38; Univ Calif, Los Angeles, MA, 42, PhD(org chem), 46. *Prof Exp:* Lab asst soil chem, Univ Calif, Los Angeles, 40-41, asst chem, 43-46; from assoc prof to prof, 46-86, EMER PROF CHEM, UNIV CALIF, DAVIS, 86- *Mem:* Am Chem Soc; Am Soc Enol; Phytochem Soc NAm. *Res:* Odor and flavor constituents of fruits and wines; grape pigments; volatile plant terpenes. *Mailing Add:* Dept Chem Univ Calif Davis CA 95616

KEPPER, JOHN C, b Baltimore, Md, Nov 7, 32; m 60; c 2. SEDIMENTARY PETROLOGY, STRATIGRAPHY. *Educ:* Franklin & Marshall Col, BS, 55; Univ Wash, MS, 60, PhD(geol), 69. *Prof Exp:* Instr geol, Cabrillo Col, 61-67; asst prof geol, State Univ NY Col Oneonta, 69-73; assoc prof geol, Univ Nev, Las Vegas, 73-80 & Reno, 80-81; explor researcher, Conoco Minerals Div, 81-82, Conoco Petrol, 82-83, Noranda Explor, 83-84; LECTR GEOL, UNIV NEV, RENO, 84- *Concurrent Pos:* Consult mineral deposits & remote sensing appln to explor. *Mem:* Soc Econ Paleont & Mineral; Am Asn Petrol Geol; Geol Soc Am. *Res:* Sedimentary petrology, especially carbonate petrology; regional stratigraphy of the Cambrian rocks in the Great Basin Region. *Mailing Add:* 512 Tara Ct Boulder City NV 89005

KEPPER, ROBERT EDGAR, b Charlestown, Mass, Apr 8, 35; m 60; c 3. AGRICULTURAL & FOOD CHEMISTRY. *Educ:* Univ Mass, Amherst, BS, 57. *Prof Exp:* Plant qual supvr, RT French Co, 60-66; chief qual control, Wm Underwood Co, 66-72; corp mgr qual control, 72-80, DIR QUAL, OCEAN SPRAY CRANBERRIES INC, 80- *Concurrent Pos:* Mem Nutrit Bd, Commonwealth Mass, 78-84; nat counr, Food, Drugs & Cosmetic Div, Am Soc Qual Control, 78-; chmn, undergrad scholarship comt, Inst Food Technologists, 84-85 & Northeast Sect, 88-89. *Mem:* Inst Food Technologists; Am Soc Qual Control (treas, Food, Drugs & Cosmetics Div, 88-89). *Res:* Quality improvement processes to meet the needs of todays world, quality responsibility understood by all. *Mailing Add:* Ocean Spray Cranberries Inc One Ocean Spray Dr Lakeville MA 02349

KEPPIE, JOHN D, b Nakuru, Kenya, Oct 3, 42. TECTONICS. *Educ:* Univ Glasgow, BSc, 64, PhD(geol), 67. *Prof Exp:* Asst prof geol, Bryn Mawr Col, 67-70; regional geologist, Geol Surv Zambia, 70-73; MGR, SURVEY SECT, MINERAL RESOURCES DIV, NOVA SCOTIA DEPT MINES & ENERGY, 74- *Concurrent Pos:* Co-leader, Terrains In Circum-Atlantic Paleozoic Orogens, Int Geol Correlation Prog, UNESCO. *Mem:* Fel Geol Soc Am; Geol Asn Canada. *Res:* Tectonics of the circum-Atlantic area. *Mailing Add:* Dept Mines & Energy PO Box 1087 Halifax NS B3J 2X1 Can

KEPPLE, PAUL C, b San Luis Potosi, Mex, Feb 6, 36; US citizen; m 62; c 2. PHYSICS. *Educ:* Univ Okla, BS, 58, MS, 61; NMex State Univ, PhD(physics), 66. *Prof Exp:* Res assoc plasma physics, Univ Md, 66-69; RES PHYSICIST, NAVAL RES LAB, 69- *Mem:* Am Phys Soc. *Res:* Plasma physics; atomic physics. *Mailing Add:* Code 4720 Naval Res Lab Washington DC 20375

KEPPLER, WILLIAM J, b Teaneck, NJ, Jan 20, 37; m 60; c 1. GENETICS, EVOLUTION. *Educ:* Univ Miami, Fla, BS, 59; Univ Ill, MS, 61, PhD(genetics), 65. *Prof Exp:* From asst prof to prof zool, Eastern Ill Univ, 65-76, asst provost, 73-76; dean arts & sch, Boise State Univ, 77-85; vchancellor, Univ Alaska, 85-88; DEAN COL HEALTH & PROF GENETICS, FLA INT UNIV, 88- *Mem:* Sigma Xi. *Res:* Cytochemistry of chromosomes. *Mailing Add:* Col Health Fla Int Univ-University Park Miami FL 33199

KEPRON, WAYNE, b Winnipeg, Man, Mar 31, 42; m 66; c 3. IMMUNOLOGY, PHYSIOLOGY. *Educ:* Univ Man, BSc & MD, 67. *Prof Exp:* ASSOC PROF INTERNAL MED, FAC MED, UNIV MAN, 75-, ASSOC PROF IMMUNOL, 78- *Concurrent Pos:* Attend physician, Respiratory Ctr, Health Sci Ctr; prin investr, Med Res Coun Group Allergy Res, Dept Immunol, Univ Man, 78- *Mem:* Am Acad Allergy; Am Thoracic Soc; Can Lung Asn. *Res:* Studies in the pathogenesis of Ige, Ige mediated asthma with specific reference to the role of local immune mechanisims in the lung; the modification of Ige metabolism with tolerogenic conjugates. *Mailing Add:* Dept Med Univ Man Winnipeg MB R3A 1R8 Can

KER, JOHN WILLIAM, b Chilliwack, BC, Aug 27, 15; m 43; c 3. FORESTRY. *Educ:* Univ BC, BASc, 41; Yale Univ, MF, 51, DF(forestry), 57. *Hon Degrees:* DSc, Univ BC, 71. *Prof Exp:* Forest ranger, BC Forest Serv, 41-45, asst forester, 45-48; asst prof forest mensuration, Univ BC, 48-53, assoc prof forest mensuration & econ, 53-61; prof forest mensuration & econ & dean, Fac Forestry, Univ NB, 61-82; RETIRED. *Concurrent Pos:* Consult, H G Acres & Agr Rehab & Develop Act, 64-65, Atlantic Develop Bd, 66-68, Royal Comn Econ State & Prospects of Nfld & Labrador, 66-67, Prov NB Land Compensation Bd, 70-71, Int Bank Reconstruct & Develop, 72-73, & Can Coun Rural Develop, 72-73; expert univ educ, Can Int Develop Agency, 73-74; Can deleg, Food & Agr Orgn, Adv Comt Forestry Educ, 75-78; chmn, Forest Mgt Task Force, NB Dept Natural Resources, 78-81; mem, Natural Sci & Eng Res Coun Can, 80-83; consult, Int Develop Res Ctr, People's Repub China, 87. *Mem:* Can Forestry Asn (vpres, 72-73); Can Inst Forestry (pres, 72-73). *Res:* Forest economics and valuation; forest measurements and biometry; university-level forestry education in both Canada and developing countries. *Mailing Add:* 760 Golf Club Rd RR3 Fredericton NB E3B 4X4 Can

KERAMIDAS, VASSILIS GEORGE, b Moudros, Greece, June 27, 38; US citizen; m 67; c 2. APPLIED PHYSICS, MATERIALS SCIENCE. *Educ:* Rockford Col, BA, 60; Univ Ill, BS, 62; John Carroll Univ, MS, 69; Pa State Univ, PhD(solid state sci), 73. *Prof Exp:* Res staff mem cadmium sulfide solar cells, Crystal Solid State Div, Harshaw Chem Co, 63-67; mem tech staff optoelectronic mat & devices, 73-80, supvr spec III-V semiconductor mat for high speed devices & integrated circuits, Bell Lab, 80-83, dist res mgr, photonic & electronic mat res, 83-85, DIV MGR, PHOTONIC & ELECTRONIC MAT RES, BELLCORE, 85- *Mem:* Mat Res Soc; Electrochem Soc; Am Phys Soc; Am Asn Crytal Growth; Metall Soc, Am Inst Mining Metall & Petrol Engrs. *Res:* Compound semiconductor light emitting sources and detectors for lightwave communications through optical fibers; compound semiconductor materials for optoelectronics; high speed devices and integrated circuits. *Mailing Add:* Bellcore 3Z-375 331 Newman Springs Rd Red Bank NJ 07701

KERBECEK, ARTHUR J(OSEPH), JR, b New York, NY, May 29, 26. CHEMICAL ENGINEERING. *Educ:* Columbia Univ, BS, 47, MS, 48, PhD(chem eng), 51. *Prof Exp:* Res engr, Titanium Metal Div, E I du Pont de Nemours & Co, Del, 52; chief metallurgist, Terrebonne Titanium Co, Can, 52-53; res engr, Ord & Inorg Chem Div, Food Mach & Chem Corp, 53-58; engr, Res Dept, Bethlehem Steel Corp, Pa, 58-68; INDUST CONSULT, 68- *Mem:* Am Chem Soc; Electrochem Soc; Am Inst Chem Engrs. *Res:* Chemical process technology; production metallurgy; corrosion; electrochemistry. *Mailing Add:* PO Box 1883 Chelan WA 98816

KERBEL, ROBERT STEPHEN, b Toronto, Ont, Apr 5, 45; m 70; c 1. CANCER. *Educ:* Univ Toronto, BS, 68; Queen's Univ, Ont, PhD(microbiol & immunol), 72. *Prof Exp:* Res fel Nat Cancer Inst Can, Chester Beatty Res Inst, London, 72-74; from asst prof to assoc prof, Dept Path, Queen's Univ, 80-91; DIR, CHAP RES DIV, STATE UNIV NY BROOK HEALTH SCI CTR, 91- *Concurrent Pos:* King George V Silver Jubilee Cancer Res fel, Nat Cancer Inst Can, 73-74, res scholar, 75-81, res assoc, 81-; mem, Grants Panel B, Nat Cancer Inst Can, 77-81; mem study sect B, NIH, 81-82; assoc ed, Inv & Metastasis, Cancer Metastasis Rev, 81- *Honors & Awards:* Wild Leitz Jr Sci Award, Exp Path, 80. *Mem:* Brit Soc Immunol; Can Soc Immunol; Am Asn Immunologists; Can Assoc Pathol. *Res:* Cancer; tumor biology; immunology; cell biology of cancer metastasis studied using membrane mutant tumor sublines; tumor progression and heterogeneity; membrane biology of activated lymphocyte and macrophage cell populations. *Mailing Add:* State Univ NY Brook Health Sci Ctr Reichmann Res Bldg 5218 2075 Bayview Ave Toronto ON M4N 3M5 Can

KERBER, ERICH RUDOLPH, b Langham, Sask, Apr 2, 26; m 56; c 2. CYTOGENETICS. *Educ:* Univ Sask, BSA, 50, MSc, 53; Univ Alta, PhD(cytogenetics), 58. *Prof Exp:* Res officer, plant breeding, 56-60, WHEAT CYTOGENETICIST, RES STA, CAN DEPT AGR, 60- *Honors & Awards:* Gold Medal, Prof Inst Pub Serv Can, 83. *Mem:* AAAS; Genetics Soc Can; Am Soc Agron; Am Soc Crop Sci. *Res:* Plant cytology and genetics; cytogenetic investigations on the transfer of rust resistance to common wheat from related species and endosperm proteins of wheat related to baking quality. *Mailing Add:* 195 Dafoe Rd Winnipeg MB R3T 2M9 Can

KERBER, RICHARD E, b New York, NY, May 10, 39; c 2. CARDIOVASCULAR DISEASES. *Educ:* Columbia Univ, AB, 60; NY Univ, MD, 64. *Prof Exp:* From asst prof to assoc prof, 71-78, PROF INT MED, COL MED, UNIV IOWA, 78- *Mem:* Am Heart Asn; fel Am Col Cardiol; Am Physiol Soc; Am Soc Clin Invest; Asn Am Physicians; Asn Univ Cardiologists. *Res:* Echocardiography; stress electrocardiography; defibrillation and cardioversion; cardiovascular pharmacology. *Mailing Add:* Dept Int Med Univ Iowa Hosp Iowa City IA 52242

KERBER, ROBERT CHARLES, b Hartford, Conn, Nov 29, 38. ORGANIC CHEMISTRY. *Educ:* Mass Inst Technol, SB, 60; Purdue Univ, PhD(org chem), 65. *Prof Exp:* Fel org chem, Purdue Univ, 65; from ast to assoc prof, 65-88, PROF CHEM, STATE UNIV NY, STONY BROOK, 88- *Concurrent Pos:* Fel, Humboldt Found, WGer, 73-74. *Mem:* AAAS; Am Chem Soc; Royal Soc Chem; NY Acad Sci. *Res:* Organo transition metal chemistry; small-ring heterocycles; organic nitrogen compounds. *Mailing Add:* Dept Chem State Univ NY Stony Brook NY 11794

KERBER, RONALD LEE, b Lafayette, Ind, July 2, 43; m 63. MECHANICAL ENGINEERING, ELECTRICAL ENGINEERING. *Educ:* Purdue Univ, BS, 65; Calif Inst Technol, MS, 66, PhD(eng sci), 70. *Prof Exp:* From asst prof to assoc prof mech eng, Mich State Univ, 69-78, assoc dir, Div Eng, 78-80, prof mech & elec eng, 78-85, assoc dean eng & dir div eng res, 80-83; dep undersecy, Dept Defense, 85-88; vpres tech bus develop, McDonnell Douglas Corp, 88-91; EXEC PRES & CHIEF EXEC OFFICER, WHIRLPOOL CORP, 91- *Concurrent Pos:* Mem tech staff, Aerospace Corp, 70-72. *Mem:* Am Soc Mech Engrs; Inst Elec & Electronics Engrs. *Res:* Theory of phase transitions and liquids, gas-surface interaction, chemical and molecular lasers. *Mailing Add:* Whirlpool Corp Monte Rd Benton Harbor MI 49022

KERBY, HOYLE RAY, b Sweetwater, Tex, Aug 9, 32; m 50; c 4. ELECTRICAL ENGINEERING. *Educ:* Tex Tech Col, BSEE, 57; Stanford Univ, MS, 60. *Prof Exp:* Res staff mem, Res Lab, Int Bus Mach Corp, 57-64, develop engr, Systs Develop Lab, 64-67, mgr, Lab Tech Opers, 67-70, mgr prod components & technol, 71-74, mgr, Explor Storage Develop, IBM Corp, San Jose, 75-80. *Mem:* Inst Elec & Electronics Engrs. *Res:* Magnetic recording technology; speech synthesis and recognition; information storage and retrieval. *Mailing Add:* IBM Corp 208-262 Harbor Dr PO Box 10501 Stanford CT 06904-2501

KERCE, ROBERT H, b Bartow, Fla, Nov 29, 25; m 46; c 3. MATHEMATICS. *Educ:* Ga Inst Tech, BME, 46; Vanderbilt Univ, MS, 57; George Peabody Col, PhD(math), 65. *Prof Exp:* From instr to assoc prof, 46-66, PROF MATH, DAVID LIPSCOMB COL, 66-, CHMN DEPT, 65- *Mem:* Math Asn Am; Nat Coun Teachers Math. *Mailing Add:* David Lipscomb Univ Nashville TN 37203

KERCHER, CONRAD J, b Yakima, Wash, June 17, 26; m 46; c 4. ANIMAL NUTRITION. *Educ:* Mont State Col, BS, 50; Cornell Univ, MS, 52, PhD(animal nutrit), 54. *Prof Exp:* Assoc animal nutrit, Cornell Univ, 50-54; head, Animal Sci Dept, 87-88, ANIMAL NUTRITIONIST, UNIV WYO, 54- *Concurrent Pos:* consult, Agency Int Develop. *Mem:* AAAS; fel Am Soc Animal Sci; Am Dairy Sci Asn; Am Registry Prof Animal Scientists; Am Forage & Grassland Coun. *Res:* Forage harvesting systems; sources of dietary fat for ruminants; alternate crops for ruminants; feed additives for cattle. *Mailing Add:* Amimal Sci Dept Univ Wyo Box 3684 Laramie WY 82071

KERCHEVAL, JAMES WILLIAM, b Rowan, Iowa, Oct 7, 06; m 29. ORGANIC CHEMISTRY. *Educ:* Iowa State Teachers Col, BA, 29; Univ Iowa, MS, 34, PhD(org chem), 39. *Prof Exp:* Teacher pub sch, Iowa, 25-37; asst, Univ Iowa, 37-39; from asst prof to prof chem, Eastern Mich Univ, 39-49; from assoc prof to prof, 49-72, EMER PROF CHEM, UNIV NORTHERN IOWA, 72- *Concurrent Pos:* Res & develop chemist, US Rubber Co, 42; capt, USAF, 43-46. *Mem:* Am Chem Soc. *Res:* Carbohydrates present in hydrolysis products of starch; separation and identification of disaccharides; ester synthesis. *Mailing Add:* 2115 Washington Cedar Falls IA 50613-2097

KERCHNER, HAROLD RICHARD, b Lewistown, Pa, Mar 5, 46; m 68; c 2. EXPERIMENTAL SOLID STATE PHYSICS. *Educ:* Harvard Univ, AB, 68; Univ Ill, MS, 72, PhD(physics), 74. *Prof Exp:* Res Assoc, Martin Marietta Energy Syst, 74-76; RES STAFF MEM, SOLID STATE PHYSICS, OAK RIDGE NAT LAB, 76- *Concurrent Pos:* Dir, Nat Low-Temp Neutron Irradation Facil, Oak Ridge Nat Lab, 85-88. *Mem:* Am Phys Soc. *Res:* Type II superconductivity; radiation effects in materials. *Mailing Add:* Solid State Div Oak Ridge Nat Lab PO Box 2008 Oak Ridge TN 37831-6061

KERDESKY, FRANCIS A J, b Wilkes-Barre, Pa, Mar 10, 53. CHEMISTRY. *Educ:* Wilkes Col, BS, 75; Univ Pa, PhD(org chem), 79. *Prof Exp:* Res assoc, Mass Inst Technol, 80-81; SR RES CHEMIST, ABBOTT LABS, NORTH CHICAGO, 81- *Mem:* Am Chem Soc; Sigma Xi. *Res:* Design and synthesis of antiarthritic drugs. *Mailing Add:* 207 Windjammer Lane Grays Lake IL 60030-2625

KEREIAKES, JAMES GUS, b Columbus, Ohio, Aug 15, 24; m 50; c 4. PHYSICS. *Educ:* Western Ky State Col, BS, 45; Univ Cincinnati, MS, 47, PhD(physics), 50; Am Bd Radiol, dipl radiol physics, 60. *Prof Exp:* Res physicist, Environ Med Br, Med Res Lab, US Army, Ky, 50-53, supvy physicist, Radiobiol Dept, 53-57, dep dir, 57-59; from asst prof to assoc prof, 59-68, PROF RADIOL PHYSICS, COL MED, UNIV CINCINNATI, 68- *Honors & Awards:* Coolidge Award, Am Asn Physicists in Med, 81, Gold Medal, 85; Gold Medal, Radiol Soc NAm, 88. *Mem:* AAAS; Am Soc Therapeut Radiol Oncol; Biophys Soc; Radiation Res Soc; Am Asn Physicists Med (pres, 69-70); Radiol Soc North Am (vpres, 81-82). *Res:* Radiation physics and biology; radiopharmaceutical dosimetry. *Mailing Add:* E555 Med Sci Bldg Univ Cincinnati Cincinnati OH 45267

KEREKES, RICHARD JOSEPH, b Welland, Ont, July 9, 40; m 78. FLUID MECHANICS, SUSPENSIONS. *Educ:* Univ Toronto, BASc, 63, MASc, 65; McGill Univ, PhD(chem eng), 70. *Prof Exp:* Scientist, Pulp & Paper Res Inst Can, 71-77, sect head, 77-83; DIR, PULP & PAPER CTR, UNIV BC, 83- *Concurrent Pos:* Hon prof chem eng, Univ BC, 78- *Mem:* Can Soc Chem Eng; Can Pulp & Paper Asn. *Res:* Fibre flocculation; mixing in pulp suspensions; pulp screening. *Mailing Add:* Chem Eng Bldg Univ BC 2075 Westbrook Mall Vancouver BC V6T 1Z2 Can

KEREN, JOSEPH, b Czech, Feb 28, 30; m 64. PHYSICS. *Educ:* Univ Melbourne, BSc, 54, MSc, 56; Columbia Univ, PhD(physics), 63. *Prof Exp:* Instr physics, Agr & Mech Col Tex, 56-57; res assoc, Australian Nat Univ, 63-64; asst prof, 65-69, ASSOC PROF PHYSICS, NORTHWESTERN UNIV, 69- *Mem:* Am Phys Soc. *Res:* Theoretical investigations of charge transfer reactions in helium-helium scattering; experimental work in elementary particles. *Mailing Add:* Dept Physics Northwestern Univ Evanston IL 60208

KERFOOT, BRANCH PRICE, b New York, NY, May 9, 25; m 65; c 1. COMPUTER SIMULATIONS. *Educ:* Yale Univ, BE, 45; Univ Mich, MSE, 47, PhD(electronics), 55; Western State Univ, JD, 87. *Prof Exp:* Ensign eng, US Navy, 45-46; AA engr, RCA Missile & Radar Dept, 49-57; prin eng, Ford Aeronutronic Div, 58-68; prin scientist, McDonnell Douglas Electronics, 68-90; RETIRED. *Concurrent Pos:* Teacher, Pasadena City Col, 58. *Mem:* Sr mem Inst Elec & Electronic Engrs; Sigma Xi; Soc Computer Simulation. *Res:* Electronic systems research and development, especially radar and communications technologies and equipment; analysis of performances. *Mailing Add:* 1420 Antigua Way Newport Beach CA 92660

KERFOOT, WILSON CHARLES, b Staten Island, NY, Mar 13, 44; m 78; c 2. LIMNOLOGY, EVOLUTIONARY ECOLOGY. *Educ:* Univ Kans, BA(zool) & BA(geol), 66; Univ Mich, PhD(zool), 72. *Prof Exp:* NSF postdoctoral limnol, Dept Zool, Univ Wash, 72-73, res assoc, 73-76; asst prof aquatic ecol, Dept Biol Sci, Dartmouth Col, 75-82, assoc prof res ecol, 82-83; vis assoc res scientist, Great Lakes Res Div, Univ Mich, 83-84, assoc res scientist, 84-90, on-leave assoc prof aquatic ecol, Sch Natural Resources, 89-; assoc prof, 89-91, PROF AQUATIC ECOL, DEPT BIOL SCI, MICH TECHNOL UNIV, 91-, CO-DIR, LAKE SUPER ECOSYSTS RES CTR, 90- *Concurrent Pos:* Vis sr scientist, Ctr Ecosysts Studies, Cornell Univ, 81-83; adj assoc prof, Dept Biol Sci, Dartmouth Col, 84-86, Dept Biol, Sch Natural Resources, Univ Mich, 85-89. *Mem:* Ecol Soc Am; Am Soc Limnol & Oceanog; Am Soc Naturalists; Int Asn Theoret & Appl Limnol; Int Asn Ecol. *Res:* Aquatic ecology of large and small lakes; zooplankton population biology and evolution; paleoecology of zooplankton; limnology; community ecology. *Mailing Add:* Dept Biol Sci Mich Technol Univ Houghton MI 49931

KERJASCHKI, DONTSCHO, b Vienna, Austria, Feb 8, 47; m 81; c 2. IMMUNOPATHOLOGY. *Educ:* Albertus Magnus Sch, Vienna, Natura, 65, Univ Vienna, MD, 72. *Prof Exp:* PROF PATH & CELL BIOL, DEPT PATH, UNIV VIENNA, 86-, DIR, DIV ULTRASTRUCT PATH & CELL BIOL, 90- *Concurrent Pos:* Vis prof cell biol, Yale Univ Sch Med, 80-88; vis res scientist cell biol, Div Cellular & Molecular Med, Univ Calif, San Diego, 89-; mem, Comn Cell Biol, Int Soc Nephrology. *Honors & Awards:* Vollhard Prize, Ger Nephrology Soc, 90. *Mem:* Int Soc Nephrology. *Res:* Molecular aspects of autoimmune diseases, especially kidney. *Mailing Add:* Dept Path Univ Vienna Wahringer Gutel 18-20 Vienna A-1090 Austria

KERKA, WILLIAM (FRANK), b Cleveland, Ohio, Apr 5, 21; m 53; c 1. MECHANICAL ENGINEERING. *Educ:* Fenn Col, BS, 48; Case Inst Technol, MS, 52. *Prof Exp:* Instr graphics, Case Inst Technol, 51-52; instr mech eng, Ore State Col, 52-54; res engr res lab, Am Soc Heating, Refrig & Air-Conditioning Engrs, 54-61; assoc dean eng, Cleveland State Univ, 61-77, assoc prof mech eng, 77-81; RETIRED. *Res:* Odor control and acoustics as related to air conditioning. *Mailing Add:* 2913 Priscilla Ave Parma OH 44134

KERKAR, AWDHOOT VASANT, b Bombay, India, Sept 20, 63; m 90. POLYMER ENGINEERING. *Educ:* Univ Bombay, BChemE, 84; Univ Pittsburgh, MS, 86; Case Western Reserve Univ, PhD(chem eng), 90. *Prof Exp:* Res asst chem eng, Univ Pittsburgh, 84-86 & Case Western Reserve Univ, 86-90; RES ENGR, W R GRACE & CO, 90- *Mem:* Am Ceramic Soc. *Res:* Products and processes for materials application such as ceramics, polymers and their composites; chemical engineering, colloid science and polymer processing to develop novel processing strategies for materials. *Mailing Add:* Res Div W R Grace & Co 7379 Rte 32 Columbia MD 21044

KERKAY, JULIUS, b Sopron, Hungary, Apr 27, 34; US citizen; c 2. CLINICAL CHEMISTRY. *Educ:* Veszprem Tech Univ, BS, 55, MS, 56; Univ Louisville, PhD(biochem), 69; Am Bd Clin Chem, dipl. *Prof Exp:* Chief chem engr, Alcohol Factory Gyor, Hungary, 56; technician, Alloys & Chem Mfg Co, Ohio, 57-58; asst in res, Cleveland Clin Found, 58-59; chief anal sect, US Army Res Inst Environ Med, 62-64; dir lab, Euclid Clin Found, 68-70; asst prof, 70-74, assoc prof & dir clin chem, 74-81, PROF CHEM & BIOL, CLEVELAND STATE UNIV, 81- *Concurrent Pos:* Adj prof, Cleveland State Univ, 68-70; speaker, Health Careers Info, Cleveland Hosp Coun, 69-; adj consult, Cleveland Clin Found, 70-85; consult, Diamond Shamrock Health Sci Labs, 74-76. sci vpres, Euclid Clin Res Found, 69-86; affil staff, St Luke's Hosp Cleveland, 77-85. *Honors & Awards:* Outstanding Contrib in Educ, Am Assoc Clin Chem, 88. *Mem:* AAAS; fel Am Asn Clin Chem; fel Am Inst Chem; NY Acad Sci; Am Chem Soc; Sigma Xi; Nat Registry Clin Chem. *Res:* protein electrophoresis; protein and steroid hormone interactions; serum constituents of mothers of down syndrome children; clinical chemistry methodology, computerization in clinical chemistry; changes in body fluid constituents of hemodialysis patients; development of radiobioassays for vitamins and isoenzymes; plasticizers and their metabolites in human organs and body fluids; awarded one US patent. *Mailing Add:* Dept Chem Cleveland State Univ Cleveland OH 44115-2440

KERKER, MILTON, b Utica, NY, Sept 25, 20; m 46; c 4. PHYSICAL CHEMISTRY. *Educ:* Columbia Univ, AB, 41, MA, 47, PhD(chem), 49. *Hon Degrees:* DSc, Lehigh Univ, 75 & Clarkson Univ, 85. *Prof Exp:* Asst chem, Columbia Univ, 46-49; from instr to prof, 49-60, chmn dept, 60-64, dean sch sci, 64-66 & 81-85, dean sch arts & sci, 66-74, THOMAS S CLARKSON PROF, CLARKSON COL TECHNOL, 74- *Concurrent Pos:* Ed-in-chief, J Colloid & Interface Sci, 65-; fel, Ford Found, 52-53; Unilever prof, Univ Bristol, 67-68; chmn, Nat Acad Sci-Nat Res Coun comt colloids & surface chem, 70-74; vis prof, Hebrew Univ & Technion, 74-75; titular mem & secy comn on colloids & surfaces, Int Union Pure & Appl Chem, 78-83. *Honors & Awards:* Kendall Award, Am Chem Soc, 71. *Mem:* Am Chem Soc; Hist Sci Soc; fel Optical Soc Am. *Res:* Light scattering; aerosols; history of science. *Mailing Add:* Sch Sci Clarkson Univ Potsdam NY 13676

KERKMAN, DANIEL JOSEPH, b Milwaukee, Wis, Sept 17, 51; m 73; c 2. CHEMISTRY. *Educ:* Johns Hopkins Univ, MA, 76; Mass Inst Technol, PhD(chem), 79. *Prof Exp:* Assoc fel, Mass Inst Technol, 79-80; MED CHEMIST, ABBOTT LABS, 80- *Mem:* Am Chem Soc. *Res:* Medicinal chemistry. *Mailing Add:* 21 Cremin Dr Lake Villa IL 60046-8864

KERKMAN, RUSSEL JOHN, b Burlington, Wis, Aug 11, 48; m 71. ELECTRICAL ENGINEERING. *Educ:* Purdue Univ, BS, 71, MS, 73, PhD(elec eng), 76. *Prof Exp:* Elec engr mach anal, Gen Elec Co, 76-80; sr proj engr, 80-86, PRIN ENGR, ALLEN-BRADLEY CO, 86- *Mem:* Inst Elec & Electronics Engrs. *Res:* Electric machine design and analysis; power systems; control systems; solid state power conditioning; AC motor drives. *Mailing Add:* Allen-Bradley Motion Control Div 8949 N Deer Brook Tr Brown Deer WI 53223

KERLAN, JOEL THOMAS, b Minneapolis, Minn, Feb 23, 40. ENDOCRINOLOGY. *Educ:* Col St Thomas, Minn, BS, 52; Univ Utah, MS, 65; Univ Mich, Ann Arbor, PhD(zool), 72. *Prof Exp:* Instr, 70-71, asst prof, 71-81, ASSOC PROF BIOL, HOBART & WILLIAM SMITH COLS, 81- *Concurrent Pos:* Vis prof, Dept Obstet & Gynec, Univ Mich, 75. *Mem:* Sigma Xi; Am Soc Zoologists. *Res:* Regulation and biosynthesis of sex steroid hormones in vertebrate testes. *Mailing Add:* Dept Biol Hobart & William Smith Col 730 S Main St Geneva NY 14456

KERLEE, DONALD D, b Ryderwood, Wash, Dec 22, 26; m 50; c 4. NUCLEAR PHYSICS, MANAGEMENT INFORMATION. *Educ:* Seattle Pac Col, BS, 51; Univ Wash, PhD(physics), 56. *Prof Exp:* Res instr physics, Univ Wash, 56; from asst prof to prof, Seattle Pac Col, 56-69, chmn dept, 62-69, dir inst res, 59-69; acad vpres, Roberts Wesleyan Col, 69-74; vpres admin, Seattle Pac Col, 74-76, dir res, 76-79, dir planning res, 79-84; RETIRED. *Concurrent Pos:* Sci fac fel, Univ Manchester, 63-64; Am Inst Physics & Am Asn Physics Teachers Regional Counsr, Washington State, 65-69. *Mem:* AAAS; Am Asn Physics Teachers; Am Phys Soc; Sigma Xi; Asn Comput Mach. *Res:* Heavy ion, nuclear and cosmic ray physics; institutional planning; forecasting models. *Mailing Add:* Seattle Pac Univ Seattle WA 98119

KERLEY, ELLIS R, physical anthropology, forensic science, for more information see previous edition

KERLEY, GERALD IRWIN, b Houston, Tex, Mar 23, 41; c 3. STATISTICAL MECHANICS, EQUATIONS OF STATE. *Educ:* Ohio Univ, BS, 63; Univ Ill, PhD(chem physics), 66. *Prof Exp:* Fel chem, Univ Ill, 66-67; staff mem phys chem, US Army Officer, 67-69, Los Alamos Nat Lab, Univ Calif, 69-84; MEM TECH STAFF, STRUCT & SOLID DYNAMICS DEPT, SANDIA NAT LAB, 84- *Mem:* Am Phys Soc. *Res:* Theory and calculation of equations of state of gases, liquids, and solids; statistical mechanics; theory of electrons in condensed matter; atomic physics; shock wave physics. *Mailing Add:* Div 1542 Sandia Nat Lab PO Box 5800 Albuquerque NM 87185

KERLEY, MICHAEL A, b Crockett, Tex, Apr 17, 41; m 85; c 3. ANATOMY. *Educ:* Stephen F Austin State Col, BS, 64; Tex A&M Univ, MS, 69, PhD(zool), 71. *Prof Exp:* ASSOC PROF BIOL, SOUTHWESTERN OKLA STATE UNIV, 71- *Concurrent Pos:* Fel, Univ Conn Health Ctr, 75-77. *Mem:* Am Asn Anatomists. *Res:* Embryonic development of mammalian dentitions. *Mailing Add:* Dept Biol Sci Southwestern Okla State Univ Weatherford OK 73096

KERLEY, TROY LAMAR, b Allen, Okla, Aug 20, 29; m 50; c 1. PHARMACOLOGY. *Educ:* Univ Okla, BS, 53, MS, 55; Purdue Univ, PhD(pharmacol), 58. *Prof Exp:* Pharmacologist, Dow Chem Co, 57-64; head biomed res dept, 64-66; dir biol sci sect, Riker Labs, Calif, 66-71; dir biol res, 3M Co, 71-73, tech dir, Riker Labs Res & Develop, 73-80, dir int new bus develop, Riker Labs, 81-84, dir new bus develop, 3M Health Care Ltd, Tokyo, 84-89, DIR TECH EVAL-LICENSING, 3M PHARMACEUT, 89- *Mem:* AAAS; Sigma Xi; Am Pharmaceut Asn; Am Found Pharmaceut Educ; Am Soc Pharmacol & Exp Therapeut. *Res:* Pharmacologic aspects of the blood-brain barrier; neuromuscular pharmacology; pharmacology and physiology of tremor and rigidity syndromes; asthmatic pharmacology. *Mailing Add:* 3M Pharmaceut 3M Ctr Bldg 225-1S-07 St Paul MN 55144-1000

KERLICK, GEORGE DAVID, b Sharon, Pa, June 24, 49. THEORETICAL PHYSICS. *Educ:* Rensselaer Polytech Inst, BS, 70; Princeton Univ, MA, 72, PhD(physics), 75. *Prof Exp:* Res assoc physics, Mont State Univ, 75; res fel physics, Alexander von Humboldt Found, Univ Cologne, 75-76 & Max-Planck Inst, Munich, 76-77; vis asst prof math, Ore State Univ, 77-78; adj asst prof physics, Univ San Francisco, 78-79; res scientist, Comput Fluid Dynamics Dept, Nielsen Eng & Res Inc, 79-83, res scientist Sterling Software, NASA Ames Res Ctr, 83-88, Computer Sci Coordr, 89; prin scientist, Tektronix Inc, 89-90; CONSULT, 90- *Honors & Awards:* Siggraph Tutorial, Asn Comput Mach, 89. *Mem:* Am Phys Soc; Asn Comput Mach; Soc Indust & Appl Math; Sigma Xi. *Res:* Scientific visualization; computer graphics; computational geometry. *Mailing Add:* 5627 SW 45th Ave Portland OR 97221

KERLIN, THOMAS W, b Charlotte, NC, Apr 7, 36; m 54; c 3. NUCLEAR ENGINEERING. *Educ:* Univ SC, BSChE, 58; Univ Tenn, MS, 59, PhD(eng sci), 65. *Prof Exp:* Res engr, Atomics Int Div, NAm Aviation, 59-61 & Oak Ridge Nat Lab, 61-66; assoc prof, 66-76, PROF NUCLEAR ENG, UNIV TENN, KNOXVILLE, 76- *Concurrent Pos:* Consult, Oak Ridge Nat Lab, 66- *Honors & Awards:* Glenn Murphy Award, Am Soc Eng Educ, 78. *Mem:* Am Nuclear Soc; Instrument Soc Am; Am Soc Eng Educ. *Res:* Instrumentation, thermometry and process simulation. *Mailing Add:* Univ Tenn Univ Tenn 212 Nuclear Eng Bldg Knoxville TN 37996

KERMAN, ARTHUR KENT, b Montreal, Que, Can, May 3, 29; m 52; c 5. THEORETICAL PHYSICS. *Educ:* McGill Univ, BSc, 50; Mass Inst Technol, PhD, 53. *Prof Exp:* Nat Res Coun Can res fel theoret physics, Calif Inst Technol, 53-54 & Inst Theoret Physics, Copenhagen, 54-55, mem res staff, 55-56; from asst prof to assoc prof, 56-64, dir, Ctr Theoret Physics, 76-83, PROF PHYSICS, MASS INST TECHNOL, 64-, DIR, LAB NUCLEAR SCI, 83- *Concurrent Pos:* Consult, Educ Serv Inc, Shell Develop Co Div, Shell Oil Co, 53-58, Argonne Nat Lab, 61-83, Los Alamos Sci Lab, 61-, Lawrence Livermore Nat Lab, 64-, Brookhaven Nat Lab, 65-81, Lawrence Berkeley Nat Lab, 75-80, Oak Ridge Nat Lab, 79- & Nat Bur Standards, 80-81; exchange prof, Guggenheim mem fel, Univ Paris, 61-62; mem, physics surv comt, Panels Nuclear Data & Heavy Ion Physics, Nat Acad Sci; vis prof physics, State Univ NY, Stony Brook, 70-71; adj prof, Brooklyn Col, City Univ New York, 71-75 & Argonne Nat Lab; foreign mem, Sci Coun Nat Inst Nuclear & Particle Physics, Nat Ctr Sci Res, Paris, France, 72-76; mem adv comt high energy, Brookhaven Nat Lab, 75, Theory Div, Los Alamos Sci Lab, 77-85 & Physics Div, 84; mem, comt sci & acad, Lawrence Livermore Nat Lab & Los Alamos Sci Lab, 81-, comt sci & technol, Exec Off Pres, White House Sci Coun, 82-85 & Argonne Nat Lab, Univ Chicago, 84-, comt nuclear sci, Dept Energy & NSF, 82-85, comt sci, Nat Res Coun, 85; chmn, Lawrence Berkeley Nat Lab, 76-78 & 81; assoc prof, Inst Nuclear Physics, Univ Paris-South, 75- *Honors & Awards:* Humboldt Sr US Scientist Award, Max Planck Inst, 85. *Mem:* Fel Am Phys Soc; fel Am Acad Arts & Sci; fel NY Acad Sci. *Res:* Theoretical nuclear physics. *Mailing Add:* Dept Physics Rm 6-305 Mass Inst Technol Cambridge MA 02139

KERMAN, R A, b Winnipeg, Manitoba, Can, May 31,43; m 67; c 2. ANALYSIS AND FUNCTIONAL ANALYSIS. *Educ:* Univ Manitoba, BA, 65, MA, 66; Univ Toronto, PhD(math), 69. *Prof Exp:* From asst prof to assoc prof math, 70-85, PROF MATH, BROCK, UNIV, 86- *Mem:* Am math soc. *Res:* Weighted norm inequalities with applications to differential equations; approximation theory in weighted lebesgue spaces; weighted convolution algebras. *Mailing Add:* Math Dept Brock Univ St Catharines ON ON L2S 3A1 Can

KERMANI-ARAB, VALI, b Bombay, India, Jan 29, 39. REGULATION OF IMMUNE SYSTEM. *Educ:* Wash State Univ, PhD(immunol & microbiol), 75. *Prof Exp:* RES SCIENTIST IMMUNOL. *Mem:* Am Immunol Asn; Am Soc Microbiologist; AAAS. *Mailing Add:* 537 Montana Ave Apt C Santa Monica CA 90403

KERMICLE, JERRY LEE, b Dundas, Ill, Mar 8, 36; m 57; c 5. CORN GENETICS, PLANT MOLECULAR GENETICS. *Educ:* Univ Ill, BS, 57; Univ Wis, MS, 59, PhD(genetics), 63. *Prof Exp:* Fel genetics & biochem, 63, from asst prof to assoc prof, 63-77, PROF GENETICS, UNIV WIS-MADISON, 77 - *Concurrent Pos:* NSF & Dept Energy grantee. *Mem:* AAAS; Genetics Soc Am. *Res:* Maize genetics, cytogenetics and development; analysis of spontaneous mutation, paramutation and complex loci. *Mailing Add:* Lab Genetics Univ Wis Madison WI 53706

KERMISCH, DORIAN, b Bucharest, Romania, Nov 13, 31; m 62; c 2. OPTICS, ELECTROMAGNETICS. *Educ:* Israel Inst Technol, BSc, 55; Polytech Inst Brooklyn, MS, 64, PhD(elec eng), 68. *Prof Exp:* Elec engr, Israeli Ministry Defense, 60-62; res fel, Polytech Inst Brooklyn, 62-66; lectr, City Col New York, 67-68; SR SCIENTIST OPTICS, XEROX CORP, 68- *Mem:* Optical Soc Am. *Res:* Theoretical investigations of blazed holograms, volume holograms and phase imaging; optical and computer image processing. *Mailing Add:* 329 Valley Green Dr Penfield NY 14526

KERN, BERNARD DONALD, b New Castle, Ind, Oct 31, 19; m 46; c 3. NUCLEAR PHYSICS. *Educ:* Univ Ind, BS, 42, MS, 47, PhD(physics), 49. *Prof Exp:* Jr physicist radar, Signal Corps, US Army, 42-43; asst nuclear physics, Metall Lab, Chicago, 43; physics, Ind Univ, 46-49; sr physicist, Nuclear Physics, Oak Ridge Nat Lab, 49-50; from asst prof to prof, 50-85, chmn, dept physics & astron, 67-69, EMER PROF PHYSICS, UNIV KY, 85- *Concurrent Pos:* Vis physicist, US Naval Radiol Defense Lab, 57-58; prof, Univ Ky Overseas Prog, Bandung Tech Inst, 61-62; vis, Stanford Univ, 71 & Inst Nuclear Physics, KFA WGer, 78. *Mem:* Am Phys Soc; Am Asn Physics Teachers. *Res:* Nuclear energy level studies with Van de Graaff accelerator; beta and gamma-ray spectroscopy; heavy ion induced reactions. *Mailing Add:* Dept Physics & Astron Univ Ky Lexington KY 40506-0055

KERN, CHARLES WILLIAM, b Middletown, Ohio, July 13, 35. THEORETICAL CHEMISTRY. *Educ:* Carnegie Inst Technol, BS, 57; Univ Minn, PhD(chem), 61. *Prof Exp:* Fel theoret chem, Dept Chem & IBM Watson Lab, Columbia Univ, 61-64; asst prof chem, State Univ NY Stony Brook, 64-66; res scientist, 66-72, mgr chem phys sect, Battelle Mem Inst, 72-76, dir Battelle Inst Prog, 74-76; prof chem, Ohio State Univ, 76-80; prog dir, 78-80 & 83-84, proj mgr CSNET, 80-83, actg dir, chem div, 84-85, SECT HEAD, PHYS CHEM & DYNAMICS, CHEM DIV, NSF, 85- *Concurrent Pos:* Adj assoc prof, Ohio State Univ, 66-71; adj prof, 71-76, acad vchmn, Dept Chem, 72-73. *Mem:* Am Phys Soc; Am Chem Soc; Sigma Xi. *Res:* Theoretical chemistry; theory of molecular structure and spectra. *Mailing Add:* Chem Div NSF 1800 G St Washington DC 20550

KERN, CLIFFORD DALTON, b Oakland, Calif, Jan 6, 28; m 51; c 3. METEOROLOGY. *Educ:* Univ Calif, Berkeley, AB, 52; Univ Calif, Los Angeles, MA, 58; Univ Wash, Seattle, PhD(atmospheric sci), 65. *Prof Exp:* weather officer, USAF, 52-58, adv weather officer, Cambridge Res Lab & Electronics Systs Div, 58-65; tech serv officer, Southeast Asia, 65-67, adv weather officer, Satellite Control Facil, Calif, 67-69, chief spec proj br, Air Force Global Weather Ctr, Offutt AFB, Nebr, 69-70 & develop activ in sci & numerical area, 70-71, staff meteorologist, Space & Missile Systs Orgn, 71-72; res supvr, Environ Transp Div, Savannah River Lab, E I du Pont de Nemours & Co, Inc, 72-78; staff engr & group leader, Atmospheric Effects Group, 78-80, SR STAFF ENGR & GROUP LEADER, PROPAGATION SCI, LOCKHEED MISSILES & SPACE C0, 80- *Mem:* AAAS; Am Meteorol Soc; Am Geophys Union; Sigma Xi. *Res:* Infrared emission of the earth and its cloud fields as seen by weather satellites; satellites considering solar interactions with the earth, its upper atmosphere and geomagnetic field; propagation of electromagnetic radiation through the atmosphere and ionosphere. *Mailing Add:* Lockheed Missiles & Space Co 0/62-47 B/076 1111 Lockheed Way Sunnyvale CA 94089-3504

KERN, CLIFFORD H, III, b New Orleans, La, Aug 30, 48; m 72; c 1. DEVELOPMENTAL GENETICS. *Educ:* Washington & Lee Univ, BS, 70; Ind Univ, MA, 72, PhD(zool), 79. *Prof Exp:* Instr genetics & biochem, DePauw Univ, 76-79, asst prof, 79-81; PROG COORDR, DIABETES UNIV, LA DEPT HEALTH & HUMAN RESOURCES, 81- *Concurrent Pos:* Vis asst prof, Ind Univ, Bloomington, 80 & 81; lectr, Ind Univ & Purdue Univ, 81. *Mem:* Soc Develop Biol; Genetics Soc Am; Sigma Xi; Am Diabetes Asn. *Res:* Diabetes; genetics and development of female-sterile mutants in Drosophila. *Mailing Add:* 1309 Richland Ave Metairie LA 70001-3634

KERN, EARL R, b Auburn, Wyo, May 22, 40; m 62; c 2. MEDICAL MICROBIOLOGY. *Educ:* Univ Utah, BS, 66, MS, 70, PhD(microbiol), 73. *Prof Exp:* Microbiologist, Dugway Proving Ground, Dept Army, 66-68; instr microbiol, 73-76, res instr, Dept Pediat, 74-76, res asst prof, Dept Microbiol, 76-78, res asst prof, Dept Pediat, 76-80, RES ASSOC PROF MICROBIOL, COL MED, UNIV UTAH, 80- *Mem:* Am Soc Microbiol; Soc Exp Biol Med; Sigma Xi; Am Asn Cancer Res. *Res:* Pathogenesis, chemotherapy and host resistance in experimental viral infections; cancer chemotherapy; immunomodulation of host resistance; herpesvirus infections. *Mailing Add:* Univ Utah Sch Med 50 N Medical Dr/Pediat Salt Lake City UT 84132

KERN, FRED, JR, b Montgomery, Ala, Sept 9, 18; m 42; c 3. INTERNAL MEDICINE, GASTROENTEROLOGY. *Educ:* Univ Ala, BA, 39; Columbia Univ, MD, 43; Am Bd Internal Med, dipl, 52. *Prof Exp:* Asst, Sch Med, Emory Univ, 43, asst path, 44; asst med, Med Col, Cornell Univ, 47-48, instr, 51; from asst prof to assoc prof, 52-65, head div gastroenterol, 59-82, PROF MED, UNIV COLO MED CTR, DENVER, 65 - *Concurrent Pos:* Res fel med, Med Col, Cornell Univ, 47-49, Ledyard Jr fel, 49-50; intern, Grady Hosp, Atlanta, 43, asst resident, 44; asst resident, NY Hosp, 47-48, provisional asst outpatients, 47 & 49; physician, Colo Gen Hosp, 52; assoc attend physician, Div Internal Med, Denver Gen Hosp, 52, chief div prev med & pub health, 52, dir gen med clin, 51-59; chmn, Am Bd Gastroenterol, 69-72; mem, Nat Arthritis, Metab & Digestive Dis Adv Coun, NIH, 78-81; pres, Digestive Dis Info Ctr, 78-80; mem, Sci Adv Bd, Nat Found Ileitis-Colitis, chmn, Grants Review Comt, 78-81. *Mem:* AAAS; Am Inst Nutrit; Am Gastroenterol Asn (vpres, 73-74, pres-elect, 74-75, pres, 75-76); hon mem Gastroenterol Soc Australia; Asn Am Physicians; master Am Col Physicians. *Res:* Medical education and gastroenterology; bile acid metabolism; intestinal absorption; pathogenesis of gallstones; effect of female sex steroid hormones on bile acid and cholesterol metabolism. *Mailing Add:* Div Gastroenterol Univ Colo Health Sch Med 4200 E Ninth Ave Box C-293 Denver CO 80262

KERN, HAROLD L, biochemistry, for more information see previous edition

KERN, JEROME, b New York, NY, Nov 2, 27; m 52; c 3. VIROLOGY. *Educ:* Brooklyn Col, BSc, 50; Ohio State Univ, MSc, 54; George Washington Univ, PhD(virol), 62. *Prof Exp:* Asst bact, Ohio State Univ, 51-54; res assoc, Smithsonian Inst, 54-56; bacteriologist, US Dept Interior, 56-57; bacteriologist, Walter Reed Army Inst Res, 57-58; bacteriologist, Nat Inst Allergy & Infectious Dis, 58-65; res microbiologist, Nat Cancer Inst, 65-66; sr scientist, Flow Labs, Inc, 66-76; RES ASSOC, AM TYPE CULT COLLECTION, 76- *Mem:* AAAS; Am Soc Microbiol; Sigma Xi. *Res:* Serological methods of virus identification; immunological relationships between human and animal viruses; virus purification. *Mailing Add:* 302 Lynn Manor Dr Rockville MD 20850-0081

KERN, JOHN PHILIP, b Springfield, Mass, Jan 3, 39; m 69. INVERTEBRATE PALEONTOLOGY. *Educ:* Univ Calif, Los Angeles, AB, 63, PhD(geol), 68. *Prof Exp:* From asst prof to assoc prof, 68-77, PROF GEOL, SAN DIEGO STATE UNIV, 77- *Mem:* Paleont Soc; Paleont Res Inst. *Res:* Paleoenvironmental studies of late Cenozoic marine invertebrates; trace fossils. *Mailing Add:* Dept Geol Sci San Diego State Univ San Diego CA 92182

KERN, JOHN W, b Mansfield, Ohio, Dec 16, 30; m 62; c 2. FUEL TECHNOLOGY, PETROLEUM ENGINEERING. *Educ:* Univ Calif, Berkeley, BS, 56, MA, 58, PhD(geophys), 60. *Prof Exp:* Phys scientist, Rand Corp, Calif, 60-64; from asst prof to prof physics, Univ Houston, 64-74, adj prof, 75-78; sr res specialist, Exxon Prod Res Co, Houston, 75-81; dir, Info & Interpretation Systs, Dresser Atlas, Houston, 81-84, staff petrophysicist, 84-87; dir training, Atlas Wireline Serv, 87-89, mgr Intepretational Support, 89-90; CONSULT, PETROPHYSICS & RESERVOIR ENG, 91- *Concurrent Pos:* Mem comn 4, Int Sci Radio Union, 62; mem working group data anal, comn II, Int Union Geod & Geophys, 63-; consult phys scientist, Rand Corp, 64-75. *Mem:* Soc Explor Geophys; Am Geophys Union; Soc Prof Well Log Analysts; Soc Petrol Engrs; Am Asn Artificial Intel. *Res:* Rock magnetism; geomagnetism; auroral and magnetospheric physics; solar and planetary physics; well log analysis; well log methods development; well log instrument theory and applications; resevoir description. *Mailing Add:* 3405 Amherst Houston TX 77005

KERN, MICHAEL DON, b Los Angeles, Calif, Nov 25, 38; m 61; c 3. AVIAN PHYSIOLOGY. *Educ:* Whittier Col, BA, 62; Wash State Univ, MS, 65, PhD(zoophysiol), 70. *Prof Exp:* USPHS trainee avian reproductive physiol, Cornell Univ, 69-71; asst prof chordate morphogenesis, Fordham Univ, 71-75; asst prof animal physiol, 76-81, ASST PROF BIOL, COL WOOSTER, 81- *Mem:* AAAS, Cooper Ornith Soc; Ecol Soc Am; Am Ornithologists Union; Sigma Xi. *Res:* Photoperiodism in birds; annual cycles of birds; avian reproductive physiology, particularly incubation, nest construction, and nesting energetics. *Mailing Add:* Dept Biol Col Wooster Wooster OH 44691

KERN, RALPH DONALD, JR, b New Orleans, La, Aug 8, 35; m 61; c 3. PHYSICAL CHEMISTRY. *Educ:* Univ Tex, Austin, BS, 57 & 60, PhD(chem), 65. *Prof Exp:* Res fel, Harvard Univ, 65-67; from asst prof to assoc prof, 67-77, PROF CHEM, UNIV NEW ORLEANS, LAKEFRONT, 77-, CHMN DEPT, 80- *Mem:* Am Chem Soc; Am Phys Soc; Combustion Inst. *Res:* Rates of gas phase reactions in shock tubes monitored by infrared emission and time-of-flight mass spectrometry. *Mailing Add:* Dept Chem Univ New Orleans Lakefront New Orleans LA 70122

KERN, ROLAND JAMES, b Bay City, Mich, Oct 29, 25; m 53; c 2. CHEMISTRY. *Educ:* Univ Mich, BS, 48; Northwestern Univ, PhD(chem), 52. *Prof Exp:* RES CHEMIST, MONSANTO CO, 52- *Concurrent Pos:* Fac, Univ Col, Wash Univ. *Mem:* Am Chem Soc. *Res:* Polymerization; polymer technology. *Mailing Add:* 50 Forest Crest Dr Chesterfield MO 63017

KERN, ROY FREDRICK, b Lewiston, Minn, Oct 25, 18; m 44; c 3. METALLURGY, MECHANICAL ENGINEERING. *Educ:* Macalester Col, BS, 39; Marquette Univ, BS, 43. *Prof Exp:* Chief metallurgist, Allis-Chalmers Mfg Co, 43-62; plant mgr, Knoxville Iron Co, 63-65; proj engr, Caterpillar Tractor Co, 65-71; OWNER, KERN ENG CO, 71- *Mem:* Fel Am Soc Metals. *Res:* Development of boron and alloy boron steels, heat treating of steel parts, designing for heat treatment, selection of steel for heat treated parts; exculusive method developed for buying steels: cost reduction-works better-costs less; gear problems; gears of all kinds and sizes. *Mailing Add:* Kern Eng Co 818 E Euclid Peoria IL 61614

KERN, WERNER, b Basel, Switz, Mar 18, 25; US citizen; m 55; c 3. CHEMICAL VAPOR DEPOSITION, SEMICONDUCTOR PROCESSING. *Educ:* Univ Basel, Cert chem, 44; Polyglot Sch Lang, Switz, dipl lang, 47; Rutgers Univ, AB, 55. *Prof Exp:* Jr chemist, Hoffmann-LaRoche, Ltd, Switz, 42-48, anal res chemist, NJ, 48-55, Dept Radioisotope Biochem, 55-57; chief chemist & radiol safety officer, Nuclear Corp Am, 58-59; engr & div health physicist, Electronics Component & Devices Div, Somerville, 59-64, fel tech staff, David Sarnoff Res Ctr, RCA Corp, 64-87; PRES, WERNER KERN ASSOC, 87-; SR SCIENTIST, MONKOWSKI-RHINE, INC, LAM RES, 88- *Concurrent Pos:* Safety coun rep, RCA Labs, RCA Corp, 70-87; chem vapor deposition course lectr, Am Vacuum Soc, 81- & Electrochem Soc, 85-; vchmn, Electrochem Soc, 81-82, chmn, 83-84, adv, 85-86. *Honors & Awards:* T C Callinan Award, Electrochem Soc Inc, 72. *Mem:* Am Chem Soc; Electrochem Soc; Sigma Xi; Am Vacuum Soc. *Res:* Semiconductor process research; chemical vapor deposition; silicon wafer cleaning technology; chemical etching of microelectronic materials; analytical process control methods; radioactive tracer applications; surface decontamination research; preparation and properties of dielectric films. *Mailing Add:* 439 Probasco Rd East Windsor NJ 08520-5518

KERN, WILLIAM H, b Nuermberg, Ger, Dec 25, 27; m; c 2. PATHOLOGY, CANCER. *Educ:* Univ Munich, MD, 52. *Prof Exp:* DIR LABS, HOSP GOOD SAMARITAN, 66- *Honors & Awards:* Papanicolaou Award, Am Soc Cytol, 87. *Mem:* Col Am Path; Am Asn Cytol; Int Acad Path; Int Acad Cytol. *Res:* cytopathology of cancer; urinary tract pathology; pathology of vascular bypass grafts. *Mailing Add:* Off Dir Labs Hosp Good Samaritan Los Angeles CA 90017

KERN, WOLFHARD, b Berlin, Ger, Feb 18, 27; m 52; c 3. HIGH ENERGY PHYSICS. *Educ:* Univ Frankfurt, BS, 48, MS, 51; Univ Bonn, PhD(physics), 58. *Prof Exp:* Asst prof physics, Univ Bonn, 58-60; res assoc, Deutsches Elektronen-Synchrotron, 60-63, scientist, 65-67; res assoc physics, Mass Inst Tech, 63-64; assoc prof, 64-65, PROF PHYSICS, SOUTHEASTERN MASS UNIV, 67- *Concurrent Pos:* Vis scientist, Max-Planck Inst, 75-76. *Mem:* Am Phys Soc; Am Asn Physics Teachers. *Res:* Experimental high energy physics. *Mailing Add:* Dept Physics Southeastern Mass Univ North Dartmouth MA 02747

KERNAGHAN, ROY PETER, b Schenectady, NY, Mar 26, 33; m 56. BIOLOGY, GENETICS. *Educ:* Dartmouth Univ, BA, 55, MA, 57; Univ Conn, PhD(genetics), 63. *Prof Exp:* NIH trainee electron micros, Col Physicians & Surgeons, Columbia Univ, 63-65; asst prof biol sci, State Univ NY Stony Brook, 65-74; PROF BIOL & CHMN DEPT, SALISBURY STATE COL, 74- *Mem:* Genetics Soc Am; Am Soc Cell Biol. *Res:* Developmental genetics. *Mailing Add:* Dept Biol Salisbury State Univ PO Box 2095 Salisbury MD 21801

KERNAN, ANNE, b Dublin, Ireland, Jan 15, 33; US citizen. PHYSICS. *Educ:* Univ Col, Dublin, BSc, 53, PhD(physics), 57. *Prof Exp:* Asst lectr physics, Univ Col, Dublin, 58-62; res physicist, Lawrence Berkeley Lab, 62-66 & Stanford Linear Accelerator Ctr, Stanford Univ, 66-67; assoc prof, 67-70, chmn dept, 73-76, PROF PHYSICS, UNIV CALIF, RIVERSIDE, 70- *Concurrent Pos:* Prin investr, Dept Energy, res contract, 69-; mem, Adv Comt Physics, NSF, 78-82, sci & educ adv comt, Lawrence Berkeley Lab, 83-87, Fermilab prog adv comt, 86-90, Dept Energy, physics adv panel, 86-90, Users Orgn SSC, 89-; counciller-at-large, Am Phys Soc, 85- *Mem:* AAAS; Am Phys Soc. *Res:* Experimental high energy physics. *Mailing Add:* Dept Physics Univ Calif 900 University Ave Riverside CA 92521-0413

KERNAN, WILLIAM J, JR, b Baltimore, Md, Oct 18, 33; m 56; c 3. PHYSICS. *Educ:* Loyola Col, Md, BS, 55; Univ Chicago, MS, 56, PhD(physics), 60. *Prof Exp:* Asst physicist, Argonne Nat Lab, 60-61; asst prof physics, NY Univ, 61-63; assoc prof, 63-66, PROF PHYSICS, IOWA STATE UNIV, 66-; SR PHYSICIST, AMES LAB, 66- *Concurrent Pos:* Guest staff mem, Brookhaven Nat Lab, 60-65; physicist, Ames Lab, 63-66; prog dir high energy physics, Ames Labs, ERDA, 75-78, assoc dir opers, Dept of Energy, 78- *Mem:* Fel Am Phys Soc. *Res:* Properties and interactions of elementary particles. *Mailing Add:* Dept Physics 12 Physics Iowa State Univ Ames IA 50011

KERNBERG, OTTO F, b Vienna, Austria, Sept 10, 28; US citizen; c 3. PSYCHOANALYSIS. *Educ:* Univ Chile, BS, 47, MD, 53. *Prof Exp:* Intern, Hosp J J Aquirre, Santiago, Chile, 53; resident psychiat, Psychiat Clin, Univ Chile, 54-57, staff mem, Dept Psychiat, 57-61, asst prof psychiat, 58-59; staff psychiatrist, C F Menninger Mem Hosp, Topeka, Kans, 61-62; staff psychiatrist, Res Dept, Menninger Found, 62-67, chief investr, Psychother Res Proj, 67-69; dir, C F Menninger Mem Hosp, 69-73; dir gen clin serv, NY State Psychiat Inst, 73-76; prof clin psychiat, Col Physicians & Surgeons, Columbia Univ, 73-76; TRAINING & SUPV ANALYST, COLUMBIA UNIV CTR FOR PSYCHOANALYTICAL TRAINING & RES, 74-; PROF PSYCHIAT, CORNELL UNIV MED COL, 76-; ASSOC CHMN & MED

DIR, NY HOSP-CORNELL UNIV MED CTR, WESTCHESTER DIV, 76- *Concurrent Pos:* Prof psychopath, Sch Social Work, Nat Health Serv, Santiago, 57-59, prof ment health, 58-59; prof ment health & prof psychol diag, Sch Psychol, Cath Univ Chile, 58-61; Rockefeller Found fel psychiat, Henry Phipps Clin, Johns Hopkins Hosp, Baltimore, 59-60; consult, Topeka State Hosp, Kans, 64-68; fac mem, Menninger Sch Psychiat & Topeka Inst Psychoanal, 64-73; resident psychiat, C F Menninger Mem Hosp, Topeka, 65-68, consult, 68-69; training analyst, Topeka Inst Psychoanal, 66-73, treas, 68-73; staff psychiatrist, Adult Outpatient Serv, Menninger Found, 68-69; assoc clin prof psychiat, Univ Kans Med Ctr, Kansas City, Kans, 71-73; mem staff, A K Rice Inst, Group Rels Conf, Washington Sch Psychiat, 72-75; vis prof, Menninger Sch Psychiat & Mass Gen Hosp, Harvard Med Sch, 74-75 & Albert Einstein Col Med, 77-; attend psychiatrist, Serv Psychiat, Presby Hosp, NY, 74-76; asst ed, J Am Psychoanal Asn, 74-77, assoc ed, 77-; fac mem, NY Psychoanal Inst, 74-81. *Honors & Awards:* Heinz Hartman Award, 72; Edward A Strecker Award, 75; William F Schonfeld Award, 82; Van Geison Award, 86. *Mem:* Fel Am Psychiat Asn; AMA; fel Am Col Physicians; Sigma Xi; fel NY Acad Med; NY Acad Sci; AAAS; Int Psychoanal Asn (vpres, 83-); Cent Neuropsychiat Hosp Asn (pres, 88-). *Res:* Diagnosis of severe personality disorders with particular reference to borderline conditions and narcissistic pathology; the process and outcome of psychoanalytic psychotherapy with severe personality disorders; psychoanalytic object relations theory, institutional dynamics, psychopathology of love relations and psychoanalytic technique. *Mailing Add:* 21 Bloomingdale Rd White Plains NY 10605-1596

KERNELL, ROBERT LEE, b Greer, SC, Mar 24, 29; m 59; c 2. NUCLEAR PHYSICS. *Educ:* Wofford Col, AB, 50; Univ SC, MS, 58; Univ Tenn, PhD(physics), 68. *Prof Exp:* Asst prof physics, Col William & Mary, 58-62; assoc prof, 67-79, PROF PHYSICS, OLD DOMINION UNIV, 79- *Mem:* Am Phys Soc; Am Asn Physics Teachers; NY Acad Sci; Sigma Xi. *Res:* Nuclear structure physics; radiation effects induced by charged particles. *Mailing Add:* Dept Physics Old Dominion Univ Norfolk VA 23529-0116

KERNER, EDWARD HASKELL, b New York, NY, Apr 22, 24; m 48; c 3. THEORETICAL PHYSICS, THEORETICAL BIOLOGY. *Educ:* Columbia Col, AB, 43; Cornell Univ, PhD(physics), 50. *Prof Exp:* Instr physics, Princeton Univ, 43-44; physicist, Manhattan eng dist, 44-45; asst prof physics, Wayne State Univ, 49-51; res assoc, Iowa State Univ, 51-53; from asst prof to assoc prof, Univ Buffalo, 53-62; PROF PHYSICS, UNIV DEL, 62- *Mem:* Fel Am Phys Soc; Soc Math Biol. *Res:* Relativistic particle dynamics; projective relativity; atomic and molecular collision theory; band structure of random lattices; theory of biological population fluctuations and of biochemical kinetics; non-linear dynamics. *Mailing Add:* Dept Physics Univ Del Newark DE 19711

KERNEY, PETER JOSEPH, b Philadelphia, Pa, Apr 7, 40; m 68; c 3. MECHANICAL ENGINEERING. *Educ:* Univ Notre Dame, BS, 62, MS, 64; Pa State Univ, PhD(mech eng), 70. *Prof Exp:* Instr mech eng, Tri-State Col, 65-66 & Pa State Univ, 66-70; asst prof, Lafayette Col, 70-72; res staff engr, Draper Lab, Mass Inst Technol, 72-77; MGR ADVAN TECHNOL, CTI-CRYOGENICS, 77- *Concurrent Pos:* Res asst, Ord Res Lab, 68-70; consult, US Army Frankford Arsenal, 71; adj prof mech eng, Tufts Univ, 77- *Mem:* Am Soc Mech Engrs; assoc mem Am Soc Eng Educ; Am Soc Heating, Refrig & Air Conditioning Engrs; Sigma Xi. *Res:* Two phase flow and turbulent jet mixing; jet penetration characteristics of a submerged steam jet; energy and momentum characteristics of liquid bath downstream of steam jet; convection and cryogenic heat transfer. *Mailing Add:* Seven Elwern Rd Arlington MA 02174

KERNIS, MARTEN MURRAY, b Chicago, Ill, Sept 21, 41; m 82; c 1. ANATOMY, TERATOLOGY. *Educ:* Roosevelt Univ, BS, 63; Univ Fla, PhD(anat sci), 68. *Prof Exp:* from asst prof to assoc prof anat, Sch Basic Med Sci, 68-76, from asst prof to assoc prof anat in obstet & gynec, Abraham Lincoln Sch Med, 68-76, from asst dean to assoc dean, Sch Basic Med Sci, 72-76; assoc prof anat, Jefferson Med Col & dean, Col Allied Health Sci, Thomas Jefferson Univ, 76-78; assoc prof anat, Sch Basic Med Sci & dep exec dean, 78-82, actg exec dean, 82-83, VDEAN, COL MED, UNIV ILL, 83- *Mem:* AAAS; Teratology Soc; Asn Am Med Cols. *Res:* Transport across the placenta during normal and abnormal embryogenesis; effects of teratogens on function; distribution of teratogens; mechanism of malformations. *Mailing Add:* Univ Ill Col Med MC 784 PO Box 6998 Chicago IL 60680

KERNKAMP, MILTON F, b Washington Co, Minn, Sept 16, 11. PLANT PATHOLOGY. *Educ:* Univ Minn, BS, 34, MS, 38, PhD(plant path), 41. *Prof Exp:* Asst pathologist, Agr & Mech Col, Tex, 35; asst, Univ Minn, 35-36, instr, 36-41; asst pathologist, USDA, Miss, 41-46; from asst prof to assoc prof, 46-56, asst dir agr exp sta, 56-61, head dept plant path, 61-72, prof PLANT PATH, 72-77, EMER PROF, UNIV MINN, ST PAUL, 77- *Mem:* AAAS; Am Soc Agron; Am Phytopath Soc; Indian Phytopath Soc. *Res:* Genetics and variability of smut fungi; diseases of forage legumes and grasses; diseases of wild rice. *Mailing Add:* 10035 W Royal Oak Dr No 100 Sun City AZ 85351

KERNS, DAVID MARLOW, b Minneapolis, Minn, Oct 7, 13. ELECTROMAGNETICS. *Educ:* Univ Minn, BEE, 35; Cath Univ Am, PhD(physics), 51. *Prof Exp:* Asst, Univ Minn, 35-39; chief elec unit, Aerial Camera Lab, Air Mat Command, USAF, Ohio, 40-42; from proj leader to asst div chief, Radio Standards Lab & Electromagnetic Fields Div, 46-71, sr res scientist, Electromagnetic Fields Div, Nat Bur Standards, 71-80; PHOTOG, 80- *Concurrent Pos:* Adj prof, Univ Colo, 61-; mem US comn, Int Sci Radio Union, 58- *Honors & Awards:* Silver Medal, US Dept Com, 60, Gold Medal, 73; Samuel Wesley Stratton Award, Nat Bur Standards, 81. *Res:* Electromagnetic theory of wave guides and wave guide junctions; scattering-matrix theory of antennas and antenna-antenna interactions. *Mailing Add:* 1365 Bear Mountain Dr Boulder CO 80303

KERPER, MATTHEW J(ULIUS), b St Louis, Mo, Apr 9, 22; m 48; c 1. CERAMICS. *Educ:* Univ Mo, BS, 43, MS, 47; George Washington Univ, MS, 59; Am Univ, MPA. *Prof Exp:* Ceramic engr, Laclede Christy Co, Mo, 47-52, Nat Bur Standards, 52-66, Off Aerospace Res, 66-70 & Air Force Systs Command, 70-72, ceramic engr, Air Force Off Sci Res, 72-88; RETIRED. *Mem:* Am Ceramic Soc; Am Inst Ceramic Engrs. *Res:* Physical properties of ceramics and glass at elevated temperatures; science administration. *Mailing Add:* 4620 N Park Ave Apt 1509E Chevy Chase MD 20815

KERR, ANDREW, JR, b Wilkinsburg, Pa, Dec 30, 14; c 3. MEDICINE. *Educ:* Colgate Univ, AB, 37; Harvard Univ, MD, 41. *Prof Exp:* Instr med, Tulane Univ, 49-50; from instr to asst prof, La State Univ, 51-55; assoc Col Med, State Univ NY Upstate Med Ctr, 55-60; assoc prof clin med, Med Col Ga, 60-61, clin prof med, 61; asst prof clin med, Sch Med & Dent, Univ Rochester, 61-67; dir, Cardiopulmonary Lab, Vet Admin Hosp, 61-67, chief med, 64-67; prof med, 78-87, EMER PROF, NORTHEAST OHIO COL MED, 87- *Concurrent Pos:* Fel med, Tulane Univ, 48-49; asst chief med, Vet Admin Hosp, Syracuse, NY, 55-60, chief, Augusta, Ga, 60-61; chmn Dept Med, Akron City Hosp, 67-76, Div Cardiol, 76-79. *Mem:* Am Fedn Clin Res. *Res:* Cardiovascular diseases. *Mailing Add:* 7431 Valley View Rd Hudson OH 44236

KERR, ANTHONY ROBERT, b Farnborough, Eng, Aug 30, 41; Australian citizen; m 74; c 1. MICROWAVE ELECTRONICS, RADIO ASTRONOMY. *Educ:* Univ Melbourne, BE, 63, MESc, 67, PhD(elec eng), 69. *Prof Exp:* Res scientist radio astron, Div Radiophys, Commonwealth Sci & Indust Res Orgn, Sydney, Australia, 69-71; electronics engr millimeter-wave electronics, Nat Radio Astron Observ, Charlottesville, Va, 71-74; physicist millimeter-wave electronics, Goddard Inst Space Studies, NASA, 74-84; SCIENTIST, NAT RADIO ASTRON OBSERV, CHARLOTTESVILLE, VA, 84- *Concurrent Pos:* Vis prof, Univ Va, 84-; consult, Macom Inc, 80-83 & NASA, 87- *Honors & Awards:* The Microwave Prize, Inst Elec & Electronics Engrs, 78; Except Eng Achievement Medal, NASA, 83. *Mem:* Fel Inst Elec & Electronics Engrs; Int Union Radio Sci; Astron Soc Australia; Sigma Xi. *Res:* Development of low-noise receivers at millimeter and submillimeter wavelengths, and their application in radio astronomy, atmospheric physics, and space communications. *Mailing Add:* Nat Radio Astron Observ 2015 Ivy Rd Charlottesville VA 22903

KERR, ARNOLD D, b Suwalki, Poland, Mar 9, 28; US citizen; m 66; c 2. ENGINEERING & ICE COVER MECHANICS, RAILROAD ENGINEERING. *Educ:* Tech Univ, Munich, Dipl Ing, 52; Northwestern Univ, MS, 56, PhD(theoret & appl mech), 58. *Prof Exp:* Engr design & anal, Hazelet & Erdal, Consult Engrs, 55; asst res scientist mech solids, Courant Inst Math & Sci, New York Univ, 58-59, from asst prof aeronaut to assoc prof aeronaut & astronaut, 59-66, prof, 66-73, dir, Lab Mech Solids, 67-73; vis prof civil eng, Princeton Univ, 73-78; PROF CIVIL ENG, UNIV DEL, 78- *Concurrent Pos:* Consult to US govt agencies & indust, 59-; pres, Inst Railroad Eng, 80-; fel, Ctr Advan Study, Univ Del, 89-90. *Mem:* Am Soc Mech Engrs; Am Railway Eng Asn. *Res:* Structural mechanics; continuously supported structures; dynamics and stability of structures; bearing capacity and dynamics of floating ice covers; railroad track analyses and technology. *Mailing Add:* Dept Civil Eng Univ Del Newark DE 19716

KERR, CARL E, b Corsicana, Tex, Sept 16, 26; m 50; c 1. MATHEMATICS. *Educ:* La Salle Col, BA, 50; Univ Del, MA, 53; Lehigh Univ, PhD(math), 59. *Prof Exp:* Asst prof math, Lafayette Col, 53-59; asst prof, Dickinson Col, 59-69; prof math, 69-76, PROF MATH & COMPUT SCI, SHIPPENSBURG STATE COL, 76- *Concurrent Pos:* Mathematician, Frankford Arsenal, 52-54; mathematician, Convair Div, Gen Dynamics Corp, 56. *Mem:* Am Math Soc; Math Asn Am. *Res:* Linear spaces; summability. *Mailing Add:* 33 Oak Lane Shippensburg PA 17257

KERR, DONALD L(AURENS), b Putnam, Conn, June 28, 43; m 66; c 1. CHEMICAL ENGINEERING. *Educ:* Worcester Polytech Inst, BS, 65; Univ Del, MChE, 68, PhD(chem eng, phys chem), 70. *Prof Exp:* Sr chemist, Photog Eng Lab, 69-75, RES ASSOC PHOTOG PROCESSES LAB, RES LABS, EASTMAN KODAK CO, 75- *Mem:* Am Inst Chem Engrs. *Res:* Mass tranfer; chemical kinetics; drying; photographic processing. *Mailing Add:* 177 Long Pond Rd Rochester NY 14612

KERR, DONALD M, JR, b Philadelphia, Pa, Apr 8, 39; m 61; c 1. RESEARCH ADMINISTRATION, GEOPHYSICS. *Educ:* Cornell Univ, BEE, 63, MS, 64, PhD(elec eng, physics), 66. *Prof Exp:* Staff mem, 66-71, group leader, 71-73, asst, Dir Off, 73-75, alt energy div leader, 75-76, dir, Los Alamos Nat Lab, Univ Calif, 79-85; dep mgr, Nev Opers, Dept Energy, 76-77, dep & actg asst secy defense progs, 77-79, dep & actg asst secy energy technol, 79; sr vpres, 85-88, exec vpres, 88-89, PRES, EG&G INC, WELLESLEY, MASS, 89- *Concurrent Pos:* Consult, Navajo Sci Comt, 74-77; mem US Army Sci Adv Panel, 75-78; chmn, comt res & develop, Int Energy Agency, OECD, 79-85; mem joint stategic target planning staff, Sci Adv Group, 81-; mem corp, Charles Stark Draper Lab, 82-, SRI Nat Security Adv Coun, 80-89, Cornell Univ Eng Col Coun, 84-; dir, Nat Asn Mfrs, 87-, Mirage Systs Inc, 88-, Resources for the Future, 90- *Mem:* Fel AAAS; Am Phys Soc; Am Geophys Union. *Res:* Nuclear weapons research and development testing; laser, heavy ion, and electron beam fusion; nuclear safeguards and security; ionospheric physics; international activities relating to nuclear technology and political, military, economic and energy affairs. *Mailing Add:* 34 Greenwood Rd Wellesley Hills MA 02181-2912

KERR, DONALD PHILIP, b Winnipeg, Man, Oct 22, 38; m 62; c 2. ATOMIC PHYSICS, MOLECULAR PHYSICS. *Educ:* Univ Man, BSc, 60, MSc, 61, PhD(physics), 65. *Prof Exp:* Nat Res Coun Can fel, Univ Giessen, 65-67; res assoc, Harvard Univ, 67-69; asst prof physics, Univ Winnipeg, 69-74, assoc prof, 74-79; CONSULT, 79- *Mem:* Can Asn Physicists. *Res:* Positron annihilation in organic liquids and solids; high resolution mass spectrometry; heavy ion scattering. *Mailing Add:* Dept Physics Univ Winnipeg 515 Portage Ave Winnipeg MB R3B 2E9 Can

KERR, DONALD R, b Chicago, Ill, Mar 12, 38; m 60; c 1. MATHEMATICS. *Educ:* Univ Ariz, BS, 60, MS, 62; Lehigh Univ, PhD(math), 67; Fla State Univ, PhD(clin psychol). *Prof Exp:* Instr math, Lafayette Col, 63-67; asst prof, State Univ NY Albany, 67-69; vis prof math, 69-71, assoc prof educ & asst dir math educ develop ctr, 71-78, acad officer & basic skills coordr math, Ind Univ, Bloomington, 78-82; CLIN PSYCHOLOGIST, PSYCHOL ASSOCS TALLAHASSEE, 87- *Mem:* Am Math Soc; Math Asn Am; Nat Coun Teachers Math. *Res:* Elementary education in mathematics and the mathematics training of elementary teachers; problem solving of children, grades 4, 5, 6; basic skills in mathematics for college students. *Mailing Add:* Psychol Assocs Tallahassee 130 Salem Ct Tallahassee FL 32301

KERR, DOUGLAS S, b Washington, DC, Nov 19, 40; m 68. COMPUTER SCIENCES. *Educ:* Yale Univ, BA, 62; Purdue Univ, MS, 64, PhD(computer sci), 67. *Prof Exp:* Asst prof, 67-71, ASSOC PROF, COMPUT & INFO SCI, OHIO STATE UNIV, 71- *Mem:* Asn Comput Mach; Inst Elec & Electronics Engrs. *Res:* Data base systems; software engineering. *Mailing Add:* Comput & Info Sci Ohio State Univ 2036 Neil Ave Columbus OH 43210

KERR, ERIC DONALD, b Gipsy, Mo, Feb 21, 30; m 71. PHYTOPATHOLOGY. *Educ:* Univ Mo, Columbia, BS, 51, MS, 60; Univ Nebr, Lincoln, PhD(plant path), 67. *Prof Exp:* Res plant pathologist, Agr Res Serv, 67; EXTEN PLANT PATHOLOGIST, PANHANDLE RES & EXT CTR, UNIV NEBR, 67- *Mem:* Soc Nematol; Am Phytopath Soc; Am Soc Sugar Beet Technol. *Res:* Control of Cercospora leaf spot on sugar beet; control of wheat disease; dry bean yield losses caused by white mold; control of nematodes on sugar beet. *Mailing Add:* 2460 Valencia Gering NE 69341

KERR, ERNEST ANDREW, b Guelph, Ont, Aug 24, 17; m 45; c 3. PLANT BREEDING. *Educ:* McMaster Univ, BA, 40; McGill Univ, MSc, 41; Univ Wis, PhD(genetics, plant path), 44. *Prof Exp:* Asst, Hort Exp Sta, Ont Ministry Agr & Food, 44-52, res assoc, 52-54, chief res scientist, Plant Breeding, 54-70, res coordr prod systs, Hort Res Inst Ont, 70-72, res scientist, 72-82; DIR PLANT BREEDING & RES, STOKES SEEDS, 83- *Mem:* Am Soc Hort Sci; Nat Sweet Corn Breeders Asn; fel Agr Inst Can; Can Soc Hort Sci; Tomato Genetics Coop; hon mem Can Phytopathological Soc. *Res:* Horticultural plant breeding, particularly tomatoes, sweet peppers and sweet corn; disease resistance; genetics and earliness in tomatoes. *Mailing Add:* Eight Eden Pl Simcoe ON N3Y 3K9 Can

KERR, FRANK JOHN, b St Albans, Eng, Jan 8, 18; m 66; c 3. RADIO ASTRONOMY. *Educ:* Univ Melbourne, BSc, 38, MSc, 40, DSc(astron), 62; Harvard Univ, MA, 51. *Prof Exp:* Res officer radiophysics, Div Radio Physics, Commonwealth Sci & Indust Res Orgn, Australia, 40-45, sr res officer radio astron, 45-55, prin res officer, 60-68; vis prof, Univ Md, 66-68, dir astron prog, 73-78, actg provost, 78-79, provost, Div Math & Phys Sci & Eng, 79-85, prof astron, 68-87, EMER PROF ASTRON, UNIV MD, COLLEGE PARK, 87- *Concurrent Pos:* Res scholar, Harvard Univ, 51-52; vis scientist, Leiden Univ, 57-58; fel, Guggenheim Found, 75; prog dir, Univ Space Res Asn, 83- *Mem:* Int Astron Union; Am Astron Soc; Royal Astron Soc; Astron Soc Australia; Australian Inst Physics. *Res:* Galactic structure; tropospheric radio propagation; moon radar; radio studies of Galaxy and Magellanic Clouds. *Mailing Add:* Astron Prog Space Sci Bldg Univ Md College Park MD 20742

KERR, GEORGE R, b Winnipeg, Man, May 15, 30; m 54; c 6. PUBLIC HEALTH NUTRITION, PEDIATRICS. *Educ:* Dalhousie Univ, MD & CM, 55. *Prof Exp:* Intern med & surg, Victoria Gen Hosp, Halifax, NS, 54-55; gen pract, 55-56; resident pediat & orthop, Gen Hosp, St Johns, Nfld, 56-57; asst resident pediat, Vancouver Gen Hosp, BC, 57-58, asst resident med, 59; instr, Univ BC, 61-62; asst prof pediat, Med Sch, Univ Ore, 62-63; from asst prof to assoc prof, Sch Med, Univ Wis-Madison, 63-71; assoc prof nutrit, Sch Pub Health, Harvard Univ, 71-77; PROF SCH PUB HEALTH, UNIV TEX HEALTH SCI CTR, 77- *Concurrent Pos:* Res fel, Vancouver Gen Hosp, BC, 59; R Samuel McLaughlin traveling fel, 59-60; res fel pediat endocrinol & metab, Med Sch, Univ Ore, 59-61; Queen Elizabeth II Can res fel, 60-61; asst dir, Ore Regional Primate Res Ctr, 62-63; res assoc, Joseph P Kennedy, Jr Mem Lab & Wis Regional Primate Res Ctr, 63-71; dir, Human Nutrit Ctr, Univ Tex, Health Sci Ctr, 77-88. *Res:* Public health maternal-child health program needs and effectiveness, nutrition and child development. *Mailing Add:* Sch Pub Health Univ Tex Health Sci Ctr PO Box 20186 Houston TX 77225

KERR, GEORGE THOMSON, b Baltimore, Md, May 7, 23; m 50; c 3. SURFACE CHEMISTRY. *Educ:* Pa State Univ, BS, 45, MS, 46, PhD(chem), 52. *Prof Exp:* Instr chem, Pa State Univ, 49-50; asst prof, Lebanon Valley Col, 50-52; res supvr, Res Div, AMP, Inc, 52-56; sr res chemist, Socony Mobil Labs, Mobil Res & Develop Corp, 56-62, res assoc, 62-75, sr res assoc, Cent Res Div Lab, 75-85; RETIRED. *Concurrent Pos:* Mem comt zeolite nomenclature, Int Union Pure & Appl Chem; mem organizing comt, Fourth Int Conf Molecular Sieves, 77. *Mem:* Int Zeolite Asn; Am Chem Soc; Int Zeolite Asn (pres, 80-83). *Res:* Synthesis and physical properties of high molecular weight hydrocarbons; kinetics of acid attack of clay minerals; synthesis and properties of crystalline zeolites. *Mailing Add:* 10 Pin Oak Dr Lawrenceville NJ 08648

KERR, HAROLD DELBERT, b Clarkton, Mo, Jan 3, 33; m 51; c 2. WEED SCIENCE. *Educ:* Univ Mo, BSc, 55, MS, 57; Wash State Univ, PhD(agron), 63. *Prof Exp:* Res agronomist, Univ Mo, Agr Res Serv, USDA, 55-57 & Wash State Univ, 58-62; res agronomist, Plant Indust Sta, Md, 62-67; asst prof weed sci, 67-71, ASSOC PROF AGRON, UNIV MO, 71- *Mem:* Weed Sci Soc Am; Am Soc Agron; Crop Sci Soc Am; Sigma Xi. *Res:* Botany of weeds and ecology of control in agronomic crops. *Mailing Add:* Six E Parkway Dr Columbia MO 65203

KERR, HUGH BARKLEY, b Maryville, Tenn, July 22, 22; m 51; c 2. MECHANICAL ENGINEERING. *Educ:* Univ Tenn, BS, 47, MS, 51. *Prof Exp:* Mech engr res & develop, Combustion Eng Co, 47-48; instr mech eng, Clemson Col, 48-51; asst prof, Univ Ala, 51-54; gen engr, Mine Detection Lab, USN, Fla, 54-55; assoc prof mech eng, Univ Miss, 55-61 & Tex Western Col, 61-62; ASSOC PROF ENG SCI & DIR, D W MATTSON COMPUT CTR, TENN TECHNOL UNIV, 62- *Concurrent Pos:* Mech engr, Int Harvester Co, 57; plant engr, Western Elec Co, 59. *Mem:* Am Soc Eng Educ. *Res:* Heat power; refrigeration and air conditioning; photoelastic stress analysis; data processing; computer sciences. *Mailing Add:* Dept Comput Sci Tenn Technol Univ Cookeville TN 38505

KERR, I LAWRENCE, b June 23, 17; m; c 4. HEALTH ADMINISTRATION. *Educ:* Univ Pa Dent Sch, DDS, 39. *Hon Degrees:* DSc, Georgetown Univ, 81. *Prof Exp:* Intern, Philadelphia Gen Hosp, 39-40, resident, Oral Surgery, 40-42; Oral surgeon, US Army Air Force, 42-46; pres, 39-47, PVT PRACT, VALLEY DENT GROUP, 63. *Concurrent Pos:* Mem Dept Health State NY Task Force Medicaid, 69; mem adv comt Dent Health, HEW, 71-72; mem, rev comt, NY-Pa Health Serv Agency, 76; mem, US Olympic Comt, comt Dent Health, 80-88. *Honors & Awards:* Jarvie-Burkhart Award; Heid Brink Award. *Mem:* Inst Med-Nat Acad Sci; Am Soc Oral & Maxillofacial Surgeons; Am Dent Asn (pres, 79-80); fel Am Dent Soc Anesthesiol. *Res:* Clinical research. *Mailing Add:* 626 Crossfield Circle Venice FL 34293-4352

KERR, JAMES S(ANFORD) S(TEPHENSON), b Vancouver, BC, Mar 9, 26; nat US; m 49; c 2. ELECTRICAL ENGINEERING, ELECTROMAGNETISM. *Educ:* Univ BC, BASc, 48; Univ Ill, MS, 49, PhD, 51. *Prof Exp:* Eng analyst elec eng, Gen Elec Co, NY, 51-56; mgr guid, tracking & eval dept, TRW Space Technol Labs, 56-65, asst lab dir antisubmarine warfare, 65-66, asst mgr, Guid & Control Opers, TRW Systs Group, 66-67, sr staff engr to gen mgr electronics systs div, 67-71, sr staff engr, Ballistic Missile Defense Prog Off, Redondo Beach, 71-76, sr staff engr spec proj, TRW Defense & Space Systs Group, 76-80, sr staff engr, Space Commun Div, 80-86, sr staff engr, Appl Technol Div, TRW Electronics Systs Group, 86-90; RETIRED. *Mem:* Sr mem Inst Elec & Electronics Engrs. *Res:* Analysis and design of space communications systems; analysis of advanced radar systems and data therefrom; management of radio guidance system and radio guidance equation development for space programs such as Mercury, Ranger and Mariner. *Mailing Add:* 1564 Prospect Pl Victoria BC V8R 5X8 Can

KERR, JAMES WILLIAM, geology, for more information see previous edition

KERR, JANET SPENCE, b New Haven, Conn; c 3. DEVELOPMENT NOVEL THERAPEUTICS. *Educ:* Beaver Col, BA, 64; Rutgers State Univ, MS, 69, PhD(physiol), 73. *Prof Exp:* Asst prof physiol, Camden Col Arts & Sci, Rutgers Univ, 73-76; NIH trainee & res assoc physiol, Sch Med, Univ Pa, 76-79; asst prof med & physiol, Rutgers Med Sch, Univ Med & Dent NJ, 79-85; SR RES ASSOC, DUPONT MERCK PHARMACEUT CO, 85- *Concurrent Pos:* Assoc mem, Grad Fac & Bur Biol Res, Rutgers Univ, 80-85; adj assoc prof, Sch Med, Univ Pa, 85-89; mem, Orgn Comt, Inflammatory Res Asn, 88-90. *Mem:* Am Physiol Soc; Am Thoracic Soc; Inflammatory Res Asn; AAAS. *Res:* Role of phospholipase A2, cytokines, and oxygen-derived free radicals in inflammation, both acute and chronic. *Mailing Add:* Du Pont Merck Pharmaceut Co PO Box 80400 E400/4223 Wilmington DE 19880-0400

KERR, JOHN M(ARTIN), b Normal, Ill, Jan 31, 34; m 56; c 4. CERAMIC & CHEMICAL ENGINEERING. *Educ:* Univ Ill, BS, 56. *Prof Exp:* Metallurgist, Oak Ridge Nat Lab, 56-61; sr engr, Babcock & Wilcox Co, 61-66, supvr nuclear ceramics, 66-71, mgr ceramics sect, 71-74, mgr nuclear fuel cycle sect, 74-81, mgr systs design & eng sect, 81-87, MGR FUELS & MAT UNIT, BABCOCK & WILCOX CO, 87- *Mem:* Fel Am Ceramic Soc; Am Soc Testing & Mat. *Res:* Ceramic bodies for waste disposal; ceramic fuel-metal compatibility studies; ceramic fuels development and development of fuels fabrication methods; nuclear ceramics; nuclear fuel cycle studies. *Mailing Add:* 1425 Brookville Lane Lynchburg VA 24502

KERR, JOHN POLK, b Little Rock, Ark, July 14, 31; m 56; c 4. ZOOLOGY, AQUATIC BIOLOGY. *Educ:* Rutgers Univ, BA, 56; Univ Calif, MS, 57; Univ Mich, PhD(zool), 62. *Prof Exp:* Asst biol, Rutgers Univ, 55-56; researcher animal behavior, Ciba Pharmaceut Co, 57-58; teaching asst zool, Univ Mich, 58-59; instr biol, Adrian Col, 61-62; res assoc fisheries, Univ Mich, 62-63; asst prof zool, Univ Ga, 63-69; assoc prof biol & marine sci, Univ WFla, 69-73; vpres & sr ecol consult, Baseline, Inc, 73-76, gen mgr & pres, 76-78; head, New South Ecosyst, Inc, 78-86; prof biol, Salem Col, 83-86; ENVIRON MGR, DEPT ENVIRON REG, STATE FLA, 86- *Concurrent Pos:* Rackham fel fisheries, Univ Mich, 62-63; vis asst prof, Ore Inst Marine Biol, Univ Ore, 65; partic, Electronics for Scientists Prof, Univ Ill, 68; faculty fel, Systs Design Prog, NASA & Am Soc Eng Educ, Auburn Univ, 70. *Mem:* Am Soc Zool; Ecol Soc Am; Am Fisheries Soc; Am Soc Ichthyol & Herpet; Am Soc Limnol & Oceanog. *Res:* Aquatic ecology; biology of vertebrates, especially fishes; ichthyology and herpetology; fisheries; marine and freshwater ecosystems, especially rivers, estuaries, wetlands, and lakes; animal behavior, especially under field conditions. *Mailing Add:* 4770 Velasquez Pensacola FL 32504

KERR, KIRKLYN M, b Green Bank, WVa, May 1, 36; m 57; c 3. VETERINARY PATHOLOGY. *Educ:* Univ WVa, BS, 61, MS, 66; Ohio State Univ, DVM, 61; Tex A&M Univ, PhD(vet path), 70; Am Col Vet Path, dipl, 68. *Prof Exp:* Vet practr, North Side Vet Clin, Carlisle, Pa, 61-62; res assoc vet microbiol & path, Univ WVa, 62-65; from instr to assoc prof vet path, Col Vet Med, Tex A&M Univ, 65-72; assoc prof vet pathobiol & dir div appl path, Col Vet Med, Ohio State Univ, 72-78; asst dean res & advan studies, Sch Vet Med & Head Vet Sci, La State Univ, 78-87; dir, Ohio Agr Res & Develop Ctr & prof poultry sci, 87-90, PROF VET PREV MED &

FAC MEM, DEPT PREV MED, COL VET MED, OHIO STATE UNIV, 91- *Mem:* Am Vet Med Asn; Int Acad Path. *Res:* Veterinary pathology; mycoplasmatacea; cancer research in animals. *Mailing Add:* Ohio State Univ ORADC 1680 Madison Ave Wooster OH 44691

KERR, MARILYN SUE, b Sumner, Ill. DEVELOPMENTAL BIOLOGY. *Educ:* Gettysburg Col, BA, 59; Duke Univ, MA, 61, PhD(zool), 66. *Prof Exp:* USPHS fel vitellogenesis, Biol Div, Oak Ridge Nat Lab, 66-69, Nat Inst Child Health & Human Develop trainee biophys, 69-70; ASST PROF BIOL, SYRACUSE UNIV, 70- *Mem:* Am Soc Zoologists; Soc Develop Biol; AAAS; Sigma Xi. *Res:* Vitellogenesis and limb regeneration correlated with studies of hemocyte origin, differentiation and functions in arthropods; hemocyanin synthesis, its isolation and characterization. *Mailing Add:* Biol Res Labs Syracuse Univ 1209 Cyman Hall Syracuse NY 13210

KERR, NORMAN STORY, b St Louis, Mo, Aug 6, 33; m 65; c 2. DEVELOPMENTAL BIOLOGY. *Educ:* Oberlin Col, AB, 54; Northwestern Univ, MS, 55, PhD(biol), 58. *Prof Exp:* Asst biol, Northwestern Univ, 54-56; from instr to prof zool, 58-76, asst dean, 66-68, assoc dean col biol sci, 69-78, PROF GENETICS & CELL BIOL, UNIV MINN, MINNEAPOLIS, 76- *Mem:* Fel AAAS; Am Soc Zool; Am Micros Soc. *Res:* Morphogenesis and genetics of the true slime molds; computer assisted instruction. *Mailing Add:* Dept Genetics & Cell Biol Univ Minn St Paul MN 55108-1095

KERR, RALPH OLIVER, b Oakland City, Ind, May 14, 26; m 46; c 4. ORGANIC CHEMISTRY. *Educ:* Ind Univ, AB, 49; Univ Ill, MS, 50, PhD(chem), 53. *Prof Exp:* Asst inorg chem, Univ Ill, 49-53; chemist org develop, Columbia Southern Chem Corp, 53-56; res assoc, Petro-Tex Chem Corp, 56-76; OWNER, INDIAN HILL PROD, 76- *Mem:* Am Chem Soc. *Res:* Reductions of hindered ketones; organic oxidations; heterogeneous oxidation catalysts; maleic catalyst development. *Mailing Add:* RR 1 Box 67 Broaddus TX 75929-0135

KERR, ROBERT LOWELL, b Dayton, Ohio, Mar 31, 36; m 91; c 3. AIRBREATHING PROPULSION, ENERGY CONVERSION & AEROSPACE POWER. *Educ:* Ohio State Univ, BS, 59; Univ Dayton, MS, 70, PhD(mech eng), 86. *Prof Exp:* Apollo-LM subsyst mgr, NASA Manned Spacecraft Ctr, Tex, 64-65; fuel cell task mgr, Aero Propulsion Lab, Wright Patterson AFB, Ohio, 59-64, develop engr, Foreign Technol Div, 65-66, aerospace engr, 66-68 & 72-73, battery task mgr, 68-72, advan develop prog mgr, 73-78, electrochem res task mgr, 78-81, ASST CHIEF SCIENTIST, AERO PROPULSION LAB, WRIGHT PATTERSON AFB, OHIO, 81- *Concurrent Pos:* Instr physics, Wilberforce Univ, 62-64; adj assoc prof mech eng, Wright State Univ, 91- *Mem:* Am Phys Soc. *Res:* Investigation of the electrochemistry of batteries and fuel cells such as nickel-cadmium, nickel hydrogen and lithium rechargeable batteries along with hydrogen-oxygen fuel cells; heat transfer in cells and fundamental electrochemical studies of lithium couples. *Mailing Add:* 5736 Mayville Dr Dayton OH 45432

KERR, ROBERT MCDOUGALL, b Harvey, Ill, Mar 22, 54; m 84; c 2. TURBULENCE, NUMERICAL METHODS. *Educ:* Univ Chicago, BA, 75, MS, 75; Cornell Univ, PhD(physics), 81. *Prof Exp:* Postdoctoral, Ames Res Ctr, NASA, 81-83 & Lawrence Livermore Nat Lab, 83-86; SCIENTIST, NAT CTR ATMOSPHERIC RES, 83- *Mem:* Am Phys Soc; Soc Indust & Appl Math. *Res:* Three-dimensional direct simulations to study isotropic turbulence, convective turbulence and vortex interactions. *Mailing Add:* Nat Ctr Atmospheric Res PO Box 3000 Boulder CO 80307

KERR, SANDRIA NEIDUS, b Youngstown, Ohio, Oct 1, 40; m 63; c 2. SCIENTIFIC PROGRAMMING. *Educ:* Col Wooster, BA, 62; Bryn Mawr Col, MA, 64; Cornell Univ, PhD(math), 71. *Prof Exp:* From asst prof to assoc prof, 71-82, PROF MATH, WINSTON-SALEM STATE UNIV, 82- *Concurrent Pos:* Collabr, Los Alamos Nat Lab, 85- *Mem:* Am Math Soc; Asn Comput Mach. *Res:* Analysis on infinite-dimensional manifolds; studies involving placement of students in math courses; scientific programming. *Mailing Add:* 1938 Fac Dr Winston-Salem NC 27106

KERR, STRATTON H, b Springfield, Mass, May 17, 24; m 48; c 2. ENTOMOLOGY. *Educ:* Univ Mass, BS, 49; Cornell Univ, PhD(entom), 53. *Prof Exp:* From asst entomologist to assoc entomologist, 53-68, ENTOMOLOGIST, UNIV FLA, 68- *Mem:* AAAS; Entom Soc Am. *Res:* Insects of turfgrass, arthropod resistance to insecticides; applied entomology. *Mailing Add:* Dept Entomol McCarty Hall Univ Fla Gainesville FL 32611

KERR, SYLVIA JEAN, b St Louis, Mo, July 2, 41. BIOCHEMISTRY. *Educ:* Smith Col, AB, 62; Columbia Univ, PhD, 67. *Prof Exp:* Asst prof surg, 70-75, ASSOC PROF BIOCHEM, UNIV COLO MED CTR, DENVER, 76- *Concurrent Pos:* Nat Cancer Inst career develop award, 71- *Mem:* Am Chem Soc; Am Soc Biol Chemists; Am Asn Cancer Res. *Res:* Biochemical control mechanisms; transfer RNA metabolism. *Mailing Add:* 4700 E Fla Ave Denver CO 80222

KERR, SYLVIA JOANN, b Detroit, Mich, June 19, 41; m 65; c 2. DEVELOPMENTAL BIOLOGY, MICROBIOLOGY. *Educ:* Carleton Col, BA, 63; Univ Minn, Minneapolis, MS, 66, PhD(zool), 68. *Prof Exp:* Asst prof biol, Augsburg Col, 68-71; res fel pharmacog, Univ Minn, 72, med fel cell biol, 74-75; asst prof biol, Anoka-Ramsey Community Col, Augsburg Col & Hamline Univ, 72-74; from asst prof to assoc prof, 76-89, PROF BIOL, HAMLINE UNIV, 89- *Mem:* Sigma Xi; AAAS. *Res:* Developmental biology-myxomycetes; planarian regeneration; population genetics-mutant allele frequencies in cats. *Mailing Add:* Dept Biol Hamline Univ St Paul MN 55104

KERR, THEODORE WILLIAM, JR, b Patterson, NJ, Nov 20, 12; m 38. ENTOMOLOGY. *Educ:* Mass State Col, BS, 36; Cornell Univ, PhD(econ entomol), 41. *Prof Exp:* Field asst, NY Exp Sta, Geneva, 35-36; asst entomologist, Cornell Univ, 37-42; sr res entomologist, US Rubber Co, 42-46; from asst res prof entomol to res prof entomol & plant path, 46-77, EMER RES PROF PLANT PATH & ENTOM, UNIV RI, 77- *Mem:* Entom Soc Am; Sigma Xi. *Res:* Biology and control of insects attacking trees, shrubs and turf. *Mailing Add:* 42 Clarke Lane Kingston RI 02881

KERR, THOMAS JAMES, b Muskogee, Okla, Oct 7, 27; m 51; c 7. MICROBIOLOGY. *Educ:* Okla A&M Univ, BS, 50; Okla State Univ, MS, 63; Univ Ga, PhD(microbiol), 76. *Prof Exp:* Asst to dir biol res, US Biol Warfare Ctr, Frederick, Md, 58-60, dep dir med res, 63-64; proj officer, US Army Test & Eval Command, Aberdeen, Md, 67-70; res asst microbiol, 71-76, DIR LAB STUDIES, UNIV GA, 76- *Mem:* Am Soc Microbiol; Sigma Xi. *Res:* Biological inhibition of fusarium moniliforme various subglutinans; the casual agent of pine pitch canker; production of single cell protein from agricultural waste products. *Mailing Add:* Athens Tech Hwy 29 St North Athens GA 30601

KERR, WARWICK ESTEVAM, b Santana de Parnaiba, Brazil, Sept 9, 22; m 56; c 7. GENETICS. *Educ:* Univ Sao Paulo, MSc, 48, PhD(genetics), 50. *Hon Degrees:* Prof Hon Causa, Univ Fed Amazonas. *Prof Exp:* Biologist genetics, Grad Sch Agr, 46-48, from asst prof to assoc prof, Univ Sao Paulo, 48-58; prof biol, Col Sci UNESP-Rio-Claro, 59-64; prof genetics, Col Med, Univ Sao Paulo, 64-75; dir, Nat Res Inst Amazon, 75-79; prof biol, Fed Univ Maranhao, Sao Luis, 81-88,; PROF GENETICS, UNIV FED UBERLANDIA, 88- *Concurrent Pos:* Rockefeller Found Fel, 51-52; Brazilian Nat Res Coun grants, USDA, 61-66; State of Sao Paulo Res Found grants, 65-67; pres, State Univ Maranhao, 87-88; hon pres, Brazilian Asn Advan Sci. *Honors & Awards:* Souzandrade Gold Medal, Univ Fed Maranhao. *Mem:* Soc Study Evolution; Brazilian Genetics Soc (pres, 64-66); Brazilian Soc Advan Sci (pres, 69-73). *Res:* Bee genetics; cytology and evolution; plant breeding. *Mailing Add:* Dept Bio-Sci Univ Fed Uberlandia Uberlandia 38400 Brazil

KERR, WENDLE LOUIS, b Harlan, Iowa, Dec 16, 17; m 43; c 2. PHARMACY. *Educ:* Univ Iowa, BS, 41, MS, 50. *Prof Exp:* From instr to asst prof, 47-61, ASSOC PROF PHARM, COL PHARM, UNIV IOWA, 61-, COORDR PHARM EXTEN SERV, 66- *Concurrent Pos:* Sta pharmacist, Univ Iowa, 46-52 & dir pharmaceut procurement, 52-66; mem, Teachers Pharm Admin & Continuing Educ. *Mem:* Am Soc Hosp Pharmacists; Am Pharmaceut Asn; Sigma XI. *Res:* Improved formulations for the oral administration of drugs; administrative pharmacy; hospital pharmacy. *Mailing Add:* Col Pharm Univ Iowa Iowa City IA 52240

KERR, WILLIAM, b Sawyer, Kans, Aug 19, 19; m 45; c 3. REACTOR SAFETY, REACTOR SHIELDING. *Educ:* Univ Tenn, BS, 42, MS, 47; Univ Mich, PhD(elec eng), 54. *Prof Exp:* Asst prof elec eng, Univ Tenn, 47-48; asst prof elec eng, 53-56, assoc prof nuclear eng, 56-58, chmn dept nuclear eng, 61-74, from assoc dir to actg dir, Mich Mem Phoenix Proj, 60-65, PROF NUCLEAR ENG, UNIV MICH, ANN ARBOR, 58-, DIR, MICH MEM PHOENIX PROJ, 65-, DIR OFF ENERGY RES, 77- *Concurrent Pos:* Proj supvr nuclear energy proj, USAID, 56-65; consult, Atomic Power Develop Assocs, 54-59, Union Carbide Nuclear Corp, 54 & USAID, 56-65; pres bd dir, Assoc Midwest Univs, 65; mem adv comt reactor safeguards, US Nuclear Regulatory Comn, 72-; mem gov task force on nuclear waste disposal, State of Mich, 76-84. *Honors & Awards:* Arthur Holly Compton Award, Am Nuclear Soc, 74. *Mem:* Fel Am Nuclear Soc; sr mem Inst Elec & Electronics Engrs; Am Soc Eng Educ; Sigma Xi. *Res:* Application of telemetering to power systems; autoradiography; nuclear reactor system dynamics; reactor control; reactor shielding; reactor safety analysis. *Mailing Add:* 3034 Phoenix Mem Lab North Campus Univ Mich Ann Arbor MI 48109

KERR, WILLIAM CLAYTON, b Steubenville, Ohio, Mar 8, 40; m 63; c 2. COMPUTER SIMULATION. *Educ:* Col Wooster, BA, 62; Cornell Univ, PhD(theoret physics), 67. *Prof Exp:* Res assoc physics, Inst Theoret Physics, Chalmers Univ Technol, Sweden, 67-68 & Solid State Sci Div, Argonne Nat Lab, 68-70; from asst prof to assoc prof, 70-83, PROF PHYSICS, WAKE FOREST UNIV, 83- *Concurrent Pos:* Vis scientist, Argonne Nat Lab, 71, 72 & 80, Chalmers Univ Technol, 74, Univ Paris, 76-77 & Los Alamos Nat Lab, 81, 86 & 87; collabr, Los Alamos Nat Lab, 84-85. *Mem:* Am Phys Soc; Am Asn Physics Teachers; Sigma Xi. *Res:* Theory of nonlinear systems; computer simulation of nonlinear, low dimensional systems of interest in statistical physics, including structural phase transitions, sine-Gordon chains and biological molecules. *Mailing Add:* Dept Physics Wake Forest Univ Box 7507 Winston-Salem NC 27109

KERREBROCK, JACK LEO, b Los Angeles, Calif, Feb 6, 28; m 53; c 3. AERONAUTICS, ASTRONAUTICS. *Educ:* Ore State Col, BS, 50; Yale Univ, MS, 51; Calif Inst Technol, PhD(mech eng), 56. *Prof Exp:* Res engr, Oak Ridge Nat Lab, 56-58; sr res fel, Calif Inst Technol, 58-60; from asst prof to prof, Mass Inst Technol, 60-75, dir, Space Propulsion Lab, 62-76, Gas Turbine Lab, 68-78, head dept, 78-81 & 83-85, assoc dean eng, 85-90, RICHARD COCKBURN MACLAURIN PROF AERONAUT & ASTRONAUT, MASS INST TECHNOL, 75- *Concurrent Pos:* Chmn, Sci & Technol Adv Group, USAF Sci Adv Bd; mem Nat Res Coun Aeronaut & Space Eng Bd; mem Am Soc Mech Engrs Turbomach Comt; NASA Adv Bd Aircraft Fuel Conserv Technol; hon prof Beijing Inst Aeronaut & Astronaut, China, 80; assoc adminr, Off Aeronaut & Space Technol, NASA, 80; mem, Nat Comn Space, 85. *Honors & Awards:* Dryden lectr, Am Inst Aeronaut & Astronaut, 80. *Mem:* Nat Acad Eng; fel Am Inst Aeronaut & Astronaut; fel Explorers Club; Am Phys Soc; sr mem Am Astronaut Soc; fel Am Acad Arts & Sci. *Res:* Aircraft propulsion; space propulsion and power generation systems; magnetohydrodynamics; nuclear rockets. *Mailing Add:* Prof Aeronaut & Astronaut Mass Inst Technol Rm 33-411 Cambridge MA 02139

KERRI, KENNETH D, b Napa, Calif, Apr 25, 34; m 58; c 2. CIVIL & SANITARY ENGINEERING. *Educ:* Ore State Univ, BS, 56, PhD(civil eng), 65; Univ Calif, Berkeley, MS, 59. *Prof Exp:* Assoc prof, 59-68, PROF CIVIL ENG, CALIF STATE UNIV, SACRAMENTO, 68- *Concurrent Pos:* Fac res award, Calif State Univ, Sacramento, 69; pres, Nat Environ Training Asn, 79-80, Asn Bds Cert, 83 & Calif Water Pollution Fedn, 83-84. *Honors & Awards:* Collection Syst Award, Water Pollution Control Fedn, 77. *Mem:* Hon mem, Water Pollution Control Fedn; Am Soc Civil Engrs; Am Water Works Asn; Am Soc Eng Educ. *Res:* Water quality economics; training manuals for operators of water and wastewater facilities. *Mailing Add:* Dept Civil Eng Calif State Univ 6000 J St Sacramento CA 95819-6025

KERRICK, DERRILL M, b Santa Cruz, Calif, Dec 27, 40; m 61; c 3. PETROLOGY, GEOCHEMISTRY. *Educ:* San Jose State Col, BS, 63; Univ Calif, Berkeley, PhD(geol), 68. *Prof Exp:* Lectr geol, Victoria Univ Manchester, 67-69; from asst prof to assoc prof petrol, 69-79, chmn geochem & mineral grad prog, 78-83, PROF PETROL, PA STATE UNIV, 79- *Concurrent Pos:* Assoc ed, Geochimica et Cosmochimica Acta. *Mem:* Fel Mineral Soc Am; Geochem Soc. *Res:* Laboratory and field investigations of metamorphic reactions and stability relations of metamorphic assemblages. *Mailing Add:* Geosci Dept Pa State Univ University Park PA 16802

KERRICK, WALLACE GLENN LEE, PHYSIOLOGY, BIOPHYSICS. *Educ:* Univ Puget Sound, Tacoma, BS, 61; Univ Wash, Seattle, PhD(physiol & biophys), 71. *Prof Exp:* Postdoctoral fel, Dept Physiol & Biophys, Univ Wash, 71-72, actg asst prof, 72-74, from asst prof to assoc prof, 74-81; assoc prof, 81-83, PROF PHYSIOL & BIOPHYS & MOLECULAR & CELLULAR PHARMACOL, UNIV MIAMI, 83- *Concurrent Pos:* Lectr, Univ Calif, San Diego, 76, Wash State Univ, 77, Univ Cincinnati, 80, Wash Univ, St Louis, 85 & Univ Calgary, 88; res grants, Muscular Dystrophy Asn, 82-91, Am Heart Asn, 82-92, NIH, 87; mem, Nat Peer Rev, Am Heart Asn, 86- *Mem:* Fedn Am Socs Exp Biol; Biophys Soc; Am Heart Asn; Sigma Xi. *Res:* Mechanism of cardiac muscle regulation by troponin; effects of weightlessness on physiological and biochemical properties of single muscle cells; author of numerous technical publications. *Mailing Add:* Dept Physiol & Biophys Sch Med Univ Miami 1600 NW Tenth Ave PO Box 016430 Miami FL 33101

KERRIDGE, KENNETH A, b London, Eng, Mar 26, 28. MEDICINAL CHEMISTRY, ORGANIC CHEMISTRY. *Educ:* Univ London, BPharm, 51, PhD(med chem), 55. *Prof Exp:* Res chemist, Smith & Nephew Res Ltd, 55-57; res chemist, Parke, Davis & Co, Eng, 57-60; sr org chemist, Arthur D Little, Inc, 60-61; sr res chemist, Armour Pharmaceut Co, 61-71; ASSOC DIR LIT SERV, BRISTOL-MYERS SQUIBB CO, 71- *Concurrent Pos:* Fel, Sch Pharm, Univ Md, 60. *Mem:* Am Chem Soc; Royal Inst Chem; fel Chem Soc; fel Pharmaceut Soc Gt Brit; Am Inst Chem. *Res:* Synthetic antibacterials, antitubercular, antifungal and antitumor agents; central nervous system stimulants and depressants; cardiovascular agents. *Mailing Add:* Bristol-Myers Squibb Co PO Box 5100 Wallingford CT 06492-7660

KERSCHNER, JEAN, b Baltimore, Md, May 31, 22. GENETICS. *Educ:* Hood Col, AB, 43; Univ Pa, PhD(zool), 50. *Prof Exp:* Chemist, E I du Pont de Nemours & Co, 43-45; lab asst, Univ Pa, 45-46; asst prof biol, Elmira Col, 50-51; histologist, Army Chem Ctr, 51-52; from asst prof to assoc prof, 52-68, PROF BIOL, WESTERN MD COL, 68- *Concurrent Pos:* Fac fel, NSF, Columbia Univ, 60-61. *Mem:* Genetics Soc Am; Sigma Xi. *Res:* X-ray induced mutations in Drosophila. *Mailing Add:* Rte 3 Box 213 Hayesville NC 28904

KERSEY, JOHN H, b Apr 22, 38. LEUKEMIA. *Educ:* Univ Minn, MD, 64. *Prof Exp:* PROF LAB MED, PATH & PEDIAT, SCH MED, UNIV MINN, 77- *Mem:* Am Asn Cancer Res; Am Soc Hemat; Am Soc Clin Investrs; Am Asn Pathologists. *Res:* Lymphoma. *Mailing Add:* Dept Lab Med & Path Univ Minn UMHC 420 Delaware St SE Minneapolis MN 55455

KERSEY, ROBERT LEE, JR, b Richmond, Va, Nov 6, 22; m 43. ANALYTICAL CHEMISTRY. *Educ:* Univ Richmond, BS, 48. *Prof Exp:* Asst proj chemist, Standard Oil Co, Ind, 48-53; chemist, 53-65, spec asst to dir res, 65-70, mgr prod develop, 70-75, dir res, 75-78, VPRES, LIGGETT & MYERS TOBACCO CO, 78-, CHIEF RES OFFICER, 81- *Mem:* Am Chem Soc; Inst Food Technol. *Res:* Tobacco and tobacco products research and development. *Mailing Add:* RR 2 No 212 Morehead City NC 28557

KERSHAW, DAVID STANLEY, b Missoula, Mont, May 20, 43; m 66; c 3. THEORETICAL PHYSICS, COMPUTER SCIENCE. *Educ:* Harvard Univ, BA, 65; Univ Calif, Berkeley, PhD(physics), 70. *Prof Exp:* Fel theoret partical physics, Stanford Linear Accelerator Ctr, 70-72 & Dept Physics & Astron, Univ Md, College Park, 72-74; PHYSICIST LASER FUSION, LAWRENCE LIVERMORE LAB, UNIV CALIF, 74- *Mem:* Am Phys Soc. *Res:* Computer simulation of the physics of laser fusion. *Mailing Add:* Lawrence Livermore Lab L477 PO Box 808 Livermore CA 94550

KERSHAW, KENNETH ANDREW, b Morecambe, Eng, Sept 5, 30; m 67; c 3. PLANT ECOLOGY, LICHENOLOGY. *Educ:* Manchester Univ, BS, 52; Univ Wales, PhD(ecol), 57, DSc, 68. *Prof Exp:* Lectr, Imp Col, Univ London, 57-63; sr lectr, Secondment to Ahmadu Bell Univ, N Nigeria, 63-65; lectr, Imp Col, Univ London, 65-68; PROF BIOL, MCMASTER UNIV, 69- *Mem:* Brit Lichen Soc; fel Royal Soc Can. *Res:* Ecology of northern plant systems with special emphasis on the interaction of microclimate and plant physiology; lichen physiology. *Mailing Add:* Dept Biol McMaster Univ 1280 Main St W Hamilton ON L8S 4L8 Can

KERSHENBAUM, AARON, b Brooklyn, NY, Oct 9, 48; m 70; c 2. ALGORITHMS, NETWORK DESIGN. *Educ:* Polytech Inst New York, BS & MS, 70, PhD(elec eng), 76. *Prof Exp:* Vpres software, Network Anal Group, 69-78; from asst prof to prof elec & comput sci, Polytech Univ, 78-90; RES MGR, IBM, 90- *Concurrent Pos:* Assoc ed, J Telecommun Networks, 83-85, Networks, 83-85; dir, Network Design Lab, Polytech Univ, 84- *Mem:* Asn Comput Mach; Inst Elec & Electronics Engrs. *Mailing Add:* IBM T J Watson Res Ctr PO Box 704 Yorktown Heights NY 10598

KERSHENSTEIN, JOHN CHARLES, b New York, NY, Sept 23, 41; m 68; c 1. PHYSICS, ELECTROOPTICS. *Educ:* Georgetown Univ, BS, 64, MS, 67, PhD(physics), 69. *Prof Exp:* Fel physics, 68-69; RES PHYSICIST, US NAVAL RES LAB, 69- *Mailing Add:* 11842 Clara Way Fairfax Station VA 22039

KERSHNER, CARL JOHN, b Lima, Ohio, Dec 15, 34; m 58; c 2. PHYSICAL CHEMISTRY, RADIOCHEMISTRY. *Educ:* Capital Univ,BS, 56; Univ Ohio, PhD(inorg chem), 61. *Prof Exp:* Sr res chemist, Monsanto Res Corp, 61-64, group leader radiochem res, 64-66, Sect mgr, 66-68, sr res specialist, 68-70, sci fel, Mound Lab, 70-79, sr sci fel, 79-85; VPRES, FEMTO-TECH, INC, 86- *Concurrent Pos:* Adj prof, Ohio State Univ, 80-81. *Mem:* Am Inst Physics; Am Chem Soc; Am Nuclear Soc; Sigma Xi. *Res:* High temperature radiochemical research; syntheses and physical property determinations; isotope separation; radioactive waste and emission control; laser photochemistry; radiation detection instrumentation. *Mailing Add:* 2123 Timberidge Circle Dayton OH 45459

KERST, A(L) FRED, b Greeley, Colo, June 3, 40; m 62; c 3. ORGANIC CHEMISTRY, PHYSICAL CHEMISTRY. *Educ:* Colo State Univ, BS, 62, MS, 63; Harvard Univ, PhD(chem), 67. *Prof Exp:* Sr res chemist, Monsanto Co, 67-68; mgr, Gates Rubber Co, 69-71; vpres res & develop, Mich Chem Corp, 71-76, Velsicol Chem Corp, 76-77; vpres res & develop, 77-84, vpres mkt, 84-89, PRES, CALGON CORP, 89- *Concurrent Pos:* Eval Panel, Nat Bur Standards Ctr Fire Res, 76-79. *Mem:* Am Chem Soc. *Res:* Chemistry of phosphorus compounds, flame retardants, polymers; agriculture pesticides and industrial biocides; water treatment chemicals. *Mailing Add:* Calgon Corp Box 1346 Pittsburgh PA 15230

KERST, DONALD WILLIAM, b Galena, Ill, Nov 1, 11; m 40; c 2. PLASMA PHYSICS. *Educ:* Univ Wis, BA, 34, PhD(physics), 37. *Hon Degrees:* ScD, Univ Wis, 61; ScD, Lawrence Col, 42; Dr, Univ Sao Paulo, Brazil, 53; ScD, Univ Ill, 87. *Prof Exp:* X-ray tube developer, Gen Elec X-ray Corp, 34-37; from instr to prof physics, Univ Ill, 38-57; tech dir, Midwestern Univ Res Asn, 53-57; thermonuclear proj, John Jay Hopkins Lab Pure & Appl Sci, Gen Atomic Div, Gen Dynamics Corp, 57-62; E M Terry prof physics, 62-80, EMER PROF, UNIV WIS-MADISON, 80- *Concurrent Pos:* Physicist, Res Lab, Gen Elec Co, 40-41 & Manhattan Dist, Los Alamos Sci Lab, Calif, 43-45. *Honors & Awards:* Comstock Prize, Nat Acad Sci, 43; Scott Award, Franklin Inst, 47; Maxwell Prize, Am Phys Soc, 84; R R Wilson Prize, Am Phys Soc. *Mem:* Nat Acad Sci; AAAS; fel Am Phys Soc; Am Acad Arts & Sci. *Res:* Nuclear physics; enriched chain reactor; electrostatic accelerator; betatron and thermonuclear research; x-ray tubes and radiography; fixed field alternating gradient accelerator developments. *Mailing Add:* Dept Physics Univ Wis Madison WI 53706

KERSTEIN, MORRIS D, b Trenton, NJ, Jan 13, 38; m 80; c 1. VASCULAR TRAUMA. *Educ:* Colgate Univ, AB, 59; Chicago Med Sch, 63. *Prof Exp:* Instr surg, Tufts Univ Sch Med, 70-71; from asst prof to assoc prof surg, Yale Univ Sch Med, 71-77; assoc prof, surg, Pritzker Sch Med, Univ Chicago, 77-79; prof surg, Tulane Univ Sch Med, 79-88, assoc dean, acad affairs & dir, grad & postgrad prog, 86-88; DEISSLER PROF & CHMN, DEPT SURG, HAHNEMANN UNIV, PHILADELPHIA, PA, 88- *Concurrent Pos:* Resident surg serv, Boston City Hosp, 69-70, chief resident, 70-71; staff surgeon, Vet Admin Hosp, West Haven, Conn, 71-77; chief, peripheral vascular surg, Michael Reese Hosp, Chicago, 77-79; surgeon, 81-; clin prof, surg, Tulane Univ Sch Med, 83- *Mem:* Am Col Surgeons; Am Heart Asn; AMA; Soc Vascular Surg; Am Surg Asn; Soc Univ Surgeons. *Res:* Prostoglandin metabolism in human vessels; vascular trauma; non-invasive laboratory. *Mailing Add:* Dept Surg MS 413 Hahnemann Univ Broad & Vine Philadelphia PA 19102-1192

KERSTEN, MILES S(TOKES), b St Paul, Minn, Aug 12, 13; m 38; c 2. ENGINEERING. *Educ:* Univ Minn, BCE, 34, MS, 36, PhD(hwys, soils), 45. *Prof Exp:* Soils engr, State Hwy Dept, Minn, 36-37; instr hwys & soils, Univ Minn, 37-44; spec investr, Hwy Res Bd, Washington, DC, 44-45; from asst prof to prof, 45-78, EMER PROF SOIL MECH, HWYS & SOILS, UNIV MINN, MINNEAPOLIS, 78- *Honors & Awards:* Harold R Peyton Award, Am Soc Civil Eng, 89. *Mem:* Hon mem Am Soc Civil Eng; Am Soc Eng Educ; Nat Asn Prof Engrs. *Res:* Thermal conductivity of soil; soil stabilization; subgrade moisture conditions and their role in flexible pavement design; airport engineering; general civil engineering; surveying and mapping. *Mailing Add:* 4300 W River Pkwy Apt 619 Minneapolis MN 55406-3662

KERSTEN, ROBERT D(ONAVON), b Carlinville, Ill, Jan 30, 27; m 50; c 2. ENGINEERING EDUCATION & WATER RESOURCES. *Educ:* Okla State Univ, BS, 49, MS, 56; Northwestern Univ, PhD(fluid mech), 61. *Prof Exp:* Hydraul engr, US Dept Interior, 49-53; res assoc civil eng, Okla State Univ, 53-56, asst prof, 56; res eng, Jersey Prod Res Co, Standard Oil NJ, 56-57; from asst prof to assoc prof eng, Ariz State Univ, 57-60, prof civil eng & chmn, 60-68; dir univ res, 68-69, dean, Col Eng, 68-87, PROF ENG, UNIV CENT FLA, 68- *Concurrent Pos:* Mem, NASA-NSF Conf Lunar Explor, 62, Flight Safety Found, 62, Col Bus Exchange Prog Found Econ Ed, 63 & NASA-Cambridge Conf Explor Mars & Venus, 65; vis scholar, Stanford Univ, 66; actg dir, Fla Solar Energy Ctr, 75; chmn bd trustees, Inst Cert Eng Technicians, 78-80; mem, Eng Accreditation Comn, 82-; Fla Bd Prof Engrs, 80-86. *Mem:* AAAS; Soc Eng Educ; Am Soc Civil Engrs; Nat Soc Prof Engrs (vpres, 84-85); Sigma Xi; Nat Coun Eng Examr. *Res:* Fluid mechanics, including fluid turbulence, turbulent diffusion, non-Newtonian flow, two phase flow; water resources engineering; applied mathematics. *Mailing Add:* Dept Civil & Environ Eng Univ Central Fla PO Box 25000 Orlando FL 32816

KERSTETTER, REX E, b Ashland, Kans, Nov 22, 38; m 60; c 2. PLANT PHYSIOLOGY. *Educ:* Ft Hays Kans State Col, BS, 60, MS, 63; Fla State Univ, PhD(plant physiol), 67. *Prof Exp:* Instr biol, Fla State Univ, 67; from asst prof to assoc prof, 67-80, PROF BIOL, FURMAN UNIV, 80- *Mem:* AAAS; Am Soc Plant Physiologists; Bot Soc Am; Sigma Xi. *Res:* Plant hormone physiology; plant tissue culture; peroxidase isoenzymes. *Mailing Add:* Dept Biol Furman Univ Greenville SC 29613

KERSTETTER, THEODORE HARVEY, b Milwaukee, Wis, Dec 16, 30; div; c 2. FISH PHYSIOLOGY, ELECTROLYTE BALANCE. *Educ:* Univ Nev, Reno, BS, 59; Wash State Univ, MS, 62, PhD(zoo physiol), 69. *Prof Exp:* Instr biol, Peninsula Col, 63-64; NIH fel, Wash State Univ, 69-70; dir marine lab

& sea grant prog, 74-79, from asst prof to assoc prof, 70-79, PROF ZOOL, HUMBOLDT STATE UNIV, 79- *Concurrent Pos:* Dir, Marine Lab & Sea Grant Prog, 74-79. *Mem:* Am Soc Zoologists. *Res:* Water and ion balance in lower vertebrates; mechanisms of ion transport through epithelia; fish physiology. *Mailing Add:* Dept Biol Humboldt State Univ Arcata CA 95521

KERSTING, EDWIN JOSEPH, b Ottawa, Ohio, Nov 4, 19; m 46; c 2. VETERINARY MEDICINE. *Educ:* Ohio State Univ, DVM, 52; Univ Conn, MS, 64. *Prof Exp:* Pvt pract, 52-61; res asst, Univ Conn, 61-62, assoc prof animal diseases, 62-65, asst dean, Col Agr & Natural Resouces & dir, Ratcliffe Hicks Sch Agr, 65-66, prof animal diseases, deal Col Agr & Natural Resources, dir, Storrs Agr Exp Sta & Coop Exten Serv, 66-88, Title XII Officer 77-88; RETIRED. *Concurrent Pos:* Consult staff, Hartford Hosp, 63-; mem adv coun, Col Vet Med, Cornell Univ, 78. *Mem:* Am Vet Med Asn; AAAS; Royal Soc Health; NY Acad Sci; Sigma Xi. *Res:* Clinical veterinary medicine. *Mailing Add:* 97 Sycamore Lane Apt C Manchester CT 06040

KERTAMUS, NORBERT JOHN, b Murray City, Utah, Oct 12, 32; m 56; c 3. FUEL & CHEMICAL ENGINEERING. *Educ:* Univ Utah, BS, 60, PhD(fuels eng), 64. *Prof Exp:* Res engr, Phillips Petrol Co, 64-66; sr res chemist nuclear fuels, Idaho Nuclear Co, 66-70; res specialist chem & combustion, Babcock & Wilcox Co, 70-75; sr process engr, C F Braun Engrs, 75-79; SR RES SCIENTIST, SOUTHERN CALIF EDISON, 79- *Mem:* Am Chem Soc; Am Inst Chem Engrs. *Res:* Conversion of coal to gases and/or liquids; hydrogen processing and catalysis; combustion. *Mailing Add:* 1027 Entrada Way Glendora CA 91740-2227

KERTESZ, ANDREW (ENDRE), b Budapest, Hungary, Oct 10, 38; US citizen; m 63; c 2. VISION, BIOMEDICAL ENGINEERING. *Educ:* McGill Univ, BEng, 63; Northwestern Univ, Evanston, MS, 66, PhD(bioeng), 69. *Prof Exp:* Asst prof elec eng, Univ Pittsburgh, 69-72; PROF ELEC ENG, PSYCHOL & BIOMED ENG, NORTHWESTERN UNIV, EVANSTON, 72-, CHMN BIOMED ENG, 83- *Concurrent Pos:* Sr res fel appl sci, Calif Inst Technol & prin investr, USPHS res grant, 70-; NIH res career award, 74; clin assoc ophthal, Evanston Hosp, 77. *Mem:* Inst Elec & Electronics Engrs; Asn Res Vision & Ophthal; Biomed Eng Soc; Optical Soc Am; Nat Eyetracking Study Group (pres, 77-). *Res:* Human binocular information processing; binocular vision. *Mailing Add:* Biomed Eng Dept 2145 Sheridan Rd Northwestern Univ Evanston IL 60208

KERTESZ, JEAN CONSTANCE, b New York, NY, Sept 3, 43. PHARMACEUTICAL CHEMISTRY. *Educ:* Northwestern Univ, BA, 63; Univ Southern Calif, PhD(pharmaceut chem), 70. *Prof Exp:* Res asst biochem, Sch Med, 65-66, res assoc biomed chem, 68-77, ASST RES PROF, SCH PHARM, UNIV SOUTHERN CALIF, 77- *Mem:* Am Chem Soc; Intra-Sci Res Found; Int Soc Magnetic Resonance; Sigma Xi. *Res:* Free radical intermediates in biological systems; utilization of electron spin resonance techniques for biomedical applications; molecular mechanisms of radiation damage; protection processes; carcinogenesis. *Mailing Add:* 2329 W 2nd St Apt No 7 Los Angeles CA 90057

KERTESZ, MIKLOS, b Budapest, Hungary, July 15, 48; m 69; c 2. CHEMISTRY, QUANTUM CHEMISTRY OF SOLIDS. *Educ:* Eotvos L Univ, Budapest, dipl, 71, Dr rer nat, 79; Hungarian Acad Sci, Cand phys sci, 78. *Prof Exp:* Res scientist, Cent Res Inst, Hungarian Acad Sci, 71-79; res fel, Quantum Theory Proj, Univ Fla, 79-80 & Dept Chem, Cornell Univ, 80; sr scientist, Cent Res Inst, Hungarian Acad Sci, 81; res assoc, Dept Chem, Cornell Univ, 82-83; PROF, DEPT CHEM, GEORGETOWN UNIV, 83- *Concurrent Pos:* Camille & Henry Dreyfus teacher-scholar, 84-89. *Mem:* Am Chem Soc; Am Phys Soc. *Res:* Theoretical solid state chemistry. *Mailing Add:* Dept Chem Georgetown Univ Washington DC 20057

KERTH, LEROY THOMAS, b Visalia, Calif, Nov 23, 28; m 50; c 4. PHYSICS. *Educ:* Univ Calif, AB, 50, PhD, 57. *Prof Exp:* Asst, 50-57, res physicist, 57-65, assoc dean col lett & sci, 66-70, PROF PHYSICS, LAWRENCE BERKELEY LAB, UNIV CALIF, 65- *Concurrent Pos:* Assoc dir & div head comput sci, 86-87, assoc lab dir gen sci, 87-89; assoc lab dir, Sci & Tech Resources, 89- *Mem:* Fel Am Phys Soc. *Res:* High energy physics; weak interactions photon positron interactions. *Mailing Add:* Dept Physics Lawrence Berkeley Lab Berkeley CA 94720

KERTZ, ALOIS FRANCIS, b Bloomsdale, Mo, Sept 15, 45; m 69; c 4. ANIMAL NUTRITION. *Educ:* Univ Mo-Columbia, BS, 67, MS, 68; Cornell Univ, PhD(animal nutrit), 74. *Prof Exp:* Nutrit officer, US Army Natick Labs, 68-69; food supply & mgt officer, US Army Depot, Sattahip, Thailand, 69-70; res asst animal nutrit, Cornell Univ, 70-73; res nutritionist, 73-75, mgr res, dairy res dept, 75-85, RES MGR, RUMINANT RES DEPT, PURINA MILLS INC, 85- *Mem:* Am Dairy Sci Asn; Am Soc Animal Sci; AAAS; Nutrit Today Soc; Am Inst Nutrit. *Res:* Efficiency of nitrogen and energy utilization by calves and dairy cattle; evaluation and utilization of common feedstuffs and other by-products. *Mailing Add:* Purina Mills Inc PO Box 66812 St Louis MO 63166

KERTZ, GEORGE J, b Bloomsdale, Mo, Dec 10, 33; m 68. MATHEMATICS, COMPUTER SCIENCE. *Educ:* Cardinal Glennon Col, BA, 55; St Louis Univ, MA, 63, PhD(math), 66; Univ Mich, MSE, 78. *Prof Exp:* Mem staff, Int Bus Mach Corp, 58-60; from asst prof to assoc prof, 66-78, PROF MATH, UNIV TOLEDO, 78- *Mem:* Asn Comput Mach; Math Asn Am. *Res:* Database. *Mailing Add:* Univ Toledo Toledo OH 43606

KERWAR, SURESH, b Madras, India, May 9, 37; US citizen; m 64; c 2. MOLECULAR BIOLOGY. *Educ:* Univ Madras, BS, 56; Univ Nagpur, MS, 58; Ore State Univ, PhD(microbiol), 64. *Prof Exp:* Res assoc biochem, Scripps Clin & Res Found, 63-67; fel, Brandeis Univ, 67-69; res assoc, Roche Inst Molecular Biol, NJ, 69-71, asst mem staff, 71-75; sr scientist, Div Metab Dis, Ciba-Geigy Inc, 75-77; GROUP LEADER, CONNECTIVE TISSUES RES SECT, LEDERLE LABS, 77- *Concurrent Pos:* Am Soc Biol Chemists travel grant, 73 & 76. *Mem:* Am Soc Biol Chemists. *Res:* Protein synthesis; regulation of connective tissue metabolism; animal models of joint diseases. *Mailing Add:* Lederle Labs Pearl River NY 10965

KERWIN, EDWARD MICHAEL, JR, b Oak Park, Ill, Apr 20, 27; m 53; c 9. ACOUSTICS. *Educ:* Mass Inst Technol, SB & SM, 50, ScD(elec eng, acoustics), 54. *Prof Exp:* Consult acoustics, Bolt Beranek & Newman, Inc, 50-54; asst elec eng, Mass Inst Technol, 50-54; consult acoust, 54-67, sr consult, INC, 67- *Concurrent Pos:* Assoc staff, Peter Bent Brigham Hosp, Boston, 72-74. *Mem:* Fel Acoust Soc Am; Soc Indust & Appl Math; sr mem Inst Elec & Electronics Engrs. *Res:* Underwater sound; noise and vibration control; vibration damping; sound radiation and transmission. *Mailing Add:* Three Legion Rd Weston MA 02193

KERWIN, JOHN LARKIN, b Quebec, Que, June 22, 24; m 50; c 8. PHYSICS. *Educ:* St Francis Xavier Univ, BSc, 44; Mass Inst Technol, MSc, 46; Laval Univ, DSc, 49. *Hon Degrees:* LLD, St Francis Xavier Univ, 70, Univ Toronto, 73, Concordia Univ, 76, Univ Alta, 83, Dalhousie Univ, 83; DSc, Univ BC, 73, McGill Univ, 74, Mem Univ & DCL, Bishops Univ, 78, Univ Ottawa, 81, Royal Mil Col, Kingston, 82, Univ Winnipeg, 83, Univ Windsor, 84, Univ Moncton, 85. *Prof Exp:* Lectr, 46, from asst prof to prof physics, 48-80, chmn dept, 61-67, vdean fac sci, 67-68, acad vrector, 69-72, rector, Laval Univ, 72-77,; pres, nat res coun Can, 80-89; PRES, CAN SPACE AGENCY, 89- *Concurrent Pos:* Mem, Defence Res Bd Can, 71-77; mem standing comt int rels, Nat Res Coun Can, 72-80; pres, Int Union Pure & Appl Physics; mem bd dirs, Can-France-Hawaii Telescope Corp, Nat Res Coun Can, 73-78; pres, Asn Univs & Cols Can, 74; mem adv coun, Order of Can, 75; Can rep, working group res & develop, Int Econ Summit, 82- *Honors & Awards:* Gov Gen Medal, 44; Laureate of Lit & Sci Competition of Prov of Que, 51; Pariseau Medal, Fr-Can Asn Advan Sci, 65; Centenary Medal, Romania, 67; Gold Medal, Can Asn Physicists, 69; Companion, Order Can, 80; Gold Medal, Can Coun Prof Engrs, 82; National Order Quebec, 88. *Mem:* Am Phys Soc; Can Asn Physicists (vpres, 53, pres, 54); Fr-Can Asn Advan Sci; fel Royal Soc Can (pres, 76); Nat Sci & Eng Coun Can (vpres, 78-80); fel Royal Soc Arts; fel AAAS; fel Am Inst Physics; Can Acad Eng (vpres, 87, pres, 89). *Res:* Mass spectrometry; atomic and molecular structure. *Mailing Add:* Can Space Agency 500 Blvd René-Léveque Quest Place Air Can 3e Etage Montreal PQ H2Z 1Z7 Can

KERWIN, RICHARD MARTIN, b West Chester, Pa, Apr 5, 22; m 57; c 3. BACTERIOLOGY. *Educ:* Dartmouth Col, AB, 47; Univ NH, MS, 49; Pa State Univ, PhD(bact), 56. *Prof Exp:* Sr res scientist bact, Wyeth Labs Div, Am Home Prod Corp, 52-75; consult, 75-85; instr, Dublin Sch, Dublin, NH, 78-88, head Sci Dept, 83-86; RETIRED. *Mem:* Am Chem Soc; Am Soc Microbiologists; Soc Indust Microbiologists; Sigma Xi. *Res:* Antibiotics; screening new antibiotics; bacterial fermentation products; microbiological production of enzymes and steroids; anti-cancer research. *Mailing Add:* Box 117 Old Dublin Rd Hancock NH 03449

KERWIN, WILLIAM J(AMES), b Portage, Wis, Sept 27, 22; m 47; c 3. ELECTRICAL ENGINEERING, PHYSICS. *Educ:* Univ Redlands, BS, 48; Stanford Univ, MS, 54, PhD(elec eng), 67. *Prof Exp:* Aeronaut res scientist, Nat Adv Comt Aeronaut & NASA, Ames Res Ctr, 48-62; head electronics dept, Stanford Linear Accelerator Ctr, 62; chief space tech br, Ames Res Ctr, NASA, 62-64, chief electronics res br, 64-69; PROF ELEC ENG, UNIV ARIZ, 69- *Concurrent Pos:* Lectr, Stanford Univ, 56-61 & 68; asst prof, San Jose State Col, 63-68; NASA res awards, 70 & 75. *Honors & Awards:* Centennial Medal, Inst Elec & Electronic Engrs, 84. *Mem:* Fel Inst Elec & Electronics Engrs. *Res:* Network and circuit theory, especially synthesis of active resistance-capacitance networks; DC-DC power conversion; switched-mode. *Mailing Add:* 1981 Shalimar Way Tucson AZ 85704

KERZMAN, NORBERTO LUIS MARIA, b Buenos Aires, Arg, Feb 1, 43. ANALYSIS, COMPLEX VARIABLES. *Educ:* Univ Buenos Aires, Lic, 66; NY Univ, PhD(math), 70. *Prof Exp:* Asst prof math, Princeton Univ, 70-73; asst prof math, Mass Inst Technol, 73-76, assoc prof, 76-79; assoc prof math, 79-82, PROF MATH, UNIV NC, CHAPEL HILL, 82- *Concurrent Pos:* Vis mem staff, Göttingen, Münster, Florence, Paris & Marseille; Sloan fel, Stockholm, Amsterdam, Zurich, 73-75. *Mem:* Am Math Soc. *Res:* Complex analysis in particular several complex variables, involving methods of partial differential equations and singular integrals; conformal mapping. *Mailing Add:* Dept Math Univ NC Chapel Hill NC 27514

KESARWANI, ROOP NARAIN, b Kanpur, India, July 6, 32; Can citizen; m 57; c 2. THEORETICAL PHYSICS. *Educ:* Univ Lucknow, BSc, 52, Hons, 53, MSc, 54, PhD(math), 57. *Prof Exp:* Asst prof math, Univ Lucknow, 54-58; assoc prof, G S Tech Inst, Indore, 58-59; lectr, Punjab Univ, India, 59-61; asst prof, Wash Univ, 61-62; assoc prof, Wayne State Univ, 62-64; assoc prof, 64-70, PROF MATH, UNIV OTTAWA, ONT, 70- *Honors & Awards:* Swami Rama Tirtha Gold Medal & Debi Sahai Misra Gold Medal, Univ Lucknow, 54. *Mem:* Am Math Soc; Math Asn Am; Indian Math Soc; Can Math Soc; Soc Indust & Appl Math; Can Appl Math Soc. *Res:* Theory of functions of a complex variable; special functions and applications to theoretical physics; integral transforms and equations; quantum mechanics. *Mailing Add:* 20 Bittern Ct Rockcliffe Park ON K1L 8J8 Can

KESHAVAN, H R, b Bangalore, India, Apr 20, 49; US citizen; m 79; c 3. GENERAL COMPUTER SCIENCES, ELECTRICAL ENGINEERING. *Educ:* Bangalore Univ, BS, 69; Southern Methodist Univ, PhD(elec eng), 77. *Prof Exp:* Asst prof, Bangalore Univ, 71-74; scientist, Rockwell Int, 77-80; sr mem tech staff, Northrop Corp, 80-85; MGR, AUTOMATION SCI LAB, NORTHROP RES, 85- *Concurrent Pos:* Vis prof, Univ Southern Calif, 89- *Mem:* Sr mem Inst Elec & Electronic Engrs. *Res:* Computer science; intelligent systems; artificial intelligence; automation. *Mailing Add:* 6902 Verde Ridge Rd Rancho Palos Verdes CA 90274

KESHAVAN, KRISHNASWAMIENGAR, b Hassan, India, June 5, 29; m 57; c 3. CIVIL ENGINEERING, ENVIRONMENTAL ENGINEERING. *Educ:* Univ Mysore, BSc, 50, BE, 55; Univ Iowa, MS, 60; Cornell Univ, PhD(sanit eng), 63. *Prof Exp:* Sect officer, Cent Pub Works Dept, Govt of India, 55-58; asst civil eng, Cornell Univ, 60-63; assoc prof, Univ Maine, Orono, 63-67; assoc prof, 67-76, head dept, 76-86, PROF CIVIL ENG,

WORCESTER POLYTECH INST, 76- *Concurrent Pos:* Water Resources Ctr res grant, Univ Maine, Orono, 65-67; Off Water Resources Res grant thermal pollution & NSF grant & chlorination, 71; co-dir, Environ Systs Study Prog, Sloan Found; sr adv, UNESCO, 75-76, consult, 76- *Mem:* Am Soc Civil Engrs; Am Soc Eng Educ; Water Pollution Control Fedn; Am Water Works Asn. *Res:* Kinetics of biological treatment of organic liquid wastes; nitrification of natural bodies of water and its effect on oxygen utilization; combined effects of thermal and organic pollution; hazardous chlorinated compounds due to chlorination. *Mailing Add:* Dept Civil Eng Worcester Polytech Inst Worcester MA 01609

KESHAVIAH, PRAKASH RAMNATHPUR, b Bangalore, India, Feb 15, 45; c 2. ARTIFICIAL ORGANS, END-STAGE RENAL DISEASE. *Educ:* Indian Inst Technol, Madras, BTech, 67; Univ Minn, MS, 70, PhD(mech eng), 74, MS, 80. *Prof Exp:* Process Planning Engr, Larsen & Toubro Ltd, India, 67-68; res asst, Dept Mech Eng, Univ Minn, 68-73; BIOMED ENGR, REGIONAL KIDNEY DIS CTR, 73-, MGR BIOENG, 76-; DIR CLIN ONCOL, BAXTER HEALTH CARE CORP. *Concurrent Pos:* Consult & mem, Artificial Kidney-Chronic Uremia Adv Comt, Nat Inst Arthritis, Metab & Digestive Dis, 78-79; prin investr, Dept Health Educ & Welfare, Food & Drug Admin, 77-81; res assoc, Dept Med, Hennepin County Med Ctr & Univ Minn, 77-81, sr res assoc, 81- *Mem:* Am Soc Artificial Internal Organs; Int Soc Artificial Organs. *Res:* Basic physiology, kinetic modeling and engineering design aspects of therapies for end-state renal disease and kinetic modeling of biological systems. *Mailing Add:* Baxter Health Care Corp 825 S 8th St Suite 722 Minneapolis MN 55404

KESHOCK, EDWARD G, b Campbell, Ohio, Mar 2, 35; m 59; c 3. FLUID MECHANICS, THERMODYNAMICS. *Educ:* Univ Detroit, BME, 58; Okla State Univ, MS, 66, PhD(mech eng), 68. *Prof Exp:* Res engr, Lewis Res Ctr, NASA, 58-64; res asst film boiling heat transfer, Okla State Univ, 64-67; asst prof mech eng, Cleveland State Univ, 67-69; assoc prof thermal eng, Old Dominion Univ, 69-77; assoc prof, 77-80, PROF MECH & AEROSPACE ENG, UNIV TENN, 80- *Concurrent Pos:* Vis scientist, Nat Sci Coun, Repub China & Nat Tsing Hua Univ, Taiwan, 74-75; vis prof, Univ Petrol & Minerals, Saudi Arabia, 83-85. *Mem:* Am Soc Mech Engrs; Am Soc Eng Educ; Am Inst Chem Engrs. *Res:* Boiling heat transfer and two-phase flow; condensation heat transfer; thermophysical properties; phase change heat transfer; multiphase flow and heat transfer; energy systems; measurement techniques in heat transfer and fluid mechanics. *Mailing Add:* Dept Mech Eng Univ Tenn Knoxville TN 37996

KESIK, ANDRZEJ B, b Warsaw, Poland, Oct 27, 30; m 64; c 2. PHYSICAL GEOGRAPHY, REMOTE SENSING. *Educ:* Marie Curie-Sklodowska Univ, MSc, 51, PhD(geog), 59; Int Training Centre Aerial Surv, Holland, dipl geomorphol, 63. *Prof Exp:* Sr asst phys geog, Marie Curie-Sklodowska Univ, 55-60, lectr, 60-64, lectr cartog & photo interpretation, 64-70; vis prof photo interpretation & geomorphol, 70-71, assoc prof photo interpretation, 71-81, PROF GEOG, UNIV WATERLOO, 76- *Concurrent Pos:* Govt training grant, Univ Amsterdam, 63; nat reporter, Comn VII, Int Soc Photogram, 68-70; Brit Coun scholar, 70. *Mem:* Can Inst Surveying; Am Soc Photogram; Can Asn Geogr; Am Asn Geogr. *Res:* Physical elements of the environment; application of remote sensing techniques to the land evaluation; geomorphological mapping. *Mailing Add:* Dept Geog Univ Waterloo Waterloo ON N2L 3G1 Can

KESKKULA, HENNO, b Tartu, Estonia, Mar 25, 26; nat US; m 52; c 2. MATERIAL SCIENCE. *Educ:* Davis & Elkins Col, BS, 49; Univ Cincinnati, MS, 51, PhD(org chem), 53. *Prof Exp:* Chemist, Dow Chem Co, 53-56, group leader, 56-68, assoc scientist, 68-82; vis scholar, 82-83, RES FEL, UNIV TEXAS, 83- *Concurrent Pos:* Sr vis res fel, Queen Mary Col, Univ London, 69-70. *Mem:* Am Chem Soc; Sigma Xi. *Res:* Elastomers; polymer chemistry; mechanical behavior of polymers; miscibility of polymers; toughened plastics. *Mailing Add:* 4159 Steck Ave #283 Austin TX 78759

KESLER, CLYDE E(RVIN), b Condit, Ill, May 7, 22; m 47; c 2. CONCRETE. *Educ:* Univ Ill, BS, 43, MS, 46. *Prof Exp:* Jr eng aide, Ill Cent RR, 46-47; res assoc theoret & appl mech, 47-48, from instr to prof theoret & appl mech, 48-82, 63-82, EMER PROF CIVIL ENG, THEORET & APPL MECH, UNIV ILL, URBANA, 82- *Concurrent Pos:* Prof civil eng, Univ Ill, Urbana, 63-82. *Honors & Awards:* Thompson Award, Am Soc Testing & Mat, 58; Lindau Award, Am Concrete Inst, 71. *Mem:* Nat Acad Eng; fel Am Soc Civil Engrs; hon mem Am Concrete Inst (pres, 67-68); Am Soc Eng Educ; hon mem Wire Reinforcement Inst. *Res:* Plain and reinforced concrete; cracking and crack control; freeze-thaw durability; fracture; creeps; fiber reinforcement; diffusion; expansion; quick-setting; history and future; United States patentee. *Mailing Add:* Dept Civil Eng Univ Ill 205 N Mathews St Urbana IL 61801

KESLER, DARREL J, b Portland, Ind, Sept 21, 49; m 73; c 2. REPRODUCTIVE ENGINEERING & BIOTECHNOLOGY. *Educ:* Purdue Univ, BS, 71, MS, 74; Univ Mo, PhD, 77. *Prof Exp:* Teaching asst acad coun & admin, Deans Off, Sch Agr, Purdue Univ, 71-74; res asst reprod & lactation physiol, Dept Dairy Husb, Univ Mo, 74-77; asst prof, 77-81, ASSOC PROF REPRODUCTIVE ENG & BIOTECHNOL, DEPT ANIMAL SCI, UNIV ILL, URBANA, 81- *Concurrent Pos:* Biochemist, Abbott Labs, Inc, 83-84; int fac mem, Int Conf Pharmaceut Sci & Clin Pharmacol, 88; mem polymeric mat, sci & eng div, Am Chem Soc. *Mem:* AAAS; Am Soc Animal Sci; Controlled Release Soc; Am Chem Soc; Sigma Xi. *Res:* Drug formulation, chemistry, delivery, metabolism and immunodetection. *Mailing Add:* Dept Animal Sci 101 Animal Genetics Lab Univ Ill Urbana IL 61801

KESLER, EARL MARSHALL, b Dunmore, WVa, Dec 3, 20; m 45; c 3. DAIRY SCIENCE. *Educ:* WVa Univ, BS, 43; Pa State Univ, MS, 48, PhD, 51. *Prof Exp:* From instr to assoc prof, 48-64, PROF DAIRY SCI, PA STATE UNIV, UNIVERSITY PARK, 64- *Mem:* AAAS; Am Soc Animal Sci; Am Dairy Sci Asn; Sigma Xi. *Res:* Dairy cattle nutrition, physiology of digestion; calf nutrition rumen metabolism and intermediary metabolism of the bovine; forage production; physiology of milk secretion; management of cows. *Mailing Add:* 534 Beaumont Dr State College PA 16801

KESLER, G(EORGE) H(ENRY), b West Terre Haute, Ind, Oct 29, 20; m 42; c 2. CHEMICAL ENGINEERING. *Educ:* Rose Polytech Inst, BS, 42; Mass Inst Technol, MS, 49, ScD, 52. *Prof Exp:* Chem engr distillation, Tex Co, 42-48; asst chem engr, Mass Inst Technol, 49-51; prin chem engr, Battelle Mem Inst, 51-55, asst div chief, 55-59; assoc dir res, Nat Steel Corp, 59-65; dept mgr, Mat Lab Dept, McDonnell Aircraft Co, St Louis, 65-77; eng consult, 77-83; RETIRED. *Mem:* Am Soc Metals. *Res:* Mass transport in spray-laden turbulent air streams; titanium production; laboratory distillation column evaluation; heat and mass transfer; steelmaking fundamentals; materials testing. *Mailing Add:* Box 255 West Terre Haute IN 47885-0255

KESLER, OREN BYRL, b Crawford Co, Ill, Aug 28, 39; m 67; c 1. ELECTRICAL ENGINEERING, APPLIED ELECTROMAGNETIC THEORY. *Educ:* Univ Ill, BS, 61, MS, 62, PhD(elec eng), 65; Univ Wis, MA, 68. *Prof Exp:* Asst prof elec eng, Univ Tex, Austin, 65-72; mem tech staff, 72-81, SR MEM TECH STAFF, ANTENNA & MICROWAVE LAB, EQUIP GROUP, TEX INSTRUMENTS, INC, 81- *Concurrent Pos:* NSF sci fac fel, 70-71. *Mem:* Sr mem Inst Elec & Electronics Engrs; Am Math Asn. *Res:* Electromagnetic field; antennas; microwaves; radar; information processing; mathematical and computational techniques for engineering analysis. *Mailing Add:* 1401 Camelia Dr Plano TX 75074

KESLER, STEPHEN EDWARD, b Washington, DC, Oct 5, 40; m 65; c 2. ECONOMIC GEOLOGY, EXPLORATION GEOCHEMISTRY. *Educ:* Univ NC, BS, 62; Stanford Univ, PhD(geol), 66. *Prof Exp:* Asst prof geol, La State Univ, Baton Rouge, 66-70; asst prof, 70-71, assoc prof geol, Univ Toronto, 71-77; PROF GEOL, UNIV MICH, 77- *Concurrent Pos:* Vis scientist, Consejo de Recursos Minerales, Mex, 75-76; assoc ed, Econ Geol, 80-90, J Geochem Expl, 85-; Fulbright Panel, Coun Int Exchange Scholars, 84-87; int lectr, Soc Econ Geologists, 89-90. *Mem:* Geol Soc Am; Soc Econ Geologists (vpres, 90-91); Asn Explor Geochemists; Geochem Soc; fel Geol Soc Am. *Res:* Tectonic and petrologic framework of ore deposition; ore deposit and exploration geochemistry; geology of Central America, Mexico and West Indies. *Mailing Add:* Dept Geol Univ Mich Ann Arbor MI 48109

KESLER, THOMAS L, b Salisbury, NC, Dec 29, 08; m 32; c 2. GEOLOGY, MINERAL EXPLORATION. *Educ:* Univ NC, Chapel Hill, BS, 29, MS, 30. *Prof Exp:* Geologist, Shell Petrol Corp, 30-31, Soil Conserv Serv, 34-36, US Geol Surv, 36-46 & Thompson-Weinman & Co, 46-52; sr geologist, Tenn Coal & Iron Div, US Steel Corp, 52-53; chief geologist, Foote Mineral Co, 53-65 & Engelhard Minerals & Chemicals Corp, 65-68; consult geologist, domestic & foreign engagements, 68-74; RETIRED. *Concurrent Pos:* Counr, Soc Econ Geologists, 55-57; mem, Beryllium Panel, Mat Adv Bd, Nat Res Coun, 57-58 & Lithium Resources Group, Comt Nuclear & Alternative Energy Systs, 76; chmn, Indust Minerals Div, Soc Mining Engrs, 58 & Southeastern Sect, Geol Soc Am, 61; pres, Carolina Geol Soc, 58; mem, Forum Earth Resources, Voice of Am, 71. *Mem:* fel Soc Econ Geologists; fel Geol Soc Am; Soc Mining Engrs. *Res:* Southern Appalachian Stratigraphy, structure and mineral resources; geology of industrial mineral deposits; author of scientific and technical papers and reports published in professional media. *Mailing Add:* PO Box 3561 Bellevue WA 98009

KESLING, ROBERT VERNON, b Cass Co, Ind, Sept 11, 17; m 42; c 3. PALEONTOLOGY. *Educ:* DePauw Univ, AB, 39; Univ Ill, MS, 41, PhD(geol), 49. *Prof Exp:* From asst prof to assoc prof geol, 49-59, assoc cur micropaleont, Mus Paleont, 49-58, dir, 66-74, PROF GEOL, UNIV MICH, ANN ARBOR, 59-, CUR, MUS PALEONT, 58- *Concurrent Pos:* Ed, J Paleont, 58-64. *Mem:* Paleont Soc; Sigma Xi; AAAS. *Res:* Living and fossil Ostracoda; middle Devonian stratigraphy; Paleozoic echinoderms. *Mailing Add:* Mus Paleont Univ Mich Ann Arbor MI 48104

KESMODEL, LARRY LEE, b Fort Worth, Tex, Mar 5, 47; m 70; c 3. SOLID STATE PHYSICS, SURFACE SCIENCE. *Educ:* Calif Inst Technol, BS, 69; Univ Tex, Austin, PhD(physics), 74. *Prof Exp:* Fel phys chem, Univ Calif, Berkeley, 73-75, staff scientist surface physics, Mat & Molecular Res Div, Lawrence Berkeley Lab, 75-78; asst prof, 78-83, ASSOC PROF PHYSICS, IND UNIV, BLOOMINGTON, 83- *Concurrent Pos:* prin investr, Off Naval Res Contract, 80- & US Dept Energy Grant, 84-; Pres, LK Technologies, 85- *Mem:* Am Phys Soc. *Res:* Experimental studies of solid surfaces and chemisorption; low-energy electron diffraction; high-resolution electron energy loss spectroscopy. *Mailing Add:* Dept Physics Swain Hall W Rm 117 Ind Univ Bloomington IN 47401

KESNER, LEO, b New York, NY, Feb 22, 31; m 54; c 3. BIOCHEMISTRY, ANALYTICAL CHEMISTRY. *Educ:* City Col New York, BS, 54; State Univ NY, PhD(biochem), 61. *Prof Exp:* Sr technician biophys, Sloan-Kettering Inst, 54; jr biochemist, 54-56, asst, 56-60, from instr to asst prof, 61-69, assoc prof, 69-89, PROF BIOCHEM, STATE UNIV NY, DOWNSTATE MED CTR, 69- *Mem:* AAAS; Am Asn Clin Chem; Am Chem Soc; NY Acad Sci; Am Soc Biol Chemists. *Res:* Relationship of acid-base balance to intermediary metabolism; nutrition; design of analytical techniques in biochemistry; protein chemistry; inborn errors of metabolism; membrane phospholipids; growth promotion; proteases; protease inhibitors. *Mailing Add:* Dept Biochem State Univ NY Downstate Med Ctr Brooklyn NY 11203

KESNER, MICHAEL H, b Pawtucket, RI, Nov 30, 45; m 69; c 1. SYSTEMATICS, MAMMALOGY. *Educ:* Northwestern Univ, BA, 69, MS, 72; Univ Mass, PhD(zool), 76. *Prof Exp:* ASSOC PROF HUMAN ANAT & COMP ANAT, IND UNIV PA, 76- *Concurrent Pos:* Grant in aid, Sigma Xi, 76, Theodore Roosevelt Grant, Am Mus Natural Hist, 76 & Univ Res Grant, Ind Univ Pa, 77, 78 & 81. *Mem:* Am Soc Mammalogists; Soc Syst Zool. *Res:* Mammalian (Rodent) functional morphology and systematics primarily of the subfamily microtinae; biogeography and systematics of insular populations of the genus Microtus from off the coast of Northeastern North America. *Mailing Add:* Dept Biol Ind Univ Pa Main Campus Indiana PA 15705

KESNER, RAYMOND PIERRE, b Oran, Algeria, Dec 19, 40; US citizen; m 65; c 2. PHYSIOLOGICAL PSYCHOLOGY, NEUROSCIENCE. *Educ:* Wayne State Univ, BS, 62; Univ Ill, MS, 64, PhD(psychol), 65. *Prof Exp:* Fel physiol, Ctr Brain Res, Rochester, NY, 65-67; from asst prof to assoc prof, 67-75, PROF PSYCHOL, UNIV UTAH, 75- *Concurrent Pos:* Fel, Ctr Advan Study Behav Sci, 71-72. *Mem:* Soc Neurosci; Psychonomic Soc. *Res:* Neurobiological mechanisms of memory. *Mailing Add:* Dept Psychol Univ Utah Salt Lake City UT 84112

KESSEL, BRINA, b Ithaca, NY, Nov 20, 25; wid. ORNITHOLOGY, VERTEBRATE ZOOLOGY. *Educ:* Cornell Univ, BS, 47, PhD(ornith), 51; Univ Wis, MS, 49. *Prof Exp:* Asst ornith & conserv, Cornell Univ, 47-48 & 49-51; asst prof biol sci, 51-54, assoc prof zool, 54-59, head dept biol sci, 57-66, dean col biol sci & renewable resources, 61-72, PROF ZOOL, UNIV ALASKA, FAIRBANKS, 59- *Concurrent Pos:* Proj dir, Ecol Invests, AEC Proj Chariot, Univ Alaska, 59-63, cur terrestrial vert mus collections, Univ mus, 72-, admin assoc acad progs, Off Chancellor, 73-80; ornithologist, invests NW Alaska pipeline, 76-81; ornithologist, Sustina Hydroelec Proj, Alaska, 80-83. *Mem:* Fel AAAS; fel Am Ornithologists Union; fel Arctic Inst NAm; Wilson Ornith Soc; Cooper Ornith Soc; Sigma Xi. *Res:* European starling in North America; biology, ecology, behavior and biogeography of Alaskan birds. *Mailing Add:* PO Box 80211 Univ Alaska Mus Fairbanks AK 99708

KESSEL, DAVID HARRY, b Monroe, Mich, Jan 8, 31; m; c 2. BIOCHEMISTRY. *Educ:* Mass Inst Technol, BS, 52; Univ Mich, MS, 54, PhD(biochem), 59. *Prof Exp:* Res assoc, Children's Cancer Res Found & asst path, Harvard Med Sch, 65-68; from asst prof to assoc prof pharmacol, Sch Med, Univ Rochester, 69-73; PROF PHARMACOL & MED, SCH MED, WAYNE STATE UNIV, 74- *Concurrent Pos:* NIH fel, Harvard Univ, 59-63; res scientist, Mich Cancer Found, 74; secy-treas, Am Soc Photobiol, 88- *Mem:* AAAS; Biochem Soc; Am Asn Cancer Res; Am Soc Pharmacol & Exp Therapeut; Am Soc Biol Chemists. *Res:* Development of anti-tumor agents; mode of action of anti-neoplastic drugs and photosensitizing agents. *Mailing Add:* Pharmacol Dept Wayne State Univ Sch Med Detroit MI 48201

KESSEL, EDWARD LUTHER, zoology, for more information see previous edition

KESSEL, QUENTIN CATTELL, b Boston, Mass, Aug 15, 38; m 60; c 2. EXPERIMENTAL ATOMIC PHYSICS, MATERIALS SCIENCE. *Educ:* Yale Univ, BS, 60; Univ Mich, MS, 62; Univ Conn, PhD(physics), 66. *Prof Exp:* Res asst physics, Univ Conn, 62-65, res assoc, 65-66; physicist, Robert J Van de Graaff Lab, High Voltage Eng Corp, 66-70; guest scientist, Inst Physics, Aarhus Univ, 70-71; asst prof, 71-73, assoc prof, 73-78, PROF PHYSICS, UNIV CONN, 78- *Concurrent Pos:* Guest prof, Univ Freiburg, Ger, 77-78. *Mem:* AAAS; fel Am Phys Soc; Europ Phys Soc. *Res:* Interaction of accelerated ions and molecules with matter. *Mailing Add:* Dept Physics Univ Conn Storrs CT 06269

KESSEL, RICHARD GLEN, b Fairfield, Iowa, July 19, 31. ZOOLOGY, ANATOMY. *Educ:* Parsons Col, BS, 53; Univ Iowa, MS, 56, PhD(zool), 59. *Prof Exp:* Marine Biol Lab, summer 57, 60, 62; from instr to asst prof anat, Bowman Gray Sch Med, 59-61; from asst prof to assoc prof, 61-68, PROF BIOL, UNIV IOWA, 68- *Concurrent Pos:* NIH fel, 60-61; Nat Inst Gen Med Sci res grant, 60-65, career develop award, 64-69; Nat Inst Child Health & Human Develop res grant, 64-69; prog dir, NIH Training grant develop biol, 66-78; NSF res grant, 68-71; assoc ed, J Exp Zool, 78-82. *Mem:* Soc Study Reproduction; Am Asn Anat; Soc Develop Biologists; Am Soc Cell Biol; Am Phys Soc; Electron Micros Soc Am. *Res:* Electron microscopic, autoradiographic, cytochemical and freeze-fracture studies on oocyte growth and differentiation; origin, structure and function of cell organelles; mechanisms of secretion; scanning electron microscopy of tissues and organs; immunocytochemistry. *Mailing Add:* Dept Biol Univ Iowa Iowa City IA 52242

KESSEL, ROSSLYN WILLIAM IAN, b London, Eng, May 5, 29; m 53; c 2. IMMUNOLOGY, MICROBIOLOGY. *Educ:* Univ London, MB & BS, 56; Rutgers Univ, PhD(microbiol), 60. *Prof Exp:* Res assoc microbiol, Rutgers Univ, 60-61, asst prof, 62-65; asst prof, Univ Mass, Amherst, 65-67; from assoc prof to prof microbiol, Univ Md Sch Med, Baltimore City, 75-86; MEM STAFF LIMECROFT, ENG, 86- *Concurrent Pos:* NSF res fel, Post-Grad Med Sch, Univ Milan, 61-62. *Mem:* AAAS; Am Soc Microbiol; Am Soc Naturalists; Reticuloendothelial Soc; Am Soc Cell Biol. *Res:* Cellular immunology; organization of biomedical knowledge; curricular organization; information transfer; pathogenesis of silicosis and other pneumoconioses; intracellular parasitism. *Mailing Add:* Limecroft Shute Hill Marlborough Devon TQ7 35F England

KESSELL, STEPHEN ROBERT, b Lewiston, Maine, May 22, 49. PLANT ECOLOGY, RESOURCE MANAGEMENT. *Educ:* Amherst Col, BA, 72; Cornell Univ, MS, 77. *Prof Exp:* Res asst plant ecol, Cornell Univ, 72-74; fire ecologist, Glacier Nat Park, Nat Park Serv, 74-76; RES SUPVR, COOP RES AGREEMENTS, USDA FOREST SERV, 76- *Concurrent Pos:* Coop biologist, Glacier Nat Park, Nat Park Serv, 72-74, consult resource modeling, 75-77; consult, Can Forest Serv, Northern Forest Res Ctr, Edmonton, 77-, div plant indust, Canberra, Australia, 77- & Nat Parks & Wildlife Serv New South Wales, Sydney, Australia, 77-; vis fel, Res Sch Biol Sci, Australian Nat Univ, 78-; invited fac, USDA Forest Serv, Nat Interagency Training Ctr, Marana, Ariz, 78- pres, Gradient Modeling, Inc, 75- *Mem:* Sigma Xi; Ecol Soc Am; Am Inst Biol Sci; AAAS. *Res:* Development and implementation of computer-based resource management and forest fire simulation models; gradient analysis and gradient modeling; habitat and niche relations of vascular plants; species diversity. *Mailing Add:* Box 329 West Perth West Australia

KESSELMAN, WARREN ARTHUR, b Newark, NJ, Apr 8, 27; m 50; c 3. ELECTROMAGNETIC COMPATIBILITY, COMMUNICATIONS SECURITY. *Educ:* Newark Col Eng, BS, 50. *Prof Exp:* Engr electron, US Army Signal Corps Eng Lab, 49-58; eng supvr electron US Army Electron Comd, 58-76; chief, Electro Magnetic Compatibility Team, electron US Army Comn Res & Develop, Comd, 76-81; chief comsec div, commun security, US Army Commun Electron Comd, 81-84; DEP DIR RES & DEVELOP, MGT CTR C3 SYST, 84- *Mem:* Fel Inst Elec & Electronics Engrs; Armed Forces Commun & Electron Asn. *Res:* Electromagnetic compatibility and interference measurement techniques, analysis methodology, standardization and interference reduction techniques. *Mailing Add:* 31 Hope Rd Tinton Falls NJ 07724

KESSELRING, JOHN PAUL, b Detroit, Mich, Mar 26, 40; m 66; c 2. ENERGY ANALYSES, HVAC EQUIPMENT. *Educ:* Univ Mich, BS, 61; Stanford Univ, MS, 62, PhD(aeronaut & astronaut sci), 68. *Prof Exp:* Res engr, Rocketdyne Div, NAm Aviation, Inc, 62-63, mem tech staff, NAm Rockwell Corp, 67-69; asst prof mech & aerospace eng, Univ Tenn, Knoxville, 69-74; mgr catalytic combustion progs, Acurex Corp, 74-80, assoc mgr combustion technol, 81-82; vpres, Alzeta Corp, 82-86; SR PROJ MGR, ELEC POWER RES INST, 86- *Concurrent Pos:* Instr aeronaut & astronaut, Stanford Univ, 77; lectr, Soviet Acad Sci, 86. *Mem:* Combustion Inst; Am Soc Mech Engrs; Sigma Xi. *Res:* Residential bldg energy analysis; HVAC equipment; heat pumps. *Mailing Add:* 279 Apricot Lane Mountain View CA 94040

KESSIN, RICHARD HARRY, b Bayonne, NJ, Feb 24, 44; c 2. DEVELOPMENTAL GENETICS, SIGNAL TRANSDUCTION. *Educ:* Yale Univ, BA, 66; Brandeis Univ, PhD(biol), 71. *Prof Exp:* asst prof, 74-80, assoc prof biol, Harvard Univ, 80-; AT DEPT ANAT & CELL BIOL, COLUMBIA UNIV. *Res:* Development and regulation of signal transduction. *Mailing Add:* Dept Anat & Cell Biol Columbia Univ 630 W 168th St New York NY 10032

KESSINGER, MARGARET ANNE, b Beckley, WVa, June 4, 41; m 71. MEDICINE. *Educ:* WVa Univ, BA, 63, MD, 67. *Prof Exp:* Asst prof to assoc prof med oncol, 72-90, PROF INTERNAL MED, UNIV NEBR MED CTR, 90- *Mem:* Am Col Physicians; Am Soc Clin Oncol; Am Asn Cancer Res; Am Asn Cancer Educ; Am Med Soc; Am Fedn Clin Res. *Res:* High dose therapy with autologous bone marrow transplantation for solid tumors; peripheral hematopietic stem cell transfusion. *Mailing Add:* 600 S 42nd St Omaha NE 68198

KESSINGER, WALTER PAUL, JR, b Corsicana, Tex, July 9, 30; m 62; c 3. MICROPALEONTOLOGY. *Educ:* Tex Technol Col, BS, 51, MS, 53; La State Univ, PhD(geol), 74. *Prof Exp:* From asst prof to assoc prof, 53-76, PROF GEOL, UNIV SOUTHWESTERN LA, 77-, HEAD DEPT, 56- *Concurrent Pos:* ARCO, 66-69, Univ Ill, 62. *Mem:* Geol Soc Am; Am Asn Petrol Geologists; Paleont Soc; Soc Econ Paleontologists & Mineralogists; Nat Asn Geol Teachers; Sigma Xi. *Res:* Ostracoda of the Comanche Series of north Texas. *Mailing Add:* 406 Orange Wood Lafayette LA 70503

KESSLER, ALEXANDER, b Vienna, Austria, Mar 19, 31; US citizen; m 53; c 4. MEDICINE, PUBLIC HEALTH. *Educ:* NY Univ, BA, 51; Columbia Univ, MD, 55; Rockefeller Univ, PhD(pop biol), 66. *Prof Exp:* Res physician, Albert Einstein Med Ctr, New York, 56-57 & Georgetown Univ Hosp, 60-61; dir spec prog res human reprod, WHO, Geneva, 66-84, dir spec leave, 84-86; RETIRED. *Res:* International public health; family planning; reproductive biology and contraceptive technology; research administration. *Mailing Add:* Four Ellerdale Close London NW3 6BE England

KESSLER, BERNARD V, b Brooklyn, NY, June 27, 28; m 57; c 2. LASERS. *Educ:* NY Univ, BA, 59; Cath Univ Am, MS, 69, PhD(physics), 72. *Prof Exp:* RES PHYSICIST, ELECTRO-OPTICS BR, NAVAL WARFARE CTR, 59- *Concurrent Pos:* Consult, Pres & Chief Exec Officer, Evgodic Processes, Ltd. *Mem:* Optical Soc Am; Am Phys Soc; Am Asn Physics Teachers. *Res:* Optics; laser mode studies; ultra low temperature thermodynamics of paramagnetic materials; adaptive optics for coherent optical systems; laser speckle phenomena; communication theory; infrared search and surveillance. *Mailing Add:* Electro-Optics Two Woodland Way Suite D Greenbelt MD 20770-1713

KESSLER, DAN, b Vienna, Austria, May 23, 24; m 53; c 3. HIGH ENERGY PHYSICS. *Educ:* Hebrew Univ, Israel, MSc, 50; Sorbonne, DesSc(physics), 54. *Prof Exp:* Res physicist, Nat Ctr Sci Res, Fr Res Coun, 51-58; sr physicist, Israel Atomic Energy Comn, 58-64; res assoc high energy physics, Univ Chicago, 64-67; PROF PHYSICS, CARLETON UNIV, 67- *Concurrent Pos:* Assoc prof, Tel-Aviv Univ, 62-64. *Mem:* Am Phys Soc; Can Asn Physicists. *Res:* Electromagnetic and nuclear interactions of cosmic-ray muons; interactions of K-mesons; mesic atoms; production of heavy mesons in P-P collisions. *Mailing Add:* Dept Physics Carleton Univ Ottawa ON K1S 5B6 Can

KESSLER, DAVID, MEDICAL ADMINISTRATION. *Prof Exp:* COMNR FOOD & DRUG, FOOD & DRUG ADMIN, 90- *Mailing Add:* Food & Drug Admin 5600 Fishers Lane Rockville MD 20857

KESSLER, DAVID PHILLIP, b Anderson, Ind, Nov 1, 34; m 57; c 4. CHEMICAL ENGINEERING, BIOENGINEERING. *Educ:* Purdue Univ, BS, 56; Univ Mich, MS, 59, PhD(chem eng), 62. *Prof Exp:* Mem res staff, Delco Remy Div, Gen Motors Corp, 52, 53, 55 & Dow Chem Co, 56-58; group leader explor develop, Procter & Gamble Co, 62-64; from asst prof to assoc prof chem eng, 64-73, asst provost, 75-80, dir, Acad Info Systs, 78-80, PROF CHEM ENG, PURDUE UNIV, 73-, HEAD, DIV INTERDISCIPLINARY ENG STUDIES, 82- *Concurrent Pos:* Mem res staff, Humble Oil & Ref Co, 65 & Phillips Petrol Co, 66; consult, Great Lakes Chem Corp, 67, Midwest Appl Sci & D&M Corp, 68, Am Oil Co, Melnor Corp, Am Filters & Araneida Corp, 69, Roper Corp, 70, Westinghouse Corp,

71, Exxon Corp, 80, Howard W Sams Div MacMillan, 86 & Citizens Gas & Coke, 86; mediator/fact finder, Ind Educ Employ Rels Bd, 74-, Ind Better Bus Bur, 82-, Mediation Farm Debt, 88- *Mem:* Am Inst Chem Engrs; Am Soc Eng Educ; Soc Prof Dispute Resolution. *Res:* Multiphase flow; transport properties in disperse media. *Mailing Add:* Dept Chem Eng Purdue Univ Lafayette IN 47907

KESSLER, DIETRICH, b Hamilton, NY, May 28, 36. CELL BIOLOGY, MOLECULAR BIOLOGY. *Educ:* Swarthmore Col, BA, 58; Univ Wis, MS, 60, PhD(zool), 64. *Prof Exp:* From asst prof to prof, dept biol, Haverford Col, 64-84; head dept biol, 84-90, PROF, DEPT BIOL, COLGATE UNIV, 84- *Concurrent Pos:* Am Cancer Soc fel, Brandeis Univ, 66-67; NSF sci fac fel, Swiss Inst Exp Cancer Res, Lausanne, 71-72; Fulbright res grant, Bonn, Ger, 80-81; vis scientist, Genetics Dept, Univ Leicester, Eng, 90-91. *Mem:* AAAS; Am Soc Cell Biol. *Res:* Molecular mechanism of amoeboid movement in Physarum polycephalum using techniques of molecular and cell biology. *Mailing Add:* Dept Biol Colgate Univ Hamilton NY 13346-1398

KESSLER, EDWIN, 3RD, b New York, NY, Dec 2, 28; m 50; c 2. METEOROLOGY. *Educ:* Columbia Univ, AB, 50; Mass Inst Technol, SM, 52, ScD(meteorol), 57. *Prof Exp:* Chief synoptic sect, Weather Radar Br, Air Force Cambridge Res Ctr, 54-61; dir atmospheric physics div, Travelers Res Ctr, 61-64; dir, Nat Severe Storms Lab, US Weather Bur, 64-86; ADJ PROF & CONSULT GEOG & METEOROL, UNIV OKLA, 86- *Concurrent Pos:* Adj prof, Univ Okla, 64-; vis prof, Mass Inst Technol, 75-76; vis lectr, McGill Univ, 80; consult, Saudi Arabia, 87. *Honors & Awards:* Cleveland Abbe Award, Am Meteorol Soc, 89. *Mem:* Fel AAAS; Am Geophys Union; fel Am Meteorol Soc; Royal Meteorol Soc; Sigma Xi. *Res:* Synthesis of varied observations and theory to improve understanding of meteorological phenomena and to develop and apply technology in the public interest. *Mailing Add:* Dept Geog Univ Okla 804 Dale Hall Tower Norman OK 73019

KESSLER, ERNEST GEORGE, JR, b Hanover, Pa, Sept 12, 40; m 62; c 4. PRECISION MEASUREMENTS, X-RAY SPECTROSCOPY. *Educ:* Shippensburg State Col, BS, 62; Univ Wis-Madison, MS, 64, PhD(physics), 69. *Prof Exp:* Nat Res Coun-Nat Bur Standards res assoc spectros, 69-71, PHYSICIST, CTR ATOMIC, MOLECULAR & OPTICAL PHYSICS, NAT INST STANDARDS & TECHNOL, 71- *Mem:* Am Phys Soc. *Res:* High resolution studies of the ionized helium spectrum; determination of the Rydberg constant from wavelength measurements on ionized helium; precise x-ray and gamma ray wavelength measurements. *Mailing Add:* A-141 Physics Bldg Nat Inst Standards & Technol Gaithersburg MD 20899

KESSLER, FREDERICK MELVYN, b Brooklyn, NY, May 15, 32; m 54; c 3. ENGINEERING ACOUSTICS, ELECTROMECHANICAL ENGINEERING. *Educ:* City Col NY, BME, 54; Rutgers Univ, MS, 67, PhD(elec eng), 71. *Prof Exp:* Jr test engr, Curtiss-Wright Corp, 54-55; sr proj mgr acoust, David Taylor Model Basin, Washington, DC, 59-61; sr develop engr eng acoust, Ingersoll-Rand Co, 61-68; fac mem elec eng, Rutgers Univ, 68-71; vpres eng acoust, Lewis S Goodfriend & Assoc, 71-73; partner, Dames & Moore, 73-80, managing partner eng acoust, 80-; AT FMK TECHNOL INC, BOUND BROOK. *Concurrent Pos:* Adj prof, Dept Mech Eng, Stevens Inst Technol, Hoboken, NJ, 78- *Mem:* Acoust Soc Am; Inst Elec & Electronics Engrs; Inst Noise Control Eng (pres, 88). *Res:* Optimization muffler design parameters using conjugate gradient search techniques. *Mailing Add:* FMK Technol Inc 31 Shady Lane PO Box 168 Bound Brook NJ 08805

KESSLER, GEORGE MORTON, b Philadelphia, Pa, July 26, 17; m 43; c 2. HORTICULTURE. *Educ:* Pa State Univ, BS, 46, MS, 47; Mich State Col, PhD(hort), 53. *Prof Exp:* From instr to assoc prof, 47-82, EMER ASSOC PROF HORT, MICH STATE UNIV, 82- *Concurrent Pos:* Ed, Fruit Varieties & Hort Digest, Am Pomol Soc, 53-72. *Mem:* Am Soc Hort Sci; Am Pomol Soc (secy-treas, 57-64, pres, 67-68). *Res:* Teaching and evaluating of fruit varieties. *Mailing Add:* 1127 Lilac East Lansing MI 48824

KESSLER, GERALD, b New York, NY, Mar 27, 30; m 52; c 4. CLINICAL CHEMISTRY. *Educ:* City Col New York, BS, 50; Univ Md, MS, 52, PhD(biochem), 54. *Prof Exp:* Clin biochemist, Albert Einstein Med Ctr, 54-57; biochemist, Technicon Instruments Corp, NY, 57-61; head clin ctr core lab, Montefiore Hosp, NY, 61-64; chief automation div, Bio-Sci Labs, Los Angeles, 65-67; DIR DIV BIOCHEM, JEWISH HOSP ST LOUIS, 67- *Concurrent Pos:* Assoc scientist, Dept Biochem, Sloan-Kettering Inst Cancer Res, 59-61; consult automation res prog, Vet Admin Hosp, Bronx, NY, 63-64; assoc prof, Sch Med, Washington Univ, 67-82, prof, 82. *Mem:* AAAS; Am Chem Soc; Am Asn Clin Chem; NY Acad Sci; Acad Clin Lab Physicians & Scientists; Nat Acad Clin Biochem. *Res:* Clinical biochemistry; methodological instrumentation research as applied to automation of analytical procedures. *Mailing Add:* 23622 Claymoor Dr Chesterfield MO 63017-7834

KESSLER, HAROLD D, b Toledo, Ohio, Dec 28, 21. TITANIUM. *Educ:* Case Inst Technol, BS, 42; Ill Inst Technol, MS, 49. *Prof Exp:* Mem staff, Nat Advan Comt Aeronaut, 43-44, Air Force, Wright Field, 44-45 & Armour Res Found, 45-54; mgr, Metall Res Lab, Titanium Metals Corp Am, 54-57 & Tech Develop Lab, 57-64; chief metallurgist, Reactive Metals Co, 64-69, vpres technol, 69-79; tech dir, DH Titanium, 79-81; mgr titanium opers, Cabot Corp, 81-86; CONSULT, 86- *Concurrent Pos:* Mem bd dirs, Am Soc Metals. *Honors & Awards:* Russ Ogden Award, Am Soc Testing & Mat, 85. *Mem:* Am Inst Mining Metall & Petrol Engrs; Am Soc Metals. *Res:* Processes for manufacture of titanium; titanium alloys. *Mailing Add:* Okes Co Inc 18010 Alyssum Dr Sun City West AZ 85375

KESSLER, IRVING ISAR, b Chelsea, Mass, Mar 22, 31; m; c 2. PREVENTIVE MEDICINE, EPIDEMIOLOGY. *Educ:* NY Univ, AB, 52; Harvard Univ, MA, 55, DrPH, 69; Stanford Univ, MD, 60; Columbia Univ, MPH, 62; Am Bd Prev Med, dipl. *Prof Exp:* Instr environ med, State Univ NY, 64-66; asst prof chronic dis, Johns Hopkins Univ, 66-69, assoc prof epidemiol, 70-72, prof, 73-78; assoc prof prev med, 70-78, chmn epidemiol & prev med, 78-88, PROF ONCOL & MED, UNIV MD, 84- *Concurrent Pos:* Fac res award, Am Cancer Soc, 72-77; med dir res, USPHS; mem exec comn, Gov Coun Toxic Substances, Md; exec dir, Md Cancer Registry, 82; scientific servicing bd, Ctr Indoor Dir Res. *Mem:* AAAS; Am Epidemiol Soc; Am Asn Cancer Res; Am Pub Health Asn; Asn Teachers Prev Med; Sigma Xi. *Res:* Epidemiological research in cancer, diabetes mellitus, birth defects and Parkinson's disease; epidemiological principles; community studies of health; gerontology; environmental and occupational health; health regulation. *Mailing Add:* Dept Epidemiol & Prev Med Sch Med Univ Md Baltimore MD 21201

KESSLER, IRVING JACK, b Brooklyn, NY, May 14, 40; c 2. MATHEMATICS. *Educ:* Brooklyn Col, BA, 62; Univ Wis, MS, 63, PhD(math), 66. *Prof Exp:* Asst prof math, Univ Mich, 66-68; from asst prof to assoc prof, Southern Ill Univ, 68-78, prof math, 78-80; RES STAFF MATH, INST DEFENSE ANALYSIS, 77-, PRES. *Concurrent Pos:* Res staff math, Inst Defense Anal, 74-75 & 77- *Mem:* Am Math Soc; Math Asn Am. *Res:* Combinatorial mathematics; number theory; applying mathematics to problems in speech recognition. *Mailing Add:* Inst Defense Analysis Thanet Rd Princeton NJ 08540

KESSLER, JOHN OTTO, b Vienna, Austria, Nov 26, 28; nat US; m 50; c 2. PHYSICS, PHYCOLOGY. *Educ:* Columbia Univ, AB, 49, MS, 50, PhD(physics), 53. *Prof Exp:* Asst physics, Columbia Univ, 48-52; physicist labs, Radio Corp Am, 52-62 & Princeton Univ, 63-65; mgr grad recruiting, RCA Corp, 65-66; PROF PHYSICS, UNIV ARIZ, 66- *Concurrent Pos:* Vis prof physics, Univ Leeds, 72-73 & T H Delft, Neth, 79; Fulbright fel, Dept Appl Math & Theoret Phys, Cambridge Univ, 83-84, Cult Ctr Algae & Protozoa, 85-, Dept Appl Math, Univ Leeds, 90-91. *Mem:* Fel AAAS; Am Phys Soc; Phycol Soc Am. *Res:* Applied phycology; biophysics; plant physiology; fluid mechanics of microorganisms; concentrative and cooperative phenomena of microrganism populations for practical applications in cell separation and phycoculture and as a means for investigating self organizing systems. *Mailing Add:* Dept Physics Bldg 81 Univ Ariz Tucson AZ 85721

KESSLER, KARL GUNTHER, b Hamburg, Germany, Aug 21, 19; US citizen; m 41; c 2. ATOMIC & MOLECULAR PHYSICS, OPTICS. *Educ:* Univ Mich, AB, 41, MS, 42, PhD(physics), 47. *Prof Exp:* Res physicist, Univ Mich, 41-48, instr, 43-48; div chief, 62-78, Dir, Ctr Basic Standards, 78-86, RES PHYSICIST, NAT BUR STANDARDS, 48-, ASSOC DIR INT AFFAIRS 86- *Honors & Awards:* Gold Medal, Dept Com, 62; Dist Service Award, Optical Soc Am, 84. *Mem:* Am Phys Soc; Optical Soc Am (pres, 69); Astron Soc Am; Int Astron Union; AAAS; Sigma Xi. *Res:* Atomic physics; atomic spectroscopy. *Mailing Add:* Nat Bur Standards A 505 Admin Bldg Gaithersburg MD 20899

KESSLER, KENNETH J, JR, b Wheeling, WVa, Mar 15, 33; m 54; c 4. PLANT PATHOLOGY. *Educ:* WVa Univ, BS, 55, MS, 57, PhD(plant path), 60. *Prof Exp:* PLANT PATHOLOGIST, NORTH CENT FOREST EXP STA, US FOREST SERV, 59- *Mem:* Am Phytopath Soc; Mycol Soc Am. *Res:* Hardwood tree diseases. *Mailing Add:* Forestry Sci Lab SIU NCent Forest Expos Sta Carbondale IL 62901

KESSLER, LAWRENCE W, b Chicago, Ill, Sept 26, 42; m 64, 85; c 7. ULTRASOUND, BIOENGINEERING. *Educ:* Purdue Univ, BS, 64; Univ Ill, Urbana-Champaign, MS, 65, PhD(elec eng), 68. *Prof Exp:* Mem res staff acoust visualization, Zenith Radio Corp, 68-74; PRES ACOUST MICROS, SONOSCAN, INC, 74- *Concurrent Pos:* Mem adv panel, Tech Electronic Prod Radiation Safety Standards Comt, Dept Health, Educ & Welfare, 72-75; nat lectr, Inst Elec & Electronics Engrs, 81-82. *Mem:* Sr mem Inst Elec & Electronics Engrs; Am Inst Ultrasound Med; fel Acoust Soc Am; Am Soc Nondestructive Testing. *Res:* Acoustic microscopy and ultrasonic visualization applications in life and materials sciences; quality assurance inspection equipment; biomedical engineering; laser scanning systems. *Mailing Add:* Sonoscan Inc 530 E Green St Bensenville IL 60106

KESSLER, NATHAN, b St Louis, Mo, Aug 19, 23; m 47; c 3. CHEMICAL ENGINEERING. *Educ:* Washington Univ, BSChE & MS, 44. *Prof Exp:* Chem engr, Biotech Resources, Inc, 44-51, tech supvr, Ohio, 51-53, sr chem eng, Ill, 53-57, chief chem engr, 57-60, dir process eng, 60-61, plant supt, 61-62, gen supt, 62-67, mem bd dirs & tech group vpres, AE Staley Mfg Co, 67-85, pres, Staley Techventures, 84-86, TECHNOL CONSULT, BIOTECH RESOURCES, INC, 85- *Concurrent Pos:* Mem eng adv bd, Rice Univ, 71 & Univ Ill, 82-87; mem bd dirs, Wastech, Inc, Tex, 70-76; mem Jr Eng & Technol Soc Bd & pres, 82-85; mem, Biotechnical Resources Bd, 82-85; mem bd, Technol Transfer Soc, 83-; White-Rodgers fel; assoc dir, Biotech Resources, Inc, 85-89; chmn, Indust Adv Comt, Mich Biotechnol Inst, 87-; mem, Indust Adv Bd, Fed Lab Consortium, 87- *Mem:* Am Inst Chem Engrs; Am Chem Soc; Technol Transfer Soc (pres, 85-88, chmn, 88-90). *Res:* Engineering, research and development; food processes; biotechnology; technology transfer. *Mailing Add:* 49 Allen Bend Dr Decatur IL 62521

KESSLER, RICHARD HOWARD, b Paterson, NJ, Dec 15, 23; m 44; c 3. MEDICINE, PHYSIOLOGY. *Educ:* Rutgers Univ, BSc, 48; NY Univ, MD, 52. *Prof Exp:* Asst chem, Rutgers Univ, 47-48; asst med, NY Univ, 52-55; from instr to assoc prof physiol, Cornell Univ, 55-68; prof med, Med Sch, Northwestern Univ, Chicago, 68-78, assoc dean Med Sch, 70-78; prof med, Pritzker Sch Med, Univ Chicago, 78-83; sr vpres, Michael Reese Hosp & Med Ctr, 78-83; PROF MED, MT SINAI SCH MED, NY, 84- *Concurrent Pos:* Life inst med res fel, 55-56; Hofheimer Found fel, 57-62; mem, Sect Renal Dis, Coun Circulation, Am Heart Asn; mem, Health Econ Adv Comt, Nat Bur Econ Res; consult, Educ Serv, Vet Admin; vpres, Kidney Found Ill; comnr, Health & Hosps Gov Comn Cook County; fel, Hastings Inst Soc, Ethics & Life Sci; comnr, Chicago Health Systs Agency. *Mem:* AAAS; Harvey Soc; Soc Exp Biol & Med; Am Physiol Soc; fel Am Col Physicians. *Res:* Cardiovascular and renal physiology; fluid balance; electrolyte transport; diuretics; renal disease; hypertension. *Mailing Add:* 349 S Wash Dr Sarasota FL 34236

KESSLER, SEYMOUR, b New York, NY, Sept 3, 28; m 53; c 2. GENETICS. *Educ:* City Col New York, BS, 60; Columbia Univ, MA, 62, PhD(zool), 65; PhD(soc-clin psychol). *Prof Exp:* Fel psychiat, 65-67, asst prof, 67-73, sr scientist, 73-74, adj prof psychiat, Sch Med, Stanford Univ, 74-75; dir & sr lectr, Univ Calif, Berkeley, 75-85; CONSULT, 85- *Mem:* AAAS; Behav Genetics Asn; Am Soc Human Genetics; Sigma Xi; Am Psychol Asn. *Res:* Behavior and psychiatric genetics; genetic counseling. *Mailing Add:* PO Box 7702 Berkeley CA 94707

KESSLER, THOMAS J, b Neptune, NJ, Nov 13, 38; m 62; c 3. GAS DYNAMICS, SYSTEMS DESIGN. *Educ:* Rutgers Univ, BS, 61, PhD(separated flow), 66; NY Univ, MME, 63. *Prof Exp:* Mem tech staff mech develop, Bell Tel Labs, Inc, 61-63; res asst mech eng, Rutgers Univ, 63-64, res asst gas dynamics, 66; mem tech staff, Bell Tel Labs, Inc, Whippany, 66-69, supvr, 69-75; AREA MGR, XEROX CORP, 75- *Mem:* Am Soc Mech Engrs; Am Soc Eng Educ; Sigma Xi. *Res:* Separated flows; cooling electronic equipment; xerographic process and design. *Mailing Add:* 30 Harwood Lane East Rochester NY 14445

KESSLER, WAYNE VINCENT, b Milo, Iowa, Jan 10, 33; m 53; c 2. BIONUCLEONICS, ENVIRONMENTAL TOXICOLOGY. *Educ:* NDak State Univ, BS, 55, MS, 56; Purdue Univ, PhD(pharmaceut chem), 59. *Prof Exp:* Asst pharmaceut chem, Purdue Univ, 56-57; asst prof, NDak State Univ, 59-60; from asst prof to assoc prof, 60-68, PROF HEALTH SCI, PURDUE UNIV, WEST LAFAYETTE, 68- *Concurrent Pos:* Lederle pharm fac award, 62; vis scientist, Am Asn Cols Pharm, 64-70; assoc prof, Sch Med, Ind Univ, 70-73. *Mem:* Fel AAAS; Am Pharmaceut Asn; Am Chem Soc; fel Acad Pharmaceut Sci; Health Physics Soc. *Res:* Inhalation toxicology of noxious substances; natural radioactivity in building materials. *Mailing Add:* Sch Health Sci Purdue Univ West Lafayette IN 47907

KESSLER, WILLIAM J(OSEPH), b Roebling, NY, Feb 28, 17; c 4. COMMUNICATIONS ENGINEERING. *Prof Exp:* Lab instr elec eng, Univ Fla, 43-45, asst prof, 45-67, asst res engr, 45-53; OWNER, W J KESSLER ASSOCS, 67- *Concurrent Pos:* Microwave, TV & commun consult, eng forensics. *Honors & Awards:* Cert Off Sci & Res & Develop & Ord Develop Award, 45. *Mem:* Nat Soc Prof Engrs; Am Electronic Soc. *Res:* Thunderstorm electricity; propagation of low frequency electromagnetic radiations; special instrumentation for radio location of thunder-storms for meteorological forecasting purposes. *Mailing Add:* 1625 SW 35th Pl Gainesville FL 32608

KESSLIN, GEORGE, organic chemistry, for more information see previous edition

KESSNER, DAVID MORTON, b New York, NY, Aug 23, 32; m 59; c 2. INTERNAL MEDICINE. *Educ:* Univ Ariz, BS, 54; Washington Univ, MD, 58; Am Bd Internal Med, dipl, 67. *Prof Exp:* Intern med, Mary Imogene Bassett Hosp, 58-59, asst resident internal med, 59-60; jr attend physician, Med Clin, Univ Ill, 61-62; second year resident internal med, Sch Med, Yale Univ, 64-65, from instr to asst prof med & epidemiol, 65-69; study dir & res assoc, Inst Med, Nat Acad Sci, 69-73; dir, Health Serv Res Off & assoc prof community med & int health, Georgetown Univ, 73-75; PROF & V CHMN COMMUNITY & FAMILY MED, UNIV MASS MED CTR, 75- *Concurrent Pos:* Fel prev med, Sch Med, Univ Ill, 60-62; fel med, Sch Med, Yale Univ, 62-64; Nat Inst Arthritis & Metab Dis fel, 63-64; attend physician, Metab Sect, Yale-New Haven Hosp, 65-69, assoc physician, Dept Med, 66-69, consult internist, Yale Psychiat Inst, 67-69; asst clin prof, Sch Med, George Washington Univ, 70-73, attend physician, Hosp, 70- *Mem:* Am Fedn Clin Res; Am Pub Health Asn; Int Epidemiol Asn; fel Am Col Physicians. *Res:* Chronic disease epidemiology; the use of epidemiology in health services research. *Mailing Add:* 17 Henshaw St Boston MA 02135

KESTEN, ARTHUR S(IDNEY), b New York, NY, Sept 10, 34; m 56; c 2. RESEARCH ADMINISTRATION. *Educ:* NY Univ, BS, 55; Univ Pittsburgh, MS, 58, PhD(chem eng), 61. *Prof Exp:* Assoc engr res & develop, Bettis Atomic Power Lab, Westinghouse Elec Corp, 55-57, engr, 57-61, sr engr, 61-63; res engr, United Technol Res Ctr, 63-65, sr res engr, 65-68, supvr kinetics & heat transfer, 68-72, prin scientist kinetics & environ sci, 72-76, mgr combustion sci, 76-77, mgr energy res, 77-81, asst dir res power indust systs technol, 81-90, ASST DIR, INDUST SYSTS & TECHNOL, UNITED TECHNOL RES CTR, 90- *Concurrent Pos:* Adj assoc prof, Rensselaer Polytech Inst, 65-; chmn, Solar Energy Res Inst Adv Bd, 84- *Mem:* Fel Am Inst Chem; Sigma Xi. *Res:* Combustion; chemical reactor analysis; transport processes. *Mailing Add:* 17 Morning Crest Dr West Hartford CT 06117

KESTEN, HARRY, MATHEMATICS. *Prof Exp:* PROF MATH, CORNELL UNIV, 65- *Honors & Awards:* Brouwer Mem Lectr, Dutch Math Soc. *Mem:* Nat Acad Sci; Am Math Soc; Inst Math Statist. *Mailing Add:* Dept Math Cornell Univ Ithaca NY 14853

KESTENBAUM, RICHARD CHARLES, b New York, NY, Apr 3, 31; m 54; c 3. BACTERIOLOGY. *Educ:* City Col New York, BS, 52; Rutgers Univ, MS, 54, PhD(bact), 59. *Prof Exp:* Asst bact, Rutgers Univ, 52-54 & 56-59; bacteriologist, 59-61, sr res microbiologist, 61-63, sect head microbiol, 63-66, sect head oral res, 66-69 & 71-75, sect head oral prod, 69-71, sr sect head oral prod & appl res, 77-78, mgr oral prod, 78-81, assoc dir personal care prod, 81-83, ASSOC DIR RES & DEVELOP PLANNING, COLGATE-PALMOLIVE CO, PISCATAWAY, 83- *Mem:* Am Soc Microbiol; Int Asn Dent Res. *Res:* Dental medicine; bacterial metabolism; microbial quality control; antimicrobial agents; oral and clinical research; oral products. *Mailing Add:* 18 Bradford Rd East Brunswick NJ 08816

KESTENBAUM, RICHARD STEVEN, b New York, NY, Mar 20, 42; m 70; c 1. NEUROPSYCHOPHARMACOLOGY, PSYCHOTHERAPY. *Educ:* NY Univ, BA, 63, PhD(psychol), 68. *Prof Exp:* Res fel psychol, NIMH, 66-68; asst prof, State Univ NY Stony Brook, 68-73; assoc dir res, Dept Psychiat, NY Med Col, 74-79; ASSOC PROF PSYCHOL & CLIN SUPVR,

TEACHER'S COL, COLUMBIA UNIV, 80- *Concurrent Pos:* Fel, Nat Inst Psychother, 74-79. *Mem:* AAAS; Am Psychol Asn; NY Acad Sci. *Res:* Neural coding of pain and neuropsychopharmacology of pain perception; neuropsychopharmacology of opiates and implications for treatment; neuropsychopharmacology of cocaine in man; psychotherapy outcome evaluation. *Mailing Add:* 142 West End Ave New York NY 10023

KESTER, ANDREW STEPHEN, b Abington, Pa, Sept 1, 32; m 60; c 2. BACTERIOLOGY. *Educ:* Pa State Univ, BS, 54; Univ Tex, PhD(bact), 61. *Prof Exp:* Res microbiologist, Miles Chem Co, Ind, 61-67; asst prof, 67-74, ASSOC PROF BIOL, NTEX STATE UNIV, 74- *Mem:* Am Soc Microbiol; Am Chem Soc. *Res:* Microbial oxidation of hydrocarbons; industrial microbiology. *Mailing Add:* Dept Biol Sci NTex State Univ Denton TX 76201

KESTER, DALE EMMERT, b Audubon, Iowa, July 28, 22; m 46; c 2. POMOLOGY. *Educ:* Iowa State Col, BS, 47; Univ Calif, MS, 49, PhD(plant physiol), 51. *Prof Exp:* Res asst, Univ Calif, 48-50, jr specialist, 51, instr pomol, 51-53, lectr, 53, from asst prof to assoc prof, 54-69, from jr pomologist to assoc pomologist, 51-69, prof pomol & pomologist, Exp Sta, 69-91; RETIRED. *Honors & Awards:* Stark Award, Am Soc Hort Sci, 80. *Mem:* AAAS; fel Am Soc Hort Sci; Int Plant Propagators Soc; Genetics Soc Am; Int Asn Plant Tissue Cult; Am Pomol Soc; Int Hort Soc. *Res:* Plant breeding, almonds, rootstocks; tissue and embryo culture of prunus species; plant propagation; somatic variation. *Mailing Add:* Dept Pomol Univ Calif Davis CA 95616

KESTER, DANA R, b Los Angeles, Calif, Jan 26, 43; m 63; c 1. CHEMICAL OCEANOGRAPHY, PHYSICAL CHEMISTRY. *Educ:* Univ Wash, BS, 64; Ore State Univ, MS, 66, PhD(oceanog), 69. *Prof Exp:* From asst prof to assoc prof, 69-76, PROF OCEANOG, UNIV RI, 76- *Concurrent Pos:* Ed, Marine Chem, 73-78, J Marine Res, 73-80. *Mem:* Am Geophys Union; Sigma Xi; AAAS. *Res:* Physical chemistry of seawater; effects of temperature and pressure on ionic equilibria; transition metal marine chemistry; oceanic chemical distributions; waste disposal in the ocean. *Mailing Add:* Grad Progs Univ RI Kingston RI 02881

KESTER, DENNIS EARL, b Eureka, Kans, Aug 21, 47; m 68; c 1. ORGANIC POLYMER CHEMISTRY. *Educ:* Col Emporia, BS, 69; Univ Ark, PhD(org chem), 75. *Prof Exp:* Res scientist polymers, coatings & eng, 74-80, supvr organic res & dir prod develop, 80-84, VPRES RES & DEVELOP, AM NAT CAN CO, 84- *Mem:* Am Chem Soc. *Res:* Polymer characterization and polymer synthesis; anionic, condensation and free radical polymerizations for the preparation of materials for specific coatings applications; radiation curable resins and coatings, adhesion; film and sheet laminating and extrusion, printing. *Mailing Add:* Am Nat Can Co PO Box 702 Neenah WI 54957

KESTEVEN, MICHAEL, b Sydney, Australia, July 30, 40. ASTRONOMY. *Educ:* Univ Sydney, BSc, 63, PhD(astron), 68. *Prof Exp:* Asst prof physics, Queen's Univ, Ont, 68-81; COMMONWEALTH SCI & INDUST RES ORGN, 81- *Mem:* Am Astron Soc; fel Royal Astron Soc; found Int Astron Soc Australia. *Res:* Radio astronomy. *Mailing Add:* Div Radiophys Commonwealth Sci & Indust Res Orgn PO Box 76 Epping N S W 2121 Australia

KESTIN, J(OSEPH), b Warsaw, Poland, Sept 18, 13; m 49; c 1. MECHANICAL ENGINEERING, THERMODYNAMICS. *Educ:* Univ Warsaw, dipl, 37; Imp Col, dipl & Univ London, PhD(thermodyn), 45. *Hon Degrees:* MA, Brown Univ, 57; DSc, Univ London, 66, Univ Claude Bernard, Lyon, France, 83. *Prof Exp:* From sr lectr to prof mech eng & head dept, Polish Univ Col, London, Eng, 44-52; prof eng, 52-83, dir Ctr Energy Studies, 76-85, RES PROF ENG, BROWN UNIV, 83- *Concurrent Pos:* US deleg, Int Conf Steam Properties, 54, 56, 63, 68, 74, 79 & 84 & Int Comn Steam Properties, 54-; tech ed, J Appl Mech, Am Soc Mech Engrs, 56-71; assoc prof, Univ Paris, 66; adv to chancellor, Univ Tehran, 68; adv working group, Int Asn Properties Steam, pres, 74-76; consult, Nat Bur Standards; assoc prof, Univ Claude Bernard, Lyon I, 74; chmn, Nat Acad Sci/Nat Res Coun Eval Panel Off Standard Ref Data Nat Bur Standards, 74-80, mem eval panel, Nat Measurement Lab, 78-; mem, Numerical Data Adv Bd, Nat Acad Sci, 76- & Mech Eng Peer Comt, Nat Acad Eng, 87-90; consult, Off Naval Res, RAND Corp; vis prof, Imperial Col, London, 82-87, Univ Md, 84-89, Univ d Bundeswehr, Ger, 90 & distinguished vis prof, Univ Del, 89; fel, Inst Advan Studies, Berlin, 84-85 & Imp Col, London, 89. *Honors & Awards:* Brit Inst Mech Engrs, Prize 49; Centennial Medal, Am Soc Mech Engrs, 80; James Harry Potter Gold Medal, 84; Humboldt Prize, 87. *Mem:* Nat Acad Eng; Am Soc Mech Eng; Brit Inst Mech Eng; foreign mem Polish Acad Sci. *Res:* Measurement of thermodynamic and transport properties; classical, irreversible and statistical thermodynamics; heat transfer; boundary layers; thermodynamics of stress and strain; geothermal energy-conversion systems; multi-phase flow. *Mailing Add:* Div Eng Brown Univ Providence RI 02912

KESTING, ROBERT E, b Jamaica, NY, Feb 8, 33; m 59; c 4. POLYMER CHEMISTRY. *Educ:* Manhattan Col, BS, 56; State Univ NY Col Forestry, Syracuse, MS, 59, PhD(chem), 61. *Prof Exp:* Res chemist, Heberlein & Co AG, Switz, 60-62 & Von Karman Ctr, Aerojet-Gen Corp Div, Gen Tire & Rubber Co, Calif, 62-65; sr scientist, Philco-Ford, 65-67; consult, 67-69; vpres polymer opers, Chem Systs, Inc, Santa Ana, 69-76, vpres, Irvine, 76-79; vpres, 79-81, PRES, PURPORE INC, DIV GELMAN SCI, 82- *Res:* Graft copolymerization on polymer substrates; structure and function of reverse osmosis membranes; new theory for the lyotropic swelling of polar polymers; developed Kesting process for the fabrication of wet-dry reversible reverse osmosis membranes; synthetic polymeric membranes. *Mailing Add:* 3220 Avenue Court E No 199 Sumner WA 98390

KESTNER, MARK OTTO, b Berea, Ohio, Dec 10, 47; m 71; c 1. INORGANIC CHEMISTRY, DUST CONTROL. *Educ:* Carnegie-Mellon Univ, BS, 69; Northwestern Univ, MS, 70, PhD(chem), 74. *Prof Exp:* Res chemist, Borg-Warner Corp, 74-78; group leader, Apollo Chem Corp, 78-80, mgr, prod develop, 80-83; pres, Chemicoal, Inc, 83-88; PRES, NATURAL ENVIRON SERV CO, 88- *Mem:* Am Chem Soc; NY Acad Sci; Sigma Xi. *Res:* Handling, storage and combustion of coal and other fossil fuel; air pollution control and chemical treatments for control of particulate and gaseous emissions; solid waste disposal. *Mailing Add:* Seven Hampshire Dr Mendham NJ 07945

KESTNER, MELVIN MICHAEL, b Wooster, Ohio, Oct 20, 45; m 71. ORGANIC CHEMISTRY. *Educ:* Heidelberg Col, BS, 67; Purdue Univ, PhD(org chem), 73. *Prof Exp:* SR CHEMIST ORG CHEM, EASTMAN KODAK CO, 73- *Mem:* Am Chem Soc. *Res:* Synthesis of compounds used in photographic products. *Mailing Add:* 590 Parma Ctr Rd Hilton NY 14615

KESTNER, NEIL R, b Milwaukee, Wis, Dec 11, 37. CHEMICAL PHYSICS. *Educ:* Univ Wis-Milwaukee, BS, 60; Yale Univ, MS, 62, PhD(theoret chem), 64. *Prof Exp:* Res assoc, Inst Study Metals, Univ Chicago, 63-64; asst prof chem, Stanford Univ, 64-66; assoc prof, 66-72, chmn freshman chem, 73-76, chmn dept, 76-81, PROF CHEM, LA STATE UNIV, BATON ROUGE, 72- *Concurrent Pos:* A P Sloan fel, 67-69; vis prof, Tel-Aviv Univ, 72; res collabr, Brookhaven Nat Lab, 81. *Mem:* Am Chem Soc; Am Phys Soc; Sigma Xi; AAAS. *Res:* Quantum chemistry; intermolecular forces; electrons in disordered media; electron transfer reactions. *Mailing Add:* Dept Chem La State Univ Baton Rouge LA 70803

KESTON, ALBERT S, RADIOACTIVE ISOTOPES. *Educ:* Yale Univ, PhD(chem), 35. *Prof Exp:* EMER PROF, MT SINAI MED SCH, 66- *Res:* Thyroid cancer diagnosis and therapy with radioactive iodine. *Mailing Add:* Five E 102nd St New York NY 10029

KETCHA, DANIEL MICHAEL, b Newark, NJ, Jan 12, 56; m; c 1. ANTITUMOR ANTIBIOTICS. *Educ:* King's Col, Wilkes-Barre, BS, 77; Temple Univ, Philadelphia, PhD(org chem), 84. *Prof Exp:* Res assoc, Dartmouth Col, Hanover, NH, 83-85, res instr chem, 85; ASST PROF CHEM, WRIGHT STATE UNIV, DAYTON, OHIO, 85- *Mem:* Am Chem Soc. *Res:* Development of novel synthetic methodologies and their subsequent application towards the synthesis of molecules of biological importance. *Mailing Add:* Dept Chem Wright State Univ Dayton OH 45435

KETCHAM, ALFRED SCHUTT, b Newark, NY, Oct 7, 24; m 46; c 6. SURGERY. *Educ:* Hobart Col, BS, 45; Univ Rochester, MD, 49; Am Bd Surg, dipl, 59. *Hon Degrees:* DSc, Hobart Col, 70. *Prof Exp:* Intern surg, US Naval Med Ctr, Bethesda, 49-50; resident, USPHS Hosps, San Francisco, 50-52, Seattle, 52-55; chief, Dept Surg, Nat Cancer Inst, 62-74, assoc sci dir clin res, Gen Labs & Clins & clin dir, Inst, 71-74; PROF SURG & CHIEF DIV ONCOL, SCH MED, UNIV MIAMI, 74- *Honors & Awards:* Meritorious Serv Medal, USPHS. *Mem:* Fel Am Col Surg; Am Asn Cancer Res; Am Radium Soc; Soc Head & Neck Surgeons; Am Surg Asn. *Res:* Cancer surgery; experimental metastases; clinical and laboratory investigation. *Mailing Add:* Dept Surg Sch Med Univ Miami PO Box 016310 Miami FL 33101

KETCHAM, BRUCE V(ALENTINE), b Wilmington, Del, Mar 17, 18; m 44. AEROSPACE ENGINEERING. *Educ:* Yale Univ, BME, 40; Univ Okla, MAE, 56. *Prof Exp:* Designer aero engines, Pratt & Whitney Aircraft Div, United Aircraft Corp, 40-47; chmn sch aero eng, Univ Okla, 53-63, dir aero res, 57-64, mem fac, 47-64; head dept aerospace eng, 64-67, dir res & develop, 66-67, dir solar energy projs, 76-83, PROF MECH & AEROSPACE ENG, UNIV TULSA, 64-, EMER PROF, 83- *Concurrent Pos:* Consult, Aero Design & Eng Co, 56-58 & Todd Eng Co, 61-, Elec Peakload Probs, 84. *Mem:* Am Soc Eng Educ; Am Inst Aeronaut & Astronaut. *Res:* Space engineering; rocket propulsion; gas turbine. *Mailing Add:* 2956 E 47th St Tulsa OK 74105

KETCHAM, ROGER, b Berea, Ohio, Sept 2, 26; m 50; c 4. ORGANIC CHEMISTRY, HETEROCYCLIC CHEMISTRY. *Educ:* Antioch Col, BS, 51; Cornell Univ, PhD(chem), 56. *Prof Exp:* From instr to prof chem & pharmaceut chem, Sch Pharm, Univ Calif, San Francisco, 69-91; RETIRED. *Concurrent Pos:* Mem, Orgn Am States Professorship, Monterrey Inst Technol & Higher Educ, 64-65; vis prof, Univ Graz, Austria, 71-72, Univ Hamburg, Ger, 79-80 & Univ Leeud, Sweden, 83. *Mem:* Am Chem Soc; Am Pharmaceut Asn; The Chem Soc; Sigma Xi. *Res:* Three-membered rings; nitrogen and sulfur heterocycles; reaction mechanisms; organo-sulfur chemistry. *Mailing Add:* Dept Pharm Chem Univ Calif San Francisco CA 94143-0446

KETCHEL, MELVIN M, b Pontiac, Mich, June 1, 22; m 58; c 3. REPRODUCTIVE PHYSIOLOGY. *Educ:* Olivet Col, AB, 48; Western Reserve Univ, MS, 49; Harvard Univ, PhD(biol), 54. *Prof Exp:* Res asst biophys chem, Harvard Univ, 54-55; res assoc cytol, Protein Found Labs, 55-56; res assoc surg, Harvard Med Sch, 56-59; staff scientist physiol, Worcester Found Exp Biol, 59-63, sr scientist, 63-65; from assoc prof to prof, Sch Med, Tufts Univ, 65-72; dir, Oak Ridge Pop Res Inst, 72-75; scientist, Human Reprod Unit, WHO, 75-77; owner, The Shelter Co, 77-81; EXEC SECY, DIV RES GRANTS, NIH, 81- *Concurrent Pos:* NSF sr fel, 71-72; prof zool, Univ Tenn, 74-75. *Res:* Hormonal aspects of pregnancy and pseudopregnancy; immunological aspects of the relationships between the fetus and its mother. *Mailing Add:* Div Res Grants Westwood Bldg Rm 2A15B NIH Bethesda MD 20205

KETCHEN, EUGENE EARL, b Miami, Fla, May 3, 21; m 51; c 3. PHYSICAL CHEMISTRY. *Educ:* Univ Miami, BS, 43; Univ Pittsburgh, PhD(phys chem), 50. *Prof Exp:* Asst, Univ Pittsburgh, 46-50; asst prof chem, Washington & Jefferson Col, 50-51; chemist, Isotope Div, 51-74, INDUST HYG CHEMIST, OAK RIDGE NAT LAB, 74- *Mem:* Am Chem Soc; Am Indust Hyg Asn; Am Acad Indust Hyg. *Res:* Radio-isotopes; industrial hygiene chemistry; industrial toxicology. *Mailing Add:* 654 Lakeshore Dr Kingston TN 37763-2010

KETCHEN, MARK B, b St Stephens, Can, Sept 15, 48. CRYOGENIC DIGITAL DEVICES. *Educ:* Mass Inst Technol, BS, 70; Univ Calif, Berkeley, MA, 71, PhD(physics), 77. *Prof Exp:* Officer, Physics Thermodynamics Reactor Oper, US Naval Nuclear Power Prog, 72-76; RES STAFF MEM & MGR, THOMAS J WATSON RES CTR, IBM, 77- *Mem:* Am Phys Soc; Inst Elec & Electronics Engrs. *Res:* Electrical design and evaluation of superconducting quantum interference devices for digital and analog applications; logic and power circuits for an ultra-high-speed cryogenic computer. *Mailing Add:* IBM T J Watson Res Ctr PO Box 218 Yorktown Heights NY 10562

KETCHIE, DELMER O, b Salisbury, NC, July 7, 32; m 59, 79; c 4. PLANT PHYSIOLOGY, BIOCHEMISTRY. *Educ:* Wash State Univ, BS, 59; Univ Idaho, MS, 61; Cornell Univ, PhD(pomol), 65. *Prof Exp:* Plant physiologist, Date & Citrus Sta, USDA, Calif, 65-67; from asst horticulturist to assoc horticulturist, 67-80, HORTICULTURIST, TREE FRUIT RES CTR, WASH STATE UNIV, 80- *Honors & Awards:* Stark Award, 73; Paul Howe Sheppard Award, 84. *Mem:* Am Soc Hort Sci; Soc Cryobiol; Int Soc Hort Sci; Sigma Xi. *Res:* Winter hardiness of deciduous fruit trees, including biochemical and physical aspects; rest dormancy of deciduous fruit trees. *Mailing Add:* 1100 N Western Wenatchee WA 98801

KETCHLEDGE, RAYMOND WAIBEL, communications, electronics; deceased, see previous edition for last biography

KETCHMAN, JEFFREY, b New York, NY, Nov 23, 42; m 62; c 2. MECHANICAL ENGINEERING, NEW PRODUCT DEVELOPMENT. *Educ:* City Col New York, BSME, 64; Ohio State Univ, MSME, 67; Columbia Univ, DrEngSci, 72. *Prof Exp:* Proj task leader, Battelle Mem Inst, 64-67; mem tech staff, Bell Labs, 67-76; dir, Eng & Res, AMF, Inc, 76-85; vpres design, eng & res, Lightolier Inc, 85-87; dir res & eng, Genlyte Group, 85-87; DIR, MECH & SAFETY ENG, INTREATY TESTING & CONSULT INC, 87- *Concurrent Pos:* Consult, new prod develop & eng mgt & forensic eng. *Mem:* AAAS; NY Acad Sci; Asn Res Dirs; Am Soc Mech Eng; Am Soc Safety Eng; Syst Safety Soc; Soc Automotive Engrs. *Res:* Optimizing radioisotope thermoelectric generators; design and development of submarine sonar systems; exercise equipment; oil-field equipment; lighting systems; product safety. *Mailing Add:* 14 Caccamo Lane Westport CT 06880

KETCHUM, GARDNER M(ASON), b Philadelphia, Pa, Oct 20, 19; m 49; c 1. FLUID MECHANICS, HEAT TRANSFER. *Educ:* Mass Inst Technol, SB, 41, SM, 44, ScD(mech eng), 49. *Prof Exp:* Instr mech eng, Mass Inst Technol, 41-48; engr gen eng lab, Gen Elec Co, 48-53; assoc prof, 53-56, chmn dept, 62-74, prof, 56-85 EMER PROF MECH ENG, UNION COL, NY, 85- *Mem:* Sigma Xi; fel Am Soc Mech Engrs; Am Soc Eng Educ. *Res:* Fluid mechanics; heat transfer; thermodynamics; technology-society interactions, particularly urban and environmental. *Mailing Add:* Dept Mech Eng Union Col Schenectady NY 12308

KETCHUM, MILO S, b Denver, Colo, Mar 8, 10. STRUCTURAL ENGINEERING. *Educ:* Univ Ill, BS, 31, MS, 32. *Hon Degrees:* DSc, Univ Colo, 76, Dr, Univ Colo, 76. *Prof Exp:* Founder & vpres, Kkbna Eng, 45-86; EMER PROF CIVIL ENG, UNIV CONN, 78- *Honors & Awards:* Turner Gold Medal, Am Concrete Inst, 72. *Mem:* Nat Acad Eng; hon mem Am Concrete Inst; hon mem Am Soc Civil Eng; Am Soc Eng Educ; fel Am Consult Eng Coun. *Mailing Add:* 165 Estes St Denver CO 80226

KETCHUM, PAUL ABBOTT, b Hyannis, Mass, Aug 11, 42; m 63; c 2. MICROBIAL PHYSIOLOGY. *Educ:* Bates Col, BS, 64; Univ Mass, PhD(microbiol), 69. *Prof Exp:* NIH fel biochem, Johns Hopkins Univ, 68-70; asst prof, 70-77, NSF grant, 71-77, ASSOC PROF BIO SCI, OAKLAND UNIV, 77- *Concurrent Pos:* Guest scientist, Univ Ga, Athens, 89-90; Mem, Nat Coun & Comt A, Am Asn Univ Professors; AAUP Comt A. *Mem:* AAAS; Am Soc Microbiol; Am Asn Univ Professors. *Res:* Inorganic nitrogen metabolism; biochemistry of nitrate reductase; role of molybdenum in inorganic nitrogen metabolism. *Mailing Add:* Dept Biol Sci Oakland Univ Rochester MI 48063

KETELLAPPER, HENDRIK JAN, b Ridderkerk, Neth, Dec 23, 25; nat US; m 51; c 3. PLANT PHYSIOLOGY. *Educ:* Univ Utrecht, BSc, 47, MSc, 51, PhD(plant physiol), 53. *Prof Exp:* Instr gen bot, Univ Utrecht, 48-51; res officer, Div Plant Indust, Commonwealth Sci & Indust Res Orgn, Canberra, Australia, 54-57; res fel biol, Calif Inst Technol, 57-64; lectr, 64-65, assoc prof, 65-69, PROF BOT, UNIV CALIF, DAVIS, 69-, ASSOC DEAN COL LETT & SCI, 67- *Concurrent Pos:* Nat Acad Sci exchange scientist, USSR, 63. *Mem:* AAAS; Bot Soc Am; Am Soc Plant Physiol; NY Acad Sci; Royal Neth Bot Soc; Sigma Xi. *Res:* Climate and plant growth and development; algal physiology; physiological ecology. *Mailing Add:* Dept Bot Univ Calif Davis CA 95616

KETHLEY, JOHN BRYAN, b Passaic, NJ, Oct 18, 42; m 68. ACAROLOGY. *Educ:* Univ Ga, BS, 64, PhD(entom), 69. *Prof Exp:* NIH trainee acarol lab, Ohio State Univ, 69-70; asst cur insects, 70-75, Head Div, 74-77 & 80-84, ASSOC CUR, HEAD DIV, FIELD MUS NATURAL HIST, 75- *Concurrent Pos:* Lectr comt evolutionary biol, Univ Chicago, 71-72, 74-; instr biol sci, Northwestern Univ, 73-; lectr acarology, Ohio State Univ, 74- *Mem:* Acarological Soc Am; Entom Soc Am; Sigma Xi. *Res:* Acarine systematics; methodologies in cladistics and phylogenetics; population ecology; functional morphology. *Mailing Add:* Div Insects Field Mus Natural Hist Chicago IL 60605-2496

KETHLEY, THOMAS WILLIAM, b Crystal Springs, Miss, Apr 3, 13; m 37; c 3. BIOLOGY. *Educ:* Emory Univ, AB, 34, MS, 35. *Prof Exp:* Drug & insecticide chemist, State Chemist's Off, Ga, 37-41; mineral chemist, State Div Mines, 41; toxicologist, Med Div, US Army, Edgewood Arsenal, Md, 41-46; from res asst prof to res assoc prof, Eng Exp Sta, Ga Inst Technol, 46-60, head bio-eng br, 60-77, res prof appl biol, 60-81, prof, 74-81, EMER PROF BIOL, GA INST TECHNOL, 81- *Concurrent Pos:* Res biologist, Ga

Inst Technol, 51-60; consult, Commun Dis Ctr, 54-55 & 57-58; consult, Chem Corps, US Army, 58- *Mem:* AAAS; Am Pub Health Asn; Am Soc Microbiol; NY Acad Sci. *Res:* Effect of ice formation on living cells; thermal properties of frozen tissues; preservation of foods by freezing; instrumental methods of analysis; gas and aerosol chamber techniques; aerobiology; water content of bacterial aerosols; aerial disinfectants. *Mailing Add:* 514 Ponce de Leon Pl Decatur GA 30030

KETLEY, ARTHUR DONALD, b London, Eng, Dec 27, 30; m; c 3. PHOTOCHEMISTRY, POLYMER CHEMISTRY. *Educ:* Univ London, BSc, 51, PhD(chem), 53. *Prof Exp:* Res assoc chem, Mass Inst Technol, 53-55; lectr, Univ Sydney, 56-57; res fel, Ga Inst Technol, 57-58; chemist, Esso Res & Eng Co, 58-59; chemist, W W Grace & Co, 59-73, mgr mat develop photopolymer systs, 73-79, dir technol prod res, 79-88; MANAGING DIR, JAPAN RES CTR, 88- *Mem:* Am Chem Soc; Int Am Photochem Soc; Soc Photog Scientists & Engrs. *Res:* Mechanisms of organic reactions; organometallic chemistry; photochemistry; development and applications of photopolymerizable materials; imaging science. *Mailing Add:* 252 Whitmoor Terr Silver Spring MD 20901

KETLEY, JEANNE NELSON, b New York, NY, Apr 18, 38. ENZYMOLOGY, DRUG METABOLISM. *Educ:* Queens Col, NY, BS, 62; Cornell Univ, MS, 67; Johns Hopkins Univ, PhD(biochem), 73. *Prof Exp:* Res assoc biochem, Cornell Univ, 67, res asst enzym & biochem, Med Col, 68-69; vis scientist develop biol, Nat Inst Dent Res, 73-74, staff fel enzyme biochem, lab biochem & metab, Nat Inst Arthritis, Metab & Digestive Dis, NIH, 74-76, sr staff fel, Nat Inst Aging, 76-77; chemist, Bur Foods, Food & Drug Admin, 77-79; exec secy, phys biochem study sect, 79-85, chief, Spec Rev Sect, 85-89, CHIEF, PHYS SCI REV SECT, DIV RES GRANTS, NIH, 89- *Concurrent Pos:* Muscular Dystrophy Asn Am fel, 73-74. *Mem:* AAAS; Am Chem Soc; Biophys Soc; Protein Soc; NY Acad Sci. *Res:* Drug metabolism; mammalian biochemical processes at the molecular level; changes in biological structure and function of proteins; physical biochemistry. *Mailing Add:* Phys Sci Sect Div Res Grants NIH Westwood Bldg Rm 203 Bethesda MD 20892

KETNER, KEITH B, b Boscobel, Wis, Feb 5, 21. GEOLOGY. *Educ:* Univ Wis, BA, 47, MA, 52, PhD(geol), 68. *Prof Exp:* Geologist & res geologist, US Geol Surv, 50-91; RETIRED. *Mem:* AAAS; fel Geol Soc Am. *Mailing Add:* US Geol Surv MS 939 Box 25046 Fed Ctr Denver CO 80225

KETOLA, H GEORGE, FISH NUTRITION. *Educ:* Cornell Univ, BS, 65, MS, 67, PhD(nutrit), 73. *Prof Exp:* RES PHYSIOLOGIST, TUNISON LAB FISH NUTRIT, US FISH & WILDLIFE SERV, 73- *Concurrent Pos:* Adj asst prof, Dept Poultry & Avian Sci, Cornell Univ, 76-91; US Fish & Wildlife Serv res grant, 84-85 & Dept Natural Resources, 91-; consult fish nutrit, feed mfrs. *Mem:* Am Inst Nutrit; Poultry Sci Asn; Am Fisheries Soc; Am Soc Animal Sci; Sigma Xi; Animal Nutrit Res Coun. *Res:* Nutritional studies to reduce nutrient discharges in effluent waters from fish hatcheries; improvement of nutritional quality of diets for fry and fingerling salmon and trout. *Mailing Add:* Tunison Lab Fish Nutrit US Fish & Wildlife Serv 3075 Gracie Rd Cortland NY 13045

KETRING, DAROLD L, b Van Nuys, Calif, Mar 14, 30; m 50; c 2. PLANT PHYSIOLOGY, BIOCHEMISTRY. *Educ:* Univ Calif, Los Angeles, BS, 63, PhD(plant sci), 67. *Prof Exp:* PLANT PHYSIOLOGIST, USDA, 67- *Mem:* Am Soc Plant Physiologists; Am Soc Agron; Am Peanut Res & Educ Asn; Plant Growth Regulator Soc Am; Sigma Xi. *Res:* Relation of ethylene and other plant hormones to plant growth, development, senescence and accompanying biochemistry; effect of environmental stress on plant growth and development. *Mailing Add:* 1933 Wildwood Dr Stillwater OK 74075

KETTEL, LOUIS JOHN, b Chicago, Ill, Nov 4, 29; m 51; c 3. PULMONARY PHYSIOLOGY, MEDICAL EDUCATION. *Educ:* Purdue Univ, BS, 51; Northwestern Univ, MD, 54, MS, 58; Am Bd Internal Med, dipl, 61; Am Bd Pulmonary Dis, dipl, 65. *Prof Exp:* Chief pulmonary dis sect, Vet Admin Res Hosp, Chicago, 62-68 & Tucson, 68-72; assoc dean, Col Med, Univ Ariz, 74-77, assoc prof, 68-73 & prof med, Sch Med, Univ Ariz, 73-87, dean col med, 77-87; VPRES, ACAD AFFAIRS ASN, AM MED COL, 88-, CLIN PROF MED, GEORGETOWN UNIV, SCH MED, 89- *Concurrent Pos:* USPHS fel, Northwestern Univ, 56-57, USPHS trainee, 60-62; asst prof, Northwestern Univ, 66-68; asst chief med serv, Vet Admin Hosp, Tucson, 72-74. *Mem:* AAAS; Am Fedn Clin Res; fel Am Col Physicians; Am Med Asn; Am Thoracic Soc. *Res:* Managing change in medical education; evaluation of techniques in medical education. *Mailing Add:* Assoc Am Med Col One Dupont Circle Washington DC 20036

KETTELKAMP, DONALD B, b Anamosa, Iowa, Jan 21, 30; m 54; c 4. ORTHOPEDIC SURGERY. *Educ:* Cornell Col, BA, 52; Univ Iowa, MD, 55, MS, 60. *Prof Exp:* From asst prof to assoc prof orthop surg, Albany Med Col, 64-68; assoc prof, Univ Iowa, 68-71; prof & chmn dept, Med Ctr, Univ Ark, Little Rock, 71-74; PROF ORTHOP SURG & CHMN DEPT, IND UNIV MED CTR, INDIANAPOLIS, 74- *Concurrent Pos:* Consult, Vet Admin Hosp, Little Rock, 71-74. *Mem:* Fel Am Acad Orthop Surg; Orthop Res Soc; fel Am Col Surg; Am Orthop Asn; Am Med Asn. *Res:* Knee and hand reconstruction and biomechanics. *Mailing Add:* 737 N Michigan Ave Suite 1150 Chicago IL 60611

KETTER, ROBERT L(EWIS), civil engineering; deceased, see previous edition for last biography

KETTERER, JOHN JOSEPH, b Philadelphia, Pa, Mar 12, 21; m 46; c 2. ZOOLOGY. *Educ:* Dickinson Col, BS, 43; NY Univ, PhD(biol), 53. *Prof Exp:* Asst biol, NY Univ, 46-52, instr, 52-53; from asst prof to prof, 53-73, head dept, 59-73, W P PRESSLY PROF BIOL, MONMOUTH COL, 73- *Mem:* AAAS; Soc Protozool; Am Soc Zool. *Res:* Invertebrate zoology with special reference to protozoa. *Mailing Add:* Dept Biol Monmouth Col 700 E Broadway Monmouth IL 61462

KETTERER, PAUL ANTHONY, b Warwick, NY, Aug 2, 41; m 62; c 2. ORGANIC CHEMISTRY. *Educ:* Syracuse Univ, AB, 64; Seton Hall Univ, MS, 73. *Prof Exp:* Chemist, Polaks Frutal Works Inc, 64-68; sr chemist, Tenneco Chem Inc, 68-73, group leader anal chem, 74-75, lab mgr polymer characterization & anal, 76-81; MGR ANAL SERV, TENNECO POLYMERS INC, FLEMINGTON, 82- *Mem:* Am Chem Soc; Soc Appl Spectros; Coblentz Soc; Sigma Xi. *Res:* Polymer characterization and analysis, determination of structure-property relationships of vinyl polymers and additives; development of analytical methods for trace analysis of air and water pollutants; organotin chemistry. *Mailing Add:* 3308 Coachman Rd Wilmington DE 19803-1946

KETTERING, JAMES DAVID, b Pekin, Ill, Mar 27, 42; m 63; c 3. MEDICAL MICROBIOLOGY. *Educ:* Andrews Univ, BA, 64; Loma Linda Univ, MS, 68, PhD(microbiol), 74. *Prof Exp:* Microbiologist, Indust Bio-Test Labs, 64-66 & Abott Labs Inc, 66; instr, 72-74, asst prof, 74-80, assoc prof microbiol, 80-89, PROF MICROBIOL, SCH MED, LOMA LINDA UNIV, 89- *Concurrent Pos:* Fel med microbiol, Calif State Dept Health, Berkeley, 75-77. *Mem:* Am Soc Microbiol; Sigma Xi; Am Asn Dent Schs; Am Asn Dent Res; Am Soc Virol. *Res:* diagnostic virology-clinical; tumor immunology; dental immunology. *Mailing Add:* 11980 Canary Ct Grand Terrace CA 92324

KETTERINGHAM, JOHN M, physical chemistry; deceased, see previous edition for last biography

KETTERSON, JOHN BOYD, b Orange, NJ, Oct 2, 34; m 61; c 3. PHYSICS. *Educ:* Univ Chicago, BS, 57, MS, 59, PhD(physics), 62. *Prof Exp:* Assoc physicist, Argonne Nat Lab, Ill, 62-72; sr physicist, 74; chmn dept, 85-90, PROF PHYSICS, NORTHWESTERN UNIV, 74-; CONSULT, ARGONNE NAT LAB, 74- *Mem:* Fel Am Phys Soc. *Res:* Low and ultra low temperature technique; properties of monomolecular films; physics of liquid crystals; properties of composition modulated structures; properties of liquid and solid helium; electronic properties of metals and Fermi surfaces; semiconducting films; nonlinear optical film. *Mailing Add:* Dept Phys & Astron Northwestern Univ Evanston IL 60201

KETTMAN, JOHN RUTHERFORD, JR, b Niles, Calif, Nov 29, 39; m 68; c 2. IMMUNOLOGY. *Educ:* Univ Calif, Berkeley, BA, 61; Ore State Univ, PhD(biochem), 68. *Prof Exp:* Asst res, Ore State Univ, 61-62, biochem, 62-64; asst res biologist, Univ Calif, San Diego, 69-72; asst prof, 73-75, assoc prof, 75-80, PROF MICROBIOL, SOUTHWESTERN MED SCH, UNIV TEX SOUTHWESTERN MED CTR, DALLAS, 80- *Concurrent Pos:* NIH trainee immunochem, Kaiser Res Found, San Francisco, 67-69; mem immunol sci study group, 80-83; Wellcome vis prof, 85. *Mem:* Am Chem Soc; Am Asn Immunologists; Am Soc Zoologists; Soc Anal Cytol; AAAS. *Res:* Cellular immunology; cell cooperation in the immune response; biology of lymphoid systems. *Mailing Add:* Microbiol Dept Univ Tex Health Sci Ctr Dallas TX 75235-9048

KETTNER, CHARLES ADRIAN, b Fredericksburg, Tex, Oct 27, 46; m 69; c 2. ENZYMOLOGY, PEPTIDE CHEMISTRY. *Educ:* Southwest Tex State Univ, BSE, 69, MA, 71; Tex A&M Univ, PhD(biochem), 74. *Prof Exp:* Asst instr chem, Southwest Tex State Univ, 71; res asst chem, Tex A&M Univ, 71-74; res collabr biochem, Brookhaven Nat Lab, 74-77, sr res assoc biochem, 77-78, from asst biologist to assoc biologist biochem, 78-80; prin investr biochem, CR&D Dept, Du Pont, 80-90, PRIN INVESTR BIOCHEM, DU PONT MERCK PHARMACEUTICALS, 91- *Mem:* Am Soc Biochem & Molecular Biol. *Res:* Design and synthesis of inhibitors of proteolytic enzymes; identification of target proteases where control of proteolysis can be theropeutically useful. *Mailing Add:* Du Pont Merck Pharmaceut E328-245 Wilmington DE 19880-0328

KETY, SEYMOUR S, b Philadelphia, Pa, Aug 25, 15; m 40; c 2. NEUROSCIENCE, PSYCHOBIOLOGY. *Educ:* Univ Pa, AB, 36, MD, 40. *Hon Degrees:* ScD, Univ Pa, 65, Loyola Univ, 69, Univ Ill, 81, Mt Sinai Sch Med, 83, Med Col Pa, 85, Georgetown Univ, 87, Wash Univ, 89, Mich Univ, 91; MD, Univ Copenhagen, 79. *Prof Exp:* Nat Res Coun fel, Harvard Med Sch, 42-43; instr pharmacol, Sch Med, Univ Pa, 43-44, assoc, 44-46, asst prof, 46-48, prof clin physiol, Grad Sch Med, 48-61; sci dir, NIMH & Nat Inst Neurol Dis & Blindness, 51-56, chief lab clin sci, NIMH, 56-67; dir psychiat res lab, Mass Gen Hosp, 67-77; dir psychiat res labs, Mailman Res Ctr, McLean Hosp, 77-83; prof psychiat, 67-82, prof, 82-83, EMER PROF NEUROSCI, HARVARD MED SCH, 83-; SR SCIENTIST, NIMH, NIH, 83- *Concurrent Pos:* Numerous hon lectureships, 51-; mem adv bd ment health res, Ford Found, 55-57; mem sci adv bd, Scottish Rite Found Res Schizophrenia, 57-, chmn, 71-; chmn biosci adv comt, NASA, 59-60; mem, President's Panel Ment Retardation, 61-62; Henry Phipps prof, Johns Hopkins Univ & psychiatrist-in-chief, Hosp, 61-62; assoc, Neurosci Res Found, 62-79; vis prof, Col France, 66-67; mem vis comt biol, Calif Inst Technol, 72-79; mem bd trustees, Rockefeller Univ, 77-85; founding ed, J Psychiat Res, 83- *Honors & Awards:* Theobald Smith Award, AAAS, 49; Max Weinstein Award, United Cerebral Palsy Res Found, 54; McAlpin Medal & Res Achievement Award, Nat Asn Ment Health, 72; Paul Hoch Award, Am Psychopath Asn, 73; Kovalenko Award, Nat Acad Sci, 73; Neurosci Award, 88; Menninger Award, Am Col Physicians, 76; Fromm-Reichmann Award, Am Acad Psychoanal, 78; Passano Award, 80; FFRP Award Psychiat Res, 80; Res Award Asn Res Nerv & Ment Dis, 80; Thomas W Salmon Medal, NY Acad Med, 82; Emil Kraepelin Medal, Max Planck Inst, 84; Mihara Award, 84; Ralph Gerard Award, Soc Neurosci, 86. *Mem:* Nat Acad Sci; Am Physiol Soc; Asn Res Nerv & Ment Dis (trustee, 63, pres, 65); Am Psychopath Asn (pres, 65); Am Acad Arts & Sci; Am Philos Soc. *Res:* Lead citrate complex, first chelation therapy of lead poisoning; circulation and metabolism of the human brain; theory of capillary-tissue exchange of diffusible tracers; brain imaging; genetics of schizophrenia; biological aspects of mental illness. *Mailing Add:* NIMH, NIH Bldg 10 Rm 4C110 Bethesda MD 20892-0001

KEUDELL, KENNETH CARSON, b Oklahoma City, Okla, May 3, 41; m 67; c 2. MICROBIOLOGY. *Educ:* Okla State Univ, BS, 63, MS, 67; Univ Mo, PhD(microbiol), 69. *Prof Exp:* Fel microbiol, Albert Einstein Col Med, 69-71; asst prof, Sch Dent Med, Wash Univ, 71-78; ASSOC PROF MICROBIOL, WESTERN ILL UNIV, 78- *Concurrent Pos:* Nat Inst Dent Res res grant, 74-76. *Mem:* Am Soc Microbiol; Sigma Xi. *Res:* Medical microbiology, with emphasis on anaerobic microorganisms. *Mailing Add:* 40 Briarwood Pl Macomb IL 61455

KEULEGAN, GARBIS, hydraulics; deceased, see previous edition for last biography

KEULKS, GEORGE WILLIAM, b East St Louis, Ill, Apr 2, 38; m 60; c 4. PHYSICAL CHEMISTRY. *Educ:* Washington Univ, AB, 60; Univ Ark, MS, 62; Northwestern Univ, PhD(chem), 64. *Prof Exp:* Res chemist, Gulf Res & Develop Co, 64-65; res assoc, Johns Hopkins Univ, 65-66; from asst prof to assoc prof chem, 66-74, assoc dean natural sci, 74-75, actg dean grad sch, 75-77, prof chem, 74-, DEAN GRAD SCH, UNIV WIS-MILWAUKEE, 77-, RES, 86- *Concurrent Pos:* Consult, Nelson Industs, Inc, 74, Atlantic Richfield Co, 80- *Mem:* AAAS; Am Chem Soc; Catalysis Soc NAm. *Res:* Heterogeneous catalysis; catalytic kinetics; synthesis of new catalysts; mechanisms of catalytic oxidation; physical characteristics of catalysis. *Mailing Add:* Dept Chem Univ Wis Milwaukee WI 53201

KEUPER, JEROME PENN, b Ft Thomas, Ky, Jan 12, 21; m 48; c 2. PHYSICS. *Educ:* Mass Inst Technol, BS, 48; Stanford Univ, MS, 49; Univ Va, PhD, 52. *Prof Exp:* Res assoc, Carnegie Inst Technol, 49-50; sr res physicist, Remington Arms Co, Inc, 52-58; sr scientist, RCA Missile Test Proj, 58-60, mgr syst anal, 60-62; PRES, FLA INST TECHNOL, 58- *Concurrent Pos:* Chmn math dept, Bridgeport Eng Inst, 54-58. *Mem:* Am Soc Eng Educ; Opers Res Soc Am; Asn Comput Mach; Sigma Xi. *Res:* Solid state, radiation and nuclear physics; computing machines; operations research; ballistics; error analysis; space technology. *Mailing Add:* President Emeritus PO Box 510394 Melbourne Beach FL 32951

KEUSCH, GERALD TILDEN, b New York, NY, Apr 30, 38; m 62, 85; c 3. INFECTIOUS DISEASES. *Educ:* Columbia Col, AB, 58; Harvard Univ, MD, 63. *Prof Exp:* Instr, Tufts-New Eng Med Ctr, 69-70; from asst prof to prof med, Mt Sinai Sch Med, 70-78; PROF MED, TUFTS-NEW ENG MED CTR, 78- *Concurrent Pos:* Mem subcomt interactions nutrit & infection, Nat Acad Sci, 71-, chmn, 76-; mem comt int nutrit progs, 75- & comn int rels, 76- *Honors & Awards:* Squibb Award, Infectious Dis Soc Am, 81. *Mem:* AAAS; Am Soc Microbiol; Am Fedn Clin Res; Infectious Dis Soc Am; NY Acad Sci; Asn Am Physicians. *Res:* Pathogenesis of enteric infections, particularly bacillary dysentery , giordiasis and cryptosporidiosis; effect of malnutrition on the immune response and host defenses. *Mailing Add:* 136 Harrison Ave Boston MA 02111

KEVAN, DOUGLAS KEITH MCEWAN, b Helsinki, Finland, Oct 31, 20; Can & UK citizen; m 43; c 3. ENTOMOLOGY. *Educ:* Univ Edinburgh, BSc Hons, 41; Imp Col Trop Agr, Trinidad, BWI, AICTA, 43; Univ Nottingham, PhD, 56. *Prof Exp:* Brit Colonial Off cadet entom, Imp Col Trop Agr, 41-43; entomologist, Kenya Dept Agr, EAfrica, 43-48; head agr zool sect, Univ Nottingham, 48-58; prof entom & chmn dept entom & plant path, Macdonald Col, 58-64,; chmn dept entom, 64-72, chmn, Lyman Entom Mus, 60-72, dir, Lyman Entom Mus & Res Lab, 72-86, EMER PROF ENTOM, MCGILL UNIV, 86- *Concurrent Pos:* Acctg sr entomologist, Uganda, 45; Tech adv locust control, E Ethiopia & Somalilands, 46-47. *Honors & Awards:* Gold Medal Outstanding Achievement, Entom Soc Can, 81. *Mem:* Fel Royal Soc Edinburgh; Am Entom Soc; fel Entom Soc Can (pres, 72-73); Entom Soc Am; fel Royal Entom Soc London; hon mem Entom Soc Finland; Can Soc Zool; Soc Syst Zool; Syst Asn; hon mem Orthopterists' Soc; Asn Appl Biol. *Res:* Taxonomy and distribution of orthopteroid insects, especially Pyrgomorphidae and North American neuroptera; biology and ecology of soil fauna, especially microarthropods; insect poetry and ethnoentomology. *Mailing Add:* Dept Entom Macdonald Campus McGill Univ, 21111 Lakeshore Rd Ste-Anne-de-Bellevue PQ H9X 1C0 Can

KEVAN, LARRY, b Kansas City, Mo, Dec 12, 38. PHYSICAL CHEMISTRY. *Educ:* Univ Kans, BS, 60; Univ Calif, Los Angeles, PhD(chem), 63. *Prof Exp:* Vis res assoc chem, Univ Newcastle, 63; instr, Univ Chicago, 63-65; from asst prof to assoc prof, Univ Kans, 65-69; prof chem, Wayne State Univ, 69-; CULLEN PROF CHEM, UNIV HOUSTON, 80- *Concurrent Pos:* Exchange fel, Czech, 69; vis scientist, Danish Atomic Energy Lab, 70; Guggenheim fel, 70-71; vis prof, Univ Utah, 71, Univ Nagoya, Japan, 76, Univ Paris, 77 & Armed Forces Tech Univ, Munich, 79, Hokkaido Univ, Japan, 87, Univ Florence, Italy, 87, 90; exchange fel USSR, Nat Acad Sci, 74, 75 & 77; vis scientist, Japan Soc Prom Sci, 76; vis comt chem, Brookhaven Nat Lab, 74-78, chmn, 78; chmn, Gordon Conf Radiation Chem, 75, Chem Div Rev Comt, Argonne Nat Lab, 82; dir, Southwest Catalysis Soc, 82-84, chmn, 86-88. *Honors & Awards:* Polish Soc Radiation Res Award, 79; Nat Honor Soc Res Award, 86; Am Chem Soc Tex Award, 86; Rector's Medal, Poland, 87; Sigma Xi Res Award, 89. *Mem:* Fel AAAS; Am Chem Soc; fel Am Phys Soc; Int Soc Magnetic Resonance; fel, Royal Soc Chem; Int Electron Paramagnetic Resonance Soc. *Res:* Electron magnetic resonance and relaxation; electron spin echo spectrometry; photoredox reactions in micelles and vesicles; ions and radicals on zeolite and oxide surfaces; peroxy and nitroxide radical probes of molecular motion; ion solvation geometry; electron localization and solvation; radiation damage in materials. *Mailing Add:* Dept Chem Univ Houston Houston TX 77204-5641

KEVAN, PETER GRAHAM, b Edinburgh, Scotland, June 17, 44; Can citizen; m 84; c 2. ENTOMOLOGY, BOTANY ECOLOGY EVOLUTION. *Educ:* McGill Univ, BSc, 65; Univ Alta, PhD(entom), 70. *Prof Exp:* Nat coordr, Int Biol Prog, Univ Alta, Can, 69-70; contract biologist, Can Wildlife Serv, Inuvik, NT, Can, 70-71; fel, Plant Res Inst, Can Agr, Ottawa, Can, 71-72; prog mgr ecol, Mem Univ Nfld, Can, 72-75; asst prof biol, Univ Colo, Colo Springs, 75-82, grad fac asst prof biol & assoc, Inst Arctic & Alpine Res, Boulder, 75-83, prin investr NSF grants, Colo Springs, 76-82; ASSOC PROF ECOL, ETOM & APICULT, BOT UNIV GUELPH, CAN, 83- *Concurrent Pos:* Prin investr, Nat Sci & Eng Res Coun grants, Univ Guelph, 83-; consult, Can Wildlife Serv, 70-71, Palm Oil Res Inst Malaysia, 83-, Can Int Develop Agency, 83-, Can Int Develop Res Ctr, 83- & Food & Agr Orgn, Rome, 83-84 & 89-90; mem bd dirs & ed, Entom Soc Ont, 86-; mem bd dirs, Entom Soc Can, 90- *Mem:* Brit Ecol Soc; Bot Soc Am; Int Bee Res Asn; Entom Soc Can; Int Asn Ecol; Sigma Xi; Fel Royal Entom Soc London. *Res:* Botanical and entomological research in co-evolutionary ecology; pollination biology; arctic and alpine ecology; apiculture and agriculture in developing countries; sustainable and ecological agriculture and development. *Mailing Add:* Dept Environ Biol Univ Guelph Guelph ON N1G 2W1 Can

KEVANE, CLEMENT JOSEPH, b Rembrandt, Iowa, May 17, 22; m 53; c 9. PHYSICS. *Educ:* Iowa State Col, BS, 48, PhD(physics), 53. *Prof Exp:* Physicist, Motorola, Inc, 53-56; assoc prof, 56-63, PROF PHYSICS, ARIZ STATE UNIV, 63- *Mem:* Am Phys Soc; Sigma Xi. *Res:* Physics of solid state, semiconductor properties; conduction of heat and electricity in refractory oxides. *Mailing Add:* 1714 S La Rosa Dr Tempe AZ 85281

KEVERN, NILES RUSSELL, b Elizabeth, Ill, May 15, 31; m 55; c 3. AQUATIC ECOLOGY, FISHERIES SCIENCE. *Educ:* Univ Mont, BS, 58; Mich State Univ, MS, 61, PhD(limnol), 63. *Prof Exp:* Limnologist, Oak Ridge Nat Lab, Union Carbide Corp, 63-66; from asst prof to assoc prof, 66-69, PROF FISHERIES & WILDLIFE & CHMN DEPT, MICH STATE UNIV, 69-, ASSOC DIR SEA GRANT PROG, 78- *Concurrent Pos:* Asst dir, Inst Water Res, Mich State Univ, 67-69. *Mem:* Am Fisheries Soc; AAAS; Wildlife Soc. *Res:* Aquatic ecology, particularly bioenergetics and mineral cycling of flowing water ecosystems; radiation ecology; fisheries. *Mailing Add:* Dept Fisheries & Wildlife Mich State Univ Nat Resources Bldg East Lansing MI 48824

KEVILL, DENNIS NEIL, b Walton-le-Dale, Eng, May 27, 35. PHYSICAL ORGANIC CHEMISTRY. *Educ:* Univ Col London, BSc, 56, PhD(chem), 60. *Hon Degrees:* DSc, London Univ, 82. *Prof Exp:* Asst lectr chem, Univ Col London, 59-60; res assoc, Univ Nebr, 60-63; from asst prof to assoc prof, 63-70, presidential res prof, 85-89, PROF CHEM, NORTHERN ILL UNIV, 70-, UNIV RES PROF, 89- *Concurrent Pos:* Grants, Petrol Res Fund, 63-64 & 65-70, NSF, 67-71, 74-75 & 84-88, NIH, 75-76 & NASA, 82; consult, Carus Chem Co, Inc, 65-68; hon res fel, Univ Col London, 75-76; vis prof, Univ Tuebingen, 83, Univ Freiburg, 87, Univ Munich, 90; Nat Acad Sci Exchange Partic, Yugoslavia, 83. *Mem:* Am Chem Soc; The Chem Soc. *Res:* Organic reaction mechanisms; nucleophilicity; reaction mechanisms in solvents of low polarity; elimination reactions; electrophilic assistance to nucleophilic substitutions; perchlorate esters; adamantane derivatives; boron-carbon compounds; chemical vapor deposition. *Mailing Add:* Dept Chem Northern Ill Univ DeKalb IL 60115

KEVLES, DANIEL JEROME, b Philadelphia, Pa, Mar 2, 39; m 61; c 2. HUMAN GENETICS, POLITICS OF SCIENCE. *Educ:* Princeton Univ, BA, 60, PhD(hist), 64. *Prof Exp:* Instr hist, Princeton Univ, 64; from assoc prof to prof hist, 64-86, KOEPFLI PROF OF HUMANITIES, CALIF INST TECHNOL, 86- *Concurrent Pos:* Vis res fel, Univ Sussex, 76; exec officer humanities, Calif Inst Technol, 78-81; vis prof, Univ Penn, 79; mem, adv comt sci autobiog, Sloan Found, 80-90, educ adv comt, Guggenheim Found, 86-; fel, Ctr Adv Study Behav Sci, 86-87. *Honors & Awards:* Nat Hist Soc Prize, 79. *Mem:* Hist Sci Soc; Am Hist Asn; Am Acad Arts & Sci; Orgn Am Historians; fel AAAS. *Res:* Social and political history of modern science, especially physics and genetics in the United States and Britain. *Mailing Add:* Humanities & Soc Sci Div Calif Inst Technol Pasadena CA 91125

KEVORKIAN, ARAM K, b Aug 27, 42; US citizen; c 4. LARGE-SCALE SCIENTIFIC COMPUTATIONS, SPARSE MATRIX COMPUTATIONS. *Educ:* Univ London, BSc, 65, PhD(appl math), 68. *Prof Exp:* Sr mathematician, Royal Dutch Shell Res Lab, Amstead, 68-80; head comput, GA Technologies, 80-86; assoc dir, Cornell Nat Supercomputer Ctr, Cornell Univ, 88-89; MEM TECH STAFF, NAVAL OCEAN SYSTS CTR, 89- *Concurrent Pos:* Vis scientist, T J Watson Res Ctr, IBM Corp, Yorktown Heights, NY, 86-87; adj prof, San Diego State Univ, 91-; sr fel, San Diego Supercomputer Ctr, 91- *Mem:* Soc Indust & Appl Math; Am Math Soc; Asn Comput Mach. *Res:* Large-scale scientific computations with emphasis on the use of graph theory to unearth structure out of large sparse matrix problems. *Mailing Add:* 935 Genter St Unit 3D La Jolla CA 92037

KEVORKIAN, JIRAIR, b Jerusalem, Palestine, May 14, 33; US citizen; m 80. APPLIED MATHEMATICS. *Educ:* Ga Inst Technol, BS, 55, MS, 56; Calif Inst Technol, PhD(aeronaut, math), 61. *Prof Exp:* Aerodynamicist, Gen Dynamics/Convair, 56-57; res fel aeronaut, Calif Inst Technol, 61-64; from asst prof to assoc prof, 64-71, PROF APPL MATH, AERONAUT & ASTRONAUT, UNIV WASH, 71- *Concurrent Pos:* Vis prof, Univ Paris, 71-72; Fulbright-Hays Award, Coun Int Exchange of Scholars, 75. *Mem:* Soc Indust & Appl Math. *Res:* Development and application of perturbation techniques. *Mailing Add:* Dept Appl Math FS-20 Univ Wash Seattle WA 98195

KEWISH, RALPH WALLACE, chemistry; deceased, see previous edition for last biography

KEY, ANTHONY W, b Edinburgh, Scotland, Mar 3, 39; m 63; c 2. PARTICLE PHYSICS. *Educ:* Aberdeen Univ, MA, 60; Oxford Univ, DPhil(physics), 64. *Prof Exp:* PROF PHYSICS, UNIV TORONTO, 73-, ASSOC CHMN, PHYSICS DEPT. *Concurrent Pos:* Psychotherapist, Pvt Pract. *Mem:* Can Asn Physicists; Brit Inst Physics. *Res:* Experimental particle physics; high energy physics using the techniques of bubble or spark chambers; higher education. *Mailing Add:* Dept Physics Univ Toronto Toronto ON M5S 1A1 Can

KEY, CHARLES R, b Oklahoma City, Okla, Aug 4, 34; m 58; c 3. PATHOLOGY, CANCER EPIDEMIOLOGY. *Educ:* Okla State Univ, BS, 56; Univ Okla, MD, 59, MS, 62, PhD(med sci), 66; Am Bd Path, dipl, 64. *Prof Exp:* Intern path, Med Ctr, Univ Okla, 59-60, resident, 60-64; pathologist, Div Air Pollution, USPHS, Ohio, 64-66; pathologist, Atomic Bomb Casualty Comn, Hiroshima, Japan, 66-69; asst prof, 69-73, assoc path, 73-83, PROF PATH, SCH MED, UNIV NMEX, 83- *Concurrent Pos:* Med dir, NMex Tumor Registry, 69- *Honors & Awards:* Div Ann Award, Am Cancer Soc, 85. *Mem:* Col Am Path; Int Acad Path; Am Asn Pathologists; Am Soc Prev Oncol. *Res:* Use of quantitative methods in morphology; pathology of radiation injury; epidemiological pathology of cancer; tumor registry. *Mailing Add:* Dept Path Med Sch Univ NMex Albuquerque NM 87131

KEY, JOE LYNN, b Troy, Tenn, Sept 10, 33; m 56; c 2. PLANT PHYSIOLOGY. *Educ:* Univ Tenn, BS, 55; Univ Ill, MS, 57, PhD(plant physiol), 59. *Prof Exp:* Asst agron, Univ Ill, 55-59, fel biochem, 59-60; NSF fel, Univ Calif, Davis, 60, asst prof, 60-62; from assoc prof to prof, plant physiol, Purdue Univ, 62-69; RES PROF BOT, UNIV GA, 69- *Concurrent Pos:* Dir, Competitive Grants Prog, Sci & Educ Admin, USDA, 78-79; vpres res, Agrigenetics Corp, Boulder, CO, 84, 85. *Res:* Biochemistry of auxin action; nucleic acid metabolism; developmental regulation in plants; RNA metabolism; control of protein synthesis; stress-regulated RNA and protein synthesis. *Mailing Add:* Dept Bot Univ Ga Athens GA 30602

KEY, MORRIS DALE, b Hobbs, NMex, May 31, 39; m 70; c 1. ENVIRONMENTAL CHEMISTRY, ANALYTICAL CHEMISTRY. *Educ:* Midwestern State Univ, BS, 64; Univ Tex, Dallas, PhD(environ geochem), 76. *Prof Exp:* Sales engr, Drew Chem Corp, 65-66; sales rep, Beckman Instruments, Inc, 66-67; co-founder, Anal Consult, Inc, 67-71; PRES, KEY LABS, INC, 71- *Mem:* Am Chem Soc; Am Inst Chemists; Am Soc Metals; Am Soc Testing & Mat; Asn Consult Chemists & Chem Engrs. *Res:* Heavy metal mobilities in sediments and water; precious metal recovery systems. *Mailing Add:* 12538 Renoir Lane Dallas TX 75230

KEYES, DAVID ELLIOT, b Brooklyn, NY, Dec 4, 56; m 80; c 2. COMPUTATIONAL FLUID DYNAMICS, PARALLEL COMPUTATION. *Educ:* Princeton Univ, BS, 78; Harvard Univ, MS, 79, PhD(appl math), 84. *Prof Exp:* Res assoc comput sci, 84-85, asst prof, 86-90, ASSOC PROF MECH ENG, YALE UNIV, 90- *Concurrent Pos:* NSF presidential young investr, 89; vis scientist, Inst Computer Applications Sci & Eng, NASA Langley Res Ctr, Hampton, Va, 90. *Mem:* Am Inst Aeronaut & Astronaut; Am Soc Mech Engrs; Combustion Inst; Int Asn Math & Computers Simulation; Soc Indust & Appl Math. *Res:* Numerical methods for partial differential equations; domain decomposition algorithms; parallel computation; computational modelling of combustion and heat transfer. *Mailing Add:* Dept Mech Eng Yale Univ PO Box 2159 New Haven CT 06520-2159

KEYES, EVERETT A, animal husbandry; deceased, see previous edition for last biography

KEYES, JACK LYNN, b St Johns, Mich, Nov 15, 41; m 68; c 2. PHYSIOLOGY. *Educ:* Linfield Col, BA, 63; Univ Ore Med Sch, PhD(physiol), 70. *Prof Exp:* Instr physiol, Med Col, Cornell Univ, 70-71; from asst prof to assoc prof physiol, Med Sch, Ore Health Sci Univ, 71-83; assoc prof, 83-88, PROF BIOL & CHMN SCI, LINFIELD COL, PORTLAND, 88- *Mem:* Assoc Am Physiol Soc. *Res:* Renal physiology; acid-base regulation; venous blood-gas composition. *Mailing Add:* Linfield Col 2255 NW Northrup St Portland OR 97210

KEYES, MARION ALVAH, IV, b Bellingham, Wash, May 11, 38; m 62; c 3. REAL-TIME MISSION CRITICAL SYSTEMS, ARTIFICIAL INTELLIGENCE. *Educ:* Stanford Univ, BS, 60; Univ Ill, MS, 68; Baldwin Wallace Col, MBA, 81. *Prof Exp:* Dir eng, Control Systs Div, Beloit Corp, 63-70; gen mgr digital systs, Div Taylor Instrument Co, 70-75; vpres eng, res & develop, Bailey Controls Co, 75-80, pres, 80-85, pres & chief exec officer, 89-90; sr vpres & group exec, McDermott Int Inc, 85-89; PRES, TRICE ENG, 90-; CHMN, DCOM CORP, 90- *Concurrent Pos:* Mem, Automation Res Coun, NSF, 70-73; dir & secy, Am Automatic Control Coun, 72-80; trustee, Baldwin Wallace Col, 83-; dir, Fact Inc. *Mem:* Tech Asn Pulp & Paper Indust; Inst Elec & Electronics Engrs; Instrument Soc Am; Am Inst Chem Engrs; Soc Mfg Engrs. *Res:* Computers; process controls; information systems; electronics; automation; holder of over 45 US patents in these fields and have 15 patents pending. *Mailing Add:* 120 Riverstone Dr Chagrin Falls OH 44022

KEYES, PAUL HOLT, b Hartford, Conn, Oct 23, 43; m 87; c 1. LIQUID CRYSTALS, PHASE TRANSITIONS. *Educ:* Rensselaer Polytech Inst, BS, 65; Univ MD, PhD (physics), 72. *Prof Exp:* Fel physics, Univ Del, 72-75, asst prof physics, Univ Mass, Boston, 75-79; from asst prof to assoc prof, Bartol Res Found, Univ Del, 79-87; ASSOC PROF, WAYNE STATE UNIV, 87- *Mem:* Am Phys Soc; Sigma Xi. *Res:* Critical phenomena; liquid crystals; phase transitions. *Mailing Add:* Wayne State Univ Detroit MI 48202

KEYES, PAUL LANDIS, b Thomasville, NC, July 7, 38; m 66; c 2. REPRODUCTIVE ENDOCRINOLOGY, PHYSIOLOGY. *Educ:* NC State Univ, BS, 60, MS, 62; Univ Ill, PhD(animal sci, physiol), 66. *Prof Exp:* Res asst prof, Dept Obstet & Gynec, Albany Med Col, 68-72; from asst prof to assoc prof path, Univ Mich, 72-75, from asst prof to assoc prof, 74-81, PROF PHYSIOL, UNIV MICH, ANN ARBOR, 84- *Concurrent Pos:* NIH fel, Med Sch, Harvard Univ, 66-68; ed, Endocrinol, 79-82; mem, Clin Sci Study Sect, NIH, 85-89; dir, Soc Study Reproduction 85-88. *Mem:* Endocrine Soc; Soc Study Reproduction; Am Physiol Soc. *Res:* Endocrine regulation of the ovary; regulation of the corpus luteum by growth factors and cytokines. *Mailing Add:* Dept Physiol 7793 Med Sci II Univ Mich Ann Arbor MI 48109-0622

KEYES, ROBERT WILLIAM, b Chicago, Ill, Dec 2, 21; m 66; c 2. PHYSICS. *Educ:* Univ Chicago, BS, 42, MS, 49, PhD(physics), 53. *Prof Exp:* Jr physicist, Argonne Nat Lab, 46-50; res physicist, Res Lab, Westinghouse Elec Corp, 53-56, adv physicist, 57-60, consult physicist, 60; PHYSICIST, IBM CORP, 60- *Concurrent Pos:* Corresp, Comments on Solid State Physics, 70-83; consult, Nat Acad Sci Physics Surv Comt, 70-72; chmn, Comt Ion Implantation, Nat Mat Adv Bd, 79, Comt Applications Physics, Am Phys Soc, 76-78 & Int Conf Heavy Doping & Metal-Insulator Transition in Semiconductors, 84; assoc ed, Rev Mod Physics, 76- *Honors & Awards:* WRG Baker Prize, Inst Elec & Electronics Engrs, 76. *Mem:* Nat Acad Eng; fel Am Phys Soc; fel Inst Elec & Electronics Engrs. *Res:* Solid state physics and its applications to electronics. *Mailing Add:* TJ Watson Res Ctr IBM Corp PO Box 218 Yorktown Heights NY 10598

KEYES, THOMAS FRANCIS, b New Haven, Conn, Sept 21, 45; m 68. THEORETICAL CHEMISTRY. *Educ:* Yale Univ, BS, 67; Univ Calif, Los Angeles, PhD(chem), 71. *Prof Exp:* NSF fel chem, Mass Inst Technol, 71-74; asst prof, 74-81, ASSOC PROF CHEM, YALE UNIV, 81- *Res:* Statistical mechanics, emphasizing theory of light scattering, dynamics of fluctuations in fluids and kinetic theory. *Mailing Add:* 15 Vista Terr New Haven CT 06515

KEYNES, HARVEY BAYARD, b Philadelphia, Pa, Dec 27, 40; m 64. MATHEMATICS. *Educ:* Univ Pa, BA, 62, MA, 63; Wesleyan Univ, PhD(math), 66. *Prof Exp:* Asst prof math, Univ Calif, Santa Barbara, 66-68; vis asst prof, 68-69, from asst prof to assoc prof, 69-78, assoc head math, 79-82, PROF MATH, UNIV MINN, MINNEAPOLIS, 78-, DIR, OUTREACH PROJS, 84- *Concurrent Pos:* Trainee math, US Naval Air Develop Ctr, Pa, 58-62; NSF grants, 67-; HEW grant, 73-75; prog dir, NSF, 82-83. *Mem:* AAAS; Am Math Soc; Fedn Am Scientists; Math Asn Am; Sigma Xi. *Res:* Topological dynamics; dynamical systems; ergodic theory; mathematics education. *Mailing Add:* Dept Math Univ Minn Minneapolis MN 55455

KEYS, ANCEL (BENJAMIN), b Colorado Springs, Colo, Jan 26, 04; m 39; c 3. PHYSIOLOGY, NUTRITION. *Educ:* Univ Calif, BA, 25, MS, 29, PhD(biol), 30; Cambridge Univ, PhD(physiol), 38. *Prof Exp:* Asst, Scripps Inst, Univ Calif, 27-30; Nat Res Coun fel, Copenhagen Univ, 30-31; Rockeffer Found Coun fel, Cambridge Univ, 31-32, lectr & demonstr physiol, 32-33; instr biochem sci, Harvard Univ, 33-36; asst prof biochem, Mayo Found, 36-37, from assoc prof to prof physiol, 37-46, prof physiol hyg, 46-75, dir lab physiol hyg, 39-75, EMER PROF PHYSIOL HYG, UNIV MINN, MINNEAPOLIS, 75- *Concurrent Pos:* Res assoc, Oceanog Int, Woods Hole, 33-34; mgr, Int High Altitude Exped, Chile, 35; sr Fulbright fel, Oxford Univ, 51-52; USPHS spec fel, 63-64; consult, UN, WHO, Food & Agr Orgn & UNESCO, 50- *Honors & Awards:* Medal Honor, Acad Finland; Medal Honor, Univ Belgrade. *Mem:* AAAS; Am Physiol Soc; Am Inst Nutrit; Am Soc Clin Nutrit; Int Soc Cardiol. *Res:* Epidemiology; cardiology. *Mailing Add:* Lab Physiol Hyg Univ Minn 611 Beacon St Minneapolis MN 55455

KEYS, CHARLES EVEREL, b Wis, May 10, 21; m 43; c 3. VERTEBRATE EMBRYOLOGY. *Educ:* Greenville Col, AB, 43; Univ Kans, PhD(zool), 52. *Prof Exp:* Asst zool, Greenville Col, 42-43; asst, Atomic Energy Proj, Univ Kans, 50-51; from asst prof to prof biol, Asbury Col, 51-56; assoc prof & chmn div sci & math, Wesleyan Col, NY, 56-60; prof biol, Asbury Col, 60-64 & Florence State Col, 64-68; prof & chmn dept, Seattle Pac Col, 68-71; PROF BIOL, UNIV N ALA, 71- *Concurrent Pos:* Asst, Univ Kans, 51-56; vis prof, Malone Col, 62; chairperson, Health Professions Adv Comt & coordr gen biol, Univ NAla. *Mem:* Sigma Xi. *Res:* Comparative development of rodents; gametogenesis and early development of fish. *Mailing Add:* Dept Biol Univ NAla Florence AL 35632

KEYS, JOHN DAVID, b Toronto, Ont, Sept 30, 22; m 45; c 2. PHYSICS. *Educ:* McGill Univ, BSc, 47, MSc, 48, PhD(nuclear physics), 51. *Prof Exp:* From asst prof to prof physics, Can Serv Col, Royal Roads, 51-58, head dept, 57-58; sr sci off, Dept Energy, Mines & Resources, Ont, 58-67, head mineral physics sect, 63-67, chief hydrol sci div, Inland Waters Br, 67-70; sci adv, Treasury Bd Secretariat, 70-71; asst vpres, Nat Res Coun Can, 71-73, vpres, 73-76; asst dep minister, Sci & Technol Sector, Dept Energy, 76-81; CONSULT, 81- *Mem:* AAAS; Am Phys Soc; Can Asn Physicists; Royal Astron Soc. *Res:* Application of radiotracers to industrial problems associated with mining; solid state physics studies applied to minerals and semiconductors. *Mailing Add:* 39 Ridean Ter Ottawa ON K1M 2A2 Can

KEYS, L KEN, b Cincinnati, Ohio, Nov 6, 39; m 61; c 3. MATERIALS SCIENCE ENGINEERING, CHEMISTRY. *Educ:* Univ Cincinnati, BS, 61; Pa State Univ, PhD(solid state sci), 65. *Prof Exp:* Res assoc, solid state sci, Pa State Univ, 65-66; prin engr, Nuclear Systs Progs Div, Gen Elec Co, 66-68; mgr advan microelectronics eng, Magnavox Co, 69-73; mgr microelectronics technol, Bell-Northern Res, 73-75; mgr, advan mfg develop, Northern Telecom Ltd, 75-77, mgr advan electronic tel develop, 77-78; mgr, tel progs, Stromberg-Carlson Corp, 77-78; prog mgr, DBX Progs & dir technol, United Technols Lexar Corp, 81-84; dir, Instrumentation Systs Ctr, Univ Wis-Madison, 84-87, assoc prof indust eng, 87-89; PROF & DEPT CHAIR INDUST & MFG SYSTS ENG, LA STATE UNIV, BATON ROUGE, 89- *Concurrent Pos:* Dir & pres-elect, Soc Eng & Mgt Systs. *Mem:* Sr mem Inst Elec & Electronics Engrs; sr mem Soc Mfg Engrs; Sigma Xi; sr mem Inst Indust Engrs; NY Acad Sci. *Res:* New technology, product development and transfer into production-market place, including follow-up support (life cycle process). *Mailing Add:* Col Eng CEBA Bldg Rm 3128 La State Univ Baton Rouge LA 70803

KEYS, RICHARD TAYLOR, b Salina, Kans, Feb 11, 31; m 55. PHYSICAL CHEMISTRY. *Educ:* Harvard Univ, AB, 53; Iowa State Col, PhD(chem), 58. *Prof Exp:* Asst chem, Iowa State Col, 53-58; res fel, Calif Inst Technol, 58-59; from asst prof to assoc prof, 59-72, PROF CHEM, CALIF STATE UNIV, LOS ANGELES, 72- *Concurrent Pos:* NIH spec fel, Univ Calif, Riverside, 66-67; vis prof, Nat Auro Univ Mex, 75-76, Univ Southern Calif, 84-85. *Mem:* Am Chem Soc; Am Phys Soc. *Res:* Chemistry of free radicals; magnetic resonance. *Mailing Add:* Dept Chem Calif State Univ 5151 State University Dr Los Angeles CA 90032-8202

KEYS, THOMAS EDWARD, b Greenville, Miss, Dec 2, 08; m 34; c 2. HISTORY OF MEDICINE. *Educ:* Beloit Col, AB, 31; Univ Chicago, MA, 34. *Hon Degrees:* ScD, Beloit Col, 72. *Prof Exp:* Order asst, Newberry Libr, Univ Chicago, 31-32; libr asst, Mayo Clin Libr, 34-35, ref librn, 35-42, librn, 46-69, sr libr consult & emer chmn, Mayo Found, 69-72, from asst prof to assoc prof hist med, Mayo Grad Sch Med, Univ Minn, 56-69, prof, 69-72, EMER PROF HIST MED, MAYO GRAD SCH MED, UNIV MINN & MAYO FOUND & MED LIBR CONSULT, MAYO CLIN LIBR, 73- *Concurrent Pos:* Hon consult, Army Med Libr, 46-50; mem, Bd Regents, Nat Libr Med, 59-62; spec lectr, Int Cong, Med Libr, Amsterdam, 69 & Int Cong Hist Med, London, 72; vis lectr, Univ Iceland Sch Med, 77, Univ Witwatersrand, Johannesburg, SAfrica & Capetown Univ, 79; lectr hist mod anesthesia, Rotterdam, Neth, 82. *Honors & Awards:* George Dock Mem lectr, Univ Calif, 66. *Mem:* Hon mem Med Libr Asn (pres, 57-58); Am Asn Hist Med; Am Soc Anesthesiol; hon mem Anesthesia Hist Asn. *Res:* History of anesthesia, cardiology, medical education and medical librarianship. *Mailing Add:* 4001 NW 19th Ave Rochester MN 55901

KEYSER, DAVID RICHARD, b Ft Wayne, Ind, Dec 5, 41; m 65; c 2. MECHANICAL ENGINEERING, APPLIED MATHEMATICS. *Educ:* Swathmore Col, BS, 63; Univ Pa, MS, 65; Eurotechnical Res Univ, PhD, 91. *Prof Exp:* Res engr fluid dynamics & flow measurement, Naval Boiler & Turbine Lab, 65-68; res engr flow & temperature measurement, Naval Ship Eng Ctr, Philadelphia, 68-72, sr proj engr fluid dynamics & control systs, 72-80; AEROSPACE ENGR FLUIDIC FLIGHT CONTROL, NAVAL AIR DEVELOP CTR, WARMINSTER, PA, 80- *Concurrent Pos:* Chmn, Govt Fluidics Coord Group, Am Soc Mech Engrs, 79-; secy, Res Comt Fluid Meters, Am Soc Mech Engrs, 78-, mem, Standards Comt, Measurement Fluid Flow, 78-, chmn, Fluid Power Control Systs Panel & vchair bd Performance Test Codes, exec comt, Fluid Power Systems & Technol Div; mem Comn A-6D Flight Controls Panel, Soc Automotive Engrs. *Mem:* Am Soc Mech Engrs; Soc Automotive Engrs. *Res:* Flight controls and flying qualities; fluidics; advanced actuation systems; unsteady flow measurements; new methods of flow measurement. *Mailing Add:* Box 1426 Southampton PA 18966

KEYSER, N(AAMAN) H(ENRY), b Philadelphia, Pa, Dec 5, 18; m 42; c 4. METALLURGY. *Educ:* Antioch Col, BS, 41; Ohio State Univ, MS, 43. *Prof Exp:* Res engr process metall, Battelle Mem Inst, 41-47; asst group leader fabrication group, Los Alamos Sci Lab, Univ Calif, 47-48; asst chief process metall, Battelle Mem Inst, 48-62; dir res, Interlake Inc, 62-80; CONSULT, 80- *Concurrent Pos:* Exec consult, Exec Serv Corps Chicago; master gardener, Coop Exten Serv, Univ Ill. *Mem:* Am Soc Metals; Am Foundrymen's Soc (secy, 48-60); Am Inst Mining, Metall & Petrol Engrs; Iron & Steel Soc. *Res:* Economic studies and new processes for metals and inorganic chemicals using elevated temperatures produced by electric and blast furnaces; making, forming and treating of steel and steel products especially in silicon and silicon alloys. *Mailing Add:* 122 W Walnut Hinsdale IL 60521-3350

KEYSER, PETER D, b Columbus, Ohio, Oct 26, 45; m 69; c 1. MEDICAL MICROBIOLOGY. *Educ:* Ohio State Univ, BS, 68, MS, 70, PhD(microbiol), 72. *Prof Exp:* Res fel microbiol, Univ Ky & Fla State Univ, 72-73; asst prof microbiol, 73-81, ASST PROF MICROBIOL & IMMUNOL, TEX COL OSTEOP MED, NTEX STATE UNIV HEALTH SCI CTR, 81-, ADJ PROF BIOL SCI, 75- *Mem:* Sigma Xi; Am Soc Microbiol. *Res:* Role of lipases in pathogenicity of gram negative organisms, particularly those species isolated from burn wound sepsis. *Mailing Add:* Dept Microbiol & Immunol Tex Col Osteopath Med Camp Bowie-Montgomery Ft Worth TX 76107

KEYT, DONALD E, b Indianapolis, Ind, Oct 30, 27; m 53; c 3. MECHANICAL ENGINEERING. *Educ:* Purdue Univ, BS, 50; Mass Inst Technol, MS, 53. *Prof Exp:* Design & res engr, Arthur D Little, Inc, 52-62; asst prof mech eng, Towne Sch Civil & Mech Eng, Univ Pa, 62-69; PROF & CHMN DEPT MECH & MFG ENG TECHNOL, SPRING GARDEN COL, 69- *Concurrent Pos:* Design consult, High Field Magnet Lab, Univ Pa, 63; lectr, 69-; design engr, plant res projs, Space Lab 1-4, 78-86, Micro-G res, 86-89; adj lectr, Dept Mech Eng & Appl Math, Univ Pa, 69- *Mem:* Am Soc Eng Educ; Am Soc Mech Engrs; Am Soc Metals. *Res:* Ultrahigh field electromagnet design, development and research; numerical solution of engineering problems; compressor design and development; instrumentation and control applied to systems; machine design; computer aided design and drafting. *Mailing Add:* 1196 Skippack Rd Harleysville PA 19438

KEYWORTH, DONALD ARTHUR, b Flint, Mich, Apr 21, 30; m 78. TECHNICAL MANAGEMENT, CATALYSIS. *Educ:* Univ Mich, BS, 51; Mich State Univ, MS, 54; Wayne State Univ, PhD(chem), 58. *Prof Exp:* Chief anal controls, Lapaco Chem Co, Mich, 51-52; anal chemist, Ethyl Corp, Mich, 54 & Wyandotte Chem Co, 58-60; asst res dir, Universal Oil Prod Co, 60-67, vpres & tech dir, Sci & Educ Serv, 67-68; mgr res, Tenneco Chem, Tenneco Inc, 68-80; mgr res, Petro-Tex Chem Corp, 80-85; GROUP RES MGR, CATALYST DIV, AKZO CHEM INC, 85- *Concurrent Pos:* Instr, Wayne State Univ, 57-60. *Honors & Awards:* Res Award, Sigma Xi, 68. *Mem:* Am Soc Testing & Mat; Am Chem Soc; Soc Appl Spectros; NY Acad Sci; Sigma Xi. *Res:* Industrial research; methods and techniques of analysis of industrial products, acetylenics synthesis and manufacturing; industrial utilization of pi-complexes; industrial catalysis and catalyst evaluation. *Mailing Add:* 5320 Dora Houston TX 77005-1818

KEYWORTH, GEORGE A, II, b Boston, Mass, Nov 30, 39; m 62; c 2. EXPERIMENTAL NUCLEAR PHYSICS. *Educ:* Yale Univ, BS, 63; Duke Univ, PhD(physics), 68. *Prof Exp:* Res assoc, Duke Univ, 68; staff mem, Los Alamos Sci Lab, 68-74, group leader, 74-78, alt physics div leader & physics div leader, 78-81, laser fusion div leader, 80-81; DIR, OFF SCI & TECHNOL POLICY & SCI ADV TO PRESIDENT, 81- *Mem:* AAAS; Am Phys Soc; Sigma Xi. *Res:* Nuclear structure problems; isobaric analogue states; polarization experiments; fission physics; neutron physics; fusion physics; science policy. *Mailing Add:* 3956 Georgetown Ct Washington DC 20007

KEYZER, HENDRIK, b Djakarta, Indonesia, Dec 7, 31; m 54; c 7. PHYSICAL CHEMISTRY, SOLID STATE CHEMISTRY. *Educ:* Univ NSW, BS, 63, PhD(chem), 66. *Prof Exp:* Res chemist, Mus Appl Arts & Sci, 62-63; lectr, Sydney Tech Col, 64-66; univ fel & NIH fel, 67, res grant, 67-68, from asst prof to assoc prof, 67-79, PROF CHEM, CALIF STATE UNIV, LOS ANGELES, 79- *Concurrent Pos:* NIMH res grant, 68-69 & 71-72; consult, Jet Propulsion Lab, 68-; reader, Victoria Univ, Wellington, 72-73; Health & Human Welfare, res grant, 76-80; Dept Educ, training grants, 81-86. *Honors & Awards:* NASA Award, 71. *Mem:* Am Chem Soc. *Res:* Physical-organic chemistry of natural compounds and compounds of biological importance; psychotropic drugs; polymer chemistry; electrochemistry; micro-analysis; natural products. *Mailing Add:* Dept Chem Calif State Univ 5151 State University Dr Los Angeles CA 90032

KEZDI, PAUL, b Hungary, Nov 13, 14; nat US; m 43, 65; c 4. CARDIOLOGY. *Educ:* Pazmany Peter Univ, Budapest, MD, 42; Am Bd Internal Med, dipl, 61. *Prof Exp:* Dir heart sta, Wesley Mem Hosp, Chicago, Ill, 54-65; prof med, Wright State Univ, 75-86; dir res, 65-67, dir, Cox Heart Inst, 67-86; assoc dean res affairs, 77-85, EMER PROF MED, WRIGHT STATE UNIV SCH MED, 86- *Concurrent Pos:* Am Heart Asn fel clin cardiol; assoc prof, Sch Med, Northwestern Univ, 62-65; mem, Med Adv Bd, Coun High Blood Pressue Res, Am Heart Asn, 63; prof med, Ind Univ, 65-72; clin prof, Ohio State Univ, 65-75; mem, Heart Prog Proj Comt, Nat Heart Lung & Blood Coun, 66-67. *Mem:* Fel Am Col Chest Physicians; fel Am Col Cardiol; Am Physiol Soc; Am Fedn Clin Res; Am Heart Asn. *Res:* Cardiac physiology; coronary artery disease; hypertension. *Mailing Add:* Cox Health Eval Serv 7076 Corporate Way Dayton OH 45459

KEZDY, FERENC J, b Budapest, Hungary, July 28, 29; m 58; c 3. BIOCHEMISTRY. *Educ:* Univ Louvain, DrSci(phys org chem), 57. *Prof Exp:* Asst phys & org chem, Univ Louvain, 57-61; assoc biochem, Northwestern Univ, 61-65; from asst prof to prof biochem, Univ Chicago, 66-88; DISTINGUISHED SCIENTIST, UPJOHN CO, 88- *Honors & Awards:* Stas-Spring Prize, Belg Royal Acad Sci, 58; Woutes Prize, Chem Soc Belg, 60. *Mem:* Am Chem Soc; AAAS; Chem Soc Fr; Soc African Insect Scientists. *Res:* Physical organic chemistry; surface biochemistry; enzymology. *Mailing Add:* 3805 Robin Lane Kalamazoo MI 49008-3148

KEZIOS, STOTHE PETER, b Chicago, Ill, Apr 8, 21; m 52; c 1. THERMODYNAMICS. *Educ:* Ill Inst Technol, BS, 42, MS, 48, PhD(mech eng), 55. *Prof Exp:* Asst instr mech eng, Ill Inst Technol, 42-44, instr, 46-47, res engr, 47-49, from asst prof to prof, 49-67, dir heat transfer lab, 55-68; PROF MECH ENG & DIR, SCH MECH ENG, GA INST TECHNOL, 67- *Concurrent Pos:* Consult govt & indust, 48-; vchmn, Nat Heat Transfer Conf, 58; consult ed, J Heat Transfer, 58-63, tech ed, 63-; secy, Int Heat Transfer Conf, 61-62, adv orgn comt, 66; Am Soc Mech Eng rep, Comt Int Rels, Eng, Joint Coun, 62-; mem sci adv panel, Eng Sect, NSF, 63-66. *Honors & Awards:* Ralph Coates Roe award, Am Soc Eng Educ, 77. *Mem:* Am Soc Mech Engrs (secy, 53-57, vpres, 73-77, pres, 77-78); Am Soc Eng Educ; Am Inst Chem Engrs; Sigma Xi. *Res:* Heat transfer; analogy between heat and mass transfer; forced convection heat transfer, especially with free jets; boiling in forced convection flow. *Mailing Add:* 1060 Winding Creek Trail NW Atlanta GA 30328

KEZLAN, THOMAS PHILLIP, b Omaha, Nebr, Aug 6, 35. MATHEMATICS. *Educ:* Univ Omaha, 57; Univ Kans, MA, 59, PhD(math), 64. *Prof Exp:* Asst prof math, Univ Mo, Kansas City, 63-64; asst prof, Univ Tex, Austin, 64-68; ASSOC PROF MATH, UNIV MO-KANSAS CITY, 68- *Mem:* Math Asn Am; Am Math Soc. *Res:* Theory of rings. *Mailing Add:* Univ Mo Kansas City MO 64110

KHABBAZ, SAMIR ANTON, b Tel Aviv, Palestine, Mar 31, 32; US citizen; m 59; c 3. MATHEMATICS. *Educ:* Bethel Col, BA, 54; Univ Kans, MA, 56, PhD, 60. *Prof Exp:* Instr math, Univ Mass, 60; asst prof, Lehigh Univ, 60-62; Off Naval Res res fel, Yale Univ, 62-63; univ res fel, 63-64; assoc prof, 64-68, PROF MATH, LEHIGH UNIV, 68- *Mem:* Am Math Soc. *Res:* Algebra and topology. *Mailing Add:* Dept Math Xmas-Saucon Hall 26 Lehigh Univ Bethlehem PA 18015

KHACHADURIAN, AVEDIS K, b Aleppo, Syria, Jan 6, 26; m 61; c 2. MEDICINE. *Educ:* Am Univ Beirut, BA, 49, MD, 53. *Prof Exp:* Resident internal med, Am Univ Beirut, 53-56, fel biochem, 56-57, from asst prof to prof biochem & internal med, Sch Med, 59-71; lectr pediat, Northwestern Univ, Chicago, 65-66, prof, 71-73; dir, Clin Res Ctr, Children's Mem Hosp, 71-73; PROF MED, UNIV MED & DENT NJ-R W JOHNSON MED SCH, 73-, CHIEF DIV ENDOCRINE, METAB DIS & NUTRIT, 73- *Concurrent Pos:* Res fel, Harvard Med Sch & Joslin Clin, Mass, 57-59. *Mem:* Am Diabetes Asn; Endocrine Soc; Am Fedn Clin Res; Am Heart Asn; NY Acad Sci; Am Inst Nutrit. *Res:* Clinical and biochemical aspects of familial hypercholesterolemia; effect of excercise on metabolic parameters in diabetes mellitus. *Mailing Add:* Dept Med Univ Med & Dent NJ R W Johnson Med Sch C N 19 New Brunswick NJ 08903

KHACHATOURIANS, GEORGE G, b 1940. TOXICOLOGY, BIOTECHNOLOGY. *Educ:* Calif State Univ, 66, BA, MA, 69; Univ British Columbia, PhD(Microbiol), 71. *Prof Exp:* Res assoc, Univ Mass, 73-74; asst prof, 74-81, PROF APPL MICROBIOL, UNIV SASK, 81- *Concurrent Pos:* Chmn, Genetics, Univ Sask, 80-83, Biotechnol, 84-86; dir, Bioinsecticide Res Lab, Univ Sask, 82- *Mem:* Am Soc Microbiol; Can Soc Microbiol; Can Biochem Soc; Can Genetics Soc; Am Entom Soc; Can Entom Soc; Am Soc Indust Microbiol. *Mailing Add:* Dept Applied Microbiol & Food Sci Univ Sask Col Agr Biotechnol Lab Saskatoon SK S7N 0W0 Can

KHACHATURIAN, NARBEY, b Teheran, Iran, Jan 12, 24; nat US; m 52; c 4. CIVIL ENGINEERING, STRUCTURAL ENGINEERING. *Educ:* Univ Ill, BS, 47, MS, 48, PhD(eng), 52. *Prof Exp:* From instr to prof, 49-89, assoc head Dept Civil Eng, 83-89, EMER PROF CIVIL ENG, UNIV ILL, URBANA, 89- *Concurrent Pos:* NSF sr fel, Univ Calif, Los Angeles, 63-64;

chmn, Ill Struct Engr Examining Comt, 71; pres, Struct Engrs Asn Ill, 89-90. *Honors & Awards:* Parmer Award, 86; Halliburton Award, 88. *Mem:* Fel Am Soc Civil Engrs; Am Soc Eng Educ; Am Concrete Inst; Nat Soc Prof Engrs; Sigma Xi. *Res:* Experimental and analytical structural engineering, especially reinforced and prestressed concrete; structural optimization; structural concrete. *Mailing Add:* PO Box 26 Philo IL 61864-0026

KHACHATURIAN, ZAVEN SETRAK, b Alleppo, Syria, Apr 15, 37; US citizen; m 63; c 1. NEUROSCIENCE. *Educ:* Yale Univ, BA, 61; Case Western Reserve Univ, PhD(neurobiol), 67. *Prof Exp:* Fel neurophysiol, Col Physicians & Surgeons, Columbia Univ, 67-69; asst prof develop neurobiol, Div Child Psychiat, Med Sch, Univ Pittsburgh, 69-72, prog dir neurophysiol, Dept Psychiat, 72-77; grants assoc, Div Res Grants, NIH, 77-78; prog dir, Neurosci Aging, Nat Inst Aging, NIH, 78-79; health policy coordr, Off Secy, Dept Health, Educ & Welfare, 79-80; spec asst dir, Physiol Aging Br, 80-81, chief, 81-85, ASSOC DIR, NEUROSCI & NEUROPSYCHOL OF AGING & DIR, OFF ALZHEIMER'S DIS RES, NAT INST AGING, NIH, 86- *Concurrent Pos:* Interim sci dir, Pittsburgh Biotechnol Ctr, 85-86; vpres res, Health Sci Univ Pittsburgh Med Ctr, 85, prof, Health Serv Admin, 86. *Mem:* Soc Neurosci. *Res:* Alzheimer's disease; neuroscience of aging; neural plasticity; science policy; calcium regulation in aging neurine. *Mailing Add:* Nat Inst Aging Bldg 31-C NIH Bethesda MD 20892

KHAIR, ABDUL WAHAB, b Kabul, Afghanistan, March 20, 41; m 69. ROCK MECHANICS, ACOUSTICS EMISSION. *Educ:* WVa Univ, BS, 67, MS, 68; Pa, State Univ, PhD(miningeng), 72. *Prof Exp:* Res assoc rock mech, Pa, State Univ, 72-75; pres, Ministry of Mines & Ind, Afghanistan, 75-79; res assoc rock mech, Pa State Univ, 79-81; assoc prof mining eng, 81-86, PROF MINING ENG, WVA UNIV, 86- *Mem:* Soc Mining Eng; Int Soc Rock Mech. *Res:* Respirable dust generation in underground coal mines, coal bump and gas outburst; subsidence due to underground excavations, monitoring, and prediction; structural stability analysis of mines and design of mine layout and support system; rock mechanics and ground control relation problems, and physical and analytical modeling of geotechnical problems. *Mailing Add:* Rm 22 White Hall Bldg WVa Univ PO Box 6070 Morgantown WV 26506

KHAIRALLAH, EDWARD A, b Beirut, Lebanon, Jan 9, 36; US citizen; m 62; c 2. ENDOCRINOLOGY, NUTRITIONAL BIOCHEMISTRY & BIOCHEMICAL TOXICOLOGY. *Educ:* Am Univ, Beirut, BSc, 56; Harvard Univ, MA, 59; Mass Inst Technol, PhD(nutrit, biochem), 64. *Prof Exp:* Instr physiol chem, Mass Inst Technol, 64-66; NIH res fel, McArdle Labs, Univ Wis, 66-67; asst prof biochem, 67-69, from assoc prof to prof biochem & biophysics, 69-81, PROF BIOL, UNIV CONN, 81- *Concurrent Pos:* Vis prof, Univ Tokushima, Japan, 81-82. *Mem:* AAAS; Am Inst Nutrit; Endocrine Soc; NY Acad Sci; Am Soc Biol Chem; Am Soc Cell Biol; Int Inst Proteolysis; Soc Toxicol; fel Japanese Biochem Soc. *Res:* Hormonal and dietary regulation of protein synthesis and degradation in mammalian cells; amino acid compartmentation; translational mechanisms regulating gluconeogenic and amino acid catabolizing enzymes; mechanisms of hepatotoxicity, lysosomes and proteolysis. *Mailing Add:* 16 Costello Circle Storrs CT 06268

KHAIRALLAH, PHILIP ASAD, b New York, NY, Feb 3, 28; m 63; c 4. PHYSIOLOGY, PHARMACOLOGY. *Educ:* Am Univ Beirut, BA, 47; Columbia Univ, MD, 51. *Prof Exp:* Am Heart Asn estab investr, Found, 58-63, mem staff res div, 63-70, sci dir res div & head dept cardiovasc res, 70-80, staff res div, 80-86, DIR ANESTHESIA RES, CLEVELAND CLIN FOUND, 86- *Concurrent Pos:* Am Heart Asn fel, Duke Univ, 51-53, Cleveland Clin Found, 56-57 & Am Univ Beirut, 57-58; adj prof, John Carroll Univ & Cleveland State Univ; mem, Coun High Blood Pressure Res, Am Heart Asn. *Mem:* Am Soc Pharmacol & Exp Therapeut; Am Physiol Soc; Am Soc Nephrology; AMA; Int Anesthesia Res Soc. *Res:* Cardiovascular research in biochemistry, physiology and pharmacology; cardiac and blood vessel contraction; medical ethics; narcotic anesthetics. *Mailing Add:* Cleveland Clin Found 9500 Euclid Ave Cleveland OH 44195-5154

KHAKOO, MURTADHA A, b Zanzibar, Tanzania, Apr 29, 53; Brit citizen; m 83; c 2. ELECTRON SCATTERING, LASER EXCITED ATOMS. *Educ:* Univ London, BSc Hons, 75, PhD(physics), 80. *Prof Exp:* Postdoctoral fel, Univ Col London, 80-81 & Univ Windsor, Ont, 84-87; vis fel, Jet Propulsion Lab, NASA-Nat Res Coun, 81-84; res asst prof mech, Univ Mo, Rolla, 87-89; ASST PROF PHYSICS, CALIF STATE UNIV, FULLERTON, 89- *Concurrent Pos:* Consult, Jet Propulsion Labs, Pasadena, 89-; reviewer, J Physics B, Phys Rev A & Phys Rev Lett. *Honors & Awards:* NASA Award, 88. *Mem:* Am Phys Soc. *Res:* Electron impact studies of gaseous targets; elastic and inelastic electron scattering from atomic and molecular targets; electron-photon coincidence studies; fragmentation of molecules, lifetime studies; spin polarized electron scattering. *Mailing Add:* Dept Physics Calif State Univ Fullerton CA 92634-9480

KHALAF, KAMEL T, b Mosul, Iraq, 1922; m 58; c 3. ZOOLOGY, MEDICAL ENTOMOLOGY. *Educ:* Univ Baghdad, BS, 44; Univ Okla, MS, 50, PhD(zool), 52. *Prof Exp:* Instr high teachers col, Univ Baghdad, 53-56, from asst prof to prof, 56-62, res prof, Iraq Natural Hist Inst, 62-63; from asst prof to assoc prof, Loyola Univ, La, 63-69, prof entom, 69-86; RETIRED. *Concurrent Pos:* NIH res grant, 65-68; acad grant fund, Loyola Univ. *Res:* Surveys of biting gnats; biology of puss caterpillar and its parasites; animal surveys; micromorphology of Arthropods Integument. *Mailing Add:* 5811 S Claiborne New Orleans LA 70125

KHALAFALLA, SANAA E, b Mit Yaish, Egypt, July 1, 24; US citizen; m 56; c 2. PHYSICAL CHEMISTRY, CHEMICAL METALLURGY. *Educ:* Cairo Univ, BSc, 44; Univ Minn, Minneapolis, MS, 49, PhD(phys chem), 53. *Prof Exp:* Demonstr chem fac sci, Cairo Univ, 44-48, from asst prof to assoc prof phys chem, 53-61; Hill Family Found res fel & res assoc, Univ Minn, 61-64; proj leader, 64-66, res supvr, Twin Cities Metall Res Ctr, Bur Mines, 66-87, CHIEF SCIENTIST, US DEPT INTERIOR, 87- *Concurrent Pos:* Vis prof, Bristol Univ, 60-61. *Honors & Awards:* Spec Act of Serv Award, Twin Cities Metall Res Ctr, US Bur Mines, 66. *Mem:* Am Inst Mining, Metall & Petrol Engrs; Am Inst Elec & Electronics Engrs. *Res:* Process and extractive metallurgy; kinetics and mechanisms of mineral reactions; electrochemistry and polarography; catalytic reactions; magneto chemistry; plasma chemistry; leaching processes; asbestos fibers; mine water and wastes; magnetic fluids; water conservation; minerals and mining. *Mailing Add:* 2551 37th Ave Minneapolis MN 55406

KHALED, MOHAMMAD ABU, b Murshidabad, India, Nov 1, 42; US citizen; m 76. BIOPHYSICS, PHYSICAL CHEMISTRY & NUTRITIONAL BIOCHEMISTRY. *Educ:* Univ Calcutta, BSc, 61; Aligarh Muslim Univ, India, MSc, 64; Univ London, PhD(biophys, chem), 75. *Prof Exp:* Lectr chem, Polytech Inst, Chittagong, Bangladesh, 64-65; lectr, Cadet Col, Rajshahi, Bangladesh, 66-71; fel, 75-77, from instr to asst prof biochem, 77-83, asst prof nutrit sci, 83-84, ASSOC PROF NUTRIT SCI, UNIV ALA, 84- *Res:* Spectroscopic approach in determining biomolecular conformations and structure-function relationships; body composition measurements. *Mailing Add:* Dept Biochem Univ Ala-Birmingham Univ Sta Birmingham AL 35294

KHALIFA, RAMZI A, b Cairo, Egypt, July 20, 40; US citizen; c 3. ACOUSTICS, METAL FABRICATING. *Educ:* Cairo Univ, BSME, 64; NJ Inst Technol, MS, 74, MS, 89. *Prof Exp:* Mfg eng mgr, HVAC, Delta Indust Co, 64-69; dir eng electronics, Edson Tool & Mfg Co, 70-89; DIR CORP MFG ACOUST, INDUST ACOUST CO, 85- *Concurrent Pos:* Teacher, Cairo Univ, 64-69; prog chmn, Precision Metal Stamping Asn, 81-85. *Res:* Developed cable enclosures with patented self-locking devices; new metal stamping tooling concepts; universal mount dies; modular correctional security ceiling system; acoustical windows; modular track wall panels and highway barrier panels; five patents. *Mailing Add:* 448 Lincoln Ave Rutherford NJ 07070

KHALIFAH, RAJA GABRIEL, b Tripoli, Lebanon, May 5, 42; nat US; m 71; c 2. CHEMISTRY, BIOCHEMISTRY. *Educ:* Am Univ Beirut, BS, 62; Princeton Univ, PhD(phys chem), 69. *Prof Exp:* Res assoc biochem, Harvard Univ, 68-70; res assoc pharmacol, Sch Med, Stanford Univ, 70-73; asst prof chem, Univ Va, 73-80; MEM STAFF, VET ADMIN MED CTR, 80- *Concurrent Pos:* Res assoc prof biochem, sch med, Univ Kans, 79- *Mem:* Am Chem Soc; Am Soc Biol Chemists; Sigma Xi. *Res:* Biophysical chemistry; kinetics and thermodynamics of protein conformation changes; enzyme kinetics and mechanism; chemical modification of active sites; nuclear magnetic resonance applications to proteins and enzymes. *Mailing Add:* Res Serv 151 Vet Admin Med Ctr 4801 Linwood Blvd Kansas City MO 64128

KHALIL, M ASLAM KHAN, b Jhansi, India, Jan 7, 50; US citizen; m 73; c 2. ATMOSPHERIC CHEMISTRY & PHYSICS. *Educ:* Univ Minn, BPhys, BA(math) & BA(psychol), 70; Va Polytech Inst & State Univ, MS, 72; Univ Tex, Austin, PhD(physics), 76; Ore Grad Ctr, MS & PhD(environ sci), 79. *Prof Exp:* Teach asst physics, Va Polytech Inst & State Univ, 70-71; grad asst math & physics, Univ Tex, Austin, 71-72, teaching asst physics, 72-73 & 76, res scientist asst, Ctr Particle Theory, 72-76; res asst, Inst Atmosphere Sci, Ore Grad Ctr, 77-79, sr res assoc, 79-80, from asst prof to assoc prof, environ sci, 80-84, prof chem, biol & environ sci, 84-87, PROF, DEPT ENVIRON SCI & ENG, ORE GRAD INST, 87- *Concurrent Pos:* Instr physics, Pac Univ, 77-78; prin investr, NSF, NASA, Environ Protection Agency & Dept Energy, 80-; owner, Andarz Co, 81- *Mem:* Am Phys Soc; Am Chem Soc; Am Geophys Union; AAAS; Air & Waste Mgt Asn. *Res:* Elementary particles with spin (theoretical physics); author or co-author of 120 publications; atmospheric physics and chemistry; models for global dispersion of trace gases; effects of man-made trace gases; long-distance transport of pollution; receptor models for urban pollution; mathematical and statistical techniques in environmental sciences; biogeochemical cycles; climate models. *Mailing Add:* Global Change Res Ctr Ore Grad Inst 19600 NW Von Neumann Dr Beaverton OR 97006-1999

KHALIL, MICHEL, b Alexandria, Egypt, June 30, 35; Can citizen; m 67; c 3. ORGANOHALOGENATED COMPOUNDS. *Educ:* Univ Alexandria, Egypt, BSc, 57, MSc, 64; Univ Laval, Can, PhD(chem), 70. *Prof Exp:* Chemist, El-Nasr Spinning, Egypt, 57-65 & Dionne Spinning, Can, 65-67; PROF CHEM, UNIV QUEBEC, RIMOUSKI, 70- *Concurrent Pos:* Vis prof chem oceanog, Univ Marie Curie, Paris, 74 & Univ Miami, 77-78; consult, Provincial Dept Environ, 82, Provincial Dept Transports, 84-85 & PCB studies, Hydro-Quebec, 84-86. *Mem:* Am Chem Soc; Can Meterol & Oceanog Soc. *Res:* Bioconcentration and contamination studies in marine and estuarine ecosystems regarding halogenated and polyaromatic hydrocarbons. *Mailing Add:* UQAR 300 Ave Ursulines Rimouski PQ G5L 3A1 Can

KHALIL, MOHAMED THANAA, b Al-Kosair, Egypt, Feb 8, 33; m 60; c 2. BIOPHYSICS, MATHEMATICS. *Educ:* Cairo Univ, BSc, 53; Univ Pittsburgh, BS, 74, PhD(biophys), 66; Univ Alexandria, MSc, 61. *Prof Exp:* Instr & res asst physics, Univ Alexandria, 53-61; res assoc biophys, Univ Pittsburgh, 61-66, res prof biophys & microbiol, 66-71; from asst prof to prof phys sci, 69-74, dean int students, 74-77, prof natural sci & eng technol, Point Park Col, 74-; AT DEPT MATH & ENG TECH. *Mem:* AAAS; Biophys Soc; NY Acad Sci; Am Chem Soc; Am Inst Physics. *Res:* Effect of laser beam on the system of polymerization of virus protein; thermodynamics of the reconstitution of virus particles in deuterium; structure and function of tobacco mosaic virus interferon and plant viruses. *Mailing Add:* Rm 403 Point Park Col Wood St & Blvd Allies Pittsburgh PA 15222

KHALIL, SHOUKRY KHALIL WAHBA, b Cairo, Egypt, Dec 7, 30; US citizen; m 64. PHARMACOGNOSY. *Educ:* Cairo Univ, BPharm & Pharm Chem, 53, MPharm, 56, PhD(pharmacog), 60. *Prof Exp:* From instr to assoc prof pharmacog, Cairo Univ, 54-68; from asst prof to assoc prof, 68-75, PROF PHARMACOG, NDAK STATE UNIV, 75- *Concurrent Pos:* Egyptian govt fel, Univ Mich, Ann Arbor, 62-63. *Mem:* Am Pharmaceut Asn; Am Soc Pharmacog; Soc Cosmetic Chemists; Sigma Xi. *Res:* Plant chemistry; analysis of medicinal plants; pharmacokinetic. *Mailing Add:* Col Pharm ND State Univ Fargo ND 58102

KHALILI, ALI A, b Ardebil, Iran, Feb 9, 32; US citizen; m 62; c 2. REHABILITATION MEDICINE. *Educ:* Tehran Univ, MD, 57. *Prof Exp:* Instr phys med & rehab, State Univ NY Downstate Med Ctr, 62-65; asst prof, Med Ctr, 65-68, dir residency prog phys med & rehab, 67-69, ASSOC PROF PHYS MED & REHAB, MCGAW MED CTR, NORTHWESTERN UNIV, 68-; DIR REHAB MED, GRANT HOSP, CHICAGO, 68-, CHMN, DEPT REHAB MED, 77- *Concurrent Pos:* Dir, Neuromuscular Diag & Res Dept, Rehab Inst Chicago, 68-70; consult physiatrist, Elmhust Mem Hosp, 68- & Rehab Inst Chicago, 70-; chmn, Comt Allied Health Professions, Grant Hosp, Chicago, 73-75; chmn med pract comt, Am Acad Phys Med & Rehab, 79-80. *Mem:* AMA; Am Acad Phys Med & Rehab; Am Cong Rehab Med; Am Asn Electromyography & Electrodiag. *Res:* Spasticity and peripheral phenol nerve block (clinical, physiological and histological aspects, including electromyographic changes); radio-linked bladder stimulation in neurogenic bladder; air splint in preprosthetic rehabilitation of lower extremity amputated limbs; sensory input discrimination in normal and hemiplegic subjects; burn rehabilitation. *Mailing Add:* Dept Rehab Med Grant Hosp Chicago 550 W Webster Chicago IL 60614

KHALIMSKY, EFIM D, b Odessa, USSR, June 23, 38; US citizen; m 62; c 1. GENERAL TOPOLOGY, GENERALIZED HOMOLOGY & HOMOTOPY THEORY. *Educ:* Pedagogical Inst, Odessa, MS, 60; Pedagogical Inst, Moscow, USSR, PhD(math), 69. *Prof Exp:* Teacher math & physics, MS, Odessa, USSR, 60-66; assoc prof math, Pedagogical Inst, Magnitogorsk, USSR, 69-72 & City Univ NY 79-80; sr res scientist, Opers Res Math Econ, Food Indust, Res & Prod Inst, 72-73; sr res scientist appl math & OS, Econ Inst, Acad Sci, Odessa, USSR, 73-77; assoc prof math & comput sci, Manhattan Col, Riverdale, NY, 80-85; assoc prof comput sci, Col Staten Island, City Univ NY, 85-89; PROF MATH, CENT STATE UNIV, 89- *Concurrent Pos:* Postdoctoral studies, Pedagogical Inst, Moscow, USSR, 69; ed-at-large, Marcel Dekker Publ Co; assoc ed, J Appl Math & Stochastic Anal. *Mem:* Am Math Soc; Soc Indust & Appl Math; Asn Comput Mach; Inst Elec & Electronics Engrs Systs, Man & Cybernetics Soc. *Res:* Defined and investigated properties of ordered topological spaces and have used them in developing the generalized homotopy and homology groups, digital topology, topological cell complexes and used those theories in cmputer graphics, systems analysis and design. *Mailing Add:* 1260 Brentwood Dr Dayton OH 45406

KHAMIS, HARRY JOSEPH, b San Jose, Calif, Dec 20, 51. LOG-LINEAR MODEL THEORY. *Educ:* Santa Clara Univ, BS, 74; Va Tech, MS, 76, PhD(statist), 80. *Prof Exp:* ASSOC PROF MATH & STATIST, WRIGHT STATE UNIV, 87- *Concurrent Pos:* Adj instr ethnic dance, dept health, phys educ & recreation, consult, 82-90, Wright State Univ, assoc dir, Statist Consult Ctr, 90-; assoc prof, dept Community health, sch med, 90- *Mem:* Am Statist Asn; Biomet Soc. *Res:* Loglinear model analysis and applications to genetic data; goodness-of-fit tests. *Mailing Add:* 535 Green Tree Pl Fairborn OH 45324

KHAN, ABDUL JAMIL, b Allahabad, India, May 5, 40; m 68; c 2. PEDIATRICS, MEDICINE. *Educ:* Univ Allahabad, BSc, 57; Univ Lucknow, MB & BS, 62; Agra Univ, DCH, 64; Panjab Univ, India, MD, 67; Am Bd Pediat, cert, 73, cert pediat nephrology, 85. *Prof Exp:* Intern, King George Med Col, Univ Lucknow, 62-63; physician med & pediat, Northern Railway Hosp, India, 64-65; teaching fel pediat, Inst Post-Grad Med, Panjab Univ, India, 66-67; registr pediat, Med Col, Aligarh Muslim Univ, India, 68-69; resident, Kings County, State Univ Hosp, Brooklyn, 69-70; fel, 71-73, chief Div Pediat Nephrology, 73-87, ASSOC DIR, DEPT PEDIAT, INNER FAITH MED CTR, BROOKLYN, 87-; PROF PEDIAT, MEHARRY MED CTR, BROOKLYN, 90- *Concurrent Pos:* From instr to asst prof clin pediat, State Univ NY Downstate Med Ctr, 73-78, assoc prof, 78- *Mem:* Am Fedn Clin Res; Soc Pediat Res; Am Soc Pediat Nephrology; Soc Exp Biol & Med; fel Am Acad Pediat. *Res:* White blood cell functions including Chemotaxis; studies on efficacy and pharmacokinetics of newer antibiotics; renal diseases in infants and children; medical education. *Mailing Add:* Inter Faith Med Ctr 1545 Atlantic Ave Brooklyn NY 11213

KHAN, ABDUL WAHEED, b Lahore, WPakistan, Apr 16, 28; Can citizen; m 60; c 2. BIOCHEMISTRY, MICROBIOLOGY. *Educ:* Univ Panjab, Pakistan, MSc, 52, PhD(biochem), 56; Manchester Col Sci & Technol, Eng, PhD(microbiol), 58. *Prof Exp:* Res fel biochem, Univ Panjab, Pakistan, 52-54, lectr org chem, 54-55; res fel microbiol, Manchester Col Sci & Technol, Eng, 55-58; fel biophys, Div Biosci, Nat Res Coun Can, 58-60, from asst res officer to assoc res officer, 60-72, sr res officer, 72-89; RETIRED. *Mem:* Inst Food Technologists; Can Biochem Soc; Can Soc Microbiol; fel Royal Soc Chem. *Res:* Conversion of biomass to fuels and chemical feed stock; anaerobic degradation of cellulose; methanogensis; meat biochemistry; effect of freezing and storage on muscle proteins; biosynthesis of cellulose; studies in bacterial cell wall components; microbiological synthesis of fat from carbohydrates; nutritive values of foods and dietary standards. *Mailing Add:* 2155 Tawney Rd Ottawa ON K1G 1C2 Can

KHAN, AKHTAR SALAMAT, b Aligarh, India, June 8, 44; m 72; c 2. SOLID MECHANICS, MECHANICAL ENGINEERING. *Educ:* Aligarh Univ, India, BS, 61, BS, 65; Johns Hopkins Univ, PhD(solid mech), 72. *Prof Exp:* Lectr mech eng, Aligarh Univ, India, 65-67; from res asst to res assoc mech, Johns Hopkins Univ, 67-73; staff engr stress anal, Arthur McKee, Cleveland, 73-74; sr staff engr, Bechtel Power Corp, 74-78; from asst prof to assoc prof, 78-84, PROF AEROSPACE, MECH & NUCLEAR ENG, UNIV OKLA, 84- *Concurrent Pos:* Ed-in-chief, Int J Plasticity. *Mem:* Fel Am Soc Mech Engrs; Am Acad Mech; Soc Exp Mech; Soc Natural Philos. *Res:* Dynamic and quasi-static behavior of metallic solids; finite amplitude wave propagation in solids; use of finite element techniques to study stresses in shell-to-shell intersections; fracture mechanics; rock mechanics. *Mailing Add:* Sch Aerospace & Mech Eng Univ Okla Norman OK 73019

KHAN, AMANULLAH RASHID, b Bhavnagar, India, Mar 1, 27; US citizen; m 52; c 2. CHEMICAL ENGINEERING. *Educ:* Univ Bombay, BS, 47; Ill Inst Technol, MS, 51. *Prof Exp:* Develop engr, Dry Freeze Corp, 51-52; res engr, Inst Gas Technol, Ill Inst Technol, 52-53; fel, French Petrol Inst, 53-54; supvr refinery processing, Attock Oil Co, 54-61; opers engr, Universal Oil Prod Co, 61-62; mgr gas opers res, 62-70, PRES, GDC, INC, INST GAS TECHNOL, ILL INST TECHNOL, 70- *Mem:* Am Chem Soc; Am Inst Chem Engrs; Am Gas Asn. *Res:* Liquefaction, storage and utilization of liquefied natural gas; gas distribution and transmission research; hydrocarbon processing. *Mailing Add:* 249 Westmoreland Dr Wilmette IL 60091-3059

KHAN, ANWAR AHMAD, b Monghyr, India, Oct 16, 34; m 67; c 2. PLANT PHYSIOLOGY, PLANT BIOCHEMISTRY. *Educ:* Univ Karachi, BS, 56, MS, 57; Univ Chicago, PhD(bot), 63. *Prof Exp:* Demonstr bot, Univ Karachi, 57-58; demonstr & lectr, Univ Sind, Pakistan, 58-60; res asst, Univ Chicago, 60-63; res assoc biochem, Mich State Univ, 63-65; from asst prof to assoc prof, 65-80, PROF SEED PHYSIOL, NY STATE AGR EXP STA, CORNELL UNIV, 80- *Concurrent Pos:* Am Seed Res Found res grants, 67-69 & 70-76; NSF traveling res fel, Univ Liege, Univ Ghent & Univ Clermont-Ferrand, 71-72; Herman Frasch Found grant agr chem, 72-76; Centre Nat de la Recherche res grant, Univ Clermont-Ferrand, France, 72; sr res fel, Agr Univ, Wageningen, Neth, 78; consult, Tenn Eastman Co, 80-81, UN Develop Prog, Pakistan, 81-85; workshop on seed sci & technol, Zhongshan Univ, Guangzhou, China, 85; vis scientist, Int Rice Res Inst, Los Baenos, Philippines,85-86. *Mem:* AAAS; Am Soc Plant Physiologists; Scand Soc Plant Physiologists; Am Soc Hort Sci; Int Plant Growth Substances Asn; Am Soc Agron; Crop Sci Soc Am. *Res:* Seed technology; seed physiology and biochemistry; molecular biology; growth regulators; stress physiology. *Mailing Add:* Dept Hort Cornell Univ State Agr Exp Sta Geneva NY 14456-0462

KHAN, ATA M, b Khan, W Pakistan, Dec 15, 41; m 69. TRANSPORTATION & TRAFFIC ENGINEERING. *Educ:* Am Univ Beirut, BEng, 63, MEng, 65; Univ Waterloo, PhD(transp planning), 70. *Prof Exp:* Asst civil eng, Am Univ Beirut, 63-65; engr, Trans Arabian Pipeline Co, 64; transp engr, De Leuw, Cather & Co, Chicago, 65-67; teaching asst transp eng, Univ Waterloo, 67-69; asst prof, 69-77, ASSOC PROF ENG, CARLETON UNIV, 77- *Concurrent Pos:* Individual supporting mem, Hwy Res Bd, Nat Res Coun-Nat Acad Sci, 68-; spec consult, N D Lea & Assocs, 71; Nat Res Coun Can fel, 68-69. *Mem:* Assoc mem Am Soc Civil Engrs; jr mem Inst Traffic Eng. *Res:* Development of transport planning methodology for the evaluation of policy and investment alternatives and for the analysis of transport subsidy policy in Canada. *Mailing Add:* Dept Civil Eng Carlton Univ Ottawa ON K1S 5B6 Can

KHAN, FAIZ MOHAMMAD, b Multan, Pakistan, Nov 1, 38; m 66; c 3. BIOPHYSICS, RADIOLOGICAL PHYSICS. *Educ:* Emerson Col, Multan, Pakistan, BSc, 57; Govt Col Lahore, MSc, 59; Univ Minn, Minneapolis, PhD(biophys), 69. *Prof Exp:* Health physicist, Radiother Inst, Mayo Hosp, Lahore, Pakistan, 60-63; instr radiol, 68-69, from asst prof to assoc prof, 69-79, PROF THERAPEUT RADIOL, UNIV MINN, MINNEAPOLIS, 79-, HEAD SECT RADIATION PHYSICS, 74- *Concurrent Pos:* Consult physicist, Vet Admin Hosp, Minneapolis, 71- *Mem:* Am Asn Physicists Med; Sigma Xi. *Res:* Radiation dosimetry and treatment techniques in radiation therapy; application of computers in radiotherapy; biological effects of radiation. *Mailing Add:* Univ Minn Hosp Radiation Box 494 Minneapolis MN 55455

KHAN, IQBAL M, b Karachi, Pakistan, Dec 27, 50; US citizen; m 80; c 4. TEACHING PHYSIOLOGY. *Educ:* Univ Karachi, BS, 69, MS, 70; Univ G06t07borg, PhD(physiol), 80. *Prof Exp:* Postdoctoral fel physiol, Univ Ill, Urbana, 80-81; res assoc physiol, Univ Ill, Chicago, 81-82, res asst prof, 81-89; ASSOC PROF & DIR IVF/ANDROLOGY LAB, DEPT OBSTET & GYNEC, MED COL GA, AUGUSTA, 90- *Concurrent Pos:* Postdoctoral fel, Ford Found, 80-81. *Mem:* Soc Study Reproduction Fertil & Steril; Endocrine Soc. *Res:* Pituitary-ovarian axis; corpus luteum function; hormonal control of luteal steroidogenesis; in vitro fertilization. *Mailing Add:* Dept Obstet & Gynec Rm CK 159 Med Col Ga 1459 Laney Walker Blvd Augusta GA 30907Ho

KHAN, JAMIL AKBER, b Hyderabad, Pakistan, Mar 17, 52; m 81; c 2. ORGANOMETALLIC CHEMISTRY. *Educ:* Univ Sind, Pakistan, BSc, 71, MSc, 73; Univ London, Eng, PhD(org chem), 79; Univ New Haven, MBA, 85. *Prof Exp:* Chemist qual control, Eastman Chem Co, 73-74; lectr chem, D J Sci Col, Pakistan, 74-76; demonstr chem, Univ Col London, Eng, 76-79; res assoc chem, Duke Univ, 79-81; res chemist, Uniroyal Inc, 81-84, res scientist, 84-85; dir mkt, Ausimont USA, 85-86, dir technol & mkt int, 86-87, prod mgr, Sales & Mkt Div, 87-90; BUS MGR, ENIMONT AM, INC, 90- *Concurrent Pos:* Secy gen, Inst Pub Affairs, Pakistan, 75-78; joint secy, Sind Lectr Asn, 76-77; mem, Presidential Task Force & Pub Educ Fund Comt, Am Inst Chemists; mem, Int Org Chem Sci Develop Network, UNESCO, bd educ, Mt Arlington, NJ; mem, Republican Presidential Task Force. *Mem:* Am Chem Soc; Royal Soc Chem; Sigma Xi; Am Inst Chemists; Chem Mkt Res Asn. *Res:* Synthesis and development of activators and catalysts, which can be used in polymerization, autoxidation of unsaturated fatty acids; synthesis of antioxidants, which can be used to inhibit the autoxidation of phospholipid biomembranes; Ziegler-Natta catalysis; organometallic chemistry; free radical chemistry polymerization; polymerization kinetics; reaction mechanism; polypeptides; bio compatible polymers; synthetic lubricants; higher performance polymers and their application to high tech industry. *Mailing Add:* Enimont Am Inc 1211 Ave of the Americas New York NY 10036

KHAN, MAHBUB R, b Dhaka, Bangladesh, Sept 11, 49, US citizen; m 76; c 4. MAGNETIC THIN FILMS, DIGITAL MAGNETIC RECORDING. *Educ:* Dhaka Univ, BSc, 69, MSc, 70; Boston Col, PhD(solid state physics), 79. *Prof Exp:* Lectr physics, Jahangirnagar Univ, Bangladesh, 72-74; grad asst, Boston Col, 74-79; teaching fel, Univ Nebr, Lincoln, 79-81; fel, Argonne Nat Lab, 81-83; sr scientist magnetic recording, Control Data Corp-MPI, 83-85;

sr scientist Magnetic Recording, Alcoa-Stolie Corp, 85-86; mgr, Process Develop, 86-89, ACTG DIR TECHNOL DEPT, SEAGATE MAGNETICS, 90- *Mem:* Am Phys Soc; Inst Elec & Electronics Engrs. *Res:* Preparation, characterization, magnetic properties, anisotropy, surface analysis, electron microscopy of magnetic thin films; longitudinal and vertical high density digital magnetic recording; metallic superlattice; optical and electro-optical properties of layered materials. *Mailing Add:* Technol Dept Seagate Magnetics 47001 Benicia St Fremont CA 94538

KHAN, MAHMOOD AHMED, b Hyderabad, India, Sept 16, 45; US citizen; m 75; c 3. FOOD SCIENCE, HOTEL RESTAURANT & INSTITUTIONAL MANAGEMENT. *Educ:* Osmania Univ, BS, 66; Andhra Pradesh Agr Univ, BS, 69; La State Univ, MS, 72, PhD(food sci), 75. *Prof Exp:* Res asst food sci, La State Univ, 71-75; asst prof foods & nutrit, Albright Col, 75-78; asst prof foods & nutrit, 78-84, assoc prof, Univ Ill, Urbana, 84-87; PROF & ASST HEAD, VA POLYTECH & STATE UNIV, 87- *Concurrent Pos:* Vis prof, Catering Res Ctr, Huddersfield Polytech, UK. *Honors & Awards:* Donald K Tessler Res Award. *Mem:* Inst Food Technologists; Am Dietetic Asn; Coun Hotel, Restaurant & Inst Educ; Nat Restaurant Asn; Am Inst Nutrit; Int Nat Acad Hospitality Res. *Res:* Food quality in food service systems; concepts of nutrition and obesity; nutritional evaluation of food processing; international food patterns; food habits; food service consultant; food snacking and eating away from home. *Mailing Add:* Dept Hotel & Restaurant Mgt VA Tech Col 13 Hillcrest Hall Blacksburg VA 24061

KHAN, MOHAMED SHAHEED, b Bloomfield, Guyana, Dec 29, 33; US citizen; m 58; c 3. PLANT PATHOLOGY, AGRICULTURE. *Educ:* Eastern Caribbean Inst, Trinidad, diplom, 57; Iowa State Univ, BSc, 64, MSc, 66, PhD(plant path & hort), 68. *Prof Exp:* Agr ext agent agr educ, Govt Brit Guyana, 57-60; agr res asst, McGill Univ, 60-63; res asst, Iowa State Univ, 63-68; plant pathologist res, Ministry Agr Govt Guyana, 69-71; assoc prof biol & physical sci, Morris Col SC, 72; EXTEN SPECIALIST & PESTICIDE COORDR PATH, ENTOM PESTICIDES, UNIV DC & USDA, 72- *Concurrent Pos:* Teacher hort & landscaping, Spingarn-Phelp's Voc Sch, Washington, DC, 73-77; teacher pesticide applicators, DC Coop Ext Serv, Washington, DC, 73-; coordr training, EPA grant, 74-; consult trop agr, Indonesia USAID, 77, consult res agronomist, JWK Int Corp, Govt Chad, Africa, 78-79, pest mgt specialist, Soma Lia, 79-80. *Mem:* Am Phytopath Soc; Am Hort Soc; Am Entom Soc; Am Chem Soc; Sigma Xi. *Res:* Mycorrhizal associates of Juglans Nigra with special emphasis on nitrogen and phosphorous uptake; pesticide screening; chemical and biological control of pests; environmental preservation. *Mailing Add:* Coop Exten Serv Univ DC 901 Newton St NE Washington DC 20017

KHAN, MOHAMMAD ASAD, b Pakistan; m 74; c 1. GEODESY, RESEARCH ADMINISTRATION. *Educ:* Univ Panjab, WPakistan, BSc, 57, MSc, 63; Univ Hawaii, PhD(geophys), 67. *Prof Exp:* Lectr geophys, Univ Panjab, WPakistan, 63-64; from asst prof to assoc prof, 67-74, PROF GEOPHYS & GEOL, HAWAII INST GEOPHYS, UNIV HAWAII, 74- *Concurrent Pos:* Sr vis scientist, Nat Acad Sci, Goddard Space Flight Ctr, NASA, 72-74; sr scientist, Comput Sci Corp, 74-76, sr consult, 76-77; adv resource surv, Gov of Pakistan, 74-76; mem, Hawaii State Environ Coun, 79-83, chmn, Environ Coun Exec Comn, 79-83 & vchmn, Environ Coun, 81-83; minister, Petrol & Natural Resources, Govt Pakistan, 83-86, chmn, Hydrocarbon Develop Inst, 84-86, chmn, Attock Oil Refinery, 84-86, cabinet mem, Nat Econ Coun, 83-86, senator, 85-86. *Mem:* Am Geophys Union; Am Geol Inst; Pakistan Asn Advan Sci. *Res:* Geophysics; Satellite altimetry; earth density modelling; geophysical, geodetic and geodynamical applications of satellites; plate tectonics; gravity and isostasy; geophysical exploration; resource surveys; geodesy; geodynamics; core-mantle boundary problems; technology transfer; technical management. *Mailing Add:* Hawaii Inst Geophys Univ Hawaii 2525 Correa Rd Honolulu HI 96822

KHAN, MOHAMMAD IQBAL, b Karachi, Pakistan, Dec 27, 50; m 80; c 2. ENDOCRINOLOGY, REPRODUCTIVE PHYSIOLOGY. *Educ:* Univ Karachi, Pakistan, BSc Hons, 69, MSc, 70; Univ Goteborg, Sweden, PhD, 80. *Prof Exp:* Fel endocrinol, Dept Physiol, Univ Goteborg, Sweden, 80 & Dept Animal Sci, Univ Ill, Urbana-Champaign, 80-81; res asst & prof endocrinol, Dept Physiol & Biophys, Med Ctr, Univ Ill, Chicago, 81-89; ASSOC PROF & DIR IVF/ANDROLOGY LABS, MED COL GA, 89- *Concurrent Pos:* Asst lectr, Dept Physiol, Sch Med, Univ Goteborg, Swed, 76-79. *Mem:* Soc Study Reprod; Scand Soc Physiologists; Swed Med Asn; Endocrine Soc; Am Fertil Soc. *Res:* Mechanism of action of gonadotropins in the maintenance and function of the corpus luteum as well as the role of prostaglandins in the regression of the corpus luteum; endocrinology. *Mailing Add:* Dept Obstet & Gynec Rm CK 159 Med Col Ga 1459 Laney Walker Blvd Augusta GA 30912

KHAN, MOHAMMED ABDUL QUDDUS, b India, Mar 10, 39; m 74; c 3. BIOCHEMICAL & ENVIRONMENTAL TOXICOLOGY, INSECTICIDE TOXICOLOGY. *Educ:* Univ Karachi, BSc, 57, MSc, 59; Univ Western Ont, PhD(zool), 64; Univ Auto de Cd Juarez, MD, 84. *Prof Exp:* Fel entom, NC State Univ, 65-67, Ore State Univ, 67-68 & Rutgers Univ, 68-69; assoc prof, 69-74, PROF BIOL, UNIV ILL, CHICAGO, 74- *Concurrent Pos:* Vis prof, Univ Wis, 70; consult Velsicol Chem Corp, Chicago, 75, NSF, 77-78, Environ Protection Agency, 80, Eastern Res Group, 85 & Continental Chemists, 85; vis scientist, Nat Inst Environ Health Sci, NIH, 75-76; vis chemist, US Environ Protection Agency, Corvallis, 80; expert toxicol, UN Develop Prog, Pakistan, 80-81; ed, J Biochem Toxicol. *Mem:* AAAS; Entom Soc Am; Am Chem Soc; Soc Environ Toxic Chem; Soc Toxicol; Am Col Toxicol. *Res:* Metabolism of insecticides, drugs and lipids; biochemistry and genetics of insecticide resistance; environmental toxicology; metabolism of xenobiotics including pesticides; in vitro detoxication and mechanisms; induction of drug metabolizing enzymes; effects on steriodogenesis. *Mailing Add:* Dept Biol Univ Ill Box 4348 Chicago IL 60680

KHAN, MOHAMMED NASRULLAH, b Hyderabad, India, Oct 11, 33; m; c 2. PHYSIOLOGY, VETERINARY SCIENCE. *Educ:* Osmania Univ, India, BVSc, 55; La State Univ, Baton Rouge, MS, 63, PhD(environ physiol), 70. *Prof Exp:* Vet, Govts Hyderabad & Andhra Pradesh, India, 55-58; asst lectr anat, State Vet Sch, Hyderabad, India, 58-61; res asst, La State Univ, Baton Rouge, 61-63, & 67-70; asst lectr reproductive physiol, Tirupati & Hyderabad Vet Cols, India, 63-67; asst prof biol, City Cols Chicago, Mayfair Col & Southwest Col, 70-72; instr microbiol, Schs Nursing, Michael Reese Hosp & Med Ctr, Chicago, 73-74; instr anat, physiol & microbiol, South Chicago Community Hosp Schs Nursing & Radiol, 72-77; assoc prof anat, physiol, microbiol & chem, Little Co Mary Hosp Sch Nursing, 75-83; assoc prof, 77-87, PROF BIOL, TRUMAN COL, 87-; PROF ANAT, PHYSIOL, RAVENSWOOD HOSP, RADIOL SCH, 90- *Concurrent Pos:* HEH Nizam fel; Ford Found res scholar. *Res:* Stress. *Mailing Add:* 2904 W Greenleaf Ave Chicago IL 60645-2916

KHAN, MUSHTAQ AHMAD, b Lyallpur, Pakistan, Dec 12, 39; US citizen; m 59; c 4. PERINATAL TOXICOLOGY, ENDOCRINOLOGY. *Educ:* Univ Punjab, Pakistan, BSc, 60; Mont State Univ, MS, 62; Wash State Univ, PhD(vet sci & physiol), 68. *Prof Exp:* Chemist steroid biochem, Syntex Res Ctr, Calif, 64-65; asst prof vet physiol, Univ Agr, Pakistan, 65-72; res assoc pediat, Med Sch, Univ Md, Baltimore, 72-74, asst prof, 74-78, asst prof path, 77-78; physiologist food additives, Div Toxicol, 78-80, res physiologist perinatal toxicol, Metab Br, 80-81, SUPVRY RES PHYSIOLOGIST & HEAD, PERINATAL TOXICOL, BUR FOODS, FOOD & DRUG ADMIN, 81- *Concurrent Pos:* Prin investr, Sch Med, Univ Md, Baltimore, 72-78; adj asst prof, Univ RI, 80- *Mem:* Am Physiol Soc; Am Inst Nutrit; Am Col Toxicol; Am Soc Vet Physiologists & Pharmacologists; World Asn Physiol, Pharmacol & Biochem. *Res:* Endocrine and nutritional factors in obesity and atherosclerosis; perinatal toxicology; perinatal nutrition and delayed effects-imprinting; cholesterol metabolism; age and sex related changes in metabolic responses. *Mailing Add:* Adminr Div Res Grants NIH Gen Med A-2 Study Sect 3543 Westwood Bldg Bethesda MD 20892

KHAN, NASIM A, b Benares, India, June 1, 38; US citizen; m 77. GENE REGULATION, MITOCHONDRIAL GENETICS. *Educ:* Univ Dacca, BSc, 58, MSc, 60; City Univ New York, PhD(genetics), 67. *Prof Exp:* Lectr bot, Univ Decca, 61-62; fel gen biol, 62-64, lectr gen biol, 64-68, instr, Genetics Lab, 68-69, from asst prof to assoc prof, 70-84, PROF GENETICS, BROOKLYN COL, CITY UNIV NEW YORK, 85- *Mem:* Sigma Xi; Am Soc Microbiol. *Res:* Construction of yeast strains exhibiting elevated levels of ethanol production using genetic selection procedures and certain recombinant DNA technique; interaction of nuclear and cytoplasmic genes in the utilization of fermentive sugars in yeast; regulation of maltase and alpha-methylglucosidose synthesis in yeast. *Mailing Add:* Dept Biol Brooklyn Col City Univ New York Brooklyn NY 11210

KHAN, PAUL, b Vienna, Austria, Nov 4, 23; nat US; m 50. FOOD SCIENCE, PRODUCT SAFETY. *Educ:* NY Univ, BA, 48; Univ Chicago, MS, 49. *Prof Exp:* Asst microbiol, Squibb Inst Med Res, Olin Mathieson Chem Co, 49-53; chief microbiologist, Food & Drug Res Lab, 53-55; dir frozen food prod lab, DCA Food Industs, Inc, 55-60, mgr cent res labs, 60-62; mgr res admin, Continental Baking Co, 62-71; dir food protection, ITT Continental Baking Co, Inc, 71-80, vpres qual & food protection, 80-85; PRES, REGU-TECH ASSOCS, INC, 85- *Concurrent Pos:* Mem, Tech Comt Food Protection-Grocery Mfg; indust liaison panel, Food Protection Comt, Nat Acad Sci-Nat Res Coun; mem, Codex Alimentarius Comn, FAO/WHO; contrib ed, Food Safety Notebook. *Mem:* Am Chem Soc; Am Asn Cereal Chemists; fel Inst Food Technologists; Asn Food & Drug Officials; Asn Res Dirs; Am Soc Qual Control. *Res:* Food preservation, technology, poisoning and protection; laboratory management; regulatory compliance; quality assurance; occupational safety and environment; nutrition. *Mailing Add:* Regu-Tech Assoc Inc 158 W Boston Post Rd PO Box 717 Mamaroneck NY 10543-0717

KHAN, RASUL AZIM, b Port Mourant, Guyana, Oct 31, 34; Can citizen; m 66; c 3. PARASITOLOGY. *Educ:* Univ Toronto, Can, BSA, 64, MSc, 66, PhD(parasitol), 69. *Prof Exp:* Asst prof, 69-74, assoc prof, 74-81, PROF BIOL, MEM UNIV NFLD, 82- *Concurrent Pos:* Res scientist, Marine Sci Res Lab, 72- *Mem:* Am Soc Parasitilogists; Am Soc Protozoologists; Can Soc Zoologists. *Res:* Studies on the effects of parasites and pollutants as causative agents of disease in commercial fish in Eastern Canada; long term effects of petroleum on fish. *Mailing Add:* Marine Sci Res Lab Mem Univ Nfld St Johns NF A1C 5S7 Can

KHAN, SAAD AKHTAR, b Dhaka, Bangladesh, May 24, 58. RHEOLOGY OF MATERIALS, POLYMER PROCESSING. *Educ:* Princeton Univ, BSE, 80; Mass Inst Technol, PhD(chem eng), 85. *Prof Exp:* Mem tech staff, AT&T Bell Labs, 85-87; MEM TECH STAFF, BELL COMMUN RES, 87- *Mem:* Soc Rheology; Polymer Processing Soc. *Res:* Rheology and physical properties of polymers, gels, ceramic, composites and foams; relationship between material microstructure and rheological/processing properties; rheological studies are complemented using techniques such as light scattering and microscopy to get all over picture. *Mailing Add:* Bell Commun Res 331 Newman Springs Rd Red Bank NJ 07701

KHAN, SEKENDER ALI, b Bogra, Bangladesh, Feb 1, 33; m 63; c 2. PLANT PATHOLOGY. *Educ:* Univ Dacca, BArg, 53, MAgr, 54; La State Univ, PhD(plant path), 59. *Prof Exp:* Sect officer, EPakistan Indust Develop Corp, 55-57; asst cane develop officer, M/S Carew & Co, 57; prof biol, Tex Col, 59-63, chmn sci & math, 60-63; actg chmn dept, 64-65, PROF BIOL, ELIZABETH CITY STATE UNIV, 64-, CHMN DEPT, 65- *Mem:* AAAS; Bot Soc Am; Bangladesh Bot Soc. *Res:* Plant hormones; plant alkaloids extraction of active chemicals from Vitex negundo L; science education; tropical vegetable plants. *Mailing Add:* Dept Biol Elizabeth City State Univ Elizabeth City NC 27909

KHAN, SHABBIR AHMED, b Dec 24, 45; m; c 1. MOLECULAR BIOLOGY, BIOCHEMISTRY. *Educ:* Bangalore Univ, India, PhD(peptide chem), 77. *Prof Exp:* Asst prof chem, Rockefeller Univ, 80-86; ASSOC PROF CHEM, WISTAR INST, 86- *Mem:* Am Soc Biochem & Molecular Biol; Am Chem Soc; AAAS. *Res:* Structure-function studies of the transacting proteins of HIV-1. *Mailing Add:* Wistar Inst 3601 Spruce St Philadelphia PA 19104

KHAN, SHAHAMAT ULLAH, b Rampur, India, Apr 25, 37; Can citizen; m 63; c 2. ENVIRONMENTAL SCIENCES. *Educ:* Agra Univ, BSc, 57; Aligarh Muslim Univ, India, MSc, 59; Univ Alta, MSc, 63, PhD(soil chem), 67. *Prof Exp:* Res asst soil chem, Univ Alta, 63-64; RES SCIENTIST ORG MATTER & ENVIRON SCI, CAN DEPT AGR, 68- *Concurrent Pos:* Ed, J Environ Sci & Health, Part B. *Mem:* Fel Royal Soc Chem; Am Chem Soc; fel Chem Inst Can. *Res:* Chemistry and reactions of organic matter in soils and waters; soil and water pollution; pesticides in soils, plants and waters; pesticide toxicology. *Mailing Add:* Four Barry Burn Ct Nepean ON K2R 1C5 Can

KHAN, SHAKIL AHMAD, b Bareilly, India; US citizen; m 74; c 2. PHYSICAL CHEMISTRY. *Educ:* Univ Karachi, Pakistan, BS, 67, MS, 68; Univ Islamabad, MPhil, 69; Northwestern Univ, Evanston, Ill, PhD(phys org chem), 74. *Prof Exp:* Fel molecular orbital theory, Univ SC, 74; asst prof spectros quantum chem, Univ Islamabad, Pakistan, 74-77; fel, Fla State Univ, 77-78, RES SPECIALIST NUCLEAR MAGNETIC RENOSANCE SPECTROS & ANALYSIS, MOBAY CHEM CORP, 78- *Mem:* Am Chem Soc; Soc Plastic Industs. *Res:* Use of nuclear magnetic renosance spectroscopy to study polymers especially polyurethanes, polycarbonates; laboratory and office automation; personal computer in chemistry. *Mailing Add:* Mobay Corp Parkway W Coating Div Bldg 8 Pittsburgh PA 15205

KHAN, SULTANA, b Dacca, Bangladesh, Dec 13, 47; m 74; c 1. SOLID STATE PHYSICS, ASTRONOMY. *Educ:* Univ Dacca, BSc, 70, MSc, 72; Univ Grenoble, Doc(solid state physics), 77. *Prof Exp:* ASSOC PROF PHYSICS, ELIZABETH CITY STATE UNIV, 78- *Concurrent Pos:* Planetarian. *Mem:* Bangladesh Phys Soc; Am Phys Soc. *Res:* Involved in a project to measure solar radiation in southeastern United States in cooperation with the Solar Energy Research Institute Department of Energy and National Oceanic and Atmospheric Administration since 1985. *Mailing Add:* Elizabeth City State Univ PO Box 845 Elizabeth City NC 27909

KHAN, WINSTON, b San Fernando, Trinidad, Mar 12, 34; US citizen; m 61; c 5. APPLIED MATHEMATICS, MATHEMATICAL PHYSICS. *Educ:* Univ London, BSc, 56, MSc, 58; Univ Birmingham, Eng, dipl, 61, PhD(mat Physics), 64. *Prof Exp:* Asst prof mat, Univ WI, London, 58-59; lectr & dir mat, Univ WI, Trinidad, 64-69; from asst prof to assoc prof mat, Univ PR, Cayey, 70-74, assoc prof, 74-82, PROF PHYSICS, UNIV PR, MAYAQUEZ, 82- *Concurrent Pos:* Exec adv mem, Nat Sci Comn, 82-83; dir, US Army grants, 82-85; coordr, US Army Mat Command, Rep, Univ PR, 84-87. *Mem:* Am Mat Soc; Am Physiol Soc; AAAS; Soc Inst & Appl Math; Int Asn Mat & Computer Modeling. *Res:* Fluid dynamics involving mathematics, physics and engineering sciences. *Mailing Add:* Univ PR Calle Uroyan AD4 Mayaquez PR 00709

KHANDELWAL, RAMJI LAL, b Dausa, India, June 2, 44; m 62; c 1. BIOCHEMISTRY. *Educ:* Univ Udaipur, India, BSc, 63; Punjab, Agr Univ, India, MSc, 66; Univ Man, PhD(biochem), 72. *Prof Exp:* Demonstr agr, Univ Udaipur, India, 63-64; res asst agr, Govt Rajasthan, India, 66-68; teaching asst biochem, Univ Man, 69-70, res asst biochem, 72-73; res biochem, Univ Calif, Davis, 73-75; asst prof oral biol, Univ Man, 75-80; assoc prof, 80-85, PROF BIOCHEM, UNIV SASK, 85- *Concurrent Pos:* Med Res Coun Can scholar, 75-80, develop grant, 81. *Mem:* Can Biochem Soc; Am Soc Biochem & Molecular Biol. *Res:* Role of insulin and cyclic adenosine monophosphate in the regulation of glycogen metabolism in mammalian systems; regulation of protein dephosphorylation in biological systems. *Mailing Add:* Dept Biochem A10-1 Health Sci Bldg Univ Sask Saskatoon SK S7N 0W0 Can

KHANDWALA, ATUL S, SCIENCE. *Prof Exp:* EXEC VPRES RES & DEVELOP, CHEMEX PHARMACEUT INC, 86- *Mailing Add:* Chemex Pharmaceut Inc PO Box 378017 Denver CO 80237

KHANG, SOON-JAI, b Seoul, Korea, Feb 26, 44; m 73; c 2. CHEMICAL ENGINEERING. *Educ:* Yonsei Univ, Korea, BE, 66; Ore State Univ, MS, 72, PhD(chem eng), 75. *Prof Exp:* Instr chem eng, Ore State Univ, 74-75; asst prof, 75-79, ASSOC PROF CHEM ENG, UNIV CINCINNATI, 80- *Concurrent Pos:* Consult, Procter & Gamble Co, 78-80, Amoco Oil, 80- & Exxon, 81. *Mem:* Am Inst Chem Engrs; Sigma Xi. *Res:* Chemical reaction engineering, including residence time distribution, mixing and catalyst deactivation; energy conversion and coal gasification; application of statistical methods for process control; mathematical modeling. *Mailing Add:* Dept Chem & Nuclear Eng Univ Cincinnati Cincinnati OH 45221

KHANNA, FAQIR CHAND, b Lyallpur, India, Jan 23, 35; m 66; c 2. NUCLEAR PHYSICS. *Educ:* Univ Panjab, India, BSc, 55, MSc, 56; Fla State Univ, PhD(physics), 62. *Prof Exp:* Lectr physics, Univ Panjab, India, 56-58; fel, Univ Iowa, 61-63 & Rice Univ, 63-65; Nat Res Coun Can fel, 66-67, from assoc to sr res physicist, 67-84, Atomic Energy Comn Labs, PROF PHYSICS, 84-, DIR, THEORETICAL PHYSICS INST, UNIV ALTA , 86- *Mem:* Fel Am Phys Soc; Can Asn Physics. *Res:* Nuclear physics, especially low and high energy physics; low energy physics study of the few nucleon problem, particularly the three nucleon system; many body problem with emphasis on nuclear physics and solid state physics; quantum liquids and solids; effective operators. *Mailing Add:* Physics Dept Univ Alberta Edmonton AB T6G 2J1 Can

KHANNA, JATINDER MOHAN, b Amritsar, India, Apr 15, 36; m 66; c 2. BIOCHEMICAL PHARMACOLOGY. *Educ:* Punjab Univ, India, BSc, 58, MSc, 60; Univ Conn, PhD(pharmacol), 64. *Prof Exp:* Res fel pharmacol, 64-65, lectr, 65-66, from asst prof to assoc prof, 66-77, PROF PHARMACOL, FAC MED, UNIV TORONTO, 77- *Concurrent Pos:* Scientist IV, Alcohol & Drug Addiction Res Found, 69-, head, Behav Pharmacol & Drug Anal Sect; mem ed bd, J Alcohol; Lederle Res fel, Univ Conn, Starrs. *Mem:* Soc Neurosci; Can Pharmacol Soc; Am Soc Pharmacol & Exp Therapeut; Res Soc Alcoholism; Int Soc Biomed Res Alcoholism; Sigma Xi. *Res:* Biochemical and behavioral mechanisms of alcohol and drug addiction. *Mailing Add:* Dept Pharmacol Univ Toronto Toronto ON M5S 1A8 Can

KHANNA, KRISHAN L, b Amritsar, India, Nov 19, 33; m 63; c 2. PHYTOCHEMISTRY, PHARMACEUTICAL CHEMISTRY. *Educ:* Univ Panjab, India, BPharm, 57, MPharm, 59; Univ Conn, PhD(pharm sci), 63. *Prof Exp:* Lectr pharmacog, Univ Panjab, India, 59-60; asst instr org chem, Univ Conn, 62-63, instr, 63-64; lectr pharmacog, Univ Panjab, India, 64-68; res assoc, Col Pharm, Univ Mich, Ann Arbor, 68-70, NIH fel toxicol, 70-71, res assoc indust toxicol, 71-74; assoc prof, Col Pharm & Pharmacol Sci, Howard Univ, 74-76; CONSULT, 76- *Mem:* AAAS; Royal Soc Chem; NY Acad Sci; Am Asn Cols Pharm. *Res:* Toxicology; drug metabolism; metabolism of foreign compounds. *Mailing Add:* US Environ Protection Agency EH-550D 401 M St SW Washington DC 20460

KHANNA, PYARE LAL, b Lahore, Mar 28, 45; US citizen; m 73; c 2. BIO-ORGANIC CHEMISTRY, SYNTHETIC CHEMISTRY & IMMUNOCHEMISTRY. *Educ:* Univ Delhi, BSc, 65, MSc, 67, PhD(chem), 70. *Prof Exp:* Res assoc natural prod, Indian Nat Sci Acad, 70-71; asst prof chem, Ramjas Col, Univ Delhi, 71-74; res assoc, Columbia Univ, 74-77; res group leader, 77-80, asst dir res, 80-86, VPRES RES & DEVELOP, MICROGENICS, 86- *Honors & Awards:* Syntex Sci Award, 84. *Mem:* Am Chem Soc; Am Asn Clin Chem; NY Acad Sci; AAAS. *Res:* Natural products isolation and synthesis; steroids; small ring compounds; fluorescent dyes; protein modifications; development of new immunoassay techniques; dipstick immunoassays; homogenous and heterogenous immunoassays; future research and development strategy. *Mailing Add:* Microgenics 2380 Bisso Lane Concord CA 94520

KHANNA, RAVI, b Kapurthala, India, Sept 27, 44; m 74. POLYMER CHEMISTRY, CHEMICAL ENGINEERING. *Educ:* Indian Inst Technol, Kanpur, BTech, 67; Mass Inst Technol, SM, 68. *Prof Exp:* Res chemist, 68-72, sr res chemist polymer chem, 72-76, res assoc, 76-77, lab head, 77-81, TECH ASST TO DIR RES, EASTMAN KODAK CO RES LABS, 81- *Mem:* Am Chem Soc; Am Inst Chem Engrs. *Res:* Investigation of the mechanism and kinetics of batch and continuous free radical polymerization. *Mailing Add:* Eastman Kodak Co 343 State St Box NJ160 Rochester NY 14650-9786

KHANNA, SARDARI LAL, b Amritsar, India, Apr 15, 37; m 63; c 2. PHYSICS. *Educ:* Panjab Univ India, BA, 56; Univ Saugar, MSc, 59, PhD(physics), 63. *Prof Exp:* Lectr physics, DAV Col, Amritsar, India, 59-60; lectr, Panjab Univ, 62-64; from asst prof to assoc prof, York Jr Col, Pa, 65-68; sci pool off, Panjab Univ, 68-69; assoc prof, 70-78, PROF PHYSICS, YORK COL, PA, 78- *Mem:* Am Phys Soc; Am Asn Physics Teachers. *Res:* Electrets; solid state physics; study of dielectrics subjected to electric and magnetic fields. *Mailing Add:* Dept Physics York Col Pa York PA 17405

KHANNA, SHADI LALL, dentistry, for more information see previous edition

KHANNA, SHYAM MOHAN, b Agra, India, May 10, 32; m 59; c 2. AUDIOLOGY, PHYSIOLOGY. *Educ:* Univ Lucknow, BS, 51; St Xavier's Col, India, DRE, 54; City Univ New York, PhD(hearing), 70. *Prof Exp:* Develop engr instrumentation, Pye Ltd, Eng, 54-55; design engr commun, Can Westinghouse Ltd, Ont, 55-58; sr engr avionics, Int Tel & Tel Labs, 58-61; adv engr commun, IBM Corp, 61-64; res assoc hearing, 64-70, from asst prof to assoc prof, 70-83, PROF OTOLARYNGOL, COL PHYSICIANS & SURGEONS, COLUMBIA UNIV, 83-, DIR RES, FOWLER MEM LABS, 83- *Concurrent Pos:* NIH res grants, 64-, prin investr, 79-; res career develop award, NINCDS, 77-82, prin investr, Prog Proj grants, 85-, mem, Commun Disorders Rev Comt. *Mem:* Sr mem Inst Elec & Electronics Engrs; fel Acoust Soc Am; Sigma Xi; Asn Res Otolaryngol; AAAS. *Res:* Physics of hearing; mechanics of the middle and inner ear; transducer action and coding in the peripheral auditory system. *Mailing Add:* Col Physicians & Surgeons Rm 11-452 Columbia Univ 630 W 168th St New York NY 10032

KHANSARI, DAVID NEMAT, b Khansar, Iran, Sept 28, 44; US citizen; m 88; c 3. IMMUNOLOGY OF AGING, LYMPHOKINE RESEARCH. *Educ:* Univ Tehran, DVM, 70; Univ Ill, MS, 78, PhD(immunol), 81. *Prof Exp:* Vet officer, Vet Med, Royal Air Force, Iran, 72-76; res assoc, Clin Immunol, Med Univ SC, 80-82, instr, 82-83; div dir, Res & Develop, Imrege, Inc, 83-85; RES DIR, RES & DEVELOP, IMMUNOLBIOL LABS, INC, 86- *Concurrent Pos:* Asst prof, Tulane Univ Med Sch, 83-85; assoc prof, NDak State Univ, 85- *Mem:* Am Asn Immunologists; Am Soc Cell Biologists; Am Soc Clin Immunol; Am Asn Vet Immunol; AAAS. *Res:* Defining effects of aging on the immune system and prevention of age associated diseases; effect of stress on immune system; trying to intervene stress induced immunosuppression by immunotherapy. *Mailing Add:* 3782 N Tenth St Fargo ND 58102

KHARAKA, YOUSIF KHOSHU, b Mosul, Iraq, May 15, 41; c 2. GEOCHEMISTRY, HYDROGEOLOGY. *Educ:* King's Col, Univ London, BSc, 63; Univ Calif, Berkeley, PhD(geol), 71. *Prof Exp:* Asst geologist explor & res dept, Ministry of Oil, Baghdad, Iraq, 63-67; asst res geologist, Univ Calif. Berkeley, 71-75; HYDROLOGIST WATER RESOURCES DIV, US GEOL SURV, 75- *Concurrent Pos:* Asst geol, Univ Baghdad, 63-64; consult explor, Mining & Metals Div, Union Carbide Corp, 74-75. *Honors & Awards:* Spec Award, Soc Econ Paleontologists & Mineralogists, 85. *Mem:* Int Asn Geochem & Cosmochem; Geochem Soc; Am Asn Petrol Geologists; Am Geophys Union; Geol Soc Am. *Res:* Geochemistry of sediments, sedimentary rocks and their associated fluids; computer modelling of water-rock interactions; membrane properties of fine grained sediments; stable isotopes. *Mailing Add:* 3385 St Michael Dr Palo Alto CA 94306

KHARASCH, NORMAN, b Poland, Sept 11, 14; nat US; m 40. CHEMISTRY. *Educ:* Univ Chicago, BS, 37, MS, 38; Northwestern Univ, PhD(chem), 44. *Prof Exp:* Instr chem, Chicago Jr Col, 38-39, Ill Inst Technol, 39-43 & Northwestern Univ, 44-46; from asst prof to prof chem, 46-66, prof biomed chem, 66-80, EMER PROF, UNIV SOUTHERN CALIF, 80-; FOUNDER & PRES, LOUIS PASTEUR FOUND, 80- *Concurrent Pos:* Sci dir, Intra-Sci Res Found, Los Angeles, 66-80. *Mem:* Am Chem Soc; Sigma Xi. *Res:* Organic sulfur compounds; mechanisms of organic reactions; chemistry of prostaglandins and related substances; free radical reactions. *Mailing Add:* PO Box 6599 Laguna Niguel CA 92677-6599

KHARE, ASHOK KUMAR, b Kanpur, India, Aug 7, 48; US citizen; m 74; c 2. FORGINGS & EXTRUSIONS. *Educ:* Agra Univ, India, BSc, 64; Indian Inst Tech, BTech, 69; Stevens Institute Tech, MS, NJ, 71. *Prof Exp:* sr prod metallurgist, dept Metall, 75-82, SR DEVELOP METALLURGIST, RES & DEVELOP CORP, NAT FORGE CO, IRVINE, PA, 82- *Mem:* AAAS; fel Am Soc Metals; Am Soc Mech Eng. *Res:* Product and process develpment for new and existing ferrous-non ferrous forgings and extrusions. *Mailing Add:* Five Leslie Blvd Warren PA 16365

KHARE, BISHUN NARAIN, b Varanasi, India, June 27, 33; m 62; c 2. PHYSICS, PHYSICAL CHEMISTRY. *Educ:* Banaras Hindu Univ, BSc, 53, MSc, 55; Syracuse Univ, PhD(physics), 61. *Prof Exp:* Res assoc, Univ Toronto, 61-62 & State Univ NY Stony Brook, 62-64; assoc res scientist, Ont Res Found, Can, 64-66; physicist, Smithsonian Astrophys Observ, Mass, 66-68; SR RES PHYSICIST, LAB PLANETARY STUDIES, CTR RADIOPHYSICS & SPACE RES, CORNELL UNIV, 68- *Concurrent Pos:* Assoc, Harvard Observ, 66-68. *Mem:* AAAS; Am Phys Soc; Am Astron Soc; Int Soc Study Origin of Life; Astron Soc India; Am Chem Soc; Int Astron Union; Sigma Xi; Planetary Soc. *Res:* Interdisciplinary research; molecular structure and spectroscopy; synthesis of organic compounds in primitive terrestrial and contemporary planetary atmospheres by photochemical reaction; hydrogen bonding among molecules of biological interest; planetary surfaces and atmospheres; interstellar and cometary chemistry. *Mailing Add:* 39 Highgate Circle Ithaca NY 14850

KHARE, MOHAN, b Varanasi, India, May 15, 42. PHYSICAL CHEMISTRY, ENVIRONMENTAL CHEMISTRY. *Educ:* Banaras Hindu Univ, BSc, 61, MSc, 63 & PhD(chem), 67. *Prof Exp:* Lectr chem, Banaras Hindu Univ, 67; res assoc & fel, Univ Md, 68-69 & Radiation Ctr, Ore State Univ, 69-70; res assoc, Cornell Univ, 70-80; at I T Enviscience Inc, 80-; AT E A ENG SCI & TECH INC, SPARKS GLENCOE, MD. *Mem:* Am Chem Soc. *Res:* Radiochemical separation of isotopes; kinetics of annealing of radiation damage; energy transfer in solids; photolysis and radiolysis of aqueous systems and interstellar molecules; water pollution characterization and environmental research; analytical methods development for trace and toxic materials, organic and inorganic, in diverse matrices. *Mailing Add:* Recro Environmental Inc 10189 Maxinest Ellicott City MD 21043-6316

KHARGONEKAR, PRAMOD P, b Indore, India, Aug 24, 56; India; m 83; c 2. CONTROL THEORY, SYSTEM THEORY. *Educ:* Indian Inst Technol, Bombay, BTech, 77; Univ Fla, Gainesville, MS, 80, PhD(elec eng), 81. *Prof Exp:* Asst prof elec eng, Univ Fla, Gainesville, 81-84; from assoc prof to prof elec eng, Univ Minn, Minneapolis, 84-89; PROF ELEC ENG & COMPUTER SCI, UNIV MICH, ANN ARBOR, 89- *Concurrent Pos:* Prin investr, Air Force Off Sci Res, Army Res Off, NSF, 81-; vis asst prof, Tex Tech Univ, Lubbock, 82; vis prof, Swiss Fed Inst Technol, Zurich, 83; consult, Honeywell Corp, 84-; prin young investr award, NSF, 85; George Taylor res award, 87. *Honors & Awards:* D Eckmann Award, 89; Axelby Award, Inst Elec & Electronics Engrs, 90, W R G Baker Prize, 91. *Mem:* Inst Elec & Electronics Engrs. *Res:* Control and systems theory; robust control, optimal control, infinite dimensional systems, adaptive control, algebraic system theory, two dimensional and multi-dimensional systems. *Mailing Add:* Dept Elec Eng & Computer Sci Univ Mich Ann Arbor MI 48109-2122

KHASNABIS, SNEHAMAY, b Dacca, India, Nov 4, 39; nat US; m 67; c 2. TRANSPORTATION ENGINEERING, SYSTEMS ANALYSIS. *Educ:* Univ Calcutta, India, BE, 62; NC State Univ, Raleigh, MCE, 70, PhD(civil eng), 73. *Prof Exp:* From asst prof to assoc prof, Wayne State Univ, Detroit, 75-82, actg chmn dept, 83-84, chmn dept, 84-87, PROF CIVIL ENG, WAYNE STATE UNIV, DETROIT, 82- *Concurrent Pos:* Transp engr, Barton-Aschman Assocs, Inc, Washington, DC, 74-75; prin investr, US Dept Transp Res Projs, Wayne State Univ, 77-91 & NSF Educ Proj, 83-86, chmn dept, 84-87; actg dir, Urban Transp Inst, Wayne State Univ, 85- *Mem:* Am Soc Civil Engrs; Inst Transp Engrs; Transp Res Bd. *Res:* Land-use models, simulation, transit station location, traffic safety and operation, transportation logistics; computer techniques for bus system planning and expansion, and financial strategies for transit system development; transit drivatization and intelligent vehicles highway systems. *Mailing Add:* Dept Civil Eng Wayne State Univ 2170 Eng Detroit MI 48202

KHATAMI, MAHIN, b Tehran, Iran, May 9, 43. PROTEIN CHEMISTRY, ENZYMOLOGY. *Educ:* Teachers Training Col, BS, 64; State Univ NY Buffalo, MS, 77; Univ Pa, PhD (molecular biol), 80. *Prof Exp:* Instr chem & physics, Teachers Training Col, Tehran, Iran, 64-69; instr biochem, Philadelphia Col Podiatric Med, 74-75; instr chem, Philadelphia Community Col, 74-75; res assoc dept physiol, Univ Va, 80-81; ASST PROF DEPT OPHTHAL, UNIV PA, 85- *Concurrent Pos:* Mem Cell Biol Grad Group, Univ Pa, 87-; sci rev, Coronary Heart Dis Res, Am Health Asst Found, 87-; ad hoc ed revs, Current Eye Res, 87- *Mem:* Asn Res Vision & Ophthal; AAAS; Soc Exp Biol & Med; Am Diabetes Asn. *Res:* Biochemistry and molecular aspects of diabetic retinopathy; role of hyperglycemia in retinal metabolism; cellular transport and metabolism of vital metabolites (e.g., sugars, ascorbate, myo-inositol); mechanism of signal transduction and receptor function; immunology; mechanisms of ocular allergic reactions. *Mailing Add:* Dept Ophthal Scheie Eye Inst Univ Pa 51 N 39 St Philadelphia PA 19104

KHATIB, HISHAM M, b Palestine, Jan 5, 36; Jordanian citizen; m 68; c 3. ENERGY & ELECTRICITY GLOBAL IMPACT, ENVIRONMENTAL EFFECTS OF ENERGY. *Educ:* Univ Ain Shams, Cairo, BSc, 59; Univ Birmingham, MSc 62; Univ London, BSc, 67, PhD(elec eng), 74. *Prof Exp:* Asst engr, Jerusalem Dist Elec Co, 59-62, engr, 62-66, chief engr, 66-73; res fel elec eng, Univ London, 73-74; dep dir gen, Jordan Elec Authority, 74-76, dir gen, 80-84; energy adv, Arab Fund, Kuwait, 76-80; minister energy, Govt Jordan, 84-89; CONSULT INT ORGN, 90- *Concurrent Pos:* Mem, Int Comt Availability Generating Plant, 82-; consult ed, Int J Elec & Energy Systs, 85- & Environ Policy, 90-; vchmn, World Energy Coun, 88-; chmn, Inst Elec Engrs Ctr, Jordan, 89- *Mem:* Fel Inst Elec & Electronics Engrs; World Energy Coun. *Res:* Global energy matters, demand and supply, effect on the environment; future of electrical energy and its environmental impact; human development and energy matters. *Mailing Add:* PO Box 925387 Amman Jordan

KHATIB-RAHBAR, MOHSEN, b Rafsandjan, Iran, Feb 21, 54; m 85. CHEMICAL ENGINEERING, NUCLEAR ENGINEERING. *Educ:* Univ Minn, BChemEng, 74; Cornell Univ, PhD(nuclear eng), 78. *Prof Exp:* mem sci staff, Brookhaven Nat Lab, 78-88, group leader accident anal group, 85-88; SR TECH ADV, US NUCLEAR REGULATORY COMN, 88- *Concurrent Pos:* Vis scientist, Germany Res Satellite, WGermany, 82; prin investr, US Nuclear Regulatory Comn; Cornell McMullin fel, 75. *Mem:* Assoc mem Am Inst Chem Engrs; Am Nuclear Soc. *Res:* Nuclear reactor dynamics and safety; heat transfer; fluid dynamics; numerical methods; probabilistic risk assessment; severe nuclear reactor accident analysis; mathematical methods for propagation of stochastic and physical uncertainty. *Mailing Add:* 7831 Whiterim Terr Brookhaven Nat Lab Rockville MD 20854

KHATRA, BALWANT SINGH, b Nabha, India, Feb 2, 45; US citizen; m 68; c 2. BIOCHEMISTRY. *Educ:* Punjab Univ, BSc Hons, 65, MSc, 67; Univ Leeds, PhD(biochem), 72. *Prof Exp:* Res fel, Clin Res Facil, Emory Univ, 72-75; res assoc, 75-78, ASST PROF, DEPT PHYSIOL, VANDERBILT UNIV, 78- *Mem:* Am Soc Biol Chemists. *Res:* Phosphoprotein phosphatases: isolation, characterization and their role in the regulation of glycogen metabolism. *Mailing Add:* Dept Anat & Physiol Calif State Univ 1250 Bellflower Blvd Long Beach CA 90840

KHATRI, HIRALAL C, b Navsari, India, Feb 6, 36; m 65; c 1. MECHANICAL ENGINEERING. *Educ:* Imp Col Eng, Addis Ababa, Ethiopia, BS, 58; Purdue Univ, MS, 63, PhD(mech eng), 66. *Prof Exp:* Engr, Imp Hwy Authority, Ethiopia, 58-61; asst prof elec eng, Mont State Univ, 65-67; assoc prof mech eng, Cath Univ Am, 67-71; mech engr, 71-80, PHYS SCIENTIST, HARRY DIAMOND LABS, 80- *Concurrent Pos:* NSF res grant, 66-68. *Mem:* Assoc mem Am Soc Mech Engrs; Inst Elec & Electronics Engrs. *Res:* Automatic control systems; optimal filter, identification and control of distributed systems; stability and sensitivity of distributed parameter systems; radar systems and signal processing. *Mailing Add:* Harry Diamond Labs 2800 Powder Mill Rd Adelphi MD 20783

KHATTAB, GHAZI M A, b Baghdad, Iraq, Nov 20, 30; c 1. POLYMER CHEMIST. *Educ:* Univ Baghad, BS, 53; Univ Ill, Urbana, MS, 58; Polytech Inst Brooklyn, PhD(polymer sci), 65. *Prof Exp:* Sr res chemist res & develop, Allied Chem Corp, 65-71; prog mgr, BiomedSci, 72-73; res assoc, GAF Corp, 73-75; sr res chemist res & develop, Allied Chem Corp, 76-80; tech dir res & develop, Wellman, Inc, 81-84; vpres, Res & Develop Corp, 84-87; RES MGR, ITW, INC, 87- *Mem:* Soc Plastics Engrs; Sigma Xi; Am Chem Soc. *Res:* New polymeric materials; engineering plastics including fluoropolymers; polyamides; polyesters and polysulfones; synthetic fibers. *Mailing Add:* ITW Inc 3650 W Lake Ave Glenview IL 60025

KHATTAK, CHANDRA PRAKASH, b Rawalpindi, Pakistan, May 19, 44; m 70; c 2. SOLID STATE PHYSICS, MATERIALS SCIENCE. *Educ:* Indian Inst Technol, Bombay, BTech, 65; State Univ NY Stony Brook, MS, 71, PhD(mat sci), 73. *Prof Exp:* Jr sci officer magnetism, DMR Lab, Hyderabad, India, 65-68; asst mat sci, State Univ NY Stony Brook, 68-73; res assoc solid state physics, Brookhaven Nat Lab, 74-75, asst physicist, 75-77; dir res & develop, 77-80, vpres technol, 80-86, sr vpres, 87-89, EXEC VPRES, CRYSTAL SYSTS INC, 89- *Honors & Awards:* IR-100 Award, 79. *Mem:* Am Ceramic Soc; Am Phys Soc; Am Asn Crystal Growth; Electrochemical Soc; Am Soc Metals. *Res:* Growth of large diameter sapphire for optical application, directional solidification of silicon crystals for photovoltaic applications and laser crystals growth by heat exchanger method; low-cost slicing of silicon by multi-ware fixed abrasive slicing technique; characterization and evaluation of silicon material for solar cells; thermodynamic evaluation of vacuum processing of silicon; single crystal growth of compound semiconductors and non-linear crystals. *Mailing Add:* Crystal Systs Inc Shetland Indust Park 27 Congress St Salem MA 01970

KHAVKIN, THEODOR, b Kharkov, Ukraine, USSR, Feb 18, 19; US citizen; m 46; c 2. HISTOLOGY, EXPERIMENTAL PATHOLOGY. *Educ:* Kharkov Med Inst, USSR, Med Doct, 41; Mil Med Acad, Leningrad, PhD cand(path), 49; USSR Acad Med Sci, SciD(path), 72. *Hon Degrees:* Med Doct Hons, Mil Med Acad, Bd, Leningrad, USSR, 45. *Prof Exp:* Mil pathologist trauma, Soviet Army, World War II, 42-44; adj path & atherosclerosis, Mil Med Acad, Leningrad, USSR, 44-49; chief pathologist, Mil Districts, USSR, 49-62; sr scientist path, Inst Exp Med, 62-79; sr scientist, 80-87, CONSULT, RES & DEVELOP LYMPHOKINE PROD, INTERFEROM SCI, INC, NEW BRUNSWICK, NJ, 88- *Concurrent Pos:* Vis prof & scientist, Rutgers State Univ, New Brunswick, NJ, 83-85; consult, Sci Testing, Inc, New Brunswick, 88-89. *Mem:* Am Asn Pathologists; hon mem Am Soc Rickettsiol & Rickettsial Dis; Am Soc Microbiol; Am Soc Exp Biol Med; Soc Leukocyte Biol. *Res:* Research and development of lymphokine production; pathology and immunomorphology of inflammatory and immune responses; pathology of infectious diseases; microscopy. *Mailing Add:* Interferom Sci Inc 783 Jersey Ave New Brunswick NJ 08901

KHAW, BAN-AN, b Bassein, Burma, July 25, 47; US citizen; m 76; c 4. MODIFICATION OF ANTIBODIES, RADIOIMMUNOSCINTIGRAPHY. *Educ:* State Univ NY, Oswego, BA, 69; Boston Col, MS, 70 & PhD(immunol), 73. *Prof Exp:* Res fel, Mass Gen Hosp & Harvard Med Sch, 73-76; instr, Harvard Med Sch, 76-78; asst biochem, Mass Gen Hosp, 76-79; asst prof, dept path, Harvard Med Sch, 78-83, dept radiol, 83-84; asst biochemist, 80-81, assoc biochemist, Dept Med, Mass Inst Technol, 81-83, asst radio chemist, Dept Radiol, 83-89; ASSOC PROF, DEPT RADIOL, HARVARD MED SCH, 84-; ASSOC RADIO CHEMIST, DEPT RADIOL, MASS GEN HOSP-EAST, 90- *Concurrent Pos:* Consult, Centocor, 82-, Vasocor, 89-; assoc prof, Dept Radiol, Harvard Med Sch, 84- *Mem:* Soc Nuclear Med; Am Heart Asn Basic Sci; Chinese Am Soc Nuclear Med (pres, 90-). *Res:* Application of antibodies in vivo diagnosis of cardiovascular diseases; modification of charge of proteins to affect changes in biodistribution and enhancement in vivo target localization; general application of radioimmunoscintigraphy. *Mailing Add:* Radiol Lab Rm 5410 Mass Gen Hosp E - 149 13th St Charlestown MA 02129

KHAWAJA, IKRAM ULLAH, b Delhi, India, Dec 25, 42; m 68. ECONOMIC GEOLOGY. *Educ:* Univ Karachi, BS, 62, MS, 63; Southern Ill Univ, MS, 68; Ind Univ, Bloomington, PhD(geol), 69. *Prof Exp:* From asst prof to assoc prof, 68-81, PROF GEOL, YOUNGSTOWN STATE UNIV, 81- *Mem:* Geol Soc Am; Sigma Xi. *Res:* Coal geology; petrology; sulphides in coal. *Mailing Add:* Dept Geol Youngstown State Univ Youngstown OH 44503

KHAYAT, ALI, b Tehran, Iran, Feb 2, 38; US citizen; m 66; c 2. AGRICULTURAL CHEMISTRY, FOOD SCIENCE. *Educ:* Univ Tehran, BS, 61; Univ Calif, Davis, MS, 64, PhD(agr chem), 68. *Prof Exp:* Res asst biochem, Univ Calif, Davis, 64-68; asst prof, Med Sch, Pahlavi Univ, Iran, 68-70; res biochemist, 70-72, assoc scientist, 72-77, sr scientist res dept, Van Camp Div, Ralston Purina Co, 77-82; ASSOC DIR RES, BEATRICE, 82- *Mem:* Am Chem Soc; Sigma Xi; Inst Food Technologists; AAAS. *Res:* Chemical modification of proteins; flavor chemistry; studies on vegetable oils; food texture; sensory and microbiological evaluation of foods. *Mailing Add:* 15484 Hamner Dr Los Angeles CA 90077-1803

KHAZAN, NAIM, b Baghdad, Iraq, Feb 15, 21; US citizen; m 52; c 2. PHARMACOLOGY. *Educ:* Col Pharm & Chem, Baghdad, PhC, 43; Hebrew Univ Jerusalem, PhD(pharmacol), 60. *Prof Exp:* Res assoc cent nerv syst pharmacol, Upjohn Co, 63-64; lectr, Hadassah Med Sch & Sch Pharm, Hebrew Univ, Jerusalem, 64, sr lectr, 66; asst prof, Med Sch, Univ Ore, 66-67; assoc prof, Columbia Univ, 67-68; assoc prof pharmacol, Mt Sinai Sch Med, 68-72; head dept pharmacol, Merrell Nat Labs, Cincinnati & assoc clin prof, Col Med, Univ Cincinnati, 72-74; prof pharmacol & toxicol, 74-80, chmn, Dept Pharmacol & Toxicol & dir, Grad Prog Pharmacol, 74-86, EMERSON PROF, PHARMACOL & TOXICOL, UNIV MD, 80- *Concurrent Pos:* Mem, Grad Coun, Univ Md, Baltimore City, 78-; NIMH grants, 67, 68-72 & 72-75; Nat Inst Drug Abuse grants, 75-78, 78-88; USPHS int fel, 68. *Honors & Awards:* Ellis Grollman Lectr Award, 90. *Mem:* Am Pharmaceut Asn; Am Soc Pharmacol & Exp Therapeut; NY Acad Sci; Soc Neurosci; Sigma Xi. *Res:* Electroencephalographic and behavioral studies of experimental drug dependence on narcotics; electroencephalographic effects of cannabinoids; pharmacology of rapid eye movement sleep; opioid multiple receptors and electroencephalograph power spectra. *Mailing Add:* Dept Pharm & Toxicol Univ Md 20 N Pine St Baltimore MD 21201

KHERA, KUNDAN SINGH, b Wadala Khurd, India, May 12, 22; Can citizen. TOXICOLOGY, TERATOLOGY. *Educ:* Punjab Univ, BSc, 44, DVM & BVSc, 52, MVSc & MSc, 56; Univ Paris, DSc, 58. *Prof Exp:* Dis invest officer, Vet Dept, Punjab Govt, India, 58-61; prof vet path, Vet Col, Punjab Univ & Govt Punjab, 61-64; SECT HEAD TERATOLOGY, HEALTH PROTECTION BR, NAT HEALTH & WELFARE, CAN, 64- *Concurrent Pos:* Fel, Col Med, Baylor Univ, 62-63. *Mem:* Soc Toxicol; Teratology Soc; Europe Teratology Soc. *Res:* Study of food additives and environmental contaminants on embryonic, postnatal and reproductive phases of mammalian development to determine admissible levels for humans. *Mailing Add:* 604-44 Emmerson Ave Ottawa ON K1Y 2L8 Can

KHO, BOEN TONG, b Tegal, Indonesia, Dec 3, 19; US citizen; m 60; c 4. ANALYTICAL CHEMISTRY. *Educ:* Univ Utrecht, BS, 42; State Univ Leiden, Drs, 47, apotheker, 49; Philadelphia Col Pharm, MS, 52; Univ Wis, PhD(pharm), 57. *Prof Exp:* Res chemist, Cent Res Labs, Gen Aniline & Film Co, Pa, 57-60; head methods develop, Merck Sharp & Dohme, 60-64; head anal res, Toms River Res Lab, Ciba Chem & Dye Co, 64-67; mgr anal res, Ciba-Agrochem Co, 67-68; asst dir anal res & develop, 68-71, DIR ANAL RES & DEVELOP, AYERST LABS, INC, 71- *Mem:* Am Chem Soc; Am Pharmaceut Asn; NY Acad Sci; Sigma Xi. *Mailing Add:* 37 Adirondack Lane Plattsburgh NY 12901-3213

KHODADAD, JENA KHADEM, b Tehran, Iran; US citizen; c 3. ULTRASTRUCTURE, MEMBRANES. *Educ:* Mt Union Col, BS, 60; Northwestern Univ, MS, 71, PhD(biol sci), 75. *Prof Exp:* Fel membrane, 77-78, instr cell biol, Dept Path, 78-79, ASST PROF, DEPT ANAT & PATH, MED COL, RUSH UNIV, 79- *Mem:* Am Soc Cell Biol; Am Asn Anatomists; AAAS. *Res:* Membrane research; biological membranes and the correlationship of ultrastructure with biochemical studies; red blood cell is the model membrane system used. *Mailing Add:* Dept Anat Med Col Rush Univ 600 S Paulina Chicago IL 60612

KHOO, TENG LEK, b Penang, Malaysia, June 1, 43; m 72; c 2. NUCLEAR PHYSICS. *Educ:* Dalhousie Univ, BSc, 65, MSc, 67; McMaster Univ, PhD(physics), 72. *Prof Exp:* Res assoc nuclear physics, Mich State Univ, 72-74, asst prof, 74-77; from asst physicist to physicist, 77-85, SR PHYSICIST, NUCLEAR PHYSICS, ARGONNE NAT LAB, 79- *Mem:* Fel Am Phys Soc. *Res:* Experimental nuclear structure physics, especially properties of deformed nuclei, effective nucleon interaction and nuclei at high angular momentum and high temperature; formation and decay of compound nucleus. *Mailing Add:* Argonne Nat Lab Bldg 203 9700 S Cass Ave Argonne IL 60439

KHOOBYARIAN, NEWTON, b Tabriz, Iran, Oct 20, 24; US citizen; c 2. VIROLOGY. *Educ:* Lafayette Col, AB, 49; Univ Ill, MS, 50; Univ Wis, PhD(med microbiol), 54; Am Bd Microbiol, dipl. *Prof Exp:* Asst biol sci, Univ Ill, 49-50; asst med microbiol, Univ Wis, 50-54; res fel microbiol, Sch Med, Ind Univ, 54-55, res assoc, 55-56, instr pediat, 56-60; actg head, Dept Microbiol & Immunol, 80-83, from asst prof to assoc prof, 60-72, PROF MICROBIOL, COL MED, UNIV ILL, CHICAGO, 72- *Concurrent Pos:* Consult, Miles Labs, Inc, 70-80. *Mem:* AAAS; Am Soc Microbiol. *Res:* Mammalian cell-virus relationship; viral genetics; chemical and viral transformation. *Mailing Add:* Dept Microbiol & Immunol Univ Ill Col Med 835 S Wolcott Chicago IL 60680

KHORANA, BRIJ MOHAN, b Multan, India, Apr 11, 39; m 67; c 3. APPLIED OPTICS, RESEARCH & MANAGEMENT. *Educ:* Univ Delhi, BSc, 58, MSc, 60; Indian Inst Technol, Kharagpur, MTech, 61; Univ Chicago, MS, 64; Case Western Reserve Univ, PhD(physics), 68. *Prof Exp:* Res assoc physics, James Franck Inst, Univ Chicago, 67-68 & Univ Rochester, 68-70; asst prof physics, Univ Notre Dame, 70-77; assoc prof physics, 77-84, PROF PHYSICS & APPL OPTICS, ROSE- HULMAN INST TECHNOL, 84-, HEAD, 80- *Concurrent Pos:* Dir, Ctr Appl Optics Studies, Rose-Hulman Inst Technol, 85- *Mem:* Am Phys Soc; fel Inst Soc Optical Eng; Optical Soc Am; Am Soc Elec Eng. *Res:* Fiber optic interferometry; fiber optic sensors; machine vision applications. *Mailing Add:* Dept Physics & Appl Optics Rose-Hulman Inst Technol Terre Haute IN 47803

KHORANA, HAR GOBIND, b Raipur, India, Jan 9, 22; m 52; c 3. ORGANIC CHEMISTRY. *Educ:* Punjab Univ, India, BSc, 43, MSc, 45; Univ Liverpool, PhD, 48. *Hon Degrees:* DSc, Univ Chicago, 67, Simon Fraser Univ, 69, Univ Liverpool, 71, Univ Wis, 76, Univ BC, 77, New Eng Col, 84. *Prof Exp:* Govt India fel, with Prof V Prelog, Swiss Fed Inst Technol, 48-49; Nuffield fel, with Prof Sir Alexander Todd, Cambridge Univ, 50-52; head org chem group, BC Res Coun & res prof fac grad studies, Univ BC, 52-59; prof & group leader, Inst Enzyme Res, Univ Wis-Madison, 60-62, prof, 62-71, Conrad A Elvejhem Prof life sci, 64-71; ALFRED P SLOAN PROF BIOL & CHEM, MASS INST TECHNOL, 70- *Concurrent Pos:* Vis prof, Rockefeller Inst, 58- *Honors & Awards:* Nobel Prize in Med, 68; Merck Award, Chem Inst Can, 58; Gold Medal, Prof Inst Can Pub Serv, 60; Dannie Heinneman Prize, 67; Lasker Found Award, 68; Louisa Gross Horwitz Prize, 68; Nat Medal Sci; Paul Kayser Int Award of Merit in Retina Res, 87. *Mem:* Nat Acad Sci; fel AAAS; Am Chem Soc; Am Soc Biol Chem; Am Acad Arts & Sci; foreign mem Royal Soc. *Res:* Peptides and proteins; chemistry of phosphate esters of biological interest; nucleic acids; chemical synthesis and structure; enzymes of nucleic acid metabolism; viruses and chemical genetics. *Mailing Add:* Dept Biol & Chem Mass Inst Technol Cambridge MA 02139

KHOSAH, ROBINSON PANGANAI, b Marondera, Zimbabwe, Jan 1, 54; US citizen. CHROMATOGRAPHY, MASS SPECTROMETRY. *Educ:* Univ Rhodesia, BS, 76; Univ Mass Amherst, MS, 81, PhD(chem), 85. *Prof Exp:* Scientist, Alcoa, 84-86, sr scientist, 86-87, staff scientist, 87-89, TECH SPECIALIST, ALCOA, 89- *Concurrent Pos:* Pres & coatings consult, Diverse Technologies, Inc, 85- *Mem:* Am Chem Soc. *Res:* Separation mechanisms under supercritical fluids; analytical chromatography; preparative-scale supercritical fluid extraction. *Mailing Add:* Alcoa Tech Ctr 100 Technical Dr Alcoa Center PA 15069-0001

KHOSHNEVISAN, MOHSEN MONTE, b 45; US citizen; m 70; c 2. ELECTROOPTICS, NONLINEAR OPTICS. *Educ:* Calif State Univ, San Jose, BS & BA, 68; Mich State Univ, MS, 71, PhD(physics), 73. *Prof Exp:* Asst prof physics, Arya-Mehr Univ Technol, Iran, 73-77; vis asst prof, Mich State Univ, Lansing, 77-78; mem tech staff, Electro-Optics Dept, 78-88, DIR OPTICS, SCI CTR, ROCKWELL INT, 88- *Concurrent Pos:* Mgr small tech groups, Infared Devices, Optical Devices & Appl Optics, dir, Optics. *Mem:* Am Phys Soc; Optical Soc Am; Soc Photo-Instrumentation Engrs. *Res:* Electro-optics; nonlinear optics; thin films. *Mailing Add:* Rockwell Int Sci Ctr 1049 Camino Dos Rios Thousand Oaks CA 91360

KHOSLA, MAHESH C, b Rajoya, Hazara, India, July 6, 25; m 57; c 3. CHEMISTRY. *Educ:* Panjab Univ, India, BS, 50, MS, 52, PhD, 57. *Prof Exp:* Sr sci officer, Cent Drug Res Inst, Lucknow, Uttar Pardesh, India, 57-69; assoc staff, 71, STAFF, CLEVELAND CLIN FOUND, 71- *Concurrent Pos:* Fel med adv bd, Coun High Blood Pressure Res, Am Heart Asn, 71-; NIH grantee, 78- *Mem:* Am Chem Soc; Am Soc Biochem & Molecular Biol; Int Soc Hypertension; Am Soc Hypertension; AAAS; Am Heart Asn; NY Acad Sci; Inter-Am Soc Hypertension; Am Peptide Soc. *Res:* Solid-phase peptide synthesis; synthesis of angiotensin II antagonists with minimum agonist activity; synthesis of tissue selective congeners of angiotensin II; hypertension; granted nine patents; author of various publications. *Mailing Add:* Health Sci Ctr REBV/NC3-147 Cleveland Clin Found 9500 Euclid Ave Cleveland OH 44195-5286

KHOSLA, RAJINDER PAUL, b Phillaur, India, July 25, 33; m 66; c 3. SOLID STATE PHYSICS. *Educ:* Univ Delhi, BSc, 53; Benaras Hindu Univ, MSc, 55; Purdue Univ, PhD(physics), 66. *Prof Exp:* Lectr physics, Govt Col, Narnaul, India, 55-56; sci asst, Nat Phys Lab India, New Delhi, 56-59; teaching & res asst, Purdue Univ, 59-66; sr physicist, 66-70, res assoc, 71-76, group leader device technol, Solid State Lab, 75-76, lab head, 76-79, sr res lab head, 79-82, asst dir, Physics Div, 82-85, DIR, MICROELECTRONICS TECHNOL DIV, EASTMAN KODAK CO, 85- *Concurrent Pos:* Lectr, Eve Sch, Univ Rochester, 72-74; vis scholar, Dept Elec Eng & Comput Sci, Univ Calif, Santa Barbara, 74-75. *Honors & Awards:* Frederick Philips Award, Inst Elec & Electronics Engrs. *Mem:* AAAS; Am Phys Soc; fel Inst Elec & Electronics Engrs. *Res:* Transport and photoconductive properties of semiconductors and solid state devices; solid state imaging devices. *Mailing Add:* Res Labs Eastman Kodak Co Bldg 81 Rochester NY 14650

KHOURY, GEORGE, molecular biology, anatomy; deceased, see previous edition for last biography

KHURANA, KRISHAN KUMAR, b New Delhi, India, June 1, 55. SPACE PHYSICS, MAGNETOSPHERIC PHYSICS. *Educ:* Delhi Univ, India, BSc, 74; Osmania Univ, India MSc, 77, PhD(appl geophys), 81; Durham Univ, UK, PhD(pure geophys), 84. *Prof Exp:* ASST RES GEOPHYSICIST SPACE PHYSICS, UNIV CALIF, LOS ANGELES, 85- *Concurrent Pos:* Prin investr, NASA, 90-92. *Mem:* Am Geophys Soc; Am Phys Soc; Am Astron Soc. *Res:* Magnetospheres of Venus, Earth, Jupiter, Saturn and Uranus; structure and motion of Jupiter's magnetosphere; spherical harmonic models of planetary magnetic fields; data processing for the Galileo mission to Jupiter. *Mailing Add:* IGPP Slichter Hall Univ Calif Los Angeles CA 90024

KHURANA, SURJIT SINGH, b Tandlianwala, Pakistan, June 15, 31; m 62; c 3. MATHEMATICS. *Educ:* Panjab Univ, India, BA, 53, MA, 55; Univ Ill, Urbana, PhD(math), 68. *Prof Exp:* Lectr math, Camp Col, Panjab Univ, India, 56-59 & Post-Grad Inst, Univ Delhi, 59-64; asst prof, 68-74, assoc prof, 74-79, PROF MATH, UNIV IOWA, 79- *Mem:* Am Math Soc. *Res:* Measure theory; functional analysis; general topology; probability theory; topological vector spaces. *Mailing Add:* Dept Math Univ Iowa Iowa City IA 52242

KHURI, NICOLA NAJIB, b Beirut, Lebanon, May 27, 33; US citizen; m 55; c 2. THEORETICAL PHYSICS. *Educ:* Am Univ, Beirut, BA, 52; Princeton Univ, PhD(physics), 57. *Prof Exp:* Asst prof physics, Am Univ, Beirut, 57-58; mem Inst Adv Study, 59-60; assoc prof physics, Am Univ, Beirut, 61-62; mem Inst Adv Study, 62-63; vis assoc prof, Columbia Univ, 63-64; assoc prof, 64-68, PROF PHYSICS, ROCKEFELLER UNIV, 68- *Concurrent Pos:* Brookhaven Nat Lab, 63-73; trustee, Am Univ Beirut, 69-; trustee, Brearley Sch, 70-79. *Mem:* Fel Am Phys Soc. *Res:* Quantum field theory; scattering theory; theory of dispersion relations and their applications; high energy particle physics. *Mailing Add:* Rockefeller Inst New York NY 10021

KHUSH, GURDEV S, b Rurki, India, Aug 22, 35; m 61; c 4. PLANT BREEDING. *Educ:* Punjab Univ, India, BSc, 55; Univ Calif, Davis, PhD(genetics), 60. *Hon Degrees:* Dr, Punjab Agr Univ, India, 87. *Prof Exp:* Asst genetics, Univ Calif, Davis, 57-61, jr res geneticist, 61-62, asst res geneticist, 62-67; plant breeder, 67-72, head, Varietal Improv Dept, 72-85, PRIN PLANT BREEDER & HEAD, DIV PLANT BREEDING, GENETICS & BIOCHEM, INT RICE RES INST, 86- *Concurrent Pos:* Vis prof, Colo State Univ, 75-76. *Honors & Awards:* Borlaung Award, 77; Japan Prize, 87; ASA Fel Award, 87; Int Agron Award, 89; Emil M Marak Int Award, 90. *Mem:* Foreign assoc, Nat Acad Sci; Genetics Soc Am; Bot Soc Am; NY Acad Sci; Indian Nat Sci Acad; Third World Acad Sci. *Res:* Cytogenetic studies of genus Secale and origin of cultivated rye and genus Lycopersicon particularly gene location, chromosome mapping and centromere location in the cultivated tomato; rice genetics and breeding. *Mailing Add:* Int Rice Res Inst PO Box 933 Manila Philippines

KHWAJA, TASNEEM AFZAL, b Pakistan, Apr 20, 36; m 65; c 2. CANCER, CHEMOTHERAPY. *Educ:* Univ Panjab, WPakistan, BSc, 55, MSc, 57; Cambridge Univ, MA, 61, PhD(synthetic nucleic acid chem), 64. *Prof Exp:* NIH res assoc grant, Univ Utah, 65-66; NIH proj assoc grant, McArdle Lab Cancer Res, Univ Wis-Madison, 66-68; res asst, Max Planck Inst Exp Med, Ger, 69-70; head dept cancer chemother, ICN Nucleic Acid Res Inst, 70-73; asst prof, 73-75, ASSOC PROF PATH, SCH MED, UNIV SOUTHERN CALIF, 76-, SR RES SCIENTIST & DIR ANIMAL TUMOR RESOURCE FACIL & PHARMACOANALYTIC FACIL, LOS ANGELES COUNTY-UNIV SOUTHERN CALIF CANCER CTR, 73- *Mem:* Am Asn Cancer Res; fel Royal Chem Soc; Am Chem Soc. *Res:* Synthesis of nucleoside antimetabolites as antitumor agents; biochemical mechanisms of drug action; use of animal tumor models. *Mailing Add:* Sch Med MOL 207 Univ Southern Calif Los Angeles CA 90033

KIANG, CHIA SZU, b Shanghai, China, Sept 9, 41; US citizen; m 68; c 2. ATMOSPHERIC SCIENCES, PHYSICS. *Educ:* Nat Taiwan Univ, BS, 62; Ga Inst Technol, MS, 64, PhD(physics), 70. *Prof Exp:* Assoc prof physics, Clark Col, 67-74; res scientist, Nat Ctr Atmospheric Res, 74-76, aerosol proj leader, 76-78; PROF GEOPHYS SCI, GA INST TECHNOL, 78- *Concurrent Pos:* Sr res assoc, Atmospheric Sci Res Ctr, State Univ NY Albany, 71-; adj prof, Dept Chem, Atlanta Univ, 74-76 & Dept Atmospheric Sci, Colo State Univ, 74-78; adj prof physics, Ga Inst Technol, 76-77. *Mem:* Am Chem Soc; Am Physics Soc; Am Meteorol Soc; Am Asn Advan Sci. *Res:* Nucleation; aerosol physics; aerosol chemistry; atmospheric chemistry; surface science; phase transition; critical phenomena; statistical physics; planetary atmosphere; environmental science and planning; natural phenomena. *Mailing Add:* Dept Geophys Ga Inst Technol Atlanta GA 30332

KIANG, DAVID TEH-MING, b Chekiang, China, Nov 13, 35; m 68; c 3. INTERNAL MEDICINE, ONCOLOGY. *Educ:* Nat Defense Med Ctr Taiwan, MB, 60; Columbia Univ, MS, 64; Univ Minn, PhD, 73. *Prof Exp:* Intern med, Beekman-Downtown Hosp, NY, 64-65, resident, 65-66; resident, Francis Delafield Hosp, 66-68; USPHS fel oncol, 68-70, instr, 70-73, asst prof, 73-81, assoc prof, 81-90, PROF MED, MED SCH, UNIV MINN, MINNEAPOLIS, 90- *Mem:* Soc Exp Biol & Med; Am Fedn Clin Res; Am Asn Cancer Res; Am Soc Clin Oncol; Sigma Xi. *Res:* Treatment of mammary cancer. *Mailing Add:* Univ Hosp Box 168 Univ Minn Minneapolis MN 55455

KIANG, NELSON YUAN-SHENG, b Wuxi, China, July 6, 29; m 57, 76; c 1. NEUROPHYSIOLOGY. *Educ:* Univ Chicago, PhB, 47, PhD(biopsychol), 55. *Hon Degrees:* MD, Univ Geneva, Switzerland, 81; MS, Harvard Univ. *Prof Exp:* Res asst otol, 57-62, DIR EATON PEABODY LAB, MASS EYE & EAR INFIRMARY, 62-; NEUROPHYSIOLOGIST, NEUROL SERV, MASS GEN HOSP, 77-; PROF, DEPT OTOL & LARYNGOL, HARVARD MED SCH, 84- *Concurrent Pos:* Res asst otol, Harvard Med Sch, 57-61, res asst otolaryngol, 61-69, sr res assoc otolaryngol & physiol, 69-; mem, Commun Sci Study Sect, Div Res Grants, NIH, 68-72, Behav & Neurosci Study Sect, 85-89; mem, Comt Hearing Bioacoust & Biomech, Nat Acad Sci-Nat Res Coun, Collegium Otorhinol-Laryngol Amiticiam Sacrum, Deafness Res Found, & Int Brain Res Orgn; staff mem, Res Lab Electronics, Mass Inst Technol, 55-, lectr elec eng, 68-, Eaton-Peabody Prof, Brain & Cognitive Sci Dept, 83. *Honors & Awards:* Beltone Award, Am Acad Arts & Sci, 68. *Mem:* AAAS; Soc Neurosci; Am Physiol Soc; fel Acoust Soc Am; Am Otol Soc; NY Acad Sci; Psychonomic Soc; Sigma Xi; Asn Res Otolaryngol; Royal Soc Med. *Res:* Physiology of auditory and other sensory systems; relation of brain to behavior. *Mailing Add:* Eaton Peabody Lab Mass Eye & Ear Infirmary Boston MA 02114

KIANG, ROBERT L, b Chungking, China, Nov 30, 39; US citizen; m 86; c 2. MECHANICAL ENGINEERING, FLUID MECHANICS. *Educ:* Nat Taiwan Univ, BS, 61; Stanford Univ, MS, 64, PhD(aeronaut eng), 70. *Prof Exp:* Res engr, Stanford Res Inst, 63-71, sr res engr, 71-87; MECH ENGR, DAVID TAYLOR RES CTR, 87- *Concurrent Pos:* Teacher, USN Acad, Johns Hopkins Univ & San Jose State Univ. *Mem:* Am Soc Mech Engrs; Am Nuclear Soc. *Res:* Fluid dynamics; dynamic modeling; nuclear safety; heat transfer; noise and vibration; friction and lubrication. *Mailing Add:* David Taylor Res Ctr Code 2721 Annapolis MD 21402-5067

KIANG, YING CHAO, b China; US citizen; m; c 2. FIBER OPTICS, LASERS. *Educ:* Nat Cheng Kung Univ, BS; Univ Ill, MS; Univ Md, PhD(elec eng). *Prof Exp:* Res mem lasers, Electronics Div, Gen Dynamics, 60-66; CHMN, OPTICAL FIBER COMMUN STEERING COMT, FIBER OPTICS, DATA SYST DIV, IBM, 67- *Concurrent Pos:* Adj prof, Polytech Univ NY, 66-69. *Mem:* Sr mem Inst Elec & Electronics Engrs; Optical Soc Am. *Res:* Analysis and experimentation on reflection induced intensity noise from laser diode and its effect on optical system performance. *Mailing Add:* Six Pasture Lane Poughkeepsie NY 12603

KIANG, YUN-TZU, b Taiwan, Feb 1, 32; US citizen; m 57; c 3. POPULATION GENETICS, PLANT BREEDING. *Educ:* Taiwan Normal Univ, BS, 56; Ohio State Univ, MA, 62; Univ Calif, Berkeley, PhD(genetics), 70. *Prof Exp:* Res asst genetics, Univ Calif, Berkeley, 67-68, teaching asst, 68-69, teaching assoc, 69-70; from asst prof to assoc prof, 70-83, PROF PLANT SCI & GENETICS, UNIV NH, 83- *Concurrent Pos:* Vis prof, Taiwan Normal Univ, 78; vis scientist, Academia Sinica, 78 & 86; assoc ed, J Hered, 87- *Mem:* AAAS; Genetics Soc Am; Am Genetic Asn; Soc Study Evolution; Bot Soc Am; Am Soc Agron; Am Soc Naturalists; Genetics Soc Can. *Res:* Genetics and evolution of economic and natural plant populations. *Mailing Add:* Dept Plant Sci Univ NH Durham NH 03824

KIBBEL, WILLIAM H, JR, b Buffalo, NY, Aug 31, 23; m 49; c 2. CHEMICAL ENGINEERING, INORGANIC CHEMISTRY. *Educ:* Case Inst Technol, BS, 44, MS, 48. *Prof Exp:* Chem eng reactor design, Buffalo Electro Chem Co, 47-52; sales mgr peroxy chem, Becco Sales Corp, 52-57, mgr mkt res, Becco Chem Div, 57-58; mgr indust appln develop, FMC Corp, 58-77, tech serv, 77-82 & tech admin, 82-85; RETIRED. *Mem:* Sigma Xi; Am Inst Chem Engrs. *Res:* Pulp and paper; textiles; pollution control; metal surface treatments. *Mailing Add:* 24 Dublin Rd Pennington NJ 08534

KIBBEY, DONALD EUGENE, mathematics; deceased, see previous edition for last biography

KIBBEY, MAURA CHRISTINE, b Pittsburgh, Pa, Mar 13, 62; m 88. EXTRACELLULAR MATRIX, METASTASIS. *Educ:* State Univ NY, Albany, BS, 83; Albany Med Col, PhD(anat & cell biol), 89. *Prof Exp:* Teaching asst gen biol, Va Polytech Inst & State Univ, 84, teaching asst histol, 85; teaching asst med gross anat & med histol, Albany Med Col, 87; RES BIOLOGIST, LAB DEVELOP BIOL, NAT INST DENT RES, NIH, 89- *Concurrent Pos:* Instr med gross anat, Georgetown Univ Med Sch, 90. *Mem:* Am Soc Cell Biol; Sigma Xi. *Res:* Investigating the biological activity of laminin, as well as its receptors and proteases, with regard to tumor cell growth and metastasis. *Mailing Add:* Lab Develop Biol Nat Inst Dent Res NIH Bldg 30 Rm 402 Bethesda MD 20892

KIBBY, CHARLES LEONARD, b Wenatchee, Wash, Jan 2, 38; m 70; c 1. CHEMICAL KINETICS, SURFACE CHEMISTRY. *Educ:* Reed Col, BA, 59; Purdue Univ, PhD, 64. *Prof Exp:* Fel chem, Harvard Univ, 63-65; res assoc, Brookhaven Nat Lab, 65-67; fel catalysis, Mellon Inst Sci, Carnegie-Mellon Univ, 67-69; res chemist, Gulf Res & Develop Co, Pittsburgh, 70-75, res chemist, Pittsburgh Energy Res Ctr, 76-77, sr res chemist, 77-80, res assoc, 81-85; SR RES ASSOC, CHEVRON RES & TECHNOL CO, RICHMOND, CA, 85- *Concurrent Pos:* Vis scientist catalysis, Inst Org Chem, Moscow, USSR, 74. *Mem:* Am Chem Soc; AAAS; Catalysis Soc. *Res:* Characterization of heterogeneous catalysts for hydrocarbon processing and chemicals production, by chemical and physical methods. *Mailing Add:* 846 Clifton Ct Benicia CA 94510

KIBENS, VALDIS, b Riga, Latvia, Oct 22, 36; US citizen; m 58, 88; c 2. AEROACOUSTICS, TURBULENCE. *Educ:* Yale Univ, BE, 57; Johns Hopkins Univ, PhD(mech), 68. *Prof Exp:* Res engr, Gen Dynamics Corp, Conn, 57-60; asst prof aerospace eng, Univ Mich, Ann Arbor, 68-74; sr scientist aeroacoust, 74-85, PRIN SCIENTIST, RES LABS, MCDONNELL DOUGLAS CORP, 85- *Concurrent Pos:* Consult & vis res scientist, Res Labs, Gen Motors Corp, Mich, 73. *Mem:* Am Phys Soc; assoc fel Am Inst Aeronaut & Astronaut; Sigma Xi. *Res:* Control of the development of turbulent structures in flow fields as the basis for technological devices to reduce aerodynamic noise, enhance mixing, and control buffeting. *Mailing Add:* 21 Cardigan Dr St Louis MO 63135

KIBLER, KENNETH G, b Peoria, Ill, Apr 15, 40; m 61; c 2. GEOPHYSICS, ENVIRONMENTAL SCIENCE. *Educ:* Univ Iowa, MS, 64, PhD(nuclear physics), 66. *Prof Exp:* Res asst nuclear physics, Univ Iowa, 62-66; res assoc, Case Western Reserve Univ, 66-69; SR RES SCIENTIST, GEN DYNAMICS-CONVAIR AEROSPACE, 69- *Concurrent Pos:* Asst prof, Tex Wesleyan Col, 69- *Mem:* AAAS; Am Phys Soc; Am Geophys Union. *Res:* Heavy-ion nuclear reactions; gravity measurement; remote sensing of earth resources; enhancement techniques for aerial or orbital images; physics of materials. *Mailing Add:* 8232 Saddlebrook Dr Ft Worth TX 76116-1417

KIBLER, RUTHANN, b Mansfield, Ohio, Dec 1, 42. IMMUNOLOGY, VIROLOGY. *Educ:* Marietta Col, BS, 64; Purdue Univ, MS, 67; Univ Calif, Berkeley, PhD(immunol), 73. *Prof Exp:* res scientist immmunol, Dept Microbiol, 76-79, asst prof microbiol & immunol, Col Med, Univ Ariz, 80-87; SR RES SCIENTIST, BIO-RAD LAB, 87- *Concurrent Pos:* mem staff, Calif State Dept Health Serv, Berkeley, 79-80. *Mem:* Am Soc Microbiol; Reticuloendothelial Soc; Sigma Xi; Am Asn Immunologists; Am Rheumatism Asn. *Res:* Infectious disease diagnostics. *Mailing Add:* 1184 Malibu Dr San Jose CA 95129-3315

KIBRICK, ANNE K, b Palmer, Mass, June 1, 19. EDUCATION. *Educ:* Boston Univ, BS, 45; Columbia Teachers Col, 48; Harvard Univ, EdD, 58. *Hon Degrees:* DLitt, St Josephs Col. *Prof Exp:* Asst educ dir, Cushing Mass Hosp, Framingham, 48-49; asst prof nursing, Simmons Col, 49-55; dir grad div nursing, Boston Univ Sch Nursing, 58-63, dean, 63-68, prof, 68-70; chmn dept nursing, Boston Col Grad Sch Arts & Sci, 70-74 & Boston State Col, 74-82; dean, 82-88, PROF, UNIV MASS, 82- *Concurrent Pos:* Mem staff, USPHS, NIH, 68-73; mem, Nat Acad Pract, 86- *Honors & Awards:* Nutting Award Leadership Nursing, Nat League Nursing. *Mem:* Nat Acad Sci; Inst Med; Nat League Nursing (pres, 71-73); fel Am Acad Nursing. *Mailing Add:* 381 Clinton Rd Brookline MA 02146

KIBRICK, SIDNEY, b Boston, Mass, Apr 2, 16; m 49; c 2. PEDIATRICS. *Educ:* Harvard Univ, AB, 38; Mass Inst Technol, PhD(bact), 43; Boston Univ, MD, 46; Am Bd Pediat, dipl, 53; Am Bd Microbiol, dipl, 65. *Prof Exp:* Res assoc bact, Sch Med, Univ Boston, 43-44; intern med, Mass Mem Hosps, Boston, 46-47; jr asst resident med, Children's Hosp Med Ctr, 50-52; asst pediat, Harvard Med Sch, 52-53, instr, 53-56, res assoc, 56-57, assoc, 57-60, asst clin prof, 60-61; assoc prof microbiol, 61-67, chief, Sect Virol, Univ Hosp, 61-68, assoc prof med, 61-81, prof microbiol, 67-81, prof pediat, 72-81, EMER PROF PEDIAT & MICROBIOL & EMER ASSOC PROF MED, SCH MED, BOSTON UNIV, 82- *Concurrent Pos:* USPHS res fel, Children's Hosp Med Ctr, Boston, 49-50, res assoc, 52-62, from asst physician to assoc physician, 52-61, consult, 61-; lectr microbiol, Sch Med, Boston Univ, 53-61; lectr pediat, Harvard Med Sch, 61-65; vis physician, Boston City Hosp & Boston Univ Hosp, 61-82; med consult, Disability Qual Br, Soc Sec Admin, HHS, 81- *Mem:* Soc Pediat Res; Infectious Dis Soc Am; Am Pediat Soc; Am Soc Microbiol; Nat Found Infectious Dis; Am Soc Virol. *Res:* Virology; infectious diseases. *Mailing Add:* Boston Univ Med Ctr 80 E Concord St Boston MA 02118

KICE, JOHN LORD, b Colorado Springs, Colo, Feb 18, 30; m 53; c 2. ORGANIC CHEMISTRY. *Educ:* Harvard Univ, AB, 50, MA, 53, PhD(chem), 54. *Prof Exp:* Sr chemist, Rohm and Haas Co, 53-56; from asst prof to assoc prof chem, Univ SC, 56-60; from assoc prof to prof, Ore State Univ, 60-70; prof & chmn dept, Univ Vt, 70-75; chmn dept, chem prof, Tex Tech Univ, 75-85; prof & chmn dept, 85-88, DEAN SCI, MATH & ENG, UNIV DENVER, 88- *Concurrent Pos:* fel, Japan Soc Prom Sci, 83; NIH spec fel, 68-69. *Mem:* Am Chem Soc; Sigma Xi. *Res:* Organic reaction mechanisms; free radical reactions; organic sulfur chemistry; organic selenium chemistry. *Mailing Add:* Dept Chem Univ Denver Denver CO 80208

KICHER, THOMAS PATRICK, b Johnsonburg, Pa, Oct 20, 37; m 62; c 3. ENGINEERING MECHANICS. *Educ:* Case West Reserve Univ, BS, 59, MS, 62, PhD(eng mech), 64. *Prof Exp:* Res engr, Aeronca Mfg Co, 59; res asst, Case Western Reserve Univ, 59-64; design engr, Douglas Aircraft Co, Inc, Calif, 64-65; asst prof eng, 65-68, assoc prof mech eng, 68-81, assoc dean sci & eng, 74-79, PROF MECH ENG, CASE WESTERN RESERVE UNIV, 81-, CHMN DEPT MECH & AEROSPACE ENG, 85- *Concurrent Pos:* Consult, 65- *Mem:* Am Inst Aeronaut & Astronaut; Am Soc Mech Engrs; Soc Exp Mech. *Res:* Computer methods of optimum design; basic phenomena of buckling of elastic systems; analysis and testing of composite materials; analysis and testing of plates and shells; failure analysis and design. *Mailing Add:* Dept Mech Eng Glennan 418 Case Inst Technol 10900 Euclid Ave Cleveland OH 44106

KICKLER, THOMAS STEVEN, b Pittsburgh, Pa, July 7, 47. HEMATOLOGY. *Educ:* W Va Univ, BA, 69, MD, 73. *Prof Exp:* Fel path, Johns Hopkins Univ, 74-76; fel med, Mayo Grad Sch, 76-78; fel hemat, Univ Rochester, 78-80; asst prof, 80-86, ASSOC PROF, JOHNS HOPKINS SCH MED, 86- *Mem:* Am Soc Hemat; Am Asn Blood Banks. *Res:* Platelet immunology. *Mailing Add:* Blood Bank Johns Hopkins Hosp 600 N Wolfe St Baltimore MD 21205

KICLITER, ERNEST EARL, JR, b Ft Pierce, Fla, June 19, 45; c 1. NEUROANATOMY, COLOR VISION. *Educ:* Univ Fla, BA, 68; State Univ NY Upstate Med Ctr, PhD(anat), 73. *Prof Exp:* Asst prof neuroanat, Col Med, Sch Basic Med Sci & asst prof physiol, Col Lib Arts & Sci, Univ Ill, Urbana-Champaign, 74-77; assoc prof, 77-84, PROF ANAT, SCH MED, UNIV PR, 84- *Concurrent Pos:* NIH fel neurol surg, Sch Med, Univ Va, 72-74; vis prof, Univ Med Sch, Pecs, Hungary, 76. *Mem:* Fel AAAS; Soc Neurosci; Am Asn Anatomists. *Res:* Comparative studies of structure and function in vertebrate visual systems; color vision and neuronal plasticity in visual systems. *Mailing Add:* c/o Inst Neurobiol Blvd del Valle 201 San Juan PR 00901

KICSKA, PAUL A, b New York, NY, Nov 23, 32; m 60; c 3. PHYSICS. *Educ:* Muhlenberg Col, BS, 58; Lehigh Univ, MS, 60, PhD(physics), 65. *Prof Exp:* Instr math, Lafayette Col, 58-60; engr, Bell Tel Labs, 60-61; instr physics, Lafayette Col, 61-63; engr, Gen Elec Co, 66-67; PROF PHYSICS, EAST STROUDSBURG STATE COL, 67- *Concurrent Pos:* Chief consult, Tech Consult Serv, 67- *Mem:* Am Phys Soc; Sigma Xi; Soc Physics Students; AAAS; Nat Fluid Power Asn. *Res:* Analytic description of vehicle collisions; accident reconstruction. *Mailing Add:* E Stroudsburg Univ E Stroudsburg Col East Stroudsburg PA 18301

KIDAWA, ANTHONY STANLEY, b Philadelphia, Pa, Apr 1, 42; m 69; c 2. PODIATRIC MEDICINE & SURGERY, ANGIOLOGY. *Educ:* Villanova Univ, BS, 64, Pa Col Pediat Med, DPM, 69. *Prof Exp:* Resident podiatric med & surg, James C Guiffre Med Ctr, Philadelphia, 69-70; chmn med, 74-84, PROF MED, PA COL PODIATRIC MED, 70-, DIR, VASCULAR LAB, 84- *Concurrent Pos:* Staff podiatrist, James C Guiuffre Med Ctr, Philadelphia, 71- & JFK Washington Div Hosp, Turnersville, 73-; consult, Bionic Instruments, Bala Cynwyd, Pa, 72-74, Pa Blue Shield, 74, Sentry Instruments, Inc, Atlanta, Ga, 74-76, Vascular Diag Instruments, Cherry Hill, 76-, NJ State Bd Med Examrs, 82 & Advan Diag, Inc, Oak Ridge, Tenn, 83-; dir res educ podiatry, JFK Washington Div Hosp, Turnersville, 76-80; vis lectr podiatry, Ohio Col Podiatric Med, Cleveland, 80-; vis lectr angiol, Logan Col Chiropractic, St Louis, 81-83; pres & dir angiol, Clindialab Inc, Cherry Hill, NJ, 84-; ed podiatry, J Am Podiatry Asn, 84- *Honors & Awards:* William J Stickel Silver Medal, Am Podiatry Asn, 82, William J Stickel Bronze Medal, 85. *Mem:* Am Podiatry Med Asn; Am Asn Hosp Podiatrists; Am Asn Col Podiatric Med; Am Soc Podiatric Angiol (pres, 83-); Am Diabetes Asn. *Res:* Effects of hypothermia and anesthetic blocks on the peripheral circulation; physiological studies for determining various parameters of circulation in the lower extremities; clinical trials for antifungal medication in the treatment of pedal fungus infections and vasodilators in the treatment of peripheral vascular diseases. *Mailing Add:* Pa Col Podiatric Med 810 Race St Philadelphia PA 19107

KIDD, BERNARD SEAN LANGFORD, b Belfast, Northern Ireland, July 7, 31; m 58; c 4. CARDIOVASCULAR PHYSIOLOGY. *Educ:* Queen's Univ Belfast, MB, BCh & BAO, 54, MD, 57. *Prof Exp:* Asst lectr physiol, Queen's Univ Belfast, 55-57; tutor & registr, Royal Victoria Hosp, Belfast, 57-60; physician & assoc scientist, Hosp for Sick Children, 61-67; assoc prof pediat physiol, Univ Toronto, 61-75; dep dir, 73-75, HARRIET LANE HOME PROF PEDIAT CARDIOL & DIR DIV, SCH MED, JOHNS HOPKINS UNIV, 75-, DIR DIV PEDIAT CARDIOL, JOHNS HOPKINS HOSPITAL. *Concurrent Pos:* Physician & assoc scientist, Hosp Sick Children, 67-75. *Mem:* Am Pediat Soc; Can Soc Clin Invest; Am Col Cardiol; Am Heart Asn; Soc Pediat Res. *Res:* Circulatory physiology, hemodynamics in congenital heart disease. *Mailing Add:* Helen B Taussig Children's Cardiac Ctr Johns Hopkins Univ Hosp 600 N Wolfe St Baltimore MD 21205

KIDD, DAVID EUGENE, b Evanston, Ill, Apr 13, 30; m 55; c 1. PHYCOLOGY. *Educ:* Ariz State Col, BS, 51; Northwestern Univ, MS, 52; Univ NH, MST, 60; Mich State Univ, PhD(bot), 63. *Prof Exp:* Sci teacher high sch, Ariz, 54-55; prof chem, Lindsey Wilson Col, 55-56; sci teacher high sch, Ariz, 56-60; asst prof natural sci, Mich State Univ, 60-67; from assoc prof to prof biol, 67-89, EMER PROF SCI EDUC & BIOL, UNIV NMEX, 89- *Concurrent Pos:* Grants, Am Philos Soc, 64-66; Sigma Xi, 65-66; NSF Undergrad Equip, 69-71; Water Resources Inst, 69-72; NSF, 71-77; instr, Sci Educ, Univ NMex, 89- *Mem:* Am Micros Soc; Am Chem Soc; Nat Sci Teachers Asn; Sigma Xi. *Res:* Taxonomy and ecology of algae in polluted ranch ponds in northern Arizona; carbon-14 primary productivity and population dynamics of phytoplankton in Elephant Butte Reservoir and Lake Powell; nutrient loading models and their modification as applied to Southwestern reservoirs. *Mailing Add:* CIMTE Univ NMex Mesa Vista Hall Albuquerque NM 87131

KIDD, FRANK ALAN, b Dodge City, Kans, July 24, 52; m 73. PLANT PHYSIOLOGY. *Educ:* Ore State Univ, BS, 74; Colo State Univ, MS, 76, PhD(tree physiol), 82. *Prof Exp:* Teaching asst forestry, Ore State Univ, 73-74; teaching asst forest biomet, econ & physiol, Colo State Univ, 74-81, res asst forest tree physiol, 74-80, instr forest ecol, 80-81; res forester, Potlatch Corp, 81-; mem staff, Dow Chem USA. *Concurrent Pos:* Consult, Repub Nat Bank, Dallas, Tex, 76-82; contributing researcher, Lightwood Res Coord Coun, US Forest Serv, 76-78. *Mem:* AAAS; Soc Am Foresters; Am Forestry Asn; Sigma Xi. *Res:* Forest tree physiology; mycorrhizal relationships; physiological consequences of silvicultural treatment of conifers. *Mailing Add:* 13916 Sandy Creek Ct Carmel IN 46032

KIDD, GEORGE JOSEPH, JR, b Grand Rapids, Mich, May 6, 34; m 56; c 3. MECHANICAL ENGINEERING. *Educ:* Northwestern Univ, BS, 56, MS, 57; Univ Tenn, PhD(eng sci), 66. *Prof Exp:* Assoc engr, Y-12 Plant, 57-58, res staff mem reactor develop, Oak Ridge Nat Lab, 58-68, develop engr, 68-77, DEPT SUPVR, OAK RIDGE GASEOUS DIFFUSION PLANT, MARTIN MARIETTA ENERGY SYSTEMS, INC, 77- *Mem:* Am Soc Mech Engrs; Am Inst Chem Engrs; Nat Soc Prof Engrs. *Res:* Heat transfer; fluid mechanics. *Mailing Add:* 120 Windham Rd Oak Ridge TN 37830

KIDD, HAROLD J, b Billings, Okla, Jan 30, 24; m 55; c 2. CYTOGENETICS, PLANT MORPHOLOGY. *Educ:* Okla State Univ, BS, 49, MS, 51; Washington Univ, PhD(bot), 56. *Prof Exp:* Instr bot, Okla State Univ, 51-53; br plant mgr, Plant Breeding, Pioneer Hi-Bred Int, Inc, 55-83; RETIRED. *Mem:* Am Genetics Asn; Sigma Xi; AAAS; Bot Soc Am; Am Inst Biol Sci. *Res:* Cultivated sorghum; plant breeding; morphology. *Mailing Add:* 1617 Dallas St Plainview TX 79072

KIDD, KENNETH KAY, b Bakersfield, Calif, Sept 5, 41; m 64. POPULATION GENETICS, HUMAN GENETICS. *Educ:* Univ Southern Calif, AB, 65; Univ Wis, MS, 67, PhD(genetics), 69; Am Bd Med Genetics, dipl, 82. *Prof Exp:* Res assoc genetics, Sch Med, Stanford Univ, 71-72; asst prof anthrop genetics, Sch Med, Washington Univ, 72-73; asst prof, 73-78, assoc prof human genetics, 78-81, assoc prof human genetics & psychiat, 81-86, PROF HUMAN GENETICS & PSYCHIATRY & BIOL, SCH MED, YALE UNIV, 86- *Concurrent Pos:* NIH fel, Univ Pavia, 69-71 & Sch Med, Stanford Univ, 71; mem, NIH Study Sect, Mammalian Genetics, 79-83; vis assoc prof, Harvard Univ, 81-82; vis scientist, Mass Inst Technol, 82; bd dir, Am Bd Med Genetics, 82-84; ed bd, J Genetics, 86- & J Genomics, 87- *Mem:* Genetics Soc Am; Soc Study Evolution; Am Soc Human Genetics; Am Asn Phys Anthrop; Behav Genetics Asn; fel AAAS. *Res:* Genetics of human behavioral disorders; genetic relationships of human populations; human gene mapping. *Mailing Add:* 333 Cedar St Yale Univ Sch Med New Haven CT 06510-8005

KIDD, RICHARD WAYNE, b Westminster, Md, June 16, 47; m 69; c 1. PHYSICAL CHEMISTRY. *Educ:* Western Md Col, BA, 69; Univ Ill, MS, 71, PhD(chem), 75. *Prof Exp:* Teaching & res asst, Univ Ill, Urbana, 69-75, res assoc chem, 75-77; res scientist, Mat Technol Sect, Battelle Mem Inst, 77-80, prin res scientist, Ceramics & Mat Processing Sect, Columbus Lab, 80-; AT SAN FERNANDO LABS, PACOIMA. *Mem:* Am Chem Soc; Sigma Xi. *Res:* Mathematical modeling and experimental determination of isotope effects in chemical reactions; mechanisms of reactions; chemical vapor deposition of coatings to enhance the properties of the substrate material. *Mailing Add:* 24542 Burr Ct New Hall CA 91321

KIDD, ROBERT GARTH, b Stockton-on-Tees, Eng, July 19, 36; Can citizen; m 59; c 3. INORGANIC CHEMISTRY. *Educ:* Univ Man, BSc, 58, MSc, 60; Univ London, PhD(chem), 62. *Prof Exp:* Asst prof, 63-69, ASSOC PROF CHEM, UNIV WESTERN ONT, 69-, ASST DEAN GRAD STUDIES, 70-. *Mem:* Chem Inst Can; Am Chem Soc. *Res:* Nature of bonding in transition metal complexes; nuclear magnetic resonance spectroscopy of inorganic compounds. *Mailing Add:* Dept Chem Univ Western Ont London ON N6A5B7 Can

KIDD, WILLIAM SPENCER FRANCIS, b Shawford, Eng, Feb 23, 47; UK citizen; m 84; c 1. GEOLOGY. *Educ:* Univ Cambridge, Eng, BA, 69, PhD(geol), 74. *Prof Exp:* Vis lectr geol, Erindale Col, Univ Toronto, 72-73; lectr, 74, asst prof, 75-81, ASSOC PROF GEOL, STATE UNIV NY, ALBANY, 81- *Mem:* Am Geophys Union; Geol Soc Am; AAAS. *Res:* Structural geology; tectonics; Appalachian geology; orogenic belts; Tibet. *Mailing Add:* Dept Geol Sci State Univ NY 1400 Washington Ave Albany NY 12222

KIDDER, ERNEST H(IGLEY), b Amiret, Minn, July 14, 12; m 39; c 2. ENGINEERING. *Educ:* Univ Minn, BAgrEng, 35; Univ Ill, MSCE, 47; US Naval Acad, CMet, 45. *Prof Exp:* Jr soil conservationist, Soil Conserv Serv, USDA, Minn, 35-39; jr hydraul engr, Univ Ill, 39-41, asst hydraul engr, 41-43, assoc hydraul engr, 46-48; proj supvr, Auburn, Ala, 48-49; from assoc prof to prof, 50-79, EMER PROF AGR ENG, MICH STATE UNIV, 79- *Honors & Awards:* Hancor Award, Am Soc Agr Engrs, 71. *Mem:* Fel Am Soc Agr Engrs. *Res:* Hydraulic, hydrologic, meteorologic, soil physical and agronomic phases of soil and water conservation. *Mailing Add:* 1709 Cahill Dr East Lansing MI 48823-4728

KIDDER, GEORGE WALLACE, JR, b Oregon City, Ore, Dec 29, 02; m 30; c 3. BIOLOGY, BIOCHEMISTRY. *Educ:* Univ Ore, AB, 26; Univ Calif, MA, 29; Columbia Univ, PhD(zool), 32. *Hon Degrees:* MA, Amherst Col, 49; ScD, Wesleyan Univ, 50. *Prof Exp:* Teacher pub sch, Ore, 26-28; instr biol, City Col New York, 29-37; asst prof, Brown Univ, 37-46; assoc prof, 46-49, Stone Prof, 49-70, EMER PROF BIOL, AMHERST COL, 70- *Concurrent Pos:* Mem corp, Marine Biol Lab, Woods Hole, 43-47; chmn, vitamins & metab, Gordon Res Confs, 57; mem, Training Grant Comt, NIH, 60-64; vis prof biochem, Univ Calif, Santa Cruz, 68. *Mem:* Fel AAAS; fel Am Acad Microbiol; hon mem Am Soc Microbiol; fel Am Acad Arts & Sci; fel Am Soc Protozoologists. *Res:* Protozoan cytology and physiology; biochemistry of microorganisms; vitamins and amino acids; purine and pyrimidine metabolism in Trypanosomatid flagellates; metabolism; nutrition; chemotherapy of cancer. *Mailing Add:* Biol Lab Amherst Col Amherst MA 01002

KIDDER, GEORGE WALLACE, III, b New York, NY, Sept 24, 34; m 57; c 2. PHYSIOLOGY, BIOPHYSICS. *Educ:* Amherst Col, AB, 56; Univ Pa, PhD(bot), 61. *Prof Exp:* Res fel, Johnson Found, Univ Pa, 61-62; res fel biophys labs, Harvard Med Sch, 62-64; asst prof biol, Wesleyan Univ, 64-73; from assoc prof to prof physiol, Univ Md Sch Dent, 73-84; PROF & CHAIR, DEPT BIOL SCI, ILL STATE UNIV, 84- *Concurrent Pos:* USPHS fel, Harvard Univ, 64. *Mem:* AAAS; Am Physiol Soc; Biophys Soc. *Res:* Gastric acid secretion; electrophysiology; active transport of ions in relation to aerobic metabolism. *Mailing Add:* Dept Biol Sci Ill State Univ Normal IL 61761

KIDDER, GERALD, b Leonville, La, Jan 2, 40; m 66; c 3. SOIL FERTILITY. *Educ:* Univ Southwestern La, BS, 61; Univ Ill, MS, 64; Okla State Univ, PhD(soil sci), 69. *Prof Exp:* Lab asst soil nitrogen, Univ Ill, 63-64; vol, Papal Vol Latin Am, US Cath Bishops' Conf, 64-66; res agronomist, Standard Fruit Co, Castle & Cooke, Inc, 69-72; mgr agr serv, 72-75; asst prof agron, 75-80, assoc prof, 80-88, PROF SOIL SCI, INST FOOD & AGR SCI, UNIV FLA, 88- *Mem:* Am Soc Agron; Int Soc Soil Sci; Soil Sci Soc Am; Sigma Xi. *Res:* Integration of agrotechnology for improved efficiency of crop production and resource utilization; soil test interpretations and crop fertilization recommendations. *Mailing Add:* 2171 McCarthy Hall Univ Fla 0151 IFAS Gainesville FL 32611-0151

KIDDER, GERALD MARSHALL, b Harlingen, Tex, Dec 26, 44; m 83; c 2. DEVELOPMENTAL CELL BIOLOGY, DEVELOPMENTAL GENETICS. *Educ:* Hiram Col, BA, 66; Yale Univ, PhD(biol), 71. *Prof Exp:* Res fel biol sci, Reed Col, Ore, 71-72; from asst prof to assoc prof, 72-89, PROF ZOOL, UNIV WESTERN ONT, 89- *Concurrent Pos:* Res scientist, Alpha Helix Exped, Honduras Reef, 77; vis res assoc radiobiol, Med Sch, Univ Calif, San Francisco, 79-80; mem, bd trustees, Soc Develop Biol, 77-80 & 86-91; vis scientist, Mass Inst Technol, 86-87; mem, Special Study Sect, Nat Inst Child Health & Human Develop, 86; distinguished res prof, Univ Western Ont, 90-91. *Mem:* AAAS; Am Soc Cell Biol; Can Soc Cell Biol; Soc Develop Biol; Can Fedn Biol Sci. *Res:* Developmental genetics of the early embryo: genetic control of morphogenesis and the early events of cell differentiation and the control of gene expression in early mammalian development. *Mailing Add:* Dept Zool Cell Sci Lab Univ Western Ont London ON N6A 5B7 Can

KIDDER, HAROLD EDWARD, physiology; deceased, see previous edition for last biography

KIDDER, JOHN NEWELL, b Boston, Mass, Apr 30, 32; m 60; c 3. PHYSICS, PHYSIOLOGICAL OPTICS. *Educ:* Calif Inst Technol, BS, 54; Duke Univ, PhD(physics), 60. *Prof Exp:* Res assoc physics, Yale Univ, 60-62; from asst prof to assoc prof, 62-74, chmn, dept physics, 84-90, PROF PHYSICS, DARTMOUTH COL, 74- *Concurrent Pos:* NSF fac fel, 71-72. *Mem:* Am Phys Soc; Am Asn Physics Teachers; Optical Soc Am; Inter-Soc Col Coun; Asn Res Vision & Ophthal; Sigma Xi. *Res:* Color science; vision; models of visual response. *Mailing Add:* Physics Dept Dartmouth NH 03755

KIDDER, RAY EDWARD, b New York, NY, Nov 12, 23; m 47; c 3. APPLIED MATHEMATICS, PHYSICS. *Educ:* Ohio State Univ, PhD(physics), 50. *Prof Exp:* Sr physicist, Calif Res Corp, 50-56; assoc div leader, Theoret Div, 56-90, ASSOC, LAWRENCE LIVERMORE LAB, UNIV CALIF, 90- *Honors & Awards:* Alexander von Humboldt Award, 89. *Mem:* Fel Am Phys Soc. *Res:* Thermonuclear physics; astrophysics; quantum electronics. *Mailing Add:* Lawrence Livermore Nat Lab Box 808 Livermore CA 94550

KIDNAY, ARTHUR J, b Milwaukee, Wis, Apr 4, 34; m 60; c 3. CHEMICAL ENGINEERING, PHYSICAL CHEMISTRY. *Educ:* Colo Sch Mines, BS, 56, DSc(chem eng), 68; Univ Colo, MS, 61. *Prof Exp:* Proj engr, Monsanto Chem Co, Mass, 56-58; res engr, Cryogenics Div, Nat Bur Standards, Colo, 59-68; from asst prof to assoc prof, Colo Sch Mines, 68-76, prof chem & petrol refining eng, 77-83, dept head chem eng, 83-90, DEAN GRAD STUDIES & RES, COLO SCH MINES, 90- *Mem:* Am Inst Chem Engrs; Am Soc Eng Educ. *Res:* Solid-vapor and liquid vapor equilibria at cryogenic temperatures; physical adsorption at cryogenic temperatures. *Mailing Add:* Dean Grad Studies & Res Colo Sch Mines Golden CO 80401

KIDWELL, ALBERT LAWS, b Auxvasse, Mo, Jan 1, 19; m 43; c 4. PETROLEUM GEOLOGY. *Educ:* Mo Sch Mines, BS, 40; Wash Univ, MS, 42; Univ Chicago, PhD(geol), 49. *Prof Exp:* Photogrammetric engr, US Coast & Geod Surv, 42-44; geologist, Mo Geol Surv, 44-47; res geologist, Carter Oil Co, 50-58; sect head, Jersey Prod Res Co, 58-65; res assoc, Esso Prod Res Co, 65-73; sr res assoc, Exxon Prod Res Co, 73-84; RETIRED. *Mem:* Geol Soc Am; Mineral Soc Am; Am Asn Petrol Geol; Soc Econ Geol. *Res:* Igneous and economic geology. *Mailing Add:* 14403 Carolcrest Dr Houston TX 77079

KIDWELL, MARGARET GALE, b Askham, Eng, Aug 17, 33; wid; c 2. EVOLUTIONARY GENETICS, DROSOPHILA. *Educ:* Nottingham Univ, BSc, 53; Iowa State Univ, MS, 62; Brown Univ, PhD(genetics), 73. *Prof Exp:* Adv officer agr, Ministry Agr, London, 55-60; assoc res, 66-70, res fel, 73-74, res assoc, 74-75, investr, 75-77, from asst prof to prof biol, Brown Univ, 77-85; PROF ECOL & EVOLUTIONARY BIOL, UNIV ARIZ, 85- *Concurrent Pos:* Assoc ed, Evolution; mem, Genetic Basis Dis Rev Comt, NIH. *Mem:* Am Genetic Asn (pres, 91); Am Soc Naturalists (vpres, 84); Genetics Soc Am; Soc Study Evolution; Sigma Xi; AAAS. *Res:* Drosophila genetics and evolution; recombination transposable elements; speciation. *Mailing Add:* Dept Ecol & Evolutionary Biol Univ Ariz Tucson AZ 85721-0001

KIDWELL, ROGER LYNN, b Nevada City, Calif, Feb 24, 38; m 61; c 1. SYNTHETIC ORGANIC CHEMISTRY. *Educ:* Chapman Col, BS, 62; Univ Southern Calif, PhD(org chem), 66. *Prof Exp:* Res chemist, 66-71, res specialist, 71-73, RES GROUP LEADER SURFACTANTS, MONSANTO CO, 73- *Mem:* Am Chem Soc; Sigma Xi. *Res:* Synthesis and process development for new surface active agents. *Mailing Add:* Monsanto Co 800 N Lindbergh Blvd St Louis MO 63167-0001

KIDWELL, WILLIAM ROBERT, b La Follette, Tenn, Sept 2, 36; m 55; c 2. CELL BIOLOGY, BIOCHEMISTRY. *Educ:* Berea Col, BA, 63; Wash Univ, PhD(biochem), 67. *Prof Exp:* Staff fel, 69-73, res biologist growth regulation, 73-75, CHIEF, CELL CYCLE REGULATION SECT, NAT CANCER INST, 75- *Concurrent Pos:* Proj officer, Breast Cancer Task Force, Nat Cancer Inst, 74- *Mem:* AAAS; Am Asn Cancer Res; Am Soc Cell Biologists. *Res:* Cell cycle regulation; rate limiting steps of the division cycle and the effect of hormones on these processes; extracellular matrix and its role in growth regulation. *Mailing Add:* Bldg 10 Rm 5B39 Nat Cancer Inst NIH Bethesda MD 20892

KIEBLER, JOHN W(ILLIAM), b Hershey, Pa, Apr 1, 28; m 81; c 3. COMMUNICATION SATELLITES, ORBIT & SPECTRUM UTILIZATION. *Educ:* Lafayette Col, BS, 50. *Prof Exp:* Electronic scientist, Nat Bur Standards, 51-52; proj engr, Off Chief Ord, 52-53; assoc engr, Appl Physics Lab, Johns Hopkins Univ, 53-58, sr engr, 58-59, prin engr, Emerson Res Labs, 59-60; head, Network Implementation Br, Goddard Space Flight Ctr, NASA HQ, 60-65, actg chief, Proj Opers Support Div, 65-66, asst chief, 66-70, proj mgr, Satellite Tracking & Data Acquistion, 70-73, sr engr, Commun Div, 73-79, sr commun engr, 79-85, head, Tech Consult Serv, 85-89, TELECOMMUN CONSULT, NASA HQ, 90- *Concurrent Pos:* mem, US Deleg World Admin Radio Conf, 79, 83, 85 & 88, Comt Radio Frequencies & Int Radio Consult Comt, Nat Acad Sci. *Honors & Awards:* Except Serv Medal, NASA, 85. *Mem:* Inst Elec & Electronics Engrs. *Res:* Communications systems; remote sensing; regulatory filings. *Mailing Add:* 14520 Dowling Dr Burtonsville MD 20866

KIEBURTZ, R(OBERT) BRUCE, b Seattle, Wash, Mar 22, 31; m 54; c 2. ELECTRICAL ENGINEERING. *Educ:* Univ Wash, BSEE, 52, MSEE, 63, PhD(elec eng), 66. *Prof Exp:* Proj engr, Gen Elec Co, 54-57; res engr, Boeing Co, 57-65, supvr, 66-67; mem tech staff, Bell Tel Labs, 67-77, ASST ENG MGR, NETWORK PLANNING & BUS SERV, AT&T CO, 77- *Mem:* AAAS; sr mem Inst Elec & Electronics Engrs; NY Acad Sci; Sigma Xi. *Res:* Ballistic missile defense; balanced magnetic circuits for logic and memory; digital processing of signals, including digital filtering. *Mailing Add:* Cramer Dr RD-1 Box C-7 Chester NJ 07930

KIEBURTZ, RICHARD B(RUCE), b Spokane, Wash, Nov 28, 33; m 59; c 2. COMPUTER SCIENCE. *Educ:* Univ Wash, BSEE, 55, MSEE, 57, PhD(elec eng), 61. *Prof Exp:* Develop engr, Gen Elec Co, 54-57; from res engr to group mgr ballistic missile defense, Boeing Co, 57-67; mem tech staff, Principles Res Lab, Bell Labs, dist eng mgr prod planning & support, 77-80, dist mgr, 80-83, eng staff mgr info systs, 83-85, SUPVR DATA NETWORKING GROUP, A T & T BELL LABS, WHIPPANY, NJ, 85- *Concurrent Pos:* Consult, CBS Labs, Conn, 62-63 & Rome Air Develop Ctr, USAF, 63-68; NSF sci fac fel, 68-69. *Mem:* AAAS; sr mem Inst Elec & Electronics Engrs; NY Acad Sci. *Res:* programming languages; distributed computing. *Mailing Add:* Oregon Grad Ctr 19600 NW Von Neumann Dr Beaverton OR 97006

KIECH, EARL LOCKETT, b Jonesboro, Ark, Jan 29, 49. ARTIFICIAL INTELLIGENCE. *Educ:* Southwestern Memphis, BS, 71; Memphis State Univ, MS, 73. *Prof Exp:* SCI PROGRAMMER, CLASPAN CORP, 73- *Concurrent Pos:* Sci programmer, Univ Tenn Space Inst, 87- *Mem:* Am Phys Soc; Sigma Xi. *Res:* Researching a program to diagnose mechanical problems or faults with the main engine in the space shuttle. *Mailing Add:* 306 Crosslake Dr Tullahoma TN 37388

KIECHLE, FREDERICK LEONARD, b Indianapolis, Ind, Mar 26, 46. CLINICAL CHEMISTRY, INSULIN ACTION. *Educ:* Evansville Col, BA, 68; Ind Univ, PhD(chem), 73, MD, 75. *Prof Exp:* Resident path, William Beaumont Hosp, 75-79; res fel clin chem, Barnes Hosp, 79-80; asst prof path, Univ Pa Sch Med, 80-83; CHIEF CLIN CHEM & CLIN PATH, WILLIAM BEAUMONT HOSP, 83-; CLIN ASST PROF PATH, WAYNE STATE UNIV SCH MED, 84- *Concurrent Pos:* Vis lectr, Nat Univ Singapore, 84; adv, Nat Comt Clin Lab Standards, 87-88; mem coun spec topics, Am Soc Clin Path, 88-91; fel, Hartford Found, 82-83. *Mem:* Am Soc Clin Pathologists; Am Asn Pathologists; Am Asn Clin Chem; Am Diabetes Asn; Col Am Pathologists; Am Fedn Clin Res. *Res:* Unraveling the temporal sequence of rapid events which occur following the binding of insulin to its receptor; membrane polarization; phospholipid metabolism; fatty acid metabolism; redox potential and impedance changes induced by insulin. *Mailing Add:* William Beaumont Hosp 3601 W 13 Mile Rd Royal Oak MI 48073

KIECKHEFER, ROBERT WILLIAM, b Milwaukee, Wis, Mar 13, 33; m 69; c 3. INSECT ECOLOGY. *Educ:* Univ Wis, BS, 55, PhD(entom), 62; Univ Minn, MS, 58. *Prof Exp:* RES ENTOMOLOGIST, NORTHERN GRAIN INSECTS RES LAB, USDA, 63- *Mem:* Entom Soc Am; Ecol Soc Am; Int Orgn Biol Control. *Res:* Aphid biology and ecology; ecology and biological control of cereal insects. *Mailing Add:* Northern Grain Insects Res RR 3 Brookings SD 57006

KIEFER, BARRY IRWIN, b Bayonne, NJ, May 16, 33; m 60. DEVELOPMENTAL BIOLOGY. *Educ:* Univ Denver, BS, 60; Univ Calif, Berkeley, PhD(zool), 65. *Prof Exp:* From asst prof to assoc prof, 65-77, PROF BIOL, WESLEYAN UNIV, 77- *Mem:* Am Soc Cell Biol. *Res:* Genetic regulation of spermiogenesis in Drosophila. *Mailing Add:* PO Box 8725 Reno NV 89507

KIEFER, CHARLES RANDOLPH, b Minneapolis, Minn, Nov 24, 47; m 73; c 1. EPITOPE MAPPING-HUMAN HEMOGLOBIN, IMMUNOPHENOTYPIC ANALYSIS- LEUKEMIA. *Educ:* Univ Cincinnati, BS, 69; Med Col Ga, PhD(microbiol), 81. *Prof Exp:* Res fel, 81-87, asst res scientist, 87-91, ASST PROF IMMUNOL, MED COL GA, 91- *Concurrent Pos:* Consult, Med Diag Technol, Inc, Augusta, Ga, 87-; prin investr, Am Heart Asn, Ga Affil, 90-91. *Mem:* Am Asn Immunologists. *Res:* residue-specific monoclonal antibodies to normal and variant human hemoglobins of clinical interest and their application in immunoassays; immunophenotypic identification of chronic lymphocytic leukemia. *Mailing Add:* Dept Immunol & Microbiol Med Col Ga Augusta GA 30912-2400

KIEFER, DAVID JOHN, b Sewickley, Pa, Oct 1, 38; m 62. MICROBIOLOGY, IMMUNOLOGY. *Educ:* Univ Pittsburgh, BS, 68; Univ Miami, PhD(microbiol), 77. *Prof Exp:* Res technician immunol, Sch Med, Univ Miami, 68-77; staff immunologist, Cordis Labs, Inc, 77-80, sr staff immunologist, 80-83, prin immunologist, 83-86; dir, 86-90, VPRES, PROD DEVELOP, DIAMEDIX CORP, 90- *Mem:* Am Soc Microbiol; Nat Comt Clin Lab Standards. *Res:* Pregnancy associated plasma proteins; immunology of pregnancy; immunosuppressive properties of pregnancy serum; enzyme-labeled immunoassays for the detection and quantitation of humoral constituents, for example antibodies to rubella, herpes simplex virus, cytomegalovirus, Human T-Cell Leukemia/Lymphoma Virus I and HIV-1, Epstein Barr Virus and autoantigens in systemic rheumatic diseases. *Mailing Add:* Diamedix Corp 2140 N Miami Ave Miami FL 33127

KIEFER, EDGAR FRANCIS, b Tsingtao, China, Sept 9, 34; US citizen; m 57, 74; c 6. ORGANIC CHEMISTRY. *Educ:* Stanford Univ, BS, 57; Calif Inst Technol, PhD(chem), 60. *Prof Exp:* Asst chem, Calif Inst Technol, 57-58; res assoc, Univ Ill, 60-61; res chemist, Chevron Res Corp, 61-62; from asst prof to assoc prof, 62-72, PROF CHEM, UNIV HAWAII, 72- *Concurrent Pos:* NSF sci fac fel, Stanford Univ, 68-69; vis scientist, Mass Inst Technol, 78 & Oxford Univ, 85-86. *Mem:* Am Chem Soc. *Res:* Physical organic and bioorganic chemistry; stereochemistry. *Mailing Add:* Dept Chem Univ Hawaii Honolulu HI 96822

KIEFER, HAROLD MILTON, b Detroit, Mich, Mar 9, 33; m 66; c 1. THEORETICAL PHYSICS. *Educ:* Wayne State Univ, PhD(physics), 69. *Prof Exp:* Personnel exam, City of Detroit, 58-62; asst prof, 69-81, ASSOC PROF PHYSICS, NORFOLK STATE COL, 81- *Mem:* Am Phys Soc; Sigma Xi. *Res:* Applications of group theory to coulomb potential problems. *Mailing Add:* 116 W Government Ave Norfolk VA 23503

KIEFER, JOHN DAVID, b Evansville, Ind, Jan 2, 40; m 64; c 5. GEOLOGIC HAZARDS, ENERGY RESOURCES. *Educ:* St Josephs Col, BA, 61; Univ Ill, MS, 65, PhD(eng geol), 70. *Prof Exp:* Instr geol, Univ Ill, 65-67; assoc prof, Eastern Ky Univ, 67-71; head, Eng Geol Div, Geotech Eng Assocs, 71-78;

eng geologist, Geol Surv Ala, 78-79, head, Water Resources Div, 79-82; ASST STATE GEOLOGIST, KY GEOL SURV, UNIV KY, 81- *Mem:* Geol Soc Am; Am Asn Petrol Geologists; Soc Econ Paleontologists & Mineralogists; AAAS; Sigma Xi. *Res:* Engineering geology and hydrogeology. *Mailing Add:* 228 Mining & Mineral Resources Bldg Ky Geol Surv Univ Ky Lexington KY 40506-0107

KIEFER, JOHN HAROLD, b New Ulm, Minn, Aug 27, 32; m 61; c 3. COMBUSTION, KINETICS. *Educ:* Univ Minn, BS, 54; Cornell Univ, PhD(phys chem), 61. *Prof Exp:* Fel, Cornell Univ, 59-60; res staff mem, Los Alamos Sci Lab, Univ Calif, 61-67; assoc prof, 67-72, PROF CHEM ENG & CHEM, UNIV ILL-CHICAGO, 72-, ACTG HEAD, 89- *Concurrent Pos:* Consult, Los Alamos Nat Lab, 67-, Ill Tool Works, 75-77; Shell res fel, Thorton, UK, 73-74; joint appointment, Argonne Nat Lab, 85- *Mem:* Combustion Inst; Am Inst Chem Eng; Am Chem Soc. *Res:* Shock tube studies of kinetics; energy transfer in high temperature gases; laser diagnostics. *Mailing Add:* Dept Chem Eng Univ Ill Box 4348 Chicago IL 60680

KIEFER, RALPH W, b Somerville, NJ, Nov 28, 34; m 59, 83; c 4. CIVIL ENGINEERING. *Educ:* Cornell Univ, BCE, 56, MS, 60, PhD(civil eng), 64. *Prof Exp:* Asst hwy engr, NJ State Hwy Dept, 58; asst civil eng, Cornell Univ, 58-62; from asst prof to assoc prof, 62-70, PROF CIVIL ENG, UNIV WIS-MADISON, 71-, ENG DIR, ENVIRON REMOTE SENSING CTR, 81- *Concurrent Pos:* Vis prof, Univ Hawaii, 70-71. *Mem:* Am Soc Civil Engrs; Am Soc Photogram. *Res:* Remote sensing of the environment; engineering applications of airphoto interpretation. *Mailing Add:* Civil & Environ Eng 1210 Eng Bldg Univ Wis Madison WI 53706

KIEFER, RICHARD L, b Columbia, Pa, Dec 14, 37; m 62; c 3. NUCLEAR CHEMISTRY. *Educ:* Drew Univ, AB, 59; Univ Calif, Berkeley, PhD(nuclear chem), 64. *Prof Exp:* Res assoc, Brookhaven Nat Lab, 63-65; asst prof, 65-68, assoc prof, 68-81, PROF CHEM, COL WILLIAM & MARY, 81- *Mem:* Am Chem Soc; Sigma Xi. *Res:* Quenching in liquid scintillation counting; irradiation studies of polymers. *Mailing Add:* Dept Chem Col William & Mary Williamsburg VA 23185

KIEFF, ELLIOTT DAN, b Philadelphia, Pa, Feb 2, 43; m 65; c 2. INFECTIOUS DISEASES, VIROLOGY. *Educ:* Univ Pa, AB, 63; Johns Hopkins Univ, MD, 66; Univ Chicago, PhD(virol), 71. *Prof Exp:* Intern, Hosp, Sch Med, Univ Chicago, 66-67; from jr resident to sr asst resident, 67-69, resident, 69-70, chief, Sect Infectious Dis, 71-87, from asst prof to prof, Sch Med, 70-87; PROF MED, MICROBIOL & MOLECULAR GENETICS, HARVARD UNIV, 87- *Concurrent Pos:* Mem, Comt Virol, Univ Chicago, 70-; fac res award, Am Cancer Soc; Carl Hartford vis prof, Washington Univ. *Mem:* Infectious Dis Soc Am; Am Soc Clin Invest; Am Asn Cancer Res; Am Soc Microbiol. *Res:* Molecular biology of animal viruses, particularly herpes viruses. *Mailing Add:* 75 Francis St Thorn Bldg 12th Floor Boston MA 02115

KIEFFER, HUGH HARTMAN, b Norwich, Conn, Oct 31, 39; c 1. PLANETARY SCIENCE, INFRARED INSTRUMENTATION. *Educ:* Calif Inst Technol, BS, 61, PhD(planetary sci), 68. *Prof Exp:* Res fel planetary sci, Calif Inst Technol, 68-69; asst prof, 69-75, assoc prof planetary sci, Univ Calif, Los Angeles, 75-78; chief, 86-90, RES GEOPHYSICIST, BR ASTROGEOLOGY, GEOL DIV, US GEOL SURV, 78- *Concurrent Pos:* Vis assoc planetary physics, Calif Inst Technol, 75-77; prin investr, Viking Infrared Thermal Mapper; coinvestr, Mariner 6 & 7 Infrared Radiometer, Mariner 9 Infrared Radiometer, Galileo Near Infrared Mapping Spectrometer, Mars Observer Thermal Emission Spectrometer; US partic scientist, Soviet Mars-94 mission; team mem, Earth Observing Syst High Resolution Imaging Spectrometer; US team mem, Earth Observing Syst Advan Spaceborne Thermal Emission Radiometer (Japan); adj prof planetary sci, Univ Calif, Los Angeles, 78-81; Nat Acad Sci, Comt Lunar & Planetary Explor, 79-83; chmn, Fourth Int Conf Mars, 89. *Honors & Awards:* Sci Achievement Medal, NASA, 77; Antarctic Serv Medal, 68. *Mem:* AAAS; Am Astron Soc; Am Meteorol Soc; Am Optical Soc; fel Am Geophys Union. *Res:* Planetary atmospheres and surfaces; spectra of the moon and planets; atmospheric condensation processes; infrared instrumentation, observations and thermal models; thermal infrared and radar observations of volcanos. *Mailing Add:* Astrogeology 2255 N Gemini Dr Flagstaff AZ 86001

KIEFFER, NAT, b Montgomery, La, July 13, 30; m 51; c 3. GENETICS. *Educ:* Univ Southwestern La, BS, 52; La State Univ, MS, 56; Okla State Univ, PhD(animal breeding), 59. *Prof Exp:* Animal geneticist, Range Livestock Exp Sta, USDA, 59-62, supt beef cattle res, 62-64; res assoc & fel molecular biol, Univ Calif, Berkeley, 64-65; asst prof, 65-69, assoc prof mammalian cytogenetics, 69-77, PROF GENETICS, ANIMAL & PLANT SCI, TEX A&M UNIV, 77- *Mem:* Genetics Soc Am; Am Soc Animal Sci. *Res:* Population genetics; beef cattle breeding; molecular biology, gene action in bacteria on molecular level; mammalian cytogenetics, beef cattle. *Mailing Add:* 1212 Winding Rd College Station TX 77840

KIEFFER, STEPHEN A, b Minneapolis, Minn, Dec 20, 35; m 58; c 4. RADIOLOGY, NEURORADIOLOGY. *Educ:* Univ Minn, BA, 56, BS, 57, MD, 59. *Prof Exp:* From instr to prof, Univ Minn, 66-74; PROF RADIOL & CHMN DEPT, STATE UNIV NY HEALTH SCI CTR, SYRACUSE, 74- *Concurrent Pos:* Nat Heart Inst cardiovasc trainee, Univ Minn, 61-62 & 64-65, Nat Inst Neurol Dis & Blindness fel neuroradiol, 66; James Picker Found scholar radiol res, 66-68; chief radiol, Minneapolis Vet Admin Hosp, 68-74; assoc ed, Year Book Diag Radiol, 81-87; Radiol, 86; panelist & subcomt chair, NIH Consensus Develop Conf Magnetic Resonance & Imaging, 87; consult to ed, Radiol, 87- *Mem:* Am Col Radiol; Am Soc Neuroradiol (pres, 78-79); Radiol Soc NAm; Am Roentgen Ray Soc; Asn Univ Radiologists. *Res:* Neuroradiology; myelography; cerebrospinal fluid circulation. *Mailing Add:* Dept Radio State Univ NY Health Sci Ctr Syracuse NY 13210

KIEFFER, SUSAN WERNER, b Warren, Pa, Nov 17, 42; m 66; c 1. VOLCANOLOGY, MINERAL PHYSICS. *Educ:* Allegheny Col, BSc, 64; Calif Inst Technol, MSc, 67, PhD(planetary sci), 71. *Hon Degrees:* DSc, Allegheny Col, 87; Dipl, USSR Acad Sci. *Prof Exp:* Res geophysicist, Univ Calif, Los Angeles, 71-73, asst prof, 73-78, assoc prof geol, 78-79; geologist, US Geol Surv, 78-90; PROF GEOL, ARIZ STATE UNIV, TEMPE, 90- *Concurrent Pos:* Alfred P Sloan res fel, 77-79. *Honors & Awards:* Mineral Soc Am Award, 80; W H Mendenhall lectr, US Geol Surv, 80; Meritorious Serv Award, Dept Interior, 87; Int Geol Spendiarov Prize, 89. *Mem:* Nat Acad Sci; Am Geophys Union; Meteoritical Soc; Sigma Xi; Geol Soc Am; Am Acad Arts & Sci. *Res:* Geological physics; high pressure geophysics and impact processes; shock metamorphism of natural materials; thermodynamic properties of minerals; mechanisms of geyser and volcano eruptions; river hydraulics. *Mailing Add:* Dept Geol Ariz State Univ Tempe AZ 85287-1404

KIEFFER, WILLIAM FRANKLIN, b Trenton, NJ, Mar 16, 15; m 40; c 2. GENERAL CHEMISTRY. *Educ:* Col Wooster, BA, 36; Ohio State Univ, MSc, 38; Brown Univ, PhD(photochem), 40. *Prof Exp:* Asst chem, Ohio State Univ, 36-38 & Brown Univ, 38-39; instr, Col Wooster, 40-42; from instr to asst prof, Western Reserve Univ, 42-46; prof, 46-80, EMER PROF CHEM, COL WOOSTER, 80- *Concurrent Pos:* Res partic, Chem Div, Oak Ridge Nat Lab, 51-52; ed J Chem Educ, Am Chem Soc, 55-67; NSF fac fel, Mass Inst Technol, 63-64; vis scholar, Stanford Univ, 69-70 & Univ Calif, Santa Cruz, 74-75, vis prof, US Naval Acad, 81. *Honors & Awards:* Award Chem Educ, Mfg Chem Asn, 65 & Am Chem Soc, 68. *Mem:* AAAS; Am Chem Soc; Am Inst Chemists; NY Acad Sci. *Res:* Photochemistry; radiation chemistry; chemical education. *Mailing Add:* 1873 Golden Rain Rd 3 Walnut Creek CA 94595

KIEFT, JOHN A, b Oak Park, Ill, Feb 27, 41; m 65; c 1. PHYSICAL CHEMISTRY, INORGANIC CHEMISTRY. *Educ:* Hope Col, BA, 63; Ill Inst Technol, PhD(chem), 68. *Prof Exp:* Chemist, Shell Chem Co, 67-70; res chemist anal chem, 70-71, supvr prod develop, 72-74, sect mgr, 74-76, sr sect mgr anal, 76-79, dept mgr res serv, Western Res Lab, 79-85, dir, regulatory affairs, Stauffer Chem Co, 85-87; MGR, REGIST W, ICI AM, 87- *Mem:* Am Chem Soc. *Res:* New product development; analytical chemistry; agricultural chemicals; research administration. *Mailing Add:* 707 Blue Serv Dr Danville CA 94526-4526

KIEFT, LESTER, b Grand Haven, Mich, Sept 18, 12; m 41; c 3. ANALYTICAL CHEMISTRY. *Educ:* Hope Col, AB, 34; Pa State Col, MS, 36, PhD(anal chem), 39. *Prof Exp:* Asst, Pa State Col, 34-37; asst prof chem, Pa State Jr Col, 37-42; from asst prof to prof, 42-81, head dept, 44-70, secy fac, 68-80, EMER PROF CHEM, BUCKNELL UNIV, 81- *Mem:* AAAS; Am Chem Soc; Nat Sci Teachers Asn. *Res:* Analytical properties of salts of the iodometallic acids. *Mailing Add:* Dept Chem Bucknell Univ Lewisburg PA 17837

KIEFT, RICHARD LEONARD, b Lewisburg, Pa, Apr 27, 45. INORGANIC CHEMISTRY, ANALYTICAL CHEMISTRY. *Educ:* Dickinson Col, BS, 67; Univ Ill, Urbana, PhD(inorg chem), 73. *Prof Exp:* Teaching res assoc chem, Tulane Univ, 73-75; from asst prof to assoc prof, 75-89, PROF CHEM, MONMOUTH COL, ILL, 89- *Honors & Awards:* Sears-Roebuck Found Award, 88. *Mem:* Am Chem Soc. *Res:* Synthesis and identification of organometallic compounds; environmental analysis. *Mailing Add:* 520 E First Ave Monmouth IL 61462

KIEHL, JEFFREY THEODORE, b Harrisburg, Pa, June 10, 52; m 80; c 1. RADIATIVE TRANSFER, CLIMATE MODELING. *Educ:* Elizabethtown Col, BS, 74; Ind Univ, MS, 77; State Univ NY, Albany, PhD(atmospheric sci), 81. *Prof Exp:* Vis scientist, 81-82, fel adv studies, 82-84, RES SCIENTIST, ATMOSPHERIC SCI, NAT CTR ATMOSPHERIC RES, 84- *Mem:* Am Meteorol Soc; Am Geophys Union. *Res:* Electromagnetic scattering from spherical and non-spherical particles; two-dimensional radiative-convective climate model; infrared transfer in the atmosphere with applications to the carbon dioxide climate problems; stratospheric modeling. *Mailing Add:* Nat Ctr Atmospheric Res PO Box 3000 Boulder CO 80307

KIEHL, RICHARD ARTHUR, b Akron, Ohio, Feb 14, 48. PHYSICAL ELECTRONICS. *Educ:* Purdue Univ, BS & MS, 70; PhD(elec eng), 74. *Prof Exp:* Asst elec eng, Sch Elec Eng, Purdue Univ, 71-74; mem tech staff, Div Solid-State Device Physics Res, Sandia Labs, 74-80; MEM TECH STAFF, EXPLOR HIGH-SPEED DEVICE GROUP, BELL LABS, 80- *Mem:* Inst Elec & Electronics Engrs; Am Phys Soc. *Res:* Microwave and optical active semiconductor devices; high-speed logic devices in compound semiconductors. *Mailing Add:* IBM Corp/T J Watson Res Ctr Rm 24-120 PO Box 218 Yorktown Heights NY 10598

KIEHLMANN, EBERHARD, b Grosshartmannsdorf, Ger, Feb 9, 37. ORGANIC CHEMISTRY. *Educ:* Univ Tuebingen, Vordiplom, 59; Univ Md, College Park, PhD(org chem), 64. *Prof Exp:* Res fel, Univ Calif, Berkeley, 64-65 & Dartmouth Col, 65-66; asst prof, 66-72, ASSOC PROF CHEM, SIMON FRASER UNIV, 72- *Mem:* Am Chem Soc; Ger Soc Chem; Chem Inst Can. *Res:* Synthesis and reactions of flavanoids. *Mailing Add:* Dept Chem Simon Fraser Univ Burnaby BC V5A 1S6 Can

KIEHN, ROBERT MITCHELL, b Oak Park, Ill, Dec 29, 29; m 58. PHYSICS. *Educ:* Mass Inst Technol, BS, 50, PhD, 54. *Prof Exp:* Staff physicist, Los Alamos Sci Lab, 54-62; assoc prof, 62-72, PROF PHYSICS, UNIV HOUSTON, 72- *Mem:* Am Phys Soc; Am Nuclear Soc. *Res:* Neutron reactor physics; hydrodynamics; thermodynamics. *Mailing Add:* Dept Physics Univ Houston University Park 4800 Calhoun Rd Houston TX 77204

KIEL, JOHNATHAN LLOYD, b Houston, Tex, Sept 4, 49; m 73; c 2. BIOCHEMICAL IMMUNOLOGY, BACTERIOLOGY. *Educ:* Tex A&M Univ, BS, 73, DVM, 74, Health Sci Ctr, PhD(microbiol & biochem), 81. *Prof Exp:* Instr vet microbiol, Sch Vet Med, Tex A&M Univ, 74-75; vet, Vet Pub Health, Grissom AFB, Ind, 75-77, res immunol, Physics Br, Radiation Sci Div, Sch Aerospace Med, Brooks AFB, Tex, 81-91, CHIEF BIOPHYS MECHANISM FUNCTION RADIO FREQUENCY, RADIATION BR, ENERGY DIV, OCCUP & ENVIRON HEALTH DIRECTORATE, USAF, 91- *Mem:* Am Vet Med Asn; Bioelectromagnetics Soc; Am Col Vet Microbiologists, 84. *Res:* Oxidative metabolism of the various cells of the immune system and how it influences the immune response and cytotoxic mechanisms; influence of radiofrequency radiation on this metabolism. *Mailing Add:* USAF AL/OEDR Brooks TX 78235-5301

KIEL, OTIS GERALD, b Wichita Falls, Tex, Feb 10, 31; m 53; c 3. ENGINEERING SCIENCES, MATHEMATICS. *Educ:* NTex State Univ, BS, 49; Univ Okla, BS, 56, MS, 57, PhD(eng sci), 63. *Prof Exp:* sect supvr, Continental Oil Co, 59-80, CHIEF RESERVOIR ENGR, CONOCO, INC, 80- *Honors & Awards:* C K Ferguson Award, Am Inst Mining, Metall & Petrol Engrs, 63. *Mem:* Am Inst Mining, Metall & Petrol Engrs; Soc Petrol Engrs. *Res:* Reservoir mechanics, mathematical modeling; reserve determination enhanced recovery projects. *Mailing Add:* 807 Soboda Ct Houston TX 77079

KIELKOPF, JOHN F, b Louisville, Ky, Aug 1, 45; m 70. ATOMIC & MOLECULAR SPECTROSCOPY, ASTROPHYSICS. *Educ:* Univ Louisville, BS & MS, 66; Johns Hopkins Univ, PhD(physics), 69. *Prof Exp:* From asst prof to assoc prof, 69-77, PROF PHYSICS, UNIV LOUISVILLE, 77- *Concurrent Pos:* Assoc res scientist, Johns Hopkins Univ, 74; vis scientist, Argonne Nat Lab, 74-75; astronomer, Observatory Paris, Meudon, 79-80; scientist in residence, Argonne Nat Lab, 81. *Mem:* Am Phys Soc; Optical Soc Am; Am Astron Soc. *Res:* High resolution atomic and molecular spectroscopy; neutral atom collisions; shape of atomic spectral lines; stellar spectroscopy; active galaxies. *Mailing Add:* Dept Physics Univ Louisville Louisville KY 40292

KIELY, DONALD EDWARD, b Waterbury, Conn, Jan 5, 38; m 63; c 3. SYNTHETIC ORGANIC CHEMISTRY. *Educ:* Fairfield Univ, BS, 60; Univ Conn, PhD(org chem), 65. *Prof Exp:* Vis asst prof org chem, Wofford Col, 65-66; staff fel, Nat Inst Arthritis & Metab Dis, 66-68; from asst prof to assoc prof, 68-77, PROF CHEM, UNIV ALA, BIRMINGHAM, 77- *Concurrent Pos:* Vis fel, Res Sch Chem, Australian Nat Univ, 74; resident prof, Staley Mfg Co, 81-82; ed, J Carbohydrate Chem, 81- *Mem:* Am Chem Soc; Sigma Xi. *Res:* Synthesis of biologically interesting carbohydrates, cyclitols and other carbocyclic and heterocyclic compounds; chemical studies related to cyclitol biosynthesis; synthetic carbohydrate chemistry; industrially related carbohydrate chemistry; synthetic carbohydrate based polymer synthesis. *Mailing Add:* 2521 Chatwood Rd Birmingham AL 35226

KIELY, JOHN ROCHE, b Berkeley, Calif, Nov 8, 06; m 40; c 5. MECHANICAL ENGINEERING, MINING ENGINEERING. *Educ:* Univ Wash, BSCE, 31. *Prof Exp:* Construct engr, Rainier Pulp & Paper Co, 24-31, supt, 31-36; resident engr, Rayonier, Inc, 37-40, asst gen supt, 40-42; mgr outfitting, subassembly & transport, Calif Shipbuilding Corp, 42-45; proj mgr, Bechtel Bros McCone Co, 45-48, mgr, Bechtel Corp, 48-51, vpres, 51-54, sr vpres, 57-67, exec vpres, 67-71, dir, 54-74, exec consult, 74-80, sr exec consult, Bechtel, Inc, 80-88; RETIRED. *Mem:* Nat Acad Eng; fel Am Inst Mech Engrs; Am Soc Civil Engrs; Am Soc Mech Eng; Sigma Xi. *Mailing Add:* PO Box 620303 Woodside CA 94062

KIELY, JOHN STEVEN, b Missoula, Mont, Oct 11, 51; m 74; c 2. PHARMACOLOGY. *Educ:* Mont State Univ, BSc, 74; NDak State Univ, PhD(organ chem), 79. *Prof Exp:* Assoc res fel, Lawrence Berkeley Lab, Univ Calif, 79-81; scientist, 81, sr scientist, 81-87, res assoc, 87-91, SR RES ASSOC, PARKE DAVIS/WARNER LAMBERT RES, WARNER LAMBERT CO, 91- *Mem:* Am Chem Soc; Sigma Xi; AAAS. *Res:* Quinolone antibacterials; cognition activators. *Mailing Add:* 4138 Sunset Ct Ann Arbor MI 48103

KIELY, LAWRENCE J, b Truxton, NY, Feb 26, 22; m 51; c 9. ANATOMY, PHYSIOLOGY. *Educ:* Niagara Univ, BS, 43, MA, 47; Columbia Univ, MA, 49, EdD(biol), 51. *Prof Exp:* From instr to assoc prof biol, 49-60, prof sci educ, 60-74, prof biol, 60-75, CHMN BIOL, NIAGARA UNIV, 75- *Concurrent Pos:* Lectr, Rosary Hill Col, 55-, grant exp psychol, 64-65; res grant human physiol, Williams Col, 65; res grant marine biol & trop ecol, Univ PR, 67; res grant sci educ, Ithaca Col, 68; res grant hist biol, Ohio State Univ, 69. *Mem:* AAAS; Am Physiol Soc; Nat Sci Teachers Asn; Nat Asn Biol Teachers. *Res:* Human anatomy, physiology and biology; science education. *Mailing Add:* Dept Biol Niagara Univ NY 14109

KIELY, MICHAEL LAWRENCE, b Springfield, Ill, June 17, 38; m 67; c 2. ANATOMY. *Educ:* Lewis Col, BS, 60; Loyola Univ, MS, 64, PhD(anat), 67. *Prof Exp:* Res assoc, Stritch Sch Med, 67, asst prof, 67-76, ASSOC PROF ANAT, SCH DENT, LOYOLA UNIV, CHICAGO, 76- *Mem:* Am Asn Anatomists; Int Asn Dent Res; Am Asn Dent Res. *Res:* Gross and oral anatomy; influence of endocrines and magnesium deficiency on the tissues of the rat incisor; morphologic variations in the human skull and cranial soft tissue. *Mailing Add:* Anatomy Dept Loyola Dental Sch 2160 S First St Maywood IL 60153

KIEN, C LAWRENCE, b Oct 5, 46; m; c 2. STABLE ISOTOPIC TRACERS, PROTEIN METABOLISM. *Educ:* Univ Cincinnati, MD, 72; Mass Inst Technol, PhD(nutrit & biochem), 77. *Prof Exp:* C E Compton prof nutrit, Sch Med, Univ WVa, 84-87; PROF PEDIAT & DIR, DIV NUTRIT, OHIO STATE UNIV 87- *Concurrent Pos:* From asst prof to assoc prof pediat & biochem, Med Col Wis, Milwaukee, 77-84. *Honors & Awards:* Future Leaders Award, Nutrit Found. *Res:* Nutrition; study of protein and energy. *Mailing Add:* Children's Hosp Rm W209 700 Children's Dr Columbus OH 43205

KIENHOLZ, ELDON W, b Moscow, Idaho, May 27, 28; m 50; c 4. ANIMAL NUTRITION. *Educ:* Manchester Col, BS, 50; Wash State Univ, BS, 52, MS, 59; Univ Wis, PhD(biochem), 62. *Prof Exp:* Instr high sch, Wash, 52-54 & 55-57; asst poultry nutrit, Wash State Univ, 57-59; res asst poultry biochem, Univ Wis, 59-61, proj asst biochem, 61-62; from asst prof to prof poultry nutrit, Colo State Univ, 62-81, prof dept animal sci, 81-88; RETIRED. *Concurrent Pos:* Nutritionist, Human Nutrit Div, USDA, Beltsville, Md, 69-70; vis prof, Dept Animal Sci, Univ Ill, 76-77. *Mem:* Poultry Sci Asn. *Res:* Mineral nutrition and metabolism; processing of legume seeds; value of foodstuffs for poultry; pollution and animal wastes. *Mailing Add:* PO Box 8 Windsor CO 80550

KIENTZ, MARVIN L, b Clovis, Calif, Jan 28, 36; m 84; c 2. BIOCHEMISTRY. *Educ:* Fresno State Col, BA, 58, MA, 61; Western Ont Univ, PhD(biochem), 66. *Prof Exp:* Teacher high sch, 59-63; res biochemist, Med Ctr, Univ Calif, San Francisco, 66-67; assoc prof, 67-74, PROF CHEM, SONOMA STATE UNIV, 74- *Concurrent Pos:* Inst dir, NSF, 71-73. *Mem:* Am Chem Soc. *Res:* Determination of the structure of proteins; protein polymorphism. *Mailing Add:* Dept Chem Sonoma State Univ Rohnert Park CA 94928

KIER, ANN B, b Littlefield, Tex, June 26, 49; m 79. PATHOLOGY, ANIMAL MEDICINE. *Educ:* Univ Tex, Austin, BA, 71; Tex A&M Univ, BS, 73, DVM, 74; Univ Mo, Columbia, PhD(path), 79; Am Col Lab Animal Med, dipl. *Prof Exp:* NIH fel, Lab Animal Med & Comp Path, Univ Mo, Columbia, 76-79, from asst prof to assoc prof path & microbiol, 84-87; assoc prof, 87-90, PROF PATH, UNIV CINCINNATI MED SCH, 90- *Concurrent Pos:* Dir, Histopath Lab, Vet Med Diag Lab, 80- *Mem:* Am Vet Med Asn; Am Asn Vet Med Col; Am Asn Lab Animal Sci. *Res:* Immunpathology, histopathology and oncogenicity of tumors; neutrophil and macrophage chemotaxis; development of a colony of coagulation factor XII deficient domestic cats; genetics; neutrophil and macrophage chemotaxis. *Mailing Add:* Dept Path Univ Cincinnati Sch Med 231 Bethesda Ave Co Cincinnati OH 45267-0529

KIER, LEMONT BURWELL, b Cleveland, Ohio, Sept 13, 30; m 53; c 5. MEDICINAL CHEMISTRY. *Educ:* Ohio State Univ, BS, 54; Univ Minn, PhD(med chem), 58. *Prof Exp:* Asst prof pharmaceut chem, Univ Fla, 58-63; assoc prof, Ohio State Univ, 63-66; sr med chemist, Columbus Labs, Battelle Mem Inst, 66-69, assoc fel med chem, 69-72; prof chem, Mass Col Pharm, 72-77; chmn, 77-87, PROF, DEPT PHARMACEUT CHEM, MED COL VA, VA COMMONWEALTH UNIV, 77- *Concurrent Pos:* Adj prof, Univ Mich, 69-72; chmn, Dept Chem, Mass Col Pharm, 72-74. *Mem:* Fel Am Asn Pharmaceut Scientists; Am Chem Soc; fel Acad Pharmaceut Sci. *Res:* Theoretical approaches to drug structure activity relationships. *Mailing Add:* Va Commonwealth Univ Box 540 MCV SAS Richmond VA 23298

KIER, PORTER MARTIN, invertebrate paleontology, for more information see previous edition

KIERAS, FRED J, CONNECTIVE TISSUE BIOCHEMISTRY, GENETIC DISEASES. *Educ:* Univ Chicago, PhD(biochem), 68. *Prof Exp:* RES SCIENTIST, NY INST BASIC RES, 71- *Mailing Add:* 414 E 65th St Apt 41 New York NY 10021

KIERBOW, JULIE VAN NOTE PARKER, b Fayetteville, Tenn, Feb 13, 25; wid. PHYSICAL CHEMISTRY, RADIOCHEMISTRY. *Educ:* Ohio State Univ, BS, 45; Univ Hawaii, MS, 48; Univ Colo, PhD(chem), 57. *Prof Exp:* Asst chem, Ohio State Univ, 44-45; res engr, Battelle Mem Inst, 45-46 & 48-50; asst, Univ Hawaii, 46-48; phys chemist, Redstone Arsenal Res Div, Rohm & Haas Co, 51-54; asst chem, Univ Colo, 54-56, res fel, 56-57; from asst prof to prof, 57-78, EMER PROF CHEM, CALIF STATE UNIV, LONG BEACH, 78- *Mem:* AAAS; Am Chem Soc. *Res:* Scintillation properties of solutions, particularly as related to structure of metal-organic compounds; use of radiotracers in development of analytical techniques. *Mailing Add:* 19038 Sombrero Circle Sun City AZ 85373-1417

KIERNAN, JOHN ALAN, b Kidderminstar, Eng, 1942; m 67; c 5. NEUROHISTOLOGY. *Educ:* Univ Birmingham, BSc, 63, MB & ChB, 66, PhD(neuroanat), 69; DSc, (anat) Univ Birmingham, Eng, 79. *Prof Exp:* House surgeon, East Birmingham Hosp, 66-67, res fel, Dept Anat, Univ Birmingham, Eng, 67-69; fel, Sidney Sussex Col, Cambridge, Eng, 69-72; asst prof, 72-75, assoc prof, 75-81, PROF, DEPT ANAT, UNIV WESTERN ONT, 81- *Concurrent Pos:* House physician, Worcester Royal Infirmary, 67; sr res fel, Sidney Sussex Col, Cambridge, 69-71, dir studies med, 70-72; univ demonstr, Dept Anat, Univ Cambridge, 71-72. *Mem:* Anat Soc Brit; Bot Soc Brit Isles; Histochem Soc; Soc Neurosci; fel Royal Microscopical Soc. *Res:* Reactions of nervous tissue to injury; histology and histochemistry; neuroanatomy. *Mailing Add:* Dept Anat Univ Western Ont London ON N6A 5C1 Can

KIERSCH, GEORGE ALFRED, b Lodi, Calif, Apr 15, 18; m 42; c 4. GEOLOGY. *Educ:* Colo Sch Mines, GE, 42; Univ Ariz, PhD(geol), 47. *Prof Exp:* Geologist, 79 Mining Co, Ariz, 46-47; instr, Mont Sch Mines, 47; geologist eng geol, CEngr, 48; geologist, Folsom Dam Proj, 49-50; supv geologist, Int Boundary & Water Comn, 50-51; asst prof geol, Univ Ariz, 51-55; asst chief explor, Southern Pac Co, 56-60; chmn dept, 65-71, prof, 60-78, EMER PROF GEOL SCI, CORNELL UNIV, 78- *Concurrent Pos:* Consult eng geologist, 53-; dir mineral resources surv, Navajo-Hopi Indian Reservations, 53-56; NSF sr fel, Tech Univ Vienna, 63-64. *Honors & Awards:* Holdredge Award, Asn Eng Geol, 65. *Mem:* Fel Geol Soc Am; Soc Econ Geol; fel Am Soc Civil Eng; hon mem Asn Eng Geologists; Int Asn Eng Geologists. *Res:* Engineering geology; nonmetallic mineral deposits; environmental geology; geomechanics; forensic geology; application of geology to planning, design and operation engineering works. *Mailing Add:* 4750 N Camino Luz Tucson AZ 85718

KIERSTEAD, HENRY ANDREW, thermodynamics; deceased, see previous edition for last biography

KIERSTEAD, RICHARD WIGHTMAN, b Fredericton, NB, Feb 17, 27; m 51, 64; c 3. ORGANIC CHEMISTRY. *Educ:* Univ NB, BSc, 48, MSc, 50; Univ London, PhD(org chem), 52. *Prof Exp:* Res assoc, Univ Toronto, 52-53 & Univ Calif, Los Angeles, 53-54; res chemist, E I du Pont de Nemours & Co, 54-55 & Harvard Univ, 55-56; sr res chemist, 56-63, group chief, 63-69, sect chief, 69-77, DIR MED CHEM, HOFFMANN-LA ROCHE, INC, 78- *Mem:* Am Chem Soc. *Res:* Synthesis of medicinal compounds. *Mailing Add:* Hoffmann-La Roche Inc Nutley NJ 07110

KIES, CONSTANCE, b Blue River, Wis, Dec 13, 34. NUTRITION SCIENCE. *Educ:* Wis State Univ, Platteville, BS, 55; Univ Wis, MS, 60, PhD(nutrit), 63. *Prof Exp:* Teacher high sch, Wis, 55-58; res asst human nutrit, Univ Wis, 60-63; from asst prof to assoc prof, 63-68, PROF HUMAN NUTRIT, UNIV NEBR, 68-,. *Concurrent Pos:* Univ res coun animal facility res grant, Univ Nebr, 65; NIH res grant, 64-69, Ross Lab grant, ADM, commodity bds, 70- *Honors & Awards:* Borden Award, Am Home Econ Asn, 73; Distinguished Serv Award, Agr & Food Chem, Am Chem Soc, 87. *Mem:* Inst Food Technologists; Am Home Econ Asn; Am Dietetic Asn; Am Chem Soc; Am Inst Nutrit; Am Soc Clin Nutrition. *Res:* Human nutrition; protein and amino acid requirements of adult humans; nutritive value of cereal and plant proteins; dietary factors as related to blood lipid components; fibre-mineral interrelationships; controlled human feeding studies. *Mailing Add:* Dept Nutrit Sci & Mgt Hospitality Univ Nebr 316 Ruth Leverton Hall Lincoln NE 68583

KIES, MARIAN WOOD, biochemistry, neurochemistry; deceased, see previous edition for last biography

KIESCHNICK, W(ILLIAM) F(REDERICK), b Dallas, Tex, Jan 5, 23; m 48, 79; c 2. CHEMICAL ENGINEERING. *Educ:* Rice Univ, BS, 47. *Prof Exp:* From jr engr to theoret oil reservoir engr, Atlantic Refining Co, Atlantic Richfield Co, 47-48, admin asst, Res Dept Admin, 48-49, sr chem engr, 49-51, supv engr, 51-54, head res sect, 54-59, asst to gen mgr explor, 59-61, mgr explor prod dist, 61-63, div mgr dists, 63-66, vpres & mgr cent region, Tex, 66-69, vpres synthetic crude & mineral opers, 69-70, vpres chem opers, 70-72, vpres & head corp planning, Atlantic Richfield Co, 72-73, exec vpres, 73-79, vchmn, 79, pres & chief opers officer, 81, pres & chief exec officer, 82-85. *Concurrent Pos:* Mem bd dirs, Atlantic Richfield Co, 73-; vchmn bd trustees, Calif Inst Technol; trustee, Carnegie Inst Wash; dir, Atlantic Richfield, TRW, First Interstate Bancorp, Pac Mutual Life Ins; chmn bd Coun Health, Safety & Environ; chmn, Biotech Group Inc. *Mem:* Am Inst Mining, Metall & Petrol Engrs; Am Inst Chem Engrs; Am Asn Petrol Geologists; Am Petrol Inst. *Mailing Add:* Atlantic Richfield Co 515 S Flower Los Angeles CA 90071

KIESEWETTER, WILLIAM BURNS, medicine; deceased, see previous edition for last biography

KIESLING, ERNST W(ILLIE), b Eola, Tex, Apr 8, 34; m 56; c 3. STRUCTURAL MECHANICS. *Educ:* Tex Tech Col, BS, 55; Mich State Univ, MS, 58, PhD(appl mech), 66. *Prof Exp:* From instr to asst prof civil eng, Tex Tech Col, 56-63; sr res engr, Struct Res Dept, Southwest Res Inst, 66-69; prof civil eng & chmn dept, 69-88, ASSOC DEAN RES, COL ENG, TEX TECH UNIV, 88-, DIR, CTR ADVAN RES & ENG, RES FOUND, 88- *Mem:* Am Soc Eng Educ; Am Soc Civil Engrs; Nat Soc Prof Engrs; Sigma Xi. *Res:* Housing; earth sheltered buildings; solar engineering. *Mailing Add:* Dept Civil Eng Tex Tech Univ Lubbock TX 79409

KIESLING, RICHARD LORIN, b Rockford, Ill, Nov 20, 22; m 47; c 4. PLANT PATHOLOGY. *Educ:* Univ Wis, BS, 49, MS, 51, PhD(plant path), 52. *Prof Exp:* From asst prof to assoc prof plant path, Mich State Univ, 52-60; PROF PLANT PATH & HEAD DEPT, NDAK STATE UNIV, 60- *Concurrent Pos:* Chmn, Nat Barley Improv Comt, 79-80. *Mem:* Am Phytopath Soc; Am Inst Biol Sci; Sigma Xi. *Res:* Nature of disease resistance; smut fungi and pathological histology; genetics of barley disease resistance. *Mailing Add:* Dept Plant Path NDak State Univ Fargo ND 58102

KIESO, ROBERT ALFRED, b Aurora, Illinois, Jan 15, 43; US citizen; m 81; c 2. CARDIOLOGY. *Educ:* Univ Ill, BA, 65, Ariz State Univ, MS, 68. *Prof Exp:* Microbiol, NASA, Planetary Quarantine Unit, 67-69; biochemist, Univ Iowa, 71-73; embryologist, Maplehurst Embryo Transplant, 73-75; RES ASST, DEPT INTERNAL MED, UNIV IOWA HOSP, 79- *Mem:* Am Heart Asn. *Res:* Conduct research in 2-D echocardiology high frequency echo and research involving defibilation. *Mailing Add:* 505 MRC Univ Iowa Cardiovasc Ctr Iowa City IA 52242

KIESS, EDWARD MARION, b Washington, DC, Mar 10, 33; m 59; c 3. PHYSICS. *Educ:* Mass Inst Technol, 55; Pa State Univ, MS, 62, PhD(physics), 65. *Prof Exp:* Asst prof physics, Lycoming Col, 65; physicist, Battelle Mem Inst, 65-67; ASSOC PROF PHYSICS, HAMPDEN-SYDNEY COL, 68- *Concurrent Pos:* Cottrell grant, 71- *Mem:* Optical Soc Am. *Res:* Brillouin scattering; Fourier spectroscopy. *Mailing Add:* Dept Physics Hampden-Sydney Col Hampden-Sydney VA 23943

KIESSLING, GEORGE ANTHONY, b New York, NY, Dec 2, 20; m 51; c 6. ELECTRICAL ENGINEERING. *Educ:* Manhattan Col, BSEE, 51; Stevens Inst Technol, MSIE, 53. *Prof Exp:* Mem tech staff, Physics Lab, Sylvania Electronics Prod, Inc, 51-54; adminr eng financial planning, RCA Corp, 54-55, mgr eng stand & serv, 56-61, staff engr, 61, mgr prod admin, 62-63, mgr prod eng prof develop, 63-69, dir prof eng serv, 69-70, dir prod safetyplans & progs, 70-85; RETIRED. *Mem:* Inst Elec & Electronics Engrs. *Res:* Product safety programs, including policy, management, systems, audit, standards, legislation and regulations; product ionizing and nonionizing radiation safety; laser safety including standards, legislation and regulations. *Mailing Add:* 421 Overhill Rd Haddonfield NJ 08033

KIESSLING, OSCAR EDWARD, b Jefferson, Wis, Apr 15, 01; m 28; c 5. MINERALOGY. *Educ:* Univ Wis, BA, 23, MA, 25; Brookings Inst, PhD(mineral econ), 27. *Prof Exp:* Mineral economist, US Bur Mines, 27-30, chief economist, Div Mineral Statist, 30-35, Div Div Mineral Prod & Econs, 35-39, chg mineral technol & output studies in coop with Works Prog Admin-Nat Res Proj, 36-39; chief mineral industs, US Bur Census, 39-41, chief basic mat, 42; secy, Mach & Allied Prod Inst, 43-47; spec indust adv, US Tariff Comn, 47-71; CONSULT & AUTHOR RESOURCE MGT, 71- *Concurrent Pos:* Mem, Harris Mem Found Round Table Conf Population Probs, 29; lectr, Am Univ, 30-31, 34-35 & 37-38; ed, Minerals Yearbook, 32-36; mem, Cent Statist Bd, US Dept Interior, 33-; spec consult, Off Prod Mgt, 41; econ analyst, 30. *Res:* Mineral economics and resources; economics of capital goods industries; cooperative development of oil pools. *Mailing Add:* 7048 Haycock Rd Falls Church VA 22043

KIEWIET DE JONGE, JOOST H A, b Leiden, Netherlands, Sept 4, 19; m 54; c 1. ASTRONOMY. *Educ:* Harvard Univ, PhD(astron), 54. *Prof Exp:* Instr, 50-54, asst prof, 54-65, assoc prof astron, Univ Pittsburgh, 66-, actg chmn, 70-75; ALLEGHENY OBSERV, PITTSBURGH. *Concurrent Pos:* Lectr, Chatham Col, 56-70; consult, J W Fecker Div, Am Optical, 60-70; actg dir, Allegheny Observ, 70-77, mem staff, 77- *Mem:* Am Astron Soc. *Res:* Astronomical navigation; stellar statistics; celestial mechanics; astronomical instrumentation. *Mailing Add:* Dept Physics 100 Allen Hall Univ Pittsburgh Main Campus Pittsburgh PA 15260

KIEWIT, DAVID ARNOLD, b Cincinnati, Ohio, Feb 25, 40; m 63; c 2. APPLIED PHYSICS, MATERIALS SCIENCE. *Educ:* Northwestern Univ, Evanston, BA, 62, PhD(mat sci), 68. *Prof Exp:* Mem tech staff, Hughes Res Labs, Hughes Aircraft Co, 67-73; mgr device develop, Gould Lab, Elec & Electronic Res, Gould Inc, 73-81; dir eng & advan develop, A C Nielsen Co, 81-90; CONSULT & REGIST PATENT AGENT, 90- *Honors & Awards:* Div Invention Awards, Hughes Aircraft Co, 70-73; Sci Achievement Award, Gould Inc, 78. *Mem:* Am Phys Soc; Inst Elec & Electronics Engrs. *Res:* Development of intelligent instrumentation systems for statistical measurements; application of pattern recognition and signal analysis methodologies to instrument development. *Mailing Add:* 2420 Seneca Palm Harbor FL 34683-2829

KIFER, EDWARD W, b Penn, Pa, Oct 31, 38; m 63; c 2. INORGANIC CHEMISTRY, PHYSICAL CHEMISTRY. *Educ:* Ind State Col, BSEd, 60; George Washington Univ, MS, 65; Carnegie-Mellon Univ, PhD(inorg chem), 69. *Prof Exp:* Teacher, Norwin Sch Syst, Pa, 60-63; asst chem, George Washington Univ, 63-65; asst res scientist phys chem, Res Dept, Koppers Co, Inc, 65-68, sr res scientist, 68-69, res group mgr physics & phys chem, 69-75, sr res group mgr physics & phys chem, Res Dept, 75-80, mgr res, Phenolic Prod Dept, 80-88; RETIRED. *Mem:* AAAS; Am Chem Soc; fel Am Inst Chemists. *Res:* Synthesis and characterization of compounds of the group IV elements, especially silicon hydrides; inorganic coordination polymer behavior; fire retardant systems for wood and plastics. *Mailing Add:* 106 Ardennes Ct Trafford PA 15146

KIFER, PAUL EDGAR, b Grove City, Pa, Aug 16, 24; m 50; c 3. FOOD SCIENCE. *Educ:* Mich State Univ, BS, 50, MS, 53, PhD(animal nutrit), 56. *Prof Exp:* Asst prof poultry, Univ Ga, 56-57; asst mgr poultry res div, Ralston Purina Co, 57-59, mgr, Spec Chows Res Div, 59-64, dir, Pet Food Res & Develop, 64-67, asst to vpres corp res, 67-69, asst dir corp res, 69-73; head dept food sci & technol, 73-83, assoc dean, int agr, 83-85, EMER PROF FOOD SCI & TECHNOL, ORE STATE UNIV, 85- *Mem:* Poultry Sci Asn; Inst Food Technologists; Am Inst Nutrit. *Res:* Nutrition as related to food science, food safety. *Mailing Add:* 3005 NW Maxine Circle Corvallis OR 97330-3721

KIFFNEY, GUSTIN THOMAS, JR, b New York, NY, May 17, 30; m 54; c 5. OPHTHALMOLOGY. *Educ:* Col Holy Cross, AB, 51; Albany Med Col, MD, 55; Am Bd Ophthal, dipl, 63. *Prof Exp:* Clin instr surg, 59-61, asst prof, 61-64, assoc prof, 64-81, CLIN ASSOC PROF OPHTHAL, SCH MED, UNIV NC, CHAPEL HILL, 81- *Concurrent Pos:* Asst attend surgeon, NC Mem Hosp, 61-64, assoc attend surgeon, 64-; attend surgeon ophthal, Watts Hosp, Durham, NC, 61- *Mem:* AMA; Am Acad Ophthal & Otolaryngol. *Res:* Experimental and clinical ocular pathology. *Mailing Add:* 67 Fennwood Lane Chapel Hill NC 27516-1607

KIGER, JOHN ANDREW, JR, b Dayton, Ohio, Feb 6, 41. GENETICS, BIOPHYSICS. *Educ:* Calif Inst Technol, BS, 63, PhD(biophys), 68. *Prof Exp:* Instr biol, Mass Inst Technol, 68-69; asst prof biochem, Ore State Univ, 69-73; asst prof, 73-76, assoc prof, 76-82, PROF GENETICS, UNIV CALIF, DAVIS, 82-, CHAIR, DEPT GENETICS, 87- *Concurrent Pos:* Am Cancer Soc fel, 69. *Mem:* AAAS; Genetics Soc Am; Soc Developmental Biol. *Res:* Biochemical and developmental genetics of Drosophila; cyclic nucleotide metabolism. *Mailing Add:* Dept Genetics Univ Calif Davis CA 95616

KIGER, ROBERT WILLIAM, b Washington, DC, Oct 4, 40; m 68; c 2. SYSTEMATIC BOTANY, HISTORY OF BIOLOGY. *Educ:* Tulane Univ, BA, 66; Univ Md, MA, 71, PhD(bot), 72. *Prof Exp:* Teacher elem schs, Montgomery County, Md, 66-67; res botanist, Smithsonian Inst, 72-73; dir develop, Sea Kal Develop Cor, 73-74; asst dir & sr res scientist, 74-77, DIR & PRIN RES SCIENTIST, HUNT INST BOT DOC, CARNEGIE MELLON UNIV, 77- *Concurrent Pos:* Res assoc, bot sect, Carnegie Mus Nat Hist, 78-; adj prof hist sci, Dept Hist, Carnegie Mellon Univ, 79-, adj prof, Dept Biol, 84- *Mem:* Linnean Soc London; Bot Soc Am; Am Soc Plant Taxonomists; Soc Study Evolution; Int Asn Plant Taxon. *Res:* Floristic and monographic study of various New World angiosperms, especially Flacourtiaceae and Talinum; history, philosophy and theory of biological systematics; evolutionary philosophy, especially in relation to theory of evolutionary mechanism and change, and to systematics; flora of North America project; botanical documentation and databanking. *Mailing Add:* Hunt Inst Carnegie Mellon Univ Pittsburgh PA 15213

KIGGINS, EDWARD M, b Stamford, Conn, Mar 26, 29; m 58; c 2. MICROBIOLOGY. *Educ:* Univ Conn, AB, 52, MS, 54, PhD(microbiol), 58. *Prof Exp:* Asst animal dis, Univ Conn, 52-57; mem staff microbiol, Abbott Labs, 58-60, group leader, 61-66, sect head, 66-67, mgr animal health res, Amdal Co Div, 67-70, dir res & develop, Agr & Vet Prod Div, 70-75, dir & prod develop, Rhodia, Hess & Clark Div, 75-80; WITH DIAMOND SHAMROCK CORP, 80- *Mem:* AAAS; Am Soc Microbiol; US Animal Health Asn; Sigma Xi; NY Acad Sci. *Res:* Veterinary microbiology. *Mailing Add:* Lubrizol Enterprises Inc 29400 Lakeland Blvd Wickliff OH 44092

KIHARA, HAYATO, b Oakland, Calif, Feb 28, 22; m 50; c 2. BIOCHEMISTRY, BIOCHEMICAL GENETICS. *Educ:* Univ Tex, BS, 44; Univ Wis, MS, 51, PhD(biochem), 52. *Prof Exp:* Chemist, US Army, 406th Med Gen Lab, Japan, 46-48; asst biochem, Univ Wis, 48-51; res scientist, Biochem Inst, Univ Tex, 51-56; asst res biochemist, Univ Calif, 56-60; res specialist, Sonoma State Hosp, Eldridge, Calif, 60-63; chief res biochemist, Pac State Hosp, 63-73; assoc res biochemist, 73-75, prof biochem, 75-87, EMER PROF BIOCHEM, UNIV CALIF, LOS ANGELES, 87-, CHIEF, BIOCHEM LAB, MENT RETARDATION RES CTR, UNIV CALIF, LOS ANGELES-LANTERMAN STATE HOSP, 73- *Concurrent Pos:* Vis prof, Sch Med, Nihon Univ, Tokyo, 80. *Mem:* Am Chem Soc; Am Soc Biol Chem; Am Soc Human Genetics; Am Asn Ment Deficiency. *Res:* Biochemistry of inborn errors of metabolism; tissue culture; enzyme purification; diagnosis of genetic disorders; heterozygote identification; prenatal diagnosis. *Mailing Add:* Res Group Lanterman Develop Ctr Univ Calif Los Angeles PO Box 100-R Pomona CA 91769

KIHN, HARRY, b Tarnow, Austria, Jan 24, 12; nat US; m 37; c 2. ELECTRONICS. *Educ:* Cooper Union, BSEE, 34; Univ Pa, MSEE, 52. *Prof Exp:* Res engr, Hygrade Sylvania Corp, 35-37; chief engr, Polytherm, Inc, 37-38; res engr, Ferris Instrument Co, 38-39; res engr, RCA Corp, Camden, 39-42 & Princeton, NJ, 42-58, group head space systs res, 58-60, staff engr to vpres res & eng, 60-70, staff tech adv corp licensing, 71-75, sr tech adv patent opers, 75-77; PRES, KIHN ASSOCS, INC, 77- *Concurrent Pos:* Mem mat adv bd, Nat Acad Sci-Nat Res Coun, 60-; mem comm sr ed, Gov Comn Sci Technol. *Honors & Awards:* Centennial Award, Inst Elec & Electronics Engr. *Mem:* AAAS; Sigma Xi; fel Inst Elec & Electronics Engrs; NY Acad Sci; Am Defense Preparedness Asn; Am Mgt Asn; Nat Soc Prof Engrs. *Res:* Electronics and systems research; television communication theory and devices; electromagnetics; radar; infrared; space physics and instrumentation; solid state circuits and devices; integrated circuits; computer design; information processing systems; medical electronics; nuclear energy and waste isolation; industrial electronics; robotics. *Mailing Add:* 30 Green Ave Lawrenceville NJ 08648

KIILSGAARD, THOR H, b Honeyville, Utah, June 10, 19; m 46; c 3. ECONOMIC GEOLOGY. *Educ:* Univ Idaho, BS, 42; Univ Calif, MA, 49. *Prof Exp:* Jr geologist, State Bur Mines & Geol, Idaho, 46-47, asst geologist, 48-49, assoc geologist, 49-51; proj geologist, US Geol Surv, 51-54, staff asst, 54-59, commodity geologist, 59-69, chief br resources res, 63-69, dep asst chief geologist, 69-72, chief, US Geol Surv-Saudi Arabia Proj, 72-76, geologist-in-chg, US Geol Surv, Wash, 76-89; RETIRED. *Concurrent Pos:* Int Coop Admin consult, Peru, 58-59 & Bolivia, 59-60; US Geol Surv adv, Iran, 63, Sudan, 77 & Yemen, 79. *Mem:* Soc Econ Geol; Geol Soc Am; distinguished mem Am Inst Mining, Metall & Petrol Engrs; Northwest Mining Asn. *Res:* Challis 2 geologic map, Idaho; Hailey 2 geologic map, Idaho; trans-Challis fault system and its control on gold and silver deposits. *Mailing Add:* US Geol Surv 656 US Court House W 920 Riverside Spokane WA 99201

KIJEWSKI, LOUIS JOSEPH, b Philadelphia, Pa, Mar 20, 36; m 75; c 1. THEORETICAL PHYSICS. *Educ:* LaSalle Col, BA, 58; Columbia Univ, MA, 61; NY Univ, PhD(physics), 67. *Prof Exp:* Asst physics, Columbia Univ, 58-61; physicist, RCA Corp, Moorestown & Princeton, 61-64; res asst atomic physics, NY Univ, 64-67, assoc res scientist, 67-68; PROF PHYSICS, MONMOUTH COL, NJ, 68- *Mem:* Am Phys Soc; Math Asn Am. *Res:* Calculations for the energy of atomic systems using density matrices. *Mailing Add:* Dept Physics Monmouth Col West Long Branch NJ 07764

KIKKAWA, YUTAKA, PATHOLOGY. *Prof Exp:* PROF & CHMN, DEPT PATH, UNIV CALIF, IRVINE, 88- *Mailing Add:* Dept Path Med Sci I D440 Univ Calif Irvine CA 92717

KIKTA, EDWARD JOSEPH, JR, b Buffalo, NY, June 11, 48. ANALYTICAL CHEMISTRY, CHROMATOGRAPHY. *Educ:* State Univ NY, Buffalo, BA, 70, PhD(anal chem), 78; Canisius Col, MS, 72. *Prof Exp:* Anal res chemist & lab supvr, 76-80, MGR ANALYTICAL SERV, AGR CHEM GROUP, FMC CORP, 80- *Honors & Awards:* Chromatography Award, Am Chem Soc, 81. *Mem:* Am Chem Soc; Am Inst Chemists; AAAS; Am Soc Testing & Mat; NY Acad Sci. *Res:* Chromographic methods and systems development; application of hplc to high resolution and high sensitivity analysis; optimization of hplc and gc systems; new hplc bonded phases. *Mailing Add:* Agr Chem Div FMC Corp US Hwy 1 PO Box 8 Princeton NJ 08543-0008

KIKUCHI, CHIHIRO, physics, nuclear engineering; deceased, see previous edition for last biography

KIKUCHI, RYOICHI, b Osaka, Japan, Dec 25, 19; US citizen; m 43; c 2. STATISTICAL MECHANICS. *Educ:* Univ Tokyo, BS, 42, PhD(physics), 51. *Prof Exp:* Res assoc physics, Univ Tokyo, 45-50; res assoc, Div Indust Coop, Mass Inst Technol, 51-53; asst prof, Inst Study Metals, Univ Chicago, 53-55; res physicist, Armour Res Found, Ill Inst Technol, 55-56; assoc prof, Wayne State Univ, 56-58; mem tech staff, Hughes Res Labs, 58-59, sr staff physicist, 59-63, sr scientist, 63-85; res prof, Math Sci Dept, Univ Wash, 85-89; ADJ PROF, MAT SCI & ENG DEPT, UNIV CALIF, LOS ANGELES, 87- *Concurrent Pos:* Consult, Lawrence Radiation Lab, Univ Calif, 63-67; adj prof, Univ Calif, Los Angeles, 75-85; vis prof, Purdue Univ,

77-, Delft Tech Univ, Neth, 80 & 81; Alexander von Humboldt fel, Max-Planck-Inst, Düsseldorf, WGer, 85, 86 & 87. *Mem:* Am Phys Soc; Phys Soc Japan; Am Inst Mining, Metall & Petrol Engrs. *Res:* Statistical mechanics of cooperative phenomena, equilibrium, and irreversible solid state physics; physical metallurgy. *Mailing Add:* Mat Sci & Eng Dept Univ Calif Los Angelas Los Angeles CA 90024

KIKUCHI, SHINYA, b Kobe, Japan, May 5, 43. FUZZY SET THEORY APPLICATION, LOGISTICS. *Educ:* Hokkaido Univ, Japan, BS, 67, MS, 69; Univ Pa, PhD(transp), 74. *Prof Exp:* Assoc, Transp Develop Assocs, Seattle, Wash, 74-77; sr proj engr, Transp Systs Div, Gen Motors Corp, 77-79, staff asst, Logistics Opers, 79-82; asst prof, 82-87, ASSOC PROF, UNIV DEL, 88- *Concurrent Pos:* Dir, Del Transp Ctr, 89- *Mem:* Am Soc Civil Engrs; Inst Transp Engrs. *Res:* Analysis of transportation systems and application of operations research; urban public transportation systems design and operation; application of fuzzy set theory; routing and scheduling of transportation system; traffic engineering. *Mailing Add:* Civil Eng Dept Univ Del Newark DE 19716

KIKUDOME, GARY YOSHINORI, b Hakalau, Hawaii, Jan 2, 25. CYTOGENETICS. *Educ:* Univ Mich, BS, 49, MS, 50; Univ Ill, PhD(bot), 59. *Prof Exp:* Asst bot, Univ Hawaii, 50-52 & Univ Ill, 54-56; biologist maize cytogenetics, Oak Ridge Nat Lab, 58-59; asst prof, 59-68, ASSOC PROF MAIZE CYTOGENETICS, UNIV MO, COLUMBIA, 68- *Mem:* Genetics Soc Am; Am Genetics Asn; Sigma Xi. *Res:* Cytogenetics of maize. *Mailing Add:* Tucker Hall Univ Mo Columbia MO 65211

KILAMBI, RAJ VARAD, b India, Feb 1, 33; m 57; c 3. FISHERIES, ZOOLOGY. *Educ:* Univ Wash, PhD(fisheries), 65. *Prof Exp:* from asst prof to assoc prof, 66-71, PROF ZOOL, UNIV ARK, FAYETTEVILLE, 77- *Mem:* Am Fisheries Soc; Sigma Xi. *Res:* Fish biology and population dynamics; reservoir fisheries. *Mailing Add:* Dept Biol Sci Univ Ark Fayetteville AR 72701

KILAMBI, SRINVASACHARYULU, BARACH ALGEBRA, TOPOLOGY OF COMPLEX MANIFOLDS. *Educ:* Univ Paris, DSc(math), 62. *Prof Exp:* Asst prof, math, Univ Md, 62-65; from asst prof to assoc prof, 65-84, PROF, MATH, UNIV MONTREAL, 84- *Mem:* Am Math Soc. *Mailing Add:* Math Dept Univ Montreal Montreal PQ H3C 3J7 Can

KILB, RALPH WOLFGANG, b Chicago, Ill, Feb 7, 31; m 59; c 2. PLASMA PHYSICS, IONOSPHERIC PHYSICS. *Educ:* Univ Nebr, BS, 52; Harvard Univ, MA, 53, PhD(chem), 56. *Prof Exp:* Asst chem, Harvard Univ, 53; phys chemist, Res Labs, Gen Elec Co, 56-60, physicist, 60-69, Tech Mil Planning Opers, 69-71; physicist, 71-73, PLASMA PHYSICS GROUP LEADER, MISSION RES CORP, 73- *Concurrent Pos:* Fel, Am Phys Soc. *Mem:* Am Chem Soc; Am Phys Soc; Am Geophys Union; NY Acad Sci; AAAS. *Res:* Microwave spectroscopy of gases; properties and structure of high polymers; thermonuclear fusion; atmospheric physics; nuclear burst effects at high altitudes; magnetohydrodynamic code simulation; collisionless magnetohydrodynamics; plasma structure in ionoshere. *Mailing Add:* Mission Res Corp PO Drawer 719 Santa Barbara CA 93102-0719

KILBERG, MICHAEL STEVEN, b Nov 11, 51; m; c 2. REGULATION OF NUTRIENT TRANSPORT. *Educ:* Univ SDak, PhD(biochem), 77. *Prof Exp:* PROF & ASSOC CHMN BIOCHEM & MOLECULAR BIOL, UNIV FLA, 80- *Mem:* Am Phys Soc; Am Soc Biochem & Molecular Biol; Am Soc Cell Biol. *Res:* Regulation of hepatic amino acid transport and transcriptional control of gene expression by amino acids. *Mailing Add:* Dept Biochem & Molecular Biol Univ Fla Box J-245 Gainesville FL 32610

KILBOURN, BARRY T, b Burton-on-Trent, Eng, Jan 21, 39; m 64; c 2. LANTHANIDE ELEMENTS. *Educ:* Oxford Univ, BA, 63, PhD(chem), 65. *Prof Exp:* CIBA res fel, Eidg Tech Hochschule, Zurich, 66-67; researcher crystallog, Imperial Chem Indust, Runcorn, Eng, 67-73, develop mat, Belgium, 73-80; DEVELOP MAT, UNOCAL/MOLYCORP, WHITE PLAINS, NY, 81- *Mem:* Royal Soc Chem; Am Chem Soc; Am Ceramic Soc; Metall Soc; Electrochem Soc. *Res:* Technology and industrial applications of yttrium and the Lanthanide elements (the Rare Earths). *Mailing Add:* Molycorp Inc 709 Westchester Ave White Plains NY 10604

KILBOURN, JOAN PRISCILLA PAYNE, b Juneau, Alaska, June 15, 36; m 61; c 2. MEDICAL MICROBIOLOGY. *Educ:* Univ Ore, BS, 58, MS, 60; Ore State Univ, PhD(microbiol), 63. *Prof Exp:* Instr bact, Univ Ore, 63-66, pediat, Med Sch, 66-68; tutor sci & math, Portland Community Col, 68-71; microbiologist, Clin Path Lab, Vet Admin Hosp, 71-74; assoc dir, ICN Med Labs, 74-76; instr, Clackamas Community Col, 78; consult microbiologist, 78-84; PRES & LAB DIR, CONSULT CLIN & MICROBIOL LAB, INC, 84- *Concurrent Pos:* Am Cancer Soc res grant, 63-65; consult, Choice, Books for Col Libraries, 64-; tutor, Portland Community Col, 71-; instr, Med Sch, Univ Ore, 73-74; instr, Portland State Univ Cont Educ Div, 77; reviewer, AAAS Science Books & Films. *Mem:* AAAS; Am Soc Clin Pathologists; Am Soc Microbiol; Nat Registry Microbiologist; Am Bd Bioanalysis. *Res:* Use of a radiation resistant microorganism as a protectant and therapeutic agent from the lethal effects of irradiation; automation of clinical medical microbiology; bacterial flora of chronic lung diseases. *Mailing Add:* 3178 SW Fairmount Blvd Portland OR 97201

KILBOURNE, EDWIN DENNIS, b Buffalo, NY, July 10, 20; m 52; c 4. MEDICINE, PUBLIC HEALTH & EPIDEMIOLOGY. *Educ:* Cornell Univ, AB, 42, MD, 44. *Hon Degrees:* DSc, Rockefeller Univ, 86. *Prof Exp:* Intern med, NY Hosp, 44-45, asst resident, 45-46; asst resident physician, Hosp Rockefeller Inst, 48-51; assoc prof med & dir, Div Infectious Dis, Tulane Univ, 51-55; assoc prof pub health, Med Col, Cornell Univ, 55-61, prof, 61-69, dir, Div Virus Res, 55-69; prof microbiol & chmn dept, 69-86, DISTINGUISHED SERV PROF, MT SINAI SCH MED, 86- *Concurrent Pos:* NIH res career award, 61-68; asst, Rockefeller Inst, 48-51; mem, Comn Influenza, US Armed Forces Epidemiol Bd, 59-71; mem, infectious dis adv comt, Nat Inst Allergy & Infectious Dis, 69-73, chmn subcomt influenza, 71-74; Virol Task Force chmn, NIH & mem, adv comt immunization pract, CDC, 76-; lectr, Harvey Soc, 78. *Honors & Awards:* R E Dyer Lectureship Award, NIH, 73; Borden Award, Asn Am Med Cols, 74; Thomas Francis Jr Mem lectr, 76; Harry F Dowling Lect Award, 76; Henry Brainerd Mem Lectr, 76; Bernard A Briody Mem Lectr, 77; Harry M Rose Mem Lectr, 80. *Mem:* Nat Acad Sci; Am Acad Microbiol; Asn Am Physicians; Am Asn Immunologists; Am Epidemiol Soc; Am Soc Clin Invest; fel NY Acad Sci. *Res:* Influenza virus genetics; host determinants of viral virulence; viral genetics and immunology; experimental epidemiology; vaccines. *Mailing Add:* Dept Microbiol Mt Sinai Sch Med One Gustave Levy Pl New York NY 10029

KILBOURNE, EDWIN MICHAEL, b New Orleans, La, Oct 1, 53; m 82; c 2. ENVIRONMENTAL EPIDEMIOLOGY, TOXICOLOGY. *Educ:* Cornell Univ, AB, 74, MD, 78; Am Bd Internal Med, cert, 84; Am Bd Prev Med, cert, 88. *Prof Exp:* Resident internal med, Univ Ala Hosps, Birmingham, 78-80 & 82-83; epidemic intel serv officer, Ctr Dis Control, Atlanta, 80-82, chief, invests sect, Spec Studies Br, Ctr Environ Health, 83-85, med epidemiologist, Foreign Assignment, 85-87, med epidemiologist, Ctr Environ Health & Injury Control, 87-89, chief, Health Studies Br, Div Environ Hazards & Health Effects, Ctr Environ Health & Injury Control, 89-90, ASST DIR SCI, EPIDEMIOL PROG OFF, CTR DIS CONTROL, 90- *Honors & Awards:* Alexander D Longmuir Award, 82. *Mem:* Am Acad Clin Toxicol; Am Pub Health Asn; fel Am Col Physicians; fel Am Col Prev Med. *Res:* Epidemiologic research regarding the adverse effects on human health of physical and chemical environmental agents; effects of heat and cold and the newly discovered toxic-oil syndrome in Spain; eosinophilia-myalgia sydrome (EMS) in the United States. *Mailing Add:* Ctr Dis Control 1600 Clifton Rd NE Mailstop F-28 Atlanta GA 30333

KILBURN, KAYE HATCH, b Logan, Utah, Sept 20, 31; m 54; c 3. INTERNAL & PREVENTIVE MEDICINE. *Educ:* Univ Utah, BS, 51, MD, 54. *Prof Exp:* Intern med, Univ Hosps, Cleveland, 54-55; resident, Univ Utah Hosps, 55-57; asst prof med, Wash Univ, 61-62; from assoc prof to prof, Med Ctr, Duke Univ, 62-73, asst prof anat, 68-73, dir, Div Environ Med, 70-73; prof med, assoc prof anat & dir, Div Pulmonary & Environ Med, Med Ctr, Univ Mo-Columbia, 73-77; prof med & community Med, Mt Sinai Sch Med, City Univ New York, 77-80; RALPH EDGINTON PROF MED, UNIV SOUTHERN CALIF SCH MED, LOS ANGELES, 80- *Concurrent Pos:* Fel cardiopulmonary physiol, Duke Univ, 57-58; Am Trudeau Soc fel, 58; USPHS fel cardiol, Brompton Hosp, Univ London, 60-61; Nat Heart Inst res grant & USPHS training grants, 62-; chief med serv, Durham Vet Admin Hosp, 63-68; consult, Vet Admin Hosps, Durham, Fayetteville & Oteen, 68-73; pres, Workers Dis Detection Serv, 86- *Mem:* Am Fedn Clin Res; Am Thoracic Soc; APS; AAP; ASCB. *Res:* Pulmonary structure and function, particularly pulmonary circulation; early detection of dysfunction; respiratory failure causing cerebral and circulatory dysfunction; pulmonary responses to environmental and occupational agents; alveolar surfactant; structure and function of cilia, mechanism of inflammation; proteolytic enzymes and antienzymes; mediators of leukocyte response; experimental pathology of lung; epidemiology of occupational diseases, byssinosis, asbestosis, and effects of exposure to formaldehyde, trichlorolthylene, welding gases and fumes; neurobehavioral toxicology of solvents, formaldehydes and PCB's; methods for measuring neurological functions in populations, nuters toxic effects of chemicals associated with birth defects associated with metals, metal chelating agents and enzyme blockers, smoking cessation intervention in blue collar workers. *Mailing Add:* 3250 Mesaloa Lane Pasadena CA 91107

KILBY, JACK ST CLAIR, b Jefferson City, Mo, Nov 8, 23; m 48; c 2. ELECTRICAL ENGINEERING. *Educ:* Univ Ill, BS, 47; Univ Wis, MS, 50. *Hon Degrees:* DSc, Univ Miami, 82, Rochester Inst Technol, 86, Univ Ill, 88, Rensselaer Polytech Inst, 89, Univ Wis, 90. *Prof Exp:* Engr, Globe-Union, Inc, Wis, 47-58; from engr to asst vpres, Tex Instruments Inc, 58-70; CONSULT, 70- *Concurrent Pos:* Distinguished prof, Tex A&M Univ, 77-85. *Honors & Awards:* Sarnoff Award, Inst Elec & Electronics Engrs, 66 & Brunetti Award, 80; Ballantine Medal, Franklin Inst, 66; Hall Minuteman Trophy, Order of Daedalians, 66; Nat Medal Sci, 70; Zworykwin Medal, Nat Acad Eng, 75; Holley Medal, Am Soc Mech Engrs, 82 & Charles Stark Draper Prize, 90; Medal Honor, Inst Elec & Electronics Engrs, 83; Nat Medal Technol, 90. *Mem:* Nat Acad Eng; fel Inst Elec & Electronics Engrs. *Res:* Monolithic integrated circuits. *Mailing Add:* 6600 LBJ Freeway Suite 4155 Dallas TX 75240

KILDAL, HELGE, b Oslo, Norway, June 23, 42; US citizen; m 67; c 3. APPLIED PHYSICS. *Educ:* Norweg Inst Technol, sivil 67; Stanford Univ, 69, PhD(appl physics), 72. *Prof Exp:* Res asst appl physics, Stanford Univ, 72-74; mem staff, Lincoln Lab, Mass Inst Technol, 74-80; mem staff, Forsvarets Forsknings Inst, 80-81; sect head, Norweg Coun Sci & Indust Res, 81-88, DIV HEAD, CTR INDUST RES, 88-; PROF, UNIV OSLO, 90- *Mem:* Optical Soc Am; Inst Elec & Electronics Engrs. *Res:* Development and applications of infrared nonlinear materials; laser spectroscopy; integrated optics. *Mailing Add:* Ctr Indust Res PO Box 124 Blindern Oslo 3 0314 Norway

KILDAY, WARREN D, b Westminster, Calif, July 10, 29; m 58; c 3. ORGANIC CHEMISTRY. *Educ:* Fresno State Col, BA, 59; Wash State Univ, PhD(org chem), 64. *Prof Exp:* Asst prof org chem, 63-67, assoc prof chem, 67-74, PROF CHEM, PEPPERDINE UNIV, 74- *Mem:* Am Chem Soc; Sigma Xi. *Res:* Papain catalyzed reactions of acylated amino acids. *Mailing Add:* Pepperdine Univ Malibu CA 90265

KILDSIG, DANE OLIN, b Oshkosh, Wis, Aug 3, 35; m 58; c 2. PHYSICAL PHARMACY. *Educ:* Univ Wis, Madison, BS, 57, PhD(phys pharm), 65. *Prof Exp:* Res scientist, Wyeth Labs, 65-66; from asst prof to assoc prof, 66-75, assoc head dept, 81-85, PROF PHYS PHARM, PURDUE UNIV, 85-, HEAD DEPT INDUST & PHYS PHARM, 85- *Mem:* Am Asn Pharmaceut Scientists; fel Am Asn Pharmaceut Scientists. *Res:* Mechanism of dissolution of solids and drug binding to protein; drug targeting; liposomes, pulmonary drug delivery. *Mailing Add:* Sch Pharm Purdue Univ Lafayette IN 47907

KILEN, THOMAS CLARENCE, b Jackson, Minn, Jan 24, 33; m 65. PLANT GENETICS. *Educ:* Univ Wis, BS, 63, MS, 66, PhD(agron, genetics), 68. *Prof Exp:* PLANT RES GENETICIST, AGR RES SERV, USDA, 67- *Mem:* Am Soc Agron; Crop Sci Soc Am; Am Genetic Asn. *Res:* Inheritance of disease resistance in soybeans; inheritance of characters modifying plant type and their effect upon seed yield; genetics of resistance to foliar feeding insects of soybeans. *Mailing Add:* PO Box 196 Agr Res Serv USDA Stoneville MS 38776

KILEY, CHARLES WALTER, b Staten Island, NY, Feb 25, 44; m 70; c 2. ICHTHYOLOGY. *Educ:* Wagner Col, BS, 66; NY Univ, MS, 69, PhD(biol, ichthyol), 73, DDS, 80. *Prof Exp:* Instr biol, 69-74, ASST PROF BIOL, WAGNER COL, 74-, CHMN DEPT, 75- *Concurrent Pos:* Adj asst prof, Staten Island Community Col, 74-75, St John's Univ, 76- & Int Med Educ, 76-; adj teaching staff, Staten Island Hosp, 81-; dir dent, NY Shipping-Port Police Union, 85- *Mem:* Am Inst Biol Sci; AAAS; Am Fisheries Soc; Am Soc Ichthyologists & Herpetologists; NY Acad Sci. *Res:* Histology and ultrastructure of immunocompetent organs in fishes; field collection and identification of fishes. *Mailing Add:* 2306 Redwood Rd Scotch Plains NJ 07076-2116

KILEY, JAMES P, b Medford, Mass, Apr 16, 52. RESPIRATORY PHYSIOLOGY, NEUROPHYSIOLOGY. *Educ:* Kans State Univ, PhD(physiol), 82. *Prof Exp:* Fel physiol, Univ NC, Chapel Hill, 82-84; HEALTH SCI ADMINR, NAT HEART, LUNG & BLOOD INST, NIH, 84- *Mem:* Am Physiol Soc; Sigma Xi. *Mailing Add:* Five Sanders Ct Gaithersburg MD 20877

KILEY, JOHN EDMUND, b New York, NY, Mar 22, 20; m 45; c 5. MEDICINE. *Educ:* Rennselaer Polytech Inst, BS, 42; Harvard Med Sch, MD, 45. *Prof Exp:* From instr to prof, Albany Med Col, 52-78; prof nephrology, Med Ctr, Univ Miss, 78-85; CONSULT, 85- *Concurrent Pos:* Fel physiol, Albany Med Col, 48-49. *Mem:* Fel Am Col Physicians; Am Fedn Clin Res. *Res:* Extracorporeal vividialysis and renal disease. *Mailing Add:* 39 Avery Circle Jackson MS 39211-2403

KILEY, LEO AUSTIN, b Boston, Mass, May 22, 18; m 44; c 2. NUCLEAR CHEMISTRY. *Educ:* Mass Inst Technol, SB, 39; Ohio State Univ, PhD(chem), 52. *Hon Degrees:* LLD, NMex State Univ, 67. *Prof Exp:* Dep dir, Atomic Warfare Directorate, Air Force Cambridge Res Ctr, 53-54, dep chief biophys div, Spec Weapons Ctr, 55-56, chief, 57-58, tech dir, Weapons Effects Tests, Field Command, Defense Atomic Support Agency, 59-60, dir, 60-63, vcomdr, Air Force Cambridge Res Labs, 63-64, comdr, 64-65, Air Force Missile Develop Ctr, 65-68, Off Aerospace Res, 68-69; gen mgr, Neutron Devices Dept, Gen Elec Co, 69-78; VPRES, LOS ALAMOS TECH ASSOC, 79- *Res:* Environmental and physical sciences; nuclear physics. *Mailing Add:* 780 Camino Pinones Santa Fe NM 87501

KILGORE, BRUCE MOODY, b Los Angeles, Calif, Mar 26, 30; m 52; c 2. FOREST ECOLOGY, FIRE ECOLOGY. *Educ:* Univ Calif, Berkeley, AB, 52, PhD(zool), 68; Univ Okla, MA, 54. *Prof Exp:* Info asst, Nature Conserv, DC, 56-57; ed, Nat Parks Mag, Nat Parks Asn, 57-60; managing ed, Sierra Club Pub & ed, Sierra Club Bull, Sierra Club, Calif, 60-65; teaching asst zool, Univ Calif, Berkeley, 63-68; res biologist, Off Chief Scientist, Nat Park Serv, 68-72, assoc regional dir prof serv, Western Region, 72-81; res proj leader, Northern Forest Fire Lab, Intermountain Forest & Range Exp Sta, US Forest Serv, Montana, 81-85; CHIEF, DIV NAT RES & RES, WESTERN REGION, NAT PARK SERV, SAN FRANCISCO, 85- *Mem:* Ecol Soc Am; Wildlife Soc; Soc Am Foresters. *Res:* Fire ecology of giant sequoia-mixed conifer forests; crown fire potential; fire history frequency; impact of prescribed burning on vegetation, fuels, and breeding birds; role of fire in wilderness/parks; understory burning in pine-larch-fir forests. *Mailing Add:* Nat Park Serv W Reg 600 Harrison St Suite 600 San Franciso CA 94107

KILGORE, DELBERT LYLE, JR, b Hutchinson, Kans, Sept 28, 42; m 66; c 2. ENVIRONMENTAL PHYSIOLOGY, RESPIRATORY PHYSIOLOGY. *Educ:* Univ Kans, BA, 64, MA, 67, PhD(physiol, cell biol), 72. *Prof Exp:* Res assoc physiol, Duke Univ, 71-73; from asst prof to assoc prof, 73-83, PROF ZOOL, UNIV MONT, 83- *Concurrent Pos:* Vis adj prof, Col Vet Med, Kans State Univ, 82-83 & 84; vis scholar, Dept Med, Univ Calif at San Diego, La Jolla, 90. *Mem:* Sigma Xi; Am Physiol Soc; Am Soc Zoologists. *Res:* Respiratory adaptations of birds to extreme environments; physiology of temperature regulation, control of respiration, acid-base balance and water balance. *Mailing Add:* Div Biol Sci Univ Mont Missoula MT 59812

KILGORE, LEE A, b Levitt, Nebr, Aug, 05. ELECTRICAL ENGINEERING. *Educ:* Univ Nebr, BS, 27; Univ Pittsburgh, MS, 29. *Hon Degrees:* DEng, Univ Nebr, 56. *Prof Exp:* Dir eng, Westinghouse Elec Co, 29-70; CONSULT, 70- *Honors & Awards:* Lamme Medal, Inst Elec & Electronics Engrs, 52. *Mem:* Nat Acad Eng; fel Inst Elec & Electronics Engrs. *Mailing Add:* 3945 Sardis Rd Murraysville PA 15668

KILGORE, LOIS TAYLOR, b Feb 9, 22; US citizen; m 44; c 2. NUTRITION. *Educ:* Miss State Univ, BS, 55, MS, 63, PhD(physiol), 68. *Prof Exp:* Med technologist, Vicksburg Hosp, 43-44 & Scales Clin, 44-46; med technologist with Dr Hunt Cleveland, 46-48; jr chemist, Petrol Prod Lab, Motor Vehicle Controller, 53-55; asst home economist, 55-69, prof, 69-84, EMER PROF HOME ECON, MISS STATE UNIV, 84- *Mem:* AAAS; Home Econ Asn Am; Am Soc Clin Path; Am Inst Nutrit; Am Dietetic Asn. *Res:* Nutrition of pre-school children; interrelationships of nutrients; international nutrition. *Mailing Add:* Route 1 Box 440A Starkville MS 39759

KILGORE, WENDELL WARREN, b Greenfield, Mo, June 21, 29; m 52; c 3. TOXICOLOGY. *Educ:* Univ Calif, AB, 51, PhD(microbiol), 59. *Prof Exp:* Asst microbiol, Univ Calif, 55-58, microbiologist, 58-59; microbiologist, Stanford Res Inst, 59-60; chmn dept, 70-77, dir Food Protection & Toxicol Ctr, 70-78, PROF ENVIRON TOXICOL, UNIV CALIF, DAVIS, 60-,

PROF PHARMACOL, SCH MED, 78- *Concurrent Pos:* Chmn, Governor's Sci Adv Panel, 87- *Mem:* AAAS; Am Soc Microbiol; Am Chem Soc; Soc Toxicol. *Res:* Environmental toxicology; toxicology of pesticides; analysis and detection of pesticides; effect of pesticides on human health. *Mailing Add:* Dept Environ Toxicol Univ Calif Davis CA 95616

KILGOUR, GORDON LESLIE, b Vancouver, BC, Apr 24, 29; m 49; c 2. BIOCHEMISTRY. *Educ:* Univ BC, BA, 51, MSc, 53; Univ Wash, PhD(biochem), 56. *Prof Exp:* Jr res biochemist, Univ Calif, Berkeley, 56-57; asst prof biochem, Mich State Univ, 57-63; from assoc prof to prof, San Fernando Valley State Col, 63-68; head dept chem, 68-74, PROF CHEM, PORTLAND STATE COL, 68- *Mem:* Am Chem Soc. *Res:* Oxidative enzymes and coenzymes; chemistry and biochemistry of phosphorylated carbohydrate compounds. *Mailing Add:* 930 OBrien Rd Lake Oswigo OR 97034

KILHAM, PETER, limnology, geochemistry; deceased, see previous edition for last biography

KILHAM, SUSAN SOLTAU, b Duluth, Minn, Jan 22, 43; wid. ECOLOGY, AQUATIC BIOLOGY. *Educ:* Eckerd Col, BS, 65; Duke Univ, PhD(zool), 71. *Prof Exp:* Res assoc microbiol, Duke Univ, 70-72; guest investr oceanog, Woods Hole Oceanog Inst, 72 & 79-80; lectr, Univ Mich, Ann Arbor, 73, asst res scientist natural resources, 73-75, asst res scientist, 75-78, assoc res scientist biol, 79-90; ASSOC PROF, DREXEL UNIV, PHILADELPHIA, 91- *Concurrent Pos:* NSF oceanog trainee, 67-70; nat lectr, Phycological Soc Am, 85-87; vis scientist, Max Planck Inst Limnol, 87-88, Max Panck Soc scholar, 88; vis assoc prof, Univ Wash, 90- *Honors & Awards:* McArthur Alumnus Award, Eckerd Col, 77. *Mem:* Am Soc Limnol & Oceanog; Phycol Soc Am; AAAS; Int Soc Limnol; Int Asn Great Lakes Res. *Res:* Aquatic ecology; freshwater phytoplankton; algal physiology and ecology; population dynamics and competition. *Mailing Add:* Dept Biosci & Biotechnol Drexel Univ Philadelphia PA 19104

KILINC, ATTILA ISHAK, b Mersin, Turkey, Feb 15, 36; m 61; c 2. GEOCHEMISTRY. *Educ:* Istanbul Univ, BS, 60; Pa State Univ, MSc, 66, PhD(geol), 69. *Prof Exp:* Res asst geochem, Pa State Univ, 64-68; res assoc, Stanford Univ, 68-70; from asst prof to assoc prof, 70-79, head dept, 75-83, PROF GEOL, UNIV CINCINNATI, 79-, HEAD DEPT, 91- *Concurrent Pos:* Asst head dept geol, Univ Cincinnati, 72-75. *Mem:* Am Geophys Union; Sigma Xi. *Res:* Geochemistry of magmatic and hydrothermal systems at high temperatures and pressures. *Mailing Add:* Geol Dept 13 Univ of Cincinnati Cincinnati OH 45221-0013

KILKSON, HENN, b Tartu, Estonia, Dec 30, 30; nat US; m 63; c 2. CHEMICAL ENGINEERING, PHYSICAL CHEMISTRY. *Educ:* Univ Colo, BS, 52, MS, 54; Cornell Univ, PhD(chem eng), 57. *Prof Exp:* Res engr, Eastern Lab, 57-59, res engr, Eng Res Lab, 60-65, sr res engr, Eng Tech Lab, 65-66, res assoc, 66-72, res fel, 72-85, sr res fel, 85-90, FEL, E I DU PONT DE NEMOURS & CO, 90- *Concurrent Pos:* Ledership team mem, Eng Res & Develop, Du Pont. *Honors & Awards:* T H Chilton Award, Am Inst Chem Engrs, 89. *Mem:* Am Chem Soc; Am Inst Chem Engrs; NY Acad Sci. *Res:* Mathematical aspects of polymerizations; chemical kinetics; chemical reactor design and stability. *Mailing Add:* Five Birch Knoll Rd Northminster Wilmington DE 19810

KILKSON, REIN, b Tartu, Estonia, Aug 1, 27; nat US. BIOPHYSICS. *Educ:* Yale Univ, BS, 53, MS, 54, PhD(physics), 56. *Prof Exp:* Mem tech staff, Bell Tel Labs, Inc, 56-58; asst prof physics, Wayne State Univ, 58-59 & biophysics, Yale Univ, 59-66; guest researcher, Dept Med Physics, Karolinska Inst, Stockholm, 64-70; vis prof chem & physics, 70-72, PROF PHYSICS, UNIV ARIZ, 72-, PROF MICROBIOL, 76- *Mem:* AAAS; Biophys Soc; Am Soc Microbiol; Sigma Xi; Am Phys Soc. *Res:* Molecular biophysics; virus structure; macromolecular arrangements in cell organelles; molecular regulation; molecular evolution; theoretical biology and biophysics; biological systems; light scattering and nerve conduction. *Mailing Add:* Dept Physics Univ Ariz Tucson AZ 85721

KILLAM, ELEANOR, b Whitefield, NH, May 18, 33. MATHEMATICS. *Educ:* Univ NH, BS, 55, MS, 56; Yale Univ, PhD(math), 61. *Prof Exp:* from asst prof to assoc prof, 60-91, EMER PROF MATH, UNIV MASS, AMHERST, 91- *Mem:* Am Math Soc; Math Asn Am. *Res:* Ring theory; banach algebras; locally m-convex algebras. *Mailing Add:* 26 Valley Lane Amherst MA 01002

KILLAM, EVA KING, b New York, NY, Nov 16, 21; m 55; c 3. PHARMACOLOGY, NEUROPHARMACOLOGY. *Educ:* Sarah Lawrence Col, AB, 42; Mt Holyoke Col, AM, 44; Univ Ill, PhD(pharmacol), 53. *Prof Exp:* Instr biol, Sarah Lawrence Col, 44-46; instr biol, Albertus Magnus Col, 46-47; asst therapeut, Col Med, NY Univ, 47-48; pharmacologist & toxicologist, Army Chem Ctr, Md, 48-51; instr pharmacol, Univ Ill, 52-53; res pharmacologist, Univ Cal, Los Angeles, 53-59; res assoc pharmacol, Sch Med, Stanford Univ, 59-68; prof in residence, 68-78, PROF PHARMACOL, SCH MED, UNIV CALIF, DAVIS, 78- *Concurrent Pos:* Ed-in-chief, J Pharmacol & Exp Therapeut, 84; pres, Western Pharmacol Soc, 84. *Honors & Awards:* Abel Award, Soc Pharm Exp Therapist, 54. *Mem:* Am Soc Pharmacol & Exp Therapeut (pres 89-90); Am Col Neuropsychopharmacol (pres, 88); Soc Exp Biol & Med; Am Epilepsy Soc. *Res:* Neuropharmacology, especially central nervous system; influence of drugs on epilepsy; eletrophysiological correlations of behavior and influence of drugs thereon. *Mailing Add:* Dept Pharmacol Sch Med Univ Calif Davis CA 95616

KILLAM, EVERETT HERBERT, b Whitefield, NH, Feb 2, 38; m; c 2. RESEARCH & DEVELOPMENT OF WASTE DISPOSAL METHODS. *Educ:* Univ NH, BS, 61; Univ Wyo, MS, 65, PhD(civil eng), 73. *Prof Exp:* Stress engr, Boeing Co, 67-68 & Earl & Wright, 73-74; instr mech, Univ Wyo, 68-72; asst prof mech, Rose-Hulman Inst Technol, 74-77; mgr res & develop, Custodis Construct Co, 77-82, engr mgr, Custodis-Cotrell, 83-88; consult,

Waste Mgt, 88-90; CONSULT, FORENSIC ENG, 90- *Concurrent Pos:* Eng mgr indust chimneys, Custodis Construct Co, 82. *Mem:* Am Soc Civil Engrs. *Res:* Dynamic and thermal effects on large industrial chimneys; new cooling tower designs using theoretical and experimental data; seperation of trash into useful end products. *Mailing Add:* 14 Fieldstone Pl Flemington NJ 08822

KILLAM, KEITH FENTON, JR, b Hollywood, Fla, Mar 2, 27; m 55; c 3. PHARMACOLOGY. *Educ:* Tufts Col, BS, 48; Univ Ill, MS, 53, PhD(pharm), 54. *Prof Exp:* Res pharmacologist, Smith Kline & French Labs, 48-50 & 54-55; NIH sr res fel, Univ Calif, Los Angeles, 55-59; prof pharmacol, Sch Med, Stanford Univ, 59-68; chmn dept, 68-81, PROF PHARMACOL, SCH MED, UNIV CALIF, DAVIS, 68- *Concurrent Pos:* Res fel neurophys, Mass Gen Hosp, 58; mem psychopharmacol study sect, NIH, 58-62; study sect, comput res, NIH, 62-65; mem, Int Brain Res Orgn, Drug Abuse Res Rev Comt, Nat Inst Drug Abuse, 82-85, chmn, 83-85. *Mem:* AAAS; Am Soc Pharmacol & Exp Therapeut; Am Col Neuropsychopharmacol; Int Soc Primatology. *Res:* Neuropharmacology; physiological mechanisms in brain; correlation between brain electrical activity and behavior and effects of psychotropic agents; substance abuse. *Mailing Add:* Dept Pharmacol Univ Calif Sch Med Davis CA 95616

KILLEBREW, FLAVIUS CHARLES, b Canadian, Tex, Apr 2, 49; m 78; c 1. MAMMALOGY. *Educ:* WTex State Univ, BS, 70, MS, 72; PhD(zool), 76. *Prof Exp:* Instr zool, Univ Ark, 72-76; ASST PROF BIOL, WTEX STATE UNIV, 76- *Concurrent Pos:* Res asst grant, Killgore Comt, 71-72, res prof grant, 76-77; res asst grant-in-aid, Sigma Xi, 71-72; res asst, Univ Ark Mus, 74-76; res prof grant, US Dept Interior, 78-79. *Mem:* Sigma Xi (vpres, 77-78, pres, 78-79); Soc Study Amphibians & Reptiles; Herpetologists League; Am Soc Ichthyologists & Herpetologists. *Res:* Systematics and ecology of reptiles. *Mailing Add:* Dept Biol WTex State Univ Canyon TX 79016

KILLEEN, JOHN, b Guam, July 28, 25; m 50; c 6. PLASMA PHYSICS, MAGNETOHYDRODYNAMICS. *Educ:* Univ Calif, AB, 49, MA, 51, PhD(math), 55. *Prof Exp:* Mathematician, Radiation Lab, Univ Calif, 50-55, Bell Tel Labs, Inc, 55 & Inst Math Sci, NY Univ, 56; mathematician, Math Radiation Lab, 57-68, PROF APPL SCI, LAWRENCE LIVERMORE LAB, UNIV CALIF, 68-, DIR, NAT MAGNETIC FUSION ENERGY COMPUT CTR, 74- *Concurrent Pos:* Ed, J Comput Physics, 68- *Mem:* Am Math Soc; Am Phys Soc; Sigma Xi. *Res:* Mathematical physics; computation; computer applications to controlled thermonuclear research. *Mailing Add:* 1528 Campus Dr Berkeley CA 94708

KILLELEA, JOSEPH R(ICHARD), nuclear engineering; deceased, see previous edition for last biography

KILLEN, JOHN YOUNG, JR, b July 20, 49. MEDICAL RESEARCH. *Educ:* Kenyon Col, AB, 71; Tufts Univ, MD, 75. *Prof Exp:* Physician, Moses Taylor Hosp, Scranton, 77-80; sr investr, Clin Invest Br, Cancer Ther Eval Prog, Div Cancer Treat, Nat Cancer Inst, 80-81, head, Med Sect, 81-84, dep chief, Clin Invest Br & prog dir, Coop Clin Trials Group, 84- 86; med dir, Whitman-Walker Clin, Washington, DC, 86-87; DEP DIR, DIV AIDS, NAT INST ALLERGY & INFECTIOUS DIS, NIH, 87- *Concurrent Pos:* Asst clin prof med, Sch Med, George Washington Univ, 86-88. *Mem:* Am Col Physicians; Am Soc Clin Oncol; Int AIDS Soc; Am Asn Physicians Human Rights. *Res:* Medicine. *Mailing Add:* Div AIDS Nat Inst Allergy & Infectious Dis NIH Control Data Bldg Rm 249P 6003 Exec Blvd Rockville MD 20892

KILLEN, ROSEMARY MARGARET, b Denver, Colo; m 79; c 1. ASTRONOMY, PHYSICS. *Educ:* Rice Univ, PhD (space physics & astron), 87. *Prof Exp:* Analyst earth resources, Lockheed Electronics Co, 74-76; engr guidance & control mechs, McDonnell Douglas Corp, 76-80; res assoc planetary atmospheres, Rice Univ, 86-89; GODDARD SPACE FLIGHT CTR, 89- *Mailing Add:* Lab Extraterrestrial Physics Goddard Space Flight Ctr NASA Code 693 Greenbelt MD 20771

KILLGOAR, PAUL CHARLES, JR, b Boston, Mass, Aug 3, 46; m 69; c 2. ELASTOMERS. *Educ:* State Col, Bridgewater, BA, 68; Mich State Univ, PhD(chem), 72. *Prof Exp:* Scientist, Cabot Corp, 68; res assoc, Mich State Univ, 72; scientist, 72-90; MGR, FUELS & LUBRICANTS DEPT, FORD MOTOR CO, 90- *Mem:* Am Chem Soc; Sigma Xi; Soc Plastic Engs. *Res:* Physical properties of elastomers; dynamic mechanical and fatigue properties; use of elastomers in automotive applications; polymer processing. *Mailing Add:* 15455 Ashurst Rd Livonia MI 48154-2603

KILLGORE, CHARLES A, b Lisbon, La, Aug 19, 34; m 54; c 3. CHEMICAL ENGINEERING, NUCLEAR PHYSICS. *Educ:* La Polytech Inst, BS, 56, MS, 63. *Prof Exp:* Chem engr, Am Oil Co, Ark, 56; chemist, Claiborne Gasoline Co, La, 59; from instr to prof chem eng, La Tech Univ, 59-76, dir, Nuclear Ctr, 63-76, assoc dean eng & dir eng res, 72-76; PRES, KILLGORE'S, INC, RADIATION & ENG CONSULTS, 76- *Concurrent Pos:* NSF sci fac fel. *Honors & Awards:* Am Soc Eng Educ Award, 70. *Mem:* Am Inst Chem Engrs; Am Soc Eng Educ; Am Nuclear Soc; Health Physics Soc; Am Asn Physicists Med. *Res:* Industrial applications of radioactive isotopes. *Mailing Add:* 506100 Oaks Dr Ruston LA 71270

KILLIAN, CARL STANLEY, b Cleveland, Ohio, Apr 13, 33. CHEMICAL PATHOLOGY, DIAGNOSTIC IMMUNOLOGY. *Educ:* Daemen Col, BS, 73; State Univ NY, Buffalo, MA, 77, PhD(chem path), 81. *Prof Exp:* Res technologist, Corning Glass Works, NY, 59-61; technologist, Hemat, Blood Banking & Urinalysis, 62-64; sr technologist, Clin Chem, 64-76, CANCER RES SCIENTIST, DIAG IMMUNOL RES & BIOCHEM, ROSWELL PARK MEM INST, BUFFALO, NY, 76- *Concurrent Pos:* Lectr, Div Grad Sch, Roswell Park Mem Inst, State Univ NY, Buffalo, 79- *Res:* Methods of detection and application of biologic tumor markers in prostate cancer. *Mailing Add:* Roswell Park Mem Inst 666 Elm St Buffalo NY 14263-0002

KILLIAN, FREDERICK LUTHER, b Lancaster, Pa, May 31, 42; m 63; c 2. POLYMER CHEMISTRY. *Educ:* Franklin & Marshall Col, AB, 63; Northwestern Univ, Ill, PhD(org chem), 67. *Prof Exp:* from res chemist to sr res chemist, 67-81, SR FINANCIAL ADV, E I DU PONT DE NEMOURS & CO, 81- *Mem:* Am Chem Soc. *Res:* Physical-organic chemistry; reaction mechanisms; polymer synthesis and conversion to synthetic fibers; fiber structure; organic polymer chemistry. *Mailing Add:* 1138 Elderon Dr Wilmington DE 19808

KILLIAN, GARY JOSEPH, b Rockville Centre, NY, Dec 6, 45; m 65; c 3. REPRODUCTIVE PHYSIOLOGY. *Educ:* Kans State Univ, BS, 67, MS, 69; Pa State Univ, PhD(reprod biol), 73. *Prof Exp:* Asst prof anat physiol, Dept Biol, Pa State Univ, 73-74, from res asst to res assoc, 72-75, asst prof reprod physiol, Dept Dairy & Animal Sci, 75-76; asst prof, 76-79, assoc prof reprod physiol, Dept Biol Sci, Kent State Univ, 79-84; PROF ANIMAL SCI, PA STATE UNIV, 84- *Mem:* Soc Study Reprod; AAAS. *Res:* Male reproductive physiology, endocrine regulation and effects of contraceptives on epididymal physiology and sperm maturation; biology of spermatozoa and sperm capacitation. *Mailing Add:* Pa State Univ Dairy Breeding Res Ctr University Park PA 16802

KILLICK, KATHLEEN ANN, b Chicago, Ill, Jan 22, 42. MICROBIAL BIOCHEMISTRY. *Educ:* Ill Inst Technol, BS, 64, MS, 66, PhD(biochem), 69. *Prof Exp:* AEC res assoc biochem, Argonne Nat Lab, 69-70; NIH trainee develop biol, Boston Biomed Res Inst, 70-72, NIH spec fel, 72-74, staff scientist develop biol, 74-; FEL RES, DEPT PHYSIOL, TUFTS UNIV, 90- *Concurrent Pos:* Instr develop biol, Harvard Univ, 75-77. *Mem:* AAAS; Am Soc Microbiol; Am Chem Soc; Am Soc Biol Chemists; Sigma Xi. *Res:* Biochemical basis of cellular differentiation in the slime mold, Dictyostelium discoideum. *Mailing Add:* Dept Physiol Sch Med Tufts Univ 136 Harrison Ave Rm 720 M&V Boston MA 02111

KILLINGBECK, STANLEY, b Blackburn, Eng, May 20, 29; US citizen. PHYSICAL CHEMISTRY. *Educ:* Blackburn Tech Col, BS, 51; Cornell Univ, MS, 54, PhD(chem), 64. *Prof Exp:* Lab asst, Walpamur Paint Co, Eng, 46-52; asst tech off, Imp Chem Indust, Eng, 52-54; asst prof, 63-71, ASSOC PROF CHEM, CENT MO STATE COL, 71- *Res:* Analytical chemistry. *Mailing Add:* Dept Chem Cent Mo State Univ Warrensburg MO 64093

KILLINGER, DENNIS K, b Boone, Iowa, Sept 23, 45; m 69; c 2. LASER REMOTE SENSING, QUANTUM OPTICS. *Educ:* Univ Iowa, BA, 67; DePauw Univ, MA, 69; Univ Mich, PhD(physics), 78. *Prof Exp:* Res physicist, Naval Avionics Fac, 68-74; res assoc, physics, Univ Mich, 74-78; mem quantum electronics staff, Mass Inst Technol, 79-81, prog mgr laser remote sensing, Lincoln Lab, 81-87; PROF PHYSICS, UNIVS FLA, 87- *Mem:* Am Phys Soc; Optical Soc Am. *Res:* Physics of new optical and laser sources, quantum electronics and non-linear optical techniques with applications toward laser remote sensing. *Mailing Add:* Phy 114 Univ SFla Tampa FL 33620

KILLINGSWORTH, LAWRENCE MADISON, b Cuthbert, Ga, Mar 9, 46; m 66; c 1. CLINICAL CHEMISTRY, PATHOLOGY. *Educ:* Emory Univ, BS, 68; Univ Fla, PhD(path, clin chem), 73. *Prof Exp:* Asst prof med & path, Sch Med, Univ NC, Chapel Hill, 73-77, assoc dir clin chem, NC Mem Hosp, 73-77, assoc dir, Radioassay Lab, 74-76; DIR CLIN CHEM & IMMUNOL LABS, SACRED HEART MED CTR, 77- *Mem:* Am Asn Clin Chem; Asn Clin Scientists. *Res:* Application of light-scattering techniques to the measurement of immunochemical reactions; application of immunochemical techniques to the clinical chemistry laboratory; study of protein physiology in health and disease. *Mailing Add:* 627 W 15th Ave Spokane WA 99203

KILLINGSWORTH, R(OY) W(ILLIAM), b Headland, Ala, Apr 8, 25; m 50; c 2. CIVIL ENGINEERING, ENGINEERING MECHANICS. *Educ:* Univ Ala, BS, 48, MS, 56. *Prof Exp:* Asst col eng, Univ Ala, 48-49; field engr, Pressure Concrete Co, 49-50, vpres, 50-52; asst dean, Col Eng, Univ Ala, 54-56, from asst prof to prof eng, 56-67, from asst dean to assoc dean, Col Eng, 63-70, dir phys planning & facil, 70-85 prof civil eng, 67-90, sr consult engr new proj, 85-80, EMER PROF CIVIL ENG, UNIV ALA, 90- *Concurrent Pos:* Gen contractor, Ala, 48-49; consult, US Army, 63- *Mem:* AAAS; Am Soc Civil Engrs; Am Soc Eng Educ; Nat Soc Prof Engrs. *Res:* Properties of materials; soil mechanics; hydrology. *Mailing Add:* 731 Capstone Ct 13th St Tuscaloosa AL 35401

KILLION, JERALD JAY, b Wichita, Kans, Oct 4, 42; c 2. IMMUNOBIOLOGY, CANCER. *Educ:* Wichita State Univ, BS, 70, MS, 71; Univ Okla, PhD(biophys), 73. *Prof Exp:* Instr biophys, 73-74, asst prof, Dept Radiol Sci, Health Sci Ctr, Univ Okla, 74-78; mem fac, Sch Med, Oral Roberts Univ, 78-81, assoc prof, Dept Physiol, 81-87; ASSOC PROF DEPT CELL BIOL, MD ANDERSON CANCER CTR, UNIV TEX, 87- *Concurrent Pos:* Affil instr, Cancer Res Prog, Okla Med Res Found, 73-74, staff scientist, 74-75, asst mem, 75- *Mem:* Sigma Xi; Am Asn Cancer Res; Am Soc Cell Biol; Biophys Soc; Tissue Cult Asn. *Res:* Membrane properties of tumor cells and tumor cell subpopulations; the influence of biological and biochemical properties of tumor on the tumor-host relationship; antigenic topography of tumor cell membranes. *Mailing Add:* 1515 Holcombe Blvd Box 173 Houston TX 77030

KILLION, LAWRENCE EUGENE, b Ross, Tex, Mar 28, 24; m 48; c 2. LASERS, COMPUTER SCIENCE. *Educ:* Baylor Univ, BA, 44; Univ Ind, MS, 48; Washington Univ, PhD(physics), 55. *Prof Exp:* Asst dir, Blast & Shock Prog, Defense Atomic Support Agency, US Dept Defense, 51-52; Nuclear Prog, 55-56, tech asst to dir defense weapons effects tests, 56-58; chief, Nuclear Qual Assurance Agency, Albuquerque Opers Off, US AEC, 58-64, sci adv electronics, US Army Electronics Proving Ground, 64-65, chief scientist, 65-67, dep asst controller info systs, US AEC, DC, 67-69, spec asst to asst gen mgr, div mil appln, 69-73; dep asst dir res & develop, 73-74; dep asst dir laser & isotope separation technol, 74-75, actg asst dir laser & isotope separation technol, Div Mil Appln, Energy Res & Develop Admin, 75-77; DIR, LASER FUSION DIV, DEPT ENERGY, 77- *Mem:* Am Phys Soc; sr mem, Inst Elec & Electronics Engrs; Am Statist Asn. *Res:* Computer controls and computation; communications electronics; laser technology for laser fusion and laser isotope separation; electron and ion beam technology and application to fusion. *Mailing Add:* Off Inertial Fusion US Dept Energy Washington DC 20545

KILLPATRICK, JOSEPH E, b Hillsboro, Ill, Feb 15, 33; m 55; c 3. ELECTRICAL ENGINEERING, PHYSICS. *Educ:* Univ Ill, BS, 55. *Prof Exp:* Res engr, Aero Res, 55-59, sr res engr, MPG Res, 59-62, supvr optics, 62-67, sect head electro-optics, Systs & Res Ctr, 67-69, mem staff, Aeronaut Div, 69-74, RES MGR, SYSTS & RES CTR, HONEYWELL INC, 74- *Mem:* Optical Soc Am; Inst Elec & Electronics Engrs. *Res:* Electronic devices and circuitry; optical scanning and detection systems; horizon scanners; sun seekers; lasers; laser devices and systems; laser gyro; frequency stability and precision control of lasers. *Mailing Add:* Honeywell Inc Minneapolis MN 55413

KILMAN, JAMES WILLIAM, b Terre Haute, Ind, Jan 22, 31; m 68; c 3. CARDIOVASCULAR SURGERY, THORACIC SURGERY. *Educ:* Ind State Univ, BS, 56; Ind Univ, Indianapolis, MD, 60; Am Bd Surg, dipl 67; Am Bd Thoracic Surg, dipl, 67. *Prof Exp:* Intern, Med Ctr, Ind Univ, Indianapolis, 60-61, resident, 61-66; from asst prof to assoc prof surg, 66-73, dir thoracic surg div, 73-81, PROF SURG, MED COL, OHIO STATE UNIV, 73- *Concurrent Pos:* USPHS fel cardiovasc surg, Med Ctr, Ind Univ, Indianapolis, 63-64; consult, Vet Admin Hosp, Dayton, Ohio, 66-; attend surgeon, Ohio State Univ & Children's Hosp, Columbus, 66- *Mem:* Am Col Surg; Am Acad Pediat; Am Col Chest Physicians; Am Col Cardiol; Am Surg Asn. *Res:* Infant cardiopulmonary bypass and peripheral blood flow. *Mailing Add:* Dept Surg Col Med Ohio State Univ 410 W Tenth Ave Columbus OH 43210

KILP, GERALD R, b Carrollton, Mo, Sept 16, 31; m 56; c 3. PHYSICAL METALLURGY, CERAMICS ENGINEERING. *Educ:* Mo Valley Col, BS, 52; Iowa State Univ, PhD(metall), 57. *Prof Exp:* Asst metall, Ames Lab, AEC, 52-57; sr engr, Atomic Power Dept, Westinghouse Elec Corp, 57-62, supvry engr, Astro-Nuclear Lab, 62-68, mgr fuel eng, Nuclear Core Opers, 68-72 & Nuclear Fuel Div, 72-80, mgr mat interactions, 80-83, ADV ENG, NUCLEAR FUEL DIV, WESTINGHOUSE ELEC CORP, 83- *Honors & Awards:* Harlan J Anderson Award, Am Soc Testing & Mat, Award of Merit. *Mem:* Am Soc Testing & Mat. *Res:* Thermoelectric and thermionic materials; graphite nuclear reactor fuels; light water reactor fuels development; nuclear waste disposal. *Mailing Add:* 890 Fredericka Dr Bethel Park PA 15102

KILP, TOOMAS, b Charleroi, Belgium, Oct 24, 48; Can citizen. PHOTOPHYSICS, PHOTOCHEMISTRY. *Educ:* Univ Toronto, BSc, 74, MSc, 75, PhD(chem), 79. *Prof Exp:* Chemist, Inmont Can Ltd, 70-72; chemist, Lumonics Res Ltd, 78-79; asst prof, Univ Notre Dame, 79-83; consult, 83-88; MGR POLYMER CHEM & MICROS, ORTECH INT, 88- *Mem:* Am Chem Soc; Can Thermal Anal Soc. *Res:* Polymer photochemistry; photophysics including degradation; polymerization; photoconductivity; intramolecular energy migration; excited state complex formation. *Mailing Add:* ORTECH Int Sheridan Park 5395 Speakman Dr Mississauga ON L5K 1B3 Can

KILPATRICK, CHARLES WILLIAM, b Wichita Falls, Tex, June 10, 44; wid; c 3. EVOLUTION, MOLECULAR SYSTEMATICS. *Educ:* Midwestern Univ, BS, 68, MS, 69; NTex State Univ, PhD(zool), 73. *Prof Exp:* Instr human biol, Midwestern Univ, 69-70; vis asst prof, Dept Biol, St Lawrence Univ, 73-74; asst prof, 74-80, ASSOC PROF, DEPT ZOOL, UNIV VT, 80- *Concurrent Pos:* Vis prof, Dept Zool, Univ Fla, 80-81; res assoc, Fla State Mus, Univ Fla, 80-81. *Mem:* Am Soc Mammalogists; Soc Study Evolution; Soc Syst Zoologists; Am Genetic Asn. *Res:* Genetic changes and evolutionary processes associated with speciation; effects of isolation on the genetic structure of populations; evolutionary relationships based upon analysis of morphology, protein electrophovesis, restriction fragments, DNA-DNA hybridization and karyology of vertebrates. *Mailing Add:* Dept Zool Univ Vt 85 S Prospect St Burlington VT 05405

KILPATRICK, DANIEL LEE, b Los Angeles, Calif, Apr 2, 51; m 79. BIOCHEMISTRY. *Educ:* Univ Calif, San Diego, BA 74; Duke Univ PhD(biochem), 80. *Prof Exp:* Postdoctoral fel, Dept Physiol Chem & Pharmacol, Roche Inst Molecular Biol, 80-81, res assoc, 82-84; sr scientist, Unigene Labs, 81-82; staff scientist, 84-89, SR SCIENTIST, WORCESTER FOUND EXP BIOL, 89- *Concurrent Pos:* Lectr, Seventh Int Cong Endocrinol, Can, 84, Tufts Univ, 86 & 88, Univ Miami & Columbia Univ, 86, Harvard Univ, 87, Univ Mass, 87, 88 & 90 & Ore Regional Primate Ctr, 90; ad hoc reviewer, Biochem Endocrinol Study Sect, 89. *Mem:* AAAS; NY Acad Sci; Am Soc Neurochem; Am Soc Biol Chemists; Endocrine Soc. *Res:* Author of numerous publications in medical journals. *Mailing Add:* Worcester Found Exp Biol 222 Maple Ave Shrewsbury MA 01545

KILPATRICK, EARL BUDDY, b Burkburnett, Tex, June 21, 20; m 56; c 6. FISH BIOLOGY. *Educ:* Univ Okla, BS, 42, MS, 49, PhD(zool), 59. *Prof Exp:* Asst zool, Univ Okla, 46-49; from asst prof to assoc prof biol, Southeastern Okla State Univ, 49-62, prof & head dept, 62-83; RETIRED. *Concurrent Pos:* Dir, NSF res partic prog, 59-60. *Mem:* AAAS. *Res:* Cytology, seasonal gonadal cycles of the fresh-water fishes. *Mailing Add:* Dept Biol Sci Southeastern Okla State Univ Durant OK 74701

KILPATRICK, JEREMY, b Fairchild, Iowa, Sept 21, 35; m 62; c 2. MATHEMATICS EDUCATION RESEARCH, MATHEMATICS CURRICULUM. *Educ:* Univ Calif, Berkeley, AB, 56, MA, 60; Stanford Univ MS, 62 PhD(educ), 67. *Prof Exp:* Teaching asst math educ, Sch Math Study Group, Stanford Univ, 61-63, res asst, 62-67; from asst prof to assoc prof math, Teachers Col, Columbia Univ, 67-75; PROF MATH EDUC, UNIV GA, 75- *Concurrent Pos:* Supvr math interns, sec educ proj, Stanford Univ, 62-63; res assoc, Sec Sch Math Curric Improv Study, Teachers Col, 69-75; vis lectr, Univ Cambridge, Eng, 73-74; guest prof, Institute Für Didactik Der Mathematics, WGer, 76; distinguished visitor, NZ Asn Res Educ, 87; distinguished scholar, San Diego State Univ, 88. *Honors & Awards:* J Wilson Award, Res Coun for Diagnostic & Prescriptive Math, 86; Fulbright Sr Lectr, Spain, 89. *Mem:* Nat Coun Teachers Math; Math Asn Am; Am Educ Res Asn; Nat Coun Supvr Math; Nat Coun Measurement Educ. *Res:* Evaluations of math curricula; studies in testing and assessment; problem solving and mathematical abilities; editing translations of Soviet studies in mathematics education; surveys of research in mathematics education. *Mailing Add:* 227 Woodlawn Ave Athens GA 30606

KILPATRICK, JOHN MICHAEL, b Feb 2, 53; m; c 3. CELL BIOLOGY, CANCER RESEARCH. *Educ:* Med Univ SC, PhD(immunol & microbiol), 81. *Prof Exp:* Postdoctoral fel, Univ Ala Birmingham, 81-84; asst prof, 85-87, ADJ ASSOC PROF IMMUNOL & MICROBIOL & LAB DIR BIOTHERAPEUT, MED UNIV SC, 87- *Mem:* AAAS; NY Acad Sci; Am Asn Immunologists; Am Diabetes Asn. *Res:* Use of biologic response modifiers in cancer therapy. *Mailing Add:* Dept Med Univ Ala THT 437 UAB Sta Birmingham AL 35294

KILPATRICK, KERRY EDWARDS, b Baltimore, Md, Mar 17, 39; m 65; c 2. INDUSTRIAL ENGINEERING, OPERATIONS RESEARCH. *Educ:* Univ Mich, Ann Arbor, BSE(mech eng) & BSE(eng math), 61, MS, 67, PhD(indust eng), 70; Harvard Univ, MBA, 63. *Prof Exp:* Methods engr, Buick Motor Div, Gen Motors Corp, 63-65; res asst indust eng, Univ Mich, Ann Arbor, 67-70; PROF INDUST & SYSTS ENG, UNIV FLA, 70-, PROF HEALTH & HOSP ADMIN, COL BUS ADMIN, 71-, PROF COMMUNITY HEALTH & FAMILY MED, 72-, DIR, HEALTH SYSTS RES DIV, J HILLIS MILLER HEALTH CTR, 72-, DIR, CTR HEALTH POLICY RES, 81- *Concurrent Pos:* Consult, World Health Orgn & Vet Admin, 77- *Mem:* Inst Mgt Sci; Opers Res Soc Am; Am Inst Indust Engrs. *Res:* Industrial and systems engineering, analysis of production and health service systems; analysis of extended function auxiliaries in health care delivery systems; evaluation of international health services delivery programs. *Mailing Add:* Dept Health Policy & Educ Admin Campus Box 7400 Univ NC 1101 McGavran-Greenberg Bldg Chapel Hill NC 27599

KILPATRICK, S JAMES, JR, b Belfast, Northern Ireland, Apr 24, 31; nat US; m 56; c 2. BIOSTATISTICS, EPIDEMIOLOGY. *Educ:* Queen's Univ Belfast, BSc, 54, MSc, 57, PhD, 60. *Prof Exp:* Asst lectr med statist, Queen's Univ Belfast, 54-58, lectr, 58-61; lectr, Aberdeen Univ, 61-65, NIH fel statist, 62-63; chmn biostatist, 65-83, prof family pract, 79-83, PROF BIOSTATIST, MED COL VA, 65- *Concurrent Pos:* NIH fel statist, Iowa State Univ, 60-61; mem, Working Comt Drug Monitoring, Aberdeen Univ, 63-65; mem, Va Health Interview Coun, 77-79; chmn, Va Health Statist Adv Coun, 78-82; mem, Oral Biol Med Study Sect, NIH, 78-80, Indoor Air Pollution Adv Comt, 85- *Mem:* Biomet Soc; Int Epidemiol Asn; Royal Statist Soc. *Res:* Statistics of epidemiological studies; risk assessment methods. *Mailing Add:* Dept Biostatist Med Col Va Richmond VA 23298-0032

KILPPER, ROBERT WILLIAM, b Houston, Tex, Oct 21, 38; m 60; c 3. TOXICOLOGY, BIOMATHEMATICS. *Educ:* Univ Houston, BS, 61, MS, 63, PhD(biophys), 67. *Prof Exp:* Asst prof biomath, radiation biol & biophys, Sch Med & Dent, Univ Rochester, 67-75, assoc prof, 75-81; TOXICOLOGIST, XEROX CORP, 81- *Mem:* Soc Toxicol. *Res:* Mathematical model analysis of physiological systems, especially the mechanical behavior of some mammalian lung models and the compartmental distribution of various compounds. *Mailing Add:* Xerox Corp 800 Phillips Rd Rochester NY 14580

KILSHEIMER, JOHN ROBERT, b Mt Vernon, NY, Sept 21, 23; m 46; c 6. AGRICULTURAL CHEMISTRY. *Educ:* Col Holy Cross, BS, 44; Fordham Univ, MS, 48; Syracuse Univ, PhD(chem), 51. *Prof Exp:* Instr chem, Fordham prep sch, 46-47; asst, Syracuse Univ, 47-48; res chemist, Polymer Div, Union Carbide Chem Co, 50-54, org-agr, 54-61; supvr agr chem, Mobil Chem Co, 61-63, mgr org chem, 63-66; sr staff adv agr prods, Esso Res & Eng Co, 66-67, head pesticides res & develop, 67-71; vpres res & develop, O M Scott & Sons Co, 71-80, sr vpres res, mfg & sales, 80-82; CONSULT, 82- *Concurrent Pos:* Mem bd dirs, Nat Agr Chem Asn, 72-81, Fertilizer Inst, 77-80. *Mem:* Am Chem Soc; Nat Agr Chem Asn; Fertil Inst. *Res:* Pesticides; fertilizers; oxidation reactions; synthetic organics; vinyl polymers; turf research. *Mailing Add:* 4141 S Atlantic Ave New Smyrna Beach FL 32169

KILSHEIMER, SIDNEY ARTHUR, b New Rochelle, NY, Oct 19, 30; wid. ORGANIC CHEMISTRY. *Educ:* Wagner Col, BS, 52; NC State Col, MS, 54; Purdue Univ, PhD(chem), 59. *Prof Exp:* Asst chem, NC State Col, 52-54; asst, Purdue Univ, 54-56, instr, 56-58; from asst prof to assoc prof, 58-72, PROF CHEM, BUTLER UNIV, 72- *Mem:* Fel AAAS; Am Chem Soc; fel The Chem Soc; fel Am Inst Chemists. *Res:* Chemical reductions; natural products. *Mailing Add:* Dept Chem Butler Univ Indianapolis IN 46208

KIM, AGNES KYUNG-HEE, b Seoul, Korea, May 17, 37; US citizen; m 65; c 3. CLINICAL PATHOLOGY, HEMATOLOGY. *Educ:* Yonsei Univ, BS, 58, MD, 62. *Prof Exp:* Asst pathologist, St Barnabas Med Col, 69-70, Lenox Hill Hosp, 71; assoc pathologist, Vet Admin Hosp, WRoxbury, Mass, 71-73; ASSOC PATHOLOGIST, NEW ENG DEACONESS HOSP & NEW ENG BAPTIST HOSP, 73-; Dept of Pathology, Harvard Med Sch. *Mem:* Fel Col Am Pathologists; Am Soc Clin Pathologists. *Mailing Add:* New England Baptist Hosp 91 Parker Hill Ave Boston MA 02120

KIM, BENJAMIN K, b Seoul, Korea, Apr, 27, 33; US citizen; m 59; c 4. TRANSDERMAL DELIVERY SYSTEMS, CONTROLLED RELEASE DELIVERY SYSTEMS. *Educ:* Seoul Nat Univ, BS, 59; Kyung Hzz Univ, MS, 69; Univ Southern Calif, PhD(pharmaceut chem), 77. *Prof Exp:* Scientist, Nat Chem Lab, 59-61; dir res & develop, Seoul Pharmaceut Co, 61-65; sect head, NIH, 65-69; res scientist, Rochelle Labs, 70-76; tech dir, Rich Life Inc, 76-80; sr res scientist, G D Searle, 80-83; res & develop dir, Nelson Res, 83-84; RES & DEVELOP VPRES, PACO RES INC, 84- *Mem:* Am Pharmaceut Asn; Am Pharmaceut Scientist Asn; Am Chem Soc. *Res:* New drug delivery systems; oral, transdermal implant and nasal delivery systems. *Mailing Add:* 226 Middle Dr Toms River NJ 08753

KIM, BORIS FINCANNON, b Commerce, Ga, Nov 19, 38; m 62; c 2. MOLECULAR SPECTROSCOPY, LASERS. *Educ:* Johns Hopkins Univ, BES, 60, PhD(physics), 67. *Prof Exp:* PRIN PROF STAFF PHYSICIST, JOHNS HOPKINS UNIV APPL PHYSICS LAB, 69- *Concurrent Pos:* Instr physics, Evening Col, Johns Hopkins Univ, 70-; prin investr, Dept Health, Educ & Welfare Res Grant, 75-81. *Mem:* Am Phys Soc. *Res:* High resolution, low temperature optical spectroscopy, and electron spin resonance studies of the class of porphyrin compounds; high temperature superconductivity; computer vision. *Mailing Add:* Johns Hopkins Univ Appl Phys Lab Johns Hopkins Rd Laurel MD 20723

KIM, BYUNG C, b Pyongyang, Korea, Nov 2, 34; US citizen; m 60; c 5. PROCESS DEVELOPMENT, PROCESS DESIGN. *Educ:* Ripon Col, AB, 56; Mass Inst Technol, BS, 58, MS, 61. *Prof Exp:* Prin chem engr, 61-66, sr chem engr, 66-79, assoc sect mgr, 79-82, PROJ MGR, COLUMBUS DIV, BATTELLE MEM INST, 82- *Mem:* Am Inst Chem Engrs; Sigma Xi. *Res:* Fluidized bed combustion; biomass gasification; bioreactors; microencapsulation; hydrothermal process; separation processes; soil remediation. *Mailing Add:* Columbus Div Battelle Mem Inst 505 King Ave Columbus OH 43201

KIM, BYUNG SUK, b Korea, Mar 20, 42; US citizen; m 67; c 2. IMMUNOLOGY. *Educ:* Seoul Nat Univ, BS, 67; Va State Univ, MS, 69; Univ Ill, PhD(microbiol), 73. *Prof Exp:* Res assoc radiation biol, Atomic Energy Res Inst, Korea, 67-68; sr res technician genetics, Univ Chicago, 69-70; sr staff assoc, Columbia Univ, 73-74; res asst prof immunol, Univ Chicago, 74-76; asst prof, 76-81, ASSOC PROF IMMUNOL, NORTHWESTERN UNIV, CHICAGO, 81- *Concurrent Pos:* Mem, study sect, NIH, 85-89. *Mem:* Am Soc Microbiol; Am Asn Immunol; Sigma Xi. *Res:* Effects of hapten on antigen presentation; expression of T-cell receptor genes; mouse model for multiple sclerosis. *Mailing Add:* Dept Microbiol & Immunol Med Sch Northwestern Univ Chicago IL 60611

KIM, CARL STEPHEN, b Los Angeles, Calif, Sept 12, 43; m 70; c 2. BIOPHYSICS, COMPUTER SCIENCE GENERAL. *Educ:* Calif State Univ, Fresno, BS, 66; Univ Calif, Davis, PhD(biophysics), 72. *Prof Exp:* Res assoc chem, Cornell Univ, 72-74; jr res scientist biophys, Roswell Park Mem Inst, 74; sr res scientist biophys, 74-77, chief biomet lab, Toxicol Ctr, Div Labs & Res, 77-79, dir, Toxicol Inst, 79-81, DIR COMPUT SYST MGT & DATA PROCESSING, NY STATE DEPT HEALTH, 81- *Concurrent Pos:* NIH fel, 74; adj asst prof physics, State Univ NY Albany, 75- *Res:* Risk assessment, molecular toxicology; elucidation of structure and function of biologically important molecules by x-ray diffraction and molecular orbital theory. *Mailing Add:* NY State Dept Health Empire State Plaza Tower Albany NY 12237

KIM, CHANGHYUN, b Seoul, Korea, Jan 1, 61; m 87; c 1. ULSI DESIGN, SOLID STATE SEMICONDUCTOR. *Educ:* Seoul Nat Univ, Korea, BS, 82, MS, 84. *Prof Exp:* Researcher dynamic random access memory design, Samsung Semiconductor Tech Co, Ltd, 84-86, asst mgr, 86-88; mgr, MOS Device Develop, Samsung Electronics Co, Ltd, 89; RES ASST, UNIV MICH, ANN ARBOR, 89- *Mem:* Inst Elec & Electronics Engrs. *Res:* Dynamic random access memory from 64K bit to 16M bit dynamic random access memory; MOS device characterization for submicron devices and reliability problems; circuit design and BiCMOS process development for sensors. *Mailing Add:* 2445 Stone Dr Ann Arbor MI 48105

KIM, CHANG-SIK, b Korea, Mar 8, 39; Korean citizen; c 2. ANTENNAS, PASSIVE MICROWAVE COMPONENTS. *Educ:* Ohio State Univ, MSc, 80; Univ Miss, PhD(electromagnetics), 83. *Prof Exp:* Radar engr, Radio Res Inst, 83-84; PRIN ENGR, PRODELIN CORP, 84- *Mem:* Inst Elec & Electronics Engrs; Korean Scientists & Engrs Asn. *Res:* Research in electromagnetics including horn antenna, reflector antenna, microwave passive components. *Mailing Add:* Prodelin Corp PO Box 368 1700 NE Calde Dr Conover NC 28613

KIM, CHARLES WESLEY, b Nashville, Tenn, Mar 20, 26; m 56; c 1. IMMUNOLOGY, PARASITOLOGY. *Educ:* Univ Calif, BA, 49; Univ NC, MSPH, 52, PhD(parasitol, bact), 56. *Prof Exp:* Instr microbiol, NY Med Col, 56-59, asst prof, 59-64; assoc scientist, 65-68, scientist, 68-70, RES COLLABR, MED DEPT, BROOKHAVEN NAT LAB, 70-; assoc prof microbiol, Health Sci Ctr, State Univ NY Stony Brook, 70-87, assoc vprovost, Grad Sch, 81-83; PROF, MICROBIOL & MED, HEALTH SCI CTR, STATE UNIV NY STONY BROOK, 87- *Concurrent Pos:* La State Univ Trop Med fel, Cent Am, 58; USPHS fel, Argonne Nat Lab & Univ Chicago, 64-65; assoc dean, Sch Basic Health Sci, Health Sci Ctr, State Univ NY Stony Brook, 72-74; assoc dean, Grad Sch, State Univ NY Stony Brook, 74-81; pres, Int Comn Trichinellosis, 88-92. *Mem:* Am Asn Immunol; Am Soc Parasitol; Am Soc Microbiol; Am Soc Trop Med & Hyg; fel Royal Soc Trop Med & Hyg; Sigma Xi. *Res:* Immune response to parasites; mechanism of immunity to parasites, including Trichinella spiralis; Cryptosporiduim; pathogenesis and chemotherapy of cryptosporidiosis; chemotherapy of trichinellosis; complement activity of Trichinella spiralis. *Mailing Add:* Div Infectious Dis State Univ NY Health Sci Ctr Stony Brook NY 11794-8153

KIM, CHUNG SUL (SUE), b Seoul, Korea, Dec 21, 32; US citizen; m 57; c 2. POLYMER CONCRETES, PROPELLANT CHEMISTRY. *Educ:* Univ Ill, BS, 55; Cornell Univ, PhD(org chem), 60. *Prof Exp:* Proj leader polymer chem, Standard Oil Ohio, 59-62; res chemist, Ga Pac Corp, 63-65; sr chem specialist, Aerojet Solid Propulsion Co, 66-73; from asst prof to assoc prof, 73-81, PROF CHEM, CALIF STATE UNIV, SACRAMENTO, 81- *Concurrent Pos:* Consult, Aerojet Solid Propulsion Co, 74-86; consult, unsaturated polyesters, vinyl esters & polymer concretes, 86- *Mem:* Am Chem Soc; Soc Advan Mat & Process Eng; Soc Plastics Engrs. *Res:* Chemistry and behavior of polymers in composite; propellant binders; polymer synthesis; bonding (coupling) agents; polymer concretes. *Mailing Add:* Dept Chem Calif State Univ 6000 J St Sacramento CA 95819

KIM, CHUNG W, b Hiroshima, Japan, Jan 8, 34; m 60; c 1. THEORETICAL HIGH ENERGY PHYSICS, THEORETICAL NUCLEAR PHYSICS. *Educ:* Seoul Nat Univ, BS, 58; Univ Ind, PhD(physics), 63. *Prof Exp:* Res assoc physics, Univ Pa, 63-66; from asst prof to assoc prof, 66-73, PROF PHYSICS, JOHNS HOPKINS UNIV, 73- *Mem:* Fel Am Phys Soc. *Res:* Nuclear and elementary particle physics; cosmology. *Mailing Add:* Dept Physics Astron Johns Hopkins Univ Baltimore MD 21218

KIM, DAE MANN, b Seoul, Korea, Apr 22, 38; m 67; c 1. ATOMIC PHYSICS, QUANTUM ELECTRONICS. *Educ:* Seoul Nat Univ, BS, 60; Yale Univ, MS, 65, PhD(physics), 67. *Prof Exp:* Res assoc physics, Mass Inst Technol, 67-69, instr, 69-70; asst prof elec eng, Rice Univ, 70-74, assoc prof, 74-; AT DEPT APPL PHYSICS & ELEC ENG, ORE GRAD CTR, BEAVERTON; Dept of Elec-Computer Eng Rice Univ. *Mem:* Am Phys Soc. *Res:* Modelocked laser pulses and their detection processes; photorefractive phase holography, dye lasers, light scattering study. *Mailing Add:* Tektronix Inc MS 59-567 PO Box 500 Beaverton OR 97077

KIM, DONG YUN, b Korea, May 6, 29; Can citizen; m 62; c 3. ELEMENTARY PARTICLE PHYSICS. *Educ:* Univ Seoul, Korea, BSc, 53; Aachen Tech Univ, Ger, PhD(theoret physics), 62. *Prof Exp:* Fel, Radiation Lab, New York Univ, 62-63; asst prof physics, Dept Physics, Mont State Univ, 63-65; ASSOC PROF PHYSICS, DEPT PHYSICS, UNIV REGINA, SASK, 65- *Concurrent Pos:* Vis scientist at var orgn including: Int Ctr Theoret Physics, Trieste, Italy; Inst Theoret Physics, Univ Heidelberg, Ger; Theory Group, Stanford Univ; Theory Group, Fermi Nat Lab; Dept Appl Math & Theoret Physics, Univ Cambridge, Eng & Bohr Inst, Denmark, 71-79. *Mem:* Can Asn Physicists; Am Phys Soc. *Res:* Theoretical nuclear and elementary particle physics. *Mailing Add:* Dept Physics & Astron Univ Regina Regina SK S4S 0A2 Can

KIM, GIHO, b Seoul, Korea, May 15, 37; m 68; c 2. CLINICAL BIOCHEMISTRY. *Educ:* Simpson Col, BA, 63, Iowa State Univ, MS, 69, PhD(biochem), 71. *Prof Exp:* Res assoc biochem, Iowa State Univ, 71-72; RES ASSOC MED RES, DOWNSTATE MED CTR, 73- *Concurrent Pos:* Clin instr, Downstate Med Ctr, 76- *Mem:* Am Chem Soc. *Res:* Biochemical effect of environmental pollutants on intestinal transport. *Mailing Add:* 81 Morewood Oaks Port Washington NY 11050

KIM, HAN JOONG, b Seoul, Korea, Nov 3, 37; m 62; c 2. METALLURGY, CERAMICS ENGINEERING. *Educ:* Seoul Nat Univ, BS, 60; San Jose State Col, MS, 66; Lehigh Univ, PhD(metall), 69. *Prof Exp:* Mem tech staff electronic mat, GTE Labs, Inc, 69-76, MGR RES & DEVELOP, GIBSON ELEC, SUBSID GTE, 76- *Mem:* Am Ceramic Soc; Am Soc Metals; Am Inst Mining, Metall & Petrol Engrs. *Res:* Oxidation of metals; glass-metal sealing; crystal growth; electrical contact materials. *Mailing Add:* GTE 40 Sylvan Rd Waltham MA 02254

KIM, HAN-SEOB, b Seoul, Korea, Sept 5, 34; m 63; c 2. PATHOLOGY. *Educ:* Seoul Nat Univ, MD, 59, MS, 62, PhD (biochem), 68. *Prof Exp:* Asst biochem, Col Med, Seoul Nat Univ, 63-67, from instr to asst prof, 67-69; from instr to asst prof, 71-84, ASSOC PROF PATH, BAYLOR COL MED, 84- *Mem:* Am Asn Pathologists; Int Acad Path; Am Soc Clin Path. *Res:* Anatomic pathology relating to cardiovascular and general pathology. *Mailing Add:* Dept Path Baylor Col Med One Baylor Plaza Houston TX 77030

KIM, HARRY HI-SOO, b Taegu, Korea, Jan 23, 22; US citizen; m 47; c 3. PATHOLOGY. *Educ:* Yonsei Univ, Korea, MD, 45; Univ Pa, MSc, 59. *Prof Exp:* Res asst path, Children's Hosp, Philadelphia, Pa, 57-58; res assoc, Sch Med, Univ Wash & Children's Orthop Hosp, Philadelphia, Pa, 57-58; res assoc, Sch Med, Univ Wash & Children's Orthop Hosp, 58-59, clin instr, 60-61; asst prof, 61-67, ASSOC PROF PATH, NY MED COL, 67-, ASSOC PATHOLOGIST, FLOWER & FIFTH AVE HOSPS, 68- *Concurrent Pos:* Fel, Grad Sch Med, Univ Pa, 56-58; assoc pathologist, Metrop Hosp, NY, 60-67; pathologist, NIH maternal & child health prog, NY Med Col Unit, 61-67, consult pathologist, 67- *Mem:* Int Acad Path; fel Am Soc Clin Path; fel Am Col Path. *Res:* Pediatric-pathology; genetics. *Mailing Add:* NYMC Metro Hosp 1901 First Ave Rm 2A47 New York NY 10028

KIM, HEE JOONG, b Seoul, Korea, Feb 10, 34; US citizen; m 63; c 2. NUCLEAR PHYSICS. *Educ:* Case Inst Technol, MS, 59, PhD(physics), 62. *Prof Exp:* PHYSICIST, PHYSICS DIV, OAK RIDGE NAT LAB, 62- *Mem:* Am Phys Soc. *Res:* Experimental nuclear physics. *Mailing Add:* Physics Div Oak Ridge Nat Lab MS 6372 PO Box 2008 Bldg 6000 Oak Ridge TN 37830

KIM, HYEONG LAK, b Korea, Jan 18, 33; nat US; m 67; c 5. NATURAL PRODUCTS CHEMISTRY. *Educ:* Seoul Nat Univ, BS, 56; St Louis Univ, MS, 68; Tex A&M Univ, PhD(biochem), 70. *Prof Exp:* Asst prof, 79-89, RES CHEMIST, TEX AGR STA, TEX, A&M UNIV, 69-, MEM GRAD FAC, DEPT VET PHYSIOL & PHARMACOL, 75-, ASSOC PROF, 89- *Mem:* Am Chem Soc; AAAS; NY Acad Sci. *Res:* Chemical constituents of poisonous plants; naturally occurring toxicants in food chain; metabolism of toxicants. *Mailing Add:* 4002 Oaklawn Bryan TX 77801

KIM, HYONG KAP, b Chungup, Korea, May 29, 30; m 63; c 3. ELECTRICAL ENGINEERING. *Educ:* Chunpuk Nat Univ, Korea, BS, 55; Univ Pa, MS, 60, PhD(elec eng), 64. *Prof Exp:* Lectr elec eng, Chunpuk Nat Univ, Korea, 55-58; res engr, Gen Elec Co, Pa, 60-61; instr elec eng, Univ Pa, 61-64, fel, 64-65; from asst prof to assoc prof, 65-73, PROF ELEC ENG, UNIV MAN, 73- *Mem:* Sr mem Inst Elec & Electronics Engrs. *Res:* Electric network analysis and synthesis. *Mailing Add:* Dept Elec Eng Univ Man Winnipeg MB R3T 2N2 Can

KIM, HYUN DJU, b Ham Buk, Korea, Jan 4, 37; US citizen; m 69; c 2. PHYSIOLOGY, BIOCHEMISTRY. *Educ:* Duke Univ, AB, 62, PhD(physiol), 68. *Prof Exp:* Muscular Dystrophy Asn of Am fel physiol, Univ Calif, Los Angeles, 69-71; Alexander von Humboldt fel, Div Med Physiol,

Aachen Tech Univ, 71-72; assoc prof physiol, Univ Ariz, 72-80; prof pharmacol & physiol, Univ Ala, Birmingham, 80-; AT DEPT PHARMACOL, UNIV MO, COLUMBIA. *Mem:* Biophys Soc; Am Soc Cell Biol; Am Physiol Soc. *Res:* Membrane transport and energy metabolism in red blood cells; protein metabolism in muscle development. *Mailing Add:* Dept Pharmacol Univ Mo Columbia MO 65212

KIM, HYUNYONG, physical chemistry, quantum chemistry, for more information see previous edition

KIM, JAE HO, b Taegu, Korea, Dec 17, 35; US citizen; m 63; c 2. RADIOBIOLOGY, RADIOTHERAPY. *Educ:* Kyung-Pook Nat Univ, Korea, MD, 59; Univ Iowa, PhD(radiobiol), 63. *Prof Exp:* Res assoc, Sloan-Kettering Inst Cancer Res, 63-68, asst prof biophys, Sloan-Kettering Div, Cornell Univ, 67-68; intern med, Montefiore Hosp, Bronx, 68-69; resident radiother, Mem Hosp, New York, 69-72; from asst prof to assoc prof radiol, 72-79, ASSOC PROF BIOPHYS, MED COL, CORNELL UNIV, 72-, PROF RADIOL, 80-; assoc mem, Sloan-Kettering Inst, 72-; AT MEM HOSP, NEW YORK. *Concurrent Pos:* Asst attend, Mem Hosp, 72-75, assoc attend, 75-77, attend, 78- *Mem:* AMA; NY Acad Sci; Radiation Res Soc; Am Asn Cancer Res; Am Radium Soc. *Res:* Cellular radio and chemo-biology; metabolic studies occurring during the cell cycle of mammalian cells in vitro; effects of ionizing radiation, metabolic inhibitors, hyperthermia on nucleic acid metabolism and cell viability in various tumors in culture. *Mailing Add:* Mem Hosp 1275 York Ave New York NY 10021

KIM, JAE HOON, b Seoul, Korea, Jan 8, 52; m 79; c 1. PHOTOTONICS & OPTOELECTRONICS, COMPOUND SEMICONDUCTOR DEVICE & PHYSICS. *Educ:* Seoul Nat Univ, BS, 76, MS, 78; Univ Fla, PhD(elec eng), 87. *Prof Exp:* Teaching asst elec eng, Seoul Nat Univ, 76-78, Univ Minn, 82-83; instr elec eng, Korea Naval Acad, 78-80, dept head, 80-81; res asst elec eng, Univ Fla, 84-87; mem tech staff, Jet Propulsion Lab, 87-89, task mgr photonics, 89-91; PRIN STAFF ENGR PHOTONICS, BOEING HIGH TECHNOL CTR, 91- *Concurrent Pos:* Prin investr, Jet Propulsion Lab, 89-91. *Mem:* Inst Elec & Electronic Engrs; Am Phys Soc; Japan Soc Appl Physics; Korea Inst Elec Engrs. *Res:* Development of optoelectronic integrated circuits for applications to neural networks, optical signal processing, and optical communications; development of integrated optoelectronic receivers and transmitters for fiber-optic avionics networks. *Mailing Add:* Boeing High Technol Ctr 7J-56 PO Box 3999 Seattle WA 98124-2499

KIM, JAI BIN, b Seoul, Korea, May 17, 34; m 60; c 4. CIVIL ENGINEERING. *Educ:* Ore State Univ, BS, 59, MS, 60; Univ Md, PhD(civil eng), 65. *Prof Exp:* Chief hwy res engr, DC Dept Transp, 64-66; from asst prof to assoc prof civil eng, 66-77, PROF CIVIL ENG & CHMN DEPT, BUCKNELL UNIV, 77- *Concurrent Pos:* Res engr, Nat Bur Standards, 76-77. *Mem:* Am Soc Testing & Mat; Am Concrete Inst; Am Soc Civil Engrs; Sigma Xi. *Res:* Structural mechanics; engineering analysis; foundation engineering; nonlinear structural analysis; pile foundations; shallow excavations; pile caps. *Mailing Add:* Dept Civil Eng Bucknell Univ Lewisburg PA 17837

KIM, JAI SOO, b Korea, Nov 1, 25; nat US; m 52; c 4. ATMOSPHERIC PHYSICS. *Educ:* Seoul Nat Univ, BSc, 49; Univ Sask, MSc, 57, PhD(physics), 58. *Prof Exp:* Instr physics, Tonga Col, Pusan, 52-53; instr, Sung Kyun Kwan Univ, Korea, 53-54; asst prof, Clarkson Univ, 58-59; from asst prof to prof, Univ Idaho, 59-67; chmn dept, 69-76, PROF ATMOSPHERIC SCI & PHYSICS, STATE UNIV NY ALBANY, 67- *Concurrent Pos:* Rep, Univ Corp Atmospheric Res, State Univ NY Albany, 71-76; consult, US Army Res Off, 78-80, Battelle Mem Inst, 77-80, NY State Environ Conserv Dept & Environ One Corp, 76-82, Norlite Corp, 82-84, Korean Studies Prog, State Univ NY-Stony Brook, 83-85, Korean Antarctic Prog, 88-; investr grants, NSF, Air for Off Sci & Res, US Army Res Off, Off Naval Res, Environ Protection Agency & NY State Environ Conserv Dept, Korea Ocean Res & Develop Inst; vis prof, Advan Inst Sci & Tech, Seoul, Korea, 83. *Mem:* Am Asn Physics Teachers; Am Geophys Union. *Res:* Upper atmospheric physics; solar-terrestrial relations; plasma physics; magneto-hydrodynamics. *Mailing Add:* Dept Atmospheric Sci State Univ NY Albany NY 12222

KIM, JEAN BARTHOLOMEW, b Philadelphia, Pa, Oct 12, 40; m 65; c 2. ORGANIC CHEMISTRY. *Educ:* Eastern Baptist Col, BA, 61; Bryn Mawr Col, PhD(chem), 68. *Prof Exp:* Asst prof chem, Haverford Col, 67-68; res assoc chem, Drexel Univ, 68-70; PROF CHEM & HEAD DEPT, EASTERN COL, 70-, VPRES & ACAD DEAN, 80- *Mem:* AAAS; Am Chem Soc; Sigma Xi; Nat Sci Teachers Asn. *Res:* Electrophilic substitution on electronic properties of porphin derivatives. *Mailing Add:* 131 Waterloo Ave Berwyn PA 19312

KIM, JIN BAI, b Sangju, Korea, June 23, 21; m 45; c 1. ALGEBRA. *Educ:* Yonsei Univ, Korea, BS, 50; Univ Chicago, MS, 56; Va Polytech Inst, PhD(math), 66. *Prof Exp:* From instr to asst prof math, Yonsei Univ, Korea, 49-61; asst prof, Mich State Univ, 65-67; from asst prof to assoc prof, 67-76, PROF MATH, WVA UNIV, 76- *Mem:* Am Math Soc; Math Asn Am. *Res:* Algebraic semigroups; linear algebra; matrices; tensors and differential manifolds. *Mailing Add:* Dept Math B25 Eiesland Hall WVa Univ Morgantown WV 26505

KIM, JINCHOON, b Choon-Chun, Korea, Mar 5, 43; US citizen; m 70; c 3. NUCLEAR FUSION, PARTICLE ACCELERATOR. *Educ:* Seoul Nat Univ, BS, 65; Univ Calif, Berkeley, MS, 68, PhD(plasma physics), 71. *Prof Exp:* Physicist, Cyclotron Corp, 71-74; res staff, Oak Ridge Nat Lab, 74-80; PRIN SCIENTIST, GEN ATOMICS, 80- *Concurrent Pos:* Vis scientist, Jet Joint Undertaking, 88; adj fac, Mesa Col, 89- *Mem:* Am Phys Soc; Am Vacumm Soc; Korean Nuclear Soc; Inst Elec & Electronics Engrs. *Res:* Physics and technology of energetic neutral beam generation; ion sources and accelerators; plasma physics and diagnostics in tokamak devices; neutron disimetry and shielding; vacuum technology. *Mailing Add:* PO Box 85608 3550 General Atomics Ct San Diego CA 92121-1194

KIM, JOHN K, b Seoul, Korea, Oct 26, 37; US citizen; m 61; c 2. ELECTRICAL ENG, MATERIAL SCIENCE. *Educ:* Ohio Univ, BS, 63; Ohio State Univ, MS, 63, PhD(elec eng), 67. *Prof Exp:* Mem tech staff electronics, Lincoln Lab, Mass Inst Technol, 66-67; sr engr solid state, Semiconductor Group, Motorola, 67-68; head sect semiconductor eng, Micro State, 68-70, head dept semiconductor res & develop, Raytheon Co, 70-75; PRES, SUPERTEK CO, 75- *Mem:* Sigma Xi; Inst Elec & Electronics Engrs; Korean Sci Eng Asn. *Res:* Analysis and development of solid state devices for high efficiency and high power generation of microwave energy. *Mailing Add:* 2231 Colby Ave Los Angeles CA 90064

KIM, JONATHAN JANG-HO, b Kwang-Ju, Korea, June 11, 32; m 58; c 1. METALLURGICAL ENGINEERING. *Educ:* Seoul Nat Univ, BS, 55; Carnegie-Mellon Univ, MS, 61: Univ Okla, PhD(metall eng), 66. *Prof Exp:* Res engr, Sci Res Lab, Ministry of Nat Defense, Korea, 55-59; process design engr, Lummus Co, 65-67; sr proj engr, Ledgemont Lab, Kennecott Copper Corp, 67-75, staff metall engr, Lexington Develop Ctr, 75-80; MGR, PROCESS TECHNOL, CARBORUNDUM CO, 80- *Mem:* Am Inst Mining, Metall & Petrol Engrs; Am Ceramic Soc. *Res:* Smelting and refining of nonferrous metals; extractive metallurgy; ceramic manufacturing processes via fusion and sintering; ceramic grains synthesis. *Mailing Add:* 79 Brandywine Dr Williamsville NY 14221

KIM, JUHEE, b Osan, Korea, Sept 13, 35; m 68; c 2. MICROBIOLOGY. *Educ:* Seoul Nat Univ, BS, 58; Cornell Univ, MS, 62, PhD(food sci & microbiol), 66. *Prof Exp:* Researcher food & nutrit, Sci Res Inst, Ministry Defense, Seoul, Korea, 58-59; from asst prof to assoc prof, 66-75, PROF MICROBIOL, CALIF STATE UNIV, LONG BEACH, 75- *Concurrent Pos:* Vis investr, Scripps Inst Oceanog, 70-71 & Wood Hole Oceanog Inst, 73; environ specialist, Southern Calif Coastal Water Res Proj, 74. *Mem:* AAAS; Am Soc Microbiol; Am Chem Soc; NY Acad Sci. *Res:* Food and industrial microbiology; public health. *Mailing Add:* Dept Microbiol Calif State Univ 1250 Bellflower Blvd Long Beach CA 90840

KIM, KE CHUNG, b Seoul, Korea, Mar 7, 34; m 64; c 2. ENTOMOLOGY, SYSTEMATICS & COEVOLUTION. *Educ:* Seoul Nat Univ, BS, 56; Univ Mont, MA, 59; Univ Minn, PhD(entom), 64. *Prof Exp:* Res fel entom, Univ Minn, 64-67, res assoc, 67-68; from asst prof to assoc prof, 68-79, PROF ENTOM & CUR, PA STATE UNIV, UNIV PARK, 79-, DIR, CTR BIODIVERSITY RES, 88- *Concurrent Pos:* Consult, Smithsonian Inst, 64-67; Nat Inst Allergy & Infectious Dis res grants, 64-; Fulbright lectr & researcher, 75; vis scientist, Atomic Energy Res Inst, Seoul, 75-76; vis prof, Seoul Nat Univ, 75-76 & Univ Heidelberg, 76; rep-at-large, Asn Systs Collections, 79-81; chmn, Coun Systs & Soc, 81-87, Coun Appl Systs, Asn Systs Collections, 82-85, Sect A, Entom Soc Am, 85, Entom Col Network, 89-90 & Int Adv Coun Biosysts Serv in Entom, 85-; dir, Ctr for Biodiversity Res, Pa State Univ. *Mem:* AAAS; Entom Soc Am; Soc Syst Zool; Am Soc Parasitol; Soc Consult Biol. *Res:* Biosystematics of the dipterous family's imuliidae, Sphaeroceridae and Tephritidae; systematics and ecology of sucking lice; ecology and evolution of ectoparasites-mammalian host relationships; biodiversity research. *Mailing Add:* Frost Entom Mus Dept Entom Pa State Univ Univ Park PA 16802

KIM, KENNETH, b Honolulu, Hawaii, Dec 24, 31. MICROBIOLOGY. *Educ:* Univ Hawaii, BA, 53; Wash Univ, MS, 60, PhD(microbiol), 64. *Prof Exp:* Instr prev med, Univ Wash, 64-71; asst prof, 71-80, ASSOC PROF CLIN PATH, BACT DIV, MED SCH, UNIV ORE, 80- *Mem:* Am Soc Microbiol; NY Acad Sci. *Res:* Relationship of diphtheria toxin to the diphtheria phage-bacterium relationship and the mode of action of toxin on cultured cells; biology of Rubella virus. *Mailing Add:* Dept Clin Path Ore Health Sci Univ 3181 S W S Jackson Pk Rd Portland OR 97201

KIM, KEUN YOUNG, b Kaesong, Korea, May 29, 28; US citizen; m 58; c 2. CHEMICAL ENGINEERING, INORGANIC CHEMISTRY. *Educ:* Seoul Nat Univ, BS, 51; Univ Wis, MS, 56, PhD(chem eng), 59. *Prof Exp:* Res investr, Inst Sci Res & Technol, Korea, 50-53; res chem engr, 62-66, res specialist, 66-70, FEL, DEPT RES & DEVELOP, DETERGENT & PHOSPHATE DIV, MONSANTO INDUST CHEM CO, 70- *Mem:* Am Inst Chem Engrs; Am Chem Soc. *Res:* Phosphates product and process research; calcium phosphates, particularly dentifrices. *Mailing Add:* 237 Ladue Lake Dr St Louis MO 63141-7412

KIM, KI HANG, b Mundok, Pyongando, Korea, Aug 5, 36; US citizen; m 63; c 2. MATHEMATICS. *Educ:* Univ Southern Miss, BS, 60, MS, 61; George Washington Univ, MP, 70, PhD(math), 71. *Prof Exp:* Instr math, Univ Hartford, 61-66; lectr, George Washington Univ, 66-68; assoc prof, St Mary's Col, Md, 68-70, assoc prof, Pembroke State Univ, 70-73; prof math, 74-87, DISTINGUISHED PROF MATH, ALA STATE UNIV, 88- *Concurrent Pos:* Managing ed, Math Social Sci, 80-; vis prof math, Portugal Inst Physics & Math, 70-74, Univ Stuttgart, 78-79, Chinese Acad Sci, 83-84; assoc ed, Future Generations Comput Systs, 83-; assoc ed, Pure Math & Applns, 90- *Mem:* Am Math Soc. *Res:* Diophantine decidability; symbolic dynamics; decision and game theory. *Mailing Add:* Box 69 Ala State Univ Montgomery AL 36101-0271

KIM, KI HONG, b Seoul, Korea; US citizen; m 68; c 2. ANALYTICAL CHEMISTRY. *Educ:* Ohio State Univ, BS, 59; Ohio Univ, MS, 62; George Washington Univ, MPH, 73, PhD(chem), 76. *Prof Exp:* Res assoc biochem, Med Ctr, Univ Ky, 63-66; res chemist, Paktron Div, Ill Tool Works Inc, 66-73, proj leader, 73-78, mgr metal systs, Emcon Div, 78-80; mat mgr, Kyocerra Int Inc, San Diego, 81-84, tech adv, 84-85; CONSULT, 85-; TECH DIR ELECROMECH, SAMSUNG, 88- *Mem:* Am Chem Soc; Soc Appl Spectros; Am Ceramic Soc. *Res:* Spectrophotometric and photopolarographic studies of organo-metallic compounds; research and development of precious metal powder and electrode ink systems for ceramic chip capacitor. *Mailing Add:* 18195 Colonnades Pl San Diego CA 92128

KIM, KI HWAN, b Pyung-Yang, Korea, May 12, 46; US citizen; m 70; c 3. COMPUTER ASSISTED MOLECULAR DESIGN. *Educ:* Yonsei Univ, Seoul, Korea, BS, 69, MS, 71; Univ Kans, Lawrence, MS, 76, PhD(med chem), 81. *Prof Exp:* Res asst, Yonsei Univ, Seoul, Korea, 69-71,; res asst, chem dept, Pomona Col, 71-73, res assoc, 76-80; res & teaching asst, med chem dept, Univ Kans, 73-76; THEORET RES CHEMIST, ABBOTT LABS, 80- *Mem:* Am Chem Soc; Drug Info Soc; Korean Scientists & Engrs Asn Am. *Res:* Quantitative correlation of chemical structure and biological activity; conformational and quantum chemical calculations; application of computer graphics to drug design; mechanism of action of biologically active compounds. *Mailing Add:* Abbott Labs D-47E AP9 Abbott Park IL 60064

KIM, KI-HAN, b Seoul, Korea, June 20, 32; US citizen; m 58; c 2. BIOCHEMISTRY. *Educ:* Univ Calif, Berkeley, BA, 57; Wayne State Univ, PhD(chem), 61. *Prof Exp:* Res assoc biochem, Wayne State Univ, 61-65 & Univ Wis-Madison, 65-67; from asst prof to assoc prof, 67-73, PROF BIOCHEM, PURDUE UNIV, 73- *Mem:* AAAS; Am Soc Biol Chem. *Res:* Mechanism of hormone action; biochemical basis of cellular differentiation. *Mailing Add:* Dept Biochem Purdue Univ West Lafayette IN 47907

KIM, KI-HYON, b Ui ju City, Korea, Apr 20, 33; m 65; c 3. MEDICAL PHYSICS, NUCLEAR MEDICINE. *Educ:* Seoul Nat Univ, BSc, 56; Univ Vienna, PhD(nuclear physics), 63. *Prof Exp:* Res off, Atomic Energy Res Inst, Seoul, 63-66; assoc prof nuclear physics, 68-72, PROF APPL PHYSICS, NC CENT UNIV, 72- *Concurrent Pos:* Resident res assoc, NASA-Langley Res Ctr, 66-68, lectr-consult, 68-; consult, Oak Ridge Assoc Univs, 71- *Mem:* Am Phys Soc; Korean Phys Soc; Am Asn Physicists in Med; Soc Nuclear Med; Korean Scientists & Engrs Am. *Res:* Nuclear spectrometry; high-flux pulsed neutron sources; biomedical applications of nuclear electronics; organ visualization by means of radiopharmaceuticals and computer based scanning. *Mailing Add:* Dept Physics NC Cent Univ Durham NC 27707

KIM, KIL CHOL, b Hyokyong Hampook, Korea, Apr 22, 19; US citizen; m 65; c 4. ANESTHESIOLOGY, PHARMACOLOGY. *Educ:* Kyung-Pook Nat Univ, Korea, MD, 44; Loyola Univ, PhD(pharm), 62. *Prof Exp:* Intern med, Edgewater Hosp, 56-57; resident, Univ Chicago Clin, 57-58; resident, Cook County Hosp, 58-59; resident, Michael Reese Hosp, 59-60; sr investr pharm, Loyola Univ, 61-66; physician, Galesbury Res Hosp, 59-60; resident med, Albert Einstein Col Med, 68-69; ASST PROF MED, MED CTR, IND UNIV, INDIANAPOLIS, 69- *Mem:* AMA; assoc Am Soc Anesthesiol; Soc Neurosci. *Res:* Mechanism of contractility depression of cardiac muscle by anesthetics through interaction; investigation of drugs which reduce sleeping time of several anesthetics. *Mailing Add:* 1100 W Michigan St Indianapolis IN 46202

KIM, KI-SOO, b Korea, Nov 6, 42; m 70; c 3. POLYMER CHEMISTRY. *Educ:* Seoul Nat Univ, BS, 65; City Univ NY, PhD(chem), 72. *Prof Exp:* Res asst chem, Atomic Energy Res Inst Korea, 66-68; assoc, Res Inst, City Univ New York, 72-74; res assoc, Eastern Res Ctr, Stauffer Chem Co, 74-87; SR RES SCIENTIST, AKZO CHEM, 88- *Mem:* Am Chem Soc; Korean Chem Soc. *Res:* Condensation and vinyl polymerization; synthesis of phosphorus and sulfur-containing polymers; engineering plastics; adhesive formulation, materials research for microelectronics, liquid crystalline polymers. *Mailing Add:* Res Ctr Akzo Chemicals Inc Dobbs Ferry NY 10522

KIM, KWANG SHIN, b Seoul, Korea, Nov 15, 37; m 65; c 2. MEDICAL MICROBIOLOGY, HORTICULTURE. *Educ:* Seoul Nat Univ, BS, 59; Rutgers Univ, New Brunswick, MS, 63, PhD(hort), 67. *Prof Exp:* Res asst hort, Rutgers Univ, New Brunswick, 65-66, res assoc, 66-67; from asst res scientist to assoc res scientist microbiol, 67-71, asst prof, 71-75, ASSOC PROF MICROBIOL, SCH MED, MED CTR, NY UNIV, 75- *Mem:* AAAS; Am Soc Microbiol; Sigma Xi; NY Acad Sci. *Res:* Ultrastructure and cytochemistry of bacteria, especially human pathogenes and their interactions; plant breeding and embryology. *Mailing Add:* Dept Microbiol NY Univ Sch Med New York NY 10016

KIM, KYEKYOON K(EVIN), b Seoul, Korea, Oct 5, 41; US citizen; m 69; c 2. COMPOUND SEMICONDUCTOR MICROELECTRONICS, ELECTROHYDRODYNAMIC SPRAYING. *Educ:* Seoul Nat Univ, BS, 66; Cornell Univ, MS, 68, PhD(appl physics), 71. *Prof Exp:* Res asst nuclear sci & appl physics, Cornell Univ, 66-71, fel appl physics, 71-72; fel phys chem, 72-74, fel elec eng, 74-76, from asst prof to assoc prof, 76-85, PROF ELEC & COMPUT ENG, MECH & INDUST ENG & NUCLEAR ENG, 85- *Concurrent Pos:* Consult, Y Div, Lawrence Livermore Lab, 78-, NASA, 80-86, Elec Power Res Inst, 81-86, Argonne Nat Lab, 86- & Lab Laser Energetics, Univ Rochester, 86-; chmn, Comt Fueling Plasma Devices, Am Vacuum Soc, 81-86. *Mem:* Am Phys Soc; Inst Elec & Electronics Engrs; Am Vacuum Soc. *Res:* Fusion technology; plasma engineering; cryogenic laser fusion targets; electrohydrodynamics; monodispersed micro-particle generation from insulators; polymers and metals; compound semiconductor microelectronics. *Mailing Add:* Dept Elec & Comput Eng Univ Ill 1406 W Green St 155 Everit Lab Urbana IL 61801

KIM, KYUNG SOO, PLANT CELL VIRUS INTERACTION, VIRUS-INDUCED INCLUSION. *Educ:* Univ Ark, PhD(plant path), 70. *Prof Exp:* PROF PLANT PATH, UNIV ARK, 70- *Res:* Plant pathology; electron microscopy. *Mailing Add:* Dept Plant Path Univ Ark Main Campus Fayetteville AR 72701

KIM, MI JA, PHYSIOLOGY, NURSING. *Educ:* Univ Ill, PhD(physiol), 75. *Prof Exp:* PROF MED SURG NURSING & DEAN, COL NURSING, UNIV ILL, CHICAGO, 78- *Mem:* Fel Am Acad Nursing. *Res:* Respiratory muscle training. *Mailing Add:* 3200 N Lake Shore Dr Apt 807 Chicago IL 60657

KIM, MOON W, b Seoul, Korea. MATHEMATICAL ANALYSIS. *Educ:* Univ NH, BA, 64; Polytech Inst Brooklyn, MS, 68, PhD(math), 69. *Prof Exp:* Teaching fel math, Polytech Inst Brooklyn, 66-67, instr, 67-69; ASSOC PROF, SETON HALL UNIV, 69- *Mem:* Am Math Soc; Math Asn Am. *Res:* Functional analysis; fixed points theorems in non-linear analysis; analytic sets in Banach spaces. *Mailing Add:* 584 White Oak Ridge Rd Short Hills NJ 07078

KIM, MYUNGHWAN, b Seoul, Korea, Feb 8, 32; US citizen; m 59; c 4. ELECTRICAL ENGINEERING. *Educ:* Univ Ala, BS, 58; Yale Univ, MEng, 59, PhD(elec eng), 62. *Prof Exp:* Elec engr, Tenn Valley Authority, 58-59; from asst prof to prof elec eng, Cornell Univ, 62-89; CONSULT, 89- *Concurrent Pos:* NSF res grants, 65-67, 72-74 & 82-84; Nat Res Coun assoc, Jet Propulsion Lab, Calif Inst Technol, 68-69, vis assoc biol, 69; NIH spec res fel, 68-70, res grant, 75-78; Res Corp grant, 71-72. *Mem:* Inst Elec & Electronics Engrs; NY Acad Sci. *Res:* Automatic control and computer systems; bioengineering; neurobiology and systems. *Mailing Add:* Seoul Nat Univ Kwanak-GU Seoul 151-742 Republic of Korea

KIM, PAIK KEE, manifolds, transformaton group; deceased, see previous edition for last biography

KIM, RHYN H(YUN), b Seoul, Korea, Feb 4, 36; m 66; c 3. MECHANICAL & CHEMICAL ENGINEERING. *Educ:* Seoul Nat Univ, BSME, 58; Mich State Univ, MS, 64, PhD(mech eng), 65. *Prof Exp:* Assoc prof mech eng, Univ NC, Charlotte, 65-76; staff engr, Off Air Qual Planning & Standards, Environ Protection Agency, 76-77, mech engr, Indust Environ Res Lab, Off Res & Develop, 77-78; PROF, COL ENG, UNIV NC, CHARLOTTE, 78- *Concurrent Pos:* Consult, Indust Environ Res Lab, Off Res & Develop, Environ Protection Agency, 78-; consult, Duke Power Co, NC, Westinghouse Steam Turbine Plant & Teledyne Allvac; bd mem, NC State Bd Examiners, Plumbing, Heating & Sprinkler's Contractors, 88- *Mem:* Am Soc Mech Engrs; Am Soc Heating, Refrig & Air-Conditioning Engrs; Instrument Soc Am. *Res:* Combustion, energy conservation and utilization; environmental emission controls; mass transfer in porous media; thermal system optimization and simulations; computer aided heat exchanger design; flow phenomenon visualization; computational fluid mechanics. *Mailing Add:* 2726 Wamath Dr Charlotte NC 28210

KIM, RYUNG-SOON (SONG), b Jeonju, Korea, Nov 24, 38; Can citizen; m 63; c 3. PHARMACOLOGY, MEDICINAL CHEMISTRY. *Educ:* Seoul Nat Univ, BSc, 61; Duquesne Univ, MSc, 63; Univ Man, PhD(chem), 76. *Prof Exp:* Teaching asst pharm, Duquesne Univ, 61-63; res asst biochem, Hahnemann Med Col, 63-64; lab instr pharm chem, 71-75, res staff pharmacol, 76-82, ASST PROF, UNIV MAN, 83-; CONSULT, INT TOXICOL CONSULTS, 87- *Concurrent Pos:* Res fel, Can Heart Found, 76-79. *Mem:* Pharmacol Soc Can. *Res:* Cardiovascular pharmacology; endogenous ligands and modulators of the digitalis receptor; chemical and biological characterization of endogenous ionophores; lipid peroxidation in biology and medicine. *Mailing Add:* Dept Pharmacol Univ Man Winnipeg MB R3E 0W3 Can

KIM, SANG HYUNG, b Kyunggi, Korea, Oct 18, 42; m 68; c 2. PHOTOGRAPHIC SCIENCE ENGINEERING, INTERFACIAL SCIENCE. *Educ:* Seoul Nat Univ, BS, 64; Univ Utah, PhD(phys chem), 71. *Prof Exp:* Asst teaching & res, Seoul Nat Univ, 66-67; res asst, Univ Utah, 67-70; fel electrochem, Univ Pa, 70-71; fel membrane biophys, Northwestern Univ, 71-74; res chemist, 74-77, sr res chemist, 79-85, RES ASSOC RES DEVELOP, RES LABS, EASTMAN KODAK CO, 86- *Mem:* Am Chem Soc; Soc Imaging Sci Tech. *Res:* Ion-selective electrodes; solution and interfacial electrochemistry; membrane biophysics; electrochemistry applied to biological and medical sciences; emulsion and polymer coatings; photographic sciences. *Mailing Add:* 19 Sutton Pt Pittsford NY 14534-1729

KIM, SANGDUK, b Seoul, Korea, June 15, 30; US citizen; m 59; c 3. ENZYMOLOGY, NEUROCHEMISTRY. *Educ:* Korea Univ Med Col, MD, 53; Univ Wis, PhD(biochem), 59. *Prof Exp:* Res assoc physiol chem, Univ Wis, 59-61; res assoc biochem, Univ Ottawa, 62-65; res assoc biochem, 65-78, assoc prof, 78-90, PROF BIOCHEM, FELS RES INST, SCH MED, TEMPLE UNIV, 90- *Mem:* Am Soc Biol Chemist; Am Assoc Cancer Res; Am Soc Neurochem; Am Chem Soc. *Res:* Biochemistry of protein methylation; enzymatic modification of protein molecule, biology of myelination. *Mailing Add:* Fels Res Inst Temple Univ Sch Med Philadelphia PA 19140

KIM, SEUNG U, b Osaka, Japan, Oct 28, 36; US citizen; m 62; c 2. NEUROPATHOLOGY, NEUROBIOLOGY. *Educ:* Univ Seoul, Korea, MD, 60; Kyoto Univ, Japan, PhD(neurobiol), 65. *Prof Exp:* Multiple Sclerosis Soc fel, Col Physicians & Surgeons, Columbia Univ, 66-69; assoc prof neurobiol, Univ Sask, Saskatoon, 70-72; asst prof neuropath, 72-75, assoc prof neuropath, 75-81, PROF NEUROL, UNIV PA, 81- *Mem:* Tissue Cult Asn; Am Asn Neuropathologists; Histochem Soc; Soc Neurosci; Am Soc Cell Biol. *Res:* Experimental neurology; neural tissue culture. *Mailing Add:* Div Neurol Dept Med Univ Hosp 2211 Wesbrook Mall Vancouver BC V6T 2B5 Can

KIM, SHOON KYUNG, b Ham Hun, Korea, Feb 29, 20; m 49; c 3. THEORETICAL CHEMISTRY. *Educ:* Osaka Univ, BS, 44; Yale Univ, PhD(phys chem), 56. *Prof Exp:* Res chemist, Inst Chem, Kyoto Univ, 44-46; instr chem, Seoul Nat Univ, 46-47, from instr to prof phys chem, 49-62; res chemist, Cent Indust Res Inst, 47-49; vis prof phys chem, Brown Univ, 62-66; prof, Univ Louisville, 66-69; PROF CHEM, TEMPLE UNIV, 69- *Concurrent Pos:* vis prof, phys chem lab, Oxford Univ, Eng, 80; vis prof, Frit* Harler Res Ctr Molecular Dynamics, Hebrew Univ, Jerusalem, Israel, 85. *Honors & Awards:* Korean Nat Sci Award, 61. *Mem:* Korean Chem Soc (secy gen, 60-61); Am Chem Soc; Am Phys Soc; Am Vacuum Soc. *Res:* Statistical mechanical theory of classical and quantal systems; chemical kinetics; theory of optical activity; theory of light scattering; Representation theory of finite and continuous groups. *Mailing Add:* 336 Ainslee Rd Huntingdon Valley PA 19006

KIM, SOO MYUNG, b Chaeryung, Korea, Oct 24, 36; m 70; c 1. SOLID STATE PHYSICS. *Educ:* Seoul Nat Univ, BS, 59; Univ NC, Chapel Hill, PhD(physics), 68. *Prof Exp:* Res assoc physics, Univ NC, Chapel Hill, 67-68; fel dept physics, Univ Guelph, 68-70; Nat Res Coun fel neutron physics br, 70-72, asst res officer, 72-76, assoc res officer, 77-84, RES OFFICER, ATOMIC ENERGY CAN RES CO, CHALK RIVER, 85- *Mem:* Am Phys Soc; Can Asn Physicists; Asn Korean Scientists & Engrs Am. *Res:* Positron annihilation in metals; defects in metals. *Mailing Add:* Neutron & Solid State Physics Br Atomic Energy Can Res Co Chalk River ON K0J 1J0 Can

KIM, SOOJA K, b Seoul, Korea; US citizen; m 85. EPIDEMIOLOGY, DIETETICS. *Educ:* Humboldt State Univ, BS, 67; Tex Woman's Univ, MS, 73, PhD(nutrit sci), 75. *Prof Exp:* Asst prof nutrit & food sci, Ga Col, 75-77; asst prof nutrit & dietetics, Bowling Green State Univ, Ohio, 77-80, from assoc prof to prof nutrit, 80-87; prog dir nutrit, Nat Inst Aging, 85-86, HEALTH SCIENTIST ADMINR NUTRIT, DIV RES GRANT, NIH, 88- *Concurrent Pos:* Nutrit consult, Med Col Ohio, Toledo, 78-80, adj prof, 80-87; vis prof, Univ Mich, 81-83. *Mem:* Am Inst Nutrit; Am Dietetic Asn. *Res:* Dietary energy consumption in relation to longevity and health; age related metabolic changes during trauma and illness; nutritional status of elderly. *Mailing Add:* NIH Westwood Bldg Rm 348 Westbard Ave Bethesda MD 20892

KIM, SOON-KYU, b Hadong, Korea, Oct 3, 32; US citizen; m 59; c 3. MATHEMATICS. *Educ:* Seoul Nat Univ, BS, 57, MS, 59; Univ Mich, PhD(math), 67. *Prof Exp:* Instr math, Seoul Univ & Kunkook Univ, 59-62; instr, Univ Ill, Urbana, 66-69; from asst prof to assoc prof, 69-83, assoc dept head, 85-88, PROF MATH, UNIV CONN, 83-, DEPT HEAD, 88- *Concurrent Pos:* Vis assoc prof, Mich State Univ, 75-76; vis lectr, Seoul Nat Univ, Korea, 83. *Mem:* Am Math Soc; Korean Math Soc; Korean Scientists & Engrs Asn Am. *Res:* Fiberings and transformation groups; fixed point theory. *Mailing Add:* Dept Math Univ Conn Storrs CT 06268

KIM, SUKYOUNG, b Korea, June 7, 54; m 82; c 2. TRIBOLOGY, CHARACTERIZATION OF MATERIALS. *Educ:* Inha Univ, Korea, BS, 80; Seoul Nat Univ, Korea, MS, 82; Alfred Univ, NY, MS, 85; Univ Vt, PhD(mat sci), 90. *Prof Exp:* Res asst diezoelec ceramic, Seoul Nat Univ, Korea, 80-82, crystal structure anal, Alfred Univ, 83-85; res asst mos devices, 85-86, aluminatool wear, 86-90, RES ASSOC DLC COATING & WEAR STUDY, UNIV VT, 90- *Concurrent Pos:* Libr asst, Cook Sci Libr, Univ Vt, 86; NSF grant, 87. *Mem:* Am Ceramic Asn; Am Crystallog Asn; Electron Micros Soc Am. *Res:* Wear, fracture and deformation of materials; diamond and DLC thin film coating; study of interfacial interaction and material surfaces; crystal structure analysis; bio-application of ceramics for hip-joint; author of various publications. *Mailing Add:* Six Ethan Allen Ave No 17 Colchester VT 05446

KIM, SUNG KYU, b Chulla Namdo, Korea, Jan 12, 39; m 68; c 4. THEORETICAL PHYSICS. *Educ:* Davidson Col, BS, 60; Duke Univ, AM, 64, PhD(physics), 65. *Prof Exp:* Asst prof, 65-70, assoc prof, 71-75, PROF PHYSICS, MACALESTER COL, 75- *Concurrent Pos:* Master teacher, Twin City Inst Talented Youth, 68-69, 83-; asst prof, Univ Calif, Irvine, 70-71; vis scholar, Fermi Inst, Univ Chicago, 80-81, Astron & Astrophys, Univ Chicago, 87-88. *Mem:* Am Phys Soc; Soc Hist Sci; Am Asn Physics Teachers. *Res:* Cosmology/astrophysics. *Mailing Add:* 3134 N Victoria St St Paul MN 55126

KIM, SUNG WAN, b Pusan, Korea, Aug 21, 40; m 66; c 2. BIOMATERIALS, PHARMACEUTICS. *Educ:* Seoul Nat Univ, BS, 63, MS, 65; Univ Utah, PhD, 69. *Prof Exp:* Res asst, Seoul Nat Univ, 63-65; res asst chem, 66-69, fel, 69, res assoc mat sci, 69-70, asst res prof mat sci & eng, 70-73, from asst prof to assoc prof pharmaceut, 73-80, PROF PHARMACEUT, COL PHARM, UNIV UTAH, 80- *Concurrent Pos:* NIH res career develop award, 77-81. *Mem:* AAAS; Am Chem Soc; Korean Chem Soc; Am Soc Artificial Internal Organs; Am Pharm Asn. *Res:* Blood compatible polymers; membrane diffusion; interface induced thrombosis; polymeric drug delivery system. *Mailing Add:* Col Pharm 4512 Jupiter Dr Skaggs Hall 206 Univ of Utah Salt Lake City UT 84124

KIM, SUNG-HOU, b Taegu, Korea, Dec 12, 37; US citizen; m 68; c 2. BIOPHYSICAL CHEMISTRY, STRUCTURAL MOLECULAR BIOLOGY. *Educ:* Seoul Nat Univ, Korea, BS, 60, MS, 62; Univ Pittsburgh, PhD(phys chem), 66. *Prof Exp:* Res assoc biol, Mass Inst Tech, 66-70, sr res scientist, 70-72; asst prof biochem, Sch Med, Duke Univ, 72-73, assoc prof, 73-78; PROF BIOPHYS CHEM, UNIV CALIF, BERKELEY, 78-, FAC SR SCIENTIST, LAWRENCE BERKELEY LAB, 79-, DIV DIR, 89- *Concurrent Pos:* Postdoctoral biophys, Dept Biol, Mass Inst Technol, Cambridge, Mass, 66-70, NIH consult, 76-80; fulbright fel, Fulbright Found, 62; Lansdowne scholar, Univ Victoria, Can, 80; ed bd, J Biol Chemistry, 79-83, Nucleic Acids Res, 83-; exchange prof, Peking Univ, Peking, China, 82; Miller res prof, Univ Calif, Berkeley, 83-84; mem, Sci Planning Comt, Nat Found Cancer Res, 83-; Guggenheim fel, New York, 85-86; vis prof, Univ Paris, France, 86; coun mem, Korean Scientists & Engrs Asn Am, 88-; chmn, Adv Comt, Ctr Korean Studies, Inst EAsian Studies, Univ Calif, Berkeley, 89-; mem, US Nat Comt Crystallog, Nat Res Coun, Nat Acad Sci, 90- *Honors & Awards:* Pres Serv Merit Award, Repub Korea, 85; fel, Found Prom Cancer Res, Nat Cancer Ctr, Tokyo, Japan, 87; Ernest O Lawrence Award, Dept Energy, 87; Javits Neurosci Investr Award, Dept Health & Human Servs, 88; Princess Takamatsu Award, Princess Takamatsu Cancer Found, Tokyo, Japan, 89. *Mem:* Am Soc Biochem & Molecular Biol; Am Crystallog Asn; Am Chem Soc; AAAS; Korean Scientists & Engrs Am; Biophys Soc; Protein Soc. *Res:* Structure and function of DNA, RNA and proteins; reviewer for several scientific journals. *Mailing Add:* Dept Chem Univ Calif Berkeley CA 94720

KIM, SUN-KEE, b Seoul, Korea, Dec 11, 37; US citizen; m 63; c 2. CELL BIOLOGY, ANATOMY. *Educ:* Yon-sei Univ, Korea, BS, 60; Univ Rochester, MS, 64, PhD(biol), 70. *Prof Exp:* Asst cancer res scientist endocrinol, Rosewell Park Mem Inst, Buffalo, 64-65; pre-doctoral trainee, NIH, 65-68; RES BIOLOGIST, VET ADMIN MED CTR, 68-, COORDR ELECTRON MICROS, RES SERV, 73- *Concurrent Pos:* Co-dir, Advan Spec Training Dent Res, Vet Admin Med Ctr, 70- 75; res assoc anat, Sch Med, Univ Mich, 70-74, asst prof anat & cell biol, 74-81, assoc prof, 81-, adj assoc prof oral biol, Sch Dent, 82-; proj dir aging studies salivary gland, Nat Inst Aging, NIH, 78-82, 87- *Mem:* Am Soc Cell Biol; Am Asn Anatomists; Int Asn Dent Res; AAAS; Sigma Xi. *Res:* Age-related changes in secretory function of exocrine glands; ultrastructure and protein synthesis in secretory cells of the salivary gland. *Mailing Add:* Res Serv Vet Admin Med Ctr 2215 Fuller Rd Ann Arbor MI 48105

KIM, TAI KYUNG, b Pyungyang, Korea, June 19, 27; m 59; c 2. PHYSICAL CHEMISTRY. *Educ:* Seoul Nat Univ, BSE, 51; Adrian Col, BS, 55; Univ Detroit, MS, 60; Duquesne Univ, PhD(phys chem), 63. *Prof Exp:* Res chemist, 63-77, ENG SPECIALIST, CHEM & METALL DIV, GTE SYLVANIA INC, 77- *Mem:* Am Chem Soc. *Res:* Metal complexation; solvent extraction; radiochemistry; molybdenum and tungsten chemistry. *Mailing Add:* RD 1 Hillcrest Towanda PA 18848

KIM, THOMAS JOON-MOCK, b Seoul, Korea, Oct 13, 36; US citizen; m 67; c 2. MECHANICAL ENGINEERING, APPLIED MECHANICS. *Educ:* Seoul Nat Univ, BS, 59; Villanova Univ, MS, 64; Univ Ill, PhD(mech), 67. *Prof Exp:* Instr mech, Univ Ill, 65-67; asst prof math, Villanova Univ, 67-68; from asst prof to assoc prof, 68-79, PROF & CHMN MECH ENG, UNIV RI, 79- *Concurrent Pos:* Consult, US Naval Underwater Systs Ctr, 77-84. *Mem:* Am Soc Mech Engrs; Am Ceramic Soc; Soc Mfg Engrs. *Res:* Solid mechanics; ceramic processing; dynamic face seals; water jet machining. *Mailing Add:* Dept Mech Eng Univ RI 105 Wales Hall Kingston RI 02881

KIM, UING W, b Masan, Korea, Nov 18, 41; US citizen; m 68; c 2. NUMERICAL MODELLING & ITS SOLUTION ON DIGITAL COMPUTER. *Educ:* Korea Mil Acad, BS, 63; Penn State Univ, MS, 73; NJ Inst Technol, PhD(civil eng), 81. *Prof Exp:* Res assoc numerical anal, Penn State Univ, 72-73; DIR COMPUTER SERV, KILLAM ASSOCS, 73- *Concurrent Pos:* Adj prof civil eng, NJ Inst Technol, 74-79; dir, Hydraul Div, Computer Appln Eng, Planning & Archit, 81-82. *Res:* Mathematical modelling and analytical solution on dissolved oxygen under three-dimensional and unsteady state flow. *Mailing Add:* 45 Old Short Hills Rd Short Hills NJ 07078

KIM, UNTAE, b Dec 16, 26. CANCER RESEARCH, TUMOR IMMUNOLOGY. *Educ:* Seoul Univ, Korea, MD, 52. *Prof Exp:* HEAD PATH RES, ROSWELL PARK MEM INST, 63- *Mem:* Am Asn Pathologists; Am Asn Immunologists; Am Asn Cancer Res; NY Acad Sci; Int Soc Differentiation; Int Soc Metastasis Res; Int Asn Breast Cancer Res. *Res:* Breast cancer research on carcinogenesis and evolution of neoplastic state; hormone dependency, immunology of tumor cells and acquisition of metastatic potential and its patterns; interaction between tumor and lymphoid cells. *Mailing Add:* Dept Path Roswell Pk Mem Inst Buffalo NY 14263

KIM, WAN H(EE), electrical engineering, for more information see previous edition

KIM, WON, b Korea, 1948. DATABASE & DISTRIBUTED SYSTEMS. *Educ:* Mass Inst Technol, BS & MS, 71; Univ Ill, Urbana-Champaign, PhD(computer sci), 80. *Prof Exp:* Prin scientist, MCC, 84-90, dir, Object-Oriented Distrib Syst Lab, 87-90; PRES, UNISQL INC, 90- *Concurrent Pos:* Chmn, Spec Interest Group Mgt Data, Asn Comput Mach, 89- *Mem:* Asn Comput Mach. *Mailing Add:* UniSQL 9390 Research Blvd Bldg II Austin TX 78759

KIM, WOO JONG, b Seoul, Korea, Jan 15, 37. MATHEMATICS. *Educ:* Seoul Nat Univ, BS, 58; Okla State Univ, MS, 60; Carnegie-Mellon Univ, MS & PhD(chem eng), 64, PhD(math), 68. *Prof Exp:* From asst prof to assoc prof appl math, 68-84, PROF APPL MATH & STATIST, STATE UNIV NY, STONY BROOK, 85- *Mem:* Am Math Soc. *Res:* Disconjugacy, oscillation and asymptotic behavior of ordinary differential equations. *Mailing Add:* Dept Appl Math & Statist State Univ NY Stony Brook NY 11794

KIM, YEE SIK, b Seoul, Korea, Apr 15, 28; US citizen; m 50; c 4. BIOCHEMISTRY, PHARMACOLOGY. *Educ:* Kans State Univ, BS, 57, MS, 60; St Louis Univ, PhD(pharmacol), 65. *Prof Exp:* Instr chem, Kans State Univ, 60; res chemist, Union Starch & Refining Co, Ill, 60-63; from asst prof to prof, 66-76, PROF PHARMACOL, ST LOUIS UNIV, 76- *Concurrent Pos:* USPHS fel biochem, State Univ NY Buffalo, 65-66; Gen res support grant & Cancer Inst res grant, 66-; USPHS res grant, 67- *Mem:* AAAS; Am Soc Pharmacol & Exp Therapeut; Am Chem Soc; NY Acad Sci. *Res:* Molecular mechanism of hormone action; enzyme reaction mechanisms; metabolisms of macromolecules; diabetic pregnancy; mechanisms of vitamin D metabolites. *Mailing Add:* Dept Pharmacol St Louis Univ Sch Med 1402 S Grand St Louis MO 63104

KIM, YEONG ELL, b SKorea; US citizen; m 61; c 2. THEORETICAL NUCLEAR PHYSICS. *Educ:* Lincoln Mem Univ, BS, 59; Univ Calif, Berkeley, PhD, 63. *Prof Exp:* Fel physics, Bell Tel Labs Inc, 63-65, Oak Ridge Nat Lab, 65-67; from asst prof to assoc prof, 67-77, PROF PHYSICS, PURDUE UNIV, 77- *Concurrent Pos:* Proj dir, Purdue Nuclear Theory Group, 71-; vis staff mem, Los Alamos Nat Lab, Univ Calif, 73-74, consult, 74-; chmn, Gordon Res Conf, 77; vis prof, Seoul Nat Univ, 79 & 80. *Honors & Awards:* US Sr Scientist Humboldt Award, 77. *Mem:* Korean Scientists & Engrs Asn Am; fel Korean Phys Soc; fel Am Phys Soc. *Res:* Quantum theory of scattering; theory of the three-nucleon systems; intermediate energy physics; theories of meson-exchange currents; parity violations in nuclear physics; photonuclear reactions; nuclear structure and reactions; quark degrees of freedom in nuclei; geodesy; geodynamics; gravitational theory; exploration geophysics. *Mailing Add:* Dept Physics Purdue Univ West Lafayette IN 47907

KIM, YEONG WOOK, b July 21, 25; m; c 3. SOLID STATE PHYSICS, ATOMIC & MOLECULAR PHYSICS. *Educ:* Seoul Nat Univ, Korea, BS, 48, MS, 50; Brown Univ, Providence, RI, PhD(physics), 61. *Prof Exp:* PROF PHYSICS, WAYNE STATE UNIV, 60- *Concurrent Pos:* Sigma Xi award, Wayne State Univ, 76; NIH cancer sci fel, 76-78. *Mem:* Fel Am Phys Soc; Am Asn Physics Teachers; Sigma Xi; Inst Elec & Electronics Engrs; Am Asn Univ Professors. *Res:* Solid State experiments; spatial pairing effects in high temperature superconductors. *Mailing Add:* Dept Physics & Astron Wayne State Univ Detroit MI 48202

KIM, YONG IL, b Seoul, Korea, June 5, 45; m 75; c 3. BIOMEDICAL ENGINEERING, NEUROMUSCULAR TRANSMISSION. *Educ:* Seoul Nat Univ, BS, 68; Cornell Univ, MS, 73, PhD(biomed eng), 77. *Prof Exp:* Systs analyst trainee, Korea Comput Ctr, 68; systs engr, Gadelius & Co, Ltd, 68-70; instr bioeng & elec systs, Cornell Univ, 77; res assoc, dept neurol & Jerry Lewis Neuromuscular Ctr, 77-79, ASST PROF BIOMED ENG & NEUROL, SCH MED, UNIV VA, 79- *Concurrent Pos:* New investr res award, NIH, 82-85; vis scientist, dept pharmacol, Univ Lund, Sweden, 83-84. *Mem:* Inst Elec & Electronics Engrs; Biomed Eng Soc; AAAS; Soc Neurosci; Korean Scientists & Engrs Am. *Res:* Neuromuscular transmission in disease states; bioelectric systems; biomedical instrumentation. *Mailing Add:* Dept Neurol & Biomed Eng Univ Va Health Sci Ctr Box 377 Charlottesville VA 22908

KIM, YONG WOOK, b Seoul, Korea, Sept 30, 38; nat US; m 66; c 3. ATOMIC PHYSICS, STATISTICAL PHYSICS. *Educ:* Seoul Nat Univ, BS, 60, MS, 62; Univ Mich, PhD(physics), 68. *Prof Exp:* From asst prof to assoc prof, 68-77, chmn dept, 84-87, PROF PHYSICS, LEHIGH UNIV, 77- *Concurrent Pos:* Lectr, USSR Acad Sci, 91. *Mem:* Fel Am Phys Soc; Sigma Xi; AAAS; Metall Soc; Electrochem Soc. *Res:* Autocorrelations and non-linear phenomena in molecular and Brownian fluctuation; transport in nonideal plasmas; optically assisted gas phase reactions; spectroscopy of light scattering from small particle suspension; electron attachment; shock waves; laser-produced plasmas; nonlinear dynamics. *Mailing Add:* Dept Physics Lehigh Univ Bldg 16 Bethlehem PA 18015

KIM, YONG-KI, b Seoul, Korea, Feb 20, 32; m 63; c 2. ATOMIC PHYSICS. *Educ:* Seoul Nat Univ, BS, 57; Univ Del, MS, 61; Univ Chicago, PhD(physics), 66. *Prof Exp:* Instr physics, Korean Air Force Acad, 57-59; res assoc, Argonne Nat Lab, 66-68, asst physicist, 68-72, physicist, 72-79, sr, physicist, 79-83; PHYSICIST, NAT BUR STANDARDS, 83- *Concurrent Pos:* Lectr, Univ Chicago, 67-68. *Mem:* Am Phys Soc; Korean Phys Soc; Radiation Res Soc. *Res:* Atomic structure theory; atomic collision theory; radiation physics. *Mailing Add:* Nat Inst Standards & Technol Bldg 221 Rm A267 Gaithersburg MD 20899

KIM, YOON BERM, b Soon Chun, Korea, Apr 25, 29; m 59; c 3. IMMUNOLOGY. *Educ:* Seoul Nat Univ, MD, 58; Univ Minn, PhD(microbiol), 65. *Prof Exp:* Intern med, Univ Hosp, Seoul Nat Univ, 58-59; from asst teaching & res to assoc teaching & res, 60-64, from instr to assoc prof microbiol, Med Sch, Univ Minn, Minneapolis, 65-73; head, lab ontogeny immune syst, Sloan-Kettering Inst Cancer Res, NY, 73-; prof biol & immunol, Grad Sch Med Sci, Cornell Univ, 73-, chmn, immunol unit, 80-; AT DEPT MICROBIOL & IMMUNOL, CHICAGO MED SCH, N CHICAGO. *Concurrent Pos:* Res fel, 60-64; USPHS career develop res award, 68-73. *Mem:* Asn Gnotobiotics (pres, 79-80); Am Asn Immunol; Am Soc Microbiol; Reticuloendothelial Soc; Am Asn Path. *Res:* Immunobiology and immunochemistry; ontogeny of the immune system; mechanism of the immune response, regulation of the immune system; tumor immunity; immunochemistry and biology of bacterial toxins and host-parasite relationships; gnotobiology. *Mailing Add:* Dept Microbiol & Immunol Chicago Med Sch 3333 Green Bay Rd N Chicago IL 60064

KIM, YOUNG BAE, b Korea, Oct 23, 22; m 51; c 3. PHYSICS. *Educ:* Univ Wash, BS, 50; Princeton Univ, PhD(physics), 54. *Prof Exp:* Asst, Princeton Univ, 51-53; res assoc, Univ Ind, 54-55; assoc prof, Univ Wash, 56-61; mem tech staff, Bell Tel Labs, Murray Hill, NJ, 61-68; prof physics & elec eng, 69-75, CHMN ELEC ENG & PROF PHYSICS, UNIV SOUTHERN CALIF, 75- *Concurrent Pos:* Guggenheim fel, Tokyo Univ, 66-67; invited prof, Korea Advan Inst Sci, Seoul, 73-74. *Mem:* Fel Am Phys Soc. *Res:* Superconductivity, superconducting magnets and technology, solid state and low temperature physics; high energy physics, energy resources management. *Mailing Add:* Dept Physics 0484 Univ Southern Calif University Park Los Angeles CA 90089

KIM, YOUNG C, b Seoul, Korea, May 25, 36; US citizen; m 65; c 3. CIVIL ENGINEERING. *Educ:* Univ Southern Calif, BS, 58, PhD(civil eng), 64; Calif Inst Technol, MS, 59. *Prof Exp:* Civil engr, Daniel, Mann, Johnson & Mendenhall, 59-61; lectr civil eng, Univ Southern Calif, 61-64; from asst prof to assoc prof, 65-73, chmn dept, 76-79, PROF CIVIL ENG, CALIF STATE UNIV, LOS ANGELES, 73-, ACTG ASSOC DEAN, SCH ENG, 78- *Concurrent Pos:* Sr lectr, Univ Southern Calif, 65-70; Res Corp res grants, 64 & 67, travel grant, 68; NSF res grant, 65; resident consult, US Naval Civil Eng Lab, 67 & 69; resident consult, Sci Eng Assocs, 67-, sr res engr, 69; vis scholar, Univ Calif, Berkeley, 71; NATO sr fel sci, Delft Technol Univ, Neth, 75; vis scientist, US-Japan Coop Sci Prog, Osaka City Univ, 76; prin investr, NSF, Res Corp & Defense Atomic Support Agency, 65-; reviewer, Sci Develop Countries Prog, NSF, 80; mem bd gov, Southern Calif Ocean Studies Consortium, 78. *Mem:* AAAS; Am Soc Civil Engrs; Int Asn Hydraul Res; Am Soc Eng Educ; Sigma Xi. *Res:* Underwater explosions; wave forces; interaction of structures and sea waves; hydraulic model studies; oil slick transport on coastal waters; wave energy absorption in coastal structures; extraction of energy from the sea. *Mailing Add:* Dept Civil Eng Calif State Univ Los Angeles CA 90032

KIM, YOUNG DUC, b Korea, Oct 28, 32; m 65; c 1. ELECTRICAL ENGINEERING. *Educ:* Newark Col Eng, MSEE, 63, ScD, 68. *Prof Exp:* RES ELECTRONIC ENGR, US NAVAL AMMUNITION DEPOT, 68- *Mem:* Inst Elec & Electronics Engrs. *Res:* Investigation of noise spectral density observed in solid state devices. *Mailing Add:* Naval Weapons Support Ctr Crane IN 47522

KIM, YOUNG JOO, b Kwangju, Dec 12, 60; m 86; c 1. BIOSEPARATION, SEPARATION PROCESS. *Educ:* Chonnam Nat Univ, Korea, BE, 83; Seoul Nat Univ, Korea, ME, 85. *Prof Exp:* Res engr chem eng, Oriental Chem Indust Res Ctr, 86, Ssangyong Oil Refinery, Korea, 87-89; ASST RESEARCHER CHEM ENG, RENSSELAER POLYTECH INST, 89- *Mem:* Am Inst Chem Engrs. *Res:* Separation and purification of various pharmaceutical proteins and biochemicals by preparative chromatography; scale-up; mathematical modeling. *Mailing Add:* RI-408 Rensselaer Polytechnic Inst Troy NY 12180

KIM, YOUNG NOK, b Seoul, Korea, Jan 30, 21; m 60; c 3. THEORETICAL PHYSICS. *Educ:* Seoul Nat Univ, BS, 47, MS, 49; Univ Birmingham, PhD(math physics), 57. *Prof Exp:* Res grants, Inst Theoret Physics, Copenhagen, Denmark, 56-58, Heidelberg, Ger, 58, Edmonton, Can, 62-63, Inst Henri Poincare, Paris, France, 58-59 & Univ Wash, 60-62; prof physics, Mem Univ, 63-64; PROF PHYSICS, TEX TECH UNIV, 64- *Mem:* Am Phys Soc; Italian Phys Soc. *Res:* Nuclear structure; collision; mesonic atoms. *Mailing Add:* Dept Physics Tex Tech Univ Lubbock TX 79409

KIM, YOUNG SHIK, b Seoul, Korea, Feb 5, 33; US citizen; c 2. MEDICINE. *Educ:* Stanford Univ, AB, 56; Cornell Univ, MD, 60. *Prof Exp:* Instr med, Med Col, Cornell Univ, 64-65; res assoc path, Stanford Univ, 65-66; asst prof biochem, New York Med Col, 67-68; from asst prof to assoc prof med, Univ Calif, San Francisco, 68-76, prof med, 76-; DIR, GI RES LAB, VET ADMIN HOSP, SAN FRANCISCO, 68- *Concurrent Pos:* Am Cancer Soc res scholar, 65-68; mem gen med study sect A, NIH, 76-79; large bowel cancer working group, Nat Cancer Inst, 85-89; clin Sci I Study Sect, NIH, 89-; assoc ed, Gastroenterol, 81-86, cancer res, 87- *Honors & Awards:* Western Gastroenterol Res Award, 78; Vet Admin Med Investr Award, 84-90; NIH Merit Award, 88. *Mem:* Am Asn Cancer Res; Am Soc Clin Invest; Am Gastroenterol Asn; Biochem Soc; Asn Am Physicians. *Res:* Glycoprotein and glycolipid chemistry, immunochemistry and metabolism of the gastrointestinal tract; biology and molecular biology of gastrointestinal cancer. *Mailing Add:* GI Res Vet Admin Hosp 4150 Clement St San Francisco CA 94121

KIM, YOUNG TAI, b Seoul, Korea, Nov 10, 30; m 55; c 3. BIOCHEMISTRY, IMMUNOLOGY. *Educ:* Seoul Nat Univ, BS, 53, MS, 57; Univ Calif, Los Angeles, PhD(plant biochem), 63. *Prof Exp:* Asst prof biochem, Kyung Hee Univ, Korea, 57-60; res assoc bot, Univ Calif, Los Angeles, 63-64; biochem specialist, Growth Sci Ctr, Int Minerals & Chem Corp, 65-70; asst prof, 70-80, ASSOC PROF MED, MED COL, CORNELL UNIV, 80- *Mem:* AAAS; Am Chem Soc; Am Inst Chemists; NY Acad Sci; Am Asn Immunologists. *Res:* Virology; molecular biology; relationship between immunological response and genetics; immune response in aging. *Mailing Add:* Dept Immunol Med Cornell Univ Med Col 1300 York Ave New York NY 10021

KIM, YUNG DAI, b Seoul, Korea, Mar 24, 36; US citizen; m 67; c 2. TUMOR IMMUNOLOGY, CLINICAL CHEMISTRY. *Educ:* Wash State Univ, BS, 61; Univ Idaho, MS, 63; Univ Minn, PhD(biophys chem), 68. *Prof Exp:* NIH fel, Northwestern Univ, 69-71; NIH res fel, Univ Pa, 71-73; sr scientist immunol, 74-88, ASSOC RES FEL, CANCER RES LAB, ABBOTT LABS, 88- *Concurrent Pos:* Vis scientist, C F Kettering Res Inst, 68-69. *Mem:* Am Asn Immunologists; Am Chem Soc; Sigma Xi; Am Asn Cancer Res. *Res:* Cancer research in immunology and immunochemistry; isolation and characterization of tumor-associated antigens; research and development of clinical diagnostic tests. *Mailing Add:* Abbott Labs Dept Cancer Res D-90C North Chicago IL 60064

KIM, YUNGKI, b Korea, Dec 11, 35; US citizen; m 61; c 2. ORGANIC CHEMISTRY, POLYMER CHEMISTRY. *Educ:* Tex Christian Univ, BA, 59; Univ Colo, Boulder, MS, 61; Ariz State Univ, PhD(org chem), 65. *Prof Exp:* Fac res assoc chem, Ariz State Univ, 64-65; res chemist, Dow Corning Corp, 65-68, sr res chemist, 68-70, res group leader, Silicone Polymers & Intermediates, 70-77, process eng mgr, 77-83, res mgr, 83-84, develop mgr, 84-86, TS&D dir, 86-89, RES & DEVELOP DIR, FRPL, DOW CORNING CORP, 89- *Mem:* Am Chem Soc; Sigma Xi. *Res:* Synthetic organic and polymer chemistry; silicone fluorosilicone and hybrid of silicone and organic/fluoro-organic leading to sealants, rubbers, resins and fluids. *Mailing Add:* Mail Stop CO-2418 Dow Corning Corp Midland MI 48640

KIM, ZAEZEUNG, b Hamhung, Korea, Feb 21, 29; US citizen; m 61; c 2. IMMUNOLOGY, ALLERGY. *Educ:* Seoul Nat Univ, MD, 60; Univ Cologne, PhD(immunol), 68. *Prof Exp:* Resident physician, Seoul Nat Univ Hosp, 61-63; resident physician, Heidelberg Univ Hosp, 63-64; res fel immunol, Max Planck Inst, 65-67; clin fel hematol, Univ Tex M D Anderson Hosp, 67-68; resident physician allergy-immunol, Temple Univ Hosp, 68-69; fel allergy-immunol, Col Med, Ohio State Univ, 69-71; from instr to assoc prof med, 72-80, CLIN ASSOC PROF, DEPT MED ALLERGY SECT, MED COL WIS, 80- *Concurrent Pos:* Travel grants, AAAS, 67 & Am Acad Allergy, 70; pvt pract. *Mem:* AAAS; Am Acad Allergy; AMA. *Res:* Modification (change) of cellular antigenicity by enzyme treatment. *Mailing Add:* 4521 N Wildwood Ave Milwaukee WI 53211

KIMBALL, ALLYN WINTHROP, b Buffalo, NY, Oct 2, 21; m 44; c 2. PUBLIC HEALTH & EPIDEMIOLOGY. *Educ:* Univ Buffalo, BS, 43; NC State Univ, PhD(statist), 50. *Prof Exp:* Exp statistician aerospace med, USAF Sch Aviation Med, 48-50; chief statist sect, Math Panel, Oak Ridge Nat Lab, 50-60; PROF BIOSTATIST, SCH HYG & PUB HEALTH, JOHNS HOPKINS UNIV, 60- *Concurrent Pos:* Chmn dept biostatist, Johns Hopkins Univ, 60-66, prof, Sch Med, 60-, chmn dept statist, 62-66, prof, Fac Arts & Sci, 62-, dean fac arts & sci, 66-70. *Mem:* Biomet Soc (treas, 55-60); Am Statist Asn. *Res:* Statistics and mathematics applied to biology and medicine. *Mailing Add:* Dept Biostatist Sch Hyg & Pub Health Johns Hopkins Univ 615 N Wolf St Baltimore MD 21205

KIMBALL, AUBREY PIERCE, b Lufkin, Tex, Oct 20, 26; m 53; c 1. DRUG METABOLISM, ENZYMOLOGY. *Educ:* Univ Houston, BS, 58, PhD(chem), 62. *Prof Exp:* Serologist, Houston Med Group, 48-50; res technician, Baylor Col Med, 52-54; fel, Stanford Res Inst, 61-62, biochemist, 62-67; from assoc prof to prof, Dept Biochem & Biophys, Univ Houston, 67-81, planning dir, Cancer Prog, 76-80; CONSULT, 80- *Concurrent Pos:* Dir, Southwest Sci Forum, 76-81, vpres, 79-80, pres, 80-81. *Mem:* AAAS; Am Chem Soc; Soc Exp Biol & Med; Am Asn Cancer Res; Am Soc Biol Chemists. *Res:* Design, synthesis and mechanism studies of carcinostatic drugs; biochemical evolution; enzyme affinity labeling. *Mailing Add:* 8849 Braesmont 222 Houston TX 77096

KIMBALL, BRUCE ARNOLD, b Aitkin, Minn, Sept 27, 41; m 66; c 3. SOIL PHYSICS, MICROMETEOROLOGY. *Educ:* Univ Minn, St Paul, BS, 63; Iowa State Univ, MS, 65; Cornell Univ, PhD(soil physics), 70. *Prof Exp:* SOIL SCIENTIST, US WATER CONSERV LAB, USDA, 70- *Mem:* Fel Am Soc Agron; fel Soil Sci Soc Am; Int Solar Energy Soc; AAAS; Int Soc Hort Sci. *Res:* Carbon dioxide effects on plants; energy relationships of greenhouses; evaporation and transpiration; soil air movement. *Mailing Add:* US Water Conserv Lab USDA-Agr Res Serv 4331 E Broadway Phoenix AZ 85040

KIMBALL, CHARLES NEWTON, b Boston, Mass, Apr 21, 11; m 51; c 2. ELECTRICAL ENGINEERING. *Educ:* Northeastern Univ, BEE, 31; Harvard Univ, MS, 32, ScD(elec eng), 34. *Hon Degrees:* DEng, Northeastern Univ, 55; ScD, Park Col, 58; LittD, Westminster Col, 78. *Prof Exp:* Res engr, Radio Corp Am, NY, 34-41; vpres & dir, Aircraft Acessories Corp, 41-47; tech dir, Res Labs Div, Bendix Aviation Corp, Mich, 48-50; pres, 50-75, chmn bd trustees, 75-79, EMER PRES, MIDWEST RES INST, 79- *Concurrent Pos:* Instr grad sch, NY Univ, 40-41; dir, Trans-World Airlines, Inc; trustee, US Comt Econ Develop, Hallmark Found & Menninger Found; chmn adv coun, Off Technol Assessment, US Cong, 79. *Mem:* Fel Inst Elec & Electronics Engrs; Sigma Xi. *Res:* Vacuum tube and circuit design; measurement and control of industrial processes; technology, transfer and assessment applications of technology to regional development. *Mailing Add:* Midwest Res Inst 425 Volker Blvd Kansas City MO 64110

KIMBALL, CHASE PATTERSON, b Ware, Mass, July 21, 32; m 58; c 4. PSYCHIATRY, INTERNAL MEDICINE. *Educ:* Brown Univ, AB, 54; State Univ NY, MD, 59. *Prof Exp:* Intern med, Univ Vt Hosps, 59-60, resident, 63-65; resident psychiat, New York Hosp, 60-61; instr psychiat med, Univ Rochester, 65-67; asst prof psychiat & med, Yale Univ, 67-72; PROF PSYCHIAT & MED & BEHAV SCI, PRITZKER SCH MED, UNIV CHICAGO, 72- *Concurrent Pos:* NIMH fel psychosom med, Univ Rochester, 65-67; lectr, Conn Acad Gen Pract, Conn Psychiat Asn, 67-; Am Heart Asn res grant, 68-71; consult, USPHS, 69-; chmn comt community health affairs & hosps, dir psychiat curric & chmn social med, Univ Chicago, 72-; lectr, Sch Med, Yale Univ, 72-; dir progs intercult med, Yale Univ & Univ Chicago, 72-; vpres, Int Col Psychosom Med, 77-79, pres-elect, 79- *Mem:* Am Psychosom Soc; Am Psychiat Soc; Am Heart Asn; NY Acad Sci; Soc Health & Human Values. *Res:* Investigation of psychosocial factors associated with illness and the adaptation of individuals to illness; medical education as a humanizing experience. *Mailing Add:* Univ Chicago Psy 950 E 59th St Chicago IL 60637

KIMBALL, CLYDE WILLIAM, b Laurium, Mich, Apr 20, 28; m 52; c 2. SOLID STATE PHYSICS. *Educ:* Mich Tech Univ, BS, 50, MS, 52; St Louis Univ, PhD(physics), 59. *Prof Exp:* Asst physicist, Argonne Nat Lab, 50-53, res assoc, 57-59; res physicist solid state physics, Autonetics Div, NAm Aviation, Inc, 59-60; mem res staff res opers, Aeronutronics Div, Ford Motor Co, Calif, 60-62; assoc physicist, Solid State Div, Argonne Nat Lab, 62-64; assoc prof physics, Northern Ill Univ, 64-68; prog dir low temperature physics, Div Mat Res, NSF, 77-78; sci & technol adv to pres, 82-88, PROF PHYSICS, NORTHERN ILL UNIV, 68-, UNIV RES PROF, 86- *Concurrent Pos:* Consult, Argonne Nat Lab, 64- *Mem:* AAAS; fel Am Phys Soc; Sigma Xi; Am Asn Physics Teachers. *Res:* Experimental reactor physics; photodisintegration cross sections; neutron cross sections and nuclear reactions; magnetic and electronic properties of metals and alloys; Mossbauer effect; superconductivity; amorphous solids; lattice properties. *Mailing Add:* Dept Physics Northern Ill Univ De Kalb IL 60115

KIMBALL, FRANCES ADRIENNE, b Oakland, Calif, May 2, 39. REPRODUCTIVE ENDOCRINOLOGY. *Educ:* Univ Calif, Berkeley, BA, 61; Calif State Univ, Chico, MA, 70; Cornell Univ, PhD(physiol), 73. *Prof Exp:* Res asst endocrinol, Reed Col, 61-68; res technician, Med Sch, Univ Calif, San Francisco, 68; instr biol, Calif State Univ, Chico, 68-70; USPHS fel, Cornell Univ, 70-73; scientist, Upjohn Co, 74-75, res scientist II, 75-80, sr res scientist III, reproductive endocrinol & fertil res, 80-85, sr proj mgr, prod devleop, 85-89, actg dir, 89-91, DIR, PROJ MGT, UPJOHN CO, 91- *Concurrent Pos:* NIH fel, Upjohn Co, 73-74. *Mem:* Am Soc Zoologists; Soc Study Reproduction; Am Soc Microbiol; Drug Info Asn. *Res:* Mechanism of hormone action in the uterus and corpus luteum; action of prostaglandins in uterine contractility; physiology of antimicrobial agents. *Mailing Add:* 2808 Ferdon Rd Kalamazoo MI 49008

KIMBALL, JOHN WARD, b Portland, Maine, 1931; m 53; c 2. BIOLOGY. *Educ:* Harvard Univ, AB, 53, AM, 70, PhD(biol), 72. *Prof Exp:* Assoc prof biol, Tufts Univ, 77-81; vis lectr biol, Harvard Univ, 82-86, 90-91; assoc prof biol, Bradford Col, 88-89. *Concurrent Pos:* Vis assoc prof, Tufts Univ, 89-91. *Mem:* Fel AAAS; Am Asn Immunologists. *Res:* Author of biology textbooks. *Mailing Add:* 89 Prospect Rd Andover MA 01810

KIMBALL, PAUL CLARK, b New London, Conn, Jan 26, 46; m 85; c 1. MOLECULAR GENETICS, VIROLOGY. *Educ:* Mass Inst Technol, BS, 68; Univ Calif, Berkeley, PhD(molecular biol), 72. *Prof Exp:* Instr microbiol, Univ Ill Med Ctr, 72-73; sr scientist tumor virol, Meloy Labs, Inc, 73-75; asst prof microbiol, Ohio State Univ & molecular virologist, Cancer Res Ctr, 75-81; prin res scientist, 81-82, sr res scientist, 83, res leader, Battelle, Columbus, 84-88; consult, Cushman, Darby & Cushman, Washington, DC, 88-90; CONSULT, FOLEY & LORDNER, ALEXANDRIA, VA, 90- *Mem:* AAAS; Am Soc Microbiol; Am Soc Virol. *Res:* Patenting of biotechnologies, especially genetric engineering products and processes; microbial production of biologicals with medical/veterinary/industrial import; use of microcomputers in biotechnology research; protein/peptide engineering. *Mailing Add:* 2801 Strauss Terr Silverspring MD 20904

KIMBALL, RALPH B, computer science, for more information see previous edition

KIMBALL, RICHARD FULLER, b Baltimore, Md, Feb 1, 15; m 40; c 2. BIOLOGY. *Educ:* Johns Hopkins Univ, AB, 35, PhD(zool), 38. *Prof Exp:* Sterling fel, Yale Univ, 38-39; instr biol, Johns Hopkins Univ, 39-43, asst prof, 43-47; sr biologist, Oak Ridge Nat Lab, 47-67 & 69-76, dir, Biol Div, 67-69, sr biologist, 69-76, assoc sect head, 76-80, consult, 80-90; RETIRED. *Concurrent Pos:* Guggenheim fel, Karolinska Inst, Sweden, 57-58; on leave, Biol Br, Div Biol & Med, AEC, 71-72; res fel, Union Carbide Corp, 79; mem, Nat Adv Environ Health Sci Coun, 80- *Mem:* AAAS; Genetics Soc Am; Radiation Res Soc; Am Soc Cell Biol; Environ Mutagen Soc; Sigma Xi. *Res:* Mutation genetics; induced mutation processes in bacteria. *Mailing Add:* 300 Laboratory Rd No 257 Oak Ridge TN 37830

KIMBEL, PHILIP, b Philadelphia, Pa, Mar 23, 25; m 53; c 3. MEDICINE, PULMONARY DISEASES. *Educ:* Temple Univ, MD, 53; Am Bd Internal Med, dipl, 61. *Prof Exp:* Resident internal med, 54-56, head pulmonary dis sect, 61-79, DIR PULMONARY FUNCTION LAB, ALBERT EINSTEIN MED CTR, 58- *Concurrent Pos:* USPHS res fel physiol & pharmacol, Grad Sch Med, Univ Pa, 56-57; res assoc, Gerontol Res Inst, Pa, 58-64; res assoc, Inst Cancer Res & Grad Sch Med, Univ Pa, 59-61; clin instr, Sch Med, Temple Univ, 59-64, assoc, 64-67, assoc prof, 67-71, prof, 71-79; clin prof med & chmn dept med, Grad Hosp, Sch Med, Univ Pa, 80- *Mem:* AAAS; fel Am Col Chest Physicians; fel Am Col Physicians; Am Thoracic Soc; Am Fedn Clin Res. *Res:* Pulmonary physiology and circulation; internal medicine; lung mechanics; rehabilitation in emphysema; experimental emphysema. *Mailing Add:* Dept Med One Graduate Plaza Philadelphia PA 19146

KIMBER, CLARISSA THERESE, b Merced, Calif, Feb 1, 29. ETHNO-MEDICINE, BIOGEOGRAPHY. *Educ:* Univ Calif, Berkeley, AB, 49; Univ Wis-Madison, MS, 62, PhD(geog), 69. *Prof Exp:* Teaching asst geog, Univ Wis-Madison, 61, res asst, 62-64; actg asst prof, Univ Calif, Riverside, 64-66, lectr, 66-67; asst prof, Calif State Col, Hayward, 67-68; from asst prof to assoc prof, 68-81, PROF GEOG, TEX A&M UNIV, 81- *Concurrent Pos:* Pron investr, Res Coun Fiscal Grant, Tex A&M Univ, 71; HEW, NIH, 71-72; co-prin investr, Fish & Wildlife Serv, US Dept Interior, 78-82; mem bd gov, Orgn Trop Studies, 81- *Mem:* Sigma Xi; Soc Women Geogrs; AAAS; Asn Am Geogrs; Orgn Trop Studies; fel Royal Geog Soc. *Res:* Island biogeography; people-plant relations as in domestication of plants and ethno-medical systems; human impacts on isolated environments. *Mailing Add:* Dept Geog Tex A&M Univ College Station TX 77843

KIMBER, GORDON, b Manchester, Eng, July 21, 32; m 57; c 1. CYTOGENETICS. *Educ:* Univ London, BSc, 54, DSc, 86; Univ Manchester, PhD(genetics), 61. *Prof Exp:* Res asst genetics, Univ Manchester, 54-58; sci officer, Plant Breeding Inst, 58-63, sr sci off, 63-67; assoc prof, 67-72, PROF AGRON, UNIV MO, COLUMBIA, 72- *Concurrent Pos:* Kellogg Found fel, 63-64; Int Atomic Energy Agency expert, Indian Agr Res Inst, New Delhi, 70-71. *Mem:* Genetics Soc Am; Brit Genetical Soc; Sigma Xi; Genetics Soc Can. *Res:* Wheat cytogenetics. *Mailing Add:* Dept Agron 220 Curtis Hall Univ Mo Columbia MO 65211

KIMBERLING, WILLIAM J, b Logansport, Ind, Oct 16, 40; m 62; c 6. GENE LINKAGE, COMMUNICATION DISORDERS. *Educ:* Ind Univ Sch Med, PhD(med genetics), 67. *Prof Exp:* Fel, Med Sch, Univ Ore, 67-71, instr med genetics, 71-72; asst prof, Med Sch, Univ Colo, 72-79; ASSOC PROF MED GENETICS, MED SCH, CREIGHTON UNIV, 79- *Concurrent Pos:* Res assoc, Boys Town Inst Hered Commun Dis, 79- *Mem:* AAAS; Am Soc Human Genetics. *Res:* Gene linkage and localization in humans. *Mailing Add:* Dept Otolaryngol Creighton Univ 2500 Cal St Omaha NE 68178

KIMBERLY, ROBERT PARKER, b New Haven, Conn; m; c 5. INTERNAL MEDICINE, RHEUMATOLOGY. *Educ:* Princeton Univ, BA, 68; Oxford Univ, BA & MA, 70; Harvard Univ, MD. *Prof Exp:* Clin assoc, Nat Inst Arthritis, Metab & Digestive Dis, NIH, 75-77; fel, Hosp Spec Surg, New York Hosp, 77-84, asst prof, 79-84; ASSOC PROF MED, CORNELL UNIV MED COL, 84- *Concurrent Pos:* Attending physician, Hosp Spec Surg, New York Hosp, 79-84, assoc attending physician, 84. *Mem:* Fel Am Col Physicians; fel Am Rheumatism Asn; Am Asn Immunologists; Am Soc Clin Invest. *Mailing Add:* Hosp Spec Surg 535 E 70th St New York NY 10021

KIMBLE, GERALD WAYNE, b Bakersfield, Calif, Apr 24, 28; m 48, 65. NUMERICAL ANALYSIS. *Educ:* Univ Calif, Berkeley, AB, 49, Univ Calif, Los Angeles, MA, 58, PhD(math), 62. *Prof Exp:* Mathematician, Nat Bur Standards Inst Numerical Anal, Univ Calif, Los Angeles, 49-51; engr, Hughes Aircraft Co, 55-58; res asst numerical anal, Univ Calif, Los Angeles, 59-61; asst prof math, Calif State Col, Long Beach, 61-63; assoc prof math & dir comput ctr, Univ Mont, 63; mathematician, TRW, Inc, Calif, 63-67; assoc prof, 67-75, EMER PROF MATH & COMPUT SCI, UNIV NEV, RENO, 75- *Mem:* Math Asn Am. *Res:* Calculus of variations; computer science. *Mailing Add:* 12360 Big Blue Rd Nevada City CA 95959

KIMBLE, GLENN CURRY, chemistry; deceased, see previous edition for last biography

KIMBLE, HARRY JEFFREY, b Floydada, Tex, Apr 23, 49. PHYSICS. *Educ:* Abilene Christian Univ, BS, 71; Univ Rochester, MA, 73, PhD(physics), 78. *Prof Exp:* Assoc sr res physicist, Gen Motors Res Labs, 77-79; PROF PHYSICS, UNIV TEX, AUSTIN, 79- *Mem:* Am Phys Soc; Optical Soc Am.

Res: Quantum optics; theory of coherence; atomic physics; infrared spectroscopy; semiconductor diode lasers; laser intracavity absorption spectroscopy; optoacoustic spectroscopy; theory of resonance fluorescence; photostatistics. *Mailing Add:* Div Physics Calif Tech MC 12-33 Pasadena CA 91125

KIMBLER, DELBERT LEE, b Whitman, WVa, Sept 8, 45; m 67. QUALITY ENGINEERING, SYSTEM SIMULATION. *Educ:* Univ SFla, BSE, 76; Va Polytech Inst & State Univ, MS, 78, PhD(indust eng & opers res), 80. *Prof Exp:* Instr indust eng & opers res, Va Polytech Inst & State Univ, 78-80; from asst prof to assoc prof indust eng, Univ SFla, 80-86; assoc prof, 86-90, PROF INDUST ENG, CLEMSON UNIV, 90- *Concurrent Pos:* Actg dept head, Indust Eng, Clemson Univ, 89-90. *Honors & Awards:* Mfg Systs Spec Citation, Inst Indust Engrs, 88. *Mem:* Sr mem Inst Indust Engrs; sr mem Am Soc Qual Control; Soc Mfg Engrs; Sigma Xi; Nat Soc Prof Engrs. *Res:* Quality engineering and quality management; discrete computer simulation; manufacturing modeling and analysis. *Mailing Add:* 104 Freeman Hall Clemson SC 29631-0920

KIMBLIN, CLIVE WILLIAM, b Stockport, Eng, Dec 25, 38; US Citizen; m 68; c 3. ARC & INTERRUPTION PHYSICS. *Educ:* Univ Liverpool, BSc, 60, PhD(elec eng), 64, Univ Pittsburgh, MSIE, 76. *Prof Exp:* Res engr arch physics, Westinghouse Res & Develop Ctr, 64-65, sr engr, 65-74, fel engr, 74-75, mgr, Power Interruption Res, 75-86; dir, Prod Develop, Holec, 86-89; DIR, TECHNOL DEVELOP, BEGEMANN, 90- *Mem:* Am Phys Soc; sr mem Inst Elec & Electronics Engrs; Inst Elec Engrs. *Res:* Physics of arcing and interruption in vacuum and air with particular reference to electrode mechanisms, properties of the inter-electrode metallic plasma and development of high power switching devices. *Mailing Add:* Bruglaan 1 3743 JB Baarn Netherlands

KIMBRELL, JACK T(HEODORE), b Peoria, Ill, Apr 23, 21; m 48; c 4. MECHANICAL ENGINEERING. *Educ:* Purdue Univ, BS, 43; Univ Mo-Columbia, MS, 47. *Prof Exp:* Prod engr, Eversharp Corp, 43-44; asst prof mech eng, Univ Mo, 47-51; sr res engr, Midwest Res Inst, 51-54; dean col eng, 69-70, chmn dept mech eng, 70-77, PROF MECH ENG, WASH STATE UNIV, 54-, ALCOA PROF, 66- *Mem:* Am Soc Eng Educ; Am Soc Mech Engrs. *Res:* Machinery design with emphasis on the kinematic synthesis of mechanisms. *Mailing Add:* 7401 Parkwood NW Albuquerque NM 87120

KIMBROUGH, JAMES W, b Eupora, Miss, Nov 7, 34; m 61; c 3. MYCOLOGY. *Educ:* Miss State Univ, BS, 57, MS, 60; Cornell Univ, PhD(mycol), 64. *Prof Exp:* Asst bacteriologist, Vet Sci Dept, Miss State Univ, 60-61, instr bot, 61-62; teaching asst mycol, Cornell Univ, 62-64; from asst prof to assoc prof, 64-76, assoc chmn dept, 74-75, actg chmn, 85-86, PROF BOT, UNIV FLA, 76- *Concurrent Pos:* Mem bd sci adv, Highlands Biol Sta; panelist, Nat Sci Found, 82-85. *Mem:* Mycol Soc Am (secy-treas, 74-77, vpres, 77-78, pres elect, 78-79, pres, 79-80); Bot Soc Am; Int Asn Plant Taxon; Brit Mycol Soc; Linnean Soc. *Res:* Developmental and taxonomic studies in the powdery mildews; morphological, developmental and taxonomic studies of operculate Discomycetes; fungi as biocontrol agents of termites. *Mailing Add:* Dept Plant Path Univ Fla Gainesville FL 32611

KIMBROUGH, THEO DANIEL, JR, b Lafayette, Ala, Sept 17, 33; m 65; c 2. PHYSIOLOGY, BIOCHEMISTRY. *Educ:* Univ Ala, BS, 55, MA, 59; Auburn Univ, PhD(zool), 65. *Prof Exp:* Teaching asst zool, Auburn Univ, 62-63, NIH fel physiol, 63-64; asst prof biol, Birmingham-Southern Col, 64-67; asst prof, 67-74, ASSOC PROF BIOL, VA COMMONWEALTH UNIV, 74- *Concurrent Pos:* NSF grant, Res Small Col Fac, Univ NC, 83. *Mem:* Fel Am Inst Chem; Soc Indust Microbiol; Sigma Xi; Am Entom Soc. *Res:* Physiology of intestinal serotonin; animal physiology; histology; physiology of intestinal serotonin; insect physiology, characterization of serotonin receptors in the cockroach digestive tract; neurophysiology, study of effects of aspartame on serotonin levels in brain and on behavior in mice. *Mailing Add:* 10300 Waltham Dr Richmond VA 23284

KIME, JOSEPH MARTIN, b Akron, Ohio, Feb 16, 17; m 41; c 3. PHYSICS. *Educ:* Univ Akron, BS, 38. *Prof Exp:* Physicist, B F Goodrich Co, 39-41 & 45-60, sect leader, New Prod Dept, 60-62, tech mgr, Bldg Prod Dept, 62-64, mgr prod res, 65-78, dir, Int Tech Admin, 78-81; CONSULT, 81- *Mem:* Am Phys Soc; Acoust Soc Am; Inst Elec & Electronics Engrs. *Res:* Underwater acoustics; application of plastics in the building industry; electrical, mechanical and acoustic properties of high polymers. *Mailing Add:* 1846 Indian Hills Trail Akron OH 44313-4790

KIMEL, JACOB DANIEL, JR, b Winston-Salem, NC, Aug 11, 37; m 65; c 4. PARTICLE PHYSICS, THEORETICAL PHYSICS. *Educ:* Univ NC, Chapel Hill, BS, 59; Univ Wis-Madison, MS, 60, PhD(physics), 66. *Prof Exp:* Res assoc physics, Univ Wis-Madison, 66; res assoc, 66-67, from asst prof to assoc prof, 67-88, PROF PHYSICS, FLA STATE UNIV, 88-, DIR GRAD AFFAIRS, PHYSICS DEPT, 89- *Mem:* Sigma Xi; Am Phys Soc. *Res:* Theoretical physics with emphasis on studies of elementary particle properties and intearctions; computational and simulational physics. *Mailing Add:* Dept Physics Fla State Univ Tallahassee FL 32306

KIMEL, WILLIAM R(OBERT), b Cunningham, Kans, May 2, 22; m 52. MECHANICAL & NUCLEAR ENGINEERING. *Educ:* Kans State Univ, BS, 44, MS, 49; Univ Wis, PhD(eng mech), 56. *Prof Exp:* Design engr, Goodyear Tire & Rubber Co, Ohio, 44-46; from instr to asst prof mech eng, Kans State Univ, 46-54; engr, US Forest Prod Lab, Univ Wis, 55-56; assoc prof mech eng, Kans State Univ, 54-59, prof nuclear eng & head dept, 58-68; prof nuclear eng, dir Eng Exp Sta & dean Col Eng, 68-86, EMER DEAN & PROF COL ENG, UNIV MO, COLUMBIA, 86- *Concurrent Pos:* Mem, Kans Gov Atomic Energy Adv Coun, 61-68, chmn, 66-68; AEC observer, UN Atomics for Peace Conf, 64 & 71; mem, Adv Comt to Off Civil Defense, 64-70, chmn, 69-70; consult, Div Nuclear Educ & Training, AEC, 66-68; mem, fel rev panel, HEW, 66-68. *Mem:* AAAS; Am Soc Eng Educ; fel Am Soc Mech Engrs; Am Nuclear Soc (vpres & pres-elect, 77-78, pres, 78-79); Nat Soc Prof Engrs (vpres-elect, 85-86); Jr Eng Tech Soc (pres, 80-81). *Res:* Engineering mechanics composite structures; nuclear reactor and shielding analysis. *Mailing Add:* Dean Emeritis Col Eng Univ Mo 900 Yale Columbia MO 65203

KIMELBERG, HAROLD KEITH, b Hertford, Eng, Dec 5, 41; nat US; m 66; c 2. BIOCHEMISTRY. *Educ:* Univ London, BS, 63; State Univ NY Buffalo, PhD(biochem), 68. *Prof Exp:* Fel, Johnson Res Found, Univ Pa, 68-69; NIH fel, Roswell Park Mem Inst, 69-70, sr cancer res scientist, 70-74; asst res prof, Roswell Park Div, State Univ NY Grad Sch, 72-74; res assoc prof neurosurg, Albany Med Col, Union Univ, 74-80, assoc prof biochem, 74-88, prof anat, 80-88, RES PROF NEUROSURG, 80-, PROF PHARMACOL & TOXICOL, ALBANY MED COL, UNION UNIV, 87- *Concurrent Pos:* Adj assoc prof biol, State Univ NY, Albany, 78-88; adj prof biol, State Univ NY, Albany; prof biochem, Albany Med Col, Union Univ, 88-; Fulbright grantee, Univ Heidelberg Frg, 88; mem, Nat Inst Neurol Dis & Stroke Study Sect, Neurol Sci, 91-94. *Mem:* Am Soc Biol Chemists; Am Soc Neurochem; Soc Neurosci. *Res:* Ion transport, electrical properties, transmitter responsiveness and enzyme properties of astroglial cells, especially using cultured cells as models; roles of astrocytes in trauma and ishemia especially in regard to cerebral cellular edema. *Mailing Add:* Div Neurosurg A-60 Albany Med Col Albany NY 12208

KIMELDORF, GEORGE S, b New York, NY, Sept 3, 40; m 64; c 2. STATISTICS, MATHEMATICS. *Educ:* Univ Rochester, AB, 60; Univ Mich, MA, 61, PhD(math), 65. *Prof Exp:* From asst prof to assoc prof math, Calif State Univ, Hayward, 65-69; from asst prof to assoc prof statist, Fla State Univ, 69-75; assoc prof, 75-76, PROF MATH SCI, UNIV TEX, DALLAS, 76- *Concurrent Pos:* Vis asst prof, Math Res Ctr, Univ Wis, 67-68, vis assoc prof, 68-69. *Mem:* Inst Math Statist; Am Statist Asn. *Res:* Mathematical statistics; probability; operations research. *Mailing Add:* Math Sci Univ Tex Dallas PO Box 830688 Richardson TX 75083-0688

KIMERLING, LIONEL COOPER, b Birmingham, Ala, Dec 2, 43; m 66; c 2. SOLID STATE SCIENCE, PHYSICS. *Educ:* Mass Inst Technol, SB, 65, PhD(mat sci), 69. *Prof Exp:* Res asst, Dept Metall & Mat Sci, Mass Inst Technol, 65-69; res physicist, Air Force Cambridge Res Labs, USAF, 69-72; mem tech staff, 72-81, HEAD, MAT PHYSICS RES DEPT, BELL LABS, 81- *Concurrent Pos:* Vis fel, Inst Study Defects Solids, 75-; adj prof phys, Lehigh Univ, 77-; lectr, Welsh Found, 79. *Mem:* AAAS; Am Inst Mining, Metall & Petrol Engrs; Am Phys Soc; Electrochem Soc; Sigma Xi. *Res:* Defects in solids: structure, electrical properties and chemical reactions; elemental and III-V semiconductor systems. *Mailing Add:* 734 Crescent Pkwy Westfield NJ 07090

KIMES, BRIAN WILLIAMS, b Tampa, Fla, Sept 7, 43; m. CANCER RESEARCH. *Educ:* Stanford Univ, BS, 66; Univ Wash, PhD(biochem), 71. *Prof Exp:* Am Cancer Soc fel, Salk Inst Biol Studies, San Diego, 71-73, res assoc, 73-75; grants assoc, Div Res Grants, NIH, 75-76, prog dir, Extramural Tumor Biol Prog, Div Cancer Resources & Ctr & Div Cancer Biol & Diag, Nat Cancer Inst, 76-79, chief, Extramural Cancer Biol Br, 79-85, assoc dir, Extramural Res Prog, 82-88, ASSOC DIR, CENTERS, TRAINING & RESOURCE PROG, DIV CANCER BIOL & DIAG, NAT CANCER INST, NIH, 89- *Concurrent Pos:* Mem, Div Cancer Biol & Diag Steering Comt, Nat Cancer Inst, 79-82, actg dir, Cancer Diag Br, 83-86. *Res:* Structure and function of ribosomes; vitro protein synthesis; vitro macromolecular synthesis; tissue culture of nerve; muscle and fibroblast cell lines; electron microscopy; cell-cell communication. *Mailing Add:* NIH Nat Cancer Inst Ctrs Training & Resources Prog Exec Plaza N Rm 300 6130 Executive Blvd Rockville MD 20892

KIMES, THOMAS FREDRIC, b Phoenixville, Pa, July 24, 28; m 52, 74; c 2. NUMERICAL ANALYSIS, MATHEMATICAL ANALYSIS. *Educ:* Ursinus Col, BS, 49; Univ Tex, MA, 56; Carnegie Inst Technol, PhD, 62. *Prof Exp:* Asst math, Univ Tex, 54-56; asst, Carnegie Inst Technol, 56-58, proj mathematician, 58-59; mathematician, Bettis Atomic Power Lab, Westinghouse Elec Corp, 59-60, sr mathematician, 60-62; from asst prof to prof math, 62-76, chmn dept, 62-73, dir interactive comput serv, 73-76, DIR, JAN TERM PROG, CHADWICK PROF MATH, 76-, CHMN DEPT, AUSTIN COL, 84- *Concurrent Pos:* Vis prof math, Furness Col, Univ Lancaster, 70; vis scholar, Cambridge Univ, 80, 87. *Mem:* Am Math Soc; Math Asn Am; Prehistoric Soc UK. *Res:* Ordinary and partial differential equations. *Mailing Add:* 2501 Turtle Creek Dr Sherman TX 75090

KIMLER, BRUCE FRANKLIN, b St Paul, Minn, Sept 23, 48; m 75; c 2. RADIATION BIOLOGY, CANCER BIOLOGY. *Educ:* Univ Tex, Austin, BA, 70, MA, 71, PhD(radiation biol), 73. *Prof Exp:* Appointee radiation biol, Argonne Nat Lab, 73-75; fel, Thomas Jefferson Univ Hosp, 75-77; from asst prof to assoc prof radiation ther, 77-84, PROF DEPT RADIATION ONCOL, UNIV KANS MED CTR, KANSAS CITY, 84- *Concurrent Pos:* Assoc scientist, Mid-Am Cancer Ctr Prog, 77-81; adj prof, Dept Radiation Biophys, Univ Kans, 77-82; adj prof, Dept Pharmacol, Toxicol & Therapeut, Univ Kans Med Ctr, 90- *Mem:* Radiation Res Soc; Am Asn Cancer Res; Cell Kinetics Soc (secy, 81-83); Am Soc Therapeut Radiol & Oncol; Sigma Xi; Int Soc Anal Cytol. *Res:* Preclinical experimental radiation oncology; cancer chemotherapeutic agents, hyperthermia and other combined modalities; flow cytometry; in utero effects of radiation. *Mailing Add:* Dept Radiation Oncol Univ Kans Med Ctr 39th & Rainbow Blvd Kansas City KS 66103

KIMLIN, MARY JAYNE, b Cresson, Pa, July 4, 24. ANALYTICAL CHEMISTRY, PHYSICAL CHEMISTRY. *Educ:* St Francis Col, Pa, BS, 48; Pa State Univ, MS, 58, PhD(anal chem), 69. *Prof Exp:* Lab asst chem, Kinetic Div, E I du Pont de Nemours & Co, 44-46; from instr to assoc prof, 48-70, PROF CHEM, ST FRANCIS COL, PA, 70- *Mem:* Sigma Xi; AAAS; Am Chem Soc. *Res:* Polarographic study of correlations between ionic diffusion coefficients and ionic strength; polarography in fused salts; polarography of copper hydroxo complexes. *Mailing Add:* 515 Ashcroft Ave Cresson PA 16630

KIMME, ERNEST GODFREY, b Long Beach, Calif, June 7, 29; m 52; c 3. MATHEMATICS, ENGINEERING. *Educ:* Pomona Col, BA, 52; Univ Minn, MA, 54, PhD(math), 56. *Prof Exp:* Instr math, Ore State Col, 55-57; mem tech staff, Bell Tel Labs, 57-63, supvr, 63-65; head appl sci dept, Collins Radio Co, 65-67, asst to vpres, 67-72; res engr, Northrop Electronics, 72-74; prin engr, Interstate Electronics, 74-79; dir spec commun prog, Gould Naval Commun Syst Div, 79-82; chief scientist & chief exec officer, Cobit, Inc, 82-84; tech staff, Gen Res Corp, 84-87; VPRES ENG, STARFIND INC, 87- *Concurrent Pos:* Vpres & dir, A S Johnston Drilling Co, 68-; prin assoc, Ameta Consult Technologists, Anaheim, Calif, 78- *Mem:* AAAS; Soc Indust & Appl Math. *Res:* Analysis, particularly probability theory, mathematical statistics, information theory and operations research; communications sciences; quantitative management theory; experimental design. *Mailing Add:* 301 N Starfire Anaheim CA 92807

KIMMEL, BRUCE LEE, b Poplar Bluff, Mo, Nov 6, 45; m 67, 87; c 2. LIMNOLOGY, AQUATIC ECOLOGY. *Educ:* Baylor Univ, BS, 67, MS, 69; Univ Calif, Davis, PhD(ecol), 77. *Prof Exp:* Proj officer oceanic biol, Ocean Sci & Technol Group, Off Naval Res, 70-71; instr chem, Div Math Sci, US Naval Acad, 71-72; asst prof zool & asst dir, biol sta, Univ Okla, 77-80; res staff mem & group leader ecosysts dynamics, Mat Transport & Fate, 80-90, PROG MGR, OFF-SITE ENVIRON RESTORATION PROG, ENVIRON SCI DIV, OAK RIDGE NAT LAB, 90- *Concurrent Pos:* Prin investr res grant, Okla Water Resources Inst, Off Water Res & Technol, Dept Interior, 78-80; prin & co-prin investr res contracts, Dept Energy, US Army CEngrs & Tenn Valley Authority, NSF Ecosysts Progs, 80-; proj dir, RCRA Facil Invest, Environ Sci Div, Oak Ridge Nat Lab, 87- *Mem:* Am Soc Limnol & Oceanog; Int Asn Theoret & Appl Limnol; Ecol Soc Am; Am Inst Biol Sci; AAAS; Sigma Xi. *Res:* Biological productivity and foodweb interactions in lakes and reservoirs; energy flow and nutrient cycling in aquatic systems; contaminant transport, fate, and effects in river-reservoir ecosystems; reservoir and lake liminology and ecology; water resources management. *Mailing Add:* Environ Sci Div Oak Ridge Nat Lab PO Box 2008 Oak Ridge TN 37831-6038

KIMMEL, CAROLE ANNE, b Lexington, Ky, Apr 26, 44; m 70; c 1. TERATOLOGY. *Educ:* Georgetown Col, BS, 66; Univ Cincinnati, PhD(anat), 70. *Prof Exp:* Fel toxicol, Col Med, Univ Cincinnati, 70-72; sect head teratology, Environ Protection Agency, 72; instr anat, Sch Med, Harvard Univ, 72-73; sr staff fel teratology, Nat Inst Environ Health Sci, 73-77, chief, Perinatal & Postnatal Eval Br, 79-84, RES PHARMACOLOGIST, DIV TERATOGENESIS RES, NAT CTR TOXICOL RES, NIH, 77-, CHIEF, PERINATAL & POSTNATAL EVAL BR, 79-; AT ENVIRON PROTECTION AGENCY. *Concurrent Pos:* Consult, Environ Protection Agency, 72-73; adj asst prof, Sch Med, Univ NC, 75-76; prog mgr, Reproductive & Develop Toxicol, Nat Toxicol Prog, 79-86; adj assoc prof, Interdis Toxicol Div, Univ Ark Med Sci, 82-89; guest worker, Food & Drug Admin, 84-; adj prof, Univ Md, 89- *Mem:* Soc Toxicol; Neurobehav Teratology Soc; Teratology Soc; Europ Teratology Soc. *Res:* The effects of developmental exposures to drugs and environmental agents on the behavior and function of offspring; test methodology in teratology; extrapolation of animal data for human risk assessment. *Mailing Add:* 401 M St SW Rm 3809 Mall USEPA RD-689 Washington DC 20460

KIMMEL, CHARLES BROWN, b New Orleans, La, May 3, 40; m 62; c 1. DEVELOPMENTAL BIOLOGY. *Educ:* Swarthmore Col, BA, 62; Johns Hopkins Univ, PhD(biol), 66. *Prof Exp:* NIH fel immunol, Salk Inst Biol Studies, 66-68; asst prof biol, 69-75, ASSOC PROF BIOL, UNIV ORE, 75- *Mem:* AAAS; NY Acad Sci; Soc Develop Biol; Soc Neurosci. *Res:* Experimental neurogenesis; vertebrate neural development. *Mailing Add:* Dept Biol Univ Ore Eugene OR 97403

KIMMEL, DONALD LORAINE, JR, b Swedesboro, NJ, Apr 15, 35; m 83; c 5. DEVELOPMENTAL BIOLOGY. *Educ:* Swarthmore Col, BA, 56; Temple Univ, MD, 60, MSc, 62; Johns Hopkins Univ, PhD(developgenetics), 64. *Prof Exp:* Asst prof, Div Med Sci, Brown Univ, 64-71; assoc prof & chmn dept, 71-78, PROF BIOL, DAVIDSON COL, 78- *Concurrent Pos:* Res fel, Calif Inst Technol, 70-71; corp mem, Bermuda Biol Sta; res scientist, Div Res NC Dept Mental Health, 78-79. *Mem:* AAAS; Am Soc Zool; Soc Develop Biol; Am Soc Cell Biol. *Res:* Web building behavior, neural control of web building and web geometry during development of orb weaving spiders. *Mailing Add:* Dept Biol Davidson Col Davidson NC 28036

KIMMEL, ELIAS, b New York, NY, Mar 7, 24; m 45; c 2. PHYSICAL CHEMISTRY. *Educ:* Univ Long Island, BS, 48; Polytech Inst Brooklyn, PhD(chem), 53. *Prof Exp:* Res chemist, Gen Chem Div, Allied Chem Corp, 52-54; consult chemist, Foster D Snell, Inc, 54-57; DIR RES & DEVELOP, TEMPIL DIV, BIG THREE INDUSTS, INC, 57- *Res:* Temperature measurement; phase-rule studies of inorganic and organic systems; surface phenomena as applied to dispersion; corrosion prevention; lubrication; properties and utilization of polymeric materials. *Mailing Add:* 2050 Audubon Ave South Plainfield NJ 07080

KIMMEL, GARY LEWIS, b Dayton, Ohio, Nov 20, 45; m 70. BIOCHEMICAL TERATOLOGY. *Educ:* Miami Univ, AB, 67; Univ Cincinnati, MS, 69, PhD(physiol), 72. *Prof Exp:* Fel steroid biochem, Worcester Found Exp Biol, 72-73; reproduction biologist, Res Triangle Inst, 73-75; RES PHYSIOLOGIST, NAT CTR TOXICOL RES, 75-, DEVELOP TOXICOLOGIST, ENVIRON PROTECTION AGENCY. *Mem:* Soc Study Reprod; NY Acad Sci. *Res:* The relationship of biochemical events and developmental alterations, specifically at the cellular and subcellular level. *Mailing Add:* 401 M St SW Rm 3809 Mall USEPA Rd-689 Washington DC 20460

KIMMEL, HOWARD S, b Brooklyn, NY, Feb 2, 38; m 64; c 1. PHYSICAL CHEMISTRY. *Educ:* Brooklyn Col, BS, 59; WVa Univ, 61; City Univ New York, PhD(phys chem), 67. *Prof Exp:* Res assoc chem, Isaac Albert Res Inst, Jewish Chronic Dis Hosp, Brooklyn, NY, 62-63; asst prof, 66-70, assoc prof, 70-80, PROF CHEM & ASSOC CHMN DEPT, NJ INST TECHNOL, 80- *Mem:* Am Chem Soc. *Res:* Vibrational spectra of inorganic coordination compounds, tin compounds, olefinic compounds, monosubstituted benzene derivatives and biochemicals; kinetics studies using infrared spectroscopy. *Mailing Add:* Dept Chem Newark Col Eng Newark NJ 07102

KIMMEL, JOE ROBERT, b DuQuoin, Ill, May 3, 22; m 47; c 4. BIOCHEMISTRY, MEDICINE. *Educ:* DePauw Univ, AB, 44; Johns Hopkins Univ, MD, 47; Univ Utah, PhD(biochem), 54. *Prof Exp:* Intern, Salt Lake Gen Hosp, 47-49; from res instr to assoc res prof biochem, Col Med, Univ Utah, 54-60, assoc prof, 60-64; prof biochem & med, Univ Kans, 64-; AT VET HOSP, KANSAS CITY. *Concurrent Pos:* USPHS fel, 60-64. *Mem:* Endocrine Soc; Am Soc Biol Chemists; Sigma Xi. *Res:* Protein chemistry; proteolytic enzymes; insulin and other pancreatic hormones. *Mailing Add:* ACOS/R & D-151 Veterans Hosp 4801 Linwood Kansas City KS 64128

KIMMEL, ROBERT MICHAEL, b Beverly, Mass, Feb 6, 43; m 65; c 3. MATERIALS ENGINEERING, POLYMER PHYSICS. *Educ:* Mass Inst Technol, BS, 64, MS, 65, Mat Engr, 67, ScD(mat eng), 68. *Prof Exp:* Res chemist, Celanese Res Co, 68-71, sr res chemist, 71-73, group leader, Celanese Plastics Co, 73-75, prod supvr, 76-79; INDUST MGR, AM HOECHST CORP, 80- *Mem:* Am Phys Soc; Fiber Soc; Sigma Xi. *Res:* Fiber physics; physics of acrylic polymers; formation structure and properties of graphite fibers; effects of high pressure on materials; nature of the glassy state; structure properties of oriented films. *Mailing Add:* 116 Bridgeton Dr Greenville SC 29615

KIMMEL, WILLIAM GRIFFITHS, b Scranton, Pa, Aug 27, 45; m 69; c 2. WATER POLLUTION BIOLOGY. *Educ:* Wilkes Col, BA, 67; Pa State Univ, MS, 70, PhD(zool), 72. *Prof Exp:* from asst prof to assoc prof, 76-82, PROF BIOL & ENVIRON SCI, 82-, CHMN, DEPT BIOL & ENVIRON SCI, CALIFORNIA UNIV, PA, 87- *Mem:* Sigma Xi; Am Fisheries Soc. *Res:* Responses of stream ecosystems to environmental stresses; aquatic macroinvertebrates and fishes as indicators of water quality. *Mailing Add:* RD 3 Box 25C Charleroi PA 15022

KIMMERLE, FRANK, b Tuebingen, Ger, Mar 27, 40; Can citizen; m 63; c 3. ELECTROCHEMISTRY. *Educ:* Univ Toronto, BS, 63, MS, 64, PhD(electrochem), 67. *Prof Exp:* NATO fel, Free Univ Brussels, 67-68; from asst prof to assoc prof chem, Univ Sherbrooke, 68-84; CHIEF ANALYTICAL CHEM, ALCAN INT LTD, 84- *Concurrent Pos:* Can Res Coun fel, 67. *Mem:* Electrochem Soc. *Res:* Thermodynamics and kinetics of adsorption and reaction of organic compounds at the electrode-electrolyte interface; electro-organic synthesis; non-aqueous battery systems. *Mailing Add:* Exp Eng Ctr Alcan Int Ltd PO Box 1500 Jonquiere PQ G7S 4L2 Can

KIMMEY, JAMES WILLIAM, b St Johns, Ore, Jan 24, 07; m 32; c 2. FOREST PATHOLOGY. *Educ:* Ore State Col, BSF, 31, MS, 32; Yale Univ, PhD(forestry), 40. *Prof Exp:* Field asst, US Forest Serv, 27; field asst div forest path, Bur Plant Indust, Agr Res Admin, USDA, 28-35, asst pathologist, 35-42, assoc pathologist, Bur Plant Indust, Soils & Agr Eng, Agr Res Admin, 42-47, pathologist, 47-53; pathologist, US Forest Serv, 53-56, sr pathologist, 56-57, chief div forest dis res, Intermountain Forest & Range Exp Sta, 57-63, proj leader, 63-65; RETIRED. *Concurrent Pos:* Consult, 65-75. *Mem:* Fel AAAS; Soc Am Foresters; Am Phytopath Soc; Am Forestry Asn; Arctic Inst NAm; Sigma Xi. *Res:* White pine blister rust in western United States and Canada; decay of wood in use; deterioration of fire-killed timber in Pacific Coast States; decay and other cull in timber stands of California and Alaska; forest disease survey in California and Alaska; dwarf mistletoes. *Mailing Add:* PO Box 19 Westport WA 98595

KIMMICH, GEORGE ARTHUR, b Cortland, NY, Dec 8, 41; m 63; c 2. BIOCHEMISTRY, CELL PHYSIOLOGY. *Educ:* Cornell Univ, BS, 63; Univ Wis-Madison, MS, 65; Univ Pa, PhD(biochem), 68. *Prof Exp:* Nat Inst Dent Res fel biophys, 68-70, from asst prof to assoc prof radiation biol & biophys, 70-83, PROF BIOCHEM, RADIATION BIOL & BIOPHYS & ASSOC CHMN BIOCHEM, SCH MED & DENT, UNIV ROCHESTER, 83- *Concurrent Pos:* Prin investr, NIH grant, 71; NIH Res Career Develop Award, 72-77; vis lectr biochem, Univ Manchester Inst Sci & Technol, Manchester, Eng, 75-76. *Mem:* Am Soc Biol Chemists; Am Physiol Soc. *Res:* Epithelial ion transport; metabolic regulation; sodium-dependent transport systems for sugars and amino acids; bioenergetics. *Mailing Add:* Dept Biochem Univ Rochester Sch Med & Dent Rochester NY 14642

KIMMINS, JAMES PETER (HAMISH), b Jerusalem, Israel, July 31, 42; m 64; c 2. FOREST ECOLOGY, ENVIRONMENTAL SCIENCES. *Educ:* Univ Wales, BSc, 64; Univ Calif, MSc, 66; Yale Univ, MPhil, 68, PhD(ecol), 70. *Prof Exp:* From asst prof to assoc prof, 69-79, PROF FOREST ECOL, DEPT FOREST SERV, UNIV BC, 79- *Concurrent Pos:* BC Govt Ecol Reserves Comt, 69-; consult, Environ Res Consults, 71-; chmn nat forest ecol working group, Can Inst Forestry, 74-76; Killam fel, Can Coun, 75-76; vis scientist, Kyoto, Japan, 82. *Honors & Awards:* Gold Medal Sci Achievement, Int Union Forest Res Orgns, 86; Gold Medal Sci Achievement, Can Inst Forestry, 87. *Mem:* Commonwealth Forestry Asn; Can Inst Forestry. *Res:* Nutrient cycling in forest ecosystems; biogeochemistry of forest land management; effects of herbicides, whole-tree logging and slashburning on system watershed nutrient budgets; forecyte and forecast ecologically-based forest management computer simulation models; ecology of forest regeneration. *Mailing Add:* Dept Forest Sci Univ BC 2357 Main Mall Vancouver BC V6T 1W5 Can

KIMMINS, WARWICK CHARLES, b London, Eng, July 20, 41; m 64; c 2. MARINE MAMMALS. *Educ:* Univ London, BSc, 62, PhD, 65. *Prof Exp:* From asst prof to assoc prof biol, 65-74, chmn dept, 81-82 & 85-90, PROF BIOL, DALHOUSIE UNIV, 74-, DEAN, FAC SCI, 90- *Concurrent Pos:* Consult res dept, Encyclop Britannica, 62-65; asst lectr, SE Essex Col Technol, Eng, 63-64. *Mem:* Soc Exp Biol & Med; Brit Asn Appl Biol; Can Soc Zool. *Res:* Marine mammals; large predator prey impact on fisheries; parasitology; population control. *Mailing Add:* Dept Biol Dalhousie Univ Halifax NS B3H 3J5 Can

KIMMONS, GEORGE H, ENGINEERING. *Prof Exp:* RETIRED. *Mem:* Nat Acad Eng. *Mailing Add:* Rte 33 Williams Rd Knoxville TN 37932

KIMOTO, MASAO, b Osaka, Japan, Aug 24, 47; m 74; c 3. IMMUNOLOGY, GENETICS. *Educ:* Osaka Univ, MD, 72, PhD(med), 81. *Prof Exp:* Res assoc immunol, Osaka Univ, 72-78, asst prof internal med, 81-85; res fel immunol, Mayo Clin, 78-81; PROF IMMUNOL, SAGA MED SCH, 86- *Mem:* Am Asn Immunol. *Res:* Structure and function of major histocompatibility complex. *Mailing Add:* Dept Immunol Saga Med Sch Nabeshima Saga 840-01 Japan

KIMOTO, WALTER IWAO, b Honolulu, Hawaii, Mar 25, 32. FOOD CHEMISTRY. *Educ:* Univ Hawaii, BA, 54; Univ Wis, PhD(org chem), 61. *Prof Exp:* Res chemist, Am Oil Co, 61-63; res chemist, Sci & Educ Admin-Agr Res, USDA, Beltsville, 63-71, res chemist, Eastern Regional Res Ctr, 71-85; RETIRED. *Mem:* Am Chem Soc; Sigma Xi. *Res:* Flavor and aroma of cured meat products. *Mailing Add:* 15790 E Alameda Pkwy Apt 4-210 Aurora CO 80017-2034

KIMPEL, JAMES FROOME, b Cincinnati, Ohio, Apr 18, 42; m 86; c 3. METEOROLOGY. *Educ:* Denison Univ, BS, 64; Univ Wis-Madison, MS, 70, PhD(meteorol), 73. *Prof Exp:* Weather officer meteorol, USAF, 64-68; res asst, Univ Wis-Madison, 68-73; asst prof, Univ Okla, 73-79, assoc dean eng, Col Eng, 78-81, dir, Sch Meteorol, 81-87, PROF METEOROL, UNIV OKLA, 86-, DEAN COL GEOSCI, 87- *Concurrent Pos:* Prin investr, Nat Oceanic & Atmospheric Admin, NSF, 74-90; pres, Appl Systs Inst, Inc, 86; mem, Adv Comt Atmospheric Sci, NSF, 83-86, chair, 89-; elected trustee, Univ Corp Atmospheric Res, 86-, chair, bd trustees, 91-; fel explor res, Elec Power Res Inst, 89- *Mem:* Fel Am Meteorol Soc; Nat Weather Asn; Am Asn Geog; Sigma Xi. *Res:* Synoptic and severe storms research for hydrometeorological applications. *Mailing Add:* Col Geosci Univ Okla Norman OK 73019

KIMSEY, LYNN SIRI, b Oakland, Calif, Feb 1, 53; m 76. SYSTEMATICS, FUNCTIONAL MORPHOLOGY. *Educ:* Univ Calif, Davis, BS, 75, PhD(entomol), 79. *Prof Exp:* Res Fel, Univ Calif, Davis, 79-80, 83-86 & Smithsonian Trop Res Inst, Panama, 81-82; lectr, dept entom, Univ Calif, Davis, 82-83; CONSULT, 83- *Mem:* AAAS; Sigma Xi. *Res:* Taxomony, behavior and physiology of a variety of insects, especially the hymenopteran families Chrysididae, Pompilidae, Apidae and Sphecidae; feeding energetics and male behavior of orchid bees and the ecology of neotropical sawflies. *Mailing Add:* Dept Entom Univ Calif Davis CA 95616

KIMURA, ARTHUR K, cancer research; deceased, see previous edition for last biography

KIMURA, EUGENE TATSURU, b Sheridan, Wyo, Sept 19, 22; m 50; c 3. PHARMACOLOGY, TOXICOLOGY. *Educ:* Univ Nebr, BS, 44, MS, 46; Univ Chicago, PhD(pharmacol), 48. *Prof Exp:* Lab instr inorg chem & microbiol, Sch Nursing, St Elizabeth Hosp, Nebr, 45-46; res pharmacologist, Nepera Chem Co, NY, 49-55; res pharmacologist, 55-60, group leader, 60-62, sect head, 62-71, mgr dept autonomic pharmacol, Sci Div, 71-72, mgr corp res develop dept, Corp Res & Exp Ther, 72-76, SR TOXICOLOGIST, ABBOTT LABS, 76-, ASSOC RES FEL, 80- *Concurrent Pos:* Fel pharmacol, Univ Chicago, 48-49; asst, Univ Nebr, 44-46. *Mem:* Am Soc Pharmacol & Exp Therapeut; Am Pharmaceut Asn; NY Acad Sci; Soc Exp Biol & Med; Soc Toxicol. *Res:* Pharmacology and toxicology of antihistamines and antiserotonins; bronchial asthma and allergy; drugs affecting ciliary motility; antiheparin agents; analeptics; antispasmodics; blood coagulants; anti-inflammatory compounds; toxicology and carcinogenicity of experimental drugs. *Mailing Add:* Abbott Labs North Chicago IL 60064

KIMURA, HIDENORI, b Tokyo, Japan, Nov 3, 41; m 71; c 3. CONTROL ENGINEERING. *Educ:* Univ Tokyo, BEng, 65, MEng, 67, DEng, 70. *Prof Exp:* Res asst control eng, Osaka Univ, 70-71, lectr, 71-73, assoc prof, Dept Control Eng, 73-87, PROF CONTROL ENG, DEPT MECH ENG, OSAKA UNIV, 87- *Mem:* Fel Soc Instrument & Control Engrs; fel Inst Elec & Electronics Engrs. *Res:* Control engineering; theory; mechanical control systems; robotics. *Mailing Add:* Dept Mech Eng Osaka Univ 2-1 Yamada-oka Suita 565 Japan

KIMURA, JAMES HIROSHI, b Kona, Hawaii, Oct 29, 44; m 75; c 2. BIOCHEMISTRY. *Educ:* Univ Hawaii, BS, 71; Case Western Reserve Univ, PhD(develop biol), 76. *Prof Exp:* Fel, Nat Inst Dent Res, NIH, 76-78, staff fel, 78-80, sr staff fel, 80-81; from asst prof to prof, Dept Biochem, 81-90, from asst prof to prof biochem, Dept Orthop Surg, Rush Presby St Luke's Med Ctr, 81-90, SR STAFF, BONE & JOINT CTR, HENRY FORD HOSP, 90- *Mem:* Orthop Res Soc; Am Soc Biochem & Molecular Biol; AAAS; Soc Complex Carbohydrates. *Res:* Biochemistry of proteoglycans and mechanisms for the control of their synthesis; organization in cartilage. *Mailing Add:* Henry Ford Hosp 1653 W Congress Pkwy Chicago IL 60612-3864

KIMURA, KAZUO KAY, b Sheridan, Wyo, Sept 16, 20; m 52; c 3. CLINICAL PHARMACOLOGY. *Educ:* Univ Wash, BS, 42; Univ Nebr, MS, 44; Univ Ill, PhD(pharmacol), 49; St Louis Univ, MD, 53; Am Bd Med Toxicol, dipl, 75. *Prof Exp:* Asst physiol & pharmacol, Univ Nebr, 43-44, instr, 44-45; asst pharmacol, Univ Ill, 45-46; from instr to asst prof, St Louis Univ, 49-54; intern, Children's Med Serv, Mass Gen Hosp, 54-55, asst resident, 55-56; from sr resident to chief resident pediat, Raymond Blank Mem Hosp Children, Des Moines, Iowa, 56-57; chief exp med br, Chem Res & Develop Labs, Army Chem Ctr, Md, 57-61; asst chief clin res div, 58-61, chief, 62; chief pharmacologist & mgr pharmacol sect, Atlas Chem Industs, Inc, 62-63; corp med dir & dir biomed res dept, 63-70; med dir & dir med res dept, ICI Am, Inc, Del, 70-73; vpres & res dir, Hazleton Labs Am, Inc, 73-75; exec vpres & med dir, Mediserve Int, Inc, 75-77; prof pharmacol & med, Sch Med, Wright State Univ, 77-88; MED DIR, WESTERN OHIO REGIONAL DRUG & POISON INFO SYST, CHILDRENS MED CTR, 80- *Concurrent Pos:* Teaching fel pediat, Harvard Med Sch, 55-56; founder & first dir, Iowa State Poison Info Ctr, 56-57; pvt pract, Md, 58-62; prof lectr, Univ Md, 60-62; consult pediat, Wilmington Med Ctr Hosps, 69-73; consult med, Vet Admin Hosp, 78-80; mem active staff, Childrens Med Ctr. *Honors & Awards:* Borden Award, 53. *Mem:* Fel AAAS; fel Am Col Clin Pharmacol (secy, 85-, vpres, 87, pres, 88-90); fel Royal Soc Med; fel Am Col Physicians; Am Soc Pharmacol & Exp Therapeut. *Res:* Striated muscle paralyzing drugs; cardiac glycosides; medical toxicology; biochemophology of muscle paralyzants; dietary fiber and refined sugars; psychopharmacology; clinical pharmacology; poison control; pediatric and general toxicology. *Mailing Add:* Vet Admin Med Ctr Environ Med Off Wright State Univ Sch Med 19 E Blossom Hill Rd Dayton OH 45449

KIMURA, KEN-ICHI, b Fushun, China, Mar 5, 33; Japan citizen; m 61; c 2. NATURAL ENERGY UTILIZATION, DESICCANT COOLING & DEHUMIDIFICATION WITH SOLAR ENERGY. *Educ:* Waseda Univ, BArch, 57, MSc, 59, DEng, 65. *Prof Exp:* From asst prof to assoc prof, 64-73, PROF ENVIRON ENG & ARCHIT, WASEDA UNIV, 73- *Concurrent Pos:* Teaching asst, Mass Inst Technol, 60-62; postdoctoral fel, Nat Res Coun, 67-69; mem bd dirs, Int Solar Energy Soc, 73-79 & Archit Inst Japan, 80-82. *Mem:* Int Solar Energy Soc; fel Am Soc Heating, Refrig & Air Conditioning Engrs. *Res:* Architectural utilization of solar energy; energy conservation in buildings; thermal comfort; indoor air quality; visual comfort in interiors; heat and moisture transfer in buildings; environmental engineering; architectural design; building science. *Mailing Add:* 13-21 Enoki-cho Tokorozawa Saitama 359 Japan

KIMURA, MINEO, b Tokyo, Japan, Nov 15, 46; m 73; c 3. CHEMICAL PHYSICS, CONDENSED MATTER PHYSICS. *Educ:* Waseda Univ, Japan, BSc, 70; Univ Tokyo, MSc, 72; Univ Alta, Can, PhD(chem physics), 81. *Prof Exp:* Asst prof atomic physics, Univ Mo, Rolla, 81-84; scientist, Joint Inst Lab Astrophys, Univ Colo & Nat Bur Standards, 84-86; PHYSICIST, ARGONNE NAT LAB, 86- *Concurrent Pos:* Adj prof, Dept Physics, Rice Univ, 86- *Mem:* Am Phys Soc; Radiation Res Soc; Sigma Xi. *Res:* Theoretical atomic physics; radiation physics and biology; theoretical condensed matter physics. *Mailing Add:* Argonne Nat Lab 9700 S Cass Ave Bldg 203 Argonne IL 60439

KIMURA, NAOKI, b Wakayama City, Japan, May 20, 22; m 47; c 3. MATHEMATICS. *Educ:* Osaka Univ, DSc, 44; Tulane Univ, PhD, 57. *Prof Exp:* Lectr math, Tokyo Inst Technol, Japan, 49-55; asst prof, Univ Wash, 58-60 & Univ Sask, 60-62; assoc prof, Univ Okla, 62-65; PROF MATH, UNIV ARK, FAYETTEVILLE, 65- *Mem:* Am Math Soc; Math Soc Japan; Swed Math Soc. *Res:* Functional analysis and general algebra. *Mailing Add:* Dept Math Univ Ark Fayetteville AR 72701

KIMURA, TOKUJI, b Osaka, Japan, Nov 14, 25; m 51; c 2. ENZYMOLOGY, BIOPHYSICS. *Educ:* Osaka Univ, BS, 50, PhD(chem), 60. *Prof Exp:* Prof chem, St Paul's Univ, Tokyo, 65-68; PROF CHEM, WAYNE STATE UNIV, 68- *Concurrent Pos:* NIH res grant, 65. *Mem:* Am Soc Biol Chemists; Am Soc Chem; Sigma Xi. *Res:* Adrenal cortex mitochondrial steroid hydroxylases. *Mailing Add:* 4209 Sedgemoor Ln Bloomfield Hills MI 48013

KINARD, FRANK EFIRD, b Newberry, SC, Jan 15, 24; m 52; c 2. RESEARCH ADMINISTRATION. *Educ:* Newberry Col, BS, 46, AB, 47; Univ NC, MS, 50, PhD(physics), 54. *Prof Exp:* Instr physics, Univ NC, 49-52; physicist tech div, Savannah River Lab, E I du Pont de Nemours & Co, 53-63, head univ rels off, 63-67; exec dir, SC Comn Higher Educ, 67-68, assoc dir, 68-87; sr assoc comnr, 87-90; CONSULT, 90- *Mem:* Am Phys Soc; Am Nuclear Soc. *Res:* Research in higher education generally; especially graduate education. *Mailing Add:* SC Comn Higher Educ 1333 Main St Suite 650 Columbia SC 29201

KINARD, FREDRICK WILLIAM, b Leesville, SC, Oct 14, 06; m 29; c 2. PHYSIOLOGY, PHYSIOLOGICAL CHEMISTRY. *Educ:* Clemson Col, BS, 27; Univ Va, MS, 32, PhD(biochem), 33; Univ Tenn, MD, 45. *Prof Exp:* Asst chem, Med Col SC, 27-30; instr biochem, Univ Va, 30-33; from instr to assoc prof, 33-53, actg dir dept physiol, 44-46, chmn grad study comt, 49-65, prof, 53-78, dean, Col Grad Studies, 65-78, EMER PROF PHYSIOL, MED UNIV SC, 78- *Mem:* AAAS; Am Soc Zool; Am Physiol Soc; Soc Exp Biol & Med. *Res:* Creatine-creatinine metabolism; phosphatase; hemopoietine; ethanol metabolism. *Mailing Add:* The Crescent Two Johnson Rd Charleston SC 29407

KINARD, W FRANK, b Greenville, SC, Nov 16, 42. NUCLEAR CHEMISTRY. *Educ:* Duke Univ, BS, 64; Univ SC, PhD(anal chem), 68. *Prof Exp:* Postdoctoral fel nuclear chem, US AEC, Fla State Univ, 68-70; res assoc chem oceanog, Univ PR, 70-72; from asst prof to assoc prof, 72-83, PROF CHEM, COL CHARLESTON, 83- *Concurrent Pos:* Sr scientist, Oak Ridge Nat Lab, 78-79; guest scientist, Lawrence Livermore Nat Lab, 83-89; vis scientist, Westinghouse Savannah River Lab, 90-91. *Mem:* Am Chem Soc; AAAS; Sigma Xi. *Res:* Research in solution chemistry of lanthanide and actinide elements; applications of ICP-MS to radiochemical problems. *Mailing Add:* Dept Chem Col Charleston Charleston SC 29424

KINARIWALA, BHARAT K, b Ahmedabad, India, Oct 14, 26; US citizen; m 53; c 2. ALGORITHMS, DATABASE SYSTEMS. *Educ:* Benares Hindu Univ, BS, 51; Univ Calif, Berkeley, MS, 54, PhD(elec eng), 57. *Prof Exp:* Actg asst prof elec eng, Univ Calif, Berkeley, 56-57; mem tech staff, Bell Tel Labs, 57-66; chmn, Dept Elec Eng, 69-75 & 78-81, PROF ELEC ENG, UNIV HAWAII, 66- *Concurrent Pos:* Comput consult, Marchant, Inc, 56-57; prog chmn, Hawaiian Int Conf Systs Sci, 67; Inst Educ Exchange Serv deleg, Popov Soc Meeting, USSR, 67, prog chmn, Int Symp Circuit Theory, 68. *Mem:* Fel Inst Elec & Electronics Engrs. *Res:* System and computer sciences. *Mailing Add:* Dept Elec Eng Univ Hawaii 2540 Dole St Holmes Hall Honolulu HI 96822

KINASEWITZ, GARY THEODORE, b New York, NY, Aug 17, 46; m 69; c 3. PULMONARY DISEASE, PULMONARY CIRCULATION. *Educ:* Boston Col, BS, 68, MEd, 69; Wayne State Univ, MD, 73. *Prof Exp:* Resident med, Univ Pa Hosp, 73-76; fel pulmonary dis, Dept Med, Cardiovasc Pulmonary Div, Univ Pa, 76-78, res assoc med, 78-79, asst prof med, 79-80; from asst prof to assoc prof, 80-88, PROF MED & PHYSIOL & BIOPHYS, LA STATE UNIV MED CTR, SHREVEPORT, 88-; PROF MED & HEAD PULMONARY DIS & CRIT CARE MED, UNIV OKLA HEALTH SCI CTR, OKLAHOMA CITY, OKLA, 88- *Concurrent Pos:* Counr cardiopulmonary dis, Am Heart Asn. *Mem:* Am Thoracic Soc; Am Col Chest Physicians; Am Fedn Clin Res; Am Col Physicians; Am Physiol Soc. *Res:* Quantitative analysis of fluid exchanges across the pulmonary capillaries of the visceral pleura; role of abnormal pulmonary vasomotor reactivity in the genesis of pulmonary hypertension; cardiopulmonary adjustments to exercise in patients with lung disease; mesothelial cell glycosaminoglycans. *Mailing Add:* Dept Med Univ Okla Health Sci Ctr PO Box 26901 Rm 35P-400 Oklahoma City OK 73190

KINBACHER, EDWARD JOHN, b Brooklyn, NY, Nov 19, 27; m 55; c 3. PLANT PHYSIOLOGY. *Educ:* Cornell Univ, BS, 49; Purdue Univ, MS, 51; Univ Calif, PhD(plant physiol), 55. *Prof Exp:* Asst, Purdue Univ, 49-51 & Univ Calif, 51-55; assoc prof plant breeding & agron, Cornell Univ, 55-63; assoc prof hort, 63-71, PROF HORT, UNIV NEBR, LINCOLN, 71- *Concurrent Pos:* Plant physiologist crops res div, Agr Res Serv, USDA, 55-62. *Mem:* Am Soc Plant Physiol; Am Soc Agron. *Res:* Heat, drought and cold resistance of horticultural crops and specifically turfgrasses. *Mailing Add:* Dept Hort Univ Nebr East Campus Lincoln NE 68583

KINCADE, PAUL W, b Moorhead, Miss, Oct 10, 44; div; c 1. IMMUNOBIOLOGY, MICROBIOLOGY. *Educ:* Miss State Univ, BS, 66, MS, 68; Univ Ala, PhD(microbiol & immunol), 71. *Prof Exp:* Res asst, Lobund Lab, Univ Notre Dame, 68-69; fel, Univ Ala, 71-72, Walter & Eliza Hall Inst, Melbourne, 72-74; assoc, Sloan Kettering Inst, 74-79, assoc mem, 79-82; assoc prof, Grad Sch Med Sci, Cornell Univ, 80-82; MEM & HEAD DEPT IMMUNOBIOL, OKLA MED RES FOUND, OKLAHOMA CITY, 82- *Concurrent Pos:* mem, Immunobiol Study Sect, 84-88; mem, Prog Comt, Am Asn Immunologists, 89- *Mem:* Am Asn Immunologists; Am Asn Pathologists; Sigma Xi; Am Soc Microbiologists; Int Soc Develop Comparative Immunologists. *Res:* Humoral immune system with particular emphasis on relationships between stem cells and B lymphocytes and utilizing normal and genetically defective animal models. *Mailing Add:* Okla Med Res Found 825 NE 13th St Oklahoma City OK 73104

KINCADE, ROBERT TYRUS, b Indianola, Miss, May 16, 41; m 68; c 3. ENTOMOLOGY, WEED SCIENCE. *Educ:* Miss State Univ, BS, 63, MS, 66, PhD(entom), 70. *Prof Exp:* Field res specialist weed control, Chevron Chem Co, 69-75, supvr herbicides, fungicides & insecticides, 75-77; FIELD RES SPECIALIST, BIOL RES INSECTICIDES, HERBICIDES, FUNGICIDES, VALENT USA, 77- *Mem:* Sigma Xi. *Res:* Development of herbicides, fungicides and insecticides for agricultural use. *Mailing Add:* Box 5008 Valent USA Greenville MS 38701

KINCAID, DENNIS CAMPBELL, b Deer Park, Wash, June 17, 44; m 75; c 3. IRRIGATION SYSTEMS. *Educ:* Wash State Univ, BS, 66; Colo State Univ, MS, 68, PhD(agr eng), 70. *Prof Exp:* AGR ENGR, AGR RES SERV, USDA, 70- *Concurrent Pos:* Affil prof, Univ Idaho, 79- *Mem:* Am Soc Agr Engrs; Am Soc Civil Engrs; Irrigation Asn; Sigma Xi. *Res:* Basic and applied research in irrigation system improvement and water management. *Mailing Add:* 3793 W 3600E Kimberly ID 83341

KINCAID, JAMES ROBERT, b Covington, Ky, Feb 11, 45; m 68; c 1. ANALYTICAL CHEMISTRY, BIOPHYSICAL CHEMISTRY. *Educ:* Xavier Univ, Ohio, BS, 70; Marquette Univ, PhD(chem), 74. *Prof Exp:* Vis fel chem, Princeton Univ, 74-78; asst prof chem, Univ Ky, 78-; AT DEPT CHEM, MARQUETTE UNIV. *Concurrent Pos:* Nat res serv award, NIH, 75-77 & Nat Cancer Inst, 77-78. *Mem:* Am Chem Soc. *Res:* Applications of Raman spectroscopy to biologically significant systems; structure and function relationships in heme proteins; role of trace metals in biochemistry and medicine. *Mailing Add:* Dept Chem Marquette Univ 1515 W Wisc Ave Milwaukee WI 53228

KINCAID, RANDALL L, PHARMACOLOGY, CALMODULIN PHOSPHODIESTERASE. *Educ:* Stanford Univ, PhD(pharmacol), 78. *Prof Exp:* RES PHARMACOLOGIST, NIH, 79- *Mailing Add:* NIAAA 12501 Washington Ave Rockville MD 20852

KINCAID, RONALD LEE, b St Joseph, Mo, Oct 29, 50; m 70; c 4. ANIMAL NUTRITION, BIOCHEMISTRY. *Educ:* Univ Mo, BS, 71, MS, 73; Univ Ga, PhD(animal nutrit), 76. *Prof Exp:* Lectr animal nutrit, Lincoln Col, Univ Canterbury, 76-77; PROF ANIMAL NUTRIT, WASH STATE UNIV, 77- *Mem:* Am Dairy Sci Asn; Am Soc Animal Sci; Am Inst Nutrit; Coun Agr Sci & Technol; Soc Exp Biol & Med. *Res:* Mineral metabolism in animals; phyto estrogens; dairy cattle nutrition. *Mailing Add:* Dept Animal Sci Wash State Univ Pullman WA 99164

KINCAID, STEVEN ALAN, b Indianapolis, Ind, July 6, 43; m 77; c 2. ANATOMY. *Educ:* Purdue Univ, BS, 65, DVM, 69, MS, 71, PhD(anat), 77. *Prof Exp:* Veterinarian, South Bend, Ind, 72-73; asst prof anat, Univ Tenn, 77-81; ASSOC PROF ANAT, PURDUE UNIV, 82- *Concurrent Pos:* Seeing Eye Found, Inc grant, 69-71; NIH fel, 71-72. *Mem:* Sigma Xi; Am Asn Vet Anatomists; Am Vet Med Asn; Am Soc Animal Sci. *Res:* Pathobiology of articular cartilage; comparative arthrology; copper metabolism. *Mailing Add:* Dept Anat & Histol Auburn Univ Auburn AL 36849-5510

KINCAID, THOMAS GARDINER, b Hamilton, Ont, Sept 18, 37; m 62; c 3. NONDESTRUCTIVE EVALUATION. *Educ:* Queens Univ Kingston, BSc, 59; Mass Inst Technol, SM, 61, PhD(elec eng), 65. *Prof Exp:* SYSTS ENGR, GEN ELEC CO, 65- *Mem:* Inst Elec & Electronics Engrs; Am Soc

Nondestructive Testing. *Res:* Investigation of methods of evaluating materials for structural defects, including ultrasonic and electromagnetic techniques and x-ray. *Mailing Add:* Col Eng Boston Univ 144 Cummington St Boston MA 02215

KINCAID, WILFRED MACDONALD, b Cornhill, Scotland, Sept 13, 18; US citizen; m 52; c 3. STATISTICS, NUMERICAL ANALYSIS. *Educ:* Univ Calif, AB, 40; Brown Univ, PhD(appl math), 46. *Prof Exp:* Instr math & eng, Brown Univ, 43-44; physicist, Nat Adv Comt Aeronaut, Langley Field, 44-46; instr math, 46-51, res mathematician, Vision Res Lab, 50-58, lectr math, 58-60, from asst prof to prof, 60-84, EMER PROF MATH, UNIV MICH, ANN ARBOR, 84- *Mem:* AAAS; Am Statist Asn; Am Math Soc; Math Asn Am; Asn Res Vision & Ophthal; Soc Sci Explor. *Res:* Numerical analysis; statistics; vision; psychophysics. *Mailing Add:* Dept Math Univ Mich Ann Arbor MI 48109-1003

KINCANNON, DONNY FRANK, b Olustee, Okla, Jan 17, 33; m 57; c 2. BIOENVIRONMENTAL ENGINEERING. *Educ:* Okla State Univ, BS, 59, MS, 60, PhD(bioeng), 66. *Prof Exp:* Sanit engr, Tex State Dept Health, 60-61; instr civil eng, Arlington State Col, 61-63; asst prof, Univ Mo, Rolla, 65-66; from asst prof to assoc prof, 66-80, PROF CIVIL ENG, OKLA STATE UNIV, 80- *Mem:* Am Soc Civil Engrs; Am Soc Eng Educ; Water Pollution Control Fedn. *Res:* Water pollution control; biological treatment; industrial wastes; solid wastes. *Mailing Add:* Dept Civil Eng Okla State Univ Stillwater OK 74075

KINCH, DONALD M(ILES), b Lexington, Nebr, Apr 13, 13; m 41; c 1. AGRICULTURAL ENGINEERING. *Educ:* Univ Nebr, BSc, 38; Univ Minn, MS, 40; Mich State Univ, PhD(agr eng), 53. *Prof Exp:* Asst agr eng, Univ Minn, 38-40; design engr, Oliver Tractor Corp, 40-45 & Climax Eng Co, 45-46; asst prof agr eng, Iowa State Univ, 46-47 & Purdue Univ, 47-50; asst, Mich State Univ, 50-53; agr engr & prof agr, 53-76, EMER PROF AGR, UNIV HAWAII, 76- *Concurrent Pos:* Fulbright res fel, Brazil, 62. *Mem:* AAAS; Am Soc Agr Engrs; Nat Soc Prof Engrs. *Res:* On-the-farm processing of food and fiber products with development of processing machinery and equipment for farm use. *Mailing Add:* 12705 SE River Rd Milwaukee WI 97222

KINCHEN, DAVID G, b Hammond, La, Mar 7, 53; m 73; c 3. MATERIALS LABORATORY TESTING & CONSULTING, STATISTICAL ANALYSIS FOR PROCESS & PROBLEM EVALUATION. *Educ:* La State Univ, BE, 80. *Prof Exp:* Tech mgr, Mat Eval Lab, Inc, 81-83; qual engr, Martin Marietta Qual Eval Lab, 83-85, sr qual engr, 85-91, GROUP ENGR, MARTIN MARIETTA QUAL ADVAN TECHNOL, MARTIN MARIETTA MANNED SPACE SYSTS, 91- *Concurrent Pos:* Prin investr, Martin Marietta Qual Eval Lab, Qual Control Performance Criteria of Aerospace Aluminum Alloys, Martin Marietta Manned Space Systs, 84-85 & Eval of Straight Line X-Ray Radiographic Indications, 88. *Mem:* Am Soc Metals Int; Am Soc Qual Control. *Res:* Laboratory identification and investigation of unknown x-ray indications in aluminum alloy weldments. *Mailing Add:* Martin Marietta Manned Space Systs PO Box 29304 MS 3773 New Orleans LA 70189

KINCL, FRED ALLAN, b Everett, Mass, Dec 13, 22; m 46; c 2. ENDOCRINOLOGY, REPRODUCTION. *Educ:* Polytech-Prague, Baccal, 41; Univ Carolinum-Prague, Dr rer Nat, 48, DSc, 67. *Prof Exp:* Jr sci officer, Nat Chem Lab, Poona, India, 49-52; res chemist, Syntex Corp, Mex & Palo Alto, Calif, 52-66; asst dir, Pop Coun, NY, 67-70; vpres, Biol Concepts, Inc, NY, 70-73; prof, 74-86, EMER PROF BIOL, COL STATEN ISLAND, CITY UNIV NY, 86- *Concurrent Pos:* Teaching fel, Univ Wis, 65; res fel, Karolinska Inst Stockholm, 66; consult, Nat Inst Child Health & Human Develop, 71-78 & Nat Inst Drug Abuse, 73-79; adj asst prof, Hahnemann Med Col, 72; adj prof, Albert Einstein Sch Med, 73; vis prof, Swiss Fed Inst Technol, Zurich, 81. *Mem:* Am Chem Soc; Endocrine Soc; Int Soc Neuroendocrinol; Royal Soc Chem; NY Acad Sci; Int Soc Res Reproduction. *Res:* Physiology and biochemistry of reproduction; drug delivery systems; methods of contraception. *Mailing Add:* Col Staten Island Colony 130 Stuyvesant Pl Staten Island NY 10301

KIND, CHARLES ALBERT, b Philadelphia, Pa, Apr 17, 17; m 44. BIOCHEMISTRY. *Educ:* Lafayette Col, BS, 39; Yale Univ, PhD(chem), 42. *Prof Exp:* Res chemist, Nat Defense Res Comt, Yale Univ, 42; from instr to asst prof chem, Univ Conn, 42-57, from assoc prof to prof zool, 57-67, asst dean, 67-68, prof biol & assoc dean Col Lib Arts & Sci, 67-77; CONSULT, 77- *Concurrent Pos:* Lalor fel, Marine Biol Lab, Woods Hole, Mass, 49-50; Am Philos Soc fel, Bermuda Biol Sta, 51; mem corp, Marine Biol Lab, Woods Hole & Bermuda Biol Sta; actg dean, Col Lib Arts & Sci, Univ Conn, 70-71. *Mem:* AAAS; Am Chem Soc; Am Inst Chemists. *Res:* Lipids of marine invertebrate animals; sterols of vegetable oils; phosphatases. *Mailing Add:* Univ Conn Box U-98 Storrs CT 06268

KIND, LEON SAUL, b Boston, Mass, Dec 26, 22. MICROBIOLOGY. *Educ:* Harvard Univ, AB, 47; Yale Univ, PhD(microbiol), 51. *Prof Exp:* From asst prof to assoc prof, Med Col SC, 51-58; asst prof, Sch Med, Univ Calif, San Francisco, 58-70; assoc prof, 70-78, PROF MICROBIOL, DALHOUSIE UNIV, 78- *Concurrent Pos:* Res grants, NIH, 53- *Mem:* Soc Exp Biol & Med; Am Soc Immunol. *Res:* Immunology; experimental allergy; sensitivity of pertussis inoculated mice to histamine; anaphylaxis. *Mailing Add:* Eight Heffler Bedford NS B4A 1N3 Can

KIND, PHYLLIS DAWN, b Sidney, Mont, July 31, 33. CELLULAR IMMUNOLOGY. *Educ:* Univ Mont, BA, 55; Univ Mich, MS, 56, PhD(bact), 60. *Prof Exp:* Fel dermat, Univ Mich, 60-63; from instr to assoc prof path, Univ Colo, 63-71; res microbiologist, Nat Cancer Inst, 71-74; assoc prof microbiol, 74-79, PROF MICROBIOL & MED, GEORGE WASHINGTON UNIV, 79-, ASSOC DIR, TISSUE TYPING LAB, 78- *Concurrent Pos:* Chairperson, Immunol Div, Am Soc Microbiol, 75-76; fel mem, evaualtion panel for NSF, 79-81; coun, immunol div, Am Soc

Microbiol, 85-88; vis prof, Nat Cheng Kung Med Col, Tainan, Taiwan, 86; mem, Drug Abuse Biomed Res Review Comt, Nat Inst Drug Abuse, 86-88; fel, Prog Evaluation Panel, 88; mem, Drug Abuse Aids Res Comt, Nat Inst Drug Abuse, 88-90. *Mem:* AAAS; Am Asn Immunol; Am Soc Microbiol; Soc Exp Biol & Med; Sigma Xi; Am Soc Histol Compatability & Immunogenetics. *Res:* Regulation of antibody synthesis. *Mailing Add:* Dept Microbiol George Washington Univ Med Ctr Washington DC 20037

KINDEL, JOSEPH MARTIN, b Barberton, Ohio, Mar 10, 43; c 2. PLASMA PHYSICS. *Educ:* Univ Akron, BS, 65; Univ Calif, Los Angeles, MS, 66, PhD(physics), 70. *Prof Exp:* Res assoc plasma physics, Princeton Univ, 70-71; mem staff controlled thermonuclear res group, Physics Div, 71-72 & laser theory group, Theoret Div, 72-75, assoc group leader laser theory, Laser Div, 75-78, head, Plasma Theory Sect, 78-79, assoc group leader, 79-80, GROUP LEADER INERTIAL FUSION & PLASMA THEORY, APPL THEORET PHYSICS DIV, LOS ALAMOS NAT LAB, 80- *Mem:* Am Phys Soc. *Res:* Space plasma physics; radio frequency heating of plasmas; nonlinear theory; controlled thermonuclear fusion; laser plasma interaction; plasma simulation studies; hydrodynamics of laser fusion. *Mailing Add:* Mission Res Corp 127 Eastgate Dr Suite 208 Los Alamos NM 87544

KINDEL, PAUL KURT, b Milwaukee, Wis, Sept 6, 34; m 61; c 3. BIOCHEMISTRY. *Educ:* Univ Wis, BS, 56; Cornell Univ, PhD(biochem), 61. *Prof Exp:* NIH fel, Max Planck Inst Cell Chem, Munich, 61-63; from asst prof to assoc prof biochem, 63-77, PROF BIOCHEM, MICH STATE UNIV, 77- *Mem:* AAAS; Am Soc Biol Chemists; Am Soc Plant Physiol. *Res:* Enzymology of plant cell wall formation; freeze inhibitor polysaccharides of winter cereals; isolation and chemical characterization of polysaccharides. *Mailing Add:* Dept Biochem Mich State Univ East Lansing MI 48824

KINDER, THOMAS HARTLEY, b Riverside, Calif, Sept 6, 43; m 70. PHYSICAL OCEANOGRAPHY. *Educ:* US Naval Acad, BS, 65; Univ Wash, MS, 74, PhD(phys oceanog), 76. *Prof Exp:* Res asst, Dept Oceanog, Univ Wash, 71-76, res assoc, 76-78; oceanogr, Naval Ocean Res & Develop Activ, 78-87, MGR, COASTAL SCI PROG, OFF NAVAL RES, 87- *Concurrent Pos:* Guest investr, Woods Hole Oceanog Inst, 84-85. *Mem:* Am Geophys Union; Am Meteorol Soc; AAAS; Sigma Xi. *Res:* Measurements of mesoscale features on shelves and in semi-enclosed seas (Bering, Caribbean, Mediterranean), especially fronts and eddies formed by flows exiting straits and flows within straits. *Mailing Add:* 8900 Cromwell Dr Springfield VA 22151-1120

KINDERLEHRER, DAVID (SAMUEL), b Allentown Pa, Oct 23, 41. MATHEMATICS. *Educ:* Mass Inst Technol, SB, 63; Univ Calif, Berkeley, PhD(math), 68. *Prof Exp:* From instr to asst prof, 68-75, PROF MATH, UNIV MINN, MINNEAPOLIS, 75- *Concurrent Pos:* Ital Govt researcher math, Advan Training Sch, Pisa, 71-72. *Res:* Partial differential equations; minimal surfaces; variational inequalities; mechanics. *Mailing Add:* Dept Math Univ Minn Inst Technol Minneapolis MN 55455

KINDERMAN, EDWIN MAX, b Cincinnati, Ohio, Aug 21, 16; m 42; c 4. NUCLEAR CHEMISTRY. *Educ:* Oberlin Col, AB, 37; Univ Notre Dame, MS, 38, PhD(phys chem), 41. *Prof Exp:* Instr chem, Univ Portland, 41-43; chemist, Radiation Lab, Univ Calif, 43-45; from asst prof to assoc prof chem, Univ Portland, 45-49; chemist, Gen Elec Co, Wash, 49-56; sr physicist, 56-57, mgr, Nuclear Physics Dept, 57-68, dir, Appl Physics Lab, 68-71, dir mkt, Phys Sci, 71-73; sr scientist, 73-75, mgr mkt Europe, 75-77, mgr nuclear & utility systs, MGR ENERGY PLANNING, SRI INT, 80- *Concurrent Pos:* Chemist, Columbia Steel Casting Co, Ore, 43; res chemist, Res & Develop Div, H J Kaiser Co, Calif, 43. *Mem:* Am Chem Soc; Am Soc Int Law, panel on nuclear energy & world order; Am Phys Soc; Am Nuclear Soc; Sigma Xi; Inst Nuclear Mat Mgt. *Res:* Nuclear and analytical chemistry of uranium and transuranic elements; radiation chemistry and radiation damage effect; nuclear materials management; energy economics and planning; non fossil energy technologies; nucler energy and weapons proliferation. *Mailing Add:* SRI Int 333 Ravenswood Ave Menlo Park CA 94025

KINDERS, ROBERT JAMES, b Feb 12, 48. TUMOR MARKERS, BREAST CANCER. *Educ:* Loyola Univ, Chicago, BS, 70, MS, 76; Kans State Univ, PhD(biol), 80. *Prof Exp:* PROJ MGR, ABBOT LAB, 85- *Concurrent Pos:* Staff, dept immunochem res, Evanston Hosp, 76-77; mem adv coun, Ctr Basic Cancer Res, Kans State Univ, 90- *Mem:* Am Asn Cancer Res. *Res:* Application of monoclonal antibodies in cancer diagnosis and therapy; use of alternative binding agents in In Vitro Diagnostics. *Mailing Add:* Dept 90C Bldg R1 Abbot Lab North Chicago IL 60064

KINDIG, NEAL B(ERT), electronics, pulmonary physiology; deceased, see previous edition for last biography

KINDLER, SHARON DEAN, b Omaha, Nebr, Apr 27, 30; m 64; c 2. ENTOMOLOGY. *Educ:* Univ Nebr, BSc, 59, PhD(entom), 67. *Prof Exp:* RES ENTOMOLOGIST, AGR RES SERV, USDA, UNIV NEBR, 64- *Mem:* Entom Soc Am; Sigma Xi. *Res:* Biology, ecology and control of sorghum and grass insects; plant resistance to sorghum and grass insects. *Mailing Add:* 5024 W Ninth St Stillwater OK 74074-1409

KINDT, GLENN W, b Alpena, Mich, Sept 10, 30; m 60; c 4. NEUROSURGERY. *Educ:* Mich State Univ, BS, 51; Pa State Univ, BS, 52; Univ Mich, MD, 59. *Prof Exp:* Intern med, Univ Mich Hosp, 59-60, resident surg, 60-61, resident neurosurg, 62-65; resident, Harvard Med Sch & Peter Bent Brigham Hosp, Boston, 61-62; instr, Med Col SC, 65-66, assoc, 66-67; asst prof, Univ Calif, Davis, 67-69; from asst prof to assoc prof neurosurg, Univ Mich Med Ctr, Ann Arbor, 69-80; MEM STAFF, DIV NEUROSURG, UNIV COLO MED CTR, 80- *Concurrent Pos:* Chief investr, US Vet Admin res grant, 66-; co-investr, NIH res grant, 67- *Mem:* AMA; Cong Neurol Surg. *Res:* Regional hypothermia and vascular insufficiency of the brain; intracerebral hematomas. *Mailing Add:* Univ Colo Health Sci 4200 E Ninth Ave C-307 Denver CO 80262

KINDT, THOMAS JAMES, b Cincinnati, Ohio, May 18, 39; m 64; c 2. BIOCHEMISTRY, IMMUNOGENETICS. *Educ:* Thomas More Col, AB, 63; Univ Ill, Urbana, PhD(biochem), 67. *Prof Exp:* NIH fel biol, City of Hope Med Ctr, 67-69, asst res scientist, 69-70; fel, Rockefeller Univ, 70-71, asst prof, 71-73, assoc prof biol, 73-77; CHIEF LAB IMMUNOGENETICS, NAT INST ALLERGY & INFECTIOUS DIS, NIH, 77- *Concurrent Pos:* Adj assoc prof med, Cornell Univ Med Col, 73-77; assoc ed, J Immunol, 73-; adv ed, J Exp Med, 77- & Immunochem, 77-; adj prof micro & pediatrics, Sch Med & Dent, Georgetown Univ, 81-; vis scientist, Inst Pasteur, Paris, 82-83. *Mem:* Am Heart Asn; Harvey Soc; Sigma Xi; Am Asn Immunol; Am Soc Biol Chemists. *Res:* Genetic determinants on immunoglobulins and histocompatibility antigens; protein and polysaccharide structure. *Mailing Add:* Chief-Lab Immunogenetics Nat Inst Health Bldg 4 Rm 213 Bethesda MD 20892

KINERSLY, THORN, b The Dalles, Ore, Sept 15, 23; m 54; c 3. DENTISTRY, DENTAL RESEARCH. *Educ:* Univ Ore, DMD, 48. *Prof Exp:* Pvt pract, 48-49; intern children's dent, Forsyth Dent Infirmary, 49-50; asst resident, Sch Med, Yale Univ & Grace-New Haven Community Hosp, 50; res fel dent, Sch Med, Yale Univ, 52-56; pvt pract, 56-64; res assoc, 64, asst prof, 64-67, ASSOC PROF DENT, DENT SCH, UNIV ORE HEALTH SCI CTR, 67- *Concurrent Pos:* Lectr, Forsyth Dent Infirmary, 52-56; res fel, Nobel Inst, 54; abstractor, Oral Res Abstr, 66-; prin investr, Educ & Res Found Prosthodont, 69; consult on-site invest res appl, Dent Sect, NIH, 70. *Mem:* Fel AAAS; Soc Exp Biol & Med; Int Asn Dent Res; Am Dent Asn; Sigma Xi. *Res:* Paper electrophoresis of saliva and salivary glands for identification of enzymes, blood group factors and anti-bacterial factors; track radioautography of teeth with Ca-45 isotopes; lasers and their relation to dentistry. *Mailing Add:* Univ Ore Dent Sch 611 SW Campus Dr Portland OR 97201

KING, A DOUGLAS, JR, b Portland, Ore, May 11, 33; m 59; c 2. FOOD MYCOLOGY, FOOD MICROBIOLOGY. *Educ:* Wash State Col, BS, 55; Univ Calif, Davis, MS, 61; Wash State Univ, PhD(food sci), 66. *Prof Exp:* Food technologist, Nalley's Fine Foods, 57-59; chemist, USDA, 61-65, res chemist, 66-72, res leader, 72-76, proj leader, 77-85, LEAD SCIENTIST, USDA, 86- *Concurrent Pos:* Vis scientist, Commonwealth Sci Indust Res Orgn, Australia, 76-77. *Mem:* Inst Food Technologists; Am Soc Microbiol; Asn Off Anal Chemists. *Res:* Food microbiology research on fruits, vegetables, tree nuts, dried fruit and meats; food mycology and methods. *Mailing Add:* 800 Buchanan St Albany CA 94710

KING, ALAN JONATHAN, b Northampton, Eng, Oct 20, 54; Can citizen. OPTIMIZATION, STOCHASTIC PROGRAMMING. *Educ:* Univ Wash, BS, 81, MS, 84, PhD(appl math), 86. *Prof Exp:* Postdoctoral fel math, Univ BC, 86-87; res scholar, Int Inst Appl Systs Anal, 87-88; RES STAFF MEM, INT BUS MACH RES, 88- *Concurrent Pos:* Actg asst prof, Dept Math, Univ Wash, 88, vis lectr, 89. *Mem:* Math Prog Soc; Opers Res Soc Am; Soc Indust & Appl Math. *Res:* Stochastic programming; applications to engineering and economic systems; optimization theory; statistical estimation; robustness and sensitivity analysis; large scale optimization. *Mailing Add:* IBM T J Watson Res Ctr PO Box 218 Yorktown Heights NY 10598

KING, ALBERT IGNATIUS, b Tokyo, Japan, June 12, 34; US citizen; m 60; c 2. BIOENGINEERING. *Educ:* Univ Hong Kong, BSc, 55; Wayne State Univ, MS, 60, PhD(eng mech), 66. *Prof Exp:* Demonstr civil eng, Hong Kong, 55-58; asst & instr eng mech, Wayne State Univ, 58-60; from instr to assoc prof, 60-76, ASSOC NEUROSURG, SCH MED, WAYNE STATE UNIV, 71-, PROF BIOENG, 76-, DISTINGUISHED PROF MECH ENG, 90- *Concurrent Pos:* NIH Career develop award. *Honors & Awards:* Charles Russ Richards Mem Award, Am Soc Mech Engrs, 80; Volvo Award, 84. *Mem:* Am Soc Eng Educ; Am Soc Mech Engrs; Am Acad Orthop Surg; Sigma Xi. *Res:* Human response to acceleration and vibration, automotive and aircraft safety; biomechanics of the spine; mathematical modelling of impact events; low back pain research. *Mailing Add:* Bioeng Ctr Wayne State Univ Detroit MI 48202

KING, ALEXANDER HARVEY, b London, Eng, July 1, 54; m 77; c 2. ELECTRON MICROSCOPY, CRYSTALLOGRAPHY. *Educ:* Univ Sheffield, BMet, 75; Univ Oxford, DPhil(metall), 79. *Prof Exp:* Harwell fel, Dept Metall & Sci Mat, Univ Oxford, 79 & Dept Mat Sci & Eng, Mass Inst Technol, 79-81; from asst prof to assoc prof, Dept Mat Sci & Eng, 81-90, assoc vprovost grad studies, 87-90, PROF, DEPT MAT SCI & ENG, STATE UNIV NY, STONY BROOK, 90-, VPROVOST GRAD STUDIES, 90- *Mem:* Inst Metall; Am Inst Mining, Metall & Petrol Engrs; Electron Microscope Soc Am; Am Soc Metals; Mat Res Soc; Am Phys Soc. *Res:* Crystal lattice defects, particularly grain boundaries; electron microscopy, electron and x-ray diffraction. *Mailing Add:* Dept Mat Sci & Eng State Univ NY Stony Brook NY 11794-2275

KING, ALFRED DOUGLAS, JR, b Portland, Ore, May 11, 33; m 59; c 2. FOOD MICROBIOLOGY. *Educ:* Wash State Univ, BS, 55, PhD(food sci), 65; Univ Calif, Davis, MS, 61. *Prof Exp:* Food technologist, Nalley's Inc, 57-59; from assoc chemist to chemist, 61-65, microbiologist, 65-72, res leader microbiol, 73-77, RES MICROBIOLOGIST, WESTERN REGIONAL RES LAB, USDA, 78- *Concurrent Pos:* Vis scientist, Commonwealth Sci & Indust Res Orgn Div Food Res, Sydney, Australia, 77-78, 89. *Mem:* Inst Food Technologists; Am Soc Microbiol; Asn Off Anal Chemists. *Res:* Microbial physiology; food sanitation and public health; wine flavor; human and microbial nutrition; food mycology. *Mailing Add:* Western Regional Res Lab USDA 800 Buchanan St Albany CA 94710

KING, ALLEN LEWIS, b Rochester, NY, Mar 27, 10; m 37; c 3. HISTORY OF PHYSICS, BIOPHYSICS. *Educ:* Univ Rochester, BA, 32, MA, 33, PhD(physics), 37. *Hon Degrees:* MA, Dartmouth Col, 48. *Prof Exp:* Instr physics, Univ Rochester, 36-37 & Rensselaer Polytech Inst, 37-42; from instr to prof, 42-75, EMER PROF PHYSICS, DARTMOUTH COL, 75-, CUR, HIST SCI APPARATUS, 71- *Concurrent Pos:* Consult, Behr-Manning Corp,

NY, 41-42 & 48-57, Tougaloo Col, 57. *Mem:* Am Phys Soc; Biophys Soc; Am Asn Physics Teachers; Optical Soc Am. *Res:* Mechanical and hydrodynamical processes in biological systems; optical measurements; thermophysics; history of physics. *Mailing Add:* Dept Physics & Astron Dartmouth Col Hanover NH 03755

KING, AMY P, b Douglas, Wyo, Dec 30, 28; m 49. ANALYSIS. *Educ:* Univ Mo, BS, 49; Wichita State Univ, MA, 60; Univ Ky, PhD, 70. *Prof Exp:* Teacher pub schs, Goddard, Kans, 56-58; teaching fel, Wichita State Univ, 58-60, instr, 60-62; asst instr, Univ Kans, 62-65; instr math, Washburn Univ, 66-67; teaching asst, Univ Ky, 67-70; from asst prof to assoc prof, 70-79, PROF MATH, EASTERN KY UNIV, 79- *Mem:* Am Math Soc; Math Asn Am; Nat Coun Teachers Math. *Res:* Complex variables; mathematics education. *Mailing Add:* 1228 Tates Creek Rd Lexington KY 40502

KING, ANN CHRISTIE, BIOCHEMISTRY, MOLECULAR BIOLOGY. *Educ:* Wash Univ, PhD(molecular biol), 79. *Prof Exp:* ASST PROF BIOCHEM, SCH MED, UNIV ILL, 82- *Res:* Mechanism of growth factor. *Mailing Add:* Div Cell Biol 3030 Cornwallis Rd Research Triangle Park NC 27709-4416

KING, ARTHUR FRANCIS, b St John's, Nfld, Feb 11, 37; m 63; c 2. GEOLOGY. *Educ:* Mem Univ, BSc, 61, MSc, 63; Univ Reading, PhD(geol), 67. *Prof Exp:* from asst prof to assoc prof, 67-80, PROF GEOL, MEM UNIV NFLD, 80-, DEPT HEAD EARTH SCI, 87- *Mem:* Geol Asn Can. *Res:* Late precambrian clastic sequences in avalon zone of Newfoundland and the Appalachian-Caledonian Orogen. *Mailing Add:* Dept Earth Sci Mem Univ St John's NF A1B 3X5 Can

KING, B(ERNARD) G(EORGE), b Kitchener, Ont, Can, Dec 22, 22; nat US; m 49; c 5. ELECTRICAL ENGINEERING. *Educ:* Univ Southern Calif, BE, 44; Univ Wis, MS, 51, PhD(elec eng), 55. *Prof Exp:* Instr elec eng, Univ Southern Calif, 44-48 & Univ Wis, 48-55; mem tech staff, Bell Tel Labs, Inc, 55-88; CONSULT. *Concurrent Pos:* Ed, Bell Syst Tech J, 81-84. *Res:* Atmospheric optical transmission; radio physics; microwave radio. *Mailing Add:* 5 Monmouth Ave Rumson NJ 07760

KING, BARRY FREDERICK, b Perham, Minn, Sept 22, 42. CELL BIOLOGY, DEVELOPMENTAL BIOLOGY. *Educ:* Univ Minn, BA, 65; Univ Nev, MS, 67; Wash Univ, PhD(anat), 70. *Prof Exp:* Asst prof anat, Med Sch, Washington Univ, 71-77; assoc prof anat, 78-82, actg chair, 87-89, PROF CELL BIOL & ANAT, MED SCH, UNIV CALIF, DAVIS, 82- *Concurrent Pos:* USPHS fel anat, Univ Wis-Madison, 70-71; human embryol & develop study sect, NIH, 85-89; res career develop award, NIH, 75-80. *Mem:* Am Asn Anat; Am Soc Cell Biol; Soc Gynecol Invest. *Res:* Cell biology of the female reproductive system, especially the functional cytology of the placenta and fetal membranes. *Mailing Add:* Dept Cell Biol & Human Anat Sch Med Univ Calif Davis CA 95616-8643

KING, BENTON DAVIS, b Hackensack, NJ, Oct 20, 19; m 45; c 2. MEDICINE. *Educ:* Haverford Col, BS, 41; Univ Pa, MD, 44; Am Bd Anesthesiol, dipl, 50. *Prof Exp:* Assoc surg, Sch Med, Univ Pa, 48-52, res assoc, Grad Sch Med, 49-52; assoc prof anesthesiol, Col Med, State Univ NY Downstate Med Ctr, 54-57, prof & chmn dept, Sch Med, State Univ NY, Buffalo, 57-69, chmn dept, 69-80, PROF ANESTHESIOL, STATE UNIV NY DOWNSTATE MED CTR, 69- *Concurrent Pos:* Head dept anesthesiol, Meyer Mem Hosp, Buffalo, 57-69, Children's Hosp, 64-69, Kings County & State Univ Hosps, 69-; consult, Vet Admin Hosp, Brooklyn, NY. *Mem:* Am Soc Anesthesiol; Asn Univ Anesthetists. *Mailing Add:* Three Bonnie Heights Flower Hill Manhasset NY 11030

KING, BETTY LOUISE, b Atlanta, Ga, Nov 26, 43; m 77; c 1. METABOLISM. *Educ:* Brandeis Univ, BA, 65; Harvard Univ, PhD(microbiol, molecular genetics), 72. *Prof Exp:* Asst prof biol, Bard Col, 72-75; asst prof biol, Skidmore Col, 76-77; from asst prof to assoc prof, 78-84, asst div chmn nat sci, 83-85, PROF BIOL, NORTHERN VA COMMUNITY COL, 78- *Res:* Genetics and regulation of primary and secondary metabolism in bacteria, fungi and plants; chemical communication; biochemistry and ecology of alkaloids; drug use and abuse. *Mailing Add:* Div Sci & Appl Technol Northern Va Community Col Alexandria VA 22311

KING, BLAKE, b Atlanta, Ga, Feb 4, 21; m 48; c 2. MECHANICAL ENGINEERING, PHYSICAL METALLURGY. *Educ:* Univ Fla, BME, 48; Calif Inst Technol, MSME, 49. *Prof Exp:* Asst prof mech eng, Univ Miami, 49-51, 53-54; sr engr, Design Integration Dept, Martin Co, Md, 54-56, unit supvr, Nuclear Div, 56-58; from assoc prof to prof mech eng, Univ Miami, 58-87; RETIRED. *Mem:* Am Soc Metals; Metall Soc; Am Soc Mech Engrs. *Res:* Dispersion-strengthened high temperature alloys; creative mechanical design. *Mailing Add:* 4102 Alhambra Circle Coral Gables FL 33146

KING, C(ARY) JUDSON, III, b Ft Monmouth, NJ, Sept 27, 34; m 57; c 3. SEPARATION PROCESSES. *Educ:* Yale Univ, BE, 56; Mass Inst Technol, SM, 58, ScD(chem eng), 60. *Prof Exp:* Asst prof chem eng, Mass Inst Technol, 59-63; from asst prof to assoc prof, Univ Calif, Berkeley, 63-69, vchmn dept, 67-72, chmn dept, 72-81, dean col chem, 81-87, PROF CHEM ENG, UNIV CALIF, BERKELEY, 69-, PROVOST, PROF SCHS & COLS, 87- *Concurrent Pos:* Dir Bayway Sta, Sch Chem Eng Pract, Mass Inst Technol, 59-61; consult, Procter & Gamble Co, 65- *Honors & Awards:* 25th Ann Inst Lectr, 73; Food, Pharmaceut & Bioeng Div Award, 75; William H Walker Award, 76, Warren K Lewis Award, 90, Am Inst Chem Engrs; George Westinghouse Award, 78, Am Soc Eng Educ; Mac Pruitt Award, Coun Chem Res, 90. *Mem:* Nat Acad Eng; Am Chem Soc; fel Am Inst Chem Engrs. *Res:* Synthesis and analysis of chemical processes; drying and concentration of foods; separation processes; water pollution abatement. *Mailing Add:* Seven Kensington Ct Kensington CA 94707-1009

KING, CALVIN ELIJAH, b Chicago, Ill, June 5, 28. MATHEMATICS. *Educ:* Morehouse Col, AB, 49; Atlanta Univ, MA, 50; Ohio State Univ, PhD(math educ), 59. *Prof Exp:* Instr math, Jackson Col, 53-55; asst instr, Ohio State Univ, 55-58; head, Dept Physics & Math, 74-79, PROF MATH, TENN STATE UNIV, 58- *Concurrent Pos:* Specialist math & head dept, Fed Adv Teachers Col, Nigeria, 62-64. *Res:* Methods of teaching remedial mathematics; teaching elementary mathematics by television; the relation of modern mathematics to traditional mathematics. *Mailing Add:* 626 N Fifth St Nashville TN 37203

KING, CHARLES C, b Kittanning, Pa, Jan 12, 33; m 56, 81; c 6. ECOLOGY, BOTANY. *Educ:* Marietta Col, BS, 54; Ohio State Univ, MSc, 56, PhD(entom), 61. *Prof Exp:* Res technician entom, Ohio Agr Res & Develop, 56-61; from asst prof to prof biol, Malone Col, 61-72; EXEC DIR, OHIO BIOL SURV, 72- *Concurrent Pos:* Consult radiation biol, Univ Wash, 62; consult dept bot, Okla State Univ, 63; consult dept geol, Univ Calgary, 68; coordr geobot conf, Malone Col, 70, NAm Prairie Conf, Ohio State Univ, 78 & Annual Conf Lepidopterists Soc, 83; mem, Orgn Biol Field Sta, Ohio Acad Sci; coordr, 38th Meeting Am Inst Biol Sci, 87. *Mem:* Am Quaternary Asn; Am Inst Biol Sci; Ecol Soc Am; Nature Conservancy. *Res:* Effects of herbicides and fungicides on honeybees, metabolism of 2, 4-D in blackjack oak; effects of 2, 4-D on nectar secretion; peat bog palynology; environmental analysis; distribution and ecology of prairies in central Ohio. *Mailing Add:* Ohio Biol Surv 484 W 12th Ave Columbus OH 43210-1292

KING, CHARLES EVERETT, b Oak Park, Ill, May 8, 34; c 2. POPULATION BIOLOGY. *Educ:* Emory Univ, AB, 58; Fla State Univ, MS, 60; Univ Wash, PhD(zool), 65. *Prof Exp:* Res assoc zool, Col Fisheries, Univ Wash, 65; instr biol, Yale Univ, 65-66; asst prof zool, Univ Ill, Urbana, 66-68; asst prof biol, Yale Univ, 68-72; from assoc prof to prof biol, Univ SFla, 72-77; chmn dept, 77-86, PROF ZOOL, ORE STATE UNIV, 77- *Concurrent Pos:* Vis prof, Shandong Col Oceanog, Peoples Repub China, 84-85 & Univ Valencia, Spain, 91-92; vis scholar, Univ Wash, 85. *Mem:* Fel AAAS; Ecol Soc Am; Am Soc Naturalists; Genetics Soc Am; Soc Study Evolution; Sigma Xi. *Res:* Laboratory and field investigation of population dynamics; interaction of genetical and ecological phenomena within populations; relation of life history characteristics to ecological adaptation; evolution of senescence. *Mailing Add:* Dept Zool Ore State Univ Corvallis OR 97331-2140

KING, CHARLES MILLER, b West Salem, Ill, Oct 2, 32; m 54; c 2. BIOCHEMISTRY, ONCOLOGY. *Educ:* Univ Ill, Urbana, BA, 54; Univ Minn, Minneapolis, PhD(biochem), 62. *Prof Exp:* Res assoc, Michael Reese Hosp & Med Ctr, 65-68, from actg dir to dir div cancer res, 65-75; asst prof med, Pritzker Sch Med, Univ Chicago, 73-75; expert, Nat Cancer Inst, Nat Ctr Toxicol Res, 75-77; assoc prof biochem, Univ Ark for Med Sci, 75-77; CHMN DEPT CHEM CARCINOGENESIS, MICH CANCER FOUND, 77-; ADJ PROF CHEM, WAYNE STATE UNIV, 82- *Concurrent Pos:* Res fel oncol, Med Sch, Univ Minn, Minneapolis, 62-63; Am Cancer Soc fel, Neth Cancer Inst, 63-65; mem, Nat Bladder Cancer Proj, Nat Cancer Inst, 73-78; mem, Amines Comt, Nat Res Coun, Nat Acad Sci, 79-80; mem, Chem Path Study Sect, NIH, 79-83; assoc ed, Cancer Res, 83-87; mem, Metab Path Study Sect, NIH, 86-90; mem, Coun Res & Clin Invest Awards, 88-91; dir carcinogenesis prog, Comprehensive Cancer Ctr Metrop Detroit, 77- *Mem:* Am Asn Cancer Res; Am Chem Soc; Am Soc Biochem & Molecular Biol; Japanese Cancer Soc; Environ Mutagen Soc. *Res:* Mechanism of action of chemical carcinogens; metabolism in covalent interaction with protein and nucleic acid. *Mailing Add:* Mich Cancer Found 110 E Warren Detroit MI 48201

KING, CHARLES O(RRIN), b Rochester, NY, Jan 9, 16; m 43; c 3. CHEMICAL ENGINEERING. *Educ:* Univ Rochester, BS, 37; Univ Mich, MS, 39, ScD(chem eng), 43. *Prof Exp:* Asst chem eng, Univ Mich, 39-41; res engr, 43-48; process develop supvr, Tenn, 48-52; asst to tech mgr, 52-56; planning engr, 56-60; process supvr, 60-61, sr engr, 61-67; asst to tech dir, Textile Fiber Dept, E I du Pont de Nemours & Co, Inc, 67-82; RETIRED. *Mem:* Am Chem Soc; Am Inst Chem Engrs; Am Inst Chem; Sigma Xi. *Res:* Solvent extraction; polyamide textile fibers. *Mailing Add:* 1514 Woodsdale Rd Wilmington DE 19809

KING, CHARLES W(ILLIS), mathematics, electrical engineering, for more information see previous edition

KING, CHERYL E, b Ont, Can, Sept 3, 54. PHYSIOLOGY. *Educ:* Queens Univ, PhD(physiol), 83. *Prof Exp:* fel, 83-86, ASST PROF PHYSIOL, QUEENS UNIV, 86- *Honors & Awards:* Queen's Nat Scholar. *Mem:* Can Phys Soc; Am Phys Soc; Sigma Xi. *Res:* Oxygen transport during hypoxia. *Mailing Add:* Sch Rehab Ther Queens Univ Kingston ON K7L 3M6 Can

KING, CHI-YU, b Nanking, China, Aug 14, 34; m 62; c 3. GEOPHYSICS. *Educ:* Univ Taiwan, BEE, 56; Duke Univ, MS, 61; Cornell Univ, PhD(appl physics), 65. *Prof Exp:* Res fel geophys, Calif Inst Technol, 65-66; asst res geophysicist, Univ Calif, Los Angeles, 66-68; geophysicist, US Geol Surv, 68-70; geophysicist, US Earthquake Mechanism Lab, Nat Oceanic & Atmospheric Admin, 70-73; GEOPHYSICIST, US GEOL SURV, 73- *Mem:* Am Geophys Union; Seismol Soc Am. *Res:* Earthquake source mechanisms and prediction; fracture of solids; heat transfer; geomagnetism; geophysical fracture phenomena. *Mailing Add:* US Geol Surv 345 Middlefield Rd Menlo Park CA 94025

KING, CRESTON ALEXANDER, JR, b San Antonio, Tex, July 9, 35; m 62; c 3. LOW TEMPERATURE PHYSICS. *Educ:* Rice Univ, AB, 58, MA, 63, PhD(physics), 65; Duke Univ, MA, 62. *Prof Exp:* Asst physics, Rice Univ, 62-64; asst prof, 66-73, ASSOC PROF PHYSICS, LOYOLA UNIV, LA, 73-, CHMN DEPT, 80- *Mem:* Am Phys Soc; Am Asn Physics Teachers. *Res:* Superconductivity; solid state physics. *Mailing Add:* Dept Physics Loyola Univ Box 74 New Orleans LA 70118

KING, DARRELL LEE, b Hall, Mont, Jan 10, 37; m 62; c 3. LIMNOLOGY. *Educ:* Mont State Col, BS, 59; Mich State Univ, MS, 62, PhD(limnol), 64. *Prof Exp:* From asst prof to prof civil eng, Univ Mo-Columbia, 64-74; PROF FISHERIES & WILDLIFE, MICH STATE UNIV, 74-, ACTG DIR, INST WATER RES, 78- *Concurrent Pos:* Consult, Ralston Purina, 64-65 & 72-73, St Louis County Water Co, 67-69 & 73-74, Campbell Soup Co, 70-71, Harland Bartholomew & Assocs, 73-74 & A P Green Refractories Co, 75. *Mem:* AAAS; Am Soc Limnol & Oceanog; Am Fisheries Soc; Water Pollution Control Fedn. *Res:* Interrelationships between physical, chemical and biological factors with specific reference to detailing interacting mechanisms which govern aquatic ecosystems. *Mailing Add:* Wildlife 13 Nat Resource Mich State Univ East Lansing MI 48824

KING, DAVID BEEMAN, b Ware, Mass, Mar 12, 37; m 61; c 2. BIOLOGY. *Educ:* Univ Mass, BS, 59, MA, 61; Ind Univ, PhD(zool), 65. *Prof Exp:* From asst prof to prof, 65-88, DR E PAUL & FRANCES H REIFF PROF BIOL, FRANKLIN & MARSHALL COL, 88- *Concurrent Pos:* Nat Inst Arthritis & Metab Dis res grant, 66-78. *Mem:* AAAS; Am Soc Zoologists; Sigma Xi. *Res:* Comparative endocrinology; hormonal regulation of growth in chickens; thyroidal influence on muscle growth and development. *Mailing Add:* Dept Biol Franklin & Marshall Col Lancaster PA 17604

KING, DAVID GEORGE, NERVE CELLS, EVOLUTION. *Educ:* Univ Calif, PhD(neurosci), 75. *Prof Exp:* ASSOC PROF HISTOL, SOUTHERN ILL UNIV, 78- *Res:* Neurobiology. *Mailing Add:* Dept Zool Southern Ill Univ Carbondale IL 62901

KING, DAVID S(COTT), b Hartford, Conn, Nov 5, 49. PHOTO-DISSOCIATION. *Educ:* Univ Pa, BA, 71, PhD(phys chem), 76. *Prof Exp:* RES CHEMIST, NAT BUR STANDARDS, 76- *Mem:* Am Chem Soc; Optical Soc Am. *Res:* State-resolved and time-resolved laser studies of the mechanisms and rates of energy flow within model molecular systems, including relaxation and bond rupture. *Mailing Add:* Nat Inst Standards & Technol Molecular Spectroscopy Div Gaithersburg MD 20899

KING, DAVID SOLOMON, b Wooster, Ohio, Jan 25, 36; m 56; c 2. ASTROPHYSICS. *Educ:* Manchester Col, BA, 60; Ind Univ, MA, 64, PhD(astrophys), 67. *Prof Exp:* From asst prof to assoc prof, 65-77, PROF ASTRON, UNIV NMEX, 77- *Concurrent Pos:* NSF res grant stellar pulsation theory, 66-77. *Mem:* Am Astron Soc. *Res:* Stellar interiors; theory of pulsating variable stars; nonlinear self-excited radial oscillations in stellar envelopes. *Mailing Add:* Dept Physics & Astron Univ NMex 800 Yale Blvd NE Albuquerque NM 87131

KING, DAVID THANE, b Wellington, NZ, Jan 16, 23; nat US; m 50; c 4. PHYSICS. *Educ:* Univ NZ, BSc, 44, MSc, 47; Bristol Univ, PhD(physics), 51. *Prof Exp:* Physicist, US Naval Res Lab, 51-55; from asst prof to assoc prof, 55-61, PROF PHYSICS, UNIV TENN, KNOXVILLE, 61- *Concurrent Pos:* Mem user's group zero gradient synchrotron Argonne Nat Lab, 62-; consult, Oak Ridge Nat Lab, 65- *Mem:* Am Phys Soc; Italian Phys Soc. *Res:* High energy physics. *Mailing Add:* Dept Physics-Astron 401 Physics Bldg Univ Tenn Knoxville TN 37996

KING, DELBERT LEO, b Kansas City, Mo, Jan 7, 34; m 57; c 7. PHYSICAL CHEMISTRY. *Educ:* Ariz State Col, BS, 56; Univ Nebr, MS, 59, PhD, 68. *Prof Exp:* Asst prof chem, Sterling Col, 59-62; asst prof, Nebr Wesleyan Univ, 62-65; assoc prof, 65-72, chmn natural sci div, 70-76, PROF CHEM, DOANE COL, 72-, DIR COMPUT CTR, 76- *Res:* Solution colorimetry. *Mailing Add:* 340 Branched Oak Rd Davey NE 68336

KING, DONALD M, b Seattle, Wash, June 5, 35; m 65; c 1. ANALYTICAL CHEMISTRY. *Educ:* Wash State Univ, BS, 57; Calif Inst Technol, PhD(chem), 63. *Prof Exp:* Asst prof chem, Calif State Col Los Angeles, 61-63; Welsh fel, Univ Tex, 63-64; chemist, Org Chem Div, E I du Pont de Nemours & Co, 64-66; asst prof, 66-69, ASSOC PROF CHEM, WESTERN WASH STATE COL, 69-, CHMN DEPT, 74- *Concurrent Pos:* NSF res grant, 68-70. *Mem:* Am Chem Soc. *Res:* Application of electroanalytical techniques to the study of reaction kinetics and mechanisms. *Mailing Add:* Dept Chem Western Washington Univ Bellingham WA 98225

KING, DONALD WEST, JR, b Cochranton, Pa, June 30, 27; m 52; c 3. PATHOLOGY. *Educ:* Syracuse Univ, MD, 49. *Prof Exp:* Resident & instr path, Col Physicians & Surgeons, Columbia Univ, 49-52; prof & chmn dept, Univ Colo, Denver, 61-67; Delafield prof path & chmn dept, Col Physicians & Surgeons, Columbia Univ, 67-82; CONSULT, 87- *Concurrent Pos:* USPHS fel, Univ Chicago, 54-55 & Carlsberg Lab, 55-56. *Mem:* Am Soc Exp Path; Am Soc Cell Biol; Human Genetics Soc; Am Asn Path; NY Acad Sci; AAAS. *Res:* Cell injury; membrane transport. *Mailing Add:* Div Biol Sci Univ Chicago 950 E 59th St Chicago IL 60637

KING, DOROTHY WEI (CHENG), b Hankow, China, Mar 3, 14; nat US; m 40, 61. NUTRITION, PHYSIOLOGY. *Educ:* Yenching Univ, China, BS, 36; Mt Holyoke Col, MA, 48; Iowa State Univ, PhD(animal nutrit), 54. *Prof Exp:* Asst biol, Yenching Univ, China, 36-37; teacher, China, 38-45; instr, Nat Med Col, Shanghai, 45-46; asst zool, Iowa State Univ, 50-51; from instr to asst prof histol, Univ Iowa, 54-67; vis prof, 67-70, prof, 70-84, EMER PROF ZOOL, NAT TAIWAN UNIV, 84- *Concurrent Pos:* Sr lectr biol, Chung Chi Col, Hong Kong, 61-62. *Mem:* AAAS; Am Asn Anatomists; Am Soc Zoologists; Am Soc Animal Sci; Am Inst Nutrit. *Res:* Effects of mitomycin C, monosodium glutamate, antihistaminic drugs and heavy metals on chick embryos. *Mailing Add:* Dept Zool Nat Taiwan Univ Taipei 107 Taiwan

KING, EDGAR PEARCE, b Oklahoma City, Okla, Nov 4, 22; m 60; c 3. MATHEMATICAL STATISTICS. *Educ:* Carnegie Inst Technol, BS, 48, MS, 49, DSc(math), 51. *Prof Exp:* Instr math, Carnegie Inst Technol, 49-51; math statistician, Nat Bur Standards, 51-53; staff res statistician, Eli Lilly & Co, 53-55, dept head opers res, 55-59, head statist res, 59-64, asst dir, Sci Serv Div, 64-67, dir, 67-69, dir corp qual assurance, 69-71, corp planning adv corp opers planning, 71-78, SCI ADV, SCI INFO SERV, ELI LILLY & CO, 78- *Mem:* Opers Res Soc Am; Am Soc Qual Control; Am Statist Asn; Inst Math Statist; Int Statist Inst. *Res:* Statistical design of experiments. *Mailing Add:* 6450 N Ewing Indianapolis IN 46220

KING, EDWARD FRAZIER, b Lake Providence, La, Dec 3, 35; m 66; c 5. INORGANIC CHEMISTRY. *Educ:* Gregorian Univ, BA, 58; Loyola Univ, BS, 65; La State Univ, New Orleans, PhD(chem), 69. *Prof Exp:* Instr, Tex Woman's Univ, 68-79, asst prof chem, 70-75; mem, fac chem & physics, Cent Tex Col, 75-89; ASST PROF & CHMN SCI DEPT, S PLAINS COL, 89- *Res:* Complexes of the first row transition metals, particularly their preparation and visible infrared spectra; halogen complexes of vanadium and copper. *Mailing Add:* Sci Dept SPlains Col 1401 College Ave Levelland TX 79336

KING, EDWARD LOUIS, b Grand Forks, NDak, Mar 15, 20; m 52; c 2. INORGANIC CHEMISTRY. *Educ:* Univ Calif, Berkeley, BS, 43, PhD(chem), 45. *Prof Exp:* Asst chem, Univ Calif, 42-44; asst chem, Manhattan Dist Proj, 44-45, res assoc, 45-46, du Pont fel & lectr chem, Harvard Univ, 46-47, instr, 47-48; from asst prof to prof, Univ Wis, 48-63; chmn dept, 70-72, prof chem, 63-86, EMER PROF, UNIV COLO, BOULDER, 86- *Concurrent Pos:* At Off Sci Res & Develop, 44; Guggenheim fel, Calif Inst Technol, 57-58; ed, Inorg Chem, Am Chem Soc, 64-68. *Mem:* Am Chem Soc. *Res:* Chemistry of chromium and other transition metals in solution; ion solvation in mixed solvents; mechanisms of reactions in solution. *Mailing Add:* Dept Chem Univ Colo Campus Box 215 Boulder CO 80309

KING, EDWARD P(ETER), b Deer Creek, Ill, May 29, 05; m 28; c 2. CHEMICAL ENGINEERING. *Educ:* Eureka Col, BS, 26; Univ Ill, MS, 27, PhD(chem eng), 30. *Prof Exp:* Instr chem, Univ Ill, 26-28; sr engr, Atlantic Ref Co, 29-35; resident chemist, Pure Oil Co, Chicago, 36-48, tech foreman, 48-53, tech supt, 53-56, oper asst, 56-60, asst mgr opers-ref, 60-65, mgr ref qual control, 65-70; CONSULT LUBRICATING OIL REF OPERS, 70- *Mem:* AAAS. *Res:* Partial oxidation in liquid phase; solvant extraction of hydrocarbons; phase equilibria in hydrocarbon systems under high pressure; technical aspects of lubricating oil manufacturing operations; refinery simulation and linear programming applications. *Mailing Add:* 107 E South St Cambridge IL 61238

KING, EDWIN WALLACE, entomology; deceased, see previous edition for last biography

KING, EILEEN BRENNEMAN, b Albany, Ore, May 13, 24; m 48; c 4. PATHOLOGY, CYTOLOGY. *Educ:* Univ Ore, BA, 46, MD, 50; Am Bd Path, cert anat path, 55. *Prof Exp:* Lectr cytol, 55-59, from instr to asst clin prof, 55-69, assoc clin prof, 69-79, CLIN PROF PATH, SCH MED, UNIV CALIF, SAN FRANCISCO, 79-, PATHOLOGIST-IN-CHG CYTOL LAB & TRAINING PROG, 60- *Concurrent Pos:* Pathologist-in-chg, Pulmonary Cytol Lab, San Francisco Gen Hosp, 55-61; pathologist & dir labs, Mary's Help Hosp, 55-66; consult, San Francisco Gen Hosp, 59-61, US Naval Hosp, Oakland, 62- & US Vet Hosp, San Francisco, 66- *Mem:* Fel Col Am Path; Am Soc Cytol (pres, 78); fel Int Acad Cytol; Soc Anal Cytol. *Res:* Evaluation of cytochemical probes for potential markers of dysplasia and carcinoma in situ of uterine cervix and cancer precursors in urinary bladder; pulmonary cytology survey study for early diagnosis of lung cancer; cytologic, histologic and histochemical methods for evaluation of radiation host response in carcinoma of the cervix; histologic and cytologic studies of breast cancer and its precursors. *Mailing Add:* Univ Calif Med Ctr San Francisco CA 94143

KING, ELBERT AUBREY, JR, b Austin, Tex, Nov 12, 35; m 57; c 2. PLANETOLOGY, SPACE GEOLOGY. *Educ:* Univ Tex, BS, 57, MA, 61; Harvard Univ, PhD(geol), 65. *Prof Exp:* Geologist, Geol & Geochem Br, Manned Spacecraft Ctr, NASA, 63-67, cur, Lunar Receiving Lab, 67-69; chmn dept, 69-74, PROF GEOL, UNIV HOUSTON, 69- *Mem:* Fel Meteoritical Soc. *Res:* Meteorites, planetary surfaces; geochemistry; mineralogy of pegmatites, petrography and composition of tektites; zoned ultramafic intrusive rocks; lunar samples. *Mailing Add:* Dept Geol Univ Houston 4800 Calhoun Rd Houston TX 77004

KING, ELIZABETH NORFLEET, b Concord, NC, May 22, 25. CELL PHYSIOLOGY. *Educ:* Randolph-Macon Woman's Col, AB, 46; Wellesley Col, MA, 51; Duke Univ, PhD(cell physiol), 63. *Prof Exp:* Instr zool, Wellesley Col, 51-53; instr biol, Woman's Col, NC, 53-56 & Bucknell Univ, 60-61; instr physiol, Vassar Col, 61-63, asst prof biol, 63-69; ASSOC PROF BIOL, WINTHROP COL, 69- *Mem:* AAAS; Am Soc Zoologists; Nat Asn Biol Teachers. *Res:* Cellular physiology; separation of cellular organelles. *Mailing Add:* Dept Biol Winthrop Col Rock Hill SC 29730

KING, ELIZABETH RAYMOND, b Halifax, NS, Dec 5, 23; US citizen. GEOLOGY. *Educ:* Smith Col, AB, 47. *Prof Exp:* Geologist, 48-51, geophysicist, 51-74, RES GEOPHYSICIST, US GEOL SURV, 74- *Mem:* AAAS; Soc Explor Geophys; Am Geophys Union; fel Geol Soc Am. *Res:* Airborne magnetometer surveys; analysis of magnetic and gravity anomalies by modeling techniques; geologic interpretation of data from geophysical studies of continents and oceans. *Mailing Add:* US Geol Surv MS 927 Reston VA 22092

KING, FRANKLIN G, JR, b Mahonoy City, Pa, Sept 23, 39; m 59; c 3. CHEMICAL ENGINEERING, BIOCHEMICAL ENGINEERING. *Educ:* Pa State Univ, BS, 61; Kans State Univ, MS, 62; Howard Univ, MEd, 76. *Hon Degrees:* DSc, Stevens Inst Technol, 66. *Prof Exp:* Res engr, Uniroyal Res Ctr, 62-65; process engr, Am Cyanamid Co, 65-66; asst prof chem eng, Lafayette Col, 66-72; from assoc prof to prof chem eng, Howard Univ, 72-85; PROF CHEM ENG, NC A&T STATE UNIV, 85- *Concurrent Pos:* Consult, Maxwell House Div, Gen Foods Corp, 67-72 & NIH, 77- *Mem:* Am Inst Chem Eng; Am Soc Eng Educ. *Res:* Modeling, simulation and control of chemical processes; development of physiological pharmacokinetic models; biochemical engineering including waste treatment and fermentation. *Mailing Add:* Dept Chem Eng NC A&T State Univ Greensboro NC 27411

KING, FREDERICK ALEXANDER, b Paterson, NJ, Oct 3, 25; div; c 2. NEUROSCIENCE. *Educ:* Stanford Univ, AB, 53; Johns Hopkins Univ, MA, 55, PhD(psychol, med sci), 56. *Prof Exp:* Instr psychol, Johns Hopkins Univ, 55-56; from instr to asst prof psychiat, Col Med, Ohio State Univ, 56-59; from asst prof to prof psychol & neurosurg, Col Med, Univ Fla, 59-69, prof neurosci & chmn dept, 69-78, dir ctr neurobiol sci, 68-78; PROF ANAT & CELL BIOL, & ASSOC DEAN, SCH MED, EMORY UNIV, 78-, DIR & RES PROF NEUROBIOL, YERKES REGIONAL PRIMATE RES CTR, 78- *Concurrent Pos:* NIMH grant, 57-60, div biol sci training grant, 65-, Nat Inst Neurol Dis & Stroke grant, 59-71, NIMH fel & vis prof, Inst Physiol, Univ Pisa, 61-62; vis scientist, Am Psychol Asn-NSF prog, 62-67 & neuroanat prog, NIH, 63-70; scientist co-dir, Ctr Neurobiol Sci, Univ Fla, 64-68; consult, Battelle Mem Inst, 56-59 & res mem psychobiol adv panel, Biol & Med Sci Div, NSF, 63-67; mem adv comt, Primate Res Ctr, NIH, 69-73; mem ed adv bd, Behav Biol, 71-; ed, Physiol & Animal Psychol, 71- & J Suppl Abstr Serv; chmn comt commun & coord, Biol Sci Training Prog, NIMH, 72-78, mem & chmn res scientist develop rev comt, 74-78; mem bd sci adv, Yerkes Regional Primate Res Ctr, 74-78; mem Nat Res Coun-Nat Acad Sci brain sci comt, 74-; chmn & mem comt educ, Soc Neurosci, 74-77; adj prof psychol, Emory Univ, 78-; mem Int Adv Bd, Nat Mus Kenya & Kenyan Inst Primate Res, 83-; chair, Comn Animals Res & Experimentation, Am Psychol Asn, 83-85, bd sci affairs, 86-88; chair, Comn Animal Res, Soc Neurosci, 87, Govt & Pub Affairs Comn, 90-93; mem bd trustees, Am Asn Accreditation Lab Animal Care, 87-90; mem bd dir, Nat Asn Biomed Res, 87-90; mem adv bd & dir, NIH, 89- *Honors & Awards:* Spec Leadership & Serv Citation, Am Psychol Asn, 84. *Mem:* Fel AAAS; fel Am Psychol Asn; Am Physiol Soc; Soc Neurosci; Int Neuropsychol Soc (secy-treas, 69-73); Asn Clin Sci. *Res:* Effects of subcortical brain lesions on motivated behavior and learning in animals; analysis of cortical functions in infrahuman primates by use of ablation and brain stimulation techniques; electrophysiological and conditioning analysis of recovery of function in the isolated forebrain. *Mailing Add:* Yerkes Primate Ctr Emory Univ Atlanta GA 30322

KING, FREDERICK JESSOP, b Niagara Falls, NY, July 4, 28; m 56; c 3. FOOD SCIENCE. *Educ:* Cornell Univ, BS, 51, MFS, 52; Mass Inst Technol, PhD, 60. *Prof Exp:* BIOCHEMIST, NORTHEAST UTILIZATION RES CTR, NAT MARINE FISHERIES SERV, NAT OCEANIC & ATMOSPHERIC ADMIN, 60- *Concurrent Pos:* Exec secy, New Eng Fisheries Inst, 70- *Mem:* Am Chem Soc; Inst Food Technologists; fel Am Inst Chemists. *Res:* Denaturation of fish proteins; irradiation preservation of fish; flavor chemistry of fish; nutritive value; fishery technology; process and product improvements; quality assurance. *Mailing Add:* Box 635 Cheshire MA 01930

KING, GAYLE NATHANIEL, b Paulding, Ohio, July 17, 48; m 71; c 2. PHYSICAL CHEMISTRY. *Educ:* Heidelberg Col, BS, 70; Univ Pac, PhD(phys chem), 77. *Prof Exp:* Instr statist, US Air Force, 70-74; asst prof chem, Rose Hulman Inst Technol, 77-81; CHIEF CHEMIST, ELF ASPHALT, 81- *Concurrent Pos:* Consult, Bituminous Mat Inc, 78- *Honors & Awards:* Emmon's Prize, Am Asn Physics Teachers. *Mem:* Am Chem Soc; Sigma Xi; Am Asn Physics Teachers. *Res:* Modified asphalt; asphalt and oil emulsions. *Mailing Add:* 113 E Lawrin Blvd Terre Haute IN 47803

KING, GENERAL TYE, b King, Ky, Apr 7, 20; m 40. MEAT SCIENCE. *Educ:* Union Col, Ky, BS, 48; Univ Ky, BS, 50, MS, 51; Tex A&M Univ, PhD(meats), 58. *Prof Exp:* Asst animal husbandman, SDak State Col, 51-53; from instr to asst prof meats, 53-60, assoc prof meats, 60-69, assoc prof animal sci, 69-76, PROF ANIMAL SCI, TEX A&M UNIV, 76-, ASST HEAD DEPT, 69- *Mem:* Am Meat Sci Asn; Am Soc Animal Sci; Inst Food Technol. *Res:* Animal science; nutrition. *Mailing Add:* Dept Animal Sci Tex A&M Univ College Station TX 77843

KING, GEORGE, III, b Tampa, Fla, Nov 12, 46; m 68; c 2. EXPERIMENTAL NUCLEAR PHYSICS. *Educ:* Talladega Col, BA, 68; Stanford Univ, MS, 74, PhD(nuclear physics), 77. *Prof Exp:* Instr physics, Albany State Col, 69-71 & Col Notre Dame, 72-73; res asst nuclear physics, Stanford Univ, 73-77; res assoc nuclear physics, Lawrence Berkeley Lab, 77-79; MEM PROF STAFF, NUCLEAR DEPT, SCHLUMBERGER DOLL RES, 79- *Concurrent Pos:* Xerox Corp fel, 74-77. *Mem:* Am Phys Soc. *Res:* Giant resonance excitation for light nuclei; relativistic heavy ion central collisions. *Mailing Add:* 92 Walnut Grove Rd Ridgefield CT 06877

KING, GERALD WILFRID, b Eng, Jan 22, 28; m 54; c 4. CHEMICAL PHYSICS. *Educ:* Univ London, BSc, 49, PhD(chem), 52. *Hon Degrees:* DSc, Univ London, 70. *Prof Exp:* Asst lectr, Univ Col, London, 54-56, lectr, 56-57; from asst prof to prof, 57-89, EMER PROF CHEM, MCMASTER UNIV, 89- *Concurrent Pos:* Hon res assoc, Univ Col, London, 65; Ed, Can J Chem, 74-79; chmn, dept chem, McMaster Univ, 79-82; bd gov, McMaster Univ, 85-88. *Honors & Awards:* Gerhard Herzberg Award, Spectros Soc Can, 81. *Mem:* Fel Chem Inst Can; Optical Soc Am; fel Royal Soc Chem; fel Royal Soc Can. *Res:* Electronic states and structures of molecules; multi-photon laser spectroscopy; optical-optical double resonance; high resolution electronic molecular spectroscopy. *Mailing Add:* Dept Chem McMaster Univ Hamilton ON L8S 4M1 Can

KING, GORDON JAMES, b Toronto, Ont, June 8, 32; m 57; c 2. REPRODUCTIVE PHYSIOLOGY, ENDOCRINOLOGY. *Educ:* Univ Toronto, DVM, 59; Univ Guelph, MS, 66, PhD(reprod physiol), 68. *Prof Exp:* Pvt pract vet med, 59-60; field veterinarian, Hamilton Dist Cattle Breeding Asn, 60-62, tech mgr, 62-65; asst prof, 68-70, assoc prof, 70-75, PROF ANIMAL SCI, UNIV GUELPH, 75- *Concurrent Pos:* Can Dept Agr res grant, 76-78; Nat Res Coun Can res grant, 76-; consult, FAO, IAEA, CIDA, IFS. *Mem:* Am Soc Animal Sci; Soc Study Reprod; Soc Study Fertil; fel Inst Biol. *Res:* Reproductive biology. *Mailing Add:* Dept Animal & Poultry Sci Univ Guelph Guelph ON N1G 2W1 Can

KING, H(ENRY) E(UGENE), b Wilmington, Va, Sept 24, 22; m 48; c 2. PSYCHOLOGY. *Educ:* Univ Richmond, BA, 42; Columbia Univ, MA, 43, PhD, 48. *Prof Exp:* Asst psychologist, NY State Psychiat Inst, 42-43 & 46-48; sr res scientist, NY State Brain Res Proj, 48-49; assoc prof psychiat & neurol, Sch Med, Tulane Univ, 49-60; prof psychol, Sch Med & Chief Serv, Western Psychiat Inst, Univ Pittsburgh, 60-80; PROF PSYCHOL, WASHINGTON & LEE UNIV, 80- *Concurrent Pos:* Lectr, Columbia Univ, 46-49; vis scientist, Charity Hosp, New Orleans, 49-60. *Mem:* Fel AAAS; Am Physiol Soc; Soc Exp Biol & Med; fel Am Psychol Asn; Am Psychopath Asn. *Res:* Psychophysiology; experimental psychopathology; motor response systems; central nervous system stimulation and ablation. *Mailing Add:* Dept Psychol Washington & Lee Univ Lexington VA 24450

KING, HAROLD, b Bedford, Ind, Aug 12, 22; m 52; c 2. SURGERY. *Educ:* Yale Univ, MD, 46. *Prof Exp:* From instr to assoc prof, 55-64, PROF SURG, MED CTR, IND UNIV, INDIANAPOLIS, 64-, DIR DIV THORACIC & CARDIOVASC SURG, 72- *Concurrent Pos:* Mem staff, Vet Admin Hosp & Indianapolis Gen Hosp. *Mem:* Soc Vascular Surg; Soc Univ Surg; Am Col Surg; Soc Thoracic Surg; Am Surg Asn. *Res:* Cardiovascular surgery. *Mailing Add:* Ind Univ Med Ctr 545 Barnhill Dr Indianapolis IN 46223

KING, HARRISS THORNTON, b Ames, Iowa, June 15, 47; m 76. NUCLEAR PHYSICS. *Educ:* Iowa State Univ, BS, 69; Stanford Univ, MS, 70, PhD(physics), 75. *Prof Exp:* Res assoc physics, Rutgers Univ, 74-77; asst prof physics, Stanford Univ, 77-80; MEM STAFF, MEASUREX CORP, 80- *Mem:* Sigma Xi; Am Phys Soc. *Res:* Hyperfine interactions; nuclear electromagnetic moments; polarization phenomena. *Mailing Add:* Measurex Corp One Results Way Cupertino CA 95014

KING, HARRY J, b Saskatoon, Sask, May 2, 34; nat US; m 57; c 4. PLASMA PHYSICS, ELECTRICAL ENGINEERING. *Educ:* Univ Sask, BS, 56, MS, 57; McMaster Univ, PhD(nuclear & atomic physics), 60. *Prof Exp:* Sr physicist, Ion Physics Corp, Burlington, 60-63; sr mem tech staff, Hughes Aircraft Co, 63-72, asst dept mgr ion & plasma sources, 72-74, assoc dept mgr power utility switches, 74-78, dept mgr high voltage devices, 78-81, proj mgr, Electron Beam Lithography, 81-85, prog mgr data link systs, 85-87, PROG MGR AUTOMOTIVE & COM PRODS, RADAR SYSTS GROUP, HUGHES AIRCRAFT CO, 87- *Mem:* Inst Elec & Electronics Engrs. *Res:* Ion and neutral beams and sources; ion propulsion; high power switches for power utilities and fusion apparatus; electron beams; data transmission systems. *Mailing Add:* Hughes Radar Systs Box 92426 Los Angeles CA 90009

KING, HARTLEY H(UGHES), b Fresno, Calif, Apr 21, 36; m 62; c 2. AERONAUTICAL SCIENCES, MECHANICAL ENGINEERING. *Educ:* Univ Calif, Berkeley, BS, 57, MS, 59, PhD(mech eng), 63. *Prof Exp:* Sr engr, Electrooptical Systs, 62-65; sr res engr, AC Electronics Defense Res Labs, Gen Motors Corp, 65-68; sr res engr, Gen Res Corp, 68-75; SR RES ENGR, EFFECTS TECHNOL, INC, 75- *Mem:* Am Inst Aeronaut & Astronaut; Inst Elec & Electronics Engrs. *Res:* Compressible fluid mechanics; wakes and separated flows; re-entry aerodynamics; flight mechanics. *Mailing Add:* General Research Corp 5383 Hollister Ave Santa Barbara CA 93111

KING, HENRY LEE, b Muleshoe, Tex, Apr 12, 21; m 48; c 4. ORGANIC CHEMISTRY, POLYMER CHEMISTRY. *Educ:* Tex Tech Col, BS, 52, MS, 54. *Prof Exp:* Res chemist, Chemstrand Res Ctr, Inc, Monsanto Co, 54-65, sr res chemist, 65-70, res specialist, 70-76, sr res specialist, 76-81; RETIRED. *Mem:* Am Chem Soc. *Res:* Synthetic polymers and fibers; polymer processes; polymer modification for specific end uses; intermediate synthesis; condensation polymers, especially polyesters. *Mailing Add:* 102 Bogue Ct Cary NC 27511

KING, HERMAN (LEE), b Grand Ledge, Mich, Feb 24, 15; m 39; c 2. ACADEMIC ADMINISTRATION. *Educ:* Mich State Univ, BS, 39; Pa State Univ, MS, 41, PhD(biochem), 42. *Prof Exp:* Res fel, Reilly Tar & Chem Corp, Ind, 39-42; instr agr & biol chem, Pa State Univ, 43; orchard mgr, Mich, 44; exten entomologist, Mich State Univ, 45-46, from asst prof to assoc prof entom, 46-59, prof & asst dean, Col Arts & Sci, 59-62, dir, Div Biol Sci, 60-62, asst dean, Col Natural Sci, 62-63, asst provost, 63-74, dir acad serv, 74-81; RETIRED. *Res:* University administration. *Mailing Add:* 1342 Marble Elans E Lansing MI 48823-2833

KING, HOWARD E, b Seattle, Wash, Oct 16, 24; m 50; c 5. ELECTRICAL ENGINEERING. *Educ:* Wash Univ, BS, 46; Univ Ill, MS, 55. *Prof Exp:* Engr, RCA Victor Div, NJ, 46-52 & Andrew Corp, Ill, 52-53; res asst electromagnetics, Univ Ill, Urbana, 53-55; mem tech staff, Ramo-Wooldridge Corp, Calif, 55-58 & Space Tech Labs Inc, 58-61; mem tech staff, 61-62, mgr electromagnetics sect, 62-70, HEAD, ANTENNAS & PROPAGATION DEPT, AEROSPACE CORP, 70- *Concurrent Pos:* Lectr, Univ Calif, Los Angeles, 62-64. *Mem:* Inst Elecs & Electronics Engrs. *Res:* Frequency modulation and television transmitters, antennas and diplexers; antennas for aircraft, reentry vehicles, spacecraft and ground installations; millimeter wavelength research in antennas and components; radar cross section measurements and analysis. *Mailing Add:* PO Box 2957 Mail Sta M4-931 Los Angeles CA 90009-2957

KING, HUBERT WYLAM, b Cardiff, Wales, Jan 20, 30; Can citizen; m 57; c 4. MATERIALS SCIENCE, ENGINEERING PHYSICS. *Educ:* Univ Birmingham, BSc, 54, PhD(metall), 56; Imp Col, London, DIC(metall), 71. *Prof Exp:* Fel metall, Univ Birmingham, 56-58; USAEC grant metal physics, Mellon Inst, 59-62; lectr phys metall, Imp Col, Univ London, 62-65, sr lectr, 65-67; vis prof metall, Univ Calif, Berkeley, 67; reader phys metall, Imp Col, Univ London, 67-70; prof eng physics, Dalhousie Univ, 71-83; PROF ENG PHYSICS, TECH UNIV, NS, 83-, HEAD ENG PHYSICS DEPT, 86- *Concurrent Pos:* UK Atomic Energy Authority & Sci Res Coun grant, Imp Col, Univ London, 62-70; co-ed comn struct reports, Int Union Crystallog, 66-; consult, USAEC, 67-69; Nat Res Coun Can grant, Dalhousie Univ, 71-83; nat sci eng res coun grant, Tech Univ NS, 83-; bd coun fed, Inf Space

Univ, 87. *Mem:* Fel Brit Inst Metall; fel Brit Inst Physics; Am Inst Mining, Metall & Petrol Engrs; Can Inst Min Metall; Can Ceram Soc; Soc Advan & Process Eng. *Res:* Structure and stability of alloy phases; effect of phase transformations on superconducting transition magnetic and ferroelectric properties of stainless steels; electronic properties of ceramics. *Mailing Add:* Dept Engr Physics Tech Univ Nova Scotia Box 1000 Halifax NS B3J 2X4 Can

KING, IVAN ROBERT, b Far Rockaway, NY, June 25, 27; c 4. ASTRONOMY. *Educ:* Hamilton Col, AB, 46; Harvard Univ, AM, 47, PhD(astron), 52. *Prof Exp:* Instr astron, Harvard Univ, 51-52; methods analyst, US Dept of Defense, 54-56; from asst prof to assoc prof astron, Univ Ill, 56-64; assoc prof, 64-66, PROF ASTRON, UNIV CALIF, BERKELEY, 66- *Mem:* Nat Acad Sci; Am Astron Soc (pres, 78-80); Int Astron Union; Am Acad Arts & Sci. *Res:* Structure of stellar systems. *Mailing Add:* Dept Astron Univ Calif Berkeley CA 94720

KING, JAMES, JR, b Columbus, Ga, Apr 23, 33; div; c 2. PHYSICAL CHEMISTRY, PHYSICS. *Educ:* Morehouse Col, BS, 53; Calif Inst Technol, MS, 55, PhD(chem, physics), 58. *Prof Exp:* Res engr electrochem, Jet Propulsion Lab, Calif Inst Technol, 56; sr res engr thermal properties, Atomics Int Div, NAm Aviation, Inc, 58-60; sr scientist, Electro-Optical Systs, Inc, 60-61; sr scientist, Jet Propulsion Lab, Calif Inst Technol, 61-69, sect mgr physics, 69-74; dir, Space Shuttle Environ Effects, Space Shuttle Prog, NASA, Washington, DC, 74-75 & Upper Atmospheric Sci Prog, 75-76; mgr, User Prog Develop Off, Jet Propulsion Lab, Calif Inst Technol, 76-78; mgr space physics, 79-80, mgr space sci & appln, 81-87, DEPT ASST LAB DIR, JET PROPULSION LAB, CALIF INST TECHNOL, 88- *Mem:* Am Chem Soc; Am Phys Soc; Sigma Xi; AAAS; Am Geophys Union. *Res:* Nuclear and electron resonance; radiation chemistry. *Mailing Add:* 1720 La Cresta Dr Pasadena CA 91103

KING, JAMES C, b Petaluma, Calif, Jan 18, 32; div; c 3. PHARMACY. *Educ:* Univ NMex, BS, 53; Univ Tex, MS, 58, PhD(pharm), 62. *Prof Exp:* Asst pharm, Univ Tex, 57-58, lectr, 59-60, asst, 60-61, spec instr, 61-62; from asst prof to prof pharm, 62-76, dir clin pharm, 71-76, PROF CLIN PHARM, UNIV PAC, 76- *Concurrent Pos:* Consult, US Army Surgeon Gen, 72- & Vet Admin Hosp, San Diego & Palo Alto, 73-80. *Mem:* Am Pharmaceut Asn; Am Soc Hosp Pharmacists. *Res:* Development of and medicament release from dermatologic medications; incompatibilities of parenteral admixtures. *Mailing Add:* Sch Pharm Univ Pac Stockton CA 95211

KING, JAMES CLAUDE, b St Joseph, Mo, Oct 2, 24; m 49; c 4. LOW TEMPERATURE PHYSICS, SOLID STATE PHYSICS. *Educ:* Amherst Col, BA, 49; Yale Univ, MS, 50, PhD(physics), 53. *Prof Exp:* Instr physics, Yale Univ, 50, asst, 50-53; mem tech staff, Bell Tel Labs, Inc, 53-65; mgr radiation physics dept, 65-68, dir appl res, 68-71, dir electrochem components & measurement systs, 71-77, dir weapons elec subsysts, 77-83, DIR MATS PROCESS ENG & FABRICATION, SANDIA LABS, 83- *Honors & Awards:* C B Sawyer Award, Sawyer Res Prod, Inc, 73. *Mem:* Fel Am Phys Soc; Sigma Xi. *Res:* Solid state physics; acoustic wave interactions with lattice defects; effects of radiation on the acoustic properties of alpha-quartz. *Mailing Add:* 7832 Academy Trail NE Albuquerque NM 87185

KING, JAMES CLEMENT, genetics; deceased, see previous edition for last biography

KING, JAMES DOUGLAS, b Welland, Ont, May 28, 34; m 58. NUCLEAR PHYSICS. *Educ:* Univ Toronto, BA, 56; Univ Sask, PhD(nuclear physics), 60. *Prof Exp:* Fel, McMaster Univ, 60-61; asst prof physics, Univ Sask, 61-64; asst prof, 65-67, assoc dean & registr, Scarborough Col, 71-74, assoc dean, 75-76, assoc prof, 67-80, PROF PHYSICS, UNIV TORONTO, 80- *Mem:* Am Asn Physics Teachers; Can Asn Physicists; Am Phys Soc; Am Astron Soc. *Res:* Nucleosynthesis; nuclear reactions; stellar reaction rates. *Mailing Add:* Dept Physics Univ Toronto Toronto ON M5S 1A7 Can

KING, JAMES EDWARD, b Escanaba, Mich, July 23, 40; m 73; c 1. PALYNOLOGY. *Educ:* Alma Col, BS, 62; Univ NMex, MS, 64; Univ Ariz, PhD(geosci), 72. *Prof Exp:* Res assoc geochronol, Univ Ariz, 71-72; cur paleobot, Ill State Mus, Springfield, 72-78, head sci sect, 78-85, asst dir sci, 85-87; DIR, CARNEGIE MUS, 87- *Concurrent Pos:* NSF res grants, 72, 74, 76 & 81; adj assoc prof geol, Univ Ill-Urbana, 78-88; res assoc, Hunt Bot Inst Carnegie, Mellon Univ, Pittsburgh, 88- *Mem:* AAAS; Am Quaternary Asn (former treas); Am Asn Stratigraphic Palynologists; Ecol Soc Am; Sigma Xi; Am Asn Mus. *Res:* Quaternary palynology, biogeography and paleoenvironments of North America; the ecology and extinction of the Pleistocene megafauna and the interaction of early man to his environments. *Mailing Add:* Dir Carnegie Mus Natural Hist 4400 Forbes Ave Pittsburgh PA 15213

KING, JAMES FREDERICK, b Moncton, NB, Apr 6, 34; m 65; c 3. ORGANIC CHEMISTRY. *Educ:* Univ NB, BSc, 54, PhD(org chem), 57. *Prof Exp:* Beaverbrook overseas scholar, Imp Col, London, 57-58; res fel chem, Harvard Univ, 58-59; from asst prof to assoc prof, 59-67, PROF CHEM, UNIV WESTERN ONT, 67- *Concurrent Pos:* Alfred P Sloan res fel, 66-68. *Honors & Awards:* Merck, Sharp and Dohme Lectr Award, 76. *Mem:* Am Chem Soc; Chem Inst Can; Royal Soc Chem. *Res:* Organic sulfur chemistry; reaction mechanisms; stereochemistry; reaction optimization. *Mailing Add:* Dept Chem Univ Western Ont London ON N6A 5B7 Can

KING, JAMES P, b Kiangsu, China, Nov 11, 33; m 61. PHYSICAL INORGANIC CHEMISTRY. *Educ:* Newberry Col, BS, 53; Loyola Univ, Ill, MS, 56; Purdue Univ, PhD(chem), 60. *Prof Exp:* PROJ LEADER, PENNWALT CHEM CORP, 59- *Mem:* Am Chem Soc; Am Soc Lubrication Eng. *Res:* Lubrication products on high temperature application; Thermochemistry of rhenium compounds; heats of formation of various rhenium compounds determined by solution calorimetry and electrochemical cell measurements; synthesis and characterization of coordination compounds and inorganic polymers. *Mailing Add:* 904 Breezewood Lane Lansdale PA 19446

KING, JAMES ROGER, b San Jose, Calif, Mar 12, 27; m 50; c 3. PHYSIOLOGICAL ECOLOGY. *Educ:* San Jose State Col, AB, 50; Wash State Univ, MS, 53, PhD(zool), 57. *Prof Exp:* Actg instr zoophysiol, Wash State Univ, 56-57; asst prof exp biol, Univ Utah, 57-60; from asst prof to assoc prof zoophysiol, 60-67, chmn dept zool, 72-78, PROF ZOOPHYSIOL, WASH STATE UNIV, 67- *Concurrent Pos:* NIH res career develop award, 63-67; Guggenheim Found fel, 69; mem adv panel environ biol, NSF, 73-76 & 84; Maytag vis prof zool, Ariz State Univ, 79; hon prof, EChina Univ, 83. *Honors & Awards:* Brewster Medal, Am Ornithologists Union, 74. *Mem:* Fel AAAS; Cooper Ornith Soc; Am Ornith Union (pres, 82-84); Am Physiol Soc; Am Soc Zool. *Res:* Avian physiology and ecology; vertebrate physiology. *Mailing Add:* Dept Zool Wash State Univ Pullman WA 99164-4220

KING, JAMES S, b Painesville, Ohio, Nov 19, 38; m 61; c 2. NEUROANATOMY. *Educ:* Taylor Univ, BS, 60; Ohio State Univ, MSc, 62, PhD(anat), 65. *Prof Exp:* From instr to assoc prof anat, 65-75, PROF ANAT, OHIO STATE UNIV, 75-, DIR, SPINAL CORD INJURY RES CTR, 77- *Concurrent Pos:* NIH gen res grant, 65-67; res fel neuroanat & electron micros, Wayne State Univ, 67-68; USPHS grant, 69-81. *Mem:* Am Asn Anat; Soc Neurosci; Sigma Xi; AAAS. *Res:* Synaptic organization and development of precerebellar nuclei. *Mailing Add:* Dept Anat 1645 Neil Ave Ohio State Univ Columbus OH 43210

KING, JANET CARLSON, b Red Oak, Iowa, Oct 3, 41; m 67; c 2. NUTRITION. *Educ:* Iowa State Univ, BS, 63; Univ Calif, Berkeley, PhD(nutrit), 72. *Prof Exp:* Dietitian, Fitzsimons Gen Hosp, Denver, 64-67; fel, 72, from asst prof to assoc prof, 72-83, PROF NUTRIT, UNIV CALIF, BERKELEY, 83-, CHAIR, DEPT NUTRIT SCI, 88- *Concurrent Pos:* Mem, Comt Nutrit Mother & Preschool Child, Nat Acad Sci, Nat Res Coun, 74-80, Nutrit Study Sect, NIH, 81-85, Comt Mil Nutrit Res, Nat Acad Sci, Nat Res Coun, 85-; consult, NIH, 79-80; mem, Comt Nutrit Status During Pregnancy & Lactation, Nat Res Coun, Nat Acad Sci, 88-, Food & Nutrit Bd, Inst Med-Nat Acad Sci, 91- *Honors & Awards:* Frances Fischer Mem Lectr, 85; Lederle Award Human Nutrit, 89. *Mem:* Am Dietetic Asn; Sigma Xi; Soc Nutrit Educ; Am Inst Nutrit; AAAS; Am Soc Clin Nutrit. *Res:* Study of the nutritional requirements for protein, energy, and trace elements in nonpregnant and pregnant women, and lactating women; zinc utilization in healthy adults. *Mailing Add:* Dept Nutrit Sci Univ Calif Berkeley CA 94720

KING, JERRY PORTER, b Dyersburg, Tenn, July 9, 35; m 62; c 2. MATHEMATICS. *Educ:* Univ Ky, BSEE, 58, MS, 59, PhD(math), 62. *Prof Exp:* Asst prof, Lehigh Univ, 62-68, assoc dean arts & sci, 79-81, dean grad sch, 81-87, PROF MATH, LEHIGH UNIV, 68- *Mem:* Math Asn Am; Am Math Soc. *Res:* Complex variables, summability. *Mailing Add:* Math Dept No 14 Lehigh Univ Bethlehem PA 18015

KING, JOE MACK, b Conroe, Tex, July 25, 44; m 67; c 1. PHYCOLOGY. *Educ:* Sam Houston State Univ, BS, 67, MA, 68; Univ Tex, Austin, PhD(bot), 71. *Prof Exp:* Res scientist aquatic ecol, Univ Tex, Austin, 68-70; asst prof biol, Univ Wis-La Crosse, 71-73; res assoc aquatic ecol, Rice Univ, 73-78; from asst prof to assoc prof, 78-81, PROF BIOL, MURRAY STATE UNIV, 81- *Concurrent Pos:* Consult, Exxon Corp, Baytown, Tex, 73-74. *Mem:* Am Soc Limnol & Oceanog; Phycol Soc Am; Int Phycol Soc; Am Inst Biol Sci. *Res:* Phytoplankton ecology; impact of nutrient enrichment on aquatic ecosystems; aquatic phytotoxicity; algal taxonomy. *Mailing Add:* Dept Biol Sci Murray State Univ Murray KY 42071

KING, JOHN, b Darlington, Eng, Nov 4, 38; m 62; c 3. PLANT PHYSIOLOGY. *Educ:* Univ Durham, BSc, 60; Univ Man, MSc, 62, PhD(bot), 66. *Prof Exp:* Asst prof biol, Bishops Univ, 64-67; from asst prof to assoc prof, Univ Sask, 67-77, head biol, 81-87, actg head chem, 88-89, PROF BIOL, UNIV SASK, 77- *Concurrent Pos:* Natural Sci & Eng res Coun Can res grant-in-aid, 64-; fel, Univ Man, 66; vis scientist, Univ Leicester, 73-74, Hawaiian Sugar Planter's Asn, 80-81, Max-Planck Inst, Cologne, 84 & Univ BC, Vancouver, 87-88, Natural Sci & Eng Res Coun Can URF Comt, 86-89, chair, 88-89 & Natural Sci & Eng Res Coun Can Plant Biol GSC, 89-92, chair, 90-91. *Mem:* Am Soc Plant Physiol; Can Soc Plant Physiol (pres, 83-84); Int Asn Plant Tissue Cult; Biol Coun Can; Can Coun Univ Biol Chmn (pres, 82-83); Int Plant Molecular Biol Asn. *Res:* Isolation, characterization and use of auxotropic mutants from plant cell cultures; genetic transformation of plant cells; isolation and use of resistance mutants in cell culture; isolation of arabidopsis metabolic mutants. *Mailing Add:* Dept Biol Univ Sask Saskatoon SK S7N 0W0 Can

KING, JOHN A(LBERT), b Columbus, Ind, Sept 6, 16; m 41; c 4. RESEARCH ADMINISTRATION, NEW PRODUCT LICENSING. *Educ:* Ind Univ, AB, 38; Univ Minn, MS, 40, PhD(org chem), 42. *Prof Exp:* Asst chem, Ind Univ, 37-38 & Univ Minn, 38-41; Merck fel, 42-43; sr chemist, Winthrop Chem Co Div, Sterling Drug, 43-46; dir chem res, Warner Lambert Pharmaceut Co, NJ, 46-57; dir res & gen mgr res div, Armour & Co, 57-60; mgr res & develop, Agr Div, Am Cyanamid Co, 60-69, assoc dir agr & pharmaceut res, 70-77, dir licensing agr prod, 78-85; RETIRED. *Concurrent Pos:* Civilian with Off Sci Res & Develop, 44. *Mem:* AAAS (vpres & chmn chem sect, 59-60); Am Chem Soc; fel Inst Chem; Sigma Xi; Soc Chem Indust. *Res:* Synthesized pharmacologically active organic compounds; mechanisms of chemical reactions; coordination of chemical research with evaluation of compounds produced; research management; research development and commercialization of pharmaceutical, nutritional and industrial chemical products; licensing of pharmaceutical and agricultural products. *Mailing Add:* 90 Battle Rd Princeton NJ 08540

KING, JOHN ARTHUR, b Detroit, Mich, June 22, 21; m 49; c 2. ZOOLOGY. *Educ:* Univ Mich, AB, 43, MS, 48, PhD, 51. *Prof Exp:* Asst genetics, Univ Mich, 47-51; USPHS fel, Jackson Mem Lab, 51-53, staff scientist, 53-60; Nat Acad Sci-Nat Res Coun fel, 60-61; from assoc prof to prof, 61-86, EMER PROF ZOOL, MICH STATE UNIV, 86- *Concurrent Pos:* Develop res career award, USPHS, 64-70. *Mem:* Am Soc Mammal; Animal Behav Soc (secy, 62-65, pres, 70). *Res:* Sociobiology; mammalian behavior; effects of early experience; behavioral evolution. *Mailing Add:* Dept Zool Mich State Univ East Lansing MI 48823

KING, JOHN EDWARD, b Columbus, Ohio, Nov 16, 39. ANATOMY, HISTOLOGY. *Educ:* Ohio State Univ, BA, 61, PhD(anat), 65. *Prof Exp:* From instr to assoc prof anat, Ohio State Univ, 65-73, assoc prof physiol optics, Col Optom, 73-91; RETIRED. *Concurrent Pos:* Asst dean, Ohio State Univ. *Mem:* AAAS. *Res:* Hematology. *Mailing Add:* 3839 Olentangy Blvd Columbus OH 43214

KING, JOHN GORDON, b London, Eng, Aug 13, 25; nat US; m 49; c 8. ATOMIC PHYSICS. *Educ:* Mass Inst Technol, SB, 50, PhD(physics), 53. *Hon Degrees:* ScD, Univ Hartford, 72. *Prof Exp:* From instr to prof physics, 53-74, FRANCIS L FRIEDMAN PROF PHYSICS, MASS INST TECHNOL, 74- *Honors & Awards:* Millikan Award, 65; Harbison Award, 71. *Mem:* Am Phys Soc; Am Asn Physics Teachers. *Res:* Studies of biological surfaces; molecule microscopy. *Mailing Add:* Mass Inst Technol 26-457 Cambridge MA 02139

KING, JOHN MCKAIN, b Boston, Mass, Jan 16, 27; m 47; c 1. VETERINARY PATHOLOGY, WILDLIFE DISEASES. *Educ:* Okla Agr & Mech Col, DVM, 55; Cornell Univ, PhD(path), 63. *Prof Exp:* Instr vet path, Cornell Univ, 59-60; asst prof, Wash State Univ, 60-62; fel path, Mellon Inst, 62-69; assoc prof, 69-80, PROF VET PATH, NY STATE COL, CORNELL UNIV, 80- *Concurrent Pos:* USPHS res fel, 60; consult, Animal Med Ctr, NY, 62- *Mem:* Am Vet Med Asn; Am Col Vet Path; Brit Vet Asn. *Res:* Veterinary pathology, especially lung and liver diseases; wildlife pathology, both gross and micropathology. *Mailing Add:* Dept Vet Path NY State Vet Col Cornell Univ Ithaca NY 14853

KING, JOHN MATHEWS, b New York, NY, Oct 25, 39; m 62; c 2. ORGANIC CHEMISTRY. *Educ:* Cornell Univ, AB, 61; Univ Mich, MS, 63, PhD, 65. *Prof Exp:* NSF fel, Calif Inst Technol, 65-66, univ fel, 66-67; sr res assoc, 67-79, mgr, Lubricating Oil Additives Div, 79-82, MGR, COM ADDITIVE PROCESS DIV, CHEVRON RES CO, STANDARD OIL CO CALIF, 82- *Mem:* Am Chem Soc. *Res:* Synthesis of organic peroxides; organic photochemistry; radiation chemistry of organic compounds; lube oil additives; organic process development. *Mailing Add:* 1194 Idylberry Rd San Rafael CA 94903

KING, JOHN PAUL, b Zena, Okla, Nov 23, 38; m 61; c 3. SOLID STATE PHYSICS. *Educ:* Cent State Col, Okla, BS, 61; Okla State Univ, PhD(solid state physics), 66. *Prof Exp:* Eng sci specialist, Missiles & Space Div, LTV Aerospace Corp, 66-68; from asst prof to assoc prof, 68-77, PROF PHYSICS, CENT STATE UNIV, OKLA, 77- *Mem:* Am Phys Soc; Sigma Xi. *Res:* Study of basic optics problems associated with imaging and interference. *Mailing Add:* 3420 Baird Dr Edmond OK 73013

KING, JOHN STUART, b Buffalo, NY, Nov 12, 27. GEOLOGY. *Educ:* Univ Buffalo, BA, 55, MA, 57; Univ Wyo, PhD(geol), 63. *Prof Exp:* Res engr petrog, Res Div, Carborundum Co, 57-60; from asst prof to assoc prof, 63-77, from actg chmn dept to chmn dept, 66-71, PROF GEOL, STATE UNIV NY, BUFFALO, 77-, ASSOC DEAN FAC NATURAL SCI & MATH, 90- *Concurrent Pos:* Consult electro minerals div, Carborundum Co, 64-66. *Mem:* Fel Geol Soc Am; Am Asn Petrol Geologists. *Res:* Igneous and metamorphic petrology and structures; structural geology; field interpretations; planetology and analog studies. *Mailing Add:* Dept Geol State Univ NY 4240 Ridge Lea Rd Amherst NY 14260

KING, JOHN SWINTON, b Detroit, Mich, Oct 31, 20; m 43; c 3. PHYSICS. *Educ:* Univ Mich, PhD(nuclear physics). *Prof Exp:* Asst, Appl Physics Lab, Johns Hopkins Univ, 42-45; res assoc reactor physics, Knolls Atomic Power Lab, Gen Elec Co, 53-56, mgr submarine adv reactor physics sub-sect, 56-59; assoc prof, 59-62, PROF NUCLEAR ENG, UNIV MICH, ANN ARBOR, 62-, CHMN DEPT, 74- *Mem:* Am Nuclear Soc; Am Phys Soc. *Res:* Neutron physics as applied to reactor design and neutron optics. *Mailing Add:* 2311 Vinewood Ann Arbor MI 48104

KING, JOHN WILLIAM, b Butler, Ind, Mar 31, 38; div; c 3. AGRONOMY, CROP SCIENCE. *Educ:* Purdue Univ, BS, 63; Univ RI, MS, 66; Mich State Univ, PhD(crop sci), 70. *Prof Exp:* Machinist, Gen Elec Co, Ind, 56-60; ASSOC PROF AGRON, UNIV ARK, FAYETTEVILLE, 70- *Mem:* Am Soc Agron; Crop Sci Soc Am. *Res:* Turf grass soils fertilization, species and variety adaptation, cultural systems, weed control and irrigation management. *Mailing Add:* Dept Agron Univ Ark Fayetteville AR 72701

KING, JONATHAN (ALAN), b Brooklyn, NY, Aug 20, 41; m 76; c 2. MOLECULAR BIOLOGY, BIOCHEMISTRY. *Educ:* Yale Univ, BS, 62; Calif Inst Technol, PhD(genetics), 67. *Prof Exp:* Assoc scientist microbial ecol, Jet Propulsion Lab, Calif Inst Technol, 67-68; fel molecular biol, Purdue Univ, 68-69; fel struct biol, Lab Molecular Biol, Brit Med Res Coun, 69-70; from asst prof to assoc prof, 70-78, PROF MOLECULAR BIOL, MASS INST TECHNOL, 78-, DIR BIOL ELECTRON MICROSCOPE FACIL, 70- *Concurrent Pos:* Childs Mem Fund Med Res res fel, 68-70; Guggenheim fel, 87; counr, Am Soc Virol; chmn, Microbiol Physiol Study Sect & Genetic Based Dis Study Sect, NIH. *Honors & Awards:* US Antarctic Serv Medal. *Mem:* Genetics Soc Am; Am Soc Microbiol; Biophys Soc; fel AAAS; Am Soc Virol; Am Soc Biochem & Molecular Biol. *Res:* Genetic control of morphogenesis; virus assembly; protein folding. *Mailing Add:* Dept Biol Mass Inst Technol Cambridge MA 02139

KING, JONATHAN STANTON, b Bristol, Tenn, Oct 30, 22; m 51; c 2. CLINICAL CHEMISTRY. *Educ:* Univ Chicago, BS, 47; Univ Tenn, PhD(biochem), 54. *Prof Exp:* Control chemist, S E Massengill Co, 47, res chemist, 48-51, res biochemist, 54-56; res assoc, Bowman Gray Sch Med, Wake Forest Univ, 56-59, instr biochem & Whitney fel, 59-62, res asst prof, 62-65, res assoc prof, 65-70; exec dir, Am Asn Clin Chem, 71-73, exec ed, Clin Chem, 69-90; RETIRED. *Concurrent Pos:* NIH res career develop award, 65-70; guest investr, Rockefeller Univ, 68-69. *Honors & Awards:* Spec Area Award, Am Asn Clin Chem, 81 & Miriam Reiner Award, 83. *Mem:* Am Asn Clin Chem. *Res:* Salicylate effect on pituitary-adrenal axis; function of adrenal ascorbic acid; enzyme adsorption by microcryst sulfas; colorimetric alkaloid estimation; metabolism of trimethylchromone; biochemistry of renal calculi; urinary macromolecular constituents; factors affecting calcium excretion; ninhydrin-positive substances in urine; taurine excretion in mongolism. *Mailing Add:* 2370 Lyndhurst Ave Winston-Salem NC 27103

KING, JOSEPH HERBERT, b Malden, Mass, Nov 16, 39; m 65; c 2. SPACE PHYSICS. *Educ:* Boston Col, PhD(physics), 66. *Prof Exp:* Lectr physics, Regis Col, Mass, 63-64; staff scientist, Aerospace Corp, 65-67; Nat Acad Sci assoc fel, 67-69, PHYSICIST, GODDARD SPACE FLIGHT CTR, NASA, 69- *Concurrent Pos:* Proj scientist, Interplanetary Monitoring Probe Satellite, NASA, 74- *Mem:* Am Geophys Union. *Res:* Spacecraft plasma and field data concerning the mass and energy coupling between the solar wind and the earth's magnetosphere. *Mailing Add:* Code 933 Goddard Space Flight Ctr NASA Greenbelt MD 20771

KING, KATHERINE CHUNG-HO, b Peiping, China, Aug 27, 37; US citizen; m. PEDIATRICS, NEONATOLOGY. *Educ:* Meredith Col, AB, 57; Bowman Gray Sch Med, MD, 62; Am Bd Pediat, dipl, 68, dipl, Neonatal & Perinatal Med, 77. *Prof Exp:* From intern pediat to resident pediat metab, Cleveland Metrop Gen Hosp, 62-66, actg dir, Newborn Serv, 74-75; from asst prof to assoc prof pediat, Sch Med, Case Western Res Univ, 71-85, assoc prof neonatology, Dept Reproductive Biol, 78-85; asst pediatrician, Cleveland Metrop Gen Hosp & co-dir, Perinatal Clin Res Ctr, 69-85, dir, Newborn Serv, 81-84; assoc prof pediat, State Univ NY, Stony Brook, 85-89; NEONATOLOGIST, SCHNEIDER CHILDREN'S HOSP, LONG ISLAND JEWISH HOSP, 85-; ASSOC PROF PEDIAT, ALBERT EINSTEIN SCH MED, 89- *Concurrent Pos:* USPHS trainee, 66-68; Cleveland Diabetes Found grant, Cleveland Metrop Gen Hosp, 69-70; sr instr, Case Western Reserve Univ, 69-71. *Mem:* Am Fedn Clin Res; Am Acad Pediat; fel Am Col Nutrit; Soc Pediat Res. *Res:* Carbohydrate metabolism of human fetus and neonatal; metabolism. *Mailing Add:* Long Island Jewish-Hillside Med Ctr Schneider Children's Hosp New Hyde Park NY 11042

KING, KENDALL WILLARD, b Pittsburgh, Pa, Feb 20, 26; div; c 2. NUTRITION. *Educ:* Va Polytech Inst, BS, 49, MS, 50; Univ Wis, PhD(biochem), 53. *Prof Exp:* From asst prof bact & biochem to prof biochem, Va Polytech Inst, 53-68, head dept biochem & nutrit, 67-68; asst vpres, Res Corp, NY, 68-77, vpres grants, 77-83; vpres, Remick Assocs, 83-85; ASSOC DEAN RES & GRAD STUDIES, WESTERN CAROLINA UNIV, 86- *Concurrent Pos:* Res assoc, Inst Nutrit Sci, Columbia Univ, 59-60; mem, US comt, Int Union Nutrit Sci, Nat Acad Sci-Nat Res Coun, 75-81; Comt Blindness Prev, Helen Keller Int; consult food & pharmaceut industs. *Mem:* AAAS; Am Chem Soc; Am Inst Nutrit; Am Soc Biol Chemists; Am Asn Cereal Chemists; Am Inst Chemists; Nat Coun Univ Res Admin. *Res:* Purification and mode of action of fungal cellulases; vuminant physiology; experimental nutrition of bacteria, insects, mammals; public health nutrition programming and evaluation in third wolrd; cellulase action; amino acid nutrition. *Mailing Add:* Western Carolina Univ 250 Robinson Bldg Cullowhee NC 28723

KING, KENNETH, JR, b Philadelphia, Pa, Dec 27, 30; m 61; c 2. BIOGEOCHEMISTRY. *Educ:* Mass Inst Technol, SB, 52, SM, 60; Columbia Univ, PhD(geol), 72. *Prof Exp:* Indust res & develop, Merck & Co, 55-64, dir res planning, 65-67; res asst, Columbia Univ, 67-72; RES ASSOC BIOGEOCHEM, LAMONT-DOHERTY GEOL OBSERV, COLUMBIA UNIV, 72- *Concurrent Pos:* Fel, Carnegie Inst Wash, 72-74. *Mem:* Sigma Xi; AAAS; Am Chem Soc; Geol Soc Am; Geochem Soc. *Res:* Proteins of modern and fossil mineralized tissues; amino acid racemization; mechanism of biomineralization. *Mailing Add:* 275 Jones Rd Falmouth MA 02540-3338

KING, L D PERCIVAL, b Williamstown, Mass, Dec 29, 06; m 79; c 2. PHYSICS. *Educ:* Univ Rochester, BS, 30; Univ Wis, PhD(exp nuclear physics), 37. *Prof Exp:* Asst physics, Univ Rochester, 30-31 & Mass Inst Technol, 31-33; asst, Univ Wis, 35-37; instr, Purdue Univ, 37-42, fel, Off Sci Res & Develop contract, 42-43; group leader, Manhattan Dist & AEC Projs, Los Alamos Sci Lab, 43-57; tech dir, US Atoms for Peace Conf, Atomic Energy Comn, Geneva, 57-58; asst div leader, Reactor Div, Los Alamos Sci Lab, 59, chmn, Rover Flight Safety Off, 60-69, res adv, 69-73; consult, 73-78; RETIRED. *Mem:* Fel AAAS; fel Am Phys Soc; fel Am Nuclear Soc. *Res:* Nuclear physics; artificial radioactivity; nuclear structure; design and construction of homogenous and fast nuclear research and test reactors; solution to flight safety problems in nuclear rocket reactors. *Mailing Add:* Rte 4 Box 16-B Santa Fe NM 87501

KING, L(EE) ELLIS, b Jamestown, NC, Aug 21, 39; m 60. TRANSPORTATION ENGINEERING, ERGONOMICS. *Educ:* NC State Univ, BS, 61; Univ Calif, Berkeley, DrEng, 67. *Prof Exp:* From asst prof to asoc prof, WVa Univ, 67-73; assoc prof, Univ Colo, Denver Ctr, 73-75; prof, Wayne State Univ, 75-76; CHMN & PROF, DEPT CIVIL ENG, UNIV NC, CHARLOTTE, 76- *Concurrent Pos:* Consult, various pvt & pub agencies. *Honors & Awards:* Walter L Huber Civil Engr Res Prize, Am Soc Civil Engrs, 73. *Mem:* Am Soc Civil Engrs; Transp Res Bd; Am Soc Eng Educ; Human Factors Soc; Nat Soc Prof Engrs; Inst Transp Engrs. *Res:* Traffic engineering problems; transportation planning; driver, vehicle and roadway interaction; driver behavior; reduced visibility performance; roadway lighting; roadway signing; pedestrian behavior; traffic signals. *Mailing Add:* Dept Civil Eng Univ NC Charlotte NC 28223

KING, LAFAYETTE CARROLL, b Marysvale, Utah, Sept 9, 14; m 37; c 5. ORGANIC CHEMISTRY. *Educ:* Utah State Col, BS, 36; Mich State Col, MS, 38, PhD(chem), 42. *Prof Exp:* Asst biol chem, Mich State Col, 36-42; res assoc, 42, from instr to prof chem, 42-85, EMER PROF CHEM, NORTHWESTERN UNIV, 85- *Concurrent Pos:* Indust consult, Food Chem, 50-; US-Japan Conf Chem Educ, Japan, 64, Berkeley, 68 & First Int Am Conf Chem Educ, Buenos Aires, 65; chmn adv coun col chem, 66-; chmn org comt, Indo-US Binat Conf Chem Educ, 69. *Honors & Awards:* Chem Educ Award, Am Chem Soc, 69. *Mem:* Fel AAAS; Am Chem Soc; Oil Chem Soc; Royal Soc Chem. *Res:* Synthesis of quaternary salts, thiazoles and selenazoles; structures of sterols; mechanism of organic reactions. *Mailing Add:* Dept Chem Northwestern Univ Evanston IL 60201

KING, LARRY DEAN, b Atlanta, Ga, Mar 10, 39; div; c 2. SOIL SCIENCE. *Educ:* Ga Inst Technol, BS, 62; Univ Ga, MS, 68, PhD(agron), 71. *Prof Exp:* Aircraft engr comput prog, Lockheed Aircraft, 62-65; fel soil sci, Univ Guelph, Ont, 71-72; agronomist agr develop, Tenn Valley Authority, 73-74; from asst prof to assoc prof, 74-88, PROF SOIL SCI, NC STATE UNIV, 88- *Mem:* Am Soc Agron. *Res:* Alternative agricultural systems; land application of municipal and industrial wastes. *Mailing Add:* Dept Soil Sci NC State Univ Box 7619 Raleigh NC 27695-7619

KING, LARRY GENE, b Prosser, Wash, June 24, 36; m 56; c 4. IRRIGATION & DRAINAGE ENGINEERING, GROUNDWATER HYDROLOGY. *Educ:* Wash State Univ, BS, 58; Colo State Univ, MS, 61, PhD(civil eng), 65. *Prof Exp:* Asst civil eng, Colo State Univ, 58-62, jr agr engr, 62; res engr hydrol, Gen Elec Co, 62-65; res scientist, Pac Northwest Labs, Battelle Mem Inst, 65, sr res engr, 65-68, res assoc, 68-69; assoc prof agr & irrig eng, Utah State Univ, 69-74; chmn dept, 79-87, PROF AGR ENG, WASH STATE UNIV, 74- *Concurrent Pos:* Consult, USAID, 84, 86-90; mem, US Nat Comt, Int Comn Irrigation & Drainage. *Mem:* Am Soc Civil Engrs; Am Soc Agr Engrs. *Res:* Irrigation; drainage; groundwater hydrology; soil and water use and conservation; water quality; soil moisture movement; erosion control; fluid flow through porous media; on-farm water management; computer modeling. *Mailing Add:* Dept Agr Eng Wash State Univ Pullman WA 99164-6120

KING, LARRY MICHAEL, b Brooklyn, NY, Nov 29, 42; m 68. MATHEMATICS. *Educ:* Brooklyn Col, BS, 63; Univ Md, College Park, MA, 66, PhD(math), 68. *Prof Exp:* Asst prof math, Univ Mass, Amherst, 68-74; asst prof, 74-79, ASSOC PROF MATH, UNIV MICH, FLINT, 79-, ASSOC DEAN, COL ARTS & SCI, 82- *Mem:* Am Math Soc; Math Asn Am. *Res:* Slices in transformation groups; topological dynamics and dynamical systems; R-isomorphisms of transformation groups; actions of non-compact semigroups; new ways of teaching calculus. *Mailing Add:* Col Arts & Sci Univ Mich Flint MI 48502-2186

KING, LEE CURTIS, b Greenville, SC, Oct 20, 54. FIBERGLASS COATINGS, TEST DEVELOPMENT. *Educ:* Univ SC, BS, 76, MEd, 80. *Prof Exp:* Technician, Cardinal Chem Co, 76-78; teacher, math, chem & phys, Midlands Tech Col, 78-80; teacher, chem, phys sci, Anderson Col, 80-82; teacher, chem, phys sci & math, N Greenville Col, 82-83; instr, chem, math, phys & astron, Tri County Tech Col, 83-84; RES CHEMIST, CLARK-SCHWEBEL FIBER GLASS, 84- *Concurrent Pos:* Salesperson, Am Indust, Lumberton, NC, 83, Real Estate Assoc, Anderson, SC, 83- *Mem:* Am Soc Mat. *Res:* Development, evaluation, fiberglass coatings based on customer needs (market or applied orientation); polymer properties and processing characteristics; test development from size preparation to end product in many fields. *Mailing Add:* 1612 Chapman Rd Anderson SC 29621

KING, LESTER SNOW, b Cambridge, Mass, Apr 18, 08; m 31; c 2. PATHOLOGY. *Educ:* Harvard Univ, AB, 27, MD, 32. *Prof Exp:* Asst path, Rockefeller Inst, Princeton, NJ, 37-40; asst & instr, Sch Med, Yale Univ, 40-42; from clin asst prof to clin prof, Col Med, Univ Ill, 46-64; sr ed, 63-73, contrib ed, J, AMA, 73-78; CONSULT, 78- *Concurrent Pos:* Res fel anat, Harvard Med Sch, 33-35; Moseley travelling fel, Inst Cancer, Madrid & Nat Hosp, London, 35-36; Nat Cancer Inst res grant, 47-49; pathologist, Fairfield State Hosp, Conn, 40-42 & Ill Masonic Hosp, 46-62; lectr path, Col Med, Univ Ill, 64-71; Boerhaave lectr, Leiden, 64; lectr, Univ Chicago, 65-; mem hist life sci study sect, NIH, 63-68, chmn, 64-66; ed, Clio Medica, 74-76; Garrison lectr, Am Asn Hist Med, 75. *Honors & Awards:* Boerhaave Medal, Leiden, 64; Welch Medal, Am Asn Hist Med, 77. *Mem:* Am Asn Hist Med (pres, 74-76). *Res:* Anatomy of the nervous system; neuropathology; cancer; virus diseases; blood brain barrier; medical history of the 17th, 18th and 19th centuries; American medicine 19th and 20th century; philosophy of medicine. *Mailing Add:* 360 Wellington Chicago IL 60657

KING, LEWIS H, b Lockeport, NS, Nov 10, 24; m 55; c 6. MARINE GEOLOGY. *Educ:* Acadia Univ, BSc, 49; Mass Inst Technol, PhD(geol), 55. *Prof Exp:* Geologist, Geol Surv Can, NS, 54-56; geochemist, Mines Br. Dept Mines & Tech Surv, Ont, 56-63; GEOLOGIST, BEDFORD INST, DEPT ENERGY, MINES & RESOURCES. 63- *Concurrent Pos:* Mem fac, Dalhousie Univ, 68- *Mem:* AAAS; fel Geol Soc Am; fel Geol Asn Can; Sigma Xi. *Res:* Geological mapping of the sea floor across the Scotian Shelf, eastern Gulf of Maine and the Grand Banks. *Mailing Add:* 50 Swanton Dr Dartmouth NS B2W 2C5 Can

KING, LLOYD ELIJAH, JR, b Mayfield, Ky, Sept 10, 39; m 68; c 3. DERMATOLOGY, ANATOMY. *Educ:* Vanderbilt Univ, BA, 61; Univ Tenn, Memphis, MD, 67, PhD(anat), 70. *Prof Exp:* Intern med, City of Memphis Hosps, 69-70; instr anat, Med Units, Univ Tenn, Memphis, 70-74, instr med, 73-76, asst prof dermat, 75-77; PROF MED & CHIEF DERMAT, VANDERBILT UNIV, 77- *Concurrent Pos:* NIH fel anat, Med Units, Univ Tenn, Memphis, 68-69; Vet Admin trainee dermat, Vet Admin Hosp, Memphis, 72-74, Vet Admin res & educ associateship, 75-; NIH spec fel, St Jude Children's Res Hosp, 74-76; resident internal med, City of Memphis Hosps, 70-71, resident dermat, 71-74; clin investr, Vet Admin Hosp, Nashville, Tenn, 77-80. *Mem:* AAAS; Am Acad Dermat; fel Am Col Physicians; Am Fedn Clin Res; AMA. *Res:* Skin diseases; cell membranes; growth factors; spider venom/bites. *Mailing Add:* Dermat Sect 3900 Vanderbilt Clin Nashville TN 37232-5227

KING, LOWELL ALVIN, b Spencer, Iowa, June 16, 32; m 54; c 3. PHYSICAL CHEMISTRY. *Educ:* Iowa State Univ, BS, 53, PhD(inorg chem), 63; Wash Univ, AM, 55. *Prof Exp:* Group leader, Mat Lab, USAF, 56-59, instr chem, USAF Acad, 59-61, res assoc, Ames Lab, Iowa State Univ, 61-63, asst prof chem, USAF Acad, 63, assoc prof, 63-64, res assoc, F J Seiler Lab, Off Aerospace Res, 64-65, assoc prof chem, 65-71, prof chem, USAF Acad, 71-80, consult, F J Seiler Lab, Air Force Systs Command, USAF, 81-83; PROF COMPUTER SCI, COLO TECH COL, 83- *Concurrent Pos:* Dir chem div, F J Seiler Lab, Off Aerospace Res, 66-68, dir chem div, F J Seiler Lab, Air Force Systs Command, 74-75, sr scientist, 75-80; prof phys sci, Europ Christian Col, Vienna, Austria, 81- *Honors & Awards:* Air Force Res & Develop Award, 70. *Mem:* Electrochem Soc; Am Chem Soc. *Res:* Electrochemistry; molten salt chemistry; measurement of physical and electrochemical properties of molten salt mixtures containing complex halo-anions. *Mailing Add:* 3945 Hill Circle Colorado Springs CO 80904

KING, LOWELL RESTELL, b Salem, Ohio, Feb 28, 32; m 60; c 2. UROLOGY. *Educ:* Johns Hopkins Univ, BA, 52, MD, 56. *Prof Exp:* Intern, Johns Hopkins Hosp, 56-57, resident, 57-61; asst prof urol, Med Sch, Johns Hopkins Univ, 63, from asst prof to assoc prof, Northwestern Univ, 63-68; prof, Univ Ill, 68-70; chmn, Div Urol, 70-77, SURGEON-IN-CHIEF, CHILDREN'S MEM HOSP, 74-; prof urol, Northwestern Univ, 70-81; PROF UROL, DUKE UNIV, 81-, HEAD, SECT PEDIAT UROL, 81- *Concurrent Pos:* Am Cancer Soc fel, 58-59; chmn dept urol, Presby-St Luke's Hosp, 68-70; mem, Am Bd Urol, 75-81. *Mem:* Am Urol Asn; fel Am Acad Pediat; fel Am Col Surgeons; Soc Pediat Urol; Am Urol Asn. *Res:* Pediatric urology. *Mailing Add:* Surgery/Box 3831 Med Ctr Duke Univ Durham NC 27710

KING, LUCY JANE, b Vandalia, Ill, Dec, 23, 32. PSYCHIATRY. *Educ:* Wash Univ, AB, 54, MD, 58; Am Bd Psychiat & Neurol, dipl psychiat, 66. *Prof Exp:* Intern, Butterworth Hosp, Grand Rapids, Mich, 58-59; asst resident psychiat, Renard Hosp, St Louis, Mo, 59-62, chief resident, 62-63; from instr to assoc prof, Sch Med, Wash Univ, 70-74; PROF PSYCHIAT & PHARMACOL, MED COL VA, VA COMMONWEALTH UNIV, 74- *Concurrent Pos:* NIMH res career develop awards neuropharmacol, 63-73. *Mem:* AAAS; fel Am Psychiat Asn; Psychiat Res Soc; Sigma Xi. *Res:* Clinical and social psychiatry; neuropharmacology. *Mailing Add:* 6012 Wheeler Lane Broad Run VA 22014

KING, LUNSFORD RICHARDSON, b Greensboro, NC, July 22, 37; m 61; c 2. MATHEMATICS. *Educ:* Davidson Col, BS, 59; Duke Univ, PhD(math), 63. *Prof Exp:* Res instr math, Univ Va, 63-64; from asst prof to assoc prof, 64-80, PROF MATH, DAVIDSON COL, 80- *Concurrent Pos:* Vis scholar, Dartmouth Col, 74-75. *Mem:* Math Asn Am; Am Math Soc. *Mailing Add:* Dept Math Davidson Col PO Box 1719 Davidson NC 28036

KING, MARVIN, b New York, NY, Apr 23, 40; m 61; c 1. OPTICAL SYSTEMS, ELECTRONIC SYSTEMS. *Educ:* City Col NY, BEE, 61; Polytech Inst Brooklyn, MSEE, 63; Columbia Univ, EngScD, 66. *Prof Exp:* Jr elec engr, ITT Fed Labs, 62-63, electrophysicist, 63-64; res asst, Electronics Res Labs, Columbia, 64-66, res engr, 66-67; sr res engr, Riverside Res Inst, 67-68, asst head electrooptics lab, 68-69, mgr optics lab, 69-76, res dir, 77-81, exec dir prog develop, 81-82, vpres res, 82-88, exec vpres, 88-90, PRES, RIVERSIDE RES INST, 90- *Concurrent Pos:* Consult, US Army Electronics Command, Ft Monmouth, NJ, 67-68, Adv Optics Ctr, Radiation, Inc, Mich, 69 & spec study sect on optics, NIH, 70; lectr, Columbia Univ Sci Honors Prog, 66-73; adj assoc prof, City Col NY 72-74. *Mem:* Inst Elec & Electronics Engrs; Optical Soc Am; Sigma Xi; Am Inst Physics; Quantum Electronics Soc; Am Inst Aeronaut & Astronaut. *Res:* Signal processing for radar and communications systems utilizing holography, photographic film, physical optics and ultrasonics; atmospheric optics; optical oceanography; infrared holography; laser radar; laser signatures; infrared acquisition and tracking systems; image processing. *Mailing Add:* Riverside Res Inst 330 W 42nd St New York NY 10036

KING, MARY MARGARET, b Oklahoma City, Okla, May 26, 46. DRUG METABOLISM, NUTRITION. *Educ:* Cent State Univ, BS, 69; Univ Okla, PhD(med physiol & biophys), 75; Okla City Univ, MBA, 84. *Prof Exp:* Instr physiol, Cent State Univ, 69; instr, Okla City Pub Schs, 69-71; fel, Okla Med Res Found, 75-76, res assoc biomembrane res, 76-77, staff scientist, 77-80; res asst prof biochem & molecular biol, Univ Okla Health Sci Ctr, 81-88, dir, Lab Animal Resources Ctr, 82-90, Sci Support Serv, 82-90, ADJ ASSOC PROF BIOCHEM & MOLECULAR BIOL, UNIV OKLA HEALTH SCI CTR, 88-, SPEC ASST TO PRES, 90- *Concurrent Pos:* Young Investr Award, NIH/Nat Inst Environ Health Sci, 78-81; consult, Nat Cancer Inst, NIH, 83-, Grad Fac, Univ Okla, Health Sci Ctr, 83-; assoc mem, Okla Med Res Found, 86- *Honors & Awards:* Young Investr Award, NIH. *Mem:* NY Acad Sci; Am Asn Cancer Res; Am Inst Nutrit; Sigma Xi; Soc Exp Biol Med. *Res:* Chemical carcinogens and dietary fat-antioxidant interactions as relates to mammary cancer; carcinogenesis; chemical carcinogens; dietary parameters; hormonal interactions in mammary gland, especially relating to the drug metabolizing system. *Mailing Add:* Okla Med Res Found President's Off 825 NE 13th St Oklahoma City OK 73104-5046

KING, MARY-CLAIRE, b Evanston, Ill, Feb 27, 46; m 73; c 1. EPIDEMIOLOGY, HUMAN GENETICS. *Educ:* Carleton Col, BA, 66; Univ Calif, Berkeley, PhD(genetics), 73. *Prof Exp:* Asst prof, 76-80, ASSOC PROF EPIDEMIOL, UNIV CALIF, BERKELEY, 80- *Concurrent Pos:* Vis prof, Univ Chile, Santiago, 73; asst prof genetics, Univ Calif, San Francisco, 74-80, assoc prof epidemiol, 80-; prin investr, genetic epidemiol of breast cancer in families, NIH grant, 79- *Mem:* Am Soc Human Genetics; Soc Epidemiol Res; Am Epidemiol Soc. *Res:* Genetics and epidemiology of breast cancer and other common chronic diseases, pedigree analysis, human and primate molecular evolution. *Mailing Add:* Behs Sch Pub Health Univ Calif Berkeley CA 94720

KING, MERRILL KENNETH, b Claymont, Del, Nov 15, 38; m 64; c 2. COMBUSTION SCIENCE. *Educ:* Carnegie Inst Technol, BS, 60, MS, 62, PhD(chem eng), 65. *Prof Exp:* Sr res engr, 64-67, head, Thermodyn & Combustion Sect, Atlantic Res Corp Div, Susquehanna Corp, 67-72, staff scientist, 72-73, head, Energy & Pollution Technol Sect, 73-74, chief kinetics & combustion group, 74-76, chief scientist res & technol, 76-84, MGR, COMBUSTION RES DEPT, ATLANTIC RES CORP, 84- *Mem:* Am Inst Aeronaut & Astronaut; Combustion Inst. *Res:* Solid propellant combustion; metals combustion; graphite oxidation and erosion; air-breathing propulsion; chemical thermodynamics; chemical kinetics; vortex flow; combustion instability. *Mailing Add:* 4634 Tara Dr Fairfax VA 22032

KING, MICHAEL DUMONT, b Kansas City, Mo, Oct 20, 49; m 72; c 2. ATMOSPHERIC RADIATION, REMOTE SENSING. *Educ:* Colo Col, BA, 71; Univ Ariz, MS, 73, PhD(atmospheric sci), 77. *Prof Exp:* Res asst, Univ Ariz, 71-77; ATMOSPHERIC SCIENTIST, GODDARD SPACE FLIGHT CTR, NASA, 78- *Concurrent Pos:* Prin investr, Earth Radiation Budget Exp Sci Team, 80-, proj scientist, 83-; chmn, Comt Atmospheric Radiation, Am Meteorol Soc, 86-88; vis prof, Dept Atmospheric Sci, Univ Wash, 86-87; assoc ed, J Atmospheric Sci; prin investr, Moderate Resolution Imaging Spectrometer Sci Team, 89-; mem, Clouds & Earth's Radiant Energy Syst Sci Team, 89- *Honors & Awards:* NASA Except Serv Medal. *Mem:* Fel Am Meteorol Soc; Am Geophys Union; Sigma Xi; AAAS. *Res:* Multiple light scattering and radiative transfer in cloud free and cloudy atmospheres; application of inversion methods to determination of aerosol size distributions; effect of clouds and aerosols on earth's radiation budget; determination of the optical thickness, particle radius and single scattering albedo of clouds from airborne measurements of scattered radiation. *Mailing Add:* 14613 Peach Orchard Rd Silver Spring MD 20905-4437

KING, MICHAEL M, b Chicago, Ill, May 10, 44; m 70; c 1. ORGANIC CHEMISTRY. *Educ:* Ill Inst Technol, BS, 66; Harvard Univ, AM, 67, PhD(org chem), 70. *Prof Exp:* Asst prof chem, NY Univ, 70-73; from asst prof to assoc prof, 73-84, PROF CHEM, GEORGE WASHINGTON UNIV, 84- *Mem:* Am Chem Soc. *Res:* Pyrroles and porphyrins; heterocycles, organofluorides, enzyme model systems. *Mailing Add:* Dept Chem George Washington Univ Washington DC 20052

KING, MICHAEL STUART, b Brackley, Eng, June 2, 31; m 62; c 3. GEOLOGICAL ENGINEERING, APPLIED GEOPHYSICS. *Educ:* Univ Glasgow, BSc, 53; Univ Calif, Berkeley, MS, 61, PhD(eng sci), 64. *Prof Exp:* Design engr, John Brown & Co, Scotland, 52-54; field engr, Iraq Petrol Co, Kirkuk, 54-59; teaching asst mineral technol, Univ Calif, Berkeley, 59-64, lectr, 64-65; sr sci officer, Geotech Div, Ministry of Technol, Eng, 65-66; prof geol sci, Univ Sask, 66-81; staff scientist, Dept Mech Eng, Univ Calif, Berkeley, 81-; MEM STAFF, DEPT MECH TECHNOL, CABOT INST ARTS & TECHNOL. *Concurrent Pos:* Mem Can Nat Comt Rock Mech. *Mem:* Assoc mem Am Geophys Union; Soc Explor Geophys; Am Inst Mining, Metall & Petrol Engrs; Brit Inst Mech Engrs; Can Inst Mining & Metall. *Res:* Geological sciences; application of ultrasonics in geology; rock mechanics; fluid flow in porous media. *Mailing Add:* Dept Mech Technol Cabot Inst Arts & Technol PO Box 1693 St Johns NF A1C 5P7 Can

KING, MORRIS KENTON, b Oklahoma City, Okla, Nov 13, 24; m 53; c 5. INTERNAL MEDICINE. *Educ:* Univ Okla, BA, 47; Vanderbilt Univ, MD, 51. *Prof Exp:* Intern med, Barnes Hosp, St Louis, Mo, 51-52, from asst resident to resident, 52-55; asst resident, Vanderbilt Hosp, Nashville, Tenn, 53-54; instr microbiol, Sch Med, Johns Hopkins Univ, 56-57; from instr to assoc prof, 57-67, assoc dean, 62-65, DEAN, SCH MED, WASH UNIV, 65-, PROF PREV MED, 67- *Concurrent Pos:* Fel microbiol, Sch Hyg & Pub Health, Johns Hopkins Univ, 55-56. *Mem:* Infectious Dis Soc Am; Cent Soc Clin Res. *Res:* Pathogenesis of fever; infectious diseases. *Mailing Add:* Dean Med Wash Univ Sch Med 660 S Euclid Ave St Louis MO 63110

KING, NICHOLAS S P, b Lafayette, Ind, Dec 3, 40; m 66; c 2. NUCLEAR STRUCTURE. *Educ:* Dartmouth Col, BA, 62; Univ NMex, MS, 64; Univ Colo, PhD(physics), 70. *Prof Exp:* Res physicist, Univ Calif, Davis, 71-77; staff physicist, 78-80, assoc group leader, 81-84, GROUP LEADER, LOS ALAMOS NAT LAB, 84- *Mem:* Am Phys Soc; Inst Elec & Electronics Engrs; Soc Photo-Optical Instrumentation Engrs. *Res:* Investigation of the reaction mechanisms of nuclear particles with nuclei and the resultant excitation modes of the nucleus; determination of plasma properties via measurements of emitted radiation. *Mailing Add:* Los Alamos Nat Lab D 406 P-15 Los Alamos NM 87545

KING, NORVAL WILLIAM, JR, b Salisbury, Md, Apr 29, 38; m 64; c 2. VETERINARY PATHOLOGY, COMPARATIVE PATHOLOGY. *Educ:* Univ Ga, DVM, 62. *Prof Exp:* Res fel, 65-67, res assoc, 68-72, prin assoc, 72-76, ASSOC PROF COMP PATH, HARVARD MED SCH, 76-, ASSOC DIR, NEW ENG REGIONAL PRIMATE RES CTR, 80- *Concurrent Pos:* Lectr, Sch Med & Vet Med, Tufts Univ, 79-; consult, Pathobiol Inc, 74-; dir, NIH Training grant, vet & comparative path, NIH, 75-; Res Award, Am Asn Lab Animal Sci, 69. *Mem:* Int Acad Path; Am Asn Pathologists; Am Col Vet Pathologists; Am Vet Med Asn; New Eng Soc Pathologists. *Res:* Ultrastructure of viruses and viral iduced lesions; pathology of the reproductive tract; animal models for human diseases. *Mailing Add:* New Eng Regional Primate Res Ctr Harvard Med Sch One Pine Hill Dr Southborough MA 01772

KING, PATRICIA A, MEMBRANE TRANSPORT, COMPARATIVE PHYSIOLOGY. *Educ:* Brown Univ, PhD(physiol), 82. *Prof Exp:* Res assoc physiol, Sch Med, Emory Univ, 84-87; AT COL MED, UNIV VT, 87- *Mailing Add:* Dept Med Given Bldg Univ Vt Col Med Burlington VT 05405

KING, PAUL HARVEY, b Ft Wayne, Ind, Sept 4, 41; m 67, 78, 82; c 4. BIOMEDICAL ENGINEERING, MECHANICAL ENGINEERING. *Educ:* Case Inst Technol, BS, 63, MS, 65; Vanderbilt Univ, PhD(mech eng), 68. *Prof Exp:* Res asst eng, Case Inst Technol, 63-65; asst prof eng, 68-72, actg chmn dept biomed eng, 71-72, prog dir biomed eng, 72-75, asst prof, Ortho & Rehab, 73-81, chmn biomed eng, 75-77, ASSOC PROF BIOMED & MECH ENG, VANDERBILT UNIV, 72-, ASST PROF ANETHESIOL DEPT, 87- *Concurrent Pos:* Sr teaching fel mech eng, Vanderbilt Univ, 65-68; researcher, Oak Ridge Assoc Univ, Med & Health Sci Div, Radiopharmaceut Develop Group, 78-79. *Honors & Awards:* Skylab Awards. *Mem:* Sigma Xi; Asn Advan Med Instrumentation; Inst Elec & Electronics Engrs. *Res:* Orthopedics research; computer analysis of electro-cardiograms; radioisotope scanning systems; biomedical data analysis; modelling and research; positron emission tomography; computer assisted monitoring in anesthesiology. *Mailing Add:* Box 1631 Sta B Nashville TN 37235-1631

KING, PERRY, JR, b West Frankfort, Ill, Sept 17, 28; m 54; c 3. RADIOCHEMISTRY. *Educ:* Univ Ill, BS, 50; Washington Univ, PhD(chem), 60. *Prof Exp:* Chemist, Mallinckrodt Chem Works, 50-53 & 60-64, sect leader, 64-71, mgr, Process Develop Radiopharmaceut, 71-72, asst dir radiopharmaceut res & develop, 73-76, DIR QUAL ASSURANCE CHEM, MALLINCKRODT, INC, 76- *Mem:* Am Chem Soc; Soc Nuclear Med; Sigma Xi. *Res:* Radiopharmaceuticals; applications of raiochemical techniques to poblems in inorganic chemistry; analytical chemistry; in vivo imaging agents. *Mailing Add:* 1722 Connemara Ballwin MO 63021

KING, PETER FOSTER, b New York, NY, Oct 7, 29; m 54; c 3. PHYSICAL CHEMSTRY. *Educ:* Ind Univ, BS, 51, MA, 52; Mass Inst Technol, DSc, 57. *Prof Exp:* Res & develop engr, Dow Chem Co, 57-65; chemist, 65-67, group leader, 67-80, sr res chemist, Res & Develop, Parker Chem Co, 80-87; RES SCIENTIST, PARKER & AMCHEM, 88- *Mem:* Electrochem Soc; Am Chem Soc; Nat Asn Corrosion Eng. *Res:* Electrochemistry; chemistry and electrochemistry of surfaces; corrosion, metal cleaning and conversion coating. *Mailing Add:* 26500 Orchard Lake Rd Farmington Hills MI 48018

KING, PETER RAMSAY, b Blackpool, UK, Nov 22, 43; m 70; c 1. COMPUTER SCIENCE. *Educ:* Univ Nottingham, BSc, 65, PhD(comput sci), 69. *Prof Exp:* Lectr math & comput sci, Univ Nottingham, 67-69; asst prof, 69-77, assoc prof, 77-80, PROF COMPUT SCI, UNIV MAN, 80- *Concurrent Pos:* Nat Res Coun Can grants, 69-72. *Mem:* Asn Comput Mach; Brit Comput Soc; Brit Inst Math & Appln. *Res:* Compiler construction for high level languages, especially Algol 68; spline interpolation; interactive problem solving. *Mailing Add:* 245 Wellington Crescent Winnipeg MB R3M 0A1 Can

KING, R MAURICE, JR, mathematics, statistics, for more information see previous edition

KING, RAY J(OHN), b Montrose, Colo, Jan 1, 33; m 64; c 2. ELECTRICAL ENGINEERING, MATERIALS EVALUATION. *Educ:* Ind Inst Technol, BS, 56 & 57; Univ Colo, Boulder, MS, 60, PhD(elec eng), 65. *Prof Exp:* Asst prof electronic eng, Ind Inst Technol, 60-61, asst prof elec eng & assoc chmn dept, 61-62; res assoc, Univ Colo, Boulder, 62-65; from assoc prof to prof elec eng, Univ Wis-Madison, 65-82; staff res engr, Lawrence Livermore Labs, 82-89; VPRES, KDC TECHNOL CORP, 83- *Concurrent Pos:* Res assoc, Univ Colo, Boulder, 60, vis lectr, 64-65; res assoc, Univ Ill, Urbana, 65; mem US nat comt, Int Sci Radio Union, Comns A, B & F, 67-; Fulbright guest prof, Tech Univ Denmark, 73-74; Erskine fel, Univ Canterbury, Christchurch, NZ, 77; prin investr, NSF, Air Force Off Sci Res grants; contracts, Naval Surface Warfare Ctr, Dept Com, Dept Agr, Naval Air Eng Ctr; admin comt mem, Antennas & Propagation Soc, 89- *Mem:* Fel Inst Elec & Electronics Engrs. *Res:* Electromagnetic wave propagation over nonuniform surfaces; microwave surface and leaky wave antennas; microwave instrumentation and measurement systems; microwave nondestructive testing; environmental effects on antenna performance; high power microwave effects; microwave evaluation of materials; electromagnetism. *Mailing Add:* KDC Technol Corp 2011 Research Dr Livermore CA 94550

KING, RAYMOND LEROY, b Burbank, Calif, Oct 4, 22; m 47; c 6. FOOD SCIENCE. *Educ:* Univ Calif, AB, 55, PhD(agr chem), 58. *Prof Exp:* Asst dairy prod, Univ Calif, 55-57, jr specialist, 57-58; asst prof dairy technol, 58-62, assoc prof food sci, 62-66, PROF FOOD SCI, UNIV MD, COLLEGE PARK, 66-, COORDR & CHMN FOOD SCI PROG, 73- *Concurrent Pos:* Consult ice cream prod, Chile, 62; consult pesticide residues in milk, 60-61; consult milk packaging, 62. *Mem:* AAAS; Am Chem Soc; Am Dairy Sci Asn; Inst Food Technol. *Res:* Mechanism of lipid oxidation in milk; characterization of milk fat globule membrane; distribution and movement of pesticides in dairy cows; food chemistry. *Mailing Add:* Dept Dairy Univ Md Col Agr College Park MD 20742

KING, REATHA CLARK, b Pavo, Ga, Apr 11, 38; m 61; c 2. INORGANIC CHEMISTRY, PHYSICAL CHEMISTRY. *Educ:* Clark Col, BS, 58; Univ Chicago, MS, 60, PhD(chem), 63; Columbia Univ, MBA, 77. *Prof Exp:* Chemist, Nat Bur Standards, 63-68; asst prof chem, 68-70, assoc prof chem & assoc dean div natural sci & math, 70-74, prof chem & assoc dean acad affairs, York Col, NY, 74-77; PRES, METROP STATE UNIV, ST PAUL, 77- *Mem:* AAAS; Am Chem Soc; Nat Orgn Prof Advan Black Chemists & Black Engrs; Sigma Xi. *Res:* Experimental study on thermochemical properties of alloys using tin solution calorimetry; heats of formation of refractory compounds; fluorine flame calorimetry at room temperature. *Mailing Add:* Gen Mills Found PO Box 1113-1E Minneapolis MN 55440

KING, RICHARD ALLEN, b Fresno, Calif, Mar 20, 39; m 63; c 2. MEDICINE, GENETICS. *Educ:* Pa State Univ, AB, 61; Jefferson Med Col, MD, 65; Univ Minn, PhD(genetics), 75. *Prof Exp:* Intern med, Univ Minn, 65-66, resident, 66-69; USPHS surgeon, Atomic Bomb Casualty Comn, Hiroshima, Japan, 69-71; from instr to assoc prof, 71-84, dir Genetics Div, 85-90, PROF MED & PEDIAT, UNIV MINN, MINNEAPOLIS, 84-, DIR, DIV GENETICS & METAB, DEPT MED & PEDIAT & INST HUMAN GENETICS, 90- *Mem:* Am Soc Human Genetics; Cent Soc Clin Res; AAAS; Am Fedn Clin Res; fel Am Col Physicians; Pan Am Soc Pigment Cell (secy, treas). *Res:* Melanin pigment metabolism and human pigment defects; genetics of common adult diseases including diabetes mellitus, arthritis, SLE and cancer. *Mailing Add:* Box 485 UMHC Univ Hosp 420 Delaware St SE Minneapolis MN 55455

KING, RICHARD AUSTIN, b Philadelphia, Pa, Mar 26, 29; m 56; c 3. PSYCHOPHYSIOLOGY, NEUROBIOLOGY. *Educ:* Univ Cincinnati, AB, 54, MA, 55; Duke Univ, PhD(psychol), 59. *Prof Exp:* From instr to asst prof psychol, Univ NC, Chapel Hill, 58-63; fel physiol, Univ Wash, 63-65; assoc prof, 65-71, PROF PSYCHOL, UNIV NC, CHAPEL HILL, 71-, ASSOC DIR NEUROBIOL, 73- *Concurrent Pos:* Vis prof psychol, Brown Univ, 71. *Mem:* Soc Neurosci; Psychonomic Soc. *Res:* Biology of memory. *Mailing Add:* Dept Psychol Univ NC Chapel Hill NC 27515

KING, RICHARD JOE, b Kansas City, Mo, Aug 30, 37; m 63; c 2. ANIMAL PHYSIOLOGY. *Educ:* Univ Mo, Columbia, BA, 59; Univ Calif, Berkeley, PhD(biophys), 70. *Prof Exp:* Asst res biophysicist, Univ Calif, San Francisco, 71-74; from asst prof to assoc prof, 74-80, PROF PHYSIOL, UNIV TEX HEALTH SCI CTR, SAN ANTONIO, 80- *Concurrent Pos:* Consult, Review Comt, NIH, 75-76, 78-79 & 80-81. *Mem:* Am Physiol Soc. *Res:* Composition and properties of pulmonary surfactant; metabolism and isolation of its associated apoproteins; interaction of lipids and proteins in pulmonary surfactant; correlation of structure function relationships; effects of nitrogens on lung cells; alterations in surfactant in chronic lung injury. *Mailing Add:* Dept Physiol Univ Tex Health Sci Ctr 7703 Floyd Curl Dr San Antonio TX 78284

KING, RICHARD WARREN, b Philadelphia, Pa, Mar 25, 25; m 47; c 4. ANALYTICAL CHEMISTRY, PHYSICAL CHEMISTRY. *Educ:* Kenyon Col, AB, 47; Univ Del, MS, 59. *Prof Exp:* Res chemist, Res & Develop Dept, 47-59, sect chief, Anal Sect, 60-69, MGR RES SERV, RES & DEVELOP DEPT, SUN OIL CO, 69- *Mem:* Am Chem Soc; fel Am Soc Testing & Mat. *Res:* Catalytic reactions of hydrocarbons; application of physical separation techniques to the study of high-boiling fractions from petroleum; gas chromatography of petroleum fractions. *Mailing Add:* 3031 Hermosa Lane Havertown PA 19083

KING, ROBBINS SYDNEY, b San Diego, Calif, Apr 23, 22; m 47; c 4. EMBRYOLOGY. *Educ:* Stanford Univ, AB, 47, PhD(biol), 54. *Prof Exp:* Res assoc, Hopkins Marine Sta, Stanford Univ, 52-54, asst prof biol, Univ, 55; instr, Menlo Col, 54-55; asst prof, Wabash Col, 55-56; asst prof biol, Calif State Univ, Chico, 56-65, chmn dept, 65-69, prof biol sci, 65-89. *Mem:* AAAS; Am Soc Mammal; Nat Sci Teachers Asn. *Res:* Experimental embryology; regeneration. *Mailing Add:* 45 Quadra Ct Chico CA 95928

KING, ROBERT (BAINTON), b Pittsburgh, Pa, Aug 26, 22; m 51; c 3. SURGERY. *Educ:* Rochester Univ, MD, 46; Am Bd Neurol Surg, dipl, 54. *Prof Exp:* Asst neuroanat, Med Sch, Wash Univ, 48; asst chief, Walter Reed Army Hosp, 49-51; from instr to asst prof neurosurg, Med Sch, Wash Univ, 51-57; chmn dept, 66-88, PROF NEUROSURG, COL MED, STATE UNIV NY HEALTH SCI CTR, 57-; MED DIR, STATE UNIV HOSP. *Concurrent Pos:* Markle scholar, 51-56; attend surg, Crouse-Irving Mem Hosp, 57 & State Univ Hosp, Syracuse, 65-; Consult, Vet Admin Hosp, 57; distinguished serv prof neurosurg, State Univ NY. *Honors & Awards:* Cushing Medal, Am Asn Neurol Surgeons. *Mem:* Neurosurg Soc Am; Am Asn Neurol Surg; Am Col Surgeons; Am Acad Neurol Surg; Soc Neurol Surg; Sigma Xi. *Res:* Neurosurgery; neurophysiology. *Mailing Add:* Col Med 766 Irving Ave Syracuse NY 13210

KING, ROBERT BRUCE, b Rochester, NH, Feb 27, 38; m 60; c 2. INORGANIC CHEMISTRY. *Educ:* Oberlin Col, BA, 57; Harvard Univ, PhD(inorg chem), 61. *Prof Exp:* Res chemist, Explosives Dept, E I du Pont de Nemours, Del, 61-62; res fel, Mellon Inst, 62-64, sr res fel, 64-66; res assoc prof chem, Univ Ga, 66-68, res prof, 68-73, actg head chem, 80-82, REGENTS' PROF CHEM, UNIV GA, 73- *Concurrent Pos:* Ed, Organometallic Syntheses, 63-; tech adv, Pressure Chem Co, Pa, 64-75; ed, J Organometallic Chem, 64-; Sloan Found fel, 67-69; fel, Japan Soc Promotion Sci, 81; consult, Los Alamos Nat Lab, 79-83. *Honors & Awards:* Am Chem Soc Awards, 71, 91. *Mem:* Am Chem Soc; Chem Soc London; Mat Res Soc. *Res:* Synthetic and spectroscopic studies on organo-metallic compounds of transition metals; molecular catalysis; organophosphous chemistry; isocyanide polymerization; chemical applications of group theory and group theory; applications of inorganic chemistry in nuclear technology. *Mailing Add:* Dept Chem Univ Ga Athens GA 30602

KING, ROBERT CHARLES, b New York, NY, June 3, 28; m 79; c 2. GENETICS. *Educ:* Yale Univ, BS, 48, PhD(zool), 52. *Prof Exp:* Scientist, Brookhaven Nat Lab, 51-56; from asst prof to assoc prof, 56-63, PROF BIOL SCI, NORTHWESTERN UNIV, 64- *Concurrent Pos:* NSF sr fels, Univ Edinburgh, 58, div entom, Commonwealth Sci & Indust Res Orgn, Canberra, Australia, 63 & sericulture exp sta, Tokyo, 70; Seoul Nat Univ, Seoul, Korea, 78 & Han Yang Univ, 79; vis investr & fel, Rockefeller Inst, 59. *Mem:* Fel AAAS; Genetics Soc Am; Am Soc Zool; Entom Soc Am; Am Soc Cell Biol (treas, 72, 73 & 74). *Res:* Developmental genetics; genetic control of oogenesis in Drosophila. *Mailing Add:* Biochem Molecular Biol & Cell Biol Dept Hogan Hall 5-130 Northwestern Univ Evanston IL 60208

KING, ROBERT EDWARD, b Zanesville, Ohio, Dec 27, 23; m 50; c 5. BIOLOGICAL CHEMISTRY, PHARMACEUTICAL CHEMISTRY. *Educ:* Ohio State Univ, BSc, 44; Univ Minn, PhD(pharmaceut chem), 48. *Prof Exp:* Res assoc, Merck Sharp & Dohme Res Labs, 48-61; emer prof indust pharm, Philadelphia Col pharm, 61-86; RETIRED. *Concurrent Pos:* Ed, J Parenteral Drug Asn, 64-78. *Mem:* Am Pharmaceut Asn; Parenteral Drug Asn. *Res:* Pharmaceutical dosage forms. *Mailing Add:* 3475 Aquetong Rd Doylestown PA 18901

KING, ROBERT WILLIAM, b Wesson, Miss, Nov 18, 29; m 59; c 2. MATHEMATICS. *Educ:* Univ Southern Miss, BS, 51, MA, 56; Fla State Univ, MS, 65, PhD(math, educ), 67. *Prof Exp:* Instr math, Copiah-Lincoln Jr Col, 56-57; asst prof, Miss Col, 59-64; asst prof, 67-69, ASSOC PROF MATH, UNIV SOUTHERN MISS, 69- *Mem:* Math Asn Am; Nat Coun Teachers Math. *Res:* Use of various modifications of programmed instruction to investigate the effects of selected social and psychological factors in the teaching and learning of mathematics. *Mailing Add:* Dept Math Univ Southern Miss Box 5045 Southern Sta Hattiesburg MS 39406

KING, ROBERT WILLIS, b Grand Haven, Mich, June 9, 61. MOLECULAR BIOLOGY, MICROBIOLOGY. *Educ:* Mich State Univ, BS, 83; Purdue Univ, PhD(microbiol), 89. *Prof Exp:* Instr microbiol, Purdue Univ, 87-88, instr med microbiol, Ind Univ Sch Med, 89; RES ASSOC, UNIV CHICAGO, 90- *Mem:* Am Soc Microbiol; AAAS. *Mailing Add:* 1617 E 50th Pl Apt 14C Chicago IL 60615

KING, ROBERT WILSON, JR, b Fayetteville, NC, Feb 8, 47; m 70; c 2. GEODESY. *Educ:* Davidson Col, BS, 70; NC State Univ, BS, 70; Mass Inst Technol, PhD(instrumentation), 75. *Prof Exp:* Res geodesist, Terrestrial Sci Div, Air Force Geophys Lab, 74-77; res assoc, 77-85, PRIN RES SCIENTIST, DEPT EARTH, ATMOS & PLANETARY SCI, MASS INST TECHNOL, 85- *Mem:* Am Geophys Union. *Res:* Use of precise extraterrestrial measurement techniques to monitor earth rotation, polar motion and crustal deformation. *Mailing Add:* Two Pinewood St Lexington MA 02173

KING, ROGER HATTON, b Barry, Wales, Dec 16, 41; m 65; c 2. PEDOLOGY, PHYSICAL GEOGRAPHY. *Educ:* Univ Wales, BSc, 63; Univ Aberdeen, MSc, 65; Univ Sask, PhD(geog), 69. *Prof Exp:* Tutor geog, Univ Wales, 69-70; from asst prof to assoc prof, 70-88, PROF GEOG, UNIV WESTERN ONT, 88- *Concurrent Pos:* Vis prof, Inst Archaeol, Oxford Univ, Eng, 77-78; consult, Parks Can, 82-; vis prof, Dept Geog, Univ Canterbury, NZ, 85. *Mem:* Can Geog Soc; Int Quaternary Asn; Geol Asn Can; Can Quaternary Asn. *Res:* Persistence of chemical residues in archaeological soils and sediments; impact of environmental stress on soil development in arctic and alpine areas; palaeoenvironmental reconstruction in the Canadian Cordillera; archaeological pottery provenance; tephras; clay mineralogy and composition. *Mailing Add:* Dept Geog Univ Western Ont London ON N6A 5B9 Can

KING, RONOLD (WYETH PERCIVAL), b Williamstown, Mass, Sept 19, 05; m 37; c 1. PHYSICS, ELECTRICAL ENGINEERING. *Educ:* Univ Rochester, AB, 27, SM, 29; Univ Wis, PhD(electrodyn), 32. *Hon Degrees:* AM, Harvard Univ, 42. *Prof Exp:* Asst physics, Univ Rochester, 28-29; asst physics, Univ Wis, 32-33, asst elec eng, 33-34, Alumni Res Found fel, 34; instr physics, Lafayette Col, 34-35, asst prof, 35-37; Guggenheim Mem Found fel, Berlin & Munich, 37-38; fac instr physics & commun eng, 38-39, from asst prof to prof, 39-72, EMER PROF APPL PHYSICS, HARVARD UNIV, 72- *Concurrent Pos:* Guggenheim Mem Found fel, 58; mem comn B, Int Sci Radio Union; consult, Raytheon Co, 74-75 & 87-91, Mitre Corp, 90; IBM distinguished scholar, Northeastern Univ, 85. *Honors & Awards:* Centennial Medal, Inst Elec & Electronics Engrs, 84. *Mem:* AAAS; fel Inst Elec & Electronics Eng; fel Am Phys Soc; fel Am Acad Arts & Sci; corresp mem Bavarian Acad Sci. *Res:* Electromagnetic theory, radiation and antennas; transmission-line theory; microwave circuits; insulated antennas, crossed antennas, antennas in dessipative media near surface; subsurface communication; electromagnetic pulses; electromagnetic surface waves; resonant antenna arrays. *Mailing Add:* 92 Hillcrest Pkwy Winchester MA 01890

KING, ROY WARBRICK, b Liverpool, Eng, July 4, 33; US citizen; m 70. MASS SPECTROMETRY, NUCLEAR MAGNETIC RESONANCE. *Educ:* Cambridge Univ, BA, 54, MA & PhD(org chem), 58. *Prof Exp:* Norman fel org chem, Hickrill Chem Res Found, NY, 58; univ fel, Iowa State Univ, 58-60, supvr instrument serv, 60-66, asst prof chem, 66-69; res assoc chem, 69-84, ASST RES SCIENTIST, UNIV FLA, 84- *Mem:* Am Chem Soc; Am Soc Mass Spectrom; Sigma Xi; Royal Soc Chem. *Res:* Organic structure determination; organic analysis; applications of physical methods to organic chemistry. *Mailing Add:* Dept Chem Univ Fla Gainesville FL 32611-2046

KING, S(ANFORD) MACCALLUM, b St Catharines, Ont, May 21, 26; US citizen; div; c 4. SOIL FERTILITY, CROP MANAGEMENT. *Educ:* Ont Agr Col, BSA, 48; Purdue Univ, MS, 50; Univ Wis, PhD(soil fertil), 56, Keller Grad Inst Mgt, MBA, 81. *Prof Exp:* Soil scientist, Stand Fruit & Steamship Co, 51-53; proj asst soil fertil, Univ Wis, 53-56; res asst, prof & sta supt, Mich State Univ, 56-61; mkt & tech serv specialist fertilizer, Int Minerals & Chem Corp, 62-70; dir agr, Develop & Resources Corp, 71-73; vpres, Taralan Corp, 73-87; vpres, Seed-Prep Corp & sr vpres, Competitive Edge Inc, 88-90; PRES, KWG INT, 90- *Concurrent Pos:* Agr proj develop & mkt tasks 18 countries, Potash Corp, Saskatchewan; Consult, 80-83; instr mkt, Northern Ill Univ, 90- *Mem:* Am Soc Agron; Soil Sci Soc Am. *Res:* Crop nutrition; crop management and production. *Mailing Add:* 36 Pine Ave Lake Zurich IL 60047-2326

KING, SHELDON SELIG, b New York, NY, Aug 28, 31. MEDICAL ADMINISTRATION. *Educ:* NY Univ, AB,52; Yale Univ, MS, 57. *Prof Exp:* Asst dir, Mt Sinai Hosp, 60-66, dir planning, 66-68; exec dir, Albert Einstein Col Med, Bronx Munic Hosp Ctr, 68-72; dir hosp & clin, Univ Hosp & assoc clin prof, Univ Calif, San Diego, 72-81; assoc vpres med affairs, exec vpres & dir, Stanford Univ Hosp, 81-85, pres, 86-89; PRES, CEDARS-SINAI MED CTR, LOS ANGELES, 89- *Concurrent Pos:* Pres, Hosp Coun San Diego & Imp County, 77; mem, Gov Coun, Sect Metrop Hosps, Am Hosp Asn, 84-89, Coun 2000 Comn, Am Podiatric Asn, 85-86, Accreditation Coun Grad Med Educ, 87-90, Bd Health Sci Policy, Inst Med & Health Leadership Coun, 89-, bd dirs, Nat Comt Qual Health Care & Vol Hosps Am, 90-; chmn bd adv, Am Bd Internal Med, 85-91. *Mem:* Inst Med-Nat Acad Sci; fel Am Col Health Care Execs; fel Am Pub Health Asn; fel Royal Soc Health; Am Hosp Asn. *Res:* Author of publications in various medical journals. *Mailing Add:* Cedars-Sinai Med Ctr 8700 Beverly Blvd PO Box 48750 Los Angeles CA 90048-0750

KING, STANLEY SHIH-TUNG, b Liau-Ning, China, Nov 12, 34; m 66; c 3. PHYSICAL CHEMISTRY, MOLECULAR SPECTROSCOPY. *Educ:* Taiwan Normal Univ, BS, 57; Drexel Univ, MS, 62; Univ Minn, Minneapolis, PhD(phys chem), 66. *Prof Exp:* RES ASSOC, DOW CHEM USA, 80- *Mem:* Am Chem Soc. *Res:* Vibrational spectroscopy; low temperature matrix isolation study; microsample analysis by vibrational spectroscopy; polymer analysis; chromatographic analysis; electromicroscopy; catalysis; analytical chemistry. *Mailing Add:* 1311 Kirkland Dr Midland MI 48640

KING, STEPHEN MURRAY, MONOCLONAL ANTIBODIES, PROTEIN SUBSTANCES. *Educ:* Univ London, Eng, PhD(cell biol), 82. *Prof Exp:* RES ASSOC, WORCESTER FOUND EXP BIOL, 83- *Mailing Add:* Cell Biol Group Worcester Found Exp Biol 222 Maple Ave Shrewsbury MA 01545

KING, TE PIAO, b Shanghai, China, Aug 21, 30; m; c 2. IMMUNOLOGY. *Educ:* Univ Calif, AB, 50, MS, 51; Univ Mich, PhD, 53. *Prof Exp:* Res assoc biochem, 53-57, asst prof, 57-63, ASSOC PROF BIOCHEM, ROCKEFELLER UNIV, 63- *Mem:* Am Soc Biol Chemists; Am Acad Allergy. *Res:* Peptides; proteins. *Mailing Add:* Dept Biochem Rockefeller Univ 1230 York Ave Box 69 New York NY 10021

KING, TERRY LEE, b Akron, Iowa, Feb 24, 45; m 71; c 3. ANALYSIS OF DATA, LINEAR MODELS. *Educ:* Westmar Col, BA, 67; Univ Iowa, MS, 69; Pa State Univ, PhD(statist), 80. *Prof Exp:* Instr math, Thiel Col, 69-71; statistician, Desmatics, Inc, 75-79; instr, Dept Math, Frostburg State Col, 79-81; assoc prof, 81-88, CHMN DEPT MATH & STATIST, NW MO STATE UNIV, 88-, PROF, 89- *Concurrent Pos:* Vchair, Mo Sect, Math Asn Am, 90-91, chair, 91- *Mem:* Am Statist Asn; Biometric Soc; Math Asn Am. *Res:* Sample size determination; linear models; data analysis. *Mailing Add:* Dept Math & Statist NW Mo State Univ Maryville MO 64468

KING, THEODORE MATTHEW, b Quincy, Ill, Feb 13, 31; m 54; c 2. OBSTETRICS & GYNECOLOGY, PHYSIOLOGY. *Educ:* Quincy Col, BS, 50; Univ Ill, Urbana, MS, 52, MD, 59; Mich State Univ, PhD(physiol), 59. *Prof Exp:* Lab asst physiol, Univ Ill, 55-59; intern surg, Presby Hosp, New York, 59-60; from resident to chief resident obstet & gynec, Sloane Hosp Women, 60-65; asst prof physiol, obstet & gynec, Sch Med, Univ Mo, 65-68, assoc prof, 68; prof & chmn dept, Albany Med Col, 68-71; PROF OBSTET & GYNEC & DIR DEPT, SCH MED, JOHNS HOPKINS UNIV, 71- *Concurrent Pos:* Macy fel, Sloane Hosp Women, 60-64, Macy fac fel obstet, 66-67; Nat Inst Child Health & Human Develop fel, 65-68. *Res:* Study of uterine contractile protein; influence of enzyme induction on animal reproduction. *Mailing Add:* Dept Obstet & Gynec Johns Hopkins Hosp Harvey 319 Baltimore MD 21205

KING, THEODORE OSCAR, b Portsmouth, Ohio, May 29, 22; m 52; c 2. TOXICOLOGY, PHARMACOLOGY. *Educ:* Univ Mich, BS, 43; Georgetown Univ, PhD(pharmacol), 49; Univ Wyo, JD, 60; Am Bd Toxicol, dipl, 81. *Prof Exp:* Pharmaceut control chemist, Wm R Warner & Co, NY, 43-45; anal chemist, Res Div, Colgate-Palmolive-Peet Co, NJ, 46; from assoc prof to prof pharmacol, Col Pharm, Univ Wyo, 49-58; dir div pharmacol, Ortho Res Found, NJ, 59-65; vpres & dir res, Bio/Dynamics, Inc, 65-71; dir, 71-76, SR DIR SAFETY EVAL, PFIZER, INC, 76- *Concurrent Pos:* WHO pub health fel, UK, 51; Fulbright res fel, State Univ Ghent, 55-56; sr pharmacologist, Johnson & Johnson Res Found, NJ, 57-59; lectr, Rutgers Univ, 58-65; assoc res prof, Col Pharmacol, Univ Conn, 80-; Intra-Acad (NAS-HAS) Exchange Scientist, Hungary, 85, 88. *Mem:* Soc Toxicol; Am Soc Pharmacol & Exp Ther; Am Chem Soc; NY Acad Sci; fel AAAS. *Res:* Drug safety evaluation; physiology of reproduction; endocrine pharmacology. *Mailing Add:* Pfizer Central Res Pfizer Pharmaceuts Groton CT 06340

KING, THOMAS CREIGHTON, b Salt Lake City, Utah, Apr 10, 28; m 52; c 4. THORACIC SURGERY, CRITICAL CARE. *Educ:* Univ Utah, BS, 50, MD, 54; Univ Mo, Kansas City, MA, 63. *Prof Exp:* From asst resident to chief resident, Univ Utah Hosps, 55-59; assoc, Univ Kans, 60, asst prof med sch, 60-64; assoc prof surg & psychol & chief training, Ctr Study Med Educ, Univ Ill Med Ctr, 64-66; from assoc prof to prof surg, Med Sch, Univ Utah, 66-73, assoc dean med sch, 66-68, acad vpres, 68-69, provost, 69-73; PROF SURG, COLUMBIA-PRESBY MED CTR, 73- *Concurrent Pos:* Fel surg, Univ Utah Hosps, 54 & 60; intern, Columbia-Presby Med Ctr, 54-55; staff surgeon, Kansas City Vet Admin Hosp, 60-64, assoc chief staff & dir res, 62-64; staff surgeon, Med Ctr, Univ Utah, 66-73; chief thoracic surg, Salt Lake City Vet Admin Hosp, 68-73; attend surg, Harlen Hosp Ctr, 73-80, Presby Hosp, NY. *Mem:* AAAS; Am Fedn Clin Res; Asn Am Med Cols; Soc Univ Surg; Am Asn Thoracic Surg; Am Col Surgeons; Am Surg Asn; Soc Int Surgeons. *Res:* Medical education and teacher training; surgical nutrition and metabolism; skin degerming and infection control; surgical intensive care. *Mailing Add:* Dept Surg Columbia-Presby Med Ctr New York NY 10032

KING, THOMAS K C, b Shanghai, China, June 1, 34; US citizen; c 2. PULMONARY PHYSIOLOGY, PULMONARY DISEASES. *Educ:* Univ Edinburgh, MB, ChB, 59, MD, 63; FRCP, 80. *Prof Exp:* Eli Lilly Int fel, Bellevue Hosp, Columbia Univ, 65-66, Polachek Found Cardiopulmonary Lab fel, 66-67; lectr med, Sch Med, Univ Hong Kong, 67-70; asst prof, 70-73, ASSOC PROF MED, COL MED, CORNELL UNIV, 73-, ASSOC PROF BIOPHYS, 75- *Concurrent Pos:* Vis prof, Univ Hong Kong, 81, Nat Defense Med Ctr, Taiwan, 87. *Honors & Awards:* Pulmonary Acad Award, Nat Heart & Lung Inst, 72. *Mem:* Med Res Soc, UK; Am Fedn Clin Res; Am Physiol Soc; Am Thoracic Soc; Am Col Chest Physicians. *Res:* The mechanism and quantitation of impaired blood gas exchange in the lungs in disease. *Mailing Add:* Col Med Cornell Univ 1300 York Ave New York NY 10021

KING, THOMAS MORGAN, b Morristown, Tenn, Aug 28, 40; m 62; c 2. PHYSICAL INORGANIC CHEMISTRY. *Educ:* Carson-Newman Col, BS, 62; Univ Tenn, PhD(chem), 66. *Prof Exp:* Res chemist, 66-68, sr res chemist, 68-69, res specialist, 69, res group leader, 69-72, com develop mgr, 72-76, dir planning & control, 77, com dir sorbates, 77-80, dir results mgt & personnel planning, 80-82, dir bus develop, 83-85, DIR TECH D&P DIV, MONSANTO INDUST CHEM CO, 86- *Mem:* Am Chem Soc. *Res:* Coordination chemistry of Cobalt-II and Nickel-II compounds; basic and applied research on phosphonate compounds; precipitation inhibition and corrosion inhibition. *Mailing Add:* 800 N Lindbergh Blvd Mail Zone 04C St Louis MO 63167

KING, TSOO E, biochemistry; deceased, see previous edition for last biography

KING, WALTER BERNARD, b Ewing, Ill, Dec 3, 00; m 36; c 4. INORGANIC CHEMISTRY. *Educ:* Univ Ill, BS, 23; Iowa State Col, MS, 24, PhD(inorg chem), 30. *Prof Exp:* From asst prof to prof, 31-71, EMER PROF CHEM, IOWA STATE UNIV, 71- *Mem:* Am Chem Soc. *Mailing Add:* 2103 Country Club Blvd Ames IA 50010-7014

KING, WENDELL L, b Fairmont, Minn, Sept 6, 40; m 61; c 4. IMPLANTABLE DEFIBRILLATORS, ANGIOPLASTY DEVICES. *Educ:* Univ Minn, BPhys, 64. *Prof Exp:* Res physicist, B F Goodrich Res, 64-68; res scientist, Owens Corning Fiberglas, 68-69; prin investr, NStar Res & Develop Inst, Univ Minn, 69-72; staff scientist, Medtronic Inc, 72-74, Div Res & Develop, 74-81, vpres & gen mgr, 81-85; CHIEF EXEC OFFICER & CHMN, ANGEION CORP, 87- *Mem:* Soc Plastics Engrs. *Res:* Implantable materials; adhesion of polymeric materials inside the body; design of implantable and disposable medica devices. *Mailing Add:* 13000 Hwy 55 Plymouth MN 55441

KING, WILLIAM CONNOR, b Newark, Ohio, Mar 10, 27; m 50; c 2. RADAR SYSTEMS, COMMUNICATION SYSTEMS. *Educ:* Denison Univ, BA, 49; Duke Univ, PhD(physics), 53. *Prof Exp:* Asst physics, Duke Univ, 49-53, res assoc, 53; res assoc radiation lab, Johns Hopkins Univ, 53-56; commun physicist, Space Sci Lab, Missile & Space Div, Gen Elec Co, 56-62, mgr space systs anal proj, 62-64, commun eng spacecraft dept, Pa, 64-67, data systs, 67-70; mgr hard point demonstration array radar prog, RCA Corp, 71-74, leader tradex programming, Missile & Surface Radar Div, 74-77; mgr, Wayland Software Eng Dept, 78-80, SR STAFF, SOFTWARE SYSTS LAB, RAYTHEON CORP, 81- *Mem:* Am Phys Soc; Asn Comput Mach; sr mem Inst Elec & Electronics Eng; Am Inst Aeronaut & Astronaut. *Res:* Microwave spectroscopy and wave propagation in ionized media; design and development of real-time computer programs for radar, missile and communication systems; communications applications of computers. *Mailing Add:* 446 Hayward Mill Rd Concord MA 01742

KING, WILLIAM DAVID, b Owensboro, Ky, Apr 9, 53; m 78. CLINICAL TOXICOLOGY, INJURY EPIDEMIOLOGY. *Educ:* DrPH, Univ Ala, Birmingham, 89. *Prof Exp:* CLIN ASST PROF, SCH NURSING, UNIV ALA, BIRMINGHAM, 79-, ASSOC PROF PEDIAT, SCH MED, 90; DIR TOXICOL, POISON CONTROL CTR, CHILDREN'S HOSP ALA, 83. *Concurrent Pos:* Adj clin fac, Sch Pharm, Samford Univ, 79-; ed, Poison Info Bull, Children's Hosp, 79-; prin investr, prescription drug ingestions in preschool-aged children, US Consumer Prod Safety Comn, 83-84. *Honors & Awards:* Child Advocacy Award, Am Acad Pediat. *Mem:* Am Asn Poison Control Ctrs; Am Acad Clin Toxicol. *Res:* Childhood injury epidemiology; adolescent parasuicide by drug ingestion; epidemiology of poison exposures; development of a regional secondary prevention system for poison injuries. *Mailing Add:* Children's Hosp Ala 1600 Seventh Ave S Birmingham AL 35233-1711

KING, WILLIAM EMMETT, JR, b Pittsburgh, Pa, July 27, 43. CHEMICAL ENGINEERING. *Educ:* Univ Pittsburgh, BS, 65; Carnegie-Mellon Univ, MS, 68; Univ Pa, PhD(chem eng), 76. *Prof Exp:* Chem engr, Esso Res & Eng Co, 66-68; engr mathematician, Cities Serv Res & Develop Co, 68-70; asst prof chem eng, Univ Md, 75-81; sr res engr, Gulf Res & Develop Co, 81-83; ASSOC PROF CHEM ENG & CHMN, DEPT CHEM ENG, BUCKNELL UNIV, 83- *Mem:* Am Inst Chem Engrs; Am Chem Soc. *Res:* Fluid-solid reactions; applied mathematics; coal conversion technology; synthetic fuels; biomedical engineering. *Mailing Add:* Dept Chem Eng Bucknell Univ Lewisburg PA 17837

KING, WILLIAM MATTERN, b Cando, NDak, Mar 12, 30; m 89; c 1. APPLIED CHEMISTRY. *Educ:* NDak State Univ, BS, 57, MS, 58. *Prof Exp:* Chemist polymers, Hooker Chem Corp, 58-60; mgr develop membranes, Envirogenics Systs Co, 60-78; res dir, Spectrum Separations/Separex, 79-90; RETIRED. *Mem:* Am Chem Soc. *Res:* Asymmetric membranes for the separation of gases and organic liquids as well as the desalination of water by reverse osmosis. *Mailing Add:* 2879 Stanbridge Ave Long Beach CA 90815-1060

KING, WILLIAM ROBERT, JR, b Los Angeles, Calif, Aug 25, 24; m 50; c 3. CHEMICAL METALLURGY. *Educ:* Calif Inst Technol, BS, 47; Univ Calif, Los Angeles, PhD(chem), 52. *Prof Exp:* Asst chem, Univ Calif, Los Angeles, 47-50; res chemist, Filtrol Corp, 52-54 & Sierra Talc & Clay Co, 54-55; sect head, Kaiser Aluminum & Chem Corp, 55-77, sr staff res chemist, 77-82; CHEM METALL CONSULT, 82- *Concurrent Pos:* Writer, chem encyclopedias. *Mem:* AAAS; Am Chem Soc. *Res:* Physical and inorganic chemistry; radiochemistry; molten salts; natural iron and aluminum minerals. *Mailing Add:* 875 Sunset Dr San Carlos CA 94070

KING, WILLIAM STANELY, b Monroe, La, June 16, 35; m 58; c 2. APPLIED PHYSICS, ENGINEERING. *Educ:* Univ Calif, Berkeley, BSME, 57; Univ Calif, Los Angeles, MS, 60, PhD(appl math, physics), 66. *Prof Exp:* Res engr, Rocket Div, Rockwell Int Corp, 57-61; staff scientist, Aerospace Corp, 61-72; sr res scientist, Rand Corp, 72-83; RES SCIENTIST, ROCKET DIV, ROCKWELL INT CORP, 83- *Concurrent Pos:* Instr math, Santa Monica Col, 73-; instr eng, Univ Calif, Los Angeles, 77-; Aerospace Corp adv study grant. *Mem:* Assoc fel Am Inst Aeronaut & Astronaut; Sigma Xi. *Res:* Fluid dynamics; numerical analysis; electromagnetic theory; laser physics; applied mathematics. *Mailing Add:* 4472 Don Milagro Dr Los Angeles CA 90008

KING, WILLIS KWONGTSU, b Shanghai, China, Sept 23, 36; m 70. COMPUTER SCIENCE, ELECTRICAL ENGINEERING. *Educ:* Darmstadt Tech Univ, Dipl Ing, 63; Univ Pa, PhD(elec eng), 69. *Prof Exp:* Res engr comput design, IBM Labs, 63-65; asst prof comput sci, 69-73, ASSOC PROF COMPUT SCI, UNIV HOUSTON, 73-, CHMN DEPT, 79- *Mem:* Asn Comput Mach; Sigma Xi; Inst Elec & Electronics Engrs. *Res:* Computer architecture; distributed computing; microprogramming. *Mailing Add:* Dept Comput Sci 3801 Cullen Blvd Houston TX 77204-3475

KING, WILTON W(AYT), b Richmond, Va, Aug 11, 37; m 58; c 4. ENGINEERING MECHANICS. *Educ:* Univ Va, BME, 59, MME, 61; Va Polytech Inst, PhD(eng mech), 65. *Prof Exp:* Instr eng mech, Va Polytech Inst, 61-64; from asst prof to assoc prof, 64-77, PROF ENG MECH, GA INST TECHNOL, 77- *Mem:* Am Soc Mech Engrs; Sigma Xi. *Res:* Vibrations; fracture mechanics. *Mailing Add:* Sch Eng Sci & Mech Ga Inst Technol Atlanta GA 30332

KINGDON, HENRY SHANNON, b Puunene, Hawaii, July 2, 34; m 57, 85; c 3. BIOCHEMISTRY, HEMATOLOGY. *Educ:* Oberlin Col, AB, 56; Western Reserve Univ, MD & PhD(biochem), 63. *Prof Exp:* Intern & resident internal med, Univ Wash, 63-65; clin assoc, Nat Heart Inst, 65-67; from asst prof to assoc prof med & biochem, Univ Chicago, 67-73; prof med & biochem, Univ NC, Chapel Hill, 73-81; med dir, Hyland Therapeut, Glendale, Calif, 81-90, vpres, 84-90, VPRES & GEN MGR, HYLAND BIOTECHNOL, HAYWARD, CALIF, 90- *Concurrent Pos:* Guggenheim fel, 72-73. *Mem:* Am Chem Soc; Am Fedn Clin Res; Int Soc Thrombosis & Haemostasis; Am Soc Hemat; Am Soc Biol Chem. *Res:* Hematology; enzymology of blood coagulation; regulation of nitrogen metabolism in microorganisms; primary structure of regulatory and coagulation enzymes. *Mailing Add:* Hyland Div Baxter Healthcare 1978 W Winton Ave Hayward CA 94545

KINGERY, BERNARD TROY, b Metter, Ga, July 16, 20; m 59; c 2. PHYSICS. *Educ:* Ga Southern Col, BS, 48; Columbia Univ, MA, 49. *Prof Exp:* Instr physics, Orange County Community Col, 50-52; supvr physics courses, Div Technol, Newark Col Eng; ASST PROF PHYSICS, NJ INST TECHNOL, 52- *Concurrent Pos:* Instr teaching of sci, Teachers Col, Columbia Univ, 57-58. *Mem:* AAAS; Am Soc Eng Educ; Am Asn Physics Teachers; Nat Sci Teachers Asn; Am Asn Univ Prof; Sigma Xi. *Res:* Science teacher education. *Mailing Add:* 92 Oakridge Ave Nutley NJ 07110

KINGERY, W(ILLIAM) D(AVID), b New York, NY, July 7, 26; div; c 4. CERAMICS, MATERIALS SCIENCE. *Educ:* Mass Inst Technol, SB, 48, ScD(ceramics), 50. *Hon Degrees:* PhD, Tokyo Inst Technol, 82; ScD, Ecole Polytechnique Federale de Lausanne, 88. *Prof Exp:* Res assoc, 49-50, from asst prof to prof, Mass Inst Technol, 50-88; PROF MAT SCI & ENG & PROF ANTHROP, UNIV ARIZ, TUCSON, 88- *Concurrent Pos:* Foreign collabr, Comn Atomic Energy, France, 64-65; vis prof, Johns Hopkins Univ, 87-88; regents fel, Smithsonian Inst, 87-88; chmn, Comt Ceramic Hist & Archaeol, Am Ceramic Soc & bd trustees, Int Acad Ceramics; mem, Mat Educ Coun, Mat Res Soc; ed-in-chief, Ceramics Int. *Honors & Awards:* Purdy Award, Am Ceramic soc, 54, John Jeppson Award, 58, Robert Sosman Mem Lectr, 73, Albert V Bleininger Award, 77, F H Norton Award, 77, Edward Orton Jr Mem Lectr, 80, Hobart M Kraner Award, 85; Wagener Lectr, Tokyo Inst Technol, 76; Kurtz Lectr, Technion, Haifa, 78. *Mem:* Nat Acad Eng; Am Chem Soc; fel Am Ceramic Soc; fel AAAS; Am Acad Arts & Sci. *Res:* Archaeological ceramics; author of more than 200 technical publications. *Mailing Add:* Dept Mat Sci & Eng Univ Ariz Tucson AZ 85721

KINGHORN, ALAN DOUGLAS, b Newcastle-upon-Tyne, UK, Aug 31, 47; m 76. PHARMACOGNOSY, PHYTOCHEMISTRY. *Educ:* Univ Bradford, UK, BPharm, 69; Univ Strathclyde, Glasgow, UK, 70; Sch Pharm, Univ London, PhD(pharmacog), 75, DSc, 91. *Prof Exp:* Anal chemist, Burroughs Wellcome Co, Dartford, UK, 70-71; teaching fel pharmacog, Sch Pharm, Univ London, UK, 71-75; postdoctoral pharmacog, Univ Miss, 75-76; res assoc, 76-77, from asst prof to assoc prof, 77-86, PROF PHARMACOG, UNIV ILL, CHICAGO, 86- *Concurrent Pos:* Ed, Photochem Anal, 89-; guest prof pharmacog, ETH-Zentrum, Zurich, Switz, 90. *Mem:* Am Chem Soc; Am Soc Pharmacog (pres, 90-91); Royal Pharmaceut Soc; Soc Econ Bot; fel Linean Soc London. *Res:* Isolation, structure elucidation, semi-synthesis, chemical analysis, and bioassay of plant secondary metabolites with interesting biological activities, especially compounds that exhibit antineoplastic, antiviral bitter-tasting, cytotoxic, insecticidal, mutagenic, skin-irritant or sweet-tasting properties. *Mailing Add:* Dept Med Chem & Pharmacog Col Pharm Univ IL 833 S Wood St Chicago IL 60612

KINGMAN, HARRY ELLIS, JR, b Ft Collins, Colo, Sept 4, 11; m 36; c 1. VETERINARY MEDICINE. *Educ:* Colo State Univ, DVM, 33. *Prof Exp:* Jr veterinarian, US Bur Animal Indust, 33-39; chief veterinarian, Wilson & Co, Inc, Ill, 39-53; asst exec secy, Am Vet Med Asn, 53-58, exec secy, 58-66; exec dir, Nat Soc Med Res, 66-78; RETIRED. *Concurrent Pos:* Mem, Nat Adv Food & Drug Coun, 65-69. *Mem:* Am Vet Med Asn (treas, 52-58). *Res:* Public health; food hygiene; physiology of reproduction of cattle. *Mailing Add:* 1707 Essex Dr Ft Collins CO 80526-1616

KINGMAN, ROBERT EARL, b Phoenix, Ariz, June 24, 38; m 59; c 4. PHYSICS. *Educ:* Walla Walla Col, BS, 61; Univ Ariz, MS, 67, PhD(physics), 71. *Prof Exp:* Instr physics, Walla Walla Col, 63-66, asst prof, 66-67; instr physics, Univ Ariz, 71; from asst prof to assoc prof, 71-79, PROF, ANDREWS UNIV, 79-, CHAIR, PHYSICS DEPT, 71- *Mem:* Am Phys Soc; Am Asn Physics Teachers; Am Asn Univ Prof; Sigma Xi; Coun Undergrad Res. *Res:* Study of super clustering of galaxies; method of describing decay of unstable quatum states; study of cosmological models based on the Robertson-Walker metric. *Mailing Add:* Physics Dept Andrews Univ Berrien Springs MI 49103

KINGREA, C(HARLES) L(EO), b Barren Springs, Va, Aug 17, 23; m 46; c 3. CHEMICAL ENGINEERING, PROGRAM MANAGEMENT. *Educ:* Va Polytech Inst, BS, 43, MS, 51, PhD(chem eng), 53. *Prof Exp:* Owner-mgr, Kingrea Milling Co, Va, 46-53; process design engr, Ethyl Corp, 53-56, asst supt mfg tech serv, 56-58, econ anal engr, 58, proj mgr, 58-60, head eng & math sci, 60-63, head spec process design assignment, Ethyl Corp, La, 63-68, gen supt alcohol opers, Tex, 68-71, gen supt opers, 71-80, mgr tech planning & proj coordr, 80-87; RETIRED. *Concurrent Pos:* Prod engr, Dallas Chem Procurement Dist, 43-44, property disposal officer, 44-46. *Mem:* Am Inst Chem Engrs. *Res:* Mass transfer operations, particularly thermal diffusion; chemical process design; chemical project management. *Mailing Add:* 12015 Oakwilde Baton Rouge LA 70810

KINGREA, JAMES I, b Philadelphia, Pa, Mar 10, 28; m 47; c 2. QUALITY ASSURANCE MANAGEMENT, SUPPLIER SOURCE ASSURANCE MANAGEMENT. *Educ:* Pa Mil Col, BSEE, 50. *Prof Exp:* Mgr, Qual Systs & Serv, Steam Div, Westinghouse Elec Corp, 69-86; mgr, Advan Develop & Eng Ctr, 86-89; DIR QUAL, ENG SYSTS CO, DIV DATRON, 89- *Concurrent Pos:* Qual consult, 70-; weld inspector, Am Weld Soc, 82- *Mem:* Sr mem Am Soc Qual Control; Am Soc Mech Engrs. *Res:* Quality discipline relating to surface texture and lay; mathematical definition of finishes; equipment used to measure different finishes; specification of surface finish requirements; generation of specific finishes. *Mailing Add:* Datron Eng Systs Div 2550 Market St PO Box 2240 Aston PA 19014-3426

KINGSBURY, CHARLES ALVIN, b Louisville, Ky, Jan 12, 35; m 58; c 5. ORGANIC CHEMISTRY. *Educ:* Iowa State Col, BS, 56; Univ Calif, Los Angeles, PhD(org chem), 60. *Prof Exp:* NSF fel, Harvard Univ, 62-63; instr org chem, Iowa State Univ, 63-67; from asst prof to assoc prof, 67-72, PROF ORG CHEM, UNIV NEBR, LINCOLN, 72- *Mem:* Royal Soc Chem; Am Chem Soc. *Res:* Stereochemistry; reaction mechanisms. *Mailing Add:* 7327 York Lane Lincoln NE 68505

KINGSBURY, DAVID THOMAS, b Seattle, Wash, Oct 24, 40; m 82. VIROLOGY, MICROBIOLOGY. *Educ:* Univ Wash, BS, 62, MS, 64; Univ Calif, San Diego, PhD(biol), 71. *Prof Exp:* Microbiologist, Naval Med Res Inst, 64-67; res fel microbiol, Am Inst Biol Sci, 67-68; from asst prof to assoc prof microbiol, Univ Calif, Irvine, 72-81; prof med microbiol & virol, Univ Calif, Berkeley, 81-86; asst dir biol behav & social sci, NSF, 84-88. *Concurrent Pos:* Am Cancer Soc Dernham fel oncol, Univ Calif, San Diego, 71-72 & 77 & NIH, 78-79; vis scientist, Scripps Clin & Res Found, La Jolla, Calif, 73-80; dir, Naval Biosci Lab, Oakland, 81-84; mem, bd regents, Nat Library Med, 84-; adj prof microbiol, George Washington Univ, 85- *Mem:* Fel AAAS; Am Soc Microbiol; Am Soc Virol; Soc Genetic Microbiol. *Res:* Oncogenic viruses; viral genetics; biochemistry of virus replication; techniques in diagnostic virology and microbiology; biochemistry and genetics of the unconventional viruses. *Mailing Add:* Dept Microbiol & Immunol George Washington Univ Med Ctr 2300 I St NW Washington DC 20037

KINGSBURY, DAVID WILSON, b Jersey City, NJ, Apr 2, 33; m 57; c 7. VIROLOGY, MOLECULAR BIOLOGY. *Educ:* Manhattan Col, BS, 55; Yale Univ, MD, 59. *Hon Degrees:* DSc, Manhattan Col, 90. *Prof Exp:* Intern path, Yale Univ, 59-60, asst resident, 60-61; from res fel to mem, 63-69, mem div virol, St Jude Hosp, Memphis, 69-88; SR SCI OFFICER, HOWARD HUGHES MED INST, 88- *Concurrent Pos:* USPHS res fel, Yale Univ, 61-63, St Jude Hosp, Memphis, 63-64, career develop award, 64-73; adj prof microbiol, Univ Tenn, Memphis, 72-85. *Mem:* Am Asn Immunol; Am Soc Microbiol; Am Soc Virol. *Res:* Negative strand RNA viruses. *Mailing Add:* Howard Hughes Med Inst 6701 Rockledge Dr Bethesda MD 20817

KINGSBURY, ELIZABETH W, b Leamington, Eng, May 8, 25; div; c 3. ONCOLOGY. *Educ:* WVa Univ, Morgantown, PhD(microbiol), 67. *Prof Exp:* Chief, Electron Micros Lab, Litton-Bionetics, 69-78; CONSULT. *Mem:* Am Soc Microbiol; Am Soc Cell Biol; AAAS; Soc Electron Micros; Sigma Xi. *Mailing Add:* 305 Washington Grove Lane Gaithersburg MD 20877

KINGSBURY, HERBERT B, b Pittsburgh, Pa, Feb 15, 34; m 56; c 3. SOLID MECHANICS, BIOMECHANICS. *Educ:* Univ Conn, BS, 58; Univ Pa, MS, 61, PhD(eng mech), 64. *Prof Exp:* Scientist, Dyna/Struct Inc, 61-64; engr, Missile & Space Div, Gen Elec Corp, 64-66; asst prof aerospace eng, Pa State Univ, 66-67; asst prof, 67-73, assoc prof aerospace eng, 73-80, PROF MECH ENG, UNIV DEL, 80- *Concurrent Pos:* Eng consult, Scott Paper Co, 70-; adj assoc prof, Sch Vet Med, Univ Pa, 78-82. *Mem:* Am Soc Mech Engrs; Am Soc Biomech; Nat Soc Prof Engrs; Am Acad Mech. *Res:* Structural mechanics; mechanics of biological structures; mechanics of porous deformable solids; structural dynamics. *Mailing Add:* Dept Mech Eng Univ Del Newark DE 19711

KINGSBURY, ROBERT FREEMAN, b Ithaca, NY, June 26, 12; m 33; c 4. ATOMIC SPECTROSCOPY. *Educ:* Bowdoin Col, BS, 34; Cornell Univ, MS, 39; Univ Pa, PhD(physics), 56. *Prof Exp:* Teacher pub schs, NY; instr sci, Mass State Teachers Col, Westfield, 42-43; instr physics, Bowdoin Col, 43, Bates Col, 44 & Univ Maine, 44-47; from instr to assoc prof, Trinity Col, Conn, 50-64; prof, 64-78, EMER PROF PHYSICS, BATES COL, 78- *Mem:* AAAS; Am Phys Soc; Am Asn Physics Teachers; Am Optical Soc; Sigma Xi. *Res:* Atomic spectra. *Mailing Add:* 65 Vale St Lewiston ME 04240

KINGSBURY, WILLIAM DENNIS, b Buffalo, NY, Nov 21, 41; m 67; c 3. ORGANIC CHEMISTRY, MEDICINAL CHEMISTRY. *Educ:* State Univ NY Buffalo, BA, 65; Wayne State Univ, PhD(chem), 70. *Prof Exp:* Chemist, Electro Refractories & Abrasives, 61-62; instr chem, Wayne State Univ, 65-67; SR CHEMIST, SMITH KLINE & FRENCH LABS, 71- *Concurrent Pos:* NIH fel, Univ Kans, 70-71. *Mem:* Am Chem Soc. *Res:* Anthelmintics; animal nutrition; heterocyclic chemistry; organo sulfur chemistry; immunochemistry; antimicrobial chemotherapy. *Mailing Add:* Smith Kline & French L421 PO Box 1539 King of Prussia PA 19406

KINGSLAKE, RUDOLF, b London, Eng, Aug 28, 03; nat US; m 29; c 1. OPTICS. *Educ:* Univ London, BSc, 24, Imp Col, MSc & dipl, 26, DSc, 50. *Hon Degrees:* DSc, Univ Rochester,86. *Prof Exp:* Optical designer, Sir Howard Grubb, Parsons & Co, Eng, 27-28; res engr, Int Standard Elec Corp, London, 28-29; from asst prof to assoc prof geom optics, 29-59, prof optics, 59-84, EMER PROF OPTICS, INST OPTICS, UNIV ROCHESTER, 84- *Concurrent Pos:* Exchange prof, Imp Col, Univ London, 36-37; optical designer, Eastman Kodak Co, 37-39, head optical design dept, 39-68. *Honors & Awards:* Ives Medal, Optical Soc Am, 73; Gold Medal, Soc Photo-optical Instrumentation Engrs, 80; Progress Medal, Soc Motion Picture & TV Engrs, 64. *Mem:* Fel & hon mem Optical Soc Am (vpres, 45-47, pres, 47-49); fel Soc Motion Picture & TV Engrs; fel Soc Photog Scientists & Engrs; fel Soc Photo-Optical Instrumentation Engrs. *Res:* Design of lenses and optical systems; measurement of aberrations; effect of aberrations on optical images; applied optics. *Mailing Add:* 56 Westland Ave Rochester NY 14618

KINGSLAND, GRAYDON CHAPMAN, b Burlington, Vt, Aug 28, 28; m 50; c 4. PLANT PATHOLOGY. *Educ:* Univ Vt, BA, 52; Univ NH, MS, 55; Pa State Univ, PhD(plant path), 58. *Prof Exp:* Res technician, Conn Tobacco Lab, 52-53; assoc pathologist, United Fruit Co, Honduras, 58-60; asst prof bot

& asst pathologist, 60-67, assoc prof, 67-84, PROF PLANT PATH & PHYSIOL, CLEMSON UNIV, 84- *Concurrent Pos:* Plant pathologist, USAID/SECID Seych Is food crop improv proj, 81-82 & 84. *Mem:* Am Phytopath Soc; Sierra Club; Wilderness Soc; Nat Audubon Soc; Sigma Xi. *Res:* Diseases of cereal grains; ecology of microflora of rhizospheres and seeds of cereal grains; chemical control of cereal grains diseases; teaching introductory and graduate phytopathology; tropical agriculture; mycology. *Mailing Add:* Dept Plant Path & Physiol Clemson Univ Clemson SC 29634-0377

KINGSLEY, HENRY A(DELBERT), b Wakefield, RI, May 5, 21; m 43; c 2. CHEMICAL ENGINEERING, PROCESS ENGINEERING. *Educ:* Univ RI, BS, 43; Yale Univ, DEng, 49. *Prof Exp:* Engr, Shell Develop Co, Calif, 49-59, supvr process eng, 59-63, dept head licensing & design eng, 64-65, dir res & develop lab, Indust Chem Div, Shell Chem Co, Tex, 65-67, mgr proj develop, Plastics & Resins Div, NY, 67-70, mgr chem & chem eng dept, Explor & Prod Res Ctr, Shell Develop Co, 70-72, mgr chem process eng, 72-74, mgr support process eng, Shell Oil Co, 74-82; consult, 82-84; PRES, COMTECH CONSULTS INC, 84- *Mem:* Am Inst Chem Engrs; Am Chem Soc; Sigma Xi. *Mailing Add:* 411 W Fair Harbor Lane Houston TX 77079-2515

KINGSLEY, JACK DEAN, b Wonewoc, Wis, Aug 10, 34; m 62; c 3. SOLID STATE DEVICES, OPTOELECTRONICS. *Educ:* Univ Wis, BSEE, 56, MSEE, 57; Univ Ill, MS, 58, PhD(physics), 60. *Prof Exp:* Physicist, Gen Elec Res & Develop Ctr, 60-71, mgr light emitting diode array prog, 71-72, mgr optoelectronics br, 72-81, mgr electronic mat br, 81-83, mgr display br, 83-85, PHYSICIST, GEN ELEC RES & DEVELOP CTR, 85- *Mem:* Am Phys Soc. *Res:* Optical spectroscopy of solids; quantum electronics; luminescence; point defects in solids; imaging and display devices. *Mailing Add:* Gen Elec Res & Develop Ctr PO Box 8 Schenectady NY 12301

KINGSLEY, MICHAEL CHARLES STEPHEN, b Harrow, UK, Oct 20, 41; Can citizen. ECOLOGY, BIOMETRICS. *Educ:* Univ Cambridge, MA, 66; Lancaster Univ, MA, 68. *Prof Exp:* Statistician, Can Forestry Serv, 71-73; statistician, Can Wildlife Serv, 73-78, biologist, 78-83; RES SCIENTIST, CAN DEPT FISHERIES & OCEANS, 83- *Mem:* Biomet Soc; Soc Marine Mammal; Arctic Inst NAm. *Res:* Ecology of marine mammals; behavior, vocalizations, population dynamics, population estimation, growth. *Mailing Add:* 71A Academy Rd Winnepeg MB R2J 1B9 Can

KINGSOLVER, CHARLES H, b Peru, Nebr, Aug 31, 14; m 41; c 4. PLANT PATHOLOGY. *Educ:* Nebr State Teachers Col, Peru, AB, 35; Iowa State Univ, MS, 39, PhD(plant path), 43. *Prof Exp:* Asst prof bot, Univ Mo, 46-51; sect chief, Biol Br, Chem Corps, Biol Warfare Labs, Md, 51-55, chief biol br II, 55-57; agr adminr, Mkt Qual Res Div, Agr Mkt Serv, USDA, 57-62; chief biol br, Crops Div, US Army Biol Labs, 62-68, chief plant path div, 68-71; dir, Plant Dis Res Lab, Northeast Region, Sci & Educ Admin-Agr Res, USDA, 71-79; CONSULT PLANT DIS RES, 79- *Concurrent Pos:* Adj prof plant path, Pa State Univ, 72- *Mem:* AAAS; Am Inst Biol Sci; Sigma Xi; Am Phytopath Soc. *Res:* Plant disease epidemiology; quantitation of disease increase and spread; predictive systems; threat potential of foreign plant disease; biological control of weeds with plant pathogens. *Mailing Add:* PO Box 337 Braddock Heights MD 21714

KINGSOLVER, JOHN MARK, b Selma, Ind, Mar 20, 25; m 48; c 2. ENTOMOLOGY. *Educ:* Purdue Univ, BS, 51; Univ Ill, MS, 56, PhD(entom), 61. *Prof Exp:* Res asst entom, Univ Ill, 54-61; res assoc, Ill Natural Hist Surv, 61-62; RES ENTOMOLOGIST, SYST ENTOM LAB, USDA, 62- *Mem:* Entom Soc Am; Am Entom Soc; Asn Trop Biol; Sigma Xi; Coleop Soc. *Res:* Taxonomy of seed-beetles (Bruchidae) of Western Hemisphere. *Mailing Add:* 429 St Lawrence Dr Silver Spring MD 20901

KINGSTON, CHARLES RICHARD, b San Diego, Calif, Apr 11, 31; m 69; c 1. FORENSIC SCIENCE. *Educ:* Univ Calif, Berkeley, BS, 59, MCriminol, 61, Dr Criminol, 64. *Prof Exp:* Lab technician, Criminalistics Lab, Sch Criminol, Univ Calif, Berkeley, 58-63, res criminalist, 63-64, asst res criminalist, 64-65; consult, NY State Identification & Intel Syst, 65-66, chief criminalistics res bur, 66-68; PROF CRIMINALISTICS, JOHN JAY COL, CITY UNIV NEW YORK, 68- *Mem:* Am Chem Soc; Am Statist Asn; Am Acad Forensic Sci. *Res:* Application of probability and statistics in criminalistics; computer applications in criminology and criminalistics. *Mailing Add:* 445 W 59th St New York NY 10019

KINGSTON, DAVID GEORGE IAN, b London, Eng, Nov 9, 38; m 66; c 3. NATURAL PRODUCTS CHEMISTRY. *Educ:* Cambridge Univ, BA, 60, PhD(org chem), 63; dipl theol, Univ London, 62. *Prof Exp:* Res fel chem, Queens' Col, Cambridge Univ, 62-66, NATO fel, 64-66; asst prof chem, State Univ NY Albany, 66-71; assoc prof, 71-77, PROF CHEM, VA POLYTECH INST & STATE UNIV, 77- *Concurrent Pos:* Res assoc, Mass Inst Technol, 63-64; mem, biomed sci study sect, NIH, 79-84; assoc ed, J Natural Prod, 83-; Div Cancer Treatment Contracts Rev Comm, NIH, 87-91; mem, Div Cancer Treatment Contracts Rev Comt, NIH, 87-92, Chmn, 89-92. *Mem:* Am Chem Soc; Royal Soc Chem; Am Soc Pharmacog (vpres, 87-88, pres, 88-89). *Res:* Natural products chemistry; structure and synthesis of biologically active natural products; organic structure determination by spectroscopic methods; mutagen structure and metabolism. *Mailing Add:* Dept Chem Va Polytech Inst & State Univ Blacksburg VA 24061-0212

KINGSTON, DAVID LYMAN, b Lansing, Mich, June 26, 30; m 54; c 3. SOLID STATE PHYSICS. *Educ:* Mich State Univ, BS, 53, MS, 55. *Prof Exp:* Res physicist, Aerospace Res Labs, 55-75, RES PHYSICIST SOLID STATE PHYSICS, AIR FORCE AVIONICS LAB, WRIGHT-PATTERSON AFB, OHIO, 75- *Mem:* Am Phys Soc. *Res:* Conducting research on the characterization of III-V compound semiconductors such as GaAs and InP using the techniques of luminescence topography and x-ray photoemission spectroscopy. *Mailing Add:* 10 E Routzong Dr Fairborn OH 45324

KINGSTON, GEORGE C, b New York, NY. POLYMER PROCESS DESIGN & DEVELOPMENT. *Educ:* Manhattan Col, BE, 70; Clarkson Col Tech, MS, 72, PhD(chem eng), 75. *Prof Exp:* Sr res engr, Monsanto Polymer Prod Co, 79-82, res specialist, 82-87, tech leader, 87-89, RES & DEVELOP MGR, MONSANTO CHEM CO, 89- *Concurrent Pos:* Instr chem eng, Clarkson Col Tech, 74-75; adj chem eng, Univ Dayton, 78-79, Univ Mass, Amherst, 82. *Mem:* Am Inst Chem Engrs; Soc Plastics Engrs. *Res:* Design and development of polymer processes; polymerization; process effects on polymer structure and properties; oxidative stability of polymers. *Mailing Add:* Monsanto Chem Co 730 Worcester St Springfield MA 01151

KINGSTON, JOHN MAURICE, b Joliet, Ill, May 25, 14; m 37; c 2. MATHEMATICS. *Educ:* Univ Western Ont, BA, 35; Univ Toronto, MA, 36, PhD(theory, abstract groups), 39. *Prof Exp:* Lectr math, Univ BC, 39-40; assoc, 40-43, from instr to asst prof, 43-59, ASSOC PROF MATH, UNIV WASH, 59-, EXEC SECY DEPT, 52- *Concurrent Pos:* Fel, NSF, 59-60. *Mem:* Nat Coun Teachers Math; Math Asn Am. *Res:* Abstract group theory. *Mailing Add:* Dept Math Univ Wash Seattle WA 98195

KINGSTON, NEWTON, b Akron, Ohio, June 6, 25; m 52; c 3. ZOOLOGY, PARASITOLOGY. *Educ:* Wayne State Univ, BA, 54, MSc, 56; Univ Toronto, PhD(zool), 62. *Prof Exp:* Asst prof biol, Detroit Inst Technol, 59-62 & Geneva Col, 62-64; NIH fel, Nat Univ Mex, 64-65; assoc prof, Geneva Col, 65-68; assoc prof, 68-76, PROF PARASITOL, UNIV WYO, 76- *Mem:* Am Soc Parasitol; Am Micros Soc; Soc Protozoologists; Wildlife Dis Asn. *Res:* Morphology of the nervous system of trematodes; life history studies of monogenetic and digenetic trematodes, cestodes; systematics of spinturnicid mites from bats; protozoan parasites domestic and wild ungulate raptors. *Mailing Add:* Microbiol/Vet Med Box 3354 Laramie WY 82070

KINGSTON, PAUL L, b Rochester, NY, Feb 12, 32; m 62; c 6. FINANCIAL MODELING, MARKETING STRATEGIES. *Educ:* Colgate Univ, BA, 54; Univ Rochester, MA, 57; Syracuse Univ, MS, 75. *Prof Exp:* Mem staff appl sci, Int Bus Mach Corp, 58-63, sr sci mkt, 63-74, consult power indust, 74-84, CONSULT ACAD MKT, IBM CORP, 84- *Concurrent Pos:* Adj lectr, Syracuse Univ, 76-85. *Mem:* Am Math Soc; Math Asn Am; Opers Res Soc Am; Inst Mgt Sci; Soc Indust & Appl Math. *Res:* Mathematical programming and financial modeling for solving business and management problems. *Mailing Add:* 179 Brookside Lane Fayetteville NY 13066

KINGSTON, ROBERT HILDRETH, b Somerville, Mass, Feb 13, 28; m 52; c 4. SOLID STATE DEVICES, OPTICS. *Educ:* Mass Inst Technol, BS, 48, MS, 48, PhD(physics), 51. *Prof Exp:* Mem staff, Transistor Res & Develop, Bell Labs, 51-52; mem, Solid State Physics Group, Mass Inst Technol, 52-61, leader, Optics & Infrared Group, 61-69, head, Optics Div, 69-72, leader, Infrared Radar Group, 72-77, sr staff, 77-84, assoc leader, Optical Communs Technol Group, 84-87, sr staff, Lincoln Lab, 87- 90; RETIRED. *Concurrent Pos:* Vis assoc prof, Stanford Univ, 64-65; ed, J Quantum Electronics, Inst Elec & Electronics Engrs, 65-70; adj prof, Mass Inst Technol, 85-90, sr lectr, 90-; auth, Detection Optical & Infrared Radiation, 78. *Honors & Awards:* Centennial Medal, Inst Elec & Electronics Engrs, 84. *Mem:* Fel Am Phys Soc; fel Optical Soc Am; fel Inst Elec & Electronics Engrs; Nat Acad Eng. *Res:* Physical principles of semiconductor devices; physics of semiconductor surfaces; magnetic resonance; solid state maser, parametric amplifiers; optical masers; non-linear optics; tuneable semiconductor lasers; infrared detectors. *Mailing Add:* Four Field Rd Lexington MA 02173-8015

KINKEL, ARLYN WALTER, b Fond du Lac, Wis, Oct 15, 29; m 55; c 3. PHARMACY. *Educ:* Univ Wis, BS, 52, MS, 57, PhD(pharm), 58. *Prof Exp:* Assoc res pharmacist, Parke-Davis Res Div, Warner-Lambert Co, 58-62, from res pharmacist to sr res pharmacist, 62-70, sect dir pharmaceut res & develop, 70-80, DIR CLIN PHARMACOKINETICS, PHARMACOKINETIC-DRUG METAB DEPT, PARKE-DAVIS RES DIV, WARNER-LAMBERT CO, 81- *Mem:* Am Pharmaceut Asn; fel Acad Pharmaceut Sci; Am Col Clin Pharmacol; fel Am Asn Pharmaceut Sci. *Res:* Biopharmaceutics; assay of blood levels and drugs; pharmacokinetics. *Mailing Add:* Parke-Davis Pharmaceut Res Div Warner Lambert Co 2800 Plymouth Rd Ann Arbor MI 48105-2430

KINLOCH, BOHUN BAKER, JR, b Charleston, SC, July 21, 34; m 61, 69; c 4. GENETICS, PLANT PATHOLOGY. *Educ:* Univ Va, BA, 56; NC State Univ, BS, 62, MS, 65, PhD(genetics), 68. *Prof Exp:* Res asst forest genetics & path, NC State Univ, 62-68; GENETICIST, PAC SOUTHWEST FOREST & RANGE EXP STA, US FOREST SERV, 68- *Mem:* Am Phytopath Soc. *Res:* Genetics of disease resistance in forest trees; population genetics of forest trees. *Mailing Add:* 1355 Queens Rd Berkeley CA 94708

KINLOCH, ROBERT ARMSTRONG, b Dumbarton, Scotland, Feb 15, 39; m 63; c 3. NEMATOLOGY. *Educ:* Glasgow Univ, BSc, 63; Univ Calif, Davis, PhD(entom, nematol), 68. *Prof Exp:* Assoc nematologist, 68-80, ASSOC PROF AGRON, AGR EXP STA, UNIV FLA, 80- *Mem:* Soc Nematol; Orgn Trop Am Nematol. *Res:* Biology and host-parasite relationships of plant parasitic nematodes; economic control of plant parasitic nematodes affecting agronomic crops. *Mailing Add:* Agr Res Ctr Univ Fla PO Box Rte 3 Jay FL 32565

KINLOUGH-RATHBONE, LORNE R, b Adelaide, S Australia, Aug 20, 38. HEMATOLOGY, ATHEROSCLEROSIS. *Educ:* Univ Adelaide, MB & BS, 61, MD, 67; McMaster Univ, PhD(med sci), 71. *Prof Exp:* Res asst, Nat Heart Found, Australia, 64-67; postdoctoral fel, Med Res Coun Can, 67-71; lectr, 71-73, from asst prof to assoc prof, 73-82, PROF PATH, MCMASTER UNIV, 82- *Concurrent Pos:* Sr res fel, Ont Heart Found, 75-80, res assoc, 80-83; consult, NIH, 77-79; mem, sci prog comt, 8th Int Cong Haemostasis & Thrombosis, 79-81, sci rev comt, Ont Heart Found, 81-84 & Coun Thrombosis, Am Heart Asn; prin investr, Med Res Coun, Can, 74- & NATO res grant, 84; chmn grad prog med sci, McMaster Univ, 81-87, actg assoc dean educ, 87-88. *Mem:* Int Soc Thrombosis & Haemostasis; Am Heart Asn; AAAS; Am Asn Pathologists; Can Soc Clin Invest; Am Soc Hemat. *Res:*

Mechanisms influencing hemostasis, thrombosis and the development of atherosclerosis; cellular and biochemical mechanisms in platelet response to stimuli; factors governing the response of blood and vessels to injury. *Mailing Add:* Dept Path McMaster Univ Med Ctr 1200 Main St W Rm 3N26 Hamilton ON L8N 3Z5 Can

KINMAN, RILEY NELSON, b Dry Ridge, Ky, Jan 25, 36; m 57; c 2. CIVIL & SANITARY ENGINEERING. *Educ:* Univ Ky, BS, 59; Univ Cincinnati, MS, 62; Univ Fla, PhD(sanit eng), 65. *Prof Exp:* Engr-in-training, Water Dept, City of Dayton, Ohio, 59-61, 62; res assoc chem & sanit sci, Univ Fla, 65-66; asst chief demonstration grants br, Fed Water Pollution Control Admin, 66-67, chief, 67-68; assoc prof civil eng, 68-73, PROF CIVIL ENG, UNIV CINCINNATI, 73-; PRES, PRISTINE, INC, 74- *Mem:* Am Soc Civil Engrs; Water Pollution Control Fedn; Am Water Works Asn; Am Chem Soc; Sigma Xi. *Res:* Treatment and ultimate disposal of hazardous wastes; research and development for control of water pollution, especially the physical, chemical, biological, physiological, economic and political aspects of water pollution. *Mailing Add:* 415 Stevenson Rd Erlanger KY 41018

KINMAN, THOMAS DAVID, b Rugby, Eng, Aug 10, 28; m 63; c 2. ASTRONOMY. *Educ:* Oxford Univ, BA, 49, MA & DPhil(astron), 53. *Prof Exp:* Dept demonstr, Univ Observ, Oxford Univ, 49-53; sci officer physics, Admirality Res Lab, Teddington, 53-54; Radcliffe travelling fel astron, Univ Observ, Oxford Univ, 54-56 & Radcliffe Observ, Pretoria, 56-59; sr sci officer, Royal Observ, Cape Town, 59-60; from asst astronr to astronr, Lick Observ, Univ Calif, Santa Cruz, 60-69; ASTRONR, KITT PEAK NAT OBSERV, 69- *Concurrent Pos:* Mem comn, Int Astron Union, 58. *Mem:* Am Astron Soc. *Res:* Large scale structure of our own and other galaxies, particularly constitution and dynamics of older stars and star clusters; quasistellar objects. *Mailing Add:* Kitt Peak Nat Observ PO Box 26732 Tucson AZ 85726

KINN, DONALD NORMAN, b Chicago, Ill. ACAROLOGY, NEMATOLOGY. *Educ:* Lawrence Col, BS, 56; Univ Wyo, MS, 62; Univ Calif, Berkeley, PhD(entom), 69. *Prof Exp:* Assoc specialist, Div Biol Control, Univ Calif, 69-70, asst res entom, 71-73; RES ENTOMOLOGIST ACAROL, SOUTHERN FOREST EXP STA, 75- *Mem:* Entom Soc Am; Entom Soc Can; Int Orgn Biol Control; Acarol Soc Am; Soc Nematologists. *Res:* Natural control of forest insects, especially bark beetles, by mites and nematodes; control of the pine wood nematode in wood products. *Mailing Add:* Southern Forest Exp Sta 2500 Shreveport Hwy Pineville LA 71360

KINNAIRD, RICHARD FARRELL, b Des Moines, Iowa, Oct 21, 12; m 44; c 2. ENGINEERING. *Educ:* Millsaps Col, BS, 34; Univ Chicago, MS, 36. *Prof Exp:* Asst, Dearborn Observ, Northwestern Univ, 36-39; optical engr, Bell & Howell Co, Chicago, 39-40; optical engr, Perkin-Elmer Corp, 40-63, sr staff engr, 63-72; RETIRED. *Mem:* Am Phys Soc; Optical Soc Am. *Res:* Design and theory of optical instruments. *Mailing Add:* RFD 1 Box 293 Ellsworth ME 04605

KINNAMON, KENNETH ELLIS, b Denison, Tex, May 28, 34; m 57; c 3. PHYSIOLOGY, RADIOBIOLOGY. *Educ:* Okla State Univ, BS, 56; Tex A&M Univ, DVM, 59; Univ Rochester, MS, 61; Univ Tenn, PhD(physiol), 71. *Prof Exp:* Res investr radiation chem, Walter Reed Army Inst Res, 59-60, chief radioisotope lab, Army Nutrit Lab, Fitzsimons Gen Hosp, Colo, 61-65, chief dept surveillance inspection, Med Dept, US Army Vet Sch, 65-68, res investr biol, 71-75; assoc prof physiol & asst dean instrnl & res support, 75-80, PROF PHYSIOL & ASSOC DEAN OPERS, UNIFORMED SERV UNIV HEALTH SCI, 80- *Mem:* Radiation Res Soc; Am Physiol Soc; Health Physics Soc; Am Vet Med Asn; Soc Exp Hemat; Sigma Xi. *Res:* Mineral metabolism; physiology of wound healing; bone marrow transplantation; secondary disease; immunology; cancer therapy; chemical radiation therapy; radiation injury therapy. *Mailing Add:* 17412 Beauvoir Blvd Rockville MD 20855

KINNARD, MATTHEW ANDERSON, b Nashville, Tenn, April 12, 36; m 62; c 2. NEUROPHYSIOLOGY, DENTAL RESEARCH. *Educ:* Tenn State Univ, BS, 57, MA, 60; Georgetown Univ, PhD(physiol), 70. *Prof Exp:* Biologist virol res, Walter Reed Army Med Ctr, 60-62; biologist brain res, NIH, 63-67; asst prof physiol, DC Teachers Col, 70-71; scientist/admin, NIH, 72-79; admin health sci specialist, Cent Off, Vet Admin, 79-85; HEALTH SCI ADMIN, NIH, NAT INST DENT RES, BETHESDA, 85- *Concurrent Pos:* Lectr, Univ DC, 71-79; consult, NIMH, 68-70, Physiol Dept, Howard Univ, 79-80; physiologist, US Civil Serv Bd Examiners, 85- *Mem:* AAAS; Am Physiol Asn; Orgn Black Scientists; Int Asn Dent Res. *Res:* Oral soft tissue diseases (oral cancer, herpes, and AIDS). *Mailing Add:* Nat Inst Dent Res Westwood Bldg Rm 509 5333 Westbard Ave Bethesda MD 20892

KINNARD, WILLIAM J, JR, b Wilmington, Del, Apr 18, 32; m 59. HIGHER EDUCATION ADMINISTRATION. *Educ:* Univ Pittsburgh, BS, 53, MS, 55; Purdue Univ, PhD(pharmacol), 57. *Prof Exp:* From asst prof to prof pharmacol, Univ Pittsburgh, 58-68; actg dean, Grad Sch, Univ Md, 74-76, dean, 76-90, prof pharmacol & dean, Sch Pharm, 68-90, actg pres, 90-91, ACTG ASSOC CHANCELLOR, UNIV MD SYST, 91- *Concurrent Pos:* Chmn bd trustees, US Pharmacopoeial Conv, 74-85; consult pharm, Surgeon Gen USAF, 84-90. *Honors & Awards:* Honor Achievement Award, Angiol Res Found, 65. *Mem:* Inst Med-Nat Acad Sci; fel AAAS; Am Phys Soc; Am Pharmaceut Asn; Am Asn Pharmaceut Scientists; Am Asn Cols Pharm (pres, 76-77); Am Coun Pharm Educ (vpres, 86-). *Res:* Health care systems and education; higher education administration. *Mailing Add:* 4000 N Charles St Baltimore MD 21218

KINNAVY, M(ARTIN) G(ERALD), b Chicago, Ill, Nov 20, 21; m 56; c 5. MECHANICAL ENGINEERING, MECHANICS. *Educ:* Ill Inst Technol, BS, 43, MS, 52. *Prof Exp:* Proj engr, Fisher Body Detroit Div, Gen Motors Corp, 43-44; asst engr, Armour Res Found, Ill Inst Technol, 48-49, assoc engr, Ill Inst Technol Res Inst, 49-50, res engr, 50-52, supvr mechanisms anal & vibration eng, 52-59; assoc dir adv res dept, Sunbeam Corp, 59-60; assoc dir eng res dept, Continental Can Co, Inc, Ill, 60-65; dir eng, Herr Equip Corp, Ohio, 65-68, vpres, 68-70; dir, 70-76, VPRES ENG, HERR-VOSS CORP, 76-, VPRES PROD DEVELOP, 87- *Mem:* Am Soc Mech Engrs; Am Soc Metals; Asn Iron & Steel Engrs. *Res:* Dynamic behavior of mechanisms, linkages and structures; vibration analysis of artillery weapons; dynamics of space vehicles; methods of cam synthesis; mill equipment; tension leveling. *Mailing Add:* Herr-Voss Corp Callery PA 16024

KINNEL, ROBIN BRYAN, b Milwaukee, Wis, Jan 18, 37; m 60; c 2. ORGANIC CHEMISTRY. *Educ:* Harvard Univ, AB, 59; Mass Inst Technol, PhD(org chem), 65. *Prof Exp:* Asst chem, Hercules Powder Co, Del, 56-57; jr chemist, Merck, Sharp & Dohme, NJ, 59-60; res assoc org chem, Stanford Univ, 64-66; from asst prof to prof org chem, 66-84, assoc dean, 73-76, SILAS D CHILDS PROF CHEM, HAMILTON COL, 86-, CHMN DEPT, 82- *Concurrent Pos:* Res assoc, Cornell Univ, 76-77, Univ Hawaii, 81-82, 86; premed adv, Univ Wis-Madison, 72-80, 88, vis prof, 79. *Mem:* Sigma Xi; AAAS; Am Chem Soc. *Res:* Organic reaction mechanisms; medium ring chemistry; chemistry of marine natural products; synthetic organic and natural products chemistry. *Mailing Add:* Dept Chem Hamilton Col Clinton NY 13323

KINNEN, EDWIN, b Buffalo, NY, Mar 9, 25; m 52; c 4. ELECTRICAL ENGINEERING. *Educ:* Univ Buffalo, BS, 49; Yale Univ, ME, 50; Purdue Univ, PhD(elec eng), 58. *Prof Exp:* Res engr, Res Lab, Westinghouse Elec Corp, 50-55; asst prof elec eng, Purdue Univ, 58-59 & Univ Minn, 59-63; assoc prof, 63-77, PROF ELEC ENG, UNIV ROCHESTER, 77-, CHAIR, 89- *Concurrent Pos:* Consult, Minneapolis-Honeywell, Wash Sci, Control Data Corp, Eastman Kodak; NIH spec fel, Westinghouse fel, Sci Res fel, Neth. *Mem:* Inst Elec & Electronics Engrs; fel Japan Soc Prom Sci; Sigma Xi; Rehab Eng Soc NAm. *Res:* Computer aided design for integrated circuits; pediatric orthoses; dynamics of blood flow; computer aided design for floorplanning, placement and routing of custom integrated circuits, based on methods of analytic optimizations; design and development of lower body orthoses for paraplegic children. *Mailing Add:* Dept Elec Eng Univ Rochester Rochester NY 14627

KINNERSLEY, WILLIAM MORRIS, b Baltimore, Md, Jan 3, 44; m 66; c 3. THEORETICAL PHYSICS. *Educ:* Rensselaer Polytech Inst, BS, 64; Calif Inst Technol, PhD(theoret physics), 69. *Prof Exp:* Fac assoc physics, Univ Tex, Austin, 68-70; Nat Acad Sci resident res assoc, Wright Patterson AFB, Ohio, 70-71; asst prof, 71-75, ASSOC PROF PHYSICS, MONT STATE UNIV, 75- *Concurrent Pos:* Vis asst prof appl math, Calif Inst Technol, 73-74; vis assoc prof comput sci, Wash State Univ, 87-88. *Res:* General relativity and gravitational radiation theory and computational physics. *Mailing Add:* Dept Physics Mont State Univ Bozeman MT 59717

KINNEY, ANTHONY JOHN, b Ilkeston, Derbyshire, UK, Sept 9, 58; m 89. PROTEIN PURIFICATION, LIPIDOLOGY. *Educ:* Sussex Univ, UK, BSc, 80; Oxford Univ, UK, DPhil, 85. *Prof Exp:* Res assoc bot, La State Univ, 83-87; res fel biochem & food sci, Rutgers Univ, 87-89; RES BIOCHEMIST BIOCHEM & BIOTECHNOL, DUPONT EXP STA, 89- *Concurrent Pos:* Res student, Agr & Food Res Coun, Letcombe Lab, 80-83; mem, Univ Col, Oxford, 80-83. *Mem:* Am Soc Biochem & Molecular Biol; Biochem Soc; NY Acad Sci; Sigma Xi. *Res:* Biochemistry, molecular biology and genetics of membrane and storage lipid metabolism in plants and yeast; genetic engineering of crop plants. *Mailing Add:* DuPont Exp Sta PO Box 80402 Wilmington DE 19880-0402

KINNEY, DOUGLAS MERRILL, b Los Angeles, Calif, Feb 24, 17; m 42; c 3. GEOLOGY. *Educ:* Occidental Col, BA, 37; Yale Univ, MS, 42, PhD(geol), 51. *Prof Exp:* Geologist, Union Oil Co, Calif, 37-40; asst, Yale Univ, 40-42; geologist, US Geol Surv, 42-56, geol map ed, 56-80; PRES, GEOL SURV ASSOC, INC, 80- *Concurrent Pos:* Vpres, NAm Comn Geol Map of World, 66-80. *Mem:* Geol Soc Am; Am Asn Petrol Geol; Asn Earth Sci Ed. *Res:* Geologic map editing, cartography and printing of colored geologic maps; geologic mapping standards and symbols. *Mailing Add:* Geol Surv Assoc Inc 5221 Baltimore Ave Bethesda MD 20816

KINNEY, EDWARD COYLE, JR, b Massillon, Ohio, Sept 27, 17; m 42. FISH BIOLOGY. *Educ:* Ohio State Univ, BSc, 41 & 46, MS, 48, PhD(hydrobiol), 54. *Prof Exp:* Asst to dir F T Stone Inst Hydrobiol, Ohio State Univ, 48-52; fisheries biologist, State Game & Fish Comn, Ga, 54-56; asst chief fish mgt, WVa Conserv Comn, 56-57, chief, 57-62; staff specialist, Bur Sport Fisheries & Wildlife, US Fish & Wildlife Serv, 62-68, chief br fishery mgt, 68-69 & br coop fishery units, 69-74, Great Lakes coordr, 74-77, chief, fishery prog staff, 78; RETIRED. *Mem:* Am Fisheries Soc; Sigma Xi. *Res:* Life history studies of fresh water fishes; sampling methods for sampling fish populations. *Mailing Add:* 807 17th St SW Massillon OH 44647-7401

KINNEY, GILBERT FORD, b Judsonia, Ark, Dec 29, 07; m 34; c 2. PHYSICAL CHEMISTRY, EXPLOSIONS. *Educ:* Ark Col, AB, 28; Univ Tenn, MS, 30; NY Univ, PhD(phys chem), 35. *Prof Exp:* Radio engr, Radio Sta WNBZ, NY, 30-32; res chemist, Titanium Pigment Co, 35; instr chem, Pratt Inst, 35-39, head instr, 39-42; radiologist, US Navy Bikini, 46; assoc prof thermodyn & phys chem, 46-50, prof chem eng, 50-71, chmn dept mat sci & chem, 60-69, EMER PROF CHEM ENG, NAVAL POSTGRAD SCH, 71- *Concurrent Pos:* Consult, Naval Weapons Ctr, China Lake, Calif & Anamet Labs, Berkeley, Calif, 71- *Mem:* Am Chem Soc; Am Soc Eng Educ. *Res:* Explosive shocks; electroplating; chemical engineering thermodynamics; applications of thermodynamics to chemical equilibria problems; plastics. *Mailing Add:* 1116 Sylvan Rd Monterey CA 93940

KINNEY, JOHN JAMES, b Dansville, NY, Aug 2, 32; m 62; c 1. SYMMETRIC FUNCTIONS IN STATISTICS. *Educ:* St Lawrence Univ, BS, 54; Harvard Univ, AMT, 56; Univ Mich, MS, 59; Iowa State Univ, PhD(statist), 71. *Prof Exp:* Instr, 55-58, asst prof math, St Lawrence Univ, 60-64; assoc prof, State Univ NY, Oneonta, 64-68; asst prof, Univ Nebr, 71-74; assoc prof, 74-79, PROF MATH, ROSE-HULMAN INST TECHNOL, 79- *Concurrent Pos:* Dupont fel, Harvard Univ, 54, Am Asn Qual Control, 58, NSF Sci Fac, 68; dir, Ind Quant Literacy Proj, 85-; chair,

Joint Comt on Curric in Probability & Statist, Am Statist Asn-Nat Coun Teachers Math, 89- *Mem:* Am Statist Asn; Sigma Xi; Soc Indust & Appl Math; Nat Counc Teachers Math. *Res:* Multivariate polykays; simulation in probability theory. *Mailing Add:* 221 Highland Rd Terre Haute IN 47802

KINNEY, JOHN MARTIN, b Evanston, Ill, May 24, 21; m 44; c 3. SURGERY. *Educ:* Denison Univ, AB, 43; Harvard Univ, MD, 46; Am Bd Surg, dipl; Am Bd Nutrit, dipl. *Prof Exp:* Surg intern, Peter Bent Brigham Hosp, 46-47; AEC-Nat Res Coun fel, Med Sch, Univ Colo, 49-52; from asst resident surgeon to chief resident surgeon, Peter Bent Brigham Hosp, 52-57, jr assoc surgeon, 58-63; assoc prof, 63-67, PROF SURG, COL PHYSICIANS & SURGEONS, COLUMBIA UNIV, 67-; DIR SURG METAB, PRESBY HOSP, 63- *Concurrent Pos:* Mead Johnson scholar, Am Col Surgeons, 56-59; Henry E Warren fel surg, Harvard Med Sch, 58-60 & 62-63; assoc attend surgeon, Presby Hosp, 63-67, attend surgeon, 67-; New York Health Res Coun career scientist award, 65; mem adv comt metabolism in trauma, US Army Med Res & Develop Command, 67; mem adv panel to comt on interplay of eng with biol & med, Nat Acad Eng, 68; chmn comt on shock, Comn Emergency Med Serv, Nat Res Coun, 69. *Mem:* Am Burn Asn; Am Asn Surg Trauma; Soc Univ Surg; fel Am Col Surgeons; Am Surg Asn. *Res:* Metabolic response to injury, burns, shock and peritonitis; gas exchange; calorimetry; energy balance; intensive care; patient monitoring; surgical nutrition. *Mailing Add:* Columbia Presby Med Ctr 630 W 168th St New York NY 10032

KINNEY, LARRY LEE, b Salem, Iowa, Oct 26, 41; m 62; c 2. ELECTRICAL ENGINEERING. *Educ:* Univ Iowa, BS, 64, MS, 65, PhD(elec eng), 68. *Prof Exp:* Asst prof elec eng, Univ Iowa, 68; PROF ELEC ENG, UNIV MINN, MINNEAPOLIS, 68- *Mem:* Inst Elec & Electronics Engrs; Asn Comput Mach. *Res:* Switching theory; computer systems. *Mailing Add:* Dept Elec Eng Univ Minn Minneapolis MN 55455

KINNEY, MICHAEL J, b Chicago, Ill, July 9, 37; m 71; c 6. INTERNAL MEDICINE, NEPHROLOGY. *Educ:* Univ Chicago, BS, 59, MD, 63. *Prof Exp:* Assoc chief, Renal Div Nephrol, US Pub Health Serv Hosp, 69-78; asst prof, State Univ NY Downstate Med Ctr, 77-78; assoc prof med & chief nephrology sect, Sch Med, Marshall Univ, 78-80; med dir, Nephrol Res & Educ Found, 80-82; MED DIR, FLA MED CARE CLIN, 89- *Mem:* Fel Am Col Physicians; fel Am Col Clin Pharmacol; Am Soc Nephrology; Soc Exp Biol & Med; Am Physiol Soc. *Res:* Hypertension; nuclear medicine. *Mailing Add:* Fla Med Care Clin 420 S Tamiami Tr Suite 302 Venice FL 34285

KINNEY, RALPH A, b Frostproof, Fla, Nov 7, 30; m 60; c 3. ELECTRICAL ENGINEERING. *Educ:* Univ Fla, BEE, 56, MSE, 58, PhD(elec eng), 67. *Prof Exp:* Res asst elec eng, Univ Fla, 57-60; scientist, Northrop Space Labs, Calif, 60-62; asst res elec eng & aerospace, Univ Fla, 64-67; assoc prof elec eng, 67-76, PROF ELEC ENG, LA STATE UNIV, BATON ROUGE, 76- *Mem:* Inst Elec & Electronics Engrs. *Res:* Wave phenomena in homogeneous media; induction heating; computer applications; field theory. *Mailing Add:* Dept Elec Eng La State Univ Baton Rouge LA 70803

KINNEY, ROBERT BRUCE, b Joplin, Mo, July 20, 37; m 61; c 3. MECHANICAL ENGINEERING. *Educ:* Univ Calif, Berkeley, BS, 59, MS, 61; Univ Minn, Minneapolis, PhD(mech eng), 65. *Prof Exp:* Sr res engr, United Aircraft Res Labs, 65-68; assoc prof, Univ Ariz, 68-78, assoc head, 81-84, prof aerospace & mech eng, 78-87. *Concurrent Pos:* Assoc tech ed, J Heat Transfer, Am Soc Mech Engrs, 73-76; Alexander von Humboldt Found vis scientist, Ger, 76; vis prof, US Mil Acad, West Point, NY, 84. *Mem:* AAAS; Am Inst Aeronaut & Astronaut; Am Soc Mech Engrs. *Res:* Energy transport in gases and liquids; dynamics of fluid flow, including unsteady viscous aerodynamics; fluid flow analogies and experimental methods. *Mailing Add:* 456 22nd Ave SE St Petersburg FL 33705

KINNEY, TERRY B, JR, b Norfolk, Mass, Sept 12, 25; m 46; c 2. POPULATION GENETICS. *Educ:* Univ Mass, BS, 55, MS, 56; Univ Minn, PhD, 63. *Prof Exp:* Res asst poultry, Univ Mass, 55-56; geneticist, Hubbard Farms Inc, NH, 56-57; instr poultry, Univ Minn, 57-62; biometrician, USDA, Md, 63-65, res geneticist, Ind, 65-69; asst dir, Animal Sci Res Div, 69-72, assoc dep adminr, NCent Region, 72-74, asst adminr livestock vet sci, 74-78, assoc adminr, 78-80, ADMINR, SCI & EDUC ADMIN-AGR RES, USDA, 80- *Mem:* Poultry Sci Asn; Sigma Xi. *Res:* Statistics; population genetic studies; administration of research relating to livestock and veterinary sciences. *Mailing Add:* US Dept Agr Rm 302 Admin Bldg Washington DC 20250

KINNIE, IRVIN GRAY, b Orlando, Fla, Apr 28, 32; m 68. COMPUTER SYSTEM DESIGN. *Educ:* US Mil Acad, BS, 53; Univ Ariz, MS, 60. *Prof Exp:* Res officer, E W Div, Army Proving Ground, Ariz, 56-58, area signal officer, Longlines Signal Battalion Pusan, Korea, 60-61, asst prof eng & info sci, US Mil Acad, 61-65, EDP specialist info sci, Allied Mil Commun Electronics Agency, Paris, 65-67, standards specialist, Mallard Proj, Ft Monmouth, NJ, 67-68, chief plans & opers, US Army Regional Commun Group, Saigon, 68-69, chief info sci, US Army Res & Develop Group, London, 69-73, dep dir res & develop, US Army Comput Systs Command, 73-74; syst analyst & engr tech planning, 74-76, engr software res, 76-77, engr mkt develop, 77-78, engr solar eng, 78-80, engr command, control & commun archit, 80-81, engr, computer systs design, command & space systs, Fed Systs Div, 81-82, nationwide network design, Entry Systs Div, 82-86, computer architect, 86-88, COMPUTER ARCHITECT, PC TECHNOL FORECASTING, IBM CORP, 88- *Mem:* Asn Comput Mach; Inst Elec & Electronics Engrs; Armed Forces Comn Electronics Asn. *Res:* Communications; electronics and information sciences; software engineering; computer systems architecture. *Mailing Add:* 540 11th Ave Boca Raton FL 33486-3461

KINNISON, GERALD LEE, b San Diego, Calif, July 16, 31; m 50, 89; c 4. PHYSIOLOGY. *Educ:* Univ Calif, Los Angeles, BS, 58, MS, 60, PhD(physics), 63. *Prof Exp:* Postdoctoral fel physics, Univ London, 63-64; res physicist, Electronics Lab, 63-67, Undersea Ctr, 67-77, RES PHYSICIST, OCEAN SYSTS CTR, USN, 77- *Concurrent Pos:* Sonar adv panel, Navsea Explor Develop Panel, Passive Sonar, 67-86; master ASW study group, Chief Naval Opers, 84- *Honors & Awards:* Arthur S Fleming Award, 68. *Mem:* Acoust Soc Am; Am Physics Teachers Soc. *Res:* Passive sonar arrays: design, instrumentation, beamforming (conventional and optimal), signal processing, display and performance prediction in various environmental noises (predicted and measured); recipient of one patent. *Mailing Add:* 3274 Trumbull St San Diego CA 92106-2421

KINNISON, ROBERT RAY, b Los Angeles, Calif, Sept 10, 34; m 59; c 2. EXTREME VALUE STATISTICS, QUALITY ASSURANCE. *Educ:* Pomona Col, BA, 56; Univ Calif Los Angeles, PhD(statist), 71. *Prof Exp:* Pharmacologist, Rexall Drug & Chem Co, 60-68; Statistician, Univ Calif, San Francisco, 71-72; statistician, US Environ Protection Agency, Las Vegas, 72-79; statistician, Battelle Pac Northwest Lab, 79-85; statistician, Desert Res Inst, 85-89; PRIN STATISTICIAN, REYNOLDS ELEC & ENG CO, LAS VEGAS, NEV, 89- *Concurrent Pos:* Assoc ed, J Simulation, 72- *Mem:* Am Statist Asn; Biomet Soc; AAAS; Sigma Xi. *Res:* Extreme value statistics; bioassay statistics; exposure-dose assessment statistics for environmental pollutants; spatial and geographic statistics; quality assurance statistics. *Mailing Add:* 846 E Pescados Dr Las Vegas NV 89123-1359

KINNMARK, INGEMAR PER ERLAND, b Sundbyberg, Sweden, Dec 8, 53; m 82; c 1. NUMERICAL METHODS OF DIFFERENTIAL, SCIENTIFIC & TECHNICAL TRASPORTATION. *Educ:* Royal Inst Technol, MS, 79; Princeton Univ, MA, 82 & PhD(water resources), 84. *Prof Exp:* Res assoc, 84-86, asst prof hydraul & numerical methods, Univ Notre Dame, 86-89; SWED SCI & TECH TRANSL, 89- *Mem:* Am Translr Asn. *Res:* Mathematical modeling of flow in shallow seas, estuaries and rivers; design, analysis and application of numerical methods for the solution of ordinary and partial differential equations. *Mailing Add:* 1860 Sherman Ave Apt 2-NE Evanston IL 60201-3732

KINO, GORDON STANLEY, b Melbourne, Australia, June 15, 28; nat US; m 57; c 1. ELECTRICAL ENGINEERING, ACOUSTICS. *Educ:* Univ London, BSc, 48, MSc, 50; Stanford Univ, PhD(elec eng), 55. *Prof Exp:* Jr scientist, Mullard Radio Valve Co, Eng, 47-51; res asst, Electronics Res Lab, Stanford Univ, 51-55, res assoc, Microwave Lab, 55; mem tech staff, Bell Tel Labs, NJ, 55-57; res assoc, Stanford Univ, 57-61, assoc prof, 61-65, assoc chmn, Elec Eng Dept, 85-88, PROF ELEC ENG, STANFORD UNIV, 65-, PROF APPL PHYSICS, 76-, ASSOC DEAN PLANNING & FACIL, 87- *Concurrent Pos:* Consult, Varian, Tex Instruments, Lockheed Aircraft Corp & Advan Res Projs Agency, 57-, Endosonics & Prometrix; Guggenheim fel, 67-68; chmn, Ultrasonics Group, Inst Elec & Electronics Engrs. *Honors & Awards:* Centennial Medal, Inst Elec & Electronics Engrs, 84. *Mem:* Nat Acad Eng; fel Inst Elec & Electronics Engrs; fel Am Phys Soc; fel AAAS; Am Optical Soc. *Res:* Electromagnetic theory; design of electron and ion guns; wave propagation in plasmas; microwave tubes; microwave acoustics; acoustic imaging; non-destructive testing; waves in solids; author of more than 400 technical publications. *Mailing Add:* Ginzton Lab Mail Code 4085 Stanford Univ Stanford CA 94305

KINOSHITA, FLORENCE KEIKO, b Salem, Ore, Aug 6, 41. TOXICOLOGY, PHARMACOLOGY. *Educ:* Univ Chicago, BS, 63, MS, 66, PhD(pharmacol), 69; Am Bd Toxicol, dipl. *Prof Exp:* From instr to asst prof pharmacol, Univ Chicago, 69-73; toxicologist, Indust Bio-Test Labs, Inc, 73-74, tech mgr toxicol, 74-77; SR TOXICOLOGIST, HERCULES, INC, 78- *Concurrent Pos:* Consult, US Environ Protection Agency, 71-72, US Fed Drug Admin, 72-73; mem, Toxicol Study Sect, NIH, 79-83 & Toxicol Data Bank Peer Rev Comt, Nat Library Med, 83-85; affil asst prof, Dept Pharmacol & Toxicol, Med Col Va, 84- *Mem:* Soc Toxicol; Am Indust Hyg Asn; Soc Exp Biol & Med; NY Acad Sci; Int Soc Ecotoxicol & Environ Safety; Int Soc Study Xenobiotics. *Res:* Interactions of drugs, pesticides and chemicals as inducers of hepatic microsomal enzyme systems; effects of organophosphorus compounds on cholinestrase and aliesterases; development of hepatic microsomal enzymes in fetal and neonatal animals. *Mailing Add:* Med Dept Hercules Inc 1313 N Market St Wilmington DE 19894

KINOSHITA, JIN HAROLD, b San Francisco, Calif, July 21, 22; m 48. BIOLOGICAL CHEMISTRY. *Educ:* Columbia Univ, AB, 44; Harvard Univ, PhD, 52. *Hon Degrees:* ScD, Bard Col, 67. *Prof Exp:* Asst chem, Bard Col, 44-46; from instr to asst prof biochem, Harvard Med Sch, 52-64, from assoc prof to prof biochem ophthal, 64-73; chief, Lab Vision Res, Nat Eye Inst, 71-81, sci dir, 81-90; CLIN PROF OPHTHAL, UNIV CALIF, DAVIS, 90- *Concurrent Pos:* Biochemist, Mass Eye & Ear Infirmary, 55-73; Friedenwald mem lectr, 65; mem visual sci study sect, NIH, 65-69. *Honors & Awards:* Proctor Medal, Asn Res Vision & Ophthal, 74. *Mem:* AAAS; Am Chem Soc; Am Soc Biol Chemists; Asn Res Vision & Ophthal. *Res:* Chemistry and metabolism of ocular tissues. *Mailing Add:* 44269 Clubhouse Dr El Macero CA 95618

KINOSHITA, KAY, b Princeton, NJ, July 17, 54. PHYSICS. *Educ:* Harvard Univ, AB & AM, 76; Univ Calif, Berkeley, PhD(physics), 82. *Prof Exp:* Res assoc, 82-84, asst prof, 84-88, ASSOC PROF PHYSICS, HARVARD UNIV, 88- *Concurrent Pos:* Sci scholar, Mary Ingraham Bunting Inst, 85-87. *Mem:* Am Phys Soc. *Res:* Elementary particle physics; weak interactions of heavy quarks; exotic heavily ionizing particles. *Mailing Add:* 42 Oxford St Cambridge MA 02138

KINOSHITA, KIMIO, b Vancouver, BC, Aug 5, 42; m 65; c 2. ELECTROCHEMISTRY. *Educ:* Univ Alta, BSc, 64; Univ Calif, Berkeley, PhD(chem), 69. *Prof Exp:* Sr res assoc phys chem, Mat Eng Res Lab, Pratt & Whitney Aircraft, 69-76; mem staff, Chem Eng Div, Argonne Nat Lab, 76-79; MEM STAFF, SRI INT, 79- *Mem:* Am Chem Soc; Electrochem Soc; Am Carbon Soc. *Res:* Corrosion; electrochemistry; carbon chemistry; catalysis. *Mailing Add:* 20644 Nancy Ct Cupertino CA 95014

KINOSHITA, SHIN'ICHI, b Osaka, Japan, June 5, 25; m 54; c 4. TOPOLOGY. *Educ:* Osaka Univ, BS, 48, PhD(math), 58. *Prof Exp:* Lectr math, North Col, Osaka Univ, 58-59; vis mem, Inst Advan Study, 59-61; res assoc, Princeton Univ, 61-62; asst prof, Univ Sask, 62-64; from assoc prof to prof, Fla State Univ, 64-84; PROF, FAC SCI, KWANSEI GAKUIN UNIV, JAPAN. *Mem:* Am Math Soc; Math Soc Japan; Sigma Xi. *Res:* Topological transformations; knot theory and its applications, fundamental group phenomena. *Mailing Add:* Kwansei Gakuin Univ Nishinomiya 662 Japan

KINOSHITA, TOICHIRO, b Tokyo, Japan, Jan 23, 25; nat US; m 51; c 3. THEORETICAL HIGH ENERGY PHYSICS. *Educ:* Univ Tokyo, BS, 47, PhD(physics), 52. *Prof Exp:* Mem, Inst Adv Study, Princeton, NJ, 52-54; fel theoret physics, Columbia Univ, 54-55; res assoc, 55-58, from asst prof to assoc prof, 58-64, PROF THEORET PHYSICS, CORNELL UNIV, 64- *Concurrent Pos:* Ford fel, European Orgn Nuclear Res, Geneva Switz, 62-63; Guggenheim Found fel, 73-74; tech adv panel univ progs, Dept Energy, 82-83; comt fundamental constants, Nat Res Coun, 84-86. *Honors & Awards:* J J Sakurai Prize, Am Phys Soc, 90. *Mem:* Nat Acad Sci; fel AAAS; fel Am Phys Soc. *Res:* Quantum field theory; quantum theory of atoms; elementary particles; symmetry law. *Mailing Add:* Lab Nuclear Studies Cornell Univ Ithaca NY 14853-5001

KINRA, VIKRAM KUMAR, b Lyallpur, India, Apr 3, 46; m 76, 88; c 1. ENGINEERING MECHANICS, MATERIALS SCIENCE. *Educ:* Indian Inst Technol, Kanpur, BTech, 67; Utah State Univ, MSc, 68; Brown Univ, PhD(eng mech), 75. *Prof Exp:* Struct eng stress anal, Northrop Corp, Hawthorne, 68-70; proj engr mech eng, Ostgaard & Assocs, Inc, Gardena, 70-71; asst & res assoc, Brown Univ, 71-75; asst prof, Univ Colo, Boulder, 75-82; assoc prof, 82-89, PROF AEROSPACE ENG & ASSOC DIR, CTR MECH COMPOSITES, TEX A&M UNIV, 89- *Concurrent Pos:* Prin investr & proj dir, NSF grants, Univ Colo, Boulder, 76-81, Off naval Res, Air Force Off Sci Res, Tex adv tech prog grants, NASA grants, IBM grant & Martin-Marietta grant; consult, Willow Water Dist, Denver, 76-77, Ponderosa Asn, Louisville, 80 & Corning Glass, 80 & Martin-Marietta Corp, 88; Halliburton prof, 86. *Honors & Awards:* Dow Outstanding Young Fac Award, 80; Ralph R Teeter Award, 82. *Mem:* Soc Exp Mech; Am Acad Mech; Am Soc Eng Educ; Am Soc Mech Engrs; Sigma Xi; Acoust Soc Am. *Res:* Damping; wave propagation; nondestructive testing and evaluation; composite materials; ultrasonics. *Mailing Add:* Dept Aerospace Eng Tex A&M Univ College Station TX 77843

KINSBOURNE, MARCEL, b Vienna, Austria, Nov 3, 31; m 65; c 4. PEDIATRIC NEUROLOGY, EXPERIMENTAL PSYCHOLOGY. *Educ:* Oxford Univ, BA, 52, MD, 55, MA, 56, DM(neuropsychol), 63. *Prof Exp:* Lectr psychol, Oxford Univ, 64-67; assoc prof pediat & neurol & lectr psychol, Med Ctr, Duke Univ, 67-74; sr staff physician, Hosp Sick Children, Toronto, 74-80; prof psychol, Univ Waterloo, 74-79; prof pediat, Univ Toronto, 74-80, prof psychol, 75-80; DIR, BEHAV NEUROL DEPT, EUNICE KENNEDY SHRIVER CTR, 80-; LECTR NEUROL, HAVARD MED SCH, 80- *Concurrent Pos:* Fel, New Col, Oxford Univ, 65-67; J Arthur Lectureship evol brain, 74; adj prof cognitive sci, Brandeis Univ, 83- *Honors & Awards:* Queen Square Prize neurol, 61. *Mem:* Fel Am Psychol Asn; Am Neurol Asn; fel Geront Soc; Int Neuropsychol Soc; Psychonomic Soc. *Res:* Human neuropsychology; developmental psychology; visual information processing; age-related changes in behavior. *Mailing Add:* 158 Cambridge St Winchester MA 01890

KINSEL, NORMA ANN, b Boston, Mass, Feb 26, 29. MICROBIOLOGY. *Educ:* Univ Va, BS, 49; Pittsburgh Univ, MS, 52, PhD(bact), 59. *Prof Exp:* Asst fel yeast chem, Mellon Inst, 52-53, asst, microbiol & micros sect, 53-56, assoc microbiologist, 56-62, fel petrol, 63-69; res biologist, Gulf Res & Develop Co, 69-73; SR MICROBIOLOGIST, ELI LILLY & CO, 74- *Concurrent Pos:* USPHS res fel, Rutgers Univ, 62-63. *Mem:* AAAS; Am Soc Microbiol; Soc Indust Microbiol. *Res:* Autotrophic iron and sulfur bacteria; petroleum microbiology; antibiotic fermentation technology. *Mailing Add:* Fermentation Res & Dev Eli Lilly & Co Lilly Res Labs Indianapolis IN 46285

KINSELLA, JOHN EDWARD, b Wexford, Ireland, Feb 22, 38; m 65; c 4. FOOD CHEMISTRY, BIOCHEMISTRY. *Educ:* Nat Univ Ireland, BS, 61; Pa State Univ, MS, 65, PhD(food biochem sci), 67. *Prof Exp:* Teacher zool, Latin & chem, CKC Onitsha, Nigeria, 61-63; from asst prof to assoc prof food sci, Cornell Univ, 67-77, prof Food Sci/Chem & chmn Dept Food Sci, 77-90, Dir, Inst Food Sci, 80-90, Liberty Hyde Bailey Prof Food Sci, 81-90, Gen Foods Distinguished Prof Food Sci, 84-90, chmn Dept Gen Foods, 84-90; DEAN, COL AGR & ENVIRON SCI, UNIV CALIF, DAVIS, 90- *Concurrent Pos:* General Foods Endowed Chair, 84. *Honors & Awards:* Outstanding Res Food Chem & Agr, Am Chem Soc, 90; Spencer Award, Am Chem Soc, 91. *Mem:* Am Inst Nutrit; AAAS; Am Chem Soc; Am Dairy Sci Asn; fel Inst Food Technol; Am Oil Chem Soc. *Res:* Essential fatty acids and prostaglandins; flavor chemistry of foods; lipid enzymology and metabolism in tissue cultures; nutrients in foods; functional properties of food proteins; protein modification; food protein from oilseeds and yeast. *Mailing Add:* Deans Off 200 Mark Hall Univ Calif Davis Davis CA 95616

KINSELLA, JOHN J, b Seneca Falls, NY, Oct 15, 26; m 49; c 8. PHYSICS. *Educ:* Univ Tex, BA, 49; Syracuse Univ, MS, 53. *Prof Exp:* Res physicist, Consol Vacuum Corp Div, Bell & Howell, 51-56; from physicist to sr physicist, Xerox Corp, 56-59, scientist, 59-62, res mgr, 62-65, mgr photoreceptor technol, 65-68, mgr xerographic consumables mfg, 68-70, mgr advan mfg eng, 70-73, mfg prog mgt, 73-78, mem res & eng tech staff, Corp Planning Consumables & Supplies, 78-81; CONSULT, 81- *Mem:* Am Vacuum Soc; Inst Elec & Electronics Engrs; Optical Soc Am; Soc Photog Sci & Eng. *Res:* Vacuum pump and gauge design development; thin film microcircuit, resistors and capacitors; evaporated silver bromide research; xerographic photoconductors. *Mailing Add:* 2846 St Paul Blvd Rochester NY 14617

KINSELLA, RALPH A, JR, b St Louis, Mo, June 4, 19; m; c 8. INTERNAL MEDICINE. *Educ:* St Louis Univ, AB, 39, MD, 43. *Prof Exp:* From instr to assoc prof, 48-70, PROF INTERNAL MED, SCH MED, ST LOUIS UNIV, 70-; physician & chief unit 2, Med Serv, St Louis City Hosp, 57-80, med dir, 80-85; MED DIR, ST LOUIS UNIV HOSP, 85- *Concurrent Pos:* Neilson fel, St Louis Univ, 47-48, Markle scholar, 48-53. *Mem:* AAAS; Endocrine Soc; NY Acad Sci; fel Am Col Physicians; Soc Exp Biol & Med; Sigma Xi. *Res:* Steroid hormone metabolism; endocrinology. *Mailing Add:* City Hosp 1515 Lafayette St Louis MO 63104

KINSER, DONALD LEROY, b Loudon, Tenn, Sept 28, 41; m 61; c 2. MATERIALS SCIENCE, ENGINEERING. *Educ:* Univ Fla, BS, 64, PhD(mat sci), 68. *Prof Exp:* From asst prof to assoc prof ceramic eng, 68-75, PROF MAT SCI, VANDERBILT UNIV, 75- *Concurrent Pos:* Co-ed, J Non Crystalline Solids, 88- *Mem:* Am Ceramic Soc; Am Soc Metals; Am Inst Mining, Metall & Petrol Engrs; Electrochem Soc; Soc Glass Technol UK. *Res:* Electrical behavior of two-phase alkali-silicate glasses; semiconducting glasses; mechanical properties of glasses; electrical behavior of glasses; radiation damage and mechanical properties of materials. *Mailing Add:* Dept Mat Eng Box 1689B Vanderbilt Univ Nashville TN 37235

KINSEY, BERNARD BRUNO, b London, Eng, June 15, 10; m; c 3. PHYSICS. *Educ:* Cambridge Univ, BA, 32, PhD, 37. *Prof Exp:* Lectr physics, Univ Liverpool, 36-39; with Telecommun Res Estab, UK, 39-42; res physicist, Atomic Energy Can, 44-54; guest scientist, Univ Calif, 54-55; res physicist, Atomic Energy Res Estab, Eng, 55-58; PROF PHYSICS, UNIV TEX, AUSTIN, 58- *Mem:* Fel Am Phys Soc; fel Royal Soc Can. *Res:* Nuclear physics. *Mailing Add:* Dept Physics RLM Bldg Univ Tex Austin TX 78712

KINSEY, DAVID WEBSTER, b Warsaw, Ind, Mar 11, 39; m 61; c 2. MATHEMATICS. *Educ:* Manchester Col, BA, 61; Univ Ariz, MS, 64; Ind Univ, PhD(math), 72. *Prof Exp:* Instr math, Carthage Col, 63-64 & Millikin Univ, 64-68; ASSOC PROF MATH, UNIV SOUTHERN IND, EVANSVILLE, 72- *Mem:* Math Asn Am; Nat Coun Teachers Math; Am Math Soc. *Mailing Add:* Univ Southern Ind 8600 University Blvd Evansville IN 47712

KINSEY, JAMES HUMPHREYS, b Xenia, Ohio, June 6, 32. APPLIED PHYSICS. *Educ:* Princeton Univ, AB, 60; Univ Md, College Park, PhD(physics), 70. *Prof Exp:* Aerospace technologist, NASA, 68-71; mem res staff x-ray diffraction, Princeton Univ, 71-73, lectr, 73-74; consult, Control Data Corp, 74-76; assoc consult, Mayo Clinic, 76-82; asst prof biophys, 79-82, advan projs physicist, Space Telescope Sci Inst, 82-88; SR SCIENTIST, APPL RES CORP, 88- *Mem:* Am Phys Soc; Am Astron Soc; Am Geophys Union; Sigma Xi; Inst Elec & Electronics Engrs; Biomed Eng Soc; Optical Soc Am. *Res:* Optical aurora; low energy cosmic rays and energetic solar particles; x-ray diffraction studies of large biological molecules; computer analysis of radiographs; x-ray and optical detector systems for biomedical imaging, computer tomography; photo therapy of cancer; imaging instrumentation. *Mailing Add:* Appl Res Corp 8201 Corporate Dr Landover MD 20785

KINSEY, JAMES LLOYD, b Paris, Tex, Oct 15, 34; m 62; c 3. PHYSICAL CHEMISTRY. *Educ:* Rice Univ, BA, 56, PhD(chem), 59. *Prof Exp:* NSF fel, Univ Uppsala, 59-60; Miller res fel, Univ Calif, Berkeley, 60-62; from asst prof to assoc prof, 62-74, chmn dept, 77-82, prof chem, Mass Inst Technol, 74-84; DEAN NAT SCI & D R BULLARD-WELCH FOUND PROF SCI, RICE UNIV, 88- *Concurrent Pos:* Alfred P Sloan res fel, 63-67; Guggenheim fel, 69-70; vis assoc prof, Univ Wis, 69-70; consult, Los Alamos Nat Lab, 74-; mem, bd chem sci & technol, Nat Acad Sci-Nat Res Coun, 80-83, co-chmn, 81-83; mem, adv comt, Army Res Off-Nat Res Coun, 81-86; assoc ed, J Chem Physics, 81-84; chmn, Am Chem Soc, Div Physics & Chemistry, 85, mem, Am Phys Soc, Div Chem, Phys Exec Comn, 86-89. *Honors & Awards:* E O Lawrence Award, US Dept Energy, 87. *Mem:* Nat Acad Sci; Am Chem Soc; fel Am Acad Arts & Scis; AAAS. *Res:* Molecular beams; intermolecular forces; chemical kinetics; scattering theory; spectroscopy; lasers. *Mailing Add:* Sch Nat Sci Rice Univ PO Box 1892 Houston TX 77251

KINSEY, JOHN AARON, JR, b Panama City, Fla, Mar 12, 39; m 62. GENETICS. *Educ:* Fla State Univ, BS, 61; Univ Tex, PhD(zool), 65. *Prof Exp:* NIH res fel genetics, Univ Wash, 65-67; asst prof, 67-80, ASSOC PROF MICROBIOL, MED CTR, UNIV KANS, 80- *Mem:* Genetics Soc Am. *Res:* Biochemical aspects of Neurospora genetics. *Mailing Add:* Dept Microbiol Univ Kans Med Ctr 39th & Rainbow Blvd Kansas City KS 66103

KINSEY, KENNETH F, b Providence, RI, Sept 14, 33. PHYSICS. *Educ:* Brown Univ, AB, 55; Univ Rochester, PhD(physics), 61. *Prof Exp:* Asst prof, Univ Rochester, 64-66; PROF PHYSICS, STATE UNIV NY COL GENESEO, 66- *Mem:* Am Phys Soc; Am Asn Physics Teachers. *Res:* Medium energy particle physics. *Mailing Add:* Dept Physics State Univ NY Col Geneseo NY 14454

KINSEY, PHILIP A, b Warsaw, Ind, Feb 10, 31; m 54; c 3. PHYSICAL CHEMISTRY. *Educ:* Manchester Col, AB, 53; Purdue Univ, PhD(chem), 57. *Prof Exp:* PROF CHEM, UNIV EVANSVILLE, 56- *Concurrent Pos:* Sabbatical, Univ Calif, Santa Barbara, 69-70, Univ Ill, Urbana, 80-81. *Mem:* Am Chem Soc. *Res:* Uses of computers in chemical education. *Mailing Add:* Dept Chem Univ Evansville 1800 Lincoln Ave Evansville IN 47722

KINSEY, WILLIAM HENDERSON, REPRODUCTION & FERTILIZATION, CELL BIOLOGY. *Educ:* Univ Wash, Seattle, PhD(anat & cell biol), 76. *Prof Exp:* ASSOC PROF GROSS ANAT & CELL BIOL, SCH MED, UNIV MIAMI, 85- *Mailing Add:* Dept Anat & Cell Biol Univ Miami Sch Med 1600 NW Tenth Ave Miami FL 33101

KINSINGER, JACK BURL, b Akron, Ohio, June 23, 25; div; c 2. PHYSICAL CHEMISTRY. *Educ:* Hiram Col, BA, 48; Cornell Univ, MSc, 51; Univ Pa, PhD(phys chem), 58. *Prof Exp:* Group leader polymer chem, Rohm & Haas Co, 51-56; from asst prof to assoc prof phys chem, Mich State Univ, 57-66, prof chem, 66-75, from assoc chmn to chmn dept chem, 65-75; dir chem div, NSF, 75-77; asst vpres res, Mich State Univ, 77-82, assoc provost, 77-82; vpres acad affairs, Ariz State Univ, 82-87; PRES & CLO, CHICAGO OSTEOP HEALTH SYST, 87- *Concurrent Pos:* Consult, Union Carbide Chem Co, 58-80; mem bd, Kirksville Osteop Col, 86-87 & Ariz State Univ Res Park, 84-87. *Mem:* Fel AAAS; Am Chem Soc; Am Phys Soc. *Res:* Micro-structure of polymers; laser light scattering spectroscopy. *Mailing Add:* Chicago Osteop Health Systems 5200 S Ellis Ave Chicago IL 60615

KINSINGER, JAMES A, b Ottumwa, Iowa, Dec 6, 44; m 66; c 3. ANALYTICAL CHEMISTRY. *Educ:* Wartburg Col, BA, 67; Univ Wis, PhD(anal chem), 72. *Prof Exp:* Asst prof chem, Tougaloo Col, 72-76; res assoc, Univ Wis, 76-77; res scientist, Warf Inst, Raltech Sci Serv & Hazelton Labs, 77-82; head, Appln & Training, Nicolet Instruments, 81-86; vpres mkt, Masstron, Inc, 87; MANAGING DIR, INDUST LABS CO, 88- *Concurrent Pos:* Head, Chem Dept, Tougaloo Col, 74-76. *Mem:* Am Chem Soc; Am Soc Mass Spectrometry; Asn Off Anal Chemists; Inst Food Technologists; Am Oil Chemists Soc; Asn Off Racing Chemists. *Mailing Add:* Indust Labs PO Box 16207 Denver CO 80216

KINSINGER, RICHARD ESTYN, b Wilmington, Del, July 23, 42; m 64; c 2. MEDICAL IMAGING PHYSICS. *Educ:* Cornell Univ, BEngPhys, 64, MEng, 65, PhD(aerospace eng), 69. *Prof Exp:* Res physicist, Res & Develop Ctr, 68-76, mgr arc interruption res, 76-81, mgr plasma technol progs, 81-82; MGR APPL SCI LAB, GEN ELEC MED SYSTS, 82- *Mem:* Inst Elec & Electronics Engrs; Am Phys Soc. *Mailing Add:* W349 S4698 Kingdom Dr Dousman WI 53118

KINSKY, STEPHEN CHARLES, b Berlin, Ger, Feb 9, 32; nat US; m 59; c 2. BIOCHEMISTRY. *Educ:* Univ Chicago, AB, 51; Johns Hopkins Univ, PhD(biochem), 57. *Prof Exp:* From instr to prof, Sch Med, Wash Univ, 59-78; PROF BIOCHEM, NAT JEWISH HOSP, 78- *Concurrent Pos:* USPHS fel, 57-59, res career develop awards, 64-70. *Mem:* Am Soc Biol Chem; Am Asn Immunol. *Res:* Membrane biochemistry; immunology. *Mailing Add:* 1035-1 Jasmine St Denver CO 80220

KINSLAND, GARY LYNN, b Eugene, Ore, June 10, 47; div; c 1. GEOPHYSICS, MINERALOGY. *Educ:* Univ Rochester, BS, 69, MS, 71, PhD(geol), 74. *Prof Exp:* Res asst geol, Univ Rochester, 74-76; vis asst prof, Ariz State Univ, 76-77; asst prof, 77-80, ASSOC PROF GEOL, UNIV SOUTHWESTERN LA, 80- *Concurrent Pos:* Prin investr, Dept Energy geopressure-geothermal energy grant, 78-79, high resolution 3-d seismic surv over geopressured resevoir, 82-85. *Honors & Awards:* A I Leversen Mem Award, Am Asn Petrol Geol, 84. *Mem:* Sigma Xi; Am Geophys Union; Soc Explor Geophysicists. *Res:* High pressure-high temperature simulation and study of earth mantle properties; geopressure-geothermal energy prospects; high pressure materials science; high-resolution 3-d seismic surveys; continental crustal structure deduced from potential field data. *Mailing Add:* Dept Geol PO Box 44530 Lafayette LA 70504

KINSLEY, HOMAN BENJAMIN, JR, b Baltimore, Md, Dec 31, 40; m 62; c 3. WET LAID NONWOVEN ENGINEERING, FIBER PHYSICS. *Educ:* Western Md Col, BA, 63; Lawrence Univ, MS, 64, PhD(chem), 67. *Prof Exp:* Res assoc, Ethyl Corp, 67-74; res chemist, Du Pont, 74-75; res assoc, James River Corp, 75-78, tech dir, 78-82, dir technol, 82-86, sr res fel, 86-91; SR RES FEL, CUSTOM PAPERS GROUP INC, 91- *Mem:* Am Chem Soc; Tech Asn Pulp & Paper Indust; AAAS; Filtration Soc. *Res:* Wet laid nonwovens; forming techniques; fibers and binders; physical properties of the resulting web structures. *Mailing Add:* 3257 Three Bridge Rd Powhatan VA 23139

KINSMAN, DONALD MARKHAM, b Framingham, Mass, May 20, 23; m 49; c 3. ANIMAL SCIENCE, MEAT SCIENCE. *Educ:* Univ Mass, BS, 49; Univ NH, MS, 51; Okla State Univ, PhD, 64. *Prof Exp:* Instr animal sci, Univ NH, 49-51 & Univ Vt, 51-52; farm mgr, Univ Mass, 52-56; from asst prof to prof, 56-88, EMER PROF ANIMAL SCI, UNIV CONN, 88- *Concurrent Pos:* Pres, New Eng Livestock Conserv, Inc, 66-78; Danforth Assoc, 79; dir, Am Soc Animal Sci, 72-74; mem, USDA Inspection Comt, Nat Acad Sci, 89-90. *Honors & Awards:* Distinguished Serv Award, Am Soc Animal Sci, 78. *Mem:* Am Soc Animal Sci; Am Meat Sci Asn (pres, 78-79); Coun Agr Sci & Technol. *Res:* Humane slaughter, reduced stress methods; evaluation and quality of meat and meat products; meat microbiology. *Mailing Add:* Dept Animal Sci Univ Conn Storrs CT 06269-4040

KINSMAN, DONALD VINCENT, b Toledo, Ohio, July 16, 43; m 69; c 2. ORGANIC CHEMISTRY. *Educ:* Univ Cincinnati, BS, 65; Pa State Univ, PhD(chem), 69. *Prof Exp:* Develop chemist household prod, Lever Bros, 69-71, group leader develop, 71-74; group leader oleo chem res, Emery Chem, 74-79, mgr, 79-87; TECH DIR, HENKEL EMERY GROUP, 87- *Mem:* Am Chem Soc; Am Oil Chemists Soc; Soc Automotive Engrs; Tech Asn Pulp & Paper Indust; Soc Tribologists & Lubrication Engrs. *Res:* Develop and administration of applications for oleo chemicals, synthetic lubricants, fatty alcohols, azelaic and pelargonic acids. *Mailing Add:* Henkel Emery Group 4900 Este Ave Cincinnati OH 45232

KINSOLVING, C RICHARD, MEDICINAL CHEMISTRY. *Educ:* Emory Univ, Phd(pharmacol), 70. *Prof Exp:* DIR, DIV BIOL & CHEM SCI DIV, PENWALT CORP, 80- *Mailing Add:* 14 E 63rd St New York NY 10021

KINSON, GORDON A, b Wood End, Eng, Sept 21, 35; m 60, 82; c 2. PHYSIOLOGY, ENDOCRINOLOGY. *Educ:* Univ Aston, BS, 57; Col Advan Technol, Birmingham, ARIC, 61; Univ Ala, Birmingham, MS, 62; Univ Birmingham, PhD(endocrinol), 67. *Hon Degrees:* FRSC (UK), London, 73. *Prof Exp:* Res asst endocrinol, Univ Ala, Birmingham, 59-63; sr res assoc, Med Sch, Birmingham Univ, 64-68; from asst prof to assoc prof, 68-79, PROF PHYSIOL, FAC MED, UNIV OTTAWA, 79- *Concurrent Pos:* Consult, Europ Orgn for Control of Circulatory Dis, 82, Comp Cancer Unit, UAB, USA, 83. *Mem:* Brit Soc Endocrinol; Royal Soc Chem; Can Physiol Soc; Am Physiol Soc. *Res:* Extrahepatic metabolism of androgen hormones; steroid hormone production by seminiferous tubules of the testis; chronic consequences of vasectomy; reninangiotensin and other factors influencing adrenocortical hormone secretion; pineal gland-adrenocortical hormone relationships; role of the pineal gland and indoles in testicular function and androgen biosynthesis; cardiac actions of anabolic androgens and estrogens; exercise and female reproductive function; steroids and Cyclosporin A in renal function. *Mailing Add:* Univ Ottawa Fac Health Sci 451 Smyth Rd Ottawa ON K1H 8M5 Can

KINSTLE, JAMES FRANCIS, b Lima, Ohio, Nov 23, 38. POLYMER CHEMISTRY. *Educ:* Bowling Green State Univ, BS, 66, MA, 67; Univ Akron, PhD(polymer sci), 70. *Prof Exp:* Res & develop polymers, Wyandotte Chem Corp, 60-63; res & develop chemist, Allied Mat Corp, 63-65; sr res scientist, Ford Motor Co Sci Res Labs, 70-72; from asst prof to assoc prof chem, Univ Tenn, Knoxville, 72-83; sr res group leader, Polaroid, 83-87; MGR CORP CHEM & POLYMER RES & DEVELOP, JAMES RIVER CORP, 87- *Concurrent Pos:* Ed-in-chief, J Radiation Curing, Technol Mkt Corp, 74-79; consult, 73-83. *Honors & Awards:* A K Doolittle Award, Am Chem Soc, 77. *Mem:* Am Chem Soc (treas, 85-87). *Res:* Polymer synthesis and characterization; homogeneous and surface reactions on polymers; cellulose and paper chemistry; radiation chemistry in polymer sciences; polymer reuse and disposal. *Mailing Add:* 2585 Millbrook Rd Appleton WI 54914

KINSTLE, THOMAS HERBERT, b Lima, Ohio, Dec 18, 36; m 58. ORGANOFLUORINE CHEMISTRY, MASS SPECTROMETRY. *Educ:* Bowling Green State Univ, BA, 58; Univ Ill, PhD(org chem), 63. *Prof Exp:* Res assoc org chem, Univ Ill, 63-64; from instr to asst prof org chem, Iowa State Univ, 64-70; assoc prof org chem, 70-74, PROF CHEM, BOWLING GREEN STATE UNIV, 74- *Concurrent Pos:* Am Chem Soc-Petrol Res Found grant, 64-; Res Corp grant, 65- *Mem:* Am Chem Soc; Am Acad Arts & Sci; Am Soc Mass Spectrometry. *Res:* Applications of mass spectrometry to organic chemical problems; synthesis and reactions of strained ring systems; structure determination; synthesis and biosynthesis of natural products; organofluorine chemistry. *Mailing Add:* 149 S Grove St Bowling Green OH 43402

KINTANAR, AGUSTIN, b New Haven, Conn, Apr 28, 58; m 89; c 1. NUCLEAR MAGNETIC RESONANCE SPECTROSCOPY, MACROMOLECULAR STRUCTURE. *Educ:* Univ Ill, Chicago, BS, 79; Univ Ill, Urbana, PhD(phys chem), 84. *Prof Exp:* Postdoctoral mem tech staff, AT&T Bell Labs, 84-85; postdoctoral, Univ Wash, 85-88; ASST PROF, IOWA STATE UNIV, 88- *Mem:* Am Chem Soc. *Res:* Nuclear magnetic resonance studies of protein and DNA structure and dynamics. *Mailing Add:* Dept Biochem & Biophys Iowa State Univ Ames IA 50011

KINTER, LEWIS BOARDMAN, b Exeter, NH, Aug 15, 50; m 73; c 3. VASOPRESSIN, WHOLE ANIMAL MODELS. *Educ:* Union Col, BS, 73; Harvard Univ, PhD(physiol), 78. *Prof Exp:* Asst dir pharmacol, 84-88, asst dir toxicol, 88-90, ASSOC DIR TOXICOL, SMITH KLINE BEECHAM PHARMACEUTICALS, 90-; SCIENTIST, SMITH KLINE & FRENCH LABS, 81- *Concurrent Pos:* Adj asst prof physiol, Univ Pa. *Mem:* Am Physiol Soc; Am Soc Nephrology; Inst Soc Nephrology. *Res:* Homeostatic mechanisms involved in control of renal function and body fluid volume and body fluid composition. *Mailing Add:* Smith Kline Beecham Pharmaceuticals L-64 Box 1539 King of Prussia PA 19406-0939

KINTNER, EDWIN E, m; c 4. ENGINEERING. *Educ:* US Naval Acad, BS; Mass Inst Technol, MA(nuclear physics) & MA(marine eng). *Prof Exp:* Exec vpres, Gen Pub Utility, Nuclear Corp, 83-90; RETIRED. *Concurrent Pos:* Chmn, Nuclear Power Div Adv Comt, Adv Light Water Reactor, Utility Steering Comt; asst dir reactor eng & dep dir, Reactors Develop Div, AEC. *Mem:* Nat Acad Eng; Am Nuclear Soc; Elec Power Res Inst. *Res:* Nuclear power. *Mailing Add:* Three Fern Hollow Rd Boonton NJ 07005

KINTNER, ROBERT ROY, b Weeping Water, Nebr, Apr 3, 28; m 52; c 3. ORGANIC CHEMISTRY. *Educ:* Iowa State Univ, BS, 53; Univ Wash, PhD(org chem), 57. *Prof Exp:* From asst prof to assoc prof, 57-65, chmn dept, 60-65, 77-85 PROF CHEM, AUGUSTANA COL SDAK, 65- *Concurrent Pos:* Petrol res fund adv sci award, Univ Wash, Seattle, 64-65; NSF fac fel, Univ Calif, Santa Cruz, 71-72; vis prof chem, Univ Nebr, Lincoln, 80-81 & vis lectr chem, Munich, Germany campus, Univ Md, 85-87; prin investr, Sioux Falls Refuse Derived Fuel Proj, 78-80; vis prof chem, Univ Calif, Santa Cruz, 87-88. *Mem:* Am Chem Soc. *Res:* Physical-organic chemistry, especially in reaction mechanisms; structure; investigation of the mechanism of the formation of 2, 3, 7, 8 Tetrachlorodibenzodioxin. *Mailing Add:* Dept Chem Augustana Col Sioux Falls SD 57191-0001

KINYON, BRICE W(HITMAN), b South Bend, Ind, May 1, 11; m 35; c 2. MECHANICAL ENGINEERING. *Educ:* Purdue Univ, BSME, 33; Oak Ridge Sch Reactor Technol, cert, 55. *Prof Exp:* Engr prod, Armstrong Cork Co, 35-37 & Hamilton Watch Co, 37-38; prod & design engr, Radio Corp Am, 38-44 & Elec Storage Battery Co, 44-47; sr design engr reactor design, Oak Ridge Nat Lab, 47-62; anal engr, Thermal Stress Consult, Combustion Eng, Inc, 62-76; RETIRED. *Honors & Awards:* Chattanooga Engr of the Year, 79. *Mem:* Am Nuclear Soc; Sigma Xi; fel Am Soc Mech Engrs; Nat Soc Prof Engrs. *Res:* Fluid flow and heat transfer for power reactors and components. *Mailing Add:* 1312 N Shady Circle Chattanooga TN 37405

KINZEL, GARY LEE, b Bremen, Ohio, Jan 18, 44; m 67; c 2. COMPUTER-AIDED DESIGN, KINEMATICS. *Educ:* Ohio State Univ, BSME, 68, MS, 69; Purdue Univ, PhD(mech eng), 73. *Prof Exp:* Researcher, Batelle Mem Inst, 68-69; res eng, Batelle Columbus Labs, 68-69; from asst prof to assoc prof, 78-87, PROF, OHIO STATE UNIV, 87- *Mem:* Sigma Xi; Am Soc Mech Engrs. *Res:* Interactive computer graphics; machine element design; biomechanics; mechanism analysis and design; design and development of CAD software. *Mailing Add:* 3160 Caris Brook Rd Columbus OH 43221

KINZEL, JERRY J, b Avon, Ohio, Apr 24, 52; m 81. MICROBIAL PHYSIOLOGY, INTERMEDIATE METABOLISM. *Educ:* Cleveland State Univ, BS, 76; Miami Univ, MS, 78, PhD(microbiol), 81. *Prof Exp:* sr assoc res scientist, 81-, SR RES SCIENTIST, INT MINERALS & CHEM CORP. *Mem:* Sigma Xi; Am Soc Microbiol. *Res:* Microbial physiology with special emphasis on genetics and enzymology. *Mailing Add:* IMC PO Box 207 Terre Haute IN 47802

KINZER, EARL T, JR, b Beckley, WVa, Apr 7, 31; m 57; c 2. THEORETICAL PHYSICS, SOLID STATE PHYSICS. *Educ:* Auburn Univ, BEP, 58, MS, 60; Univ Va, PhD(physics), 62. *Prof Exp:* From asst prof to assoc prof physics, Univ Ala, Tuscaloosa, 61-67; ASSOC PROF PHYSICS, AUBURN UNIV, 67-, ACTG HEAD PHYSICS DEPT, 86- *Res:* Theory of the exchange of energy between gas particles and solids; operational solution of partial difference equations. *Mailing Add:* Dept Physics Auburn Univ Main Campus Auburn AL 36849

KINZER, H GRANT, b Grandfield, Okla, July 22, 37; m 57, 81; c 5. ENTOMOLOGY. *Educ:* Okla State Univ, BS, 59, MS, 60, PhD(entom), 62. *Prof Exp:* From asst prof to assoc prof entom, 64-76, PROF ENTOM, NMEX STATE UNIV, 76- *Mem:* Entom Soc Am; Am Registery Prof Entomologists; Can Entom Soc; Am Mosquito Control Asn. *Res:* Biology, physiology and control of livestock insects; biology and physiology of bark beetles. *Mailing Add:* Box 3 BE Las Cruces NM 88003

KINZER, ROBERT LEE, b Grandfield, Okla, June 23, 41; m 67; c 2. GAMMA RAY & X-RAY ASTRONOMY. *Educ:* Univ Okla, BS, 62, MS, 66, PhD(elem particle physics), 67. *Prof Exp:* Nat Acad Sci resident res fel cosmic ray physics, 67-69, RES PHYSICIST, US NAVAL RES LAB, 69- *Mem:* Am Phys Soc; AAAS; Sigma Xi. *Res:* Cosmic ray physics; experimental elementary particle physics; experimental gamma-ray astronomy and x-ray astronomy. *Mailing Add:* Code 4153 Naval Res Lab Washington DC 20375

KINZEY, BERTRAM Y(ORK), JR, b Rutland, Mass, Sept 25, 21; m 44; c 2. ARCHITECTURAL ACOUSTICS. *Educ:* Va Polytech Inst, BS, 42, MS, 43. *Prof Exp:* Naval architect, Norfolk Navy Yard, 43-45; archit draftsman & struct engr, Baskervill & Son, Va, 45-47; from asst prof to assoc prof archit eng, Va Polytech Inst, 47-59; from assoc prof to prof, 59-85, EMER PROF ARCHIT, UNIV FLA, 85- *Concurrent Pos:* Mem comt archit & acoust, Am Guild Organists, 52-56; mem comn archit, Nat Coun Churches, 54-, chmn joint comt church archit & music, 56-; consult archit acoust, 60- *Mem:* Fel Acoust Soc Am; Am Inst Archit; Am Soc Heating, Refrig & Air-Conditioning Engrs; Nat Coun Acoust Consults. *Res:* Heating, lighting, acoustics and sanitation; tests on wood box columns to determine formulas for design; thermal performance of ventilated building skins. *Mailing Add:* 212 SW 42nd St Gainesville FL 32607-2769

KINZEY, WARREN GLENFORD, b Orange, NJ, Oct 31, 35; m 57, 83; c 3. ANATOMY, PHYSICAL ANTHROPOLOGY. *Educ:* Univ Minn, BA, 56, MA, 58; Univ Calif, Berkeley, PhD(anat), 64. *Prof Exp:* Asst prof zool, Univ Calif, Davis, 63-67, asst prof anat, Sch Med, 67-68; planning officer regional med progs, 67-68; assoc prof, 70-82, chmn, 84-87, PROF ANTHROP, CITY COL NEW YORK, 83- *Concurrent Pos:* Asst res anatomist, Nat Ctr Primate Biol, 64-65; Wenner-Gren Found Anthrop Res Conf grant, 65; Nat Inst Child Health & Human Develop res grant, 65-67; Wenner-Gren res grant, Belg, 71; Explor Club res grant, Peru, 74; Earthwatch res grants, Peru, 75, 76, 77, 80 & 83; NSF res grants, 80, 81, 88, 91, prog dir phys anthrop, 88-90; consult ed, Am J Primatology, 84-; dir, Prog Community Health Educ, Div Social Sci, City Col, 87-88; assoc ed, Am J Phys Anthrop, 86-90. *Mem:* Fel AAAS; Am Asn Phys Anthrop; Am Asn Anat; Am Soc Mammal; Int Primatol Soc; fel Am Anthrop Asn. *Res:* Comparative morphology and evolution of primates; dental anthropology; new world primates; primate ecology. *Mailing Add:* Dept Anthrop City Col New York New York NY 10031-9198

KINZIE, JEANNIE JONES, b Gt Falls, Mont, Mar 14, 40; m 65; c 1. RADIOTHERAPY. *Educ:* Mont State Univ, BS, 61; Washington Univ, MD, 65. *Prof Exp:* Intern surg, Univ NC, 65-66; resident therapeut radiol, Washington Univ, 68-71, instr radiol, 71-73; asst prof, Med Col Wis, 73-74; asst prof radiol, Univ Chicago, 75-78, assoc prof, 78-80; assoc prof radiation oncol, Wayne State Univ, 80-85; PROF RADIOL & DIR RADIATION ONCOL, UNIV COLO, 85- *Concurrent Pos:* Am Cancer Soc advan clin fel, 71-74; consult radiol, Homer G Phillips Hosp, St Louis, 71-73, Vet Hosp, Wood, Wis & West Allis Mem Hosps, 73-74, Children's Hosp Mich, Detroit Receiving Hosp, Hutzel Hosp & Harper Grace Hosps, Detroit, 80-85; consult, Rose Med Ctr, Denver Gen Hosp & Denver Vet Hosp, 85-; Consult, FDA, 87. *Mem:* Fel Am Col Radiol; AMA; Am Soc Therapeut Radiol & Oncol; Am Soc Clin Oncol; AAAS; Soc Head & Neck Surgeons. *Res:* Patterns of care in Hodgkins disease; pediatric cancer treatment; combination treatment of head and neck cancer; combination radiation/chemotherapy studies for numerous types of cancer. *Mailing Add:* 3221 Interlocken Dr Box A031 4200 E Ninth Ave Evergreen CO 80439

KINZIE, ROBERT ALLEN, III, b Santa Cruz, Calif, June 7, 41. ZOOLOGY. *Educ:* Univ Hawaii, MS, 66; Yale Univ, PhD(biol), 70. *Prof Exp:* Fel, Univ Ga, 70-71; asst prof zool, 71-79, ASSOC PROF ZOOL, UNIV HAWAII, 75- *Concurrent Pos:* Consult, Upjohn Drug Co, 70- *Mem:* Ecol Soc Am; Soc Study Evolution; Soc Syst Zool; Am Soc Limnol & Oceanog. *Res:* Coral reef ecology; symbiosis. *Mailing Add:* Dept Zool Univ Hawaii 2538 The Mall Honolulu HI 96822

KINZLY, ROBERT EDWARD, b North Tonawanda, NY, July 4, 39; m 63; c 2. ELECTROOPTICS, OPTICAL ENGINEERING. *Educ:* Univ Buffalo, BA, 61; Cornell Univ, MS, 64. *Prof Exp:* Physicist, Cornell Aeronaut Lab, Inc, 63-67, sect head & br head, Calspan Corp, 67-75; PRES, SCIPAR, INC, 75- *Concurrent Pos:* Mem, Am Stand Inst working group microdensity. *Mem:* Optical Soc Am; Asn Old Crows. *Res:* Camouflage; image evaluation; atmospheric optics; remote sensing; visual perception; planetology; computer science; electro-optical warfare; optical countermeasures; target acquisition; image processing. *Mailing Add:* SCIPAR Inc PO Box 400 Buffalo NY 14221

KIPHART, KERRY, b South Haven, Mich, Apr 21, 49. INORGANIC PHYSICAL & CHEMICAL TREATMENT. *Educ:* Ill Wesleyan Univ, BA, 71; Univ Notre Dame, MS, 80, PhD(environ eng), 84. *Prof Exp:* Develop engr, Endicott, NY, 83-87, ENVIRON ENGR, IBM CORP, POUGHKEEPSIE, NY, 87- *Mem:* Am Chem Soc; Int Asn Impact Assessment. *Res:* Physical and organizational structures ensuring responsible chemical management, spill prevention and environmentally compatible products. *Mailing Add:* 121 Dusinberre Rd Gardiner NY 12525

KIPLING, ARLIN LLOYD, b Melfort, Sask, Dec 15, 36; m 63; c 3. SOLID STATE PHYSICS. *Educ:* Univ Sask, BEng, 58; McGill Univ, MSc, 61; Univ Exeter, PhD(physics), 67. *Prof Exp:* Engr, Northern Elec Co, Que, 58-59; physicist, Noranda Res Ctr, 63-64; PROF PHYSICS, CONCORDIA UNIV, 67- *Mem:* Can Asn Physicists. *Res:* Semiconductors at low temperatures. *Mailing Add:* Dept Physics Concordia Univ Sir G Williams 1455 DeMaisonneuve Blvd W Montreal PQ H3G 1M8 Can

KIPLINGER, GLENN FRANCIS, b Indianapolis, Ind, Sept 29, 30; m 53; c 4. CLINICAL PHARMACOLOGY. *Educ:* Butler Univ, BS, 53; Univ Mich, PhD(pharmacol), 58; Univ Tex, Galveston, MD, 67. *Prof Exp:* Asst pharmacol, Boston Univ Sch Med, 58-62 & Univ Tex, Galveston, 62-67; clin pharmacologist, Eli Lilly & Co, Indianapolis, 67-72, dir toxicol, 72-75, vpres res, 75-77; managing dir, Lilly Res Ctr, UK, 77-80; vpres res & develop, Ortho Pharmaceut, Johnson & Johnson, 80-88 & R W Johnson Pharmaceut Res Inst, 88-89; CONSULT, 89- *Concurrent Pos:* Trustee, Butler Univ, Indianapolis, 89- *Mem:* Am Soc Pharmacol & Exp Therapeut; Am Soc Clin Pharmacol & Therapeut. *Res:* Cardiovascular pharmacology; clinical pharmacology of analgesics and of marijuana; research and development management. *Mailing Add:* 1633 S Lodge Dr Sarasota FL 34239

KIPNIS, DAVID MORRIS, b Baltimore, Md, May 23, 27; m 53; c 3. INTERNAL MEDICINE, ENDOCRINOLOGY. *Educ:* Johns Hopkins Univ, AB, 45, MA, 49; Univ Md, MD, 51. *Prof Exp:* Intern med, Johns Hopkins Hosp, 51-52; from jr asst resident to sr asst resident, Duke Univ Hosp, 52-54; chief resident, Univ Md Hosp, 54-55; asst prof biochem in med, Wash Univ, 57-62, dir, Clin Res Ctr, 60-87, from assoc prof to prof, 62-73, BUSCH PROF MED & CHMN DEPT, MED, WASH UNIV, 73- *Concurrent Pos:* Am Col Physicians fel biochem, Sch Med, Wash Univ, 55-56; Markle scholar, 56-61; mem sci adv bd, USAF; mem Endocrinol Study Sect, NIH; mem, Nat Pituitary Agency; mem & chmn, Nat Diabetes Adv Bd, 77-81. *Honors & Awards:* Oppenheimer Award, Endocrine Soc. *Mem:* Nat Acad Sci; Inst Med-Nat Acad Sci; Asn Am Physicians; Endocrine Soc; Biochem Soc; Am Soc Clin Invest; Am Acad Arts & Sci. *Res:* Hormonal control of carbohydrate and protein metabolism. *Mailing Add:* Dept Med Wash Univ Sch Med St Louis MO 63110

KIPOUROS, GEORGES JOHN, production of light metals, electrodeposition of refractory metals, for more information see previous edition

KIPP, EGBERT MASON, b Angola, Port WAfrica, Nov 27, 14; m 35; c 3. RESEARCH ADMINISTRATION. *Educ:* Iowa Wesleyan Col, BS, 34; Boston Univ, MS, 35; Pa State Univ, PhD(phys chem), 39. *Hon Degrees:* DSc, Iowa Wesleyan Col, 61. *Prof Exp:* Asst chem, Pa State Univ, 35-39; res chemist, Aluminum Co Am, 39-44, asst chief phys chem div, 44-47, exec secy oils & lubricants comt, 46-57, chief lubricants div, 47-57; dir res & develop, Foote Mineral Co, 57-59; asst to mgr prod develop, Sun Oil Co, 59-60, mgr basic res, 60-62, assoc dir res & develop, 62-70; INDEPENDENT CONSULT TECH MGT, 70- *Concurrent Pos:* Consult, chem & metals indust, 71-; eval agent, Off Energy Related Inventions, Nat Bur Standards; chmn, tech comt K, Am Soc Test & Mat, vchmn, Pittsburgh dist; chmn, Am Inst Chemists, Philadelphia & pres-elect, Pa Inst Chemists; mem, Res Mgt Group Philadelphia, pres elect, 74 & pres, 75-76; plenary lectr, 3rd Int Conf Metal Working Lubricants, Esslingen, Ger, 82. *Mem:* Am Chem Soc; fel Am Soc Lubrication Eng (pres, 46); fel Am Inst Chemists (treas, 70); Sigma Xi; Am Soc Testing & Mat. *Res:* Lubrication engineering sciences; research and development management and organization; energy, market and commercial development; air, soil and water conservation; technology audits; author of one book and 16 publications; holder of 11 patents. *Mailing Add:* 745 Thomas St State College PA 16803

KIPP, HAROLD LYMAN, b Graig, Nebr, Feb 19, 03; m 28; c 6. MECHANICAL ENGINEERING. *Educ:* Univ Nebr, BSc, 31, MSc, 34. *Prof Exp:* Instr heat power, Univ Kans, 36-37; instr, Sch Mines, Univ SDak, 37-38; from assoc prof to prof, Tex Tech Col, 38-47; prof, 47-73, asst chmn dept mech eng, 66-73, EMER PROF HEAT POWER, UNIV KANS, 73- *Concurrent Pos:* Dir ground eng, Midland AFB, 44; Propelle Lab Eng, Wright Patterson AFB, 46; ground eng officer staff, Gen Atkinson 2nd AFB, Ca, 46; ground eng officer, Omaha, 46. *Mem:* Am Soc Mech Engrs; Am Soc Eng Educ. *Res:* Thermodynamics; heat power; heat transfer. *Mailing Add:* 1647 University Dr Lawrence KS 66044

KIPP, JAMES EDWIN, b Detroit, Mich, June 15, 53; m 86; c 1. CHEMICAL KINETICS, PHARMACEUTICAL FORMULATIONS. *Educ:* Albion Col, BA, 75; Univ Mich, MS, PhD(chem), 83. *Prof Exp:* Sr res assoc, Baxter Travenol, 83-85; res scientist, 85-90, SR RES SCIENTIST, BAXTER HEALTHCARE CORP, 90- *Mem:* Am Chem Soc; Am Asn Pharmaceut Scientist. *Res:* Photochemistry of strained bicyclic alkenes; mathematical modeling of nonisothermal decomposition of drugs in pharmaceutical formulations; prediction of pharmaceutical shelf life; variable pH experiments for rapid preformulation screening; use of NMR spectrometry in equilibrium constant determinations. *Mailing Add:* Baxter Healthcare Corp Pharmaceut Res & Develop Rte 120 & Wilson Rd Round Lake IL 60073

KIPP, RAYMOND J, b Ossian, Iowa, Dec 7, 22; m 49; c 2. SANITARY ENGINEERING. *Educ:* Marquette Univ, BS, 51; Univ Wis, MS, 57, PhD(sanit eng, bact), 65. *Prof Exp:* from asst prof civil eng to dean Col Eng, 57-87, PROF CIVIL ENG, MARQUETTE UNIV, 69- *Concurrent Pos:*

Chmn, Milwaukee Metrop Sewerage Dist, 78-79. *Mem:* Water Pollution Control Fedn; Am Acad Environ Engrs; Am Soc Civil Engrs; Nat Soc Prof Engrs; Am Soc Eng Educ; Am Water Works Asn. *Res:* Industrial wastes; toxic and hazardous wastes. *Mailing Add:* Col Eng Marquette Univ 1515 W Wisconsin Ave Milwaukee WI 53233

KIPPAX, DONALD, chemistry and chemical engineering; deceased, see previous edition for last biography

KIPPENBERGER, DONALD JUSTIN, b St Louis, Mo, Feb 25, 47; m 73; c 2. PHYSICAL CHEMISTRY. *Educ:* Univ St Thomas, BS & BA, 70; Sam Houston State Univ, MS, 72; Tex A&M Univ, PhD(chem), 82. *Prof Exp:* Chemist path residents, Gorgas Army Hosp, 77-80 & Tex A&M Univ, 80-82; chemist off chem readiness, Edgewood Arsenal, 82-87; mem staff, Weisbbaden Forensic Toxicol Drug Testing Lab, Ger, 87-90; MEM STAFF, OFF ARMY SURGEON GEN, 90- *Concurrent Pos:* Lectr, Fayetteville Univ, 74-77; consult chem, Commander XVIII Airborne Corp, 74-77 & Gov, Canal Zone, 77-80; co-investr, Heart Enzyme Group, Walter Reed Army Inst, 76-77; asst prof, Canal Zone Col, 77-80. *Mem:* Am Inst Chemists. *Res:* Membrane memitic chemistry: characterization of polymeric vesicles utilizing proton and carbon nuclear magnetic resonance, fluorescense and absorption techniques; surface chemistry: characterization of catalytically active inorganic complexes in zeolites and ambevlytes using electron spin resonance techniques. *Mailing Add:* 625 Shore Dr Joppatowne MD 21085

KIPPENHAN, C(HARLES) J(ACOB), b Middle Amana, Iowa, Nov 8, 19; m 41; c 2. MECHANICAL ENGINEERING. *Educ:* Univ Iowa, BSME, 40, MSME, 46, PhD(mech eng), 48. *Prof Exp:* Jr mech engr, Stanley Eng Co, 40; instr mech eng, Univ Iowa, 41-42; from asst prof to assoc prof, Univ Wash, St Louis, 48-54, prof & head dept, 54-63; chmn dept, 63-73, prof mech eng, 63-84, EMER PROF MECH ENG, UNIV WASH, SEATTLE, 84- *Concurrent Pos:* Mech eng consult; adj prof archit, Univ Wash, Seattle. *Mem:* Am Soc Heating, Refrigerating & Airconditioning Engrs; Am Soc Mech Engrs; Am Soc Eng Educ. *Res:* Building energy systems analysis, simulation; energy conservation and management; convective and radiation heat transfer; thermal properties; transient techniques. *Mailing Add:* 3908 Northeast 38th St Seattle WA 98105

KIPPS, THOMAS CHARLES, b Eureka, Utah, Feb 28, 23; m 48; c 4. MATHEMATICS. *Educ:* Univ Calif, Berkeley, AB, 49, MA, 50, PhD(math), 57. *Prof Exp:* Instr math & physics, Univ Santa Clara, 53-55; asst math, Univ Calif, Berkeley, 55-56; from instr to assoc prof, 56-67, chmn dept, 69, PROF MATH, CALIF STATE UNIV, FRESNO, 67- *Mem:* Am Math Soc; Math Asn Am; Sigma Xi. *Res:* Double integral problems in the calculus of variations; methods of linear algebra in numerical analysis. *Mailing Add:* 1459 E Portals Ave Fresno CA 93710

KIPROV, DOBRI D, b Sofia, Bulgaria, May 1, 49; US citizen; c 1. IMMUNOLOGY. *Educ:* Med Acad, Bulgaria, MD, 74. *Prof Exp:* Instr pathol, Sackler Sch Med, Israel, 74-77; resident, Mt Sinai Hosp, Case Western Reserve Univ, 77-79; clin res fel, Mass Gen Hosp, Harvard Med Sch, 79-81; clin res fel, 81-82, DIR, PLASMAPHERESIS, DEPT CELLULAR IMMUNOL, CHILDREN'S HOSP, UNIV CALIF, SAN FRANCISCO, 82- *Concurrent Pos:* Training & Res Award, Nat Inst Allergies & Infectious Dis; investr, Nat Inst Allergies & Infectious Dis. *Mem:* AMA; Am Soc Clin Pathol; Am Col Pathologists; World Apheresis Asn; World Med Asn. *Res:* Basic immunologic defects in autoimmune diseases, including renal diseases; evaluation of lymphocyte subsets from peripheral blood and tissues using moncolonal antibodies; effect of plasmapheresis, lymphocytapheresis and immunosuppressive therapy on the immunologic system of patients with autoimmune diseases; clinical correlation with immunologic assays. *Mailing Add:* Calif Pac Med Ctr Rm 308 3700 Calif St San Francisco CA 94118

KIRBER, MARIA WIENER, b Prague, Czech, Feb 19, 17; nat US; m 43; c 2. MICROBIOLOGY, VIROLOGY. *Educ:* Univ Prague, MUC, 38; Univ Pa, MS, 41, PhD(bact), 42; Am Bd Microbiol, dipl. *Hon Degrees:* ScD, Med Col Pa, 73. *Prof Exp:* Asst bact, 41-43, from instr prof, prof microbiol, 62-72, EMER PROF VIROL & MICROBIOL, MED COL PA, 72- *Concurrent Pos:* Nat Coun Combat Blindness grant, Med Col Pa, 58-59, 62-63 & 66-67 & Nat Soc Prev Blindness grant, 68-70; mem res staff, Children's Hosp, Philadelphia, 51-52. *Honors & Awards:* Christian R & Mary E Lindback Award, 71. *Mem:* Fel Am Acad Microbiol; Am Soc Microbiol. *Res:* Antigenic structure of hemolytic streptococci; complement fixing antigens of influenza viruses; mouse brain tissue culture; experimental viral infections and autoimmune reactions of the eye; experimental mycobacterial eye infection. *Mailing Add:* Juniper Dr Lakeville CT 06039

KIRBY, ALBERT CHARLES, b Baton Rouge, La, Mar 31, 41; m 61; c 3. PHYSIOLOGY. *Educ:* La State Univ, Baton Rouge, BS, 62, MS, 63; Univ Ill, Urbana, PhD(physiol), 67. *Prof Exp:* NIH sr fel physiol, Univ Wash, 67-69; asst prof, 69-75, ASSOC PROF PHYSIOL, CASE WESTERN RESERVE UNIV, 75-, ASSOC DEAN, 84- *Concurrent Pos:* NIH res grants, Case Western Reserve Univ, 69- *Mem:* Soc Gen Physiol; Biophys Soc; Am Physiol Soc. *Res:* Excitation-contraction coupling and contractile proteins in denervated mammalian muscle. *Mailing Add:* Dept Physiol Case Western Reserve Univ Sch Med Cleveland OH 44106

KIRBY, ANDREW FULLER, b Washington, DC, Feb 15, 55; m 81; c 1. ELECTRONIC SPECTROSCOPY, COORDINATION CHEMISTRY. *Educ:* Col Holy Cross, AB, 77; Duke Univ, PhD(phys chem), 81. *Prof Exp:* Fel assoc, Univ Va, 81-82; phys scientist, Off Sci & Weapons Res, 85-87, res scientist br chief, 87-90, DIV CHIEF MAT RES, OFF RES & DEVELOP, CIA, 91- *Mem:* Am Chem Soc; Sigma Xi. *Res:* Detailed electronic and molecular structures of solid-state lanthanide complexes; advanced spectroscopic measurement techniques. *Mailing Add:* PO Box 3313 Falls Church VA 22043-3313

KIRBY, BRUCE JOHN, b Toronto, Ont, Nov 19, 28; m 58; c 2. APPLIED MATHEMATICS. *Educ:* Univ Toronto, BA, 50, MA, 51; Univ London, PhD, 67. *Prof Exp:* Teaching fel math, Univ Toronto, 51-53; lectr, Univ Liverpool, 53-54; lectr, 54-60; from asst prof to assoc prof, 60-69, PROF MATH, QUEEN'S UNIV, ONT, 69- *Mem:* Soc Indust & Appl Math. *Res:* Mathematics of control theory. *Mailing Add:* Dept Math & Statist Queen's Univ Kingston ON K7L 3N6 Can

KIRBY, CONRAD JOSEPH, JR, b Opelousas, La, Aug 2, 41; m 68; c 2. ECOLOGY. *Educ:* Univ Southwestern La, BS, 64; La State Univ, MS, 67, PhD(marine sci), 71. *Prof Exp:* Asst prof biol, Univ Southeastern La, 67-72; res ecologist, 72-74, SUPVRY ECOLOGIST, WATERWAYS EXP STA, 74- *Mem:* Ecol Soc Am; Nat Estuarine Res Fedn; Gulf Estuarine Res Fedn. *Res:* Impacts and mitigation of construction activities in wetlands; development and management of wetlands and other ecosystems. *Mailing Add:* Waterways Exp Sta ER 3909 Halls Ferry Rd Vicksburg MS 39180

KIRBY, EDWARD PAUL, b Ithaca, NY, Sept 28, 41; m 66. BIOCHEMISTRY, HEMATOLOGY. *Educ:* Univ Rochester, BS, 63; Case Western Reserve Univ, PhD(biochem), 68. *Prof Exp:* NSF resident res associateship, Naval Med Res Inst, Bethesda, Md, 68-70; NIH fel, Univ Wash, 70-71; from asst prof to assoc prof, 71-85, PROF BIOCHEM, TEMPLE UNIV, 85-; PROF THROMBOSIS RES, TEMPLE UNIV, 85- *Concurrent Pos:* Mem coun thrombosis, Am Heart Asn; Res career develop award, 76-80. *Mem:* Am Soc Biol Chemists; AAAS; NY Acad Sci. *Res:* Blood coagulation; protein chemistry; structure and function of von Willebrand factor and factor XII. *Mailing Add:* Dept Biochem Temple Univ Health Sci Ctr Philadelphia PA 19140

KIRBY, HILLIARD WALKER, b Asheville, NC, June 12, 49; m 81. INTEGRATED PEST MANAGEMENT. *Educ:* Univ NC, BS, 71; NC State Univ, MS, 74, PhD(plant path), 81. *Prof Exp:* Agr exten agent, NC Agr Exten Serv, 74-78; ASST PROF PLANT PATH, COL AGR, UNIV ILL, 81- *Mem:* Am Phytopath Soc. *Res:* Effects of conservation tillage on plant disease development. *Mailing Add:* Dept Plant Path N-531 Turner Hall Univ Ill 1102 S Goodwin Ave Urbana IL 61801

KIRBY, JAMES M, b Corona, Calif, Jan 23, 01; m 23; c 2. EARTH SCIENCES, APPLIED GEOPHYSICS. *Educ:* Stanford Univ, BA, 24. *Prof Exp:* Geologist, Standard Oil Co Calif, 25-42; chief geologist, Calif Standard Co, Alta, Can, 42-46; explor supvr, Calif Co, Denver, Colo, 46-49; res geologist, Richmond Petrol Co, Ltd, 52-57; prog mgr explor operations, West Australian Petrol Party Ltd, 57-62; mgr explor operations, Am Overseas Petrol Ltd, Nigeria, 62-64; CONSULT, 64- *Mem:* Am Assoc Petrol Geologists; Geological Soc Am. *Res:* Geological mapping, domestic and foreign; direction of the exploration activities of others. *Mailing Add:* 301 White Oak Dr No 253 Santa Rosa CA 95409

KIRBY, JAMES RAY, b Goldsboro, NC, Oct 22, 33; m 62; c 3. ANALYTICAL CHEMISTRY. *Educ:* E Carolina Col, AB & BS, 55; Duke Univ, MA, 57, PhD(chem, physics), 60. *Prof Exp:* From res chemist to sr res chemist, 60-68, res specialist, 68-73, sr res specialist, Monsanto Co, 73-; MGR CHEM ANALYSIS, IBM, RESEARCH TRIANGLE PARK, 84- *Mem:* Am Chem Soc. *Res:* Chelate chemistry; polymer characterization. *Mailing Add:* IBM Dept E62 Bldg 061 PO Box 12195 Research Triangle Park NC 27709

KIRBY, JON ALLAN, flight software; deceased, see previous edition for last biography

KIRBY, KATE PAGE, b Washington, DC, Dec 5, 45; m 77; c 4. ELECTRON & ATOMIC PHYSICS. *Educ:* Harvard, AB, 67; Univ Chicago, MS, 68, PhD(chem physics), 72. *Prof Exp:* RES PHYSICIST, SMITHSONIAN ASTROPHYS OBSERV, 74-; ASSOC DIR, CTR FOR ASTROPHYS, ATOMIC & MOLECULAR PHYSICS DIV, 88- *Concurrent Pos:* Lectr, Astron Dept, Harvard, 74- *Mem:* Fel Am Phys Soc; Int Astron Union; Am Astron Soc. *Res:* Molecular structure and properties; radiation and molecular interaction; atomic and molecular processes in astrophysics; chemical physics. *Mailing Add:* Ctr Astrophys 60 Garden St Cambridge MA 02138

KIRBY, MARGARET LOEWY, b Ft Smith, Ark, June 5, 46; m 71; c 2. ANATOMY, PHARMACOLOGY. *Educ:* Manhattanville Col, AB, 68; Univ Ark, PhD(anat), 72. *Prof Exp:* Teaching asst anat, Univ Ark, 69-71; asst prof gross anat, Cent Univ Ark, 72-74; from asst prof to prof, 77-88, REGENTS PROF ANAT, MED COL GA, 88- *Concurrent Pos:* Prin investr, Ga Heart Asn, 78-79, Nat Inst Drug Abuse, 79-81 & NIH, 81-89; Nat Heart, Lung & Blood Adv Coun, 85-91. *Mem:* Am Asn Anatomists; Sigma Xi; Soc Neurosci; Soc Develop Biol. *Res:* Effects of neurotransmitters on embryonic development; neural crest; heart development; molecular biology of development; pediatric cardiology. *Mailing Add:* Dept Anat Med Col Ga Augusta GA 30912

KIRBY, MAURICE J(OSEPH), systems & electrical engineering; deceased, see previous edition for last biography

KIRBY, PAUL EDWARD, b Washington, DC, Mar 12, 49; m 71; c 4. CELL BIOLOGY. *Educ:* Mt St Mary's Col, Md, BS, 71; Cath Univ Am, MS, 74, PhD(cell biol), 78. *Prof Exp:* Res scientist in vitro mutagenesis, EG&G Mason Res Inst, 78-80, chief, Mammalian Mutagenesis Sect, 80-82; dir oper, Microbiol Asn, 82-84; VPRES, DIR OPER, SITEK RES LABS, 84- *Concurrent Pos:* Prin investr, Nat Cancer Inst; consult, 80- *Mem:* AAAS; Environ Mutagen Soc; Tissue Cult Asn; Genetic Toxicol Asn; Genetic & Environ Mutagen Soc. *Res:* In vitro carcinogenesis and mutagenesis in mammalian cells, especially the L5178Y mouse lymphoma mutagenesis assay as a tool for the screening of compounds for mutagenic potential; invitro toxicity test development. *Mailing Add:* 14007 Castlebar Dr Glenwood MD 21738

KIRBY, RALPH C(LOUDSBERRY), b Washington, DC, July 21, 25; m 50; c 1. EXTRACTIVE METALLURGY, MINERAL ENGINEERING. *Educ:* Cath Univ Am, BChE, 50. *Prof Exp:* Chem engr, Metall Res Ctr, 50-54, metall engr, 55-58, proj leader, 59-65, staff metallurgist, Div Metall, 66-70, sr staff metallurgist, 71-72, chief, Div Metall, 72-76, asst dir metall, 76-79, DIR, MIN RESOURCES TECHNOL, US BUR MINES, 79- *Concurrent Pos:* Bd dirs, Metall Soc, 76- *Honors & Awards:* Gold Medal, Am Inst Chem, 50; Invention Award, US Bur Mines, 6. *Mem:* Am Inst Mining, Metall & Petrol Engrs; Am Inst Chem Engrs; AAAS; Am Soc Metals; Sigma Xi. *Res:* Metallurgical research; research and development management. *Mailing Add:* 116 Southwood Ave Silver Spring MD 20901

KIRBY, RICHARD C(YRIL), b Galesburg, Ill, Nov 22, 22; m 44; c 7. ELECTRICAL ENGINEERING. *Educ:* Univ Minn, BEE, 51. *Prof Exp:* Telegrapher, West Union Tel Co, Minn, 40-42; asst chief engr radio broadcast, KFEQ Inc, Mo, 46-48; res worker & proj leader radio propagation studies, 48-55, chief, Ionospheric Res Sect, 55-57, asst chief, Radio Propagation Physics Div, 57-59, chief Radio Systs div, Nat Bur Standards, 59-65; dir, Ionospheric Telecommun Lab, 65-68, dir Inst Telecommun Sci, 68-71, assoc dir, Off Telecommun, US Dept Com, 71-74; DIR, INT RADIO CONSULT COMT, GENEVA, SWITZ, 74- *Concurrent Pos:* Asst, Univ Minn, 41-42; mem, Interdept Radio Adv Comt, Exec Off President, 59-60; US Dept Com fel sci & technol, 65-66; chmn, Commun Technol Group, 69-71; adj prof, Univ Denver, 69-74. *Honors & Awards:* US Dept Com, Gold Medal Award, 56. *Mem:* AAAS; Sigma Xi; fel Inst Elec & Electronics Engrs; Int Union Radio Sci; fel Radio Club Am. *Res:* Telecommunications; radio wave propagation; antennas; information transmission; ionospheric studies. *Mailing Add:* Int Radio Consult Comt Two Rue Varembe 1211 Geneva 20 Switzerland

KIRBY, ROBERT EMMET, b Stowe Twp, Pa, Feb 27, 21; m 44; c 4. ANALYTICAL CHEMISTRY. *Educ:* Univ Pittsburgh, BS, 49; Carnegie Inst Technol, MS, 52; Univ Ariz, PhD(anal chem), 61. *Prof Exp:* Res chemist, Shell Chem Co, 52-57; salesman, William C Buchanan Co, 57-58; instr, Tex Col Arts & Indust, 58-59; sr res chemist, Colgate-Palmolive Co, 62-63; asst prof, 63-70, ASSOC PROF CHEM, QUEENS COL, NY, 70- *Mem:* Am Chem Soc; Royal Chem Soc. *Res:* Electroanalytical methods; chelate chemistry; photoelectron spectroscopy. *Mailing Add:* 1435 W Chapala Dr Tucson AZ 85704

KIRBY, ROBION C, b Chicago, Ill, Feb 25, 38; m 82; c 2. MATHEMATICS. *Educ:* Univ Chicago, BS, 59, MS, 60, PhD(math), 65. *Prof Exp:* Asst prof math, Univ Calif, Los Angeles, 65-69, PROF MATH, UNIV CALIF, BERKELEY, 71- *Concurrent Pos:* Dep dir, Math Sci Res Inst, 85- *Honors & Awards:* Veblen Prize Geom, 71. *Mem:* Am Math Soc. *Res:* Topology, specifically topology of manifolds, differential and combinatorial topology. *Mailing Add:* Dept Math Univ Calif Berkeley CA 94720

KIRBY, ROGER D, b Lansing, Mich, June 1, 42; m 64; c 1. PHYSICS. *Educ:* Mich State Univ, BS, 64; Cornell Univ, PhD(physics), 69. *Prof Exp:* Res assoc physics, Cornell Univ, 68-69 & Univ Ill, 69-71; from asst prof to assoc prof, 71-81, PROF PHYSICS, UNIV NEBR, LINCOLN, 81- *Mem:* Am Phys Soc. *Res:* Lattice dynamics; Raman scattering from solids; far infrared optical properties of solids; transport properties of solids. *Mailing Add:* Behlen Lab Physics Univ Nebr Lincoln NE 68588

KIRBY, WILLIAM M M, medicine, for more information see previous edition

KIRCH, LAWRENCE S, b Philadelphia, Pa, Dec 7, 32; m 55; c 3. ORGANIC CHEMISTRY. *Educ:* Temple Univ, AB, 55; Univ Cincinnati, MS, 57, PhD(chem), 58. *Prof Exp:* Sr res chemist, 58-73, proj leader, 73-81, SECT MGR, ROHM & HAAS CO, 81- *Mem:* Am Chem Soc. *Res:* Organic reactions at high pressures; catalysis; reaction kinetics and mechanisms; organometallic compounds; coordination chemistry; reactions of acetylenes and of carbon monoxide; oxidation of organic compounds; chlorination, sulfonation and dehydrogenation of organic compounds; development of new chemical processes and optimization of existing processes. *Mailing Add:* 871 Oriole Lane Huntingdon Valley PA 19006

KIRCH, MURRAY R, b Philadelphia, Pa, Oct 11, 40; m 65; c 2. PROGRAMMING METHODOLOGY. *Educ:* Temple Univ, AB, 62; Lehigh Univ, MS, 64, PhD(math), 68. *Prof Exp:* Instr math, Lehigh Univ, 65-68; asst prof, State Univ NY Buffalo, 68-72; assoc prof math, 72-83, assoc prof info sci & math, 83-84, PROF COMPUT SCI & MATH, STOCKTON STATE COL, 84- *Concurrent Pos:* MacArthur distinguished vis prof arts & sci, New Col, Univ S Fla, 84-85; asst dir, Inst Retraining in Comput Sci, Clarkson Univ, 85- *Mem:* Am Math Soc; Math Asn Am; Soc Indust & Appl Math; Am Statist Asn; Asn Comput Mach; Inst Elec & Electronics Engrs. *Res:* Point-set topology; functional analysis; numerical methods for nonlinear systems; mathematical analysis of gambling and risk taking; algorithms and data structure; programming methodology; computer science theory; computer science education; artificial intelligence and expert systems; software engineering. *Mailing Add:* Dept Info Stockton State Col Pomona NJ 08240

KIRCH, PATRICK VINTON, b Honolulu, Hawaii, July 7, 50; div. ARCHEOLOGY, PALEOECOLOGY. *Educ:* Univ Pa, BA, 71; Yale Univ, MPhil, 74, PhD(anthrop), 75. *Prof Exp:* Assoc anthropologist, Bernice P Bishop Mus, Hawaii, 74-75, anthropologist, 73-85, head, Div Archaeol, 83-84; dir, Burke Mem Wash State Mus, Univ Wash, 85-87, from assoc prof to prof anthrop, 85-89; RES ASSOC, BERNICE P BISHOP MUS, HAWAII, 84-; PROF ANTHROP, UNIV CALIF, BERKELEY, 87- *Concurrent Pos:* Prin investr, NSF grants, Bernice P Bishop Mus, 76- & var other grants, 84-91; affil assoc prof, Univ Hawaii, 82-84. *Mem:* Nat Acad Sci; Sigma Xi; Asn Field Archaeol; Soc Am Archaeol; Am Anthrop Asn. *Res:* Archaeology, paleoecology and biogeography of the oceanic region. *Mailing Add:* Dept Anthrop Univ Calif Berkeley CA 94720

KIRCHANSKI, STEFAN J, cell biology, hematology, for more information see previous edition

KIRCHBERGER, MADELEINE, b Buffalo, NY. PHYSIOLOGY. *Educ:* Hunter Col, BA, 60; Columbia Univ, MA, 62, PhD(cell physiol), 66. *Prof Exp:* Res fel med, Mass Gen Hosp & Sch Med, Harvard Univ, 66-69; instr physiol, 70-71, assoc, 71-74, asst prof, 74-78, ASSOC PROF PHYSIOL, BIOPHYS & MED, MT SINAI SCH MED, 78- *Mem:* Am Physiol Soc; Biophys Soc; Am Soc Biol Chem; Int Soc Heart Res. *Res:* Regulation of cardiac contractility by catecholamines; calcium transport in biological membranes. *Mailing Add:* Dept Physiol & Biophys Mount Sinai Sch Med One Gustave L Levy Pl New York NY 10029

KIRCHER, HENRY WINFRIED, organic chemistry; deceased, see previous edition for last biography

KIRCHER, JOHN FREDERICK, b Athens, Ohio, Jan 31, 29; m 51; c 4. RADIATION CHEMISTRY, MATERIALS SCIENCES. *Educ:* Ohio Univ, BS, 50, MS, 51; Syracuse Univ, PhD(chem), 56. *Prof Exp:* Design & develop engr, Radio Corp Am, 55-57; proj leader, Battelle Mem Inst, 57-60, assoc div chief chem physics res, 60-65, fel chem physics res, 65-70, div chief chem physics res, 70-73, sr chemist, dept physics, 73-76, sr chemist, Energy & Environ Technol Dept, 76-78, prog mgr, 78-80, waste package dept mgr, 80-82, Syst Anal Dept Mgr, Off Nuclear Waste Isolation, 82-89, PROG MGR, ENERGY SYSTS GROUP, BATTELLE MEM INST, 89- *Mem:* AAAS; Am Chem Soc; fel Am Inst Chem; Sigma Xi. *Res:* Radiation chemistry of inorganic gas phase reactions; polymers and other organic systems; radiation dosimetry; surface chemistry; vacuum techniques; gas kinetics and electrical discharges; flames; nuclear waste disposal. *Mailing Add:* Battelle Energy Systs Group 505 King Ave Columbus OH 43201-2693

KIRCHER, MORTON S(UMMER), b Rome, NY, May 3, 17; m 42; c 3. CHEMICAL ENGINEERING. *Educ:* Univ Rochester, BS, 38. *Prof Exp:* Chem engr, Hooker Chem Corp, NY, 38-51, res supvr, 51-58, mgr electrochem & inorg res, 58-59, sr engr electrochem develop, 59-61; gen mgr, Dryden Chem Ltd, 61-71, pres, 67-71; INDUST CONSULT ELECTROCHEM, MORTON S KIRCHER, 71- *Concurrent Pos:* Civilian, Off Sci Res & Develop, AEC; gen mgr, BC Chem Ltd, 65-71. *Mem:* Electrochem Soc; Am Chem Soc; Am Inst Chem Eng. *Res:* Electrochemistry of caustic-chlorine cells, fluorine cells, chlorate, perchlorate and chloralkali cells including diaphragm type; development of membrane diaphragm, cells. *Mailing Add:* 1111 Granada St Clearwater FL 34615

KIRCHGESSNER, JOSEPH L, b Wheeling, WVa, June 23, 32; m 56; c 3. PHYSICS, ELECTRICAL ENGINEERING. *Educ:* Le Moyne Col, NY, BS, 54; Cornell Univ, MS, 56. *Prof Exp:* Mem prof tech staff, Princeton Penn Accelerator, 56-65, sr tech staff, 65-70, div head, 68-70; ACCELERATOR PHYSICIST, CORNELL ELECTRON SYNCHROTRON, 70- *Concurrent Pos:* Del, Int Accelerator Conf, Italy, 65, Mass, 67 & USSR, 69; mem prog comn, US Nat Particle Accelerator Conf, 66 & 67; consult, Princeton Penn Accelerator, 71. *Mem:* Am Phys Soc. *Res:* Accelerator design and development. *Mailing Add:* Wilson Sync Lab Cornell Univ Ithaca NY 14853

KIRCHHEIMER, WALDEMAR FRANZ, b Schneidemuhl, Ger, Jan 11, 13; nat US; m 45. MICROBIOLOGY. *Educ:* Univ Giessen, MD, 47; Univ Wash, PhD, 49. *Prof Exp:* Res physician, King County Tuberc Hosp, Seattle, Wash, 42-46; res assoc tuberc, Univ Wash, 46-47, instr microbiol, Sch Med, 48-49; from asst prof to assoc prof bact, Med Sch, Northwestern Univ, 49-56; dep safety dir & med bacteriologist, Ft Detrick, Md, 56-61; mem res staff, Inst Allergy & Infectious Dis, 61-62; chief, Microbiol Sect, USPHS Hosp, 62-64, Lab Br, 65-67 & Lab Res Dept, 67-71, chief, Lab Res Br, 71-83; RETIRED. *Concurrent Pos:* Clin prof bact trop dis & med parasitol, La State Univ Med Ctr, New Orleans. *Honors & Awards:* Medal Award Super Serv, Dept Health Educ & Welfare, 71, Medal Award Distinguished Serv, 77. *Mem:* Am Soc Microbiol. *Res:* Immunology; mechanism of tuberculin-type sensitivity; medical bacteriology; cell culture of leprosy bacillus; animal transmission of leprosy. *Mailing Add:* 3121 Alki Ave Seattle WA 98116

KIRCHHOFF, WILLIAM HAYES, b Chicago, Ill, Nov 27, 36; m 58; c 4. PHYSICAL CHEMISTRY. *Educ:* Univ Ill, BS, 58; Harvard Univ, MA, 61, PhD(chem physics), 63. *Prof Exp:* NATO fel phys chem, Inorg Chem Lab, Oxford Univ, 62-63; Nat Res Coun res assoc chem physics, Nat Bur Standards, 64-66, physicist, 66-72, phys sci adminr affair & water measurement, 72-78, dept dir, Ctr Thermodynamics & Molecular Sci, 78-80, chief off environ measurements, 80-82, physicist, 83-87; TECH MGR CHEM PHYSICS, CHEM SCI DIV, OFF ENERGY RES, US DEPT ENERGY, 87- *Mem:* AAAS; Am Phys Soc; Am Chem Soc. *Res:* Statistical analysis of physical chemistry data including spectroscopy and thermodynamics; computer modelling of physical chemical systems. *Mailing Add:* Div Chem Sci US Dept Energy ER-141 GTN Washington DC 20585

KIRCHMAYER, LEON K, b Milwaukee, Wis, July 24, 24; m 50; c 2. ELECTRICAL POWER ENGINEERING. *Educ:* Marquette Univ, BS, 45; Univ Wis, MS, 47, PhD, 50. *Prof Exp:* Mgr power syst oper invests, Gen Elec Co, 56-58, mgr syst generation anal eng, 58-63, mgr syst planning & control, 63-77, mgr advan syst technol & planning, 77-84; RETIRED. *Concurrent Pos:* Dir ESIN, SAm consult firm; mem, Int Conf Large High Tension Elec Systs. *Mem:* Nat Acad Eng; fel Am Soc Mech Engrs; fel Inst Elec & Electronics Engrs; Opers Res Soc Am; Nat Soc Prof Engrs. *Res:* Computer applications; system technology; power generation. *Mailing Add:* 11 Rivercrest Dr Rexford NY 12148

KIRCHNER, ERNST KARL, b San Francisco, Calif, June 18, 37; m 60; c 3. BULK ACOUSTIC WAVES IN CRYSTALS, MICROWAVE DELAY DEVICES. *Educ:* Stanford Univ, BS, 59, MS, 60, PhD(elec eng), 63. *Prof Exp:* Res physicist, US Army Res & Develop Activ, Ft Huachuca, Ariz, 63-65; mem tech staff, Teledyne MEC, Palo Alto, Calif, 65-72, proj engr, 72-79, staff engr, 79-81, mgr, Mountain View, Calif, 81-82, opers mgr, 82-83, sr mgr, 83-84; mgr eng, 84-87, dir eng, 87-88, VPRES BUS DEVELOP, TELEDYNE MICROWAVE, MOUNTAIN VIEW, CALIF, 88-, VPRES, DELAY DEVICE PROD LINE, 90- *Concurrent Pos:* Teacher, Univ Ariz,

63-65. *Mem:* Inst Elec & Electronics Engrs; Am Phys Soc. *Res:* Acoustic wave propagation in single crystalline materials and their applications; magneto elastic waves in ferrites and acousto-optic interactions. *Mailing Add:* 41 Ashfield Rd Atherton CA 94027

KIRCHNER, FREDERICK KARL, b New YOrk, NY, May 10, 11; m 38; c 1. ORGANIC CHEMISTRY, INFORMATION SCIENCE. *Educ:* Maryville Col, AB, 34; Univ Tenn, MS, 35; Ohio State Univ, PhD(org chem), 38. *Prof Exp:* Asst prof chem, Bridgewater Col, 38-43; res chemist, Winthrop Chem Co, NY, 43-45; sr res chemist & group leader, Sterling-Winthrop Res Inst, 45-66, dir coord, 66-76; RETIRED. *Mem:* Am Chem Soc; fel NY Acad Sci; fel Am Inst Chem. *Res:* Synthesis of pharmaceuticals; scientific information research. *Mailing Add:* 1714 Glen Echo Rd Nashville TN 37215

KIRCHNER, H(ENRY) P(AUL), b Buffalo, NY, Sept 9, 23; m 50; c 3. CERAMICS. *Educ:* Cornell Univ, BME, 47; Pa State Univ, PhD(ceramics), 55. *Prof Exp:* Prod engr, Carborundum Co, 47-51; proj engr, Corning Glass Works, 54-57; prin ceramic engr, Aeronaut Lab, Cornell Univ, 57-60, asst head mat dept, 60-63; tech dir, Linden Labs, Inc, 63-66, vpres, 66-68; pres, Ceramic Finishing Co, 68-88; CONSULT, 88- *Mem:* Fel Am Ceramic Soc; Nat Inst Ceramic Engrs. *Res:* Physical and chemical properties of ceramic materials, especially refractories, abrasives and dielectrics; processes to improve strength; fractography; failure analysis; fracture mechanics. *Mailing Add:* 700 S Sparks St State College PA 16801

KIRCHNER, JAMES GARY, b Detroit, Mich, Sept 11, 38; m; c 5. PETROLOGY, GEOLOGY. *Educ:* Wayne State Univ, BS, 60, MS, 62; Univ Iowa, PhD(geol), 71. *Prof Exp:* Geologist petrol, Gulf Oil Corp, 62-66; PROF GEOL, ILL STATE UNIV, 69- *Mem:* Geol Soc Am; Nat Asn Geol Teachers; Sigma Xi. *Res:* Igneous geology of the northern Black Hills in South Dakota. *Mailing Add:* Dept Geog & Geol Ill State Univ Normal IL 61761

KIRCHNER, JOHN ALBERT, b Waynesboro, Pa, Mar 27, 15; m 47; c 5. OTOLARYNGOLOGY. *Educ:* Univ Va, MD, 40. *Hon Degrees:* MS, Yale Univ, 52. *Prof Exp:* Instr otolaryngol, Johns Hopkins Hosp, 48-49, resident, 49; from asst prof to prof, 51-85, EMER PROF OTOLARYNGOL, SCH MED, YALE UNIV, 85- *Concurrent Pos:* Commonwealth Fund fel otolaryngol, Royal Col Surg, 63-64; jr attend otolaryngologist, Children's, Deaconess, Christ & Good Samaritan Hosps, 49-50; assoc surgeon, Grace-New Haven Community Hosp, 51-; mem, otolaryngol post-grad training comt, NIH, 56-59. *Honors & Awards:* Mosher Award, 58; Casselbery Award, Am Laryngol Asn, 66, Newcomb Award, 69, DeRoaldes Medal, 85; Semon lectr, Univ London, 81. *Mem:* Am Laryngol Asn (pres, 79); Am Laryngol, Rhinol & Otol Soc (pres, 81-82); Am Acad Ophthal & Otolaryngol; fel Am Col Surg; Am Soc Head & Neck Surg (pres, 76); Ger Soc Otolaryngol. *Res:* Physiology of the larynx and pharynx; pathology of laryngeal cancer. *Mailing Add:* Yale-New Haven Med Ctr Box 3333 333 Cedar St New Haven CT 06510

KIRCHNER, RICHARD MARTIN, b San Francisco, Calif, Dec 1, 41. INORGANIC CHEMISTRY, STRUCTURAL CHEMISTRY. *Educ:* Univ Calif, Berkeley, AB, 64; Calif State Univ, San Jose, MS, 66; Univ Wash, PhD(chem), 71. *Prof Exp:* Fel chem, Northwestern Univ, 71-73; from asst prof to assoc prof, 73-87, PROF CHEM, MANHATTAN COL, 87- *Concurrent Pos:* Consult, Tarrytown Tech Ctr, Union Carbide Corp, 74-; res collabr chem, Brookhaven Nat Lab, 75-85. *Mem:* Am Crystallog Asn; Am Chem Soc; Am Phys Soc; Sigma Xi. *Res:* Preparation and characterization of transition metal complexes with unusual ligands; x-ray crystallography; structural characterization of micro-crystaline molecular sieves. *Mailing Add:* Dept Chem Manhattan Col Bronx NY 10471

KIRCHNER, ROBERT P, b Orange, NJ, Jan 21, 39; m 62; c 2. MECHANICAL ENGINEERING. *Educ:* Newark Col Eng, BS, 62, MS, 64; Rutgers Univ, PhD(mech eng), 68. *Prof Exp:* From asst instr to assoc prof, Newark Col Eng, 62-76, PROF MECH ENG, NJ INST TECHNOL, 76- *Concurrent Pos:* Consult, Solar Energy Syst Design. *Mem:* Am Soc Mech Engrs; Int Solar Energy Soc. *Res:* Thermodynamics; fluid mechanics; heat transfer; solar energy; energy conservation. *Mailing Add:* Dept Mech Eng NJ Inst Technol 323 High St Newark NJ 07102

KIRCHOFF, WILLIAM F, b New Rochelle, NY, Apr 28, 29; m 54; c 4. MICROBIOLOGY. *Educ:* Fordham Univ, BS, 51; Purdue Univ, MS, 57, PhD(microbiol), 59; Seton Hall Univ, JD, 72. *Prof Exp:* Sr microbiologist, Nuodex Prod Co Div, Tenneco Corp, 59-60; res microbiologist biochem res & develop, Chas Pfizer & Co, 60-65; tech coordr drug regulatory affairs, Hoffmann-La Roche, Inc, 65-72; dir res, Drug Regulatory Affairs, Sandoz Inc, 72-74, asst dir, Short Term Res & Develop, 74-77, consult, 77-79; SR ATTY, WARNER-LAMBERT CO, 79- *Mem:* Am Soc Microbiol; NY Acad Sci; Sigma Xi; Am Bar Asn. *Res:* Microbial fermentation; antibiotics; enzymes. *Mailing Add:* 284 Roseland Ave Essex Fells NJ 07021

KIRCZENOW, GEORGE, Australian citizen. THEORY OF SEMICONDUCTORS, THEORY OF INTERCALATION COMPOUNDS. *Educ:* Univ Western Australia, BSc, 70; Oxford Univ, D. Phil(theoret physics), 74. *Prof Exp:* Asst prof physics, Boston Univ, 79-83; assoc prof, 83-87, PROF PHYSICS, SIMON FRASER UNIV, 87- *Concurrent Pos:* Prin investr, NSF grant, 81-83; US Dept Energy, 83 & Natural Sci & Eng Res Coun, Can, 84- *Mem:* Am Phys Soc; Can Asn Physicists. *Res:* Theoretical solid state physics; semiconductors; superconductors; intercalation compounds; surface physics including scanning tunneling microscopy theory. *Mailing Add:* Dept Physics Simon Fraser Univ Burnbary BC V5A 1S6 Can

KIRDANI, RASHAD Y, b Cairo, Egypt, Oct 20, 29; US citizen; m 58; c 2. ORGANIC CHEMISTRY, BIOCHEMISTRY. *Educ:* Cairo Univ, BSc, 50, MSc, 55; Univ Buffalo, PhD(org chem), 61. *Prof Exp:* Trainee steroid training prog, Worcester Found Exp Biol & Clark Univ, 60-61; res assoc steroid chem, Worcester Found Exp Biol, 61-62, staff scientist steroid chem & biochem,

62-65; asst res prof chem, 72-86, ASSOC RES PROF BIOCHEM, STATE UNIV NY, BUFFALO, 86-; PRIN CANCER RES SCIENTIST, ROSWELL PARK MEM INST, 65- *Mem:* AAAS; Endocrine Soc; NY Acad Sci; Sigma Xi. *Res:* Chemistry and biochemistry of steroid hormones and interferons. *Mailing Add:* Roswell Park Mem Inst 666 Elm St Buffalo NY 14203

KIREMIDJIAN, ANNE SETIAN, b Sofia, Bulgaria, Aug 11, 49; US citizen; m 72; c 1. EARTHQUAKES. *Educ:* Columbia Univ, BS, 72; Stanford Univ, MS, 73, PhD(civil eng), 77. *Prof Exp:* From asst prof to assoc prof, 78-91, PROF STRUCT ENG, DEPT CIVIL ENG, STANFORD UNIV, 91- *Concurrent Pos:* Chmn, Risk Comt, Am Soc Civil Engrs, 85-; prin investr, NSF, mem adv comt, 88- *Mem:* Am Soc Civil Engrs; Seismol Soc Am; Int Asn Struct Safety & Reliability; Earthquake Eng Res Inst. *Res:* Earthquake occurrence modelings for long term forecasting; ground motion characterization from earthquake vibrations; damage estimation and forecasting models; application of stochastic processes to earthquake engineering; structural reliability analysis models. *Mailing Add:* 14210 Berry Hill Ct Los Altos Hills CA 94022

KIRK, ALEXANDER DAVID, b London, Eng, Apr 17, 34; Can citizen; c 3. INORGANIC CHEMISTRY, PHYSICAL CHEMISTRY. *Educ:* Univ Edinburgh, BSc, 56, PhD(chem), 59. *Prof Exp:* Res fel chem, Univ BC, 59-61; from asst prof to assoc prof, 61-71, chmn dept, 74-79, PROF CHEM, UNIV VICTORIA, BC, 71- *Concurrent Pos:* Humbolt fel, 68-69; vis prof, Univ Sussex, Sch Molecular Sci, 75-76; res assoc, Univ Southern Calif, Los Angeles, 80; actg dir, Can Ctr Picosecond Laser Flash Photolysis, Concordia Univ, Montreal, 83. *Mem:* Fel Chem Inst Can; Can Asn University Teachers; InterAm Photochemical Soc. *Res:* Photochemistry; inorganic photochemistry and luminescence; photochemistry and luminescence of coordination compounds. *Mailing Add:* Dept Chem Univ Victoria Victoria BC V8W 2Y2 Can

KIRK, BEN TRUETT, b Natchitoches, La, Oct 1, 42; m 66. PLANT PATHOLOGY. *Educ:* La Polytech Inst, BS, 64; La State Univ, MS, 66, PhD(plant path), 68. *Prof Exp:* From asst prof to assoc prof bot, Southeastern La Univ, 68-87. *Mem:* Am Phytopath Soc. *Res:* Fungal ultrastructure; systemic fungicides. *Mailing Add:* 103 Robinwood Dr Hammond LA 70403

KIRK, BILLY EDWARD, b Robinson, Ill, May 5, 27; div; c 2. VIROLOGY. *Educ:* Univ Ill, BS, 49; Ohio State Univ, MSc, 55, PhD(microbiol), 57. *Prof Exp:* Sr bacteriologist, Eli Lilly & Co, Ind, 57-62; instr, Univ Mich, 62-64; asst prof, 64-73, ASSOC PROF MICROBIOL, WVA UNIV, 73- *Mem:* AAAS; Am Soc Microbiol; Brit Soc Gen Microbiol; Tissue Cult Asn; Sigma Xi. *Res:* Virology; biology and pathogenic role of defective viruses; virus persistence. *Mailing Add:* 914 Guyasuta Lane Pittsburgh PA 15215

KIRK, DALE E(ARL), b Payette, Idaho, July 2, 18; m 39; c 5. AGRICULTURAL ENGINEERING. *Educ:* Ore State Univ, BS, 42; Mich State Univ, MS, 54. *Prof Exp:* Asst, Agr Eng, 41-42, asst agr engr, 42-44 & 46-54, assoc prof agr eng, 54-63, actg head dept, 70-71 & 80-81, PROF AGR ENG & AGR ENGR, ORE STATE UNIV, 63-, EMER PROF, 83- *Mem:* Fel Am Soc Agr Engrs; Sigma Xi. *Res:* Food engineering; processing and handling agricultural products; agricultural machine design. *Mailing Add:* Dept Bioresource Eng Ore State Univ Corvallis OR 97331

KIRK, DANIEL EDDINS, b Rocky Mount, NC, Feb 19, 24; m 46; c 4. BIOLOGY. *Educ:* Furman Univ, BS, 48; Univ NC, MA, 50; Emory Univ, PhD, 57. *Prof Exp:* Asst prof biol, Furman Univ, 50-57; PROF BIOL, CATAWBA COL, 57- *Mem:* Am Soc Parasitol; Am Micros Soc. *Res:* Helminthology; reptilian blood flukes. *Mailing Add:* Catawba Col Box 343 Salisbury NC 28144

KIRK, DAVID BLACKBURN, b Lock Haven, Pa, Nov 18, 21; m 44; c 2. MATHEMATICS. *Educ:* Haverford Col, BS, 43; Univ Pa, MA, 48. *Prof Exp:* Instr statist, Univ Pa, 47-48; methods analyst electronics comt, Mutual Benefit Life Ins Co, NJ, 48-56; sr res mathematician, Res Div, Curtiss-Wright Corp, Pa, 56-59; res mathematician, Willow Run Labs, Univ Mich, 59-66; head statist & res prog, Data Processing Div, Educ Testing Serv, Princeton NJ, 66-67; opers res scientist, 67-74; dir statist serv, Univ City Sci Ctr, 74-77; RETIRED. *Concurrent Pos:* Part-time teaching comput sci & numerical anal, Rutgers Univ, 69-70 & quant methods & statist, Rider Col, 70-74. *Mem:* Am Math Soc; Math Asn Am; Am Statist Asn; Asn Comput Mach; Sigma Xi. *Res:* Application of digital computers to mathematical, engineering and statistical research problems. *Mailing Add:* 1914 Yardley Rd Yardley PA 19067

KIRK, DAVID CLARK, b Newark, NJ, May 19, 24; m 53; c 3. PHYSICAL CHEMISTRY. *Educ:* Lehigh Univ, BS, 44; Polytech Inst New York, MS, 51; Univ Iowa, PhD(chem), 53; Furman Univ, MBA, 79. *Prof Exp:* Chem engr, Am Dyewood Co, 46-48; chemist, Merck & Co, 48-50; res chemist, Hercules Powder Co, 53-59; dir fundamental res, Ecusta Paper Div, Olin Corp, 59-69, dir res & develop, 69-76; TECH DIR, ALLIED PAPER DIV, SCM CORP, 76- *Mem:* Am Chem Soc; Sigma Xi; Tech Asn Pulp & Paper Indust. *Res:* Organic chemistry; kinetics, photo and surface chemistry of protective coatings; elastomers and papers. *Mailing Add:* Rte 3 River Ridge Box 158 Brevard NC 28712

KIRK, DAVID LIVINGSTONE, b Clinton, Mass, Mar 19, 34; m 58; c 1. DEVELOPMENTAL BIOLOGY. *Educ:* Northeastern Univ, AB, 56; Univ Wis, MS, 58, PhD(biochem), 60. *Prof Exp:* Res technician, Lovett Mem Lab, Mass Gen Hosp, Boston, 52-56; res asst, Univ Wis, 56-60; sr res chemist, Biol Res Labs, Colgate-Palmolive Co, 60-62; res assoc develop biol, Univ Chicago, 62-65, asst prof, 65-69; assoc prof, 69-79, PROF, BIOL, WASH UNIV, 79- *Concurrent Pos:* Dean Grad Sch, Wash Univ, 79. *Mem:* AAAS; Am Chem Soc; Soc Develop Biol; Soc Cell Biol; Sigma Xi. *Res:* Developmental biochemistry and genetics; analysis of genetic, cytological and molecular basis of cell determination and cytodifferentiation in simple eukaryotes. *Mailing Add:* Dept Biol Wash Univ St Louis MO 63130

KIRK, DONALD EVAN, b Baltimore, Md, Apr 4, 37; m 62; c 3. ELECTRICAL ENGINEERING. *Educ:* Worcester Polytech Inst, BS, 59; Naval Postgrad Sch, MS, 61; Univ Ill, PhD(elec eng), 65. *Prof Exp:* Instr elec eng, Naval Postgrad Sch, 59-62; teaching asst, Univ Ill, 62-63, instr, 63-64; from asst prof to assoc prof, Naval Postgrad Sch, 65-76, chmn dept, 76-83, prof elec eng, 76-87; assoc dean eng, 87-90, PROF ELEC ENG, SAN JOSE STATE UNIV, 90- *Concurrent Pos:* Vis staff scientist, Lincoln Lab, Mass Inst Technol, 81-82. *Mem:* Inst Elec & Electronics Engrs; Am Soc Eng Educ; Sigma Xi. *Res:* Signal processing. *Mailing Add:* 3369 Trevis Way Carmel CA 93923

KIRK, DONALD WAYNE, b Carleton Place, Ont, Aug 18, 34; m 58; c 3. CIVIL ENGINEERING. *Educ:* Queen's Univ, BSc, 56, MSc, 65, PhD(struct eng), 69. *Prof Exp:* from lectr to assoc prof, 62-75, head dept civil eng, 77-84, PROF STRUCT, ROYAL MIL COL CAN, 75-; DEAN CAN FORCES MIL COL, 84- *Honors & Awards:* Duggan Medal, Eng Inst Can, 65. *Mem:* Am Concrete Inst; Eng Inst Can; Am Soc Eng Educ. *Res:* Ultimate strength of reinforced concrete slab and girder systems; analysis and design of beams containing web openings; slab column connections. *Mailing Add:* Can Forces Mil Col Royal Mil Col Kingston ON K7K 5L0 Can

KIRK, IVAN WAYNE, b Lark, Tex, Jan 25, 37; m 60; c 2. COTTON PRODUCTION, COTTON PROCESSING. *Educ:* Tex Tech Univ, BS, 59; Clemson Univ, MS, 61; Auburn Univ PhD(agr eng), 68. *Prof Exp:* Res agr engr, Agr Res Serv, USDA, Lubbock, 60-65, Auburn, 65-67, Lubbock, 67-71, lab dir, NMex, 71-77, assoc dir, New Orleans, 77-80, actg dir, 80-82, dir, Southern Regional Res Ctr, 82-87, AGR ENGR, PESTICIDE MGT RES, AGR RES SERV, USDA, COLLEGE STATION, TEX, 87- *Concurrent Pos:* Instr & asst prof, Dept Agr Engr, Tex Tech Univ, 63-65. *Honors & Awards:* Arthur S Fleming Award, 75. *Mem:* AAAS; Am Soc Agr Engrs; Coun Agr Sci & Technol. *Res:* New and improved methods, and machinery for cotton production, harvesting, and ginning; improved technology for aerial application of pesticides. *Mailing Add:* 231 Scoates Hall Tex A&M College Station TX 77843

KIRK, JAMES CURTIS, b Hubbard, Tex, May 10, 21; m 44; c 5. ORGANIC CHEMISTRY. *Educ:* Baylor Univ, BS, 44; Ohio State Univ, PhD(chem), 49. *Prof Exp:* Analyst, Pan Am Ref Corp, 44-46; asst chem, Ohio State Univ, 46-49; from assoc res chemist to sr res chemist, Continental Oil Co, 49-53, res group leader, 53-55, supvry res chemist, 55-57; dir res, Petrol Chem, Inc, 57-60; dir, Petrochem Res Div, Res & Develop Dept, 60-66, dir, Environ Conserv, Res & Eng Dept, 66-67, gen mgr, 67-75, VPRES RES & DEVELOP DEPT, CONOCO INC, 75- *Mem:* Am Chem Soc; Soc Petrol Engrs. *Res:* Hydrocarbon oxidation; lubricating oil additives; surface active agents; reaction mechanisms; polymerication; research administration. *Mailing Add:* 1308 Arronimink Circle Austin TX 78746

KIRK, JOE ECKLEY, JR, b Houston, Tex, May 17, 39; m 67; c 3. MATHEMATICS. *Educ:* Sam Houston State Univ, BA, 60; Univ Tex, Austin, MA, 62, PhD(math), 67. *Prof Exp:* Opers res analyst, US Arms Control & Disarmament Agency, 67-69; asst prof math, Univ Wyo, 69-74; asst prof math, Univ Tenn, Chattanooga, 74-76, assoc prof, 76-80; assoc prof math, 80-88, PROF MATH, SAM HOUSTON STATE UNIV, 88- *Mem:* Asn Comput Mach; Am Math Soc; Math Asn Am. *Res:* Complex analysis; function theory. *Mailing Add:* Dept Math Sam Houston State Univ Huntsville TX 77341

KIRK, JOHN GALLATIN, b Wilmington, Ohio, Oct 21, 38. SOLAR PHYSICS, AEROSPACE SCIENCES. *Educ:* Amherst Col, AB, 60; Univ Mich, AM, 62, PhD(astron), 66. *Prof Exp:* Jr astronr, Kitt Peak Nat Observ, 66-69; asst prof astron, Univ Toledo, 69-74; staff scientist, Comput Sci Corp, 74-79; sr analyst, Electronics Div, Gen Dynamics Corp, 79-80, MEM PROF STAFF, GEODYNAMICS CORP, 80- *Honors & Awards:* Tech Innovation Award, NASA, 81. *Mem:* Am Astron Soc; Sigma Xi; Am Geophys Union Inst Navig. *Res:* Physics of the solar atmosphere; physical geodesy; satellite orbit management. *Mailing Add:* 325 Palisades Dr Santa Barbara CA 93109

KIRK, KENT T, b 1940. MICROBIAL BIOCHEMISTRY, OXIDATIVE PROCESSES. *Educ:* La Polytech Inst, BS, 62; NC State Univ, MS, 64, PhD(biochem & plant path), 68. *Prof Exp:* Postdoctoral polymer chem, NC State Univ, 67-68; postdoctoral org chem, Chalmers Univ Technol, Sweden, 68-69; res scientist, 70-80, proj leader, 80-85, DIR, INST MICROBIAL & BIOCHEM TECHNOL, FOREST SERV, USDA, 85-; USDA PROF, DEPT BACT, UNIV WIS-MADISON, 82- *Concurrent Pos:* Vis prof, Kyoto Univ, Japan, 79-80; chmn, Gordon Res Conf Chem & Mat Natural Resources, 82; co-organizer, Int Conf Biotechnol in Pulp & Paper Indust, 89; consult lignin biodegradation & applications of bio-ligninolytic systs, industs, univs & res insts. *Honors & Awards:* William H Aiken Prize, Tech Asn Pulp & Paper Indust, 86; Marcus Wallenberg Prize, Sweden, 85. *Mem:* Nat Acad Sci; fel Int Acad Wood Sci (secy-treas, 85-90); Am Soc Microbiol; Am Chem Soc; Am Soc Biochem & Molecular Biol; Tech Asn Pulp & Paper Indust. *Res:* Biochemistry and physiology of wood decomposition by fungi, industrial application of fungi and enzymes; author of over 100 articles in scientific journals. *Mailing Add:* Forest Prod Lab Univ Wis Madison WI 53706

KIRK, MARILYN M, b Bridgeport, Nebr, May 8, 27; m 58; c 1. DEVELOPMENTAL BIOLOGY. *Educ:* Univ Nebr, BS, 48; Univ Wis, MS, 54, PhD(nutrit, biochem), 56. *Prof Exp:* Asst nutrit, Univ Nebr, 48-52; asst prof foods & nutrit, Sch Home Econ, Univ Wis, 56-60; res assoc, Am Meat Inst Found, Ill, 63-64; res assoc, Dept Biol, Univ Chicago, 65-69; RES ASSOC, DEPT BIOL, WASH UNIV, 69- *Res:* Biochemical studies of development and cytodifferentiation in simple eukaryotes. *Mailing Add:* Dept Biol Wash Univ Lindell & Skinker St Louis MO 63130

KIRK, PAUL WHEELER, JR, b Jacksonville, Fla, Feb 23, 31; m 58; c 2. MYCOLOGY, BACTERIOLOGY. *Educ:* Univ Richmond, BS, 57, MS, 61; Duke Univ, PhD(bot), 66. *Prof Exp:* Asst prof biol, Western Carolina Col, 65-66; asst prof bot, Va Polytech Inst, 66-70; assoc prof biol, 71-77, asst dean sci & health professions, 73-78, PROF BIOL, OLD DOMINION UNIV, 77-,. *Concurrent Pos:* Consult med microbiol & pre-health prof adv. *Mem:* Mycol Soc Am; Sigma Xi; Nat Asn Adv Health Prof. *Res:* Marine ascomycetes and deuteromycetes. *Mailing Add:* 1213 Kittery Dr Virginia Beach VA 23464

KIRK, R(OBERT) S(TEWART), b Chicago, Ill, Nov 2, 22; m 58; c 2. CHEMICAL ENGINEERING. *Educ:* Ill Inst Technol, BS & MS, 43; Univ Wis, PhD(chem eng), 48. *Prof Exp:* Asst prof chem eng, Univ Wis, 48-55; res engr, Calif Res Corp, 55-58, group supvr thermal recovery, 58-64, sr res engr, Chevron Res Co, Standard Oil Co Calif, 64-66; ASSOC PROF CHEM ENG, UNIV MASS, AMHERST, 66- *Mem:* Am Chem Soc; Am Inst Chem Engrs. *Res:* Process design and evaluation; chemical kinetics and reactor design; thermal methods of secondary recovery of crude oil. *Mailing Add:* PO Box 355 Leverett MA 01054-0355

KIRK, ROBERT WARREN, b Stamford, Conn, May 20, 22; m 49; c 3. VETERINARY MEDICINE. *Educ:* Univ Conn, BS, 43; Cornell Univ, DVM, 46; Am Col Vet Internal Med, dipl & cert internal med & dermat. *Prof Exp:* Pvt pract, 46-50; from asst prof to assoc prof, 52-74, chmn dept small animal med & surg & dir small animal hosp, 69-77, PROF MED, STATE UNIV NY VET COL, CORNELL UNIV, 74- *Concurrent Pos:* Fel, Sch Med, Univ Colo, 60-61; NSF sci fac fel, Sch Med Stanford Univ, 67-69; mem grants adv bd, Seeing Eye Found, NY, 70-73; Evelyn Williams fel & vis scholar, Univ Sydney, Australia, 74; pres, Am Col Vet Internal Med, 74-76; vis prof, Sch Med Stanford Univ, 75; pres & chmn bd regents, Am Col Vet Internal Med; trustee, Seeing Eye Found, 77- *Honors & Awards:* Fido Award, Am Animal Hosp Asn, 64; Gaines Medal, 67. *Mem:* Am Vet Med Asn; Am Animal Hosp Asn; Am Col Vet Dermat (pres). *Res:* Clinical medicine and dermatology therapeutics. *Mailing Add:* 84 Turkey Hill Rd Ithaca NY 14850

KIRK, ROGER E, b Princeton, Ind, Feb 23, 30; m 83. EXPERIMENTAL DESIGN. *Educ:* Ohio State Univ, BS, 51, MA, 52, PhD(psychol), 55. *Prof Exp:* Sr psychoacoust engr, Baldwin Piano & Organ Co, 55-58; from asst prof to assoc prof, 58-64, PROF PSYCHOL, BAYLOR UNIV, 64-, DIR, INST GRAD STATIST, 91-; PRES, RES CONSULTS, 82- *Concurrent Pos:* Postdoctoral statist, Univ Mich, 71; vis prof, Seinan Gaukin Univ, Fukuoka, Japan, 73-74; dir, Behav Statist Prog, Baylor Univ, 76-; assoc ed, J Educ Statist, 76-89; adv ed statist, Contemp Psychol, 81-85; mem-at-large div five exec coun, Am Psychol Asn, 89-91. *Mem:* Fel Am Psychol Asn; Am Statist Asn; Psychometric Soc; Human Factors Soc; fel Am Psychol Soc. *Res:* Statistical methodology; author of various publications. *Mailing Add:* Psychol Dept Baylor Univ Waco TX 76798-7334

KIRK, THOMAS BERNARD WALTER, b Denver, Colo, June 13, 40; m 74; c 2. HIGH ENERGY PHYSICS. *Educ:* Univ Colo, Boulder, BS, 62; Univ Wash, MS, 64, PhD(physics), 67. *Prof Exp:* Fel physics, Harvard Univ, 67-69, from asst prof to assoc prof, 69-73; assoc prof physics, Univ Ill, Urbana, 73-76; head neutrino dept, Fermi Nat Accelerator Lab, 76-81, Tev II proj mgr, physics res div head, 81-89; DIR HEP DIV, ARGONNE NAT LAB, 89- *Concurrent Pos:* Mem prog adv comt, Fermilab, 70-72; consult, Dept Eng, Can Sci Coun, 84- *Mem:* Fel Am Phys Soc. *Res:* Experimental investigation of fundamental particle processes in strong, electromagnetic and weak interactions at high energies. *Mailing Add:* Argonne Nat Lab 9700 S Cass Ave Bldg 362 Argonne IL 60439-4815

KIRK, VERNON MILES, economic entomology, for more information see previous edition

KIRK, WILBER WOLFE, b Brownsville, Pa, Sept 21, 32; m 54; c 4. MARINE CORROSION, CONSULTING. *Educ:* Otterbein Col, BS, 54; Ohio State Univ, MS, 59. *Prof Exp:* Engr, Bettis Atomic Power Lab, Westinghouse Corp, 58-62; engr, 62-67, pres, 68-90, SR TECH ADV, LAQUE CTR CORROSION TECHNOL, INCO, LTD, 91- *Concurrent Pos:* Chmn, Offshore Technol Conf, 77; chmn, indust panel, Nat Sea Grant Prog, 81-83 & Subcomt Atmospheric Corrosion, Am Soc Testing & Mat, 86- *Mem:* Nat Asn Corrosion Engrs; Am Soc Testing & Mat; Minerals, Metals & Mat Soc; fel Am Soc Metals Int. *Res:* Materials in marine environments, including steels, stainless steels, copper alloys, nickel alloys, aluminum and titanium alloys; corrosion and protection. *Mailing Add:* Rte 1 Box 248D Ivanhoe NC 28447-9801

KIRK, WILEY PRICE, b Joplin, Mo, July 24, 42; m 64; c 2. CONDENSED MATTER PHYSICS, NANOSTRUCTURES & LOW TEMPERATURE PHYSICS. *Educ:* Wash Univ NY, BA, 64; State Univ NY, Stony Brook, MS, 67, PhD(physics), 70. *Prof Exp:* Jr res assoc physics, Brookhaven Nat Lab, 67-69; instr, State Univ NY Stony Brook, 69-70; fel, Univ Fla, 70-71, interim asst prof, 71-73; asst prof, 73-75; from asst prof to prof physics, 75-84, PROF PHYSICS & ELEC ENG, TEX A&M UNIV, 86- *Concurrent Pos:* Tech collabr, Brookhaven Nat Lab, 69-70; consult, Sci Instruments, Inc, 70-75; Nalorac Cryog Corp, 80-84, Tex Instruments, Inc, 83-; dir, Ctr Nanostruct Mat & Quantum Device Fabrication, 90- *Mem:* AAAS; Am Phys Soc; Sigma Xi; Am Vacuum Soc; Mat Res Soc; Inst Elec & Electronics Engrs. *Res:* 2-D charge transport and magnetoconduction; quantum Hall effect; nanostructures and electron-beam patterning; transport in mesoscopic systems and super-ethics; molecular beam epitaxy; quantum effect devices; thermodynamic, magnetic and nuclear magnetic resonance properties of materials; quantum crystals of helium; methods of low temperature thermometry; cryogenic and superconducting devices; low temperature thermoelectric studies; millikelvin-thermocouple-thermometry; superconducting quantum interference detector and pulsed nuclear magnetic resonance techniques; high temperature superconductivity; magnetic surface effects. *Mailing Add:* Eng-Physics Bldg Tex A&M Univ College Station TX 77843-4242

KIRK, WILLIAM ARTHUR, b Montour Falls, NY, Oct 3, 36; m 59; c 3. MATHEMATICS. *Educ:* DePauw Univ, AB, 58; Univ Mo, MA, 60, PhD(math), 62. *Prof Exp:* Asst prof math, Univ Calif, Riverside, 62-67; assoc prof, 67-70, PROF MATH, UNIV IOWA, 70-, CHMN DEPT, 85- *Mem:* Am Math Soc; Math Asn Am. *Res:* Metric and geodesic geometry; functional analysis. *Mailing Add:* Dept Math Univ Iowa Iowa City IA 52242

KIRK, WILLIAM LEROY, b Charleston, Miss, Aug 29, 30; m 53; c 3. NUCLEAR ENGINEERING. *Educ:* US Naval Acad, BS, 52; Air Force Inst Technol, MS, 57. *Prof Exp:* Staff mem, Nuclear Rocket Div, Los Alamos Nat Lab, 61-67, asst div leader, 67-73, assoc div leader energy technol, 77-78, alt div leader, 78-81, prin asst div leader, 81-83, dep div leader, Energy Div, 83-87, DEP DIV LEADER, NUCLEAR TECHNOL & ENG DIV, LOS ALAMOS NAT LAB, 87- *Res:* Energy research and development, including nuclear safeguards and safety, nuclear system development and other energy systems. *Mailing Add:* Energy Div Los Alamos Nat Lab MS-E561 PO Box 1663 Los Alamos NM 87545

KIRKALDY, J(OHN) S(AMUEL), b Victoria, BC, May 13, 26; m 52; c 3. PHYSICAL METALLURGY. *Educ:* Univ BC, BASc, 49, MASc, 51; McGill Univ, PhD(physics), 53. *Prof Exp:* Res assoc physics, McGill Univ, 53-54, asst prof metall eng, 54-57; from asst prof to assoc prof metall, 57-63, chmn dept, 62-66, PROF METALL, MCMASTER UNIV, 63-, STEEL CO CAN CHAIR METALL, 66- *Mem:* Am Soc Metals; Am Inst Mining, Metall & Petrol Engrs; Can Asn Physicists; Can Inst Mining & Metall; Sigma Xi. *Res:* Application of thermodynamics of irreversible processes to metallurgy. *Mailing Add:* McMaster Univ Eng Bldg 1280 Main St Hamilton ON L8S 4M1 Can

KIRKBRIDE, CHALMER GATLIN, b Tyrone, Okla, Dec 27, 06; wid. PETROLEUM ENGINEERING. *Educ:* Univ Mich, BSE & MSE, 30; Beaver Col, ScD, 59. *Hon Degrees:* DSc, Beaver Women's Col; DEng, Drexel Univ, Widner Univ. *Prof Exp:* Petrol scientist, Fed Energy Off, 74-75; sci adv, Energy Res Develop Admin, 75-77; dir, Kirkbride Assoc, 63-70, pres, 76-89; RETIRED. *Mem:* Nat Acad Eng; Am Inst Chem Eng; emer mem Am Chem Soc. *Mailing Add:* 4000 Massachusetts Ave NW Suite 805 Washington DC 20016

KIRKBRIDE, CLYDE ARNOLD, b Los Angeles, Calif, Mar 14, 24; m 44; c 5. VETERINARY BACTERIOLOGY. *Educ:* Okla State Univ, DVM, 53; SDak State Univ, MS, 70. *Prof Exp:* Vet practr, 53-63; asst prof vet med, Col Vet Med, Kans State Univ, 63-67; from instr to assoc prof, 67-82, PROF VET MED, ANIMAL DIS RES & DIAG LAB, SDAK STATE UNIV, 82- *Concurrent Pos:* Vet investr officer, NZ Ministry Agr & Fisheries, 74-75; pres, Am Leptospirosis Res Conf, 81; pres-elect, Western Vet Conf, 88. *Mem:* Am Vet Med Asn; Am Asn Vet Lab Diagnosticians; US Animal Health Asn. *Res:* Water intoxication in cattle; relationship of milking machine function to mastitis in cattle; diseases affecting reproduction in animals; fetal serology in diagnosis of bovine abortion; swine tuberculosis; diagnosis of leptospirosis; nonclassified anaerobic bacterium that causes abortion in sheep. *Mailing Add:* 2015 Iowa St Brookings SD 57006

KIRKBRIDE, JOSEPH HAROLD, JR, b St Louis, Mo, Feb 4, 43; m 75; c 2. TAXONOMIC BOTANY. *Educ:* St Louis Univ, BA, 66, MS(R), 68; City Univ New York, PhD(biol), 75. *Prof Exp:* Assoc cur bot, Smithsonian Inst, 75-79; adj prof, Univ Brasilia, 79-84; BOTANIST, AGR RES SERV, USDA, 84- *Concurrent Pos:* Res assoc, Smithsonian Inst, 79-; consult, Interamer Inst Coop Agr, 83; mem bd adv, Int Ctr Trop Ecol, 85- *Mem:* Am Asn Plant Taxonomists; Int Asn Plant Taxon; Asn Trop Biol; Sigma Xi; Bot Soc Brasil. *Res:* Taxonomic revision of plants cultivated on American farms; taxonomic revision of selected neotropical Rubiaceae. *Mailing Add:* Agr Res Serv USDA Beltsville MD 20705

KIRKBRIDE, L(OUIS) D(ALE), b Morris, Ill, Oct 18, 32; m 57; c 3. MEDICAL INSTRUMENTATION. *Educ:* Carnegie Inst Technol, BS, 54, MS & PhD(metall), 57. *Prof Exp:* Develop engr nuclear mat, Knolls Atomic Power Lab, Gen Elec Co, 57-60, mgr core mat develop, 60-61; group leader power reactor mat, Los Alamos Sci Lab, 61-66; proj analyst, Gen Elec Res & Develop Ctr, Schenectady, 66-68, mgr clin equip develop, Med Develop Oper, 68-70, mgr bus develop & strategic planning, Med Systs Bus Div, Gen Elec Co, 70-71, mgr nuclear diag, Med Systs Div, 71-74; gen mgr, Diag Div, J T Baker, 74-81; vpres & gen mgr, Lab Prod Div, Litton Bionetics, 81-85; VPRES, BIONETICS LAB PROD, ORGANON TEKNIKA CORP, 85-, VPRES, MKT. *Mem:* Am Nuclear Soc; Instrument Soc Am; fel Am Inst Chemists. *Res:* Semiconductor materials; radioimmunoassay; clinical chemistry; hematology; immunology; enzyme immunoassay. *Mailing Add:* 12712 Waterman Dr Raleigh NC 27614

KIRKEMO, HAROLD, b Birney, Mont, Oct 16, 15; m 41; c 2. GEOLOGY. *Educ:* Univ Wash, BS, 38; Indust Col Armed Forces, dipl, 63. *Prof Exp:* Mine geologist, Anaconda Copper Mining Co, Mont, 41-45; explor geologist, Thailand, 46-47, Pac Northwest & BC, 47-52; geologist defense minerals explor prog, US Geol Surv, 52-53, chief, Colo Plateau Uranium Explor Prog, 54-56, geologist minerals resources sect & resources res br, DC, 56-69, chief officer minerals explor, 69-79; RETIRED. *Mem:* Soc Econ Geologists; Am Inst Mining, Metall & Petrol Eng; Geol Soc Am. *Res:* Economic, mining and structural geology. *Mailing Add:* 638 Duquesna Dr Sun City Center FL 33573

KIRKENDALL, ERNEST OLIVER, b East Jordon, Mich, July 6, 14; m 38, 78; c 3. DIFFUSION IN SOLID STATE METALS. *Educ:* Wayne State Univ, BS, 34; Univ Mich, MSE, 35, DSc(metall eng), 38. *Prof Exp:* From instr to asst prof chem eng, Wayne State Univ, 37-46; asst secy, Am Inst Mining & Metall Engrs, 46-55; gen secy, Am Inst Mining, Metall & Petrol Engrs, 55-63; secy & gen mgr, United Eng Trustees, 63-65; metall engr, Am Iron & Steel Inst, 65-66, asst vpres, 66-68, vpres, 68-79; RETIRED. *Concurrent Pos:* Secy, Eng Found, 63-65; adj prof, Univ DC, 82-85. *Mem:* Iron & Steel Soc; Minerals Metals & Mat Soc; Am Soc Metals Int; hon mem Inst Metals. *Res:* Kirkendall Effect, difference in the diffusion rates of copper and zinc in copper plated alpha brass causing the interface to move. *Mailing Add:* 5100 Fillmore Ave No 909 Alexandria VA 22311-5047

KIRKENDALL, THOMAS DODGE, b Columbus, Ohio, Sept 8, 37; m 61; c 3. ANALYTICAL CHEMISTRY, SPECTROSCOPY. *Educ:* Colby Col, BA, 61. *Prof Exp:* Asst physics, Middlebury Col, 61-62; res scientist, Machlett Labs, Raytheon Co, 62-69; tech staff mem, 69-81, mgr, Dept Anal Chem & Failure Anal, 81-89, MGR, DEPT SEMICONDUCTOR RELIABILITY & QUAL ASSURANCE, 89- *Honors & Awards:* NASA ATS-6 Propagation Exp Award, Commun Satellite Corp, 74 & Centimeter Wave Beacon Award, 76; Outstanding Mem of the Year, Soc Appl Spectros, 80. *Mem:* Soc Appl Spectros (treas, 74-76, chmn, 77-78); Microbeam Anal Soc; Am Chem Soc; Electron Microscopy Soc Am. *Res:* Physics and failure mechanisms of solid state devices; energy conversion and storage; surface analysis and characterization. *Mailing Add:* Dept Analytical Chem & Failure Analysis COMSAT Labs Clarksburg MD 20871

KIRKENDALL, WALTER MURRAY, b Louisville, Ky, Mar 31, 17; m 48; c 10. INTERNAL MEDICINE. *Educ:* Univ Louisville, MD, 41. *Prof Exp:* Asst anat, Univ Louisville, 38-39; from asst to assoc internal med, Univ Iowa, 49-51, asst prof, 51-52, clin assoc prof, 52-58, from assoc prof to prof, 58-71; chmn dept, 72-76, PROF MED, UNIV TEX MED SCH, HOUSTON, 72-, DIR HYPERTENSION DIV, 76- *Concurrent Pos:* Chief med serv, Vet Admin Hosp, Iowa City, 52-58; dir cardiovasc res lab, Univ Iowa, 58-70, dir renal-hypertension-electrolyte div, 58-71. *Mem:* AAAS; AMA; Am Heart Asn; Am Soc Internal Med; Am Soc Nephrology; Sigma Xi. *Res:* Clinical investigation of kidney disease and hypertension. *Mailing Add:* Dept Internal Med Univ Tex Health Sci Ctr Med Sch PO Box 20708 Houston TX 77225

KIRKHAM, DON, b Provo, Utah, Feb 11, 08; m 39; c 3. SOIL PHYSICS, LAND DRAINAGE. *Educ:* Columbia Univ, BA, 33, MA, 34, PhD(physics), 38. *Hon Degrees:* Dr Agr, Royal Agr Univ, Ghent, Belgium, 63. *Prof Exp:* Instr & asst prof math & physics, Utah State Univ, Logan, 37-40; civilian scientist, US Navy, 40-46; from assoc prof to prof soils & physics, 46-59, CURTISS DISTINGUISHED PROF AGR, SOILS & PHYSICS, IOWA STATE UNIV, AMES, 59-, EMER PROF SOILS & PHYSICS, 78- *Concurrent Pos:* Res prof, Fulbright-Hays grantee, Holland, 50-51 & Belg, 57-58, Guggenheim award, 57-58 & Ford Found appointee, Egypt, 61; consult, TAMS dam construct, eng co, Turkey, 59 & Orgn Am States, Arg, 65; dir, Iowa State Water Resources Inst Res, 64-73; vis prof, Hohenheim-Stuttgart Agr Univ, Ger, 82; Lectr, People's Repub China, 85. *Honors & Awards:* Stevenson Award, Soil Sci Soc Am, 52; Wolf Prize Agr, 84. *Mem:* Fel Am Phys Soc; fel Am Soc Agron; fel & hon mem Soil Sci Soc Am; hon mem Int Soil Tillage Res Orgn; Am Geophys Union; Math Asn Am; Int Soil Sci Soc. *Res:* Application of basic physics and mathematics to soil problems of agriculture, for the design of land drainage systems and other physical land management practices to increase soil fertility and maintain soil environment. *Mailing Add:* 2109 Clark Ave Ames IA 50010

KIRKHAM, M B, b Cedar Rapids, Iowa. DROUGHT PHYSIOLOGY. *Educ:* Wellesley Col, BA; Univ Wis-Madison, MS; PhD(bot). *Prof Exp:* NSF fel, Inst Environ Studies, Univ Wis-Madison; plant physiologist, US Environ Protection Agency, Cincinnati, Ohio; plant physiologist, Univ Mass, Amherst; crop physiologist, Okla State Univ, Stillwater; PLANT PHYSIOLOGIST, KANS STATE UNIV, MANHATTAN, 80- *Concurrent Pos:* Vis lectr, evapotranspiration, China, 85, Italy, 89; vis scholar, Harvard Univ, 90; vis scientist, Dept Sci & Indust Res, Palmerston North, 91. *Honors & Awards:* Travel Award, Soil Sci Soc Am, 90. *Mem:* Am Soc Plant Physiol; Am Meteorol Soc; fel Soil Sci Soc Am; fel Am Soc Agr; Int Soc Hort Sci; fel AAAS; Crop Sci Soc Am. *Res:* Plant-soil-water relationships; uptake of trace elements by plants; effect on plants of elevated levels of carbon dioxide. *Mailing Add:* 1420 McCain Lane 244 Manhattan KS 66506

KIRKHAM, WAYNE WOLPERT, veterinary science, microbiology; deceased, see previous edition for last biography

KIRKHAM, WILLIAM R, pathology, biochemistry; deceased, see previous edition for last biography

KIRKIEN-RZESZOTARSKI, ALICJA M, b Lodz, Poland; m 73. PHYSICAL CHEMISTRY. *Educ:* Polish Univ Col, London, MChEng, 51; Univ London, PhD(phys org chem), 55. *Prof Exp:* From asst prof to assoc prof phys chem, Univ Col West Indies, 56-61; assoc prof, Univ West Indies, 61-65; assoc prof, 65-69, PROF & CHAIR PHYS CHEM, TRINITY COL DC, 69- *Concurrent Pos:* Hon res fel, Univ Col, Univ London, 71-72; res assoc, George Washington Univ, 84. *Mem:* Royal Chem Soc; fel Royal Inst Chem; Am Chem Soc. *Res:* Physical organic chemistry; kinetics of reactions in solutions and in the gas-phase, kinetic isotope effects; organic mass spectrometry, effect of chemical structure on ionization potentials; fragmentation patterns; high performance liquid chromatography; history of science. *Mailing Add:* 407 Buckspur Ct Millersville MD 21108-1764

KIRKLAND, GORDON LAIDLAW, JR, b Troy, NY, June 4, 43; m 66; c 1. MAMMALOGY, VERTEBRATE ECOLOGY. *Educ:* Cornell Univ, BS, 65; Mich State Univ, MS, 68, PhD(ornith), 69. *Prof Exp:* From asst prof to assoc prof, 69-78, Cur, Vert Mus, 71-89, PROF BIOL, SHIPPENSBURG UNIV, 78-, DIR, VERT MUS, 89- *Concurrent Pos:* Res collabr, Dept Vert Zool, Smithsonian Inst, 78-88; res assoc, Sect Mammals, Carnegie Mus, 73-; vis prof, Mountain Lake Biol Sta, Univ Va, 87; pres, Pa Chap, Wildlife Soc, 89-90. *Mem:* Am Soc Mammal (secy-treas, 80-86, dir, 86-89); Wildlife Soc; Am Soc Naturalists; Soc Syst Zoologists; Ecol Soc Am; fel AAAS. *Res:* Mammalian ecology, systematics, and natural history; boreal and sub-boreal ecology; ecology of vertebrates on disturbed ecosystems; mammalian zoogeography. *Mailing Add:* Vert Mus Box F-207 Shippensburg Univ Shippensburg PA 17257

KIRKLAND, JAMES T, b Mt Kisco, NY, Mar 27, 43. ENGINEERING GEOLOGY. *Educ:* Syracuse Univ, BA, 66; State Univ NY, MS, 69, PhD(geol). *Prof Exp:* Asst prof geol, Univ Tex, Arlington, 75-79; asst prof, Univ Mo, Kans City, 79-81; consult & sr geologist, Core Lab, Dallas, Tex, 81-83; consult geologist, 83-87; CONSULT GEOLOGIST, SCHNABEL ENG ASSOC, BETHESDA, MD, 87- *Mem:* Am Inst Prof Geologists; fel Geol Soc Am; Sigma Xi; Asn Eng Geologists. *Res:* Expansive soils and geomorphology. *Mailing Add:* 6625 Fairfax Chevy Chase MD 20815

KIRKLAND, JERRY J, b Elk City, Okla, May 18, 36; m 57; c 4. MICROBIOLOGY. *Educ:* Northwestern State Col, Okla, BS, 58; Okla State Univ, MS, 61, PhD(microbiol), 64. *Prof Exp:* MICROBIAL PHYSIOLOGIST, PROCTER & GAMBLE, 64- *Mem:* Am Soc Microbiol. *Res:* Inducible enzyme formation in microorganisms and their role in dental plaque; microbiology of skin; etiology of acne; toxic shock syndrome; etiology of acne and development of anti-acne products; quality control of manufacture of acne product and etiology of acne; the organisms of dental plaque and dental plaque formation; co-author of numerous articles. *Mailing Add:* Sharon Woods Health & Beauty Tech Ctr Procter & Gamble Co 11511 Reed Hartman Hwy Cincinnati OH 45241

KIRKLAND, JOSEPH JACK, b Winter Garden, Fla, May 24, 25; m 49, 83; c 5. ANALYTICAL CHEMISTRY. *Educ:* Emory Univ, AB, 48, MS, 49; Univ Va, PhD(chem), 53. *Hon Degrees:* DSc, Emory Univ, 74. *Prof Exp:* Chemist, Exp Sta, Hercules Powder Co, 49-51; from res chemist to sr res chemist, 53-61, res assoc, 61-69, res fel, 69-82, DUPONT FEL, E I DU PONT DE NEMOURS & CO, 82- *Concurrent Pos:* Adj prof chem, Univ Del, 80, 82. *Honors & Awards:* Chromatography Award, Am Chem Soc, 72; Stephen Dal Nogare Award in Chromatography, Chromatography Forum, 73; Anachem Award, 79; Torbern Bergman Medal, Analytic Chem, Swed Chem Soc, 82. *Mem:* Am Chem Soc. *Res:* Liquid chromatography; field flow fractionation. *Mailing Add:* Cent R&D Exp Sta Bldg 228 E I Du Pont De Nemours & Co Wilmington DE 19880-0228

KIRKLAND, WILLIS L, b Galesburg, Ill, Aug 8, 44. CANCER RESEARCH. *Educ:* Univ Kans, PhD(physiol & cell biol), 73. *Prof Exp:* Asst prof, 80-85, ASSOC PROF BIOL, MT MERCY COL, 85- *Mem:* Am Soc Cell Biol; Am Soc Col Sci Teachers. *Mailing Add:* Dept Biol Mt Mercy Col 1330 Elmhurst Dr NE Cedar Rapids IA 52402

KIRKLIN, JOHN W, b Muncie, Ind, Aug 5, 17; c 3. CARDIOVASCULAR SURGERY. *Educ:* Univ Minn, BA, 38, Harvard Univ, MD, 42; Am Bd Surg, dipl, 50; Am Bd Thoracic Surg, dipl, 50. *Hon Degrees:* DMed, Univ Munich, 61; DSc, Hamline Univ, 66. *Prof Exp:* Surgeon, Mayo Clin & Grad Sch Med, 50-66, from instr to assoc prof surg, 51-60, prof, 60-66, chmn dept, 64-66; surgeon-in-chief Hosps & Clin, 66-88, PROF SURG, MED COL, UNIV ALA, BIRMINGHAM, 66- *Concurrent Pos:* Mem bd gov, Mayo Clin, 65-66; mem surg study sect B, NIH, 64-67 & chmn Sect A, 68-70, mem comt standardization of classification congenital heart dis, 67, mem adv comt artificial heart-myocardial infarction prog, 67-69 & adv comt crippled children serv regional prog; chmn ad hoc comt to consider appln for subspec bd pediat surg, Am Bd Surg, 66-73. *Mem:* Nat Acad Sci; Am Col Surgeons; Am Surg Asn; Am Asn Thoracic Surg; Am Col Cardiol. *Mailing Add:* Dept Surg Univ Ala Med Ctr Birmingham AL 35294

KIRKLIN, PERRY WILLIAM, b Ellwood City, Pa, Feb 28, 35; m 56; c 3. PETROLEUM CHEMISTRY. *Educ:* Westminster Col, BS, 57; Univ Minn, Minneapolis, PhD(phys chem), 64. *Prof Exp:* Group leader anal res, Rohm and Haas Co, Pa, 64-70; ASSOC CHEM & PROJ LEADER, AVIATION FUELS RES, MOBIL RES & DEVELOP CORP, PAULSBORO, NJ, 70- *Mem:* Am Chem Soc; Nat Orgn Black Chemists & Chem Engrs. *Res:* Electron nuclear double resonance studies of hole center in magnesium oxide single crystals; catalyst characterization, especially metal function of catalysts using x-ray and chemisorption techniques; exploratory process research; aviation fuels. *Mailing Add:* 1860 Hillside Rd Southampton PA 18966

KIRKMAN, HENRY NEIL, JR, b Jacksonville, Fla, Sept 14, 27; m 50; c 4. MEDICINE, PEDIATRICS. *Educ:* Ga Inst Technol, BS, 47; Emory Univ, MS, 50; Johns Hopkins Univ, MD, 52; Am Bd Pediat, dipl, 60. *Prof Exp:* Intern pediat, Johns Hopkins Hosp, 52-53; resident, Vanderbilt Univ Hosp, 55-57; res investr, Nat Inst Arthritis & Metab Dis, 58-59; asst prof pediat, Sch Med, Univ Okla, 59-65; PROF PEDIAT, SCH MED, UNIV NC, CHAPEL HILL, 65- *Concurrent Pos:* Nat Inst Arthritis & Metab Dis fel metab enzymes, 57-58; Markle scholar, 61. *Honors & Awards:* Mead Johnson Award, 67. *Mem:* Am Pediat Soc; Am Soc Biol Chem. *Res:* Metabolic and enzymatic disturbances in children; human biochemical genetics. *Mailing Add:* Dept Pediat Univ NC Sch Med Chapel Hill NC 27599-7220

KIRKPATRICK, CHARLES HARVEY, b Topeka, Kans, Nov 5, 31; m 59; c 3. ALLERGY, IMMUNOLOGY. *Educ:* Univ Kans, BA, 54, MD, 58. *Prof Exp:* Asst med, Med Ctr, Univ Colo, 62-63, instr, 63-65; asst prof, Med Ctr, Univ Kans, 65-68, assoc prof, 68; sr investr & head, Sect Allergy & Hypersensitivity, Lab Clin Invest, Nat Inst Allergy & Infectious Dis, 68-79; DIR, DIV ALLERGY & CLINICAL IMMUNOL, DEPT MED, NAT JEWISH HOSP, 79-; PROF, UNIV COLO, 79- *Concurrent Pos:* Fel allergy & immunol, Univ Colo, 62-65. *Mem:* AAAS; Am Acad Allergy; Am Soc Clin Invest; Am Asn Immunol; Am Col Physicians. *Res:* Mechanisms of cellular immunity and the role of cellular immunity to resistance to infectious diseases and neoplasia; methods of correcting diseases associated with abnormal cellular immunity. *Mailing Add:* Dept Med Nat Jewish Hosp 1400 Jackson St Denver CO 80206-1997

KIRKPATRICK, CHARLES MILTON, b Greensburg, Ind, Jan 1, 15; m 39; c 2. WILDLIFE ECOLOGY. *Educ:* Purdue Univ, BS, 38; Univ Wis, MA, 40, PhD(zool), 43. *Prof Exp:* Asst zool, Univ Wis, 38-41; from instr to prof wildlife mgt, Agr Exp Sta, Purdue Univ, 41-81; RETIRED. *Concurrent Pos:* Ed, J Wildlife Mgt, Wildlife Soc, 59-62. *Mem:* Wilson Ornith Soc; hon mem Wildlife Soc. *Res:* Wildlife physiology and ecology. *Mailing Add:* 800 State Lafayette IN 47901-1744

KIRKPATRICK, DIANA (RORABAUGH) M, b Washington, DC, Mar 24, 44. QUALITY ASSURANCE AUDITING. *Educ:* George Washington Univ, BS, 67, PhD(phys chem), 72. *Prof Exp:* Forensic scientist chem, Bur Alcohol, Tobacco & Firearms, 72, res analyst, 73-74; res analyst, Consumer Prod Safety Comn, 74-77; prof leader thermal insulation & lab accreditation, Nat Bur Standards, 77-83; mgr, Prod Assurance Div Off Qual Assurance, Bur Engraving & Printing, Washington, DC, 83-85; res assoc, Paffenbarger Res Ctr, Am Dent Asn, Nat Bur Standards, Gaithersburg, Md, 85-88; CONSULT, 88- *Res:* Quality assurance testing programs; dental composites and bonding materials. *Mailing Add:* 1670 Jefferson Blvd Hagerstown MD 21742

KIRKPATRICK, E(DWARD) T(HOMSON), b Cranbrook, BC, Jan 15, 25; m 48; c 4. MECHANICAL ENGINEERING. *Educ:* Univ BC, BASc, 47; Carnegie Inst Technol, MS, 56, PhD, 58. *Prof Exp:* Test engr, Gen Elec Co, Can, 47; sales engr, F D Bolton, Ltd, 47-51, dist mgr, 51-53, sales mgr, 53-54; from instr to asst prof mech eng, Carnegie Inst Technol, 54-58; asst prof, Univ Pittsburgh, 58-59; prof & chmn dept, Univ Toledo, 59-64; dean, Col Appl Sci, Rochester Inst Technol, 64-71; pres, Wentworth Inst Technol, 71-90; RETIRED. *Mem:* Fel Am Soc Eng Educ; Am Soc Mech Engrs; Sigma Xi. *Res:* Conduction heat transfer; numerical analysis and digital computer technology. *Mailing Add:* 40 Radcliffe Rd Weston MA 02193

KIRKPATRICK, EDWARD SCOTT, b Wilmington, Del, Dec 12, 41. SOLID STATE PHYSICS. *Educ:* Princeton Univ, AB, 63; Harvard Univ, PhD(physics), 69. *Prof Exp:* Res assoc physics, James Franck Inst, Univ Chicago, 69-71; STAFF MEM, RES DIV, IBM CORP, 71- *Concurrent Pos:* Consult, Lincoln Labs, Mass Inst Technol, 65-66, Argonne Nat Labs, AEC, 69-71; vis assoc prof, State Univ NY, Stony Brook, 77; exchange prof, Ecole Normale Superieure, Paris France, 78- *Honors & Awards:* Am Phys Soc Prize Indust Appln Physics, 87. *Mem:* Fel Am Phys Soc; fel AAAS. *Res:* Magnetic order and excitations in disordered materials, transport in low-mobility materials; optimization and pattern recognition using techniques of statistical physics, computer design and architecture. *Mailing Add:* IBM Res Ctr Yorktown Heights NY 10598

KIRKPATRICK, FRANCIS HUBBARD, b Laurel Hill, NC, Nov 7, 43; m 69; c 1. BIOTECHNOLOGY, SEPARATIONS TECHNOLOGY. *Educ:* Harvard Col, BA, 64; Stanford Univ, PhD(biophys), 70. *Prof Exp:* Postdoctoral fel, Wash State Univ, 69-71; postdoctoral fel, Sch Med, Univ Rochester, 72-74, asst prof biophys, 74-80; lab mgr, Pall Corp, 80-84; TECH DIR, BIOPROD DEPT, MARINE COLLOIDS DIV, FMC CORP, 84- *Concurrent Pos:* SBIR study sect genetics, NIH, 87- *Mem:* Biophys Soc; Optical Soc Am; Am Soc Cell Biol; Am Chem Soc; Electrophoresis Soc; Am Soc Biochem & Molecular Biol. *Res:* Development of innovative products for research and analysis in biotechnology, life sciences and medicine. *Mailing Add:* 191 Thomaston St Rockland ME 04841

KIRKPATRICK, JAMES W, b Monmouth, Ill, Jan 9, 36; m 55; c 3. TECHNICAL TRAINING. *Educ:* Western Ill Univ, BS, 58, MSEduc, 60; Univ Ill, PhD(anal chem), 70. *Prof Exp:* Teacher high schs, Ill, 58-65; asst, Univ Ill, 65-69, from asst prof to assoc prof chem, Western Ill Univ, 69-78; mem staff, Inst Gas Technol, 78-79; MEM STAFF, VARIAN ASSOCS, PALO ALTO, 79- *Concurrent Pos:* Mem, Region 5 Lab, Environ Protection Agency, 76-78. *Mem:* AAAS; Am Chem Soc; Am Soc Training & Develop. *Res:* Quantitative determination of nitrogen isotope ratios; spectrometric determination of nitrogen content; spectrometric analysis of nonmetals. *Mailing Add:* 642 Bryson Ave Palo Alto CA 94306

KIRKPATRICK, JAY FRANKLIN, b Quakertown, Pa, Feb 24, 40; m 66. ANIMAL PHYSIOLOGY. *Educ:* East Stroudsburg State Col, BS, 62, MS, 64; Cornell Univ, PhD(physiol), 71. *Prof Exp:* Teacher biol, Quakertown High Sch, Pa, 62-63 & Pennsburg High Sch, Yardley, Pa, 64-65; asst prof & chmn dept, Bucks County Community Col, 65-67; assoc prof animal physiol, 70-76, dean, Sch Arts & Sci, 76-85, ASSOC PROF PHYSIOL, EASTERN MONT COL, 85- *Concurrent Pos:* Fel, Col Vet Med, Univ Pa, 73. *Honors & Awards:* Burlington Northern Found Res Award, 86. *Mem:* Soc Study Reproduction; Soc Exp Biol & Med. *Res:* Comparative mammalian reproduction, especially species indigenous to hostile environments, such as the pika and wild horses; chemical fertility control in wild and feral species. *Mailing Add:* Dept Biol Sci Eastern Mont Col Billings MT 59101

KIRKPATRICK, JOEL BRIAN, b Odessa, Tex, Feb 19, 36; c 4. NEUROPATHOLOGY. *Educ:* Rice Univ, BA, 58; Wash Univ, MD, 62. *Prof Exp:* Instr path, Wash Univ, 65-67; asst prof pharmacol, Rutgers Univ, New Brunswick, 68-70; assoc prof path, Univ Ariz, 70-72; assoc prof path & neurol, Univ Tex Health Sci Ctr, Dallas, 72-78, prof, 78-80; PROF PATH, BAYLOR COL MED, 81- *Concurrent Pos:* Am Cancer Soc fel, 64-65; NIH spec fel, 67-68; Nat Inst Neurol Dis & Stroke grant, 70; with Vet Admin, NIH, 71-, mem, Study Sect Path A, 87-91. *Mem:* Soc Neurosci; Am Asn Neuropath; Am Asn Pathologists. *Res:* Video enhancement microscopy; quantitative analysis of cerebral cortex; dementia and trauma. *Mailing Add:* Dept Path Methodist Hosp MS 205 6565 Fannin Houston TX 77030

KIRKPATRICK, JOEL LEE, b Abilene, Tex, June 21, 36; m 57, 91; c 2. ORGANIC CHEMISTRY, PESTICIDE CHEMISTRY. *Educ:* Abilene Christian Col, BS, 58; Univ Tex, MS, 60; Univ Ill, PhD(org chem), 69. *Prof Exp:* Med chemist, Smith Kline & French Labs, 61-65; sr res chemist, Gulf Oil Chem Co, 69-79; res assoc, Mobil Chem Co, 79-81; sr scientist, Velsicol Chem Corp, 81-86; res mgr, Sandoz, Ltd, 86-89, SR MGR, SANDOZ CROP PROTECTION CORP, 89- *Mem:* AAAS; Am Chem Soc. *Res:* Synthetic organic chemistry; nitrogen containing heterocycles; structure activity relationships, particularly pesticides; pheromones insect growth regulants; organophosphorus chemistry. *Mailing Add:* Sandoz Crop Protection Corp 1300 E Touhy Des Plaines IL 60018

KIRKPATRICK, LARRY DALE, b Lewiston, Idaho, Feb 11, 41; div; c 3. PHYSICS, PHYSICS EDUCATION. *Educ:* Wash State Univ, BS, 63, Mass Inst Technol, PhD(physics), 68. *Prof Exp:* Res assoc physics, Mass Inst Technol, 68-69; asst prof, Univ Wash, 69-74; from asst prof to assoc prof, 74-85, PROF PHYSICS, MONT STATE UNIV, 85- *Concurrent Pos:* Vis assoc prof physics, Kans State Univ, 83-84; coach, US Physics Team, 88- *Honors & Awards:* Distinguished Serv Citation, Am Asn Physics Teachers, 82. *Mem:* Am Asn Physics Teachers; Nat Sci Teachers Asn. *Res:* Physics education; use of computers in physics education; textbook writing for general physics. *Mailing Add:* Dept Physics Mont State Univ Bozeman MT 59717

KIRKPATRICK, MARK ADAMS, b New York, NY, Apr 20, 56. ZOOLOGY. *Educ:* Harvard Univ, BA, 78; Univ Wash, PhD(zool), 83. *Prof Exp:* Miller fel zool, Miller Inst Basic Res Sci, Univ Calif, Berkeley, 83-85; ASST PROF ZOOL, UNIV TEX, AUSTIN, 85- *Mem:* Soc Study Evolution; Am Soc Naturalists; Ecol Soc Am. *Res:* Theoretical population genetics; evolution of mating systems; sexual selection; morphology; development. *Mailing Add:* Dept Zool Univ Tex Austin TX 78712

KIRKPATRICK, R JAMES, b Schnectady, NY, Dec 31, 46; m 85; c 2. NUCLEAR RESONANCE SPECTROSCOPY. *Educ:* Cornell Univ, AB, 68; Univ Ill, PhD(geol), 72. *Prof Exp:* Sr res geologist, Exxon Prod Res Co, 72-73; res fel geophys, Harvard Univ, 73-75; asst res geologist, DSDP, Scripps Inst Oceanog, 76-78; asst prof geol to assoc prof geol, 78-83, PROF GEOL, UNIV ILL, URBANA-CHAMPAIGN, 83-, HEAD, DEPT GEOL, 88- *Concurrent Pos:* Mem var comts, Deep Sea Drilling Proj/Joint Oceanog Insts Deep Earth Sampling, 76-78 & Int Mineral Asn, 82-; prin investr, NSF & other grants, 77-; fel, Churchill Col, Cambridge Univ, 85-; assoc ed, Am Mineralogist, 87-91. *Mem:* Fel Mineral Soc Amer; Am Geophys Union; Am Ceramic Soc. *Res:* Use of nuclear magnetic resonance spectroscopy to investigate minerals, glasses, and cements; kinetics of geochemical reactions and igneous petrology. *Mailing Add:* Dept Geol Univ Ill Urbana-Champaign Urbana IL 61801

KIRKPATRICK, RALPH DONALD, b Jonesboro, Ind, Feb 10, 30; m; c 4. WILDLIFE ECOLOGY. *Educ:* Ball State Univ, BS, 53; Univ Ariz, MS, 57; Okla State Univ, PhD(zool), 64. *Prof Exp:* Game biologist, State Dept Conserv, Ind, 54-55, game res biologist, 56-59; asst prof biol, Taylor Univ, 59-60; res cur div birds, Smithsonian Inst, 64-65; asst prof zool, Ind Univ, 65-67; from asst prof to assoc prof, 67-73, PROF BIOL, BALL STATE UNIV, 73- *Concurrent Pos:* Consult, Pac Proj div birds, Smithsonian Inst, 65-, Aquatic Control, Inc, Seymour, Ind, 71- & Upper Wabash Resource Ctr, Huntington Col, Huntington, Ind, 75- *Mem:* Wildlife Soc; Am Soc Mammal; Wilson Soc. *Res:* Ecology of Indiana wildlife; population dynamics of mammals, including rodents and house cats on coral atolls. *Mailing Add:* Dept Biol Ball State Univ Muncie IN 47306

KIRKPATRICK, ROBERT JAMES, b Schenectady, NY, Dec 31, 46; m 68, 85; c 2. MINERALOGY-PETROLOGY, GEOCHEMISTRY. *Educ:* Cornell Univ, AB, 68; Univ Ill, PhD(geol), 72. *Prof Exp:* Sr res geologist, Exxon Prod Res Co, 72-73; res fel geophys, Harvard Univ, 73-76; asst res scientist, Deep Sea Drilling Proj, Scripps Inst Oceanog, Univ Calif, San Diego, 76-78; from asst prof to assoc prof, 78-83, PROF GEOL, UNIV ILL, URBANA, 83- , DEPT HEAD, 88- *Concurrent Pos:* Ed, Deep Sea Drilling Proj Initial Reports Legs, 46, 55 & 75-78; NSF & other grants, 77-; consult, 78-; US Rep Int Mineral Asn Crystal Growth Comn, 81-; fel, Churchill Col, Cambridge, Eng, 85; prog chmn, Int Mineral Asn Meeting, 86. *Mem:* Am Geophys Union; fel Am Mineral Soc. *Res:* Igneous petrology; processes of crystallization of igneous rocks, NMR spectroscopy of solids, structure of amorphous materials, crystal physics and chemistry, rates and mechanisms of geologic processes. *Mailing Add:* Dept Geol Univ Ill 254 Natural Hist Bldg 1301 W Green St Urbana IL 61801

KIRKPATRICK, ROY LEE, b Fairview, WVa, Apr 16, 40; m 61; c 2. NUTRITIONAL ECOLOGY, REPRODUCTIVE PHYSIOLOGY. *Educ:* WVa Univ, BS, 62; Univ Wis, MS, 64, PhD(reproductive physiol, endocrinol), 66. *Prof Exp:* Res asst reproductive physiol, Univ Wis, 62-64, instr, 64-66, asst prof animal sci, 69-71; asst prof wildlife physiol, Va Polytech Inst & State Univ, 66-69, from asst prof to assoc prof, 72-77, prof wildlife sci, 77-89, T H JONES PROF FISHERIES & WILDLIFE, VA POLYTECH INST & STATE UNIV, 89- *Mem:* Wildlife Soc; Am Soc Mammal; Wildlife Dis Asn. *Res:* Environmental influences on reproduction and mortality of wildlife, particularly effects of nutrition. *Mailing Add:* Dept Fisheries & Wildlife Va Polytech Inst & State Univ Blacksburg VA 24061

KIRKPATRICK, THEODORE ROSS, b Kalispell, Mont, Aug 29, 53. STATISTICAL MECHANICS, CONDENSED MATTER THEORY. *Educ:* Univ Calif, Los Angeles, BS, 77; Rockefeller Univ, PhD(theoret physics), 81. *Prof Exp:* Res assoc, 81-83, ASST PROF THEORET STATIST MECH, INST PHYS SCI & TECHNOL, UNIV MD, COLLEGE PARK, 83- *Concurrent Pos:* Presidential Young Investr Award, NSF, 84. *Mem:* Am Phys Soc. *Res:* Condensed matter theory and statistical mechanics; physics of disordered solids and liquids. *Mailing Add:* Inst Phys Sci & Technol Univ Md College Park MD 20742

KIRKSEY, AVANELLE, b Mulberry, Ark, Mar 23, 26. NUTRITION, BIOCHEMISTRY. *Educ:* Univ Ark, BS, 47; Univ Tenn, MS, 50; Pa State Univ, PhD(nutrit), 61. *Prof Exp:* Assoc prof home econ, Ark Polytech Col, 50-55; res asst nutrit, Pa State Univ, 56-59; from assoc prof to prof, 61-85, MEREDITH DISTINGUISHED PROF NUTRIT, PURDUE UNIV, 85- *Concurrent Pos:* Prog dir, Nutrit Collab Res Support Prog, Egypt, Kenya & Mex. *Honors & Awards:* Borden Award. *Mem:* Am Inst Nutrit. *Res:* Vitamin B-6 metabolism; nutrition in pregnancy and development; human lactation; international nutrition. *Mailing Add:* Dept Foods & Nutrit Purdue Univ West Lafayette IN 47907

KIRKSEY, DONNY FRANK, b Aberdeen, Miss, Apr 13, 48; m 70. PHARMACOLOGY. *Educ:* Delta State Univ, BS, 70; Univ Miss, PhD(pharmacol), 76. *Prof Exp:* Fel, Duke Univ, 76-78; asst prof biomed sci, Ohio Univ, 78-80; clin res scientist, Burroughs Wellcome, 80-86; pres, Clindar, Inc, Durham, NC, 86-87; ASSOC DIR CLIN RES, GLAXO, INC, 87- *Concurrent Pos:* Neurosci fel, NIMH, 76-78; grant, NC Heart Asn, 77-78; NIH fel, Nat Inst Drug Abuse, 78. *Mem:* Sigma Xi. *Res:* Investigations of pre and postsynaptic neuronal mechaniisms in the central monoaminergic systems and pharmacological manipulation of those systems by drugs of abuse. *Mailing Add:* 4100 Thetford Rd Durham NC 27707

KIRKSEY, HOWARD GRADEN, JR, b Memphis, Tenn, June 19, 40; c 3. SCIENCE EDUCATION. *Educ:* Mid Tenn State Univ, BS, 61; Auburn Univ, PhD(phys chem), 66. *Prof Exp:* Assoc prof, 69-79, PROF CHEM, MEMPHIS STATE UNIV, 79-, CHMN DEPT, 82- *Concurrent Pos:* Staff scientist phys sci group, Boston Univ, 70-72. *Mem:* AAAS; Am Chem Soc. *Res:* Chemical education, especially development of text and laboratory teaching materials. *Mailing Add:* Dept Chem Memphis State Univ Memphis TN 38152

KIRKWOOD, CHARLES EDWARD, JR, b Richmond, Va, Oct 10, 13; m 42; c 2. MATHEMATICAL SCIENCES. *Educ:* Lynchburg Col, AB, 35; Univ Ga, MS, 37. *Prof Exp:* Teacher pub sch, Ga, 36-37; from instr to asst prof math, Clemson Univ, 37-48, assoc prof math sci, 48-79; RETIRED. *Concurrent Pos:* Assoc prof elec eng, Comput Ctr, Clemson Univ, 51-52, comput analyst, 64-70, mgr prog, 70-75; vis assoc prof comput sci & math sci, Clemson Univ, 79-86. *Res:* Dielectric properties of ceramic materials; electrical properties of cotton; thermoconductivity of felts. *Mailing Add:* Wren St Clemson SC 29631

KIRKWOOD, JAMES BENJAMINE, b Beulah, Ky, Jan 22, 24; m 53. ZOOLOGY, RADIATION BIOLOGY. *Educ:* West Ky State Col, 48; Univ Louisville, MS, 52, PhD(zool), 62. *Prof Exp:* Proj leader fishery biol, Ky Dept Fish & Wildlife Resources, 52-57; US Bur Com Fisheries, 57-60, prog supvr invert biol, 62-64, prog leader, 64-68; tech coordr bio-environ studies, Battelle-Columbus, 68-73, mgr, W F Clapp Labs, 73-75; coastal ecosysts activ leader, US Fish & Wildlife Serv, Region IV, 75-89; RETIRED. *Mem:* AAAS; Am Fisheries Soc; Sigma Xi; Nat Shellfish Asn; Int Acad Fishery Scientists. *Res:* Ecology and ichthyology of Kentucky teleost fishes; life history and ecology of Pacific salmon; biology and population dynamics of shellfish species in Gulf of Alaska and Bering Sea. *Mailing Add:* 34 N Country Club Dr Crystal River FL 32629-5358

KIRKWOOD, SAMUEL, b Edmonton, Alta, June 14, 20; US citizen; m 47; c 3. BIOCHEMISTRY. *Educ:* Univ Alta, BS, 42; Univ Wis, MS, 44, PhD(biochem), 47. *Prof Exp:* Biochemist, Camp Detrick, Md, 47; asst res officer, Nat Res Coun Can, 48; from assoc prof to prof chem, McMaster Univ, 50-56; assoc prof, 56-62, PROF BIOCHEM, COL BIOL SCI, UNIV MINN, ST PAUL, 62- *Mem:* Sigma Xi. *Res:* Intermediary metabolism; route of synthesis of thyroxin; enzymology of carbohydrates. *Mailing Add:* 9819 152nd St N Hugo MN 55038

KIRMSE, DALE WILLIAM, b Alva, Okla, July 9, 38; m 62; c 4. COMPUTER AIDED DESIGN, ENERGY SYSTEMS. *Educ:* Okla State Univ, BS, 60; Iowa State Univ, MS, 63, PhD(chem eng), 64. *Prof Exp:* Res assoc chem eng, Univ Fla, 64-65, asst prof, 65-67; mgr asst, Parma Tech Ctr, Union Carbide Corp, 67-69; asst prof, 69-80, ASSOC PROF CHEM ENG, UNIV FLA, 80- *Concurrent Pos:* Reynolds Smith & Hill, 75-76. *Mem:* Am Soc Qual Control; Am Inst Chem Engrs. *Res:* Statistical process quality control and reliability; mathematical modeling and computer methods; stochastic systems and Monte Carlo techniques; knowledge base export systems; computer aided process design; energy systems analysis and design. *Mailing Add:* Dept Chem Eng Univ Fla Gainesville FL 32611

KIRMSER, P(HILIP) G(EORGE), b St Paul, Minn, Dec 17, 19; m 42; c 2. APPLIED MATHEMATICS, ENGINEERING. *Educ:* Univ Minn, BChE, 39, MS, 44, PhD, 58. *Prof Exp:* Instr, Kans State Col, 42-44; mech engr, US Naval Ord Lab, 46-48; instr, Univ Minn, 49-54; from assoc prof to appl mech, Kans State Univ, 54-75, head dept, 62, prof, 75-90, EMER PROF ENG & MATH, KANS STATE UNIV, 90- *Concurrent Pos:* Consult, Phillips Petrol Co, Bayer & McElrath & Boeing Co, Digital Equip Co; vis lectr, Soc Indust & Appl Math, 75. *Mem:* Am Math Soc; Math Asn Am; Soc Indust & Appl Math; Neth Royal Inst Eng. *Res:* Partial differential equations of engineering dealing with heat flow, vibrations and stresses; dynamics and motion of artificial satellites; analog computers; simulation; approximation; automatic controls; industrial processes; analysis of data; co-inventor of Chinese word-processing and typing system. *Mailing Add:* Dept Elec & Comput Eng Kans State Univ Manhattan KS 66506

KIRON, RAVI, b Shimoga, Karnataka, India, Mar 4, 59; m 89. PROTEIN PURIFICATION, RECEPTOROLOGY. *Educ:* Bombay Univ, India, BS, 79, MS, 81; Indian Inst Sci, Bangalore, India, PhD(biochem), 86. *Prof Exp:* Res fel, Cornell Univ Med Col, 82-86, postdoctoral, 86-89, asst prof med, 89-91, asst prof biochem, 90-91; SR RES SCIENTIST, PFIZER CENT RES, 91- *Concurrent Pos:* Sr res fel, US Dept Agr, 82-85, Indian Inst Sci, 85-86; young leadership award, Am Biograph Inst, 86; young investr award, Eastern Hypertension Soc, 88. *Mem:* AAAS; Harvey Soc; Am Soc Hypertension. *Res:* Biochemistry of renin angiotensin system; characterization of angiotensin II receptors; immunologic analysis and cloning of gene for receptor; renin and prorenin study and analysis. *Mailing Add:* Pfizer Central Res Eastern Point Rd Box 161 Rm B-205 Groton CT 06340

KIRSCH, DONALD R, b Newark, NJ, Apr 28, 50; m 74. MOLECULAR BIOLOGY. *Educ:* Rutgers Col, BA, 72; Princeton Univ, MA, 74, PhD(biol), 78. *Prof Exp:* Instr pharmacol, Rutgers Med Sch, 78-81; res investr molecular biol, Squibb Inst Med Res, 81-82, sr res investr, 82-88, res group leader, 83-88; PRIN RES SCIENTIST, AGRIC RES DIV, AM CYANAMID CO, 88- *Concurrent Pos:* Adj asst prof, Rutgers Med Sch, 82-83. *Mem:* Genetics Soc Am; Am Soc Microbiol; Sigma Xi. *Res:* Genome organization and gene expression in lower eucaryotes. *Mailing Add:* 152 Terhune Rd Princeton NJ 08540

KIRSCH, EDWIN JOSEPH, b Hoboken, NJ, Aug 25, 24; m 45; c 3. MICROBIOLOGY. *Educ:* Mich State Univ, BS, 49; Purdue Univ, MS, 55, PhD(microbiol), 58. *Prof Exp:* Biologist, Lederle Labs, Am Cyanamid Corp, 50-53, sr res scientist, 57-63; assoc prof sanit eng microbiol, 63-70, PROF ENVIRON ENG, PURDUE UNIV, 70- *Mem:* Am Soc Microbiol; Soc Indust Microbiol; Water Pollution Control Fedn. *Res:* Microbiology of wastewater purification; microbial interactions; environmental microbiology; biodegradation. *Mailing Add:* 1616 Sheridan Rd West Lafayette IN 47906

KIRSCH, FRANCIS WILLIAM, b Wheeling, WVa, Aug 27, 25; m 61; c 2. CHEMISTRY. *Educ:* Univ Del, BChE, 45, MChE, 47; Univ Pa, PhD(chem), 52. *Prof Exp:* Instr inorg qual anal, Univ Pa, 46-50; assoc res chemist, Houdry Process Corp, 50-59, proj dir, 59-64; proj dir, Sun Oil Co, 64-67, res assoc, Explor Res Div, 67-72; DIR, CTR ENERGY MGT & ECON DEVELOP, UNIV CITY SCI CTR, PHILADELPHIA, 73- *Concurrent Pos:* Consult, Gov Energy Coun Pa, 79-, Pa Pub Util Comn, 77-78, World Bank, 80 & NMex State Govt. *Mem:* AAAS; Am Chem Soc. *Res:* Catalysis, heterogeneous and homogeneous; basic and process research and development: petroleum, chemicals, edible oils; industrial energy conservation; offshore oil and gas production; science policy; economic analysis and evaluation of manufacturing processes. *Mailing Add:* Univ City Sci Ctr 3624 Market St Philadelphia PA 19104

KIRSCH, JACK FREDERICK, b Detroit, Mich, Aug 14, 34; m 62; c 2. BIOCHEMISTRY. *Educ:* Univ Mich, BS, 56; Rockefeller Inst, PhD(biochem, cytol), 61. *Prof Exp:* Jane Coffin Childs fel biochem, Brandeis Univ, 61-63; Helen Hay Whitney fel biophys, Weizmann Inst, 63-64; from asst prof to prof biochem, 64-89, PROF BIOCHEM & MOLECULAR BIOL, UNIV CALIF, BERKELEY, 89- *Concurrent Pos:* Guggenheim fel, Max Planck Inst Biophys Chem, 71-72; vis prof, Univ Basel, 79-80. *Mem:* Am Chem Soc; Fedn Am Socs Exp Biol; AAAS. *Res:* Mechanism of enzyme action; genetic engineering. *Mailing Add:* Molecular Biol Dept Barker Hall Univ of Calif Berkeley CA 94720

KIRSCH, JOSEPH LAWRENCE, JR, b Indianapolis, Ind, Aug 20, 42; m 65; c 2. PHYSICAL CHEMISTRY. *Educ:* Butler Univ, BS, 64; Univ Ill, MS, 66, PhD(phys chem), 68. *Prof Exp:* Asst prof inorg chem, Fairleigh Dickinson Univ, 68-70; asst prof, 70-74, assoc prof, 74-81, PROF PHYS CHEM, BUTLER UNIV, 81- *Mem:* Am Chem Soc; Sigma Xi. *Res:* Study of chemical bonding in molecules through the use of chemical spectroscopy. *Mailing Add:* Dept Chem Butler Univ Indianapolis IN 46208

KIRSCH, LAWRENCE EDWARD, b Newark, NJ, Feb 24, 38. PHYSICS, COMPUTER SCIENCE. *Educ:* Columbia Univ, AB, 60; Rutgers Univ, MS, 62, PhD(physics), 65. *Prof Exp:* Res assoc physics, Nevis Labs, Columbia Univ, 64-66; asst prof, 66-72, ASSOC PROF PHYSICS, BRANDEIS UNIV, 72-, DIR COMPUT CTR, 70- *Mem:* Am Phys Soc; Inst Elec & Electronics Engrs; Asn Comput Mach. *Res:* High energy and particle physics. *Mailing Add:* Dept Physics Brandeis Univ Waltham MA 02254

KIRSCH, MILTON, b Montreal, Que, Jan 16, 23; nat US; m 47; c 3. CHEMISTRY, ENVIRONMENTAL SYSTEMS. *Educ:* McGill Univ, BSc, 42, PhD(chem), 45. *Prof Exp:* Res assoc rubber chem, McGill Univ, 45; asst prof chem, Univ Man, 46-49 & Univ BC, 49-56; sr res chemist, E I du Pont de Nemours & Co, 57-63; mem tech staff, Rocketdyne Div, Rockwell Inst, 63-78, sr staff scientist, Environ Monitoring & Serv Ctr, 78-84, combustion engr, Enrviron Monitoring & Serv Inc, 84-85; CONSULT, 85- *Mem:* Royal Soc Chem; fel Am Inst Chem; Sigma Xi. *Res:* Detonation velocity; cross-section for absorption of thermal neutrons; radioactive tracers; chemical oceanography; colloid and surface chemistry; propellants; rheology; environmental chemistry; wastewater treatment; environmental technology. *Mailing Add:* 20414 Haynes St Canoga Park CA 91306

KIRSCH, WOLFF M, b St Louis, Mo, Mar 2, 31; m 55; c 4. NEUROSURGERY, BIOCHEMISTRY. *Educ:* Wash Univ, BA, 51, MD, 55. *Prof Exp:* Intern med, NC Mem Hosp, Univ NC, 55-56; asst resident gen surg, Barnes Hosp, Wash Univ, 59-60 & neurosurg, 61-62, chief resident & instr, 64; asst prof neurosurg, Med Sch, Univ Colo, Denver, 65-68, assoc prof, 68-, dir & chmn training prog & chmn div, 71-; AT DEPT SURG, UNIV NMEX, ALBUQUERQUE. *Concurrent Pos:* Fel neurol med, NC Mem Hosp, Univ NC, 56-57; fel neuroanat, Sch Med, Wash Univ, 60, res fel neuro-pharmacol, 63 & 65; attend, Vet Admin Hosp, Denver, 65-; consult, Fitzsimons Gen Hosp, Aurora, Colo, 69- *Mem:* Am Acad Neurol; Asn Acad Surg; Am Asn Neurol Surg; Int Soc Neurochem; Am Col Surg. *Res:* Experimental biology of brain tumors. *Mailing Add:* Loma Linda Univ Sch Med Loma Linda CA 92354

KIRSCHBAUM, H(ERBERT) S(PENCER), b Cleveland, Ohio, Feb 6, 20; m 46; c 4. ELECTRICAL ENGINEERING. *Educ:* Cooper Union, BS, 42; Univ Pittsburgh, MS, 46; Carnegie Inst Technol, PhD(elec eng), 53. *Prof Exp:* Engr, Westinghouse Elec Corp, 42-46; assoc prof elec eng, Ohio State Univ, 47-57; div consult, Systs Div, Battelle Mem Inst, 57-59, dept consult eng physics, 59-69; mgr systs eng, Info Systs Lab, Westinghouse Elec Co, 69-82; RETIRED. *Mem:* Inst Elec & Electronics Engrs. *Res:* Systems engineering; control systems; process control. *Mailing Add:* RD 2 Box 384C Asheville NC 28805

KIRSCHBAUM, JOEL BRUCE, b Palo Alto, Calif, Aug 29, 45; m 74; c 2. MOLECULAR BIOLOGY, GENETICS. *Educ:* Pomona Col, BA, 67; Harvard Univ, PhD(molecular biol), 72. *Prof Exp:* Fel molecular biol, H H Whitney Found, 73-75; sr researcher, Univ Geneva, 75-77; instr neuropath, Harvard Med Sch, 77-81; supvr, genetic eng div, Stauffer Chem Co, 81-85; dir res & develop, Codon, 85-90; ACTG VPRES, OCTAMER, 91- *Concurrent Pos:* Teaching fel, Harvard Univ, 69; asst instr bact genetics, Cold Spring Harbor Lab, 71; res assoc, Children's Hosp Med Ctr, 77-81; res fel, Med Found Inc, 78-80; prin investr, Am Cancer Soc, 79-81; consult, biotechnol & pharmaceut, 90- *Mem:* AAAS. *Res:* Molecular basis for the control of gene expression; biochemistry. *Mailing Add:* 6132 Johnston Dr Oakland CA 94611

KIRSCHBAUM, JOEL JEROME, b New York, NY, Nov 23, 35; m 60; c 2. ANALYTICAL CHEMISTRY, BIOCHEMISTRY. *Educ:* City Col New York, BS, 57; Rutgers Univ, PhD(biochem), 63. *Prof Exp:* RES LEADER, ANALYTIC RES & DEVELOP DIV, BRISTOL-MYERS SQUIBB PHARM RES INST, 64- *Concurrent Pos:* Mem grants comt, Am Found Scholarly Res & mem, Directorate of Inst Motivated Behav, Belle Mead, NJ. *Mem:* Am Chem Soc; Am Soc Biochem & Molecular Biol; Am Asn Pharm Scientists. *Res:* Analyses of drugs; lycanthropy; high pressure liquid chromatography; association and dissociation of proteins, enzymes and antibiotics. *Mailing Add:* Analytic Res & Develop Div Bristol-Myers Squibb Pharm Res Inst One Squibb Dr New Brunswick NJ 08903-0191

KIRSCHBAUM, THOMAS H, b Minneapolis, Minn, Apr 22, 29; m 83; c 2. OBSTETRICS & GYNECOLOGY. *Educ:* Univ Minn, BA, 50, BS, 51, MD, 53. *Prof Exp:* Resident physician obstet & gynec, Univ Minn Hosps, 56-59; asst prof, Col Med, Univ Utah, 59-64; assoc prof, Med Ctr, Univ Calif, Los Angeles, 64-71; prof obstet, gynec & reproductive biol & chmn dept, Col Human Med, Mich State Univ, 71-83; PROF OBSTET & GYNEC, COL MED, UNIV SOUTHERN CALIF, 85- *Concurrent Pos:* Consult, Rand Corp, Calif, 64-71; spec expert, Nat Inst Child Health & Human Develop, 83-85; assoc ed, Yearbk Obstet-Gynec, 87-; Nat & Child Health Res Comt, NIH, 86- *Mem:* AAAS; Perinatal Res Soc; Soc Gynec Invest; Am Gynec & Obstet Soc; Am Col Obstet-Gynec. *Res:* Fetal physiology and maternal-fetal interrelationships. *Mailing Add:* Women's Hosp Rm 5K40 1240 N Mission Rd Los Angeles CA 90033

KIRSCHENBAUM, DONALD (MONROE), biochemistry; deceased, see previous edition for last biography

KIRSCHENBAUM, LOUIS JEAN, b Washington, DC, Apr 17, 43; m 64; c 2. TRANSITION METAL SOLUTION CHEMISTRY, KINETICS & MECHANISMS. *Educ:* Howard Univ, BS, 65; Brandeis Univ, MA, 67, PhD(chem), 68. *Prof Exp:* Lectr chem, Brandeis Univ, 68-69; asst Naval Ord Lab, 69-70; from asst prof to assoc prof, 70-83, PROF CHEM, UNIV RI, 83- *Concurrent Pos:* Vis prof, Ben Gurion Univ Negev, Israel, 78-79. *Mem:* Sigma Xi; Am Chem Soc. *Res:* Chemistry of transition metals in uncommon oxidation states; kinetics and mechanisms of metal ion oxidation-reduction and complexation reactions; rapid reaction techniques; kinetics of chemical analysis. *Mailing Add:* Dept Chem Univ RI Kingston RI 02881

KIRSCHENBAUM, SUSAN S, b Washington, DC, Sept 15, 43; m 64; c 2. ENGINEERING PSYCHOLOGY, HUMAN DECISION MAKING. *Educ:* George Washington Univ, BA, 65; Univ RI, MA, 75 & 83, PhD(exp psychol), 85. *Prof Exp:* Teacher, Boston Sch Dept, 65-67; dir, South Kingstown Orgn Laymen Educ, 74-77; dir educ serv, South County Community Action, 77-78; lectr eng as foreign lang, Ben Gurion Univ, Negev, 78-79; spec instr psychol, Univ RI, 81-85; ENG PSYCHOLOGIST, NAVAL UNDERWATER SYSTS CTR, 85- *Concurrent Pos:* Personnel psychologist, Naval Underwater Systs Ctr, 84-85; adj asst prof, Univ RI, 85- *Mem:* Am Psychol Asn; Asn Appl Exp & Eng Psychologists; Am Psychol Soc; Sigma Xi. *Res:* Information management for submarine combat control; human information gathering and usage for situation understanding and decision making in complex environments; apply findings to design of command decision aids. *Mailing Add:* Naval Underwater Systs Ctr Code 2212 Bldg 1171-1 Newport RI 02841

KIRSCHNER, LEONARD BURTON, b Chicago, Ill, Nov 12, 23; m 50; c 4. PHYSIOLOGY. *Educ:* Univ Ill, BS, 44, MS, 47; Univ Wis, PhD(physiol), 51. *Prof Exp:* Nat Found Infantile Paralysis res fel, Copenhagen Univ, 51-53; from instr to assoc prof, 53-65, PROF ZOOL, WASH STATE UNIV, 65- *Mem:* Am Physiol Soc; Soc Gen Physiol; Am Soc Zoologists; Sigma Xi. *Res:* Active transport of solutes and water; invertebrate excretory organs. *Mailing Add:* Dept Zool Wash State Univ Pullman WA 99164

KIRSCHNER, MARC WALLACE, b Chicago, Ill, Feb 28, 45; m 68; c 3. BIOCHEMISTRY, CELL BIOLOGY. *Educ:* Northwestern Univ, BA, 66; Univ Calif, Berkeley, PhD(biochem), 71. *Prof Exp:* NSF fel develop biol, Univ Calif, Berkeley, 71-72; asst prof, Princeton Univ, 72-77, prof biochem, 77-78; PROF BIOCHEM, UNIV CALIF, SAN FRANCISCO, 78- *Concurrent Pos:* NIH res career develop award, 75-80. *Honors & Awards:* Lounsberg Award, Nat Acad Sci, 91. *Mem:* Nat Acad Sci; Am Soc Biol Chemists; Am Soc Cell Biol; Am Acad Arts & Sci. *Res:* Mechanism of microtubule assembly, regulation of mitosis and regulation of cell division in amphibian eggs; biophysical studies of macromolecules; embryonic induction. *Mailing Add:* Dept Biochem & Biophys Univ Calif Sch Med San Francisco CA 94143-0448

KIRSCHNER, MARVIN ABRAHAM, b Brooklyn, NY, Mar 5, 35; m 57; c 3. INTERNAL MEDICINE, ENDOCRINOLOGY. *Educ:* Albert Einstein Col Med, MD, 59. *Prof Exp:* Med intern, Bronx Munic Hosp Ctr, 59-60, asst resident, 60-61, resident med, 64-65; clin assoc endocrinol, Endocrine Br, Nat Cancer Inst, 61-64, fel reproductive endocrinol, Karolinska Inst, Stockholm, 65-66; sr investr, Endocrinol Br, 66-69; dir med, Newark Beth Israel Med Ctr, 69; assoc prof, 69-72, PROF MED, NJ MED SCH, 73- *Concurrent Pos:* Consult, Res Comt, East Orange Vet Admin Hosp, 71-; mem, Breast Cancer Task Force-Epidemiol, Nat Cancer Inst, 74-78, prog proj comt, 80- *Mem:* Am Fedn Clin Res; Endocrine Soc; Am Asn Cancer Res. *Res:* Androgen and estrogen metabolism; hirsutism; endocrine tumors; breast cancer; obesity management. *Mailing Add:* Dept Med Univ Med & Dent NS 100 Bergen St Newark NJ 07103

KIRSCHNER, ROBERT H, b Philadelphia, Pa, Oct 30, 40; m 65; c 3. FORENSIC PATHOLOGY. *Educ:* Washington & Jefferson Col, BA, 62; Jefferson Med Col, MD, 66. *Prof Exp:* Instr path, Univ Chicago, 69-71; asst chief path dept, USPHS Hosp, 71-73; asst prof path, 73-78, Ben Horwich scholar med sci, 75-78, ASSOC PROF PATH, UNIV CHICAGO, 78-; DEP CHIEF MED EXAMR, COOK CO, 87- *Concurrent Pos:* Am Cancer Soc clin fel path, Univ Chicago, 69-70; dep med examer,78-86; med adv comt, Lincoln Park Zoo, Chicago; AAAS forensic consult, Govt Arg, 85; mem, Comt Sci Freedom & Responsibility, AAAS, 87-; comt mem, Child Abuse & Neglect, Am Acad Pediat. *Mem:* AAAS; Am Soc Cell Biol; Sigma Xi; Am Acad Forensic Sci; NY Acad Sci; Nat Asn Med Examiners (bd dir, 86-); Physicians Human Rights (bd dir, 88-); Am Acad Pediat. *Res:* Sudden cardiac death; child abuse and neglect; trauma and injury; forensic documentation of human rights abuses. *Mailing Add:* Inst Forensic Med 2121 Harrison St Chicago IL 60612

KIRSCHNER, RONALD ALLEN, b New York, NY, Jan 18, 42; m 64; c 2. LASER-TISSUE INTERACTION, OPTICS & IMAGING. *Educ:* New York Univ, BA, 62; Philadelphia Col Osteopath Med, DO, 66, MSc, 72. *Prof Exp:* Dir, head & neck serv, Main Navy Dispensary, 68-70, Neurosensory Unit, 73-76, clin assoc prof, head & neck, Philadelphia Col Osteopath Med, 75-85, pres, Suburban Ear, Nose & Throat, Group Ltd, 76-, chmn, head & neck, Suburban Gen Hosp, 76-, exec dir, Inst Appl Laser Surg, 80-, chmn, Div Surg, Suburban Gen Hosp, 83-89, clin prof, head & neck, 85-89, PROF & CHMN, HEAD & NECK & FACIAL PLASTIC SURG, PHIL COL OSTHEOPATH MED, 90- *Concurrent Pos:* Vpres, The Courtlandt Group Inc, 79-84, design consult, Pilling Corp, 81-, exec vpres, The Courtlandt Group, 84-85, design consult, Inframed Corp, 84-87, guest ed, Surgical Clins North Am, 84, design consult, Sigma Dynamics, 87-, contrib ed, Photonics Spectra, 87-, pres, Kirschner Design Group Inc, 88- *Mem:* Laser Assoc Am; NY Acad Sci; Sigma Xi; Laser Inst Am; Am Soc Lasers Med Surv. *Res:* Laboratory and clinical research regarding lasers of various wavelengths and their interaction with tissue, delivery systems, related imaging systems and development of new instruments. *Mailing Add:* Kirschner Design Group Inc Two Bala Plaza Suite Plaza 13 Baladyhwya PA 19004

KIRSCHNER, STANLEY, b Brooklyn, NY, Dec 17, 27; m 50; c 2. INORGANIC CHEMISTRY. *Educ:* Brooklyn Col, BS, 50; Harvard Univ, AM, 52; Univ Ill, PhD(chem), 54. *Prof Exp:* Res chemist inorg chem, Monsanto Chem Co, 51; from asst prof to assoc prof, 54-60, vchmn dept, 61-64, actg chmn, 64-65, PROF CHEM, WAYNE STATE UNIV, 60- *Concurrent Pos:* Res fels, Res Corp, 55-57, Chattanooga Med Co, 56, NSF, 58, 65-67 & 73-77, sr fel, 63-64 & Ford Found, 69; NIH grant, 62-65; Fulbright scholar, 63-64 & 84; perm secy, Int Conf Coord Chem, 66-89; vis prof, Univ London, 63-64, Univ Sao Paulo, Brazil, 69, Polytech Univ Timisoara, Romania, 73, Univ Florence, Italy, 76, Tohoku Univ, Sendai, Japan, Inst Chem, Cluj, Romania, 78, Polytech Inst, Lisbon, & Univ Porto, Portugal, 84; chmn, comt educ, 81-83, Div Chem Educ, Am Chem Soc, 86- & bd dirs, 88-92; ed, Inorg Synthesis, bd dirs, Am Chem Soc, 88-92; adv bd, Seaborg Ctr for Teaching & Learning of Sci & Math. *Honors & Awards:* Fac Res Award, Sigma Xi, 74; Heyrovsky Medal, Czech Acad Sci, 78; Catalyst Award, Mfg Chem Asn, 84. *Mem:* fel AAAS; Am Chem Soc; fel NY Acad Sci; Chem Soc; Brazilian Acad Sci; fel Indian Chem Soc; fel Japan Soc Prom Sci. *Res:* Structure and stereochemistry of organosilicon and complex inorganic compounds; rotatory dispersion and circular dichroism of asymmetric coordination compounds; physiologically important complex inorganic compounds and their biological activity; application of computer techniques to the storage and retrieval of chemical literature references; optical properties of fast-racemizing complexes; the pfeiffer effect; chem education. *Mailing Add:* Dept Chem Wayne State Univ Detroit MI 48202

KIRSCHSTEIN, RUTH LILLIAN, b Brooklyn, NY, Oct 12, 26; m 50; c 1. PATHOLOGY. *Educ:* Long Island Univ, AB, 47; Tulane Univ, MD, 51. *Hon Degrees:* DSc, Mt Sinai Sch Med, 84; LLD, Atlanta Univ, 85; DSc, Med Col Ohio, 86. *Prof Exp:* Pathologist, Div Biologics Stand, NIH, 57-60, chief, sect path, Lab Virol Immunol, 60-65 & asst chief, 62-65, chief, Lab Path, 65-72, asst dir, Div Biologics Stand, 71-72; dep dir, Bur Biologics, Food & Drug Admin, 72-73, dep assoc comnr sci, 73-74; DIR, NAT INST GEN MED SCI, NIH, 74- *Concurrent Pos:* Mem,expert comt poliomyelitis, WHO, 65 & 67 & 71; chairperson, NIH Grants Peer Rev Study Team, 75-76, PHS Task Force Women's Health Issues, 83-84, PHS Coord, Comt Women's Health Issues, 85- *Mem:* Am Asn Pathologists; Am Asn Immunologists; Am Soc Microbiol. *Res:* Pathology and pathogenesis of viral diseases; poliomyelitis,; oncogenic viruses; viral vaccines; scientific peer review; scientific research administration; carcinogenesis. *Mailing Add:* Nat Inst Gen Med Sci NIH Westwood Bldg Rm 926 5333 Westbard Ave Bethesda MD 20892

KIRSHENBAUM, ABRAHAM DAVID, b New York, NY, July 8, 19; m 43; c 2. PHYSICAL INORGANIC CHEMISTRY, EXPLOSIVES. *Educ:* City Col New York, BS, 41; Polytech Inst Brooklyn, MS, 45. *Prof Exp:* Res scientist, S A M Labs, Columbia Univ, 41-46; res chemist, Houdry Process Corp, Pa, 46-48 & Standard Oil Develop Co, 48-49; res supvr, Res Inst, Temple Univ, 49-68; res chemist & proj leader, US Army Armament Munitions & Chem Command, 68-88; RETIRED. *Mem:* Am Chem Soc; Sci Res Soc Am; Sigma Xi; NAm Thermal Anal Soc. *Res:* Exchange reactions; lubrication oil additives; catalysis; inorganic fluorine chemistry; high vacuum techniques; gaseous burning velocity, calorimetric studies; high temperature physical measurements; rare gas chemistry; pyrotechnics and delays; propellants and explosives. *Mailing Add:* 7234 Fairfax Dr Tamarac FL 33321-4303

KIRSHENBAUM, GERALD STEVEN, b New York, NY, Dec 6, 44; m; c 3. POLYMER CHEMISTRY, PLASTICS RESEARCH. *Educ:* Case Inst Technol, BS, 66; Polytech Inst Brooklyn, MS, 70, PhD(polymer chem), 71. *Prof Exp:* Res chemist, Celanese Plastics Co, 71-74; sr res chemist, Soltex Polymer Corp, 74-75; sr chemist, Union Carbide Corp, Bound Brook, 76-78; develop assoc, supvr & mgr, Celanese Eng Resins Co, 78-89; MGR PROD SAFETY, HOECHST CELANESE, 90- *Concurrent Pos:* Asst ed, Polymer News, 72-75, ed, 75- *Mem:* Am Chem Soc; Soc Plastics Engrs. *Res:* Polyolefins; high-density polyethylene; polyethylene catalysis; polyethylene terephthalate for container application and plastics for packaging; regulatory affairs; author of several books; engineering plastics; product safety. *Mailing Add:* 10 Byron Lane Fanwood NJ 07023

KIRSHENBAUM, ISIDOR, b New York, NY, June 22, 17; m 47; c 4. CHEMISTRY. *Educ:* City Col New York, BS, 38; Columbia Univ, MA, 39, PhD(chem), 42. *Prof Exp:* Asst instr phys chem, Columbia Univ, 40-42, res scientist, 42-45; res assoc, 45-68, sr res assoc, 68-77, sci adv, Exxon Res & Eng Co, 77-85; PVT CONSULT,86- *Concurrent Pos:* Consult, AEC, 47-55. *Mem:* AAAS; Am Chem Soc. *Res:* Oxidation reactions; patents and research analysis; polymer properties; petrochemical petroleum processes; heterogeneous catalysis; separation and physical properties of isotopes. *Mailing Add:* 5601 Kennedy Blvd West New York NJ 07093

KIRSHNER, HOWARD STEPHEN, b Bryn Mawr, Pa, July 11, 46; m 69; c 2. NEUROLOGY. *Educ:* Williams Col, BA, 68; Harvard Med Sch, MD, 72. *Prof Exp:* Intern med, Mass Gen Hosp, 72-73; staff assoc lab perinatal physiol, Nat Inst Neurol & Commun Dis, 73-75; resident & clin instr, Mass Gen Hosp, 75-78; assoc prof, 78-87, PROF NEUROL, SCH MED, VANDERBILT UNIV, 87- *Concurrent Pos:* Consult, Nashville Gen Hosp, 78- & Mid Tenn Ment Health Inst, 78-; vchmn, dept neurol, sch med & assoc dir, Rehab Unit, Vanderbilt Univ; stroke coun, AMA. *Mem:* Am Acad Neurol; Am Neurol Asn; Nat Aphasia Asn; Acad Aphasia. *Res:* Aphasia; higher cortical functions; cerebrovascular disease. *Mailing Add:* Dept Neurol Vanderbilt Univ Sch Med Nashville TN 37240

KIRSHNER, NORMAN, b Wilkes-Barre, Pa, Sept 21, 23; m 61; c 3. BIOCHEMISTRY. *Educ:* Univ Scranton, BS, 47; Pa State Univ, MS, 51, PhD(biochem), 52. *Prof Exp:* Res asst physiol, Univ Rochester, 52-54; res asst, Yale Univ, 54; res asst, 55-57, assoc biochem, 56-59, from asst prof to assoc prof, 59-70, chmn dept pharmacol, 77, PROF BIOCHEM, DUKE UNIV, 70- *Concurrent Pos:* USPHS sr res fel, 59-62, career develop award, 62-; mem, NINCDS Prog Proj Rev Comt, 75-79, Coun Sci Adv, Roche Inst Molecular Biol, 79; ed, Molecular Pharmacol, 77. *Mem:* Am Soc Biol Chem; Am Soc Pharmacol & Exp Therapeut; Am Soc Neurochem. *Res:* Metabolism of Catecholamines; mechanism of storage and release of neurotransmitters. *Mailing Add:* Med Ctr Box 3813 Duke Univ Durham NC 27710

KIRSHNER, ROBERT PAUL, b Long Branch, NJ, Aug 15, 49; m 70; c 2. ASTRONOMY. *Educ:* Harvard Col, AB, 70; Calif Inst Technol, PhD(astron), 75. *Prof Exp:* Res assoc astron, Kitt Peak Nat Observ, 74-76; from asst prof to prof astron, Univ Mich, 76-85, dir, McGraw Hill Observ, 80-85; PROF ASTRON, HARVARD UNIV, 85- *Concurrent Pos:* Users comt, Kitt Peak Nat Observ, 78-81, Cerro-Tololo Inter-Am Observ, 79-82; telescope allocation comt, Kitt Peak Nat Cerro-Tololo Inter Am Observ, 80-82; comt, Space Astron & Astrophys, 82-85; sci adv comt, Nat New Technol Telescope, 83-84; vis comt mem, Asn Univ Res Astron, 83-86; chmn, NSF subcomt Large Optical-IR Telescopes, 85-86; Alfred P Sloan fel, 79-82; coun mem, Am Astron Soc, 85-; vis comt, Mt Wilson & Las Camponas Observ; vis comt, Space Telescope Sci Inst. *Honors & Awards:* Bowdon Prize, Harvard Col, 70. *Mem:* Am Astron Soc; Int Astron Union; Am Phys Soc; AAAS. *Res:* Extragalactic supernovae and galactic supernova remnants; dynamics and luminosity of galaxies; large scale structure in the universe. *Mailing Add:* Dept Astron Harvard Univ 60 Garden St Cambridge MA 02138

KIRSNER, JOSEPH BARNETT, b Boston, Mass, Sept 21, 09; m 34; c 1. INTERNAL MEDICINE, GASTROENTEROLOGY. *Educ:* Tufts Col, MD, 33; Univ Chicago, PhD(biol sci) & Am Bd Internal Med, dipl, 42. *Prof Exp:* From asst prof to prof med, 35-74, dean med affairs & chief staff, Univ Hosp, 71-76, LOUIS BLOCK DISTINGUISHED SERV PROF MED, UNIV CHICAGO, 74- *Concurrent Pos:* Mem nat adv coun, Nat Inst Arthritis & Metab Dis; chmn, Adv Group, Nat Comn Digestive Dis, 78; assoc ed, Advances Internal Med, 70- *Honors & Awards:* Friedenwald Medal, Am Gastroenterol Asn; John Phillips Mem Award, Am Col Physicians; George Howell Coleman Medal, Inst Med, Chicago; Rudolph Schindler Award, Am Soc Gastrointestinal Endoscopy. *Mem:* Am Soc Clin Invest; Am Soc Gastrointestinal Endoscopy (secy-treas, Gastroscopic Soc, 42-48, pres, 49-50); fel AMA; Am Gastroenterol Asn (treas & pres, 65-66); mastership Am Col Physicians. *Res:* Gastroenterology, especially peptic ulcer, cancer, inflammatory diseases including regional enteritis and ulcerative colitis; protein metabolism; hepatic disease; immunological mechanisms in gastrointestinal disease. *Mailing Add:* Dept Med Univ Chicago Chicago IL 60637

KIRST, HERBERT ANDREW, b St Paul, Minn, Sept 22, 44. ORGANIC CHEMISTRY. *Educ:* Univ Minn, BS, 66; Harvard Univ, PhD(org chem), 71. *Prof Exp:* Fel org chem, Calif Inst Technol, 71-73; sr chemist, 73-77, res scientist, 77-83, SR RES SCIENTIST, ELI LILLY & CO, 83- *Mem:* Am Chem Soc; AAAS. *Res:* Structure determination and chemical modification of new antibiotics and other fermentation products. *Mailing Add:* Dept M539 Fermentation Prod Lilly Res Lab Indianapolis IN 46285

KIRSTEN, EDWARD BRUCE, b New York, NY, Jan 28, 42; m 63; c 2. NEUROPHARMACOLOGY, CELL PHYSIOLOGY. *Educ:* Fairleigh Dickinson Univ, BS, 62; NY Univ, MS, 66; City Univ New York, PhD(biol), 69. *Prof Exp:* Assoc, Col Physicians & Surgeons, Columbia Univ, 71-72, asst prof pharmacol, 72-77; DIR CLIN PHARMACOL, KNOLL PHARMACEUT CO, 77- *Concurrent Pos:* NIH fel pharmacol, Col Physicians & Surgeons, Columbia Univ, 69-71, res career develop award, 72-77. *Mem:* AAAS; assoc Am Physiol Soc; Am Soc Pharmacol & Exp Therapeut; Sigma Xi. *Res:* Vestibular neuropharmacology; antiarrhythmic drugs; muscle physiology. *Mailing Add:* Knoll Pharmaceut Co 30 N Jefferson Rd Whippany NJ 07981

KIRSTEN, WERNER H, b Leipzig, Ger, Oct 29, 25; US citizen; m 60; c 3. PATHOLOGY. *Educ:* Univ Frankfurt, MD, 54. *Prof Exp:* Asst path, Path Inst, Frankfurt, 53-55; intern med, Englewood Hosp, 55-56; resident path, 56-59, from instr to assoc prof, 59-66, assoc prof pediat, 66-68, prof path & pediat, 68-88, chamn dept path, Univ Chicago,72-88; ASSOC DIR, FREDERICK CANCER RES & DEV CTR, 88- *Concurrent Pos:* USPHS career develop award, 60-68. *Mem:* AAAS; Am Asn Path & Bact; Am Soc Exp Path; Am Asn Cancer Res. *Res:* Experimental pathology; cancer research; oncogenic virus, especially leukemia. *Mailing Add:* Assoc Dir Nat Cancer Inst Frederick Cancer Res & Dev Ctr PO Box B Frederick MD 21702-1201

KIRSTEUER, ERNST, invertebrate zoology, for more information see previous edition

KIRTLEY, JOHN ROBERT, b Palo Alto, Calif, Aug 27, 49; m 73; c 1. SUPERCONDUCTIVITY, MICROSCOPY. *Educ:* Univ Calif, Santa Barbara, BA, 71, PhD(physics), 76. *Prof Exp:* Post doctoral, Univ Pa, 76-77, res asst prof, 77-78; RES STAFF MEM, IBM RES DIV, 78- *Mem:* Fel Am Phys Soc. *Res:* Solid state physics in planar structures and using a scanning tunneling microscope; collective phenomena such as superconductivity, quantum hall effect at low temperatures. *Mailing Add:* T J Watson Res Ctr Yorktown Heights NY 10598

KIRTLEY, MARY ELIZABETH, b Mansfield, Ohio, Aug 27, 35. BIOCHEMISTRY. *Educ:* Univ Chicago, BA, 56; Smith Col, MA, 58; Western Reserve Univ, PhD(biochem), 64. *Prof Exp:* Lab instr chem, Smith Col, 58-59; res assoc biochem, Brookhaven Nat Lab, 64-65; res assoc, Univ Calif, Berkeley, 65-66; from asst prof to prof, 66-81, RES PROF BIOL CHEM, SCH MED, UNIV MD, BALTIMORE, 83- *Concurrent Pos:* Proj dir, Biochem Prog, NSF, 82-83; dir, Off Grad Studies, Dickinson Col, 83- *Mem:* Am Chem Soc; Am Soc Biol Chem; Biophys Soc. *Res:* Enzyme mechanisms; biochemical regulation; enzyme structure and function; enzyme-membrane interactions. *Mailing Add:* Dickinson Col High & College Sts Carlisle PA 17013

KIRTLEY, THOMAS L(OYD), b Salmon, Idaho, Nov 16, 18; m 43; c 2. CHEMICAL ENGINEERING. *Educ:* San Jose State Col, AB(chem) & AB(physics & math), 40; Calif Inst Technol, MS, 42. *Prof Exp:* Tech shift supvr, Ammonia Prod, Hercules Powder Co, 42-44, shift supvr, Rocket Powder Prod, 44-45, serv supvr, Chem Cotton Prod, 45-56, asst chief chemist, 56-60, chem supvr, 60-65, chem cotton tech coordr, 65-69, chem cotton supt, 69-75, supt eng develop, 75-77, SUPT ENVIRON ENG, HERCULES INC, 77- *Mem:* Am Chem Soc. *Res:* Cellulose chemistry; chemical cotton; cellulose ethers. *Mailing Add:* 306 Sherwood Dr Hopewell VA 23860

KIRTLEY, WILLIAM RAYMOND, b Crawfordsville, Ind, May 30, 14; m 40; c 2. CLINICAL MEDICINE. *Educ:* Wabash Col, AB, 36; Northwestern Univ, MB, 40, MD, 41. *Prof Exp:* Staff physician, Eli Lilly & Co, 47-56, physician in chg, Diabetes Res Lab Clin Res, 53-70, sr physician, 56-61, head, Clin Med Dept, 61-65, asst dir, Clin Res Div, 63-65, dir, Med Res Div, 65-72, dir, Lilly Lab Clin Res, 72-78; assoc prof med, Sch Med, Ind Univ, 73-78; RETIRED. *Concurrent Pos:* Assoc med, Wishard Mem Hosp, 49-, chief, Diabetes Clin, 53-70; assoc med, Sch Med, Ind Univ, 52-61, asst prof, 61-73; abstr ed, Diabetes, Am Diabetes Asn, 50-68. *Honors & Awards:* Banting Medal, Am Diabetes Asn, 71. *Mem:* Am Soc Clin Pharmacol & Therapeut; Endocrine Soc; AMA; Am Diabetes Asn. *Res:* Insulin modifications; metabolism; diabetes mellitus; glucagon. *Mailing Add:* 40 N Port Royal Dr Hilton Head SC 29928

KIRTMAN, BERNARD, b New York, NY, Mar 30, 35; m 58; c 2. PHYSICAL CHEMISTRY. *Educ:* Harvard Univ, PhD(phys chem), 61. *Prof Exp:* Fel chem, Univ Wash, 60-62; asst prof chem, Univ Calif, Berkeley, 62-65, from asst prof to assoc prof chem, Univ Calif, Santa Barbara, 65-74, PROF THEORET PHYS CHEM, UNIV CALIF, SANTA BARBARA, 74- *Res:* Theoretical chemistry; application of quantum mechanics to electronic structure of atoms and molecules, intermolecular interactions and molecular dynamics. *Mailing Add:* Dept Chem Univ Calif Santa Barbara CA 93106

KIRWAN, ALBERT DENNIS, JR, b Louisville, Ky, Nov 29, 33; m 56; c 3. PHYSICAL OCEANOGRAPHY. *Educ:* Princeton Univ, AB, 56; Tex A&M Univ, PhD(phys oceanog), 64. *Prof Exp:* Res asst oceanog, Tex A&M Univ, 59-64; from asst prof to assoc prof, NY Univ, 64-70; prog dir phys oceanog, Off Naval Res, 70-72; AT DEPT MARINE SCI, UNIV SFLA. *Mem:* AAAS; Am Meteorol Soc; Am Geophys Union. *Res:* Air-sea interaction, general circulation of oceans; physics of fluids; engineering science. *Mailing Add:* Dept Oceanog Old Dominion Univ Norfolk VA 23520-0276

KIRWAN, DONALD FRAZIER, b Oklahoma City, Okla, Nov 9, 37; m 59; c 5. ENERGY, ENERGY EDUCATION. *Educ:* Univ Mo-Columbia, BS, 63, MS, 64, PhD(physics), 69. *Prof Exp:* From instr to asst prof, 67-74, dir, Off Energy Educ, 79, PROF PHYSICS, UNIV RI, 81-; MGR, EDUC DIV, AM INST PHYSICS, 88- *Concurrent Pos:* Consult, Fed Emergency Mgt Agency, 80-, Argonne Nat Labs, 82-, Int Atomic Energy Agency, 83-, Environ Protection Agency, 84-; ed, The Physics Teacher, 85-; at large mem exec bd, Soc Physics Students, 85-88, exec bd, Am Asn Physics Teachers, 85-; energy leader, Conf Sci & Technol Educ & Future Human Needs, 85; chmn, 81 Int Conf on Energy Educ. *Mem:* AAAS; Am Phys Soc; Am Asn Physics Teachers; Nat Sci Teachers Asn. *Res:* Fundamental interactions; few-nucleon systematics; excitation mechanisms of the nucleus; physics pedagogy; radiological accident assessment; radiological emergency preparedness. *Mailing Add:* Dept Physics Univ RI Kingston RI 02881

KIRWAN, WILLIAM ENGLISH, b Louisville, Ky, Apr 14, 38; m 60; c 2. MATHEMATICAL ANALYSIS. *Educ:* Univ Ky, BA, 60; Rutgers Univ, MS, 62, PhD(math), 64. *Prof Exp:* Instr math, Rutgers Univ, 63-64; from asst prof to assoc prof math, Univ Md, College Park, 64-72, chmn dept, 77-81, vchancellor acad affairs, 81-88, actg pres, 88-89, PROF MATH, UNIV MD, COLLEGE PARK, 72-, PRES, 89- *Concurrent Pos:* Vis lectr, Royal Holloway Col, Univ London, 66-67; prog dir, NSF, 75-76. *Mem:* Math Asn Am; Am Math Soc; Sigma Xi. *Res:* Functions of one complex variable, particularly extremal properties of conformal and quasiconformal mappings of the unit disc. *Mailing Add:* President's Off Univ Md College Park MD 20742

KIRWIN, GERALD JAMES, b Lowell, Mass, Feb 11, 29; m 55; c 2. ELECTRICAL ENGINEERING, APPLIED MATHEMATICS. *Educ:* Northeast Univ, BSEE, 52; Mass Inst Technol, MSEE, 55; Syracuse Univ, PhD(elec eng), 68. *Prof Exp:* Mem tech staff, Bell Tel Labs, 55-56; assoc prof elec eng, Merrimack Col, 56-64; instr, Syracuse Univ, 64-68; prof, Univ Maine, Portland-Gorham, 68-73; PROF ELEC ENG, SCH ENG, UNIV NEW HAVEN, 73- *Mem:* Inst Elec & Electronics Engrs. *Res:* Nonlinear systems; circuit theory and design; optimal network design, electrical engineering education. *Mailing Add:* Univ New Haven PO Box 1306 New Haven CT 06516

KIRZ, JANOS, b Budapest, Hungary, Aug 11, 37; US citizen; m 88; c 1. SYNCHROTRON RADIATION, X- RAY OPTICS. *Educ:* Univ Calif, Berkeley, BA, 59, PhD(physics), 63. *Prof Exp:* Nat Acad Sci-Nat Res Coun fel physics, Saclay Nuclear Res Ctr, France, 63-64; physicist, Lawrence Radiation Lab, Univ Calif, 64-68; assoc prof, 68-72, assoc chmn, 84-85, PROF PHYSICS, STATE UNIV NY, STONY BROOK, 72- *Concurrent Pos:* Lectr, Univ Calif, Berkeley, 67; Sloan Found fel, 70-72; visitor, Lab Molecular Biophys, Oxford Univ, 72-73; Guggenheim fel, 85-86. *Mem:* Fel AAAS; fel Am Phys Soc; Optical Soc Am. *Res:* X-ray microscopy. *Mailing Add:* Dept Physics State Univ NY Stony Brook NY 11794-3800

KISCHER, CLAYTON WARD, b Des Moines, Iowa, Mar 2, 30; m 64; c 2. EMBRYOLOGY, CELL BIOLOGY IN SURGICAL RESEARCH. *Educ:* Univ Omaha, BS, 53; Iowa State Univ, MS, 60, PhD(embryol), 62. *Prof Exp:* Cytologist, Col Med, Univ Nebr, 53-56; teacher high sch, Nebr, 56-58; asst prof zool, Ill State Univ, 62-63; resident res assoc biol & med, Argonne Nat Lab, 63; asst prof zool, Iowa State Univ, 63-64; chief sect electron micros, Southwest Found Res & Educ, 66-67; asst prof, 67-70, assoc prof, Univ Tex Med Br; vis prof surg biol, 76-77, dir Sem Lab, 77-85, ASSOC PROF ANAT, UNIV ARIZ COL MED, 77- *Concurrent Pos:* Fel biochem, Univ Tex M D Anderson Hosp, 64-66; res consult to chief staff, Shriners Burns Inst, Galveston, 70-73. *Mem:* AAAS; Soc Develop Biol; Am Soc Cell Biol; Electron Micros Soc Am. *Res:* Induction organogenesis; ultrastructural changes during morphogenesis; organ culture; tissue culture; electron microscopy; hypertrophic scarring and fibronectin; microvessels in wound healing. *Mailing Add:* Dept Anat Univ Ariz Col Med Tucson AZ 85724

KISER, DONALD LEE, b Keokuk, Iowa, Jan 2, 33; m 53; c 3. ANALYTICAL CHEMISTRY. *Educ:* Iowa State Univ, BS, 59; Ind State Univ, MS, 61; Univ Iowa, PhD(anal chem), 64. *Prof Exp:* Anal chemist, Com Solvents Corp, Ind, 59-61; sr anal chemist, 64-73, mgr anal develop, 73-83, MGR INTELLECTUAL PROPERTY, GRAIN PROCESSING CORP, 83-, VPRES, 85- *Concurrent Pos:* Coun mem, Am Chem Soc, 73-; mem, bd trustees, Group Ins Plans, Am Chem Soc, 87- *Mem:* Am Chem Soc; Asn Off Anal Chemists; Am Asn Cereal Chemists; Licensing Execs Soc. *Res:* Automated methods of analysis; protein and amino acid analysis; alcohol congeners by gas chromatography; carbohydrate molecular weight profiles by liquid chromatography. *Mailing Add:* Grain Processing Corp 1600 Oregon Ave Box 349 Muscatine IA 52761

KISER, KENNETH M(AYNARD), b Detroit, Mich, Nov 28, 29; m 54; c 5. CHEMICAL ENGINEERING. *Educ:* Lawrence Tech Univ, BS, 51; Univ Cincinnati, MS, 52; Johns Hopkins Univ, DS(chem eng), 56. *Prof Exp:* Asst, Univ Cincinnati, 51-52; res assoc, Inst Coop Res, Johns Hopkins Univ, 52-56; res assoc, Chem Dept, Res Lab, Gen Elec Co, 56-64; adj staff, Rensselaer Polytech Inst, 62-64; from asst prof to assoc prof, 64-80, PROF CHEM ENG, STATE UNIV NY, BUFFALO, 80-, ASSOC DEAN ENG, 78- *Concurrent Pos:* Consult. *Mem:* Am Inst Chem Engrs; Am Soc Eng Educ; Sigma Xi. *Res:* Turbulent transport; non-Newtonian fluids; fluid mechanics in the human body; air and water pollution. *Mailing Add:* Eng Off State Univ NY 412 Bonner Hall Buffalo NY 14260

KISER, LOLA FRANCES, b Selmer, Tenn, Dec 6, 30. MATHEMATICS. *Educ:* Memphis State Univ, BS, 52; Univ Ga, MA, 54; Univ Ala, Tuscaloosa, PhD(math), 71. *Prof Exp:* Instr math, Univ Ga, 54-55; from asst prof to assoc prof, 55-71, PROF MATH, BIRMINGHAM-SOUTHERN COL, 71- *Mem:* Math Asn Am. *Res:* Complex analysis; differential equations. *Mailing Add:* Dept Math Box A-32 Birmingham-Southern Col Birmingham AL 35254

KISER, ROBERT WAYNE, b Rock Island, Ill, Apr 26, 32; m 54; c 3. INORGANIC CHEMISTRY. *Educ:* St Ambrose Col, BA, 53; Purdue Univ, MS, 55, PhD, 58. *Prof Exp:* From asst prof to prof inorg chem, Kans State Univ, 57-67; chmn chem dept, 68-72, PROF CHEM & DIR MASS SPECTROMETRY CTR, UNIV KY, 67- *Mem:* Am Soc Mass Spectrometry; Mass Spectros Soc Japan; Am Chem Soc; Am Phys Soc; The Chem Soc. *Res:* Exited states of negative ions; energetics and thermochemistry of ionic species; artificial intelligence in mass spectrometry; mass spectrometry and molecular structures. *Mailing Add:* Dept Chem CP-9 Univ Ky Lexington KY 40506

KISH, VALERIE MAYO, b Paintsville, Ky, Nov 28, 44. CELL BIOLOGY. *Educ:* Univ Ky, BS, 65; Ind Univ, MA, 66; Univ Mich, PhD(cell biol), 73. *Prof Exp:* Res asst, Inst Cancer Res, Philadelphia, 66-69; res assoc, Worcester Found Exp Biol, Shrewsbury, 73-76; PROF BIOL, HOBART & WILLIAM SMITH COLS, 76- *Mem:* Am Soc Cell Biol; Am Soc Plant Physiol; Sigma Xi. *Res:* Interaction of proteins with RNA in eukaryotic cells and the role these interactions play in the regulation of gene expression. *Mailing Add:* Dept of Biol Hobart & William Smith Cols Geneva NY 14456

KISHEL, CHESTER JOSEPH, b Cleveland, Ohio, Oct 12, 15; m 39; c 3. INDUSTRIAL & HUMAN FACTORS ENGINEERING. *Educ:* Fenn Col, BSME, 44, BSIE, 47; Case Inst Technol, MSIE, 54, PhD(eng admin), 59. *Prof Exp:* Asst chief prod eng, Iron Fireman Mfg Co, 43-45; prod mgr, A W Hecker Co, 45-46; vpres mfg, Kramic Corp, 46-48; from asst prof to assoc prof mech eng, Fenn Col, 48-57; spec lectr eng admin, Case Inst Technol, 57-59; assoc prof & dir comput ctr, Fenn Col, 59-62; contract prof comput indust eng, Mich State Univ/USAID, Polytech Sch, Sao Paulo, Brazil, 62-64; prof mech & indust eng, Cleveland State Univ, 65-70, dir, Comput Ctr, 65-66, prof indust eng & actg chmn dept, 70-71, SR PROF INDUST ENG, CLEVELAND STATE UNIV, 71- *Concurrent Pos:* Polio Found grant, Highland View Hosp, 60-61. *Mem:* Am Inst Indust Engrs; Soc Mfg Eng; Asn Comput Mach; Am Arbit Asn. *Res:* Neuroanatomy and neurophysiology in desision making aspects of organization theory; electromyographical study-hypotheses of muscular control of human hand languages and techniques of digital computing; numerical controls. *Mailing Add:* 17512 Greenbrier Dr Strongsville OH 44136

KISHI, KEIJI, b Tokyo, Japan, Mar 31, 30; m 63; c 2. RESEARCH ADMINISTRATION & MANAGEMENT. *Educ:* Keio Univ, BA, 52, Dr(elec eng), 69. *Prof Exp:* Sr researcher, Toshiba Corp, 66-73, res fel, 73-79, dep dir, 79-84, gen mgr, 84-86; gen mgr planning, 86-87, vpres & gen mgr, Silicon Div, 87-89, VPRES & GEN MGR, RES & DEVELOP CTR, TOSHIBA CERAMIC CO, 87- *Concurrent Pos:* Mem, Workshop Ministry State Sci & Technol, 85-88. *Mem:* Fel Inst Elec & Electronics Engrs. *Res:* Thyristor or converter for hvdc, power devices and their application; optoelectronics application for high voltage measurement. *Mailing Add:* Toshiba Ceramics Co Ltd 26-2 Nishi-Shinjuku 1-Chrome Shinjuku-Ku Tokyo 163 Japan

KISHI, YOSHITO, b Nagoya, Japan, April 13, 37; m 63; c 2. MEDICINAL CHEMISTRY. *Educ:* Nagoya Univ, BS, 61; Harvard Univ, MA, 74 PhD (chem), 66. *Prof Exp:* Instr chem, Nagoya Univ, 66-69, assoc prof, 69-74; prof, 74-82, MORRIS LOEB PROF CHEM, HARVARD UNIV, 82- *Concurrent Pos:* Fel, Harvard Univ, 66-68, vis prof, 72-73; Arthur C Cope scholar award, 88, Javits neurosci invest award, 88. *Honors & Awards:* Japan Chem Soc Prize, 67; Am Chem Soc Award Creative Work Synthetic Org Chem, 80; Harrison Howe Award, 81. *Mem:* Am Chem Soc; Chem Soc Japan; Swiss Chem Soc; Am Acad Arts & Sci. *Res:* Total synthesis of complex natural products typified by the completed works of neurotoxins, metabolites of microorganisms and polyether, ansamycin and antitumor antibiotics. *Mailing Add:* Dept Chem Harvard Univ 12 Oxford St Cambridge MA 02138

KISHIMOTO, YASUO, b Osaka-Shi, Japan, Apr 11, 25; m 49; c 4. MOLECULAR BIOLOGY, GENETICS. *Educ:* Kyoto Univ, BS, 48, PhD(pharmaceut chem), 56. *Prof Exp:* Res chemist, Osaka Gas Co, 48-50; lectr pharm, Kyushu Univ, 50-54; asst prof, Shizuoka Col Pharm, 54-61; from asst to assoc res biochemist, Ment Health Res Inst, Univ Mich, Ann Arbor, 62-67; mem staff, Div Chem Res, G D Searle & Co, Ill, 67-69; asst biochemist, Mass Gen Hosp, 69-70, assoc biochemist neurol serv, 70-76; from assoc prof to prof, Sch Med, Johns Hopkins Univ, 76-88; ADJ PROF, DEPT NEUROSCI SCH MED, UNIV CALIF, SAN DIEGO, 88- *Concurrent Pos:* Res asst biochem, Med Sch, Northwestern Univ, 57-59; sr investr, Eunice Kennedy Shriver Ctr Ment Retardation, 69-76; assoc neurol, Harvard Med Sch, 69-76; dir biochem res, John F Kennedy Inst, 76-88. *Mem:* Am Soc Biochem & Molecular Biol; Int Soc Neurochem; AAAS; Am Soc Neurochem; Japanese Biochem Soc. *Res:* Structures and metabolism of brain lipids; myelination, demyelination. *Mailing Add:* Ctr Molecular Genetics 0634 J Dept Neurosci Univ Calif San Diego Sch Med La Jolla CA 92093

KISHK, AHMED A, b Ashmoun, Egypt, Dec 9, 54; m 83; c 2. APPLIED ELECTROMAGNETICS, NUMERICAL SOLUTIONS OF ELECROMAGNETIC PROBLEMS. *Educ:* Cairo Univ, Egypt, BS, 77; Ain Shams Univ, Egypt, BS, 80; Univ Manitoba, Can, MS, 83, PhD(elec eng), 86. *Prof Exp:* Teaching & res asst elec eng, Cairo Univ, 77-81; teaching & res asst elec eng, Univ Man, Can, 81-85, res assoc, 85-86; ASSOC PROF ELEC ENG, UNIV MISS, 86- *Mem:* Inst Elec & Electronics Engrs; Sigma Xi. *Res:* Numerical solutions of electromagnetic problems and antenna design, especially antenna feeds, microstrip antennas, ground station antennas and mobile satellite antennas; antenna design. *Mailing Add:* Dept Elec Eng Univ Mississippi University MS 38677

KISHORE, GANESH M, b Hunsur, India, Sept 26, 53; US citizen; m 76; c 2. PLANT GENETIC ENGINEERING, BIOCHEMISTRY. *Educ:* Univ Mysore, BSc, 70, MSc, 72; Indian Inst Sci, PhD(biochem), 76. *Prof Exp:* Sr res biochem, Monsanto Corp Res Labs, 80-82, res specialist, 82-86, assoc fel, 87-89, res mgr, 89-90, MGR, MONSANTO CORP RES LABS, 90- *Mem:* Am Soc Biochem & Molecular Biol; Am Soc Plant Physiol. *Res:* Plant biochemistry and plant genetics engineering; genetic engineering of herbicide tolerance, metabolism of herbicides and crop quality improvement. *Mailing Add:* 15354 Grantley Dr Chesterfield MO 63017

KISHORE, GOLLAMUDI SITARAM, b Madras, Tamilnadu, India, Feb 22, 45; m 71; c 2. SIALOGLYCOCONJUGATES IN CANCER, NUTRITION & CANCER. *Educ:* Univ Madras, BS, 64, MS, 66; Indian Inst Sci, Bangalore, PhD(biochem), 72. *Prof Exp:* Res fel, Okla Med Res Found, 73-76, res assoc, 76-78; res assoc, McArdle Lab Cancer Res, Univ Wis-Madison, 78-80; CANCER RES SCIENTIST III, DEPT BREAST SURG, ROSWELL PARK MEM INST, BUFFALO, NY, 81-; ASST PROF BIOCHEM, STATE UNIV NY, BUFFALO, 84- *Concurrent Pos:* Res assoc, Univ Okla Health Sci Ctr, Okla, 73-78. *Mem:* Am Asn Cancer Res; Am Soc Biochemists; Int Asn Vitamin & Nutrit Oncol; Sigma Xi. *Res:* Metabolism of sialoglycoconjugates in cancer; biochemical studies of skin carcinogenesis. *Mailing Add:* Dept Breast Surg Roswell Park Mem Inst Buffalo NY 14263

KISILEVSKY, ROBERT, b Montreal, Can, Dec 19, 37; m 67; c 3. BIOCHEMISTRY, PATHOLOGY. *Educ:* McGill Univ, BSc, 58, MD, CM, 62; Univ Pittsburgh, PhD(biochem), 69; FRCP(C), 72; Am Bd Path, dipl, 72. *Prof Exp:* from asst prof to assoc prof, 70-79, PROF PATH, QUEEN'S UNIV, 79-, HEAD, 86- *Concurrent Pos:* Asst pathologist, Kingston Gen Hosp, 70-86, pathologist-in-chief, 86-; from asst prof to assoc prof biochem, Queen's Univ, 71-90, prof, 90-; mem, Grant Comt, Med Res Coun, 74-77, sci officer, 77-83; mem, Nat Cancer Inst, 88-91. *Mem:* Int Acad Path; Am Asn Pathologists; Can Asn Pathologists; Can Biochem Soc. *Res:* Pathogenetic mechanisms of Amyloidosis and its relationship to Alzheimer's disease. *Mailing Add:* Dept Path Queen's Univ Kingston ON K7L 3N6 Can

KISKIS, JOSEPH EDWARD, JR, b Lynwood, Calif, Oct 2, 47. THEORETICAL PHYSICS. *Educ:* Univ Calif, Davis, BS, 69; Stanford Univ, MS, 71, PhD(physics), 74. *Prof Exp:* Fel, Mass Inst Technol, 74-76 & Los Alamos Sci Lab, 76-77; mem, Inst Advan Study, 77-78; Oppenheimer fel, Los Alamos Nat Lab, 78-80; PROF PHYSICS, UNIV CALIF, DAVIS, 80- *Res:* High energy physics; quantum field theory. *Mailing Add:* Dept Physics Univ Calif Davis CA 95616

KISLIUK, PAUL, b Philadelphia, Pa, Feb 22, 22; m 50; c 4. PHYSICS. *Educ:* Queen's Col, NY, BA, 43; Columbia Univ, MS, 47, PhD(physics), 52. *Prof Exp:* Mem tech staff, Bell Tel Labs, 52-62; dept head, 62-66, SR SCIENTIST, AEROSPACE CORP, 66- *Mem:* Am Phys Soc. *Res:* Microwave spectroscopy; contact physics; surface physics; lasers; solid state spectroscopy. *Mailing Add:* M5-665 Aerospace Corp PO Box 92957 Los Angeles CA 90009

KISLIUK, ROY LOUIS, b Philadelphia, Pa, Aug 4, 28; m 54; c 2. BIOCHEMISTRY. *Educ:* Queen's Col, NY, BS, 50; Yale Univ, MS, 52; Western Reserve Univ, PhD(biochem), 56. *Prof Exp:* Vis scientist, Nat Inst Arthritis & Metab Dis, 58-60; from asst prof to assoc prof pharmacol, 60-71, assoc prof biochem, 71-72, PROF BIOCHEM, SCH MED, TUFTS UNIV, 72- *Concurrent Pos:* Nat Found Infantile Paralysis fel biochem, Oxford Univ, 56-58; prog dir biochem, NSF, Washington, DC, 72-73. *Mem:* Am Soc Biol Chem; Am Chem Soc; Am Soc Pharmacol & Exp Therapeut; Am Soc Microbiol; Am Soc Cancer Res. *Res:* Folate enzymes, coenzymes and antimetabolites. *Mailing Add:* Dept Biochem Tufts Univ Sch Med Boston MA 02111

KISMAN, KENNETH EDWIN, b Sudbury, Ont, Nov 18, 46; m 70; c 2. PETROLEUM ENGINEERING. *Educ:* Queens Univ, BS, 68; Univ Toronto, MS, 70, PhD(molecular physics), 74. *Prof Exp:* Res officer hyperbaric biophys, Defence & Civil Inst Environ Med, Can, 74-79, res supvr reservoir eng, 79-84, prin reservoir engr, 84-79; OIL SANDS TECH & RES AUTHORITY, PRIN RESERVOIR ENG, 84- *Mem:* Soc Petrol Eng; Can Inst Mining & Metall. *Res:* Engineering studies for heavy oil petroleum recovery processes, including field pilots, numerical simulation and direction of lab studies. *Mailing Add:* 5720 Buckboard Rd NW Calgary AB T3A 4R3 Can

KISPERT, LOWELL DONALD, b Faribault, Minn, June 9, 40; m 89; c 7. SOLID STATE CHEMISTRY, RADIATION CHEMISTRY. *Educ:* St Olaf Col, BA, 62; Mich State Univ, PhD(chem), 66. *Prof Exp:* Fel electron spin resonance, Varian Assocs, Calif, 66-67; radiation chem, Mellon Inst, 68; from asst prof to prof, 68-80, RES PROF CHEM, UNIV ALA, 80- *Honors & Awards:* Burnum Award, 88. *Mem:* Am Chem Soc; Am Phys Soc; Sigma Xi; Int Soc Magnet Resonance. *Res:* Free radicals as produced in irradiated organic single crystals by electron spin resonance; electron nuclear double resonance and electron-electron double resonance of paramagnetic single crystals, anions and cations in solution; radiation chemistry; conducting polymers; photosynthesis; solid-state photochemistry; polymer batteries, role of carotenoids in plant photosynthesis; coal porosity. *Mailing Add:* Chem Dept Box 870336 Univ Ala Tuscaloosa AL 35487-0336

KISS, KLARA, b Budapest, Hungary, Aug 28, 30; m 51; c 2. PHYSICAL CHEMISTRY, POLYMER CHEMISTRY. *Educ:* Budapest Tech Univ, dipl Chemiker, 54, PHd, 82. *Prof Exp:* Jr chemist, Res Ctr Org & Polymer Chem, Budapest, 49-51; res chemist, Res Ctr Telecommun, 55-56; res & develop chemist, Filmfabrik AGFA, EGer, 56; res chemist, Dom Dyeing & Printing Co, Can, 57-59; sr res assoc phys chem & polymer sci, Horizons, Inc, Ohio, 61-65; res assoc electron micros, Case Western Reserve Univ, 65-67; res chemist, GAF Corp, 67-70; sr res assoc, Stauffer Chem Co, 70-87; SR RES SCIENTIST, AKZO CHEMICALS INC, 87- *Mem:* Soc Appl Spectros; Asn Hungarian Chemists; Ger Chem Soc; Am Crystallog Asn; Microbeam Analysis Soc. *Res:* Kinetics of redox bulk polymerization of acrylates; photopolymerization; controlled, ultrafine particle size ferroelectrics; epitaxial crystallization of polymers; identification of wairakite single crystals by electron diffraction; microbeam analysis; mechanical properties relationship in materials. *Mailing Add:* Akzo Chem Inc Livingstone Ave Dobbs Ferry NY 10522

KISSA, ERIK, b Abja, Estonia, Apr 7, 23; nat US; m 52; c 2. COLLOID CHEMISTRY, TEXTILE CHEMISTRY. *Educ:* Karlsruhe Univ, dipl, 51; Univ Del, PhD(chem), 56. *Prof Exp:* Anal chemist, 51-56, res chemist, 56-67, sr res chemist, 67-74, res assoc, 74-86, sr res assoc, Du Pont Chem Dept, 86-90, RES FEL, E I DU PONT DE NEMOURS & CO, INC, 90- *Concurrent Pos:* UN Indust Develop Orgn consult, Atira Textile Res Inst, Ahmedabad, India, 78 & 79, Shanghai Dye Res Inst, China, 82, Korea Res Inst Chem Technol, 86, 87 & 88. *Mem:* AAAS; Am Chem Soc; fel Am Inst Chem; Int Asn Colloid & Interface Scientists; Fiber Soc; Am Oil Chemists Soc. *Res:* Colloid and surface chemistry; association colloids; dispersions; emulsions adsorption; detergency; surfactants fluorinated surfactants dyes; textile chemicals; physical chemistry of dyeing and stain resistance; polymers; analytical chemistry; surface chemistry of fibers. *Mailing Add:* 1436 Fresno Rd Wilmington DE 19803

KISSANE, JOHN M, b Oxford, Ohio, Mar 30, 28; m 51; c 5. MEDICINE. *Educ:* Univ Rochester, AB, 48; Wash Univ, MD, 52. *Prof Exp:* Asst path, 52-53, from instr to assoc prof, 53-68, PROF PATH & PROF PATH IN PEDIAT, SCH MED, WASH UNIV, 68- *Concurrent Pos:* Intern, Barnes Hosp, 52-53, chief resident, 53-54, asst pathologist, Barnes & Assoc Hosps, 58-; Nat Found Infantile Paralysis res fel, Sch Med, Wash Univ, 54-55 & 57-58, med alumni scholar, 70-71; mem sect renal dis, Coun Circulation, Am Heart Asn. *Mem:* AMA; Am Asn Path & Bact; Am Soc Exp Path; Histochem Soc; Int Acad Path. *Res:* Pathology; pediatric pathology; quantitative histochemistry of nervous system; kidney. *Mailing Add:* Dept Path & Pediat Wash Univ Sch Med 660 S Euclid St Louis MO 63110

KISSEL, CHARLES LOUIS, b Chicago, Ill, Aug 5, 47; m 70; c 2. INDUSTRIAL ORGANIC CHEMISTRY. *Educ:* Univ Calif, Irvine, BA, 69; Univ Calif, Santa Barbara, PhD(chem), 73. *Prof Exp:* Group leader chem res, 73-84, dir res, Magna Corp, 84-85; sr res chemist, Unocal Corp, 85-90; PRIN OWNER, CNC DEVELOP, 90- *Mem:* Am Chem Soc; Sigma Xi; AAAS; Nat Asn Corrosion Engrs; Tech Asn Pulp & Paper Indust. *Res:* Treating mechanisms and synthesis of industrial biocides, corrosion inhibitors, scale inhibitors, emulsion breakers and water clarifiers, zero formaldehyde textile binders for non-wovens, as well as specialty chemicals; special emphasis on acrolein chemistry; sol-gel chemistry; nonionic surfactants. *Mailing Add:* 2856 Skywood Circle Anaheim CA 92804

KISSEL, DAVID E, b Vanderburg Co, Ind, Aug 10, 43; m 66; c 3. SOIL CHEMISTRY, SOIL FERTILITY. *Educ:* Purdue Univ, BS, 65; Univ Ky, MS, 67, PhD(soil acidity), 69. *Prof Exp:* From asst prof to assoc prof soil chem, Tex A&M Univ, 69-77, assoc prof & asst dir for res, Blackland Res Ctr, 77-78; PROF AGRON, KANS STATE UNIV, 78- *Concurrent Pos:* Assoc ed, J Environ Qual, 75-78; ed, Soil Sci Am J. *Mem:* Am Soc Agron; fel Soil Sci Soc Am. *Res:* Plant nutrition; soil acidity; movement of water and nitrate in soils; nitrogen and phosphorus fertilizer use efficiency; ammonia volatilization; nitrogen mineralization. *Mailing Add:* Dept Agron 3111 Plant-Sci Bldg Univ Ga Athens GA 30602

KISSEL, JOHN WALTER, b St Louis, Mo, Dec 12, 25; m 53; c 4. PHARMACOLOGY. *Educ:* Wash Univ, AB, 48; St Louis Col Pharm, BS, 51; St Louis Univ, MS, 55; Univ Mich, PhD(pharmacol), 58. *Prof Exp:* Asst prof pharmacog & pharmacol, Col Pharm, Univ Fla, 56-57; assoc sr pharmacologist, Mead Johnson & Co, 57-59, group leader cent nerv syst pharmacol, 59-62, sect leader, 62-69; from pharmacologist to sr pharmacologist, 69-71, Am Cyanamid Co, mgr drug eval, 71-73, clin res assoc med res dept, Cyanamid Int, 73-75, clin res assoc, clin res dept, Lederle Labs, 75-88, sr clin res assoc, 88-91; RETIRED. *Mem:* Fel AAAS; Am Soc Pharmacol & Exp Therapeut; NY Acad Sci; Am Soc Clin Pharmacol Therapeut; Drug Info Asn. *Res:* Central nervous system pharmacology, especially opiate tolerance and addiction; spinal cord reflexes; muscle relaxants; effect of drugs on behavior; clinical evaluation of psychotherapeutic agents; antibiotics; anti-cancer therapy. *Mailing Add:* 115 Sherwood Dr Ramsey NJ 07446

KISSEL, THOMAS ROBERT, b Chicago, Ill, Sept 26, 47; m 72. ANALYTICAL CHEMISTRY, CLINICAL ANALYSIS. *Educ:* Univ Notre Dame, BS, 69; Univ Wis, PhD(anal chem), 74. *Prof Exp:* RES CHEMIST, EASTMAN KODAK CO, 74- *Mem:* Sigma Xi; Electrophoresis Soc. *Res:* Electrochemistry, specifically ion selective electrodes. *Mailing Add:* 200 Willowood Dr Rochester NY 14612

KISSEL, WILLIAM JOHN, b New York, NY, Mar 12, 41; m 66; c 2. POLYMER PROPERTIES. *Educ:* City Col New York, BS, 62; State Univ NY, Buffalo, PhD(org chem), 68. *Prof Exp:* SR RES CHEMIST, AMOCO CHEM CO, 67- *Mem:* Am Chem Soc; Am Soc Testing Mat. *Res:* Polymer evaluation; polymer stability. *Mailing Add:* 1355 Old Dominion Rd Naperville IL 60540

KISSELL, KENNETH EUGENE, b Ohio, June 28, 28; div; c 2. SPACE OBSERVATORIES, SPACE SURVEILLANCE. *Educ:* Ohio State Univ, BSc, 49, MSc, 58, PhD, 69. *Prof Exp:* Res assoc rocket res lab, Res Found, Ohio State Univ, 48-51; instrumentation physicist, Propulsion Br, Flight Res Lab, Wright-Patterson AFB, 51-57, physicist, Appl Math Lab, Aeronaut Res Lab, 58-59, chief, Gen Physics Res Lab, Aerospace Res Labs, Off Aerospace Res, 59-61, physicist, 61-66, dir, 66-72, br chief, Surveillance Br, Air Force Avionics Lab, 72-75, sr scientist, Reconnaissance & Weapon Delivery Div, Air Force Avionics Lab, 76-80; staff scientist laser systs anal, Rocketdyne Div, Rockwell Int, 80-83; prin staff mem, Optical Astron, BDM Corp, 83-87; SR RES ASSOC PHYSICS & ASTRON, UNIV MD, 87- *Concurrent Pos:* Solar eclipse experimenter, var expeds, 54-73; ed measurement applns, Trans, Instrument Soc Am, 62-68; vis prof, Arcetri Astrophys Observ, Italy, 69; lectr, Wright State Univ, 70-71; assoc dir, Aerospace Instrumentation Div ISA, 78-83; instr physics, Sinclair Community Col, 80; vis astronomer, Kitt Peak Nat Observ, 81. *Mem:* Am Astron Soc; Am Geophys Union; sr mem Instrument Soc Am; Int Astron Union; fel Royal Astron Soc; Optical Soc Am. *Res:* Rocket exhaust temperature measurement; automated telescopes; satellite photometry; photoelectric imaging devices; stellar spectroscopy in near infrared; solar eclipse observation; optical scattering; super luminous stars; Cepheid variables; laser systems; space telescope optical systems and science instruments. *Mailing Add:* Dept Physics & Astron Box 122 Univ Md 1328 Physics Bldg College Park MD 20742

KISSEN, ABBOTT THEODORE, b New York, NY, Nov 24, 22; m 46; c 2. PHYSIOLOGY. *Educ:* Brooklyn Col, BA, 50; Ohio State Univ, MA, 52, PhD(zool), 56. *Prof Exp:* Res assoc physiol, Ohio State Univ, 57-59, asst prof, 59-61; RES PHYSIOLOGIST, AEROSPACE MED RES LAB, WRIGHT-PATTERSON AFB, 61- *Concurrent Pos:* Am Heart Asn fel, 57-58. *Mem:* Aerospace Med Asn; Am Physiol Soc; Sigma Xi. *Res:* Physiological effects of acceleration stresses encountered or anticipated in aerospace flight. *Mailing Add:* 311 Passage Way Osprey FL 34229

KISSIN, BENJAMIN, b Philadelphia, Pa, July 17, 17; m 50; c 1. INTERNAL MEDICINE. *Educ:* Columbia Univ, BS, 38; Long Island Col Med, MD, 41; Am Bd Internal Med, dipl, 51. *Prof Exp:* Assoc prof, 60-68, dir div alcoholism & drug dependence, 70-81, PROF PSYCHIAT, STATE UNIV NY DOWNSTATE MED CTR, 68- *Concurrent Pos:* Dir alcohol clin, Kings County Hosp, 56-67, attend physician, 67-; dir psychosom clin & assoc physician med, Jewish Hosp, Brooklyn, 56-71, consult physician, 71-81. *Mem:* Fel Am Col Physicians; NY Acad Sci. *Res:* Psychosomatic medicine, especially alcoholism; autonomic nervous system and the endocrines. *Mailing Add:* 525 E 86th St Apt 5A New York NY 10028

KISSIN, G(ERALD) H(ARVEY), b New York, NY, May 7, 14; m 41; c 4. ELECTROCHEMISTRY, PROCESS METALLURGY. *Educ:* City Col New York, BS, 35; Univ Mich, MS, 36, PhD(phys chem), 44. *Prof Exp:* Asst phys chem, Univ Mich, 35-37, asst electrochem, 37-39, asst metall, Sch Dent, 39-41; res assoc chem, Nat Defense Res Comt contract, Cornell Univ, 42; res engr, Am Smelting & Refining Co, NJ, 42-46; res asst prof chem eng, Ga Inst Technol, 46-48, res assoc prof, 48-49; chem eng consult, 49-50; head, Finishing & Electrochem Appln Br, Dept Metall Res, Kaiser Aluminum & Chem Corp, 50-62, tech asst to dir, 62-65, tech mgr, 65-68, mgr fabrication & appln res dept, Aluminum Div, 68-72, mgr appln res dept, Metals Div, Kaiser Aluminum & Chem Corp, 72-79; CONSULT METALL ENGR, 79- *Concurrent Pos:* US delegate, Int Standards Orgn, 70-82; mem bd dir, Mont Energy Res & Develop Inst, 82-89; regist prof eng, 67- *Mem:* Fel AAAS; Am Chem Soc; Electrochem Soc; Am Soc Metals; Am Electroplaters' Soc; Sigma Xi. *Res:* Raman spectroscopy; electrode potential phenomena; electrodeposition; dry battery technology; non-ferrous process metallurgy; electrolytic capacitors; finishing of aluminum; primary fabrication processes; corrosion. *Mailing Add:* PO Box 8606 Spokane WA 99203-0606

KISSIN, STEPHEN ALEXANDER, b Ithaca, NY, Apr 11, 42; m 71; c 2. MINERALOGY, ECONOMIC GEOLOGY. *Educ:* Univ Wash, BS, 64; Pa State Univ, MS, 68; Univ Toronto, PhD(geol), 74. *Prof Exp:* Aerospace scientist, Goddard Space Flight Ctr, NASA, 67-68; engr, Siemens AG, WGer, 69; from asst prof to assoc prof, 75-87, PROF GEOL, LAKEHEAD UNIV, 87- *Concurrent Pos:* Fel, McMaster Univ, Hamilton, 73; Nat Res Coun fel, Dept Energy, Mines & Resources, Ottawa, 74-75; Assoc Comt Meteorites, Nat Res Coun Can, 81-91; vis res prof, dept chem & geol, Ariz State Univ, 81-82. *Mem:* Geol Asn Can; Mineral Asn Can; Mineral Soc Am; AAAS; Meteoritical Soc; Sigma Xi. *Res:* Mineralogy and crystal chemistry of sulfides; genesis of ore deposits; meteoritics. *Mailing Add:* Dept Geol Lakehead Univ Thunder Bay ON P7B 5E1 Can

KISSINGER, DAVID GEORGE, b Reading, Pa, July 26, 33; m 55; c 2. TAXONOMY, DATABASE SYSTEMS. *Educ:* Columbia Union Col, BA, 54; Univ Md, MS, 55, PhD(entom), 57; Univ Calif, MPH, 58. *Prof Exp:* Asst entom, Univ Md, 55-57; prof biol & head dept, Oakwood Col, 58-60; prof biol & head dept, Atlantic Union Col, 60-72; prof epidemiol, Loma Linda Univ, 72-83; CONSULT; MGR DATABASE, LOMA LINDA UNIV, 86- *Concurrent Pos:* NIH fel, 74-76. *Mem:* Asn Comput Mach. *Res:* Taxonomy of new world apionidae (coleoptera); application of computer techniques to taxonomic problems. *Mailing Add:* Univ Computing Loma Linda Univ Loma Linda CA 92354

KISSINGER, HOMER EVERETT, b Ottawa, Kans, Aug 29, 23; m 48; c 4. METAL PHYSICS. *Educ:* Kans State Col, BS, 49, MS, 50. *Prof Exp:* Phys sci aide, Nat Bur Standards, 48, physicist, 50-60; asst physics, Kans State Col, 49-50; sr scientist, Gen Elec Co, 60-65; sr res scientist, Battelle-Northwest, 65-87; RETIRED. *Mem:* Am Crystallog Asn; Sigma Xi. *Res:* Crystallography of radiation damage, phase transformations and x-ray diffraction methods; crystal chemistry of solid nuclear waste forms and containments. *Mailing Add:* 1733 Horn Ave Richland WA 99352

KISSINGER, JOHN CALVIN, b Shamokin, Pa, June 8, 25; m 50; c 2. MICROBIOLOGY. *Educ:* Bucknell Univ, BS, 49, MS, 50. *Prof Exp:* Chemist, Campbell Soup Co, 50-51; asst mgr biol standards, Sharp & Dohme, 53-55; supvr fermentation, Grain Processing Corp, 55-56; res microbiologist, Eastern Regional Res Ctr, USDA, 57-83; RETIRED. *Concurrent Pos:* Assoc, Nat Maple Syrup Coun, 67-75. *Honors & Awards:* Medal, Fedn Sewage & Indust Wastes Asn, 57. *Mem:* Fel Asn Off Anal Chem; Int Asn Milk, Food & Environ Sanit. *Res:* Industrial waste treatment; food microbiology; meat and meat products; maple products. *Mailing Add:* 320 Clearspring Rd Lansdale PA 19664

KISSINGER, PAUL BERTRAM, b New York, NY, Mar 30, 30; m 57; c 2. MAGNETIC RESONANCE. *Educ:* Albright Col, BS, 52; Northwestern Univ, MS, 54; Rutgers Univ, PhD(physics), 61. *Prof Exp:* Physicist, Gen Elec Co, 56; instr physics, Rutgers Univ, 59-60; from asst prof to assoc prof, 60-70, assoc dir, Develop Acad Progs, 73-74, chmn dept, 81-84, PROF PHYSICS & ASTRON, DEPAUW UNIV, 70- *Concurrent Pos:* Vis lectr, Univ Colo, 64; vis investr, Woods Hole Oceanog Inst, 67-68; NSF lectr, Munich, Ger, 68-69 & Lima, Peru, 70; physicist, Biol Warfare Labs, 54-56, res, 56-62; consult, IBM Corp, 67, Dept State Schs Latin Am, 70 & 84, United Ministries Higher ed, 79-82, Dept Defense Schs Europe, 77- & Danforth Assocs, 81-87; res assoc, Univ Va, 82, Midwest Assoc in Higher Ed, 88-; solar exclipse exped, Nat Geog soc, 73, US Naval Obs, 83. *Mem:* AAAS; Am Asn Physics Teachers; Am Phys Soc; Sigma Xi; Nat Sci Teachers Asn. *Res:* Electron spin resonance. *Mailing Add:* Dept Physics DePauw Univ Greencastle IN 46135-0037

KISSINGER, PETER THOMAS, b Staten Island, NY, Dec 19, 44; m 78; c 2. CHEMISTRY. *Educ:* Union Col, BS, 66; Univ NC, PhD(chem), 70. *Prof Exp:* Res assoc chem, Univ Kans, 70-72; asst prof chem, Mich State Univ, 72-75; asst prof, 75-76, assoc prof, 77-82, PROF CHEM, PURDUE UNIV, 82- *Concurrent Pos:* Pres, Bioanal Systs Inc, 74- *Mem:* Am Chem Soc; Am Asn Clin Chemists; AAAS; Am Asn Mass Spectrometry; Sigma Xi; Soc Electronanal Chem (pres, 87-89). *Res:* Trace organic analysis using chromatographic and electrochemical techniques; metabolic pathways of aromatic compounds; neurochemistry; organic redox reactions; chemical instrumentation. *Mailing Add:* 111 Lorene Pl West Lafayette IN 47906

KISSLING, DON LESTER, b St Louis, Mo, Jan 29, 34; m 59; c 2. PALEOECOLOGY, SEDIMENTOLOGY. *Educ:* Mo Sch Mines, BS, 58; Univ Wis, MS, 60; Ind Univ, PhD(geol), 67. *Prof Exp:* Res geologist, Superior Oil Co, 60-62; from instr to assoc prof paleont, State Univ NY, Binghamton, 65-74, assoc prof geol, 74-82; OWNER & PRES, JACKALOPE GEOLOGICAL LTD, 84- *Mem:* Geol Soc Am; Soc Econ Paleontologists & Mineralogists; Paleont Soc. *Res:* Paleoecology of Paleozoic corals and bioherma1 fossil assemblages; ecology of modern corals; Paleozoic sedimentary environments and recent carbonate sediments. *Mailing Add:* 406 Welch Ave Berthoud CO 80005

KISSLINGER, CARL, b St Louis, Mo, Aug 30, 26; m 48; c 5. SEISMOLOGY. *Educ:* St Louis Univ, BS, 47, MS, 49, PhD(geophys), 52. *Prof Exp:* From instr to prof geophys & geophys eng, St Louis Univ, 49-72, chmn dept earth & atmospheric sci, 63-72; dir, Coop Inst Res Environ Sci, 72-79, PROF GEOL SCI, UNIV COLO, BOULDER, 72- *Concurrent Pos:* UNESCO expert in seismol & chief tech adv, Int Inst Seismol & Earthquake Eng, Tokyo, 66-67; consult, US Dept Energy, 69-78; chmn comt seismol, Nat Acad Sci-Nat Res Coun, 70-72, mem US Geodynamics Comt, 75-78; mem earth sci adv panel, NSF, 71-74; mem, US Nat Comt, Int Union Geod & Geophys, 74-, mem bur, 75-83, vpres, 83-; mem earthquake hazards reduction adv group, Off Sci

Technol Policy, 77-78; mem earthquake studies adv panel, US Geol Surv, 77-82, chmn, 81-82; mem comt scholarly communication with People's Repub China, Nat Acad Sci, 78-81, chmn Subcomt on Earthquake Res, 84-88; fel, Univ Colo, Boulder, 79-; mem comt adv, US Geol Surv, 83-88. *Honors & Awards:* Alexander von Humboldt Found US Sr Scientist Award, 79; Commemorative Medal, USSR Acad Sci, 85. *Mem:* Fel AAAS; Soc Explor Geophys; Seismol Soc Am (pres, 72-73); fel Am Geophys Union (foreign secy, 74-84); fel Geol Soc Am; NY Acad Sci; corresp mem Austrian Acad Sci. *Res:* Generation of seismic waves by explosions and earthquakes; propagation of elastic waves in layered systems; earthquake prediction. *Mailing Add:* Coop Inst Res Environ Sci Univ Colo Campus Box 216 Boulder CO 80309-0216

KISSLINGER, FRED, b St Louis, Mo, Nov 19, 19; m 45; c 3. METALLURGICAL ENGINEERING. *Educ:* Mo Sch Mines, BS, 42; Univ Cincinnati, MS, 45, PhD(metall eng), 47. *Prof Exp:* From instr to assoc prof metall eng, Ill Inst Technol, 47-64; assoc prof, 64-69, PROF METALL ENG, UNIV MO-ROLLA, 69-; PARTNER, ASKELAND, KISSLINGER & WOLF, METALL ENG CONSULTS, 73- *Mem:* Am Soc Metals; Am Inst Mining, Metall & Petrol Engrs. *Res:* Thermodynamics; heat treating. *Mailing Add:* Dept Metall Eng Univ Mo Rolla MO 65401

KISSLINGER, LEONARD SOL, b St Louis, Mo, Aug 15, 30; m 56. THEORETICAL PHYSICS. *Educ:* St Louis Univ, BS, 51; Ind Univ, MS, 52, PhD(physics), 56. *Prof Exp:* From instr to prof physics, Case Western Reserve Univ, 56-68; PROF PHYSICS, CARNEGIE-MELLON UNIV, 68- *Concurrent Pos:* Res Corp fel, Bohr Inst, Copenhagen, Denmark, 58-59; res assoc, Mass Inst Technol, 66-67; vis staff mem, Los Alamos Sci Lab, 69- *Mem:* Am Phys Soc; Sigma Xi. *Res:* Nuclear models and structure; many-body problem; particle physics. *Mailing Add:* Dept Physics Carnegie-Mellon Univ Pittsburgh PA 15213

KISSMAN, HENRY MARCEL, b Graz, Austria, Sept 9, 22; nat US; m 56; c 2. ORGANIC CHEMISTRY, COMPUTER SCIENCES. *Educ:* Sterling Col, BS, 44; Univ Cincinnati, MS, 48; Univ Rochester, PhD(org chem), 50. *Prof Exp:* Sr asst scientist org chem, NIH, 50-52; res chemist, Lederle Labs Div, Am Cyanamid Co, 52-62, dept head tech info, 62-67; dir sci info facility, US Food & Drug Admin, DC, 67-70; ASSOC DIR SPECIALIZED INFO SERVS, US NAT LIBR MED, 70- *Concurrent Pos:* Chmn comt study environ qual info progs in fed govt, Off Sci & Technol; mem adv bd, Chem Abstr Serv, 74-77; chmn, Toxicol Info Subcomt, US Dept Health, Educ & Welfare, 73- *Honors & Awards:* Super Serv Award, US Dept Health, Educ & Welfare, 73; Dirs Award, NIH, 85. *Mem:* AAAS; Am Chem Soc; Am Soc Info Sci; Soc Toxicol. *Res:* Ethylenimine chemistry; amino acids; carbohydrates; nucleosides; steroids; tetracyclines; chemical documentation; development and national operation of online databases in toxicology such as toxline, chemline, hazardous substances data bank. *Mailing Add:* Specialized Info Serv Nat Libr Med 8600 Rockville Pike Bethesda MD 20894

KISSMEYER-NIELSEN, ERIK, b Silkeborg, Denmark, Oct 22, 22; m 62; c 2. FOOD SCIENCE, BOTANY. *Educ:* Royal Vet & Agr Col, Copenhagen, BS, 48; Cornell Univ, MS, 60; Univ Wis, PhD(food sci), 64. *Prof Exp:* Trainee food technol, Am Scand Soc, 48-50; inspector, R T French Co, NY, 50-53; asst mgr, Grimstrup Coop Starch & Dehydration Plant, Denmark, 53-58; asst food sci, Cornell Univ, 58-60 & Univ Wis, 60-63; asst prof biochem & food sci, Univ Del, 63-66; INT ADV, AGR & FOOD INDUST, 66- *Concurrent Pos:* Agr & food indust adv, UN Indust Develop Orgn & World Bank, 70-73; regional agr & food indust adv, UN Econ Comn for Africa, 73-75, Agr & Food Indust Adv Int, 75- *Mem:* Fel AAAS; Inst Food Technol. *Mailing Add:* 65 Crabtree Rd Concord MA 01742

KIST, JOSEPH EDMUND, b Buffalo, NY, Aug 11, 29; div; c 1. MATHEMATICS. *Educ:* Univ Buffalo, BA, 52; Purdue Univ, MS, 54, PhD(math), 57. *Prof Exp:* Instr math, Purdue Univ, 57; asst prof, Wayne State Univ, 57-59 & Pa State Univ, 59-62; vis assoc prof, Purdue Univ, 62-63; assoc prof, Pa State Univ, 63-66; assoc prof, 66-67, PROF MATH, NMEX STATE UNIV, 67- *Mem:* Fel AAAS; Am Math Soc; Math Asn Am; Sigma Xi. *Res:* Functional analysis; lattices; semigroups. *Mailing Add:* Dept Math NMex State Univ Las Cruces NM 88003

KISTER, JAMES MILTON, b Cleveland, Ohio, June 29, 30; m 56, 78; c 1. MATHEMATICS. *Educ:* Wooster Col, AB, 52; Univ Wis, AM, 56, PhD, 59. *Prof Exp:* Res asst, Los Alamos Sci Lab, 53-55; from instr to assoc prof, 59-66, chmn dept, 71-73, PROF MATH, UNIV MICH, ANN ARBOR, 66- *Concurrent Pos:* Fel, Off Naval Res, Univ Va, 60-61; mem, Inst Advan Study, 62-64; vis prof, Univ Calif, Los Angeles, 67; vis fel, Clare Hall, Cambridge Univ, Eng, 70; vis mem, Institut des Hautes Etudes, France, 74; vis fel Wolfson Col, Oxford Univ, Eng, 77 & 85-86; ed, Duke Math J, 72-75; ed, Mich Math J, 76-78, managing ed, 77-78, 83-85 & 86-88. *Mem:* Am Math Soc. *Res:* Topology; isotopies; transformation groups; manifolds. *Mailing Add:* Dept Math Univ Mich Ann Arbor MI 48109

KISTIAKOWSKY, VERA, b Princeton, NJ, Sept 9, 28; div; c 2. ELEMENTARY PARTICLE PHYSICS. *Educ:* Mt Holyoke Col, AB, 48; Univ Calif, PhD(chem), 52. *Hon Degrees:* DSc, Mt Holyoke Col, 78. *Prof Exp:* Scientist, USN Radiol Defense Lab, 52-53; Berliner fel physics, Radiation Lab, Univ Calif, 53-54; res assoc, Columbia Univ, 54-57, instr, 57-59; from asst prof to adj assoc prof, Brandeis Univ, 59-65; scientist, Lab Nuclear Sci, 65-69, sr res scientist dept physics, 69-72, PROF PHYSICS, MASS INST TECHNOL, 72- *Mem:* Fel AAAS; fel Am Phys Soc; Asn Women Sci (pres, 82-84). *Res:* Observational astrophysics. *Mailing Add:* 24-522 Mass Inst Technol Cambridge MA 02139

KISTLER, ALAN L(EE), b Laramie, Wyo, Nov 26, 28; m 55; c 3. FLUID DYNAMICS. *Educ:* Johns Hopkins Univ, BE, 50, MS, 52, PhD(aeronaut), 55. *Prof Exp:* Res group supvr, Jet Propulsion Lab, Calif Inst Technol, 57-61; assoc prof, Yale Univ, 61-65; fluid physics sect mgr, Jet Propulsion Lab, Calif Inst Technol, 65-69; PROF MECH ENG & ASTRONAUT, TECHNOL

INST, NORTHWESTERN UNIV, EVANSTON, 69- *Mem:* Am Phys Soc; Am Soc Mech Engrs; AAAS. *Res:* Diffusion in turbulent flow fields; turbulence in compressible media; mechanics of wakes and separated flows. *Mailing Add:* 3241 Park Pl Evanston IL 60201

KISTLER, MALATHI K, b India, Oct 29, 44. CHEMISTRY. *Educ:* Indian Inst Sci, PhD(biochem), 70. *Prof Exp:* RES ASSOC PROF, DEPT CHEM, UNIV SC, 76- *Mem:* Endocrine Soc; Cell Biol Soc. *Mailing Add:* Dept Chem Univ SC Columbia SC 29208

KISTLER, RONALD WAYNE, b Chicago, Ill, May 18, 31; m 57; c 2. GEOLOGY. *Educ:* Johns Hopkins Univ, BA, 53; Univ Calif, Berkeley, PhD(geol), 60. *Prof Exp:* GEOLOGIST, US GEOL SURV, 60- *Concurrent Pos:* Vis prof, Northwestern Univ, 71. *Mem:* Chinese Soc Mineral Petrol & Geochemistry; Geol Soc Am; Am Geophys Union. *Res:* Structural geology; geochronology. *Mailing Add:* Geol Surv 345 Middlefield Rd Menlo Park CA 94025

KISTLER, WILSON STEPHEN, JR, b Newport News, Va, Mar 1, 42; m 76; c 2. BIOCHEMISTRY, REPRODUCTIVE BIOLOGY. *Educ:* Princeton Univ, AB, 64; Harvard Univ, PhD(biochem), 70. *Prof Exp:* Res assoc, Dept Microbiol & Molecular Genetics, Sch Med, Harvard Univ, 70-71, Ben May Lab Cancer Res, Univ Chicago, 71-75; from asst prof to assoc prof, 75-86, PROF CHEM, UNIV SC, 86- *Mem:* Am Soc Biol Chemists; Am Chem Soc. *Res:* Study of changes in basic nuclear proteins accompanying mammalian spermatogenesis and the regulation of protein synthesis by androgenic steroid hormones; gene analysis using recombinant DNA. *Mailing Add:* 226 Banbury Rd Columbia SC 29210

KISTNER, CLIFFORD RICHARD, b Cincinnati, Ohio, Dec 16, 36; m 61; c 1. INORGANIC CHEMISTRY. *Educ:* Carthage Col, AB, 59; Univ Iowa, MS, 62, PhD(chem), 63. *Prof Exp:* Instr chem, Univ Iowa, 63, asst prof, 63-64; from asst prof to assoc prof, 64-68, PROF CHEM & CHMN DEPT, UNIV WIS, LA CROSSE, 68- *Mem:* AAAS; Am Chem Soc; Sigma Xi. *Res:* Inorganic coordination chemistry. *Mailing Add:* N1596 Skyline Blvd La Crosse WI 54601

KISTNER, DAVID HAROLD, b Cincinnati, Ohio, July 30, 31; m 57; c 2. ENTOMOLOGY. *Educ:* Univ Chicago, AB, 52, SB, 56, PhD(zool), 57. *Prof Exp:* Asst termites, Univ Chicago, 53-54, asst comp anat, 55, field zoologist comp anat & arthropods, 56-57; instr biol, Univ Rochester, 57-59; instr, 59-60, from asst prof to assoc prof, 60-67, PROF BIOL, CALIF STATE UNIV, CHICO, 67- *Concurrent Pos:* NSF grants, 58-59, 60-71 & 72-88; Guggenheim Mem Found fel, 65-66; hon res assoc, Div Insects, Field Mus Natural Hist, Chicago, 67 & Atlantica Ecol Res Sta, Rhodesia; dir, Shinner Inst Study Interrelated Insects, 68-72. *Mem:* AAAS; Entom Soc Am; Soc Study Evolution; Soc Syst Zool; Am Soc Zool; fel Explorers Club. *Res:* Systematics, evolution; zoogeography and behavior of myrmecophilous and termitophilous insects; systematics of Staphylinidae. *Mailing Add:* Dept Biol Calif State Univ Chico CA 95929-0515

KISTNER, OTTMAR CASPER, b New York, NY, Mar 22, 30; m 59; c 3. PHOTO NUCLEAR PHYSICS, HEAVY ION NUCLEAR REACTIONS. *Educ:* Polytech Inst Brooklyn, BS, 52; Columbia Univ, PhD(physics), 59. *Prof Exp:* Asst physicist, 59-63, assoc physicist, 63-66, PHYSICIST, BROOKHAVEN NAT LAB, 66- *Concurrent Pos:* Guest physicist, Max Planck Inst Nuclear Physics, 69-70; Weizmann Inst Sci, 80. *Mem:* Sigma Xi; fel Am Phys Soc. *Res:* Experimental nuclear physics; nuclear structure and hyperfine interactions by beta and gamma ray spectroscopy; on line spectroscopy with heavy ion reactions; Mossbauer effect; medium energy (300 MeV) photonuclear research and facility development of Laser-Electron-Gamma-Source at BNL NSLS. *Mailing Add:* 14 Sands Ln Port Jefferson NY 11777

KISVARSANYI, EVA BOGNAR, b Budapest, Hungary; US citizen; m 56; c 1. GEOLOGY, PETROLOGY. *Educ:* Univ Mo-Rolla, BS, 58, MS, 60. *Prof Exp:* Res geologist, 59-89, ASST DIR, MO DEPT NATURAL RES, GEOL SURV, 89- *Concurrent Pos:* Mem, Working Group Precambrian Correlation Cent Interior Region US, Int Union Geol Sci, 76-84. *Mem:* Mineral Soc Am; Geol Soc Am; Soc Econ Geologists. *Res:* Precambrian geology of the St Francois Mountains and vicinity in southeast Missouri; geology and structure of buried Precambrian basement, midcontinent region; mineral resources of Missouri. *Mailing Add:* Mo Dept Natural Resources PO Box 250 Rolla MO 65401

KISVARSANYI, GEZA, b Tokay, Hungary, Feb 23, 26; US citizen; m 56; c 1. ECONOMIC GEOLOGY. *Educ:* Eotvos Lorand, Budapest, MS, 52; Univ Mo, Rolla, PhD(geol), 66. *Prof Exp:* Asst prof geol, Eotvos Lorand, Budapest, 52-55; chief geologist, Ore Mining & Develop Co, Hungary, 55-56; explor geologist, Bear Creek Mining Co Div, Kennecott Copper Corp, 57-62; from instr to assoc prof, 62-81, PROF GEOL, UNIV MO, ROLLA, 81- *Concurrent Pos:* Chmn, Int Geol Conf; consult, gold explor. *Mem:* Geol Soc Am; Soc Econ Geol. *Res:* Hydrothermal ore deposits; Mississippi Valley-type lead-zinc deposits; iron-titanium ore deposits; geotectonics of the midcontinent; remote sensing, radar, landsat, of geologic structures, precious and base metal deposits, prophyry copper and molybdenum deposits; author and editor of four books and over 160 papers and reports of investigations. *Mailing Add:* Dept Geol & Geophys Univ Mo Rolla MO 65401

KISZENICK, WALTER, b New York, NY, Apr 1, 18; m 43; c 2. PHYSICS, ELECTRICAL ENGINEERING. *Educ:* Brooklyn Col, BA, 39; Polytech Inst Brooklyn, MS, 47, PhD(physics), 54. *Prof Exp:* Jr metallurgist, NY Naval Shipyard, 41-43; jr scientist, Los Alamos Sci Lab, 44-46; res asst phosphors, Polytech Inst Brooklyn, 46-52; physicist, Freed Radio Co, 52-53; from instr to asst prof physics, 53-64, ASSOC PROF PHYSICS & NUCLEAR ENG, POLYTECH INST NY, 62- *Mem:* Am Phys Soc; Electron Micros Soc Am; Sigma Xi. *Res:* Dielectric properties of phosphors; physical properties of ice. *Mailing Add:* Dept Physics Polytech Inst NY 333 Jay St Brooklyn NY 11201

KIT, SAUL, b Passaic, NJ, Nov 25, 20; m 45; c 3. BIOCHEMISTRY, CANCER. *Educ:* Univ Calif, AB, 48, PhD(biochem), 51. *Prof Exp:* Nat Cancer Inst fel biochem, Chicago, 51-52, Nat Found Infantile Paralysis fel, 52; res biochemist, Univ Tex M D Anderson Hosp & Tumor Inst, 53-55; asst prof biochem, Col Med, Baylor Univ, 56-57; from asst prof to assoc prof, Post-Grad Sch Med, Univ Tex, 57-61; vis prof virol, 62, PROF BIOCHEM & HEAD DIV BIOCHEM VIROL, BAYLOR COL MED, 62- *Concurrent Pos:* Assoc biochemist, Univ Tex M D Anderson Hosp & Tumor Inst, 57-60, biochemist & chief Sect Nucleoprotein Metab, 61-62; Nat Inst Arthritis & Infectious Dis res career award, 63-88; mem cancer virol panel, USPHS, 61-62, consult, 71-; chairperson, Pathobiol Chem Study Sect, NIH, 75-79; mem Cancer Virol Panel, Nat Cancer Inst, USPHS, 61-62; sci consult, molecular biol dept, Miles Lab, Elkhart, Ind, 69-72; sci adv bd mem, Am Genetics Int, Inc, Denver, Colo, 81-84; distinguished vis prof, dept microbiol, La Trobe Univ, Australia, 82; mem deleg, US-USSR exchange virol, 67; chmn sci adv bd, Novagene, Inc, Houston, 82- *Mem:* Am Soc Biol Chemists; Am Soc Cell Biol (treas, 65-67, pres, 71); Am Chem Soc; Am Soc Virol; Am Soc Microbiol; Am Asn Cancer Res; corresp mem Arg Soc Virol. *Res:* Molecular biology; biochemical virology; biochemistry of cancer; nucleic acids; genetically engineered vaccines; author of over 250 research publications. *Mailing Add:* Div Biochem Virol Baylor Col Med Houston TX 77030

KITABCHI, ABBAS E, b Tehran, Iran, Aug 28, 33; US citizen; m; c 4. INTERNAL MEDICINE, ENDOCRINOLOGY & DIABETES. *Educ:* Cornell Col, BA, 54; Univ Okla, MS, 56, PhD(med sci), 58, MD, 65. *Prof Exp:* Res assoc biochem, Okla Med Res Found, 60-61, biochemist, 61-65, sr investr, 65-66; instr med, Univ Wash, 66-68; from asst prof to assoc prof med, 68-73, assoc prof biochem, 68-73, PROF MED & BIOCHEM, UNIV TENN CTR FOR HEALTH SCI, 73-, CHIEF DIV ENDOCRINOL & METAB & DIR CLIN RES CTR, 73- *Concurrent Pos:* Fel biochem, Okla Med Res Found, 58-60; NIH spec fel endocrinol, Univ Wash, 66-68; assoc chief staff res & chief endocrinol & metab labs, Vet Admin Hosp, 68-73, assoc chief metab, Med Serv, 73-; chief Diabetes & Endocrinol Clin, City of Memphis Hosp; consult, Baptist Hosp; chief endocrinol & attend physcian, Univ Tenn Hosp; vis scientist, Va Mason Res Ctr & Univ Wash, Seattle, 90-91; NIH spec fel, 66-68. *Mem:* Fel Am Col Physicians; Asn Am Physicians; Am Fedn Clin Res; Am Diabetes Asn; Endocrine Soc; Am Soc Biochem & Molecular Biol; Am Soc Clin Investrs; Am Inst Nutrit. *Res:* Mechanism of action of steroids and pancreatic hormones at molecular level; pathogenesis and treatment of diabetes. *Mailing Add:* Univ Tenn Memphis 951 Court Ave Memphis TN 38163

KITAHATA, LUKE MASAHIKO, b Jan 12, 25; m; c 3. ANESTHESIA, PAIN & NEUROSURGICAL ANESTHESIA. *Educ:* Tokyo Imperial Univ, Japan, MD, 47. *Prof Exp:* PROF ANESTHESIOL, SCH MED, YALE UNIV, 73- *Mailing Add:* PO Box 3333 New Haven CT 06510

KITAI, REUVEN, b Johannesburg, SAfrica, Oct 4, 24; m 52; c 3. ELECTRICAL ENGINEERING. *Educ:* Univ Witwatersrand, BSc, 44, MSc, 48, DSc(elec eng), 62. *Prof Exp:* Lectr elec eng, Univ Witwatersrand, 47-55, sr lectr, 55-64; from assoc prof to prof, 65-88, EMER PROF ELEC ENG, MCMASTER UNIV, 88- *Concurrent Pos:* Vis lectr, Eng Labs, Cambridge Univ, 53-54 & Dept Elec Eng, Imp Col, Univ London, 61-62; vis res officer, Standard Telecommun Labs, Ltd, Eng, 61 & Brown Boveri Co, Switz, 77; pres, Instrumentation & Measurement Soc, Inst Elec & Electronics Engrs, 82. *Mem:* Sr mem Inst Elec & Electronics Engrs. *Res:* Instrumentation. *Mailing Add:* Dept Elec Eng McMaster Univ Hamilton ON L8S 4L7 Can

KITAIGORODSKII, SERGEI ALEXANDER, b Moscow, USSR, Sept 13, 34; m 69; c 2. PHYSICAL OCEANOGRAPHY. *Educ:* Inst Physics Atmosphere Acad Sci, USSR, PhD(geophysics), 60; Inst Oceanology Acad Sci, DSc, 68. *Prof Exp:* Head lab res, Inst Oceanog Acad Sci, USSR, 68-78; lectr, Inst Phys Oceanog, Univ Copenhagen, 79-85; PROF OCEANOG, DEPT EARTH & PLANETARY SCI, JOHNS HOPKINS UNIV, 80- *Res:* Oceanic turbulence and its modelling; physics of air-sea interaction; wave motions in the ocean; geophysical fluid dynamics. *Mailing Add:* Dept Earth & Planetary Sci Johns Hopkins Univ Baltimore MD 21218

KITAY, JULIAN I, b Kearny, NJ, Aug 29, 27; m 73; c 2. INTERNAL MEDICINE, PHYSIOLOGY. *Educ:* Princeton Univ, AB, 49; Harvard Med Sch, MD, 54. *Prof Exp:* Intern med, Grace-New Haven Hosp, Conn, 54-55; asst resident, Beth Israel Hosp, Mass, 55-56; instr, Col Physicians & Surgeons, Columbia Univ, 58-59; from asst prof to assoc prof internal med, 59-70, from asst prof to assoc prof physiol, 61-70, prof internal med & physiol, Sch Med, Univ VA, 70-78; assoc dean curricular affairs, 78-84, PROF INTERNAL MED, PHYSIOL & BIOPHYSICS, UNIV TEX MED BR, GALVESTON, TEX, 78-, ASST VPRES & ASSOC DEAN ACAD AFFAIRS, 84- *Concurrent Pos:* Commonwealth Fund fel, Col Physicians & Surgeons, Columbia Univ, 56-58; USPHS res career develop award, 61-70; asst physician, Med Serv, Presby Hosp, NY, 58-59; attend physician, Univ Hosp, Univ Va, 59-78, head div endocrinol & metab, Dept Internal Med, 70-78; mem neuroendocrinol panel, Int Brain Res Orgn. *Mem:* Endocrine Soc; Am Fedn Clin Res; Am Soc Clin Invest; Am Physiol Soc; Soc Exp Biol & Med. *Res:* Endocrine physiology; clinical aspects of endocrine disease. *Mailing Add:* Off Dean Med Univ Tex Med Br Galveston TX 77550

KITAZAWA, GEORGE, b San Jose, Calif, May 2, 17; m 43; c 3. CHEMISTRY, WOOD SCIENCE. *Educ:* Univ Calif, BS, 40; State Univ NY, MS, 44, PhD(wood technol), 47. *Prof Exp:* Technician, Guayule Rubber Res, War Relocation Auth, Calif, 42-43; indust res fel, State Univ NY Col Forestry, Syracuse, 44-47; res wood technologist, Casein Co Am, NY, 47; wood technologist, Timber Eng Co, 47-50; res assoc, State Univ NY Col Forestry, Syracuse, 50-53; res physicist, Gillette Safety Razor Co, 53-56; group leader, Cent Res Lab, Borden Chem Co, Philadelphia, 56-57, asst lab head, 57-59, lab head, 59-73; mgr, Wood Sci Group, Koppers Co, 73-78, forest prod res sect, 78-80; CONSULT & TECH INTERPRETER, 80- *Concurrent Pos:* Industrial liaison hydrostatic and ultrasonic pipe testers & coke oven mach for steel indust. *Mem:* AAAS; Am Chem Soc; Asn Asian Studies. *Res:* Wood preservatives and fire retardants; polymer characterization; analytical chemistry; adhesives; instrumentation; sonic and ultrasonic nondestructive testing; surface chemistry and physics; wood physics; steel industry machinery. *Mailing Add:* 926 Harvard Rd Monroeville PA 15146-4303

KITCHELL, JAMES FREDERICK, b Gary, Ind, July 20, 42; m 77; c 2. AQUATIC ECOLOGY, BIOENERGETICS. *Educ:* Ball State Teachers Col, BS, 64; Univ Colo, PhD(biol), 70. *Prof Exp:* Proj assoc ecol, Inst Environ Studies, 70-72, asst scientist, 72-74, asst prof zool, 74-77, ASSOC PROF ZOOL, UNIV WIS-MADISON, 77- *Concurrent Pos:* Scientist, Smithsonian Inst Proj, Skadar Lake, Yugoslavia, 72-77; vis scientist, Nat Marine Fisheries Serv, Honolulu, 73-78. *Mem:* Am Fisheries Soc; Ecol Soc Am; Int Soc Limnol; AAAS; Am Inst Biol Sci. *Res:* Application of ecosystem models; predator-prey interactions; trophic ecology. *Mailing Add:* Dept Zool 151 Noland Hall Univ Wis 1050 Bascom Mall Madison WI 53706

KITCHELL, JENNIFER ANN, b Zanesville, Ohio, May 25, 45. PALEOBIOLOGY, EVOLUTIONARY THEORY. *Educ:* Univ Wis, BS, 68, MS, 71, PhD(geol), 78. *Prof Exp:* res assoc, Univ Wis-Madison, 78-80, asst scientist geol & zool, 80-; AT MUS PALEONT, UNIV MICH, ANN ARBOR. *Mem:* Sigma Xi; Soc Econ Paleontol & Mineralogists; Soc Oceanog & Limnol; Paleont Soc; Geol Soc Am. *Res:* Paleobiology; coevolution; taxonomic diversification; predator-prey interactions; siliceous sedimentation; paleolimnological studies of predator-prey interactions. *Mailing Add:* Mus Paleont Univ Mich Ann Arbor MI 48109

KITCHELL, RALPH LLOYD, b Waukee, Iowa, July 9, 19; m 47; c 3. VETERINARY ANATOMY. *Educ:* Iowa State Univ, DVM, 43; Univ Minn, PhD(human anat), 51. *Prof Exp:* Instr vet bact, Kans State Col, 43; instr vet med, Univ Minn, St Paul, 47-51, from assoc prof to prof vet anat, 51-64, head dept, 51-64; dean vet med, Kans State Univ, 64-66; prof vet anat, Iowa State Univ, 66-72, dean col vet med & dir vet med res inst, 66-71; prof vet anat, 72-90, EMER PROF, SCH VET MED, UNIV CALIF, DAVIS, 90- *Concurrent Pos:* USPHS res fel, Swed & Eng, 57-58; vis prof, Sch Vet Med, Univ Calif, Davis, 71-72; res fel & vis prof, Royal Vet Col, Edinburgh, Scotland, 76; vis prof, Sch Vet Med, Massey Univ, Palmerstone North, New Zealand. *Honors & Awards:* Small Animal Res Award, Rallston-Purina, 85. *Mem:* NY Acad Sci; Am Vet Med Asn; Am Asn Vet Anat (secy, 51, pres, 54); Am Asn Anatomists; Am Vet Neurol Asn; Soc Neurosci. *Res:* Veterinary neuroanatomy; sexual and fetal physiology; somatosensory physiology. *Mailing Add:* Dept Anat Univ Calif Sch Vet Med Davis CA 95616

KITCHEN, HYRAM, biochemistry, hematology; deceased, see previous edition for last biography

KITCHEN, SUMNER WENDELL, b Somerville, Mass, Sept 17, 21. NUCLEAR SCIENCE, NUCLEAR ENGINEERING. *Educ:* Oberlin Col, BA, 43; NY Univ, PhD(physics), 51. *Prof Exp:* Physicist, physics of metals, Frankford Arsenal, 43-46; instr physics, NY Univ, 46-50; group leader accelerator design, E O Lawrence Radiation Lab, 50-54; res assoc reactors physics, Knolls Atomic Power Lab, 54-56, mgr, 56-78, sr physicist, reactor physics, 78-79, mgr, Qual Assurance Eng, 79-90, CONSULT QUAL ASSURANCE, KNOLLS ATOMIC POWER LAB, 90- *Concurrent Pos:* Consult, Res & Develop Coun, Dept Defense, 48-49. *Mem:* Am Phys Soc; Am Nuclear Soc; NY Acad Sci; Sr Mem Am Soc Qual Control. *Res:* Reactor and criticality safety; reactor physics; gas discharges; accelerators. *Mailing Add:* 1352 Stanley Lane Schenectady NY 12309

KITCHENS, CLARENCE WESLEY, JR, b Panama City, Fla, Nov 8, 43; m 66; c 2. ARMOR-ANTI-ARMOR MECHANICS, BLAST DYNAMICS. *Educ:* Va Polytech Inst & State Univ, BS, 66, MS, 68; NC State Univ, PhD(eng mech), 70. *Prof Exp:* Eng asst, Atlantic Res Corp, Va, 62-65; res engr, Aberdeen Res & Develop Ctr, 70-72, aerospace engr, 72-77, asst to dir, 77-78, leader fluid dynamics anal team, 78-80, chief, blast dynamics br, 80-82, chief, penetration mech br, 82-85, actg chief, 85-86, CHIEF, TERMINAL BALLISTICS DIV, BALLISTIC RES LAB, US ARMY, 86- *Concurrent Pos:* Fel US Army Ballistic Res Lab; fel US Army Lab Command. *Honors & Awards:* Fire Power Award, Am Defense Preparedness Asn. *Mem:* Assoc fel Am Inst Aeronaut & Astronaut; Am Defense Preparedness Asn; Asn US Army; Sr Execs Asn. *Res:* Numerical computations in gas dynamics; experimental techniques in fluid dynamics; numerical analysis; computer applications in engineering; blast loading and response; flight mechanics; penetration mechanics; armor/anti-armor research. *Mailing Add:* Terminal Ballistics Div Ballistic Res Lab Aberdeen Proving Ground MD 21005

KITCHENS, THOMAS ADREN, b Amarillo, Tex, Oct 31, 35; m 58; c 3. CONDENSED MATTER PHYSICS. *Educ:* Rice Inst, BA, 58, Rice Univ, MA, 60, PhD(physics), 63. *Prof Exp:* Staff mem physics, Los Alamos Sci Lab, 63-65; physicist, Brookhaven Nat Lab, 65-75; liaison physicist, Off Naval Res, London, 75-76; alternate group leader, Los Alamos Sci Lab, 76-80, group leader physics, 78-80; staff assoc, Div Mat Res, NSF, 80-82; SCI COMPUT STAFF, DEPT ENERGY, 82- *Concurrent Pos:* Sr res fel, Univ Sussex, Gr Brit, 70-71. *Mem:* Fel Inst Physics UK; fel Am Inst Physics; Asn Comput Mach; AAAS. *Res:* Ultralow temperature physics and condensed matter research utilizing neutron and light scattering; high performance computing. *Mailing Add:* Dept Energy ER-7 1000 Independence Ave Washington DC 20585

KITCHENS, WILEY M, b Porth Arthur, Tex, Jan 6, 44; m 70; c 3. WETLANDS ECOLOGY, ESTUARINE ECOLOGY. *Educ:* Lamar Univ, BS, 66; Miami Univ, MA, 70; NC State Univ, PhD(zool), 78. *Prof Exp:* Fel coastal ecol, Univ SC, 71-73, res assoc salt marsh ecol, 73-79; ecologist, Nat Coastal Ecosysts Team, US Fish & Wildlife Serv, 79-85; LEADER & ECOLOGIST, FLA COOP FISH & WILDLIFE RES UNIT, 85- *Concurrent Pos:* Adj asst prof, La State Univ, 82-85; adj assoc prof, Univ Fla, 85- *Mem:* Soc Wetlands Scientists; Wildlife Soc. *Res:* Ecosystems approach to wetlands ecology; simulation modelling; use of geographic information systems and remote sensing technologies; landscape dynamics through coupling of simulation and GIS models. *Mailing Add:* Fla Coop Fish & Wildlife Res Unit Univ Fla 117 Newins-Ziegler Hall Gainesville FL 32611

KITCHIN, JOHN FRANCIS, b Greenwood, Miss, May 6, 53. RELIABILITY & LIFE TESTING. *Educ:* Univ Southern Miss, BS, 76; Fla State Univ, MS, 78, PhD(statist), 80. *Prof Exp:* Asst prof statist, Purdue Univ, 80-81; mem tech staff, Bell Labs, 81-84, mem tech staff, Commun Res, 84-85, tech mgr, Bell Commun Res, 85-; RES SCIENTIST, DIGITAL EQUIP. *Mem:* Am Statist Asn; Inst Math Statist; Inst Elec & Electronics Engrs; Reliability Soc. *Res:* Development and comparison of statistical estimators of reliability; development of methods of reliability prediction for complex systems. *Mailing Add:* Digital Equip Co 77 Reed Rd Hudson MA 01749

KITE, FRANCIS ERVIN, b Galesburg, Ill, Dec 8, 18; m 42; c 2. CEREAL CHEMISTRY. *Educ:* Knox Col, AB, 40; Univ Iowa, MS, 42, PhD(phys chem), 48. *Prof Exp:* Res chemist, Corn Prods Co, 48-71; sr res chemist, CPC Int, 71-84; RETIRED. *Mem:* Am Chem Soc; Am Asn Cereal Chemists. *Res:* Raman spectroscopy; molecular structure; chemistry of corn sugar and its derivatives; starch and starch derivatives. *Mailing Add:* 332 Nuttall Rd Riverside IL 60546

KITE, JOSEPH HIRAM, JR, b Decatur, Ga, Nov 11, 26; m 70. IMMUNOLOGY, BACTERIOLOGY. *Educ:* Emory Univ, AB, 48; Univ Tenn, MS, 54; Univ Mich, PhD(bact), 59. *Prof Exp:* Res assoc, 58-59, from instr to assoc prof, 59-72, PROF MICROBIOL, SCH MED, STATE UNIV NY, BUFFALO, 72- *Concurrent Pos:* Am Heart Asn adv res fel, 63-67. *Mem:* AAAS; Am Soc Microbiol; Tissue Cult Asn; Am Asn Immunol; NY Acad Sci. *Res:* Autoimmune thyroiditis, cell-mediated immune reactions. *Mailing Add:* Dept Microbiol State Univ NY Sch Med Buffalo NY 14214

KITHIER, KAREL, b Prague, Czechoslovakia, Dec 6, 30; m 61; c 1. IMMUNOCHEMISTRY, COMPARATIVE PATHOLOGY. *Educ:* Charles Univ, MD, 62, PhD(biochem), 67. *Prof Exp:* Res scientist immunochem, Res Inst Child Develop, Charles Univ, 67-68, Mich Cancer Found, Detroit, 72-74; res assoc, Child Res Ctr Mich, 68-71; assoc head, div clin chem, Detroit Res Hosp & Univ Health Ctr, 74-88, div head immunol, 80-88; asst prof chem & immunochem, 74-78, ASSOC PROF PATH, DEPT PATH, SCH MED, WAYNE STATE UNIV, DETROIT, 78-; MED DIR, SPECIAL CHEM, DAMON CLIN LAB, DETROIT MED CTR, 89- *Concurrent Pos:* Staff pathologist immunol, Vet Admin Med Ctr, Allen Park, 78- *Mem:* Am Asn Cancer Res; Am Asn Immunol; Am Asn Clin Chem; Nat Acad Clin Biochem; NY Acad Sci. *Res:* Proteins of blood and tissues in health and disease; proteins of fetuses and cancer patients; development and pathology of proteins and related substances. *Mailing Add:* 540 E Canfield St Rm 9231 Detroit MI 48201

KITOS, PAUL ALAN, b Saskatoon, Sask, May 31, 27; m 52; c 7. BIOCHEMISTRY. *Educ:* Univ BC, BSA, 50, MSA, 52; Ore State Univ, PhD(chem), 56. *Prof Exp:* Chemist, E I du Pont de Nemours & Co, 56-59; asst prof biochem, 59-62, from asst prof to assoc prof comp biochem & physiol, 62-69, actg chmn dept, 69-71, PROF BIOCHEM, UNIV KANS, 69- *Concurrent Pos:* NIH fel, dept microbiol, Harvard Med Sch, 71-72; sr scientist, Mid Am Cancer Ctr, 75-79; Fogarty fel, Inst d'Embryologie, Nogent-Sur-Marne, France, 83. *Honors & Awards:* Amoco Award, 74. *Mem:* AAAS; Am Chem Soc; Tissue Cult Asn; Am Soc Biol Chem & Mol Biol. *Res:* Metabolism in animal cells; developmental biology; effects of organophosphorus insecticides on avian embryos; biochemical basis of some birth defects; teratogenesis. *Mailing Add:* Dept Biochem Univ Kans Lawrence KS 66045

KITSON, JOHN AIDAN, b Victoria, BC, Feb 14, 27; m 54; c 2. FOOD SCIENCE. *Educ:* Univ BC, BA, 49; Ore State Univ, MSc, 54. *Prof Exp:* Food technologist, Sun Rype Prod Ltd, BC, 49-50; food technologist prod & process develop, Can Dept Agr, 50-64; food technologist & vis scientist, Eng & Develop Lab, USDA, Calif, 64-65; food technologist prod & process develop, Can Dept Agr, 65-71, head, Food Processing Sect, 71-80, assoc dir, Res Sta, Agr Can, 80-83; RETIRED. *Honors & Awards:* Prix Industs Award, 68 & 70; W J Eva Award for Indust Serv, 77. *Mem:* Inst Food Technologists; Can Inst Food Technologists; fel Inst Food Sci & Technol (UK). *Res:* Research and development of processes, products and equipment for fruit and vegetable processing industry. *Mailing Add:* RR 4 Fite 104 Summerland BC V0H 1Z0 Can

KITSON, ROBERT EDWARD, b Ashtabula, Ohio, Aug 9, 18; m 42; c 2. POLYMER CHEMISTRY. *Educ:* Mt Union Col, BS, 40; Purdue Univ, MS, 42, PhD(anal chem), 44. *Prof Exp:* Res chemist, Rayon Tech Div, 43-44, Ammonia Chem Div, 43-52 & Dacron Res Div, 52-64, SUPVR, DACRON TEXTILE RES DIV, E I DU PONT DE NEMOURS & CO, INC, 65- *Mem:* Am Chem Soc. *Res:* Texturing of textile fibers; physical chemistry of high polymers. *Mailing Add:* 322 Hampton Rd Wilmington DE 19803

KITTAKA, ROBERT SHINNOSUKE, b Los Angeles, Calif, Sept 23, 34; m 62; c 3. FOOD SCIENCE, MICROBIOLOGY. *Educ:* Univ Ill, BS, 57, MS, 59, PhD(food sci), 64. *Prof Exp:* Microbiologist, Swift & Co, 64-69; mgr microbiol, 69-71, MICROBIOL & QUAL ASSURANCE DIR, FOOD RES, CENT SOYA CO, INC, 71- *Mem:* Inst Food Technol; Am Soc Microbiol; Asn Milk, Food & Environ Sanitarians; Soc Indust Microbiol; Brit Soc Appl Bact. *Res:* Food microbiology; public health and spoilage microbiology as related to food products and processes; development, implementation and auditing of quality assurance programs in food processing systems to assure compliance with governmental regulations and corporate standards for cost efficient operations. *Mailing Add:* 5231 Chippewa Trail Ft Wayne IN 46804

KITTEL, CHARLES, b New York, NY, July 18, 16; m 38; c 3. PHYSICS. *Educ:* Cambridge Univ, BA, 38; Univ Wis, PhD(physics), 41. *Prof Exp:* Physicist, Naval Ord Lab, Washington, DC, 40-42; opers analyst, US Fleet, 43-45; res assoc physics, Mass Inst Technol, 45-46; res physicist, Bell Tel Labs, 47-50; vis assoc prof, 50, prof, 51-78, EMER PROF PHYSICS, UNIV CALIF, BERKELEY, 79- *Concurrent Pos:* Guggenheim fel, 46, 57 & 64. *Honors & Awards:* Buckley Prize, 57; Oersted Medal, 79. *Mem:* Nat Acad Sci; Am Acad Arts & Sci. *Res:* Solid state physics; theory of ferromagnetism; mathematical physics. *Mailing Add:* Dept Physics Univ Calif Berkeley CA 94720

KITTEL, J HOWARD, b Ritzville, Wash, Oct 9, 19; m 43; c 4. TECHNOLOGY TRANSFER. *Educ:* Wash State Univ, BS, 43. *Prof Exp:* Aeronaut res scientist, Nat Aeronaut & Space Agency, 43-51; metal engr, 51-61, mgr adv fuels, 74-79, mgr nuclear waste res & develop, 79-85, SR METALLURGIST, ARGONNE NAT LAB, 61-, TECHNOL TRANSFER SPECIALIST, 89- *Concurrent Pos:* Mem, Mat Adv Bd, Nat Acad Sci, 56-58; US delegate, UN Geneva Conf, 58-64. *Mem:* Fel Am Nuclear Soc; Sigma Xi; Scientists & Engrs Secure Energy. *Res:* Technology transfer. *Mailing Add:* Argonne Nat Lab 9700 S Cass Ave Argonne IL 60439

KITTEL, PETER, b Mt Vernon Dist, Va, Mar 23, 45; m 72; c 1. PHYSICS, CRYOGENICS. *Educ:* Univ Calif, Berkeley, BS, 67; Univ Calif, San Diego, MS, 69; Univ Oxford, Eng, DPhil(physics), 74. *Prof Exp:* Res assoc & adj asst prof physics, Univ Ore, 74-78; res assoc radiol, Stanford Univ, 78; Nat Res Coun assoc cryog, 78-80, RES SCIENTIST & TEAM LEADER, AMES RES CTR, NASA, 80- *Honors & Awards:* Medal for Except Eng Achievement, NASA, 90. *Mem:* Am Phys Soc; AAAS. *Res:* Low temperature physics; far infrared spectroscopy; stocastic processes and applications of cryogenics in space. *Mailing Add:* MS 244-10 NASA-Ames Res Ctr Moffett Field CA 94035-1000

KITTELBERGER, JOHN STEPHEN, b Palmerton, Pa, Mar 14, 39; m 63. PHYSICAL CHEMISTRY. *Educ:* Hamilton Col, AB, 61; Princeton Univ, AM, 63, PhD(phys chem), 66. *Prof Exp:* Instr chem, Princeton Univ, 65-66; res assoc, Mass Inst Technol, 66-68; asst prof, Amherst Col, 68-73; scientist, 73-81, MGR, XEROX CORP, 81- *Concurrent Pos:* Consult, surface sci & powder technol. *Mem:* Am Phys Soc; Am Chem Soc. *Res:* Structural chemistry; surface science; reprographic science; materials science; powder technology; paper science. *Mailing Add:* Xerox Corp 800 Phillips Rd Webster NY 14580

KITTELSON, DAVID BURNELLE, b Pelican Rapids, Minn, Mar 12, 42; m 70. MECHANICAL & CHEMICAL ENGINEERING. *Educ:* Univ Minn, Minneapolis, BSc, 64, MSc, 66; Cambridge Univ, PhD(chem eng), 72. *Prof Exp:* asst prof, 70-76, assoc prof, 76-80, PROF MECH ENG, UNIV MINN, 80- *Concurrent Pos:* Overseas fel, Churchill Col, Cambridge Univ, Cambridge, England, 85-86. *Honors & Awards:* Teeter Award, Soc Automotive Engrs, 73, Arch T Colwell Merit Award, 78 & 83. *Mem:* Sigma Xi; Am Chem Soc; Am Soc Mech Engrs; Soc Automotive Engrs. *Res:* Electronic engine control; diesel engine combustion and emissions; combustion generated aerosols. *Mailing Add:* Dept Mech Eng 111 Church St SE Minneapolis MN 55455

KITTILA, RICHARD SULO, b Ely, Minn, July 12, 17; m 49. ORGANIC CHEMISTRY. *Educ:* Univ Minn, BChem, 41; Duke Univ, MA, 44, PhD(chem), 49. *Prof Exp:* Jr chemist, Shell Develop Co, 41-42; sr res chemist, Biochem Dept, E I du Pont de Nemours & Co, Inc, 49-82; RETIRED. *Mem:* Am Chem Soc. *Res:* Organic synthesis; agricultural chemicals; dimethylformamide; agricultural chemical formulations. *Mailing Add:* 3730 Mill Creek Rd Hockessin DE 19707

KITTING, CHRISTOPHER LEE, b Monroe, Mich, May 23, 53. ECOLOGY OF MARINE POPULATIONS. *Educ:* Univ Calif, Irvine, BS, 74; Stanford Univ, PhD(biol sci), 79. *Prof Exp:* Res assoc, Stanford Med Sch, Hopkins Marine Sta, 78-79; res biologist, Marine Sci Inst, Univ Calif, Santa Barbara, 79; ASST PROF MARINE STUDIES & ZOOL, PORT ARANSAS MARINE LAB, UNIV TEX, Austin, 79-; DEPT MARINE BIOL, CALIF STATE UNIV, HAYWARD. *Concurrent Pos:* Vis teaching asst, WI Lab, Fairleigh Dickinson Univ, 75, vis investr, 76; teaching asst, Dept Biol Sci, Hopkins Marine Sta, Stanford Univ, 77, fel, 74-79. *Mem:* Ecol Soc Am; Am Soc Limnol & Oceanog; Sigma Xi; NY Acad Sci. *Res:* Advancing ecological theory using specialized natural history studies, especially marine invertebrates, algae, foraging and competition; field experiments using close-up listening and visual records of activities on semi-isolated surfaces of shallow rocks, pilings, oyster reefs and seagrass blades. *Mailing Add:* Dept Biol Sci Calif State Univ Hayward CA 94542

KITTINGER, GEORGE WILLIAM, b Toledo, Ohio, Dec 10, 21; m 47. BIOCHEMISTRY. *Educ:* Northwestern Univ, BS, 48, MS, 50; Univ Ore, PhD(biochem), 53. *Prof Exp:* Instr chem, Univ Ore, 53-56; res chemist, Procter & Gamble Co, 56-58; sr biochemist, May Inst Med Res, 58-64; assoc prof biochem, Med Sch, Univ Ore, 65-67; scientist, Ore Regional Primate Res Ctr, 65-79; prof biochem, Med Sch, Univ Ore, 67-79; RETIRED. *Concurrent Pos:* Fel chem, Univ Ore, 53-56; asst prof, Col Med, Univ Cincinnati, 59-64. *Mem:* Am Chem Soc; Endocrine Soc; Soc Exp Biol & Med. *Res:* Corticosteroid biosynthesis and metabolism; mechanism of action of steroid hormones; maternal-fetal-endocrine relationships. *Mailing Add:* 4885 SW 152nd Beaverton OR 97007

KITTLE, CHARLES FREDERICK, b Athens, Ohio, Oct 24, 21; m 45, 71; c 4. SURGERY. *Educ:* Ohio Univ, BA, 42; Univ Chicago, MD, 45; Univ Kans, MS, 50; Am Bd Surg, dipl; Am Bd Thoracic Surg, dipl. *Hon Degrees:* LLD, Ohio Univ, 67. *Prof Exp:* Chief lab serv, Brentwood Vet Admin Hosp, Los Angeles, 47-48; from instr to assoc prof surg, Sch Med, Univ Kans, 50-66; prof surg & chief sect throacic & cardiovasc surg, Univ Chicago, 66-73; PROF SURG & HEAD SECT THORACIC SURG, RUSH MED COL, 73-; DIR, RUSH CANCER CTR, 78. *Concurrent Pos:* Consult, Oak Ridge Inst Nuclear Studies, 50-57 & Vet Admin Hosps, Kans & Mo, 53-66; Am Cancer Soc Clin fel, 50-52; Murdock fel, 51; Markle scholar, 53-58; mem, Bd Thoracic Surg, 66-77. *Mem:* Soc Univ Surg (pres, 66-67); Am Col Surgeons; Soc Thoracic Surgeons; Am Asn Thoracic Surgeons; Int Cardiovasc Soc (secy, 65-71); Am Surg Assoc; Soc Clin Surg. *Res:* Cardiovascular hemodynamics; extracorporeal circulation; clinical trials. *Mailing Add:* 1725 W Harrison St Chicago IL 60612

KITTLITZ, RUDOLF GOTTLIEB, JR, b Waco, Tex, Apr 19, 35; m 66; c 4. CHEMICAL ENGINEERING. *Educ:* Univ Miss, BSChemE, 57. *Prof Exp:* Engr, Du Pont Co, 57-62, res engr, 62-68, sr res engr, 68-87, RES ASSOC, DU PONT CO, 87- *Concurrent Pos:* Adj prof, Univ Tenn Chattanooga, 80-82; regional dir, Am Soc Qual Control, 86-91, exec regional dir, 87-91, dir at large, 91-93. *Honors & Awards:* W G Hunter Award, Am Soc Qual Control, 89. *Mem:* Fel Am Soc Qual Control; Am Statist Asn. *Mailing Add:* 917 N Atlanta Circle Seaford DE 19973-1131

KITTO, GEORGE BARRIE, b Wellington, NZ, July 31, 37; m 62; c 3. BIOCHEMISTRY. *Educ:* Victoria Univ, BSc, 61, MSc, 62; Brandeis Univ, PhD(biochem), 66. *Prof Exp:* Biochemist, Wellington Pub Hosp, 60-61; asst prof, 66-71, PROF CHEM, UNIV TEX, AUSTIN, 71-; RES SCIENTIST, CLAYTON FOUND BIOCHEM INST, 66- *Concurrent Pos:* Vis prof, Univ Calif, Berkeley, 78, Duke Univ, 86; consult, Am Cyanamid, Dell Corp, Aquanautics Corp, Whatman, Inc. *Mem:* AAAS; Am Chem Soc; Royal Soc Chem; assoc mem NZ Inst Chem; Royal Soc NZ; Australian Biochem Soc. *Res:* Enzyme structure and taxonomy; evolution of protein structure; multiple molecular forms of enzymes; immobilized enzymes. *Mailing Add:* Dept Chem Welch 4 260 Univ Tex Austin TX 78712

KITTREDGE, CLIFFORD PROCTOR, b Lowell, Mass, June 30, 06; wid; c 2. ENGINEERING. *Educ:* Mass Inst Technol, BS, 29; Tech Hochsch, Munich, DrTechWissen(fluid mech), 32. *Prof Exp:* Asst mech eng, Mass Inst Technol, 29-30, instr, 32-33; instr power eng, Thayer Sch Civil Eng, Dartmouth Col, 33-34; engr, Freeman Trust Estate, Providence, 34-36; asst prof theoret & appl mech, Univ Ill, 36-41; develop engr, Underwater Sound Lab, Div War Res, Columbia Univ, 41-45; assoc prof mech eng, 45-71, EMER ASSOC PROF MECH ENG, PRINCETON UNIV, 71- *Concurrent Pos:* Consult engr, Metrop Water Dist Southern Calif, Los Angeles. *Mem:* Am Soc Mech Engrs. *Res:* General fluid mechanics; hydraulic machinery. *Mailing Add:* 763 Princeton-Kingston Rd Princeton NJ 08540

KITTRELL, BENJAMIN UPCHURCH, b Kittrell, NC, Oct 25, 37; m 58; c 2. TOBACCO, SOYBEANS. *Educ:* NC State Univ, BS, 60, MEd, 69, Phd(crop sci), 75. *Prof Exp:* Teacher voc agr, Vance Co, NC, 60-63 & Wake Co, NC, 63-65; supt, res sta, NC State Univ, 65-68, agronomist tobacco, 68-75; asst prof agron, Univ Ga, 75-78; from assoc prof to prof agron, Clemson Univ, 78-87; RESIDENT DIR, PEE DEE RES & EDUC CTR, 87- *Mem:* Am Soc Agron. *Res:* Tobacco production management including plant density, leaf area index, fertilization, sucker control, harvest, curing, disease control; soybean production management. *Mailing Add:* Pee Dee Res & Educ Ctr Clemson Univ Rte One Box 531 Florence SC 29501-9603

KITTRELL, JAMES RAYMOND, b Akransas City, Kans, Oct 28, 40; m 60; c 4. POLYMER CHEMISTRY. *Educ:* Okla State Univ, BS, 62; Univ Wis, MS, 63, PhD(chem eng), 66. *Prof Exp:* From res engr to sr res engr, Chevron Res Co, 66-69; oper asst, Standard Oil Co Calif, 69-70; prof chem eng, Univ Mass, Amherst, 70-80; PRES, KSE, INC, 80- *Concurrent Pos:* NSF fel & instr, Univ Wis, 66. *Mem:* Am Inst Chem Engrs; Am Chem Soc. *Res:* Petroleum refining; polymers; reactor design and analysis; catalyst deactivation; environmental processes; bioengineering. *Mailing Add:* PO Box 368 Amherst MA 01004

KITTRICK, JAMES ALLEN, b Milwaukee, Wis, Aug 4, 29; m 53; c 2. SOIL MINERALOGY. *Educ:* Univ Wis, BS, 51, MS, 53, PhD, 55. *Prof Exp:* From asst prof to assoc prof, 55-67, PROF SOILS, WASH STATE UNIV, 67- *Honors & Awards:* Fel Am Soc Agron. *Mem:* Soil Sci Soc; Clay Minerals Soc. *Res:* Mineral stability, weathering. *Mailing Add:* Dept Agron & Soils Wash State Univ Pullman WA 99164

KITTS, DAVID BURLINGAME, b Oswego, NY, Oct 27, 23; m 45; c 2. VERTEBRATE PALEONTOLOGY, GEOLOGY. *Educ:* Univ Pa, AB, 49; Columbia Univ, PhD(zool), 53. *Prof Exp:* Instr biol, Amherst Col, 53-54; from asst prof to assoc prof geol, Univ Okla, 54-62, assoc prof geol & hist sci, 62-66, David Ross Boyd prof geol & hist sci, 66-80, head cur, Dept Geol, Stoval Mus, 68-80; at Dept Geol Physics, Univ Okla, Norman, 80-87; RETIRED. *Concurrent Pos:* Vis fel, Princeton Univ, 64-65. *Mem:* Soc Vert Paleontol; Philos Sci Asn. *Res:* Historical geology; Cenozoic mammals and stratigraphy; philosophy of geology and evolutionary theory. *Mailing Add:* 6559 Stonecroft Terr Santa Rosa CA 95409

KITTSLEY, SCOTT LOREN, b Port Washington, Wis, Feb 17, 21; m 46. PHYSICAL CHEMISTRY. *Educ:* Univ Wis, BS, 42; Case Western Reserve Univ, MS, 44, PhD(phys chem), 45. *Prof Exp:* Asst chem, Case Western Reserve Univ, 42-45; from instr to prof, 45-81, chmn, dept, 57-62, EMER PROF CHEM, MARQUETTE UNIV, 82- *Mem:* Am Chem Soc. *Res:* Chemical thermodynamics; solutions of nonelectrolytes. *Mailing Add:* 3838 N Oakland Apt 169 Shorewood WI 53211-2258

KITZ, RICHARD J, b Oshkosh, Wis, Mar 25, 29; m 54; c 1. ANESTHESIOLOGY, ENZYMOLOGY. *Educ:* Marquette Univ, BS, 51, MD, 54; Harvard Univ, MA, 69; Am Bd Anesthesiol, dipl, 62. *Prof Exp:* Surg intern, Columbia-Presby Med Ctr, 54-55, surg resident, 57-58, resident anesthesiol, 58-60, instr, 60-61, from asst prof to assoc prof anesthesiol, 62-69; prof, 69-70, HENRY ISAIAH DORR PROF ANESTHESIA, HARVARD MED SCH, 70- *Concurrent Pos:* NIH spec res fel, Columbia-Presby Med Ctr, 61-62 & Karolinska Inst, Stockholm, 68; NIH spec res fel, Karolinska Inst, Stockholm, 68; anesthetist-in-chief, Mass Gen Hosp, 69-; consult, Air Force Surg Gen, 70-80, Brigham & Women's Hosp & Beth Israel Hosp, Boston, 70-; ed-in-chief, J Clin Anesthesia; co-dir, Harvard-Mass Inst Technol Div Health Sci & Technol, 85-; consult anesthesiol, Dept Navy, 69-71 & Surgeon Gen, USAF, 70-79. *Honors & Awards:* Golden Emblem Award, Finnish Soc Anesthesiologists, 77. *Mem:* Inst Med-Nat Acad Sci; AMA; Am Soc Anesthesiol; fel Am Col Anesthesiol; NY Acad Med; Am Chem Soc; Am Soc Pharmacol & Exp Therapeut; AAAS. *Res:* Basic sciences as related to anesthesiology with emphasis on uptake and distribution of anesthetic agents and enzymology; design, synthesis and testing of novel compounds used as

molecular probes (active site investigations) and drugs (short- acting, non-depolarizing neuromuscular blocking agents, anticholinesterases); design, constructing and testing of new anesthesia delivery systems and monitoring devices for care of the critically ill patient in operating rooms and intensive care units; author of two books. *Mailing Add:* Dept Anesthesia Mass Gen Hosp Boston MA 02114

KITZES, ARNOLD S(TANLEY), b Boston, Mass, Sept 21, 17; m 42; c 2. NUCLEAR ENGINEERING, CHEMICAL ENGINEERING. *Educ:* City Col New York, BChE, 39; Univ Minn, MChE, 41, PhD(chem eng), 47. *Prof Exp:* Instr chem eng, Univ Minn, 41-42, res assoc, 45-47, chief chemist, Sangamon Ord, Ill, 42-45; group leader, Oak Ridge Nat Lab, 48-57; mgr test eng, Atomic Power Div, Westinghouse Elec Corp, 57-72, adv engr, Nuclear Serv Div, 72-86; RETIRED. *Res:* Reactor technology; environment; spray drying and explosives; design of high pressure equipment; heat transfer; fluid flow; waste management; decontamination; technical administration. *Mailing Add:* 2740 Mount Royal Rd Pittsburgh PA 15217

KITZES, GEORGE, b Denver, Colo, May 28, 19; m 42; c 3. BIOCHEMISTRY, PHARMACOLOGY. *Educ:* City Col New York, BS, 41; Univ Wis, MS, 42, PhD(chem), 44. *Prof Exp:* Res biochemist, White Labs, NJ, 44-49 & Vet Admin, 49-51; res biochemist indust toxicol, USAF, 51-57, asst chief, Physiol Br, Aeromed Lab, 57-60, chief, Physiol Div, Aerospace Med Res Labs, Aero-Med Div, Wright-Patterson AFB, Ohio, 60-66; health sci adminr & gastroenterol prog dir, Extramural Prog, Nat Inst Arthritis, Metab & Digestive Dis, 66-81; CONSULT, 81- *Concurrent Pos:* Consult, 81- *Mem:* AAAS; Am Gastroenterol Asn. *Res:* Gastrointestinal physiology and biochemistry; nutrition; aviation physiology and toxicology; aero-space life support sciences. *Mailing Add:* 10201 Grosvenor Pl Rockville MD 20852

KITZES, LEONARD MARTIN, b New York, NY, Apr 10, 41; m 67; c 2. NEUROPHYSIOLOGY. *Educ:* Univ Calif, Los Angeles, BA, 62; Univ Calif, Irvine, PhD(psychobiol), 70. *Prof Exp:* Fel neurophysiol, Univ Wis, 70-73, res assoc, 73-74; res neurophysiol neurol, 74-77, ASST PROF NEUROPHYSIOL, DEPT ANAT, UNIV CALIF, IRVINE, 77- *Concurrent Pos:* Nat Inst Neurol Dis & Stroke res grant, 78. *Res:* Auditory neurophysiology with primary interests in responses of single inferior colliculus neurons to binaural stimulation and functional and structural development of the brainstem audiroty system. *Mailing Add:* Dept Anat Univ Calif Col Med Irvine CA 92717

KITZKE, EUGENE DAVID, b Milwaukee, Wis, Sept 2, 23; m 46; c 4. ENVIRONMENTAL HEALTH, TECHNICAL MANAGEMENT. *Educ:* Marquette Univ, BS, 45, MS, 47. *Prof Exp:* Instr biol, Marquette Univ, 46-47; from asst prof to assoc prof, St Thomas Aquinas Col, Grand Rapids, Mich, 47-51; res mgr biol, S C Johnson & Son, Inc, 57-76, vpres, Corp Res & Develop, 76-81; PRES, OAK CRETE BLOCK CORP, 82-; OWNER, DANEL ENTERPRISES. *Concurrent Pos:* Instr microbiol, St Mary's Nursing Sch, Grand Rapids, Mich, 47-51; asst clin prof environ med, Dept Environ Med, Med Col Wis, 73-82. *Mem:* AAAS; Sigma Xi; Hist Sci Soc. *Res:* Aspects, theory and practice of creative technical management. *Mailing Add:* 616 Aspen St South Milwaukee WI 53172

KITZMILLER, JAMES BLAINE, b Toledo, Ohio, June 30, 18; m 74; c 4. GENETICS, MEDICAL ENTOMOLOGY. *Educ:* De Sales Col, BS, 39; Univ Mich, MS, 41, PhD(genetics), 48. *Prof Exp:* Entomologist, Toledo Mus Sci, 39-41; asst zool, Univ Mich, 41-42; from instr to prof, 48-74, EMER PROF ZOOL, UNIV ILL, URBANA, 74-; AT FLA MED ENTOMOL LAB. *Concurrent Pos:* Fulbright fel, Univ Pavia, Italy, 53; NIH fel, Univ Cagliari, Italy & Johannes Gutenberg Univ, Ger, 65-66; consult, WHO & Pan Am Health Orgn, 53-, NIH, 58- *Honors & Awards:* Meritorious Serv Award, Am Mosquito Control Asn, 78; John Belkin Memorial Award, 86. *Mem:* AAAS; Genetics Soc Am; Soc Study Evolution; Am Entom Soc; Am Mosquito Control Asn. *Res:* Genetics and cytogenetics of mosquitoes, especially Anophelines; evolutionary cytogenetics, polymorphism and cytotaxonomy. *Mailing Add:* Fla Med Entomol Lab 200 Ninth St SE Vero Beach FL 32962

KIUSALAAS, JAAN, b Tartu, Estonia, June 23, 31; m 59; c 3. MECHANICS. *Educ:* Univ Adelaide, BE, 56; Northwestern Univ, MS, 59, PhD(mech), 62. *Prof Exp:* Plant design engr, Chrysler Australia Ltd, 56-57, prod design & test engr, 57-58, prod develop engr, Eng Div, Chrysler Corp, 59-60; res fel, Mat Res Ctr, Northwestern Univ, 62-63; from asst prof to assoc prof, 63-74, PROF ENG MECH, PA STATE UNIV, 74- *Concurrent Pos:* Sr resident res assoc, Marshall Space Flight Ctr, NASA, 71-72. *Mem:* Am Soc Mech Engrs; Am Acad Mech. *Res:* Finite elements; optimal structural design; structural stability. *Mailing Add:* Dept Eng Sci Pa State Univ Main Campus University Park PA 16802

KIVEL, BENNETT, theoretical physics; deceased, see previous edition for last biography

KIVELSON, DANIEL, b New York, NY, July 11, 29; m 49; c 2. CHEMICAL PHYSICS. *Educ:* Harvard Univ, AB, 49, MA, 50, PhD(chem physics), 53. *Prof Exp:* Instr physics, Mass Inst Technol, 53-55; from asst prof to assoc prof chem, 55-63, chmn dept, 75-78, PROF CHEM, UNIV CALIF, LOS ANGELES, 63- *Concurrent Pos:* Guggenheim fel, 59; Res Corp spec grant, 59; Sloan fel, 61-65; NSF sr fel, 65-66; consult, NAm Sci Ctr, 63-64, Advan Res Projs Agency, 65-66 & NSF, 76-77; assoc ed, J Chem Phys, 68-71 & Molecular Physics, 75-85. *Mem:* Fel Am Phys Soc; Am Chem Soc. *Res:* Microwave spectroscopy; molecular structure; nuclear magnetic resonance; electron spin resonance; light scattering; theory of liquids; supercooled liquids and glasses. *Mailing Add:* Dept Chem Univ Calif Los Angeles CA 90024

KIVELSON, MARGARET GALLAND, b New York, NY, Oct 21, 28; m 49; c 2. SPACE PHYSICS. *Educ:* Radcliffe Col, AB, 50, AM, 51, PhD(physics), 57. *Prof Exp:* Adj asst prof physics, 67-72, res geophysicist, Inst Geophys & Planetary Physics, 67-80, adj assoc prof, 72-73, assoc prof in residence, 75-77, prof geophys & space physics in residence, 77-80, chmn dept earth & space sci, 84-87, PROF, UNIV CALIF, LOS ANGELES, 80- Concurrent Pos: Consult, Rand Corp, 55-71; scholar, Radcliffe Inst Independent Study, 65-66; fel, John Simon Guggenheim Mem Found, 73-74; overseer, Harvard Col, 77-83. Mem: AAAS; Am Geophys Union; Am Phys Soc; Am Astron Soc. Res: Magnetospheric physics; plasma physics; particles and fields in the magnetospheres of Earth and Jupiter; interplanetary magnetic fields. Mailing Add: Inst Geophys & Planetary Physics Univ Calif Los Angeles CA 90024-1567

KIVENSON, GILBERT, b Pittsburgh, Pa, Dec 5, 20. ENGINEERING INSTRUMENTATION. Educ: Carnegie Inst Technol, BS, 42; Univ Pittsburgh, MS, 47. Prof Exp: Rubber Reserve Co fel anal distillation, Mellon Inst, 47-50, develop engr, 50-51; chem engr, US Steel Co, 51-53; consult instrumentation, 53-55; chem engr, Atomic Power Dept, Westinghouse Elec Corp, 55-62, sr engr, Res & Develop Ctr, 62-68; sr engr, Electro-Optical Systs Div, Xerox Corp, 68-70; sr engr, Liquid Metals Eng Ctr, Atomics Int, Chatsworth, 71-73; sr engr, C F Braun Co, Calif, 73-75, J B Lansing, Northridge, Calif, 80-81; CONSULT & PATENT AGENT, 75- Concurrent Pos: Lectr, Pa Technol Inst & Northrop Inst Technol. Mem: Am Chem Soc. Res: Electronic instruments; stroboscopes; control systems; transducers; process control instrumentation. Mailing Add: 22030 Wyandotte St Canoga Park CA 91303

KIVER, EUGENE P, b Cleveland, Ohio, Feb 26, 37; m 64; c 3. GEOMORPHOLOGY. Educ: Case Western Reserve Univ, BA, 64; Univ Wyo, PhD(glacial geol, geomorphol), 68. Prof Exp: Chmn dept, 71-74, from asst prof to assoc prof, 71-77, PROF GEOL, EASTERN WASH UNIV, 77-, CHMN DEPT, 90- Mem: Am Quaternary Asn; Geol Soc Am; Nat Speleol Soc. Res: Pleistocene and neoglacial history of alpine regions of western US; general geomorphology; volcanism in the Cascade Mountains; geology of the national parks; quaternary geology of NE Washington. Mailing Add: Dept Geol Eastern Wash Univ Cheney WA 99004

KIVIAT, ERIK, b New York, NY, June 9, 47; m 82. WETLAND & HUMAN ECOLOGY. Educ: Bard Col, BS, 76; State Univ NY, New Paltz, MA, 79. Prof Exp: From instr to asst prof natural hist, 73-78, dir, field sta, 72-78, RES ASSOC ECOL, BARD COL, 78-, ASST PROF ENVIRON STUDIES, GRAD SCH ENVIRON STUDIES, 88-; EXEC DIR, HUDSONIA LTD, 88- Concurrent Pos: Consult, Hudsonia Ltd, 78-, ecologist, 81- Mem: Torrey Bot Club; Ecol Soc Am; Estuarine Res Fedn; Soc Conserv Biol; Soc Wetland Scientists; Soc Study Amphibians & Reptiles. Res: Wetland ecology and management; vertebrates; vascular plants; vegetation change; habitat ecology; rare species conservation; introduced species ecology; cultural ecology. Mailing Add: Hudsonia Ltd Bard Col Field Sta Annandale NY 12504

KIVIAT, FRED E, b New York, NY, May 16, 40. OPERATIONS RESEARCH. Educ: City Col, New York, BS, 62; Univ Pittsburgh, PhD(chem), 68, MS, 78. Prof Exp: Sr res chemist, Gulf Res & Develop Co, 68-83, dir, anal instrumentation & process comput, 83-85; automation consult, ICI Americas, 86; RES ENGR, DUPONT, 87- Res: Determination of properties of heterogeneous catalysts via spectroscopic techniques; mathematical modelling of physio-chemical processes; development and implementation of process control strategies in refining and chemical facilities; development and implementation of product quality management systems for chemical facilities. Mailing Add: 400 Woodland Rd 400 Woodland Rd Seaford DE 19973

KIVILAAN, ALEKSANDER, b Jaarja, Estonia, July 20, 06; nat US; m 35; c 3. PLANT PHYSIOLOGY. Educ: Univ Tartu, Estonia, BS, 32, MS, 35; Univ Berlin, dipl, 38; Mich State Univ, PhD(plant path, bot), 57. Prof Exp: Asst plant path, Univ Tartu, Estonia, 32-36, res instr pomol, 39-44; lectr dendrol, Baltic Univ, Ger, 48-50; plant propagator, Mt Arbor Nurseries, Iowa, 50-53; asst plant path, Mich State Univ, 54-56, res instr plant physiol, 57, asst prof, 58-60, assoc prof 60-78, emer assoc prof plant physiol, 78-; RETIRED. Mem: Am Phytopath Soc; Am Soc Plant Physiol. Res: Physiology of fungi, parasitism, plant growth and plant cell wall. Mailing Add: 1167 Marigold Ave East Lansing MI 48823

KIVIOJA, LASSI A, b Finland, Mar 29, 27; US citizen; m 64; c 2. GEODESY. Educ: Univ Helsinki, BS, 51, MS, 52; Ohio State Univ, PhD, 63. Prof Exp: Res asst, Int Isostatic Inst, Helsinki, Finland, 49-52; res assoc geod gravity, Ohio State Univ, 55-63; prof, 64-90, EMER PROF, PURDUE UNIV, 90- Concurrent Pos: Instr, Dept Geod Sci, Ohio State Univ, 59-62; geodesist, US Coast & Geod Surv, 70-71; res scientist, Defense Mapping Agency, 79-80 & 86-87. Mem: Am Geophys Union; Am Cong Surv & Mapping. Res: Gravity anomalies isostasy; geodetic and astro-geodetic instruments; geodesic lines on the ellipsoid; the vertical mirror and its applications; mean sea level; mercury leveling instruments; hydrostatic leveling on land; precise astro-azimuths by mercury leveling of a theodolite. Mailing Add: 60 Blackfoot Ct Lafayette IN 47905

KIVISILD, H(ANS) R(OBERT), b Tartu, Estonia, July 19, 22; m 47; c 4. CIVIL ENGINEERING. Educ: Royal Inst Technol, Swed, CE, 46, DEng(hydraul eng), 54. Prof Exp: Engr, City of Stockholm, Swed, 46-48 & VBB, 48-50; with Found Co Can, Ltd, Montreal, 50-53; designing engr, Found Can Eng Corp Ltd, Montreal, 54-56, dist engr, Vancouver, 57-63, chief hydraul engr, Toronto, 64-65, chief civil engr, 65-70, dir res & develop, 70-73, vpres, 73-75; vpres & mgr, western opers & dir, Fenco Consults Ltd, Calgary, 75-81; vpres, Lavalin Inc, Calgary, 82-87; PRIN, HRK CONSULT INC, CALGARY, 87- Concurrent Pos: Consult, Belg Govt, 56, Tech Assistance Opers, UN, 61, Spec Fund, 63, Food & Agr Orgn, 64; mem snow & ice subcomt, Nat Res Coun Can, 65-75, mem cold regions res comt, 77-79. Honors & Awards: Eng Medal, Asn Prof Engrs Ont, 76; Queen Elizabeth II Silver Jubilee Medal, 77; Award Merit Innovation, Manning Awards, 85. Mem: Fel Can Soc Civil Eng; Int Asn Hydraul Res; Marine Technol Soc; fel Eng Inst Can. Res: Oceanography and hydraulics of ice-covered waters; arctic development. Mailing Add: 1420 Premier Way SW Calgary AB T2T 1L9 Can

KIVLIGHN, HERBERT DANIEL, JR, b Mineola, NY, Apr 10, 31; m 54; c 3. PHYSICAL CHEMISTRY, INORGANIC CHEMISTRY. Educ: Hofstra Univ, BA, 54; Pa State Univ, PhD(phys chem), 58. Prof Exp: Res chemist, Corning Glass Works, 57-58, res supvr, 58-61; res scientist, Am Stand Res Labs, 61-63; res scientist, 63-69, STAFF SCIENTIST, GRUMMAN AEROSPACE CORP, 69- Concurrent Pos: Mem, comt aerospace transparencies, Am Soc Test & Mats. Mem: AAAS; fel Am Inst Chemists; Am Ceramic Soc; Am Chem Soc; Sigma Xi. Res: Photochromic glass response under high excitation intensities; effect on the kinetics of darkening and fading; laser countermeasures research related to flash blindness phenomena. Mailing Add: PO Box 262 Long Island Bethpage NY 11714-0262

KIVLIGHN, SALAH DEAN, b Iowa City, Iowa, July 12, 57; m 89. DRUG DEVELOPMENT & DESIGN NOVEL THERAPEUTICS AGENTS. Educ: Iowa State Univ, BS, 79; Univ Houston, PhD(cardiovasc oharmacol), 89. Prof Exp: Postdoctoral fel physiol, Dept Physiol, Univ Miss Med Ctr, Jackson, 87-88, instr physiol, 88-90; SR RES PHARMACOLOGIST, MERCK, SHARP & DOHME RES LABS, WEST POINT, PA, 90- Concurrent Pos: Referee ed, Heart & Circulatory & Res & Integ, Am J Physiol, 88- Mem: Am Soc Hypertension; Am Physiol Soc; Am Soc Nephrol. Res: Design and development of novel therapeutic agents for the treatment of hypertension and heart failure. Mailing Add: Merck Sharp & Dohme Res Lab WP26-265 West Point PA 19486

KIVNICK, ARNOLD, b Philadelphia, Pa, June 30, 23; m 46; c 3. CHEMICAL ENGINEERING, APPLIED MATH. Educ: Univ Pa, BS, 43, PhD(chem eng), 51. Prof Exp: Asst, SAM Labs, Columbia Univ, 43-44; jr chem engr, Tenn Eastman Co, 44-46; res assoc chem eng, Univ Ill, 50-52; sr chem engr, Tech Div, Pennsalt Chem Corp, 52-60, group leader, 60-68, mgr, Process Develop Dept, 68-72; liaison engr, Cent Eng Dept, Pennwalt Corp, 72-88; RETIRED. Concurrent Pos: Adj prof chem eng, Univ Pa, 80- Mem: Am Inst Chem Engrs. Res: Process development; applied statistics; cost estimation; process economics. Mailing Add: 308 Glenway Rd Philadelphia PA 19118

KIYASU, JOHN YUTAKA, b San Francisco, Calif, Dec 25, 27; m 54; c 4. BIOCHEMISTRY. Educ: Univ Calif, Berkeley, BA, 50, MA, 51, PhD(physiol), 55. Prof Exp: Fel biochem, Univ Chicago, 56-57, instr & res assoc, 57-60; asst prof biochem & asst res prof psychiat, Sch Med & Dent, Univ Rochester, 60-63; from asst prof to assoc prof chem, Adelphi Univ, 63-70; dir div biochem & asst dir lab, dept path, Roosevelt Hosp, 70-84; ASSOC CLIN PROF PATH, COL PHYSICIANS & SURGEONS, COLUMBIA UNIV, 72- Concurrent Pos: USPHS fel, 55-57; NIH grant, Adelphi Univ; res biochemist & mem staff, Dept Path & Labs, Div Biochem, Meadowbrook Hosp, East Meadow, NY, 67-69, dir lab systs, 69-70. Mem: Am Oil Chem Soc; Am Chem Soc; fel Am Inst Chem; NY Acad Sci; Am Asn Clin Chem. Res: Biosynthesis of phospholipids and liponucleotides; lipid enzymology and control mechanisms. Mailing Add: Dept Clin Labs Brookdale Hosp Med Ctr Linden Blvd & Brockdale Plaza Brooklyn NY 11212

KIZER, DONALD EARL, b Benton Co, Iowa, Oct 12, 21; m 42; c 6. BIOCHEMISTRY. Educ: Upper Iowa Univ, BS, 47; Purdue Univ, MS, 50; Univ NC, PhD(bact), 54. Prof Exp: Asst dairy, Purdue Univ, 48-50; asst animal indust, Univ NC, 52-54; res assoc, 64-60, head biochem pharmacol sect, Biomed Div, Samuel Roberts Noble Found, Inc, 60-85; RETIRED. Mem: AAAS; Am Chem Soc; Am Soc Microbiol; Soc Exp Biol & Med; Am Asn Cancer Res. Res: Biochemistry of carcinogenesis; biochemical changes in pre-cancerous tissues. Mailing Add: 825 H St NW Ardmore OK 73401

KIZER, JOHN STEPHEN, b Charleston, WVa, Jan 8, 45; m 70; c 6. NEUROENDOCRINOLOGY. Educ: Princeton Univ, AB, 66; Duke Univ, MD, 70. Prof Exp: Intern med, Johns Hopkins Hosp, Baltimore, 70-71, resident, 71-72; res assoc neuroendocrinol, Lab Clin Sci, Nat Inst Ment Health, 72-75; from asst prof to assoc prof, 75-85, PROF MED & PHARMACOL, SCH MED, UNIV NC, CHAPEL HILL, 86- Concurrent Pos: Assoc dir, Biol Sci Res Ctr, 85-; res sci career develop award, 77-88. Mem: Am Soc Clin Invest. Res: Investigation of protein neurotransmitters; post-translational processing of protien neurotransmitters. Mailing Add: Biol Sci Res Ctr Sch Med Univ NC Chapel Hill NC 27514

KJAR, RAYMOND ARTHUR, b Farnam, Nebr, Feb 27, 38; m 67; c 1. ELECTRONICS ENGINEERING. Educ: Univ Nebr, BSEE, 60; Iowa State Univ, MS, 62, PhD, 64. Prof Exp: Res engr, Naval Res Lab, Off Naval Res, 64-68; res engr, R & E Div, Rockwell Int, 68-76, supvr, Electronics Div, IC Tech, 76-80, mgr, Mil IC Prod, 80-90, DIR ADVAN PROCESS TECHNOL, ROCKWELL DIGITAL COMMUN DIV, ROCKWELL INT, 90- Mem: Inst Elec & Electronics Engrs. Res: Semiconductor devices; integrated circuits; reliability physics; semiconductor surfaces. Mailing Add: Rockwell Int 4311 Jamboree Rd Newport Beach CA 92660-8902

KJELDAAS, TERJE, JR, b Oslo, Norway, Oct 24, 24; US citizen; m 50; c 2. THEORETICAL PHYSICS. Educ: Polytech Inst Brooklyn, BS, 48; Columbia Univ, AM, 49; Univ Pittsburgh, PhD, 59. Prof Exp: Res engr, Res Labs, Westinghouse Elec Corp, 49-59; from asst prof to assoc prof, 59-63, head dept, 77-80, PROF PHYSICS, POLYTECH INST NY, 63- Concurrent Pos: Consult, indust labs. Mem: Am Phys Soc. Res: Solid state theory; atomic-molecule-electron interactions. Mailing Add: Dept Physics NY Polytech Inst 333 Jay St Brooklyn NY 11201

KJELDGAARD, EDWIN ANDREAS, b Brush, Colo, Sept 14, 39; m 65; c 2. APPLIED CHEMISTRY, METROLOGY. Educ: St Olaf Col, BA, 61; Univ Colo, PhD(chem), 66. Prof Exp: Staff mem, 66-70, div supvr, 70-84, SR MEM TECH STAFF, SANDIA LABS, 84- Res: Organic fluorine chemistry; fluorinated cyclobutenes; explosive chemistry; coordination compounds; Risk analysis. Mailing Add: Sandia Labs Div 6321 Box 5800 Albuquerque NM 87185

KJELDSEN, CHRIS KELVIN, b Stockton, Calif, Apr 26, 39; m 62; c 3. PHYCOLOGY. *Educ:* Univ of the Pac, BA, 60, MS, 62; Ore State Univ, PhD(bot), 66. *Prof Exp:* Instr bot, Ore State Univ, 62-66; from asst prof to assoc prof, 66-73, chmn dept, 72-75, PROF BIOL, SONOMA STATE UNIV, 73- *Concurrent Pos:* NSF, DOE, grants for sci educ, 69-; Sonoma County Planning Comn, 74-76; Sabbatical leave, Univ NWales, 76; IPA assignment, US Dept Energy, Educ Progs, Washington, DC, 78-79; Sonoma County Hazardous Mat Comn, 85- *Honors & Awards:* Mosser Award, Ore State Univ, 66. *Mem:* Sigma Xi (vpres, 67). *Res:* Marine algae; physiological ecology and taxonomy; Sacramento San Jouquin Delta habitar analysis. *Mailing Add:* Dept Biol Sonoma State Univ 1801 E Cotati Ave Rohnert Park CA 94928

KJELGAARD, WILLIAM L, b Lindley, NY, Aug 27, 20; m 50; c 5. AGRICULTURAL ENGINEERING. *Educ:* Pa State Univ, BS, 50, MS, 53. *Prof Exp:* Instr agr eng, WVa Univ, 51-52; from instr to asst prof, 52-60, ASSOC PROF AGR ENG, PA STATE UNIV, 60-; RETIRED. *Mem:* Am Soc Agr Engrs. *Res:* Mechanization and processing of forage crops. *Mailing Add:* 1311 Circleville Rd State College PA 16801

KJELSBERG, MARCUS OLAF, b Mayville, NDak, Dec 27, 32; m 62; c 2. BIOSTATISTICS. *Educ:* Concordia Col, Moorhead, Minn, BA, 52; Univ Minn, MA, 55, PhD(biostatist), 62. *Prof Exp:* Instr biostatist, Sch Med, Tulane Univ, 57-60; asst prof biostatist & res assoc epidemiol, Sch Pub Health, Univ Mich, Ann Arbor, 61-66; assoc prof, 66-75, div head, biomet, 72-87, PROF BIOSTATIST, SCH PUB HEALTH, UNIV MINN, MINNEAPOLIS, 75- *Concurrent Pos:* Mem, US Nat Comt Vital & Health Statist, 73-77; prin investr, Mult Risk Factor Intervention Ctr, 72- *Mem:* Am Pub Health Asn; Am Statist Asn; Am Heart Asn; Biometric Soc; Pop Asn Am. *Res:* Statistical epidemiology; health statistics; clinical trials methodology. *Mailing Add:* Biostat Box 197 Mayo Bldg Univ of Minn Minneapolis MN 55455

KJONAAS, RICHARD A, b Minot, NDak, Mar 20, 49; m 79; c 2. ORGANIC CHEMISTRY. *Educ:* Valley City State Col, BS, 71; Purdue Univ, PhD(chem), 78. *Prof Exp:* Teacher high sch, NDak, 71-73; fel, Ohio State Univ, 78-79; asst prof org chem, Ft Hays State Univ, 79-83; asst prof, 83-87, ASSOC PROF ORG CHEM, IND STATE UNIV, 87- *Mem:* Am Chem Soc; Sigma Xi. *Res:* Organic synthesis via transition metals. *Mailing Add:* Dept Chem Ind State Univ Terre Haute IN 47809

KLAAS, ERWIN EUGENE, b Batchtown, Ill, Aug 23, 35; m 69; c 3. WILDLIFE BIOLOGY. *Educ:* Univ Mo, BS, 56; Univ Kans, MA, 63, PhD(zool), 70. *Prof Exp:* From asst prof to assoc prof biol, Rockhurst Col, 65-71; assoc prof, 75-80, PROF WILDLIFE BIOL, IOWA STATE UNIV, 80-; RES BIOLOGIST, US FISH & WILDLIFE SERV, 71- *Mem:* Am Ornith Union; Cooper Ornith Soc; Wilson Ornith Soc; Wildlife Soc; Sigma Xi. *Res:* Population ecology of birds; effects of environmental pollution on wild bird populations; management of habitat for optimal utilization by game and non-game species. *Mailing Add:* Iowa Coop Fish & Wildlife Res Unit Iowa State Univ Ames IA 50011

KLAAS, NICHOLAS PAUL, b Kieler, Wis, June 25, 25; m 49; c 4. FUEL TECHNOLOGY, PETROLEUM ENGINEERING. *Educ:* Loras Col, BA, 45; Univ Notre Dame, PhD(org chem), 48. *Prof Exp:* Chemist, Rohm and Haas Co, 48, actg prod mgr, 49-51; tech asst to sales develop mgr, Minn Mining & Mfg Co, 52, mkt specialist, 52-53, mgr oil indust prod, 53-59, mgr res & develop, 60-65; exec vpres, Wyomissing Corp, Reading, 65-71, dir, 68-71, chief operating officer, 70-71; vpres com develop, GAF Corp, 71-74, group vpres chem group, 74-77; gen mgr spec chem div, Ga-Pac Corp, 77; from exec vpres to pres, 77-84, gen mgr, J T Baker Chem Co, 77-84; PRES, INST INNOVATIVE ENTERPRISE, INC, 85- *Mem:* AAAS; Am Chem Soc. *Res:* Organic chemical intermediates and polymers; chemicals derived from acetylene; surfactants; dyes; pigments; carbonyl iron powders; textile auxiliaries; felts; filters; electronic, laboratory and agricultural chemicals; ceramics; materials science; roofing granules; building, photographic and quarry products; abrasives and adhesives; herbicides; ion exchange; soils and soil science; horticulutre; natural products. *Mailing Add:* 51 Hoot Owl Terrace Kinnelon NJ 07405-2409

KLAASEN, GENE ALLEN, b Holland, Mich, Feb 3, 41; m 63; c 3. MATHEMATICS. *Educ:* Hope Col, BA, 63; Univ Nebr, Lincoln, MA, 65, PhD(math), 68. *Prof Exp:* Asst prof math, Univ Nebr, Lincoln, 68-69; asst prof, 69-74, ASSOC PROF MATH, UNIV TENN, KNOXVILLE, 74- *Mem:* Am Math Soc; Math Asn Am. *Res:* Boundary value problems for ordinary differential equations. *Mailing Add:* Calvin Col 3201 Burton St Grand Rapids MI 49506

KLAASSEN, CURTIS DEAN, b Ft Dodge, Iowa, Nov 23, 42; m 64; c 2. PHARMACOLOGY, TOXICOLOGY. *Educ:* Wartburg Col, BA, 64; Univ Iowa, MS, 66, PhD(pharmacol), 68. *Prof Exp:* From instr to assoc prof, 68-77, PROF PHARMACOL & TOXICOL, MED CTR, UNIV KANS, 77- *Concurrent Pos:* Burroughs Wellcome scholar, toxicol, 82-87. *Honors & Awards:* Achievement Award, Soc Toxicol, 76. *Mem:* Soc Toxicol (pres, 90-91); Am Soc Pharmacol & Exp Therapeut; Sigma Xi. *Res:* Biliary excretion of drugs and toxicants. *Mailing Add:* Dept Pharmacol Univ Kans Med Ctr Kansas City KS 66103

KLAASSEN, DWIGHT HOMER, b Weatherford, Okla, Aug 15, 36; m 57; c 3. BIOCHEMISTRY. *Educ:* Tabor Col, BA, 58; Kans State Univ, MS, 61, PhD(biochem), 65. *Prof Exp:* Asst instr biochem, Kans State Univ, 63-64; assoc prof, 64-67, PROF CHEM, UNIV WIS-PLATTEVILLE, 67- *Concurrent Pos:* Coordr Coop Educ & Internships, 77-81; Assoc Dean, Student Affairs, 81-84, asst chancellor, Univ Relations, 84- *Mem:* Am Sci Affil; Sigma Xi. *Res:* Binding of sulfur-containing azo dyes related to dimethylaminoazobenzene to rat liver proteins; comparative study of mitochondrial proteins involving amino acid composition and solubility. *Mailing Add:* Univ Relations Univ Wis One Univ Plaza Platteville WI 53818

KLAASSEN, HAROLD EUGENE, b Hillsboro, Kans, Apr 18, 35; m 56; c 2. ECOLOGY, FISH BIOLOGY. *Educ:* Tabor Col, BA, 57; Kans State Univ, MS, 59; Univ Wash, Seattle, PhD(aquatic ecol), 67. *Prof Exp:* Fishery biologist, Univ Wash, Seattle, 59; ASSOC PROF BIOL, KANS STATE UNIV, 67- *Mem:* Am Fisheries Soc; Sigma Xi. *Res:* Fisheries management; fish distribution and production; aquaculture. *Mailing Add:* Div Biol Kans State Univ Manhattan KS 66506

KLABUNDE, KENNETH JOHN, b Madison, Wis, May 30, 43; m 67; c 3. ORGANIC CHEMISTRY, INORGANIC CHEMISTRY. *Educ:* Augustana Col, BA, 65; Univ Iowa, PhD(org chem), 69. *Prof Exp:* Res assoc org chem, Pa State Univ, 69-70; from asst prof to prof org chem, Univ NDak, 70-79; PROF CHEM & HEAD DEPT, KANS STATE UNIV, 79- *Concurrent Pos:* Grants, Res Corp, 70-72, Univ NDak, 70-72, Petrol Res Fund, 71-74 & 79-85; grants from NSF, 72-91, US Dept Energy, 74-80, Indust, 79-88, Army Res Off, 84-91 & Naval Res Off, 85-88. *Honors & Awards:* Sigma Xi Res Award, 77, 87. *Mem:* Am Chem Soc; Sigma Xi. *Res:* Organic-Inorganic: reactive intermediates such as metal atom chemistry, carbonmonosulfide chemistry, organometallic synthesis, thin film materials; metal oxide surface chemistry; use of metal vapors and other reactive species as synthetic reagents; adsorbents. *Mailing Add:* Dept Chem Kans State Univ Manhattan KS 66506

KLABUNDE, RICHARD EDWIN, b Pasadena, Calif, Oct 7, 48; m 68; c 4. CARDIOVASCULAR PHYSIOLOGY, PHARMACOLOGY. *Educ:* Pepperdine Univ, BS, 70; Univ Ariz, PhD(physiol), 75. *Prof Exp:* Am Heart Asn fel physiol, Univ Ariz, 75-76; fel, Nat Heart Lung & Blood Inst, Univ Calif, San Diego, 76, asst res physiologist, Pharmacol Div, 77-78; asst prof, Dept Physiol, WVa Univ Med Ctr, 78-85; SR GROUP LEADER, CARDIOVASC PHARMACOL, ABBOTT LABS, 85- *Concurrent Pos:* Vis assoc prof, Dept Biomed Sci, Univ Ill Col Med, Rockford, 86-; adj assoc prof, Dept Physiol, Chicago Med Sch, 89- *Mem:* Am Physiol Soc; Microcirculatory Soc. *Res:* Mechanisms of blood flow regulation, particularly in skeletal muscle during exercise and following short periods of ischemia; EDRF and blood flow regulation in vivo. *Mailing Add:* Dept Pharmacol D-46R AP9 Abbott Labs Abbott Park IL 60064

KLACSMANN, JOHN ANTHONY, b West New York, NJ, Oct 6, 21; m 44; c 3. ORGANIC CHEMISTRY. *Educ:* Yale Univ, BS, 42, MS, 44, PhD(org chem), 47. *Prof Exp:* Lab instr, Yale Univ, 42-44 & 46-47; dept supvr, Carbide & Carbon Chems Corp, Tenn, 44-46; res chemist, Fabrics & Finishes Dept, Newburgh Lab, E I Du Pont de Nemours & Co, Inc, 47-49, supvr, 49-53, res mgr, 53-56, res mgr, Marshall Lab, 56-58, lab dir, Exp Sta, Del, 58-59, asst dir res, Fabrics & Finishes Dept, Res Div, 60-64, asst dir mkt, Automative & Indust Prods, 64-66, gen sales mgr consumer prods, 66-67, dir mkt, Consumer Prod Div, 67-69, dir, Finishes Mkt Div, Fabrics & Finishes Dept, 69-71, dir, Finishes Div, 71-73, vpres & gen mgr, Fabrics & Finishes Dept, 73-75, vpres & gen mgr, Int Dept, 75-78; exec vpres, Cleansites Inc, 85-86; RETIRED. *Mem:* Am Chem Soc; AAAS; Sigma Xi. *Res:* Finishes; coated fabrics; polymers. *Mailing Add:* 5160 Bridlewood Ct Marsh Landing Ponte Vedra Beach FL 32082

KLAFTER, RICHARD D(AVID), b New York, NY, Aug 5, 36; m 59; c 2. CONTROLS, ROBOTICS. *Educ:* Mass Inst Technol, SB, 58; Columbia Univ, MSEE, 59, EE, 63; City Univ New York, PhD(optimal control), 69. *Prof Exp:* Lectr elec eng, City Col New York, 59-64 & 65-67; from asst prof to assoc prof elec eng, Drexel Univ, 67-84; PROF ELEC ENG, TEMPLE UNIV, 84- *Concurrent Pos:* Assoc proj dir, NASA-Am Soc Eng Educ, 71; proj dir cardiac pacemakers, NSF, 73-75. *Mem:* Sigma Xi; Inst Elec & Electronics Engrs; Soc Mfg Engr; Robotics & Automation Soc (vpres). *Res:* Mobile robots, optimal trajectory control of robots, tactile sensing, sensored and nonsensored robot navigation. *Mailing Add:* 607 Park Lane Wyncote PA 19095

KLAGER, KARL, b Vienna, Austria, May 15, 08; nat US; m 38; c 1. ORGANIC CHEMISTRY, PROPELLANTS & EXPLOSIVES. *Educ:* Univ Vienna, PhD(chem), 34. *Prof Exp:* Instr, Univ Vienna, 32-34; asst prod engr, Neuman Bros, Roumania, 34-36; res chemist, Chinoin, AG, Hungary, 36-38 & I G Farben, Ger, 39-48; res & develop chemist, Off Naval Res, 49; mgr solid propellant develop, Aerojet-Gen Corp, 50-67, mgr res & tech opers, 67-69, asst gen mgr, Solid Rocket Div, 69-71, vpres-dir opers, Aerojet Solid Propulsion Co, 71-73, TECH CONSULT, AEROJET-GEN CORP, 73- *Concurrent Pos:* Tech consult, US govt, 73-, Aerojet Gen Corp, 73-, Cordova Chem Co, 80-85, Martin Marietta Aerospace, 75-79, Teledyne Coast Proseal, 73-76, Coalcon Co, NY, 78-79, Artek Burling Game, Calif, 83, Brea Chem Co, 60-61, Ger Ministry Defense, 66-67, Aerochemie, Italy, 70, Dinippon Cellulose/Daicel, Japan, 64-86, Serv Poudre, France, 70, Australian Dept Defense Labs, Adelaide, 84, Essex Chem Co, 73-77. *Honors & Awards:* James H Wyld Propulsion Award, Am Inst Aeronaut & Astronaut, 72; Chem Pioneer Award, Am Inst Chemists, 78; Austrian Honor Cross, Sci & Art Medal, 89. *Mem:* Fel Am Inst Aeronaut & Astronaut; fel Am Inst Chemists; Am Chem Soc; Sigma Xi. *Res:* Acetylene derivatives; hydrogenation; cyclooctatetraene; organic preparative and catalytic chemistry; rocket fuels; chemical rocket propellants; synthesis and composition; manufacturing methods; combustion properties; safety characteristics. *Mailing Add:* 4110 Riding Club Lane Sacramento CA 95864-1649

KLAHR, CARL NATHAN, b Pittsburgh, Pa, July 3, 27; m 53; c 3. APPLIED PHYSICS. *Educ:* Carnegie Inst Technol, BS & MS, 48, MS & DSc(physics), 50. *Prof Exp:* Physicist, Res Labs, Westinghouse Elec Corp, 50-52; physicist & proj mgr, Nuclear Develop Corp Am, 52-57; proj mgr & sr physicist, Tech Res Group, Inc, 57-61; sr assoc, 61-67, PRES, FUNDAMENTAL METHODS ASSOCS, INC, 67- *Concurrent Pos:* Lectr, Columbia Univ, 53-58. *Mem:* Am Phys Soc; Am Nuclear Soc; Opers Res Soc Am; Inst Mgt Sci; Inst Elec & Electronics Engrs. *Res:* Hypervelocity physics; space vehicle technology; semiconductor device design and technology; solid state physics; operations research; electromagnetic radiation and quantum electronics. *Mailing Add:* Fundamental Methods Assoc Inc 678 Cedar Lawn Ave Lawrence NY 11559

KLAHR, PHILIP, b Mar 7, 46; US citizen. COMPUTER SCIENCE. *Educ:* Univ Mich, BS, 67; Univ Wis, MS, 69, PhD(comput sci), 75. *Prof Exp:* Sr analyst, Syst Develop Corp, 72-78; dir, Info Processing Systs Prog, Rand Corp, 78-86; VPRES PROF SERV, INFERENCE CORP, 86- *Mem:* Asn Comput Mach; Cognitive Sci Soc; Am Asn Artificial Intel; Inst Elec & Electronics Engrs. *Res:* Artificial intelligence research in knowledge-based systems; rule-based modeling, simulation, languages; explanation techniques, man-machine interfaces,; deductive question-answering, problem solving, learning, planning; abstraction, data-base management and cognitive modeling. *Mailing Add:* 3315 Colbert Ave Los Angeles CA 90066

KLAHR, SAULO, b Santander, Colombia, June 8, 35; US citizen; m 65; c 2. NEPHROLOGY. *Educ:* Col de Santa Librada, BS, 54; Nat Univ Colombia, MD, 59. *Prof Exp:* USPHS trainee, 61-63, from instr to assoc prof, 63-72, PROF MED & DIR RENAL DIV, SCH MED, WASHINGTON UNIV, 72- *Concurrent Pos:* Asst physician, Barnes Hosp, St Louis, Mo, 66-72, assoc physician, 72-75, physician, 75-; established investr, Am Heart Asn, 68-73; mem adv comt, Artificial Kidney, Chronic Uremia Prog, Nat Inst Arthritis & Metab Dis, 70-78; chmn med adv bd & mem bd dirs; Kidney Found Eastern Mo & Metro East, 73-74; mem fel comt, Nat Kidney Found, 77-81 & chmn, 80-81; assoc ed, J Clin Invest, 77-82. *Mem:* NY Acad Sci; Am Physiol Soc; Biophys Soc; Am Soc Clin Invest; Asn Am Physicians. *Res:* Hormonal control of ion transport across isolated membranes; studies on the functional and metabolic alterations produced by kidney disease; intermediary metabolism of the kidney. *Mailing Add:* Sch Med-Nephrology Wash Univ Sch Med 660 S Euclid Ave St Louis MO 63110

KLAIBER, FRED WAYNE, b Lafayette, Ind, Oct 7, 40; m 64; c 2. STRUCTURAL ENGINEERING. *Educ:* Purdue Univ, BSCE, 62, MSCE, 64, PhD(struct eng), 68. *Prof Exp:* Res engr, Caterpillar Tractor Co, 68; assoc prof, 68-80, PROF CIVIL ENG, IOWA STATE UNIV, 68- *Concurrent Pos:* ACI fel, 87. *Honors & Awards:* Raymond C Reese Res Prize, 78. *Mem:* Am Soc Civil Engrs; Am Railway Eng Asn; Am Concrete Inst. *Res:* Bridge rehabilitation and strengthening; prestressed folded plate theory; study of bridges behavior. *Mailing Add:* Dept Civil & Construct Eng Iowa State Univ Ames IA 50011

KLAIBER, GEORGE STANLEY, b Toledo, Ohio, Nov 20, 16; m 44; c 2. PHYSICS. *Educ:* Univ Buffalo, BA, 38; Univ Ill, MA, 41, PhD(physics), 43. *Prof Exp:* Asst physics, Univ Buffalo, 38-39; from asst to instr, Univ Ill, 39-44; res physicist, Gen Elec Co, 44-47; from asst prof to assoc prof physics, Univ Buffalo, 47-60; consult physicist, Wurlitzer Co, 60-67; RETIRED. *Concurrent Pos:* Independent consult physicist, 67- *Mem:* Am Phys Soc. *Res:* Acoustics and vibrations. *Mailing Add:* 2504 Colvin Blvd Tonawanda NY 14150

KLAIN, GEORGE JOHN, TOXICOLOGY, METABOLISM. *Educ:* Univ Ill, PhD(biochem), 59. *Prof Exp:* SUPVR CHEM RES, LETTERMAN ARMY INST RES, 74- *Mailing Add:* Dept Cutaneous Hazards Letterman Army Inst Res Presidio of San Francisco CA 94129-6800

KLAINER, ALBERT S, b Chelsea, Mass, Oct 29, 35; m 57; c 3. INTERNAL MEDICINE, INFECTIOUS DISEASES. *Educ:* Mass Inst Technol, BSc, 57; Tufts Univ, MD, 61. *Prof Exp:* Assoc prof med, Col Med, Ohio State Univ, 71-72; prof med & infectious dis, Sch Med, WVa Univ, 72-75; PROF MED, MED SCH, RUTGERS UNIV, 75-; CHMN DEPT MED, MORRISTOWN MEM HOSP, 75-; PROF CLIN MED, COL PHYSICIANS & SURGEONS, COLUMBIA UNIV, 80- *Concurrent Pos:* Grant infectious dis & internal med, New Eng Med Ctr Hosps, 63-64 & 65-66. *Honors & Awards:* Hull Award, AMA, 71, Morrissey Award, 72 & Bronze Medal, 72. *Mem:* Infectious Dis Soc Am; fel Am Col Physicians; Am Fedn Clin Res. *Res:* Scanning electron microscopy; infectious diseases, antibiotic pharmacology and AIDS research. *Mailing Add:* 315 W 70th St NEW York NY 10023

KLAINER, STANLEY M, b Chelsea, Mass, Apr 11, 30; m 52; c 6. INSTRUMENTATION, PHYSICAL & ANALYTICAL CHEMISTRY. *Educ:* Clark Univ, BA, 52, MA, 55, PhD, 59. *Prof Exp:* Res coordr qual control, Martin Div, Martin Marietta Corp, 57-59; chemist, Res Labs, Bendix Corp, 59-60; sr chemist, Nat Res Corp, 60-61; sr chemist, Tracerlab Div, Lab for Electronics, Inc, 61-63, spec projs mgr instruments, 63-64, from res & develop mgr to res mgr, 64-67; staff chemist, Block Eng, Inc, 67-70, mgr anal systs dept, 70-78; dep group leader geosci, Lawrence Berkeley Lab, 78-81, group mgr, 81-82; TECH CONSULT LASER SPECTROSCOPY, 78-; PRES, ST&E, INC, 82- *Concurrent Pos:* Joseph F Donnelly fel, 55. *Mem:* Am Chem Soc; Soc Appl Spectros; Soc Photo-Optical Engrs; Combustion Inst. *Res:* Development of optical, RF and microwave spectrometers and nuclear instruments; analysis of trace atmospheric constituents, ablation, encapsulation, special quality control analytical techniques and ultracentrifugation; interferometry; Raman, remote Raman, micro Raman and infrared spectroscopy; nuclear quadrupole resonance; micro particle analysis; chemical transport in natural systems; fiber optical chemical sensors for environmental and chemical measurements; medical, biological diagnostics and process control; surface chemistry and structure studies. *Mailing Add:* 2063 Sutton Way Henderson NV 89014

KLAMKIN, MURRAY S, b Brooklyn, NY, Mar 5, 21. MATHEMATICS. *Educ:* Cooper Union, BChE, 42; Polytech Inst Brooklyn, MS, 47. *Hon Degrees:* DrMath, Univ Waterloo, 83. *Prof Exp:* From instr to assoc prof math, Polytech Inst Brooklyn, 48-56; prin staff mathematician, Res & Adv Develop Div, Avco Corp, 56-62; prof interdisciplinary studies & res in eng, State Univ NY, Buffalo, 62-64; vis prof math, Univ Minn, 64-65; prin res scientist, Sci Lab, Ford Motor Co, 65-75; prof appl math, Univ Waterloo, 74-76; chmn dept, 76-81, PROF MATH, UNIV ALTA, 76- *Concurrent Pos:* Problem ed, Soc Indust & Appl Math Review, 59-; mem, Nat High Sch Math Contest Comt, 72-85; mem, Can Olympiad Comt, 74-83; coach, US Int Math Olympiad Team, 75-84. *Mem:* AAAS; Am Math Soc; Math Asn Am; Soc Indust & Appl Math. *Res:* Applied mathematics; geometry; heat conduction and radiation. *Mailing Add:* Univ Alta Ctr Acad Bldg Rm 698 Edmonton AB T6G 2G1 Can

KLAND, MATHILDE JUNE, b Chicago, Ill, June 6, 16; m 50; c 2. ENVIRONMENTAL CHEMISTRY, FOSSIL FUEL CHEMISTRY. *Educ:* Univ Chicago, BS, 39; Northwestern Univ, PhD(org chem), 48. *Prof Exp:* Sr chemist, Distillation Prods, Inc, NY, 39-41; sr chemist, Revere & La Ord Divs, 41-43; asst prof chem, Goucher Col, 48-49; E I du Pont fel, Ohio State Univ, 49-51; res assoc, Boston Univ, 51-52; writer & abstractor, 52-54 & 56-58; res assoc, Med Sch, Tufts Univ, 54-56; staff scientist, Lawrence Berkeley Lab, Univ Calif, 58-82; CONSULT, HEALTH EFFECTS CHEMS, 82- *Concurrent Pos:* Univ fel, Northwestern Univ, 44-45, Allied chem & dye fel, 45-46; Dupont postdoctoral fel, Ohio State Univ, 49-51; writer, environ chem-toxicol, 86- *Mem:* AAAS; Am Chem Soc; NY Acad Sci. *Res:* Reactions of styrene oxide; structure of styrene oxide dimers; molecular structure and spectra of organic compounds; abnormal bimolecular reactions of furfuryl chloride; grignard reactions; radiation chemistry of peptides; water pollution monitoring of bio-parameters and organics; toxic effects of pollutants, food additives, pesticides and drugs; use of structure-toxicity/carcinogenicity relationships in prediction of health effects; fossil fuel chemistry; teratogenicity of pesticides and other environmental pollutants. *Mailing Add:* 3678 Hastings Ct Lafayette CA 94549

KLANDERMAN, BRUCE HOLMES, b Grand Rapids, Mich, Feb 27, 38; m 60; c 3. ORGANIC CHEMISTRY,ENVIRONMENTAL SCIENCES. *Educ:* Calvin Col, AB, 59; Univ Ill, MS, 61, PhD(org chem),63. *Prof Exp:* Res chemist, 63-64, sr res chemist, 64-68, res assoc, 68-74, tech assoc, Synthetic Chem Div, 75, dept head, 76-77, govt regulations coordr, Gen Mgt, 78-80, dir, Environ Tech Serv, Kodak Park Div, 81-84; dir, Environ Tech Serv, 85-86, occup health lab, 87, DIR, ENVIRON TECH SERV, HEALTH ENVIRON LABS, EASTMAN KODAK CO, 88- *Mem:* Soc Photog Scientists & Engrs; Am Chem Soc. *Res:* Benzyne chemistry; liquid crystals; organic semiconductors; aliphatic diazonium chemistry; anthracene and triptycene chemistry. *Mailing Add:* 130 Bunker Hill Dr Rochester NY 14625

KLANFER, KARL, b Vienna, Austria, Oct 10, 04; m 46; c 1. CHEMICAL ENGINEERING. *Educ:* Inst Technol, Vienna, Austria, ChemE, 27. *Prof Exp:* Chief chemist, Wiener Leather Indust, Vienna, Austria, 27-34; supt, E Traub Co, Prague, Czech, 35-39; res chemist, Beardmore & Co Ltd, Acton, Ont, Can, 40-46; res dir, Cortume Carioca, Brazil, 46-50; res dir, AR Clarke & Co Ltd, Toronto, Can, 50-56; tech dir, A C Lawrence Leather Co, Peabody, 56-61; RETIRED. *Concurrent Pos:* Lectr, Univ Exten, Vienna, 30-34; consult chem eng, 61- *Mem:* Sr mem Am Chem Soc; Am Leather Chem Asn; fel Chem Inst Can; Royal Soc Chem; Am Leather Chem Asn; Can Soc Chem Eng; Soc Leather Technologists, London. *Res:* Chemistry and technology of leather and tanning; author or coauthor of 40 publications; microanalysis; chemistry of chrom-complexes; chemistry of fats & oils; instrumental process control. *Mailing Add:* 18 Colgate Rd Marblehead MA 01945

KLAPMAN, SOLOMON JOEL, b Chicago, Ill, Apr 25, 12; m 37; c 2. MATHEMATICAL PHYSICS. *Educ:* Univ Chicago, BS, 32, PhD(physics), 40; Univ Mich, MS, 35. *Prof Exp:* Math physicist, Jensen Mfg Co, Ill, 37-40; physicist, Frankford Arsenal, Pa, 40-41, Utah Radio Prods Co, Ill, 41-46, Indust Res Prods Co, 47-49, E I Guthman & Co, 49-50 & Admiral Corp, 51-52; sr electronics engr, Chicago Midway Labs, 52-55 & Lockheed Missile Systs, 55-56; sr scientist, Hughes Aircraft Co, Culver City, 56-73; adj prof civil eng, Univ Southern Calif, 73-84. *Concurrent Pos:* Instr evening div, Ill Inst Technol, 42-50; lectr, Roosevelt Col, 50-55, Mt St Mary Col, 63-64 & Immaculate Heart Col, 64-65. *Mem:* Am Phys Soc; Acoust Soc Am; Am Asn Physics Teachers. *Res:* Acoustics; magnetism; infrared. *Mailing Add:* 7842 E Lakeview Trail Orange CA 92669

KLAPPER, CLARENCE EDWARD, embryology; deceased, see previous edition for last biography

KLAPPER, DAVID G, b New York, NY, Mar 15, 44. MICROBIOLOGY, IMMUNOLOGY. *Educ:* Tulane Univ, BS, 65; Univ Fla, PhD(microbiol & immunol), 72. *Prof Exp:* Prof microbiol & immunol, Rockefeller Univ, Univ Tex Southwestern Med Sch; PROF MICROBIOL & IMMUNOL, MED SCH, UNIV NC. *Mailing Add:* Dept Microbiol & Immunol Med Sch Univ NC CB No 7290 Chapel Hill NC 27599

KLAPPER, GILBERT, b Wichita, Kans, Sept 29, 34; m 59; c 3. PALEONTOLOGY. *Educ:* Stanford Univ, BS, 56; Univ Kans, MS, 58; Univ Iowa, PhD(geol), 62. *Prof Exp:* Paleontol, Shell Oil Co, La, 58-59; NSF fel & res assoc, Ill State Geol Surv, 62-63; res paleontol, Res Ctr, Pan Am Petrol Corp, 63-68; assoc prof, 68-73, PROF GEOL, UNIV IOWA, 73- *Concurrent Pos:* Vis prof, Ore State Univ, 78. *Mem:* Paleont Res Inst; Paleont Soc; Soc Econ Paleont & Mineral; Brit Paleont Asn; Ger Paleont Soc. *Res:* Micropaleontology, especially research in conodonts; biostratigraphy, Silurian, Devonian and Mississippian. *Mailing Add:* Dept Geol Univ Iowa Iowa City IA 52242

KLAPPER, JACOB, b Ulanow, Poland, Sept 17, 30; US citizen; m 58; c 2. COMMUNICATIONS SYSTEMS. *Educ:* City Col New York, BEE, 56; Columbia Univ, MS, 58; NY Univ, EngScD, 65. *Prof Exp:* Elec engr, Columbia Broadcasting Syst, 52-56; lectr elec eng, City Col New York, 56-59; proj engr, Fed Sci Corp, 59-60; sr mem tech staff, Adv Commun Labs, Radio Corp Am, 60-65, sr proj mem tech staff, 65-67; assoc prof elec eng, Newark Col Eng, 67-71; PROF, NJ INST TECHNOL, NEWARK, 71-, CHMN ELEC ENG DEPT, 86- *Concurrent Pos:* Consult, various orgn. *Honors & Awards:* Region One Award, Inst Elec & Electronics Engrs, 86. *Mem:* Sr mem Inst Elec & Electronics Engrs; Commun Soc. *Res:* Electrical communication; systems and techniques with emphasis on phase-locked loops and FM systems. *Mailing Add:* NJ Inst Technol 323 King Blvd Newark NJ 07102

KLAPPER, MICHAEL H, b Berlin, Ger, June 10, 37; US citizen; m 60; c 2. BIOCHEMISTRY, BIOPHYSICS. *Educ:* Harvard Univ, AB, 58; Univ Rochester, MS, 59; Univ Calif, PhD(biochem), 64. *Prof Exp:* Fel chem, Northwestern Univ, 64-66; from asst prof to assoc prof, 66-86, PROF CHEM,

OHIO STATE UNIV, 86- *Concurrent Pos:* Co-dir, Nat Ctr Sci Teaching & Learning, 90. *Mem:* Biophys Soc; Am Soc Biol Chemists; Protein Soc; Am Chem Soc; AAAS. *Res:* Physical biochemistry of enzyme structure; theoretical and experimental studies of enzyme catalysis; long range electron transfer in polypeptides and proteins; fast radical reactions. *Mailing Add:* Chem Dept Ohio State Univ 120 W 18th Ave Columbus OH 43210

KLAPPROTH, WILLIAM JACOB, JR, b Springfield, Ohio, Aug 2, 20; m 45; c 2. POLYMER CHEMISTRY. *Educ:* Wittenberg Col, AB, 42; Univ Chicago, MS & PhD(org chem), 49. *Prof Exp:* Res chemist, Stamford Res Labs, Am Cyanamid Co, 49-54, group leader, Warners Plant, 54-57, group leader catalytic & gen process improv, Bridgeville Plant, 57-59, mgr acids & miscellaneous chems develop, 59-63; tech dir, Bridgeville Plant, Koppers Co, Inc, 63-67, sr proj scientist, Monroeville Res Ctr, Arco Polymers Inc, 67-70, sr group mgr, 70-78, prin scientist, Arco Chem Co, 79-81; RETIRED. *Concurrent Pos:* Consult, 81-83. *Mem:* Am Chem Soc. *Res:* Research and development in polymers, especially polystyrene, high pressure polyethylene and related copolymers. *Mailing Add:* 3576 Logans Ferry Rd Murrysville PA 15668

KLARFELD, JOSEPH, b Poland, Dec 22, 35; c 2. THEORETICAL PHYSICS. *Educ:* Israel Inst Technol, BSc, 59, MSc, 62; Yeshiva Univ, PhD(physics), 69. *Prof Exp:* Instr physics, Israel Inst Technol, 58-61; res assoc, Israel Atomic Energy Comn, 61-63; asst prof, 69-73, asst chmn dept, 69-76, dep chmn dept, 76-78, ASSOC PROF PHYSICS, QUEENS COL, NEW YORK, 74-; ASSOC PROF, GRAD CTR, CITY UNIV NY, 80- *Concurrent Pos:* Vis assoc prof, Tel Aviv Univ, 78-79; res fel, City Univ New York, 78-79. *Mem:* Asn Math Physics; Am Phys Soc; NY Acad Sci; Int Soc Gen Relativity & Gravitation. *Res:* Quantization of general relativity; foundations of quantum field theory; relativistic astrophysics. *Mailing Add:* Dept Physics Queens Col 65-30 Kissena Blvd Flushing NY 11367-0904

KLARMAN, HERBERT E, b Chmielnick, Poland, Dec 21, 16. MEDICAL ADMINISTRATION. *Educ:* Columbia Univ, AB, 39; Univ Wis, MA, 46, PhD, 46. *Prof Exp:* asst dir, Hosp Coun Greater NY, 49-51, assoc dir, 52-62; mem, Health Serv Res Study Sect, NIH, 62-66; mem fac, Johns Hopkins Univ, 62-69, prof pub health admin & political econ, 65-69; prof environ med & commun health, Downstate Med Ctr, 69-70; prof pub admin, Grad Sch Pub Admin, New York Univ, 70-82; NY State Health Adv Coun, 76-83; RETIRED. *Concurrent Pos:* asst dir, NY State Hosp Study, 48-49; Guggenheim fel, 48-49. *Honors & Awards:* Norman A Welch Award, 65. *Mem:* Fel AAAS; Inst Med Nat Acad Sci; Pub Health Asn; Am Econ Asn. *Mailing Add:* One E University Pkwy Baltimore MD 21218

KLARMAN, KARL J(OSEPH), b Scotia, NY, Mar 18, 22; c 2. ELECTRICAL ENGINEERING. *Educ:* Union Col, BS, 44; Columbia Univ, MS, 47. *Prof Exp:* Jr engr, Carl L Norden, Inc, 44; lectr elec eng, Union Univ, NY, 45-46; proj engr, Avion Instrument Corp, 46-47; proj engr, Eclipse-Pioneer Div, Bendix Aviation Corp, 47-51; plant mgr, Electro Tec Corp, 51-53, dir eng, 53-55, vpres, 55-57; sect head, Sanders Assocs, 57-59; chief prod engr, Precision Prod Dept, Northrop, 59-61; eng scientist, Defense Electronic Prod Div, Radio Corp Am, 61-74, eng scientist, Govt & Com Div, RCA Corp, 71-74; CONSULT, 74- *Mem:* Sigma Xi; Nat Soc Prof Engrs; Inst Elec & Electronics Engrs. *Res:* Gyroscopic instruments, design, development and production; design, development, test and production of inertial and other electro-mechanical instruments and systems. *Mailing Add:* 20 Kipling St Nashua NH 03062

KLARMAN, WILLIAM L, b Moweaqua, Ill, Sept 21, 35; c 3. RESEARCH ADMINISTRATION. *Educ:* Eastern Ill Univ, BS, 57; Univ Ill, MS, 60, PhD(plant path), 62. *Prof Exp:* From asst prof to prof plant path, Univ Md, 62-80; dept head plant path, Okla State Univ, 80-84; dept head plant path, 84-90, INTERIM VCHANCELLOR RES, NC STATE UNIV, 90- *Concurrent Pos:* Fulbright prof, 74-75; counr, Am Phytopath Soc, 76-80; plant path, USDA, 80. *Mem:* Am Phytopath Soc; AAAS. *Mailing Add:* Holladay Hall Box 7003 NC State Univ Raleigh NC 27695-7003

KLARMANN, JOSEPH, b Berlin, Ger, Jan 16, 28; m 57; c 2. PHYSICS. *Educ:* Hebrew Univ, Israel, MSc, 54; Univ Rochester, PhD(physics), 58. *Prof Exp:* Res assoc & instr, Univ Rochester, 57-58, instr, 58-61; from asst prof to assoc prof, 61-74, PROF PHYSICS, WASH UNIV, 74- *Mem:* Int Astron Union; AAAS; Am Phys Soc. *Res:* Cosmic ray astrophysics. *Mailing Add:* Dept Physics Wash Univ St Louis MO 63130

KLASINC, LEO, b Zagreb, Yugoslavia, May 20, 37; m 61; c 2. MOLECULAR SPECTROSCOPY, ATMOSPHERIC CHEMISTRY. *Educ:* Univ Zagreb, dipl, 60, PhD(chem), 63; Ruder Boskovic Inst, Zagreb, PhD(phys chem), 68. *Prof Exp:* Postdoctoral fel theoret & radiation chem, Nuclear Res Ctr, Karlsruhe, 66-68,; from asst prof to assoc prof, 71-79, PROF FAC SCI, UNIV ZAGREB, 79- *Concurrent Pos:* Res asst phys chem, Ruder Boskovic Inst, Zagreb, 61-63 & 64-68, res assoc, 68-72, sr res assoc, 72-77, sr scientist, 77-; vis prof, dept chem, La State Univ. *Mem:* Am Phys Soc; Int Soc Quantum Biol; World Asn Theoretical Org Chemist; Europ Photochemical Asn. *Res:* Electronically excited states of molecules and ions spectroscopy, quantum chemistry, photochemistry and chemical kinetics; photochemical processing in the atmosphere, photosmog and troposphere ozone formation. *Mailing Add:* Dept Chem La State Univ Baton Rouge LA 70803

KLASNER, JOHN SAMUEL, b Flint, Mich, June 22, 35; m 64; c 4. STRUCTURAL GEOLOGY, APPLIED GEOPHYSICS. *Educ:* Mich State Univ, BS, 57, MS, 64; Mich Technol Univ, PhD(geol), 72. *Prof Exp:* Geophys engr, Geophys Serv Inc, 57-62; geophysicist, Standard Oil Co Calif, 64-69; from asst prof to assoc prof geol, 72-79, chmn dept, 74-78, PROF GEOL, WESTERN ILL UNIV, 80-; GEOLOGIST ECON GEOL, US GEOL SURV, 72- *Concurrent Pos:* Assoc dir, Honors Prog, Western Ill Univ, 90- *Mem:* Sigma Xi; Geol Soc Am; Soc Explor Geophysicists; Geol Asn Can; Am Geophys Union; Nat Asn Geol Teachers. *Res:* Economic and Precambrian geology of Northern Michigan; structural geology and tectonics of Northern Michigan and Wisconsin; geophysical archeology; regional geophysics North and South Dakota. *Mailing Add:* Dept Geol Western Ill Univ Macomb IL 61455

KLASS, ALAN ARNOLD, b Russia, Aug 13, 07; nat US; m 39; c 2. ANATOMY, SURGERY. *Educ:* Univ Man, BA, 27, MD, 32; FRCS(E), 37; FRCS(C), 43. *Hon Degrees:* LLD, Univ Man, 74. *Prof Exp:* Assoc prof surg & assoc prof anat, Univ Man, 37-; RETIRED. *Concurrent Pos:* Assoc surgeon, Winnipeg Gen Hosp, 38- & Misericordia Gen Hosp, 38-; pres, Medico-Legal Soc Man, 63-65; chmn, Govt Comt Cent Drug Purchasing & Distrib. *Mem:* Can Surg Soc (past secy); Can Fedn Biol Soc; Can Asn Anat; Medico-Legal Soc (pres); hon mem Can Soc Vascular Surgeons. *Res:* Pilo-nidal sinus; acute mesenteric occlusion; fatigue fractures tibia; engineering and scientific principles; professionalism and professional integrity. *Mailing Add:* 594 Oak St Winnipeg MB R3M 3R6 Can

KLASS, DONALD LEROY, b Waukegan, Ill, July 23, 26; c 3. ORGANIC CHEMISTRY. *Educ:* Univ Ill, BS, 51; Harvard Univ, AM, 52, PhD(org chem), 55. *Prof Exp:* Res chemist, Standard Oil Co, Ind, 54-55 & Am Can Co, 56-59; dir process & prod res div, Pure Oil Co, Ill, 59-65; asst dir basic res, 65-69, asst res dir, 69-76, dir basic res, 76-77, dir eng & sci res, 77-79, asst vpres, 79-80, VPRES EDUC, INST GAS TECHNOL, ILL INST TECHNOL, 80- *Concurrent Pos:* consult energy & chem prod; tech ed, J Solar Energy Eng, 83-; bd mem & pres, Biomass Energy Res Asn, 85-; pres, Entech, 88- *Honors & Awards:* Nat Lubricating Grease Inst Award, 66; Mem Award, Foote Chem Co, 66; Richard A Glenn Award, Bituminous Coal Res, Inc, 76. *Mem:* Am Chem Soc; Biomass Energy Res Asn (pres, 85-); Am Inst Chem Eng. *Res:* Research and education administration; gas processing; petrochemicals; refining; catalysis; fermentation; waste treatment; pollution control; gasification-liquefaction of wastes, biomass & fossil fuels; energy supplies. *Mailing Add:* 25543 W Scott Rd Barrington IL 60010

KLASS, MICHAEL R, b July 31, 49. EXPERIMENTAL BIOLOGY. *Educ:* Univ Wis-Madison, BS, 71; Univ Wyo, Laramie, PhD(cell biol), 74. *Prof Exp:* Res asst, Dept Zool, Univ Wyo, 72-74; res assoc, Dept Molecular, Cellular & Develop Biol, Univ Colo, 74-79; from asst prof to assoc prof biol, Univ Houston, 79-86; lab head biochem genetics, 86-90, INTERIM DIR, CORP MOLECULAR BIOL, ABBOTT LABS, 90- *Concurrent Pos:* Res career develop award, NIH, 84; postdoctoral fel develop genetics, Univ Colo, 74-79. *Res:* Aging; numerous publications. *Mailing Add:* Abbott Labs 93D AP9A-3 Abbott Park IL 60064

KLASSEN, DAVID MORRIS, b Clovis, NMex, June 15, 39; m 65; c 2. INORGANIC CHEMISTRY, SPECTROSCOPY. *Educ:* Univ Tex, El Paso, BS, 61; Univ NMex, PhD(phys chem), 67. *Prof Exp:* Teaching asst, Univ NMex, 61-62; NATO res fel, Inst Phys Chem, Frankfurt, WGer, 66-67; res assoc, Univ NC, Chapel Hill, 67-69; from asst prof to assoc prof, 69-77, PROF CHEM, MCMURRY COL, 77- *Mem:* Am Chem Soc; Sigma Xi; Sci Res Soc Am. *Res:* Synthesis, bonding and electronic structure of transition-metal complexes; luminescence of ruthenium and osmium complexes. *Mailing Add:* Dept Chem McMurry Univ Abilene TX 79697

KLASSEN, J(OHN), b Waterloo, Ont, Jan 9, 28; m 50; c 4. CHEMICAL ENGINEERING. *Educ:* Queen's Univ, Ont, BSc, 48, MSc, 49; Univ Wis, PhD(chem eng), 54. *Prof Exp:* Jr res officer, chem eng, Nat Res Coun Can, 49-53, asst res officer, 53-55; res engr, 55-57, res supvr, 57-59, tech supt, 59-62, tech mgr, 62-66, asst works mgr, 66-69, mgr, Cent Res Lab, Maitland, 69-71, tech mgr, 71-75, SR ENG SPECIALIST, DU PONT CAN, 75- *Mem:* Can Soc Chem Engrs; fel Chem Inst Can. *Res:* Commercial processing of industrial chemicals. *Mailing Add:* Champaign Reg Col Saint Lawrence Campus 790 Neree-Tremblay St Santa Fe PQ G1V 4K2 Can

KLASSEN, LYNELL W, b Gossel, Kans, Jan 24, 47; m 67; c 4. RHEUMATOLOGY, TRANSPLANT IMMUNOLOGY. *Educ:* Tabor Col, Hillsboro, Kans, AB, 69; Univ Kans, Kansas City, MD, 73. *Prof Exp:* Resident internal med, Univ Iowa Hosps & Clins, 73-75; res assoc immunol, Arthritis & Rheumatism Br, NIH, 75-77; chief resident internal med, Univ Iowa Hosps & Clins, 77-78, asst prof, 78-82; assoc prof rheumatology & immnuol, 82-90, PROF & VCHMN INTERNAL MED, UNIV NEBR MED CTR, 90-; CHIEF ARTHRITIS SERV RHEUMATOLOGY, OMAHA VET ADMIN, 82- *Concurrent Pos:* Chmn, Sci Rev Comt, Nat Inst Alchol Abuse, Alcoholism, 89-; mem, Educ Coun, Am Col Rheumatology, 89- *Mem:* Am Col Physicians; Am Asn Immunol. *Res:* Mechanisms of hematopoietic allograft rejection; pathophysiology of graft-versus-host disease; use of cytotoxic therapy in non-malignant diseases. *Mailing Add:* 600 S 42nd St Omaha NE 68198-3332

KLASSEN, NORMAN VICTOR, b Winnipeg, Man, Nov 6, 33; m 61; c 3. RADIATION CHEMISTRY, DOSIMETRY. *Educ:* McGill Univ, BSc, 54, PhD(chem), 57; Univ Col, London, PhD(chem), 61. *Prof Exp:* Fel, Nat Res Coun Can, 61-63; fel phys chem, Mellon Inst, 63-66; MEM RES STAFF, INST NAT MEASUREMENT STANDARDS, NAT RES COUN CAN, 66- *Mem:* Royal Soc Chem; fel Chem Inst Can; Radiation Res Soc. *Res:* Radiation chemistry; pulse radiolysis; dosimetry. *Mailing Add:* Inst Nat Measurement Standards Nat Res Coun Can M-35 Ottawa ON K1A 0R6 Can

KLASSEN, RUDOLPH WALDEMAR, b Hanna, Alta, Sept 30, 28; m 67; c 4. GEOLOGY. *Educ:* Univ Alta, BSc, 59, MSc, 60; Univ Sask, PhD(geol), 65. *Prof Exp:* RES SCIENTIST QUATERNARY GEOLOGY, GEOL SURV CAN, 65- *Concurrent Pos:* Lectr, Univ of Calgary, 74-86. *Mem:* Geol Asn Can; Am Quaternary Asn. *Res:* Quaternary stratigraphy and geomorphology; reports and maps on Quaternary geology of Manitoba, Southern Saskotchewom, Northern Northwest Territoris and Southern Yukon published mainly by geological survey of Canada; current studies of tertiory and Quaternary geomorshology and stratigreshy of Cypress Lake and Wood Mountain alias Southwestern Saskotchewan. *Mailing Add:* Geol Surv Can 3303 33rd St NW Calgary AB T2L 2A7 Can

KLASSEN, WALDEMAR, b Vauxhall, Alta, Dec 28, 35; US citizen; m 64; c 1. ENTOMOLOGY, GENETICS. *Educ:* Univ Alta, BSc, 57, MSc, 59; Univ Western Ont, PhD(zool), 63. *Prof Exp:* USPHS res assoc zool, Univ Ill, 63-65; res geneticist, Metab & Radiation Res Lab, Entom Res Div, 65-67, leader insect physiol & metab sect, 67-70, asst to dep adminr plant sci & entom, 70-72, STAFF SCIENTIST PEST MGT, AGR RES SERV, USDA, 72- *Honors & Awards:* Dipl & Medal, Int Cong Plant Protection, 75. *Mem:* AAAS; Entom Soc Am; Am Chem Soc; Genetics Soc Can; Sigma Xi. *Res:* Dispersal of mosquitoes; inheritance of resistance to insecticides in mosquitoes; cytogenetics of mosquitoes; chemosterilization of insects; program planning in entomological research; insect population dynamics. *Mailing Add:* 10900 Fleetwood Dr Beltsville MD 20705

KLATSKIN, GERALD, internal medicine; deceased, see previous edition for last biography

KLATT, ARTHUR RAYMOND, b Hamilton, Tex, June 10, 43; m 80; c 5. PLANT BREEDING. *Educ:* Tex Tech Univ, BS, 66; Colo State Univ, MS, 68, PhD(plant breeding & genetics), 69. *Prof Exp:* Plant breeder, 69-79, assoc dir, Wheat Prog Int Maize & Wheat Improv Ctr, 79-87, ASST DEAN, INT PROG, OKLA STATE UNIV, 88- *Mem:* Am Soc Agron; Crop Sci Soc Am; Am Genetic Asn; AAAS; Weed Sci Soc Am. *Res:* Genetics and environmental factors affecting drought tolerance; genetics and environmental influence on dormancy, vernalization and winterhardiness; incorporation of horizontal resistance; breeding adapted high yielding winter and spring wheat varieties. *Mailing Add:* Int Prog Div Agr Okla State Univ 139 Agr Hall Stillwater OK 74078

KLATT, GARY BRANDT, b Milwaukee, Wis, Nov 22, 39; m 64; c 2. MATHEMATICS. *Educ:* Case Western Reserve Univ, BS, 61; Univ Wis, MS, 62, PhD(math), 69. *Prof Exp:* Instr math, Marquette Univ, 64-65; from asst prof to assoc prof, 67-75, PROF MATH, UNIV WIS, WHITEWATER, 75- *Mem:* Math Asn Am; Nat Coun Teachers Math. *Res:* Theory of rings and modules. *Mailing Add:* 145 N Fremont St Whitewater WI 53190

KLATT, LEON NICHOLAS, b Underhill, Wis, Aug 28, 40; m 61; c 2. CHEMICAL INSTRUMENTATION. *Educ:* Univ Wis, Oshkosh, BS, 62; Univ Wis, Madison, PhD(anal chem), 67. *Prof Exp:* Chemist, Dow Chem Co, Mich, 62-63; asst, Univ Wis-Madison, 63-66; asst prof chem, Southern Ill Univ, 67-69 & Univ Ga, 69-74; RES STAFF MEM, OAK RIDGE NAT LAB, 74- *Mem:* Am Chem Soc; Sigma Xi; AAAS; SAS. *Res:* Instrumentation for remote analyses; on-line control of instruments with small computers; multi variable data reduction systems; application of small computers to analytical problems; process monitor and control systems. *Mailing Add:* Oak Ridge Nat Lab PO Box 2008 Oak Ridge TN 37831-6306

KLATTE, EUGENE, b Indianapolis, Ind, Mar 19, 28; m 50; c 4. RADIOLOGY. *Educ:* Ind Univ, AB, 49, MD, 52. *Prof Exp:* Resident radiol, Univ Calif, 55-57; from instr to assoc prof, Sch Med, Ind Univ, 58-62; prof & chmn dept, Sch Med, Vanderbilt Univ, 62-71; prof, 71-80, DISTINGUISHED PROF & CHMN DEPT, SCH MED, IND UNIV, INDIANAPOLIS, 80- *Concurrent Pos:* Picker res scholar, Sch Med, Ind Univ, 57-58; consult, Vet Admin Hosp, Nashville, Tenn, 62-71; clin prof, Meharry Med Col, 64-71. *Mem:* Am Col Radiol; Asn Univ Radiol; Radiol Soc NAm; Soc Pediat Radiol; AMA. *Res:* Diagnostic radiology, especially cardiovascular radiology. *Mailing Add:* Ind Univ Med Ctr UH X-98 926 W Michigan St Indianapolis IN 46202

KLATZO, IGOR, NEUROPATHOLOGY. *Prof Exp:* HEAD CEREBROVASC PATHOPHYSIOL SECT, NAT INST NEUROL DISORDERS & STROKE, NIH, 91- *Mailing Add:* NIH Nat Inst Neurol Disorders & Stroke Cerebrovasc Pathophysiol Br Bldg 36 Rm 4004 9000 Rockville Pike Bethesda MD 20892

KLAUBER, MELVILLE ROBERTS, b San Diego, Calif, Aug 9, 33; m 53; c 6. PUBLIC HEALTH & EPIDEMIOLOGY. *Educ:* Stanford Univ, AB, 54, MS, 56, PhD(statist), 64. *Prof Exp:* From asst prof to prof family & community med & chief, Div Biostatist, Col Med, Univ Utah, 67-78. *Concurrent Pos:* Adj assoc prof math, Univ Utah, 72-76, adj prof math, 76-78; prin statistician, Sch Med, Univ Calif, La Jolla, 78-81; adj prof community & family med, 81- *Mem:* Inst Math Statist; Am Statist Asn; Biomet Soc. *Res:* Biostatistics; statistical methods for the medical sciences, especially epidemiology. *Mailing Add:* Dept Community & Family Med Univ Calif Sch Med Code 0816 La Jolla CA 92093-0816

KLAUBERT, DIETER HEINZ, b Ger, Dec 15, 44; Can citizen; m 69; c 2. MEDICINAL CHEMISTRY. *Educ:* Univ Alta, BSc, 67; Mass Inst Technol, PhD(org chem), 71. *Prof Exp:* Fel org chem, Univ Calif, Berkeley, 71-73; RES CHEMIST, WYETH LABS INC, AM HOME PROD CORP, 73- *Mem:* Am Chem Soc; The Chem Soc; Sigma Xi. *Res:* Synthesis of novel compounds of pharmaceutical interest. *Mailing Add:* RD Five Box 402 Flemington NJ 08822

KLAUDER, JOHN RIDER, b Reading, Pa, Jan 24, 32; m 53, 80; c 5. THEORETICAL PHYSICS, APPLIED MATHEMATICS. *Educ:* Univ Calif, Berkeley, BS, 53; Stevens Inst Technol, MS, 56; Princeton Univ, MA, 57, PhD(physics), 59. *Prof Exp:* Head theoret physics dept, Bell Tel Labs, 66-67 & 69-71, head solid state spectros dept, 71-76, mem tech staff, 53-88; PROF, DEPT PHYSICS & MATH, UNIV FLA, 88- *Concurrent Pos:* Vis assoc prof, Univ Bern, 61-62; adj prof, Rutgers Univ, 65; prof, Syracuse Univ, 67-68; vis prof, Univ Bern, 80, Gakushuin Univ, 82 & 84, Univ Trento, 88, Imperial Col, 88 & 90. *Mem:* Fel Am Phys Soc; fel AAAS. *Res:* Solid state physics; quantum optics; quantum field theory; fundamentals of quantum theory. *Mailing Add:* Dept Physics & Math Univ Florida Gainesville FL 32611

KLAUNIG, JAMES E, b Newark, NJ, May 27, 51. ENVIRONMENTAL TOXICOLOGY, CHEMICAL CARCINOGENESIS. *Educ:* Ursinus Col, BS, 73; Montclair State Col, MA, 76; Univ Md, PhD(path), 80. *Prof Exp:* Lab scientist, dept path, Univ Md, Baltimore, 76-80; from instr to asst prof, 80-86, ASSOC PROF PATH, DEPT PATH, MED COL OHIO, TOLEDO, 86-; ASSOC PROF, DEPT PHARMACOL, 87- *Concurrent Pos:* Adj fac mem,

W Alton Jones Cell Sci Ctr, Lake Placid, NY, 81; mem grad fac, Med Col, Ohio Grad Sch, 82-; prin investr on grants, US Environ Protection Agency, NIH & US Army, 82-; adj prof, Ctr Photochemical Sci, Dept Chem, Bowling Green State Univ, Ohio; vis scientist, Chem Indust Inst Toxicol, Res Triangle Park, NC, 90-91. *Mem:* Am Asn Cancer Res; Soc Toxicol Pathologists; Soc Toxicol; Sigma Xi; Tissue Cult Asn; Am Col Toxicol. *Res:* Environmental toxicology and carcinogenesis; liver tumor promotion; hepatic carcinogenesis; cell pathology; liver cell isolation and tissue culture; hepatotoxicology. *Mailing Add:* Dept Path Health Educ Bldg Rm 263 Med Col Ohio 3000 Arlington Ave Toledo OH 43699

KLAUS, E(LMER) ERWIN, b Neffsville, Pa, Apr 19, 21; m 45; c 2. PETROLEUM REFINING, TRIBOLOGY. *Educ:* Franklin & Marshall Col, BS, 43; Pa State Univ, MS, 46, PhD(chem), 52. *Prof Exp:* Res asst, 43-47, instr petrol chem, 47-52, from asst prof to prof, 52-83, EMER PROF CHEM ENG, PA STATE UNIV, 83- *Concurrent Pos:* Consult, US Dept Defense, 55; mem mechanisms comt, Off Naval Res Mech Failures Prevention Group; coun mem, Nat Bur Standards Mech Failures Prev Group, 70-, coun chmn, 76-80; assoc ed, J Lubrication Technol, 68-80; mem nat res coun comt, US Army Basic Res, 77-80; Fenske fac fel chem eng, 79-83, emer fel, 83-; consult, Nat Inst Standards & Technol, 83-89. *Honors & Awards:* Nat Award, Am Soc Lubrication Engrs, 76; Capt Alfred E Hunt Award, Am Soc Lubrication Engrs, 80; Mayo D Hersey Award, Am Soc Mech Engrs, 82, Innovative Res Award, 88; Al Sonntag Award, Soc Tribologists & Lubrication Engrs, 91. *Mem:* Am Chem Soc; fel Soc Tribologist & Lubrication Engrs; fel Am Inst Chem; Am Inst Chem Engrs; fel Am Soc Mech Engrs; Am Soc Testing Mats. *Res:* Boundary lubrication; metal corrosion; petroleum refining; physical and chemical properties of lubricants; use of surfactants and polymers in enhanced oil recovery; surface chemical effects; high temperature tribology; tribology of ceramics; vapor phase lubrication. *Mailing Add:* Dept Chem Eng 108 Fenske Lab Pa State Univ University Park PA 16802

KLAUS, EWALD FRED, JR, b Needville, Tex, Oct 22, 28; m 64. ZOOLOGY, ENTOMOLOGY. *Educ:* Univ Tex, BA, 52, MA, 58; Tex A&M Univ, PhD(entom), 65. *Prof Exp:* From asst prof to assoc prof, 64-74, PROF BIOL, ETEX STATE UNIV, 74- *Concurrent Pos:* Fac res grant, 67-68, NSF fel, Col Sci Improv Prog, 71-72. *Res:* Biology and taxonomy of mosquitoes; genetics of insecticide resistance in cotton insects; insecticide residue studies; food sanitation; aquatic ecology and water quality studies. *Mailing Add:* Dept Biol Sci ETex State Univ Commerce TX 75428

KLAUS, RONALD LOUIS, b Brooklyn, NY, Apr 23, 40; m 68; c 2. CHEMICAL ENGINEERING. *Educ:* Rensselaer Polytech Inst, BChE, 60, PhD(chem eng), 67. *Prof Exp:* Res engr combustion, Jet Propulsion Lab, 63-65 & 63-70; asst prof chem eng, Univ Pa, 70-76; MINISTER. *Concurrent Pos:* Resident res assoc fel, Jet Propulsion Lab, 68-70. *Mem:* Am Inst Chem Engrs. *Res:* Computer-aided design in chemical engineering; thermodynamics of liquid mixtures; computer-based estimation of thermodynamic properties; numerical methods; design and implementation of computer-based problem-oriented languages for engineering. *Mailing Add:* 607 S 48th St Philadelphia PA 19143

KLAUS, SIDNEY N, dermatology, for more information see previous edition

KLAUSMEIER, ROBERT EDWARD, b Evansville, Ind, June 6, 26; m 51; c 4. MICROBIOLOGY, CHEMISTRY. *Educ:* Univ Ind, AB, 51, MA, 53; La State Univ, PhD(bact), 58. *Prof Exp:* Asst bact, Univ Ind, 51-53; res bacteriologist, Army Chem Corps, Ft Detrick, Md, 53-55; asst prof bact, Southwestern La Inst, 55-57; asst, La State Univ, 57-58; microbiologist, Weapons Qual Eng Ctr, Naval Weapons Support Ctr, 58-80, chemist, 80-89; RETIRED. *Mem:* AAAS; Am Soc Microbiol; Soc Indust Microbiol. *Res:* Microbial deterioration of materials, particularly explosives and synthetic polymers; microbial physiology; enzymology; economically and environmentally effective demilitarization of ammunition. *Mailing Add:* 4111 Granhaven Dr Bloomington IN 47401

KLAUSTERMEYER, WILLIAM BERNER, b San Pedro, Calif, Nov 30, 39; m 62; c 3. ALLERGY, IMMUNOLOGY. *Educ:* Univ Calif, Los Angeles, BA, 62; Univ Cincinnati, MD, 66. *Prof Exp:* Intern, Wadsworth Vet Admin Hosp, 66-67, resident, 67-68, fel pulmonary, 68-69; fel allergy & immunol, Nat Jewish Hosp, 69-70; allergist, USAF Med Ctr, Wright Patterson Air Force Base, 70-72; CHIEF ALLERGY & IMMUNOL, WADSWORTH VET ADMIN HOSP, LOS ANGELES, 72-; PROF MED & CONSULT ALLERGY & RESPIRATORY DIS, UNIV CALIF, LOS ANGELES, 72- *Mem:* Fel Am Acad Allergy; Am Thoracic Soc; fel Am Col Allergy; Am Fedn Clin Res. *Res:* Pharmacologic and immunologic aspects of allergic and respiratory disease; nasal and bronchial provocation testing; diagnostic and therapeutic approaches to patients with severe steroid dependent asthma; role of troleandomycin, gold salts and methotrexate in the prevention of corticosteroid complications. *Mailing Add:* Allergy & Immunol WIIIR West Los Angeles Vet Admin Med Ctr Los Angeles CA 90073

KLAVANO, PAUL ARTHUR, b Valley, Wash, Nov 30, 19; m 45; c 4. PHARMACOLOGY. *Educ:* State Col Wash, BS, 41, DVM, 44. *Prof Exp:* Instr vet anat, Wash State Univ, 44-45, vet physiol & pharmacol, 45-48, from asst prof to prof vet pharmacol, 48-83, chmn dept, 52-72; RETIRED. *Concurrent Pos:* Chemist, Wash Horse Racing Comn, 42-46. *Mem:* Am Col Vet Pharmacol & Therapeut; Am Soc Vet Physiol & Pharmacol (secy, 53-54, pres, 64-65); NY Acad Sci; Am Soc Vet Anesthesiol. *Res:* Anesthesia of domestic animals. *Mailing Add:* South East 1125 Kamiaken Pullman WA 99163

KLAVERKAMP, JOHN FREDERICK, b Sauk Rapids, Minn, Aug 6, 41; m 65; c 4. PHARMACOLOGY, TOXICOLOGY. *Educ:* Univ Minn, BS, 64; Univ Wash, MS, 70, PhD(pharmacol), 72. *Prof Exp:* Res scientist, 73-77, res mgr, 77-78, RES SCIENTIST, FISHERIES & MARINE SERV, FRESHWATER INST, 78- *Concurrent Pos:* Fel, Wash State Univ, Toxicol

Lab, 72-73; adj prof, Dept Zool, Univ Man, 77- *Res:* Acidification of freshwater; biochemical mechanisms of tolerance in fish; cadmium toxicology; cardiovascular-respiratory physiology of fish; embryology of fish; heavy metal toxicology; mercury toxicology; selenium toxicology; organophosphate insecticides. *Mailing Add:* Box 1 Group 9 RR 1 Dugald MB R0E 0K0 Can

KLAVINS, JANIS VILBERTS, b Latvia, May 6, 21; nat US; m 50; c 4. MEDICINE. *Educ:* Univ Kiel, MD, 48, PhD, 59; Am Bd Path, cert anal path, 57, cert clin path, 59; Am Bd Nutrit, cert, 68. *Prof Exp:* Demonstr path, Sch Med, Western Reserve Univ, 54-55, from instr to sr instr, 55-60, asst prof, 60; from assoc prof to prof, Med Ctr, Duke Univ, 60-65, dir sch cytotech, 63-65; clin prof path, State Univ NY Downstate Med Ctr, 65-71; prof path, State Univ NY Stony Brook, 71-77; DIR DEPT PATH, LONG ISLAND JEWISH MED CTR-QUEENS HOSP CTR, 70-; CHMN DEPT PATH, CATH MED CTR, 77- *Concurrent Pos:* Asst pathologist, Marymount Hosp, Garfield Heights, Ohio, 55-57; cytologist-in-chg, Cleveland Metrop Gen Hosp, 55-60, assoc pathologist, 58-60; chief lab serv, Vet Admin Hosp, Durham, NC, 61-63; pathologist-in-chief, Brooklyn-Cumberland Med Ctr, 65-70; adj prof biol, Fac Grad Arts, Long Island Univ, 68-; clin prof path, Col Physicians & Surgeons, Columbia Univ, 69-; prof lectr, State Univ NY Downstate Med Ctr, 71-77, prof path, 77-85; prof path, Med Col, Cornell Univ, 85- *Mem:* Fel AAAS; Am Soc Cytol; Am Asn Pathologists; Col Am Path; Asn Clin Scientists; Int Acad Path; Biochem Soc; Col Am Path; Am Inst Nutrit; AMA; Int Asn Tumor Marker Oncologists. *Res:* Iron metabolism; pathology of iron excess; effects of antimetabolites, particularly entionine; embryonic and specific proteins in malignant neoplasms; pathology of amino acid excess; tumor markers. *Mailing Add:* 5 Broadmoor Rd Scarsdale NY 10583

KLAWE, WITOLD L, b Piotrkow Trybunalski, Poland, June 9, 23; US citizen; m 55; c 1. FISH BIOLOGY. *Educ:* Univ Toronto, BA, 53, MA, 55. *Prof Exp:* Jr scientist, 55-56, scientist, 56-61, SR SCIENTIST, INTER-AM TROP TUNA COMN, SCRIPPS INST OCEANOG, 61- *Concurrent Pos:* Mem working groups, Expert Panel Facilitation Tuna Res, Food & Agr Orgn, UN, 65, consult, Food & Agr Orgn, UN & SPacific Comn. *Honors & Awards:* Gold Insignia of the Order of Merit Polish People's Repub, 88. *Mem:* Am Inst Fishery Res Biol; AAAS. *Res:* Early life history of scombroid fishes; general marine biology; fishery oceanography; statistics on global catches of tunas. *Mailing Add:* Inter-Am Trop Tuna Comn Scripps Inst Oceanog La Jolla CA 92093

KLAY, ROBERT FRANK, b Ft Benton, Mont, Jan 2, 30; m 59; c 3. ANIMAL NUTRITION. *Educ:* Mont State Univ, BS, 52; Wash State Univ, MS, 58; Univ Minn, Minneapolis, PhD(nutrit), 64. *Prof Exp:* Asst animal sci, Wash State Univ, 56-58, asst prof, 61-64; res fel animal genetics, Commonwealth Sci & Indust Res Orgn, Australia, 58-59; res asst animal nutrit, Univ Minn, Minneapolis, 59-61; NUTRITIONIST, MOORMAN MFG CO, 64- *Mem:* Am Soc Animal Sci; Sigma Xi; ARPAS. *Res:* Protein and amino acid digestion and availability; ruminant and nonruminant nutrition. *Mailing Add:* Moorman Mfg Co Res Dept Quincy IL 62301

KLAYMAN, DANIEL LESLIE, b New York, NY, Feb 28, 29; m 58; c 2. MEDICINAL CHEMISTRY. *Educ:* Columbia Univ, BS, 50; Rutgers Univ, MS, 52 & 54, PhD(org chem), 56. *Prof Exp:* Fulbright grant chem, Sch Trop Med, Calcutta, 56-57; asst prof chem, Hofstra Col, 58-59; res chemist med chem, Dept Radiobiol, 59-63, res chemist med chem, Div Med Chem, 63-78, RES CHEMIST MED CHEM, DIV EXP THERAPEUT, WALTER REED ARMY INST RES, 78- *Concurrent Pos:* Partic, US-Soviet Health Exchange, Moscow, 69; lectr, US-India Exchange Scientists, 72. *Mem:* Am Chem Soc. *Res:* Isolation and structural investigation of alkaloids and terpenes; organic sulfur, selenium and nitrogen chemistry; synthesis of antimalarial, antiradiation, antibacterial, and antiviral compounds. *Mailing Add:* Div Exp Therapeut Walter Reed Army Inst Res Washington DC 20307-5100

KLEBAN, MORTON H, b Brooklyn, NY, Oct 23, 31; m 55; c 3. SOCIAL GERONTOLOGY, STATISTICS & MEASUREMENT. *Educ:* City Col New York, BBA, 53; State Univ Iowa, MA, 55; Univ NDak, PhD(exp psychol), 60. *Prof Exp:* Clin psychologist, Norristown State Hosp, Pa, 60-64; res psychologist, Off Ment Health, State of Pa, 64-66; MED RES SCIENTIST, NORRISTOWN STATE HOSP, 66-; SR RES PSYCHOLOGIST, PHILADELPHIA GERIAT CTR, 66- *Mem:* Am Psychol Asn; Geront Soc. *Res:* Design, statistical applications and computer technology in applied gerontological research. *Mailing Add:* Philadelphia Geriat Ctr 5301 Old York Rd Philadelphia PA 19141

KLEBANOFF, P(HILIP) S(AMUEL), b New York, NY, July 21, 18; m 50; c 3. FLUID MECHANICS. *Educ:* Brooklyn Col, BA, 39, Dr Eng, Hokkaid Univ, Japan, 79. *Prof Exp:* Asst phys aide, 41-42, jr physicist, 42-44, from asst physicist to physicist, 44-57, aeronaut res engr, 57-61, physicist, 61-69, chief aerodynamics sect & asst chief fluid mech, Mech Div, 69-75, chief, Fluid Mech Sect, Mech Div, 75-78, sr scientist, 78-83, consult, 83-85, GUEST SCIENTIST, NAT BUR STANDARDS, 85- *Concurrent Pos:* Mem, US Nat Comt Theoret & Appl Mech, 70-74, hydromechanics comt, Naval Sea Systs Command, 74-81, boundary layer transition study group, 70-82; indust prof adv comt, aerospace eng, Penn State Univ, 70-75; guest worker, Nat Bur Standards, 85- *Honors & Awards:* Naval Ord Develop Award; Gold Medal, US Dept of Com, 75; Fluid Dynamics Prize, Am Phys Soc, 81. *Mem:* Nat Acad Eng; fel Am Inst Aeronaut & Astronaut; fel AAAS; fel Am Phy Soc (vchmn, div fluid dynamics, 68, chmn exec comt, 69). *Res:* Flow instability; magnetohydrodynamics; anemometry; turbulence; boundary layers. *Mailing Add:* Fluid Eng Div Nat Bur Standards Washington DC 20234

KLEBANOFF, SEYMOUR J, b Toronto, Ont, Feb 3, 27; m 51; c 2. INFECTIOUS DISEASES, BIOCHEMISTRY. *Educ:* Univ Toronto, MD, 51; Univ London, PhD(biochem), 54. *Prof Exp:* Intern, Toronto Gen Hosp, 51-52; lectr path chem, Univ Toronto, 54-57; guest investr & asst physician, Rockefeller Inst, 57-59, res assoc, 59, asst prof, assoc physician & radiation protection officer, 59-62; assoc prof, 62-68, PROF MED, SCH MED, UNIV WASH, 68- *Concurrent Pos:* NIH res career develop award, 64-68. *Mem:* Nat Acad Sci; Am Soc Clin Invest; Infectious Dis Soc Am; Asn Am Physicians; Am Soc Biol Chem; Endocrine Soc. *Res:* Role of granulocytes in host defense; role of enzyme peroxidase in biological processes; microbicidal activity of peroxidases. *Mailing Add:* Dept Med Univ Wash Sch Med Seattle WA 98195

KLEBANOV, IGOR ROMANOVICH, b Mar 29, 62; US citizen; m 91. STRING THEORY, QUANTUM GRAVITY. *Educ:* Mass Inst Technol, SB, 82; Princeton Univ, PhD(physics), 86. *Prof Exp:* Res assoc, Stanford Linear Accelerator Ctr, 86-89; ASST PROF PHYSICS, PRINCETON UNIV, 89- *Concurrent Pos:* Alfred P Sloan Res Fel, 91-; Presidential young investr, NSF, 91- *Res:* Theoretical high energy physics; low energy properties of quantum chromodynamics; string theory and quantum theory of gravity. *Mailing Add:* Dept Physics Jadwin Hall Princeton Univ Princeton NJ 08544-0708

KLEBBA, PHILLIP E, b Ypsilanti, Mich, Dec 27, 51; m 82; c 3. MONOCLONAL ANTIBODIES, MEMBRANE BIOLOGY. *Educ:* Univ Notre Dame, BS, 75; Univ Calif, Berkeley, PhD(biochem), 81. *Prof Exp:* Postdoctoral fel mcd microbiol, Sch Med, Stanford Univ, 81-82; postdoctoral fel microbiol & immunol, Univ Calif, Berkeley, 82-84; asst prof microbiol, Univ Notre Dame, 84-88; ASST PROF MICROBIOL, MED COL WIS, 86- *Concurrent Pos:* Prin investr, USPHS, 86-90; Consult, Chevron Chem Co, 86-89, Ensys, Inc, 88-90 & Thymax Corp, 90- *Mem:* Sigma Xi; Am Soc Microbiol; AAAS. *Res:* Bacterial pathogens shield themselves from the mammalian immune system by the barrier properties of molecules that reside in their outer membrane; structure of outer membrane proteins and lipopoly saccharide and the interactions of these two molecules with each other and the immune system. *Mailing Add:* Dept Microbiol Med Col Wis 8701 Watertown Plank Rd Milwaukee WI 53226

KLEBE, ROBERT JOHN, b Philadelphia, Pa, Oct 26, 43. CELL ADHESION PROTEINS. *Educ:* Johns Hopkins Univ, BA, 65; Yale Univ, PhD(biol), 70. *Prof Exp:* From asst prof to assoc prof human genetics, Grad Sch Biomed Sci, Univ Tex Med Br, Galveston, 76-81; assoc prof, 81-86, PROF ANAT, GRAD SCH BIOMED SCI, UNIV TEX HEALTH SCI CTR, 86- *Concurrent Pos:* Jane Coffin Child Mem Fund fel, Salk Inst, 70-72. *Mem:* Am Soc Human Genetics; Am Soc Cell Biol. *Res:* Somatic cell genetics; biochemistry of cell adhesion; developmental genetics; biochemical mechanism of cell adhesion: the analysis of the binding of fibronectin, laminim and other cell adhesion proteins to cell surface molecules. *Mailing Add:* Dept Cell & Struct Biol Univ Tex Health Sci Ctr San Antonio TX 78284-7762

KLEBER, EUGENE VICTOR, b Cleveland, Ohio, July 7, 20; m 42, 64, 78; c 5. CHEMISTRY, ENVIRONMENTAL PROGRAMS. *Educ:* Univ Calif, Los Angeles, AB, 40, MA, 41; Univ Wis, PhD(chem), 43. *Prof Exp:* Res chemist, Sharples Chem, Inc, Mich, 43-45, Golden Bear Oil Co, Calif, 45-46 & Lockheed Aircraft Corp, 46-47; pres, Res Chem, Inc, 47-59, gen mgr res chem div, Nuclear Corp Am, 59-62; staff asst, Atomics Int Div, Rockwell Int Corp, 62-78; phys scientist, Fed Energy Regulatory Comn, Dept Energy, 78-85; RETIRED. *Concurrent Pos:* Res chemist, Coast Paint & Chem Co, 47 & Lockheed Aircraft Corp, 51-56. *Mem:* Am Chem Soc. *Res:* Nuclear fuels and materials; organic synthesis; fine chemical manufacturing; production and use of purified rare earth oxides and metals; energy and environmental programs. *Mailing Add:* 406 Upper Wood Way Burnsville MN 55337

KLEBER, HERBERT DAVID, b Pittsburgh, Pa, June 19, 34; div; c 3. PSYCHIATRY, DRUG ABUSE. *Educ:* Dartmouth Col, BA, 56; Jefferson Med Col, MD, 60. *Hon Degrees:* MA, Yale Univ, 75. *Prof Exp:* Resident psychiat, Sch Med, Yale Univ, 61-64; chief receiving serv, USPHS Hosp, Lexington, Ky, 64-66; from asst prof to assoc prof, 66-75, PROF PSYCHIAT, SCH MED, YALE UNIV, 75-; DEP DIR DEMAND REDUCTION, OFF NAT DRUG CONTROL POLICY, EXEC OFF OF THE PRES, 68- *Concurrent Pos:* Consult, Nat Inst Drug Abuse, 71- & Nat Acad Sci, 73-; fund prize res psychiat, Am Psychiat Asn, 81. *Honors & Awards:* Gold Medal, Am Psychiat Asn, 75. *Mem:* Fel Am Psychiat Asn; Am Col Neuropsychopharmacol; Am Acad Psychiat in Alcoholism & Addiction. *Res:* Treatment of drug dependence; etiological aspects of drug abuse. *Mailing Add:* Off Nat Drug Control Policy Exec Off of the Pres Washington DC 20500

KLEBER, JOHN WILLIAM, b Warsaw, Ill, Jan 1, 23; m 45; c 5. BIOPHARMACEUTICS. *Educ:* Duquesne Univ, BS, 43; Univ Minn, PhD(pharmaceut chem), 49. *Prof Exp:* Asst prof pharm, Univ Buffalo, 49-52, assoc prof, 52-60; sr biologist, Eli Lilly & Co, 60-70, sr anal chemist, 70-75, sr formulations chemist, 75-81; RETIRED. *Concurrent Pos:* USPHS trainee grant steroid biochem, Univ Utah, 59-60. *Mem:* Am Chem Soc. *Res:* Development of animal health care products. *Mailing Add:* 216 N Lynn St Seymour IN 47274

KLECKA, MIROSLAV EZIDOR, b Yoakum, Tex, Nov 9, 21; m 45. CHEMICAL ENGINEERING. *Educ:* Univ Tex, BS, 43, MS, 46, PhD(chem eng), 48. *Prof Exp:* Res asst chem eng, Univ Tex, 45-46, res assoc, 46, instr, 46-47; sr res engr, Shell Oil Co, 47-66, staff res engr, 66-77, STAFF RES ENGR, SHELL DEVELOP CO, HOUSTON, 77- *Mem:* Am Inst Chem Engrs. *Res:* Petroleum refining design and evaluation; phase equilibria and separation processes; operations research and computer calculations systems design; process control systems. *Mailing Add:* 14215 Swiss Hill Dr Houston TX 77077

KLECKNER, ALBERT LOUIS, veterinary microbiology; deceased, see previous edition for last biography

KLEE, CLAUDE BLENC, CALMODULIN, PROTEIN-PROTEIN INTERACTION. *Educ:* Univ Marsailles, France, MD, 59. *Prof Exp:* HEAD MACROMOLECULAR INTERACTIONS, LAB BIOCHEM, NAT CANCER INST, 74- *Mailing Add:* Nat Cancer Inst NIH Bldg 37 Rm 4E28 Bethesda MD 20892

KLEE, GERALD D'ARCY, b New York, NY, Jan 29, 27; m 50; c 5. PSYCHIATRY, MEDICINE. *Educ:* Harvard Med Sch, MD, 52; Am Bd Psychiat & Neurol, dipl, 59. *Prof Exp:* Sr asst surgeon, USPHS, 53-54; res assoc psychiat, Sch Med, Univ Md, 56-58, from asst prof to assoc prof, 58-67, dir div outpatient psychiat, Psychiat Inst, 58-67; prof psychiat, Sch Med, Temple Univ, 67-70; LECTR PSYCHIAT, SCH MED, JOHNS HOPKINS UNIV, 76- *Concurrent Pos:* Pvt pract, psychiat, 71-; med dir, USPHS. *Mem:* AMA; fel Am Psychiat Asn. *Res:* Psychopharmacology; psychotherapy; community psychiatry; epidemiology. *Mailing Add:* 28 Allegheny Ave Suite 1300 Towson MD 21204

KLEE, LUCILLE HOLLJES, b Baltimore, Md, Dec 8, 24; m 59; c 2. BIOCHEMISTRY. *Educ:* Bryn Mawr Col, AB, 46, AM, 47, PhD(chem), 51. *Prof Exp:* Instr, Barnard Col, Columbia Univ, 51-56, class adv, 54-56; res chemist, Toni Co, Div Gillette Co, 56-57; res assoc, Brandeis Univ, 57-59, lectr chem, 58-69; sci coord, Douglas County Bd Educ, 69-70; assoc prof chem, Lowell State Col, 70-71; assoc prof chem, 71-74, assoc prof, 74-78, PROF SCI EDUC, W GA COL, 78- *Mem:* AAAS; Nat Sci Teachers Asn. *Res:* Science education. *Mailing Add:* 24 Forest Dr Carrollton GA 30117

KLEE, VICTOR LA RUE, JR, b San Francisco, Calif, Sept 18, 25; m 45, 85; c 4. MATHEMATICS. *Educ:* Pomona Col, BA, 45, Univ Va, PhD(math), 49. *Hon Degrees:* DSc, Pomona Col, 65; Dr, Univ Liege, 84. *Prof Exp:* Instr math, Univ Va, 47-48, asst prof, 49-53; from asst prof to assoc prof, 53-57, PROF MATH, UNIV WASH, 57-, PROF APPL MATH, 76- *Concurrent Pos:* Nat Res Coun fel, Inst Advan Study, 51-52; vis assoc prof, Univ Calif, Los Angeles, 55-56 & Univ Western Australia, 79; Sloan res fel, Univ Wash, 56-58 & 60-61; NSF sr fel, Copenhagen Univ, 58-59, Sloan res fel, 59-60; consult, Boeing Sci Res Labs, 63-69, Rand Corp, 66-69, Holt, Rinehart & Winston, 66-76, E I du Pont de Nemours & Co, Inc, 68-72, IBM Corp, 72 & W H Freeman, 76-; Sigma Xi nat lectr, 69; vis prof, Univ Colo, 71; adj prof comput sci, Univ Wash, 74-; vis prof, Univ Victoria, 75; fel, Ctr Advan Study Behav Sci, 75-76; trustee, Conf Bd Math Sci, 72-73; Guggenheim fel & Von Humboldt awardee, Univ Erlangen-Nurnberg, 80-81; mem, Math Sci Res Inst, 85-86; sr fel, Inst Math & Its Applns, 87. *Honors & Awards:* Pres & Visitor's Res Prize, Univ Va, 52; L R Ford Award, 72; C B Allendoerfer Award, Math Asn Am, 80; Vollum Award, Reed Col, 82; Barrows Award, Pomona Col, 88. *Mem:* Fel AAAS; Am Math Soc (assoc secy, 55-58); Math Asn Am (1st vpres, 68-69, pres-elect, 70, pres, 71-72); Soc Indust & Appl Math; Opers Res Soc Am. *Res:* Convex sets; mathematical programming; combinatorial mathematics; design and analysis of algorithms; functional analysis; point-set topology. *Mailing Add:* Dept Math/Comp Sci Univ Wash Seattle WA 98195

KLEEMAN, CHARLES RICHARD, b Los Angeles, Calif, Aug 19, 23; m 45; c 3. PHYSIOLOGY, METABOLISM. *Educ:* Univ Calif, BS, 44, MD, 47. *Prof Exp:* Rotating internship, San Francisco City Hosp, 47-48; asst resident path, Mallory Inst, Boston City Hosp, 48-49; from intermediate resident to sr resident med, Newington Vet Admin Hosp, 49-51; from instr to asst prof med, Metab Sect, Sch Med, Yale Univ, 53-56; from assoc clin prof to assoc prof, Sch Med, Univ Calif, Los Angeles, 56-64, prof, Sch Med & dir div med, Cedars-Sinai Med Ctr, 64-74; prof med & chief dept, Hadassah Med Sch, Hebrew Univ, Israel, 74-75; prof med & chief, Div Nephrology, 75-77, EMER PROF MED & EMER CHIEF, DIV NEPHROLOGY, SCH MED, UNIV CALIF, LOS ANGELES, 77-; SCI DIR, RES INST, CEDAR SINAI MED CTR, 86-, DIR, DIV NEPHROL, 86- *Concurrent Pos:* Fel metab, Newington Vet Admin Hosp, 50-51; Upjohn-Endocrine Soc scholar, Univ Col, Univ London, 60-61; chief metab sect, Vet Admin Hosp, Los Angeles, 56-60, consult, 62-; dir div med, Mt Sinai Hosp, Los Angeles, 61-; mem sci adv bd, Nat Kidney Dis Found; vis prof, Univ Queensland, 66; consult artificial kidney, Chronic Uremia Prog & Kidney Dis Control Prog, NIH, 67-; vis prof, Beilinson Hosp, Tel-Aviv Univ & Med Sch, Hadassah-Hebrew Univ, 68; vis prof, St Francis Hosp, Honolulu, 69; chmn, Internal Med Sect, Nat Bd Med Exam, 71; dir, Ctr Health Enhancement Educ & Res, Cedar Sinai Med Ctr, 71-86. *Mem:* Inst Med-Nat Acad Sci; AMA; Am Physiol Soc; Am Soc Clin Invest; Endocrine Soc. *Res:* Renal physiology; electrolyte and water metabolism; nephrology. *Mailing Add:* Dept Med Cedar Sinai Med Ctr 8700 Beverly Los Angeles CA 90048

KLEEN, HAROLD J, b Nebr, July 2, 11; m 34; c 2. GEOLOGY. *Educ:* Univ Nebr, BSc, 33. *Prof Exp:* Geologist, Skelly Oil Co, 37-44, dist geologist, Okla, 44-49; div mgr, Cent US, Kerr-McGee Oil Industs, Inc, 49-53, chief geologist, 53-58, explor mgr, 58-61, geol adv to pres, 61-67, mgr mineral explor, 67-68, explor asst to chief exec off, Kerr-McGee Corp, 68-69, vpres minerals explor, 69-74; CONSULT, 74- *Honors & Awards:* Sigma Xi. *Mem:* Am Inst Prof Geol. *Res:* Investment economics. *Mailing Add:* 6229 Smith Blvd Oklahoma City OK 73112

KLEENE, STEPHEN COLE, b Hartford, Conn, Jan 5, 09; m 42, 78; c 4. MATHEMATICAL LOGIC. *Educ:* Amherst Col, AB, 30; Princeton Univ, PhD(math), 34. *Hon Degrees:* ScD, Amherst Col, 70. *Prof Exp:* Instr math, Princeton Unv, 30-31, asst, 34-35; from instr to asst prof, Univ Wis, 35-41; assoc prof, Amherst Col, 41-42; from assoc prof to prof math, Univ Wis-Madison, 46-64, chmn dept, 57-58 & 60-62, numerical analysis, 62-63, dean, Col Letters & Sci, 69-74, Cyrus C MacDuffee prof math, 64-74, Cyrus C MacDuffee prof math & comput sci, 74-79, EMER PROF MATH & COMPUT SCI & EMER DEAN LETT & SCI, UNIV WIS-MADISON, 79- *Concurrent Pos:* Mem, Inst Advan Study, 39-40 & 65-66; Guggenheim fel, 49-50; vis prof, Princeton Univ, 56-57; mem math div, Nat Res Coun, 57-58, chmn designate div math sci, 69-72; NSF grant, Univ Marburg, 58-59; pres, Int Union Hist & Philos Sci, 61, pres div logic, methodology & philos sci, 60-62; actg dir math res ctr, US Army, 66-67. *Honors & Awards:* Steele Prize, Am Math Soc, 83; Nat Medal Sci, 90. *Mem:* Nat Acad Sci; Am Math Soc; Asn Symbolic Logic (vpres, 41-42 & 47-49, pres, 56-58); Am Acad Arts & Sci. *Res:* Recursive functions; author of mathematics textbook. *Mailing Add:* 1514 Wood Lane Madison WI 53711

KLEHR, EDWIN HENRY, environmental chemistry; water chemistry; deceased, see previous edition for last biography

KLEI, HERBERT EDWARD, JR, b Detroit, Mich, May 5, 35; m 59; c 5. CHEMICAL ENGINEERING. *Educ:* Mass Inst Technol, BS, 57; Univ Mich, MS, 58 & 59; Univ Conn, PhD(chem eng), 65. *Prof Exp:* Res engr, Chas Pfizer & Co, 59-63; instr, 64-65, from asst prof to assoc prof, 65-78, PROF CHEM ENG, SCH ENG, UNIV CONN, 78- *Concurrent Pos:* Vis prof, US Military Acad, 85. *Honors & Awards:* Ralph Teetor Award, Soc Automotive Engrs, 78. *Mem:* Am Inst Chem Eng; Am Chem Soc; Catalysis Soc NAm. *Res:* Water pollution control, including biological kinetics and reactor design, membrane polarization and process control. *Mailing Add:* Dept Chem Eng Univ Conn Main Campus U-139 191 Auditorium Storrs CT 06268

KLEI, THOMAS RAY, b Detroit, Mich, Dec 11, 42; m 65; c 1. PARASITOLOGY, IMMUNOLOGY. *Educ:* Northern Mich Univ, BS, 65; Wayne State Univ, PhD(biol & parasitol), 71. *Prof Exp:* NIH fel parasitol, Sch Vet Med, Univ Ga, 71-73; asst prof biol & zool, Millersville State Col, 73-75; from asst to assoc prof, 77-82, PROF PARASITOL, SCH VET MED, LA STATE UNIV, 82- *Concurrent Pos:* Prin investr, WHO & USDA Coop States Res Study grants, 77-; consult vet par asitol. *Mem:* Am Soc Parasitologists; Am Soc Trop Med & Hyg; AAAS; Am Asn Vet Parasitol, (pres, 87); Wildlife Dis Asn. *Res:* Immunologic and pathologic responses of vertebrate hosts to parasitic animals; parasitic diseases of horses. *Mailing Add:* Dept Vet Microbiol & Parasitol La State Univ Baton Rouge LA 70803

KLEIER, DANIEL ANTHONY, b Louisville, Ky, Aug 19, 45; m 68; c 2. PHYSICAL CHEMISTRY, THEORETICAL CHEMISTRY. *Educ:* Bellarmine Col, Ky, BA, 67; Univ Notre Dame, PhD(chem), 71. *Prof Exp:* Woodrow Wilson teaching intern & asst prof chem, Va State Col, 70-72; res fel, Harvard Univ, 72-75; asst prof chem, Williams Col, 75-81; chemist, Shell Develop Co, 81-86; RES ASSOC, DUPONT CO, 86- *Concurrent Pos:* NSF fel, Harvard Univ, 72-73, Am Cancer Soc fel, 74-75; vis staff mem, Los Alamos Nat Lab, 78-79. *Mem:* Am Chem Soc; Sigma Xi. *Res:* Theoretical investigations of chemical bonding and potential energy surfaces; nuclear magnetic resonance studies of stereodynamic processes; computer aided design of crop protection chemicals. *Mailing Add:* 31 Johnston Dr Elkton MD 21921

KLEIMAN, DEVRA GAIL, b New York, NY, Nov 15, 42; div; c 2. ETHOLOGY, REPRODUCTIVE BIOLOGY. *Educ:* Univ Chicago, BS, 64; Univ Col, Univ London, PhD(zool), 69. *Prof Exp:* Res asst biopsychol, Univ Chicago, 64-65; res asst reproductive biol, Wellcome Inst Comp Physiol, Zool Soc London, 65-69; NIH fel develop, Inst Animal Behav, Rutgers Univ, 69-71; reproduction zoologist, 72-79, head zool res, 79-83, asst dir animal progs 83-84, ASST DIR RES, NAT ZOOL PARK, SMITHSONIAN INST, 84- *Concurrent Pos:* Adj asst prof psychol, Rutgers Univ, 70-71; res assoc, Smithsonian Inst, 70-72; adj assoc prof, George Washington Univ, 73-76 & Univ Md, 79-81; adj prof, George Mason Univ, 80- & Univ Md, 82- *Honors & Awards:* Ann WISE Award, 87; Fel, Animal Behav Soc; Fel, AAAS; Distinguished Achievement Award, Soc Cons Biol, 88. *Mem:* Animal Behav Soc (secy, 77-81 & pres, 81-82); Am Asn Zool Parks & Aquariums; Am Soc Mammalogists; Sigma Xi; Am Inst Biol Sci; AAAS. *Res:* Social behavior and social organization of mammals; mammalian reproductive strategies. *Mailing Add:* Zool Res Nat Zool Park Smithsonian Inst Washington DC 20008

KLEIMAN, HERBERT, b New York, NY, Oct 1, 33; m 60; c 1. PHYSICS. *Educ:* Mass Inst Technol, BS, 54; Purdue Univ, MS, 57, PhD(physics), 61. *Prof Exp:* NSF fel physics, Univ Calif, Berkeley, 61-63, asst prof, 63-66; MEM STAFF, LINCOLN LAB, MASS INST TECHNOL, 66- *Res:* High resolution spectroscopy; atomic structure and atomic spectra; quantum optics and photon correlation studies; physical optics. *Mailing Add:* Lincoln Labs Mass Inst Tech Lexington MA 02173

KLEIMAN, HOWARD, b New York, NY, Apr 15, 29; m 56; c 3. ALGEBRA, NUMBER THEORY. *Educ:* NY Univ, BA, 50, MS, 61; Columbia Univ, MA, 54; King's Col, Univ London, PhD(math), 69. *Prof Exp:* Teacher, New York City Bd Educ, 55-56 & Bur Educ Physically Handicapped, 56-67; from asst prof to assoc prof, 67-78, PROF MATH, QUEENSBOROUGH COMMUNITY COL, 78- *Mem:* Am Math Soc. *Res:* Bounds for all solutions of a new class of diophantine equations; a sharp threshold algorithm for obtaining a Hamilton circuit in a random digraph; algorithms for randomly chosen regular cubic graphs and randomly chosen G(N,3) graphs in the definition of Friozo and Fenner. *Mailing Add:* 188-83 85th Rd Hollis NY 11423

KLEIMAN, MORTON, b Kansas City, Mo, Mar 8, 16; m 40; c 2. ORGANIC CHEMISTRY. *Educ:* Univ Mich, BS, 37, MS, 38; Univ Chicago, PhD, 42. *Prof Exp:* DuPont fel, Univ Chicago, 41-42, Coman fel, 42-43; res chemist, Velsicol Corp, 43-46, head org res dept, 46-51, dir res, 51-53; pres & tech dir, Chemley Prods co, 53-82; PRES & TECH DIR, M KLEIMAN ASSOCS, 53- *Mem:* Sigma Xi; Am Chem Soc. *Res:* Syntheses of various medicinals; insecticides and fungicides; reactions such as liquid ammonia, Grignard, elimination, carboxylation, chlorination, Diels-Alder and redistribution; structure proofs; ionic and free radical mechanisms; synthetic resins; organo-metallics; industrial organic chemicals. *Mailing Add:* 2827 W Catalpa Ave Chicago IL 60625

KLEIN, ABEL, b Rio de Janeiro, Brazil, Jan 16, 45; m 83; c 2. MATHEMATICAL PHYSICS. *Educ:* Univ Brazil, BS, 67; Inst Pure & Appl Math, MS, 68; Mass Inst Technol, PhD(math), 71. *Prof Exp:* Instr math, Inst Pure & Appl Math, 67; teaching asst, Mass Inst Technol, 68-71; actg asst prof, Univ Calif, Los Angeles, 71-72; instr, Princeton Univ, 72-74; from asst prof to assoc prof, 74-82, PROF MATH, UNIV CALIF, IRVINE, 82- *Mem:* Am Math Soc; Int Asn Math Physics. *Res:* mathematical physics; functional analysis. *Mailing Add:* Dept Math Univ Calif Irvine CA 92717

KLEIN, ABRAHAM, b Brooklyn, NY, Jan 10, 27; m 50; c 2. THEORETICAL PHYSICS. *Educ:* Brooklyn Col, BA, 47; Harvard Univ, MA, 48, PhD(physics), 50. *Prof Exp:* Asst physics, Harvard Univ, 47-49, instr, 50-52; assoc prof, 55-58, PROF PHYSICS, UNIV PA, 58- *Concurrent Pos:* NSF sr fel, 61-62; Alfred P Sloan Found fel, 61-63; vis prof, Univ Paris, 61-62, Princeton Univ, 69-70, Univ Tsukuba, Japan, 81, Yale Univ, 83, Tech Univ, Munich, 84 & Univ Frankfurt, 87 & 88; consult, Res Inst Advan Study, Martin Marietta, 59 & Gen Dynamics-Convair, 60; Guggenheim fel, 75; vis scientist, Ctr Theoret Physics, Mass Inst Technol, 75-76; assoc, Nat Ctr Sci Res, France, 84. *Honors & Awards:* A von Humboldt Sr Scientist Award, 87. *Mem:* Fel Am Phys Soc; Am Asn Physics Teachers. *Res:* Quantum electrodynamics; meson theory of nuclear forces; theory of scattering; many body problem; quantum field theory; theory of nuclear structure. *Mailing Add:* Dept Physics Univ Pa Philadelphia PA 19104-6396

KLEIN, ALBERT JONATHAN, b Dayton, Ohio, Nov 16, 44; m 65; c 3. TOPOLOGY. *Educ:* Ohio State Univ, BSc, 66, MS, 67, PhD(math), 69. *Prof Exp:* Asst prof, 69-74, assoc prof math, 74-80, assoc prof, 80-83, PROF MATH & COMPUT SCI, YOUNGSTOWN STATE UNIV, 83- *Mem:* Math Asn Am; Asn Comput Mach; Sigma Xi. *Res:* Fuzzy topology; digital topology. *Mailing Add:* Dept Math & Comput Sci Youngstown State Univ Youngstown OH 44555

KLEIN, ANDREW JOHN, b Wilkes-Barre, Pa, Dec 31, 51; m 78. TRACE ORGANIC ANALYSIS, ENVIRONMENTAL ANALYSIS. *Educ:* King's Col, BS, 73; Univ Wis, PhD(chem), 78. *Prof Exp:* Res assoc, Univ Wis, 78-80; sr res chemist, 80-82, res specialist, 82, res group leader, 82-85, sr group leader, 85-86, MGR, REGULATORY AFFAIRS, MONSANTO CO, 86- *Mem:* Am Chem Soc; AAAS; Asn Ground Water Scientists & Engrs. *Res:* Trace organic analysis in environmental matrices and natural water; analysis of xenobiotics in animal tissues. *Mailing Add:* Monsanto Agr Co 800 N Lindbergh Blvd C25F St Louis MO 63167

KLEIN, ATTILA OTTO, b Subotica, Yugoslavia, July 10, 30; nat US; m 52; c 3. PLANT PHYSIOLOGY. *Educ:* Brooklyn Col, BA, 53; Ind Univ, PhD(plant physiol), 59. *Prof Exp:* USPHS fel biochem, Yale Univ, 59-61; asst biochemist, Conn Agr Exp Sta, 61-62; asst prof, 62-67, chmn dept, 68-70, ASSOC PROF BIOL, BRANDEIS UNIV, 67- *Res:* Cellular and plant physiology; developmental biochemistry of leaves; light-induced metabolic oscillations. *Mailing Add:* Dept Biol Brandeis Univ Waltham MA 02254

KLEIN, AUGUST S, b Newton, Mass, Aug 31, 24; m 67; c 3. PHYSICS, CHEMISTRY. *Educ:* Williams Col, BA, 48; Harvard Univ, MS, 50. *Prof Exp:* Physicist, Atomic Power Div, Westinghouse, 50-58; west coast mgr, High Voltage Eng Corp, 59-64; gen mgr, spec prod div, Tech Measurement Corp, 64-67; pres, Nuclear Equip Corp, 67-84; PRES, ION IMPLANTATION CORP, 85- *Concurrent Pos:* Chmn, Greater Silicon Valley Implant Users Group; bd dir, Northern Calif Sect, Am Vacuum Soc. *Mem:* Electron Micros Soc Am; Am Inst Mining Engrs. *Res:* Energy dispersive x-ray fluorescent analysis; x-ray crystallography; ion implantation. *Mailing Add:* 160 La Questa Way Woodside CA 94062

KLEIN, BARBARA P, b New York, NY, Dec 30, 36; m 56; c 2. FOOD CHEMISTRY. *Educ:* Cornell Univ, BS, 57, MS, 59; Univ Ill, PhD(foods & nutrit), 74. *Prof Exp:* Res asst food chem, Col Home Econ, Cornell Univ, 57-58; res asst, Col Agr, Univ Ill, Urbana, 66-68, teaching asst, 69-72, from asst prof to assoc prof, 74-85, div chair, 85-90, PROF FOODS & NUTRIT FOOD SCI, COL AGR, UNIV ILL, URBANA, 85- *Honors & Awards:* Borden Award, Am Home Econ Asn, 88. *Mem:* Inst Food Technologists; Am Home Econ Asn; Am Chem Soc; Am Dietetics Asn; Am Asn Cereal Chemists. *Res:* Alterations in food quality that occurs during storage and processing of foods for human consumption; specifically sensory, biochemical and nutritional assessment of home and commercially processed foods. *Mailing Add:* Food & Nutrit 274 Bevier Hall Univ Ill Urbana IL 61801

KLEIN, BENJAMIN GARRETT, b Durham, NC, Jan 24, 42; m 71; c 2. MATHEMATICAL ANALYSIS. *Educ:* Univ Rochester, BA, 63; Yale Univ, MA, 65, PhD(Ergodic theory), 68. *Prof Exp:* Lectr math, NY Univ, 67-68, asst prof, 69-71; from asst prof to assoc prof, 71-85, PROF MATH, DAVIDSON COL, 85- *Concurrent Pos:* Consult, NC Dept Public Instr; vpres, Col Western Region NC, Coun Teachers Math, 89. *Honors & Awards:* Thomas Jefferson Award, 90. *Mem:* Am Math Soc; Math Asn Am. *Res:* General mathematics. *Mailing Add:* Dept Math Davidson Col Box 1719 Davidson NC 28036

KLEIN, BERNARD, b New York, NY, Sept 16, 14; m 42; c 2. ORGANIC CHEMISTRY. *Educ:* Brooklyn Col, BS, 34; Polytech Inst Brooklyn, PhD(chem), 50. *Prof Exp:* Chemist, Bethel Hosp, NY, 36-41; biochemist, Jewish Sanitarium & Hosp Chronic Dis, 41-42; res chemist, Warner Inst, NY, 46-48 & Harlem Hosp Cancer Found, 48; biochemist, US Vet Admin Hosp, Bronx, 48-67; biochemist, Res Div, Hoffmann-La Roche, Inc, 67, group chief clin chem, 67-71, asst dir, Dept Diag Res, 71-80; prof, 81-85, EMER PROF, DEPT LAB MED, EINSTEIN COL MED, BRONX, NY, 85- *Concurrent Pos:* Consult chemist, Area Reference Lab. *Honors & Awards:* Van Slyke Award, Am Asn Clin Chemists, 69, Ames Award, 75. *Mem:* Am Chem Soc; Am Asn Clin Chem. *Res:* Pyrazine chemistry; automated biochemical analyses. *Mailing Add:* 129 Patton Blvd New Hyde Park NY 11040-1726

KLEIN, CERRY M, b Kansas City, Mo, Dec 11, 55; m 80; c 4. OPTIMIZATION, COMPUTER-AIDED DESIGN & MANUFACTURING. *Educ:* North West Mo State Univ, BS, 77; Purdue Univ, MS, 80, PhD(indust eng), 83. *Prof Exp:* Teacher math, Consol Sch Dist No 1, Kansas City, Mo, 77-78; asst calculus, Purdue Univ, 78-82, reseacher, 82-83; systs analyst, Nisus Corp, Indianapolis, Ind, 80-83; ASST PROF INDUST ENG, UNIV MO, COLUMBIA, 84- *Concurrent Pos:* Consult, 3M Co, 86 & McDonnell Douglas Corp, 87; ed, Comput & Info Systs, 87-88; prin investr, grant, Off Naval Res, 88 & McDonnell Douglas Corp, 88. *Mem:* Opers Res Soc Am; Soc Indust & Appl Math; Math Prog Soc; Inst Indust Eng; Sigma Xi. *Res:* Dynamic programming; combinatiorial optimization; design and analysis of heuristics; submodular functions; manufacturing processes; decision analysis; applications of operations research techniques and interior methods for mathematical programming. *Mailing Add:* Indust Eng 121 Elec Eng Univ Mo Columbia MO 65211

KLEIN, CHRISTOPHER FRANCIS, b Los Angeles, Calif, Sept 11, 43; m 73; c 2. LASERS, ELECTRO-OPTICS. *Educ:* Calif State Univ, Long Beach, BS, 67, MS, 72. *Prof Exp:* Mem res & develop staff electro-optic design, Autonetics Div, NAm Rockwell, 67-72; mem res & develop staff laser design, Hughes Aircraft Co, 72-75; RES LASER SPECTROSCOPY, LASER SYSTMS DESIGN & ANALYSIS, AEROSPACE CORP, 75- *Concurrent Pos:* Instr physics, El Camino Col, 76-77. *Mem:* Soc Photo-Optical Instrumentation Engrs. *Res:* Laser spectroscopy; laser design; laser damage to optical materials. *Mailing Add:* Aerospace Corp MY-980 2350 E El Segundo Blvd El Segundo CA 90245

KLEIN, CLAUDE A, b Strasbourg, France, Nov 4, 25; nat US; m 50; c 1. PHYSICS. *Educ:* Univ Paris, EE, 51, PhD(physics), 55. *Prof Exp:* Asst to mgr, Mil Dept, French AEC, 55-57; prin scientist, Res Div, Raytheon Co, 57-88; CONSULT SCIENTIST, 89- *Concurrent Pos:* Vis lectr, Univ Lyons, 53-54, Univ Paris, 54-56 & Univ Lowell, 62-63. *Mem:* Fel Am Phys Soc; Inst Elec & Electronics Engrs. *Res:* Solid state physics; lasers and infrared; systems engineering. *Mailing Add:* Churchill Lane Lexington MA 02173

KLEIN, CORNELIS, b Haarlem, Holland, Sept 4, 37; US citizen; m 60; c 2. MINERALOGY, PETROLOGY. *Educ:* McGill Univ, BSc, 58, MSc, 60; Harvard Univ, PhD(geol), 65. *Prof Exp:* Res assoc geol, Harvard Univ, 63-65, lectr, 65-69, assoc prof mineral, 69-72; prof mineral, Ind Univ, Bloomington, 72-84; PROF GEOL, UNIV NMEX, 84- *Concurrent Pos:* Allston Burr sr tutor & asst dean, Harvard Col, 66-70; assoc ed, Am Mineralogist, 77-82, Precambrian Res, 83-, & Can Mineralogist, 89-91; Guggenheim fel, 78; John Simon Guggenheim fel, 78; mem, Precambrian Paleobiol Res Group, Univ Calif Los Angeles, 79- *Mem:* Fel Mineral Soc Am; fel Geol Soc Am; Mineral Asn Can; Microbeam Anal Soc; fel AAAS. *Res:* Precambrian iron formation; chemical, optical and x-ray properties of amphiboles; minerals in meteorites; mineralogy and petrology of lunar rocks; electron probe analysis of minerals. *Mailing Add:* Dept Geol Univ NMex Albuquerque NM 87131

KLEIN, DALE EDWARD, b Cooper Co, Mo, July 6, 47; m 71. NUCLEAR ENGINEERING. *Educ:* Univ Mo, Columbia, BS, 70, MS, 71, PhD(nuclear eng), 77. *Prof Exp:* Design Engr, Procter & Gamble Co, 70-72; teaching & res asst nuclear eng, Univ Mo, Columbia, 73-77; asst prof, 77-82, DIR, NUCLEAR ENG TEACHING PROG, UNIV TEX, AUSTIN, 78-, ASSOC PROG MECH ENG,82-G PROG, 78- *Concurrent Pos:* Engr, Gen Atomic Co, 74; dep dir, Ctr Energy Studies, 86- *Honors & Awards:* Eng Found Award, Univ Tex, Austin, 79 & 82; Young Eng Yr, Travis Chap, Tex Soc Prof Eng, 82. *Mem:* Am Soc Mech Engrs; Am Nuclear Soc; Nat Soc Prof Engrs. *Res:* Thermal analysis of nuclear shipping containers; heat transfer augmentation for flow over rough surfaces; liquid metal flows through a packed bed under the influence of a transverse magnetic field. *Mailing Add:* Dept Mech Eng Univ Tex Austin TX 78712

KLEIN, DAVID C, b New York, NY, May 11, 40. NEUROENDOCRINOLOGY. *Educ:* Cornell Univ, AB, 62; Rice Univ, PhD(biol), 68. *Prof Exp:* Res asst endocrinol, Cornell Univ, 61 & phys biol, 62; res asst biophys cytol, Rockefeller Univ, 62-64; lab instr gen biol, endocrinol & radioisotope methodology, Rice Univ, 64-66; fel pharmacol, Univ Rochester Sch Med & Dent, 67-69; sr staff fel, Sect Physiol Controls, Lab Biomed Sci, 71-73, physiologist, 73-77, CHIEF NEUROENDOCRINOL SECT, LAB DEVELOP NEUROBIOL, NAT INST CHILD HEALTH & HUMAN DEVELOP, NIH, 77- *Concurrent Pos:* Rice fel biol, 64; Nat Inst Dent Res trainee, 65-67; pres, Nat Inst Child Health & Human Develop Assembly Scientists, 74-75; chmn, Nat Inst Health Child Care Adv Comt, 75-76. *Mem:* AAAS; Sigma Xi; Tissue Cult Soc; Endocrine Soc; Am Soc Pharmacol & Exp Therapeut; Int Soc Neurochem; Am Soc Neurochem. *Res:* Biochemical signal transduction, using pinealocyte as an experimental model; the neural regulation of gene expression; the molecular basis of the biochemical "AND" gate. *Mailing Add:* Lab Develop Neurobiol Nat Inst Child Health & Human Develop NIH Bldg 36 Rm 4A07 Bethesda MD 20892

KLEIN, DAVID HENRY, b Milwaukee, Wis, May 28, 33; m 54; c 2. ANALYTICAL METHODS DEVELOPMENTS. *Educ:* Albion Col, BA, 54; Case Western Reserve Univ, PhD, 59. *Prof Exp:* Instr chem, Calif Inst Technol, 59-60; asst prof, Los Angeles State Col, 60-64; from assoc prof to prof chem, Hope Col, 64-81, chmn dept, 69-73; proj scientist, Parke Davis Co, 81-85; ANALYSIS GROUP LEADER, ADAMANTECH INC, 85- *Concurrent Pos:* NSF fel, Scripps Inst Onceanog, 68-69; vis scientist, Oak Ridge Nat Lab, 73-74. *Mem:* AAAS; Am Chem Soc. *Res:* Kinetics of nucleation; precipitation and co-precipitation; mercury and other heavy metals in the environment; geochemistry and marine chemistry; analysis of pharmaceutical materials; analysis of perfluorocarbons. *Mailing Add:* 4615 Buckingham Lane Carlsbad CA 92008-6402

KLEIN, DAVID JOSEPH, b Los Angeles, Calif, Aug 3, 22; m 45; c 2. RADIOLOGICAL PHYSICS. *Educ:* Calif Inst Technol, BS, 43, PhD(physics), 51. *Prof Exp:* ASST CLIN PROF RADIOL, MED SCH, UNIV SOUTHERN CALIF, 74- *Concurrent Pos:* Asst, Calif Inst Technol, 46-51; res engr, NAm Aviation, Inc, 51-52, sr engr, 53-58, sr physicist, 59-63, mem tech staff, Autonetics Div, NAm Rockwell Corp, 63-71; consult, Advan Info Methods, Inc, 71-72; asst radiation physicist, Los Angeles County-Univ Southern Calif Med Ctr, 72-88. *Mem:* Am Asn Physicists Med; Am Phys Soc. *Res:* Solid state physics; reactor fuels; X-ray diffraction; electron diffraction and microscopy; radiation effects; semiconductors; dosimetry; diagnostic radiological physics; image quality; resolution; modulation transfer function; diagnostic quality assurance. *Mailing Add:* 2339 Kenilworth Ave Los Angeles CA 90039

KLEIN, DAVID ROBERT, b Fitchburg, Mass, May 18, 27; c 3. MAMMALIAN ECOLOGY. *Educ:* Univ Conn, BS, 51; Univ Alaska, MS, 53; Univ BC, PhD(zool), 63. *Prof Exp:* Biologist, US Fish & Wildlife Serv, 55-59; biologist, Alaska Dept Fish & Game, 59-61, res dir, 61-62; DIR, ALASKA COOP WILDLIFE RES UNIT & PROF WILDLIFE ECOL,

UNIV ALASKA, 62- *Concurrent Pos:* NSF inst grant, 63-64; Bur Sport Fisheries & Wildlife grant, 64-65 & Bur Land Mgr, 65-67; vis res biologist, Kalo Game Biol Sta, Denmark, 67; vis prof, Univ Oslo, 71-72, Univ Pretoria, 83. *Mem:* AAAS; Wildlife Soc; Arctic Inst NAm; Soc Range Mgt; Am Soc Mammalogists. *Res:* Foraging dynamics of arctic ungulates; man's impact on the environment. *Mailing Add:* Alaska Coop Wildlife Res Unit Univ Alaska Fairbanks AK 99775-0990

KLEIN, DAVID XAVIER, organic chemistry; deceased, see previous edition for last biography

KLEIN, DEANA TARSON, b Chicago, Ill, Jan 7, 25; m 47. MYCOLOGY. *Educ:* Univ Chicago, BS, 47, MS, 48, PhD(bot), 52. *Prof Exp:* Res asst, Food Res Inst, Univ Chicago, 52-53; res asst chem embryol, Columbia Univ, 54-55, USPHS res fel, Dept Zool, 55-57; USPHS res fel microbiol & immunol, Albert Einstein Col Med, 57-58, from instr to asst prof, 58-66; asst prof biol sci, Hunter Col, 66-67; assoc prof, 67-71, PROF BIOL, ST MICHAEL'S COL, VT, 71- *Mem:* Am Soc Plant Physiologists; Bot Soc Am. *Res:* Physiology microorganisms and stress in angiosperm. *Mailing Add:* Dept Biol St Michael's Col Colchester VT 05439

KLEIN, DIANE M, b Chicago, Ill, Mar 21, 48. CIRCULATORY SHOCK. *Educ:* Univ Ill, PhD(physiol), 78. *Prof Exp:* Asst prof, 78-83, ASSOC PROF PHYSIOL, LOYOLA UNIV, 83- *Mem:* Am Physiol Soc; Circulatory Shock Soc. *Mailing Add:* Dept Physiol Loyola Univ Sch Nursing 2160 S First Ave Bldg 131 S Maywood IL 60153

KLEIN, DOLPH, b New York, NY, May 2, 28; m 56; c 4. CLINICAL MICROBIOLOGY. *Educ:* City Col New York, BS, 50; Rutgers Univ, PhD(microbiol), 61; Am Bd Med Microbiol, dipl, 77. *Prof Exp:* Med technologist, Manhattan Gen Hosp, 50-51; lab asst microbiol, New York City Dept Health, 51-53; bacteriologist, Beth Israel Hosp, NY, 53-54; chief med technologist, Lakeside Hosp, Copiague, 54-57; res asst biophys, Sloan-Kettering Inst Cancer Res, 57-58; res asst agr microbiol, Rutgers Univ, 58-61; res microbiologist, Monsanto Co, Mo, 61-62, sr res microbiologist, Monsanto Res Corp, Mass, 62-63; res assoc biophys chem, Purdue Univ, 63-67; asst prof biochem, Univ Minn, Minneapolis, 67-71, asst prof microbiol & asst dir, Diag Microbiol Lab, Med Ctr, 72-74; dir, Clin Microbiol Lab, Dulce Hosp, 74-89, ASSOC PROF MICROBIOL, MED CTR, DUKE UNIV, 74-,. *Concurrent Pos:* Vis prof, Univ Groningen, Neth, 82-83 & Univ Amsterdam, Neth, 91. *Mem:* Am Soc Microbiol. *Res:* Applications of biotechnology for rapid detection and identification of pathogenic bacteria; bacterial dissimilation of streptomycin; evaluation of antimicrobial agents and their modes of action; physicochemical basis of biological stability in structural proteins. *Mailing Add:* Dept Microbiol & Immunol Duke Univ Med Ctr Box 2929 Durham NC 27710

KLEIN, DONALD ALBERT, b Bridgeport, Conn, Sept 11, 35; m 56; c 4. MICROBIOLOGY. *Educ:* Univ Vt, BS, 57, MS, 61; Pa State Univ, PhD(microbiol), 66. *Prof Exp:* Asst qual control, Nat Dairy Prod Corp, Vt, 57-58; instr food microbiol, Univ Vt, 58-61; res asst microbiol, Pa State Univ, 62-66; asst prof, Ore State Univ, 67-70; from asst prof to assoc prof, 70-78, PROF MICROBIOL, COLO STATE UNIV, 78- *Concurrent Pos:* Vis prof, Univ Kiel, 75 & Univ Copenhagen, 78. *Mem:* Am Soc Microbiol; Am Soc Agron. *Res:* Mined land reclamation microbiology; rhizosphere microbiology; microbial transformation of hydrocarbons and pesticides; soil ecology. *Mailing Add:* Dept Microbiol Colo State Univ Ft Collins CO 80523

KLEIN, DONALD FRANKLIN, b New York, NY, Sept 4, 28; c 5. PSYCHIATRY, PSYCHOPHARMACOLOGY. *Educ:* Colby Col, BA, 47; State Univ NY Downstate Med Ctr, MD, 52; Am Bd Psychiat & Neurol, dipl psychiat, 59. *Prof Exp:* Intern, USPHS Hosp, Staten Island, NY, 52-53; resident, Creedmoor State Hosp, 53-54 & 56-58; asst surg & staff psychiatrist, USPHS Hosp, Lexington, Ky, 54-56; res assoc psychiat, Creedmoor Inst Psychobiol Studies, 57-59; res assoc, Hillside Hosp, 59-64, dir res, 65-70, med dir for eval, 70-71, dir dept psychiat, res & eval, Long Island Jewish-Hillside Med Ctr, 72-76; prof psychiat, Col Med, State Univ NY, Stony Brook, 72-76; DIR RES, NY STATE PSYCHIAT INST, 76-; PROF PSYCHIAT, COL PHYSICIANS & SURGEONS, COLUMBIA UNIV, 78-; CONSULT, NAT INST DRUG ADMIN, 90- *Concurrent Pos:* Pvt pract, 56-; candidate, New York Psychoanal Inst, 57-61; USPHS ment health career investr, 61-64, sr staff psychiatrist, 65; NIMH grants, 61-; mem Hofheimer Prize Bd, Am Psychiat Asn, 69-75 & task force methadone & narcotic antagonist eval, 71-73; adj prof psychol, Queens Col, City Univ NY, 69-; psychiatrist-in-chief, Queen Hosp Ctr, 70-71, full attend psychiatrist, 72-85; mem clin pharmacol study sect, NIMH, 71-75 & Neuropharmacol Adv Comt, Food & Drug Admin, 71; chmn, Comt Res & Pub, Long Island Jewish-Hillside Med Ctr, 72-75, consult, 76-85; vis prof psychiat, Univ Auckland, NZ, 75, Albert Einstein Col Med, 76-77; lectr, Columbia Univ, 76-78; chmn, Res Adv Coun, Texas Dept Mental Health & Mental Retardation, 83-85; pres, Nat Found Depressive Illness, 83-; mem, Sci Adv Bd, Nat Depressive & Manic Depressive Asn, 86-, chmn, Clin Comt, 88-; consult, ADAMHA, 88-89, sr sci adv, 89-90. *Honors & Awards:* A E Bennett Neuropsychiat Res Award, 64; Res Award, Nat Asn Pvt Psychiat Hosps, 65 & 71; Samuel W Hamilton Award, Am Psychopath Asn, 80; William R McAlpin Award, Res Achievement, 88; Gold Medal Award, Soc Biol Psychiat, 90. *Mem:* Fel Am Col Neuropsychopharmacol (pres, 81); fel Am Psychiat Asn; Am Psychopath Asn (treas, 72, pres, 78); Int Neuropsychol Soc; fel Royal Col Psychiat; AAAS. *Res:* Diagnosis and drug treatment of psychiatric disorders; psychiatric case studies, treatment, drugs and outcome; age of onset of drug abuse in psychiatric inpatients; phobic anxiety syndrome complicated by drug dependence and addiction. *Mailing Add:* NY State Psychiat Inst 722 W 168 New York NY 10032

KLEIN, DONALD LEE, b Brooklyn, NY, Dec 19, 30; m 52; c 6. INORGANIC CHEMISTRY. *Educ:* Polytech Inst Brooklyn, BSCh, 52; Univ Conn, MS, 56, PhD(chem), 59. *Prof Exp:* Engr, Sylvania Elec Prod, Inc, 52-54; asst, Univ Conn, 54-55, asst instr, 55-58; mem tech staff, Bell Tel Labs, Inc, 58-67; sr chemist, Gen Technol Div, IBM Corp, 67-87; CONSULT, 87- *Concurrent Pos:* Adj lectr, Rochester Inst Technol & Dutchess Community Col, 87- *Mem:* Am Chem Soc; Electrochem Soc; Sigma Xi. *Res:* Electrochemistry; photochemistry; semiconductor materials and processing. *Mailing Add:* Four Carnelli Ct Poughkeepsie NY 12603

KLEIN, DOUGLAS J, b Portland, Ore, Nov 8, 42; m 66; c 2. QUANTUM CHEMISTRY, MOLECULAR PHYSICS. *Educ:* Ore State Univ, BSc, 64; Univ Tex, Austin, MA, 67, PhD(chem), 69. *Prof Exp:* Instr chem, Univ Tex, Austin, 69; Air Force Off Sci Res-Nat Res Coun fel, Princeton Univ, 69-70, univ fel, 70-71; asst prof physics, Univ Tex, Austin, 71-78; from asst prof to assoc prof, 79-87, PROF CHEM, TEX A&M UNIV, GALVESTON, 88- *Concurrent Pos:* Vis asst prof, Rice Univ, Houston, 79; vis sci officer, Off Naval Res, 84. *Mem:* Am Chem Soc; Am Phys Soc; AAAS. *Res:* Theoretical models for molecules and solids, with special emphasis on correlation effects and group-theoretic methods. *Mailing Add:* Dept Marine Sci Tex A&M Univ Galveston TX 77553-1675

KLEIN, EDMUND, b Vienna, Austria, Oct 22, 21; Can citizen; m 52; c 5. MEDICINE, DERMATOLOGY. *Educ:* Univ Toronto, BA, 47, MD, 51. *Prof Exp:* Nat Res Coun fel, Univ Labs, Harvard Med Sch, 51-52, Children's Cancer Res Found fel, Children's Med Ctr, 52-53, res assoc, 53-55, assoc clin path, 54-56; asst resident, Mass Gen Hosp, 58-59, res assoc, 59-60; asst prof med, Sch Med, Tufts Univ, 59-60, asst prof dermat, 60-61; assoc prof exp path, Grad Sch, 62-70, RES PROF MED DERMAT, SCH MED, STATE UNIV NY, BUFFALO, 70-; CHIEF DERMAT, ROSWELL PARK MEM INST, 61- *Concurrent Pos:* Clin & res fel dermat, Mass Gen Hosp, 56-58, teaching fel, 58-59; consult, Children's Cancer Res Found, Children's Med Ctr & Acute Leukemia Task Froce, Nat Cancer Inst; mem adv comt biol effects of optical masers, US Army Med Res & Develop Command, Off Surgeon Gen, DC. *Honors & Awards:* Lasker Immunol Award, 72; Founders' Award Immunol, 75. *Mem:* Am Asn Cancer Res; Am Soc Exp Path; Soc Invest Dermat; Am Acad Dermat; Am Soc Clin Pharmacol & Therapeut; Sigma Xi. *Res:* Biological effects of lasers; lipid transport; hemorrhagic diathesis; cutaneous neoplasms. *Mailing Add:* 1331 N Forest Rd Williamsville NY 14221

KLEIN, EDWARD LAWRENCE, b Roscoe, Pa, Feb 17, 36; m 63; c 4. WOOD ENERGY, FOREST PRODUCTS. *Educ:* Pa State Univ, BS, 58, MS, 61; Baylor Univ, MBA, 63; La State Univ, PhD(forestry mkt), 68. *Prof Exp:* Asst dist forester, Md Dept of Forestry, 58-60; instr in charge mkt res, La State Univ, 65-68; mkt analyst, US Forest Serv, Princeton, WVa, 68-69; supvr econ sect, Tenn Valley Auth, 69-73; mfg mgr, Indust Wood & Pallet Co, 73-75; staff asst to dir, Div Forestry, Tenn Valley Auth, Norris, 75-77, proj leader wood energy, 77-79; dir mkt, Enerco Assocs, Langhorne, PA, 79-86; DIR TIMBER OPERS, BROOKE INT, 91- *Mem:* Soc Am Foresters; Forest Prod Res Soc. *Res:* Economics of new plant sites and construction; production of saw mills and pallet plants; utilization of biomass as an energy source. *Mailing Add:* 1022 Olive St Coatesville PA 19320

KLEIN, ELENA BUIMOVICI, b Bucharest, Romania, Nov 12, 30; m 62; c 1. MICROBIOLOGY, INFECTIOUS DISEASES. *Educ:* Univ Bucharest, MD, 53. *Prof Exp:* Sr res scientist microbiol, Cantacuzino Inst, Bucharest, 53-72; res scientist virol, Bellevue Hosp, NY Univ, 72-73; ASST PROF PEDIAT, ROOSEVELT HOSP, COLUMBIA UNIV, 73-; DIR, VIRUS LAB, ST LUKE'S-ROOSEVELT HOSP CTR, 80- *Mem:* Romanian Soc Infectious Path; Am Soc Microbiol; fel Soc Infectious Dis. *Res:* Epidemiology of diptheria; genetics of enteroviruses; epidemiology of poliomyelities; congenital rubella; cell mediated immunity in viral infections; viral vaccines. *Mailing Add:* 27 Nob Ct New Rochelle NY 10804

KLEIN, ELIAS, b Leipzig, Ger, Oct 26, 24; m 48; c 3. PHYSICAL CHEMISTRY. *Educ:* Tulane Univ, MS, 52, PhD(phys chem), 54. *Prof Exp:* Res chemist, Southern Regional Res Lab, USDA, 54-55, invest head, 55-58; res chemist phys chem, Courtaulds, Inc, Ala, 58-60, sect head, 60-62, mgr res dept, 62-64, dir res & develop, 64-67; dir phys chem, Gulf Southern Res Inst, 67, dir Lake Pontchartrain Lab, 67-81; ASSOC PROF ENG, SCH CHEM ENG, 82-, PROF MED, UNIV LOUISVILLE SCH MED, 84- *Concurrent Pos:* Consult, Kalvar Corp, 52-55; adj prof, Loyola Univ, La, 69-; founder & pres, NAm Membrane Soc. *Mem:* Am Chem Soc; Sci Res Soc Am; Am Soc Artificial Internal Organs. *Res:* Kinetics; cellulose and fiber chemistry; membrane transport; hemodialysis; reverse osmosis. *Mailing Add:* 5517 Hempstead Rd Louisville KY 40207

KLEIN, FRANCIS MICHAEL, b Wilkes Barre, Pa, Nov 1, 41; m 64; c 5. ORGANIC CHEMISTRY. *Educ:* King's Col, Pa, BS, 63; Univ Notre Dame, PhD(org chem), 67. *Prof Exp:* NIH fel, Iowa State Univ, 67-68; asst prof, 68-73, asst dean, 78-79, ASSOC PROF CHEM, CREIGHTON UNIV, 73-, CHMN DEPT, 87- *Mem:* Am Chem Soc. *Res:* Organic photochemistry; reaction mechanisms, especially stereochemical factors; molecular orbital calculations of reaction energetics; electrophilic addition reactions. *Mailing Add:* Dept Chem Creighton Univ Omaha NE 68178-0104

KLEIN, GEORGE DEVRIES, b s'Gravenhage, Neth, Jan 21, 33; US citizen; m 82. SEDIMENTOLOGY, BASIN ANALYSIS. *Educ:* Wesleyan Univ, BA, 54; Univ Kans, MA, 57; Yale Univ, PhD(geol), 60. *Prof Exp:* Part-time geologist, State Geol Surv, Kans, 55-56; asst instr, Univ Kans, 56-57; lab instr, Yale Univ, 57-58 & 59-60; res geologist, Sinclair Res, Inc, 60-61; asst prof geol, Univ Pittsburgh, 61-63; asst prof, Univ Pa, 63-66, assoc prof, 66-69; assoc prof, 70-72, PROF GEOL, UNIV ILL, URBANA, 72- *Concurrent Pos:* Vis fel, Oxford Univ, 69; vis assoc prof geol & geophys, Univ Calif, Berkeley; vis prof oceanog, Ore State Univ, 74, Seoul Nat Univ, 80, Univ Tokyo, 83; vis exchange prof geophys sci, Univ Chicago, 79-80; assoc, Ctr Adv Study, Univ Ill, 74 & 83; sr res fel, Japan Soc Promotion Sci, 83; sr Fulbright res fel, Vrije Univ, Amsterdam, 89. *Mem:* AAAS; Geol Soc Am; Am Asn Petrol Geol; Soc Econ Paleontol & Mineral; Int Asn Sedimentol; Am Geophys Union; Soc Explor Geophysicists. *Res:* Recent sediments; sedimentary and sandstone petrology; basin analysis; marine geology; turbidites; sedimentation on tidal flats; tidalites; back arc and cratonic basins; Deep-ocean sediment transport; petroleum sandstone reservoir prediction and diagenesis. *Mailing Add:* 245 Natural Hist Bldg Univ Ill 1301 W Green St Urbana IL 61801-2999

KLEIN, GERALD I(RWIN), b Brooklyn, NY, Sept 22, 28; m 48; c 4. ELECTRICAL ENGINEERING. *Educ:* Cooper Union, BEE, 48; Polytech Inst Brooklyn, MEE, 53. *Prof Exp:* Asst head, Radar & Microwave Electronics Sect, Naval Mat Lab, NY, 48-55; chief, Oscillators & Amplifiers Sect, Evans Signal Lab, NJ, 55-58; mgr, Microwave Tubes Sect, Electronic Tube Div, Westinghouse Elec Corp, 58-65, mgr microwave tech lab, Aerospace Div, Md, 65-68; vpres eng & mfg, Solitron Microwave, 68-70; ADV ENGR, WESTINGHOUSE ELEC CORP, 70- *Mem:* Sr mem Inst Elec & Electronics Engrs. *Res:* Microwave electronics; electromagnetic theory; microwave plasmas; electron tubes. *Mailing Add:* Herley Indust Ten Industry Dr Lancaster PA 17603

KLEIN, GERALD WAYNE, b Seattle, Wash, Mar 28, 39; m 58; c 3. POLYMER CHEMISTRY. *Educ:* Seattle Pac Col, BS, 60; Yale Univ, PhD(org chem), 65. *Prof Exp:* NSF fel, Gothenburg Univ, 64-65; asst prof chem, Univ Wyo, 65-69; NIH spec res fel & vis prof, Columbia Univ, 69-70; sr res chemist, 70-77, RES ASSOC, EASTMAN KODAK CO, 77- *Mem:* Am Chem Soc. *Res:* Novel synthetic polymers as vehicles, mordants and chemically active components of photographic systems; photographic science; polymeric films. *Mailing Add:* Eight Sutherland St Pittsford NY 14534

KLEIN, GORDON LESLIE, b New York, NY, Aug 26, 46; m 73; c 1. GASTROENTEROLOGY, PEDIATRICS. *Educ:* Columbia Univ, BA, 67; Albert Einstein Col Med, MD, 71; Univ Calif, Los Angeles, MPH, 80. *Prof Exp:* Intern & resident pediat, Stanford Univ Med Ctr, Calif, 71-74; fel nutrit, Sch Med, Johns Hopkins Univ, 76-78; fel gastroenterol, Med Ctr, Univ Calif, Los Angeles, 78-80; adj asst prof pediat, 80-82; asst prof pediat, Med Ctr, & adj asst prof nutrit, Sch Pub Health, Tulane Univ, 82-84; chief serv, Pediat Gastroenterol & Nutrit, City Hope Nat Med Ctr, 84-86; ASSOC PROF PEDIAT & NUTRIT, UNIV TEX MED BR, GALVESTON, 86- *Concurrent Pos:* Res affil, Wadsworth Med Ctr, Vet Admin, Calif, 80-82; clin assoc prof pediat, Sch Med, Univ Southern Calif, 84-; US Pharmacopeia Gen Comt of Rev, 90-95; Am Soc Parenteral & Enteral Nutrit Tech Adv Group on Parenteral Nutrit, 90- *Mem:* Am Soc Bone & Mineral Res; Am Soc Clin Nutrit; Am Gastroenterol Asn; Am Fedn Clin Res; Soc Pediat Res. *Res:* Investigation of abnormalities in calcium and bone metabolism; study of aluminum contamination of parenteral solution and toxicity to bone and other organs. *Mailing Add:* Pediat Dept Univ Tex Med Br Galveston TX 77550-2776

KLEIN, HAROLD GEORGE, b Jersey City, NJ, Mar 14, 29; m 65. VERTEBRATE ZOOLOGY. *Educ:* Cornell Univ, BS, 53, MS, 54, PhD(vert zool), 58. *Prof Exp:* Instr biol, Swarthmore Col, 57-58; asst prof zool, Pa State Univ, 59-62; asst prof, 62-65, ASSOC PROF BIOL, STATE UNIV NY COL PLATTSBURGH, 65- *Mem:* Ecol Soc Am; Am Soc Mammal; Animal Behav Soc. *Res:* Ecological research on vertebrate animals, particularly mammals. *Mailing Add:* Dept Biol Sci State Univ NY Col Plattsburgh NY 12901

KLEIN, HAROLD PAUL, b New York, NY, Apr 1, 21; m 42; c 2. MICROBIAL PHYSIOLOGY, EXOBIOLOGY. *Educ:* Brooklyn Col, BA, 42; Univ Calif, Berkeley, PhD(bact, biochem), 50. *Prof Exp:* Chemist, US Civil Serv, 42-43; instr bact, Armstrong Jr Col, 46-47; asst, Univ Calif, 47-50; res fel, Am Cancer Soc, Mass Gen Hosp, 50-51; from instr to asst prof microbiol, Univ Wash, Seattle, 51-55; from asst prof to prof biol, Brandeis Univ, 55-63, chmn dept, 56-63; chief exobiol div, Ames Res Ctr, NASA, 63-64, dir life sci, 64-84; SCIENTIST-IN-RESIDENCE, SANTA CLARA UNIV, 84- *Concurrent Pos:* Am Cancer Soc fel, 50; vis prof, Univ Calif, 60-61; NSF sr fel, 63; mem, Joint US-USSR Space Med & Biol Working Group, 71-; leader biol team, Viking Mission to Mars, 75-77; investr, US-USSR Cosmos Flights, 75 & 79; mem space sci bd, Nat Acad Sci, 84-89. *Honors & Awards:* NASA Medal for Except Sci Achievement, 77; Co-recipient, Cleveland-Newcomb Award, AAAS, 77. *Mem:* AAAS; Am Soc Microbiol; Am Chem Soc; Am Soc Biol Chem; Int Acad Astronaut. *Res:* Microbial metabolism; formation of adaptive enzymes; lipid synthesis; space biology; exobiology. *Mailing Add:* Biol Dept Santa Clara Univ Santa Clara CA 95053

KLEIN, HARVEY GERALD, b New York, NY, Oct 22, 30; m 83; c 1. PHARMACEUTICS, MEDICAL INSTRUMENTATION. *Educ:* City Col New York, BS, 53; NY Univ, MS, 57; Purdue Univ, PhD(chem), 61. *Prof Exp:* Chemist plastic additives res, Cent Res Div, Am Cyanamid Co, Stamford, Conn, 61-63; sr mkt analyst, Com Develop Div, Wayne, NJ, 63-66, tech rep & liaison, Washington, DC, 66-67; drug & health care financial analyst, R W Pressprich & Co, 67-69, Wertheim & Co, 69-70 & Andresen & Co, 71-72; pres, Klein Assocs, 73-80, PRES, KLEIN BIOMED CONSULTS, 81- *Mem:* Am Chem Soc; Am Inst Ultrasound Med; Int Soc Optical Eng; Am Soc Echocardiography. *Res:* Medical diagnostic instrumentation and pharmaceutical development. *Mailing Add:* 215 W 90th St New York NY 10024

KLEIN, HERBERT A, b Milwaukee, Wis, Mar 28, 36; m 73; c 2. NUCLEAR MEDICINE. *Educ:* Columbia Univ, AB, 56, MD, 60; Harvard Univ, MA, 68, PhD(biochem), 75; Am Bd Nuclear Med, cert, 74. *Prof Exp:* ASSOC DIR, DIV NUCLEAR MED, DEPT RADIOL, SCH MED, UNIV PITTSBURGH, 80-, DIR, NUCLEAR MED EDUC, 84-; MEM, MED STAFF, PRESBY-UNIV HOSP, PITTSBURGH, 80- *Concurrent Pos:* Prin investr, Wechsler Res Found Grant, 84-85. *Mem:* Soc Nuclear Med; Harvey Soc. *Res:* Computer analysis of radionuclide studies of esophageal function; use of radioiodine in the diagnosis and treatment of thyroid carcinoma. *Mailing Add:* Div Nuclear Med Presby-Univ Hosp DeSoto at O'Hara St Pittsburgh PA 15213

KLEIN, HOWARD JOSEPH, b Kokomo, Ind, July 5, 41; m 64; c 5. METALLURGICAL ENGINEERING. *Educ:* Purdue Univ, BS, 63; Univ Ala, MS, 65; Univ Tenn, PhD(metall eng), 69. *Prof Exp:* Engr, Haynes Stellite Div, Cabot Corp, 69-71, sr engr process metall, 72-73, group leader process metall & ceramics, 73-75, sect mgr, 75-77, dir technol, 77-79, oper mgr high technol, 79-82, dir inventory mgt & qual control, CWP Div, 82-83, dir opers, 83-85, dir & gen mgr technol, Cabot Corp, 85-86; VPRES, HAYNES INT INC, 87- *Concurrent Pos:* Mem, Comt Electroslag Remelting & Plasma Melting Technol, Nat Mat Adv Bd, 74-75 & Comt Joint Coop Electro Metall, US State Dept, 76-; mem comt review US-USSR Agreement Coop Fields Sci & Technol, Nat Acad Sci, 77; chmn tech bd, Am Soc Metals, 85-87; dir, Metals Property Coun, 87. *Honors & Awards:* Von Karman Award, Theodore Von Karman Mem Found, 74; IR 100, Indust Res, 75 & 77. *Mem:* Fel Am Soc Metals Int; Am Inst Mining, Metall & Petrol Engrs; Am Vaccum Soc. *Res:* Development and processing of nickel and cobolt base alloys for application in the aerospace corrosion and wear resistant areas; primary and secondary processing technologies. *Mailing Add:* Haynes Int Inc 1020 W Park Ave PO Box 9013 Kokomo IN 46901

KLEIN, IMRICH, b Kosice, Czech, Sept 14, 28; US citizen; m 50; c 2. PLASTICS & CHEMICAL ENGINEERING. *Educ:* Israel Inst Technol, BS, 56; Case Inst Technol, MS, 58, PhD(chem eng), 59. *Prof Exp:* Res asst, Israel Inst Technol, 55-56; chem engr, Israel Govt, 56-57; res engr, E I du Pont de Nemours & Co, 59-61 & Esso Res & Eng Co, 61-63; sr res engr, Eng Res Cent, Western Elec Co, 63-68; PRES, SCI PROCESS & RES, INC, 68- *Mem:* Am Chem Soc; Am Inst Chem Engrs; Soc Plastics Engrs; Am Inst Chemists. *Res:* Plastics extrusion and processing; development of computerized physical models; computer technology and programming; numerical analysis; polymerization kinetics; design and analysis of experiments; multicomponent phase equilibria; semicrystalline polyolefins; separation processes; chemical process development. *Mailing Add:* 70 S Adelaide Ave Highland Park NJ 08904

KLEIN, JAMES H(ENRY), b Boston, Mass, Oct 12, 20; m 55; c 4. CHEMICAL ENGINEERING. *Educ:* Mass Inst Technol, BS & MS, 43. *Hon Degrees:* ScD, Mass Inst Tech, 50. *Prof Exp:* Maintenance engr, Union Oil Co, 43-46; engr, Quinton Engrs Ltd, 46-47; res assoc, Mass Inst Technol, 47-50, engr & bus mgr, Lexington Proj, 48, instr, 48-49; proj engr, Am Res & Develop Co, 50-51; tech dir, Inst Inventive Res, Div Southwest Res Inst, 51-53; tech consult, State of Mass, 53-56; mem tech staff, Thompson Ramo-Wooldridge, 56-57; dir eng, Stanley Aviation Corp, 58; pres, Klein Aerospace, Inc, 66-88; pres, Jelcar, Inc, 88-89; consult engr, 58-90; RETIRED. *Concurrent Pos:* Consult, Baird-Atomic, 50-56 & Southwest Res Inst, 53-56; dir res, Cook Batteries, 59-63; vpres res, Frost Eng Develop Corp, 61-64; vpres, Aerospace Eng Sales Co, 61-70. *Mem:* Am Chem Soc; Electrochem Soc; Am Inst Chem Engrs; Am Inst Physics; Air Pollution Control Asn. *Res:* Heat and mass transfer; photosynthesis; energy storage and conversion; nuclear engineering; medical instrumentation; air pollution; bio-engineering; development and commercialization of new ideas and inventions. *Mailing Add:* PO Box 990 Breckenridge CO 80424

KLEIN, JAN, b Opava, Czech, Jan 18, 36; m 69; c 3. BIOLOGY, IMMUNOGENETICS. *Educ:* Charles Univ, Prague, MS, 58; Czech Acad Sci, PhD(genetics), 64. *Prof Exp:* Res assoc genetics, Inst Exp Biol & Genetics, Prague, 64-65; res assoc, Sch Med, Stanford Univ, 68-69; from asst prof to assoc prof, Univ Mich, Ann Arbor, 69-74; from assoc prof to prof microbiol, Univ Tex Health Sci Ctr, Dallas, 74-78; DIR DEPT IMMUNOGENETICS, MAX PLANCK INST BIOL, 78-; DISTINGUISHED RES PROF, DEPT MICROBIOL & IMMUNOL, UNIV MIAMI SCH MED, MIAMI, FL, 87- *Concurrent Pos:* Levere K Purcell & Truman St John Mem fels genetics, Sch Med, Stanford Univ, 65-66; NIH grant; Nat Inst Dent Res grant, 69; managing ed, Immunogenetics; mem immunobiol study sect, NIH. *Honors & Awards:* Elisabeth Goldschmidt Mem lectr; Rabbi Schacknai Mem Prize, Transplantation Soc. *Mem:* Am Asn Immunol; Transplantation Soc. *Res:* Immunogenetics; cellular immunology. *Mailing Add:* Dept Immunogenetics Corrensstrasse 42 Tübingen 1 7400 Germany

KLEIN, JERRY ALAN, b Neenah, Wis, Apr 19, 45; m 66; c 2. CHEMICAL ENGINEERING. *Educ:* Univ Wis, BS, 67; Princeton Univ, PhD(chem eng), 72. *Prof Exp:* Develop engr, 72-78, GROUP LEADER ENVIRON CONTROL TECHNOL, CHEM TECHNOL DIV, OAK RIDGE NAT LAB, 78- *Res:* Environmental control technology for coal conversion processes. *Mailing Add:* K-25 Area Bldg 1037 Mail Stop 7358 Oak Ridge Nat Lab Oak Ridge TN 37831

KLEIN, JOHN PETER, b Milwaukee, Wis, Dec 30, 50. RELIABILITY, COMPETING RISKS. *Educ:* Univ Wis, Milwaukee, BA, MS, 75; Univ Mo, PhD(statist), 80. *Prof Exp:* Asst math, Univ Wis, Milwaukee, 74-75, statist, Univ Mo, 75-80; asst prof, 80-86, ASSOC PROF STATIST, OHIO STATE UNIV, 86- *Concurrent Pos:* Researcher, Oak Ridge Nat Labs, 76-77; biostatistician, Ohio State Comprehensive Cancer Ctr, 81- *Mem:* Biomet Soc; Am Statist Asn; Inst Mat Statist; Soc Clin Trials. *Res:* Survival analysis. *Mailing Add:* Cockins Hall Rm 138 1958 Neil Ave Columbus OH 43210

KLEIN, JOHN SHARPLESS, b Ossining, NY, Sept 9, 22; m 63; c 2. APPLIED MATHEMATICS. *Educ:* Haverford Col, BS, 43; Mass Inst Technol, SM, 49; Univ Mich, PhD, 59. *Prof Exp:* Instr math, Williams Col, 49-51, Oberlin Col, 54-55 & Case Inst Technol, 55-56; asst prof, Univ RI, 56-58 & Wilson Col, 59; assoc prof, Lafayette Col, 59-60, Wilson Col, 60-63 & Monmouth Col, NJ, 63-64; head dept, 64-74, PROF MATH, HOBART & WILLIAM SMITH COLS, 64- *Mem:* Soc Indust & Appl Math; Math Asn Am. *Res:* Integral transforms. *Mailing Add:* 11 Norway Maple Dr Geneva NY 14456

KLEIN, LARRY L, b Chicago, Ill, Jan 24, 53. ORGANIC CONDUCTORS, CARBOHYDRATE CHEMISTRY. *Educ:* Ill Inst Technol, BS, 75; Mich State Univ, PhD(chem), 80. *Prof Exp:* NIH fel chem, Harvard Univ, 80-82; ASST PROF CHEM, TEX A&M UNIV, 82- *Mem:* Am Chem Soc. *Res:* Organic synthesis of natural products. *Mailing Add:* Abbott Labs Dept 47N AP9A Abbott Park IL 60064

KLEIN, LEROY, b Newark, NJ, Oct 1, 26; m 58; c 3. BIOCHEMISTRY, MEDICAL RESEARCH. *Educ:* Syracuse Univ, BA, 50; Boston Univ, MA, 52, PhD(biochem), 58; Case Western Reserve Univ, MD, 65. *Prof Exp:* From instr to assoc prof biochem in orthop surg, 63-77, from sr instr to asst prof biochem, 65-71, PROF BIOCHEM IN ORTHOP & MACROMOLECULAR SCI, CASE WESTERN RESERVE UNIV, 77-, ASSOC PROF BIOCHEM, 71- *Concurrent Pos:* Res fel biochem, Case Western Reserve Univ, 58-60, res fel orthop, 60-63, instr orthop surg, 63-65; Kappa Delta res award, 63, res grants, 63- *Mem:* Am Fedn Clin Res; Am Soc Biol Chem; Orthop Res Soc; Geront Soc; Am Soc Bone & Mineral Res; Sigma Xi. *Res:* Connective tissue metabolism and diseases; collagen and mineral turnover in experimental and metabolic bone diseases, aging, wound healing. *Mailing Add:* 511 Wearn Bldg Case Western Med Cleveland OH 44106

KLEIN, LEWIS S, b Youngstown, Ohio, Sept 2, 32; m 60. THEORETICAL PHYSICS. *Educ:* Union Col, NY, BS, 54; Yale Univ, MS, 55, PhD(physics), 58. *Prof Exp:* Fulbright fel & res physicist, Nat Ctr Sci Res, France, 58-59; instr physics, Northwestern Univ, 59-60; sr res staff, Nat Bur Stand, 60-65; assoc prof, 65-70, PROF PHYSICS, HOWARD UNIV, 70- *Mem:* Am Phys Soc. *Res:* Field theory; statistical mechanics; plasma physics. *Mailing Add:* Dept Physics & Astron Howard Univ Washington DC 20059

KLEIN, LUELLA, b Walker, IA, 24. GYNECOLOGY, MATERNAL FETAL MEDICINE. *Educ:* Univ Iowa Sch Med, MD, 49. *Prof Exp:* Rotating intern, 49-50, jr asst resident med, 50-51, jr asst resident surg, 51-52, asst resident Ob-Gyn, 52-54, resident Ob- Gyn, 54-55; instr Ob-Gyn, Western Residency, 55; clin instr, 56-67, assoc prof, 67-73, PROF GYNECOL & OBSTET, EMORY UNIV, 73-, CHMN, 86- *Concurrent Pos:* Sr Fulbright Res Scholar Univ London, 56; consult, Ga Dept Publ Health, 58-62; staff, Piedmont Hosp, Ga Byst Hosp, Crawford Hosp, Atlanta, 60- *Mem:* Inst Med-Nat Acad Sci; Am Med; Am Col Obstet & Gynecol; Am Med Women's Asn; Am Bd Obstet & Gynec. *Mailing Add:* 69 Butler St SE Atlanta GA 30303

KLEIN, MARTIN J(ESSE), b New York, NY, June 25, 24; m 80; c 4. HISTORY OF PHYSICS. *Educ:* Columbia Univ, AB, 42, AM, 44; Mass Inst Technol, PhD(physics), 48. *Prof Exp:* Asst physics, Columbia Univ, 42-44, physicist, Underwater Sound Reference Lab, 44-45; mem, Opers Res Group, Washington, DC, 45-46; res assoc physics, Mass Inst Technol, 46-49; from instr to prof, Case Inst Technol, 49-67; Guggenheim fel, 67-68, prof, 67-73, EUGENE HIGGINS PROF HIST PHYSICS, YALE UNIV, 73- *Concurrent Pos:* Nat Res fel, Dublin Inst Advan Studies, 52-53; Guggenheim fel, Inst Lorentz, Leiden, 58-59; mem, Inst Advan Study, 72; Van der Waals prof, Univ Amsterdam, 74; vis prof, Rockefeller Univ, 75 & Harvard, 89-90. *Mem:* Nat Acad Sci; fel AAAS; fel Am Phys Soc; Hist Sci Soc; Acad Int Hist Sci; Am Acad Arts & Sci. *Res:* History of modern physics; statistical mechanics. *Mailing Add:* Dept Hist Sci Yale Univ Box 2036 New Haven CT 06520-2036

KLEIN, MAX, b New Bedford, Mass, Feb 5, 25; m 85; c 3. CHEMICAL PHYSICS, THERMODYNAMICS. *Educ:* Univ Mass, BS, 48; Univ Md, PhD(physics), 62. *Prof Exp:* Electronics engr comput memory, Nat Bur Standards, 50-55, physicist thermodynamics, 55-63; physicist chem physics, Weizmann Inst Sci, 63-65; physicist thermodynamics, Nat Bur Standards, 65-67, sect chief, 67-77, supvry physicist, 77-81; SR SCIENTIST THERMODYNAMICS, GAS RES INST, 81- *Concurrent Pos:* Mem fac grad sch, NIH, 66-76; mem adv comt grad sch, Dept Agr, 67-72. *Honors & Awards:* Dept Silver Medal, US Dept Com. *Mem:* Am Phys Soc; Am Inst Chem Engrs; AAAS; Sigma Xi; Am Chem Soc; Am Ceramic Soc; NY Acad Sci. *Mailing Add:* c/o GRI 8600 W Bryn Mawr Ave Chicago IL 60631

KLEIN, MELVIN PHILLIP, b Denver, Colo, July 27, 21; m 60; c 2. BIOPHYSICS, BIOPHYSICAL SPECTROSCOPY. *Educ:* Univ Calif, Berkeley, AB, 52, PhD(biophys), 59. *Prof Exp:* Physicist, Radiation Lab, 52-59, biophysicist, Lawrence Radiation Lab, 59-69, BIOPHYSICIST & ASSOC DIR, CHEM BIODYN LAB, LAWRENCE BERKELEY LAB, UNIV CALIF, 69- *Concurrent Pos:* Vis mem tech staff, Bell Tel Labs, Murray Hill, NJ, 60-61; mem biophys & biophys chem study sect, Div Res Grants, NIH, 69-74; mem adv comt, Stable Isotopes Resource, Los Alamos Sci Labs, 75; mem exec comt, Stanford Magnetic Resonsance Lab, Stanford Univ, 76; sabbatical vis, Biophys Group, Ecole Polytech, Palaiseau & Synchrotron Radiation Lab, Univ Paris, 76-77; John Simon Guggenheim Mem Found, 76; chmn, Gordon Res Conf Magnetic Resonance, 77. *Mem:* Biophys Soc; Am Phys Soc. *Res:* Nitrogen fixation; photosynthesis; magnetic resonance spectroscopy; x-ray spectroscopy with synchrotron radiation; membranes. *Mailing Add:* 1140 Keith Ave Berkeley CA 94708

KLEIN, MICHAEL, b Chicago, Ill, Mar 5, 46; m 80; c 2. MEDICINAL CHEMISTRY, MASS SPECTROMETRY. *Educ:* Univ Ill, BS, 67, PhD(med chem), 71. *Prof Exp:* Appointee chem, Chem Div, Argonne Nat Lab, Ill, 72-74; sr res chemist mass spectrometry, liquid chromatography & forensic drug chem, Testing & Res Lab, 74-78, SR CHEMIST, DRUG CONTROL SECT, DRUG ENFORCEMENT ADMIN, US DEPT JUSTICE, 78- *Concurrent Pos:* Presidential internship, Argonne Nat Lab, Ill, 72-73. *Honors & Awards:* Spec Achievement Award, Drug Enforcement Admin, US Dept Justice, 78. *Mem:* Am Chem Soc. *Res:* Identification and quantitation of trace impurities in illicit drugs for forensic purposes; techniques applied include organic synthesis, mass spectrometry, high and resolution, gas and liquid chromatography, nuclear magnetic resonance, infrared; drug control decisions based on actual and potential abuse. *Mailing Add:* 7819 Moorland Lane Bethesda MD 20814

KLEIN, MICHAEL JOHN, b Ames, Iowa, Jan 19, 40; m 62; c 3. RADIO ASTRONOMY. *Educ:* Iowa State Univ, BS, 62; Univ Mich, MS, 66, PhD(astron), 68. *Prof Exp:* Asst res engr, Radio Astron Lab, Univ Mich, 63-64; Nat Res Coun-NASA resident res assoc, 68-69, sr scientist radio astron, Space Sci Div, 69-73, MEM TECH STAFF, SPACE SCI DIV, JET PROPULSION LAB, CALIF INST TECHNOL, 73-; JPC PROJ MGR, SEARCH FOR EXTRA-TERRESTRIAL INTEL (SETI), 81- *Mem:* Am Inst Elec & Electronics Eng; Am Astron Soc; Int Astron Union; Int Union Radio Sci; Am Inst Aeronaut & Astronaut. *Res:* Measuring and interpretating radio frequency emission of galactic and extragalactic radio sources and solar system planets and satellites; apply radio astronomy techniques to the Search for Extraterrestrial Intelligence (SETI). *Mailing Add:* 348 Camino Del Sol South Pasadena CA 91030

KLEIN, MICHAEL LAWRENCE, b London, Eng, Mar 13, 40; m 62; c 2. CONDENSED MATTER PHYSICS. *Educ:* Bristol Univ, BSc, 61, PhD(theoret chem), 64. *Prof Exp:* Ciba Found fel physics, Univ Genoa, 64-65; Imp Chem Industs fel theoret chem, Bristol Univ, 65-67; res assoc physics, Rutgers Univ, 67-68; pr res off chem, Nat Res Coun Can, 68-87; PROF CHEM, UNIV PA, 87- *Concurrent Pos:* Fel World Trade, IBM, San Jose, 70; prof, Univ Paris, 75; prof chem, McMaster Univ, Hamilton, Ont, 77; fel, Japan Soc Prom Sci, 82, fel, Commoner Trinity Col, Cambridge, UK, 85; Louis Néel prof, Ecole Normale Superierre, Lyon, France, 88. *Mem:* Am Phys Soc; Royal Soc Chem; Can Inst Physics; fel Chem Inst Can; fel Royal Soc Can. *Res:* Computer simulation studies in physical chemistry; solid state physics. *Mailing Add:* Dept Chem Univ Pa Philadelphia PA 19104-6323

KLEIN, MICHAEL TULLY, b Wilmington, Del, Mar 15, 55; m; c 3. MODELLING, APPLIED & INDUSTRIAL CHEMISTRY. *Educ:* Univ Del, BChE, 77; Mass Inst Technol, ScD, 81. *Prof Exp:* Instr chem eng, Mass Inst Technol, 80; from asst prof to assoc prof chem eng, 81-89, assoc dean, Col Eng, 87-88, DIR, CTR CATALYTIC SCI & TECHNOL, UNIV DEL, 88-, PROF CHEM ENG, 89- *Concurrent Pos:* Consult, var oil & chem cos; lectr, var univ & orgn, 82-91; prin young investr, Outstanding Young Men Am, NSF, 85 & 88. *Mem:* Am Inst Chem Engrs; Am Chem Soc. *Res:* Chemical reaction engineering of complex mixtures, including resid upgrading, hydroprocessing, and applied catalysis; author of over 80 technical papers. *Mailing Add:* Dept Chem Eng Univ Del Newark DE 19716

KLEIN, MICHAEL W, b Teglas, Hungary, Mar 29, 31; US citizen; m 55; c 2. SOLID STATE PHYSICS, STATISTICAL MECHANICS. *Educ:* Univ Colo, BS, 56; Cornell Univ, PhD(physics), 62. *Prof Exp:* Asst physics, Cornell Univ, 56-61; res physicist, Lincoln Labs, Mass Inst Technol, 61-62 & Sperry Rand Res Ctr, Mass, 62-68; assoc prof physics, Wesleyan Univ, 68-71; assoc prof physics, Bar-Ilan Univ, Israel, 71-77; vis prof, Physics Dept, Univ Ill, 77-79; head dept, 79-84, PROF PHYSICS, WORCESTER POLYTECH INST, 79- *Concurrent Pos:* Vis assoc prof, Brandeis Univ, 67-68. *Mem:* Am Phys Soc. *Res:* Theory of magnetism; solid state physics; statistical mechanics; plasma physics; glassy and amorphous systems. *Mailing Add:* Dept Physics Worcester Polytech Inst Worcester MA 01609

KLEIN, MILES VINCENT, b Cleveland, Ohio, Mar 9, 33; m 56; c 2. PHYSICS. *Educ:* Northwestern Univ, BS, 54; Cornell Univ, PhD(physics), 61. *Prof Exp:* NSF fel physics, Stuttgart Tech Univ, 61-62; from asst prof to assoc prof, 62-69, PROF PHYSICS, UNIV ILL, URBANA, 69-, DIR, SCI & TECHNOL CTR SUPERCONDUCTIVITY, 89- *Honors & Awards:* Frank Isakson Prize, Am Phys Soc, 90. *Mem:* Sr mem, Inst Elec & Electronics Engrs; Optical Soc Am. *Res:* Raman scattering in solids; optical properties of solids. *Mailing Add:* Loomis Lab Univ Ill 1110 W Green Urbana IL 61801

KLEIN, MILTON M, b New York, NY, Apr 19, 17; m 51; c 1. FLUID DYNAMICS. *Educ:* NY Univ, MS, 50, PhD(physics), 56. *Prof Exp:* Physicist aerodyn, Nat Adv Comt Aeronaut, 42-47, aeronaut res scientist, 48-50; instr physics, NY Univ, 50-55; physicist, Gen Elec Co, 55-61; physicist nuclear debris studies, Geophys Corp Am, 61-63, re-entry physics, Lincoln Lab, 63-64 & plasma physics & hydrodyn, GCA Corp, 64-67; RES PHYSICIST, AIR FORCE CAMBRIDGE RES LABS, BEDFORD, 67- *Concurrent Pos:* Lectr, Grad Sch, Univ Pa, 57-58. *Mem:* Am Phys Soc. *Res:* Fluid dynamics; diffusion; turbulence; heat transfer; interaction of buoyant turbulent jets with air; dissipation of fog by heat; cloud physics; meteorology. *Mailing Add:* 54 Burlington St Lexington MA 02173

KLEIN, MORTON, b New York, NY, Aug 9, 25; m 49; c 2. INDUSTRIAL ENGINEERING, OPERATIONS RESEARCH. *Educ:* Duke Univ, BS, 46; Columbia Univ, MS, 52, EngScD, 57. *Prof Exp:* Ord engr, Picatinny Arsenal, US Army, 50-54; from instr to assoc prof indust eng, 56-69, chmn dept, 82-85, PROF INDUST ENG & OPERS RES, COLUMBIA UNIV, 69- *Concurrent Pos:* Consult Ind & Govt; ed, Mgt Sci, 70-77. *Mem:* Opers Res Soc Am; Inst Mgt Sci; Am Inst Indust Eng. *Res:* Network flows; production planning; research and publications on production planning; scheduling early cancer detection examinations and network flows. *Mailing Add:* Dept Indust Eng Columbia Univ Main Div New York NY 10027

KLEIN, MORTON, b Philadelphia, Pa, Nov 30, 14; m 36; c 2. VIROLOGY. *Educ:* Univ Pa, BS, 38, MS, 40, PhD(immunol), 42. *Prof Exp:* Asst instr bact, Univ Pa, 40-42; res assoc, Dept Med, Univ Chicago, 42-43; res assoc, Virus Infections & Chemother, Sch Med, Univ Pa, 43-48; asst prof bact, Jefferson Med Col, 48-50; assoc prof, 50-57, PROF MICROBIOL, SCH MED, TEMPLE UNIV, 57- *Mem:* Fel AAAS; Am Soc Microbiol; fel Am Acad Microbiol. *Res:* Inactivation of viruses; immunology of viral infections; role of macrophages in resistance to infection. *Mailing Add:* Dept Microbiol Temple Univ Sch Med Philadelphia PA 19140

KLEIN, MORTON JOSEPH, b Chicago, Ill, Feb 26, 28; m 53; c 3. INORGANIC CHEMISTRY. *Educ:* Univ Ill, BS, 48; Ill Inst Technol, PhD(chem), 53. *Prof Exp:* Assoc chemist catalysis, 53-54, res chemist org chem, 54-56, asst supvr propellant res, 56-59, supvr, 59-62, asst dir chem res, 62-65, dir appl chem, 65-69, dir chem res, 69-75, dir chem & chem eng res, 75-77, VPRES RES OPERS, IIT RES INST, 77- *Mem:* Am Chem Soc; Am Inst Aeronaut & Astronaut; Am Inst Chem; Int Ozone Asn (pres, 75-79). *Res:* Synthesis and evaluation of new high energy materials; research management. *Mailing Add:* IIT Res Inst Ten W 35th St Chicago IL 60616

KLEIN, NATHAN, b New York, NY, July 29, 31; m 52; c 5. PHYSICAL CHEMISTRY. *Educ:* City Col New York, BS, 51; Columbia Univ, MA, 52; Univ Del, PhD(chem), 67. *Prof Exp:* Res asst biochem, Sloan-Kettering Inst Cancer Res, 52; chemist, Edgewood Arsenal, 55-66, group leader radiation chem, US Army Nuclear Defense Lab, 66-74, GROUP LEADER, US ARMY BALLISTIC RES LABS, 74- *Mem:* Am Chem Soc; Sigma Xi; fel Am Inst Chem. *Res:* Physical chemistry of aqueous solutions; reactions rates of high energy compounds; chemical kinetics of propellants and explosives; high temperature, high pressure reaction studies. *Mailing Add:* US Army Ballistic Res Labs Aberdeen Proving Ground MD 21005

KLEIN, NELSON HAROLD, b New York, NY, Mar 6, 42; m 66; c 2. PHYSICS. *Educ:* Drexel Univ, BS, 64, MS, 68, PhD(physics), 72. *Prof Exp:* Assoc prof physics, 72-78, CHMN DEPT SCI, BUCKS COUNTY COMMUN COL, 78-, PROF PHYSICS, 80- *Mem:* Am Asn Physics Teachers. *Mailing Add:* Dept Sci Bucks County Commun Col Swamp Rd Newtown PA 18940

KLEIN, NORMAN W, b San Francisco, Calif, Feb 6, 31. REPRODUCTIVE TOXICOLOGY. *Educ:* Univ Calif, PhD(nutrit), 60. *Prof Exp:* PROF ANIMAL GENETICS & NUTRIT & MOLECULAR CELL BIOL, UNIV CONN, 76- *Mailing Add:* Ctr Environ Health Univ Conn 3636 Horsebarn Rd Exten Storrs CT 06268

KLEIN, PAUL ALVIN, b Weehawken, NJ, Feb 1, 41; m 63; c 2. VIROLOGY, IMMUNOLOGY. *Educ:* Rutgers Univ, New Brunswick, BA, 63; Univ Fla, PhD(med sci), 67. *Prof Exp:* PROF PATH, COL MED, UNIV FLA, 69- *Mem:* AAAS; Am Asn Immunol; Reticuloendothelial Soc; Sigma Xi. *Res:* Viral immunology; autoimmunity; diabetes; cancer biology. *Mailing Add:* Dept Path Univ Fla Col Med Gainesville FL 32601

KLEIN, PETER DOUGLAS, b Elmhurst, Ill, Nov 30, 27; m 50, 68; c 3. ANALYTICAL CHEMISTRY, BIOCHEMISTRY. *Educ:* Antioch Col, BS, 48; Wayne State Univ, MS, 50, PhD(physiol chem), 54. *Prof Exp:* From asst biochemist to sr biochemist, Div Biol Med Res, Argonne Nat Lab, 69-72; prof med, Univ Chicago, 72-80; mem staff, Div Biol & Med Res, Argonne Nat Lab, 76-80; PROF PEDIAT, BAYLOR COL MED, 80-; DIR, STABLE ISOTOPE LAB, CHILDREN'S NUTRIT RES CTR, 80- *Mem:* Am Soc Mass Spectrometry; Am Asn Study Liver Dis; Am Soc Biol Chemists; Am Gastroenterol Asn; Am Soc Clin Nutrit; Am Pediat Asn. *Res:* Application of stable isotope tracer methodology and mass spectrometry to clinical research in nutrition and gastroenterology; development of non-invasive diagnostic and functional assessments in infants; pregnant and lactating women. *Mailing Add:* Children's Nutrit Res Ctr 1100 Bates St Houston TX 77030

KLEIN, PHILIPP HILLEL, b New York, NY, Sept 14, 26; m 53; c 3. PHYSICAL INORGANIC CHEMISTRY. *Educ:* Syracuse Univ, BS, 48, MS, 51, PhD(phys chem), 53. *Prof Exp:* Asst chem, Syracuse Univ, 48-49, res asst, 51-52; res assoc, Knolls Atomic Power Lab, Gen Elec Co, 52-56, phys chemist, Electronics Lab, 56-61; mem sci staff, Sperry Rand Res Ctr, 61-66; head dielec mat sect, Electronics Res Ctr, NASA, Mass, 66-70; head crystals & pure mat sect, 70-73, head, dielec mat sect, 73-87, RES CONSULT, ELECTRONIC MAT, USN RES LAB, 87- *Mem:* Fel Am Inst Chem; Sigma Xi; Inst Elec & Electronics Engrs; Am Ceramic Soc; Am Phys Soc; Am Asn Crystal Growth. *Res:* Effects of nuclear radiation on gases and dielectric solids; thermoelectricity; compound semiconductors; thermal properties of solids; electronically and optically active solids; crystal growth. *Mailing Add:* 2017 Hillyer Pl NW Washington DC 20009-1005

KLEIN, RALPH, b Pittsburgh, Pa, Jan 24, 18; wid; c 2. PHYSICAL CHEMISTRY. *Educ:* Carnegie Inst Technol, BS, 38; Univ Minn, MS, 40; Univ Pittsburgh, PhD(chem), 50. *Prof Exp:* Phys chemist, US Bur Mines, 38-39, Chem Warfare Serv, 40-41, US Bur Mines, 46-56, Olin Mathieson Chem Corp, 56-60 & Melpar, Inc, 60-61; chief surface chem sect, Nat Bur Standards, Washington, DC, 61-74, sr scientist, 74-85; RETIRED. *Concurrent Pos:* Lady Davis Fel, Technion, Israel, 85-86. *Mem:* Am Chem Soc; Am Phys Soc; Sigma Xi. *Res:* Low temperature chemistry; surface science. *Mailing Add:* 11 Moriah St No 4 Beer Sheva Israel

KLEIN, RICHARD JOSEPH, b Lugoj, Romania, June 11, 26; US citizen; m 57; c 1. MICROBIOLOGY, EXPERIMENTAL MEDICINE. *Educ:* Inst Med, Bucharest, Romania, MD, 53, DSc(virol), 67. *Prof Exp:* From instr to asst prof microbiol, Inst Med, Bucharest, 49-54; sr res scientist, Cantacuzino Inst, Bucharest, 51-72; assoc prof microbiol, Sch Dent, 77-89, ASSOC PROF MICROBIOL, SCH MED, MED CTR, NY UNIV, 72- *Mem:* Am Soc Microbiol; Brit Soc Gen Microbiol; NY Acad Sci. *Res:* Virology; experimental chemotherapy; antiviral drugs; epidemiology; genetics of enteroviruses; herpes simplex virus infections. *Mailing Add:* NY Univ Med Ctr 550 First Ave New York NY 10016

KLEIN, RICHARD LESTER, b Hempstead, NY, Nov 6, 29; m 53; c 2. PHARMACOLOGY, CELL PHYSIOLOGY. *Educ:* Hofstra Univ, BA, 51, MA, 52; Vanderbilt Univ, PhD(biol, chem), 57. *Prof Exp:* Instr pharmacol, Vanderbilt Univ, 58-59; from asst prof to assoc prof, 59-66, prof pharmacol, 66-81, PROF PHARMACOL & TOXICOL, MED CTR, UNIV MISS, 81- *Concurrent Pos:* Nat Heart Inst fel pharmacol, Vanderbilt Univ, 57-59; NIH career prog award, 62-72; vis prof, Biophys Lab, Wenner-Gren Inst, 63-64; vis prof, Karolinska Inst, 69-70. *Mem:* Am Soc Pharmacol & Exp Therapeut; Electron Micros Soc Am; Soc Neurosci; Sigma Xi. *Res:* Sympathetic nerve and catecholamine storage visicles, including permeability, enzyme activity, composition, histochemistry and ultrastructure; human serum dopamine beta-hydroxylase and sympathetic homeostasis. *Mailing Add:* Dept Pharmacol Univ Miss Med Ctr Jackson MS 39216

KLEIN, RICHARD M, b Philadelphia, Pa, Nov 10, 37. INORGANIC CHEMISTRY, ORGANIC CHEMISTRY. *Educ:* Williams Col, BA, 59; Univ Ill, MS, 62, PhD(polymer chem), 63. *Prof Exp:* Mem staff sales develop spec prod, Rohm and Haas Co, 63-65, mgr sales develop, Pa, 65-66, mgr int

opers sales develop, 66-68; asst managing dir, Triton Chem SAfrica, 68-69; asst to pres int opers, Tanatex Chem Corp, NJ, 69-70; pres, Ionac Chem Co, 70-78; group vpres chem, 78-84, PRES & CHIEF EXEC OFFICER, SYBRON CHEM, INC, 84- *Concurrent Pos:* Pres, Gamlen Chem Co NAm, 75-78. *Mem:* Am Chem Soc. *Res:* Chemical management and development. *Mailing Add:* Sybron Chem Inc Birmingham Rd PO Box 66 Birmingham NJ 08011

KLEIN, RICHARD M, b Chicago, Ill, Mar 17, 23; m 47. PLANT PHYSIOLOGY. *Educ:* Univ Chicago, BS, 47, MS, 48, PhD(bot), 51. *Prof Exp:* Am Cancer Soc fel bot, Univ Chicago, 51-53; res assoc, NY Bot Garden, 53-55, assoc cur, 55-57, Alfred H Caspary cur plant physiol, 57-67; PROF BOT, UNIV VT, 67- *Honors & Awards:* UVM Biolog Sci Univ Scholar. *Mem:* Am Soc Plant Physiol; Bot Soc Am; Am Soc Photobiol. *Res:* Growth and development of plants; photobiology; physiological ecology of forest decline. *Mailing Add:* Dept Bot Univ Vt Burlington VT 05405

KLEIN, RICHARD MORRIS, b Brooklyn, NY, Apr 26, 42; m 64; c 2. MATERIALS SCIENCE, GLASS SCIENCE. *Educ:* State Univ NY Col Ceramics, Alfred, BS, 63, PhD(ceramic sci), 67. *Prof Exp:* mem tech staff, 66-, actg res mgr, 81-82, PRIN INVESTR, 75-, DEPT MGR, GTE LABS, INC, 82- *Concurrent Pos:* Mem Nat Bd Adv, Rose-Hulman Inst Technol, Terre-Haute, Ind, 87- *Mem:* Am Ceramic Soc; Nat Inst Ceramic Engrs; Soc Glass Technol; Sigma Xi. *Res:* Management of research on components and techniques for optical communications, including optical fibers, semiconductor devices, optical amplification, and optical distribution techniques. *Mailing Add:* GTE Labs Inc 40 Sylvan Rd Waltham MA 02254

KLEIN, ROBERT HERBERT, b New York, NY, Dec 5, 32; m 61; c 3. PHYSICS. *Educ:* Columbia Univ, AB, 53; Carnegie Inst Technol, PhD(physics), 63. *Prof Exp:* Asst physics, Carnegie Inst Technol, 53-55; assoc scientist, Avco Res & Adv Develop Corp, 56-57; solid state physicist, Electronic Res Directorate, Air Force Cambridge Res Ctr, Mass, 57-59; asst theoret physics, Carnegie Inst Technol, 59-63; res scientist, Courant Inst, NY Univ, 63-65; res assoc physics, Case Inst Technol, 65-67; dir First Col, 75-78, 84-90, ASSOC PROF PHYSICS, CLEVELAND STATE UNIV, 67- *Mem:* Am Phys Soc. *Res:* Theoretical physics. *Mailing Add:* First Col Dept Physics Cleveland State Univ Cleveland OH 44115

KLEIN, ROBERT MELVIN, b New York, NY, Dec 17, 49; m 75; c 3. ANATOMY, CELL BIOLOGY. *Educ:* Queens Col, City Univ New York, BA, 70; NY Univ, MS, 73, PhD(anat), 74. *Prof Exp:* from asst prof to assoc prof, 81-86, PROF & CHMN DEPT ANAT, MED CTR, UNIV KANS, 87- *Concurrent Pos:* Assoc scientist, Mid-Am Cancer Ctr, 75-; NIH Travel Awards, 77-; fel, Marquette Univ, 74-75; prin investr grants, NIH, 78-88 & Am Heart Asn 80-82. *Mem:* Am Asn Anatomists; Cell Kinetics Soc; Soc Develop Biol; NY Acad Sci; Am Soc Cell Biol; Sigma Xi. *Res:* Influence of growth factors on tooth eruption and development; influences of autonomic nervous system on growth and differentiation of neonatal digestive system. *Mailing Add:* Dept Anat Univ Kans Med Ctr 39th & Rainbow Kansas City KS 66103

KLEIN, RONALD, b New York, NY, July 25, 43; m 65; c 1. OPHTHALMOLOGY. *Educ:* Brooklyn Col, BS, 65; NY Univ, MD, 69; Univ NC, MPH, 73. *Prof Exp:* From asst prof to assoc prof, 78-85, PROF OPHTHAL, SCH MED, UNIV WIS-MADISON, 85- *Mem:* Am Epidemiol Soc; Soc Epidemiol; Am Col Epidemiol; Asn Res Vision & Ophthal; AMA; Am Diabetes Asn. *Res:* Epidemiological studies on ocular and systemic complications of diabetes mellitus; age-related ocular diseases, macular degeneration and cataract. *Mailing Add:* Dept Ophthal Univ Wis 600 Highland Ave Madison WI 53792

KLEIN, RONALD DON, biotechnology, for more information see previous edition

KLEIN, SHERWIN JARED, b Toledo, Ohio, Jan 13, 19; m 43; c 2. PSYCHOPHYSIOLOGY. *Educ:* Western Reserve Univ, AB, 41; Univ Pa, MA, 47, PhD, 51. *Prof Exp:* Asst psychol, Western Reserve Univ, 40-41, instr, 46-51; proj scientist, Air Crew Equip Lab, Naval Air Eng Ctr, 51-65; coordr, 65-74, PROF PSYCHOL, WRIGHT STATE UNIV, 65- *Concurrent Pos:* Assoc & lectr, Grad Sch Educ, Univ Pa, 59-65. *Mem:* Fel AAAS; Am Psychol Asn; Psychonomic Soc; Soc Psychophysiol Res. *Res:* Neurophysiology; biophysics; electrophysiological correlates of behavior, especially quantitative measures of stress; relationships of electrophysiological patterns to performance in mental and motor work. *Mailing Add:* 10438 E Cedar Wax Wing Ct Sun Lake AZ 85248

KLEIN, SIGRID MARTA, b Koenigsberg, Ger, May 1, 32. MICROBIOLOGY, BIOCHEMISTRY. *Educ:* Univ Kiel, Staatsexamen, 57; Brigham Young Univ, MS, 61, PhD(microbiol), 64. *Prof Exp:* Res asst biochem, Sloan-Kettering Lab, 58-59; res assoc microbiol, Brigham Young Univ, 64-67; fel, Charles F Kettering Labs, 67070; res assoc biochem, Brigham Young Univ, 70-81; chemist, Becton Dickinson, 82-87; QUAL ASSURANCE MGR, MURDOCK INT, SPRINGVILLE, 87- *Mem:* Am Soc Microbiol. *Res:* Organization of photosynthetic membranes; intermediary metabolism in blue-green algae; immunodiagnostics. *Mailing Add:* 972 W 200 N Provo UT 84601

KLEIN, V(ERNON) A(LFRED), b Marion, Tex, Sept 10, 18; m 50; c 2. CHEMICAL ENGINEERING. *Educ:* Univ Tex, BSChE, 40, MSChE, 42. *Prof Exp:* Instr chem eng, Univ Tex, 41-42; asst supt org chem prod, Dow Chem Co, 42-44, res & develop engr & group leader org res, 44-51, tech specialist, 51-56, sr tech specialist, 56-60, consult, 60-63, systs specialist, 63-68, sr process specialist, 68-81; RETIRED. *Mem:* Am Chem Soc; Am Inst Chem Engrs; Sigma Xi. *Res:* High pressure vapor-liquid equilibrium; physical chemistry. *Mailing Add:* 63 Plantation Ct Lake Jackson TX 77566

KLEIN, VLADISLAV, b Brezolupy, Czech, Feb 7, 29; m 56; c 1. AERONAUTICS, FLIGHT DYNAMICS. *Educ:* Mil Acad Czech, dipl ing, 49, CSc, 62; Cranfield Inst Technol, PhD(flight dynamics), 74. *Prof Exp:* Res scientist flight dynamics, Aeronaut Res & Test Inst, Prague-Letnany, Czech, 54-70; sr res officer, Cranfield Inst Technol, 70-75; res scientist, 75-78, AT DEPT ENG, GEORGE WASHINGTON UNIV, 78- *Mem:* Soc Natural Philosophy; Am Inst Aeroanut & Astronaut. *Res:* Flight dynamics; flight test data analysis; system identification; control theory. *Mailing Add:* Dept Eng George Washington Univ 2121 Eye St NW Washington DC 20052

KLEIN, WILLIAM, b Philadelphia, Pa, Apr 1, 43; m 67; c 2. STATISTICAL MECHANICS. *Educ:* Temple Univ, BS, 65, PhD(physics), 72. *Prof Exp:* Res scientist physics, Univ Cologne, 74-76; asst prof, 76-81, ASSOC PROF PHYSICS, BOSTON UNIV, 81- *Res:* Mathematics and physics of phase transitions. *Mailing Add:* Dept Physics Boston Univ 590 Commonwealth Boston MA 02215

KLEIN, WILLIAM ARTHUR, b St Paul, Minn, Aug 30, 29; m 52; c 4. ORGANIC CHEMISTRY. *Educ:* Col St Thomas, BS, 51; Univ Md, PhD(org chem), 56. *Prof Exp:* Sr chemist, Minn Mining & Mfg Co, 56-61, res supvr, 61-64, res mgr, 64-67, tech mgr, 67-69, TECH DIR DECORATIVE PROD DIV, 3M CO, ST PAUL, 69- *Mem:* Am Chem Soc. *Res:* Phenolic resins; addition polymers; exocyclic dienes; coated abrasives; decorative films. *Mailing Add:* 2775 Lexington Ave N No 321 St Paul MN 55113

KLEIN, WILLIAM RICHARD, b Grayling, Mich, July 8, 37; m 56; c 4. ACOUSTICS. *Educ:* Cent Mich Univ, AB, 59; Mich State Univ, MS, 62, PhD(physics), 64; Pepperdine Univ, MBA, 80. *Prof Exp:* Asst prof, Southern Ill Univ, 66-69; from assoc prof to prof physics, Murray State Univ, 69-76; MGR RES & ENG, TIDELAND SIGNAL CORP, 76- *Mem:* Int Asn Dent Res; Acoust Soc Am. *Res:* Interaction of light with ultrasound; acoustic birefringence; marine navigational aids; non-linear acoustics. *Mailing Add:* 7210 Pebblemill Lane Houston TX 77086

KLEINBERG, ISRAEL, b Toronto, Ont, May 1, 30; m 55; c 4. ORAL BIOLOGY. *Educ:* Univ Toronto, DDS, 52; Univ Durham, PhD(physiol & biochem), 58; Royal Co Dentists Can, FRCD(C), 69. *Hon Degrees:* DSc, Univ Man, 83. *Prof Exp:* Demonstr, Univ Durham, 55-58; from asst prof to prof biochem, Univ Man, 58-73; PROF & CHMN, DEPT ORAL BIOL & PATH, STATE UNIV NY STONY BROOK, 73- *Concurrent Pos:* Mem assoc comt dent res, Nat Res Coun Can, 59-60, exec mem, 60-65; consult, Nat Inst Dent Res, NIH, 74-81; mem, NY State Health Res Coun, 81-87. *Honors & Awards:* Can Centennial Medal, Govt Can, 67. *Mem:* Int Asn Dent Res; Am Asn Dental Res; Am Soc Microbiol; Am Asn Oral Biol. *Res:* Metabolism of the dental bacterial plaque; plaque formation; peptide growth factors in plaque and other microbial flora control, saliva composition and its oral microbial effects; microchemical and oral diagnostic techniques; biomedical instrumentation development; gingival crevice fluid and its relation to oral and systemic disease. *Mailing Add:* Dept Oral Biol & Path State Univ NY Stony Brook NY 11794

KLEINBERG, JACOB, b Passaic, NJ, Feb 14, 14; m 42; c 2. INORGANIC CHEMISTRY. *Educ:* Randolph-Macon Col, BS, 34; Univ Ill, MS, 37, PhD(chem), 39. *Prof Exp:* Asst chem, Univ Ill, 37-39; asst prof, James Millikin Univ, 40-43; assoc prof, Col Pharm, Univ Ill, 43-46; from asst prof to assoc prof, 46-51, chmn dept, 63-70, PROF CHEM, UNIV KANS, 51- *Mem:* Am Chem Soc. *Res:* Reactions in non-aqueous solvents; unfamiliar oxidation states of the elements. *Mailing Add:* Dept Chem Univ Kans Lawrence KS 66044

KLEINBERG, ROBERT LEONARD, b San Francisco, Calif, Aug 3, 49. GEOPHYSICAL INSTRUMENTATION, PHYSICS OF POROUS MEDIA. *Educ:* Univ Calif, Berkeley, BS, 71; Univ Calif, San Diego, PhD(physics), 78. *Prof Exp:* Postdoctoral fel, Exxon Res & Eng Co, 78-80; res physicist, 80-85, prog leader electromagnetics, 85-88, SR RES SCIENTIST, SCHLUMBERGER-DOLL RES, 88- *Mem:* Am Phys Soc. *Res:* Ultrasonic, electromagnetic, nuclear magnetic resonance, and gravimetric instrumentation for in situ characterization of subsurface geologic formations; physics of porous media. *Mailing Add:* Schlumberger-Doll Res Old Quarry Rd Ridgefield CT 06877

KLEINBERG, WILLIAM, b New York, NY, Jan 24, 11; m 41; c 3. PHYSIOLOGY. *Educ:* NY Univ, BS, 32, MS, 35; Princeton Univ, PhD(physiol, endocrinol), 43. *Prof Exp:* Asst biol, NY Univ, 32-33, teaching fel, 33-36, asst, 36-41; instr biol, Princeton Univ, 41-44; instr pharmacol, Col Physicians & Surgeons, Columbia Univ, 44-45; asst prof & Upjohn fel, Princeton Univ, 45-46; dir, Princeton Labs, Inc, 47-74, CONSULT, PRINCETON LAB PROD CO, 74-77. *Concurrent Pos:* Res assoc, Princeton Univ, 46-50, vis biologist, 50-65, sr vis biologist, 65-76. *Mem:* Endocrine Soc; NY Acad Sci; Reticuloendothelial Soc. *Res:* Blood; protein hormones; enzymes; immunological diagnostics; development of diagnostic tests. *Mailing Add:* 50 Woodland Dr Princeton NJ 08540

KLEINER, ALEXANDER F, JR, b New York, NY, June 18, 42; m 65; c 2. MATHEMATICS. *Educ:* Univ St Thomas, Tex, BA, 64; Tex A&M Univ, MS, 66, PhD(math), 69. *Prof Exp:* Instr math, Tex A&M Univ, 66-69; from asst prof to assoc prof, 69-79, PROF MATH, DRAKE UNIV, 79-, CHMN DEPT, MATH & COMPUT SCI, 84- *Mem:* Math Asn Am; Am Math Soc; Asn Comput Mach. *Res:* Summability theory; mathematics of political processes; graph theory. *Mailing Add:* Dept Math & Comput Sci Drake Univ Des Moines IA 50311

KLEINER, BEAT, applied statistics, for more information see previous edition

KLEINER, SUSAN MALA, b Cleveland, Ohio, Oct 17, 57; m 81. SPORTS NUTRITION, CARDIOVASCULAR NUTRITION. *Educ:* Hiram Col, BA, 79; Case Western Res Univ, MS, 82, PhD(nutrit), 87. *Prof Exp:* Vis prof nutrit, Univ NC, Greensboro, 87-88; asst res prof nutrit, Dept Med, div nutrit, Prev Approach Cardiol & assoc dir, Sarah W Stedman Ctr Nutrit Studies, Dept Med, Duke Univ Med Ctr, 88-91; SPORTS NUTRIT CONSULT, CLEVELAND BROWNS/CLEVELAND CAVALIERS, 91- *Concurrent Pos:* Young investr award, Am Col Nutrit, 87; Columnist, Physician & Sportsmed J, 88- & Exec Health Report Newslett, 89-; adj prof, Dept Nutrit, Sch Med, Case Western Reserve Univ, 91- *Mem:* Am Dietetic Asn; Am Col Sports Med; fel Am Col Nutrit; Nat Strength & Conditioning Asn. *Res:* Nutritional requirements of strength training and muscle building and bodybuilding; writings on nutrition and health. *Mailing Add:* 7133 Deepwood Dr Russel Township OH 44022

KLEINER, WALTER BERNHARD, b Plainfield, NJ, Mar 1, 18; m 44; c 4. ELECTROCHEMISTRY. *Educ:* Rutgers Univ, BS, 39; Ohio State Univ, PhD(electrochem), 46. *Prof Exp:* Anat & control chemist, Calco Chem Div, Am Cyanamid Co, NJ, 39-40; asst chem, Ohio State Univ, 40-43; res engr, Battelle Mem Inst, 44; res & develop electrochemist, Manhattan Proj, Dayton, 45; res electrochemist, Am Smelting & Refining Co, 47-52; from asst prof to prof chem, Essex County Col, 74-85; RES DIR, KLEINER ELECTROCHEM CO, 52- *Concurrent Pos:* Instr chem, Rutgers Univ, 51-52; res dir, Spiral Glass Pipe Co, 53-55; res electrochemists, Hanson-Van Winkle-Munning Co, 58-61, res dir, Electrochem Mach Div, 61-63; mem, Am Electroplaters Soc Res Comt, Nat Bur Standards; res dir, Ionic Mach Co, 63-65; asst prof chem, Upsala Col, 65-66; lectr chem, Newark Col Eng, 66-67; res dir, View-Formall Co, BC, 67. *Mem:* AAAS; Am Chem Soc; Electrochem Soc; Am Electroplaters Soc; Am Soc Metals. *Res:* Electrochemical machining; electroplating-electrodeposition. *Mailing Add:* 1845 First St Dunellen NJ 08812-1340

KLEINERMAN, JEROME, b Pittsburgh, Pa, July 7, 24; m 44; c 3. PATHOLOGY, PHYSIOLOGY. *Educ:* Univ Pittsburgh, BS, 43, MD, 46; Am Bd Path, dipl, 52. *Prof Exp:* Demonstr path, Sch Med, Case Western Reserve Univ, 51-52, from instr to assoc prof, 52-72; assoc dir, St Luke's Hosp, 57-64, head, Dept Path Res & Clin Path, 65-70, assoc dir med res, 70-80, dir, Div Path Res, 76-80; prof path, Sch Med, Case Western Reserve Univ, 72-80, dir dept, 76-80; prof & chmn, Dept Path, Mt Sinai Med Sch, 80-86; DIR, DEPT PATH, METRO HEALTH MED CTR, 86- *Concurrent Pos:* Res fel physiol, Grad Sch Med, Univ Pa, 50-51; Am Heart Asn res fel, Cleveland Metrop Gen Hosp, 52-54, asst path, 52-63, vis assoc path, 62; lectr, Sch Med, Univ Pittsburgh, 63-; attend pulmonary dis, Vet Admin Hosp, 65-; consult path, Saranac Lab, Trudeau Found; mem path B study sect NIH, 65- *Mem:* Am Soc Exp Path; fel Am Soc Clin Path; Am Asn Path & Bacteriologists; Am Heart Asn; Col Am Path. *Res:* Experimental pathology; pulmonary emphysema and physiology; cerrbral blood flow; experimental renal disease; electron microscopy. *Mailing Add:* Metro Health Med Ctr 3395 Scranton Rd Cleveland OH 44109

KLEINFELD, A M, b New York, NY, Feb 6, 41. MEMBRANE BIOPHYSICS. *Educ:* Univ Wis, BA, 62; Rutgers Univ, PhD(nuclear physics), 68. *Prof Exp:* Privat dozent physics, Univ Cologne, Ger, 68-76, consult physics, Weigman Inst, 70; Yale Univ, 72; assoc prof, physiol biophys, Harvard Univ, 76-87; MEM MED BIOL INST, CALIF, 87-; SR SCIENTIST, LIDAK PHARMACEUT, LA JOLLA, CALIF, 88- *Mem:* Am Phys Soc; Biophys Soc. *Res:* Research in fatty acids transport and interaction with immune cells. *Mailing Add:* Med Bio Inst 11077 N Torrey Pines Rd La Jolla CA 92037

KLEINFELD, ERWIN, b Vienna, Austria, Apr 19, 27; nat US; m 68. ALGEBRA,COMBINATORICS & FINITE MATHEMATICS. *Educ:* City Col New York, BS, 48; Univ Pa, MA, 49; Univ Wis, PhD(math), 51. *Prof Exp:* Instr math, Univ Chicago, 51-53; from asst prof to prof, Ohio State Univ, 53-62; prof, Syracuse Univ, 62-68; PROF MATH, UNIV IOWA, 68- *Concurrent Pos:* Vis lectr, Yale Univ, 56-57; partic conf algebra, Bowdoin Col, 57; res assoc, Cornell Univ, 58; vis lectr, Univ Calif, Los Angeles, 59, Stanford Univ, 60, Inst Defense Anal, 61-62 & Agency Int Develop Educ, India, 64-65; consult ed, Charles E Merrill Publ Co, 63-; vis prof, Emory Univ, 76-77, Univ Hawaii, 67-68; mucia consult, Univ Indonesia, Jakarta, 85-86; ed, J Algebra, 64-; prof, ITM/Mucia, Prog Malaysia, 88-89. *Mem:* Am Math Soc. *Res:* Algebra and the foundations of projective geometry. *Mailing Add:* Dept Math Univ Iowa Iowa City IA 52242

KLEINFELD, IRA H, b New York, NY, Apr 5, 47; m 71; c 2. ENGINEERING ECONOMY. *Educ:* Columbia Univ, BS, 67, MS, 69, Eng ScD(indust eng), 74. *Prof Exp:* Instr math, John Jay Col Criminal Justice, City Univ New York, 70-72; from instr to asst prof quant methods, Sch Bus, Hofstra Univ, 72-76; from asst prof to assoc prof, 76-83, PROF INDUST ENG & CHMN, DEPT INDUST ENG & COMPUT SCI, SCH ENG, UNIV NEW HAVEN, 83- *Mem:* Inst Indust Engrs; Inst Mgt Sci; Soc Mfg Engrs; Am Soc Eng Educ. *Res:* Engineering economy; use of computers in industrial engineering. *Mailing Add:* Dept Indust Eng & Comput Sci Univ New Haven 300 Orange Ave West Haven CT 06516

KLEINFELD, MARGARET HUMM, b St Louis, Mo, Apr 7, 38; m 59, 68; c 2. MATHEMATICS. *Educ:* Univ Rochester, BA, 60; Syracuse Univ, MS, 63, PhD(math), 65. *Prof Exp:* Asst prof math, Syracuse Univ, 65-68; asst prof, 68-75, ASSOC PROF MATH, UNIV IOWA, 76- *Mem:* Am Math Soc; Asn Women in Math. *Res:* Non associative ring theory; algebra. *Mailing Add:* Dept Math Univ Iowa Iowa City IA 52242

KLEINFELD, RUTH GRAFMAN, b New York, NY, Feb 9, 28; m 48, 73. CELL BIOLOGY, REPRODUCTIVE BIOLOGY. *Educ:* Brooklyn Col, BS, 49; Univ Wis, MA, 51; Univ Chicago, PhD(cell biol), 53. *Prof Exp:* Res assoc prev med, Yale Univ, 56-57; res assoc path, Ohio State Univ, 57-62; assoc prof pharmacol, State Univ NY Syracuse, 62-70; assoc prof anat, 70-72, chmn, 80-83, PROF ANAT & REPRODUCTIVE BIOL, UNIV HAWAII, MANOA,

72- *Concurrent Pos:* USPHS fel physiol, Ohio State Univ, 53-55, Mary S Muellhaupt scholar, 55-56; USPHS res career develop award pharmacol, State Univ NY Syracuse, 62-70; mem NIH Reproductive Biol Study Sect. *Mem:* Histochem Soc; Am Soc Cell Biol; Am Asn Anat; Soc Develop Biol. *Res:* cell replication and cytodifferentiation; decidualization and placentation in pregnancy; cytochemistry and cell-fine structure. *Mailing Add:* Dept Anat & Reprod Biol Univ Hawaii Manoa Honolulu HI 96822

KLEINHENZ, WILLIAM A, mechanical engineering, metallurgy; deceased, see previous edition for last biography

KLEINHOFS, ANDRIS, b Dobele, Latvia, Dec 25, 37; US citizen; m 65; c 2. GENETICS, AGRONOMY. *Educ:* Univ Nebr, Lincoln, BS, 58, MS, 64, PhD(genetics), 67. *Prof Exp:* Instr genetics, Univ Nebr, Lincoln, 65-67; from asst prof to assoc prof, 67-77, PROF GENETICS, WASH STATE UNIV, 77- *Mem:* AAAS; Genetics Soc Am; Sigma Xi. *Res:* Genetics and biochemistry of nitrate reduction in plants; chloroplast development; uptake and fate of exogenous DNA by plants; mechanisms of chemical mutagenesis. *Mailing Add:* Dept Agron Wash State Univ Pullman WA 99163

KLEINHOLZ, LEWIS HERMANN, b New York, NY, May 18, 10. PHYSIOLOGY. *Educ:* Colby Col, BS, 30; Harvard Univ, MA, 35, PhD(biol), 37; US Army Sch Aviation Med, dipl, 42. *Hon Degrees:* DSc, Colby Col, 63, Reed Col, 85. *Prof Exp:* Instr biol, Colby Col, 30-33; asst, Harvard Univ, 35-37, Sheldon traveling fel, Plymouth & Naples Biol Stas, 37-38, instr biol, 40-41; proj engr, Aero-Med Lab, Wright Field, 45-46; from asst prof to assoc prof, 46-50, PROF BIOL, REED COL, 50- *Concurrent Pos:* Res investr, Marine Biol Lab, Woods Hole, 33-67, dir invert zool, 50-54, trustee, 50-; asst, Radcliffe Col, 36-37; instr, Cambridge Jr Col, 38-42; Jones Scholar, Long Island Biol Sta, 40; Guggenheim fel, Harvard Univ, 45-46; Fulbright fel, 51-52; NSF fac fel, 58; consult, NIH, 60-65; vis res prof, Med-Kem Inst, Univ Lund, 61-62; mem US Nat Comt, Int Union Biol Sci Comt Int Biol Stas, 62-; mem, Sci Adv Bd, Wesleyan Univ, 64-67; Porter scholar, Bermuda Biol Sta; adv comt & pub adv bd, Naples Zool Sta; consult, US Off Educ, 75-76; distinguished vis prof, Univ W Fla, 83. *Mem:* Am Soc Zool; Soc Gen Physiol; NY Acad Sci. *Res:* Biochemistry; physiology of endocrine systems of vertebrates and invertebrates; physiology of nervous system and chromatophores; physiology and specificity of neurosecretory hormones. *Mailing Add:* Dept Biol Reed Col Portland OR 97202

KLEINKOPF, GALE EUGENE, b Twin Falls, Idaho, Oct 2, 40; c 1. PLANT PHYSIOLOGY, POTATO SCIENCE. *Educ:* Univ Idaho, BS, 63; Univ Calif, Davis, PhD(plant physiol), 70. *Prof Exp:* Chemist, Aerojet Gen Corp, 63-64; res assoc agron, Univ Calif, Davis, 64-70, fel plant sci, 70-72; asst prof plant ecol, Univ Calif, Los Angeles, 72-75; assoc prof, 75-82, PROF CROP PHYSIOL, UNIV IDAHO, 82- *Mem:* AAAS; Crop Sci Soc Am; Am Soc Plant Physiol. *Res:* Carbon and nitrogen cycling in some crop species as affected by environmental stress. *Mailing Add:* 3793 N 3600 E Univ Idaho Kimberly ID 83341

KLEINKOPF, MERLIN DEAN, b Macomb, Ill, Feb 1, 26; div; c 4. GEOPHYSICS. *Educ:* Monmouth Col, Ill, BS, 49; Univ Mo-Rolla, BS, 51; Columbia Univ, PhD(geol), 55. *Prof Exp:* Seismic comput geologist, Atlantic Refining Co, 51-52; geologist-geophysicist, Standard Oil Co Calif, 55-57, geologist, 57-58, lead geophysicist gravity & magnetics, 58-62 & 64-65, dist explor geologist, 62-64, prof specialist gravity & magnetics, 65-66; res geophysicist, Br Regional Geophys, US Geol Surv, Colo, 66-70, dep asst to chief geologist, 70-72, RES GEOPHYSICIST, US GEOL SURV, 72- *Concurrent Pos:* Leader, US Deleg Geophysicists to USSR, 71. *Mem:* Geol Soc Am; Am Inst Mining, Metall & Petrol Engrs; Soc Explor Geophysicists; Am Asn Petrol Geologists; Soc Econ Geologists; Sigma Xi. *Res:* Gravity and magnetic model studies using electronic computer; regional geophysical studies of Belt Basin, Northwestern Montana; geophysical studies of porphyry copper, Sonora, Mexico; geophysical studies of Arabian Shield, Saudi Arabia; geophysical studies of the Idaho batholith and associated mineral deposits; geological survey of Bangladesh. *Mailing Add:* US Geol Surv MS 964 Box 25046 Denver CO 80225

KLEINMAN, ARTHUR MICHAEL, b New York, NY, Mar 11, 41; m 65; c 2. PSYCHIATRY, MEDICAL ANTHROPOLOGY. *Educ:* Stanford Univ, AB, 62, MD, 67; Harvard Univ, MA, 74. *Prof Exp:* Lectr anthrop, Harvard Univ, 74-76; clin instr psychiat, Mass Gen Hosp & Harvard Med Sch, 75-76; assoc prof & adj assoc prof, Univ Wash, 76-79, prof psychiat & adj prof anthrop, 79-82; PROF MED ANTHROP & PSYCHIAT, MED SCH & FAC ARTS & SCI, HARVARD UNIV, 82- *Concurrent Pos:* Intern med, Yale-New Haven Hosp, Yale Univ, 67-68; resident psychiat, Mass Gen Hosp, 72-75; Dupont Warren fel, Harvard Med Sch, 74-75 & Milton Fund fel, 75-76; Found Fund Res Psychiat fel, 74-76; prin investr, NIMH res grant, Univ Wash, 78-79, NSF res grant, 83-86; ed-in-chief, Cult, Med & Psychiat, 76-86; Rockefeller Found grant, 83-86; NIMH training grant, 84-; Carnegie Corp grant, 89- *Honors & Awards:* Wellcome Medal Med Anthrop, Royal Anthrop Inst, 80. *Mem:* Inst Med-Nat Acad Sci; fel AAAS; fel Am Anthrop Asn; Soc Med Anthrop; fel Am Psychiat Asn. *Res:* Medical anthropology; depression; cross cultural psychiatry; pain and disability; therapeutic relationships and indigenous healing; China; anthropology of suffering. *Mailing Add:* 330 William James Hall Harvard Univ Cambridge MA 02138

KLEINMAN, CHEMIA JACOB, b Sandomierz, Poland, Feb 1, 32; US citizen; m 55; c 2. PHYSICS. *Educ:* Yeshiva Univ, BA, 53; NY Univ, MS, 56, PhD(physics), 65. *Prof Exp:* Physicist, Mat Lab, Brooklyn Naval Shipyard, 56-57; engr microwave res, Ford Instrument Co, 57-60; lectr physics, City Col New York, 60-64; PROF PHYSICS, LONG ISLAND UNIV, 64-, CHMN DEPT, 80- *Concurrent Pos:* Instr, Yeshiva Univ, 54-55, 58-60; consult, Budd-Lewyt Corp, 56-57; res fel, NY Univ, 65; NASA fel, Goddard Space Flight Ctr, 66; NSF fel, Univ Colo, Boulder, 67; assoc res scientist, NY Univ, 69, consult, 71. *Mem:* Am Phys Soc. *Res:* Atomic and electromagnetic scattering and bound state problems. *Mailing Add:* Dept Physics Long Island Univ Brooklyn NY 11201

KLEINMAN, HYNDA KAREN, b Boston, Mass, May 20, 47; m 68; c 2. CELL BIOLOGY, CONNECTIVE TISSUE RESEARCH. *Educ:* Simmons Col, BS, 69; Mass Inst Technol, MS, 71, PhD(nutrit biochem), 73. *Prof Exp:* Res fel, Med Sch, Tufts Univ & Vet Admin Hosp, 73-75; res chemist, 75-85, CHIEF CELL BIOL SECT, NAT INST DENT RES, NIH, 85- *Honors & Awards:* Dirs Award, NIH, 85; Doren Kamp-Zbinden Award, 87. *Mem:* Soc Biol Chemists; Soc Cell Biol; Soc Complex Carbohydrates; Tissue Cult Asn. *Res:* Structure and function of extracellular matrices and their role in development and in diseases. *Mailing Add:* Nat Inst Dent Res NIH Bldg 30 Rm 407 Bethesda MD 20892

KLEINMAN, JACK G, b New York, NY, Feb 8, 44; m 66; c 2. ACID-BASE PHYSIOLOGY, RENAL DISEASES. *Educ:* NY Univ, MD, 68. *Prof Exp:* Asst chief, Renal Dis Sect, Zabloski Vet Admin Med Ctr, 76-86; assoc prof, 80-89, PROF MED, MED COL, UNIV WIS, 89-; CHIEF, RENAL DIS SECT, ZABLOSKI VET ADMIN MED CTR, 86- *Mem:* Am Physiol Soc; Am Fedn Clin Res; Am Soc Nephrology. *Res:* Cell biology of renal transport mechanisms involved in cell pH regulation and acid-base transport, intestinal acid-base transport; pathophysiology of kidney stone disease. *Mailing Add:* Renal Dis Sect Zabloski Vet Admin Ctr 5000 W National Ave Milwaukee WI 53295

KLEINMAN, KENNETH MARTIN, b Brooklyn, NY, Aug 10, 41; m 82; c 2. PSYCHOPHYSIOLOGY. *Educ:* Grinnell Col, BA, 62; Wash Univ, St Louis, MA, 64, PhD(psychol), 67. *Prof Exp:* Res fel, Dept Psychiat, Wash Univ Sch Med, 62-66; Instr, Dept Psychiat, Univ Mo Sch Med, 67-69; from asst prof to assoc prof, 69-78, PROF PSYCHOL, SOUTHERN ILL UNIV, EDWARDSVILLE, 78-, CHMN DEPT, 79- *Concurrent Pos:* Res consult, St Louis Vet Admin Med Ctr, 69-; res fel psychiat, Charing Cross Hosp Med Sch, London, Eng, 77. *Mem:* Sigma Xi; Soc Psychophysiol Res; Am Psychol Soc. *Res:* Brain-behavior relationships; behavioral med; effects of stress on physiology and behavior; research design and statistical analysis; program evaluation. *Mailing Add:* Dept Psychol Southern Ill Univ Edwardsville IL 62026-1121

KLEINMAN, LEONARD, b New York, NY, July 25, 33; m 57; c 2. SOLID STATE PHYSICS. *Educ:* Univ Calif, Los Angeles, BA, 55, MA, 56; Univ Calif, Berkeley, PhD(physics), 60. *Prof Exp:* Res assoc physics, Univ Chicago, 60-61; asst prof, Univ Pa, 61-64; assoc prof, Univ Southern Calif, 64-67; PROF PHYSICS, UNIV TEX, AUSTIN, 67- *Mem:* Fel Am Phys Soc. *Res:* Energy band theory; theory of electronhypherphonon interactions; semiconductor superlattices; covalent bonding and theory of cohesive energies; electron gas theory; theory of metal surfaces. *Mailing Add:* Dept Physics Univ Tex Austin TX 78712

KLEINMAN, LEONARD I, b Brooklyn, NY, June 29, 35; m 61; c 3. PHYSIOLOGY, NEONATOLOGY. *Educ:* Columbia Col, AB, 56; State Univ NY, MD, 60. *Prof Exp:* From intern to resident pediat, Mass Gen Hosp, 60-62; from asst prof to assoc prof, 66-75, PROF PHYSIOL & PEDIAT, COL MED, UNIV CINCINNATI, 75- *Concurrent Pos:* Res fel physiol, Harvard Univ, 62-65; Fulbright res scholar, Univ Milan, 65-66. *Mem:* Am Physiol Soc; Soc Pediat Res; Am Pediat Soc; Soc Exp Biol Med; AAAS. *Res:* Respiratory, renal and neonatal physiology. *Mailing Add:* Dept Pediat State Univ NY Health Sci Ctr Stony Brook NY 11794

KLEINMAN, MICHAEL THOMAS, b Brooklyn, NY, Mar 8, 42; m 65; c 2. ENVIRONMENTAL & OCCUPATIONAL HEALTH, ANALYTICAL CHEMISTRY. *Educ:* City Univ NY, BS, 65; Polytechnic Inst Brooklyn, MS, 71; NY Univ, PhD(environ health), 77. *Prof Exp:* Radiochemist, US Atomic Energy Comn, 63-65, physical scientist, 65-72; asst res scientist, NY Univ Med Ctr, 72-77; dir Aerosol Lab, Rancho Los Amigos Hosp, 77-82; ASSOC PROF COMMUNITY & ENVIRON MED, UNIV CALIF, IRVINE, 82- *Concurrent Pos:* Consult, Con Ed NY, 74-76, Equitable Environ Asn, 75-77, Intersoc Comt,75-77 & Nat Resources Defense Coun, 76-77; consult, 74-; guest lectr, Univ Calif, Irvine, 79-82; mem, Adv Coun S Coast Air Qual Mgt Dist, 85. *Mem:* Air Pollution Control Asn; AAAS; Sigma Xi; Am Indust Hyg Asn. *Res:* Health effects of pollutant aerosols and gases in humans and animals; chemical alterations of airborne pollutants; development of methods for the generation and characterizations of air pollutants; industrial hygiene. *Mailing Add:* 3492 Lotus St Irvine CA 92714

KLEINMAN, RALPH ELLIS, b New York, NY, July 27, 29; m 55; c 2. APPLIED MATHEMATICS. *Educ:* NY Univ, BA, 50; Univ Mich, MA, 51; Delft Univ Technol, PhD(appl math), 61. *Prof Exp:* Res asst math, Univ Mich, Ann Arbor, 51-53, res assoc, 55-58, from assoc res mathematician to res mathematician, 59-68; assoc prof, 68-72, PROF MATH, UNIV DEL, 72- *Concurrent Pos:* Danish Nat Found Tech Sci grant, Lab Electromagnetic Theory, Tech Univ Denmark, 65-66; vis prof math, Univ Strathclyde, 72 & 82; Nat Res Coun sr resident res assoc, Air Force Cambridge Res Labs, 74-75; vis scientist, David Taylor Naval Ship Res & Develop Ctr, 82 & Naval Res Lab, 86; vis prof, Delft Univ Technol, 87. *Mem:* Am Math Soc; Edinburgh Math Soc; Gesellschaft Angewandte Math & Mech; Soc Indust & Appl Math; Inst Elec & Electronics Engrs; Int Sci Radio Union. *Res:* Classical electromagnetic theory; propagation and scattering of electromagnetic and acoustic waves; boundary value problems; integral equations; partial differential equations; special functions. *Mailing Add:* Dept Math Univ Del Newark DE 19716

KLEINMAN, ROBERT L P, b New York, NY, Dec 22, 51; m 76; c 2. GEOMICROBIOLOGY, BIOGEOCHEMISTRY. *Educ:* Pa State Univ, BS, 74; Princeton Univ, MA, 76, PhD(water resources) 79. *Prof Exp:* Group supvr environ eng, Bur Mines, 79-83, RES SUPVR ENVIRON TECHNOL, BUR MINES, PITTSBURGH RES CTR, 83- *Concurrent Pos:* Chair, Eastern sect, Am Soc Surface Mining & Reclamation, 88. *Honors & Awards:* Five Star Award, Pollution Eng Mag; Stewardship Award Sci & Technol, US Dept Interior. *Mem:* Am Soc Surface Mining & Reclamation; Int Mine Water Asn. *Res:* ameliorate the environmental impacts of mining activities; fires in abandoned mines and acid mine draininge. *Mailing Add:* Bur Mines PO Box 18070 Pittsburgh PA 15236

KLEINMAN, ROBERTA WILMA, b New York, NY, Oct 10, 42. ORGANIC CHEMISTRY, CONSERVATION CHEMISTRY. *Educ:* Barnard Col, Columbia Univ, AB, 64; Rutgers Univ, NB, PhD(org chem), 69. *Prof Exp:* NIH fel chem, Rutgers Univ, NB, 69-70, instr, 70-71, fel, 71-72; asst prof chem, Univ Mich-Dearborn, 72-79; Lectr, Univ Mich-Ann Arbor, 79-82; CONSERV CHEMIST, PANHANDLE-PLAINS HIST MUS, TEX, 82- *Concurrent Pos:* Consult, Henry Ford Mus, Dearborn, 79-80. *Mem:* Am Chem Soc; AAAS. *Res:* Carbene additions to steroid analogues; micelle formation and catalysis; synthetic polynucleotides as catalysts of enzymatic ractions; nucleic acid interactions; identification of natural dyes on textiles; stability studies of natural dyes. *Mailing Add:* Dept Chem Lock Haven Univ Lock Haven PA 17745

KLEINMANN, DOUGLAS ERWIN, b Chicago, Ill, July 11, 42; m 70; c 1. ASTRONOMY, PHYSICS. *Educ:* Rice Univ, BA, 64, PhD(space sci), 69; Mass Inst Technol, SM, 80. *Prof Exp:* Fel astron, Rice Univ, 68-70; astronomer, Smithsonian Astrophys Observ, Smithsonian Inst, 70-79; PROG MGR, HONEYWELL ELECTRO-OPTICS, 80- *Concurrent Pos:* Lectr, Harvard Col Observ, Harvard Univ, 71-79; res affil, Mass Inst Technol, 73-79; mem infrared instrument definition team, Space Telescope, NASA, 73-77. *Mem:* Am Astron Soc; Int Astrophys Union; Am Optical Soc. *Res:* Infrared devices, infrared astronomy; instrumentation using infrared devices. *Mailing Add:* 15 Hastings Rd Lexington MA 02173

KLEINROCK, LEONARD, b New York, NY, June 13, 34; m 54; c 2. COMPUTER SCIENCE. *Educ:* City Col New York, BEE, 57; Mass Inst Technol, SMEE, 59, PhD(elec eng), 63. *Prof Exp:* Asst engr, Photobell Co, NY, 51-57; res asst, Servomechanism Lab, Mass Inst Technol, 57-58, res asst, Electronics Res Lab, 58-61, staff mem, Lincoln Lab, 63; from asst prof to assoc prof, 63-70, PROF COMPUT SCI, UNIV CALIF, LOS ANGELES, 70- *Concurrent Pos:* Consult, Babcock Electronics Corp, Calif & Beckman Instruments Inc, 64, Jet Propulsion Lab, Calif Inst Technol, 64-, TRW Systs Group, 65-, Solid State Radiation, Inc, 65-, Cubic Corp, 66, Magnavox Res Labs, 67, Assoc Comput Mach, 67-, Technol Serv Corp, 68-, US Army Comput Syst Command, 71- & Off Emergency Preparedness, Exec Off President, 68-; prin investr, Advan Res Projs Agency, Dept Defense Contract, 69-; pres, Linkabit Corp, 68-69; chief exec officer, Technol Transfer Inst, 76-; mem adv coun sci & eng, City Col New York; mem, sci adv comt, IBM, Comput Sci & Technol Bd, Nat Res Coun; Guggenheim Found fel; Marconi Int Fel, 86. *Honors & Awards:* Lanchester Prize, 76; L M Ericsson Prize, 82; Marconi Int Fel, 86; ACM Sigcomm Award, 90. *Mem:* Nat Acad Eng; Opers Res Soc Am; fel Inst Elec & Electronics Engrs. *Res:* Communication theory; queueing theory; computer systems modeling and analysis; local area and computer networks; performance evaluation; distributed systems. *Mailing Add:* Comput Sci Dept 3732 Boelter Hall Univ Calif Los Angeles CA 90024-1596

KLEINROCK, MARTIN CHARLES, b Boston, Mass, Aug 21, 58; m 86; c 2. PLATE TECTONIC PROCESSES, OCEANIC CRUST GENERATION & DEFORMATION. *Educ:* Univ Calif, Santa Barbara, BA, 81; Univ Calif, San Diego, MS, 84, PhD(earth sci), 88. *Prof Exp:* Teaching asst field geol & optical mineral, Scripps Inst Oceanog, 82-83, res asst marine geol & geophys, 81-86; vis prof colleague marine geol & geophys, Hawaii Inst Geophys, 86-87, asst geophysicist, 88; postdoctoral investr, 88-89, ASST SCIENTIST MARINE GEOL & GEOPHYS, WOODS HOLE OCEANOG INST, 89- *Concurrent Pos:* Earle C Anthony fel, Univ Calif, San Diego, 85-86. *Mem:* Am Geophys Union; Geol Soc Am; Oceanog Soc. *Res:* Plate tectonics; plate boundary processes; generation and deformation of oceanic lithosphere; mid-ocean ridge tectonics; evolution of seafloor morphology; hotspot processes; seafloor survey instruments; author of various publications. *Mailing Add:* Dept Geol & Geophys Clark 240 Woods Hole Oceanog Inst Woods Hole MA 02543

KLEINSCHMIDT, ALBERT WILLOUGHBY, b Clinton, Iowa, Mar 20, 13; m 43; c 3. ORGANIC CHEMISTRY. *Educ:* Iowa State Univ, BS, 35; Purdue Univ, PhD(org chem), 41. *Prof Exp:* Res chemist, Cent Soya Co, Inc, 40-44, Beatrice Foods Inc, Ill, 44-47 & Am Maize Prod Co, Inc, 47-58; res chemist, J R Short Milling Co, 58-63, lab mgr, 63-67, tech dir, 67-69, vpres, 69-79; RETIRED. *Mem:* Am Chem Soc; Am Oil Chem Soc; Am Asn Cereal Chem; Inst Food Technol; Am Soc Brewing Chem. *Res:* Oil; fats; carbohydrates. *Mailing Add:* 452 King St Oviedo FL 32765

KLEINSCHMIDT, ERIC WALKER, b Indianapolis, Ind, Aug 10, 55; m 86; c 2. COATING & LAMINATING PROCESSES, ADHESIVE & COATING FORMULATION. *Educ:* Purdue Univ, BS, 78; Univ Phoenix, MBA, 91. *Prof Exp:* Process engr, Thermark Div, Avery Int, 78-80; process engr, Circuit Mat Div, Rogers Corp, 80-82, mat engr, Bus Prod Div, 82-85; sales & mkt mgr electronics prod, Gila River Prods, Inc, 85-87; TECH DIR, COURTAULDS PERFORMANCE FILMS, 87- *Mem:* Am Chem Soc; Am Inst Chem Engrs; Soc Advan Mat & Process Engrs; Soc Mfg Engrs. *Res:* Technology transfer of materials technology into industrial use; polymer science in application as films manufactured via a variety of processes; one United States patent. *Mailing Add:* 856 W Kiva Ave Mesa AZ 85210

KLEINSCHMIDT, R STEVENS, b Boston, Mass, Oct 8, 25; wid; c 4. CIVIL ENGINEERING, WATER POWER ENGINEERING. *Educ:* Harvard Univ, AB, 49, SM, 51, ScD(civil eng), 58. *Hon Degrees:* Dr Community Develop, Unity Col, Maine, 76. *Prof Exp:* Res engr, Harvard Univ, 50-58; asst engr water supply & sewage disposal, Camp Dresser & McKee, Mass, 58-59; asst prof hydraul & sanit eng, Northeastern Univ, 59-62; hydraul engr, Great Northern Paper Co, Maine, 62-66; independent consult engr, 66-70; partner, Kleinschmidt & Dutting Consult Engrs, 70-80, chmn bd & sr vpres, 80-85, pres, 85-, VPRES, KLEINSCHMIDT ASSOCS. *Honors & Awards:* Herschel Prize, 51. *Mem:* Am Soc Civil Engrs; Am Consult Engrs Coun; Sigma Xi. *Res:* Hydraulics as applied to sanitary engineering. *Mailing Add:* RFD 4 Box 134 A Ellsworth ME 04605

KLEINSCHMIDT, ROGER FREDERICK, b New York, NY, May 12, 19; m 45; c 3. ORGANIC CHEMISTRY, PETROLEUM TECHNOLOGY. *Educ:* Lehigh Univ, BS, 40; Columbia Univ, PhD(org chem), 44. *Prof Exp:* Asst chem, Columbia Univ, 41-44; res chemist, Interchem Corp, NY, 44 & Gen Aniline & Film Corp, 46-52; group leader org chem, Phillips Petrol Co, 52-57, sr group supvr, 57-59, sect mgr org chem synthesis, 59-65, mgr hydrocarbon chem br, 65-68, vpres res & develop, Phillips Sci Corp, 68-71, licensing rep, Phillips Petrol Co, 71-74, PLANNING CONSULT NATURAL RESOURCES, PHILLIPS PETROL CO, 74-, PROG MGR TERTIARY RECOVERY PETROL, ENERGY RES & DEVELOP ADMIN-PHILLIPS PETROL CO PROJ, 75-, MEM STAFF CORP PLANNING NORWEG INDUST DEVELOP, 81- *Concurrent Pos:* Instr, Exten, Okla State Univ, 53- *Mem:* Soc Petrol Engrs. *Res:* Synthetic organic chemicals; acetylene chemistry; polymerization; pressure reactions; petrochemicals from olefins and diolefins; naphthenes; non-aromatic cyclics; organometallic catalysts and intermediates; industrial organic chemistry; enhanced recovery of oil; micellar/polymer recovery methods; exploration and production planning, economics and budgets. *Mailing Add:* 1827 SE Hampden Rd Bartlesville OK 74006

KLEINSCHMIDT, WALTER JOHN, b Wabash Co, Ill, Apr 11, 18; m 42; c 1. BIOCHEMISTRY. *Educ:* Ind Univ, BS, 40; Univ Minn, MS, 49, PhD(biochem), 50. *Prof Exp:* Sr res biochemist cell & molecular biol, Lilly Res Labs, 50-66, res scientist, Biol Res Div, 66-83; RETIRED. *Mem:* AAAS; Am Chem Soc; Am Soc Biol Chem. *Res:* Antiviral agents; interferon; viruses; virus inhibition; aging; nucleic acids. *Mailing Add:* 9732 Trilobi Dr Indianapolis IN 46236-9704

KLEINSCHUSTER, JACOB JOHN, b Northampton, Pa, July 5, 43; m 70; c 3. ORGANIC POLYMER CHEMISTRY. *Educ:* Va Mil Inst, BS, 64; Pa State Univ, MS, 66, PhD(chem), 72. *Prof Exp:* Res chemist, E I du Pont de Nemours & Co, Inc, 72-74, sr res chemist, 74-75, supvr org polymer chem, 75-78, sr supvr, 79-82, tech supt, 82-83, mfg supt, 83-84, tech mgr, 84-85, worldwide tech mgr, 85-89, TECH DIR, E I DU PONT DE NEMOURS & CO, INC, 89- *Mem:* Am Chem Soc. *Res:* High performance organic industrial fibers, and apparel and spandex fibers. *Mailing Add:* Six Twin Turns Lane Chadds Ford PA 19317-9347

KLEINSCHUSTER, STEPHEN J, III, b Bath, Pa, June 3, 39; m 66; c 2. DEVELOPMENTAL BIOLOGY, IMMUNOTHERAPY. *Educ:* Colo State Univ, BS, 63, MS, 66; Ore State Univ, PhD(zool), 70. *Prof Exp:* Fel develop biol, Univ Chicago, 71; asst prof biol, Metrop State Col, Denver, 71-73; affil prof bot & plant path, Colo State Univ, 73, assoc prof anat, 75-77, chmn exec comt anat, 76-77; actg head, 80-81, DIR ANIMAL TUMOR PROG, ANIMAL, DAIRY & VET SCI, UTAH STATE UNIV, 77-, PROF & HEAD, ANIMAL, DAIRY & VET SCI DEPT, 81-; DEAN, AGR & NAT RESOURCES, COOK COL, RUTGERS UNIV. *Concurrent Pos:* Consult, NASA, 71-77; prin investr, NIH, Cancer Immunoprophylaxis Contracts, 77-81 & Immunother Procurement Contracts, 75-81; Impact Rev Group, Dept Agr, State Utah, 77; surg oncol res group, LDS Hosp, Salt Lake City, Utah, 79-; organizer & dir, Vet Sci Tissue Cult Facil, Utah State Univ, 79- *Mem:* AAAS; Am Asn Anatomists; Am Asn Vet Anatomists; NY Acad Sci; Am Soc Animal Sci. *Res:* Cancer biology; immunotherapy and immunoprophylaxis; molecular biology of development and morphogenesis. *Mailing Add:* Cook Col Rutgers Univ PO Box 231 New Brunswick NJ 08903

KLEINSMITH, LEWIS JOEL, b Detroit, Mich, Apr 13, 42; m 64; c 2. CELL BIOLOGY, BIOCHEMISTRY. *Educ:* Univ Mich, BS, 64; Rockefeller Univ, PhD(life sci), 68. *Prof Exp:* From asst prof to assoc prof zool, Univ Mich, Ann Arbor, 68-74; vis prof biochem, Univ Fla, Gainesville, 74-75; PROF BIOL SCI, UNIV MICH, ANN ARBOR, 75- *Concurrent Pos:* Lectr vis biologists prog, Am Inst Biol Sci, 69-71; Guggenheim fel, 74-75. *Honors & Awards:* Henry Russel Award, 71. *Mem:* Am Inst Biol Sci; Sigma Xi; Am Soc Cell Biol; NY Acad Sci; Am Soc Biol Chemists. *Res:* Biochemistry of cell nucleus; role of nuclear proteins in regulating gene function; nucleoprotein chemistry and function; biochemical regulatory mechanisms; regulation of normal and malignant cell growth. *Mailing Add:* Div Biol Sci Krause Nat Sci Bldg Univ Mich Ann Arbor MI 48109

KLEINSPEHN, GEORGE GEHRET, b Middlebury, Vt, Mar 27, 24; div; c 2. ORGANIC CHEMISTRY. *Educ:* Colgate Univ, AB, 44; Johns Hopkins Univ, AM, 47, PhD(chem), 51. *Prof Exp:* Jr chemist, Clinton Eng Works-Tenn Eastman Corp, 44-46; jr instr, Johns Hopkins Univ, 46-49, res assoc & univ fel, 51-56; sr res assoc & treas, Monadnock Res Inst, 56-59; res assoc, Johns Hopkins Univ, 60; chemist, US Army Ballistic Res Labs, 60-63, chief, Org Chem Sect, Chem Br, 63-67; dept chair, 79-82, PROF CHEM, HOOD COL, 67-, WHITAKER PROF CHEM, 83- *Concurrent Pos:* USPHS fel, NIH, 51-52; consult, US Army Ballistic Res Labs, 68; fel, DuPont, 50 & Beneficial-Hodson, 84. *Mem:* AAAS; Am Chem Soc; NY Acad Sci; Sigma Xi. *Res:* Nitrogenous heterocyclic compounds, especially pyrroles and porphyrins; organic substances of high nitrogen content. *Mailing Add:* Dept Chem Hood Col Frederick MD 21701

KLEINSTEUBER, TILMANN CHRISTOPH WERNER, b Berlin, Ger, July 16, 34. PHYSICS, PHYSICAL CHEMISTRY. *Educ:* Univ Hamburg, BSc, 56; Univ Munich, PhD(phys chem), 61. *Prof Exp:* Res assoc phys chem, Univ Munich, 61-63; res assoc, Amherst Col, 63-64, asst prof chem, 64-65; asst prof, 65-68, ASSOC PROF PHYSICS, KING'S COL, PA, 68-, CHMN DEPT, 65- *Mem:* Am Asn Physics Teachers; Sigma Xi. *Res:* Calorimetry at high and low temperatures; thermodynamics of metals and alloys; surface chemistry; gas chromatography. *Mailing Add:* Dept Physics King's Col Wilkes-Barre PA 18711

KLEINZELLER, ARNOST, b Ostrava, Czech, Dec 6, 14; m 43; c 2. CELL PHYSIOLOGY. *Educ:* Univ Brno, Czech, MD, 38; Univ Sheffield, PhD(biochem), 42. *Hon Degrees:* DSc, Czech Acad Sci, 59; MA, Univ Pa, 73. *Prof Exp:* Med Res Coun grant biochem, Cambridge Univ, 43-44; head lab cell metab, Czech Inst Health, 46-48; assoc prof fermentation chem &

head dept, Prague Tech Univ, 48-52; assoc prof biochem, Charles Univ, Prague, 52-55; head lab cell metab, Czech Acad Sci, 56-66; vis prof physiol, Univ Rochester, 66-67; prof, 67-85, EMER PROF PHYSIOL, SCH MED, UNIV PA, 85- Concurrent Pos: Rockefeller fel, Cambridge Univ, 41-42; mem, Acad Leopoldina, 66-; ed, Biochem Biophys Acta & Current Topics in Membranes & Transport, 69-; mem & secy, US Nat Comt Physiol Sci, 76-; Fogarty Sr Int Fel, 80. Mem: Biophys Soc; Am Physiol Soc; Am Soc Cell Biol; Soc Gen Physiol; Brit Biochem Soc. Res: Intermediate metabolism; transport of electrolytes and sugars across cell membranes. Mailing Add: Dept Physiol Univ Pa Sch Med Philadelphia PA 19104

KLEIS, JOHN DIEFFENBACH, b Hamburg, NY, Feb 1, 12; m 51; c 3. PHYSICS, METALLURGY. Educ: Univ Buffalo, BA, 32, MA, 33; Yale Univ, PhD(physics), 36; Harvard Bus Sch, AMP, 57. Prof Exp: Physicist elec contacts, Fansteel, Inc, 36-57, vpres res refractory metals, 57-63, vpres & gen mgr elec & electronic prod, 63-69; vpres res elec contacts, Sterndent Corp, 69-70, pres, Cooper Div, 70-78, vpres & technol dir precious metals group, 78-82; CONSULT TECH DIR. Concurrent Pos: Electronics consult, 83- Mem: Soc Automotive Engrs; fel Am Soc Testing & Mat; Inst Elec & Electronics Engrs. Res: Solid state, areas relating to bonding procedures and bonding, interconnect materials; precious metals, metallurgy of Al and AliSi materials. Mailing Add: Stern-Leach 262 Broad St North Attleboro MA 02761

KLEIS, ROBERT W(ILLIAM), b Martin, Mich, Nov 30, 25; m 49; c 2. AGRICULTURAL ENGINEERING. Educ: Mich State Univ, BS, 49, MS, 51, PhD(agr eng), 57. Prof Exp: Instr agr eng, Mich State Univ, 49-51; instr, Univ Ill, 51-53, asst prof, 53-56; prof & head dept, Univ Mass, Amherst, 57-66; prof & chmn dept, Univ Nebr, Lincoln, 66-67, assoc dir, Agr Exp Sta, 67-83, dean int progs, 76-84, exec dean int affairs, 84-90; RETIRED. Concurrent Pos: Dir, Midamerica Int Agr Consortium, Inc, 76-; consult, Indust, 50- & Int Agr Develop, 75-; dir, Int Collab Res Support Prog, 78-; exec dir, Bd Int Food & Agr, Washington, DC, 85-87. Mem: Fel Am Soc Agr Engrs; Am Soc Eng Educ; Nat Asn Univ Dirs Int Progs; Agr Exp Sta Dir Asn; Sigma Xi. Res: Materials handling systems for agriculture; product processing; food and feed preservation; farm operations mechanization. Mailing Add: 6520 Sumner Lincoln NE 68506

KLEIS, WILLIAM DELONG, b Akron, Ohio, Feb 26, 24; m 64; c 4. ATMOSPHERIC PHYSICS. Educ: Univ Colo, AB, 49; Univ Chicago, MS, 52. Prof Exp: With USAF, 52-68, chief forecaster, Weather Detachment, Eng, 52-56, climat officer, March AFB, Calif, 56-61, chief forecaster, Turkey, 61-62, chief sci serv br, Andrews AFB, Md, 64-68; PROG DEVELOP SCIENTIST, PROG OFF, ENVIRON RES LABS, NAT OCEANIC & ATMOSPHERIC ADMIN, 69- Mem: Am Meteorol Soc; Am Geophys Union; Royal Meteorol Soc. Res: Aeronomy; upper atmospheric physics and chemistry; solar-terrestrial physics; meteorology; remote sensing of the atmosphere. Mailing Add: 3095 Heidelberg Dr Boulder CO 80303

KLEITMAN, DANIEL J, b New York, NY, Oct 4, 34; m 64; c 3. MATHEMATICS. Educ: Cornell Univ, AB, 54; Harvard Univ, AM, 55, PhD, 58. Prof Exp: NSF fel physics, Copenhagen Univ, 58-59 & Harvard Univ, 59-60; asst prof, Brandeis Univ, 60-66; assoc prof, 66-69, head dept, 79-84, PROF MATH, MASS INST TECHNOL, 69- Concurrent Pos: Consult, Nuclear Regulatory Comn, Gen Acct Off, 73-81; managing ed, SIAM J Algebraic & Discrete Methods, 75-82; ed, J Networks. Mem: Am Math Soc; Oper Res Soc Am; Soc Indust & Appl Math; Am Acad Arts & Sci; NY Acad Sci. Res: Combinatorial mathematics; graph theory; numeration and optimization; applications to operations research. Mailing Add: Mass Inst Technol Rm 2-347 Cambridge MA 02139

KLEITMAN, DAVID, b New York, NY, June 28, 31; m 52; c 3. SOLID STATE PHYSICS. Educ: Cornell Univ, BA, 52; Purdue Univ, MS, 53, PhD(physics), 58. Prof Exp: Group head labs & head display res, David Sarnoff Res Ctr, RCA, 57-67; vpres & dir res & develop, Signetics Corp, 67-78, vpres res, 78-80; pres, Protracoa, 80-91; dir technol, Branson/IPC, 81-82; RETIRED. Mem: Inst Elec & Electronics Engrs; AAAS. Res: Semiconductors; radiation damage; luminescence; optical amplification; color television system and display system management; integrated circuits, their materials, processing devices, circuits design and applications in computers, communications and consumer and industrial systems; semiconductor devices for integrated circuits, microwave, power, computer, instrumentation, optical and novel applications and systems; computer software; plasma processing. Mailing Add: 12387 Stonebrook Dr Los Altos Hills CA 94022

KLEITSCH, WILLIAM PHILIP, b Cincinnati, Ohio, July 12, 12. SURGERY. Educ: Univ Ill, BS, 34, MD & MS, 37; Am Bd Surg, dipl. Prof Exp: Instr surg, Col Med, Univ Ill, 38-42; from asst instr to assoc prof, Col Med, Univ Nebr, Omaha, 55-66; assoc prof, Sch Med, Creighton Univ, 55-66; CHIEF SURG, VET ADMIN HOSP, PHOENIX, 66- Concurrent Pos: Lectr, Col Dent, Univ Nebr, Omaha, 47-50; chief surg serv, Vet Admin Hosp, Nebr, 46-66; mem staff, Good Samaritan Hosp; consult, Phoenix Indian Med Ctr. Mem: AAAS; Am Thoracic Soc; AMA; Asn Mil Surg US; Nat Tuberc & Respiratory Dis Asn; Sigma Xi. Mailing Add: 2201 E Solano Dr Phoenix AZ 85016

KLEKOWSKI, EDWARD JOSEPH, JR, b Brooklyn, NY, Oct 24, 40. BOTANY, GENETICS. Educ: NC State Univ, BS, 62, MS, 64; Univ Calif, Berkeley, PhD(bot), 68. Prof Exp: Asst prof, 68-73, ASSOC PROF BOT, UNIV MASS, AMHERST, 73- Mem: Bot Soc Am. Res: Pteridology; genetic and evolutionary studies of homosporous ferns. Mailing Add: Dept Bot Univ Mass Amherst MA 01003

KLEMA, ERNEST DONALD, b Wilson, Kans, Oct 4, 20; m 53; c 2. NUCLEAR PHYSICS. Educ: Univ Kans, AB, 41, AM, 42; Rice Inst, PhD(physics), 51. Prof Exp: Jr scientist, Los Alamos Sci Lab, 43-46; sr physicist nuclear physics, Oak Ridge Nat Lab, 50-56; assoc prof nuclear eng, Univ Mich, 56-58; prof nuclear & sci eng, Northwestern Univ, 58-68, chmn

dept eng sci, 60-67; dean col eng, 68-73, prof, 68-88, EMER PROF ENG SCI, TUFTS UNIV, 88-, EMER DEAN, COL ENG, 88- Concurrent Pos: Adj prof int politics, Fletcher Sch Law & Diplomacy, 73-83. Mem: Fel Am Phys Soc; fel Am Nuclear Soc; sr mem Inst Elec & Electronics Engrs. Res: Angular correlations of gamma rays; fission cross sections; empirical nuclear models; semiconductor detectors; science and technology policy. Mailing Add: 105 Anderson Hall Tufts Univ Medford MA 02155

KLEMANN, LAWRENCE PAUL, b Cincinnati, Ohio, Aug 13, 43; m 63; c 2. ORGANOMETALLIC CHEMISTRY, PHYSICAL CHEMISTRY. Educ: Univ Mass, BS, 65, PhD(chem), 69. Prof Exp: Staff chemist, Exxon Res & Eng Co, 69-86; RES ASSOC, NABISCO BRANDS, INC, 86- Mem: Am Chem Soc; AAAS; Am Oil Chemists' Soc. Res: Investigations of the solution properties of polyamine chelated lithium compounds; investigation of alkali metal organic electrolytes; lipid chemistry; homogeneous and heterogeneous catalysis involving carbon monoxide and hydrogen; surfactant mineral interactions; synthetic organic and natural products chemistry; fats and oils; food science and technology; agricultural and food chemistry. Mailing Add: 196 Tanglewood Dr Somerville NJ 08876

KLEMAS, VICTOR V, b Klaipeda, Lithuania, Nov 29, 34; US citizen; m 60; c 3. OPTICAL PHYSICS, MARINE STUDIES. Educ: Mass Inst Technol, BS, 57, MS, 59; Univ Brunswick, PhD(optical physics), 65. Prof Exp: Mgr optical physics & space explor, Space Div, Gen Elec Co, 59-71; assoc prof, 71-80, PROF MARINE STUDIES, UNIV DEL, 80-, DIR REMOTE SENSING CTR, 75-, DIR APPL OCEAN SCI PROG, 81- Concurrent Pos: Fel, Gen Elec Co, 63; consult, Environ Protection Agency, NASA, NSF & AID, UNESCO, United Nations, Ecuador, Peru, Costa Rica, Panama, India, Sri Lanka, Egypt, Korea, France & Ger; mem Comn Natural Resources, Nat Acad Sci, 75-78; mem, Ocean Policy Comt, Nat Acad Sci; mem adv comts, NASA, 75-; prog mgr, Scientists & Engrs Econ Develop, NSF, 77-78; mem, Man & the Biosphere, UNESCO, 77-; mem, Comt Earth Studies, Nat Acad Sci, 88- Honors & Awards: Achievement Medal, Korean Advan Inst Sci & Merit Award, India Remote Sensing Agency, 78. Mem: Inst Elec & Electronics Engrs; Asn Am Geographers; Am Geophys Union; Am Soc Photogram. Res: Management of coastal resources; remote sensing of environment, especially physical and biological coastal processes; third world resources development; geographic information systems. Mailing Add: Col Marine Studies Univ Del Newark DE 19711

KLEMCHUK, PETER PAUL, b Oakville, Conn, Oct 31, 28; m 49; c 5. ORGANIC CHEMISTRY. Educ: Mass Inst Technol, BS, 50; Rutgers Univ, MS, 56, PhD(org chem), 57. Prof Exp: Chemist, Merck & Co, Inc, 50-56; chemist, Esso Res & Eng Co, 57-59; chemist, Stauffer Chem Co, 59-60; SR RES FEL, CIBA-GEIGY CORP, 60- Mem: AAAS; Am Chem Soc; Soc Plastics Engrs. Res: Polymer stabilization; polymer additives; polymer degradation; degradable polymers; recyclable polymers. Mailing Add: Ciba-Geigy Corp 444 Sawmill River Rd Ardsley NY 10502

KLEMENS, PAUL GUSTAV, b Vienna, Austria, May 24, 25; m 50; c 2. PHYSICS, THERMAL CONDUCTIVITY. Educ: Univ Sydney, BSc, 46, MSc, 48; Oxford Univ, DPhil(theoret physics), 50. Prof Exp: Prin res officer, Nat Standards Lab, Sydney, Australia, 50-59; physicist, Westinghouse Res Labs, 59-64, mgr transport properties, Solids Dept, 64-67; chmn dept, 70-74, PROF PHYSICS, UNIV CONN, 67- Mem: Fel Am Phys Soc; fel Brit Inst Physics & Phys Soc. Res: Theoretical solid state and low temperature physics, particularly thermal conductivity of solids and other non-equilibrium and transport properties; ultrasonic attenuation, properties of composites; laser welding and surface modification. Mailing Add: Dept Physics Univ Conn Storrs CT 06269

KLEMENT, VACLAV, b Pilsen, Czech, May 7, 35; m 67; c 2. RADIATION ONCOLOGY, MICROBIOLOGY. Educ: Charles Univ, Prague, MD, 59; Czech Acad Sci, Prague, PhD(biol), 64. Prof Exp: Vis scientist viral oncol, Nat Inst Allergy & Infectious Dis, NIH, 67-68; res fel pediat, 68-69, ASSOC PROF RADIATION ONCOL & MICROBIOL, SCH MED, UNIV SOUTHERN CALIF, 79- Mem: Am Asn Cancer Res; Int Asn Comp Res Leukemia & Related Dis; Am Soc Therapeut Radiol & Oncol. Res: Viral carcinogenis and tumor biology; radiation oncology; oncogenes. Mailing Add: Sch Med 2025 Zonal Ave Los Angeles CA 90033

KLEMENT, WILLIAM, JR, b Chicago, Ill, Sept 30, 37. MATERIALS SCIENCE. Educ: Calif Inst Technol, BS, 58, PhD(eng sci), 62. Prof Exp: Asst res geophysicist, Inst Geophys & Planetary Physics, Univ Calif, Los Angeles, 62-64; Miller res fel physics, Univ Calif, Berkeley, 64-66; asst prof eng, 66-67, ASSOC PROF ENG, UNIV CALIF, LOS ANGELES, 67- Concurrent Pos: NATO fel, Royal Inst Technol, Sweden, 63; Guggenheim Mem Found fel, Australian Nat Univ, 68-69; Ford Found Prog, Univ Chile, 73; vis scientist, Nat Phys Res Lab, Pretoria, SAfrica, 74-76 & Inorg Chem Lab, Oxford Univ, 79. Honors & Awards: co-recipient, Int Prize New Ma, 80. Res: Phase transformations; archaeological, ethnographic and historical materials. Mailing Add: Dept Mat Sci Univ Calif 405 Hilgard Ave Los Angeles CA 90024

KLEMER, ANDREW ROBERT, b St Clair, Pa, June 4, 42; m 63; c 4. PHYSIOLOGICAL ECOLOGY, LIMNOLOGY. Educ: La Salle Col, BA, 64; Univ Minn, PhD(ecol), 73. Prof Exp: Consult limnol, Dept Sci & Indust Res, NZ, 73-74; vis scientist algal physiol, Cawthron Inst, 74-75; ASST PROF BIOL & ENVIRON SCI, STATE UNIV NY, PURCHASE, 76- Concurrent Pos: Nat Res Adv Coun NZ, res fel, 73-75; State Univ NY Res Found fac res fel, 77 & 79. Mem: Phycol Soc Am; Am Soc Limnol & Oceanog; Int Asn Theoret & Appl Limnol; AAAS; Sigma Xi. Res: Factors that limit the distribution and affect the community structure of phytoplankton; physiological mechanisms involved in responses to those factors. Mailing Add: Dept Biol Univ Minn 305 Life Sci Bldg Duluth MN 55812

KLEMM, DONALD J, b Detroit, Mich, Jan 13, 38. AQUATIC BIOLOGY, AQUATIC ECOLOGY & TOXICOLOGY. *Educ:* Valley City State Col, BS, 63; Eastern Mich Univ, MS & SpecS, 70; Univ Mich, Ann Arbor, PhD(fisheries), 74. *Prof Exp:* Res assoc malacol, Mollusk Div, Mus Zool, Univ Mich, Ann Arbor, 72-74; RES AQUATIC BIOLOGIST, ENVIRON MONITORING & SUPPORT LAB, US ENVIRON PROTECTION AGENCY, 74- *Mem:* Soc Environ Toxicol & Chem; Int Asn Theoret & Appl Limnol; Am Soc Testing & Mat; NAm Benthol Soc; Brit Freshwater Biol Asn; Am Fisheries Soc. *Res:* Stream and lake ecology; ecology of polluted waters; invertebrate and fish zoology; macroinvertebrates and fish methodology, parasitology and toxicology; systematics and ecology of freshwater fish, macroparasites, insects, mollusks, annelids, aquatic oligochaetes and Hirudinea of the world. *Mailing Add:* PO Box 44090 Cincinnati OH 45244

KLEMM, JAMES L, b South Bend, Ind, Oct 30, 39; m 81; c 2. APPLIED MATHEMATICS, COMPUTER SCIENCE. *Educ:* Univ Chicago, BS, 61; Purdue Univ, MS, 63; Mich State Univ, PhD(eng mech), 70. *Prof Exp:* Asst math, Purdue Univ, 61-65; asst prof, Ind Univ Pa, 65-67; asst, Dept Metall, Mech & Mat Sci, Mich State Univ, 67-69, res assoc, 70; asst prof eng sci, Univ Cincinnati, 70-77; sr publ, NCR Corp, 77-80, sr systs analyst, 80-82, consult analyst, 82-91, CONSULT ENG, QUAL ASSURANCE, NCR CORP, 91- *Concurrent Pos:* Partic, NSF Inst Appl Math & Mech, Mich State Univ, 67. *Mem:* Sigma Xi; AAAS. *Res:* St Venant boundary value problems in two and three dimensional theories of classical elasticity; computer applications in manufacturing. *Mailing Add:* NCR Corp 3245 Plattsprings Rd West Columbia SC 29169

KLEMM, LEROY HENRY, b Maple Park, Ill, July 31, 19; m 45; c 3. ORGANIC CHEMISTRY. *Educ:* Univ Ill, BS, 41; Univ Mich, MS, 43, PhD(org chem), 45. *Prof Exp:* Res chemist, Am Oil Co, Tex, 44-45; fel, Univ Res Found, Ohio State Univ, 46; instr chem, Harvard Univ, 46-47; from instr to asst prof, Ind Univ, 47-52; from asst prof to assoc prof, 52-63, PROF CHEM, UNIV ORE, 63- *Concurrent Pos:* Fel, Guggenheim Mem Found, Med Res Coun Labs, London & Swiss Fed Inst Tech, Zurich, 58-59; vis prof, Univ Cincinnati, 65-66; Fulbright-Hays res fel & NATO res grant, Aarhus Univ, Denmark & Univ Groningen, Neth, 72-73; vis prof, La Trobe Univ & sr assoc, Univ Melbourne, Australia, 79-80; vis prof, Univ Queensland, Australia, 86. *Mem:* Fel AAAS; Am Chem Soc; Int Soc Heterocyclic Chem; Sigma Xi; Am Asn Univ Professors. *Res:* Synthesis of carbocyclic and heterocyclic compounds; organic reactions; biologically active compounds; chromatography; heterogeneous catalysis. *Mailing Add:* Dept Chem Univ Ore Eugene OR 97403

KLEMM, REBECCA JANE, b Bloomington, Ind, Feb 21, 50. STATISTICS, OPERATIONS RESEARCH. *Educ:* Miami Univ, BS, 71, Iowa State Univ, MS, 73, PhD(statist), 76. *Prof Exp:* asst prof statist, Temple Univ, 76-80; MEM FAC STATIST, SCH BUS ADMIN, GEORGETOWN UNIV, 80- *Concurrent Pos:* US Dept Energy fac fel, Sch Bus, Am Assembly Col, 78-79. *Mem:* Am Statist Asn; Opers Res Soc Am; Inst Mgt Sci. *Res:* Statistical education; constrained least squares, econometric model building. *Mailing Add:* 1785 Massachusetts Ave NW 5th Fl Washington DC 20036

KLEMM, RICHARD ANDREW, b Bloomington, Ind, Mar 13, 48; m 80; c 2. THEORETICAL SOLID STATE PHYSICS. *Educ:* Stanford Univ, BS, 69; Harvard Univ, MA, 72, PhD(physics), 74. *Prof Exp:* Fel, Stanford Univ, 74-76; asst prof physics, Ames Lab, Iowa State Univ, 76-81, assoc prof, 81; staff physicist, 82-84, sr staff physicist, Exxon Res & Eng Co, 84-86; VIS SCIENTIST, AMES LAB, US DEPT ENERGY, DEPT PHYSICS, IOWA STATE UNIV, 88- *Concurrent Pos:* Vis scientist, Univ, BC, 78 & 79, Univ Hamburg, 80; vis prof physics, Univ Calif, La Jolla, 86-88. *Mem:* Am Phys Soc. *Res:* Theory of condensed matter involving superconductivity; p-wave superconductivity; lower dimensional conductors; charge-density waves and spin-glasses; high temperatur superconductors. *Mailing Add:* Ames Lab US Dept Energy/Dept Physics Iowa State Univ Ames IA 50011

KLEMM, ROBERT DAVID, b Youngstown, Ohio, Sept 13, 29; div; c 1. VERTEBRATE MORPHOLOGY. *Educ:* Capital Univ, BS, 57; Ohio Univ, MS, 59; Southern Ill Univ, PhD(vert zool), 64. *Prof Exp:* Asst prof biol, Capital Univ, 64-67; asst prof anat, Col Vet Med, Kans State Univ, 67-70; asst prof, Mich State Univ, 70-72, invests leader, Avian Anat Invests, USDA, 70-72; assoc prof, 72-79, PROF ANAT, DEPT ANAT & PHYSIOL, COL VET MED, KANS STATE UNIV, 79- *Concurrent Pos:* Guest prof, Institut fur Anat u Cytobiol Der Justus-Liebig Univ Giessen WGer, 79-80. *Mem:* Am Asn Anatomists; Am Asn Vet Anatomists; Am Soc Zoologists. *Res:* Gross, light and electron microscopy studies of vertebrate structure with special reference to normal and diseased lungs of domestic animals. *Mailing Add:* Dept Anat & Physiol Col Vet Med Kans State Univ Manhattan KS 66506

KLEMM, WALDEMAR ARTHUR, JR, b Elgin, Ill, July 10, 34; m 59; c 3. SILICATE CHEMISTRY. *Educ:* Univ Calif, Riverside, BA, 56; Ore State Univ, MS, 67. *Prof Exp:* Res chemist propellant chem, Lockheed Propulsion Co, 59-66; group leader cement chem, Tech Ctr, Am Cement Corp, 66-70; assoc specialist geochem, Inst Geophys & Planetary Physics, Univ Calif, 70-72; sr res scientist, Gen Portland Inc, 72-75; sr scientist, Martin Marietta Labs, 75-83; MGR, CENT PROCESS LAB, SOUTHDOWN, INC, 83- *Concurrent Pos:* Vchmn solid-liquid interactions cement hydration, Gordon Res Conf, 75-76; trustee, Cements Div, Am Ceramic Soc, 87-90. *Mem:* fel Am Ceramic Soc; Am Chem Soc. *Res:* High-temperature silicate chemistry; cement clinkering reactions; admixture interactions; expansive cements; cement hydration. *Mailing Add:* Southdown Inc Box 937 Victorville CA 92393

KLEMM, WILLIAM ROBERT, b South Bend, Ind, July 24, 34; m 57; c 2. ANIMAL PHYSIOLOGY. *Educ:* Auburn Univ, DVM, 58; Univ Notre Dame, PhD(biol), 63. *Prof Exp:* NIH fel, 60-63; assoc prof physiol & pharmacol, Iowa State Univ, 63-66; assoc prof, 66-70, PROF, TEX A&M UNIV, 70- *Mem:* Am Physiol Soc; Soc Neurosci. *Res:* Animal hypnosis; theta rhythm; brain stem functions; animal electroencephalography; psychopharmacology. *Mailing Add:* Dept Vet Anat & Pub Health Tex A&M Univ College Station TX 77843

KLEMME, HUGH DOUGLAS, b Belmond, Iowa, Jan 24, 21; m 43; c 3. PETROLEUM GEOLOGY. *Educ:* Coe Col, AB, 42; Princeton Univ, MA, 48, PhD(geol), 49. *Prof Exp:* Regional geologist, Standard Oil Co, 49-51; staff geologist, Am Overseas Petrol, Ltd, 51-58, mgr explor, 58-63, asst chief geologist, 63-69; vpres explor, Lewis G Weeks Assocs, Ltd, Westport, 69-76; sr vpres, Weeks Petrol Corp, Westport, 76-79; PRES, GEO BASINS LTD, 80-; CONSULT, PETROL GEOLOGIST, 82- *Mem:* Geol Soc Am; hon mem Am Asn Petrol Geologists; AAAS; Am Geophys Union; Am Petrol Inst. *Res:* Regional petroleum geology; regional tectonics; basin studies; petroleum formation, migration and accumulation. *Mailing Add:* RR 1 Box 179-B Bondville VT 05340

KLEMMEDSON, JAMES OTTO, b Ft Collins, Colo, Aug 20, 27; m 52; c 4. SOIL SCIENCE, FOREST & RANGE ECOLOGY. *Educ:* Univ Calif, Berkeley, BS, 50, PhD(soil sci), 59; Colo State Univ, MS, 53. *Prof Exp:* Soil conservationist, Soil Conserv Serv, USDA, 50-51; instr forestry, Colo State Univ, 51-52, res asst range mgt, 52-53; instr, Mont State Univ, 53-55; res asst forestry, Univ Calif, Berkeley, 55-56, soils & plant nutrit, 56-59; range scientist, Int Forest & Range Exp Sta, USDA, 59-66; prof range & forestry, 66-88, PROF RANGE & WATERSHED MGT, UNIV ARIZ, 66-, RES SCIENTIST, AGR EXP STA, 80- *Concurrent Pos:* Charles Bullard forest res fel, Harvard Univ, 74-75; NATO/Hienemann Found grant, study vis chair soil sci, Univ Munich, Ger, 83; vis scientist, Swiss Fed Inst Forest, Snow & Landscape Res, Bermensdorf, Switz, 82-83, 89-90. *Mem:* Soc Am Foresters; Soc Range Mgt; Soil Sci Soc Am; Am Soc Agron. *Res:* Soil-plant-nutrient relations in forest, range and shrub ecosystems; ecology of forest and range ecosystems. *Mailing Add:* Sch Renewable Natural Resources Univ Ariz Tucson AZ 85721

KLEMMER, HOWARD WESLEY, b Sask, Can, 22; nat US; m 60; c 1. MICROBIOLOGY. *Educ:* Univ Sask, BS, 49, MS, 50; Univ Wis, PhD(bact), 54. *Prof Exp:* Microbiologist, Pineapple Inst Hawaii, 54-63; microbiologist, Pac Biomed Res Ctr, Univ Hawaii, 63-75, proj dir, Community Studies Pesticides, 65-75; RETIRED. *Mem:* AAAS; Am Soc Microbiol; Am Phytopath Soc; Am Chem Soc; fel Am Inst Chem; Sigma Xi. *Res:* Sanitary and industrial microbiology; environmental pollution. *Mailing Add:* 4915 Kalanianaole Honolulu HI 96821

KLEMOLA, ARNOLD R, b Pomfret, Conn, Feb 20, 31. ASTRONOMY. *Educ:* Ind Univ, AB, 53; Univ Calif, Berkeley, PhD(astron), 62. *Prof Exp:* Res asst astron, Yale Univ, 61-63, res staff astronr, 63-67; from asst res astronr to assoc res astronr, 67-84, RES ASTRONR, LICK OBSERV, UNIV CALIF, SANTA CRUZ, 84- *Mem:* Am Astron Soc; Int Astron Union; Astron Soc Pac. *Res:* Photographic astrometry. *Mailing Add:* UCO/Lick Observ Univ Calif Santa Cruz CA 95064

KLEMOLA, TAPIO, b Pori, Finland, July 20, 34; m 55; c 2. MATHEMATICS. *Educ:* Univ Helsinki, MS, 56, PhD(math), 59. *Prof Exp:* Asst math, Univ Helsinki, 56-58; Aaltonen Saatio & Govt Finland grant, Paris, 59; actg asst prof, Univ Oulu, 59-60; asst prof, Univ Windsor, 60-62; NSF grant, Inst Advan Study, 62-63; lectr & res assoc, Johns Hopkins Univ, 63-64; vis prof, 64-65, ASSOC PROF MATH, UNIV MONTREAL, 65- *Concurrent Pos:* Swiss Govt grant, Swiss Fed Inst Technol, 71-72. *Mem:* Am Math Soc; Can Math Cong. *Res:* Complex manifolds and spaces; differential operators on manifolds. *Mailing Add:* Dept Math Univ Mont CP 6128 Sta A Montreal PQ H3C 3J7 Can

KLEMPERER, FRIEDRICH W, RHEUMATOLOGY. *Educ:* Harvard Univ, MD, 37. *Prof Exp:* Prof med, Sch Med, Syracuse Univ, 65-78; RETIRED. *Mailing Add:* PO Box 487 Saranac Lake NY 12983

KLEMPERER, MARTIN R, b New York, NY, June 26, 31; m 59; c 3. HEMATOLOGY, ONCOLOGY. *Educ:* Dartmouth Col, AB, 53; NY Univ, MD, 57. *Prof Exp:* Instr pediat, Harvard Med Sch, 65-67, assoc, 67-69, asst prof, 69-70, tutor med sci, 67-70; assoc prof pediat, Sch Med, Univ Rochester, 70-74, prof med, 71-74, prof pediat & med, 74-; PROF, DEPT PEDIAT, MARSHALL UNIV. *Concurrent Pos:* Res fel pediat, Harvard Med Sch, 63-65; fel hemat & med, Children's Hosp Med Ctr, Boston, Mass, 63-65, asst med, 65-66, res assoc immunol & hemat, 66-68, assoc med, immunol & hemat, 68-70; sr assoc pediatrician, Strong Mem Hosp, Med Ctr, Univ Rochester, 70- *Mem:* Soc Pediat Res; Am Fedn Clin Res; Int Soc Exp Hemat; Am Soc Hemat; NY Acad Sci. *Res:* Hereditary and acquired defects of the serum complement system; role of the complement system in inflammation; therapy of childhood malignancies. *Mailing Add:* All Childrens Hosp 801 Sixth St S St Petersburg FL 33701

KLEMPERER, WALTER GEORGE, b Saranac Lake, NY, Apr 2, 47; m 77; c 2. MATERIALS CHEMISTRY. *Educ:* Harvard Univ, BA, 68; Mass Inst Technol, PhD(chem), 73. *Prof Exp:* From asst prof to prof chem, Columbia Univ, 73-81; PROF CHEM, UNIV ILL, URBANA-CHAMPAIGN, 81- *Concurrent Pos:* Alfred P Sloan Found fel, 76; Camille & Henry Dreyfus Found grant, 78; Guggenheim fel, 80. *Mem:* Am Chem Soc; The Chem Soc. *Res:* Inorganic chemistry; materials chemistry of oxides; polyoxuanion chemistry; sul-gel chemistry; cement chemistry; zeolite chemistry. *Mailing Add:* Dept Chem Univ Ill Urbana IL 61801

KLEMPERER, WILLIAM, b New York, NY, Oct 6, 27; m 49; c 3. PHYSICAL CHEMISTRY. *Educ:* Harvard Univ, AB, 50; Univ Calif, PhD, 54. *Prof Exp:* Instr chem, Univ Calif, 54; from instr to assoc prof, 54-65, PROF CHEM, HARVARD UNIV, 65- *Honors & Awards:* John Price Wetherill Medal, Franklin Inst, 78; The Irving Langmuir Award in Chem Physics, Am Soc Phys, 80; Evans Lectr, Ohio State Univ, 81; Pratt Lectr, Univ Va, 84; Rollefson Lectr, Univ Calif, Berkeley; Flygare Mem Lectr, Univ Ill, Urbana, 85; Oesper Lectr, Univ Cincinnati, 87; Kolthoff Lectr, Univ Minn, 87; Mary E Kapp Lectr, Va Commonwealth Univ, 87; Linus Pauling Distinguished Lectrs, Ore State Univ, 88; Harry Emmett Gunning Lectrs, Univ Alta, 88; Fritz London Mem Lectr, Duke Univ, 89; Hinshelwood Lectr, Oxford Univ, Eng, 89; Bomem Michelson Award, Coblentz Soc, 90; Neckers Lectr, Southern Ill Univ, 90. *Mem:* Nat Acad Sci; fel Am Chem Soc; Am Acad Arts & Sci; Am Phys Soc. *Res:* Molecular structure; molecular spectroscopy. *Mailing Add:* Dept Chem Harvard Univ Cambridge MA 02138

KLEMPNER, DANIEL, b Brooklyn, NY, June 4, 43. PHYSICAL CHEMISTRY, POLYMER SCIENCE. *Educ:* Rensselaer Polytech Inst, BS, 64; Williams Col, MS, 68; State Univ NY Albany, PhD(phys chem), 70. *Prof Exp:* Engr, Sprague Elec Co, Mass, 64-68; vis scientist, Univ Mass, Amherst, 70-72; MEM FAC, POLYMER INST, UNIV DETROIT, 72- *Mem:* AAAS; Am Chem Soc; Am Inst Chem Engrs; Am Phys Soc; Am Inst Chem; Soc Plastics Engrs; Fedn Coatings Socs; Nat Forensic Ctr. *Res:* Electrical properties of materials; inter-penetrating polymer networks; high pressure effects on polymers; x-ray diffraction studies of polymers; morphological and viscoelastic studies on polymers; theories of fusion and blending of polymers, flammability; polyurethanes of all types. *Mailing Add:* Polymer Technologies Univ Detroit 4001 W McNichols Rd Detroit MI 48221

KLEMPNER, MARK STEVEN, b Utica, NY, Jan 18, 49; m 79; c 3. INFECTIOUS DISEASES. *Educ:* Cornell Univ, MD, 73. *Prof Exp:* Intern med, Mass Gen Hosp, 73-74; resident, 74-75; clin assoc, NIH, 75-78; asst prof med, Tufts Univ, 78-82; assoc prof med, 83-88, PROF MED, NEW ENG MED CTR, 89- *Concurrent Pos:* Prin investr, NIH, Nat Inst Allergy & Infectious Dis & US Army Res Command, 79-; mem, Nat Inst Allergy & Infectious Dis study sect & sub-spec Bd, Am Bd Internal Med, 88-; vis lectr, Brazil, Austria, Italy, Sweden, Switz, UK & Denmark; vis prof, Boston Univ, 87. *Mem:* Fel Am Clin Res; Am Soc Clin Invest; Infectious Dis Soc Am; AAAS. *Res:* Interactions of infectious agents with host cells; cell activation for eradicating pathogens. *Mailing Add:* Dept Med Div Exp Med New Eng Med Ctr Boston MA 02111

KLEMS, GEORGE J, b Brno, Czech, May 4, 36; US citizen; m 68; c 2. METALLURGY, TECHNICAL MARKETING & SALES. *Educ:* Harvard Univ, AB, 58; Ill Inst Technol, MS, 61; Case Western Reserve Univ, PhD(metall & mat sci), 71. *Prof Exp:* Res asst solid state physics & mat sci, Ill Inst Technol, 58-61 & x-ray crystallog, 61-64; res metallurgist, Res Ctr, Repub Steel Corp, 64-73; mkt develop metallurgist, Molycorp, Inc, 73-76; prod metallurgist high strength steels, Steel Group, Flat Rolled Div, Repub Steel Corp, 76-84; div metallurgist, Flat Rolled Div, 84-85, staff metallurgist, Flat Rolled & Coated Prods, Prod Develop Div, Res Ctr, 85-86, MARKET DEVELOP ENGR, AUTOMOTIVE DEVELOP GROUP, LTV STEEL CO, 86- *Mem:* Am Soc Metals Int; Metall Soc Am Inst Mining & Metall Engrs; Soc Automotive Engrs; Sigma Xi; Am Soc Testing & Mat; Am Welding Soc; Soc Mfg Engrs. *Res:* Phase transformations; alloy development; sheet steel formability. *Mailing Add:* 32840 Ledge Hill Dr Solon OH 44139-1917

KLEMS, JOSEPH HENRY, b Cincinnati, Ohio, July 14, 42; m 67; c 2. ENERGY CONSERVATION. *Educ:* Univ Chicago, SB, 64, SM, 65, PhD(physics), 70. *Prof Exp:* Res assoc physics, Lab Nuclear Studies, Cornell Univ, 70-73; asst res physicist, Univ Calif, Davis, 73-78; STAFF SCIENTIST, LAWRENCE BERKELEY LAB, 78- *Mem:* Am Phys Soc; AAAS; Sigma Xi. *Res:* Energy-efficient windows and lighting systems; solar energy; measurement of energy flows through architectural windows under realistic conditions; contrast mechanisms for X-ray microscopy. *Mailing Add:* Appl Sci Div Lawrence Berkeley Lab Berkeley CA 94720

KLENIN, MARJORIE A, b Lancaster, Pa. MECHANICS. *Educ:* Swarthmore Col, BA, 65; Univ Pa, MS, 66, PhD(physics), 70. *Prof Exp:* Res assoc solid state physics, Inst Max von Lane-Paul Langevin, 70-72 & Univ Saarlandes, 72-74; guest scientist, Brookhaven Nat Lab, 74-76; sr res assoc reactor safety, 76-77; asst prof, 77-80, ASSOC PROF SOLID STATE PHYSICS, NC STATE UNIV, 80- *Concurrent Pos:* Vis asst prof solid state physics, State Univ NY, Stony Brook, 74-76; guest scientist, Max Plank Inst Solid State Physics Res, 85- *Mem:* Am Phys Soc; Sigma Xi. *Res:* Structural modeling and growth dynamics of bond-directed disordered compounds; covalent semiconductors; equilibrium dynamic properties; complex orientational ordering as mediated by quadrupolar couplings. *Mailing Add:* Dept Physics NC State Univ Box 8202 Raleigh NC 27650

KLENKE, EDWARD FREDERICK, JR, b New York, NY, May 22, 16; m 38; c 2. CHEMICAL ENGINEERING. *Educ:* Newark Col Eng, BS, 40. *Prof Exp:* Sr supvr, Kankakee Ord Works, 41-44; chem engr, Manhattan Proj, Univ Chicago, Hanford, Wash, 44-45; res chem & engr, pigments dept, E I du Pont de Nemours & Co, Inc, 45-64, sr res engr, 64-69, res supvr, 69-74, coordr new facil, 74-76; safety, health & environ coord, 77-81; MEM, INT EXEC SERV CORPS, 85- *Concurrent Pos:* Eng consult, 46-51 & 81- *Mem:* Am Inst Chem Engrs. *Res:* Colored pigments research; process design and development. *Mailing Add:* Ten Rigg Ct Basking Ridge NJ 07920

KLENKNECHT, KENNETH S(AMUEL), b Washington, DC, July 24, 19; m 47; c 3. AEROSPACE & AERONAUTICAL ENGINEERING, TECHNICAL MANAGEMENT. *Educ:* Purdue Univ, BS, 42. *Prof Exp:* Proj engr, NASA Lewis Res Ctr, 42-51; head opers eng sect & aeronaut res scientist, NASA Flight Res Ctr, Calif, 51-59, mem, Space Task Group, Langley Field, Va, 59-61, tech asst to dir, NASA Johnson Space Ctr, 61-62, mgr, Proj Mercury, 62-63, dep mgr, Gemini Prog, 63-67, mgr, Command & Serv Modules, Apollo Spacecraft Prog, 67-70, mgr, Skylab Prog, 70-74, dir flight opers, 74-76, asst mgr, Orbiter Proj, 76-77, dep assoc adminr space transp systs, Europ Opers, NASA HQ, Washington, DC, 77-79, asst mgr, Orbiter Proj & vehicle mgr, Orbiter 102, NASA Johnson Space Ctr, 79-81, sr space transport syst tech adv, Martin Marietta Denver Aerospace, Colo, 81-84, dir, design-to-cost productivity, Space Sta Proj, 84-88, dir, Zenith Star Proj, 88-89; RETIRED. *Honors & Awards:* John J Montgomery Award, Nat Soc Aerospace Prof, 63; W Randolph Lovelace, II Award, Am Astronaut Soc, Inc, 75. *Mem:* Fel Am Astronaut Soc; Int Acad Astronaut; assoc fel Am Inst Aeronaut & Astronaut. *Res:* Space science. *Mailing Add:* 825 Front Range Rd Littleton CO 80120

KLENS, PAUL FRANK, b Scranton, Pa, July 21, 18; m 47; c 4. MICROBIOLOGY. *Educ:* Syracuse Univ, AB, 40, MS, 42, PhD(microbiol), 51. *Prof Exp:* Asst bot, Syracuse Univ, 40-42, instr bact & mycol, 45-51; chemist & mat engr, Carrier Corp, NY, 42-45; chief germicides unit, QM Res

& Develop, US Dept Army, 51-53; chief microbiol lab, Nuodex Prods Co, NJ, 54-58; from assoc prof to prof, 58-86, dean arts & sci, 66-74, EMER PROF BIOL, LOCK HAVEN UNIV, 86- *Mem:* Am Soc Microbiol; Soc Indust Microbiol. *Res:* Physiology of fungi; microbiological deterioration; industrial microbiology; water pollution studies. *Mailing Add:* Box 405 RD 3 Mill Hall PA 17751-9520

KLENSIN, JOHN, b Tucson, Ariz, Feb 1, 45. COMPUTER SCIENCE. *Educ:* Mass Inst Technol, BS, 67, PhD(computer appln & use polit sci), 79. *Prof Exp:* PRIN RES SCIENTIST ARCHIT, MASS INST TECHNOL, 78- *Concurrent Pos:* Lectr, Dept Polit Sci, Mass Inst Technol, 80-; chmn, Comt X3J1, Am Nat Standards, 84-; dir secretariat, Int Network Food Data Syst, 89- *Mem:* Asn Comput Mach; Inst Elec & Electronics Engrs Computer Soc; Am Statist Asn; Int Asn Statist Comput. *Mailing Add:* Mass Inst Technol Rm N52-457 Cambridge MA 02139

KLEPCZYNSKI, WILLIAM J(OHN), b Philadelphia, Pa, Apr 16, 39; m 61; c 2. ASTRONOMY. *Educ:* Univ Pa, AB, 61; Georgetown Col, MA, 64; Yale Univ, PhD(astron), 69. *Prof Exp:* Astronr, Nautical Almanac Off, 61-71; ASTRONR, TIME SERV DIV, US NAVAL OBSERV, 71- *Concurrent Pos:* Pres, Inst Navig, 88-89. *Mem:* AAAS; Am Astron Soc; Am Inst Navig; Int Astron Union. *Res:* Planetary motion; masses of the planets; motion of minor planets; observations of minor planets; eclipsing variable stars. *Mailing Add:* US Naval Observ Washington DC 20392-5100

KLEPINGER, LINDA LEHMAN, b Hammond, Ind, Mar 27, 41. BIOLOGICAL ANTHROPOLOGY. *Educ:* Ind Univ, AB, 63; Univ Kans, MPhil, 71, PhD(anthrop), 72. *Prof Exp:* Asst prof, 72-79, ASSOC PROF ANTHROP, UNIV ILL, URBANA, 79- *Mem:* Am Anthrop Asn; Am Asn Phys Anthropologists; AAAS; Paleopath Asn. *Res:* Biological relationships of prehistoric populations; dental pathology; paleopathology of New and Old World populations; paleodemography of New World populations; chemical analyses of archaeological bone; bone biology. *Mailing Add:* Dept Anthrop Univ Ill 607 S Matthews Ave Urbana IL 60801

KLEPPA, OLE JAKOB, b Oslo, Norway, Feb 4, 20; m 48; c 2. PHYSICAL CHEMISTRY. *Educ:* Norwegian Tech Univ, ChE, 46; Dr techn(chem), 56. *Prof Exp:* Instr, Inst Study Metals, Univ Chicago, 49-50; res supvr, Dept Chem & Metall, Norweg Defense Res Estab, 50-51; asst prof chem, Inst Study Metals, 52-57, assoc prof, Inst Study Metals & Dept Chem, 57-62, chmn calorimetry conf, 66-67, assoc dir, James Franck Inst, 68-71, dir, 71-77, PROF, DEPT CHEM, 62-, PROF, DEPT GEOPHYS SCI, UNIV CHICAGO, 68- *Concurrent Pos:* Consult, Argonne Nat Lab; Alexander von Humboldt Award, 83-84; vis prof, Japan Soc Prom Sci, 75, Univ Paris, Orsay, 77. *Honors & Awards:* Huffman Mem Award, 82. *Mem:* Am Chem Soc; Am Inst Mining, Metall & Petrol Eng; Am Ceramic Soc; Norweg Chem Soc; fel AAAS; Am Soc Metals; Soc Norweg Engrs. *Res:* Thermodynamics; thermochemistry; electrochemistry; chemical and physical metallurgy; solid state chemistry; fused salts. *Mailing Add:* James Franck Inst Univ Chicago 5640 Ellis Ave Chicago IL 60637

KLEPPER, DAVID LLOYD, b New York, NY, Jan 25, 32. ARCHITECTURAL ACOUSTICS, ELECTROACOUSTICS. *Educ:* Mass Inst Technol, BS, 53, MS, 57. *Prof Exp:* Sr consult acoust, Bolt Beranek & Newman, Inc, 57-71; PRES ACOUST, KLEPPER MARSHALL KING ASSOC LTD, 71- *Mem:* Fel Acoust Soc Am; fel Audio Eng Soc; US Inst Theatre Technol; Inst Noise Control Eng. *Mailing Add:* 142 E 16th St New York NY 10003

KLEPPER, ELIZABETH LEE (BETTY), b Memphis, Tenn, Mar 8, 36. PLANT PHYSIOLOGY. *Educ:* Vanderbilt Univ, BA, 58; Duke Univ, AM, 63, PhD(bot), 66. *Prof Exp:* Teacher high sch, Tenn, 60-61; teaching asst bot, Duke Univ, 63-64; res scientist, Div Irrig Res, Commonwealth Sci & Indust Res Orgn Griffith, Australia, 66-68; asst prof bot, Auburn Univ, 68-72; res scientist, Battelle Northwest Labs, 72-74; sr res scientist, Ecosysts Dept, 74-76; mem staff & supvry plant physiologist, 76-85, RES LEADER, AGR RES SERV, USDA, 85- *Concurrent Pos:* Tech ed, Crop Sci, 90-; adv ed, Irrig Sci, 87- *Mem:* Fel AAAS; fel Soil Sci Soc Am; Bot Soc Am; Am Soc Plant Physiol; fel Am Soc Agron; Sigma Xi; fel Crop Sci Soc Am. *Res:* Environmental plant physiology including water relations and stress; root growth and uptake of water; cereal developmental history and modelling of cereal yield. *Mailing Add:* Agr Res Serv USDA PO Box 370 Pendleton OR 97801

KLEPPER, JOHN RICHARD, b Dayton, Ohio, Sept 20, 47; m 69; c 1. ULTRASONIC IMAGING, PATTERN RECOGNITION. *Educ:* Ohio State Univ, BS, 69; Wash Univ, MS, 75, PhD(physics), 80. *Prof Exp:* Res asst, Biomed Comput Lab, Wash Univ, 77-80; res bioengr, 80-81, DIR, DEPT PHYS SCI, INST APPL PHYSIOL & MED, 81-, EXEC DIR, 87- *Concurrent Pos:* Affil asst prof elec eng, Univ Wash, 82-87. *Mem:* Inst Elec & Electronics Engrs; Eng in Med & Biol Soc; Am Inst Ultrasound Med. *Res:* Development of computer aided ultrasonic imaging systems for use in medical diagnosis; ultrasonic tissue characterization, through application of computed tomographic techniques; blood flow analysis, through Doppler shift measurements. *Mailing Add:* Dept Phys Sci Inst Appl Physiol & Med 701 16th Ave Seattle WA 98122

KLEPPNER, ADAM, b New York, NY, June 5, 31; m 58. MATHEMATICAL ANALYSIS. *Educ:* Yale Univ, BS, 53; Univ Mich, MA, 54; Harvard Univ, PhD(math), 60. *Prof Exp:* From asst prof to assoc prof, 61-68, PROF MATH, UNIV MD, COLLEGE PARK, 68- *Concurrent Pos:* Vis prof, Univ Colo, 70-71 & Univ Calif, Berkeley, 75. *Mem:* Am Math Soc. *Res:* Group representations; functional analysis. *Mailing Add:* Dept Math Univ Md College Park MD 20742

KLEPPNER, DANIEL, b New York, NY, Dec 16, 32; m 58; c 3. PHYSICS. *Educ:* Williams Col, BA, 53; Cambridge Univ, BA, 55; Harvard Univ, PhD(physics), 59. *Prof Exp:* Res fel physics, Harvard Univ, 59-60, from instr to asst prof, 60-66; assoc prof, 66-73, head, dept physics, Div Atomic, Plasma & Condensed Matter Physics, 76-79, PROF PHYSICS, MASS INST TECHNOL, 74-, LESTER WOLFE PROF, 85-, ASSOC DIR, RES LAB ELECTRONICS, 87- *Concurrent Pos:* Alfred P Sloan Found fel, 62-64; mem, Comt Atomic, Molecular & Optical Physics, Nat Res Coun, 73-76 & 80-85; chmn, Div Atomic, Molecular & Optical Physics, Am Phys Soc, 83-84; mem bd physics & astron, Nat Acad Sci, 87-90. *Honors & Awards:* Davisson-Germer Prize, Am Phys Soc, 85, Julius Edgar Lilienfeld Prize, 90. *Mem:* Nat Acad Sci; fel AAAS; fel Am Phys Soc; fel Am Acad Arts & Sci. *Res:* Atomic physics; redetermination of the Rydberg constant; quantum chaos; studies of hydrogen in the microkelvin regime; ultra precise laser spectroscopy. *Mailing Add:* Dept Physics Mass Inst Technol Cambridge MA 02179

KLEPSER, HARRY JOHN, b Buffalo, NY, Mar 10, 08; m 36; c 3. STRATIGRAPHY, GEOLOGY. *Educ:* Syracuse Univ, AB, 32, MA, 33; Ohio State Univ, PhD(geol), 37. *Prof Exp:* Asst geol, Ohio State Univ, 34; assoc prof geol & geog, Capital Univ, 36-43 & SDak State Col, 43-44; geologist, US Geol Surv, Ky, 44-45; from asst prof to prof, 46-78, head dept, 61-72, EMER PROF GEOL, UNIV TENN, KNOXVILLE, 78- *Mem:* AAAS; Geol Soc Am; Am Asn Petrol Geol. *Res:* Fluorspar deposits; lower Mississippian rocks of the highland rim in southern Kentucky and Tennessee; Chattanooga shale of Tennessee. *Mailing Add:* 301 Engert Rd Knoxville TN 37922-3614

KLERER, JULIUS, b New York, NY, July 19, 28; m 61. PHYSICAL CHEMISTRY, SOLID STATE CHEMISTRY. *Educ:* NY Univ, BA, 49, MS, 55, PhD(phys chem), 58. *Prof Exp:* Instr phys chem, Brooklyn Col, 57-58; group leader, Radio Corp Am, 58-60; supvry mem tech staff, Bell Tel Labs, 60-67; assoc prof, 67-80, PROF CHEM ENG, COOPER UNION, 80- *Mem:* Electrochem Soc; Royal Soc Chem; NY Acad Sci; Sigma Xi. *Res:* Solid state chemistry concerned with thin films of oxides and metals, their preparation and properties both on semiconductors and semiconducting oxides. *Mailing Add:* 29 Grove St New York NY 10014

KLERER, MELVIN, b New York, NY, Feb 17, 26; m 51; c 2. COMPUTER SCIENCE. *Educ:* NY Univ, BA, 48, MS, 50, PhD(theoret physics), 54. *Prof Exp:* Tutor physics, City Col New York, 52-53, instr, 54-57; sr res assoc, dir comput & data processing facil & head, Comput Sci Prog, Hudson Labs, Columbia Univ, 57-67; prof indust eng, NY Univ, 67-73; vis scientist, Weizman Inst Sci, 73-74; PROF COMPUT SCI, POLYTECH INST NY, 74- *Concurrent Pos:* Nat lectr, Asn Comput Mach, 67; chmn, sci comput, Hudson Labs Columbia Univ, 57-67, Joint Users Group, Asn Comput Mach, 66-68, comt social implications, Inst Elec & Electronics Engrs, 70-71, Eval Panel, Nat Res Coun, 76; reviewer comput res, Int Fedn Info Processing Soc & Am Fedn Info Processing Soc; ed, Asn Comput Mach Comput Reviews. *Mem:* AAAS; Asn Comput Mach; Inst Elec & Electronics Engrs; Am Asn Artificial Intel; Am Asn Ling. *Res:* Automating the programming process for scientific-engineering-mathematical application programming and in user-oriented computer languages and terminal design. *Mailing Add:* Dept Comput Sci Polytech Univ 333 Jay St Brooklyn NY 11201

KLERLEIN, JOSEPH BALLARD, b Baltimore, Md, Dec 16, 48; m 70; c 4. FINITE MATHEMATICS. *Educ:* Furman Univ, BS, 70; Vanderbilt Univ, PhD(math), 75. *Prof Exp:* Instr, 74-75, from asst prof to assoc prof, 75-85, PROF MATH, WESTERN CAROLINA UNIV, 85-, DEPT HEAD, 90- *Mem:* Am Math Soc; Math Asn Am; Sigma Xi. *Res:* Graph theory especially the study of traversability in cayley color graphs; line graphs for directed graphs. *Mailing Add:* Dept Math Western Carolina Univ Cullowhee NC 28723

KLERMAN, GERALD L, b New York, NY, Dec 20, 28. PSYCHIATRY. *Educ:* Cornell Univ, AB, 50; NY Univ, MD, 54. *Prof Exp:* Asst prof psychiat, Med Sch, Yale Univ, 65-70; prof psychiat, Harvard Med Sch, 70-77; adminr, Alcohol, Drug Abuse & Ment Health Admin, 77-81; PROF PSYCHIAT, CORNELL UNIV, 85-, ASSOC CHMN RES, 85- *Concurrent Pos:* Dir, Conn Ment Health Ctr, 67-69; supt, Erich Lindemann Ment Health Ctr, Boston, 70-76; dir, Cobb Psychiat Res Lab, Mass Gen Hosp, 76-77; consult, AMA, 67-77, NIMH, 61-77, Med Lett, 68-77. *Honors & Awards:* Hofheimer Prize, Am Psychiat Asn, 69. *Mem:* Inst Med-Nat Acad Sci; AMA; Am Psychiat Asn; Am Psychopath Asn; AAAS. *Mailing Add:* Dept Psychiat Payne Whitney Clin Cornell Univ 525 E 68th St New York NY 10021

KLERMAN, LORRAINE VOGEL, b New York, NY, July 10, 29; div; c 4. PUBLIC HEALTH. *Educ:* Cornell Univ, BA, 50; Harvard Univ, MPH, 53, DrPH, 62. *Prof Exp:* Mem staff vol health agencies, NJ, 52-56; fac assoc res, Florence Heller Grad Sch Advan Studies Social Welfare, Brandeis Univ, 62-65, lectr, 71-73, assoc prof pub health, 73-82, prof pub health, 82-84; policy studies, Dept Health, Educ & Welfare, Washington, DC, 78-80; asst prof pub health, Yale Univ, 65-70, res assoc, 70-71, consult, 71-73, prof & head pub health, Div Health Serv Admin, 84-87, DIR, MASTERS PROG, DEPT EPIDEMIOL & PUB HEALTH, SCH MED, YALE UNIV, 90- *Concurrent Pos:* Mem, Res Rev Comt, Nat Inst Alcohol Abuse & Alcoholism, 73-76, Health Serv Res Rev Subcomt, Nat Ctr Health Serv Res & Health Care Technol Assessment, 86-90, Expert Panel Content Prenatal Care, Pub Health Serv, 86-89, Priority Expert Panel A (Low Birthweight), Nat Ctr Nursing Res, 88-89, Prog Develop Bd, Am Pub Health Asn, 90-; mem ed adv comt, Family Planning Perspectives, 90-94; chairperson, Maternal & Child Health Sect, Am Pub Health Asn, 86-88, New Haven Family Alliance; consult, Comt Study Prev Low Birthweight, Inst med; mem, Comt Study Outreach Prenatal Care, Inst Med, sci adv panel & bd dirs, Alan Guttmacher Inst, coun adv, Nat Ctr Children in Poverty, tech adv comt, Community Childhood Hunger Identification Proj, adv comt, Primary Care Asst & Accountability Proj, Asn Maternal & Child Health Progs, prof adv comt, Conn Child Health Access Proj. *Mem:* Fel Am Pub Health Asn. *Res:* Maternal and child health; handicapped children; child welfare; adolescent parenting. *Mailing Add:* Dept Epidemiol & Pub Health Yale Univ Sch Med 60 College St PO Box 3333 New Haven CT 06510-8034

KLESIUS, PHILLIP HARRY, b Bryn Mawr, Pa, Mar 1, 38; m 70; c 2. PARASITOLOGICAL RESEARCH. *Educ:* Fla Southern Col, BS, 61; Northwestern State Univ, MS, 63; Univ Tex, Austin, PhD(microbiol), 66. *Prof Exp:* Asst prof, microbiol, Univ Tex, Austin, 68-69; asst prof, microbiol, Univ Ariz, 69-72; asst chief, Ctr Dis Control, Ft Collins, Colo, 72-73; RES LEADER & MICROBIOLOGIST, ANIMAL PARASITE RES LAB, AUBURN, AGR RES SERV, US DEPT AGR, 73- *Concurrent Pos:* Adj prof, Dept Pathobiol, Col Vet Med, Auburn Univ, 73-; vis prof vet microbiol, Col Vet Med, Tuskegee Univ, 74-; assoc prof, Med Univ SC, 76-; comn mem patent biotech, Agr Res Serv, 88- *Mem:* Am Asn Vet Immunologists (secy-treas, 88-91); Am Asn Immunologists; Am Soc Microbiologists. *Res:* Immunological and parasitological research on control of internal parasites of food animals; immune system of cultured catfish. *Mailing Add:* Animal Parasite Res Lab USDA Agr Res Serv PO Box 952 Auburn AL 36830

KLESSIG, DANIEL FREDERICK, b Fond du Lac, Wis, Feb 24, 49. MOLECULAR BIOLOGY, BIOCHEMISTRY. *Educ:* Univ Wis-Madison, BS, 71; Univ Edinburgh, BSc, 73; Harvard Univ, PhD(molecular biol, biochem), 78. *Prof Exp:* Fel, Cold Spring Harbor Lab, 78, staff scientist tumor virol, 79-80; mem fac, Dept Cell & Molecular Biol, Univ Utah, 80-85; PROF MOLECULAR BIOL & ASSOC DIR WAKSMAN INST, 85- *Concurrent Pos:* Marshall scholar, 71-73, Searle scholar, 82-85, McKnight scholar, 83-86. *Res:* Control of gene expression in plants and in animal cells and their viruses; molecular and cellular biology; virus research. *Mailing Add:* Waksman Inst Rutgers Univ PO Box 759 Piscataway NJ 08855

KLESTADT, BERNARD, b Buren, Ger, Jan 31, 25; nat US; m 56; c 1. ELECTRICAL ENGINEERING. *Educ:* Columbia Univ, BS, 49, MS, 50, Univ Southern Calif, PhD(elec eng), 58. *Prof Exp:* Elec engr, Aircraft Radiation Lab, Wright Air Develop Ctr, 49; asst proj engr, Sperry Gyroscope Co, 50; mem tech staff, Systs Develop Labs, Hughes Aircraft Co, 50-58, sr staff eng, 58-62, sr scientist, 62-63, asst mgr flight control systs dept, 63-66, mgr missile control systs dept, 66-69, sr scientist, Space & Commun Group, 69-76, sr scientist, Missiles Systs Group, 76-81, mgr, Control Systs Dept, 81-86, chief scientist, 87-88, prog mgr, Automotive Controls Eng, 88-89; TECH CONSULT, 90- *Concurrent Pos:* Lectr, Univ Southern Calif, 56-58. *Mem:* Sigma Xi; Inst Elec & Electronics Engrs; NY Acad Sci. *Res:* Servomechanisms; circuit theory; guidance and control systems; automatic computation; space flight development. *Mailing Add:* 56-845 Merion La Quinta CA 92253

KLETSKY, EARL J(USTIN), b Springfield, Mass, July 22, 30; m 58; c 1. ELECTRICAL ENGINEERING. *Educ:* Mass Inst Technol, BS, 51, MS, 53; Syracuse Univ, PhD(elec eng), 61. *Prof Exp:* Torchiana fel & res engr elec eng, Univ Delft, 55-56; elec engr, Gen Electronics Labs, 56-57; from instr to prof elec eng, 57-82, asst dir, Lab Sensory Commun, 64-74, coordr bioeng prog, 73-82, ASST DEAN, COL ENG, SYRACUSE UNIV, 82- *Concurrent Pos:* Admin dir, Inst Sensory Res, Syracuse Univ, 74-80. *Mem:* Inst Elec & Electronics Engrs; Am Soc Eng Educ. *Res:* Biosimulation; analog and digital simulation of sensory systems; modeling of sensory information processing. *Mailing Add:* Col Eng Syracuse Univ 223 Link Hall Syracuse NY 13244-1240

KLETT, JAMES ELMER, b Cincinnati, Ohio, May 20, 47. LANDSCAPE HORTICULTURE, NURSERY PRODUCTION. *Educ:* Ohio State Univ, BS, 69; Univ Ill, MS, 71, PhD(hort), 74. *Prof Exp:* Res asst ornamental hort, Univ Ill, 69-72, teaching asst, 72-74; asst prof, ornamental hort, SDak State Univ 74-77, assoc prof, 77-79; ASSOC PROF ORNAMENTAL HORT, COLO STATE UNIV, 79- *Concurrent Pos:* Water Qual Inst grant, SDak State Univ, 76, new chem prod grants; numerous nat & indust res grants, landscape hort. *Mem:* Am Soc Hort Sci; Am Hort Soc; Int Plant Propagators Soc; Sigma Xi; Int Soc Arboriculture. *Res:* Herbaceous and woody ornamental plant evaluation research; water utilization studies with landscape plants; herbicide research with container nursery crops and landscape management studies. *Mailing Add:* Dept Hort Colo State Univ Ft Collins CO 80523

KLETZIEN, ROLF FREDERICK, b Beloit, Wis, Dec 15, 46; m 69; c 2. BIOCHEMISTRY. *Educ:* Univ Wis, BS, 70, PhD(oncol), 74. *Prof Exp:* Fel biochem, Princeton Univ, 74-75, Harvard Univ, 75-77; ASSOC PROF BIOCHEM, WVA UNIV, 77- *Mem:* Am Soc Cellular Biol. *Res:* Regulation of cellular function and metabolism. *Mailing Add:* Dept Biochem Sch Med WVa Univ Morgantown WV 26506

KLEVANS, EDWARD HARRIS, b Roaring Spring, Pa, Oct 13, 35; m 59; c 2. PLASMA PHYSICS, RESEARCH ADMINISTRATION. *Educ:* Pa State Univ, BS, 57; Univ Mich, MS, 58, PhD(nuclear eng), 62. *Prof Exp:* Sr scientist, Jet Propulsion Lab, Calif Inst Technol, 62-66; from asst prof to assoc prof, 66-76, assoc dean res, 80-84, PROF NUCLEAR ENG, PA STATE UNIV, 76-, DEPT HEAD, 87- *Concurrent Pos:* Physicist, Off Fusion Energy, US Dept Energy, 84-85. *Mem:* Am Phys Soc; Am Nuclear Soc; fel AAAS. *Res:* Plasma physics; thermonuclear engineering. *Mailing Add:* 103 W Marylyn Ave State College PA 16801

KLEVAY, LESLIE MICHAEL, b Chicago, Ill. NUTRITION MEDICAL & HEALTH SCIENCES, ENVIRONMENTAL & PUBLIC HEALTH & EPIDEMIOLOGY. *Educ:* Univ Wis-Madison, BS, 56, MD, 60; Harvard Univ, MS, 62. *Hon Degrees:* DSc, Harvard Univ, 65. *Prof Exp:* Teaching asst chem, Univ Wis-Madison, 55-57; intern med, St Louis City Hosp, Mo, 60-61; from instr to asst prof internal med, Col Med, Univ Cincinnati, 65-72, from asst prof to assoc prof environ health, 65-72; RES MED OFFICER, HUMAN NUTRIT RES CTR, AGR RES SERV, USDA, 72- *Concurrent Pos:* Asst, Wash Univ, 60-61; consult, Off Int Res, NIH, 67, Ky Dept Health, 68-69, Div Chronic Dis Progs, Health Serv & Ment Health Admin, 69- & Nat Ctr Health Statist, 70; assoc prof & prof internal med, Univ NDak, 72, mem, attending med staff, 76-; Joseph Goldberger vis prof clin nutrit, 76; consult, Nat Heart & Lung Inst, 76; adv, Scientific Rev Comt, Human Nutrit Res Coun, Ont, 80; adv, Nutrit Comt, Am Acad Pediat, 81; mem Comt Clin Issues Health Dis, Am Soc Clin Nutrit; tech adv, Comt Sci & Educ Res Grants Prog,

USDA, 83; chmn, Spec Study Sect, USPH, NIH & DHHS, 86. *Mem:* AAAS; Am Fedn Clin Res; Soc Exp Biol & Med; Am Inst Nutrit; Am Soc Clin Nutrit. *Res:* Experimental atherosclerosis; epidemiology of ischemic heart disease; metabolism of metallic trace elements; definition of nutritional requirement; interrelationships of nutrients; nutritional aspects of the human environment; mammalian metabolism of insecticides; nutritional problems of underdeveloped countries. *Mailing Add:* USDA ARS Human Nutrit Res Ctr PO Box 7166 Univ Sta Grand Forks ND 58202

KLEVECZ, ROBERT RAYMOND, b Stratford, Conn, Feb 8, 39; m 61; c 2. CELL BIOLOGY, MOLECULAR BIOLOGY. *Educ:* Ga Inst Technol, BS, 62; Univ Tex, PhD(cell biol), 66. *Prof Exp:* SR RES SCIENTIST, CITY OF HOPE MED CTR, 67- *Concurrent Pos:* Fel, Yale Univ, 66, Nat Cancer Inst res fel enzyme chem, 66-67. *Mem:* AAAS; Am Soc Cell Biol. *Res:* Cellular regulatory mechanisms; cellular clocks and oscillators; control of growth and division in mammalian cells in culture; periodic gene function and the temporal organization of RNA and enzyme synthesis. *Mailing Add:* Dept Biol City Hope Med Ctr Duarte CA 91010

KLEVEN, STANLEY H, b Dawson, Minn, June 24, 40; m 60; c 3. VETERINARY MICROBIOLOGY, AVIAN MEDICINE. *Educ:* Univ Minn, St Paul, BS, 63, DVM, 65, PhD(microbiol), 70. *Prof Exp:* Pvt pract, 65-66; instr vet microbiol, Univ Minn, 66-67, res fel, 67-70; from asst prof to assoc prof med microbiol, 70-78, head dept Avian Med, 73-82, PROF AVIAN MED & MED MICROBIOL, POULTRY DIS RES CTR, UNIV GA, 78- *Honors & Awards:* Upjohn Achievement Award, Am Asn Avian Pathologists,80; Am Feed Indust Award, Am Vet Med Asn, 85. *Mem:* AAAS; Am Vet Med Asn; Am Asn Avian Path; World Vet Poultry Asn; Poultry Sci Asn. *Res:* Respiratory infections of poultry; avian mycoplasmosis. *Mailing Add:* Dept Avian Med Univ Ga 953 College Station Rd Athens GA 30605

KLEYN, DICK HENRY, b Heemstede, Neth, Oct 29, 29; US citizen; m 62; c 1. DAIRY SCIENCE, BIOCHEMISTRY. *Educ:* Ohio State Univ, BS, 53; Cornell Univ, MS, 56, PhD(dairy sci), 60. *Prof Exp:* Asst prof dairy sci, Univ Fla, 59-60; food technologist, Gen Foods Corp, 60-62; exten specialist dairy tech, Ohio State Univ, 62-63; assoc res specialist food sci, 63-80, PROF FOOD SCI, RUTGERS UNIV, NEW BRUNSWICK, 80- *Concurrent Pos:* Fel, Asn Off Anal Chemists, 87. *Mem:* Am Dairy Sci Asn; Inst Food Technol; Sigma Xi; Asn Off Anal Chemists. *Res:* Chemistry of milk; freezing point studies; factors affecting chemical composition; analytical methods for estimating phosphatase activity in milk; yogurt-digestibility and manufacture. *Mailing Add:* 357 Riva Ave Milltown NJ 08850

KLIBANOV, ALEXANDER M, b Moscow, USSR, July 15, 49; US citizen; m 72; c 1. APPLIED ENZYMOLOGY, BIOTECHNOLOGY. *Educ:* Moscow Univ, MS, 71, PhD(chem enzym), 74. *Prof Exp:* Res chemist, Moscow Univ, 74-77; res assoc, Univ Calif, San Diego, 78-79; asst prof appl biochem & Henry L Doherty prof, 79-83, from assoc prof to prof appl biochem, 83-88, PROF CHEM, MASS INST TECHNOL, 88- *Honors & Awards:* Leo Friend Award, Am Chem Soc, Ipatieff Prize. *Mem:* Am Chem Soc; Am Soc Biochem & Molecular Biol; AAAS. *Res:* Mechanisms of protein inactivation; stabilization of proteins; immobilized enzymes and cells; enzymes as catalysts in organic syntheses; biocatalysis in extreme environments. *Mailing Add:* Rm 16-209 Mass Inst Technol Cambridge MA 02139

KLICK, CLIFFORD C, b Strausstown, Pa, Aug 31, 18; m 47; c 6. SOLID STATE PHYSICS. *Educ:* Muhlenberg Col, AB, 39; Harvard Univ, MA, 47; Carnegie Inst Technol, ScD(physics), 49. *Prof Exp:* Asst elec eng, Mass Inst Technol, 41-42; assoc physicist, Radiation Lab, Johns Hopkins Univ, 42-45; physicist, Naval Res Lab, 49-52, head luminescent mat sect, 53-67, supt mat sci div, 67-77; mem staff, Off Naval Res, London, 77-79; RETIRED. *Concurrent Pos:* Consult, US Army, 45. *Mem:* Fel Am Phys Soc; Sigma Xi. *Res:* Color centers and luminescent centers in solids. *Mailing Add:* 5355 Nevada Ave NW Washington DC 20015

KLICKA, JOHN KENNETH, b Chicago, Ill, Dec 9, 33; m 54; c 7. COMPARATIVE ENDOCRINOLOGY. *Educ:* Northern Ill Univ, BS, 57, MS, 58; Univ Ill, Urbana, PhD(physiol & endocrinol), 62. *Prof Exp:* From instr to prof physiol & biochem, Wis State Univ, Oshkosh, 62-79; RES ASSOC, VET ADMIN HOSP, UNIV MINN, 79- *Concurrent Pos:* NIH training prog fel, Med Sch, Univ Minn, 65-67; career develop award environ toxicol, Nat Inst Environ Health Sci, NIH, 81-84. *Mem:* AAAS; Am Soc Zool; Am Physiol Soc; NY Acad Sci; Sigma Xi. *Res:* Mechanism by which estrogens act to induce renal tumors in Syrian golden hamsters; chemical carcinogenesis. *Mailing Add:* Vet Admin Med Ctr Burnsville MN 55337

KLIEBENSTEIN, JAMES BERNARD, b Dodgeville, Wis, June 1, 47; m 67; c 3. ECONOMICS OF LIVESTOCK PRODUCTION, ANIMAL HEALTH MANAGEMENT. *Educ:* Univ Wis-Platteville, BS, 69; Univ Ill, MS, 70, PhD(agr econ), 72. *Prof Exp:* Asst prof, Northwest Mo State Univ, 72-74; from asst prof to prof, Univ Mo, 74-86; PROF, IOWA STATE UNIV, 86- *Concurrent Pos:* Vis prof, Univ Wis-Madison, 82-83; mem eval comt, Nat Animal Health Monitoring Syst, 87-88; respondent, Off Technol Assessment, US Cong, 89-90, consult, 90-91. *Mem:* Am Agr Econ Asn; Am Soc Farm Managers & Rural Appraisers. *Res:* Economics of livestock production; economics of animal health management and food safety, primary focus at the product production level. *Mailing Add:* 1620 Buchanan Dr Ames IA 50010

KLIEFORTH, HAROLD ERNEST, b San Francisco, Calif, July 6, 27; m 54; c 2. METEOROLOGY. *Educ:* Univ Calif, Los Angeles, BA, 49, MA, 51. *Prof Exp:* Res meteorologist, Univ Calif, Los Angeles, 51-56; field dir flight group meteorol, Air Force Cambridge Res Labs, Edwards Air Force Base, Calif, 58-61, chief, Exp Meteorol Br, 61-65; RES PROF ATMOSPHERIC SCI, DESERT RES INST, UNIV NEV, RENO, 65- *Honors & Awards:* Paul Tuntland Mem Res Award, Soaring Soc Am, 54. *Mem:* AAAS; Am Meteorol Soc; Royal Meteorol Soc; Sigma Xi. *Res:* Mountain meteorology; air flow over mountains; mountain lee waves; meso-scale meteorology; synoptic meteorology and climatology; severe storms; snow water resources; macrophysics of clouds. *Mailing Add:* Desert Res Inst Lab Atmo Stead Facil Reno NV 89506

KLIEGER, PAUL, b Milwaukee, Wis, Oct 26, 16; m 42; c 1. CIVIL ENGINEERING. *Educ:* Univ Wis, BS, 39. *Prof Exp:* Mat inspector & state engr, Wis, 39; engr farm planning, Soil Conserv Serv, 39-41; sr res engr, 41-60, mgr field res sect, 60-63 & concrete res sect, Portland Cement Asn, 63-71, dir, concrete mat res dept, 71-86; CONSULT, CONCRETE & CONCRETE MAT, 86- *Concurrent Pos:* Mem, Hwy Res Bd & US Comt on Large Dams, Am Concrete Inst. *Honors & Awards:* Award of Merit, Am Soc Testing & Mat, 75, Frank Richart Award, 77; Hon Mem, Am Concrete Inst. *Mem:* Am Soc Testing & Mat; Am Concrete Inst. *Res:* Cement and concrete. *Mailing Add:* 2050 Valencia Dr Northbrook IL 60062

KLIEJUNAS, JOHN THOMAS, b Sheboygan, Wis, May 4, 43; m 68; c 2. PLANT PATHOLOGY. *Educ:* Univ Wis-Stevens Point, BS, 65; Univ Minn, MF, 67; Univ Wis-Madison, PhD(plant path), 71. *Prof Exp:* Fel plant path, Univ Wis-Madison, 71-72; jr plant pathologist, Univ Hawaii, Hilo, 72-75, asst plant pathologist, 75-79; plant pathologist, 79-87, SUPVR PLANT PATHOLOGIST, PAC SOUTHWEST REGION, US FOREST SERV, 87- *Mem:* Am Phytopath Soc. *Res:* Epidemiology and control of forest nursery disease and other diseases of Californian forest trees. *Mailing Add:* Forest Pest Mgt 630 Sansome St San Francisco CA 94111

KLIEM, PETER O, b Berlin, Ger, May 13, 38; US citizen; m 62; c 3. ANALYTICAL CHEMISTRY, PHOTOGRAPHIC CHEMISTRY. *Educ:* Bates Col, BS, 60; Northeastern Univ, MS, 65. *Prof Exp:* Asst scientist to sr scientist, 60-66, dept mgr to sr develop mgr, 66-75, div vpres negative res & develop, 75-77, asst corp vpres res, 77-80, VPRES RES, POLAROID CORP, 80- *Concurrent Pos:* Sr vpres & dir res & eng, Polaroid Corp. *Mem:* Nat Acad Sci; Soc Photog Sci & Eng; Indust Res Inst; Am Chem Soc. *Res:* Electronic imaging, medical diagnostic research and development; general management, photographic research and development. *Mailing Add:* 158 Sherburn Circle Weston MA 02193

KLIER, KAMIL, b Prague, Czech, Mar 21, 32; m 61; c 2. PHYSICAL CHEMISTRY. *Educ:* Charles Univ, Prague, dipl chem, 54; Czech Acad Sci, CSc(phys chem), 61. *Prof Exp:* Res fel surface phys chem, Inst Phys Chem, Czech Acad Sci, 54-57, asst, 57-61, res scientist, 61-67; vis prof physics & chem of solids & surfaces, 67-68, res assoc prof, 68-73, PROF CHEM, LEHIGH UNIV, 73-, DIR, CATALYSIS LAB, 75-, ASSOC DIR, CTR SURFACE RES, 78- *Concurrent Pos:* Int Atomic Energy Agency fel radiation chem surfaces, Wantage Res Labs, Eng, 59-60; consult catalysis, spectros & separation processes. *Mem:* Am Chem Soc; Sigma Xi. *Res:* Physics and chemistry of solids; surface chemistry; chemisorption; catalysis. *Mailing Add:* Ctr Surface Coatings Lehigh Univ Bethlehem PA 18015

KLIEWER, JOHN WALLACE, b Lanigan, Sask, Jan 20, 24; US citizen; m 54; c 4. MEDICAL ENTOMOLOGY, ENVIRONMENTAL PROTECTION. *Educ:* Bethel Col, Kans, BA(biol), 50; Univ Utah, MS(Invert Zool), 52; Univ Kans, PhD(entom), 62. *Prof Exp:* Asst prof biol sci, Bethel Col, Kans, 53-56; sr vector control specialist & proj leader, Calif State Dept Pub Health, 60-67; proj officer, Aedes Aegypti Eradication Prog, 67-69, sr res entomologist, Malaria Prog, Ctr Dis Control, USPHS, 69-72; pesticides specialist, Pesticide Prog, Environ Protection Agency, 72-90; CONSULT, 90- *Concurrent Pos:* Consult, WHO, 66-, AID, 83; adj prof prev med, Med Univ SC, 74-83. *Honors & Awards:* Spec Achievement Awards, Environ Protection Agency, 87 & 88. *Mem:* Entom Soc Am; Am Mosquito Control Asn; Am Registry Prof Entomologists; Soc Vector Ecologists. *Res:* Ecology and behavior of insects, mites and ticks; public health and agricultural uses of pesticides; environmental aspects of pest control. *Mailing Add:* 9805 Meadow Knoll Ct Vienna VA 22181-3213

KLIEWER, KENNETH L, b Mountain Lake, Minn, Dec 31, 35; m 59; c 3. THEORETICAL SOLID STATE PHYSICS. *Educ:* Univ Minn, BS, 57, MSEE, 59; Univ Ill, PhD(physics), 64. *Prof Exp:* From asst prof to assoc prof, Iowa State Univ, 63-69, prof physics, 69-81; assoc dir phys res, Argonne Nat Lab, 81-86; DEAN, SCH SCI, PURDUE UNIV, 86- *Concurrent Pos:* From assoc physicist to sr physicist, Ames Lab, US Dept Energy, 63-81, prog dir solid state physics, 74-78, assoc dir sci & technol, 78-81, mem staff, Off Basic Eng Sci, Off Energy Res, 79-80; guest prof, Univ Hamburg, Ger, 72-73; Free Univ, Berlin, 74 & Fritz-Haber Inst, Berlin, 75; vis scientist, Rockwell Int Sci Ctr, Thousand Oaks, Calif, 76. *Mem:* Am Phys Soc; AAAS; Sigma Xi. *Res:* Optical properties of solids, particularly metals; lattice dynamics; surface physics; photoemission; optics. *Mailing Add:* Dean Sch Sci Math Sci Bldg Purdue Univ West Lafayette IN 47907

KLIEWER, WALTER MARK, b Escondido, Calif, Dec 10, 33; m 62; c 1. BIOCHEMISTRY, PLANT PHYSIOLOGY. *Educ:* Calif State Polytech Col, BS, 55; Cornell Univ, MS, 58, PhD(agron), 61. *Prof Exp:* Scientist, Soil Conserv Serv, USDA, 55; fel, Ore State Univ, 61-63; asst biochemist, 63-68, assoc biochemist, 68-74, BIOCHEMIST, UNIV CALIF, DAVIS, 74- *Concurrent Pos:* Pres, Am Soc Viticult & Enol, 82. *Mem:* Am Soc Plant Physiol; Am Soc Enol; Am Soc Hort Sci; Sigma Xi. *Res:* Effect of environment on fruit quality and growth and development of grapevines; organic acid, amino acid and carbohydrate metabolism of grapevines; translocation; photosynthesis; plant growth regulators; fruit coloration; mineral nutrition; vineyard canopy management. *Mailing Add:* Dept Viticult & Enol Univ Calif Davis CA 95616

KLIGER, DAVID SAUL, b Newark, NJ, Nov 3, 43; m 79. PHYSICAL CHEMISTRY. *Educ:* Rutgers Univ, BS, 65; Cornell Univ, PhD(phys chem), 70. *Prof Exp:* NIH res fel phys chem, Harvard Univ, 70-71; PROF CHEM, UNIV CALIF, SANTA CRUZ, 71- *Concurrent Pos:* Petro Res Fund-Am Chem Soc res grant, Univ Calif, Santa Cruz, 71-74; NIH res grant, 73-; NSF res grant, 76- *Mem:* Am Soc Photobiol; Biophys Soc; Am Chem Soc. *Res:* Molecular spectroscopy of electronically excited states; spectroscopic studies of visual pigments; photochemistry; biophysics. *Mailing Add:* Div Natural Sci Univ Calif Santa Cruz CA 95064

KLIGMAN, ALBERT MONTGOMERY, b Philadelphia, Pa, Mar 17, 16; m 42; c 3. DERMATOLOGY. *Educ:* Pa State Univ, BS, 39; Univ Pa, PhD(bot), 42, MD, 47; Am Bd Dermat & Syphilol, dipl, 51. *Prof Exp:* Dir res, J B Swayne Co, Pa, 39-44; intern, N Div, Albert Einstein Med Ctr, 47-48; resident dermat, Univ Hosp, 48-51, prof, Div Grad Med, Hosp Univ Pa, 58-72, from instr to assoc prof, 48-57, PROF DERMAT, SCH MED, UNIV PA, 57- *Mem:* AAAS; Soc Invest Dermat; Soc Exp Biol & Med; AMA; Am Acad Dermat; Sigma Xi. *Res:* Medical mycology; dermatologic allergy. *Mailing Add:* Dermat Dept Univ Pa Clin Res Bldg 219 Philadelphia PA 19104

KLIGMAN, RONALD LEE, b Philadelphia, Pa, Aug 20, 40; m 68; c 2. STATISTICAL MECHANICS, ACOUSTICS. *Educ:* Temple Univ, BA, 62; Am Univ, MS, 67, PhD(physics), 68. *Prof Exp:* Asst prof physics, Sweetbriar Col, 68-69; physicist acoust, Naval Ship Res & Develop Ctr, 69-71; asst prof physics, Robert Col, 71-72; RES PHYSICIST STATIST MECH & ACOUST, NAVAL SURFACE WEAPONS CTR, 72- *Mem:* Am Phys Soc. *Res:* Statistical mechanics applied to phase transitions in solids; transport theory in ionized media; wave propagation and scattering, acoustic and electromagnetic. *Mailing Add:* 1394 Canterbury Way Rockville MD 20854

KLIJANOWICZ, JAMES EDWARD, b Baltimore, Md, Sept 24, 44; m 75; c 1. ORGANIC CHEMISTRY. *Educ:* Loyola Col, Md, BS, 66; Carnegie-Mellon Univ, MS, 69, PhD(org chem), 71. *Prof Exp:* Sr res chemist, 70-78, res lab head, Chemiphotog Systs Lab, 78-81, RES LAB HEAD, PHOTOG MECHANISMS LAB, EASTMAN KODAK CO, 81- *Mem:* AAAS; Am Chem Soc; Sigma Xi. *Res:* Mechanisms of photographic chemical reactions. *Mailing Add:* Seven Millstone Ct Pittsford NY 14534

KLIMAN, ALLAN, b Boston, Mass, Dec 20, 33; m 56; c 2. INTERNAL MEDICINE. *Educ:* Harvard Univ, AB, 54, MD, 58. *Prof Exp:* Res assoc org chem, Harvard Univ, 53-54, res assoc endocrinol, 55; intern, Beth Israel Hosp, 58-59; chief, Clin Ctr Blood Bank, NIH, 59-61; instr, 61-71, ASST CLIN PROF MED, HARVARD MED SCH, 71-; CHIEF HEMAT-ONCOL DEPT, SPAULDING REHAB HOSP, 74- *Concurrent Pos:* Med dir, Mass Red Cross Blood Prog, 64-74; clin assoc med, Mass Gen Hosp. *Mem:* Am Fedn Clin Res; Am Soc Hemat. *Res:* Hematology; blood transfusion; serum hepatitis; automation of laboratory procedures; plasmapheresis; cancer chemotherapy; clinical pharmacology. *Mailing Add:* 40 Newton St Brookline MA 02146

KLIMAN, GERALD BURT, b Boston, Mass, July 28, 31; m 60; c 2. ROTATING MACHINES, ELECTRONIC DRIVES. *Educ:* Mass Inst Technol, SB, 55, SM, 59, ScD(elec eng), 65. *Prof Exp:* Teaching asst elec eng, Mass Inst Technol, 57-61, instr, 61-65; asst prof, Rensselaer Polytech Inst, 65-71; electromagnetic engr advan propulsion equip prog, Transp Technol Ctr, 71-75, prin engr, Fast Breeder Reactor Dept, 75-77, ELEC ENGR, CORP RES & DEVELOP, GEN ELEC CO, 77- *Concurrent Pos:* Lectr, Diag, Elec Power Res Inst. *Mem:* Inst Elec & Electronics Engrs; Am Phys Soc. *Res:* Magnetohydrodynamics; Alfven waves in wave guides with non uniform magnetic fields; flow in pipes; electromagnetic pumps; electrical machines, materials and electronic drive systems; diagnostics. *Mailing Add:* Corp Res & Develop Gen Elec Co Schenectady NY 12301

KLIMAN, HARVEY LOUIS, b Boston, Mass, May 28, 42; m 66. PHYSICAL CHEMISTRY, POLYMER PHYSICS. *Educ:* Boston Univ, AB, 63; Princeton Univ, MA, 66, PhD(phys chem), 70. *Prof Exp:* SR RES CHEMIST POLYMER CHEM & PHYSICS, FIBERS DEPT, FIBERS & COMPOSITES DEVELOP CTRS, E I DU PONT DE NEMOURS & CO, INC, 69- *Mem:* AAAS; Am Chem Soc; Sigma Xi. *Res:* High pressure physical chemistry of solutions; hydrophobic interactions in biochemical macromolecules; physical chemistry and physics of fiber forming polymers; textile yarn process engineering; computer modelling; composites; 2 United States patents. *Mailing Add:* 48 H Webb Rd Chadds Ford PA 19317

KLIMCZAK, WALTER JOHN, b New Haven, Conn, Nov 17, 16; m 54; c 5. MATHEMATICS. *Educ:* Yale Univ, PhD(math), 48. *Prof Exp:* Instr, Yale Univ, 43-47; instr & asst prof math, Univ Rochester, 47-51; from asst prof to prof, 51-74, SEABURY PROF MATH & NATURAL PHILOS, TRINITY COL, CONN, 74- *Mem:* Am Math Soc; Math Asn Am. *Res:* Differential operators of infinite order. *Mailing Add:* 66 Johnson St Newington CT 06111

KLIMEK, JOSEPH JOHN, b Wilkes-Barre, Pa, Sept 14, 46; m 71; c 1. INFECTIOUS DISEASES, HOSPITAL EPIDEMIOLOGY. *Educ:* Princeton Univ, AB, 68; Pa State Univ, MD, 72. *Prof Exp:* Intern & resident med, 72-74, fel infectious dis, 74-76, chief epidemiol, 76-88, assoc dir infectious dis & asst dir med, 76-88, assoc dir med, 88-90, dir, Aids Prog, 85-90, DIR DEPT MED, HARTFORD HOSP, CONN, 90-; ASSOC DIR, DEPT MED, UNIV CONN SCH MED, 90-, PROF MED, 90- *Concurrent Pos:* From asst prof to assoc prof med, Sch Med, Univ Conn, 77-90; sr ed, Am J Infection Control, 82-; lectr, Merck Sharpe & Dohme Vaccine, 82-; med ed, Asepsis, Infection Control Forum, 83-; bd dir, Asn Practr Infection Control, 78-82. *Honors & Awards:* Lange Award, Pa State Univ Col Med, 72; Am Red Cross Commun Award, 88. *Mem:* Fel Am Col Physicians; fel Infectious Dis Soc Am; Am Soc Microbiol; Soc Hosp Epidemiologists Am; Am Pub Health Asn; AAAS; Asn Practr Infection Control. *Res:* Hospital epidemiology and infection control; antibiotic pharmacokinetics; acquired immunodeficiency syndrome. *Mailing Add:* Dept Med Hartford Hosp 80 Seymour St Hartford CT 06115

KLIMISCH, RICHARD L, b Yankton, SDak, Jan 1, 38; m 62; c 2. CATALYSIS, ENVIRONMENTAL HEALTH. *Educ:* Loras Col, BS, 60; Purdue Univ, PhD(org chem), 64. *Prof Exp:* Res chemist, Explosives Dept, Exp Sta, E I du Pont de Nemours & Co, Inc, 64-67; sr res chemist, 67-71, supvry res chemist, Fuels & Lubrication Dept, 71-73, asst dept head, Phys Chem Dept, 73-75, DEPT HEAD, ENVIRON SCI DEPT, GEN MOTORS RES LABS, 75- *Mem:* Am Chem Soc; Sigma Xi. *Res:* Catalysis; air pollution; surface chemistry; atmospheric chemistry. *Mailing Add:* 43 Fairford Rd Grosse Point Shore MI 48236-2617

KLIMKO, EUGENE M, b Youngstown, Ohio, Mar 13, 39. MATHEMATICS. *Educ:* Ohio State Univ, BS, 61, MS, 64, PhD(math), 67. *Prof Exp:* Sr res eng, NAm Aviation, Inc, 62-65; asst prof math, 67-75; MEM FAC, DEPT MATH, STATE UNIV NY, BINGHAMTON, 73- *Mem:* Am Math Soc; Inst Math Statist. *Res:* Application of ratio ergodic theorems to Glivenko-Cantelli theorem and to convergence of information ratios. *Mailing Add:* Dept Math Sci State Univ NY Binghamton NY 13901

KLIMPEL, GARY R, INTERFERON, CYTOTOXIC EFFECTOR CELLS. *Educ:* Univ Ariz, PhD(microbiol), 76. *Prof Exp:* ASSOC PROF MICROBIOL & IMMUNOL, MED BR, UNIV TEX, 80- *Mailing Add:* Dept Microbiol Univ Tex Med Br Galveston TX 77550

KLIMPEL, RICHARD ROBERT, b Billings, Mont, Sept 23, 39; m 58; c 2. MINERAL PROCESSING, ENGINEERING MATHEMATICS. *Educ:* NDak State Univ, BS, 61, MS, 62; Pa State Univ, PhD(mat sci), 64. *Prof Exp:* Teaching asst math, NDak State Univ, 61-62; res assoc fuel sci, Pa State Univ, 62-64; res mgr math, 64-77; SR RES SCIENTIST ENG, DOW CHEM CO, 77- *Concurrent Pos:* Lectr math, Saginaw Valley State Col, 65-75; adj prof mineral eng, Pa State Univ, 78- *Honors & Awards:* Robert H Richards Award, Am Inst Mining Engrs, 88. *Mem:* Sigma Xi; Am Inst Mining Engrs; Am Inst Chem Engrs; Am Chem Soc. *Res:* Mining chemicals; math modeling of engineering processes; operations research; engineering research in grinding, flotation, solids separation and other related particulate handling processes. *Mailing Add:* 4805 Oakridge Dr Midland MI 48640

KLIMSTRA, PAUL D, b Erie, Ill, Aug 25, 33; m 57; c 2. MEDICINAL CHEMISTRY. *Educ:* Augustana Col, BA, 55; Univ Iowa, MS, 57, PhD, 59. *Prof Exp:* Res chemist, G D Searle & Co, 59-70, from asst dir to dir chem res, 70-73, dir preclin res & develop, 73-74, vpres NAm Preclin Res & Develop, 74-86, sr vpres, World-Wide Preclin Develop, 86-89, EXEC VPRES, SCI & TECHNOL, G D SEARLE & CO, 89- *Concurrent Pos:* Mem, Intra-Sci Res Found & Indust Res Inst. *Mem:* Sigma Xi; Am Pharmaceut Asn; NY Acad Sci; Am Soc Pharmacol & Exp Therapeut; Pharmaceut Mfrs Asn. *Res:* Steroids; heterocyclics. *Mailing Add:* G D Searle & Co 4901 Searle Pkwy Skokie IL 60077

KLIMSTRA, WILLARD DAVID, b Erie, Ill, Dec 25, 19; m 42; c 3. WILDLIFE MANAGEMENT, VERTEBRATE ECOLOGY. *Educ:* Blackburn Col, BA, 39; Maryville Col, BA, 41; Iowa State Univ, MS, 48, PhD(econ zool), 49. *Prof Exp:* Student work mgr, Blackburn Col, 38-39; asst instr bot, Univ NC, 41-42, lectr, Sch Pub Health, 42; foreman consol, Nat Munitions Corp, 42-45; asst coop wildlife res unit, Iowa State Univ, 46-47, res assoc, 47-49; from asst prof to assoc prof, 49-59, actg dir, Coal Utilization & Res Ctr, 76-77, prof, 59-83, dist prof, 83-85, dir, 51-87, dir grad studies in zool, 73-83, EMER DISTINGUISHED PROF ZOOL, SOUTHERN ILL UNIV, CARBONDALE, 85-, EMER DIR COOP WILDLIFE RES LAB, 87- *Concurrent Pos:* Wildlife consult, Ill Natural Hist Surv, 52 & Nat Pest Control Asn; professional scientist, Ill Nat Hist Surv, 80-86. *Honors & Awards:* Kaplan Res Award, Sigma Xi, 74; Aldo Leopold Award, Wildlife Soc, 88. *Mem:* Fel AAAS; Am Soc Zool; Wildlife Soc (pres, 73-74); hon mem Nat Pest Control Asn; NY Acad Sci; hon mem Wildlife Soc; Am Inst Biol Sci; Am Soc Mammologist; Sigma Xi; Soil Conserv Soc Am; Nat Wildlife Refuge Asn. *Res:* Life history, ecology and management of vertebrates; ecology of disturbed lands by mineral extraction. *Mailing Add:* Coop Wildlife Res Lab Southern Ill Univ Carbondale IL 62903

KLINCK, HAROLD RUTHERFORD, b Gormley, Ont, Sept 24, 22; m 51; c 5. AGRONOMY, PLANT BREEDING. *Educ:* Ont Agr Col, BSA, 50; McGill Univ, MSc, 52, PhD, 55. *Prof Exp:* Lectr, 54-56, from asst prof to assoc prof, 56-71, prof, 71-88, EMER PROF AGRON, MACDONALD COL, MCGILL UNIV, 88- *Mem:* Am Soc Agron; Can Soc Agron (secy-treas, 55-58, pres elect, 65-66, pres, 66-67); fel Agr Inst Can. *Res:* Oat and barley breeding. *Mailing Add:* Dept Plant Sci Macdonald Col McGill Univ 21111 Lakeshore Rd Ste Anne de Bellevue PQ H9X 1C0 Can

KLINCK, ROSS EDWARD, b Kitchener, Ont, Dec 1, 38; m 63; c 1. PHYSICAL & ORGANIC CHEMISTRY. *Educ:* Univ Western Ont, BSc, 60, PhD(chem), 65. *Prof Exp:* Res assoc physics, Duke Univ, 64-66; asst prof chem, Univ Conn, 66-71; assoc prof chem, Urbana Col, 71-76, coordr sci, 74-76; assoc prof & chmn, Sci Div, 78-83, asst dean, 84-86, PROF CHEM, ADIRONDACK COMMUNITY COL, 79-, EXEC ASST TO PRES, 89- *Concurrent Pos:* Vis researcher chem, Univ Western Ont, 72-77 & 86-87. *Mem:* Am Chem Soc; Chem Inst Can. *Res:* High resolution nuclear magnetic resonance spectroscopy related to conformational studies and barriers to rotation; homoenolization of cyclic ketones; MNDO and molecular mechanics calculations. *Mailing Add:* Adirondack Community Col Glens Falls NY 12801

KLINE, BERRY JAMES, b Mont Alto, Pa, Jan 9, 41; m 63; c 2. CHROMOTOGRAPHY. *Educ:* Philadelphia Col Pharm, BS, 62; Temple Univ, MS, 65; Univ Wis, PhD(pharm), 68. *Prof Exp:* Sr analyst, Vick Divisions Res, Richardson-Merrell, Inc, 68-69; Sr scientist, CIBA Pharmaceut Co, 69-72, sr staff scientist, CIBA-Geigy Pharmaceut, 72-76; assoc prof & dir, anal serv, Sch Pharm, Med Col Va, 76-83; asst dir, 83-90, DIR, ANALYTICAL RES & DEVELOP, SQUIBB INST MED RES, 90- *Concurrent Pos:* Consult, NIH-Nat Cancer Inst. *Mem:* Am Pharmaceut Asn; Am Asn Pharmaceut Scientists; Am Chem Soc; fel Am Found, Pharmaceut Educ; Sigma Xi. *Res:* Homogeneous solution kinetics; complexation interactions; pharmaceutical analysis, especially gas and high performance liquid chromatography. *Mailing Add:* Bristol-Myers Squibb PO Box 191 New Brunswick NJ 08903-0191

KLINE, BRUCE CLAYTON, b Grand Rapids, Mich, June 22, 37; m 63; c 3. MOLECULAR BIOLOGY, MICROBIOLOGY. *Educ:* Aquinas Col, BS, 59; Mich State Univ, MS, 66, PhD(microbiol), 68. *Prof Exp:* Microbiologist, Mich Dept Health, 60-63; instr microbial genetics, Mich State Univ, 68; asst prof biochem, Univ Tenn, Knoxville, 71-75; assoc prof microbiol, 84-87,

PROF BIOCHEN & MOLECULAR BIOL, GRAD SCH MED, MAYO CLIN, ROCHESTER, MINN, 87-, CONSULT, DEPT BIOCHEM & MOLECULAR BIOL, 75- *Concurrent Pos:* NIH fel, Univ Calif, San Diego, 68-70; assoc prof microbiol, Univ Minn, 80-84. *Mem:* Am Soc Microbiol; fel Am Acad Microbiol. *Res:* Mechanism and control of plasmid maintenance in procaryotic organisms; control DNA replication, Domain analysis in proteins. *Mailing Add:* Dept Molecular Biol & Med Mayo Med Sch 200 First St SW Rochester MN 55902

KLINE, CHARLES HOWARD, b Pittsfield, Mass, Oct 22, 18; m 47; c 4. PHYSICAL CHEMISTRY, MARKETING & BUSINESS STRATEGY IN CHEMICALS. *Educ:* Princeton Univ, AB, 40, PhD(phys chem), 44. *Prof Exp:* Supvr, Chem & Metall Prog, Chem Div, Gen Elec Co, 46-49, mgr prod planning, 49-52; mgr chem develop div, Climax Molybdenum Co, 52-56; sci dir, Shulton, Inc, 56-59; pres, Charles H Kline & Co, Inc, 59-86, adv to bd, 86-88; PRES, PANGRAPHION INC, 89- *Concurrent Pos:* Dir, Rosario Resources Corp, 68-80. *Honors & Awards:* Mem Award, Chem Mkt Res Asn, 87. *Mem:* Int Asn Bus Res & Corp Develop; Am Chem Soc; Soc Chem Indust; Com Develop Asn; Chem Mkt Res Asn. *Res:* Marketing and strategy in the international chemical industry. *Mailing Add:* 389 Ski Trail Kinnelon NJ 07405

KLINE, DANIEL LOUIS, b Philadelphia, Pa, Dec 25, 17; m 45; c 3. PHYSIOLOGY. *Educ:* Purdue Univ, BS, 42; Columbia Univ, PhD(physiol), 46. *Prof Exp:* Instr physiol, Col Physicians & Surgeons, Columbia Univ, 45; instr, Long Island Col Med, 46-47; nat res coun fel physiol chem, Sch Med, Yale Univ, 47-49, asst prof, 49-52, from asst prof to assoc prof physiol, 52-66; from chmn dept, prof, 66-88, EMER PROF PHYSIOL, COL MED, UNIV CINCINNATI, 88. *Concurrent Pos:* Guggenheim fel, 58-59. *Honors & Awards:* Harold Lamport Award, NY Acad Sci, 79. *Mem:* AAAS; Sigma Xi; Soc Exp Biol & Med; Int Soc Thrombosis & Haemostasis; Am Physiol Soc. *Res:* Purification of plasminogen and plasmin; mechanism of activation. *Mailing Add:* Dept Physiol Univ Cincinnati Col Med Cincinnati OH 45267

KLINE, DAVID G, b Philadelphia, Pa, Oct 13, 34; m 58; c 3. NEUROSURGERY. *Educ:* Univ Pa, AB, 56, MD, 60. *Prof Exp:* From intern to resident surg, Univ Mich, Ann Arbor, 60-62; res investr neurosurg, Walter Reed Gen Hosp & Inst Res, 62-64; teaching assoc, Univ Mich & res investr, Kresge Neurosurg Labs, 64-67; instr surg & neurosurg, 67-68, asst prof neurosurg, 68-70, assoc prof surg & neurosurg, 70-71, assoc prof neurosurg, 71-73, prof surg & neurosurg, 73-76, chmn div, 71-76, PROF NEUROSURG & CHMN DEPT, SCH MED, LA STATE UNIV, NEW ORLEANS, 76- *Concurrent Pos:* Vis investr, Delta Regional Primate Ctr, 67-; vis surgeon, Charity Hosp, New Orleans, 67-; mem staff, Southern Baptist Hosp, Hotel Dieu, Touro Infirmary & Ochsner Clin & Found, 67-; consult, Keesler AFB, 69-, USPHS Hosp, 71- & Vet Admin Hosp, 74; secy, Am Bd Neurol Surg. *Honors & Awards:* Frederick A Coller Award, Am Col Surgeons, 67. *Mem:* Cong Neurol Surg; Asn Acad Surg; Am Asn Neurol Surg; Soc Univ Surgeons; Res Soc Neurosurgeons; Soc Neurol Surg (treas, 87-92); Am Bd Neurol Surg (secy, 78-83, chmn, 83-84); Southern Neurol Surg Soc (secy, 75-78, pres, 85-86). *Res:* Peripheral nerve injuries and their repair; computer utilization for neurosurgical research; hepatic encephalopathy. *Mailing Add:* Dept Neurosurg La State Univ Sch Med 1542 Tulane Ave New Orleans LA 70112

KLINE, DONALD EDGAR, b DuBois, Pa, Aug 28, 28; m 49; c 3. PHYSICS. *Educ:* Pa State Univ, BS, 51, MS, 53, PhD(physics), 55. *Prof Exp:* Instr eng mech, Pa State Univ, 54-55, asst prof physics, 55-56, vis physicist, Nuclear Reactor Facil, 56-57; staff res physicist, HRB-Singer, Inc, 57-61; from assoc prof to prof nuclear eng, 61-68, PROF MAT SCI, PA STATE UNIV, 68- *Concurrent Pos:* Consult, Jet Propulsion Lab, Calif Inst Technol, NASA, HRB-Singer, Inc, Avco Corp, Pfaudler, Hershey Med Ctr & NETCO. *Mem:* Am Phys Soc; Am Nuclear Soc; Am Soc Eng Educ. *Res:* Radiation effects; dosimetry; polymer physics; polymer impregnated concrete, wood, biomaterials; composite polymer systems. *Mailing Add:* Forest Res Lab Pa State Univ University Park PA 16802

KLINE, EDWARD SAMUEL, b Philadelphia, Pa, June 26, 24; m 50; c 2. BIOCHEMISTRY. *Educ:* Univ Pa, AB, 48; George Washington Univ, MS, 55, PhD(biochem), 61. *Prof Exp:* Clin bacteriologist, Grad Hosp, Univ Pa, 50-51; chemist, Publicker Indust, Inc, 51-52; bacteriologist, Ft Detrick, 52-54; bacteriologist, Walter Reed Army Inst Res, 54-57; biochemist, Armed Forces Inst Path, 57-61; fel, Dept Chem, Ind Univ, 61-63; asst prof, 63-68, ASSOC PROF BIOCHEM, MED COL VA, VA COMMONWEALTH UNIV, 68- *Mem:* AAAS; Sigma Xi. *Res:* Metabolic control mechanisms; metabolic effects of alcohol. *Mailing Add:* Dept Biochem Va Commonwealth Univ Sch Med MCV Sta Box 565 Richmond VA 23298

KLINE, EDWIN A, animal science; deceased, see previous edition for last biography

KLINE, FRANK MENEFEE, b Cumberland, Md, May 14, 28; m 53; c 2. PSYCHIATRY. *Educ:* Univ Md, BS, 50, MD, 52. *Prof Exp:* Intern med, Cincinnati Gen Hosp, 52-53; psychiat resident, Brentwood Vet Admin Hosp, Los Angeles, 55-58; consult, E Los Angeles Probation Off, 60-63; psychiat consult, Univ High Sch & Francis Blend Sch Blind, Los Angeles, 66-67; regional chief, W Cent Ment Health Serv, Los Angeles County, 67-68; assoc dir, Psychiat Outpatient Dept, Los Angeles County-Univ Southern Calif Med Ctr, 68-77; assoc prof psychiat, Sch Med, Univ Southern Calif, 74-78; CHIEF PSYCHIAT, LONG BEACH VET MED CTR, 77-; PROF PSYCHIAT & VCHMN DEPT, UNIV CALIF, IRVINE, 78- *Concurrent Pos:* Pvt pract, 58-; instr, Exten Div, Southern Calif Psychoanal Inst, 64-67, instr, Inst, 67-79; fac mem, Psychother Group, Los Angeles Ctr, 67-69; reviewer, JAPA & J Neuropsychiat. *Res:* Evaluation of psychotropic drugs; training of psychiatric residents and evaluation of the best methods for accomplishing this; group psychotherapy, particularly as a device for maintaining competence in practicing psychotherapists; historical evolution of psychoanalytic and psychodynamic theory. *Mailing Add:* Long Beach Vet Med Ctr 5901 E Seventh St Long Beach CA 90822

KLINE, GORDON MABEY, b Trenton, NJ, Feb 9, 03; wid; c 1. PLASTICS CHEMISTRY, POLYMER ENGINEERING. *Educ:* Colgate Univ, AB, 25; George Washington Univ, MS, 26; Univ Md, PhD(chem), 34. *Prof Exp:* Res chemist, State Dept Health, NY, 26-28 & Picatinny Arsenal, Dept of War, 28-29; chief plastics sect, Nat Bur Standards, 29-52, chief, Div Polymers, 52-63; consult, 64-69; sci editor, 69-90; RETIRED. *Concurrent Pos:* Tech ed, Mod Plastics, 36-90; ed dir & consult, Mod Plastics Encycl, 36-60; chmn, Fed Specifications Plastics Tech Comt, 41-54; tech investr, US Army, Europe, 45; chmn tech comt plastics & US deleg, Int Standardization Orgn, var US & foreign countries, 51-90; hon secy, Plastics & High Polymers Sect, Int Union Pure & Appl Chem, 51-59, from vpres to pres, 59-67. *Honors & Awards:* Gold Medal, Dept Com, 53; Am Soc Testing & Mat Award, 54; Rosa Award, Nat Bur Standards, 65; Am Nat Standards Inst Award, 87; Plastics Hall of Fame, 73. *Mem:* Am Chem Soc; Am Soc Testing & Mat; Soc Plastics Eng; Soc Plastics Indust; fel Am Inst Chem. *Res:* Plastics; adhesives; polymers; pioneer in research on testing methods and properties of polymers and plastics and in the preparation and adoption of engineering standards for polymeric materials and products. *Mailing Add:* 3063 Donnelly Dr Apt C-318 Lantana FL 33462-6405

KLINE, IRA, b Plainfield, NJ, July 7, 24; m 45; c 2. CANCER. *Educ:* Am Univ, BS, 48; George Washington Univ, MS, 50, PhD, 57. *Prof Exp:* Biologist, Nat Cancer Inst, 48-57; head dept cancer chemother & res & develop, Microbiol Assocs, Inc, 57-65, asst dir, 65-75; expert consult, Nat Cancer Inst, NIH, 75-78, health sci adminr, Div Res Grants, 78-84; RETIRED. *Concurrent Pos:* Consult biologist, Microbiol Assocs, Inc, 55-57. *Mem:* Soc Exp Biol & Med; Am Asn Cancer Res; NY Acad Sci; Am Soc Pharmacol & Exp Therapeut. *Res:* Scientific responsibility of review of grants related to cancer experimental therapeutics. *Mailing Add:* 7800 Winterberry Pl Bethesda MD 20817

KLINE, IRWIN KAVEN, b Canton, Ohio, Mar 18, 31; m 56; c 4. MEDICINE, PATHOLOGY. *Educ:* Columbia Univ, AB, 53; Western Reserve Univ, MD, 57; Am Bd Path, dipl, 62. *Prof Exp:* Intern, Mt Sinai Hosp, Cleveland, 57-58; resident path, Michael Reese Hosp, Chicago, 58-63, USPHS trainee, 60-63; instr, Univ Ill Col Med, 63-64; asst prof, Sch Med & assoc pathologist, Hosp, Boston Univ, 64-66; clin assoc path, Harvard Med Sch, 66-68; asst pathologist, Mass Gen Hosp, 68-69; from clin assoc prof to clin prof path, Sch Med, Temple Univ, 69-79; CHMN PATH, LANKENAU HOSP, 69-; PROF PATH, JEFFERSON MED COL, THOMAS JEFFERSON UNIV, PA, 79- *Concurrent Pos:* Pathologist & chief anat path, Cambridge City Hosp, 66-68. *Mem:* Int Soc Lymphology; fel Col Am Pathologists; Am Soc Clin Path; Am Asn Path; fel Am Col Cardiologists. *Res:* Cardiac disease, principally infections and immunologic myocarditis and the effect of the obstructed cardiac lymphatics. *Mailing Add:* Dept Path Lankenau Hosp 100 E Lancaster Ave Wynnewood PA 19096

KLINE, JACOB, b Boston, Mass, Aug 3, 17; m 57; c 3. BIOMEDICAL ENGINEERING. *Educ:* Mass Inst Technol, BSc, 42, MSc, 51; Iowa State Univ, PhD, 62. *Prof Exp:* Electronics engr, Int Tel & Tel Co, 42-46 & Continental TV & Electronic Co, 46-48; asst elec eng, Mass Inst Technol, 48-51, res engr, 51-52; from asst prof to assoc prof, Univ RI, 52-66; coordr biomed eng prog, 62-66, dir, Med Instrumentation Lab, 70-78, PROF BIOMED ENG & DIR DEPT, UNIV MIAMI, 66- *Concurrent Pos:* Consult, 48-; NASA-Am Soc Eng Educ fel, Stanford Univ & NASA Ames Res Ctr, 65, 66. *Mem:* Asn Advan Med Instrumentation; Inst Elec & Electronics Engrs; Am Soc Artificial Organs; NY Acad Sci; Sigma Xi. *Res:* Biomedical engineering and electronic instrumentation as applied to medical electronics; artificial hearts; patient safety problems. *Mailing Add:* PO Box 248294 Coral Gables FL 33124

KLINE, JENNIE KATHERINE, b Boston, Mass, Jan 15, 50. EPIDEMIOLOGY. *Educ:* Univ Chicago, BA, 72; Columbia Univ, MS, 74, PhD(epidemiol), 77. *Prof Exp:* SR RES SCIENTIST, NY STATE PSYCHIAT INST, 75- *Concurrent Pos:* Adj assoc prof public health, Sch Pub Health & Gertrude H Sergievsky Ctr, Columbia Univ, 85- *Mem:* Soc Epidemiol Res; Int Epidemiol Asn; Am Pub Health Asn. *Res:* Epidemiology of fetal defects and spontaneous abortions; mental retardation; prenatal HIV infection. *Mailing Add:* 722 W 168th St 630 W 168th St New York NY 10032

KLINE, JERRY ROBERT, b Minneapolis, Minn, May 20, 32; m 54; c 5. SOIL CHEMISTRY, ANALYTICAL CHEMISTRY. *Educ:* Univ Minn, BS, 57, MS, 60; Univ Minn, PhD(soil sci), 64. *Prof Exp:* Res assoc neutron activation appl to soils, Argonne Nat Lab, 64-65; assoc scientist, PR Nuclear Ctr, 65-66, dir terrestrial ecol proj, 66-68; ecologist, Radiol & Environ Res Div, Argonne Nat Lab, 68-74; sr land use analyst, US Nuclear Regulatory Comn, 74-76, sect leader, 76-80; ADMIN JUDGE, ATOMIC SAFETY & LICENSING BD, US NUCLEAR REGULATORY COMN, 80- *Concurrent Pos:* Adj prof, Univ Ill, Chicago Circle, 72-76. *Mem:* AAAS; Sigma Xi; Nature Conservancy. *Res:* Terrestrial ecology; trace elements in environmental systems; water relationships in soil-plant systems. *Mailing Add:* 13624 Middlevale Lane Silver Spring MD 20906

KLINE, KENNETH A(LAN), b Chicago, Ill, July 11, 39; m 60; c 4. ENGINEERING MECHANICS, MECHANICAL ENGINEERING. *Educ:* Univ Minn, BS, 61, PhD(eng mech), 65. *Prof Exp:* Sr res engr, Esso Prod Res Co, Standard Oil Co, NJ, 65-66; assoc prof, 66-73, PROF MECH ENG, 73-, CHMN MECH ENG, WAYNE STATE UNIV, 86- *Concurrent Pos:* Prin investr, NSF res grants, 67-68, 69-71, 72-75, 76-78 & 80-82; coprin investr, Dept Energy res grant, 77-79; prin investr, Gm res grant, 80-81, Ford res grants, 84, TACOM res grant, 84-86. *Honors & Awards:* Sr US Scientist Award, Alexander von Humboldt-Stiftung, 72-73. *Mem:* Soc Rheology; Soc Automotive Engrs; Am Soc Mech Engrs; Sigma Xi; Am Inst Astronaut & Aeronaut. *Res:* Computer-aided structural analysis; boundary integral method of structural analysis; optimal design; structural dynamics, system identification. *Mailing Add:* Mech Eng Dept Wayne State Univ 2105 Engineering St Detroit MI 48202

KLINE, LARRY KEITH, b Buffalo, NY, Oct 20, 39; m 61; c 3. MOLECULAR BIOLOGY. *Educ:* Valparaiso Univ, BS, 61; Pa State Univ, MS, 65; State Univ NY Buffalo, PhD(biochem), 70. *Prof Exp:* Asst cancer res scientist, Roswell Park Mem Inst, 65-67; NIH fel, Yale Univ, 69-71; asst prof, 71-74, ASSOC PROF BIOL SCI, STATE UNIV NY COL BROCKPORT, 74- *Mem:* AAAS; Sigma Xi. *Res:* Nucleic acid biosynthesis and function in mammalian cells. *Mailing Add:* Dept Biol Sci State Univ NY Col Brockport NY 14420

KLINE, MORRIS, b New York, NY, May 1, 08; m 39; c 3. APPLIED MATHEMATICS. *Educ:* NY Univ, BSc, 30, MSc, 32, PhD(math), 36. *Prof Exp:* Instr math, Wash Sq Col, NY Univ, 30-36; asst, Inst Adv Study, 36-38; instr math, NY Univ, 38-42; physicist & radio engr, Signal Corps Eng Labs, NJ, 42-45; from asst prof to prof, 45-76, dir, Div Electromagnetic Res, Courant Inst Math Sci, 46-66, chmn math dept, 59-70, EMER PROF MATH, WASH SQ COL, NY UNIV, 76- *Concurrent Pos:* Lectr, New Sch Soc Res, 39-40 & Hunter Col, 40-41; consult, Reeves Instrument Corp, NY, 45-56; vis prof, Stanford Univ, 58, 61 & 66; Fulbright lectr, Ger, 58-59; Guggenheim fel, 58-59; vis distinguished prof, Brooklyn Col, City Univ New York, 74-76; assoc ed, Arch Hist Exact Sci, 70-85. *Mem:* Am Math Soc; Math Asn Am. *Res:* Topology; electromagnetic theory; ultrahigh frequency radio theory; pedagogy; history of mathematics. *Mailing Add:* 1024 E 26th St Brooklyn NY 11210

KLINE, RALPH WILLARD, b Omaha, Nebr, Sept 23, 17; m 46; c 2. FOOD SCIENCE. *Educ:* Univ Omaha, AB, 39; Iowa State Univ, PhD(poultry prod technol), 45. *Prof Exp:* mem staff, Food Res Div, Armour & Co, 42-80. *Concurrent Pos:* Consult, 80- *Mem:* Am Chem Soc; Poultry Sci Asn; Inst Food Technologists. *Res:* Technology of egg and poultry products. *Mailing Add:* 4940 E Laurel Lane Scottsdale AZ 85254

KLINE, RAYMOND MILTON, b St Louis, Mo, Feb 25, 29; m 51; c 4. ELECTRICAL ENGINEERING. *Educ:* Univ Mo-Rolla, BS, 51; Iowa State Univ, MS, 54; Purdue Univ, PhD(elec eng), 62. *Prof Exp:* Systs engr, Sperry Gyroscope Co, NY, 54-57; sr systs engr, Aircraft Div, McDonnell Aircraft Co, Mo, 57-59; instr elec eng, Purdue Univ, 59-62; from asst prof to assoc prof, 62-77, PROF ELEC ENG, WASH UNIV, 77- *Mem:* Inst Elec & Electronics Engrs; Am Soc Eng Educ; Asn Comput Mach; Sigma Xi. *Res:* Design and application of information processing systems including digital computers, switching theory, especially areas in the field of artificial intelligence, pattern recognition and learning machines; image processing. *Mailing Add:* 14824 Ralls Dr Bridgeton MO 63044

KLINE, RICHARD WILLIAM, b Philadelphia, Pa, Dec 33, 42. POLYMER RHEOLOGY, ENERGY ENGINEERING. *Educ:* Mass Inst Technol, SB, 64, SM, 65, PhD(chem eng), 70. *Prof Exp:* res engr, Milliken, Inc, 70-75; group leader, Lockwood Greene Engrs, 75-76; SR DEVELOP ENGR, CRYOVAC DIV, W R GRACE & CO, 76- *Concurrent Pos:* Consult, Am Hoechst, 78-79, Batchelder-Blasius, Inc, 77-81, City Landrum, 80. *Mem:* Am Inst Chem Engrs; Soc Plastics Engrs; Sigma Xi. *Res:* Simulation and modelling of plasticating extrusion, including extruder screws and dies; development of rheological theory. *Mailing Add:* 929 Campbell St Williamsport PA 17701

KLINE, ROBERT JOSEPH, b Minocqua, Wis, Nov 6, 21; m 55; c 5. INORGANIC CHEMISTRY. *Educ:* Univ Wis, BS, 47, PhD(chem), 53. *Prof Exp:* From asst prof to prof, 53-88, EMER PROF INORG CHEM, OHIO UNIV, 88- *Mem:* Am Chem Soc. *Res:* Colloidal electrolytes; coordination compounds; actinide elements. *Mailing Add:* Dept Chem Ohio Univ Athens OH 45701-2979

KLINE, RONALD ALAN, b Wilkes-Barre, Pa, June 28, 52; m 81. ENGINEERING. *Educ:* Johns Hopkins Univ, BES, 74, MSE, 75, PhD(mech & mat sci), 78. *Prof Exp:* Res asst, Johns Hopkins Univ, 72-78; sr res scientist, Gen Dynamics Corp, 78-79; sr res engr, Gen Motors Res Labs, 79-82; ASSOC PROF MECH ENG, 82- *Mem:* Adhesion Soc Am; Am Soc Nondestructive Testing. *Res:* Mechanical behavior of fiber reinforced composite materials and adhesively bonded composite joints. *Mailing Add:* Depy Aerospace & Mech Eng Univ Okla 865 Asp Blvd Norman OK 73019

KLINE, STEPHEN JAY, b Los Angeles, Calif, Feb 25, 22; m 46; c 3. MECHANICAL ENGINEERING. *Educ:* Stanford Univ, BS, 43, MS, 49; Mass Inst Technol, ScD, 52. *Prof Exp:* Res analyst turbomach, Aerophys Lab, NAm Aviation, Inc, 46-48; asst mech eng, Stanford Univ, 48-50; from instr to asst prof, Mass Inst Technol, 50-52; from asst prof to assoc prof mech eng, 52-57, dir, Thermosci Div, 61-73, PROF MECH ENG, STANFORD UNIV, 57- *Concurrent Pos:* Consult, Gen Elec Co, 57- & Gen Motors Co, 57-81. *Honors & Awards:* Melville Medal, Am Soc Mech Engrs, 59, Fluids Eng Award, 75; George Stephenson Medal, Brit Inst Mech Engrs, 66. *Mem:* Nat Acad Eng; Am Soc Mech Engrs; AAAS; Nat Asn Sci Technol & Soc. *Res:* Internal flow; thermodynamics; innovation. *Mailing Add:* Dept Mech Eng Rm 500K Stanford Univ Stanford CA 94305-3030

KLINE, TONI BETH, b Los Angeles, Calif, Aug 23, 50; m 76. NEUROSCIENCES, PHARMACY. *Educ:* Univ Calif, Berkeley, AB, 73; Univ Calif, San Francisco, MS, 76; Univ Ala, Birmingham, PhD(chem), 80. *Prof Exp:* Teaching asst org chem, Dept Pharmaceut Chem, Univ Calif, 74-76; teaching asst org chem, Dept Chem, Univ Ala, 76-78, res asst, 76-79, teaching asst biochem, 78; res assoc, Dept Chem, Ore State Univ, 79-80; res assoc, Dept Chem, State Univ NY, 80-82, lectr org chem, 81-82; ASST PROF CHEM, DEPT PHARMACOL, MT SINAI MED SCH, 82- *Mem:* AAAS; Am Chem Soc. *Res:* Use of organic chemistry in investigating chamsims of drug actions and natural product synthesis; biogenesis, pharmacognosy and ecology; structure activity relationships of all biologically active compounds. *Mailing Add:* Biogen 14 Cambridge Ctr Cambridge MA 02142

KLINE, VIRGINIA MARCH, b Cleveland, Ohio, Jan 26, 26. PLANT ECOLOGY, VEGETATION MANAGEMENT. *Educ:* Univ Wis, BS, 47, MS, 75, PhD(bot), 76. *Prof Exp:* STAFF ECOLOGIST, UNIV WIS ARBORETUM & LECTR BOT, UNIV WIS, 76- *Mem:* Am Inst Biol Sci; Ecol Soc Am; Sigma Xi. *Res:* Community ecology; prairie, temperate forest, wetland; relationship of climate and geology to vegetation; succession; management of natural vegetation. *Mailing Add:* Arboretum Univ Wis 1207 Seminole Hwy Madison WI 53711

KLINEDINST, KEITH ALLEN, b York, Pa, Nov 8, 44; m 74; c 2. CHEMICAL VAPOR DEPOSITION, ELECTROCHEMISTRY. *Educ:* Franklin & Marshall Col, BA, 66; Stanford Univ, MS, 70, PhD(chem), 72. *Prof Exp:* Res assoc, Advan Fuel Cell Res Lab, Pratt & Whitney Aircraft, United Technol Corp, 72-76; MEM TECH STAFF MAT SCI LAB, GTE LABS INC, GEN TEL & ELECTRONICS CORP, 76- *Concurrent Pos:* Woodrow Wilson fel,. *Mem:* Sigma Xi; Am Chem Soc; Electrochem Soc. *Res:* Fuel cell electrochemistry; lithium batteries; heterogeneous catalysis; porous electrode research and development; luminescence; chemical vapor deposition. *Mailing Add:* GTE Labs Inc 40 Sylvan Rd Waltham MA 02254

KLINEDINST, PAUL EDWARD, JR, b York, Pa, Dec 29, 33; m 67; c 2. ORGANIC CHEMISTRY. *Educ:* Lehigh Univ, BS, 55; Univ Calif, Los Angeles, PhD(chem), 59. *Prof Exp:* NSF fel chem, Harvard Univ, 59-60; from asst prof to assoc prof, 60-69, chair chem, 83-88, PROF CHEM, CALIF STATE UNIV, NORTHRIDGE, 69-, ASSOC DEAN SCI & MATH, 88- *Mem:* Am Chem Soc; Sigma Xi. *Res:* Organic reaction mechanisms; salt effects and ion pairs in solvolysis and related reactions. *Mailing Add:* Sch Sci & Math Calif State Univ Northridge CA 91330

KLINEFELTER, HARRY FITCH, clinical medicine; deceased, see previous edition for last biography

KLINENBERG, JAMES ROBERT, b Chicago, Ill, June 17, 34; m 59; c 4. INTERNAL MEDICINE, RHEUMATOLOGY. *Educ:* Johns Hopkins Univ, AB & AM, 55; George Washington Univ, MD, 59. *Prof Exp:* From asst prof to assoc prof, 66-73, PROF MED, UNIV CALIF, LOS ANGELES, 73-, VCHMN DEPT, 72-, ASST DEAN, SCH MED, 80-; SR VPRES MED AFFAIRS, CEDARS-SINAI MED CTR, 85- *Concurrent Pos:* Consult, Wadsworth Vet Admin Hosp & Sepulveda Vet Admin Hosp; consult, Calif Regional Med Progs, Inland Area VI, 66-72; chmn, Calif State Arthritis Coun, 74-78; fel, Arthritis Found, 66-70, clin scholar; attend physician, Cedars-Sinai Med Ctr, 72-, dir, Dept Med, 72-85; mem, US Pharmacopeia Adv Panel Number 1 Allergy, Immunol & Connective Tissue Dis, 75-; mem bd trustees, Arthritis Found, 78-; chmn, Nat Arthritis Adv Bd, 81-87, Arthritis Found, 91- *Mem:* Fel Am Col Physicians; Am Rheumatism Asn; Am Fedn Clin Res; Asn Prog Dirs in Internal Med; Am Fedn Clin Res. *Res:* Clinical investigation in purine metabolism and gout. *Mailing Add:* Cedars-Sinai Med Ctr 8700 Beverly Blvd Los Angeles CA 90048

KLING, GERALD FAIRCHILD, b Lewisburg, Pa, Dec 12, 41; m 64; c 1. SOIL SCIENCE. *Educ:* Purdue Univ, BS, 68; Cornell Univ, MS, 73, PhD(soil sci), 74. *Prof Exp:* ASST PROF SOIL SCI, ORE STATE UNIV, 74- *Mem:* Am Soc Agron; Soil Sci Soc Am; Int Soc Soil Sci; Sigma Xi. *Res:* Quantification of the dynamic soil system so that predictions can be made regarding the probable effects of various land use changes on the system. *Mailing Add:* Dept Soil Sci Ore State Univ Corvallis OR 97331

KLING, OZRO RAY, b Peru, Ind, May 3, 42; m 66; c 2. REPRODUCTIVE ENDOCRINOLOGY, BIOLOGY. *Educ:* Butler Univ, BS, 65; Ind Univ, Bloomington, PhD(zool), 69. *Prof Exp:* NIH fel, Div Steroid Res, Ohio State Univ, 69-70; asst prof gynec & obstet & adj asst prof physiol and biophys, Sch Med, 70-75, assoc prof gynec & obstet & adj assoc prof physiol & biophys, Sch Med, 75-84, PROF GYNEC & OBSTET, PHYSIOL & BIOPHYS, & ZOOL, HEALTH SCI CTR, UNIV OKLA, 84-, ASSOC DEAN GRAD COL & ASST VPROVOST, RES ADMIN, 87- *Concurrent Pos:* Ford Found res fel, Human Reproductive Endocrinol Res Unit, Karolinska Inst, Stockholm, 74-75. *Mem:* Endocrine Soc; Am Soc Primatologists; Soc Study Reproduction; Soc Gynec Invest; Am Soc Zool. *Res:* Reproductive biology and physiology; factors regulating ovarian function; endocrine regulation of pregnancy and fetal development. *Mailing Add:* Dept Gynec & Obstet Univ Okla PO Box 26901 Oklahoma City OK 73190

KLINGBEIL, WERNER WALTER, b Onoway, Alta, June 19, 32; m 66; c 2. APPLIED MECHANICS, APPLIED MATHEMATICS. *Educ:* Univ Alta, BSc, 54; Col Aeronaut, Eng, dipl, 56; Brown Univ, SM, 64, PhD(appl math), 66. *Prof Exp:* Stress engr, Avro Aircraft Ltd, Can, 56-59; res engr, Allied Res Assocs, Inc, 59-61; res scientist, 66-71, sr res scientist, 71-82, RES ASSOC, RES CTR, UNIROYAL, INC, 82- *Mem:* Sigma Xi. *Res:* Stress analysis and design of engineering structures; deformation and flow behavior of polymeric materials; finite elasticity; viscoelasticity; composite materials; tire mechanics. *Mailing Add:* 9744 Shenandoah Dr Brecksville OH 44141-2834

KLINGBEIL, WILLIAM GENE, b Wichita, Kans, Sept 17, 16; m 41; c 4. PEDIATRICS, HEMATOLOGY. *Educ:* Univ Wichita, AB, 38; Washington Univ, MD, 43. *Prof Exp:* From instr to assoc prof pediat, Sch Med, Washington Univ, 48-60; PROF PEDIAT & CHMN DEPT, SCH MED, WVA UNIV, 60- *Concurrent Pos:* Consult, US Army, Ft Leonard Wood, Mo, 47-60 & Div Crippled Children's Serv, State of Mo, 47-60; vis prof, Sch Med, Univ Ankara, 57-58; consult, US Air Force Hosp, Scott AFB, 58-60. *Mem:* Am Pediat Soc; Am Acad Pediat; Am Fedn Clin Res. *Res:* Hematology research and cancer chemotherapy in children. *Mailing Add:* Med Ctr Morgantown WV 26505

KLINGE, ALBERT FREDERICK, b Dudleytown, Ind, May 8, 23; m 53; c 2. AGRICULTURAL ENGINEERING. *Educ:* Purdue Univ, BS, 52, MS, 55; Univ Calif, Los Angeles, PhD(eng hydraul), 66. *Prof Exp:* Asst eng, Purdue Univ, 52-55; assoc, Univ Calif, Los Angeles, 55-62, lectr, 62-63; from assoc prof to prof, 65-, EMER PROF AGR ENGR, UNIV MAINE, ORONO. *Mem:* Am Soc Agr Engrs; Am Soc Eng Educ. *Res:* Hydraulics, water and soil resource management; waste management. *Mailing Add:* 108 Forest Ave Orona ME 04473

KLINGEBIEL, ALBERT ARNOLD, b Hinton, Iowa, Oct 1, 10; m 37; c 4. SOIL SCIENCE. *Educ:* Iowa State Univ, BS, 36, MS, 37. *Prof Exp:* Res asst range reseeding, Intermt Forest & Range Exp Sta, US Forest Serv, 37-38; guest lectr, Univ Ill, 60; pres, Salut Corp, 76-84; soil scientist, Soil Conserv Serv, USDA, 38-42, dir training, 42-46, state soil scientist, 46-52, Ill state dir soil & water mgt res, Agr Res Serv, 52-54, from asst dir to dir soil surv interpretations, Soil Conserv Serv, 54-73, CONSULT, AGR RES SERV, USDA, 86- *Concurrent Pos:* Soils consult, Int Bank Reconstruct & Develop, Mex, 73-75, Econ Res Serv, USDA, DC, 75-76, Nat Res Coun, DC, 75, Remote Sensing Inst, SDak State Univ, 77-81, Mexico DF, 78-79, Soil Conserv Serv, 82-89 & World Bank, 83. *Mem:* Fel Am Soc Agron; fel Soil Sci Soc Am; Am Soc Planning Off; Int Soc Soil Sci; Soil Conserv Soc Am. *Res:* Soil classification and interpretation; soil and water management; land use planning; remote sensing; author of over 50 scientific articles. *Mailing Add:* 2413 Countryside Dr Silver Spring MD 20904

KLINGELE, HAROLD OTTO, b Niagara Falls, NY, Aug 4, 37. ANALYTICAL CHEMISTRY, ORGANIC CHEMISTRY. *Educ:* Mass Inst Technol, BSc, 59; Yale Univ, MS, 61; Cornell Univ, PhD(org chem), 65. *Prof Exp:* Instr pharmacol, Univ Louisville, 65-66, asst prof, 66-71; PRES, HOK ASSOCS, 71- *Concurrent Pos:* Sr res assoc, Dept Chem, State Univ NY, Buffalo, 73-75; mgr, treas & chem consult, Peninsula Chem Anal Ltd, 76-78; chem consult & chem analysis, HOK Assoc, 71-82; vis indust chemist, Chem Dept, Canisius Col, Buffalo, NY, 80; pres, Soap Factory Stores, Inc, 80. *Mem:* Am Chem Soc; Royal Soc Chem. *Res:* Forensics; analytical method development; industrial problems involving chemistry; drug analysis; organophosphorous heterocycles; carcinogens; toxicology; environmental chemistry; organic synthesis. *Mailing Add:* 505 Meadowbrook Dr Lewiston NY 14092

KLINGEMAN, PETER C, b Evanston, Ill, May 31, 34; m 57; c 2. HYDRAULIC ENGINEERING, HYDROLOGY. *Educ:* Northwestern Univ, BS, 57, MS, 59; Univ Calif, Berkeley, PhD(civil eng), 65. *Prof Exp:* Asst prof civil eng, NDak State Univ, 59; res engr, Univ Calif, Berkeley, 62-64; Ford Found Prog vis prof hydraul eng, Cath Univ Chile, 64-66; from asst prof to assoc prof, 66-75, PROF CIVIL ENG, ORE STATE UNIV, 75-; ACTG DEPT HEAD CIVIL ENG, ORE WATER RESOURCES RES INST, 89- *Concurrent Pos:* Dir civil eng, Ore Water Resources Res Inst, 75-89. *Honors & Awards:* Hilgard Hydraul Prize, Am Soc Civil Eng, 83. *Mem:* Am Soc Civil Eng; Am Geophys Union; Int Asn Hydraul Res; Am Soc Engrs. *Res:* Planning development and management of river basins and estuaries, including hydraulics, hydrology, sediment transport, problem analysis, impact assessment and related aspects of water resources development. *Mailing Add:* Dept Civil Eng Ore State Univ Corvallis OR 97331

KLINGEN, THEODORE JAMES, b St Louis, Mo, Oct 7, 31; m 58; c 2. PHYSICAL INORGANIC CHEMISTRY. *Educ:* St Louis Univ, BS, 53, MS, 55; Fla State Univ, PhD(chem), 62. *Prof Exp:* Nuclear res officer, Res Div, Spec Weapons Ctr, US Air Force, NMex, 55-57; analyst chem, McDonnell Aircraft Corp, 57-58; fel, Fla State Univ, 58-60, asst, 60-62; res scientist, Res Div, McDonnell Aircraft Corp, 62-64; from asst prof to assoc prof, 64-70, dir, Ctr Radiation Res, 72-74, PROF CHEM, UNIV MISS, 70-, DIR OFF ENVIRON SAFETY, 85- *Concurrent Pos:* Grants, US Dept of Energy & NSF; Am conf govt indust hygienists, Am Biol Safety Asn. *Mem:* Am Chem Soc; Am Phys Soc; Am Nuclear Soc; Health Physics Soc. *Res:* Radiation chemistry of plastic crystals; radiation induced polymerization of organo-substituted carboranes; environmental chemistry. *Mailing Add:* Environ Safety Off Univ Miss University MS 38677

KLINGENBERG, JOSEPH JOHN, b Bellevue, Ky, Nov 16, 19; wid; c 5. ANALYTICAL CHEMISTRY, MANDELIC ACIDS OF ZIRCONIUM. *Educ:* Xavier Univ, Ohio, BS, 41; Univ Cincinnati, MS, 47, PhD(chem), 49. *Prof Exp:* From instr to prof, 49-86, EMER PROF CHEM, XAVIER UNIV, OHIO, 86- *Concurrent Pos:* Vis lectr, Univ Cincinnati, 66, 69 & 75; sci adv, Food & Drug Admin, 67-72; vis lectr, Univ Ky, 76 & 77; vis scientist, Va Polytech Inst & State Univ, 81-82. *Mem:* Am Chem Soc. *Res:* Chemistry of zirconium; mandelic acid derivatives. *Mailing Add:* 51 Pleasant Ridge Ave Ft Mitchell KY 41017

KLINGENER, DAVID JOHN, b Meadville, Pa, Sept 4, 37; m 60; c 1. ZOOLOGY. *Educ:* Swarthmore Col, BA, 59; Univ Mich, MA, 61, PhD(zool), 64. *Prof Exp:* Instr zool & cur mammals, Mus Zool, Univ Mich, 63-64; asst prof, 64-70, ASSOC PROF ZOOL, UNIV MASS, AMHERST, 70- *Mem:* Am Soc Mammal; Soc Study Evolution; Am Soc Zool; Soc Vert Paleont. *Res:* Comparative anatomy and paleontology of bats and rodents. *Mailing Add:* Dept Zool Univ Mass Amherst MA 01003

KLINGENSMITH, GEORGE BRUCE, b Pittsburgh, Pa, Dec 6, 34; m 60; c 2. PHYSICAL & ORGANIC CHEMISTRY, POLYMER CHEMISTRY & ENGINEERING. *Educ:* Univ Pittsburgh, BSc, 57, PhD(phys-org chem), 63. *Prof Exp:* Fel phys & org chem, Pa State Univ, 63-64; res supvr, Shell Chem Co Woodbury, 66-74, sr staff res chemist, Shell Develop Co, 74-90; DIR RES & DEVELOP, HUNTSMAN POLYPROPYLENE CORP, 90- *Mem:* Am Chem Soc; Sigma Xi; Soc Advan Mat & Process Eng. *Res:* Reactions and physical properties of aromatic systems; solvent effects; crystallization and crystal structure of polymers; nuclear magnetic resonance spectroscopy. *Mailing Add:* 1222 N Bay Shore Dr Virginia Beach VA 23451

KLINGENSMITH, MERLE JOSEPH, b Grenora, NDak, Mar 27, 32; m 59; c 2. PLANT PHYSIOLOGY. *Educ:* Wheaton Col, Ill, BS, 54; Univ Mich, MS, 56, PhD(bot), 59. *Prof Exp:* Lab asst bact, Fla State Univ, 54-55; asst bot, Univ Mich, 55-56; vis asst prof bot & bact, Ohio Wesleyan Univ, 59-60; asst prof bot, Colgate Univ, 60-65; from asst prof to assoc prof, 65-76, PROF BIOL, ROCHESTER INST TECHNOL, 76- *Mem:* Fel AAAS; Am Soc Hort Sci; Am Sci Affiliation. *Res:* Plant tissue culture; exogenous growth regulators; radiation effects on plant growth. *Mailing Add:* Dept Biol Rochester Inst Technol Rochester NY 14623-0887

KLINGENSMITH, RAYMOND W, b Pittsburgh, Pa, Mar 21, 31; m 53, 84; c 2. NUCLEAR PHYSICS. *Educ:* Hanover Col, AB, 53; Miami Univ, MA, 55; Ohio State Univ, PhD(elastic scattering), 63. *Prof Exp:* Physicist, Westinghouse Elec Corp, 56-57; prin physicist, Battelle Mem Inst, 57-63, proj leader nuclear physics, 63-68, prog mgr strategic technol, 68-74, supvr, Hot Lab, 74-77, mgr & assoc sect mgr nuclear mat technol, Battelle-Columbus, 77-81, proj mgr, Off Nuclear Waste Isolation, 81-88, PROJ MGR, BATTELLE COLUMBUS, 88- *Res:* Low energy scattering as related to nuclear structure physics; nuclear weapons effects; strategic technology; nuclear materials technology; hot cell technology and operations; research management; nuclear waste repository development; licensing. *Mailing Add:* Battelle Columbus Div 505 King Ave Columbus OH 43201

KLINGER, ALLEN, b New York, NY, Apr 2, 37; m 88; c 5. PATTERN ANALYSIS, ENGINEERING ELECTRONICS. *Educ:* Cooper Union, BEE, 57; Calif Inst Technol, MS, 58; Univ Calif, Berkeley, PhD(elec eng), 66. *Prof Exp:* Mem tech staff electronics, Hughes Aircraft Co, 57; electronics engr elec systs, ITT Labs, 58-59; electronics systs engr comput systs, Syst Develop Corp, 59-62; sr res engr electronics systs, Jet Propulsion Lab, 64-65, consult, 78; researcher math, Rand Corp, 65-67; PROF COMPUT SCI & ENG, UNIV CALIF, LOS ANGELES, 67- *Concurrent Pos:* Consult, Syst Develop Corp, 67-68 & 78-, res dept, Gateways Hosp, 71-72, Rand Corp, 67-69 & 72-73, Int Bank Reconstruct & Develop, 74; sr radar systs specialist, Litton Industs, 68-69; prin investr, NSF, 68-71 & Air Force Off Sci Res, 70-77; chmn, Conf Data Struct Pattern Recognition & Comput Graphics, 74-75; pres, Data Structure & Display Co, 76-; consult, Los Angeles Unified Sch Dist, 76-78, radiol dept, Long Beach Mem Hosp, 77-78, IBM Los Angeles Sci Ctr, 78-79, US Army Eng Topogr Labs, 78-80, Aerospace Corp, 80-87, Edo Corp, World Bank, 80-81, Ayres-Sowell Assocs, 87, Comput Technol Assocs, 90; res travel fel, USSR, Nat Acad Sci, 82-83 & 85-86; SAIC chmn, panel Soviet image pattern recognition; Fulbright fel, 90-91. *Mem:* Fel Inst Elec & Electronics Engrs; Pattern Recognition Soc; Classification Soc; Am Asn Artificial Intel. *Res:* Computer vision and neural networks; image data bases; human computer interaction; allocation of unreliable units; composite views from tomography; biomedical wave forms; data analysis. *Mailing Add:* 3531-C Boelter Hall Univ Calif Los Angeles CA 90024-1596

KLINGER, HAROLD P, b Brooklyn, NY, July 20, 29; m 59. GENETICS. *Educ:* Harvard Univ, BA, 52; Univ Basel, MD, 59, PhD, 63. *Prof Exp:* Res asst neurosurg, New York Med Col, 45-52; demonstr anat, Univ Basel, 55-57, from second asst to first asst, 59-61, dir cytogenetics res unit, 61-63; from asst prof to assoc prof anat & genetics, 63-72, PROF GENETICS, ALBERT EINSTEIN COL MED, 72- *Concurrent Pos:* Ed, Cytogenetics & Cell Genetics, 60-; NIH career develop award, 65-74; mem adv comt, Pop Coun, Rockefeller Univ, 71- *Mem:* Genetics Soc Am; Am Soc Human Genetics; Am Asn Phys Anthrop; NY Acad Sci; Swiss Anat Soc. *Res:* Cytogenetics; role of chromosomal aberrations in human development; somatic cell genetics; gene regulation and interaction in normal and malignant cells. *Mailing Add:* Dept Pediat & Genetics Albert Einstein Col Med 1300 Morris Park Ave Bronx NY 10461

KLINGER, LAWRENCE EDWARD, b Chicago, Ill, Nov 18, 29; m 53; c 4. BACTERIOLOGY. *Educ:* Loyola Univ, Ill, BS, 51; Ill Inst Technol, MS, 53. *Prof Exp:* Res chemist, Swift & Co, 52-54, asst to dir labs, 54-55, div head, Res Labs, 55-59, asst to vpres res, 59-61, div head, Res Labs, 61-62, gen mgr new prod develop dept, 62-68, dir planning & acquisitions, Swift Chem Co, 68-69, plant mgr, 69-70, sr admin asst, 70-71, dir pub responsibility, 71-77, dir qual assurance, 78-84; mgr qual & anal serv, Beatrice Refrig Foods, 84-85, vpres, Beatrice Meats Inc, 85-86; dir qual & anal regulatory servs, Swift-Eckrich, Inc, 86-91; RETIRED. *Concurrent Pos:* Dir, Food Update, 75-79; steering comt, Nutrit Planning Conf, Food & Drug Admin, 75-76; mem, Coun Agr Sci & Technol. *Mem:* Inst Food Technol (treas, 57 & 58); Am Soc Qual Control. *Res:* New products development; nutrition education; food safety. *Mailing Add:* 1218 Indian Trail Hinsdale IL 60521

KLINGER, THOMAS SCOTT, b Kalamazoo, Mich, May 5, 55; m 87. MARINE BIOLOGY, PHYSIOLOGICAL ECOLOGY. *Educ:* Macalester Col, BA, 75; Univ SFla, MA, 79, PhD(biol), 84. *Prof Exp:* Teaching asst biol, Univ SFla, 76-83, adj lectr, 83-84; instr biol, Pasco-Hernando Community Col, 84; adj prof biol, St Leo Col, 84-85; asst prof, 85-90, ASSOC PROF BIOL, BLOOMSBURG UNIV, 90- *Concurrent Pos:* Dir, Marine Sci Consortium, 86-, vpres acad affairs, 88- *Mem:* Am Soc Zoologists; AAAS; Sigma Xi. *Res:* Physiological ecology and nutritional physiology of marine invertebrate animals, primarily echinoderms; feeding, digestion, and energetics. *Mailing Add:* Dept Biol & Allied Health Sci Bloomsburg Univ Bloomsburg PA 17815

KLINGER, WILLIAM RUSSELL, b Columbia City, Ind, Feb 9, 39; m 60; c 1. MATHEMATICS. *Educ:* Taylor Univ, BS in Ed, 61; Ohio State Univ, MSc, 67, PhD(math, educ), 73. *Prof Exp:* Teacher math, Marion Community Schs, Ind, 61-68; instr, Ohio State Univ, 68-73; asst prof, 73-74, ASSOC PROF MATH, MARION COL, 74-, HEAD DEPT, 73- *Concurrent Pos:* Mem assoc fac, Ind Univ, Kokomo, 73- *Mem:* Math Asn Am. *Res:* Necessary and sufficient conditions for continuity in metric spaces and topological spaces. *Mailing Add:* Taylor Univ Upland IN 46989

KLINGHAMMER, ERICH, b Kassel, Ger, Feb 28, 30; m 58; c 1. ETHOLOGY, PSYCHOLOGY. *Educ:* Univ Chicago, AB, 58, PhD(psychol), 62. *Prof Exp:* From instr to asst prof psychol, Univ Chicago, 63-68; ASSOC PROF PSYCHOL, PURDUE UNIV, LAFAYETTE, 68- *Concurrent Pos:* Pres, NAm Wildlife Park Found, 72-; sci ed, Grzimek's Animal Life Encycl; consult, animal behavior. *Mem:* AAAS; Am Ornith Union; Animal Behav Soc. *Res:* Ethology; imprinting; effects of early experience on adult behavior; behavior mechanisms in canids development and motivation; predator-prey interactions in wolves and bison; applied ethology. *Mailing Add:* Dept Psychol Sci Purdue Univ Lafayette IN 47907

KLINGHOFFER, JUNE F, b Philadelphia, Pa, Feb 12, 21; m 47; c 1. INTERNAL MEDICINE. *Educ:* Univ Pa, BA, 41; Woman's Med Col Pa, MD, 45; Am Bd Internal Med, cert, 51; Spec Bd Rheumatology, cert, 76. *Prof Exp:* Intern, Albert Einstein Med Ctr, 45, resident internal med, 45-47; fel path, 47-48, clin asst med, 48-50, dir student health serv, 48-51, from instr to prof, 50-87, ETHEL RUSSELL MORRIS PROF MED, MED COL PA, 87- *Honors & Awards:* Commonwealth Citation, Commonwealth Bd of Med Col Pa, 73. *Mem:* Am Med Women's Asn; Asn Women Sci; fel Am Col Physicians; AMA; Am Col Rheumatology; Asn Am Med Col. *Mailing Add:* Med Col Pa 3300 Henry Ave Philadelphia PA 19129

KLINGLER, EUGENE H(ERMAN), b Ft Wayne, Ind, Sept 3, 32; m 54; c 6. ELECTROMECHANICAL ENGINEERING. *Educ:* Ind Inst Technol, BSEE, 53; NMex State Univ, MSEE, 57; Carnegie Inst Technol, PhD(elec eng), 61. *Prof Exp:* Servomech engr, Bell Aircraft Corp, 53-55; instr, NMex State Univ, 57; proj engr, Carnegie Inst Technol, 57-61; staff engr, Space Tech Labs, 61-62; chief electronics engr, Fairchild Camera & Instrument Corp, 62; sr mem tech staff, Northrop Space Labs, 62-63; mgr eng res lab, NAm Aviation Inc, Okla, 63-65; chmn dept elec eng, Ind Inst Technol, 65-69; prof & chmn dept, Univ Detroit, 69-70; PRES & CHMN BD, EUGENE KLINGLER INC, 70- *Concurrent Pos:* Instr, Ind Inst Technol, 58. *Mem:* Inst Elec & Electronics Engrs. *Res:* Synthesis of artificial dielectric materials by means of control of electric and magnetic losses as a function of frequency. *Mailing Add:* 5045 Charing Cross Rd Bloomfield MI 48013

KLINGMAN, DARWIN DEE, b Dickinson, NDak, Feb 5, 44; m 64. OPERATIONS RESEARCH, MATHEMATICAL STATISTICS. *Educ:* Wash State Univ, BA, 66, MA, 67; Univ Tex, Austin, PhD, 69. *Prof Exp:* Teaching asst, Wash State Univ, 66-67; teaching assoc, 68-69, from asst prof to assoc prof oper res & computer sci, 69-76, PROF OPER RES & COMPUTER SCI, UNIV TEX, AUSTIN, 76-; DIR, CTR BUS DECISION ANALYSIS, 83- *Concurrent Pos:* Consult, Tex Water Develop Bd, 71-72, Farah Mfg Co, 72-, Mathematica, Inc, 72-73 & Univac, 73-76, Southland, 82, Citgo, 83-85; Hugh Roy Cullen Centennial Chair, 85-, dir, Info Syst Mgt Prog, 85-; assoc ed, Naval Res Logistics Quart, 82-; vpres, Inst Mgt Sci, 81-83, Tex rep, Math Asn Am, 80-83; mem bd dir, Timix Publ, 87-, Decision Anal & Res Inst, 78-82, Anal, Res & Comput, Inc, 75-85. *Honors & Awards:* Franz Edelman Award Mgt Sci, Int Mgt Sci. *Mem:* Oper Res Soc Am; Math Prog Soc; Asn Comput Mach. *Res:* Mathematical programming and computational algorithms and comparisons. *Mailing Add:* Dept Mgt Sci & Statist Univ Tex Austin TX 78712

KLINGMAN, DAYTON L, b Neosho Falls, Kans, Feb 10, 13; m 41; c 4. AGRONOMY. *Educ:* Univ Nebr, BSc, 38, PhD(agron), 54; Purdue Univ, MSc, 42. *Prof Exp:* Asst agronomist & asst prof agron, Univ Wyo, 42-48; agronomist, Crops Res Div, USDA, 48-52, sr agronomist & coordr weed invests, 53-56, leader weed invests-grazing lands, 57-72, chief, Turfgrass Lab, 72-74, chief, Field Crops Lab, 74-76 & Weed Sci Lab, 76-83; RETIRED. *Mem:* AAAS; Am Soc Agron; Weed Sci Soc Am; Soc Range Mgt. *Res:* Technique study of use of cages in pasture research; small grain improvement; weed control in pastures and field crops. *Mailing Add:* 407 Russell Ave Gaithersburg MD 20877

KLINGMAN, GERDA ISOLDE, b Berlin, Ger, May 6, 24; US citizen; m 53; c 1. PHARMACOLOGY. *Educ:* Fordham Univ, BS, 52; Med Col Va, PhD(pharmacol), 56. *Prof Exp:* Res assoc, Dept Pharmacol & Physiol, Med Sch, Duke Univ, 55-57; instr, Dept Pharmacol & Exp Therapeut, Sch Med, Johns Hopkins Univ, 57-61; from instr to assoc prof, 61-73, PROF BIOCHEM PHARMACOL, STATE UNIV NY, BUFFALO, 73- *Concurrent Pos:* Career develop award, NIH, 62-67. *Mem:* Am Soc Pharmacol & Exp Therapeut. *Res:* Neuropharmacology; neurochemistry, adrenergic nervous system and catecholamine metabolism; acute and chronic tolerance; drug dependence; nerve growth factor and nerve growth factor antiserum. *Mailing Add:* Dept Biochem Pharmacol State Univ NY Buffalo N Campus 449 Hochestetter Hall Buffalo NY 14260

KLINGMAN, JACK DENNIS, b Johnson City, NY, Apr 21, 27; m 53; c 1. BIOCHEMISTRY. *Educ:* Syracuse Univ, BA, 51; Med Col Va, MS, 53; Duke Univ, PhD(biochem), 58. *Prof Exp:* Res assoc pharmacol, Med Col Va, 53; asst biochem, Duke Univ, 54-55; res assoc neurochem, Johns Hopkins Univ, 58-61; from instr to assoc prof, 61-73, PROF BIOCHEM, STATE UNIV NY, BUFFALO, 73- *Concurrent Pos:* Mem fac, Dept Biochem, Sch Med, Monash Univ, Australia, 71-72; Hayes-Fulbright fel, 71-72. *Mem:* AAAS; Am Chem Soc; Neurochem Soc; Am Soc Biol Chem; Sigma Xi; Int Neurochem Soc. *Res:* N15-ethanolamine metabolism; purification and mechanism of renal glutaminase; C14-glucose metabolism in surviving superior cervical ganglion; biochemical events in excitation; phospholipid and amino acid metabolism. *Mailing Add:* Dept Biochem State Univ NY Buffalo NY 14214

KLINGSBERG, CYRUS, b Philadelphia, Pa, Nov 12, 24; m 50. SOLID STATE CHEMISTRY, CERAMICS. *Educ:* Univ Pa, BA, 48; Bryn Mawr Col, MA, 49; Pa State Univ, PhD(geo chem), 58. *Prof Exp:* Res mgr, G F Pettinos Inc, 50-51; petrologist, Simonds Abrasive Co, 51-54; asst geochem, Pa State Univ, 54-57; res chemist, Corning Glass Works, NY, 57-59; ceramist, Off Naval Res, 59-63, liaison scientist ceramics, London, 63-64, ceramist, 64-66; exec secy, Comt Radioactive Waste Mgt, Nat Acad Sci-Nat Res Coun, 68-75; sr res assoc, Arhco, 76-77; geologist, US Dept Energy, 77-87; RETIRED. *Concurrent Pos:* Vis prof, Japan Soc Prom Sci, 75. *Mem:* Fel Am Ceramic Soc; Mineral Soc Am; Sigma Xi; AAAS. *Res:* Solid state chemistry of ceramics, minerals and ionic solids; synthesis and characterization of crystalline phases; management of radioactive wastes. *Mailing Add:* 1318 Deerfield Dr State Col PA 16803-2208

KLINGSBERG, ERWIN, b Philadelphia, Pa, Mar 13, 21; m 45; c 3. ORGANIC CHEMISTRY. *Educ:* Univ Pa, BS, 41; Univ Rochester, PhD(chem), 44. *Prof Exp:* Res chemist, Schering Corp, 44-46; res chemist, Am Cyanamid Co, 46-65, prin res scientist, 65-81; RETIRED. *Concurrent Pos:* Assoc prof, City Col New York, 63-64; Brit Govt traveling fel, 72; vis prof, Univ Caen, France, 73 & Univ Sci Tech Lang, Montpellier, France, 73-74. *Mem:* Fel AAAS; Am Chem Soc. *Res:* Dyestuffs; anthraquinone derivatives; heteroaromatic compounds of nitrogen and sulfur. *Mailing Add:* 1597 Deer Path Mountainside NJ 07092

KLINK, JOEL RICHARD, b Nevada, Ohio, June 28, 35; m 59; c 2. ORGANIC CHEMISTRY. *Educ:* Ohio State Univ, BS, 57, PhD(chem), 64. *Prof Exp:* Instr chem, Ohio Northern Univ, 61-63; from asst prof to assoc prof, 63-71, PROF CHEM, UNIV WIS-EAU CLAIRE, 71-, CHMN DEPT, 78-83, 91- *Mem:* Am Chem Soc. *Res:* Reactions of diazoalkenes. *Mailing Add:* Dept Chem Univ Wis Eau Claire WI 54701

KLINK, WILLIAM H, b Chicago, Ill, Sept 29, 37; m 59. THEORETICAL PHYSICS. *Educ:* Univ Mich, BA, 59; Johns Hopkins Univ, PhD(physics), 64. *Prof Exp:* From asst prof to assoc prof, 65-77, PROF PHYSICS & ASTRON, UNIV IOWA, 77- *Concurrent Pos:* Fulbright grant, Univ Heidelberg, 64-65. *Mem:* Am Phys Soc. *Res:* Elementary particle physics, primarily using group theory. *Mailing Add:* Dept Physics & Astron Univ Iowa Iowa City IA 52240

KLINKE, DAVID J, b Detroit, Mich, Feb 27, 32; m 64; c 3. ORGANIC CHEMISTRY. *Educ:* Mich State Univ, BS, 54, PhD(org chem), 63. *Prof Exp:* Teacher jr high sch, Mich, 54-55 & high sch, 55-59; res chemist petrol additives, Jackson Lab, 63-66, prod supvr miscellaneous org intermediates, Chambers Works, 66-68, supvr mgt training & personnel develop, 68-71, prod supvr dyes, 71-74, sr supvr mat distrib, 74-80, supt safety, Environment, Protection, 80-86, SR QUAL MGR, E I DU PONT DE NEMOURS & CO, 86- *Mem:* Am Chem Soc. *Res:* Thiophene chemistry; organo-metallics. *Mailing Add:* RR Three Box 228 Woodstown NJ 08098

KLINMAN, JUDITH POLLOCK, b Philadelphia, Pa, Apr 17, 41; div; c 2. BIOCHEMISTRY, PHYSICAL ORGANIC CHEMISTRY. *Educ:* Univ Pa, AB, 62, PhD(org chem), 66. *Prof Exp:* Fel phys org chem, Isotopes Dept, Weizmann Inst, 66-67; assoc, Inst Cancer Res, 68-70, res assoc biochem, 70-72, asst mem, 72-77, assoc mem, 77-78; assoc prof, 78-82, PROF CHEM, UNIV CALIF, BERKELEY, 82- *Concurrent Pos:* Asst prof med biophys, Univ Pa, 74-78; Guggenheim fel, 88. *Mem:* Am Chem Soc; Am Soc Biochem & Molecular Biol. *Res:* Mechanism and regulation of enzyme action. *Mailing Add:* Dept Chem Univ Calif Berkeley CA 94720

KLINMAN, NORMAN RALPH, b Philadelphia, Pa, Mar 23, 37; m 78; c 2. IMMUNOLOGY. *Educ:* Haverford Col, AB, 58; Jefferson Med Col, MD, 62; Univ Pa, PhD(microbiol), 65. *Prof Exp:* Fel immunol, Univ Pa, 62-66, Weizmann Inst, 66-67 & Nat Inst Med Res, London, 67-68; from asst prof to assoc prof microbiol, Sch Med, Univ Pa, 68-75, prof path, 75-78; MEM STAFF, SCRIPPS CLIN & RES FOUND, 78- *Concurrent Pos:* NIH res fel, 62-63; Helen Hay Whitney Found res fel, 63-66; Am Cancer Soc res scholar, 66-68; adj prof, Univ Calif, San Diego, 79- *Mem:* Am Asn Immunol; Am Asn Exp Pathologists. *Res:* Structure, activity and synthesis of antibody. *Mailing Add:* Dept Immunol IMM-16 Scripps Clin & Res Found 10666 N Torrey Pines Rd La Jolla CA 92037

KLINTWORTH, GORDON K, b Ft Victoria, Rhodesia, Aug 4, 32; US citizen; m 57; c 3. PATHOLOGY, ANATOMY. *Educ:* Univ Witwatersrand, BSc, 54, MB, BCh, 57, BSc(Hons), 61, PhD(anat), 66. *Prof Exp:* Intern med & surg, Johannesburg Hosp, 58-59, sr house physician, psychiat, 59-60, registr, Neurol & Neurosurg, 60-61; assoc, 64-66, from asst prof to assoc prof, 66-73, PROF PATH, MED CTR, DUKE UNIV, 73-, PROF OPHTHAL, 81- *Concurrent Pos:* Louis B Mayer scholar, 72; distinguished prof, Duke Univ. *Honors & Awards:* Zimmerman Award. *Mem:* AAAS; Am Asn Pathologists; Sigma Xi; Int Soc Neuropath; Tissue Cult Asn; NY Acad Sci; Int Acad Path; Am Acad Ophthal. *Res:* Diseases of the eye and nervous system; infectious diseases; secondary effects of increased intracranial pressure; human genetics and diseases of the cornea. *Mailing Add:* Dept Path Duke Univ Med Ctr Durham NC 27706

KLINZING, GEORGE ENGELBERT, b Natrona Heights, Pa, Mar 22, 38; m 69; c 2. CHEMICAL ENGINEERING. *Educ:* Univ Pittsburgh, BS, 59; Carnegie Inst Technol, MS, 61, PhD(chem eng), 63. *Prof Exp:* From asst prof to assoc prof, 63-81, PROF CHEM ENG, UNIV PITTSBURGH, 81-, ASSOC DEAN RES, 87- *Concurrent Pos:* Consult, Univ Develop Proj, Ecuador, 63-66; hon prof, Cent Univ Ecuador, 66; continuing ed lectr, Prev Transport, Am Inst Chem Engrs. *Mem:* Fel Am Inst Chem Engrs; Am Soc Eng Educ. *Res:* Solid/gas flow systems; electrostatics; mass transfer in partially miscible systems; molecular hydrogen permeation; micrographic analysis of particles. *Mailing Add:* 5121 Beeler St Pittsburgh PA 15217

KLIONSKY, BERNARD LEON, b Binghamton, NY, Oct 8, 25; m 50; c 4. MEDICINE, PATHOLOGY. *Educ:* Harvard Univ, AB, 47; Hahnemann Med Col, MD, 52; Am Bd Path, dipl, 57. *Prof Exp:* Nat Cancer Inst trainee path, Med Ctr, Univ Kans, 53-55, Am Cancer Soc clin fel, 55-57, fel path, 56-57, from instr to assoc prof, 56-61; assoc prof, 61-70, PROF PATH, SCH MED, UNIV PITTSBURGH, 71- *Mem:* Am Cancer Soc. *Res:* Intrauterine fetal growth retardation; yellow hyaline membranes. *Mailing Add:* Rm 711B Scaife Univ Pittsburgh Sch Med Pittsburgh PA 15261

KLIORE, ARVYDAS J(OSEPH), b Kaunas, Lithuania, Aug 5, 35; US citizen; m 60; c 2. PLANETARY SCIENCE, ATMOSPHERIC PHYSICS. *Educ:* Univ Ill, BS, 56; Univ Mich, MS, 57; Mich State Univ, PhD(elec eng), 62. *Prof Exp:* Engr, Armour Res Found, Ill Inst Technol, 57-59; instr elec eng, Mich State Univ, 61-62; sr res engr, 62-64, res specialist, 64-66, res scientist, 66-87, SR RES SCIENTIST, JET PROPULSION LAB, CALIF INST TECHNOL, 87- *Concurrent Pos:* Lectr, Univ Calif, Los Angeles, 63-64. *Honors & Awards:* Exceptional Sci Achievement Medal, NASA, 72. *Mem:* AAAS; Am Astron Soc; Am Geophys Union; Comt Space Res; Int Astron Union. *Res:* Space astronomy; radio propagation experiments to measure planetary atmospheres; spacecraft radio propagation experiments to study the atmospheres and ionospheres of planets and their satellites. *Mailing Add:* Jet Propulsion Lab 4800 Oak Grove Dr Pasadena CA 91109

KLIOZE, OSCAR, b Baltimore, Md, Jan 2, 19; m 43; c 3. PHARMACEUTICAL CHEMISTRY. *Educ:* George Washington Univ, BS, 40; Va Polytech Inst, BS, 44; Univ Md, PhD(pharmaceut chem), 49. *Prof Exp:* Jr chemist, Baltimore Paint & Color Works, Inc, 40-41 & Bur Plant Indust, USDA, 41-42 & 46; jr biochemist, Manhattan Proj, US Army Engrs, 44-46; res assoc biochem, Northwestern Univ, 49-50; res chemist pharmaceut chem, Chas Pfizer & Co, Inc, 50-54, res supvr, 54-58; dir prod develop, A H Robins Co, Inc, 58-60 & prod develop & qual control, 60-64, dir, 65-81, vpres pharm res & anal serv, 81-84; RETIRED. *Concurrent Pos:* Lectr, Med Col Va, 66-75. *Mem:* Am Chem Soc; Am Pharmaceut Asn; Am Inst Chem; Parenteral Drug Asn. *Res:* Relationship of chemical structure to biological activity; pharmaceutical research and development; physiological effects of radiant energy; plant biochemistry; protein synthesis. *Mailing Add:* Two High Stepper Ct Apt 203 Baltimore MD 21208

KLIP, DOROTHEA A, b The Hague, Netherlands, Sept 27, 21; m 55; c 4. FUNCTIONAL ANALYSIS, APPLIED MATHEMATICS. *Educ:* State Univ Utrecht, Dr(theoret physics), 62. *Prof Exp:* Asst prof, physiol, 63-73, ASSOC PROF PHYSIOL & BIOPHYS, UNIV ALA, BIRMINGHAM, 73-, ASST PROF INFO SCI, 71- *Concurrent Pos:* Reviewer, NSF, Inst Elec & Electronics Engrs & J Comput Appl Math. *Mem:* AAAS; Asn Comput Mach; Sigma Xi; Soc Indust & Appl Math; Math Asn Am. *Res:* Design and implementation of algorithms for the solution of nonlinear (polynomial) equations; symbolic algebraic manipulation by computer. *Mailing Add:* Dept Physiol & Biophys Univ Ala Birmingham AL 35294

KLIP, WILLEM, b Rotterdam, Neth, Nov 26, 17; US citizen; m 55; c 4. BIOPHYSICS. *Educ:* Univ Utrecht, MD, 45, PhD(bact), 51, PhD(theoret physics), 55, DSc(physics), 62. *Prof Exp:* Staff mem of Dr H C Burger, Dept Med Physics, Univ Utrecht, 53-58; PROF MED PHYSICS, DEPT PHYSIOL MED & PHYSICS, UNIV ALA, BIRMINGHAM, 58- *Res:* Medical and theoretical physics. *Mailing Add:* Dept Physiol Med & Physics VH 421 Univ Ala Univ Sta Birmingham AL 35294

KLIPHARDT, RAYMOND A(DOLPH), b Chicago, Ill, Mar 18, 17; m 45; c 5. ENGINEERING SCIENCES. *Educ:* Ill Inst Technol, BS, 38, MS, 48. *Prof Exp:* Instr graphics & math, NPark Col, 38-43; asst math, Ill Inst Technol, 43-44; asst civil eng, Northwestern Univ, 45-46, from asst prof to assoc prof eng graphics, 46-58, from assoc prof to prof eng sci, 58-70, prof, 70-87, chmn dept, 78-87, EMER PROF ENG SCI & APPL MATH, NORTHWESTERN UNIV, 87- *Concurrent Pos:* Campus coordr, Khartoum Proj, USAID; consult, Appl Math Div, Argonne Nat Lab. *Mem:* AAAS; Am Soc Eng Educ; Asn Comput Mach; Am Acad Mech. *Res:* Abstract geometry; computer automation. *Mailing Add:* Dept Eng Sci & Appl Math Northwestern Univ Evanston IL 60201

KLIPPEL, JOHN HOWARD, b Warren, Ohio, Oct 15, 44; m 67; c 2. MEDICAL RESEARCH. *Educ:* Univ Cincinnati, MD, 70; Bowling Green State Univ, BA, 66; Am Bd Internal Med, cert, 74. *Prof Exp:* Sr investr, Arthritis & Rheumatism Br, 76-87, CLIN DIR, NAT INST ARTHRITIS & MUSCULOSKELETAL & SKIN DIS, BETHESDA, 87- *Concurrent Pos:* Borden res award, Univ Cincinnati, 70; clin asst prof med, Med Ctr, Georgetown Univ, 85- *Mem:* Am Col Physicians; Am Col Rheumatology. *Res:* Numerous publications; medicine. *Mailing Add:* Nat Inst Arthritis Musculoskeletal & Skin Dis Intramural Res Prog 9000 Rockville Pike Bldg 10 Rm 9N228 Bethesda MD 20892

KLIPPLE, EDMUND CHESTER, b Cuero, Tex, July 5, 06; m 28; c 3. MATHEMATICAL ANALYSIS. *Educ:* Univ Tex, BA, 26, PhD(pure math), 32. *Prof Exp:* Instr math, Univ Tex, 26-27; instr, San Antonio Jr Col, 27-29; instr, Univ Tex, 29-35; from instr to prof, 35-71, EMER PROF MATH, TEX A&M UNIV, 71- *Mem:* AAAS; Am Math Soc; Math Asn Am; Sigma Xi. *Res:* Point set theory; real variables; Laplace transformations; spaces in which there exist contiguous points. *Mailing Add:* Rte 7 PO Box 1450 Bryan TX 77802

KLIPSTEIN, DAVID HAMPTON, b New York, NY, July 25, 30; m 55, 72; c 8. ENGINEERING MANAGEMENT, ALTERNATIVE ENERGY UTILIZATION. *Educ:* Princeton Univ, BSE, 52; Mass Inst Technol, SM, 56, ScD, 63. *Prof Exp:* Res engr, Am Cyanamid Corp, 51-54; dir, Bound Brook Sta, Sch Chem Eng Practice, Mass Inst Technol, 58-60; mkt rep, Union Carbide Chem Corp, 62-69, mkt develop mgr, Develop Div, 69-70, prod mgr acrylate monomers & polymers, 70-71, mkt mgr, Trade Paint Intermediates, 71-72; bus develop mgr, Res Cottrell Inc, 72-73, vpres planning & develop oper, Air Pollution Control Group, 73-74, vpres particulate opers, 74-76, dir advan technol corp develop, Res Cottrell Inc, 76-80, dir biphase energy systs, 80-; VPRES CORP DEVELOP, BIOSYM TECHNOL. *Concurrent Pos:* Mem, Environ Adv Comt, Fed Energy Admin, 74-76. *Mem:* Am Chem Soc; Am Inst Chem Eng; Geothermal Resources Coun. *Res:* Commercial development; optimization of combustion processes, precombustion fuel cleaning, high efficiency energy conversion systems, load leveling controls. *Mailing Add:* Biosym Technol 10065 Barnes Canyon Rd San Diego CA 92121-2777

KLIPSTEIN, FREDERICK AUGUST, b Greenwich, Conn, June 5, 28; m 65; c 3. MEDICINE. *Educ:* Williams Col, BA, 50; Columbia Univ, MD, 54; Am Bd Internal Med, dipl, 63; Am Bd Clin Nutrit, dipl, 67. *Prof Exp:* From intern to asst resident, Med Serv, Presby Hosp, NY, 54-56; NIH trainee, Col Physicians & Surgeons, Columbia Univ, 58-59, instr med, 61-63, from asst prof to assoc prof, 63-68; Postgrad Med Sch, London, Eng, 59-60; chief med resident, Francis Delafield Hosp, New York, 60-61; assoc prof, 68-72, PROF MED & MICROBIOL, SCH MED & DENT, UNIV ROCHESTER, 72- *Concurrent Pos:* Asst physician, Presby Hosp, NY, 60-68; Am Cancer Soc advan clin fel, 61-63; clin asst vis physician, First Med Div, Bellevue Hosp, 63-68; consult, Greenwich Hosp, Conn, 63-68 & Harlem Hosp, 66-68; vis physician, Francis Delafield Hosp, 66-68; physician, Strong Mem Hosp, Med Ctr, Univ Rochester; dir trop malabsorption unit, Univ Rochester-Univ PR, San Juan, 70-73; assoc prof, Sch Med, Univ PR, San Juan, 70-73. *Mem:* Am Gastroenterol Asn; Am Soc Hemat; Am Soc Clin Nutrit; fel Am Col Gastroenterol; Am Soc Microbiol; fel Am Col Med. *Res:* Diarrheal disorders; tropical malabsorption. *Mailing Add:* Dept Med Univ Rochester Med Ctr Rochester NY 14642

KLIR, GEORGE JIRI, b Prague, Czech, Apr 22, 32; m 62; c 2. HISTORY & PHILOSOPHY OF SCIENCE, MATHEMATICS GENERAL. *Educ:* Tech Univ, Prague, MSEE, 57; Czech Acad Sci, PhD(comput sci), 64. *Prof Exp:* Res asst, Res Inst Telecommun, Prague, 51-52; lectr, Charles Univ, 62-64; lectr elec eng, Univ Baghdad, 64-66; lectr comput sci, Univ Calif, Los Angeles, 66-68; assoc prof elec eng, Fairleigh Dickinson Univ, 68-69; from assoc prof to prof systs sci, 69-84, CHMN DEPT, SCH ADVAN TECHNOL, STATE UNIV NY, BINGHAMTON, 76-, DISTINGUISHED PROF, T J WATSON SCH, 84- *Concurrent Pos:* Ed, Czech Acad Sci, Prague, 62-63; IBM Systs Res Inst fel, 69; Ed-in-chief, Int J Gen Systs, Am Soc Cybernet, 74; Neth Inst Advan Studies fel, 75-76 & 82-83; Japan Soc Prom Sci fel, 80. *Honors & Awards:* Advancing Gen Systs Res Award, Neth Soc Systs Res, 76; Outstanding Contribution to Systs Res & Cybernet Award, Austrian Soc Cybernet Studies. *Mem:* Sr mem Inst Elec & Electronics Engrs (pres); Philos Sci Asn; Soc Gen Systs Res (pres, 81); Cognitive Sci Soc; Int Fed Systs Res (pres, 80-84). *Res:* Switching and automata theory; logical design of digital computers; digital codes; cybernetic methodology; general systems theory and methodology; logical design of computers; computer architecture; discrete mathematics; information theory; knowledge engineering; expert systems. *Mailing Add:* Dept Systs Sci TJ Watson Sch State Univ NY Binghamton NY 13901

KLITGAARD, HOWARD MAYNARD, b Harlan, Iowa, Oct 16, 24; m 45; c 5. PHYSIOLOGY. *Educ:* Univ Iowa, BA, 49, MS, 50, PhD(physiol), 53. *Prof Exp:* Instr physiol, Univ Iowa, 51-53; from instr to prof, 53-78, asst chmn dept, 61-66, vchmn dept, 67-78, ADJ PROF, MED COL WIS, 78- *Concurrent Pos:* Consult, Vet Admin Hosp, Wood, Wis, 57-89; chmn basic sci, Marquette Univ Sch Dent, 78-90. *Mem:* AAAS; Endocrine Soc; Am Physiol Soc; Soc Exp Biol & Med; Int Asn Dent Res. *Res:* Physiology and biochemistry of the thyroid hormone, endocrines and metabolism; radioisotope methodology. *Mailing Add:* 604 N 16th St Milwaukee WI 53233

KLITZMAN, BRUCE, b Dayton, Ohio, Nov 4, 51; m 80; c 2. MICROCIRCULATION, PLASTIC & RECONSTRUCTION SURGERY. *Educ:* Duke Univ, BSE, 74; Univ Va, PhD(physiol), 79. *Prof Exp:* Res assoc microcirculation, Univ Ariz, 79-81; from asst prof to assoc prof physiol, La State Univ Med Ctr, Shreveport, 82-85; ASST MED RES PROF PLASTIC SURG & PHYSIOL (CELL BIOL), DUKE UNIV MED CTR, 85-, DIR, PLASTIC SURG RES LABS, 85- *Concurrent Pos:* Young investr award, European Soc Microcirculation, 80; mem, Mem Comt, Microcirculatory Soc, 82-85; vis prof, Univ Manchester, UK, 85; vis scientist, Burroughs-Wellcome Found, 85; study sect reviewer, NIH, 85-; assoc ed, J Reconstructive Microsurg, 88- *Honors & Awards:* First Prize Investr, Plastic Surg Educ Found, 88. *Mem:* Microcirculatory Soc; Am Physiol Soc; Plastic Surg Res Coun; Am Heart Asn; European Soc Microcirculation; Soc Biomat. *Res:* Regulation of microcirculation and oxygenation of tissue; adaptation of microcirculation to different environments; hyperbaric physiol; biomaterials; microvascular prostheses; soft tissue implants; pressure sore prevention. *Mailing Add:* Duke Univ Med Ctr Box 3906 Durham NC 27710

KLIVINGTON, KENNETH ALBERT, b Cleveland, Ohio, Sept 23, 40; m 68; c 1. NEUROSCIENCE, COGNITIVE SCIENCE. *Educ:* Mass Inst Technol, SB, 62; Columbia Univ, MS, 64; Yale Univ, PhD(neurosci), 67. *Prof Exp:* Res engr electronics, Electronics Res Lab, Columbia Univ, 62-64; asst res neuroscientist, Univ Calif, San Diego, 67-68; dir res urban design, Fisher-Jackson Assocs, 68-69; prog officer sci, Alfred P Sloan Found, 69-81; vpres res & dev, Electro-Biol Inc, 81-84; ASST PRES SCI PLANNING, SALK INST, LA JOLLA, CALIF, 84- *Concurrent Pos:* Vis scientist, Univ Calif, San Diego, 73; consult, Nat Res Coun, 75-77; vis comt, Dept Psychol, Mass Inst Technol, 81-86; fel, Fetzer Inst, 90- *Mem:* AAAS; Soc Neurosci; Bioelectromagnetics Soc; Cognitive Sci Soc. *Res:* Electromagnetic properties of biological tissues; neural correlates of behavior; neural information processing. *Mailing Add:* Salk Inst 10010 N Torrey Pines Rd La Jolla CA 92037-1099

KLIWER, JAMES KARL, b Abilene, Kans, Dec 17, 28; m 63. PHYSICS. *Educ:* Univ Colo, BS, 57, MS, 59, PhD, 63. *Prof Exp:* Asst, Nuclear Physics Lab, Univ Colo, 57-63; res assoc, 63-65, from asst prof to assoc prof, 65-75, PROF PHYSICS, UNIV NEV, RENO, 75- *Mem:* Sigma Xi. *Res:* Atomic and nuclear spectroscopy. *Mailing Add:* Dept Physics Univ Nev Reno NV 89557

KLOBUCHAR, RICHARD LOUIS, b Chicago, Ill, Oct 15, 48; m 71. NAVAL ANALYSIS. *Educ:* Univ Ill, BS, 70; Carnegie-Mellon Univ, MS, 72, PhD(chem), 75. *Prof Exp:* Res assoc nuclear chem, Brookhaven Nat Lab, 75-77; mem prof staff, Ctr Naval Analyses, 77-81; DIR ADV TECHNOL, INC, 81- *Mem:* Am Chem Soc; Am Nuclear Soc; Am Phys Soc; Sigma Xi. *Res:* Scientific analysis of naval weapons systems; analytical support of fleet activities; applications of positronium chemistry; high energy nuclear reactions. *Mailing Add:* 758 Suffolk Lane Virginia Beach VA 23452

KLOBUKOWSKI, MARIUSZ ANDRZEJ, b Wroclaw, Poland, Mar 6, 48; Can citizen; m 84; c 1. MOLECULAR STRUCTURES & PROPERTIES. *Educ:* N Copernicus Univ, Torun, Poland, BSc, 71, PhD(physics), 78. *Prof Exp:* Asst prof chem, N Copernicus Univ, Poland, 78-81; I W Killam fel, Univ Alta, 80-83, res assoc, 83-88, programmer analyst comput sci, 88-89, ASST PROF CHEM, UNIV ALTA, 89- *Mem:* Chem Inst Can. *Res:* Development and use of accurate Gaussian basis sets for the studies of molecular structure and properties; calculations of the molecular structure and properties of molecules in their excited electronic states; development of parallel algorithms. *Mailing Add:* Dept Chem Univ Alta Edmonton AB T6G 2G2

KLOCK, BENNY LEROY, b Wash, DC, Oct 29, 34; m 57, 76; c 3. DIGITAL MAPPING, RESEARCH ADMINISTRATION. *Educ:* Cornell Univ, BA, 56, MS, 60; Georgetown Univ, PhD(astron), 64. *Prof Exp:* Tech asst dir six-inch transit circle div, 60-69, dir, Northern Transit Circle Div, 69-76, chief instrumentation br, US Naval Observ, 76-84; geodist, Defense Mapping Agency, 84-85, phys scientist, 85-89; SR CONSULT, ADV MAPPING CONCEPTS, 89- *Honors & Awards:* NSF int grant, 74. *Mem:* Am Astron Soc; Int Astron Union; Am Geophys Union. *Res:* Design and development of transit circle instrumentation; microcomputer systems; determination of star positions; automation of telescopes; electro-optics system design; advanced weapon system requirements for digital mapping data; digital mapping. *Mailing Add:* 4509 Bayside Dr Milton FL 82583-8423

KLOCK, GLEN ORVAL, b Portland, Ore, Aug 26, 37; m 58; c 4. FOREST SOILS. *Educ:* Ore State Univ, BS, 59, PhD(soil physics), 68; Iowa State Univ, MS, 63. *Prof Exp:* Res assoc soil physics, Ore State Univ, 64-67; prin res soil scientist, Pac Northwest Forest & Range Exp Sta, USDA Forest Serv, 68-82; PRIN SCIENTIST, WESTERN RESOURCES ANALYSIS, 82- *Concurrent Pos:* Pres, Western Resources Analysis,. *Mem:* Am Soc Agron; Soil Sci Soc Am; Soil Conserv Soc Am; Int Soil Sci Soc; Am Forestry Assoc. *Res:* Water resource and plant nutrient management for maintaining and enhancing the productivity of forest ecosystems in the western United States; use of image processing for development of geographic information system data bases for natural resources management. *Mailing Add:* Western Resources Analysis 2113 Sunrise Circle Wenatchee WA 98801

KLOCK, HAROLD F(RANCIS), b Miami Beach, Fla, Mar 21, 29; m 55; c 3. ELECTRICAL ENGINEERING. *Educ:* Northwestern Univ, BS, 52, MS, 54, PhD(elec eng), 56. *Prof Exp:* Lectr elec eng, Northwestern Univ, 56; asst prof, Case Western Reserve Univ, 56-62; prof lectr, 62-64; systs engr, Bailey Meter Co, Ohio, 64-66; PROF ELEC ENG, OHIO UNIV, 66- *Concurrent Pos:* Consult, Reliance Elec & Mfg Co, Nat Cash Register Co & Curtiss-Wright Corp. *Mem:* Inst Elec & Electronics Engrs; Asn Comput Mach; Soc Indust & Appl Math. *Res:* Feedback control systems; switching theory. *Mailing Add:* ECE Dept Stocker Ctr Athens OH 45701

KLOCK, JOHN W, b Orange, NJ, Nov 12, 28; m 53; c 2. SANITARY & CIVIL ENGINEERING. *Educ:* Southern Calif Univ, BE, 51; Univ Calif, Berkeley, MS, 56, PhD(sanit eng), 60. *Prof Exp:* PROF ENG, ARIZ STATE UNIV, 60- *Concurrent Pos:* Consult, Ariz Health Planning Authority, USPHS, 51-55, Commun Dis Ctr, 60-, Off Surgeon Gen, 61- & Honeywell Corp, 80-84. *Mem:* Am Water Works Asn; Water Pollution Control Fedn. *Res:* Communicable disease control; water pollution; waste water reclamation. *Mailing Add:* 2626 N 58th Pl Scottsdale AZ 85257

KLOCKE, FRANCIS J, CARDIOLOGY. *Prof Exp:* DIR, FEINBERG CARDIOVASC RES INST & PROF MED, MED CTR, NORTHWESTERN UNIV, 91- *Mailing Add:* Northwestern Univ 303 E Chicago Ave Chicago IL 60611

KLOCKE, ROBERT ALBERT, b Buffalo, NY, Oct 4, 36; c 3. PULMONARY DISEASES, PULMONARY PHYSIOLOGY. *Educ:* Manhattan Col, BS, 58; State Univ NY Buffalo, MD, 62. *Prof Exp:* Res asst prof med, 70-71, from asst prof to assoc prof med, 71-78, from asst prof to assoc prof physiol, 76-81, PROF MED, STATE UNIV NY BUFFALO, 78-, PROF PHYSIOL, 81- *Concurrent Pos:* Chief pulmonary lab, Walter Reed Gen Hosp, Washington, DC, 63-66; mem attend staff, E J Meyer Mem Hosp, Buffalo, 70-; chief pulmonary div, dept med, 77- *Mem:* Am Physiol Soc; Am Thoracic Soc; Am Fedn Clin Res. *Res:* Pulmonary gas exchange, particularly the rates of chemical reactions of carbon dioxide and oxygen in blood. *Mailing Add:* Dept Med & Physiol State Univ NY 462 Grider St Buffalo NY 14215

KLOEPFER, HENRY WARNER, human genetics; deceased, see previous edition for last biography

KLOET, WILLEM M, b Neth. NUCLEAR PHYSICS. *Educ:* Univ Utrecht, PhD(theoret physics), 73. *Prof Exp:* Res assoc theoret physics, Inst Fisica Teorica, Sao Paulo, 68-70, Univ Md, 73-75 & Los Alamos Sci Lab, 75-77; asst prof, 77-82, ASSOC PROF THEORET PHYSICS, RUTGERS UNIV, 82- *Mem:* Am Phys Soc; AAAS. *Res:* Theoretical nuclear physics. *Mailing Add:* Dept Physics & Astron Rutgers Univ Serin Physics Lab Frelinghuysen Rd Piscataway NJ 08854

KLOETZEL, JOHN ARTHUR, b Cambridge, Mass, Mar 21, 41; m 62; c 4. CELL BIOLOGY, PROTOZOOLOGY. *Educ:* Univ Southern Calif, BA, 62; Johns Hopkins Univ, PhD(biol), 67. *Prof Exp:* NIH fel biol, Univ Colo, 67-70; asst prof, 70-75, ASSOC PROF BIOL, UNIV MD BALTIMORE COUNTY, 75- *Concurrent Pos:* Fel, Alexander von Humboldt Found, WGer, 78; Vis Assoc Prof, Biochem, Johns Hopkins Univ Sch Med, 87. *Mem:* Soc Protozool; Am Soc Cell Biol. *Res:* Fine-structural aspects of cellular function, development and differentiation; Morphogenesis and post-conjugant development in ciliated protozoans, form and function of the ciliate cytoskeleton. *Mailing Add:* Dept Biol Sci Univ Md Baltimore County Catonsville MD 21228

KLOETZEL, MILTON CARL, b Detroit, Mich, Aug 28, 13; m 38; c 4. ORGANIC CHEMISTRY. *Educ:* Univ Mich, BS, 34, PhD(org chem), 37. *Prof Exp:* Du Pont fel chem, Univ Mich, 37-38; instr, Harvard Univ, 38-41; from asst prof to assoc prof, DePauw Univ, 41-45; from asst prof to prof, 45-58, dean, Grad Sch, 58-68, vpres res & grad affairs, 67-70, acad vpres, 70-75, EMER ACAD VPRES, UNIV SOUTHERN CALIF, 75- *Mem:* Am Chem Soc. *Res:* Chemistry of polycyclic and heterocyclic compounds; Diels-Alder reaction; chemistry of nitroparaffins. *Mailing Add:* 425 Ena Rd Apt 1106A Honolulu HI 96815

KLOHS, WAYNE D, BIOCHEMISTRY, CELL BIOLOGY. *Educ:* Ind State Univ, PhD(cell biol), 77. *Prof Exp:* SR SCIENTIST, WARNER-LAMBERT CO, 83- *Mailing Add:* Dept Chemother Warner-Lambert Co 2800 Plymouth Rd Ann Arbor MI 48105

KLOKHOLM, ERIK, b Nykobing, Denmark, Mar 13, 22; US citizen; m 43. SOLID STATE PHYSICS. *Educ:* Mass Inst Technol, BS, 51; Temple Univ, PhD(physics), 60. *Prof Exp:* From res asst to head struct & metals br, Labs Res & Develop, Franklin Inst, 51-59; res physicist, Moorehead Patterson Res Ctr, Am Mach & Foundry Co, 59-61; assoc prof physics, State Univ NY Col Ceramics, Alfred Univ, 61-62; res staff mem, Thomas J Watson Res Ctr, 62-73, RES PROJ MGR, MFR RES LABS, DATA SYSTS DIV, IBM CORP, 73- *Mem:* NY Acad Sci. *Res:* Structure and properties of solids, particularly thin metallic films; crystallographic aspects of solid state physics. *Mailing Add:* 64 Willard Terr Stamford CT 06903

KLOMBERS, NORMAN, b New York, NY, Jan 28, 23; m 55; c 2. ADMINISTRATION. *Educ:* New York Col Podiat Med, DPM, 44. *Prof Exp:* Fac mem & clin instr, New York Col Podiat Med, 44-54; dir div sci affairs, 78-80, exec dir, Am Podiatric Med Asn, 80-90; EXEC DIR, ANXIETY DIS AM, 91- *Concurrent Pos:* Practr podiatric med, New York, NY, 44-68; dir peer rev activ & dir prof serv, Podiatry Soc State NY, 69-78; mem, adv comm Health & Hosp Corp, New York, 70-77 & Health Comt, City Long Beach, NY, 76. *Mem:* Am Col Foot Orthopedists; Am Bd Podiatric Orthopedics; Nat Acad Pract; Acad Podiatric Med. *Mailing Add:* 6000 Executive Blvd Rockville MD 20852

KLOMP, EDWARD, b Detroit, Mich, Oct 18, 30; m 59; c 3. MECHANICAL ENGINEERING, FLUID MECHANICS. *Educ:* Wayne State Univ, BS, 52, MS, 53. *Prof Exp:* Res engr, 53-59, assoc sr res engr, 59-65, sr res engr, 65-77, staff res engr, 77-80, SR STAFF RES ENGR, RES LABS, GEN MOTORS CORP, WARREN, 80- *Concurrent Pos:* Instr, Wayne State Univ, 55-65. *Mem:* Am Soc Mech Engrs; Soc Automotive Engrs; Sigma Xi. *Res:* Fluid mechanics relating to turbomachinery and internal combustion engines; author of 7 publications; 43 US patents. *Mailing Add:* 36237 Acton Dr Mt Clemens MI 48043

KLOMPARENS, KAREN L, b E Lansing, Mich, Sept 17, 50. PLANT SCIENCE, ELECTRON OPTICS. *Educ:* Mich State Univ, BS, 72, MS, 74, PhD(bot & electron optics), 77. *Prof Exp:* Asst prof, 80-85, ASSOC PROF ELECTRON OPTICS, DEPTS BOT & PLANT PATH ENTOM, MICH STATE UNIV, 85-, DIR, CTR ELECTRON OPTICS, 80- *Mem:* Electron Micros Soc Am; Am Phytopath Soc; AAAS; Am Inst Biol Sci; Mycol Soc Am. *Res:* Ultrastructural and analytical electron microscopy methods relevant to plant science; applications to plant host-pathogen-vector relationships, fungal morphology and spore development. *Mailing Add:* B5 Pesticide Res Ctr Mich State Univ E Lansing MI 48824-1311

KLONTZ, EVERETT EARL, b Akron, Ohio, Sept 28, 21; m 42; c 3. PHYSICS. *Educ:* Kent State Univ, BS, 42; Univ Ill, MS, 43; Purdue Univ, PhD(physics), 52. *Prof Exp:* Asst physics, Univ Ill, 42-44; instr, Bowling Green State Univ, 44; asst, 46-52, res assoc & asst prof, 52-62, ASSOC PROF PHYSICS, PURDUE UNIV, WEST LAFAYETTE, 62- *Mem:* Am Phys Soc; Am Asn Physics Teachers. *Res:* Effects of high energy particle irradiations on physical properties of crystals. *Mailing Add:* Dept Physics Purdue Univ West Lafayette IN 47907-1396

KLOPATEK, JEFFREY MATTHEW, b Milwaukee, Wis, Dec 5, 44; m 84; c 2. ECOLOGY, BOTANY. *Educ:* Univ Wis-Milwaukee, BS, 71, MS, 74; Univ Okla, PhD(bot), 78. *Prof Exp:* Res assoc ecol, Okla Biol Surv, 73-76; res ecologist, Environ Sci Div, Oak Ridge Nat Lab, 76-81; PROF & RES ECOLOGIST, DEPT BOT, ARIZ STATE UNIV, 81- *Concurrent Pos:* Consult, Elec Power Res Inst, AEC-Energy Res & Develop Admin, 73-74 & Forest Serv, USDA, 81-; lectr, Univ Tenn, 80-81; chmn, Municipal Planning Comn, Farragut, Tenn, 80-81; vis scientist, US Environ Protection Assoc, 90; Fulbright Scholar, 90-91. *Mem:* Ecol Soc Am; Int Asn Ecol; Am Inst Biol Sci; Soc Wetland Sci; Interdisciplinary Group Ecol Develop & Energy; Soil Sci Soc Am; Int Soc Ecol Model; Am Soc Surface Mining & Reclamation. *Res:* Nutrient cycling; ecosystem analysis; ecosystem restoration; wetland ecology; integration of natural and cultural systems; landscape ecology; microbial processes; succession and disturbance. *Mailing Add:* Dept Bot Ariz State Univ Tempe AZ 85287-1601

KLOPFENSTEIN, CHARLES E, b Los Angeles, Calif, July 6, 40; m 63. PHYSICAL ORGANIC CHEMISTRY, CHEMICAL INSTRUMENTATION. *Educ:* Univ Ore, BA, 62, PhD(chem), 66. *Prof Exp:* Asst prof chem, 66-80, DIR LABS, UNIV ORE, 66-, ASSOC PROF, 80- *Concurrent Pos:* NATO res fel, Lab Org Chem, Swiss Fed Inst Technol, 66-67. *Mem:* Am Chem Soc. *Res:* Synthesis of aromatic heterocyclic compounds; calculation of physical properties of aromatic compounds; computer analysis of physical data. *Mailing Add:* Dept Chem Univ Ore Eugene OR 97403

KLOPFENSTEIN, KENNETH F, b Mt Pleasant, Iowa, Mar 13, 40; m 61; c 3. MATHEMATICS. *Educ:* Iowa Wesleyan Col, BA, 61; Colo State Univ, MS, 63; Purdue Univ, PhD(math), 67. *Prof Exp:* Instr math, Wabash Col, 66-67; asst prof, 67-73, ASSOC PROF MATH, COLO STATE UNIV, 73- *Mem:* Math Asn Am; Am Math Soc. *Res:* Hilbert space; operator theory; mathematics education. *Mailing Add:* Dept Math & Statist Colo State Univ Ft Collins CO 80523

KLOPFENSTEIN, WILLIAM ELMER, b Paris, Ohio, Dec 23, 35; m 59; c 3. BIOCHEMISTRY. *Educ:* Pa State Univ, BS, 58, MS, 61, PhD(biochem), 64. *Prof Exp:* Asst technologist food res, Gen Foods Corp, 58; instr biochem, Pa State Univ, 60-64; from asst prof to prof biochem, Kansas State Univ, 64-88, assoc biochemist, Agr Exp Sta, 72-86, chmn, grad biochem group, 77-86; PROF BIOCHEM & CHMN CHEM DEPT, WESTERN ILL UNIV, 88- *Mem:* Am Chem Soc; Sigma Xi; Am Soc Biol Chemists; Am Oil Chem Soc. *Res:* Structure and function of lipids; physical properties of lipids; binding of lipids to proteins; use of lipids as alternative fuels. *Mailing Add:* Dept Chem Western Ill Univ Macomb IL 61445

KLOPFER, PETER HUBERT, b Berlin, Ger, Aug 9, 30; m 55; c 3. ZOOLOGY. *Educ:* Univ Calif, Los Angeles, AB, 52; Yale Univ, PhD(zool), 57. *Prof Exp:* Head sci dept, Windsor Mountain Sch, Mass, 52-53 & 55-56; USPHS fel, Cambridge Univ, 57-58; from asst prof to assoc prof zool, 58-68, PROF ZOOL & DIR FIELD STA ANIMAL BEHAV, DUKE UNIV, 68- *Concurrent Pos:* Nat Inst Ment Health career develop award, 65; Alexander von Humboldt Prize, 79-80. *Honors & Awards:* NIMH Res Scientist Award, 70. *Mem:* Fel AAAS; Ecol Soc Am; fel Animal Behav Soc; Int Soc Res Aggression. *Res:* Behavior and ecology, especially analysis of the development of species-specific behavior in birds and mammals; maternal-filial relations and aggression. *Mailing Add:* Dept Zool Duke Univ Durham NC 27706

KLOPMAN, GILLES, b Brussels, Belg, Feb 24, 33; m 57. CHEMISTRY. *Educ:* Free Univ Brussels, Lic es Sci, 56, Dr es Sci, 60. *Prof Exp:* Res assoc org chem, Cyanamid Europ Res Inst, 60-67; assoc prof chem, 67-69, dean, math & sci, 86-88, PROF CHEM, CASE WESTERN RESERVE UNIV, 69-, CHMN, CHEM DEPT, 81- *Concurrent Pos:* Welch fel, Univ Tex, 65-66; vpres, Biofor Inc. *Honors & Awards:* Stas-Spring Award, Belg Chem Soc. *Mem:* Am Chem Soc; The Chem Soc; Swiss Chem Soc; Belg Chem Soc; Am Asn Univ Professors; Sigma Xi. *Res:* Applied theoretical organic chemistry; chemical reactivity; nucleophilic reactivity; quantum mechanical calculation of large organic molecules; quantitative structure activity relationship of pharmacological and of carcinogenic molecules. *Mailing Add:* Dept Chem Case Western Reserve Univ Cleveland OH 44106

KLOPOTEK, DAVID L, b Green Bay, Wis, Jan 11, 42; m 63; c 2. ORGANIC CHEMISTRY. *Educ:* St Norbert Col, BA, 64; Utah State Univ, PhD(chem), 68. *Prof Exp:* Res assoc chem, E I du Pont de Nemours & Co, Inc, 67-68; from asst prof to assoc prof, 68-82, PROF CHEM, ST NORBERT COL, 83- *Concurrent Pos:* NSF grant, 71-73; Res Corp grant, 72-74; vis prof, Dartmouth Univ, 83; summer fac fel, Lewis Res Ctr, Cleveland, Ohio, 85 & 86, NASA, Am Soc Environ Educ; NASA grants, polyimide res, 86-91. *Mem:* Am Chem Soc. *Res:* Chemistry of compounds containing nitrogen-fluorine bonds; reactivity of fluoronitrene with nucleophiles; diamines for polyimides. *Mailing Add:* Dept Chem St Norbert Col DePere WI 54115

KLOPP, CALVIN TREXLER, b Atlantic City, NJ, Dec 7, 12; c 3. SURGERY. *Educ:* Swarthmore Col, BA, 34; Harvard Univ, MD, 38. *Prof Exp:* Intern surg, Boston City Hosp, Mass, 39-40; rotating intern med, Reading Hosp, Pa, 40-41; asst resident, Mem Hosp, New York, 41-44; from asst clin prof to assoc prof, 46-60, Warwick prof surg, Sch Med, 60-76, med dir, Univ Clin, 68-76, EMER PROF SURG, SCH MED, GEORGE WASHINGTON UNIV, 76- *Concurrent Pos:* Consult, var hosps, 46- *Mem:* Am Radium Soc; Am Asn Cancer Res; Am Col Surgeons; James Ewing Soc; Am Thyroid Asn; Sch Head & Neck Surg; Southern Surg Asn. *Res:* Cancer. *Mailing Add:* 4443 64th Ave Dr W Bradenton FL 33507

KLOPPEL, THOMAS MATHEW, b Denver, Colo, Oct 27, 50; m 76; c 2. CELL BIOLOGY. *Educ:* Colo State Univ, BS, 72, MS, 74; Purdue Univ, PhD(cell biol), 79. *Prof Exp:* Res biologist, Vet Admin Med Ctr, Denver, 81-88; asst prof biochem, Univ Colo Sch Med, 81-88; RES SCIENTIST, CORTECH, INC, 88- *Concurrent Pos:* Fel, Am Cancer Soc, 80. *Mem:* Am Soc Cell Biol. *Res:* Examination of potential anti-inflammatory drugs for clinical use; role of membrane receptors in intracellular vesicle trafficking. *Mailing Add:* Cortech Inc 6840 N Broadway Denver CO 80221

KLOS, EDWARD JOHN, b Hamilton, Ont, June 27, 25; m 49; c 2. PLANT PATHOLOGY. *Educ:* Ont Agr Col, BSAg, 50; Cornell Univ, PhD(plant path), 54. *Prof Exp:* Asst plant path, Cornell Univ, 50-54; from asst prof to assoc prof, 54-67, PROF PLANT PATH, MICH STATE UNIV, 67-, CHMN DEPT, 80- *Mem:* Am Phytopath Soc. *Res:* Effect of Erwinia herbicola on fire blight of pome fruits; population studies of Erwinia amylovora; resistance or tolerance of Venturia inaequalis to fungicides; control of tree fruit diseases by chemicals and other means. *Mailing Add:* 166 Plant Biol-Bot Mich State Univ East Lansing MI 48824

KLOS, WILLIAM ANTON, b Houston, Tex, Aug 14, 36; m 63; c 1. ELECTRICAL ENGINEERING. *Educ:* Univ Houston, BS, 63 & 64, PhD(elec eng), 69. *Prof Exp:* Res asst elec eng, Univ Houston, 66-69; prin engr, Lockheed Electronics Co, Tex, 69-70; assoc prof, 70-80, PROF ELEC ENG, UNIV SOUTHWESTERN LA, 80-, HEAD DEPT, 70- *Mem:* Nat Soc Prof Engrs; Acoust Soc Am; Am Soc Eng Educ; Am Geophys Union; Inst Elec & Electronics Engrs; Sigma Xi. *Res:* Electromagnetic wave propagation; radar cross-section and radar systems. *Mailing Add:* PO Box 43890 Lafayette LA 70504

KLOSE, JULES ZEISER, b St Louis, Mo, Aug 7, 27; m 58; c 4. ATOMIC PHYSICS, VACUUM ULTRAVIOLET RADIOMETRY. *Educ:* Wash Univ, St Louis, AB, 49; Univ Rochester, MS, 53; Cath Univ Am, PhD(physics), 58. *Prof Exp:* Physics aid, US Naval Gun Factory, 48; instr, Dunford Sch, Mo, 49; asst physics, Univ Rochester, 49-53; from asst prof to assoc prof, US Naval Acad, 53-61; physicist, Nat Bur Standards, 61-88, GUEST SCIENTIST, NAT INST STANDARDS & TECHNOL, 88- *Concurrent Pos:* Res assoc & lectr, Univ Mich, 60-61. *Mem:* Am Phys Soc; Optical Soc Am; Sigma Xi; Coun Optical Radiation Measurements. *Res:* Vacuum ultraviolet radiometry; measurement of atomic lifetimes and transition probabilities; ultrasonics; thermal relaxation in gases; cosmic rays; calibration of space instrumentation. *Mailing Add:* Nat Inst Standards & Technol Gaithersburg MD 20899

KLOSE, THOMAS RICHARD, b Adelaide, Australia, Apr 20, 46; m 74; c 2. ORGANIC CHEMISTRY. *Educ:* Univ Adelaide, BSc, 67, Hons, 68, PhD(org chem), 72. *Prof Exp:* Fel org chem, Res Inst Med & Chem, 72-74 & Mass Inst Technol, 74-75; RES CHEMIST ORG CHEM, EASTMAN KODAK CO RES LABS, 75- *Mem:* Am Chem Soc. *Res:* Synthesis of novel dyes and pigments for use in non-silver imaging systems. *Mailing Add:* 19 Sandpiper Hill Fairport NY 14450

KLOSEK, RICHARD C, b Olyphant, Pa, Feb 18, 33; m 56; c 3. MICROBIOLOGY. *Educ:* Univ Scranton, BS, 54; St John's Univ, MS, 56, PhD(microbiol), 60. *Prof Exp:* From asst prof to assoc prof, 60-74, PROF MICROBIOL, FAIRLEIGH DICKINSON UNIV, 74- *Concurrent Pos:* Res grants, Fairleigh Dickinson Univ, 61-62 & 71; res grant, Jomol Pharmaceut Corp, 63-64, consult, 65; mem, Smithsonian Inst, 74. *Mem:* Fel AAAS; Am Acad Microbiol; Soc Protozool; Sigma Xi. *Res:* Protozooan nutrition and cellular chemistry, especially pathways associated with carbohydrate, protein and lipid metabolism; isolation and functional aspects of chemotherapeutic agents utilized in bacterial, fungal and viral diseases. *Mailing Add:* 25 Palmer Dr Wayne NJ 07470

KLOSNER, JEROME M, b New York, NY, Mar 23, 28; m 65; c 3. APPLIED MECHANICS, STRUCTURAL DYNAMICS. *Educ:* City Col New York, BCE, 48; Columbia Univ, MS, 50; Polytech Inst Brooklyn, PhD(appl mech), 59. *Prof Exp:* Sr stress analyst, Repub Aviation Corp, 52-56; sr scientist appl mech, Res & Advan Develop Div, Avco Corp, 56; from res assoc to assoc prof, 56-67, PROF APPL MECH, POLYTECH INST NY, 67- *Concurrent Pos:* Consult, Res & Advan Develop Div, Avco Corp, Gen Appl Sci Labs, Fed Trade Comn, Res Ctr, Hazeltine Corp, Ingersoll-Rand Corp & Technautics Corp; consult, Weidlinger Assocs, Consult Engrs, 76-; mem comt on recommendations US Army basic sci res, Nat Res Coun, 76-79, & 85-88. *Mem:* Am Soc Mech Engrs; assoc fel Am Inst Aeronaut & Astronaut; Soc Rheol; fel Am Soc Civil Engrs. *Res:* Structural Dynamics; hydroelasticity; acoustic radiation; elastodynamics. *Mailing Add:* Prof Appl Mech Polytech Univ Rte 110 Farmingdale NY 11735

KLOSTERMAN, ALBERT LEONARD, b Cincinnati, Ohio, Oct 22, 42; m 64; c 4. SOLID GEOMETRIC MODELING, PRODUCT DEFINITION DATA BASE. *Educ:* Univ Cincinnati, BSME, 65, MSME, 68, PhD(mech eng), 71. *Prof Exp:* Instr mech eng, Univ Cincinnati, 65-70; proj mgr, 70-72, mem tech staff, 72-73, dir tech staff, 73-78, vpres & gen mgr, 78-83, SR VPRES, CHIEF TECH OFFICER & GEN MGR, STRUCT DYNAMICS RES CORP, 83- *Concurrent Pos:* Adj assoc prof, Univ Cincinnati, 72- *Mem:* Am Soc Mech Engrs; Asn Comput Mach. *Res:* System dynamics; experimental modal analysis; solid geometric modeling; product definition data base for mechanical design; mechanical computer aided engineering (MCAE) technology. *Mailing Add:* 5444 Forest Ridge Circle Milford OH 45150

KLOSTERMAN, HAROLD J, b Mooreton, NDak, Jan 11, 24; m 46; c 7. BIOCHEMISTRY. *Educ:* NDak State Univ, BS, 46, MS, 49; Univ Minn, PhD(biochem), 55. *Hon Degrees:* DSc, NDak State Univ, 90. *Prof Exp:* From asst to assoc chemist, 46-57, prof biochem & chmn dept, 57-88, EMER PROF, NDAK STATE UNIV, 88- *Mem:* Am Chem Soc. *Res:* Isolation and characterization of natural products. *Mailing Add:* Dept Biochem NDak State Univ Fargo ND 58102

KLOSTERMEYER, EDWARD CHARLES, b Omaha, Nebr, Feb 25, 19; m 41; c 2. ENTOMOLOGY. *Educ:* Univ Nebr, BSc, 40, MSc, 42; State Col Wash, PhD, 52. *Prof Exp:* Asst entom, Univ Nebr, 40-42; asst, Univ Calif, 46-47; asst entomologist, Irrig Exp Sta, 47-58, assoc entomologist, 58-62, entomologist, 62-81, prof, 75-81, EMER PROF ENTOM, WASH STATE UNIV, 81- *Mem:* Entom Soc Am. *Res:* Field crop insect control; insect pollination; bee behavior. *Mailing Add:* 1915 Benson Ave Prosser WA 99350

KLOSTERMEYER, LYLE EDWARD, b Oakland, Calif, Dec 4, 44. ENTOMOLOGY. *Educ:* Wash State Univ, BS, 68; NDak State Univ, MS, 74; Univ Nebr, PhD, 78. *Prof Exp:* Res asst dept entom, NDak State Univ, 70-73 & Univ Nebr, 73-78; asst prof entom, Dept Agr Biol, Univ Tenn, 78-80, asst prof entom, Dept Entom & Plant Path, 80-85; OWNER, LYLE'S PEST CONTROL & ENTOM SERV, 85- *Mem:* Entom Soc Am; Sigma Xi. *Mailing Add:* PO Box 1167 Prosser WA 99350

KLOTMAN, PAUL, NEPHROLOGY, CARDIOLOGY. *Educ:* Ind Univ, MD, 76. *Prof Exp:* ASST PROF NEPHROLOGY & DIR, DUKE HYPERTENSION CTR, MED CTR, DUKE UNIV, 82- *Res:* Renal transplantation; prostaglandin metabolism; hypertension. *Mailing Add:* Vet Admin Med Ctr 508 Fulton St Rm B3000 Durham NC 27705

KLOTS, CORNELIUS E, b Rochester, NY, Oct 19, 33; m 59; c 4. UNIMOLECULAR REACTIONS, VAN DER WAALS MOLECULES. *Educ:* Haverford Col, BS, 55; Harvard Univ, PhD(phys chem), 59. *Prof Exp:* Res assoc chem, Fla State Univ, 61-64; STAFF SCIENTIST CHEM PHYSICS, OAK RIDGE NAT LAB, 64- *Concurrent Pos:* Ford Found prof, Physics Dept, Univ Tenn, 66-69; vis prof, Univ Paris-Sud, 81-82. *Mem:* Fel Am Phys Soc; Sigma Xi. *Res:* Properties of reactions in small isolated aggregates of matter. *Mailing Add:* Oak Ridge Nat Lab Bldg 4500S MS 6125 PO Box 2008 Oak Ridge TN 37831-6125

KLOTZ, ARTHUR PAUL, b Milwaukee, Wis, Sept 28, 13; m 41; c 4. CLINICAL MEDICINE. *Educ:* Univ Chicago, SB & MD, 38. *Prof Exp:* Asst in med, Univ Chicago, 49-51, instr, 51-54; from asst prof to prof med, Med Ctr, Univ Kans, 54-75, dir, Div Gastroenterol, 54-75; dir gastrointestinal lab, Boswell Mem Hosp, Sun City, 75-86, dir res Biogerontol Res Inst, 86-88; RETIRED. *Concurrent Pos:* Consult, Menorah Hosp, Vet Hosp, Kansas City, Mo & Wadsworth, Kans; mem, Gastroenterol Res Group; res collabr, Brookhaven Nat Lab, NY; pvt pract clin gastroenterol, 75-; ed, Boswell Hosp Proc, 80-88. *Mem:* Am Soc Gastrointestinal Endoscopy; Soc Nuclear Med; Am Gastroenterol Asn; Am Physiol Soc; Am Col Physicians; Am Geriat Soc. *Res:* Gastric secretions; pancreatic function; liver disease; ulcerative colitis; small bowel absorption in humans by perfusion technique. *Mailing Add:* 365 Woodland Dr Sedona AZ 86336

KLOTZ, EUGENE ARTHUR, b Fredericksburg, Iowa, June 25, 35; m 57; c 2. MATHEMATICS. *Educ:* Antioch Col, BS, 58; Yale Univ, PhD(math), 65. *Prof Exp:* Actg instr math, Yale Univ, 62-63; instr, 63-69, assoc prof, 69-77, PROF MATH, SWARTHMORE COL, 77- *Concurrent Pos:* NSF sci fac fel,

74; prin investr, prog math educ using info technol, NSF-Nat Inst Educ, 81-82. *Mem:* Am Math Soc; Math Asn Am; Asn Comput Mach; Asn Develop Comput Based Instrnl Systs; Sigma Xi. *Res:* Real-time microcomputer color graphics units for mathematics instruction, using video arcade technology; social science mathematics. *Mailing Add:* Dept Math Swarthmore Col Swarthmore PA 19081

KLOTZ, IRVING MYRON, b Chicago, Ill, Jan 22, 16; m 47, 66. PHYSICAL BIOCHEMISTRY. *Educ:* Univ Chicago, SB, 37, PhD(chem), 40. *Prof Exp:* Asst chem, Univ Chicago, 37-39; Abbott res assoc, 40, Nat Defense Res Comt assoc, 41-42, from instr to prof, 42-50, Morrison prof, 63-86, EMER PROF CHEM, NORTHWESTERN UNIV, EVANSTON, 86-; WINZLER PROF BIOCHEM, UNIV ILL MED SCH, 88- *Concurrent Pos:* Lalor fel, 47-48; chmn biophys & biophys chem, Study Sect, NIH, 63-66; mem corp & trustee, Marine Biol Lab, Woods Hole; distinguished vis prof, Univ Buffalo, 67, Univ Calif, Davis, 82, Ohio State Univ, 89. *Honors & Awards:* Reilly Lectr, Univ Notre Dame; Mack Lectr, Ohio State Univ; Barton Lectr, Univ Okla; Gooch-Stephens Lectr, Baylor Univ; Midwest Award, Am Chem Soc, 70; Welch Lectr, Univ Tex, 73; Winzler Lectr, Fla State Univ, 77; Steiner Lectr, Oberlin, 81; Shaw Lectr, Univ SDak, 82; Watkins Lectr, Wichita State Univ, 83; Bull Lectr, Iowa, 85; Shrage Lectr, Univ Ill, 88. *Mem:* Nat Acad Sci; fel Am Acad Arts & Sci; Am Chem Soc; Am Soc Biol Chem; fel Royal Soc Med. *Res:* Structure and function of proteins and polymers; spectroscopy; biochemical energetics; thermodynamics. *Mailing Add:* Dept Chem Northwestern Univ 2145 Sheridan Rd Evanston IL 60208

KLOTZ, JAMES ALLEN, b Milwaukee, Wis, Oct 18, 22; m 26; c 3. PETROLEUM ENGINEERING, APPLIED MATHEMATICS. *Educ:* Northwestern Univ, BS, 44; Univ Mich, MS, 47. *Prof Exp:* Engr, Manhattan Proj, US Naval Res Lab, 44-46; res engr, Oil Field Res Div, Calif Res Corp, 47-58; petrol engr, Arabian Am Oil Co, 58-60; res technologist, Pure Oil Co, 60-62, sr res technologist, 62-65; eng assoc, 66-70, supvr, Arctic Ocean & reservoir eng res, 70-81, SUPVR WELLBORE MECH RES, UNION OIL CO CALIF, 81- *Concurrent Pos:* Mem tech adv comt, Sucker Rod Pumping Res, Inc, 61-64; mem drilling domain comt, Div Prod, Am Petrol Inst, 61-64, mem adv comt fundamental res origin & recovery of petrol, 65, mem tech adv comt, Ocean Margin Drilling Proj, 79-81, mem prod res adv comt, 81- *Mem:* AAAS; Soc Petrol Engrs; Am Inst Chem Engrs. *Res:* Oil well drilling; reservoir engineering; thermal secondary recovery; application of computer methods to oil fild engineering; application of operations research methods to oil field engineering; offshore and arctic engineering for oil production. *Mailing Add:* 2981 Lakeview Dr Fullerton CA 92635

KLOTZ, JEROME HAMILTON, b Loma Linda, Calif, June 21, 34; m 56; c 2. BIOSTATISTICS. *Educ:* Univ Calif, Berkeley, AB, 56, PhD(statist), 60. *Prof Exp:* Lectr math & statist, McGill Univ, 60-61; asst prof statist, Univ Calif, Berkeley, 61-62; asst prof, Harvard Univ, 62-65; assoc prof, 65-69, PROF STATIST, UNIV WIS-MADISON, 69- *Concurrent Pos:* Consult statistician clin oncol, Univ Wis, 72-; prof statist, Cent Oncol Group, 72- & Wis Clin Cancer Ctr, 73-; prof, Ohio State Univ, 81-82. *Mem:* Fel Inst Math Statist; Am Statist Asn; Biomet Soc. *Res:* Nonparametric methods; computer techniques; components of variance; biostatistical methods. *Mailing Add:* Dept Statist Univ Wis 1210 W Dayton St Madison WI 53706

KLOTZ, JOHN WILLIAM, b Pittsburgh, Pa, Jan 10, 18; m 42; c 8. GENETICS, PHILOSOPHY OF SCIENCE. *Educ:* Univ Pittsburgh, MS, 40, PhD(genetics), 47; Concordia Sem, MDiv, 41. *Prof Exp:* Lab asst, Univ Pittsburgh, 40; instr, Concordia Collegiate Inst, 41-43; prof sci, Bethany Lutheran Col, 43-45; from assoc prof to prof biol, physiol & nature study, Concordia Teachers Col, 45-59; prof natural sci, Concordia Sr Col, 59-74; dir grad studies, 77-88 PROF PRACTICAL THEOL, CONCORDIA SEM, 74-,. *Concurrent Pos:* Registr, Bethany Lutheran Col, 44-45. *Mem:* AAAS; Am Genetic Asn; Nat Sci Teachers Asn; Nat Asn Biol Teachers. *Res:* Genetics of Habrobracon and Mormoniella; ecology; physiology; lethals, semi-lethals and an inversion in Habrobracon juglandis. *Mailing Add:* 6417 San Bonita St Louis MO 63105

KLOTZ, LOUIS HERMAN, b Elizabeth, NJ, May 21, 28; m 66; c 2. STRUCTURAL ENGINEERING, GEOTECHNICAL ENGINEERING. *Educ:* Pa State Univ, BSCE, 51; NY Univ, MCE, 56; Rutgers Univ, New Brunswick, PhD(civil eng), 67. *Prof Exp:* Struct engr, firms, NY & NJ metrop area, 51-56; civil engr, Ebasco Int Corp, New York, 56-58, construct proj engr defense electronic prod, Missile & Surface Radar Div, Radio Corp Am , NJ, 58-59; res assoc civil eng, Univ Ill, Urbana, 59-61; consult engr, Ohio & NJ, 61-65; asst prof, 65-69, actg chmn dept, 69-71, chmn dept, 71-73, assoc prof, 69-86, EMER ASSOC PROF, CIVIL ENG, UNIV NH, 86- *Concurrent Pos:* Ed, Energy Sources, Promises & Probs, 80; pres, Durham Inst, 80-85; consult engr, 85-, Univ NH, 87-; spec projs dir, ASCE Hq, NY, 86-87. *Mem:* AAAS; Am Soc Civil Engrs; Am Soc Eng Educ; Int Asn Bridge & Struct Engrs; NY Acad Sci. *Res:* applications of linear graph system; mathematical models and computer applications in structures, manufacturing processes and soil mechanics; manufacturing processes and soil mechanics; small hydro-power technical and economic analyses; forensic analysis of structural problems. *Mailing Add:* 90 Mainmast Circle New Castle NH 03854-0204

KLOTZ, LYNN CHARLES, b Trenton, NJ, Nov 25, 40; c 1. PHYSICAL BIOCHEMISTRY. *Educ:* Princeton Univ, AB, 65; Univ Calif, San Diego, PhD(chem), 71. *Prof Exp:* Res asst molecular biol, Princeton Univ, 61-62; asst prof biochem, Harvard Univ, 71-74, assoc prof, 74-79; with BioTechnica Int, 81-89. *Concurrent Pos:* Vis lectr, Princeton Univ, 79-81; Carilla & Henry Dreyfus Teacher Scholar Grant, 75. *Mem:* AAAS; Sigma Xi; Am Chem Soc. *Res:* Physical studies of DNA and chromosomes; evolution of DNA and chromosomes; design of DNA probe diagnostic kits. *Mailing Add:* PO Box 2748 Cambridge MA 02238

KLOTZ, RICHARD LAWRENCE, b Philadelphia, Pa, Jan 4, 50; m 75; c 3. BIOLOGY. *Educ:* Denison Univ, BS, 72; Univ Conn, MS, 75, PhD(bot), 79. *Prof Exp:* From asst prof to assoc prof, 79-89, PROF BIOL, STATE UNIV NY, CORTLAND, 89- *Mem:* Am Soc Limnol & Oceanog; AAAS; Sigma Xi; NAm Benthol Soc. *Res:* Phosphorus influence on stream ecosystems. *Mailing Add:* Dept Biol Sci State Univ NY Box 2000 Cortland NY 13045

KLOTZBACH, ROBERT J(AMES), b New York, NY, Aug 27, 22; m 46; c 1. CHEMICAL ENGINEERING. *Educ:* Fordham Univ, BS, 43. *Prof Exp:* Develop engr, Oak Ridge Nat Lab, 46-48, design engr, 48-53, design problem leader, 53, chmn long range planning group, Chem Technol Div, 53-55; proj engr, Union Carbide Nuclear Div, 55-57, mgr eng dept, 57-65, mgr, Union Carbide Mining & Metals Div, 65-69, asst dir eng, 69, dir eng, 69-72, dir technol, Union Carbide Metals Div Union Carbide Corp, 73-85; RETIRED. *Mem:* AAAS. *Res:* Solvent extraction; power reactor fuel reprocessing; mining and milling; ion exchange. *Mailing Add:* 5140 Dana Dr Lewiston NY 14092

KLOUDA, MARY ANN ABERLE, b Peoria, Ill, Jan 8, 37; m 62; c 4. PHYSIOLOGY. *Educ:* Col Notre Dame, Calif, BA, 58; Loyola Univ, Ill, PhD(physiol), 64. *Prof Exp:* Res assoc physiol, Loyola Univ, Ill, 64-65; from instr to asst prof, Univ Mass, Amherst, 65-71; lectr, Col of Our Lady of the Elms, 74-79, asst prof, 79-82, assoc prof biol, 82-87; INSTR BIOL SCI, CALIF STATE UNIV, SACRAMENTO, 87- *Mem:* Assoc Am Physiol Soc; Sigma Xi. *Res:* Cardiac response to sympathetic stimulation; effects of cardiac sympathectomy. *Mailing Add:* 4137 Hancock Dr Sacramento CA 95821

KLOWDEN, MARC JEFFREY, b Chicago, Ill, June 6, 48; m 70; c 2. MEDICAL ENTOMOLOGY. *Educ:* Univ Ill, Chicago Circle, BS, 70, MS, 73, PhD(biol), 76. *Prof Exp:* Res asst biol, Univ Ill, Chicago Circle, 70-73, teaching asst biol, 73-76; res assoc, dept entom, Univ Ga, 76-81; PROF ENTOM, UNIV IDAHO, MOSCOW, 81- *Honors & Awards:* Sigma Xi Res Awards, 80, 82 & 88. *Mem:* AAAS; Entom Soc Am; Sigma Xi; Soc Vector Ecologists; Am Mosquito Control Asn; Am Soc Zool. *Res:* Physiology of mosquito behavior. *Mailing Add:* Dept Entom Univ Idaho Moscow ID 83843

KLUBA, RICHARD MICHAEL, b Altoona, Pa, July 16, 47; m 81. FOOD SCIENCE, ANALYTICAL CHEMISTRY. *Educ:* Pa State Univ, BS, 69, MS, 73; Cornell Univ, PhD(food sci), 77. *Prof Exp:* Res assoc, 77-80, DIR RES, TAYLOR WINE CO INC, 80- *Mem:* Inst Food Technol; Am Soc Enologists; Sigma Xi. *Res:* Wine chemistry; analytical instrumentation. *Mailing Add:* 78 Edward Dr Eureka MO 63025

KLUBEK, BRIAN PAUL, b Buffalo, NY, Apr 21, 48; m 72; c 3. SOIL MICROBIOLOGY, MICROBIAL ECOLOGY. *Educ:* Colo State Univ, BS, 71; Ore State Univ, MS, 74; Utah State Univ, PhD(microbiol ecol), 77. *Prof Exp:* Res assoc microbiol, NC State Univ, 77-78; asst prof, 78-84, ASSOC PROF SOIL MICROBIOL, SOUTHERN ILL UNIV, CARBONDALE, 84- *Concurrent Pos:* Comt mem, Pesticide Waste Mgt Task Force, Ill Environ Protection Agency, 81-82; prin investr, Ill Inst Environ Qual, 81-83, Ill Soybean Prog Operating Bd, 81-84, NCent Regional Pesticide Impact Assessment Prog, 83-84, Dept Energy, 84-89, Nat Agr Chem Asn, 85-86 & Ill Dept Energy & Natural Resources, 87-88; co-investr, Ctr Res on High Sulfur Coal, Ill Dept Energy & Natural Resources. *Mem:* Am Soc Microbiol; Am Soc Agron; Soil Sci Soc Am; Sigma Xi. *Res:* Biological sulfur oxidation; disposal of waste pesticide solutions; pesticide decomposition; microbial desulfurization of coal; pesticide movement in soil; reclamation of mine lands. *Mailing Add:* Dept Plant & Soil Sci Southern Ill Univ Carbondale IL 62901-4415

KLUBES, PHILIP, b Brooklyn, NY, June 23, 35; m 64; c 2. PHARMACOLOGY. *Educ:* Queens Col, NY, BS, 56; Univ Minn, MS, 59, PhD(biochem), 62. *Prof Exp:* Res fel bact, Harvard Med Sch, 62-63; res assoc microbiol, Sch Med, Univ Southern Calif, 63-64, instr, 64-65; asst res prof, 65-70, asst prof, 70-73, assoc prof, 73-79, PROF PHARMACOL, MED CTR, GEORGE WASHINGTON UNIV, 79- *Concurrent Pos:* Res fel, USPHS, 62-63 & Bank Am-Giannini Med Found, 63-64. *Mem:* Am Asn Cancer Res; Am Soc Pharmacol & Exp Therapeut; AAAS. *Res:* Studies on the mechanism of action and metabolism of drugs. *Mailing Add:* Dept Pharmacol George Washington Univ Med Ctr Washington DC 20037

KLUCAS, ROBERT VERNON, b Montevideo, Minn, Oct 19, 40; m 66; c 2. MICROBIAL BIOCHEMISTRY, PLANT BIOCHEMISTRY. *Educ:* SDak State Col, BS, 62; Univ Wis, MS, 64, PhD(biochem), 67. *Prof Exp:* Res asst biochem, Univ Wis, 62-67; res assoc plant physiol, Ore State Univ, 67-69; from asst prof to assoc prof, 69-81, PROF BIOCHEM, UNIV NEBR, LINCOLN, 81- *Concurrent Pos:* NIH fel, Ore State Univ, 67-68. *Mem:* Am Soc Plant Physiol; Am Soc Microbiol. *Res:* Biological nitrogen fixation. *Mailing Add:* Dept Biochem Univ Nebr Lincoln NE 68583-0718

KLUEH, RONALD LLOYD, b Ferdinand, Ind, Oct 23, 36; m 59; c 2. ALLOY DEVELOPMENT, MECHANICAL PROPERTIES STUDIES. *Educ:* Purdue Univ, BS, 61; Carnegie-Mellon Univ, MS, 64, PhD(metall), 66. *Prof Exp:* Res staff, corrosion, 66-71, sr res staff, mech properties, 71-80, SR RES STAFF, ALLOY DEVELOP, OAK RIDGE NAT LAB, 80- *Mem:* Am Soc Metals. *Res:* New steels for fusion-reactor applications; published over 120 papers. *Mailing Add:* Oak Ridge Nat Lab PO Box 2008 Oak Ridge TN 37831

KLUENDER, HAROLD CLINTON, b Baraboo, Wis, Jan 28, 44; m 69; c 2. ORGANIC CHEMISTRY, MEDICINAL CHEMISTRY. *Educ:* Univ Wis-Stevens Point, BS, 66; Univ Wis-Madison, MS, 68; Wesleyan Univ, PhD(org chem), 71. *Prof Exp:* Res assoc org chem, Pharmaceut Dept, Univ Wis, 72-73; RES SCIENTIST & SUPVR MED CHEM, NATURAL PROD LAB, 73-, MILES DELEGATE, MILES LABS, INC. *Concurrent Pos:* NIH fel org chem, Harvard Univ, 70-72. *Mem:* Am Chem Soc; AAAS. *Res:* Prostanoids, chemistry and biological activity. *Mailing Add:* Miles Inc Res Ctr 400 Morgan Lane West Haven CT 06516

KLUEPFEL, DIETER, b Zurich, Switz, Oct 7, 30; Can & Swiss citizen; m 59; c 2. MICROBIOLOGY, BIOCHEMISTRY. *Educ:* Swiss Fed Inst Technol, Dipl Sc nat, 54, Dr Sc nat, 56. *Prof Exp:* Res asst microbiol, Swiss Fed Inst Technol, 54-56, res assoc biochem, 56-57; Nat Res Coun Can fel, 57-58; res scientist microbiol, Lepetit SPA, Milan, Italy, 59-61, head lab biochem, 61-65; sr res scientist microbiol, Ayerst Labs, 65-70, res assoc, 70-75; RES PROF, INST A FRAPPIER, UNIV QUE, MONTREAL, 75-, DEPT

HEAD, APPL MICROBIOL, 89- Concurrent Pos: Lectr, Univ Montreal, 70-78; adj prof, Univ Concordia, Montreal, 78-87. Mem: Soc Indust Microbiol; Am Soc Microbiol. Res: Microbial metabolism; biosynthesis of natural products; biodegradation and bioconversion; isolation of secondary metabolites and antibiotics. Mailing Add: Inst Armand-Frappier Univ Que Ville de Laval PQ H7N 4Z3 Can

KLUESSENDORF, JOANNE, b Milwaukee, Wis, Apr, 8, 49. PALEONTOLOGY. Educ: Univ, Wis-Milwaukee, 71-75; Ore State Univ, Corvallis, 75-79; Univ Ill, Urbana-Champaign, BS, 83, MS, 86. Prof Exp: Res asst, Ore State Univ, 75-79; res asst, Ill State Geol Surv, Champaign, 80-83; RES ASST & TEACHING ASST, DEPT GEOL, UNIV ILL, 83-; CUR, GREENE GEOL MUS, UNIV WIS-MILWAUKEE, 84- Concurrent Pos: Sci consult, Milwaukee Pub Mus, 80-83; leader, Silurian reef and general geology fiel trips for nat soc, petrol geologists and educ orgn. Mem: Geol Soc Am; Paleont Soc Found; Sigma Xi; Am Asn Petrol Geologists; Nat Asn Geol Teachers; Soc Econ Paleont & Mineralogists. Res: Preservation of lagerstatten; genesis of oolitic iron stones; evolution and depositional environment of Silurian reefs; gastropod and polyplacophoran systematics and paleoecology; author and co-author of numerous books and articles on geology. Mailing Add: 116 W McHenry St Urbana IL 61801

KLUETZ, MICHAEL DAVID, b Wausau, Wis, June 20, 48. BIOPHYSICAL CHEMISTRY. Educ: Univ Wis-Madison, BS, 71; Univ Ill, Urbana, PhD(phys chem), 75. Prof Exp: Asst prof chem, Univ Idaho, 75-, asst prof biochem, 80-; MEM STAFF, CARGILL INC. Mem: Am Chem Soc; Biophys Soc. Res: Biophysical, particularly magnetic resonance, studies of enzyme systems which are responsible for the degradation of several physiologically important polyamines and histamine in both plant and animal systems. Mailing Add: 5435 Joyce St Maple Plain MN 55359-9632

KLUG, AARON, b Aug 11, 26; m; c 2. BIOLOGICAL CHEMISTRY. Educ: Univ Wit Watersrand, Cape Town & Cambridge. Prof Exp: Dir virus struct res group, Birkbeck Col, 58-61; DIR STUDIES, PETERHOUSE COL, UNIV CAMBRIDGE, 62-; DIR, LAB MOLECULAR BIOL, MED RES COUN, 86- Honors & Awards: Nobel Prize Chem, 82. Mem: Foreign assoc Nat Acad Sci; Am Acad Arts & Sci. Mailing Add: Lab Molecular Biol Med Res Coun Hills Rd Cambridge CB2 2QH England

KLUG, DENNIS DWAYNE, b Milwaukee, Wis, Aug 22, 42; m 62; c 1. PHYSICAL CHEMISTRY. Educ: Univ Wis-Milwaukee, 54; Univ Wis-Madison, PhD(phys chem), 68. Prof Exp: Fel, 68-70, RES CHEMIST, NAT RES COUN CAN, 70- Mem: Sigma Xi. Res: Experimental and theoretical studies of lattice vibrations in crystalline and disordered solids; raman and infrared spectroscopy, high pressure techniques and instrumentation. Mailing Add: Steacie Inst Molecular Sci Nat Res Coun Can Ottawa ON K1A 0R6 Can

KLUG, MICHAEL J, b Milwaukee, Wis, Mar 7, 41; m 69. MICROBIOLOGY, ECOLOGY. Educ: SDak State Univ, BS, 63; Univ Iowa, MS, 66, PhD(microbiol), 69. Prof Exp: NIH res fel microbiol, Univ Ill, Urbana, 69-70; mem staff, 70-80, ASSOC PROF, W K KELLOGG BIOL STA, MICH STATE UNIV, 80- Mem: AAAS; Am Chem Soc; Am Soc Microbiol; Am Inst Biol Sci. Res: Ecology and metabolism of heterotrophic bacteria in natural waters and insects. Mailing Add: W K Kellogg Biol Sta-Micro Mich State Univ Hickory Corners MI 48060

KLUG, WILLIAM STEPHEN, b Parkersburg, WVa, Sept 2, 41; div; c 3. BIOLOGY. Educ: Wabash Col, BA, 63; Northwestern Univ, PhD(develop genetics), 68. Prof Exp: Instr biol, Wabash Col, 63-65, asst prof, 68-73; assoc prof biol & chmn dept, 73-78, PROF BIOL, TRENTON STATE COL, 79-, CHMN DEPT, 81- Mem: Soc Develop Biol; Sigma Xi. Res: Developmental genetics in the ovarian system of Drosophila melanogaster; co-author one book on genetics. Mailing Add: Dept Biol Trenton State Col Trenton NJ 08650

KLUGE, ARNOLD GIRARD, b Glendale, Calif, July 27, 35; m 59; c 2. VERTEBRATE ZOOLOGY. Educ: Univ Southern Calif, BA, 57, MS, 60, PhD(biol), 64. Prof Exp: Lectr embryol, Univ Southern Calif, 64; asst prof comp anat, San Fernando Valley State Col, 64-65; asst prof comp anat & embryol, 65-66, assoc prof biol, 74-76, UNIV MICH, ANN ARBOR, 66-, PROF BIOL, 76- Concurrent Pos: Res asst, Los Angeles Ment Health Asn grant, 58; NSF grants, 59 & 69-71; USPHS grant, 60-61; Fulbright scholar, Australia, 61-62; Am Philos Soc grant, 69; Guggenheim fel, 71-72; NSF fel, 60, Sigma Xi & Sci Res Soc Am fel, 63. Mem: AAAS; Am Soc Ichthyol & Herpet; Soc Study Evolution; Soc Syst Zool. Res: Evolution; numerical taxonomy; herpetology. Mailing Add: 4042a Nat Sci Biol Univ Mich Ann Arbor MI 48109

KLUGE, JOHN PAUL, b St Louis, Mo, July 7, 37; m 58; c 4. VETERINARY PATHOLOGY, COMPARATIVE PATHOLOGY. Educ: Univ Mo, BS & DVM, 62; Iowa State Univ, MS, 65; George Washington Univ, PhD(comp path), 68; Am Col Vet Path, dipl, 70. Prof Exp: Res vet, Nat Animal Dis Lab, 62-68; from assoc prof to prof path, 68-75, chmn dept, 75-90, PROF VET PATH, IOWA STATE UNIV, 75- Concurrent Pos: Consult, Agr Res Serv & Animal & Plant Health Inspection Serv, USDA. Mem: Am Asn Vet Lab Diagnosticians; Am Vet Med Asn; US Animal Health Asn; Conf Res Workers Animal Dis; Am Col Vet Path; Intermountain Vet Med Asn. Res: Comparative pathology of infectious diseases and neoplasms. Mailing Add: Dept Vet Path Iowa State Univ Ames IA 50011-1250

KLUGER, MATTHEW JAY, b Brooklyn, NY, Dec 14, 46; m 67; c 2. PHYSIOLOGY. Educ: Cornell Univ, BS, 67; Univ Ill, MS, 69, PhD(zool), 70. Prof Exp: NIH fel, Yale Univ & J B Pierce Found Lab, 70-72; from asst prof to assoc prof, 72-81, PROF PHYSIOL, MED SCH, UNIV MICH, ANN ARBOR, 81- Mem: Am Physiol Soc; Am Soc Zoologists; Soc Exp Biol Med; Sigma Xi. Res: Temperature regulation and bioenergetics; evolution and adaptive value of fever; host responses to infection; role of monokines and cytokines in regulation of body temperature and food intake. Mailing Add: Dept Physiol Univ Mich Med Sch Ann Arbor MI 48109

KLUGER, RONALD H, b Newark, NJ, Dec 22, 43; m 69; c 2. ORGANIC CHEMISTRY, BIOCHEMISTRY. Educ: Columbia Univ, AB, 65; Harvard Univ, AM, 66, PhD(chem), 69. Prof Exp: NIH fel biochem, Brandeis Univ, 69-70; asst prof chem, Univ Chicago, 70-74; from asst prof to assoc prof, 74-81, PROF CHEM, UNIV TORONTO, 81-, ASSOC CHAIR, 89- Concurrent Pos: Sloan Found fel, 73. Honors & Awards: Merck Sharp & Dohme lectr, Chem Inst Can, 83; Labatt Award, Can Soc Chem, 90. Mem: Chem Inst Can; Am Chem Soc. Res: Mechanisms of biochemical catalysis and related organic reaction mechanisms; thiamin, biotin and enzymes; enzyme inhibitors based on mechanistic analysis; functional group interactions and reactive intermediates; biotechnology of drug and agrichemical design. Mailing Add: Dept Chem Univ Toronto Toronto ON M5S 1A1 Can

KLUGHERZ, PETER D(AVID), b Brooklyn, NY, Feb 25, 42; m 62; c 4. CHEMICAL ENGINEERING. Educ: Cornell Univ, BChE, 63, PhD(chem eng), 69. Prof Exp: SR CHEMIST, RES LABS, ROHM AND HAAS CO, SPRING HOUSE, PA, 68- Mem: Am Chem Soc; Am Inst Chem Engrs; Catalysis Soc. Res: Heterogeneous catalytic oxidation; monomer process research. Mailing Add: 760 Killdeer Lane Huntingdon Valley PA 19006

KLUIBER, RUDOLPH W, b Chicago, Ill, Feb 20, 30; m 55. INORGANIC CHEMISTRY. Educ: Univ Ill, BS, 50; Columbia Univ, AM, 52; Univ Wis, PhD(chem), 54. Prof Exp: Chemist, Plastics Div, Union Carbide Corp, 54-64; res assoc, Princeton Univ, 64-66; from asst prof to assoc prof, 66-71, PROF CHEM, RUTGERS UNIV, NEWARK, 71- Mem: Am Chem Soc. Res: Metal chelate compounds. Mailing Add: Dept of Chem Rutgers Univ Newark NJ 07102

KLUKSDAHL, HARRIS EUDELL, b Bismarck, NDak, Mar 4, 33; m 89; c 3. CHEMISTRY, CATALYSIS. Educ: Western Wash State Col, BA, 54; Univ Wash, PhD(inorg chem), 58. Prof Exp: Res chemist, E I du Pont de Nemours & Co, 58-60; res chemist, Chevron Res Co, 60-67, sr res chemist, 67-70, sr res assoc, 70-89, RES SCIENTIST, CHEVRON RES & TECHNOL CO, 89- Mem: AAAS. Res: Heterogeneous catalysis; petroleum refining processes; extractive metallurgy. Mailing Add: Chevron Res & Technol Co 100 Chevron Way Richmond CA 94802

KLUMPAR, DAVID MICHAEL, b Jacksonville, Fla, Apr 14, 43; m 64; c 2. SPACE PHYSICS. Educ: Univ Iowa, BA, 65, MS, 68; Univ NH, PhD(physics), 73. Prof Exp: Res assoc physics, Univ NH, 72-74; res assoc space physics, Inst Phys Sci, Univ Tex, Dallas, 74-77,; res scientist, Ctr Space Sci, 78-84, res scientist, 84-90, STAFF SCIENTIST, LOCKHEED PALO ALTO RES LABS, 90- Mem: Am Geophys Union; assoc mem Sigma Xi; Am Phys Soc. Res: Investigations of the low energy charged particle environment in the earth's near magnetosphere and interactions between the magnetosphere and ionosphere, especially at high latitudes in the auroral region. Mailing Add: Dept 91-20 Bldg 255 Lockheed Palo Alto Res Labs 3251 Hanover St Palo Alto CA 94304

KLUMPP, THEODORE GEORGE, b New York, NY, May 15, 03; m 34; c 6. INTERNAL MEDICINE. Educ: Princeton Univ, BS, 24; Harvard Univ, MD, 28. Hon Degrees: DSc, Philadelphia Col Pharm, 43, New Eng Col Pharm, 61 & Albany Med Col, 64; LLD, Univ Chattanooga, 60. Prof Exp: Intern, Peter Bent Brigham Hosp, 29-30; asst resident physician, Lakeside Hosp, Cleveland, 30-32; instr & asst clin prof internal med, Med Sch, Yale Univ, 32-36; chief drug div, Food & Drug Admin, Fed Security Agency, Washington, DC, 36-41; dir drugs, food & phys ther & secy coun pharm & chem, AMA, Ill, 41-42; pres, Winthrop Labs Div, Sterling Drug, Inc, 42-70, chmn, 70-73, mem bd dirs & vpres, Sterling Drug, Inc, 60-73; RETIRED. Concurrent Pos: Assoc physician, New Haven Hosp, 32-36, chief hemat clin & dir med lab, 33-36; adj clin prof, Med Sch, George Washington Univ, 40-41; attend physician, Gallinger Munic Hosp, 41; mem pharmaceut mfrs indust adv comt & penicillin producers indust adv comt, War Prod Bd, 42-47; chmn bd gov, Nat Vitamin Found, 47-56; dir, Sterwin Chem, Inc, 49-70; chmn task force on handicapped, Off Defense Mobilization, Washington, DC, 51-52; pres, Nat Pharmaceut Coun, Inc, 53-55; chmn med serv task force, Hoover Comn on Orgn Exec Br Govt, 53-55; mem study comt fed aid to pub health, Comn Inter-Govt Rels, Washington, DC, 54; mem, Nat Adv Coun Voc Rehab, 55-59; consult surgeon, Chesapeake & Ohio RR, 56; mem bd vis, Sch Pub Health, Harvard Univ, 58-64 & Med & Dent Schs, 64-70; mem med adv comt, Off Voc Rehab, 60-62; mem health resources adv comt, Off Emergency Planning, Exec Off of President, 62-69; mem, NY State Voc Rehab Coun, 67-73; secy, Nat Fund Med Educ, 68-69, pres, 69-71, chmn, 71-75, vpres, 75-80; mem bd trustees, Brooklyn Col Pharm, Long Island Univ, 68-77, chmn, 74-77; mem comt on aging, Coun Med Serv, AMA; mem, Gov Coun Rehab, NY & Found Trop Med; mem bd trustees, Affil Cols & Univs, Inc, 72-; med consult, President's Coun Phys Fitness & Sports, 73-; assoc ed, Med Times, 73-; mem bd trustees, Human Resources Sch, NY, 74- & Arnold & Marie Schwartz Col Pharm & Health Sci, Long Island Univ, 77-; chmn bd dirs, Nat Asn Human Develop, 74-80. Mem: AAAS; fel Am Soc Clin Invest; fel Am Col Physicians; fel AMA. Res: Therapeutics; materia medica; medical pharmacology; hematology; physiological chemistry. Mailing Add: Box 72 Rte 6 Charlottsville VA 22902

KLUN, JEROME ANTHONY, b Ely, Minn, May 4, 39; div; c 3. ENTOMOLOGY. Educ: Univ Minn, Duluth, BS, 61; Iowa State Univ, PhD(entom), 65. Prof Exp: Res asst, Iowa State Univ, 61-65; RES SCI & EDUC ADMIN-AGR RES, USDA, 65- Concurrent Pos: Assoc prof entom, Iowa State Univ, 69-77. Honors & Awards: Superior Serv Award, USDA, 82. Mem: AAAS; Am Chem Soc; Entom Soc Am. Res: Insect sex pheromone chemistry, behavior, biosynthesis, and neurohormones. Mailing Add: USDA Agr Res Serv BARC-West Bldg 007 Beltsville MD 20705

KLUNDT, IRWIN LEE, b Pasco, Wash, Aug 7, 36; m 59; c 3. ORGANIC CHEMISTRY. Educ: State Col Wash, BS, 58; Mont State Univ, MS, 59; Wayne State Univ, PhD(org chem), 63. Prof Exp: Chemist, Detroit Inst Cancer Res, 63; sr res scientist, Pac Northwest Labs, Battelle Mem Inst, 65-66; proj leader, Aldrich Chem Co, 66, group leader, 66-71, biochem tech mgr, 71-73, tech serv mgr, 73-74, vpres, Aldrich Chem Co Inc, 74-90, vpres, Sigma-Aldrich Corp, 75-83, dir, Aldrich Chem Co Inc, 78-90, dir, Sigma Chem Co, 78-90; AT EARTH TECHNOL CORP, 90- Mem: Am Chem Soc; NY Acad Sci. Res: Aliphatic and alicyclic chemistry; small ring compounds; carbohydrates. Mailing Add: 2320 Kevenauer Dr Brookfield WI 53005

KLURFELD, DAVID MICHAEL, b New York, NY Feb 22, 51; m 73; c 2. ATHEROSCLEROSIS, TUMOR PROMOTION. Educ: Cornell Univ, BS, 72; Med Col Va, MS, 75, PhD(path), 77. Prof Exp: Res fel, 77-79, res assoc, 79-82, asst prof, 82-87, ASSOC PROF, WISTAR INST ANAT & BIOL, 87- Concurrent Pos: USPHS fel, 77-79; mem, Coun Arteriosclerosis, Am Heart Asn, 87-; course dir, Nutrit Res Tech, Univ Pa Sch Med, assoc prof, nutrit surg, 89- Honors & Awards: Nutrit Res Award, J Nutrit Res, 82. Mem: Fel Am Col Nutrit; Am Heart Asn; Am Inst Nutrit; NY Acad Sci; Soc Exp Biol & Med; Sigma Xi; Am Asn Pathologists. Res: The relationship of atherosclerosis to immune system, macrophages and nutrition; nutrition and tumor promotion; dietary fiber. Mailing Add: Wistar Inst 36th St & Spruce St Philadelphia PA 19104

KLUS, JOHN P, b Goodman, Wis, June 13, 35; m 61; c 4. NEW PRODUCT DEVELOPMENT. Educ: Mich Technol Univ, BS, 57, MS, 61; Univ Wis, PhD(civil eng), 65. Prof Exp: Appraiser, Am Appraisal Co, 57-58; res engr designer, Eng & Res Develop Labs, Va, 58-59; instr, Mich Technol Univ, 60-61; instr drawing, Univ Wis, 61-62; struct designer, Warzyn Eng Co, 62; from instr to assoc prof struct, 62-70, chmn dept eng & prof struct, 70-80, PROF DEVELOP, UNIV WIS-MADISON, 80- Concurrent Pos: Fulbright scholar, Finland, 66-67; chmn, Working Group Continuing Educ Engrs, UNESCO, 73-; gen chmn, 1st World Conf Continuing Eng Educ, 79; Fulbright Researcher, Finland, 85. Honors & Awards: Leonardo da Vinci Medal, 87. Mem: AAAS; Am Soc Civil Engrs; Am Soc Eng Educ; Nat Soc Prof Engrs. Res: Continuing education research; new product development. Mailing Add: Dept Eng Prof Develop Univ Wis-Exten 432 N Lake St Madison WI 53706-1498

KLUSKENS, LARRY F, PATHOLOGY. Educ: Univ Chicago, PhD(path), 75, MD, 76, Am Bd Path, cert. Prof Exp: Resident, Univ Chicago, dir, cellular immunol; dir cytol, Univ Iowa; AT RUSH-PRESBY-ST LUKES MED CTR. Mailing Add: Dept Path Rush-Presby-St Luke's Med Ctr 1653 W Congress Pkwy Chicago IL 60612

KLUSMAN, RONALD WILLIAM, b Batesville, Ind, June 16, 41; m 64; c 2. GEOCHEMISTRY. Educ: Ind Univ, Bloomington, BS, 64, MS, 67, PhD(geochem), 69. Prof Exp: Instrumental analyst geochem, Ind Geol Surv, 64-67; asst prof, Purdue Univ, 69-72; assoc prof, 72-77, PROF GEOCHEM, COLO SCH MINES, 77-, PROF CHEM, 80- Mem: Geol Soc Am; Soc Environ Geochem & Health. Res: Trace elements in geological and environmental systems; instrumental analysis; computer applications in geology. Mailing Add: Dept Chem & Geochem Colo Sch Mines Golden CO 80401

KLUSS, BYRON CURTIS, b Luzerne, Iowa, May 25, 28. CELL BIOLOGY. Educ: Univ Iowa, BA, 49, MS, 55, PhD(zool), 57. Prof Exp: Asst zool, Univ Iowa, 53-57; instr, Albion Col, 57-59; from asst prof to assoc prof biol, 59-66, dir spec progs, 70-75, prof biol, 66-80, PROF ZOOL, CALIF STATE UNIV, LONG BEACH, 80- Concurrent Pos: Lalor Found fel, 58; Fulbright fel, Assiut, 65-66. Mem: Am Soc Zool; Am Soc Cell Biol; Electron Micros Soc Am. Res: Cytology; bioluminescence; electron microscopy. Mailing Add: Dept Biol Calif State Univ Long Beach CA 90840

KLUTCHKO, SYLVESTER, b Wilkes-Barre, Pa, Sept 2, 33; m 62; c 3. SYNTHETIC ORGANIC CHEMISTRY, ORGANIC CHEMISTRY. Educ: Pa State Univ, BS, 55. Prof Exp: From asst scientist to scientist org chem, 55-75, SR SCIENTIST ORG CHEM, WARNER-LAMBERT CO, INC, ANN ARBOR, 75- Mem: Am Chem Soc. Res: Synthesis of agents that affect the renin-angiotensin system; ie angiotensin converting enzyme (ACE) inhibitors and renin inhibitors; invention of antihypertensive ACE inhibitor quinapril; preparation of antiallergy chromones by novel synthetic procedure; study of ageneral rearrangement of 3-substituted chromones. Mailing Add: 5143 Pratt Rd Scio Twp Ann Arbor MI 48103

KLUTE, ARNOLD, b Galein, Mich, Sept 24, 21; m 48; c 4. AGRONOMY. Educ: Mich State Col, BS, 47, MS, 48; Cornell Univ, PhD(soil physics), 51. Prof Exp: Res engr, Schlumberger Well Surv Corp, 51-53; from asst prof to prof agron, Univ Ill, 53-70; PROF SOILS, COLO STATE UNIV, 70-; RES LEADER IRRIG & SOIL-PLANT-WATER RELS, AGR RES SERV, USDA, FT COLLINS, 78- Concurrent Pos: US Salinity Lab, Riverside, Calif, 60; soil scientist, Sci & Educ Admin-Agr Res, USDA, 70-78. Honors & Awards: Soil Sci Award, Am Soc Agron, 65. Mem: Am Soc Agron; Soil Sci Soc Am; Am Geophys Union; Sigma Xi. Res: Investigations of the transport of water, gases, heat, and solutes in soils. Mailing Add: 1112 Parkwood Dr Ft Collins CO 80525

KLUVER, J(OHAN) W(ILHELM), b Monaco, Nov 13, 27; m 64; c 2. COMMUNICATIONS. Educ: Royal Inst Technol, Sweden, Engr, 51; Univ Calif, MS, 55, PhD, 57. Prof Exp: Engr TV develop, Thomson-Houston Co, France, 53-54; asst prof elec eng, Univ Calif, 57-58; mem tech staff, Bell Tel Labs, 58-68; PRES, EXPS IN ART & TECHNOL, INC, 68- Mem: Assoc Am Phys Soc; Assoc Inst Elec & Electronics Engrs; Optic Soc Am; Soc Info Display. Res: Microwave electronics; electron dynamics; electron devices; microwave tubes; gas lasers; optics. Mailing Add: 69 Apple Tree Row Berkeley Heights NJ 07922

KLYCE, STEPHEN DOWNING, b Arlington, Mass, Oct 30, 42; m 64; c 3. PHYSIOLOGY. Educ: Univ Mass, BS, 64; Yale Univ, PhD(physiol), 71. Prof Exp: Res assoc ocular physiol, Yale Univ, 71-72; res assoc, 72-75, sr res assoc ocular physiol, Stanford Univ, 75-79; PROF OPHTHALMOL, MED SCH, LA STATE UNIV, 79- Concurrent Pos: Co-prin investr res grant, Nat Eye Inst, NIH, 72-79, prin investr res grant, 80- & corneal dis panel, Nat Eye Adv

Coun; consult, Vet Admin Hosp, Palo Alto, Calif, 76-79, 80-81; trustee, Asn Res Vision & Ophthal; consult, Allergan Med Optics, 84- Mem: Asn Res Vision & Ophthal; Biophys Soc; Am Physiol Soc; Int Soc Eye Res; Int Soc Contact Lens Res. Res: Physiology and biophysics of membrane transport and permeability in epithelial tissues; corneal retractive surgery research. Mailing Add: Dept Opthal La State Univ Sch Med Eye Ctr 2020 Gravier St Suite B New Orleans LA 70112

KLYMKOWSKY, MICHAEL W, CELL BIOLOGY. Educ: Calif Inst Technol, PhD(biophysics), 80. Prof Exp: ASST PROF CELL BIOL, UNIV COLO, 83- Res: Intermediate filaments; early xenopus development. Mailing Add: Dept Molecular, Cellular & Develop Biol Univ Colo Campus Box 347 Boulder CO 80309

KMAK, WALTER S(TEVEN), b Garfield, NJ, June 5, 28; m 59; c 4. CHEMICAL ENGINEERING. Educ: Pa State Univ, BS, 49; Univ Ill, MS, 50, PhD, 56. Prof Exp: Res chem engr, Esso Res & Eng Co, 55-64, sr engr, 64-74, eng assoc, 74-79, SR ENG ASSOC, EXXON RES & ENG CO, 79- Mem: Am Chem Soc; Am Inst Chem Engrs. Res: Chemical reaction kinetics; catalytic reforming; engineering computer applications. Mailing Add: 18 Chiplou Lane Scotch Plains NJ 07076

KMETEC, EMIL PHILIP, b Carlinville, Ill, Sept 29, 27; m 55; c 5. BIOLOGICAL CHEMISTRY. Educ: Univ Chicago, MS, 53; Univ Wis, PhD(plant physiol), 57. Prof Exp: Asst bot, Univ Chicago, 50-52; asst plant physiol, Iowa State Univ, 52-53; asst, Univ Wis, 53-57, res assoc, 57; res assoc biochem, Sch Med, La State Univ, 57-60; sr instr biochem & pediat, Sch Med, Case-Western Reserve Univ, 60-61, asst prof, 61-64; from assoc prof to prof biol, 64-75, asst vpres acad affairs, 79-84, PROF BIOL CHEM, SCH MED & COL SCI & ENG, WRIGHT STATE UNIV, 75- Concurrent Pos: Vis prof, Okayama Univ of Sci, Japan, 84. Mem: AAAS; Am Chem Soc; NY Acad Sci; Sigma Xi. Res: RNA metabolism and protein synthesis; extracellular Matrix and basement membranes. Mailing Add: 2172 Crabtree Dr Dayton OH 45431

KMETZ, JOHN MICHAEL, b Johnstown, Pa, Jan 14, 43; m 66; c 3. HISTOLOGY, CYTOCHEMISTRY. Educ: Pa State Univ, BS, 64, PhD(physiol), 68. Prof Exp: Asst prof biol, Pa State Univ, 68-73; res biol, Sci Unlimited Res Found, 73-78; ASST PROF BIOL SCI, KEAN COL NJ, 78- Mem: NY Acad Sci; Am Inst Biol Sci. Res: Quantitative histochemistry and cytophotometry. Mailing Add: Dept Biol Kean Col NJ Union NJ 07083

KMIECIK, JAMES EDWARD, b New Waverly, Tex, Feb 11, 36; m 59; c 1. ORGANIC CHEMISTRY. Educ: St Edwards Univ, BS, 56; Univ Tex, MA, 60, PhD(chem), 61. Prof Exp: Res chemist, Cities Serv Res & Develop Co, 61-64; sr res chemist, Columbia Carbon Co, 64-66; sr res chemist, Com Develop Div, Jefferson Chem Co, 66-68, proj chemist, 68-70, mgr amine prod develop, Mkt Dept, 70-77, mgr new prod develop, 77-80; BUS MGR, TEXACO CHEM CO, 80- Mem: Soc Aerospace Mat & Process Engrs; Am Chem Soc. Res: Synthesis of organic nitrogen heterocyclic compounds; reactions of carbon monoxide with organic compounds; reactions of aliphatic and aromatic nitro compounds. Mailing Add: Texaco Chem Co PO Box 27707 Houston TX 77227

KNAAK, JAMES BRUCE, b Milwaukee, Wis, Aug 20, 32; m 58; c 3. BIOCHEMISTRY, TOXICOLOGY. Educ: Univ Wis, BS, 54, MS, 57, PhD(biochem, dairy husb), 62. Prof Exp: Res asst dairy husb, biochem & entom, Univ Wis, 56-61; Union Carbide Indust fel biochem, Mellon Inst, 61-66, sr res fel, 67; sr res chemist, Niagara Chem Div, FMC Corp, 67-71; group leader agr chem, CIBA-Geigy Corp, 71-73; staff toxicologist, Calif Dept Food & Agr, 73-85; staff toxicologist, Calif Dept Health Serv, 86-89; PROD STEWARDSHIP SCIENTIST, OCCIDENTAL CHEM CORP, 90- Mem: AAAS; Am Chem Soc; Soc Toxicol; NY Acad Sci; fel Am Inst Chem. Res: Toxicology and metabolism of organophosphate and carbamate insecticides; urea and s-triazine herbicides; metabolism of industrial chemicals; biochemical pharmacology; dermal dose response and dermal absorption studies; environmental monitoring; behavioral and biochemical pharmacology; risk assessment; PBPK modeling; agricultural field worker safety studies; indoor safety studies. Mailing Add: Occidental Chem Corp 360 Rainbow Blvd 5 Niagara Falls NY 14302

KNABE, GEORGE W, JR, b Grand Rapids, Mich, June 29, 24; m 54; c 4. PATHOLOGY. Educ: Univ Md, MD, 49. Prof Exp: Fel path, Cleveland Clin Found, 50-51; resident path, Henry Ford Hosp, Detroit, 53-54; chief lab serv, Vet Admin Ctr, Dayton, Ohio, 55-57; med ed adv, Int Coop Admin, 57-59; asst prof path & chief clin lab, Sch Med, Puerto Rico, 59-60; prof path & chmn dept, Sch Med, Univ SDak, 60-72, dean sch med, 67-72; assoc dean clin affairs, 72-75, PROF PATH, SCH MED, UNIV MINN, DULUTH, 72- Concurrent Pos: Mem staff, Va Regional Med Ctr, Virginia, Minn, 77- Mem: Am Med Asn; Am Soc Clin Pathologists; Col Am Pathologists; Int Acad Pathologists. Res: Infectious disease. Mailing Add: Va Regional Med Ctr 901 Ninth St N Virginia MN 55792

KNACKE, ROGER FRITZ, b Stuttgart, Ger, June 22, 41; US citizen; m 72; c 2. ASTROPHYSICS, ASTRONOMY. Educ: Univ Calif, Berkeley, BA, 63, PhD(physics), 69. Prof Exp: Fel astron, Lick Observ, Univ Calif, 70-71; from asst prof to assoc prof, 71-79, PROF ASTRON, DEPT EARTH SCI, STATE UNIV NY, STONY BROOK, 79- Concurrent Pos: Vis scientist, Max Plank Inst Nuclear Physics, Heidelberg, Ger, 78-85. Mem: Am Astron Soc; Int Astron Union. Res: Interstellar matter; infrared astronomy; planetary atmospheres; comets. Mailing Add: Dept Earth Sci State Univ NY Stony Brook NY 11794-2100

KNAEBEL, KENT SCHOFIELD, b Cincinnati, Ohio, Aug 20, 51; m 73; c 3. CHEMICAL ENGINEERING. Educ: Univ Ky, BSChE, 73; Univ Del, MChE, 78, PhD(chem eng), 80. Prof Exp: Chem engr, Tenn Eastman Co; instr chem eng, Univ Del, 79-80; asst prof, 80-86, ASSOC PROF CHEM ENG, OHIO STATE UNIV, 86- Concurrent Pos: Vis scientist, Brookhaven

Nat Lab, 81, Sch Areospace Med, 84. *Mem:* Am Chem Soc; Am Inst Chem Engrs. *Res:* Separation process: cyclic sorption, including both gas and liquid phase versions. *Mailing Add:* Dept Chem Eng Ohio State Univ 140 W 19th Ave Columbus OH 43210

KNAFF, DAVID BARRY, b New York, NY, June 5, 41; m 62; c 1. PHOTOBIOLOGY. *Educ:* Mass Inst Technol, BS, 62; Yale Univ, MS, 63, PhD(chem), 66. *Prof Exp:* Biochemist, Dept Cell Physiol, Univ Calif, Berkeley, 66-76; assoc prof chem, 76-80, PROF CHEM, TEX TECH UNIV, 80- *Mem:* Biophys Soc; Am Soc Photobiol; Am Soc Plant Physiol; AAAS; Am Soc Biol Chemists; Sigma Xi. *Res:* Electron transport in plants and photosynthetic bacteria with emphasis on the roles of cytochromes and iron-sulfur proteins. *Mailing Add:* 3902 54th St Lubbock TX 79413

KNAGGS, EDWARD ANDREW, b Oak Park, Ill, July 28, 22; m 47; c 2. ORGANIC CHEMISTRY. *Educ:* YMCA Col, BS, 45; Ill Inst Technol, MS, 53. *Prof Exp:* Pilot plant technician, Glue Div, Swift & Co, 42; asst chemist, Inst Gas Technol, 42-45; res chemist, Ninol Labs, Inc, 45-49, chief chemist & plant engr, 49-58; assoc tech dir, 58-62, asst to gen mgr, 65-69, DIR RES & DEVELOP, STEPAN CHEM CO, 62-, TECH DIR INDUST CHEM DIV, 69- *Mem:* AAAS; Am Chem Soc; Am Oil Chemists' Soc; Water Pollution Control Fedn; Am Water Works Asn. *Res:* Organic synthesis; organic sulfur compounds; desulfurization of gas and petroleum; SO3 sulfation and sulfonation; surface-active agents. *Mailing Add:* 715 Colwyn Terr Deerfield IL 60015

KNAKE, ELLERY LOUIS, b Gibson City, Ill, Aug 26, 27; m 51; c 2. WEED SCIENCE, AGRONOMY. *Educ:* Univ Ill, BS, 49, MS, 50, PhD(agron), 60. *Prof Exp:* Teacher, High Sch, Ill, 50-56; instr plant sci, Voc Agr Serv, 56-60, from asst prof to assoc prof agron, 60-64, PROF AGRON, UNIV ILL, URBANA, 69- *Concurrent Pos:* Assoc ed, Agron J, 76-78; partic, East-West Center Confs, Honolulu, 76 & 77; mem comt integrated pest mgt, Off Technol Assessment, 79; weed sci rep, People to People Prog, Peoples Rep China, 83; mem bd dirs, Coun Agr Sci & Technol, 84-, task force, Ecol Impacts Fed Conserv & Cropland Reduction Progs, 88-90; bd dirs, Weed Sci Soc Am, 72-75 & 86-89, N Cent Weed Sci Soc, 69-72 & 85-88, pres, 71. *Honors & Awards:* Crops & Soils Mag Award, Am Soc Agron, 67; Outstanding Extension Worker Award, Weed Sci Soc Am, 72; Ciba-Geigy Award for Outstanding Contrib to Agr, 72; Educr Award, Midwest Agr Chem Asn, 75; Exten Educ Award, Am Soc Agron, 78; Sustained Excellence Award, Ill Coop Exten Serv, 83; Super Serv Award, US Dept Agr, 83. *Mem:* Fel Weed Sci Soc Am (pres, 74); fel Am Soc Agron; Crop Sci Soc Am; Am Agr Ed Asn; Sigma Xi. *Res:* Competitive effects of giant foxtail; cultivation versus chemical weed control; herbicide incorporation; improving effectiveness of pre-emergence herbicides; site of herbicide uptake; weed control for conservation tillage and conservation acreage reserve; herbicide performance; effect of herbicides on crops. *Mailing Add:* N323 Turner Hall Univ Ill 1102 S Goodwin Ave Urbana IL 61801

KNAP, JAMES E(LI), b Denver, Colo, Oct 5, 26; m 49; c 6. INVESTMENT RECOVERY, CHEMICAL ENGINEERING. *Educ:* Univ Colo, BS, 49; Univ Ill, MS, 51, PhD(chem eng), 53. *Prof Exp:* Asst chem eng, Ill, 49-50; chem engr, Process Develop Lab, Carbide & Carbon Chem Co, 52-55; group leader, Res & Develop Dept, Union Carbide Chem Div, 55-68, asst mgr, Invest Recovery Dept, 68-74, gen mgr, Invest Recovery Dept, 74-86, pres, Invest Recovery Asn, 83-85, Union Carbide Corp; INVEST RECOVERY ASN CONSULT,86-, EMER PRES,87- *Concurrent Pos:* Pres, Invest Recovery Asn, 83-85. *Honors & Awards:* Bronze Medal & Chromium Plating Award, Am Electroplaters Soc, 67. *Mem:* Am Chem Soc; Am Inst Chem Engrs; Am Electroplaters Soc; Nat Soc Prof Engrs. *Res:* High pressure reactions; reaction kinetics; reactions of carbon monoxide; organometallic reactions; vapor plating; unit operations. *Mailing Add:* 120 Pine Cone Dr Huddleston VA 24104-2824

KNAPCZYK, JEROME WALTER, b Chicago, Ill, Sept 3, 38. CORROSION SCIENCE, ADHESION. *Educ:* Benedictine Col, BS, 60; Univ Mass, PhD(chem), 64. *Prof Exp:* Postdoctoral, Univ NC, 69, Duke Univ, 70, Univ Mass, 71 & 72; prof chem, Johnson State Col, 74-78, chmn, Div Sci & Math, 74-79; postdoctoral fel chem, Univ Mass, Amherst, 79-80. *Concurrent Pos:* Vis prof chem, Univ Mass, Amherst, 75 & 76. *Mem:* Am Chem Soc; Sigma Xi; Mat Res Soc. *Res:* Design of transparent, infrared reflecting films for glazing applications; mechanism of corrosion and transmission loss in optical stacks containing silver; adhesion modification of plastic substrates. *Mailing Add:* Monsanto Chem Co 730 Worcester St Springfield MA 01151

KNAPHUS, GEORGE, b McCallsburg, Iowa, Aug 31, 24; m 47; c 4. PLANT PATHOLOGY. *Educ:* Univ Northern Iowa, BA, 49; Iowa State Univ, MS, 51, PhD(plant path), 64. *Prof Exp:* Prin & teacher, High Sch, Iowa, 58-62; from instr to assoc prof bot, 62-72, PROF BOT, IOWA STATE UNIV, 72-, PROF SEC EDUC, 80- *Mem:* Am Phytopath Soc; Bot Soc Am; Am Inst Biol Sci; Nat Sci Teacher Asn. *Res:* Fungistasis of soil fungi, taxonomy and physiology of mushrooms. *Mailing Add:* Dept Bot Iowa State Univ 353 Bessay Ames IA 50011

KNAPKA, JOSEPH J, b Benton Pa, Jan 27, 35. PHYSIOLOGY. *Educ:* Univ Tenn, PhD(animal sci), 67. *Prof Exp:* NUTRITIONIST, DIV RES SERV, NIH, 67- *Mem:* Am Inst Nutrit; NY Acad Sci; Am Asn Lab Sci. *Res:* Animal nutrition. *Mailing Add:* Res Serv Div Bldg 14A Rm A109 NIH 900 Rockville Pike Bethesda MD 20205

KNAPP, ANTHONY WILLIAM, b Morristown, NJ, Dec 2, 41; m 63; c 2. LIE GROUPS, REPRESENTATION THEORY. *Educ:* Dartmouth Col, BA, 62; Princeton Univ, MA, 64, PhD(math), 65. *Prof Exp:* C L E Moore instr math, Mass Inst Technol, 65-67; from asst prof to prof math, Cornell Univ, 67-90; PROF MATH, STATE UNIV NY, STONY BROOK, 86- *Concurrent Pos:* Mem, Inst Advan Study, NJ, 68-69, 75-76 & 82-83; invited address, Int Congress Math, 74; res assoc, Princeton Univ, 71; prof d'échange, Univ de Paris-Sud, Orsay, 72; vis assoc prof, Rice Univ, 73; vis scholar, Univ Chicago, 81 & 83; prof associé, 82, prof invité, Univ Paris Seventh, 87; vis prof, Univ di Trento, Italy, 84, Tata Inst, Bombay, India, 88, Univ New SWales, Australia, 89, Univ Montréal, 90; foreign expert, Hunan Normal Univ, Changsha, PR China, 88. *Mem:* Am Math Soc. *Res:* Representations of semi-simple Lie groups. *Mailing Add:* Dept Math State Univ NY Stony Brook NY 11794-3651

KNAPP, CHARLES FRANCIS, b Evansville, Ind, Mar 28, 40; m 68; c 3. BIOENGINEERING. *Educ:* St Procopius Col, BA, 62; Univ Notre Dame, BS, 63, MS, 65, PhD(aerospace eng), 68. *Prof Exp:* From asst prof to prof mech eng, 68-85, PROF, BIOMED ENG CTR, UNIV KY, 85- *Concurrent Pos:* Prin investr grants, NIH, NASA, Air Force Off Sci Res; co-investr, NIH grants. *Mem:* Inst Elec & Electronics Engrs. *Res:* Frequency response characteristics of cardiovascular regulation; cardiovascular changes during oscillatory lower-body negative pressure; custom designed surgical implants and aids from CT scans; blood rheology. *Mailing Add:* Wenner-Gren Biomed Eng Ctr Univ Ky Lexington KY 40506

KNAPP, CHARLES H, b New York, NY, June 8, 31; m 55; c 4. ELECTRICAL ENGINEERING. *Educ:* Univ Conn, BSEE, 53, PhD(elec eng), 62; Yale Univ, ME, 56. *Prof Exp:* Eng trainee, RCA Victor Div, Radio Corp Am, 53; assoc engr, Res Div, Int Bus Mach Corp, 56-57; from instr to assoc prof, 57-74, PROF ELEC ENG, UNIV CONN, 74- *Concurrent Pos:* Consult, Elec Boat Div, Gen Dynamics Corp, 61-78, Unimation Inc, 80, US Navy Underwater Systs Ctr, 84- *Mem:* Inst Elec & Electronics Engrs. *Res:* Automatic control; estimation and identification; communications and signal processing. *Mailing Add:* Dept Elec Eng Box U-157 Univ Conn Storrs CT 06269

KNAPP, DANIEL ROGER, b Evansville, Ind, July 29, 43; m; c 2. PHARMACOLOGY. *Educ:* Univ Evansville, BA, 65; Ind Univ, Bloomington, PhD(org chem), 69. *Prof Exp:* NIH fel, Univ Calif, Berkeley, 69-70; asst prof exp med, Col Med, Univ Cincinnati, 71-72; from asst prof to assoc prof, 72-84, PROF PHARMACOL, MED UNIV SC, 84- *Mem:* Am Chem Soc; Am Soc Mass Spectrometry; Am Soc Pharmacol & Exp Therapeut; The Protein Soc. *Res:* Organoanalytical chemistry; mass spectrometry; protein structure. *Mailing Add:* Med Univ SC 171 Ashley Ave Charleston SC 29425

KNAPP, DAVID ALLAN, b Cleveland, Ohio, Feb 25, 38; m 62; c 1. PHARMACY, DRUGS & PUBLIC POLICY. *Educ:* Purdue Univ, BS, 60, MS, 62, PhD(pharm admin), 65. *Prof Exp:* From asst prof to assoc prof pharm admin, Col Pharm, Ohio State Univ, 64-71; assoc prof, 71-72, chmn, 73-79, dir, grad studies, 79-81, assoc dean, 81-83, PROF PHARM ADMIN, SCH PHARM, UNIV MD, 72-, CHMN, PHARM PRACT ADMIN, 87- *Concurrent Pos:* Vis scholar, Sch Pub Health, Univ Mich, 70-71; researcher, Nat Ctr Health Serv Res, Dept Health & Human Serv, 78; consult, Am Pharm Asn, 84-85; Schering-Plough scholar in residence, Am Asn Col Pharm, 86-87; mem bd dirs, Am Asn Col Pharm, 86- *Honors & Awards:* E H Volwiler Res Award, Am Asn Col Pharm, 86. *Mem:* Fel AAAS; fel Am Pharmaceut Asn; fel Am Pub Health Asn; Am Asn Col Pharm; Am Soc Hosp Pharm. *Res:* Applications of the social and administrative sciences to the drug component of medical care; cost and quality control methods in third party drug programs; drugs and public policy. *Mailing Add:* 11318 Cushman Rd Rockville MD 20852

KNAPP, DAVID EDWIN, b El Paso, Tex, July 9, 32; m 54; c 4. NUCLEAR PHYSICS. *Educ:* Calif Inst Technol, BS, 53; Univ Rochester, MA, 59, PhD(physics), 61. *Prof Exp:* Res scientist, Long Beach Div, Douglas Aircraft Co, Inc, 54-55, consult, 55-57, res scientist, 57-58, res scientist, Missile & Space Systs Div, 61-62, chief nuclear res br, 62-64, asst chief scientist nuclear dept, 64-69, chief scientist, Donald W Douglas Labs, Richland, Wash, 69-74, chief scientist, 74-76, PRIN STAFF ENGR, McDONNELL DOUGLAS ASTRONAUTICS CO, 76- *Mem:* AAAS; Am Phys Soc; Sigma Xi. *Res:* Elementary particle physics; energy conversion; nuclear power sources and propulsion. *Mailing Add:* 814 Market St Emporia KS 66801

KNAPP, EDWARD ALAN, b Salem, Ore, Mar 7, 32; m 54; c 4. PHYSICS. *Educ:* Pomona Col, BA, 54; Univ Calif, PhD(physics), 58. *Hon Degrees:* DSc, Pomona Col, 84, Bucknell Univ, 84. *Prof Exp:* Group leader, 59-68, asst div leader, 68-72, assoc div lab, Medium Engery Physics Div, 72-76, alt div leader, Physics Div, 76-77, div leader, Accelerator Technol Div, Los Alamos Sci Lab, Univ Calif, 78-82; dir, NSF, Washington, DC, 82-84; pres, Univs Res Asn, Washington, DC, 85-89; DIR, LOS ALAMOS MESON PHYSICS FACIL, LOS ALAMOS NAT LAB, 90- *Concurrent Pos:* Consult, Sci Applications, Inc & EMI Ther Systs, Inc, 71-; sr fel, NSF, 82- *Honors & Awards:* David Barrows Awards, Pomona Col, 88. *Mem:* AAAS; Am Phys Soc; Sigma Xi; Inst Elec & Electronics Engrs. *Res:* Medical application of accelerators and accelerator produced particles to cancer therapy; application of particle accelerators; high energy nuclear physics; photomeson processes and pi meson interactions; high energy linear accelerators, microwave cavities and related electromagnetic phenomena; applied physics; scientific administration. *Mailing Add:* Los Alamos Nat Lab MS-H850 Los Alamos NM 87544

KNAPP, FRANCIS MARION, b Caldwell, Idaho, Oct 17, 24; div; c 3. CARDIOVASCULAR PHYSIOLOGY, NEUROPHYSIOLOGY. *Educ:* Col Idaho, AB, 49; Univ Southern Calif, MS, 55, PhD(physiol), 60. *Prof Exp:* Asst physiol, Sch Med, Univ Southern Calif, 54-59; res assoc, Thudichum Lab, State Res Hosp, Galesburg, Ill, 61-64; from asst prof to assoc prof physiol & biol, Duquesne Univ, 64-70; chmn, Dept Biol, Stetson Univ, 70-77; dir acad affairs, Pa State Univ, New Kensington, 78-83; chmn, Dept Biol, 83-91, PROF BIOL, STETSON UNIV, 91- *Concurrent Pos:* NIH fel, Karolinska Inst, Sweden, 60-61. *Mem:* AAAS; Am Physiol Soc; Am Inst Biol Sci; Microcirc Soc; Am Soc Zool; Sigma Xi. *Res:* Cerebro-vascular and peripheral blood flow problems; neurophysiology and central nervous system; behavioral studies and drug action. *Mailing Add:* Dept Biol Stetson Univ De Land FL 32720-3756

KNAPP, FRED WILLIAM, b Princeton, Ill, Oct 14, 28; m 58; c 2. ENTOMOLOGY. *Educ:* Univ Ill, BS, 56; Kans State Univ, MS, 58, PhD(entom), 61. *Prof Exp:* Asst entom, Kans State Univ, 56-58, asst instr, 58-60, instr, 60-61; from asst prof to assoc prof, 61-71, PROF ENTOM, UNIV KY, 71- *Concurrent Pos:* Consult, USAID, pub health & indust, 61-85; entom adv, Agr Ctr Northeast, Thailand, 68-70; consult pesticide indust, 71-; pres, Am Registry Prof Entomologist, 87; pres, Ky Vector Control Asn & NCent Br, Entom Soc Am. *Honors & Awards:* Distinguished Med Vet Entomologist, 86; C V Riley Award, Entom Soc Am. *Mem:* Entom Soc Am; Am Mosquito Control Asn; Thailand Agr Soc; Sigma Xi. *Res:* Medical and veterinary entomology; insecticides; application methods and residues; pest management; integrated control of insects affecting man and animals. *Mailing Add:* Dept Entom Univ Ky Lexington KY 40546-0091

KNAPP, FREDERICK WHITON, b Danbury, Conn, Mar 19, 15; m 48. FOOD CHEMISTRY. *Educ:* Univ Calif, Davis, BS, 35, MS, 56, PhD(agr chem), 60. *Prof Exp:* Asst biochemist, Univ Fla, 60-67, assoc prof food sci & assoc biochemist, Inst Food & Agr Sci, 67-77; RETIRED. *Mem:* Inst Food Technologists. *Res:* Use and control of enzymes in food processing; protein recovery from animal by-products. *Mailing Add:* Dept Entom Univ Ky Lexington KY 40506

KNAPP, GAYLE, b Norwich, NY, July 31, 49. MOLECULAR BIOLOGY. *Educ:* Barnard Col, AB, 71; Univ Ill, PhD(biochem), 77. *Prof Exp:* Fel molecular biol, Dept Chem, Univ Calif, San Diego, 77-81; asst prof micromolecular biol, Dept Microbiol, Univ Ala, Birmingham, 81-88; ASST PROF, DEPT CHEM & BIOCHEM, UTAH STATE UNIV, 88- *Mem:* Am Chem Soc; Sigma Xi; Am Soc Microbiol; NY Acad Sci; AAAS. *Res:* Biosynthesis of eukaryotic (yeast) tRNA's, in particular those which arise through splicing of intron-containing RNA precursors; nucleic acid structure and how altering structure affects the biological activity of the nucleic acid. *Mailing Add:* Dept Chem & Biochem Utah State Univ Logan UT 84322-0300

KNAPP, GILLIAN REVILL, b 1944; m 68. RADIO ASTRONOMY. *Educ:* Univ Edinburgh, BSc, 66; Univ Md, PhD(astron), 72. *Prof Exp:* Asst astron, Univ Md, 66-72, teaching assoc, 72-74; res fel astron, 74-80, VIS ASSOC RADIO ASTRON, CALIF INST TECHNOL, 80- *Mem:* Royal Astron Soc; Am Astron Soc. *Res:* Interstellar microwave spectroscopy. *Mailing Add:* Dept Astrophys Sci Princeton Univ Princeton NJ 08544

KNAPP, GORDON GRAYSON, b Miami, Ariz, Nov 26, 30; m 55; c 4. ORGANIC CHEMISTRY. *Educ:* Ore State Col, BS, 52; Univ Wis, MS, 53, PhD(chem), 57. *Prof Exp:* Res chemist, Ethyl Corp, 56-63, res supvr, 66-69, res assoc, 70-85, sr res assoc, 85-89, RES ADV, ETHYL CORP, 89- *Mem:* Am Chem Soc. *Res:* Polymer research, especially epoxies and polyurethanes; hydrometallurgy; industrial organic chemicals. *Mailing Add:* Ethyl Corp PO Box 14799 Baton Rouge LA 70898

KNAPP, HAROLD ANTHONY, JR, technical administration; deceased, see previous edition for last biography

KNAPP, JOHN WILLIAMS, b Huntington, WVa, Dec 9, 32; m 57; c 3. CIVIL & SANITARY ENGINEERING. *Educ:* Va Mil Inst, BS, 54; Johns Hopkins Univ, MSE, 62, PhD(sanit eng, water resources), 65. *Prof Exp:* Admin asst, Chesapeake & Potomac Tel Co, 54; off engr, Concrete Pipe & Prod Co, 58-59; instr civil eng, Va Mil Inst, 59-61; res asst sanit eng, Johns Hopkins Univ, 61-64; from asst prof to assoc prof, 64-68, head dept, 66-71, PROF CIVIL ENG, VA MIL INST, 68- *Mem:* Am Soc Civil Engrs; Am Water Works Asn; Water Pollution Control Fedn; Nat Soc Prof Engrs. *Res:* Urban hydrology; economics and systems analysis; water supply and treatment; waste treatment and disposal; radioactive waste disposal. *Mailing Add:* Va Mil Inst Lexington VA 24450

KNAPP, JOSEPH LEONCE, JR, b New Boston, Tex, Nov 6, 37; m 57; c 2. AGRICULTURAL CHEMISTRY, INTEGRATED PEST MANAGEMENT. *Educ:* Miss State Univ, BS, 60, PhD(entom), 65; Kans State Univ, MS, 62. *Prof Exp:* Scientist host plant resistance entom, USDA, 62-65; supvr field entom, Int Minerals & Chem Co, 65; entomologist, Upjohn Co, 69-70, plant scientist, 70-77; mem staff, 77-80, PROF ENTOM & NEMATOL, UNIV FLA, 80- *Concurrent Pos:* consult, IPM, Grenada, Honduras, Egypt, Israel. *Mem:* Entom Soc Am; Am Registry Prof Entom. *Res:* Field research and development of insecticides, fungicides, herbicides and plant growth regulators for use on a wide range of agronomic crops. *Mailing Add:* Univ Fla CREC 700 Experiment Station Rd Lake Alfred FL 33850

KNAPP, KARL, mechanical engineering, physics, for more information see previous edition

KNAPP, KENNETH T, b Jacksonville, Fla, June 9, 30; m 54; c 3. ENVIRONMENTAL CHEMISTRY. *Educ:* Univ Fla, BS, 54, PhD(chem), 60. *Prof Exp:* Chemist, Res Div, Procter & Gamble Co, 60-63, chemist, Foods Div, 63-65; chemist, Int Latex & Chem Corp, 65-67; chemist, Southern Res Inst, 67-70; head anal sect, Vick Chem Co, Mt Vernon, NY, 70-71; chief nonmetal sect, Div Atmos Surveillance, 71-73, chief, Particulate Emissions Res Sect, 73-80, chief, Stationary Sources Emissions Res Br, 80-88, CHIEF, MOBILE SOURCES EMISSIONS RES BR, ATMOSPHERIC RES & EXPOSURE ASSESSMENT LAB, US ENVIRON PROTECTION AGENCY, RESEARCH TRIANGLE PARK, NC, 88- *Mem:* Am Chem Soc. *Res:* Instrumental analysis, especially gas chromatography, x-ray analysis; infrared spectroscopy, ultraviolet and visible spectroscopy, isolation and identification of naturally occurring compounds; measurement and characterization of air pollutants from source emissions. *Mailing Add:* 317 Glasgow Rd Cary NC 27511

KNAPP, LESLIE W, b Port Byron, NY, Nov 17, 29; m 57; c 2. ICHTHYOLOGY. *Educ:* Cornell Univ, BS, 52, PhD(vert zool), 64; Univ Mo, MA, 58. *Prof Exp:* Supvr vert, 63-68, dep dir, 68-72, supvr vert, oceanog sorting ctr, 72-81, dir, 81-88, SUPVR VERT, OCEANOG SORTING CTR, SMITHSONIAN INST, 88- *Mem:* Am Soc Ichthyol & Herpet; Soc Syst Zool. *Res:* Systematic ichthyology, particularly the families Percidae and Platycephalidae. *Mailing Add:* Oceanog Sorting Ctr Smithsonian Inst Washington DC 20560

KNAPP, MALCOLM HAMMOND, b Orange, NJ, Sept 20, 39; m 70; c 2. INDUSTRIAL ORGANIC CHEMISTRY. *Educ:* Rutgers Univ, BS, 61. *Prof Exp:* Chemist electrochem, Nuodex Div, Heyden Newport Corp, 64-66, sr chemist organometallics synthesis, Nuodex Div, Tenneco Chem, Inc, 67-70, lab mgr, 71-80, mgr lubricants, Res & Develop Dept, Tenneco Chem, Inc, 80-82 & Nuodex Inc, 83-85; mgr, Lubricants Res & Develop, Hüls Am, 85-91; CONSULT, 91- *Mem:* Soc Automotive Engrs; Am Soc Testing & Mat; Soc Tribologists & Lubrication Engrs. *Res:* Development of synthetic lubricants, novel base fluids, additives; test development. *Mailing Add:* 316 Raymond Ct Bridgewater NJ 08807

KNAPP, PETER HOBART, b Syracuse, NY, June 30, 16; m 56; c 6. PSYCHOSOMATICS, SCHIZOPHRENIA. *Educ:* Harvard Univ, BA, 37, MD, 41. *Prof Exp:* Instr psychiat, Sch Med Harvard Univ, 46-47 & Sch Med, Boston Univ, 47-51; from asst prof to assoc prof, 51-59, PROF PSYCHIAT, SCH MED, BOSTON UNIV, 59-; ATTEND PHYSICIAN PSYCHIAT, UNIV HOSP, BOSTON, 50- *Concurrent Pos:* Career investr, NIMH, 55-60, mem rev comt, 70-80 & 81-84; fel, Ctr Adv Study Behav Sci, 64-65. *Mem:* Am Psychiat Asn; Am Psychosom Soc (pres, 69-70); Am Psychoanal Asn; Am Acad Psychoanal; Am Col Psychanalysts. *Res:* Psychoanalytic and psychobiologic study of emotions with special reference to quantitative study of language as evidence for emotional processes in health and disease; especially allergic psychoimmunologic disorders; emotion. *Mailing Add:* 85 E Newton St M-810 Boston MA 02118

KNAPP, ROBERT HAZARD, JR, b Boston, Mass, May 18, 44; m; c 3. THEORETICAL PHYSICS, TECHNOLOGY STUDIES. *Educ:* Harvard Col, BA, 65; Oxford Univ, PhD(theoret physics), 68. *Prof Exp:* Res physicist, Carnegie-Mellon Univ, 68-70; lectr, Calif State Polytech Col, 70-72; mem fac physics, Evergreen State Col, 72- & asst acad dean, 76-79; res asst, Univ Col, London, 80; CONSULT, 80- *Concurrent Pos:* Mem, Nat Fac Humanities Arts & Sci, 87- *Honors & Awards:* Burlington Northern Award, 86. *Mem:* Am Phys Soc; Soc Values Higher Educ; AAAS. *Res:* Philosophy of education; energy and transportation; design of college-level interdisciplinary studies; physics and natural history. *Mailing Add:* Evergreen State Col Olympia WA 98505

KNAPP, ROBERT LESTER, b Keokuk, Iowa, Nov 17, 21; m 42; c 4. ORGANIC CHEMISTRY. *Educ:* Brown Univ, ScB, 43. *Prof Exp:* Jr res chemist, Naugatuck Chem Div, US Rubber Co, 46-48, sr res chemist plastics, 48-52, sr group leader, 52-57, sect mgr Kralastic res & develop, 57-60, prod supt, Synthetic Rubber Plant, 60-63, sales mgr Kralastic, 64-66; dir mkt, Uniroyal, Inc, 66-69, group mgr plastics res & develop, Uniroyal Chem Div, 69-76, dir bus develop & planning, 77-81, dir, Int Bus Develop, 81. *Concurrent Pos:* Consult, 81- *Mem:* Am Chem Soc; Soc Plastics Eng. *Res:* High polymers, both synthesis and physical behavior, particularly polyesters and gum plastics. *Mailing Add:* 189 Alder Lane PO Box 65 North Falmouth MA 02556-0065

KNAPP, ROGER DALE, b Natchez, Miss, Sept 6, 43. NUCLEAR MAGNETIC RESONANCE SPECTROSCOPY. *Educ:* Miss State Univ, BS, 65; Univ Houston, PhD(chem), 74. *Prof Exp:* ASST PROF, BAYLOR COL MED, 81- *Mem:* Am Soc Biochem & Molecular Biol. *Res:* Biopolymer structure by nuclear magnetic resonance; computer, instrument interface. *Mailing Add:* Dept Med Baylor Col Med 1200 Moursund Ave Houston TX 77030

KNAPP, ROY M, b Gridley, Kans, May 20, 40; m 62; c 3. PETROLEUM RESERVOIR ENGINEERING, MATHEMATICAL SIMULATION OF PETROLEUM PRODUCTION PROCESSES. *Educ:* Univ Kans, BS, 63, MS, 69, DEng, 73. *Prof Exp:* Dir opers res, Northern Nat Gas Co, 64-71; res asst, Ctr for Res Inc, 71-73; from asst prof to assoc prof petrol eng, Univ Tex, Austin, 73-78; assoc prof, 79-83, dir, Sch Petrol & Geol Eng, 79-88, PROF PETROL ENG, UNIV OKLA, 80- *Concurrent Pos:* Distinguished lectr, Soc Petrol Engrs, 80-81; mem res comt, Nat Inst Petrol & Eng Res, 83-84; mem bd dirs, US Nat Comt for World Energy Conf, 83-86. *Mem:* Soc Petrol Engrs of Am Inst Mining, Metall & Petrol Engrs; Am Soc Eng Educ; Am Petrol Inst; Int Asn Math & Comput Simulation. *Res:* Development and application of computer simulators for petroleum reservoirs and ground water hydrology; use of microorganisms for enhanced oil recovery. *Mailing Add:* Dept Geol Eng Univ Okla Main Campus Norman OK 73019

KNAPP, THEODORE MARTIN, b Berkeley, Calif, Sept 2, 47; div; c 2. EMERGENCY MEDICINE. *Educ:* Univ Calif, Berkeley, BA, 69; Univ Tenn, Knoxville, PhD(physiol psychol), 73. *Prof Exp:* Vis lectr & res assoc psychol, Univ Houston, 73-74; res assoc psycho-physiol, Baylor Col Med, 74-75; fel, Tex Res Inst Ment Sci, 75; assoc neuropsychologist, Midwest Res Inst, 75-81; asst prof allied health, Univ Kans, 81-86; RETIRED. *Concurrent Pos:* Consult ed, Physiol & Behav & Psychophysiol, 75- *Mem:* Soc Neurosci; Am Soc Allied Health Prof. *Res:* Learning styles, biofeedback, physiological reeducation, issues in emergency medical training. *Mailing Add:* 3561 E 17th No A San Francisco CA 94110-1002

KNAPP, WILLIAM ARNOLD, JR, b Atlanta, Ga, Oct 4, 25; m 50; c 3. VETERINARY PHARMACOLOGY, TOXICOLOGY. *Educ:* Univ Ga, DVM, 51, MS, 64. *Prof Exp:* Asst prof vet med & surg, Univ Ga, 51-52; pvt practice, 52-54; asst prof physiol & pharmacol, Sch Vet Med, Univ Ga, 54-62; dir res, Morris Res Labs, Inc, 62-65; assoc dir & res coordr, Toxicol Div, Hazleton Labs, Inc, 65-68, pres, Hazleton Res Animals, Inc, 68-71; dir,

Animal Sci-Prod Div, Flow Labs Inc, Rockville, Md, 71-78, pres, Flow Res Animals Inc, 71-77; vpres, Flow Labs Inc, McLean, Va, 76-83; CONSULT VET PHARMACEUT, AGRICHEM & RELATED INDUST, W A KNAPP ASSOC, 83- *Mem:* Am Soc Vet Physiol & Pharmacol; Vet Med Asn; Am Asn Lab Animal Sci; Indust Vet Asn (pres, 69-70); Am Col Vet Toxicologists. *Res:* veterinary pharmacology; drug evaluations; toxicology; nutrition; research administration and general management. *Mailing Add:* 3212 Queens Rd Raleigh NC 27612-6233

KNAPP, WILLIAM JOHN, ceramic engineering, materials science; deceased, see previous edition for last biography

KNAPPE, LAVERNE F, b Ellsworth, Wis, Jan 8, 22; m 44; c 3. MECHANICAL ENGINEERING, APPLIED MECHANICS. *Educ:* Univ Minn, BME, 44, MSME, 47, PhD(mech eng), 53. *Prof Exp:* Mech engr, Barber Colman Co, 47-50; instr mech eng, Univ Minn, 50-53; res engr, Am Mach & Foundry Co, 53-55; consult, Booz, Allen & Hamilton, Inc, 55-57; mgr mech anal lab, Int Bus Mach Corp, 57-70, SR ENGR, IBM CORP, 70- *Concurrent Pos:* Consult, Gen Mills, Inc, 51-53. *Mem:* Am Soc Mech Engrs; Soc Exp Stress Anal; NY Acad Sci. *Res:* Research and development of computer-aided mechanical design systems, including engineer-computer communication, analytical design procedures, system modeling and design optimization. *Mailing Add:* 501 SW 17th St Rochester MN 55902

KNAPPENBERGER, HERBERT ALLAN, b Reading, Pa, May 24, 32; m 57; c 3. INDUSTRIAL ENGINEERING, APPLIED STATISTICS. *Educ:* Pa State Univ, BS, 57, MS, 60; NC State Univ, PhD(exp statist), 66. *Prof Exp:* Apprentice draftsman, Textile Mach Works, Pa, 50-54; instr indust eng, Pa State Univ, 58-60; from instr to asst prof, NC State Univ, 60-68; from assoc prof to prof, Univ Mo-Columbia, 68-77; MEM STAFF, DEPT INDUST ENG, WAYNE STATE UNIV, 77-, CHMN OPERS RES, 81- *Concurrent Pos:* Mem health serv res training comt, Nat Ctr Health Serv Res & Develop. *Mem:* Am Inst Indust Engrs; Am Statist Asn; Opers Res Soc Am; Am Soc Eng Educ. *Res:* Health care systems design; patient scheduling systems design; automated radiology systems design; resource allocation on large systems. *Mailing Add:* Dept Indust Eng Wayne State Univ 5950 Cass Ave Detroit MI 48202

KNAPPENBERGER, PAUL HENRY, JR, b Reading, Pa, Sept 5, 42; m 63; c 2. SCIENCE ADMINISTRATION, ASTRONOMY. *Educ:* Franklin & Marshall Col, AB, 64; Univ Va, MA, 66, PhD(astron), 68. *Prof Exp:* Chmn dept astron, Fernbank Sci Ctr, Atlanta, 68-72; DIR, SCI MUS VA, 73- *Concurrent Pos:* Instr astron, Emory Univ & adj prof, Ga State Univ, 70-72; asst prof, Va Commonwealth Univ, 73-; adj assoc prof, Univ Richmond, 74-81; pres, Asn Sci & Technol Ctrs, 85; councilman, Nat Mus Act. *Mem:* Am Astron Soc; AAAS; Am Asn Mus; Int Coun Mus. *Res:* Astronomical interferometry; astronomical applications of image converters and intensifiers; development of educational activities in astronomy; design and evaluation of interactive exhibits in science; education in science museums. *Mailing Add:* Sci Mus Va 2500 W Broad St Richmond VA 23220

KNASTER, TATYANA, b Moscow, USSR, Sept 11, 33; US citizen; m 56; c 1. DURABILITY OF CONCRETE, STRUCTURAL DESIGN. *Educ:* Moscow Inst Transp Eng, BS, 56, MS, 58, PhD(civil eng), 65. *Prof Exp:* Struct engr, Design Inst, Moscow, 56-61; prof civil eng, Moscow Inst Transp Eng, 66-70; sr res scientist concrete durability, Res Inst Concrete & Reinforced Concrete, Moscow, 70-76; res assoc struct res, Univ Southern Calif, 76-77; struct engr, struct design, Ruthroff & Englekirk, Co, 77-78 & Day & Zimmerman, Inc, 78-79; CHAIRPERSON CIVIL ENG, PA INST TECHNOL, 79- *Concurrent Pos:* Part-time lectr, Moscow Inst Transp Eng, 65-66, Calif State Univ, 77-78. *Mem:* Am Concrete Inst. *Res:* Durability of building materials & structures, primarily concrete in severe weather conditions and in aggressive mediae. *Mailing Add:* Pa Inst Technol 800 Manchester Ave Media PA 19063

KNATTERUD, GENELL LAVONNE, b Minot, NDak. BIOSTATISTICS. *Educ:* Macalester Col, BA, 52; Univ Minn, MS, 59, PhD(biostatist), 63. *Prof Exp:* Asst biochemist, Pillsbury Mills Res Labs, 52-53; teaching asst anat, Univ Minn, 54-56, statistician, Biostatist Div, 56-57, sr statistician, 58-60, instr, Sch Pub Health, 60-62; anal statistician, Off Biomet, Consult Sect, NIMH, 63-64; asst prof epidemiol & biostatist, Pakistan Med Res Ctr, Univ Md, 66-67, from asst prof to assoc prof, Inst Int Med, 67-72, from assoc prof to prof, 72-84, RES PROF EPIDEMIOL & PREV MED, UNIV MD, BALTIMORE, 85- *Concurrent Pos:* Mem nat cancer adv comt, Nat Bladder-Prostate Cancer Projs, 72-74; vpres, Md Med Res Inst, 74-; mem lipid metab adv comt, Nat Heart & Lung Inst, 75-83. *Mem:* AAAS; Am Diabetes Asn; Am Pub Health Asn; Am Statist Asn; Biomet Soc. *Res:* Design, methods and applications of clinical trials; epidemiology of cardiovascular disease and diabetes. *Mailing Add:* 40 Bouton Green Baltimore MD 21210

KNAUER, BRUCE RICHARD, b New York, NY, Nov 24, 42. PHYSICAL-ORGANIC CHEMISTRY. *Educ:* Cooper Union, BChE, 63; Cornell Univ, MS, 65, PhD(chem), 69. *Prof Exp:* USAF Off Aerospace Res fel, Univ Ga, 68-70; asst prof, 70-78, chmn chem dept, 82-85, ASSOC PROF ORG CHEM, STATE UNIV NY COL ONEONTA, 78- *Mem:* Am Chem Soc. *Res:* Electron spin resonance; nitroxide free radicals; reaction mechanisms. *Mailing Add:* Dept Chem State Univ NY Col Oneonta NY 13820-4015

KNAUER, THOMAS E, EXPERIMENTAL BIOLOGY. *Prof Exp:* ATTY ENVIRON LAW, MCSWEENEY, BURTCH & CRUMP, 90- *Mailing Add:* McSweeney Burtch & Crump 11 S 12th St Richmond VA 23212

KNAUFF, RAYMOND EUGENE, b Venus, Pa, July 22, 25; m 49; c 1. ENDOCRINE BIOCHEMISTRY. *Educ:* Capital Univ, BS, 47; Univ Mich, MS, 49, PhD(biol chem), 52. *Prof Exp:* Chem technician, Barneby-Cheney Eng Co, 44; org chemist, Dow Chem Co, 45, anal chemist, 46, biochemist toxicol, 47; biochemist, Univ Mich Hosp, 47-49, asst biol chem, Med Sch, 49-50; endocrinologist, Upjohn Co, 51-55; head bioanal dept, G D Searle Co, 56-57; asst prof biochem, Univ Mich, 57-61; assoc prof biochem, Sch Med, Temple Univ, 61-74; prof biochem & chmn dept, 74-89, EMER PROF & CHMN, PHILADELPHIA COL OSTEOP MED, 89- *Concurrent Pos:* Dir res, Cystic Fibrosis Res Inst, 61-67. *Mem:* AAAS; Am Chem Soc; Am Inst Chemists; NY Acad Sci. *Res:* Biological chemistry; endocrine biochemistry; bioanalytical chemistry; protein and amino acid chemistry and metabolism; cystic fibrosis; nutritional biochemistry; eicosanoid biochemistry. *Mailing Add:* 37 Meade Rd Ambler PA 19002

KNAUFT, DAVID A, b Evergreen Park, Ill, May 10, 51; m 78; c 1. PLANT BREEDING, FARMING SYSTEMS. *Educ:* Univ Wis-Madison, BS, 73; Cornell Univ, PhD(plant breeding), 77. *Prof Exp:* Vis instr genetics, Agron & Soils Dept, Clemson Univ, 77-78; from asst prof to assoc prof, 78-90, PROF PLANT BREEDING & GENETICS, AGRON DEPT, UNIV FLA, 90- *Mem:* Am Soc Agron; Am Genetic Asn; Crop Sci Soc Am. *Res:* Genetic factors important in the improvement of cultivated peanuts, including genetic stability, response to stress environments, disease resistance and intercropping. *Mailing Add:* Agron Dept Univ Fla 304 Newell Hall Gainesville FL 32611-0311

KNAUS, EDWARD ELMER, b Leroy, Sask, Jan 7, 43; m 74. MEDICINAL CHEMISTRY. *Educ:* Univ Sask, BSP, 65, MSc, 67, PhD(pharmaceut chem), 70. *Prof Exp:* Med Res Coun Can fel chem, Tex A&M Univ, 70-71 & Univ BC, 71-72; asst prof, 72-80, PROF MED CHEM, UNIV ALTA, 80- *Honors & Awards:* McNeill Res Award. *Mem:* Am Chem Soc; Can Pharmaceut Asn; Chem Inst Can. *Res:* Synthesis of new nitrogen heterocycles and diagnostic agents; structure-activity studies; drug design. *Mailing Add:* Fac Pharm & Pharmaceut Sci Univ Alta Edmonton AB T6G 2N8 Can

KNAUS, RONALD MALLEN, b San Jose, Calif, June 9, 37; m 60; c 2. RADIOECOLOGY, RADIOBIOLOGY. *Educ:* San Jose State Univ, AB, 60, MA, 62; Ore State Univ, PhD(radiation biol), 71. *Prof Exp:* Teacher, Fremont Union High Sch Dist, 60-65 & Fresno City Col, 65-68; researcher biochem, Ore State Univ, 68-71; prof biol, Univ Tex, Arlington, 71-75; PROF & RESEARCHER RADIOECOL & RADIOBIOL, LA STATE UNIV, 75- *Concurrent Pos:* Consult, Comp Planning Inst, Dallas, 76- & City & Parish, East Baton Rouge, 77-81; prin investr, Lake Restoration Proj, City Baton Rouge, 77-, study marsh sediments, US Dept Interior & US Geol Surv; neutron activation anal biol sci, US Dept Energy; southeast regional dir, Sigma Xi. *Mem:* AAAS; Sigma Xi. *Res:* Investigation into the behavior of stable metal and rare earth tracers in the lotic environment; establish stable, activable tracers as soil horizon markers in fresh, brackish, and saltwater marshlands. *Mailing Add:* Nuclear Sci Ctr La State Univ Baton Rouge LA 70803

KNAUSENBERGER, WULF H, b Vienna, Austria, May 3, 43; US citizen; m 67; c 4. ELECTRONICS ENGINEERING. *Educ:* Pa State Univ, BS, 65, PhD(solid state sci), 69. *Prof Exp:* Res assoc, Pa State Univ, 69-70; mem tech staff, 70-78, TECH SUPVR, BELL LABS, 78- *Concurrent Pos:* Assoc ed, CHMT trans, Inst Elec & Electronics Engrs, 85- *Mem:* Inst Elec & Electronics Engrs; Int Electronics Packaging Soc. *Res:* Analysis of electronic system design, electronic packaging and interconnection system design. *Mailing Add:* AT&T Bell Labs 8C-017 Whippany NJ 07981

KNAUSS, JOHN ATKINSON, b Detroit, Mich, Sept 1, 25; m 54; c 2. OCEANOGRAPHY. *Educ:* Mass Inst Technol, BS, 46; Univ Mich, MA, 49; Univ Calif, Los Angeles, PhD(oceanog), 59. *Prof Exp:* Oceanographer, Navy Electronics Lab, 47-48; oceanogr, Off Naval Res, 49-51; res staff, Scripps Inst Oceanog, 51-53; oceanogr, Off Naval Res, 53-54; oceanogr, Scripps Inst Oceanog, 55-62; dean, Grad Sch Oceanog, Univ RI, 62-, provost marine affairs, 69-; AT JOINT OCEANOG INST DEEP EARTH SAMPLING, WASHINGTON, DC. *Concurrent Pos:* Mem, President's Comn Marine Sci, Eng & Resources, 67-68; chmn, Ocean Sci Comt, Nat Acad Sci, 71-73, Ocean Policy Comt, Nat Acad Sci, 72-, Univ Nat Oceanog Lab Syst, 74-75; mem, Nat Adv Comt Oceans & Atmosphere, 77-, chmn, 81- *Honors & Awards:* Sea Grant Asn Award, 74. *Mem:* Am Geophys Union; AAAS; Am Meteorol Soc. *Res:* Ocean circulation; law of the sea; marine affairs. *Mailing Add:* 3910 18th St NW Washington DC 20011

KNAVEL, DEAN EDGAR, b Windber, Pa, Sept 5, 24; m 47; c 3. HORTICULTURE. *Educ:* Pa State Univ, BS, 54; Univ Del, MS, 56; Mich State Univ, PhD(hort), 59. *Prof Exp:* From asst prof to assoc prof hort, 59-78, PROF HORT, UNIV KY, 78- *Mem:* Fel Am Soc Hort Sci. *Res:* Breeding, nutrition and minimum tillage of vegetable crops. *Mailing Add:* Dept Hort Univ Ky N318 Agr Sci Ctr N Lexington KY 40506

KNAZEK, RICHARD ALLAN, b Cleveland, Ohio, Mar 23, 42; m 67; c 2. MEDICINE, ENGINEERING. *Educ:* Case Inst Technol, BS, 62; Lehigh Univ, MS, 64; Ohio State Univ, MD, 69; Am Bd Internal Med, dipl, 74. *Prof Exp:* Engr cryogenic res, Air Prod & Chem, 62-63; engr plastics develop, E I du Pont de Nemours & Co, Inc, 63-65; intern med, Duke Hosp, 69-70, resident, 70-71; INVESTR MED, NIH, 71- *Concurrent Pos:* Vis fac, W A Jones Cell Sci Ctr, 73-76; vis lectr, Univ Toronto & Mass Gen Hosp, 74; contract officer, Breast Cancer Task Force, Nat Cancer Inst, 74-80; PhD thesis adv, Univ Del, 76-77. *Mem:* Endocrine Soc; Am Asn Cancer Res; Soc Exp Biol Med. *Res:* Developer of artificial capillary cell culture technique to grow solid organs in vitro; study of the control of prolactin receptors in liver and mammary cancer. *Mailing Add:* Nat Cancer Inst NIH Bldg 10 Rm 10N262 9000 Rockville Pike Bethesda MD 20892

KNEALE, SAMUEL GEORGE, b Tulsa, Okla, Dec 13, 21; m 45; c 2. MATHEMATICS. *Educ:* Univ Kans, AB, 47, MA, 48; Harvard Univ, PhD(math), 53. *Prof Exp:* Consult math, Philco Corp, Pa, 51-56, Gen Elec Co, 56-59 & Avco Corp, Ohio, 59-61; PRIN SCIENTIST MATH, OPERS RES INC, 61- *Mem:* Am Math Soc; Soc Indust & Appl Math; Math Asn Am; Opers Res Soc Am. *Res:* Applied mathematics, including probability theory and statistics, game theory, systems analysis, and other aspects of operations research. *Mailing Add:* 4111 Cheney Pl Wilmington NC 28412

KNEBEL, HARLEY JOHN, b Iowa City, Iowa, Nov 10, 41; m 69; c 2. GEOLOGICAL OCEANOGRAPHY, MARINE GEOLOGY. *Educ:* Univ Iowa, BA, 65; Univ Wash, MS, 67, PhD(oceanog), 72. *Prof Exp:* Res asst oceanog, Univ Wash, 65-67; oceanogr, Nat Oceanic & Atmospheric Admin, Atlantic Oceanog & Meteorol Labs, 67-69; res assoc, Univ Wash, 69-73; OCEANOGR, BR ATLANTIC MARINE GEOL, US GEOL SURV, 73-, ASSOC BR CHIEF, 85- *Concurrent Pos:* Texaco fel oceanog, Univ Wash, 71-72; mem adv bd, Geol Dept, Univ Iowa. *Mem:* AAAS; Am Geophys Union; fel Geol Soc Am; Sigma Xi. *Res:* Sedimentology; estuarine, nearshore, and continental shelf sedimentary processes and stratigraphy; statistics applied to geological oceanography; mass movements of sediments on continental slopes; clay mineralogy; submarine canyon development. *Mailing Add:* Br Atlantic Marine Geol US Geol Surv Woods Hole MA 02543

KNECHT, CHARLES DANIEL, b Halethorpe, Md, Mar 22, 32; wid; c 2. VETERINARY SURGERY, NEUROLOGY. *Educ:* Univ Pa, VMD, 56; Univ Md, College Park, BS, 60; Univ Ill, Urbana, MS, 66; Am Col Vet Surgeons, dipl, 68; Am Col Vet Internal Med, dipl & cert neurol, 74. *Prof Exp:* Assoc vet, Broad St Vet Hosp, Richmond, Va, 56; assoc vet, Wertz Mem Animal Hosp, Pittsburgh, 58-59; assoc vet, Towson Vet Hosp, Md, 59-64; from instr to asst prof vet med & surg, Col Vet Med, Univ Ill, Urbana, 64-68, assoc prof vet surg, 68-70; prof med & surg, Col Vet Med, Univ Ga, 70-72; prof & chief of surg, Sch Vet Sci & Med, Purdue Univ, West Lafayette, 72-79; PROF & HEAD, DEPT SMALL ANIMAL SURG & MED, COL VET MED, AUBURN UNIV, 79- *Concurrent Pos:* Abstractor, Chirurgia Veterinaria, WGer, 66-68 & Auburn Univ, 79-; mem grad fac, Univ Ill, Urbana, 68-70, Univ Ga, 71-72, Purdue Univ, 72-79, Auburn Univ, 80-; abstractor, J World & Europ Vet Surgeons, 68-72; pres neurol specialty, Am Col Vet Internal Med, 83-86; pres & chmn bd, Am Col Vet Surg, 88-90; pres elect, Am Asn Vet Clinicians, 90-91; assoc ed, Vet Med Report, 87-90. *Honors & Awards:* Norden Award, 76; Gaines Award, 82. *Mem:* Am Vet Med Asn; Am Col Vet Surg (pres, 88-89); Am Asn Vet Neurol (pres, 74-75); Am Asn Vet Clinicians; Am Animal Hosp Asn. *Res:* Orthopedic surgery; neurosurgery; electrodiagnostics. *Mailing Add:* Dept Small Animal Surg & Med Col Vet Med Auburn Univ Auburn AL 36849

KNECHT, DAVID JORDAN, b Elgin, Ill, June 2, 30; m 57; c 1. MAGNETOSPHERIC PHYSICS. *Educ:* Univ Ill, BS, 51, MS, 52; Univ Wis, PhD(physics), 58. *Prof Exp:* Proj assoc nuclear physics, Univ Wis, 58-59; proj officer physics div, Air Force Spec Weapons Ctr, 59-61, res physicist spec proj div, 61-63; sci dir space physics br, Air Force Weapons Lab, 63-64; res physicist, Space Physics Lab, Air Force Cambridge Res Labs, 64-75; PHYSICIST, SPACE PHYSICS DIV, GEOPHYS LAB, 75- *Mem:* Am Geophys Union. *Res:* Magnetospheric substorms and other disturbances using magnetic measurements made by ground networks and spacecraft; interactions of orbiting vehicles with space plasmas using spectroscopic observations made on space shuttle. *Mailing Add:* 56 South Rd Bedford MA 01730

KNECHT, LAURANCE A, b Elgin, Ill, Mar 16, 32; m 66; c 1. ANALYTICAL CHEMISTRY, PHYSICAL CHEMISTRY. *Educ:* Univ Ill, BS, 53; Univ Minn, PhD(anal chem), 59. *Prof Exp:* Instr chem, Iowa State Univ, 60-63; asst prof, Univ Cincinnati, 63-68; from assoc prof to prof chem, Marietta Col, 68-84; AT NC SCH SCI & MATH, 85- *Mem:* AAAS; Am Chem Soc. *Res:* Electroanalytical techniques. *Mailing Add:* NC Sch Sci & Math Durham NC 27705

KNECHT, WALTER LUDWIG, b Ludwigsburg, Ger, Feb 2, 09; US citizen; m 35; c 3. PHYSICS, CHEMISTRY. *Educ:* Univ Munich, MS, 32; Univ Berlin, PhD(quantum electronics), 34. *Prof Exp:* Res physicist, Fernseh A G, Berlin, 34-37; dir res high vacuum sci & microwave electronics, Flugfunk Forschungs Inst, Munich, 38-45; dir div head electronic mat, Dept Econ, Munich, 45-47; sect chief electron tube & molecular electronics, Air Force Aeronaut Systs Div, Ohio, 47-60; sr scientist, Air Force Avionics Lab, Dayton, 61-79; supvr physicist & tech area mgr, Air Force Mat Lab, Dayton, 79-85; RETIRED. *Concurrent Pos:* Mem, Armed Serv Electron Device Comt, Defense Dept, 49-60; consult, indust, univs & govt agencies, 50 -; mem adv group dir defense res, 54-62; chmn, Nat Conf Aerospace Electronics, 63; progressive space studies & curriculum, 85 - *Honors & Awards:* Lilienthal Award, 38; Air Force Res & Pub Awards, 55-67. *Mem:* Inst Elec & Electronics Engrs. *Res:* Laser components designed for satellite space laser communication systems; integrated optical circuits for compact electrooptical systems and progressive fiber optics; material purification; processing and development for military systems, including missles and high energy lasers; gallium aluminum arsenide lasers designed for electrooptical communication systems operating at room temperature; initiation of research program on materials processing in space, including high sensitivity detector materials. *Mailing Add:* 1616 Mercer Ct Yellow Springs OH 45387

KNECHTLI, RONALD (C), b Geneva, Switz, Aug 14, 27; nat US; m 53; c 3. PHYSICS. *Educ:* Swiss Fed Inst Technol, Dipl, 50, PhD(elec eng), 55. *Prof Exp:* Res engr, Brown Boveri & Co, Switz, 50-51; asst, Mass Inst Technol, 51-52; res engr, Brown Boveri & Co, 52-53; res engr, Res Labs, Radio Corp Am, 53-58; sr scientist, Res Labs, Hughes Aircraft Co, 58-86; RETIRED. *Honors & Awards:* Outstanding Work Res Award, RCA Lab; L A Hyland Patent Award. *Mem:* Am Inst Aeronaut & Astronaut; Am Phys Soc; Sigma Xi; Inst Elec & Electronics Engrs. *Res:* Photovoltaic and electrochemical devices. *Mailing Add:* 22929 Ardwick St Woodland Hills CA 91364

KNEE, DAVID ISAAC, b New York, NY, July 13, 34; m 83; c 2. TEACHING, TEACHER TRAINING. *Educ:* City Col New York, BS, 56; Mass Inst Technol, PhD(math), 62. *Prof Exp:* Mathematician, Arde Assocs, NJ, 56; instr math, Columbia Univ, 62-65; asst prof, 65-69, ASSOC PROF MATH, HOFSTRA UNIV, 69- *Concurrent Pos:* Dir, Teacher Training Inst, Hofstra, 86- *Mem:* Math Asn Am; Nat Coun Teachers Math. *Res:* Algebra; mathematics education; representation theory; mathematical linguistics. *Mailing Add:* Dept Math Hofstra Univ Hempstead NY 11550

KNEE, TERENCE EDWARD CREASEY, b Brussels, Belg, Apr 20, 32; m 56; c 2. PHYSICAL ORGANIC CHEMISTRY. *Educ:* Trinity Col, Dublin, BASc, 53; Mass Inst Technol, PhD(chem), 56. *Prof Exp:* Instr chem, Franklin Tech Inst, 55-56; res chemist, DuPont Co Can, 57-60, res supvr, E I du Pont de Nemours & Co, Inc, 60-70, sr supvr, 70-72, tech supt, 72-76, tech supt, Chattanooga Res & Develop Sect, 76-85; CONSULT, 85- *Mem:* Am Chem Soc. *Res:* Synthetic textile fibers; polymer chemistry; reaction mechanisms. *Mailing Add:* 228 Masters Rd Hixson TN 37343

KNEEBONE, LEON RUSSELL, b Bangor, Pa, May 28, 20; m 45; c 3. MYCOLOGY. *Educ:* Pa State Univ, BS, 42, PhD(bot), 50. *Prof Exp:* Asst bot, 47-50, from asst prof to prof, 50-78, EMER PROF BOT & PLANT PATH, PA STATE UNIV, 78- *Concurrent Pos:* Consult mushroom indust, int consult, nat govt Australia, Jamaica, Haiti & Dominician Repub; founder & gen chmn, mushroom indust short course, Pa State Univ, 56-78. *Mem:* AAAS; Bot Soc Am; Mycol Soc Am; Am Phytopath Soc; Am Inst Biol Sci; Am Mushroom Inst; Mushroom Grower Asn UK; Mushroom Grower Asn Can; Mushroom Grower Asn Australia; Int Soc for mushroom Sci; Sigma Xi. *Res:* Mushroom culture, especially spawn and strain development, diseases; edible fungi. *Mailing Add:* 628 Fairway Rd State College University Park PA 16803

KNEEBONE, WILLIAM ROBERT, b Eveleth, Minn, July 11, 22; m 48; c 3. AGRONOMY, PLANT BREEDING. *Educ:* Univ Minn, BS, 47, MS, 50, PhD(plant genetics), 51. *Prof Exp:* Asst grass breeding, Univ Minn, 47-50; res agronomist, Okla Agr Exp Sta & Crop Res Div Agr Res Serv, USDA, 51-63; PROF AGRON, UNIV ARIZ, 63- *Concurrent Pos:* Consult revegetation & golf course maintainence. *Mem:* Fel AAAS; Am Soc Agron; Coun Agr Sci & Technol; Soc Econ Bot; Sigma Xi. *Res:* Breeding; genetics; seed production; factors involved in stand establishment, vigor and spread of forage and turf grasses; water use by grasses. *Mailing Add:* 2491 N Camino De Oeste Tucson AZ 85745

KNEECE, ROLAND ROYCE, JR, b Tifton, Ga, Oct 15, 39; m 63. MATHEMATICS, OPERATIONS RESEARCH. *Educ:* Ga Inst Technol, BS, 61, MS, 62; Univ Md, PhD(math), 70. *Prof Exp:* Mem staff opers res, Inst Defense Anal, Arlington, 67-74; MEM STAFF, ACQUISITION & LOGISTICS, OFF SECY DEFENSE, 74-; OPER RES ANALYST, DEFENSE DEPT. *Concurrent Pos:* Prof lectr, Am Univ, 69- *Mem:* Am Math Soc; Asn Comput Mach. *Res:* Operator theory; functional analysis; strictly singular operators; computer technology. *Mailing Add:* 9235 Georgetown Pike Great Falls VA 22066

KNEIB, RONALD THOMAS, b Pittsburgh, Pa, Jan 29, 51. ECOLOGY. *Educ:* Pa State Univ, BS, 72; Univ NC, Chapel Hill, MA, 76, PhD(ecol), 80. *Prof Exp:* Res assoc, 81-83, asst res scientist, 84-86, ASSOC RES SCIENTIST, UNIV GA MARINE INST, 86- *Concurrent Pos:* Adj asst prof, Zool Dept, Univ Ga, 85-86, adj assoc res scientist, 86-; vis assoc prof, Col Marine Studies, Univ Del, 86. *Mem:* Ecol Soc Am; Am Fisheries Soc; Estuarine Res Fedn; Am Soc Zoologists; Am Soc Ichthyologists & Herpetologists; Sigma Xi. *Res:* Ecological interactions within and between populations of estuarine fishes and invertebrates; processes affecting recruitment and population dynamics in salt marshes. *Mailing Add:* UGA Marine Inst Sapelo Island GA 31327

KNEIP, G(EORGE) D(EWEY), JR, b Cleveland, Ohio, Jan 18, 25; m 49; c 2. METALLURGY. *Educ:* Case Inst Technol, BS, 45, MS, 48; Mass Inst Technol, ScD, 52. *Prof Exp:* Proj engr, S K Wellman Co, 50-53; head metallurgist, Carbide & Carbon Chem Co, 53-62; tech dir, Supercon Div, Nat Res Corp, Mass, 62-66; SR SCIENTIST, INSTRUMENT DIV, VARIAN ASSOCS, 66- *Mem:* Am Phys Soc; Am Soc Metals; Sigma Xi. *Res:* Physics of metals. *Mailing Add:* 24 Palm Court Menlo Park CA 94025

KNEIP, THEODORE JOSEPH, b St Paul, MN, Dec 20, 26; div; c 6. ANALYTICAL CHEMISTRY, ENVIRONMENTAL CHEMISTRY. *Educ:* Univ Minn, BCh, 50; Univ Ill, MS, 52, PhD, 54. *Prof Exp:* Chemist, Mallinckrodt Chem Works, 54-58, head, Anal Res Lab, 59-61, asst mgr uranium div, Anal Lab, 61-63, mgr, Lab Supply Res, 63-66; asst dir, 67-70, dep dir, 71-83, DIR, LAB ENVIRON STUDIES, NY UNIV MED CTR, 84- *Concurrent Pos:* Chmn & ed, CLSP subcomt biol monitoring-manual of method, Am Pub Health Asn,86-88; consult, US NIH, US Environ Protection Agency, Off Tech Assessment-Cong, Nat Oceanic & Atmospheric Admin, States of NY & NJ, NY City Dept Health & var industs; mem, environ health subcomt, NY Acad Med, 84-; mem, Intersoc Comt Methods Air Sampling & Anal, 87- *Mem:* NY Acad Sci; AAAS; Inst Standards Orgn (secy, 75-83); Am Chem Soc; Am Pub Health Asn; Am Indust Hyg Asn; Sigma XI. *Res:* Analysis, wet and instrumental; sampling and evaluation of natural and polluted environmental systems, air, water, biota, hazardous wastes; toxicological studies in aquatic and mammalian species. *Mailing Add:* NY Univ Med Ctr Inst Environ Med 550 First Ave New York NY 10016

KNELLER, WILLIAM ARTHUR, b Cleveland, Ohio, Apr 7, 29; m 51; c 4. ORGANIC PETROLOGY, ECONOMIC GEOLOGY. *Educ:* Miami Univ, AB, 51, MS, 55; Univ Mich, PhD(econ geol), 64. *Prof Exp:* Asst prof geol, Eastern Mich Univ, 56-59; prof geol & chmn dept, 61-85, dir Eitel Inst Silicate Res, 76-85, EMER PROF GEOL, UNIV TOLEDO, 89-, EMER DIR, EITEL INST SILICATE RES, 89- *Mem:* Fel Geol Soc Am; Soc Econ Paleontologists & Mineralogists; Soc Mining Eng; NAm Soc Thermal Anal. *Res:* Coal characterization for industrial use, coal petrology and petrography; industrial mineralogy; economic geology of industrial rocks and minerals; characterization of waste materials for industrial use; geochemistry of chert; concrete petrology. *Mailing Add:* 7027 Hickory Ridge Rd Sylvania OH 43560-1108

KNERR, REINHARD H, b Pirmasens, Ger, Feb 18, 39; m 68; c 4. MICROWAVES, LIGHTWAVE. *Educ:* Tech Univ Aachen, BS, 60; Enseeht, Toulouse France, dipl eng, 62; Lehigh Univ, MS, 64, PhD(elec eng), 68. *Prof Exp:* Mem tech staff, 68-79, SUPVR, AT&T BELL LABS, 79- *Concurrent*

Pos: Mem tech staff microwave, 68-79, supvr fiber optic components, 79-81, supvr integrated optics, 81-84, supvr lightwave technol; mem admin comt, Inst Elec & Electronics Engrs-Microwave Theory & Tech Soc, 80-, pres, 86, lectr, 88- *Mem:* Fel Inst Elec & Electronics Engrs. *Res:* Fiber optics; lightwaves; microwave circulators; power amps; integrated optics; local area networks; data interfaces. *Mailing Add:* AT&T Bell Labs 9999 Hamilton Blvd Breinigsville PA 18031

KNEVEL, ADELBERT MICHAEL, b St Joseph, Minn, Oct 20, 22; m 50; c 5. MEDICINAL CHEMISTRY. *Educ:* NDak State Univ, BS, 52, MS, 53; Purdue Univ, PhD(med chem), 57. *Prof Exp:* Instr pharmaceut chem, NDak State Univ, 53-54; from instr to assoc prof med chem, 54-65, asst dean, Sch Pharm & Pharm Sci, 68-75, PROF MED CHEM, SCH PHARM & PHARM SCI, PURDUE UNIV, WEST LAFAYETTE, 65-, ASSOC DEAN, 75- *Mem:* AAAS; Am Chem Soc; Am Asn Pharm Scientists; fel Am Asn Pharm Sci. *Res:* Methods development for drugs; drug metabolites in biological systems; studies of mechanism of drug action. *Mailing Add:* 62 Thise Ct West Lafayette IN 47905

KNIAZUK, MICHAEL, b Wilkes-Barre, Pa, June 12, 14; m 36. ELECTRONICS. *Educ:* NY Univ, BEE, 41. *Prof Exp:* Lab asst, 32-33, technician, 33-37, res assoc, 37-43, asst dir, Res Labs, 46-60, MGR BIOELECTRONICS LAB, MERCK INST THERAPEUT RES, 60- *Mem:* AAAS; Am Phys Soc; Inst Elec & Electronics Engrs. *Res:* Design and development of electronic instruments for biological research. *Mailing Add:* Merck & Co Inc Res PO Box 2000 Rahway NJ 07065

KNIAZZEH, ALFREDO G(IOVANNI) F(RANCESCO), b New York, NY, July 31, 38; m 68; c 2. MECHANICAL ENGINEERING, PHYSICAL CHEMISTRY. *Educ:* Mass Inst Technol, BS, 59, MS, 61, PhD(mech eng), 66. *Prof Exp:* Physicist, Electronics Res Ctr, NASA, 66-70; from scientist to sr scientist, 70-78, asst lab mgr, 78-84, PROD DEVELOP MGR, POLAROID CORP, 84- *Mem:* Am Phys Soc; Am Vacuum Soc; Electrochem Soc; Mat Res Soc. *Res:* Thin films; physical electronics; mechanics of materials; photochemistry; chemical kinetics; thermodynamics; radiation transport; fluid mechanics; direct energy conversion; electrochemistry; optics; solid state physics; electronics engineering. *Mailing Add:* 76 Prince St West Newton MA 02165

KNICKLE, HAROLD NORMAN, b Boston, Mass, Jan 6, 36; m 63; c 2. CHEMICAL ENGINEERING. *Educ:* Univ Mass, BSME, 62; Rensselaer Polytech Inst, MS, 65, PhD(nuclear eng), 69. *Prof Exp:* Engr nuclear eng, Knolls Atomic Power Lab, Gen Elec Co, 62-66; ASSOC PROF CHEM ENG, UNIV RI, 69- *Concurrent Pos:* Res prof, Pittsburgh Energy Technol Ctr, 77-80. *Mem:* Am Chem Soc; Am Inst Chem Engrs; Am Soc Eng Educ; Sigma Xi. *Res:* Mass transfer including azeotropic and extractive distillation, gas absorption, and leaching; heat transfer including single and two phase flow and insulation properties; design including mass transfer and heat transfer equipment; multiphase flow. *Mailing Add:* 99 Vaughn Ave Warwick RI 02886

KNIEBES, DUANE VAN, b Marquette, Mich, May 17, 26; m 50; c 3. ANALYTICAL CHEMISTRY, RESEARCH ADMINISTRATION. *Educ:* Mich State Univ, BS, 48; Ill Inst Technol, MS, 54. *Prof Exp:* Asst chemist, 49-51, supvr, Instrumental Anal Lab, Inst Gas Technol, 51-54, head Anal Div, 54-57, asst res dir, 57-62, assoc dir, 62-69, dir opers, 69-75, asst vpres educ serv, 75-84; CONSULT, 84- *Honors & Awards:* Merit Award, Am Soc Testing & Mat; Gold Merit Award, Am Gas Asn. *Mem:* Am Chem Soc; Am Soc Testing & Mat; Am Gas Asn. *Res:* Odorization of natural and liquid propane gases; analysis of gaseous fuels; management of research and research facilities; design and operation of engineering education and technician training programs in natural gas technology. *Mailing Add:* 4612 Hampshire St Boulder CO 80301-4211

KNIEF, RONALD ALLEN, b Hinsdale, Ill, Oct 8, 44; m 83. REACTOR & FUEL FACILITY SAFETY, EDUCATION & TRAINING & RISK MANAGEMENT. *Educ:* Albion Col, BA, 67; Univ Ill, PhD(nuclear eng), 72. *Prof Exp:* Sr physicist, Combustion Eng, Inc, 72-74; from asst to assoc prof nuclear eng, Univ NMex, 74-80; mgr plant training, Three Mile Island, GPU Nuclear Corp, 80-83, mgr educ develop, 83-85, co-chair, prog safety comn, 85-86, mgr corp training, 86-87, staff consult, 87-90; PRIN CONSULT, ERC ENVIRON & ENERGY SERV CO, 90- *Concurrent Pos:* Adj prof, Univ Hartford, 72-74; consult, Sandia Nat Lab, 74-81, Babcock & Wilcox Naval Nuclear Fuel Div & Westinghouse Goco Nuclear Safety Comt; adj prof, Univ NMex & Pa State Univ, 80- *Honors & Awards:* Achievement Award, Am Nuclear Soc, Nuclear Criticality Safety Div, 83. *Mem:* Am Nuclear Soc; Sigma Xi; Inst Nuclear Mat Mgt; AAAS. *Res:* Nuclear reactor technology and safety; nuclear fuel facility safety and risk management; education and training development. *Mailing Add:* PO Box 7010 Mechanicsburg PA 17055-7010

KNIEVEL, DANIEL PAUL, b West Point, Nebr, Jan 29, 43; m 65; c 2. CROP PHYSIOLOGY, STRESS PHYSIOLOGY. *Educ:* Univ Nebr, Lincoln, BS, 65; Univ Wis-Madison, MS, 67, PhD(agron, biochem), 68. *Prof Exp:* Crop physiologist plant sci, Univ Wyo, 68-72; asst prof, 72-75, ASSOC PROF CROP PHYSIOL, PA STATE UNIV, UNIVERSITY PARK, 75- *Concurrent Pos:* NSF grant, 69-72; crop sci adv, Pa State Univ-USAID, Grad Sch Develop Proj Arg, 72-74; Postgrad Develop Proj Sri Lanka, 85; vis scientist, Sea Educ Admin, USDA, 80-81; assoc ed, Agron J, 78-83; bd dirs, Coun Agr Sci & Technol, 87-; mem, rev panel, USDA-SBIR grant prog, 88, topic mgr, 89. *Mem:* Am Soc Agron; Crop Sci Soc Am; Am Soc Plant Physiol; Coun Agr Sci & Technol; Sigma Xi; Soil Sci Soc Am. *Res:* Physiology of assimilate transport in plants; physiology of crop response to environmental stress; computer simulation of crop growth and development; phloem unloading and sink metabolism. *Mailing Add:* Dept Agron Pa State Univ University Park PA 16802

KNIFFEN, DONALD AVERY, b Kalamazoo, Mich, Apr 27, 33; m 52; c 3. ASTROPHYSICS, GAMMA-RAY ASTRONOMY. *Educ:* La State Univ, BS, 59; Wash Univ, St Louis, MA, 60; Cath Univ Am, PhD, 67. *Prof Exp:* ASTROPHYSICIST, GODDARD SPACE FLIGHT CTR, NASA, 60- *Concurrent Pos:* Lectr, Univ Md, 78-87; proj scientist, Gamma Ray Observ, 79-; vis scientist, CSIRO Div Radiophys, Australia, 87-88. *Mem:* AAAS; Am Astron Soc; Am Phys Soc; Am Geophys Union. *Res:* Galactic and solar cosmic rays, including both charged particles and gamma rays; trapped radiation, pulsars. *Mailing Add:* Code 662 Goddard Space Flight Ctr Greenbelt MD 20771

KNIGGE, KARL MAX, b Brooklyn, NY, July 17, 26; m 48; c 3. NEUROENDOCRINOLOGY. *Educ:* Rutgers Univ, BS, 50; Univ Mich, PhD(anat), 53. *Prof Exp:* Instr anat, Univ Pittsburgh, 53-55; asst prof, Univ Calif, Los Angeles, 55-59; from assoc prof to prof, Univ Cincinnati, 59-65; prof anat & chmn dept, Sch Med & Dent, 65-79, PROF & DIR, NEUROENDOCRINE UNIT, UNIV ROCHESTER, 80- *Mem:* Endocrine Soc; Am Physiol Soc; Am Asn Anatomists; Int Soc Neuroendocrinol; Soc Neurosci. *Res:* Neuroendocrinology. *Mailing Add:* 2561 Clover Rochester NY 14618

KNIGHT, ALAN CAMPBELL, b Hartford, Conn, Nov 2, 22; m 48; c 2. POLYMER SCIENCE. *Educ:* Ore State Col, BS, 48; Univ Calif, Berkeley, PhD(chem), 50. *Prof Exp:* Chemist, E I du Pont de Nemours & Co, Inc, Wilmington, 50-55, asst tech supt mfg, 55-56, tech supt, 56-58, sr supvr res & develop, 58-67, res assoc, 67-73, res assoc, Polymer Prod Dept, Wash Lab, 73-86; RETIRED. *Mem:* Am Chem Soc. *Res:* Reaction kinetics and mechanisms; polymer synthesis; degradation mechanisms; stabilization; relation of structure to properties; manufacture of heavy organic chemicals; applied mathematics; computer applications; applied physical theory. *Mailing Add:* Rte 1 Box 231 Parkersburg WV 26101

KNIGHT, ALLEN WARNER, b Grand Rapids, Mich, Feb 7, 32; m 55; c 4. AQUATIC ECOLOGY, WATER POLLUTION. *Educ:* Western Mich Univ, BS, 59; Mich State Univ, MS, 61; Univ Utah, PhD(zool), 65. *Prof Exp:* Asst prof entom & zool, Mich State Univ, 65-68; from asst prof to assoc prof, 68-76, PROF HYDROBIOL, UNIV CALIF, DAVIS, 76- *Concurrent Pos:* Consult, Stanford Res Inst, 70-75; Calif State Water Resources Control Bd, 76-81 & Wickland Oil Co, 81-; mem, Inst Ecol, Univ Calif. *Mem:* AAAS; Ecol Soc Am; Am Soc Limnol & Oceanog; Inst Soc Limnol; Entom Soc Am. *Res:* Pollution ecology; effect of environmental factors and pollutants on aquatic life; aquaculture, culture of freshwater prawn; clam (Corbicula); growth metabolism studies of aquatic life; hydrobiology. *Mailing Add:* Dept Land Air & Water Resources Univ Calif Davis CA 95616

KNIGHT, ANNE BRADLEY, b Durham, NC, July 7, 42. BIOCHEMISTRY, PHYSIOLOGY. *Educ:* Univ NC, Chapel Hill, BA, 64, PhD(biochem), 72. *Prof Exp:* Res technician membrane transport, Med Sch, Univ NC, 64-67, res asst, 71-72; res assoc erythrocyte metab, Dept Physiol, Duke Univ, 72-75; instr anat & physiol, NC Cent Univ, 75; ASST PROF ANAT & PHYSIOL, WINTHROP COL, 75- *Concurrent Pos:* NIH res fel, Duke Univ, 72-74. *Mem:* Sigma Xi; Biophys Soc; AAAS. *Res:* Erythrocytes; comparative metabolism and transport; cellular volume regulation; interaction of active and passive transport with metabolism. *Mailing Add:* 201 MC CLG St Bennettsville SC 29512

KNIGHT, ARTHUR ROBERT, b St John's, Nfld, Feb 24, 38; m 59, 86; c 8. PHOTOCHEMISTRY. *Educ:* Mem Univ Nfld, BSc, 58, MSc, 60; Univ Alta, PhD(chem), 62. *Prof Exp:* Fel chem, Univ Alta, 62-64; from asst prof to assoc prof chem, Univ Sask, 64-76, head dept, 76-81, dean, Col Art & Sci, 81-90, PROF CHEM, UNIV SASK, 76-, ACTG ASSOC VPRES ACAD, 90- *Mem:* Fel Can Inst Chem. *Res:* Reactions of radicals produced in photolytic decompositions; primary process studies in photolyses; photochemistry and photophysics of sulfur containing compounds. *Mailing Add:* Off Vpres Acad Univ Sask Saskatoon SK S7N 0W0 Can

KNIGHT, BRUCE L, b Kansas City, Mo, Jan 4, 42; m 64; c 2. CHEMICAL & PETROLEUM ENGINEERING. *Educ:* Univ Kans, BS, 64; Univ Colo, MS, 65, PhD(chem eng), 69. *Prof Exp:* Res engr, Denver Res Ctr, 69-75, sr petrol engr, 75-77, environ coordr, 77-81, SR RES ENGR, MARATHON OIL CO, 81- *Mem:* Am Inst Chem Engrs; Soc Petrol Engrs; Am Petrol Inst; Sigma Xi. *Res:* Cryogenic heat transfer through porous media; tertiary oil recovery processes; production logging. *Mailing Add:* Marathon Oil Co PO Box 269 Littleton CO 80160

KNIGHT, BRUCE WINTON, (JR), b Milwaukee, Wis, Dec 4, 30; m 73; c 2. BIOPHYSICS, APPLIED MATHEMATICS. *Educ:* Dartmouth Col, BS, 52. *Prof Exp:* Staff mem, Los Alamos Sci Lab, 55-61; MEM FAC, ROCKFELLER UNIV, 61- *Res:* Neurophysiology of vision; applied theoretical physics. *Mailing Add:* Dept Biophys Rockefeller Univ 1230 York Ave New York NY 10021

KNIGHT, CHARLES ALFRED, b Chicago, Ill, Mar 28, 36; m 62. CLOUD PHYSICS. *Educ:* Univ Chicago, MS, 57, PhD(geol), 59. *Prof Exp:* Res scientist arctic ice, Univ Wash, 59-61; prog scientist cloud physics, Lab Atmospheric Sci, 62-74, SR SCIENTIST, NAT CTR ATMOSPHERIC RES, 76- *Concurrent Pos:* Mem fac, Colo State Univ, 69-, Univ Colo, 73- & Univ Wyo, 77- *Honors & Awards:* Publications Prize, Nat Ctr Atmospheric Res, 70. *Mem:* Am Meteorol Soc; Am Geophys Union; Am Asn Crystal Growth; Soc Cryobiology; Glaciol Soc. *Res:* Structure of hail; hail formation in relation to severe storm structure; hail suppression; ice crystal nucleation and growth. *Mailing Add:* Nat Ctr Atmospheric Res Boulder CO 80307

KNIGHT, CLIFFORD BURNHAM, b Rockville, Conn, Jan 6, 26; m 56, 80; c 2. INSECT ECOLOGY. *Educ:* Univ Conn, BA, 50, MA, 52; Duke Univ, PhD(invert ecol), 57. *Prof Exp:* Instr biol, East Carolina Univ, 56-57, from asst prof to assoc prof zool, 57-64, dir grad studies biol, PROF ZOOL, EAST CAROLINA UNIV, 64-, DIR UNDERGRAD STUDIES BIOL, 89- *Mem:*

Ecol Soc Am; Nat Audubon Soc; Am Inst Biol Sci; Nat Wildlife Fed; Am Entom Soc. *Res:* Ecology of Collembola in forest communities of North Carolina; benthic invertebrate estuarine ecology. *Mailing Add:* Dept Biol ECarolina Univ Greenville NC 27858-4353

KNIGHT, DAVID BATES, b Louisville, Ky, Sept 23, 39; m 65; c 2. ORGANIC CHEMISTRY. *Educ:* Univ Louisville, BS, 61; Duke Univ, MA, 63, PhD(org chem), 66. *Prof Exp:* Vis res assoc chem, Ohio State Univ, 66-67 & 77-78; from asst prof to assoc prof, 67-82, PROF CHEM, UNIV NC, GREENSBORO, 82- *Mem:* AAAS; Am Chem Soc. *Res:* Protium-deuterium exchange in hydrocarbons; chemistry of fulvenes. *Mailing Add:* Dept Chem Univ of NC Greensboro NC 27412

KNIGHT, DENNIS HAL, b Clear Lake, SDak, Dec 24, 37; m 67; c 2. PLANT ECOLOGY. *Educ:* Augustana Col, SDak, BA, 59; Univ Wis-Madison, MS, 61, PhD(bot), 64. *Prof Exp:* Instr bot, with Peace Corps, Loja, Ecuador, 64-66; from asst prof to assoc prof, 66-79, PROF BOT, UNIV WYO, 79- *Mem:* AAAS; Ecol Soc Am; Am Inst Biol Sci; Soc Range Mgt; Sigma Xi. *Res:* Ecology of Great Plains and Rocky Mountain vegetation; tropical plant ecology; impact of vegetation structure on ecosystem function. *Mailing Add:* Dept Bot Univ Wyo Laramie WY 82071

KNIGHT, DOUGLAS MAITLAND, b Cambridge, Mass, June 8, 21; m 42; c 4. EDUCATIONAL ADMINISTRATION. *Educ:* Yale, AB, 42, MA, 44, PhD, 46. *Hon Degrees:* LLD, Ripon Col, Knox Col, Davidson Col, 63, Univ NC, 65, Emory Univ, 65, Ohio Wesleyan Univ, 70, Centre Col, 73; LHD, Lawrence Univ, 64. *Prof Exp:* Instr English, Yale, 46-47, asst prof, 47-53; pres, Lawrence Col, Appleton, Wis, 54-63 & Duke Univ, Durham, NC, 63-69; div vpres educ develop, RCA, New York, 69-71; educ serv, 71-72; staff vpres educ & community relations, 72-73; consult, 73-75; pres, Iran, 71-72; dir, 71-73; PRES SOCIAL, ECON & EDUC DEVELOP, INC, 73-; DIR & PRES, QUESTAR CORP, 76- *Concurrent Pos:* Morse res fel, 51-52; bd dirs, Woodrow Wilson Nat Fel Found, 59-, chmn, 82; US deleg, SEATO Conf Asian Univ, Pres, Pakistan, 61, Nat Comn Sci & Eng Manpower, 65 & Nat Comn, UNESCO, 65-67; mem, Corp Mass Inst Technol, 65-70; chmn, Nat Adv Comn Libr, 66-68; adv, Imp Orgn Social Serv Govt Iran, 70-77; chmn, Near East Found, 75-, Int Sch Serv, 75-81, Solebury Sch, 75-83, Nat Comn Higher Educ Issues, 81-; trustee, Questar Libr Sci & Art, 81- *Res:* Author and editor of seven books and over 100 articles. *Mailing Add:* 68 Upper Creek Rd Stockton NJ 08559

KNIGHT, DOUGLAS WAYNE, b Batavia, NY, Oct 7, 38; m 61; c 3. COMPUTER SCIENCE. *Educ:* Ariz State Univ, BS, 61, MS, 69, PhD(elec eng), 75. *Prof Exp:* Geophysicist seismol, Shell Oil Co, 61-67; fac assoc programming, Ariz State Univ, 71-73; contractor, Govt Electronics Div, Motorola, 73-74; programming suprvr, Trans Test Ctr, Dynalectron Corp, 74-79; prog mgr, Kentron Int, 79; dept chmn, 80-90, ASSOC PROF, COMPUTER SCI TECHNOL FAC, UNIV SOUTHERN COLO, 91- *Concurrent Pos:* Referee, Fed Info Processing Standard, 73; owner, Computerland, Colorado Springs, 78-81. *Mem:* Data Processing Mgt Asn; Asn Comput Mach. *Res:* Microprocessing operating systems; real-time dynamic structures; role playing simulations. *Mailing Add:* Univ Southern Colo 2200 N Bonforte Blvd Pueblo CO 81001

KNIGHT, FRANK B, b Chicago, Ill, Oct 11, 33; m 70; c 3. MATHEMATICS, MATHEMATICAL STATISTICS. *Educ:* Cornell Univ, BA, 55; Princeton Univ, PhD(math), 59. *Prof Exp:* Res asst math, Univ Minn, 59-60, from instr to asst prof, 60-63; from asst prof to assoc prof, 63-69, PROF MATH, UNIV ILL, URBANA, 69- *Res:* Probability theory; continuous time stochastic processes. *Mailing Add:* Dept Math 121C Altgeld Hall Univ Ill Urbana IL 61801

KNIGHT, FRED BARROWS, b Waterville, Maine, Dec 12, 25; m 45; c 3. FOREST ENTOMOLOGY. *Educ:* Univ Maine, BSF, 49; Duke Univ, MF, 50, DF(forest entom), 56. *Prof Exp:* Entomologist, Bur Entom & Plant Quarantine, NC, 50-51 & Colo, 51-54; entomologist, Forest Insect & Dis Lab, US Forest Serv, Colo, 54-60; from assoc prof to prof forestry, Univ Mich, 60-72, chmn dept, 66-70; dir, Sch Forest Resources, Univ Maine, Orono, 72-83, assoc dean, 83-86, prof, 72-90, dean, 86-90, EMER PROF & DEAN, FOREST RESOURCES, UNIV MAINE, ORONO, 91- *Concurrent Pos:* Vis prof, Sch Forestry, Univ Canterbury, 70; from vpres to pres, Forestry Res Orgn, Asn State Col & Univ, 76-80; assoc dir, Maine Agr Exp Sta, Univ Maine, Orono, 76-83 & 86-90. *Mem:* Fel AAAS; fel Soc Am Foresters; Entom Soc Am; Soil & Water Conserv Soc; Ecol Soc Am; Forestry Hist Soc. *Res:* Forest insect population, biological and silvicultural control; silviculture; forest ecology. *Mailing Add:* Col Forest Resources Univ Maine Orono ME 04469

KNIGHT, FRED G, b Fargo, NDak, May 23, 20; div; c 2. EARTH SCIENCES. *Educ:* Colo Sch Mines, Geol Engr, 42. *Prof Exp:* Computer-party chief, Seismograph Serv Corp, 42-46; geophysicist, 47-53, dist geologist, Gulf & West Coasts, 53-61, adv sr staff geologist, 61-65, div explor mgr, 65-67, coord mgr explor, US & Can, 67-72, ASSOC RES DIR EXPLOR, MARATHON OIL CO, 72- *Mem:* Am Asn Petrol Geologists; Soc Explor Geophysicists; Sigma Xi. *Mailing Add:* 6624 S Prescott Littleton CO 80120

KNIGHT, GLENN B, IMMUNOLOGY. *Prof Exp:* STAFF RESEARCHER, LAHEY CLIN MED CTR, 88- *Mailing Add:* Dept Immunol Res & Molecular Biol Lahey Clin Med Ctr 41 Mall Rd Burlington MA 01805

KNIGHT, GORDON RAYMOND, electrooptics, electrical engineering, for more information see previous edition

KNIGHT, HOMER TALCOTT, b Rochelle, Ill, June 2, 23; m 48; c 2. ANALYTICAL CHEMISTRY. *Educ:* Northern Ill State Teachers Col, BS, 47; Colo State Col, MA, 48; Univ Ill, MS, 50; Univ Wis, PhD(chem), 52. *Prof Exp:* Chemist, Fansteel Metall Corp, Ill, 52-57, asst dir res, 57-63, tech dir,

63-65; tech asst to gen mgr, Gen Instruments Corp, SC, 65; mgr, Newport Facil, Elpac, Inc, 65-68; asst prof, 69-74, ASSOC PROF ANALYTICAL CHEM, UNIV WIS-PARKSIDE, 74- *Mem:* Am Chem Soc; Soc Appl Spectros. *Mailing Add:* Univ Wis-Parkside PO Box 2000 Kenosha WI 53141-2000

KNIGHT, JAMES ALBERT, JR, b La Grange, Ga, Oct 16, 20; m 48; c 3. ORGANIC CHEMISTRY. *Educ:* Wofford Col, BS, 42; Ga Inst Technol, MS, 44; Pa State Univ, PhD(chem), 50. *Prof Exp:* From assoc prof to assoc prof chem, Ga Tech Res Inst, Ga Inst Technol, 50-58, from res assoc prof to res prof, 58-85; RETIRED. *Concurrent Pos:* Consult, Indust Firms, 70-, Nat Bur Standards, 80-81 & UNFAD, 84-85. *Mem:* AAAS; Forest Prod Res Soc; Am Chem Soc; Int Union Pure & Appl Chem. *Res:* Pyrolytic and carbon technologies utilizing agricultural and forestry materials; radiation chemistry of organic systems; synthetic organic chemistry; gas and liquid chromatography. *Mailing Add:* 2117 Kodiak Dr NE Altanta GA 30345

KNIGHT, JAMES ALLEN, b St George, SC, Oct 20, 18; m 63; c 1. PSYCHIATRY. *Educ:* Wofford Col, AB, 41; Duke Univ, BD, 44; Vanderbilt Univ, MD, 52; Tulane Univ, MPH, 62. *Prof Exp:* Intern, Grady Mem Hosp, Atlanta, Ga, 52-53; asst resident pediat & obstet, Duke Univ Hosp, 53-54; instr psychiat, Sch Med, Tulane Univ, 55-58; asst prof, Col Med, Baylor Univ, 58-61, asst dean, 60-61; assoc prof, Sch Med, Tulane Univ, 61-63; prof psychiat, Union Theol Sem, NY, 63-64; prof psychiat & assoc dean, Sch Med, Tulane Univ, 64-74; dean, Col Med, Tex A&M Univ, 74-77; PROF PSYCHIAT & MED ETHICS, LA STATE UNIV SCH OF MED, NEW ORLEANS, 78- *Concurrent Pos:* Resident, Tulane Serv, Charity Hosp, New Orleans, 55-58, chief resident, 57-58. *Mem:* Am Psychiat Asn; Acad Psychoanal; Group Advan Psychiat. *Res:* Interrelationships of religion and psychiatry; suicide; motivation; psychosomatic medicine; ethics and human values; medical student maturation. *Mailing Add:* La State Univ Sch Med 1542 Tulane Ave New Orleans LA 70112-2822

KNIGHT, JAMES MILTON, b Jacksonville, Fla, Feb 20, 33; m 67; c 2. THEORETICAL PHYSICS. *Educ:* Spring Hill Col, BS, 54; Univ Md, PhD(physics), 60. *Prof Exp:* Res assoc & instr physics, Univ Md, 60-61; res assoc, Duke Univ, 63-65; assoc prof physics, 65-77, PROF PHYSICS & ASTRON, UNIV SC, 77- *Mem:* Am Phys Soc. *Res:* Quantum optics; chaotic dynamics; symmetry properties; phase transitions. *Mailing Add:* Dept Physics Univ SC Columbia SC 29208

KNIGHT, JAMES WILLIAM, b Alexandria, La, Nov 27, 48. REPRODUCTIVE PHYSIOLOGY. *Educ:* Univ Southwestern La, BS, 70; Univ Fla, MS, 72, PhD(reprod physiol), 75. *Prof Exp:* Fel reprod physiol, Univ Mo, 75-76; PROF REPROD PHYSIOL, VA POLYTECH INST & STATE UNIV, 76- *Concurrent Pos:* Sabbatical, Fed Repub of Ger, 88-89. *Mem:* Am Soc Animal Sci; Soc Study Reprod; AAAS. *Res:* Conceptus-maternal interrelationships; uterine protein secretions; placental function; endocrinology of gestation. *Mailing Add:* Dept Animal Sci Va Polytech Inst & State Univ 3160 Animal Sci Bldg Blacksburg VA 24061

KNIGHT, JERE DONALD, b Deer River, Minn, July 2, 16; m 81. PHYSICAL CHEMISTRY. *Educ:* St John's Univ, Minn, BS, 38; Univ Minn, PhD(phys chem), 48. *Prof Exp:* Instr, St John's Prep Sch, Minn, 38-40; asst chem, Univ Minn, 41-42; asst, Nat Defense Res Comt Chicago, 42-43, jr chemist, Metall Lab, 43; assoc chemist, Clinton Labs, Tenn, 43-45; instr chem, Univ Ill, 48, res assoc physics, 48-49; mem staff, 49-80, fel, Los Alamos Nat Lab, 81-82; CONSULT, 82 - *Concurrent Pos:* Vis researcher, Brookhaven Nat Lab, 54-55. *Honors & Awards:* Clark Medal, Am Chem Soc, 80. *Mem:* Am Chem Soc; fel Am Phys Soc. *Res:* Nuclear chemistry; nuclear structure and reactions; chemical and nuclear studies with mesons. *Mailing Add:* PO Box 90637 Columbia SC 29290-0637

KNIGHT, JOHN C (IAN), b Musselburgh, Scotland, June 16, 26; US citizen; m 63; c 1. UNDERWATER ACOUSTICS, OPERATIONS RESEARCH. *Educ:* Univ Edinburgh, BSc, 50, PhD(physics), 53. *Prof Exp:* Asst physics, Univ Edinburgh, 52-54; sr sci officer, Admiralty, Eng, 54-58, prin sci officer, Home Fleet, 58-59; mem staff opers res, Saclant Anti-Submarine Warfare Res Ctr, Italy, 59-63; prin sci officer, Ministry of Defence, Eng, 63-65; sr staff mem, Opers Res Inc, Md, 65-68; sr staff & dep dir, John D Kettelle Corp, Va, 68-70; head, Systs Anal Group, Acoust Div, Naval Res Lab, 70-80; prin scientist, EG & G Wash Anal Serv Ctr, 80; mem prin staff, Summit Res Corp, 80-88; RETIRED. *Concurrent Pos:* Consult comt undersea warfare, Nat Acad Sci, 67-70; staff scientist, Commander Oceanog Syst Atlantic, 74-77. *Mem:* Acoust Soc Am; Opers Res Soc Am; Sigma Xi. *Res:* Beta and gamma ray spectroscopy; military operations research; analysis of naval system performance. *Mailing Add:* 3403 Fessenden St NW Washington DC 20008

KNIGHT, KATHERINE LATHROP, b Jackson, Mich, May 13, 41. BIOCHEMISTRY, IMMUNOLOGY. *Educ:* Elmira Col, BA, 62; Ind Univ, PhD(chem), 66. *Prof Exp:* Res assoc, 66-68, from asst prof to assoc prof, 68-75, PROF MICROBIOL, UNIV ILL MED CTR, CHICAGO, 75- *Concurrent Pos:* NIH res career develop award, 70-75. *Mem:* AAAS; Am Chem Soc; Am Asn Immunologists. *Res:* Immunochemistry; immunogenetics; protein chemistry. *Mailing Add:* Dept Microbiol Loyola Univ Chicago Stritch Sch Med Maywood IL 60153

KNIGHT, LARRY V, b Pocatello, Idaho, Mar 13, 35; m 58; c 8. PHYSICS. *Educ:* Brigham Young Univ, BS, 58, MS, 59; Stanford Univ, PhD(physics), 65. *Prof Exp:* Res assoc physics, Stanford Univ, 65-69; asst prof physics, 69-77, assoc prof, 77-80, PROF PHYSICS & ASTRON, BRIGHAM YOUNG UNIV, 80- *Concurrent Pos:* Mem tech staff, Hewlett-Packard Co, 64-69; vpres, Holograf Corp, 70-72; consult, Lawrence Livermore Lab, 75-; pres, Moxtek Corp, 86-; ed, J X-ray Sci & Technol. *Mem:* Am Phys Soc. *Res:* Low temperature physics; fundamental constants; magnetic resonance; electron spectroscopy; quantum electronics; holography; laser fusion; plasma and x-ray physics. *Mailing Add:* Dept Physics & Astron Brigham Young Univ Provo UT 84602

KNIGHT, LEE H, JR, b Westville, Fla, July 8, 28; m 60; c 1. MECHANICAL ENGINEERING. *Educ:* Univ SC, BS, 57; Ga Inst Technol, MS, 62. *Prof Exp:* Res asst, Ga Inst Technol, 57-69, asst res engr, 60-62; sr engr, Lockheed Ga Co, 62-64; br head mech eng, Ga Inst Technol, 64-67; ASST DIR SERV, SKIDAWAY INST OCEANOG, 67- *Mem:* Oceanog Soc. *Res:* Heat and mass transfer; underwater propulsion systems; electromechanical equipment for oceanographic research. *Mailing Add:* Skidaway Inst Oceanog PO Box 13687 Savannah GA 31406

KNIGHT, LON BISHOP, JR, b Milledgeville, Ga, Apr 24, 44; m 66; c 2. PHYSICAL CHEMISTRY. *Educ:* Mercer Univ, BS, 66; Univ Fla, PhD(chem), 70. *Prof Exp:* Res assoc phys chem, Univ Fla, 70-71; asst prof, 71-75, ASSOC PROF CHEM, FURMAN UNIV, 75- *Mem:* Am Chem Soc. *Res:* Study of reactions of metal atoms in the gas phase at low temperatures; metallic transport mechanisms; ESR matrix isolation of high temperature species. *Mailing Add:* Dept Chem Furman Univ Poinsett Hwy Greenville SC 29613

KNIGHT, LUTHER AUGUSTUS, JR, b Clarendon, Ark, Dec 19, 30; m 55; c 2. AQUATIC BIOLOGY. *Educ:* Ark State Col, BS, 57; Univ Miss, MS, 61, PhD(biol), 69. *Prof Exp:* Teacher, High Sch, Ark, 56-57 & Mo, 57-64; asst prof & asst res prof, 68-72, assoc prof, 72-81, PROF BIOL & CUR ZOOL MUS, UNIV MISS, 81- *Mem:* Am Soc Limnol & Oceanog; Am Micros Soc; Sigma Xi. *Res:* Taxonomy, ecology and distribution of Rotifera; water quality as related to freshwater plankton and biological productivity in reservoirs; trace elements in aquatic organisms. *Mailing Add:* Tre 15 13 Larhonda Dr Oxford MS 38655

KNIGHT, LYMAN COLEMAN, b Wooster, Ohio, Nov 2, 15; m 37; c 3. MATHEMATICS. *Educ:* Col Wooster, BS, 37; Kent State Univ, MA, 41; Univ Pittsburgh, DEd(math educ), 58. *Prof Exp:* Teacher pub schs, Ohio, 37-40; instr math & band, Northeastern Okla Jr Col, 40-42; from asst prof to assoc prof, 42-59, PROF MATH, MUSKINGUM COL, 59-, PROF COMPUT SCI, 81- *Concurrent Pos:* Consult, Nat Defense Educ Act. *Mem:* AAAS; Am Math Soc; Math Asn Am; Am Statist Asn. *Res:* Mathematical education; training of elementary and secondary teachers; undergraduate curriculum in mathematics in liberal arts colleges. *Mailing Add:* 156 Lakeside Dr New Concord OH 43762

KNIGHT, PATRICIA MARIE, b Schnectady, NY, Jan 25, 52. MEDICAL DEVICE RESEARCH & DEVELOPMENT. *Educ:* Ariz State Univ, BS, 74, MS, 76; Univ Utah, PhD(biomed eng), 83. *Prof Exp:* Teaching & res asst, Ariz State Univ, 74-76; proj engr, Am Med Optics, 76-77; res asst & PhD cand, Univ Utah, 79-83; mgr mat res, Am Med Optics, 83-87; dir mat res, 87-88, dir res, 88-91, VPRES RES & DEVELOP, ALLERGAN MED OPTICS, 91- *Mem:* Soc Biomat; Asn Res in Vision & Ophthal; Am Chem Soc; Biomed Eng Soc; Soc Women Engrs. *Res:* Develop products for ophthalmic surgery; polymer research; optical and mechanical engineering; biological science. *Mailing Add:* 9701 Jeronimo Irvine CA 92718

KNIGHT, PAUL R, b Mechanicsburg, Pa, June 27, 47; m; c 2. ANESTHESIOLOGY. *Educ:* Pa State Univ, MD & PhD(med microbiol), 73. *Prof Exp:* Assoc prof, 82-86, PROF ANESTHESIA, UNIV MICH HOSPS, 86- *Mem:* Am Soc Microbiol; Asn Univ Anesthesiologists; Am Soc Anesthesiol. *Res:* Anesthetic action and its effects and on cellular function and viral replication. *Mailing Add:* 1066 N Wagner Ann Arbor MI 48103

KNIGHT, SAMUEL BRADLEY, chemistry; deceased, see previous edition for last biography

KNIGHT, STEPHEN, b San Mateo, Calif, Feb 24, 38; m 59; c 3. EXPERIMENTAL PHYSICS. *Educ:* Beloit Col, BS, 59; Yale Univ, MS, 60, PhD(physics), 64. *Prof Exp:* Mem tech staff, Bell Tel Labs, 64-68, supvr explor device tech group, 68-71, supvr optical mat properties group, 71-72, supvr displays group, 72-73, supvr opto-isolators group, 73-78, supvr govt systs support, 78-80, supvr technol options, 80-81, head electronics technol planning, 81-84, div mgr, AT&T Corp Hq, 83-86, dir, AT&T Bell Labs, Kelly Educ & Training Ctr, 86-90, DIR, SEMATECH TECHNOL TRANSFER, AT&T BELL LABS, 90- *Concurrent Pos:* Assoc ed, Inst Elec & Electronics Engrs Trans Electron Devices, 68-79, ed, 79-83, pub chmn, 83-86. *Mem:* Am Phys Soc; sr mem Inst Elec & Electronics Engrs; Electron Devices Soc. *Res:* Low temperature physics, turbulent superfluid helium; semiconductor physics, bulk negative resistance; light emitting diodes; magnetic bubble materials; integrated circuit planning; high performance technology evaluation. *Mailing Add:* AT&T Bell Lab Rm 2A225 555 Union Blvd Allentown PA 18103

KNIGHT, VERNON, b Osceola, Mo, Sept 6, 17; m 46; c 4. MEDICINE. *Educ:* William Jewell Col, AB, 39; Harvard Univ, MD, 43; Am Bd Internal Med, dipl. *Prof Exp:* Asst prof med, Med Col, Cornell Univ, 53-54; assoc prof, Sch Med, Vanderbilt Univ, 54-59; clin dir, Nat Inst Allergy & Infectious Dis, Md, 59-66; PROF MED & CHMN DEPT MICROBIOL & IMMUNOL, BAYLOR COL MED, 66-, CO-DIR, CTR BIOTECHNOL & PROF, BIOTECHNOL, MED, MICROBIOL & IMMUNOL. *Concurrent Pos:* Mem, Nat Adv Allergy & Infectious Dis Coun; prof consult, US Army Med Res Inst of Infectious Dis, Ft Detrick. *Mem:* Soc Exp Biol & Med; Am Soc Clin Invest; Am Clin & Climat Asn; Am Col Physicians. *Res:* Infectious disease. *Mailing Add:* Dept Microbiol Baylor Col Med One Baylor Plaza Houston TX 77030

KNIGHT, WALTER DAVID, b New York, NY, Oct 14, 19; m 45, 72; c 3. SOLID STATE PHYSICS, ATOMIC & MOLECULAR PHYSICS. *Educ:* Middlebury Col, AB, 41; Duke Univ, MA, 43, PhD(physics), 50. *Hon Degrees:* DSc, Middlebury Col, 76, Fed Polytech Sch, Lausanne, Switz, 83. *Prof Exp:* Instr physics, Duke Univ, 43-44; instr, Trinity Col, 46-50; from asst prof to assoc prof, 50-61, asst dean col lett & sci, 59-61, assoc dean, 61-63, dean, 67-72, PROF PHYSICS, UNIV CALIF, BERKELEY, 61- *Concurrent Pos:* Alfred P Sloan fel, 56-59; Guggenheim fel, 61; Miller fel, 79; vis res fel,

Clarendon Lab, Oxford, UK, 83-84; Christensen fel, St Catharine's Col, Oxford, 83; vis distinguished prof, Niels Bohr Inst, 88; vis prof, Fed Polytech Sch, Lausanne, Switz, 80. *Honors & Awards:* Sackler Distinguished Lectr Chem, Univ Tel Aviv, 89. *Mem:* Nat Acad Sci; fel AAAS; fel Am Phys Soc; Am Acad Arts & Sci; Europ Phys Soc; Sigma Xi. *Res:* Study of metal microclusters by molecular beam methods. *Mailing Add:* Dept Physics Univ Calif Berkeley CA 94720

KNIGHT, WALTER REA, b Cortland, Ohio, Apr 25, 32; m 54; c 3. PSYCHOBIOLOGY. *Educ:* Baldwin-Wallace Col, AB, 54; Pa State Univ, MS, 56, PhD(psychol), 61. *Prof Exp:* Instr psychol, 56-57, asst prof, 57-58, 60-65, assoc prof, 65-68, PROF PSYCHOL & BIOL, HIRAM COL, 68- *Concurrent Pos:* NIMH spec res training fel, Univ Fla, 64-66. *Mem:* AAAS; Animal Behav Soc; Psychonomic Soc; Sigma Xi. *Res:* Neural, hormonal and early experience factors in animal social behavior and social stress. *Mailing Add:* Dept Biol Hiram Col Hiram OH 44234

KNIGHT, WILBUR HALL, b Denver, Colo, May 8, 21; m 41; c 1. OIL & GAS EXPLORATION & DEVELOPMENT. *Educ:* Univ Wyo, BA, 40, MA, 41. *Prof Exp:* Geologist oil & gas, Union Prod Co, 41-48, dist geologist, 41-55; chief geologist oil & gas, Larco Drilling Co, 55-59; CONSULT GEOLOGIST, 59- *Concurrent Pos:* Mem, div prof affairs cert bd, Am Assoc Petrol Geol, 75-; instr, Millsaps Col, 79-81. *Mem:* Fe Geol Soc Am; Am Asn Petrol Geologists; Soc Petrol Engr; Soc Independent Prof Earth Scientists; Am Inst Prof Geologists; Soc Explor Geophysicists. *Res:* Oil and gas exploration and development. *Mailing Add:* L100B Capital Towers Bldg Jackson MS 39201

KNIGHT, WILLIAM ALLEN, JR, b St Louis, Mo, Oct 5, 14; m 41; c 3. GASTROENTEROLOGY. *Educ:* Drury Col, AB, 36; St Louis Univ, MD, 40; Am Bd Int Med cert, 51. *Prof Exp:* Asst int med, 47-48, instr, 48-49, sr instr, 49-51, asst prof med, 51-55, dir int med, Housestaff & Residency Training Prog, 50-59, DIR, DIV GASTROENTEROL, SCH MED, ST LOUIS UNIV, 54-, ASSOC PROF MED, 55- *Concurrent Pos:* Area consult int med, US Vet Admin, 56-60; dir int med, St Louis City Hosp, St Louis Univ Med Serv, 48-55 & Firmin Desloge Hosp, 55-58; dir, Dept Internal Med, St Mary's Health Ctr, 65-80, emer dir, Cancer Res & Treatment Ctr, 81- *Mem:* Am Col Gastroenterol; Am Soc Gastrointestinal Endoscopy; Am Gastroenterol Asn; Am Asn Study Liver Dis; fel Am Col Phys; Sigma Xi. *Res:* Qualitative and quantitative analysis of pancreatic secretions, proteins and electrolytes in the normal and diseased pancreas to aid in the diagnosis of pancreatic disease. *Mailing Add:* 1035 Bellevue Suite 502 St Louis MO 63117

KNIGHT, WILLIAM ERIC, b Lacon, Ala, July 24, 20; m 44; c 1. PLANT BREEDING. *Educ:* Ala Polytech Inst, BS, 42, MS, 48; Pa State Univ, PhD(agron), 51. *Prof Exp:* Teacher, Vet Voc Agr, State Dept Educ, Fla, 46-47; asst soils res, Ala Polytech Inst, 47-48; res agronomist plant breeding & genetics, USDA, Miss State Univ, 51-86, from adj sci & educ admin, 61-86; RETIRED. *Concurrent Pos:* Mem, South Pasture & Forage Crop Improvement Conf. *Mem:* Am Soc Agron; Crop Sci Soc Am; Am Genetic Asn; AAAS; Sigma Xi; Coun Agr Sci Technol. *Res:* Genetic and plant breeding in annual Trifolium species; plant improvement by breeding. *Mailing Add:* Box 272 Miss Mississippi State MS 39762

KNIGHT, WILSON BLAINE, b Aug 20, 55. PROTEIN CHEMISTRY, ENZYMOLOGY. *Educ:* Univ Va, BA; Univ Md, College Park, PhD(biochem), 83. *Prof Exp:* Res assoc, Univ Wis, 83-87; sr biochemist, 87-89, RES FEL, MERCK, SHARPE & DOHME RES LAB, 89- *Mem:* AAAS; NY Acad Sci; Sigma Xi. *Res:* Enzyme mechanisms; mechanism of inhibition of elastases; mechanism of signal peptidase; enzyme inhibitors. *Mailing Add:* Dept Enzym Bldg 80Y-150 Merck Sharpe & Dohme Res Labs PO Box 2000 Rahway NJ 07060

KNIGHTEN, JAMES LEO, b Lafayette, La, Apr 1, 43; m 70; c 2. ELECTROMAGNETICS, ELECTROMAGNETIC PULSE ENGINEERING. *Educ:* La State Univ, BS, 65, MS, 68; Iowa State Univ, PhD(elec eng), 76. *Prof Exp:* Asst prof electromagnetics, Iowa State Univ, Ames, 76-77; staff engr, 77-80, MGR ELECTROMAGNETICS, IRT CORP, 80- *Mem:* Inst Elec & Electronics Engrs. *Res:* Nuclear electromagnetic pulse effects on systems; numerical methods; nuclear electromagnetic pulse hardening of ground launched Cruise missile; assessment of electromagnetic pulse coupling to ground based command, control and communications facilities, tactical systems, aircraft and missiles. *Mailing Add:* Survivability SVC Grp S-Cubed 3020 Callan Rd San Diego CA 92121

KNIGHTEN, ROBERT LEE, b Marshfield, Ore, May 7, 40; m 63; c 2. COMPILERS, CATEGORICAL ALGEBRA. *Educ:* Mass Inst of Technol, BS, 62, PhD(math), 66. *Prof Exp:* Mathematician, Navy Electronics Lab, Calif, 57-64; instr math, Mass Inst Technol, 64-66 & Univ Chicago, 66-68; asst prof, Univ Ill, Chicago Circle, 68-71; from asst prof to assoc prof math, Univ PR, 76-83; SYST CONSULT, SOFTECH, INC, MASS, 83- *Mem:* AAAS; Am Math Soc; Math Asn Am; Soc Indust & Appl Math; Asn Comput Mach; Inst Elec & Electronics Engrs. *Res:* Compilers; semantics of programming languages; applications of category theory to commutative algebra; algebraic geometry and homotopy theory. *Mailing Add:* 40 Dartmouth St Watertown MA 02172

KNIGHTS, JOHN CHRISTOPHER, b Felixstowe, Eng, July 2, 47. PHYSICS. *Educ:* Univ Sussex, BSc, 68; Univ Cambridge, MA & PhD(physics), 72. *Prof Exp:* Imp Chem Industs fel, Univ Cambridge, 72-73; RES SCIENTIST PHYSICS, PALO ALTO RES CTR, XEROX CORP, 73- *Concurrent Pos:* Res fel, Sidney Sussex Col, Cambridge, Eng, 72-73. *Mem:* Am Phys Soc. *Res:* Transport properties of amorphous and crystalline semiconductors. *Mailing Add:* 2607 Bryant St Palo Alto CA 94306

KNIKER, WILLIAM THEODORE, b Seguin, Tex, Aug 30, 29; div; c 4. ALLERGY, IMMUNOLOGY. *Educ:* Univ Tex, BA, 50, MD, 53; Am Bd Pediat, dipl, 59; Conjoint Bd Allergy & Immunol, dipl, 74. *Prof Exp:* Intern, Henry Ford Hosp, Detroit, 53-54; resident pediat, Univ Tex Med Br, 56-58; chief resident, Med Ctr, Univ Ark, 58-59, from instr to asst prof pediat, 59-62, from asst prof to assoc prof pediat & path, 65-69; from assoc prof to prof pediat & microbiol, 69-88, PROF PEDIAT, MICROBIOL & INTERNAL MED, UNIV TEX MED SCH, SAN ANTONIO, 88- *Concurrent Pos:* Res fel infectious dis, Am Thoracic Soc, 59-62; res fel, Div Exp Path, Scripps Clin & Res Found, Calif, 62-65; asst dir clin res unit, Univ Ark Med Ctr, 65-69, dir sect immunol-allergy, 69-; NIH res career develop award, 68-69; mem bd regents, Am Col Allergy & Immunol, 81-84. *Honors & Awards:* Stanley Jaros Lectr, Am Asn Clin Immunol & Allergy, 77; Bela Schick Award & Lectr, Am Col Allergy & Immunol, 84. *Mem:* Am Asn Path; Am Asn Immunologists; Am Col Allergy & Immunol; Am In Vitro Allergy & Immunol Soc (vpres); Am Acad Allergy & Immunol. *Res:* Immunopathology, mechanisms of hypersensitivity and immunological diseases; immunochemistry of mycobacterial antigens; primate immunology; pediatrics. *Mailing Add:* Dept Pediat Univ Tex Health Sci Ctr 7703 Floyd Curl Dr San Antonio TX 78284

KNILL, RONALD JOHN, b Chicago, Ill, Feb 20, 35; m 58; c 3. GEOMETRIC TOPOLOGY. *Educ:* Marquette Univ, BS, 56; Univ Notre Dame, MS, 60, PhD(math), 62. *Prof Exp:* NSF fel, Univ Calif, Berkeley, 62-63; from asst prof to assoc prof, 63-74, PROF MATH, TULANE UNIV, 74- *Concurrent Pos:* Res assoc, Col France, 69-70; vis fel, Univ Warwick, 78; vis scholar, Specola Vaticana, 84; vis prof, Col Holy Cross, 85-86. *Mem:* Am Math Soc; Math Asn Am; Math Soc France; Sigma Xi. *Res:* Low dimensional topology; gravilation and harmonic maps; fixed point theory; dynamical systems. *Mailing Add:* Dept Math Tulane Univ New Orleans LA 70118

KNIPE, DAVID MAHAN, b Lancaster, Ohio, Aug 6, 50; m 73; c 2. VIROLOGY, CELL BIOLOGY. *Educ:* Case Western Reserve Univ, BA, 72; Mass Inst Technol, PhD(cell biol), 76. *Prof Exp:* Fel, 76-79, from asst prof to assoc prof, 79-89, PROF VIROL, DEPT MICROBIOL, HARVARD MED SCH, 89- *Concurrent Pos:* Fac res award, Am Cancer Soc, 84-; mem, Clin Sci Study Sect, NIH, 85-89 & Virol Study Sect, 90- *Mem:* Am Soc Microbiol; Am Soc Virol; AAAS. *Res:* How herpes simplex virus replicates in and interacts with its host cell; molecular biological and genetic approaches to study how viral and cellular proteins function within the cell nucleus. *Mailing Add:* Dept Microbiol & Molecular Genetics Harvard Med Sch 200 Longwood Ave Boston MA 02115

KNIPE, RICHARD HUBERT, b Salmon, Idaho, Sept 12, 27; m 57; c 3. CHEMICAL PHYSICS. *Educ:* Calif Inst Technol, BS, 50; Duke Univ, PhD(physics), 54. *Prof Exp:* Physicist, 54-57, aeronaut power plant res engr, 57-62, phys chemist, 62, RES PHYSICIST, NAVAL WEAPONS CTR, 62- *Mem:* Am Phys Soc; Am Chem Soc. *Res:* Quantum theory of molecules; gas phase chemical kinetics and mechanism; performance analysis of flash lamps pumped dye lasers. *Mailing Add:* 1121 Lucille Ct Ridgecrest CA 93555

KNIPFEL, JERRY EARL, b Calgary, Alta, June 30, 41; m 66; c 2. NUTRITION & BIOCHEMISTRY, RESEARCH ADMINISTRATION. *Educ:* Univ Sask, BSA, 65, MSc, 67; McGill Univ, PhD(nutrit), 73. *Prof Exp:* Chemist, Health Protection Br, Nutrit Res Div, Health & Welfare Can, 67-69, res scientist, 69-73; prog specialist, Forage Prod & Utilization Sect, 84-87, prog dir Prarie Region Res Br, 84-87, RES SCIENTIST, FORAGE PROD & UTILIZATION SECT, RES STA, AGR CAN, 73-84, 87-, SECT HEAD, 89- *Concurrent Pos:* Asst ed, Can J Animal Sci, 77-80, assoc ed, 80-83; lectr, Dept Biol, Univ Regina, 78; consult, Topline Feeds, Inc, Swift Current, 74; mem, Exp Comt Animal Nutrit, 81-91, Exp Comt Forage Crops, 79-82, prog rev comt, Agr Can Western Region, 80; sci authority, Res Br, Agr Can, Contract Res Progs, 79-; asst prof biol, Univ Regina, 81 & 90, lectr & dept chmn, 91; adj prof, Univ Sask, 87-92; dir & rep, Can Feed Info Ctr, 86-, chmn exec comt, Int Network Feed Info Ctrs, 88-; mgr, Res Br Prarie Region Contract Res & ERDA progs, 85-90; coordr, Forage Commodity Coord Directorate Agr Can, 86; mem Nat Res Coun task force on feed composition, 89- *Mem:* Can Soc Animal Sci; Nutrit Soc Can; Can Asn Lab Animal Sci; Am Chem Soc; Inst Food Technologists; NY Acad Sci. *Res:* Nutrition in relation to fetal development and post partum performance; nitrogen metabolism in ruminants; evaluation of nutritional quality and nutrient availability; improvement in nutritional value of roughage. *Mailing Add:* Forage Prod & Utilization Sect Res Sta Agr Can Swift Current SK S9H 3X2 Can

KNIPLING, EDWARD FRED, b Port Lavaca, Tex, Mar 20, 09; m 34; c 5. ENTOMOLOGY. *Educ:* Tex A&M Univ, BS, 30; Iowa State Univ, MS, 32, PhD(entom), 47. *Hon Degrees:* DSc, Catawba Col, 62, NDak State Univ, 70, Clemson Univ, 70, Fla State Univ, 75. *Prof Exp:* Field asst, Bur Entom & Plant Quarantine, USDA, 30, from jr entomologist to assoc entomologist, 31-42, sr entomologist, Off Sci Res & Develop Contract, 42-46, prin entomologist, 46-53, dir, Entom Res Div, Agr Res Serv, 53-71, sci adv, 71-73; COLLABR, USDA, 73- *Honors & Awards:* Medal, Typhus Comn; Hoblitzelle Nat Award, 60; Nat Medal Sci, 66; Agr Res Serv, Scientist Hall of Fame, USDA, 85. *Mem:* Nat Acad Sci; Entom Soc Am. *Res:* Biology, ecology, population dynamics and control of insects. *Mailing Add:* 2623 Military Rd Arlington VA 22207

KNIPMEYER, HUBERT ELMER, b Sharon, Conn, Nov 7, 29; m 52; c 4. ORGANIC CHEMISTRY, RESEARCH ADMINISTRATION. *Educ:* Mass Inst Technol, SB, 51; Univ Ill, PhD(chem), 57. *Prof Exp:* Res chemist, Cent Res Dept, 56-60, tech investr, Film Dept, 60-61, res chemist, 61-62, staff scientist, 62-63, tech rep, 64-65, group mgr, 65-69, res mgr, Ohio, 69-71, prod mgr, 71-74, tech mgr, Film Dept, 74-75, mgr mkt develop & customer serv, 76, lab supt, 77, res mgr, plastic prod & resins dept & prog mgr corp automotive develop, corp plans dept, 78-82, mgr membranes, 83-84, TECH MGR POLYMERS PRODUCTS DEPT, E I DU PONT DE NEMOURS & CO, INC, 85- *Mem:* Am Chem Soc; NAm Membrane Soc. *Res:* Physical organic chemistry; organic synthesis; heterocyclic organic chemistry; polyimide, polyester and polyolefin chemistry; membranes. *Mailing Add:* 1017 Weldin Circle Wilmington DE 19803

KNIPP, ERNEST A, (JR), b Houston, Tex, Oct 10, 29; m 79. COMPUTER SCIENCES, GENERAL. *Educ:* Rice Univ, BS, 50, MS, 52; Yale Univ, PhD(chem eng), 58. *Prof Exp:* Fel chem, Univ NC, 57-59, Rice Univ, 60; asst prof chem eng, Mich Col Mining & Technol, 59; res chem engr, Humble Oil & Ref Co, 61-63; sr res chem engr, Esso Res & Eng Co, 63-65, res specialist, 68-71; CONSULT, 71- *Mem:* Am Chem Soc; Am Inst Chem Engrs; Asn Comput Mach; Royal Soc Chem. *Res:* Computer applications. *Mailing Add:* PO Box 3041 Houston TX 77253-3041

KNIPPLE, WARREN RUSSELL, b Johnstown, Pa, May 28, 34; m 59; c 2. INDUSTRIAL & PETROLEUM CHEMISTRY, REFRACTORY METALS. *Educ:* Univ Pittsburgh, BS, 59; Case Western Reserve Univ, MS, 61, PhD(chem), 68. *Prof Exp:* Sr chemist, Res Dept, Standard Oil Co (Ohio), 59-68; sr res chemist, Chase Brass & Copper Co, Inc, 68-69, prod mgr chem, 69-70, chem opers mgr, 70-72, gen mgr, Cleveland Refractory Metals Div, 72-85; REGIONAL SALES MGR, CLIMAX SPECIALTY METALS, 86- *Mem:* Am Chem Soc; AAAS; Am Mgt Asn. *Res:* Organic synthesis; organometallic research; heterogeneous catalysis; peroxide chemistry; aromatic substitution; rhenium chemistry; ion exchange and solvent extraction technology; molybdenum; tungsten chemistry. *Mailing Add:* Climax Specialty Metals 21801 Tungsten Rd Cleveland OH 44117

KNISELEY, RICHARD NEWMAN, b Wichita, Kans, Jan 23, 30; m 51; c 2. SPECTROCHEMISTRY, ANALYTICAL INSTRUMENTATION. *Educ:* Univ Kansas City, BA, 51; Iowa State Univ, MS, 54, PhD, 71. *Prof Exp:* Res assoc spectros, Inst Atomic Res, 55-59, from assoc chemist to chemist, 59-74, sr chemist, Ames Lab, Dept Energy, 74, asst dir environ progs, Iowa State Univ, 79-88; CONSULT, 88- *Mem:* Optical Soc Am; Soc Appl Spectros; Microbeam Anal Soc; Sigma Xi. *Res:* High temperature, stable molecules; materials science; instrument design and development; inductively coupled plasma excitation sources; determination of trace metals in environmental and biomedical samples; microbiological air quality; photoacoustic spectroscopy. *Mailing Add:* Ames Lab Dept Energy Iowa State Univ Ames IA 50011

KNISELY, WILLIAM HAGERMAN, b Houghton, Mich, Feb 3, 22; m 47; c 5. ANATOMY. *Educ:* Univ Chicago, PhB, 47, BS, 50; Med Col SC, MS, 52, PhD, 54. *Prof Exp:* Asst anat, Med Col SC, 49-50 & 51-54; from instr to asst prof med & from instr to assoc prof anat, Duke Univ, 54-59; prof anat & chmn dept, Univ Ky, 59-63; dir inst biol & med, Mich State Univ, 63-70; vchancellor health affairs, Univ Tex Syst, 70-73, prof anat & cell biol, 70-75, asst to chancellor health affairs, 73-75; vpres acad affairs, 75, PRES, MED UNIV SC, 75-, PROF ANAT, 75- *Concurrent Pos:* Univ res fel med, Duke Univ, 54-57, Am Heart Asn fel, 55-57, USPHS sr res fel, 57-59; mem, Nat Adv Coun Educ for Health Prof, 68-71, chmn policy plan comn, 69-71, vchmn, Study Comn on Dietetics, 69-71; comnr, Navajo Health Authority, 72-76; mem adv comt, Regional Health Serv Res Inst, Univ Tex Health Sci Ctr San Antonio, 73-75 & Univ Tex Austin Ctr Social Work, 74-75; mem bd dirs, Holy Cross Hosp, 73-75; consult, Regions IX & X, Dept HEW, 73-79 & Dept Med & Surg, Vet Admin, 74-; assoc mem, Inst Soc, Ethics & Life Sci, 74-75; gen adj prof, Union Grad Sch, Yellow Springs, Ohio, 74- *Mem:* AAAS; Geront Soc; Am Heart Asn; Am Asn Anatomists; Microcirc Soc. *Res:* Anatomy, physiology, pharmacology and pathology of small blood vessels, especially the lung. *Mailing Add:* PO Box 26901 BSB Rm 553 Oklahoma City OK 73190

KNISKERN, VERNE BURTON, b Negaunee, Mich, Oct 16, 21; m 42; c 3. PARASITOLOGY. *Educ:* Univ Mich, BS, 47, MS, 48, PhD(zool), 50. *Prof Exp:* From assoc prof to prof, Eastern Ill Univ, 50-86; RETIRED. *Mem:* Am Micros Soc; Am Soc Parasitol; Wildlife Dis Asn. *Res:* Parasitology; protozoology; genetics; histology; malacology; medical biology. *Mailing Add:* 1531 Division St Charleston IL 61920

KNITTEL, MARTIN DEAN, b Torrington, Wyo, Dec 19, 32; m 56; c 3. MICROBIOLOGY. *Educ:* Willamette Univ, BS, 55; Ore State Univ, MS, 62, PhD(microbiol), 65. *Prof Exp:* Bacteriologist, Ore Agr Div, 55-56; med technologist, Doctors' Clin, 58-59; aquatic biologist, Ore Fish Comn, 62-63; sr res microbiologist, Norwich Pharmacal Co, 65-66; microbiologist, Pac Northwest Water Lab, Fed Water Pollution Control Admin, 66-69; sr engr, Jet Propulsion Lab, 69-71; res microbiologist, Pac Northwest Water Lab, Environ Protection Agency, 71-74, res microbiologist, Western Fish Toxicol Sta, 74-88; RETIRED. *Mem:* AAAS; Am Soc Microbiol. *Res:* Physiology of the bacterium Sphaerotilus natans as is related to its growth in polluted streams; microbiology of waste treatment; detection of pathogenic organisms in waste water; biochemistry of waste water treatment; effect of stress on disease in fish. *Mailing Add:* 2244 Kiger Inland Dr Corvallis OR 97333

KNIZE, RANDALL JAMES, b Tacoma, Wash, Feb 4, 53. LASER COOLING, OPTICAL COMPUTING. *Educ:* Univ Chicago, BA & MS, 75; Harvard Univ, MA, 76, PhD(physics), 81. *Prof Exp:* Res asst, Harvard Univ, 76-81; res physicist, Princeton Univ, 80-88; ASST PROF, UNIV SOUTHERN CALIF, 88- *Mem:* Am Phys Soc; Am Vacuum Soc. *Res:* Fundamental atomic physics; optical pumping. *Mailing Add:* Dept Physics Univ Southern Calif Los Angeles CA 90089

KNOBEL, LEROY LYLE, b Valley City, NDak, July 10, 45. GROUND WATER, WATER-ROCK INTERACTION. *Educ:* Univ Wash, BS, 72; George Washington Univ, MS, 79. *Prof Exp:* Hydrologist, Peace Corps, 73-75; HYDROLOGIST, US GEOL SURV, 77- *Honors & Awards:* Presidential Citation, Dominican Repub, 75; Serv Award, Peace Corps, 75. *Mem:* Geochem Soc; Am Geophys Union. *Res:* Geochemical, microbial and physical processes controlling the carbon dioxide cycle and water-rock interactions in the Snake River Plain aquifer, Idaho. *Mailing Add:* PO Box 2230 MS 4148 Idaho Falls ID 83403-2230

KNOBELOCH, F X CALVIN, b Tell City, Ind, Aug 5, 25; m 53; c 3. SPEECH PATHOLOGY, AUDIOLOGY. *Educ:* Ind Univ, BSEd, 49; Univ Fla, PhD(speech path, audiol), 59. *Prof Exp:* Assoc prof speech & dir speech & hearing clin, Univ Miss, 59-60; asst chief speech path & audiol, Vet Admin Regional Off, Louisville, Ky, 60-62, chief speech path & audiol, Winston-Salem, NC, 62-66 & Durham, NC, 66-69; assoc dir, Biol Sci Res Ctr, Univ NC, Chapel Hill, 69-85, co-dir interdisciplinary training & head, Commun Disorders Ctr Develop & Learning, 69-90; ASSOC PROF, DIV SPEECH, HEARING & LANG, NC CENTRAL UNIV, DURHAM, 90- *Concurrent Pos:* Clin assoc prof, 77-85, clin prof, Inst Speech & Hearing Sci & Dept Pediat, Univ NC, Chapel Hill, 85-90; assoc prof, Shaw Univ, 72-79; adj assoc prof, Speech, Lang & Auditory Path Prog, East Carolina Univ, 72-80. *Honors & Awards:* Hons, NC Speech-Hearing-Lang Asn, 83. *Mem:* Am Speech, Language & Hearing Asn; fel Am Speech & Hearing Asn. *Res:* Hearing and language development in high risk and normal infants. *Mailing Add:* Div Speech Hearing & Lang NC Central Univ PO Box 19776 Durham NC 27707

KNOBIL, ERNST, b Berlin, Ger, Sept 20, 26; nat US. ENDOCRINOLOGY, PHYSIOLOGY. *Educ:* Cornell Univ, BS, 48, PhD(zool), 51. *Hon Degrees:* Dr, Univ Bordeaux, France, 80; ScD, Med Col Wis, 83. *Prof Exp:* Milton res fel, Biol Res Lab, Sch Dent Med, Harvard Univ, 51-53, instr physiol, Med Sch, 53-55, assoc, 55-57, asst prof, 57-61; Richard Beatty Mellon prof physiol & chmn dept, Sch Med, Univ Pittsburgh, 61-81, dir, Ctr Res Primate Reprod, 74-81; dean, Med Sch, 81-84, H WAYNE HIGHTOWER PROF PHYSIOL & DIR, LAB NEUROENDOCRINOL, UNIV TEX HEALTH SCI CTR, HOUSTON, 81- *Concurrent Pos:* Markle scholar acad med, 56-61; mem, Human Growth & Develop Study Sect, NIH, 64-66, chmn, Reproductive Biol Study Sect, 66-68, mem, primate res ctr adv comt, 69-73 & Corpus Luteum Panel, Contraceptive Develop Br, 69-71; mem adv coun, Inst Lab Animal Resources, Nat Acad Sci, 66-69; mem physiol test comt, Nat Bd Med Examr, 70-74; mem liaison comt med educ, AMA-Am Asn Med Cols, 74; consult, Pop Off, Ford Found, 74-75, Uniformed Serv Univ Health Sci, 75 & Human Reproduction Unit, WHO, 76-; mem comt priorities & planning, Fedn Am Socs Exp Biol, 74; mem pop res comt, Ctr Pop Res, Nat Inst Child Health & Human Develop, 74-77; mem bd adv gynec, Alza Corp, 76-78; mem med adv bd, Nat Pituitary Agency, 80-83; ed-in-chief, Sect Endocrinol & Metab, Am J Physiol, 79-82; ed, Ann Rev Physiol, 74-77; mem adv comt, Searle Scholars Prog, 81-82; mem planning comt for develop endocrinol & phys growth, Nat Inst Child Health & Human Develop, 84-86; mem, steering comt info sci & med educ, Asn Am Med Col, 84 -86, ad hoc comt fac pract, 86; mem, Expert Adv Panel Biol Standarization, WHO, 87-, exec coun & distinguished serv mem, Asn Am Med Cols & Comt Int Orgn & Progs, Nat Res Coun. *Honors & Awards:* Ciba Award, Endocrine Soc, 61; Bowditch Lectr, Am Physiol Soc, 65; Gregory Pincus Mem Lectr, Laurentian Hormone Conf, 73; Upjohn Lectr, Am Fertil Soc, 74; Kathleen M Osborn Mem Lectr, Sch Med, Univ Kans, 74; Hopkins-Md Lectr, 74; Karl Paschkis Lectr, Philadelphia Endocrine Soc, 75; Transatlantic Lectr, Soc Endocrinol Gt Brit, 79; Lawson Wilkins Lectr, Endocrine Soc, 80; Bard Lectr, Sch Med, Johns Hopkins Univ, 81; Herbert M Evans Lectr, Univ Calif, 81; Fred Conrad Koch Award, Endocrine Soc, 82; Carl G Hartman Award, Soc Study Reproduction, 83; Potter Lectr, Thomas Jefferson Univ, 85; Axel Munthe Award, 85. *Mem:* Nat Acad Sci; Am Physiol Soc (pres, 78-79); Int Soc Endocrinol (pres, 84-88); Endocrine Soc (pres, 76-77); hon mem Hungarian Acad Sci. *Res:* Physiology of the pituitary gland; reproductive physiology. *Mailing Add:* Lab Neuroendocrinol Univ Tex Med Sch PO Box 20708 Houston TX 77225

KNOBLER, CAROLYN BERK, b New Brunswick, NJ, Jan 6, 34. STRUCTURAL CHEMISTRY, INORGANIC CHEMISTRY. *Educ:* George Washington Univ, BS, 55; Pa State Univ, PhD(inorg chem), 59. *Prof Exp:* Res asst geol, Calif Inst Technol, 62-64; asst res chemist, 64-84, ASSOC RES CHEMIST, DEPT CHEM & BIOCHEM, UNIV CALIF, LOS ANGELES, 84- *Concurrent Pos:* Res chemist, Pierre & Marie Curie Univ, 83-84 & 87, Univ Canterbury, 78-79, Univ Leiden, 70-71 & 60-61; vis lectr dept chem, Univ Calif, Los Angeles, 75; Nat Sci Found fel chem, Univ Amsterdam, 59-60. *Mem:* Am Crystallog Asn. *Res:* Structural characterization (single crystal x-ray crystallography) of metal clusters, organometallic compounds and metallacarboranes; conformational changes in uncomplexed host and in host-guest complexes. *Mailing Add:* Dept Chem & Biochem Univ Calif 405 Hilgard Ave Los Angeles CA 90024

KNOBLER, CHARLES MARTIN, b Newark, NJ, June 1, 34; m 57; c 2. PHYSICAL CHEMISTRY. *Educ:* NY Univ, BA, 55; Leiden Univ, Netherlands, PhD(molecular physics), 61. *Prof Exp:* Res assoc phys chem, Ohio State Univ, 61-62; fel chem eng, Calif Inst Technol, 62-64; from asst prof to assoc prof, 64-77, PROF CHEM, UNIV CALIF, LOS ANGELES, 77- *Mem:* Am Chem Soc; fel Am Phys Soc; Sigma Xi. *Res:* kinetics of phase transitions; statistical mechanics of complex fluids. *Mailing Add:* 2023 Malcom Ave Los Angeles CA 90025

KNOBLER, ROBERT LEONARD, b New York, NY, Nov 10, 48; m 79; c 3. NEUROIMMUNOLOGY, NEUROVIROLOGY. *Educ:* City Col New York, BS, 69; State Univ NY, Brooklyn, MD & PhD(anat), 75. *Prof Exp:* Intern med & psychiat, Kings County Hosp Ctr, State Univ NY, 75-76, resident neurol, 76-79; FEL IMMUNOPATH, SCRIPPS CLIN & RES FOUND, 79-, CLIN COORDR, MULTIPLE SCLEROSIS CLIN, GEN CLIN RES CTR, 81-; AT DEPT NEUROL, JEFFERSON MED COL, PHILADELPHIA. *Concurrent Pos:* Co-dir, Multiple Sclerosis Comprehensive Clin Ctr; dir, Reflex Sympathetic Dystrophy Clin. *Honors & Awards:* Roland P Mackay Award, Am Acad Neurol. *Mem:* Am Acad Neurol; Am Asn Anatomists; Sigma Xi; Soc Neurosci; Am Asn Neuropathologists. *Res:* Viral and immune mechanisms of multiple sclerosis and reflex sympathetic dystrophy; ultrastructure of neurocellular relationships; clinical neurophysiology. *Mailing Add:* Dept Neurol Jefferson Med Col 1025 Walnut St Philadelphia PA 19107

KNOBLOCH, EDGAR, b Praha, Czechoslovakia, Mar 30, 53; UK citizen; m 87. STABILITY THEORY, NONLINEAR DYNAMICS. *Educ:* Cambridge Univ, BA, 74; Harvard Univ, AM, 75, PhD(astron), 78. *Prof Exp:* Res asst astron, Harvard Univ, 76-78; from asst prof to assoc prof, 78-87, PROF PHYSICS, UNIV CALIF, BERKELEY, 87- *Concurrent Pos:* Jr fel, Harvard Soc Fels, 78-80; Alfred P Sloan res fel, 80-84. *Mem:* Am Phys Soc. *Res:* Astrophysical fluid dynamics; magnetohydrodynamics (dynamo theory); stochastic processes; stellar dynamics; bifurcation theory; nonlinear dynamics. *Mailing Add:* Dept Physics Univ Calif Berkeley CA 94720

KNOBLOCH, HILDA, b New York, NY, Dec 14, 15; div. PEDIATRICS, DEVELOPMENTAL DISABILITIES. *Educ:* Barnard Col, Columbia Univ, BA, 36; NY Univ, MD, 40; Johns Hopkins Univ, MPH, 51, DrPH, 55; Am Bd Pediat, dipl, 47; Am Bd Prev Med, dipl, 54. *Prof Exp:* Asst clin child develop, Sch Med, Yale Univ, 45-46; clin asst child guid clin, Mt Sinai Hosp, NY, 47-49; pediat consult, Maternity & Newborn Div, New York City Health Dept, 49-50; res assoc, Maternal & Child Health Div, Sch Hyg & Pub Health, Johns Hopkins Univ, 51-55, asst prof, 55; from assoc prof to prof pediat, Col Med, Ohio State Univ, 55-66, asst prof psychiat, 58-66; prof pediat, Univ Ill Col Med, 66-67; prof, Mt Sinai Sch Med, 67-70; prof, 72-82, EMER PROF PEDIAT, ALBANY MED COL, 82- *Concurrent Pos:* Pvt pract, NY, 47-49; asst clin vis pediatrician, Bellevue Hosp, NY, 47-49; dir clin child develop, Children's Hosp, Colo, 56-64, dir div child develop, 64-66; dir, Child Develop Div, Dept Pediat, Mt Sinai Hosp, NY; med specialist, NY State Off Ment Retardation & Develop Disabilities, 70-81. *Mem:* Fel Soc Res Child Develop; fel Am Acad Pediat; Am Pediat Soc; Am Acad Cerebral Palsy; Sigma Xi. *Res:* Developmental assessment and infant neurology, especially etiologic factors in neuropsychiatric disabilities of childhood. *Mailing Add:* 230 E Oglethorpe Ave Savannah GA 31401

KNOBLOCH, IRVING WILLIAM, b Buffalo, NY, Mar 1, 07; m 34; c 3. BOTANY. *Educ:* Univ Buffalo, BA, 30, MA, 32; Iowa State Univ, PhD(bot), 42. *Prof Exp:* Asst, Univ Buffalo, 28-30; dir bot gardens, City of Buffalo, 31-33; wildlife technician, US Fish & Wildlife Serv, 33-37; instr bot, Iowa State Univ, 42-43; asst prof bot, Univ Buffalo, 43-45; from asst prof to prof natural sci, 45-59, prof bot, 59-76, EMER PROF BOT, MICH STATE UNIV, 76- *Mem:* Bot Soc Am; Am Fern Soc(vpres, 65-66, pres, 66-69); Am Inst Biol Sci; Am Asn Univ Professors. *Res:* Structure of economic plants; cytology of plants; morphology; agrostology; hybrids and evolution; pteridophytes. *Mailing Add:* 6104 Brookhaven East Lansing MI 48824

KNOBLOCH, JAMES OTIS, b Thibodaux, La, Jan 9, 20; m 50; c 3. ORGANIC CHEMISTRY. *Educ:* La State Univ, BS, 41; Univ Notre Dame, MS, 47, PhD(chem), 49. *Prof Exp:* Inspector, Radford Ord Works, 41-42; shift supvr acid plant, Pa Ord Works, US Rubber Co, 42-43; chem engr, Magnesium Plant, Mathieson Alkali Works, 43-44; chemist, Synthetic Rubber Plant, Firestone Tire & Rubber Co, 44-45; chemist, Whiting Res Labs, Standard Oil Co, Ind, Amoco Chem Corp, 49-59, sr res scientist, 59-70, sr res chemist, 70-82; CONSULT, 82- *Concurrent Pos:* Adj prof, Ill Benedictine Col, 82-88. *Mem:* Am Chem Soc. *Res:* Structure of petroleum sulfonic acids; reactions of ozonides; synthesis of trifluoromethyl olefins; reduction of acetylenic glycols; oxidation studies; aromatic acid halogenations; polyanhydrides; fire retardants; catalytic hydrogenation of aromatic polycarboxylic acids. *Mailing Add:* 7S 242 Green Acres Dr Naperville IL 60540

KNOBLOCK, EDWARD C, biochemistry; deceased, see previous edition for last biography

KNOCHE, HERMAN WILLIAM, b Stafford, Kans, Nov 15, 34; m 55; c 2. PHYTOPATHOLOGY. *Educ:* Kans State Univ, BS, 59, PhD(biochem), 63. *Prof Exp:* From instr to assoc prof biochem & nutrit, 62-73, PROF, DEPT BIOCHEM, UNIV NEBR, 73-, HEAD DEPT, 74- *Concurrent Pos:* Consult, Physicians Path Lab, Univ Nebr, 66-70; hon res assoc & NIH fel, Harvard Univ, 71-72; grants, NSF, Nebr Wheat Comn & AEC, 71-72, NSF, 80-83; consult, Norden Labs, Inc, 73- & Vet Hosp, Lincoln, Nebr, 75- *Mem:* Am Chem Soc; Am Soc Biol Chemists. *Res:* Biochemistry and structure of lipids; structures and activity of plant toxins; radioactive tracer methodology. *Mailing Add:* Dept Biochem Univ Nebr Lincoln NE 68583-0718

KNOCKEMUS, WARD WILBUR, b Des Plaines, Ill, Jan 29, 34; m 64; c 3. INORGANIC CHEMISTRY. *Educ:* Knox Col, BA, 55; Pa State Univ, MS, 58; Univ Nebr, PhD, 69. *Prof Exp:* From asst prof to assoc prof chem, Morningside Col, 61-70; asst prof chem, Behrend Campus, Pa State Univ, 70-74; PROF CHEM & CHMN DEPT, HUNTINGDON COL, 74- *Concurrent Pos:* Consult, Great Lakes Res Inst, Pa, 70-; res consult, Wesley Indust Inc, Mobile Ala, 87-, electrochem corrosion res, NASA, Marshall Space Flight Ctr, 85, 86. *Mem:* Am Chem Soc; Space Studies Inst. *Res:* Chemistry of metal-organic chelates; hydrolysis of oxymolybdenum chelates; adduct type compounds of octamolybdic acid; elecrochemical corrosion studies. *Mailing Add:* Dept Chem Huntingdon Col Montgomery AL 36106-2148

KNODEL, ELINOR LIVINGSTON, b New York, NY; m 80; c 2. BIOCHEMISTRY, CELLCULTURE TECHNOLOGY. *Educ:* Columbia Univ, AB, 69; Yale Univ, MS, 72; Univ Conn, PhD(biochem), 76. *Prof Exp:* Res fel, Univ Conn, 74-76; res fel neuroendocrinol, Rockefeller Univ, 76-77; res fel neurochem, Mayo Clin, 78-80; process chemist, 80-85, res & develop chemist, 86-90, SR TECH WRITER, CLIN SYSTS DIV, E I DU PONT DE NEMOURS & CO, 90- *Mem:* AAAS; Tissue Cult Asn; Soc Neurosci; Am Chem Soc. *Res:* Biochemistry of neural transmission in the brain and its regulation by hormones and drugs; automated clinical chemistry test development; cellculture technology; monoclonal antibody production. *Mailing Add:* E I du Pont de Nemours & Co Concord Plaza Quillen Bldg Wilmington DE 19810

KNODEL, RAYMOND WILLARD, b Butte, NDak, June 10, 32; m 59; c 3. MATHEMATICS EDUCATION. *Educ:* Minot State Col, BS, 55; Univ Northern Colo, MA, 60, DEduc(math educ), 70. *Prof Exp:* Instr sci, Ashley High Sch, NDak, 55-56; instr math, Mandan Sr High Sch, 56-59 & Anaconda Sr High Sch, Mont, 60-61; PROF MATH, BEMIDJI STATE COL, 61- *Res:* Team teaching mathematics and arithmetic methods to prospective elementary school teachers. *Mailing Add:* Dept Math & Comput Sci Bemidji State Univ Bemidji MN 56601

KNODT, CLOY BERNARD, b Hastings, Minn, Feb 20, 17; m 40; c 3. DAIRY SCIENCE. *Educ:* Univ Minn, BS, 40, PhD(dairy husb), 44; Univ Conn, MS, 42. *Prof Exp:* Lab technician, Univ Minn, 37-40, asst physiol, 41-44; from instr to asst prof, Cornell Univ, 44-45; from assoc prof to prof dairy prod, Pa State Col, 46-54; cattle specialist, Gen Mills, Inc, 54-56; dir res, Burrus Mills, Inc, 57; res farm, Cargill, Inc, 57-61; head dept nutrit res, Squibb Inst Med Res, 61-62; DIR RES & DEVELOP, A O SMITH HARVESTORE PROD, INC, 62- *Mem:* AAAS; Am Chem Soc; Am Soc Animal Sci; Soc Exp Biol & Med; Am Dairy Sci Asn. *Res:* Nutrition, housing and management of farm livestock; agricultural sciences. *Mailing Add:* 26573 W Taylor St Barrington IL 60010-2731

KNOEBEL, LEON KENNETH, b Shamokin, Pa, Dec 7, 27. PHYSIOLOGY. *Educ:* Pa State Univ, BS, 50, MS, 52; Univ Rochester, PhD(physiol), 55. *Prof Exp:* Instr physiol, Univ Rochester, 55; from asst prof to assoc prof, 55-70, PROF PHYSIOL, SCH MED, IND UNIV, INDIANAPOLIS, 70- *Mem:* AAAS; Am Physiol Soc. *Res:* Gastrointestinal digestion and absorption of lipid. *Mailing Add:* Dept Physiol & Biophys Ind Univ Sch Med Indianapolis IN 46207

KNOEBEL, SUZANNE BUCKNER, b Ft Wayne, Ind, Dec 13, 26. INTERNAL MEDICINE, CARDIOLOGY. *Educ:* Goucher Col, AB, 48; Ind Univ, Indianapolis, MD, 60. *Prof Exp:* From asst prof to prof, 66-77, HERMAN C & ELLNORA D KRANNERT PROF MED, SCH MED, IND UNIV, INDIANAPOLIS, 77- *Mem:* Fel Am Col Cardiol; Asn Univ Cardiologists; Am Fedn Clin Res; Am Heart Asn. *Res:* Myocardial blood flow; arrhythmias in coronary artery disease; computer analysis of cardiovascular data. *Mailing Add:* Krannert Inst Med Dept Ind Univ Med Ctr 1001 W Tenth St Indianapolis IN 46202

KNOECHEL, EDWIN LEWIS, b Milwaukee, Wis, June 15, 31; m 53; c 2. PHARMACY. *Educ:* Univ Wis, BS, 53, MS, 55, PhD(pharm), 58. *Prof Exp:* Res assoc prod res & develop, Upjohn Co, 58-65, head mat inspection, control div, 65-90; CONSULT, 90- *Mem:* Am Pharmaceut Asn. *Res:* Control activities associated with Food and Drug Administration, especially good manufacturing practices, self inspection, raw material evaluation, contract processor inspections, determination of mesh analysis, bulk volume, dissolution rates, surface area and particle size distributions, solids technology and in-process testing. *Mailing Add:* 1201 Turwill Lane Kalamazoo MI 49007

KNOEDLER, ELMER L, b Gloucester, NJ, Feb 12, 12; m 41. CHEMICAL ENGINEERING. *Educ:* Cornell Univ, ME, 34; Columbia Univ, MS, 36, PhD(chem eng), 52. *Prof Exp:* Mem tech staff, Atlantic Refining Co, Pa, 34-35; asst supt, Davis Emergency Equip Co, NJ, 36-37; develop engr & supt iron powder prod, Metals Disintegrating Co, 39-41; sr proj engr, Sheppad T Powell, 41-58; SR PROJ ENGR & PARTNER, SHEPPARD T POWELL ASSOCS, 58- *Mem:* Am Chem Soc; Am Soc Mech Eng; Inst Elec & Electronics Engrs; Am Inst Chemists; Am Inst Chem Eng. *Res:* Industrial water; boiler feedwater; corrosion; industrial waste waters. *Mailing Add:* 513 Little John Hill Sherwood Forest MD 21405

KNOEFEL, PETER KLERNER, b New Albany, Ind, Aug 4, 06; m 53. HISTORY & PHILOSOPHY OF SCIENCE, PHARMACOLOGY. *Educ:* Univ Wis, BA, 27, MA, 28; Harvard Univ, MD, 31. *Prof Exp:* Res fel pharmacol, Nat Res Coun, 31-33; assoc pharmacol, Vanderbilt Univ, 33-35; from asst prof to prof, 35-68, chmn dept, 41-66, EMER PROF PHARMACOL, UNIV LOUISVILLE, 69- *Concurrent Pos:* Vis prof, Inst & Mus Hist & Sci Firenze, Italy, 68-81. *Mem:* Emer mem Am Chem Soc; Am Asn Hist Med; Hist Sci Soc; Int Soc Toxicol. *Res:* Classical studies of venomous serpents. *Mailing Add:* 800 S Fourth St Apt 1306 Louisville KY 40203-2146

KNOEPFLER, NESTOR B(EYER), b New Orleans, La, Oct 1, 18; m 43; c 3. CHEMICAL ENGINEERING, CHEMISTRY. *Educ:* Tulane Univ, BE, 40. *Prof Exp:* Lab asst, Southern Cotton Oil Co, 40-41; asst mgr, Southern Photocraft, 45-50 fel, Nat Cottonseed Prod Asn, Southern Regional Res Lab, USDA, 50-52, asst chem engr, 52-55, assoc chem engr, 55-56, tech asst to dir, 56-58, asst to dir, 58-60, res chem engr, 60-66, head, Cotton Prod Invests Eng & Develop Lab, 66-73, res leader, textile chem eng, eng & develop, 73-79, CONSULT CHEM ENGR, USDA, 79- *Mem:* Am Oil Chem Soc; Sigma Xi; Am Asn Textile Chem & Colorists. *Res:* Processing of Southern grown agricultural products, especially oilseeds, fruits, vegetables, pine gum; mechanical processing and chemical finishing of cotton; author of 129 tech articles & 9 patents. *Mailing Add:* 17 Jennifer Court Mandeville LA 70448

KNOERR, KENNETH RICHARD, b Milwaukee, Wis, Sept 2, 27; m 52; c 3. FORESTRY. *Educ:* Univ Idaho, BS, 52; Yale Univ, MF, 55, PhD, 61. *Prof Exp:* Res forester meteorol & asst proj leader snow physics studies, Pac Southwest Forest & Range Exp Sta, US Forest Serv, 56-61, proj leader forest microclimate studies, Cent States Forest Exp Sta, 61; asst prof forest climat, 61-67, assoc prof forest meteorol, 67-72, assoc prof biometeorol, 67-81, prof forest meteorol, 72-81, assoc prof bot, 76-81, PROF FORESTRY & ENVIRON STUDIES, SCH FORESTRY, DUKE UNIV, 81- *Concurrent Pos:* Mem, Nat Acad Sci Adv Comt Climat, US Weather Bur, 65- *Mem:* AAAS; Soc Am Foresters; Am Meteorol Soc; Am Geophys Union; Int Asn Sci Hydrol; Sigma Xi. *Res:* Micrometeorology and microclimatology of forests related to surface energy balance; evapotranspiration and watershed management. *Mailing Add:* Sch Forestry Duke Univ Durham NC 27706

KNOKE, JAMES DEAN, b Des Moines, Iowa, Mar 22, 41; m 69; c 2. BIOSTATISTICS. *Educ:* Univ Iowa, BA, 63; Stanford Univ, MS, 65; Univ Calif, Los Angeles, PhD(biostatist), 70. *Prof Exp:* Opers res analyst, Autonetics Div, NAm Rockwell Corp, 65-68; asst prof biomet, Case Western Reserve Univ, 71-78; assoc prof biostatist, Univ NC, 78-87; CONSULT, 87- *Mem:* Am Statist Asn; Biomet Soc. *Res:* Statistical methodology and biomedical applications. *Mailing Add:* One Peach Leaf Ct Gaithersburg MD 20878

KNOKE, JOHN KEITH, b Detroit, Mich, Mar 31, 30; m 56; c 2. VIROLOGY, EPIPHYTOLOGY. *Educ:* Univ Wis, BS, 52, MS, 59, PhD(entom), 62. *Prof Exp:* Entomologist, Cacao Entom, Inter-Am Inst Agr Sci, 60-63; proj assoc, Univ Wis-Madison, 63, asst prof, 63-67; RES ENTOMOLOGIST, AGR RES SERV, USDA, OHIO AGR RES & DEVELOP CTR, 67- *Concurrent Pos:* Collab, Univ Wis-Madison, 60-63; adj assoc prof entom, Ohio Agr Res & Develop Ctr & Ohio State Univ, 67- *Mem:* Entom Soc Am; Am Phytopath Soc. *Res:* Control of insects attacking vegetables, especially through the use of systemic insecticides; study of all phases of entomology relative to production of Theobroma cacao L; epidemiology of virus diseases of corn; vectors of maize viruses. *Mailing Add:* Dept Entom Ohio State Univ Main Campus 103 Bot/Zool Bldg Columbus OH 43210

KNOLL, ALAN HOWARD, b St Joseph, Mich, Apr 16, 31; m 52; c 6. CIVIL ENGINEERING, SOFTWARE SYSTEMS. *Educ:* Univ Mich, BSc, 52, MSc, 56. *Prof Exp:* Res engr, Alcoa Labs, 56-67, group leader, 67-70, sect head, 70-75, mgr sci & bus systs, 75-77 & mgr elec prods div, 77-80; gen mgr, REA Magnet Wire Co, 80-82, vpres technol, 82-86; INDEPENDENT CONSULT, 86- *Mem:* Sigma Xi; Am Soc Civil Engrs. *Res:* Structural use of aluminum; application of digital computation to structural and other engineering problems; new products and test methods for overhead electrical transmission and magnet wire. *Mailing Add:* 15 Cambridge Dr Frankenmuth MI 48734-9779

KNOLL, ANDREW HERBERT, b West Reading, Pa, Apr 23, 51; m 74; c 2. PALEONTOLOGY, GEOLOGY. *Educ:* Lehigh Univ, BA, 73; Harvard Univ, AM, 74, PhD(geol), 77. *Prof Exp:* Asst prof, Oberlin Col, 77-81; assoc prof biol, 82-85, PROF BIOL & CUR BOT MUS, HARVARD UNIV, 85- *Concurrent Pos:* Assoc ed, Palebiol, 80-, Precambrian Res, Rev Paleobot Palynology, 87-, Trends in Ecol & Evolution, 87-, Am J Sci, 91-; mem, Comt Planetary Biol & Chem Evolution, US Space Sci Bd, 80-88 & Space Sci Bd Task Group, Major Directions in Space Sci: 1995-2015, 85-88; prin investr, NSF, 80-; mem NRC Bd Earth Sci, 87-89, NASA Exobiol Adv Comt, 88-, chmn, Int Stratig Comn Working Group on terminal Precambrian Syst, 88-; Guggenheim fel, 87; vis fel, Australian Nat Univ, 87-, Gonville & Caius Col, Cambridge, 91; vis scientist, Australian Bur Mineral Resources, 87- *Honors & Awards:* Walcott Medal, NAS, 87; Schuchbert Award, Paleontol Soc, 87. *Mem:* Nat Acad Sci; Bot Soc Am; Soc Econ Paleontologists & Mineralogists; Sigma Xi; Paleont Soc; fel Am Acad Arts & Sci; fel Geol Soc Am. *Res:* Precambrian biological evolution; evolution of land plants; Precambrian sedimentary geology. *Mailing Add:* Bot Mus Harvard Univ Cambridge MA 02138

KNOLL, GLENN F, b St Joseph, Mich, Aug 3, 35; m 57; c 3. NUCLEAR ENGINEERING. *Educ:* Case Western Reserve Univ, BS, 57; Stanford Univ, MS, 59; Univ Mich, PhD(nuclear eng), 63. *Prof Exp:* From asst prof to assoc prof, 62-72, PROF NUCLEAR ENG, UNIV MICH, ANN ARBOR, 72-, CHMN DEPT, 79- *Concurrent Pos:* Fulbright travel grant & vis scientist, Nuclear Res Ctr, Karlsruhe, Ger, 65-66; consult to var indust orgn, 82- *Honors & Awards:* Glenn Murphy Award, Am Soc Eng Educ, 79; Fel Award, Am Nuclear Soc, 83. *Mem:* Am Nuclear Soc; Am Phys Soc; Inst Elec & Electronics Engrs; Am Soc Eng Educ. *Res:* Neutron spectroscopy; radiation detection and measurements; radioisotope imaging; medical instrumentation; neutron cross sections. *Mailing Add:* 3891 Waldenwood Ann Arbor MI 48105

KNOLL, HENRY ALBERT, b Englewood, NJ, Dec 10, 22; m 47; c 2. VISION. *Educ:* Univ Rochester, BS, 44; Ohio State Univ, MS, 48, PhD(physiol optics), 50. *Prof Exp:* Asst physiol optics, Ohio State Univ, 46-49; asst prof optics, Los Angeles Col Optom, 50-52, assoc prof geomet optics, 52-53, dean 53-55; from asst res biophysicist to assoc res biophysicist, Med Ctr, Univ Calif, Los Angeles, 55-58; res engr, 58-59, dept head, Lens Prod Develop & Instrument Res, 60-62 & Biophys Res & Develop, 62-66, SR SCIENTIST, BIOPHYS RES & DEVELOP, BAUSCH & LOMB INC, 66- *Mem:* AAAS; Optical Soc Am; Am Acad Optom. *Res:* Pupillary size changes associated with accommodation and convergence; refractive state of the eye in the absence of optical stimuli to accommodation; point source thresholds of the human eye; optics of ophthalmic lenses; infrared communication systems; contact lens research and development. *Mailing Add:* PO Box 755 Brunswick ME 04011

KNOLL, JACK, b Ashland, Wis, Feb 17, 24; m 48; c 1. ANIMAL PHYSIOLOGY. *Educ:* Mich State Univ, BS, 50, MS, 59, PhD(physiol), 62. *Prof Exp:* Asst physiol, Mich State Univ, 57-61; from asst prof to assoc prof, 62-76, prof, 76-81, EMER PROF BIOL, UNIV NEV, RENO, 81- *Mem:* AAAS; Sigma Xi. *Res:* Ion transport across natural biological membranes; fish physiology. *Mailing Add:* Rte 2 Box 271 Mason WI 54856

KNOLL, KENNETH MARK, b Pittsburgh, Pa, Apr 10, 41; m 63; c 2. GLACIAL GEOLOGY, PETROLEUM GEOLOGY. *Educ:* Antioch Col, BA, 64; Univ Wash, MS, 67; Univ Kans, PhD(glacial geol), 73. *Prof Exp:* Instr geol, Univ Wash, 68; asst prof geol, Winona State Col, 69-71; geologist, Shell Oil Co, 73-78; geologist, Rocky Mountain Div, Koch Explor Co, 78-80; GEOL MGR, FRONTIER GROUP, SOHIO PETROL CO, 80- *Mem:* Geol Soc Am; Sigma Xi; Am Asn Petrol Geologists. *Res:* Petroleum exploration geology of Rocky Mountains using seismic, magnetics and geophysical well logging techniques. *Mailing Add:* 443 E Gaywood Houston TX 77079

KNOLLENBERG, ROBERT GEORGE, b Mattoon, Ill, Aug 28, 39; m 66; c 3. INSTRUMENTATION. *Educ:* Eastern Ill Univ, BS, 61; Univ Wis-Madison, MS, 64, PhD(cloud physics), 67. *Prof Exp:* Res asst, Univ Wis-Madison, 64-66; scientist, NCAR Res Aviation, Colo, 67-69; asst prof geophys sci, Univ Chicago, 69-72; vis prof atmospheric sci, Colo State Univ, 73-75; PRES, PARTICLE MEASURING SYSTS, INC, 72-, CHIEF RES EXEC, 76- *Mem:* Am Meteorol Soc; Am Chem Soc; Am Inst Physics; Inst Elec & Electronics Engrs; Optical Soc Am. *Res:* Particle physics research and instrumentation development applied to the development of particle size spectrometers. *Mailing Add:* Particle Measuring Systs Inc 1855 S 57th Ct Boulder CO 80301

KNOLLMAN, GILBERT CARL, b Cleveland, Ohio, Mar 14, 28; m 59; c 4. MATERIAL SCIENCE, ULTRASONICS. *Educ:* Ga Inst Technol, BS, 49, MS, 50, PhD(physics), 61. *Prof Exp:* Instr physics, Ga Inst Technol, 49-50, res physicist, Eng Exp Sta, 50-62, asst prof math, Inst, 52-60; res scientist, Res Labs, Lockheed Missiles & Space Co, 62-63, head, Hydrospace Physics Lab, 62-70, staff scientist, 63-64, sr staff scientist, 64-, SR MEM, LOCKHEED PALO ALTO RES LABS, 66-, DIR, ADV ACOUST LAB, 71- *Concurrent Pos:* NSF grant, 60-61; consult, Ga Tech Res Inst, 62-65 & Lockheed Calif Co, 64-67; fel, Stanford Univ, 65-66; consult, Saratoga Systs, 70-76; mem Lockheed Res Comt, 70-71, chmn, 72-73; staff, Lockheed Eng Sci, 72-74; consult, Naval Weapons Lab, 73-; staff, Lockheed Mat Sci, 74- *Mem:* Fel AAAS; fel Am Phys Soc; Am Asn Physics Teachers; fel Am Inst Physics; fel NY Acad Sci; Res Soc Am; Acoust Soc Am; Sigma Xi. *Res:* Quantum field theory; quantum and statistical mechanics; many-body theory; superconductivity and superfluidity; liquid state physics; theoretical acoustics and hydrodynamics; orbital and wave mechanics; electromagnetic theory; viscoelasticity; ultrasonics; underwater acoustics and electronics; material sciences; nondestructive test and evaluation; ocean science. *Mailing Add:* 705 Charleston Ct Palo Alto CA 94303

KNOLLMUELLER, KARL OTTO, b Regensburg, Ger, July 12, 31; US citizen; m 68. INDUSTRIAL ORGANIC CHEMISTRY. *Educ:* Univ Munich, MS, 57, PhD(chem), 60.. *Prof Exp:* Res chemist, Diversey Corp, Ill, 60; from res chemist to sr res chemist, 60-74, res assoc, 74-84, SR RES ASSOC, OLIN CORP, CONN, 78- *Mem:* Am Chem Soc; AAAS; Sigma Xi. *Res:* Inorganic and organic phosphorus compounds; sequestration agents; metal treatment chemicals; high temperature stable polymers; functional fluids; chemicals for electronics; photoresists; chlorine/hypochlorites; sulfur chemistry; hydrosulfite bleaching/reductions; pulp & paper chemistry. *Mailing Add:* 28 Apple Tree Lane Hamden CT 06518

KNOP, CHARLES M(ILTON), b Chicago, Ill, Feb 18, 31; m 77. ELECTROMAGNETIC ENGINEERING, ELECTRICAL ENGINEERING. *Educ:* Ill Inst Technol, BSEE, 54, MSEE, 60, PhD(elec eng), 63. *Prof Exp:* Mem tech staff antenna res, Hughes Aircraft Co, 54-55; res asst, Princeton Univ, 55-56; asst engr, Armour Res Found, Ill, 56-58, assoc engr, 58-60; sr engr, Res Lab, Systs Div, Bendix Corp, 60-61; asst dir res & develop, Hallicrafters Co, 61-64; mem sr staff, Nat Eng Sci Co, Calif, 64-65; assoc dir antenna design, Andrew Corp, 65-67, res consult, 67-68; independent consult electrodyn, 68-70; mgr res & develop, 70-76, dir res & develop, 76-80, CHIEF SCIENTIST & DIR, ANTENNA RES, ANDREW CORP, 80- *Concurrent Pos:* Lectr, Dept Elec Eng, Ill Inst Technol, 66-72. *Mem:* fel Inst Elec & Electronics Engrs; Am Phys Soc. *Res:* Electromagnetic wave radiation, propagation, scattering and diffraction as related to antennas and communication systems; waveguiding systems. *Mailing Add:* Andrew Corp 10500 W 153rd St Orland Park IL 60462

KNOP, CHARLES PHILIP, b Detroit, Mich, May 23, 27; m 52; c 9. INORGANIC CHEMISTRY. *Educ:* Aquinas Col, BS, 52; Mich State Univ, PhD(chem), 58. *Prof Exp:* Chemist, Haviland Prod Co, 52-53; asst, Mich State Univ, 53-58; res asst prof, 65-70, chmn dept, 69, assoc prof, 70-76, PROF CHEM, GRAND VALLEY STATE COL, 76-, CHMN DEPT, 74- *Mem:* Am Chem Soc; Sigma Xi. *Res:* Pollution abatement; plating wastes recovery. *Mailing Add:* Dept Chem Grand Valley State Col Allendale MI 49401

KNOP, HARRY WILLIAM, JR, b Chicago, Ill, June 20, 20; m 50. PHYSICS. *Educ:* Ripon Col, BA, 42; Univ Wis, PhD(physics), 48. *Prof Exp:* Res physicist, Photo Prods Dept, E I DuPont De Nemours & Co, 49-56, res supvr, 56-60, sales tech supvr, 60-65, field sales mgr, 65-66, mgr, Rochester Res & Develop, 66-69, res mgr, Exp Sta Lab, 69-73, dir Du Pont Photo Prods, Japan, 73-75, Lab dir, Exp Sta Lab, 76-83; RETIRED. *Mem:* Am Phys Soc; Soc Motion Picture & TV Eng; Sigma Xi. *Res:* Physics of photography, solid state physics. *Mailing Add:* 313 Brockton Rd Sharpley Wilmington DE 19803

KNOP, OSVALD, b Kurim, Czech, July 11, 22; nat Can; m 51; c 1. INORGANIC CHEMISTRY, SOLID STATE CHEMISTRY. *Educ:* Masaryk Univ, Czech, BS, 46; Laval Univ, DSc(phys chem), 57. *Prof Exp:* Asst, Calif Inst Technol, 49; lectr indust chem, Dept Chem Eng, NS Tech Col, 50-53, from asst prof to assoc prof chem, 53-64; from assoc prof to prof, 64-90, EMER PROF CHEM, DALHOUSIE UNIV, 90-, HARRY SHIRREFF PROF CHEM RES, 81- *Mem:* Fel Chem Inst Can; Royal Soc Chem. *Res:* Structural inorganic and solid state chemistry; computer-simulation methods; application of combinatorial analysis and graph theory to chemistry and physics. *Mailing Add:* Dept Chem Dalhousie Univ Halifax NS B3H 4J3 Can

KNOPF, DANIEL PETER, b Louisville, Ky, July 28, 16; m 44; c 4. CHEMISTRY. *Educ:* Univ Louisville, AB, 38, MS, 40, JD, 68. *Prof Exp:* Night supt, 42-45, plant mgr, 45-52, mgr labs, 52-66, dir labs, Brown-Forman Distillers Corp, 66-81; CONSULT, 81- *Honors & Awards:* Distinguished Serv Award, Distillers Feed Res Coun. *Mem:* Am Chem Soc; Am Soc Testing & Mat. *Mailing Add:* Distillery & Flavor Consult 118 Beechwood Rd Louisville KY 40207

KNOPF, FRITZ L, b Aurora, Ohio, June 6, 45; c 3. WILDLIFE ECOLOGY, ORNITHOLOGY. *Educ:* Hiram Col, Ohio, BA, 67; Utah State Univ, MS, 73, PhD(wildlife ecol), 75. *Prof Exp:* Instr, Utah State Univ, 75-76; asst prof wildlife ecol, Okla State Univ, 76-79; res wildlife biologist, 80-81, PROJ LEADER, US FISH & WILDLIFE SERV, 82- *Concurrent Pos:* Ed, Wildlife Soc Bull, vols, 11-13, 83-85. *Honors & Awards:* Douglas L Gilbert Award for Outstanding Prof Achievement. *Mem:* Wildlife Soc; Cooper Ornith Union; Am Ornithologists Union; Wilson Ornith Soc; Soc Conserv Biol. *Res:* Habitat preference and utilization by wild birds; landscape ecology, management of landscapes; conservation of biodiversity. *Mailing Add:* Nat Ecol Res Ctr US Fish & Wildlife Serv 4512 McMurray Ave Ft Collins CO 80525-3400

KNOPF, PAUL M, b Trenton, NJ, Apr 4, 36; m 58; c 3. BIOLOGY. *Educ:* Mass Inst Technol, PhD(molecular biol), 62. *Prof Exp:* PROF MED SCI, BROWN UNIV, 77- *Concurrent Pos:* Fulbright fel, 78-79 & Fogarty fel, 86-87. *Mem:* AAAS; Am Asn Immunologists; Am Soc Trop Med & Hyg; Soc Neurosci. *Res:* Development of vaccine against human schistosomiasis; central nervous system/immune system interactions. *Mailing Add:* Div Biol & Med Brown Univ Box G-B616 Providence RI 02912

KNOPF, RALPH FRED, b Muskegon, Mich, Mar 26, 26; m 54; c 3. INTERNAL MEDICINE. *Educ:* Univ Mich, BS, 51, MD, 54; Am Bd Internal Med, dipl, 63. *Prof Exp:* Intern med, Virginia Mason Hosp, Seattle, Wash, 54-55, resident internal med, 55-56; resident internal med, 56-58, resident ophthal, 58-59, fel internal med, 59-62, from instr to assoc prof, 62-73, PROF INTERNAL MED, MED CTR, UNIV MICH, ANN ARBOR, 73- *Mem:* Am Fedn Clin Res; Am Diabetes Asn; Endocrine Soc. *Res:* Diabetes mellitus; inter-relationship between carbohydrate, protein and lipid metabolism and insulin glucagon and growth hormone secretion. *Mailing Add:* 3820 Taubman Ctr Univ Mich Hosps Univ Mich Ann Arbor MI 48109

KNOPF, ROBERT JOHN, b West New York, NJ, Apr 18, 32; m 83; c 3. ORGANIC CHEMISTRY, POLYMER CHEMISTRY. *Educ:* Gettysburg Col, BA, 54; Princeton Univ, MA, 56, PhD(org chem), 57. *Prof Exp:* Chemist, Union Carbide Corp, SCharleston, 57-66, group leader res & develop, 66-70, res scientist, 70-75, prod develop mgr, 74-76, sr develop scientist, Ethylene Oxide-Glycol Div, 75-86; RETIRED. *Concurrent Pos:* Consult alkoxylation chem, Union Carbide, 86-90. *Mem:* Am Chem Soc. *Res:* Flame retardants for plastics; chemistry of isocyanates; polyurethane foams, coatings and elastic fibers; aldol condensation chemistry; condensation polymerizations in solution; polyethers; hydrogel polymers; photocure coatings intermediates; surfactant intermediates; ethylene oxide derivatives; catalysis of ethoxylation reactions. *Mailing Add:* 2657 Lakeview Dr St Albans WV 25177

KNOPKA, W N, b Buffalo, NY, Dec 1, 38; m 65; c 3. FIBER TECHNOLOGY, WIRE TECHNOLOGY. *Educ:* Canisius Col, BS, 61; Seton Hall Univ, MS, 63, PhD(org chem), 65. *Prof Exp:* Sr res chemist, Cent Res & Develop, FMC Corp, NJ, 65-69, group leader, Fiber Div, Pa, 69-73, prod mgr, 73-76, agr chem, 76-77; dir fabric develop, Goodyear Tire & Rubber Co, 77-79, dir elastomer & chem res, 79-83; vpres corp develop, Interchem Inc, 86-88; PRES, INTERCHEM DEVELOP CO, 88- *Mem:* AAAS; Asn Comput Mach; Indust Res Inst. *Res:* Specialty organic chemical intermediates; fiber processing; wire process technology; condensation and addition polymers; synthetic rubbers; rubber chemicals. *Mailing Add:* Interchem Develop Co 2859 Paces Ferry Rd Suite 600 Atlanta GA 30339

KNOPOFF, LEON, b Los Angeles, Calif, July 1, 25; m 61; c 3. GEOPHYSICS, PHYSICS. *Educ:* Calif Inst Technol, BS, 44, MS, 46, PhD(physics), 49. *Prof Exp:* From asst prof to assoc prof physics, Miami Univ, 48-50; from res assoc to assoc res geophysicist, Inst Geophys, 50-56, assoc prof geophys, 56-57, assoc dir, Inst Geophys & Planetary Physics, 72-86, PROF GEOPHYS, UNIV CALIF, LOS ANGELES, 57-, PROF PHYSICS, 59-, RES MUSICOLOGIST, 61- *Concurrent Pos:* Mem earth sci panel, NSF, 59-62; sr fel NSF, Cambridge Univ, 60-61; prof, Calif Inst Technol, 62-63; secy gen, Int Upper Mantle Comt & chmn US comt, 63-71; vis prof, Technische Hochschule, Karlsruhe, Ger, 66; chmn comt math geophys, Int Union Geophys & Geophysicists, 71-82; vis prof, Harvard Univ, 72 & Univ Chile, Santiago, 73; mem US nat comt, Int Union Geophys & Geophysicists, 73-77; Guggenheim fel, 76-77; mem educ adv bd, Guggenheim Found, 89-; distinguished geophys lectr, Tex A&M Univ, 90. *Honors & Awards:* Harold Jeffreys Lectr, Royal Astron Soc, 77; Emil Wiechert Medal, Ger Ger Geophys Soc, 78; Gold Medal, Royal Astron Soc, 79; Sidney Chapman Mem Lectr, Univ Alaska, 88; Medal, Seismol Soc Am, 90. *Mem:* Nat Acad Sci; fel Am Acad Arts & Sci; Am Phys Soc; hon mem Seismol Soc Am; fel Am Geophys Union; fel AAAS; Royal Astron Soc. *Res:* Elastic wave propagation; theoretical and observational seismology; acoustics of solids; physics and chemistry of deep interior of earth; physics of high pressures; systematic musicology; author or co-author of over 300 scientific papers and publications; theory of earthquakes. *Mailing Add:* Inst Geophys & Planetary Physics Univ Calif Los Angeles CA 90024

KNOPP, JAMES A, b Grand Rapids, Mich, Oct 26, 40; m 62; c 2. BIOPHYSICAL CHEMISTRY. *Educ:* Carleton Col, BA, 62; Univ Ill, PhD(biophys chem), 67. *Prof Exp:* Investr biol div, Oak Ridge Nat Lab, 67-69, biophysicist, 67-69; asst prof, 69-74, ASSOC PROF BIOCHEM, NC STATE UNIV, 74- *Mem:* Am Chem Soc; Am Soc Biol Chemists. *Res:* Physical chemistry of proteins; fluorescence techniques in biochemistry; video image microscopy. *Mailing Add:* Dept Biochem PO Box 7622 NC State Univ Raleigh NC 27695-7622

KNOPP, MARVIN ISADORE, b Chicago, Ill, Jan 4, 33; m 57; c 4. NUMBER THEORY, MATHEMATICAL ANALYSIS. *Educ:* Univ Ill, BS, 54, AM, 55, PhD(math), 58. *Prof Exp:* Asst math, Univ Ill, 54-58; res mathematician, Space Tech Labs, 58-59; fel math, NSF Inst Advan Study, 59-60; from asst prof to prof math, Univ Wis, Madison, 60-70; prof math, Univ Ill, Chicago Circle, 70-76; prof math, Bryn Mawr Col, 88-89; PROF MATH, TEMPLE UNIV, 76- *Concurrent Pos:* Res grants, NSF, 60-89, NSA, 90-; mathematician, Nat Bur Standards, 63-64; vis prof, Math Inst, Univ Basel,

Switz, 68-69, Ohio State Univ, 79; visitor, Inst Advan Study, 75, 78, mem, 88. *Mem:* Am Math Soc. *Res:* Construction of automorphic forms; uniformization and Riemann surfaces; Eichler cohomology of automorphic forms; rational period functions of automorphic integrals and quadratic forms; Fourier coefficients of modular forms of small weight; Mellin transforms of modular integrals. *Mailing Add:* Temple Univ Philadelphia PA 19122

KNOPP, ROBERT H, MEDICINE. *Prof Exp:* DIR, NORTHWEST LIPID RES CLIN, 78-; PROF MED, SCH MED, UNIV WASH, 82- *Mailing Add:* Sch Med Univ Wash 326 Ninth Ave Rm 465-ZA36 Seattle WA 98104

KNOPP, WALTER, b Ostrava, Czech, Oct 22, 22; US citizen; m 51; c 2. PSYCHIATRY. *Educ:* Univ Heidelberg, MD, 50; Am Bd Psychiat & Neurol, dipl, 61. *Prof Exp:* Intern med, Univ Heidelberg Hosp, 50, resident pediat, 51; toxicologist, Europ Lab, US Army Med Ctr, Ger, 52-54; intern, Glens Falls Hosp, NY, 54-55; psychiat resident, Springfield State Hosp, Sykesville, Md, 55-58, chief serv, Men's Group, 58-60; from instr to prof, 85, EMER PROF PSYCHIAT, COL MED, OHIO STATE UNIV, 85- *Concurrent Pos:* Attend staff physician, Ohio State Univ Hosps, 60-85, coordr preclin educ, 66-85, premed educ, 73-85, asst prof, Sch Soc Work, 64-67, adv, Grad Sch, 65-85, asst prof phys med, Univ, 66-67; consult, Vet Admin Hosp, Chillicothe, Ohio, 61-85, Vet Admin Ment Hyg Clin, Columbus, 65-76 & Athens Ment Health Ctr, 79-85; dipl mem psychiat Pan-Am Med Asn, 65-75. *Mem:* Fel Am Psychiat Asn; Soc Neurosci; NY Acad Sci; Int Col Neuropsychopharmacol. *Res:* Bridging the gaps between neuro-sciences, human behavior and patient oriented therapeutic research and between behavioral sciences, biological sciences and educational technology and research. *Mailing Add:* 4829 Sherry Lane Ft Myers FL 33908-2025

KNOPPERS, ANTONIE THEODOOR, b Kapelle, Neth, Feb 27, 15; nat US; m 39; c 4. PHARMACOLOGY. *Educ:* Univ Amsterdam, MD, 39; Univ Leyden, PharD, 41. *Hon Degrees:* DSc, Worcester Polytech Inst, 65. *Prof Exp:* First asst, Pharmacol Inst, Univ Amsterdam, 40-43; dir pharmacol, Amsterdamsche Chininefabriek, 43-49; mem managing bd combination Amsterdamsche, Bandoengsche en Nederlandsche Kininefabrieken, 50-52; mgr med serv, Merck-NAm, Inc, 52-53, dir med serv, Merck Sharp & Dohme Int Div, 53-55, dir sci activities, 55, vpres & gen mgr, 55-57, pres, 57-67, from sr vpres to pres, Merck & Co, 67-74, vchmn, 74-75; RETIRED. *Concurrent Pos:* Mem, Malaria Comn Neth, 51-53; mem, Coun Foreign Rels; dir, Centucor Inc; mem bd dirs, Neth-Am Found; vchmn, Salk Inst. *Mem:* Fel NY Acad Sci. *Res:* Physiology and pharmacology of the regulation of the body temperature; chemotherapy of malaria, especially chemoresistance; cardiovascular pharmacology. *Mailing Add:* Seven Obow Lane Summit NJ 07901

KNORR, DIETRICH W, b Waidhofen, Austria, July 3, 44; m 68; c 3. FOOD TECHNOLOGY. *Educ:* Univ Agr, Austria, dipling, 71, Dr, 74. *Prof Exp:* Asst prof, dept food technol, Univ Agr, Vienna, 74-78; assoc prof food processing, 78-84, PROF FOOD PROCESSING & BIOTECHNOL, UNIV DEL, 84- *Concurrent Pos:* Vis prof food sci, Western Regional Res Ctr-USDA, 76-77 & Cornell Univ, 78; vis prof food sci, Asn Biotechnol Res, Braunschweig, Fed Rep Ger, 85-86; ed, J Food Biotechnol. *Mem:* Am Chem Soc; Inst Food Technologist; Am Acad Appl Sci; NY Acad Sci; Asn Austrian Food & Biotechnologists. *Res:* Food processing; biopolymers; plant cell culture, immobilization, enzyme technology; food technology. *Mailing Add:* Dept Food Sci Alison Hall Univ Del Newark DE 19716

KNORR, GEORGE E, b Munich, Ger, Jan 29, 29; m 60; c 3. PLASMA PHYSICS. *Educ:* Munich Tech Univ, Vordiplom, 51, dipl, 54; Univ Munich, PhD(physics), 63. *Prof Exp:* Physicist, Philips Corp, C H F Mueller AG, 55-58; res asst, Max Planck Inst Physics, 58-65; res fel plasma physics, Princeton Univ, 63-64; res assoc, Inst Plasma Physics, Garching, Ger, 65-66; asst prof & res assoc physics, Univ Calif, Los Angeles, 66-67; assoc prof physics & astron, 67-74, PROF PHYSICS, UNIV IOWA, 74- *Mem:* Am Phys Soc. *Res:* Radiation from plasmas; confinement and instabilities of plasmas; wave propagation, non-linear effects, numerical methods and computer simulation of plasmas. *Mailing Add:* Dept Physics & Astron Univ Iowa Iowa City IA 52240

KNORR, PHILIP NOEL, b Mitchell, Nebr, Apr 9, 16; m 64. FOREST ECONOMICS, RESOURCE MANAGEMENT. *Educ:* Univ Calif, BS, 38; Duke Univ, MF, 40; Univ Minn, PhD(forest econ), 63. *Prof Exp:* Timber supvr & mapper, US Forest Serv, 38-39, jr forester, Southern Forest Exp Sta, New Orleans, 41; asst res forester, Weyerhaeuser Co, 46-48; asst prof forest mgt, Ore State Univ, 48-51; from asst to instr forest mgt, Univ Minn, 54-59; assoc prof forestry, 59-66, PROF FORESTRY, UNIV ARIZ, 66-, RES SCIENTIST, AGR RES STA, 76- *Honors & Awards:* Ford-Bartlett Award, Am Soc Photogram, 66. *Mem:* Fel AAAS; Soc Am Foresters; Am Econ Asn; Am Agr Econ Asn; Am Soc Photogram; Sigma Xi. *Res:* Decision making in forest management; remote sensing, including photo-interpretation; decision making in renewable resources policy. *Mailing Add:* 1333 E Indian Wells Rd Tucson AZ 85718

KNORR, THOMAS GEORGE, b Buffalo, NY, Apr 14, 32; m 56; c 5. SOLID STATE PHYSICS. *Educ:* Canisius Col, BS, 53; Case Inst Technol, MS, 55, PhD(physics), 58; Univ Detroit, MA, 78. *Prof Exp:* Asst, Case Inst Technol, 53-58; from instr to asst prof physics, Univ Dayton, 58-60; sr physicist, Battelle Mem Inst, 60-65; assoc prof, 65-77, PROF PHYSICS, WHEELING COL, 77- *Mem:* Am Phys Soc; Am Inst Physics; Am Asn Physics Teachers. *Res:* Thin films; defect properties and structure; radiation damage; linguistics; surface structures. *Mailing Add:* 18 Bae Mar Pl Wheeling WV 26003

KNOSPE, WILLIAM H, b Oak Park, Ill, May 26, 29; m 54; c 3. HEMATOLOGY. *Educ:* Univ Ill, Urbana, AB, 51, BS, 52; Univ Ill Med Ctr, MD, 54; Univ Rochester, MS, 62. *Prof Exp:* Chief med serv, US Army Hosp, Berlin, Ger, 58-61; attend physician, Walter Reed Gen Hosp, Washington, DC, 63-64, asst chief hemat serv, 64-66; assoc prof med, Univ Ill Col Med, 69-72; assoc dir hemat sect & chief radio hemat lab, Presby-St Luke's Hosp, 67-74, dir clin hemat sect, 74-82, ELODIA KEHM PROF & DIR HEMAT, RUSH PRESBY ST LUKE'S MED CTR, 86-; PROF MED, RUSH MED COL, 74- *Concurrent Pos:* Investr radiation biol, Walter Reed Army Inst Res, Washington, DC, 62-64 & investr hemat, 64-66; asst prof med, Col Med, Univ Ill, 67-69 & assoc prof med, Rush Med Col, 71-74; attend staff & attend physician, Presby-St Luke's Hosp, 67-; prin investr, Southeastern Cancer Study Group, 69-80 & Polycythemia Vera Study Group, 78-; vis prof med, dept hematol, Univ Basel, Switz, 80-81, Free Univ Berlin, WGermany, 81, Nat Tiwan Univ Sch Med, Taipei, Taiwan, 85 & McGill Cancer Ctr, McGill Univ, Montreal, 87; vis prof med, McGill Cancer Ctr, McGill Univ, Montreal, Can, 87. *Mem:* Fel Am Col Physicians; Am Soc Hemat; Am Fedn Clin Res; Radiation Res Soc; Sigma Xi; Int Soc Exp Hemat. *Res:* Radiation effects upon bone marrow; role of sinusoidal microcirculation in aplastic anemias; regulation of hematopoietic stem cells; clinical investigation of leukemia and lymphomas; role of hematopoietic stroma in hematopoiesis. *Mailing Add:* Rush-Presby-St Luke's Med Ctr 1653 W Congress Pkwy Chicago IL 60612-3864

KNOTEK, MICHAEL LOUIS, b Norfolk, Nebr, Nov 23, 43; m 69; c 3. SOLID STATE PHYSICS, SURFACE PHYSICS. *Educ:* Iowa State Univ, BS, 66; Univ Calif, Riverside, MS, 69, PhD(physics), 72. *Prof Exp:* Physicist amorphous semiconductors, Naval Weapons Ctr, 70-72; mem tech staff superionic conductors & surface physics, Sandia Lab, 73-85, supvr, Surface Physics Div, 79-85; chmn, Nat Synchrotron Light Source, Brookhaven Nat Lab, 85-89; SR SCI DIR, PAC NORTHWEST LABS, 89-, DIR, MOLECULAR SCI RES CTR & ENVIRON & MOLECULAR SCI LAB, 89- *Concurrent Pos:* Consult, Bourns Inc, 70. *Mem:* Am Vacuum Soc; Sigma Xi. *Res:* Transport properties of amorphous and disordered semiconductors; transport properties of superionic conductors, especially related to surface and interface properties; surface studies of photocatalytic properties of rutile and related materials; surface science; electron and photon stimulated desorption. *Mailing Add:* One Sprout Rd Richland WA 99352

KNOTH, WALTER HENRY, JR, b New York, NY, Feb 18, 30; m 53; c 2. HETEROPOLYANIONS, BORON HYDRIDES. *Educ:* Syracuse Univ, BS, 50; Pa State Univ, PhD(chem), 54. *Prof Exp:* Res chemist, E I du Pont de Nemours & Co, Inc, 53-85; RETIRED. *Mem:* Sigma Xi. *Res:* Organo-silicon and organoboron chemistry; boron hydrides; nitrogen complexes; transition metal chemistry. *Mailing Add:* Box Six Mendenhall PA 19357-0006

KNOTT, DONALD MACMILLAN, b Boston, Mass, Oct 20, 19; m 43; c 4. INDUSTRIAL ORGANIC CHEMISTRY. *Educ:* Mass Inst Technol, SB, 41, PhD(org Chem), 47. *Prof Exp:* With Gen Elec Co, 47-54; with Chas Pfizer & Co, 54-63, chem consult, 63-68; pres, Chemconsul Inc, 68-88; PRES, KNOTT ASSOCS, 88- *Mem:* Am Chem Soc; Com Develop Asn; Tech Asn Pulp & Paper Indust. *Res:* Commercial chemical development; pulp and paper technology; wood science and technology. *Mailing Add:* 9790 NE Benton St Newport OR 97365

KNOTT, DOUGLAS RONALD, b New Westminster BC, Nov 10, 27; m 50; c 4. CYTOLOGY. *Educ:* Univ BC, BSA, 48; Univ Wis, MS, 49, PhD, 52. *Prof Exp:* From asst prof to assoc prof, 52-65; prof & head dept, 65-75, prof crop sci, 75-88, ASSOC DEAN RES, UNIV SASK, 88- *Concurrent Pos:* Res Adv, Zambia-Can Wheat Proj. *Mem:* Fel Am Soc Agron; Genetics Soc Can; fel Agr Inst Can; Can Soc Agron; Crop Sci Soc Am; Sigma Xi. *Res:* Genetics and cytogenetics of rust resistance in wheat; transfer of resistance to wheat from its relatives; wheat breeding. *Mailing Add:* Col Agr Univ Saskatchewan Saskatoon SK S7N 0W0 Can

KNOTT, FRED NELSON, b Oxford, NC, July 18, 33; m 55; c 4. ANIMAL NUTRITION. *Educ:* NC State Col, BS, 55; NC State Univ, MS, 62; Va Polytech Inst, PhD(animal nutrit), 68. *Prof Exp:* Res asst, NC State Univ, 55-56, agr exten agent, 56-67, assoc prof, 72-79, exten dairy specialist, 57-65, specialist in chg, Exten Dairy Husb, 82, EXTEN DAIRY SPECIALIST DAIRY HUSB, NC STATE UNIV, 67-, PROF, 79- *Concurrent Pos:* Mem, Exten Serv Team, VI, 78. *Mem:* Am Dairy Sci Asn; Sigma Xi. *Res:* Utilization of urea by dairy cattle as a supplement to protein nutrition. *Mailing Add:* 821 Ravenwood Dr Raleigh NC 27606

KNOTT, JOHN RUSSELL, b Chicago, Ill, Nov 6, 11; m 35; c 3. NEUROPHYSIOLOGY. *Educ:* Univ Iowa, BA, 35, MA, 36, PhD, 38. *Prof Exp:* Res asst psychol, 38-41, asst prof, 41-48, from assoc prof to prof psychiat, 48-74, EMER PROF PSYCHIAT, UNIV IOWA, 74-; PROF NEUROL, TUFTS UNIV, 77- *Concurrent Pos:* Vpres, Inst Fedn Socs Electroencephalog & Clin Neurophysiol, 57-61; head div electroencephalog & neurophysiol, Univ Psychopath Hosp, 58-74; mem neurol A study sect, NIH, 67-71, mem neurol prog proj rev comt B, 72-76; vis prof psychiat, La State Univ, 69-71; prof neurol, Sch Med, Boston Univ, 75-77; consult electroencephalography, 80- *Mem:* Am Electroencephalog Soc (pres, 57). *Res:* Neurophysiological bases of behavior. *Mailing Add:* 701 Oaknoll Dr Apt 5438 Iowa City IA 52246

KNOTT, ROBERT F(RANKLIN), chemical engineering, organic chemistry, for more information see previous edition

KNOTTS, GLENN RICHARD, pharmacology, public health, for more information see previous edition

KNOUS, TED R, b Ely, Nev, May 11, 49; m 69; c 2. FUNGAL PHYSIOLOGY & TOXINS, PLANT TISSUE CULTURE. *Educ:* Univ Nev, BS, 72, MS, 74; Univ Minn, PhD(plant path), 79. *Prof Exp:* Teaching fel plant path, Univ Minn, 79; ASST PROF DIS PHYSIOL, DEPT PLANT SCI, UNIV NEV, 79- *Concurrent Pos:* Consult, Sierra Biotechnol, 82-; mem, Admin Comt, Nat Plant Pest Surv & Detection Prog, 83- & Dis & Pathogen Physiol Comt, Am Phytopath Soc, 84-; assoc ed, Plant Dis, Am Phytopath Soc, 84- *Mem:* Am Phytopath Soc; Am Soc Plant Physiologists; AAAS; Sigma Xi. *Mailing Add:* Res Promo Serv Bowman Hall No 400 Univ Wis-Stout Menomonie WI 54751

KNOWLER, LLOYD A, actuarial science, statistics; deceased, see previous edition for last biography

KNOWLES, AILEEN FOUNG, b China, Aug 9, 42; US citizen; m 69; c 2. BIOCHEMISTRY. *Educ:* Nat Taiwan Unvi, BS, 63; Univ Calif, Riverside, PhD(biochem), 68. *Prof Exp:* Fel biochem, Cornell Univ, 68-70, res assoc, 72-77; scientist, Pub Health Res Inst, New York, 70-72; from res biochemist to ASSOC PROF, BIOCHEM, CANCER CTR, UNIV CALIF, 77- *Mem:* Am Soc Biochem & Molecular Biol. *Res:* Biophysics and biochemsitry of membrane-bound enzymes; ATP hydrolyzing enzymes of tumor membranes. *Mailing Add:* Dept Biol Northeastern Univ 360 Huntington Ave Boston MA 02115

KNOWLES, BARBARA B, b New York, NY, Feb 27, 37; div; c 2. GENETICS, CELL BIOLOGY. *Educ:* Middlebury Col, AB, 58, Ariz State Univ, MS, 63, PhD(zool), 65. *Prof Exp:* Res asst drosophila genetics, Ariz State Univ, 62-65; res fel genetics, Univ Calif, Berkeley, 65-66; from res asst to res assoc, 67-76, assoc prof, 77-83, PROF, WISTAR INST ANAT & BIOL, UNIV PA, 83-, WISTAR PROF PATH, LAB MED & MICROBIOL, 84- *Concurrent Pos:* Career develop award, Nat Inst Allergy & Infectious Dis, 75; mem, Univ Pa Immunol Grad Group, 77-; consult, Can Info Dissemination Serv, Am Can Soc; mem, Can Res Manpower Review Bd, 80-83; vis sr scientist, Cold Spring Harbor Lab, 87-88. *Mem:* Am Soc Human Genetics; Genetics Soc Am. *Res:* Immunogenetics; genetic control of human cell surface molecules; murine immune response genes to tumor specific antigens; cell surface molecules of preimplantation stage mouse embryos; HBV and hepatocellular carcinoma. *Mailing Add:* Wistar Inst Anat & Biol Univ Pa 36th St Spruce Philadelphia PA 19104

KNOWLES, CECIL MARTIN, b Newton, Miss, Jan 6, 18; m 48; c 2. ORGANIC CHEMISTRY. *Educ:* Miss Col, BA, 39; Univ Tex, MA, 41, PhD(org chem), 43. *Prof Exp:* Instr chem, Univ Tex, 39-42; chemist, GAF Corp, NY, 43-51, mgr tech serv & com develop, 51-67; vpres & dir res, Trylon Chem, Inc, 68-72; vpres & tech dir, chem specialties groups, Emery Indust, Inc, 72-83; RETIRED. *Res:* Surface active agents; organic chemical specialties. *Mailing Add:* 1204 Edwards Rd Greenville SC 29615

KNOWLES, CHARLES ERNEST, b Ogden, Utah, Mar 7, 37; m 76; c 6. PHYSICAL OCEANOGRAPHY. *Educ:* Univ Utah, BS, 60; Tex A&M Univ, MS, 67, PhD(phys oceanog), 70. *Prof Exp:* Res scientist phys oceanog, Tex A&M Univ, 69-70; asst prof, 70-76, ASSOC PROF PHYS OCEANOG, NC STATE UNIV, 76- *Concurrent Pos:* Consult, NUC Corp, 75. *Mem:* Am Geophys Union. *Res:* Wind wave generation and dissipation in deep and finite depth water; wave current interaction; non-ideal wind wave generation and parametergation of directonal wave spectrum; tributary esturarine circulation dynamics. *Mailing Add:* Dept Earth Sci NC State Univ Main Campus Box 8208 Raleigh NC 27695-8208

KNOWLES, CHARLES OTIS, b Tallassee, Ala, Feb 1, 38; m 59; c 3. ENTOMOLOGY, PESTICIDE TOXICOLOGY. *Educ:* Auburn Univ, BS, 60, MS, 62; Univ Wis, PhD(entom), 65. *Prof Exp:* From asst prof to prof, 65-74, PROF ENTOM, UNIV MO, COLUMBIA, 74- *Concurrent Pos:* Vis sr scientist, Commonwealth Sci & Indust Res Orgn, Long Pocket Labs, Brisbane, Australia, 72. *Mem:* Entom Soc Am; Am Chem Soc; Soc Toxicol; Soc Environ Toxicol & Chem; Int Soc Study Xenobiotics. *Res:* Toxicology of insecticides; comparative insect and mite biochemistry; mode of action and metabolism of acaricides; environmental impact of pesticides. *Mailing Add:* Dept Entom 1-87 Agr Bldg Univ Mo Columbia MO 65211

KNOWLES, DAVID M, b Saginaw, Mich, July 22, 27; m 45; c 2. GEOLOGY. *Educ:* Mich Technol Univ, BS, 54, MS, 55; Columbia Univ, PhD(geol), 67. *Prof Exp:* Chief geologist, Can Javelin Ltd, 54-69; ASSOC PROF GEOL, LAKE SUPERIOR STATE COL, 69- *Mem:* Geol Soc Am; Geol Asn Can; Can Inst Mining & Metall; Sigma Xi. *Res:* Structural geology of Labrador Trough formations near Wabush Lake; superposed folds, Hudsonian events and Grenville Province events. *Mailing Add:* 905 McCandles St Sault Ste Marie MI 49783

KNOWLES, FRANCIS CHARLES, b Akron, Ohio, Sept 11, 41; m 69; c 2. BIOPHYSICAL CHEMISTRY. *Educ:* Univ Southern Calif, BA, 63; Univ Calif, Riverside, PhD(biochem), 68. *Prof Exp:* Lectr physiol & biochem, Mt Sinai Sch Med, New York, 68-72; res fel biochem, Cornell Univ, 72-77; SCIENTIST BIOCHEM, SCRIPPS INST OCEANOG, 77- *Mem:* Am Soc Plant Physiol; Am Chem Soc; Scand Soc Plant Physiol. *Res:* Biophysical chemistry; evolution of cooperative mechanisms of dioxygen transport proteins; arsenic biochemistry and mechanisms of arsenic toxicity; regulation of photosynthetic carbon dioxide fixation; pathways of photorespiration. *Mailing Add:* Seven Montchanin Ct Montchanin DE 19710

KNOWLES, HAROLD LORAINE, b Chicago, Ill, Aug 21, 05; m 31; c 1. PHYSICS. *Educ:* Phillips Univ, BA, 26; Univ Kans, PhD(physics), 31. *Prof Exp:* Asst instr physics, Univ Kans, 27-31; from instr to prof physics, 31-54, prof phys sci, Univ Fla, 54-72, head dept, 54-67, EMER PROF PHYS SCI, UNIV FLA, 72- *Concurrent Pos:* Supvr sig corps contract proj, War Res Lab, Fla Eng & Indust Exp Sta, 44-45. *Mem:* Am Phys Soc; Am Asn Physics Teachers. *Res:* Dielectric constant measurements; direction finder for atmospherics; physical sciences in general education. *Mailing Add:* 2805 NW 83rd St Apt C-106 Gainesville FL 32606

KNOWLES, HARROLD B, b Berkeley, Calif, July 28, 25; m 49; c 2. ELECTROMAGNETISM. *Educ:* Univ Calif, Berkeley, AB, 47, MA, 51, PhD(physics), 57. *Prof Exp:* Res asst oceanog, Univ Wash, 48-50; res asst physics, Univ Calif, Berkeley, 51-57; sr exp physicist, Lawrence Radiation Lab, Univ Calif, Livermore, 57-61; res assoc, Yale Univ, 61-64; from assoc prof to prof physics, Wash State Univ, 64-80; mem prin staff, BDM Corp, 80-83; MEM, TECH STAFF VI, ROCKETDYNE, 84- *Concurrent Pos:* Vis scientist, Los Alamos Sci Lab, 73-74, vis staff mem, 74-80; consult, Lawrence Berkeley Lab, 78, BDM Corp, 83-84, Los Alamos Sci Lab, 83-85, Rensselaer Polytech Inst, 83-85; res assoc, Air Force Weapon Lab, 80. *Mem:* Am Phys Soc. *Res:* Nuclear physics; accelerator design; protection against radiation hazards; radiological physics; charged particle optics, track detectors; free electron lasers. *Mailing Add:* PO Box 5225 Chatsworth CA 91313-5225

KNOWLES, JAMES KENYON, b Cleveland, Ohio, Apr 14, 31; m 52; c 3. APPLIED MATHEMATICS & MECHANICS. *Educ:* Mass Inst Technol, SB, 52, PhD(math), 57. *Hon Degrees:* DSc, Nat Univ Ireland, 85. *Prof Exp:* Instr math, Mass Inst Technol, 57-58; from asst prof to assoc prof, 58-65, PROF APPL MECH, CALIF INST TECHNOL, 65- *Mem:* fel Am Soc Mech Engrs; fel Am Acad Mech; Soc Indust & Appl Math. *Res:* Mathematical problems in continuum mechanics. *Mailing Add:* Thomas Lab Calif Inst Technol Pasadena CA 91125

KNOWLES, JEREMY RANDALL, b Apr 28, 35; m 60; c 3. CHEMISTRY. *Educ:* Magdalen Col, MA. *Prof Exp:* Res assoc, Calif Inst Tech, 61-62; fel & tutor, Wadham Col, Oxford, 62-74; lectr chem, Univ Oxford, 66-74; AMORY HOUGHTON PROF CHEM & BIOCHEM, HARVARD UNIV, 74-, DEAN, FAC ARTS & SCI, 91- *Concurrent Pos:* Vis prof, Yale Univ, 69, 71; Sloan vis prof, Harvard Univ, 73; Newton-Abraham vis prof, Oxford Univ, 83. *Honors & Awards:* Charmian Medal, Royal Soc Chem, 80; Prelog Medal, 89; Bader Award, Am Chem Soc, 89, Arthur Cope Scholar Award, 89. *Mem:* Nat Acad Sci; fel Am Acad Arts & Sci; Am Philos Soc; fel Royal Soc London. *Res:* Biorganic chemistry. *Mailing Add:* 44 Coolidge Ave Cambridge MA 02138

KNOWLES, JOHN APPLETON, III, b Portchester, NY, Nov 20, 35; m 84; c 2. DRUG METABOLISM. *Educ:* Middlebury Col, AB, 58; Ariz State Univ, PhD(anal chem), 66. *Prof Exp:* Res chemist, Orchem Dept, Chambers Works, E I du Pont de Nemours & Co, 66-68; res scientist, 68-78, MGR DRUG KINETICS SECT, WYETH LABS, AM HOME PROD CORP, 78- *Mem:* AAAS; Am Chem Soc. *Res:* Analysis of drugs and their metabolites. *Mailing Add:* 402 Herritage Dr Harlasville PA 19438

KNOWLES, JOHN WARWICK, b Toronto, Ont, Dec 9, 20; m 43; c 5. NUCLEAR PHYSICS. *Educ:* Univ Toronto, BA, 43; McGill Univ, PhD(physics), 47. *Prof Exp:* Res physicist nuclear physics, Nat Res Coun, 43-45; asst res officer gen physics, 47-54, assoc res officer, 54-58, SR RES OFFICER NUCLEAR PHYSICS, ATOMIC ENERGY CAN LTD, 58- *Mem:* Fel Am Phys Soc; Can Asn Physicists. *Res:* Low energy nuclear physics, crystal diffraction of neutron capture x-rays; precision measurements of reference x-rays; photo fission and related photo nuclear reactions. *Mailing Add:* Two Alexander Pl Box 736 Deep River ON K0J 1P0 Can

KNOWLES, PAULDEN FORD, genetics; deceased, see previous edition for last biography

KNOWLES, RICHARD JAMES ROBERT, b McPherson, Kans, Aug 2, 43; m 70; c 1. BIOPHYSICS. *Educ:* St Louis Univ, HBS, 65; Cornell Univ, MS, 69; Polytechnic Univ, PhD(physics), 79. *Prof Exp:* Chief med physicist, Long Island Col Hosp, 77-81; dir, Radiation Physics Lab, Downstate Med Ctr, 81-82; SR MED PHYSICIST, NY HOSP-CORNELL MED CTR, 82- *Concurrent Pos:* Clin asst prof radiol, Col Med, State Univ NY Downstate Med Ctr, 80-82; adj asst prof physics, Dept Natural Sci, York Col, City Univ NY, 82; asst prof, 82-89, ASSOC PROF PHYSICS IN RADIOL, CORNELL UNIV MED COL, 89- *Mem:* Am Phys Soc; Soc Nuclear Med; Health Physics Soc; Am Asn Physicists Med; Soc Magnetic Resonance Med; Soc Photo-Optical Instrumentation Engrs. *Res:* Medical image formation including MRI, CT, radioisotopes, and ultrasound; computerized image processing and analysis; data processing including artificial intelligence; data communications including PACS. *Mailing Add:* Dept Radiol NY Hosp Cornell Med Ctr 525 E 68th St New York NY 10021

KNOWLES, RICHARD N, b Wilmington, Del, Aug 8, 35; div; c 3. ORGANIC CHEMISTRY. *Educ:* Oberlin Col, BA, 57; Univ Rochester, PhD(chem), 60. *Prof Exp:* Res chemist, 60-73, res supvr, 73, develop supvr, Indust Chem Dept, 73-75, tech supt, 75, prod supt, 76-77, mfg mgr, 78-80, asst plant mgr, 80-83, Niagara, 83-87, PLANT MGR, E I DU PONT DE NEMOURS & CO, INC, BELLE, WVA, 87- *Concurrent Pos:* Vpres, SPUR; bd mem, Nat Inst Chem Studies. *Mem:* Am Chem Soc; Sierra Club; Nat Audubon Soc. *Res:* Herbicides, fungicides, insecticides, pharmaceutical agents, azo catalysts and flame retardants; colloidal silica; production of organic and inorganic chemicals in volumes ranging from lab scale to bulk commodities. *Mailing Add:* Agr Prod 901 W DuPont Ave Belle WV 25015

KNOWLES, ROGER, b Halifax, Eng, July 7, 29; m 63; c 2. MICROBIOLOGY. *Educ:* Univ Birmingham, BSc, 53; Univ London, PhD, 57. *Prof Exp:* From asst prof to prof, 57-71, PROF MICROBIOL, MACDONALD COL, MCGILL UNIV, 71- *Concurrent Pos:* Dept chmn, Macdonald Col, McGill Univ, 70-74 & 79-87. *Honors & Awards:* Can Soc Microbiol Award, 82. *Mem:* Soil Sci Soc Am; Am Soc Microbiol; Can Soc Microbiol; fel Royal Soc Can. *Res:* Soil and aquatic microbiology; forest soils; nitrogen fixation; denitrification; nitrification; methane metabolism. *Mailing Add:* Dept Microbiol Macdonald Campus McGill Univ 21111 Lakeshore Rd Ste Anne de Bellevue Quebec PQ H9X 1C0 Can

KNOWLES, STEPHEN H, b New York, NY, Feb 28, 40; m 65; c 2. IONOSPHERIC PHYSICS, RADIO ASTRONOMY. *Educ:* Amherst Col, BA, 61; Yale Univ, PhD(astron), 68. *Prof Exp:* Astronr, US Naval Observ, 61, ASTRONR, US NAVAL RES LAB, 61- *Mem:* AAAS; Int Union Radio Sci; Am Astron Soc; Int Astron Union; Am Geophys Union. *Res:* Radar astronomy; celestial mechanics; radio spectroscopy; very long baseline interferometry; ionospheric physics. *Mailing Add:* Naval Space Surveillance Systs Ctr Code 7134B Dalgren VA 22448

KNOWLTON, CARROLL BABBIDGE, JR, b Nashua, NH, Oct 11, 26; m 50; c 2. ENTOMOLOGY. *Educ:* Amherst Col, BA, 50; Cornell Univ, MS, 58, PhD(entom), 61. *Prof Exp:* From asst prof to assoc prof zool, 61-70, prof, 70-81, PROF BIOL, ORANGE COUNTY COMMUNITY COL, 81-, CHMN, DEPT BIOL & HEALTH SCI, 70- *Mem:* AAAS; Am Inst Biol Sci; Entom Soc Am; Lepidop Soc; Soc Syst Zool; Sigma Xi. *Res:* Mechanisms of survival in the Arthropoda, especially the Insecta. *Mailing Add:* Four Robalene Dr Goshen NY 10924

KNOWLTON, DAVID A, b Washington, DC, June 20, 38; m 83; c 2. CHEMISTRY. *Educ:* Capital Univ, BS, 61; Ohio State Univ, MSc, 67, PhD(biochem), 69. *Prof Exp:* Instr, Ohio State Univ, 70-71, asst prof, 71-73; researcher, Battelle Mem Inst, 73-76; res chemist, Gunning Refractories Inv, 77-81, dir new prof develop, 82-83; MGR RES & DEVELOP, QUIGLEY CO, INC, 84- *Concurrent Pos:* Consult, Cent Labs, State Ohio & Consolidated Biomed Labs, 71-73. *Mem:* Sigma Xi; Am Chem Soc; Am Ceramic Soc. *Res:* Development of carbon-bonded refractory, especially products for use in the manufacture of iron and steel. *Mailing Add:* BMI Refractories Inc PO Box 267 South Webster OH 45682

KNOWLTON, FLOYD M(ARION), b Milan, Ind, Jan 18, 18; m 41; c 3. CHEMICAL ENGINEERING. *Educ:* Purdue Univ, 39. *Prof Exp:* From process operator to res chemist, Joseph E Seagram & Sons, Inc, 39-42; supvr, Chem Control Div, Pa Ord Works, 42-43; develop engr, Naugatuck Chem Div, US Rubber Co, 43-45; develop engr, Bristol Labs Div, Bristol-Myers Co, 45-50, dept head, Chem Develop Pilot Plant, 50-58, mgr chem develop, 58-67, dir develop, 67-75, dir develop spec projs, Indust Div, 75-81; RETIRED. *Mem:* Am Chem Soc; Am Inst Chem Engrs. *Res:* Industrial fermentations; recovery processes; organic syntheses. *Mailing Add:* 1509 Browning Lane Bloomington IN 47401

KNOWLTON, FREDERICK FRANK, b Springville, NY, Nov 24, 34; c 4. WILDLIFE RESEARCH. *Educ:* Cornell Univ, BS, 57; Mont Stat Col, MS, 59; Purdue Univ, PhD(ecol, physiol), 64. *Prof Exp:* Proj biologist, Mont Fish & Game Dept, 59; lectr biol, Univ Mo, Kansas City, 64; WILDLIFE BIOLOGIST, US FISH & WILDLIFE SERV, 64- *Concurrent Pos:* Vis assoc prof, Cornell Univ, 71; assoc prof wildlife sci, Utah State Univ, 72- *Mem:* Wildlife Soc; Am Soc Mammal; Wildlife Dis Asn; Nat Audubon Soc. *Res:* Dynamics and mechanisms of natural vertebrate populations; especially mammalian physiology and phenomenoon of predation; ungulates; the larger carnivores. *Mailing Add:* 1398 N 1720 E Logan UT 84321

KNOWLTON, GEORGE FRANKLIN, entomology; deceased, see previous edition for last biography

KNOWLTON, GREGORY DEAN, b Santa Barbara, Calif, Jan 6, 46; m 83; c 1. CHEMISTRY OF PROPELLANTS, PYROTECHNICS & EXPLOSIVES. *Educ:* San Jose State Univ, BS, 74, MS, 76; Ariz State Univ, PhD(chem), 82. *Prof Exp:* Res geochemist, Lawrence Livermore Nat Lab, 75 & 76; res assoc, Ariz State Univ, 76-79; anal lab mgr, Commerce Metal Refiners, 79-80; SR CHEMIST & PRIN INVESTR, TALLEY DEFENSE SYSTS, 80- *Concurrent Pos:* Instr chem, Phoenix Col, 78-80. *Mem:* Am Chem Soc; fel Am Inst Chemists; Int Pyrotechnics Soc; Sigma Xi; Am Defense Preparedness Asn. *Res:* Pyrolysis of complex salts and organometallics; adsorption and desorption of gases and liquids inzeolites; determination of light elements in refractory materials; inorganic azide research; development, analysis and testing of propellants, pyrotechnics and explosives. *Mailing Add:* 615 W Summit Cir Chandler AZ 85224

KNOWLTON, NANCY, b Evanston, Ill, May 30, 49; m 83; c 1. SYSTEMATICS, MARINE BIOLOGY. *Educ:* Harvard Univ, AB, 71; Univ Calif, Berkeley, PhD(zool), 78. *Prof Exp:* NATO fel, Univ Liverpool & Cambridge Univ, Eng, 78-79; from asst prof to assoc prof biol, Yale Univ, 79-84; RES SCIENTIST, SMITHSONIAN TROP RES INST, 85- *Concurrent Pos:* Ed, Am Scientist; panelist, Animal Learning & Behav, NSF. *Mem:* Animal Behav Soc; Soc Study Evolution; Ecol Soc Am; AAAS; Soc Systematic Zool; Am Soc Naturalists. *Res:* Evolution of aggression; coral reef ecology; crustaceans; anemone commensals; sibling species. *Mailing Add:* Smithsonian Trop Res Inst Miami FL 34002-0011

KNOWLTON, ROBERT CHARLES, b Gardner, Mass, Aug 3, 29; m 52; c 1. CHEMICAL ENGINEERING. *Educ:* Northeastern Univ, BS, 52; Newark Col Eng, MS, 58. *Prof Exp:* Appl Engr, Worthington Corp, 52-55; serv engr, Eng Dept, E I du Pont de Nemours & Co, Inc, 55-58, res engr, Textile Fibers Dept, 58-65, sr res engr, 65-75, sr res assoc, Textile Fibers Dept, 75-87; RETIRED. *Mem:* Am Inst Chem Engrs; Textured Yarn Asn Am. *Res:* Synthetic fibers; spinning, drawing, and processing of polyester fibers; texturing of continuous filament yarns. *Mailing Add:* 1904 Hanson Rd Kinston NC 28501

KNOWLTON, ROBERT EARLE, b Summit, NJ, Oct 14, 39; m 63; c 2. INVERTEBRATE ZOOLOGY, MARINE BIOLOGY. *Educ:* Bowdoin Col, AB, 60; Univ NC, Chapel Hill, PhD(zool), 70. *Prof Exp:* Asst zool, Univ NC, Chapel Hill, 60-64; res asst fisheries, Inst Marine Sci, 64-65; from instr to asst prof biol, Bowdoin Col, 65-72; asst prof, 72-75, asst dean, 80-85, ASSOC PROF BIOL, GEORGE WASHINGTON UNIV, 75- *Concurrent Pos:* Teaching fel zool, Univ NC, Chapel Hill, 62-63; vis lectr biol, Univ Southern Maine, 72-80. *Mem:* AAAS; Am Inst Biol Sci; Am Soc Zool. *Res:* Autecology of decapod larvae; plankton ecology; sound production of Crustacea. *Mailing Add:* Dept Biol Sci George Washington Univ Washington DC 20052

KNOX, ARTHUR STEWART, b Charlestown, Mass, Jan 10, 03; m 50. GEOLOGY, PALYNOLOGY. *Educ:* Tufts Col, BS, 28; Harvard Univ, AM, 30; Tufts Univ, MEd, 39. *Prof Exp:* Control chemist, Cities Serv Refining Co, 28-29; instr geol, mineral, paleontol, geog & astron, Tufts Col, 29-38; geologist, US Geol Surv, 39-40; res assoc, Harvard Univ, 40-42; peat technologist, Champion Peat Co, 43-44; petrol geologist, Continental Oil Co, 44; topog engr, Topog Br, US Geol Surv, 44-52, cartogr, 52-58, supvry cartogr, 57-62, geologist, Mil Geol Br, 62-65, cartogr, Geog Names, 65-73; RETIRED. *Concurrent Pos:* Asst cur, Barnum Mus, Tufts Col, 30-38; lectr & dean, Ferry Beach Nature Inst, 39-41; photogeologist, Alaska Br, US Geol Surv, 59; partic, First Int Conf Palynol, 62; staff cartogr, US Bd Domestic Geog Names, 65-73. *Honors & Awards:* Knox Peak, US Bd Geog Names, 62. *Mem:* Fel AAAS; fel Geol Soc Am; fel Am Geog Soc; Am Soc Photogram; Am Asn Stratigraphic Palynologists; Am Polar Soc; Antarctican Soc. *Res:* Investigations of the geology of downtown Washington, DC; numerous publications on geology and paleobotany. *Mailing Add:* 2006 Columbia Rd NW Washington DC 20009

KNOX, BRUCE E, b Binghamton, NY, Aug 4, 31; m 53; c 3. MATERIALS SCIENCE. *Educ:* Rensselaer Polytech Inst, BS, 53; Syracuse Univ, MS, 58; Pa State Univ, PhD(fuel technol), 63. *Prof Exp:* Res asst, Syracuse Univ, 56-57; res asst shock tube chem, Pa State Univ, 57-59 & 60-62, instr geochem, 63, asst prof solid state technol, 63-67, solid state sci, 67-68 & mat sci, 68-69, ASSOC PROF MAT SCI, PA STATE UNIV, 69-, ASST DIR, MAT RES LAB, 75- *Concurrent Pos:* Vis prof, Pohang Inst Sci & Engr, Pohang, Korea, 88; assoc prof, Sci Technol & Soc. *Mem:* AAAS; fel Am Inst Chemists; Am Chem Soc; Am Soc Eng Educ; Am Phys Soc; Am Vacuum Soc; NY Acad Sci; Sigma Xi; Nat Asn Sci; Technol & Soc. *Res:* Mass spectrometry; vapor species of solid materials; laser-solid interaction; thin films; characterization of materials; chemical kinetics; trace elements in disease; materials science and engineering education; auger electron spectrometry; ion scattering spectrometry. *Mailing Add:* Mat Res Lab Pa State Univ University Park PA 16802

KNOX, BURNAL RAY, b Pineville, Mo, Mar 29, 31; m 55; c 3. FLUVIAL GEOMORPHOLOGY, SPELEOLOGY. *Educ:* Univ Ark, BS, 53, MS, 57; Univ Iowa, PhD(geol), 66. *Prof Exp:* Explor geologist, Gulf Oil Corp, 56-58; teacher high sch, Kans, 59-62; from asst prof to assoc prof, 65-76, PROF GEOL SCI, SOUTHEAST MO UNIV, 76- *Mem:* Geol Soc Am; Nat Asn Geol Teachers; Am Sci Affil; Sigma Xi. *Res:* Geomorphic evolution of Ozarks and Mississippi-Ohio rivers confluence area; Karst geomorphology and speleology of southeast Missouri; correlation of landforms with quaternary events, particularly climatic changes as controls of landscape evolution. *Mailing Add:* Dept Earth Sci Southeast Mo State Univ Cape Girardeau MO 63701

KNOX, CHARLES KENNETH, b Minneapolis, Minn, Nov 19, 38. NEUROPHYSIOLOGY, ENGINEERING. *Educ:* Univ Minn, Minneapolis, BS, 61, MS, 62, PhD(physiol), 69. *Prof Exp:* Asst prof, 69-76, ASSOC PROF PHYSIOL, UNIV MINN, MINNEAPOLIS, 76- *Concurrent Pos:* NIH res fel, Nobel Inst Neurophysiol, Stockholm, 69-70; NIH res grant, 73-76 & 76-79. *Mem:* Am Physiol Soc; Soc Neurosci; NY Acad Sci. *Res:* Neural control of respiration. *Mailing Add:* Dept Physiol 1000 Ingerson Rd St Paul MN 55126

KNOX, DAVID LALONDE, b Chicago, Ill, Sept 3, 30; m 58; c 3. OPHTHALMOLOGY. *Educ:* Baylor Univ, MD, 55. *Prof Exp:* ASSOC PROF OPHTHAL, SCH MED, JOHNS HOPKINS UNIV, 62-, ASST DEAN ADMIS, 76- *Res:* Neuro-medical ophthalmology. *Mailing Add:* Johns Hopkins Hosp Baltimore MD 21205

KNOX, ELLIS GILBERT, b Sterling, Ill, Mar 25, 28; m 48; c 2. NEW CROP INTRODUCTION. *Educ:* Univ Ill, BS, 49, MS, 50; Cornell Univ, PhD(soils), 54. *Prof Exp:* From asst prof to prof soils, Ore State Univ, 54-73; pedologist, Aero Serv Corp, Bogota, Colombia, 73-75; exec officer soil & land use Technol, Inc, 75-85; SOIL SCIENTIST, SOIL CONSERV SERV, USDA, 85- *Concurrent Pos:* Soil scientist, Soil Conserv Serv, USDA, 62-63; consult, Interam Inst Agr Sci, Turrialba, Costa Rica, 66; tech officer, Food & Agr Org UN, Turrialba, Costa Rica, 69-70. *Mem:* AAAS; Soil Sci Soc Am; Sigma Xi; Soil & Water Conserv Soc; Am Soc Agron. *Res:* Soil classification and mapping; soil survey interpretations; land resource and land use evaluation; analysis of crop production-marketing-consumption systems; biomass crops; development of crop systems. *Mailing Add:* 1040 Piedmont Rd Lincoln NE 68510

KNOX, FRANCIS STRATTON, III, b Wilmington, Del, Jan 28, 41; m 65; c 2. PHYSIOLOGY, BIOMEDICAL ENGINEERING. *Educ:* Brown Univ, BA, 63; Iowa State Univ, MS, 66; Univ Ill, PhD(physiol & biomed eng), 71. *Prof Exp:* Grad teaching asst zool, Iowa State Univ, 63-66; USPHS trainee biomed eng, Med Ctr, Univ Ill, 66-70; chief bioinstrumentation br, US Army Aeromed Res Lab, 70-73; asst prof, Med Ctr, La State Univ, Shreveport, 73-76, mem Grad Fac, 74-80, assoc prof physiol & biophys, 76-80; res physiologist & chief crew biotechnol br, Biomed Appln Res Div, Aeromed Res Lab, 80-86, supvry res physiologist, spec proj, 87-88, SUPVRY RES PHYSIOLOGIST & CHIEF CREW, LIFE SUPPORT BR, US ARMY, 88- *Concurrent Pos:* Consult physiol & bioeng, US Army Aeromed Res Lab, Ft Rucker, Ala, 73-80; affil asst prof bioeng, La Tech Univ, 75- *Mem:* Sigma Xi; Am Burn Asn; Inst Elec & Electronics Engrs; Soc Neurosci; Aerospace Med Asn; assoc fel, Aerospace Med Asn. *Res:* Quantitative physiology; systems analysis of physiological systems; use of biomedical instrumentation and computers to study physiological systems and to create economical, comprehensive diagnostic and patient monitoring systems; physiological effects of protective clothing under operational conditions. *Mailing Add:* Attn Comdr Sgrd-CB US Army Aeromed Res Lab Biomed Appln Div PO Box 577 Ft Rucker AL 36362

KNOX, FRANKLIN G, b Rochester, NY, Dec 20, 37. RENAL PHYSIOLOGY. *Educ:* State Univ NY, Buffalo, MD & PhD(physiol), 65. *Prof Exp:* DEAN, MAYO MED SCH, 82- *Mem:* Am Physiol Soc (pres, 86-87); Am Soc Nephrol; Am Heart Asn; Am Asn Physicians; Am Soc Clin Invest; Am Fedn Clin Res. *Mailing Add:* Dept Physiol Mayo Med Sch Rochester MN 55905

KNOX, GAYLORD SHEARER, b Bangkok, Thailand, Oct 18, 23; US citizen; m 46; c 3. MEDICINE, RADIOLOGY. *Educ:* Tulane Univ, MD, 51; Am Bd Radiol, cert, 58. *Prof Exp:* Intern gen med, Charity Hosp, New Orleans, 51-52; pvt pract, La, 52-53; physician, US Army, Ft Bliss, Tex, 53-55, resident radiol, Walter Reed Army Hosp, 55-58, chief radiol serv, Hosp, Bad Canstatt, Ger, 58-61; assoc prof radiol, Univ Okla, 61-65; chief, Dept Radiol, Baltimore City Hosps, 65-79; RADIOLOGIST, GREATER LAUREL BELTSVILLE HOSP, 79- *Concurrent Pos:* Consult, Vet Admin Hosp, Oklahoma City, 61-65 & Perry Point, Md, 65-; asst prof, Sch Med, Johns Hopkins Univ, 65-; assoc prof, Sch Med, Univ Md, Baltimore City, 65-; active consult, Sinai Hosp Baltimore, 66-; pres, Chesapeake Physicians, 73-78, Chesapeake Casualty Ins Co, Denver, 74-78. *Mem:* Radiol Soc NAm; fel Am Col Radiol; AMA; Soc Nuclear Med; Am Inst Ultrasound Med. *Res:* Clinical radiology, particularly skeletal and visceral changes in the aging process and vascular changes in aging. *Mailing Add:* 14201 Laurel Pike Dr Suite 106 Greater Laurel Beltsville Hosp Laurel MD 20707

KNOX, JACK ROWLES, b Hot Springs, Ark, Feb 8, 29; m 53; c 2. POLYMER SCIENCE. *Educ:* La State Univ, BS, 50, MS, 52; Univ Del, PhD(phys chem), 63. *Prof Exp:* Res chemist, E I du Pont de Nemours & Co, 52-60; res chemist, Avisun Corp, 60-61, group leader polymer struct, 61-63; mgr polymer physics, Continental Can Co, 63-65; sect leader polymer properties, Avisun Corp, 65-70; res assoc, 70-80, SR RES ASSOC, AMOCO CHEM CO, 80- *Mem:* Soc Rheology (pres, 76-77); N Am Thermal Anal Soc; Am Phys Soc. *Res:* Thermal analysis; polymer rheology, structure and analysis; mechanical properties. *Mailing Add:* 1564 Swallow St Naperville IL 60565

KNOX, JAMES CLARENCE, b Platteville, Wis, Nov 29, 41; m 64; c 2. GEOMORPHOLOGY, PHYSICAL GEOGRAPHY. *Educ:* Univ Wis-Platteville, BS, 63; Univ Iowa, PhD(geog), 70. *Prof Exp:* From asst prof to assoc prof geog, Univ Wis-Madison, 68-76, chmn dept, 83-86, dir, Ctr Geol Anal, Inst Environ Studies, 73-78, PROF GEOG, UNIV WIS-MADISON, 77- *Concurrent Pos:* Assoc ed, Geog Annals, 78-81, Counr Am Quaternary Asn, 78-; mem, US Nat Comt Int Quaternary Union, 82-91 secy, 86-91 vchmn & chmn, Quaternary Geol & Geomorphology Div, Geol Soc Am, 85-88; US deleg & corresp mem, Int Geomorphology Assoc, 86-90; Geog & Regional Sci Panel, NSF, 88-90, Continental Hydrol Panel, 91-; assoc ed, Geol Soc Am Bull, 91- *Honors & Awards:* Honors Award, Asn Am Geogr, 90. *Mem:* Am Quaternary Asn; fel AAAS; Geol Soc Am; Soil Conserv Soc Am; Asn Am Geogr. *Res:* Fluvial geomorphology; paleoclimatology and paleohydrology of the Quaternary; effects of present-day climate variation and land use on stream flow characteristics and sedimentation problems; Quaternary landscape evolution of upper Miss Valley; erosion, transportation and storage of sediment in river systems; hydrology and water resources. *Mailing Add:* Dept Geog 234 Sci Hall Univ Wis Madison WI 53706

KNOX, JAMES L(ESTER), b Youngstown, Ohio, July 30, 19; m 46; c 5. ELECTRICAL ENGINEERING. *Educ:* Univ Tenn, BS, 42; Univ Mich, MSE, 54; Ohio State Univ, PhD(elec eng), 62. *Prof Exp:* Instr elec eng, Univ Tenn, 42-43; engr, Gen Elec Co, 43-46; Am Baptist For Mission Soc missionary assigned as instr physics, Univ Shanghai, 47-48; asst prof eng, Cent Philippine Univ, 48-50, tech dir eng, Radio Sta DYSR, 50-51, assoc prof, Cent Philippine Univ, 51-59, prof & dean, 62-65; instr elec eng, Ohio State Univ, 60-61, res assoc, Res Found, 61-62; assoc prof, 65-69, prof elec eng, Mont State Univ, 69-; at dept elec eng, Univ Petrol & Minerals, Saudi Arabia; RETIRED. *Concurrent Pos:* Prof elec eng, Univ Petrol & Minerals, Dhahran, Saudi Arabia, 73-75 & 76-78. *Mem:* Inst Elec & Electronics Engrs; Instrument Soc; Am Soc Eng Educ; Sigma Xi. *Res:* Electromagnetics; electric power; instrumentation; community acoustics; remote sensing. *Mailing Add:* Dept Elec Eng 603 S 7th Ave Bozeman MT 59715

KNOX, JAMES RUSSELL, JR, b Bonne Terre, Mo, May 28, 41; m 65; c 2. PHYSICAL BIOCHEMISTRY, X-RAY CRYSTALLOGRAPHY. *Educ:* Univ Mo, Rolla, BS, 63; Boston Univ, PhD(phys chem), 67. *Prof Exp:* NIH Fellowship, Oxford Univ, 66-69; res assoc biophysics, Dept Molecular Biochem, Yale Univ, 69-70; assoc prof, 70-80, PROF BIOPHYS, DEPT MOLECULAR & CELL BIOL, UNIV CONN, 80- *Concurrent Pos:* vis prof, Harvard Biol Labs, 77; NIH study sect, biophysical chem, 78, spec study sect, 87-88; consult, Hoffman-LaRoche Co, 85-87, Eli Lilly & Co, 88; external adv comt, Crystallographic Data Anal Ctr, Purdue Univ, 88- *Honors & Awards:* Author Award, Am Chem Soc, 82. *Mem:* AAAS; Am Chem Soc; Am Crystallog Asn; Sigma Xi; Biophys Soc. *Res:* Enzyme structure and function by means of x-ray analysis; penicillin-binding proteins; interactive computer graphics; beta-lactamases and bacterial cell-wall synthesizing enzymes. *Mailing Add:* Molecular & Cell Biol Dept Univ Conn U-125 Storrs CT 06269

KNOX, JOHN MACMURRAY, b Sheboygan, Wis, Nov 21, 46; m 69; c 2. PHYSICS. *Educ:* Gustavus Adolphus Col, BA, 68; Univ Wyo, MS, 77, PhD(physics), 81. *Prof Exp:* Res fel & asst prof, Inst Paper Chem, 81-84, ASSOC PROF PHYSICS, IDAHO UNIV, 84- *Mem:* Am Phys Soc; Am Vac Soc; Am Asn Physics Teachers; Health Physics Soc. *Res:* Interaction of 2 Mev ion beams with materials, analysis, primarily using particle induced x-ray emission and backscattered ion spectroscopy; gas phase electron capture; negative ion and optically modified mass spectra. *Mailing Add:* Physics Dept Idaho State Univ Pocatello ID 83209

KNOX, KENNETH L, b Winnipeg, Man, Sept 18, 20; US citizen; m 42; c 3. CHEMICAL ENGINEERING. *Educ:* Univ Saskatchewan, BE, 42, MSc, 46; Columbia Univ, PhD(chem eng), 49. *Prof Exp:* Res engr, Yerkes Res & Develop Lab, 48-52, develop supvr, Cellophane Plants, NY, Iowa & Kans, 52-60, tech supt Cellophane Lab, Kans, 60-63, RES ASSOC, CIRCLEVILLE RES & DEVELOP LAB, E I DU PONT DE NEMOURS & CO INC, NY, 63- *Mem:* Am Chem Soc; Am Inst Chem Engrs. *Res:* Industrial research on polyester film. *Mailing Add:* 327 Meadow Lane Circleville OH 43113

KNOX, KERRO, b Philadelphia, Pa, June 17, 24; m 49; c 4. INORGANIC CHEMISTRY. *Educ:* Yale Univ, BS, 45, PhD(phys chem), 50; Cambridge Univ, PhD(phys chem), 52. *Prof Exp:* From instr to assoc prof chem, Univ NC, 51-56; mem tech staff, Bell Tel Labs, 56-63; assoc prof chem, Case Western Reserve Univ, 63-69; PROF CHEM, CLEVELAND STATE UNIV, 69- *Mem:* Am Crystallog Asn. *Res:* X-ray crystallography. *Mailing Add:* Dept Chem Univ Mass Amherst MA 01003

KNOX, KIRVIN L, b Sayre, Okla, Aug 9, 36; m 58; c 3. NUTRITION. *Educ:* Fresno State Col, BS, 58; Colo State Univ, MS, 60; Univ Calif, PhD(physiol, nutrit), 64. *Prof Exp:* Lab supt, Escalon Packers Inc, 60; asst prof animal sci, Colo State Univ, 64-65, from asst prof to assoc prof animal sci & physiol, 65-72, dir metab lab, 65-72; PROF NUTRIT SCI & HEAD DEPT, UNIV CONN, 72- *Concurrent Pos:* Co-investr, NIH res grant, 65-68 & prin investr, 66-69; Am Cancer Soc res grant, 66-68; sabbatical leave, Univ Calif, Berkeley, 78-79. *Mem:* Am Dairy Sci Asn; Am Soc Animal Sci; Am Inst Nutrit; NY Acad Sci; AAAS. *Res:* Comparative nutrition as related to energy and vitamin metabolism; behavioral response to nutrition. *Mailing Add:* 170 David Rd Storrs CT 06268

KNOX, LARRY WILLIAM, b Mishawaka, Ind, Nov 10, 42; m 64; c 2. PALEONTOLOGY. *Educ:* Ind Univ, AB, 65, AM, 71, PhD(geol), 74. *Prof Exp:* From asst prof to assoc prof, 74-83, PROF GEOL, TENN TECHNOL UNIV, 83- *Mem:* Geol Soc Am; Paleont Soc; Paleont Asn; Soc Econ Paleontologists & Mineralogists. *Res:* Paleoecology and biostratigraphy of paleozoic ostracodes. *Mailing Add:* Box 5125 Tenn Technol Univ Cookeville TN 38505

KNOX, ROBERT ARTHUR, b Washington, DC, Jan 15, 43; m 66; c 2. PHYSICAL OCEANOGRAPHY. *Educ:* Amherst Col, AB, 64; Mass Inst Technol-Woods Hole Oceanog Inst, PhD(oceanog), 71. *Prof Exp:* Res assoc, Mass Inst Technol, 71-73; res asst oceanogr, 73-81, assoc res oceanogr, 81-86, acad adminr, 80- 90, RES OCEANOGR, SCRIPPS INST OCEANOG, 86- *Mem:* Sigma Xi; Am Meteorol Soc; Am Geophys Union. *Res:* Equatorial ocean dynamics and circulation; structure and dynamics of oceanic mixed layer; acoustic sensing of ocean circulation. *Mailing Add:* Scripps Inst Oceanog A-030 Univ Calif at San Diego La Jolla CA 92093

KNOX, ROBERT GAYLORD, b Wash, DC, Oct 24, 56; m 78. QUANTITATIVE FOREST ECOLOGY, VEGETATION SCIENCE. *Educ:* Princeton Univ, AB, 78; Univ NC, PhD(bot), 87. *Prof Exp:* Computer programmer cancer epidemiol, Univ Tex Syst Cancer Ctr, 78-82; Huxley res instr evolution, Dept Biol, Rice Univ, 87-90, sr res assoc ecol, Dept Ecol & Evolutionary Biol, 90-91; RES SCIENTIST FOREST ECOL, BIOSPHERIC SCI BR, UNIVS SPACE RES ASN, NASA/GODDARD SPACE FLIGHT CTR, 91- *Concurrent Pos:* Vis lectr, Dept Ecol & Evolutionary Biol, Rice Univ, 90, sr res assoc, 91- *Mem:* Ecol Soc Am; Int Asn Veg Sci; Bot Soc Am; Am Inst Biol Sci; Sigma Xi; Torrey Bot Club. *Res:* Plant community structure and distribution; methods of vegetation analysis; ecological theory; physiological ecology of resource capture and limitation; computer modeling of individuals, communities and ecosystems; inference from long term studies and from spatial pattern; philosophy of biology. *Mailing Add:* Biospheric Sci Br NASA-Goddard Space Flight Ctr Code 923-0 Greenbelt MD 20771

KNOX, ROBERT SEIPLE, b Franklin, NJ, July 13, 31; m 54; c 3. BIOPHYSICS, CONDENSED MATTER. *Educ:* Lehigh Univ, BS, 53; Univ Rochester, PhD(physics, optics), 58. *Prof Exp:* Res assoc physics, Univ Ill, 58-59, res asst prof, 59-60; from asst prof to assoc prof, 60-68, chmn dept, 69-74, PROF PHYSICS, UNIV ROCHESTER, 68- *Concurrent Pos:* Consult, Solid State Sci Div, Argonne Nat Lab, 59-69, Naval Res Lab, 60-70; NSF sr fel, Univ Leiden, 67-68; fel, Japanese Soc Prom Sci, Kyoto, Japan, 79; Dean, Univ Col, Univ Rochester, 82-85. *Mem:* Fel Am Phys Soc; Am Soc Photobiol; Am Asn Physics Teachers; Biophys Soc. *Res:* Optical and electrical properties of ionic and molecular crystals; theory of photosynthesis, picosecond spectroscopy. *Mailing Add:* Dept Physics & Astron Univ Rochester Rochester NY 14627-0011

KNOX, WALTER ROBERT, b Childress, Tex, July 18, 26; m 48; c 3. ORGANIC CHEMISTRY, PETROCHEMISTRY. *Educ:* Baylor Univ, BS, 47; Univ Iowa, MS, 49, PhD(org chem), 50. *Prof Exp:* Control chemist, Fibreboard Prod, Inc, 47; res chemist, Gen Aniline & Film Corp, 50-52; chemist, Hydrocarbons & Polymers Div, Monsanto Co, 52-54, sr res chemist, 54-56, group leader, 56-62, mgr res, 62-70, dir res, Petrochem Div, Monsanto Polymers & Petrochem Co, 70-80, TECHNOL DIR SPEC ASST, MONSANTO INT CO, 80- *Mem:* Am Chem Soc; Catalysis Soc. *Res:* Catalysts and catalytic interconversions, especially with hydrocarbons; monomer synthesis; surfactant synthesis; general organic synthesis. *Mailing Add:* 2118 Babler Ridge Lane Glencoe MO 63038

KNOX, WILLIAM JORDAN, b Pomona, Calif, Mar 21, 21; m 48; c 4. NUCLEAR PHYSICS. *Educ:* Univ Calif, BS, 42, PhD(physics), 51. *Prof Exp:* Asst chem, Metall Lab, Univ Chicago, 42-43; from jr chemist to chemist, Clinton Labs, Oak Ridge, 43-46; jr technologist, Hanford Eng Works, 44-45; chemist, Radiation Lab, Univ Calif, 46, physicist, 47-51; asst prof physics, Yale Univ, 51-53; physicist, AEC, Washington, DC, 53-55 & consult, 55-56; asst prof physics, Yale Univ, 55-59; assoc prof physics, Univ Calif, Davis, 60-66, chmn dept, 63-66; actg dir Crocker Nuclear Lab, 66-67 & 78-79; chmn dept, 71-75 & 80-83, PROF PHYSICS, UNIV CALIF, DAVIS, 66- *Concurrent Pos:* Vis sci, Cambridge Univ, 67-68, Europ Ctr Nuclear Res, 73-74; sr Fulbright Hayes Prog res fel France, Comt Int Exchange Persons, 73-74; vis scientist, Lawrence Berkeley Lab, 81-86. *Mem:* Fel Am Phys Soc; Am Asn Physics Teachers. *Res:* Nuclear reactions; nuclear structure; plutonium and fission product chemistry; particle production. *Mailing Add:* Dept Physics Univ Calif Davis CA 95616

KNUCKLES, JOSEPH LEWIS, b Lumberton, NC, Mar 17, 24. PARASITOLOGY. *Educ:* NC Cent Univ, BS, 48, MS, 50; Univ Conn, PhD(parasitol), 59. *Prof Exp:* Instr biol, Bishop Col, 50-52, chmn dept sci, 51-52; instr biol & math, 56-59, coordr biol, 59-67, asst to acad dean, 71, dir summer sch, 71, actg head, Div Sci & Math, 73-75, actg head, Div Arts Scis, 75-76, chmn, Dept Biol & Phys Sci, 67-78, coodr, Area Biol, 78-80 PROF BIOL, FAYETTEVILLE STATE UNIV, 59- *Concurrent Pos:* Sigma Xi grant, 58; consult, NC Teachers Asn, 59; NC Acad Sci grant, 60; NSF stipend, 64; dir coop prog, Fayetteville State Univ, 70, dir consortium prom acad excellence & dir sci improv & exp biol prog; Title III grants, 72-76; NIH res grant, 82-86. *Mem:* AAAS; Am Entom Soc; Am Soc Parasitol; Am Asn Univ Professors; Am Pub Health Asn; Am Inst Biol Sci. *Res:* Transmission of disease agents by flies; general parasitology. *Mailing Add:* Dept Life Sci Fayetteville State Univ PO Box 14339 Fayetteville NC 28301

KNUDSEN, DENNIS RALPH, b Warren, Minn, July 22, 43; m 65; c 2. PHYSICAL CHEMISTRY. *Educ:* NDak State Univ, BS, 65, PhD(phys chem), 70. *Prof Exp:* Res chemist, 70-86, PHYS SCIENTIST, US NAVAL SURFACE WARFARE CTR, 86- *Mem:* Am Chem Soc. *Res:* Thermodynamics of solutions; determination of chemical compounds in air; chemical instrument development, combustion; fiber optics in adverse environments. *Mailing Add:* US Naval Surface Warfare Ctr Dahlgren VA 22448-5000

KNUDSEN, ERIC INGVALD, b Palo Alto, Calif, Oct 7, 49; m 75. NEUROSCIENCE, BIOLOGY. *Educ:* Univ Calif, Santa Barbara, BA, 71, MA, 73; Univ Calif, San Diego, PhD(neurosci), 76. *Prof Exp:* Res fel neurosci, Calif Inst Technol, 76-80; MEM FAC, DEPT NEUROSCI, STANFORD SCH MED, 80- *Mem:* Acoust Soc Am; AAAS; Soc Neurosci; Sigma Xi. *Res:* Neurophysiology, anatomy and ethology related to the evolution of the auditory system, specifically encoding of space by the auditory system. *Mailing Add:* Dept Neurobiol Stanford Univ Sch Med Stanford CA 94305

KNUDSEN, HAROLD KNUD, b San Francisco, Calif, Aug 6, 36; m 58; c 2. ELECTRICAL ENGINEERING. *Educ:* Univ Calif, Berkeley, BS, 58, MS, 60, PhD(elec eng), 62. *Prof Exp:* Staff mem data systs anal, Lincoln Lab, Mass Inst Technol, 62-66; assoc prof, 66-74, PROF ELEC ENG, UNIV NMEX, 74- *Mem:* Inst Elec & Electronics Engrs; Sigma Xi. *Res:* Data systems analysis; system theory, especially the application of theories of optimization. *Mailing Add:* 6605 Loftus Ave NE Albuquerque NM 87109

KNUDSEN, J(AMES) G(EORGE), b Youngstown, Alta, Mar 27, 20; nat US; m 47; c 2. CHEMICAL ENGINEERING. *Educ:* Univ Alta, BS, 43, MS, 44; Univ Mich, PhD(chem eng), 50. *Prof Exp:* From asst prof to assoc prof chem eng, Ore State Univ, 49-57, asst dean, 59-71, assoc dean, 71-81, PROF CHEM ENG, ORE STATE UNIV, 57- *Concurrent Pos:* Nat Sci sr fel, Cambridge Univ, 61-62; eng consult. *Honors & Awards:* Founders Award, Am Inst Chem Engrs, 77. *Mem:* Am Inst Chem Engrs (pres, 81); Am Chem Soc; Sigma Xi. *Res:* Fluid mechanics; heat transfer; relationship between these processes; applied mathematics. *Mailing Add:* Dept Chem Eng Ore State Univ Corvallis OR 97331-2702

KNUDSEN, JOHN R, b Brooklyn, NY, July 12, 16; m 42; c 2. APPLIED MATHEMATICS. *Educ:* NY Univ, BS, 37, PhD(math), 51. *Prof Exp:* Instr math, NY Univ, 39-51, from asst prof to prof, 51-72, asst dean & budget officer, 66-72; mem tech staff, Educ Ctr, Bell Tel Labs, 72-83; RETIRED. *Concurrent Pos:* Consult, Bell Tel Labs, NJ, 57-72, 83-87 & Univ Bangalore, India, 68. *Mem:* Am Math Soc; Math Asn Am. *Res:* Self-study instructional techniques. *Mailing Add:* 10105 Jupiter Hills Dr Austin TX 78747-1312

KNUDSEN, KAREN ANN, b June 30, 43; m; c 1. CELL ADHESION, MYOGENESIS. *Educ:* Univ Pa, PhD(biochem), 77. *Prof Exp:* SR SCIENTIST, LANKENAU MED RES CTR, 85- *Mem:* AAAS; Am Women sci; Am Soc Cell Biol. *Mailing Add:* Dept Cell Biol Lankenau Med Res Ctr Lancaster Ave-W of City Line Philadelphia PA 19151

KNUDSEN, RICHARD CARL, VIROLOGY, IMMUNOLOGY. *Educ:* Univ Ariz, PhD(microbiol), 71. *Prof Exp:* Lead scientist, 76-91, CHIEF, PLUM ISLAND ANIMAL DIS CTR, 91- *Mailing Add:* Biosafety Br Ctr Dis Control 1600 Clifton Rd NE Atlanta GA 30333

KNUDSEN, WILLIAM CLAIRE, b Provo, Utah, Dec 12, 25; m 48; c 4. PHYSICS. *Educ:* Brigham Young Univ, BS, 50; Univ Wis, MS, 52, PhD(physics), 54. *Prof Exp:* Res physicist geophys, Calif Res Corp Div, Standard Oil Co, Calif, 54-62; res physicist geophys, Lockheed Aircraft Corp, Lockheed Res Lab, 62-67; staff scientist, 67-84; PRES, KNUDSEN GEOPHYS RES, 84- *Mem:* Am Geophys Union. *Res:* Low temperature physics; exploration geophysics; ground water hydrology; planetary atmospheres; planetary ionospheres. *Mailing Add:* Knudsen Geophys Res 18475 Twin Creeks Rd Monte Sereno CA 95030

KNUDSON, ALFRED GEORGE, JR, b Los Angeles, Calif, Aug 9, 22; m 76; c 3. MEDICINE, GENETICS. *Educ:* Calif Inst Technol, BS, 44, PhD(biochem, genetics), 56; Columbia Univ, MD, 47. *Prof Exp:* Chmn dept pediat, City of Hope Med Ctr, Calif, 56-62, chmn dept biol, 62-66; prof pediat & assoc dean, Health Sci Ctr, State Univ NY Stony Brook, 66-69; prof biol & pediat & assoc dir educ, Univ Tex M D Anderson Hosp & Tumor Inst, Houston, 69-70; prof med genetics & dean, Univ Tex Grad Sch Biomed Sci, 70-76; dir, 76-83, SR MEM, INST CANCER RES, FOX CHASE CANCER CTR, 76- *Honors & Awards:* Mott prize, Gen Motors Cancer Res Found, 88; Medal of Honor, Am Cancer Soc, 89. *Mem:* Nat Acad Sci; fel AAAS; Am Asn Cancer Res; Am Pediat Soc; Am Soc Human Genetics (pres, 78); Asn Am Physicians; Int Soc Pediat Oncol. *Res:* Cancer genetics; medical genetics; tumor suppressor genes. *Mailing Add:* 7701 Burholme Ave Inst Cancer Res Fox Chase Philadelphia PA 19111

KNUDSON, ALVIN RICHARD, b Minneapolis, Minn, Aug 17, 34; m 60; c 3. RADIATION EFFECTS. *Educ:* Catholic Univ, AB, 54; Johns Hopkins Univ, PhD(physics), 60. *Prof Exp:* Res asst nuclear physics, Johns Hopkins Univ, 57-58 & 59-60; res physicist, 60-67, head charged particle reactions sect, 67-70, head mat anal sect, 70-85, head radiation effects sect, 85-88, HEAD VAN DE GRAAFF APPLICATIONS SECT, US NAVAL RES LAB, 88- *Mem:* Am Phys Soc. *Res:* Study of charged particle radiation effects in microelectronics; use of high energy ion beams for materials analysis. *Mailing Add:* US Naval Res Lab Code 4673 Washington DC 20375-5000

KNUDSON, DOUGLAS MARVIN, b Anoka, Minn, June 11, 36; m 57; c 2. FOREST RECREATION, TROPICAL SILVICULTURE. *Educ:* Colo State Univ, BS, 59, MS, 60; Purdue Univ, PhD(forest econ), 65. *Hon Degrees:* Dr, Fed Univ Vicosa, Brazil, 68. *Prof Exp:* Forester, Purdue Univ-Agr Univ Brazil, 60-62, asst prof forestry, 65-67; assoc prof forestry, 69-82, PROF FORESTRY, PURDUE UNIV, WEST LAFAYETTE, 82- *Concurrent Pos:* Wilderness planner, Nat Park Serv, 70; res assoc, US Forest Serv, 79; consult corp engrs, US Forest Serv, 80-81; vis scientist, Bogor Agr Univ, Indonesia, 88; resident adv, Fuelwood Res Prog, Dominican Repub, 83-87. *Mem:* Soc Am Foresters; Nat Recreation & Park Asn; Int Asn Torch Clubs (pres, 78-79); Nat Asn Interpretation. *Res:* Outdoor recreation economics and planning. *Mailing Add:* Dept Forestery & Natural Resources Purdue Univ West Lafayette IN 47907

KNUDSON, GREGORY BLAIR, b Salina, Kans, Aug 9, 46; m 72; c 2. GENE CLONING, PATHOGENIC MICROBIOLOGY. *Educ:* Calif State Univ, Fullerton, BA, 69, MA, 71; Univ Calif, Riverside, PhD(genetics), 77. *Prof Exp:* Res biologist, Univ Calif, Riverside, 77-78; res microbiologist, 78-81, RES MICROBIOLOGIST GS-12, US ARMY MED RES INST INFECTIOUS DIS, FT DETRICK, MD, 81- *Concurrent Pos:* Lectr molecular biol, Calif State Col, San Bernardino, 78; instr recombinant DNA technol, Hood Col, Frederick, Md, 80-81. *Mem:* Genetics Soc Am; AAAS; Am Soc Microbiol; NY Acad Sci. *Res:* Established technological base for genetic analysis of pathogens of military importance; genetic engineering of bacterial for the production of medically useful gene products; toxin plasmids. *Mailing Add:* Dept Path Area Microbiol Letterman Army Med Ctr Presidio CA 94129

KNUDSON, RONALD JOEL, b Chicago, Ill, Feb 22, 32; m 71. PULMONARY PHYSIOLOGY. *Educ:* Yale Univ, BS, 53; Northwestern Univ, MD, 57. *Prof Exp:* Intern, Chicago Wesley Mem Hosp, 57-58; fel surg, Ochsner Found Hosp, New Orleans, 59-63; fel thoracic surg, Overholt Thoracic Clin, Boston, 63-64; res fel pulmonary physiol, Boston Univ Med Ctr, 64-66 & Sch Pub Health, Harvard Univ, 66-68; asst prof physiol & med, Sch Med, Yale Univ, 68-70; assoc prof internal med, 70-75, PROF INTERNAL MED, COL MED, UNIV ARIZ, 75-, ASSOC DIR DIV RESPIRATORY SCI, 74- *Concurrent Pos:* Dir respiratory serv, Ariz Med Ctr, Tucson, 74- *Mem:* Am Physiol Soc; Am Thoracic Soc; AAAS; Am Col Chest Physicians. *Res:* Respiratory mechanics, airway dynamics; respiratory structure and function relationships. *Mailing Add:* Div Respiratory Sci Univ Ariz Med Col Tucson AZ 85724

KNUDSON, VERNIE ANTON, b Olsburg, Kans, Sept 18, 32; m 54; c 3. LIMNOLOGY. *Educ:* Bethany Col, BS, 54; Ft Hays Kans State Col, MS, 59; Okla State Univ, PhD(zool), 70. *Prof Exp:* Instr biol, Bethany Col, 59-60; instr, Dodge City Col, 60-63; partic, Acad Yr Inst, Univ Ore, 64-65; asst prof fisheries & wildlife, Mich State Univ, 66-69; asst prof biol sci, 71-77, asst prof, 77-80, ASSOC PROF NATURAL RESOURCES TECHNOL, LAKE SUPERIOR STATE COL, 80- *Mem:* Am Inst Biol Sci; Am Chem Soc. *Res:* Water quality; eutrophication; nutrient removal by algae. *Mailing Add:* Dept Biol & Chem Lake Superior State Univ Sault Ste Marie MI 49783

KNUDTSON, JOHN THOMAS, b Charleston, SC, July 3, 45; m 84; c 2. PHYSICAL CHEMISTRY. *Educ:* Colo Col, BS, 67; Columbia Univ, MS, 69, PhD(chem), 72. *Prof Exp:* Res assoc chem, Univ Utah, 72-74 & Univ Wis, 74-75; from asst prof to assoc prof chem, Northern Ill Univ, 75-85; MEM TECH STAFF, AEROSPACE CORP, 85- *Mem:* Am Phys Soc; Am Chem Soc. *Res:* Laser spectroscopy; spectral radiometery. *Mailing Add:* Aerospace Corp M5/747 PO Box 92957 Los Angeles CA 90009

KNULL, HARVEY ROBERT, b Thorsby, Alta, Sept 15, 41; m 65; c 2. BIOCHEMISTRY. *Educ:* Univ Alta, BSc, 63; Univ Nebr, MS, 65; Pa State Univ, PhD(biochem), 70. *Prof Exp:* Fel neurochem, Dept Biochem, Mich State Univ, 70-73; asst prof oral biol, Univ Man, 73-78, assoc prof, 78-80; assoc prof, 80-87, PROF BIOCHEM, UNIV NDAK, 87- *Mem:* Int Soc Neurochem; Am Soc Neurochem; Soc Complex Carbohydrates; Can Biochem Soc; Can Fedn Biol Sci; Am Soc Biol Chemists; Sigma Xi. *Res:* Neurochemistry; axonal transport; brain energy metabolism; compartmentation of glycolytic enzymes; proteoglycans; carbohydrate metabolism. *Mailing Add:* Dept Biochem Univ NDak Sch Med Grand Forks ND 58202

KNUTH, DONALD ERVIN, b Milwaukee, Wis, Jan 10, 38; m 61; c 2. ALGORITHMS, DIGITAL TYPOGRAPHY. *Educ:* Case Inst Technol, BS & MS, 60; Calif Inst Technol, PhD(math), 63. *Hon Degrees:* Numerous from US & foreign univs, 80-88. *Prof Exp:* From asst prof to prof math, Calif Inst Technol, 63-68; prof, 68-77, Fletcher Jones prof comput sci & elec eng, 77-91, PROF ART COMPUTER PROG, STANFORD UNIV, 90- *Concurrent Pos:* Consult, Burroughs Corp, 60-68; staff mathematician, Commun Res Div, Inst Defense Anal, 68-69; Guggenheim Found fel, 72. *Honors & Awards:* G M Hopper Award, 71; A M Turing Award, Asn for Comput Mach, 74; J B Priestley Award, 81; W McDowell Award, 80 & Comput Pioneer Award, Inst Elec & Electronics Engrs, 82; Software Systs Award, Asn Comput Mach, 86; Steele Prize, Am Math Soc, 86; New York Acad of Sci Award, 87; Franklin Medal, 88. *Mem:* Nat Acad Sci; Nat Acad Eng; Soc Indust & Appl Math; Asn Comput Mach; Math Asn Am; Brit Comput Soc; hon mem Inst Elec & Electronics Engrs. *Res:* Analysis of algorithms; combinatorial theory; programming languages; history of computer science; typography. *Mailing Add:* Comput Sci Dept Stanford Univ Stanford CA 94305

KNUTH, ELDON L(UVERNE), b Luana, Iowa, May 10, 25; m 73; c 4. MOLECULAR DYNAMICS, COMBUSTION. *Educ:* Purdue Univ, BS, 49, MS, 50; Calif Inst Technol, PhD(aeronaut eng), 53. *Prof Exp:* Group leader aerothermodyn, Aerophys Develop Corp, 53-56; assoc res engr, 56-58, assoc prof, 59-68, gen develop, Heat Transfer & Fluid Mech Inst, 59, head molecular-beam lab, 61-88, head, Chem, Nuclear & Thermal Div, 63-65, chmn, Dept Energy & Kinetics, 69-75, PROF ENG, UNIV CALIF, LOS ANGELES, 65- *Concurrent Pos:* Consult, Marquardt Aircraft Corp, 58-60, Jet Propulsion Lab & TRW Inc, 79-; Alexander von Humboldt Found fel, 75. *Honors & Awards:* Award, Am Inst Aeronaut & Astronaut, 50. *Mem:* Am Phys Soc; Am Inst Aeronaut & Astronaut; Am Inst Chem Engrs; Combustion Inst. *Res:* Combustion; thermodynamics and statistical mechanics; transport phenomena and properties; free-molecule flows and molecular beams. *Mailing Add:* Dept Chem Eng 5531 Boelter Hall Univ Calif Los Angeles CA 90024

KNUTSON, CARROLL FIELD, b Santa Monica, Calif, Mar 14, 24; m 48; c 4. GEOSCIENCES SCIENCE. *Educ:* Stanford Univ, BS, 50, MS, 51; Univ Calif, Los Angeles, PhD(geol), 59. *Prof Exp:* Reservoir engr, Continental Oil Co, 51-54, from prod engr to sr prod engr, 54-58, sr pet, 58-61, res group leader, 61-66, res assoc, 66-67; chief geologist, Cer Geonuclear Corp, 67-74; consult geologist, C F Knutson & Assocs, 74-76; vpres, C K Geoenergy, 76;

PRES & BD DIRS, C K M RESOURCES, 75-; SR SCIENTIST, EG&G IDAHO, INC. *Concurrent Pos:* Environ dir, Western Oil Shale Corp, 75-76. *Mem:* Geol Soc Am; Am Asn Petrol Geol; Soc Petrol Engrs; Am Geophys Union; Soc Independent Prof Earth Scientists; Sigma Xi. *Res:* Formation evaluation; rock mechanics; subsurface nuclear engineering geology; hydrology, environmental geology; development of energy with minimum adverse environmental impact. *Mailing Add:* EG&G Idaho Inc PO Box 1625 Idaho Falls ID 83415-2107

KNUTSON, CHARLES DWAINE, b Milbank, SDak, Sept 23, 34; m 55; c 4. COMPUTER SCIENCES. *Educ:* SDak Sch Mines & Technol, BS, 56; Brown Univ, PhD(physics), 62. *Prof Exp:* Asst instr math, SDak Sch Mines & Technol, 54-55, asst instr math & physics, 55-56; math asst, Sperry Rand Univac, 55; lab asst, Knolls Res Lab, Gen Elec Co, 56; res asst, Brown Univ, 56-61; sr res physicist, Cent Res Labs, 61-68, SR RES SPECIALIST, COMPUT-ASSISTED RES & DEVELOP, CENT RES LABS, MINN MINING & MFG CO, 68- *Res:* Energy systems, psychology, systems modeling and artificial intelligence; computer science; education technology; nuclear magnetic resonance; solid state physics computer applications. *Mailing Add:* Cent Res Lab 208-1 3M Co 3M Ctr St Paul MN 55144

KNUTSON, CLARENCE ARTHUR, JR, b Minot, NDak, June 18, 37; m 59; c 2. ORGANIC CHEMISTRY. *Educ:* Concordia Col, Moorhead, Minn, BA, 59; NDak State Univ, MS, 61. *Prof Exp:* RES CHEMIST, NORTHERN REGIONAL RES CTR, AGR RES SERV, USDA, 61- *Mem:* Am Chem Soc; Am Asn Cereal Chemists. *Res:* Carbohydrate chemistry; composition structure and properties of cereal polysaccharides; quantitative analytical methods. *Mailing Add:* 5927 Tampico Dr Peoria IL 61614

KNUTSON, DAVID W, b Minneapolis, Minn, Feb 12, 41; c 4. NEPHROLOGY. *Educ:* Univ Minn, MD, 67. *Prof Exp:* PROF NEPHROLOGY & CHIEF, MILTON S HERSHEY MED CTR, PA STATE UNIV, 85- *Mem:* Am Soc Nephrology; Int Soc Nephrology; Nat Kidney Found; Am Asn Immunol. *Res:* Immunology. *Mailing Add:* Div Nephrol Dept Med Milton S Hershey Med Ctr Pa State Univ PO Box 850 Hershey PA 17033

KNUTSON, KENNETH WAYNE, b Williams, Minn, Feb 11, 32; m 57; c 2. PLANT PATHOLOGY. *Educ:* Univ Minn, BS, 54, MS, 56, PhD(plant path), 60. *Prof Exp:* Asst plant pathologist, Univ Idaho, 60-64; EXTEN ASSOC PROF, COLO STATE UNIV, 64- *Mem:* Potato Asn Am. *Res:* Potato diseases and other cultural problems; soil-borne fungi; viruses. *Mailing Add:* 1116 Morgan Ft Collins CO 80521

KNUTSON, LLOYD VERNON, b Ottawa, Ill, July 4, 34; m 57; c 2. BIOLOGICAL CONTROL, TAXONOMY. *Educ:* Macalester Col, BA, 57; Cornell Univ, MS, 59, PhD(limnol), 63. *Prof Exp:* Res assoc entom, Cornell Univ, 63-68; res entomologist, Syst Entom Lab, 68-73, dir, Biosystematics & Beneficial Insects Inst, 73-88, SPEC ASST, BIOSYST & BIOCONTROL, PLANT SCI INST, AGR RES CTR, AGR RES SERV, USDA, BELTSVILLE, 88- *Concurrent Pos:* Sci cooperator, Royal Inst Natural Sci, Belgium, 65-; resident ecologist, Smithsonian Inst, 71-72; dir, Systematic Biol Prog, NSF, 83-84. *Mem:* Entom Soc Am (pres, 88); Am Soc Parasitologists; Soc Syst Zool (treas, 71-74); Ecol Soc Am; Am Registry Prof Entomologist (pres, 87). *Res:* Taxonomy and biology of Diptera, especially malacophagous and entomophagous groups; phylogeny of Sciomyzoidea; biological control of pest molluscs; taxonomic services; immigrant arthropods; computer applications to taxonomy and biocontrol. *Mailing Add:* Biol Control Weeds Lab-Europe Am Embassy-Agr APO New York NY 09794-0007

KNUTSON, LYNN D, b Red Wing, Minn, Aug 22, 46; m 68. POLARIZATION PHENOMENA IN NUCLEAR REACTIONS, FEW-BODY SYSTEMS. *Educ:* St Olaf Col, BA, 68; Univ Wis-Madison, MA, 70, PhD(physics), 73. *Prof Exp:* From asst prof to assoc prof, 77-85, PROF PHYSICS, UNIV WIS-MADISON, 85- *Mem:* Fel Am Phys Soc. *Res:* Polarization effects in nuclear reactions and scattering at low and intermediate energies with emphasis on few-body systems; wave functions of A-2 and 3 nuclei; tests of charge symmetry; production of polarized beams and targets. *Mailing Add:* Physics Dept Univ Wis 1150 University Ave Madison WI 53706

KNUTSON, ROGER M, b Montevideo, Minn, Jan 3, 33; m 57; c 5. BOTANY, PLANT LIFE HISTORIES. *Educ:* St Olaf Col, BA, 57; Mich State Univ, MS, 61, PhD(plant path), 65. *Prof Exp:* From asst prof to assoc prof, 64-74, PROF BIOL, LUTHER COL, IOWA, 74- *Concurrent Pos:* NSF sci fac fel, Univ Ga, 71-72. *Mem:* AAAS; Am Inst Biol Sci. *Res:* Plant thermoregulation. *Mailing Add:* Dept Biol Luther Col Decorah IA 52101

KNUTSON, VICTORIA P, BIOCHEMISTRY, ENDOCRINOLOGY. *Educ:* Univ Minn, PhD(biochem), 81. *Prof Exp:* ASST PROF PHARMACOL, SCH MED, HEALTH & SCI CTR, UNIV TEX, 83- *Mailing Add:* Dept Pharmacol Univ Tex Med Sch PO Box 20708 Houston TX 77225

KNUTTGEN, HOWARD G, b Yonkers, NY, May 5, 31; m 61; c 2. APPLIED PHYSIOLOGY. *Educ:* Springfield Col, BS, 52; Pa State Univ, MS, 53; Ohio State Univ, PhD(phys educ), 59. *Prof Exp:* Instr phys educ, Ohio State Univ, 54-59; asst prof anat & physiol, Boston Univ, 61-65, assoc prof physiol, 65-71, assoc dean, 75-80, PROF PHYSIOL, COL ALLIED HEALTH PROFESSIONS, BOSTON UNIV, 71-, CHMN, DEPT HEALTH SCI, 80- *Concurrent Pos:* Ed-in-chief, Med & Sci in Sports, 74- *Mem:* AAAS; Am Col Sports Med; Am Physiol Soc. *Res:* Human performance; exercise physiology; muscle metabolism. *Mailing Add:* Ctr Sports Med Pa State Univ Greenberg Sports Complex University Park PA 16802

KNYCH, EDWARD THOMAS, b Chicago, Ill, Oct 8, 42; m 65; c 3. PHARMACOLOGY. *Educ:* Loyola Univ, Chicago, BS, 64; Creighton Univ, MS, 66; WVa Univ, PhD(pharmacol), 70. *Prof Exp:* Fel pharmacol, Univ Wis, 70-72; ASST PROF PHARMACOL, UNIV MINN, 72- *Mem:* AAAS; Sigma Xi. *Res:* Mechanism of polypeptide hormone action and effect of drugs of abuse on hormone release. *Mailing Add:* Dept Pharmacol Univ Minn Sch Med Duluth MN 55812

KNYSTAUTAS, EMILE J, Can citizen. ATOMIC PHYSICS. *Educ:* Univ Montreal, BSc, 65; Univ Conn, MS, 67, PhD(physics), 69. *Prof Exp:* From asst prof to assoc prof, 71-82, PROF PHYSICS, LAVAL UNIV, 82- *Concurrent Pos:* Foreign guest worker, Nat Bur Standards, Washington, DC, 78-79. *Mem:* Can Asn Physicists; Mat Res Soc. *Res:* Atomic spectroscopy; ion implantation; accelerator technology. *Mailing Add:* Dept Phys Laval Univ Quebec PQ G1K 7P4 Can

KO, CHE MING, b Szechuan, China, Jan 7, 43; US citizen; m 73; c 2. HEAVY-ION PHYSICS, MEDIUM-ENERGY PHYSICS. *Educ:* Tunghai Univ, Taiwan, BSc, 65; McMaster Univ, Can, MSc, 68; State Univ NY, Stony Brook, PhD(physics), 73. *Prof Exp:* Res assoc, McMaster Univ, 73-74; vis scientist nuclear physics, Max Planck Inst, 74-77; res assoc, Mich State Univ, 77-78; staff physicist, Lawrence Berkeley Lab, 78-80; from asst prof to assoc prof, 80-88, PROF PHYSICS, TEX A&M UNIV, 88- *Concurrent Pos:* Vis scientist, Oak Ridge Nat Lab, 84-85. *Mem:* Am Phys Soc. *Res:* Theoretical nuclear physics, especially the studies of transport processes and particle productions in heavy-ion reactions. *Mailing Add:* Physic Dept Tex A&M Univ Col Sta TX 77843-4242

KO, CHIEN-PING, b Taipei, Taiwan, July 5, 48; US citizen; m 75; c 2. DEVELOPMENTAL NEUROBIOLOGY. *Educ:* Nat Taiwan Univ, BS, 70; Washington Univ, PhD(physiol & biophys), 75. *Prof Exp:* Res fel, dept anat, Univ Colo, 75-78, Nat Inst Neurol & Commun Dis & Stroke, NIH, 78-81; asst prof, 81-87, ASSOC PROF, DEPT BIOL SCI, UNIV SOUTHERN CALIF, 87- *Honors & Awards:* Res Career Develop Award, NIH, 83-87. *Mem:* Soc Neurosci; Am Soc Cell Biol. *Res:* Membrane structure and function of synapses; development and plasticity of synaptic connections; freeze-fracture electron microscopy. *Mailing Add:* Dept Biol Sci Univ Southern Calif Los Angeles CA 90089-2520

KO, EDMOND INQ-MING, b Hong Kong, July 8, 52. CHEMICAL CATALYSIS, SOLID STATE CHEMISTY. *Educ:* Univ Wis, Madison, BS, 74; Stanford Univ, MS, 75, PhD(chem eng), 80. *Prof Exp:* From asst prof to assoc prof, 80-88, PROF CHEM ENG, CARNEGIE-MELLON UNIV, 88- *Concurrent Pos:* Vis assoc prof, Univ Calif, Berkeley, 87-88. *Mem:* AAAS; Am Inst Chem Engrs; Am Chem Soc; Sigma Xi. *Res:* Synthesis and characterization of heterogeneous catalysts; adsorption and reaction of gases on solid surfaces; semiconductor processing. *Mailing Add:* Dept Chem Eng Carnegie-Mellon Univ Pittsburgh PA 15213

KO, FRANK K, b Canton, China, Aug 5, 47; c 2. TEXTILE MATERIALS ENGINEERING, TEXTILE STRUCTURAL COMPOSITES. *Educ:* Philadelphia Col Textiles & Sci, BS, 70; Ga Inst Technol, MS(textile eng & polymer, plastic eng), 71, PhD(textile engr), 77. *Prof Exp:* Res engr, textile engr, Ga Inst Technol, 72-73; from asst prof to assoc prof textile eng, Philadelphia Col Textile & Sci, 76-84; assoc prof, 84-90, PROF MAT ENG, DREXEL UNIV, 90-, DIR FIBROUS MAT RES LAB, 84- *Concurrent Pos:* Adj prof, Temple Univ, 80; chmn, student chapters comt, Soc Advan Mat & Processing Eng, 85-; advan composites roadmap team mem, Aerospace Indust Asn, 88; comt mem, Soc Advan Mat & Process Eng, 88-; mem, sci bd Am Composites Technol Inc, 88- *Honors & Awards:* Distinguished Achievement Award, Fiber Soc. *Mem:* Fel Soc Advan Mat & Process Eng; Am Ceramic Soc; AAAS; Am Soc Mech Eng. *Res:* Fibrous materials ranging from textile surgical implants to textile structural composites; textile structural mechanics; 3-D composites. *Mailing Add:* 29W-202 Drexel Univ 31st & Chestnut Sts Philadelphia PA 19104

KO, H(SIEN) C(HING), b Formosa, Apr 28, 28; m 55; c 3. ELECTRICAL ENGINEERING, RADIO ASTRONOMY. *Educ:* Nat Taiwan Univ, BS, 51; Ohio State Univ, MSc, 53, PhD(elec eng), 55. *Prof Exp:* Asst, Radio Wave Res Labs, Formosa, 51-52; res asst, Radio Observ, Ohio State Univ, 52-55, from instr to assoc prof elec eng, 55-63, asst dir, Radio Observ, 57-66, PROF ELEC ENG & ASTRON, OHIO STATE UNIV, 63-, CHMN, DEPT ELEC ENG, 77- *Mem:* Int Union Radio Sci; fel Inst Elec & Electronics Engrs; Am Asn Eng Educ. *Res:* Space physics; electromagnetic theory and antennas; electronics and communications. *Mailing Add:* Dept Elec Eng Ohio State Univ 2015 Neil Ave Columbus OH 43210

KO, HON-CHUNG, b Canton, China, June 27, 37; m 68; c 1. PHYSICAL CHEMISTRY. *Educ:* Chung Chi Col, BS, 59; Univ Va, MS, 62; Carnegie Inst Technol, PhD(phys chem), 64. *Prof Exp:* Res chemist, Rocket Power Res Lab, Maremont Corp, 63-67 & Space Sci Inc, 67-68; res assoc chem, Univ Chicago, 69-70, Univ Pittsburgh, 70-72 & Univ Lethbridge, 72-74; RES CHEMIST, ALBANY RES CTR, BUR MINES, 74- *Mem:* Sigma Xi; Am Chem Soc. *Res:* Thermochemistry; thermodynamics; calorimetry; molten salts; solvent effects; high temperature chemistry; low-temperature heat capacities; chlorination of metals; vapor-liquid equilibria; instrumental analysis. *Mailing Add:* Albany Res Ctr Bur Mines Albany OR 97321-3554

KO, HON-YIM, b Hong Kong, China, Jan 18, 40; m 64; c 2. SOIL & ROCK MECHANICS. *Educ:* Univ Hong Kong, BSc, 62; Calif Inst Technol, MS, 63, PhD(civil eng), 66. *Prof Exp:* Res fel eng, Calif Inst Technol, 66-67; from asst prof to assoc prof, 67-75, dept chmn, 83-90, PROF CIVIL ENG, UNIV COLO, BOULDER, 75- *Concurrent Pos:* Consult, Jet Propulsion Lab, Calif Inst Technol, 67-69, prin investr, NSF, Air Force Off Sci Res, US Bur Reclamation & US Bur Mines res grants, 67-; consult, Martin Marietta Corp, 70-, Sandia Corp, 79-81, Exxon Prod Res, 81-, Earth Technol Corp, 84- *Honors & Awards:* Huber Res Prize, Am Soc Civil Engrs, 79. *Mem:* Am Soc Civil Engrs; Am Soc Eng Educ; Soc Exp Stress Anal. *Res:* Fundamental mechanical properties of soil, rock and other geological materials, and the analysis of engineering problems in geotechnics; centrifugal modeling of geotechnical structures; geotechnical engineering. *Mailing Add:* Dept Civil Eng Univ Colo Boulder CO 80309

KO, HOWARD WHA KEE, biophysics, for more information see previous edition

KO, LI-WEN, b Taipei, Taiwan, Jan 31, 49. EXPERIMENTAL BIOLOGY. *Educ:* Taiwan Univ, Vet Med; Univ Wash, MS; Ohio State Univ, PhD(exp path). *Prof Exp:* ASST PROF PATH, MED SCH, CORNELL UNIV, NY; RES SCIENTIST, BURKE MED RES INST, WHITE PLAINS, NY. *Concurrent Pos:* Vis scientist, Univ Miss, Kansas City; assoc prof, Nat Yeng-Ming Med, Taipei. *Mem:* Am Soc Path Biol; Soc Exp Biol & Med. *Mailing Add:* Burke Rehab Ctr 785 Mamaroneck Ave White Plains NY 10605

KO, PAK LIM, b Hong Kong, Mar 4, 37; Can citizen; m 64; c 2. TRIBOLOGY. *Educ:* Univ Strathclyde, BSc, 63; Univ BC, MASc, 65, PhD(tribology), 70. *Prof Exp:* Fel mech eng, Univ BC, 69-70; res engr, Chalk River Nuclear Labs, Atomic Energy Can Ltd, 70-84; RES ENGR, NAT RES COUN CAN, 84- *Concurrent Pos:* Adj prof, dept mech eng, Univ BC, 84- *Mem:* Inst Mech Engrs, Eng; Inst Mech Engrs; Am Soc Mech Eng. *Res:* Friction and friction-induced vibration mechanisms; impact and fretting wear studies; flow-induced vibration and tube fretting wear in steam generators and heat exchangers. *Mailing Add:* Nat Res Coun Can 3650 Wesbrook Mall Vancouver BC V6S 2L2 Can

KO, WEN HSIUNG, b Fukien, China, Apr 12, 23; US citizen; m 57; c 4. BIONICS. *Educ:* Nat Amoy Univ, BS, 46; Case Inst Technol, MS, 56, PhD(elec eng), 59. *Prof Exp:* Engr, Taiwan Telecommun Admin, 46-54; from asst prof to assoc prof elec eng, 59-67, assoc prof surg, Sch Med, 64-70, actg dir, Eng Design Ctr, 70-71, dir, Electronics Design Ctr, 71-83, PROF ELEC & BIOMED ENG, CASE WESTERN RESERVE UNIV, 67- *Concurrent Pos:* Mem biomed eng training comt, NIH, 66-70, mem oviduct panel, Contraceptive Develop Br, Ctr Pop Res, 71; NIH fel, Sch Med, Stanford Univ, 67-68; mem, NASA Life Sci Prog Space Sci Bd, Nat Acad Sci, 69-70; reviewer, NIH, NASA, NSF & Inst Elec & Electronics Engrs. *Honors & Awards:* Cecon Award, Electronics Rep Asn, 70. *Mem:* Fel Inst Elec & Electronics Engrs; Electronics & Biol Eng; Int Soc Biotelemetry; Biomed Eng Soc. *Res:* Microelectronic instrumentation and technology; medical instrumentation; implant electronic transducers, telemetry and stimulators; solid state sensors and actuators. *Mailing Add:* 1356 Forest Hills Blvd Cleveland OH 44118

KO, WEN-HSIUNG, b Chao Chow, Taiwan, May 14, 39; m 68; c 2. PLANT PATHOLOGY, SOIL MICROBIOLOGY. *Educ:* Nat Taiwan Univ, BS, 62; Mich State Univ, PhD(plant path), 66. *Prof Exp:* From res assoc to assoc prof, 66-76, PROF PLANT PATH, UNIV HAWAII, HILO, 76- *Honors & Awards:* Ruth Allen Award, Am Phytopath Soc, 84. *Mem:* Fel Am Phytopath Soc. *Res:* Ecology of soil-borne diseases; general soil microbiology; fungal physiology. *Mailing Add:* Agr Exp Sta Univ Hawaii 461 Lanikaula St Hilo HI 96720

KO, WINSTON TAI-KAN, b Shanghai, China, Apr 5, 43; m 70; c 2. HIGH ENERGY PHYSICS. *Educ:* Carnegie Inst Technol, BS, 65; Univ Pa, MS, 66, PhD(physics), 71. *Prof Exp:* Asst res physicist, 70-72, from asst prof to assoc prof, 72-82, PROF PHYSICS, UNIV CALIF, DAVIS, 82- *Concurrent Pos:* Assoc, Europ Coun Nuclear Res, 85-86. *Mem:* Am Phys Soc. *Res:* Experimental high energy physics; elementary particle physics; computer application to physics experiments. *Mailing Add:* Dept Physics Univ Calif Davis CA 95616

KOBALLA, THOMAS RAYMOND, JR, b Manchester, NH, June 28, 54. ATTITUDE CHANGE. *Educ:* East Carolina Univ, BS, 76, MA, 78; Pa State Univ, PhD, 81. *Prof Exp:* Asst prof, Pikeville Col, 81-; AT DEPT CONTINUING EDUC. *Res:* Systematic design of attitude change paradigms in changing attitudes toward science. *Mailing Add:* Dept Curric & Instr Univ Tex Austin TX 78712

KOBATA, AKIRA, b Nemuro, Japan, Mar 17, 33; m 60; c 3. SYNTHETIC ORGANIC & NATURAL PRODUCTS CHEMISTRY, MEDICAL SCIENCE. *Educ:* Univ Tokyo, BS, 56, MS, 58, PhD(biochem), 62. *Prof Exp:* Staff mem, Res Inst, Takeda Chem Indust Co, 58-67; prof, Kobe Univ Sch Med, 71-83; PROF, INST MED SCI, UNIV TOKYO, 82-, DIR MGT, 90- *Concurrent Pos:* Vis assoc, Nat Inst Arthritis & Metab Dis, NIH, 67-69, vis scientist, 69-71; Fogarty scholar-in-residence, Fogarty Int Ctr, 85-87; nat rep, Int Glycoconjugata Asn, 87-; dir, Japanese Soc Protein Eng, 88- & Japanese biochem Soc, 89- *Mem:* Am Soc Biol Chemists; AAAS. *Res:* Structure and function of the sugar chains of glycoprotein; clinical application of glycoconjugate research. *Mailing Add:* Inst Med Sci Univ Tokyo 4-6-1 Shirokanedai Minato-ku Tokyo 108 Japan

KOBAYASHI, ALBERT S(ATOSHI), b Chicago, Ill, Dec 9, 24; m; c 3. SOLID MECHANICS. *Educ:* Univ Tokyo, BS, 47; Univ Wash, MS, 52; Ill Inst Technol, PhD(mech eng), 58. *Prof Exp:* Tool engr, Konishiroku Photo Indust, Japan, 47-50 design engr, Ill Tool Works, 53-55; res engr exp stress anal, Armour Res Found, Ill Inst Technol, 55-58; from asst prof to prof, 58-88, BOEING PENNELL PROF STRUCT ANALYSIS, UNIV WASH, 88- *Concurrent Pos:* Mem staff, Boeing Co, Wash, 58-75; assoc ed, J Appl Mech, 77-84, Trans Japan Soc Composite Mats, 74-; honoree, Int Conf Dynamic Fracture Mech, San Antonio, Tex, 84. *Honors & Awards:* F G Tatnall Award, Soc Exp Stress Anal, 73, B J Lazan Award, 81, R E Peterson Award, & William Murray Medal, 83. *Mem:* Nat Acad Eng; fel Am Soc Mech Engrs; fel Soc Exp Mech; Am Ceramic Soc. *Res:* Fracture mechanics; experimental stress analysis; theories of elasticity; theory of structures and dynamic response of structures; author of over 350 publications. *Mailing Add:* Dept Mech Eng Univ Wash Seattle WA 98195

KOBAYASHI, F(RANCIS) M(ASAO), b Seattle, Wash, Nov 19, 25; m 63; c 3. ENGINEERING MECHANICS. *Educ:* Univ Notre Dame, BS, 47, MS, 48, ScD(eng mech), 53. *Prof Exp:* Asst prof eng mech, 48-58, assoc prof eng sci, 58-59, 60-64, asst vpres res & sponsored progs, 68-71, PROF ENG SCI, UNIV NOTRE DAME, 64-, ASST VPRES ADVAN STUDIES, RES & SPONSORED PROGS, 71- *Concurrent Pos:* Asst dir eng sci prog, NSF, 59-60. *Res:* Fluid mechanics; wave resistance; systems engineering; operations research. *Mailing Add:* Univ Notre Dame Notre Dame IN 46556

KOBAYASHI, GEORGE S, b San Francisco, Calif, Nov 25, 27; m 56; c 4. MYCOLOGY, BIOCHEMISTRY. *Educ:* Univ Calif, Berkeley, BS, 52; Tulane Univ, PhD(microbiol), 63. *Prof Exp:* Chemist enol, Roma Wine Co, Calif, 52; sr lab technician mycol, Sch Pub Health, Univ Calif, Berkeley, 52-59; from instr to assoc prof, 64-77, PROF MYCOL, SCH MED, WASHINGTON UNIV, 77- *Concurrent Pos:* Assoc dir, Microbiol Labs, Barnes Hosp, St Louis, 65-; assoc ed, Cutaneous Path, 74-; consult, Bur of Biologics, Food & Drug Admin, 74-76; mem NIH study sect, Nat Inst Allergy & Infectious Dis, 78-84. *Mem:* AAAS; Int Soc Human & Animal Mycol; NY Acad Sci; Med Mycol Soc Am; Infectious Dis Soc; Am Soc Microbiol. *Res:* Immunology and biochemistry of medically important fungi. *Mailing Add:* Dermat Div Box 8123 Washington Univ St Louis MO 63110

KOBAYASHI, HISASHI, b Tokyo, Japan, June 13, 38; m 63. RESEARCH MANAGEMENT, UNIVERSITY ADMINISTRATION. *Educ:* Univ Tokyo, BS, 61, MS, 63; Princeton Univ, MA, 66, PhD(elec eng) 67. *Prof Exp:* Res staff mem, IBM T J Watson Res Ctr, 67-71, mgr syst measurement-modeling, 71-73, sr mgr systs anal, 75-75, 77-79, sr mgr, VLSI design, 81-82; dir, IBM Japan Sci Inst, 82-86; SHERMAN FAIRCHILD PROF ELEC ENG & COMPUT SCI & DEAN ENG & APPL SCI, PRINCETON UNIV, 86- *Concurrent Pos:* vis asst prof, Syst Sci Dept, Univ Calif, Los Angeles, 69-70, vis prof, Info Sci Dept, Univ Hawaii, 75, Technische Hochschule Darmstadt, WGer, 79-80 & int prof comput sci, Free Univ Brussells, Belgium, 80; consult, ALOHA systs proj, 75; consult prof, Comput Systs Lab, Stanford Univ, 76. *Honors & Awards:* Humboldt Prize, 79; Silver Core, Int Fed Info Processing, 80. *Mem:* Fel Inst Elec & Electronics Engrs; Asn Comput Mach. *Res:* Radar signal design; detection and estimation theory; data transmission theory; seismic signal processing; image date compression; magnetic recording theory; optical communication; communication networks; queueing theory. *Mailing Add:* Dean Sch Eng & Appl Sci Princeton Univ Princeton NJ 08544

KOBAYASHI, KAZUMI, b Fukuyama-City, Japan, Feb 16, 52. PHARMACEUTICAL SCIENCE. *Educ:* Kyoto Univ, Japan, BS, 74, MS, PhD(pharmaceut sci), 83. *Prof Exp:* Asst lectr & instr, Dept Biochem, Niigata Col Pharm, Japan, 79-83; postdoctoral assoc, Dept Chem, Mass Inst Technol, 83-85; res fel, Dept Molecular Biol, Mass Gen Hosp & Dept Genetics, Med Sch, Harvard Univ, 85-87; STAFF SCIENTIST II & GROUP LEADER PROTEIN CHEM, CAMBRIDGE NEUROSCI, INC, 87- *Concurrent Pos:* Prin investr, NIH, 88-89 & 90-92. *Mem:* Am Soc Biochem & Molecular Biol; Am Chem Soc; Protein Soc; AAAS. *Res:* Isolation of novel neuroactive compounds from natural products; author of 11 technical publications. *Mailing Add:* Cambridge Neurosci Inc One Kendall Sq Bldg 700 Cambridge MA 02139

KOBAYASHI, NOBUHISA, b Osaka, Japan, May 4, 50; m; c 2. COASTAL ENGINEERING, HYDRODYNAMICS. *Educ:* Kyoto Univ, Japan, BCE, 74, MCE, 76; Mass Inst Technol, PhD(civil eng), 79. *Prof Exp:* Res asst coastal eng, Mass Inst Technol, 78-79; sr engr ocean eng, Brian Watt Assocs, Inc, Houston, 79-81; from asst prof to assoc prof, 81-91, PROF COASTAL ENG, UNIV DEL, 91- *Concurrent Pos:* Chmn, Tidal Hydraul Comt, Am Soc Civil Eng, 86-87 & Task Comt Sea Level Rise & Its Effects, 87-90; mem, Waves & Wave Forces Comt, Am Soc Civil Engrs, 87-; vis lectr, Disaster Prevention Res Inst, Kyoto Univ, 88-; assoc dir, Ctr Appl Coastal Res, Univ Del, 89-; asst ed, J Waterway, Port, Coastal & Ocean Eng, Am Soc Civil Engrs, 90- *Mem:* Am Soc Civil Engrs; Am Geophys Union; Int Asn Hydraul Res; Japan Soc Civil Engrs. *Res:* Interaction of wind waves with coastal structures; wave mechanics in surf and swash zones; sediment transport mechanics in nearshore region; prediction of oil spills and their effects; numerical prediction of tsunami run-up. *Mailing Add:* Dept Civil Eng Univ Del Newark DE 19716

KOBAYASHI, RIKI, b Webster, Tex, May 13, 24; c 4. CHEMICAL ENGINEERING, CHEMISTRY. *Educ:* Rice Inst, BS, 44; Univ Mich, MSE, 47, PhD(chem eng), 51. *Prof Exp:* From asst prof to assoc prof, 51-65, prof, 65-67, LOUIS CALDER PROF CHEM ENG, RICE UNIV, 67- *Concurrent Pos:* Res engr, Continental Oil Co, Okla; consult, appl thermodynamics. *Honors & Awards:* Katz Lectureship, Univ Mich, 75; Katz Award, 85. *Mem:* Fel Am Inst Chem Engrs; fel Am Inst Chemists; Am Inst Physics, Am Inst Mining & Metal Engrs; Am Chem Soc. *Res:* Thermodynamic and transport properties of fluids and solids, particularly at advanced pressures; cryogenic temperatures to moderately high temperatures; approx 190 papers in learned journals. *Mailing Add:* Dept Chem Eng Rice Univ PO Box 1892 Houston TX 77251

KOBAYASHI, ROGER HIDEO, b Honolulu, Hawaii, May 21, 47; m; 74; c 2. PEDIATRIC IMMUNOLOGY & ALLERGY, PEDIATRIC RHEUMATOLOGY. *Educ:* Univ Nebr, Lincoln, BA, 69, MD, 75; Univ Hawaii, MS, 75; Am Bd Pediat, cert pediat, 80; Am Bd Allergy & Immunol, cert allergy & immunonol, 81. *Prof Exp:* Resident pediat, Sch Med, Univ Southern Calif, 75-77; clin fel pediat immunol, Sch Med, Univ Calif, Los Angeles, 77-78, res fel immunol, 78-79; asst prof pediat & microbiol, Univ Nebr Med Ctr, 80-84, assoc prof pediat, path & microbiol, 84-88; ASSOC PROF PEDIAT, UNIV CALIF, LOS ANGELES CTR HEALTH SCI, 88- *Mem:* Soc Microbiol; Am Fedn Clin Res; Am Acad Pediat; Am Acad Allergy & Immunol. *Res:* Neutrophil modulation of natural killer cell activity; viral-neutrophil interaction with special interests adverse effects of viral infections of neutrophil function; treatment of asthma in infants and young children; use of intravenous gammaglobulin in children. *Mailing Add:* Dept Allergy & Immunol Univ Nebr Med Ctr Bloomfield Hills Omaha NE 68105

KOBAYASHI, SHIRO, b Gotsu, Japan, Feb 21, 24; m 61. MECHANICAL ENGINEERING. *Educ:* Univ Tokyo, BS, 46; Univ Calif, Berkeley, MS, 57, PhD(mech eng), 60. *Prof Exp:* Asst prof indust eng, 61-65, assoc prof mech eng, 65-68, PROF MECH ENG, UNIV CALIF, BERKELEY, 68- *Concurrent Pos:* Battelle vis prof, Ohio State Univ, 67-68; E A Taylor vis prof, Univ Birmingham, 70. *Honors & Awards:* Blackall Mach Tool & Gage Award, 63; Gold Medal Award, Soc Mfg Engrs, 83. *Mem:* Nat Acad Eng; Am Soc Mech Engrs; Japan Soc Tech Plasticity; Soc Mfg Engrs. *Res:* Materials processing, machining and forming. *Mailing Add:* Dept Mech Eng Univ Calif Berkeley CA 94720

KOBAYASHI, SHOSHICHI, b Kofu, Japan, Jan 4, 32; m 57; c 2. MATHEMATICS. *Educ:* Univ Tokyo, BS, 53; Univ Wash, PhD(math), 56. *Prof Exp:* Mem staff, Inst Advan Study, 56-58; res assoc math, Mass Inst Technol, 58-60; asst prof, Univ BC, 60-62; from asst prof to assoc prof, 62-66, chmn dept, 78-81, PROF MATH, UNIV CALIF, BERKELEY, 66- *Concurrent Pos:* A P Sloan fel, 64-66; lectr, Univ Tokyo, 65; vis prof, Univ Mainz, 66, Univ Bonn, 69 & 78 & Mass Inst Technol, 70; assoc ed, Duke J Math, 70-80; ed, J Differential Geometry, 73-85; Guggenheim fel, 77-78; vis prof, Univ Tokyo, 81; ed, Int J Math, 90- *Honors & Awards:* Geom Prize, Math Soc Japan, 87. *Mem:* Am Math Soc; Math Soc Japan; Math Soc France; Swiss Math Soc. *Res:* Differential geometry and functions of several complex variables. *Mailing Add:* Dept Math Univ Calif Berkeley CA 94720

KOBAYASHI, YUTAKA, b San Francisco, Calif, Mar 11, 24; m 54, 82; c 3. BIOCHEMISTRY. *Educ:* Iowa State Col, BS, 46, MS, 50; Univ Iowa, PhD(biochem), 53. *Prof Exp:* Res assoc, Rheumatic Fever Res Inst, Chicago, Ill, 53-57; sr scientist, Worcester Found for Exp Biol, 57-74; mgr, Appln Lab, New Eng Nuclear Corp, 74-82; DuPont, 82-85; RETIRED. *Concurrent Pos:* Consult, biotechnol, 85- *Mem:* Am Chem Soc; Am Soc Pharmacol & Exp Therapeut; Am Soc Biol Chemists; Sigma Xi. *Res:* Intermediary metabolism of amino acids and biogenic amines; amine oxidases. *Mailing Add:* 60 Audubon Rd Wellesley MA 02181

KOBE, DONALD HOLM, b Seattle, Wash, Jan 13, 34. QUANTUM PHYSICS. *Educ:* Univ Tex, Austin, BS, 56; Univ Minn, Minneapolis, MS, 59, PhD(physics), 61. *Prof Exp:* Vis asst prof physics, Ohio State Univ, 61-63; Fulbright lectr, Nat Taiwan Univ & Taiwan Norm Univ, 63-64; vis scientist, Quantum Chem Inst, Univ Uppsala, 64-66; vis asst prof physics, H C Oersted Inst, Copenhagen Univ, 66-67 & Northeastern Univ, 67-68; assoc prof, 68-75, PROF PHYSICS, UNIV N TEX, 75- *Concurrent Pos:* Fulbright lectr/researcher Instituto de Física Teórica, Sao Paulo, Brazil, 88-89. *Mem:* AAAS; Am Phys Soc; Am Sci Affil; Am Asn Physics Teachers. *Res:* Quantum theory of many-particle systems; applications to superfluid helium; quantum theory of radiation; interaction of electromagnetic radiation and matter; geometrical phase in quantum theory. *Mailing Add:* Dept Physics Univ North Texas Denton TX 76203

KOBER, C(ARL) L(EOPOLD), b Vienna, Austria, Nov 22, 13; nat US; m 42; c 2. MECHANICAL ENGINEERING. *Educ:* Vienna Univ, PhD, 35; Vienna Tech Univ, Dr phil habil, 38. *Prof Exp:* From tech asst to vpres prod, Elin, Inc, Austria, 36-40; chief radar dept, GEMA GmbH, Ger, 40-45; mgr, Secowerk, 48-49; consult, Wright Air Develop Ctr, Wright-Patterson Air Force Base, Ohio, 49-55; tech dir, Tech Div, Gen Mills, Inc, Minn, 55-58; vpres and oper opers, Crosley Div, Avco Corp, 58-61; dir manned space opers, Denver Div, Martin Marietta Corp, 61-73; PRES, DENVER MINERAL EXPLOR CORP, 73- *Concurrent Pos:* Prof, Colo State Univ, 68-79. *Mem:* Inst Elec & Electronics Engrs; AAAS. *Res:* Ordnance; missile detection; plasma physics; data processing; remote sensing geology. *Mailing Add:* 605 Front Range Rd Littleton CO 80120

KOBER, EHRENFRIED H, organic chemistry, for more information see previous edition

KOBERNICK, SIDNEY D, b Montreal, Que, May 7, 19; nat US; m 41; c 3. PATHOLOGY. *Educ:* McGill Univ, BSc, 41, MD, CM, 43, MSc, 49, PhD(exp path), 51; Am Bd Path, cert path anat, 53, cert clin path, 59. *Prof Exp:* Asst prof path, McGill Univ, 51-52; CLIN PROF PATH, WAYNE STATE UNIV, 53-; DIR LABS, SINAI HOSP, DETROIT, 52- *Concurrent Pos:* Adj prof med technol, Wayne State Univ. *Mem:* Soc Exp Biol & Med; Am Asn Path & Bact; fel Am Soc Clin Path; fel Col Am Pathologists; Electron Micros Soc Am. *Res:* Experimental atherosclerosis; tissue hypersensitivity; morphological pathology. *Mailing Add:* 8319 W Country Club Dr N Sarasota FL 33580

KOBERSTEIN, JEFFREY THOMAS, b Milwaukee, Wis, Sept 27, 52; m 75; c 2. POLYMER CHEMISTRY, CHEMICAL ENGINEERING. *Educ:* Univ Wis, BS, 74; Univ Mass, PhD(chem eng), 79. *Prof Exp:* Fel, Ctr Res Macromolecules, 79-80; asst prof chem eng, Princeton Univ, 80-86; assoc prof, 86-89, PROF CHEM ENG, UNIV CONN, 89- *Concurrent Pos:* Vis asst prof, Univ Wis, 81; vis res scientist, IBM, 85. *Honors & Awards:* Doolittle Award, Am Chem Soc, 84. *Mem:* Am Chem Soc; Am Phys Soc; NAm Thermal Anal Soc; Soc Plastics Engrs. *Res:* Polymer morphology; structure property relationships in block copolymers; polymer-polymer interfaces; microphase separation; small angle x-ray, neutron and light scattering; polymer blends; compatibility. *Mailing Add:* Dept Chem Eng U-136 Univ Conn Storrs CT 06268

KOBILINSKY, LAWRENCE, b New York, NY, Nov 7, 46; m 71; c 1. IMMUNOLOGY, BIOCHEMISTRY. *Educ:* City Univ New York, BS, 69, MA, 71, PhD(biol), 77. *Prof Exp:* Res asst biophys, Columbia Presbyterian Med Ctr, 69-70; lectr biol, City Univ New York, 70-71; Brooklyn Col, 72-74; Hunter Col, 74-75 & John Jay Col Criminal Justice, 75-77; res fel immunol, Sloan Kettering Inst Cancer Res, 77-80. *Concurrent Pos:* Adj asst prof, John Jay Col Criminal Justice, City Univ New York, 77- *Mem:* Sigma Xi; AAAS; NY Acad Sci; Am Chem Soc; Am Acad Forensic Sci; Am Asn Immunol. *Res:* Indentification of individuals by DNA profiling analysis. *Mailing Add:* Dept Forensic Sci John Jay Col Criminal Justice 445 W 59th St New York NY 10019

KOBISKE, RONALD ALBERT, b New London, Wis, Feb 2, 38; m 58; c 3. ADMINISTRATION, PHYSICS LABORATORY DEVELOPMENT IN NUCLEAR SPECTROSCOPY. *Educ:* Ind Inst Technol, BS(physics) & BS(math), 62; Highlands Univ, MS, 64; Univ Wis Milwaukee, PhD(physics), 76. *Prof Exp:* PROF & CHMN, PHYSICS DEPT, MILWAUKEE SCH ENG, 63- *Mem:* Am Asn Physics Teachers. *Res:* General relativity; teaching and laboratory development efforts. *Mailing Add:* Milwaukee Sch Eng Box 644 Milwaukee WI 53201

KOBLICK, DANIEL CECIL, b San Francisco, Calif, May 13, 22; m 60; c 2. PHYSIOLOGY. *Educ:* Univ Calif, AB, 44; Univ Ore, PhD(biol), 57. *Prof Exp:* Instr biol, Univ Ore, 57-58; asst prof zool, Univ Mo, 58-59; USPHS fel, Univ Calif, Berkeley, 59-60; asst res physiologist, 60-63; ASSOC PROF PHYSIOL, ILL INST TECHNOL, 63- *Concurrent Pos:* Lectr physiol, Univ Calif, 60-61. *Mem:* Fel AAAS; Am Physiol Soc; Biophys Soc; Soc Gen Physiol; Sigma Xi. *Res:* Ion transport; bioelectricity; invertebrate physiology. *Mailing Add:* Dept Biol Ill Inst Technol Chicago IL 60616

KOBLICK, IAN, b San Francisco, Calif, July 12, 39; m 62; c 2. DIVING TECHNOLOGY. *Prof Exp:* PRES, MARINE EDUC, MARINE RESOURCES DEVELOP FOUND, 84- *Mailing Add:* PO Box 787 Key Largo FL 33037

KOBLINSKY, CHESTER JOHN, b Hartford, Conn, March 25, 48; m 77; c 1. SATELLITE REMOTE SENSING, SCIENTIFIC DATA ANALYSIS. *Educ:* Reed Col, BA, 71; Ore State Univ, PhD(oceanog), 79. *Prof Exp:* Oceanographer, US Environ Protection Agency, 75-76; res assoc, Ore State Univ, 79; res fel, Scripps Inst Oceanog, 79-81, asst res oceanographer, 82-83; OCEANOGRAPHER, GODDARD SPACE FLIGHT CTR, NASA, 83- *Concurrent Pos:* Adj assoc prof, Univ Colo, 87- *Honors & Awards:* Spec Achievement Award, NASA, 85; Except Scientific Achievement Medal, NASA, 90. *Mem:* Am Geophys Union; Inst Elec & Electronics Engrs; Am Meteorol Soc; Oceanog Soc. *Res:* Physical oceanography: general ocean circulation, models, observations; satellite remote sensing; airborne remote sensing. *Mailing Add:* Goddard Space Flight Ctr NASA 926 Greenbelt MD 20771

KOBLUK, DAVID RONALD, b Feb 4, 49; Can citizen. PALEOECOLOGY, MARINE GEOLOGY. *Educ:* McGill Univ, BSc, 71, MSc, 73; McMaster Univ, PhD(geol), 76. *Prof Exp:* ASST PROF PALEONT, UNIV TORONTO, 77- *Concurrent Pos:* Fel, Mem Univ Nfld, 76-77; Nat Res Coun Can oper grant, 77-78; Earthwatch res grants, 77, 78 & 79; Connaught grant, Univ Toronto, 78. *Mem:* Am Asn Petrol Geologists; Can Soc Petrol Geologists; Soc Econ Paleontologists & Mineralogists; Paleont Asn; fel Explorers Club. *Res:* Ecology of modern and ancient coral reefs; processes of marine biological erosion; cavity-dwelling marine organisms; invertebrate evolution and community ecology. *Mailing Add:* Dept Geol Erindale Col Univ Toronto 3359 Mississauga Rd N Mississauga ON L5L 1C6 Can

KOBRIN, ROBERT JAY, b New York, NY, Nov 4, 37; m 69; c 3. CHEMICAL INSTRUMENTATION. *Educ:* City Col New York, BS, 60; Univ Del, PhD(phys chem), 65. *Prof Exp:* Jr chemist anal chem, Sonneborn Chem & Refining Co, 60-61; sr chem, 61-73, res assoc, 74-84, RES CONSULTANT, ADV AUTOMATION & DATA SYSTS, MOBIL RES & DEVELOP CORP, 84- *Mem:* Asn Comput Mach; Sigma Xi; Am Chem Soc. *Res:* Laboratory automation; information systems and computer networks; analytical instrumentation. *Mailing Add:* Paulsboro Lab Mobil Res & Develop Corp Paulsboro NJ 08066

KOBRINE, ARTHUR, b Chicago, Ill, Oct 9, 43; m 69; c 2. NEUROSURGERY. *Educ:* Northwestern Univ, BS, 64, MD, 68; George Washington Univ, PhD(physiol), 79. *Prof Exp:* Resident neurosurg, Walter Reed Hosp, 70-73; asst chief, 73-75; asst prof, 75-77, assoc prof, 77-79, PROF NEUROSURG, GEORGE WASHINGTON UNIV, 79- *Mem:* Soc Neurol Surgeons; Am Phys Soc; Am Bd Neurol Surg. *Mailing Add:* 2440 M St NW Washington DC 20037

KOBSA, HENRY, b Vienna, Austria, May 4, 29; nat US; m 80; c 2. PHYSICAL CHEMISTRY. *Educ:* Univ Vienna, PhD(chem), 56. *Prof Exp:* Res chemist, Dacron Res Lab, E I Du Pont de Nemours & Co, Del, 56-58, res chemist, Pioneering Res Lab, 58-61, sr res chemist, 61, res assoc, 61-62, res supvr, 62-64, res supvr, Benger Res Lab, Va, 64-65, tech supvr, Lycra Tech Sect, Va, 65-66, sr supvr, 66-68, sr supvr, May Plant Tech Sect, 68-69, res mgr, Dacron Res Lab, 69-73, tech supt, Indust Tech Sect, Kinston plant, NC, 73-77, res fel, Pioneering Res Lab, 77-81, sr res fel, 81-89, E I DU PONT DE NEMOURS & CO, DEL, 89- *Mem:* Am Chem Soc; Am Inst Chem Engrs; AAAS; Am Asn Textile Chemists & Colorists; Int Soc Optical Eng. *Res:* Photochemistry of polymers and dyes; energy transfer phenomena; polymer physics; polymerization kinetics; diffusion. *Mailing Add:* 111 Greenspring Rd Greenville DE 19807

KOBURGER, JOHN ALFRED, b Queens, NY, Apr 20, 31; m 52; c 4. FOOD MICROBIOLOGY. *Educ:* Kans State Univ, BS, 59, MS, 60; NC State Col, PhD(food microbiol), 62. *Prof Exp:* Res asst, Res Labs, Nat Dairy Prod Corp, 56-57; from asst prof to assoc prof food microbiol, WVa Univ, 62-69; prof food microbiol, Dept Food Sci, Univ Fla, 69-86. *Mem:* Am Chem Soc; Am Dairy Sci Asn; Am Soc Microbiol; Inst Food Technol. *Res:* Physiology of microorganisms important to the food industry. *Mailing Add:* PO Box 58 Suwannee FL 32692

KOBYLNYK, RONALD WILLIAM, b Calgary, Alta, Aug 19, 42. LASER ENTOMOLOGY. *Educ:* Univ Calgary, BSc, 63; Univ Guelph, Ont, MSc, 65, PhD(entom), 72. *Prof Exp:* Eval officer insecticides, Agr Can, 70-80; MEM STAFF, PESTICIDE CONTROL BR, MINISTRY ENVIRON, VICTORIA, 80- *Mem:* Entom Soc Can. *Res:* Effects of laser radiation on insects. *Mailing Add:* Pesticide Mgt Br Ministry Environ 4th Floor 737 Courtnay St Victoria BC V8V 1X5 Can

KOCAOGLU, DUNDAR F, b Turkey, June 1, 39; US citizen; m 68; c 1. ENGINEERING MANAGEMENT, MULTICRITERIA DECISION MAKING. *Educ:* Robert Col, Turkey, BS, 60; Lehigh Univ, MS, 62; Univ Pittsburgh, MS, 72, PhD(opers res), 76. *Prof Exp:* Struct engr, Modjeski & Masters, 62-63; proj engr, United Engrs & Constructors, 63-66, consult proj engr, 69-71; partner, Tekser Consult Co, 66-69; res asst indust eng, Univ Pittsburgh, 71-74, vis asst prof mgt, 74-76, assoc prof opers res, Indust Eng & Eng mgt & dir eng mgt proj, 76-87; pres, Technol Mgt Assoc, 73-87; PROF & DIR, ENG MGT PROG, PORTLAND STATE UNIV, 87- *Concurrent Pos:* Consult, Tokten Prog, UN, 79-80, 87; pres, Col Eng Mgt, Inst Mgt Sci, 79-81; publ dir, Eng Mgt Soc, Inst Elec & Electronics Engrs, 82-85, ed-in-chief, Trans Eng Mgt, 85- *Honors & Awards:* Centennial Medal, Inst Elec & Electronics Engrs, 84. *Mem:* Am Soc Eng Educ; Inst Elec & Electronics Engrs; Am Soc Civil Engrs; Am Soc Eng Mgt; Inst Mgt Sci. *Res:* Engineering management; hierarchical decision modeling; decision support systems for project management, strategic planning, manpower analysis and strategic management; operations research; resource optimization; quantification of expert judgements; technological innovations; risk assessment; technology management; conflict resolution. *Mailing Add:* Eng Mgt Prog Sch Eng & Appl Sci Portland State Univ Portland OR 97207

KOCATAS, BABUR M(EHMET), b Istanbul, Turkey, Apr 7, 27; m 55; c 1. CHEMICAL ENGINEERING. *Educ:* Robert Col, Istanbul, BS, 47; Univ Tex, MS, 53, PhD(chem eng), 62. *Prof Exp:* Design engr, Union Carbide Chem Co, 56-57; lectr math, Univ Tex, 57-62; sr res engr, Monsanto Co, 62-76, sr res group leader, 76-85; CONSULT, 85- *Mem:* Am Chem Soc; Am Inst Chem Engrs; Sigma Xi. *Res:* Process design and development; micro pilot planting; distillation; fuel cells; mathematical modeling; systems engineering; air pollution control. *Mailing Add:* Dilhayat Sok 22 13 Camlik Etiler Istanbul Turkey

KOCH, ALAN R, b St Louis, Mo, Oct 6, 30; m 56; c 3. PHYSIOLOGY. *Educ:* Univ Mich, BS, 51; Columbia Univ, PhD(pharmacol), 55. *Prof Exp:* Asst pharmacol, Columbia Univ, 53-55; res instr, Col Med, Univ Utah, 55-57; from res instr to res asst prof to asst prof physiol, Univ Wash, 59-65; ASSOC PROF ZOOPHYSIOL, MED SCH, WASH STATE UNIV, 66- *Concurrent Pos:* USPHS fel, 55-58. *Mem:* Biophys Soc; Am Physiol Soc. *Res:* Renal physiology; ion transport in brain; active ion transport. *Mailing Add:* Dept Zool Wash State Univ Pullman WA 99163

KOCH, ARTHUR LOUIS, b St Paul, Minn, Oct 25, 25; m 47; c 2. THEORETICAL BIOLOGY, BACTERIAL PHYSIOLOGY. *Educ:* Calif Inst Technol, BS, 48; Univ Chicago, PhD(biochem), 51. *Prof Exp:* Res assoc & instr, Univ Chicago, 51-52 & 53-56; from asst prof to assoc prof biochem, Col Med, Univ Fla, 56-63, prof biochem & microbiol, 63-67; PROF MICROBIOL, IND UNIV, BLOOMINGTON, 67- *Concurrent Pos:* Assoc scientist, Argonne Nat Lab, 52-56; Guggenheim fel, 60-61 & 81-82. *Mem:* Am Soc Microbiol; Am Chem Soc; Am Soc Biol Chemists; Biophys Soc; Genetics Soc Am. *Res:* Enzyme and haploid evolution; active transport systems; microbial growth physiology; microbial response to toxic and antibiotic substances. *Mailing Add:* Dept Biol Ind Univ Bloomington IN 47405

KOCH, CARL CONRAD, b Cleveland, Ohio, Oct 19, 37; m 65; c 2. RAPID SOLIDIFICATION, ALLOY BEHAVIOR. *Educ:* Case Inst Technol, BS, 59, MS, 61, PhD(metall), 64. *Prof Exp:* NSF fel metall, Univ Birmingham, Eng, 64-65; staff scientist superconducting mat, Oak Ridge Nat Lab, 65-70, group leader, 70-83; PROF MAT SCI & ENG, NC STATE UNIV, 83- *Concurrent Pos:* Vis scientist, AERE, Harwell, Eng, 71-72; lectr, Univ Tenn, Knoxville, 71; secy, Alloy-phases Comt, Metall Soc, 81-83, chmn, 83-85. *Honors & Awards:* Metall & Ceramics Award, US Dept Energy, 80; IR 100 Award, Indus Res & Develop Mag, 83. *Mem:* Fel Am Phys Soc; fel AAAS; Mat Res Soc; Am Soc Metals; Am Inst Mining, Metall & Petrol Engrs. *Res:* Materials science; rare earth alloy behavior; superconducting materials; fluxoid pinning; amorphous superconductors; non-equilibrium processing; rapid solidification; vapor deposition; mechanical alloying; intermetallic compounds; solid state amorphization. *Mailing Add:* 1713 Lookout Point Ct Raleigh NC 27612

KOCH, CARL FRED, b Washington, DC, July 13, 32; m 78; c 1. PALEOECOLOGY, BIOSTRATIGRAPHY. *Educ:* Univ Md, BS, 57, MS, 61; George Washington Univ, PhD(geol), 77. *Prof Exp:* Engr, Appl Physics Lab, Johns Hopkins Univ, 58-61; Rixon Electronics Inc, 62-67 & Seismic Data Anal Ctr, 68-77; asst prof, 78-82, ASSOC PROF GEOL, OLD DOMINION UNIV, 82- *Concurrent Pos:* Geologist, US Geol Surv, 77-; res assoc, Smithsonian Inst, 81- *Mem:* Geol Soc Am; Paleont Soc; Int Paleont Inst. *Res:* Geologic data for restricted time intervals of large geographic extent; biosphere history; quantitative paleoecology and biostatigraphy using upper cretaceous molluscs. *Mailing Add:* Dept Geosci Old Dominion Univ 5215 Hampton Blvd Norfolk VA 23508

KOCH, CARL MARK, b Orefield, Pa, Apr 29, 44; m 66; c 3. WATER RESOURCES. *Educ:* Univ Del, BSCE, 66; Univ Pa, MSCE, 67, PhD(water resources), 72. *Prof Exp:* Environ engr, Reentry & Environ Syst Div, Gen Elec, 70-74; consult engr, United Engrs & Constructors, 74-76; ENVIRON ENGR, GREELEY & HANSEN, 76- *Concurrent Pos:* Hydraul engr, US Army CEngr, 66; Lectr water resources, Univ Del, 73; chmn, Gen Elec Corp Inter-Div Panel Waste Disposal & Pollution Control, 73-74; consult, Engrs Energy & Environ, 74-77. *Mem:* Water Pollution Control Fedn; Am Soc Civil Engrs; Am Acad Environ Engrs. *Res:* Biological treatment of municipal and industrial wastewaters and sludge residues; water resources, hydrological investigations and water quality modeling; environmental assessment and impact evaluations. *Mailing Add:* 1919 Gravers Lane Wilmington DE 19810

KOCH, CHARLES FREDERICK, b Tarrytown, NY, Mar 23, 32. MATHEMATICS. *Educ:* Union Col, BS, 53; Univ Ill, MS, 57, PhD(math), 61. *Prof Exp:* Instr math, Univ Minn, 61-64; asst prof math, Kans State Univ, 64-66; ASST PROF MATH, SOUTHERN ILL UNIV, CARBONDALE, 66- *Mem:* Am Math Soc; Math Asn Am; Sigma Xi. *Res:* Summability of sequences and series. *Mailing Add:* Dept Math Southern Ill Univ Carbondale IL 62901

KOCH, CHRISTIAN BURDICK, wood science, forestry; deceased, see previous edition for last biography

KOCH, DAVID GILBERT, b Milwaukee, Wis, Aug 6, 45; m 74; c 3. INFRARED ASTRONOMY, GAMMA-RAY ASTRONOMY. *Educ:* Univ Wis-Madison, BS, 67; Cornell Univ, MS, 71, PhD(physics), 72. *Prof Exp:* Sr scientist x-ray astron, Am Sci & Eng, 72-76, staff scientist, 76-77; astrophysicist infrared astron, Astrophys Observ, Smithsonian Inst, 77-89; ASTROPHYSICIST, NASA AMES RES CTR, 88- *Concurrent Pos:* Assoc, Harvard Col Observ, 77-89. *Mem:* AAAS; Am Astron Soc; Am Inst Aeronaut & Astronaut; Int Astron Union. *Res:* Infrared, x-ray and gamma-ray astrophysics with particular emphasis on spaceborne instrumentation and computer aided data reduction. *Mailing Add:* NASA Ames Res Ctr MS 245-6 Moffett Field CA 94035

KOCH, DAVID WILLIAM, b Frankfort, Kans, Nov 22, 42; m 66. AGRONOMY. *Educ:* Kans State Univ, BS, 64, MS, 66; Colo State Univ, PhD(agron), 71. *Prof Exp:* Asst prof, 71-77, ASSOC PROF CROP PHYSIOL, UNIV NH, 77- *Mem:* Am Soc Agron. *Res:* Minimum tillage establishment of forage crops; forage crop management; methods of forage conservation. *Mailing Add:* Ext Agronomist Plant Sci Div Univ Wyo PO Box 3354 Univ Sta Laramie WY 82071

KOCH, DONALD LEROY, b Dubuque, Iowa, June 3, 37; m 62; c 3. STRATIGRAPHY, GROUNDWATER GEOLOGY. *Educ:* State Univ Iowa, BS, 59, MS, 67. *Prof Exp:* Res geologist, 59-71, chief subsurface geol, 71-75, asst state geologist, 75-80, state geologist & dir, Iowa Geol Surv, 80-86, STATE GEOLOGIST & BUR CHIEF, GEOL SURV BUR, 86- *Mem:* Sigma Xi. *Res:* Carbonate petrology and carbonate hydrology, defining the relationship of primary and secondary porosity to parameters of water availability and water quality. *Mailing Add:* 123 N Capitol Geol Surv Bureau/IDNR Iowa City IA 52242

KOCH, ELIZABETH ANNE, b Toronto, Ohio, Oct 8, 36; m 73. BIOCHEMICAL GENETICS, ELECTRON MICROSCOPY. *Educ:* Mt Union Col, BS, 58; Northwestern Univ, PhD(genetics), 64. *Prof Exp:* Teacher gen sci & eng, Ely Jr High Sch, Elyria, Ohio, 58-59; instr gen biol & genetics, Hope Col, 63-64; res assoc electron microscopy & genetics, Northwestern Univ, 64-69; asst prof, 69-74, ASSOC PROF BIOCHEM & GENETICS, CHICAGO MED SCH, 74- *Concurrent Pos:* Fel, Cancer Inst, NIH, 64-66; lectr, Eve Div, Northwestern Univ, 65-72. *Mem:* Am Soc Cell Biol; Genetics Soc Am; AAAS; Sigma Xi; Am Asn Univ Prof. *Res:* Ultrastructural localization of proteins using immune chemical methods; oogenesis. *Mailing Add:* 3333 Greenbay Rd N Chicago IL 60064

KOCH, FREDERICK BAYARD, b St Paul, Minn, Aug 15, 35; m 63; c 3. PHYSICAL METALLURGY. *Educ:* Carleton Col, BA, 57; Univ Minn, Minneapolis, MS, 62; Northwestern Univ, PhD(mat sci), 67. *Prof Exp:* Mem staff, Sci Lab, Ford Motor Co, 60-62; MEM TECH STAFF, BELL TEL LABS, 67- *Mem:* Am Inst Metall Engrs; Electrochem Soc. *Res:* Discrete wiring methods for circuit board development. *Mailing Add:* 184 Kent Place Blvd Summit NJ 07901

KOCH, GARY MARLIN, b Pottsville, Pa, Nov 7, 41. ORNAMENTAL HORTICULTURE. *Educ:* Pa State Univ, BS, 63, MS, 65, PhD(bot), 70. *Prof Exp:* Asst prof ornamental hort, 70, assoc prof plant sci, 70-77, PROF PLANT SCI, CALIF STATE UNIV, FRESNO, 77- *Mem:* Am Soc Hort Sci. *Res:* Plant materials and usage; commercial floriculture; floral design; turfgrass production. *Mailing Add:* Dept Plant Sci Calif State Univ Fresno CA 93740

KOCH, GEORGE SCHNEIDER, JR, b Washington, DC, Oct 30, 26; m 73; c 3. ECONOMIC GEOLOGY, STATISTICS. *Educ:* Harvard Univ, SB, 48, PhD(econ geol), 55; Johns Hopkins Univ, MA, 49. *Prof Exp:* Geologist, US Geol Surv, 48-52; chief geologist, Minera Frisco, SA, Mex, 52-56; asst prof geol, Ore State Univ, 56-62; res geologist, US Bur Mines, 62-71; PROF GEOL, UNIV GA, 71- *Concurrent Pos:* Consult, firms in mineral indust & US govt. *Mem:* Soc Econ Geol; Am Inst Mining, Metall & Petrol Eng; Int Asn Math Geol. *Res:* Statistical analysis of geological data; exploration for and evaluation of mineral deposits; precious and nonferrous metal and uranium deposits. *Mailing Add:* Dept Geol 133 GGS Univ Ga Athens GA 30602

KOCH, HEINZ FRANK, b Berlin, Ger, June 21, 32; US citizen; m 58; c 2. PHYSICAL ORGANIC CHEMISTRY, FLUORINE CHEMISTRY. *Educ:* Haverford Col, BS, 54, MS, 56; Cornell Univ, PhD(org chem), 60. *Prof Exp:* Res chemist, Univ Calif, Berkeley, 60-62; res chemist, Plastics Dept, E I Du Pont de Nemours & Co, Inc, 62-65; from asst prof to assoc prof, Ithace Col, 65-70, chmn, Dept Chem, 67-79, Div Fluorine Chem, 77, PROF ORG CHEM, ITHACA COL, 70- *Concurrent Pos:* NSF fac fel, Univ Calif, Berkeley, 71-72, vis prof, 72-73; vis prof, Univ Grenoble, France, 79 & Univ Auckland, NZ, 79-80. *Mem:* Am Chem Soc (secy-treas, Div Fluorine Chem, 74-75); Royal Chem Soc; NY Acad Sci; Sigma Xi. *Res:* Physical organic studies of reaction mechanisms, particularly fluorohalocarbon chemistry; studies of 1, 2 elimination reactions, carbanions and isotope effects. *Mailing Add:* Dept Chem Ithaca Col Ithaca NY 14850

KOCH, HENRY GEORGE, b Mt Holly, NJ, May 22, 48; m 68; c 2. MEDICAL ENTOMOLOGY. *Educ:* Okla State Univ, BS, 71, MS, 74; NC State Univ, PhD(entom), 77. *Prof Exp:* RES ENTOMOLOGIST, LONE STAR TICK RES LAB, SCI & EDUC ADMIN-AGR RES, USDA, 77- *Concurrent Pos:* Adj assoc prof entom, Okla State Univ, 78- *Mem:* Entom Soc Am; Coun Agr Sci & Technol; Am Registry Prof Entomologists. *Res:* Acarology, tick ecology and biology; host suitability; acarid susceptibility. *Mailing Add:* 100 Wedgewood Dr Poteau OK 74953

KOCH, HERMAN WILLIAM, b New York, NY, Sept 28, 20; m 45; c 5. PHYSICS. *Educ:* Queens Col, NY, BS, 41; Univ Ill, MS & PhD(physics), 44. *Prof Exp:* Res physicist, Univ Ill, 44-45 & Clinton Labs, Oak Ridge, 45-46; asst prof res nuclear physics, Univ Ill, 46-49; physicist, High Energy Radiation Sect, Nat Bur Stand, 49-62, chief, Radiation Physics Div, 62-66; dir, Am Inst Physics, 66-67; RETIRED. *Concurrent Pos:* Chmn, US Nat Comt, Comt Data Sci & Technol, past-chmn, Copy Clearance Ctr; past-pres, Nat Fedn Abstracting & Indexing Serv. *Mem:* Fel Am Phys Soc; fel Optical Soc Am; Am Asn Physics Teachers; fel AAAS; Acoust Soc Am. *Res:* Nuclear physics research on 20 million electron volt, 50 million electron volt and 180 million electron volt electron accelerator; development of high-energy x-ray spectrometer; Bremsstrahlung production; radiation physics research from 50 kiloelectron volts to 180 million electron volts; studies of science information flow. *Mailing Add:* Am Inst Physics 335 E 45th St New York NY 10017

KOCH, HOWARD A(LEXANDER), b Evanston, Ill, June 15, 22; m 43; c 2. CHEMICAL ENGINEERING. *Educ:* Northwestern Univ, BS, 43, MS, 46, PhD(chem eng), 49. *Prof Exp:* Asst chem engr, Northwestern Univ, 46, res supvr, 46-49; proj leader, Reservoir Mechs, Atlantic Refining Co, 49-59, sr reservoir engr, 59-61, div reservoir engr, 61-63, mgr, Block 31 Unit Opers, 63-67, dist mgr, Rocky Mt-Mid Continent Dist, 67-69, res eng mgr, Atlantic Richfield Co, 69-73, vpres eng, Arco Oil & Gas Co, 73-84. *Mem:* Am Inst Chem Engrs; Am Inst Mining, Metall & Petrol Engrs. *Res:* Unit operations; gas absorption; reservoir mechanics; fluid flow and mass transfer. *Mailing Add:* 4529 Crooked Lane Dallas TX 75229

KOCH, J FREDERICK, b Berlin, Ger, June 1, 37; US citizen; m 85; c 4. SEMICONDUCTOR PHYSICS. *Educ:* NY Univ, BA, 58; Univ Calif, Berkeley, PhD(physics), 62. *Prof Exp:* Asst prof physics, Univ Calif, Berkeley, 62-63; from asst prof to prof physics, Univ Md, College Park, 63-73; PROF PHYSICS, TECH UNIV MUNCHEN, 73- *Mem:* Fel Am Phys Soc; Deutsche Phys Gesellschaft. *Mailing Add:* Dept Physics E16 Tech Univ Munchen Garching 8046 Germany

KOCH, KAY FRANCES, b Tremont, Ill, June 18, 36; m 71. ORGANIC CHEMISTRY. *Educ:* Univ Ill, BS, 58; Univ Calif, Berkeley, PhD(org chem), 62. *Prof Exp:* Instr chem, Wellesley Col, 61-63, asst prof, 63-66; sr org chemist, Res Labs, 66-77, head microbiol & fermentation res, 77-80, head phys chem res, 80-88, MGR, SCI INFO SERV, ELI LILLY CO, 88- *Concurrent Pos:* Nat Inst Gen Med Sci spec fel, 64-65. *Mem:* Am Chem Soc. *Res:* Photochemistry of highly conjugated cyclic organic and other organic compounds; biosynthesis of quinones in insects and study of defensive secretions of insects; fermentation products chemistry; antibiotics, especially amino glycosides. *Mailing Add:* The Lilly Res Labs Lilly Corp Ctr Indianapolis IN 46285-0725

KOCH, LEONARD JOHN, b Chicago, Ill, Mar 30, 20; c 1. NUCLEAR POWER. *Educ:* Ill Inst Technol, BS, 43; Univ Chicago, MBA, 68. *Prof Exp:* Var mgt positions, Argonne Nat Lab, 48-72; mgr nuclear projs, Ill Power Co, 72-76, vpres, 76-83; RETIRED. *Mem:* Nat Acad Eng; fel Am Nuclear Soc. *Mailing Add:* One E Desert Sky Rd No 16 Tucson AZ 85737

KOCH, MELVIN VERNON, b Chicago, Ill, June 12, 40. PHARMACEUTICAL CHEMISTRY. *Educ:* St Olaf Col, BA, 62; Univ Iowa, MS, 64, PhD(med chem), 67. *Prof Exp:* Res chemist, 67-69, proj leader fine organics, 69-71, group leader, 71-74, GROUP LEADER PHARMACEUT CHEM, DOW CHEM USA, 74- *Mem:* Am Chem Soc; Sigma Xi. *Mailing Add:* 5300 Sturgeon Creek Pkwy Midland MI 48640

KOCH, PETER, b Missoula, Mont, Oct 15, 20; m 50. WOOD SCIENCE & TECHNOLOGY. *Educ:* Mont State Univ, BS, 42; Univ Wash, PhD(wood sci), 54. *Hon Degrees:* DSc, Univ Maine, 80. *Prof Exp:* Asst to pres, Stetson-Ross Mach Co, Wash, 46-52; consult engr, 52-55; assoc prof wood sci, Mich State Univ, 55-57; vpres & dir, Champlin Co, NH, 57-62; res & writing, 62-63; chief wood scientist & proj leader, Southern Forest Exp Sta, Forest Serv, USDA, Pineville, 63-82, res scientist, Intermountain Res Sta, Missoula, 82-84; PRES, WOOD SCI LABS INC, 84- *Concurrent Pos:* Adj prof wood & paper sci, NC State Univ, 73-; mem, Comt Renewable Resources for Indust Mat, Nat Res Coun, 75; adj prof, Wood Sci, Univ Mont, 83-; distinguished affil prof, Wood Sci, Univ Idaho, 83; concurrent prof, Univ Nanjing, 86- *Honors & Awards:* John Scott Award, 73; Distinguished Serv Award, Soc Wood Sci & Technol, 87. *Mem:* Am Soc Mech Engrs; Nat Soc Prof Engrs; Wood Sci & Technol; Forest Prod Res Soc (pres, 72-73); fel Int Acad Wood Sci. *Res:* Wood machining processes, wood conversion processes; improved utilization of southern pines, southern hardwoods and lodgepole pine; author of more than 200 papers and three major texts. *Mailing Add:* Wood Sci Lab Inc 942 Little Willow Creek Rd Corvallis MT 59828

KOCH, PETER M, b Washington, DC, Feb 11, 45; m; c 2. ATOMIC PHYSICS, DYNAMICS. *Educ:* Univ Mich, BS, 66; Yale Univ, MP, 69, PhD(physics), 74. *Prof Exp:* Asst prof physics, Yale Univ, 76-82; from asst prof to assoc prof, 82-89, PROF PHYSICS, STATE UNIV NY, STONY BROOK, 89- *Concurrent Pos:* A P Sloan Found fel, 78-82; Alexander von Humboldt Found Sr US Scientist Award, 89-90. *Mem:* Am Inst Physics; Am Phys Soc; Sigma Xi. *Res:* Experimental study of highly excited simple atoms (hydrogen and helium) in intense electromagnetic fields for understanding quantum dynamics in classically chaotic systems, ie, quantum chaology; atomic laser and synchrotron radiation spectroscopy; rf electric discharges. *Mailing Add:* Dept Physics SUNY at Stony Brook Stony Brook NY 11794-3800

KOCH, RICHARD, b NDak, Nov 24, 21; m 44; c 5. PEDIATRICS. *Educ:* Univ Rochester, MD, 51; Univ Calif, AB, 58. *Prof Exp:* From instr to prof pediat, 55-75, head div child develop, 65-75, PROF CLIN PEDIAT, SCH MED, UNIV SOUTHERN CALIF, 76-; DIR, MATERNAL PKU NAT COL LAB STUDY, 68- *Concurrent Pos:* Dir, Child Develop Clin, Los Angeles Children's Hosp, 55-75, mem, Div Med Genetics, 76-80, actg head, 80-; dir, Regional Ctr for Developmentally Disabled; dep dir chg ment health, develop, disabilities & drug abuse, Calif State Dept Health, 66-76. *Mem:* Acad Pediat; Am Asn Ment Deficiency. *Res:* Mental retardation in children; metabolic diseases. *Mailing Add:* Childrens Hospital 4614 Sunset Blvd Los Angeles CA 90027

KOCH, RICHARD CARL, b Pittsburgh, Pa, Aug 10, 30; m 47; c 4. MEDICINAL CHEMISTRY. *Educ:* Cornell Univ, BA, 52; Yale Univ, PhD(org chem), 57. *Prof Exp:* Res chemist, Res & Eng Div, Monsanto Chem Co, 57-59; chemist, 59-66, proj leader, 66-68, sect mgr, 68-72, asst dir, 72-76, DIR, CENT RES DIV, PFIZER, INC, 76- *Mem:* Am Chem Soc; Sigma Xi; Soc Environ Toxicol & Chem. *Res:* Discovery and development of animal health drugs. *Mailing Add:* Cent Res Div Pfizer Inc Groton CT 06340

KOCH, RICHARD MONCRIEF, b Hutchinson, Kans, Jan 29, 39. GEOMETRY. *Educ:* Harvard Univ, AB, 61; Princeton Univ, PhD(math), 64. *Prof Exp:* Instr math, Univ Pa, 64-66; from asst prof to assoc prof, 66-84, PROF MATH, UNIV ORE, 84- *Mem:* Am Math Soc; Math Asn Am. *Res:* Differential geometry, particularly pseudogroups. *Mailing Add:* Dept Math Univ Ore Eugene OR 97403

KOCH, ROBERT B, b St Paul, Minn, May 14, 23; m 45; c 5. BIOCHEMISTRY. *Educ:* Univ Minn, BS, 48, PhD(biochem), 52. *Prof Exp:* Chief chem & microbiol br, Qm Food & Container Inst, US Army, 57-61; res sect head, Honeywell Inc, 61-65, staff scientist, 65-73; prof biochem, Miss State Univ, 73-88; RETIRED. *Mem:* Am Soc Biol Chem; Sigma Xi; NY Acad Sci. *Res:* Enzyme purification and kinetics; enzymology of subcellular particles and organelles; biochemical mechanism of olfaction; properties of adenoxine triphosphatase system; soybean lipoxygenase isozymes. *Mailing Add:* Dept Biochem Miss State Univ Mississippi State MS 39762

KOCH, ROBERT HARRY, b York, Pa, Dec 19, 29; m 59; c 4. ASTRONOMY. *Educ:* Univ Pa, AB, 51, MA, 55 & PhD(astron), 59. *Prof Exp:* Instr astron, Amherst & Mt Holyoke Cols, 59-60, from asst prof to assoc prof, Joint Dept Astron, Amherst, Mt Holyoke & Smith Cols & Univ Mass, 60-66; assoc prof astron, Univ NMex,66-67; assoc prof, 67-69, actg chmn dept, 68-73, PROF ASTRON, UNIV PA, 69- *Mem:* AAAS; Am Astron Soc; Int Astron Union. *Res:* Photoelectric photometry, polarimetry, visible band and ultraviolet spectroscopy and evolution of eclipsing variable stars. *Mailing Add:* 210 Roberts Rd Ardmore PA 19003

KOCH, ROBERT JACOB, b Chicago, Ill, Apr 17, 26; m 57; c 2. MATHEMATICS. *Educ:* Tulane Univ, La, PhD(math), 53. *Prof Exp:* From asst prof to assoc prof, 53-62, PROF MATH, LA STATE UNIV, BATON ROUGE, 62- *Mem:* Am Math Soc. *Res:* Topological semigroups. *Mailing Add:* Dept Math La State Univ Baton Rouge LA 70803

KOCH, ROBERT MILTON, b Sioux City, Iowa, May 15, 24; m 46; c 3. ANIMAL BREEDING. *Educ:* Mont State Univ, BS, 48; Iowa State Univ, MS, 50, PhD(animal breeding, genetics), 53. *Prof Exp:* From asst prof to assoc prof animal husb, 50-59, chmn dept, 59-66, PROF ANIMAL SCI, UNIV NEBR, 59- *Concurrent Pos:* Supt, Ft Robinson Beef Cattle Res Sta, 54-57. *Honors & Awards:* Animal Breeding & Genetics Award, Am Soc Animal Sci, 76; Pioneer Award, Beef Improvement Fed, 79. *Mem:* Fel AAAS; fel Am Soc Animal Sci. *Res:* Beef cattle breeding; population genetics. *Mailing Add:* Animal Sci Dept Lincoln NE 68583-0908

KOCH, RONALD JOSEPH, b Cincinnati, Ohio, June 30, 39; m 61; c 4. PHYSICS, MATERIALS SCIENCE. *Educ:* Xavier Univ, Ohio, BS, 61; Johns Hopkins Univ, PhD(physics), 69. *Prof Exp:* From res spectrochemist to sr res spectrochemist, 69-75, sr res chemist, 75-78, sr res physicist, 78-83, RADIATION SAFETY OFFICER, ARMCO INC, 78-, SR STAFF PHYSICIST, 83- *Mem:* Am Iron & Steel Inst; Am Soc for Testing & Mat. *Res:* X-ray physics and diffraction; auger spectroscopy; surface analysis; radiation safety. *Mailing Add:* Armco Inc 703 Curtis St Middletown OH 45043

KOCH, RONALD N, b Pittsburgh, Pa, Aug 19, 41; m 63; c 3. FLOW MEASUREMENT. *Educ:* Carnegie-Mellon Univ, BSME, 63; Univ Pittsburgh, MBA, 83. *Prof Exp:* Develop engr, 63-76, MGR ENG, ROCKWELL INT, 76- *Honors & Awards:* Indust Prod Award, Soc Petrol Engrs, 84. *Mem:* Am Water Works Asn; Am Soc Sanit Engrs; Asn Mech Engrs; Int Water Supply Asn. *Res:* New product development activities for equipment in the field of flow measurement. *Mailing Add:* 400 N Lexington Ave Pittsburgh PA 15208

KOCH, RUDY G, b Richland Center, Wis, Sept 26, 39; m 61; c 3. SCIENCE EDUCATION, PLANT TAXONOMY. *Educ:* Concordia Teacher's Col, Nebr, BSEd, 61; Univ Mo-Kansas City, MA, 63; Ore State Univ, MS, 65; Okla State Univ, EdD, 68, Univ Nebr, PhD(bot), 75. *Prof Exp:* Asst prof biol, St John's Col, Kans, 63-67; asst cur plant taxon, Nebr State Mus, Lincoln, 67-68; Asst prof, 68-77, ASSOC PROF BIOL, UNIV WIS, SUPERIOR, 68-, CHMN DEPT, 77- *Mem:* Am Soc Plant Taxon; Am Bryol & Lichenological Soc; Nat Asn Biol Teachers; Int Asn Plant Taxon. *Res:* Curriculum development in biology; systematic studies of Bidens in North Cental United States; flora of Northwestern Wisconsin. *Mailing Add:* Univ WI LaCrosse Cowley Hall Rm 345 LaCrosse WI 54061

KOCH, STANLEY D, b Cleveland, Ohio, Feb 10, 24; m 58; c 3. ORGANIC CHEMISTRY, POLYMER CHEMISTRY. *Educ:* Univ Ill, BA, 47; Cornell Univ, PhD(org chem), 50. *Prof Exp:* Fel, Univ Chicago, 50-51; res chemist, Jackson Lab, E I Du Pont de Nemours & Co, 51-54 & Glendale Plaskon Lab, Barrett Div, Allied Chem Corp, 54-57; sr res chemist, Boston Lab, Monsanto Res Corp, 57-59, group leader, 59-64, res mgr, 64, Dayton Lab, 65-69; dept head, Horizons Res, Inc, 69-73; sec head, photo cure, Dwight P Joyce Res Ctr, Glidden-Durkee Div, SCM Corp, 74-77; develop assoc, Washington Res Ctr, W R Grace & Co, 77-78; MGR POLYMERIC COATINGS RES, ENTHONE, INC, 79- *Mem:* Am Chem Soc; Inst Interconnecting & Packaging Electronic Circuits. *Res:* UV-curable coatings, including those used in the electronics industry; radiation curing; herbicides. *Mailing Add:* PO Box 1900 Enthone-OMI Inc New Haven CT 06508-1900

KOCH, STEPHEN ANDREW, b Jamaica, NY, Nov 19, 48; m 75. INORGANIC CHEMISTRY. *Educ:* Fordham Univ, BS, 70; Mass Inst Technol, PhD(chem), 75. *Prof Exp:* Assoc inorg chem, Tex A&M Univ, 75-77 & Cornell Univ, 78; asst prof, 78-84, ASSOC PROF CHEM, STATE UNIV NY, 84- *Mem:* Am Chem Soc. *Res:* Structural, electronic and reactivity properties of transition metal compound; bioinorganic chemistry; catalysis. *Mailing Add:* Dept Chem State Univ NY Stony Brook NY 11794-3400

KOCH, STEPHEN DOUGLAS, b New York, NY, Dec 16, 40; m 68; c 2. PLANT TAXONOMY. *Educ:* Swarthmore Col, BA, 62; Univ Mich, MS, 64, PhD(plant taxon), 69. *Prof Exp:* Instr bot, Duke Univ, 67-68; asst prof, NC State Univ, 68-73, PROF-INVESTR BOT, POSTGRAD COL, CHAPINGO, MEX, 73- *Honors & Awards:* Sistema Nacional de Investigadores, 84- *Mem:* AAAS; Am Soc Plant Taxonomists; Int Asn Plant Taxonomists; Mex Bot Soc; Sigma Xi; Torrey Bot Club. *Res:* Grass systematics. *Mailing Add:* Centro Botanica Colegio de Postgraduados Chapingo Edo 56230 Mexico

KOCH, TAD H, b Mt Vernon, Ohio, Jan 1, 43; m. ORGANIC CHEMISTRY. *Educ:* Ohio State Univ, BS, 64, Iowa State Univ, PhD(org photochem), 68. *Prof Exp:* from asst prof to assoc prof, 68-80, chair, 83-86, PROF ORG CHEM, UNIV COLO, BOULDER, 80- *Concurrent Pos:* Grants, Petrol Res Fund, 68-72, 77-79 & 83-86, Res Corp, 69-70, Gen Med Sci Inst, 71-77, Nat Cancer Inst, 78-91, Nat Heart Lung Blood Inst, 87-89, NSF, 86-91, Army Res Off, 80-88. *Mem:* Am Chem Soc; Am Asn Cancer Res; Am Soc Photobiol; AAAS. *Res:* Mechanistic and synthetic photochemistry; free radical chemistry; bioorganic chemistry; lasers. *Mailing Add:* Dept Chem & Biochem Univ Colo Boulder CO 80309-0215

KOCH, THEODORE AUGUR, b Schenectady, NY, Oct 21, 25; m 52; c 5. PHYSICAL CHEMISTRY, CATALYSIS. *Educ:* St Michael's Col, BS, 46, MS, 47; Univ Pa, PhD(chem), 52. *Prof Exp:* Instr chem, Univ Vt, 46 & Drexel Inst, 47-51; CHEMIST, PETROCHEM DEPT, E I DU PONT DE NEMOURS & CO, INC, 52- *Mailing Add:* 600 Cheltenham Rd Wilmington DE 19808-1805

KOCH, THOMAS L, b Boston, Mass, July 13, 55; m 79; c 2. OPTOELECTRONICS, OPTICAL FIBER COMMUNICATIONS. *Educ:* Princeton Univ, AB, 77; Calif Inst Technol, PhD(appl physics), 82. *Prof Exp:* Mem tech staff, Electronic Device Res Dept, 82-87, supvr, Photonic Circuits Res Dept, 87-89, HEAD, OPTOELECTRONICS RES DEPT, AT&T BELL LABS, 89- *Concurrent Pos:* Mem prog comts for numerous int & nat conf, 85-; distinguished lectr award, Laser & Electroptics Soc, Inst Elec & Electronics Engrs, 90, mem bd gov, 91-93. *Mem:* Sr mem Inst Elec & Electronics Engrs Laser & Electroptics Soc; fel Optical Soc Am. *Res:* Semiconductor lasers and optical fiber communications; single-frequency and tunable lasers; device fabrication, dynamic characteristics for high speed transmission; integration of semiconductor guided-wave optoelectronic components to form photonic integrated circuits. *Mailing Add:* AT&T Bell Labs MS 4E-338 Crawfords Corner Rd Holmdel NJ 07733-1988

KOCH, THOMAS RICHARD, b Strasburg, Pa, Oct 9, 44; m 68; c 3. CLINICAL CHEMISTRY, ANALYTICAL CHEMISTRY. *Educ:* Lebanon Valley Col, BS, 66; Univ Md, PhD(anal Chem), 70; Am Bd Clin Chem, dipl, 76. *Prof Exp:* Trainee clin chem, State Univ NY, Buffalo, 70-72; clin chemist, St Joseph Hosp, 72-75; asst prof path & assoc dir clin chem, 75-82, ASSOC PROF PATH & DIR CLIN CHEM, SCH MED, UNIV MD, 82- *Concurrent Pos:* Consult, Vet Admin Hosp, 76-80 & Food & Drug Admin, USPHS, 77- *Mem:* Am Asn Clin Chem. *Res:* Trace elements in human disease; bilirubin measurement. *Mailing Add:* Md Med Labs 1901 Sulphur Spring Rd Baltimore MD 21227

KOCH, WALTER THEODORE, b Orwigsburg, Pa, Jan 4, 23; m 46; c 3. ORGANIC CHEMISTRY. *Educ:* Albright Col, BS, 44; Rutgers Univ, MS, 50, PhD(org chem), 51. *Prof Exp:* Asst gen chem, Rutgers Univ, 47-50; res chemist, Merck & Co, Inc, 51-53 & Am Viscose Corp, 53-63; res chemist, 63-66, sect leader, FMC Corp, 66-78; RETIRED. *Mem:* Am Chem Soc. *Res:* High polymers; cellophane, coatings; thermoplastic films. *Mailing Add:* 4 E Langhorne Ave Havertown PA 19083

KOCH, WILLIAM EDWARD, b York, Pa, Nov 22, 33; m 61; c 2. ANATOMY. *Educ:* Univ Pa, AB, 56, AM, 59; Stanford Univ, PhD(biol), 62. *Prof Exp:* From instr to asst prof anat, Sch Med, Yale Univ, 62-68; assoc prof, 68-75, PROF ANAT, SCH MED, UNIV NC, CHAPEL HILL, 68-, ADJ PROF ZOOL, 77- *Mem:* AAAS; Am Soc Zoologists; Am Asn Anatomists; Soc Develop Biol. *Res:* Study of embryonic tissue interacting and differentiating in vitro. *Mailing Add:* Dept Anat Univ NC Sch Med Chapel Hill NC 27514

KOCH, WILLIAM FREDERICK, b Oak Park, Ill, Mar 11, 50; m 75; c 2. PH MEASUREMENTS, ION CHROMATOGRAPHY RESEACH. *Educ:* Loyola Univ, Chicago, BS, 72; Iowa State Univ, MS, 74, PhD(anal chem), 75. *Prof Exp:* Res fel, 75-77, res chemist, 77-79, group leader, 79-88, DEP DIV CHIEF, NAT INST STANDARDS & TECHNOL, 88- *Concurrent Pos:* Lectr chem, Montgomery Col, Rockville, Md, 83-; bd dirs, Nat Comt Clin Lab Standards. *Mem:* Am Chem Soc; Soc Electroanal Chem; Sigma Xi; Am Soc Testing & Mat; Nat Comt Clin Lab Standards. *Res:* Responsible for the nations pH standards; ion chromatography; conductivity; coulometry; potentiometry; voltammetry; standard reference materials; acid rain. *Mailing Add:* 20468 Watkins Meadow Dr Germantown MD 20876

KOCH, WILLIAM GEORGE, b Forsyth, Mont, May 16, 24; m 51; c 1. PHYSICAL CHEMISTRY. *Educ:* Univ Notre Dame, BS, 47; Mont State Univ, MA, 53. *Prof Exp:* Asst chemist, Great Western Sugar Co, Colo, 54-55; asst prof, 55-58, assoc prof, 59, chmn dept, 74-77, PROF CHEM, UNIV NORTHERN COLO, 59- *Mem:* AAAS; Am Chem Soc. *Res:* Carbon isotope effects in decarboxylation reactions; effect of deuterium on carbon isotope effects; kinetics in the reaction between organolithium compounds and ether. *Mailing Add:* Dept Chem Div Sci Univ Northern Colo Greeley CO 80631

KOCH, WILLIAM JULIAN, b Durham, NC, May 17, 24; m 47; c 4. MYCOLOGY. *Educ:* Univ NC, MA, 50, PhD(bot, plant physiol, zool), 55. *Prof Exp:* From asst to assoc prof, 47-74, PROF BOT, UNIV NC, CHAPEL HILL, 74- *Concurrent Pos:* Vis investr, Mich Biol Sta, 56, Highlands Biol Sta, 57, 58 & Int Bot Cong, Can, 59. *Mem:* AAAS; Bot Soc Am; Mycol Soc Am; Am Soc Plant Taxon; Electron Micros Soc Am. *Res:* Culture, comparative morphology, sexuality and mobility of fungus of reproductive cells; fungi parasitic on algae. *Mailing Add:* Dept Bot Univ NC Chapel Hill NC 27515

KOCHAKIAN, CHARLES DANIEL, b Haverhill, Mass, Nov 18, 08; m 40; c 1. ENDOCRINOLOGY, BIOCHEMISTRY. *Educ:* Boston Univ, AB, 30, AM, 31; Univ Rochester, PhD(physiol chem), 36. *Prof Exp:* Instr physiol, Sch Med & Dent, Univ Rochester, 36-40, assoc, 40-44, from asst prof to assoc prof endocrinol, 44-51; prof res biochem, Sch Med, Univ Okla, 51-57; prof physiol & biophys, 57-61, prof biochem & prof & dir exp endocrinol, 61-79, prof physiol & biophys, 64-75, EMER PROF BIOCHEM, MED & DENT SCHS, UNIV ALA, BIRMINGHAM, 79- *Concurrent Pos:* Assoc, Jackson Mem Lab, 46-49; mem panel hormones, Comt Growth, Nat Res Coun, 49-51, mem panel appraisers handbk biol data, 49-52; vis Claude Bernard prof, Inst Exp Med Surg, Univ Montreal, 50; assoc dir, Okla Med Res Found, 51-53, head dept biochem & endocrinol, 51-57, coordr res, 53-55; consult, Dent Sch Comt, Okla Dent Asn, 54-57; consult, Univ Tex M D Anderson Hosp & Tumor Inst Houston, 56; actg coordr res, Univ Ala, Birmingham, 60-61; consult, Div Int Med Educ, Asn Am Med Cols, 64-70; mem panel drugs for metab disturbances, Drug Efficacy Study, Nat Acad Sci-Nat Res Coun, 67-69; mem metab & endocrine eval comt, Vet Admin Hosp, 69-72; mem nomenclature comt, Int Union Physiol Sci, 69-73. *Honors & Awards:* Claude Bernard Medal, Univ Montreal, 50; Medal, Osaka Endocrine Soc, 62; Charles D Kochakian Award, Endocrinol & Nutrit, Sch Med & Dent, Univ Rochester, 85. *Mem:* AAAS; Am Chem Soc; Am Physiol Soc; Am Soc Biol Chemists; Endocrine Soc. *Res:* Steroid biochemistry; protein; carbohydrate and fat metabolism; enzymes; hormones; nucleic acids; mechanisms of anabolic action of steroid hormones. *Mailing Add:* 3617 Oakdale Rd Birmingham AL 35223

KOCHAN, IVAN, b Ukraine, Aug 20, 23; nat US; m 49; c 3. IMMUNOLOGY. *Educ:* Univ Man, BSc, 53, MSc, 55; Stanford Univ, PhD(med microbiol), 58; Am Bd Med Microbiol, dipl. *Prof Exp:* Res assoc, Stanford Univ, 58-59; assoc prof microbiol, Baylor Univ, 59-61, prof & chmn dept, 61-67; prof microbiol, Sch Med, Wright State Univ, 74-77; PROF MICROBIOL, MIAMI UNIV, 67- *Mem:* AAAS; Am Asn Immunologists; Reticuloendothelial Soc; Am Soc Microbiol; Am Thoracic Soc; fel Am Trudeau Soc; Am Acad Microbiol. *Res:* Study of transferrin-iron-mycobactin interplay in host-parasite relationship; nutritional immunity; role of fatty acids in cellular immunity and immunological diseases. *Mailing Add:* Dept Microbiol Miami Univ Oxford OH 45056

KOCHAN, ROBERT GEORGE, b Prince Albert, Sask, Oct 25, 49; m 72; c 4. EXERCISE PHYSIOLOGY, BIOCHEMISTRY. *Educ:* Univ Sask, BA, 71; Univ Toledo, PhD(exercise physiol), 78. *Prof Exp:* Teaching asst, Univ Sask, 71-73 & Univ Toledo, 73-77; fel biochem, Med Col Ohio, 77-79, res assoc, 79-80; asst prof exercise physiol, Univ Wis-Madison, 80-87. *Concurrent Pos:* Res fel, Juvenile Diabetes Found, 77-79. *Mem:* Am Col Sports Med. *Res:* Exercise metabolism; glycogen synthesis; insulin action mechanism; diabetes control. *Mailing Add:* 814 Lewis Ct Madison WI 53715

KOCHAN, WALTER J, b Plainfield, NJ, July 15, 22; m 48; c 1. PLANT PHYSIOLOGY. *Educ:* Utah State Univ, BS, 50, MS, 52; Rutgers Univ, PhD(plant physiol), 55. *Prof Exp:* Asst horticulturist, Univ Idaho, 55-57; asst plant biochemist, Div Indust Res, Wash State Univ, 57; from asst horticulturist to assoc horticulturist, 57-70, RES PROF HORT, EXP STA, UNIV IDAHO, 70- *Mem:* Am Soc Plant Physiologists; Sigma Xi. *Res:* Mineral nutrition of plants; post-harvest physiology of tree fruits. *Mailing Add:* Plant Soil/Ent Sci Dept Univ Idaho Moscow ID 83843

KOCHANOWSKI, BARBARA ANN, b Beaver, Pa, 1957. EXPERIMENTAL BIOLOGY. *Educ:* Pa State Univ, BS, 79; Univ Ill, Champaign, MS, 81, PhD(nutrit), 84. *Prof Exp:* Res scientist health care, 85-90, SECT HEAD, HEALTH CARE, PROCTER & GAMBLE CO, 91- *Mem:* Am Inst Nutrit. *Mailing Add:* Procter & Gamble Co 11511 Reed Hartman Hwy Cincinnati OH 45241

KOCHAR, HARVINDER K, b Punjab, India, Jan 15, 53. MUSCLE DISEASES. *Educ:* Punjab Univ, BS, 74, MSc, 75. *Prof Exp:* Res technologist, 77-80, SR RES TECHNOLOGIST, RES DEPT, MED CTR, NORTHWESTERN UNIV, 80- *Mem:* Histochem Soc Am; Int Histochem Soc. *Mailing Add:* 9325A Jamison Ave Philadelphia PA 19115

KOCHEN, MANFRED, b Vienna, Austria, July 4, 28; nat US; m 54; c 2. INFORMATION SCIENCE, COGNITIVE SCIENCE. *Educ:* Mass Inst Technol, BS, 50; Columbia Univ, MA, 51, PhD(appl math), 55. *Prof Exp:* Asst, Spectros Lab, Mass Inst Technol, 49-50; mathematician aeroelasticity res, Biot & Arnold Co, 50-52; lectr math, Columbia Univ, 52-53; mem staff, Electronic Comput Proj, Inst Advan Study, 53-55; Ford fel math models in behav sci, Princeton Univ & Harvard Univ, 55-56; staff mathematician, Thomas J Watson Res Ctr, Int Bus Mach Corp, 56-58, mem res tech staff, 58-60, mgr info retrieval proj, 60-63, exchange vis expert at Euratom, Italy, 63-64; assoc prof math biol, 65-69, PROF INFO SCI & URBAN/REGIONAL PLANNING & RES MATHEMATICIAN, MENT HEALTH RES INST, 69-, ADJ PROF COMPUT INFO SYSTS, GRAD SCH BUS ADMIN, 82- *Concurrent Pos:* Consult, Paul Rosenberg Assocs, 53-55, RCA Res Lab, 65, United Aircraft Corp, 66-69, Rand Corp, 68- & Sci Ctr Berlin, 78-; assoc ed, Behav Sci, 68-70, J Asn Comput Mach, 72- & managing ed, Human Systs Mgt, 79-; nat lectr, Asn Comput Mach, 69; hon res assoc, Harvard Univ, 73-74; pres, Wise Found, 75-; vis prof, Rockefeller Univ, 80-81; chmn, sociotechnol systs area in planning PhD prog, UTEP, Ment Health Res Inst, 83-; prin investr grants, NIH & NSF, 85- *Mem:* Am Math Soc; Am Soc Info Sci; Fedn Am Scientists; Am Phys Soc; fel AAAS; NY

Acad Sci; Inst Elec & Electronics Engrs. *Res:* Information systems and organization of knowledge; models for information-seeking behavior, problem representation solving, cognitive learning processes; decentralization theory; science of science; social planning; artificial intelligence; decision support systems. *Mailing Add:* 2026 Devonshire Ann Arbor MI 48104

KOCHEN, SIMON BERNARD, b Antwerp, Belg, Aug 14, 34; nat US. MATHEMATICS. *Educ:* McGill Univ, BSc, 54, MSc, 55; Princeton Univ, MA, 56, PhD(math), 58. *Prof Exp:* Demonstr physics, McGill Univ, 52-53; asst lectr math, Princeton Univ, 57-58; Nat Res Coun Can res assoc & asst prof, Univ Montreal, 58-59; from asst prof to prof, Cornell Univ, 59-67; PROF MATH, PRINCETON UNIV, 67- *Concurrent Pos:* Guggenheim fel, 62-63; mem, Inst Advan Study, 66-67. *Honors & Awards:* Cole Prize, Am Math Soc, 67. *Mem:* Am Math Soc; Asn Symbolic Logic. *Res:* Mathematical logic. *Mailing Add:* Dept Math Princeton Univ Princeton NJ 08544

KOCHER, BRYAN S, b July 3, 48; m 71; c 2. COMPUTER SCIENCE. *Educ:* Moravian Col, AB, 70; Univ Mass, MS, 75. *Prof Exp:* Proj leader, Com Union, 75-77; syst mgr, Raytheon Data Systs, 77-80; sr software eng, Cullinane, 80-82; prod mgr, Interactive Data Corp, 82-85; dept mgr, Data Resources Inc, 85-86; CONSULT PROJ MGR, CONSULTS MGT DECISIONS, INC, 86- *Concurrent Pos:* Chmn, Boston Chap, Asn Comput Mach, 78-80; regional vpres, 80-88. *Mem:* Asn Comput Mach (pres, 88-90). *Mailing Add:* CMD Inc One Main St Cambridge MA 02142

KOCHER, CARL A, b Seattle, Wash, Feb 14, 42; m 68; c 3. EXPERIMENTAL ATOMIC PHYSICS. *Educ:* Univ Calif, Berkeley, AB, 63, PhD(physics), 67. *Prof Exp:* Fel, Oxford Univ, 67-68; vis scientist, Mass Inst Technol, 68-69; lectr, Columbia Univ, 69-73; asst prof, 73-78, ASSOC PROF PHYSICS, ORE STATE UNIV, 78- *Concurrent Pos:* Postdoc Fel, NSF, 67-69. *Mem:* Sigma Xi; Am Phys Soc; Am Asn Physics Teachers; Fedn Am Scientists; Union Concerned Scientists. *Res:* Atomic collisions; radiative and autoionization processes; high Rydberg states; surface physics; computer instrumentation. *Mailing Add:* Dept Physics Ore State Univ Corvallis OR 97331-6507

KOCHER, CHARLES WILLIAM, b Johnson City, NY, May 16, 32; m 54; c 3. NUCLEAR PHYSICS. *Educ:* Harpur Col, BA, 54; NC State Col, MS, 56; Ind Univ, PhD(nuclear physics), 61. *Prof Exp:* Res assoc physics, Solid State Physics Group, Brookhaven Nat Lab, 60-62; sr staff scientist, Phys Res Lab, Budd Co, 62-65; scientist, 65-67, SR RES PHYSICIST, ANALYSIS LAB, DOW CHEM USA, 67- *Mem:* AAAS; Am Chem Soc; Am Phys Soc; Sigma Xi. *Res:* Nuclear reactors and decay schemes; Mossbauer effect; solid state physics, corrosion testing; metalorganic chemistry. *Mailing Add:* 907 Deerfield Ct Midland MI 48640-2708

KOCHER, DAVID CHARLES, b Washington, DC, Nov 9, 41; m 77. PHYSICS, ENVIRONMENTAL SCIENCES. *Educ:* Univ Md, BS, 63; Univ Wis-Madison, MS, 65, PhD(physics), 70. *Prof Exp:* Res assoc physics, Univ Birmingham, 70-71; res assoc, 71-76, RES ASSOC PHYSICS & ENVIRON SCI, OAK RIDGE NAT LAB, 76- *Mem:* Am Phys Soc; Am Nuclear Soc; Health Physics Soc. *Res:* Development of models and data bases for the assessment of health and safety impacts on man from energy production technologies. *Mailing Add:* Oak Ridge Nat Lab PO Box 2008 Oak Ridge TN 37831

KOCHER, HARIBHAJAN S(INGH), b Lyall Pur, WPakistan, Sept 29, 34; m 63; c 2. FLUID MECHANICS. *Educ:* Okla State Univ, BS, 57; Purdue Univ, MS, 59; Mich State Univ, PhD(fluid mech), 63. *Prof Exp:* Lectr mech eng, Indian Inst Technol, New Delhi, 62-63; develop engr, Civilian Atomic Power Dept, Can Gen Elec Co, 63-64; sr res physicist, Delco Prod Div, Gen Motors Corp, NY, 64-66; scientist, Xerox Corp, 66-82; RES ASSOC, EASTMAN KODAK CO, ROCHESTER, NY, 82- *Concurrent Pos:* Adj asst prof, Univ Rochester, 76- *Mem:* Am Soc Mech Engrs; Inst Elec & Electronics Engrs. *Res:* Thermal properties of metals; heat transfer in nuclear reactors and electrical motors; boundary layer studies; fluid flow instability; air bearing technology; heat transfer in xerographic systems; photographic film process fluid mechanics. *Mailing Add:* 56 Leonard Circle Penfield NY 14526

KOCHERLAKOTA, KATHLEEN, b Pittsburgh, Pa, Mar 31, 38; m 62; c 2. STATISTICS. *Educ:* Muskingum Col, BS, 60; Johns Hopkins Univ, MS, 63, DSc(statist), 69. *Prof Exp:* Lectr math, Muskingum Col, 62-63; consult statist, Dept Community Med, Univ Western Ont, 65-70; sessional lectr statist, 67-69, form asst prof to assoc prof, 69-87, PROF, UNIV MAN, 87- *Mem:* Biomet Soc; Am Statist Asn. *Res:* Applied multivariate analysis and analysis of discrete data. *Mailing Add:* Dept Statist Univ Man Winnipeg MB R3T 2N2 Can

KOCHERLAKOTA, SUBRAHMANIAM, b Bangalore, India, Feb 3, 35; m 62; c 2. MATHEMATICAL STATISTICS. *Educ:* Univ Col Sci, Benares, India, BSc, 54, MSc, 57; Inst Agr Res Statist, dipl, 63; ScD, Johns Hopkins, 64. *Prof Exp:* Jr res fel statist, Inst Agr Res Statist, 57-58, sr res fel, 58-59; investr, Rockefeller Found, India, 59-60; asst prof math, Univ Western Ont, 64-66; assoc prof, 66-70, PROF STATIST, UNIV MAN, 70- *Concurrent Pos:* Nat Res Coun operating grants pure & appl math, 66- *Mem:* Inst Math Statist; Am Statist Asn; fel Royal Statist Soc; Int Statist Inst. *Res:* Multivariate analysis; distribution theory; statistical tests of significance; applied probability theory; non-normality. *Mailing Add:* Dept Statist Univ Man Winnipeg MB R3T 2N2 Can

KOCHERT, GARY DEAN, b Louisville, Ky, Oct 12, 39; m 63; c 2. BOTANY. *Educ:* Ind Univ, AB, 63, PhD(microbiol), 67. *Prof Exp:* From asst prof to assoc prof, 67-78, PROF BOT & CHMN DEPT, UNIV GA, 78- *Res:* Molecular genetics and systematics of rice, peanuts, and bamboo. *Mailing Add:* 330 Crestwood Dr Athens GA 30605

KOCHHAR, DEVENDRA M, b Sailkot, India, Mar 10, 38; m 62; c 2. ANATOMY, EMBRYOLOGY. *Educ:* Punjab Univ, BSc, 58, MSc, 59; Univ Fla, PhD(anat), 64. *Prof Exp:* Instr anat, Univ Fla, 64-65; vis scientist, Karolinska & Wenner-Grens Insts, Stockholm, Sweden, 65-66; vis scientist, Strangeways Res Lab, Cambridge, 66-67; guest investr, Rockefeller Univ, 67-68; from asst prof to assoc prof anat, Univ Iowa, 68-71; from assoc prof to prof, Univ Va, 71-76; PROF ANAT, JEFFERSON MED COL, 76- *Concurrent Pos:* Consult pharmaceut indust, Environ Protection Agency & Food & Drug Admin. *Honors & Awards:* Warkany Award, 90. *Mem:* Am Soc Cell Biol; Am Asn Anatomists; Teratology Soc (pres, 82-83); Soc Develop Biol. *Res:* Experimental teratology; development of skeletal system; congenital abnormalities of limb; collagen genes in development; retinoids in health and disease. *Mailing Add:* Dept Anat Jefferson Col Med 1020 Locust St Philadelphia PA 19107

KOCHHAR, MAN MOHAN, b Lahore, Pakistan, Sept 14, 32; US citizen; m 54; c 3. MEDICINAL CHEMISTRY, BIOCHEMICAL TOXICOLOGY. *Educ:* Punjab Univ, India, BS, 53; Univ Tex, Austin, MS, 61, PhD(med chem), 64. *Prof Exp:* Chief chemist, Dr Nayer Chem Works, India, 54-55; med rep, Geigy Pharmaceut, India, 55-58; asst med chem, Univ Tex, Austin, 59-63; spec instr, Col Pharm, 63-64; from asst prof to assoc prof pharm & pharmaceut chem, Auburn Univ, 64-75, prof toxicol, 75-81, prof pharmacol & toxicol, 81-83; RES SCI, OFF STANDARDS-FDA, 83- *Concurrent Pos:* Lederle fac awards, 64 & 65; grant-in-aid, Auburn Univ, 65-68; Nat Inst Drug Abuse award, 73; Dept Ment Health award, State of Ala, 74; dir, Drug Screening Training Prog, Auburn Univ & Drug Anal Lab, 73- *Mem:* Am Acad Clin Toxicol; Am Asn Clin Chem; Am Pub Health Asn; Can Acad Clin Anal Toxicol. *Res:* Structure-activity relationships among psychotropic and antineoplastic agents; biochemical approach to toxicology including analytical toxicology. *Mailing Add:* Div Bioequivalence Off Standards-FDA 5600 Fishers Ln Rockville MD 20852

KOCHHAR, RAJINDAR KUMAR, b Nurmahal, India, Aug 1, 22; US citizen; m 54; c 2. POLYMER CHEMISTRY. *Educ:* Panjab Univ, India, BS, 45; Univ Delhi, MS, 48; Univ Tex, Austin, PhD(phys org chem), 65; Univ Mo-Kansas City, MBA, 69. *Prof Exp:* Asst res med, Lady Hardinge Med Col, Delhi, 55-58; res chemist, Gulf Res & Develop Co, 64-68, sr res chemist, 68-77, mgr polymerization res, 77-81, dir process chem, plastics div, 81-83; CHEM CONSULT, 83- *Res:* Color and chemical constitution; dyes for nylon; research on medicinal plants; fixed oils; polymer synthesis and characterization; polyolefin development; polypropylene catalysts. *Mailing Add:* 610 Brenwick Ct Katy TX 77450

KOCHI, JAY KAZUO, b Los Angeles, Calif, May 17, 27; m 59; c 3. ORGANIC CHEMISTRY. *Educ:* Univ Calif, Los Angeles, BS, 49; Iowa State Univ, PhD(chem), 52. *Prof Exp:* Instr org chem, Harvard Univ, 52-55; NIH fel, Cambridge Univ, 55-56; vis asst prof, Iowa State Univ, 56; chemist, Shell Develop Co, 57-62; from assoc prof to prof chem, Case Western Reserve Univ, 62-69; prof chem, Ind Univ, Bloomington, 69-74, Earl Blough prof chem, 74-84; ROBERT WELCH DISTINGUISHED PROF CHEM, UNIV HOUSTON, UNIV PARK, 84- *Honors & Awards:* James Flack Norris Award, Am Chem Soc, 81; A C Cope Scholar Award, Am Chem Soc, 88; A V Humboldt Sr Scientist Award, 87. *Mem:* Nat Acad Sci; Am Chem Soc; Royal Soc Chem. *Res:* Mechanisms of organic reactions catalyzed by metal complexes; application of metal complexes to organic synthesis; electron-transfer and charge-transfer processes in organic chemistry; photochemistry of organometallic compounds; application of electron spin resonance spectroscopy to organic and organometallic free radicals and to the mechanism of homolytic reactions; time-resolved spectroscopy of inactive intermediates. *Mailing Add:* Univ Houston Univ Park 4800 Calhoun Rd Houston TX 77204-5641

KOCHMAN, RONALD LAWRENCE, b Rome, NY, Apr, 7, 46; m 68; c 2. CLINICAL RESEARCH, GASTROENTEROLOGY. *Educ:* Pa State Univ, BS, 68; Northeastern Univ, MS, 74. *Prof Exp:* Res biologist, 75-79, res assoc, 79-82, res investr, 82-86, sr res assoc, 86-88, ASST DIR, G D SEARLE & CO, SUBSID MONSANTO CO, 88- *Mem:* Am Soc Pharmacol & Exp Therapeut; Am Soc Neurochem. *Res:* Clinical trails of anti-ulcer, analgesic and anxiolytic medications; neurochemistry of memory and learning; neurotransmitters and drug receptors. *Mailing Add:* G D Searle & Co 4901 Searle Pkwy Skokie IL 60077

KOCHMAN, STANLEY OSCAR, b New York, NY, July 18, 46; m 69; c 7. MATHEMATICS. *Educ:* Kenyon Col, AB, 66; Univ Chicago, MS, 67, PhD(math), 70. *Prof Exp:* Gibbs instr math, Yale Univ, 70-72; asst prof, Purdue Univ, West Lafayette, 72-77; from asst prof to assoc prof, Univ Western Ont, 77-85; PROF MATH, YORK UNIV, 85- *Concurrent Pos:* Lectr, Mass Inst Technol, 73-74. *Mem:* Am Math Soc; Can Math Soc. *Res:* Algebraic topology, especially homology operations; Cobordism theory and stable homology. *Mailing Add:* York Univ 4700 Keele St Downsview ON M3J 1P3 Can

KOCHWA, SHAUL, b Vienna, Austria, Apr 30, 15; nat US; m 40; c 1. BIOCHEMISTRY, IMMUNOCHEMISTRY. *Educ:* Hebrew Univ Jerusalem, MSc, 40, PhD(immunol), 49; Harvard Univ, MPH, 53. *Prof Exp:* Chief chemist, Gordon Co, Israel, 43-45; med lab dir, Bikur Cholim Hosp, Jerusalem, 45-47; assoc dir immunochem, Rogoff Med Res Inst, Beilinson Hosp, 55-58; sr res assoc, Mt Sinai Hosp, 59-66; assoc prof med, 66-72, prof path, 72-85, prof med, 79-83, EMER PROF MED, MT SINAI SCH MED, 85- *Mem:* Am Asn Immunologists; Am Chem Soc; Harvey Soc; Am Soc Hemat; NY Acad Sci. *Res:* Physicochemistry and purification of proteins, toxins, antibodies and enzymes; protein-protein interaction and complex formation. *Mailing Add:* 67-22 Harrow St Forest Hills NY 11375

KOCH-WESER, DIETER, b Kassel, Ger, July 13, 16; nat US; m 50; c 2. EXPERIMENTAL PATHOLOGY, PREVENTIVE MEDICINE. *Educ:* Univ Sao Paulo, MD, 43; Northwestern Univ, MS, 50, PhD, 56. *Prof Exp:* Asst med, Hosps & Clin, Sao Paulo, Brazil, 44-47; asst path, Hektoen Inst

Med Res, Cook County Hosp, Chicago, 49-51; from instr to asst prof med, Grad Sch Med, Univ Chicago, 51-56; assoc prof, Sch Med, Western Reserve Univ, 57-62; chief, Latin-Am Off, NIH, Brazil, 62-64; assoc prof, Sch Pub Health, 64-71, PROF PREV & SOCIAL MED, MED SCH, HARVARD UNIV, 71-, ASSOC DEAN INT PROGS, 67- *Honors & Awards:* Couto Prize, Acad Med, Brazil, 45. *Mem:* Am Soc Clin Invest; Am Thoracic Soc; AMA; Am Col Chest Physicians; Sigma Xi. *Res:* Isotope and biochemical studies in tuberculosis, hypersensitivity and liver diseases. *Mailing Add:* 19 Standish Rd Wellesley MA 02181

KOCH-WESER, JAN, b Berlin, Ger, Oct 30, 30; US citizen. INTERNAL MEDICINE, CLINICAL PHARMACOLOGY. *Educ:* Univ Chicago, AB, 50; Harvard Med Sch, MD, 54. *Prof Exp:* Intern med, Mass Gen Hosp, 54-55, asst resident, 55-56, resident, 59; from instr to assoc prof pharmacol, Harvard Med Sch, 62-75; assoc dir res ctr & chief med res, 75-76, dir res ctr & vpres res, Merrell Int, 76-; AT F HOFFMANN LA ROCHE & CO, SWITZ. *Concurrent Pos:* Assoc physician & chief hypertension & clin pharmacol unit, Mass Gen Hosp, 66-75; mem med adv bd, Coun High Blood Pressure Res, Am Heart Asn; mem sci adv bd, Pan Am Health Orgn; mem pharmacol-toxicol rev comt & prog comt, Nat Inst Gen Med Sci; mem bd trustees, US Pharmacopeial Coun; USPHS res fel pharmacol, Harvard Med Sch, 60-61, Burroughs Wellcome scholar clin pharmacol, 66-71; USPHS spec res fel, 62 & grant, 64-75. *Mem:* Am Col Cardiol; Am Fedn Clin Res; Am Soc Clin Pharmacol & Therapeut; Am Soc Pharmacol & Exp Therapeut; Cardiac Muscle Soc. *Res:* Cardiovascular physiology and pharmacology; clinical pharmacology and human therapeutics; adverse drug reactions; drug metabolism and pharmacokinetics; antiarrhythmic, antihypertensive and anticoagulant drugs. *Mailing Add:* F Hoffmann La Roche & Co Grenzacherstr 124 Bldg 52 Rm 1601 Basel 4002 Switzerland

KOCIBA, RICHARD JOSEPH, b Harbor Beach, Mich, Apr 8, 39; m 66; c 2. VETERINARY PATHOLOGY, TOXICOLOGY. *Educ:* Mich State Univ, BS, 64, DVM, 66, MS, 69, PhD(path), 70. *Prof Exp:* Practr, Milford Vet Clin, 66-67; instr anat, Mich State Univ, 67-68, NIH fel path, 68-70; res pathologist, 70-80, SR ASSOC SCIENTIST, TOXICOL RES LAB, DOW CHEM CO, 80- *Concurrent Pos:* Adj asst prof path, Mich State Univ, 81- *Mem:* Am Vet Med Asn; Am Col Vet Pathologists; Soc Toxicol; Soc Pharmacol & Environ Pathologists; Am Bd Toxicol. *Res:* Design, conduction, evaluation and interpretation of research in the area of acute and chronic toxicity, with special emphasis on carcinogenesis and pathology. *Mailing Add:* Toxicol Res Lab 1803 Bldg Dow Chem Co Midland MI 48640

KOCKS, U(LRICH) FRED, b Dusseldorf, Ger, Nov 25, 29; nat US; m 54; c 4. MATERIALS SCIENCE. *Educ:* Univ Gottingen, dipl physics, 54; Harvard Univ, PhD(appl physics), 59. *Hon Degrees:* DrTech, Tampere Univ Tech, 82. *Prof Exp:* Lectr & res fel, Harvard Univ, 59-61, asst prof, 61-65; sr scientist, Argonne Nat Lab, 65-83; FEL, LOS ALAMOS NAT LAB, 83- *Concurrent Pos:* Vis prof, Munich Tech Univ, 64, Aachen Tech Univ, 78 & McMaster Univ, 78; Humboldt award, Fed Repub Ger, 79. *Honors & Awards:* Japan Soc Promotion Sci Sr Award, 85. *Mem:* The Metall Soc; Am Soc Mat; Int Soc Mat; fel Am Inst Mining Eng. *Res:* Mechanics and thermodynamics of solids; defects in crystals; strengthening mechanisms; kinetics of plasticity and creep; textures, constitutive relations. *Mailing Add:* 902 Paseo de Lacuma Santa Fe NM 87501

KOCOL, HENRY, b Chicago, Ill, July 16, 37; m 71; c 1. HEALTH PHYSICS. *Educ:* Loyola Univ, Chicago, BS, 58; Purdue Univ, Lafayette, MS, 61. *Prof Exp:* Radiochemist, Nat Bur Standards, 61-64 & USPHS, 64-71; mgr x-ray control & regulation, Wash State Dept Social & Health Serv, Seattle, 79-82; res chemist, Rockville, Md, 71-73, regional radiation control, Philadelphia, 73-77, regional radiol health rep, 77-79, FED-STATE LIAISON, FOOD & DRUG ADMIN, SEATTLE, WASH, 82- *Mem:* Health Physics Soc. *Res:* The explanation of scientific concepts and facts, especially in the field of radiation safety, to lay audiences, to enable societal decisions to be based on fact rather than hype. *Mailing Add:* Radiation Health Br 1232 A St Sacramento CA 95816

KOCON, RICHARD WILLIAM, b Fall River, Mass, Apr 18, 42; m 68; c 2. CLINICAL CHEMISTRY, MEDICAL LABORATORY SCIENCE. *Educ:* Southeastern Mass Univ, BS, 70; Providence Col, PhD(chem), 73. *Prof Exp:* Org Chem, Rhode Island Hosp, 74-77; LAB DIR, DAMON MED LAB, INC, 79- *Mem:* Am Asn Clin Chem. *Res:* Clinical chemistry, specifically clinical application of radioimmunoassay procedures and enzyme-linked immunosorbent blocking assay techniques. *Mailing Add:* 408 Oakland Pkwy Franklin MA 02038

KOCSIS, JAMES JOSEPH, b Barberton, Ohio, Aug 13, 20; m 52; c 3. PHARMACOLOGY, TOXICOLOGY. *Educ:* Ohio State Univ, BA, 43; Univ Chicago, MS, 52, PhD(pharmacol), 56. *Prof Exp:* From instr to assoc prof, 56-74, PROF PHARMACOL, JEFFERSON MED COL, 74- *Mem:* Am Chem Soc; Am Soc Pharmacol & Exp Therapeut; Soc Toxicol; Sigma Xi. *Res:* Bioassay; drug metabolism. *Mailing Add:* 306 N Woodstock Dr Cherry Hill NJ 08034

KOCUREK, MICHAEL JOSEPH, b New York, NY, Jan 6, 43; m 67; c 2. PAPER SCIENCE, ENGINEERING. *Educ:* State Univ NY, BS, 64, MS, 67; Syracuse Univ, PhD(paper sci & eng), 71. *Prof Exp:* Assoc prof paper sci, 70-80, PROF PAPER SCI & ENG, UNIV WIS-STEVENS POINT, 80-, CHMN DEPT, 70- *Concurrent Pos:* Mem acad adv coun, Tech Asn Pulp & Paper Indust, 71-, mem prof develop oper coun, 76-, chmn continuing educ div, 76-78, mem US-Can joint textbook comt, 78-, instr intro to pulp & paper tech, 75- *Mem:* Tech Asn Pulp & Paper Indust. *Res:* Wood and pulping chemistry; paper and fiber physics. *Mailing Add:* Dept Paper Sci & Forestry Univ Wis Stevens Point WI 54481

KOCURKO, MICHAEL JOHN, b Orange, Calif, Jan 28, 45; m. GEOLOGY, SEDIMENTOLOGY. *Educ:* Midwestern Univ, BS, 66; Univ Wis-Milwaukee, MS, 68; Tex Tech Univ, PhD(geol), 72. *Prof Exp:* Explor geologist, Union Oil Co Calif, 68-69 & 72-75; asst prof geol, Tulane Univ, 75-79; PROF GEOL, MIDWESTERN STATE UNIV, WICHITA FALLS, TEX, 79- *Concurrent Pos:* Chmn, Geol Dept, Midwestern State. *Mem:* Soc Econ Paleontologists & Mineralogists; Geol Soc Am. *Res:* Application of modern carbonate depositional environments and post-depositional history to the paragenesis of carbonate rocks; taxonomy of fossil Octocorallia. *Mailing Add:* Dept Geol Midwestern State Univ Wichita Falls TX 76308

KOCZAK, MICHAEL JULIUS, b New York, NY, Apr 29, 44; m 68; c 2. COMPOSITE MATERIALS, MATERIALS PROCESSING. *Educ:* Polytech Inst NY, BS, 65; Univ Pa, MS, 67, PhD(metall & mat sci), 69. *Prof Exp:* Lectr, mat sci, State Univ NY, 69-70; 1st lieutenant, US Army Corp Eng, 70; from asst to assoc prof mat eng, 71-81, PROF MAT ENG, DREXEL UNIV, 81- *Concurrent Pos:* Guest researcher, Solid State Physics, Brookhaven Nat Lab, 69-70; vis fel, Univ Surrey, 75-76; sr scientist, Off Naval Res, Tokyo, 82-83 & London, 89-91. *Mem:* Sigma Xi; Metall Soc; Am Soc Metals; Am Ceramic Soc; Am Powder Metall Inst. *Res:* Composite materials; powder metallurgy with emphasis on structure, property processing relationships; thermoplastic metal matrix and ceramic matrix composite materials. *Mailing Add:* Dept Mat Eng Drexel Univ Philadelphia PA 19104

KODA, ROBERT T, b Watsonville, Calif, June 18, 33; m 59; c 2. PHARMACOLOGY. *Educ:* Univ Southern Calif, PharmD, 61, PhD(pharmaceut chem), 68. *Prof Exp:* Asst prof, 68-74, fel, 69, assoc prof pharmaceut chem & assoc dean, 75-85, ASSOC PROF PHARMACEUT, UNIV SOUTHERN CALIF, 85- *Concurrent Pos:* Fel acad admin internship prog, Am Coun Educ, 71. *Mem:* Am Pharmaceut Asn; Sigma Xi; AAAS. *Res:* Pharmacokinetics of drugs of current clinical interest; percutaneous absorption. *Mailing Add:* Sch Pharm Univ Southern Calif Los Angeles CA 90033

KODAMA, ARTHUR MASAYOSHI, b Honolulu, Hawaii, Dec 17, 31; m 59; c 1. PHYSIOLOGY. *Educ:* Washington Univ, BA, 54; Univ Calif, Berkeley, PhD(physiol), 63, MPH, 78. *Prof Exp:* Res physiologist, Univ Calif, Berkeley, 63-77; asst prof physiol, 78-80, MEM FAC, DEPT PUB HEALTH SCI, SCH PUB HEALTH, UNIV HAWAII, 80- *Mem:* AAAS; Am Physiol Soc. *Res:* Environmental physiology and space biology; occupational health and industrial hygiene. *Mailing Add:* Dept Pub Health Sci Sch Pub Health Univ Hawaii 1960 East-West Rd Honolulu HI 96822

KODAMA, GOJI, b Sakai City, Japan, Dec 2, 27; m 57; c 2. BORON HYDRIDE CHEMISTRY, METALLABORANE CHEMISTRY. *Educ:* Tokyo Inst Technol, BE, 51; Univ Mich, Ann Arbor, MS, 52, PhD(chem), 58. *Prof Exp:* From res assoc to instr chem, Univ Mich, 58-60; res fel, Harvard Univ, 60-61; from asst prof to prof, Tokyo Sci Univ, 61-69; assoc res prof, 69-80, RES PROF CHEM, UNIV UTAH, 80- *Mem:* Am Chem Soc. *Res:* Reactions of boron hydrides with various bases; transition metal-boron hydride complexes. *Mailing Add:* Dept Chem Univ Utah Salt Lake City UT 84112

KODAMA, HIDEOMI, b Tokyo, Japan, Oct 9, 31; m 59; c 3. SOILS & SOIL SCIENCE. *Educ:* Tokyo Univ Educ, BSc, 56, MSc, 58, DSc(mineral), 61. *Prof Exp:* Japan Soc Promoting Sci fel & lectr, Int Christian Univ, Tokyo, 61-62; Nat Res Coun Can fel, 62-64; res scientist, Soil Res Inst, Can Dept Agr, 64-77, res scientist, Chem & Biol Res Inst & head Mineral Anal Serv, 78-86, RES SCIENTIST & PROJ LEADER, LAND RESOURCE RES CTR, AGR CAN, 86- *Concurrent Pos:* Vis scientist, Nat Ctr Sci Res, Orleans, France, 69-70; Japan Soc Prom Sci, 84; assoc ed, Mineral Asn Can, 83-86, Clay Minerals Soc, 86-, consult ed, Soil Sci, 90-; fel, Japan Soc Prom Sci, 61-62, Nat Res Coun Can, 62-64. *Mem:* Mineral Soc Am; Clay Minerals Soc; fel Can Soc Soil Sci; Int Asn Study Clay (treas, 78-85); Mineral Soc Japan; Mineral Soc UK; Am Soc Soil Sci; Clay Sci Soc Japan. *Res:* Structure and genesis of interstratified clay minerals; fine structure analysis of layer silicates; crystal chemistry of silicate minerals; interactions between clay minerals and soil organic matter; intercalations of clay materials; characterization of soils; non-crystalline inorganic soil components. *Mailing Add:* Soil Mineral Land Resource Res Ctr Agr Can Ottawa ON K1A 0C6 Can

KODAMA, JIRO KENNETH, b Reedley, Calif, Mar 4, 24; m 51; c 5. PHARMACOLOGY, TOXICOLOGY. *Educ:* Univ Calif, AB, 51, MS, 55, PhD(pharmacol), 57. *Prof Exp:* Instr pharmacol, Med Sch & toxicol, Sch Pub Health, Univ Calif, 57-59; from asst dept chief toxicol & pharmacol to sr pharmacologist, Hazleton Labs, Inc, 59-63; pharmacologist, Shell Develop Co, 63-66, supvr pharmacol dept, 66-68, vis Shell scientist, Tunstall Lab, Shell Res Ltd, Eng, 69-70, staff toxicologist, Agr Div, Shell Chem Co, 70-77; staff toxicologist, 77-80, TECH LIAISON REP, CHEVRON CORP, CALIF, 81- *Mem:* AAAS; Am Soc Pharmacol & Exp Therapeut; Soc Toxicol. *Res:* Drug research and development; pharmacotoxic characterization of chemical warfare agents; mechanisms of toxic actions of organophosphorus chemicals and cytotoxic alkylating agents; toxicology and safety evaluation of industrial and agricultural chemicals; preclinical evaluation of pharmaceuticals; forensic toxicology; technical management. *Mailing Add:* Three Corwin Dr Alamo CA 94507

KODAMA, ROBERT MAKOTO, b Kauai, Hawaii, May 30, 32; m 64; c 2. CELL PHYSIOLOGY. *Educ:* Univ Hawaii, BA, 55; Univ Ill, PhD(physiol), 67. *Prof Exp:* Phys sci aide, US Fish & Wildlife Serv, 55-56; med technologist, Mt Sinai Hosp, Chicago, Ill, 59; asst cell physiol, Med Ctr, Univ Ill, 60-62, physiol, 66-67; from asst prof to assoc prof, 67-76, PROF BIOL, DRAKE UNIV, 76- *Mem:* Sigma Xi. *Res:* Biological transport; endocrinology; electron microscopy. *Mailing Add:* 3830 Twana Dr Des Moines IA 50310

KODAVANTI, PRASADA RAO S, b Dharmavaram, AP, India, Aug 1, 54; m 87. TOXICOLOGY. *Educ:* Andhra Univ, India, BS, 74, MS, 76; SVUPG Univ, India, PhD(toxicol), 81. *Prof Exp:* Scientist off, aquatic toxicol, SVUPG Univ, India, 82-83, asst neurotoxicol, Univ Miss Med Ctr, Jackson, 83-84; res assoc hepatotoxicol, 84-89, RES ASST PROF NEUROTOXICOL, UNIV MISS MED CTR, JACKSON, 89- *Mem:* Sigma Xi; Soc Toxicol. *Res:* Neuro, hepato and pulmonary toxicity by chemicals and drugs. *Mailing Add:* Dept Pharmacol & Toxicol Univ Miss Med Ctr 2500 N State St Jackson MS 39216

KODITSCHEK, LEAH K, microbial ecology; deceased, see previous edition for last biography

KODRES, UNO ROBERT, b Tartu, Estonia, May 21, 31; nat US; m 59; c 2. COMPUTER SCIENCE. *Educ:* Wartburg Col, BA, 54; Iowa State Univ, MS, 56, PhD(math), 58. *Prof Exp:* Staff mathematician, Prod Develop Lab, Int Bus Mach Corp, 58-63; assoc prof math, 63-82, actg chmn comput sci group, 75-76, PROF COMPUT SCI, NAVAL POSTGRAD SCH, 82- *Concurrent Pos:* Consult, IBM Corp, 67-69 & 84-85, Collins Radio Co, 67-69. *Mem:* Math Asn Am; Asn Comput Mach; Soc Indust & Appl Math; Sigma Xi. *Mailing Add:* Dept Comput Sci Naval Postgrad Sch Monterey CA 93943

KODRICH, WILLIAM RALPH, b Cooperstown, NY, Aug 26, 33; m 60. ECOLOGY. *Educ:* Hartwick Col, BA, 55; Univ Pittsburgh, PhD(biol), 67. *Prof Exp:* Assoc prof, 67-74, PROF BIOL, CLARION STATE COL, 74- *Mem:* Ecol Soc Am; Am Soc Mammalogists. *Res:* Physiological rates of small mammals living freely in their natural environments; relative thyroid release rates of 131-I of mammals living at different altitudes; bioenergetics of mammals in natural environments. *Mailing Add:* Biology Dept Clarion State Coll Clarion PA 16214

KOE, B KENNETH, b Astoria, Ore, Apr 15, 25; m 55; c 2. NEUROCHEMISTRY. *Educ:* Reed Col, BA, 45; Univ Wash, MS, 48; Calif Inst Technol, PhD(chem), 52. *Prof Exp:* Res fel chem, Calif Inst Technol, 52-54; assoc org chemist, Southwest Res Inst, 54-55; res chemist, 55-74, sr res investr, 74-79, RES ADV, PFIZER INC, 79- *Mem:* Am Soc Pharmacol & Exp Therapeut; Soc Neurosci; Am Col Neuropsychopharmacol. *Res:* Neurochemistry of psychotherapeutic drugs (antidepressants, antipsychotics, anxiolytics); psychopharmacology; neurotransmitters; receptor binding. *Mailing Add:* Dept Pharmacol Pfizer Inc Groton CT 06340

KOEDERITZ, LEONARD FREDERICK, b St Louis, Mo, Aug 21, 46; m 68; c 3. RESERVOIR SIMULATION, TRANSIENT PRESSURE ANALYSIS. *Educ:* Univ Mo, BS, 68, MS, 69, PhD(petrol eng), 70. *Prof Exp:* Sr engr, Atlantic Richfield, 70-74, proj dir, 74-75; assoc prof, 75-80, PROF PETROL ENG, UNIV MO, ROLLA, 80- *Concurrent Pos:* Head petrol eng dept, Univ Mo, Rolla, 79-80 & 89-91, dept chmn, 80-81. *Mem:* Soc Petrol Engrs. *Res:* Reservoirs imulation; transient pressure analysis; advanced reservoir applications in petroleum engineering. *Mailing Add:* Dept Petrol Eng Univ Mo 119 McNutt Hall Rolla MO 65401

KOEGLE, JOHN S(TUART), b Rochester, Pa, Jan 8, 26; m 50; c 5. CHEMICAL ENGINEERING. *Educ:* Purdue Univ, BChE, 48; Kans State Col, MS, 49; Ohio State Univ, PhD(chem eng), 51. *Prof Exp:* Res chem engr, Monsanto Co, 51-56, res design engr, 56-61, res group leader, 61-63, pilot plant dir marine colloids, 63-68, dir eng marine colloids, 68-70, dir develop marine colloids, Maine, 70-71; dir eng, Technol, Res & Develop, Inc, 71-74; sr process eng, Pedco, 74-77; dir process eng, Velsicol, 77-80; SR PROCESS ENG & PROJ MGR, PEDCO, 80- *Mem:* Am Chem Soc; Am Inst Chem Engrs. *Res:* Plant design, process and product development. *Mailing Add:* 9302 Gregg Dr West Chester OH 45069

KOEHL, GEORGE MARTIN, optics, for more information see previous edition

KOEHL, WILLIAM JOHN, JR, b Newport, Ky, July 27, 35; m 60. FUEL SCIENCE, AIR POLLUTION. *Educ:* Xavier Univ, Ohio, BS, 55, MS, 57; Univ Ill, PhD(org chem), 60. *Prof Exp:* Sr res chemist, Cent Res Div Lab, Socony Mobil Oil Co Inc, NJ, 60-68; sr res chemist, 68-76, supv chemist, 76-83, sr res assoc, Prod Res & Technol Serv Div, 83-88, SCIENTIST, MOBIL RES & DEVELOP CORP, 88- *Mem:* Am Chem Soc; Soc Automotive Engrs; Air & Waste Mgt Asn. *Res:* Electrochemical synthesis of organic compounds; automotive fuels and exhaust emissions. *Mailing Add:* Res Dept Mobil Res & Develop Corp Paulsboro NJ 08066

KOEHLER, ANDREAS MARTIN, b Weimar, Ger, Jan 21, 30; US citizen; m 52; c 3. MEDICAL PHYSICS, ACCELERATOR PHYSICS. *Educ:* Harvard Univ, BS, 50. *Prof Exp:* Proj engr mech design, Hesse-Eastern Corp, 51-53; tech assoc accelerator physics, 53-61, asst dir, 61-77, DIR ACCELERATOR MED PHYSICS, CYCLOTRON LAB, HARVARD UNIV, 77- *Res:* Radiation therapy using beams of charged particles; radiation physics and dosimetry of protons; radiography using protons and alpha particles; proton activation analysis; accelerator designs for medical applications. *Mailing Add:* Harvard Cyclotron Lab 44 Oxford St Cambridge MA 02138

KOEHLER, CARLTON SMITH, b Holyoke, Mass, July 20, 32; m 54; c 3. ECONOMIC ENTOMOLOGY. *Educ:* Univ Mass, BS, 54; Cornell Univ, MS, 56, PhD(econ entom), 58. *Prof Exp:* Res assoc econ entom, Cornell Univ, 58-59; asst prof entom & plant path, 59-62; lectr, Univ Calif, Berkeley, 62-69; from asst entomologist to entomologist, 62-75, prof entom, 69-75; prof entom & head dept, Ore State Univ, 75-76; EXTEN ENTOMOLOGIST, UNIV CALIF, BERKELEY, 76- *Concurrent Pos:* Spec field staff mem, Rockefeller Found, 68-69; consult, Int Rice Res Inst, 70-71 & Rockefeller Found, 70-; Ford Found travel fel, 71. *Mem:* Entom Soc Am. *Res:* Biology; ecology; control of insects and mites of importance in urban environments. *Mailing Add:* 201 Wellman Hall Univ of Calif Berkeley CA 94720

KOEHLER, DALE ROLAND, b Milwaukee, Wis, Oct 13, 32; m 55; c 4. RADIATION EFFECTS ON QUARTZ. *Educ:* Auburn Univ, BS, 54, MS, 55 & Univ Ala, PhD(physics), 64. *Prof Exp:* Physicist, Signal Eng Labs, NJ, 55 & Army Ballistic Missile Agency, Ala, 57-58; physicist, Phys Sci Lab, Army Missile Command, 58-64, chief radiation physics br, 64-67; mgr advan res lab, Bulova Watch Co, 67-77; PHYSICIST, SANDIA LABS, 77- *Res:* Ionizing radiation effects on quartz crystal resonators, primarily on frequency and acoustic loss changes; development of quartz purification and radiation hardness assurance technologies; quartz transducer development. *Mailing Add:* Sandia Labs Org 2533 Albuquerque NM 87185

KOEHLER, DON EDWARD, b Urbana, Ill, May 10, 42. PLANT PHYSIOLOGY. *Educ:* Univ Ill, BS, 64; Purdue Univ, MS, 67; Mich State Univ, PhD(biochem), 72. *Prof Exp:* Fel develop biol, Univ Chicago, 72-74; fel plant physiol, Univ Calif, Riverside, 74-77; asst prof plant sci, Tex A&M Univ, 77-82; PLANT PHYSIOLOGIST, DEPT PESTICIDE REGULATION, CALIF ENVIRON PROTECTION AGENCY, 82- *Mem:* Am Soc Plant Physiologists; AAAS; Western Soc Weed Sci; Western Plant Growth Regulators Soc. *Res:* Hormonal control of enzyme induction and developmental processes in plants. *Mailing Add:* Dept Pesticide Regulation Calif Environ Protection Agency 1220 N St Rm 8400 Sacramento CA 95814

KOEHLER, DONALD OTTO, pure mathematics; deceased, see previous edition for last biography

KOEHLER, FRED EUGENE, b Naylor, Mo, Jan 25, 23; m 47; c 4. SOIL FERTILITY. *Educ:* Univ Mo, BS, 43, MS, 50, PhD(soils), 51. *Prof Exp:* Soil scientist, USDA & asst agronomist, Univ Nebr, 51-57; assoc soil scientist, 58-66, soil scientist & prof, 66-88, EMER PROF SOILS, WASH STATE UNIV, 88- *Mem:* Am Soc Agron; Soil Sci Soc Am; Sigma Xi. *Res:* Soil fertility and soil chemistry. *Mailing Add:* Dept Agron & Soils Wash State Univ Pullman WA 99164-6420

KOEHLER, HELMUT A, b Berlin, Ger, Mar 10, 33; US citizen; m 69. NUCLEAR PHYSICS, SOLID STATE PHYSICS. *Educ:* Gen Motors Inst, BSE, 61; Univ Mich, MSE, 64, PhD(nuclear sci), 68. *Prof Exp:* PHYSICIST, LAWRENCE LIVERMORE LAB, UNIV CALIF, 68- *Concurrent Pos:* Tech consult serv, 68- *Mem:* Am Phys Soc. *Res:* Lasers and new laser systems and their appropriate application; neutron imaging and measurement systems. *Mailing Add:* Sauerbrchst 6 Heidenheim D7920 Germany

KOEHLER, HENRY MAX, b Offenbach, Ger, Oct 13, 31; nat US; m 57. SCIENTIFIC INFORMATION, TRANSLATION. *Educ:* Roosevelt Univ, BS, 52, MS, 57, MBA, 62. *Prof Exp:* Sr water chemist, Water Bur, Chicago, 53-54; asst biochem, Med Sch, Northwestern Univ, 54-55; asst chem, Roosevelt Univ, 55-56, sr chemist, Div Chem, 56-64; ed supvr, Oral Res Abstr, Am Dent Asn, 64-72, ed updates, 70-79, ed, Oral Res Abstr, 74-79; CONSULT ED, 80- *Concurrent Pos:* Mem, State of Ill Weather Modification Bd, 74-77. *Mem:* Fel AAAS; fel Am Inst Chemists; Am Chem Soc; Am Med Writers' Asn; Am Translr Asn. *Res:* Abstract preparation; editing; translation. *Mailing Add:* 211 E Chicago Ave No 1200 Chicago IL 60611

KOEHLER, JAMES K, b Darmstadt, Ger, June 7, 33; US citizen; m 57; c 5. CELL BIOLOGY. *Educ:* Univ Ill, BS, 55; Univ Calif, Berkeley, MS, 58, PhD(biophys), 61. *Prof Exp:* Asst prof physics, NMex Highlands Univ, 62-63; from asst prof to assoc prof, 63-75, PROF BIOL STRUCT, UNIV WASH, 75- *Concurrent Pos:* NIH fel, Swiss Fed Inst Technol, 61-62; vis lectr, Dept Anat, Univ Malaya, 71-72. *Mem:* Electron Micros Soc Am; Am Asn Anat; Am Soc Cell Biologists; Am Soc Study Reproduction. *Res:* Fine structure of cells and tissues; cryobiology. *Mailing Add:* Dept Biol Struct Univ Wash Seattle WA 98195

KOEHLER, JAMES STARK, b Oshkosh, Wis, Nov 10, 14; m 40; c 2. PHYSICS. *Educ:* Wis State Teachers Col, BEd, 35; Univ Mich, PhD, 40. *Prof Exp:* Rackham fel, Univ Mich, 40-41; Westinghouse res fel, 41-42; instr physics, Carnegie Inst Technol, 42-46, assoc prof, 46-50; assoc prof, 50-53, PROF PHYSICS, UNIV ILL, URBANA, 53- *Concurrent Pos:* Guggenheim fel, 57; mem solid state adv comt, Oak Ridge Nat Lab. *Mem:* Fel Am Phys Soc. *Res:* Effects of internal rotation on molecular spectra; plastic deformation of solids; radiation damage; point defects produced by quenching, irradiation and ion bombardment. *Mailing Add:* Dept Physics Univ Ill 1110 W Green St Urbana IL 61801

KOEHLER, LAWRENCE D, b Grand Rapids, Mich, Feb 19, 32; m 60; c 3. DEVELOPMENTAL BIOLOGY, CELL BIOLOGY. *Educ:* Otterbein Col, BS, 54; Mich State Univ, PhD(zool), 60. *Prof Exp:* From asst prof to assoc prof, 60-68, PROF BIOL, CENT MICH UNIV, 68-, CHMN DEPT, 75- *Concurrent Pos:* NIH spec res fel, Inst Molecular Evolution, Univ Miami, 68-69. *Mem:* Am Soc Zoologists; Am Soc Cell Biologists; Int Soc Develop Biol; Electron Micros Soc Am. *Res:* Gametogenesis and fertilization, including ultrastructural changes in gametes and early zygotes. *Mailing Add:* Dept Biol Cent Mich Univ Mt Pleasant MI 48859

KOEHLER, MARK E, b Dayton, Ohio, July 6, 49; m 70; c 1. INSTRUMENTATION. *Educ:* Univ Dayton, BS, 71; Wright State Univ, MS, 73; Case Western Reserve Univ, PhD(chem), 78. *Prof Exp:* Chemist, Glidden Coatings & Resins, Div SCM Corp, 77-80, sect leader, 80-83; GROUP LEADER, GLIDDEN CO-ICI, 77- *Mem:* Am Chem Soc. *Res:* Development and computer interfacing of laboratory instrumentation; areas of computer applications in chemistry; scientific computing and analog and digital electronics. *Mailing Add:* 6715 Winona Circle Middleburg Heights OH 44130

KOEHLER, P RUBEN, b Berlin, Ger, Apr 29, 31; US citizen; c 2. RADIOLOGY. *Educ:* Univ Bern, MD, 56. *Prof Exp:* Assoc radiol, Albert Einstein Med Ctr, Philadelphia, Pa, 61-62; instr, Sch Med, Temple Univ, 62-64; from asst prof to prof, Sch Med, Washington Univ, 64-70; PROF RADIOL, COL MED, UNIV UTAH, 70-, CHIEF, DIV DIAG RADIOL, 80- *Mem:* Am Col Radiol; Am Roentgen Ray Soc; Radiol Soc NAm; Asn Univ Radiologists; Int Soc Lymphology. *Res:* Lymphology; visceral arteriography. *Mailing Add:* Univ Utah Col Med Salt Lake City UT 84132

KOEHLER, PHILIP EDWARD, b Kansas City, Mo, Mar 30, 43; m 66; c 2. FOOD SCIENCE. *Educ:* Emporia Kans State Col, BS, 65; Okla State Univ, PhD(biochem), 69. *Prof Exp:* From asst prof to assoc prof, 69-80, PROF FOOD SCI, UNIV GA, 80- *Mem:* Am Chem Soc; Inst Food Technologists; Sigma Xi. *Res:* Food safety and toxicology; food colorants; flavor chemistry. *Mailing Add:* Dept Food Sci Univ Ga Athens GA 30601

KOEHLER, PHILIP GENE, b Doylestown, Pa, July 21, 47; m 74. ENTOMOLOGY. *Educ:* Catawba Col, AB, 69; Cornell Univ, PhD(entom), 72. *Prof Exp:* From asst prof to assoc prof exten entom, 75-84, PROF ENTOM, UNIV FLA, 84- *Concurrent Pos:* Proj leader, USDA Household Insect Res Proj. *Honors & Awards:* Hon mem Fla Pest Control Asn. *Mem:* Am Mosquito Control Asn; Entom Soc Am; Am Registry Prof Entomologists; Sigma Xi. *Res:* Cockroach and flea research; household and structural pest management. *Mailing Add:* 214 Newell Hall Univ Fla Gainesville FL 32611

KOEHLER, RAYMOND CHARLES, BRAIN BLOOD FLOW. *Educ:* State Univ NY, Buffalo, PhD(physiol), 78. *Prof Exp:* ASST PROF ANESTHESIOL, JOHNS HOPKINS HOSP, 80- *Mailing Add:* Dept Anesthesiol/Critical Care Med Blalock 1404 Johns Hopkins Hosp 601 N Wolfe St Baltimore MD 21205

KOEHLER, RICHARD FREDERICK, JR, b New York, NY, Mar 27, 45; m 69; c 2. PHYSICS, ELECTRICAL ENGINEERING. *Educ:* Mass Inst Technol, BS, 67; Stanford Univ, MS, 68, PhD(elec eng), 72. *Prof Exp:* Assoc scientist, Xerox Corp, 72-75, scientist, 75-78, sr scientist, 78-80, AREA MGR, XEROX CORP, 80- *Mem:* Soc Photog Scientists & Engrs. *Res:* Physics and materials of the xerographic system. *Mailing Add:* 15 Woodrose Dr Webster NY 14580

KOEHLER, THOMAS RICHARD, b Toledo, Ohio, Aug 8, 32; m 61. PHYSICS. *Educ:* Seattle Univ, BS, 54; Calif Inst Technol, PhD(physics), 60. *Prof Exp:* Physicist, Aeronutronic Div, Ford Motor Co, 59-60; STAFF PHYSICIST, SAN JOSE RES LAB, IBM CORP, 60- *Mem:* Am Phys Soc. *Res:* Theoretical solid state and low temperature physics. *Mailing Add:* K32-803-D IBM Res 650 Harry Rd San Jose CA 95120

KOEHLER, TRUMAN L, JR, b Allentown, Pa, Apr 9, 31; m 54; c 3. STATISTICS. *Educ:* Muhlenberg Col, BS, 52; Rutgers Univ, MS, 57. *Prof Exp:* Engr qual control, Sylvania Elec Prod Inc, 52-57; statistician, Am Cyanamid Co, 57, head qual control sect, Org Chem Div, 57-62, mgr systs anal, 62-66, mfg mgr, Org Pigments Dept, 66-68, dir mkt, Pigments Div, 68-70, mgr titanium dioxide dept, 70-77, dir planning, 77-79, gen mgr, Spec Prods Dept, 79-80; exec vpres & chief oper officer, Sodyeco Div, Martin Marietta Chem, 81-84; pres & chief exec officer, 84-87, GROUP VPRES CHEM, SANDOZ CORP, 87- *Concurrent Pos:* Lectr, Rutgers Univ, 57-; partic, NSF TV Prog, Pursuit of Perfection, 65-; dir, Nat Asn Mfrs & Chem Mfrs Asn. *Mem:* Fel Am Soc Qual Control; Am Statist Asn; Am Inst Chem Engrs; Nat Asn Mfrs; Chem Mfrs Asn. *Res:* Design and analysis of experimental programs; numerical analysis and computing. *Mailing Add:* 5222 Winding Brook Rd Charlotte NC 28226

KOEHLER, WILBERT FREDERICK, b Braddock, Pa, Feb 20, 13; m 36; c 2. OPTICAL PHYSICS. *Educ:* Allegheny Col, BS, 33; Cornell Univ, MA, 34; Johns Hopkins Univ, PhD(physics), 49. *Prof Exp:* Pub sch instr, Pa, 34-36; instr, Punahou Acad, Hawaii, 36-43; instr physics, Johns Hopkins Univ, 46-48; assoc prof, Naval Postgrad Sch, 48-51; sect head phys optics, Naval Ord Test Sta, 51-54, res scientist, 54-58, head physics div, 58-61; asst dean, 61-62, dean progs, 62-76, EMER DEAN & DISTINGUISHED EMER PROF PHYSICS, NAVAL POSTGRAD SCH, 76- *Concurrent Pos:* Mem, Naval Officer Prof Develop Study Group, 74. *Mem:* Am Phys Soc; fel Optical Soc Am; Am Asn Physics Teachers; Am Soc Eng Educ. *Res:* Multiple-beam interferometry; optical constants; surface smoothness; specific heats of gases; solid state physics. *Mailing Add:* 52 Alta Mesa Circle Monterey CA 93940

KOEHLER, WILLIAM HENRY, b Houston, Tex, Feb 17, 39. INORGANIC CHEMISTRY. *Educ:* Southern Methodist Univ, BS, 60, MS, 62; Univ Tex, Austin, PhD(chem), 69. *Prof Exp:* Instr chem, Southern Methodist Univ, 61-63; res scientist, Tracor, Inc, Tex, 68-69; asst prof, 69-74, ASSOC PROF CHEM, TEX CHRISTIAN UNIV, 74-, ACTG DEAN GRAD SCH, 78-, VCHANCELLOR ACAD AFFAIRS, 80- *Concurrent Pos:* Vpres, Tex Christian Univ Res Found. *Mem:* Nat Coun Univ Res Adminr; Am Chem Soc. *Res:* Raman spectroscopy; reflection and transmission spectroscopy; characterization of metal-ammonia solutions; synthesis and reaction mechanisms in nonaqueous solvents. *Mailing Add:* Chem Dept Tex Christian Univ Ft Worth TX 76129

KOEHMSTEDT, PAUL LEON, b Seattle, Wash, Sept 21, 23; m 56; c 3. INJECTION ELECTRICAL GROUNDING, CHEMICAL AGENT DECONTAMINATION. *Educ:* Ore State Univ, BS, 49, MS, 51. *Prof Exp:* Chemist, Gen Elec Co, 50-53; res engr mat, Boeing Airplane Co, 53-67; RES SCIENTIST MAT, BATTELLE PAC NORTHWEST LABS, 67- *Mem:* Am Chem Soc; Nat Asn Corrosion Engrs. *Res:* Innovative applications towards solving material and methods problems involving chemical processes, corrosion, chemical and radiochemical decontamination, preparation and testing. *Mailing Add:* 2021 Davison Richland WA 99352

KOEHN, ENNO, b Flushing, NY, April 29, 36; m 67; c 2. PROJECT MANAGEMENT SYSTEMS, OPTIMAL PRODUCTIVITY & COST FACTORS. *Educ:* City Univ NY, BCE, 58; Columbia Univ, MS, 60; NY Univ, MCE, 65; Wayne State Univ, PhD(civil eng), 75. *Prof Exp:* Res engr struct eng, NAm Rockwell, 58-59; asst prof eng, Long Island Univ, 60-66; educ specialist continuing educ, Int Bus Mach, 66-67; from assoc prof to prof, civil eng, Ohio Northern Univ, 67-79; assoc prof civil eng, Purdue Univ, 79-84; CHAIR, DEPT CIVIL ENG, LAMAR UNIV, 84- *Concurrent Pos:* Prin investr, Stanford Univ/NSF grants, 64-66, NSF grants, 70-72, US Army Construct Eng Res Lab, 83-88; fel, Mass Inst Technol, 72, Univ Mich/NSF Construct Res Sem, 72 & 82, Univ Pa, Dept Energy Sem, 76, fel, NASA-ASEE, Stanford Univ Inst, 77; sr civil engr, Bechtel Corp, 78-81; chmn, Social & Environ Concerns Comt, Am Soc Civil Engrs, 79-87. *Honors & Awards:* Pres citation, Am Soc Civil Engrs, 83. *Mem:* Fel Am Soc Civil Engrs; Sigma Xi; Am Asn Cost Engrs; Nat Soc Prof Engrs; Am Soc Eng Educ. *Res:* International productivity in design and construction systems; probabilistic and pre-design cost estimating; optimal productivity of small and medium size companies; weather related productivity factors in construction; national infrastructure/rehabilitation costs; application of fuzzy sets to construction systems. *Mailing Add:* Civil Eng Dept Lamar Univ PO Box 10024 Beaumont TX 77710

KOEHN, PAUL V, b Bristol, Conn, Jan 10, 31; m 63; c 3. BIOCHEMISTRY. *Educ:* Bates Col, BS, 52; Cent Mo State Col, MSEd, 58; Univ Conn, PhD(biochem), 64. *Prof Exp:* Jr chemist, Am Cyanamid Co, 52-55; instr gen chem, Cent Mo State Col, 56-58; instr biochem, Univ Conn, 63-64; assoc prof biochem, 67-70, head dept, 69-70, fac res grant, 71, chmn dept, 70-82, PROF BIOCHEM, STATE UNIV NY COL ONEONTA, 69- *Concurrent Pos:* NY State Res Found grant, 68-70; fac grants, State Univ NY, 74 & 77. *Mem:* Am Chem Soc; Sigma Xi; NY Acad Sci. *Res:* Synthesis of phosphopeptides and structural studies of proteins. *Mailing Add:* Dept Chem State Univ NY Col Oneonta NY 13820

KOEHN, RICHARD KARL, b Niles, Mich, Aug 25, 40; m 83; c 3. POPULATION GENETICS. *Educ:* Western Mich Univ, BA, 63; Ariz State Univ, PhD(genetics), 67. *Prof Exp:* Trainee immunol, Univ Kans, 67, asst prof zool, 67-70; vis scientist genetics, Aarhus Univ, 70-71 & 76-77; assoc prof ecol & evolution & provost biol sci, 71-80, dean biol sci, 78-88, PROF ECOL & EVOLUTION, STATE UNIV NY STONY BROOK, 78-, DIR CTR BIOTECHNOL, 83- *Concurrent Pos:* NSF & NIH grants; NATO sr sci fel, 70; assoc ed, J Soc Study Evolution, 75-77, Molecular Biol Evolution, 83- & J Exp Marine Biol Ecol, 84-; George C Marshall fel, Denmark, 76-77; ed, Marine Biol Lett, 78-84; Guggenheim Fel, 88-89; bd dirs, Boyce Thompson Inst Plant Res, 88-, Long Island Forum Technol, 88-, Asn Biotechnol Co, 89-; vpres, develop, Orgn Trop Studies, 88-; chmn, Coun Biotechnol Ctrs, 89-91. *Mem:* AAAS; Am Soc Naturalists; Genetics Soc Am; Soc Study Evolution (pres, 85); Linnean Soc; Europ Soc Comp Physiol Biochem. *Res:* Evolutionary genetics and physiological energetics of natural populations, particularly marine invertebrates; protein function, structure and adaptation. *Mailing Add:* Dept Ecol & Evolution State Univ NY Stony Brook NY 11794-5200

KOELLA, WERNER PAUL, b Zurich, Switz, Apr 13, 17; nat US; m 55; c 3. NEUROPHYSIOLOGY. *Educ:* Univ Zurich, MD, 42. *Prof Exp:* Resident neurosurg, Univ Zurich, 43-45, resident physiol, 45-48, head asst dept, 48-51; res assoc neurophysiol, Univ Minn, 51-52, assoc prof, 52-55; mem staff, Worcester Found Exp Biol, 57-68; chmn dept pharmacol & med, Robapharm Ag, 68-70; sr mem staff, Lab Neurophysiol, Ciba-Geigy Ltd, 70-82; RETIRED. *Concurrent Pos:* Prof affil, Clark Univ, 57-, Boston Univ, 59 & Univ Berne, 70. *Mem:* AAAS; Am Physiol Soc; Am Soc Pharmacol & Exp Therapeut; fel Am Col Neuropsychopharmacol; NY Acad Sci. *Res:* Cerebellum; vestibular apparatus; subcortical-cortical relationships; sleep; organization of autonomic functions; neuropharmacology. *Mailing Add:* Buchenstrasse 1 CH-4104 Oberwil BL304329 Switzerland

KOELLE, GEORGE BRAMPTON, b Philadelphia, Pa, Oct 8, 18; m 54; c 3. PHARMACOLOGY. *Educ:* Philadelphia Col Pharm & Sci, BSc, 39; Univ Pa, PhD(pharmacol), 46; Johns Hopkins Univ, MD, 50. *Hon Degrees:* DSc, Philadelphia Col Pharm & Sci, 65; Dr Med, Univ Zurich, 72. *Prof Exp:* Bioassayist, LaWall & Harrisson, Pa, 39-42; asst prof pharmacol, Columbia Univ, 50-52; prof, Grad Div Med, Univ Pa, 52-65, dean, 57-59, prof pharmacol & chmn dept, 59-81, distinguished prof, 81-89, EMER PROF, SCH MED, UNIV PA, 89- *Concurrent Pos:* Consult, pharmaceut indust, 51-, Philadelphia Gen Hosp, 53-71, Valley Forge Army Hosp, 54-70, Army Chem Corps, 56-61, Philadelphia Naval Hosp, 57-; Hachmeister lectr, Med Sch, Georgetown Univ, 52; Merck lectr, Med Sch, McGill Univ, 55; vis lectr, Philadelphia Col Pharm & Sci, 55-57, trustee, 62- 86; consult, Study Sect Pharmacol & Exp Therapeut, NIH, 58-62, chmn, 65-68, bd sci counr, Nat Heart Inst, 60-64, chmn eval clin ther adv comt, Nat Inst Neurol Dis & Stroke, 68-70, mem, Nat Adv Neurol Dis & Stroke Coun, 70-74; ed, Pharmacol Rev, 59-62; spec lectr, Univ London, 62; vis lectr, Biophys Inst, Univ Brazil, 62; Guggenheim fel & vis prof, Inst Physiol, Univ Lausanne, 63-64; secy-gen, Int Union Pharmacol, 66-69, vpres, 69-72; prof & actg chmn dept pharmacol, Med Sch, Pahlavi Univ, Iran, 69-70; P K Smith lectr, George Washington Univ, 71; vis prof, Polish Acad Sci, 79; delegate, Soviet Acad Sci, 89. *Honors & Awards:* Abel Prize & Borden Award, 50; Sollmann Award, 90. *Mem:* Nat Acad Sci; fel AAAS (vpres, 71); Am Soc Pharmacol & Exp Therapeut (pres, 65-66); Brit Pharmacol Soc; Soc Neurosci. *Res:* Neuropharmacology; histochemistry; electron microscopy; anticholinesterase agents; neurohumoral transmitters; neurotrophic factors. *Mailing Add:* Dept Pharmacol Univ Pa Sch Med Philadelphia PA 19104-6084

KOELLE, WINIFRED ANGENENT, b Soerakarta, Indonesia, Mar 26, 26; m 54; c 3. ESSENTIAL HYPERTENSION, CARDIAC DRUGS. *Educ:* Wellesley Col, BA, 48; Columbia Univ, MD, 52; Am Bd Internal Med, cert, 79. *Prof Exp:* Teaching assoc pharmacol, Grad Sch Med & Sch Med, 56-72, teaching assoc med, Sch Med, 70-76, ASST PROF PHARMACOL, SCH MED, UNIV PA, 72-, ASST PROF MED, 76- *Concurrent Pos:* Chief, Intensive Care Unit, Taylor Hosp, Ridley Park, Pa, 67-69; asst prof pharmacol, Pahlavi Univ, Shiraz, Iran, 69-70; co-chief & chief, med clin, Philadelphia Gen Hosp, Univ Pa Serv, 73-77; vis prof, Sch Med, Free Fac, Lille, France, 76-, Mahidol Univ, Bangkok, Thailand, 78 & St George's Univ, St Vincent, WI, 84; deleg, US Pharmacopeia, Univ Pa, 80-85. *Mem:* Am Soc Pharmacol & Exp Therapeut; fel Am Col Clin Pharmacol; Sigma Xi. *Res:* Metabolism of catecholamines by ocular tissue; histochemistry and life cycles of cholinesterases; anticholinesterase agents; mechanisms of release of acetylcholine; neurotrophic factors. *Mailing Add:* Dept Pharmacol & Med Univ Pa Sch Med Philadelphia PA 19104

KOELLER, RALPH CARL, b Chicago, Ill, Aug 9, 33; div; c 3. MECHANICAL ENGINEERING, CONTINUUM MECHANICS. *Educ:* Ill Inst Technol, BS, 57, MS, 59, PhD(mech), 63. *Prof Exp:* Res & teaching asst mech eng, Ill Inst Technol, 59-62; from instr to prof, Univ Colo, 62-84; prof mech eng, Univ Wis, 84-89; PRES, MECH DESIGN GROUP, 89- *Concurrent Pos:* Univ Colo Fac fel, Univ Calif, Berkeley, 65-66; consult, Colo Instruments Inc, 69, Dow Chem Co, 70-71, Dieterich Standards Corp, 73 & Hewlett-Packard Co, 74; resident fac fel, Am Soc Eng Educrs, 74-75; vis asst prof, Cornell Univ, 76-77; consult, Ponderosa Assoc, 78-88. *Mem:* Am Soc Mech Engrs; Sigma Xi. *Res:* Deformation of materials with memory; design of a flexible rbotic arm. *Mailing Add:* 960 Stonebridge No 6 Platteville WI 53818-2078

KOELLING, DALE DEAN, b Great Bend, Kans, May 8, 41; m 68; c 2. SOLID STATE SCIENCE. *Educ:* Kans State Univ, BS, 63; Mass Inst Technol, PhD(physics), 68. *Prof Exp:* Res assoc, Northwestern Univ, 68-72; physicist, 72-87, SR PHYSICIST MAT SCI DIV, ARGONNE NAT LAB, 87- *Concurrent Pos:* Vis prof, Northern Ill Univ, 85-87. *Mem:* Am Phys Soc; Mat Res Soc. *Res:* Electronic structure and resulting properties primarily in metallic or semiconducting actinide, rare-earth, or transition element materials. *Mailing Add:* Mat Sci Div Argonne Nat Lab Argonne IL 60439-4845

KOELLING, MELVIN R, b Sullivan, Mo, July 18, 37; m 59. FORESTRY. *Educ:* Univ Mo, BS, 59, MS, 61, PhD(bot), 64. *Prof Exp:* Assoc plant physiol, Northeast Forest Exp Sta, USDA, 64-67; from asst prof to assoc prof, 67-77, PROF FORESTRY & EXTEN SPECIALIST, MICH STATE UNIV, 77- *Mem:* AAAS; Am Inst Biol Sci; Ecol Soc Am; Soc Am Foresters. *Res:* Botany; improvement of sugar maple with respect to sap production; maple sap physiology. *Mailing Add:* 126 Nat Resources Forestry Mich State Univ East Lansing MI 48824

KOELSCHE, CHARLES L, b Seattle, Wash, Sept 23, 06; m 44. INORGANIC CHEMISTRY, ANALYTICAL CHEMISTRY. *Educ:* Univ Southern Calif, AB, 28, MS, 30; Ind Univ, EdD(sci educ), 53. *Prof Exp:* Pub sch teacher sci & math, 29-42; Fed Security Agency inspector, US Food & Drug Admin, 42-44; assoc prof chem & chmn dept, Univ Alaska, 44-47; assoc prof, Ariz State Univ, 47-52; assoc prof phys sci, Wis State Col, Eau Claire, 52-57; prof sci educ & chem, Univ Toledo, 57-58; specialist for sci, US Off Educ, 58-59; spec asst to dir off sci personnel, Nat Acad Sci-Nat Res Coun, 59-60; prof sci educ, Univ Ga, 60-74; consult sci educ, 74-75; chmn dept sci, Athens Acad, 75-81; RETIRED. *Concurrent Pos:* NSF Summer & In-serv Insts Sci Teachers, Univ Ga, 61-73; vis prof chem, Embry-Riddle Aeronaut Univ, Prescott, Ariz, 82, 83 & 85- *Res:* Chemistry; physics; physical science; science education. *Mailing Add:* 116 Cedar Ct Forsyth GA 31029-9131

KOELSCHE, GILES ALEXANDER, b Ashland, Ore, Sept 3, 08; m 30; c 2. INTERNAL MEDICINE. *Educ:* Pac Union Col, BS, 30; Loma Linda Univ, MD, 31; Univ Minn, MS & PhD(med), 35. *Prof Exp:* Fel, Mayo Found, Univ Minn, 31-35; asst, Mayo Grad Sch Med, Univ Minn, 35-37, from instr to assoc prof, 37-74, consult, Div Med, Mayo Clin, 37-74, emer assoc prof clin med, 74,84; RETIRED. *Mem:* Emer fel Am Acad Allergy; emer fel Am Col Allergists (pres, 60-61). *Res:* Newer therapy for asthma, hay fever and perennial allergic rhinitis; management of chronic urticaria; problems of immunologic tolerance and organ transplantation; auto-immune diseases; immunology in relation to cancer. *Mailing Add:* 13825 Crown Point Sun City AZ 85351

KOELTZOW, DONALD EARL, b Clovis, NMex, May 9, 44; m 65; c 3. BIOCHEMISTRY, ORGANIC CHEMISTRY. *Educ:* NMex Inst Mining & Technol, BS, 66; Univ Ill, MS, 68, PhD(biochem), 70. *Prof Exp:* Fel med microbiol, Stanford Univ, 70-71; assoc res scientist pharmacol, Univ Iowa, 77-78; asst prof, 71-77, ASSOC PROF CHEM, LUTHER COL, 77-, CHMN DEPT, 78- *Concurrent Pos:* US Army med res grant, 73-75. *Mem:* Am Chem Soc; Am Soc Microbiol; AAAS; Midwest Asn Chem Teachers Lib Arts Cols. *Res:* Structure and function of membrane components, particularly carbohydrates and lipids. *Mailing Add:* USDA FGIS Tech Ctr PO Box 20285 Kansas City MO 64195

KOELZER, VICTOR A, b Seneca, Mo, May 3, 14; m 50; c 1. HYDRAULIC ENGINEERING. *Educ:* Univ Kans, BS, 37; Univ Iowa, MS, 39. *Prof Exp:* Jr engr, US Geol Surv, Iowa, 38-40; jr asst engr, Corps Engrs, Ky, 40-42; from sr engr to head engr, Bur Reclamation, Washington, DC, 46-48 & 54-56 & Colo, 48-54; from hydraul engr to vpres, Harza Eng Co, Ill, 56-69; chief eng & environ sci, Nat Water Comn, 69-76; pres, Eng Farms Inc, 76-84; RETIRED. *Concurrent Pos:* Mem comt water, Nat Acad Sci, 65-67; teacher, Colo State Univ, Ft Collins, 72-; consult, 72-88. *Honors & Awards:* Julian Hinds Award, Am Soc Civil Engrs, 75. *Mem:* Am Soc Civil Engrs; Am Geophys Union; Sigma Xi; Am Water Resources Asn. *Res:* Hydrology; sedimentation; hydropower; water resources; development and operating of irrigation land. *Mailing Add:* 1801 Sheely Dr Ft Collins CO 80526-1941

KOEN, BILLY VAUGHN, b Graham, Tex, May 2, 38; m 67; c 2. NUCLEAR ENGINEERING. *Educ:* Univ Tex, Austin, BA, 60, BS, 61; Mass Inst Technol, SM, 62, ScD(nuclear eng), 68; Saclay Nuclear Res Ctr, France, dipl eng, 63. *Prof Exp:* From asst prof to assoc prof, 68-82, PROF MECH ENG, UNIV TEX, AUSTIN, 82- *Concurrent Pos:* Foreign collabr, French Atomic Energy Comn, 71-72. *Mem:* Am Nuclear Soc; Am Soc Eng Educ; NY Acad Sci. *Res:* Nuclear reactor kinetics; engineering education. *Mailing Add:* Dept Mech Eng Univ Tex ETC 5160 Austin TX 78712-1063

KOENG, FRED R, b Wilmington, Del, Aug 6, 41; m 63; c 1. ORGANIC CHEMISTRY. *Educ:* Franklin & Marshall Col, AB, 63; Northwestern Univ, PhD(org chem), 70. *Prof Exp:* Chemist, Rohm and Haas Co, 65-67; sr res chemist, 70-80, RES ASSOC, EASTMAN KODAK CO, 80- *Mem:* AAAS; Soc Photog Scientists & Engrs; Sigma Xi. *Res:* Photographic systems research. *Mailing Add:* 93 Damsen Rd Rochester NY 14612

KOENIG, CHARLES JACOB, b St Marys, Ohio, June 3, 11; m 52; c 5. CERAMIC ENGINEERING. *Educ:* Ohio State Univ, BCerE, 32, MS, 33, PhD(ceramic eng), 35, CE, 48. *Prof Exp:* Res engr, Eng Exp Sta, 35-41, prof, 56-73, EMER PROF CERAMIC ENG, OHIO STATE UNIV, 73-; SR RES ENGR, EDWARD ORTON JR CERAMIC FOUNDATION, 73- *Concurrent Pos:* Dir res, Am Nepheline Corp, 45-66 vpres res & dir, 60-64, pres, 64-66; dir, Indust Minerals Can, Ltd, 47-65, vpres res, 60-66; trustee, Whiteware Div, Am Ceramic Soc, 63-66; consult, Anchor Hocking Corp, 73-76, Gorham Div, Textron, Inc, 76-78, Sumerbank Yarimca Seramik, Yarimca, Turkey, 81 & Haeng-Nam Sa Co, Mokpo, Korea, 85. *Mem:* Fel Am Ceramic Soc; Am Soc Testing & Mat; Nat Inst Ceramic Engrs. *Res:* Electrical porcelain; dinnerware; tile; glass; porcelain enamel; sanitary porcelain; refractories; multiple fluxes for promoting glassy phases in whiteware bodies; effect of furnace atmosphere in firing ceramics; pyrometric cones. *Mailing Add:* 1521 Guilford Rd Columbus OH 43221

KOENIG, CHARLES LOUIS, b Yonkers, NY, Oct 11, 11; c 4. PHYSICAL CHEMISTRY. *Educ:* NY Univ, BS, 32, PhD(chem), 36. *Prof Exp:* Sr chemist, Solvay Process Co, NY, 36-45; res chemist, Lithaloys Corp, 45-46; chief res br, AEC, 46-47; chmn dept chem & chem eng res, Armour Res Found, 47-49; asst dir res, Stanford Res Inst, 50-51; vpres, Southwest Res Inst, 51-56; OWNER, LOUIS KOENIG RES, 56- *Concurrent Pos:* Sect ed, Chem Abstr, 49-; adv, Saline Water Conservation Prog, Secy Interior, 52-56; adv, Advan Waste Treatment Prog, USPHS, 60-63. *Mem:* Sigma Xi; Am Chem Soc. *Res:* Phase relations of aqueous systems; water resources; waste disposal; cost engineering; market research; economics. *Mailing Add:* Louis Koenig Res 26890 Sherwood Forest San Antonio TX 78258

KOENIG, DANIEL RENE, b Rouen, France, Oct 6, 36; US citizen; div; c 2. NUCLEAR ENGINEERING, COMPUTER GRAPHICS. *Educ:* Univ Calif, Berkeley, BS, 59, MS, 65, PhD(eng sci), 66. *Prof Exp:* Physicist, Defense Atomic Support Agency, 65-67; physicist, Los Alamos Sci Lab, Univ Calif, 69-85; SOFTWARE ENG, TIME ARTS, INC, SANTA ROSA, CALIF, 86- *Concurrent Pos:* Vis physicist, Ctr Nuclear Studies, France, 67-68, Saclay, 79-80. *Honors & Awards:* Teller Award, Am Nuclear Soc, 66. *Mem:* Am Nuclear Soc. *Res:* Surface physics phenomena such as thermionic emission and surface ionization; detection and theoretical transport of neutron, gamma and x-ray radiations; application of heat pipes to solar energy; design of fast-spectrum nuclear reactors for space applications; design of small solar thermodynamic engines. *Mailing Add:* Time Arts Inc 1425 Corporate Center Pkwy Santa Rosa CA 95407

KOENIG, EDWARD, b New York, NY, Nov 10, 28; m 53; c 3. NEUROBIOLOGY, NEUROCHEMISTRY. *Educ:* Franklin & Marshall Col, BA, 56; Univ Pa, PhD(physiol), 61. *Prof Exp:* From asst prof to assoc prof, 63-75, PROF PHYSIOL, STATE UNIV NY BUFFALO, 75- *Concurrent Pos:* Res career prog award, Nat Inst Neurol Dis & Stroke, 68-73. *Mem:* AAAS; Am Physiol Soc; Am Soc Neurochem; Int Soc Neurochem; Soc Neurosci. *Res:* Cellular biology of the neuron as related to (1) central and local regulation of synthesis of axonal proteins and (2) structure, function and organizational regulation of axonal cytoskeleton; microchemistry; microanalysis. *Mailing Add:* Dept Physiol 321 Cary Hall State Univ NY Buffalo NY 14214

KOENIG, ELDO C(LYDE), b Marissa, Ill, Oct 17, 19; m 50; c 4. ELECTRICAL ENGINEERING. *Educ:* Wash Univ, St Louis, BS, 43; Ill Inst Technol, MS, 49; Univ Wis, PhD(elec eng), 56. *Prof Exp:* Test engr, Allis Chalmers Mfg Co, 44-46, engr, 46-52, supvr comput lab, 54-57, engr in chg eng anal, 57-62; assoc prof numerical anal, 62-64, ASSOC PROF COMPUT SCI, ASSOC PROF INSTR RES LAB EDUC & ASSOC DIR SYNNOETICS LAB, SCH EDUC, UNIV WIS-MADISON, 64- *Honors & Awards:* Nobel Prize, 51. *Mem:* AAAS; Am Math Soc; Soc Indust & Appl Math; Asn Comput Mach; assoc Inst Elec & Electronics Engrs. *Res:* Engineering and mathematical analysis and research for computers; systems and design; intelligent properties of systems. *Mailing Add:* 35005 W Fairview Rd Oconomowoc WI 53066

KOENIG, HAROLD, b New York, NY, Mar 16, 21; m 45; c 2. NEUROLOGY. *Educ:* Rutgers Univ, BS, 42; Chicago Med Sch, MD, 47; Northwestern Univ, MS, 45; Univ Pa, PhD(anat), 48. *Prof Exp:* Res assoc anat, Med Sch, Univ Wash, 47; from instr to asst prof, Univ Pa, 47-49; from asst prof to assoc prof, Chicago Med Sch, 49-54; from asst prof to assoc prof neurol, 57-63, PROF NEUROL, NORTHWESTERN UNIV, CHICAGO, 63-; CHIEF NEUROL, VET ADMIN LAKESIDE MED CTR, 57- *Concurrent Pos:* Resident physician, Univ Chicago Clin, 55; mem, Vet Admin Prog Comt Psychiat, Neurol & Psychol, 61-63; sr physician, Vet Admin, 71-75. *Mem:* Am Soc Cell Biol; Histochem Soc; Sigma Xi; NY Acad Sci; Am Acad Neurol. *Res:* Metabolic bases of neurological disease; nucleic acid and protein metabolism in nervous system; effects of nucleic acid antimetabolites; experimental neuropathology; lysosomes and other storage particles; neurochemistry; neurohistochemistry; electron microscopy. *Mailing Add:* Neurol Serv Vet Admin Lakeside Med Ctr Northwestern Univ Med Sch 333 E Huron St Chicago IL 60611

KOENIG, HERMAN E, b Marissa, Ill, Dec 12, 24; m 49; c 3. ELECTRICAL ENGINEERING. *Educ:* Univ Ill, BS, 47, MS, 49, PhD(elec eng), 53. *Prof Exp:* Asst prof elec eng, Univ Ill, 53-54 & 55-56 & Mass Inst Technol, 54-55; assoc prof, 56-59, chmn dept elec eng & systs sci, 69-75, asst vpres indust asst, 85, PROF ELEC ENG, MICH STATE UNIV, 59-, DIR SYSTS SCI PROG, 67-, DIR CTR ENVIRON QUAL, 75-, OFF VPRES RES & GRAD STUDIES, 80- *Concurrent Pos:* Consult, Reliance Elec & Eng Co, Ohio, 51-54 & Lear Siegler, Inc, Mich, 63-65. *Mem:* Am Soc Eng Educ; Soc Eng Sci; Inst Elec & Electronics Engrs. *Res:* Theory of electrical networks and other physical systems; transportation, business and other socio-economic systems; operations research; industrialized ecosystem design and management; energy and energy resources. *Mailing Add:* 4733 Mohican Lane Okemos MI 48864

KOENIG, INGE RABES, physical chemistry, for more information see previous edition

KOENIG, JACK L, b Cody, Nebr, Feb 12, 33; m 53; c 4. POLYMER CHEMISTRY, PHYSICAL CHEMISTRY. *Educ:* Yankton Col, BA, 56; Univ Nebr, MS, 58, PhD(chem), 60. *Prof Exp:* Mem staff, Plastics Dept, E I du Pont de Nemours & Co, 59-63; from asst prof to assoc prof chem, 63-70, PROF MACROMOLECULAR SCI, CASE WESTERN RESERVE UNIV, 70- *Concurrent Pos:* Consult, El Tech, Charden, Ohio & 3M Co, 59-63; NSF, US Army, US Navy res grants, 65-88. *Honors & Awards:* Res Award, Sigma Xi. *Mem:* Am Chem Soc; Am Phys Soc; Soc Appl Spectros. *Res:* Spectroscopy of polymeric materials. *Mailing Add:* Dept Macro Sci Case Western Reserve Univ Univ Circle Cleveland OH 44106

KOENIG, JAMES BENNETT, b New York, NY, Nov 25, 32; c 3. STRUCTURAL GEOLOGY. *Educ:* Brooklyn Col, BS, 54, Ind Univ, MA, 56; USN Postgrad Sch, dipl meteorol, 58. *Prof Exp:* Ground water geologist, US Geol Surv, St Paul, Minn, 55-56; jr mining geologist, Calif Div Mines, San Francisco, 56-57, asst geologist, 60-63, supvry geologist, 65-72; ensign & lieutenant, USN Postgrad Sch, Monterey & USN Weather Res Facil, Norfolk, 57-60; asst geologist, Calif Div Mines & Geol, 60-63, supvry geologist, 65-72; PRES, GEOTHERMEX INC, 73- *Concurrent Pos:* Instr econ geog, Col William & Mary, 59-60; student geol & seismol, Univ Nev, 63-65; instr geol, Univ Calif, Berkeley, 68-70; consult, UN Geothermal Explor, Ethiopia & El Salvador, 71, Weyerhaeuser Co & Pac Power & Light Co, 71-72; mem int working group, Int Geothermal Asn, 86-; mem bd dir, Geothermal Resource Coun, 75-, pres-elect, 87. *Mem:* Fel Geol Soc Am; Am Geophys Union; Int Asn Volcanology. *Res:* Exploration, drilling and development of geothermal resources as energy source, on behalf of electric utilities, international lender and donor agencies, major and independent oil/gas/mining companies, landowners, turbine manufacturers, government agencies in US and abroad (Japan, Costa Rica, Kenya, Philippines, etc); exploration for geothermal energy; feasibility assessments of energy resources. *Mailing Add:* Geothermex Inc 5221 Cent Ave Suite 201 Richmond CA 94804

KOENIG, JAMES J(ACOB), b Le Mars, Iowa, Sept 4, 18; m 51; c 4. CHEMICAL ENGINEERING. *Educ:* Iowa State Col, BS, 39. *Prof Exp:* Foreman, Procter & Gamble Mfg Co, Kans, 40-41, gen foreman & tech supvr, Tenn, 41-44; asst dir prod div, NY Opers Off, USAEC, 46-47, asst area mgr, Mo, 47-52, chief, Fla Field Off, 52-56; res engr, Aluminum Co Am, 56-67, sr res engr, 67-77; MEM STAFF, JAMES J KOENIG CONSULT, INC, 81- *Mem:* Sigma Xi. *Res:* Explosives loading; uranium extraction and metallurgy; alumina chemicals. *Mailing Add:* 129 Country Club Acres Belleville IL 62223

KOENIG, JANE QUINN, b Seattle, Wash, Sept 16, 35; c 2. RESPIRATORY PHYSIOLOGY. *Educ:* Univ Wash, BS, 59, MS, 61, PhD(physiol psychol), 63. *Prof Exp:* Fel neurophysiol, Med Sch, Stanford Univ, 63-65; asst prof, Med Sch, 66-70, vis scientist, 70-71, actg asst prof physiol, Dept Zool, 72-73, ASSOC PROF, DEPT ENVIRON HEALTH, SCH PUB HEALTH & COMMUNITY MED, UNIV WASH, 74- *Concurrent Pos:* Consult, Clear Air Sci Adv Comt, Environ Protection Agency. *Mem:* AAAS; Fedn Am Scientists; Union Concerned Scientists; Am Thoracic Soc; Air Pollution Control Asn; Am Pub Health Asn. *Res:* Effects of acute exposures to air pollutants upon respiratory physiology in human volunteers especially susceptible individuals. *Mailing Add:* Dept Environ Health SC-34 Univ Wash Seattle WA 98195

KOENIG, JOHN HENRY, b St Marys, Ohio, Apr 9, 09; m 42; c 5. CERAMIC ENGINEERING. *Educ:* Ohio State Univ, BChE, 31, MSc, 36, PhD(ceramic eng), 38. *Prof Exp:* Res chemist, Gen Elec Co, Mass, 31-35; res engr, Hall China Co, Ohio, 38-42; dir sch ceramics, NJ Ceramic Res Sta, 45-70, adj prof ceramics, 70, EMER PROF, RUTGERS UNIV, 70- *Concurrent Pos:* Pres, VI Int Glass Cong, 62-65; consult, Int Exec Serv Corps, S Korea, Arg, Mex, Colombia & Turkey. *Honors & Awards:* Jeppson Award, Am Ceramic Soc, 63; Award Merit, Am Soc Testing & Mat, 69. *Mem:* Hon mem Am Ceramic Soc; fel Am Soc Testing & Mat; Ceramic Educ Coun (pres, 55-56). *Mailing Add:* 1967 H-Rd Delta CO 81416

KOENIG, JOHN WALDO, b Newark, NJ, July 19, 20; m 50; c 1. PALEONTOLOGY. *Educ:* Columbia Univ, BS, 47; Univ Kans, MS, 51. *Prof Exp:* Sci illusr, Kans State Geol Surv, 47-51; geol engr, Phillips Petrol Co, 51-54; geologist, Mo Geol Surv & Water Resources, 54-65, Continental Oil Co, 65-66 & Mo Geol Surv, 66-67; tech ed, Univ Mo, Rolla, 67-85; RETIRED. *Concurrent Pos:* Lectr art, Univ Mo-Rolla, 74-80. *Mem:* Asn Earth Sci Ed. *Res:* Invertebrate paleontology; Bryozoa and Crinoidea; Mississippian stratigraphy. *Mailing Add:* 1319 Woodlawn Dr Rolla MO 65401

KOENIG, KARL E, b Washington, DC, Dec 27, 47; m. AMINO ACID SYNTHESIS, HERBICIDE SYNTHESIS. *Educ:* Univ Tex, Austin, BS, 70; Univ Southern Calif, PhD(chem), 74. *Prof Exp:* Fel chem, Univ Calif, Los Angeles, 74-76; sr res chemist, asymmetric catalysis, Monsanto Corp Res Labs, 76-79, res specialist, Corp Res & Develop Biomed Prog, 79-81, sr res specialist, Nutrit Chem Div, 81-86, RES GROUP LEADER, MONSANTO AGR CHEM CO, 86- *Mem:* Am Chem Soc; AAAS. *Res:* Homogeneous catalysis; drugs based on low molecular weight polyelectrolytes; asymmetric synthesis; nutritional chemicals; selective complexation of transition metals; organosilicon chemistry; growth promotants; weed control agents. *Mailing Add:* Monsanto Corp 800 N Lindbergh Blvd 03C St Louis MO 63167

KOENIG, KARL JOSEPH, b Milwaukee, Wis, Jan 9, 20; m 59; c 1. GEOLOGY. *Educ:* Univ Ill, BS, 41, MS, 46, PhD, 49. *Prof Exp:* Stratigrapher, Shell Oil Co, Tex, 49-55; ASSOC PROF GEOL, TEX A&M UNIV, 55- *Mem:* Soc Econ Paleont & Mineral; Geol Soc Am; Am Asn Petrol Geol. *Res:* Miocene stratigraphy and paleontology; sedimentation and clay mineralogy. *Mailing Add:* Dept Geol Tex A&M Univ College Station TX 77843

KOENIG, LLOYD RANDALL, b St Louis, Mo, July 17, 29; m 55; c 2. METEOROLOGY. *Educ:* Washington Univ, BSChE, 50; Univ Chicago, MS, 59, PhD(meteorol), 62. *Prof Exp:* Instr chem eng, USN Postgrad Sch, 50-53; res asst meteorol, Univ Chicago, 57-59, res assoc, 60-62; chief atmospheric sci br, Missile & Space Systs Div, Douglas Aircraft Co, 62-66; phys scientist, Rand Corp, Santa Monica, 66-79; assoc prog dir meteorol, NSF, 79-80; SR SCI OFFICER, WORLD METEOROL ORG, GENEVA, SWITZ, 80- *Mem:* Am Geophys Union; Am Meteorol Soc; Sigma Xi. *Res:* Cloud physics, including natural and artificial precipitation mechanisms; scavenging, effects of atmospheric processes on the atmosphere. *Mailing Add:* 258 Notteargenta Rd Pacific Palisades CA 90272-3110

KOENIG, MICHAEL EDWARD DAVISON, b Rochester, NY, Nov 1, 41; m 80; c 2. INFORMATION MANAGEMENT. *Educ:* Yale Univ, BA, 63; Univ Chicago, MS, 68, MBA, 70; Drexel Univ, PhD(info sci), 82. *Prof Exp:* Mgr info serv, Pfizer, Inc, Groton, Conn, 70-74; dir prod opers, Inst Sci Info, Philadelphia, Pa, 74-77, dir develop, 77-78; vpres opers, Swets NAm, Berwyn, Pa, 78-80; assoc prof info systs, Sch Libr Serv, Columbia Univ, 83-85; vpres info mgt, Tradenet Inc, 85-88; DEAN, GRAD SCH LIBR & INFO SCI, ROSARY COL, 88- *Concurrent Pos:* Adj fac, Grad Sch Bus, Columbia Univ, 83-85, adj prof, Sch Libr Serv, 85-88. *Mem:* Am Soc Info Sci; Asn Comput Mach; AAAS; Spec Libr Asn; Soc Social Study Sci. *Res:* Bibliometrics; relationship between information technology and productivity; research productivity and the information environment. *Mailing Add:* 320 E 42nd St New York NY 10017

KOENIG, MILTON G, b Moberly, Mo, Aug 23, 27; m 56; c 3. THERMODYNAMICS. *Educ:* Wayne State Univ, BSME, 56, MSME, 57. *Prof Exp:* From instr to assoc prof, 56-89, EMER PROF MECH ENG, WAYNE STATE UNIV, 89- *Mem:* Soc Automotive Eng. *Res:* Thermodynamics and its applications; automotive design; vehicle dynamics and handling. *Mailing Add:* Dept Mech Eng Wayne State Univ Detroit MI 48202

KOENIG, NATHAN HART, b NH, Apr 18, 15; m 40; c 3. TEXTILE CHEMISTRY. *Educ:* Univ Chicago, BS, 37; Calif Inst Technol, PhD(chem), 50. *Prof Exp:* Chem lab instr, Cent YMCA Col, Chicago, 37-41; asst chem, US Regional Res Lab, USDA, Ill & Calif, 41-42; explosives inspector, Green River Ord Plant, 42-43; sr res chemist, Shell Develop Co, 50-54; chemist, US Regional Res Labs, Pa, 54-57, chemist fiber sci, Western Regional Res Ctr, 57-81; RETIRED. *Res:* Proton-olefin complexes; isolation of trace amounts of insecticides; sulfur derivatives of fats; wool chemistry. *Mailing Add:* 824 Ramona Ave Albany CA 94706

KOENIG, PAUL EDWARD, b Gallup, NMex, May 30, 29; m 50; c 8. ORGANIC CHEMISTRY. *Educ:* Univ Ariz, BS, 50, MS, 52; Univ Iowa, PhD(chem), 55. *Prof Exp:* Chemist, Ethyl Corp, 55-58; from asst prof to prof chem, La State Univ, Baton Rouge, 58-83, vchancellor acad affairs, 70-81. *Concurrent Pos:* Asst head dept chem, La State Univ, Baton Rouge, 63-67, assoc dean grad sch, 67-70, dean, Runnels Sch, 84-; consult chemist, 83- *Mem:* Am Chem Soc. *Res:* Organic reaction mechanisms; physical organic chemistry; organic synthesis; reactions of metal nitrides with organic compounds; structure of tertiary amides. *Mailing Add:* 2006 Cherrydale Ave Baton Rouge LA 70808

KOENIG, SEYMOUR HILLEL, b Manchester, NH, July 16, 27; m 47; c 2. BIOPHYSICS. *Educ:* Columbia Univ, BS, 49, MA, 50, PhD, 52. *Prof Exp:* Asst physics, Columbia Univ, 49-51; mem staff, Watson Res Lab, IBM Corp, 52-64, from asst dir to dir, 64-70, dir gen sci, Res Ctr, 70-71, STAFF MEM, PHYS SCI DEPT, WATSON RES CTR, IBM CORP, 71- *Concurrent Pos:* From adj asst prof to adj prof, Dept Elec Eng, Columbia Univ, 57-68, lectr, Dept Art Hist & Archeol, 70-76, adj prof, 76- consult, Physics Div, Los Alamos Sci Lab, 59-; mem gov coun, Am Phys Soc, 70-74. *Mem:* Fel Am Phys Soc; Biophys Soc; Sigma Xi; Am Soc Biol Chemists; NY Acad Sci. *Res:* Low temperature electrical transport in semiconductors and semi-metals; inelastic neutron scattering by solids; biophysics of proteins; nuclear magnetic relaxation in protein solutions; protein-water interactions; laser light scattering from macromolecule and virus solutions. *Mailing Add:* IBM T J Watson Res Ctr PO BOX 218 Yorktown Heights NY 10598

KOENIG, THOMAS W, b Kansas City, Mo, Feb 11, 38; m 61; c 2. ORGANIC CHEMISTRY. *Educ:* Southern Methodist Univ, BS, 59; Univ Ill, MS, 61, PhD(chem), 63. *Prof Exp:* From asst prof to assoc prof, 63-74, PROF CHEM, UNIV ORE, 74- *Mem:* Am Chem Soc. *Res:* Mechanisms of organic reactions. *Mailing Add:* Dept Chem Univ Ore Eugene OR 97401

KOENIG, VIRGIL LEROY, biochemistry; deceased, see previous edition for last biography

KOENIGSBERG, ERNEST, b New York, NY, Apr 15, 23; m 55; c 2. OPERATIONS RESEARCH, MANAGEMENT SCIENCE. *Educ:* NY Univ, BA, 48; Iowa State Univ, PhD(theoret physics), 53. *Prof Exp:* Sr physicist, Midwest Res Inst, 53-55; group leader, EMI Eng Develop, 55-57; sect head opers res, Midwest Res Inst, 57-58; mgr mgt serv, Touche, Ross, Bailey & Smart, 58-61; mgr tech serv, CEIR Inc, 61-64; prof indust, Univ Pa, 64-65; vpres, Matson Res Corp, 65-69; sr vpres & tech dir, Manalytics Inc, 69-72; SR LECTR, SCHS BUS ADMIN, UNIV CALIF, BERKELEY, 72- *Concurrent Pos:* Vis lectr, Stanford Univ, 60; vis lectr, Univ Calif, Berkeley, 61-63, grad sch bus admin, 63-64, lectr, 66-72; mem comt future port develop, Maritime Transp Res Bd, Nat Acad Eng, 74-75. *Mem:* Fel Royal Statist Soc; Opers Res Soc; Inst Mgt Sci (vpres, 61-65). *Res:* Application of operations research to business, commercial and non-military government problems; queue theory; inventory theory; linear programming; transportation, distribution and energy development. *Mailing Add:* Schs Bus Admin Univ Calif Berkeley CA 94720

KOEPF, ERNEST HENRY, b Bruceville, Tex, Jan 23, 12; m 38; c 2. CHEMICAL ENGINEERING. *Educ:* Univ Tex, BS, 34, MS, 36, PhD(chem eng), 39. *Prof Exp:* Chem engr, Atlantic Refining Co, 39-52; vpres & gen mgr, Texas City Chem, Inc, 52-54; admin coordr, Crude Oil Producing Dept, Atlantic Refining Co, 54-55; mgr res & tech servs, Core Labs, Inc, 55-65, gen mgr Francorelab, 65-68, vpres res & tech servs, 68-78, vpres res & develop, 78-82. *Concurrent Pos:* Engr, AAAS, Interstate Oil Compact Comn Res Comt & Am Petrol Inst; pres, Ocean Pollution Control, Inc, 71-72, Ecol Audits, Inc, 72-78, P-V-T, Inc, 75-78 & Syndrill Carbide Diamond Co, 80-82; consult, 82- *Mem:* Am Chem Soc; Am Inst Mining, Metall & Petrol Engrs; Soc Independent Prof Earth Scientists. *Res:* High pressure phase behavior of hydrocarbons; petroleum reservoir operation; physical properties of oil reservoir rock and their contained fluids; distribution and flow of hydrocarbons in porous media; environmental protection. *Mailing Add:* 3607 Greenbrier Dr Dallas TX 75225

KOEPFINGER, J L, b Sewickley, Pa, May 6, 25; m; c 6. ELECTRICAL ENGINEERING. *Educ:* Univ Pittsburgh, BS, 49, MS, 53. *Prof Exp:* DIR, SYST & RES ELEC UTILITIES, DUQUESNE LIGHT CO, 85- *Honors & Awards:* Steinmetz Award. *Mem:* Fel Inst Elec & Electronics Engrs. *Mailing Add:* Duquesne Light Co 301 Grant St 19-5 Pittsburgh PA 15279

KOEPFLI, JOSEPH B, b Los Angeles, Calif, Feb 5, 04; m 35; c 2. ORGANIC CHEMISTRY. *Educ:* Stanford Univ, BA, 24, MA, 25; Oxford Univ, DPhil(chem), 28. *Prof Exp:* Instr pharmacol, Sch Med, Johns Hopkins Univ, 30-32; res assoc chem, 32-72, EMER SR RES ASSOC, CALIF INST TECHNOL, PASADENA, 72- *Res:* Alkaloids; phytohormones; antimalarials. *Mailing Add:* 580 Freehaven Dr Santa Barbara CA 93108

KOEPKE, BARRY GEORGE, b Detroit, Mich, Oct 27, 37. CERAMICS. *Educ:* Univ Ill, BS, 60, MS, 62; Iowa State Univ, PhD(metall), 68. *Prof Exp:* Res engr metall, Rocketdyne Div, NAm Aviation, 62-64; ceramics prog dir, NSF, 79-81; scientist, Honeywell Corp Res Ctr, Honeywell Inc, 64-79, prog mgr, Syst & Res Ctr, 81-83, res & develop, Ceramics Ctr, 83-85, opers mgr, 85-88, OPERS MGR & COMPONENTS, ALLIANT TECHSYSTS, 88- *Concurrent Pos:* Adj prof mat sci, Univ Minn, 76-; dir Ceramics Prog Div Mat Res, NSF, 80-81. *Mem:* fel Am Ceramic Soc; Am Soc Metals; Sigma Xi. *Res:* Studies of the mechanical properties and fracture behavior of ceramic materials, the nature and extent of surface damage introduced into dielectrics by machining and polishing; piezo electric ceramics; production of ceramics with tailored microstructures by unique processing techniques. *Mailing Add:* Alliant Techsysts 5121 Winnetka Ave N New Hope MN 55428

KOEPKE, GEORGE HENRY, b Toledo, Ohio, Jan 1, 16; m 40; c 2. MEDICINE. *Educ:* Univ Toledo, BS, 45; Univ Cincinnati, MD, 49; Am Bd Phys Med & Rehab, dipl, 55. *Prof Exp:* Intern, Toledo Hosp, 49-50; resident, Univ Mich Hosp, 50-52, instr phys med & rehab, Med Sch, Univ, 52-53; pvt pract, 53-54; from asst prof to prof phys med & rehab, Med Ctr, Univ Mich, Ann Arbor, 54-76; RETIRED. *Concurrent Pos:* Consult, Vet Admin Hosp, Ann Arbor, 55-75, Lapeer State Home & Training Sch, 64- & Mary Free Bed Hosp, Grand Rapids, 71-76; mem staff, Saginaw Community Hosp, St Mary's Hosp, Saginaw Gen Hosp & St Luke's Hosp; chmn, Am Bd Phys Med & Rehab, 76; Am Bd Phys Med & Rehab; emer mem prof adv coun, United Cerebral Palsy Asn; Am Bd Electrodiagnostic Med. *Mem:* Fel Am Acad Phys Med & Rehab; Am Asn Electromyog & Electrodiag; Am Acad Orthop Surg; Am Cong Rehab Med; AMA. *Res:* Physical medicine and rehabilitation; electromyography and prosthetics. *Mailing Add:* 2222 S Mainrson Findlay OH 45840

KOEPKE, JOHN ARTHUR, b Milwaukee, Wis, Mar 25, 29; m 55; c 4. CLINICAL PATHOLOGY, HEMATOLOGY. *Educ:* Valparaiso Univ, BA, 51; Univ Wis, MD, 56; Marquette Univ, MS, 64. *Prof Exp:* Instr path, Marquette Univ, 58-60; from asst prof to assoc prof, Univ Ky, 61-70; assoc clin prof med technol, Col Med, Univ Iowa, 70-71, prof path & vchmn dept, 72-79; PROF PATH & MED DIR, CLIN HEMAT LABS, DUKE UNIV MED CTR, 80- *Concurrent Pos:* From asst pathologist to assoc pathologist, Univ Ky Hosp, 61-71; attend pathologist, Vet Admin Hosp & consult, USPHS, 63-71; vis scientist, Karolinska Inst, Sweden, 67-68; dir lab, Lexington Clin, 71-72; attend pathologist, Univ Iowa Hosp & Clins, 72-; chief lab serv, Vet Admin Hosp, Iowa City, 72-78; vis colleague, Royal Postgrad Med Sch, London, Eng. *Mem:* Col Am Path; Am Soc Hemat; Am Soc Clin Path; Am Asn Blood Banks. *Res:* Blood coagulation; flow cytometry; immunohematology; quality assurance systems. *Mailing Add:* Duke Univ Med Ctr PO Box 2929 Durham NC 27710-2929

KOEPNICK, RICHARD BORLAND, b Dayton, Ohio, Feb 5, 44. SEDIMENTARY PETROLOGY. *Educ:* Univ Colo, BA, 67; Univ Kans, MS, 69, PhD(geol), 76. *Prof Exp:* Asst prof, Dept Geol, Williams Col, 75-77; res geologist, Mobil Field Res Lab, 77-84, assoc geol res, 85-86, RES ASSOC GEOL, DALLAS RES LAB, MOBIL RES DEVELOP CORP, 86- *Mem:* Soc Econ Paleontologists & Mineralogists; Am Asn Petrol Geologists; Sigma Xi. *Res:* Application of strontium isotope analysis to stratigraphic and diagenetic studies; diagenesis of carbonate rocks and sandstones. *Mailing Add:* 4249 Southcrest Rd Dallas TX 75229

KOEPP, LEILA H, b Haifa, Israel, July 7, 45; US citizen; m 69; c 2. SCIENCE EDUCATION. *Educ:* Messiah Col, BA, 68; NTex State Univ, MS, 70; Univ Med & Dent NJ, PhD(microbiol), 81. *Prof Exp:* Instr microbiol, Montclair State Col, 74-76, instr biol, anat & physiol, Fairleigh Dickson Univ, Madison, 76-77; asst med microbiol & researcher microbiol, Univ Med & Dent NJ, 78-81; ASSOC PROF BIOL, ANAT & PHYSIOL, BLOOMFIELD COL, NJ, 81- *Concurrent Pos:* Bacteriologist clin microbiol, genetics, immunol & gen biol, Overlook Hosp, Summit, NJ, 75-81. *Mem:* Am Soc Microbiol. *Res:* Molecular basis for the biological activity of the slime glycolipoprotein of Pseudomonas aeruginosa; isolation and identification of fungi from stone monuments of the New York Metropolitan Art Museum. *Mailing Add:* Dept Math & Natural Sci Bloomfield Col Bloomfield NJ 09003

KOEPP, STEPHEN JOHN, b Los Angeles, Calif, Apr 24, 46; m 69; c 2. ZOOLOGY, CYTOPATHOLOGY. *Educ:* Messiah Col, BA, 68; NTex State Univ, MS, 70, PhD(biol), 73. *Prof Exp:* Asst prof, 73-78, assoc prof biol, 79-85, PROF BIOL, MONTCLAIR STATE COL, 85- *Concurrent Pos:* Environ consult. *Mem:* Electron Micros Soc Am; Am Col Toxicol, 80- *Res:* Histopathologic and cytopathologic response of aquatic fauna following toxic exposure to heavy metals. *Mailing Add:* Dept Biol Montclair State Col Upper Montclair NJ 07043

KOEPPE, DAVID EDWARD, plant biochemistry, plant physiology; deceased, see previous edition for last biography

KOEPPE, JOHN K, b Beaver Dam, Wis, July 20, 44. INSECT PHYSIOLOGY, ENDOCRINOLOGY. *Educ:* Hope Col, BA, 67; Tulane Univ, PhD(biol), 71. *Prof Exp:* Fel insect physiol, Northwestern Univ, 71-73; res fel biochem, Roche Inst Molecular Biol, 73-75; ASST PROF ZOOL, UNIV NC, CHAPEL HILL, 75- *Mem:* Am Soc Zoologists; Entom Soc Am. *Res:* Molecular analysis of the mode of action of the juvenile hormone during insect vitellogenesis. *Mailing Add:* Dept Zool Univ NC Chapel Hill NC 27514

KOEPPE, MARY KOLEAN, b Holland, Mich, July 7, 55; m 78. ENVIRONMENTAL TOXICOLOGY. *Educ:* Hope Col, BA, 77; Univ Wis, Madison, MS, 79, PhD(entomol), 83. *Prof Exp:* SECT RES CHEMIST, E I DU PONT NEMOURS & CO INC, 83- *Mem:* Am Chem Soc. *Res:* Metabolism and environmental fate of new agricultural chemicals in plants, animals and soil; elucidation of the metabolic pathway of agricultural chemicals in plants and animals; pesticide metabolism in plants and animals. *Mailing Add:* DuPont Co Exp Sta Bldg 402/5328 PO Box 80402 Wilmington DE 19880-0402

KOEPPE, OWEN JOHN, b Cedar Grove, Wis, May 29, 26; m 50; c 3. BIOCHEMISTRY. *Educ:* Hope Col, AB, 49; Univ Ill, MS, 51, PhD(biochem), 53. *Prof Exp:* Asst chem, Univ Ill, 49-51, asst biochem, 51-52; USPHS res fel, Univ Minn, 53-55; from asst prof to assoc prof biochem, Sch Med, Univ Mo, Columbia, 55-61, chmn dept, 68-73, prof, 61-, provost acad affairs, 73-80; provost, 80-87, PROF BIOCHEM, KANS STATE UNIV, 80- *Mem:* Am Chem Soc; Am Soc Biochem & Molecular Biol. *Res:* Mechanism of enzyme action; peptide bond synthesis. *Mailing Add:* Dept Biochem Kans State Univ Manhattan KS 66506

KOEPPE, ROGER E, II, b Champaign, Ill, July 1, 49. CHEMISTRY, BIOCHEMISTRY. *Educ:* Haverford Col, BA, 71; Calif Inst Technol, PhD(chem & biochem), 76. *Prof Exp:* NIH postdoctoral fel struct biol, Sch Med, Stanford Univ, 76-79; from asst prof to assoc prof, 79-87, PROF, DEPT CHEM & BIOCHEM, UNIV ARK, 87- *Concurrent Pos:* Guest asst scientist, Brookhaven Nat Lab, 80-86; vis assoc, Calif Inst Technol, 85-86. *Honors & Awards:* Harold Lamport Lectr, Cornell Univ, 87. *Mem:* Fedn Am Socs Exp Biol. *Res:* Mechanism of ion transport through membrane channels; physicochemical studies of proteins and nucleic acids; author of 39 technical publications. *Mailing Add:* Dept Chem & Biochem Univ Ark 115 Chem Bldg Fayetteville AR 72701

KOEPPE, ROGER ERDMAN, b Amoy, China, May 2, 22; m 47; c 5. BIOCHEMISTRY. *Educ:* Hope Col, AB, 44; Univ Ill, MS, 47, PhD(biochem), 50. *Prof Exp:* Asst, Univ Ill, 46-48 & 50-51; res assoc, Univ Tenn, 51-52, from instr to assoc prof chem, 52-59; from assoc prof to prof, 59-90, head dept, 63-90, EMER PROF & HEAD, BIOCHEM, OKLA STATE UNIV, 90- *Concurrent Pos:* NIH sr fel, Univ Pa, 66-67; Sigma Xi lectr, 74. *Mem:* AAAS; Am Chem Soc; Am Soc Biol Chemists; Brit Biochem Soc. *Res:* Metabolism, including enzymology, of acetate, mannose, pyruvate and glutamate in mammalian brain and liver; neurochemistry. *Mailing Add:* Dept Biochem Okla State Univ Stillwater OK 74074

KOEPPEN, BRUCE MICHAEL, b Oct 7, 51; m; c 2. ELECTROPHYSIOLOGY, ION TRANSPORT. *Educ:* Univ Chicago, MD, 77; Univ Ill, PhD(physiol), 80. *Prof Exp:* ASSOC PROF MED & PHYSIOL, HEALTH CTR, UNIV CONN, 82- *Mem:* Am Soc Nephrology; Am Physiol Soc; Am Biophys Soc; Int Soc Nephrol; Soc Gen Physiol. *Mailing Add:* Dept Med Sch Med Univ Conn Health Ctr Farmington CT 06030

KOEPPEN, ROBERT CARL, botany, research administration, for more information see previous edition

KOEPPL, GERALD WALTER, b Chicago, Ill, Dec 4, 42; m; c 2. CHEMICAL PHYSICS. *Educ:* Ill Inst Technol, BSc, 65, PhD(chem), 69. *Prof Exp:* NIH res fel chem, Harvard Univ, 69-70, res fel, 70; from instr to assoc prof, 71-88, PROF CHEM, QUEENS COL, NY, 89- *Concurrent Pos:* Proj dir res grant, Res Found City Univ New York, 71-89; Sloan Found fel, 75-79. *Mem:* Am Chem Soc; NY Acad Sci; Am Phys Soc; Fedn Am Scientists. *Res:* Classical mechanical trajectory studies of the statistical theories of chemical reaction dynamics. *Mailing Add:* Dept Chem & Biochem Queens Col City Univ NY Flushing NY 11367

KOEPSEL, WELLINGTON WESLEY, b McQueeney, Tex, Dec 5, 21; m 50; c 3. ELECTRICAL ENGINEERING. *Educ:* Univ Tex, BS, 44, MS, 51; Okla State Univ, PhD(elec eng), 60. *Prof Exp:* Res scientist elec eng, Univ Tex, 48-51; res engr aerophys lab, NAm Aviation, Inc, 51; asst prof, Southern Methodist Univ, 51-56 & Okla State Univ, 56-58; assoc prof, Southern Methodist Univ, 58-59, Univ NMex, 60-63 & Duke Univ, 63-64; head dept, 64-76, prof elec eng, Kans State Univ, 64-; AT MUTRONIL SYSTS. *Mem:* Sr mem Inst Elec & Electronics Engrs; Nat Soc Prof Engrs. *Res:* Feedback control systems; microcomputer and digital control systems. *Mailing Add:* PO Box 459 Port Aransas TX 78373

KOEPSELL, PAUL L(OEL), b Canova, SDak, June 17, 30; m 52; c 3. CIVIL ENGINEERING. *Educ:* SDak State Univ, BS, 52; Univ Wash, Seattle, MS, 54; Okla State Univ, PhD, 65. *Prof Exp:* Stress analyst, Aircraft Structures, Boeing Airplane Co, 52-57; asst prof civil eng, 58, assoc prof, 59-65, dir res

& data processing, 65-76, PROF CIVIL ENG, SDAK STATE UNIV, 67-, DIR COMPUT CTR, 76-, HEAD, DEPT COMPUT SCI, 81- Mem: Am Soc Civil Engrs; Am Concrete Inst; Nat Soc Prof Engrs (vpres, 79-81); Asn Comput Mach. Res: Structural components; approximate analysis of structures; application of matrix methods to structural analysis. Mailing Add: Box 2219 Univ Station Brookings SD 57007

KOERBER, GEORGE G(REGORY), b Akron, Ohio, Aug 30, 24; m 51; c 3. ELECTRICAL ENGINEERING. Educ: Hiram Col, BA, 48; Purdue Univ, MS, 50, PhD(phys chem), 52. Prof Exp: Mem tech staff component develop, Bell Tel Labs, Inc, 52-56; assoc prof mech, Rensselaer Polytech Inst, 56-58; assoc prof eng mech, Mich Col Mining & Technol, 58-59; assoc prof theoret & appl mech, Iowa State Univ, 60-61, from assoc prof to prof elec eng, 61-84; RETIRED. Mem: Inst Elec & Electronics Engrs. Res: Applied mathematics; properties of solids. Mailing Add: 206 Ridge View Dr Sequim WA 98382

KOERBER, WALTER LUDWIG, b Berlin, Ger, Aug 3, 12; US citizen; m 37; c 1. BIOCHEMISTRY, MICROBIOLOGY. Educ: Univ Frankfurt, PhD(org chem, bact), 36. Prof Exp: Fel, Sch Hyg & Trop Med, Univ London, 36-37; asst head dept bact, E R Squibb & Sons, 39-42 & Mfg Pilot Plant, 42-44, head microbiol develop, Squibb Inst Med Res, 44-51, tech adv, Squibb-Rome, 52-53, tech dir, Squibb-Sao Paulo, 54-55, dir res & develop, 56-57, dir & regional coordr, 57-58, dir int res, Squibb Inst Med Res, 58-63, coordr, 63-70, dir, 65-69, dir sci affairs, Squibb Europe Ltd, 70-77; consult, Revlon Health Care Group, 78-84; RETIRED. Concurrent Pos: Consult to Badger Ltd, London, Govt Pakistan, 64. Mem: Am Chem Soc; Am Asn Immunol; Am Soc Microbiol. Res: Fermentations; biologicals; natural products; development of pharmaceuticals. Mailing Add: Dr Lechner-Weg 17 Mondsee A 5310 Austria

KOERING, MARILYN JEAN, b Brainerd, Minn, Jan 7, 38. ANATOMY, REPRODUCTIVE PHYSIOLOGY. Educ: Col St Scholastica, BA, 60; Univ Wis-Madison, MS, 63, PhD(anat), 67. Prof Exp: Res asst chem, Col St Scholastica, 60-61; instr anat, Univ Wis, 63-64; res trainee reproductive physiol, Ore Primate Res Ctr, 66-67; NIH fel, Primate Res Ctr, 67-68, proj assoc, Univ Wis- Madison, 68-69; asst prof, 69-73, assoc prof, 73-79, PROF ANAT, MED CTR, GEORGE WASH UNIV, 79- Concurrent Pos: Vis scientist, div biol, Calif Inst Technol, 76, Jones Inst Reproductive Med, E Va Med Sch, Norfolk, 85-; guest worker, Pregnancy Res Br, Nat Inst Child Health & Human Develop, 77-85; mem primate res adv bd, NIH, 78-82. Mem: AAAS; Am Asn Anatomists; Electron Micros Soc Am; Soc Study Reproduction. Res: Cyclic changes in ovarian morphology as observed in light and electron microscopy; correlation of reproductive morphology with physiology. Mailing Add: Dept Anat George Washington Univ Med Ctr Washington DC 20037

KOERKER, FREDERICK WILLIAM, b Milwaukee, Wis, June 9, 13; m 37; c 3. INORGANIC CHEMISTRY. Educ: Univ Wis, BS, 34, MS, 36. Prof Exp: Asst chem, Univ Wis, 34-36; analyst, Main Lab, 36-37, chemist, Chlorine Dept, 37-41, asst supt, 41-53, tech expert electrochem planning, 53-62, tech expert, Chem Dept, 62-68, mgr qual standards, 68-78, consult, Dow Chem Co, 78-82; RETIRED. Mem: Am Chem Soc; Electrochem Soc. Res: Electrolytic production of chlorine, caustic and allied products; specifications and methods of analysis for general chemicals. Mailing Add: Judson Park Retirement Community 23600 Marine View Dr S Apt 233 Des Moines WA 98198

KOERKER, ROBERT LELAND, b Saginaw, Mich, Jan 10, 43. PHARMACOLOGY. Educ: Kalamazoo Col, BA, 65; Emory Univ, PhD(pharmacol), 70. Prof Exp: Asst biologist microbiol, Biochem Res Lab, Dow Chem Co, 63-64, asst chemist, Dept Chem Res Lab, 65; NIH res fel pharmacol, Univ Colo Med Ctr, 70-73; USPHS res fel, Emory Univ, 73-74; asst prof pharmacol, La State Univ Med Ctr, 74-80; ASSOC PROF PHARMACOL, MED SCH, WRIGHT STATE UNIV, 80- Mem: Tissue Cult Asn; Sigma Xi; AAAS; Am Soc Pharmacol & Exp Therapeut; Soc Toxicol. Res: Toxicity of aldehydes, alcohols, organo-mercurials and other agents in cultured mouse neuroblastoma cells; characterization of uptake and storage mechanisms in cultured mouse neuroblastoma cells. Mailing Add: Sch Med Wright State Univ Dayton OH 45435

KOERNER, E(RNEST) L(EE), b Cleveland, Ohio, Mar 17, 31; m 53; c 6. CHEMICAL ENGINEERING. Educ: Univ Dayton, BChE, 53; Iowa State Col, MS, 55, PhD(chem eng), 56. Prof Exp: Asst chem eng, Ames Lab, AEC, Iowa State Col, 53-56; res engr & sect leader extractive processes, Union Carbide Metals Co, 57-59; res specialist, Monsanto Co, 59-67; sr res specialist, Kerr-McGee Corp, 67, sr res group leader, 67-70; pres, Technol Res & Develop Inc, 70-83; PRES, TECHRAD INC, 83- Mem: Am Inst Chem Engrs; Am Inst Mining, Metall & Petrol Engrs; Nat Soc Prof Engrs. Res: Extractive metallurgical processes; liquid-liquid extraction; hazard wastes treatment; energy recovery from solid wastes; biological and enzyme treatment of waste waters. Mailing Add: 12721 St Andrews Terrace Oklahoma City OK 73120-8807

KOERNER, HEINZ, aquatic biology, ecology, for more information see previous edition

KOERNER, JAMES FREDERICK, b Charles City, Iowa, June 30, 29; m 58; c 2. BIOCHEMISTRY, NEUROSCIENCE. Educ: Iowa State Col, BS, 50, PhD(biochem), 56. Prof Exp: Res assoc biochem, Iowa State Col, 50-52, asst, 52-56; USPHS fel, Mass Inst Technol, 56-58, res assoc, 58-61; from asst prof to assoc prof, 61-72, PROF BIOCHEM, SCH MED, UNIV MINN, MINNEAPOLIS, 72- Mem: AAAS; Am Soc Biol Chemists; Soc Neurosci; Am Soc Neurochem. Res: Neurochemistry; acidic amino acids as excitatory neurotransmitters; glutamate metabolism. Mailing Add: Dept Biochem Univ Minn Med Sch 435 Delaware St S E Minneapolis MN 55455

KOERNER, ROBERT M, b Philadelphia, Pa, Dec 2, 33; m 59; c 3. SOIL MECHANICS. Educ: Drexel Inst Technol, BSCE, 56, MSCE, 63; Duke Univ, PhD(soil mech), 68. Prof Exp: Engr & supt, Conduit & Found Corp, 56-60; engr analyst, Dames & Moore, 60-62; engr & supt, J J Skelly, Inc, 62-63; instr, Pa Mil Col, 64-65; NSF teaching intern, Duke Univ, 65-67, instr, part-time, 67-68; from asst prof to assoc prof, 68-76, PROF CIVIL ENG, DREXEL UNIV, 76- Mem: Am Soc Civil Engrs; Am Soc Eng Educ. Res: Foundation engineering; particle mechanics; powder metallurgy. Mailing Add: Dept Civil Eng Drexel Univ 32nd & Chestnut St Philadelphia PA 19104

KOERNER, T J, b Rochester, NY, June 15, 54; m 78; c 3. FACILITATE PEER REVIEW. Educ: Univ Toronto, Ont, Can, BSc, 76; Univ Cincinnati, PhD(develop biol), 85. Prof Exp: Postdoctoral researcher, Columbia Univ, New York, 82-85; asst med res prof, Duke Univ Med Ctr, 85-88; SCI PROG DIR, AM CANCER SOC, 88- Mem: Am Soc Biochem & Molecular Biol; Genetics Soc Am; Am Asn Cancer Res; Am Soc Microbiol; AAAS. Res: Administration of the peer review of applications and oversight of awards in the areas of molecular biology, immunology, and virology. Mailing Add: Res Dept Am Cancer Soc 1599 Clifton Rd NE Atlanta GA 30329-4251

KOERNER, THEODORE ALFRED WILLIAM, JR, b Waco, Tex, July 30, 47; m 71; c 2. BLOOD TRANSFUSION MEDICINE, CELL MEMBRANE BIOCHEMISTRY. Educ: La State Univ, BS, 70, PhD(biochem), 75; Tulane Univ, MD, 78. Prof Exp: Intern path, Yale-New Haven Hosp, 78-79, resident physician, 79-81; fel lab med, Yale Univ Sch Med, 81-82; asst prof path, Tulane Univ Sch Med, 82-86; asst prof, 86-88, ASSOC PROF PATH, UNIV IOWA COL MED, 88- Concurrent Pos: Acad Clin Lab Physicians & Scientists young investr award, 82; adj asst prof, Dept Biochem, Tulane Univ Sch Med & med dir, Blood Ctr & Apheresis Serv, Tulane Univ Hosp, 82-86; asst med dir, DeGowin Blood Ctr, Univ Iowa Hosp & Clins, 86-88, assoc med dir, 88-; legal expert, blood-borne acquired immune deficiency syndrome, Law Off Meis & Waite, San Francisco, 89- Mem: Am Chem Soc; Am Asn Blood Banks; Am Soc Biochem & Molecular Biol; Am Asn Pathologists; Am Soc Hemat. Res: Structure, function, immunology and pathophysiology of blood platelet membrane glycoproteins, glycolipids and phospholipids; oligosaccharide and lipid chemistry; high pressure liquid chromatography and multidimensional nuclear magnetic resonance spectroscopy; transfusion medicine; storage and survival of platelets; transfusion-transmitted acquired immune deficiency syndrome and other infectious diseases. Mailing Add: Dept Path Med Res Ctr 143 Univ Iowa Col Med Iowa City IA 52242

KOERNER, WILLIAM ELMER, b Neenah, Wis, Nov 3, 23; m 47; c 2. PHYSICAL CHEMISTRY, ANALYTICAL CHEMISTRY. Educ: Univ Wis, BS, 46, PhD(phys chem), 49. Prof Exp: Res phys chemist, 49-54, group leader phys chem, 54-64, mgr, Phys Sci Ctr, 64-80, dir phys sci, 80-86, FEL PROG COORDR, MONSANTO CO, 86- Concurrent Pos: Dual tech ladder consult & coordr, 80- Mem: AAAS; Soc Appl Spectros; Am Chem Soc; Sigma Xi. Res: Chemical reaction kinetics; thermochemistry; physical analytical chemistry. Mailing Add: 5642 Murdoch Ave St Louis MO 63109

KOERTING, LOLA ELISABETH, b Munich, Ger, Jan 31, 24; US citizen; m 53; c 2. GENETICS, CYTOLOGY. Educ: Munich Tech Univ, BS, 47, MS, 49, PhD(biol, agr), 53. Prof Exp: Res asst plant genetics, Munich Tech Univ, 48-53; chief histol & ultrasonics, New Eng Inst Med Res, 54-55; res assoc forest genetics, Sch Forestry, Yale Univ, 55-58; res fel cytogenetics, New Eng Inst Med Res, 65-73; CONSULT GENETICIST, KITCHAWAN RES LAB, BROOKLYN BOT GARDEN, NY, 75- Mem: AAAS; Environ Mutagen Soc; Bot Soc Am; Sigma Xi; Tissue Cult Asn. Res: Cytogenetics; induced chromosome abnormalities and mutations; tissue culture. Mailing Add: PO Box 551 Ridgefield CT 06877

KOESTER, CHARLES JOHN, b Niagara Falls, NY, Jan 26, 29; m 53; c 4. OPHTHALMIC OPTICS, MICROSCOPY. Educ: Carnegie Inst Technol, BS, 50; Univ Rochester, PhD(physics, optics), 55. Prof Exp: Asst, Univ Rochester, 50-55; physicist, Am Optical Co, 55-58; res assoc, Nat Bur Standards, 58-59; physicist, Am Optical Corp, 59-65, appl res mgr, 65-75, dir res, Sci Instrument Div, 75-77; asst prof biophys ophthal, Columbia Univ Col Physicians & Surgeons, 78-84, assoc prof clin biophys ophthal, 84-91; RETIRED. Concurrent Pos: Lectr ophthal, Columbia Univ, 70-; indust rep, Food & Drug Admin Ophthalmic Device Classification Panel, 74-78; mem bd dirs, Optical Soc Am, 74-76. Mem: Optical Soc Am; Asn Res Vision & Ophthal; Am Acad Ophthal; Am Soc Cataract Refractive Surg; Am Acad Optom. Res: Ophthalmic instruments and microscopes; interference, polarizing, and confocal microscopes; image enhancement in fiber optics; laser photocoagulation; microscopy of the cornea; intraocular lenses. Mailing Add: 60 Kent Rd Glen Rock NJ 07452

KOESTER, JOHN D, BEHAVIOR, BIOPHYSIOLOGY. Educ: Columbia Univ, PhD(physiol), 71. Prof Exp: ACTG DIR, CTR NEURO-BIOL & BEHAV, SCH MED, COLUMBIA UNIV, 74- Mailing Add: NY State Psychiat Inst Columbia Univ 722 W 168 St New York NY 10032

KOESTER, LOUIS JULIUS, JR, experimental high energy physics, nuclear physics; deceased, see previous edition for last biography

KOESTERER, MARTIN GEORGE, b Rochester, NY, July 2, 33; m 58; c 4. MICROBIOLOGY. Educ: Univ Rochester, AB, 55; Syracuse Univ, MS, 57. Prof Exp: Res supvr sterilization, Wilmot Castle Co, 59-64; sr microbiologist & prog mgr sterilization & planetary quarantine, Valley Forge Space Technol Ctr, Gen Elec Co, 65-75; mgr microbiol res, Ethicon Inc, 75-77; CONSULT INDUST, 77-; MEM STAFF MICROBIOL & STERILIZATION TECHNOL, WYETH LABS, INC, 78- Concurrent Pos: Spec lectr, Grad Sch Environ Sci, Drexel Univ, 69-70; guest lectr, Contamination Control Sem, Rochester Inst Technol, 71-75; consult microbial contamination control & sterilization, 69- Honors & Awards: NASA Tech Brief, 72. Mem: Am Soc Microbiol; Inst Environ Sci; Soc Indust Microbiol; Am Soc Testing & Mat. Res: Resistance of bacterial spores to dry heat, moist heat, irradiation, various chemical agents; bioburden and microbial contamination control as it pertains to good manufacturing practices on pharmaceutical, biomedical devices and products; biological indicator development; sterilization development and validation. *Mailing Add:* Wyeth Labs, Inc PO Box 565 West Chester PA 19380

KOESTLER, ROBERT CHARLES, b Elizabeth, NJ, Oct 31, 32; m 58; c 3. PESTICIDE CHEMISTRY. *Educ:* Cornell Univ, BA, 54; Univ NC, Chapel Hill, PhD(org chem), 61. *Prof Exp:* Res chemist, Am Viscose Div, FMC Corp, 61-65; sr res chemist, Pennwalt Corp, 65-77, proj leader, 77-83, res scientist, 83-86; SR RES SCIENTIST, ATOCHEM NAM, 86- *Concurrent Pos:* Mem, Gov Bd, Int Symposium Controlled Release Bioactive Mat. *Mem:* Am Chem Soc; Int Controlled Release Soc (treas). *Res:* Organometallics; films and coatings; organic synthesis; microencapsulation and controlled release technology. *Mailing Add:* 2004 Pebblestone Court College Station TX 77845

KOESTNER, ADALBERT, b Hatzfeld, Rumania, Sept 10, 20; US citizen; m 51; c 2. VETERINARY PATHOLOGY. *Educ:* Univ Munich, DMV, 51; Ohio State Univ, MSc, 57, PhD, 59. *Prof Exp:* Res assoc bact, Vet Col, Univ Munich, 51-52; from instr to assoc prof, 55-64, chmn, Dept Vet Pathobiol, 72-81, PROF VET PATH, OHIO STATE UNIV, 64-; CHMN PATH, MICH STATE UNIV, 81- *Mem:* AAAS; Am Asn Pathologists; Soc Neurosci; Am Vet Med Asn; Am Asn Cancer Res; Am Asn Neuropath; Am Col Vet Path. *Res:* Comparative neuropathology; comparative and experimental oncology. *Mailing Add:* Dept Path A622 E Fee Hall Mich State Univ East Lansing MI 48824

KOETHE, SUSAN M, b San Diego, Calif, Sept 4, 45; m 68; c 1. IMMUNOLOGY. *Educ:* San Diego State Col, BS, 67; Harvard Univ, PhD(immunol), 74. *Prof Exp:* NIH fel, 74-76, asst prof, 76-81, ASSOC PROF PATH, MED COL WIS, 81- *Concurrent Pos:* Chmn, Milwaukee Immunol Group, 78-82. *Mem:* Am Asn Immunol. *Res:* Immunoregulation in myasthenia gravis and multiple sclerosis. *Mailing Add:* Med Col Wis 8700 W Wisconsin Ave Med Col Wis 8701 W Watertown Plank Rd Milwaukee WI 53226

KOETKE, DONALD D, b Chicago, Ill, Dec 12, 37; m 59; c 3. NUCLEAR STRUCTURE. *Educ:* Concordia Col, Ill, BS, 59; Northwestern Univ, Ill, MS, 63, PhD(physics), 68. *Prof Exp:* Assoc prof physics, Concordia Col, Ill, 67-77; PROF PHYSICS, VALPARAISO UNIV, 77- *Concurrent Pos:* Vis scientist, Argonne Nat Lab, 69 & 71-82; Los Alamos Nat Lab, 82-; consult, Int Atomic Energy Agency, 80- *Mem:* Am Phys Soc; Am Asn Physics Teachers; Sigma Xi. *Res:* Experiments in muon and neutrino physics done at Los Alamos; low energy nuclear cross-section measurements relative to solar neutrino production. *Mailing Add:* Dept Physics Valparaiso Univ Valparaiso IN 46383

KOETSCH, PHILIP, b Vanceburg, Ky, Nov 25, 35; m 57; c 4. RELIABILITY ANALYSIS. *Educ:* Univ SC, BS, 59, MS, 60. *Prof Exp:* Engr, Westinghouse, 60-63; sr engr, Lockhead, 63-66; consult, 66-70; supvr, Honeywell, 70-71; mgr, Nat Cash Register, 71-76; chief engr, ACDC Electronics, 76-80; VPRES ENG, POWERTEC, 80- *Mem:* Inst Elec & Electronics Engrs; Am Elec Asn. *Res:* Solid state power conversion technology developments ranging from 5-100,000 watts involving semiconductor and high-frequency magnetics in several architectures; off-line switchmode converter design. *Mailing Add:* Astec America Inc 401 Jones Rd Oceanside CA 92054

KOETZLE, THOMAS F, b Brooklyn, NY, Oct 15, 43; m 67; c 2. CHEMICAL CRYSTALLOGRAPHY, NEUTRON AND X-RAY DIFFRACTION. *Educ:* Harvard Univ, BA, 64, MA, 65, PhD(chem), 70. *Prof Exp:* Res assoc chem, Brookhaven Nat Lab, 70-73, assoc chemist, 73-75, chemist, 75-90, SR CHEMIST, BROOKHAVEN NAT LAB, 90- *Concurrent Pos:* AEC fel, 70-71; NIH fel, 71-73; prin investr, The Protein Data Bank, 75-; mem, US Nat Comt Crystallog, 84-86 & 88-90; chmn, Neutron Scattering Spec Interest Group, Am Crystallog Asn, 83-84; ed, Molecular Struct Biol, 88- *Mem:* AAAS; NY Acad Sci; Am Chem Soc; Am Crystallog Asn. *Res:* Applications of neutron and x-ray diffraction to the analysis of molecular structure and chemical bonding; macromolecular structure databases. *Mailing Add:* Dept Chem Brookhaven Nat Lab Upton NY 11973-5000

KOEVENIG, JAMES L, b Postville, Iowa, Mar 18, 31; m 54, 85; c 2. BOTANY, BIOLOGY. *Educ:* Univ Iowa, BA, 55, PhD(sci educ & bot), 61; State Col Iowa, MA, 57. *Prof Exp:* Elem sch teacher, 55-56; res assoc, Univ Iowa, 61; asst prof zool, San Diego State Col, 61-62; res consult, Biol Sci Curric Study, Univ Colo, 62-64; from assoc prof to prof bot & biol, Univ Kans, 64-72; PROF BIOL, UNIV CENT FLA, 72- *Concurrent Pos:* Vis lectr, Univ Colo, 63-64; mem eval panel, Comn Undergrad Educ in Biol, 65; NSF sci fac fel, Princeton Univ, 67-68; United Nations Educ Sci & Cult Orgn Panel on Short Biol Films, 64; Fac Res Partic, Savannah River Ecol Lab, 78. *Mem:* AAAS; Bot Soc Am; Mycol Soc Am; Nat Asn Biol Teachers; Am Inst Biol Sci; Sigma Xi. *Res:* Development of nematodes; plant response to nematodes; plant growth and development; science visual aid and evaluation. *Mailing Add:* Dept Biol Univ Cent Fla Box 25000 Orlando FL 32816

KOFF, BERNARD LOUIS, b Huntington, NY, Mar 24, 27. GAS TURBINE. *Educ:* Clarkson Univ, BS, 51; New York Univ, MS, 58. *Prof Exp:* Test engr, Gen Elec, 51-52; design engr, Fairchild, 52-56; design engr, Curtis Wright, 56-58; design engr, Evendale, 58-62, sr engr, 62-65, mgr, 65-73, gen mgr, 73-75, chief engr, Gen Elec, 75-80; sr vpres, Govt Prod Div, Pratt & Whitney, 80-83, Eng Div, 83-87 & Govt Eng Bus, 87-90, EXEC VPRES, GROUP ENG & TECHNOL, PRATT & WHITNEY, 90- *Concurrent Pos:* Mem, Sci Adv Bd, USAFm 86-, Space Div Adv Bd, 88- *Honors & Awards:* R Tom Sawyer Award, Am Soc Mech Engr, 88; Theodore von Karman Award, Air Force Asn, 88; Reed Aeronaut Award, 90. *Mem:* Nat Acad Eng; fel Am Industr Arts Asn; fel Am Soc Mech Engrs; fel Am Inst Aeronaut & Astronaut; fel Soc Automotive Engrs. *Res:* Gas turbine industry; author of 11 technical publications; awarded 13 patents. *Mailing Add:* Pratt & Whitney PO Box 109600 West Palm Beach FL 33410-9600

KOFF, RAYMOND STEVEN, b Brooklyn, NY, June 11, 39; m 60; c 2. INTERNAL MEDICINE, GASTROENTEROLOGY. *Educ:* Adelphi Col, BA, 58; Albert Einstein Col Med, MD, 62; Am Bd Internal Med, dipl. *Prof Exp:* Intern med, Barnes Hosp, Washington Univ, 62-63, asst resident, 63-64; teaching fel, Tufts Univ & Lemuel Shattuck Hosp, 64-65, res fel, 65-66; clin & res fel, Mass Gen Hosp, Harvard Med Sch, 66-68, res fel, 68-69; from asst prof to assoc prof, 69-78, PROF, SCH MED, BOSTON UNIV, 79- *Concurrent Pos:* NIH trainee gastroenterol, Mass Gen Hosp, 66-69; clin investr, Vet Admin, 69-72, chief hepatology sect, 73-86, chmn med, Framingham Union Hosp, 86- *Mem:* Soc Epidemiol Res; Am Asn for Study Dis of Liver; fel Am Col Physicians; Am Gastroenterol Asn; Int Asn Study Liver. *Res:* Viral hepatitis; drug hepatotoxicity; chronic hepatitis. *Mailing Add:* Framingham Union Hosp 115 Lincoln St Framingham MA 01701

KOFFLER, DAVID, b New York, NY, Mar 28, 34; c 3. IMMUNOPATHOLOGY. *Educ:* State Univ NY, MD, 58. *Prof Exp:* Res fel, Mt Sinai Hosp, 59-60; NIH trainee, 60-63; instr path, Col Physicians & Surgeons, Columbia Univ, 63-66; from asst prof to assoc prof path, Mt Sinai Sch Med, 66-69, asst dean acad affairs, 66-72, prof, 69-76; PROF PATH & DIR LAB MED, HAHNEMANN MED COL & HOSP, 76- *Concurrent Pos:* NIH spec res fel, 63-66; asst attend pathologist, Mt Sinai Hosp, 63-71, attend pathologist, 71-; guest investr, Rockefeller Univ, 66-78, adj prof, 78- *Mem:* Am Soc Clin Invest; Am Asn Path; Am Soc Exp Path; Am Asn Nephrology; Am Rheumatism Asn; Am Asn Immunol. *Res:* Mechanisms of tissue injury in SLE. *Mailing Add:* Dept Path Hahnemann Univ Broad & Vine Sts Philadelphia PA 19102-1192

KOFFLER, HENRY, b Vienna, Austria, Sept 17, 22; nat US; m 46. MICROBIOLOGY. *Educ:* Univ Ariz, BS, 43; Univ Wis, MS, 44,PhD(bact), 47. *Prof Exp:* From asst prof to assoc prof bact, Purdue Univ, West Lafayette, 47-52, coordr res, 49-59, prof biol, 52-74, asst to dean grad sch, 57-59, asst dean, 59-60, head dept biol sci, 59-75, F L Hovde Distinguished prof, 74-75; prof biochem & microbiol & vpres acad affairs, Univ Minn, Minneapolis, 75-79; prof biochem & microbiol, Univ Mass, Amherst, 79-82; PROF BIOCHEM & MICROBIOL & PRESIDENT, UNIV ARIZ, 82- *Concurrent Pos:* Guggenheim fel, Sch Med, Case Western Reserve Univ, 53-54; mem, Comn Undergrad Educ in Biol Sci, 66-69, vchmn, 66-67, chmn, 67-69; mem, Purdue Res Found, 67-; consult-examr, NCent Asn Cols, 67-; mem, 2nd-7th Int Cong Biochem, Paris, Brussels, Vienna, Moscow & Tokyo; mem, 6th-8th & 10th Int Cong Microbiol, Rome, Stockholm, Montreal & Mexico City; mem, 9th & 11th Int Bot Cong, Montreal & Seattle; mem, 1st-3rd Int Biophys Cong, Stockholm, Vienna & Boston; mem, 5th Int Cong Electron Micros, Philadelphia, 16th Int Zool Cong, Washington, DC, 4th Int Cong Chemother, Washington, DC, 24th Int Cong Physiol Sci, 1st Int Cong Bact, Jerusalem & 1st Int Cong Int Asn Microbiol Soc, Tokyo. *Honors & Awards:* Eli Lilly & Co Award, 57. *Mem:* Am Soc Biol Chemists; Biophys Soc; Am Soc Microbiol; fel Am Acad Microbiol; Am Soc Cell Biologists. *Res:* Biosynthesis of carbohydrates; chemistry, biosynthesis and mechanism of action of antibiological peptides; structure and biosynthesis of flagellin and bacterial flagella; self-assembly of macromolecular structures; molecular bases for biological stability. *Mailing Add:* 6880 Firenze Dr Tuscon AZ 85704

KOFFMAN, ELLIOT B, b Boston, Mass, May 7, 42. COMPUTER SCIENCE EDUCATION, ARTIFICIAL INTELLIGENCE. *Educ:* Mass Inst Technol, SBEE & SMEE, 64, PhD(eng), 67. *Prof Exp:* Assoc prof, 74-77, PROF COMPUTER SCI, TEMPLE UNIV, 77- *Mem:* Asn Comput Mach. *Mailing Add:* Dept Computer & Info Sci Temple Univ 038-24 Broad & Montgomery Philadelphia PA 19122

KOFFYBERG, FRANCOIS PIERRE, b Eindhoven, Neth, Nov 17, 34; m 59; c 3. SOLID STATE SCIENCE. *Educ:* Free Univ, Amsterdam, Drs, 59. *Prof Exp:* Nat Res Coun Can fel, 59-62; res chemist, Corning Glass Works, NY, 62-65; assoc prof chem, 65-67, actg head physics, 67-68, assoc prof, 68-74, PROF PHYSICS, BROCK UNIV, 74- *Mem:* AAAS; Am Asn Physics Teachers; Can Asn Physicists. *Res:* Semiconductivity of oxides and glasses; photo-electronic properties. *Mailing Add:* Dept Physics Brock Univ St Catharines ON L2S 3A1 Can

KOFLER, RICHARD ROBERT, b Milwaukee, Wis, July 4, 35; m 59; c 3. PHYSICS. *Educ:* Marquette Univ, BS, 58; Univ Wis, MS, 60, PhD(elem particle physics), 64. *Prof Exp:* Res assoc physics, Univ Wis, 64-65; asst prof, 65-69, assoc prof physics, Univ Mass, Amherst, 69-; AT LAWRENCE BERKELEY LAB. *Mem:* Am Inst Physics; Am Phys Soc. *Res:* High energy elementary particle physics. *Mailing Add:* Bldg 50B-6208 Lawrence Berkeley Lab Berkeley CA 94720

KOFOID, MELVIN J(ULIUS), b Portland, Ore, July 16, 10; m 41; c 2. ELECTRICAL ENGINEERING. *Educ:* Ore State Col, BS, 33, MS, 35, PhD(elec eng), 42. *Prof Exp:* Elec & hydraul engr, Bingham Pump Co, Ore, 35-37; res engr lamp div, Westinghouse Elec Co, NJ, 37, res lab, Pa, 38-46; assoc prof elec eng, Ore State Col, 46-50; res specialist, 50-58, STAFF MEM, SCI RES LABS, BOEING AIRPLANE CO, 58- *Mem:* Fel Inst Elec & Electronics Engrs; Am Phys Soc. *Res:* Plasma physics; gaseous electrical conductors; dielectrics; high voltage insulation. *Mailing Add:* 18625 Beverly Rd Seattle WA 98166

KOFORD, JAMES SHINGLE, b Cheyenne, Wyo, July 26, 38. VLSI COMPUTER AIDED DESIGN. *Educ:* Stanford Univ, BS, 59, MS, 60, PhD(elec eng), 64. *Prof Exp:* Res asst mem tech staff, Stanford Electronics Lab, 60-64; proj engr, IBM Components Div, 64-66; mem tech staff, Fairchild Semiconductor Corp, 66-69, sr mem tech staff, 69-73; vpres develop, Packet Commun, Inc, 73-75; mgr, Network Develop Lab, Boeing Comput Servs, 75-81; VPRES COMPUT AIDED DESIGN, LSI LOGIC INC, 81- *Mem:* Inst Elec & Electronics Engrs. *Res:* Adaptive pattern-recognition systems, speech recognition, threshold elements, adaptation algorithms; computer-aided design for microelectronic circuits, logic simulation, graphic data processing. *Mailing Add:* 1601 McCarthy Blvd Milpitas CA 95035

KOFRANEK, ANTON MILES, b Chicago, Ill, Feb 5, 21; m 42; c 2. FLORICULTURE. *Educ:* Univ Minn, BS, 47; Cornell Univ, MS, 49, PhD, 50. *Prof Exp:* Asst, Cornell Univ, 47-50; from asst prof to prof floricult, Univ Calif, Los Angeles, 50-67; prof environ hort, Univ Calif, Davis, 67-90; RETIRED. *Concurrent Pos:* Vis prof, Hebrew Univ, Israel, 72-73 & 80; mem AID Proj, Egypt, 79-83; mem, Food & Agr Orgn, UN, India, 85. *Honors & Awards:* Am Carnation Soc Res Award, 74. *Mem:* Am Soc Hort Sci; Am Orchid Soc. *Res:* Photoperiod and temperature in floricultural plants; post harvest physiology; floriculture crop production. *Mailing Add:* Dept Environ Hort Univ Calif Davis CA 95616

KOFRON, JAMES THOMAS, JR, b Petersburg, Va, Mar 11, 28; m 60; c 4. PHOTOGRAPHIC CHEMISTRY. *Educ:* Univ Notre Dame, BS, 52; Mass Inst Technol, PhD(org chem), 56. *Prof Exp:* Res assoc chem, 56-75, SR RES ASSOC, RES LABS, EASTMAN KODAK CO, 75- *Mem:* AAAS; Am Chem Soc; Soc Photog Scientists & Engrs. *Res:* Reaction mechanisms in chemistry of photographic processes. *Mailing Add:* 123 El Mar Dr Rochester NY 14616

KOFRON, WILLIAM G, b Petersburg, Va, Aug 9, 34. ORGANIC CHEMISTRY. *Educ:* Univ Notre Dame, BS, 56; Univ Rochester, PhD(chem), 61. *Prof Exp:* Res assoc chem, Duke Univ, 60-62; fel, Columbia Univ, 62-63; sr chemist, Med Chem Dept, Geigy Res Labs, NY, 63-65; from asst prof to assoc prof, 65-76, PROF CHEM, UNIV AKRON, 76- *Mem:* Am Chem Soc. *Res:* Chemistry of carbanions; heterocyclic chemistry. *Mailing Add:* Dept Chem Univ Akron Akron OH 44325-3601

KOFSKY, IRVING LOUIS, b New York, NY,; m 69; c 3. PHYSICS. *Educ:* Syracuse Univ, BA, 45, PhD(physics), 52. *Prof Exp:* Instr physics, Syracuse Univ, 47-51; asst prof, Smith Col, 52-56; physicist, Tech Opers, Inc, 57-68; PRES & TECH DIR, PHOTOMETRICS INC, 68- *Mem:* Am Phys Soc; Am Asn Physics Teachers. *Res:* Extensive showers in cosmic radiation; weapons effects; gaseous electronics; atmospheric optics; photometry, image analysis and scanning theory. *Mailing Add:* PhotoMetrics Inc Four Arrow Dr Woburn MA 01801

KOFT, BERNARD WALDEMAR, b Hammonton, NJ, Nov 21, 21; m 44; c 5. BACTERIOLOGY. *Educ:* Rutgers Univ, BS, 43; Univ Pa, MS, 47, PhD(bact), 50. *Prof Exp:* From instr to asst prof bact, Jefferson Med Col, 50-57; from asst prof to prof, 57-87, EMER PROF MICROBIOL, RUTGERS UNIV, 87- *Mem:* Am Soc Microbiol; fel NY Acad Sci; Sigma Xi. *Res:* Bacterial nutrition; metabolism; vitamin synthesis; cellulose degradation. *Mailing Add:* 196 Hardenburg Lane East Brunswick NJ 08816

KOGA, ROKUTARO, b Nagoya, Japan, Aug 18, 42; US citizen; m 81; c 2. ASTROPHYSICS. *Educ:* Univ Calif, Berkeley, BA, 66; Univ Calif, Riverside, PhD(physics), 74. *Prof Exp:* Physicist, Berkeley Sci Labs, 66-69; fel, Univ Calif, Riverside, 74; res assoc astrophys, Case Western Reserve Univ, 74-76, sr res assoc astrophys, 77-78, asst prof physics, 79-80; RES PHYSICIST SPACE & ASTROPHYS, AEROSPACE CORP, 81- *Mem:* Am Phys Soc; Sigma Xi; NY Acad Sci; Am Geophys Union. *Res:* Measurements of heavy ions in space using satellite based sensors; the effects of cosmic rays on microcircuits in space; gamma-ray astronomy. *Mailing Add:* Space Sci Lab Aerospace Corp PO Box 92957 M2-259 Los Angeles CA 90009-2957

KOGA, TOYOKI, b Japan, Apr 1, 12; div; c 2. FOUNDATIONS OF QUANTUM PHYSICS. *Educ:* Univ Tokyo, MS, 37. *Hon Degrees:* DSc, Univ Tokyo, 48. *Prof Exp:* Asst prof aeronaut, Nagoya Univ, 40-48, prof mech eng & appl physics, 48-59; res scientist, Eng Ctr, Univ Southern Calif, 59-63; prof mech eng, Univ NC, 63-64; mem prof staff, TRW Systs, Inc, 67-69; RES & WRITING, 69- *Concurrent Pos:* Fulbright sr res fel, Calif Inst Technol, 55-56; vis prof, Univ Calif, 56-59 & Grad Ctr, Polytech Inst Brooklyn, 64-67. *Mem:* Am Phys Soc. *Res:* Gas dynamics; kinetic theory of gases; plasma physics; kinetic theory of quantum mechanical systems; revision of quantum mechanics; theory of elementary particles; superconductivity. *Mailing Add:* 3061 Ewing Ave Altadena CA 91001

KOGAN, MARCOS, b Rio de Janeiro, Brazil, June 9, 33; m 53; c 2. ENTOMOLOGY, ECOLOGY. *Educ:* Univ Rural do Rio de Janeiro, BS, 61; Univ Calif, Riverside, PhD(entom), 69. *Prof Exp:* Entomologist res, S Cent Inst Agr Res, Rio de Janeiro, 61-63; biologist res, Inst Oswaldo Cruz, Rio de Janeiro, 63-66; res fel entom, Univ Calif, Riverside, 66-69, res assoc, 69; assoc entomologist res, Ill Natural Hist Surv, Urbana, 69-76; assoc prof agr entom, Univ Ill, Urbana, 73-77, prof entom & agr entom, 77-90; entomologist, Ill Natural Hist Surv, Urbana, 76-90; PROF ENTOM & DIR, INTEGRATED PLANT PROTECTION CTR, ORE STATE UNIV, CORVALLIS, 91. *Concurrent Pos:* Consult soybean entom Brazil, 74-78, Korea, 78-79; mem sci deleg to Repub China, 81; ed comt, Ann Rev Entom, 84-88, Entomologia Experimentalis et Applicata, 85-90, Thomas Say Pub Entom Soc Am, 88-91. *Honors & Awards:* ASA/ICI-Americas Soybean Res Award, 86. *Mem:* Entom Soc Am; Brazilian Entom Soc; Sigma Xi; Am Chem Soc. *Res:* Management of soybean insect pests; soybean resistance to insects; insect plant interactions; nutrition of phytophagous insects; bionomics of Strepsiptera; international cooperation in soybean entomology and crop protection; chemical ecology. *Mailing Add:* Integrated Plant Protection Ctr Ore State Univ Corvallis OR 97330

KOGEL, MARCUS DAVID, public health, academic administration; deceased, see previous edition for last biography

KOGELNIK, H W, b Graz, Austria, June 2, 32; m 64; c 3. LASERS, COMMUNICATIONS. *Educ:* Vienna Tech Univ, Dipl Ing, 55, Dr Tech, 58; Oxford Univ, DPhil(electromagnetic theory), 60. *Prof Exp:* Asst prof electronics, Inst High frequency Electronics, Vienna, Austria, 55-58; Brit Coun scholar, Oxford Univ, 58-60; mem staff electronics res, Bell Labs, 61-67, head coherent optics, res dept, 67-76, dir, electronics res lab, 76-83, DIR, PHOTONICS RES LAB, AT&T BELL LABS, 83- *Mem:* Nat Acad Eng; Am Phys Soc; fel Inst Elec & Electronics Engrs; fel Optical Soc Am; AAAS. *Res:* Lasers; integrated optics; optical communication. *Mailing Add:* Crawford Hill Lab AT&T Bell Labs Box 400 Holmdel NJ 07733

KOGER, JOHN W, b Florence, Ala, Aug 20, 40. PHYSICAL METALLURGY. *Prof Exp:* PROG MGR WASTE MINIMALIZATION, MARTIN MARIETTA ENERGY SYSTS INC, 90- *Mailing Add:* Martin Marietta Energy Systems, Inc Y-12 Plant Bldg 9202 MS 8097 Oak Ridge TN 37831

KOGER, MARVIN, b Colgate, Okla; m 38; c 4. ANIMAL BREEDING, GENETICS. *Educ:* NMex Col, BS, 39; Kans State Univ, MS, 41; Univ Mo, PhD(physiol), 43. *Prof Exp:* From instr to assoc prof animal husb, NMex Col, 43-51; prof animal husb & animal geneticist, 51-80, PROF ANIMAL SCI, AGR EXP STA, UNIV FLA, 80- *Mem:* Am Soc Animal Sci; Am Dairy Sci Asn. *Res:* Nutritional deficiencies of sorghums; effects of hyperthyroidism; genetics of cattle and sheep; physiology of reproduction. *Mailing Add:* 1764 NW 17th Lane Gainesville FL 32605

KOGON, IRVING CHARLES, b Brooklyn, NY, Aug 8, 23; m 48; c 2. POLYMER CHEMISTRY. *Educ:* Brooklyn Col, BA, 48, MA, 51; Polytech Inst Brooklyn, PhD(chem), 54, Univ Wis, PhD, 54. *Prof Exp:* Asst org chem, Brooklyn Col, 49-51; res assoc, Univ Wis, 53-54; res assoc, Polymer Prod Dept, Exp Sta, E I Du Pont De Nemours & Co, Inc, 54-82; RETIRED. *Concurrent Pos:* Consult, polyurethane indust, 82- *Mem:* AAAS; Am Chem Soc; Polyurethane Mfg Asn; Soc Plastics Indust. *Res:* Synthesis of antispasmodics, antihistamines and local anesthetics; heterocyclic vinyl monomers; mechanism of organic reactions; chemistry of isocyanates and polyurethanes; new urethane curatives and synthetic rubbers; polymer chemistry. *Mailing Add:* 1420 Drake Rd Green Acres Wilmington DE 19803

KOGOS, L(AURENCE), b Boston, Mass, July 24, 29; m 51; c 2. CHEMICAL ENGINEERING. *Educ:* Northeastern Univ, BS, 51. *Prof Exp:* Chem engr, Sawyer-Tower, Inc, 51-52, chief res engr, 52-53, dir tech sales, 55, tech dir, 56-58, gen mgr, 58-59; exec vpres, Farrington Texol Corp, 59-63; gen mgr, Dynamic Coaters, Inc, 63-70; dir opers, Plymouth Rubber Co, Inc, 70-72; pres, Polymeric Fabricants Div, Whittaker Corp, 72-73; pres, Plastics Div, W R Grace & Co, 73-79; vpres & gen mgr, Roper Plastics, Inc, NY, 80-82, TLB Plastics, Brewster, NY, 82-84; chief exec officer, Ouimet group, Brockton, Mass, 84-90; CONSULT, 90- *Mem:* Am Chem Soc; Am Soc Plastics Engrs; Am Inst Chem Engrs. *Res:* Application of protective and decorative coatings to fabrics; plastics molding, extrusion and forming. *Mailing Add:* Nine Pioneer Circle Sharon MA 02067-2724

KOGUT, JOHN BENJAMIN, b Brooklyn, NY, Mar 6, 45; m 85; c 1. THEORETICAL PHYSICS. *Educ:* Princeton Univ, BA, 67; Stanford Univ, MS, 68, PhD(physics), 71. *Prof Exp:* Assoc physics, Inst Advan Study, 71-73 & Tel Aviv Univ, 73; res assoc, Cornell Univ, 73-74, from asst prof to assoc prof, 74-78; PROF PHYSICS, UNIV ILL, 78- *Concurrent Pos:* Sloan Found fel & NSF grant, Cornell Univ, 76-78; NSF grant, Univ Ill, 78-; Guggenheim fel, 88-89. *Mem:* Fel Am Phys Soc; Comt Concerned Scientists. *Res:* Theory of elementary particles; field theory and statistical mechanics. *Mailing Add:* Dept Phys Loomis Lab Univ Ill 1110 W Green St Urbana IL 61801

KOGUT, MAURICE D, b Brooklyn, NY, July 7, 30; m 59; c 3. PEDIATRICS, ENDOCRINOLOGY. *Educ:* NY Univ, BA, 51, MD, 55. *Prof Exp:* Intern pediat, Bellevue Hosp, New York, 55-56, resident, 56-57; chief resident, Children's Hosp, Los Angeles, 59-60, USPHS fel endocrinol, 60-62; from instr to assoc prof, Sch Med, Univ Southern Calif, 62-73, prof pediat, 73-80, assoc head dept, 75-80, clin prof med, 80-; head div endocrinol & metab & prog dir, Clin Res Ctr, Children's Hosp, 71-; AT DEPT PEDIAT, WRIGHT STATE UNIV, DAYTON, OHIO. *Mem:* Endocrine Soc; Am Diabetes Asn; Am Acad Pediat; Soc Pediat Res; Am Pediat Soc. *Res:* Carbohydrate metabolism in idiopathic hypoglycemia; growth hormone and insulin metabolism in hypopituitarism; the role of circulating insulin and glucagon in children with genetic predisposition to diabetes mellitus; uric acid metabolism. *Mailing Add:* Childrens Med Ctr One Childrens Plaza Dayton OH 45404

KOH, EUNSOOK TAK, b Seoul, Korea, May 3, 36; US citizen; m 61; c 3. NUTRITIONAL SCIENCE. *Educ:* Seoul Nat Univ, Korea, BS, 58; Univ Md, MS, 70, PhD(nutrit sci), 73. *Prof Exp:* Res asst, Univ Md, College Park, 68-70; res asst, USDA, Agr Res Serv, Nutrit Inst, 70-73, res assoc, 73-74; assoc prof, Alcorn State Univ, Miss, 74-81; prof, Univ Okla, Norman, 81-89; PROF, DEPT CLIN DIETETICS, COL ALLIED HEALTH, UNIV OKLA, OKLAHOMA CITY, 89- *Concurrent Pos:* Prin investr, Alcorn State Univ, 74-81, Univ Okla, 81-91, USDA, Agr Res Serv, Nutrit Inst, 87-88 & Hallym Univ, 87; vis prof, Hallym Univ, Kangwon Do, Korea, 87 & USDA, Agr Res Serv, Nutrit Inst, Beltsville, 87-88. *Mem:* Am Inst Nutrit; Am Dietetic Asn; Nutrit Educ Soc. *Res:* Nutrition survey; carbohydrates and lipid metabolism; copper and fructose interaction; interaction of fructose, magnesium deficiency and sex hormone on nephrocalcinosis; mechanism of the interaction on nephrocalcinosis. *Mailing Add:* Dept Clin Dietetics Col Allied Health Univ Okla PO Box 26901 Oklahoma City OK 73190

KOH, EUSEBIO LEGARDA, b Manila, Philippines, Oct 4, 31; Can citizen; m 58; c 4. DISTRIBUTION THEORY, INTEGRAL TRANSFORMATIONS & OPERATIONAL CALCULUS. *Educ:* Univ Philippines, BS, 54; Purdue Univ, Ind, MS, 56; Univ Birmingham, UK, MSc, 61; State Univ NY, Stony Brook, PhD(appl math), 67. *Prof Exp:* Res eng, Int Harvester Co, Ill, 56-57; asst prof, 59-64, dept head mech eng, Univ Philippines, 63-64; asst prof math, Univ SC, 67-68, Univ Sask, 68-70; assoc prof, 70-75, dept head, 77-79, PROF MATH & STATIST, UNIV REGINA, 75- *Concurrent Pos:* Dir, DCCD Eng Corp, Philippines, 62-64; guest prof, Technische Hochschule Dermstadt, Ger, 75-76; prof, Univ Petrol & Minerals, Saudi Arabia, 79-82; Nat Sci Eng Res Coun res grant, Can, 71-; travel fel, Nat Sci Eng Res Coun, Ger, 75-76. *Mem:* Am Math Soc; Math Asn Am; Soc Indust & Appl Math; Can Math Soc; Can Appl Math Soc. *Res:* Extension of integral transformations to generalized functions; development of operational calculus by algebraic approach; association of variables technique; functional equations in distributions. *Mailing Add:* Dept Math & Statist Univ Regina Regina SK S4S 0A2 Can

KOH, KWANGIL, b Seoul, Korea, July 8, 31; m 58; c 3. MATHEMATICS. *Educ:* Auburn Univ, BS, 59, MS, 60; Univ NC, PhD(math), 64. *Prof Exp:* From instr to assoc prof, 64-68, PROF MATH, NC STATE UNIV, 68- *Mem:* Am Math Soc; Math Asn Am. *Res:* Algebra; theory of rings. *Mailing Add:* Dept Math NC State Univ Raleigh NC 27607

KOH, P(UN) K(IEN), b China, Jan 31, 14; nat US; m 67; c 2. PHYSICAL METALLURGY. *Educ:* Mass Inst Technol, DSc, 39. *Prof Exp:* Head, Metall Sect, Stand Oil Co, Ind, 45-47; assoc dir res & staff scientist, Allegheny Ludlum Steel Corp, 48-60; res engr, Bethlehem Steel Corp, 60-66; prof mat sci, 66-76, EMER PROF MECH ENG, TEX TECH UNIV, 81- *Concurrent Pos:* Lectr, Ill Inst Technol, 46-47, Univ Pittsburgh, 53-58 & Stevens Inst Technol, 65-66; hon assoc dir, Metal Indust Res Inst, Kaosiung, Taiwan; tech adv, Indust Technol Res Inst, Taiwan; consult, Nuclear Power Div, Taiwan Power Co, Repub China. *Mem:* Fel Am Inst Chemists. *Res:* X-ray diffraction; electron microprobe; metal physics; polymers; environmental engineering. *Mailing Add:* 3318 24th St Lubbock TX 79410

KOH, ROBERT CY, b Shanghai, China, May 23, 38; m 61; c 1. FLUID MECHANICS, APPLIED MATHEMATICS. *Educ:* Calif Inst Technol, BS, 60, MS, 61, PhD(appl mech, math), 64. *Prof Exp:* Res fel eng, Calif Inst Technol, 64-65; mem tech staff, Nat Eng Sci Co, 65-66; sr scientist, Tetra Tech Inc, 66-72; MEM STAFF, CALIF INST TECH, 72- *Mem:* Int Asn Hydraul Res. *Res:* Fluid mechanics; applied mathematics. *Mailing Add:* 212 S Marengo Ave Pasadena CA 91101

KOH, SEVERINO LEGARDA, b Manila, Philippines, Jan 8, 27; m 52; c 5. THEORETICAL & APPLIED MECHANICS, THERMAL SCIENCES. *Educ:* NY Univ, BS, 50; Nat Univ, Manila, BS, 52; Pa State Univ, MS, 57; Purdue Univ, PhD(eng sci), 62. *Prof Exp:* Meteorologist-in-chg marine unit, Weather Bur, Manila, 48-54; res asst hydrodyn lab, Johns Hopkins Univ, 54-55; instr eng mech, Pa State Univ, 55-57; instr eng sci, Purdue Univ, 57-59; res assoc viscoelasticity, Gen Tech Corp, 59-61; mech engr, Major Appliance Lab, Gen Elec Co, 61-62; from asst prof to prof aeronaut & eng sci, Purdue Univ, 62-73, prof mech eng, 73-80, asst head, Div Interdisciplinary Eng Studies, 77-80, head, Dept Eng, 80-81; PROF & CHMN MECH & AEROSPACE ENG, WVA UNIV, 81- *Concurrent Pos:* Mem, President's Fact-Finding Comt, Philippines, 53-54; res assoc, B G Bantegui & Assocs, Manila, 53-54; scientist-consult, Gen Tech Corp, 62-; Standard Oil Co (Ind) Found teaching award, 67; vis prof & res assoc, Clausthal Tech Univ, 68-69; vis prof, Univ Karlsruhe, 69; Humboldt vis prof, Univ Bonn, 74-75; Balik scientist, Philippines, 76; consult, 3IE, Inc, 77- & Batelle Northwest, 78; dir, The Eng Sci Perspective, 70-73 & 78-80, ed-in-chief, 76- *Mem:* Soc Eng Sci (secy, 63-68); Am Acad Mech; Am Soc Eng Educ; Am Soc Mech Engrs; Soc Rheology. *Res:* Elasticity; viscoelasticity; fluid dynamics; rheology of nonlinear materials; composite materials; testing of materials; sandwich structures; micromechanics; solar energy systems; geotechnical engineering problems; heat and mass transfer. *Mailing Add:* Col Eng Univ Md Baltimore County Baltimore MD 21228

KOHAN, MELVIN IRA, b Boston, Mass, Mar 11, 21; m 43; c 4. ORGANIC POLYMER CHEMISTRY. *Educ:* Harvard Univ, AB, 42; Univ Ill, PhD(chem), 50. *Prof Exp:* Chemist, Dept Electrochem, 42-44 & 46-47, chemist, Dept Plastics, 50-62, sr res chemist, 62-74, res assoc, Dept Plastics, 74-80, res assoc, Polymer Products Dept, Exp Sta, E I du Pont de Nemours & Co, Inc, 80-82; RETIRED. *Concurrent Pos:* Consult, Eng Thermoplastics, 83-; adj prof, Drexel Univ. *Mem:* Am Chem Soc; Soc Plastic Engrs; Sigma Xi. *Res:* Polymer chemistry; plastics engineering; nylon plastics technology; publications, patents. *Mailing Add:* 1913 Longcome Dr Wilmington DE 19810

KOHANE, THEODORE, b New York, NY, Apr 20, 23; m 55; c 2. PHYSICS. *Educ:* City Col New York, BS, 44; Rutgers Univ, PhD(physics), 53. *Prof Exp:* Physicist, Nat Adv Comt Aeronaut, 44-46; asst physics & instr, NY Univ, 46-48; asst & fel, Rutgers Univ, 48-53; MEM RES STAFF, RAYTHEON RES DIV, 53- *Mem:* Am Phys Soc; Optical Soc Am. *Res:* Nuclear magnetic resonance; magnetic and electrical properties of ferrites; microwaves; optical properties of solids. *Mailing Add:* Raytheon Res Div 131 Spring St Lexington MA 02173

KOHEL, RUSSELL JAMES, b Omaha, Nebr, Nov 30, 34; m 57; c 3. PLANT GENETICS. *Educ:* Iowa State Univ, BS, 56; Purdue Univ, MS, 58, PhD(plant breeding), 59. *Prof Exp:* RES GENETICIST COTTON, USDA, AGR RES SERV, 59- *Mem:* Am Soc Plant Physiologists; fel Am Soc Agron; Am Genetic Asn; Genetics Soc Am. *Res:* Qualitative and quantitative genetics of the cotton plant. *Mailing Add:* Rte 5 Box 805 College Station TX 77845

KOHIN, BARBARA CASTLE, b Providence, RI, Dec 11, 32; m 59; c 3. MOLECULAR PHYSICS. *Educ:* Col William & Mary, BS, 53; Univ Md, MS, 56, PhD(physics), 60. *Prof Exp:* Res assoc molecular physics, Cath Univ Am, 59-61; physicist theoret physics, Inst Battelle, Geneva, Switz, 61-62; instr physics, Mass State Col Worcester, 64-67; asst prof physics, Clark Univ, 67-68; assoc dir, Off Spec Studies, Col of the Holy Cross, 78-85 & 86-87, actg dir, 85-86. *Concurrent Pos:* Res assoc, Mass Inst Technol, 73-74. *Res:* Quantum chemistry; solid state physics; elementary and atomic physics. *Mailing Add:* 11 Berwick St Worcester MA 01602

KOHIN, ROGER PATRICK, b Chicago, Ill, Mar 2, 31; m 59; c 3. PHYSICS. *Educ:* Univ Notre Dame, BSEE, 53; Univ Md, PhD(physics), 61. *Prof Exp:* Scientist physics, Battelle Mem Inst, Geneva, Switz, 61-62; asst prof 62-67, chmn dept 74-76 & 85-87, ASSOC PROF PHYSICS, CLARK UNIV 67- *Concurrent Pos:* Vis scientist, Inst J Stefan, Ljubljana, Yugoslavia, 68-69; Indo-Am fel, Indian Inst Tech-Kanpur, 76-77; Vis Prof, Univ Nairobi, Kenya, 87-88. *Mem:* Am Phys Soc; Optical Soc Am. *Res:* Electron-spin resonance spectroscopy; radiation damage of solids; experimental ferroelectric materials; organic and inorganic free radicals; computer simulation. *Mailing Add:* Dept Physics Clark Univ Worcester MA 01610

KOHL, A(RTHUR) L(IONEL), b Ont, Can, Aug 21, 19; nat US; m 43; c 3. GAS PURIFICATION, NUCLEAR ENGINEERING. *Educ:* Univ Southern Calif, BE, 43, MS, 47. *Prof Exp:* Res engr, Turco Prod, Inc, 42-44 & 46-47; chief chem eng res, Fluor Corp Ltd, 47-60; group leader process develop, Atomics Int, 60-68, proj engr & proj mgr advan develop, 68-78, prog mgr fossil energy, Rocketdyne Div, Rockwell Int, 78-89; CONSULT ENGR, 89- *Honors & Awards:* Outstanding Achievement Award, Am Inst Chem Engrs, 66; Tech Achievement Award, Engrs Joint Coun, 67. *Mem:* Am Inst Chem Engrs. *Res:* Gas purification; process equipment; saline water conversion; chemical process development; nuclear reactor fuels and materials; nuclear reactor component development; coal conversion; paper mill black liquor gasification. *Mailing Add:* 22555 Tiara St Woodland Hills CA 91367

KOHL, DANIEL HOWARD, b Cleveland, Ohio, July 30, 28; m 50; c 4. PLANT PHYSIOLOGY. *Educ:* Univ Calif, Berkeley, BS, 60; Washington Univ, PhD(molecular biol), 65. *Prof Exp:* Asst prof bot, 65-70, assoc prof, 70-79, sr fel, Ctr Biol Natural Systs, 71-81, PROF BIOL, WASHINGTON UNIV, 79- *Mem:* Am Soc Plant Physiol; Soil Sci Soc Am. *Res:* N isotope distribution in various components of N cycle; N fixation biochemistry. *Mailing Add:* Dept Biol Wash Univ St Louis MO 63130

KOHL, FRED JOHN, b Cleveland, Ohio, Jan 1, 42. PHYSICAL CHEMISTRY. *Educ:* Case Inst Technol, BS, 63; Case Western Reserve Univ, PhD(chem), 68. *Prof Exp:* RES CHEMIST & MGR, MAT SCI SPACE PROJ, LEWIS RES CTR, NASA, 68-, BR CHIEF. *Mem:* Am Chem Soc; Am Soc Mass Spectrometry; Combustion Inst. *Res:* Hot corrosion of superalloys; oxidation of metals; high temperature vaporization and thermodynamics; mass spectrometry; oxidation/vaporization processes; high temperature chemistry; vaporization of refractories; materials science experiments in space; combustion process related to corrosion; microgravity science and applications. *Mailing Add:* 27521 Laurell Lane North Olmsted OH 44070

KOHL, HARRY CHARLES, JR, b St Louis, Mo, Aug 6, 19; m 41; c 1. FLORICULTURE, PLANT PHYSIOLOGY. *Educ:* Univ Ill, BS, 40, MS, 48; Cornell Univ, PhD(hort), 50. *Prof Exp:* Assoc exten specialist floricult, Rutgers Univ, 50-53; asst prof, Univ Calif, Los Angeles & asst plant physiologist, 57-62; prof floricult & plant physiologist, 62-77, prof & chairperson, Plant Phys Grad Group, 77-80, EMER PROF ENVIRON HORT, UNIV CALIF, DAVIS, 80- *Mem:* Am Soc Plant Physiol; Am Soc Hort Sci; Bot Soc Am; Int Soc Soil Sci. *Res:* Control of plant growth and differentiation; preharvest and post-harvest physiology of flowers; mineral translocation; root aeration; salinity tolerance. *Mailing Add:* 1113 Halifax Ave Davis CA 95618

KOHL, JEROME, b Montreal, Que, Mar 13, 18; nat US; m 45; c 2. CHEMICAL ENGINEERING. *Educ:* Calif Inst Technol, BS, 40; NC State Univ, MS, 75. *Prof Exp:* Chem & proj engr, Avon Refinery, Tidewater Assoc Oil Co, Calif, 43-46, asst supt, 46-48; chem engr, Tracerlab, Inc, 48-51, sect leader, Mobile Radiochem Lab, 51-53, chief engr, 53-58, mgr eng & develop, 58-60; coordr spec prod, Gen Atomic Div, Gen Dynamics Corp, 60-64; mkt mgr, Oak Ridge Tech Enterprises, 65-69; sr nuclear eng exten specialist & lectr nuclear eng, 69-88, EMER NUCLEAR ENG SPECIALIST & LECTR, NC STATE UNIV, 88- *Concurrent Pos:* Instr & lectr, Univ Calif, Berkeley & San Diego, 47-64; lectr, Univ Delft, 56 & int lectr on waste reduction, mgt & minimization hazardous waste, NC State Univ. *Mem:* Am Nuclear Soc; Am Inst Chem Engrs; Am Electroplating Soc; Soc Photog Educ. *Res:* Radiation monitoring instrumentation; industrial applications of radioisotopes; measurement of nuclear radiations; energy conservation; co-generation; management and minimization of hazardous waste; waste reduction. *Mailing Add:* Dept Nuclear Eng NC State Univ Box 7909 Raleigh NC 27695-7909

KOHL, JOHN C(LAYTON), b New York, NY, June 22, 08; m 35; c 2. CIVIL ENGINEERING. *Educ:* Univ Mich, BSE, 29. *Hon Degrees:* MA, Univ Pa, 73. *Prof Exp:* From draftsman to inspector construct, Cincinnati Union Terminal Co, 29-30; instr civil eng, Carnegie Inst Technol, 31-37; asst dir cent develop lab, Pittsburgh Plate Glass Co, 37-38; res & develop engr, Pittsburgh Corning Corp, 38-44 & 46; from asst prof to prof civil eng, Univ Mich, 46-66, dir transp inst, 55-66; exec secy div eng, Nat Acad Sci-Nat Res Coun, 66-68; sr assoc, Wilbur Smith & Assocs, 68-70; comnr, NJ Dept Transp, 70-74; prof, 74-76, EMER PROF CIVIL & URBAN ENG, UNIV PA, 76- *Concurrent Pos:* Consult, Am Buslines, Inc, 48, Haugh & Keenan Transfer & Storage Co, 49 & Fruehauf Trailer Co, 50; asst adminr, US Housing & Home Finance Agency, DC, 61-66; chmn, Tristate Regional Planning Comn, 70-71; vchmn, Gov Comn Transit Financing, 71-73; mem transp res bd, Nat Acad Sci-Nat Res Coun; mem, Transp Res Forum; vis sr fel, Dept Civil Eng, Princeton Univ, 76-81. *Mem:* Am Soc Civil Engrs; Am Soc Traffic & Transp; hon mem Inst Transp; hon mem Am Pub Works Asn. *Res:* Transportation engineering and economics; urban transportation planning; rail network analysis. *Mailing Add:* 200 Aspen Ct Irving TX 75062

KOHL, JOHN LESLIE, b Zanesville, Ohio, Apr 27, 41; m 65; c 2. EXPERIMENTAL ATOMIC PHYSICS, SOLAR PHYSICS. *Educ:* Muskingum Col, BS, 63; Univ Toledo, MS, 66, PhD(physics), 69. *Prof Exp:* Res fel physics, 69-72, res assoc physics, Harvard Col Observ, 72-76, ASTROPHYSICIST & LECTR, SMITHSONIAN ASTROPHYS OBSERV, HARVARD UNIV, 76- *Concurrent Pos:* Nat Sci Found fel. *Mem:* Am Phys Soc; Am Astron Soc; Int Astron Union; Am Geophys Union. *Res:* Experimental studies of atomic and molecular processes needed to understand astrophysical and laboratory plasmas; experimental studies of solar wind generation using space instrumentation. *Mailing Add:* 160 Lawsbrook Rd Concord MA 01742

KOHL, PAUL ALBERT, b Buffalo, NY, Aug 6, 52; m 74; c 2. ELECTROCHEMISTRY, PHYSICAL CHEMISTRY. *Educ:* Bethany Col, BS, 74; Univ Tex, PhD(chem), 78. *Prof Exp:* Chemist, Nuclear Radiation Develop, 74; CHEMIST, AT&T BELL LABS, 78- *Honors & Awards:* Weston Award, Electrochem Soc, 77; Ayres Award, Univ Tex, 78. *Mem:* Am Chem

Soc; Electrochem Soc. *Res:* Chemical and electrochemical reactions involved in the processing of semiconductor materials for the development of microelectronic devices. *Mailing Add:* 130 Cameron Glen Dr Atlanta GA 30328

KOHL, ROBERT A, b Harvey, Ill, Jan 22, 36; m 57; c 4. SOIL PHYSICS, IRRIGATION. *Educ:* Purdue Univ, BS, 58; Utah State Univ, MS, 60, PhD(soils, irrig), 63. *Prof Exp:* Agr missionary, Lutheran Mission, Nigeria, 63-66; res soil scientist, Snake River Conserv Res Ctr, Agr Res Serv, USDA, Idaho, 67-74; assoc prof, 75-87, PROF PLANT SCI, SDAK STATE UNIV, 87- *Mem:* Soil Sci Soc Am; Am Soc Agron. *Res:* Water management; sprinkler irrigation. *Mailing Add:* Dept Plant Sci SDak State Univ Brookings SD 57007

KOHL, SCHUYLER G, b Philadelphia, Pa, Feb 22, 13; m 43; c 1. MEDICINE. *Educ:* Univ Md, BS, 36, MD, 40; Columbia Univ, MS, 52, DrPH, 54; Am Bd Obstet & Gynec, dipl, 51. *Prof Exp:* Asst obstet, Univ Md, 42-49, instr, 49-50; res assoc obstet & gynec, 50-51, from asst prof to assoc prof, 51-62, from asst dean to assoc dean, 58-71, PROF & VCHMN OBSTET & GYNEC, STATE UNIV NY DOWNSTATE MED CTR, 62- *Concurrent Pos:* Consult, Nat Inst Neurol Dis & Stroke, mem hosp facil res study sect, 58-60, mem human ecol study sect, 60-65; lectr, Columbia Univ, 59-84; consult, Pan Am Health Orgn, 60-80. *Mem:* Asn Planned Parenthood Physicians; Soc Gynec Invest; Am Pub Health Asn; AMA; Asn Am Med Cols. *Res:* Application of statistical methods to clinical practice. *Mailing Add:* Dept Obstet & Gynec State Univ NY Health Sci Ctr Brooklyn NY 11203

KOHL, WALTER H(EINRICH), b Kitzingen, Bavaria, Ger, Jan 22, 05; nat US; m 32; c 1. ELECTRONICS. *Educ:* Dresden Tech Univ, Dipl Ing, 28, Dr Ing, 30. *Prof Exp:* Demonstr physics, Dresden Tech Univ, 28-30; develop engr, Rogers Electronic Tubes, Toronto, 31-40, proj engr, 41-43, chief engr, vpres & dir, 44-45; sect head vacuum tube lab, Collins Radio Co, Iowa, 46-48, consult to dir res, 49-52; res assoc electronics res lab, Stanford Univ, 52-58; sr eng specialist, Mt View Components Lab, Sylvania Elec Prod, Inc, 58-61; consult mat & technol for electron devices, 62-66; sr scientist microwave lab, Electronics Res Ctr, NASA, 66-68, chief univ affairs off, 68-70; consult mat & technol for electron devices, 70-78; RETIRED. *Concurrent Pos:* Lectr, Univ Toronto, 35-40; consult, Stanford Res Inst, 56-58 & Off Asst Secy Defense, 57. *Mem:* AAAS; fel Inst Elec & Electronics Engrs; fel Am Ceramic Soc; Sigma Xi; Am Soc Metals. *Res:* Materials and techniques for electron devices. *Mailing Add:* 3210 Wisconsin Ave NW Apt 801 Washington DC 20016-3836

KOHLAND, WILLIAM FRANCIS, b Chester, Pa, May 13, 25; m 56; c 1. PETROLOGY, ATMOSPERIC SCIENCE. *Educ:* Bucknell Univ, AB, 51; Univ Tenn, Knoxville, MS, 52, PhD(earth sci), 69. *Prof Exp:* Asst prof geol & earth sci, Edinboro State Col, 59-67; PROF GEOL & EARTH SCI, MID TENN STATE UNIV, 67- *Concurrent Pos:* Consult & field geologist. *Mem:* Geol Soc Am; Nat Asn Geol Teachers; Soil Sci Soc Am. *Res:* Petrology of St Francois NIT; metasomatic changes in rocks; mineral identification; minerology of Unakite. *Mailing Add:* Dept Geol & Geog Geol Mid Tenn State Univ Box 416 Murfreesboro TN 37132

KOHLBRENNER, PHILIP JOHN, b South Bend, Ind, Nov 17, 31; m 57; c 6. TECHNICAL REGULATORY AFFAIRS. *Educ:* Purdue Univ, BS, 53; State Univ NY Col Forestry, Syracuse Univ, PhD(org chem), 58. *Prof Exp:* Res chemist synthetic org chem, Cowles Chem Co, 57-58; res chemist, Am Cyanamid, 58-64, group leader synthetic org chem, 64-76, dept head basic pharmaceut, Lederle Labs Div, 76-80, assoc dir, 80-87, dir, Regulatory Affairs-Global Tech Support, Med Res Div, 87-89, DIR, REGULATORY AFFAIRS INT, AM CYANAMID, 89- *Mem:* Am Chem Soc; Am Asn Pharm Scientists. *Res:* Synthetic organic chemistry. *Mailing Add:* Nine Borger Pl Pearl River NY 10965

KOHLER, BRYAN EARL, b Heber City, Utah, June 9, 40; m 60; c 3. PHYSICAL CHEMISTRY. *Educ:* Univ Utah, BA, 62; Univ Chicago, PhD(chem), 67. *Prof Exp:* Fermi Inst res fel chem, Univ Chicago, 67; NSF fel, Calif Inst Technol, 67-68; from asst prof to assoc prof, Harvard Univ, 69-75; assoc prof, Wesleyan Univ, 75-77, prof chem, 77-, chmn dept, 80-; AT DEPT CHEM, UNIV CALIF, RIVERSIDE. *Concurrent Pos:* Grants, Am Chem Soc, Harvard Univ, 69-70, Advan Res Proj Agency, 69-, NIH, 69- & NSF, 72-; Sloan Found fel, 74-76; Alfred P Sloan Found fel, 74; vis fel, Joint Inst Lab Astrophysics, 78; Alexander von Humboldt fel, 79. *Mem:* Am Phys Soc; NIH; Am Phys Soc; NY Acad Sci; Biophys Soc; Sigma Xi. *Res:* Investigation of the electronic structure of molecules and molecular crystals using the techniques of magnetic resonance and optical spectroscopy; electronic properties of biomolecules; dynamics of excitation. *Mailing Add:* Dept Chem Univ Calif Riverside CA 92521

KOHLER, CARL, b Hamilton, Ont, June 24, 30; m 50; c 4. MARINE BIOLOGY, ICHTHYOPLANKTON. *Educ:* McMaster Univ, BA, 53; McGill Univ, MSc, 56, PhD(zool), 60. *Prof Exp:* Asst conservationist, Royal Bot Gardens, Ont, 51; from asst technician to technician, Biol Sta, Fisheries Res Bd Can, 53-56, from asst scientist to sr scientist, 56-67, head groundfish prog, 67-73, head fishery biol sect, 73-85; RETIRED. *Concurrent Pos:* Demonstr, McGill Univ, 55-56 & 58-59. *Mem:* Am Fisheries Soc. *Res:* Fishery biology and biostatistics; fishery management. *Mailing Add:* 316 Prince of Wales St Andrews NB E0G 2X0 Can

KOHLER, CONSTANCE ANNE, b Flushing, NY, Jan 9, 43; m 80; c 2. PHARMACOLOGY, BIOCHEMISTRY. *Educ:* St John's Univ, NY, BS, 65; Univ Calif, San Francisco, PhD(pharmacol), 73. *Prof Exp:* Fel pharmacol, Roche Inst Molecular Biol, NJ, 72-74; res pharmacologist, 74-80, SR RES SCIENTIST, LEDERLE LABS, DIV AM CYANAMID, NY, 80- *Mem:* AAAS; Am Heart Asn; Am Chem Soc; NY Acad Sci; Sigma Xi; Soc Exp Biol & Med. *Res:* Allergy and asthma; platelet biochemistry; phospholipid, platelet-activating factor, arachidonic acid metabolism and pharmacology; mammalian cell culture; growth factors; hormones, drugs and signal transduction. *Mailing Add:* Med Res Div Am Cyanamid Co Middletown Rd Pearl River NY 10965

KOHLER, DONALD ALVIN, b Rainier, Ore, Oct 29, 28; m 59; c 2. X-RAY PHYSICS, PLASMA PHYSICS. *Educ:* Univ Ore, BS, 51, MS, 52; Calif Inst Technol, PhD(physics), 59. *Prof Exp:* Res assoc nuclear physics, Stanford Univ, 59-62, res assoc, 62-63, lectr, 62-65; res scientist, 65-73, staff scientist, 73-79, SR STAFF SCIENTIST, LOCKHEED PALO ALTO RES LAB, 79- *Mem:* AAAS; Am Phys Soc; Inst Elec & Electronics Engrs. *Res:* X-ray physics and instrumentation; laser-plasma interaction; experimental low-energy nuclear physics, particularly of the light nuclei; cosmology and astrophysics; elementary particle physics; weak interactions. *Mailing Add:* Lockheed Palo Alto Res Lab Bldg 203 Dept 91-10 Palo Alto CA 94304-1191

KOHLER, ELAINE ELOISE HUMPHREYS, pediatric endocrinology; deceased, see previous edition for last biography

KOHLER, ERWIN MILLER, b Cincinnati, Ohio, June 24, 30; m 54; c 2. VETERINARY MICROBIOLOGY. *Educ:* Ohio State Univ, DVM, 55, MS, 63, PhD(microbiol, immunol), 65. *Prof Exp:* Vet, Winchester Animal Hosp, Va, 55-62; from asst prof to assoc prof, 65-73, PROF INFECTIOUS DIS DOMESTIC ANIMALS, VET SCI DEPT, OHIO AGR RES & DEVELOP CTR, 73-, CHMN DEPT, 76- *Mem:* Am Vet Med Asn; Am Soc Microbiol; Conf Res Workers Animal Diseases; Am Asn Swine Practitioners. *Res:* Studies of colibacillosis of gnotobiotic and conventional swine; studies of the oral immunization of sows as an aid in the prevention of neonatal enteric colibacillosis of pigs. *Mailing Add:* 2853 Mara Loma Circle Wooster OH 44619

KOHLER, GEORGE OSCAR, b Milwaukee, Wis, Apr 9, 13; m 40; c 3. BIOCHEMISTRY, BIOCHEMICAL ENGINEERING. *Educ:* Univ Wis, BS, 34, MS, 36, PhD(biochem), 38. *Prof Exp:* Asst, Univ Wis, 34-39, Cerophyl Labs grant, 38-39; assoc dir res, Cerophyl Labs, Inc, 39-50, vpres & dir res, 50-54, dir res, Alfalfa Dehydration & Milling Co, 54-55; pres, Cerophyl Labs, Inc, 55-56; sr exec serv res leader, Western Regional Res Ctr, Sci & Educ Admin-Agr Res, USDA, 56-81; OWNER, G O KOHLER & ASSOC, 81- *Mem:* Am Inst Nutrit; Am Chem Soc; Inst Food Technologists; Am Asn Cereal Chemists; Poultry Sci Asn. *Res:* Chicken and guinea pig nutrition; hormone assay and synthesis; vitamin assay; isolation of compounds from natural materials; process development; amino acid analysis; protein isolates from leaves and oilseeds. *Mailing Add:* PO Box 454 Inverness CA 94937

KOHLER, HEINZ, b Duisburg, Ger, Sept 11, 39; m 65; c 2. PROTEIN CHEMISTRY, IMMUNOLOGY. *Educ:* Univ Munich, MD, 65. *Prof Exp:* Res fel, Max Planck Inst Biochem, 65-67; res assoc, Div Biol Sci, Ind Univ, 67-70; asst prof, Dept Path, 70-74, assoc prof, depts path & biochem, Univ Chicago, 74-81; Dept Molecular Immunol, Roswell Park Mem Inst, Buffalo, NY, 81-87; dir res, 87-89, SR SCI FEL, IDEC PHARMACEUT CORP, LA JOLLA, CALIF, 89- *Concurrent Pos:* Res career develop award, USPHS, 73; adj prof, Dept Path, Univ Calif, San Diego, 87-90; adj mem, San Diego Regional Cancer Ctr, La Jolla, 90- *Mem:* Fedn Am Soc Exp Biol. *Res:* Relationship of function and structure of proteins; regulation of immune response. *Mailing Add:* IDEC Pharmaceut Corp 11099 N Torrey Pines Rd Suite 160 La Jolla CA 92037

KOHLER, MAX A, b Lincolnville, Kans, Sept 6, 15. HYDROLOGY, GEOPHYSICS. *Educ:* Univ NMex, BS, 39. *Prof Exp:* Hydrologist, 41-51, Nat Weather Serv, chief hydrologist, 51-71, assoc dir, 71-73; RETIRED. *Concurrent Pos:* Pres, Comn Hydrol, World Meteorol Orgn, 60-68. *Honors & Awards:* Hydrol Prize, Int Am Hydrol Sci, 86. *Mem:* Nat Acad Eng; fel Am Geophys Union; fel Am Meteorol Soc. *Mailing Add:* 402 Dennis Ave Silver Spring MD 20901

KOHLER, PETER, b Brooklyn, NY, July 18, 38; m 59; c 4. ENDOCRINOLOGY, CELL CULTURE. *Educ:* Univ Va, Charlottesville, BA, 59; Duke Univ, MD, 63. *Prof Exp:* Intern med, Duke Hosp, 63-64, fel endocrinol, 64-65; clin assoc, Nat Cancer Inst, NIH, 65-67, sr investr, Nat Inst Child Health & Develop, 68-73; prof med & cell biol & chief endocrinol, Baylor Col Med, 73-77; prof & chmn med, Univ Ark, Little Rock, 77-86, interim dean, Col Med, 86; dean sch med, Health Sci Ctr, Univ Tex, 86-88; PRES, ORE HEALTH SCI UNIV, 88- *Concurrent Pos:* Head endocrinol serv, Nat Inst Child Health & Develop, NIH, 72-73; prof med & cell biol & chief endocrionol, Baylor Col Med, 73-77; mem, NIH Endocrinol Study Sect, 81-85, chmn, 84-85; mem endocrinol bd, Am Bd Internal Med, 83-; mem, NICHD Bd Sci Counrs, 86- *Honors & Awards:* Qual Award, NIH, 69, 71. *Mem:* Am Fedn Clin Res; Am Soc Clin Invest; Asn Am Physicians; Am Diabetes Asn; Endocrine Soc; Sigma Xi; Am Soc Cell Biol. *Res:* Regulation of cell function and pituitary pathophysiology. *Mailing Add:* Ore Health Sci Univ 3181 SW Sam Jackson Park Rd Portland OR 97201

KOHLER, PETER FRANCIS, b Milwaukee, Wis, Apr 14, 35; m 62; c 3. IMMUNOLOGY. *Educ:* Princeton Univ, AB, 57; Columbia Univ, MD, 61. *Prof Exp:* From asst prof to prof med, Univ Colo Med Ctr, 67-81, head div clin immunol, 75-81; PROF MED, SCH MED, TULANE UNIV, 85-; CHIEF, CHARITY HOSP, LA, 85- *Mem:* Am Soc Clin Invest; Am Asn Immunologists; Am Acad Allergy. *Res:* Immunopathogenic mechanisms in disease; complement; immunology of hepatitis B virus infection. *Mailing Add:* Dept Med Tulane Univ Med Sch 1430 Tulane Ave New Orleans LA 70211

KOHLER, R RAMON, b Midway, Utah, July 24, 31; m 56; c 7. SPEECH PATHOLOGY, AUDIOLOGY. *Educ:* Brigham Young Univ, BS, 58, MS, 61; Univ Utah, PhD(speech path), 67. *Prof Exp:* Speech pathologist mobile unit, Mont Soc Crippled Children, 59; teacher elem, Idaho Falls Pub Schs, 60-62; speech pathologist, Daggett, Hinckley & Newberry Pub Schs, 62-65 & Salt Lake City Pub Schs, 66-68; asst prof, Univ Wyo, 68-71, prof speech path, 71-89. *Concurrent Pos:* Consult & team mem, Wyo Cleft Palate Eval Team, 68- *Mem:* Am Cleft Palate Asn; Am Speech & Hearing Asn. *Res:* Velopharyngeal closure in cleft palate children and its effect upon speech; non-verbal aspects of stuttering. *Mailing Add:* 777 S 400 E No 70 St George UT 84770

KOHLER, ROBERT HENRY, b Philadelphia, Pa, Apr 25, 33. PHYSICS. *Educ:* Mass Inst Technol, BS, 55, PhD(physics), 60. *Prof Exp:* Res assoc exp physics, Columbia Univ, 60-63; asst prof physics, NY Univ, 63-65; vis asst prof, Rutgers Univ, 65-66; PROF PHYSICS, STATE UNIV NY COL BUFFALO, 66- *Mem:* Am Phys Soc; Am Asn Physics Teachers. *Res:* Lasers and quantum electronics; optical pumping. *Mailing Add:* Dept Physics State Univ Col Buffalo 1300 Elmwood Ave Buffalo NY 14222

KOHLER, SIGURD H, b Uppsala, Sweden, Dec 1, 28; wid. NUCLEAR PHYSICS. *Educ:* Univ Uppsala, Fil Kand, 51, Fil Mag, 52, Fil Lic, 56, Fil Dr(theoret physics), 59. *Prof Exp:* Asst, Inst Meteorol, Univ Uppsala, 50-53 & Inst Theoret Physics, 53-57; fel theoret physics, Cern, Geneva, Switz, 57-59; res assoc, Cornell Univ, 59-60; asst res physicist, Univ Calif, Los Angeles, 60-61; spec res, AEC Sweden, Uppsala, 61-63; asst res physicist, Univ Calif, San Diego, 63-65; vis assoc prof physics, Rice Univ, 65-68; PROF PHYSICS, UNIV ARIZ, 68- *Mem:* Am Phys Soc. *Res:* Nuclear theory; many body problems; properties of nuclear matter. *Mailing Add:* Dept Physics Univ Ariz Tucson AZ 85721

KOHLHAW, GUNTER B, b Elbing, Ger, May 5, 31; m 59. BIOCHEMISTRY. *Educ:* Univ Freiburg, MS, 59, PhD(biochem), 62. *Prof Exp:* Res asst gen biochem, Univ Freiburg, 62-64; NATO fel intracellular regulation, 64-66, from asst prof to assoc prof, 66-73, PROF BIOCHEM, PURDUE UNIV, WEST LAFAYETTE, 73- *Mem:* Genetics Soc Am; Am Soc Biol Chemists; Ger Soc Biol Chem. *Res:* Gene structure function; metabolic regulation; analysis of the structure-function relationship of structural and regulatory genes in yeast; with special attention to the regulation of gene expression by upstream elements. *Mailing Add:* Dept Biochem Purdue Univ West Lafayette IN 47907

KOHLHEPP, SUE JOANNE, b Kittanning, Pa, July 15, 39. ANALYTICAL BIOCHEMISTRY. *Educ:* WVa Wesleyan Col, BS, 61; Pa State Univ, MS, 63, PhD(biophys), 69. *Prof Exp:* Teacher physics & phys sci, Marple-Newtown Sch Dist, Pa, 63-66; res assoc clin chem, St Anthony Hosp, Louisville, Ky, 69-74; res assoc biochem, Ore State Univ, 75-77; RES ASSOC, GILBERT RES LAB, PROVIDENCE MED CTR, PORTLAND, 77- *Concurrent Pos:* Assoc prof biochem, Catherine Spalding Col, 71. *Mem:* AAAS; Biophys Soc; Am Soc Microbiol; NY Acad Sci; fel Am Inst Chemists. *Res:* Identification and quantitation of metabolic products of the anti-tumor agent 1-2-chloroethyl-3-cyclohexyl-1-nitrosourea in rats, monkeys and humans as a means of deducing the mechanisms of action of the drug and decreasing its toxicity; infectious diseases such as molecular mechanism of aminoglycoside renal toxicity and subcellular distribution of gentamicin in renal cortical tissue; mechanism of action of toxins from Clostridium difficila. *Mailing Add:* 11704 NE 70th Ave Vancouver WA 98686

KOHLI, DILIP, b Kanpur, India, July 22, 47; m 77. MECHANICAL ENGINEERING. *Educ:* Indian Inst Technol, Kanpur, BS, 69, MS, 71; Okla State Univ, PhD(mech eng), 73. *Prof Exp:* Res asst mech eng, Okla State Univ, 71-73, assoc, 74; assoc & instr, Rensselaer Polytech Inst, 74-75; asst prof, 76-79, ASSOC PROF MECH ENG, UNIV WIS-MILWAUKEE, 79- *Concurrent Pos:* Consult, Procter & Gamble, 76-, Burroughs Corp, 77- & Control Data Corp, 78-; vis scientist, Univ Fla, Gainesville, 78- *Mem:* Am Soc Mech Engrs; Am Soc Eng Educ. *Res:* Kinematics and dynamics of machinery, vibrations, rotor dynamics and machine elements; robotics and manipulators. *Mailing Add:* Dept Mech Eng Univ Wis Milwaukee WI 53201

KOHLI, JAI DEV, b Jullundur City, India, Dec 27, 18; m 46; c 3. MEDICAL PHARMACOLOGY. *Educ:* Glancy Med Sch, Amritsar, India, Dipl, 42; Univ Chicago, MS, 51; Univ Man, PhD(pharmacol), 65. *Prof Exp:* Res fel pharmacol, Indian Coun Med Res, 42-50; jr sci officer, Coun Sci & Indust Res, India, 51-59, sr sci officer, 59-61; asst prof, Univ Man, 61-65; res scientist pharmacol, Food & Drug Directorate, Can, 65-70; asst dir, Indust Toxicol Res Ctr, 70-75; assoc prof, 75-78, RES PROF PHARMACOL, UNIV CHICAGO, 79- *Concurrent Pos:* Fulbright fel, 50; Wellcome res fel, 61. *Mem:* Sigma Xi; Am Soc Pharmacol & Exp Therapeut; Pharmacol Soc Can; Indian Pharmacol Soc; NY Acad Sci. *Res:* Receptor pharmacology; medicinal plants; cardiovascular pharmacology; dopamine and analogs. *Mailing Add:* Dept Pharmacol 947 E 58th St Chicago IL 60637-4931

KOHLMAN, DAVID L(ESLIE), b Houston, Tex, Oct 13, 37; m 59; c 2. AEROSPACE ENGINEERING. *Educ:* Univ Kans, BS, 59, MS, 60; Mass Inst Technol, PhD(aeronaut, astronaut), 63. *Prof Exp:* Res engr, Boeing Co, 63-64; from asst prof to assoc prof, Univ Kans, 64-70, chmn dept, 67-72, prof aerospace eng, 70-81, dir, flight res lab, 81-82; pres, 82-88, CHMN BD, KOHLMAN SYSTEMS RES INC, 88-; PRES, KOHLMAN AVIATION CORP, 77- *Concurrent Pos:* Consult, Centron Corp, 66-75, Beech Aircraft Corp, 69-70, Bell Helicopter Co, 70, Cessna Aircraft Co, 74 & 78-, NASA, 75-77, Gates Learjet, 78-, Piaggio, 78 & Singer-Link, 81-88; mem, Flight Mech Panel, NAtlantic Treat Org-Adv Group Aeronaut Res & Develop, 81-85. *Mem:* Assoc fel Am Inst Aeronaut & Astronaut; Soc Automotive Engrs. *Res:* Aerodynamic design of aircraft; aircraft stability and control; flight simulation; aircraft ice protection systems; flight testing. *Mailing Add:* PO Box 49533 Colorado Springs CO 80949

KOHLMAYR, GERHARD FRANZ, b Klagenfurt, Austria, Nov 30, 30; m 63. MATHEMATICAL PHYSICS, APPLIED MATHEMATICS. *Educ:* Graz Tech Univ, BS, 51, PhD(theoret physics), 59. *Prof Exp:* Sci asst, Darmstadt Tech Univ, 59-60; fel, Von Humboldt Found, 60-61; staff scientist, 61-71, sr appl mathematician, Pratt & Whitney Aircraft, 71-74, FOUNDER, MATHMODEL CONSULT BUR, 74-, FOUNDER, MATHMODEL PRESS, 79- *Concurrent Pos:* Adj asst prof, Rensselaer Polytech Inst, 61-66. *Mem:* Am Math Soc. *Res:* Acoustical duct lining theory; inconsistency of Zermelo-Fraenkel set theory; neutron transport theory; mathematical foundation of electrodynamic theory; elementary particle theory; negative solution of Hilbert's second problem; absolute invalidity of Hilbert's program; transient heat transfer; numerical operational calculus; generalized functions. *Mailing Add:* 80 Founders Rd Glastonbury CT 06033-3608

KOHLMEIER, RONALD HAROLD, b Craig, Nebr, Oct 16, 36; m 63; c 3. VETERINARY PHYSIOLOGY, NUTRITIONAL PHYSIOLOGY. *Educ:* Univ Nebr, BS, 59; Iowa State Univ, PhD(ruminant nutrit), 66, DVM, 68. *Prof Exp:* Farmer, 59; eng change notice coordr, RCA Missile Div, 60; asst nutritionist, Iowa State Univ, 60-66, res assoc nutrit & physiol, 66-68, asst prof, 68-70, assoc prof nutrit & physiol & res ruminant nutritionist & vet, 70-73; res ruminant nutritionist & vet, Agr Res Serv, USDA & assoc prof animal sci, Univ Nebr, 73-75; mgr tech serv processing group, 75-80, mgr feed sci servs, The Andersons, 80-; AT FEED FORTIFIERS INC; TECH DIR ANIMAL NUTRIT, SOYBEAN ASN, 88- *Concurrent Pos:* Nutrit consult for Dr Richard Hubbard, Gowrie Vet Serv, 68-73. *Mem:* Am Soc Animal Sci; Am Asn Bovine Practioners; Am Vet Med Asn; Am Asn Swine Practitioners; Sigma Xi. *Res:* Animal nutrition and usage of feed additives; animal nutrition and disease interrelationships. *Mailing Add:* Am Soybean Asn PO Box 27300 St Louis MO 63141

KOHLMEYER, JAN JUSTUS, b Berlin, Ger, Mar 15, 28; US citizen; m. MYCOLOGY. *Educ:* Univ Berlin, Dr rer nat(bot), 55. *Prof Exp:* Res asst mycol, Fed Inst Mat Testing, Berlin-Dahlem, Ger, 56-59 & Bot Mus, 60-64; asst prof, Inst Fisheries Res, 64-69; assoc prof, 69-74, PROF INST MARINE SCI, UNIV NC, 74- *Concurrent Pos:* Res assoc, Univ Wash, 59-60 & Duke Univ, 63-64. *Mem:* Mycol Soc Am; Brit Mycol Soc; Am Inst Biol Sci; Ger Soc Mycol; Sigma Xi. *Res:* Marine mycology; taxonomy of fungi; animal-fungus relationships; phytopathology. *Mailing Add:* Inst Marine Sci Univ NC Morehead City NC 28557

KOHLS, CARL WILLIAM, b Rochester, NY, Mar 14, 31. DISCRETE APPLIED MATHEMATICS. *Educ:* Univ Rochester, AB, 53; Purdue Univ, MS, 55, PhD(math), 57. *Prof Exp:* Res asst math, Purdue Univ, 57; instr, Columbia Univ, 57-58; asst prof, Univ Ill, 58-61; from asst prof to prof, 61-91, EMER PROF MATH, SYRACUSE UNIV, 91- *Concurrent Pos:* Res assoc & vis asst prof, Univ Rochester, 60-61; translr, Am Math Soc Russian Transl Proj, 88-91. *Mem:* Am Math Soc; Math Asn Am. *Res:* The study of regulatory systems using Boolean methods. *Mailing Add:* 215 Carnegie Bldg Syracuse Univ Syracuse NY 13244-1150

KOHLS, DONALD W, b Minneapolis, Minn, Oct 21, 34; m 62; c 2. EXPLORATION GEOLOGY. *Educ:* Carleton Col, BA, 56; Univ Minn, MS, 58, PhD(geol), 61. *Prof Exp:* Res scientist, NJ Zinc Co, 64-74, gen mgr & asst to pres, NJ Zinc Explor Co, 74-76; VPRES EXPLOR, GOLD FIELDS MINING CORP, 76-, MEM BD DIRS, 79- *Mem:* Am Asn Petrol Geologists; Am Inst Mining, Metall & Petrol Engrs; Geol Soc Am; Soc Econ Paleont & Mineral; Soc Econ Geologists. *Res:* Economic geology; petrology; mineralogy; geochemistry; field mapping. *Mailing Add:* Gold Fields Mining Co 1687 Cole Blvd Golden CO 80401

KOHLS, ROBERT E, b Portage, Wis, Mar 15, 31; m 54; c 4. VETERINARY PARASITOLOGY, ENTOMOLOGY. *Educ:* Univ Wis, BS, 53, MS, 55, PhD, 58. *Prof Exp:* Proj asst, Dept Vet Sci, Univ Wis, 58; dir res, Specifide, Inc, Ind, 58-59; vet parasitologist, Upjohn Co, 59-68; chief parasitol, 68-73, chief avian prod develop, 73-75, chief feed additives, 75-77, Norwich Pharmacal Co; chief parasitol, W Agro Chem Inc, 77-84; Prod mgr, 84, ANIMAL HEALTH CONSULT, VET DIV, BRISTOL LAB, 85- *Concurrent Pos:* Consult & pres animal health, Grey Fox Ltd, 85- *Mem:* Am Soc Parasitol. *Res:* Insect taxonomy; Diptera and Coleoptera; prophylactic worm control in cattle and sheep using phenothiazine; internal parasite control, especially prophylaxis. *Mailing Add:* RR 04 Box 317 Norwich NY 13815

KOHLSTAEDT, KENNETH GEORGE, b Indianapolis, Ind, May 10, 08; m 35; c 2. MEDICINE. *Educ:* Ind Univ, BS, 30, MD, 32. *Hon Degrees:* DSc, Ind Univ, 77. *Prof Exp:* Intern, Indianapolis Gen Hosp, 32-33, resident med, 33-34, resident neuropsychiat, 34-36; assoc, 37-47, from asst prof to prof med, Sch Med, Ind Univ, Indianapolis, 47-78, spec asst to dean, 74-77; RETIRED. *Concurrent Pos:* Asst supt, Indianapolis Gen Hosp, 36-44, med dir, 45; dir, Lilly Lab Clin Res, 45-60, dir clin res div, Eli Lilly & Co, 54-60, exec dir med res, 60-64, vpres med res, 64-73; chmn med adv comt, Coun High Blood Pressure Res, Am Heart Asn; mem drug res bd, Nat Acad Sci. *Honors & Awards:* Henery Elliot Award, Am Soc Clin Pharmacol & Therapeut, 78. *Mem:* Am Soc Clin Pharmacol & Therapeut (vpres, 74-75); master Am Col Physicians; Soc Exp Biol & Med; Am Physiol Soc; fel AMA. *Res:* Research administration; clinical pharmacology; cardiovascular research. *Mailing Add:* 1430 Paseo De Marcia Palm Springs CA 92264

KOHLSTEDT, SALLY GREGORY, b Ypsilanti, Mich, Jan 30, 43; m 66; c 2. HISTORY OF SCIENCE. *Educ:* Valparaiso Univ, BA, 65; Mich State Univ, MA, 66; Univ Ill, PhD(Am hist), 72. *Prof Exp:* Asst prof hist, Simmons Col, 71-75; from asst prof to assoc prof hist, Syracuse Univ, 75-88; PROF HIST SCI, UNIV MINN, 89- *Concurrent Pos:* Fel, Smithsonian Inst, 70-71; mem, US Nat Comt Int Union Hist & Philos Sci, 78-81; mem, adv comt, US Nat Archive Records Serv, 79-81; sr fel, Fulbright Found, Australia, 83; res fel, Woodrow Wilson Int Ctr, 86; consult, NY Acad Sci, 86-; NASA Hist Adv Comt, 84-87; chair, Sect L, AAAS, 86-87; Smithsonian Inst Sr Fel, 87. *Mem:* Hist Sci Soc (secy, 78-81); AAAS; Orgn Am Historians; Am Hist Asn. *Res:* History of the institutional development of scientific activity in the United States; professional origins of scientific societies; inclusion of women in scientific organizations; role of museums in connecting science and the public. *Mailing Add:* 107 Walter Libr Univ Minn Minneapolis MN 55455

KOHMAN, TRUMAN PAUL, b Champaign, Ill, Mar 8, 16; m 45; c 3. ASTRONOMY. *Educ:* Harvard Univ, AB, 38; Univ Wis, PhD(inorg & anal chem), 43. *Prof Exp:* Asst chem, Univ Wis, 38-42; res assoc metall lab, Univ Chicago, 42-44 & 45-46; chemist, Hanford Eng Works, Wash, 44-45; res assoc, Argonne Nat Lab, 46; fel chem, Inst Nuclear Studies, Univ Chicago, 46-48; from asst prof to prof, 48-81, EMER PROF CHEM, CARNEGIE-MELLON UNIV, 81- *Concurrent Pos:* NSF fel, Max-Planck Inst Chem, 57-58; vis prof, Indian Inst Technol, Kanpur, 62-63. *Honors & Awards:* Am Chem Soc Award, 62. *Mem:* Fel AAAS; Am Chem Soc; fel Am Phys Soc;

Geochem Soc; Am Astron Soc; Meteoritical Soc. *Res:* Artificial and natural radioactivity; nuclear reactions; geochronometry; meteorites; high-energy astronomy instrumenutation. *Mailing Add:* Dept Chem Carnegie-Mellon Univ Pittsburgh PA 15213

KOHN, ALAN JACOBS, b New Haven, Conn, July 15, 31; m 59; c 4. ZOOLOGY. *Educ:* Princeton Univ, AB, 53; Yale Univ, PhD(zool), 57. *Prof Exp:* Res assoc zool, Marine Lab, Univ Hawaii, 54-56; Anderson fel, Bingham Oceanog Lab, Yale Univ, 58; asst prof zool, Fla State Univ, 58-61; from asst prof to assoc prof, 61-67, PROF ZOOL, UNIV WASH, 67- *Concurrent Pos:* Biologist, Yale Exped to Seychelles Islands, 57-58; partic, Int Indian Ocean Exped, 63; sr vis res assoc, Smithsonian Inst, 67, res assoc, 85-, sr post-doctorial fel, 90; vis prof, Univ Hawaii, 68; Guggenheim Found fel, 74; adj cur malacol, Thomas Burke Mem Wash State Mus, 71-; adj prof, Inst Environ Studies, Quaternary Res Inst, 78-; prog officer, NSF, 85-86. *Mem:* Fel AAAS; Ecol Soc Am; Am Soc Zool (treas, 71-73); Am Soc Naturalists; Soc Syst Zool; Am Malacological Union (pres, 83); Am Soc Limnol & Oceanog; Brit Ecol Soc; Pac Sci Assoc; Int Soc Reef Studies. *Res:* Ecology, systematics and paleobiology of marine mollusks; coral reefs. *Mailing Add:* Dept Zool Univ Wash Seattle WA 98195

KOHN, ERWIN, b Vienna, Austria, Aug 23, 23; nat US; m 49; c 6. PHYSICAL CHEMISTRY, POLYMER CHEMISTRY. *Educ:* Univ Ill, BS, 48; Univ Notre Dame, MS, 50; Univ Tex, PhD(chem), 56. *Prof Exp:* From res chemist to sr res chemist, Monsanto Co, Tex, 55-62, res specialist, 62-66; assoc prof chem, Southwestern Okla State Univ, 66-68; assoc prof polymer chem, NDak State Univ, 68-72, dir NSF prog, 71-72; SR PROJ SCIENTIST, DEVELOP DIV, MASON & HANGER CO, 72- *Mem:* Fel Am Inst Chemists; Am Chem Soc; Am Phys Soc; AAAS. *Res:* Physical and physical-organic chemistry; analytical chemistry; liquid and gel permeation chromatography; explosives analysis; kinetics and mechanisms; Ziegler-Natta polymerization; structure of polyolefins; polymer characterization; organometallic chemistry; kinetic isotope effects; supercritical fluid chromatography. *Mailing Add:* Develop Div Mason & Hanger Co PO Box 30020 Amarillo TX 79177

KOHN, FRANK S, b Bristol, Pa, June 21, 42; m 82; c 2. ORGANIZATIONAL BEHAVIOR SCIENCES. *Educ:* NJ State Col Trenton, BA, 69; Drexel Univ, Philadelphia, Pa, MS, 72; Univ Wis-Madison, PDD, 79. *Prof Exp:* Virologist, NJ Dept Health, 63-69; from asst microbiologist to assoc microbiologist pharmaceut, Schering Plough Corp, 69-72, scientist, 76-78, dir oper pharm & biol, 78-82; dir mfg pharm & biol, Am Home Prod, Ft Dodge Labs, 82-86; DIR MFG PHARM, SANOFI INC, SANOFI ANIMAL HEALTH, 86- *Concurrent Pos:* Consult, Bristol Sanitation Co; instr org behavior, Iowa Cent Community Col, 82-87; consult org development, Mussic & Assoc. *Honors & Awards:* H Burrows Award, Franklin Sch Sci, 61. *Mem:* Am Soc Microbiologists; Parenteral Drug Asn; Am Soc Pharmaceut Engrs; AAAS; Am Acad Microbiol. *Res:* Pharmaceutical microbiology; environmental microbiology in clean rooms; infectious diseases; leptospirosis lab methods development; management to technical staff; interpersonal relationship; value engineering as applied to the pharmaceutical industry. *Mailing Add:* N332 Twin Lakes RR 1 Manson IA 50563

KOHN, FRED R, EXPERIMENTAL BIOLOGY. *Prof Exp:* PHARMACOLOGIST, XOMA CORP, 89- *Mailing Add:* Xoma Corp 2910 Seventh St Berkeley CA 94710

KOHN, GUSTAVE K, b Syracuse, NY, Feb 12, 10; m 50; c 3. CHEMISTRY. *Educ:* NY Univ, BS, 30. *Prof Exp:* Control chemist, Ortho Div, Chevron Chem Co, Standard Oil Co, Calif, 46-51, res chemist, 51-54, group leader org synthesis, 54-56, chief res chemist, 56-62, mgr cent res labs, 62-70, sr res scientist, 70-75; pesticide prod adv to govt India, UN Indust Develop Orgn, New Delhi, 75-76; res dir, Zoecon Corp, 76-78, sr scientist, 79-90, mgr licensing & technol, Sandor Corp, 84-90; RETIRED. *Concurrent Pos:* Coun, Am Chem Soc, 79-, immediate past chmn, agrochem div. *Mem:* AAAS; Am Chem Soc; Entom Soc Am; Am Inst Biol Sci. *Res:* Synthesis of biologically active and agriculturally useful compounds; organophosphate insecticides; halo-organic fungicides; plant and insect growth regulators. *Mailing Add:* 198 Pine Lane Los Altos CA 94022

KOHN, HAROLD WILLIAM, b Newark, NJ, Nov 9, 20; m 57; c 4. ENVIRONMENTAL SCIENCE. *Educ:* Univ Mich, Ann Arbor, BS, 43; Syracuse Univ, PhD(chem), 53. *Prof Exp:* Asst chem, Syracuse Univ, 48-53; res engr, Battelle Mem Inst, 52-53; chemist, Oak Ridge Nat Labs, 47-48 & 54-73; staff scientist environ, Ohio Environ Protection Agency, 73-86; RETIRED. *Concurrent Pos:* Vis lectr, Univ Calif, Berkeley, 63-64; vis prof, Dickinson Col, 71-72. *Mem:* Am Chem Soc; Health Physics Soc; Am Nuclear Soc. *Res:* Effects of ionizing radiations on heterogeneous catalysts; radiation chemistry of surfaces; molten salt chemistry; power plant siting and productivity; environmental impacts. *Mailing Add:* 147 Chatham Rd Columbus OH 43214

KOHN, HENRY IRVING, b New York, NY, Aug 19, 09; m 61; c 2. RADIATION EPIDEMIOLOGY, RADIATION BIOLOGY. *Educ:* Dartmouth Col, AB, 30; Harvard Univ, PhD(physiol), 35, MD, 46. *Prof Exp:* Gen Educ Bd fel, cell biol, Univ Stockholm, 35-36, Univ Cambridge, 36-37; instr & asst prof physiol & pharmacol, Med Sch, Duke Univ, 37-43; clin prof exp radiol, Univ Calif, San Francisco, 53-63; Fuller-Am Cancer Soc prof radiol, 63-68, Gaiser prof, 68-76, EMER PROF RADIATION BIOL, SCH MED, HARVARD UNIV, 76- *Concurrent Pos:* Sci secy, adv comt biol & med, US AEC, 57-61; dir, Shields Warren Radiation Lab, New Eng Deaconess Hosp, Boston, 64-79, Ctr Human Genetics, Med Sch, Harvard Univ, 71-76; Chair, Bikini Atoll Rehab Comt, US Dept Interior, 84-88; referee, Rongelap Reassessment Proj, Repub Marshall Islands, 88-90. *Res:* Radiation biology: epidemiology, toxicology, environmental. *Mailing Add:* 1203 Shattuck Ave Berkeley CA 94709

KOHN, HERBERT MYRON, b Chicago, Ill, Feb 24, 35; m 57; c 2. NEUROPSYCHOLOGY, ELECTROENCEPHALOGRAPHY. *Educ:* Univ Ill, BA, 58; Roosevelt Univ, MA, 60; Ill Inst Technol, PhD(psychol), 65. *Prof Exp:* Med res assoc, Ill State Psychiat Inst, 60-67; dir, Darrow Mem Lab, Inst Juv Res, 68-70; res scientist, Ill State Pediat Inst, 71-72; asst prof, 72-75, ASSOC PROF PSYCHIAT, RUTGERS MED SCH, COL MED & DENT NJ, 75-, CHIEF NEURODIAG LAB, 72- *Concurrent Pos:* Lectr, Roosevelt Univ, 65-72; asst prof, Ill Inst Technol, 67-72 & Abraham Lincoln Sch Med, Univ Ill, 68-72; lectr, Northeastern Ill State Univ, 70-72 & Univ Ill, Chicago Circle, 72; assoc psychobiol, Grad Fac, Rutgers Univ, 72-78; adj assoc prof, Grad Sch of Appl & Prof Psychol, Rutgers Univ, 76- *Mem:* Am Psychol Asn; Psychonomic Soc; Int Neuropsychol Soc. *Res:* Neural bases of human behavior; primate behavior; vision and effects of early brain damage. *Mailing Add:* Dept Psychiat Univ Med & Dent Robert Wood Johnson Sch 675 Hoes Lane Piscataway NJ 08854

KOHN, JACK ARNOLD, b Trenton, NJ, July 17, 25; m 51; c 2. MATERIALS RESEARCH, CRYSTALLOGRAPHY. *Educ:* Univ Mich, BS, 47, MS, 48, PhD(mineral), 50. *Prof Exp:* Asst mineral, Univ Mich, 48-50, res assoc, 50-51; mineralogist, Electrotech Lab, US Bur Mines, 51-55; physicist & dep dir solid state sci div, US Army Electronics Res & Develop Command, Ft Monmouth, 55-69, dep dir inst explor res, 69-71, dep dir technol, 71-73, dir electronic mat res, Electronics Technol & Devices Lab, 74-85. *Concurrent Pos:* Consult mineralogist, US Bur Mines, 50-51. *Honors & Awards:* Prize, Army Sci Conf, 59, 62 & 70. *Mem:* Fel AAAS; fel Mineral Soc Am; Am Crystallog Asn. *Res:* Crystallography of electronic and magnetic materials; twinning; polymorphism; polytypism; general x-ray crystallography. *Mailing Add:* 65 Wigwam Rd Locust NJ 07760

KOHN, JAMES P(AUL), b Dubuque, Iowa, Oct 31, 24; m 58; c 3. CHEMICAL ENGINEERING. *Educ:* Univ Notre Dame, BS, 51; Univ Mich, MSE, 52; Univ Kans, PhD, 56. *Prof Exp:* Chem engr, Reilly Tar & Chem Corp, 46-51; from asst prof to assoc prof, 55-64, PROF CHEM ENG, UNIV NOTRE DAME, 64- *Concurrent Pos:* Consult, Eng Enterprises; dir, Solar Lab for Thermal Appln, 74- *Honors & Awards:* D L Katz Award, Gas Processors Asn, 88. *Mem:* AAAS; Am Chem Soc; Am Inst Chem Engrs. *Res:* Heterogeneous phase equilibrium; applied thermodynamics; unsteady state diffusion; physical properties; molecular transport. *Mailing Add:* Dept Chem Eng Univ Notre Dame Notre Dame IN 46556

KOHN, JOSEPH JOHN, b Prague, Czech, May 18, 32; nat US; m 66; c 3. MATHEMATICS. *Educ:* Mass Inst Technol, BS, 53; Princeton Univ, MA, 54, PhD(math), 56. *Prof Exp:* Instr math, Princeton Univ, 56-57; mem, Inst Advan Study, 57-58; from asst prof to prof math, Brandeis Univ, 58-68; PROF MATH, PRINCETON UNIV, 68-, CHMN DEPT, 74- *Concurrent Pos:* Ed, Transactions of Am Math Soc, Advances Math & Annals Math. *Honors & Awards:* Steele Prize, Am Math Soc, 79. *Mem:* Nat Acad Sci; Am Acad Arts & Sci; Am Math Soc. *Res:* Several complex variables; partial differential equations. *Mailing Add:* Dept Math Princeton Univ Princeton NJ 08540

KOHN, KURT WILLIAM, b Austria, Sept 14, 30; nat US; m 56; c 2. CHEMICAL PHARMACOLOGY. *Educ:* Harvard Univ, AB, 52, PhD(biochem), 66; Columbia Univ, MD, 56. *Prof Exp:* Clin assoc, Nat Cancer Inst, 57-59, SR INVESTR, NAT CANCER INST, 59-, CHIEF LAB MOLECULAR PHARMACOL, 68- *Mem:* Am Chem Soc; Am Asn Cancer Res; Am Asn Biol Chemists. *Res:* Effects of chemotherapeutic agents on structure and function of deoxyribonucleic acid. *Mailing Add:* 11519 Gainsborough Rd Potomac MD 20854

KOHN, LEONARD DAVID, b New York, NY, Aug 1, 35; m 62; c 2. BIOCHEMICAL PHARMACOLOGY. *Educ:* Columbia Univ, BA, 57, MD, 61. *Prof Exp:* Intern med, Columbia Presby Med Ctr, 61-62, asst resident, 62-63, sr resident, 63-64; res assoc, Lab Biochem & Metab, 64-66, med res officer, Lab Biochem Pharmacol, 66-74, CHIEF SECT BIOCHEM OF CELL REGULATION, LAB BIOCHEM PHARMACOL, NAT INST ARTHRITIS, METAB & DIGESTIVE DIS, 74- *Concurrent Pos:* Vis prof, Dept Med, Univ Liege, Belg, 70-71. *Mem:* Am Soc Biol Chemists. *Res:* Mechanism by which hormones interact with membrane components to elicit functional responses; enzymatic conversion of precursors of collagen to collagen; enzymes concerned with solute transport across membranes. *Mailing Add:* Sect Cell Regul NIDDK Bldg 10 Rm 9B13 NIH Bethesda MD 20892

KOHN, MICHAEL, b Budapest, Hungary, June 18, 34; US citizen; m 55; c 1. BIOMEDICAL ENGINEERING, NEUROPHYSIOLOGY. *Educ:* City Univ New York, BEE, 60, MEE, 68; NY Univ, PhD(elec eng), 74. *Prof Exp:* Res engr, 57-68, DIR, BIOENG DEPT, ROCKLAND RES INST, 68- *Concurrent Pos:* Consult, Mnemotron Corp, 61-62. *Res:* Development of biomedical instrumentation; analysis of electrophysiological data. *Mailing Add:* Nathan Kline Inst Orangeburg NY 10962

KOHN, MICHAEL CHARLES, b Brooklyn, NY, July 29, 41; m 71. BIOCHEMISTRY, THEORETICAL CHEMISTRY. *Educ:* Mass Inst Technol, BS, 64; Univ SC, PhD(chem), 70. *Prof Exp:* Technician chem, Gen Latex & Chem Co, 62; consult, BB Chem Co, 62-63; fel, Univ Tex, Austin, 69-71; chemist, Naval Undersea Res & Develop Ctr, 71-73; sr res investr & adj assoc prof, Univ, Pa, 74-84; at Dept Physiol, Duke Med Ctr, 84-91. *Concurrent Pos:* Nat Res Coun grant, Naval Undersea Res & Develop Ctr, 71-73. *Mem:* Am Chem Soc; Sigma Xi; Soc Comput Simulation; NY Acad Sci. *Res:* Valence force field calculations of strain energy; molecular orbital theory; statistical mechanics of polymer solutions; biomedical computer models; sensitivity analysis; graph-theoretical analysis of metabolic networks. *Mailing Add:* Dept Comput & Info Sci Univ Pa Philadelphia PA 19104

KOHN, WALTER, b Vienna, Austria, Mar 9, 23; nat US; m 78; c 3. PHYSICS. *Educ:* Univ Toronto, BA, 45, MA, 46; Harvard Univ, PhD(physics), 48. *Hon Degrees:* LLD, Univ Toronto, 67; DSC, Univ Paris, 80, Brandeis Univ, 81; DPhil, Hebrew Univ, Jerusalem, 81; DSc, Queens Univ, Can, 86. *Prof Exp:*

Instr physics, Harvard Univ, Cambridge, Mass, 48-50; prof, Carnegie Mellon Inst, Pittsburgh Pa, 50-60; prof, 60-81, chmn dept, San Diego, 61-63, dir, Inst Theoret Physics, Santa Barbara, 79-84, PROF PHYSICS, UNIV CALIF, SANTA BARBARA, 84- Concurrent Pos: Nat Res Coun fel, Copenhaagen, 50-51; Oersted fel, 51-52; sr NSF fel, Imperial Col, 58; NSF sr fel, Univ Paris, 67; Guggenheim fel, 63; vis prof, Superior Normal School, Paris, 63-64 & Hebrew Univ, Jerusalem, 70; Battelle distinguished vis prof, Univ Wash, 74; mem solid state sci panel, Nat Acad Sci; ed, J Non-Metals & J Physics & Chem of Solids; vis scholar, Univ Pa, Univ Mich, Univ Wash, Seattle, Univ Paris, Univ Copenhaagen, Univ Jerusalem, Imperial Col, London, ETH, Zurich, 58-85; consult, Westinghouse Res Lab, 53-57, Bell Tel Labs, 53-66, Gen Atomic, 60-72, IBM, 78; mem, Brookhaven Nat Labs, Argonne Nat Labs, Oak Ridge Nat Labs, Int Adv Comt Strongly Interacting Plasmas. Honors & Awards: Oliver E Buckley Prize, 60; Davisson Germer Prize, 77; Nat Medal Sci, President US, 88. Mem: Nat Acad Sci; fel Am Acad Arts & Sci; fel Am Phys Soc; fel AAAS. Res: Theory of solids; surface physics; collision theory. Mailing Add: Dept Physics Univ Calif Santa Barbara CA 93106

KOHNHORST, EARL EUGENE, b Louisville, Ky, Apr, 15, 47; m 72; c 1. PROCESS & ENGINEERING AUTOMATION, CHEMICAL & SENSORY RESEARCH. Educ: Univ Louisville, BChE, 70, MChE, 71. Prof Exp: Process engr, Brown & Williamson Tobacco Corp, 71-76, mgr, Develop Ctr, 76-79, dir mfg planning, 79-80, dir mfg planning & eng, 80-83, VPRES RES DEVELOP & ENG, BROWN & WILLIAMSON TOBACCO CORP, 83- Mem: Am Inst Chem Engrs. Res: Catalytic conversion of nitric oxides using rare earth catalysts; determining mechanisms and kinetic rate equations. Mailing Add: Brown & Williamson Tobacco Corp 1500 Brown & Williamson Tower Louisville KY 40202

KOHNKE, HELMUT, b Rostov, Russia, Aug 6, 01; nat US; m 36; c 2. AGRONOMY. Educ: Univ Berlin, BSc, 25, DrAgr, 26; Univ Alta, MSc, 32; Ohio State Univ, PhD(soils), 34. Prof Exp: Agr agronomist, Ger, 26-27; instr Ger, Univ Alta, 29-30; soil surveyor, Soil Conserv Serv, USDA, 34-35, chg soil dept, Northern Appalachian Watershed Exp Sta, 35-39, Ind Agr Hydrol Studies, 39-43; soil scientist, 43-70, EMER PROF AGRON, PURDUE UNIV, WEST LAFAYETTE, 70- Concurrent Pos: Exec dir comt sci & soc, Ind Acad Sci, 70-72; soil conserv consult UN, Food & Agr Orgn, Bulgaria, 72; consult soil scientist, Corn Prod Syst, Inc, Ill, 75. Mem: Fel AAAS; fel Am Soc Agron; Soil Sci Soc Am. Res: Soil fertility; physics; conservation; hydrology; run-off chemistry. Mailing Add: 208 Forest Hill Dr W Lafayette IN 47906

KOHOUT, FREDERICK CHARLES, III, b Flint, Mich, June 19, 40; m 61; c 3. LUBRICATION SCIENCE, ENVIRONMENTAL SCIENCE. Educ: Mich State Univ, BS, 62; Pa State Univ, PhD(phys chem), 66. Prof Exp: Mem staff, Cent Res Div, 66-73, mem staff, Prod Res Div, 73-81, mgr comp res & environ serv, 81-88, MGR & RES CONSULT, RES SER DIV, MOBIL RES & DEVELOP, PAULSBORO LAB, 88- Mem: Am Chem Soc; Am Soc Lubrication Engrs. Res: Development of marine diesel lubricants, gear lubricants and greases; analytical chemistry of petroleum streams and products; environmental science; groundwater. Mailing Add: 305 Seneca Dr Wenonah NJ 08090

KOHRMAN, ARTHUR FISHER, b Cleveland, Ohio, Dec 19, 34; m 55; c 4. PEDIATRICS. Educ: Univ Chicago, BA & BS, 55; Western Reserve Univ, MD, 59. Prof Exp: NIH trainee & spec fel pediat, Stanford Univ, 65-68; from asst prof to prof, Col Human Med, Mich State Univ, 68- 81, assoc dean educ prog & prof med educ res & develop, 77-80; PROF & ASSOC CHMN DEPT PEDIAT, PRITZKER SCH MED, UNIV CHICAGO, 81-; DIR, LA RABIDA CHILDREN'S HOSP & RES CTR, 81- Concurrent Pos: Prof biol sci, Col Div, Univ Chicago, 85- Mem: Lawson Wilkins Prod Endocrinol Soc; AAAS; Am Acad Pediat; Am Pediat Soc; Soc Pediat Res. Res: Childhood chronic disease and health policy; developmental endocrinology and biochemistry; effects of environmental agents on human development. Mailing Add: La Rabida Children's Hosp & Res Ctr E 65th St at Lake Michigan Chicago IL 60649

KOHRT, CARL FREDRICK, b Normal, Ill, Dec 18, 43; m 62; c 3. SURFACE SCIENCE, PHOTOGRAPHIC IMAGING SCIENCES. Educ: Furman Univ, BS, 65; Univ Chicago, PhD(phys chem), 71; Mass Inst Technol, MS, 91. Prof Exp: Fel, James Frank Inst, Univ Chicago, 70-71; sr scientist, Res Labs, Eastman Kodak Co, 71-76, lab head, Color-Photog Res Div, 76-82, asst div dir, Instant Photog Res Div, 82-83, analyst, Corp Strategic Planning Off, 83, asst exec vpres, Corp Staff, 84-85, div dir, Hybrid Imaging Systs Div, 85-87, group lab dir, Photog Res Labs, 87-90; Sloan fel, Sloan Sch Mgt, Mass Inst Technol, 90-91. Mem: Soc Photographic Scientists & Engrs. Res: Focus on heterogeneous catalysis or thermal catalysis to generate dyes or other photographically useful species; developed photographic quality sublimation thermal imaging systems and other digital imaging systems. Mailing Add: Eastman Kodak Co Rochester NY 14650-1713

KOHUT, ROBERT JOHN, b Cannonsburg, Pa, Nov 19, 43. POLLUTION EFFECTS ON VEGETATION, VEGETATION STRESS. Educ: Pa State Univ, BS, 65, MS, 72, PhD(plant path), 75. Prof Exp: Res fel, Dept Plant Path, Univ Minn, 75-77; plant pathologist, Environ Res & Technol, 77-80; RES ASSOC, BOYCE THOMPSON INST, CORNELL UNIV, 80- Concurrent Pos: Affil fac, Dept Plant Path, Colo State Univ, 78-80; comt mem, Colo Gov Air Pollution Tech Working Comt, 79-80, Toxic Substances Subcomt, Environ Protection Agency Sci Adv Bd, 80- Mem: Am Phytopath Soc; Air Pollution Control Asn; Sigma Xi. Res: Field and laboratory research evaluating the effects of air pollutants on growth and yield of agricultural crops, trees and on native plants and plant communities. Mailing Add: 214 Eastern Heights Dr Ithaca NY 14850

KOIDE, FRANK T, b Honolulu, Hawaii, Dec 25, 35; c 2. BIOMEDICAL & ELECTRONICS ENGINEERING. Educ: Univ Ill, BSEE, 58; Clarkson Univ, MS, 61; Univ Iowa, PhD(physiol), 66. Prof Exp: Engr res div, Collins Radio Co, 59-61; asst prof elec eng, physiol & biomed eng, Iowa State Univ, 66-68; prin res scientist, Life Sci Div, Technol Inc, Tex, 68-69; assoc prof elec eng, 69-74, PROF ELEC ENG & PHYSIOL, UNIV HAWAII, 74- Concurrent Pos: Instr, Cedar Rapids Adult Educ, 60-61; consult, Collins Radio Co, 61-63; NASA-Am Soc Eng Educ fac fel, 67; consult, Shared Clin Eng Servs Hawaii & Acupuncture Asn Hawaii, 74-; external examr, Chinese Univ Hong Kong, 77- Mem: Sigma Xi; Asn Advan Med Instrumentation; Inst Elec & Electronics Engrs; AAAS. Res: Application of engineering techniques in solution of biomedical problems; membrane physiology; electrophysiology; nerve; bioinstrumentation; nutrition. Mailing Add: Dept Elec Eng Univ Hawaii 2540 Dole St Honolulu HI 96822

KOIDE, ROGER TAI, b Berkeley, Calif, Dec 14, 57; m 79; c 2. PLANT PHYSIOLOGICAL ECOLOGY. Educ: Pomona Col, BA, 80; Univ Calif, Berkeley, PhD(bot), 84. Prof Exp: Postdoctoral res affil plant ecol, Stanford Univ, 84-86; ASST PROF PLANT ECOL, PA STATE UNIV, 86- Concurrent Pos: NSF presidential young investr award, 87. Mem: Ecol Soc Am; Bot Soc Am; Am Soc Plant Physiologists; AAAS. Res: Plant physiological ecology; nutrient ecology; reproductive biology; mycorrhizal symbiosis. Mailing Add: Dept Biol Pa State Univ 208 Mueller University Park PA 16802

KOIDE, SAMUEL SABURO, b Honolulu, Hawaii, Oct 6, 23; m 60; c 3. BIOCHEMISTRY, INTERNAL MEDICINE. Educ: Univ Hawaii, BS, 45; Northwestern Univ, MD, 53, MS, 54, PhD(biochem), 60. Prof Exp: Assoc, Sloan-Kettering Inst, 60-65, asst prof biochem, 64-65; asst dir biomed div, 65-70, ASSOC DIR & SR SCIENTIST, CTR FOR BIOMED RES, POP COUN, ROCKEFELLER UNIV, 70- Concurrent Pos: Asst prof, Cornell Univ, 61-65; Nat Inst Arthritis & Metab Dis career develop award, 63-65. Honors & Awards: Joseph A Capps Prize Med Res, 58. Mem: Biochem Soc; Endocrine Soc; Am Col Physicians; Am Soc Biol Chemists; Soc Exp Biol & Med; Am Col Physicians. Res: Signal transduction system in genetics; immunobiology of sperm; causation of unexplained infertility; reproductive biology. Mailing Add: Biomed Div Pop Coun Rockefeller Univ New York NY 10021

KOIKE, HIDEO, b Hilo, Hawaii, Mar 10, 21; m 48; c 3. PHYTOPATHOLOGY. Educ: Univ Hawaii, BA, 44; Kans State Univ, MS, 51, PhD(bact), 56. Prof Exp: Asst, Kans State Univ, 49-51; asst pathologist, Exp Sta, Hawaiian Sugar Planters Asn, 52-54; asst, Kans State Univ, 54-56; assoc pathologist, Exp Sta, Hawaiian Sugar Planters Asn, 57-66; res microbiologist sugarcane & sweet sorghum invest, Tobacco & Sugar Crops Res Br, Crops Res Div, Agr Res Serv, Univ PR, Gurabo, 66-69; res plant pathologist, US Sugarcane Field Lab, USDA, 69-83; CONSULT, SUGARCANE PATH, 84- Mem: Fel AAAS; Am Phytopath Soc; Int Soc Plant Path; Sigma Xi; Int Soc Sugarcane Technol. Res: Sugarcane pathology. Mailing Add: 43 Alamo Dr Houma LA 70360

KOIKE, THOMAS ISAO, b Watsonville, Calif, July 27, 27; m 55; c 2. PHYSIOLOGY. Educ: Univ Calif, Berkeley, AB, 51, PhD(physiol), 58. Prof Exp: Jr res physiologist, Univ Calif, Berkeley, 58-61; USPHS fel animal physiol, Univ Calif, Davis, 61-63, asst specialist physiol, 63-64, asst res physiologist, 64-65; from asst prof to assoc prof, 65-78, PROF PHYSIOL, MED CTR, UNIV ARK, LITTLE ROCK, 78- Concurrent Pos: Co-prin investr grants, Nat Inst Arthritis & Metab Dis, 63-65; NIH grant, 66-72. Mem: AAAS; Am Physiol Soc; Soc Exp Biol & Med; NY Acad Sci; Am Asn Univ Professors; Sigma Xi. Res: Regulation of body fluids. Mailing Add: Dept Physiol Univ Ark Med Ctr Little Rock AR 72205

KOIRTYOHANN, SAMUEL ROY, b Washington, Mo, Sept 11, 30; m 52; c 3. ANALYTICAL CHEMISTRY. Educ: Univ Mo, BS, 53, MS, 58, PhD(agr chem), 66. Prof Exp: Chemist, Oak Ridge Nat Lab, 59-63; from instr to asst prof, 63-70, assoc prof agr chem, 70-80, chmn chem dept, 84-90, PROF CHEM, UNIV MO, COLUMBIA, 80- Mem: Am Chem Soc; Soc Appl Spectros. Res: Determination of trace elements in biological and agricultural materials using spectroscopic and other instrumental methods. Mailing Add: 123 Chem Univ Mo Columbia MO 65211

KOISTINEN, DONALD PETER, b Lake Norden, SDak, Nov 19, 27; m 59; c 3. METAL PHYSICS. Educ: Univ Mich, BS, 52; Wayne State Univ, MS, 58. Prof Exp: Res physicist, 52-58, sr res physicist, 58-69, supvry res physicist metal physics, 69-82, SR RES SCIENTIST, RES LABS, GEN MOTORS CORP, 82- Mem: Am Phys Soc; Sigma Xi; Metall Soc; Am Inst Mining, Metall & Petrol Engrs. Res: Mechanics of large-scale plasticity in metals; strain hardening; precipitation; crystalline deformation mechanisms and transformations; fatigue in metals; surface hardening techniques. Mailing Add: 603 Dorchester Apt Y Rochester Hills MI 48307

KOIVO, ANTTI J, b Vaasa, Finland, Apr 9, 32; m 69; c 2. ELECTRICAL ENGINEERING, ROBOTICS BIOENGINEERING. Educ: Inst Technol, Finland, dipl eng, 56; Indiana Univ, MS; Cornell Univ, PhD(elec eng), 63. Prof Exp: Design engr, Oy Stroemberg Ab, Finland, 57-59; from asst prof to assoc prof, 64-78, PROF ELEC ENG, PURDUE UNIV, WEST LAFAYETTE, 78- Mem: Sr mem Inst Elec & Electronics Engrs; Sigma Xi. Res: Robotics, info techn for sensors-based robots, multiple manipulators; application of system theory and pattern recognition to biomedical problems. Mailing Add: Dept Elec Eng Purdue Univ West Lafayette IN 47907

KOIZUMI, CARL JAN, b Reno, Nev, Jan 7, 43; m 68; c 1. NUCLEAR GEOPHYSICS, SOLID STATE PHYSICS. Educ: Univ Nev, Reno, BS, 65, MS, 73, PhD(physics), 77; Ariz State Univ, MS, 67. Prof Exp: Res geophysicist, Bendix Field Eng Corp, 77-81; res physicist, Austin Res Ctr, Gearhart Indust, 82-86; staff scientist, Rockwell Hanford, 86-87; PRIN SCIENTIST, WESTINGHOUSE HANFORD, 87- Concurrent Pos: Mem, Borehole Sensors Task Group, Am Soc Testing & Mat, 79-86; mem, Spectral

Gamma-Ray Calibration Comt, Am Petrol Inst, 82-90. *Mem:* Am Phys Soc; Soc Prof Well Log Analysts. *Res:* Calibration of devices used for uranium detection; gamma-ray logging theory; neutron logging theory; Mossbauer spectroscopic studies of hydrides of intermetallic compounds; modeling responses of nuclear logging tools with radiation transport calculations; characterization of nuclear waste disposal sites by nuclear logging. *Mailing Add:* 315 Spokane St Richland WA 99352

KOIZUMI, KIYOMI, b Kobe, Japan, Sept 4, 24; m 54; c 1. AUTONOMIC NERVOUS SYSTEM, NEUROENDOCRINOLOGY. *Educ:* Tokyo Women's Med Col, MD, 47; Wayne State Univ, MS, 51; Kobe Med Col, PhD(physiol), 57. *Prof Exp:* Fel physiol, State Univ NY Downstate Med Ctr, 51-52, from instr to asst prof, 52-60; vis lectr, Kobe Med Col, 60-61; assoc prof, 63-70, PROF PHYSIOL, STATE UNIV NY DOWNSTATE MED CTR, 70- *Concurrent Pos:* Hon res fel, Aberdeen Univ, 62; NIH res grants, 55-; NSF grants, 74-78; vis scientist, Univ Heidelberg, WGer, 71, Tokyo Metrop Inst Geront, 76, Semmelweis Med Univ, Hungary, 79-80 & vis prof, Univ Occup Health, Japan, 84 & 89-90; assoc ed, J Autonomic Nerv Syst; sr int fel, Fogarty Ctr, NIH, 79-80. *Honors & Awards:* Medal Hon, Semmelweis Med Univ, Hungary, 79. *Mem:* Soc Neurosci; Int Brain Res Orgn; Am Physiol Soc; Harvey Soc; fel NY Acad Sci; Sigma Xi. *Res:* Neurophysiology; neuroendocrinology. *Mailing Add:* Dept Physiol State Univ NY Health Sci Ctr Brooklyn NY 11203

KOJIMA, HARUO, b Japan, May 18, 45; m 70; c 2. LOW TEMPERATURE PHYSICS. *Educ:* Univ Calif, Los Angeles, BS, 68, MS, 70, PhD(physics), 72. *Prof Exp:* Adj asst prof, Univ Calif, Los Angeles, 72-73; res assoc, Univ Calif, San Diego, 73-75; from asst prof to assoc prof, 75-87, PROF PHYSICS, RUTGERS UNIV, 87- *Mem:* Am Phys Soc. *Res:* Experimental investigation of superfluid phases of helium at ultra low temperatures. *Mailing Add:* Dept Physics Rutgers Univ New Brunswick NJ 08903

KOJOIAN, GABRIEL, b Providence, RI, Dec 11, 27; m 53; c 2. ASTRONOMY, RADIO ASTRONOMY. *Educ:* Brown Univ, BSc, 52; Univ Calif, Berkeley, DPhil(physics), 66. *Prof Exp:* Head theoret div physics, Div Lab Electronics, Tracerlab, 66-67; Nat Acad Sci res fel, NASA-Ames Res Ctr, 67-69; lectr, Dept Physics, Univ Mass, Amherst, 69-71; staff scientist astron, Northeast Radio Observ, Mass Inst Technol, 72-73; assoc prof, Pahlavi Univ, Shiraz, Iran, 75-76; vis assoc prof, 76-80, ADJ ASSOC PROF PHYSICS & ASTRON, UNIV WIS-EAU CLAIRE, 80- *Concurrent Pos:* Invited guest, Am Acad Sci, Soviet Socialist Repub, 76; exchange scientist, Nat Acad Sci & Soviet Acad Sci, 78. *Mem:* Am Astron Soc; AAAS; Sigma Xi. *Res:* Radio-continuum measurements of galactic and extragalactic objects. *Mailing Add:* Dept Physics & Astron Univ Wis Eau Claire WI 54701

KOK, LOKE-TUCK, b Ipoh, Malaysia, Nov 10, 39; m 66; c 1. ENTOMOLOGY, BIOLOGICAL CONTROL. *Educ:* Univ Malaya, BAgrSc Hons, 63, MAgrSc, 65; Univ Wis-Madison, PhD(entom), 71. *Prof Exp:* Tutor, Univ Malaya, 63-65, from asst lectr to lectr, 65-71; from asst prof to assoc prof, 78-82, PROF ENTOM, VA POLYTECH INST & STATE UNIV, 82- *Concurrent Pos:* Res scholar, Int Rice Res Inst, Philippines, 64; from res asst to res assoc, Univ Wis-Madison, 68-71. *Honors & Awards:* Nat Agr Recognition Award, Outstanding Contrib to Agr, Entom Soc Am, 88. *Mem:* Entom Soc Can; Entom Soc Am; Weed Sci Soc Am; Int Orgn Biol Control. *Res:* Biological control of insect and weed pests of forage and field crops in Virginia, with special emphasis on the control of Carduus thistles using introduced beneficial insects; pest management of cruciferous crop pests. *Mailing Add:* Dept Entom Va Polytech Inst & State Univ Blacksburg VA 24061

KOKALIS, SOTER GEORGE, b E Chicago, Ind, Jan 29, 36. INORGANIC CHEMISTRY. *Educ:* Purdue Univ, BSc, 58; Univ Ill, MSc, 60, PhD(inorg chem), 62. *Prof Exp:* Asst inorg chem, Univ Ill, 59-62; asst prof, Wash Univ, 62-64 & Univ Ill, Chicago Circle, 64-67; assoc prof, Chicago State Univ, 67-69; ASSOC PROF INORG CHEM, WILLIAM RAINEY HARPER COL, 69- *Concurrent Pos:* Consult, Col Bd Advan Placement Chem Exam, Income Tax Planning. *Mem:* AAAS; Am Chem Soc; Royal Soc Chem; Am Col Sports Med. *Res:* Synthesis and chemical properties of inorganic ring structures; analysis of electron delocalization in heterocyclic compounds; phosphonitrilic compounds and their applications. *Mailing Add:* Dept Chem William Rainey Harper Col Palatine IL 60067-7398

KOKAME, GLENN MEGUMI, b Waimea, Hawaii, July 7, 26; m 53; c 2. SURGERY, THORACIC SURGERY. *Educ:* Univ Hawaii, BA, 50; Tulane Univ, BS, 52, MD, 55; Am Bd Surg, dipl, 62; Am Bd Thoracic Surg, dipl, 63. *Prof Exp:* From instr to asst prof surg, Sch Med, Tulane Univ, 55-67; asst prof, 67-71, ASSOC PROF SURG, SCH MED, UNIV HAWAII, 71- *Concurrent Pos:* Am Cancer Soc adv clin fel, 64-66. *Mem:* Fel Am Col Surgeons; AMA; Am Asn Cancer Res; Am Soc Clin Oncol. *Res:* Regional chemotherapy of cancer; hyperbaric oxygenation in medicine; immunology of cancer; vascular surgery; heterotransplantation of human cancer and tissue culture; transplantation of organs. *Mailing Add:* 321 Kuakini St Suite 3007 Honolulu HI 96817

KOKAS, ESZTER B, PHYSIOLOGY. *Educ:* Univ Budapest, MD, 27. *Prof Exp:* Prof physiol, Univ NC; RETIRED. *Mailing Add:* Highland Hills Apt No 4C Carrboro NC 27510

KOKATNUR, MOHAN GUNDO, b Belgaum, India, Mar 19, 30; m 63; c 2. CLINICAL BIOCHEMISTRY, NUTRITION. *Educ:* Univ Poona, BS, 51; Univ Nagpur, BS, 53; Univ Ill, Urbana, PhD(food sci, biochem, nutrit), 59. *Prof Exp:* Res assoc food sci & lipids, Univ Ill, Urbana, 59-61; Coun Sci & Indust Res Pool fel biochem & nutrit, Cent Food Res Inst, Mysore, India, 61-63; res assoc nutrit biochem, Univ Ill, Urbana, 63-66; asst prof, 66-72, ASSOC PROF PATH, LA STATE UNIV MED CTR, NEW ORLEANS, 72-; ASSOC DIR, CLIN CHEM LAB, CHARITY HOSP, NEW ORLEANS, 78- *Concurrent Pos:* La Heart Asn sr res grant-in-aid, 67-69; mem coun arteriosclerosis, Am Heart Asn, 68. *Mem:* Am Soc Clin Nutrit; Am

Inst Nutrit; Am Asn Clin Chemists; Soc Exp Biol Med. *Res:* Lipid chemistry, biochemistry and metabolism; importance of lipids and nutrition in atherosclerosis; lipids and atherosclerosis; clinical chemistry methodology; vitamin E deficiency and fat oxidation. *Mailing Add:* Dept Path La State Univ Med Ctr New Orleans LA 70112

KOKENGE, BERNARD RUSSELL, b Dayton, Ohio, Dec 7, 39; m 59; c 2. INORGANIC CHEMISTRY. *Educ:* Univ Dayton, BS, 61; Ohio Univ, PhD(inorg chem), 66. *Prof Exp:* Lab technician, Wright-Patterson AFB, summers 60 & 61; sr res chemist, Mound Lab, Monsanto Res Corp, Miamisburg, Ohio, 65-66, group leader inorg chem & isotopic fuels, 66-72, plutonium processing mgr, 72-77, mgr nuclear technol, 77-82, dir, Nuclear Oper Dept, 82-85, assoc dir & mound dir, Advan Devices Dept, 85-86; chmn, Gen Studies Dept, Ky Christian Col, Grayson, Ky, 86-88, vpres, Strategic Planning & Prog Develop, 88-90; CONSULT, EG&G ROCKY FLATS, INC, 90-; PRES, TECH & MGT CONSULT, BRK ASSOCS, INC, 90- *Concurrent Pos:* Chmn, Dept Energy Mgt Team for Galileo & Ulysses RTG space mission progs; Achievement Awards, Significant Overall Prog Contrib, Dept Energy. *Mem:* Am Chem Soc. *Res:* Synthesis of various compounds of plutonium for use as isotopic fuels; high temperature vapor pressure of various plutonium-oxide compounds; management of plutonium fuel fabrication and nuclear waste treatment facilities; patent on plutonium-238 isotopic fuels. *Mailing Add:* 5233 S Clayton Rd Farmersville OH 45325

KOKESH, FRITZ CARL, b Minneapolis, Minn, Jan 12, 43; m 69; c 2. POLYMER CHARACTERIZATION. *Educ:* Lewis Univ, BSc, 65; Ohio State Univ, PhD(org chem), 69. *Prof Exp:* NIH fel, Harvard Univ, 69-71; asst prof biochem, Univ Guelph, 72-77; res chemist, 77-86, supvr, 86-90, TECHNOL PLANNER, PHILLIPS PETROL CO, 90- *Mem:* Am Chem Soc; Soc Plastics Engrs. *Res:* Characterization of thermoplastic polymers; polymer structure and processability. *Mailing Add:* Res Ctr Phillips Petrol Co Bartlesville OK 74004

KOKJER, KENNETH JORDAN, b Beatrice, Nebr, Feb 27, 41; div; c 2. ELECTRICAL ENGINEERING, COMPUTER SCIENCE. *Educ:* Nebr Wesleyan Univ, BS, 63; Univ Ill, Urbana-Champaign, MS, 67, PhD(biophys), 70. *Prof Exp:* From asst prof to assoc prof elec eng, Univ Alaska, Fairbanks, 70-87; COMPUTER CONSULT, COGNITECH, 88- *Concurrent Pos:* Vis scientist, Tohoku Univ, Sendai, Japan, 81. *Mem:* Inst Elec & Electronics Engrs; Comput Soc; Instrument Soc Am; Nat Soc Prof Engrs. *Res:* Applications of computers to real time support of biological research laboratories; computer based instrumentation. *Mailing Add:* Cognitech PO Box 80907 Fairbanks AK 99708-0907

KOKKINAKIS, DEMETRIUS MICHAEL, b Heraklion, Crete, March 5, 50; m; c 2. CARCINOGENESIS, DNA REPAIR & DAMAGE. *Educ:* Nat Univ Athens, BS, 73; Pa State Univ, MS, 75; WVa Univ, PhD(biochem), 77. *Prof Exp:* Fel med biochem, Sch Med, Tex Tech Univ, 78-80; fel, 80-81, assoc path, 81-85, RES ASST PROF, MED SCH, NORTHWESTERN UNIV, 85- *Concurrent Pos:* Chairman, IACUC, 90- *Mem:* Am Asn Cancer Res; NY Acad Sci; AAAS. *Res:* Chemistry of carcinogens; metabolism of carcinogens by target organs; carcinogen mediated DNA damage and cellular mechanisms of its repair. *Mailing Add:* Dept Path Northwestern Univ Med Sch 303 E Chicago Ave Chicago IL 60611

KOKKO, JUHA PEKKA, NEPHROLOGY. *Educ:* Emory Univ, MD & PhD(phys chem), 64. *Prof Exp:* ASA G CANDLER PROF & CHMN, DEPT MED, EMORY UNIV SCH MED, ATLANTA, GA, 86- *Mem:* Am Soc Clin Invest; Am Asn Physicians; Am Fedn Clin Res. *Mailing Add:* Dept Medicine 1364 Clifton Rd NE Suite F410 Atlanta GA 30322

KOKNAT, FRIEDRICH WILHELM, b Muenster, Germany, Feb 19, 38; m 64; c 2. INORGANIC ANALYSIS, X-RAY CRYSTALLOGRAPHY. *Educ:* Univ Giessen, BS, 59, MS, 63, PhD(chem), 65. *Prof Exp:* Instr chem, Univ Giessen, 64-66; fel, Iowa State Univ, 66-69; instr, Boone Jr Col, 68-69; asst prof, 69-74, assoc prof, 74-80, PROF CHEM, YOUNGSTOWN STATE UNIV, 80- *Concurrent Pos:* Consult, Tri-State Labs, Inc, 83- *Mem:* Am Chem Soc; Am Crystallog Asn; Sigma Xi. *Res:* Structural inorganic chemistry; transition metal cluster compounds; phase relationships and stabilization of low oxidation states by formation of complexes and double salts. *Mailing Add:* Dept of Chem Youngstown State Univ Youngstown OH 44555

KOKOROPOULOS, PANOS, environmental engineering, physical chemistry, for more information see previous edition

KOKOSKI, CHARLES JOSEPH, b Chicopee Falls, Mass, June 2, 27; m 52; c 3. PHARMACY. *Educ:* Univ Md, BS, 51, MS, 53, PhD(pharm), 56. *Prof Exp:* From asst prof to assoc prof pharm, George Washington Univ, 56-64; biochemist, Food & Drug Admin, Dept Health & Human Serv, Washington, DC, 64-77; CHIEF, DIV TOXICOL, STANDARDS & MONITORING BR, FOOD & DRUG ADMIN, 77- *Mem:* AAAS; Am Pharmaceut Asn; Soc Toxicol; Sigma Xi. *Res:* Toxicology; pharmaceutical and cosmetic product development. *Mailing Add:* 4504 Maple Ave Halethorpe MD 21227

KOKOSZKA, GERALD FRANCIS, b Meriden, Conn, Sept 26, 38; m 61; c 3. PHYSICAL INORGANIC CHEMISTRY. *Educ:* Univ Conn, BA, 60; Univ Md, MS, 64, PhD(chem physics), 66. *Prof Exp:* Res scientist, Inorg Chem Sect, Nat Bur Standards, 61-68; bd dirs, State Univ NY, Res Found, 84-91; from asst prof to assoc prof, 68-73, PROF CHEM, STATE UNIV NY COL PLATTSBURGH, 73- *Concurrent Pos:* State Univ NY Res Found grants, 68 & 70; Res Corp grant, 69; Petrol Res Found grants, 70, 71-73 & 74-90. *Mem:* Am Chem Soc; Am Phys Soc; Royal Chem Soc. *Res:* Electron spin resonance of metal complexes, free radicals, minerals, low-dimensional systems and biochemical systems. *Mailing Add:* Dept Chem State Univ NY Col Arts & Sci Plattsburgh NY 12901

KOKOTAILO, GEORGE T, b Willingdon, Alta, June 21, 19; US citizen; m 53; c 2. SOLID STATE PHYSICS. *Educ:* Univ Alta, BSc, 41, MSc, 48; Temple Univ, PhD(physics), 55. *Prof Exp:* Physicist, Ont Res Found, 41-42 & Defense Indust Ltd, 42-44; physicist, Nat Res Coun Can, 44-45, sr res physicist, Socony Mobil Oil Co, 48-60, res assoc, Mobil Res & Develop Corp, 60-82; VIS PROF PHYSICS, UNIV GUELPH, 83- *Concurrent Pos:* Adj prof physics, Drexel Inst Technol, 58- *Honors & Awards:* Sci Award, Am Chem Soc, 76; Alexander von Humboldt Sr US Scientist Award, 85. *Mem:* Fel Am Phys Soc; Am Chem Soc; Electron Micros Soc Am; Am Crystallog Asn; AAAS. *Res:* Radiowave propagation; cloud chamber physics; rubber physics; x-ray spectroscopy; x-ray absorption fine structure; crystal structure; anomolous transmission of x-rays and electrons; chemistry and structure of zeolites; solid state nuclear magnetic resonance. *Mailing Add:* 98 N American Woodbury NJ 08096

KOKTA, BOHUSLAV VACLAV, b Brno, Czech, Apr 15, 40. COMPOSITES OF THERMOPLASTICS, EXPLOSION PULPING. *Educ:* Univ Chem Technol, Pardubice, BS, 60, MSc, 62, Acad Sci, Prague, PhD, 67. *Prof Exp:* Sr res chemist, Res Inst of Macromolecular Chem, Brand, Czech, 62-67; fel, reverse osmosis, Syracuse Univ, 67-71; sr res chemist, wood fibers, polymers, Consolidated Bathhurst Ltd, Grand Mere, Que, 69-71; res assoc, polymers, 71-72, PROF, WOOD CHEM, QUEBEC UNIV, 72- *Honors & Awards:* Bates Prizes, Can Pulp & Paper Asn, 83 & 88. *Mem:* Can Inst Can; Am Chem Soc; Can Pulp & Paper Asn. *Res:* Composites of thermoplastics reinforced with wood fibers; grafting of thermoplastics with lignocellulosic materials; explosion pulping (ultra high yield pulp for paper); bleaching of ultra-high yield pulps. *Mailing Add:* Univ Quebec Pates & Papiers Recherche Ctr CP 500 Trois Rivieres PQ G9A 5H7 Can

KOKTA, MILAN RASTISLAV, b Brno, Czech, Mar 22, 41; m 70. SOLID STATE CHEMISTRY. *Educ:* Inst Chem Technol, Pardubice, MS, 68; Newark Col Eng, DESc, 72. *Prof Exp:* Staff chemist inorg chem res, Lachema, Czech, 65-68; mem tech staff, Bell Labs, 72-73; staff chemist res, Allied Chem Corp, 73-77; STAFF SCIENTIST, UNION CARBIDE CORP, 77- *Mem:* Am Chem Soc; Sigma Xi; Am Asn Crystal Growth. *Res:* Liquid phase epitaxy; crystal chemistry of oxide and chalcogenide compounds; relation between structure and physical properties; magnetism; phase relations in oxide systems with respect to crystal growth; crystal growth of electro-optical materials. *Mailing Add:* 1906 SE 331St Ave Washougal WA 98671

KOLAIAN, JACK H, b Troy, NY, July 22, 29; m 60; c 3. COLLOID CHEMISTRY, COAL GASIFICATION. *Educ:* Cornell Univ, BS, 56; Purdue Univ, MS, 58, PhD(clay chem), 60. *Prof Exp:* Asst, Purdue Univ, 56-60; res assoc, Cornell Univ, 60; sr chemist, Bellaire Lab, Texaco, Inc, 60-64, group leader chem res, 64-68, res chemist, 68-73, group leader catalysts-refining, Beacon Res Lab, Chem Prod Develop, 74-79, coordr, Cool Water Coal Gasification Prog, 80-82, MGR GASIFICATION DEVELOP, TEXACO, INC, 82- *Mem:* Am Chem Soc; Am Inst Chem Engrs. *Res:* Oil production; catalysts; petroleum refining; coal gasification. *Mailing Add:* Texaco Inc 2000 Westchester Ave White Plains NY 10650

KOLAKOWSKI, DONALD LOUIS, b Chicago, Ill, Jan 7, 44. PSYCHOMETRICS, HUMAN QUANTITATIVE GENETICS. *Educ:* Knox Col, BA, 66; Univ Chicago, MA, 67, PhD(measurement & statist), 70. *Prof Exp:* Asst prof biobehav sci, Univ Conn Health Ctr, 70-74, asst prof behav sci, 74-82. *Concurrent Pos:* Prin investr, NIH & NIMH grants, Univ Conn, 73-82. *Honors & Awards:* Res Career Develop Award, Nat Inst Dental Res, 75-80. *Mem:* Am Soc Human Genetics; Behav Genetics Asn; Psychometric Soc; Soc for Study Social Biol; Am Educ Res Asn. *Res:* Inheritance of mental traits, cranio-facial structures, disease susceptibility, and their measurement in diverse human populations; human behavioral genetics; dental anthropology. *Mailing Add:* 15 Prentiss St Cambridge MA 02140

KOLAR, JOHN JOSEPH, b Raynesford, Mont, June 14, 22; m 55; c 2. PLANT BREEDING. *Educ:* Mont State Col, BS, 50, MS, 52; Iowa State Col, PhD(plant breeding), 55. *Prof Exp:* Asst agronomist, 56-69, assoc prof agron & assoc agronomist, 69-77, RES PROF AGRON, UNIV IDAHO, 77- *Mem:* Crop Sci Soc Am; Am Soc Agron; Sigma Xi; Coun Agr Sci & Technol. *Res:* Bean breeding and production. *Mailing Add:* 892 Sunrise Blvd N Twin Falls ID 83301

KOLAR, JOSEPH ROBERT, JR, b Chicago, Ill, Sept 26, 38; m 72; c 1. VETERINARY VIROLOGY. *Educ:* Southern Ill Univ, Carbondale, BA, 65, MA, 68, PhD(microbiol), 72. *Prof Exp:* Res assoc dent med, Dent Res Ctr, Univ NC, 72-73; prod mgr virus, Armour-Baldwin Labs, 73-74; res scientist vet virol, 74-77, res dir, Fromm Labs, Inc, Salisbury Labs, 77-84; virus group leader, Beecham Labs, 84-85; VIRUS PROD MGR, BIOLOGICS CORP, 86-; VIRUS RES MGR, FERMENTA ANIMAL HEALTH, 87- *Concurrent Pos:* Consult com poultry oper, 84. *Mem:* Am Tissue Cult Asn. *Res:* Applied in development of veterinary viral vaccines. *Mailing Add:* 5642 Blackwell Dr Omaha NE 68137-2471

KOLAR, OSCAR CLINTON, b Los Angeles, Calif, Sept 26, 28; m 82; c 3. NUCLEAR CRITICALITY SAFETY, PHYSICS. *Educ:* Univ Calif, Los Angeles, BA, 49; Univ Calif, PhD(physics), 55. *Prof Exp:* Sr physicist, Lawrence Livermore Lab, Univ Calif, 55-87. *Mem:* Am Phys Soc; Sigma Xi; Am Nuclear Soc; Am Asn Physics Teachers; AAAS; Am Soc Safety Engrs. *Res:* Nuclear physics, especially nuclear reactions; reactor physics, including criticality hazards evaluation; geophysics; seismology. *Mailing Add:* 7595 NW McDonald Pl Corvallis OR 97330

KOLASA, KATHRYN MARIANNE, b Detroit, Mich, July 26, 49; m 83. NUTRITION, ANTHROPOLOGY. *Educ:* Mich State Univ, BS, 70; Univ Tenn, Knoxville, PhD(food sci), 74. *Prof Exp:* Test kitchen home economist, Kellogg Co, 71; from asst prof to assoc prof community nutrit, Mich State Univ, 80-82; prof & chairperson food & nutrit, 83-86, PROF & SECT HEAD, NUTRIT EDUC & SERV SECT, DEPT FAMILY MED, E CAROLINA UNIV SCH MED, 86- *Concurrent Pos:* Res assoc, Home Learning Ctr Res Proj, Off Educ, 74-75; Kellogg Nat Leadership fel, 85-88; mem, Comt Nutrit Anthrop. *Mem:* Am Inst Nutrit; Soc Nutrit Educ (pres, 84-85); Inst Food Technologists; Am Dietetic Asn; Soc Teachers Family Med. *Res:* Interactions of nutrition and culture upon the health of the individual and family in the US and the developing world; nutrition in medicine. *Mailing Add:* Family Med Dept E Carolina Univ Greenville NC 27858-4354

KOLAT, ROBERT S, b Bay City, Mich, May 8, 31; m 54; c 4. ANALYTICAL CHEMISTRY. *Educ:* Mich State Univ, BS, 58; Iowa State Univ, PhD(phys chem), 61. *Prof Exp:* Res chemist, Am Cyanamid Co, 61-65; res chemist, 65-73, RES MGR, DOW CHEM CO, MIDLAND, 73- *Mem:* AAAS; Am Chem Soc. *Res:* Chelation; bomb calorimetry; aerosol research. *Mailing Add:* 3370 Parkway Dr Bay City MI 48706

KOLATA, DENNIS ROBERT, b Rockford, Ill, June 9, 42; m 63; c 2. STRATIGRAPHY, INVERTEBRATE PALEONTOLOGY. *Educ:* Northern Ill Univ, BS, 68, MS, 70; Univ Ill, PhD(geol), 73. *Prof Exp:* Geologist explor & develop, Texaco Inc, 73-74; assoc geologist, 74-80, GEOLOGIST, ILL STATE GEOL SURV, 80- *Mem:* Paleontol Soc; Paleont Asn; Geol Soc Am. *Res:* Stratigraphy and paleontology of Paleozoic rocks in the Eastern Interior of North America. *Mailing Add:* 2314 Brookshire Champaign IL 61821

KOLATA, JAMES JOHN, b Milwaukee, Wis, Dec 26, 42; m 67; c 2. HEAVY-ION REACTION MECHANISMS. *Educ:* Marquette Univ, BS, 64; Mich State Univ, MS, 66, PhD(physics), 69. *Prof Exp:* Res assoc physics, US Naval Res Lab, 69-70 & Univ Pittsburgh, 70-72; asst physicist, Brookhaven Nat Lab, 72-73, assoc physicist, 73-76, physicist, 76-77; assoc prof, 77-84, PROF PHYSICS, UNIV NOTRE DAME, 84- *Concurrent Pos:* Vis prof, Ctr Nuclear Res, Strasbourg, France, 78; chmn steering comt, Nat Superconducting Cyclotron Facil, Mich State Univ, 78-79 and 90-91; vis scientist, Argonne Nat Lab, 83-84 & chmn User's Group Exec Comt, Atlas facil, 86-87. *Mem:* AAAS; Am Phys Soc; Sigma Xi. *Res:* Nuclear reaction mechanisms; nuclear fusion of heavy ions. *Mailing Add:* Dept Physics Univ Notre Dame Notre Dame IN 46556

KOLATTUKUDY, P E, b Kerala, India; US citizen. GENE EXPRESSION, RESEARCH ADMINISTRATION. *Educ:* Univ Madras, BSc, 57; Univ Kerala, BEd, 59; Ore State Univ, PhD(chem), 64. *Prof Exp:* Asst biochemist, Conn Agr Exp Sta, 64-69; from assoc prof to prof biochem, Wash State Univ, 69-80, dir, fel & prof, Inst Biochem, 80-86; DIR & PROF, OHIO STATE BIOTECH CTR, OHIO STATE UNIV, 86- *Concurrent Pos:* Mem, Physiol Chem Study Sect, NIH, 84-88. *Mem:* Am Soc Plant Physiologists; Am Soc Biochem & Molecular Biol; Am Soc Advan Sci. *Res:* Structure and function of genes and enzymes involved in lipid metabolism; gene expression in plant-fungus interaction; plant genes involved in defense against pathogens; gene expression in differentiation and genetic dificiency diseases. *Mailing Add:* Ohio State Biotechnol Ctr 206 Rightmire Hall 1060 Carmack Rd Columbus OH 43210-1002

KOLB, ALAN CHARLES, b Hoboken, NJ, Dec 14, 28. PHYSICS. *Educ:* Ga Inst Technol, BS, 49; Univ Mich, MS, 50, PhD(theoret physics), 55. *Prof Exp:* Supt plasma physics div, Naval Res Lab, 55-70, pres & chief exec officer, 70-78, CHMN & CHIEF EXEC OFFICER, MAXWELL LABS INC, 78- *Concurrent Pos:* Adj prof, Univ Md, College Park, 68-70; vis prof, Cath Univ Am, 65-68. *Mem:* Fel Am Phys Soc; NY Acad Sci. *Res:* Plasma physics and controlled thermonuclear research; theoretical and experimental spectroscopy; hydrodynamics and very high Mach number shock waves; electron beam research; high voltage engineering; laser development. *Mailing Add:* 9244 Balboa Ave San Diego CA 92123

KOLB, BRYAN EDWARD, b Calgary, Alta, Nov 10, 47; m. NEUROPSYCHOLOGY, BEHAVIORAL NEUROSCIENCE. *Educ:* Univ Calgary, BSc, 68, MSc, 70; Pa State Univ, PhD(psychol), 73. *Prof Exp:* Fel psychol, Univ Western Ont, 73-75; med res coun fel, Montreal Neurol Inst, 75-76; from asst prof to assoc prof, 76-83, PROF PSYCHOL, UNIV LETHBRIDGE, 83- *Concurrent Pos:* Dept chair, Univ Lethbridge, 87-90. *Mem:* Soc Neurosci; fel Am Psychol Asn; fel Can Psychol Asn; Am Psychol Soc. *Res:* Frontal lobe function in mammals; recovery of function following brain damage. *Mailing Add:* Dept Psychol Univ Lethbridge Lethbridge AB T1K 3M4 Can

KOLB, CHARLES EUGENE, JR, b Cumberland, Md, May 21, 45; m 65; c 2. CHEMICAL KINETICS, ATMOSPHERIC CHEMISTRY. *Educ:* Mass Inst Technol, SB, 67; Princeton Univ, MA, 68, PhD(phys chem), 71. *Prof Exp:* Sr res scientist, 71-75, dir, Ctr Chem & Environ Physics, 77-80, tech dir, Appl Sci Div, 80-81, vpres & dir, Appl Sci Div, 81-85, PRIN RES SCIENTIST, AERODYNE RES INC, 75-, PRES & CHIEF EXEC OFFICER, 85- *Concurrent Pos:* Hon res fel atmospheric chem, Ctr Earth & Planetary Physics, Harvard Univ, 76-85; res affil, Spectros Lab, Mass Inst Technol, 81-; mem, Nat Acad Sci-Nat Res Coun Comt Atmospheric Chem, 87-89; chmn-elect, Northeastern Sect, Am Chem Soc, 90, chair, 91. *Mem:* Combustion Inst; Am Chem Soc; Am Phys Soc; fel Optical Soc Am; Am Geophys Union; AAAS. *Res:* Experimental and theoretical studies of inelastic energy exchange in hyperthermal molecular collisions; chemistry and physics of trace atmospheric species; chemical kinetics and spectroscopy of combustion and gas lasers. *Mailing Add:* Aerodyne Res Inc 45 Manning Rd Billerica MA 01821

KOLB, DORIS KASEY, b Louisville, Ky, Aug 4, 27; m 48; c 3. ORGANIC CHEMISTRY. *Educ:* Univ Louisville, BS, 48; Ohio State Univ, MSc, 50, PhD(chem), 53. *Prof Exp:* Chemist info res, Standard Oil Co, (Ind), 53-57; assoc prof chem & head dept, Corning Community Col, 61-62, chemist, 59-62; prof chem, Ill Cent Col, 67-86; PROF CHEM, BRADLEY UNIV, 86- *Honors & Awards:* Chem Mfrs Award, 81. *Mem:* Am Chem Soc. *Res:* Sugars; fatty acid solubility; petroleum chemistry; plastics; chemical education. *Mailing Add:* 7309 N Edgewild Dr Peoria IL 61614-2113

KOLB, EDWARD WILLIAM, b New Orleans, La, Oct 2, 51; m 72; c 3. COSMOLOGY, SUPERNOVAE. *Educ:* Univ New Orleans, BS, 73; Univ Tex, PhD(physics), 78. *Prof Exp:* Fel astrophysics, Calif Inst Technol, 78-80; J Robert Opphenheimer res fel, Los Alamos Nat Lab, 80-81, mem staff astrophys, 80-82; HEAD, FERMILAB ASTROPHYS, 83- *Concurrent Pos:* Prof, Dept Astron & Astrophys, Enrico Fermi Inst, Univ Chicago, 83- *Mem:* Fel Am Phys Soc; Am Astron Soc. *Res:* Application of particle physics to the study of the early universe; cosmology; neutrino processes in supernovae; weak interactions. *Mailing Add:* Theoret Astrophys Fermilab Batavia IL 60510

KOLB, FELIX OSCAR, b Vienna, Austria, Nov 12, 21; nat US; m 66; c 2. MEDICINE. *Educ:* Univ Calif, AB, 41, MD, 43; Am Bd Internal Med, cert endocrinol & metab. *Prof Exp:* Asst med, Univ Calif, 46-49; asst, Mass Gen Hosp, 50-51; asst, 51-53, from clin instr to asst clin prof, 52-59, assoc clin prof & assoc res physician, 59-68, res physician, Metab Unit, 53-59 & 68-81, CLIN PROF MED, SCH MED, UNIV CALIF, SAN FRANCISCO, 68- *Mem:* Endocrine Soc; AMA; Am Diabetes Asn; fel Am Col Physicians; Am Fedn Clin Res; Am Soc Bone & Mineral Res. *Res:* Metabolic bone disease; renal tubular disorders and renal stones, including cystinuria. *Mailing Add:* Nine Starboard Ct Mill Valley CA 94941

KOLB, FREDERICK J(OHN), JR, b Rochester, NY, May 7, 17; m 42; c 4. RECORDING MEDIA. *Educ:* Mass Inst Technol, SB, 38, SM, 39, ScD(chem eng), 47. *Prof Exp:* Chem engr, Eastman Kodak Co, 42-50, sr, sect supvr, 54-67, tech assoc, 67-73, proj coordr, 73-82, res assoc, 82-86; RETIRED. *Concurrent Pos:* Instr, Univ Rochester, 45. *Honors & Awards:* Samuel L Warner Medal, Soc Motion Picture & TV Engrs; Tech Achievement Award, Acad Motion Picture Arts & Sci. *Mem:* AAAS; Am Chem Soc; Soc Motion Picture & TV Engrs; Am Inst Chem Engrs; Inst Elec & Electronics Engrs; Soc Photog Scientists & Eng; British Kinematograph Sound & TV Soc; Sigma Xi. *Res:* Physical and chemical properties of photographic and magnetic media; theory and practice of magnetic and photographic recording systems, especially from the viewpoint of information theory; production of cellulose ester and polyester films; cine film manufacture and applications; effects of radiation on motion-picture films; development manufacture and applications of magnetic recording media; storage and retrieval of audio and visual information; international standardization; image recording and processing in film and video systems. *Mailing Add:* 211 Oakridge Dr Rochester NY 14617-2511

KOLB, KENNETH EMIL, b Louisville, Ky, Jan 21, 28; m 48; c 3. ORGANIC CHEMISTRY. *Educ:* Univ Louisville, BS, 48; Ohio State Univ, PhD(chem), 53. *Prof Exp:* Chemist, Nat Distillers, 48, Standard Oil Co, Ind, 53-58 & Corning Glass Works, 58-65; CHEMIST, BRADLEY UNIV, 65- *Mem:* Am Chem Soc; Soc Plastics Eng; Royal Soc Chem. *Res:* Electro-organic chemistry; electrophilic bromination and aklylation; iodine organic complexes; furan chemistry. *Mailing Add:* Dept Chem Bradley Univ Peoria IL 61625

KOLB, LAWRENCE COLEMAN, b Baltimore, Md, June 16, 11; m 37; c 3. PSYCHIATRY. *Educ:* Trinity Col, Dublin, BA, 32; Johns Hopkins Univ, MD, 34; Am Bd Psychiat & Neurol, dipl, 42. *Prof Exp:* Intern med, Strong Mem Hosp, NY, 34-35, intern surg, 35-36; asst dispensary neurologist, Sch Med, Johns Hopkins Univ, 36-38, instr neurol, 39-41; resident psychiatrist, Milwaukee Sanitarium, 41-42; dir res, Div Ment Hyg, USPHS, 46-49; consult, Mayo Clin, 49-54; comnr, NY State Dept Ment Hyg, 75-77; prof, 54-75, EMER PROF PSYCHIAT, COL PHYSICIANS & SURGEONS, COLUMBIA UNIV, 76- DISTINGUISHED PHYSICIAN, US VET ADMIN, 78- & PROF PSYCHIAT, ALBANY MED COL, 78- *Concurrent Pos:* Fel, Sch Med, Johns Hopkins Univ, 36-38; Markle Found fel, Nat Hosp, London, 38; consult, USN, Washington, DC, 46-49, USPHS, 54-62 & NIMH; res assoc, Wash Sch Psychiat, 47-49; assoc prof, Univ Minn, 49-53; chmn dept psychiat & dir psychiat serv, Presby Hosp, New York, 54-74, pres med bd, 62-64, trustee, 71-73; dir, NY State Psychiat Inst, New York, 54-74; dir & mem bd dirs, Res Fedn Ment Hyg Inc, 54-75, pres & chmn bd, 60-75; mem comt, Navy Med Res, Nat Res Coun, 56-59; dir, Am Bd Psychiat & Neurol, 60-68, pres, 68; assoc comnr res, NY State Dept Ment Hyg, 68-69; ed, Yearbk Psychiat & Appl Ment Health, 71-; pres adv bd, PR Inst Psychiat, 72. *Honors & Awards:* Henry Wisner Miller Mem Award; Joan Plehn Award Humane Serv, Ment Health Asn New York & Bronx Counties, 72. *Mem:* Am Acad Arts & Sci; Am Neurol Asn; hon fel Royal Col Psychiat; Asn Res Nerv & Ment Dis (pres, 59; Am Psychiat Asn (pres, 68). *Res:* Psychiatry and psychoanalysis; neurology. *Mailing Add:* Van Wies Point Rd Glenmont NY 12077

KOLB, LEONARD H, b Chicago, Ill, Feb 2, 13; m; c 4. SURGERY, ONCOLOGY. *Educ:* Univ Ill, BS, 35, MD, 38. *Prof Exp:* Intern, Mt Sinai Hosp, Chicago, 37-39, resident surg, 39-40, ASSOC PROF SURG, RUSH MED SCH & MT SINAI HOSP, 48- *Concurrent Pos:* Sr attend surgeon, Mt Sinai Hosp, 46- *Mem:* Fel Am Col Surgeons; AMA; Am Asn Cancer Res; Am Asn Cancer Educ; Soc Clin Oncol. *Res:* Cancer. *Mailing Add:* 600 Woodrow St Unit J Columbia SC 29205

KOLB, VERA, b Belgrade, Yugoslavia, Feb 5, 48; div. ORGANIC CHEMISTRY, MEDICINAL CHEMISTRY. *Educ:* Univ Belgrade, BS, 71, MS, 73; Southern Ill Univ, Carbondale, PhD(org chem), 76. *Prof Exp:* Fel, Univ Res Found, La Jolla, Calif, 77-78; res assoc & instr chem, Southern Ill Univ, Carbondale, 78-81, adj asst prof, 81-85; assoc prof, 85-90, PROF CHEM, UNIV WIS-PARKSIDE, 90- *Concurrent Pos:* Fulbright travel grant, 73-76; tour speaker, Am Chem Soc, 80; mem, task force occup safety & health, Am Chem Soc, 80-; NIH/Nat Inst Drug Abuse grant, 84-85. *Mem:* Am Chem Soc; Sigma Xi; Serbian Chem Soc; fel Am Soc Biochem & Molecular Biol. *Res:* Reaction mechanisms; steroid chemistry; carbanion chemistry; electron transfer reactions; conformational analysis; mechanism of action of morphine agonists and antagonists; intra-red frequencies of semicarbazones. *Mailing Add:* Dept Chem Univ Wis-Parkside Kenosha WI 53141

KOLBECK, RALPH CARL, b Wausau, Wis, Sept 2, 44; m 66; c 2. PHYSIOLOGY, BIOCHEMISTRY. *Educ:* Univ Minn, BA, 66, PhD(physiol, biochem), 70. *Prof Exp:* NIH Res fel cardiac physiol, Univ Minn, Minneapolis, 70-73; instr med, Med Col Ga, 73-77, asst prof med & asst dir hemodynamic res, 77-80, lectr physiol, 77-89, ASSOC PROF MED & DIR PULMONARY RES, MED COL GA, 80-, ASSOC PROF PHYSIOL, 89- *Concurrent Pos:* Ga Heart Asn grant, 74-76; Gen Res Support grant, 74-76, 80-82, 84-85, 88-89; NIH grant, 75-78. 87-90; Am Heart Asn grant, 76-78, 87-88; Ga Heart Asn Investr, 77-81; Am Lung Asn grant, 82-83, 87-91. *Mem:* Am Fed Clin Res; Soc Exp Biol Med; Am Physiol Soc; Am Heart Asn; Sigma Xi; Biophys Soc. *Res:* Calcium uptake by mammalian myocardium; subcellular calcium localization in mammalian myocardium; smooth muscle contractility; fatigue of skeletal muscle. *Mailing Add:* 1120 15th St Med Col Georgia Augusta GA 30912

KOLBER, HARRY JOHN, b Buffalo, NY, June 5, 18; c 2. CHEMISTRY. *Educ:* Hamilton Col, BS, 40; Haverford Col, MS, 41; Northwestern Univ, PhD(chem), 43. *Prof Exp:* Proj engr, Naval Res Lab, Washington, DC, 43-44; res chemist, NY, 45-50, res supvr, Del, 50-66, TECH SERV MGR, E I DU PONT DE NEMOURS & CO, INC, 66- *Concurrent Pos:* Mem body armour comt, Nat Res Coun. *Mem:* Sigma Xi. *Res:* Thermodynamics of dehydration of alcohols; crystallography; corrosion chemistry; polymer chemistry; textile research. *Mailing Add:* 807 22nd St Wilmington DE 19802

KOLBEZEN, MARTIN (JOSEPH), b Pueblo, Colo, Apr 16, 14; m 53. PESTICIDE CHEMISTRY. *Educ:* Colo State Univ, BS, 39; Univ Utah, MS, 41, PhD(org chem), 50. *Prof Exp:* Nat Defense Res Coun fel, Iowa State Univ, 41-42; asst, Univ Utah, 39-41, 47-48; anal chemist, US Bur Mines, Utah, 42-44; asst insect toxicologist, 50-56, asst chemist, Plant Path, 56-58, assoc chemist, 58-64, chemist & lectr, 64-81, EMER CHEMIST & LECTR, PLANT PATH, CITRUS EXP STA, UNIV CALIF, RIVERSIDE, 81- *Mem:* Am Chem Soc. *Res:* Chemistry and mode of action of pesticides; residue analysis and development of methods of analysis; climatic and biological breakdown of pesticides; soil fumigation measurements and techniques. *Mailing Add:* 4721 Monroe St Riverside CA 92504

KOLBYE, ALBERT CHRISTIAN, JR, b Philadelphia, Pa, Feb 15, 35. PUBLIC HEALTH. *Educ:* Harvard Col, AB, 57; Temple Univ, MD, 61; Johns Hopkins Univ, MPH, 65; Univ Md, JD, 66. *Prof Exp:* Intern med, Univ Hosps, Madison, Wis, 62; resident physician, Div Chronic Dis, USPHS, 62-65, chief field staff, Heart Dis Control Prog, 65-67, assoc dir sci, Nat Ctr Smoking & Health, 67-68, staff dir & exec secy, Comn Pesticides & Environ Health, Dept Health, Educ & Welfare, 69, dir off standards & compliance, Consumer Protection & Environ Health Serv, USPHS, 69-70, dep dir, Bur Foods, 70-72, assoc dir sci, Food & Drug Admin, 72-82, asst surgeon gen, USPHS, 71-82; RETIRED. *Concurrent Pos:* Consult. *Mem:* Fel Am Pub Health Asn; fel Am Col Prev Med; fel Am Col Legal Med; fel Am Acad Clin Toxicol; Am Acad Forensic Med. *Res:* Epidemiology of heavy metal toxicity, halogenated hydrocarbons; environmental contaminants; epidemiology of cancer and carcinogenesis; medicolegal aspects of clinical and epidemiological research; medicolegal aspects of malpractice. *Mailing Add:* 7313 Helmsdale Rd Bethesda MD 20817

KOLCHIN, ELLIS ROBERT, b New York, NY, Apr 18, 16; m 40; c 2. MATHEMATICS. *Educ:* Columbia Univ, AB, 37, PhD(math), 41. *Prof Exp:* Instr math, Hofstra Col & lectr Barnard Col, 40-41; Nat Res fel, Inst Advan Study, 41-42; from instr to assoc prof, 46-58, prof, 58-76, Adrian prof, 76-86, EMER PROF MATH, COLUMBIA UNIV, 86- *Concurrent Pos:* Guggenheim fel, 54-55 & 61-62; NSF sr fel, 60-61; vis prof, Bucknell Univ, 86-87, Rutgers Univ, 87-88. *Mem:* AAAS; Am Acad Arts & Sci; Am Math Soc. *Res:* Differential algebra and algebraic groups. *Mailing Add:* Dept Math Columbia Univ New York NY 10027

KOLDER, HANSJOERG E, b Vienna, Austria, Nov 29, 26; c 3. OPHTHALMOLOGY, PHYSIOLOGY. *Educ:* Univ Vienna, MD, 50. *Prof Exp:* Asst physiol, Univ Vienna, 51-59, docent, 59; vis asst prof, Emory Univ, 59-63, assoc prof, 63-68; assoc prof, 68-73, PROF OPHTHAL, UNIV IOWA, 73- *Concurrent Pos:* Europ Coun res fel aviation med, Karolinska Inst, Sweden, 58 & 61. *Mem:* AAAS; Am Acad Ophthal & Otolaryngol; Am Physiol Soc. *Res:* Aviation and sensory physiology; ophthalmic electrodiagnosis; cataract management. *Mailing Add:* Dept Ophthal Univ Iowa Iowa City IA 52242

KOLDEWYN, WILLIAM A, b Ogden, Utah, Apr 23, 42; m 67; c 2. PHYSICS, SYSTEM DESIGN MANAGEMENT. *Educ:* Weber State Col, BS, 67; Wesleyan Univ, PhD(physics), 76. *Prof Exp:* Electro-mech engr, Xytex Corp, 74-75; staff physicist, Off Prod Div, IBM, 78-80; dir eng, Scientech Inc, 83-87; sr mem tech staff, Ball Aerospace Div, 75-78, prin mem staff, 80-83, SR SYST MGR, BALL CORP, 88- *Concurrent Pos:* Consult syst design, 82- *Mem:* Am Phys Soc; Laser Inst Am. *Res:* Remote sensing; high precision control systems; high accuracy measurements of physical constants; laser power and energy measurements. *Mailing Add:* 933 Columbia Pl Boulder CO 80303

KOLDOVSKY, OTAKAR, b Olomouc, Czech, Mar 31, 30; m 71. DEVELOPMENTAL PHYSIOLOGY, GASTROENTEROLOGY. *Educ:* Charles Univ, Prague, MD, 55; Czech Acad Sci, Prague, PhD(develop physiol), 62. *Hon Degrees:* MA, Univ Pa, 74. *Prof Exp:* Scientist nutrit biochem & develop physiol, Inst Physiol, Czech Acad Sci, 56-68; res assoc, Dept Pediat, Stanford Univ, 68-69; from asst prof to prof pediat, Univ Pa, 69-79; PROF PEDIAT & PHYSIOL, UNIV ARIZ, 80- *Concurrent Pos:* Vis scientist, Dept Pediat, Stanford Univ, 65 & Dept Biochem, Univ Lund, Sweden, 67-68. *Honors & Awards:* Spec Award, Czech Acad Sci, 67; Nutrit Award, Am Acad Pediat, 86. *Mem:* Am Inst Nutrit; Am Physiol Soc; Am Pediat Soc; Am Gastroenterol Soc; Perinatal Res Soc. *Res:* Role of hormonal and dietary factors in expression of normal developmental patterns of gastrointestinal functions; milk-borne hormones and their role for the neonate. *Mailing Add:* Dept Pediat Univ Ariz 1501 N Campbell St Tuscon AZ 85724

KOLEK, ROBERT LOUIS, b Pittsburgh, Pa, Feb 5, 36; m 68; c 2. PLASTICS TECHNOLOGY, MATERIALS SCIENCE. *Educ:* Univ Pittsburgh, BS, 58, MS, 61. *Prof Exp:* Chemist fiber glass, PPG Industs, 61-68; FEL SCIENTIST PLASTICS, WESTINGHOUSE ELEC CO, 68- *Mem:* Soc Plastics Engrs; Am Chem Soc; Am Asn Testile Technol. *Res:* Reinforced plastics; composite material; textile technology. *Mailing Add:* Hendy Ave Sunnyvale CA 94086

KOLENBRANDER, HAROLD MARK, b Sibley, Iowa, Oct 7, 38; m 58; c 3. METABOLISM, ENZYMOLOGY. *Educ:* Cent Col, Iowa, BA, 60; Univ Iowa, PhD(biochem), 64. *Prof Exp:* From asst prof to prof chem, Cent Col, Iowa, 64-71; asst to pres, Grand Valley State Col, 71-72, dean, Col Planning, 72-75; prof chem, provost & dean, Cent Col, Iowa, 75-86; PRES, MT UNION COL, 86- *Concurrent Pos:* USPHS grant amino acid metab, 65-68, spec fel, 69-70; vis scientist, Case Western Reserve Univ, 70. *Mem:* Royal Soc Chem; Biochem Soc; Sigma Xi. *Res:* Histidine metabolism and the associated enzymes. *Mailing Add:* Mt Union Col Alliance OH 44601

KOLER, ROBERT DONALD, b Casper, Wyo, Feb 14, 24; m 45; c 2. MEDICAL GENETICS, HEMATOLOGY. *Educ:* Univ Ore, Md, 47; Am Bd Internal Med, dipl, 55. *Prof Exp:* Intern, Med Sch, Univ Ore, 47-48, resident hemat, 48-49; instr basic sci, Med Dept Res & Grad Sch, US Army, 49-50, chief gen med, 181st Gen Hosp, 50-51; resident med, Med Sch, Univ Ore, 51-53, clin assoc med & hemat, 53-56, from asst prof to assoc prof, 56-64, head div hemat & exp med, 64-67, head div med genetics, 67-81, chmn dept med genetics, 81-87, EMER PROF MED & HEMAT, MED SCH, UNIV ORE, 64-, ASSOC VPRES ACAD AFFAIRS, 88- *Concurrent Pos:* USPHS res fel & hon res asst, Univ Col, Univ London, 60-61. *Mem:* Am Soc Hemat; Am Fedn Clin Res; fel Am Col Physicians; Am Soc Human Genetics; Int Soc Hemat. *Res:* Medical and human genetics; characterization of hemoglobin and red cell enzymes. *Mailing Add:* 3181 SW Sam Jackson Ore Health Sci Univ Portland OR 97201

KOLESAR, EDWARD S, b Canton, Ohio, June 24, 50; m 76; c 2. MICROELECTRONICS, SOLID STATE PHYSICS. *Educ:* Univ Akron, BSEE, 73; Midwestern Univ, MBA, 76, Air Force Inst Technol, MSEE, 78; Univ Tex-Austin, PhD(elec eng), 85. *Prof Exp:* Staff engr, USAF Elec Syst Div, Hanscom AFB, Mass, 73-77, design engr, Sch Aerospace Med, Brooks AFB, Tex, 78-82; ASSOC PROF ELEC ENG, AIR FORCE INST TECH, WPAFB, OHIO, 85- *Concurrent Pos:* Co-op eng student, Hoover Co, N Canton, Ohio, 71-72; consult, USAF Sci Adv Bd, Washington, DC, 81, Johns Hopkins Univ, Sch Hyg & Pub Health, Baltimore, MD, 83, Ardex Inc, Austin Tex, 85, Foreign Technol Div, Wright-Patterson AFB, Ohio, 86- & EG&G Mound Appl Technol Lab, Mat Div, Miamisburgh, OH, 88- *Honors & Awards:* H V Nobel Award, Inst Elec & Electronics Engrs, 88; Outstanding Engr & Scientist Award, Eng & Sci Found, E W Kettering Ctr, Dayton, Ohio, 90. *Mem:* Sr mem, Inst Elec & Electronics Engrs; Am Soc Eng Educ; Sigma Xi. *Res:* Design and development of microelectronic sensors to detect environmentally sensitive chemical compounds and to facilitate robotics potential, eg, sensors having a tactile sense. *Mailing Add:* 3630 Navara Dr Beavercreek OH 45431-3121

KOLESAR, PETER JOHN, b New York, NY, Nov 25, 36; c 2. OPERATIONS RESEARCH, STATISTICAL ANALYSIS. *Educ:* Queens Col, NY, BA, 59; Columbia Univ, BSIE, 59, MS, 61, PhD(opers res), 64. *Prof Exp:* Systs analyst appl statist, Procter & Gamble Co, 59-61; lectr opers res, Imp Col, Univ London, 64-65; asst prof, Columbia Univ, 65-70; assoc prof, Univ Montreal, 70-71; sr analyst, Rand Corp, 71-72; assoc prof comput sci, City Col New York, 72-75; assoc prof, 75-77, PROF MGT SCI, COLUMBIA UNIV, 77- *Concurrent Pos:* Consult, Rand Corp, 72-, NY State, 73-, Mt Sinai Hosp, 74-80, New York, 80-81, Citibank, 81-84, Int Paper, 84 & Alcoa, 88. *Honors & Awards:* Lanchester Prize, Opers Res Soc Am, 75; NATO Syst Sci Prize, 76. *Mem:* Opers Res Soc Am; Inst Mgt Sci; fel AAAS; Am Statist Assoc; Am Soc Qual Control. *Res:* Quality management and control application of operations researches; applied optimization; probability and statistics particularly in litigation, clinical trial and public systems analysis. *Mailing Add:* Uris Hall Columbia Univ New York NY 10027

KOLESAR, PETER THOMAS, b Bridgeport, Conn, Oct 14, 42; m 65; c 2. GEOCHEMISTRY, PETROLOGY. *Educ:* Rensselaer Polytech Inst, BS, 66, MS, 68; Univ Calif, Riverside, PhD(geol), 73. *Prof Exp:* Fel isotope geochem, Inst Geophys & Planetary Physics, Univ Calif, Riverside, 73-74; asst prof, 74-80, ASSOC PROF GEOL, UTAH STATE UNIV, 80- *Concurrent Pos:* Fel, US Geol Surv, Water Resources Div, 83-91. *Mem:* Sigma Xi; Soc Econ Paleontologists & Mineralogists; Geochem Soc; Nat Asn Geol Teachers; Int Asn Geochem & Cosmochem. *Res:* Deciphering carbonate rocks; their original depositional environments, the changes which they have undergone (diagenesis) and the chemistry of fluids responsible for those changes; investigation of groundwater resources. *Mailing Add:* Dept Geol Utah State Univ Logan UT 84322-4050

KOLESARI, GARY LEE, b Milwaukee, Wis, Aug 5, 48; m 73. TERATOLOGY. *Educ:* Univ Wis-Milwaukee, BS, 71; Med Col Wis, MS, 73, PhD(anat), 76, MD, 77. *Prof Exp:* Asst prof anat & teratology, 78-81, ASST ADJ PROF, DEPT ANAT, MED COL WIS, 81- *Mem:* Teratology Soc; AMA. *Res:* Teratology, environmental and abuse drug related. *Mailing Add:* Dept Anat Med Col Wis 8701 Waterton Plank Rd Milwaukee WI 53226

KOLESKE, JOSEPH VICTOR, b Stratford, Wis, Jan 23, 30; m 51; c 2. POLYMER CHEMISTRY, RADIATION CHEMISTRY. *Educ:* Univ Wis, BS, 58; Inst of Paper Chem, MS, 60, PhD, 63. *Prof Exp:* Sr res scientist, 63-77, res assoc, chemicals & plastics, 77-83, corp res fel, polymer sci, solvents & coatings mat, Union Carbide Corp, 83-88; CONSULT, 88- *Mem:* Am Chem Soc. *Res:* Polymer physical chemistry; high solids; polyurethane; powder; water-borne and cationic free radical radiation cure coatings. *Mailing Add:* 1513 Brentwood Rd Charleston WV 25314

KOLFF, WILLEM JOHAN, b Leiden, Holland, Feb 14, 11; nat US; m 37; c 5. EXPERIMENTAL MEDICINE, CLINICAL MEDICINE. *Educ:* State Univ Leiden, MD, 38; State Univ Groningen, PhD, 46. *Hon Degrees:* DSc, Allegheny Col, 60, Tulane Univ, 75, Univ L'Aquila, Italy, 81, Univ Twente, Neth, 86 & Univ Athens, 88; MD, Univ Turin, 69, Univ Rostock, 75 & Univ Bologna, 77. *Prof Exp:* Asst path anat, State Univ Leiden, 34-36; asst med, State Univ Groningen, 38-41; head dept, Munic Hosp, Kampen, 41-50; prof clin invest, Educ Found, Cleveland Clin Found, 50-67, mem staff res div, 50-63, mem surg div & head dept artificial organs, 58-67; DISTINGUISHED PROF MED & SURG & RES PROF, COL MED, UNIV UTAH, 67- *Concurrent Pos:* Pvt docent, Med Sch, State Univ Leiden, 49-51; hon mem, Europ Dialysis & Transplant Asn, Europ Renal Asn & Europ Soc Artificial Organs, 86. *Honors & Awards:* Landsteiner Silver Medal, Neth Red Cross, 42; Frances Amory Award, Am Acad Arts & Sci, 48; Addingham Gold Medal, Univ Leeds, 62; K Award, Nat Kidney Dis Found, 63; Oliver Sharpey Prize, Royal Col Physicians, 63; Cameron Prize, Univ Edinburgh, 64; Gairdner Prize, Gairdner Found, Can, 66; 1st Gold Medal, Neth Surg Soc, 70; Ubbo Emmius Medal, State Univ Groningen, 70; Leo Harvey Prize, 72; Austrian Gewerbeverein's Wilhelm-Exner Award, 80; Jean Hamburger Award, Int Soc Nephrology, 87. *Mem:* Nat Acad Eng; fel Am Col Physicians; Am Heart Asn; hon fel AMA; Am Physiol Soc; hon mem Austrian Soc Nephrology; AAAS. *Res:* Kidney transplantation; application of heart-lung machines; development of artificial heart inside the chest; avoidance of thrombiosis on plastics; development of blood oxygenators; new types of artificial kidneys and dialysis techniques; development of techniques for organ preservation for transplantation and visual prosthesis; development of artificial hearts, valves and cardiac assist devices. *Mailing Add:* Dept Surg Artificial Organs Univ Utah Salt Lake City UT 84112

KOLHOFF, M(ARVIN) J(OSEPH), b Goodland, Ind, Oct 22, 15; m 37; c 2. ENGINEERING. *Educ:* Purdue Univ, BSEE, 39. *Prof Exp:* Mem staff, Gen Elec Co, NY, 39-42, requisition engr control eng div, Locomotive & Car Equip Dept, Pa, 42-46, admin asst, 46-51, proj engr, 51, asst mgr lab, 51-53, mgr, 53-56, chmn opers res & synthesis study, 56-59, prog planning engr gen eng lab, NY, 59-60, consult eng applns, 60-63, mgr design eval, Gen Purpose Control Dept, 63-66, adminr modern eng course, NY, 66-71, staff assoc tech res, Corp Tech Staff, Gen Elec Co, Conn, 71-79; sr partner, Kolhoff Assocs, Fla, 80-88; RETIRED. *Mem:* Inst Elec & Electronics Engrs; Nat Soc Prof Engrs. *Res:* Technical resources and technological implications of legislative and regulatory issues related to environment, safety and other protection of consumers and public. *Mailing Add:* Kolhoff Assocs 342 N 14th St Quincy FL 32351

KOLI, ANDREW KAITAN, b Bombay, India, Aug 1, 25; US citizen; m 58; c 2. ORGANIC CHEMISTRY. *Educ:* Univ Bombay, BSc, 55; Howard Univ, MS, 64, PhD(chem), 68. *Prof Exp:* Develop chemist res, Dow Chem Co, 55-61; res asst, Howard Univ, 61-66, res assoc, 66-67, instr, Dept Pharm, 67-68; PROF CHEM, SC STATE COL, 68- *Mem:* Am Chem Soc; Sigma Xi; Indian Chem Soc; Am Soc Microbiol. *Res:* Organic synthesis and environmental pollution. *Mailing Add:* SC State Col Box 1633 Orangeburg SC 29117

KOLIN, ALEXANDER, b Odessa, Russia, Mar 12, 10; nat US; m 51. BIOPHYSICS. *Educ:* Prague Ger Univ, PhD(physics), 34. *Prof Exp:* Res fel biophys, Reese Hosp, Chicago, 35-37; physicist, Mt Sinai Hosp, NY, 38-41; res assoc sch eng, Columbia Univ, 41-46, instr physics, 44-45; asst prof, NY Univ, 45-46; asst prof, Univ Chicago, 46, actg chmn col physics, 47-50, chmn, 50-53, assoc prof, 53-55; assoc res biophysicist, 56-57, from assoc prof to prof, 57-76, EMER PROF BIOPHYS, SCH MED, UNIV CALIF, LOS ANGELES, 77- *Concurrent Pos:* Res fel, Med Sch, NY Univ, 41-42; instr physics, City Col NY, 41-44. *Honors & Awards:* John Scott Medal, 65; Albert F Sperry Medal, 67; Humboldt Award, 77; Founders Award, Electrophoresis Soc, 80. *Mem:* AAAS; Biophys Soc; Am Phys Soc; Sigma Xi; Am Physiol Soc; hon mem Electrophoresis Soc. *Res:* Photoelectric effects; Geiger counters; gas discharges; biophysics of circulation of blood; electromagnetic measurement of fluid flow and turbulence; isoelectric focusing; electromagnetophoresis; electrophoresis; cell electrophoresis; studies of vasomotion and blood flow. *Mailing Add:* Jules Stein Inst Univ Calif 1000 Stein Pl Los Angeles CA 90024-1771

KOLINER, RALPH, b New York, NY, Mar 20, 17; m 42; c 1. CIVIL ENGINEERING. *Educ:* Cooper Union, BChE, 39; Univ Pa, MS, 48, PhD(civil eng), 56. *Hon Degrees:* DSc, Villanova Univ, 85. *Prof Exp:* Naval architect, Philadelphia Naval Shipyard, 40-46; instr mech eng, Drexel Inst, 46; instr & asst prof civil eng, Univ Pa, 46-57; from assoc prof to prof civil eng, Villanova Univ, 57-83; RETIRED. *Concurrent Pos:* Consult, 46-; lectr & coordr, Rutgers Univ, 53-67. *Honors & Awards:* Lindback Award, Villanova Univ, 69 & Farrell Award, 81. *Mem:* Am Soc Eng Educ; Am Soc Civil Engrs; Am Concrete Inst. *Res:* Reinforced concrete; fluid and applied mechanics; engineering materials. *Mailing Add:* 2210 Cambridge Rd Broomall PA 15202

KOLIS, STANLEY JOSEPH, DRUG METABOLISM. *Educ:* St Peter's Col, BS, 67; Rutgers Univ, MS, 70. *Prof Exp:* Sr scientist, 80-86, RES SCIENTIST, HOFFMAN-LA ROCHE INC, 86- *Mem:* Am Chem Soc; Am Soc Pharmacol & Exp Therapeut. *Res:* Drug metabolism. *Mailing Add:* Hoffman-La Roche Inc 340 Kingsland St Nutley NJ 07110

KOLIWAD, KRISHNA M, b Byadgi, India, Feb 27, 38; m 67; c 2. MICROELECTRONICS, SEMICONDUCTORS. *Educ:* Karnatak Univ, India, BSc, 58, MSc, 60; Rensselaer Polytech Inst, MS, 64; Cornell Univ, PhD(mat sci), 67. *Prof Exp:* Res assoc mat sci, Cornell Univ, 67; res assoc solid state physics, Univ Md, College Park, 67-68; res assoc mat sci, Cornell Univ, 68-70; mem tech staff semiconductor, Tex Instruments, Inc, 70-75; MEM TECH STAFF, JET PROPULSION LAB, CALIF INST TECHNOL, 75- *Res:* Development of low cost silicon crystal growth technology for terrestrial solar energy application; photovoltaic devices. *Mailing Add:* 429 Paulette Pl La Canada CA 91011

KOLKA, MARGARET A, b Bay City, Mich, Aug 19, 52. TEMPERATURE REGULATION. *Educ:* Ind Univ, PhD(exercise physiol), 80. *Prof Exp:* Fel physiol, Ind Univ, 80-83; RES PHYSIOLOGIST, INST ENVIRON, US ARMY RES LABS, 83- *Mem:* Am Physiol Soc; Sigma Xi; AAAS. *Res:* Evaluation of environmental shessors on the physiological and biophysical properties of heat exchange in humans; skin blood flow; local sweating. *Mailing Add:* Thermal Physiol & Med USARIEM Natick MA 01760

KOLLAR, EDWARD JAMES, b Forest City, Pa, Mar 3, 34; m 63; c 5. ORAL BIOLOGY, EMBRYOLOGY. *Educ:* Univ Scranton, BS, 55; Syracuse Univ, MS, 59, PhD(zool), 63. *Prof Exp:* Instr zool, Univ Chicago, 63-66, asst prof biol, 66-67, asst prof anat, 67-71; assoc prof, 71-75, actg head dept, 85-86, PROF ORAL BIOL, SCH DENT MED, UNIV CONN, FARMINGTON, 76-, ASSOC DEAN ACAD AFFAIRS, 88- *Concurrent Pos:* Vis fac, W Alton Jones Cell Sci Ctr, Lake Placid, NY, 71-75; mem educ comt, Tissue Cult Asn, 74-78; bd dir, cranio-facial group, Int Asn Dent Res, 78-81; nat bd exam comt, Am Dent Asn, 78-83; ed in chief, Arch Oral Biol; vpres & prog dir, Craniofacial Group, Asn Dent Res, 81-82. *Honors & Awards:* Quantrell Teaching Award, Univ Chicago, 68, Ryerson Fac Fel, 69; Fogerty Int Fel, 78, Nat Acad Sci Exchange Fel, 78; Isaac Schour Mem Award, Int Asn Dent Res, 81. *Mem:* Int Asn Develop Biol; Soc Develop Biol; Am Asn Anatomists; Int Soc Differentiation. *Res:* Experimental studies of tooth and skin development with special reference to the etiology of craniofacial defects. *Mailing Add:* Dept Bio Struct & Function Univ Conn Health Ctr Farmington CT 06032-9984

KOLLEN, WENDELL JAMES, b Adrian, Mich, Feb 22, 35; m 55; c 4. SURFACE PHYSICS, POLYMER SCIENCE. *Educ:* Hope Col, AB, 64; Clarkson Col Technol, MS, 67, PhD(physics), 69. *Prof Exp:* Res physicist, 69-74, sr physicist, Tech Ctr, Owens-Ill, Inc, 74-87; ADJ RES PROF, UNIV TOLEDO, 87- *Mem:* Am Phys Soc; Am Vacuum Soc; Am Chem Soc; NY Acad Sci; Sigma Xi. *Res:* Vacuum ultramicrogravimetry; gas-solid interactions; chemical physics; high temperature corrosion; catalysis; gas and vapor transport in polymers; plastic barrier packaging, surface properties of plastics. *Mailing Add:* Univ Toledo Polymer Inst Toledo OH 43606-3390

KOLLER, CHARLES RICHARD, b North Manchester, Ind, Nov 16, 20; m 44; c 5. TEXTILE FIBERS, NONWOVEN FABRICS. *Educ:* Manchester Col, AB, 43; Purdue Univ, MS, 48, PhD(chem), 50. *Prof Exp:* Asst, Purdue Univ, 46-49; from res chemist to sr res chemist, E I Du Pont De Nemours & Co, 50-62, res supvr, 62-67, res assoc, 67-82; CONSULT, C R KOLLER ASSOC, 82- *Concurrent Pos:* Consult, Int Exec Serv Corp, 89. *Mem:* Am Chem Soc; Sigma Xi; fel Am Inst Chem; Fiber Soc. *Res:* Textile and inorganic fibers; nonwoven fabrics; condensation and vinyl polymerization; nitroparaffins; polymer and textile chemistry; fiber and textile engineering; fiber reinforced composites; engineered fabrics design and evaluation; biopolymers. *Mailing Add:* 317 Cecil St North Manchester IN 46962

KOLLER, EARL LEONARD, b Brooklyn, NY, Dec 8, 31; m 56; c 2. PHYSICS. *Educ:* Columbia Univ, AB, 52, MA, 58, PhD(physics), 59. *Hon Degrees:* MEng, Stevens Inst Technol, 73. *Prof Exp:* Asst physics, Columbia Univ, 52-59; from instr to assoc prof, 59-69, PROF PHYSICS, STEVENS INST TECHNOL, 69- *Honors & Awards:* Ottens Res Award, 63. *Mem:* Am Phys Soc; Sigma Xi; Am Asn Univ Prof. *Res:* High energy nuclear physics, especially particle physics; investigation of strange particle and pi meson properties; K meson decays; pi-p, K-p, p-p and p-d interactions using the Fermilab 30 inch bubble chamber hybrid system; neutronino interactions using bubble chamber techniques. *Mailing Add:* Dept Physics Stevens Inst Technol Hoboken NJ 07030

KOLLER, GLENN R, b Buffalo, NY, Nov 25, 51; m; c 3. GEOSTATISTICS. *Educ:* State Univ NY, Buffalo, BA, 73; Syracuse Univ, MS, 76, PhD(geol), 78. *Prof Exp:* Geologist, Savannah River Lab, Dept Energy, 78-80; MATH GEOLOGIST, AMOCO PROD RES, 80- *Mem:* Sigma Xi. *Res:* Mathematical and statistical manipulation of geologic data in the area of petroleum exploration; numerous patents. *Mailing Add:* 6642 S 67 E Ave Tulsa OK 74133

KOLLER, LOREN D, b Pomeroy, Wash, June 16, 40; m 63; c 3. NUTRITION, VETERINARY MEDICINE. *Educ:* Wash State Univ, DVM, 65; Univ Wis, MS, 69, PhD(path), 71. *Prof Exp:* Pvt vet pract, 65-66; capt, US Army Med Univ, 66-68; res assoc path, Vet Sci Dept, Univ Wis, 68-71; head diag & comt path, Animal Sci & Technol Br, Nat Inst Environ Health Sci, Research Triangle Park, NC, 71-72; res assoc, Sch Vet Med, Ore State Univ, 72-76, assoc prof, 76-78; asst dean, Vet Med, Univ Idaho, 78-81, from assoc prof to prof, 78-85, assoc dean, 81-85; PROF & DEAN, COL VET MED, ORE STATE UNIV, CORVALLIS, 85- *Concurrent Pos:* Prin investr & co-investr numerous grants, NIH, Environ Protection Agency, Health, Educ & Welfare Dept, US Dept Agr, Food & Drug Admin, 75-90; ed, Am J Vet Res, 76, 86, 87 & 89, J Am Vet Med Asn, 80-81 & 84-86, J Reticuloendothelial Soc, 82-85, Can J Comp Med, 84-85, J Toxicol & Environ Health, 84-85, J Clin Toxicol, 91; grants, Dow Chem, 80-81, Merck, Sharp & Dohme, 81-82, Idaho Beef Coun, 84-85, Warner Lambert Co, 87 & Pew Found, 88-89; mem, Pub Relations & Standards, Soc Toxicol, 81-84, prog comt, 84-87, immunotoxicol steering comt, 84-85, immunotoxicol spec sect, vpres, 85, pres, 86, Animals in Res, chmn, 89-91; adm adv, WRCC-46 Ram Epididymitis Regional Res Prog, 82-; adv ed, Int J Immunopharmacol, 87-89; mem subcomt, Immunotoxicol, Nat Acad Sci, 89-90, comt toxicol, Nat Res Coun, 90-93. *Mem:* Soc Toxicol; Soc Toxicol Pathologists; Am Vet Med Assoc; Am Asn Vet Immunolgists; Acad Vet & Comp Toxicologists; Asn Vet Med Col. *Res:* Effect of drugs and chemicals on immunity; toxicological, pathological and immunological studies of toxic substances; effect of environmental contaminants on tumor growth and immunity; development of immunopharmacology/toxicology procedures; development of polycythemia-vera-like condition in rats; selenium responsive diseases in livestock; pathology service; extensive publications. *Mailing Add:* Col Vet Med Ore State Univ Corvallis OR 97331-4801

KOLLER, NOEMIE, b Vienna, Austria, Aug 21, 33; US citizen; m 56; c 2. NUCLEAR PHYSICS, SOLID STATE PHYSICS. *Educ:* Columbia Univ, BA, 53, MA, 55, PhD(physics), 58. *Prof Exp:* Fel physics, Columbia Univ, 58-60; from asst prof to assoc prof, 60-70, PROF PHYSICS, RUTGERS UNIV, NEW BRUNSWICK, 70- *Mem:* Fel Am Phys Soc. *Res:* Spectroscopy of low lying nuclear levels; study of hyperfine interactions at nuclei in ionized atoms or in solids. *Mailing Add:* Dept Physics Rutgers Univ New Brunswick NJ 08903

KOLLER, ROBERT DENE, b Sidney, Nebr, Mar 7, 45; m 66; c 1. PHYSICAL CHEMISTRY, STATISTICS. *Educ:* Univ Nebr, BS, 67, PhD(theoret chem), 72. *Prof Exp:* Anal chemist, Com Solvents Corp, 67 & 68; fel theoret chem, Mellon Inst, Carnegie Mellon Univ, 72-73; SR SCIENTIST CHEM, ROHM & HAAS CO, 73- *Mem:* Am Chem Soc; Soc Automotive Engrs. *Res:* The application of computer and statistical techniques for analysis of chemical problems arising in industrial research; research and development of oil additives. *Mailing Add:* Rohm & Haas Co 727 Norristown Rd Spring House PA 19477

KOLLIG, HEINZ PHILIPP, b Bonn, Ger, May 19, 28; US citizen; m 51; c 2. ANALYTICAL CHEMISTRY, CLINICAL CHEMISTRY. *Educ:* Univ Bonn, Ger, BS, 50; Fla Inst Technol, BS, 69. *Prof Exp:* Chemist clin chem, Univ Bonn, Ger, 51-57; chemist power plant chem, City Bonn, Ger, 57-59; res chemist, Univ Ala, 59-60; anal chemist microanal, Southern Res Inst, 60-68; anal chemist environ health, TWA Kennedy Space Ctr, Fla, 68-71; RES CHEMIST WATER, US ENVIRON PROTECTION AGENCY, 71- *Mem:* Am Chem Soc. *Res:* Ecology. *Mailing Add:* 1050 Dogwood Hill NW Watkinsville GA 30677

KOLLMAN, PETER ANDREW, b Iowa City, Iowa, July 24, 44; m 70; c 2. THEORETICAL CHEMISTRY, BIOPHYSICAL CHEMISTRY. *Educ:* Grinnell Col, BA, 66; Princeton Univ, PhD(chem), 70. *Prof Exp:* NATO fel theoret chem, Cambridge Univ, 70-71; asst prof, 71-76, assoc prof, 76-80, PROF CHEM & PHARMACEUT CHEM, SCH PHARM, UNIV CALIF, SAN FRANCISCO, 80- *Concurrent Pos:* Career develop award, Nat Inst Gen Med Sci, 74. *Mem:* Am Chem Soc; Am Phys Soc; Sigma Xi. *Res:* Application of quantum mechanics and molecular mechanics to intermolecular interactions and to structure activity relationships in biological systems. *Mailing Add:* Dept Pharm/Chem 926S Univ Calif San Francisco CA 94122

KOLLMORGEN, G MARK, b Bancroft, Nebr, June 23, 32; m 54; c 4. CELL BIOLOGY, IMMUNOLOGY. *Educ:* Univ Iowa, BA, 57, MA, 60, PhD(radiation biol), 63. *Prof Exp:* Instr biol sci, Univ Iowa, 58-60, instr radiation biol, 61-63; resident res assoc, Argonne Nat Lab, 63-65, asst biologist, 65-66; asst mem, Okla Med Res Found, 66-69, assoc mem, 69-76; from asst prof to assoc prof, 66-74, PROF RADIOL, SCH MED, UNIV OKLA, 74-; MEM OKLA MED RES FOUND, 76- *Mem:* Am Asn Cancer Res; Am Soc Cell Biol; Soc Exp Biol & Med; Tissue Cult Asn; Sigma Xi. *Res:* Effects of dietary fat on tumor incidence and immune responses; serum factors which inhibit immune responses; influence of prostaglandins on tumor growth and transplantability; effects of products derived from cyclooxygenase and lipoxygenase pathways on function of natural killer cells. *Mailing Add:* 825 NE 13th St Okla Med Res Found Oklahoma City OK 73104

KOLLROS, JERRY JOHN, b Vienna, Austria, Dec 29, 17; nat US; m 42; c 2. EMBRYOLOGY, CELL BIOLOGY. *Educ:* Univ Chicago, SB, 38, PhD(zool), 42. *Prof Exp:* Asst zool, Univ Chicago, 40-42, neurosurg, 43-45, toxicity lab, 45, instr zool, Col, 45-46; assoc, 46-47, from asst prof to assoc prof, 47-57, actg chmn, 54-55, chmn dept, 55-77, prof zool, Univ Iowa, 57-88; EMER PROF ZOOL, UNIV IOWA, 57- *Concurrent Pos:* Consult, Am Col Dictionary & Random House Dictionary of the English Language; consult div inst progs, NSF, 64-66; mem cell biol study sect, NIH, 60-64, biol sci training rev comt, NIMH, 66-70; Comn Undergrad Educ Biol Sci, 67-71, chmn, 69-71. *Mem:* AAAS; Am Soc Cell Biol; Soc Develop Biol; Am Asn Anat; Am Soc Zool (treas, 59-62). *Res:* Control of skin gland development and segregation of skin regions; development of behavior in Amphibia; experimental embryology of amphibian central nervous system; amphibian metamorphosis; regeneration of Amphibia. *Mailing Add:* Dept Biol Univ Iowa Iowa City IA 52242

KOLM, HENRY HERBERT, b Vienna, Austria, Sept 10, 24; nat US; m 53; c 4. MAGNETISM. *Educ:* Mass Inst Technol, SB, 50, PhD(physics), 54. *Prof Exp:* Asst low temperature physics, Mass Inst Technol, 50-54, mem res staff, Lincoln Lab, 54-60, sr scientist & lectr, Dept Aeronaut & Astronaut, Nat Magnet Lab, 60-82; pres, Electromagnetic Launch Res Inc, 82-89; DIR, DIEZO ELEC PROD INC, 81-; PRES, MAGNEPLANE INT INC, 89- *Mem:* Am Phys Soc; Am Inst Aeronaut & Astronaut; Inst Elec & Electronic Engrs. *Res:* Hydrodynamics of liquid helium; semiconductor surface physics; cyclotron resonance in solids; design of pulsed and continous high-field solenoid magnets; superconductivity; magnetic separation; magnetic levitation and propulsion of high speed vehicles; applications of magnetism; piezoelectricity; electromagnetic launch technology. *Mailing Add:* Weir Meadow Rd Wayland MA 01778

KOLMAN, BERNARD, b Havana, Cuba, July 4, 32; US citizen; m 86; c 2. MATHEMATICS. *Educ:* Brooklyn Col, BS, 54; Brown Univ, ScM, 56; Univ Pa, PhD(math), 65. *Prof Exp:* Prin mathematician, Univac Div, Sperry Rand Corp, 57-64; from asst prof to prof, 64-76, PROF MATH, DREXEL UNIV, 76- *Mem:* Am Math Soc; Math Asn Am; Soc Indust & Appl Math; Asn Comput Mach. *Res:* Lie algebras; operations research. *Mailing Add:* Dept Math Drexel Univ 32nd & Chestnut St Philadelphia PA 19104

KOLMEN, SAMUEL NORMAN, b Brownsville, Tex, Mar 20, 30; m 54; c 2. PHYSIOLOGY. *Educ:* Univ Tex, BA, 54, PhD(physiol), 57. *Prof Exp:* From asst prof to prof physiol, Univ Tex Med Br, Galveston, 58-75, head div physiol, Shriners Burns Inst, 68-70, res coordr, 70-75; prof physiol & chmn dept, Sch Med, Wright State Univ, 75-84, asst dean sci & eng med, 80-84; prof physiol & assoc dean acad affairs, Hahnemann Univ Sch Med, 84-89, DIR MED EDUC & RES, MERCY HOSP PITTSBURGH, 89- *Concurrent Pos:* Kempner fel med, London Hosp Med Col, Univ London, 57-58; mem coun thrombosis, Am Heart Asn, res & rev comt, Ohio affil, 81-84; consult, Nat Bd Med Examr, Philadelphia, 89. *Mem:* AAAS; Soc Exp Biol & Med; Am Physiol Soc; Brit Biochem Soc; Microcirculatory Soc; Sigma Xi. *Res:* Adsorptive phenomena related to biological processes; fibrinogen metabolism, storage and distribution; lymphatic circulation; microcirculation; burn physiopathology. *Mailing Add:* 256 Sweet Gum Rd Pittsburgh PA 15238

KOLMES, STEVEN ALBERT, b Poughkeepsie, NY, Sept 17, 54; m 87; c 1. ERGONOMIC EFFICIENCY, BEHAVIORAL TOXICOLOGY. *Educ:* Ohio Univ, BS, 76; Univ Wis, MS, 78, PhD(zool), 84. *Prof Exp:* Asst prof, 84-89, ASSOC PROF BIOL, HOBART & WILLIAM SMITH COLS, 89- *Concurrent Pos:* Assoc ed, J Apicult Res, 89-91. *Mem:* Animal Behav Soc; Entom Soc Am; Int Bee Res Asn; Int Union Study Social Insects. *Res:* Ethology of invertebrates, specifically how ecological considerations have evolutionary shaped different aspects of behavior; concentrations are social insect division of labor, and behavioral aspects of pesticide resistance in spider mites. *Mailing Add:* Dept Biol Hobart & William Smith Cols Geneva NY 14456

KOLOBIELSKI, MARJAN, b Warsaw, Poland, Aug 17, 15; nat US; m 48; c 1. PETROLEUM CHEMISTRY, FUEL SCIENCE. *Educ:* Univ Lodz, MPhil, 48; Univ Paris, PhD(phys sci), 54. *Prof Exp:* Mem res staff, French Nat Ctr Sci Res, Normal Sch, Paris, 49-55; fel org chem, Northwestern Univ, 55-56; res fel, Mellon Inst, 56-63; res chemist, Borden Chem Co, 63-64; res chemist, Coating & Chem Lab, Aberdeen Proving Ground, 64-74, RES CHEMIST, US ARMY MOBILITY EQUIP RES & DEVELOP CTR, 74- *Mem:* Am Chem Soc; Combustion Inst; Chem Soc Fr; Am Inst Chemists; Am Soc Testing & Mat; Sigma Xi. *Res:* Development of new products, fuels and lubricants. *Mailing Add:* 6710 Sherwood Rd Baltimore MD 21239

KOLODNER, IGNACE IZAAK, b Warsaw, Poland, Apr 12, 20; nat US; div; c 3. MATHEMATICS, NONLINEAR PROBLEMS. *Educ:* Univ Warsaw, BA, 37; Univ Grenoble, Dipl, 40, NY Univ, PhD(math), 50. *Prof Exp:* Asst, Col Eng, NY Univ, 47-48, asst, Inst Math Sci, 48-51, res assoc, 51-53, sr scientist, 53-56; prof math, Univ NMex, 56-64; head dept, 64-71, prof, 64-90, EMER PROF MATH, CARNEGIE-MELLON UNIV, 90- *Concurrent Pos:* Instr, Wash Sq Col, NY Univ, 48-51; lectr, Stevens Inst Tech, 50-53, nat lectr, Soc Indust & Appl Math, 60-61 & 80-83; consult, Underwater Mine Comt, Nat Res Coun, 55, Courant Inst Math Sci, 56-60, 69; Sandia Corp, 56-68 & Lawrence Radiation Lab, 60-67; vis mem, Math Res Ctr, Univ Wis-Madison, 62; sch math study group, Stanford Univ, 65; Sussman vis prof, Israel Inst Technol, Haifa, 72-73; adj prof, Univ Pittsburgh, 76-77 & 78-79; grants, NSF, 57-60, 62-65, 71-78, OOR, 57-61, NONR, 65-69. *Honors & Awards:* Fulbright lectr, Republic Univ, Montevideo, Uruguay, 67; SIAM Nat lectr, 60-61. *Mem:* Am Asn Univ Profs; Am Math Soc; Math Asn Am; Soc Indust & Appl Math; Soc Natural Philos. *Res:* Differential equations; integral equations; mathematical physics. *Mailing Add:* Dept Math Carnegie-Mellon Univ Pittsburgh PA 15213

KOLODNER, PAUL R, b Morristown, NJ, Dec 16, 53; m 80. NONLINEAR DYNAMICS & PATTERN FORMATION. *Educ:* Princeton Univ, AB, 75; Harvard Univ, AM, 77, PhD(physics), 80. *Prof Exp:* MEM TECH STAFF, AT&T BELL LABS, 80- *Mem:* Am Phys Soc. *Res:* Nonlinear dynamics and pattern formation, specifically, convection in binary fluids. *Mailing Add:* AT&T Bell Labs Rm 1E-446 600 Mountain Ave Murray Hill NJ 07974-0636

KOLODNER, RICHARD DAVID, b Morristown, NJ, Apr 3, 51. MOLECULAR BIOLOGY. *Educ:* Univ Calif, Irvine, BS, 71, PhD(biol), 75. *Hon Degrees:* BSc, Harvard Univ, 88. *Prof Exp:* Res fel molecular biol, 75-78, from asst prof to assoc prof, 79-88, PROF BIOCHEM & MOLECULAR PHARMACOL, SCH MED & DANA-FARBER CANCER INST, HARVARD UNIV, 88- *Concurrent Pos:* Fel, Cystic Fibrosis Found, 75-76 & NIH, 76-78; jr fac res award, Am Cancer Soc, 81-83, fac res award, 88-89. *Mem:* Am Soc Biol Chemists; Am Soc Microbiol; Genetics Soc Am. *Res:* Enzymatic and molecular mechanism of genetic recombination in procaryotes and eucaryotes; DNA structure. *Mailing Add:* Dana-Farber Cancer Inst 44 Binney St Boston MA 02215

KOLODNY, GERALD MORDECAI, b Brookline, Mass, Apr 22, 37; m 64; c 3. RADIOLOGY. *Educ:* Harvard Univ, AB, 58; Northwestern Univ, Chicago, MD, 62; Am Bd Radiol, dipl, 67; Am Bd Nuclear Med, dipl, 74. *Prof Exp:* Intern med, Stanford Univ Med Ctr, 62-63; resident radiol, Mass Gen Hosp, 63-66; Picker Found fel biol, Mass Inst Technol, 66-69; asst radiologist, Mass Gen Hosp, 69-75, assoc radiologist & dir, Radiol Res Lab, 75-79; from instr to asst prof, 69-75, ASSOC PROF RADIOL, HARVARD MED SCH, 75-; DIR, DIV NUCLEAR MED, BETH ISRAEL HOSP, 79- *Concurrent Pos:* Picker Found grant, 69-71; res assoc biochem, Huntington Labs, Harvard Univ, 69-70, assoc chmn curric, Div Med Sci, 74- *Mem:* Am Soc Cell Biol; Tissue Cult Asn; Inst Elec & Electronics Engrs; Soc Nuclear Med; Radiol Soc NAm. *Res:* Gene regulation; RNA biochemistry; cell to cell communication; contact inhibition; electronics and computers in medicine; nuclear medicine; cellular radiation biology. *Mailing Add:* Div Nuclear Med Beth Israel Hosp 330 Brookline Ave Boston MA 02215

KOLODNY, NANCY HARRISON, b Brooklyn, NY, Mar 30, 44; m 64; c 3. PHYSICAL BIOCHEMISTRY. *Educ:* Wellesley Col, BA, 64; Mass Inst Technol, PhD(phys chem), 69. *Prof Exp:* Dean class of 76, 72-74, dir, Sci Ctr, 74-77, from asst prof to assoc prof, 69-85, PROF CHEM, WELLESLEY COL, 85- *Concurrent Pos:* Radcliffe Inst scholar, 70-72; res fel med, Mass Gen Hosp, Boston, 71-72; res assoc & lectr ophthal, Mass Eye & Ear Infirmary, Harvard Med Sch, Boston, 84- *Mem:* Am Chem Soc; Sigma Xi. *Res:* Electron spin resonance spectroscopy of charge transfer complexes in solution; nuclear magnetic resonance spectroscopy of protein-nucleic acid interactions; magnetic resonance imaging and spectroscopy of ocular disorders. *Mailing Add:* 124 Dartmouth St West Newton MA 02165

KOLODZIEJ, BRUNO J, b Chicago, Ill, Aug 27, 34; m 58; c 3. MICROBIOLOGY, MICROBIAL PHYSIOLOGY. *Educ:* Northern Ill Univ, BSEd, 58; Northwestern Univ, MS, 60, PhD(biol), 63. *Prof Exp:* NIH fel microbial physiol, Univ Chicago, 63-65; res assoc, Albert Einstein Med Ctr, Pa, 65-66; from asst prof to assoc prof microbiol, Ohio State Univ, 66-91. *Mem:* AAAS; Am Soc Microbiol; Sigma Xi. *Res:* Elucidation and characterization of bacterial cell surface components with emphasis on membrane binding-transport proteins associated with sugar and amino acid transport and structure and function of exocellular capsule material. *Mailing Add:* Dept Microbiol Ohio State Univ 484 W 12th Ave Columbus OH 43210-1292

KOLODZIEJSKI, LESLIE ANN, b Ft Leonard Wood, Mo, July 31, 58; m 79. ELECTRONIC MATERIALS, OPTOELECTRONIC DEVICES. *Educ:* Purdue Univ, BSc, 83, MS, 84, PhD(elec eng), 86. *Prof Exp:* Asst prof elec eng, Purdue Univ, 86-88; ASST PROF ELEC ENG, MASS INST TECHNOL, 88- *Concurrent Pos:* NSF presidential young investr & Off Naval Res young investr, 87. *Mem:* Am Phys Soc; Mat Res Soc; Optical Soc Am; Inst Elec & Electronics Engrs. *Res:* Fabrication of thin film semiconductors, such as zinc-selenium and gallium-arsenic, which are layered to form sophisticated heterostructures. *Mailing Add:* Mass Inst Technol Rm 13-3061 77 Massachusetts Ave Cambridge MA 02139

KOLODZY, PAUL JOHN, b Akron, Ohio, Aug, 1959; m 86; c 1. NEURAL NETWORKS, SMART SENSORS. *Educ:* Purdue Univ, BS, 83; Case Western Reserve Univ, MS, 84, PhD(chem eng), 86. *Prof Exp:* Staff mem, opto-radar systs, 86-89, ASST GROUP LEADER, NEURAL NETWORKS, MASS INST TECHNOL LINCOLN LAB, 89- *Concurrent Pos:* Co-chair, Simulation Panel, DARPA Neural Network Study, 88-89. *Res:* Develop and exploit neural network techniques for active and passive optical sensors; design and model optical sensor systems; advanced simulation technology. *Mailing Add:* Opto-Radar Systs Group MIT Lincoln Lab 244 Wood St Lexington MA 02173

KOLOPAJLO, LAWRENCE HUGH, b Steubenville, Ohio, Dec 27, 50; m 80; c 1. PHYSICAL-ANALYTICAL CHEMISTRY, CHEMICAL KINETICS. *Educ:* Muskingum Col, BS, 74; Pa State Univ, MS, 78; Wester Mich Univ, PhD(chem), 82. *Prof Exp:* Chemist, Dow Chem Co, 76-77; qual control, Allied Paper Co, SCM, 78-80; res assoc, State Univ NY, Buffalo, 82-84; asst prof phys chem, Claflin Col, 84-85; ASST PROF PHYS & ANALYTICAL CHEM, MARIETTA COL, 85- *Concurrent Pos:* Instr, Univ SC, 85. *Mem:* Am Chem Soc. *Res:* Coordination chemistry; kinetics and mechanisms of complex formation and ligand exchange reactions carried out in solution. *Mailing Add:* 716 S Blanchard Ct Findlay OH 45840

KOLP, BERNARD J, b Caroll, Iowa, Oct 20, 28; m 52; c 3. AGRONOMY, PLANT BREEDING. *Educ:* Iowa State Univ, BS, 54; Kans State Univ, MS, 55, PhD, 58. *Prof Exp:* Asst prof agron, Univ Wyo, 57-70, prof plant breeding, 70-86; RETIRED. *Res:* Drought resistance and emergence of wheat; nitrate content in oats; winter hardiness in wheat. *Mailing Add:* 1808 Ord Laramie WY 82070

KOLSKI, THADDEUS L(EONARD), b Chicago, Ill, Nov 29, 28; m 55; c 2. INORGANIC CHEMISTRY. *Educ:* Ill Inst Technol, BS, 50; St Louis Univ, MS, 54, PhD(inorg chem), 57. *Prof Exp:* Res chemist, E I du Pont de Nemours Co, Inc, 56-85; RETIRED. *Concurrent Pos:* Ed, High-Solids Coatings, 79-81. *Res:* Oxidation of niobium and niobium alloys; anodic characteristics of tantalum and niobium; electrolytic capacitators; surface and boron chemistry; white and colored pigments. *Mailing Add:* 1116 Graylyn Rd Chatham Wilmington DE 19803

KOLSKY, HARWOOD GEORGE, b Portland, Ore, Jan 18, 21; m 42; c 4. PHYSICS, COMPUTER SCIENCE. *Educ:* Univ Kans, BS, 43, MS, 47; Harvard Univ, PhD(physics), 50. *Prof Exp:* Asst instr physics, Univ Kans, 46-47; mem staff, Weapons Div & Theoret Div, Los Alamos Sci Lab, 50-52, assoc group leader, Hydrodyn Group, Theoret Div, 52-57; sr planning rep prod planning, IBM Sci Ctr, IBM Corp, 57-59, proj coordr stretch comput, 59, asst mgr, 438 L Proj, Omaha, 59-60, mgr, Systs Sci Res Lab, San Jose, 61-62, spec proj, Adv Systs Develop Lab, 62-64, univ prog, Palo Alto, 64-66, mgr, atmospheric physics dept, 66-69, IBM fel, 69-86; PROF COMPUT ENG, UNIV CALIF, SANTA CRUZ, 85- *Mem:* Am Phys Soc; Inst Elec & Electronics Engrs; Sigma Xi. *Res:* Digital computer application and design; compressible fluid hydrodynamics; nuclear moments by molecular beam technique; numerical meteorology. *Mailing Add:* 181 Seacliff Dr E Aptos CA 95003

KOLSKY, HERBERT, b London, Eng, Sept 22, 16; m 45; c 3. PHYSICS, MECHANICS. *Educ:* Imp Col, Univ London, BSc, 37; Univ London, PhD(phys chem), 40, DSc(physics), 57. *Hon Degrees:* DSc, Swiss Fed Inst Technol, 84. *Prof Exp:* Head dept physics, Akers Res Labs, Imp Chem Industs Ltd, 46-56; Fulbright vis prof eng, Brown Univ, 56-58; sr prin sci officer, Ministry Supply, UK, 58-60; PROF APPL PHYSICS, BROWN UNIV, 60- *Concurrent Pos:* Vis prof, Imperial Col, 68, METU, Ankara, 68, Oxford Univ, 74, Berkeley, 78 & ETH, Zurich, 78-79 & 80. *Honors & Awards:* Worcester Reed Warner Medal. Am Soc Mech Engrs, 83. *Mem:* Fel Acoust Soc Am; fel Brit Inst Physics; Brit Soc Rheol (pres, 59-60); Am Soc Mech Engrs; fel Acad Mech. *Res:* Stress waves in solids; viscoelasticity; rate of strain effects in metals; dynamic fracture phenomena; experimental techniques; composite materials. *Mailing Add:* Div Appl Math Brown Univ Providence RI 02912

KOLSRUD, GRETCHEN SCHABTACH, b Schenectady, NY, Jan 9, 39. GENETICS, HUMAN FACTORS ENGINEERING. *Educ:* McGill Univ, BSc, 60; Johns Hopkins Univ, MA, 63, PhD(psychol), 66. *Prof Exp:* Prin res scientist physiol psychol, Systs & Res Ctr, Honeywell Inc, 66-67; sr syst scientist human factors eng, Serendipity Inc, 67-71; staff scientist, BioTechnol Inc, 71-73; prog mgr, NASA, 73-74; prog mgr, Transp Prog, US Cong, 74-76, asst to dir, New & Emerging Technol, 76-78, actg group mgr, Health Group, prog mgr, biol appl, 79-89, SR ASSOC, OFF TECHNOL ASSESSMENT,

US CONG, 78- Concurrent Pos: Consult, Flight Mgt Systs, 67 & Nat Hwy Traffic Safety Admin, 71. Mem: Fel AAAS; Human Factors Soc (secy & treas, 74-75); NY Acad Sci; Sigma Xi. Res: Science policy; emerging technologies in biological sciences and in medicine; relationships between society and technology. Mailing Add: 114 Roberts Ct Alexandria VA 22314

KOLSTAD, GEORGE ANDREW, b Elmira, NY, Dec 10, 19; m 44; c 3. NUCLEAR PHYSICS, EARTH SCIENCES. Educ: Bates Col, BS, 43; Yale Univ, PhD(physics), 48. Prof Exp: Asst physics, Wesleyan Univ, 43-44; res assoc radar, Harvard Univ, 44-45; spec asst physics, Yale Univ, 45-46, asst, 46-47, instr, 48-50; with physics & math prog, Div Res, US Atomic Energy Comn, 50-52, head, 52-73; sr physicist & head geosci prog, Div Phys Res, US Energy Res & Develop Admin, 73-77, sr physicist & head geosci prog, Off Energy Res, Div Eng & Geosci, US Dept Energy, 77-90; RETIRED. Concurrent Pos: Guest staff mem, Inst Theoret Physics, Copenhagen, 56-57; trustee, Bates Col, 58-64 & Laytonsville Elem Sch, 66-71; mem bd dirs, Sandy Springs Friends Sch, 71-77; mem, European-Am Nuclear Data Comt, 60-73; US del, Int Nuclear Data Comt, Int Atomic Energy Agency, 63-73, chmn, 70-72; mem, Fed Coun Sci & Technol, Ad Hoc Comt Int Geodyn Proj, 73-78; liaison mem, Geophys Res Bd, Nat Acad Sci, 79-82; Geophysics Study Comn, 78-90; liaison mem, Bd Earth Sci, 79-86; mem, Interagency Coord Group Continental Sci Drilling, DUE, NSF & US Geol Surv, Dept Energy, 84-90; liaison mem, US Geodynamics Comt, 73-90; mem, Bd Energy & Natural Resources, 87-90. Honors & Awards: DOSECC Award Outstanding Contrib Continental Sci Drilling, 90. Mem: Fel Am Phys Soc; Am Geophys Union; Sigma Xi. Res: Mass spectroscopy; radar countermeasures; a linear accelerator for the production of high intensity gamma rays and neutrons; earth sciences and solar-terrestrial physics. Mailing Add: Oak Hill Farm 7920 Brink Rd Laytonsville MD 20882

KOLTHOFF, IZAAK MAURITS, b Almelo, Holland, Feb 11, 94; nat US. ANALYTICAL CHEMISTRY. Educ: State Univ Utrecht, PhD(chem), 18. Hon Degrees: DSc, Univ Chicago, 54; Dr, Univ Groningen, 64, Brandeis Univ, 74, Hebrew Univ, Jerusalem, 74 & Univ Ariz, 85. Prof Exp: Conservator, Pharmaceut Inst, State Univ Utrecht, 17-27, privat docent appl electrochem, 24-27; prof & head div, 27-62, EMER PROF ANALYTICAL CHEM, UNIV MINN, MINNEAPOLIS, 62- Concurrent Pos: Hon prof, Lima & La Plata; consult, Phillips Petrol Co; chmn comt anal chem & mem comt Fulbright scholars, Nat Res Coun; vpres, Int Union Pure & Appl Chem. Honors & Awards: Comdr, Order of Orange-Nassau, 47; Nichols Medal, Am Chem Soc, 49, Fisher Award, 50 & Minn Award, 60; Charles Medal, Charles Univ, Prague; Willard Gibbs Medal, 64; Polarographic Medal, Brit Polarographic Soc, 64; first recipient, Kolthoff Gold Medal, Acad Pharmaceut Sci, 67; Olin-Palladium Medal Award, Electrochem Soc, 81; Hanus Medal, Czechoslovak Chem Soc. Mem: Nat Acad Sci; Am Pharmaceut Asn; Int Union Pure & Appl Chem; hon mem Royal Soc Chem; Neth Acad Sci; Belg Acad Sci; AAAS; Neth Chem Soc. Res: Macromolecular compounds, physical chemistry; acid-base indicators; theory of acids and bases; volumetric analysis; potentiometric, conductometric and amperometric titrations; polarography; properties of crystalline precipitates; kinetics and mechanism of emulsion polymerization; nonaqueous chemistry; proteins; pharmaceutical chemistry. Mailing Add: 740 River Dr Apt 236 St Paul MN 55116

KOLTUN, DANIEL S, b Brooklyn, NY, Dec 7, 33; m 56; c 2. THEORETICAL PHYSICS, NUCLEAR PHYSICS. Educ: Harvard Col, AB, 55; Princeton Univ, PhD(physics), 61. Prof Exp: Res assoc physics, Princeton Univ, 60-61; NSF vis fel nuclear physics, Weizmann Inst, 61-62 & Inst Theoret Phys, Copenhagen, 62; res assoc physics, 62-63, from asst prof to assoc prof, 63-74, PROF PHYSICS, UNIV ROCHESTER, 74- Concurrent Pos: Res assoc ctr theoret physics, Mass Inst Technol, 69-70; Alfred P Sloan res fel, 69-71; vis prof, Tel Aviv Univ, 76-77 & J S Guggenheim fel, 76-77, Lady Davis vis prof, Hebrew Univ, 85; vis scientist, Mass Inst Technol, 84; assoc ed, Phys Rev C, 78-80, Phys Rev Letters, 79-81. Mem: Fel Am Phys Soc. Res: Theoretical nuclear spectroscopy; many-body theory; interaction of nuclei with mesons; scattering theory; quarks in nuclei. Mailing Add: Dept Physics & Astron Univ of Rochester Rochester NY 14627

KOLTUN, STANLEY PHELPS, b Bogalusa, La, Mar 5, 25; m 55; c 3. CHEMICAL ENGINEERING. Educ: La State Univ, BSChE, 48. Prof Exp: Chem engr, Proc Design Unit, 56-59, cost engr, Cost & Design Unit, 59-63, proj leader, Food Prod Invests, 63-70, res chem engr, 70-76, actg res leader, Oilseed Prod Res, 76-82, RES LEADER, FOOD & FEED ENGR RES UNIT, SOUTHERN REGIONAL RES CTR, USDA, 82- Mem: Am Inst Chem Engrs; Nat Soc Prof Engrs; Am Oil Chemists' Soc; Asn Off Anal Chemists; Sigma Xi; Inst Food Technologists. Res: Detoxification and inactivation of aflatoxin contaminated oilseeds; oilseed solvent extraction; food dehydration; utilization of oilseed proteins as human food; oilseed meals; cottonseed; peanuts; sweetpotatoes; human nutrition; food safety. Mailing Add: 5601 Avron Blvd Metairie LA 70003

KOLTUN, WALTER LANG, b New York, NY, Apr 23, 28; m 62; c 2. BIOPHYSICAL CHEMISTRY. Educ: Mass Inst Technol, BS, 48, PhD(biochem), 52. Prof Exp: Asst biol, Mass Inst Technol, 51-52, res assoc, 52-53; asst prof biochem, Sch Med, Univ Va, 55-56; res assoc, Med Col, Cornell Univ, 56-59; consult biophys, Univ Calif, Berkeley, 59-61; staff mem, Sci Resources Planning Off, NSF, 61-64, prog dir molecular biol, 64-65; spec asst, Off Vpres & Secy & Inst Secy for Found, Mass Inst Technol, 65-68; dir prog advan study, Bolt, Beranek & Newman, 68-70; ASST DIR RESOURCES, HARVARD-MASS INST TECHNOL HEALTH PROG, 70- Mem: AAAS; Fedn Am Sci; Am Soc Biol Chem; Sigma Xi. Res: Structure, function and interaction of macromolecules, particularly proteins. Mailing Add: 76 Goodnough Rd Brookline MA 02167

KOLYER, JOHN M, b East Williston, NY, June 30, 33; m 60; c 4. ORGANIC CHEMISTRY, POLYMER CHEMISTRY. Educ: Hofstra Col, BA, 55; Univ Pa, PhD(org chem), 60. Prof Exp: Technician pesticides, Olin Mathieson Chem Corp, NY, 55-56; res chemist, FMC Corp, NJ, 60-62; res chemist,

Thompson Chem Co, Mass, 62-63; sr res chemist, Plastics Div, Allied Chem Corp, 64-65, group leader, 65-67, tech supvr, Morristown, 67-71; MEM TECH STAFF, AUTONETICS DIV, ROCKWELL INT CORP, 73- Concurrent Pos: Mem, Lepidoptera Found, 65-73. Mem: Am Chem Soc; Lepidop Soc; SAMPE; EOS/ESD Asn. Res: Preparation processes for plastics additives and monomers; polymerization processes and fabrication methods; polymer modifications; lepidopterological research; ESD (electrostatic discharge) control; author of one publication. Mailing Add: 885 Sea Gull Lane Apt B-311 Newport Beach CA 92663

KOMAI, HIROCHIKA, ANESTHESIOLOGY. Educ: Univ Calif, Berkeley, PhD(biochem), 67. Prof Exp: ASSOC SCIENTIST, UNIV WIS, 80- Mailing Add: Dept Anesthesiol Univ Wis Health Sci Ctr 600 Highland Ave Madison WI 53792

KOMANDURI, RANGA, materials engineering, for more information see previous edition

KOMAR, ARTHUR BARAWAY, b Brooklyn, NY, Mar 26, 31; m 52; c 2. THEORETICAL PHYSICS. Educ: Princeton Univ, AB, 52, PhD(physics), 56. Prof Exp: Fel, Scandinavian-Am Found, Inst Theoret Physics, Denmark, 56-57; asst prof physics, Syracuse Univ, 58-60, assoc prof, 60-63; dean, Belfer Grad Sch Sci, 69-78; assoc prof, 63-66, PROF PHYSICS, YESHIVA UNIV, 66-; RES PROF PHYSICS, NY UNIV, 84- Concurrent Pos: Prog dir, Gravitational Physics, NSF, 82-83, 86-87. Res: General relativity and quantum field theory. Mailing Add: 2621 Palisade Ave New York NY 10463

KOMAR, PAUL D, b Grand Rapids, Mich, Dec 2, 39; m 62; c 1. OCEANOGRAPHY, MARINE GEOLOGY. Educ: Univ Mich, BA, 61, MS, 62 & 65; Univ Calif, San Diego, PhD(oceanog), 69. Prof Exp: NATO fel, St Andrews Univ, Scotland, 69-70; asst prof, 70, assoc prof, 70-78, PROF OCEANOG, ORE STATE UNIV, 78- Mem: Soc Econ Paleont & Mineral; Am Geophys Union; Geol Soc Am. Res: Transport of sand on beaches; mechanics of sediment transport. Mailing Add: Col Oceanog Ore State Univ Corvallis OR 97331

KOMARKOVA, VERA, b Pisek, Czech, Dec, 25, 42. PLANT ECOLOGY. Educ: Charles Univ, Czech, MSc, 64; Univ Colo, PhD(biol), 76. Prof Exp: RES ASSOC, INST ARCTIC & ALPINE RES, UNIV COLO, BOULDER, 76-, ASST PROF, DEPT ENVIRON POP & ORGANISMIC BIOL, 79- Concurrent Pos: Prin investr, 79- Mem: AAAS; Am Inst Biol Sci; Ecol Soc Am; Int Soc Vegetation Sci. Res: Phytosociology; photogeography; vegetation mapping; methods of vegetation analysis; environment-vegetation relationship; community development; vegetation and environmental data management; effects of perturbation. Mailing Add: Leyson Am Sch CH-1854 Leyson Switzerland

KOMARMY, JULIUS MICHAEL, b Franklin, NJ, Oct 19, 26; m 52; c 4. PHYSICAL CHEMISTRY, ORGANIC CHEMISTRY. Educ: Southwest Mo State Col, BS, 49; Univ Ark, MS, 52, PhD, 58. Prof Exp: Asst chem, Univ Ark, 49-52, instr, 52-54, 56-67; instr, Flint Jr Col, 57-61; Supvr chem, Res Dept, AC Spark Plug Div, Gen Motors Corp, 61-65, Supvr chem res, 65-66, sr staff res scientist, 66-86; CATALYTIC CONVERTER COLLETORS INC, 87- Mem: AAAS; Am Chem Soc; Soc Automotive Eng; Am Soc Testing & Mat; Soc Info Displays. Res: Automotive applications of plastics and elastomers; fuels and lubricants and filtration processes; catalyst development for automotive exhaust environmental controls; materials development for fabrication of large liquid crystal displays. Mailing Add: 1636 Miller Rd Flint MI 48503

KOMARNENI, SRIDHAR, b Komarneni Varipalem, India, Sept 26, 44; US citizen; m 79; c 1. MATERIALS RESEARCH, CLAY MINERALOGY. Educ: Andhra Pradesh Agr Univ, India, BSc, 68; Indian Agr Res Inst, MSc, 70; Univ Wis, Madison, PhD(soils), 73. Prof Exp: Proj assoc, Univ Wis, Madison, 73-76; proj assoc solid state sci, 76-78, res assoc, 78-81, sr res assoc, 81-84, assoc prof, 84-87, PROF CLAY MINERAL, PA STATE UNIV, 87- Mem: AAAS; Am Soc Agron; Soil Sci Soc Am; Clay Minerals Soc; Am Ceramic Soc; Sigma Xi. Res: Crystal chemistry of clay minerals and zeolites; new materials preparation and characterization; sol-gel chemistry; nuclear and hazardous waste disposal. Mailing Add: Mat Res Lab Penn State Univ University Park PA 16802

KOMIAK, JAMES JOSEPH, b Chicago, Ill, Oct 16, 53; m; c 1. MMIC & MIC TECHNOLOGY-CIRCUIT DESIGN, T-R MODULE & SYSTEM ARCHITECTURES. Educ: Cornell Univ, BS, 74, MS, 76, PhD(elec eng), 78. Prof Exp: Sr assoc, staff & adv eng, antenna/receiver develop, IBM Fed Systs, 78-81, develop engr & mgr, 81-83; sr engr, 83-87, PRIN STAFF, MICRO/ MM-WAVE TECHNOL, GE ELECTRONICS LAB, 87- Concurrent Pos: Instr, IBM Systs Eng Course, 81 & GE Eng Course, 83; adj prof elec eng, State Univ NY, Binghamton, 83. Mem: Sr mem Inst Elec & Electronics Engrs. Res: Technical; management; consulting in microwave; millimeter-wave technology; system architectures; T/R module/MMIC/MIC circuit design; design of mil-spec analog, digital, RF, microwave sub-systems; holder of 2 US patents; author of 32 publications. Mailing Add: 26 Lakeshore Rd Lansing NY 14882

KOMINEK, LEO ALOYSIUS, b Chicago, Ill, Apr 11, 37; m 59; c 4. MICROBIOLOGY, BIOCHEMISTRY. Educ: St Joseph's Col, Ind, BS, 59; Univ Ill, PhD(microbiol), 64. Prof Exp: Res scientist microbiol, 72-78, res mgr, 78-79, SR SCIENTIST, UPJOHN CO, 79- Mem: Am Soc Microbiol; Am Chem Soc; Sigma Xi; Am Acad Microbiol; NY Acad Sci. Res: Biochemical aspects of bacterial sporulation; microbial metabolism; antibiotic biosynthesis; steroid and sterol bioconversions. Mailing Add: 2209 Hickory Point Dr Portage MI 49081

KOMINZ, DAVID RICHARD, b Rochester, NY, Apr 2, 24; div; c 3. PROTEIN CHEMISTRY. *Educ:* Univ Rochester, MD, 47; Harvard Univ, BA, 50. *Prof Exp:* Intern, Gorgas Hosp, CZ, 47-48; res fel phys chem lab, Harvard Univ, 50-51; from sr asst surgeon to med dir, Nat Inst Arthritis & Metab Dis, 51-65, chief, NIH Pac Off, Tokyo, 66-68, med dir, Nat Inst Arthritis, Metab & Digestive Dis, 68-76, chief sect bioenergetics, Lab Biophys Chem, 70-76; assoc dir, Nat Bladder Cancer Proj, 76-78; RES PROF PHYSIOL, UNIV MASS MED SCH, 78- *Mem:* AAAS; Am Chem Soc; Am Soc Biol Chem; Biophys Soc. *Res:* Amino acid analysis; comparative biochemistry of muscle proteins; protein modifications; interactions of muscle proteins. *Mailing Add:* Dept Physiol 55 N Lake Ave Worcester MA 01605

KOMISARUK, BARRY RICHARD, b New York, NY, Apr 4, 41; m 61; c 2. PSYCHOBIOLOGY, NEUROPHYSIOLOGY. *Educ:* City Univ New York, BS, 61; Rutgers Univ, PhD(neuroendocrinol), 65. *Prof Exp:* NIMH fel neuroendocrinol, Univ Calif, Los Angeles, 65-66; asst prof zool, 66-68, assoc prof, 68-72, PROF ZOOL, RUTGERS UNIV, NEWARK, 72- *Concurrent Pos:* NIMH res grant, Rutgers Univ, Newark, 66-79, NIMH Res Scientist Develop Award, 69-79; NSF res grant, Rutgers Univ, Newark, 79- *Mem:* AAAS; Soc Neurosci; Endocrine Soc; Am Physiol Soc; Int Soc Psychoneuroendocrinol. *Res:* Neurophysiological bases of species characteristic, hormonally influenced behavior, analgesic mechanisms. *Mailing Add:* Dept Biol Sci Rutgers Univ Newark NJ 07102

KOMISKEY, HAROLD LOUIS, CENTRAL NERVOUS SYSTEM, TOXICOLOGY. *Educ:* Univ Wis, PhD(pharmacol), 75; Am Bd Toxicol, dipl, 81. *Prof Exp:* SYST PROF PHARMACOL, COL MED, UNIV ILL, 83- *Mailing Add:* 4200 Arkansas Ave Kenner LA 70065

KOMKOV, VADIM, b Moscow, USSR, Aug 18, 19; m 46; c 5. OPTIMIZATION, CONTROL OF NONLINEAR SYSTEMS. *Educ:* Warsaw Polytech Inst, dipl ing mech eng, 48; Univ Utah, PhD(math), 64. *Prof Exp:* Asst prof mech eng, Univ Utah, 57-64; vis assoc prof math, MRC, Univ Wis, 64-65; assoc prof math, Fla State, 65-69; prof math, Tex Tech, 69-77; ed, Math Rev, 77-80; prof & head math, WVa Univ, 80-83 & Winthrop Col, 83-87; PROF & HEAD MATH, AIR FORCE INST TECHNOL, 87- *Concurrent Pos:* Consult, Southwest Res Inst, 70-75, US Army Arm Com, 72-77 & Univ Cincinnati, 80-82; vis prof, Polish Acad Sci, Warsaw, 80. *Mem:* Soc Indust & Appl Math; Math Asn Am. *Res:* Applied mathematics-variational methods; theoretical classical mechanics; continuum mechanics; engineering optimization; sensitivity of mechanical and structural systems to design changes. *Mailing Add:* PO Box 175 Dayton OH 45402

KOMM, HORACE, b Russia, Dec 30, 16; nat US; m 47; c 3. MATHEMATICS. *Educ:* Univ Buffalo, BA, 37; Univ Mich, MA, 38, PhD(math), 42. *Prof Exp:* Structures engr, Curtiss-Wright Corp, 42-44, asst to chief mathematician, Res Lab, 44-46; instr math, Univ Rochester, 46-48, asst prof, 48-52; assoc prof, Univ of the South, 52-53; asst prof, Rensselaer Polytech Inst, 53-62; assoc prof, 63-71, chmn dept, 71-77, PROF MATH, HOWARD UNIV, 71- *Mem:* Am Math Soc; Math Asn Am. *Res:* Dimension of partially ordered sets; general and algebraic topology. *Mailing Add:* 5130 Wickett Terr Bethesda MD 20814

KOMMEDAHL, THOR, b Minneapolis, Minn, Apr 1, 20; m 51; c 3. PLANT PATHOLOGY. *Educ:* Univ Minn, BS, 45, MS, 47, PhD(plant path), 51. *Prof Exp:* With bur plant indust, USDA, 43-46; instr, Agr Exp Sta, Univ Minn, 46-51; asst prof, Agr Exp Sta, Ohio State Univ, 51-53; from asst prof to prof, 53-90, EMER PROF PLANT PATH, UNIV MINN, ST PAUL, 90- *Concurrent Pos:* Assoc ed, Am Phytopath Soc, 50-52, ed-in-chief, 64-67, coordr publ, 78-84, sci adv, 84-; consult, botanist & taxonomist, Div Plant Indust, State Dept Agr, Dairy & Food, Minn, 54-60; Guggenheim fel, Waite Agr Res Inst, Australia, 61-62; Fulbright fel, Iceland, 68; consult, McGraw-Hill Co; counr, Int Soc Plant Path, 72-78, secy gen & treas, 83-88; ed, Int Newslett Plant Path, 83-93; consult, Sci Mus Minn, 90- *Honors & Awards:* Award of Excellence, Weed Sci Soc Am, 66; Distinguished Serv Award, Am Phytopath Soc, 84; E C Stakman Award, 90. *Mem:* Fel AAAS; Am Inst Biol Sci; fel Am Phytopath Soc (vpres, 69, pres, 71); Mycol Soc Am; Int Soc Plant Path (treas, 83-93); Bot Soc Am; NY Acad Sci; Weed Sci Soc Am; Coun Biol Eds. *Res:* Flax and corn diseases; weed ecology; root diseases and ecology of root-infecting fungi; biological control root diseases; Fusarium species. *Mailing Add:* Dept Plant Path 495 Borlaug Hall Univ Minn St Paul MN 55108

KOMOREK, MICHAEL JOESPH, JR, b Buffalo, NY, July 14, 52; m 82. NUCLEAR MEDICINE, RADIATION SAFETY. *Educ:* State Univ NY, Buffalo, BA, 74, BS, 77. *Prof Exp:* Res radioactive gases, Nuclear Sci & Technol Ctr, State Univ NY, Buffalo, 72-77, asst health physicist, 78-82; health physicist/nuclear engr, West Valley Demonstration Proj, 83; corp health physicist, Syncor Int Radiopharm, 83-84; PRES & RADIATION SAFETY OFFICER, ALARA MGT CO, 79-; RADIATION SAFETY OFFICER, ROSWELL PARK CANCER INST, 88- *Concurrent Pos:* Health physicist, bd dirs, Elma Nuclear Consults, 84- *Mem:* Health Physics Soc; Am Asn Physicists in Med; Laser Inst Am; Soc Nuclear Med; Am Indust Hyg Asn. *Res:* Nuclear medicine related research for new imaging compounds which may be used for imaging of diseases related to cancer. *Mailing Add:* 80 Pinewood Trail East Aurora NY 14052

KOMORIYA, AKIRA, STRUCTURE-FUNCTION OF GROWTH FACTORS & THEIR RECEPTORS. *Educ:* Duke Univ, PhD(phys chem), 77. *Prof Exp:* Prin investr, Biotechnol Res Ctr, Meloy Lab, Inc, 84-88; SR STAFF FEL, DIV CYTOKINE BIOL, FOOD & DRUG ADMIN, BETHESDA, MD, 88- *Mailing Add:* 10605 Pine Haven Terr Rockville MD 20852

KOMORNICKI, ANDREW, b Louth, Eng, Oct 23, 48; US citizen. THEORETICAL CHEMISTRY. *Educ:* Univ Wis-Milwaukee, BS, 70; State Univ NY Buffalo, PhD(theoret math), 74. *Prof Exp:* Res assoc, Univ Tex, Austin, 73-74 & Univ Rochester, NY, 74-76; Nat Res Coun assoc, Ames Res Ctr, NASA, Moffett Field, Calif, 76-78; res scientist & pres, Polyatomics Res Inst, 78-; AT AMES RES CTR, NASA, MOFFETT FIELD, CALIF. *Concurrent Pos:* Samuel B Silbert fel, 72-73; Nat Res Coun fel, 76; mem, Bd Dir, Molecular Res Inst, Atherton, Calif, 79. *Mem:* Am Chem Soc; Am Phys Soc. *Res:* Theoretical chemistry; molecular quantum mechanics; molecular spectroscopy, infrared and raman vibrational intensities; dynamics of chemical reactions; author or coauthor of over 40 publications. *Mailing Add:* Polyatomics Res Inst 1101 San Antonio Rd No 420 Mountain View CA 94043

KOMOROSKI, RICHARD ANDREW, b St Louis, Mo, Feb 4, 47. PHYSICAL CHEMISTRY, ANALYTICAL CHEMISTRY. *Educ:* St Louis Univ, BS, 69; Ind Univ, PhD(phys chem), 73. *Prof Exp:* Fel chem, Fla State Univ, 73-76; sr res chemist, Diamond Shamrock Corp, 76-; AT RES & DEVELOP CTR, B F GOODRICH CO. *Mem:* Am Chem Soc; Royal Soc Chem; Soc Appl Spectros; Sigma Xi. *Res:* Nuclear magnetic resonance spectroscopy; structure of synthetic and biological macromolecules. *Mailing Add:* Radiol Dept Slot 582 Univ Ark Med 4301 W Marham Little Rock AR 72205

KOMPALA, DHINAKAR S, b Madras, India, Nov 20, 58; US citizen; m 83; c 1. BIOPROCESS ENGINEERING, RECOMBINANT MICROBIAL & MAMMALIAN CELL CULTURE. *Educ:* Indian Inst Technol, Madras, BTech, 79; Purdue Univ, MS, 82, PhD(chem eng), 84. *Prof Exp:* asst prof, 85-91, ASSOC PROF CHEM ENG, UNIV COLO, BOULDER, 91- *Concurrent Pos:* Presidential Young Investr, NSF, 88; vis assoc, Calif Inst Technol, 91-92. *Mem:* Am Inst Chem Engrs; Am Chem Soc; Am Soc Eng Educ; Soc Indust Microbiol; Int Soc Anal Cytometry. *Res:* Development of optimal operating strategies and novel bioreactor designs for maximizing bioprocesses or heterologous protein expressions in recombinant microbial and mammalian cultures; metabolic modeling; recombinant DNA; biotechnology. *Mailing Add:* Dept Chem Eng Univ Colo Boulder CO 80309-0424

KON, MARK A, PARTIAL DIFFERENTIAL OPERATIONS. *Educ:* Cornell Univ, BA, 74; Mass Inst Technol, PhD(math), 79. *Prof Exp:* Irvine lectr, Univ Calif, Irvine, 80; asst prof, 81-88, ASSOC PROF MATH, BOSTON UNIV, 88- *Concurrent Pos:* Vis asst prof math & comput sci, Columbia Univ, 85-87. *Mem:* Am Math Soc; Int Asn Math Physics; AAAS. *Res:* Mathematical physics, primarily in quantum statistical mechanics; partial differential equations, primarily the Schrödinger equation; complexity of approximately solved problems. *Mailing Add:* Dept Math Boston Univ Boston MA 02215

KONAT, GREGORY W, b Mar 6, 47; Danish citizen. MOLECULAR BIOLOGY, BIOCHEMISTRY. *Educ:* Univ Warsaw, MSc, 69; Univ Odense Med Sch, PhD(med sci), 75. *Prof Exp:* Res teaching asst biochem, Dept Neurochem, Med Res Ctr, Polish Acad Sci, Warsaw, 69-71; res scientist biochem-neurochem, Neurochem Inst, Copenhagen, 71-81; res assoc prof neurochem, Inst Med Physiol & Neuropath, Univ Copenhagen, 81-83; res scientist neurochem, Dept Neurobiol & Anat, Med Sch, Univ Tex, Houston, 83-84; asst prof neurochem, 84-87, ASSOC PROF NEUROCHEM & MOLECULAR BIOL, DEPT NEUROL MED, UNIV SC, CHARLESTON, 87-; PROF MOLECULAR NEUROBIOL, DEPT ANAT, SCH MED, WVA UNIV, MORGANTOWN, 89- *Concurrent Pos:* Adj prof molecular neurobiol, Dept Pediat, Sch Med, WVa Univ, Morgantown, 90- *Mem:* Int Soc Neurochem; Int Soc Develop Neurosci; Soc Complex Carbohydrates; Am Soc Neurochem; Am Soc Biochem & Molecular Biol. *Res:* Mechanisms of myelin gene expression; myelinogenesis and myelin disorders; intracellular processing of membrane proteins. *Mailing Add:* Dept Anat Sch Med WVa Univ 4052 HSN Morgantown WV 26506

KONDE, ANTHONY JOSEPH, b Passaic, NJ, Jan 15, 12; m 40; c 1. PHYSICAL CHEMISTRY. *Educ:* Trenton State Col, BS, 34; Columbia Univ, MA, 37; Fordham Univ, PhD, 63. *Prof Exp:* Sci teacher & chmn high schs, NJ, 35-42; from instr to prof chem, Univ Col, St John's Univ, NY, 42-62, chmn dept, 46-62; prof chem, Pace Univ, 62-70, Harold Blancke prof, 70-80, chmn dept, 64-80; RETIRED. *Concurrent Pos:* Instr chem, St Francis Col, NY & Panzer Col, 58-61. *Mem:* Fel Am Inst Chemists; Am Chem Soc; NY Acad Sci; AAAS. *Res:* Chemical instrumentation; chemistry education. *Mailing Add:* 22 Great Hill Terrace Short Hills NJ 07078

KONDO, EDWARD SHINICHI, b Victoria, BC, Sept 5, 39; m 70; c 1. FOREST PATHOLOGY. *Educ:* Univ Toronto, BScF, 64, MScF, 66, PhD(plant path), 70. *Prof Exp:* Res officer, Can Forestry Serv, Dept Environ, 69-71, res scientist forest path, 71-83, dir, Forest Insect & Dis Surv, 83-88 & Biorational Control Agents Prog, 88-90, DIR GEN, FOREST PEST MGT INST, CAN FORESTRY SERV, 90- *Concurrent Pos:* Adj prof, Forestry Fac, Univ Toronto, 72-74. *Res:* Vascular wilt tree diseases; Dutch elm disease; urban forestry; tree and fungus physiology, mycology, chemotherapy of tree diseases. *Mailing Add:* Can Forestry Serv Box 490 Sault Ste Marie ON P6A 5M7 Can

KONDO, NORMAN SHIGERU, b Honolulu, Hawaii, Oct 30, 41; m 71; c 1. ORGANIC CHEMISTRY. *Educ:* Univ Hawaii, BA, 63; Univ Calif, Riverside, PhD(chem), 67. *Prof Exp:* Res assoc biophys, Argonne Nat Lab, 71-73; asst prof chem, Fed City Col, 73-77; ASSOC PROF CHEM, UNIV DC, 77- *Concurrent Pos:* NIH fel, Johns Hopkins Univ, 67-69 & fel, 69-71; NIH grant, Fed City Col, Univ DC, 75- *Mem:* Am Chem Soc; Biophys Soc. *Res:* Synthesis and conformational studies on nucleic acid constituents. *Mailing Add:* 7135 Red Horse Tavern Lane Springfield VA 22153

KONDO, YOJI, b Hitachi, Japan, May 26, 33; US citizen; m 65; c 3. ASTRONOMY, ASTROPHYSICS. *Educ:* Tokyo Univ Foreign Studies, BA, 58; Univ Pa, MS, 63, PhD(astron), 65. *Prof Exp:* Nat res assoc astron & space sci, Goddard Space Flight Ctr, NASA, 65-68, astronr, 68-69, chief, Astrophys Sect, Johnson Space Ctr, 69-77, ASTROPHYSICIST, GODDARD SPACE FLIGHT CTR, NASA, 78-, PROJ SCIENTIST, INT ULTRAVIOLET EXPLORER SATELLITE OBSERV, 82-, PROJ

SCIENTIST, EXTREME ULTRAVIOLET EXPLORER, 88- *Concurrent Pos:* Mem adj grad fac, Univ Houston, 68-74, adj prof, 74-77; from adj assoc prof to adj prof, Univ Okla, 71-77; ed, Earth & Extraterrestrial Sci, 74-79 & Comments Astrophys, 79-; consult, NASA Hq, 78-; vis prof, Inst Space & Astronaut Sci, Tokyo, 83; pres comt 44 astron from space, Int Astron Union, 85-88 & comt 42 close binary stars, 91- *Mem:* AAAS; Am Astron Soc; Int Astron Union. *Res:* Astronomical observations from space; interacting close binary stars; interstellar medium; active galactic nuclei. *Mailing Add:* Lab Astron & Solar Physics Code 680 Goddard Space Flight Ctr Greenbelt MD 20771

KONDO, YOSHIO, zoology; deceased, see previous edition for last biography

KONDRA, PETER ALEXANDER, b Mikado, Sask, July 30, 11; m 39; c 3. POULTRY GENETICS. *Educ:* Univ Man, BSA, 34, MSc, 43; Univ Minn, PhD(poultry genetics), 53. *Prof Exp:* Poultry inspector, Man Dept Agr, 36-40; asst poultry, Univ Man, 40-43, mgr hatchery, 44-45; asst poultry specialist, Man Dept Agr, 45-46; from asst prof to assoc prof poultry sci, Univ Man, 46-64, prof poultry sci, 64-78; RETIRED. *Concurrent Pos:* Mem poultry breeding comt, Can Dept Agr, 53-; exchange scientist, Acad Sci, USSR, 64, adv, Thailand 68-69, Brazil, 74 & Costa Rica, 78. *Mem:* Poultry Sci Asn; World Poultry Sci Asn; Genetics Soc Can; Agr Inst Can. *Res:* Poultry breeding biology; incubation; housing. *Mailing Add:* 60 Purdue Winnipeg MB R3T 3C7 Can

KONECCI, EUGENE B, b Chicago, Ill, Jan 7, 25; c 2. MEDICAL PHYSIOLOGY, RESOURCE MANAGEMENT. *Educ:* Roosevelt Univ, BS, 48; Univ Bern, PhD(med physiol), 50. *Prof Exp:* Res scientist, US Air Force Sch Aviation Med, 50-56; chief physiol & toxicol & develop eng inspector, I G Norton AFB, Calif, 56-57; chief life sci, Douglas Aircraft Co, 57-62; dir human factor systs, Off Advan Res & Tech Hqs, NASA, Washington, DC, 62-64, sr prof staff mem manned space flight, spacecraft & missions, Nat Aeronaut & Space Coun, Exec Off Pres, 64-66; PROF MGT, AEROSPACE ENG & CLIN MED & KLEBERG KING RANCH PROF INTERDISCIPLINARY RES, UNIV TEX, AUSTIN, 66- *Concurrent Pos:* Mem staff, US Air Force Command & Staff Sch, Maxwell AFB, Ala, 52; chmn bioastronaut comt, Int Astronaut Fedn, 62-65; academician, Int Acad Astronaut, Paris, 65-; mem bd dirs, Appl Devices Corp, New York, 67-71, SysteMed, Inc, Newport Beach, Calif, 68-71, San Jacinto Savings Asn, 70- & Applied Solar Energy Corp, 79-; founder & pres, Inteck Assocs, 68-71; co-founder, Eugenics Inc, 69-73 & Amerigenics Inc, 75-77. *Honors & Awards:* John Jeffries Award, 64; Outstanding Dir Award, 83. *Mem:* Am Astronaut Soc (pres, 68-69); assoc fel Am Inst Aeronaut & Astronaut; Macro-Eng Soc Am. *Res:* Research and development management. *Mailing Add:* Dept Mgt Univ Tex Austin TX 78712

KONECKY, MILTON STUART, b Omaha, Nebr, July 29, 22; m 48; c 2. ORGANIC CHEMISTRY. *Educ:* Creighton Univ, BS, 44, MS, 48; Univ Ill, PhD(chem), 58. *Prof Exp:* Chemist, Omaha Grain Exchange Labs, Nebr, 47-50; chemist pesticide chem res br, Entom Res Div, USDA, Md, 50-54; sr chemist, 57-60, res assoc, 61-63, sect head, 63-69, sr staff adv, 69-87, SR RES ASSOC, EXXON RES & ENG CO, 88- *Mem:* AAAS; Sigma Xi; Am Chem Soc; NY Acad Sci. *Res:* Agricultural chemicals; petrochemicals; olefin and diolefin utilization; biodegradation methods for and synthesis of detergents; industrial, trade sales and specialty resins for surface coatings; heterogeneous catalysis; information research and analysis. *Mailing Add:* Dryden Rd PO Box 307 Pottersville NJ 07979

KONECNY, JAN, physical biochemistry, for more information see previous edition

KONEN, HARRY P, b Dayton, Ky, Sept 18, 40; m 71; c 2. NUMERICAL ANALYSIS. *Educ:* St Thomas Univ, BA, 62; Tex A&M Univ, MS, 65, PhD(math), 67. *Prof Exp:* Instr math, San Jacinto Col, 66-69; from asst prof to assoc prof, 69-83, PROF MATH, SAM HOUSTON STATE UNIV, 83- *Mem:* Soc Indust & Appl Math; Math Asn Am; Asn Comput Mach; Sigma Xi. *Res:* Fifth-order Runge-Kutta methods for the numerical solution of differential equations. *Mailing Add:* Box 2194 Huntsville TX 77341-2194

KONETZKA, WALTER ANTHONY, b Pa, Sept 8, 23; m 49; c 4. MICROBIAL PHYSIOLOGY. *Educ:* Univ Md, BS, 50, MS, 52, PhD(bact), 54. *Prof Exp:* Asst, Univ Md, 50-51; fel, Walter Reed Army Med Ctr, 52-54; microbiologist, Merck & Co, 54-55; from asst prof to prof bact, Ind Univ, Bloomington, 55-66; prof biol sci, Univ Md, Baltimore County, 66-68; PROF MICROBIOL, IND UNIV, BLOOMINGTON, 68- *Mem:* AAAS; Am Soc Microbiol. *Res:* Microbial metabolism of environmentally important molecules; magnetotaxis in bacterial cells. *Mailing Add:* Dept Biol Jordan Hall Ind Univ Bloomington IN 47405

KONG, ERIC SIU-WAI, b Hong Kong, Jan 14, 53; US citizen; m 74; c 2. MATERIALS SCIENCE, POLYMER PHYSICS. *Educ:* Univ Calif, Berkeley, BA, 74; Rensselaer Polytech Inst, MSc, 76, PhD(polymer phys), 78. *Prof Exp:* Res assoc, Va Polytech Inst, 78-79; res scientist, Ames Res Ctr, NASA, 79-83; mem tech staff, Sandia Nat Labs, 83-84 & Hewlett Packard Labs, 84-86; sr res specialist, Swedlow Inc, 86-87; biomat tech adv, Mentor Corp, 87-89; fac med chem, Univ Calif, San Francisco, 89-90; SR RES CHEMIST, BECTON DICKINSON, 91- *Concurrent Pos:* Chmn, San Francisco Chap, Soc Plastics Engrs, 83-84; consult, Elec Power Res Inst, Palo Alto, 91- *Mem:* Fel Am Inst Chemists; NY Acad Sci; Am Chem Soc; Am Phys Soc; Soc Polymer Sci Japan. *Res:* Polymer chemistry research and development where there is a close tie between basic science and applied engineering. *Mailing Add:* 936 Bluebonnet Dr Sunnyvale CA 94086

KONG, JIN AU, b Kiangsu, China, Dec 27, 42; US citizen; m 70; c 2. ELECTRICAL ENGINEERING. *Educ:* Nat Taiwan Univ, BS, 62; Chiao Tung Univ, MS, 65; Syracuse Univ, PhD(elec eng), 68. *Prof Exp:* Res engr elec eng, Syracuse Univ, 68-69; asst prof, 69-73, assoc prof, 73-80, PROF

ELEC ENG, MASS INST TECHNOL, 80- *Concurrent Pos:* Vinton Hayes fel eng, Mass Inst Technol, 69-71; consult remote sensing technol, Off Tech Coop, UN, 77-79, Raytheon Co, Lincoln Lab & Hughes Aircraft Co, 80. *Mem:* Inst Elec & Electronics Engrs; Am Phys Soc; Optical Soc Am; Am Geophys Union; Am Soc Eng Educ; Int Union Radio Sci. *Res:* Electromagnetic Wave Theory; author of 5 books, 123 refereed journal articles and 125 conference papers. *Mailing Add:* 26-305 Mass Inst Technol Cambridge MA 02139

KONG, YI-CHI MEI, b Boston, Mass, Feb 2, 34; c 2. IMMUNOLOGY, MICROBIOLOGY. *Educ:* Wellesley Col, BA, 55; Univ Mich, MS, 57, PhD(microbiol), 61. *Prof Exp:* Res assoc, Univ Mich, 60-61; asst res bacteriologist, Naval Biol Lab, Univ Calif, Berkeley, 61-66; from asst prof to assoc prof, 66-77, PROF IMMUNOL & MICROBIOL, SCH MED, WAYNE STATE UNIV, 77- *Concurrent Pos:* Mem immunol sci study sect, NIH, 74-77 & bacteriol mycol study sect, 81-84; ed, Infection & Immunity, 78-83, Clin Immunol & Immunopath, 87-92. *Honors & Awards:* Fogarty Int Sr Fel, NIH, 88; Found Lectr, Am Soc Microbiol, 87. *Mem:* AAAS; Transplantation Am Soc Microbiol; Am Asn Immunologists; Am Thyroid Asn. *Res:* Mechanisms of immunologic tolerance; transplantation antigens and immunity; immunogenetic and cellular control of autoimmunity; effect of adjuvants. *Mailing Add:* Dept Immunol & Microbiol Wayne State Univ Sch Med Detroit MI 48201

KONHAUSER, JOSEPH DANIEL EDWARD, b Ford City, Pa, Oct 5, 24; m 48; c 1. MATHEMATICS. *Educ:* Pa State Univ, BS, 48, MA, 51, PhD(math), 63. *Prof Exp:* Instr math, Pa State Univ, 50-55; sr engr, HRB-Singer, Inc, 55-61, staff mathematician, 61-65; assoc prof math, Univ Minn, Minneapolis, 65-68, assoc dir col geom proj, Minnemath Ctr, 66-68; assoc prof, 68-70, chmn dept, 69-81, PROF MATH, MACALESTER COL, 70- *Concurrent Pos:* Ed, Pi Mu Epsilon J, 84- *Mem:* Am Math Soc; Math Asn Am; The Math Asn. *Res:* Applied mathematics; geometry; biorthogonal polynomial sets. *Mailing Add:* Dept Math & Comput Sci Macalester Col St Paul MN 55105

KONHEIM, ALAN G, b Brooklyn, NY, Oct 17, 34; m 57; c 2. MATHEMATICS. *Educ:* Polytech Inst Brooklyn, BEE, 55, MS, 57; Cornell Univ, PhD(math), 60. *Prof Exp:* Res staff mem math, Thomas J Watson Res Lab, IBM Corp, 60-82, res staff mem, IBM Res Lab, Switz, 70-71; PROF COMPUTER SCI, UNIV CALIF, SANTA BARBARA, 82- *Mem:* Am Math Soc; Math Asn Am; Soc Indust & Appl Math; Asn Comput Mach. *Res:* Probability theory; harmonic analysis. *Mailing Add:* Dept Computer Sci Univ Calif Santa Barbara CA 93106

KONIECZNY, STEPHEN FRANCIS, GENE EXPRESSION, DIFFERENTIATION. *Educ:* Brown Univ, PhD(biol), 82. *Prof Exp:* SYST PROF BIOL, PURDUE UNIV, 84- *Mailing Add:* Dept Biol Sci Purdue Univ West Lafayette IN 47907

KONIG, RONALD H, b Albany, NY, Aug 12, 32; m 76; c 2. STRUCTURAL GEOLOGY, ECONOMIC GEOLOGY. *Educ:* St Lawrence Univ, BS, 54; Cornell Univ, MS, 56, PhD(geol), 59. *Prof Exp:* From asst prof to assoc prof, 59-71, chmn dept geol, 71-80, PROF, UNIV ARK, FAYETTEVILLE, 71- *Mem:* Geol Soc Am; Soc Econ Geol. *Res:* Areal geologic mapping; geologic investigation of mineral deposits. *Mailing Add:* Dept Geol Univ Ark Fayetteville AR 72701

KONIGSBACHER, KURT S, b Switz, Sept 15, 23; nat US; m 45; c 2. ORGANIC CHEMISTRY, BIOCHEMISTRY. *Educ:* Dartmouth Col, BA, 44; Swiss Fed Inst Technol, DrSc Tech, 49. *Prof Exp:* Group leader org chem, Foster D Snell, Inc, 49-50; group leader, Evans Res & Develop Corp, 50-57, assoc develop mgr, 57-60, develop mgr, 60-63, vpres, 63-69, sr vpres, 68-69; vpres, Foster D Snell, Inc, Booz, Allen & Hamilton, Inc, 69-74; vpres opers, William T Thompson Co, 74-75; vpres, Herbert V Shuster, Inc, 75-89; PRES, K-BACH CONSULTS, INC, 89- *Concurrent Pos:* Guest lectr, NY Univ, 55. *Mem:* Am Chem Soc; Am Inst Chem; Inst Food Technol; NY Acad Sci; Am Soc Testing & Mat. *Res:* Biochemistry of foods and food products; product development; pharmaceuticals; market research; enzymes; dehydrated and intermediate moisture compressed foods; canless sterilization of foods; performance testing; cosmetics; health and beauty aids. *Mailing Add:* 128 Dogwood Lane Stamford CT 06903

KONIGSBERG, ALVIN STUART, b New York, NY, Apr 28, 43; div; c 2. ATMOSPHERIC SCIENCE, BIOMETEOROLOGY. *Educ:* City Col NY, BS, 63; Syracuse Univ, MS, 65, PhD(physics), 69. *Prof Exp:* Res assoc physics, Atmospheric Sci Res Ctr, State Univ NY, Albany, 65; teaching asst, Syracuse Univ, 63-68; from asst prof to assoc prof, physics, 68-77, chmn dept, 77-79, dir innovative studies, 79-82, ASSOC PROF GEOL SCI, STATE UNIV NY COL NEW PALTZ, 82- *Concurrent Pos:* Aeronaut & Space res fel, NASA & Am Asn Eng Educ, 73 & 74. *Mem:* AAAS; Am Water Resources Asn. *Res:* Oxidant pollution; condensation nuclei studies; light scattering instrumentation; alternate energy systems. *Mailing Add:* Dept Geol Sci State Univ of NY Col New Paltz NY 12561

KONIGSBERG, IRWIN R, b Brooklyn, NY, May 6, 23; m 54; c 2. DEVELOPMENTAL BIOLOGY. *Educ:* Brooklyn Col, BA, 48; Johns Hopkins Univ, PhD(biol), 52. *Prof Exp:* Jr instr biol, Johns Hopkins Univ, 49-51; asst prof, Lab Chem Embryol, Sch Med, Univ Colo, 52-58; biologist, Geront Br, NIH, Baltimore City Hosps, 58-61; staff mem, Dept Embryol, Carnegie Inst Washington, 61-66; prof, 66-77, COMMONWEALTH PROF BIOL, UNIV VA, CHARLOTTESVILLE, 77- *Concurrent Pos:* Assoc prof, Johns Hopkins Univ, 64-66; instr, Embryol Training Prog, Marine Biol Lab, 66-68; mem molecular biol study sect, Div Res Grants, NIH, 66-70; res assoc Dept Embryol, Carnegie Inst, Washington, 68- *Mem:* AAAS; Am Asn Anatomists; Am Soc Zoologists; Am Soc Cell Biol; Soc Develop Biol; Sigma Xi. *Res:* Regulation of proliferation and differentiation of embryonic muscle cells in culture; regeneration of adult muscle fibers; properties of myoblasts from regenerating normal and dystrophic muscle. *Mailing Add:* Dept Biol Univ Va Charlottesville VA 22901

KONIGSBERG, MOSES, b Montreal, Que, Can, Sept 21, 12; nat US; m 39; c 2. ORGANIC CHEMISTRY. *Educ:* Ohio State Univ, AB, 35, PhD(org chem), 39. *Prof Exp:* Atlas Powder Co fel, Ohio State Univ, 39-40; res chemist, Nat Starch Prod, Inc, NY, 40-46; res chemist & vpres, Polymer Industs, Inc, 46-54, vpres res & develop, 54-58, commercial develop, 58-61; vpres & tech dir, Hudson Industs Corp, 61-84; CONSULT, 84- *Mem:* AAAS; Am Chem Soc; NY Acad Sci. *Res:* Natural and synthetic polymers; industrial adhesives. *Mailing Add:* 40019 Village 40 Camarillo CA 93012

KONIGSBERG, WILLIAM HENRY, b New York, NY, Apr 5, 30; m 56; c 1. BIOCHEMISTRY. *Educ:* Rensselaer Polytech Inst, BSc, 52; Columbia Univ, PhD(chem), 56. *Prof Exp:* Asst prof biochem, Rockefeller Inst, 58-64; assoc prof biochem, 64-74, assoc prof molecular biophys & biochem & human genetics, 74-77, PROF MOLECULAR BIOPHYS & BIOCHEM & HUMAN GENETICS, YALE UNIV, 77- *Res:* Protein chemistry; structure of hemoglobin; structure and function of proteins, peptides and natural products; antibodies; virus proteins; mechanism of T4 DNA replication. *Mailing Add:* Dept Biochem Yale Univ Sch Med New Haven CT 06520

KONIJN, HENDRIK SALOMON, b Amsterdam, Netherlands, Mar 17, 18; nat US. ECONOMICS. *Educ:* Columbia Univ, MA, 42; Univ Calif, PhD(statist), 54. *Prof Exp:* Jr staff mem, Nat Bur Econ Res, NY, 41-42; statistician, Combined Shipping Adj Bd, DC, 42-45; res analyst, Off Strategic Servs, 45 & Off Far Eastern Affairs, US Dept State, 45-47, statist consult econ res, Univ Calif, 53-54; lectr agr econ, Univ Calif Berkeley, 54-56; sr lectr econ statist, Sydney, 56-61; prof, City Col New York, 63-65; PROF STATIST, TEL-AVIV UNIV, 65- *Concurrent Pos:* Vis assoc prof, Cowles Found, Yale Univ, 61-62; vis prof statist, Univ Minn, 62-63 & Univ BC, 88. *Mem:* Fel AAAS; fel Am Statist Asn; Am Econ Asn; Inst Math Statist; Royal Statist Soc. *Res:* Statistical methodology; econometric studies. *Mailing Add:* Dept Statist Tel-Aviv Univ Tel Aviv 69978 Israel

KONIKOFF, JOHN JACOB, b Philadelphia, Pa, May 1, 21; m 44; c 2. BIOMEDICAL ENGINEERING, CLINICAL PHARMACOLOGY. *Educ:* Drexel Univ, BSME, 55; Union Grad Sch, PhD(biomed eng sci), 72. *Prof Exp:* Proj engr, US Naval Air Mat Ctr, 42-46, specialist chem & pyrotechnics, US Naval Air Develop Ctr, 47-49; sr mech engr, Philco Corp, 49-50; chief develop engr, Thermal Res & Eng Corp, 50-56; mgr phys biol opers, Space Sci Lab, Gen Elec Co, 56-64, consult scientist, Res & Eng Dept, 65-70; res instr, dept orthop surg, Jefferson Med Col, Thomas Jefferson Univ, 68-71, res asst prof physiol & orthop surg, 71-75, consult orthop surgeon, 71-85, res assoc prof, 76-87; RES ASSOC PROF, DEPT SURG, RUTGERS UNIV, 84- *Concurrent Pos:* Deleg, Int Astronaut Fedn Cong, Athens, Greece, 65 & New York, 68, Int Fedn Med & Biol Eng, Stockholm, 67; consult prof, La State Univ, 69-70; consult scientist, Gen Elec Co, 70-73; consult, coop wildlife res unit, Pa State Univ, 71-73; consult scientist, 78-; clin res dir, Hoffman-Roche Inc, 73-85. *Honors & Awards:* Aerospace Appln Award, Am Inst Aeronaut & Astronaut, 70; Breakthrough 60 Award, Gen Elec Co, 60. *Mem:* AAAS; Am Soc Clin Pharm & Therapeut; Soc Biomed Eng; Sigma Xi; NY Acad Sci. *Res:* Role bioelectric potentials play in mediating homeostasis, particularly studies related to wound repair enhancement and tissue regeneration; development of artificial tendons; biomedical instrumentation; investigation of safety and efficacy of new and investigational drugs in man. *Mailing Add:* 32 Fenton Dr Short Hills NJ 07078

KONIKOW, LEONARD FRANKLIN, b Far Rockaway, NY, Jan 26, 46; m 66; c 2. HYDROLOGY, GEOLOGY. *Educ:* Hofstra Univ, BA, 66; Pa State Univ, MS, 69, PhD(geol), 73. *Prof Exp:* Geologist, Geraghty & Miller Inc, 66; instr geol, Hofstra Univ, 66; HYDROLOGIST, US GEOL SURV, 72- *Concurrent Pos:* Assoc ed, Water Resource Res; chmn, hydrol prog, Am Geophys Union; Birdsall distinguished lectr, Geol Soc Am, 85-86. *Mem:* Geol Soc Am; Am Geophys Union. *Res:* Transport and dispersion of solutes in flowing ground water. *Mailing Add:* 11316 Myrtle Lane Reston VA 22091

KONING, ROSS E, b Adrian, Mich, Oct 5, 53; m 81; c 3. PLANT GROWTH REGULATION, FLOWER PHYSIOLOGY. *Educ:* Univ Mich, BS, 75, MS, 76, PhD(bot), 81. *Prof Exp:* Asst prof bot, Rutgers Univ, 81-87; asst prof, 87-89, ASSOC PROF BOT, EASTERN CONN STATE UNIV, 89- *Concurrent Pos:* Spec ed, Am Jour Bot, 83-86; ed, Physiol Sect, Bot Soc Am, 85- *Mem:* Bot Soc Am; Am Soc Plant Physiologist; Plant Growth Regulators Soc Am. *Res:* Developmental physiology of flowering, particularly growth of flower parts and hormonal and environmental cues involved in the timing mechanisms. *Mailing Add:* 141 Oak St Willimantic CT 06226

KONINGSTEIN, JOHANNES A, b Velsen, Netherlands, Nov 30, 33; Can citizen; m 59; c 2. CHEMISTRY, CHEMICAL PHYSICS. *Educ:* Univ Amsterdam, Drs & Dr(chem), 59. *Prof Exp:* Fel, Nat Res Coun Can, 59-61; mem res staff, Bell Tel Labs, 62-65; from asst prof to assoc prof, 65-72, PROF CHEM, CARLETON UNIV, 72- *Concurrent Pos:* Vis prof, Nat Res Coun Can, 69- *Honors & Awards:* Isac Walton Killam Res fel, 85-87. *Mem:* Am Phys Soc; Chem Inst Can. *Res:* Molecular and atomic research; electronic Raman spectroscopy. *Mailing Add:* Dept Chem Carleton Univ Ottawa ON K1S 5B6 Can

KONISHI, FRANK, b Ft Lupton, Colo, Dec 2, 28; m 50; c 3. NUTRITION. *Educ:* Colo State Univ, BS, 50, MS, 52; Cornell Univ, PhD(animal nutrit), 58. *Prof Exp:* Asst, Colo State Univ, 50-52 & Cornell Univ, 52-54, 57-58; radiobiologist, US Naval Radiol Defense Lab, 58-61; assoc prof nutrit, Southern Ill Univ, Carbondale, 61-65, chmn dept, 65-77, prof sch med, 74-83, prof nutrit, 66-83; RETIRED. *Concurrent Pos:* Adj prof, Univ Colo, Boulder, 84- *Mem:* Am Inst Nutrit; Am Dietetic Asn; NY Acad Sci; Sigma Xi. *Res:* Nutritional dietary surveys; obesity; energy metabolism. *Mailing Add:* 2736 Winding Trail Place Boulder CO 80304

KONISHI, MASAKAZU, b Kyoto, Japan, Feb 17, 33. NEUROBIOLOGY. *Educ:* Hokkaido Univ, BS, 56, MS, 58; Univ Calif, Berkeley, PhD(zool), 63. *Prof Exp:* Alexander von Humboldt Found fel, 63-64; Int Brain Res Orgn-UNESCO fel, 64-65; asst prof zool, Univ Wis, 65-66; from asst prof to assoc prof biol, Princeton Univ, 70-75; prof biol, 75-79, BING PROF BEHAV BIOL, CALIF INST TECHNOL, 79- *Honors & Awards:* Newcomb Cleveland Prize, AAAS, 78; Elliot Couse Award, Am Ornithol Union, 83; F O Schmitt Prize, 87; Int Prize, Biol, Japanese Asn Prom Sci, 90. *Mem:* Nat Acad Sci; AAAS; Am Soc Zoologistss; Am Soc Naturalists; Acoust Soc Am; Soc Neurosci; Am Acad Arts & Sci; Int Soc Neuroethology (pres, 87-89). *Res:* Behavior and neurobiology. *Mailing Add:* Calif Inst Technol Div Biol 216-76 Pasadena CA 91125

KONISKY, JORDAN, b Providence, RI, Apr 8, 41; m 67; c 2. MICROBIAL PHYSIOLOGY. *Educ:* Providence Col, BA, 63; Univ Wis, Madison, PhD(genetics), 68. *Prof Exp:* Res assoc genetics, Univ Wis, Madison, 68; NIH fel molecular biophys & biochem, Yale Univ, 68-70; from asst prof to assoc prof, 70-81, head, Dept Microbiol, 84-89, PROF MICROBIOL, UNIV ILL, URBANA, 81-; DIR, SCH LIFE SCI, 89- *Concurrent Pos:* Career develop award, NIH, 75. *Mem:* Am Soc Microbiol; AAAS; fel Am Acad Microbiol. *Res:* regulation of gene expression in bacteria; functions of bacterial membrane; physiology and molecular biology of methanogenic bacteria. *Mailing Add:* Dept Microbiol Univ Ill 407 S Goodwin Ave Urbana IL 61801

KONIZER, GEORGE BURR, b Wilmington, Del, Dec 24, 42; m 66; c 2. PHYSICAL ORGANIC CHEMISTRY. *Educ:* Univ Del, BA, 64, MBA, 76; Univ SC, PhD(chem), 69. *Prof Exp:* Res chemist, Dacron Res Lab, 69-71 & Textile Res Lab, 71-75, sr res chemist, Textile Res Lab, 75-78, supvr res & develop orlon, 78-81, SUPVR RES & DEVELOP NYLON, E I DU PONT DE NEMOURS & CO, INC, 81- *Concurrent Pos:* Fel, State Univ NY Col Forestry, Syracuse Univ, 68-69. *Mem:* Am Chem Soc; Sigma Xi. *Res:* Fiber research. *Mailing Add:* 1601 Chatham Rd Waynesboro VA 22980

KONKEL, DAVID ANTHONY, b Washington, DC, Feb 20, 48; m 70. MOLECULAR GENETICS, CELL BIOLOGY. *Educ:* Boston Col, BS, 70; Mass Inst Technol, PhD(biochem), 77. *Prof Exp:* Asst biochem, Dept Biol, Mass Inst Technol, 70-77; staff fel, Lab Molecular Genetics, Nat Inst Child & Human Develop, 77-80; ASST PROF CELL BIOL, DEPT HUMAN GENETICS & CELL BIOL, UNIV TEX MED BR GALVESTON, 80- *Concurrent Pos:* Prin investr grants, NSF, 80-83, NIH, 82-85 & Tex Heart Asn, 85-86. *Mem:* AAAS; Sigma Xi; Am Soc Cell Biol. *Res:* Recombinant DNA technology to study the structure and regulation of genes encoding mouse and chicken myoglobin; characterization of a new gene family related to ras oncogenes. *Mailing Add:* Dept Human Biol Chem & Genetics Univ Tex Med Br Stop F-43 Galveston TX 77550

KONKEL, PHILIP M, b Brush, Colo, May 5, 12; m 36. PETROLEUM GEOLOGY. *Educ:* Univ Wyo, BA, 34, MA, 35. *Prof Exp:* Geologist, Ohio, 40-55, asst chief geologist, 55-61, geologist, Tulsa Div, 61-65, explor mgr, 65-74, EXPLOR CONSULT, 74- *Mem:* Am Asn Petrol Geologists; Soc Explor Geophys. *Res:* Petroleum exploration. *Mailing Add:* 3227 Quebec Ave Tulsa OK 74135

KONNERTH, KARL LOUIS, b Mt Pleasant, Pa, Aug 15, 32. MEDICAL ELECTRONICS, BIO-MEDICAL ENGINEERING. *Educ:* Carnegie Inst Technol, BS, 54, MS, 55, PhD(elec eng), 61. *Prof Exp:* Asst prof elec eng, Carnegie Inst Technol, 61-64; staff res engr, Thomas J Watson Res Ctr, 64-74, mgr, I/O Systs, IBM Res Div, 74-77, MGR ADVAN TECHNOL, IBM BIOMED SYSTS, IBM CORP, 77- *Honors & Awards:* Outstanding Innovation Award, IBM Corp, 76. *Mem:* Sr mem Inst Elec & Electronics Engrs; Asn Advan Med Instrumentation. *Res:* Investigation of new types of bio-medical instrumentation. *Mailing Add:* IBM Corp T J Watson Res Ctr PO Box 218 Yorktown Heights NY 10598

KONO, TETSURO, b Tokyo, Japan, May 17, 25; m 61; c 3. PHYSIOLOGY, BIOCHEMISTRY. *Educ:* Univ Tokyo, BA, 47, PhD(anal chem), 58. *Prof Exp:* Instr agr chem, Univ Tokyo, 47-58 & 60-63; from asst prof to prof physiol, 63-85, PROF MOLECULAR PHYSIOL & BIOPHYS, SCH MED, VANDERBILT UNIV, 85- *Mem:* Am Soc Biochem; Am Diabetes Asn. *Res:* Mechanism of insulin action. *Mailing Add:* Dept Molecular Physiol & Biophys Vanderbilt Univ Sch Med Nashville TN 37232

KONOPINSKI, VIRGIL J, b Toledo, Ohio, July 11, 35; m 64; c 3. SAFETY, INDUSTRIAL HYGIENE. *Educ:* Univ Toledo, BS, 56; Pratt Inst, MS, 60; Bowling Green State Univ, MBA, 71. *Prof Exp:* Asst to dir environ control, Owens Corning Fiberglas, Toledo, 67-72; chief exec officer, Midwest Environ Mgt, 72-73; staff specialist, Williams Brothers Waste Control, Tulsa, 73-75; dir IH & RH, Ind State Bd Health, Indianapolis, 75-87; exec vpres, Asbestos Technol, Indianapolis, 87-89; consult pvt pract, Zionsville, 89-90; sr consult, Occusafe, Wheeling, 90-91; REGIONAL SAFETY ENGR, US POSTAL SERV, CHICAGO, 91- *Concurrent Pos:* Consult pvt pract, 75-87. *Mem:* Am Soc Safety Engrs; Am Indust Hyg Asn; Am Conf Govt Indust Hygienists. *Res:* Occupational health and safety; formaldehyde; carbon monoxide; carbon dioxide; pesticides; indoor air; mercury; sampling strategies and techniques; long range planning and forecasting. *Mailing Add:* 14 Fairfield Lane Cary IL 60013

KONOPKA, ALLAN EUGENE, b Chicago, Ill, Feb 26, 50; m 73; c 3. MICROBIOLOGY. *Educ:* Univ Ill, Urbana, BS, 71; Univ Wash, MS, 73, PhD(microbiol), 75. *Prof Exp:* Res assoc microbiol, Univ Wis-Madison, 75-77; asst prof, 77-83, ASSOC PROF BIOL, PURDUE UNIV, 83- *Mem:* Am Soc Microbiol; Am Soc Limnol & Oceanog; Phycol Soc Am. *Res:* Microbial ecology; physiological ecology of planktonic blue-green algae; buoyancy regulation by procaryotic microorganisms. *Mailing Add:* Dept Biol Sci Purdue Univ West Lafayette IN 47907

KONOPKA, RONALD J, b Cleveland, Ohio, Oct 19, 47. NEUROBIOLOGY. *Educ:* Univ Dayton, BS, 67; Calif Inst Technol, PhD(biochem), 72. *Prof Exp:* Fel biol, Stanford Univ, 72-74; asst prof biol, Calif Inst Technol, 74-82; AT DEPT BIOL, CLARKSON UNIV, POTSDAM, NY, 83- *Concurrent Pos:* NSF fel, 72-73; mem sci adv bd, Found Res Hereditary Dis, 72-81; Helen Hay Whitney fel, 73-74. *Mem:* Genetics Soc Am; Soc Neurosci. *Res:* Circadian rhythm; behavior genetics of Drosophila. *Mailing Add:* Dept Biol Clarkson Univ Potsdam NY 13676

KONORT, MARK D, b Sliven, Bulgaria, Nov 20, 18; nat US; m 52; c 2. ORGANIC CHEMISTRY. *Educ:* Univ Toulouse, Dipl Ing Chim, 40; Rutgers Univ, MS, 50, PhD(org chem), 52. *Prof Exp:* Chemist, Matam Corp, 42-44; develop chemist, R J Prentiss & Co, 47; asst, Rutgers Univ, 48-52; sr res chemist, Lever Bros Co, 52-56, prin res chemist, 56-59, res assoc, 59-63, sr res assoc, 63-72, res scientist, 72-81, sr res scientist, 81-84; RETIRED. *Mem:* Am Chem Soc; Sigma Xi. *Res:* Lubricating and cutting oils; insecticides; substituted phenanthrenes and hydrophenanthrenes; detergents; edible products; organic synthesis. *Mailing Add:* 90 Morris Ave Haworth NJ 07641

KONOWALOW, DANIEL DIMITRI, b Cleveland, Ohio, Apr 28, 29; m 78. THEORETICAL CHEMISTRY. *Educ:* Ohio State Univ, BS, 53; Univ Wis, PhD(chem), 61. *Prof Exp:* Chemist, Plastics Dept, E I Du Pont de Nemours & Co, Del, 60-62; asst dir, Theoret Chem Inst, Univ Wis, 62-65; from asst prof to assoc prof, State Univ NY, Binghamton, NY, 65-89; SR SCIENTIST, UNIV DAYTON RES INST, 87- *Concurrent Pos:* Vis prof, Uppsala Univ, Sweden, 71-72; vis scientist, Nat Bur Standards, Gaithersburg, Md, 78-79; sr res assoc, Nat Res Coun, Ballistics Res Lab, Aberdeen Proving Ground, Md, 85-86. *Mem:* Am Chem Soc; AAAS; Am Phys Soc. *Res:* Calculation and correlation of physical properties of atoms and molecules by quantum-mechanical methods; intramolecular and intermolecular interactions; theoretical spectroscopy. *Mailing Add:* OLAC PL RKFE Edwards AFB CA 93523-5000

KONRAD, DUSAN, b Brno, Czech, Jan 7, 35; US citizen; m 72; c 2. ELECTROCHEMISTRY. *Educ:* Masaryk Univ, Czech, MS, 57; Czech Acad Sci, PhD(chem), 62. *Prof Exp:* Instr inorg chem, Masaryk Univ, Czech, 57-62; res scientist phys chem, J Heyrovsky Inst Polarography, Czech Acad Sci, 63-66; res chemist, Govt Assay Off, Czech, 67-68; sr res fel, Rudjer Boskovic Inst, Yugoslavia, 68-69; res fel chem, Calif Inst Technol, 70-71; sr scientist electrochem, Technol Ctr, ESB Inc, Yardley, Pa, 71-78; MEM TECH STAFF, TEX INSTRUMENTS, INC, DALLAS, 78- *Mem:* Electrochem Soc. *Res:* Electrochemical instrumentation; automatized data taking and processing; electrode impedance in Laplace plane analysis; electrochemistry of lead-acid cell; non-stoichiometric oxide electrodes; porous electrodes; computer software systems. *Mailing Add:* Tex Instruments Inc MS 944 PO Box 655-012 Dallas TX 75265

KONRAD, GERHARD T(HIES), b Konigsberg, Ger, Feb 23, 35; US citizen; m 64; c 2. ELECTRICAL ENGINEERING. *Educ:* Univ Mich, BSE, 57, MSE, 60, PhD(elec eng), 69. *Prof Exp:* Asst res engr, Univ Mich, 60-61, assoc res engr, 61-69; staff mem, Lincoln Lab, Mass Inst Technol, 69-72; staff mem, 72-77, HEAD KLYSTRON DEPT, STANFORD LINEAR ACCELERATOR CTR, 77- *Concurrent Pos:* Consult, Litton Indust, 79- & Valvo, Ger, 78- *Mem:* Inst Elec & Electronics Engrs; Am Phys Soc; Nat Soc Prof Engrs; Sigma Xi. *Res:* Electron devices; microwave circuits; electron optics; high-voltage techniques; vacuum techniques; plasma physics; electromagnetic theory. *Mailing Add:* 787 Kirkcrest Rd Danville CA 94526

KONRADI, ANDREI, b Prague, Czech, July 22, 31; US citizen; m 57; c 5. SPACE PHYSICS. *Educ:* Franklin & Marshall Col, BS, 54; Univ Rochester, PhD(nuclear physics), 62. *Prof Exp:* Aero space technologist, Goddard Space Flight Ctr, NASA, 61-70, AERO SPACE TECHNOLOGIST, JOHNSON SPACE CTR, NASA, 70- *Mem:* Am Geophys Union; Am Inst Aeronaut & Astronaut; Am Phys Soc; NY Acad Sci; AAAS. *Res:* Study of dynamic processes in the magnetosphere; space plasma simulation; magnetospheric physics; solar-terrestrial interations. *Mailing Add:* NASA Johnson Space Ctr Houston TX 77058

KONSLER, THOMAS RHINEHART, b Henderson, Ky, Apr 17, 25; m 54; c 9. HORTICULTURE. *Educ:* Univ Ky, BS, 55; NC State Univ, MS, 57, PhD(exp statist), 61. *Prof Exp:* From asst prof to prof, 61-88, EMER PROF HORT SCI, MOUNTAIN HORT CROPS RES, NC STATE UNIV, 88- *Mem:* Am Soc Hort Sci. *Res:* Cultural practices and plant breeding with vegetable crops and American Ginseng. *Mailing Add:* 805 Oakland St Hendersonville NC 28739

KONSTAM, AARON HARRY, b Bronx, NY, Aug 11, 36; m 61; c 1. PROGRAMMING LANGUAGES, EXPERT SYSTEMS. *Educ:* Polytech Inst Brooklyn, BS, 57; Pa State Univ, PhD(phys org chem), 61. *Prof Exp:* Instr chem, Brooklyn Col, 61-62; fel & res assoc phys org chem, Israel Inst Technol, 62-64; sr res chemist, Monsanto Res Corp, 65-69; dir comput ctr & assoc prof math, Lindenwood Cols, 69-72; assoc prof, 72-86, PROF COMPUT SCI, TRINITY UNIV, 86- *Concurrent Pos:* Treas, 77-80, vpres, Vanguard Systs Corp, 80-83. *Mem:* Asn Comput Mach; Inst Elec & Electronics Engrs; Asn Computational Ling. *Res:* Artificial intelligence, expert systems and programming languages. *Mailing Add:* Trinity Univ 715 Stadium Dr San Antonio TX 78212

KONTOS, HERMES A, b Lefka, Cyprus, Dec 13, 33; US citizen; m 60; c 2. MEDICINE, PHYSIOLOGY. *Educ:* Nat Univ Athens, MD, 58; Med Col Va, PhD(physiol), 67; Am Col Physicians, dipl, 69. *Prof Exp:* From instr to assoc prof, 64-72, PROF MED, MED COL VA, 72-, CHMN DIV CARDIO PULMONARY MED, 81- *Concurrent Pos:* USPHS res career develop award, 67-72; Markle scholar acad med, 69-74. *Mem:* Am Fedn Clin Res; Am Heart Asn; Am Physiol Soc; Am Soc Clin Invest; Sigma Xi. *Res:* Circulatory physiology and pathophysiology. *Mailing Add:* Med Col Va Box 281 Richmond VA 23298

KONTRAS, STELLA B, b Newport News, Va, June 28, 28; m 47; c 3. PEDIATRICS, GENETICS. *Educ:* Ohio State Univ, BA, 48, MA, 49, MD, 53. *Prof Exp:* Intern med, 54, instr pediat, 54-60, asst prof pediat & anat, 60-66, assoc prof pediat, 66-69, PROF PEDIAT, COL MED, OHIO STATE UNIV, 69- *Concurrent Pos:* Resident pediat, Ohio State Univ, 55 & 56, resident path, 57-58; consult, Ohio Dept Health, 62-70; NIH spec fel cancer, Ohio State Univ, 63-64; dir, Med Genetics Ctr, Children's Hosp, Columbus, 65-71, vchmn, Dept Pediat, 80. *Mem:* Soc Pediat Res; Am Soc Hemat; Am Soc Human Genetics. *Res:* Hematology. *Mailing Add:* Children's Hosp 700 Children's Dr Columbus OH 43205

KONYA, CALVIN JOSEPH, b Cleveland, Ohio, June 23, 43. MINING ENGINEERING, BLASTING. *Educ:* Mo Sch Mines, BS, 66; Univ Mo-Rolla, MS, 68 & 70, PhD(mining eng), 72. *Hon Degrees:* Dr, Univ Miskolc, Hungary. *Prof Exp:* From asst prof to assoc prof mining eng, WVa Univ, 71-78; assoc prof, 78-81, chmn dept, 81-85, prof mining eng, 81-88, ADJ PROF, OHIO STATE UNIV, 88- *Concurrent Pos:* Mgr tech servs, Precision Blasting Servs, 73-; exchange scientist, Nat Acad Sci, 75; pres, Hydrocarbon Fuels, 83-87; dir, Ohio Mining & Mineral Resources Res Inst; pres, Precision Blasting Syst, 87; adj prof, Ohio Univ, 88-; adj prof, John Caroll Univ, 88- & Univ Miskolc, 88- *Mem:* Soc Explosives Engrs (pres, 75-77, exec dir, 74-87); Am Inst Mining, Metall & Petrol Engrs. *Res:* Rock mechanics; explosives engineering; definition of mechanisms of rock fragmentation by blasting both for production rounds and for controlled blasting techniques; development of practical equations for prediction of blast design variables in the field. *Mailing Add:* PO Box 189 Montville OH 44064

KONZ, STEPHAN A, b Milwaukee, Wis, Nov 25, 33; m 58; c 5. ERGONOMICS. *Educ:* Univ Mich, BS, 56, MBA, 56; Univ Iowa, MS, 60; Univ Ill, PhD(indust eng), 64. *Prof Exp:* Indust engr, Westinghouse Elec Corp, 56-57 & Collins Radio Co, 58-60; instr mech & indust eng, Univ Ill, 60-64; from asst prof to assoc prof, 64-69, PROF INDUST ENG, KANS STATE UNIV, 69- *Honors & Awards:* Fitts Award, Human Factors Soc. *Mem:* Inst Indust Engrs; Human Factors Soc; Am Indust Hyg Asn; Am Soc Heat, Refrig & Air-Conditioning Engrs. *Res:* Ergonomics, especially design of industrial jobs; heat stress; Inspection; hand tools. *Mailing Add:* Dept Indust Eng Kans State Univ Manhattan KS 66502

KONZAK, CALVIN FRANCIS, b Devils Lake, NDak, Oct 17, 24; div; c 2. GENETICS, PLANT BREEDING. *Educ:* NDak Agr Col, BS, 48; Cornell Univ, PhD, 52. *Prof Exp:* Assoc geneticist, Dept Biol, Brookhaven Nat Lab, 51-57; assoc prof & assoc agronomist, 57-62, PROF AGRON & AGRONOMIST, WASH STATE UNIV, 62-, PROF GENETICS, 66- *Concurrent Pos:* USPHS sr fel, 65-66; spec adv plant breeding & genetics sect, Joint Food & Agr Orgn-Int Atomic Energy Agency Div Food & Agr, Vienna, 65-67; Food & Agr consult, Crop Res & Introd Ctr, Izmir, Turkey, 71; sci adv, Plant Breeding & Genetics Sect, Joint Food & Agr Orgn-Int Atomic Energy Agency Div Atomic Energy Food & Agr, 73-74 & 82-83. *Mem:* AAAS; Genetics Soc Am; fel Am Soc Agron; Crop Sci Soc Am; Radiation Res Soc. *Res:* Breeding semidwarf, hard red and soft white spring wheat, durum wheat oats for improved yield, and tolerance in semidwarf oats to barley yellow dwarf virus; genetics of reduced height and other traits in wheat and oats; induction of useful mutations in wheat and oats; development and application of electronic data capture and management systems for plant breeding; agronomic crop research. *Mailing Add:* Dept Crop & Soil Sci Wash State Univ Pullman WA 99164-6420

KONZELMAN, LEROY MICHAEL, b Jersey City, NJ, May 27, 36; m 60; c 3. ORGANIC CHEMISTRY. *Educ:* St Peter's Col, NJ, BSc, 58; Seton Hall Univ, MSc, 64, PhD(org chem), 66. *Prof Exp:* Chemist, Schering Corp, NJ, 60-65; res chemist, 66-72, group leader dyes & intermediates dept, 72-74, chief chemist dyes & intermediates mfg dept, 74-79, chief chemist pharmaceut mfg dept, Am Cyanamid Co, Bound Brook, 79-82; mgr, color res & develop, Inmont Corp, Hawthorne, NJ, 82-83; DIR BUS DEVELOP, CHEM DYNAMICS CORP, SOUTH PLAINFIELD, NJ, 84- *Mem:* Am Chem Soc; NY Acad Sci; Sigma Xi; AAAS. *Res:* Process and product development of organic intermediates for pharmaceutical and agricultural products and a variety of specialty organic intermediates; marketing and product development for organic intermediates for use in the pharmaceutical, agricultural and specialty fields. *Mailing Add:* 61 Elm St Florham Park NJ 07932

KONZO, SEICHI, b Tacoma, Wash, Aug 2, 05; m 32; c 2. MECHANICAL ENGINEERING. *Educ:* Univ Wash, BS, 27; Univ Ill, MS, 29. *Prof Exp:* Asst mech eng, 27-32, res assoc, 32-34, res asst prof, 34-36, res assoc prof, 36-40, res prof, 40-47, prof, 47-71, assoc head dept, 63-70, EMER PROF MECH ENG, UNIV ILL, URBANA, 71- *Concurrent Pos:* Consult, Ill State Geol Surv; coordr eng teachers prog, Ford Found; eng consult, 71- *Honors & Awards:* F Paul Anderson Medal, Am Soc Heat, Refrig & Air-Conditioning. *Mem:* Am Soc Heat, Refrig & Air-Conditioning Engrs. *Res:* Heating and heat transfer; air conditioning; fluid flow; combustion. *Mailing Add:* Small Homes Coun-Bldg Res Coun Univ of Ill Champaign IL 61820

KOO, BENJAMIN, b Shanghai, China, Apr 4, 20; US citizen; wid. STRUCTURAL ENGINEERING, APPLIED MECHANICS. *Educ:* St John's Univ, BS, 41; Cornell Univ, MS, 42, PhD(struct eng), 46. *Prof Exp:* Struct engr, 42-55; engr concrete & found sect, M H Treadwell Co, 56-61; proj engr, Am Car & Foundry Div, ACF Industs, 61-65; prof civil eng, 65-90, EMER PROF ENG, UNIV TOLEDO, 90- *Concurrent Pos:* NSF, NASA & Dept Transp Res Awards. *Mem:* Am Soc Civil Engrs; Am Soc Eng Educ; Am Concrete Inst; Sigma Xi. *Res:* Structural reliability in reinforced concrete members and frames; trailer train freight car patent; cushioned underframe system patent; structural stability; structural analysis and design; structural reliability. *Mailing Add:* Dept Civil Eng Univ Toledo Toledo OH 43606

KOO, DAVID CHIH-YUEN, b Bangkok, Thailand, Jan 1, 51. OBSERVATIONAL COSMOLOGY, EVOLUTION OF GALAXIES. *Educ:* Cornell Univ, AB, 72; Univ Calif, Berkeley, MA, 74, PhD(astron), 81. *Prof Exp:* Postdoctoral fel, Dept Terrestrial Magnetism, Carnegie Inst Wash, 81-83, sr res fel, 83-84; postdoctoral fel, Space Telescope Sci Inst, 84-86, asst astronr, 86-87; ASST ASTRONR/ASST PROF ASTRON, LICK OBSERV, UNIV CALIF, SANTA CRUZ, 88- *Concurrent Pos:* Vis fel, Sci & Eng Res Coun, UK, 83; NSF presidential young investr award, 88- *Mem:* Int Astron Union; Am Astron Soc; Astron Soc Pac; Int Asn Pattern Recognition. *Res:* Faint optical imaging and spectroscopy of distant galaxies, quasars, and radio sources; examination of the age, size, texture, and shape of the universe; probe very large-scale distributions of galaxies. *Mailing Add:* 220 Dickens Way Santa Cruz CA 95064

KOO, DELIA WEI, b Hankow, China, May 14, 21; US citizen; m 43; c 3. MATHEMATICS, STATISTICS. *Educ:* St John's Univ, China, BA, 41; Radcliffe Col, AM, 42, PhD(eng philol), 47; Mich State Univ, MA, 54. *Prof Exp:* Instr math, Mich State Univ, 55-56; lectr, Douglass Col, Rutgers Univ, 56-57; instr, Mich State Univ, 57-58; from asst prof to assoc prof, 65-77, PROF MATH, EASTERN MICH UNIV, 77- *Mem:* Math Asn Am; Inst Math Statist; Economet Soc. *Res:* Author of 2 books. *Mailing Add:* 4554 Sequoia Trail Okemos MI 48864

KOO, KEE P, b Hong Kong, Mar 30, 49; UK citizen; m 87. FIBER-OPTIC SENSORS, LASER PHYSICS. *Educ:* Univ Ill, Chicago, BS, 71; Case Western Reserve Univ, MS, 75, PhD(elec eng), 77. *Prof Exp:* Res assoc, Brookhaven Nat Lab, 77-78 & John Carroll Univ, 78-80; RES PHYSICIST, NAVAL RES LAB, 80- *Mem:* Inst Elec & Electronics Engrs; Optical Soc Am. *Res:* Fiber-optic sensors research with emphasis in interferometric sensors, especially in magnetic field sensing; fiber-optic magnetometer, gradiometer. *Mailing Add:* Naval Res Lab Code 6570 Washington DC 20375

KOO, PETER H, b Shanghai, China; US citizen; m 67; c 2. CANCER IMMUNOLOGY & BIOLOGY. *Educ:* Univ Wash, BA, 64; Univ Md, PhD(biochem), 70. *Prof Exp:* Fel immunol, Johns Hopkins Univ, 71-74; staff fel, NIH, 74-75; asst prof oncol, Johns Hopkins Univ, 75-77; asst prof, 77-83, ASSOC PROF IMMUNOL & MICROBIOL, COL MED, NORTHEASTERN OHIO UNIV, 83- *Concurrent Pos:* Assoc prof, Dept Chem, Kent State Univ 78-, asst prof, Dept Biol, 80-; prin investr & grantee, Nat Cancer Inst, NIH, 78-82, Am Cancer Soc, 78-79, 81-82 Cystic Fibrosis Care Fund & Pediat of Akron, Inc, 78-79, 81-82, Nat Sci Found, 84-87, Mefcom Found Funds, 82-, United Way, 82-83 & 89-90, Oncol Fund Akron Gen Develop Fund, 90-91; bd dir, Am Cancer Soc, Portage County, 81-; chmn, Prof Educ Comt, Am Cancer Soc, Portage Co, 85-89, vpres, 87-89. *Mem:* Sigma Xi; NY Acad Sci; AAAS; Am Asn Immunologists; Am Chem Soc. *Res:* Structure and function of Alpha-2 macroglobulins; development of nonspecific tumor cytotoxic soluble factors in serum; development of memory B lymphocytes; nerve growth factor regulation and catabolism. *Mailing Add:* Dept Microbiol & Immunol Northeastern Ohio Univ Col Med 4209 St Rte 44 Rootstown OH 44272

KOO, ROBERT CHUNG JEN, b Shanghai, China, Mar 20, 21; nat US; m 49; c 3. POMOLOGY. *Educ:* Cornell Univ, BS, 44; Univ Fla, MS, 50, PhD(fruit crops), 53. *Prof Exp:* Interim asst biochemist, Citrus Exp Sta, Univ Fla, 53-57, from asst horticulturist to assoc horticulturist, 57-68, prof & horticulturist, Agr Res & Educ Ctr, 69-90, EMER PROF, AGR RES & EDUC CTR, UNIV FLA, 90- *Honors & Awards:* Presidential Gold Medal Award, Fla State Horticult Soc, 65; Res Award, Fla Fruit & Vegetable Asn, 75. *Mem:* Am Soc Hort Sci; Am Agron Soc. *Res:* Plant nutrition of citrus; irrigation and water management. *Mailing Add:* Citrus Res & Educ Ctr Univ of Fla Lake Alfred FL 33850

KOO, SUNG IL, NUTRITION, LIPID BIOCHEMISTRY. *Educ:* Clemson Univ, PhD(nutrit), 76. *Prof Exp:* Assoc prof, 78-88, PROF BIOCHEM, ORAL ROBERTS UNIV, 89-; PROF FOODS & NUTRIT, KANS STATE UNIV, 90- *Mailing Add:* Kans State Univ Justin Hall Manhattan KS 66506

KOO, TED SWEI-YEN, fish biology; deceased, see previous edition for last biography

KOOB, ROBERT DUANE, b Graettinger, Iowa, Oct 14, 41; m 60; c 7. PHYSICAL CHEMISTRY. *Educ:* Univ Northern Iowa, BA, 62; Univ Kans, PhD(chem), 67. *Prof Exp:* Instr high sch, Iowa, 63-64; res assoc chem, Univ Kans, 67; from asst prof to assoc prof, 67-72, chmn dept, 73-77, dir, Water Resources Res Inst, 74-85, dean, Col Sci & Math, 81-84, PROF CHEM, NDAK STATE UNIV, 72-, VPRES ACAD AFFAIRS, 85-, INTERIM PRES, 87- *Mem:* Am Chem Soc; Sigma Xi. *Res:* Radiation chemistry; photochemistry; mass spectrometry. *Mailing Add:* 1219 Fourth Ave S Moorhead MN 56560

KOOB, ROBERT PHILIP, b Philadelphia, Pa, Jan 3, 22; m 54; c 3. PHOTOCHEMISTRY, PHYSICAL CHEMISTRY. *Educ:* Villanova Col, BS, 43; Univ Pa, MS, 47, PhD(chem), 49. *Prof Exp:* Asst instr, Univ Pa, 43-44, 46-49; asst prof chem, Villanova Col, 49-55; from asst prof to assoc prof, 55-59, PROF CHEM, ST JOSEPH'S COL, PA, 59- *Concurrent Pos:* Res chemist, E I du Pont de Nemours & Co, 47 & Eastern Lab, Dept Agr, 59, 60, 62 & 63; asst prof, Rosemont Col, 51-52. *Mem:* Am Chem Soc; Sigma Xi. *Res:* Phase and reaction rate studies; effects of solvent on reaction; extensions of the Fries Rearrangement; organic photochemistry. *Mailing Add:* Dept Chem St Joseph's Univ City Line Ave 54th St Philadelphia PA 19131

KOOBS, DICK HERMAN, b Hinsdale, Ill, July 22, 28; m 55. PATHOLOGY, BIOLOGICAL CHEMISTRY. *Educ:* Andrews Univ, BA, 50; Loma Linda Univ, MD, 55; Univ Calif, Los Angeles, PhD(biol chem), 65; Am Bd Path, dipl, 66. *Prof Exp:* Intern, Robert B Green Mem Hosp, San Antonio, Tex, 55-56; resident physician path, White Mem Hosp, Los Angeles, 56-59; asst prof, 65-78, ASSOC PROF PATH, SCH MED, LOMA LINDA UNIV, 78- *Mem:* Int Acad Path. *Res:* Experimental and molecular pathology, particularly pulmonary disease. *Mailing Add:* Dept Path Loma Linda Univ Sch Med Loma Linda CA 92354

KOOH, SANG WHAY, b Seoul, Korea, Oct 5, 30; m 56; c 3. PEDIATRICS. *Educ:* Yan-Sei Univ, MD, 55; Univ Toronto, PhD(physiol), 67; Am Acad Pediat, dipl, 61; FRCP, 68. *Prof Exp:* Fel pediat, Michael Reese Hosp & Med Ctr, 60-62; sr res fel, Res Inst, Hosp Sick Children, Toronto, 62-67; ASSOC PROF, DEPT PHYSIOL & DEPT PEDIAT, UNIV TORONTO, 68- *Concurrent Pos:* Sr staff physician, Hosp Sick Children, 68- *Mem:* Soc Pediat Res; Can Soc Clin Investigation. *Res:* Metabolism bone diseases in children; metabolism of vitamin D in human and in experimental animals; regulation of bone mineralization. *Mailing Add:* 555 University Ave Toronto ON M5G 1X8 Can

KOOIJ, THEO, b Dordrecht, Neth, Nov 29, 33; m 56; c 3. ACOUSTIC SIGNAL PROCESSING, ELECTRONICS ENGINEERING. *Educ:* Delft Univ Technol, BSc, 58, MSc, 61; Cath Univ Am, PhD(elec eng), 77. *Prof Exp:* Jr res scientist, Saclant ASW Res Ctr, La Spezia, Italy, 61-65; sr res scientist & teamleader underwater acoustics, 65-68; sci adv sonar interpretation, US Naval Ship Res & Develop Ctr, Washington, DC, 68-74, head target physics br, Ocean Sci Dept, Naval Underwater Systs Ctr, New London, Conn, 74-76; tech dir, Acoust Res Ctr, 76-78, PROG MGR, TACTICAL TECHNOL OFF, DEFENSE ADVAN RES PROJS AGENCY, 78- *Concurrent Pos:* Lectr, Am Univ, 68-69. *Mem:* Acoust Soc Am; Netherlands Royal Inst Eng; Am Soc Cybernet; Inst Elec & Electronics Engrs. *Res:* Theoretical, model and computer simulated, and full scale experimental research in detection and classification of underwater targets; design and development of digital sonar signal processing systems. *Mailing Add:* Bolt Beranek & Newman Labs Inc 1300 N 17th St Suite 400 Arlington VA 22209

KOOMEN, MARTIN J, b Bristol, NY, Dec 30, 17; m 46; c 2. SOLAR PHYSICS. *Educ:* Univ Rochester, BS, 40, MS, 43. *Prof Exp:* Instrument inspector, Bausch & Lomb Optical Co, 41-42; res physicist, Univ Rochester, 42-46; res physicist, US Naval Res Lab, 46-82; RES PHYSICIST, US NAVAL RES LAB & SACHS/FREEMAN INC, 82- *Mem:* Optical Soc Am; Am Astron Soc; Am Geophys Union; AAAS. *Res:* Night vision; light emission from the upper atmosphere (night airglow); solar physics; design of rocket and satellite-borne instrumentation for study of the night airglow and the sun. *Mailing Add:* 5194 Dungannon RD Fairfax VA 22030

KOONCE, ANDREA LAVENDER, b Denver, Colo, Dec 31, 51. FOREST PATHOLOGY, TROPICAL FORESTRY. *Prof Exp:* Researcher, US Forest Serv, 78; teacher pathol & ecol, Ore State Univ, 75-78, res asst, 75-78; asst prof ecol, bot, pathol & head genetic res, Nat Sch Forestry Sci, 81-; AT DEPT FORESTRY, UNIV WIS, STEVENS POINT. *Concurrent Pos:* Consult, Lew Roth Forest, 80-81. *Mem:* Soc Am Foresters; Soc Trop Foresters. *Res:* Tree improvement of tropical pines, hardwoods and legumes; tree disease-fire interactions; tree physiology. *Mailing Add:* USDA Forest Serv Riverside Fire Lab 4955 Canyon Crest Dr Riverside CA 92507

KOONCE, KENNETH LOWELL, b Lake Charles, La, Sept 6, 39; m 62; c 3. EXPERIMENTAL STATISTICS. *Educ:* Univ Southwestern La, BS, 61; La State Univ, MS, 63; NC State Univ, PhD(animal genetics), 68. *Prof Exp:* Instr animal sci, NC State Univ, 67; from asst prof to assoc prof, 67-76, head dept, exp statist, 82-89, PROF EXP STATIST, LA STATE UNIV, BATON ROUGE, 76-, ASST DIR, AGR EXP STA, 89- *Mem:* Biomet Soc; Am Soc Animal Sci; Am Soc Info Sci; Am Statist Asn. *Mailing Add:* Dept Exp Statist La State Univ Baton Rouge LA 70803

KOONCE, SAMUEL DAVID, b Titusville, Fla, Nov 26, 15; m 40; c 3. TECHNICAL INTELLIGENCE. *Educ:* Oberlin Col, AB, 36; Ohio State Univ, MS, 40, PhD(physiol chem), 43. *Prof Exp:* Tech sales, Innis Speiden & Co, 36-39; teaching asst physiol chem, Ohio State Univ, 40-43; res chemist, Innis Speiden & Co, Boyce Thompson Inst, 43-45 & Distillation Prod, Inc, 45-47; mkt researcher, Jefferson Chem Co, 47-52, mkt develop mgr, 52-55; mkt res mgr, Com Develop Div, Am Cyanamid Co, 55-61; mgr com develop, Lummus Co, 61-68; mgr mkt, Ugine Kuhlmann Am, Inc, 68-74, dir technol div, 74-78, vpres, 78-83; sr consult, Atochem, Inc, 83-86; sr consult, Uranium Pechiney, 80-; RETIRED. *Mem:* Am Chem Soc; Sigma Xi; Chem Mkt Res Asn (pres, 54-55); Com Develop Asn; AAAS. *Res:* Commercial development of ethylene oxide derivatives as surfactants; transfer of plastics technology on an international basis; role of chlorofluorocarbons in the chemistry and dynamics of the earth's atmosphere; industrial development of magnetic fusion. *Mailing Add:* 255 Hempstead Rd Ridgewood NJ 07450

KOONG, LING-JUNG, EXPERIMENTAL BIOLOGY. *Prof Exp:* DEPT HEAD ANIMAL SCI, ORE STATE UNIV, 91- *Mailing Add:* Agr Exp Sta Ore State Univ Corvallis OR 97331

KOONIN, STEVEN ELLIOT, b Brooklyn, NY, Dec 12, 51; m 75; c 3. THEORETICAL NUCLEAR PHYSICS. *Educ:* Calif Inst Tech, BS, 72; Mass Inst Technol, PhD(physics), 75. *Prof Exp:* Asst prof theoret physics, 75-78, assoc prof physics, 78-81, PROF THEORET PHYSICS, CALIF INST TECHNOL, 81- *Concurrent Pos:* Consult, Lawrence Berkeley Lab, Lawrence Livermore Lab, Los Alamos Sci Lab & Oak Ridge Nat Lab, 77-; res fel, Niels Bohr Inst, 76-77; Alfred P Sloan Found res fel, 77-79. *Honors & Awards:* Humboldt Sr Scientist Award, 85. *Mem:* Fel Am Phys Soc; fel AAAS. *Res:* Nuclear reaction models; heavy ion physics. *Mailing Add:* Dept Physics Calif Inst Technol 1201 E California Blvd Pasadena CA 91125

KOONS, CHARLES BRUCE, b Oklahoma City, Okla, Nov 14, 29; m 56; c 3. ORGANIC CHEMISTRY. *Educ:* Southern Ill Univ, BS, 51; Univ Minn, PhD(org chem), 58. *Prof Exp:* Res chemist, Jersey Prod Res Co, Div Standard Oil Co, NJ, Exxon Prod Res Co, 58-64, sr res chemist, 64-75, res assoc, 75-83, res adv, 83-89; CONSULT, 89- *Mem:* Am Chem Soc. *Res:* Kinetics of aromatic substitution reactions; organic geochemistry, involving studies of the origin, migration and accumulation of petroleum; environmental chemistry, involving the fate of petroleum in the marine environment. *Mailing Add:* 10835 Saint Marys Lane Houston TX 77079-3619

KOONS, DAVID SWARNER, b Fresno, Calif, June 17, 30; div; c 2. CHEMICAL ENGINEERING, PETROLEUM ENGINEERING. *Educ:* Calif Inst Technol, BS, 52, MS, 55; Univ Colo, PhD(chem eng), 60. *Prof Exp:* Process engr, Texaco Inc, 52-54; res asst, Exp Sta, Univ Colo, 60; sr res technologist, Mobil Oil Co Inc, 60-77; staff engr, Mobil Corp, 77-86; RETIRED. *Mem:* Am Chem Soc; Am Inst Chem Engrs; Am Inst Mining, Metall & Petrol Engrs. *Res:* Simultaneous fluid flow; heat transfer and reaction kinetics of processes for recovering oil from underground reservoirs; pneumatic conveying of solids and petroleum refining. *Mailing Add:* 30343 Arena Dr Evergreen CO 80439

KOONS, DONALDSON, b Seoul, Korea, Aug 23, 17; US citizen; m 44; c 4. GEOMORPHOLOGY. *Educ:* Columbia Univ, AB, 39, AM, 41, PhD(geol), 45. *Hon Degrees:* DSc, Col of Wooster, 74, Unity Col, 76. *Prof Exp:* Instr geol & geog, Carleton Col, 42-43; lectr geol, Columbia Univ, 46; asst prof, WVa Univ, 46-47; from asst prof to prof geol, Colby Col, 47-75, Dana prof, 75-82, head dept, 47-82; RETIRED. *Concurrent Pos:* Comnr, Maine Dept of Conserv, 73-75. *Honors & Awards:* Huddleston Medal. *Mem:* Fel AAAS; fel Geol Soc Am. *Res:* Areal geology; dynamic geomorphology; geology of Colorado Plateau; Pleistocene Glaciation of Maine. *Mailing Add:* RFD 1 Box 5810 Oakland ME 04963

KOONTZ, FRANK P, b Baltimore, Md, Nov 13, 32; m 60; c 3. CLINICAL MICROBIOLOGY, PARASITOLOGY. *Educ:* Univ Md, BS, 58, MS, 60, PhD(biochem, microbiol), 62; Am Bd Med Microbiol, dipl. *Prof Exp:* NIH fel, Oxford Univ, 62-64; asst prof prev med, 64-70, asst dir, State Hygienic Lab, 67-77, prin bacteriologist, 73-76, assoc prof, dept path, 76-79, ASSOC PROF PREV MED & ENVIRON HEALTH, UNIV IOWA, 70-, PROF, DEPT PATH, 80-, DIR CLIN MICROBIOL, UNIV HOSP, 76- *Mem:* Am Soc Microbiol; fel Am Pub Health Asn. *Res:* Automation and rapid methods in clinical microbiology; blood culture techniques; evaluation of antimicrobial agents. *Mailing Add:* Univ Iowa Iowa City IA 52242

KOONTZ, HAROLD VIVIEN, plant physiology, horticulture, for more information see previous edition

KOONTZ, PHILIP G, PHYSICS. *Prof Exp:* RETIRED. *Mailing Add:* 2525 Taft Dr No 906 Boulder CO 80302

KOONTZ, WARREN WOODSON, JR, b Lynchburg, Va, June 10, 32; m 57; c 2. UROLOGY. *Educ:* Va Mil Inst, BA, 53; Univ Va, MD, 57. *Prof Exp:* From intern to resident surg, New York Hosp, 57-62, resident urol, 62-66; from instr to asst prof, Med Col Va, 66-69; asst prof, Harvard Univ, 69-70; prof urol & chmn dept, Med Col Va, 70-; AT DEPT SURG, VA COMMONWEALTH UNIV. *Concurrent Pos:* Asst urologist, Mass Gen Hosp, 69-70; consult, McGuire Vet Admin Hosp, 70- & Portsmouth Naval Hosp, 71- *Mem:* Am Col Surgeons; Am Urol Asn; Soc Pediat Urol; Soc Univ Urol. *Res:* Pediatric urology; urinary tract cancer. *Mailing Add:* Med Col Va PO Box 118 Richmond VA 23298

KOOP, CHARLES EVERETT, b Brooklyn, NY, Oct 14, 16; wid; c 3. SURGERY. *Educ:* Dartmouth Col, AB, 37; Cornell Univ, MD, 41; Univ Pa, ScD(med), 47. *Hon Degrees:* LLD, Eastern Baptist Col, 60; MD, Univ Liverpool, 68; LHD, Wheaton Col, 73. *Prof Exp:* From asst instr to instr surg, Univ Pa, 42-47, assoc, 47-48, from asst prof to assoc prof, 48-59, prof pediat surg, 59-89, prof pediat, 71-89; CHMN, NAT SAFE KIDS CAMPAIGN, WASHINGTON, DC, 89- *Concurrent Pos:* Surgeon-in-chief, Children's Hosp, Philadelphia, 48-81; consult, US Naval Hosp, 64-; ed, J Pediat Surg; dep asst secy health, USPHS, 80-89, surg gen, 81-89 & dir, Off Int Health, 82; chmn, bd trustees, Nat Mus Health & Med Found & Bd Int Health, Inst Med. *Honors & Awards:* Denis Brown Gold Medal, Brit Asn Pediat Surgeons; William E Ladd Gold Medal, Am Acad Pediat; Order of Duarte, Sanchez & Mella; Medal, Legion of Honor, France, 80. *Mem:* Inst Med-Nat Acad Sci; Am Surg Asn; fel Am Col Surgeons; Brit Asn Pediat Surg; Am Pediat Surg Asn; fel Am Acad Pediat; Royal Col Surgeons Eng; Soc Univ Surgeons. *Res:* Pediatric surgical techniques; neo-natalogy; childhood tumors. *Mailing Add:* Nat Safe Kids Campaign 111 Michigan Ave NW Washington DC 20010-2970

KOOP, DENNIS RAY, ENZYMOLOGY, METABOLISM. *Educ:* Northwestern Univ, PhD(biochem), 79. *Prof Exp:* ASST PROF BIOCHEM, UNIV MICH, 83- *Mailing Add:* Dept Environ Health Sci Case Western Reserve Univ Sch Med 2119 Abington Rd Cleveland OH 44106

KOOP, JOHN C, b Myitkyina, Burma, Mar 6, 19; US citizen; m 43; c 8. STATISTICS. *Educ:* Univ Rangoon, BSc, 42; NC State Univ, PhD(statist), 58. *Prof Exp:* Chief labor statist, Directorate Labour, Govt Burma, 48-58; mem, Int Labour Off, Switzerland, 59-60; vis asst prof exp statist, NC State Univ, 60-61, vis assoc prof, 61-65, assoc prof, 65-66; sr adv sampling in agr & head res training, Dominion Bur Statist, Can, 66-70; sr group scientist, statist sci admin, Res Triangle Inst, 70-81; CONSULT, 82- *Mem:* Fel Am Statist Asn; Royal Statist Soc; Burma Res Soc; Int Asn Survey Statisticians; Int Statist Inst. *Res:* Sampling theory for finite universes derived on basis of axioms; theory of ratio estimation; theory of functional relationships for a finite universe; unified theory of estimation for sample surveys taking into account response and measurement errors; statistical inference; demographic study of minority community in Burma. *Mailing Add:* 3201 Clark Ave Raleigh NC 27607

KOOPMAN, KARL FRIEDRICH, b Honolulu, Hawaii, Apr 1, 20. MAMMALOGY. *Educ:* Columbia Univ, BA, 43, MA, 45, PhD(zool), 50. *Prof Exp:* Instr biol, Middletown Collegiate Ctr, 49-50 & Queens Col, NY, 52-58; asst cur, Acad Natural Sci, Philadelphia, 58-59 & Chicago Natural Hist Mus, 59-61; from asst cur to cur, 61-85, EMER CUR, AM MUS NATURAL HIST, 85- *Honors & Awards:* Newberry Prize, 49; Gerrit S Miller Jr Award, 77; Hartley H T Jackson Award, 88. *Mem:* Am Soc Mammal; Nature Conservancy. *Res:* Systematic mammalogy; taxonomy and zoogeography of bats; paleontology and zoogeography of West Indian mammals. *Mailing Add:* Dept Mammal 79th St & Central Park W New York NY 10024

KOOPMAN, RICHARD J(OHN) W(ALTER), b St Louis, Mo, June 24, 05; m 34; c 1. ELECTRICAL ENGINEERING. *Educ:* Univ Mo, BS, 28, PhD, 42; Yale Univ, MS, 33. *Prof Exp:* Test engr, Gen Elec Co, 28-30; instr elec eng, Yale Univ, 30-32 & Mich Col Mining & Technol, 35-37; from asst prof to assoc prof, Univ Kans, 37-43; head electro-mech sect, Cornell Aeronaut Lab, 43-46; from assoc prof to prof elec eng, Wash Univ, 46-72, chmn dept, 49-65, Samuel C Sachs prof, 72-73, emer prof, 73-; RETIRED. *Concurrent Pos:* Consult chmn, Nat Comt Aerospace Instrumentation, 59-60; consult var utility co, testing labs & ins co, 78- *Mem:* Am Soc Eng Educ; fel Inst Elec &

Electronics Engrs. *Res:* Electrical machinery; servo motors; response of instruments and systems; causes and effects of electrical failure; direct simulation of electrical machinery; investigations of electrical damage in transformers, distribution systems, etc; history of electrical engineering. *Mailing Add:* 2201 St Clair Brentwood MO 63144

KOOPMANS, HENRY SJOERD, b Washington, DC, Jan 28, 44; m 90; c 3. NUTRITION, ENERGY BALANCE. *Educ:* Harvard Col, BA, 66; Univ Calif, San Diego, PhD(physiol psychol), 72. *Prof Exp:* Res assoc psychol, Univ Calif, San Diego, 72-73; from asst prof to assoc prof, Columbia Univ, 73-83; assoc prof, 83-86, PROF MED PHYSIOL, UNIV CALGARY, 86- *Concurrent Pos:* Res assoc, Obesity Ctr, St Luke's-Roosevelt Hosp, 73-83; adj prof, Rockefeller Univ, 78-79; vis colleague, Hammersmith Hosp, London, 79-81. *Mem:* Neurosci Soc; Int Union Physiol Sci. *Res:* Internal control of food intake, energy expenditure and body weight. *Mailing Add:* Dept Physiol Univ Calgary Calgary AB T2N 4N1 Can

KOOPMANS, LAMBERT HERMAN, b Chicago, Ill, July 23, 30; m 55; c 4. MATHEMATICAL STATISTICS. *Educ:* San Diego State Col, AB, 52; Univ Calif, PhD, 58. *Prof Exp:* Asst statist, Univ Calif, 52-56, assoc biostatist, 57-58; mem staff, Sandia Corp, 58-64; assoc prof, Univ NMex, 64-68, chmn dept, 69-74, prof math & statist, 68-89; RETIRED. *Concurrent Pos:* Consult, Sandia Corp, 64-72, Westinghouse Corp, 65-67, civil eng fac & dept path, Univ NMex, 76-77, diabetes proj, 78-82 & Vet Admin Hosp, Albuquerque, 78-79; sabbatical leave fac math, Univ Calif, Santa Cruz, 71-72; vis prof statist, Princeton Univ, 75. *Mem:* Fel Am Statist Asn; Biomet Soc; fel Inst Math Statist. *Res:* Data analysis; time series analysis. *Mailing Add:* 7400 Dellwood Rd NE Albuquerque NM 87110

KOOPOWITZ, HAROLD, b East London, SAfrica, Sept 10, 40; m 69; c 1. NEUROPHYSIOLOGY, INVERTEBRATE ZOOLOGY. *Educ:* Rhodes Univ, SAfrica, BSc, 62, MSc, 64; Univ Calif, Los Angeles, PhD(zool), 68. *Prof Exp:* Asst prof, 68-75, ASSOC PROF BIOL, UNIV CALIF, IRVINE, 75- *Mem:* Brit Soc Exp Biol; Am Soc Zool; Soc Gen Physiol; Soc Neurosci. *Res:* Organization of flatworm nervous systems; electrophysiology of vision in insect eyes. *Mailing Add:* Dept Ecol Univ Calif Irvine CA 92717

KOOSER, ROBERT GALEN, b Mankato, Minn, July 23, 41; m 85; c 2. PHYSICAL CHEMISTRY. *Educ:* St Olaf Col, BA, 63; Cornell Univ, PhD(chem), 68. *Prof Exp:* From asst prof to assoc prof, 68-84, PROF CHEM, KNOX COL, ILL, 84- *Concurrent Pos:* Vis assoc prof, Dartmouth Col, 83-84. *Mem:* Am Chem Soc; Am Phys Soc; Sigma Xi. *Res:* Electron spin resonance, primarily spin relaxation processes of organic and inorganic singlet systems in solution. *Mailing Add:* O E South St-46 Knox Col Galesburg IL 61401

KOOSIS, PAUL, b Los Angeles, Calif, Apr 20, 29. MATHEMATICAL ANALYSIS. *Educ:* Univ Calif, Berkeley, BA, 50, PhD(math), 54. *Prof Exp:* Instr math, Univ Mich, 54-55; asst instr math sci, NY Univ, 55-57, 59-60; Fulbright fel to France, 57-58; NSF fels, 58-59, 60-61; from asst prof to assoc prof, Fordham Univ, 62-63; from asst prof to assoc prof, 63-70, PROF MATH, UNIV CALIF, LOS ANGELES, 70- *Mem:* Am Math Soc; Math Soc France. *Res:* Classical harmonic analysis; complex variable theory; theory of approximation. *Mailing Add:* Dept Math McGill Univ 805 Sherbrooke St W Montreal PQ H3A 2K6 Can

KOOSTRA, WALTER L, microbiology; deceased, see previous edition for last biography

KOOTSEY, JOSEPH MAILEN, b Houston, Tex, Sept 3, 39; m 61; c 2. PHYSIOLOGY. *Educ:* Pac Union Col, BA, 60; Brown Univ, ScM, 64, PhD(physics), 66. *Prof Exp:* Asst prof physiol & biophys, Loma Linda Univ, 67-69; prof biophys, Andrews Univ, 76-79; asst prof physiol & pharmacol, 71-76, res assoc prof physiol, 79-84, ASSOC PROF PHYSIOL & RES ASSOC PROF COMPUT SCI, DUKE UNIV, 84-, DIR NAT BIOMED SIMULATION RESOURCE, 84- *Concurrent Pos:* Bank Am-Giannini Found grant, Loma Linda Univ, 65-67; NIH spec fel, Duke Univ, 69-71; pres, Simulation Resources, Inc, 91- *Mem:* AAAS; Am Physiol Soc; Biophys Soc; Asn Comput Mach; Am Heart Asn; Inst Elec & Electronics Engrs. *Res:* Cardiac electrophysiology; computer simulation in physiology; medical education; utilization of computer simulation to reassemble complex biological systems; ion regulation and electrical activity in cardiac muscle cells. *Mailing Add:* Simulation Resources Inc One University Pl Suite 250 Durham NC 27707

KOOYMAN, GERALD LEE, b Salt Lake City, Utah, June 16, 34; m 62; c 2. COMPARATIVE PHYSIOLOGY. *Educ:* Univ Calif, Los Angeles, AB, 57; Univ Ariz, PhD(zool), 66. *Prof Exp:* Res asst antarctic seal studies, Univ Ariz, 63-66; NSF fel, Anat & Physiol of Marine Mammals, London Hosp Med Col, Eng, 66-67; res physiologist, 67-68, from asst res physiologist to assoc res physiologist, 68-78, RES PHYSIOLOGIST, SCRIPPS INST OCEANOG, UNIV CALIF, SAN DIEGO, 78- *Concurrent Pos:* Sci fel Zool Soc London. *Honors & Awards:* AAAS. *Mem:* AAAS; Am Soc Zool; Am Physiol Soc; Sigma Xi; Explorers Club. *Res:* Behavior and physiology of diving in aquatic birds and mammals, especially pressure effects; comparative respiratory physiology and anatomy of vertebrates. *Mailing Add:* PRL A-004 Scripps Inst Oceanog La Jolla CA 92093

KOOZEKANAI, SAID H, b Mash-had, Iran, Mar 15, 33; m 65. ELECTRICAL ENGINEERING, BIOMEDICAL ENGINEERING. *Educ:* Univ Tehran, Electro-Mech Eng, 56; Brown Univ, PhD(elec eng), 61; Univ Dayton, MS, 69. *Prof Exp:* Sr res scientist, Raytheon Res Div, Mass, 61-65; from asst prof to assoc prof, 66-76, PROF ELEC ENG, OHIO STATE UNIV, 76- *Mem:* Am Inst Physics; Am Phys Soc; Inst Elec & Electronics Engrs. *Res:* Quantum electronics; lasers; antennas and propagation. *Mailing Add:* Dept Elec Eng 205 Electronics Hall Ohio State Univ Main Campus Columbus OH 43210

KOPAL, ZDENEK, b Litomysl, Czech, Apr 4, 14; nat US; m 38; c 3. ASTRONOMY. *Educ:* Charles Univ, Prague, DSc(astron), 37. *Hon Degrees:* MSc, Univ Manchester, 55; DSc, Univ Patras, 71, Univ Krakow, Poland, 74. *Prof Exp:* Res assoc astron, Harvard Univ, 40-42; mem staff, Mass Inst Technol, 42-45, res assoc appl math, 45-47, assoc prof, 48-51, mem court gov, 64-67, prof astron & head dept, 51-81, EMER PROF, VICTORIA UNIV MANCHESTER, 81- *Concurrent Pos:* Mem, Solar Eclipse Exped, Japan, 36; pres comn 42, Int Astron Union, 48-55; res assoc, Harvard Col Observ, 46-51, lectr, 49-51; mem, Nat Adv Comt Aeronaut, 49-52; founding ed, Icarus, 62-69, ed-in-chief, Astrophys & Space Sci, 68-, ed, Moon J, 70-; vpres, Int Found Pic-du-Midi, 62, pres, 78. *Honors & Awards:* von Neumann Medal, Univ Brussels, 66; Gold Medal, Czech Acad Sci, 69; Copernicus Medal, Univ Krakow, 74; Gold Medal, Japanese Govt, 88. *Mem:* Am Astron Soc; fel Royal Astron Soc; Int Acad Astronaut; Explorers Club; foreign mem Acad Athens (Greece); hon mem Indian Astron Soc. *Res:* Theory of close binary systems; mathematical astronomy; numerical analysis; history of science. *Mailing Add:* Dept Astron Victoria Univ Manchester Manchester M13 9PL England

KOPANSKI, JOSEPH J, b Cleveland, OH, Feb 8, 60. SEMICONDUCTOR PROCESSING, INTEGRATED CIRCUIT ENGINEERING. *Educ:* Case Western Reserve Univ, BS, 82, MS, 85. *Prof Exp:* ELEC ENGR, NAT INST STANDARDS & TECHNOL, 85- *Mem:* Inst Elec & Electronics Engrs; Electrochem Soc. *Res:* Electrical characterization of semiconductor materials, processes and circuits; development of processes for integrated circuit fabrication. *Mailing Add:* Bldg 225 Rm A305 Nat Inst Standards & Technol Gaithersburg MD 20899

KOPCHICK, JOHN J, b Punxsutawney, Pa, Nov 2, 50; m. MOLECULAR BIOLOGY, ZOOLOGY. *Educ:* Ind Univ Pa, BS, 72, MS, 75; Univ Tex, PhD(virol-biomed sci), 80. *Prof Exp:* Teaching asst microbiol, genetics & gen biol, Ind Univ Pa, 72-75; med technologist, Bellaire Gen Hosp, 76-80; Am Cancer Soc postdoctoral fel, Roche Inst Molecular Biol, 80-82; sr res biochemist, Dept Biochem Genetics, Merck Inst Therapeut Res, Merck Sharp & Dohme Res Labs, 82-84, res fel, 84-85, group leader molecular biol, Dept Animal Drug Discovery, 85-86; PROF MOLECULAR BIOL, ZOOL & BIOMED SCI & DIR, MOLECULAR BIOL DEPT, EDISON ANIMAL BIOTECHNOL CTR, OHIO UNIV, 87- *Concurrent Pos:* Milton & Lawrence H Goll eminent scholar endowed prof molecular & cellular biol, Ohio Univ, 87-; mem, Child Health & Human Develop Grant Rev Comt, NIH, 88 & 91, Adv Comt Emerging Agr Technologies, Off Technol Assessment, 89-90, Competitive Res Grants Prog, USDA, 90-91; sci adv, DNX, Princeton, NJ. *Mem:* Sigma Xi; Am Soc Microbiol; AAAS; Am Soc Biochem & Molecular Biol. *Res:* Molecular cloning of DNA molecules encoding wildtype and in vitro mutated growth hormones, growth hormone receptors, hypothalmic regulatory proteins and neuropeptides; cloning of efficient eucaryotic expression vectors for the production of proteins in cultured mammalian cells; structure-function studies of growth hormone employing transgenic animals; author of more than 60 publications. *Mailing Add:* Edison Animal Biotechnol Ctr Ohio Univ Athens OH 45701

KOPCHIK, RICHARD MICHAEL, b Punxsutawney, Pa, Apr 29, 41; m 64; c 2. POLYMERIC CHEMICAL REAGENTS. *Educ:* Carnegie-Mellon Univ, BS, 63; Univ Rochester, PhD(org chem), 68. *Prof Exp:* Res chemist polymers, 69-80, sr res assoc, 80-86, TECH MGR, ROHM & HAAS CO, 86- *Mem:* Am Chem Soc; Sigma Xi; Soc Plastics Eng. *Res:* Organic chemistry; free radical reactions; chemical modification of polymers; organophosphorous chemistry; continuous preparation and processing of polymers; polymer chemistry; polymeric sorbents; engineering plastics. *Mailing Add:* 1335 Stephen Way Southampton PA 18966-4349

KOPE, ROBERT GLENN, b Reedley, Calif, Aug 5, 53; m 76; c 2. STOCK ASSESSMENT, OPTIMAL MANAGEMENT. *Educ:* Calif State Univ, BA, 80; Univ Calif, Davis, PhD(ecol), 87. *Prof Exp:* Postgrad researcher, Univ Calif, Davis, 87-89; FISHERY BIOLOGIST, NAT MARINE FISHERIES SERV, 89- *Concurrent Pos:* Lectr, Univ Calif, Davis, 89. *Mem:* Resource Modeling Asn; AAAS; Pac Fisheries Biologists. *Res:* Optimal management of renewable resources; stock assessment; spatial population modeling. *Mailing Add:* Nat Marine Fisheries Serv 3150 Paradise Dr Tiburon CA 94920

KOPECEK, JINDRICH, b Strakonice, Czech, Jan 27, 40; m 85; c 1. POLYMERIC DRUG DELIVERY SYSTEMS, BIOCOMPATIBILITY OF POLYMERS. *Educ:* Inst Chem Technol, Prague, MS, 61; Czech Acad Sci, Prague, PhD(polymer chem), 65, DSc, 90. *Prof Exp:* Res sci officer polymer synthesis, Czech Acad Sci, Inst Macromolecular Chem, 65-72, head lab med polymers, 72-88; postdoctoral fel chem eng, Nat Res Coun Can, 67-68; PROF BIOENG & PHARMACEUT, UNIV UTAH, 89- *Concurrent Pos:* Prin investr numerous grants, 69-; mem, Comt New Polymers, Ministry Health, Czech, 76-86; vis prof, Univ Paris-Nord, 83 & Univ Utah, 86-88; co-dir, Ctr Controlled Chem Delivery, Univ Utah, 86-; bd gov, Controlled Release Soc, 88-91. *Honors & Awards:* Sci Award, Chem Sect, Czech Chem Soc, 72, 75, 77, 78 & 85; Sci Award, Presidiums Czech & USSR Acad Sci, 77; Barre's Lectr, Univ Montreal, 90. *Mem:* Am Chem Soc; Biomat Soc; AAAS; Am Asn Pharmaceut Scientists; Am Asn Cancer Res; Controlled Release Soc. *Res:* Biocompatibility and biodegradation of polymers; tailor-made synthesis of bioactive, biorecognizable polymers; targetable polymeric anticancer drug carriers; hydrogels for oral delivery of peptides and proteins. *Mailing Add:* Dept Bioeng 2480 MEB Univ Utah Salt Lake City UT 84112

KOPECKY, KARL RUDOLPH, b Hradec Kralove, Czech, Oct 5, 32; US citizen; m 63; c 2. ORGANIC CHEMISTRY. *Educ:* Iowa State Col, BS, 54; Univ Calif, Los Angeles, PhD, 59. *Prof Exp:* Instr chem, Univ Calif, Los Angeles, 59; NIH fel, Calif Inst Technol, 59-61; from asst prof to assoc prof, 61-77, PROF CHEM, UNIV ALTA, 77- *Mem:* Am Chem Soc; Chem Inst Can. *Res:* Thermal reactions of styrene; radicals from chiral sources; peroxide reactions; chemiluminescent compounds; reactions of stable radicals. *Mailing Add:* Dept Chem Univ Alta Edmonton AB T6G 2G2 Can

KOPELMAN, JAY B, b New York, NY, Feb 24, 39; m 90. INTERNATIONAL ENERGY SCIENCES. *Educ:* Rensselaer Polytech Inst, BS, 60; Northwestern Univ, PhD(physics), 65. *Prof Exp:* Res assoc physics, Univ Colo, Boulder, 64-66, asst prof, 66-74, asst dean grad sch, 68-74; mgr energy modeling prog, Stanford Res Inst, 74-78; mgr, spec studies, 78-82, MGR, INTERNAT ACTIVITIES & EXEC ASST VPRES, ELEC POWER RES INST, 82- *Mem:* AAAS; Am Phys Soc. *Res:* Energy technology; energy economics; modelling. *Mailing Add:* Elec Power Res Inst PO Box 10412 Palo Alto CA 94303

KOPELMAN, RAOUL, b Vienna, Austria, Oct 21, 33; US citizen; m 55; c 3. PHYSICAL CHEMISTRY. *Educ:* Israel Inst Technol, BS, 55, dipl eng, 56, MSc, 57; Columbia Univ, PhD(chem), 60. *Prof Exp:* Res assoc chem, Harvard Univ, 60-62; lectr, Israel Inst Technol, 62-64; res fel, Calif Inst Technol, 64-65, sr res fel, 65-66; from asst prof to assoc prof, 66-71, PROF CHEM, UNIV MICH, ANN ARBOR, 71-, PROF PHYSICS, 91- *Concurrent Pos:* Fulbright award, 57; spec res fel, NIH, 72-73, Fogarty int fel, 79; sr fel, NATO, 76; Nat Res Serv Award, NIH, 87-88; Fulbright Fel, 87-88. *Mem:* Am Phys Soc; Am Chem Soc; Biophys Soc; Mat Res Soc. *Res:* Excitation dynamics in molecular and biomimetic aggregates, excitons and phonons in disordered materials, membranes and photosynthetic units; 1.5 K time-resolved high-resolution tunable-laser microspectroscopy; supercomputer simulations of critical phenomena, transport and chemical kinetics; heterogeneous kinetics in low-dimensional and fractal domains; near-field optical microscopy and DNA mapping. *Mailing Add:* Dept Chem Univ Mich Ann Arbor MI 48109-1055

KOPF, ALFRED WALTER, b Buffalo, NY, June 21, 26; m 49; c 4. MEDICINE, ONCOLOGY. *Educ:* Cornell Univ, BA, 48, MD, 51; NY Univ, MS, 55; Am Bd Dermat, dipl, 57. *Prof Exp:* Intern, Cleveland City Hosp, Ohio, 51-52; clin resident dermat & syphil, Skin & Cancer Unit, Univ Hosp, 53-54, organizer oncol sect, 54, asst dermat & syphil, Post-Grad Med Sch & Univ Hosp, 54-55, instr, Post-Grad Med Sch, 57-59, asst clin prof, 59-61; from asst prof to prof 61-83, CLIN PROF DERMAT, SCH MED, NY UNIV, 83- *Concurrent Pos:* Pvt pract dermat, New York, 55-; from asst attend to assoc attend dermat, Univ Hosp, NY Univ Med Ctr, 58-63, assoc attend med staff, 63-64, attend, 64-; from co-ed to sr ed, Yearbook Dermat, 63-70; mem bd dirs, Inst Dermat Commun & Educ, 63-87; mem bd dirs, Am Acad Dermat, 66, Rudolf L Baer Found Skin Dis, Inc, 76, Am Dermat Asn, 81; med adv bd, Skin Cancer Found, 82; chmn bd dirs, Int Found Dermat, 87. *Mem:* AAAS; Soc Invest Dermat; Am Acad Dermat (pres, 80); Am Dermat Asn (treas); AMA. *Res:* Cutaneous oncology, especially neoplasms of the melanocyte including clinical studies on pigmented nevi and malignant melanoma. *Mailing Add:* Dept Dermat NY Univ Med Ctr 562 First Ave New York NY 10016

KOPF, PETER W, b Philadelphia, Pa, Apr 23, 44; m 70; c 2. PHYSICAL CHEMISTRY, POLYMER CHEMISTRY. *Educ:* Rutgers Col, AB, 66; Univ Rochester, PhD(phys chem), 70. *Prof Exp:* chemist, Res & Develop Dept, Union Carbide Corp, Bound Brook, 70-85; CONSULT, 85- *Mem:* Am Chem Soc; Sigma Xi. *Res:* Polymer physical chemistry; polymer microstructural analysis; stable and transient free radicals; engineering thermoplastics; computer assisted calculations and simulations; polymer synthesis; thermoset reaction mechanisms; surface chemistry; dynamic mechanical analysis; polymers in coatings and adhesives applications; new specialty chemicals. *Mailing Add:* Arthur D Little Inc Acorn Park Cambridge MA 02140

KOPF, RUDOLPH WILLIAM, b Munich, Ger, Sept 10, 22; US citizen; m 49; c 2. STRATIGRAPHY, STRUCTURAL GEOLOGY. *Educ:* Univ Buffalo, BA, 50, MA, 52. *Prof Exp:* Instr geol, Univ Buffalo, 50-51; oceanogr, US Navy Hydrographic Off, 52; geologist, US Geol Surv, 52-56; geol engr, USAEC, 56-61; geologist, US Geol Surv, 61-82, 84; CONSULT GEOLOGIST, 84- *Concurrent Pos:* WCoast rep & ed, Geol Names Comt, US Geol Surv, 61-80, lectr, 82- *Mem:* fel Geol Soc Am. *Res:* Mechanics of thrust faulting; development of fault breccia, clastic pipes and dikes, cryptoexplosion structures, diatremes and mud volcanoes; uranium-vanadium deposits, stratigraphy, and structure of parts of Colorado plateaus, basin and range province; origin of auriferous placers in California; source of diamonds in California. *Mailing Add:* 129 E Empire St Grass Valley CA 95945

KOPFLER, FREDERICK CHARLES, b New Orleans, La, Aug 14, 38; m 61; c 4. ENVIRONMENTAL CHEMISTRY. *Educ:* Southeastern La Col, BS, 60; La State Univ, MS, 62, PhD(food sci), 64. *Prof Exp:* Res chemist milk proteins, Agr Res Serv, USDA, 64-66, supvry chemist, 66-71; supvry chemist, Water Supply Progs Div, US Environ Protection Agency, 71-79, chief, Chem & Statist Support Br, 79-87, chief, Exposure Assessment & Pharmacokinetics Sect, Health Effects Res Lab, 87-89, CHIEF SCIENTIST, GULF OF MEX PROG, US ENVIRON PROTECTION AGENCY, 89- *Concurrent Pos:* Adj asst prof, Univ Ala, 68-73. *Mem:* Sigma Xi; AAAS; Am Water Works Asn; Am Chem Soc; Soc Environ Toxicol & Chem. *Res:* Bioassay directed isolation, fractionation and identification of mutagenic organic chemicals from water; effects of toxic substances on estuarine ecosystems. *Mailing Add:* Gulf Mex Prog US EPA Bldg No 1103 Rm 202 Stennis Space Ctr NSTL Station MS 39529

KOPIN, IRWIN J, b New York, NY, Mar 27, 29; m 52; c 3. INTERNAL MEDICINE, PHARMACOLOGY. *Educ:* McGill Univ, BSc, 51, MD, 55. *Prof Exp:* Intern med, Boston City Hosp, 55-56, resident, 56-57; res assoc, NIH, 57-60; resident med, Columbia-Presby Med Ctr, 60-61; actg chief, Sect Med, Lab Clin Sci, 61-63, actg chief, Lab Clin Sci, 68-69, chief, sect med, 63-83, chief, Lab Clin Sci, 69-83, assoc dir, clin res, NIMH, 82-83, DIR, IRP, NAT INST NEUROL & COMMUN DIS & STROKE, NIH, 83- *Mem:* AAAS; Asn Am Physicians; Am Soc Biol Chemists; Am Soc Clin Invest; Am Soc Pharmacol & Exp Therapeut. *Res:* Biochemical pharmacology. *Mailing Add:* IRP Nat Inst Neurol & Commun Dis & Stroke Bldg 10 5N-214 NIH 9000 Rockville Pike Bethesda MD 20892

KOPITO, RON RIEGER, b Haifa, Israel, Dec 21, 54; US citizen; m 87; c 2. BIOCHEMISTRY, NEUROSCIENCES. *Educ:* Bowdoin Col, AB, 76; Mass Inst Technol, PhD(biochem), 82. *Prof Exp:* NIH postdoctoral fel, Mass Inst Technol & Whitehead Inst, 82-86; ASST PROF BIOL SCI, STANFORD UNIV, 87- *Concurrent Pos:* Lucille P Markey scholar biomed sci, 85; Basil O'Connor starter scholar res award, 89; NSF presidential young investr award, 89; asst prof, Dept Molecular & Cell Physiol, Stanford Univ Sch Med, 91- *Mem:* Soc Gen Physiologists; Am Soc Cell Biol; Sigma Xi. *Res:* Molecular physiology; membrane transport; cytoskeleton-membrane interactions. *Mailing Add:* Dept Biol Sci Stanford Univ Stanford CA 94305

KOPLIK, JOEL, b Brooklyn, NY, Oct 31, 48. THEORETICAL PHYSICS, FLUIDS. *Educ:* Cooper Union, BS, 69; Univ Calif, Berkeley, PhD(physics), 74. *Prof Exp:* Res assoc physics, Columbia Univ, 74-76; mem, Inst Advan Study, 76-79; researcher, Ecole Normal Sup, Paris, France, 77-78; MEM PROF STAFF, SCHLUMBERGER-DOLL RES, 79- *Mem:* Am Phys Soc; Soc Petrol Engrs. *Res:* Theoretical physics of transport in random systems, particularly fluid flow, diffusive phenomena and electrical properties of porous media; pattern selection in non-equilibrium growth processes such as multi-fluid displacement and solidification. *Mailing Add:* Levich Inst T202 City Col NY Steinman 202 New York NY 10031

KOPLOW, JANE, b Ulm, Germany, Mar 15, 48; US citizen. MICROBIOLOGY, MOLECULAR BIOLOGY. *Educ:* Univ Wis, BA, 70; Univ Pa, MS, 73, PhD(molecular biol), 77. *Prof Exp:* Fel virol, Sch Med, Wash Univ, 78-79, fel immunol & membranes, 79-81, fel plasmids, Dept Biol, 82-84; CONSULT, 84- *Concurrent Pos:* fel virol, Sch Med, Wash Univ, 78-79, fel immunol & membranes, 79-81, fel plant pathogen plasmids, Dept Biol, Wash Univ, 82-84. *Res:* Synthesis of microbial membranes; defense mechanisms of pathogens. *Mailing Add:* 7146 Tulane St University City MO 63130

KOPLOWITZ, JACK, b Lenger, USSR, Mar 12, 44; US citizen; m 82; c 1. PATTERN RECOGNITION, IMAGE ANALYSIS. *Educ:* City Col New York, BEE, 67; Stanford Univ, MEE, 68; Univ Colo, PhD(elec eng), 73. *Prof Exp:* Mem tech staff Elec Eng, Bell Tele Labs, 67-70; ASSOC PROF ELEC & COMPUTER ENG, CLARKSON UNIV, 73- *Concurrent Pos:* Secy, Info Theory Group, Inst Elec & Electronics Engrs, 81-83; assoc ed, Inst Elec & Electronics Engrs Trans Info Theory, 83-87 & Pattern Recognition, 90-; consult, Teltech, Inc, 89- *Mem:* Sigma Xi; sr mem Inst Elec & Electronics Engrs. *Res:* Pattern recognition; image analysis; graphics; information theory; digital encoding of bilevel images; subpixel reconstruction of image edge information; estimation of shape characteristics of digitized shapes. *Mailing Add:* Elec & Comput Eng Dept Clarkson Univ Potsdam NY 13699

KOPLYAY, JANOS BERNATH, b Budapest, Hungary, June 24, 24; US citizen; m 58; c 4. APPLIED MATHEMATICS, COMPUTER SCIENCE. *Educ:* Royal Hungarian Air Force Acad, BS, 43; Royal Hungarian Polytech Inst, PhD(elect & mech eng), 49; Northwestern Univ, MA, 64, PhD(math, statist), 66. *Prof Exp:* Elec engr, Hungarian Utilities, Budapest, 48-49; chief engr electronic mech, Ygnis AG, Switz, 49-51; asst chief engr mech eng, Sociedade Paulista de Inst Gerais, Brazil, 51-55; engr design, Rochester & Goodell Engrs Inc, 56-63; asst prof math & comput sci, Northwestern Univ, 65-68; SCI ADV MATH & COMPUT SCI, AEROSPACE MED DIV, USAF SYSTS COMMAND, 68- *Concurrent Pos:* Consult, US Armed Serv & Friendly Allied Nations, 74-; charter mem, USAF Tech Adv Bd, 75-; prof math, San Antonio Col, 75- *Mem:* Math Asn Am; Mil Testing Asn; Am Statist Asn; Asn Comput Mach; AAAS. *Res:* Mathematics; aerospace medicine; computer science. *Mailing Add:* Dept Math San Antonio Col 1300 San Pedro Ave San Antonio TX 78284

KOPP, EUGENE H(OWARD), b New York, NY, Oct 1, 29; m 50; c 3. ELECTRICAL ENGINEERING. *Educ:* City Col New York, BEE, 50, MEE, 53; Univ Calif, Los Angeles, PhD(eng), 65. *Prof Exp:* Proj engr, Polarad Electronics Corp, 50-53 & Kaye-Halbert Corp, 53-54; proj engr, Precision Radiation Instruments, Inc, 54-56, chief engr, 56-58; from asst prof to prof eng, Calif State Col, Los Angeles, 58-73; dean sch eng, 67-73; vpres acad affairs, West Coast Univ, 73-79; sr scientist, 80-85, MGR RES & DEVELOP, HUGHES AIRCRAFT CO, 85- *Concurrent Pos:* Res fel, Univ Leeds, 66-67; adj fac, Univ Calif, Los Angeles, 79- *Honors & Awards:* Excellence in Eng Educ Award, Western Elec Co, Inc, 67. *Mem:* Sr mem Inst Elec & Electronics Engrs. *Res:* Microwave components and antennas; satellite communications. *Mailing Add:* PO Box 1351 South Pasadena CA 91031-1351

KOPP, JAY PATRICK, b Buffalo Center, Iowa, 38. PHYSICS. *Educ:* Loras Col, BS, 59; Univ Wis, MS, 61; Northwestern Univ, PhD(physics), 68. *Prof Exp:* Instr physics, Loras Col, 61-64; res assoc, Solid State Physics Lab, Swiss Fed Inst Technol, 67-69; ASST PROF PHYSICS, LORAS COL, 69- *Mem:* Am Phys Soc; Am Asn Physics Teachers; Swiss Phys Soc. *Res:* Experimental solid state physics using nuclear magnetic resonance and magnetization measurements to study magnetic properties of rare earth systems, principally indirect exchange mechanisms. *Mailing Add:* Dept Physics & Eng Loras Col 1450 Alta Vista Dubuque IA 52001

KOPP, MANFRED KURT, b Koenigsberg, Ger, Mar 8, 32; US citizen; m 58; c 1. INSTRUMENTATION. *Educ:* Univ Buenos Aires, Argentina, BSEE, 62; Univ Tenn, BSEE, 70, MSEE, 75. *Prof Exp:* Develop engr nuclear instrumentation, Comision Nac de Energia Atomica, Buenos Aires, Arg, 56-67; RES ENGR INSTRUMENTATION, OAK RIDGE NAT LAB, 67- *Mem:* Inst Elec & Electronics Engrs; Sigma Xi. *Res:* Low noise electronics; position sensitive proportional counters; radiation detectors, and basic measurement science. *Mailing Add:* Ordela Inc 139 Valley Court Oak Ridge TN 37830

KOPP, OTTO CHARLES, b Brooklyn, NY, July 22, 29; m 54; c 4. GEOLOGY. *Educ:* Univ Notre Dame, BS, 51; Columbia Univ, MA, 55, PhD(geol), 58. *Prof Exp:* Res asst, Columbia Univ, 55-58; from asst prof to assoc prof, 58-68, PROF GEOL, UNIV TENN, KNOXVILLE, 68- *Concurrent Pos:* Consult, Oak Ridge Nat Lab, 59-77, res partic, 77- *Honors & Awards:* Centennial of Sci Award, Univ Notre Dame, 65; Distinguished Prof Award, Am Fedn of Mineral Soc, 76. *Mem:* Fel Geol Soc Am; Soc Appl Coal Sci; Sigma Xi; Soc Sedimentary Geol; Am Inst Mining, Metall & Petrol Engrs; Nat Asn Geol Teachers (exec secy, 82-84). *Res:* Mineralogy; cathodoluminescence microscopy; petrology; coal geology; differential thermal analysis. *Mailing Add:* Dept Geol Sci Univ Tenn Knoxville TN 37996-1410

KOPP, RICHARD E, b Brooklyn, NY, July 12, 31; m 53; c 4. SYSTEMS ANALYSIS, CONTROL THEORY. *Educ:* Polytech Inst Brooklyn, BEE, 53, MEE, 57, DEE(control theory), 61. *Prof Exp:* Res engr comput, Grumman Aircraft Eng Corp, 53-57, group leader comput res, Grumman Corp, 57-63, sect head, 63-74, dir syst sci, Corp Res Ctr, 74-90, dir, sci adv bd, 90-91; CONSULT, 91- *Concurrent Pos:* Adj prof, Polytech Inst Brooklyn, 61- *Mem:* Inst Elec & Electronic Engrs; Am Inst Aeronaut & Astronaut. *Res:* Electronics; control theory; astrodynamics; computing; marine sciences; mathematics; data processing; image processing; filtering; simulation. *Mailing Add:* Grumman Aerospace Corp Five Oyster Bay Rd A08-35 Bethpage NY 11714

KOPP, ROGER ALAN, b Detroit, Mich, Feb 17, 40; m 62; c 3. SOLAR PHYSICS, LASER FUSION. *Educ:* Univ Mich, BS, 61; Harvard Univ, MA, 63, PhD(astron), 68. *Prof Exp:* Staff scientist, High Altitude Observ, Nat Ctr Atmospheric Res, 66-76; STAFF MEM, LOS ALAMOS NAT LAB, 76- *Concurrent Pos:* Vis scientist, Max Planck Inst Physics & Astrophys, 71-72, 79-80; guest prof, Univ Florence, 88. *Mem:* Am Astron Soc; Am Geophys Union; Int Astron Union. *Res:* Heating of the solar corona; origin and dynamics of the solar wind; structure of the chromosphere-corona transition region; laser fusion target design and experiments; laser-plasma interactions. *Mailing Add:* Los Alamos Nat Lab MS F645 PO Box 1663 Los Alamos NM 87545

KOPPA, RODGER J, b Oak Park, Ill, June 23, 36; m 57; c 2. HUMAN FACTORS, REHABILITATIVE ENGINEERING. *Educ:* Univ Tex, Austin, BA, 58, MA, 60; Tex A&M Univ, PhD(indust eng), 79. *Prof Exp:* Engr, LTV Aerospace Corp, 61-67; specialist, Gen Elec Co, 67-72; asst res psychologist, 73-79, ASSOC RES ENGR, TEX TRANSP INST, 79-; ASSOC PROF INDUST ENG, DEPT INDUST ENG, TEX A&M UNIV, 82- *Concurrent Pos:* Lectr, Dept Indust Eng, Tex A&M Univ, 79-82; chair spec task force, Transp Vehicle Res, Transp Res Bd Nat Res Coun-Nat Acad Sci, 80-81; mem, Adaptive Devices Standards Comt, Soc Automotive Engrs, 88-; chair, Educr Prof Group, Human Factors Soc, 89-91. *Mem:* Fel Human Factors Soc; Inst Indust Engrs; Soc Automotive Engrs. *Res:* Design and evaluation of automotive adaptive equipment for disabled drivers; transportation human factors; highway safety; visibility and driver information presentation; job performance aid design. *Mailing Add:* 1214 N Ridgefield Circle College Station TX 77840

KOPPEL, GARY ALLEN, b Cleveland, Ohio, Aug 8, 43; m 66; c 2. ORGANIC CHEMISTRY, IMMUNOLOGY. *Educ:* Case Western Reserve Univ, 65; Univ Pittsburgh, PhD(org chem), 69. *Prof Exp:* NIH fel org chem, Columbia Univ, 69-70; res scientist org chem, 70-80, RES ASSOC, ELI LILLY & CO, 80- *Concurrent Pos:* Mobay fel, 66-68; Nat Cancer Inst fel, 69-70. *Mem:* Sigma Xi; AAAS; Am Asn Immunologist; Am Chem Soc; NY Acad Sci; Clin Immunol Soc. *Res:* Synthesis of natural products; development of new cephalosporins; new synthetic methods in the synthesis of penicillins and cephalosporins; biochemistry; molecular biology. *Mailing Add:* 7823 Sunset Ln Indianapolis IN 46260

KOPPEL, LOWELL B, b Chicago, Ill, Sept 13, 35; m 57; c 2. CHEMICAL ENGINEERING. *Educ:* Northwestern Univ, BS, 57, PhD(chem eng), 60; Univ Mich, MSE, 58. *Prof Exp:* Instr chem eng, Calif Inst Technol, 60-61; from asst prof to prof chem eng, Purdue Univ, 61-85; SR CONSULT, SET POINT, INC, HOUSTON, TEX, 85- *Concurrent Pos:* Consult, Argonne Nat Lab, 62- *Mem:* Am Chem Soc; Am Inst Chem Engrs. *Res:* Process control; transport phenomena; applied mathematics; process simulation and optimization. *Mailing Add:* Set Point Inc 14701 St Mary's Lane Houston TX 77079

KOPPELMAN, ELAINE, b Brooklyn, NY, Mar 28, 37; m 70. MATHEMATICS. *Educ:* Brooklyn Col, BA, 57; Yale Univ, MA, 59; Johns Hopkins Univ, PhD(hist sci), 69. *Prof Exp:* From instr to asst prof, 61-74, ASSOC PROF MATH, GOUCHER COL, 74- *Mem:* Math Asn Am; Hist Sci Soc. *Res:* History of modern mathematics, particularly the development of algebra during the nineteenth and twentieth centuries. *Mailing Add:* Dept Math Goucher Col Towson MD 21204

KOPPELMAN, LEE EDWARD, b New York City, NY, May 19, 27; m 48; c 4. ENVIRONMENTAL SCIENCE. *Educ:* City Col, New York, BEE, 50; Pratt Inst Grad Sch Architecture, IASP, 64; NY Univ Grad Sch, DPA, 70. *Hon Degrees:* LD, Long Island Univ, 78. *Prof Exp:* PROF PLANNING & RESOURCE MGT, STATE UNIV NY, STONY BROOK, 67-, LEADING PROF & DIR, CTR REGIONAL POLICY STUDIES, 88- *Concurrent Pos:* Dir planning, Suffolk County Planning Comn, 60-; exec dir regional planning, Long Island Regional Planning Bd, 65-; appointee, Coastal Zone Mgt Adv Coun, Nat Oceanic & Atmospheric Admin, 73-75; Nat Shoreline Erosion Adv Panel, US Army, 74-81; adj prof, Grad Sch Environ Sci & Forestry, Syracuse Univ, 75-; consult, US Dept Housing & Urban Develop, 75-78; UN Off Ocean Econ & Technol, 81. *Mem:* Am Planning Asn; Sigma Xi; Am Inst Architect. *Res:* Integraton of coastal zone sciences and the regional planning process, including pollution studies of surface waters and the institutional management mechanisms required for coastal development. *Mailing Add:* Two Dune Ct Setauket NY 11733

KOPPELMAN, RAY, b Chicago, Ill, Aug 25, 22; m 46; c 2. BIOCHEMISTRY. *Educ:* Univ Chicago, BS, 44, PhD(biochem), 52. *Hon Degrees:* DSc, Univ Osteop Med & Health Sci. *Prof Exp:* Asst biochem, dept biochem, Univ Chicago, 47-52, instr biochem & biol, 52-56, from asst prof to assoc prof, 56-70, assoc dean, Col Biol Sci, 65-67; head, NSF Liaison Staff, New Delhi, 67-69; vpres res, WVa Univ, 70-82; provost, 83-90, EMER PROVOST & PROF BIOCHEM, UNIV OSTEOP MED & HEALTH SCI, 90- *Concurrent Pos:* Consult-eval, NCent Asn, 64-75 & Am Osteop Asn, 71-; mem, Comt Undergrad Educ Biol, 65-67, Biol Grad Rec Exam Comt, 72-76. *Mem:* AAAS; Am Soc Biol Chemists; Am Inst Biol Sci; Sigma Xi. *Res:* Medical education administration. *Mailing Add:* Univ Osteop Med & Health Sci 3440 Grand Ave Des Moines IA 50312

KOPPENAAL, THEODORE J, b Milwaukee, Wis, Dec 19, 31; m 54; c 3. METALLURGY. *Educ:* Univ Wis, BS, 54; Univ Ill, MS, 58; Northwestern Univ, PhD(metall), 61. *Prof Exp:* Asst metallurgist, Argonne Nat Lab, Ill, 62-66, assoc metallurgist, 66-67; supvr phys metall, Aeronutronic Div, Ford Aerospace & Commun Corp, 67-79; dir eng, Heavy Metals Div, Aerojet Ordnance Co, 79-81; PRES, KOPPENAAL & ASSOC, 81- *Mem:* Fel Am Soc Metals. *Res:* Technical consultant; technical marketing for diversified programs related to materials science and engineering. *Mailing Add:* Koppenaal & Assoc 21095 Whitebark Mission Viejo CA 92692

KOPPENHEFFER, THOMAS LYNN, b Harrisburg, Pa, May 23, 42; m 67; c 2. BIOCHEMISTRY. *Educ:* Bloomsburg State Col, BS, 64; Williams Col, MA, 66; Boston Univ, PhD(biol), 70. *Prof Exp:* Asst prof biol, Boston Univ, 70-71; assoc surg, Harvard Med Sch, 72-73; asst prof biol, Williams Col, 73-79; from asst prof to assoc prof, 79-88, PROF BIOL, DEPT BIOL, TRINITY UNIV, 88- *Concurrent Pos:* Res fel surg, Harvard Med Sch, 71-72. *Mem:* Am Asn Immunol; Am Soc Zoologists; Int Soc Develop & Comp Immunol. *Res:* Biology of vertebrate complement systems. *Mailing Add:* Dept Biol Trinity Univ 715 Stadium Dr San Antonio TX 78284

KOPPERL, SHELDON JEROME, b Cleveland, Ohio, Sept 11, 43; m 67; c 2. HISTORY OF SCIENCE. *Educ:* Case Inst Technol, BS, 65; Univ Wis, Madison, PhD(chem, hist sci), 70. *Prof Exp:* Asst prof hist sci, 70-72, asst prof health sci, 72-75, assoc prof, 75-81, COORDR HIST SCI PROG, GRAND VALLEY STATE UNIV, 73-, PROF, SCH HEALTH SCI, 81- *Mem:* Am Chem Soc; Hist Sci Soc; Soc Hist Technol; Sigma Xi. *Res:* Historical studies in inorganic, physical and organo-metallic chemistry and medicine, chiefly since 1800; studies in the history of art and science, chiefly Renaissance and Baroque. *Mailing Add:* Sch Health Sci Grand Valley State Univ Allendale MI 49401

KOPPERMAN, RALPH DAVID, b New York, NY, Feb 17, 42; div; c 2. SET-THEORETIC TOPOLOGY, MATHEMATICAL LOGIC. *Educ:* Columbia Col, AB, 62; Mass Inst Technol, PhD(math), 65. *Prof Exp:* Lectr math, Boston Univ, 63-65; asst prof, Univ RI, 65-67; from asst prof to assoc prof, 67-83, PROF, CITY COL NEW YORK, 84- *Mem:* Am Math Soc; Asn Symbolic Logic; Math Asn Am. *Res:* topological rings and modules; generalized metric spaces and uniform spaces; general topology applied to computer graphics; spaces of ideals. *Mailing Add:* Dept Math City Col New York New York NY 10031

KOPPLE, KENNETH D(AVID), b Philadelphia, Pa, Oct 21, 30; m 60. PEPTIDE CHEMISTRY, NUCLEAR MAGNETIC RESONANCE. *Educ:* Mass Inst Technol, SB, 51, PhD(chem), 54. *Prof Exp:* Instr org chem, Univ Chicago, 54-56, asst prof, 56-62; res chemist, Gen Elec Co Res Lab, 62-65; from assoc prof to prof chem, Ill Inst Technol, 65-85, chmn dept, 82-85; DIR PHYS & STRUCT CHEM, SMITH KLINE & FRENCH LABS, 85- *Concurrent Pos:* J S Guggenheim Found fel, Lab Chem Biodynamics, Univ Calif, Berkeley, 64-65; NIH res career develop award, 70-75; for res guest, SNAM Progetti Laboratori Ricerche di Base, Rome, Italy, 74. *Mem:* Am Chem Soc; fel AAAS; Am Soc Biochem & Molecular Biol; Royal Soc Chem. *Res:* Peptide chemistry, synthesis and spectroscopic determination of conformation; nuclear magnetic resonance. *Mailing Add:* Smith Kline Beecham Pharm L940 PO Box 1539 King of Prussia PA 19406

KOPPLIN, J(ULIUS) O(TTO), b Appleton, Wis, Feb 6, 25; m 50; c 4. ELECTRICAL ENGINEERING, MATERIALS SCIENCE. *Educ:* Univ Wis, BS, 49; Purdue Univ, MS, 54, PhD(elec eng), 59. *Prof Exp:* Corrosion engr, Northern Ind Pub Serv Co, 49-53; instr elec eng, Purdue Univ, 54-58; asst prof, Univ Ill, 58-61; vis asst prof, Mass Inst Technol, 61; assoc prof, Univ Ill, 62-68; prof elec eng & head dept, Univ Tex, El Paso, 68-75; PROF ELEC ENG & CHMN DEPT, IOWA STATE UNIV, 75- *Mem:* Inst Elec & Electronics Engrs; Am Soc Eng Educ. *Res:* Superconductivity; electric and magnetic properties of materials; liquid and solid surface phenomena. *Mailing Add:* Dept Elec Eng Iowa State Univ Ames IA 50011

KOPRIWA, BEATRIX MARKUS, histology, for more information see previous edition

KOPROWSKA, IRENA, b Warsaw, Poland, May 12, 17; nat US; m 38; c 2. PATHOLOGY, CYTOLOGY. *Educ:* Warsaw Med Sch, MD, 39. *Prof Exp:* Intern med, Villejuif Lunatic Asylum, France, 40; asst pathologist, Rio de Janeiro City Hosps, 42-44; res asst & asst pathologist, Med Col, Cornell Univ & New York Hosp, 45-46; res asst appl immunol, Pub Health Res Inst, City of New York, 46-47; asst pathologist, New York Infirmary, 47-49; res fel & assoc anat, Med Col, Cornell Univ, 49-54; asst prof path, State Univ NY Downstate Med Ctr, 54-57; from assoc prof to prof, Hahnemann Med Col, 57-70; PROF PATH & DIR CYTOL SERV, HEALTH SCI CTR, TEMPLE UNIV, 70- *Concurrent Pos:* Res fel, Med Col, Cornell Univ, 49-54; USPHS res grants, 54-; Runyon Mem Fund grant, 55-56; Am Cancer Soc grant, 58-61; lectr, France, Poland, India & Iran, 52-; consult, WHO, 62- *Mem:* AAAS; Am Soc Cytol; Am Asn Cancer Res; Am Soc Exp Path; Am Soc Clin Pathologists. *Res:* Studies of progressive morphologic cellular changes, especially neoplastic progression in human beings, mice and in tissue culture systems. *Mailing Add:* 334 Fairhill Rd Ardmore PA 19140

KOPROWSKI, HILARY, b Warsaw, Poland; nat US; m 38; c 2. BIOLOGY. *Educ:* Univ Warsaw, MD. *Hon Degrees:* Numerous from US & foreign Univs. *Prof Exp:* Mem staff, Yellow Fever Res Serv, Rio de Janeiro, Brazil, 40-44; mem staff res div, Am Cyanamid Co, 44-46; asst dir viral & rickettsial res, Lederle Lab, Pearl River, 46-57; PROF MICROBIOL, FAC ARTS & SCI & WISTAR PROF RES MED, UNIV PA, 57-, DIR, WISTAR INST ANAT & BIOL, 57- *Concurrent Pos:* Consult, Nat Cancer Inst, NIH & USPHS, 62-70; mem expert comt on rabies, WHO, Switz; co-ed, Methods in Virol, Viruses & Immunity; Current Topics in Microbiol & Immunol, 65-; Fulbright scholar, Max Planck Inst Physiol of Behav, Germany, 71; Alexander Von Humbolt Sr Scientist Award, Max Planck Inst, WGermany. *Honors & Awards:* Comdr, Order of Merit; Chevalier, Order of the Royal Lion, Belg; Alvarenga Prize; Polish Millennium Award, Alfred Jurzykowski Found, 66; Felix Wankel Tierschutz Prize, 79. *Mem:* Nat Acad Sci; NY Acad Med; NY Acad Sci (pres, 59); AAAS. *Res:* Cell biology, virology, and immunology; cancer; vaccine against poliomyelitis, hog cholera and rabies. *Mailing Add:* Wistar Inst 36th & Spruce Sts Philadelphia PA 19104

KORACH, KENNETH STEVEN, b Buffalo, NY, Nov 26, 46; m 70; c 2. STEROID HORMONE ACTION, STEROID RECEPTOR PROTEINS. *Educ:* Augusta Col, Ga, BA, 69; Med Col Ga, PhD(endocrinol), 74. *Prof Exp:* Res fel endocrinol, Sch Med, Harvard Univ, 74-75, Ford res fel, 75-76; staff fel, Nat Inst Environ Health Sci, 76-78, sr staff fel, 78-80, res endocrinologist, 80-84, SR RES ENDOCRINOLOGIST, NAT INST ENVIRON HEALTH SCI, 85-, CHIEF, RECEPTOR BIOL SECT, 87- *Concurrent Pos:* Guest lectr, Dept Biochem, NC State Univ, 81-, adj prof, 90-; adj prof, lab reproductive biol, Univ NC Med Sch, 81-, adj prof, pharmacol, 89-; chmn, Radiation Safety Comt & Res Prod Subcomt, Nat Inst Environ Health Sci, 85-, Arts Graphics Comt, 87- & Res Assoc Prom Comt, 90-; transatlantic lectr, Soc Toxicol, 85; consult, Glaxo Res Inst, 90- *Mem:* Endocrine Soc; Sigma Xi. *Res:* Estrogen hormone action; investigations of estrogen receptor proteins, using a structure-activity approach; determination of intracellular sites of action; tissue responses in protein synthesis and DNA-RNA induction. *Mailing Add:* Lab Reproductive & Develop Toxicol Nat Inst Environ Health Sci PO Box 12233 Research Triangle Park NC 27709

KORACH, MALCOLM, b New York, NY, Apr 25, 22; m 46; c 3. ORGANIC CHEMISTRY. *Educ:* Yale Univ, BS, 42, PhD(chem), 48. *Prof Exp:* Asst, Manhattan Proj, Columbia Univ, 43 & Oak Ridge Inst Nuclear Studies, 44-46; res chemist, org group leader & asst dir res, 49-70, dir res, 70-74, MGR RES, CHEM DIV, PPG INDUSTS, 74- *Mem:* Am Chem Soc. *Res:* Heavy organic chemicals; chlorinated organics; hydrogen peroxide and its utilization. *Mailing Add:* 1745 Brookwood Dr Akron OH 44313

KORAN, LORRIN MICHAEL, b Los Angeles, Calif, Apr 4, 40; m 67; c 2. OBSESSIVE-COMPULSIVE DISORDERS. *Educ:* Harvard Col, BA, 62; Harvard Med Sch, MC, 66; Am Bd Psychiat & Neurol, cert, 73. *Prof Exp:* From asst prof to assoc prof, psychiat, State Univ NY, Stony Brook, 72-77; assoc prof, 79-84, PROF PSYCHIAT, STANFORD UNIV, 84- *Concurrent Pos:* Dir, med student educ psychiat, State Univ NY, Stony Brook, 73-77; dir, residency training prog, Stanford Univ, Calif, 79-81; med dir, CMU, Stanford Hosp, Stanford Calif, 80- *Mem:* Fel Am Psychiat Asn; Int Asn Study Pain. *Res:* Relationships between physical and mental disorders; treatment of obsessive-compulsive disorder; mental health policy issues. *Mailing Add:* Dept Psychiat TD114 Stanford Univ Med Ctr Stanford Univ Sch Med Stanford CA 94305

KORAN, ZOLTAN, b Hungary, May 27, 34; Can citizen; m 68; c 4. FOREST PRODUCTS, PULP & PAPER SCIENCE & TECHNOLOGY. *Educ:* Univ BC, BSc, 59, MF, 61; Syracuse Univ, PhD(forestry), 64. *Prof Exp:* Asst prof forestry, Univ N H, 63-64; res scientist pulp & paper, Pulp & Paper Res Inst Can, 65-68; asst prof forestry, Univ Toronto, 68-73; res scientist, Que Indust Res Ctr, 73-76; PROF ENG, UNIV QUE, 76- *Mem:* Forest Prod Res Soc; Int Asn Wood Anat; Can Pulp & Paper Asn; Micros Soc Can; Tech Asn Pulp & Paper Indust. *Res:* Anatomy and ultrastructure of wood, bark, fiber, pulp and paper; thermomechanical pulping; pulp and paper properties; forest products and utilization; composite boards; wood finishing and impregnation; x-ray, light and electron microscopic studies; material science engineering. *Mailing Add:* 3845 Limoges Trois-Rivieres PQ G8Y 5N1 Can

KORANT, BRUCE DAVID, b Brooklyn, NY, Aug 9, 43; m 69; c 1. VIROLOGY. *Educ:* Brooklyn Col, BS, 65; Pa State Univ, MS, 67, PhD(microbiol), 69. *Prof Exp:* BIOCHEMIST VIROL, CENT RES DEPT, EXP STA, E I DU PONT DE NEMOURS & CO, INC, 69- *Concurrent Pos:* NATO fac, 78, 82 & 84. *Mem:* Am Soc Microbiol; Am Soc Virol; Soc Interferon Res. *Res:* Animal virology; bacteriophages; virus structure and replication; effects of viruses on cells; protein chemistry; proteolytic enzymes; interferon mechanism. *Mailing Add:* Cent Res Dept Exp Sta Bldg 328 E I Du Pont de Nemours & Co Inc Wilmington DE 19898

KORANYI, ADAM, b Szeged, Hungary, July 13, 32; US citizen. MATHEMATICS. *Educ:* Univ Szeged, dipl, 54; Univ Chicago, PhD(math), 59. *Prof Exp:* Instr math, Harvard Univ, 59-60; asst prof, Univ Calif, Berkeley, 60-64; vis asst prof, Princeton Univ, 64-65; assoc prof math, Belfer Grad Sch, Yeshiva Univ, 65-68, prof, 68-79; PROF MATH, WASH UNIV, 79- *Mem:* Am Math Soc. *Res:* Symmetric spaces; Lie groups; theory of functions of several complex variables. *Mailing Add:* Herbert H Lehman Col City Univ New York Bronx NY 10468

KORC, MURRAY, b Ger, Apr 3, 47; m; c 3. ENDOCRINOLOGY. *Educ:* Albany Med Col, MD, 74. *Prof Exp:* From asst prof to assoc prof, Dept Internal Med, Health Sci Ctr, Univ Ariz, 81-89, assoc prof internal med & biochem, 87-89; PROF & CHIEF, DIV ENDOCRINOL & METAB, UNIV CALIF, IRVINE, 89- *Mem:* Am Col Physicians; AAAS; Am Soc Cell Biol; Am Diabetes Asn; Endocrine Soc; Am Fedn Clin Res. *Mailing Add:* Dept Med Div Endocrinol & Metab Univ Calif Med Sci I C240 Irvine CA 92717

KORCEK, STEFAN, b Trnava, Czech, May 28, 34; m 65; c 2. PHYSICAL ORGANIC CHEMISTRY, LUBRICANTS CHEMISTRY. *Educ:* Slovak Tech Univ, Bratislava, MS, 57, PhD(chem, chem eng & fuel technol), 67. *Prof Exp:* Assoc prof chem kinetics & reactors design, Dept Chem & Technol Petrol, Slovak Tech Univ, 57-68; fel phys org chem, Div Chem, Nat Res Coun, Ottawa, Can, 68-70; sr res scientist, 71-76, prin res scientist assoc, 76-81, STAFF SCIENTIST LUBRICANT CHEM & PHYS ORG CHEM, RES FUELS & LUBRICANTS DEPT, FORD MOTOR CO, DEARBORN, MICH, 81- *Concurrent Pos:* Vis res off, Div Chem, Nat Res Coun, Ottawa, Can, 70-71. *Honors & Awards:* F R McFarland Award, Soc Automotive Engrs, 81. *Mem:* Am Chem Soc; Soc Automotive Engrs. *Res:* Kinetics and mechanisms of autooxidation and inhibited oxidation of organic substrates in liquid phase at elevated temperatures; mechanism of action of antioxidants; automotive lubricants, their chemistry and degradation in service; reactions of lubricant antioxidant additives in engines; author or coauthor of over 40 publications. *Mailing Add:* 4778 Crestview Ct Birmingham MI 48010

KORCHAK, ERNEST I(AN), b Opava, Czech, Feb 15, 34; Australian citizen; m 59; c 3. POLYMER CHEMISTRY, TECHNICAL MANAGEMENT. *Educ:* Univ Melbourne, BChE, 57; Mass Inst Technol, SM, 61, ScD(chem eng), 64. *Prof Exp:* Res engr chem eng, Imp Chem Industs, Australia & NZ, 58-59; chem engr, Halcon Int, Inc, 64-67; sales exec, Sci Design Co, Inc, 67-71, vpres & gen mgr, Halcon Catalyst Industs, 71-75, pres, Halcon Res & Develop Corp, 75-80, pres, Sci Design Co, 81-86; chmn, Riverside Polymer Systs Inc, 86-90; PRES, PERFORMANCE COATINGS CORP, 90- *Concurrent Pos:* Pres, Chem Indust Asn, 84-85; chmn, Chemtech Mgt Ltd, 86- *Mem:* Am Chem Soc; Am Inst Chem Engrs. *Res:* Gas flow and turbulence, application to packed beds; process research and development of organic chemical processes, special emphasis on kinetics and separations. *Mailing Add:* 1118 Old Gulph Rd Bryn Mawr PA 19010-1649

KORCHAK, HELEN MARIE, PHOSPHOLIPID METABOLISM. *Educ:* Tufts Univ, PhD(physiol), 62. *Prof Exp:* ASSOC PROF EXP MED & DIR RES, DIV RHEUMATOLOGY, NY UNIV MED CTR, 84- *Res:* Neutrophil activation. *Mailing Add:* Immunol Sect Rm 8113 Univ Pa Children's Hosp Philadelphia 34th & Civic Ctr Rd Philadelphia PA 19104

KORCHIN, LEO, b Brooklyn, NY, July 1, 14; m; c 2. ORAL & MAXILLOFACIAL SURGERY. *Educ:* Cornell Univ, BS, 36; NY Univ, DDS, 41; Georgetown Univ, MS, 54; Am Bd Oral & Maxillofacial Surg, dipl; FACD; FICD. *Prof Exp:* Chief oral surg sect, Army Hosp, US Army, Ft Jay, NY, 48-52, instr oral surg & asst prof mil sci & tactics, Sch Dent, Georgetown Univ, 52-54, chief oral surg sect, Rodriguez Army Hosp, PR, 54-57, chief oral surg br, Dent Detachment, Ft Devens, Mass, 57-62, chief dent clin & oral surg sect, 97th Gen Hosp, Frankfurt, Germany, 62-65, chief dent serv & oral surg & dir dent intern training prog, Martin Army Hosp, Ft Benning, Ga, 65-67; assoc prof oral surg, 67-72, dir dept surg sci, 72-75, asst dean for clin instr, 79-82, assoc dean, 84-86, PROF ORAL & MAXILLOFACIAL SURG, SCH DENT, UNIV PR, SAN JUAN, 72- *Concurrent Pos:* Fel oral path, Armed Forces Inst Path, 75-76. *Honors & Awards:* Novice Award, Int Asn Dent Res, 54. *Mem:* Am Soc Oral & Maxillofacial Surgeons; Am Acad Oral Path; Am Dent Asn; Int Asn Oral & Maxillofacial Surgeons. *Res:* Effects of starch sponge implanted in bone. *Mailing Add:* Univ PR Sch Dent San Juan PR 00936

KORCHYNSKY, M(ICHAEL), b Kiev, Ukraine, Apr 11, 18; nat US; m 51; c 3. PHYSICAL METALLURGY, MATERIALS SCIENCE. *Educ:* Tech Univ Lviv, Ukraine, Dipl Ing, 42. *Prof Exp:* Asst metall, Tech Univ Lviv, 42-44; chief engr, US Army, Ger, 45-50; res metallurgist phys metall, Metals Res Labs, Union Carbide Metals Co, 51-60, tech supvr, Tech Dept, 60-61; res supvr alloy & high strength steels, Jones & Laughlin Steel Corp, 61-65, asst dir res new prod develop, Graham Res Lab, 65-70, dir prod res, 70-73; dir alloy develop, Umetco Minerals Corp, Union Carbide Corp, 73-86; PRIN, KORCHYNSKY & ASSOCS, CONSULTS METALL, 86- *Concurrent Pos:* Sr fel, Union Carbide Corp, 79. *Honors & Awards:* Am Iron & Steel Inst Medal, 65; Howe Mem lectr, Asn Inst Mining, Metall & Petrol Eng, 83; W H Eiseman Award, Am Soc Metals, 84 & E C Bain Award, 86. *Mem:* Fel Am Soc Metals; Iron & Steel Soc-Am Inst Mining, Metall & Petrol Engrs; Soc Automotive Engrs; Wire Asn Int. *Res:* Physical metallurgy of steels; materials for high-temperature service; nuclear fuels; alloy design and development; technology and application of high-strength, low-alloy (microalloyed) steels; management of industrial research and product development; technological marketing. *Mailing Add:* c/o Stratcor Tech Sales Inc Twin Towers Off Bld Steubenville Pike 4955 Pittsburgh PA 15205

KORDA, EDWARD J(OHN), b Duluth, Minn, Nov 17, 18; m 45; c 3. MICROSCOPY IMAGE ANALYSIS, CERAMICS. *Educ:* Univ Minn, BMetE, 47; Stevens Inst Technol, MS, 51. *Prof Exp:* Lab asst chem, Duluth Jr Col, 38-39; lab asst metallog, Univ Minn, 39-41, scientist metall res, 47-48; mem res staff metall, Manhattan Proj, 44-46; proj engr metall eng, Curtiss Wright Corp, NJ, 48-49 & 50-56; prof metall eng, Drexel Inst, 56-65; dir ed & res, Del Sci Labs, Inc, 65; electron microscopist, Corning Glass Works, 65-66, sr res scientist, 66-75, res supvr, 75-79 & 81-83; vis prof, Atlanta Univ, 79-81; RES PROF, ATLANTA UNIV, 83- *Concurrent Pos:* Laboratorian, Am Steel & Wire Co, Minn, 39-41; res assoc metall, Stevens Inst Technol, 49-50, instr, 49-51; prof, Elmira Col, 66-76. *Mem:* Am Soc Testing & Mat; Am Soc Metals; Electron Micros Soc Am; Am Ceramic Soc; Sigma Xi. *Res:* Electron microscopy and optics; metallography; physical metallurgy; solid state physics, x-ray analysis; ceramography; optical microscopy; image analysis; electron diffraction. *Mailing Add:* 3913 Jericho Rd Tucker GA 30084-7412

KORDA, PETER E, b Budapest, Hungary, Dec 5, 31; US citizen; m 54; c 4. ENGINEERING MECHANICS, STRUCTURAL ENGINEERING. *Educ:* Budapest Tech Univ, Dipl Eng, 54; Ohio State Univ, PhD(eng mech), 64. *Prof Exp:* Struct designer, Indust Bldg Design Off, Hungary, 54-56, Livesley & Henderson, Eng, 56-57 & Dominion Bridge Co, Ltd, Can, 57-60; asst proj design, Sch Archit, Ohio state Univ, 60-61, res assoc & instr eng mech, 61-63, from asst prof to prof, 64-81; PRES, KORDA/NEMETH ENG, 85- *Concurrent Pos:* Consult engr, Miller & Korda, 64-67, pres, Korda Eng Co Consult Engrs, 67-85. *Mem:* Am Soc Civil Engrs; Concrete Inst Am; Int Asn Shell Struct; Nat Soc Prof Engrs; Am Soc Eng Educ. *Res:* Shallow shell theory with computer applications; dynamic stability and structural damping. *Mailing Add:* 5544 Dublin Rd Dublin OH 43017

KORDAN, HERBERT ALLEN, b St Louis, Mo, Apr 10, 26; m 49; c 3. DEVELOPMENTAL PHYSIOLOGY, PLANT MORPHOGENETICS. *Educ:* Univ Calif, Los Angeles, BA, 55, MS, 58, PhD(plant sci), 61. *Prof Exp:* Fel tumorigenesis, Cedars of Lebanon Hosp, Los Angeles, 61-62; res staff mem plant physiol & consult, Stanford Res Inst, 63; asst res plant physiologist, Univ Calif, Los Angeles, 63-65; asst prof biol, Mt St Mary's Col, Calif, 64-66; assoc prof, Gustavus Adolphus Col, 66-67; UNIV LECTR BOT, UNIV BIRMINGHAM, 68- *Concurrent Pos:* USPHS fel, Univ Leeds, 67-68. *Mem:* Soc Exp Biol; fel Bot Soc Edinburgh; Sigma Xi. *Res:* Comparative cytology; morphology and physiology of growing and non-growing plant cells; investigations on effects of organic and inorganic central nervous system depressant chemical agents on pollen and seed germination and subsequent developmental behavior. *Mailing Add:* Dept Plant Biol Univ Birmingham PO Box 363 Birmingham England

KORDESCH, KARL VICTOR, b Vienna, Austria, Mar 18, 22; nat US; m 46; c 4. ELECTROCHEMISTRY, BATTERIES & FUEL CELLS. *Educ:* Univ Vienna, PhD(chem), 48. *Hon Degrees:* Dr techn hc, Tech Univ Vienna, 90. *Prof Exp:* Asst & lectr chem, Chem Inst, Univ Vienna, 46-48, asst prof, 48-53; chem engr, Signal Corps Eng Labs, NJ, 53-55; res chemist & group leader, Develop Dept, 55-70, sr res assoc, Parma Res Lab, Union Carbide Corp, 70-74, corp res fel, Battery Prod Div, 74-77; PROF, TECH UNIV GRAZ, AUSTRIA, 77-; DIR, INST INORG CHEM TECHNOL, 77-; CONSULT ELECTROCHEM, 77- *Concurrent Pos:* Vpres technol, Battery Technol, Inc, Can, 88-; secy gen, Int Electrochem Soc, 81. *Honors & Awards:* Nat Energy Award, Austria, 81; Vittorio De Nora-Diamond Shamrock Award, Electrochem Soc, 87; E Schroedinger Prize, Austrian Acad Sci, 90- *Mem:* Am Chem Soc; Electrochem Soc; Austrian Chem Soc; Int Electrochem Soc (vpres, 86). *Res:* Electrochemical systems; batteries, especially with alkaline electrolytes; hydrogen-oxygen fuel cells; carbon electrodes; test and control instruments; electronic circuitry; technical management. *Mailing Add:* Tech Univ Graz Stremayrgasse 16 Graz A-8010 Austria

KORDOSKI, EDWARD WILLIAM, b New Britain, Conn, Aug 15, 54; m 77. CONTINUOUS CHEMICAL PROCESS DEVELOPMENT. *Educ:* King's Col, BS, 77; Univ Md, PhD(org chem), 82; Monmouth Col, MBA, 86. *Prof Exp:* Teaching asst org chem, Univ Md, 77-79 & res asst, 79-82; process develop chemist, CIBA-GEIGY Corp, Basel, Switzerland, 82-84, sr process develop chemist, Specialty Chem & Dyes, 84-87, sr chemist, 87-88, staff chemist, Baton Rouge, La, RES & DEVELOP GROUP LEADER DYESTUFFS & CHEMS DIV, CIBA-GEIGY CORP, BATON ROUGE, LA, 89- *Mem:* Am Chem Soc. *Res:* Developing and using state-of-the-art, proto-type, automated, continuous small reactor technology and analytical monitoring and feedback control equipment to produce quality specialty chemicals, intermediates and dyes. *Mailing Add:* 838 High Lake Dr Baton Rouge LA 70810-4341

KORDOVA, NONNA, b Krasnodar, USSR; Can citizen; m 45. RICKETTSIAL DISEASES. *Educ:* Charles Univ, Prague, MD, 45; Czech Acad Sci, PhD(med virol), 60. *Prof Exp:* Res assoc pediat, Children's Hosp, Komensky Univ, 45-54, asst prof, 54-56; res assoc rickettsioses, Inst Virol, Czech Acad Sci, 56-60, sr scientist, 60-66, chief lab, 66-68; res assoc, Univ Kans, 68-69; assoc prof, 70-77, PROF, MED COL, UNIV MAN, 77- *Honors & Awards:* Recognition Dipl, Czech Acad Sci, 65. *Mem:* Can Soc Microbiol; Am Soc Microbiol; Can Pub Health Asn. *Res:* Pathogenesis of chlamydial diseases; parasite-host interactions at the cellular and subcellular level. *Mailing Add:* 501-71 Roslyn Rd Winnipeg MB R3L 0G2 Can

KORDYBAN, EUGENE S, b Ukraine, May 20, 28; m 53; c 6. FLUID MECHANICS. *Educ:* Univ Detroit, BME, 54; State Univ NY Buffalo, MS, 60, PhD(mech eng), 69. *Prof Exp:* Engr, Linde Div, Union Carbide Corp, 54-64, sr engr, 64-69; asst prof, 69-72, assoc prof civil eng, 72-81, PROF MECH ENG, UNIV DETROIT, 81- *Concurrent Pos:* NSF res grant, 70-72. *Mem:* Am Soc Mech Engrs; Am Acad Mech; AAAS. *Res:* Polyphase flow, especially slug flow, basic fluid mechanics, flow visualization techniques and flow measurement. *Mailing Add:* Dept Mech Eng Univ of Detroit Detroit MI 48221

KORECKY, BORIVOJ, b Prague, Czech, Sept 9, 29; c 2. MEDICAL PHYSIOLOGY. *Educ:* Charles Univ, Prague, MD, 55; Czech Acad Sci, PhD, 61. *Prof Exp:* From asst prof to assoc prof path physiol, Charles Univ, Prague, 55-66; assoc prof, 66-71, PROF PHYSIOL, UNIV OTTAWA, 71- *Concurrent Pos:* Med Res Coun Can res fel, 63-64. *Mem:* Can Physiol Soc; Am Physiol Soc. *Res:* Cardiovascular and respiratory physiology. *Mailing Add:* Dept Physiol Univ Ottawa Sch Med 451 Smyth Rd Ottawa ON K1H 8M5 Can

KOREIN, JULIUS, b New York, NY, Sept 27, 28; m 57; c 3. NEUROLOGY. *Educ:* NY Univ, BA, 49, MD, 53. *Prof Exp:* Intern, Maimonides Hosp, Brooklyn, 53-54; asst resident neurol, Mt Sinai Hosp, 54-55; asst & chief resident, NY Univ-Bellevue Hosp Ctr, 55-57; from asst prof to assoc prof, 61-72, assoc dir anal & comput methodology, Dept Radiol, 68-72, PROF NEUROL, MED CTR, NY UNIV, 72- *Concurrent Pos:* Fel, Mt Sinai Hosp, 53-54, asst attend, 59-70, spec trainee, 60-61; spec trainee, NY Univ-Bellevue Hosp Ctr, 59-60; vis asst, Bellevue Hosp, 59-68, vis assoc, 68-72, attend, 72-, dir EEG, 61-70, chief, 70-; attend physician, Vet Admin Hosp, Manhattan, 61-73, consult, 73-87; asst attend, NY Univ Hosp, 61-72, attend, 71-; consult, Gen Elec Corp, 66-67, Children's Bur, Dept Health, Educ & Welfare, 66-71, Int Info Processing, 67-70, Nat Inst Neurol Dis & Stroke, 71-74; proj dir, Health Res Coun Grants, City of New York, 63-65 & Nat Cancer Inst, 66-70; prin investr, USV Pharmaceut Corp, 63-72, Warner-Lambert Inst, 66-67, Nat

Ctr Health Serv Res & Develop, 66-70 & Hoffmann-La Roche Labs, 69-70; co-investr, Nat Inst Neurol Dis & Stroke, 65-67, proj dir & vchmn study cerebral death, 71-72; consult, sensory feedback ther, Int Ctr Disabled Res & Rehab Ctr, 72-84; assoc ed, Am Soc Cybernet Forum, 74-80; mem adv bd, Int J Neurosci, 79-; adv, President's Comn Study Ethical Problems Med, Biomed & Behav Res, Washington DC, 80-81; adv bd, Neurosci Info Ctr, Upjohn Co, 78-84; various educ video presentations, 67-84; mem, Sci & Soc Comt, NY Acad Sci, 84-86; chmn, Biomed Ethics Comt, Bellevue Hosp, 85-; career scientist award, New York City Health Res Coun, 66-72. *Honors & Awards:* Bronze Award, Diag of Brain Death, Int Film & TV Fest, NY, 83. *Mem:* AAAS; Am Neurol Asn; Am Acad Neurol; Asn Res Nerv & Ment Dis; Am Electroencephalographic; NY Acad Sci. *Res:* Computer applications in capture, storage, retrieval and analysis of narrative medical data for the purpose of patient care and clinical research; sensory feedback therapy in neuromuscular disorders; electroencephalography and behavior, including computer analysis of the electroencephalogram, effects of drugs on the electroencephalogram and behavior, diagnosis of brain death; pathophysiology and treatment of segmental torsion dystonia; models of neurophysiological structures; ontogenesis of cerebral function in the human fetus; aquatic medicine; electro physiological probes to analyse movement of cortical and subcortical origins. *Mailing Add:* Dept Neurol NY Univ Med Ctr New York NY 10016

KORENBROT, JUAN IGAL, b Mexico City, Mex, Nov 29, 47; m 72; c 2. BIOPHYSICS. *Educ:* Johns Hopkins Univ, MA, 71, PhD(biophys), 72. *Prof Exp:* Res assoc biophys, Johns Hopkins Univ, 71-72; res assoc physiol, 72-73, lectr, Univ Calif, Los Angeles, 73-74; from asst prof to assoc prof, 74-85, PROF PHYSIOL, UNIV CALIF, SAN FRANCISCO, 85- *Concurrent Pos:* Vis scientist biochem, Nat Polytech Inst, 73. *Mem:* Soc Neurosci; Soc Gen Physiol. *Res:* Molecular mechanisms of ion transport; mechanisms of function of Rhodopsins and phototransduction; photoreceptor development. *Mailing Add:* Dept Physiol Univ Calif Sch Med 513 Parnassus Ave San Francisco CA 94143

KORENMAN, STANLEY G, b New York, NY, Jan 21, 33; m 56; c 3. ENDOCRINOLOGY, BIOCHEMISTRY. *Educ:* Princeton Univ, AB, 54; Columbia Univ, MD, 58. *Prof Exp:* Intern, Second Div, Bellevue Hosp & Mem Hosp, New York, 58-59, asst resident med, 50-61; clin assoc, Endocrinol Br, Nat Cancer Inst, 61-63, med officer & sr investr, 63-66; from asst prof to assoc prof med, Sch Med, Univ Calif, 66-70; prof med & biochem & chief endocrinol div, Sch Med, Univ Iowa, 70-74; prof med & chmn dept, 74-89, ASSOC DEAN, SAN FERNANDO VALLEY PROG, UNIV CALIF, LOS ANGELES, 81-, ASSOC DEAN EDUC DEVELOP, SCH MED, 89- *Concurrent Pos:* Collab investr, Lab Chem Biol, Inst Arthritis & Metab Dis, 64-65; clin instr, Med Ctr, George Washington Univ, 65-66; coordr regional med prog, Dept Med, Harbor Gen Hosp, 68-70, dir clin res ctr, 69-70; mem, Reproductive Biol Study Sect, 70-73; mem, Breast Cancer Task Force, 72; chmn dept med, San Fernando Valley Prog; chief med, Vet Admin Med Ctr, Sepulveda. *Mem:* Am Soc Clin Invest; fel Am Col Physicians; Am Fedn Clin Res; Endocrine Soc; Asn Am Physicians. *Res:* Molecular mechanisms of hormone action; clinical reproductive endocrinology and impotence. *Mailing Add:* 924 Westwood Blvd Los Angeles CA 90024

KORENMAN, VICTOR, b Brooklyn, NY, Feb 5, 37; m 68; c 1. THEORETICAL CONDENSED MATTER PHYSICS. *Educ:* Princeton Univ, AB, 58; Harvard Univ, MA, 59, PhD(physics), 66. *Prof Exp:* Res assoc, 65-67, from asst prof to assoc prof, 67-79, PROF PHYSICS, UNIV MD, 79- *Concurrent Pos:* Fel, Alfred P Sloan Found, 71. *Mem:* AAAS; Am Phys Soc; Fedn Am Scientists. *Res:* Theory of itinerant ferromagnetism. *Mailing Add:* Dept Physics Univ Md College Park MD 20742-4111

KORENSTEIN, RALPH, b Havannah, Cuba, Dec 6, 51; US citizen; m 76. INORGANIC CHEMISTRY, SOLID STATE CHEMISTRY. *Educ:* Polytech Inst Brooklyn, BS, 73; Brown Univ, PhD(chem), 77. *Prof Exp:* Mem tech staff chem, Tex Instruments Inc, 76-; AT RES DIV, RAYTHEON CO, LEXINGTON, MASS. *Mem:* Am Chem Soc. *Res:* Crystal growth of oxides by liquid phase epitaxy; synthesis of new inorganic compounds; thin film technology. *Mailing Add:* Res Div Raytheon Co 131 Spring St Lexington MA 02173

KORETZ, JANE FAITH, b New York, NY, Aug 12, 47. IMAGE ANALYSIS, COMPUTER MODELING. *Educ:* Swarthmore Col, BA, 69; Univ Chicago, PhD(biophysics), 74. *Prof Exp:* Adj asst prof human physiol, Kean Col, NJ, 77; from asst prof to assoc prof, 77-90, PROF, DEPT BIOL, RENSSELAER POLYTECH INST, 90-; ADJ ASSOC PROF, SCH PUB HEALTH, STATE UNIV NY, ALBANY, NY, 88- *Concurrent Pos:* Vis scientist, Cell Biophysics Unit, Med Res Coun, London, 74-76; res affil, dept physiol, NJ Med Sch, 76-77; Fulbright scholar, Univ Oxford, 90-91; vis prof, Open Univ, 91. *Honors & Awards:* Henry Fukui Mem Travel Award, 89. *Mem:* Biophys Soc (coun, 88-91); Inst Elec & Electronic Engrs; Optical Soc Am; Sigma Xi; Asn Women Sci; Asn Res Vision & Ophthal; Am Soc Biochem & Molecular Biol; Int Soc Eyes Res. *Res:* Characterization of native reconstituted l-crystallin assemblies; computer-based modelling of human and rhesus monkey visual accommodation; etiology of presbyopia. *Mailing Add:* Ctr Biophys & Dept Biol Rensselaer Polytech Inst Troy NY 12180-3590

KOREVAAR, JACOB, b Netherlands, Jan 25, 23; nat US; c 8. MATHEMATICS. *Educ:* Univ Leiden, PhD(math), 49. *Hon Degrees:* Dr, Univ Gothenburg, 78. *Prof Exp:* Asst math, Delft Univ Technol, 44-46, prof, 51-53; res assoc, Math Ctr, Univ Amsterdam, 47-49; from asst prof to prof, Univ Wis, 53-64; chmn dept, Univ Calif, San Diego, 71-73, prof math, 64-74; PROF, MATH INST, UNIV AMSTERDAM, 74-, DIR, 80- *Concurrent Pos:* Mem Nat Sci Found fel comt, 64-66; vis prof math, Univ Amsterdam, 74-76. *Honors & Awards:* Reynolds Teaching Award, 58. *Mem:* AAAS; Am Math Soc; London Math Soc; Math Asn Am; Soc Indust & Appl Math. *Res:* Approximation; complex analysis; distributions; Fourier analysis; Tauberian theorems. *Mailing Add:* Fac Math & Comput Sci Univ Amsterdam Amsterdam Netherlands

KORF, RICHARD E, b Geneva, Switz, Dec 7, 56; US citizen. ARTIFICIAL INTELLIGENCE, HEURISTIC SEARCH. *Educ:* Mass Inst Technol, BS, 77; Carnegie Mellon Univ, MS, 80, PhD(computer sci), 83. *Prof Exp:* Asst prof computer sci, Columbia Univ, 83-85; asst prof, 85-88, ASSOC PROF COMPUTER SCI, UNIV CALIF, LOS ANGELES, 88- *Concurrent Pos:* Int Bus Mach fac develop award, 85; NSF presidential young investr, 86-; assoc ed, Inst Elec & Electronics Engrs Pattern Anal & Mach Intel, 87-89. *Mem:* Am Asn Artificial Intel. *Res:* High-level problem solving in artificial intelligence; algorithms. *Mailing Add:* Computer Sci Dept Univ Calif Los Angeles CA 90024

KORF, RICHARD PAUL, b Bronxville, NY, May 28, 25; m 59; c 4. MYCOLOGY, TAXONOMY. *Educ:* Cornell Univ, BSc, 46, PhD(mycol), 50. *Prof Exp:* Asst plant path, Cornell Univ, 47-50, from asst prof to assoc prof, 51-60, prof & chmn, Theatre Arts, 85-86, PROF MYCOL, CORNELL UNIV, 60-, PROF BOT, 82- *Concurrent Pos:* Lectr, Glasgow Univ, 50-51; Fulbright res prof & NSF sr fel, Yokohama Nat Univ, Japan, 57-58; chmn, Nomenclature Secretariat, Int Mycol Asn, 72-77; co-ed, Mycotaxon J, 74-91; adj prof, Copenhagen Univ, 78. *Mem:* Mycol Soc Am (secy-treas, 65-68, vpres, 68-69, pres, 70-71); Brit Mycol Soc; Mycol Soc France; Mycol Soc Japan; Int Asn Plant Taxon. *Res:* Taxonomic mycology; taxonomy of discomycetes; life histories and genetics of ascomycetes; botanical nomenclature; fungi of Asia, Neotropics and Macaronesia. *Mailing Add:* Plant Path Herbarium Cornell Univ Ithaca NY 14850

KORFHAGE, ROBERT R, b Syracuse, NY, Dec 2, 30; m 55; c 4. MATHEMATICS, COMPUTER SCIENCE. *Educ:* Univ Mich, BSE, 52, MS, 55, PhD(math), 62. *Prof Exp:* Engr comput res lab, United Aircraft Corp, 52-54; asst prof math, NC State Col, 60-62; from asst prof to assoc prof math & comput sci, Purdue Univ, 62-70; dir, Comput Sci/Opers Res Ctr, 70-72, actg chmn, Comput Sci & Eng, 81-83, prof comput sci, Southern Methodist Univ, 70-87; chmn dept, 87-89, PROF INFO SCI, UNIV PITTSBURGH, 87- *Concurrent Pos:* Consult, Proj Comput in Eng Educ, Univ Mich, 62, Eli Lilly & Co, 64-66, Indianapolis Hosp Develop Asn, 65-66, Los Alamos Sci Lab, 65-76, Alpha Systs, Inc, 70-72, Xerox Corp, 77-78, On-Line Data, Inc, 78-84 & IBM, 82-; Fulbright prof, 73, 75. *Mem:* Sigma Xi; Asn Comput Mach; Am Soc Info Sci; Inst Elect & Electronic Engrs Comput Soc. *Res:* Finite mathematical structures; logic and algorithms; non-numeric uses of computers; information retrieval; library information systems; graph theory; visual languages. *Mailing Add:* DIS Sch Libr & Info Sci Univ Pittsburgh Pittsburgh PA 15260

KORFMACHER, WALTER AVERILL, b St Louis, Mo, Nov 6, 51; m 74; c 2. MASS SPECTROMETRY, ENVIRONMENTAL ANALYSIS. *Educ:* St Louis Univ, BSCh, 73; Univ Ill, Urbana, MS, 75, PhD(anal chem), 78. *Prof Exp:* Teaching asst, Chem Div, Univ Ill, Urbana, 73-75, res asst, 75; res asst, Chem Div, Colo State Univ, Ft Collins, 76-78; CHEMIST, NAT CTR TOXICOL RES, FOOD & DRUG ADMIN, 78- *Mem:* Sigma Xi; Am Chem Soc; AAAS; Am Soc Mass Spectrometry; Soc Appl Spectros. *Res:* Environmental analytical methods, particularly in trace organic quantitative methods; development of capillary gas chromatography combined with atmospheric pressure ionization mass spectrometry and chemical ionization mass spectrometry; thermospray mass spectometry and tandem mass spectrometry. *Mailing Add:* Div Biochem Toxicol Nat Ctr Toxicol Res Jefferson AR 72079

KORGEN, BENJAMIN JEFFRY, b Duluth, Minn, Jan 6, 31; m 59; c 3. PHYSICAL OCEANOGRAPHY. *Educ:* Univ Minn, BS, 56; Univ Mich, MA, 58; Ore State Univ, PhD(phys oceanog), 69. *Prof Exp:* Asst prof phys oceanog, Univ N C Chapel Hill, 69-74; writing & consult, 74-78; OCEANOGR, US NAVAL OCEANOG OFF, 78- *Concurrent Pos:* Adj assoc prof, Tulane Univ, 78- *Mem:* AAAS; Am Geophys Union; Am Soc Limnol & Oceanog; Geol Soc Am; Oceanography Soc. *Res:* Seiches and related phenomena; near-bottom processes; interdisciplinary estuarine and harbor studies; terrestrial heat flow; marine geophysics. *Mailing Add:* 219 Loop Dr Slidell LA 70458

KORGES, EMERSON, b Victoria, Tex, Aug 6, 11; m 42; c 1. ELECTRICAL ENGINEERING, PHYSICS. *Educ:* Tex Col Arts & Indust, BS, 31, MS, 42; Colo State Univ, MEE, 57. *Prof Exp:* Instr eng, 42-47, from asst prof to assoc prof, 47-57, PROF ELEC ENG, TEX A&I UNIV, 57- *Concurrent Pos:* Consult, Tex A&I Univ, 57- *Mem:* Am Soc Eng Educ; Inst Elec & Electronics Engrs. *Res:* Corrosion and performance of copper-to-aluminum and aluminum-to-aluminum non tension electrical connectors. *Mailing Add:* Dept Elec Eng Emer Tex A&I Univ PO Box 2275 Kingsville TX 78363

KORIN, AMOS, b Rehovoth, Israel, Sept 11, 44; m 67; c 3. MEMBRANE SCIENCE, SEPARATION TECHNOLOGY. *Educ:* Technion Israel Inst Technol, BSc, 67; Weizmann Inst Sci, MSc, 72, PhD(polymer chem), 78. *Prof Exp:* Sr proj mgr, Israel Atomic Energy Comn, 67-73; proj engr, Weizmann Inst Sci, 73-78; dept head water distillation, Mehorot Water Co, 78-79; dir membrane develop, Gelman Sci, Inc, 79-83; independent consult, 83-85; dir technol, Stan Ohio, 85-87; DIR TECHNOL, W R GRACE, 87- *Concurrent Pos:* Lectr chem & physics, Col Eng, Tel Aviv, 73-77; consult, Amplast Co, Israel, 78-79. *Mem:* Israel Inst Chem Eng; Am Chem Soc; Filtration Soc. *Res:* Novel polymer systems; polymeric ultra filtration and microporous membranes; separation systems and their application in industrial and medical fields. *Mailing Add:* 16 Mountain View Weston CT 06883

KORIN, BASIL PETER, b Oxford, Conn, Sept 15, 32; m 59; c 1. MATHEMATICAL STATISTICS. *Educ:* Univ Conn, BA, 57; Stanford Univ, MS, 60; George Washington Univ, PhD(statist), 67. *Prof Exp:* Mathematician, Lockheed Missiles & Space Co, 57-60; math statistician, US Bur Census, 60-61; from instr to assoc prof math & statist, 61-74, PROF MATH & STATIST, AMERICAN UNIV, 74- *Mem:* Inst Math Statist; Am Statist Asn; Biometric Soc. *Res:* Statistics; multivariate analysis. *Mailing Add:* Dept Math & Statist Am Univ Washington DC 20016

KORINEK, GEORGE JIRI, b Jicin, Czech, July 8, 27; m 58. PHYSICAL CHEMISTRY. *Educ:* Univ BC, MSc, 54, PhD(metall), 56. *Prof Exp:* Fel, Nat Res Coun Can, 56-57; proj leader, Metals Res Lab, Union Carbide Metals Co, 57-61; group leader, Rare Metals Div, Ciba Ltd, Switz, 61-65, mgr, Rare Metals Dept, Ciba Corp, 65-70; managing dir, 70-74, PRES, H C STARCK, INC, 74- *Mem:* Am Chem Soc; Electrochem Soc; Am Soc Metals; Am Inst Mining, Metall & Petrol Eng. *Res:* Physical chemistry of extractive metallurgy; catalysis; hydrometallurgy; refractory metals; Ta capacitors. *Mailing Add:* H C Starck Inc 280 Park Ave New York NY 10017

KORITALA, SANBASIVAROA, b India, Apr 10, 32; nat US; m 61; c 2. LIPID CHEMISTRY. *Educ:* Andhra Univ, India, BS, 52; Nagpur Univ, BS, 54; Ohio State Univ, MS, 57, PhD(physiol chem), 60. *Prof Exp:* Res assoc & instr physiol chem, Ohio State Univ, 61-63; RES CHEMIST, NORTHERN REGIONAL RES CTR, AGR RES SERV, USDA, 63- *Concurrent Pos:* Abstr ed, Am Oil Chemists' Soc, 75- *Mem:* Am Chem Soc; Am Oil Chemists' Soc. *Res:* Selective hydrogenation of vegetable oils to improve their flavor stability and to modify their functional properties for application in margarines, cooking and other food uses; preparations of selective and active catalysts for hydrogenation of animal and vegetable oils. *Mailing Add:* Nat Ctr Agr Utilization Res Agr Res Serv USDA Peoria IL 61604

KORITNIK, DONALD RAYMOND, b Rock Springs, Wyo, Feb 28, 46; m 80; c 2. REPRODUCTIVE ENDOCRINOLOGY, ATHEROSCLEROSIS. *Educ:* Univ Wyo, BS, 68, MS, 73, PhD(animal sci), 77. *Prof Exp:* Fel, Reprod Endocrinol Prog, Univ Calif, San Francisco, 77-80; ASST PROF COMP MED, BOWMAN GRAY SCH MED, WAKE FOREST UNIV, 80- *Mem:* Endocrine Soc; Soc Study Reprod. *Res:* Regulation of lipid and carbohydrate metabolism by reproductive hormones; binding of steroid hormones to serum binding proteins and steroid receptors during pregnancy, puberty and contraceptive steroid usage; endocrine risk factors in atherosclerosis. *Mailing Add:* 1850 Fannwood Circle Winston-Salem NC 27107

KORITZ, GARY DUANE, b DeKalb, Ill, May 18, 44; m 68; c 2. VETERINARY MEDICINE, PHARMACOLOGY. *Educ:* Univ Ill, Urbana, BS, 66, DVM, 68, PhD(vet pharmacol), 75. *Prof Exp:* Clinician pvt pract, Dundee Animal Hosp, 68-70; from asst prof to assoc prof, 759-85, PROF PHARMACOL, COL VET MED, UNIV ILL, URBANA, 85- *Concurrent Pos:* NIH fel, Univ Ill, Urbana, 73-74. *Mem:* Am Vet Med Asn; Am Col Vet Toxicol; Am Acad Vet Pharmacol & Therapeut; Am Soc Vet Physiologists & Pharmacologists. *Res:* Comparative pharmacology including therapeutics, pharmacokinetics and drug disposition. *Mailing Add:* Dept Vet Biosci Univ Ill 3619 Vet Basic Sci Bldg Urbana IL 61801

KORITZ, SEYMOUR BENJAMIN, b Boston, Mass, Nov 25, 21; m 51; c 4. BIOCHEMISTRY. *Educ:* Univ Mass, BS, 44; Univ Wis, PhD(biochem), 51. *Prof Exp:* Res assoc, Ohio State Univ, 52-53; staff biochemist, Worcester Found Exp Biol, 53-59; from asst prof to prof biochem, Sch Med, Univ Pittsburgh, 59-68; PROF BIOCHEM, MT SINAI SCH MED, 68- *Concurrent Pos:* Am Cancer Soc fel, Brussels, 51-52. *Mem:* Am Soc Biol Chemists. *Res:* Mode of action of adrenocorticotropic hormone. *Mailing Add:* Dept Biochem Mt Sinai Sch Med Fifth Ave at 100th St New York NY 10029

KORKEGI, ROBERT HANI, b Milan, Italy, Dec 3, 25; m 46; c 2. AEROSPACE VEHICLES, SPACE SYSTEMS. *Educ:* Lehigh Univ, BS, 49; Calif Inst Technol, MS, 50, PhD(aerospace), 54. *Prof Exp:* Res assoc, Eng Ctr, Univ Southern Calif, 54-57; tech dir teaching & res, von Karman Inst Fluid Dynamics, 57-64; dir res & tech admin, Hypersonic Res Lab, USAF, 64-76 & int res & develop, Adv Group Aerospace Res & Develop, NATO, 76-79; vis prof teaching & res, George Washington Univ, 79-81; dir, Aerospace Bd, Nat Res Coun, Nat Acad Sci, 81-90; CONSULT, AEROSPACE RES & DEVELOP, 91- *Concurrent Pos:* Mem, Adv Subcomt Fluid Mech, NASA, 67-71, US mem, Fluid Dynamics Panel, AGARD/NATO, 69-76 & bd dir, von Karman Inst Fluid Dynamics, 76-79; mem, Sci Comt Study Group, 81 & Steering Comt, Space Explor Study, Am Inst Aerospace & Astronaut, 90; mem, Study Comt, Hypersonic Technol, Nat Res Coun, 87-88 & Aeronaut Technol, 90-92; prof, Univ Md, College Park, 90- *Honors & Awards:* Pub Serv Medal, NASA, 88. *Mem:* Fel Am Inst Aeronaut & Astronaut. *Res:* Aerodynamics of supersonic and hypersonic flows including two-and three-dimensional shock interactions with viscous flows; supersonic and hypersonic ground test facilities; aerospace policy issues; author of 60 publications. *Mailing Add:* 4418 Springdale St NW Washington DC 20016

KORMAN, N(ATHANIEL) I(RVING), b Providence, RI, Feb 23, 16; m 41; c 2. ELECTRONICS, COMPUTER SCIENCE. *Educ:* Worcester Polytech Inst, BS, 37; Mass Inst Technol, MS, 38; Univ Pa, PhD(elec eng), 58. *Prof Exp:* Student engr, RCA Corp, 38-40, develop engr, 40-44, eng group leader, 44-48, eng group supvr, 48-50, adminr radar systs activities, 50-52, mgr develop eng group, 52-54, mgr systs eng group, 54, asst chief engr, 54-56, chief systs, 57-58, dir advan mil systs, 58-63, dir tech progs, 63-65, chief engr graphic systs div, 65-66, dir med electronics plans & progs, 66-69; PRES, VENTURES RES & DEVELOP GROUP, 69- *Mem:* Fel Inst Elec & Electronics Engrs. *Res:* Advanced development of frequency modulation transmitters; microwave and waveguide components for radar and television; development of microwave studio-to-transmitter link; fire control radar; frequency modulation techniques; waveguide techniques; systems engineering; color science; patentee in field. *Mailing Add:* 108 Yucca Lane Placitas NM 87043

KORMENDY, JOHN, b Graz, Austria, June 13, 48; Can citizen; m 87. ASTRONOMY. *Educ:* Univ Toronto, BSc, 70; Calif Inst Technol, PhD(astron), 76. *Prof Exp:* Parisot fel astron, Univ Calif, Berkeley, 76-78; sr vis fel, Inst Astron, Cambridge, 78 & 80; staff mem, Dominion Astrophys Observ, 80-89; AT UNIV HAWAII, 90- *Concurrent Pos:* Fel astron, Kitt Peak Nat Observ, 78-79; sr fel, Inst Astron, Univ Cambridge, Eng, 78-80. *Honors & Awards:* Muhlmann Prize, Astron Soc Pac, 88. *Mem:* Am Astron Soc; Int Astron Union; Astron Soc Pac; Royal Astron Soc; Can Astron Soc. *Res:* Extragalactic observational astronomy, with particular emphasis on the structure of normal and peculiar galaxies; theoretical dynamics of the structure of galaxies; astronomical image processing. *Mailing Add:* Inst Anat Univ Hawaii 2680 Woodlawn Dr Honolulu HI 96822

KORN, ALFRED, b Long Island City, NY, July 19, 30. STRUCTURAL & CIVIL ENGINEERING. *Educ:* Purdue Univ, BS, 52; Univ Ill, Urbana, MS, 61; Wash Univ, DSc(appl mech, struct), 67. *Prof Exp:* Designer, Bell Aircraft Corp, NY, 52-53; engr, Sverdrup & Parcel Eng Co, Mo, 55-59 & 61-63; lectr, Washington Univ, 63-66; asst prof civil eng, Univ Ky, 67-69; from assoc prof to prof, 69-90, EMER PROF CIVIL ENG, SOUTHERN ILL UNIV, EDWARDSVILLE, 90- *Mem:* Am Concrete Inst; Am Soc Civil Engrs; Int Asn Bridge & Struct Eng; Prestressed Concrete Inst. *Res:* Structural mechanics; elastic and inelastic frame stability; plastic design; numerical and computer analysis of structures; structural optimization. *Mailing Add:* Dept Cvil Eng Sch Eng Southern Ill Univ Edwardsville IL 62026

KORN, DAVID, b Providence, RI, Mar 5, 33; m 55; c 3. PATHOLOGY, MOLECULAR BIOLOGY. *Educ:* Harvard Univ, BA, 54, MD, 59. *Prof Exp:* Res assoc biochem, Nat Inst Arthritis & Metab Dis, 61-63, staff mem, 63-68, staff pathologist, NIH, 64-68; prof path & chmn dept, 68-84, DEAN, SCH MED, STANFORD UNIV, 84-, VPRES, 86- *Concurrent Pos:* chmn, Nat Cancer Adv Bd, 84- *Mem:* AAAS; Am Soc Biochem & Molecular Biol; Asn Am Path; Am Soc Cell Biol; Int Acad Path. *Res:* Biochemistry; nucleic acid biochemistry; regulation of gene expression. *Mailing Add:* Sch Med Stanford Univ 17121 Medical Ctr Stanford CA 94305

KORN, EDWARD DAVID, b Philadelphia, Pa, Aug 3, 28; m 50; c 2. BIOCHEMISTRY, CELL BIOLOGY. *Educ:* Univ Pa, AB, 49, PhD(biochem), 54. *Prof Exp:* Asst instr, Dept Physiol Chem, Univ Pa, 49-51, Harrison fel, 51-52, Damon Runyon fel, 52-53; Damon Runyon fel, Lab Cellular Physiol & Metab, Nat Heart Inst, 53-54, asst scientist & sr asst scientist, Nat Heart & Lung Inst, 54-56, res chemist, Lab Cellular Physiol & Metab & Lab Biochem, Sect Cellular Physiol, 56-69, head, Lab Biochem, Sect Biochem & Ultrastruct, Nat Heart & Lung Inst, dep sci dir, 82-88, actg sci dir, 88-89, CHIEF, LAB CELL BIOL & HEAD, SECT CELLULAR BIOCHEM & ULTRASTRUCT, NAT HEART, LUNG & BLOOD INST, 74-, SCI DIR, 89- *Concurrent Pos:* Vis scientist, Biochem Dept, Cambridge Univ, Eng, 58-59; fac, FAES Grad Prog, NIH, 66-76; prof, FAES Grad Prog, Johns Hopkins Univ, 66-77; vis scientist, Inst Animal Physiol, Cambridge, Eng, 69-70; assoc ed, J Biol Chem, 77-; mem bd dirs, Found Advan Educ Sci, 77-, treas, 80-82, vpres, 82-84, pres, 84-86; rep, Am Soc Biol Chemists to AAAS, 84-; mem, Centennial Comt, NIH, 87, Educ Comt, 89-, H-1 Waiver Comt, 89-, AIDS Loan Repayment Comt, 89, Facil Planning Group, 89-; chair, Comt Guidelines Conduct Sci Res, 90. *Honors & Awards:* Super Serv Award, USPHS, 80; Mider Lectr, 85; Presidential Meritorious Exec Rank Award, 87. *Mem:* Nat Acad Sci; Am Soc Biochem & Molecular Biol; Biophys Soc; Am Soc Cell Biol. *Res:* Pinocytosis and phagocytosis; cell motility; cytoplasmic actin and myosin; author of 10 books and author or co-author of over 240 publications. *Mailing Add:* Lab Cell Biol Nat Heart Lung & Blood Inst Bethesda MD 20892

KORN, GRANINO A(RTHUR), b Berlin, Ger, May 7, 22; nat US; m 48; c 2. COMPUTER SCIENCE, ELECTRICAL ENGINEERING. *Educ:* Brown Univ, BA, 42, PhD(physics), 48; Columbia Univ, MA, 42. *Prof Exp:* Proj engr, Sperry Gyroscope Co, 47-48; head anal group, Airplane Div, Curtiss-Wright Corp, 48-49; staff engr, Lockheed Aircraft Co, 49-52; INDUST CONSULT, 52-; prof elec eng, Univ Ariz, 57-83; GA & TM Korn Industrial Consults, 84- *Concurrent Pos:* Consult, Nat Acad Sci, Chile, 61; mem, NIH-Nat Adv Res Coun, 78-79. *Honors & Awards:* Sr Sci Award, Soc Comput Simulation, 68; Humboldt Prize, Humboldt Found, WGer, 76. *Mem:* Int Asn Math-Simulation; fel Inst Elec & Electronics Engrs; Soc Comput Simulation. *Res:* Desire computer systems for simulation; microdare desire laboratory-automation software; mini-microcomputer system design; DESIRE simulation system; neural-network simulations. *Mailing Add:* 6801 Opatas St Tucson AZ 85715

KORN, JOSEPH HOWARD, b Augsburg, Ger, Jan 31, 47; US citizen; m 71; c 4. RHEUMATOLOGY, CELL BIOLOGY. *Educ:* City Col NY, BS, 68; Columbia Univ, MD, 72. *Prof Exp:* Asst prof med & immunol, Med Univ SC, 77-78; from asst prof to assoc prof, 78-89, PROF MED, SCH MED, UNIV CONN, 89- *Concurrent Pos:* Vis prof, Weitzmann Inst Sci, Israel, 85-86; assoc chief staff res & develop, Vet Admin Med Ctr, Newington, Conn, 85- *Mem:* Am Asn Immunologists; Am Col Rheumatology; Am Soc Clin Invest; NY Acad Sci; AAAS. *Res:* Immunobiology of connective tissue; pathogenesis of scleroderma; fibroblast biology. *Mailing Add:* Vet Admin Med Ctr Bldg 5 555 Willard Ave Newington CT 06111

KORN, ROY JOSEPH, b Chicago, Ill, July 25, 20; m 55; c 4. MEDICAL ADMINISTRATION. *Educ:* Northwestern Univ, BS, 42, MD, 46. *Prof Exp:* Intern med, Wesley Mem Hosp, Chicago, Ill, 45-46; resident internal med, Vet Admin Hosp, Hines, Ill, 49-52; staff physician, Vet Admin Hosp, Omaha, Nebr, 52-53; from asst chief to chief med serv, West Side Vet Admin Hosp, Chicago, Ill, 53-62; adv & prof med, Chiengmai Med Sch, Thailand, 62-64; chief of staff, Vet Admin Hosp, Indianapolis, 65-72; PROF MED, ABRAHAM LINCOLN SCH MED, UNIV ILL, 72-; CHIEF OF STAFF, VET ADMIN WEST SIDE HOSP, CHICAGO, 72- *Concurrent Pos:* Instr, Univ Nebr, 52-53; from asst prof to assoc prof, Col Med, Univ Ill, 55-64; prof, Chicago Med Sch, 64-65; clin prof med, Sch Med, Ind Univ-Purdue Univ, Indianapolis, 65-72. *Mem:* Fel Am Col Physicians. *Res:* Liver disease. *Mailing Add:* 516 N Lincoln Hinsdale IL 60521

KORNACKER, KARL, b Chicago, Ill, Oct 14, 37; m 60, 79; c 2. COGNITIVE SCIENCE. *Educ:* Mass Inst Technol, BS, 58, PhD(neurophysiol), 62. *Prof Exp:* Res assoc biol, Mass Inst Technol, 62-68; asst prof biophys, 68-69, ASSOC PROF, OHIO STATE UNIV, 69- *Mem:* AAAS; Cognitive Sci Soc. *Res:* Cognitive physiology. *Mailing Add:* Genetics 963 Biol Sci Bldg Ohio State Univ Main Campus Columbus OH 43210

KORNBERG, ARTHUR, b Brooklyn, NY, Mar 3, 18; m 43; c 3. BIOCHEMISTRY. *Educ:* City Col New York, BS, 37; Univ Rochester, MD, 41. *Prof Exp:* Intern, Strong Mem Hosp, Rochester, 41-42; asst surgeon to med dir, Nat Inst Arthritis & Metab Dis, NIH, 42-53, chief enzyme & metab sect, 47-53; prof microbiol & head dept, Sch Med, Wash Univ, 53-59; prof, 59-88, chmn dept, 59-69, EMER PROF, DEPT BIOCHEM, STANFORD UNIV SCH MED, 88- *Concurrent Pos:* Mem bd gov, Weizmann Inst; sci adv, Div Schering-Plough, Inc, DNAX Res Inst Molecular & Cellular Biol, Regeneron Pharmaceuticals; mem bd dirs, Xoma Corp. *Honors & Awards:* Nobel Prize in Med & Physiol, 59; Paul-Lewis Award, Am Chem Soc, 51; Nat Medal Sci, Royal Soc, 79. *Mem:* Nat Acad Sci; Am Philos Soc. *Res:* Enzymatic studies of DNA replication, membrane biochemistry. *Mailing Add:* Dept Biochem Sch Med Stanford Univ Stanford CA 94305

KORNBERG, FRED, b Lemberg, Poland, Jan 28, 36; US citizen; m 58; c 3. ELECTRONICS ENGINEERING. *Educ:* NY Univ, BSEE, 58, MSEE, 59. *Prof Exp:* Teacher, Col Eng, NY Univ, 58-59; staff scientist, Res Div, NY Univ, 58-59; vpres, Radio Eng Labs, Dynamics Corp Am, 59-69 & Nardcom Corp, 69-71; PRES & CHIEF EXEC OFFICER, COMTECH TELECOMMUN CORP, 71-, TECHNETRONIC DATA SYSTS INC, 85- & OCTAGON COMMUN CORP, 90- *Mem:* Sr mem Inst Elec & Electronics Engrs; sr mem Armed Forces Commun Eng Asn. *Mailing Add:* 17 Palatine Ct Syosset NY 11791

KORNBERG, ROGER DAVID, b St Louis, Mo, Apr 24, 47; m; c 1. BIOCHEMISTRY. *Educ:* Harvard Univ, BA, 67; Stanford Univ, PhD(chem), 71. *Prof Exp:* Mem sci staff cell biol, Med Res Coun Lab Molecular Biol, Cambridge, Eng, 74-75; asst prof biol chem, Harvard Med Sch, 76-78; PROF CELL BIOL, SCH MED, STANFORD UNIV, 78-, CHMN, 84- *Honors & Awards:* Eli Lilly Award, 80; Passano Award, 81. *Res:* Structure and transcription of chromosomes. *Mailing Add:* Dept Cell Biol Fairchild D123 Stanford Univ Stanford CA 94305

KORNBERG, THOMAS B, b Washington, DC, Nov 10, 48. CELL BIOLOGY, DEVELOPMENTAL BIOLOGY. *Educ:* Columbia Col, New York, BA, 70; Columbia Univ, PhD(biochem), 73. *Prof Exp:* Res assoc biochem, Princeton Univ, 73-75; res assoc develop biol, Med Res Coun Lab Molecular Biol, Cambridge Univ, UK, 75-76; mem staff, Molecular Biol Inst, Univ Calif, Los Angeles, 76-77; ASST PROF BIOCHEM & BIOPHYS, UNIV CALIF, SAN FRANCISCO, 78- *Res:* Genetic and biochemical description of the cellular events which govern determination in higher organisms. *Mailing Add:* Dept Biochem & Biophys Univ Calif Med Sch 513 Parnassus Ave San Francisco CA 94143

KORNBLITH, CAROL LEE, b Chicago, Ill, Sept 6, 45. PHYSIOLOGICAL PSYCHOLOGY, NEUROBIOLOGY. *Educ:* Univ Mich, AB, 66, AM, 68; Calif Inst Technol, PhD(biol), 72. *Prof Exp:* Fel psychol, Princeton Univ, 72-74; asst prof, Univ NC, 74-80, interdisciplinary fel neurosci, 80-81; assoc prof psychol, Ill State Univ, 81-83; assoc prof psychol, Oberlin Col, 83-84; MED ED, SECT PUBLS, MAYO FOUND, 84- *Mem:* Soc Neurosci; Sigma Xi; AAAS. *Res:* Development and function of sexually dimorphic brain regions in the rat as revealed by deoxyglucose autoradiography and the development of feeding behavior and its relation to reinforcement. *Mailing Add:* 15200 NW Eighth Ave Rochester MN 55901

KORNBLITH, LESTER, b Chicago, Ill, Apr 27, 17; m 40; c 3. NUCLEAR ENGINEERING. *Educ:* Mass Inst Technol, SB, 38. *Prof Exp:* Chief engr, Enrico Fermi Inst Nuclear Studies, Univ Chicago, 47-55; mgr reactor tech oper, Vallecitos Atomic Lab, Gen Elec Co, 56-63; asst dir reactors, Div Compliance, US Atomic Energy Comn, 63-67, asst dir tech progs, 67-72, mem atomic safety & licensing bd panel, US Nuclear Regulatory Comn, 72-79; vpres & prin engr, Nat Nuclear Corp, 79-86; CONSULT ENGR, 79- *Mem:* Fel Am Nuclear Soc; Inst Elec & Electronics Engrs. *Res:* Design, construction and operation of nuclear reactors and accelerators. *Mailing Add:* 1611 Clover Creek Dr Apt 136 Sarasota FL 34231-8922

KORNBLUM, NATHAN, b New York, NY, Mar 22, 14; m 47; c 4. ORGANIC CHEMISTRY. *Educ:* NY Univ, BS, 35, MS, 37; Univ Ill, PhD(org chem), 40. *Prof Exp:* Hall res instr, Oberlin Col, 40-42; Nat Res Coun fel chem, Harvard Univ, 42-43; from asst prof to assoc prof, 43-53, PROF CHEM, PURDUE UNIV, 53- *Concurrent Pos:* Fulbright sr scholar, Univ Col, London, 52-53; Guggenheim mem fel, Swiss Fed Inst Technol, 53, NSF fel, 64-65, vis prof org chem, 73-74; vis prof, Nat Ctr Sci Res, Thias, France, 75, Univ Marseilles, 78 & Univ Kyoto, Japan, 79, 85. *Honors & Awards:* Alexander von Humboldt Award, 88. *Mem:* Am Chem Soc. *Res:* Reaction mechanisms and their application to synthetic organic chemistry. *Mailing Add:* Dept Chem Purdue Univ West Lafayette IN 47907

KORNBLUM, RONALD NORMAN, b Chicago, Ill, Dec 5, 33. PATHOLOGY. *Educ:* Univ Calif, Los Angeles, BA, 55, MD, 59. *Prof Exp:* Resident gen path, Santa Clara County Hosp, 62-66; resident neuropath, Md Dept Ment Hyg, 66-67; fel forensic path, Md Postmortem Exam, 67-68; asst med examr, state med exam off, Md, 68-73; MED EXAMR, VENTURA COUNTY, CALIF, 73- *Concurrent Pos:* Lectr pub health admin, Johns Hopkins Univ, 69- *Mem:* Am Acad Forensic Sci; Am Soc Clin Pathologists; Col Am Pathologists. *Res:* Forensic pathology; investigation into causes of sudden death in infancy syndrome; investigation of craniocerebral injuries and shock in relation to cerebral anoria. *Mailing Add:* 1104 N Mission Los Angeles CA 90033

KORNBLUM, SAUL S, b Far Rockaway, NY, Feb 24, 34; m 58; c 2. PHYSICAL PHARMACY, PHYSICAL CHEMISTRY. *Educ:* Brooklyn Col Pharm, BS, 55; Columbia Univ, MS, 57; Rutgers Univ, PhD(pharm, phys chem), 63. *Prof Exp:* Instr chem, Newark Col Eng, 59-61; asst prof phys pharm, Brooklyn Col Pharm, Long Island Univ, 62-66; sr scientist, Sandoz Pharmaceut, 66-67, mgr, 67-73, assoc sect head prod develop & clin prod, Sandoz Inc, 73-85; PRES, S S KORNBLUM ASSOCS, 85- *Concurrent Pos:* CIBA res grant, 61-62. *Honors & Awards:* Lunsford Richardson Award, 63. *Mem:* Am Acad Pharmaceut Sci; NY Acad Sci; Parenteral Drug Asn. *Res:* Solid state kinetics; dissolution of poorly water-soluble drugs; sustained-release dosage forms; pharmaceutical dosage form design and evaluation; preformulation stability evaluation for new drugs; troubleshooting and reformulation. *Mailing Add:* 144 Short Hills Ave Springfield NJ 07081

KORNBLUTH, RICHARD SYD, b Kansas City, Mo, Sept 14, 48; m 85; c 2. INFECTIOUS DISEASES. *Educ:* Harvard Col, BA, 70; NY Med Col, MD, 75; Columbia Univ, PhD(path), 83; Am Bd Internal Med, cert, 78, cert pulmonary dis, 80. *Prof Exp:* Res asst, Dept Surg, Children's Hosp, Boston, 73-74; intern & resident, Dept Med, Mt Sinai Hosp, 75-78; res fel, Cardiopulmonary Res Lab, Col Physicians & Surgeons, Columbia Univ, 78-83, instr clin med, Pulmonary Div, 81-83; res assoc, Dept Immunol, Res Inst Scripps Clin, La Jolla, Calif, 83-86; asst clin prof, 86-89, ASST ADJ PROF MED, DIV INFECTIOUS DIS, DEPT MED, SCH MED, UNIV CALIF, SAN DIEGO, 90- *Concurrent Pos:* Fel, Pulmonary Div, Dept Med, Col Physicians & Surgeons, Columbia Univ, 78-81, vis physician, Columbia-Presby Med Ctr, 81-83; Am Lung Asn fel, 78-80; Parker B Francis Found fel, 80-83; attend physician, Emergency Dept, Elmhurst City Hosp, Queens, NY, 81-83 & Clairemont Community Hosp, San Diego, 83-86; prin investr, USPHS, NIH, Nat Heart, Lung & Blood Inst, 89-94, Am Found AIDS Res, 90-91 & USPHS, NIH, Nat Inst Allergy & Infectious Dis, 91-94. *Mem:* Am Thoracic Soc; Am Asn Pathologists; Am Asn Immunologists; AAAS; Reticuloendothelial Soc; Physicians Social Responsibility. *Res:* Human immunodeficiency virus; acquired immunodeficiency syndrome; macrophage immunobiology; cytokines; apoptosis; author of more than 20 technical publications. *Mailing Add:* Infectious Dis Sect 111F Vet Admin Med Ctr 3350 La Jolla Village Dr San Diego CA 92161

KORNEGAY, ERVIN THADDEUS, b Faison, NC, Mar 16, 31; m 56; c 3. ANIMAL SCIENCE. *Educ:* NC State Univ, BS, 53, MS, 60; Mich State Univ, PhD(animal nutrit), 63. *Prof Exp:* Asst agr agent, NC State Univ, 56-59; asst, Mich State Univ, 59-63; asst res prof, Animal Nutrit, Rutgers Univ, 63-67; assoc prof, 67-73, PROF ANIMAL SCI, VA POLYTECH INST & STATE UNIV, 73- *Concurrent Pos:* Travel fel, Nat Feed Ingredients Asn, 83. *Honors & Awards:* Res Award, Nutrit, Am Feed Mfg Asn, 82; Gustav Bohstedt Mineral Award, Am Soc Animal Sci, 86, Animal Mgt Award, 90. *Mem:* Am Soc Animal Sci; NY Acad Sci; Am Inst Nutrit; Can Soc Animal Sci. *Res:* Nutrition, environment and immune response; evaluation of feedstuffs for swine; fiber and mineral utilization; sow management and nutrition; artificial rearing of baby pigs; mineral availability and interactions. *Mailing Add:* Dept Animal Sci Va Polytech Inst & State Univ Blacksburg VA 24061

KORNEL, LUDWIG, b Jaslo, Poland, Feb 27, 23; m 52; c 2. ENDOCRINOLOGY, BIOCHEMISTRY. *Educ:* Wroclaw Univ, MD, 50; Univ Birmingham, PhD(endocrinol, steroid biochem), 58. *Prof Exp:* Intern med, surg, gynec & pediat, Wroclaw Univ Hosp, 49-50; from intern to resident med, Hadassah Univ Hosp, Jerusalem, 50-54; asst physician & instr, Hadassah Med Sch, Hebrew Univ, Israel, 54-55; lectr med, Univ Birmingham, 56-57; asst physician med & community health, Hadassah Univ Hosp & Community Health Ctr, Jerusalem, 57-58; from asst prof to prof med, Med Ctr, Univ Ala, 61-67, assoc prof biochem, 65-67, dir steroid sect & consult endocrinol, 63-67; DIR STEROID UNIT, RUSH-PRESBY-ST LUKE'S MED CTR, 67-, SR ATTEND PHYSICIAN & SR SCIENTIST, 71-; PROF MED & BIOCHEM, RUSH MED COL, 70- *Concurrent Pos:* Res fel hemat, Hosp Broussai, Univ Paris, 51-52; Brit Coun res scholar med & steroid chem, Univ Birmingham, 55-57; res fel endocrinol & metab, Med Ctr, Univ Ala, Birmingham, 58-59; USPHS trainee, Inst Steroid Biochem, Univ Utah, 59-61; hon vis prof, Polish Acad Sci, Warsaw, 65; prof med, Col Med, Univ Ill, 67-71; vis prof, Kanazawa Univ, Japan, 73, 82 & 88 & Inst Hypertension, Tel-Hashomer Med Ctr, Univ Tel Aviv, Israel, 90; nat corresp, Fedn Am Socs Exp Biol, 75-; mem bd dirs, Nat Acad Clin Biochem, 82-86; co-ed, Yearbook Endocrinol, 85-90. *Honors & Awards:* Physicians Recognition Award, Am Med Asn, 69, 73, 76 & 81. *Mem:* AAAS; Am Fedn Clin Res; Endocrine Soc; Sigma Xi; Cent Soc Clin Res; Am Asn Univ Prof; fel Am Soc Clin Pharmacol & Therapeut; Am Physiol Soc; fel Royal Soc Health; fel Nat Acad Clin Biochem. *Res:* Metabolism and mechanism of action of steroidal hormones, especially relation of corticosteroids to mechanism of arterial hypertension; mineralocorticoid receptors in arterial walls and hypertension; role of mineralocorticoids in mechanism of hypertension; control by steroids of transmembrene ionic fluxes in vascular smooth muscle; co-author of encyclopedia on human biology. *Mailing Add:* Steroid Unit Rush-Presby St Lukes Med Ctr 1653 W Congress Pkwy Chicago IL 60612

KORNET, MILTON JOSEPH, b East Chicago, Ind, Dec 31, 35; m 62; c 3. PHARMACEUTICAL CHEMISTRY, ORGANIC CHEMISTRY. *Educ:* Purdue Univ, BS, 57; Univ Ill, PhD(pharmaceut chem), 63. *Prof Exp:* Chemist, Abbott Labs, 57-59; res assoc org synthesis, Northwestern Univ, 62-63; asst prof, 63-67, ASSOC PROF, PHARMACEUT CHEM, UNIV KY, 67- *Mem:* Am Chem Soc. *Res:* Heterocyclic organic chemistry; medicinal chemistry; chemistry of hydrazines. *Mailing Add:* Col Pharm Univ Ky Lexington KY 40536-0082

KORNETSKY, CONAN, b Portland, Maine, Feb 9, 26; m 49; c 2. PSYCHOLOGY, PSYCHOPHARMACOLOGY. *Educ:* Univ Maine, BA, 48; Univ Ky, MS, 51, PhD(psychol), 52. *Prof Exp:* Res scientist, NIMH, 52-59; assoc prof, 59-62, PROF PSYCHIAT & PHARMACOL, SCH MED, BOSTON UNIV, 62- *Concurrent Pos:* NIH sr res fel, Boston Univ, 59-62, NIH res scientist award, 62-70, NIMH res scientist award, 70-78, Nat Inst Drug Abuse res scientist award, 83-88; mem psychopharmacol study sect, NIH, 62-67, mem clin psychopharmacol res rev comt, NIMH, 67-71; mem comt tobacco habituation, Am Cancer Soc, 66-70; mem panel behav modification drugs for hyperkinetic children, Dept Health, Educ & Welfare, 71; mem merit rev bd neurobiol, Vet Admin, 72-76; mem psychopharmacol agents adv comt, Food & Drug Admin, 73-77; mem biomed rev comt, Nat Inst Drug Abuse, 80-84; pres, psychopharmacol div, Am Psychol Asn, 85-86. *Mem:* Am Soc Pharmacol & Exp Therapeut; Am Psychol Asn; Am Col Neuropsychopharmacol; Int Col Neuropsychopharmacol; Psychonomic Soc; Soc Neurosci. *Res:* Neurobehavioral buses for the rewarding effects of abused substances, pain & analgesia; behavioral and neuropsychological studies of the action of antipsychotic and analgesic drugs. *Mailing Add:* Seven Rumford Rd Lexington MA 02173

KORNFEIL, FRED, b Vienna, Austria, Feb 14, 24; nat US; m 53. PHYSICAL CHEMISTRY, ELECTROCHEMISTRY. *Educ:* Univ Vienna, MS, 50, PhD(chem), 53. *Prof Exp:* Chemist, Power Sources Div, Electronics Technol & Devices Lab, US Army Electronics Command, Ft Monmouth, 53-59, phys scientist, Explor Res Div E, 59-71, phys scientist, Power Sources Area, 71-78; RETIRED. *Mem:* Am Chem Soc. *Res:* Fuel cells; battery test techniques; kinetics of electrode processes. *Mailing Add:* Seven Lakes Box 2035 West End NC 27376

KORNFELD, EDMUND CARL, b Philadelphia, Pa, Feb 24, 19; m 45. MEDICINAL CHEMISTRY. *Educ:* Temple Univ, AB, 40, AM, 42; Harvard Univ, MA, 44, PhD(org chem), 46. *Hon Degrees:* DSc, Temple Univ, 64. *Prof Exp:* Res chemist, Off Sci Res & Develop Contract, Harvard Univ, 45; res chemist, 46-65, res adv, Eli Lilly & Co, 65-83; RETIRED. *Mem:* Am Chem Soc. *Res:* Rubber chemistry; organic structural determination; synthetic organic medicinals; organic chemicals development; medicinal chemistry of indol derivatives and ergot alkaloids. *Mailing Add:* 3550 Bay Rd S Dr Indianapolis IN 46240

KORNFELD, LOTTIE, b Vienna, Austria, Feb 8, 25; US citizen. IMMUNOBIOLOGY. *Educ:* Col Wooster, BA, 45; Ohio State Univ, MS, 47; Univ Chicago, PhD(microbiol), 60. *Prof Exp:* Asst bact, Ohio State Univ, 45-47; bacteriologist, Viral & Rickettsial Res Div, Lederle Labs, Am Cyanamid Co, 47-54; res asst bact, Univ Mich, 54-55; res asst bact, Dept Med, Univ Chicago, 55-60, res assoc, Dept of Med & Argonne Cancer Res Hosp, 60-61; res fel, Dept Exp Path, Scripps Clin & Res Found, Calif, 61-63; res microbiologist, US Naval Radiol Defense Lab, 63-69 & Letterman Army Inst Res, 69-72; microbiologist, Div Biomed & Environ Res, US Atomic Energy Comn, 72-74; health scientist adminr, NIH, 74-85; Coordr, Univ-Wide AIDS Res Prog, Univ Calif, Berkeley, 85-89; RETIRED. *Concurrent Pos:* Lectr, Dept Microbiol, San Francisco State Col, 69-71. *Mem:* AAAS; Am Soc Microbiol; Radiation Res Soc; Reticuloendothelial Soc; Am Asn Immunologists. *Res:* Immunology; host resistance; effects of irradiation on host-parasite relationship; science administration. *Mailing Add:* 508 Tampico Dr Walnut Creek CA 94598

KORNFELD, MARIO O, b Zagreb, Yugoslavia, July 9, 27; m 56; c 1. NEUROPATHOLOGY. *Educ:* Univ Zagreb, MD, 53, ScD, 64. *Prof Exp:* Staff pathologist, Inst Path, Gen Hosp, Zagreb, 59-64; instr neuropath, Col Physicians & Surgeons, Columbia Univ, 67-68; asst prof, 68-70, assoc prof path & neuropath, 70-80, PROF PATH, SCH MED, UNIV NMEX, 80- *Concurrent Pos:* Trainee & fel, Col Physicians & Surgeons, Columbia Univ, 64-67; staff pathologist, Bernalillo County Med Ctr, Albuquerque, 70-; attend neuropathologist, Vet Admin Hosp, 70- *Honors & Awards:* Matthew T Moore Award, Am Asn Neuropath; Weil Award, Am Asn Neuropath. *Mem:* Assoc Am Asn Neuropathologists; Am Asn Pathologists; Asn Res Neuropath Ment Dis. *Res:* Histopathology of inner ear and temporal bone; ultrastructural aspects of neurolipidoses, peripheral nervous system diseases and astroglia in metabolic encephalopathies; morphometry of secretion in pituitary adenonomas. *Mailing Add:* Dept Path Univ NMex Sch Med Albuquerque NM 87196

KORNFELD, ROSALIND HAUK, b Dallas, Tex, Aug 2, 35; m 59; c 3. OLIGOSACCHARIDE STRUCTURE, GLYCOPROTEIN SYNTHESIS. *Educ:* George Wash Univ, BS, 57; Wash Univ, St Louis, PhD(biochem), 61. *Prof Exp:* Staff fel, Nat Inst Arthritis & Metab Dis, NIH, 63-65; from res instr to res assoc prof, 65-78, assoc prof biochem, div hemat-oncol & assoc prof, 78-81, PROF BIOCHEM, DIV HEMAT-ONCOL, DEPT MED, SCH MED & PROF, DEPT BIOL CHEM, WASH UNIV, 81-, COORDR, GRAD TRAINING PROG, 84- *Concurrent Pos:* Mem, Comt Cancer Immunobiol, Nat Cancer Inst, NIH, 75-78, & Physiol Chem Study Sect, 80-83. *Mem:* Am Soc Hemat; Am Soc Biochem & Molecular Biol. *Res:* Biosynthesis and structural analysis of the oligosaccharides on glycoproteins and the role of mannosidases in oligosaccharide processing. *Mailing Add:* Dept Med & Biochem Box 8125 Sch Med Wash Univ 660 S Euclid St Louis MO 63110

KORNFELD, STUART ARTHUR, b St Louis, Mo, Oct 4, 36; m 59; c 3. HEMATOLOGY. *Educ:* Dartmouth Col, AB, 58; Washington Univ, MD, 62. *Prof Exp:* Res asst, biochem dept, Wash Univ Sch Med, 58-62; intern ward med, Barnes Hosp, 62-63; res assoc, Nat Inst Arthritis & Metab Dis, NIH, 63-65; asst resident ward med, Barnes Hosp, 65-66; from instr to asst prof med, Sch Med, Wash Univ, 66-70, from asst prof to assoc prof biochem, 68-72, dir, Div Oncol, 73-76, PROF MED, SCH MED, WASH UNIV, 72-, PROF BIOCHEM & CO-DIR, DIV HEMAT & ONCOL, 76- *Concurrent Pos:* Fac res assoc, Am Cancer Soc, 66-71; NIH res career develop award, 71-76; counr, Am Soc Clin Invest, 72-75; mem, Cell Biol Study Sect, NIH, 74-77, Bd Sci Counselors, Nat Inst Arthritis, Diabetes & Digestive & Kidney Dis, 83-87, Sci Rev Bd, Howard Hughes Med Inst, 86- & Bd Sci Adv, Jane Coffin Childs Mem Fund Res, 87-; assoc ed, J Clin Invest, 77-81, ed, 81-82; assoc ed, J Biol Chem, 82-87. *Honors & Awards:* Borden Award, 62; Jubilee Lectr & Harden Medallist, Biochem Soc, 89; Passano Award, 91. *Mem:* Nat Acad Sci; Inst Med-Nat Acad Sci; Am Soc Clin Invest; Am Soc Hemat; Am Soc Biol Chemists; Asn Am Physicians (secy, 86-); Am Acad Arts & Sci; Am Chem Soc; Sigma Xi. *Res:* Studies of the structure, biosynthesis and function of glycoproteins, especially those which are found on the surface of normal and malignant cells; targeting of newly synthesized acid hydroloses to lysosomes; author of 145 publications. *Mailing Add:* Sch Med Wash Univ St Louis MO 63110

KORNFIELD, A(LFRED) T(HEODORE), b Philadelphia, Pa, Oct 12, 18; m 44; c 4. ENVIRONMENTAL PHYSIOLOGY. *Educ:* Univ Pa, Philadelphia, AB, 42; Ohio State Univ, Columbus, PhD(physiol), 53. *Prof Exp:* Res physicist, Fire Control Develop Lab, Philadelphia, US Army, 42-44; serv res scientist, Med Field Res Lab, US Navy, Camp Lejeune, NC, 44-46, res physicist, Aeromed Equip Lab, Philadelphia, 46-50; Am Heart Asn res fel physiol, Ohio State Univ, Columbus, 50-53; asst prof physiol, med & grad fac, St Louis Univ, 53-55; sr analyst opers res, Combat Opers Res Group, Ft Monroe, Va, 55-57; chief human factors spaceship life support, Space Flight Div, Bell Aerosysts, Buffalo, 57-59, asst to dir res mgt, res & eng labs, 59-60; mgr biosci, Allied Res Assocs, Boston, 60-61; pres, Biosearch Co Boston, Philadelphia, 61-66, 80-82; health sci adminr bioeng & biophys, NIH, 67-69; fac assoc physiol, dept surg, Med Sch, Harvard Univ, 70-72; RETIRED. *Concurrent Pos:* Personal serv contract, Artificial Heart Prog, Nat Heart & Lung Inst, NIH, 69-70; consult, Off Secy Defense, Adv Res Projs Agency, US Army Res Off, Durham & Land War Labs, Aberdeen, 64-66, Booz-Allen Appl Res, Bethesda, 66-67, 73-74, Geomet Corp, 73, Foster D Snell, 74; prin investr, Off Naval Res contract, combined stresses in occup environ health, via Biosearch Co, 81-83; sci consult, 71- *Mem:* Biomed Eng Soc; Biophys Soc; Inst Elec & Electronics Engrs; Am Soc Artificial Internal Organs; Sigma Xi. *Res:* Chemical signalling in living systems; biosensor analogs for design; combined stresses in occupational environments; isolated organ perfusion; biological signalling and control. *Mailing Add:* 306 Revere Rd No 16 Drexel Hill PA 19026

KORNFIELD, IRVING LESLIE, b Jacksonville, NC, July 16, 45; m 68; c 2. EVOLUTIONARY BIOLOGY. *Educ:* Syracuse Univ, AB, 68; State Univ NY, Stony Brook, NY, 72, PhD(ecol), 74. *Prof Exp:* Fel, Smithsonian Inst, 74-75; res collabr, Dept Genetics, Hebrew Univ, 75-76; assoc prof zool, 77-85, PROF ZOOL, UNIV MAINE, 85- *Concurrent Pos:* Assoc, Danforth Found, 80- *Mem:* Am Soc Ichthyologists & Herpetologists; Genetics Soc Am; Soc Study Evolution. *Res:* Evolutionary genetics of fishes; biochemical systematics. *Mailing Add:* Dept Zool Univ Maine Orono ME 04473

KORNFIELD, JACK I, b New York, NY. SATELLITE METEOROLOGY. *Educ:* City Col New York, BS, 61; Northeastern Univ, MS, 63; Univ Wis, PhD(meteorol), 73. *Prof Exp:* Sr analyst, Space Res Corp, 73-74; sr mathematician, GTE, 74-76; sr scientist, Agr Res Inst Israel, 76-77; sr colorimetrist, SCI, Inc, Tex, 77-78; sr scientist, Systs & Appl Sci Corp, 78-79; mem advan prog staff, OAO Corp, 79-80; assoc prog dir meteorol, NSF, 81-83; tech ed, Bull Am Meteorol Soc, 83-89; CONSULT, 89- *Concurrent Pos:* Freelance ed & publ, 89. *Mem:* Am Meteorol Soc; Am Geophys Union; Sigma Xi; AAAS. *Res:* Remote sensing to extract parameters of the earth's land, ocean and atmosphere systems; climate and hydrological modeling; colorimetry applied to the display and analysis of information; balistic analysis; laser propagation through the atmosphere; general theory of remote sensing. *Mailing Add:* 175 Freeman St Suite 115 Brookline MA 02146

KORNGOLD, ROBERT, GRAFT VS HOST DISEASE, BONE MARROW TRANSPLANTATION. *Educ:* Univ Pa, PhD(immunol), 79. *Prof Exp:* Asst prof, Wistar Inst Anat & Biol, 81-87; ASSOC PROF, MICROBIOL & IMMUNOL, JEFFERSON MED COL, 87- *Mailing Add:* Dept Microbiol & Immunol Jefferson Med Col 1020 Locust St Philadelphia PA 19107

KORNGUTH, STEVEN E, b New York, NY, Dec 1, 35; m 58; c 2. BIOCHEMISTRY. *Educ:* Columbia Univ, BA, 57; Univ Wis, MA, 59, PhD(biochem), 61. *Prof Exp:* Res scientist neurochem, NY State Psychiat Inst, 61-63; from asst prof to assoc prof, 63-72, PROF NEUROL & PHYSIOL CHEM, UNIV WIS-MADISON, 72- *Mem:* Am Soc Biol Chemists. *Res:* Magnetic resonance contrast agents; antigenic properties of such proteins; synaptic complexes, isolation and chemical properties; paraneoplastic disorders. *Mailing Add:* Dept Neurol & Physiol Chem Univ Wis Med Ctr Madison WI 53706

KORNHAUSER, ALAIN LUCIEN, b Beaurepaire, France, June 12, 44; US citizen; m 65. ASTRODYNAMICS, TRANSPORTATION. *Educ:* Pa State Univ, BS, 65, MS, 67; Princeton Univ, MA, 69, PhD(aerospace sci), 71. *Prof Exp:* Res asst cavitation, Ord Res Lab, 67; asst prof astrodyn, Univ Minn, Minneapolis, 71-77; assoc prof, 77-78, PROF, DEPT CIVIL ENG, PRINCETON UNIV, 78-, DIR, TRANSPORTATION PROG, 76- *Concurrent Pos:* Consult, Princeton Univ 71- & Optimal Data Co, 71- *Honors & Awards:* R T Knapp & Melville Medal, Am Soc Mech Eng, 70. *Mem:* Am Soc Mech Eng; Am Inst Aeronaut & Astronaut; Am Astronaut Soc; Sigma Xi. *Res:* Optimal space flight; cavitation; urban transportation; computer graphics; freight railroad operations and planning analysis. *Mailing Add:* Dept Civil/Geo Eng E 414 Engr Quadrangle Princeton NJ 08540

KORNHAUSER, ANDRIJA, b Zagreb, Yugoslavia, Feb 5, 30; US citizen; m 78. TOXICOLOGY, BASIC MEDICAL SCIENCES. *Educ:* Univ Zagreb, Yugoslavia, BSci, 54, PhD(biochem), 62. *Prof Exp:* Res assoc, Sch Med, Univ Frankfurt, Germany, 64-66; assoc prof, Rudjer Boskavic Inst, Univ Zagreb, 66-70; mem fac, Sch Med, Harvard Univ & Sch Dent Med, 70-78; res biologist, res & mgt, div toxicol, 78-80, CHIEF, DERMAL & OCULAR BR, DIV TOXICOL, CFSAN, FOOD & DRUG ADMIN, 80- *Concurrent Pos:* Lectr oral path, Sch Dermal Med, Harvard Univ, 78-; adj assoc prof dermat, Sch Med, George Washington Univ, 78- *Mem:* AAAS; Am Asn Photobiol; Soc Investigative Dermat Inc; NY Acad Sci; hon fel Skin Cancer Found. *Res:* Cutaneous toxicol; phototoxicity; carcinogenesis; photocarcinogenesis; photomedicine; molecular toxicology; pharmacology; protection against phototoxicity and carcinogenesis by dietary antioxidants; development of animal models for clinical studies. *Mailing Add:* Food & Drug Admin 200 C St SW Washington DC 20204

KORNHAUSER, EDWARD T(HEODORE), electrical engineering; deceased, see previous edition for last biography

KORNICKER, LOUIS SAMPSON, b Brooklyn, NY, May 23, 19; m 51; c 3. GEOLOGY. *Educ:* Univ Ala, BS, 41 & 42; Columbia Univ, MA, 54, PhD, 58. *Prof Exp:* Prod supvr trinitrotoluene, Tech Invest Group, Hercules Powder Co, 42-44; sr process engr & pilot plant supt, Cities Serv Ref Co, 44-47; treas & plant supt, Uncle Sam Chem Co, Inc, 47-54; asst, Columbia Univ, 54-57; asst dir, Inst Marine Sci, Univ Tex, 57-60; geologist, Off Naval Res, Chicago, 60-61; from assoc prof to prof oceanog, Tex A&M Univ, 61-64; assoc cur, 64-67, CUR DIV CRUSTACEA, US NATURAL HIST MUS, SMITHSONIAN INST, 67- *Concurrent Pos:* Adj prof biol, George Washington Univ, 70- *Mem:* Soc Syst Zool; Crustacean Soc. *Res:* Marine geology; micropaleontology; paleoecology; ecology; ostracodes; coral reefs; ostracoda systematics and ecology. *Mailing Add:* 10400 Lake Ridge Dr Oakton VA 22124

KORNICKER, WILLIAM ALAN, b New York, NY, July 24, 56. ROCK-WATER INTERACTIONS, GEOCHEMICAL MODELING. *Educ:* Old Dominion Univ, BS, 78, MS, 80; Tex A&M Univ, PhD(oceanog), 88. *Prof Exp:* Postdoctoral fel chem & environ eng, McMaster Univ, 88-91. *Mem:* AAAS; Am Chem Soc; Am Geophys Union; Am Soc Limnol & Oceanog; Geochem Soc; Sigma Xi. *Res:* Thermodynamics and kinetic control of mineral formation, dissolution and solute transport in low temperature environments. *Mailing Add:* 10400 Lake Ridge Dr Oakton VA 22124-1511

KORNMAN, BRENT D, b Dothan, Ala, Sept 20, 56; m 81; c 2. ARTIFICIAL INTELLIGENCE. *Educ:* Univ Md, BS, 78. *Prof Exp:* Staff programmer, PAR Technol Corp, 79-82; ADV PROGRAMMER, IBM CORP, 82- *Mem:* Am Asn Artificial Intel. *Res:* Expert system applications; automated plan construction; knowledge representation languages; knowledge base design and development techniques; knowledge base verification. *Mailing Add:* IBM Corp 100 Lake Forest Blvd Gaithersburg MD 20877

KORNREICH, HELEN KASS, b Newark, NJ, Sept 4, 31; m 65. PEDIATRICS, RHEUMATOLOGY. *Educ:* Rutgers Univ, BS, 52; Hahnemann Med Col, MD, 56. *Prof Exp:* From instr to asst prof, 63-70, ASSOC PROF PEDIAT, SCH MED, UNIV SOUTHERN CALIF, 70- *Concurrent Pos:* Arthritis & Rheumatism Found fel pediat rheumatology, Childrens Hosp, Los Angeles, Calif, 63-65. *Mem:* Am Rheumatism Asn; Am Acad Pediat. *Res:* Connective tissue diseases of childhood; medical education. *Mailing Add:* 4650 Sunset Blvd Los Angeles CA 90027

KORNREICH, PHILIPP G, b Vienna, Austria, Nov 4, 31; US citizen; m 60; c 3. SOLID STATE PHYSICS, SOLID STATE MICROWAVE DEVICES. *Educ:* Carnegie Inst Tech, BS, 62; Univ Pa, PhD(elec eng), 67. *Prof Exp:* Sr res engr thin film technol res, Sperry Rand Univac, 60-66; res assoc solid state physics res, Univ Pa, 66-67; from asst prof to assoc prof, 67-78, PROF ELEC ENG, SYRACUSE UNIV, 78- *Concurrent Pos:* Consult, Gen Elec Co; consult & co-founder, DEFT Labs; vis prof, Technon Israel Inst Technol, 80-81; consult, Electronic Device Reliability Group, RADC, Rome, NY, 82-; res remote optical sensing & light frequency electronic devices optical comput, US Air Force Photonics Lab, RADC, 88- *Mem:* Am Phys Soc; Inst Elec & Electronics Engrs; Sigma Xi; AAAS; Soc Photo-Optical Instrumentation Engrs; Int Soc Optical Eng. *Res:* Phonon microwave oscillator; variable delay magnetic strip line; directional dependence of photoconductivity; direct electronic fourier transforms of images; systems with delay and memory; vibrational modes of superlattices; ultra high speed electron devices for both microwave and very high speed integrated circuits applications; three dimensional integrated circuits; light frequency devices for optical computing. *Mailing Add:* Dept Elec & Comput Eng Syracuse Univ Syracuse NY 13244

KORNSTEIN, EDWARD, b New York, NY, Sept 7, 29; m 58. ELECTRO-OPTICS, ENGINEERING MANAGEMENT. *Educ:* NY Univ, BA, 51; Drexel Inst Technol, MS, 54. *Prof Exp:* Physicist optics, Radio Corp Am, 51-57 & Phys Res Lab, Boston Univ, 58; consult optics, 59-60; physicist, Radio Corp Am, 60-70; vpres, Optel Corp, 70-72; pres, Kortron Consults, 72-80; vpres, Object Recognition Systs, Inc, 77-86; PRES, OBJECT RECOGNITION SYSTS AUTOMATION, INC, 87- *Mem:* Soc Motion Picture & TV Engrs; Optical Soc Am; Inst Elec & Electronics Engrs; Soc Info Display. *Res:* Infrared optical and detection systems; aerial reconnaissance and data processing systems; physical optics; laser devices and systems; electro-optical displays; electronic digital timepieces; pattern recognition; machine vision systems. *Mailing Add:* Ten Channing Way RD 1 Cranbury NJ 08512

KORNYLAK, ANDREW T, innovative house design, new conveyor concepts; deceased, see previous edition for last biography

KOROBKIN, IRVING, b New York, NY, Oct 18, 25; m 47; c 4. PHYSICS, SYSTEMS ANALYSIS. *Educ:* City Col New York, BME, 45; Columbia Univ, BS, 48; Univ Md, PhD(physics), 60. *Prof Exp:* Instr physics, City Col New York, 47-48; instr mech eng, Syracuse Univ, 48-51; res scientist & adminr fluid dynamics, US Naval Ord Lab, 51-61; sr systs analyst, IBM CORP, 61-68; OPERS RES ANALYST MIL OPERS RES, NAVAL SURFACE WEAPON CTR, 68- *Concurrent Pos:* Consult, Missile & Space Vehicle Dept, Gen Elec Co, 56-59; assoc prof lectr, George Washington Univ, 57-66. *Honors & Awards:* Meritorious Civil Serv Award, Naval Ord Lab, 57. *Mem:* Sigma Xi; assoc fel Am Inst Aeronaut & Astronaut; AAAS. *Res:* High speed fluid dynamics; reentry physics; nuclear weapons effects; military systems analyst with emphasis on strategic warfare. *Mailing Add:* 8510 Hunter Creek Trail Potomac MD 20854

KOROL, BERNARD, b Chicago, Ill, Feb 2, 29; m 52; c 3. PSYCHOPHARMACOLOGY. *Educ:* Roosevelt Col, BS, 49; Univ Chicago, MS, 52; McGill Univ, PhD(pharmacol), 56. *Prof Exp:* Res pharmacologist, Smith Kline & French Labs, 56-58 & Chas Pfizer & Co, Inc, 58-61; group leader pharmacol, Geigy Res Labs, 61-64; asst prof physiol & pharmacol & chief pharmacol sect, Sch Med, Univ Mo-St Louis, 64-69; assoc prof psychiat, Sch Med, St Louis Univ & supvr psychopharmacol, St Louis Vet Admin Hosp, 69-87; RES ADMIN, ENQUAY PHARMACEUT ASSOCS, 87- *Res:* Physiology and pharmacology of mental illness. *Mailing Add:* Enquay 2840 NW Second Ave Boca Raton FL 33431

KOROLY, MARY JO, b Philadelphia, Pa, Jan 28, 43; c 1. CELL BIOLOGY. *Educ:* Bryn Mawr Col, PhD(biochem), 69. *Prof Exp:* Asst prof cell biol, Bryn Mawr Col, 72-77; Harvard Univ, 77-79; ASST PROF CELL BIOL, UNIV FLA, 79- *Mem:* AAAS; Am Soc Cell Biol; Am Women Sci; Am Soc Protozool; Am Soc Biol Chemists; Sigma Xi. *Mailing Add:* Dept Biochem & Molecular Biol Univ Fla JHMHC Box 245 Gainesville FL 32610

KOROS, AURELIA M CARISSIMO, b Boston, Mass, Aug 28, 34; m 57; c 5. IMMUNOLOGY, CELL BIOLOGY. *Educ:* Radcliffe Col, AB, 56; Univ Pittsburgh, MS, 60, PhD(microbiol), 65. *Prof Exp:* Res asst cell physiol, Sch Med, Harvard Univ, 56-58; Am Cancer Soc Inst res grant immunol, Sch Med, Univ Pittsburgh, 65-66, from instr to asst prof microbiol, 66-73, res asst prof obstet & gynec, 73-75, res asst prof path, 75-76; res assoc, Allegheny Gen Hosp, Pittsburgh, 77-78, asst biologist, Cancer Res Unit, 78-80; mem staff, Allegheny County Health Dept, 80-83; MEM STAFF, GRAD SCH PUB HEALTH, UNIV PITTSBURGH, 81- *Concurrent Pos:* NIH grants, 66-74, NCI grants, 80-83; Health Res & Serv Found grants, 69 & 70, Cancer Fedn, Inc, 83- & Candle Found; FIDIA fel neurosci, 88. *Mem:* AAAS; Am Soc Microbiol; Am Asn Immunologists; NY Acad Sci; Am Asn Cancer Res; Int Asn Study of Lung Cancer; Clin Immunol Soc; Am Soc Hemat; Marine Biol Asn UK. *Res:* Elucidation of the mechanism by which antigen and antibody regulate the proliferation of antibody-producing cells; immunological relations in maternal-fetal interactions; immunoregulation in tumor models; biology of lung cancer; evolutionarily conserved antigens on lung cancer cells; sea urchin coelomocytes. *Mailing Add:* 154 Maple Heights Rd Pittsburgh PA 15232

KOROS, PETER J, b Berlin, Ger, July 14, 32; US citizen; m 57; c 5. METALLURGY, MATERIALS SCIENCE. *Educ:* Drexel Univ, BS, 54; Mass Inst Technol, SM, 55 & ScD, 58. *Prof Exp:* Res engr & sr res engr, Jones & Laughlin Steel Corp, 58-63, res supvr steelmaking, 63-65, chief process metallurgist, Qual Control Div, 65-75, dir process metall res, 75-78, dir res spec projs, 78-80, mgr process develop & qual control, 80-82, sr res assoc, 82-84; SR RES CONSULT, LTV STEEL CORP, 84- *Concurrent Pos:* Chmn, Iron & Steel Div, Am Inst Mining, Metall & Petrol Engrs, 69-70, bd dirs, 74, chmn prog comt, 5th Int Iron & Steel Cong, Iron & Steel Soc, 86; mem, US Bur Mines, Generic Technol Res Ctr Pyrometall, 82- & Chmn, 84-85; adv bd, NSF Ctr Iron & Steel Res, Carnegie-Mellon Univ, 85-; Am Iron & Steel Inst, Steel Iniative Task Force mem, 85- *Honors & Awards:* Toy Award, 62 & McKune & Herty Mem Awards, 63, Am Inst Mining, Metall & Petrol Engrs; Jalmet Award, Jones & Laughlin Steel Corp, 63; Silver Medal, Am Iron & Steel Inst, 69, Gold Medal, 77; Design & Appln Award, Int Magnesium Asn, 78. *Mem:* Am Iron & Steel Inst; fel Am Inst Mining, Metall & Petrol Engrs; fel Am Soc Metals Int. *Res:* Process and quality control in steel production; physical chemistry of iron and steelmaking; applied research in steelmaking, coal and cokemaking development. *Mailing Add:* LTV Steel Co Technol Ctr 6801 Brecksville Rd Independence OH 44131-5099

KOROS, WILLIAM JOHN, b Omaha, Nebr, Aug 31, 47; m 70. POLYMER SCIENCE & ENGINEERING. *Educ:* Univ Tex, Austin, BS, 69, MS, 75, PhD(chem eng), 77. *Prof Exp:* Engr polymer processing, E I Du Pont de Nemours & Co, 69-73; asst prof, 77-80, ASSOC PROF CHEM ENG, NC STATE UNIV, 80- *Concurrent Pos:* Prin investr dual mode sorption & transport in glassy polymers, NSF grant, 77-79, 80-82; Army Res Off grant, 80-83; res award, Sigma Xi, 80. *Mem:* Am Inst Chem Eng; Sigma Xi. *Res:* Sorption and transport of low molecular weight compounds such as gases, solvents, monomers and additives in the polymeric solid state. *Mailing Add:* PO Box 200789 Austin TX 78720

KOROSTOFF, EDWARD, b Philadelphia, Pa, Feb 25, 21; m 51; c 3. DENTAL MATERIALS. *Educ:* Univ Pa, BS, 41, MS, 50, PhD(metall eng), 61. *Prof Exp:* Lectr dent mat sci, Univ Pa, 63-65, from asst prof to assoc prof biomat, Sch Dent Med, Sch Med & Col Eng & Appl Sci, 65-75, PROF RESTORATIVE DENT, SCH DENT MED, UNIV PA, 75- *Concurrent Pos:* USPHS career develop award, Univ Pa, 66; chmn med-dent mat comt, Metall Soc, 69-71. *Mem:* Soc Biomat; Acad Dent Mat; Int Asn Dent Res; AAAS; Am Asn Univ Prof. *Res:* Stress generated electrical potentials in bone and dentin; viscoelastic properties of bone and dentin; electric stimulation of bone remodeling. *Mailing Add:* LRSM Bldg 33/Walnut Univ Pa Philadelphia PA 19104

KOROTEV, RANDALL LEE, b Green Bay, Wis, May 15, 49; m 74. GEOCHEMISTRY, ANALYTICAL CHEMISTRY. *Educ:* Univ Wis-Madison, BS, 71, PhD(chem), 76. *Prof Exp:* Proj assoc soil sci, Univ Wis-Madison, 76-79; SR RES SCIENTIST GEOCHEM, WASHINGTON UNIV, 79- *Concurrent Pos:* Mem, Lunar & Planetary Sample Team, NASA, 82-85. *Mem:* Geochem Soc; Am Ornithologists Union; Am Geophys Union; Sigma Xi; Meteoritical Soc. *Res:* Factors affecting the distribution of elements in geologic and environmental systems; geochemistry of lunar soils and rocks; chemical analysis by neutron activation. *Mailing Add:* Dept Earth & Planetary Sci PO Box 1169 Wash Univ St Louis MO 63130

KORPEL, ADRIANUS, b Rotterdam, Neth, Feb 18, 32; m 56; c 1. OPTICS, ACOUSTICS. *Educ:* Delft Technol Univ, MSEE, 56, PhD, 69. *Prof Exp:* Res engr commun, Postmaster Gen Dept, Melbourne, Australia, 56-60; div chief laser appln, Zenith Radio Corp, 60-73, dir res eng physics, 73-77; PROF ELEC ENG, UNIV IOWA, 77- *Honors & Awards:* Alexander von Humboldt Award, 84. *Mem:* Acoust Soc Am; fel Inst Elec & Electronics Engrs; fel Optical Soc Am; Soc Photog Instrumentation Eng; foreign assoc mem Royal Acad Belg. *Res:* Information and communication theory; microwaves; laser optics; acoustic holography and microscopy; acousto-optics; nonlinear waves; optical metrology and microscopy. *Mailing Add:* Dept Elec & Comput Eng Univ Iowa Iowa City IA 52242

KORPER, SAMUEL, AGING RESEARCH. *Prof Exp:* ASSOC DIR, PLANNING ANALYSIS & INT ACTIV, NAT INST AGING, NIH, 87- *Mailing Add:* NIH Nat Inst Aging Off Int Activ Bldg 31 Rm 2C12 Bethesda MD 20892

KORPMAN, RALPH ANDREW, b New York, NY, Aug 9, 52. HEMATALOGY, MEDICAL INFORMATION SCIENCE. *Educ:* Loma Linda Univ, BA, 71, MD, 74; Claremont Grad Sch, CEM, 78. *Prof Exp:* Intern, Med Ctr, 74-75, resident path, 75-78, fel hemat, 78-79, from asst prof to assoc prof hemat & path, 83-87, PROF PATH & LAB MED, LOMA LINDA UNIV, 87-, DIR LABS, 79- *Concurrent Pos:* Dir, Med Data Corp,

76-81; consult, Technician Instruments Corp, 78-81; mem, comput adv comt, finance comt & chmn, govt rels, Am Soc Clin Path, 78-; sci adv, HBO & Co, 81-83; pres & chmn, Health Data Sci, 83- *Honors & Awards:* Sheard-Sanford Award, Am Soc Clin Pathologists, 76. *Mem:* Fel Col Am Pathologists; fel Am Soc Clin Path; NY Acad Sci; fel Am Col Physician Execs. *Res:* Characterization of cellular membranes, especially red blood cells, laboratory quality control, applications of computers to medical care and instrument design and evaluation. *Mailing Add:* PO Box 548 Loma Linda CA 92354

KORR, IRVIN MORRIS, b Philadelphia, Pa, Aug 24, 09; m 39; c 1. PHYSIOLOGY, NEUROSCIENCES. *Educ:* Univ Pa, BA, 30, MA, 31; Princeton Univ, PhD(cellular physiol), 35. *Hon Degrees:* DSc, Kirksville Col Osteop Med, 76; DOsteop Educ, Col Osteop Med Pac, 82. *Prof Exp:* Asst instr physiol, Princeton Univ, 32-33; instr, Col Med, NY Univ, 36-43; sr physiologist, Signal Lab, US War Dept, Ft Monmouth, NJ, 43-44; physiol investr wound ballistics, Princeton Univ, 45; from prof & chmn div physiol sci to distinguished prof & dir prog neurobiol, Kirksville Col Osteop Med, 45-75, emer prof physiol, 75-78; PROF MED EDUC, TEX COL OSTEOP MED, 78- *Concurrent Pos:* Procter fel, Princeton Univ, 35-36; investr, Aviation Res Labs, Columbia Univ, 42-43; prof biomech, Col Osteop Med, Mich State Univ, 75-78. *Honors & Awards:* Robert A Kistner Award, Am Asn Col Osteop Med, 83. *Mem:* AAAS; Am Physiol Soc; Soc Exp Biol & Med; Soc Neurosci; Am Soc Neurochem; Harvey Soc; Am Inst Biol Sci. *Res:* Bioluminescence; oxidation-reduction potentials; cellular metabolism; renal physiology; aviation and climatic physiology; human spinal reflexes; referred pain mechanisms; interchange between somatic and autonomic nervous systems; trophic functions of nerves. *Mailing Add:* 740 Oakwood Trail Ft Worth TX 76112

KORRINGA, JAN, b Heemstede, Netherlands, Mar 31, 15; m 43; c 3. THEORETICAL PHYSICS, GEOPHYSICS. *Educ:* Delft Univ Technol, DSc, 42. *Prof Exp:* From asst to instr physics, Delft Univ Technol, 41-46; from lectr to sr lectr, Univ Leiden, 46-53; from assoc prof to prof physics, Ohio State Univ, 53-80; sr res assoc, Chevron Oil Field Res Co, 80-86; RETIRED. *Concurrent Pos:* Guggenheim fel, 63; vis prof, Univ Besancon, 63 & Univ Paris, 68; consult, Chevron Oil Field Res Co, 55-80 & Union Carbide Nuclear Co, 57-80. *Mem:* Fel Am Phys Soc; Netherlands Phys Soc. *Res:* Statistical physics; metals physics; theory of solids; theory of heterogeneous materials. *Mailing Add:* 620 Mystic Way Laguna Beach CA 92651

KORSCH, BARBARA M, b Jena, Ger, Mar 30, 21; US citizen; wid; c 1. PEDIATRICS. *Educ:* Smith Col, BA, 41; Johns Hopkins Univ, MD, 44; Am Bd Pediat, cert, 50. *Prof Exp:* Asst resident, Bellevue Hosp, 45, Mary Imogene Basset Hosp, 46, New York Hosp, 47; fel Inst Child Develop, New York Hosp, 48-49; asst pediat, Med Col, Cornell Univ, 49-50, from instr to assoc prof, 50-61; assoc clin prof prev med, Sch Med, Univ Calif, Los Angeles, 61-64; assoc prof, 64-69, PROF PEDIAT, SCH MED, UNIV SOUTHERN CALIF, LOS ANGELES, 69- *Concurrent Pos:* Asst outpatient pediatrician, NY Hosp, 49-50, asst attend pediatrician, 50-55, clin dir, Pediat Outpatient Dept, 50-61, assoc attend pediatrician, 55-61; pediat consult, Dept Health, NY, 49-51, Hosp Spec Surg, 55-61, Gen Pediat, Childrens Hosp Los Angeles, 61-65, Med Ctr, Univ Southern Calif, 69-74; coordr, Pediat Rehab Prog, Nat Found Infantile Paralysis, 53-61; pediat dir, Observ Clin Children Los Angeles, 61-64; assoc attend pediatrician, Cedars Lebanon Hosp, 61-; dir, Introd Clin Med & Res & Training Rehab, Sch Med, Univ Southern Calif, 69-74, consult, 74-; vis prof, numerous US & foreign univs, 73-89; hon staff mem, Dept Pediat, Cedars-Sinai Med Ctr, 76-; staff, Div Gen Pediat, Childrens Hosp Los Angeles, 81-91; chair, Coun Am Pediat Soc, 89. *Honors & Awards:* George Armstrong Lectr, Ambulatory Pediat Asn, 73; Katherine D McCormick Distinguished Lectr, Standford Univ, 77; Kathy Newman Mem Lectr, Tulane Univ, 87; C Anderson Aldrich Award, Am Acad Pediat, 88. *Mem:* Inst Med-Nat Acad Sci; Am Acad Pediat; Sigma Xi; Am Pediat Soc; Soc Behav Pediat (pres, 85); Soc Pediat Res. *Res:* Doctor-patient communication; health care delivery; psychosocial aspects of pediatrics including growth and development; medical education; comprehensive care of patients with chronic illness; high risk infants, transition from hospital to home; author of numerous technical publications. *Mailing Add:* Div Gen Pediat Childrens Hosp 4650 Sunset Blvd Los Angeles CA 90027

KORSCH, DIETRICH G, b Waren-Mueritz, Germany, Nov 30, 37; US citizen; m 66; c 2. ASTRONOMY. *Educ:* Univ Tubingen, Germany, BS, 63, MS, 65, PhD, 69. *Prof Exp:* Asst to pres, Univ Tubingen, 66-68, res asst appl optics, 68-70; staff engr optics, Bendix Aerospace Systs, 70-73; eng consult, Sperry Rand Support Serv, 73-76; vpres sci, TAI Corp, 77-81; OPTICAL SCI CONSULT, KORSCH OPTICS, INC, 81- *Honors & Awards:* Six Cert Recognition, NASA. *Mem:* Optical Soc Am; Soc Optical Eng. *Res:* Design and analysis of optical systems, primarily in the area of large space optics; development of design and optimization methods for all- reflective imaging systems from near-normal to grazing incidence. *Mailing Add:* 10111 Bluff Dr Huntsville AL 35803

KORSH, JAMES F, b Philadelphia, Pa, June 16, 38; m 62; c 3. COMPUTER SCIENCE, OPERATIONS RESEARCH. *Educ:* Univ Pa, BS, 60, PhD(comput sci), 66; Univ Ill, MS, 62. *Prof Exp:* Asst prof comput sci, Univ Pa, 66-71; sr res fel, Calif Inst Technol, 71-72; assoc prof, 72-76, PROF COMPUT SCI, TEMPLE UNIV, 77-, CHMN, 75-78 & 89- *Concurrent Pos:* Chmn, CIS Dept, Temple Univ, 75-78. *Mem:* Asn Comput Mach. *Res:* Quantitative methods in computer systems; analysis of algorithms; data structures. *Mailing Add:* Dept Comput & Info Sci Temple Univ Philadelphia PA 19122

KORSLUND, MARY KATHERINE, THERAPEUTIC NUTRITION. *Educ:* Univ Nebr, Lincoln, PhD(nutrit), 72. *Prof Exp:* ASSOC PROF HUMAN NUTRIT & FOOD, VA POLYTECH INST & STATE UNIV, 64- *Mailing Add:* Va Polytech Inst 2990 Telestar Ct Falls Church VA 22042

KORSMEYER, STANLEY JOEL, INTERNAL MEDICINE, IMMUNOLOGY. *Educ:* Univ Ill, MD, 76. *Prof Exp:* Sr Investigator, Nat Cancer Inst, Bethesda, Md, 79-86; ASSOC PROF INTERNAL MED & IMMUNOL, HOWARD HUGHES MED INST, SCH MED, WASH UNIV, 86- *Mailing Add:* Dept Med Microbiol & Immunol Wash Univ Sch Med 660 S Euclid Box 8045 St Louis MO 63110

KORSON, ROY, b Philadelphia, Pa, Oct 24, 22; m 46. PATHOLOGY. *Educ:* Univ Pa, AB, 43; Jefferson Med Col, MD, 47; Am Bd Path, dipl, 56. *Prof Exp:* Asst, Univ Vt, 50-51, asst prof, 51-52 & 54-57, assoc prof, 57-67, actg chmn dept, 74, PROF PATH, COL MED, UNIV VT, 67- *Concurrent Pos:* Nat Cancer Inst res fel, Columbia Univ, 48-49 & Col Med, Univ Vt, 49-50; USPHS sr res fel, 58-63; resident, Mary Fletcher Hosp, Burlington, Vt, 51-52. *Mem:* AAAS; Am Asn Pathologists; Col Am Pathologists; Int Acad Path; Sigma Xi; Am Soc Cytol. *Res:* Cytology; histopathology; histochemistry. *Mailing Add:* Med Alumni Bldg Univ Vt Col Med Burlington VT 05405-0068

KORSRUD, GARY OLAF, b Peterborough, Ont, Mar 23, 42; m 65; c 3. ANTIBIOTIC RESIDUE ANALYSIS. *Educ:* Univ Sask, BSA, 64, MSc, 66; Univ Calif, Davis, PhD(nutrit), 70. *Prof Exp:* Res asst, Univ Sask, 64-66; teaching asst animal sci, Univ Calif, Davis, 66-67, res asst, 67-70; res scientist, Health & Welfare Can, 70-77; RES SCIENTIST, AGR CAN, 77- *Mem:* Agr Inst Can; Can Soc Animal Sci; Can Soc Nutrit Sci; Soc Toxicol Can; Am Col Vet Toxicologists. *Res:* Antibiotic residue analysis research; nutritional and biochemical aspects of veterinary toxicology; human carbohydrate nutrition research and advising; lipid nutrition; detection and assessment of chemically induced liver damage. *Mailing Add:* Health Animals Lab Agr Can 116 Vet Rd Saskatoon SK S7N 2R3 Can

KORST, DONALD RICHARDSON, b Janesville, Wis, July 17, 24; m 48; c 3. INTERNAL MEDICINE. *Educ:* Univ Wis, MD, 48. *Prof Exp:* Resident Internal med, Univ Hosp, Univ Wis-Madison; chief radioisotope serv & hematologist, Ann Arbor Vet Admin Hosp, 55-65; assoc prof med, Sch Med, 65-70, asst dean educ, 71, dir independent study prog, 72, coordr educ, Dept Med, 74, prof & asst dean educ admin, 70-79, head sect gen internal med, Univ Wis-Madison, 77-79; dir educ internal med, Madison Gen Hosp, 65-79; PROF MED, SCH MED, BOSTON UNIV, 79-, CHIEF, SECT GEN INTERNAL MED, 79- *Concurrent Pos:* Fel hemat, Med Sch, Univ Wis; asst prof internal med, Sch Med, Univ Mich, 55-65; consult, St Joseph Mercy Hosp, Ann Arbor, Mich, 58-65. *Mem:* AAAS; Am Soc Nuclear Med; Am Soc Hemat; AMA; Am Fedn Clin Res. *Res:* General medicine and health care; medical education. *Mailing Add:* Dept Med/Pub Health Box 266 Cape Porpoise ME 04014

KORST, HELMUT HANS, b Vienna, Austria, Jan 4, 16; US citizen; m 42; c 4. GAS DYNAMICS, PROPULSION. *Educ:* Vienna Tech Univ, Dipl Ing, 41, Dr tech Sc, 47. *Prof Exp:* Res engr, Maschinenfabrik Augsburg-Nurnberg AG, Ger, 41-45; asst prof mech eng, Vienna Tech Univ, 45-48; vis lectr gas dynamics, 48-49, from assoc prof to prof mech eng, 49-84, head dept mech & indust eng, 62-74, EMER PROF MECH ENG, UNIV ILL, URBANA, 84- *Concurrent Pos:* Vis prof, Kans State Univ, 50, Va Polytech Inst, 54 & Vienna Tech Univ, 57; design specialist, Gen Dynamics Convair, Ft Worth, 55; sr fel, NSF, 57; consult, Gen Elec Co, 59; propulsion specialist, Rocketdyne Div, NAm Aviation, 60 & 65-68; owner, H H Korst engrs consult, Urbana, Ill, 56-; consult, Adv Group Aeronaut Res & Develop, NATO, 64 & US Army Missile Command, 71-; res chair naval air power, Navy Postgrad Sch, Monterey, Calif, 79; Ebaugh chair prof mech eng, Univ Fla, 84. *Mem:* Fel Am Soc Mech Engrs; fel Am Inst Aeronaut & Astronaut; Am Soc Eng Educ; Sigma Xi. *Res:* Internal and external aerodynamics; jet and rocket propulsion; heat transfer. *Mailing Add:* Three Eton Ct Champaign IL 61820

KORST, JAMES JOSEPH, b Joliet, Ill, Nov 24, 31; m 60; c 3. ORGANIC CHEMISTRY. *Educ:* Univ Ill, BS, 53; Dartmouth Col, MA, 55; Univ Wis, PhD(org chem), 59. *Prof Exp:* Chemist, Chas Pfizer & Co, Inc, 59-70, supvr, 70-71, mgr qual control, 71-73, OPERS MGR, QUALITY CONTROL, PFIZER INC, 73- *Mem:* Am Chem Soc. *Res:* Structures of steroid intermediates; tetracycline chemistry; quality control aspects of organic chemicals and pharmaceuticals; quality control management. *Mailing Add:* Qual Control Bldg 157 Pfizer Inc Groton CT 06340

KORST, WILLIAM LAWRENCE, b Joliet, Ill, Mar 23, 22; m 54; c 4. INORGANIC CHEMISTRY, PHYSICAL CHEMISTRY. *Educ:* Univ Chicago, PhB, 46, SB, 47, SM, 49; Univ Southern Calif, PhD, 56. *Prof Exp:* Asst res chemist, Univ Calif, 56-57; asst prof chem, Polytech Inst Brooklyn, 57-58; sr res chemist, Atomics Int Div, NAm Aviation, Inc, 58-59, res specialist, 59-67; instr chem, Los Angeles City Col, 69-70; from instr to asst prof, 70-75, assoc prof, 75-80, PROF CHEM, WEST LOS ANGELES COL, 80- *Concurrent Pos:* Fel US Atomic Energy Comn, Univ Southern Calif, 52-53; Fulbright scholar, Univ Amsterdam, 54-55; vis prof, Tech Univ Vienna, 78. *Mem:* Am Chem Soc. *Res:* X-ray diffraction and crystal structures; solid-state chemistry; heavy metal hydrides; high-vacuum and high-temperature techniques; atmospheric chemistry. *Mailing Add:* 7106 Quartz Ave Canoga Park CA 91306-3636

KORSTAD, JOHN EDWARD, b Woodland, Calif, July 4, 49; m 72; c 4. LIMNOLOGY, AQUACULTURE. *Educ:* Calif Lutheran Col, BA & BS, 72; Calif State Univ, Hayward, MS, 80; Univ Mich, Ann Arbor, MS, 79, PhD(zool), 80. *Prof Exp:* Teaching asst, Calif Lutheran Col, 70-71, Calif State Univ, Hayward, 72-74 & Univ Mich, Ann Arbor, 75-79; PROF BIOL, ORAL ROBERTS UNIV, 80- *Concurrent Pos:* Asst geologist, Cities Serv Oil Co, Alaska, 71; res asst, Calif State Univ, Hayward, 73; asst consult, Univ Calif, Davis, 74; asst limnologist, Great Lakes Res Div, Univ Mich, Ann Arbor, 75 & res asst, 75 & 80; vis scientist, Sintef Ctr for Aquacult, Trondheim, Norway, 87-88; col acad dir, Okla Acad Sci, 83-86. *Mem:* Am Soc Limnol & Oceanog; Sigma Xi; Ecol Soc Am; Great Plains Limnologist; World Aquacult Soc. *Res:* Ecology; limnology, particularly in phytoplankton-zooplankton interactions, nutrient regeneration, life history of zooplankton, scanning electron microscopy of zooplankton; aquaculture, particularly live feed with rotifers and Artemia; Tilapia. *Mailing Add:* Dept Biol Oral Roberts Univ Tulsa OK 74171

KORT, MARGARET ALEXANDER, b Jerusalem, Jordan, Jan 16, 28; US citizen. HISTOLOGY, CELL BIOLOGY. *Educ:* Georgetown Col, BS, 58; Univ Louisville, MS, 60; Univ Northern Colo, EdD, 68. *Prof Exp:* Instr biol, Coe Col, 61-63; prof biol, Southwest Baptist Col, 67-90; RETIRED. *Mem:* AAAS; Nat Sci Teachers Asn. *Res:* Acid phosphatase patterns in the involuting rat uterus. *Mailing Add:* 1623 W Northwood Bolivar MO 65613

KORTANEK, KENNETH O, b Chicago, Ill, Nov 13, 36; c 1. OPERATIONS RESEARCH, SYSTEMS ANALYSIS. *Educ:* Northwestern Univ, BSBA, 58, MA, 59, PhD(eng sci), 64. *Prof Exp:* Asst prof appl math & indust adminr, Univ Chicago, 65-66; assoc prof opers res, Cornell Univ, 66-69; PROF MATH SCIENCES, CARNEGIE-MELLON UNIV, 69- *Concurrent Pos:* Vis prof, Col Eng, Va Polytech Inst & State Univ, 79 & Univ NC, 81; pres, Kwel Corp, 81; mem, Int Symp Semi-Infinite Prog & Applns, 81. *Mem:* Opers Res Soc Am; Inst Mgt Sci; Am Math Soc; Economet Soc; Soc Indust & Appl Math. *Res:* Linear programming; duality theory in mathematical programming; applications to engineering plasticity design; equilibrium theory in economic systems; theory and applications of semi-infinite programming and design of telecommunications networks. *Mailing Add:* Dept Mgt Sci Univ Iowa Col Bus Admin Iowa City IA 52242

KORTE, WILLIAM DAVID, b Chicago, Ill, Oct 11, 37; m 64; c 3. ORGANIC CHEMISTRY. *Educ:* Northwestern Univ, BA, 60; Univ Mich, MS, 62; Univ Calif, Davis, PhD(chem), 66. *Prof Exp:* From asst prof to assoc prof, 66-75, chmn dept, 77-80, PROF CHEM, CALIF STATE UNIV, CHICO, 75- *Concurrent Pos:* Am Chem Soc-Petrol Res Fund res grants, 70-72; NSF grant, 83, 87; US Army res assoc, IPA, 87-89. *Mem:* AAAS; Am Chem Soc. *Res:* Stereochemistry; organometallic reaction mechanisms; organic analysis. *Mailing Add:* Dept Chem Calif State Univ Chico CA 95929

KORTELING, RALPH GARRET, b Madanapalle, SIndia, Jan 2, 37; US citizen; m 61; c 2. NUCLEAR CHEMISTRY. *Educ:* Hope Col, AB, 58; Univ Calif, Berkeley, PhD(chem), 63. *Prof Exp:* Fel chem, Carnegie Inst Technol, 62-63, asst prof, 63-65; from asst prof to assoc prof, 65-81, PROF CHEM, SIMON FRASER UNIV, 81- *Mem:* Am Phys Soc. *Res:* High energy nuclear reactions. *Mailing Add:* Dept Chem Simon Fraser Univ Burnaby BC V5A 1S6 Can

KORTH, GARY E, b Tremonton, Utah, Feb 27, 38; m 61; c 5. METALLURGY, MATERIALS SCIENCE. *Educ:* Univ Utah, BS, 63, PhD(metall), 68. *Prof Exp:* Test lab engr, Gen Dynamics-Convair, 63-64; MAT RES SCIENTIST, IDAHO NAT ENG LAB, 68- *Mem:* Am Soc Mat Int. *Res:* Elevated temperature fatigue and creep fatigue; mechanical properties; neutron irradiation effects of metals; rapidly solidified metals technology; dynamic consolidation of rapidly solidified metal powders using explosives. *Mailing Add:* RR 2 No 168 Blackfoot ID 83221

KORTIER, WILLIAM E, physics, nuclear engineering, for more information see previous edition

KORTRIGHT, JAMES MCDOUGALL, b Huntington, NY, Apr 3, 27; m 52; c 2. MEDICAL PHYSICS, RADIOLOGICAL PHYSICS. *Educ:* Cornell Univ, AB, 49; Purdue Univ, MS, 53, PhD(physics), 63. *Prof Exp:* From instr to asst prof, Temple Univ, 62-66; assoc prof physics, Rose-Hulman Inst Technol, 66-72; physicist, Radiol Sci Dept, Calif Col Med, Univ Calif, Irvine, 72-73; physicist, St Francis Hosp, Lynwood, Cal, 73-74; MED PHYSICIST, RADIATION CALIBRATION CO, 74- *Mem:* Am Asn Physicists Med. *Res:* X-ray diffraction; radiation damage; semiconductor properties; gamma ray scattering; radiological and health physics. *Mailing Add:* 436 E Hoover Ave Orange CA 92667-4820

KORWEK, ALEXANDER DONALD, b Madison, Ill, Feb 20, 32; m 75; c 4. MODERN MANAGEMENT TECHNIQUES, FINANCIAL MANAGEMENT TECHNIQUES. *Educ:* Wash Univ, St Louis, Mo, BSBA, 62; Univ Utah, Salt Lake City, MBA, 67. *Prof Exp:* Asst secy & asst treas, Hoechst Hystron Fibers Inc, 66-72; vpres finance, Reeves/Teletape Corp, 72-76; chief finance off & bus mgr, Queens Col, City Univ New York, 77-79; managing dir, Am Soc Civil Engrs, 79-81; secy & gen mgr, United Eng Trustees, Inc, 81-90; PRIN, A D KORWEK CONSULTS, 75-77 & 90- *Concurrent Pos:* Exec secy, Eng Found, 81-90; secy, Eng Socs Libr, 81-90, John Fritz Medal Bd, 81-90, Frank F Aplan Bd Award, 89-90. *Mem:* Coun Eng & Sci Soc Execs; NY Acad Sci; Am Soc Civil Engrs. *Res:* Management and cost containment in the plant environment inclusive of all interfaces with corporate and administrative functions; role of systems in all phases of modern business. *Mailing Add:* 27 Cool Water Ct Palm Coast FL 32137-8330

KORWIN-PAWLOWSKI, MICHAEL LECH, b Warsaw, Poland, Apr 10, 41; Can citizen; m 74; c 1. ELECTRICAL ENGINEERING. *Educ:* Warsaw Tech Univ, MS, 63; Univ Waterloo, Can, PhD(elec eng), 74. *Prof Exp:* Sr scientist elec eng, Inst Electron Technol, Polish Acad Sci, 63-69; res & teaching asst elec eng, dept elec eng, Univ Waterloo, Can, 69-74; product line mgr rectifiers, Erie Technol Prod, Can, 74-78; vpres & chief engr, Nat Semiconductors Ltd, 78-82; dir eng, Gen Instrument Taiwan, 82-87; MGR DEVELOP ENG, POWER SEMICONDUCTOR DIV, GEN INSTRUMENT CORP, 87- *Res:* Silicon rectifiers; transient voltage suppressors; semiconductor process technology; semiconductor devices. *Mailing Add:* Gen Instrument Corp 600 W John St Hicksville NY 11802

KORY, MITCHELL, b Brooklyn, NY, Jan 6, 14; m 43; c 2. BIOLOGICAL SCIENCE, MEDICINE. *Educ:* Univ Calif, Los Angeles, AB, 42, PhD(physiol bact), 53. *Prof Exp:* Instr bact, Univ Kans, 46-51; sr res biochemist res labs, 53-63, mgr pub info, 63-66, MGR MED EDUC SERV, ELI LILLY & CO, 66- *Mem:* Am Soc Biol Chemists; Brit Biochem Soc. *Res:* Infectious diseases including effects on host physiology and biochemistry; role and nature of host defenses; microbiology; antibiotic pharmacokinetics; antibiotic mechanisms of action and resistance including genetic aspects; therapeutic and ecological considerations of antibiotic development. *Mailing Add:* 129 Willow Rd Greenfield IN 46140

KORY, ROSS CONKLIN, b Petersburg, Va, Sept 17, 18; m 47; c 3. MEDICINE, PHYSIOLOGY. *Educ:* Columbia Univ, AB, 38, MD, 42. *Prof Exp:* Asst path, Sch Med, Emory Univ, 47-48; instr med, Sch Med, Vanderbilt Univ, 49-53; from asst prof to assoc prof, Med Col Wis, 54-60, prof clin res, 60-72; assoc chief of staff & chief pulmonary function lab, Wood Vet Admin Hosp, Milwaukee, 54-72; asst dean, 72-75, PROF MED, COL MED, UNIV S FLA, 72- *Concurrent Pos:* Vis lectr, Univ Valle, Colombia, 63; attend staff, St Joseph's & Univ Community Hosps, Tampa, Fla, 72-; chief of staff, Tampa Vet Admin Hosp, 72-75; med dir respiratory serv, Tampa Gen Hosp, 75-86, consult occup lung dis, 86- *Mem:* Am Physiol Soc; fel Am Col Physicians; fel Am Col Chest Physicians; Thoracic Soc. *Res:* Pulmonary physiology; obstructive pulmonary disease; sputum viscosity; electron microscopy of lung and sputum; non-obstructive pulmonary over-inflation; phonopneumography; occupational lung disease. *Mailing Add:* 1300 Crystal Dr PH-11 Arlington VA 22202-3234

KORYTNYK, WALTER, medicinal chemistry, biochemistry; deceased, see previous edition for last biography

KOS, CLAIR MICHAEL, b Washington, Iowa, Aug 6, 11; m 36; c 3. OTOLOGY. *Educ:* Univ Nebr, BSc & MD, 37; Harvard Univ, dipl, 39. *Prof Exp:* Intern, Bishop Clarkson Mem Hosp, Omaha, Nebr, 37-38; resident, Mass Eye & Ear Infirmary, 39-41; assoc surg, Div Otolaryngol, Duke Univ, 46-47; from asst prof to prof otolaryngol, Col Med, Univ Iowa, 47-60; dir, Iowa Found Otol, 60-80; pres, Otol Med serv, PC, 71-80; RETIRED. *Concurrent Pos:* Consult, Surgeon Gen, US Air Force Hosp, Washington, DC, 41-70, emer consult, 70-; US Air Force mem, Nat Res Coun; mem bioacoust, Nat Acad Sci-Nat Res Coun; dir, Am Bd Otolaryngol; exec secy-treas, Am Acad Ophthal & Otolaryngol, 69-78; exec vpres, Am Acad Otolaryngol, 78 & 79. *Honors & Awards:* Gold Medal, Am Cong Rehab Med, 51; Award, Am Acad Ophthal & Otolaryngol, 54. *Mem:* Am Laryngol, Rhinol & Otol Soc; Am Geriat Soc; Am Otol Soc (past pres); AMA; Am Col Surgeons. *Res:* Physiology of hearing. *Mailing Add:* One Knollwood Lane Iowa City IA 52240

KOS, EDWARD STANLEY, b Chicago, Ill, Aug 10, 28; m 52; c 2. MICROBIOLOGY. *Educ:* Loyola Univ, Ill, BS, 50; Marquette Univ, MS, 52; Univ Ill, PhD(microbiol), 58. *Prof Exp:* Actg instr life sci, Univ Calif, Riverside, 57-58; instr microbiol, Col Med, Univ Ill, 58-60; prof biol & chmn dept, Parsons Col, 60-61; assoc prof, 61-69, PROF BIOL, ROCKHURST COL, 69- *Mem:* Am Soc Microbiol; Sigma Xi. *Res:* Nutrition and metabolism of bacteria; Melanin pigmentation in Azotobacter chrococcum. *Mailing Add:* 5926 McGee St Kansas City MO 64113

KOS, JOSEPH FRANK, b Montreal, Que. SOLID STATE PHYSICS. *Educ:* Univ Waterloo, BSc, 62; Univ Ottawa, PhD(physics), 67. *Prof Exp:* Asst prof physics & astron, Regina Campus, Univ Sask, 67-71, assoc prof, 71-74; assoc prof, 74-81, PROF PHYSICS & ASTRON, UNIV REGINA, 81- *Mem:* Can Asn Physicists; Electrochem Soc Am; Solar Energy Soc Can Inc. *Res:* Measurements of electron transport properties of metals; design and construction of gravitational antenna; photoelectrochemical conversion of solar energy to electricity or to production of hydrogen gas. *Mailing Add:* Dept Physics & Astron Univ Regina Regina SK S4S 0A2 Can

KOSAI, KENNETH, b Spokane, Wash, July 27, 44; m 62; c 1. SEMICONDUCTOR DEVICE PHYSICS, INFRARED DETECTORS. *Educ:* Calif Inst Technol, BS, 66; Univ Southern Calif, MSEE, 68, PhD(elec eng), 73. *Prof Exp:* Mem tech staff, Philips Labs, Briarcliff Manor, NY, 73-81; MEM TECH STAFF, SANTA BARBARA RES CTR, 81- *Concurrent Pos:* Vis scientist, Dept Solid State Physics, Univ Lund, Sweden, 79. *Mem:* Inst Elec & Electronics Engrs; Am Phys Soc; Sigma Xi. *Res:* Device physics of semiconductor heterojunctions and infrared detectors; semiconductor device modeling. *Mailing Add:* 234 Old Ranch Dr Goleta CA 93117

KOSAK, ALVIN IRA, b New York, NY, Feb 29, 24; m 58; c 3. ORGANIC CHEMISTRY. *Educ:* City Col New York, BS, 43; Ohio State Univ, PhD(org chem), 48. *Prof Exp:* Res chemist, Socony-Vacuum Oil Co, 43-45; asst chem, Ohio State Univ, 45-46, asst instr, 48; Jewett fel, Harvard Univ, 48-49; asst prof, Univ Cincinnati, 49-52; asst prof indust med, NY Univ, 52-56, assoc prof chem, 56-62, chmn dept, 62-65, head, All-Univ Dept, 65-77, actg dean fac arts & sci, 77-78, dir grad study chem, 84-87, PROF CHEM, NY UNIV, 62- *Concurrent Pos:* USPHS spec fel, Univ Zurich, 62. *Mem:* Fel AAAS; Am Chem Soc; fel NY Acad Sci. *Res:* Thiophene chemistry; natural products; polynuclear hydrocarbons. *Mailing Add:* Dept Chem Rm 514 NY Univ Four Washington Pl New York NY 10003

KOSAK, JOHN R, b Wilmington, Del, May 18, 30; m 57; c 3. INDUSTRIAL ORGANIC CHEMISTRY. *Educ:* Univ Del, BS, 51, MS, 52; Mich State Univ, PhD(org chem), 57. *Prof Exp:* Instr org & gen chem, Ferris State Col, 55-56; sr chemist, 57-80, RES ASSOC, E I DU PONT DE NEMOURS & CO, INC, 80- *Mem:* Am Chem Soc; Catalysis Soc. *Res:* Catalysis; catalytic hydrogenation. *Mailing Add:* 103 Willowspring Rd Wilmington DE 19807

KOSANKE, ROBERT MAX, b Park Ridge, Ill, Sept 4, 17; m 41; c 2. PALEOBOTANY. *Educ:* Coe Col, BA, 40; Univ Cincinnati, MS, 42; Univ Ill, PhD(paleobot), 52. *Prof Exp:* Lab instr geol, Coe Col, 39-40; asst bot, Univ Cincinnati, 40-43; asst, Ill Geol Surv, 43, from asst geologist to geologist, 43-63; GEOLOGIST, US GEOL SURV, 63- *Concurrent Pos:* Assoc prof bot, Univ Ill, 59-63. *Honors & Awards:* Cady Award, Geol Soc Am, 89. *Mem:* AAAS; fel Geol Soc Am; Bot Soc Am; Soc Econ Geol; Paleont Soc. *Res:* Spore studies of coal beds of Pennsylvania age; Pennsylvanian stratigraphy and paleobotany. *Mailing Add:* US Geol Surv MS 919 Box 25046 Denver Fed Ctr Denver CO 80225

KOSANOVICH, ROBERT JOSEPH, b Monroe, Mich, Sept 27, 38; m 60; c 3. MATHEMATICS. *Educ:* Eastern Mich Univ, BS, 60, MA, 62; Univ Detroit, MA, 63; Mich State Univ, PhD, 72. *Prof Exp:* PROF MATH, FERRIS STATE COL, 65-, HEAD DEPT, 75- *Mailing Add:* Ferris State Univ Ferris State Col 901 South St Street Big Rapids MI 49307

KOSARAJU, S RAO, b Cadhra Pradesh, India, Feb 20, 43. COMPUTER THEORY. *Educ:* Univ Pa, PhD(comput sci eng) 69. *Prof Exp:* PROF, JOHN UNIV, 77- *Mem:* Asn Comput Mach; Soc Int Appl Math; Instr Elec & Electronics Engrs. *Mailing Add:* Dept Comput Sci New Eng Bldg Johns Hopkins Univ Baltimore MD 21218

KOSARIC, NAIM, b Sarajevo, Yugoslavia, Sept 27, 28; Can citizen; m 55; c 2. WATER POLLUTION, BIOTECHNOLOGY. *Educ:* Univ Zagreb, dipl, 55; Univ Western Ont, PhD(biochem), 69. *Prof Exp:* Process engr & lab head, Iron, Coke Oven & Steelworks, Zenica, Yugoslavia, 56-59; process mgr, Pulp & Paper Indust, Maglay, 59-61; sr process eng, petrochem, Organic Chem Indust, Zagreb, Yugoslavia, 61-65; sr res eng, Bur Res & Partic Ministry, Rabat, Morocco, 65-66; res asst, Dept Biochem, Univ Western Ont, 66-69; from asst prof to assoc prof, 69-79, PROF CHEM & BIOCHEM ENG, FAC ENG, SCI, UNIV WESTERN ONT, 79- *Concurrent Pos:* Chmn chem & biochem eng, Univ Western Ont, 77-80 & 84-85; vis prof, Swiss Fed Inst Technol, Switz, 75-76, Inst Fermentation & Biotechnol, Tech Univ, Ger, 83, Inst Microbiol, Czech Acad Sci, Prague, 75 & 84, Inst Biochem Technol & Microbiol Tech Univ, Vienna, Austria, 84 & Fed Inst Biotechnol, Fed Repub Ger, 84; expert, CIDA consult, Agr & Food Eng, Unicamp, Compinas, Brazil, 79, 81 & 83; vis scientist, USSR Acad Sci, Moscow, 57-58 & Europ Nuclear Ctr, Mol, Belgium, 60; vis engr, Nippon Kokan Chem, Tokyo-Kawasaki, Japan, 62; consult, indust & govt, 70- *Mem:* Am Inst Chem Engrs; NY Acad Sci; Am Oil Chemists Soc; Can Assoc Water Pollution Res & Control; Can Inst Food Sci Technol; Can Soc Chem Eng. *Res:* Development of new processes and products in biochemical and food engineering; industrial wastewater treatment; biotechnology; microbial facts and oils; biosurfactants; fuel alcohol; anaerobic digestion of industrial pollutants; microbial detoxification; microbial proteins; economics in biotechnology. *Mailing Add:* Dept Chem & Biochem Eng Univ Western Ont London ON N6A 5B9 Can

KOSASKY, HAROLD JACK, b Winnipeg, Man, Oct 19, 27; m 55; c 3. GYNECOLOGY. *Educ:* Univ Man, BA, 48, MD, 53; FRCS, 60; Am Bd Obstet & Gynec, dipl, 64. *Prof Exp:* Rotating intern, Deer Lodge Vet & Grace Hosps, Winnipeg, Can, 52-53; resident gen surg, Colonel Belcher Hosp, Calgary, 53-54; resident psychiat, Warren State Hosp, Warren, Pa, 55-56; asst resident obstet & gynec, Chicago Lying-In Hosp, Univ Chicago, 56-58, sr res, 58-59; from asst prof to assoc prof, Sch Med, Univ Louisville, 61-65; INSTR OBSTET & GYNEC, HARVARD MED SCH, 66- *Concurrent Pos:* Exchange fel, Univ Durham, 59-60; dir dept obstet & gynec, Cambridge Hosp, 68-70, jr assoc surgeon, Peter Bent Brigham Hosp, 66-80, obstetrician & gynecologist, Boston Hosp Women, 66-80; consult, Jordan Hosp, 69-; active staff, Brigham & Women's Hosp, 80- *Mem:* Fel Am Col Obstetricians & Gynecologists; fel Am Col Surgeons; Asn Profs Gynec & Obstet; Royal Col Obstetricians & Gynecologists. *Res:* Endocrinology; gynecologic surgery; infertility; clinical gynecology. *Mailing Add:* 25 Boylston St Chestnut Hill MA 02167

KOSBAB, FREDERIC PAUL GUSTAV, b Berlin, Ger, Mar 29, 22; US citizen; m 51. PSYCHIATRY, INTERNAL MEDICINE. *Educ:* Univ Berlin, MD, 45; Am Bd Psychiat & Neurol, dipl, 63; Okla State Univ, MA, 89. *Prof Exp:* Intern, Army Hosps & Refugee Infirmary, Friedland, WGer, 45-46; resident internal med, Dist Hosp, Hannoversch Muenden, WGer, 46-48 & Evangel Hosp, Goettingen-Weende, WGer, 48-51; pvt pract internal med, 51-55; rotating intern, Swed Covenant Hosp, Chicago, Ill, 56-57; staff physician I, Psychiat Serv, Manteno State Hosp, Ill, 57-58; resident psychiat, Col Med, Univ Nebr, 58-59; staff physician II, Psychiat Serv, Northern State Hosp, Sedro-Woolley, Wash, 59-60; resident psychiat, Sch Med, Univ Wash, 60-61 & Northern State Hosp, Sedro-Woolley, Wash, 61-62; clin instr psychiat, Med Sch, Univ Ore, 62-64; from asst prof to prof psychiat, Med Col Va, Va Commonwealth Univ, 64-73, dir residency training in psychiat, 66-69, from actg chmn dept to assoc chmn dept, 69-73; med dir, E Plains Ment Health Ctr, 74-77; chief psychiat servs, Hampton, Va Med Ctr, 77-82; prof psychiat & chmn dept, 82-86, EMER PROF BIOL MED & PSYCHIAT & CHMN DEPT, SCH MED, ORAL ROBERTS UNIV, TULSA, 87- *Concurrent Pos:* Consult, WGer Vet Admin, Landau, 52-56 & Residency Training Prog, East State Hosp, Williamsburg, Va, 68-73; sr staff psychiatrist & unit med dir, Ore State Hosp, Salem, 62-63; pvt pract psychiat, 63-; mem, Med Col Va Hosps, 64-73, bd gov, City Faith Med & Res Ctr, 82-, Med Pract Coun, Conf Med & Res Ctr, 82-; attend & consult, McGuire Vet Admin Hosp, Richmond, Va, 64-73; chmn, Comt Postgrad Training in Psychiat, 67-70; mem dean's comt, Richmond Vet Admin Hosp, 69-70; prof dir NIMH grant, 69-73; prof, Dept Psychiat & Behav Sci, Eastern Va Med Sch, 77-82; chief, Dept Behav Med & Psychiat, City Faith Med & Res Ctr, Tulsa, Okla, 82-86. *Mem:* Fel Am Psychiat Asn; hon fel Arbeitsgemeinschaft F Katathymes Bilderleben, WGer; AMA. *Res:* Contribution to the problem of superfetation and superfecundation in twins; camptocormia in the female; introduction of a buddy system for hospitalized geriatric patients; symbol formation; affective imagery and its didactic uses in psychiatry; teaching and learning in medical school. *Mailing Add:* PO Box 701677 Tulsa OK 74170-1677

KOSCHIER, FRANCIS JOSEPH, b New York, NY, June 16, 50. TOXICOLOGY, PHARMACY. *Educ:* Bard Col, AB, 72; Univ Miss, PhD(pharmacol), 76. *Prof Exp:* Res asst prof, State Univ NY, Buffalo, 76-79; sr toxicologist, Food & Drug Res Labs, 79-80; sr toxicologist, Am Cyanamid Corp, 80-83; mgr toxicol, Ciba-Geigy Corp, 83-89; SR TOXICOL CONSULT, ARCO, 89- *Mem:* Soc Toxicol; Am Soc Pharmacol & Exp Therapeut; Soc Environ Toxicol & Anal Chem. *Res:* Health and environmental risk assessment of petroleum products and synthetic chemicals. *Mailing Add:* Dept Environ Protection ARCO 515 S Flower St Los Angeles CA 90071

KOSCHMIEDER, ERNST LOTHAR, b Danzig, Ger, May 1, 29; m 62; c 2. FLUID DYNAMICS. *Educ:* Univ Bonn, dipl physics, 58, Dr rer nat(physics), 63. *Prof Exp:* Res assoc, Univ Chicago, 65-67; from asst prof to assoc prof, 68-85, PROF, UNIV TEX, AUSTIN, 86- *Concurrent Pos:* Res fel, Harvard Univ, 63-65; Consult, Apollo XVII Convection Exp, Lockheed Space & Missiles Co, 71-74; mem, Ctr Statist Mech, Univ Tex, Austin, 72-; sr vis fel, Nat Ctr Atmospheric Res, Colo, 73-74; vis sci, Ctr Nuclear Studies, Saclay, France, 81. *Mem:* Am Phys Soc; Am Acad Mech. *Res:* Hydrodynamic stability. *Mailing Add:* E Cockrell Jr Hall 9-126 Univ Tex Austin TX 78712

KOSCO, JOHN C(ARROLL), b Du Bois, Pa, Sept 20, 32; m 56; c 6. METALLURGY. *Educ:* Univ Notre Dame, BS, 54; Princeton Univ, MSE, 56; Pa State Univ, PhD(metall), 58. *Prof Exp:* Res metallurgist, Stackpole Carbon Co, 58-66, chief engr metals, 66-67, dir metall res, 67-71; DIR POWDER METALL RES, KEYSTONE CARBON CO, 71- *Mem:* Am Soc Metals; Am Inst Mining, Metall & Petrol Engrs; Am Chem Soc. *Res:* Powder metallurgy of ferrous and non-ferrous materials, high temperature materials and electrical contacts; electrical ceramics and thermoelectric materials; coatings and metal joining. *Mailing Add:* Keystone Carbon Co Powder Metall Res Dept St Marys PA 15857

KOSEL, GEORGE EUGENE, b Rochester, NY, July 22, 23; m 50; c 3. ELECTROPHOTOGRAPHY. *Educ:* Cornell Univ, AB, 44; Univ Rochester, MS, 51. *Prof Exp:* Res assoc biochem, Atomic Energy Comn, Univ Rochester, 47-50; chief biochemist dental res, Passaic Gen Hosp, NJ, 50-54; mgr res graphic arts, Philip A Hunt Chem Corp, 54-67, mgr res, Electrostatic Div, 67-70, asst dir res, 70-75, dir basic chem res, 75-81, consult, electrophotography, 82-83; CHIEF CHEMIST, AM GAS & CHEM CO, LTD, 83- *Mem:* AAAS; Am Chem Soc. *Res:* Fluoride metabolism; solid state and physical chemistry; powders and liquid developers for electrophotography; chemicals for non-destructive testing. *Mailing Add:* 181 North Ave Park Ridge NJ 07656

KOSEL, PETER BOHDAN, b Northeim, Ger, Aug 20, 46; Australian citizen; m 73; c 2. MICROELECTRONIC DEVICES. *Educ:* Univ Sydney, Australia, BSc, 68; Univ New South Wales, Australia, PhD(elec eng), 76. *Prof Exp:* Prof officer elec eng, Univ NSW, 73-80; ASSOC PROF ELEC ENG, UNIV CINCINNATI, 80- *Mem:* Inst Elec & Electron Engrs; Sigma Xi. *Res:* High speed charge-coupled devices in gallium arsenite and fabrication technology of compound semiconductor devices; computer aided design and simulation of signal processing devices and circuits; submicron-line lithography and very-large-scale integration pattern generation. *Mailing Add:* Elec Comput Eng Dept ML No 30 Univ Cincinnati Cincinnati OH 45221

KOSERSKY, DONALD SAADIA, b Waterbury, Conn, Oct 16, 32; m 60. PHARMACOLOGY. *Educ:* Univ Conn, BA, 57, MS, 68; Univ of the Pac, PhD, 71. *Prof Exp:* Res assoc, Sch Med, Univ NC, Chapel Hill, 71-73; asst prof, Northeastern Univ, 73-77, assoc prof pharmacol, 77-81; ASSOC PROF PHARMACOL & COORDR GRAD PROG, MASS COL PHARM & ALLIED HEALTH SCI, 81- *Mem:* AAAS; Am Soc Pharmacol & Exp Therapeut; Neurosci Soc. *Res:* Autonomic and central nervous system pharmacology; pharmacology of addiction and drugs of abuse; classical pharmacology. *Mailing Add:* Mass Col Pharm & Allied Health Sci 179 Longwood Ave Boston MA 02115

KOSH, JOSEPH WILLIAM, b Hempstead, Tex, Sept 30, 40. NEUROPHARMACOLOGY. *Educ:* Univ Tex, BS, 64, MS, 67; Univ Colo, PhD(pharmacol), 71. *Prof Exp:* From asst prof to assoc prof pharmacol, 71-80, AT DEPT PHARMACOL, UNIV SC, COLUMBIA. *Mem:* AAAS; Sigma Xi. *Res:* Cardiovascular and neuropharmacology; pharmacology of gamma-aminobutyric acid intermediates and relation to convulsive threshold. *Mailing Add:* Dept Pharmacol Univ SC Main Campus Columbia SC 29208

KOSHEL, RICHARD DONALD, b Argo, Ill, Feb 1, 36; m 62, 80; c 2. NUCLEAR PHYSICS. *Educ:* Univ Ill, BS, 58, MS, 59; Univ Kans, PhD(theoret nuclear physics), 63. *Prof Exp:* From asst prof to prof physics, 63-85, assoc dean, Col Arts & Sci, Ohio Univ, 80-85; PROF PHYSICS, 85-, ASSOC PROVOST RES & DEAN GRAD STUDIES, ILL STATE UNIV, 85- *Concurrent Pos:* Vis prof, Fla State Univ, 69-70 & Univ Md, 78-79. *Mem:* Am Phys Soc; Am Asn Physics Teachers. *Res:* Theoretical nuclear physics and numerical analysis; nuclear structure using many body techniques and nuclear reactions; non-linear system. *Mailing Add:* Grad Sch Miss State Univ PO Box G Mississippi State MS 39762

KOSHER, ROBERT ANDREW, b Key West, Fla, Mar 1, 45; m 68; c 3. DEVELOPMENTAL BIOLOGY. *Educ:* Wilkes Col, BA, 67; Temple Univ, PhD(biol), 72. *Prof Exp:* NIH fel anat, Sch Med, Univ Pa, 72-74; asst prof, 74-80, ASSOC PROF ANAT, UNIV CONN HEALTH CTR, 80- *Mem:* Soc Develop Biol; Am Soc Zoologists; Am Asn Anatomists. *Res:* The role of extracellular matrix components in tissue interactions and other developmental processes; the control of somite chondrogenesis by extracellular matrix components produced by the embryonic notochord and spinal cord. *Mailing Add:* Dept Anat Univ Conn Sch Med Farmington CT 06032

KOSHI, JAMES H, b Agate, Colo, June 13, 19; m 77. DAIRY SCIENCE, ANIMAL SCIENCE. *Educ:* Colo Agr & Mech Col, BS, 48; Univ Minn, PhD(dairy sci), 55. *Prof Exp:* Dairy specialist & prof, Univ Hawaii, 55-74; gen mgr agr, Micronesian Develop Co, Tinian, 75-77; consult dairy prod, Hawaiian Agron Co Int, Iran, 77-78; gen mgr, 50th State Diary Farmers Coop, 79-82, consult, 83-85; RETIRED. *Concurrent Pos:* Consult & prof, Kasetsart Univ Thailand, 62-65. *Mem:* Am Dairy Sci Asn; Sigma Xi. *Res:* All areas of dairy cattle management and milk production. *Mailing Add:* 2333 Kapiolani Blvd Apt 2011 Honolulu HI 96826-4444

KOSHLAND, DANIEL EDWARD, JR, b New York, NY, Mar 30, 20; m 45; c 5. BIOCHEMISTRY. *Educ:* Univ Calif, BS, 41; Univ Chicago, PhD(biochem), 49. *Hon Degrees:* PhD, Weizman Inst Sci, 84, DSc, Carnegie-Mellon Univ, 85, LLD, Simon Fraser Univ, 86. *Prof Exp:* Anal chemist, Shell Chem Co, 41-42; asst, Manhattan Proj, Univ Chicago, 42-44, group leader, 44-46; fel, Harvard Univ, 49-51; assoc biochemist, Brookhaven Nat Lab, 51-54, biochemist, 54-56, sr biochemist, 56-65; PROF BIOCHEM, UNIV CALIF, BERKELEY, 65- *Concurrent Pos:* Affil, Rockefeller Univ, 58-65; mem panel, USPHS, 59-64; vis fel, All Souls Col, Oxford, 72-73; fel, Guggenheim Found, 72-73; ed-in-chief, Sci Mag, 85- *Honors & Awards:* T Duckett Jones Award, Helen Hay Whitney Found, 77; Distinguished Lectr Award, Soc Gen Physiol, 78; Pauling Award & Edgar Fahs Smith Award, Am Chem Soc, 79; Rosentiel Award, Brandeis Univ, 84; Waterford Prize, Scripps Clin & Res Found, 84; Nat Medal Sci, 90; Merol Award, Am Soc Biochem

& Molecular Biol, 91. *Mem:* Nat Acad Sci; AAAS; Am Chem Soc; Am Soc Biol Chemists (pres, 73-74); Japanese Biochem Soc; Royal Swed Acad Sci; Am Acad Arts & Sci; Am Philos Soc. *Res:* General principles of enzymology and regulatory control; understanding of memory and sensory processes; correlation of protein structure and function. *Mailing Add:* Dept Biochem Univ Calif Berkeley CA 94720

KOSHLAND, MARIAN ELLIOTT, b New Haven, Conn, Oct 25, 21; m 45; c 5. MOLECULAR BIOLOGY. *Educ:* Vassar Univ, BA, 42; Univ Chicago, MS, 43, PhD(bact), 49. *Prof Exp:* Asst, Cholera Proj, Off Sci Res & Develop, Chicago, 43 & 44- 45; asst, Comn Air Borne Dis, Colo, 43-44; jr chemist, Atomic Bomb Proj, Manhattan Dist, Tenn, 45-46; from assoc bacteriologist to bacteriologist, Brookhaven Nat Lab, 53-65; from assoc res immunologist to res immunologist, 65-70, chmn Dept Microbiol & Immunol, 82-89, PROF IMMUNOL, DIV IMMUNOL, UNIV CALIF, BERKELEY, 70- *Concurrent Pos:* Fel bact & immunol, Harvard Univ, 49-51; mem, Nat Sci Bd, NSF, 76-82, coun, Nat Acad Sci, 85-88, Comn Life Sci, Nat Res Coun, 89; vis prof, Cancer Ctr, Mass Inst Technol, 79 & 85-86. *Honors & Awards:* R E Dyer Lectr, NIH, 88; Excellence in Sci Award, Fedn Am Soc Exp Biol, 89. *Mem:* Nat Acad Sci; Am Asn Immunologists (pres, 82-83); Am Soc Biol Chemists; Sigma Xi; Am Acad Microbiologists. *Res:* Mechanism of antibody biosynthesis; lymphokine regulation of immunoglobin gene expression; mechanisms of lymphokine signaling of B lymphocytes. *Mailing Add:* Dept Molecular & Cell Biol LSA439 Univ Calif Berkeley CA 94720

KOSHY, K THOMAS, b Kerala, India, Sept 22, 24; m 50; c 2. PHARMACY, PHARMACEUTICAL CHEMISTRY. *Educ:* Kerala Univ, India, BSc, 43; Benares Hindu Univ, BA, 48; Univ Iowa, MS, 58, PhD (pharm & pharmaceut chem), 60. *Prof Exp:* Mfg chemist, Sterling Pharmaceut, India, 48-49; jr sci asst, Inspectorate of Gen Stores Lab, 49-51; med serv rep, Parke Davis & Co, Ltd, 51-56; asst col pharm, Univ Iowa, 60-61; sr res pharmacist, Miles Labs, Inc, 61-66; SR RES SCIENTIST, UPJOHN CO, 66- *Mem:* Acad fel Am Pharmaceut Asn; Am Chem Soc. *Res:* Analytical methods development for drugs and pharmaceuticals; kinetic studies and stability testing of pharmaceuticals; residue analysis in plants and animals; metabolism in plants and animals; photolysis of pesticides and herbicides. *Mailing Add:* Agr Prod Div Upjohn Co Kalamazoo MI 49001

KOSIBA, WALTER LOUIS, b Braddock, Pa, Feb 13, 21. PHYSICAL CHEMISTRY. *Educ:* Canisius Col, BS, 43; Ohio State Univ, MSc, 49, PhD(chem), 51. *Prof Exp:* Res assoc, S A M Labs, Columbia Univ, 44-45; chemist, Phys Chem, Uranium, Tenn Eastman Corp, 45-46; res chemist, Phys Chem Solids, Vitro Corp Am, 51-53; assoc physicist, Brookhaven Nat Lab, 53-58; mem res staff, Gen Atomic Div, Gen Dynamics Corp, 58-61; consult, European Atomic Energy Community, Belgium, 61-63; sr scientist, Nuclear Dept, Douglas Aircraft Co, Inc, 64-66 & Aerospace Corp, 66; asst to dir, NAm Rockwell Sci Ctr, 66-70; SPECIALIST, LA JOLLA RADIOCARBON & TRITIUM LAB, UNIV CALIF, SAN DIEGO, 71- *Concurrent Pos:* Consult, Int Atomic Energy Agency, Austria, 61. *Mem:* AAAS; Am Chem Soc; Sigma Xi. *Res:* Materials sciences; radiation effects; solid state chemistry; radiocarbon dating. *Mailing Add:* 3920 Ingraham St Apt 118 San Diego CA 92109

KOSIER, FRANK J, b Lansing, Mich, July 2, 34; m 52; c 2. MATHEMATICS. *Educ:* Mich State Univ, BS, 56, MS, 57, PhD(math), 60. *Prof Exp:* Instr math, Univ Calif, Berkeley, 60-61 & Univ Wis, 61-63; asst prof, Syracuse Univ, 63-64 & Univ Wis-Madison, 64-66; assoc prof, 66-69, PROF MATH, UNIV IOWA, 69- *Mem:* Am Math Soc; Math Asn Am. *Res:* Non-associative rings. *Mailing Add:* Dept Math Univ Iowa Iowa City IA 52240

KOSIEWICZ, STANLEY TIMOTHY, b Chicago, Ill, July 21, 44; m 89; c 1. NUCLEAR CHEMISTRY, NUCLEAR WASTE MANAGEMENT. *Educ:* Univ Ill, BS, 67; Univ Wis, MS, 69, PhD(anal chem), 73. *Prof Exp:* Process engr chem eng, Olin Corp, 68-69; independent consult, 86-89; staff mem anal chem, 73-86, STAFF SCIENTIST, NUCLEAR WASTE MGT, LOS ALAMOS NAT LAB, UNIV CALIF, 89- *Mem:* Am Chem Soc; AAAS. *Res:* Transuranium radioactive waste degradation; nuclear waste management; trace element geochemistry and archaeology; automation of analytical instrumentation. *Mailing Add:* Group HSE-7 MS J594 Los Alamos Nat Lab Los Alamos NM 87544

KOSIKOWSKI, FRANK VINCENT, b Torrington, Conn, Jan 10, 16; m 44; c 1. FOOD SCIENCE. *Educ:* Univ Conn, BS, 39; Cornell Univ, MS, 41, PhD(dairy chem), 44. *Prof Exp:* Instr, 42-44, res assoc, 44-45, from asst prof to prof, 45-86, EMER PROF FOOD SCI, CORNELL UNIV, 86- *Concurrent Pos:* Consult, Food & Agr Orgn, UN, 52-; Fulbright res scholar, France, 55; State Dept exchange scholar, Ireland, 59; mem expert adv comt food hyg, WHO, 70. *Honors & Awards:* Borden Award, Am Dairy Sci Asn, 55 & Pfizer Award, 60; Int Award, Inst Food Technol, 83. *Mem:* AAAS; Am Chem Soc; Am Dairy Sci Asn; Inst Food Technol. *Res:* Chemistry and bacteriology of milk and cheese products, especially flavor reactions, analytical methods and enzyme activity; foods from fermentations; international food development. *Mailing Add:* Dept Food Sci Stocking Hall Cornell Univ Ithaca NY 14853

KOSINSKI, ANTONI A, b Warsaw, Poland, May 25, 30; div; c 1. MATHEMATICS. *Educ:* Univ Warsaw, PhD, 56. *Prof Exp:* Asst prof inst math, Polish Acad Sci, 56-59 & Univ Calif, Berkeley, 59-62; mem Inst Adv Study, 62-64; assoc prof math, Univ Calif, Berkeley, 64-66; PROF MATH, RUTGERS UNIV, 66- *Mem:* Am Math Soc. *Res:* Topology and differential topology. *Mailing Add:* Dept Math Rutgers Univ New Brunswick NJ 08903

KOSINSKI, ROBERT JOSEPH, b Montclair, NJ, Jan 8, 49; m 89; c 2. POPULATION ECOLOGY, LIMNOLOGY. *Educ:* Seton Hall Univ, BS, 72; Rutgers Univ, PhD(ecol), 77. *Prof Exp:* Asst prof biol, Tex A&M Univ, 77-84; ASSOC PROF, BIOL PROG, CLEMSON UNIV, 84- *Concurrent Pos:* Grants, Environ Protection Agency, Nat Sci Found, Dept Educ. *Mem:* AAAS; Ecol Soc Am; Am Soc Limnol & Oceanog; Nat Asn Biol Teachers;

Nat Sci Teachers Asn. *Res:* stream ecology; primary productivity in streams; effects of pesticides in streams; computer modeling of antigenic variation of trypanosome infections; use of computers as teaching tools. *Mailing Add:* Biol Prog Clemson Univ 330 Long Hall Clemson SC 29634-1902

KOSKI, RAYMOND ALLEN, b Corvallis, Ore, Aug 21, 51; m 79; c 2. GROWTH FACTORS, ONCOGENES. *Educ:* Stanford Univ, BS, 73; Yale Univ, MPhil, 75, PhD(biol), 78. *Prof Exp:* Fel molecular biol, Inst Molecular Biol II, Univ Zurich, 78; fel, Dept Microbiol, Univ Geneva, 78-79; fel, Dept Genetics, Univ Wash, 79-81; RES SCIENTIST, AMGEN INC, 81- *Res:* Regulation of cellular proliferation by growth factors and proto- oncogenes. *Mailing Add:* Amgen Inc Amgen Ctr Thousand Oaks CA 91320-1789

KOSKI, WALTER S, b Philadelphia, Pa, Dec 1, 13; m 40; c 4. PHYSICAL CHEMISTRY. *Educ:* Johns Hopkins Univ, PhD(phys chem), 42. *Prof Exp:* Res chemist, Hercules Powder Co, 42-43; group leader, Los Alamos Sci Lab, 44-47; assoc prof phys chem, 47-55, chmn dept, 55-74, PROF CHEM, JOHNS HOPKINS UNIV, 55-, BERNARD N BAKER PROF CHEM, 74- *Concurrent Pos:* Physicist, Brookhaven Nat Lab, NY, 47-48; consult chem corps, US Army, 49- *Mem:* Am Chem Soc; fel Am Phys Soc. *Res:* Radioactive and stable isotopes as tracers; chemistry of boron hydrides; electron and nuclear magnetic resonance; mass spectroscopy; nuclear chemistry; ion-molecule reactions; reactive scattering of ions: mechanism of drug action. *Mailing Add:* Dept Chem Johns Hopkins Univ 3400 N Charles St Baltimore MD 21218

KOSKO, ERYK, aerospace engineering; deceased, see previous edition for last biography

KOSKY, PHILIP GEORGE, b London, Eng,; m; c 2. MATERIALS SCIENCE. *Educ:* Univ London, BSc; Univ Calif, Berkeley, MS, PhD(chem eng). *Prof Exp:* Res asst chem eng, Univ Calif, Berkeley; sr sci officer, Harwell Nat Lab, Eng; chem engr, Res & Develop Ctr, Gen Elec; assoc prof mech eng & mech, Lehigh Univ; mgr fuel sci unit, STAFF CHEM ENGR, GE RES & DEVELOP CTR, 77- *Concurrent Pos:* Adj assoc prof, Union Col NY. *Mem:* Am Chem Soc. *Res:* Applied polymer chemistry; kinetics; chemical vapor deposition (CVD) and materials science. *Mailing Add:* GE Res & Develop Ctr PO Box 8 Schenectady NY 12301-0008

KOSLOW, JULIAN ANTHONY, b Los Angeles, Calif, May 14, 47; m 79; c 2. FISHERIES OCEANOGRAPHY. *Educ:* Harvard Univ, BA, 69; Univ Wash, BA, 73; Univ Calif, San Diego, PhD(biol oceanog), 80. *Prof Exp:* Res asst biol oceanog, Scripps Inst Oceanog, Univ Calif, San Diego, 74-79; asst prof Fisheries Oceanog, Oceanog Dept, Dalhousie Univ, 80-88; res assoc, dept Fisheries & Oceans, Halifax, 88-89; SR RES SCIENTIST, DIV FISHERIES, COMMONWEALTH SCI & IND RES ORGN, 90- *Concurrent Pos:* Hon res assoc, Zool Dept, Univ West Indies, Jamaica, 86. *Honors & Awards:* E W Fager Award, Scripps Inst Oceanog, 80; Chapman-Schaefer Award, Marine Technol Soc, 75. *Res:* Deep water fisheries biology; oceanography; management; physical oceanography and plankton behavior; the regulation between stock and recruitment in fish populations; interactions with biological and climatic change and larval ecology; effect of fisheries management on fishing communities; reef fish ecology and management. *Mailing Add:* Div Fisheries Commonwealth Sci & Ind Res Orgn Marine Lab GPO Box 1538 Hobart Tasmania 7001 Australia

KOSLOW, STEPHEN HUGH, b New York, NY, Oct 14, 40; m 62; c 2. PHARMACOLOGY, PSYCHOPHARMACOLOGY. *Educ:* Columbia Univ, BS, 62; Univ Chicago, PhD, 67. *Prof Exp:* Fel pharmacol, Karolinska Inst, Stockholm, Sweden, 68-69; NIMH staff fel, Lab Preclin Pharmacol, St Elizabeth's Hosp, Washington, DC, 70-73; chief unit neurobiol & appl mass spectrometry, Lab Preclin Pharmacol, 73-75, chief, neurosci res, Biol Res Sect, Clin Res Br, 75-81, chief, Div Extramural Res, 81-85, chief, Neurosci Res Br, Div Basic Sci, 85-89, DIR, DIV BASIC BRAIN & BEHAV SCI, NIMH, 89- *Concurrent Pos:* Dir, Clin Res Br Collab Prog Psychol Depression, NIMH, 75-, Presidential Comn Mental Health, Special Asst Res Panel, 78; med adv bd, Tourette Syndrome Asn. *Honors & Awards:* Meritorious Achievement Award, Alcohol Drug Abuse & Ment Health Admin, 79, 85. *Mem:* Am Soc Pharmacol & Exp Therapeut; Soc Neurosci; Am Col Neuropsychopharmacol; Soc Biol Psychiat; Am Soc Neurochem; Collegium Internationale Neuro-Psychopharmacologicum. *Res:* Neuropharmacology and psychopharmacology; depression and schizophrenia; neurotransmitters; metabolites and central nervous system function; neuroendocrinology. *Mailing Add:* NIMH Parklawn Bldg Rm 11-103 5600 Fishers Lane Rockville MD 20857

KOSLOWSKY, VERNON THEODORE, b Leamington, Ont, Can, Dec 15, 53; m 77; c 3. NUCLEAR PHYSICS. *Educ:* Univ Waterloo, BSc 77; Univ Toronto, MSc, 78, PhD(physics), 83. *Prof Exp:* Res fel, GSI, Darmstadt, Fed Repub Germany, 83-84; ASST RES OFFICER, CHALK RIVER LABS, ONT, CAN, 85- *Mem:* Can Asn Physicists. *Res:* Experimental nuclear physics; use of accelerated heavy ions as probes of the nucleus; weak interaction and the nucleus; nuclei far from stability. *Mailing Add:* Chalk River Labs Chalk River ON K0J 1J0 Can

KOSMAHL, HENRY G, b Wartha, Ger, Dec 14, 19; US citizen; m 43; c 3. ELECTRON PHYSICS. *Educ:* Dresden Tech Univ, MS, 43; Darmstadt Tech Univ, DS(electron physics), 49. *Prof Exp:* Asst prof physics, Darmstadt Tech Univ, 49-51; res physicist, AEG-Telefunken Res Ctr, Ger, 52-56; head power amplifier, Electron Lab, US Army, 56-62; head power amplifier, Lewis Res Ctr, NASA, 62-84; CONSULT, ELECTRON DYNAMICS DIV, HUGHES AIRCRAFT, 84- *Concurrent Pos:* Consult, Aeronaut Systs & Space Div, US Air Force, Dayton, 62-84, Westinghouse Elec Defense Div, 82-83. *Honors & Awards:* Sci Achievement Medal, NASA, 74; Technol Adv Award, 77, CECON Centennial Award, Inst Elec & Electronics Engrs, 83. *Mem:* Fel Inst Elec & Electronics Engrs. *Res:* Interaction of charged particles with waves and matter. *Mailing Add:* NASA Lerc Analex Corp Space Commun Corp 21000 Brookpark Rd Cleveland OH 44135

KOSMAN, DANIEL JACOB, b Chicago, Ill, Nov 29, 41; m 64; c 2. BIOCHEMISTRY. *Educ:* Oberlin Col, BA, 63; Univ Chicago, PhD(phys org chem), 68. *Prof Exp:* Res assoc biophys, Univ Hawaii, 68-69; Cornell Univ res assoc molecular biol, Dept Chem, Cambridge Univ & Med Res Coun Lab of Molecular Biol, Cambridge, Eng, 69-70; from asst prof to assoc prof, 70-81, PROF BIOCHEM, STATE UNIV NY BUFFALO, 81- *Mem:* AAAS; Am Chem Soc. *Res:* Mechanism of enzyme action; enzyme modification; bioinorganic chemistry; protein biosynthesis. *Mailing Add:* Dept Biochem 140 Farber Hall State Univ NY Buffalo NY 14214

KOSMAN, MARY ELLEN, b Denver, Colo, Mar 27, 26; m 53; c 1. PHARMACOLOGY. *Educ:* Univ Denver, BA, 50, MA, 51; Northwestern Univ, PhD(physiol psychol), 53. *Prof Exp:* Res assoc neuropharmacol, Col Med, Univ Ill, Chicago, 53-55 & 62-65, sr scientist, Div Drugs, AMA, 69-90; RETIRED. *Concurrent Pos:* Fel pharmacol, Univ Ill, Chicago, 66-69. *Mem:* Am Soc Clin Pharmacol Therapeut; AAAS. *Res:* Cardiovascular-renal pharmacology; ocular pharmacology. *Mailing Add:* 1029 Deerfield Pl Highland Park IL 60035

KOSMAN, WARREN MELVIN, b Chicago, Ill, Mar 23, 46; m 70; c 1. CHEMICAL PHYSICS. *Educ:* Valparaiso Univ, BS, 67; Univ Chicago, MS, 69, PhD(chem physics), 74. *Prof Exp:* Instr chem, Valparaiso Univ, 69-70, instr math, 70-71; asst prof chem, Ohio State Univ, 74-77; from asst prof to assoc prof, 77-89, PROF CHEM, VALPARAISO UNIV, 89- *Mem:* Am Chem Soc. *Res:* Molecular spectroscopy and ab initio molecular orbital calculations of atoms and small molecules. *Mailing Add:* Dept Chem Valparaiso Univ Valparaiso IN 46383

KOSMATKA, JOHN BENEDICT, b Milwaukee, Wis, Aug 24, 56; m 88. STRUCTURAL DYNAMICS, COMPOSITE MATERIALS. *Educ:* Univ Wis-Madison, BS, 78; Univ Mich, MS, 80; Univ Calif, Los Angeles, PhD(aerospace eng), 86. *Prof Exp:* Engr, Aerospace Corp, El Segundo, Calif, 80-82; sr engr, TRW Corp, Redondo Beach, Calif, 82-86; asst prof mech eng, Va Polytech Inst, Blacksburg, 86-89; ASST PROF AEROSPACE STRUCT, DEPT APPL MECH, UNIV CALIF, 89- *Concurrent Pos:* Fac fel, Ames Res Ctr, NASA, Moffettfield, Calif, 88, Langley Res Ctr, Hampton, Va, 89 & Newport News Shipbuilding, San Diego, Calif, 90. *Mem:* Am Inst Aeronaut & Astronaut; Am Soc Mech Engrs; Am Helicopter Soc. *Res:* Structural dynamic and aeroelastic analysis of advanced composite helicopter, tilt-rotor, and turbo-propeller blades; hybrid composite materials that have reduced vibration behavior using passive and/or active techniques; author of various publications. *Mailing Add:* Dept Appl Mech & Eng Sci Univ Calif San Diego CA 92093-0411

KOSOW, DAVID PHILLIP, b Jersey City, NJ, Mar 15, 36; m 58; c 2. ENZYMOLOGY, BLOOD COAGULATION. *Educ:* Antioch Col, BS, 58; Va Polytech Inst, MS, 60, PhD(biochem & nutrit), 62. *Prof Exp:* Asst prof biochem, Va Polytech Inst, 62-63; Am Cancer Soc fel, Oak Ridge Nat Lab, 63-65; neurochemist, Philadelphia Gen Hosp, 65-66; res assoc, Inst Cancer Res, 66-70, sr res assoc, 70-73; res scientist, 73-77, sr res scientist, 77-81, asst dir, 81-84, COORDR RES & DEVELOP, AM RED CROSS BLOOD SERV, 85- *Concurrent Pos:* Fel, Fogarty Int Ctr, Oxford, UK, 80-81; adj assoc prof, dept biol, Cath Univ Am, 84- *Mem:* Am Soc Biol Chemists; Am Chem Soc; Sigma Xi; Int Soc Thrombosis & Haemostasis. *Res:* Regulation and mechanism of plasma coagulation factors; development of blood plasma derivatives for clinical use; inactivation of viruses in plasma derivatives. *Mailing Add:* Alpha Therapeut Corp 5555 Valley Blvd Los Angeles CA 90032

KOSOWER, EDWARD MALCOLM, b Brooklyn, NY, Feb 2, 29; m 61; c 2. BIOPHYSICAL ORGANIC CHEMISTRY. *Educ:* Mass Inst Technol, SB, 48; Univ Calif, Los Angeles, PhD(chem), 52. *Prof Exp:* NIH res fel org chem, Univ Basel, 52-53 & Harvard Univ, 53-54; asst prof chem, Lehigh Univ, 54-56; from instr to asst prof, Univ Wis, 56-61; from assoc prof to prof chem, State Univ NY, Stony Brook, 61-72, adj prof, 72-; AT DEPT CHEM, TEL-AVIV UNIV, ISRAEL. *Concurrent Pos:* Alfred P Sloan fel, 60-64; NSF fel, Weizmann Inst Sci, Israel, 68-69; prof, Tel-Aviv Univ, 72-; John Simon Guggenheim fel, 77-78; vis prof, Univ Calif, San Diego, 77, Univ Calif, Berkeley, 78, Kyoto, Japan, 78, Mass Inst Technol, 83 & Bologna Italy, 87. *Honors & Awards:* Weizmann Prize, 77; Kolthoff Award, 84; Lemburg lectr, Australian Acad Sci, 91. *Mem:* Am Chem Soc; Royal Soc Chem; Am Soc Biochem; Soc Neurosci; Israel Chem Soc; fel AAAS. *Res:* Charge-transfer spectra; pyridinium ion chemistry; solvent effects on spectra; stable free radicals; molecular medicine; neurophysiology; glutathione in chemistry, biochemistry, biology and medicine; fluorescence mechanisms; membrane mobility agents; mechanism of cell fusion; bimanes (diazabicyclo(3.3.0) octadienediones); sodium channel and acetylcholine receptor models; mechanism of fast intramolecular electron transfers; molecular basis learning and memory. *Mailing Add:* Dept Chem Tel-Aviv Univ Ramat-Aviv Tel-Aviv Israel

KOSOWSKY, DAVID I, b New York, NY, Feb 27, 30; c 3. HEALTH SCIENCES, ELECTRONICS. *Educ:* City Col New York, BEE, 51; Mass Inst Technol, SM, 52, ScD(network theory), 55. *Prof Exp:* Res asst & staff mem, Res Lab Electronics, Mass Inst Technol, 51-55; dir crystal div, Hermes Electronics Co, 55-60; vpres, Itek Electro-Prod Co, 60-61; PRES, DAMON CORP, 61-, CHMN & CHIEF EXEC OFFICER, 83- *Concurrent Pos:* Trustee, New Eng Aquarium, 68- & Univ Hosp Boston, 70-; vchmn, Childrens Hosp Med Ctr, 76-83, chmn, 83. *Mem:* Inst Elec & Electronics Engrs; Sigma Xi; AAAS; NY Acad Sci. *Res:* Network theory; statistical theory of communication; crystal filters; voltage controlled crystal oscillators; spectrum analyzers; health service delivery systems; medical and electronic instrumentation. *Mailing Add:* 403D Dedham St Newton MA 02159

KOSS, DONALD A, b Dodge Co, Minn; m 64; c 3. METALLURGY. *Educ:* Univ Minn, BS, 60; Yale Univ, MS, 62, PhD(metall), 65. *Prof Exp:* Res assoc, Pratt & Whitney Aircraft, 65-70; from assoc prof to prof metall eng, Mich Technol Univ, 70-85; PROF METALS SCI & ENG & CHMN DEPT, PA STATE UNIV, 86- *Concurrent Pos:* NSF fel, Los Alamos Nat Lab, 78-79. *Mem:* Am Soc Metals; Metall Soc; Mat Res Soc. *Res:* Processing, deformation, and fracture of high performance alloys. *Mailing Add:* Pa State Univ 208 Steidle Bldg University Park PA 16802

KOSS, LEOPOLD GEORGE, b Danzig, Poland, Oct 2, 20; nat US; m; c 3. PATHOLOGY, CYTOLOGY. *Educ:* Univ Bern, MD, 46. *Prof Exp:* Asst path, St Gallen, Switz, 47 & Long Island Col Med, 49; instr, Col Med, State Univ NY Downstate Med Ctr, 50-52; from assoc dir to dir cytol, Mem Hosp Cancer & Allied Dis, 52-60, from asst attend pathologist to assoc attend pathologist, 53-60, chief cytol serv & attend pathologist, 60-70; prof path, Jefferson Med Col, 70-73; PROF PATH & CHMN DEPT, ALBERT EINSTEIN COL MED & CHMN DEPT PATH, MONTEFIORE HOSP & MED CTR, 73- *Concurrent Pos:* From asst to assoc, Sloan-Kettering Inst, 53-60, head secy cytopath, 60-70; from asst prof to assoc prof, Sloan-Kettering Div, Med Col, Cornell Univ, 54-70; vis pathologist, James Ewing Hosp, 60-68; consult, NY State Dept Health, 62-; pathologist-in-chief, Sinai Hosp Baltimore, Inc, 70-73; mem, Ger Acad Sci, 89. *Honors & Awards:* Wien Award, 61; Goldblatt Award, 62; Papanicolaou Award, Am Soc Cytol, 66; Stewart Award, 84; Vandenberge-Hill Award, 84. *Mem:* Fel Am Soc Clin Pathologists; Am Asn Pathologists & Bacteriologists; Am Soc Cytol (pres, 62); fel Col Am Path; Int Acad Path; fel Int Acad Cytol; Soc Surg Oncol. *Res:* Cytology and pathology of cancer. *Mailing Add:* Montefiore Hosp & Med Ctr 111 E 210th St Bronx NY 10467

KOSS, MICHAEL CAMPBELL, b Ann Arbor, Mich, Sept 24, 40. PHARMACOLOGY. *Educ:* NY Univ, BA, 66; Columbia Univ, PhD(pharmacol), 71. *Prof Exp:* From asst to assoc prof, 71-81, PROF PHARMACOL, COL MED, UNIV OKLA, 81- *Mem:* Asn Res Vision & Ophthal; Soc Neurosci; Am Soc Pharmacol & Exp Therapeut; Sigma Xi. *Res:* Neuropharmacology; neurophysiology; brain stem regulatory mechanisms; autonomic nervous system. *Mailing Add:* Dept Pharmacol Univ Okla PO Box 26901 Oklahoma City OK 73190

KOSS, VALERY ALEXANDER, b Dnepropetrovsk, USSR, Aug 4, 41; US citizen; m 67; c 1. MATHEMATICAL PHYSICS. *Educ:* Leningrad Polytech Inst, USSR, MS, 64; Phys Tech Inst, Leningrad, PhD(math physics), 72. *Prof Exp:* Sr scientist, 82-88, CONSULT SCIENTIST, TECH CTR, BOC GROUP, INC, 88- *Res:* Modeling of various phenomena pertaining to chemistry, vacuum technologies and optics. *Mailing Add:* 1056 Carteret Rd Bridgewater NJ 08807

KOSSIAKOFF, ALEXANDER, b St Petersburg, Russia, June 26, 14; US citizen; m 39; c 2. PHYSICAL CHEMISTRY, SYSTEM ENGINEERING. *Educ:* Calif Inst Technol, BS, 36; Johns Hopkins Univ, PhD(chem), 38. *Prof Exp:* Fel, Calif Inst Technol, 38-39; instr chem, Cath Univ, 39-42; tech aide, Nat Defense Res Comt, 42-43; dep dir res, Allegheny Ballistics Lab, George Washington Univ, 44-46; physicist, Appl Physics Lab, Johns Hopkins Univ, 46-48, asst dir, 48-61, assoc dir, 61-66, dep dir, 66-69, dir, 69-80, EMER DIR & CHIEF SCIENTIST, APPL PHYSICS LAB, JOHNS HOPKINS UNIV, 80-, CHAIR, TECH MGT, SCH ENG, 80- *Concurrent Pos:* Consult tech adv panel aeronaut, Defense Dept, 54-58; mem panel launching & handling comt guided missiles, Res & Develop Bd, 48-52; Carmrand comt, Nat Planning Asn, 62-73; mem Gov's Sci Adv Coun, 79- *Mem:* Fel Am Inst Chemists; AAAS. *Res:* Prediction and determination of molecular structure; relation between molecular structure and physical and chemical properties; mechanism of chemical reactions; mechanism of neural processes; administration of research; computer languages; computer aided instruction for handicapped children. *Mailing Add:* Appl Physics Lab Johns Hopkins Rd Laurel MD 20723

KOSSLER, WILLIAM JOHN, b Charleston, SC, Mar 26, 37; m 61; c 3. PHYSICS. *Educ:* Mass Inst Technol, BS, 59; Princeton Univ, PhD(physics), 64. *Prof Exp:* Staff mem nuclear physics, Mass Inst Technol, 64-66, asst prof physics, 66-69; from asst prof to assoc prof, 69-78, PROF PHYSICS, COL WILLIAM & MARY, 78- *Mem:* Am Phys Soc. *Res:* Experimental nuclear and solid state physics. *Mailing Add:* Dept Physics Col William & Mary Williamsburg VA 23185

KOSSMANN, CHARLES EDWARD, b Brooklyn, NY, Apr 20, 09; m 46; c 2. MEDICINE. *Educ:* NY Univ, BS, 28, MD, 31, MedScD, 38. *Prof Exp:* House physician, Bellevue Hosp, 31-33; asst med, Heart Sta, Univ Hosp, Univ Mich, 34; asst, Sch Med, NY Univ, 34-38, from instr to prof, 38-67, head cardiovasc sect, 64-67; chmn, Div Circulatory Dis, 67-74, prof med, 67-76, EMER PROF, COL MED, UNIV TENN, MEMPHIS, 76- *Concurrent Pos:* Chief peripheral vascular dis clin, NY Univ, 36-41, from assoc attend physician to attend physician, 49-67; asst clin vis physician, Bellevue Hosp, 34-40, from asst vis physician to vis physician, 40-67, consult physician, 68-, chief adult cardiac clin, 40-56; adj physician, Lenox Hill Hosp, 37-46, assoc physician, 46-49, attend physician, 49-64, consult physician, 64-67. *Honors & Awards:* Distinguished Scientist Award, NAm Soc Pacing & Electrophysiol, 86. *Mem:* Fel AAAS; Asn Am Physicians; Asn Univ Cardiol; master Am Col Physicians; fel AMA. *Res:* Cardiovascular diseases; physiology of circulation; aviation medicine. *Mailing Add:* Div Cardiovascular Dis Univ Tenn Col Med Memphis TN 38163

KOSSOY, AARON DAVID, b New York, NY, Aug 19, 36; m 70. ORGANIC CHEMISTRY. *Educ:* City Col New York, BS, 58; Polytech Inst Brooklyn, PhD(org chem), 66. *Prof Exp:* Anal chemist, Trubek Labs, Inc, 58-61; fel chem, Univ Calif, Berkeley, 66-69; sr anal chemist, 69-80, RES SCIENTIST, ELI LILLY RES LABS, 80- *Mem:* Am Chem Soc; Sigma Xi. *Res:* Chemical and physical properties of organic compounds; spectroscopic characterization of organic compounds; organic synthesis and structure determination. *Mailing Add:* Eli Lilly & Co Lilly Corp Ctr Indianapolis IN 46285

KOSSUTH, SUSAN, b Boston, Mass, Apr 28, 46. PLANT PHYSIOLOGY, GENETICS. *Educ:* Colo State Univ, BS, 68, MS, 71; Yale Univ, MS, 72, MPhil, 73, PhD(tree physiol-genetics), 74. *Prof Exp:* Consult, Fla Citrus Comn, 74-76; asst prof tree physiol, Univ Ark, 76-77; asst res scientist, Univ Fla, 77-78; PROJ LEADER, US FOREST SERV, 79- *Concurrent Pos:* Adj asst prof, Univ Fla, 74-76; prin investr, Weyerhaeuser Corp, Eli-Lilly Co, Ark Kraft Co, Southern Regional Educ Bd, 76; co-prin investr, Fla Citrus Comn, 77-79. *Mem:* Am Soc Plant Physiol; Sigma Xi; Soc Am Foresters; Am Forestry Asn; Plant Growth Regulator Soc Am. *Res:* Reproductive physiology and breeding and improvement of Southern pines; vegetative propagation of pines; flowering in pines; early genetic testing of pines; effects of ultraviolet light on plants. *Mailing Add:* Fruit Crops Dept 1119 HS-PP Bldg Univ Fla Gainesville FL 32611

KOSTANT, BERTRAM, b New York, NY, May 24, 28; m 49, 68; c 5. GEOMETRIC QUANTIZATION. *Educ:* Purdue Univ, BS, 50; Univ Chicago, MS, 51, PhD(math), 54. *Prof Exp:* NSF fel, Inst Advan Study, 53-54, mem, 54-56; from asst prof to prof math, Univ Calif, Berkeley, 56-63; PROF MATH, MASS INST TECHNOL, 63- *Concurrent Pos:* Higgins lectr, Princeton Univ, 55-56; mem, Miller Inst Basic Res, 58-59; Guggenheim fel, Paris, France, 59-60; prof, Oxford Univ, Tel Aviv Univ & Paris, France, 74-75, 81-82. *Mem:* Nat Acad Sci; Am Acad Arts & Sci; Am Math Soc. *Res:* Operator theory; Lie groups; representation theory; differential geometry; mathematical physics. *Mailing Add:* Dept Math Mass Inst Technol Cambridge MA 02139

KOSTELNICEK, RICHARD J, b Chicago, Ill, May 16, 42; m 67. ELECTRICAL ENGINEERING. *Educ:* Univ Ill, Urbana, BS, 64, MS, 65, PhD(elec eng), 69. *Prof Exp:* SR RES ASSOC, ESSO PROD RES CO, 69- *Mem:* AAAS; Soc Explor Geophys; Inst Elec & Electronics Engrs. *Res:* Antennas; plasma physics; wave propagation in inhomogeneous media; geoscience. *Mailing Add:* 609 Bayou Crest Dr Dickinson TX 77539

KOSTENBADER, KENNETH DAVID, JR, b Allentown, Pa, May 6, 41; m 77. MICROBIOLOGY, FOOD SCIENCE & TECHNOLOGY. *Educ:* Albright Col, BS, 64. *Prof Exp:* Microbiologist I salmonella, Pa Dept Health, Philadelphia, 65-66; res asst, 66-67; res specialist, 67-87, SR RES VIROL, FOOD RES INST, UNIV WIS-MADISON, 87- *Concurrent Pos:* Consult, WHO Collab Ctr on Food Virol, Food Res Inst, Univ Wis-Madison, 75- *Mem:* AAAS; Am Soc Microbiol. *Res:* Occurrence, transmission and detection of animal and human viruses in food, water and wastewater. *Mailing Add:* Food Res Inst Dept Food Microbiol & Toxicol Univ Wis 1925 Willow Dr Madison WI 53706

KOSTENBAUDER, HARRY BARR, b Danville, Pa, Apr 9, 29. PHARMACY. *Educ:* Phila Col Pharm, BSc, 51; Temple Univ, MSc, 53; Univ Wis, PhD(pharm), 56. *Prof Exp:* Asst pharm, Temple Univ, 51-53 & Univ Wis, 55; from asst prof to prof, Temple Univ, 56-68; PROF PHARM & ASSOC DEAN RES, COL PHARM, UNIV KY, 68- *Mem:* Am Chem Soc; Am Pharmaceut Asn; NY Acad Sci; fel Acad Pharmaceut Sci (pres, 71-72); fel Am Asn Pharmaceut Scientists. *Res:* Drug binding by macromolecules; drug stability; pharmacokinetics. *Mailing Add:* Col Pharm Univ Ky Lexington KY 40506

KOSTER, DAVID F, b Houston, Tex, Nov 3, 36; m 59; c 5. PHYSICAL CHEMISTRY, SPECTROSCOPY. *Educ:* St Thomas Univ, BA, 59; Tex A&M Univ, MS, 63, PhD(chem), 65. *Prof Exp:* Chemist, Diamond Alkali Co, 59-60; res fel, Mellon Inst, 64-67; asst prof, 67-71, assoc prof 71-81, PROF CHEM, SOUTHERN ILL UNIV, CARBONDALE, 81- *Mem:* Am Chem Soc; Soc Appl Spectros; Sigma Xi. *Res:* Infrared laser induced reactions; nuclear magnetic resonance; Raman and infrared spectroscopy; structure and conformation studies of polyatomic molecules. *Mailing Add:* Dept Chem Southern Ill Univ Carbondale IL 62901

KOSTER, GEORGE FRED, b New York, NY, Apr 9, 27; m 51; c 3. PHYSICS. *Educ:* Mass Inst Technol, SB, 48, PhD(physics), 51. *Prof Exp:* Res assoc, 51-52, Lincoln Lab, 52-55, from asst prof to assoc prof, 56-64, PROF PHYSICS, MASS INST TECHNOL, 64- *Concurrent Pos:* Guggenheim fel, 55-56. *Mem:* Am Phys Soc. *Res:* Theoretical physics including theory of solids and molecular theory. *Mailing Add:* Dept Physics Mass Inst Technol Cambridge MA 02139

KOSTER, ROBERT ALLEN, b Grand Rapids, Mich, July 12, 41; m 63; c 2. ORGANIC CHEMISTRY. *Educ:* Hope Col, AB, 63; Univ Mich, Ann Arbor, MS, 65, PhD(chem), 68. *Prof Exp:* Res chemist, 68-70, proj leader, Org Chem Prod Res, 70-80, res leader, Styrene Plastics, 80-82, RES ASSOC, DOW CHEM USA, 82- *Res:* Carbonium ion chemistry; reaction mechanisms via kinetic studies; process development on fine organic chemicals; chemistry of 2.2.1 bicyclic systems; polymer process development. *Mailing Add:* 400 Hollybrook Dr Midland MI 48640

KOSTER, RUDOLF, PHARMACOLOGY. *Educ:* State Univ Iowa, PhD(embryol & endocrinol), 41. *Prof Exp:* Sr Pharmacologist, Burrough's Wellcome Co, Research Triangle Park, NC, 53-78. *Mailing Add:* 232 Hayes Rd Chapel Hill NC 27514

KOSTER, W(ILLIAM) P(FEIFFER), b Fords, NJ, Apr 18, 29; m 54; c 4. METALLURGICAL ENGINEERING. *Educ:* Rutgers Univ, BS, 50; Univ Cincinnati, MS, 51, PhD(metall eng), 53. *Prof Exp:* Staff mem, 53-57, vpres, 57-78, DIR METALL ENG, METCUT RES ASSOC INC, 57-, PRES, 78- *Honors & Awards:* Gold Medal, Soc Mfg Engrs. *Mem:* Fel Am Soc Metals; fel Soc Adv Mat & Process Eng; fel Soc Mfg Eng. *Res:* Mechanical engineering. *Mailing Add:* Metcut Res Assocs 3980 Rosslyn Dr Cincinnati OH 45209-1196

KOSTER, WILLIAM HENRY, b Teaneck, NJ, Apr 20, 44; m 68; c 1. SYNTHETIC ORGANIC CHEMISTRY. *Educ:* Colby Col, BA, 66; Tufts Univ, PhD(chem), 72. *Prof Exp:* postdoc fel, Squibb Inst Med Res, 71-72, res investr, 72-77, sr res investr, 77-80, group leader synthetic antibact agents & nat prods, 80-83, sect head, 83-84, dir, dept chem, Infectious & Metabolic Dis, 84-87, exec dir, 87-90; VPRES, DIV CHEM, BRISTOL-MYERS SQUIBB PHARM RES INST, PRINCETON, NJ, 90- *Mem:* Am Chem Soc; Am Soc Microbiol. *Res:* Synthetic organic chemistry; bioorganic chemistry; mechanism-based design of new antibacterials/antifungals; antiviral agents; cardiovascular agents and inhibitors of cholesterol biosynthesis; natural products isolation; structure elucidation; semi-synthetic modification. *Mailing Add:* Bristol-Myers Squibb PO Box 4000 Princeton NJ 08543-4000

KOSTER VAN GROOS, AUGUST FERDINAND, b Leeuwarden, Neth, Jan 9, 38; m 71. GEOCHEMISTRY. *Educ:* Univ Leiden, BSc, 58, MS, 62, PhD(exp petrol). 66. *Prof Exp:* Res assoc, Goddard Space Flight Ctr, NASA, 66-68; asst prof petrol, State Univ Utrecht, 68-70; asst prof, 70-75, ASSOC PROF GEOL SCI, UNIV ILL, CHICAGO CIRCLE, 75- *Mem:* AAAS. *Res:* Genesis of carbonatite, experimental work in synthetic systems containing carbon dioxide and water at elevated pressure and temperature; salt-silicate-water systems; studies of liquid immiscibility occurring in rocks; partitioning of minor elements in multi- phase systems; mantle metapomatsm crystal development. *Mailing Add:* Dept Geol Sci-MC 186 Univ Ill Box 4348 Chicago IL 60680

KOSTIC, NENAD M, b Belgrade, Yugoslavia, Nov 18, 52; m 76; c 2. BIOINORGANIC CHEMISTRY, ORGANOMETALLIC CHEMISTRY. *Educ:* Univ Belgrade, Yugoslavia, dipl, 76; Univ Wis-Madison, PhD(inorg chem), 82. *Prof Exp:* Teaching & res asst, Univ Wis-Mad, 78-82; res fel chem, Calif Inst Technol, 82-84; asst prof chem, 84-89, ASSOC PROF CHEM & ADJ ASSOC PROF BIOCHEM, IOWA STATE UNIV, 89- *Honors & Awards:* Presidential Young Investr Award, NSF, 88. *Mem:* Am Chem Soc; Serbian Chem Soc. *Res:* Bioinorganic chemistry; stereochemistry; electron-transfer reactions. *Mailing Add:* Dept Chem Iowa State Univ Ames IA 50011

KOSTINER, EDWARD S, b New York, NY, Feb 25, 40; m 60; c 2. SOLID STATE CHEMISTRY. *Educ:* Tufts Univ, BS, 60; Polytech Inst Brooklyn, PhD(inorg chem), 66. *Prof Exp:* Asst prof, Cornell Univ, 66-72; assoc prof, 72-77, PROF CHEM, UNIV CONN, 77-, HEAD DEPT, 85- *Mem:* Am Chem Soc; NY Acad Sci; Am Asn Crystal Growth; Mineral Soc Am; Am Crystallog Asn. *Res:* Crystal growth; crystal and structural chemistry of apatites and other halophosphates and orthophosphates; Mossbauer effect spectroscopy. *Mailing Add:* Dept Chem Univ Conn U-60 Storrs CT 06269-3060

KOSTISHACK, DANIEL F(RANK), b Pittsburgh, Pa, Mar 25, 40; m 66; c 2. ELECTRICAL ENGINEERING, SOLID STATE PHYSICS. *Educ:* Carnegie Inst Technol, BS, 63, MS, 65; Carnegie-Mellon Univ, PhD(elec eng), 68. *Prof Exp:* mem res staff, 67-81, GROUP LEADER, LINCOLN LAB, MASS INST TECHNOL, 81- *Res:* Solid-state and high frequency devices and circuits; solid-state imaging devices and electro-optical systems. *Mailing Add:* Lincoln Lab Box 73 Mass Inst Technol Lexington MA 02173

KOSTIUK, THEODOR, b Plauen, Ger, Aug 12, 44; m 70. SPACE PHYSICS. *Educ:* City Col New York, BS, 66; Syracuse Univ, PhD(physics), 73. *Prof Exp:* Nat Acad Sci resident res assoc, 73-74, space scientist, Infrared Astron Br, 74-83, head Molecular Astrophys Sect, 83-84, SPACE SCIENTIST, PLANETARY SYSTS BR, GODDARD SPACE FLIGHT CTR, NASA, 85- *Concurrent Pos:* Discipline leader, Auroral Discipline, Int Jupiter Watch. *Mem:* Am Phys Soc; Optical Soc Am; Soc Photo-Optical Instrument Engrs; AAAS; Am Astron Soc Div Planetary Sci. *Res:* Atmosphere of planets, comets, the sun, stars and the earth's stratosphere using ultra-high resolution infrared spectroscopy; infrared heterodyne spectroscopy; discovery of the first natural laser (carbon dioxide on Mars); molecular spectroscopy; hydrocarbon chemistry on outer planets; global circulation on Mars and Venus; dynamics and planetary oscillations on Jupiter; aurorae and infrared emission on Jupiter. *Mailing Add:* Code 693 NASA Goddard Space Flight Ctr Greenbelt MD 20771

KOSTKOWSKI, HENRY JOHN, b Garwood, NJ, May 16, 26; m 47, 76; c 3. SPECTRORADIOMETRY. *Educ:* Johns Hopkins Univ, PhD(physics), 54. *Prof Exp:* Molecular physicist, Nat Inst Standards & Technol, 54-56, physicist high temperature, 56-58, supvry physicist, 58-65, chief radiation thermometry sect, 65-71, chief optical radiation sect, 71-81; PROPRIETOR SPECTRORADIOMETRY CONSULT, 81- *Concurrent Pos:* Consult, Inst Advan Study, 54, 55 & 57. *Honors & Awards:* Gold Medal Award, US Dept Com, Edward Bennett Rosa Award. *Mem:* Fel Optical Soc Am. *Res:* Spectroradiometry; optical pyrometry; spectral line intensity measurements; physical optics. *Mailing Add:* Rte 1 Box 69 Charlotte Hall MD 20622-9705

KOSTOFF, MORRIS R, b Jamestown, NDak, Dec 2, 33; m 55; c 5. NUCLEAR PHYSICS, ACOUSTICS. *Educ:* Pac Lutheran Univ, BS, 62; Univ Tex, Austin, PhD(physics), 67. *Prof Exp:* Res asst nuclear physics, Ctr Nuclear Studies, Univ Tex, Austin, 63-66; engr-scientist, Sci & Systs Div, Tracor, Inc, 66-70, sr scientist, 70-79, prin scientist, Anal & Appl Res Div, Appl Sci Group, 79- 82, DIR ACOUST WARFARE PROG, ANAL & APPL RES DIV, TRACOR, INC, 82- *Concurrent Pos:* Sen Warren G Magnuson Scholarship, 60-61; Nat Defense Educ Act fel, 63-66; asst prof physics, Southwestern Univ, Tex, 71-74. *Mem:* Am Phys Soc. *Res:* Spin polarization measurements for elastic and inelastic proton scattering; systems analysis and simulation of signal processors for sonar systems; propagation of acoustic waves in water medium; operator interactive; realtime computer simulation of state-of-the-art sonar systems for evaluating mission effectiveness in a countermeasure environment. *Mailing Add:* 11513 Juniper Ridge Dr Austin TX 78759-3845

KOSTOPOULOS, GEORGE, B Athens, Greece, Dec 23, 39; US citizen; m; c 3. ELECTRICAL & COMPUTER ENGINEERING. *Educ:* Pacific States Univ, Los Angeles, BS, 62; Ariz State Univ, Tempe, MS, 70, PhD(elec & comput eng), 71; Calif State Polytech Univ, Pomona, MS, 77. *Prof Exp:* Assoc prof, Dept Elec Eng, Calif State Polytech Univ, Pomona, 75-78; prof & head, Dept Comput Eng, Nat Inst Elec Eng, Boumerdes, Algeria, 78-82; prof, Dept Elec & Comput Eng, Fla Inst Technol, Melbourne, 82-85; assoc prof elec & comput eng, Fla Atlantic Univ, 84-90; ENG CONSULT, 90- *Concurrent Pos:* Electronic engr, Blass Antenna Electronics Corp, NY, 62-64 & Maxson Electronics Corp, East Islip, NY, 64-66; sr engr-consult, Comput Applns-Sys Consults Inc, NY, Ariz, Calif, 66-72; staff engr, Singer-Librascope, Glendale, Calif, 72-73; prin scientist, Textron Corp, Belmont, 73; sr engr, Raytheon Co, Goleta, 73-74; sr pron engr, Honewell Inc, West Covina, 74-75; hon dean, Atom Col Eng, Athens Greece. *Mem:* Inst Elec & Electronics Engrs; Am Soc Eng Educ; Inst Strategic Studies, Athens. *Res:* Chinese language text processing; holder of one patent on Chinese character generation; new Chinese language dictionary structure; study to thermal printhead extension life; algorithm development for binary root and log computations; logical symmetric functions; microprocessor implementation of traffic control algorithms; Arabic language word processing; electrical vehicle performance simulation; author of numerous articles, textbooks and other publications. *Mailing Add:* 6714 Sweet Maple Lane Boca Raton FL 33433

KOSTREVA, DAVID ROBERT, b Milwaukee, Wis, Aug 14, 45; m 75; c 4. ANESTHESIOLOGY. *Educ:* Univ Wis, Milwaukee, Ba, 72; Med Col Wis, MS, 74, PhD(physiol), 76. *Prof Exp:* Fel, Am Heart Asn, 76-77 & Nat Heart, Lung and Blood Inst, NIH, 77-78; asst prof physiol & anesthesia, 78-81, ASSOC PROF ANESTHESIOL & PHYSIOL, MED COL WIS, 81- *Concurrent Pos:* Chmn, Ad Hoc Study Sect, NIMH, 81. *Honors & Awards:* Henry Pickering Bowditch lectr, 83; Res Career Develop Award, NIH, 82, Young Cardiovasc Investr Award, 79. *Mem:* Am Physiol Soc; Soc Neurosci; Sigma Xi; Soc Exp Biol & Med; Int Soc Heart Res. *Res:* Neural control of circulation and respiration in adult and fetal dogs, cats and monkeys, rabbits and chickens using afferent and efferent recording techniques and brain and heart mapping studies of visceral reflexes using the carbon-fourteen-deoxyglucose technique. *Mailing Add:* Res Serv 151 Vet Admin Med Ctr Milwaukee WI 53295

KOSTREVA, MICHAEL MARTIN, b Pittsburgh, Pa, May 9, 48; m 71; c 3. MATHEMATICS. *Educ:* Clarion State Col, BA, 71; Rensselaer Polytech Inst, MS, 73, PhD(math), 76. *Prof Exp:* Asst prof math, Univ Maine, Orono, 76-78; res scientist, Gen Motors Res Labs, 78-84; prin mem tech staff, GTE Labs, Inc, 84-86; mem tech staff, Alphatech, Inc, 86; assoc prof, 86-89, PROF MATH, CLEMSON UNIV, 89- *Concurrent Pos:* Consult, Gen Motors Corp, 87-, Gillette Res Inst, 90-; pres, Systematica Inc, Clemson, 89- *Mem:* Soc Indust & Appl Math; Am Math Soc; Opers Res Soc Am; Math Programming soc; Inst Elec & Electronics Engrs; AAAS. *Res:* Complementarity theory of mathematical programming, multiple objective programming, game theory, lubrication theory, scheduling, circadian rhythms. *Mailing Add:* Dept Math Sci Clemson Univ Clemson SC 29634

KOSTROUN, VACLAV O, b Brasov, Rumania, Dec 30, 38; US citizen; m 63; c 2. ATOMIC PHYSICS. *Educ:* Univ Wash, BSc, 61, MSc, 63; Univ Ore, PhD(physics), 68. *Prof Exp:* Res assoc & lectr, 68-70, asst prof, 70-77, ASSOC PROF APPL & ENG PHYSICS, CORNELL UNIV, 77- *Mem:* AAAS; Am Phys Soc. *Res:* Interactions of highly charged ions with atoms at lee V energies; production of low energy, very highly charged ions; theoretical atomic physics. *Mailing Add:* Ward Reactor Lab Cornell Univ Ithaca NY 14853

KOSTRZEWA, RICHARD MICHAEL, b Trenton, NJ, July 22, 43; m 65; c 10. PHARMACOLOGY, NEUROSCIENCE. *Educ:* Philadelphia Col Pharm & Sci, BS, 65, MS, 67; Univ Pa, PhD(pharmacol), 71. *Prof Exp:* Res pharmacologist, Vet Admin Hosp, New Orleans, 71-75; asst prof physiol, La State Univ Med Ctr, New Orleans, 75-78; assoc prof, 78-84, PROF PHARMACOL, E TENN STATE UNIV, 84- *Concurrent Pos:* Asst prof pharmacol, Tulane Univ Med Ctr, New Orleans, 74-75; prin investr, NIH grant, 75-81, March Dimes grant, 77-79 & Am Parkinson's Dis Asn grant, 77-78; res award, E Tenn State Univ Found, 81. *Mem:* Am Soc Pharmacol & Exp Therapeut; Neurosci Soc; Int Brain Res Orgn. *Res:* Development of monoaminergic neurons; dopamine receptors and behavior; Parkinson's disease; neurotoxins; psychopharmacology. *Mailing Add:* Col Med ETenn State Univ Johnson City TN 37614-0002

KOSTYNIAK, PAUL J, b Schenectady, NY, April 8, 47; m 70; c 3. TOXICOLOGY. *Educ:* St John Fisher Col, BS, 70; Univ Rochester, PhD(toxicol), 75. *Prof Exp:* Fel radiation, biol & biophys, Univ Rochester, 75-77; asst prof, 77-84, ASSOC PROF PHARMACOL, STATE UNIV NY, BUFFALO, 84-, DIR TOXICOL RES CTR, 85- *Concurrent Pos:* Reviewer, Toxicol & Appl Pharmacol, J Pharmacol & Exp Therapeut, J Appl Toxicol, Archives Biochem Biophys & Toxicol Letters; speaker, Gordon Conf, 87; dipl, Am Bd Toxicol. *Mem:* Soc Toxicol; Sigma Xi; Am Chem Soc; AAAS; NY Acad Sci. *Res:* Toxicology of heavy metals, antidote development and the metabolism and toxicity of organofluroic compounds; mechanisms of metal transport and disposition; role of endogenous thiols in the elimination of toxic metal pollutants; developing new in vitro models of nephrotoxicity and animal models of neurotoxicity; mechanisms of degradation of polymer films. *Mailing Add:* Dept Pharmacol & Therapeut State Univ 111 Farber Hall Buffalo NY 14214

KOSTYO, JACK LAWRENCE, b Elyria, Ohio, Oct 1, 31; m 53; c 2. PHYSIOLOGY, ENDOCRINOLOGY. *Educ:* Oberlin Col, AB, 53; Cornell Univ, PhD(zool), 57. *Hon Degrees:* MD, Univ Goteborg, 78. *Prof Exp:* From asst prof to prof physiol, Duke Univ, 59-68; prof physiol & chmn dept, Emory Univ, 68-79; chmn dept, 79-85, PROF PHYSIOL, MED SCH, UNIV MICH, 79-; ASSOC DIR, MICH DIABETES RES & TRAINING CTR, 86- *Concurrent Pos:* Nat Res Coun fel, Harvard Med Sch, 57-59; Lederle med fac award, 61-64; mem endocrinol study sect, NIH, 67-71; chmn educ comt, Am Physiol Soc, 70-76, mem coun, 79-82, rep to coun Acad Socs, Asn Am Med Cols, 81-89, chmn, Coun Endocrinol & Metabolism Sect, 90-91; vis foreign scientist, Swed Med Res Coun, 72; mem physiol test comt, comprehensive part II comt, 86-, Nat Bd Mem Examr, 74-77; mem Com Med Physiol; Int Union Physiol Sci; ed-in-chief, endocrinol, 78-82; pres, Asn Chmn Dept Physiol, 79-80; sect ed, Endocrinol, Ann Rev Physiol, 82-86; mem sci adv comt, Searle Scholars Prog, 82-85; mem admin bd, Coun Acad Socs, Am Asn Med Cols, 83-86. *Honors & Awards:* Ernst Oppenheimer Mem Award, Endocrine Soc, 69. *Mem:* Am Physiol Soc; Endocrine Soc; Sigma Xi; Am Diabetes Asn. *Res:* Mechanism of action of pituitary growth hormone; relationship between structure and functions of growth hormone; nature of growth hormone in blood. *Mailing Add:* Dept Physiol Univ Mich Med Sch 1335 E Catherine Ann Arbor MI 48109

KOSTYRKO, GEORGE JURIJ, b Ukraine, May 9, 37; US citizen. CIVIL ENGINEERING. *Educ:* City Col New York, BChE, 57; Univ Mich, Ann Arbor, MSE, 58; Sacramento State Col, MS, 63; Univ Calif, Davis, PhD(civil eng), 69. *Prof Exp:* Develop engr, Air Prod Inc, Pa, 57; Aerojet-Gen Corp, Calif, 58-61, sr res engr, 61-68; ASSOC PROF CIVIL ENG, SACRAMENTO STATE COL, 68-, HEAD PROG APPL MECH, 71- *Concurrent Pos:* NSF grant, Sacramento State Col, 71-72. *Mem:* Sigma Xi. *Res:* Detection of static and dynamic stresses in solids and structures by means of acoustic wave propagation, holography and photoelasticity. *Mailing Add:* 1721 Cathay Way Sacramento CA 95825

KOSTYU, DONNA D, b Ashtabula, Ohio, Oct 17, 47; m 70; c 2. IMMUNOLOGY. *Educ:* Duke Univ, PhD(microbiol & immunol), 79. *Prof Exp:* Res assoc, 81-86, ASST RES PROF IMMUNOL, SCH MED, DUKE UNIV, 86- *Mem:* Am Asn Immunologists; Am Soc Histocompatibility & Immunogenetics. *Res:* Research focuses on the immunogenetics of the HLA supergene, the human major histocompatibility complex. *Mailing Add:* Dept Microbiol & Immunol Duke Univ Med Ctr Box 3010 Durham NC 27710

KOSZALKA, THOMAS R, b Rochester, NY, Jan 25, 27; m 54; c 3. BIOCHEMISTRY. *Educ:* Univ Rochester, BA, 50, PhD, 59. *Prof Exp:* From instr to asst prof biochem, Sch Med & Dent, Univ Rochester, 59-65; assoc prof radiol, 65-70, assoc prof biochem, 67-75, PROF RADIOL, JEFFERSON MED COL, 70-, PROF BIOCHEM, 75-, PROF PEDIAT, 87- *Concurrent Pos:* Assoc dir, Eleanor Roosevelt Res Labs & dir, Harry Bock Labs, 65- *Res:* Developmental biochemistry; teratology. *Mailing Add:* Stein Res Lab Jefferson Med Col Philadelphia PA 19107

KOSZTARAB, MICHAEL, b Bucharest, Romania, July 7, 27; US citizen; m 53; c 1. ENTOMOLOGY. *Educ:* Hungarian Univ Agr Sci, HortE, 51; Ohio State Univ, PhD(entom), 62. *Prof Exp:* Exten asst, Hungarian State Bur Plant Protection, 47-50; asst prof hort entom, Hungarian Univ Agr Sci, 51-56; consult entomologist, Insect Control & Res Inc, Md, 57-58, asst dir res, 59-60; assoc prof entom, 62-68, dir, Ctr Systs Collections, 87-91, PROF ENTOM, VA POLYTECH INST & STATE UNIV, 68-; FOUNDING DIR, MUS NATURAL HIST, 90- *Concurrent Pos:* Planning Comt chmn, Nat Biol Surv Proj, 84-88. *Mem:* Entom Soc Am; Soc Syst Zool; AAAS; hon mem Hungarian Entom Soc; Am Inst Biol Sci. *Res:* Systematics and biology of scale insects (Homoptera Coccoidea) in North America and Europe. *Mailing Add:* Dept Entom VA Polytech Inst & State Univ Blacksburg VA 24061-0319

KOT, PETER ALOYSIUS, b Stanley, Wis, Jan 13, 32; m 58; c 6. CARDIOVASCULAR PHYSIOLOGY. *Educ:* Marquette Univ, MS, 56, MD, 57. *Prof Exp:* Intern med, Med Ctr, Georgetown Univ, 57-58, resident, 58-60, instr physiol, 60-64, instr med, 63-64, from asst prof to assoc prof physiol, 64-76, asst prof med, 64-69, PROF PHYSIOL, MED SCH, GEORGETOWN UNIV, 76- *Concurrent Pos:* Fel coun circulation, Am Heart Asn, 63, investr, 64-69; lectr physiol, US Naval Dent Sch, Bethesda, Md, 66-71; lectr, US Army Inst Dent Res, 68-71. *Mem:* AAAS; Am Fedn Clin Res; Am Physiol Soc; Soc Exp Biol & Med; Am Heart Asn. *Res:* Cardiovascular physiology, especially hemodynamic effects of the prostaglandins and their precursors; radiation injury. *Mailing Add:* Georgetown Univ Med Sch 3900 Reservoir Rd NW Washington DC 20007

KOT, RICHARD ANTHONY, b Syracuse, NY, May 22, 41; m 61; c 3. METALLURGY, MATERIALS SCIENCE. *Educ:* LeMoyne Col, BS, 64; Syracuse Univ, MS, 67, PhD(solid state sci), 69. *Prof Exp:* Res metallurgist, 69-74, supvr, 74-75, sect chief, 75-78, res adv, 78-80, sr res adv, 80-81, asst div head metall, Repub Steel Res Ctr, 81-86; dir res & develop, Touchstone Res Lab, 86-88; MGR MAT TECH, RES & DEVELOP, WORTHINGTON INDUST, 89- *Mem:* Am Soc Metals; Am Inst Mining, Metall & Petrol Engrs; Sigma Xi; NY Acad Sci. *Res:* Physical metallurgy; plastic deformation; recrystallization. *Mailing Add:* Worthington Indust 7407 Worthington-Galena Rd Columbus OH 43085

KOTANSKY, D(ONALD) R(ICHARD), b Hinsdale, Ill, July 28, 39; m 62; c 3. FLUID MECHANICS, AERODYNAMICS. *Educ:* Gen Motors Inst, BME, 62; Mass Inst Technol, SM, 62, MechE, 64, ScD(fluid mech), 66. *Prof Exp:* Asst prof mech eng, Purdue Univ, 65-68; sr & proj propulsion engr, Ft Worth Div, Gen Dynamics Corp, 68-70; sr group engr propulsion 70-73, sr scientist, McDonnel Douglas Res Lab, 73-77, sect chief, 77-80, BR CHIEF TECHNOL, AERODYN, MCDONNELL AIRCRAFT CO, MCDONNELL DOUGLAS CORP, 80- *Concurrent Pos:* Vis Caterpillar prof, Bradley Univ, 66-67; consult, Allison Div, Gen Motors Corp, Ind, 67-68. *Mem:* Assoc fel Am Inst Aeronaut & Astronaut; Am Soc Mech Engrs; Sigma Xi. *Res:* Theoretical and experimental investigations of external and internal aerodynamics, including laminar and turbulent flows, boundary layers, rotational and secondary flows, jet and multiple jet dominated flows, unsteadiness, acoustics and shock boundary layer interaction. *Mailing Add:* 14575 Appalachian Tr Chesterfield MO 63017-1901

KOTB, MALAK Y, b Cairo, Egypt, July 20, 53; m 86. BIOCHEMISTRY, IMMUNOLOGY. *Educ:* Ain Shams Univ, Cairo, Egypt, BS, 74; Univ Tenn, Memphis, PhD(biochem), 82. *Prof Exp:* Instr, Aim Shams Univ, 74; res assoc, Duke Univ Med Ctr, 82-85; asst prof, Div Infectious Dis, Dept Med, 86-90, ASST PROF, DEPT SURG & DIR SURG IMMUNOL, UNIV TENN, MEMPHIS, 90- *Concurrent Pos:* Rotary Int Educ fel, 77; mem res serv, Va Med Ctr, Memphis, 86-; consult, Nat Inst Child Health & Develop. *Mem:* Am Soc Biochemists & Molecular Biologists; Sigma Xi. *Res:* Synthesis and metabolism of S-Adenosylmethonine; biochemical regulation of T lymphocyte differentiation; mechanism of pathogenesis of poststreptococcal autoimmune diseases; role of superantigenesis autoimmunity. *Mailing Add:* Va Med Ctr 1030 Jefferson Ave No 151 Memphis TN 38104

KOTCH, ALEX, b Edwardsville, Pa, Aug 18, 26; m 52; c 4. SCIENCE & ACADEMIC ADMINISTRATION. *Educ:* Pa State Col, BS, 46, MS, 47; Univ Ill, PhD(org chem), 50. *Prof Exp:* Asst org chem, Pa State Col, 46-47 & Univ Ill, 47-49; Fulbright fel, Delft Tech Univ, 50-51; Little fel, Mass Inst Tech, 51-52; res chemist, Cent Res Dept, Exp Sta, E I du Pont de Nemours & Co, 52-54, Org Chem Dept, Jackson Lab, 54-59; assoc prog dir chem, Nat Sci Found, 59-63, prog dir org chem, 63-65; chief biosci div, Off Saline Water, US Dept Interior, 65-66; staff assoc, Sci Develop Eval Group, Div Instnl Progs, Nat Sci Found, 66-67; prof chem & assoc chmn dept, Univ Wis-Madison, 67-77; asst dir info, educ & int progs, Solar Energy Res Inst, 77-78, spec asst to dir, 78-79, br chief, Analyt Univ Progs, 78-81, prin mgr, Univ Res & Storage Progs, 81-82; PROD CHEM & DIR, OFF RES & PROG DEVELOP, UNIV NDAK, 82- *Concurrent Pos:* Consult-exam, N Cent Asn Cols & Schs, 69-, comnr-at-large, 84-88; bd dir, Assoc Western Univs, 82-, mem exec comt bd dir, 88-91. *Mem:* AAAS; Am Chem Soc; Nat Coun Univ Res Adminrs; Soc Res Adminrs; Sigma Xi. *Res:* Synthetic organic chemistry; polymers; heterocyclics; fluorescent whitening agents; dyes; science research and academic administration. *Mailing Add:* Off Res-Prog Develop Univ NDak Box 8138 University Sta Grand Forks ND 58202

KOTCHER, EMIL, medical microbiology; deceased, see previous edition for last biography

KOTCHOUBEY, ANDREW, b Florence, Italy, Mar 31, 38; US citizen; m 68; c 5. COMPUTER SCIENCE, APPLIED MATHEMATICS. *Educ:* Stevens Inst Technol, ME, 59; Columbia Univ, MA, 61, PhD(appl math), 66. *Prof Exp:* Supvr comput installation, Watson Sci Comput Lab, IBM Corp, 60-62, res asst appl math, Watson Lab, 62-66, sr staff mem appl math & comput, 66-69; dir info syst, Interway Corp, 69-71; pres, subsidiary I/W Data Systs, Inc, 71-73; vpres, Automatech Graphics Corp, 73-83; MGR, TRAIN SMITH COUNSEL, INC, 83- *Concurrent Pos:* Assoc grad facs math, Columbia Univ, 67-68, adj asst prof, 68-69. *Mem:* AAAS; Asn Comput Mach; Sigma Xi; Soc Indust & Appl Math. *Res:* Calculations in atomic and molecular physics; mathematical physics; numerical analysis. *Mailing Add:* 50 E 96th St New York NY 10128

KOTHMANN, MERWYN MORTIMER, b Castell, Tex, Jan 30, 40; m 62; c 3. RANGE SCIENCE, RANGE MANAGEMENT. *Educ:* Tex A&M Univ, BS, 61, PhD(range sci), 68; Utah State Univ, MS, 63. *Prof Exp:* Res asst range nutrit, Utah State Univ, 61-64; res asst range nutrit, 64-67; from asst prof to assoc prof range mgt, Tex Agr Exp Sta, 67-79, PROF RANGE SCI, TEX A&M UNIV, 79- *Mem:* Soc Range Mgt; Am Soc Animal Sci; Am Forage & Grassland Coun. *Res:* Simulation of natural vegetation and livestock responses to various grazing management systems; nutrition of range livestock and botanical and chemical characteristics of diets of grazing animals. *Mailing Add:* Dept Rangeland Ecol & Mgt Tex A&M Univ College Station TX 77843

KOTHNY, EVALDO LUIS, b Buenos Aires, Argentina, Oct 6, 25; US citizen; m 60; c 2. AIR POLLUTION, GEOCHEMISTRY. *Educ:* Univ Buenos Aires, MS, 55, PhD(chem), 64. *Prof Exp:* Plant chemist, Coplan Br, US Rubber Co, Argentina, 55-56; res chemist, Buenos Aires, 56-57; asst anal instrumentation, Univ Buenos Aires, 57-60 & 61-63; asst specialist qual control, Monsanto, Argentina, 60-61; sr specialist, Gen Elec, Argentina, 61-64; res chemist Air Indust Hyg Lab, 64-87, Sanitation & Radiation Lab, 87-90, RES CHEMIST, ENVIRON LAB ACCREDITATION PROG, CALIF STATE DEPT HEALTH SERV, 90- *Concurrent Pos:* Intersoc Comt, Am Pub Health Asn, 66-83, chmn subcomt 3, 71-74 & 76-79; consult, Nat Res Coun, Nat Acad Sci, 74-76, Div Med Scis, 74, 75, 76; abstractor, chem abstrs, 67-71. *Mem:* Am Chem Soc; Asn Explor Geochem. *Res:* Industrial inorganic preparative chemistry; trace inorganic analysis; geochemistry; environmental chemistry; geochemistry of noble metals; nitrogen oxides analysis; geochemical cycle of mercury; platinum and palladium in the environment; exploration of noble metals; biogeochemistry of palladium. *Mailing Add:* 3016 Stinson Circle Walnut Creek CA 94598-3621

KOTHS, JAY SANFORD, b Taylor, Mich, July 22, 26; m 47; c 3. FLORICULTURE. *Educ:* Mich State Univ, BS, 48; Purdue Univ, MS, 50; Univ Mass, PhD, 67. *Prof Exp:* Instr floricult, Purdue Univ, 48-50; greenhouse mgr, Kemble-Smith Co, Iowa, 50-53; asst greenhouse mgr, A Washburn & Sons, Ill, 53-54; gen mgr, A Weiler Greenhouse, Wis, 54-55; prof, 55-86, EMER PROF FLORICULTURE, UNIV CONN, 86- *Concurrent Pos:* Consult greenhouse mgt. *Honors & Awards:* Extension Award A&A Fel, Am Soc Hort Sci. *Mem:* Am Soc Agron; fel Am Soc Hort Sci; Soil Sci Soc Am; Int Soc Hort Sci; Am Hort Soc. *Res:* Automation of greenhouse microclimate; greenhouse crop fertility control; biological control of soilborne diseases; pollution effects on soil nitrification. *Mailing Add:* Dept Plant Sci Univ Conn Box U-67 Storrs CT 06268

KOTHS, KIRSTON EDWARD, b La Fayette, Ind, Dec 24, 48; m 85. PROTEIN ENGINEERING, PROTEIN CHEMISTRY OF PHARMACEUTICALS. *Educ:* Amherst Col, BA, 71; Harvard Univ, PhD(biochem & molecular biol), 79. *Prof Exp:* Scientist, 79-84, mgr protein chem, 82-84, SR SCIENTIST & DIR PROTEIN CHEM, CETUS CORP 84-, SR DIR RES, 89- *Mem:* AAAS. *Res:* Characterization of rare human proteins with therapeutic potential; development of cloned human proteins for clinical use; protein engineering through mutagenesis; pharmacokinetics of proteinaceous pharmaceuticals. *Mailing Add:* Cetus Corp 1400 53rd St Emeryville CA 94608

KOTICK, MICHAEL PAUL, b Buffalo, NY, Dec 28, 40; m 65; c 2. RECOMBINANT DNA, MEDICINAL CHEMISTRY. *Educ:* State Univ NY Buffalo, BS, 62, PhD(med chem), 68; Ind Univ, MS, 81. *Prof Exp:* Res asst med chem, Sch Pharm, State Univ NY Buffalo, 63-68; fel org chem, Walker Labs, Sloan-Kettering Inst Cancer Res, 68-69; res scientist, Molecular Biol Dept, Miles Labs, 69-75, sr res scientist, Chem Dept, 75-81, prin res scientist, biotechnol group, 81-88, supvr, recombinant DNA & prin staff scientist, Food Ingredients Div, 88-90, MGR, PROPRIETARY SERV, MILES PHARMACEUT DIV, MILES LABS, 90- *Mem:* AAAS; Am Chem Soc. *Res:* Chemistry of oligonucleotides, nucleosides, carbohydrates, narcotic drugs; medicinal chemistry; recombinant DNA technology; microbiology, molecular biology; research and resource management. *Mailing Add:* Res Serv Miles Inc 400 Morgan Lane West Haven CT 06516-4175

KOTILA, PAUL MYRON, b Hancock, Mich, Oct 14, 50. AQUATIC ECOLOGY, ENTOMOLOGY. *Educ:* Mich Technol Univ, BS, 72, MS, 74; Univ Wis-Madison, PhD(entom), 78. *Prof Exp:* asst prof biol, Allegheny Col, 78-; asst prof environ studies, St Lawrence Univ, 86-88; ASSOC PROF BIOL, FRANKLIN PIERCE COL, 88- *Mem:* AAAS; Am Fisheries Soc; Entom Soc Am; NAm Benthological Soc; Sigma Xi. *Res:* Effects of impoundments, toxicants and other disturbances on stream insects; ecology of aquatic invertebrates. *Mailing Add:* Dept Natural Sci Franklin Pierce Col College Rd MH 106 Rindge NH 03461

KOTIN, LEON, mathematics; deceased, see previous edition for last biography

KOTIN, LEONARD, b New York, NY, June 3, 32. PHYSICAL CHEMISTRY. *Educ:* Queens Col, BS, 54; Harvard Univ, AM, 55, PhD(chem physics), 60. *Prof Exp:* Res assoc chem, Inst Study Metals, Chicago, 59-61; asst prof, Wash Univ, 61-65; ASST PROF CHEM, UNIV ILL, CHICAGO CIRCLE, 65- *Concurrent Pos:* Res assoc, Nat Acad Sci-Nat Res Coun, 59-61. *Mem:* AAAS; Am Chem Soc; NY Acad Sci. *Res:* Equilibrium and transport properties of synthetic and biological macromolecules; polyelectrolytes; thermodynamics and statistical mechanics of condensed phases. *Mailing Add:* Dept Chem MC 111 Univ Ill Box 4348 Chicago IL 60680

KOTIN, PAUL, b Chicago, Ill, Aug 13, 16; m 70; c 2. PATHOLOGY. *Educ:* Univ Ill, BS, 37, MD, 40; Am Bd Path, dipl, 53. *Prof Exp:* From instr to prof path, Univ Southern Calif, 51-60, Paul Peirce prof, 60-62; chief carcinogenesis studies br, Nat Cancer Inst, 62-63, sci dir etiology, 64-66; dir div environ health sci, Nat Inst Environ Health Sci, 66-69, dir, Inst, 69-71; prof path, vpres health sci & dean sch med, Temple Univ, 71-74; sr vpres health, safety & environ, Johns-Manville Corp, 74-81; CONSULT PATHOLOGIST, 81- *Concurrent Pos:* Res fel path, Sch Med, Univ Southern Calif, 49-50, NSF sr fel, 59-60; med microbiologist, Los Angeles County Gen Hosp, 50-51, attend staff pathologist, 51-62. *Honors & Awards:* Sappington Lectr, Am Occup Med Asn, 80; Knudsen Award, 81; Gehrmann Lectr, Am Acad Occup Med, Nashville, 81. *Mem:* Am Asn Cancer Res; Am Asn Pathologists & Bacteriologists; fel Col Am Pathologists. *Res:* Mechanisms of carcinogenesis; experimental cancer production; environmental factors in cancer; air pollution; teratogenesis. *Mailing Add:* 4505 S Yosemite #339 Denver CO 80237

KOTLARSKI, IGNACY ICCHAK, b Warsaw, Poland, July 29, 23; US citizen. MATHEMATICS. *Educ:* Univ Warsaw, MA, 52; Wroclaw Univ, PhD(math), 61; Warsaw Tech Univ, Docent, 67. *Prof Exp:* Lectr math & statist, Planning & Statist Acad, Warsaw, 50-53; asst sampling inspection, Math Inst, Polish Acad Sci, 53-54; lectr math, Warsaw Tech Univ, 54-68; vis prof, Rome Univ, 68-69 & Univ Md, College Park, 69; PROF MATH & STATIST, OKLA STATE UNIV, 69- *Concurrent Pos:* Mem staff sampling inspection, Polish Stand Comt, 50-53; lectr math, Army Tech Acad, Warsaw, 53-59. *Mem:* Inst Math Statist. *Res:* Characterization problems in probability; mathematical modeling. *Mailing Add:* Dept Math & Statist Okla State Univ Stillwater OK 74078

KOTLER, DONALD P, b New Brunswick, NJ, Sept 30, 47; m 73; c 2. GASTROENTEROLOGY, CLINICAL IMMUNOLOGY. *Educ:* Rutgers Univ, BS, 69; Albert Einstein Col Med, MD, 73. *Prof Exp:* House officer internal med, Bronx Munic Hosp Ctr, 73-76; fel gastroenter, Hosp Univ Pa, 76-78, asst prof med, Univ Pa, 78-79; assoc prof med, COLUMBIA COL, 87- *Concurrent Pos:* Assoc attend physician, St Lukes Roosevelt Hosp Ctr, 79- *Mem:* AAAS; NY Acad Sci; Am Gastroenterol Asn; Am Fedn Clin Res. *Res:* Currently engaged in research to define, describe and control the gastrointestinal and nutritional complications of the acquired immunodeficiency syndrome. *Mailing Add:* GI Div St Lukes Roosevelt Hosp Ctr 421 W 113 St New York NY 10025

KOTLIAR, ABRAHAM MORRIS, b Brooklyn, NY, Oct 8, 26; m 55; c 4. PHYSICAL CHEMISTRY, POLYMER PHYSICS. *Educ:* Adelphi Col, BA, 49; Polytech Inst Brooklyn, PhD(chem), 55. *Prof Exp:* Res assoc & fel chem, Duke Univ, 55-56; chemist radiation effects, US Naval Res Lab, 56-60; chemist polymer physics, Esso Res & Eng Co, 60-64; group leader, Allied Chem Corp, 64-66, sr scientist, 66-69, sr res assoc, 69-88; ASSOC PROF, UNIV GA, 88- *Mem:* Am Chem Soc; Am Phys Soc; Soc Rheology; Sigma Xi. *Res:* Solution properties; molecular weight distributions; random processes; rheology and mechanical properties of plastics. *Mailing Add:* 112 Skyview Ct Athens GA 30606

KOTNIK, LOUIS JOHN, organic chemistry, chemical engineering; deceased, see previous edition for last biography

KOTOVYCH, GEORGE, b Jan 3, 41; Can citizen; m 74; c 4. BIOPHYSICAL CHEMISTRY. *Educ:* Univ Man, BSc, 63, MSc, 64, PhD(phys chem), 68. *Prof Exp:* Nat Res Coun Can fel bio-phys chem, Lawrence Radiation Lab, Univ Calif, Berkeley, 68-69; from asst prof to assoc prof, 70-89, PROF CHEM, UNIV ALTA, 89- *Mem:* Chem Inst Can; Am Chem Soc. *Res:* Application of nuclear magnetic resonance techniques to the study of biological systems; structure and conformation of polypeptides, tubulin, collagen telopeptides, and bradykinin analogs. *Mailing Add:* Dept Chem Univ Alta Edmonton AB T6G 2G2 Can

KOTT, EDWARD, b Toronto, Ont, Mar 25, 39. ZOOLOGY. *Educ:* Univ Toronto, BA, 60, PhD(ecol), 65. *Prof Exp:* Lectr zool, Lakehead Col, 63-65; assoc scientist fisheries res, Bedford Inst Oceanog, 65-69; ASST PROF BIOL, WATERLOO LUTHERAN UNIV, 69- *Mem:* Am Soc Mammalogists; Soc Syst Zool. *Res:* Mammalian and fish population ecology. *Mailing Add:* Dept Biol Sci Wilfrid Laurier Univ Waterloo ON N2L 3C5 Can

KOTTAS, HARRY, b Milligan, Nebr, Oct 24, 10; m 38; c 2. MECHANICAL ENGINEERING. *Educ:* Univ Nebr, BSc, 32, MSc, 33. *Prof Exp:* Mech engr food processing, Roberts Dairy Co, Nebr, 33-36 & Swift & Co, 36-37; chief mech eng div, Nat Adv Comt Aeronaut, Ohio, 37-52; chief tech panels, Redstone Arsenal, Ala, 52-56; asst dir eng, AK Div, Avco Mfg Corp, Ind, 56-59; chief spec prod eng, Curtiss-Wright Corp, 59-60; chief engr, Tuthill Spring Co, Ill, 60-62; mgr eng, Int Staple & Mach Co, Pa, 62-64; mfg mgr, Am Device Mfg Co, 64-69; prof design & drafting technol, Lake Land Col, 69-76; PRES, K-SERV, 76- *Res:* Product-market characteristics; mobile vehicles materials handling; industrial noise phenomena; tillage components; fluid and solid metal flow phenomena; automated packaging; engineering and industrial human factors; manufacturing optimization. *Mailing Add:* 403 S Randall St Steeleville IL 62288-2002

KOTTCAMP, EDWARD H, JR, b York, Pa, July 12, 34; c 3. METALLURGY & PHYSICAL METALLURGICAL ENGINEERING. *Educ:* Lehigh Univ, BS, 56, MS, 57, PhD(metallurgical eng & mat sci), 60. *Prof Exp:* Vpres res, Bethlehem Steel Corp, 82-85, sr vpres, 85-86, exec vpres, 86-87; GROUP VPRES, SPS TECHNOLS, INC, 87- *Concurrent Pos:* Prof, Lehigh Univ Col Eng. *Honors & Awards:* William Sparagan Award for Outstanding Res, 73; William Eisenman Award, Am Soc Metals, 88. *Mem:* Fel Am Soc Metals; Indust Res Inst; Am Iron & Steel Inst; Welding Res Coun. *Res:* cold extrusion of steels; pressure vessel design; research management and innovation; high-strength steels; pressure vessels; microstructure; fracutre; metal forming; author of numerous technical articles. *Mailing Add:* SPS Technols Inc PO Box 1000 Newtown PA 18940

KOTTER, F(RED) RALPH, b Salt Lake City, Utah, Dec 8, 15; m 49; c 5. ELECTRICAL MEASUREMENTS, HIGH VOLTAGE PHENOMENA. *Educ:* Univ Utah, BSc, 37; George Washington Univ, AM, 40; Mass Inst Technol, ScD, 55. *Hon Degrees:* ScD, Mass Inst Technol, 55. *Prof Exp:* Physicist, Nat Bur Standards, 37-47; from instr to asst prof elec eng, Mass Inst Technol, 47-54; physicist, Nat Bur Standards, 55-81; CONSULT, 82- *Mem:* Inst Elec & Electronics Engrs. *Res:* Precise electrical measurements; high voltage measurements. *Mailing Add:* 12921 Crisfield Rd Silver Spring MD 20906

KOTTKE, BRUCE ALLEN, b Blue Earth, Minn, Jan 22, 29; m 79; c 2. EXPERIMENTAL PATHOLOGY, INTERNAL MEDICINE. *Educ:* Hamline Univ, BS, 51; Univ Minn, Minneapolis, MD, 54, PhD, 62. *Prof Exp:* Consult, Mayo Found & Clin, 62-71, from asst prof to assoc prof med, 62-76, PROF MED, MAYO GRAD SCH MED, UNIV MINN, 76- *Concurrent Pos:* Fel int med, Mayo Found, 57-61; mem coun arteriosclerosis, Am Heart Asn, mem coun atherosclerosis, mem coun thrombosis, fel coun circulation. *Mem:* Am Heart Asn; Am Fedn Clin Res; Sigma Xi. *Res:* Atherosclerosis; cholesterol metabolism; bile acid metabolism. *Mailing Add:* Mayo Clin Rochester MN 55905

KOTTKE, FREDERIC JAMES, b Hayfield, Minn, May 26, 17; m 39; c 3. PHYSICAL MEDICINE & REHABILITATION. *Educ:* Univ Minn, BS & MS, 41, PhD(physiol), 44, MD, 45; Am Bd Phys Med & Rehab, dipl, 49. *Prof Exp:* Asst physiol, FUniv Minn, , 41-44, from asst prof to prof, 47-86, dir div, 49-52, head dept, 52-82, EMER PROF PHYS MED & REHAB, UNIV MINN, MINNEAPOLIS, 86- *Concurrent Pos:* Baruch fel phys med, Univ Minn, Minneapolis, 46-47; mem, Am Bd Phys Med & Rehab, 55-70, chmn, 64-70; consult, Minneapolis Vet Admin Hosp, 56; mem, Minn Gov Adv Comt Voc Rehab, 56-60; mem exec comt, prog chmn & vpres, Int Cong Phys Med, 60; mem med adv comt, Off Voc Rehab, 60-67; mem Med Res Study Sect, Voc Rehab Admin, 61-63; secy & mem bd dirs & mem expert med comt, Am Rehab Found, 64; mem, Minn State Bd Health, 64-67, Med Adv Comt, Social & Rehab Serv, 68-69 & Coun Cerebrovasc Dis & Coun Clin Cardiol, Am Heart Asn, 70-83. *Honors & Awards:* Frank H Krusen Award, Am Acad Phys Med & Rehab, 79; Sidney Licht lectr, Univ Pa, 79 & Ohio State Univ, 81; Lewis Leavitt Mem lectr, Baylor Univ Med Sch, 82. *Mem:* Fel AMA; fel Am Cong Phys Med & Rehab (vpres, 54-58, pres elect, 58-59, pres, 59-60); Am Acad Phys Med & Rehab (pres-elect, 77, pres, 78); Int Soc Rehab Disabled; hon mem Columbian Soc Phys Med & Rehab; hon mem Mex Acad Surg; hon mem Brazilian Acad Rehab Med; hon mem Venezuelan Soc Phys Med & Rehab; hon mem Neth Soc Phys Med & Rehab. *Res:* Circulation; neuromuscular diseases; poliomyelitis; rehabilitation; work of the heart. *Mailing Add:* 2741 Drew Ave S Minneapolis MN 55416

KOTTLOWSKI, FRANK EDWARD, b Indianapolis, Ind, Apr 11, 21; m 45; c 3. ECONOMIC GEOLOGY, COAL GEOLOGY. *Educ:* Ind Univ, AB, 47, AM, 49, PhD(econ geol), 51. *Prof Exp:* Asst geologist econ geol, State Geol Surv, Ind, 46-51; econ geologist, NMex Bur Mines & Mineral Res, 51-66, asst dir & sr geologist, 66-73, dir & state geologist, 73-91, EMER DIR STATE GEOLOGIST, NMEX BUR MINES & MINERAL RES, 91- *Concurrent Pos:* Asst, Ind Univ, 47-48, instr, 50; fac assoc, NMex Inst Mining & Technol, 54-73, adj prof, 74-; ed, Am Asn Petrol Geologists, 71-75; chmn, Nat Acad Sci Codes Comt, 80-81, Nat Acad Sci Comre Comt, 82-83; pres, Energy Minerals Div, Am Asn Petrol Geologists, 87-88. *Honors & Awards:* Distinguished Serv Award, Am Asn Petrol Geologists; Public Serv Award, Am Inst Prof Geologists, 86. *Mem:* AAAS; Soc Econ Geol; fel Geol Soc Am; Soc Econ Paleont & Mineral; hon mem Am Asn Petrol Geologists; Asn Am State Geologists (pres, 85-86). *Res:* Coal geology; Pennsylvanian and Permian stratigraphy; Cenozoic sediments and volcanic rocks; industrial minerals and rocks; areal mapping in Indiana, New Mexico and Montana; measuring stratigraphic sections. *Mailing Add:* 703 Sunset Dr Socorro NM 87801

KOTTMAN, CLIFFORD ALFONS, b San Diego, Calif, Aug 3, 42; m 66; c 3. MATHEMATICS. *Educ:* Loyola Univ, Los Angeles, BS, 64; Univ Iowa, MS, 66, PhD(math), 69. *Prof Exp:* Asst prof math, La State Univ, 69-70; asst prof math, Ore State Univ, 70-75, assoc prof, 75-77; mathematician, Defense Mapping Agency, 77-90; EXEC MGR, INER GRAPH CORP, 90- *Concurrent Pos:* Consult in non-destructive testing. *Mem:* AAAS; Am Math Soc; Math Asn Am; Am Soc Photogram. *Res:* Functional analysis; Banach spaces; photogrammetry. *Mailing Add:* 6614 Rockland Dr Clifton VA 22024

KOTTMAN, ROY MILTON, b Thornton, Iowa, Dec 22, 16; m 41; c 4. ANIMAL BREEDING. *Educ:* Iowa State Univ, BS, 41, PhD(animal breeding), 52; Univ Wis, MS, 48. *Hon Degrees:* LLD, Col Wooster, 72. *Prof Exp:* Asst, Univ Wis, 47-48; asst prof animal husb, Iowa State Univ, 46-47, 49-52, assoc prof, 52-54, prof & assoc dean agr, 54-58; dean col agr, Forestry & Home Econ & dir, Agr Exp Sta, WVa Univ, 58-60; dean, Col Agr & Home Econ & dir Ohio Agr Res & Develop Ctr, Ohio State Univ, 60-82, dir Coop Exten Serv, 64-82, vpres, Agr Admin, 82; RETIRED. *Concurrent Pos:* Asst to dean, Iowa State Univ, 50-51, asst dean, 51-54; mem bd dirs, Swift Independent Packing Co, 81-85; mem bd trustees, Farm Found, 78-88; actg assoc dir, Nev Agr Exp Sta, 82-83; consult to chancellor, Univ PR, 84-85. *Honors & Awards:* Soil Conserv Soc of America's Honor Award, 74. *Mem:* Sigma Xi; fel Am Soc Animal Sci. *Res:* Population genetics; genetic improvement of swine, beef cattle and sheep. *Mailing Add:* Col Agr Ohio State Univ 2120 Fyffe Rd Columbus OH 43210

KOTTMEIER, PETER KLAUS, b Munich, Ger, Feb 1, 28; m 56; c 4. SURGERY. *Educ:* Univ Munich, MD, 51, Ohio State Univ, MMSc, 60. *Prof Exp:* Asst instr surg, State Univ NY Downstate Med Ctr, 57-60; instr, Ohio State Univ, 60-61; from asst prof to assoc prof, 67-70, PROF SURG, STATE UNIV NY DOWNSTATE MED CTR, 70-, DIR PEDIAT SURG SERV, UNIV HOSP, 67-, DIR PEDIAT SURG SERV, KINGS COUNTY HOSP, BROOKLYN, 62- *Mem:* Fel Am Acad Pediat; fel Am Col Surgeons; fel Am Pediat Surg Asn. *Res:* Pediatric surgery. *Mailing Add:* Dept Surg State Univ NY Downstate Med Ctr Brooklyn NY 11203

KOTULA, ANTHONY W, b Holyoke, Mass, June 12, 29; m 57; c 2. FOOD SCIENCE. *Educ:* Univ Mass, BS, 51, MS, 54; Univ Md, PhD(food sci), 64. *Prof Exp:* Proj leader, 54-67, invests leader, 67-71, SUPVRY RES FOOD TECHNOLOGIST, CHIEF MEAT SCI RES LAB, ANIMAL SCI INST, AGR RES SERV, USDA, 71- *Honors & Awards:* Res Award, Poultry Sci Res Asn, 67; Signal Serv Award, Am Meat Sci Asn, 83; Meat Res Award, Am Soc Animal Sci, 88. *Mem:* Poultry Sci Asn; Inst Food Technologists; World Poultry Sci Asn; Am Meat Sci Asn; Am Soc Animal Sci; Sigma Xi. *Res:* Maintaining and improving quality of animal products. *Mailing Add:* 4310 Howard Rd Beltsville MD 20705

KOTVAL, PESHO SOHRAB, b Nagpur, India, Aug 31, 42; US citizen; m 65; c 2. MEDICAL ENGINEERING, MANAGEMENT SCIENCE. *Educ:* Univ Nagpur, BSc, 60; Univ Sheffield, MMet, 62, PhD(phys metall), 65; Pace Univ, MBA, 77; NY Med Col, MD, 83; Nat Bd Med Examr, dipl, 84; Am Bd Radiol, dipl, 87. *Prof Exp:* Scientist, res assoc & mgr superally metall, Stellite Div, Union Carbide Corp, 66-70; vis scientist metall, Res Inst Advan Studies, 70-71; sr group leader metals & ceramics, Corp Res Lab, Union Carbide Corp, 71-78, res mgr mat sic, Med Prod Div, 78-80; res physician radiol, 83-87, ASSOC PROF RADIOL, NY MED COL, 87-, ASSOC PROF SURG, 90- *Concurrent Pos:* Fel, Sheffield Univ, 65-66; adj prof physics, Ind Univ, 67-68; adj prof mgt econ, Pace Univ, 77- *Honors & Awards:* Coatings Award, Am Soc Metals, 73. *Mem:* Fel Am Soc Metals; Brit Inst Metallurgists; AMA; Radiol Soc NAm; Am Inst Ultrasound Med. *Res:* Superalloys for high temperature gas turbines and corrosion resistance; powder metallurgy; crystal growth; process development; low cost solar cells; medical instruments; blood flow technology. *Mailing Add:* Eight Verne Pl Hartsdale NY 10530

KOTYK, MICHAEL, b Ford City, Pa, Mar 10, 29; m 52; c 5. METALLURGY, CERAMICS. *Educ:* Pa State Univ, BS, 54, MS, 56; NC State Univ, PhD(metall, ceramics), 68. *Prof Exp:* Instr metall, Pa State Univ, 54-56; sr technologist, US Steel Corp, 56-63; assoc dir metall & ceramics div, US Army Res Off, 63-68; sect supvr, Appl Res Lab, 68-73, div chief sheet prod res, 73-82, div chief basic res div, 82-84, DIV MGR TECH SERV, 84-, TECHNOL COORDR, PROD TECHNOL, US STEEL CORP, 84- *Mem:* Am Soc Metals; Am Inst Mining, Metall & Petrol Engrs; fel Am Chem Soc; Iron & Steel Inst Japan. *Res:* Formability of sheet steels; gases in metals; physical and mechanical properties of ferrous alloys; phase equilibria studies; productions sheet steel products. *Mailing Add:* US Steel Res Monroeville PA 15146

KOTZ, ARTHUR RUDOLPH, b Eau Claire, Wis, Feb 21, 33; m 55; c 3. SOLID STATE ELECTRONICS. *Educ:* Univ Minn, BA, 55; Univ Wis, MS, 62, PhD(solid state physics), 66. *Prof Exp:* Jr physicist, 55-57, sr physicist, 57-58, supvr phys res, 58-60, proj leader, 60-61, res specialist, 66-68, sr res specialist, 68-70, mgr electronic imaging group, 70-73, CORP SCIENTIST, 3M CO, 73- *Mem:* Am Phys Soc; Soc Photog Sci & Eng. *Res:* Electrical transport properties of organic semiconductors; electron beam recording; gas discharge devices; photoeffects in solids, including photoconductivity, photovoltaic effect and photoemission; electrophotography; electronic imaging; electrography; reprography; electronic printing. *Mailing Add:* 5826 S Hobe Lane St Paul MN 55110

KOTZ, JOHN CARL, b Massillon, Ohio, June 27, 37; m 61; c 2. INORGANIC CHEMISTRY, ORGANOMETALLIC CHEMISTRY. *Educ:* Wash & Lee Univ, BS, 59; Cornell Univ, PhD(inorg chem), 64. *Prof Exp:* NIH fel chem, Manchester Col Sci & Technol, Eng, 63-64 & Ind Univ, 64-65; asst prof, Kans State Univ, 65-70; prof chem, State Univ NY Col Oneonta, 70-87; DISTINGUISHED TEACHING PROF, DEPT CHEM, STATE UNIV NY, ONEONTA, 87- *Concurrent Pos:* Fulbright lectr & res scholar, Portugal, 79. *Mem:* Am Chem Soc; Royal Soc Chem. *Res:* Synthetic organometallic chemistry; electrochemistry of organometallic compounds. *Mailing Add:* Dept Chem State Univ NY Col Oneonta NY 13820-1381

KOTZ, SAMUEL, b Harbin, China, Aug 28, 30; m 63; c 3. MATHEMATICAL STATISTICS, APPLIED PROBABILITY. *Educ:* Hebrew Univ, Israel, MSc, 56; Cornell Univ, PhD(math statist), 60. *Prof Exp:* Instr math, 56-58, lectr, Bar-Ilan Univ, Israel, 60-62; res assoc, Inst Statist, Univ NC, 62-63; sr res fel indust eng, 63-64, assoc prof, Univ Toronto, 64-67; prof math, Temple Univ, 67-79; PROF STATIST, UNIV MD, 79- *Concurrent Pos:* Assoc ed, J Am Statist Asn; Distinguished vis prof, Bucknell Univ, 77, Guelph Univ, 86; co-ed Encycl Statist Sci; adv prof, Harbin Polytech Inst. *Honors & Awards:* Wolfowitz Prize, 83. *Mem:* Am Math Soc; fel Am Statist Asn; fel Inst Math Statist; Intern Statist Inst. *Res:* Information theory; statistical distribution theory and methodology; scientific terminology; probabilistic models with special applications to business and engineering. *Mailing Add:* Dept Mgt & Statist Univ MD College Park MD 20742

KOTZEBUE, KENNETH LEE, b San Antonio, Tex, Dec 4, 33; m 54; c 3. ELECTRICAL ENGINEERING. *Educ:* Univ Tex, BS, 54; Univ Calif, Los Angeles, MS, 56; Stanford Univ, PhD(elec eng), 59. *Prof Exp:* Sr engr, Tex Instruments, Inc, 58-59; mem tech staff elec eng, Watkins-Johnson Co, 59-63, dept head solid state devices res & develop, 63-64; assoc prof, 64-68, PROF ELEC ENG, UNIV CALIF, SANTA BARBARA, 68- *Mem:* Inst Elec & Electronics Engrs. *Res:* Microwave solid-state device electronics. *Mailing Add:* 4737 Woodview Dr Santa Rosa CA 95405

KOTZIG, ANTON, mathematics; deceased, see previous edition for last biography

KOUBA, DELORE LOREN, b Lincoln, Nebr, Apr 18, 19; m 71. CHEMICAL ENGINEERING, ORGANIC CHEMISTRY. *Educ:* Univ Nebr, BSc, 41. *Prof Exp:* Anal chemist, Smokeless Powder Plant, Hercules, Inc, NJ, 41-42, lab supvr, 42, chief chemist, 42-43, anal chemist, Res & Develop Res Ctr, Del, 43-46, explosives chemist, 46-50, res chemist, 50-78, sr res chemist, Res & Develop Res Ctr, Hercules, Inc, Del, 79-82; RETIRED. *Mem:* Am Chem Soc. *Res:* Smokeless powder testing; high explosives; semi-plant nitration; oxidation of aromatic compounds and hazardous chemicals evaluation; synthetic lubricants. *Mailing Add:* 1808 Windermere Ave Wilmington DE 19804

KOUBEK, EDWARD, b Bayshore, NY, July 25, 37; m 63; c 2. INORGANIC CHEMISTRY. *Educ:* State Univ NY Albany, BS, 59; Brown Univ, PhD(chem), 64. *Prof Exp:* Fel, Bell Tel Labs, NJ, 63-64; from asst prof to assoc prof, 67-75, PROF CHEM, US NAVAL ACAD, 75- *Mem:* Am Chem Soc. *Res:* Kinetics and mechanisms of inorganic reactions. *Mailing Add:* Dept Chem US Naval Acad Annapolis MD 21402

KOUCKY, FRANK LOUIS, JR, b Chicago, Ill, June 24, 27; m 49; c 4. MINERALOGY, GEOCHEMISTRY. *Educ:* Univ Chicago, MS, 53, PhD(geol), 56. *Prof Exp:* Instr phys sci, Navy Pier, Univ Ill, 51-55; from instr to asst prof, Mont Sch Mines, 55-57; asst prof & dir field camp, Univ Ill, 57-71; from asst prof to assoc prof, Univ Cincinnati, 60-71; PROF GEOL, COL WOOSTER, 71- *Concurrent Pos:* Assoc, Danforth Found, 68; res assoc, Mass Inst Technol, 78 & 83-; NEH fel, 88. *Mem:* Am Mineral Soc; Geol Soc Am; Am Asn Petrol Geol; Geochem Soc; Clay Mineral Soc; Soc Econ Geologists; Sigma Xi. *Res:* X-ray crystallography; sulfide and sulfosalt minerals; geology of Wyoming and Montana; Precambrian geology; ancient technology related to mining and smelting; archaeological geology of Cyprus, Israel and Jordan. *Mailing Add:* Dept Geol Col Wooster Wooster OH 44691

KOUL, HIRA LAL, b Srinagar, India, May 27, 43. MATHEMATICAL STATISTICS. *Educ:* Univ Jammu & Kashmir, India, BA, 62, Univ Poona, MA, 64; Univ Calif, Berkeley, PhD(math statist), 67. *Prof Exp:* Asst, Univ Calif, Berkeley, 65-67; asst prof statist, 68-72, assoc prof, 72-77, PROF STATIST & PROBABILITY, MICH STATE UNIV, 77- *Mem:* Fel Inst Math Statist. *Res:* Nonparametric statistics; inference on stochastic processes; reliability theory and survival analysis. *Mailing Add:* A413 Wells Hall Statist Mich State Univ East Lansing MI 48824

KOUL, MAHARAJ KISHEN, b Srinagar, India, Sept 10, 41; m; c 2. MATERIALS SCIENCE, METALLURGY. *Educ:* Univ Jammu & Kashmir, BSc, 59; Banaras Hindu Univ, BSc, 63; Mass Inst Technol, PhD(mat sci), 68. *Prof Exp:* Metall asst, Union Carbide India Pvt Ltd, 63-65; res asst mat sci, Mass Inst Technol, 65-68, fel, 68-69; res scientist, Res & Develop, Mining & Metals Div, Union Carbide Corp, 69-70, proj engr, New Prod Develop, 70-76; mgr, Steel Res & Develop, Foote Mineral Co, 76-79; exec vpres, Div Indian Metals & Ferro Alloys Ltd, Newmont Mining Co, 79-80; sr res scientist, Johnson & Johnson, 80-82; VPRES & GEN MGR, ATLANTIC METALS CORP, Philadelphia, PA. *Mem:* Am Inst Mining, Metall & Petrol Engrs; Am Soc Metals; Iron & Steel Soc. *Res:* Electron microscopic investigation of phase transformation and deformation behavior in Beta-isomorphous titanium alloys; strengthening mechanisms and their application to the development of high strength-low alloy steels; dissolution kinetics of solids in liquid metals; thermodynamics and its application to metallurgical phenomenon; boron steel developments; deoxidation, desulfurization and sulfide modification in steel; dental alloy development; mold powders for casting of steel. *Mailing Add:* 136 E Delaware Ave Pennington NJ 08534

KOUL, OMANAND, b Kashmir, India, Feb 17, 47; m 71; c 2. NEUROCHEMISTRY, GLYCOCONJUGATES. *Educ:* Banaras Hindu Univ, India, MSc, 68, PhD(zool), 74. *Prof Exp:* Lectr physiol & biochem, Banaras Hindu Univ, India, 70-74; asst prof genetics & biol, Govind Ballabh Pant Univ Agr & Technol, India, 75-76; res assoc biochem, Eunice Kennedy Shriver Ctr Mental Retardation, 76-82, SCIENTIST, E K SHRIVER CTR, 87-; RES ASSOC, HARVARD MED SCH, 84-86, 88- *Concurrent Pos:* Res fel, Dept Neurol, Mass Gen Hosp, 76-82, asst dir biochem, Dept Neurol, 86- *Mem:* Am Soc Neurochem; Soc Gerantol. *Res:* Brain function in health and disease; enzymology of lipids; metabolism of glycolipids in animals and cell cultures; myelin biosynthesis during development; regulation of glycosylation in tissues. *Mailing Add:* 130 Madison Ave Arlington MA 02174

KOULOURIDES, THEODORE I, b Preveza, Greece, Sept 11, 25; US citizen; m 56; c 3. DENTISTRY, ORAL BIOLOGY. *Educ:* Nat Univ Athens, Dent Surgeon, 50; Univ Rochester, MS, 58; Univ Ala, DMD, 60. *Prof Exp:* From asst prof to assoc prof, 60-69, PROF DENT, MED CTR, UNIV ALA, BIRMINGHAM, 69-; SR SCIENTIST, INST DENT RES, 71- *Concurrent Pos:* Fel pedodontics, Guggenheim Dent Clin, New York, 55; fel, Eastman Dent Dispensary, Rochester, NY, 55-56; USPHS res career develop award, 63-68. *Mem:* Am Dent Asn; Am Col Dent; Sigma Xi; Int Dent Fedn; Int Asn Dent Res. *Res:* Biological mineralization, especially factors involved in dental caries and calculus formation. *Mailing Add:* Sch Dent Box 46 Univ Station Univ Ala Birmingham AL 35294

KOUNOSU, SHIGERU, b Tokyo, Japan, Aug 23, 28; m 61; c 2. HIGH ENERGY PHYSICS, THEORETICAL PHYSICS. *Educ:* Fukushima Univ, Japan, BEd, 51; Univ Pa, MS, 63, PhD(physics), 65. *Prof Exp:* Res assoc physics, Princeton Univ, 65-67; asst prof, 67-70, ASSOC PROF PHYSICS, UNIV LETHBRIDGE, 70- *Mem:* Am Phys Soc. *Res:* Elementary particle physics. *Mailing Add:* Dept Physics Univ Lethbridge Lethbridge AB T1K 3M4 Can

KOURANY, MIGUEL, b Panama City, Panama, Sept 16, 24; div; c 4. MICROBIOLOGY, PUBLIC HEALTH. *Educ:* Iowa State Col, BS, 50; Loyola Univ, Chicago, MS, 53; Univ Mich, Ann Arbor, MPH, 54, PhD(epidemiol sci), 63. *Prof Exp:* Dir, Pub Health Lab, 54-63, chief, Bact Dept, 63-83, DIR, GORGAS MEM LAB, 90-; MINISTRY HEALTH, DEPT TECH, PANAMA, 83- *Concurrent Pos:* Consult, Pan Am Health Orgn Lab Serv in var countries, 71 -; supv ad honoratium, Pub Health Lab Serv, Ministry Health, 63-83; mem, Epert Adv Panel Health Lab Serv, WHO, 67 -; mem, Pan Am Health Org Sci Adv Comt to Zoonosis Ctr, Argentina, 74 - *Mem:* Am Soc Trop Med & Hyg; AAAS; Panamanian Soc Microbiol & Parasitol (pres, 68, 69 & 78); Panamanian Acad Med & Surg. *Res:* Intracellular infections; etiological agents of diarrheal disease; ecology of vibrio parahaemolyticus; zoonosis in Panama. *Mailing Add:* Lab Central de Salud Apartado 1474 Panama 1 Panama

KOURI, DONALD JACK, b Hobart, Okla, July 25, 38; m 65; c 2. THEORETICAL CHEMISTRY. *Educ:* Okla Baptist Univ, BA, 60; Univ Wis, MS, 62, PhD(phys chem), 65. *Prof Exp:* Instr chem & physics, Okla Baptist Univ, 62-63; res assoc physics & mem joint inst lab astrophys, Univ Colo, 65-66; asst prof chem, Midwestern Univ, 66-67; from asst prof to prof, 67-88, DISTINGUISHED PROF CHEM & PHYSICS, UNIV HOUSTON, 88- *Concurrent Pos:* Fel, A P Sloan Found, 72-; Weizmann Inst fel, 73; J S Guggenheim fel, 78-79; fel Inst Advan Studies, Hebrew Univ, Jerusalem, 78-79. *Honors & Awards:* US Sr Scientist Award, Alexander von Humboldt Found, 73. *Mem:* Fel Am Phys Soc; Am Chem Soc; Am Asn Physics Teachers. *Res:* Theoretical research on quantum mechanical scattering phenomena; reactive and nonreactive molecular collisions; approximations for inelastic and reactive collisions. *Mailing Add:* Dept Chem Univ Houston Houston TX 77204-5641

KOUSHANPOUR, ESMAIL, b Teheran, Iran, June 9, 34; US citizen; m 78; c 4. PHYSIOLOGY, BIOPHYSICS. *Educ:* Columbia Univ, AB, 58; Mich State Univ, MS, 61, PhD(physiol), 63. *Prof Exp:* Asst prof, 63-68, ASSOC PROF PHYSIOL, MED SCH, NORTHWESTERN UNIV, ILL, 68-, ASSOC PROF ANESTHESIA, 82- *Concurrent Pos:* Nat Heart Inst fel, 65-; vis prof, Heidelberg Univ, West Germany, 83-84; sr Fulbright prof, 83-84. *Mem:* AAAS; Am Physiol Soc; NY Acad Sci; Am Heart Asn. *Res:* Mathematical and experimental analyses of the cardiovascular and renal regulators; mechanism of the baroceptor process in the carotid sinus; role of carotid sinus in renal hypertension. *Mailing Add:* Dept Physiol Northwestern Univ Med Sch Chicago IL 60611

KOUSKOLEKAS, COSTAS ALEXANDER, b Thessaloniki, Greece, May 10, 27; m 58; c 2. ENTOMOLOGY. *Educ:* Univ Thessaloniki, Dipl agr, 51; Univ Mo-Columbia, MS, 58; Univ Ill, Urbana, PhD(entom), 64. *Prof Exp:* Teacher agron, Am Farm Sch, Thessaloniki, 54-56; res assoc agr entom, Natural Hist Surv & Agr Exp Sta, Univ Ill, Urbana, 62-63; consult, Doxiadis Assocs Int, Athens, Greece, 64-65; sr res officer entom, Benaki Phytopath Inst, Athens, 65-67; ASSOC PROF ENTOM, AUBURN UNIV, 67- *Mem:* Entom Soc Am; Int Orgn Biol Control. *Res:* Biology and control of insects of ornamentals and vegetables; integrated pest management. *Mailing Add:* Dept Zool-Entom 331 Funchess Hall Auburn Univ Auburn AL 36830

KOUSKY, VERNON E, b Detroit, Mich, Nov 2, 43; m 73; c 3. DYNAMIC METEOROLOGY. *Educ:* Pa State Univ, BS, 65, MS, 67; Univ Wash, PhD(atmospheric sci), 70. *Prof Exp:* Asst prof meteor, Univ Utah, 70-75; prof collabr, Inst Astron & Geophys, Univ Sao Paulo, 75-77; assoc researcher & researcher, Inst Space Res, Brazil, 77-83; RES METEOROLOGIST, CLIMATE ANAL CTR, 84- *Mem:* Am Meteorol Soc. *Res:* Synoptic meteorology; diagnostic study of wave motions in the tropical stratosphere; severe local storms; jetstream formation; tropopause deformation; atmospheric teleconnections; tropical meteorology; climate anomalies. *Mailing Add:* NOAA-NMC Climate Anal Ctr Rm 605 WWB Washington DC 20233

KOUTS, HERBERT JOHN CECIL, b Bisbee, Ariz, Dec 18, 19; m 42; c 2. NUCLEAR ENERGY, NUCLEAR REACTOR SAFETY. *Educ:* La State Univ, BS, 41, MS, 46; Princeton Univ, PhD(physics), 52. *Prof Exp:* Assoc physicist, Brookhaven Nat Lab, 50-51, asst group leader, 51-52, group leader reactor physics, 52-58, sr scientist & assoc div head, 58-73; dir, Div Reactor Safety Res, AEC, 73-75; dir, Off Nuclear Regulatory Res, US Nuclear Regulatory Comn, 75-76; chmn, Dept Nuclear Energy, Brookhaven Nat Lab, 77-87, sr scientist, 88-89; MEM, DEFENSE NUCLEAR FACIL SAFETY BD, US GOVT, 89- *Concurrent Pos:* Mem adv comt reactor safeguards, AEC, 62-66, chmn, 65; mem, Europ-Am Comt Reactor Physics, Europ Nuclear Energy Agency, 62-68; mem, Mayor's Tech Adv Comt on Radiation, New York, 69-73; chmn, Nuclear Adv Comt, Hall of Sci, New York, 69-73; prin adv reactor safety, NY State Atomic & Space Develop Authority, 69-73; mem, Int Nuclear Safety Adv Group, chmn, 88-91. *Honors & Awards:* E O Lawrence Award, AEC, 63 & Distinguished Serv Award, 75; Distinguished Serv Award, US Nuclear Regulatory Comn, 76; Theos Thompson Award, Am Nuclear Soc. *Mem:* Nat Acad Eng; fel Am Nuclear Soc; Int Atomic Energy Agency; Int Nuclear Safety Soc Group. *Res:* Elementary particle physics; shielding and physics of nuclear reactors. *Mailing Add:* 249 S Country Rd Brookhaven NY 11719

KOUTSKY, JAMES A, b Cleveland, Ohio, Dec 1, 39. CHEMICAL ENGINEERING, POLYMER SCIENCE. *Educ:* Case Inst Technol, BS, 61, MS, 63, PhD(polymer sci), 66. *Prof Exp:* From asst prof to assoc prof, 66-77, PROF CHEM ENG, UNIV WIS-MADISON, 77- *Concurrent Pos:* Du Pont Young Fac res grant, 68-69. *Mem:* Am Phys Soc; Am Inst Chem Engrs; Am Chem Soc; Soc Plastics Engrs. *Res:* Solid state characterization of macromolecules by electron and optical microscopy, electron and x-ray diffraction and differential thermal analysis. *Mailing Add:* Chem Eng Dept Univ Wis 3016 Engr Bldg Madison WI 53706

KOUVEL, JAMES SPYROS, b Jersey City, NJ, May 23, 26; m 53; c 2. SOLID STATE PHYSICS. *Educ:* Yale Univ, BEng, 46, PhD(phys & elec eng), 51. *Prof Exp:* Res engr, Microwave Devices, Fed Telecommun Labs, NJ, 47-48; res fel physics, Univ Leeds, 51-53; res fel solid state physics, Harvard Univ, 53-55; physicist, Res & Develop Ctr, Gen Elec Co, 55-69; PROF PHYSICS, UNIV ILL, CHICAGO, 69- *Concurrent Pos:* Guggenheim fel, 67-68; vis scientist, Atomic Energy Res Estab, Harwell, Eng, 67-68; consult, Argonne Nat Lab, 69-89, mem rev comts, Solid State Sci & Mat Sci Div, 70-72, vis scientist, 73-74; mem, Mat Res Adv Comt, NSF, 80-82, eval panels, Nat Res Coun, 81-85; vis prof, Univ Paris, Orsay, 81. *Mem:* Fel Am Phys Soc; fel AAAS. *Res:* magnetic materials; critical phenomena; phase transitions; superconductors. *Mailing Add:* Dept Physics Univ Ill Chicago IL 60680

KOUZEL, BERNARD, b New York, NY, Aug 21, 20; m 48; c 3. CHEMICAL ENGINEERING. *Educ:* City Col New York, BChE, 41; Univ Southern Calif, MSChE, 63. *Prof Exp:* Chem engr, Air Prod, Inc, Pa, 48-54; tech ed, Rocketdyne Div NAm Aviation, Inc, Calif, 54-55; develop engr, Res Dept, Union Oil Co Calif, 55-63, sr develop engr, 63-69, eng assoc, sci & technol div, 69-83, sr engr assoc, 83-85; RETIRED. *Concurrent Pos:* Mem tech data subcomt, Refining Div, Am Petrol Inst; mem phys data comm, Gas Processors Asn; rev tech papers, J Chem & Eng Data. *Mem:* Am Inst Chem Engrs; Am Petrol Inst; GAs Processors Asn. *Res:* Methods for prediction and correlation of physical properties; development of computer calculation procedures for process design; Stretford technology for treatment of sulfur-plant tail gas; hydrogen sulfide abatement procedures for geothermal energy systems. *Mailing Add:* 1048 E Brookdale Pl Fullerton CA 92631

KOUZES, RICHARD THOMAS, b Arlington, Va, July 8, 47; m 70; c 2. NUCLEAR PHYSICS. *Educ:* Mich State Univ, BS, 69; Princeton Univ, MA, 72, PhD(physics), 74. *Prof Exp:* Sr systs analyst, Univ Comput Co, 70-71; res assoc nuclear physics, Cyclotron Facil, Ind Univ, Bloomington, 75-76; res staff & lectr, 76-87, SR RES PHYSICIST & LECTR, PRINCETON UNIV, 87. *Mem:* Am Phys Soc; Sigma Xi; Inst Elec & Electronics Engrs. *Res:* Solar neutrinos. *Mailing Add:* 138 Philips Dr Princeton NJ 08540

KOVAC, JEFFREY DEAN, b Cleveland, Ohio, May 29, 48; m 73; c 2. PHYSICAL CHEMISTRY. *Educ:* Reed Col, BA, 70; Yale Univ, MPhil, 72, PhD(chem), 74. *Prof Exp:* Res assoc chem, Mass Inst Technol, 74-76; asst prof, 76-83, ASSOC PROF CHEM, UNIV TENN, 83- *Concurrent Pos:* Consult, Oak Ridge Nat Lab, 84-88. *Mem:* AAAS; Am Phys Soc; Am Chem Soc. *Res:* Statistical mechanics of polymers and simple fluids; equilibrium and non equilibrium thermodynamics; rubber elasticity; structure and formation of coal; computer simulation. *Mailing Add:* Dept Chem Univ Tenn Knoxville TN 37996-1600

KOVACH, EUGENE GEORGE, b Irvington, NJ, May 18, 22; m 50; c 5. ORGANIC CHEMISTRY, SCIENCE ADMINISTRATION. *Educ:* Wayne State Univ, BS, 43, MS, 44; Harvard Univ, MA, 48, PhD, 49. *Prof Exp:* Res tutor, Harvard, 46-49; instr, Univ Fla, 49-50, asst prof, 51-54; asst prof, Colgate Univ, 50-51; sci adv, US Naval Forces, Germany, 54-57; chem prog, Nat Sci Found, 57-59; asst sci adv, Int Sci & Tech Affairs, US Dept State, 59-65, actg dir, Off Gen Sci Affairs, 65-70; dep asst secy gen for sci affairs, NATO, 70-76; with Div of Policy Res, NSF, 76-78; mem staff, Off Technol Policy, 78-80, dir, 80-82, CONSULT, OFF ADVAN TECHNOL, US DEPT STATE, 83- *Mem:* AAAS; Am Chem Soc; Ger Chem Soc; Sigma Xi. *Res:* Structure of natural products; chelate compounds; theoretical organic chemistry; science education and administration; international relations. *Mailing Add:* Off Advan Technol US Dept State Washington DC 20520

KOVACH, JACK, b Rices Landing, Pa, Mar 23, 40; m 65; c 2. GEOLOGY. *Educ:* Waynesburg Col, BSc, 62; Ohio State Univ, MSc, 67, PhD(geol), 74. *Prof Exp:* From asst prof to assoc prof, 68-82, PROF GEOL, MUSKINGUM COL, 83- *Concurrent Pos:* Res assoc, Nat Res Coun, Nat Acad Sci, 79-80; assoc, US Geol Surv, Denver. *Res:* Strontium isotope geochemistry and rubidium-strontium geochronology; biogeochemistry of nonmarine mollusk shells; composition of atmospheric precipitation; Silurian stratigraphy and paleontology; biogeochemistry and isotopic composition of conodonts. *Mailing Add:* Dept Geol Muskingum Col New Concord OH 43762

KOVACH, LADIS DANIEL, b Budapest, Hungary, Nov 21, 14; nat US; c 3. MATHEMATICS. *Educ:* Case Inst Technol, BS, 36, MS, 48; Western Reserve Univ, MA, 40; Purdue Univ, PhD(math), 51. *Prof Exp:* Elec draftsman, Picker X-ray Corp, 37-40; chief elec draftsman, Am Shipbldg Co, 41-44; sr elec designer, Ohio Crankshaft Co, 44-48; instr math, Purdue Univ, 48-51; design specialist, Douglas Aircraft Co, 51-61; prof math & head dept math & physics, Pepperdine Col, 58-68, PROF MATH, NAVAL POST-GRAD SCH, 68- *Concurrent Pos:* Instr, Case Inst Technol, 46-48; vis lectr, Pepperdine Col, 57-58. *Mem:* Math Asn Am; Soc Computer Simulation; Sigma Xi. *Res:* Nonlinear differential equations; algebra; teacher education. *Mailing Add:* County Col Morris A 304 Randolph NJ 07869

KOVACH, STEPHEN MICHAEL, petroleum chemistry, for more information see previous edition

KOVACHICH, GYULA BERTALAN, b Budapest, Hungary, Mar 27, 36; m 64, 90; c 2. NEUROCHEMISTRY. *Educ:* Haverford Col, BA, 62; Univ Pa, PhD(pharmacol), 75. *Prof Exp:* Res asst neurochem, Ciba-Geigy Corp, 64-69; instr pharmacol, 75-77, res assoc, 77-80, asst prof, Dept Pharmacol, Sch Med, Univ Pa, 80-85; RES SCIENTIST, VA MED CTR, 85- *Mem:* Soc Neurosci; Int Brain Res Orgn. *Res:* Oxygen toxicity of the central nervous system; regulation of pyruvate dehydrogenase complex; interaction of ascorbic acid with biological systems; mechanism of action of antidepressants. *Mailing Add:* Neuropsycho Pharmacol E-151 Vet Admin Med Ctr Unit Woodland Ave Philadelphia PA 19104

KOVACIC, JOSEPH EDWARD, b Youngstown, Ohio, Apr 4, 30; div; c 5. ORGANIC CHEMISTRY. *Educ:* Univ Ohio, BS, 52, MS, 53. *Prof Exp:* Chemist nylon res labs, E I Du Pont de Nemours & Co, 55-56; teaching asst chem, Fla State Univ, 56-57; res chemist, Resinous Prod Lab, Dow Chem Co, 57-60, org chemist, Chem-Physics Res Lab, 60-63; actg head analytic sect, Silicone Div, Stauffer Chem Co, 63-65; sr develop engr, Adv Tech Div, Sperry Corp, 65-68, group supvr, Mat & Process Eng, Analytic Serv Lab, 68-69, prin chem engr & group supvr, Chem & Analytic Serv Lab, 69-83; SUPVR, CHEM & ANALYTIC SERV & POLYMER MAT LABS, UNISYS CORP, 83- *Mem:* fel Am Inst Chem. *Res:* Infrared spectroscopy; infrared spectra of chelates; polymer chemistry; scanning electron microscopy; gas chromatography. *Mailing Add:* 1887 Silver Bell Rd No 314 Eagan MN 55122

KOVACIC, PETER, b Wylandville, Pa, Aug 1, 21; m 46; c 6. ORGANIC CHEMISTRY. *Educ:* Hanover Col, AB, 43, DSc, 64; Univ Ill, PhD(chem), 46. *Prof Exp:* Asst org chem, Mass Inst Technol, 46-47; instr, Columbia Univ, 47-48; res chemist, E I du Pont de Nemours & Co, 48-55; from asst prof to prof chem, Case Inst Technol, 55-68; PROF CHEM, UNIV WIS-MILWAUKEE, 68- *Mem:* Am Chem Soc. *Res:* N-Halamines; bridgehead imines; rearrangements; nitrenium ions; polymerization of aromatic nuclei; charge transfer and oxy radicals in living systems. *Mailing Add:* Dept Chem Univ Wis Milwaukee WI 53201

KOVACS (NAGY), HANNA, b Szeged, Hungary, Oct 31, 19; US citizen; m 50; c 2. ORGANIC CHEMISTRY. *Educ:* Univ Szeged, PhD(org chem), 45. *Prof Exp:* Res assoc org chem, Univ Szeged, 44-46, physiol, 46-50; res assoc org chem, Univ Budapest, 50-56; res assoc bact, Univ Basel, 57 & Detroit Inst Cancer Res, 58-59; res assoc peptide chem, St John's Univ, NY, 59-63; res chemist, Naval Appl Sci Lab, Brooklyn, 63-70; clin chemist, Mt Sinai Hosp, NY, 70-85; RETIRED. *Mem:* Nat Acad Clin Biochem; Am Chem Soc; NY Acad Sci; Sigma Xi; Am Asn Clin Chem. *Res:* Author or coauthor of twenty-eight publications in the field of peptide, heterocyclic, polymer, medicinal and clinical chemistry. *Mailing Add:* 639 S Grand Ave Pasadena CA 91105

KOVACS, BELA A, b Nagykoros, Hungary, Aug 28, 21; Can citizen; m 52. PHARMACOLOGY, ALLERGY. *Educ:* Med Univ Szeged, MD, 46; Univ London, DrPhil(pharmacol), 61. *Prof Exp:* From asst prof to assoc prof pharmacol, Med Univ Szeged, 49-56; asst prof, 61-64, assoc prof pharmacol & exp med, 64-69, ASSOC PROF EXP MED, McGILL UNIV, 69-; SCI ADV, DEPT NAT HEALTH & WELFARE, FOOD & DRUG DIRECTORATE, 69- *Concurrent Pos:* Res fel org chem, Univ Basel, 56-57; res fel pharmacol, Nat Inst Med Res, London, 57-61; lectr, Sch Pharm, Univ London, 59-61. *Mem:* Am Soc Pharmacol & Exp Therapeut; Pharmacol Soc Can; Brit Pharmacol Soc; Am Col Clin Pharmacol & Therapeut; Can Soc Immunol. *Res:* Histamine and antihistaminics; inflammation; gastric secretion; pulmonary edema. *Mailing Add:* Dept Health & Welfare Can 300 Driveway No 9D Ottawa ON K1S 3M6 Can

KOVACS, BELA VICTOR, b Tiszaors, Hungary, Nov 9, 30; m 64; c 2. ACOUSTICS TESTING. *Educ:* Wayne State Univ, BSME,65; Univ Conn, MSMET, 69. *Prof Exp:* Design engr, Luster Corp of Can, 58-60; res engr, Ford Sci Res Lab, 62-87; VPRES, ATMOSPHERE FURN CO TECH CTR, 87- *Concurrent Pos:* Adj prof, Univ Mich & Wayne State Univ, 85; consult, Hentschel Instruments. *Honors & Awards:* Tech Achievement Award, Ford Motor Co, 86. *Mem:* Am Foundrymens Soc; Am Instit Mining & Metall Eng; Am Soc Non Distructive Testing; Am Soc Metals. *Res:* Solid state thermodynamics, crystallography acoustical properties of metals and composites cast iron metallurgy. *Mailing Add:* 12238 Newburgh Rd Livonia MI 48150

KOVACS, CHARLES J, b Fairfield, Conn, Apr 7, 41. EXPERIMENTAL BIOLOGY. *Educ:* Siena Col, BS, 63; St John's Univ, MS, 65, PhD(microbiol & biochem, 69. *Prof Exp:* USPHS fel, Nat Cancer Inst, NIH, 69-71; res instr, Dept Med, Hahnemann Med Col, Philadelphia, Pa, 71-72; instr, Div Radiobiol & Biophys, Sch Med, Univ Va, Charlottesville, 72-75, asst prof, Dept Pediat, 75-76; assoc scientist, Cancer Res Unit, Div Radiation Oncol, Allegheny Gen Hosp, Pittsburgh, Pa, 76-79, sr scientist, Cancer Res Labs, 79-80; assoc prof, Dept Radiol, Div Radiation Oncol, Col Med, Univ SAla, Mobile, 80-81; assoc prof & dir, Radiation Oncol Labs, Dept Radiol, Sect Radiother, Bowman Gray Sch Med, Winston-Salem, NC, 81-85; PROF RADIATION ONCOL & DIR, DIV RADIATION BIOL & ONCOL, RADIATION ONCOL CTR, SCH MED, E CAROLINA UNIV,

GREENVILLE, NC, 85- *Concurrent Pos:* Res assoc, Brookhaven Nat Lab, AEC, 700-71; prin investr or co-prin investr grants, NIH, 83-86, 84-86 & 90-; consult, NIH, Vet Admin, Mariculture, Inc, NC Biotechnol Ctr. *Mem:* Sigma Xi; AAAS; Am Asn Cancer Res; Am Soc Cell Biol; Int Soc Exp Hemat; Am Soc Clin Immunol. *Mailing Add:* Dept Radiation Biol & Oncol Sch Med E Carolina Univ Moye Blvd Greenville NC 27858

KOVACS, EVE MARIA, b Budapest, Hungary, Apr 13, 25; m 52. MEDICINE. *Educ:* Univ Szeged, MD, 52. *Prof Exp:* Lectr pharmacol, Univ Szeged, 52-54, lectr internal med, Univ Clin, 54-55, asst prof, 55-56; pharmacologist, Geigy AG, Switz, 57-58; lectr, 61-64, ASST PROF PHARMACOL, McGILL UNIV, 64 -; SCI ADV, DEPT NAT HEALTH & WELFARE, FOOD DIRECTORATE, 70 - *Concurrent Pos:* Cancer res fel, Dept Pharmacol, Univ London, 58-61. *Mem:* Pharmacol Soc Can; Am Soc Pharmacol & Exp Therapeut; Int Soc Biochem Pharmacol; Can Med Asn. *Res:* Cancer immunology; allergy; histamine; histamine metabolites; gastric secretion. *Mailing Add:* 300 Driveway No 9D Ottawa ON K1S 3M6 Can

KOVACS, EVE VERONIKA, b Melbourne, Australia, Nov 12, 54. COMPUTER SCIENCE. *Educ:* Univ Melbourne, BSc, 76, PhD(physics), 80, dipl comput sci, 80. *Prof Exp:* Vis Scientist, Stanford Linear Accelerator Ctr, 80-81; res assoc. Rockefeller Univ, 81-; AT ARGONNE NAT LAB. *Mem:* Am Phys Soc. *Res:* Monte Carlo simulations of lattice guage theories with particular emphasis on finite size effects and the interquark potential. *Mailing Add:* HEP Argonne Nat Lab 9700 S Cass Ave Argonne IL 60439

KOVACS, JULIUS STEPHEN, b Trenton, NJ, Aug 20, 28; m 56; c 2. THEORETICAL PHYSICS. *Educ:* Lehigh Univ, BS, 50; Ind Univ, MS, 52, PhD, 55. *Prof Exp:* Asst prof physics, Univ Toledo, 54-55; res assoc, Ind Univ, 55-56; from asst prof to assoc prof, 56-68, PROF PHYSICS, MICH STATE UNIV, 68-, ASSOC CHMN DEPT, 77- *Res:* Meson physics; elementary particles. *Mailing Add:* Physics & Astron Bldg 106 Mich State Univ East Lansing MI 48824

KOVACS, KALMAN T, b Szeged, Hungary, July 11, 26; Can citizen; m 62. ENDOCRINOLOGY, ELECTRON MICROSCOPY. *Educ:* Univ Szeged, Hungary, MD, 50; Univ Liverpool, PhD(path), 66; FCAP & FRCP(C), 73; FRCPath, 80, DSc 66. *Prof Exp:* Demonstr & lectr, Dept Path, Univ Szeged, 50-54, sr lectr, Dept Med, 54-68; vis scientist exp med, Univ Montreal, 68-71; asst prof, 71-80, PROF PATH, UNIV TORONTO, 80-; PATHOLOGIST, ST MICHAEL'S HOSP, TORONTO, 71- *Concurrent Pos:* Res fel path, Docent Univ Szeged, 60 & Crosby res fel, Univ Liverpool, 64-65. *Honors & Awards:* Hungarian Acad Sci Award, 68. *Mem:* Int Acad Path; Am Path Soc; Can Micros Soc; US Endocrine Soc. *Res:* Morphologic study of endocrine glands, especially human pituitaries and pituitary adenomas; correlation of structural features with secretory activity. *Mailing Add:* Dept Pathol Univ Toronto St Michael's Hosp 30 Bond St Toronto ON M5B 1W8 Can

KOVACS, KIT M, b Iserloln, Ger, Nov 7, 56; Can; m 81. PARENTAL INVESTMENT STRATEGISTS, MATING SYSTEMS. *Educ:* York Univ, Toronto, HBSC, 79; Lakehead Univ, Thunder Bay, MSC, 82; Univ Guelph, PhD(zool), 86. *Prof Exp:* Post doctorate fel zool, NSERC & NATO, 86-87; ASST PROF BIOL, UNIV WATERLOO, 87- *Concurrent Pos:* Res assoc, La Vie Wildlife Res Assoc Ltd, 82- *Mem:* Marine Mammal Soc; Can Zool Soc; Am Ornithol Union; Animal Behav Soc. *Res:* Behavioral ecology, evolution, mating systems, parental investment, pinnipeds. *Mailing Add:* Biol Dept Univ Waterloo Waterloo ON N2L 3C1 Can

KOVACS, MIKLOS I P, b Budapest, Hungary, Feb 1, 36; Can citizen; m 61; c 2. ANALYTICAL CHEMISTRY. *Educ:* Univ Keszthely, Hungary, BS, 60; Univ Budapest, BSc, 64; Univ Guelph, MSc, 69; Univ Man, PhD(biochem), 74. *Prof Exp:* Teaching fel biochem, Univ Sask, 74-75; res scientist marine lipids, fisheries & oceans res, 75-79, RES SCIENTIST GEN CHEM, AGR CAN RES INST, 79- *Mem:* Am Asn Cereal Chemists. *Res:* Wheat quality; interaction of protein, starch and lipids. *Mailing Add:* 67 Fordhem Bay Winnipeg MB R3T 3B8 Can

KOVACS, SANDOR J, JR, b Budapest, Hungary, Aug 17, 47; US citizen. CARDIOLOGY. *Educ:* Cornell Univ, BS, 69; Calif Inst Technol, MS, 72, PhD(theoret physics), 77; Univ Miami, MD, 79. *Prof Exp:* Res asst theoret physics, Calif Inst Technol, 71-77; res fel, dept med, 82-85, ASST PROF CARDIOL & RADIOL, DEPT INTERNAL MED, WASHINGTON UNIV, 85- *Concurrent Pos:* Med consult & lectr, Nat Asn Underwater Instrs, 74- *Mem:* Sigma Xi; Int Soc Gen Relativity & Gravitation; AAAS; Am Col Physicians; Am Col Cardiol; Am Bd Internal Med; Am Physiol Soc. *Res:* Noninvasive cardiological diagnostic methods including cardiac electrophysiology, arrythmia detection and analysis; biophysics. *Mailing Add:* Three Buckhammon Pl St Louis MO 63124

KOVAL, CARL ANTHONY, b York, Pa, June 28, 52; div; c 1. CHEMISTRY. *Educ:* Juniata Col, BS, 74; Calif Inst Technol, PhD(chem), 79. *Prof Exp:* Fel, Purdue Univ, 78-80; asst prof chem, 80-87, ASSOC PROF, UNIV COLO, 87- *Mem:* Am Chem Soc. *Res:* Electrochemistry at semiconductor electrodes; steric inhibition of exothermic redox reactions; facilitated transport of molecules across liquid membranes. *Mailing Add:* Dept Chem Univ Colo Campus Box 215 Boulder CO 80302

KOVAL, CHARLES FRANCIS, b Ashland, Wis, May 10, 38; m 57; c 3. OUTREACH PROGRAMMING, LYME DISEASE EDUCATION. *Educ:* Northland Col, BA, 60; Univ Wis, MS, 63, PhD(entom), 66. *Prof Exp:* Res asst, Univ Wis-Madison, 60-65, from instr to assoc prof, 65-73, exten entomologist, 65-80, dir, Univ Exp Farms, 80-83, dean, Wis Coop Exten Serv, 83-87, chmn dept, 88-90, PROF ENTOM, UNIV WIS-MADISON, 73- *Mem:* Entom Soc Am; AAAS. *Res:* Insect management on turf, landscape plants and greenhouse crops with emphasis on integrated pest management strategies; urban forestry; development of extension outreach programs. *Mailing Add:* Dept Entom 237 Russell Labs Madison WI 53706

KOVAL, DANIEL, b Fitchburg, Mass, Nov 28, 22; m 45; c 2. MATHEMATICS. *Educ:* Worcester Polytech Inst, BS, 44; Boston Univ, AM, 52, PhD(math), 65. *Prof Exp:* Physicist radiation lab, Mass Inst Technol, 44-46; asst prof appl math & physics, Atlantic Union Col, 46-60; assoc prof math, Columbia Union Col, 60-71; chmn dept, 71-77, PROF MATH, PAC UNION COL, 71- *Mem:* Math Asn Am. *Res:* Partial differential equations. *Mailing Add:* Dept Physics Pac Union Col Angwin CA 94508

KOVAL, LESLIE R(OBERT), b Rochester, NY, Jan 12, 33; m 56; c 3. ENGINEERING MECHANICS, STRUCTURAL ACOUSTICS. *Educ:* Univ Rochester, BS, 55; Cornell Univ, MS, 57, PhD(eng mech), 61. *Prof Exp:* McMullen fel, Cornell Univ, 55-56, instr eng mech, 58-61; mem tech staff, Ramo-Wooldridge Corp, 57-58; mem tech staff, TRW Systs Group, 61-66, staff engr, TRW Systs, Inc, Calif, 66-69; US Agency Int Develop vis prof, Fed Univ Rio de Janeiro, 69-70; assoc prof mech eng, 71-76, PROF MECH & AEROSPACE ENG, UNIV MO-ROLLA, 76-, ASSOC CHMN MECH ENG, 85- *Concurrent Pos:* Consult, Fed Systs Div, Int Bus Mach, Inc, 59 & Lockheed-Calif Co, 78-85; lectr, Univ Southern Calif, 62-69; mem tech staff, Litton Ship Systs, Calif, 71. *Mem:* Am Soc Mech Engrs; Acoust Soc Am; Am Acad Mech; Am Inst Aeronaut & Astronaut. *Res:* Vibrations and dynamic response of shell structures; liquid sloshing in rigid and flexible tanks; shimmy of aircraft landing gears; liquid behavior in low-gravity environments; acoustics; structure-borne noise propagation. *Mailing Add:* Dept Mech & Aeronaut Eng Univ Mo Rolla MO 65401

KOVAL, THOMAS MICHAEL, b Brownsville, Pa, Nov 20, 50; m 84; c 3. CELL BIOLOGY, RADIATION BIOLOGY. *Educ:* Pa State Univ, BS, 72; Ohio State Univ, MS, 74, PhD(zool), 76. *Prof Exp:* Nat Res Serv award fel physiol & biophys, Univ Ill, 76-77; res assoc cancer res, Allegheny Gen Hosp, 77-79; from asst prof to assoc prof radiation ther & nuclear med, Hahnemann Med Col, 79-82; assoc res prof radiol, George Wash Univ Sch Med, 82-88; assoc staff scientist, Nat Coun Radiation Protection & Measurements, 82-86, staff scientist, 87-88; SR ASSOC CONSULT, DIV RADIATION ONCOL, MAYO CLIN, 88-, PROF ONCOL, 88- *Concurrent Pos:* Partic, NATO Advan Study Inst, Italy, 78. *Mem:* Radiation Res Soc; Am Soc Cell Biol; Tissue Cult Asn; Sigma Xi; Am Soc Photobiol; Am Asn Cancer Res; AAAS. *Res:* Cell and molecular biology; oncogenesis, aging, DNA repair mechanisms; cell differentiation; radiation biology; mechanisms of radioresistance of cultured lepidopteran insect cells. *Mailing Add:* Div Radiation Oncol Mayo Clin Rochester MN 55905

KOVALAK, WILLIAM PAUL, b Detroit, Mich, Apr 12, 46; m 70; c 3. AQUATIC BIOLOGY. *Educ:* Eastern Mich Univ, BS, 67; Univ Mich, MS, 69, PhD(fisheries), 75. *Prof Exp:* Asst prof biol, Allegheny Col, 75-78; ASST PROF AQUATIC BIOL, UNIV MICH, DEARBORN, 78- *Concurrent Pos:* Biol systs scientist, Detroit Edison, 79- *Mem:* NAm Benthological Soc. *Res:* Behavioral ecology of stream insects; ecology of Great Lakes fishes. *Mailing Add:* Dept Nat Sci Univ Mich 4901 Evergreen Rd Dearborn MI 48128

KOVALY, JOHN J, b McKeesport, Pa, June 12, 28; m 57; c 2. RADAR, MISSILE SYSTEMS. *Educ:* Muskingum Col, BS, 50; Univ Ill, MS, 53. *Prof Exp:* Teaching asst, Physics Dept, Univ Ill, 50-51, res eng, Control Systs Lab, 51-56; lectr, Northeastern Univ, 56-58; adv res eng, Sylvana Electronics Corp, 58-65; CONSULT ENG, RAYTHEON CO, 65- *Concurrent Pos:* Lieutenant, US Navy, 56-58; lectr, UCLA, 76-86. *Mem:* Fel Inst Elec & Electronics Engrs (pres,72). *Res:* Synthetic aperture radar, a class of high resolution radar which obtains fine angular resolution by coherent processing of backscattered doppler histories; member of which did much of the original research and development on synthetic arrays, built the first flyable system and made first synthetic aperture radar map. *Mailing Add:* Raytheon Co Missile Sysst Div 350 Lowell St Andover MA 01810

KOVAR, FREDERICK RICHARD, b Cleveland, Ohio, Sept 20, 33; m 62; c 4. NUCLEAR PHYSICS, PLASMA PHYSICS. *Educ:* John Carroll Univ, BS, 55; Wash Univ, St Louis, MA, 57, PhD(physics), 63. *Prof Exp:* Instr physics, St Bonaventure Univ, 59-61; SR PHYSICIST & PROJ MGR, LAWRENCE LIVERMORE LAB, UNIV CALIF, 63-; SPEC SCI ADV TO ASST SECY DEFENSE ATOMIC ENERGY, 85- *Concurrent Pos:* Consult, Bradford Components Co, NY, 60. *Mem:* Am Phys Soc. *Res:* Hydrodynamics, strategic analysis, and nuclear energy. *Mailing Add:* 1078 Hacienda Dr Walnut Creek CA 94598

KOVAR, JOHN ALVIS, b Ennis, Tex, Nov 30, 32; div; c 2. SOIL MORPHOLOGY, SOIL FERTILITY. *Educ:* Tex Tech Univ, BS, 56; Tex A&M Univ, MS, 63; Iowa State Univ, PhD(agron), 67. *Prof Exp:* Soil scientist, Soil Conserv Serv, USDA, 53-56 & 59-60; res asst soil anal, Tex A&M Univ, 60-62; res assoc soil surv, Iowa State Univ, 62-67; area dir, 67-86, REGIONAL DIR, TENN VALLEY AUTHORITY, 86- *Concurrent Pos:* Fel, Welder Wildlife Found. *Mem:* Sigma Xi; Am Soil Sci Soc; Am Soc Agron; Coun Agr Sci & Technol; Soil & Water Conserv Soc; Int Soc Soil Sci. *Res:* Soil morphology and genesis; soil fertility. *Mailing Add:* 17330 Preston Rd Suite 209D Dallas TX 75252-5728

KOVATCH, GEORGE, b Scranton, Pa, Feb 20, 34; m 68; c 4. ELECTRONICS & SYSTEMS ENGINEERING. *Educ:* Princeton Univ, BSE, 55; Cornell Univ, MS, 60, PhD, 62. *Prof Exp:* Engr electronics, Gen Elec Co, 55-60; commun officer, US Air Force, 56-57; instr control eng, Cornell Univ, 60-62; sr eng specialist control & guid systs, Martin Co, 62-64; lab chief control & info systs, NASA, 64-67, dept dir, Off Control Theory & Appln, 67-70; proj mgr, 70-76, chief, transp indust anal br, 76-81, chief, indust anal & productivity div, 81-83, CHIEF, UNIV RES & TECHNOL INNOVATION OFF, TRANSP SYSTS CTR, US DEPT TRANSP, 83- *Concurrent Pos:* Mem vis sci staff, Res Inst Adv Studies, Md, 62-64; vis lectr, Drexel Inst, 63-64, Brown Univ, 68-70. *Honors & Awards:* Bronze Medal, US Dept Transp. *Mem:* sr mem Inst Elec & Electronics Engrs. *Res:* Analysis and synthesis of automatic control and guidance systems utilizing modern control theory and techniques; analysis of intermodal transportation systems including new urban systems; automotive fuel economy studies; automotive industry analysis; university research and small business innovative research programs management. *Mailing Add:* Three Saw Mill Pond Rd Hingham MA 02043

KOVATS, ANDRE, b Budapest, Hungary, May 7, 97; US citizen; m 24. MECHANICAL ENGINEERING. *Educ:* Tech Univ Budapest, dipl(mech & elec eng), 22. *Prof Exp:* Chief engr, turbines & pumps, Rateau Soc, Paris, 47-54; chief engr, Foster Wheeler Corp, Livingston, NJ, 54-70; RETIRED. *Honors & Awards:* Silver Medal, Arpad Acad Sci (Can), 83. *Mem:* Fel Am Soc Mech Engrs; fel AAAS; NY Acad Sci; Fedn Am Scientists. *Res:* Pumps and compressors; calculation and construction of pumps, fans and compressors. *Mailing Add:* 13 Baker Rd Livingston NJ 07039

KOVESI-DOMOKOS, SUSAN, b Budapest, Hungary, Aug 16, 39; US citizen; m 67. ELEMENTARY PARTICLE PHYSICS. *Educ:* Eotvos Lorand Univ, dipl physics, 63. *Prof Exp:* Res asst theoret physics, Eotvos Lorand Univ, 62-63; res assoc, Cent Res Inst Physics, Budapest, 63-68; assoc res scientist, 69-74, asst prof theoret physics, 74-79, assoc prof physics, 79-82, PROF THEORET PHYSICS, JOHNS HOPKINS UNIV, 82- *Concurrent Pos:* Vis sci consult, Rutherford Lab, Eng, 73; vis scientist, Europ Orgn Nuclear Res, Switz, 75-76, Univ Florence, Italy, 83 & Stanford Linear Acceleration Ctr, 84; vis sci staff mem, Deutsches Electronen-Synchrotron, Hamburg, Ger, 76. *Mem:* Ital Phys Soc; Europ Phys Soc; Am Math Soc. *Res:* Strong interactions of elementary particles at high energy; critical phenomena. *Mailing Add:* Dept Physics & Astron Johns Hopkins Univ Baltimore MD 21218

KOVITZ, ARTHUR A(BRAHAM), b Detroit, Mich, Aug 6, 28; m 57; c 2. FLUID DYNAMICS. *Educ:* Univ Mich, BSE, 50, MS, 51; Princeton Univ, PhD(aeronaut eng), 57. *Prof Exp:* Rocket res engr, Bell Aircraft Corp, NY, 51-52; res assoc, Princeton Univ, 57, asst dir proj Squid, 57-58; from asst prof to assoc prof, 58-69, actg chmn dept mech eng & astronaut sci, 71-73, PROF MECH ENG, NORTHWESTERN UNIV, EVANSTON, 69- *Concurrent Pos:* Consult, Aeronaut Res Assocs, Princeton, 57-58, Bendix Aviation Corp, 60, Am Mach & Foundry, 62, Argonne Nat Labs, 79-80 & 81-82, Southern Conf Eng Educ, 84, Universal Energy Systs, 85, AFOSR, 86-87 & Vislaase Corp, 88- *Mem:* Am Phys Soc. *Res:* Heat transfer; fluid mechanics. *Mailing Add:* Dept Mech Eng Technol Inst Northwestern Univ Evanston IL 60208

KOVNER, JACOB L, forest biometry; deceased, see previous edition for last biography

KOW, LEE-MING, HYPOTHALAMUS, NEUROPEPTIDES. *Educ:* Calif Inst Technol, PhD(neurophysiol), 72. *Prof Exp:* SR RES ASSOC, ROCKEFELLER UNIV, 72- *Mailing Add:* Dept Neurobiol Rockfeller Univ 1230 York Ave New York NY 10021

KOWAL, CHARLES THOMAS, b Buffalo, NY, Nov 8, 40; m 68; c 1. ASTRONOMY. *Educ:* Univ Southern Calif, BS, 63. *Prof Exp:* Res asst astron, Calif Inst Technol, 63-65 & Univ Hawaii, 65-66; res asst, Calif Inst Technol, 66-75, assoc scientist, 76-77, senior astron, 78-81, mem prof staff, 81-85; STAFF SCIENTIST, COMPUTER SCI CORP, SPACE TELESCOPE SCI INST, 86- *Honors & Awards:* James Craig Watson Medal, Nat Acad Sci, 79. *Mem:* Am Astron Soc; Int Astron Union. *Res:* Supernovae; planetary satellites; asteroids; comets. *Mailing Add:* Comput Sci Corp Space Telescope Sci Inst Homewood Campus Baltimore MD 21218

KOWAL, GEORGE M, b July 6, 38; US citizen; m 63; c 4. NUCLEAR ENGINEERING, MECHANICAL ENGINEERING. *Educ:* Univ Detroit, BS, 61; Pa State Univ, MS, 64. *Prof Exp:* Nuclear eng radiation protection, Elec Boat Div, Gen Dynamics Corp, 64-67; nuclear proj engr gen anal, 67-73; DEPT MGR APPL ENG ANAL, GILBERT ASSOCS INC, 73- *Concurrent Pos:* Mem indust prof adv coun, PaState Univ, 73-77; instr, Reading Area Community Col, 76-; adj assoc prof, Drexel Univ, 76- *Mem:* Am Nuclear Soc. *Res:* Analysts associated with nuclear power generation, especially nuclear safety, shielding, heaalth physics, fuel management, licensing, regulation and emergency core cooling systems. *Mailing Add:* 1512 Colony Dr Wyomissing PA 19610

KOWAL, JEROME, b New York, NY, Mar 16, 31; m 58; c 2. INTERNAL MEDICINE, BIOCHEMISTRY. *Educ:* Tufts Univ, BS, 52; Johns Hopkins Univ, MD, 56. *Prof Exp:* Steroid trainee, Worcester Found Exp Biol, 62-63; from asst prof to assoc prof med, Mt Sinai Sch Med, 65-70; assoc prof, Sch Med, Case Western Reserve Univ, 70-74, assoc dean vet affairs, 77-84; chief staff, Cleveland Vet Admin Med Ctr, 77-84; dir, Geriatrics Ctr Clin Assessment, Educ & Res, 84, Div Geriatric Med, Case Western Reserve Univ, 84; ASSOC CHIEF STAFF, GERIATRIC EXTENDED CARE, CLEVELAND VA MED CTR, 84- *Concurrent Pos:* Fel endocrinol, Mt Sinai Sch Med, 60-61; fel molecular biol, Albert Einstein Col Med, 63-65; chief med serv, Cleveland Vet Admin Hosp, 73-77. *Mem:* Endocrine Soc; Am Soc Clin Invest; Am Soc Biol Chemists; fel NY Acad Sci; Am Geriat Soc; Geront Soc Am. *Res:* Mechanisms of hormone and enzyme action; biochemical regulation of adrenal cells. *Mailing Add:* Dept Med Case Western Reserve Univ 2040 Adelbert Rd Cleveland OH 44106

KOWAL, NORMAN EDWARD, b Paterson, NJ, Sept 3, 37; m 62; c 3. WASTEWATER TREATMENT, COMPUTER SIMULATION. *Educ:* New York Univ, BA, 58; Duke Univ, MA, 60, PhD(plant ecol), 66; WVa Univ, MD, 77. *Prof Exp:* Instr bot, Univ Philippines, 63-65; asst & assoc prof biol, Clark Col, 65-67; fel systs ecol, Oak Ridge Nat Lab, 67-68; fel entom, Univ Ga, 68-69; asst prof biol, WVa Univ, 69-73; RES MED OFFICER, HEALTH EFFECTS RES LAB, US ENVIRON PROTECTION AGENCY, 77- *Mem:* AAAS; Soc Comput Simulation. *Res:* Health effects of nonconventional municipal wastewater treatment, including land treatment, wastewater aquaculture, and land application of sludge; human exposure to and health effects of cadmium; computer simulation of pharmacokinetics; environmental exposure assessment. *Mailing Add:* Environ Criteria & Assessment Off US Environ Protection Agency Cincinnati OH 45268

KOWAL, ROBERT RAYMOND, b Paterson, NJ, Apr 23, 39. SYSTEMATIC BOTANY, BIOMETRY. *Educ:* Cornell Univ, BA, 60, PhD(plant taxon & ecol), 68. *Prof Exp:* Fel biomath, Dept Exp Statist, NC State Univ, 67-69; vis asst prof biol, Kans State Univ, 69-71; asst prof, 71-76, ASSOC PROF BOT, UNIV WIS-MADISON, 76- *Concurrent Pos:* vis asst prof biol, Kans State Univ, 69-71. *Mem:* AAAS; Bot Soc Am; Soc Study Evolution; Am Soc Plant Taxon; Am Inst Biol Sci. *Res:* Systematics of Senecio aureus and allied species; multivariate analysis, especially canonical analysis, as a tool in plant systematics; cytology of asteraceae tribe senecioneae. *Mailing Add:* Dept Bot Univ Wis Madison WI 53706

KOWALAK, ALBERT DOUGLAS, b Portsmouth, Va, Aug 14, 36; div; c 2. PHYSICAL INORGANIC CHEMISTRY, BIOINORGANIC CHEMISTRY. *Educ:* Col William & Mary, BS, 58; Va Polytech Inst, MS, 63, PhD(chem), 65. *Prof Exp:* Teacher chem high sch, Va, 58-59; rubber chemist, O'Sullivan Rubber Corp, 60; teacher math high sch, Va, 60-61; res asst, Air Force Off Sci Res, 63; instr inorg chem, Rose Polytech Inst, 65-67; asst prof, 67-71, ASSOC PROF CHEM, LOWELL TECHNOL INST, 71-, CHMN DEPT, 77- *Honors & Awards:* Fulbright lectr, Univ Repub, Montevideo, Uruguay. *Mem:* Am Chem Soc. *Res:* Kinetics of the arsenic-chromium reaction in various buffer solutions; synthesis coordination compounds. *Mailing Add:* Dept Chem Univ Lowell One University Ave Lowell MA 01854

KOWALCZYK, JEANNE STUART, b Atlanta, Ga, Dec 22, 42; m 82; c 3. BIOCHEMISTRY. *Educ:* Jacksonville State Univ, BS, 65, MS, 66; Auburn Univ, PhD(zool), 72. *Prof Exp:* Teacher French, Calhoun County Bd Educ, Ala, 65-67; instr biol, Jacksonville State Univ, 67-68; teaching asst zool, Auburn Univ, 68-72; res assoc biochem, 72-73; asst prof biol & head dept, Belmont Abbey Col, 73-78; ASSOC PROF BIOL, UNIV SC, SPARTANBURG, 78- *Mem:* AAAS; Sigma Xi; Am Soc Parasitologists; Am Inst Biol Sci. *Res:* Immunological phenomena associated with trichostrongylid parasitism and ecological factors in the distrubution of pathogenic Naegleria Fowleri. *Mailing Add:* Dept Math & Sci Univ SC Spartanburg SC 29303

KOWALCZYK, LEON S(TANISLAW), b Motycz, Poland, May 3, 08; m 33. CHEMICAL ENGINEERING. *Educ:* Warsaw Sch Eng, dipl, 31, DSc(eng), 36. *Prof Exp:* Dep to chief, Res Sta, Polish State Spirit Monopoly, 35-39; asst prof chem eng & dep to head dept, Polish Univ Col Eng, 47-50; from assoc prof to prof, 50-73, chmn dept, 55-73, EMER PROF CHEM ENG, UNIV DETROIT, 73- *Concurrent Pos:* Dir, Pub Health Serv (IR), 56- *Mem:* Am Inst Chem Engrs; Inst Chem Engrs, UK. *Res:* Reaction kinetics; chemical reactor design. *Mailing Add:* 540 Valley Dr Bonita Springs FL 33923

KOWALENKO, CHARLES GRANT, b Saskatoon, Sask, May 14, 46; m 71; c 2. SOIL FERTILITY, SOIL BIOCHEMISTRY. *Educ:* Univ Sask, BSA, 68, MSc, 70; Univ BC, PhD(soil sci), 74. *Prof Exp:* Soils adv, Sri Lanka-Can Dry Zone Res & Develop Proj, Kandy, 82-83; res scientist, Soil Res Inst, 74-78, RES SCIENTIST, AGR CAN RES STA, 78- *Concurrent Pos:* Assoc ed, Can J Soil Sci, 80-82 & 88-89; ed, Can J Soil Sci, 90-93. *Mem:* Can Soc Soil Sci; Int Soc Soil Sci; Agron Soc Am; Soil Sci Soc Am; Agr Inst Can-BC Inst Agrologists. *Res:* Studies on the nutrient requirements of a wide range of crops including forages, vegetables, and fruit; primary specializationin nitrogen and sulfur but also concerned with entire range of nutrients, both macro and micro. *Mailing Add:* Agr Can Res Sta Box 1000 Agassiz BC V0M 1A0 Can

KOWALEWSKI, EDWARD JOSEPH, b Mt Carmel, Pa, Apr 21, 20; m 42; c 3. FAMILY MEDICINE. *Educ:* Gettysburg Col, BS, 42; George Washington Univ, MD, 45. *Prof Exp:* Pvt pract, 33-71; PROF FAMILY MED & CHMN DEPT, SCH MED, UNIV MD, BALTIMORE, 72- *Honors & Awards:* Clarence E Shaffner Award, 71. *Mem:* Am Acad Family Physicians (pres, 69-70); Soc Teachers Family Med. *Res:* Teaching of family medicine; core content of family medicine. *Mailing Add:* Dept Family Med Univ Md Sch Med 22 S Greene St Baltimore MD 21201

KOWALIK, JANUSZ SZCZESNY, b Krzemieniec, Poland, Feb 28, 34; US citizen; m 59; c 1. EXPERT SYSTEMS, HIGH SPEED COMPUTING. *Educ:* Gdansk Tech Univ, MSc, 57; Polish Acad Sci, Dr Techn Sc, 61. *Prof Exp:* Head comput ctr, Cent Shipbuilding Design Off, Poland, 61-64; res fel, Royal Norweg Coun Sci & Indust Res, 64-66; res fel comput sci, Inst Advan Studies, Australian Nat Univ, 66-67; sr specialist & mgr math anal, Boeing Comput Serv, Inc, 67-73; mem fac, dept comput sci, Sir George Williams Univ, 73-74; dir systs & comput & prof comput sci, Wash State Univ, 74-83; MGR, SCI COMPUT & ANALYSIS, BOEING CO, 83-; AFFIL PROF COMPUT SCI, UNIV WASH, SEATTLE, 85- *Concurrent Pos:* Consult. *Mem:* Am Asn Artificial Intel; Asn Comput Mach. *Res:* Parallel computation; artificial intelligence; knowledge based systems; coupling numerical and symbolic computation; supercomputing. *Mailing Add:* 16477-107 PL NE Bothell WA 98011

KOWALIK, VIRGIL C, b Sinton, Tex, Feb 8, 32; m 59. MATHEMATICS. *Educ:* St Mary's Univ, Tex, BS, 53; Univ Tex, MA, 59, PhD(math), 66. *Prof Exp:* Instr math & physics, St Edward's Univ, 61-63; from asst prof to assoc prof, 65-69, chmn dept, 66-80, PROF MATH, TEX A&I UNIV, 69- *Mem:* Am Math Soc; Am Soc Eng Educ; Math Asn Am. *Res:* Uniqueness and existence theorems for differential equations in complex space, application of functional analysis techniques to these theorems; real variables; functional analysis; generalized derivatives. *Mailing Add:* 839 W Ave G Kingsville TX 78363

KOWALSKI, BRUCE RICHARD, b Chicago, Ill, Mar 7, 42; m 74; c 2. ANALYTICAL CHEMISTRY. *Educ:* Millikin Univ, BA, 65; Univ Wash, PhD, 69. *Prof Exp:* Chemist, Shell Develop Co, Emerville, Calif, 69-71 & Houston, Tex, 71-72; asst prof, Colo State Univ, 72-73; from asst prof to assoc prof, 75-77, PROF, UNIV WASH, 78-, DIR, CTR PROCESS ANALYTICAL CHEM, 83-, DISTINGUISHED PROF ANALYTICAL

CHEM, 91- *Concurrent Pos:* Chemist, Lawrence Livermore Lab, Univ Calif, 71-72 & consult, 72-; mem, Dir Res Appln, NSF. *Honors & Awards:* Res Award, Eli Lilly Res Lab, 76; Alexander von Humboldt Award, 80; Coun Chem Res Award, 87. *Mem:* Pattern Recognition Soc: Am Chem Soc; AAAS; NY Acad Sci; Chemometrics Soc. *Res:* Chemometrics-the development of novel mathematical approaches for improving the measurement process; application of pattern recognition and other multivariant analysis methods to chemical data; process analytical chemistry including non invasive chemical analysis. *Mailing Add:* Chem Dept BG-10 Univ Wash Seattle WA 98195

KOWALSKI, CHARLES JOSEPH, b Chicago, Ill, May 8, 38; m 62; c 3. STATISTICS, BIOMETRICS. *Educ:* Roosevelt Univ, BS, 62; Mich State Univ, MS, 65; Univ Mich, Ann Arbor, PhD(biostatist), 68. *Prof Exp:* Asst prof dent, Sch Dent, 68-74, ASSOC PROF DENT, SCH DENT, UNIV MICH, ANN ARBOR, 74-, ASST DIR STATIST RES LAB, 71-; DIR BIOMET LAB, DENT RES INST, 68- *Concurrent Pos:* Consult, Statist Res Lab, 68-71, Nat Football League, 69 & Parke, Davis & Co, 70. *Mem:* Am Statist Asn; Biomet Soc; Inst Math Statist; Int Asn Dent Res; Am Asn Phys Anthrop. *Res:* Multivariate statistical analysis, especially as applied to biomedical research; problems in growth and development; sequential and time series analysis. *Mailing Add:* Sch Dent Rm B399 Univ Mich 1011 N University Ave Ann Arbor MI 48109

KOWALSKI, CONRAD JOHN, b Chicago, Ill, July 9, 47; m 68; c 2. CARBANION CHEMISTRY, SYNTHETIC METHODOLOGY. *Educ:* Mass Inst Technol, SB, 68; Calif Inst Technol, MS, 71, PhD(chem), 74. *Prof Exp:* NIH fel, Columbia Univ, 74-76; asst prof chem, Univ Notre Dame, 76-82; asst dir, Smith Kline Beecham, 82-85, assoc dir, 85-86, dir, 86-88, GROUP DIR, SYNTHETIC CHEM, SMITH KLINE BEECHAM, 88- *Concurrent Pos:* Founder & ed, Synthetic Pathways J, 81-84. *Mem:* Am Chem Soc; AAAS; Sigma Xi. *Res:* Development of new reactions and reactive intermediates for organic synthesis; devising syntheses of pharmaceutical products. *Mailing Add:* 1724 Jennings Way Paoli PA 19301

KOWALSKI, DAVID FRANCIS, b Chester, Pa, Feb 20, 47; m 81; c 2. DNA ENZYMOLOGY, DNA STRUCTURE. *Educ:* LaSalle Col, BA, 68; Purdue Univ, PhD(chem), 74. *Prof Exp:* Chemist, USDA Eastern Regional Lab, 68; asst, Purdue Univ, 69-73; RES SCIENTIST BIOCHEM, ROSWELL PARK MEM INST, 74- *Mem:* Am Soc Biol Chemists. *Res:* Structure, reactivity and functions of supercoiled DNA; occurrence, properties and functions of DNA topoisomerses; purification and characterization of mung bean nuclease. *Mailing Add:* Dept Molecular & Cell Biol Roswell Park Mem Inst 666 Elm St Buffalo NY 14263

KOWALSKI, DONALD T, b Dearborn, Mich, Mar 23, 38; c 3. MYCOLOGY. *Educ:* Univ Mich, BS, 60, MS, 61, PhD(bot), 64. *Prof Exp:* From asst prof to assoc prof, 64-73, PROF BIOL, CALIF STATE UNIV, CHICO, 73- *Concurrent Pos:* NSF res grants, 65-75. *Mem:* Mycol Soc Am; Am Bryol & Lichenol Soc; Brit Mycol Soc; Sigma Xi. *Res:* Developmental and cytological studies in the Ascomycetes and taxonomy of Myxo mycetes; biosystematics of Myxomycetes; lichen distribution. *Mailing Add:* Dept Biol Chico State Col Chico CA 95927

KOWALSKI, KENNETH L, b Chicago, Ill, July 24, 32; m 60; c 2. FINITE-TEMPERATURE FIELD THEORY, SCATTERING THEORY. *Educ:* Ill Inst Technol, BS, 54; Brown Univ, PhD(physics), 63. *Prof Exp:* Aeronaut res scientist, Nat Adv Comt Aeronaut, 54-56; res assoc physics, Brown Univ, 62; res assoc, 62-63, from asst prof to assoc prof, 63-73, chmn dept, 71-76, PROF PHYSICS, CASE WESTERN RESERVE UNIV, 73- *Concurrent Pos:* Vis prof, Inst Theoret Physics, Univ Leuven, 68-69; exec officer dept physics, Case Western Reserve Univ, 70-71, chmn dept, 71-76; scientist-in-residence, Argonne Nat Lab, 86-87. *Mem:* Am Phys Soc. *Res:* Properties of field theories and elementary particle interactions at finite temperature; cosmological applications. *Mailing Add:* Dept Physics Case Western Reserve Univ Cleveland OH 44106

KOWALSKI, LUDWIK, b Warsaw, Poland, Oct 24, 31; m 67. NUCLEAR PHYSICS, NUCLEAR CHEMISTRY. *Educ:* Warsaw Tech Univ, ME, 55; Univ Paris, MS, 62, PhD(nuclear physics), 63; Kean Col NJ, MA, 85. *Prof Exp:* Res assoc nuclear chem, Columbia Univ, 64-69; assoc prof, 69-78, PROF PHYSICS, MONTCLAIR STATE COL, 78- *Concurrent Pos:* Teaching & using VAX/VMS Comput Simulations. *Res:* Experimental nuclear physics; high energy fission; nuclear reactions at low energies; heavy ion nuclear reactions; application of semiconductor detectors and mica track detectors for nuclear research. *Mailing Add:* Dept Physics & Earth Sci Montclair State Col Upper Montclair NJ 07043

KOWALSKI, RICHARD, b Boston, Mass, April 8, 40; m. MATHEMATICS. *Educ:* Northeastern Univ, BS, 62; Case Inst Technol, MS, 63, PhD(math), 67. *Prof Exp:* Group mgr, 75-81, SR PROJ LEADER, ARINC RES CORP, 81- *Concurrent Pos:* Mem, Reliability Soc Admin Comt, Inst Elec & Electronics Engrs, 84-, ed, Trans on Reliability, 86-87. *Mem:* Armed Forces Commun & Electronics Asn; Math Asn Am; Sigma Xi; Inst Elec & Electronics Engrs (secy, 89-90, treas, 91-). *Res:* Software quality assurance and reliability. *Mailing Add:* ARINC Res Corp 2551 Riva Rd Annapolis MD 21401

KOWALSKI, STANLEY BENEDICT, b Wishart, Sask, Feb 23, 35; m 61; c 2. NUCLEAR PHYSICS. *Educ:* Univ Sask, BEng, 57, MSc, 58; Mass Inst Technol, PhD(physics), 63. *Prof Exp:* Res physicist, 63-64, asst prof physics, 64-77, SR RES SCIENTIST, MASS INST TECHNOL, 64- *Mem:* Am Phys Soc. *Res:* Photonuclear reactions; accelerator physics. *Mailing Add:* Dept Physics 26/427 Mass Inst Technol Cambridge MA 02139

KOWALSKI, STEPHEN WESLEY, b Bayonne, NJ, June 24, 31; m 55, 71; c 6. INORGANIC CHEMISTRY, SCIENCE EDUCATION. *Educ:* Fairleigh Dickinson Univ, BS, 53; NY Univ, MA, 54, PhD(sci educ), 64. *Prof Exp:* Instr, Upsala Col, 53-54 & NY Univ, 54-55; teacher high sch, NJ, 55-56; chmn, Physics-Geosci Dept, 68-72, PROF SCI, MONTCLAIR STATE COL, 56- *Concurrent Pos:* Res chemist & consult, Shulton, Inc, NJ, 53-56; guest lectr, Upsala Col, 54-65 & Fairleigh Dickinson Univ, 55-64, res chemist, Hoffmann-La Roche, 56-67, consult, sr assoc, Danforth Found, 62-; coordr-supvr, Summer Sci Insts, AID, India, 66 & 67; coordr & supvr sci & math, MA in Teaching Prog, 68-69; consult, NSF & Memory Flavors, Inc. *Mem:* AAAS; Am Chem Soc; Nat Sci Teachers Asn. *Res:* Consumer testing; polyethylene permeability; synthetic flavor derivatives; chromatography; consumer science. *Mailing Add:* Dept Physics-Geosci Montclair State Col Upper Montclair NJ 07043

KOWALSKI, TADEUSZ, b Kutno, Poland, Nov 2, 22; US citizen; m 51; c 2. ENGINEERING, HYDRODYNAMICS. *Educ:* Glasgow Univ, BSc, 44; Stevens Inst Technol, MS, 63; Univ Waterloo, PhD(mech eng, hydrodyn), 69. *Prof Exp:* Res asst ship hydrodyn, Brit Ship Res Asn, 47-49; lectr mech eng, McGill Univ, 49-51; res engr ship res, Davidson Lab, Stevens Inst Technol, 60-63; asst prof eng, US Naval Acad, 63-66; lectr mech eng, Univ Waterloo, 66-69; PROF OCEAN ENG, UNIV RI, 69- *Mem:* Fel Royal Inst of Naval Architects; Soc Naval Archit & Marine Eng. *Res:* Ship hydrodynamics-ship model research on novel propulsion, drag reducing and motion reducing systems; pressure and velocity measurements in liquids; application of drag reducing agents to waterborne craft; environmental measurements in the coastal zone; fishing gear; novel method of acid rain prevention. *Mailing Add:* Dept Ocean Eng Univ RI Kingston RI 02881

KOWALSKY, ARTHUR, b Utica, NY, Nov 16, 23. BIOPHYSICAL CHEMISTRY. *Educ:* Clarkson Col Technol, BS, 47; Univ Chicago, MS, 50, PhD(chem), 54. *Prof Exp:* Res assoc, Brookhaven Nat Lab, 54-56 & Univ Minn, 58-62; res assoc, Johnson Found, Univ Pa, 62-63, asst prof, 63-69; assoc scientist, Papanicolaou Cancer Res Inst, 69-71; assoc prof biophys, Albert Einstein Col Med, 71-78; PROG DIR BIOPHYS, NSF, 78- *Concurrent Pos:* Res fel physiol chem, Univ Minn, 56-58. *Mem:* Am Chem Soc; Am Soc Biol Chemists; Biophys Soc; AAAS. *Res:* Protein structure; nuclear magnetic resonance; mechanism of ion and electron transfer. *Mailing Add:* Nat Sci Found 1800 G St NW Washington DC 20550

KOWANKO, NICHOLAS, b Charkov, Ukraine, June 7, 34; div; c 2. ORGANIC CHEMISTRY. *Educ:* Univ Adelaide, BSc, 56, PhD(org chem), 61. *Prof Exp:* Teacher high sch, Australia, 57; Fulbright travel grant to US, 60; res assoc chem, Univ Calif, Berkeley, 61; fel, Univ Minn, 61-62, asst prof, 62-64; sr chemist cent res labs, Minn Mining & Mfg Co, 64-68; chmn dept chem, 69-73, assoc prof, 68-77, PROF CHEM, MOORHEAD STATE UNIV, 77- *Concurrent Pos:* Instr chem exten div, Univ Minn, 62-68. *Mem:* Am Chem Soc; Royal Soc Chem; Royal Australian Chem Inst. *Res:* Catalysis and desulfurization of organic compounds by metals; structure and synthesis of natural products, biosynthesis of natural products; direct fluorination studies. *Mailing Add:* Dept Chem Moorhead State Univ Moorhead MN 56560

KOWARSKI, A AVINOAM, b Tel-Aviv, Israel, Dec 30, 27; m 50; c 2. PEDIATRICS, ENDOCRINOLOGY. *Educ:* Hebrew Univ, MD, 55. *Prof Exp:* Asst physician, Hadassah Univ Hosp, Israel, 55-62, chief physician, 65-67; from instr to asst prof, 67-72, ASSOC PROF PEDIAT, SCH MED, JOHNS HOPKINS UNIV, 72-; PROF & DIR, DIV PEDIAT ENDOCRINOL, SCH MED, UNIV MD, 81- *Concurrent Pos:* Fel pediat endocrinol, Sch Med, Johns Hopkins Univ, 62-65. *Mem:* Endocrine Soc; Am Pediat Soc; Soc Pediat Res; Am Fedn Clin Res. *Res:* Human metabolism of hormones in healthy and diseased children and adults; growth hormone; diabetes; hypoglycemia; hypertension. *Mailing Add:* Pediat Endocrinol BRB 10-047 Sch Med Univ Md Baltimore MD 21201

KOWARSKI, CHANA ROSE, b Kaunas, Lithuania, June 1, 29; US citizen; m 50; c 2. PHARMACEUTICS. *Educ:* Sch Pharm, Switz, BS, 53; Sch Pharm, Israel, PhD(pharm chem), 62. *Prof Exp:* Chief pharmacist, RAFA Labs, Israel, 53-59; teaching fel, Sch Pharm, Israel, 57-62; fel phys chem, Hebrew Univ, Jerusalem, 66-67; fel, 63-65, vis prof, 67-69, assoc prof, 69-75, PROF PHARM, TEMPLE UNIV, 75- *Honors & Awards:* Lederle Res Award, 76. *Mem:* Am Pharmaceut Asn; Am Pharmaceut Soc; Sigma Xi; Am Asn Cols Pharm. *Res:* Absorption and bioavailability of drugs using the nonthrombogenic continuous withdrawal method; exemplary subjects include sulfamthiazole, sulfaethylthiadiazole, aspirin; insulin glucagon. *Mailing Add:* Dept Pharmaceut Temple Univ Health Sci Campus-Broad & Ont Philadelphia PA 19122

KOWEL, STEPHEN THOMAS, b Philadelphia, Pa, Nov 20, 42; m 70; c 3. ELECTRICAL ENGINEERING, APPLIED OPTICS. *Educ:* Univ Pa, BSEE, 64, PhD(elec eng), 68; Polytech Inst Brooklyn, MSEE, 66. *Prof Exp:* Assoc elec eng, Moore Sch, Univ Pa, 68-69; from asst prof to assoc prof, Syracuse Univ, 69-79, prof elec eng, 79-84; prof, Dept Elec Eng & Comput Sci, Univ Calif, Davis, 84-90, vchmn, 86-90, dir, Organized Res Prog Polymeric Thin Film Systs, 88-90; PROF & CHMN, DEPT ELEC & COMPUT ENG, UNIV ALA, HUNTSVILLE, 90- *Concurrent Pos:* Prin investr, Syracuse Univ; grants & res contracts, NSF, US Army Night Vision & Electrooptics Lab, US Air Force Rome Air Develop Ctr, 71-; consult, Electronics Lab, Gen Elec Co, 76-84; vis prof, Nat Res & Resource Facil, Submicron Struct, Cornell Univ, 82-83, Sch Elec Eng; vpres, Deft Labs, Inc, 76-84. *Honors & Awards:* Centennial Medal, Inst Elec & Electronics Engr, 84. *Mem:* Sr mem Inst Elec & Electronics Engrs; Sigma Xi; AAAS; Soc Photo-Optical Instrumentation Engrs; Optical Soc Am. *Res:* Acoustooptics and electrooptics; optical imaging with surface acoustic waves; optical and electronic applications of polymers. *Mailing Add:* Dept Elec & Comput Eng Univ Ala Huntsville Huntsville AL 35899

KOWERT, BRUCE ARTHUR, b Fredericksburg, Tex, Feb 11, 42. PHYSICAL CHEMISTRY. *Educ:* Univ Tex, Austin, BS, 64, PhD(chem), 71. *Prof Exp:* Res assoc phys chem, Phys Chem Inst, Univ Basel, 71-73 & Univ Calif, Los Angeles, 73-75; asst prof phys chem, Mich State Univ, 75-77; mem fac chem, 77-79, ASSOC PROF CHEM, ST LOUIS UNIV, 79- *Mem:* Am Chem Soc; Am Phys Soc. *Res:* Spin relaxation, molecular motion in liquids, electron transfer reactions, and the electronic structure of organic ion radicals employing electron spin resonance. *Mailing Add:* Box U-136 Inst Mat Sci Univ Conn Storrs CT 06268

KOWKABANY, GEORGE NORMAN, b Jacksonville, Fla, Sept 16, 23. ORGANIC CHEMISTRY. *Educ:* Univ Fla, BS, 47; Yale Univ, MS, 49, PhD(chem), 51. *Prof Exp:* Fel carbohydrate res, Ohio State Univ, 50-52; chemist, Nat Bur Standards, 52-53; from instr to asst prof, 53-60, assoc prof org chem, Cath Univ Am, 60-86; RETIRED. *Concurrent Pos:* NIH spec fel, Univ Ferrara, 63-64; vis assoc prof, Med Sch, Univ Miami, 70-71; res chemist, USDA, Beltsville, Md, 80-81. *Mem:* Fel AAAS; fel Am Inst Chemists; Am Chem Soc. *Res:* Paper chromatography; separation of amino acids and carbohydrates; structures of polysaccharides; enzymology. *Mailing Add:* 9252 San Jose Blvd No 604 Jacksonville FL 32257-5576

KOWLES, RICHARD VINCENT, b Ivanhoe, Minn, May 9, 32; m 56; c 5. GENETICS. *Educ:* Winona State Col, BS, 54, MS, 63; St Mary's Col Minn, MS, 67; Univ Minn, St Paul, PhD(genetics), 72. *Prof Exp:* Teacher high schs, Minn, 54-68; instr biol, Univ Wis-River Falls, 71-72, asst prof, 72-74; assoc prof, 74-77, PROF BIOL, ST MARY'S COL, MINN, 77- *Concurrent Pos:* vis res prof, Univ Minn, 83-84. *Mem:* Genetics Soc Am; Am Genetic Asn; Soc Study Evolution; Radiation Res Asn. *Res:* Chromosome aberrations; supernumerary chromosomes in maize; molecular cytogenetics of endosperm in maize. *Mailing Add:* Dept Biol St Mary's Col Winona MN 55987

KOWLESSAR, O DHODANAND, b India. MEDICINE. *Educ:* Univ Rochester, PhD(med), 55. *Prof Exp:* Resident internal med, Cornell Hosp, NY, fel; asst prof, Med Sch, Cornell Univ, dir med, 64, dir gastroneurol, 66; PROF & ASSOC CHMN, DEPT MED, THOMAS JEFFERSON UNIV, 87- *Mem:* Sigma Xi; Am Soc Clin Nutrit; Inst Nutrit. *Mailing Add:* Dept Med Jefferson Med Col Thomas Jefferson Univ 1025 Walnut St Philadelphia PA 19107

KOWOLENKO, MICHAEL D, b July 23, 55. IMMUNOTOXICOLOGY. *Educ:* Northeastern Univ, BS, 78, MS, 81, PhD(med lab sci), 86. *Prof Exp:* Teaching asst, Northeastern Univ, 83-85; Nat Inst Environ Health Sci fel, Albany Med Col, 87-88, asst res prof, Dept Microbiol & Immunol & Dept Med, 88-89; MGR IMMUNOTOXICOL, DEPT INVESTIGATIVE TOXICOL, BRISTOL-MYERS SQUIBB PHARMACEUT RES INST, 89- *Mem:* Am Asn Immunologists; Soc Toxicol. *Res:* Investigative toxicology; author of more than 20 technical publications. *Mailing Add:* Investigative Toxicol Bristol-Myers Squibb Pharmaceut Res Inst PO Box 4755 Syracuse NY 13221-4755

KOYAMA, TETSUO, b Tokyo, Japan, Oct 9, 35. SYSTEMATIC BOTANY, ECONOMIC BOTANY. *Educ:* Tokyo Univ, BSc, 56, MA, 58, PhD(bot), 61. *Prof Exp:* Vis assoc prof bot, Ryukyus Univ, 58-59; asst prof, Tokyo Univ, 61-63; res assoc taxon, NY Bot Garden, 63-64, assoc cur, 64-67, cur taxon, 67-78, sr cur & dir, Asiatic Progs, 78-90; PROF, COL AGR, NIHON UNIV, 89- *Concurrent Pos:* Lectr, Fac Arts & Sci, Nippon Univ, 61; assoc prof, Tamagawa Univ, 61-62, vis prof, Sch Agr, 74-; Nat Res Coun Can res fel, 61-63; adj prof, City Univ New York, 71-; corresp mem, Am Mus Natural Hist, 74-; vis prof, Bot Inst, Aarhus Univ, Denmark, 77-78; mem bd dirs, Asian Vegetable Res & Develop Ctr, 84-88; mem, consult comt, Plant Genetic Resources, Japan Agency Sci & Technol, 84-; tech consult, Japanese Hort Soc, 85-; mem forum, Bioscience, Nomura Res Inst, Tokyo, 85; adv, Japan Flower Soc, 85-; counr, Japan Asn Int Garden & Greenery Exposition, 87-91; adj prof, Grad Sch, City Univ NY, 90- *Mem:* Am Soc Plant Taxon; Asn Trop Biol (secy-treas, 65-67); Bot Soc Japan; Int Asn Plant Taxonomists; Japanese Soc Trop Agr. *Res:* inventory studies of economic plants, with special emphasis on the tropics and subtropics; Asia botany; exploitation and development of new crops and new material for plant industries; organization and production of botanical and horticultural exhibitions and events. *Mailing Add:* PO Box 366 Tuckahoe NY 10707-0366

KOZAK, ANTAL, b Tiszapuspoki, Hungary, May 22, 36; Can citizen; m 63; c 2. FOREST BIOMETRICS. *Educ:* Univ BC, BSF, 59, MF, 61, PhD(biomet), 63. *Hon Degrees:* DSc, Sopron, Hungary, 89. *Prof Exp:* Res asst data processing, Univ BC, 62-63; res off statist, Can Dept Forestry, 63-65; from asst prof to assoc prof, 65-72, PROF FAC FORESTRY, UNIV BC, 72-, ASSOC DEAN, 78- *Concurrent Pos:* Vis lectr, Univ BC, 63-65. *Mem:* Am Statist Asn; Biomet Soc; Can Inst Forestry. *Res:* Application of statistics for forestry problems; development of estimating systems for forest inventory; taper equations; biomass equations. *Mailing Add:* Fac Forestry Univ BC Vancouver BC V6T 1W5 Can

KOZAK, GARY S, b Pittsburgh, Pa, June 13, 38; m 57; c 3. ANALYTICAL CHEMISTRY, PHYSICAL CHEMISTRY. *Educ:* Ind Univ, BS, 60; Univ Ariz, PhD(chem), 63. *Prof Exp:* Sr assoc chemist, 63-64, staff chemist, 64-66, proj chemist, 66-68, proj mgr & develop chemist, 68-69, mgr PhD recruitment progs, 69-71, mgr educ & sci rels, IBM World Trade Corp, 71-74, dir sci & contrib progs, IBM Europe, Paris, 74-78, dir spec univ prog, 78-81, prog dir tech personnel resources, 81-86, PROG DIR, TECH INTERCHANGE PROG, IBM CORP, 87- *Mem:* AAAS; Am Chem Soc; World Wildlife Fund. *Res:* Kinetic studies with electrogenerated halogens; fluorescence; photosensitive polymers; epoxy resins and laminates. *Mailing Add:* IBM Corp Old Orchard Rd Armonk NY 10504

KOZAK, JOHN JOSEPH, b Cleveland, Ohio, Sept 14, 40; m 69; c 3. CHEMICAL PHYSICS, BIOPHYSICAL CHEMISTRY. *Educ:* Case Inst Technol, BS, 61; Princeton Univ, PhD(chem), 65. *Prof Exp:* NIH fel chem, Free Univ Brussels, 65-67; res assoc, Univ Chicago, 67-68; from asst prof to prof chem, Univ Notre Dame, 76-88; DEAN & PROF CHEM, FRANKLIN COL ARTS & SCI, UNIV GA, 88- *Concurrent Pos:* Chmn, Prog Unified Sci, Univ Notre Dame, 70-; vis prof, Free Univ Brussels, 75, Ecole Polytechnique Federale de Lausanne, 78. *Mem:* Am Chem Soc; Sigma Xi. *Res:* Interaction of radiation and matter; investigations of liquid dissolved state; theory of phase transitions; studies on nature of irreversibility; reaction-diffusion theory. *Mailing Add:* Dean's Off Franklin Col Arts & Sci Univ Ga Rm 101 Athens GA 30602

KOZAK, LESLIE P, b Dauphin, Manitoba, Oct 28, 40. BIOLOGY. *Educ:* Univ Notre Dame, PhD(biochem), 69. *Prof Exp:* SR STAFF SCIENTIST, THE JACKSON LAB, 70- *Res:* Molecular genetics of mammals. *Mailing Add:* The Jackson Lab Bar Harbor ME 04609

KOZAK, MARILYN SUE, BIOCHEMISTRY OF PROTEIN SYNTHESIS. *Educ:* Johns Hopkins Univ, PhD(microbiol), 72. *Prof Exp:* PROF BIOL, UNIV PITTSBURGH, 85- *Mailing Add:* Dept Biol A-234 Langley Hall Univ Pittsburgh Main Campus Pittsburgh PA 15260

KOZAK, SAMUEL J, b Peabody, Mass, Apr 13, 31; m 59; c 2. GEOLOGY. *Educ:* Bates Col, BS, 54; Brown Univ, MS, 58; Univ Iowa, PhD(geol), 61. *Prof Exp:* Asst prof, 61-70, PROF GEOL, WASHINGTON & LEE UNIV, 70- *Mem:* Geol Soc Am; Nat Asn Geol Teachers. *Res:* Structural geology; igneous and metamorphic petrology; geology of the Central Appalachians. *Mailing Add:* Dept Geol Washington & Lee Univ Lexington VA 24450

KOZAK, WLODZIMIERZ M, b Warsaw, Poland, May 7, 27; m 74; c 1. VISUAL PHYSIOLOGY, PSYCHOPHYSICS. *Educ:* Univ Lodz, MS, 51; Univ Sydney, PhD(visual electrophysiol), 64; Polish Acad Sci, DSc(visual electrophysiol), 66. *Prof Exp:* Asst prof neurophysiol, Nencki Inst Exp Biol, Univ Lodz, 46-56, assoc prof & sr scientist, Nencki Inst Exp Biol, Polish Acad Sci, Warsaw, 56-64, head, Lab Electrophysiol, 64-67 & Lab Afferent Systs, 67-68; vis assoc res prof visual physiol, State Univ NY Buffalo, 68-70; assoc prof, 70-73, PROF PHYSIOL & BIOENG, CARNEGIE-MELLON UNIV, 74- *Concurrent Pos:* Recipient habilitation grant, Div Natural Sci, Polish Acad Sci, 65-66; Brit Coun visitor, Gt Brit, 65, Polish Acad Sci & USSR Acad Sci visitor, USSR, 65, Karolinska Inst visitor, Sweden, 66, visitor, Sch Med, Johns Hopkins Univ, Chile, 70 & NSF visitor, Japan, 78; Wellcome Trust fel, Inst Ophthal, Univ London, 68; United Health Found Western NY grant & Res Found grant, State Univ NY Albany, 69-70; Scaife Fund grant & Ford Found grant, 73-75, 88, NSF grant, 76-77 & Juv Diabetes Found grant, 78-80 & 85-87, Diabetes Res & Educ Found grant, 89-90; lectr, Univ Lodz, 52-55 & Warsaw Tech Univ, 65; sr scientist dept med, Shadyside Hosp, Pittsburgh, 75-; Rockefeller Found fel, 59-60, vis fel, Australian Nat Univ, Canberra, 79; NAS exchange scholar, Hungary, 84; Pfizer, Inc grant, 84; res scholar, Monash Univ, Australia, 86; Hewlett-Packard, Inc grant, 87; fel, Ophthalmic Res Inst Australia, Univ Sydney, 60-63. *Honors & Awards:* Sci Award, Div Natural Sci, Polish Acad Sci, 55. *Mem:* AAAS; Soc Neurosci; Int Brain Res Orgn; Polish Inst Arts & Sci Am; Asn Res Vision & Ophthal; Int Soc Clin Electrophysiol Vision; Juv Diabetes Found Int. *Res:* Electrophysiology and conditioning of salivary secretion; plasticity and memory traces of spinal cord reflexes; eye optics; electrophysiology of retina and visual pathway; oscillatory components of electroretinograms and evoked potentials; electroretinograms in diabetic retinopathy; coding of brightness and color information in eye and brain; subjective color sensations; computer Fourier analysis; neurophysiology; aldose reductase inhibition in diabetes mellitus; nuclear magnetic resonance imaging of the eye; integrity of blood retinal barrier in diabetes mellitus. *Mailing Add:* Biomed Eng Prog Doherty Hall 2313 Carnegie-Mellon Univ Pittsburgh PA 15213-3890

KOZAM, GEORGE, b Union City, NJ, Mar 28, 24; m 53. ANATOMY, PATHOLOGY. *Educ:* NY Univ, BA, 45, MS, 46, PhD(human anat), 50, DDS, 53. *Prof Exp:* Asst biol, NY Univ, 46-47, instr anat, Dent Col, 47-50, instr bact, 53-54; from asst prof to assoc prof, 58-71, PROF ANAT, COL MED & DENT NJ, 71- *Concurrent Pos:* Vis asst prof path, Dent Col, Fairleigh Dickinson Univ, 64-65. *Mem:* Am Dent Asn; fel Am Acad Oral Path; NY Acad Sci; Int Asn Dent Res. *Res:* Capillary fragility; circulation in dental pulp; respiration of rat and rabbit dental pulp; effects of local anesthetics on the respiration of dental pulp; research on trigeminal nerve; effect of eugenol on nerve transmission and oral mucous membranes. *Mailing Add:* Dept Anat Univ Med & Dent NJ Med Sch 185 S Orange Ave Newark NJ 07103

KOZARICH, JOHN WARREN, b Jersey City, NJ, June 20, 49; m 85. BIOLOGICAL CHEMISTRY, BIOCHEMISTRY. *Educ:* Boston Col, BS, 71; Mass Inst Technol, PhD(biol chem), 75. *Prof Exp:* NIH fel biochem, Harvard Univ, 74-77; from asst prof to prof pharmacol, Yale Univ, 77-84; PROF CHEM & BIOCHEM, UNIV MD, 84-; PROF, AGR BIOTECH CTR, MD BIOTECH INST, 87-; VPRES RES & DEVELOP, ALKERMES INC, CAMBRIDGE, MASS, 89- *Concurrent Pos:* Bioorg & Natural Prod Study Sect, NIH, 83-87; Am Cancer Soc Fac Res Award, 83-88. *Honors & Awards:* Pfizer Award in Enzyme Chemistry, Am Chem Soc, 88. *Mem:* Am Chem Soc; Sigma Xi; Am Soc Biol Chemists. *Res:* Design of enzyme inhibitors; mechanisms and stereochemistry of enzyme action; chemistry and biochemistry of modified nucleosides; mechanisms of drug induced DNA degradation. *Mailing Add:* Dept Chem & Biochem Univ Md College Park MD 20742

KOZAWA, AKIYA, b Japan, Jan 2, 28; m 53; c 2. ELECTROCHEMISTRY. *Educ:* Nagoya Univ, DrEng(electrochem), 59. *Prof Exp:* Instr appl chem, Nagoya Univ, 52-59, asst prof, 59-62; asst prof electrochem, Western Reserve Univ, 63-64; sr res assoc, Union Carbide Corp, 64-74, corp res fel, Battery Tech Ctr, 74-89; RETIRED. *Concurrent Pos:* Res assoc, Duke Univ, 56-57 & Western Reserve Univ, 57-59; Pres Int Battery Mat Asn 83- *Honors & Awards:* Takei Award, Electrochem Soc Japan, 83. *Mem:* Electrochem Soc; Int Battery Mat Asn (pres, 83-). *Res:* Batteries, fuel cells; reliability. *Mailing Add:* 31304 Floweridge Dr Rancho Palos Verdes CA 90274-6240

KOZEK, WIESLAW JOSEPH, b Poniatowka, Poland, Feb 6, 39; US citizen; m 71; c 3. IMMUNOLOGY, ULTRASTRUCTURE. *Educ:* Canisius Col, BS, 61; Tulane Univ, MS, 67, PhD(parasitol), 69. *Prof Exp:* Fel, Dept Microbiol, Univ Chicago, 69-71, Dept Immunol & Med Microbiol, Univ Fla, Gainesville, 71-72; asst res parasitologist, Calif Primate Res Ctr, Univ Calif, Davis, 73-77; SCIENTIST, INT COLLABR INFECTIOUS DIS RES PROG, TULANE UNIV, CALI, COLOMBIA, 77- *Concurrent Pos:* Adj assoc prof, Dept Trop Med, Tulane Sch Pub Health & Trop Med, 80-; prin investr human filariasis, Int Collabr Infectious Dis Res Prog, Cali, Colombia, 80- *Mem:* Am Soc Trop Med & Hyg; Royal Soc Trop Med & Hyg; Am Soc Parasitologists; Sigma Xi; AAAS. *Res:* Medical helminthology; immunology, morphology, ultrastructure, animal models, and host-parasite relationship of filariae; epidemiology of human filariases in Colombia; culture of helminth cells; characterization of intracellular microorganisms of filarids; host-parasite relationships of trichinella spiralis. *Mailing Add:* Microbiol Med Sci Campus Univ PR GPO Box 5067 San Juan PR 00936

KOZEL, THOMAS RANDALL, b Ft Dodge, Iowa, Jan 31, 46. MEDICAL MYCOLOGY, MEDICAL BACTERIOLOGY. *Educ:* Univ Iowa, BA, 67, MS, 69, PhD(microbiol), 71. *Prof Exp:* Instr microbiol, Univ Iowa, 69-70; from asst prof to assoc prof, 71-82, dir med admis, 72-76, PROF & CHMN MICROBIOL, UNIV NEV, RENO, 82- *Concurrent Pos:* vis assoc prof, Rockefeller Univ, 80-81. *Mem:* Am Soc Microbiol; Harvey Soc; Sigma Xi. *Res:* Cellular and molecular mechanisms of infection and resistance in systemic mycoses. *Mailing Add:* Sch Med Sci Univ Nev Reno NV 89507

KOZELKA, ROBERT M, b Minneapolis, Minn, July 20, 26; m 50; c 4. STATISTICS. *Educ:* Univ Minn, BA, 47, MA, 48; Harvard Univ, PhD(math), 53. *Prof Exp:* From instr to asst prof math, Tufts Univ, 49-53; asst prof, Univ Nebr, 53-57; from asst prof to assoc prof, 57-66, PROF MATH, WILLIAMS COL, 66- *Concurrent Pos:* Vis assoc prof math, Univ NC, Chapel Hill, 63-64; vis prof anthrop & sociol, Univ Tex, Austin, 70-71. *Mem:* Inst Math Statist; Am Statist Asn; Math Asn Am. *Res:* Applications of mathematics and statistics to behavioral science. *Mailing Add:* 602 Croom Ct Chapel Hill NC 27514

KOZIAR, JOSEPH CLEVELAND, b Baltimore, Md, Jan 6, 46; m 68; c 1. POLYMER CHEMISTRY, ORGANIC CHEMISTRY. *Educ:* Johns Hopkins Univ, BA, 68, PhD(org chem), 75. *Prof Exp:* Res chemist process develop, Diamond Shamrock Corp, 69-71; sr res chemist plastics & coatings, 75-90, GROUP LEADER, EMULSION PROCESS GROUP, ROHM & HAAS CO, 90- *Mem:* Am Chem Soc; AAAS. *Res:* Polymer synthesis and characterization; monomer synthesis; organic photochemistry; polymer process research. *Mailing Add:* Rohm and Haas Res Labs PO Box 219 Bristol PA 19007

KOZICKI, WILLIAM, b Kenora, Ont, June 11, 31; m 63; c 1. THERMODYNAMICS. *Educ:* Univ Toronto, BASc, 53, MASc, 57; Calif Inst Technol, PhD(thermodyn), 62. *Prof Exp:* Process engr, Textile Fibres Div, Du Pont of Can, 53-55; res fel thermodyn, Calif Inst Technol, 61-62; from asst prof to assoc prof, 62-71, PROF CHEM ENG, UNIV OTTAWA, 71-, ASSOC DEAN ENG, 76- *Mem:* Sigma Xi. *Res:* Transport phenomena: rheology and flow of complex systems with particular emphasis on characterization of polymer adsorption and its role in improved oil recovery, turbulent drag reduction and as filtration aid. *Mailing Add:* Dept Chem Eng Univ Ottawa Ottawa ON K1N 6N5 Can

KOZICKY, EDWARD LOUIS, b Elberon, NJ, Feb 11, 18; m 41; c 3. WILDLIFE MANAGEMENT. *Educ:* Univ Maine, BS, 41; Pa State Col, MS, 42, PhD(zool), 48. *Prof Exp:* Chief res, State Div Fish & Game, NJ, 48; leader, Co-op Wildlife Res Unit, Iowa State Col, 48-56; DIR CONSERV DEPT, WINCHESTER-WESTERN DIV, OLIN CORP, 56- *Concurrent Pos:* Dir, Wildlife Legis Fund. *Mem:* Wildlife Soc (pres, 69-70); Am Forestry Asn. *Res:* Life history, ecology and management of game birds and mammals; the development, evaluation and improvement of game animal census techniques; development and promotion of shooting preserves. *Mailing Add:* 817 Southmoore Dr Godfrey IL 62035

KOZIK, EUGENE, b Duquesne, Pa, Sept 22, 24; m 56; c 2. COMPUTER SCIENCE, OPERATIONS RESEARCH. *Educ:* Univ Pittsburgh, BS, 49, ML, 50, PhD, 60. *Prof Exp:* Engr, Gulf Oil Corp, 48-50; tech adminstr res & develop, Wright Air Develop Ctr, 53-57; mgt sci consult, Univ Pittsburgh, 59-60; mgr planning & controls, Gen Dynamics Corp, 60-61; prog mgr mgt sci, Opers Res, Inc, 61-62; dir adv studies, Burroughs Corp, 62-66; mgr info sci, Gen Elec Co, 66-70; PRES, KOZIK & ASSOCS, 70- *Concurrent Pos:* Lectr, Duquesne Univ, 60, Univ Rochester, 61 & Pa State Univ, 64-; mem, Int Comt Sci Mgt, Hist Eval Res Orgn, McLean, Va, 62-, comput comt, Am Inst Planners, 67- & urban info & measurement comt, Nat Acad Sci, 68- *Res:* Management and information science; intergrated management system; data management; computer technology. *Mailing Add:* 38 Rabbit Run Rd Malvern PA 19355

KOZIKOWSKI, ALAN PAUL, b Menominee, Mich, Oct 27, 48; m 75; c 2. ORGANIC CHEMISTRY, NEUROCHEMISTRY. *Educ:* Univ Mich, BS, 70; Univ Calif, Berkeley, PhD(org chem), 74. *Prof Exp:* NIH fel org chem, Harvard Univ, 74-76; asst prof, 76-80, assoc prof & Camille & Henry Dreyfus teacher scholar, 80-83, ALFRED P SLOAN FEL ORG CHEM, UNIV PITTSBURGH, 76-, PROF CHEM, 84- *Concurrent Pos:* Fel, Japan Soc for Promoting Sci, 84; prof behav neurosci, 88. *Honors & Awards:* Cope Award, 82. *Mem:* Am Chem Soc; The Chem Soc; Sigma Xi; Soc Neurosci. *Res:* Synthetic organic chemistry; synthesis of alkaloids and carbohydrates; organometallics; neuroscience. *Mailing Add:* Dept Chem Univ Pittsburgh Pittsburgh PA 15260

KOZIKOWSKI, BARBARA ANN, b Chicago, Ill, Jan 20, 54. PHYSICAL CHEMISTRY. *Educ:* Loyola Univ, BS, 75; Univ Ill, Chicago Circle, MS, 77, PhD(phys chem), 81. *Prof Exp:* STAFF SCIENTIST, PROCTER & GAMBLE CO, 81- *Mem:* Am Chem Soc; Soc Appl Spectroscopy. *Res:* Electronic structure of heavy transition metal complexes by means of cryogenic absorption; magnetic circular dichroism; emission and two-proton excitation techniques. *Mailing Add:* PO Box 398707 Cincinnati OH 45239

KOZINSKI, ANDRZEI, b Poland, Oct 1, 25; m 49; c 1. BIOCHEMISTRY. *Educ:* Univ Warsaw, MD, 50, PhD(biochem), 56. *Prof Exp:* Res assoc, State Inst Health, Poland, 50-55; asst prof, Inst Biochem, Polish Acad Sci, 55-57; res assoc, Virus Lab, Univ Calif, 57-58; asst prof, Inst Microbiol, Rutgers Univ, 58-59; assoc prof biochem, 62-68, PROF HUMAN GENETICS, UNIV PA, 68- *Concurrent Pos:* NIH fel biochem, Johns Hopkins Univ, 59-62. *Res:* Biochemistry of DNA replication; structure of phage chromosome. *Mailing Add:* Dept Human Genetics Univ Pa Philadelphia PA 19104

KOZIOL, BRIAN JOSEPH, b Gardner, Mass, Aug 24, 51; m. CLINICAL BIOCHEMISTRY, CLINICAL NUTRITION. *Educ:* Univ Mass, Amherst, BS, 73; Univ Calif, Los Angeles, MS, 77, PhD (exp & clin nutrit), 84. *Prof Exp:* Res asst biochem, Univ Mass, Amherst, 69-73; res asst physiol, dept

kinesiol, 73-78, DIR, LIPID LAB, DIV NUTRIT, SCH PUB HEALTH, UNIV CALIF, LOS ANGELES, 78-, RES ASSOC LIPID METAB, DIV NUTRIT, SCH PUB HEALTH & CTR HEALTH ENHANCEMENT, 84-, ASST PROF BIOCHEM & NUTRIT, DEPT MED, DIV CLIN NUTRIT, 84-, DIR, LIPID-HORMONE LAB, 85-; ASSOC PROF BIOL, UNIV BRIDGEPORT, 84- *Concurrent Pos:* Adj lectr biochem, dept chem & biochem & NIH scholar, Sch Pub Health, Univ Calif, Los Angeles, 84; consult-lectr, Northrop Aircraft Div, 82-; consult, Calif Museum Sci & Ind, 83-; mem, Jonsson Comprehensive Cancer Ctr, 85- *Mem:* Am Inst Nutrit; Am Soc Clin Nutrit; Sigma Xi; AAAS; Am Coun Sci & Health. *Res:* Effect(s) of both the quantity and quality of dietary fat on the incidence of breast tumors in humans and experimental animals; possible link between dietary fat, the breast tissue hormonal milieu and breast cancer development. *Mailing Add:* 5892 Camphor Ave Westminster CA 92683

KOZLIK, ROLAND A, b Hackensack, NJ, Mar 31, 21. METALLURGY. *Educ:* Columbia Univ, BS, 43; Univ Ky, MS, 44. *Prof Exp:* RETIRED. *Mailing Add:* 301 Sunrise Ridge Dr Spruce Pine NC 28777

KOZLOFF, EUGENE NICHOLAS, b Teheran, Iran, Sept 26, 20; nat US; m 44; c 1. ZOOLOGY. *Educ:* Univ Calif, AB, 42, MA, 46, PhD(zool, protozool), 50. *Prof Exp:* Asst zool, Univ Calif, 44, lectr micros technol, 45; from instr to prof biol, Lewis & Clark Col, 45-66, chmn dept, 60-66; PROF ZOOL, UNIV WASH, 66- *Concurrent Pos:* Guggenheim fel, 53-54; vis prof, Inst Marine Biol, Univ Ore, 57-60 & 64; vis prof, Friday Harbor Labs, Univ Wash, 61 & 62, resident assoc dir, 66-73; vis prof, Pac Marine Sta, 63; dir, NSF Inst Col Teachers, Univ Ore, 64. *Mem:* Marine Biol Asn UK; Western Soc Naturalists. *Res:* Cytology, morphology and taxonomy of protozoa; commensal ostracodes; acoel and rhabdocoel Turbellaria; orthonectid Mesozoa; development of kinorhynchs. *Mailing Add:* Dept Zool Univ Wash Seattle WA 98195

KOZLOFF, LLOYD M, b Chicago, Ill, Oct 15, 23; m 47; c 4. VIROLOGY, MOLECULAR BIOLOGY. *Educ:* Univ Chicago, BS, 43, PhD(biochem), 48. *Prof Exp:* Res assoc biochem, Univ Chicago, 49-52, from asst prof to prof, 52-64; prof microbiol, Univ Colo Med Ctr, Denver, 64-80, chmn dept, 66-76, assoc dean fac affairs, 76-79; PROF MICROBIOL & DEAN, GRAD DIV, UNIV CALIF, SAN FRANCISCO, 81- *Concurrent Pos:* Mem virol & rickettsiology study sect, NIH, 63-68; ed, J Virol, 66-74; vchmn, Virol Sect, Am Soc Microbiol, 74-75, chmn, 75-76; Found Microbiol lectr, 75-76. *Mem:* Hon fel AAAS; Am Soc Microbiol; Am Soc Biol Chemists; Am Chem Soc. *Res:* Virus structure, function and assembly; reactions during viral invasion. *Mailing Add:* Dean Grad Div Univ Calif San Francisco CA 94143

KOZLOWSKI, ADRIENNE WICKENDEN, b Hackensack, NJ, Apr 26, 41; m 68, 77. INORGANIC CHEMISTRY, CHEMICAL INFORMATION. *Educ:* MacMurray Col, AB, 62; Univ Conn, MS, 64, PhD(chem), 68. *Prof Exp:* Res asst phys chem, Univ Conn, 68, fel biol sci, 69-70; from asst prof to assoc prof, 70-82, PROF CHEM, CENT CONN STATE UNIV, 82- *Concurrent Pos:* Lectr, Univ Copenhagen, 69; vis educr, Chem Abstrs Serv, 84-85. *Mem:* Am Chem Soc; Sigma Xi. *Res:* Coordination compounds; inorganic structural chemistry; information retrieval, computerized searching by inorganic structure. *Mailing Add:* Dept Chem Cent Conn State Univ New Britain CT 06050

KOZLOWSKI, BETTY ANN, b Dothan, Ala, Dec 14, 43; m 78; c 1. NUTRITION. *Educ:* Ala Col, BS, 65; Univ Tenn, Knoxville, PhD(nutrit), 70. *Prof Exp:* Res asst nutrit, Univ Tenn, Knoxville, 69-70; asst prof, Auburn Univ, 70-74; ASSOC PROF NUTRIT, DEPT HUMAN NUTRIT & FOOD MGT & CHIEF NUTRIT, NISONGER CTR MENTAL RETARDATION & DEVELOP DISABILITIES, OHIO STATE UNIV, 74- *Mem:* Sigma Xi. *Res:* Nutritional needs, and ways of meeting them in persons with developmental disabilities. *Mailing Add:* Human Nutrit 265 Campbell Hall Ohio State Univ Main Campus Columbus OH 43210

KOZLOWSKI, DON ROBERT, b St Louis, Mo, Dec 5, 37; m 60; c 3. AVIONICS. *Educ:* Univ St Louis, BS, 59; Washington Univ, St Louis, MS, 67. *Prof Exp:* Sr engr, McDonnell Aircraft Co, McDonnell Douglas Corp, 59-62; mgr prog develop, Electronic Specialty Co, 62-64; vpres, Aerospace Systs Corp, 64-65; sect mgr advan reconnaissance systs, 65-72, sr prog engr, 72-80, CHIEF PROG ENGR ADVAN ENG, MCDONNELL AIRCRAFT CO, MCDONNELL DOUGLAS CORP, 80- *Concurrent Pos:* Dir & consult, USAF/Air Force Systs Command Offensive Air Support Mission Anal, 74-76. *Mem:* Inst Elec & Electronics Engrs; Am Soc Photogram; Am Inst Aeronaut & Astronaut; Am Defense Preparedness Asn. *Res:* Avionics, displays and data processing systems for reconnaissance and intelligence; communications and electronic warfare; aircraft systems design. *Mailing Add:* Mcdonnell Douglas Box 516 66-2n St Louis MO 63136

KOZLOWSKI, GERALD P, b Grand Rapids, Mich, Dec 24, 42. NEUROENDOCRINOLOGY. *Educ:* Aquinas Col, BS, 64; Mich State Univ, MS, 67; Univ Ill, Urbana-Champaign, PhD(anat), 71. *Prof Exp:* Technician histopath, Mich State Univ, 64-65; asst instr anat, 65-66; res assoc, Univ Mo-Columbia, 67-68; instr, Univ Ill, Urbana-Champaign, 68-70; teaching fel, Sch Med & Dent, Univ Rochester, 71-73; asst prof anat, Col Vet Med & Biomed Sci, Colo State Univ, 73-76, assoc prof, 76-78; assoc prof neurobiol & anat, Univ Tex Health Sci Ctr, Houston, 78-80; ASSOC PROF, DEPT PHYSIOL, SOUTHWESTERN MED CTR, DALLAS, TEX, 80- *Concurrent Pos:* NIH grant, 74-92; ed, Histochem. *Mem:* Am Asn Anatomists; Int Soc Neuroendocrinologists; Biol Stain Comn; Soc Neurosci; Res Soc Alcoholism. *Res:* Effects at alcohol on central nervous system; vasopressin and oxytocin; scanning and high-voltage electron microscopy of the median eminence; immunocytochemistry of HIV receptor; neuroimmunology; light and election microscopic immunocytochemistry for visualization of releasing-hormines and neuropeptides of the hypothalamus. *Mailing Add:* Dept Physiol Southwestern Med Ctr Univ Tex 5323 Harry Hines Blvd Dallas TX 75235-9040

KOZLOWSKI, LESTER JOSEPH, b Chicago, Ill, Aug 31, 53; m 83; c 2. HIGH DENSITY INFRARED FOCAL PLANE ARRAYS DESIGN, ULTRA LOW NOISE FOCAL PLANE ARRAYS DESIGN & TEST. *Educ:* Univ Ill, Chicago, BSEE, 75, MSEE, 77. *Prof Exp:* Tech asst elec eng, Univ Ill, Chicago, 75-77; sr scientist, Hughes Aircraft Missile Systs Group, 78-87; MEM TECH STAFF, ROCKWELL INT SCI CTR, 87- *Mem:* Inst Elec & Electronics Engrs Electronic Devices Soc. *Res:* Design, development and characterization of infrared focal plane arrays for a variety of applications including military and astronomy. *Mailing Add:* Rockwell Int 1049 Camino Dos Rios Thousand Oaks CA 91360

KOZLOWSKI, ROBERT H, b Duquesne, Pa, May 17, 28; m 51; c 2. ORGANIC CHEMISTRY. *Educ:* St Mary's Col Calif, BS, 50; Northwestern Univ, PhD(chem), 55. *Prof Exp:* Res chemist, 55-60, sr res chemist, 60-64, supvry res chemist, 64-66, sr res assoc, Petrol Process Res & Develop, Chevron Res Co Div, 66-76, litigation support coordr, Secy's Dept, 76-77, litigation support mgr, Secy's Dept, Standard Oil Co Calif, 78-84, litigation support mgr, 85-88, MGR, ADMIN SUPPORT, LAW DEPT, CHEVRON CORP, 88- *Concurrent Pos:* Consult, Kenwood Vineyards, 70-86. *Mem:* Am Chem Soc; Am Soc Enol. *Res:* Petroleum processing; hydrocarbon reactions and mechanisms. *Mailing Add:* 41 Sutter St #1300 San Francisco CA 94104

KOZLOWSKI, THEODORE R, b Niagara Falls, NY, Dec 21, 37; m 61; c 5. PHYSICAL INORGANIC CHEMISTRY. *Educ:* Niagara Univ, BS, 59; Rensselaer Polytech Inst, PhD(phys inorg chem), 63; Harvard Univ, PMD, 72. *Prof Exp:* Sr chemist, Corning Glass Works, 63-64, res chemist, 66-70, mgr, indust prod develop, 70-73, mgr, sunglass proj, France, 73-76, portfolio mgr, consumer prod, 76-82, develop mgr, elec & electronic prod, 82-84, develop mgr, tech & elec prod, 84-87, DIR, TECH PROD DEVELOP, CORNING INC, 87- *Mem:* AAAS; Am Chem Soc; Sigma Xi; Nat Geog Soc; Int Soc Hybrid Microelectronics. *Res:* Glass-molten salt interactions; high strength glasses and glass ceramics by ion exchange from molten salts; high strength materials; product development of photochromic ophthalmic and sunglass products; consumer tableware ovenware clear glass ceramics, hybrid PC boards; low dielectric materials, strong glass ceramics, epoxy products, specialty glasses, LCD display glasses, magnetic disk substrates, dental glass ceramics, sheet, pressing and tubing processes. *Mailing Add:* Corning Glass Works Sullivan Sci Park Corning NY 14831

KOZLOWSKI, THEODORE THOMAS, b Buffalo, NY, May 21, 17; m 54. PLANT PHYSIOLOGY, FOREST BIOLOGY. *Educ:* Syracuse Univ, BS, 39; Duke Univ, MA, 41, PhD(plant physiol), 47. *Hon Degrees:* DSc, Univ Louvain, Belgium, 78; State Univ NY. *Prof Exp:* Asst, Duke Univ, 46; from asst prof to prof bot, Univ Mass, 47-58, head dept, 50-58; prof, 58-72, chmn dept, 61-64, dir biotron, 77-87, A J RIKER PROF FORESTRY, UNIV WIS-MADISON, 72- *Concurrent Pos:* Vis prof, Univ Pa, 54; vis scientist, Soc Am Foresters, 63, 66, 68, 69 & 70; Fulbright sr res scholar & exchange lectr, Oxford Univ, 64-65; assoc ed, Am Midland Naturalist, 65-71, Can J Forest Res, 70-76; Int Shade Tree Conf res fel, 69, 70 & 71-; external PhD thesis examr, Australian Nat Univ, Univ Western Australia, Univ Ibadan, Nigeria, Sri Venkateswara Univ, India, Univ WI Found; ed, Physiol-Ecol Book Series, Acad Press; res collabr, US Forest Serv; consult, UN Food & Agr Orgn, Nat Park Serv, Oak Ridge Nat Lab, Stanford Res Inst, NSF, Fed Forest Res Sta, Brazil, Univ BC; sr distinguished res prof forestry, Wis Alumni Res Found, 84-; A J Riker prof forestry, Univ Wis-Madison, 72- *Honors & Awards:* Auth Award, Int Shade Tree Conf, 71; Barrington Moore Res Award Biol Sci, Soc Am Foresters, 74; George Lamb Lectr, Univ Nebr, 74; Arboricult Res Award, Inst Soc Arboricult, 76; George S Long Lectr, Univ Washington, 78; Merit Award, Bot Soc Am, 84; Merit Award, Int Soc Arborcult, 87. *Mem:* Soc Am Foresters; Am Soc Plant Physiol; Bot Soc Am; Ecol Soc Am; Am Inst Biol Sci; Sigma Xi; Scand Soc Plant Physiologists; Int Soc Arboricult; hon mem Finnish Forestry Soc; hon mem Polish Bot Soc. *Res:* Physiology of woody plants; plant water relations; physiological ecology; effects of environmental stresses on plant growth. *Mailing Add:* Environ Studies Prog Univ Calif Santa Barbara CA 93106

KOZMA, ADAM, b Cleveland, Ohio, Feb 2, 28; wid; c 2. RADAR SYSTEMS, ELECTROOPTICS. *Educ:* Univ Mich, BSE, 52, MSE, 64; Wayne State Univ, MSEM, 61; Univ London, PhD(elec eng), 68 & dipl, Imp Col, 69. *Prof Exp:* Design engr, US Broach Co, Mich, 51-56, sales engr, 56-58; asst mech & electrooptical design, Inst Sci & Technol, Univ Mich, Ann Arbor, 58-61, res assoc, 61-63, assoc res engr, 63-65, res engr & asst head optics group, Radar & Optics Lab, 65-69; gen mgr, Electrooptics Ctr, Radiation Div, Harris, Inc, 69-73; sr res engr & mgr, Electromagnetics & Electronics Dept, Environ Res Inst, Mich, 73-75, mgr tech staff, 75-76, vpres & dir, Radar & Optics Div, 75-85, vpres, Corp Develop, 85-86; vpres & dir Depense Electronics Eng Div, Syracuse Res Corp, 86-88; HEAD, ADVAN SYSTS DEPT, MITRE CORP, 89- *Concurrent Pos:* Consult, Conductron Corp, 65-66, IBM Systs Develop Div, 66-67, UK Atomic Weapons Estab, 67-68 & Radiation Inc, Fla, 68-69; on leave, Imp Col, Univ London, 66-68, acad visitor & lectr, 67-68; consult phys sci directorate, USAMRDEL, MICOM, Redstone, Ala, 74-78; co-chmn & lectr, Synthetic Aperture Radar Intensive Course, Col Eng, Univ Mich, Ann Arbor, 80- *Honors & Awards:* Ordnance Medal, Avionics Sect, Am Defense Preparedness Asn. *Mem:* Fel Optical Soc Am; fel Inst Elec & Electronics Engrs; Am Defense Preparedness Asn; Sigma Xi; Soc Photog & Instrument Engrs. *Res:* Coherent optics with application to signal processing and optical correlation; holography; application of lasers and holography to storage and retrieval; speckle effects in coherent systems; synthetic aperture radar systems and applications. *Mailing Add:* Infrared & Electro-Optics Systs Mitre Corp MSK 208 Burlington Rd Bedford MA 01730

KOZUB, RAYMOND LEE, b Ladysmith, Wis, June 16, 40; m 65; c 2. NUCLEAR PHYSICS. *Educ:* Univ of Wis-River Falls, BS, 62; Mich State Univ, MS, 64, PhD(physics), 67. *Prof Exp:* Asst prof physics, Tex A&M Univ, 67-71, res scientist, Cyclotron Inst, 71-72; res assoc chem, Columbia Univ, 72-74; asst prof physics, Queen's Univ, Kingston, Ont, 74-77; assoc prof, 77-80, PROF PHYSICS, TENN TECHNOL UNIV, 80-, DEPT CHMN, 86- *Concurrent Pos:* Prin investr sponsored res, Dept Energy, 78- *Mem:* Am Phys Soc; Sigma Xi; Am Asn Phys Teachers. *Res:* Nuclear structure studies; transfer reactions; stripping reactions to unbound final states; isobaric analog states; nuclear lifetime measurements; heavy ion reactions; neutron-rich nuclei; rare electron capture processes. *Mailing Add:* Dept Physics Tenn Technol Univ Cookeville TN 38505

KRA, IRWIN, b Poland, Jan 5, 37; US citizen; m 61; c 3. MATHEMATICS. *Educ:* Polytech Inst Brooklyn, BS, 60; Columbia Univ, MA, 64, PhD(math), 66. *Prof Exp:* C L E Moore instr math, Mass Inst Technol, 66-68; from asst prof to assoc prof, 68-71, chmn dept, 75-81, PROF MATH, STATE UNIV NY, STONY BROOK, 72- *Concurrent Pos:* Guggenheim Found fel, 70-71; actg chmn dept math, State Univ NY Stony Brook, 70-71, actg provost, Div Math Sci, 71-72; vis Israel, Chile, Eng, Japan & China; adv prof, Fudan Univ, Shargai, China. *Mem:* Am Math Soc. *Res:* One complex variable, particularly moduli of Riemann surfaces and Kleinian groups. *Mailing Add:* Dept Math State Univ NY Stony Brook NY 11794-3651

KRAABEL, JOHN STANFORD, MECHANICAL ENGINEERING. *Educ:* Univ Colo, BS, 69; Univ Calif, Davis, MS, 77, PhD(mech eng), 79. *Prof Exp:* MEM TECH STAFF, SANDIA NAT LABS, 86- *Mem:* Am Soc Mech Engrs; Am Inst Aeronaut & Astronaut. *Res:* Fluid mechanics; heat transfer. *Mailing Add:* Sandia Nat Labs PO Box 969 Livermore CA 94551

KRAAKEVIK, JAMES HENRY, b Chicago, Ill, Feb 18, 28; m 50; c 5. ATMOSPHERIC PHYSICS. *Educ:* Wheaton Col, Ill, BS, 48; Univ Md, PhD(physics), 57. *Prof Exp:* Physicist, US Naval Res Lab, 48-54, res sect head, 54-58; chair, Physics Dept, 60-64 & 70-81, from asst prof to prof physics, 70-84, DIR, BILLY GRAHAM CTR, WHEATON COL, ILL, 84- *Concurrent Pos:* Teacher, Titcombe Col, Nigeria, 64-66, prin, 66-67; consult, US Naval Res Lab, 58-73, Ill State Water Surv, 61-64 & Coronet Instr Films, 62-65; educ secy, Sudan Interior Mission, 69-70, educ consult, 70-, dir, res & ministry, 81-84; educ consult, Ministry Educ, Sudan, 73- *Mem:* Am Asn Physics Teachers; AAAS; Am Meteorol Soc. *Res:* Electrical properties of atmosphere and relationship with meteorology; electrical characteristics of upper atmosphere and relationship with radiation; conduction of electricity through gases; characteristics of sub-micron particles in the atmosphere. *Mailing Add:* Billy Graham Ctr Wheaton Col Wheaton IL 60187

KRAATZ, CHARLES PARRY, BOTULINUM TOXIX, MUSCLE POTENTIALS. *Educ:* Univ Cincinnati, PhD(physiol), 36. *Prof Exp:* prof pharmacol, Med Col, Jefferson Univ, 47-72; CONSULT, 72- *Mailing Add:* 329 S Norwinden Dr Apt B Springfield PA 19064

KRAAY, GERRIT JACOB, b Amsterdam, Neth, Oct 14, 35; m 63; c 4. GENETICS. *Educ:* State Agr Univ Wageningen, BSc, 60, MSc, 63, PhD, 67. *Prof Exp:* Res scientist, Found for Blood Group Res, Wageningen, Netherlands, 63; res scientist, Dept Animal Sci, State Agr Univ Wageningen, 64-67; asst prof vet bact, Univ Guelph, 67-69, asst prof biomed sci, 69-72; HEAD BLOOD TYPING SECT, ANIMAL DIS RES INST EAST, CAN DEPT AGR, 72- *Mem:* Royal Dutch Soc Agr Sci; Int Soc Animal Bloodgroup Res; Genetics Soc Can. *Res:* Population genetics of blood groups and serum-protein polymorphisms in animals; immuno-reproduction. *Mailing Add:* Bovine Blood Testing Lab Saskatoon Res Coun 30 Campus Dr Annex Saskatoon SK S7N 0X1 Can

KRABACHER, BERNARD, b Cincinnati, Ohio, Dec 25, 25. PHYSICAL ORGANIC CHEMISTRY. *Educ:* Univ Cincinnati, Chem Eng, 49, PhD(phys org chem), 61. *Prof Exp:* Chemist, Emery Industs, Inc, 49-57, group leader ozone res, 61-63; from assoc prof to prof chem, WVa State Col, 63-90, chmn dept, 76-82; RETIRED. *Mem:* Am Chem Soc; fel Am Inst Chemists. *Res:* Reactions of organic compounds with cobalt carbonyls and ozone; preparation of unusual compounds; reaction mechanisms. *Mailing Add:* PO Box 1000 Institute WV 25112-1000

KRABBE, GREGERS LOUIS, b Roskilde, Denmark, Jan 5, 20; nat US; m 55; c 2. MATHEMATICAL ANALYSIS. *Educ:* Univ Calif, AB, 49, MS, 51, PhD, 54. *Prof Exp:* Assoc prof math, Purdue Univ, 54-60 & Yale Univ, 60-61; NATO fel, Univ Rennes, 61-62; assoc prof math Purdue Univ, 62-68, Prof, 68-88; RETIRED. *Mem:* Am Math Soc. *Res:* Algebraic operational calculus, as applied to lumped systems; theory of linear operators; electrical engineering, linear networks, signal and system analysis. *Mailing Add:* 779 Spruce St Berkeley CA 94707

KRABBENHOFT, HERMAN OTTO, b Detroit, Mich, July 15, 45; m 76; c 2. ORGANIC CHEMISTRY. *Educ:* Wayne State Univ, BS, 70; Univ Mich, MS, 71, PhD(chem), 74. *Prof Exp:* Staff chemist, Gen Elec Plastics, 82-87, STAFF CHEMIST, GEN ELEC CORP RES & DEVELOP CTR, 76-81, 88- *Concurrent Pos:* NIH grant chem, Univ Calif, Berkeley, 75-76. *Mem:* Am Chem Soc. *Res:* Structure and mechanism in organic chemistry; organic synthesis. *Mailing Add:* Gen Elec Corp/Res & Develop PO Box 8 Schenectady NY 12301

KRABBENHOFT, KENNETH LOUIS, b Page, NDak, Feb 24, 31; m 55; c 3. MICROBIOLOGY. *Educ:* Univ Valparaiso, BA, 53; NDak State Univ, MS, 56; Ore State Univ, PhD(microbiol), 65. *Prof Exp:* Instr biol, Mankato State Col, 58-62; asst prof, NMex State Univ, 65-67; assoc prof, Mankato State Univ, 67-70, prof, 70-; MEM FAC, WAYNE STATE UNIV, MICH. *Concurrent Pos:* NASA res grant, 65-67; NSF res grant, 69-71. *Mem:* AAAS; Am Soc Microbiol; Inst Food Technologists; Sigma Xi. *Res:* Mechanisms of radiation resistance in microorganisms. *Mailing Add:* Am Bd Radiol 2301 W Big Beaver Rd No 625 Troy MI 48064

KRABEC, CHARLES FRANK, JR, electrooptics, for more information see previous edition

KRACHER, ALFRED, b Vienna, Austria, Sept 21, 45; m 74; c 2. METEORITICS, COSMOCHEMISTRY. *Educ:* Univ Vienna, PhD(chem), 74. *Prof Exp:* Res asst chem, Univ Vienna, 72-74, res asst petrol, 74-76; staff scientist mineral & petrol, Mus Natural Hist, Vienna, 76-81; researcher earth sci, Inst Meteoritics, Univ NMex, 81-82; ASST SCIENTIST & MICROPROBE SPECIALIST, IOWA STATE UNIV, 84- *Concurrent Pos:* Researcher chem, Inst Geophysics, Univ Calif, Los Angeles, 77-78; counr, Meteoritical Soc, 85-86. *Mem:* Meteoritical Soc; AAAS; Am Geophys Union; Geochem Soc. *Res:* Petrology and composition of meteorites; computer application to petrologic problems; theory of science. *Mailing Add:* Dept Geol Sci Iowa State Univ Ames IA 50011-3212

KRACKOV, MARK HARRY, b Brooklyn, NY, June 2, 32; m 54; c 2. ORGANIC CHEMISTRY, FLUORINE CHEMISTRY. *Educ:* Univ Calif, Berkeley, BS, 55; Ore State Univ, PhD(org chem), 62. *Prof Exp:* Instr chem, Ore State Univ, 61-62; USPHS fel org chem, Sch Med, Yale Univ, 62-65; RES CHEMIST, JACKSON LAB, ORG CHEM DEPT, E I DU PONT DE NEMOURS & CO, 65- *Mem:* Am Chem Soc; AAAS; Sigma Xi; Catalysis Soc. *Res:* Synthesis and physical properties of heterocyclic nitrogen, sulfur and selenium compounds; chemical synthesis of polynucleotides; labelling compounds via neutron activation; photopolymerization; reverse osmosis membranes; organic chemical process development; catalysis; fluoroaromatic chemistry. *Mailing Add:* Chem & Pigments Dept Jackson Lab E I du Pont de Nemours & Co Wilmington DE 19898

KRAELING, ROBERT RUSSELL, b Pittsburgh, Pa, Aug 22, 42; m 62; c 5. ANIMAL SCIENCE. *Educ:* Univ Md, BS, 64, MS, 67; Iowa State Univ, PhD(animal sci physiol reprod), 70. *Prof Exp:* Res asst, Swine Res Br, Animal Husb Res Div, Agr Res Serv, USDA, 64-66, agr res technician, 66-67, agr res scientist, 67; res assoc, Iowa State Univ, 67-70; res animal physiologist, Animal Physiol & Genetics Inst, Reproduction Lab, Md, 70-74, Animal Prod Lab, Russell Agr Res Ctr, Athens, 74-77, SUPVR RES PHYSIOLOGIST, ANIMAL PHYSIOL RES UNIT, RUSSELL RES CTR, AGR RES SERV, USDA, 77- *Concurrent Pos:* Adj assoc prof, Dept Animal & Dairy Sci, Univ Ga, 74-, mem grad fac, 79- *Honors & Awards:* Animal Physiol & Endocrinol Res Award, Am Soc Animal Sci, 90. *Mem:* Am Soc Animal Sci; Soc Study Reproduction; Sigma Xi. *Res:* Determining the physiological and endocrinolgoical factors which control puberty, ovulation, corpus luteum function and the post-partum interval in swine and cattle and the effects of environment and management systems. *Mailing Add:* Richard B Russell Agr Res Ctr USDA PO Box 5677 Athens GA 30613

KRAEMER, DUANE CARL, b Willow, Wis, Oct 27, 33; m 60; c 2. REPRODUCTIVE PHYSIOLOGY, MEDICINE. *Educ:* Univ Wis, BS, 55; Tex A&M Univ, MS, 60, BS, PhD(physiol of reprod) & DVM, 66. *Prof Exp:* Asst scientist, Southwest Found Res & Educ, 66-75; assoc prof, 75-77, PROF VET PHYSIOL & PHARMACOL, COL VET MED, TEX A&M UNIV, 77- *Mem:* Soc Study Reprod; Sigma Xi; Am Vet Med Asn; Am Soc Animal Sci; Am Asn Lab Animal Sci. *Res:* Reproductive gamete physiology; contraceptive development and testing. *Mailing Add:* Col Vet Med Physiol/Pharmacol Vet Tex A&M Univ College Station TX 77843

KRAEMER, HELENA CHMURA, b Derby, Conn, July 10, 37; m 62; c 2. BIOSTATISTICS. *Educ:* Smith Col, BA, 58; Stanford Univ, PhD(statist), 63. *Prof Exp:* Actg asst prof, 64-69, res assoc, 69-72, from asst prof to assoc prof, 72-86, PROF BIOSTATIST, DEPT PSYCHIAT & BEHAV SCI, SCH MED, STANFORD UNIV, 86- *Concurrent Pos:* Lectr, Div Biostatist, Dept Community & Prev Med, Stanford Univ, 71- *Mem:* Fel Am Statist Asn; Psychomet Soc; fel Am Statist Asn. *Res:* Development of statistical methods for use in clinical and behavioral research, with particular emphasis on correlational methods. *Mailing Add:* Dept Psychiat & Behav Sci Sch Med Stanford Univ Stanford CA 94305

KRAEMER, J HUGO, forestry; deceased, see previous edition for last biography

KRAEMER, JOHN FRANCIS, b St Louis, Mo, June 20, 41. ORGANIC CHEMISTRY, POLYMER CHEMISTRY. *Educ:* St Louis Univ, BS, 63; Loyola Univ, MS, 65, PhD(org chem), 68. *Prof Exp:* Org chemist, Int Minerals & Chem Corp, 68-86. *Mem:* Am Chem Soc. *Res:* Organic synthesis; synthesis of biologically active compounds. *Mailing Add:* RR 21 Box 147 Terre Haute IN 47802

KRAEMER, KENNETH H, b Newark, NJ, June 22, 43; m 65; c 4. DNA REPAIR, CARCINOGENESIS. *Educ:* Brown Univ, BS, 65; Tufts Univ, MD, 69, diplomat, Nat Bd Med Examiners, 70; Am Bd Internal Med, 73; Am Bd Dermat, 76. *Prof Exp:* Med intern & resident, Harlem Hosp Ctr, NY, 69-71; clin assoc dermat, Dermat Br, NIH, 71-74; resident dermat, Univ Miami Sch Med, 74-76; res scientist, Lab Molecular Carcinogenesis, NIH, 76- *Concurrent Pos:* Surgeon, US Pub Health Serv, NIH, 71-74, sr surgeon, 74-86, med dir, 86- *Mem:* Am Acad Dermat; Soc Invest Dermat; Am Soc Photobiol; Am Soc Clin Invest. *Res:* Cancer-prone human genetic disease, xerodorma pigmentosum, familial malignant melanoma or displastic nevus syndrome, atoxin telangiectasia; DNA repair; photo carcinogenesis. *Mailing Add:* Nat Cancer Inst Molecular Carcinogenesis Lab Bldg 37 Rm 3E24 Bethesda MD 20892

KRAEMER, LOUISE MARGARET, b New York, NY, Dec 26, 10. PHYSICAL CHEMISTRY, BIOCHEMISTRY. *Educ:* Univ Pa, AB, 43; Univ Chicago, PhD(chem), 49. *Prof Exp:* Asst engr, Brown Instrument Co, 43-45; asst chem, Univ Chicago, 45-49; res assoc med, Univ Minn, 49-51; assoc chemist, Argonne Nat Lab, 51-53; asst prof natural sci, Univ Chicago, 53-59, assoc prof phys sci, 59-64; prof chem, New Col, 64-65; prof, 66-76, EMER PROF PHYS SCI, UNIV COLO, BOULDER, 76- *Concurrent Pos:* Prof, Dept Biol Sci, Univ Ark. *Mem:* AAAS; Am Chem Soc. *Res:* Enzymes in wheat germ; porphyrin chemistry; color centers in alkali halides. *Mailing Add:* Dept Zool Sci & Eng Univ Ark Rm 718 Fayetteville AR 72701

KRAEMER, LOUISE RUSSERT, b Milwaukee, Wis, Dec 17, 23; m 50; c 4. MALACOLOGY. *Educ:* Marquette Univ, BS, 45; Univ Mich, Ann Arbor, MS, 47, PhD(malacol), 66. *Prof Exp:* Asst zool & fisheries, Univ Mich, Ann Arbor, 46-48; asst prof zool, 48-50, instr, 56-58 & 59-66, from asst prof to assoc prof, 66-77, PROF ZOOL, UNIV ARK, FAYETTEVILLE, 77- *Mem:* AAAS; Am Malacol Union (pres, 81-82); Am Soc Zoologists; Sigma Xi; Animal Behav Soc. *Res:* Functional morphology; behavior of freshwater mollusks; nervous systems and reproductive systems of Lampsilis and Corbicula; macrobenthic communities in lotic systems. *Mailing Add:* Dept Zool 632 Sci-Eng Bldg Univ Ark Fayetteville AR 72701

KRAEMER, PAUL MICHAEL, b Philadelphia, Pa, Mar 19, 30. CELL BIOLOGY. *Educ:* Univ Colo, BA, 57; Tulane Univ, MPH, 59, DrPH, 61; Univ Pa, PhD(microbiol), 64. *Prof Exp:* Group leader exp pathol, 79-81, STAFF MEM CELLULAR BIOL, LOS ALAMOS NAT LAB, 64- *Concurrent Pos:* Fel, Wistar Inst, 61-64; ed, J Cellular Physiol; cellular physiol study sect, NIH, 77-81, Am Cancer Soc, 87-90. *Mem:* Am Soc Biol Chemists; Am Soc Cell Biol; Am Soc Exp Path. *Res:* Mammalian cell surface complex carbohydrates; chromosome changes in cancer; tumor biology. *Mailing Add:* Cell & Molecular Biol Group MS/M888 Los Alamos Nat Lab Los Alamos NM 87545

KRAEMER, ROBERT WALTER, b Philadelphia, Pa, Jan 27, 35; m 60, 87; c 3. EXPERIMENTAL HIGH ENERGY PHYSICS. *Educ:* La Salle Col, BA, 57; Johns Hopkins Univ, PhD(physics), 62. *Prof Exp:* Instr physics, Johns Hopkins Univ, 61-62, res assoc, 62-64; res assoc, 64-65, from asst prof to assoc prof, 65-74, PROF PHYSICS, CARNEGIE-MELLON UNIV, 74- *Mem:* Am Phys Soc. *Res:* High energy experimental nuclear physics. *Mailing Add:* Dept Physics Carnegie-Mellon Univ Pittsburgh PA 15213

KRAEUTER, JOHN NORMAN, b Glen Gardner, NJ, Mar 26, 42; m 70; c 2. BIOLOGICAL OCEANOGRAPHY, MARINE ECOLOGY. *Educ:* Fla State Univ, BA, 64; Col William & Mary, MA, 66; Univ Del, PhD(biol sci), 71. *Prof Exp:* Res fel biol sci, Marine Inst, Univ Ga, 71-73; res assoc, Skidaway Inst Oceanog, Ga, 73-74; asst marine scientist Va Inst Marine Sci, 74-80, asst prof, 74-80, assoc prof marine sci, Univ Va & Col William & Mary, 81-82, assoc marine scientist, Va Inst Marine Sci, 81-82; Baltimore Gas & Elec, 82-87; ASSOC DIR, FISHERIES & AQUACULT TECH CTR, RUTGERS UNIV, 87- *Concurrent Pos:* bus mgr, J Estuaries, Estuarine Res Fedn, 78- *Mem:* Atlantic Estuarine Res Soc (pres, 77-79); Estuarine Res Fedn (treas, 82, secy, 87-88); Malacol Soc London; AAAS; Am Malacol Union; Am Fisheries Soc; Nat Shellfisheries Asn. *Res:* Systematics and ecology of scaphopod mollusks; zoogeography of the western Atlantic marine invertebrates; benthic infaunal ecology; aquaculture. *Mailing Add:* 722 Jonathan Hoffman Rd Cold Springs NJ 08204

KRAFFT, GEOFFREY ARTHUR, b Enid, Okla, May 14, 58. ACCELERATOR PHYSICS. *Educ:* Rutgers Univ, BA, 78; Univ Calif, Berkeley, MA, 80, PhD(physics), 86. *Prof Exp:* STAFF SCIENTIST, CONTINUOUS ELECTRON BEAM ACCELERATOR FACIL, 86- *Mem:* Am Phys Soc. *Res:* Theory of particle accelerators; collective effects in intense particle beams. *Mailing Add:* CEBAF 12000 Jefferson Ave Newport News VA 23606-1909

KRAFFT, JOSEPH MARTIN, b Alexandria, Va, Jan 13, 23. PHYSICS. *Educ:* Cath Univ Am, Washington, DC, BME, 43, PhD, 51. *Prof Exp:* Various positions teaching, eng, patent search & writing, DC, 41-49; physicist res & develop terminal ballistics, dynamic plastic flow & fracture fatigue mech, 48-70, head, Mech Mat Br, 70-81, CONSULT, STRUCT INTEGRITY BR, NAVAL RES LAB, WASHINGTON, DC, 81- *Concurrent Pos:* Tech ed, Trans, Am Soc Mech Engrs, J Eng Mat & Technol, 78-81. *Mem:* Fel Am Soc Metals; fel Am Soc Testing & Mat; Am Soc Mech Engrs. *Mailing Add:* 1709 Oakcrest Dr Alexandria VA 22302

KRAFFT, MARIE ELIZABETH, b Washington, DC, Aug 15, 56; m 84. TOTAL SYNTHESIS OF NATURAL PRODUCTS, ORGANOMETALLIC CHEMISTRY. *Educ:* Va Polytech Inst & State Univ, BA, 79, MS, 80, PhD(org chem), 83. *Prof Exp:* ASST PROF, FLA STATE UNIV, 85- *Concurrent Pos:* NIH res fel, Columbia Univ, 83-85. *Mem:* Am Chem Soc; Sigma Xi. *Res:* Synthetic organic and organometallic chemistry; natural products synthesis; synthetic methodology. *Mailing Add:* Dept Chem Fla State Univ Tallahassee FL 32306-3006

KRAFSUR, ELLIOT SCOVILLE, m 66; c 1. ENTOMOLOGY. *Educ:* Univ Md, BS, 62, MS, 64; Univ London PhD(zool), 72. *Prof Exp:* Res asst entomol, Dept Entomol, Univ Md, 62-64; ensign lieutenant entomol, US Navy, 64-69; fel zool, US Nat Sci Found, 70-72; Oxford Univ, 72-73; res entomologist, Animal, Plant & Health Inspection Serv, US Dept Agr, 74-76; from asst prof to assoc prof, 76-85, PROF ENTOM, IOWA STATE UNIV, 85- *Concurrent Pos:* Res med entomologist, USN, 64-69; prin invstr, US Army Biol labs, 64-67; prin invstr, US Navy Med Res Unit, Ethiopia, 67-69; consult, Commonwealth Inst & Indust Res Org, New Guinea & Australia. *Mem:* Entom Soc Am; Soc Study Evolution; Royal Entom Soc London. *Res:* Population ecology and genetics of synanthropic flies with special reference to age, breeding structure and phenology; epidemiology of arthropod borne disease. *Mailing Add:* Dept Entomol Iowa State Univ Ames IA 50011

KRAFT, ALAN M, b Passaic, NJ, May 24, 25; m 51; c 2. PSYCHIATRY. *Educ:* Chicago Med Sch, MD, 51. *Prof Exp:* Staff psychiatrist, Vet Admin Hosp, Denver, Colo, 55-57; chief psychiatrist, Ment Health Ctr Am, 58-61; dir, Ft Logan Ment Health Ctr, 61-67; PROF PSYCHIAT & CHMN DEPT, ALBANY MED COL, 67- *Concurrent Pos:* Fel psychiat, Menninger Sch Psychiat, 52-55; consult, Vet Admin Hosp, Albany, NY, 71-; dir, Capital Dist Psychiat Ctr, Albany, 67-79. *Mem:* Am Psychiat Asn. *Res:* Treatment of chronic schizophrenia; program evaluation. *Mailing Add:* 47 New Scotland Ave Albany NY 12208

KRAFT, ALLEN ABRAHAM, b New York, NY, 1923; m 47; c 2. FOOD TECHNOLOGY. *Educ:* Cornell Univ, BS, 47, MS, 49; Iowa State Col, PhD(food technol), 53. *Prof Exp:* Asst food technol & bact, Iowa State Col, 49-53; asst poultry prod technologist, Animal & Poultry Husb Res Br, Agr Res Serv, USDA, 53-59; from asst prof to assoc prof, 59-72, PROF FOOD TECHNOL, IOWA STATE UNIV, 72-, PROF MICROBIOL, 81- *Mem:* Fel AAAS; Poultry Sci Asn; Inst Food Technologists; Am Soc Microbiol; World Poultry Sci Asn. *Res:* Microbiology and technology of meat and poultry products. *Mailing Add:* Dept Food Technol G62FD Tech Lab Iowa State Univ Ames IA 50011

KRAFT, CHRISTOPHER COLUMBUS, JR, b Phoebus, Va, Feb 28, 24; m 50; c 2. AEROSPACE ENGINEERING. *Educ:* Va Polytech Inst, BS, 44. *Hon Degrees:* DEng, Ind Inst Technol, 66 & St Louis Univ, 67. *Prof Exp:* Aeronaut res engr, Langley Aeronaut Lab, Nat Adv Comt Aeronaut, Va, 45-48, space task group, NASA, 58-59, supvry aeronaut res engr, 59-61, asst chief flight opers div, 61-62, chief div, 62-63, dir flight opers, 63-69, dep dir ctr, 69-72, dir, Lyndon B Johnson Space Ctr, 72-82; CONSULT, ROCKWELL INT, HOUSTON, 82- *Honors & Awards:* Arthur S Flemming Award, 63; Louis W Hill Award, Am Inst Aeronaut & Astronaut, 70; Space Flight Award, Am Astronaut Soc, 70. *Mem:* Nat Acad Eng; fel Am Inst Aeronaut & Astronaut; fel Am Astronaut Soc. *Res:* Aerospace engineering. *Mailing Add:* 14919 Village Elm St Houston TX 77062

KRAFT, DAVID WERNER, b Worms, Ger, Apr 21, 33; nat US; m 58; c 3. PHYSICS. *Educ:* City Col New York, BS, 54; Pa State Univ, PhD(physics), 59. *Prof Exp:* Res physicist, Pa State Univ, 59-60; sr physicist, Philips Labs Div, NAm Philips Co, Inc, 60-64 & Electronics Systs Div, Loral Corp, 64; sr res physicist, Cent Res Div, Am Cyanamid Co, 65; res scientist, Hudson Labs, Columbia, 65-68; from assoc prof to prof physics, Cooper Union, 68-76; vis prof opers res, Grad Sch Bus Admin, NY Univ, 76-77; dir Manpower Placement Div & Sr Staff Adv Planning, Am Inst Physics, 77-79; DEP EXEC SECY, AM PHYS SOC, 79- *Concurrent Pos:* Consult electronic systs div, Loral Corp, 65; vis scientist & consult, Gen Tel & Electronics Labs, 69-70; vis scientist, NY Univ, 72-74 & Philips Labs Div, NAm Philips Co, 75; consult, NY Tel Co, 76-77 & Am Inst Physics, 77; adj prof opers res, Grad Sch Bus Admin, NY Univ, 69- *Mem:* Am Phys Soc; Am Asn Physics Teachers; NY Acad Sci; AAAS. *Res:* Ultrasonic absorption; scattering of elastic waves; magnetic, optical and elastic properties of solids; atmospheric physics; operations and economic analysis; scientific manpower utilization; science administration. *Mailing Add:* Elec Eng Dept Tech Bldg Col Sci Eng Univ Bridgeport Bridgeport CT 06601

KRAFT, DONALD HARRIS, b Omaha, Nebr, Dec 21, 43; m; c 2. INFORMATION RETRIEVAL, FUZZY SET THEORY. *Educ:* Purdue Univ, BS, 65, MS, 66, PhD(indust eng), 71. *Prof Exp:* Asst prof libr & info serv, Univ Md, 70-75; vis asst prof librarianship, Univ Calif, Berkeley, 75-76; assoc prof, 76-82, chmn, 85-91, PROF COMPUTER SCI, LA STATE UNIV, 82- *Concurrent Pos:* Adj prof libr & info sci, La State Univ, 83-; mem, NAm Fuzzy Info Processing Soc Coun, 90-92; ed, J Am Soc Info Sci. *Honors & Awards:* K S Fu Award, NAm Fuzzy Info Processing Soc, 86. *Mem:* Asn Comput Mach; Inst Elec & Electronics Engrs Computer Soc; Asn Libr & Info Sci Educ; Am Soc Info Sci. *Res:* Use of fuzzy set theory to model generalized Boolean information retrieval mechanisms; use of operations research to model and evaluate ranked retrieval output. *Mailing Add:* Dept Computer Sci La State Univ Baton Rouge LA 70803-4020

KRAFT, DONALD J, b Strasburg, NDak, Nov 9, 36; m 60; c 4. PLANT PHYSIOLOGY. *Educ:* NDak State Univ, BS, 59, PhD(plant physiol), 68. *Prof Exp:* Teacher high sch, NDak, 61-62, chmn dept sci, 62-63; asst prof biol, St Mary's Col Minn, 68-69; asst prof, 69-72, assoc prof, 72-79, PROF, BEMIDJI STATE UNIV, 79- *Concurrent Pos:* Prin investr grants, NIH, 68-69 & Minn State Col Bd, 69-70 & 73-74. *Mem:* AAAS; Am Soc Plant Physiologists; Sigma Xi. *Res:* Biochemistry of seed germination as a means to eliminate noxious weeds through natural components of seeds rather than through use of sprays. *Mailing Add:* Dept Biol Bemidji State Univ Bemidji MN 56601

KRAFT, EDWARD MICHAEL, b Cincinnati, Ohio, Nov 13, 44; m 71; c 1. AERODYNAMICS, WIND TUNNEL TECHNOLOGY. *Educ:* Univ Cincinnati, BS, 68; Univ Tenn, MS, 72, PhD(aerodyn eng), 75. *Prof Exp:* Res asst, Space Inst, Univ Tenn, 68-69; proj engr, Sverdrup/Aro, Inc, 69-72, res engr, 72-78, engr supv, 78-80; BR MGR, CALSPAN FIELD SERV, INC, 81- *Concurrent Pos:* Asst prof, Space Inst, Univ Tenn, 77- *Mem:* Am Inst Aeronaut & Astronaut. *Res:* Wind tunnel wall interference, including subsonic, transonic and vertical/short take off and landing theories and development of the adaptive wall concept. *Mailing Add:* Calspan AEDC Opers Arnold AFB TN 37389

KRAFT, GERALD F, b Salinas, Calif, Feb 22, 28; m 48, 79; c 4. ENTOMOLOGY, SCIENCE EDUCATION. *Educ:* San Jose State Col, BA, 54; Wash State Univ, MS, 56; Ore State Univ, PhD(entom), 62. *Prof Exp:* Asst prof zool, 61-66, chmn dept biol, 71-74 & 77-85, ASSOC PROF BIOL, WESTERN WASH UNIV, 66- *Concurrent Pos:* City of Bellingham grant, 62-64; dir inst freshwater studies, Western Wash State Col, 64-68; US Dept Interior grant, 66-; res assoc, Univ Calif, Berkeley, 68-69. *Mem:* Entom Soc Am; Nat Acad Advising Asn. *Res:* General entomology; aquatic insects. *Mailing Add:* Dept Biol Western Wash Univ Bellingham WA 98225

KRAFT, IRVIN ALAN, b Huntington, WVa, Nov 20, 21; m 51; c 4. PSYCHIATRY. *Educ:* NY Univ, MD, 49. *Prof Exp:* Asst prof psychiat, 57-61, asst prof pediat, 58-61, assoc prof psychiat & pediat, 61-77, CLIN PROF PSYCHIAT, BAYLOR COL MED, 77-; PROF MENT HEALTH, UNIV TEX SCH PUB HEALTH HOUSTON, 75- *Concurrent Pos:* Med dir, Tex Inst Family Psychiat, 64-79. *Mem:* Fel Am Psychiat Asn; fel Am Acad Psychoanal; fel Am Acad Child Psychiat; life fel Am Orthopsychiat Asn. *Res:* Child psychiatry; psychoanalysis. *Mailing Add:* 2423 Gramercy Houston TX 77030

KRAFT, JOAN CREECH, b Washington, DC, June 8, 43; c 3. RETINOIC ACID, DYSMORPHOGENESIS. *Educ:* Univ Pa, BS, 65; Free Univ Berlin, Ger, dipl biol, 75, Dr rer nat (biol), 77. *Prof Exp:* Lab asst molecular biol, Free Univ Berlin, 76-81, teaching asst biol, 79-83; sr res scientist, Inst Toxicol & Embryopharmacol, Berlin, 84-88; SR RES FEL TERATOLOGY, DEPT PHARMACOL, UNIV WASH, 90- *Concurrent Pos:* Postdoctoral res protein biosynthesis, Inst Biochem & Molecular Biol, Free Univ Berlin & Max Planck Inst Molecular Genetics, 78-79. *Mem:* Teratology Soc. *Res:* Teratology; pharmacokinetics, metabolism and placental transfer of the potent teratogen accutane. *Mailing Add:* Dept Pharmacol SJ-30 Univ Wash Seattle WA 98195

KRAFT, JOHN CHRISTIAN, b Schwenksville, Pa, Nov 15, 29; div; c 2. HOLOCEN GEOLOGY, ARCHAEOLOGICAL GEOLOGY. *Educ:* Pa State Univ, BS, 51; Univ Minn, MS, 52, PhD(micropaleont), 55. *Prof Exp:* Geologist, Shell Can Ltd, 55-61, div stratigrapher, 61-64; from asst prof to assoc prof, 64-69, chmn dept, 69-84, PROF GEOL, UNIV DEL, 69-, H FLETCHER BROWN PROF GEOL & MARINE STUDIES, 83- *Honors & Awards:* Archaeol Geol Award, Geol Soc Am, 87. *Mem:* fel AAAS; Am Asn Petrol Geologists; fel Geol Soc Am; Soc Econ Paleont & Mineral; Am Inst Prof Geologists; Archaeol Inst Am. *Res:* Geology of coasts; Holocene sedimentary environments; Ordovician and Holocene Ostracoda; archaeological geology. *Mailing Add:* Dept Geol Univ Del Newark DE 19716

KRAFT, JOHN M, b Gary, Ind, July 14, 38; m 64; c 2. PLANT PATHOLOGY. *Educ:* Ariz State Univ, BSc, 60; Univ Minn, St Paul, MS, 62; Univ Calif, Riverside, PhD(plant path), 66. *Prof Exp:* Res asst plant path, Univ Minn, St Paul, 60-62 & Univ Calif, Riverside, 62-66; RES PLANT PATHOLOGIST, CROP RES DIV, AGR RES SERV, USDA, 66- *Concurrent Pos:* Mem grad fac, Dept Plant Path, Wash State Univ & Dept Plant Sci, Univ Idaho. *Mem:* Am Phytopath Soc; Am Soc Agron & Soil Sci; Pisum Genetics Asn; Sigma Xi; Nat Pea Improv Asn. *Res:* Soil-borne diseases of peas and their etiology, biology of the fungi, their control and the nature and inheritance of resistance when found; control of soil-borne diseases of peas; breeding for root disease resistance in peas. *Mailing Add:* Irrigated Agr Res & Exten Ctr RR 2 Box 2953 A Prosser WA 99350

KRAFT, KENNETH J, b Dows, Iowa, Mar 3, 30; m 68; c 1. INVERTEBRATE ECOLOGY. *Educ:* Bemidji State Col, BS, 52; Univ NDak, MS, 53; Univ Minn, PhD(entom), 58. *Prof Exp:* Instr biol, Univ Minn, 56-58; asst prof, Moorhead State Col, 58-59 & Bemidji State Col, 59-61; asst prof, 61-64, ASSOC PROF BIOL, MICH TECHNOL UNIV, 64- *Mem:* Entom Soc Am. *Res:* Ecology of aquatic insects. *Mailing Add:* Dept Biol Sci Mich Technol Univ Houghton MI 49931

KRAFT, LISBETH MARTHA, b Vienna, Austria, May 16, 20; nat US. PATHOLOGY, RADIOBIOLOGY. *Educ:* Cornell Univ, BS, 42, DVM, 45. *Prof Exp:* Asst parasitol, Cornell Univ, 45-46 & nutrit, Harvard Univ, 46; bacteriologist, NY State Dept Health, 47-49; asst, Yale Univ, 49-51, from instr to asst prof, 51-55; asst prof, NY Univ, 55-57; res assoc path & vet, Sch Med, Yale Univ, 57-61; asst dir, New York City Bur Labs & assoc mem, Pub Health Res Inst, 61-65; staff scientist, Bioquest Div, Becton Dickinson Co, 65-66; res vet, Oak Ridge Assoc Univs, 66-68; owner, L M Kraft Assocs, 68-74; specialist, Space Sci Lab, Univ Calif, Berkeley, 74-75; assoc scientist dept physics, Univ San Francisco, 75-77; RES SCIENTIST, AMES RES CTR, NASA, 77- *Concurrent Pos:* Consult, Sloan-Kettering Inst Cancer Res, NY, 59-61. *Honors & Awards:* Griffin Award, Am Asn Lab Animal Sci, 72; Charles River Prize, Am Vet Med Asn, 81. *Mem:* AAAS; Am Soc Microbiol; Am Vet Med Asn; Am Asn Lab Animal Sci; Am Col Lab Animal Med; Sigma Xi. *Res:* Central nervous system effects of cosmic ray (homogeneous differential equation) particles as applicable to manned spaceflight safety standards; investigations on health status of animals in spaceflight research; laboratory animal medicine and science. *Mailing Add:* 2101B Fallen Leaf Lane Los Altos CA 94022

KRAFT, PATRICIA LYNN, b Somerville, NJ. TOXICOLOGY. *Educ:* Rutgers Col Agr & Environ Sci, BS, 72; Mass Inst Technol, PhD(toxicol), 79. *Prof Exp:* Sr toxicologist, Best Foods Res & Eng Ctr, CPC North Am, 80-83; GROUP MGR, LIFE SCI, PEPSI CO, VALHALLA, NY, 83- *Concurrent Pos:* Chmn, Toxicol & Safety Evaluation Div, 85-86. *Mem:* Am Inst Nutrit; Am Col Toxicol; Inst Food Technologists; Sigma Xi. *Res:* Nutritional biochemical mechanisms of tumorigenesis; effect of diet on maintenance of health and progression of disease, hunger and satiety. *Mailing Add:* Pepsi Co 100 Stevens Ave Valhalla NY 10595

KRAFT, R(ALPH) WAYNE, b Collingswood, NJ, Jan 14, 25; m 48; c 4. SYSTEMS THEORY, PHYSICAL METALLURGY. *Educ:* Lehigh Univ, BS, 48; Univ Mich, MS, 56, PhD(metall eng), 58. *Hon Degrees:* Dr, Allentown Col St Francis de Sales, 84. *Prof Exp:* Metallurgist, Am Brake Shoe Co, 48-54; instr metall, Univ Mich, 54-58; group leader mat res, United Aircraft Corp, 58-62; from assoc prof to prof metall eng, 62-67, N J Zinc Prof, 67-78, PROF METALL ENG, LEHIGH UNIV, 78- *Mem:* Fel AAAS; fel Am Soc Metals; Soc Gen Systs Res. *Res:* General systems theory and the philosophy of science and religion; analysis of cosmic processes using thermodynamic and information theory concepts; evolutionary ideas of thinkers such as Teihard de Chardin. *Mailing Add:* 645 Bierys Bridge Rd Bethlehem PA 18017

KRAFT, ROBERT PAUL, b Seattle, Wash, June 16, 27; m 49; c 2. ASTROPHYSICS. *Educ:* Univ Wash, BS, 47, MS, 49; Univ Calif, Berkeley, PhD(astron), 55. *Prof Exp:* Instr math & astron, Whittier Col, 49-51; NSF fel, Mt Wilson & Palomar Observs, 55-56; asst prof astron, Ind Univ, 56-58 & Univ Chicago, 58-59; mem staff, Mt Wilson & Palomar Observs, 60-67; actg dir, 68-70, 71-73 & 80-81, PROF ASTRON & ASTRONR, LICK OBSERV, UNIV CALIF, SANTA CRUZ, 67-, DIR, LICK OBSERV, 81-, DIR, UNIV CALIF OBSERV, 88- *Concurrent Pos:* Warner Prize lectr, Am Astron Soc, 62; chmn bd studies astron & astrophys, Univ Calif, Santa Cruz, 68-70 & 78-80; vis fel, Univ Colo, 70; Fairchild scholar, Cal Inst Technol, 80. *Mem:* Nat Acad Sci; Am Astron Soc (pres, 74-76); Am Acad Arts & Sci; Int Astron Union (vpres, 82-88); fel AAAS. *Res:* Stellar spectroscopy; galactic structure. *Mailing Add:* Lick Observ Univ Calif Santa Cruz CA 95064

KRAFT, SUMNER CHARLES, b Lynn, Mass, Aug 21, 28; m 63; c 3. INTERNAL MEDICINE, GASTROENTEROLOGY. *Educ:* Tufts Col, BS, 48; Boston Univ, AM, 49; Univ Chicago, MD, 55; Am Bd Internal Med, dipl, 62, Am Bd Gastroenterol, dipl, 65. *Prof Exp:* From instr to assoc prof, 59-73, PROF MED & COMT IMMUNOL, SCH MED, UNIV CHICAGO, 74- *Concurrent Pos:* USPHS spec res fel, 61-66, USPHS res career develop award, 67-71; res fel, Div Allergy, Immunol & Rheumatol, Scripps Clin & Res Found, La Jolla, Calif, 64-66; vis affil prof med, Uniformed Serv, Univ Health Sci, Bethesda, 79- *Honors & Awards:* William Beaumont Award, 77. *Mem:* AAAS; Am Asn Immunologists; Am Col Physicians; Am Fedn Clin Res; Am Gastroenterol Asn; Am Soc Gastrointestinal Endoscopy; Sigma Xi. *Res:* Gastrointestinal immunolgy. *Mailing Add:* Dept Med Univ Chicago Box 400 5841 S Maryland Ave Chicago IL 60637

KRAFT, WALTER H, b Newark, NJ, Dec 31, 38; m 59; c 3. MARKETING. *Educ:* Newark Col Eng, BS, 62, MS, 65; NJ Inst Technol, PhD(civil eng), 75. *Prof Exp:* FROM ASST ENG TO PARTNER & SR VPRES, EDWARDS & KELCEY, INC, 62- *Concurrent Pos:* Adj prof, NJ Inst Technol & Polytech Inst, NY; chairperson, Urban Transp Div, Am Soc Civil Engrs, 77-79, Nat Transp Policy Comt, 80-82, comt AIE03 Intermodal Transfer Facil, Nat Res Coun, Transp Res Bd, 82-85; mem, comt A3A10 Hwy Capacity & Qual Serv, 80-85; lectr, Carnegie-Mellon Univ, St Johns Univ, Int Conf Traffic Eng & Planning, Beijing, Peoples Repub China, 87 & Sino-Am- British Urban Transp Planning Sem, Beijing, Peoples Repub China, 88. *Honors & Awards:* Robert Ridgward Award, Am Soc Civil Engrs, 62, Frank Masters Award, 82; Ivor S Wisepart Transp Eng Award & Distinguished Serv Award, Inst Transp Engrs, 86. *Mem:* Fel Inst Transp Engrs (pres, 88); Am Soc Civil Engrs. *Res:* Civil engineering; traffic engineering and planning; author of numerous technical articles. *Mailing Add:* Edwards & Kelcey Inc 70 S Orange Ave Livingston NJ 07039

KRAFT, WILLIAM GERALD, b Evansville, Ind, July 15, 44; m 77. BACTERIAL PATHOGENESIS, MICROBIAL PHYSIOLOGY. *Educ:* Purdue Univ, BS, 66; Univ Wash, MS, 72; Ind Univ, PhD(microbiol), 77. *Prof Exp:* res eng chem, Dow Chem Co, 66-68; sr res microbiologist, 77-82, group leader, 82-84, SECT HEAD, NORWICH-EATON PHARMACEUT, 84- *Mem:* Am Soc Microbiol. *Res:* Physiology and pathogenesis of campylobacters; thermal resistance of aerobic bacilli spores and define sterilization cycles for pharmaceutical products; microbial ecology of natural and synthetic substances; develop novel antibacterials. *Mailing Add:* 23 Brown Ave Norwich NY 13815

KRAH, DAVID LEE, b Ashland, Pa, Jan 23, 56. VIROLOGY, CELL BIOLOGY. *Educ:* Pa State Univ, BS, 77; Hahnemann Med Col, MS, 80; Hahnemann Univ, PhD(microbiol & immunol), 82. *Prof Exp:* Fel, Rockefeller Univ, 82-85, res assoc, 85-86; sr res scientist, Schering Corp, 86-88; sr res virologist, 88-91, RES FEL, MERCK SHARP & DOHME RES LABS, MERCK & CO, INC, 91- *Mem:* AAAS; Am Soc Microbiol; Am Soc Virol; Soc Gen Microbiol. *Res:* Development and characterization of live virus vaccines; measurement of humoral immune responses following viral infection or vaccination; interactions of viruses with cell membrane receptors. *Mailing Add:* Dept Cellular & Molecular Biol Merck Sharp & Dohme Res Labs West Point PA 19486

KRAHENBUHL, JAMES LEE, b Appleton, Wis, Oct 7, 42; m 65; c 1. INFECTIOUS DISEASES, TUMOR IMMUNOLOGY. *Educ:* Univ Wis-Madison, BS, 64, MS, 67, PhD(med microbiol), 70. *Prof Exp:* Fel, Palo Alto Med Res Found, 70-71, sr res assoc immunol & infectious dis, 72-79; CHIEF, LEPROSY RES UNIT, USPHS HOSP, SAN FRANCISCO, 79- *Concurrent Pos:* Fel med, Med Ctr, Stanford Univ, 70-71, res assoc, 72-78; Pub Health Serv Res Career Develop Award, Nat Inst Allergy & Infectious Dis, 74-79. *Mem:* AAAS; Am Asn Immunologists; Soc Exp Biol & Med; Am Soc Microbiol; Int Leprosy Asn. *Res:* Mechanisms of host resistance to intracellular pathogens and tumors. *Mailing Add:* Gillis Long Hansons Dis Ctr Carville LA 70721

KRAHL, MAURICE EDWARD, b Cambridge City, Ind, Sept 17, 08; m 32, 67. ENDOCRINOLOGY, BIOCHEMISTRY. *Educ:* Depauw Univ, AB, 29; Johns Hopkins Univ, PhD(chem), 32. *Hon Degrees:* LLD, Univ Toronto, 71. *Prof Exp:* Res chemist, Eli Lilly & co, 33-44; instr pharmacol, Col Physiol & Surg, Columbia Univ, 44-46; asst prof to assoc prof pharmacol & biochem, Wash Univ, St Louis, 46-53; prof physiol, Univ Chicago, 53-69; vis prof, 67-69, chmn & prof, 69-77, EMER PROF PHYSIOL, STANFORD UNIV, 77- *Concurrent Pos:* Vis prof physiol, Univ Rio Grande Do Sul, Brazil, 59; consult, endocrine study sect, NIH, 61-65 & pvt found, 77-; vis prof biochem, Monash Univ, Australia, 66-77; trustee, Marine Biol Lab, Woods Hole. *Honors & Awards:* Banting Lectr, Eng Biabetic Asn, 50. *Mem:* Am Soc Biol Chemists; Am Soc Pharmocol & Exper Therapeut; Soc Gen Physiologists; Am Diabetes Asn. *Res:* Physical chemistry of drugs and hormones; etiology of diabetes mellitus. *Mailing Add:* 2783 W Casas Cir Tucson AZ 85741

KRAHL, NAT W(ETZEL), structural engineering; deceased, see previous edition for last biography

KRAHMER, ROBERT LEE, b Forest Grove, Ore, Dec 28, 32; m 57; c 2. FOREST PRODUCTS. *Educ:* Ore State Univ, BS, 58, MS, 60; State Univ NY Col Forestry, Syracuse Univ, PhD(wood prod eng), 62. *Prof Exp:* Instr forest prod, 59-60, asst prof, forest res labs, 62-67, assoc prof, 67-77, PROF FOREST PROD, ORE STATE UNIV, 77- *Mem:* Forest Prod Res Soc; Soc Wood Sci & Technol; Int Asn Wood Anat. *Res:* Light and electron microscope studies of fine structure of wood; variability of anatomical properties of wood. *Mailing Add:* Forest Prod Dept Ore State Univ Corvallis OR 97331

KRAHN, ROBERT CARL, b Minneapolis, Minn, Dec 1, 41; m 78; c 2. ORGANIC CHEMISTRY. *Educ:* Univ Minn, BChE, 63; Univ Wash, PhD(org chem), 68. *Prof Exp:* Res chemist, Org Chem Dept, 68-73, process chemist, 73-78, sr chemist, Chem, Dyes & Pigments Dept, Jackson Lab, E I du Pont de Nemours & Co, Inc, 78-82; HEALTH CONSULT, 82- *Mem:* Am Chem Soc; Sigma Xi; AAAS. *Res:* Emulsion polymerization; monomer synthesis; flurochemicals; surfactant chemistry; ethoxylation; textile finishing. *Mailing Add:* 17 Polaris Dr N Star Newark DE 19711

KRAHN, THOMAS RICHARD, b Swiftcurrent, Sask, May 23, 43; m 67; c 2. HORTICULTURE. *Educ:* Univ Alta, BSc, 67; Mich State Univ, MSc, 73. *Prof Exp:* Exten specialist, 67-72, head lab serv, 72-83, DIR, ALTA HORT RES CTR, 83- *Mem:* Agr Inst of Can. *Res:* Post harvest physiology, mainly storage and handling practices for vegetables, potatoes, and nursey crops. *Mailing Add:* Alta Spec Crops & Hort Res Ctr Bag Service 200 Brooks AB T0J 0J0 Can

KRAHNKE, HAROLD C, b Beloit, Wis, Oct 12, 07; m 31; c 2. PHYSIOLOGICAL CHEMISTRY, PHARMACY. *Educ:* Univ Wis, BS, 40, MS, 41. *Prof Exp:* Pharmacist, Retail Pharm, 26-36; res asst physiol chem, Univ Wis, 40-41; res chemist, Lakeside Labs, Inc, 41-47, chief control chemist, 47-52, chief pharmaceut div, 52-62; dir pharmaceut dept, Lakeside Labs Div, Colgate-Palmolive Co, 62-73; RETIRED. *Mem:* Am Chem Soc; Am Pharmaceut Asn; Acad Pharmaceut Sci. *Res:* Pharmaceutical research and product development; formulation and manufacturing procedures; synthesis of organic medicinal compounds; quality control of pharmaceuticals. *Mailing Add:* 6770 N Yates Rd Milwaukee WI 53217

KRAHULA, JOSEPH L(OUIS), b Czech, July 22, 23; nat US; m 59; c 2. MECHANICS. *Educ:* Rensselaer Polytech Inst, BME, 46, MS, 50; Univ Ill, PhD(mech), 52. *Prof Exp:* Instr mech, Rensselaer Polytech Inst, 46-50; asst, Univ Ill, 50-52; from asst prof to assoc prof, 52-67, PROF MECH, HARTFORD GRAD CTR, RENSSELAER POLYTECH INST, 67- *Mem:* Assoc fel Am Inst Aeronaut & Astronaut; Int Asn Bridge & Struct Engrs. *Res:* Vibrations and elasticity. *Mailing Add:* Dept Mech Hartford Grad Ctr 275 Windsor St Hartford CT 06120

KRAICER, JACOB, b Toronto, Ont, Oct 28, 31; m 57; c 2. ENDOCRINOLOGY, NEUROENDOCRINOLOGY. *Educ:* Univ Toronto, BA, 54, MD, 58, PhD(physiol), 62. *Prof Exp:* Rotating intern med, Toronto Gen Hosp, 58-59; res assoc physiol, Banting & Best Dept Med Res, Univ Toronto, 61-62; from asst prof to prof physiol, Queen's Univ, Ont, 64-84; PROF & CHMN PHYSIOL, UNIV WESTERN ONT, 84- *Concurrent Pos:* Med Res Coun Can fel, Endocrinol Lab, Fac Med, Laval Univ, 62-63 & Animal Morphol Lab, Free Univ Brussels, 63-64; Med Res Coun Can scholar, 64-66; ed, Can J Physiol & Pharmacol, 81- *Honors & Awards:* Sarrazin Lectr, Can Physiol Soc, 83. *Mem:* Int Soc Neuroendocrinol; Am Physiol Soc; Endocrine Soc; Can Physiol Soc (pres, 85-86); Can Soc Endocrinol & Metab. *Res:* Mechanism and related peptides and of regulation of corticotrophin and growth hormone secretion in the adenohypophysis. *Mailing Add:* Dept Physiol Univ Western Ont London ON N6A 5C1 Can

KRAICER, PERETZ FREEMAN, b Toronto, Ont, Aug 15, 32; m 58, 80; c 5. REPRODUCTIVE PHYSIOLOGY, ENDOCRINOLOGY. *Educ:* Univ Toronto, BA, 55; Weizmann Inst Sci, Israel, PhD(reproductive physiol), 60. *Prof Exp:* Res asst exp biol, Weizmann Inst Sci, 60-62, res assoc, 62-65, sr scientist, 65-69; sr lectr zool, Tel Aviv Univ, 65-68, assoc prof, 69-76, dir, Soferman Inst Fertil, 69-82, PROF ENDOCRINOL, TEL AVIV UNIV, 77-, PROF ZOOL, 77- *Concurrent Pos:* Vis scientist, Biomed Div, Pop Coun, NY, 75-76; vis prof obstet & gynec, Univ Med & Dent, NJ, 82-83. *Mem:* Israel Soc Study Fertil (secy, 75-); Int Soc Res Reproduction; Brit Soc Study Fertil; Israel Endocrine Soc. *Res:* Comparative biology of reproduction, ovulation, fertilization, ovum implantation and decidual cell grow th and differentiation. *Mailing Add:* Dept Zool George S Wise Ctr Life Sci Life Sci Tel Aviv Univ Ramat Aviv 69978 Israel

KRAICHNAN, ROBERT HARRY, b Philadelphia, Pa, Jan 15, 28; m 54; c 1. STATISTICAL MECHANICS. *Educ:* Mass Inst Technol, BS, 47, PhD, 49. *Prof Exp:* Mem, Inst Advan Study, 49-50; mem tech staff, Bell Tel Labs, Inc, 50-52; res assoc, Electronics Res Lab, Columbia Univ, 52-56; inst math sci, NY Univ, 56-59, sr res scientist, 59-62; INDEPENDENT CONSULT, 62- *Concurrent Pos:* Consult, Naval Res Lab, 57-59 & NASA, 61-; assoc physics, Woods Hole Oceanog Inst, 60-; res affiliate, Mass Inst Technol, 63- *Mem:* Fel Am Phys Soc; Acoust Soc Am; Italian Phys Soc. *Res:* Quantum and classical statistical mechanics; random processes; turbulence, quantum field and relativity theory; acoustics. *Mailing Add:* 303 Potrillo Dr Los Alamos NM 87544

KRAIG, ELLEN, b Ft Worth, Tex, Feb 9, 53; m 78; c 1. MOLECULAR IMMUNOGENETICS. *Educ:* Univ Denver, BS, 75; Brandeis Univ, PhD(biol), 81. *Prof Exp:* Teaching fel biol, Calif Inst Technol, 80-83; ASST PROF CELL BIOL, HEALTH SCI CTR, UNIV TEX, 83- *Mem:* Am Asn Immunologists; Sigma Xi. *Res:* Use of recombinant DNA approaches to elucidate the molecular bases of immune response regulation. *Mailing Add:* 4923 E Beverly Mae San Antonio TX 78229

KRAIHANZEL, CHARLES S, b New Bedford, Mass, Sept 6, 35; m 57; c 5. INORGANIC CHEMISTRY. *Educ:* Brown Univ, ScB, 57; Univ Wis, MS, 59, PhD(chem), 62. *Prof Exp:* From asst prof to assoc prof, 62-70, PROF CHEM, LEHIGH UNIV, 70- *Mem:* AAAS; Am Chem Soc. *Res:* Syntheses, reactions, nature of bonding and physical properties, of organosilicon, organophosphorous and transition metal compounds; molecular modeling of inorganic and organometallic compounds. *Mailing Add:* Dept Chem Seeley G Mudd Bldg 6 Bethlehem PA 18015

KRAIMAN, EUGENE ALFRED, b Philadelphia, Pa, Apr 11, 29; m 56. POLYMER CHEMISTRY. *Educ:* Univ Pa, BS, 50; Univ Ill, MS, 51, PhD(chem), 53. *Prof Exp:* Res chemist, Union Carbide Plastics Co, 53-56 & Hooker Chem Co, 56-58; supvr, Plastics Div, Nopco Chem Co, 58-61; res sect head, Sun Chem Corp, 61-68; PRES, POLYMER SYSTS CORP, 68- *Mem:* Am Chem Soc. *Res:* Polyurethanes; organic chemicals; elastomers; coatings; adhesives; ultraviolet radiation cured systems. *Mailing Add:* 7922 Sandpoint Blvd Orlando FL 32819

KRAINES, DAVID PAUL, b Chicago, Ill, Mar 7, 41; m 64; c 2. TOPOLOGY. *Educ:* Oberlin Col, AB, 61; Univ Calif, Berkeley, MA, 63, PhD(math), 65. *Prof Exp:* Instr math, Mass Inst Technol, 65-67; asst prof, Haverford Col, 67-70, actg chmn, 68-69; guest prof, Aarhus Univ, 70-71; asst prof, 71-73, ASSOC PROF MATH, DUKE UNIV, 73- *Mem:* Am Math Soc. *Res:* Algebraic topology. *Mailing Add:* Dept Math Duke Univ Durham NC 27706

KRAINTZ, LEON, b Johnstown, Pa, Oct 3, 24; m 49; c 3. PHYSIOLOGY, ENDOCRINOLOGY. *Educ:* Harvard Univ, AB, 50; Rice Inst, MA, 52, PhD(biol), 54. *Prof Exp:* Res scientist exp med, Univ Tex, M D Anderson Hosp & Tumor Inst, 51-52, from instr to assoc prof physiol, Dent Br, 54-62; prof biol, Rice Univ, 62-64; from assoc prof to prof oral biol, 64-87, head dept, 69-81, EMER PROF ORAL BIOL & HON PROF PHYSIOL, UNIV BC, 87- *Concurrent Pos:* USPHS spec fel physiol, Howard Florey Inst Exp Physiol, Univ Melbourne, 68-69; asst, Rice Inst, 51-52; vis instr, Col Med, Baylor Univ, 56-63; vis lectr, Univ St Thomas, Tex, 57-63; vis prof, Dental Sci Inst, Houston, 81-82. *Mem:* Fel AAAS; Endocrine Soc; Am Physiol Soc; Can Physiol Soc; Soc Exp Biol Med; fel NSF. *Res:* Radioisotopic techniques in biology and medicine; protein hormones; mineral metabolism; salivation. *Mailing Add:* 6478 Dunbar St Vancouver BC V6N 1X6 Can

KRAITCHMAN, JEROME, b New York, NY, Mar 5, 26; m 57; c 3. INSTRUMENTATION, SURFACE SCIENCE. *Educ:* Syracuse Univ, AB, 48; Columbia Univ, AM, 50, PhD(chem physics), 54. *Prof Exp:* Asst, Columbia Univ, 49-53; res physicist, Res Labs, Westinghouse Elec Corp, 53-66; res physicist, Glass Res Ctr, PPG Industs, Inc, 66-75; supvr, Cent Res Facil, 76-80, ASST DIR, MAT RES LAB, CARNEGIE-MELLON UNIV, 80- *Mem:* Am Phys Soc. *Res:* Materials science; microwave spectroscopy; molecular structure; dielectrics; semiconductors; physics and chemistry of surfaces; thin films; adhesion. *Mailing Add:* 2409 Collins Rd Pittsburgh PA 15235

KRAJCA, KENNETH EDWARD, b Witchita Falls, Tex, April 1, 44; m 67; c 2. ADHESIVES & ADHESION, COATINGS TECHNOLOGY. *Educ:* Midwestern State Univ, BS, 67; Univ Fla, PhD(chem), 72. *Prof Exp:* Res scientist, Union Camp Corp, 72-76; res chemist, SCM Corp, 76-80, sect head, 80-84, tech mgr, 84; group leader, 84-86, DEVELOP MGR, CHEM DIV, UNION CAMP CORP, 86- *Mem:* Am Chem Soc; Adhesion Soc. *Res:* Design and develop tall oil-based products for the adhesives industry (from thermoplastic polyamides and epoxy curing agents to rosin-based tackifying resins). *Mailing Add:* 110 N Cromwell Rd Savannah GA 31410

KRAJEWSKI, JOHN J, b Chicago, Ill, Mar 27, 31; m 59; c 2. SYNTHETIC ORGANIC CHEMISTRY, POLYMER CHEMISTRY. *Educ:* Loyola Univ Ill, BS, 53, MS, 54; Carnegie Inst Technol, PhD(org chem), 58. *Prof Exp:* Res chemist, Swift & Co, 58-60, div head, 60-62; res chemist, Int Minerals & Chem Corp, 62-67, synthetic org specialist, 67-70; res chemist, De Soto, Inc, 70-73, tech mgr, 73-79, mgr polymer develop, 79-87, mgr, New Venture Res, 87-89; POLYMER & COATINGS CONSULT, 90- *Concurrent Pos:* Polymer chem instructor, De Paul Univ, 90; adv res proj selection & mgt, dir, Innovative Prob Solving Serv. *Mem:* Am Chem Soc. *Res:* Organic synthesis; polymer chemistry; natural products; plant growth regulants; fungicides; photopolymerization; organic photoconductors; chemical coatings; laser curing of resins. *Mailing Add:* 932 Valley Stream Dr Wheeling IL 60090

KRAJINA, VLADIMIR JOSEPH, b Slavice, Czech, Jan 30, 05; Can citizen; m 30; c 2. PLANT ECOLOGY, PLANT TAXONOMY. *Educ:* Charles Univ, Prague, ScD(bot), 27; FCIF, 76. *Hon Degrees:* LLD, Notre Dame Univ, 73; ScD, Univ BC, 82. *Prof Exp:* Asst, Bot Inst, Charles Univ, 25-33, docent habil plant taxon & ecol, 34-45, prof bot & head, Div Plant Sociol & Ecol, Bot Inst, 45-48; spec lectr, 49-51, from asst prof to prof, 54-73, hon prof, 73-82, EMER PROF BOT, UNIV BC, 82- *Concurrent Pos:* Res fel, Yale Univ & Univ Honolulu, 29-30 & Charles Univ & Masaryk Nat Res Coun, 37; vis prof, Univ Hawaii, 61-62; hon assoc, Bernice P Bishop Mus, 63-72; mem & chmn, BC Ecol Reserves Comt, 68-75; chmn ecol subcomt, Standing Comt Pac Bot, 71-75; mem adv comt, Ecol Reserves Act, 71-84, asn comt ecol reserves, Nat Res Coun Can, 75-84; assoc ed, Syesis, 73-84. *Honors & Awards:* George Lawson Medal, Can Bot Asn, 72; Douglas H Pimlott Award, Can Nat Fedn, 82; Can Forestry Sci Achievement, 85; Award of Merit, David Douglas Soc, 90; Order of the White Lion, Czech Govt, 90. *Mem:* AAAS; Bot Soc Am; Ecol Soc Am; Am Soc Plant Taxon; Soc Am Foresters; Can Bot Asn; fel Linnean Soc London; Am Bryol & Lichenol Soc; fel Can Inst Forestry. *Res:* Forest autecology and synecology; grassland; reforestation; ecological classification; plant taxonomy of Central Europe, Pacific North America, Western Canadian Arctic, and Hawaiian Islands; taxonomy of vascular plants, bryophytes and lichens; experimental forest tree nutrition; conservation of nature, including ecological reserves in Canada, especially British Columbia. *Mailing Add:* Dept Bot Univ BC Vancouver BC V6T 2B1 Can

KRAKAUER, HENRY, b Jaworzno, Poland, May 31, 39; US citizen; m 74; c 2. BIOMETRICS-BIOSTATISTICS. *Educ:* Yeshiva Univ, BA, 60, BHL, 60; NY Univ, MD, 64; Yale Univ, PhD(chem), 68. *Prof Exp:* From asst prof to assoc prof chem, Wash State Univ, 68-79; chief, Genetics & Transplantation Biol Br, Nat Inst Allergy & Infectious Dis, 79-83; med adv, Off Med Rev, Health Care Financing Admin, 85-88, dir, Off Prog Assessment & Info, 88-91, MED ADV, HEALTH STANDARDS & QUAL BUR, HEALTH CARE FINANCING ADMIN, 91- *Concurrent Pos:* Res assoc prof, Sch Med, Univ Md, 87- *Honors & Awards:* Commendation Medal, USPHS, 81, Outstanding Serv Medal, 88, Meritorious Serv Medal, 91; Distinguished Serv Award, Dept Health & Human Serv, 89. *Mem:* Am Soc Biochem & Molecular Biol; Am Asn Immunologists; Biophys Soc; Am Chem Soc; Biomet Soc; Sigma Xi. *Res:* Epidemiologic analysis of medical practice. *Mailing Add:* Health Standards & Qual Bur-Health Care Financing Admin 6325 Security Blvd Baltimore MD 21207

KRAKAUER, TERESA, b China; US citizen. BIOCHEMISTRY, CELL BIOLOGY. *Educ:* Wash State Univ, BSc(chem) & BSc(biochem), 71; Iowa State Univ, PhD(biochem), 75. *Prof Exp:* Res assoc biochem, Wash State Univ, 75-78; NIH staff fel biochem, Nat Inst Arthritis, Metab & Digestive Dis, Nat Inst Dent Res, 78-80, NIH STAFF FEL, 80-; MICROBIOLOGIST, DEPT PATHOGENESIS & IMMUNOL DIS ASSESSMENT DIV, US ARMY MED RES INST INFECTIOUS DIS. *Mem:* Am Chem Soc; Sigma Xi; Biophys Soc. *Res:* Molecular biology; cell surface proteins; molecular basis of immunogenicity; developmental biology; gene transfer. *Mailing Add:* Dis Assessment Div USAMRIID Ft Detrick Frederick MD 21702-5011

KRAKOFF, IRWIN HAROLD, b Columbus, Ohio, July 20, 23; m 84; c 3. MEDICINE. *Educ:* Ohio State Univ, BA, 43, MD, 47; Am Bd Internal Med, dipl. *Prof Exp:* Res fel, Sloan-Kettering Inst Cancer Res, 53-54, from asst to assoc, 54-61, from asoc mem to mem, 69-76, chief, Div Chemother Res, 70-72, chief med oncol serv, 70-74, head, Lab Clin Chemother & Pharmacol, 73-76, Mem Hosp for Cancer & Allied Dis, assoc chmn, Dept Med, attend physician & chief clin chemother serv, 74-76; dir, Vt Regional Cancer Ctr, 76-83; prof med & pharmacol, Univ Vt, 77-83; PROF MED PHARMACOL & HEAD, DIV MED, UNIV TEX M D ANDERSON CANCER CTR, 83- *Concurrent Pos:* Spec fel, Mem Hosp Cancer & Allied Dis, 53-55, clin asst, 55-58, from asst attend physician to physician, 58-76, chief, Med Oncol Serv, 69-74-; res assoc, Med Col, Cornell Univ, 55-58, from asst prof to prof med, 58-76. *Honors & Awards:* Sloan Cancer Res Award, 65. *Mem:* Harvey Soc; Am Asn Cancer Res; Am Col Physicians; Am Soc Clin Oncol; Am Soc Pharmacol & Exp Therapeut. *Res:* Cancer chemotherapy. *Mailing Add:* Anderson Cancer Ctr 1515 Holcombe Blvd Houston TX 77030

KRAKOW, BURTON, b Brooklyn, NY, Feg 12, 28. PHYSICAL CHEMISTRY, SPECTROSCOPY. *Educ:* City Col New York, BS, 49; Brooklyn Col, MA, 58; Mass Inst Technol, PhD(phys chem), 62. *Prof Exp:* Jr engr extractive metall, Metall Lab, Sylvania Elec Prod, 54-57; sr phys chemist molecular spectros, Control Instrument Div, Warner & Swasey Co, 62-69; sr prin develop engr, Honeywell Inc, 69-73; res engr, ARO Inc, 73-78; PROJ MGR ADV TECHNOL, NY STATE ENERGY RES & DEVELOP AUTHORITY, 78- *Mem:* Am Chem Soc; Am Phys Soc; Coblentz Soc. *Res:* Infrared spectroscopy; radiant heat transfer; spectroscopic pyrometry; infrared instrumentation; solar energy conversion. *Mailing Add:* NY State Energy Res & Agency Bldg 2 Empire State Plaza Albany NY 12223

KRAKOW, JOSEPH S, b New York, NY, Dec 23, 29; m 55; c 2. BIOCHEMISTRY, MOLECULAR BIOLOGY. *Educ:* Univ Mich, BS, 55; Yale Univ, PhD(pharmacol), 61. *Prof Exp:* USPHS fel biochem, NY Univ, 61-63; assoc res biochemist, Space Sci Lab, Univ Calif, Berkeley, 63-71, lectr, Dept Med Physics, 64-71; PROF BIOL SCI, HUNTER COL, 71- *Concurrent Pos:* Biochem Study Sect, NIH, 75-, chmn, 78. *Mem:* Am Chem Soc; Am Soc Biol Chem. *Res:* Nucleic acids; enzymology. *Mailing Add:* Dept Biol Hunter Col 695 Park Ave New York NY 10021

KRAKOWER, GERALD W, b Brooklyn, NY, Nov 14, 29; m 59; c 6. ORGANIC CHEMISTRY. *Educ:* Yeshiva Univ, BA, 51; Columbia Univ, MA, 53; Wayne State Univ, PhD(org chem), 58. *Prof Exp:* USPHS fel, Weizmann Inst Sci, Israel, 58-59; sr res chemist, Squibb Inst Med Res, Olin Mathieson Chem Corp, NJ, 59-68; ASSOC PROF CHEM, BAR-ILAN UNIV, ISRAEL, 68- *Mem:* Royal Soc Chem. *Res:* Structure and stereochemistry of natural products; steroid chemistry; conformational analysis. *Mailing Add:* Dept Chem Bar-Ilan Univ Ramat-Gan Israel

KRAKOWSKI, FRED, b Zuoz, Switz, July 31, 27; m 58; c 1. MATHEMATICS. *Educ:* Swiss Fed Inst Technol, DSc(math), 57. *Prof Exp:* From instr to asst prof math, Univ Calif, Davis, 57-67; assoc prof, Sacramento State Col, 67-70, PROF MATH, CALIF STATE UNIV, SACRAMENTO, 70- *Mem:* Am Math Soc. *Res:* Algebra. *Mailing Add:* Dept Math Calif State Univ-6000 J St Sacramento CA 95819

KRAL, ROBERT, b Highland Park, Ill, Feb 28, 26; m 57; c 1. PLANT TAXONOMY. *Educ:* NC State Col, BS, 52; Fla State Univ, PhD(bot), 59. *Prof Exp:* Asst bot, Fla State Univ, 55-58; instr, Northeast La State Col, 58-59; asst prof, Va Polytech Inst, 59-65; assoc prof, LA Tech, 62-65; from asst prof to assoc prof biol, 65-72, PROF BIOL, VANDERBILT UNIV, 72- *Res:* Vascular plant taxonomy; floristics of southeastern coastal plain; studies in Annonaceae, Cyperaceae and Xyridaceae; flora of Alabama and Tennessee; flora of North America. *Mailing Add:* Dept Gen Biol Vanderbilt Univ Box 1705 Sta B Nashville TN 37240

KRALL, ALBERT RAYMOND, b Eaton, Ohio, July 23, 22; m 45, 67; c 8. METABOLISM OF PB CA & MG. *Educ:* Univ Wis, BS, 50, MS, 52, PhD(biochem), 53. *Prof Exp:* Asst biochem, Univ Wis, 50-53; assoc biochemist, Biol Div, Oak Ridge Nat Lab, 53-55; assoc plant physiol, Univ Minn, 55-56; sr biochemist, Res Inst Advan Studies, Md, 56-60; from asst prof to assoc prof psychiat & biochem, Sch Med, Univ Miami, 60-65; assoc prof biochem psychiat, Sch Med, Univ NC, 65-69; prof, 69-87, EMER PROF BIOCHEM, MED UNIV SC, 87- *Concurrent Pos:* NIH career develop award, 61-65 & 66-69; vis prof, Dept Zool, Univ Col, Dublin, 79-80; Fogarty sr int fel, USPHS, 79-80; consult, Bula Mining Ltd, Dublin, 80. *Mem:* Brit Biochem Soc; Am Soc Cell Biol; Am Chem Soc; Sigma Xi; Am Soc Biochem & Molecular Biol. *Res:* Biochemistry of calcium and magnesium; calcium transport in mitochondria and its regulation; biochemistry and toxicology of lead; carbon monoxide metabolism. *Mailing Add:* Dept Biochem Med Univ SC Charleston SC 29425

KRALL, ALLAN M, b Bellefonte, Pa, Feb 25, 36; m 58; c 4. MATHEMATICS. *Educ:* Pa State Univ, BS, 58; Univ Va, MA, 60, PhD(math), 63. *Prof Exp:* From asst prof to assoc prof, 63-71, PROF MATH, PA STATE UNIV, 71- *Concurrent Pos:* Grants, NASA, 65-69 & US Air Force, 77-79. *Mem:* Math Asn Am; Am Math Soc. *Res:* Differential operators. *Mailing Add:* 845 N Thomas St State College PA 16803

KRALL, HARRY LEVERN, b York, Pa, June 13, 07; m 34, 63; c 2. MATHEMATICS. *Educ:* Gettysburg Col, BS, 27, MS, 28; Brown Univ, PhD(math), 32. *Prof Exp:* Instr math, Gettysburg Col, 27-28 & Brown Univ, 28-32; from instr to assoc prof, Pa State Col, 33-48; prof, 48-76, EMER PROF MATH, PA STATE UNIV, UNIVERSITY PARK, 76- *Mem:* Am Math Soc; Math Asn Am. *Res:* Analysis; differential equations; special functions. *Mailing Add:* 500 E Marylyn Ave H 129 State College PA 16801-4454

KRALL, JOHN MORTON, b Bellefonte, Pa, July 28, 38; div; c 2. BIOSTATISTICS, CANCER RESEARCH. *Educ:* Pa State Univ, BA, 60; Univ Iowa, MS, 62, PhD(statist), 69. *Prof Exp:* Mathematician, Comput Br, NIH, 62-65; asst prof biomet, Univ Tex MD Anderson Hosp & Tumor Inst Houston, 69-70; asst prof pub health & prev med, WVa Univ, 70-73, from assoc prof to prof biostatist, 73-82; SR BIOSTATISTICIAN, AM COL RADIOL, PHILADELPHIA, 82- *Mem:* Biomet Soc; Am Statist Asn; Soc Clin Trials. *Res:* Development of methodology for medical statistical applications; study of factors affecting survival. *Mailing Add:* Am Col Radiol 1101 Market St 14th Floor Philadelphia PA 19107

KRALL, NICHOLAS ANTHONY, b Kansas City, Kans, Feb 16, 32; m 54, 85; c 6. PHYSICS. *Educ:* Univ Notre Dame, BSc, 54; Cornell Univ, PhD(theoret physics), 59. *Prof Exp:* Mem res staff solid state physics, RCA Labs, NJ, 54; mem res staff theoret physics, John Jay Hopkins Lab Pure & Appl Sci, Gen Atomic Div, Gen Dynamics Corp, 59-67, asst mgr theory, Gen Atomic Controlled Fusion Res Prog, 67; prof physics, Univ Md, College Park, 67-73; vis res prof physics, Univ Calif, San Diego, 73-74; dir lab appl plasma studies, Sci Applns Inc, 74-78, vpres, 77-78; vpres, Jaycor, San Diego, 78-87; COFOUNDER, KRALL ASSOCS, 88- *Concurrent Pos:* Dir joint prog plasma physics, Naval Res Lab, Univ Md, 71-73; Guggenheim Found fel, 73-74; chmn, APS/DPP, 81; chmn, Fusion Power Asn, 82-84. *Mem:* Fel Am Phys Soc; Sigma Xi; AAAS. *Res:* Controlled thermonuclear fusion; plasma physics; high energy nuclear physics; electron scattering; application of dispersion relation technique to atomic physics; magnetohydrodynamics; plasma stability theory; laser system modeling; Raman cell modeling. *Mailing Add:* Krall Assocs 1070 America Way Del Mar CA 92014

KRAMAN, STEVE SETH, b Chicago, Ill, Aug 30, 44; m 71; c 3. PULMONARY MEDICINE. *Educ:* Univ PR, BS, 67, MD, 73. *Prof Exp:* Asst prof, 78-84, ASSOC PROF MED, UNIV KY, 84-; CHIEF OF STAFF, VET ADMIN MED CTR, LEXINGTON, KY, 86- *Mem:* Am Col Chest Physicians; Am Thoracic Soc; Inst Elec & Electronics Engrs; Acoust Soc Am; Am Physiol Soc. *Res:* Determination of the acoustic properties of the respiratory system including lung sounds and transmitted sounds. *Mailing Add:* Vet Admin Med Ctr 11 Cooper Dr Lexington KY 40511

KRAMER, AARON R, b New York, NY, Apr 26, 32; m 60; c 3. MECHANICAL ENGINEERING, INSTRUMENTATION. *Educ:* State Univ NY Maritime Col, BME, 54; City Col New York, MME, 63. *Prof Exp:* Appl engr instr, Bailey Meter Co, 56-63; assoc prof eng, 63-71, PROF ENG, STATE UNIV NY, MARITIME COL, 71- *Concurrent Pos:* Consult, Simulation Autodyn Inc, 67- & Stone & Webster Eng Corp, 73-; Inst Environ Sci res grant, 68-69; Maritime Admin, US Dept Commerce grant, 69-73; prof eng, NY, NJ. *Mem:* Instrument Soc Am; Am Soc Mech Engrs; Am Soc Eng Educ. *Res:* Automatic control design and analysis; simulation of mechanical, chemical processes and energy management techniques with analogue, digital and hybrid computers; instrumentation and data collection systems and analysis. *Mailing Add:* Dept Engr State Univ NY Maritime Col Fort Schuyler New York NY 10465

KRAMER, ALFRED WILLIAM, JR, b Astoria, NY, Jan 19, 30; m 59; c 2. MORPHOLOGICAL BIOLOGY. *Educ:* Fordham Univ, BS, 56, MS, 59. *Prof Exp:* Jr pharmacologist cardiovasc physiol, Res Labs, Burroughs Wellcome & Co, 56-59; RES BIOLOGIST EXP PATH, LEDERLE LABS, AM CYANAMID CO, 59- *Concurrent Pos:* Instr, Bronx Community Col, 61-72, asst prof, 72- *Res:* Cardiovascular pharmacology of anti-arhythmia compounds; normal and pathological morphology and technical data of laboratory animals; development of practical techniques for use in experimental pathology. *Mailing Add:* 206 Lake Rd Valley Cottage NY 10989

KRAMER, BARNETT SHELDON, b Baltimore, Md, July 29, 48; m 72; c 1. MEDICAL ONCOLOGY, CANCER PREVENTION-PUBLIC HEALTH. *Educ:* Univ Md, MD, 73. *Prof Exp:* Intern, Washington Univ, 73-74, resident, 74-75; from asst prof to assoc prof med oncol, Univ Fla, 78-86; clin assoc oncol, Nat Cancer Inst, 75-78; assoc prof, 86-89, PROF, UNIFORMED SERV UNIV HEALTH SCI, 89-; SR INVESTR MED ONCOL, NAT CANCER INST, 86- *Mem:* Am Asn Cancer Res; Am Soc Clin Oncologists; Asn Clin Trials. *Res:* Infections in febrile neutropenic cancer patients; lung cancer; new drug development; cancer prevention and control. *Mailing Add:* Early Detection & Community Oncol Prog Nat Cancer Inst Bethesda MD 20892

KRAMER, BERNARD, b New York, NY, Nov 12, 22; m 46; c 2. SOLID STATE PHYSICS. *Educ:* City Col New York, BS, 42; NY Univ, PhD(physics), 52. *Prof Exp:* Physicist, Fed Tel & Radio Corp, 42-47; instr, Brooklyn Col, 47-49; res asst, NY Univ, 49-52, res assoc, 52-65, res scientist, 65-69; lectr, Hunter Col, City Univ New York, 53-55, from asst prof to prof physics, 55-86, chmn dept physics & astron, 60-71; ADJ PROF, FAIRLEIGH DICKINSON UNIV, 86- *Concurrent Pos:* Vis prof, Munich Technol Univ, 59-60; vis fel, Princeton Univ, 73-74; sci collab, Brookhaven Nat Lab, 80-81; vis scholar, Univ Del, 81. *Mem:* Am Phys Soc. *Res:* Luminescence; photoconductivity; photovoltaic effects. *Mailing Add:* 115 Carnation St Bergenfield NJ 07621

KRAMER, BRADLEY ALAN, b Manhattan, Kans, May 30, 58; m 78; c 3. PRODUCTION PLANNING & CONTROL, ARTIFICIAL INTELLIGENCE APPLICATIONS IN MANUFACTURING. *Educ:* Kans State Univ, BS, 80, MS, 81, PhD(indust eng), 85. *Prof Exp:* Asst prof indust eng, 85-91, assoc dir res, Ctr Res Computer Controlled Automation, 88-90, ASSOC PROF INDUST ENG, KANS STATE UNIV, 91- *Concurrent Pos:* Dow outstanding young fac award, Am Soc Eng Educ, 90. *Mem:* Soc Mfg Engrs; Am Soc Eng Educ; Inst Indust Engrs. *Res:* Manufacturing engineering; integration of the production process from product design to manufacturing to assembly with focus in production planning, scheduling and control. *Mailing Add:* Dept Indust Eng Kans State Univ 227 Durland Manhattan KS 66506-5101

KRAMER, BRIAN DALE, b Pottsville, Pa, Nov 17, 42; m 67. PHYSICAL ORGANIC CHEMISTRY. *Educ:* Pa State Univ, BS, 64; Harvard Univ, PhD(chem), 68. *Prof Exp:* Res chemist, Cent Res Div, 68-73, sr res chemist, 73-76, res supvr, New Enterprises Res Div, 76-77, res scientist, Chem Sci Div & Mat Sci Div, 77-82, proj leader paper chem, 80-87, res assoc, Mat Sci Div, 82-87, dir labs, Res Ctr, 87-89, VPRES TECHNOL, PAPER TECHNOL GROUP, HERCULES INC, 89- *Mem:* AAAS; Am Chem Soc; Tech Asn Pulp Paper Indust. *Res:* Kinetics and mechanisms of 1, 2 cycloaddition reactions; photochemical generation of reactive intermediates; Ziegler polymerization of alpha-olefins; physical and chemical characterization of organic polymers; radiation chemistry; chemical additives for paper. *Mailing Add:* Three Boysenberry Dr Ramsey Ridge Hockessin DE 19707

KRAMER, BRUCE MICHAEL, b New York City, NY, July 23, 49. MECHANICAL ENGINEERING. *Educ:* Mass Inst Technol, SB & SM, 72, PhD(mech eng), 79. *Prof Exp:* From asst prof to assoc prof mech eng, Mass Inst Technol, 79-85; PROF MECH ENG, GEORGE WASHINGTON UNIV, 85- *Concurrent Pos:* Chmn, Zoom Telephonics, Inc, Boston, 76-; consult, 79- *Honors & Awards:* Blackall Award, Am Soc Mech Eng, 82; F W Taylor Medal, Int Inst Prod Eng Res, 84; R F Bunshah Award, Int Conf Metal Coating, 86. *Mem:* Am Soc Mech Engrs; Am Soc Metals; Sigma Xi. *Res:* Machining automation and tool material development; tribology and wear theory. *Mailing Add:* Dept Civil & Environ Eng George Washington Univ Washington DC 20052

KRAMER, CAROLYN MARGARET, b Chicago, Ill, Mar 12, 53. HIGH TECHNOLOGY CERAMICS. *Educ:* Univ Ill, Champaign, BS, 74; Univ Calif, Berkeley, MS, 75; Univ Calif, Davis, PhD(mat sci & eng), 80. *Prof Exp:* Staff mem, Sandia Nat Lab, 73-81; postdoctoral, Nat Acad Sci, Nat Res Coun, 81-82; ceramic engr, Naval Res Lab, 82-83; mat scientist, Advan Technol Lab, Gillette, 83-84, mgr & group mgr mat res, Boston Res & Develop Lab, 84-89; DIR RES & DEVELOP, CERAMCO, INC, 89- *Mem:* AAAS; Soc Women Engrs; corp mem Am Ceramic Soc; Int Asn Dent Res; Am Soc Metals. *Res:* Uses of high technology ceramics. *Mailing Add:* 312 Highland Ave Moorestown NJ 08057

KRAMER, CHARLES EDWIN, b Lancaster, Pa, Apr 1, 47; m 69; c 2. POLYMER SYNTHESIS. *Educ:* Franklin & Marshall Col, BS, 69; Northeastern Univ, MS, 71, PhD(chem), 75. *Prof Exp:* Res chemist, Celanese Corp, 74-77; sr res assoc, 77-81, ASST DIR RES, ALBANY INT CORP, 81- *Mem:* Am Chem Soc; Sigma Xi. *Res:* Polymer chemistry; monomer and polymer synthesis; polymer flammability; membrane science; polymer blends; polymer structure-property relationships. *Mailing Add:* Kramer E-C 19 Churchill Rd Franklin MA 02038

KRAMER, CHARLES LAWRENCE, b Leavenworth, Kans, Apr 4, 28; m 51; c 2. MYCOLOGY. *Educ:* Univ Kans, BA, 50, MS, 53, PhD(bot), 57. *Prof Exp:* Asst prof biol, Western Ill Univ, 57-58; from asst prof to assoc prof, 58-73, PROF BOT, KANS STATE UNIV, 73- *Concurrent Pos:* Grants, USPHS, 58- & NSF, 67- *Mem:* Am Mycol Soc; Int Asn Plant Taxon; Brit Mycol Soc. *Res:* Kansas fungi, especially parasitic forms; taxonomy of Taphinales; aeromycology. *Mailing Add:* Ackert Hall Kans State Univ Manhattan KS 66506

KRAMER, CLYDE YOUNG, statistics; deceased, see previous edition for last biography

KRAMER, DAVID, metallurgy, for more information see previous edition

KRAMER, DAVID BUCKLEY, b Turtle Creek, Pa, Oct 21, 27; m 76; c 2. MECHANICAL ENGINEERING, COMPUTER SCIENCE. *Educ:* Univ Pittsburgh, BS, 50; Carnegie-Mellon Univ, MS, 53; Univ Md, PhD(mech eng), 75. *Prof Exp:* Res engr analog comput, Westinghouse Elec Corp, 50-52; res engr process control, E I du Pont de Nemours & Co, Inc, 53-61; RES ENGR OPERS RES, ORI, INC, 61- *Mem:* Am Soc Mech Engrs; Opers Res Soc Am. *Res:* Acoustical holography; nonlinear programming; nuclear weapons effects; inventories; scheduling. *Mailing Add:* ARC Prof Serv Group 5501 Backlick Rd Springfield VA 22151

KRAMER, EARL SIDNEY, b Chippewa Falls, Wis, Nov 13, 40. MATHEMATICS. *Educ:* Wis State Univ, Eau Claire, BS, 62; Univ Mich, MS, 64 & 66, PhD(math), 69. *Prof Exp:* Temp lectr, Univ Birmingham, 69-70; from asst prof to assoc prof, 70-82, PROF MATH, UNIV NEBR, 82- *Mem:* Am Math Soc; Math Asn Am. *Res:* Existence of various combinatorial structures. *Mailing Add:* Dept Math Univ Nebr Lincoln NE 68588-0323

KRAMER, EDWARD J(OHN), b Wilmington, Del, Aug 5, 39; m 63; c 2. MATERIALS SCIENCE. *Educ:* Cornell Univ, BChE, 62; Carnegie-Mellon Univ, PhD(metal & mat sci), 67. *Prof Exp:* NATO fel metall, Oxford Univ, 66-67; from asst prof to prof, 67-88, SAMUEL B ECKERT PROF MATH SCI & ENG, CORNELL UNIV, 88- *Concurrent Pos:* Vis scientist, Argonne Nat Lab, 74-75; Gauss prof, Acad Wissenschaften, Göttingen, 79; vis prof, Polytech Sch Fed Lausanne, 82. *Honors & Awards:* High Polymer Physics Prize, Am Phys Soc, 85. *Mem:* Nat Acad Eng; Am Chem Soc; Mat Res Soc; Soc Plastic Engrs; AAAS; Am Phys Soc. *Res:* Surfaces, interfaces, diffusion, deformation and fracture of polymeric materials. *Mailing Add:* Dept Mat Sci & Eng Cornell Univ Ithaca NY 14853

KRAMER, ELIZABETH, b Milwaukee, Wis, June 7, 18. DIETETICS, BIOCHEMISTRY. *Educ:* Alverno Col, BSE, 43; De Paul Univ, MS, 48; St Louis Univ, PhD(chem), 54; Mt Mary Col, BS, 81. *Prof Exp:* Instr biol & math, high sch, Ill, 43-47; instr biol, Alverno Col, 47-49, assoc prof chem & chmn dept, 54-69, prof chem, 69-79; dietitian, Vet Admin Med Ctr, Milwaukee, Wis, 81-85; CONSULT DIETITIAN, VILLA CLEMENT NSG HOME, MILWAUKEE, 85 - *Concurrent Pos:* Fel biochem, Univ Iowa, 69-70. *Mem:* Am Dietetic Asn; Sigma Xi; Soc Nutrit Educ. *Res:* Studies of drug binding, especially aspirin, salicylates and D-tubocurarine to purified proteins of human blood as well as to serum and plasma using fluorometric and gel filtration techniques. *Mailing Add:* 9020 W Morgan Ave Milwaukee WI 53228

KRAMER, ELMER E, b New Orleans, La, Sept 22, 14; m 47; c 2. PATHOLOGY. *Educ:* Tulane Univ, BS, 35, MD, 38; Am Bd Obstet & Gynec, dipl. *Prof Exp:* Intern, Hotel Dieu Hosp, New Orleans, La, 38-39, resident, 39-40; asst obstet & gynec, Med Col, Cornell Univ, 46-49, from instr to prof obstet & gynec, 49-70, prof clin path, 69-85, prof obstet & gynec, 70-85; assoc dir, 79-85, EMER PROF CLIN PATH & OBSTET & GYNEC & HON CONSULT, NY HOSP, 85- *Concurrent Pos:* Intern, NY Lying-In-Hosp, 46, from asst resident to resident, 46-50, asst pathologist, 50-56, pathologist, 56-; asst attend obstetrician & gynecologist, NY Hosp, 50-53, attend pathologist, 69-; consult, Payne Whitney Psychiat Clin, 52-56, from assoc attend obstetrician & gynecologist to attend obstetrician & gynecologist, 53-57; consult, Lenox Hill Hosp, 74-79; Am Bd Qual Assurance & Utilization Rev Physicians, 78-; Am Col Utilization Rev Physicians, 78- *Mem:* AMA; fel Am Col Surgeons; Am Col Obstetricians & Gynecologists. *Res:* Obstetrical and gynecological pathology. *Mailing Add:* NY Hosp 525 E 68 St New York NY 10021

KRAMER, FRANKLIN, b Brooklyn, NY, Mar 6, 23; m 51; c 2. PROCESS & EQUIPMENT DEVELOPMENT. *Educ:* City Col New York, BS, 44; Polytech Inst Brooklyn, MS, 47. *Prof Exp:* Res chem engr, Cent Res Labs, Gen Foods Corp, 44-53, tech supt, Atlantic Gelatin Div, 53-59; mgr res & develop, Walter Baker Chocolate Co, 59-62 & Cracker Jack Co Div, Borden Co, 62-65; mgr equip & process develop, Kitchens of Sarah Lee, 65-68; vpres mfg & eng, La Touraine-Bickford's Foods Inc, 68-73; vpres mfg, Seapak Div, W R Grace & Co, 73-75; prin scientist process eng, Cent Res Div, Gen Foods Corp, 75-88; PRES, FREMARK CO, FOOD PROCESS CONSULTS, 88- *Concurrent Pos:* Adj prof food sci, Rutgers Univ, 88-; chmn, Tech Achievement Awards, Gen Foods Corp. *Mem:* Sr mem Am Chem Soc; fel Am Inst Chem Engrs; Inst Food Technol; Sigma Xi. *Res:* Development of food processes from lab bench scale through commercialization; technical management and teaching. *Mailing Add:* 132 Holbrook Rd Briarcliff Manor NY 10510

KRAMER, FRED RUSSELL, b New York, NY, July 7, 42; m 65; c 2. MOLECULAR BIOLOGY, BIOCHEMISTRY. *Educ:* Univ Mich, BS, 64; Rockefeller Univ, PhD(molecular biol), 69. *Prof Exp:* Am Cancer Soc fel, Inst Cancer Res, Col Physicians & Surgeons, Columbia Univ, 69-71, res assoc molecular biol, 71-72, instr, 72-73, asst prof, human genetics & develop, 73-80, sr res assoc, 80-83, res scientist, 83-86; MEM & CHMN, DEPT MOLECULAR GENETICS, PUB HEALTH RES INST, 86-; RES PROF, DEPT MICROBIOL, NY UNIV MED SCH, 87- *Concurrent Pos:* Adv panalist, Nat Sci Found Biochem, 85; prin investr, NIH grant, 84-, Am Cancer Soc grant, 84-86, 89, Am Cancer Soc grant, 84-86, NSF grant, 87-90. *Mem:* NY Acad Sci; Am Soc Molecular Biol & Biol Chem; Am Soc Microbiologists; Am Asn Univ Prof; Sigma Xi. *Res:* Evolution and synthesis of nucleic acids in vitro; computer applications in molecular biology; morphology, physiology, genetics and evolution of replicatable nucleic acids. *Mailing Add:* Pub Health Res Inst 455 1st Ave New York NY 10016

KRAMER, GEORGE MORTIMER, b Brooklyn, NY, May 15, 29; m 51; c 2. PHYSICAL ORGANIC CHEMISTRY. *Educ:* Queen's Col NY, BS, 51; Univ Pa, MS, 55, PhD(phys chem), 57. *Prof Exp:* Chemist, Frankford Arsenal, Pa, 51-52; chemist, Process Res Div, Esso Res & Eng Co, 57-64, Baytown Res & Develop Div, Tex, 65-66; res assoc, Cent Basic Res Lab, 66-69, res assoc, 69-78, SR RES ASSOC, CORP RES LAB, EXXON RES & ENG CO, NJ, 78- *Mem:* Am Chem Soc. *Res:* Pressure-volume-temperature behavior of gases; equation of state; surface chemistry; thermochemical data; acid catalyzed alkylation and isomerization; catalysis; hydride transfer reactions; free radical reactions; carbonium ion rearrangement mechanisms and acid characterization; uranium chemistry. *Mailing Add:* 36 Arden Ct Berkeley Heights NJ 07922

KRAMER, GERALD M, b Gloucester, Mass, June 11, 22; m 45; c 2. PERIODONTOLOGY, DENTISTRY. *Educ:* Tufts Univ, DMD, 44; Reisman Clin, cert, 52; Am Bd Periodont, dipl. *Prof Exp:* Instr periodont, Sch Dent Med, Univ Pa, 55-60; from asst prof to assoc prof, 63-68, PROF PERIODONT & CHMN DEPT, SCH GRAD DENT, BOSTON UNIV, 68- *Concurrent Pos:* Head periodont sect, Reisman Dent Clin, Beth Israel Hosp, Boston & Boston Univ Hosp, 68- *Mem:* AAAS; Am Dent Asn; fel Am Col Dent; Am Acad Periodont; hon mem Periodont Soc SAfrica. *Res:* Clinical periodontics and oral medicine. *Mailing Add:* 90 Humphrey Swampscott MA 01907

KRAMER, GISELA A, b Nov 29, 36; wid. REGULATION OF PROTEIN SYNTHESIS, PHOSPHORYLATION. *Educ:* Univ Keil, Ger, PhD(biochem), 67. *Prof Exp:* RES SCIENTIST, UNIV TEX, AUSTIN, 74- *Mem:* Am Soc Biochem & Molecular Biol. *Mailing Add:* Dept Chem & Biochem Clayton Found Biochem Inst Univ Tex Austin TX 78712

KRAMER, HENRY HERMAN, b New York, NY, Aug 19, 30; m 59; c 3. NUCLEAR MEDICINE, DIAGNOSTIC MEDICINE. *Educ:* Columbia Univ, BA, 52, MA, 53; Univ Ind, PhD(phys chem), 60. *Prof Exp:* Res chemist nuclear methods anal, Union Carbide Corp, Nuclear Res Ctr, 60-65, group leader, Nucleonics Res & Develop, 66-73, sr group leader, Tarrytown Tech Ctr, 73-76, mgr, Sterling Forest Res Ctr, 76-78; vpres res & develop, Medi

Physics Inc, 78-89; CONSULT, 90- *Mem:* AAAS; Am Chem Soc; Am Nuclear Soc; Soc Nuclear Med; Am Col Nuclear Physicians. *Res:* Radiochemicals; radiodiagnostics; nuclear methods of analysis; clinical microbiology; educational aids; biomedical significance of trace elements; nucleonics in industry and ore body exploration; diagnostic medicine. *Mailing Add:* 3911 Campolindo Dr Moraga CA 94556

KRAMER, IRVIN RAYMOND, b Baltimore, Md, Sept 18, 12; m 35. METALLURGY. *Educ:* Johns Hopkins Univ, BS, 35, MS, 47, DE, 51. *Prof Exp:* Chemist, Am Radiator & Standard Sanit Mfg Corp, Md, 35-38; metallurgist & chief spec alloys sect, Naval Res Lab, 38-46, metallurgist, phys scientist & head mech & mat br, Off Naval Res, 46-51; asst to pres, Horizons Titanium Corp, NY, 51-53; vpres, Mercast Corp, NY, 53-55; chief mat res, Martin Co, 55-68 & Metals Sci Res Dept, Martin Marietta Corp, 68-71, mgr independent progs, 71-74; tech adv, Naval Ship Res & Develop Ctr, 75-81; RES PROF, UNIV MD, 81- *Concurrent Pos:* Mem, heat resistant mat comt, Nat Adv Comt Aeronaut, metall panel, Res & Develop Bd, Dept Nat Defense & comt ship steel, Nat Res Coun, 44; pres, Severn Tech Soc, 78. *Mem:* Fel Am Soc Metals; Am Inst Mining, Metall & Petrol Engrs; Sigma Xi; Severn Tech Soc. *Res:* Hardenability of steels; effects of alloying elements on mechanical and physical properties of metals; iron-manganese-nickel alloys; age-hardening steels; dynamic behavior of metals and phenomena of plastic deformation; surface effect as related to flow and fracture of metals; fatigue; stress corrosion; high temperature behavior of metals. *Mailing Add:* 3031 Fallstaff Rd Apt 502 Baltimore MD 21209

KRAMER, J DAVID R, JR, b Bayonne, NJ, Oct 29, 35; m 66; c 3. ELECTRICAL ENGINEERING. *Educ:* Univ Pa, BSEE, 57; Mass Inst Technol, MS, 58, ScD(elec eng), 64. *Prof Exp:* Mem tech staff, Mitre Corp, 64-69, group leader, 69-79, assoc dept head, 79-84; CONSULTING SCIENTIST, 84- *Concurrent Pos:* Lectr, Northeastern Univ, 65-68. *Mem:* Inst Elec & Electronics Engrs. *Res:* Systems optimization; signal design and processing; operations research. *Mailing Add:* 26 Fairbanks Rd Lexington MA 02173

KRAMER, JAMES M, MOLECULAR BIOLOGY. *Prof Exp:* ASSOC PROF LAB MOLECULAR BIOL, UNIV ILL, 90- *Mailing Add:* Lab Molecular Biol M-C 067 Univ Ill PO Box 4348 Chicago IL 60680

KRAMER, JAMES RICHARD, b Marine City, Mich, Oct 27, 31; m 55; c 4. GEOCHEMISTRY, AQUATIC CHEMISTRY. *Educ:* Mass Inst Technol, BS, 53; Univ Mich, MS, 54, PhD(geol), 58. *Prof Exp:* Asst, Res Lab, Carter Oil Co, 54; instr geol, Univ Mich, 57-58; fel, Nat Res Coun Can, Western Ont Univ, 58-59, lectr, 59-61, asst prof, 61-63; res assoc, Univ Mich, 63-64; from asst prof to assoc prof, Syracuse Univ, 64-68; assoc prof, 68-71, PROF GEOL, MCMASTER UNIV, 71- *Mem:* Am Geol Soc; Mineral Soc Am; Geochem Soc; Am Asn Petrol Geologists; Am Chem Soc. *Res:* Physical chemistry of carbonate minerals; limnological investigation of the Great Lakes; sedimentation and facies analysis of the Proterozoic sediments of Canada; aquatic chemistry of shield lakes. *Mailing Add:* Dept Geol McMaster Univ Hamilton ON L8S 4M1 Can

KRAMER, JERRY MARTIN, b Bronx, NY, Dec 16, 42; m 70; c 2. CHEMICAL PHYSICS. *Educ:* Univ Calif, Berkeley, BS, 65; Univ Chicago, PhD(chem), 71. *Prof Exp:* Res assoc chem, Case Western Reserve Univ, 71-72; MEM TECH STAFF, GTE LABS, 72- *Mem:* Am Phys Soc; Am Chem Soc. *Res:* Arc and discharge physics optogalvanic spectroscopy laser chemistry; low energy ion-electron excitation of inorganic phosphors; photodissociation of gaseous ions; chemical kinetics. *Mailing Add:* GTE Labs 40 Sylvan Rd Waltham MA 02254

KRAMER, JOHN J(ACOB), b Pittsburgh, Pa, July 9, 31; m 56; c 3. PHYSICAL METALLURGY, MATERIAL SCIENCE. *Educ:* Carnegie Inst Technol, BS, 53, MS & PhD(metall), 56. *Prof Exp:* Sr engr, magnetic mat develop sect, Westinghouse Elec Corp, 56-60, res metallurgist, res lab, 60-63, supv metallurgist, 63-64, sect mgr, 64-65, adv metallurgist, 65; assoc prof elec eng, 65-69, PROF ELEC ENG, UNIV DEL, 69- *Concurrent Pos:* Phys metallurgist, ballistics res lab, Aberdeen Proving Grounds, Md, 57. *Mem:* Am Soc Metals; Am Inst Mining, Metall & Petrol Engrs; Sigma Xi. *Res:* Application of thermodynamics; surfaces; grain and crystal growth; solid state reactions; magnetic and electrical properties of solids. *Mailing Add:* 410 Arbour Dr Arbour Park Newark DE 19713

KRAMER, JOHN KARL GERHARD, b Bololo, Congo, Oct 6, 39; Can citizen; m 68; c 1. BIOCHEMISTRY, ORGANIC CHEMISTRY. *Educ:* Univ Man, BSc, 63, MSc, 65; Univ Minn, Minneapolis, PhD(biochem), 68. *Prof Exp:* Hormel fel, Univ Minn, Austin, 68-70; Nat Res Coun Can fel, Univ Ottawa, 70-71; RES SCIENTIST, ANIMAL RES CTR, RES BR, AGR CAN, OTTAWA, 71- *Concurrent Pos:* Eastman Kodak fel, 64. *Honors & Awards:* CSP Canola Res Award, 84. *Mem:* Am Oil Chemists Soc. *Res:* Lipid chemistry, biochemistry and nutrition; pesticide metabolism in animals. *Mailing Add:* Animal Res Ctr Res Br Agr Can Ottawa ON K1A 0C6 Can

KRAMER, JOHN MICHAEL, b Nov 22, 41; US citizen. CONTINUUM MECHANICS, RHEOLOGY. *Educ:* Univ Wis, BS, 63, MS, 64, PhD(eng mech), 69. *Prof Exp:* Asst prof mech eng, Lamar Univ, 69-71, 72-74; res fel, Rheology Res Ctr, Univ Wis, 71-72; MECH ENGR, ARGONNE NAT LAB, 74- *Concurrent Pos:* Guest worker, Nat Bur Standards, US Dept Com, 73 & 74. *Res:* Engineering problems requiring an interdisciplinary approach in the general areas of continuum mechanics, materials behavior and heat transfer; behavior of materials and structures in hostile stress, temperature and irradiation environments. *Mailing Add:* Reactor Eng Div Bldg 207 Argonne Nat Lab 9700 S Cass Ave Argonne IL 60439

KRAMER, JOHN PAUL, b Elgin, Ill, Mar 13, 28; div; c 2. INSECT PATHOLOGY. *Educ:* Beloit Col, BS, 50; Univ Mo, MS, 52; Univ Ill, PhD(entom), 58. *Prof Exp:* Asst res prof entom, NC State Col, 58-59; asst entomologist econ entom, Ill Natural Hist Surv, 59, assoc entomologist, 59-65; from assoc prof to prof, 65-90, EMER PROF INSECT PATH, CORNELL UNIV, 90- *Concurrent Pos:* NIH res grant, 59-72; lectr, 8th Int Cong Microbiol, Montreal, 62, dept biol sci, Northwestern Univ, 64, 2nd Int Conf Protozool, London, 65, Dept Biol, Ithaca Col, 81, Dept Biol, State Univ NY, Cortland, 83, Col Biol Sci, Ohio State Univ, 84; consult, Environ Biol Unit, WHO, Geneva, 62-; mem trop med & parasitol study sect, NIH, 66-69; NSF vis insect pathologist, Japan, 67; NSF, fel, 67; mem, Eval Panel Life Sci, Nat Res Coun, 69-; NIH-Off Naval Res res grant microbiol, 71-74; vis biologist, Arctic Health Res Ctr, Inst Arctic Biol, Alaska, 72 & WHO res agreement, 78-; res grant, WHO, 79-82, US Dept Agr, 80-81; vis prof entom, Ohio State Univ, Columbus, 84. *Mem:* Soc Invert Path; NY Entom Soc. *Res:* Infectious diseases of insects, especially those caused by microsporidians and entomophthorans; ecology of microsporidians; epidemiology of diseases of insects. *Mailing Add:* Dept Entom 3142 Comstock Hall Cornell Univ Ithaca NY 14853

KRAMER, JOHN WILLIAM, b Dearborn, Mich, Aug 17, 35; m 59; c 2. CLINICAL PATHOLOGY. *Educ:* Mich State Univ, BSc, 58, DVM, 60, MSc, 68; Univ Calif, Davis, PhD(comp path), 72; Am Col Vet Pathologists, dipl & cert vet clin path. *Prof Exp:* Vet, NZ Dept Agr, 60-64; adv clin path, Mich State Univ, Nsukka, Nigeria, 64-66, asst instr, Univ, 66-68; trainee, Col Vet Med, Univ Calif, Davis, 68-72; assoc prof, 72-77, PROF VET CLIN SURG & MED, COL VET MED, WASH STATE UNIV, 77- *Mem:* Am Vet Med Asn; Am Soc Vet Clin Pathologists; Am Col Vet Pathologists; Am Asn Clin Chem. *Res:* Pathophysiology of diabetes mellitus. *Mailing Add:* Dept Vet Clin Surg & Med Col Vet Med Wash State Univ Pullman WA 99164-6610

KRAMER, KARL JOSEPH, b Evansville, Ind, Aug 20, 42; m 66; c 2. ENTOMOLOGY, BIOCHEMISTRY. *Educ:* Purdue Univ, BS, 64; Univ Ariz, PhD(chem), 71. *Prof Exp:* Res assoc biochem, Univ Chicago, 71-74; RES CHEMIST BIOCHEM, USDA, 74- *Concurrent Pos:* NIH fel, 71; from asst prof to assoc prof biochem, Kans State Univ, 74-82, prof, 82- *Mem:* Am Soc Biol Chemists; Am Chem Soc; Entom Soc Am; Am Inst Biol Sci; AAAS. *Res:* Insect biochemistry; endocrinology; physiology. *Mailing Add:* US Grain Mkt Res Lab 1515 College Ave Manhattan KS 66502

KRAMER, MARTIN A, b Ellenville, NY, Oct 20, 41; m 63; c 2. HEAVY ION PHYSICS. *Educ:* Columbia Univ, BA, 63, MA, 64, PhD(physics), 69. *Prof Exp:* Res assoc physics, Enrico Fermi Inst, Univ Chicago, 69-71; physicist, Brookhaven Nat Lab, 71-73; PROF PHYSICS, CITY COL NEW YORK, 73- *Mem:* Am Phys Soc; Am Asn Avian Pathologists. *Res:* Designing, building, running and analyzing experiments in high energy and heavy ion physics. *Mailing Add:* Brookhaven Lab Bldg 510A Upton NY 11973-5000

KRAMER, MILTON, b Chicago, Ill, Nov 11, 29; c 4. PSYCHIATRY. *Educ:* Univ Ill, BS, 50, BS, 52, MD, 54; Am Bd Psychiat & Neurol, dipl & cert psychiat, 61. *Prof Exp:* Assoc dir res, Dept Psychiat, 72-80, PROF PSYCHIAT, SCH MED, UNIV CINCINNATI, 72-, DIR, DREAM & SLEEP LAB & SLEEP DISORDERS CTR, 80- *Concurrent Pos:* Pvt practr psychiat, Cincinnati, 60-; clinician psychiat, Outpatient Dept, Cincinnati Gen Hosp, 61- & attend staff psychiatrist, 65-; consult, Coun Drugs, AMA, 62-; dir psychiat res, Vet Admin Hosp, Cincinnati, 63- & asst chief, Dept Psychiat, 64-80; Upjohn Co grant, 70-72; mem, Therapeut Care Comt, Group Advan Psychiat, 70-; res investr, Wm S Merrell Co, 70-72 & Upjohn, 70-; proj dir sonic boom res data anal, Fed Aviation Admin, 70-72; prin investr, US Vet Admin, 75-; mem, Ohio Ment Health & Ment Retardation Adv Bd, 71-75; mem staffs, Christian R Holmes Hosp, Jewish Hosp & Good Samaritan Hosp. *Mem:* Fel Am Psychiat Asn; AMA; Asn Psychophysiol Study Sleep (mem exec comt, 71); sci assoc Am Acad Psychoanal; Am Col Psychiat; Sigma Xi. *Res:* Psychology and psychophysiology of dreaming; drugs and sleep. *Mailing Add:* 3900 Rose Hill Ave Apt 1102B Cincinnati OH 45229

KRAMER, MORTON, b Baltimore, Md, Mar 21, 14; m 39; c 4. BIOSTATISTICS, EPIDEMIOLOGY. *Educ:* Johns Hopkins Univ, AB, 34, ScD, 39. *Prof Exp:* Asst biostatist, Sch Hyg & Pub Health, Johns Hopkins Univ, 37-38; instr prev med, Col Med, NY Univ, 38; statistician, State Dept Health, NY, 39-40; asst prof biostatist, Sch Trop Med, Univ PR & statistician, Insular Dept Health, San Juan, 40-42; econ analyst, US Dept Treas, DC, 42-43; assoc biostatist, Sch Med, Western Reserve Univ, 43-46; chief info & res, Off Int Health Rels, USPHS, DC, 46-49, chief biomet br, NIMH, 49-75, dir div biometry & epidemiol, 75-76; prof dept ment hyg, 76-84, EMER PROF, SCH HYG & PUB HEALTH, JOHNS HOPKINS UNIV, 84- *Concurrent Pos:* Consult ment health unit, WHO, 59-; mem expert panel health statist, WHO, 61-84; vis scientist, Dept Pub Health Admin, London Sch Hyg & Trop Med & Soc Med Res Unit Med Res Coun, Eng, 68-69; mem, Task Panel, President's Comn Ment Health, 77-78, Adv Comn Secy Health on Health Res Studies Three Mile Island Disaster, 79-85, Selection Panel Multidisciplinary Dept Geriatrics & Gerontol, Govt Ont, Can, 87. *Honors & Awards:* Superior Serv Award, Dept HEW, 62, Distinguished Serv Award, 74; Rema Lapousse Award, Am Pub Health Asn, 73; WHO Health for All Medal, 87. *Mem:* Inst Med Nat Acad Sci; fel Am Pub Health Asn; fel Am Statist Asn; fel Am Orthopsychiat Asn; Am Epidemiol Soc; Am Col Epidemiol; fel Am Psychiat Asn. *Res:* Epidemiology of mental disorders; application of biostatistical and epidemiologic methods to planning mental health and related human services and evaluating their effectiveness; classification of mental disorders. *Mailing Add:* Dept Ment Hyg Johns Hopkins Univ Sch Hyg & Pub Health Baltimore MD 21205

KRAMER, NICHOLAS WILLIAM, agronomy, genetics, for more information see previous edition

KRAMER, NOAH HERBERT, b New York, NY, Apr 10, 24; m 54; c 3. ELECTRICAL ENGINEERING. *Educ:* Mich State Univ, BS, 47, MS, 49, PhD(elec eng), 51. *Prof Exp:* Asst, Mich State Univ, 47-49, instr, 49-51; engr, Int Bus Mach Corp, 51-57 & Stelma, Inc, 57-70; mgr transmission planning, Int Tel & Tel Corp, 73-78; pres, N H Kramer & Assocs, 78-79; mem staff, Am Satellite Corp, 79-80; TECH MGR, LITTON DATA COMMAND SYSTS, 80- *Mem:* Inst Elec & Electronics Engrs. *Res:* Communication systems and equipment; digital data transmission; digital techniques; solid state electronics; computer peripherals. *Mailing Add:* 24410 Victory Blvd Woodland Hills CA 91367

KRAMER, NORMAN CLIFFORD, b New York, NY, Aug 16, 28; m 54; c 5. INTERNAL MEDICINE, IMMUNOLOGY. *Educ:* The Citadel, BS, 48; George Washington Univ, MS, 50, MD, 54; Am Bd Internal Med, dipl, 63. *Prof Exp:* Consult biochem, Vet Admin Hosp, Martinsburg, WVa, 50-53; from instr to prof, 60-85, EMER PROF INT MED, GEORGE WASHINGTON UNIV, 85- *Concurrent Pos:* USPHS res fel, 59-60, USPHS res career develop award, 61-66; dir, Washington Regional Histocompatability Typing Lab; dir, Hemopheresis Serv, George Washington Univ Med Ctr. *Mem:* Fel Am Col Physicians; Am Fedn Clin Res; Am Asn Clin Histocompatibility Test; Am Soc Artificial Internal Organs; Int Soc Nephrology. *Res:* Pathophysiology and immunology of diseases of the kidney. *Mailing Add:* 14505 Hollyhock Way Burtonsville MD 20866

KRAMER, PAUL ALAN, b Hartford, Conn, July 22, 42; m 64, 87; c 2. PHYSICAL PHARMACY, BIOPHARMACEUTICS. *Educ:* Rensselaer Polytech Inst, BChE, 64; Univ Wis, MS, 66, PhD(pharm), 68. *Prof Exp:* Res biochemist, Walter Reed Army Inst Res, 68-71; asst prof phys pharm, Purdue Univ, 71-76; PROF PHARM & ASSOC PROF LAB MED, UNIV CONN, 76-, ASST PROF PEDIAT, 86- *Mem:* Am Asn Pharm Scientists; Am Col Clin Pharmacol; Am Soc Clin Pharmacol & Therapeut. *Res:* Pharmacokinetics of drugs in the elderly; drug dispositon in the neonate; drug delivery systems. *Mailing Add:* Dept Pharm Univ Conn Health Ctr Farmington CT 06032

KRAMER, PAUL JACKSON, b Brookville, Ind, May 8, 04; m 31; c 2. PLANT PHYSIOLOGY. *Educ:* Miami Univ, AB, 26; Ohio State Univ, MS, 29, PhD(plant physiol), 31. *Hon Degrees:* LittD, Miami Univ, 66; DSc, Univ NC, 66 & Ohio State Univ, 72; Dr, Univ Paris, 75. *Prof Exp:* Asst bot, Ohio State Univ, 28-31; from instr to prof, 31-54, Duke prof, 54-74, EMER PROF BOT, DUKE UNIV, 74- *Concurrent Pos:* Mem agr bd, Nat Res Coun, 58-60, mem comt wartime modification biol teaching; prog dir regulatory biol, NSF, 60-61, consult, 60-65, mem divisional comt, Div Biol & Med, 62-65, chmn, 64-65; chmn phytron bd, Duke Univ & NC State Univ, 62-78; mem comt int biol proj, Nat Acad Sci-Nat Res Coun, 63-64; mem comt agr sci, USDA, 65-68; vis comt, Harvard Univ, 65-71; mem nat comt, Int Bot Cong, 65-71; bd trustees, Biol Abstr, 66-71, pres, 71; chmn comt to rev US Int Biol Prog, Nat Acad Sci-Nat Res Coun, 73-75; consult, Sarah P Duke Gardens, 74-78; vis prof, Univ Tex, 76; Walker-Ames vis prof, Univ Wash, 77. *Honors & Awards:* Soc Am Foresters Award, 61; Charles Reid Barnes Life Mem Award, Am Soc Plant Physioloists, 67. *Mem:* Nat Acad Sci; AAAS; Am Philos Soc; Bot Soc Am (vpres, 59, pres, 64); Am Soc Plant Physiologists (vpres, 43, pres, 45); hon mem Chinese Soc Forestry. *Res:* Plant and soil water relations and absorption of water by plants; physiology of woody plants; effects of environmental stress on plants; controlled environments for plant research; use of nuclear magnetic resonance imaging in research on plants. *Mailing Add:* Dept Bot Duke Univ Durham NC 27706-7758

KRAMER, PAUL ROBERT, b Montclair, NJ, Nov 17, 35; m 64; c 2. PHYSICS, INSTRUCTIONAL TECHNOLOGY. *Educ:* Cornell Univ, BA, 57; Rutgers Univ, MS, 59, PhD(physics), 66. *Prof Exp:* Instrumentation specialist physics, State Univ NY Stony Brook, 64-66, asst prof, 66-70; assoc prof, State Univ NY, 70-75, chmn dept, 77-81, dean acad serv, 81-84, PROF PHYSICS, AGR & TECH COL, STATE UNIV NY, FARMINGDALE, 75- *Concurrent Pos:* Dir, Comput Assisted Instr, State Univ NY Stony Brook, 67-68, proj coordr, Instr Resources Ctr, 68-70; chmn, Acad Comput Planning Comt, SUNY, 81-; treas & sci mkt consult, Safety Corp Am, Huntington, NY, 84-87; fac access, Comput Technol Adv Comt, State Univ NY, 89- *Res:* Instructional technology using computers and other media; improvement of instruction in physics and other fields; consumer product safety. *Mailing Add:* State Univ NY Agr & Tech Col Farmingdale NY 11735

KRAMER, PHILIP, b Chicago, Ill, Jan 26, 15; m 44. MEDICINE. *Educ:* Univ Chicago, MD, 39; Am Bd Internal Med, dipl, 52; Am Bd Gastroenterol, dipl, 58. *Prof Exp:* Intern, Cook County Hosp, Chicago, 39-40; resident, St Joseph Hosp, Marshfield, Wis, 41-42; fel med, Evans Mem Hosp & Mass Mem Hosp, 46-53; chief clin gastroenterol, Univ Hosp. 74-80; from instr to prof, 51-85, EMER PROF MED, SCH MED, BOSTON UNIV, 85- *Concurrent Pos:* Mem, gastrointestinal subspecialty bd, Am Bd Internal Med; consult, Boston & Framingham Union Hosp, 55-; vis physician, Mass Mem Hosp, Univ Hosp, 56-85, pres, med & dent staff, 75-76; assoc prof, State Univ NY, Upstate Med Ctr, 58; chief gastroenterol, Syracuse Vet Admin Hosp, 58; sr physician, Univ Hosp, 85- *Mem:* Capt Am Fedn Clin Res; Am Gastroenterol Asn; AMA; Am Col Physicians; Am Physiol Soc. *Res:* Esophageal and small intestinal motility in health and disease; metabolic studies on ileostomized subjects to evaluate effect of varying sodium loads, hormones, drugs and foods. *Mailing Add:* Univ Hosp Boston Univ Med Ctr 88 E Newton St Boston MA 02118

KRAMER, RAYMOND ARTHUR, b Buffalo, NY, Dec 7, 29; m 55. ANALYTICAL CHEMISTRY. *Educ:* Canisius Col, BS, 54; Rensselaer Polytech Inst, PhD(anal chem), 59. *Prof Exp:* Technician, Aluminum Co Am, NY, 48-54; asst, Rensselaer Polytech Inst, 54-58; phys chemist, Alcoa Res Labs, 58-67, ANALYTICAL CHEMIST, ALCOA TECH CTR, ALUMINUM CO AM, 67- *Mem:* Am Chem Soc; Soc Appl Spectros; Am Soc Testing & Mat. *Res:* Neutron activation analysis; x-ray diffraction and fluorescence; characterization of ultra-pure aluminum and gallium; emission spectroscopy; aluminum in fusion reactors. *Mailing Add:* 4377 Frederick Dr Lower Burrell PA 15068

KRAMER, RAYMOND EDWARD, b Warren, Ohio, Feb 2, 19; m 47; c 3. ELECTRICAL ENGINEERING, PHYSICS. *Educ:* Heidelberg Col, BSc, 43; Case Inst Technol, MSc, 50. *Prof Exp:* Asst prof, 50-54, chmn dept, 54-78, assoc prof, 54-78, PROF ELEC ENG, YOUNGSTOWN STATE UNIV, 78- *Concurrent Pos:* Consult elec engr, US Steel Corp, 51 & ARC Res Inc, 51-53; consult develop engr, Westinghouse Elec Corp, 54 & 55 & Ohio Bell Tel, 61. *Mem:* Inst Elec & Electronics Engrs; AAAS. *Res:* Spark machining; ferromagnetic domains and computer cores; particle physics; electrical, kinetic and quantum properties of fundamental particles; gravity waves. *Mailing Add:* Dept Elec Eng Youngstown State Univ 410 Wick Ave Youngstown OH 44555

KRAMER, RICHARD ALLEN, MOLECULAR BIOLOGY. *Educ:* Yale Univ, PhD(biochem), 75. *Prof Exp:* Res investr, Dept Molecular Genetics, Hoffman-Laroche, Inc, 81-89; DEPT MOLECULAR BIOL & ENZYMOL, CIBA-GEIGY PHARMACEUT, 89- *Res:* Yeast gene expression; recombinant DNA technology. *Mailing Add:* Dept Molecular Biol & Enzymol Ciba-Geigy Pharmaceut 556 Morris Ave 842 Summit Nutley NJ 07901

KRAMER, RICHARD MELVYN, b Brooklyn, NY, Dec 20, 35; m 57; c 3. HERBICIDES, PLANT GROWTH REGULATORS. *Educ:* Polytech Inst Brooklyn, BChE, 57, MChE, 60, PhD(chem eng), 63; St Louis Univ, MBA, 71. *Prof Exp:* Res assoc electrodialysis, bioferm div, Int Minerals & Chem Co, 62-63, develop engr, 63-65, sr process engr, 65-66; sr res engr, 66-73, sr res group leader, 73-79, res mgr, 79-82, RES MGR OPERS, MONSANTO CO, LATIN AM, 82- *Mem:* Am Chem Soc. *Res:* Herbicide and plant growth regulator research and development; pesticide residue research; pesticide formulation and environmental science. *Mailing Add:* 800 N Lindbergh Blvd Monsanto Res Ctr St Louis MO 63167

KRAMER, ROBERT, b Boston, Mass, Apr 25, 27; m 50; c 4. ELECTRICAL & SYSTEMS ENGINEERING. *Educ:* Mass Inst Technol, SB, 49, SM, 52, ScD(elec eng), 59. *Prof Exp:* Proj engr, Servomech Lab, Mass Inst Technol, 49-62, lectr elec eng, Univ, 59-62, staff engr, Lincoln Lab, 62-69; sr consult, Harrington, Davenport & Curtis, Inc, 69-71; STAFF ENGR, LINCOLN LAB, MASS INST TECHNOL, 71- *Mem:* Inst Elec & Electronics Engrs; Optical Soc Am. *Res:* Optical and data systems. *Mailing Add:* Lincoln Lab N356 PO Box 73 Lexington MA 02173

KRAMER, SHELDON J, b Chicago, Ill, Aug 17, 38; m 62; c 3. CHEMICAL ENGINEERING. *Educ:* Univ Ill, Urbana, BSChE, 60; Princeton Univ, MAChE, 62, PhD(chem eng), 66. *Prof Exp:* Chem engr, Gul Res & Develop Co, 65-68; sr proj chem engr, Res & Develop Dept, 68-77, SR RES ENGR, AMOCO RES CTR, 77- *Mem:* Am Inst Chem Eng; Am Chem Soc; Sigma Xi. *Res:* Petroleum refining. *Mailing Add:* Amoco Oil Co H-1 PO Box 3011 Naperville IL 60566-7011

KRAMER, SHERMAN FRANCIS, b Elcho, Wis, Nov 15, 28; m 59; c 4. PHARMACY, PHARMACOLOGY. *Educ:* Univ Wis, BS, 50, PhD(pharm), 60. *Prof Exp:* Res assoc, Upjohn Co, 60-66, sect head, 66-90; RETIRED. *Mem:* Am Pharmaceut Asn; Acad Pharmaceut Sci. *Res:* Pharmaceutical product research and development. *Mailing Add:* 9701 Oakview Portage MI 49002

KRAMER, STANLEY PHILLIP, b Baltimore, Md, Oct 7, 23; m 62; c 2. MEDICINAL CHEMISTRY, PHARMACOLOGY. *Educ:* Univ Md, BS, 49, PhD(med chem), 55. *Prof Exp:* Res assoc cancer chemother, Sinai Hosp Baltimore, Inc, 55-68; sci instrument specialist, 68-71; sr sci instrument specialist, Florence Agreement Prog, 71-74; PROG MGR, FLORENCE AGREEMENT STAFF, US DEPT COM, 76- *Concurrent Pos:* Res assoc, Sch Med, Johns Hopkins Univ, 55-60, asst surg, 60-68. *Mem:* AAAS; Am Chem Soc; Am Asn Cancer Res; Electron Micros Soc Am; Nat Soc Med Res; Sigma Xi. *Res:* Cancer chemotherapy; toxicology; enzymology; biochemistry; alkylating agents; esterase; lipase; amidase; trypsin; clinical chemistry. *Mailing Add:* 3204 Burnbrook Lane Baltimore MD 21207

KRAMER, STANLEY ZACHARY, b Philadelphia, Pa, Sept 10, 21; m 41; c 1. NEUROPHARMACOLOGY. *Educ:* Univ Pa, AB, 52, PhD(physiol), 58. *Prof Exp:* Asst instr physiol, Univ Pa, 53-58; instr, Vassar Col, 58-60; instr, NY Med Col, 60-64, asst prof physiol, 64-67; from assoc prof to prof biol, Seton Hall Univ, 67-86; RETIRED. *Res:* Effects of drugs on brain electrical activity and behavior; neurophysiology. *Mailing Add:* 1801 J F Kennedy Blvd Philadelphia PA 19103

KRAMER, STEPHEN LEONARD, b Philadelphia, Pa, July 22, 43; m; c 2. EXPERIMENTAL HIGH ENERGY PHYSICS, ACCELERATOR PHYSICS. *Educ:* Drexel Inst Technol, BS, 66; Purdue Univ, MS, 67, PhD(physics), 71. *Prof Exp:* Res asst high energy physics, 71-74, asst physicist, 74-81, PHYSICIST, ARGONNE NAT LAB, 81- *Honors & Awards:* Lark Horovitz Award; George Tautfest Award. *Mem:* Am Phys Soc. *Res:* Synchrotron radiation sources; accelerator physics; radiographic imaging; applications of particle accelerators to medical research; high energy accelerator system design and application; nuclear instrumentation. *Mailing Add:* Brookhaven Nat Lab Nat Synchrotron Light Source Bldg 725C Upton NY 11973

KRAMER, STEVEN DAVID, b Lakewood, NJ, Aug 27, 48. LASER PHYSICS, NONLINEAR OPTICS. *Educ:* Cornell Univ, AB, 70; Harvard Univ, AM, 71, PhD(physics), 76. *Prof Exp:* Res fel, Harvard Univ, 72-76, teaching fel, 74-75; res staff, Oak Ridge Nat Lab, 76-87; RES STAFF, INST DEFENSE ANALYSIS, 87- *Concurrent Pos:* Consult, IDA, 85-. *Honors & Awards:* IR-100 Award, 84 & 87. *Mem:* Am Phys Soc; Optical Soc Am; Archeol Soc Am; AAAS; Inst Elec & Electronics Engrs. *Res:* Optics; optical sensors; nonlinear optics; spectroscopy; trace analysis. *Mailing Add:* Inst Defense Analyses 1801 N Beauregard St Alexandria VA 22311

KRAMER, THEODORE TIVADAR, b Novi-Sad, Yugoslavia, Jan 4, 28; US citizen; m 57; c 3. VETERINARY MICROBIOLOGY, IMMUNOLOGY. *Educ:* Nat Vet Sch, Alfort, France, DVM, 52; Univ Strasbourg, dipl, 53; Colo State Univ, MSc, 63, PhD(microbiol), 65; Am Col Vet Microbiol, dipl, 70. *Prof Exp:* Res off microbiol, Can Dept Agr, 57-60; jr pathologist, Colo State Univ, 60-65; asst prof microbiol, Univ Col, Nairobi, Kenya, 65-67; assoc prof vet microbiol, Western Vet Col Med, Univ Sask, 67-70; prof microbiol & head dept, Sch Vet Med, Auburn Univ, 71-80; PROF VET MICROBIOL & PREV MED, COL VET MED, IOWA STATE UNIV, 80- *Mem:* Am Soc Microbiol; Am Asn Immunologists; Am Vet Med Asn. *Res:* Experimental colibacillosis in piglets; immunoglobulins of bovine colostrum; vaccine against bovine vibriosis; maternal immunity and the newborn; cell-mediated immunity to infectious diseases of animals; salmonellosis. *Mailing Add:* Dept Vet Microbiol & Prev Med Col Vet Med Iowa State Univ Ames IA 50011

KRAMER, TIM R, b Garden Plain, Kans, Mar 1 ,43. MICROBIOLOGY. *Educ:* Okla Univ, MA, PhD(med sci), 73. *Prof Exp:* Postdoctoral sr fel, Sloan Cancer Inst, NY, 73-76; mem proj, Sch Med, St Louis Univ & Fai-Land Anemia & Malnutrit Ctr, Chiang-Mai Univ, 76-78; RES BIOLOGIST, NIH, 79- *Mem:* Soc Exp Biol; Inst Nutrit. *Mailing Add:* Beltsville Human Nutrit Res Ctr Bldg 307 Rm 215 Beltsville MD 20705

KRAMER, WILLIAM GEOFFREY, b Pittsburgh, Pa, Sept 16, 48; m 73. PHARMACY, PHARMACOKINETICS. *Educ:* Univ Pittsburgh, BS, 71; Ohio State Univ, PhD(pharm), 76. *Prof Exp:* From asst prof to assoc prof pharmaceut, Col Pharm, Univ Houston, 76-86; mgr pharmacokinetics, Schering-Plough, 86-89; ASSOC DIR CLIN RES, BOEHRINGER MANHEIM PHARMACEUT, 89- *Concurrent Pos:* Res assoc, Inst Cardiovasc Studies, Col Pharm, Univ Houston, 77-86. *Mem:* Am Asn Pharm Scientists; Drug Info Asn; AAAS; Sigma Xi; Am Soc Clin Pharmacol & Therapeut. *Res:* Pediatric pharmacokinetics; effects of disease and other abnormal conditions on drug pharmacokinetics; application of computers in pharmacokinetic data analysis and dosage regimen design. *Mailing Add:* Boehringer Mannheim Pharmaceut Co 15204 Omega Dr Rockville MD 20850

KRAMER, WILLIAM J, b Coldwater, Ohio, Oct 13, 19. ORGANIC CHEMISTRY, ANALYTICAL CHEMISTRY. *Educ:* Univ Fribourg, Lic es Sci, 52, ScD(chem), 53. *Hon Degrees:* ScD, Univ Fribourg, 53. *Prof Exp:* From instr to assoc prof, 53-68, PROF CHEM, ST JOSEPH'S COL IND, 68-, CHMN DEPT, 77- *Mem:* Am Chem Soc. *Res:* Reactivity of methyl groups in substituted benzene rings; history and philosophy of science. *Mailing Add:* Dept Chem St Joseph's Col Rensselaer IN 47979

KRAMER, WILLIAM S, b Butte, Nebr, Jan 10, 22; m 44; c 4. DENTISTRY. *Educ:* Univ Nebr, BSc, 46, DDS, 48, MSc, 54; Am Bd Pedodont, dipl. *Prof Exp:* Instr operative dent, Univ Nebr, Lincoln, 48-52, prof operative dent, 54-58, chmn dept, pedodont, 58-80, prof pedodont, 80-87; RETIRED. *Concurrent Pos:* Dir, Dent Asst Utilization Prog; past examr, Am Bd Pedodont, exec secy, 74- *Mem:* Am Acad Pedodont (pres, 78-); Int Asn Dent Res. *Res:* Clinical studies on local anesthetic solutions; morphology of the primary dentition; physical properties of gold foil; ultrasonic sterilization; pedodontic failures. *Mailing Add:* Dept Pedodont Col Dent Univ Nebr Lincoln NE 68583

KRAMERICH, GEORGE L, b Aliquippa, Pa, Nov 26, 29; m 54; c 3. CONTROL & ELECTRICAL ENGINEERING. *Educ:* Fla State Univ, BS, 63, MS, 64; Case Western Reserve Univ, PhD(control eng), 70. *Prof Exp:* Teaching asst eng sci, Fla State Univ, 63-64; teaching asst elec eng, Case Western Reserve Univ, 64-69; from asst prof to assoc prof, 69-77, PROF ELEC ENG, CLEVELAND STATE UNIV, 77- *Concurrent Pos:* Consult, Gen Elec Lighting Res Lab, 70-, Chemstress Consults, 77-, Gould Instrument Div, 77- & Ohio Legis Serv Comn, 78-79. *Mem:* Sr mem Instrument Soc Am; Inst Elec & Electronics Engrs; Am Soc Eng Educ. *Res:* Economic and management decision making applied to the evaluation of advanced process control technology; computer simulation. *Mailing Add:* Dept Engr Technol Cleveland State Univ Euclid Ave at E 24th Cleveland OH 44115

KRAMISH, ARNOLD, b Denver, Colo, June 6, 23; m 52; c 2. NUCLEAR PHYSICS, INTERNATIONAL RELATIONS. *Educ:* Univ Denver, BS, 45; Harvard Univ, MA, 47. *Prof Exp:* Mass spectroscopist, Oak Ridge Nat Lab, 44-45; physicist, Los Alamos Sci Labs, 45-46; staff physicist, AEC, 47-51 & physics dept, Rand Corp, Calif, 51-69; adj prof int studies, Univ Miami, 69-73; US sci liaison attache, UNESCO, 70-73; counr sci & technol, US Mission, Orgn Econ Coop & Develop, 74-76; sr scientist, Res & Develop Assocs, Arlington, 76-81; TECHNOL CONSULT, 81- *Concurrent Pos:* Consult, Int Bank Reconstruct & Develop, 58; fel, Coun For Rels, 58-59; consult, NSF, 59-62; prof in residence, Univ Calif, Los Angeles, 65-66; Guggenheim fel, 66-67; fel, Woodrow Wilson Ctr, Smithsonian Inst, 82-83; Rockefeller scholar, Bellagio, Italy, 84. *Res:* Fission physics; applied nuclear energy; political and economic implications of nuclear energy; research and development policy and planning; space and strategic defense systems; international energy policy. *Mailing Add:* 2065 Wethersfield Ct Reston VA 22091

KRAMP, ROBERT CHARLES, b Alexandria, Va, Aug 2, 42; m 65; c 2. RADIATION BIOLOGY, ENDOCRINOLOGY. *Educ:* Univ Md, BS, 64; Univ Okla, MS, 69; Univ Tenn, PhD(radiation biol), 73. *Prof Exp:* Fel endocrinol, Inst Clin Biochem, 73-75; res instr, Vanderbilt Univ, 75-78; asst prof biol, Va Polytech Inst & State Univ, 78-83; RETIRED. *Mem:* Am Diabetes Asn; Europ Asn Study Diabetes. *Res:* Diabetes; transplantation of pancreatic islet tissue in mice; experimental and genetic diabetes in rodents. *Mailing Add:* Rte 3 Box 116A Floyd VA 24091

KRAMPITZ, LESTER ORVILLE, b Maple Lake, Minn, July 9, 09; m 32; c 1. BACTERIOLOGY, MICROBIOLOGY. *Educ:* Macalester Col, BA, 31; Iowa State Col, PhD(bact), 42. *Hon Degrees:* DSc, Macalester Col, 58. *Prof Exp:* Asst, Rockefeller Inst, 42-43; asst prof bact, Indust Sci Res Inst, Iowa

State Col, 43-46; assoc prof biochem, 46-48, prof microbiol & dir dept, 48-79, EMER PROF, SCH MED, CASE WESTERN RESERVE UNIV, 79- *Concurrent Pos:* Fulbright res scholar, Univ Munich, 55-56; mem biochem study sect, NIH, 54-58, mem bact study sect, 59-64, mem res career award comt, Nat Inst Gen Med Sci, 64-68, mem microbiol training comt, 70-74. *Mem:* Nat Acad Sci; hon mem Am Soc Microbiol; Am Soc Photobiol; Am Acad Microbiol; NY Acad Sci; Sigma Xi. *Res:* Metabolism of bacteria using isotopes; antivitamin studies; hydrogen function of biophotolysis of water; mode of action of thiamin diphosphate. *Mailing Add:* Dept Microbiol Case Western Reserve Univ Sch Med Cleveland OH 44106

KRANBUEHL, DAVID EDWIN, b Madison, Wis, Apr 16, 43; m 66; c 2. PHYSICAL CHEMISTRY, POLYMER PHYSICS. *Educ:* DePauw Univ, BA, 65; Univ Wis, PhD(chem), 69. *Prof Exp:* Res chemist polymers, Nat Bur Standards, 69-70; PROF CHEM, COL WILLIAM & MARY, 70- *Concurrent Pos:* Nat Acad Sci fel, 69-70; consult, Nat Bur Standards, 70-, Union Carbide, Gen Elec, US Steel Corp, McDonnell Douglas Corp. *Mem:* Am Chem Soc. *Res:* Physical properties of polymers; dielectric phenomena; molecular dynamics in the liquid and glassy state; materials science engineering. *Mailing Add:* Dept Chem Col William & Mary Williamsburg VA 23185

KRANC, GEORGE M(AXIMILIAN), b Lodz, Poland, Feb 1, 20; nat US. ELECTRICAL ENGINEERING. *Educ:* St Andrews Univ, BSc, 44; Columbia Univ, MS, 53, DEng Sc(elec eng), 56. *Prof Exp:* Radio engr, Jewel Radio Co, 49-51; asst elec eng, Columbia Univ, 51-53, from instr to assoc prof, 53-62; vis prof, Polytech Inst Brooklyn, 62-63; assoc prof, 63-71, PROF SCH ENG, CITY COL NEW YORK, 71- *Concurrent Pos:* Consult, Gen Appl Sci Labs, 57 & Norden Labs, 58-59; sci ed, Scripta Technica Inc, 63-71. *Mem:* Inst Elec & Electronics Engrs; Sigma Xi. *Res:* Control systems theory, particularly sampled data systems and optimal controls. *Mailing Add:* 25 5th Ave New York NY 10003

KRANC, STANLEY CHARLES, b Peoria, Ill, Sept 29, 42. MECHANICAL ENGINEERING, CHEMICAL PHYSICS. *Educ:* Northwestern Univ, BS, 64, PhD(mech eng), 69. *Prof Exp:* Asst prof eng sci, Fla State Univ, 67-71; from asst prof to assoc prof, 71-78, PROF ENG, UNIV S FLA, 71- *Mem:* Am Inst Aeronaut & Astronaut; Newcomen Soc; Am Soc Mech Engrs. *Res:* Gas dynamics; plasma physics; combustion; two phase flow. *Mailing Add:* Dept Civil Eng & Mech Col Eng Univ SFla Tampa FL 33620

KRANE, KENNETH SAUL, b Philadelphia, Pa, May 15, 44; m 66; c 1. EXPERIMENTAL NUCLEAR PHYSICS. *Educ:* Univ Ariz, BS, 65; Purdue Univ, MS, 67, PhD(physics), 70. *Prof Exp:* Res assoc physics, Los Alamos Sci Lab, 70-72 & nuclear chem, Lawrence Berkeley Lab, 72-74; from asst prof to assoc prof, 74-84, PROF & CHMN PHYSICS, ORE STATE UNIV, 84- *Mem:* Fel Am Phys Soc; Am Asn Physics Teachers. *Res:* Angular distributions and correlations of gamma rays; nuclear spectroscopy; nuclear physics at ultralow temperatures; beta decay. *Mailing Add:* Dept Physics Ore State Univ Corvallis OR 97331

KRANE, STANLEY GARSON, b New York, NY, Feb 16, 37. CELL BIOLOGY, MOLECULAR BIOLOGY. *Educ:* City Col New York, BS, 57; Mich State Univ, MS, 58; Calif Inst Technol, PhD(biochem), 66. *Prof Exp:* Res fel biochem, Brandeis Univ, 66-67; asst prof biol, Univ Mass, Boston, 68-75; asst prof, 75-80, ASSOC PROF BIOL, FITCHBURG STATE COL, 80- *Mem:* AAAS; Am Inst Biol Sci. *Res:* Mutagenesis of microorganisms. *Mailing Add:* Dept Biol Fitchburg State Col 160 Pearl St Fitchburg MA 01420

KRANE, STEPHEN MARTIN, b New York, NY, July 15, 27; m 52; c 4. MEDICINE, BIOCHEMISTRY. *Educ:* Columbia Col, AB, 46; Columbia Univ, MD, 51; Am Bd Internal Med, dipl, 58. *Hon Degrees:* AM, Harvard Univ, 68; MD, Univ Geneva, Switz, 89. *Prof Exp:* Asst, Harvard Med Sch, 55-59, instr, 59-60, assoc, 60-63, from asst prof to prof, 63-87, PERSIS, CYROS & MARLOW B HARRISON PROF MED, HARVARD MED SCH, 87-; PHYSICIAN, MASS GEN HOSP, 69- *Concurrent Pos:* Fel med, Harvard Med Sch, 53-55; fel, Sch Med, Wash Univ, 56; Guggenheim fel, Oxford Univ, 73-74. *Honors & Awards:* Geigy Rheumatism Prize, 77; Heberden Medal, London, 80; Kleruperer Medal, 90. *Mem:* Am Soc Clin Invest; Endocrine Soc; Asn Am Physicians; Am Col Rheumatology; Am Fedn Clin Res; fel AAAS; Am Soc Biol Chemists; Am Soc Bone & Mineral Res. *Res:* Connective tissue biology and metabolism; internal medicine and rheumatology; transport mechanisms. *Mailing Add:* Dept Med Harvard Med Sch 25 Shattuck St Boston MA 02115

KRANIAS, EVANGELIA GALANI, ION TRANSPORT, PHOSPHORYLATION REGULATION. *Educ:* Northwestern Univ, PhD(biochem), 74. *Prof Exp:* ASSOC PROF PHOSPHORYLATION, COL MED, UNIV CINCINNATI, 78- *Mailing Add:* Dept Pharmacol & Cell Biophys Univ Cincinnati Med Ctr 231 Bethesda Ave Cincinnati OH 45267

KRANICH, WILMER LEROY, b Philadelphia, Pa, Nov 20, 19; m 50; c 3. CHEMICAL ENGINEERING, SYNTHETIC FUELS. *Educ:* Univ Pa, BS, 40; Cornell Univ, PhD(chem eng), 44. *Prof Exp:* Instr chem eng, Cornell Univ, 41-44; asst prof, Princeton Univ, 46-48; from assoc prof to prof, Worcester Polytech Inst, 48-85, head, dept, 58-75, dean grad studies, 75-85; ADJ PROF CHEM ENG, DUKE UNIV, 85- *Concurrent Pos:* Consult, Arthur D Little, Inc, 49-74 & Norton Co, 64-68; res scientist, Hungarian Acad Sci, Budapest, 83. *Mem:* Am Chem Soc; Am Soc Eng Educ; Am Inst Chem Engrs; Sigma Xi. *Res:* Process development; chemical kinetics. *Mailing Add:* 6503 Falconbridge Rd Chapel Hill NC 27514

KRANNICH, LARRY KENT, b Pekin, Ill, Sept 5, 42; m 74; c 2. INORGANIC CHEMISTRY. *Educ:* Ill State Univ, BS, 63, MS, 65; Univ Fla, PhD(inorg chem), 68. *Prof Exp:* Asst chem, Ill State Univ, 63-65; asst, Univ Fla, 65-68, res asst, 68; asst prof, Univ Miss, 68-69; from asst prof to assoc prof, 69-76, actg chmn dept, 74-75, PROF & CHMN DEPT CHEM, UNIV ALA, BIRMINGHAM, 76. *Concurrent Pos:* Vis prof, Tech Univ Vienna, 69. *Mem:* Am Chem Soc; Sigma Xi. *Res:* Chemistry of the arsenic-nitrogen bond; chemistry of chloramine and its reaction with group V bases; reactivity of boranes with arsenic-nitrogen and phosphorus-nitrogen containing bases. *Mailing Add:* Dept Chem Univ Ala Birmingham AL 35294

KRANTZ, ALLEN, b New York, NY, Jan 25, 40; m 79; c 3. BIOCHEMISTRY. *Educ:* City Col New York, BS, 61; Yale Univ, MS, 62, PhD(chem), 67. *Prof Exp:* Fel, Univ Reading, 67-68; from asst prof to assoc prof org chem, 68-81, ADJ ASSOC PROF, DEPT PHARMACOL SCI, STATE UNIV NY STONY BROOK, 77-; RES DIR, SYNTEX CAN INC, 81- *Concurrent Pos:* Petrol Res Fund grant, 68-71; Res Corp grant, 68-; State Univ NY Res Found fac fel & grant in aid, 69-70; NATO sr fel, 75; NSF grants, 74, 76 & 77; NIH grant, 77-81; adj prof, Dept Chem, Univ Guelph & Univ Toronto, 85- *Mem:* Am Chem Soc; Royal Soc Chem; Chem Inst Can. *Res:* Drug design; photochemistry of matrix isolated species; mechanism of reactions of heterocycles; bio-organic chemistry; design of enzyme inhibitors. *Mailing Add:* Syntex Inc 2100 Syntex Ct Mississauga ON L5N 3X4 Can

KRANTZ, DAVID S, b New York, NY, Feb 9, 49; m 82; c 2. MEDICAL PSYCHOLOGY, BEHAVIORAL MEDICINE. *Educ:* City Col New York, BS, 71; Univ Tex, Austin, PhD(psychol), 75. *Prof Exp:* Asst prof psychol, Univ Southern Calif, 75-78; from asst prof to assoc prof, 78-87, PROF MED PSYCHOL, UNIFORMED SERV UNIV HEALTH SCI, 87- *Honors & Awards:* Health Psychol Ann Award, Am Psychol Asn, 81, Early Career Sci Award, 82. *Mem:* fel Am Psychol Asn; Am Psychosomatic Soc; Soc Psychophysiol Res; Acad Behav Med Res; fel Am Psychol Soc. *Res:* Behavioral and psychophysiological factors in cardiovascular disorders; psychological stress. *Mailing Add:* Dept Med Psychol Uniformed Serv Univ Health Sci 4301 Jones Bridge Rd Bethesda MD 20889-4799

KRANTZ, GERALD WILLIAM, b Pittsburgh, Pa, Mar 12, 28; m 55; c 3. ENTOMOLOGY. *Educ:* Univ Pittsburgh, BSc, 51; Cornell Univ, PhD, 55. *Prof Exp:* From asst prof to assoc prof, 55-65, PROF ENTOM, ORE STATE UNIV, 65- *Concurrent Pos:* Microzoologist, Am Quintana Roo Exped, 65, zoologist & dep leader, Exped II, 68; sr res scientist, CSIRO, Pretoria, South Africa, 79; prog officer, Systematic Biol, NSF, Washington, DC, 84-85; vis res prof, Nat Mus Natural Hist, Paris, 87; adv bd, High Desert Museum, Bend Ore, 74-; exec comt, Int Congress Acarology, 82-; exp & appl Acarology, Amsterdam, 84- *Mem:* Entom Soc Am; Acarology Soc Am (mem, Gov Bd, 71-76, chmn, 75-76); Sigma Xi. *Res:* Systematics and behavior of Acari diversi; systematics and behavior of mites (acari), with emphasis on insect associates and marine forms. *Mailing Add:* Dept Entom Ore State Univ Corvallis OR 97331

KRANTZ, KARL WALTER, b Waterbury, Conn, May 9, 18; m 42; c 2. ORGANIC CHEMISTRY, POLYMER CHEMISTRY. *Educ:* Univ Conn, BS, 39, MS, 40; Stanford Univ, PhD(chem), 51. *Prof Exp:* Asst chem, Stanford Univ, 40-41; instr, Univ Conn, 42-43; res chemist, E I du Pont de Nemours & Co, 45-50; res chemist, 50-60, SPECIALIST SILICONE RES TECHNOL, SILICONE PROD DEPT, GEN ELEC CO, 60- *Mem:* Am Chem Soc; fel Am Inst Chemists. *Res:* Aliphatic diamines; local anesthetics; fluorocarbons; catalytic oxidation of hydrocarbons; silicones. *Mailing Add:* 16009 S Arlington Dr Seneca SC 29678

KRANTZ, KERMIT EDWARD, b Oak Park, Ill, June 4, 23; m 46. MEDICINE. *Educ:* Northwestern Univ, BS, 45, BM & MS, 47, MD, 48; Am Bd Obstet & Gynec, dipl. *Hon Degrees:* LittD, William Woods Col, 71. *Prof Exp:* Asst zool, Northwestern Univ, 43, resident anat, Med Sch, 44-47; intern obstet & gynec, New York Lying-In-Hosp, Cornell Univ, 47-48; asst resident, New York Lying-In-Hosp, Cornell Univ, 47-48; from instr to asst prof, Univ Vt, 51-55; asst prof, Sch Med, Univ Ark, 55-59; dean clin affairs, Univ Kans Med Ctr, 72-74, assoc to exec vchancellor facil develop, Med Ctr, 74-83, prof & chmn, Dept Gynec & Obstet, 56-91, PROF ANAT, UNIV KANS MED CTR, 63-, UNIV DISTINGUISHED PROF, 91- *Concurrent Pos:* NY Acad Med Bowen-Brooks fel, New York Hosp, 48-50; res fel, Col Med, Univ Vt, 50-51; Markle scholar, Sch Med, Univ Kans, 57-62; asst dir, Div Maternal & Child Health & Welfare, Vt State Dept Health, 51-55, civilian Nat Consult, Surgeon Gen, US Air Force, 64-; pres, Int Family Planning Res Asn, Inc, 75-76. *Mem:* AAAS; Am Asn Anatomists; found fel Am Col Obstetricians & Gynecologists; Am Med Writers Asn; Soc Med Consults Armed Froces (pres, 90-91). *Res:* Human placenta; anatomy and physiology; female anatomy, urethra, bladder, vagina, uterus, tubes and ovaries; renal function in pregnancy. *Mailing Add:* Univ Kans Med Ctr Kansas City KS 66103

KRANTZ, REINHOLD JOHN, b Bradford, Pa, Aug 12, 15; m 37; c 4. ORGANIC CHEMISTRY. *Educ:* Greenville Col, AB, 36; Univ Ill, MS, 37; Mich State Col, PhD(org chem), 47. *Prof Exp:* Control chemist, Kendall Refrig Co, 37-38; instr chem & math, Moberly Jr Col, 38-41 & chem, Mich State Col, 41-47; from asst prof to prof, 47-81, EMER PROF CHEM, UNIV REDLANDS, 81- *Concurrent Pos:* Lectr, Pakistan Univ, 58-59; chemist, Off Naval Res, 51-; dir div sci & math, Univ Redlands, 61-73; mem bd, Redlands-Highland-Yucaipa Resource Conserv Dist, 69- *Mem:* AAAS; Am Chem Soc; Sigma Xi. *Res:* Fragmentation of tertiary alcohols; organosilicon compounds; reaction of alkenes, 1-alkenes and formaldehyde. *Mailing Add:* 1310 Monterey Redlands CA 92373

KRANTZ, SANFORD B, b Chicago, Ill, Feb 6, 34; m 58; c 4. INTERNAL MEDICINE, HEMATOLOGY. *Educ:* Univ Chicago, AB & BS, 55, MD, 59. *Prof Exp:* Intern med, Univ Chicago Hosps, 60-61, asst resident, 61-62, res assoc, 63-64; asst prof med, Univ Chicago Hosps & Argonne Cancer Res Hosp, 65-68; asst chief hemat serv, Clin Ctr, NIH, 68-70; assoc prof, 70-75, PROF MED & DIR HEMAT, SCH MED, VANDERBILT UNIV, 75-; CHIEF HEMAT UNIT, VET ADMIN HOSP, NASHVILLE, 70- *Concurrent Pos:* USPHS fel, Univ Chicago Hosps, 62-64; NATO fel biochem, Univ Glasgow, 64-65; Leukemia Soc scholar, 65-68. *Honors & Awards:* Joseph A Capps Prize, Inst Med Chicago, 64. *Mem:* AAAS; Am Soc Clin Invest; Am Fedn Clin Res; Am Soc Hemat; Asn Am Physicians; Int Soc Exp Hematol. *Res:* Erythropoietin; erythropoietic diseases; polycythemia and red cell aplasia; Friend virus polycythemia in mice. *Mailing Add:* 838 Rodney Dr Nashville TN 37205

KRANTZ, STEVEN GEORGE, b San Francisco, Calif, Feb 3, 51; m 74. SEVERAL COMPLEX VARIABLES. *Educ:* Univ Calif, Santa Cruz, BA, 71; Princeton Univ, PhD(math), 74. *Prof Exp:* Asst prof math, Univ Calif, Los Angeles, 74-81; assoc prof, 81-84, prof math, Pa State Univ, University Park, 84-87; PROF MATH, WASHINGTON UNIV, ST LOUIS, 86- *Concurrent Pos:* NSF res fel, 75- *Mem:* Am Math Soc. *Res:* Function theory on pseudoconvex domains in complex n-space; harmonic analysis of Euclidean spaces, real function theory, differentiability of functions, and interpolation theory. *Mailing Add:* Dept Math Washington Univ St Louis MD 63130

KRANTZ, WILLIAM BERNARD, b Freeport, Ill, Jan 27, 39; m 68; c 1. POLYMERIC MEMBRANES, GEOMORPHOLOGY. *Educ:* St Joseph's Col, Ind, BA, 61; Univ Ill, Urbana, BS, 62; Univ Calif, Berkeley, PhD(chem eng), 68. *Prof Exp:* Asst prof, 68-77, assoc prof, 77-79, PROF CHEM ENG, UNIV COLO, BOULDER, 79- *Concurrent Pos:* Consult, Dow Chem Co, Mich, 69-71; Fulbright-Hays lectr, Istanbul Tech Univ, 74-75; NSF, NATO sr fel, Univ Essex, Eng, 75; consult, Laramie Energy Technol Ctr, Dept Energy, 76-; dir, Thermodyn & Mass Transfer Prog, NSF, 77-78; mem area adv comt, US Coun Int Exchange Scholar, Int Commun Agency, 77-80; consult, US Dept Com, 79-80; Fulbright-Hayes sr res fel, Aachen Tech Univ, WGer, 81-82; Nat res lectr, Sigma Xi, 84-86; Guggenheim fel, Univ Oxford, Eng, 88-89; consult, Bend Res, 90- *Honors & Awards:* Spec Achievement & Outstanding Performance Awards, NSF, 78; George Westinghouse Award, Am Soc Eng Educ, 80. *Mem:* fel AAAS; Am Inst Chem Engrs; Am Soc Eng Educ; Am Chem Soc; Sigma Xi; NAm Membrane Soc. *Res:* Polymeric membrane morphology; self-organization in geophysical processing; materials science in low-gravity; global change in polar and sub polar regions. *Mailing Add:* Dept Chem Eng Univ Colo Campus Box 424 Boulder CO 80309-0424

KRANZ, EUGENE FRANCIS, b Toledo, Ohio, Aug 17, 33; c 6. AERONAUTICAL ENGINEERING. *Educ:* St Louis Univ, BS. *Prof Exp:* Flight test engr, McDonnell Aircraft Co, 54-55; supvr carrier flight test maintenance & checkout, Holloman AFB, NMex, 58-60; flight opers dir shuttle prog, 80-83, FLIGHT DIR GEMINI, APOLLO & SKYLAB MISSIONS, MANNED SPACECRAFT CTR, NASA, 64-, CHIEF FLIGHT CONTROL DIV, 69-, DIR MISSION OPERS, NASA, JOHNSON SPACE CTR, 83- *Concurrent Pos:* Flight controller, Mercury Missions. *Honors & Awards:* Lawrence Sperry Award, Am Inst Aeronaut & Astronaut, 67. *Mem:* Fel Am Astronaut Soc. *Mailing Add:* NASA Johnson Space Ctr NASA Rd 1 Houston TX 77058

KRANZER, HERBERT C, b New York, NY, Apr 10, 32; m 58; c 3. APPLIED MATHEMATICS, CONSERVATION LAWS. *Educ:* NY Univ, BA, 52, PhD(math), 57. *Prof Exp:* Asst math, NY Univ, 52-57, assoc res scientist, 57-58, instr, 58-59; assoc prof, 59-63, PROF MATH, ADELPHI UNIV, 63- *Concurrent Pos:* Consult, Los Alamos Sci Lab, 56-68; NSF sr fel, 66; vis prof, Imp Col London, 66-67; vis scholar, Columbia Univ, 76-77; consult, FONAR Corp, 83- *Honors & Awards:* Putnam Awards, 51 & 52. *Mem:* Am Math Soc; Soc Indust & Appl Math; Math Asn Am; AAAS. *Res:* Magnetohydrodynamics; numerical analysis; Weiner-Hopf problems; conservation laws; magnetic resonance imaging. *Mailing Add:* Dept Math & Comput Sci Adelphi Univ Garden City NY 11530

KRANZLER, ALBERT WILLIAM, b Bismarck, NDak, July 11, 16; m 39; c 2. MATHEMATICS. *Educ:* Univ NDak, BS, 37; Univ Minn, MS, 50. *Prof Exp:* Prin pub sch, NDak, 37-41; teacher, SDak, 41-42; instr training div, Sioux Falls Army Air Force Sch, 42-43; prin pub sch, Colo, 43-45; from assoc prof to prof math, SDak State Univ, 45-81, actg head dept, 61-68; RETIRED. *Mem:* Math Asn Am; Am Math Soc. *Res:* Reorganization of high school mathematics curriculum. *Mailing Add:* 808 Christine Ave No 206 Brookings SD 57006-3903

KRAPCHO, ANDREW PAUL, b Alden, Pa, Mar 6, 32; m 58; c 3. ORGANIC CHEMISTRY. *Educ:* Pa State Univ, BS, 53; Harvard Univ, MA, 57, PhD(chem), 58. *Prof Exp:* Instr chem, Smith Col, 57-59; res fel, Pa State Univ, 59-60; from asst prof to assoc prof, 60-67, PROF CHEM, UNIV VT, 67- *Concurrent Pos:* Fulbright scholar, France, 68-69. *Mem:* Am Chem Soc. *Res:* Chemistry of thiones and photochemistry of cyclic ketones; physical-organic chemistry; metal-amine reductions; solvolytic studies of spirane systems; sesquiterpene syntheses; bivalent carbon species. *Mailing Add:* Dept Chem Univ Vt Burlington VT 05405

KRAPF, GEORGE, b Millvale, Pa, July 20, 22; m 46. ANALYTICAL CHEMISTRY. *Educ:* Univ Pittsburgh, BS, 44. *Prof Exp:* Instr chem, sec schs, WPa, 44-50; res chemist, US Steel Res Labs, 50-70, sr res chemist, 70-83; RETIRED. *Concurrent Pos:* Consult, 83- *Mem:* Am Chem Soc. *Res:* Polarography, thermal analysis and second phase analysis in steels; twenty publications on polargraphy and thermal analysis. *Mailing Add:* Thompson Manor 307 Russell St Pittsburgh PA 15209-1613

KRAPU, GARY LEE, b Oakes, ND, Mar 12, 44; m 67, 85; c 3. WILDLIFE RESEARCH, ANIMAL ECOLOGY. *Educ:* NDak State Univ, BS, 66; Iowa State Univ, MS, 68, PhD(animal ecol), 72. *Prof Exp:* RES SCIENTIST, NORTHERN PRAIRIE WILDLIFE RES CTR, US FISH & WILDLIFE SERV, 71- *Mem:* Wildlife Soc; Ecol Soc Am; Am Ornithologist's Union; Wilson Soc. *Res:* Ecological aspects of waterfowl reproduction; sandhill crane biology; feeding ecology and nutrition; reproductive physiology; lipid storage; bioenergetics; marsh ecology. *Mailing Add:* Northern Prairie Wildlife Res Ctr PO Box 2096 Jamestown ND 58402-9736

KRASAVAGE, WALTER JOSEPH, b Luzerne, Pa, Mar 12, 33; m 55; c 4. INDUSTRIAL TOXICOLOGY. *Educ:* King's Col, BS, 55; Univ Rochester, MS, 63. *Prof Exp:* Technician parasitol, Merck Inst Therapeut Res, 55-56; sr res assoc, Atomic Energy Proj, Dept Radiation Biol & Biophys, Sch Med & Dent, Univ Rochester, 58-65; SR TOXICOLOGIST INDUST TOXICOL, HEALTH SAFETY & HUMAN FACTORS LAB, EASTMAN KODAK CO, 65-, MGR REPROD & DEVELOP TOXICOL, HEALTH & ENVIRON LABS, EASTMAN KODAK CO, 65- *Concurrent Pos:* Instr biol, Rochester Inst Technol, 63-65. *Mem:* Soc Toxicol; Teratol Soc; Environ Mutagen Soc. *Res:* Subchronic and chronic toxicology of industrial chemicals, especially reproduction and embryo-fetotoxicity. *Mailing Add:* 288 Shorecliff Dr Rochester NY 14612

KRASHES, DAVID, b Brooklyn, NY, Jan 31, 25; m 56; c 1. METALLURGY. *Educ:* Rensselaer Polytech Inst, BS, 49, MS, 52, PhD(metall), 58. *Prof Exp:* Res assoc metall, Rensselaer Polytech Inst, 53-54; mem staff, Nuclear Metals, Inc, 55-57; assoc prof, Worcester Polytech Inst, 57-65; PRES, MASS MAT RES, INC, 65-; PRES, LEHIGH TESTING LABS, 72- *Concurrent Pos:* Consult, Wyman-Gordon Co, 59- & Reed Rolled Thread Die Co, 60-; dir, Richard D Brew Co; pres, Conn Metall, Inc, 81- *Mem:* Fel Am Soc Metals (treas, 71-73, vpres, 80-81, pres, 81-82); Am Inst Mining, Metall & Petrol Engrs; Am Soc Testing & Mat; Am Foundry Soc. *Res:* Failure analysis of metals and mechanical products; solving industrial manufacturing problems relating to materials; fabrication; microscopy; economic studies. *Mailing Add:* 106 Rhodes Rd Princeton MA 01541

KRASHIN, BERNARD R(OBERT), b Buffalo, NY, Nov 9, 18; m 46; c 2. METALS. *Educ:* Western Reserve Univ, BS, 41. *Prof Exp:* Chemist, Cosma Labs Co, 41-43, dir metals labs, 43-45, chief chemist & asst tech dir, 45-50; vpres, Colton Chem Co Div, Air Reduction Co, Inc, 50-56, pres, 56-62; pres, Macco Chem Co Div, Glidden Co, 62-64, vpres & gen mgr, 64-67, vpres opers, Glidden-Durkee Div, 67-70, vpres & gen Mgr, Macco Adhesives Group, SCM Corp, 71-86; RETIRED. *Mem:* Am Chem Soc; Am Ord Asn; Am Inst Chem. *Res:* Fungicides; vinyl resins; emulsions, wax and synthetic resin; process for fusion of bronze to steel; fungicide for ropes, nets and twine; radiological physics; synthetic resins; analytic chemistry. *Mailing Add:* 22150 Shaker Blvd Shaker Heights OH 44122

KRASKIN, KENNETH STANFORD, b Kearny, NJ, Dec 28, 29; m 54; c 2. MICROBIOLOGY. *Educ:* Rutgers Univ, BS, 51, MS, 55, PhD(bact), 57. *Prof Exp:* Lab instr bact, Rutgers Univ, 54-57; microbiologist, Rohm & Haas Co, 57-70; head microbiol, Personal Prod Co, Johnson & Johnson, 71-77, dir appl res, 77-88; TECH CONSULT, 88- *Mem:* Am Soc Microbiol; Sigma Xi; Soc Indust Microbiol. *Res:* Antibiotics; enzyme fermentations; sterility; disinfectants; biodegradation; vaginal microbiology and physiology. *Mailing Add:* 14 N Garden Terr Milltown NJ 08850

KRASNA, ALVIN ISAAC, b New York, NY, June 23, 29; m 55; c 3. BIOCHEMISTRY. *Educ:* Yeshiva Col, BA, 50; Columbia Univ, PhD(biochem), 55. *Prof Exp:* Res worker, 54-56, from instr to assoc prof, 56-70, PROF BIOCHEM, COLUMBIA UNIV, 70- *Concurrent Pos:* Guggenheim Mem Found fel, 62-63. *Mem:* AAAS; Am Chem Soc; Harvey Soc; Am Soc Biol Chemists; Am Soc Microbiol; Am Soc Photobiol. *Res:* Bioconversion of solar energy; regulation of biosynthesis of enzymes. *Mailing Add:* Col Physicians & Surgeons Columbia Univ New York NY 10032

KRASNER, JEROME L, b St Louis, Mo, Feb 25, 40; m 62, 80; c 6. MANAGEMENT. *Educ:* Wash Univ, BS, 61, MS, 64; Boston Univ, PhD(medsci), 69; Nichols Col, MBA, 87. *Prof Exp:* Chmn Dept Biomed Eng & assoc prof healthsci, physiol & biomed eng, Boston Univ, 69-74; pres, Clinco, Inc, 75-81; exec vpres, Plasmedics, Inc, 81-85; vpres res & develop, Clin Develop Corp, 83-89; DEPT HEAD & PROF, DEPT ELEC ENG TECHNOL, WENTWORTH INST TECHNOL, 90- *Concurrent Pos:* Pres, Biocybernetics, Inc, 69-75; dir, Ctr Med Sci, Carnegie-Mellon Inst Res, 77-79 & Ctr Clin Eng, Wentworth Inst Technol. *Honors & Awards:* Cert of Commendation, NASA, 69. *Mem:* Inst Elec & Electronics Engrs; Asn Advan Med Instrumentation; Sigma Xi; NY Acad Sci; AAAS. *Res:* Design, development and international marketing of clinical instrumentation and medical devices. *Mailing Add:* 638 Main St Ashland MA 01721

KRASNER, JOSEPH, b Buffalo, NY, Jan 10, 26; m 53; c 2. BIOCHEMISTRY. *Educ:* Univ Buffalo, BS, 48, EdM, 50, MA, 63, PhD(biochem), 65. *Prof Exp:* High sch teacher, 50-51; asst cancer res scientist, Roswell Park Mem Inst, 51-61; res assoc, Children's Hosp, Buffalo, NY, 65-66; assoc res prof pediat, 66 79, ASSOC RES PROF OBSTET & GYNEC, STATE UNIV NY-BUFFALO, 79-; CANCER RES SCIENTIST II, ROSWELL PARK MEM INST, BUFFALO, 79- *Concurrent Pos:* Dir core labs, Children's Hosp, NY, 66-72. *Mem:* AAAS; Am Chem Soc; NY Acad Sci; Am Asn Clin Chemists; Am Soc Pharmacol & Exp Therapeut; Sigma Xi; Am Soc Biochem & Molecular Biol. *Res:* Biochemical changes during mammalian development and the effect of endogenous and exogenous compounds on development; drug-protein interactions during development and in pathological situations; physical biochemical techniques as used to study the antibody combining site. *Mailing Add:* 60 Snughaven Tonawanda NY 14150

KRASNER, ROBERT IRVING, b Providence, RI, Dec 3, 29; m 64; c 2. BACTERIOLOGY. *Educ:* Providence Col, BS, 51; Boston Univ, AM, 52, PhD(biol), 56. *Prof Exp:* From instr to assoc prof, 58-65, PROF BIOL, PROVIDENCE COL, 65- *Concurrent Pos:* Mem, La State Univ Sch Med Interam Training Prog Trop Med in Cent Am, 62, adv coun clin labs, RI Dept Health, 62- & US Army Biol Labs, Ft Detrick, Md, 65-66; vis prof, Sch Med, Georgetown Univ, 69-71. *Mem:* AAAS; Am Soc Microbiol. *Res:* Medical bacteriology; host-parasite relationships. *Mailing Add:* Dept Biol Providence Col River Ave & Eaton St Providence RI 02918

KRASNER, SOL H, b St Louis, Mo, June 14, 23; m 58; c 4. PHYSICS. *Educ:* Univ Calif, Los Angeles, AB, 57, MA, 48; Univ Chicago, PhD(physics), 55. *Prof Exp:* Jr physicist, Argonne Nat Lab, 49-53; nuclear physicist, Off Naval Res, 55-63; prof lectr, 63-71, ASSOC PROF PHYSICS, UNIV CHICAGO, 70-, DEAN STUDENTS, DIV PHYS SCI, 63- *Concurrent Pos:* Asst to the chmn & dept counselor, Univ Chicago, 76- *Mem:* Am Phys Soc; Am Asn Physics Teachers. *Res:* Nuclear reactor physics; scientific administration. *Mailing Add:* Dept Physics Univ Chicago 5801 Ellis Ave Chicago IL 60680

KRASNEY, JOHN ANDREW, b Long Beach, Calif, Nov 29, 40; m 64, 75; c 5. CARDIOVASCULAR PHYSIOLOGY. *Educ:* Elmhurst Col, BS, 62; Univ Wis-Madison, PhD(physiol), 66. *Prof Exp:* From instr to assoc prof physiol, Albany Med Col, 67-74; assoc prof, 74-83, PROF PHYSIOL, STATE UNIV NY, BUFFALO, 83- *Concurrent Pos:* Nat Heart Inst fel physiol, Univ Wis-Madison, 66-67; mem, Coun Cardiopulmonary Dis & Coun Circulation, Am Heart Asn; animal care & experimentation comt, Am Physiol Soc, 82-85. *Mem:* Am Physiol Soc; Can Physiol Soc; Am Heart Asn; Undersea Med Soc. *Res:* Regulation of circulation; respiration; renal function and blood volume during environmental stresses including chronic hypoxia, water immersion and exercise; neuroendocrine and mechanisms in control of fluid and electrolyte balance; cerebral circulation in chronic hypoxia; mechanisms of high-altitude cerebral edema and acute mountain sickness; ethanol on cerebral blood flow; cardiac function and arterial blood pressure. *Mailing Add:* Dept Physiol State Univ NY Schs Med & Dent 124 Sherman Hall Buffalo NY 14214

KRASNO, LOUIS RICHARD, b Chicago, Ill, Sept 2, 14; m 40; c 1. MEDICINE. *Educ:* Northwestern Univ, BS, 36, MS, 37, PhD(physiol), 39, MD, 44. *Prof Exp:* Asst physiol, Northwestern Univ, 36-39; instr, Chicago City Jr Col, 39-40; instr, Med Sch, Northwestern Univ, 40-47; asst prof clin sci, Univ Ill, 47-57; ASST PROF MED, STANFORD UNIV, 57-; DIR CLIN RES, UNITED AIR LINES, 57- *Concurrent Pos:* Practicing physician, Ill, 47-57. *Mem:* Soc Exp Biol & Med; fel Int Col Angiol; fel AMA; fel Aerospace Med Asn; fel Am Col Cardiol; Sigma Xi. *Res:* Aviation medicine; physiology; cardiovascular medicine. *Mailing Add:* 149 Flying Cloud Isle Foster City CA 94404

KRASNOW, FRANCES, b New York, NY, Oct 16, 94; m 30; c 1. BIOCHEMISTRY. *Educ:* Columbia Univ, BS & AM, 17, PhD(bact, chem, biochem), 22. *Hon Degrees:* LHD, Jewish Theol Sem of America, 74. *Prof Exp:* Asst biochem, Col Physicians & Surgeons, Columbia Univ, 19-22, instr, 22-32, Rhein-Levy Res Fund fel, 20-28; asst dir & head fundamental sci, Sch Dent Hyg, Guggenheim Dent Clin, 32-44, dir res, 44-52; RES CONSULT, 52- *Concurrent Pos:* Consult biochemist, Dept Dermat, Skin & Cancer Unit, NY Postgrad Med Sch & Hosp, Columbia Univ, 23-44; fel, Lehn-Fink Res Fund, NJ, 25-26; consult biochemist, NY State Labor Dept, 29; spec consult, NY Bur Dent Info, 40; res dir, Universal Coatings, Inc, 52-72; specialist clin chem, Am Bd Clin Chem, 53- *Mem:* AAAS; Am Chem Soc; fel Am Inst Chemists; Int Asn Dent Res (ed, vpres, pres, 34-60); assoc fel NY Acad Med; fel NY Acad Sci. *Res:* Skin disease; syphilis; cholesterol; phospholipids; biochemistry of saliva; caries; place of nutrition in dentistry; correlation between metabolic inorganic-organic levels in blood, saliva, urine and tooth conditions. *Mailing Add:* 595 Columbus Ave New York NY 10024

KRASNOW, MARVIN ELLMAN, b Chicago, Ill, Apr 27, 24; m 49; c 5. PHYSICAL CHEMISTRY. *Educ:* Ohio State Univ, PhD(chem), 52. *Prof Exp:* Res assoc electron scattering, Ohio State Univ, 52-53; res chemist polyethylene, Visking Corp, 53-56; mgr chem, Physics & Petrol Lab, Inland Testing Labs, 56-59; dir govt res & develop, Hallicrafters Co, 59-64; coordr indust rels, Univ Ill, Urbana, 85-89, asst dean, Col Eng, 59-64; ASST TO DEAN, COL ENG, ARIZ STATE UNIV, PHOENIX, 89- *Mem:* Inst Elec & Electronics Engrs; Am Soc Eng Educ. *Res:* Electron scattering by gases; investigation of structure of polyethylene; interaction of electromagnetic radiation with matter; evaluation of fuels and lubricants; quantum phenomena. *Mailing Add:* 2526 E Taxidea Way Phoenix AZ 85044

KRASNY, HARVEY CHARLES, b Highpoint, NC, July 27, 45. DRUG METABOLISM, PHARMACOKINETICS. *Educ:* Lynchburg Col, BS, 67; Univ NC, MS, 69, PhD(biochem), 76. *Prof Exp:* SR RES SCIENTIST, WELLCOME RES LABS, BURROUGHS WELLCOME CO, 69- *Mem:* Am Soc Clin Pharmacol & Therapeut; Am Soc Pharmacol & Exp Therapeut; Soc Toxicol; Sigma Xi; NY Acad Sci. *Res:* Drug metabolism and the pharmacokinetic disposition of nucleic acid antagonist in animals and in man. *Mailing Add:* 120 Woodbridge Lane Chapel Hill NC 27514

KRASS, ALLAN S(HALE), b Milwaukee, Wis, May 16, 35. THEORETICAL PHYSICS, SCIENCE POLICY. *Educ:* Cornell Univ, BS, 58; Stanford Univ, PhD(theoret physics), 63. *Prof Exp:* Res assoc physics, Univ Iowa, 62-64; lectr, Univ Calif, Santa Barbara, 64-65, asst prof, 65-72; lectr, Princeton Univ, 72-73; vis lectr, Open Univ Gr Brit, 73-74; assoc prof, 74-79, PROF PHYSICS & SCI POLICY, HAMPSHIRE COL, 81- *Concurrent Pos:* Consult, Off Technol Assessment, US Cong, 76-; NSF fac fel, 76-77; vis researcher, Stockholm Instrnl Peace Res Inst, 80-81, 83-84; sr arms analyst, Union Concerned Scientists, 85-89. *Res:* Elementary particle physics; theoretical high energy physics; science policy, especially arms control, energy and environmental. *Mailing Add:* Sch Natural Sci Hampshire Col Amherst MA 01002

KRASSNER, JERRY, b Brooklyn, NY, Feb 23, 53. ELECTRO-OPTICS, PHENOMENOLOGY. *Educ:* State Univ NY, Stony Brook, BS, 74; Univ Rochester, MA, 76, PhD(physics & astron), 83. *Prof Exp:* Res asst, Univ Rochester, 74-77; staff scientist, 77-85, lab head, 85-90, DIR WASH TECHNOL OFF, GRUMMAN CORP, 90- *Concurrent Pos:* Dir, Infrared Search & Track Working Group, Nat Security Indust Asn, 85, Navy Red Stripe Comt, 86; fel selectee, Indust Res Inst, White House, 90; mem, Am Astron Soc Comt Manpower & Employment, 87-90, Steering Comt, Nat Symposium on Sensors & Sensor Fusion, 89-90 & 91, Advan Space Sensor Technol Assessment Panel, USAF/Am Inst Aeronaut & Astronaut, 88. *Honors & Awards:* Cert Merit, Nat Security Indust Asn, 86. *Mem:* Nat Security Indust Asn; Soc Photo-Optical Instrumentation Engrs; Am Astron Soc; Am Defense Preparedness Asn. *Res:* Theoretical and experimental characterization of naturally occurring electro-optical backgrounds; development and evaluation of new sensor concepts, primarily for aerospace applications; infrared and radio astronomy. *Mailing Add:* Grumman Corp 1000 Wilson Blvd Suite 2800 Rosslyn VA 22209

KRASSNER, STUART M, b New York, NY, Aug 21, 35; m 86; c 2. PARASITOLOGY. *Educ:* Brooklyn Col, BS, 57; Johns Hopkins Univ, ScD(parasitol), 61. *Prof Exp:* NIH fel, int coop med res & training prog, Johns Hopkins Univ-Sch Trop Med, Univ Calcutta, 61-62; guest investr & NIH res trainee, Rockefeller Univ, 62-65; instr invert zool, Hunter Col, 64-65; asst prof organismic biol, 65-69, assoc prof develop & cell biol, 69-73, from vchmn to chmn dept, 69-84, assoc dean grad div, 74-76, biol sci, 77-80 & grad studies & res, 84-85, actg dean grad studies & res, 85-89, JOINT PROF DEVELOP & CELL BIOL & MED MICROBIOL, UNIV CALIF, IRVINE, 73- *Mem:* AAAS; Am Soc Parasitol; Soc Protozool; Am Soc Trop Med Hyg. *Res:* Immune responses in hemoflagellate infections; control of transformation in parasitic hemoflagellates. *Mailing Add:* Dept Develop & Cell Biol Univ Calif Irvine CA 92717

KRATOCHVIL, BYRON, b Osmond, Nebr, Sept 15, 32; m 60; c 4. ANALYTICAL CHEMISTRY. *Educ:* Iowa State Univ, BS, 57, MS, 59, PhD(anal chem), 61. *Prof Exp:* Instr chem, Univ Wis, Madison, 61-62, asst prof, 62-67; assoc prof, 67-71, PROF CHEM, UNIV ALTA, 71-, CHAIR, 89- *Concurrent Pos:* Guest worker, Nat Bur Standards, Washington, DC, 80-81; bd dirs, Chem Inst Can, 77-80; adv bd, Can J Chem, 82-85, anal ed, 85-88, sr ed, 88- *Honors & Awards:* Fisher Award, Can Soc Chem. *Mem:* Fel AAAS; Am Chem Soc; fel Chem Inst Can. *Res:* Metal complex studies; metal ion speciation; analysis using nonaqueous solvents; clinical analysis; sampling for chemical analysis. *Mailing Add:* Dept Chem Univ Alta Edmonton AB T6G 2G2 Can

KRATOCHVIL, CLYDE HARDING, b Racine, Wis, Aug 3, 23; m 44; c 5. PHYSIOLOGY, BIOCHEMISTRY. *Educ:* Univ Wis-Madison, BS, 50, MD, 52, PhD(physiol, biochem), 56. *Prof Exp:* Intern, US Air Force, William Beaumont Army Hosp, Tex, 52-53, officer-in-chg dept physiol & biophys, Sch Aviation Med, Randolph AFB, 55-59, proj officer biosci div, Europ Off, Off Aerospace Res, Brussels, Belg, 59-61, chief biosci div, 61-63, comdr, 6571st Aeromed Res Lab, Holloman AFB, NMex, 63-68 & 6570th Aerospace Med Res Labs, Wright Patterson AFB, 68-70; group mgr, Med Serv, Upjohn Co, 70-74, dir clin res, Europe, 74-88; ACTG DIR, FAMILY HEALTH CTR, 91- *Concurrent Pos:* Flight controller for Proj Mercury, 59-62; consult, Manned Spacecraft Ctr, NASA, 62-; clin prof prev med, Ohio State Univ, 69-71. *Mem:* AAAS; AMA; fel Aerospace Med Asn; Transplantation Soc; Int Acad Aviation & Space Med. *Res:* Aerospace medical research; immunohematology; use of primates in medical research; protein chemistry; renal physiology; central nervous system function in high stress environments; circadian rhythms. *Mailing Add:* Health Int Ltd 6403 Lifeolier St PO Box 3145 Kalamazoo MI 49003

KRATOCHVIL, JIRI, b Prague, Czech, June 11, 44; m 72; c 1. ELECTROCHEMISTRY, BIOMEDICAL SCIENCES. *Educ:* Southampton Univ, PhD(electrochem), 72. *Prof Exp:* Fel electrochem, Univ Okla, 72-73; lectr physiol, St Thomas' Hosp Med Sch, London, 73-75; res fel, Webb-Waring Lung Inst, Denver, 75-76; sr chemist electrochem, Beckman Instruments Inc, 77-79; mgr res & develop, Critikon Inc, Salt Lake City, 79-83; PRES, IE SENSORS INC, SALT LAKE CITY, UTAH, 84- *Mem:* Fel The Chem Soc; Am Chem Soc. *Res:* Ion-selective electrodes; electrochemistry of membranes; polarography; biomedical transducers; semiconductor technology; semiconductor packaging. *Mailing Add:* 3551 S Canyon Way Salt Lake City UT 84106

KRATOHVIL, JOSIP, b Morovic, Yugoslavia, Feb 26, 28; m 52; c 2. COLLOID CHEMISTRY, POLYMER PHYSICAL CHEMISTRY. *Educ:* Univ Zagreb, BS, 52, PhD(chem), 54. *Prof Exp:* Asst chem, Med Sch, Univ Zagreb, 52-59; res fel, Nat Res Coun Can, 59-60; res assoc, 60-64, from asst prof to assoc prof, 64-67, PROF CHEM, CLARKSON COL TECHNOL, 67-, DIR, INST COLLOID & SURFACE SCI, 81- *Mem:* AAAS; Am Chem Soc; Fine Particle Soc; NY Acad Sci; Sigma Xi. *Res:* Coagulation and stability of colloids; light scattering; physical biochemistry; polymer chemistry; micellar systems; bile salts; polyelectrolytes and macromolecules in solutions. *Mailing Add:* Dept Chem Clarkson Univ Potsdam NY 13676

KRATTIGER, JOHN TRUBERT, b Denison, Tex, Aug 30, 16; wid; c 1. MATHEMATICS. *Educ:* Austin Col, BA, 38; Southern Methodist Univ, MA, 39; Univ Okla, EdD, 58. *Prof Exp:* Asst math, Southern Methodist Univ, 39-40; teacher pub sch, Tex, 40-41; asst prof, Col Ozarks, 41-44; instr, Univ Okla, 46-48; from assoc prof to prof, 48-84, vpres student serv, 75-84, EMER PROF MATH, SOUTHEASTERN OKLA STATE UNIV, 84- *Mem:* Math Asn Am. *Res:* Educational guidance. *Mailing Add:* 1729 Asberry Dr Durant OK 74701-2415

KRATZ, HOWARD RUSSEL, b Mattoon, Wis, Nov 2, 16; m 42; c 2. EXPERIMENTAL PHYSICS. *Educ:* Ripon Col, AB, 38; Univ Wis, PhD(physics), 42. *Prof Exp:* Asst physics, Univ Wis, 38-40; asst spectros, Princeton Univ, 40-42; res assoc, Metall Lab, Univ Chicago, 42-44, Los Alamos Sci Lab, 44-46, Northwestern Univ, 46, Res Lab, Gen Elec Co, 46-59, Gen Atomic Co, 59-72; sr res scientist, S-Cubed, 72-79; RETIRED. *Mem:* Am Phys Soc. *Res:* Ultraviolet and infrared spectroscopy; thermal conduction and transfer; plasma physics; explosion phenomena; accelerator development; nuclear weapons effects; instrumentation. *Mailing Add:* 2620 Kanuga Pines Dr Hendersonville NC 28739

KRATZEL, ROBERT JEFFREY, b New York, NY, Feb 5, 49; m 75. IMMUNOHEMATOLOGY, MICROBIOLOGY. *Educ:* Hofstra Univ, BA, 71; State Univ NY, Buffalo, MA, 73, PhD(microbiol), 77. *Prof Exp:* Trainee lab med, Erie County Lab, E J Meyer Mem Hosp, Buffalo, 77; dir tech serv, 78-81, SCI DIR BUFFALO REGION, AM RED CROSS BLOOD SERV, 81- *Concurrent Pos:* Clin instr, Dept Microbiol, State Univ NY, Buffalo, 77-81; clin asst prof, Dept Microbiol, State Univ NY Buffalo, 81-; mem bd dir, Blood Bank Asn NY State Inc. *Mem:* Am Asn Blood Banks; Am Soc Microbiol; Am Soc Histocompatibility & immunogenetics; Am Coun Transplantation. *Res:* The chemical characterization and localization of blood group antigens on blood and tissue cells; health care management. *Mailing Add:* Buffalo Regional Red Cross Blood Ctr 786 Delaware Ave Buffalo NY 14209

KRATZER, D DAL, b Amazonia, Mo, Dec 16, 37; m 63; c 2. ANIMAL BREEDING, STATISTICS. *Educ:* Univ Mo, BS, 59; Iowa State Univ, MS, 64, PhD(animal breeding), 65. *Prof Exp:* Asst animal breeding, Iowa State Univ, 59-62, res assoc animal breeding & comput sci, 62-65, asst prof animal sci & comput sci, 65-68; from asst prof to prof animal sci & statist, Univ Ky, 68-77; BIOSTATISTICIAN, UPJOHN CO, 78- *Mem:* Am Soc Animal Sci; Biomet Soc. *Res:* Behavior of domestic animals. *Mailing Add:* Upjohn Co 700 Portage Rd 7927-190-38 Kalamazoo MI 49001

KRATZER, FRANK HOWARD, b Baldwinsville, NY, Jan 24, 18; m 46; c 3. NUTRITION. *Educ:* Cornell Univ, BS, 40; Univ Calif, PhD(animal nutrit), 44. *Prof Exp:* Asst poultry husb, Univ Calif, 40-43, res assoc, 43-44; assoc prof, Colo Agr & Mech Col, 44-45; asst prof, Col Agr & Environ Sci, 45-49, from assoc prof to prof poultry husb, 49-83, chmn dept, 76-81, EMER PROF AVIAN SCI, COL AGR & ENVIRON SCI, UNIV CALIF, DAVIS, 83- *Concurrent Pos:* NSF fel, Nat Inst Res Dairying, Reading, Eng, 59-60; guest prof, Justus Liebig Univ, Giessen, Germany, 68-69; vis prof, Univ Sydney, 75-76, Fed Univ of Rio Grande do Sol, Porto Alegre, Brasil, 82. *Honors & Awards:* Nat Turkey Fedn Res Award, 49; Am Feed Mfrs Res Award, 60; CPC Res Award, 73. *Mem:* Am Chem Soc; Soc Exp Biol & Med; fel Am Inst Nutrit; fel Poultry Sci Asn; Biochem Soc; fel Am Asn Adv Sci. *Res:* Nutrition of poultry amino acid requirements of chickens and turkeys; vitamin needs and function; minerals and mineral availability; growth inhibitors. *Mailing Add:* Dept Avian Sci Univ Calif Davis CA 95616

KRATZER, REINHOLD, b Kaaden, CSR, Nov 14, 28. POLYMER CHEMISTRY. *Educ:* Univ Munich, Dr rer nat(inorg chem), 60. *Prof Exp:* Res asst inorg chem, Univ Southern Calif, 60-62; res chemist, Naval Ord Lab, Corona, Calif, 62-64; sr scientist, MHD Res Inc, Hercules Powder Co, 64-66; spec mem adv tech staff, Marquardt Corp, 66, mgr chem res, 66-70; MGR, CHEM DEPT, ULTRASYSTS INC, 70- *Mem:* Am Chem Soc; Royal Soc Chem; Ger Chem Soc; AAAS; NY Acad Sci. *Res:* Hydrides of low atomic weight elements and their Lewis base adducts; organometallic chemistry of these elements; phosphonitriles; arc and glow discharge processes; degradation and flammability of polymers; mechanism of acid formation in coal mines; corrosion and oxidation inhibition in lubricating fluids; fluids-seals interactions; preceramic polymers and ceramics. *Mailing Add:* 1425 Seacrest Dr Corona Del Mar CA 92625

KRATZKE, THOMAS MARTIN, b Seattle, Wash, Oct 29, 53; m 79; c 3. STATISTICS, COMPUTER SCIENCE. *Educ:* Pac Lutheran Univ, BS, 75; Wash State Univ, MA, 78; Univ Ill, Urbana-Champaign, PhD(math), 88. *Prof Exp:* Adv mem tech staff, Boeing Computer Serv, 78-80; statistician, Rockwell Hanford Opers, 80-82; teaching & res asst, Math Dept, Univ Ill, 82-88; ANALYST, METRON, INC, 88- *Mem:* Soc Indust & Appl Math; Discrete Math Activ Group. *Res:* Bayesian updating algorithm for use in anti-submarine warfare on different computers. *Mailing Add:* 2373 Old Trail Dr Reston VA 22091

KRAUEL, DAVID PAUL, b Kitchener, Ont, Nov 20, 44; m 69; c 2. TURBULENT DIFFUSION, WIND-WAVE MODELLING. *Educ:* McMaster Univ, BSc, 66; Dalhousie Univ, MSc, 69; Liverpool Univ, PhD(phys oceanog), 72. *Prof Exp:* Res scientist coastal oceanog, Bedford Inst Oceanog, 66-74; asst prof, 74-79, head physics dept, 81-88 ASSOC PROF PHYS OCEANOG, ROYAL ROADS MIL COL, 79-,DEAN GRAD STUDIES, 88- *Concurrent Pos:* Consult, 79- *Mem:* Am Geophys Union; Can Meteorol & Oceanog Soc; Estuarine & Brackish-Water Sci Asn. *Res:* Modelling turbulent diffusion in coastal waters and wind-wave hindcast modelling; estuarine circulation; the effects of flux of wave energy on coastlines and coastal structures. *Mailing Add:* Royal Roads Mil Col FMO Victoria BC V0S 1B0 Can

KRAUS, ALFRED ANDREW, JR, b Richmond, Calif, May 24, 25; m 49; c 4. PHYSICS. *Educ:* Mass Inst Technol, BS, 49; Calif Inst Technol, PhD(physics), 53. *Prof Exp:* Asst, Calif Inst Technol, 50-52; res assoc nuclear physics, Rice Univ, 53-55; from res assoc to instr physics, Univ Pac, 55-61, chmn dept, 56-61; from assoc prof to prof, NMex Highlands Univ, 61-64, chmn dept physics & math, 63-64; phys sci proj dir, Killgore Res Ctr, West Tex State Univ, 64-65 & 67-68, dir, 65-67, prof physics, 64-68; OWNER, CANYON RES CO, 68- *Concurrent Pos:* Radiol physicist, San Joaquin Gen Hosp, 57-60. *Mem:* Am Phys Soc. *Res:* Nuclear physics; computers; relativistic astrophysics. *Mailing Add:* 10313 Betts Dr NE Albuquerque NM 87112

KRAUS, ALFRED PAUL, b Vienna, Austria, June 24, 16; nat US; m 44; c 2. MEDICINE. *Educ:* Univ Chicago, MD, 41. *Prof Exp:* Intern, Michael Reese Hosp, Ill, 41-42, asst resident & resident internal med, 42-44, resident dept hemat res, 48; chief hemat sect, Vet Admin Hosp, Ala, 49; asst chief, Kennedy Vet Admin Hosp, Tenn, 50-52; from asst prof to prof, 53-81, chief sect hemat, 64-81, dir, Ctr Res & Serv Sickle Cell Dis, 74-79, EMER PROF MED, COL MED, UNIV TENN, MEMPHIS, 81- *Concurrent Pos:* Consult hematologist, Baptist Mem Hosp, 53-, Le Bonheur Children's Hosp, Methodist Hosp & St Joseph's Hosp; vis asst prof, Univ Indonesia, 55-56; investr natural hist sickle cell dis, 79- *Mem:* AAAS; Am Soc Hemat; fel AMA; fel Am Col Physicians; Am Fedn Clin Res. *Res:* Hematology; sickle cell disease; abnormal hemoglobins; hemorrhagic diseases; red cell enzymes. *Mailing Add:* 1597 Peabody Ave Memphis TN 38104

KRAUS, ARTHUR SAMUEL, b New York, NY, Aug 2, 25; m 46; c 3. EPIDEMIOLOGY. *Educ:* City Col New York, BS, 49; Columbia Univ, MS, 53; Univ Pittsburgh, ScD(biostatist), 58. *Prof Exp:* Biostatistician, NY State Dept Health, 50-57; chief, Div Statist, Res & Rec, Md State Dept Health, 58-62; asst dir, Off Res, New York City Dept Health, 62-65; head dept biostatist, Montefiore Hosp & Med Ctr, Bronx, NY, 65-66; prof biostatist, Dept Community Health & Epidemiol, 66-85, PROF & COORDR GRAD STUDIES, QUEEN UNIV ONT, 85- *Concurrent Pos:* Consult biostatist, Ont Dept Health, 66- *Mem:* Fel Am Pub Health Asn; Am Heart Asn; Can Pub Health Asn; Soc Epidemiol Res. *Res:* Epidemiologic and health care studies, particularly regarding the elderly and conditions which are disabling to them, such as stroke, dementia, incontinence, deafness and depression. *Mailing Add:* Dept Commun Hlth & Epidemiol Queen's Univ Kingston ON K7L 3N6 Can

KRAUS, C RAYMOND, ENGINEERING. *Prof Exp:* CONSULT, CONSULTING COMM ENG INC. *Mailing Add:* Consulting Comm Eng Inc 845 Mt Moro Rd Villanova PA 19085

KRAUS, ERIC BRADSHAW, b Liberec, Czech, Mar 22, 12; m 42; c 3. METEOROLOGY. *Educ:* Charles Univ, Prague, PhD(geophysics), 46. *Hon Degrees:* PhD, Univ Liege, Belg. *Prof Exp:* Sr res officer, Div Radiophysics, Commonwealth Sci & Indust Res Orgn, Australia, 46-49; authority meteorologist, Snowy Mt, Hydro-Elec Authority, 52-61; sr scientist, Woods Hole Oceanog Inst, 61-66; prof, Univ Miami, 66-77, Div Atmospheric Sci, 69-77, dir, Coop Inst Marine & Atmospheric Studies, 77-81, emer prof meteorol & phys oceanog, 81-; sr res assoc, Coop Inst Res Environ Sci, 81-89; RETIRED. *Concurrent Pos:* Lectr, Univ Sydney, 47-51; mem, Australian Nat Comt Geophys & Geod, 49-55; convener sub-comt oceanog, 52-55; chief, UN tech assistance mission, Nairobi, Kenya, 55-56; mem panel water resources develop, World Meteorol Orgn, 55-56; adj prof, Yale Univ, 61-63; dir, NATO Atmospheric Studies Inst, Urbino Italy 75, trustee, Univ Corp Atmospheric Res, 74-80, dir NATO Atmospheric Res Inst, Bonas, France, 81; vis prof, Paris, France, 81, Monash Univ, Melbourne, Australia, 83, Univ Liège, Belgium, 87, India Nat Inst Oceanog Soc, Indian, 88. *Mem:* Fel Am Meteorol Soc; Am Geophys Union; fel Royal Meteorol Soc. *Res:* Physical oceanography; dynamic climatology and climatic change; air-sea interaction. *Mailing Add:* 322 Jacobson Circle Pagosa Springs CO 81147

KRAUS, GEORGE ANDREW, b Buffalo, NY, June 28, 50; m 85; c 1. ORGANIC CHEMISTRY. *Educ:* Univ Rochester, BS, 72; Columbia Univ, PhD(chem), 76. *Prof Exp:* From asst prof to assoc prof, 76-86, PROF, DEPT CHEM, IOWA STATE UNIV, 86- *Concurrent Pos:* DuPont young fac grant, 76-78; 3M Young Fac grant, 81-82. *Mem:* AAAS; Am Chem Soc; fel The Chem Soc. *Res:* Active in the development of new synthetic methods and the application of these methods to the total synthesis of natural products; interests include kinetic anions, photochemistry and thermal chemistry; research on antiprotozoan diseases, suicide enzyme inhibitors and antiretroviral drugs; 107 publications. *Mailing Add:* Dept Chem Iowa State Univ Ames IA 50011

KRAUS, HUBERT ADOLPH, b Graz, Austria, Nov 28, 07; nat US; m 29; c 2. CHEMISTRY, PHYSICS. *Educ:* Drexel Inst, Philadelphia, Pa, BS, 28; City Col NY, MS, 34. *Prof Exp:* Consult protective coatings, Durok Bldg Mat, Inc, Hastings-on-Hudson, NY, 59-81; assoc prof chem, Burlington Community Col, 71-73; RETIRED. *Mem:* Am Chem Soc; AAAS; Am Inst Chemists; Int Union Pure & Appl Chem; NY Acad Sci. *Res:* Protective coating components and products; the twelve basic physical charges. *Mailing Add:* 503 Wayne Dr Cinnaminson NJ 08077

KRAUS, JAMES ELLSWORTH, b Rocky Ford, Colo, Nov 19, 09; m 35; c 1. HORTICULTURE. *Educ:* Colo State Col, BS, 32; Univ Wis, MS, 34; Cornell Univ, PhD(veg crops), 40. *Hon Degrees:* Prof, Univ Ecuador, SAm, 56; DSc, Univ Idaho, 83. *Prof Exp:* Asst hort, Univ Wis, 33-34; asst veg crops, Cornell Univ, 39-40; asst physiologist, Bur Plant Indust, USDA, 36-41; assoc horticulturist, Aberdeen br exp sta, Idaho, 41-44; plant breeder, Calif Packing Corp, Ill, 44-45; assoc horticulturist, Univ Idaho, 45-47, horticulturist, Agr Exp Sta, 48-49, assoc dir, 49-55, dir, Agr Exp Sta & Agr Exten Serv & dean, Col Agr, 55-72, prof hort & head dept, 48-49, prof plant sci, 49-72, EMER DEAN COL AGR & EMER PROF PLANT SCI, UNIV IDAHO, 72- *Mem:* Sigma Xi. *Res:* Culture and physiology of potatoes; genetics and breeding hybrid onions; culture and physiology of freezing and canning crops; vegetable seed production. *Mailing Add:* 718 E First St Moscow ID 83843

KRAUS, JESS F, b Los Angeles, Calif, Apr 4, 36; m 57; c 5. EPIDEMIOLOGY, ENVIRONMENTAL HEALTH. *Educ:* Sacramento State Col, BA, 59, MS, 63; Univ Calif, Berkeley, MPH, 64; Univ Minn, Minneapolis, PhD(environ epidemiol), 67. *Prof Exp:* Instr epidemiol & environ health, Univ Minn, 67-68; adj asst prof, Univ Cincinnati, 68-69; from asst prof to assoc prof community health, Sch Med, Univ Calif, Davis, 69-80, prof, 80-, AT DEPT PUB HEALTH, UNIV CALIF, LOS ANGELES. *Concurrent Pos:* Chief environ epidemiol, USPHS, 68-69, epidemiologist, Bur Environ Mgt, 69-71; fel coun epidemiol, Am Heart Asn. *Mem:* Am Pub Health Asn; Soc Epidemiol Res; Asn Teachers Prev Med. *Res:* Design and execution of community and epidemiologic research involving the interrelationship of man with his physical environment. *Mailing Add:* Comm Health Univ Calif Davis Davis CA 94616

KRAUS, JOHN DANIEL, b Ann Arbor, Mich, June 28, 10; m 41; c 2. ELECTRICAL ENGINEERING, ASTRONOMY. *Educ:* Univ Mich, BS, 30, MS, 31, PhD(physics), 33. *Prof Exp:* Asst physics, Univ Mich, 31-32, res assoc, Dept Eng Res, 34-35, res physicist, Dept Physics, 36-37; res physicist, Physicist Res Co, 37-38; independent res & consult, Ann Arbor, 38-40; physicist, Naval Ord Lab, 40-43; res assoc, Radio Res Lab, Harvard Univ, 43-46; from assoc prof to prof elec eng, 46-71, TAINE G MCDOUGAL PROF ELEC ENG & ASTRON, OHIO STATE UNIV, 71-, DIR, RADIO OBSERV, 52- *Honors & Awards:* Sullivant Medal, Inst Elec & Electronics Engrs, 78, Edison Medal, 84, Edison Medal, 85, Heinrich Hertz Medal, 90. *Mem:* Nat Acad Eng; Am Phys Soc; Am Astron Soc; Inst Elec & Electronics Engrs. *Res:* Electromagnetic theory; antennas; radio astronomy; author of several textbooks. *Mailing Add:* Radio Observ Ohio State Univ Columbus OH 43210

KRAUS, JOHN FRANKLYN, b Brooklyn, NY, Nov 12, 29; m 57; c 5. FOREST GENETICS. *Educ:* Univ Mich, BSF, 53, MF, 56; Univ Minn, PhD(forestry), 66. *Prof Exp:* Res forester, Southeastern Forest Exp Sta, US Forest Serv, 56-64, plant geneticist, 64-72, plant geneticist, Ga Forestry Ctr, 72-86; RETIRED. *Mem:* AAAS; Soc Am Foresters; Sigma Xi. *Res:* Breeding improved strains of southern pines. *Mailing Add:* 1211 Timberlane Dr Macon GA 31210

KRAUS, JON ERIC, b Cambridge, Mass, May 16, 51; m 77; c 1. FUNCTIONAL ANALYSIS, OPERATOR ALGEBRAS. *Educ:* Univ Calif, Santa Barbara, BA, 72; Univ Calif, Berkeley, MA, 75, PhD(math), 77. *Prof Exp:* Hill res instr math, 77-79, asst prof, 79-85, ASSOC PROF MATH, STATE UNIV NY, BUFFALO, 85- *Mem:* Am Math Soc. *Res:* Operator algebras including von Neumann, C-algebras and reflexive algebras; noncommutative dynamical systems. *Mailing Add:* Dept Math State Univ NY 106 Diefendorf Hall Buffalo NY 14214

KRAUS, KENNETH WAYNE, b Waterloo, Iowa, Oct 20, 35; m 56; c 4. ORGANIC CHEMISTRY. *Educ:* Loras Col, BS, 57; Univ Calif, Berkeley, PhD(chem), 60. *Prof Exp:* From asst prof to assoc prof, 60-72, chmn dept, 65-69 & 70-71, PROF CHEM, LORAS COL, 72- VPRES ACAD AFFAIRS, 87- *Concurrent Pos:* NSF grant, 64-66; lectr, Dept Chem, Calif State Col Long Beach, 69-70. *Mem:* Am Chem Soc. *Res:* Mechanism and use of the reaction of organocadmium reagents with acid chlorides, dipole moments, syntheses and structure proof. *Mailing Add:* Dept Chem Loras Col Dubuque IA 52001

KRAUS, LORRAINE MARQUARDT, b Suffern, NY, Sept 6, 22; m 44; c 2. BIOCHEMISTRY. *Educ:* Mt Mary Col, BS, 43; Univ Tenn, MS, 52, PhD, 56; Memphis Col Art, BFA, 82. *Prof Exp:* Res technician, Dept Endocrinol & Metab Dis, Michael Reese Hosp, 43-44, 48; instr chem, Univ Indonesia, 55-56; res assoc, 57-60, from asst prof to assoc prof, 60-72, chairperson dept biochem, 82-84, PROF BIOCHEM, CTR HEALTH SCI, UNIV TENN, MEMPHIS, 72- *Concurrent Pos:* Mem, Blood Dis & Resources Adv Comt, Blood Div, Nat Heart, Lung & Blood Inst, 79-83. *Mem:* Int Soc Hemat; Am Chem Soc; Am Soc Hemat; Am Soc Human Genetics; Am Soc Biol Chemists; Sigma Xi. *Res:* Biosynthesis of abnormal human hemoglobins; erythropoietin; tissue culture of hemic cells; immunochemistry; chemistry of sickle cell disease; carbanoylation of amino acids and proteins in renal diseases. *Mailing Add:* Dept Biochem Univ Tenn Ctr Health Sci 800 Madison Ave Memphis TN 38163

KRAUS, OLEN, b Berwick, Pa, Apr 7, 24; m 46; c 2. PHYSICS. *Educ:* Pa State Univ, BS, 50; Mich State Univ, MS, 52, PhD(physics), 65. *Prof Exp:* Physicist, Nat Bur Standards, 55-62; chmn dept physics, Univ SDak, 62-67; chmn dept, 67-77, PROF PHYSICS, UNIV NDAK, 67-, ASSOC DEAN, COL ARTS & SCI, 77- *Mem:* Am Phys Soc; Am Asn Physics Teachers. *Res:* Nuclear magnetic resonance; quantum mechanics; mathematical physics; electron resonance. *Mailing Add:* Physics 214 Witmer Hall Univ NDak PO Box 8008 Grand Forks ND 58202

KRAUS, SAMUEL, b Irvington, NJ, Mar 15, 25; m 54; c 3. AEROTHERMODYNAMICS, FLUID MECHANICS. *Educ:* Rensselaer Polytech Inst, BAeroEng, 44, MAeroEng, 49. *Prof Exp:* Jr engr, preliminary aeronaut design, propeller div, Curtiss-Wright Corp, 46-47; aeronaut res engr, Ames Res Ctr, NASA, 49-62; res specialist, space div, 62-67, lead engr, S-II aerothermodynamics, 67-69; lead engr, 69-, mem tech staff, Shuttle Aerodynamic Loads, 69-85; prin eng specialist, gas dynamics, shuttle aerodyn, aero sci, Space Transp Systs Div, Rockwell Int Corp, 85-90; RETIRED. *Mem:* Assoc fel Am Inst Aeronaut & Astronaut; Sigma Xi. *Res:* Experimental and theoretical research in aerothermodynamics; planning, development and utilization of corporate experimental facilities; preflight prediction and postflight verification of aerospace vehicle environment; ignition overpressure; wake recirculation; venting of aerospace vehicles; free molecular flow forces. *Mailing Add:* 6108 Monero Dr Rancho Palos Verdes CA 90274

KRAUS, SHIRLEY RUTH, b New York, NY, Dec 24, 19; m 46; c 2. CLINICAL PHARMACOLOGY. *Educ:* Hunter Col, BA, 40; Cornell Univ, MA, 42; Univ Ill, PhD(physiol), 46. *Prof Exp:* Asst dept exp biol, Am Mus Natural Hist, NY, 40-41; hematologist, Jewish Mem Hosp, 41; asst biol & chem, Adelphi Col, 42-43, instr, Sch Nursing, 43; asst dept zool & physiol, Univ Ill, 43-46; high sch teacher, 46-47; physiologist, Gastroenterol Res Lab, Mt Sinai Hosp, 47-48; biochemist, Cancer Res Found, Harlem Hosp, 48-50; lectr physiol, Col Dent, NY Univ, 50-51; pharmacologist, Cancer Res & Metab Unit, Mt Alto Hosp, Washington, DC, 51-53; instr pharmacol, Sch Med, Howard Univ, 55-56; assoc prof pharmacol, Long Island Univ, prof, Brooklyn Col Pharm, 65-75; prof pharmacol & physiol, 75-82, dir, Div Pharmacotherapeut & Health Sci, 79-82, EMER PROF PHARMACOL & PHYSIOL, ARNOLD & MARIE SCHWARTZ COL PHARM & HEALTH SCI, LONG ISLAND UNIV, 82- *Concurrent Pos:* instr, Eve Sch, Brooklyn Col, 47-48; assoc prof, State Univ NY Downstate Med Ctr, 69-70; vis fel clin pharmacol, Cornell Univ Med Col, 77-78; mem, Instnl Rev Bd, Clin Drug Investr Inc, 81; vis prof physiol & pharmacol, NY Col Pediat Med, 83-84. *Mem:* Soc Exp Biol & Med; Endocrine Soc; Am Physiol Soc; Am Soc Pharmacol & Exp Therapeut; fel Am Soc Clin Pharmacol & Therapeut. *Res:* Alloxan diabetes-anaphylaxis and granuloma pouch formation; pituitary-adrenal stress response in alloxan diabetes; anti-estrogenic action of B glycyrrhetinic acid; hyperthermia on blood platelets in male rats; corticosterone and adrenocorticotropic hormone in alloxan diabetic rats. *Mailing Add:* 13901 Coolidge Ave Jamaica NY 11345

KRAUS, WILLIAM LUDWIG, b Augsburg, Ger, Aug 12, 22; nat US; m 56; c 4. CARDIOLOGY. *Educ:* Harvard Med Sch, MD, 52. *Prof Exp:* Intern med, Roosevelt Hosp, New York, 52-53, asst resident, 53-55; mem med staff, Baylor Univ Hosp, 59-60; DIR CARDIAC LAB, ST PAUL HOSP, 60- *Concurrent Pos:* Life Inst Med Res Fund fel, Harvard Med Sch, 57-59; asst, Peter Bent Brigham Hosp, Boston, 57-59; mem med staff, Gaston Hosp, 59-60 & Parkland Mem Hosp, 59-; clin prof med, Univ Tex Health Sci Ctr, 59- *Mem:* Fel Am Col Cardiol; fel Am Col Physicians; Am Fedn Clin Res. *Res:* Cardiovascular physiology. *Mailing Add:* 5939 Harry Hines Blve Suite 600 Dallas TX 75235

KRAUSCHE, DOLORES SMOLENY, b Cleveland, Ohio, Jan 27, 42. PHYSICS, ELECTRICAL ENGINEERING. *Educ:* Univ Fla, BS, 65, MS, 67, BSEE, 67, PhD(physics, astron), 75. *Prof Exp:* Res asst radioastron, 65-75, fel physics, 76-77, RES PHYSICIST, UNIV FLA, 78- *Concurrent Pos:* Interim engr, Electronic Commun Inc, 67; prin staff mem, Oper Res Inc, 78- *Mem:* Am Astron Soc; assoc mem Sigma Xi; assoc mem Inst Atmospheric Optics & Remote Sensing. *Res:* Electromagnetic phenomena relating to engineering problems and astronomical research. *Mailing Add:* 805 E University Ave Gainesville FL 32601

KRAUSE, DALE CURTISS, b Wichita, Kans, Dec 27, 29; c 2. OCEANOGRAPHY, MARINE GEOLOGY. *Educ:* Calif Inst Technol, BS, 52; Univ Calif, Los Angeles, MS, 57; Univ Calif, San Diego, PhD, 61. *Prof Exp:* Mining geologist, Cerro de Pasco Corp, Peru, 52-54; res asst marine geol, Univ Calif, 56-61; asst res prof oceanog, Grad Sch Oceanog, Univ RI, 62-66, from assoc prof to prof, 66-73, actg prof, 73-; dir, Div Marine Sci, UNESCO, 73-; RETIRED. *Concurrent Pos:* Consult, various industs; NSF fel, NZ Oceanog Inst, 61-62; exchange fel, Nat Acad Sci-USSR Acad Sci, Moscow, 67-68; fel, Cambridge Univ, UK, 68-69; res oceanographer, Nat Oceanic & Atmospheric Admin, Miami, 72-73. *Mem:* Geol Soc Am; AAAS; Am Geophys Union. *Res:* Origin and evolution of the sea floor, ocean basins and continents. *Mailing Add:* 19 Villa Croix Nivert Paris 75015 France

KRAUSE, DANIEL, JR, b Sudbury, Mass, Feb 21, 45; m 75. CHEMICAL OCEANOGRAPHY. *Educ:* Univ Mass, BS, 66, PhD(chem), 72. *Prof Exp:* RES ASSOC OCEANOG, AMHERST COL, 72- *Mem:* Sigma Xi; Am Phys Soc; Am Geophys Union. *Res:* Mass spectrometric and gasometric studies of gases dissolved in water. *Mailing Add:* 734 Bay Rd Amherst MA 01002

KRAUSE, DAVID, b Eggenfelden, WGermany, May 10, 50; US citizen; m 75; c 3. IN VITRO CELL BIOLOGY. *Educ:* Atlantic Union Col, SLancaster, Mass, BA, 72; Andrews Univ, Berrien Springs, Mich, MA, 78; Cath Univ Am, PhD(cell biol), 82. *Prof Exp:* Sr lab tech, Inst Infectous Dis, US Army Res, 73-74; teaching asst biol, Cath Univ Am, 78-82; fel res assoc cell biol, Uniformed Serv Univ Health Sci, 82-84, res instr, 85-86; SR RES ASSOC, CATH UNIV AM, 86- *Mem:* AAAS; Tissue Cult Asn; Am Soc Cell Biol; Int Soc Interferon Res. *Res:* Cell differentiation; in vitro alternatives to animal testing; mechanism of action of interferon. *Mailing Add:* 903 Helena Dr Silver Spring MD 20901

KRAUSE, DAVID WILFRED, b Medicine Hat, Alta, Feb 15, 50; m 78; c 2. ANATOMY. *Educ:* Univ Alta, BSc, 71, MSc, 76; Univ Mich, PhD(geol), 82. *Prof Exp:* Assoc prof anat, State Univ NY, Stony Brook, 82-87; AT MUS PALEONT, UNIV MICH, 88- *Honors & Awards:* Anna M Jackson Award, Am Soc Mammalogists, 81. *Mem:* Am Soc Mammalogists; Paleont Soc; Sigma Xi; Soc Syst Zool; Soc Vert Paleont. *Res:* Evolution of late Mesozoic and early Cenozoic vertebrates, particularly mammals, and the form and function of the mammalian dentition and postcranial skelton. *Mailing Add:* Dept Anat Sci Health Sci Ctr State Univ NY Stony Brook NY 11794-8081

KRAUSE, ELIOT, b New York, NY, June 7, 38; m 59; c 3. POPULATION GENETICS. *Educ:* Cornell Univ, BS, 60; Purdue Univ, MS, 63, PhD(genetics), 68. *Prof Exp:* Res asst pop genetics, Purdue Univ, 60-65; instr, Seton Hall Univ, 65-68, grad biol adv, 77-88, dept chmn, 88-89, ASST PROF BIOL, SETON HALL UNIV, 68- *Mem:* AAAS; Genetics Soc Am; Biomet Soc; Am Genetic Asn; Sigma Xi. *Res:* Cytogenetic research with various mutagens using human lymphocytes as detected by sister chromatic exchange; selection of quantitative traits in Tribolium castaneum using genotype-environment, nutrition, interactions; study of cannibalism and competitive factors which affect population size in Tribolium; study of fragile sites in cytogenetics. *Mailing Add:* Dept Biol Seton Hall Univ South Orange NJ 07079

KRAUSE, ERNST HENRY, physics; deceased, see previous edition for last biography

KRAUSE, EUGENE FRANKLIN, b Kenosha, Wis, Apr 7, 37; m 59; c 2. MATHEMATICS. *Educ:* Univ Wis, BS, 59, MA, 60, PhD(math), 63. *Prof Exp:* From instr to assoc prof, 63-76, PROF MATH, UNIV MICH, ANN ARBOR, 76- *Mem:* Nat Coun Teachers Math; Math Asn Am. *Res:* Mathematics education. *Mailing Add:* Dept Math Univ Mich Main Campus Ann Arbor MI 48109

KRAUSE, HELMUT, forest soils, for more information see previous edition

KRAUSE, HELMUT G L, b Koenigsberg, Ger, Nov 10, 11; US citizen; m 55; c 1. LOW DENSITY AERODYNAMICS, THERMODYNAMICS. *Educ:* Albertus Univ, Ger, PhD(astron), 38. *Prof Exp:* Sci res asst, Koenigsberg Univ Observ, 37-44; res scientist, Inst Ballistics, Ger Air Force Acad, 44-45 & Carl Zeiss Optical Factory, 45-47; res asst, Inst Theoret Physics, Univ Jena, 47 & Hamburg Univ Observ, 47-48; tech physicist, Glycerine & Aliphatic Acid Factory, 49-50; dep chief, Astronaut Res Inst, Univ Stuttgart, 51-54; sr res scientist, Inst Physics Jet Propulsion, 54-57; spec asst space sci, Army Ballistic Missile Agency, Ala, 57-60; sci adv studies off, Marshall Space Flight Ctr, NASA, 60-65, sci adv to dir aero-astrodyn lab, 65-74; mgr bur anal res, 75-78; aerospace scientist, Marshall Space Flight Ctr, NASA, 78-89; RETIRED. *Honors & Awards:* Ernst Heinckel Space Flight Award, 51. *Mem:* Assoc fel Am Inst Aeronaut & Astronaut; sr mem Am Astronaut Soc; Am Astron Soc; fel Brit Interplanetary Soc; Am Soc Rocket & Space Flight. *Res:* Rocket ballistics, space mechanics and astronautical sciences; first-order perturbation theory used for Explorer I and Vanguard I; author or coauthor of 50 scientific publications. *Mailing Add:* 2718 Briarwood Dr Huntsville AL 35801

KRAUSE, HERBERT FRANCIS, b Woodbury, NJ, Mar 10, 42; m 70; c 2. ATOMIC PHYSICS, CHEMICAL PHYSICS. *Educ:* Drexel Univ, BS, 65; Univ Pittsburgh, PhD(physics), 71. *Prof Exp:* SR RES PHYSICIST, OAK RIDGE NAT LAB, 71- *Concurrent Pos:* Pvt computer consult. *Mem:* Am

Phys Soc; Am Asn Physics Teachers; AAAS; Sigma Xi. *Res:* Atomic and molecular beam research; thermal-high energies; molecular dynamics studies involving inner and outer shell excited species; high energy atomic physics and particle solid interactions. *Mailing Add:* 1068 W Outer Dr Oak Ridge TN 37830

KRAUSE, HORATIO HENRY, b St Paul, Minn, Oct 11, 18; m 50; c 2. INORGANIC CHEMISTRY. *Educ:* St Mary's Col Minn, BS, 39; Univ Minn, PhD(chem), 55. *Prof Exp:* Teacher pvt sch, 39-43; instr chem & physics, De LaSalle Mil Acad, 43-48; asst, Univ Minn, 49-54; prin chemist, 55-57, proj leader chem, 57-62, sr chemist, 62-70, prin res scientist, 70-80, SR RES SCIENTIST, BATTELLE MEM INST, 80- *Mem:* AAAS; Am Chem Soc; Am Soc Mech Eng; Nat Asn Corrosion Engrs; Sigma Xi. *Res:* High temperature chemistry, deposits and corrosion in boilers and incinerators; molten salts; solid propellants; air pollution; sulfur oxides. *Mailing Add:* Battelle Mem Inst 505 King Ave Columbus OH 43201-2693

KRAUSE, IRVIN, b New York, NY, July 18, 32; m 53; c 3. MECHANICAL ENGINEERING. *Educ:* City Col New York, BME, 54; Columbia Univ, MS, 55; NY Univ, EngScD(mech eng), 60. *Prof Exp:* Lectr graphics, City Col New York, 54-59; asst mech eng, NY Univ, 59-60; res scientist, Res Div, American-Standard Corp, 60-63; assoc prof mech eng & dir mat labs, Fairleign Dickinson Univ, 63-66; chief res & develop, Diehl Div, Singer Co, 66-70, mgr eng, Indust Prod Div, 70-76; dir eng, Acushnet Co, 76-78; mgr, Mfg Technol, Arthur D Little, Inc, 78-81, tech dir, Comput Integrated Mfg, 81-85; dir, Comput Integrated Mfg, 85-86 MANAGING PARTNER, CTR MFG TECHNOL, COOPERS & LYBRAND, 86- *Concurrent Pos:* NSF fel, 59; consult, Army Res Off NC, 64- *Mem:* AAAS; Am Soc Mech Engrs; Sigma Xi; Soc Mfg Engrs. *Res:* Visco-elastic behavior of polymeric materials; fracture in brittle materials; kinematics and mechanism synthesis; tunnel diode accelerometers; laser system metrology; electric motors and controls; automated production systems; computerization and automation as applied to the design and manufacturing activities in industry. *Mailing Add:* One Seaward Lane One Post Office Sq South Dartmouth MA 02748

KRAUSE, JOSEF GERALD, b Kearny, NJ, Mar 21, 42; m 63; c 2. ORGANIC CHEMISTRY. *Educ:* Hobart Col, BS, 63; Northeastern Univ, PhD(org chem), 67. *Prof Exp:* NSF res fel org chem, Univ Mass, 67-68; PROF CHEM, NIAGARA UNIV, 68- *Concurrent Pos:* Sigma Xi res grant-in-aid, Niagara Univ, 68-69. *Mem:* Am Chem Soc. *Res:* Organic nitrogen compounds; organic synthesis; reaction mechanisms; bicyclic ring systems. *Mailing Add:* Dept Chem Niagara University NY 14109

KRAUSE, LEONARD ANTHONY, b Hartford, Conn, May 13, 25; m 55, 68; c 4. BIOCHEMISTRY. *Educ:* Univ Conn, BA, 50, MS, 51; Univ Cincinnati, ScD(indust health, med), 62; Am Bd Indust Hyg, dipl. *Prof Exp:* Instr physiol & zool, Univ Conn, 49-51; biochemist-indust hygienist, Bur Labs, Conn, 51-55, sr indust hygienist, 56-59; Virginia, chief indust hyg-air pollution, Resources Res, DC, 61-62; dir environ hyg, Nat Insts Health, 62-63; mgr environ hyg serv, 63-75, DIR ENVIRON HYG & TOXICOL, OLIN CORP, 75- *Concurrent Pos:* Instr environ sci, Middlesex Coun Col, 66-75; dir, Chem Indust Inst Toxicol. *Mem:* AAAS; Am Indust Hyg Asn; Air Pollution Control Asn; Am Pub Health Asn; Chem Indust Inst Toxicol. *Res:* Physiology of invertebrates; aspects of metal fume fever; use of vaporphase chromatography; insecticides and effects on human metabolism. *Mailing Add:* 19 Wellsweep Rd Branford CT 06405

KRAUSE, LLOYD O(SCAR), b Hamburg, Wis, Oct 23, 18; m 42; c 2. ELECTRICAL ENGINEERING. *Educ:* Rose-Hulman Inst Technol, BS, 40; Syracuse Univ, MEE, 64, PhD(elec eng), 66. *Prof Exp:* Test engr, Gen Elec Co, 40-41, prog engr, 41-43, develop engr, 43-47, proj engr, 47-52, asst sect engr, 52-53, mgr elec eng, 53-63, consult engr, Electronics Lab, 63-67; tech adv, Autonetics Div, NAm Rockwell Corp, 67-76; systs analyst, Space Div, 76-84, PRIN ENG SPECIALIST, SATELLITE SYSTS DIV, ROCKWELL INT CORP, 84- *Honors & Awards:* Coffin Award, Gen Elec Co, 53. *Mem:* Sigma Xi; Electronic Industs Asn; Inst Elec & Electronics Engrs; Nat Soc Prof Engrs. *Res:* Radio frequency and microwave radiators; antennas; transmission lines and networks; solid state microwave; paramagnetic amplifiers; phase shifters; switches; ferrites; ferroelectrics; ground screens; electronic systems; computer reliability; correlation loops; low angle radiation; satellite sensors; navigation error analysis; adaptive arrays and noise filtering; coding and data compression; probability estimation and system simulation. *Mailing Add:* Satellite Systs Div Rockwell Int 2600 Westminster Blvd PO Box 3644 Seal Beach CA 90740-7644

KRAUSE, LUCJAN, b Poznan, Poland, Jan 8, 28; nat Can; m 50; c 6. PHYSICS, ATOMIC & MOLECULAR COLLISIONS. *Educ:* Univ London, BSc, 51, DSc(physics), 68; Univ Toronto, MA, 53, PhD(physics), 55. *Hon Degrees:* DSc, Copernicus Univ, Torun, Poland, 83. *Prof Exp:* Assoc prof physics, Mem Univ, 55-58 & Assumption Univ, 58-63; prof physics & head dept, Univ Windsor, 63-83; RETIRED. *Concurrent Pos:* Hon res fel, Univ Col, Univ London, 70-71; adj prof eng sci, Wayne State Univ, Detroit, 72-; fel, Churchill Col, Cambridge. *Mem:* Fel Am Phys Soc; Can Asn Physicists; fel Brit Inst Physics; Optical Soc Am. *Res:* Laser spectroscopy of atoms and molecules, sensitized fluorescence and quenching, lifetimes of excited atomic and molecular states. *Mailing Add:* 890 Bartlet Dr Windsor ON N9G 1V4 Can

KRAUSE, MANFRED OTTO, b Stuttgart, Ger, Mar 11, 31; m 63. ATOMIC PHYSICS, CHEMICAL PHYSICS. *Educ:* Univ Stuttgart, Diplom Phys, 57; Max Planck Inst, Dr rer nat, 60. *Prof Exp:* Sr scientist mass spectrometry, William H Johnston Lab, Inc, Md, 60-63; SR SCIENTIST ELECTRON SPECTROMETRY, OAK RIDGE NAT LAB, 63- *Concurrent Pos:* Exchange prof, Lab Curie, Paris France, 75; Humboldt awardee, Stiftung, Ger, 75-76. *Mem:* AAAS; Sigma Xi; fel Am Phys Soc. *Res:* Transuranic chemistry; electron spectrometry; x-ray analysis; photoionization; atomic and molecular physics. *Mailing Add:* Oak Ridge Nat Lab Box 2008 4500N Oak Ridge TN 37831-6201

KRAUSE, MARGARIDA OLIVEIRA, b Lisbon, Portugal, Jan 13, 31; Can citizen; m 56; c 3. CELL BIOLOGY, GENETICS. *Educ:* Univ Lisbon, BSc, 53; Univ Wis, MSc, 57, PhD(cell biol), 60. *Prof Exp:* Res assoc cell biol, Univ Wis, 60-61 & Univ Toronto, 63-66; res assoc, 66-70, assoc prof, 70-76, PROF CELL BIOL, UNIV NB, 76- *Concurrent Pos:* Res grants, Banting Res Found, 65 & 67, Med Res Coun Can, 65-66, Nat Res Coun Can, 67-78, Nat Cancer Inst Can, 76-79 & 84-87, Cancer Res Soc Inc, 88-90, Nat Sci & Eng Res Coun Can, 78-91 & Fed Centres Excellence Prog in Insect Biotech Network, 90-94; mem, Can Nat Comt Int Union Biol Soc, 80-83, chmn, 83-86, mem int sci & technol affairs, 81-87; mem Nat Sci & Eng Res Coun Can grant selection comts, 80-86; exchange scientists, Int Prog, NRC, 84-87; co-chmn, Prog Comt Int Congress Cell Biol, 88. *Mem:* Biol Coun Can; Am Soc Cell Biol; Can Soc Cell Biol (pres, 84-86); Int Cell Cycle Soc. *Res:* Role of chromosomal proteins and small nuclear RNA in gene expression; chromatin template activity and structure-cellcycle control and cancer transformation. *Mailing Add:* Dept Biol Univ NB Fredericton NB E3B 6E1 Can

KRAUSE, PAUL CARL, JR, b Reynolds, Nebr, Jan 27, 32; m 53; c 4. ELECTRICAL ENGINEERING. *Educ:* Univ Nebr, BS, 56 & 57, MS, 58; Univ Kans, PhD(elec eng), 61. *Prof Exp:* Instr elec eng, Univ Kans, 58-61; res elec engr, Allis-Chalmers Mfg Co, 61-62; asst prof elec eng, Univ Wis-Milwaukee, 62-65; assoc prof, Univ Wis-Madison, 65-70; PROF ELEC ENG, PURDUE UNIV, 70- *Concurrent Pos:* Consult, Allis-Chalmers Mfg Co, 63-66. *Mem:* Sr mem Inst Elec & Electronics Engrs; Am Soc Eng Educ. *Res:* Electric machines, power systems and control systems; hybrid computer applications in analysis of systems. *Mailing Add:* Dept Elec Eng Purdue Univ West Lafayette IN 47907

KRAUSE, PAUL FREDERICK, b Racine, Wis, July 30, 45; m 70; c 1. PHYSICAL CHEMISTRY. *Educ:* Dubuque Univ, BS, 67; Univ Iowa, PhD(chem), 72. *Prof Exp:* Res asst chem, Univ Iowa, 68-72; res assoc, Univ Pittsburgh, 72-73; res assoc, 73-74, instr, 74, teaching fel & vis asst prof chem, Miami Univ, 74-77; ASST PROF CHEM, UNIV CENT ARK, 77- *Mem:* Sigma Xi. *Res:* Molecular spectroscopy and its use for structural considerations, particularly in the solid state. *Mailing Add:* Dept Chem Univ Cent Ark Conway AR 72032

KRAUSE, PETER JAMES, b Denver, Colo, Mar 17, 45; m 75; c 3. PEDIATRICS, PEDIATRIC INFECTIOUS DISEASES. *Educ:* Williams Col, BA, 67; Tufts Univ Sch Med, MD, 71. *Prof Exp:* Intern/resident pediat, New Haven Hosp, Yale Univ, 71-73; resident pediat, Med Ctr, Stanford Univ, 73-74; physician pediat, US Army, Bad Kreuznach, WGer, 74-76; res fel pediat infectious dis, Med Ctr, Univ Calif, Los Angeles, 76-79; PHYSICIAN PEDIAT INFECTIOUS DIS, HARTFORD HOSP, SCH MED, UNIV CONN, 79-, PROF PEDIAT, 91- *Concurrent Pos:* Chief pediat infectious dis, Hartford Hosp, Sch Med, Univ Conn, 79-, from asst prof to assoc prof pediat, 79-91. *Mem:* Am Soc Microbiol; fel Infectious Dis Soc Am; Am Fedn Clin Res; Sigma Xi; AAAS. *Res:* Epidemiology, pathogenesis, immunology, diagnosis and treatment of human babesiosis; neutrophil function including: neutrophil function in neonates, neutrophil subsets, therapy of neutrophil dysfunction and neutrophil cytokine production. *Mailing Add:* Dept Pediat Hartford Hosp Hartford CT 06115-0729

KRAUSE, RALPH A(LVIN), b San Francisco, Calif, Nov 11, 09; m 34, 76; c 4. ENGINEERING. *Educ:* Univ Calif, BA, 32. *Prof Exp:* Biophys res, Inst Exp Biol, Univ Calif, 30-33; radio engr, Remler Co, 32-37; electronic res engr, Instrumentation Res & Develop Dept, Calif Res Corp, 37-41; electronics res, Off US Secy Navy, 41-45; sci br, Off Naval Res, 45-46; asst dir, Lab Nuclear Sci & Eng, Mass Inst Technol, 46-47; asst to pres, Raytheon Mfg Co, Mass, 47-48; dir res, Stanford Res Inst, 48-54, assoc dir, 54-64; dir dept appl sci, UNESCO, 64-68; int consult, Res Anal Corp, 68-69; RETIRED. *Concurrent Pos:* Sr eng consult, Brookhaven Nat Lab & Res & Develop Bd. *Mem:* AAAS; Am Soc Naval Engrs; Am Nuclear Soc; Solar Energy Soc; Inst Elec & Electronics Engrs. *Res:* Laboratory organization and research administration; electronic engineering; electrical masking of nerves; electro-encephalographic equipment design; radio speech input design; acoustic engineering; magnetostrictive pressure gauges; nuclear research administration; health physics and nuclear instrumentation; radar; loran; radio and radar counter measures. *Mailing Add:* 550 Battery St Apt 2109 San Francisco CA 94111

KRAUSE, RALPH M, b New York, NY, Nov 23, 31; m 60; c 2. MATHEMATICS. *Educ:* Harvard Univ, BA, 53, MA, 54, PhD(math), 59. *Prof Exp:* Asst prof math, Univ Ill, 58-60 & Ill Inst Technol, 60-62; PROG DIR, NAT SCI FOUND, 62- *Mem:* Am Math Soc; Math Asn Am. *Res:* Topology. *Mailing Add:* NSF Topology & Foundations 1800 G St NW Washington DC 20550

KRAUSE, REGINALD FREDERICK, b Moyers, WVa, July 4, 12; m 43. BIOCHEMISTRY. *Educ:* WVa Univ, AB, 35, BS, 36, MS, 39; Univ Rochester, PhD(biochem), 42; Univ Vt, MD, 51. *Prof Exp:* Asst pharmacol, Univ Rochester, 39-42; instr, Univ Vt, 42, res assoc biochem, Col Med, 45-51; prof biochem & chmn dept, 51-77, EMER PROF BIOCHEM, MED CTR, WVA UNIV, 77- *Mem:* Am Chem Soc. *Res:* Lipid biochemistry; aviation physiology; microchemical determinations of vitamins; biochemistry of vitamin A and carotene; biochemical study of lipids of bone marrow in normal and anemic animals; lipid metabolism in atherosclerosis. *Mailing Add:* 901 Hawthorne Ave Morgantown WV 26505

KRAUSE, RICHARD MICHAEL, b Marietta, Ohio, Jan 4, 25. MICROBIOLOGY, IMMUNOLOGY. *Educ:* Marietta Col, AB, 47; Western Reserve Univ, MD, 52. *Hon Degrees:* DSc, Marietta Col, 78, Sch Med & Dent, Univ Rochester, 79, Med Col Ohio, 81, Hahnemann Med Col & Hosp, 82; LLD, Thomas Jefferson Univ, 82. *Prof Exp:* Intern med, Barnes Hosp, 52-53, asst resident, 53-54; asst, Rockefeller Inst, 54-57, from asst prof to assoc prof, 57-62; from assoc prof to prof, Sch Med, Wash Univ, 62-66; from assoc to prof microbiol & immunol & sr physician, Univ Hosp, Rockefeller Univ, 66-75; dir, Nat Inst Allergy & Infectious Dis, 75-84, asst surgeon gen, 77-84; Woodruff prof med & dean, Sch Med, Emory Univ, 84-88; SR SCI ADV,

FOGARTY INT CTR, NIH, BETHESDA, MD, 89- *Concurrent Pos:* Mem coun rheumatic fever & congenital heart dis & mem coun epidemiol, Am Heart Asn, mem res comt, 63-66; mem, Comn Streptococcal & Staphylococcal Dis, Armed Forces Epidemiol Bd, 63-72; chmn, Allergy & Immunol A Study Sect, NIH, 66-70; consult & mem, Coccal Expert Comt, WHO, 67- & mem steering comt, Biomed Sci Working Group, 78-83; mem bd dirs, NY Heart Asn, 67-73; mem infectious dis adv comt, Nat Inst Allergy & Infectious Dis, 70-74; mem bd dirs, Royal Soc Med Found, Inc, 71-77, treas, 73-75; chmn, Bd Int Health, Inst Med, Nat Acad Sci, 84-88; chmn, Middle East Res Prog, Inst Med, 89. *Honors & Awards:* Robert Koch Gold Medal, 85; Humboldt Award, 87. *Mem:* Inst Med-Nat Acad Sci; Am Col Allergists; Am Soc Clin Invest; Harvey Soc; Asn Am Physicians; Am Acad Allergy; Am Acad Microbiol; fel AAAS; Am Asn Immunologists; Am Epidemiol Soc. *Res:* Pathogenesis and epidemiology of streptococcal diseases; immunochemistry; immunogenetics; antibody structure and mechanisms that generate antibody diversity; studies on streptococcal antigens; author or co-author of over 150 publications and 3 books. *Mailing Add:* Fogarty Int Ctr Bldg 31 Rm B2C 39 NIH 9000 Rockville Pike Bethesda MD 20892

KRAUSE, RONALD ALFRED, b Boston, Mass, Oct 30, 31; m 52, 77; c 2. INORGANIC CHEMISTRY. *Educ:* Ohio State Univ, BSc, 56, PhD(chem), 59. *Prof Exp:* Res scientist, Am Cyanamid Co, Conn, 59-62; from asst prof to assoc prof, 67-77, PROF CHEM, UNIV CONN, 78- *Concurrent Pos:* Consult, Arco, 78-81; guest prof, Univ Copenhagen, Denmark, 68-69, 76, 83 & 90. *Mem:* Am Chem Soc; Nat Speleol Soc; Danish Chem Soc. *Res:* Synthesis, reactions, structure, spectra and photochemistry of coordination compounds. *Mailing Add:* Dept Chem Univ Conn Storrs CT 06268

KRAUSE, SONJA, b St Gall, Switz, Aug 10, 33; nat US; m 70. PHYSICAL CHEMISTRY, POLYMER CHEMISTRY. *Educ:* Rensselaer Polytech Inst, 54; Univ Calif, PhD(phys chem), 57. *Prof Exp:* Res chemist, Rohm & Haas Co, Pa, 57-64; US Peace Corps vol, Lagos Univ, Nigeria, 64-65 & Gondar Health Col, Ethiopia, 65-66; asst prof chem, Univ Southern Calif, 66-67; from asst prof to assoc prof, 67-78, PROF CHEM, RENSSELAER POLYTECH INST, 78- *Concurrent Pos:* Mem coun, Gordon Res Conf, 81-84; sabbatical leave, Inst charles Sadron, Strasbourg, France, 87. *Mem:* AAAS; Am Chem Soc; fel Am Phys Soc; Biophys Soc; NY Acad Sci. *Res:* Dilute solution properties of polymers; block copolymers; polymer compatibility; transient electric birefringence; muscle proteins; membranes. *Mailing Add:* Dept Chem Rensselaer Polytech Inst Troy NY 12180-3590

KRAUSE, THOMAS OTTO, b Grand Rapids, Mich, May 5, 44. THEORETICAL PHYSICS, NUCLEAR PHYSICS. *Educ:* Mass Inst Technol, BS, 66; Ohio State Univ, PhD(physics), 73. *Prof Exp:* Vis asst prof, Dept Physics, Ohio State Univ, 73-76; ASST PROF, DEPT PHYSICS, TOWSON STATE UNIV, 76- *Mem:* Am Phys Soc; Sigma Xi. *Res:* Scattering theory, especially nuclear, with nonlocal potentials; gravitation; theoretical astrophysics. *Mailing Add:* Dept Physics Towson State Univ Towson MD 21204

KRAUSE, WILLIAM JOHN, b Glasgow, Mont, Mar 24, 42; m 67; c 2. ANATOMY, HISTOLOGY. *Educ:* Augustana Col, BA, 64; Univ Iowa, MS, 66; Univ Mo-Columbia, PhD(anat), 69. *Prof Exp:* Lectr anat, Monash Univ, 69-71; from asst prof to assoc prof, 71-83, PROF ANAT, UNIV MO-COLUMBIA, 83- *Concurrent Pos:* Vis prof, Univ Western Australia, Perth, 91. *Mem:* Am Asn Anatomists; Anat Soc Gt Brit & Ireland. *Res:* Postnatal development of respiratory, urinary and digestive systems; biology of Brunner's glands. *Mailing Add:* Dept Anat Univ Mo Columbia MO 65202

KRAUS-FRIEDMANN, NAOMI, b Budapest, Hungary, July 4, 33; div; c 1. PHYSIOLOGY. *Educ:* Hebrew Univ, Jerusalem, MSc, 60, PhD(biochem), 65. *Prof Exp:* Res assoc, Columbia Univ, 65-66; res assoc physiol, Vanderbilt Univ, Nashville, 66-68; instr biochem, Sch Med, Univ Pa, 68-74; from asst prof to assoc prof, 74-86, PROF PHYSIOL, SCH MED, UNIV TEX, HOUSTON, 86 - *Concurrent Pos:* Vis prof, Eidgenossische Tech Hochschule, Zurich, 81-82. *Mem:* Am Physiol Soc; Am Soc Biol Chemists. *Res:* Hormonal regulation of gluconeogenesis; role of calcium and other ions in regulation of metabolic processes, microsomal Ca2 sequestration. *Mailing Add:* Sch Med Univ Tex PO Box 20708 Houston TX 77025

KRAUSHAAR, JACK JOURDAN, b Newark, NJ, Sept 6, 23; m 51; c 3. NUCLEAR PHYSICS. *Educ:* Lafayette Col, BS, 44; Syracuse Univ, MS, 48, PhD, 52. *Prof Exp:* Asst physics, Syracuse Univ, 46-50; res assoc nuclear spectros, Brookhaven Nat Lab, 51-53; instr physics, Stanford Univ, 53-56; from asst prof to prof, 56-88, EMER PROF PHYSICS, UNIV COLO, BOULDER, 88- *Concurrent Pos:* Fulbright award, Free Univ, Amsterdam, 67-68; fac fel, Tri-Univ Meson Facil, Univ BC, Vancouver, 78-79; vis prof, Osaka Univ, 85. *Mem:* AAAS; fel Am Phys Soc; Fedn Am Sci. *Res:* Nuclear reactions and spectroscopy; pi meson interactions and scattering; energy and environmental problems in the United States; author. *Mailing Add:* Dept Physics Univ Colo Campus Box 390 Boulder CO 80309-0390

KRAUSHAAR, WILLIAM LESTER, b Newark, NJ, Apr 1, 20; m 80; c 3. PHYSICS. *Educ:* Lafayette Col, BS, 42; Cornell Univ, PhD(physics), 49. *Prof Exp:* Physicist, Nat Bur Stand, 42-45; res assoc, Mass Inst Technol, 49-51, from asst prof to prof physics, 51-65; prof, 65-80, max mason prof, 80-85, EMER PROF PHYSICS, UNIV WIS-MADISON, 85- *Mem:* Nat Acad Sci; fel Am Acad Arts & Sci; fel Am Phys Soc. *Res:* High energy astrophysics; space science; cosmic rays. *Mailing Add:* Chamberlin Hall Dept Physics Univ Wis Madison WI 53706

KRAUSKOPF, JOHN, b New York, NY, Mar 30, 28; m 52; c 3. PSYCHOPHYSIOLOGY. *Educ:* Cornell Univ, AB, 49; Univ Tex, PhD, 53. *Prof Exp:* Asst psychol, Cornell Univ, 49-50; asst, Univ Tex, 50-52, res assoc, 52-53; USPHS fel, Brown Univ, 56-57, asst prof, 57-59; asst prof, Rutgers Univ, 59-62; res assoc, Univ Md, 62-64; res scientist, Inst Behav Res, 64-66; mem tech staff, Bell Labs, 66-87; DEPT PSYCHOL, NY UNIV, 88- *Concurrent Pos:* Vis asst prof, Bryn Mawr Col, 59-60; mem vision comt, Armed Forces-Nat Res Coun, 60- *Mem:* Optical Soc Am. *Res:* Vision; visual perception. *Mailing Add:* Dept Psychol Ctr Neural Sci NY Univ Six Washington Pl 8th Fl New York NY 10003

KRAUSKOPF, KONRAD BATES, b Madison, Wis, Nov 30, 10; m 36; c 4. GENERAL EARTH SCIENCES. *Educ:* Univ Wis, AB, 31; Univ Calif, PhD(chem), 34; Stanford Univ, PhD(geol), 39. *Hon Degrees:* DSc, Univ Wis-Milwaukee, 71. *Prof Exp:* Instr chem, Univ Calif, 34-35; actg instr phys sci, Dept Geol, 35-39, from asst prof to assoc prof geol, 39-50, prof, 50-76, EMER PROF GEOCHEM, STANFORD UNIV, 76- *Concurrent Pos:* Fulbright & Guggenheim fels, Norway, 52-53; NSF fac fel, Ger, 60-61; chmn bd, Radioactive Waste Mgt, Nat Acad Sci, 81-85. *Honors & Awards:* Day Medal, Geol Soc Am, 61; Goldschmidt Medal, Geochem Soc, 82; Ian Campbell Medal, Am Geol Inst, 84. *Mem:* Nat Acad Sci; Geol Soc Am (pres, 67); Am Geol Inst (pres, 64); Geochem Soc (pres, 70); Soc Econ Geologists; Am Philos Soc; Am Geophys Union. *Res:* Petrology of igneous and metamorphic rocks; physical chemistry of ore solutions; trace elements in sea water and in sedimentary rocks. *Mailing Add:* Dept Geol Stanford Univ Stanford CA 94305-2115

KRAUSMAN, PAUL RICHARD, b Washington, DC, Nov 17, 46; m 66; c 2. WILDLIFE ECOLOGY. *Educ:* Ohio State Univ, BS, 68; NMex State Univ, MS, 71; Univ Idaho, PhD(wildlife sci), 76. *Prof Exp:* Res asst environ alteration, Aeromed Res Lab, NMex, 68-71; res asst, Environ Res Lab & Radiation Lab, Brooks AFB, Tex, 71-72; asst prof wildlife ecol, Auburn Univ, 76-78; asst prof wildlife ecol & asst res assoc, 78-81, ASSOC PROF WILDLIFE ECOL, UNIV ARIZ, 81- *Concurrent Pos:* Welder wildlife fel ecol, 72-76. *Mem:* Wildlife Soc; Am Soc Mammalogists; Soc Range Mgt. *Res:* Ungulate ecology. *Mailing Add:* Sch Renewable Natural Resources Univ Ariz Dept Natural Resources AZ 85721

KRAUSS, ALAN ROBERT, b Chicago, Ill, Oct 3, 43; m 65; c 1. PHYSICS. *Educ:* Univ Chicago, BS, 65; Purdue Univ, MS, 69, PhD(physics), 72. *Prof Exp:* Res assoc surface physics, James Franck Inst, Univ Chicago, 71-74; asst physicist, 74-80, PHYSICIST SURFACE PHYSICS, ARGONNE NAT LAB, 80- *Concurrent Pos:* Assoc ed, Appl Physics Lett; adj prof, Univ Wis. *Mem:* Am Phys Soc; Am Vacuum Soc; Sigma Xi; Mat Res Soc. *Res:* Surface physics and chemistry; sputtering; secondary ion emission and ion-bombardment phenomena; applications to thermonuclear fusion devices. *Mailing Add:* 24461 W Blvd De John Naperville IL 60564

KRAUSS, BEATRICE HILMER, b Honolulu, Hawaii, Aug 4, 03. PLANT PHYSIOLOGY, ETHNOBOTANY. *Educ:* Univ Hawaii, BS, 26, MS, 30. *Prof Exp:* From asst plant physiologist to plant physiologist, Pineapple Res Inst Hawaii, 26-68; lectr ethnobot & res affil pineapple physiol, 68-73, RES AFFIL HAWAIIAN ETHNOBOT, LYON ARBORETUM, UNIV HAWAII, MANOA, 74- *Honors & Awards:* Award Merit, Am Asn Bot Garden & Arbor. *Mem:* Fel AAAS; Sigma Xi; fel Asn Trop Biol. *Res:* Morphology and anatomy of pineapple plant; pineapple nutrition, especially micronutrients; Hawaiian ethnobotany. *Mailing Add:* 2437 Parker Pl Honolulu HI 96822

KRAUSS, GEORGE, b Philadelphia, Pa, May 14, 33; m 60; c 4. PHYSICAL METALLURGY, METALLURGICAL ENGINEERING. *Educ:* Lehigh Univ, BS, 55; Mass Inst Technol, MS, 58, ScD, 61. *Prof Exp:* Mem staff div sponsored res, Mass Inst Technol, 61-62; NSF fel, Max-Planck Inst Iron Res, 62-63; from asst prof to prof metall, Lehigh Univ, 63-75, dir electron micros lab, 69-75; Amax Found prof, 75-90, PROF PHYS METALL, COLO SCH MINES, 90-, DIR, ADVAN STEEL PROCESSING & PROD RES CTR, 84- *Concurrent Pos:* Ed, J Heat Treating, 78-82 & Prof Engr Pa & Co; pres, Int Fedn Heat Treatment, 88-90. *Honors & Awards:* Adolf Martens Medal, 90. *Mem:* Am Inst Mining, Metall & Petrol Engrs; fel Am Soc Metals; Electron Micros Soc Am. *Res:* Mechanical and fracture behavior of steels; microstructural characterization by light and electron microscopy; failure analysis; author of over 170 publications; principles of heat treatment of steel. *Mailing Add:* Dept Metall Eng Colo Sch Mines Golden CO 80401

KRAUSS, HERBERT HARRIS, b Philadelphia, Pa, June 13, 40; m 65; c 2. CLINICAL PSYCHOLOGY, REHABILITATION PSYCHOLOGY. *Educ:* Pa State Univ, BS, 61, MS, 63; Northwestern Univ, PhD(psychol), 66. *Prof Exp:* Asst prof psychiat & psychol, Univ Kans Med Ctr, 66-67 & Col Med, Ohio State, 67-69; assoc prof psychol, Univ Ga, 69-71; PROF PSYCHOL, HUNTER COL, CITY UNIV NY, 71- *Concurrent Pos:* Adj assoc psychologist, Payne Whitney Clin, NY Hosp, 79-; adj assoc prof psychol & psychiat, Cornell Med Sch, 79-; dir res, Int Ctr Disabled, 83- *Mem:* Int Orgn Study Group Tensions; Am Psychol Asn; NY Acad Sci; Am Cong Rehab Med; Sigma Xi. *Res:* Psycho-social etiology and treatment of behavioral abnormalities; social construction of reality. *Mailing Add:* Six Downing Ct Irvington NY 10533

KRAUSS, JONATHAN SETH, b Brooklyn, New York, May 25, 45; m 72; c 2. HEMATOLOGY, COAGULATION. *Educ:* Cornell Univ, AB, 66; Univ Fla, Gainesville, MD, 70. *Prof Exp:* Intern med, Med Col Va, 70-71; gen med officer, US Naval Reserve, 71-73; resident path, NC Mem Hosp, 73-78; asst prof, 78-83, ASSOC PROF PATH, MED COL GA, 83-, DIR, HEMAT & HEMOSTATIS LAB, 78- *Concurrent Pos:* Fel path, Univ NC, Chapel Hill, 75-76. *Mem:* Am Col Physicians; Col Am Pathologists; Soc Hematopath; Am Soc Hematol; Asn Clin Scientists; Soc Hemophilia. *Res:* Von Willebrand factor antigen in body fluids; glycosylated hemoglobin determination hemolysis; measurement of fetal hemoglobin; granulocytic fragments in sepsis. *Mailing Add:* Dept Path BIH 222B Med Col Ga 1120 15th St Augusta GA 30912-3620

KRAUSS, LAWRENCE MAXWELL, b New York, NY, May 27, 54; Can & US citizen; m 80; c 1. PARTICLE & ASTROPHYSICS INTERFACE. *Educ:* Carleton Univ, BSc, 77; Mass Inst Technol, PhD(physics), 82. *Prof Exp:* Jr fel, Harvard Soc Fels, 82-85; asst prof, 85-88, ASSOC PROF PHYSICS & ASTRON, YALE UNIV, 88- *Concurrent Pos:* Vis scientist, Smithsonian Ctr Astrophys, Harvard Univ, 85-88, Boston Univ, 85-86; assoc physics dept, Harvard Univ, 87-88; Nat Comt Lectureships, Sigma Xi, 88; pres young investr award, Nat Sci Found, 86; nat lectr, Sigma Xi, 91- *Honors & Awards:* First Prize Award, Gravity Res Found, 84. *Mem:* Am Phys Soc; Sigma Xi.

Res: The interface of particle physics and astrophysics and cosmology; particle physics phenomenology; field theory; ultrasensitive detection and particle physics. *Mailing Add:* Sloane Lab Dept Physics Yale Univ New Haven CT 06511

KRAUSS, ROBERT WALLFAR, b Cleveland, Ohio Dec 27, 21; m 47; c 2. BIOCHEMISTRY, ECOLOGY. *Educ:* Oberlin Col, BA, 47; Univ Hawaii, MS, 49; Univ Md, College Park, PhD(bot), 51. *Prof Exp:* Asst bot, Univ Hawaii, 47-49; asst bot, Univ Md, College Park, 49-51, res assoc plant physiol, 51-55, from asst prof to prof, 55-73, head dept bot, 64-73; dean col sci, Ore State Univ, 73-80; exec dir, Fedn Am Soc Biol, 79-90; VIS SR SCIENTIST, CALIF INST TECHNOL, JPL, 90- *Concurrent Pos:* Res fel, Carnegie Inst, 51-55; biologist, Coastal Studies Inst, La State Univ, 58-59; mem, Nat Res Coun, 59-60; sr res affil, Chesapeake Biol Lab, 68-73; mem bd dirs, Ed Projs Inc, 69-; consult, US Air Force Sch Aviation Med, NASA & NSF; mem corp, Marine Biol Lab, Woods Hole, Mass. *Honors & Awards:* Darbaker Award, Bot Soc Am, 56; Presidents Leadership Award, Am Inst Biol Sci, 74; Achievement Awards, NASA, 76 & 89. *Mem:* Phycol Soc Am (pres, 64); Bot Soc Am; Am Soc Plant Physiol; Am Inst Biol Sci (secy-treas, 63-69, vpres, 72, pres, 73); Sigma Xi. *Res:* Algal physiology and biochemistry; science policy. *Mailing Add:* Calif Inst Technol JPL Wash Off 600 Missouri Ave Suite 510 Washington DC 20024

KRAUSS, RONALD, b New York, NY, May 12, 43; m 69; c 2. LIPOPROTEIN METABOLISM. *Educ:* Harvard Univ, BA, 64, MD, 68. *Prof Exp:* Intern, Boston City Hosp, 68-70; clin assoc, NIH, 70-73, sr investr, 73-74; asst clin prof med, Univ Calif, San Francisco, 74-82; staff scientist, 76-84, SR SCIENTIST, LAWRENCE BERKELEY LAB, 84- *Concurrent Pos:* Assoc adj prof med, Univ Calif, San Francisco, 82-; dir, Endocrine & Metab Serv, Alta Bates Hosp, 86-89; head, Molecular Med Res Prog, Donner Lab, Lawrence Berkeley Lab, 89-; fel, Arteriosclerosis Coun, Am Heart Asn. *Mem:* Am Fed Clin Res; Am Soc Clin Invest; Am Diabetes Asn; Am Inst Nutrit. *Mailing Add:* Donner Lab Lawrence Berkeley Lab Univ Calif Berkeley CA 94720

KRAUSZ, ALEXANDER STEPHEN, b Budapest, Hungary, Sept 16, 24; Can citizen; m 49. MATERIALS SCIENCE, MECHANICAL ENGINEERING. *Educ:* Budapest Tech Univ, BSc, 51; Queen's Univ, Ont, MSc, 59; Univ Toronto, PhD(metall), 65. *Prof Exp:* Mgr mfg, Gamma Instrument Co, Hungary, 49-52; res off plastic deformation, Nat Res Coun Can, 59-70; assoc prof mech eng, Univ Ottawa, 70-72, prof & chmn dept, 72-81, prof mech eng & dir, Eng Mgt Prog, 81-86, EMER PROF MECH ENG, UNIV OTTAWA, 90- *Concurrent Pos:* Assoc ed, J Eng Mat & Technol; regional ed, Int J Fracture. *Mem:* Am Soc Metals; Eng Inst Can; fel Can Soc Mech Eng. *Res:* Fracture mechanics; deformation kinetics, thermally activated plastic flow and fracture and deformation processes in manufacturing; product design. *Mailing Add:* Dept Mech Eng Univ Ottawa Ottawa ON K1N 6N5 Can

KRAUSZ, STEPHEN, b Salford, Eng, Aug 4, 50; US citizen; m 72; c 2. SCIENCE EDUCATION, ANIMAL PHYSIOLOGY. *Educ:* City Univ New York, BSc, 71; Hebrew Univ, Jerusalem, Israel, MSc, 73, PhD(physiol), 77. *Prof Exp:* Res fel, Sch Med, Univ Calif, Los Angeles, 77-78; asst prof physiol, Col Med, Howard Univ, 78-83; LAB COORDR GEN SCI, HILLEL ACAD, DENVER, COLO, 83- *Mem:* Sigma Xi; Am Physiol Soc. *Res:* Respiratory, cardiovascular and endocrinological responses of mammals to extremes of environment. *Mailing Add:* 1376 Utica Denver CO 80204

KRAUT, EDGAR A, b Cleveland, Ohio, May 4, 34; m 80. PHYSICS. *Educ:* Univ Calif, Los Angeles, AB, 56, MA, 57, PhD(physics), 62. *Prof Exp:* MEM TECH STAFF, SCI CTR, ROCKWELL INT CORP, 67- *Mem:* AAAS; Am Phys Soc; Inst Elec & Electronics Engrs; Soc Indust Appl Math. *Res:* Theoretical and mathematical physics; wave propagation; physics of semiconductor surfaces and interfaces; heterojunctions; energy bands; device modeling. *Mailing Add:* Rockwell Int Sci Ctr PO Box 1085 Thousand Oaks CA 91360

KRAUT, JOSEPH, b New York, NY, Dec 5, 26; m 53; c 3. PHYSICAL BIOCHEMISTRY. *Educ:* Bucknell Univ, BS, 50; Calif Inst Technol, PhD(phys chem), 54. *Prof Exp:* From instr to asst prof biochem, Univ Wash, 53-62; assoc prof chem, 62-66, actg chmn dept, 72-73, PROF CHEM, UNIV CALIF, SAN DIEGO, 66- *Concurrent Pos:* Fel, Howard Hughes Med Inst, 55-60. *Honors & Awards:* Keilin Medal, Brit Biochem Soc, 80. *Mem:* Nat Acad Sci; Am Chem Soc; Am Crystallog Asn; Am Soc Biol Chem; AAAS. *Res:* Structure, function and evolution of biological macromolecules; x-ray diffraction crystallography. *Mailing Add:* Dept Chem Univ Calif La Jolla CA 92093

KRAUTER, ALLAN IRVING, b Newark, NJ, Oct 15, 41; m 68; c 3. MECHANICAL ENGINEERING. *Educ:* Stevens Inst Technol, ME, 63; Stanford Univ, MS, 64, PhD(mech eng), 68. *Prof Exp:* Asst prof mech eng, Cornell Univ, 68-74; sr consult engr & mgr, Technol Dept, Shaker Res Corp, 75-81; PROG MGR, MECH SYSTS, CARRIER CORP, 81- *Mem:* Am Soc Mech Engrs; Soc Automotive Engrs. *Res:* Vibrations and dynamics of mechanical systems; simulation of mechanical and economic system behavior. *Mailing Add:* Res Div Carrier Corp Carrier Pkwy Syracuse NY 13221

KRAUTHAMER, GEORGE MICHAEL, b Ger, Sept 14, 26; nat US; div; c 6. NEUROSCIENCE. *Educ:* City Col New York, BS, 51, MA, 52; NY Univ, PhD(psychol), 59. *Prof Exp:* Asst psychophysiol, Sch Med, NY Univ, 57-59, instr psychol, 59-60; res assoc, Univ Paris, 63-66; asst prof, Col Physicians & Surgeons, Columbia Univ, 67-69; assoc prof, 69-79, PROF ANAT, ROBERT WOOD JOHNSON MED SCH, UNIV MED & DENT NJ, 79- *Concurrent Pos:* USPHS res fel, Ctr Study Physiol of Cent Nerv Syst, Univ Paris, 60-63; res assoc, Hillside Hosp, Glen Oak, NY, 59-60; lectr psychol, City Col New York, 59-60; asst to exec secy, Int Brain Res Orgn-UNESCO, 65-68. *Mem:* AAAS; Soc Neurosci. *Res:* Electrophysiology and neuroanatomy of brain;

behavior correlates of brain function; effects of brain injury; perception and intersensory relationships; electroencephalography; drug effects. *Mailing Add:* Dept Neurosci & Cell Biol Robert Wood Johnson Med Sch UMDNJ 675 Hues Lane Piscataway NJ 08554

KRAUTZ, FRED GERHARD, b Cottbus, Ger; US citizen; m 62; c 2. FIBER GLASS REINFORCEMENTS, FIBER GLASS COMPOSITES. *Educ:* Univ Cincinnati, ChE, 61, MS, 66. *Prof Exp:* Engr, Cincinnati Milacron, 61-63, consult, 63-65; sr engr, Owens-Corning Fiberglas Corp, 65-72, supvr reinforcements & mats, 72-78, mgr reinforcement & tires, 78-81, mgr composite prod, 81-83, tech support mgr, 83-87; res & develop dir, Certainteed Corp, 87-90, dir res & develop, 90-91, VPRES RES & DEVELOP, VETROTEX CERTAINTEED CORP, 91- *Mem:* Soc Plastics Engrs; Am Inst Chem Engrs. *Res:* Fiberglass products for all thermosetting and thermoplastic composites for Vetrotex Certainteed fiberglass reinforcements division. *Mailing Add:* Vetrotex Certainteed Corp 4515 Allendale Rd Wichita Falls TX 76310

KRAVITZ, EDWARD, science editing, for more information see previous edition

KRAVITZ, EDWARD ARTHUR, b New York, NY, Dec 19, 32; m 58; c 2. BIOCHEMISTRY, NEUROBIOLOGY. *Educ:* City Col New York, BS, 54; Univ Mich, PhD(biochem), 59. *Prof Exp:* Nat Heart Inst fel biochem, 59-60; Nat Inst Neurol Dis & Blindness res fel neurophysiol & neuropharmacol, 60-61, instr neurophysiol & neuropharmacol, 61-63, assoc, 63-66, dir prog neurosci, 82-90, from asst prof to assoc prof, 66-69, prof neurobiol, 69-86, GEORGE PACKER BERRY PROF NEUROBIOL, HARVARD MED SCH, 86- *Concurrent Pos:* USPHS spec fel, 61-64, career develop award, 66-71; mem, Bd Trustees & Exec Comt, Marine Biol Lab, dir neurobiol course, 75-79; mem, Governing Coun, Inst Med, 91-93; co-founder, Neurobiol Dis Teaching Workshops, Soc Neurosci. *Honors & Awards:* Flexner Lectr, Univ Pa, 72; Krantz Lectr, Univ Md, 75; Magnes Mem Lectr, Hebrew Univ Med Sch, Jerusalem, 81; Snider Lectr, Univ Toronto, 87; Lang Lectr, Marine Biol Lab, 83; Schmitt Lectr, Univ Pa, 91; Von Humboldt Award, 91. *Mem:* Inst Med Nat Acad Sci; Soc Neurosci; Am Acad Arts & Sci; Am Soc Biol Chemists; NY Acad Sci. *Res:* Biochemical studies on single physiologically identified nerve cells; identification of gamma-aminobutyric acid and other neurotransmitters in the lobster nervous system; amines, peptides neurohormones and behavior in lobsters. *Mailing Add:* Dept Neurobiol Harvard Med Sch 220 Longwood Ave Boston MA 02115

KRAVITZ, HENRY, b Poland, Oct 18, 18; Can citizen; m 42; c 1. PSYCHIATRY, PSYCHOANALYSIS. *Educ:* McGill Univ, BA, 45, MD, CM, 49; dipl psychiat, 54; Royal Col Physicians & Surgeons Can, cert psychiat, 54; FRCP(C). *Prof Exp:* Assoc prof, 67-72, PROF PSYCHIAT, MCGILL UNIV, 72-, ACTG CHMN, DEPT PSYCHIAT, 85-; chmn psychiat, 67-88, EMER CHIEF, JEWISH GEN HOSP & DIR, INST COMMUNITY & FAMILY PSYCHIAT, 88- *Concurrent Pos:* Training analyst, Can Psychoanal Inst, 62-, assoc dir, 68-, dir, 80-; chmn psychiat, Jewish Gen Hosp, 67-88; chmn bd examiners, Royal Col Physicians & Surgeons, 78-82, chmn sect psychiat, 82- *Mem:* Fel AAAS; fel Am Col Psychiat; fel Am Psychiat Asn; Can Psychoanal Soc (pres, 68-71); fel Am Col Psychoanalysts. *Res:* Theoretical and practical considerations for unwed mothers; use of methadone and other substitute therapies in drug addiction; psychiatric education. *Mailing Add:* Dept Psychiat Jewish Gen Hosp Montreal PQ H3T 1E2 Can

KRAVITZ, JOSEPH HENRY, b Nanticoke, Pa, Aug 14, 35; m; c 2. MARINE GEOTECHNIC, PROGRAM MANAGER. *Educ:* Syracuse Univ, BS, 57; George Washington Univ, MS, 75, MPH, 77, PhD(geol), 83. *Prof Exp:* Res asst, Geol Dept, Yale Univ, 61-64; marine geologist, Lamont-Doherty Geol Observ, Columbia Univ, 64-65; oceanographer, US Naval Oceanog Off, 65-71, head, Geol Lab, 71-78, actg head, marine geol & geophys br, 78; sr geologist, outer continental shelf environ assessment, Nat Ocean & Atmospheric Admin, 78-80, actg dir, Marine Ecosysts Anal Div, Off Marine Pollution Assessment, 80-82, hq staff geologist, Ocean Assessment Div, 82-84, sr oceanogr, prog develop & coord staff, Off Oceanic & Atmospheric Res, Nat Oceanic & Atmospheric Admin, 84; sci off, 84-86, PROG MGR, MARINE GEOL & GEOPHYS PROG, OFF NAVAL RES, 86- *Concurrent Pos:* Res assoc, Inst Artic & Alpine Res, 81-86; assoc prof lectr geol, George Wash Univ, 87- *Mem:* Fel Geol Soc Am; fel Artic Inst NAm; fel Explor Club; Soc Econ Paleontologists & Mineralogists. *Res:* High resolution seismic profiling; geotechnical analysis of sediments and its relation to depositional environments. *Mailing Add:* 9216 Orchard Brook Dr Potomac MD 20854

KRAVITZ, LAWRENCE C, b New York, NY, July 27, 32; m 58; c 3. PHYSICS, ELECTRONICS. *Educ:* Kans Univ, BS, 54; Air Force Inst Technol, MS, 55; Harvard Univ, PhD(physics), 63. *Prof Exp:* Physicist solid state, Corp Res & Develop Ctr, Gen Elec Co, 63-71, mgr display prog, 72-73; dir electronics, 73-78, dir, Air Force Off Sci Res, 78-81; dir res, Bendix Advan Technol Ctr, 81-; ALLIED SIGNAL. *Mem:* Inst Elec & Electronics Engrs. *Res:* Solid state science. *Mailing Add:* 7128 Wolftree Lane Rockville MD 20852

KRAWETZ, ARTHUR ALTSHULER, b Chicago, Ill, Oct 30, 32. ANALYTICAL CHEMISTRY, PHYSICAL CHEMISTRY. *Educ:* Northwestern Univ, BS, 52; Univ Chicago, SM, 53, PhD(chem), 55. *Prof Exp:* Vpres, 54-74, PRES, PHOENIX CHEM LAB, INC, CHICAGO, 74- *Mem:* Am Chem Soc; Am Soc Testing & Mat; Am Inst Chem; Royal Soc Chem; Nat Fire Protection Asn. *Res:* Fuel and lubricant technology; spontaneous ignition; flammability; air and water pollution; forensic chemistry; differential thermal analysis; solution chemistry; thermodynamics; molecular spectroscopy; industrial hygiene; safety; hydraulic fluids; protective coatings; rubber and plastic. *Mailing Add:* 1010 Isabella St Evanston IL 60201

KRAWETZ, STEPHEN ANDREW, b Fort Frances, Ont, Sept 17, 55; m 77; c 1. HUMAN GENOME INITIATIVE, MEDICAL GENETICS & BIOTECHNOLOGY. *Educ:* Univ Toronto, BSc, 77 & PhD(biochem), 83. *Prof Exp:* Occas teacher math, music & sci, Scarborough Bd Educ, 76-77; lab demonstr biochem, Univ Toronto, 77-81; postdoctoral fel, Dept Med Biochem, Univ Calgary, 83-89; asst prof res, Dept Molecular Biol & Genetics & Ctr Molecular Biol, 89-90, ASST PROF, DEPT MOLECULAR BIOL & GENETICS, WAYNE STATE UNIV, 90- *Concurrent Pos:* Brit Columbia Children's Hosp res fel, 84-; Alberta Heritage Found med res fel, 84-89; Biotechnol Consult, 85-; prin investr Comput Video Expert Systs for Biochemical Appl, 87; cofounder & med res dir, Genetic Imaging Inc, 87-88. *Honors & Awards:* Intelli Genetics Computer Appln Award, 88. *Mem:* Can Biochem Soc; AAAS; NY Acad Sci. *Res:* Control of development and differentiation; expression of elastic-tissue genes; human genome initiative; gene expression during spermatogenesis; computer assisted sequence analysis; molecular diagnostic probes. *Mailing Add:* Dept Molecular Biol & Genetics Biol Wayne State Univ 2228 Scott Hall 540 E Canfield Detroit MI 48201

KRAWIEC, STEVEN STACK, b Corvallis, Ore, Nov 4, 41; m 65; c 2. MOLECULAR BIOLOGY. *Educ:* Brown Univ, AB, 63; Yale Univ, PhD(microbiol), 68. *Prof Exp:* Trainee, Univ Wis-Madison, 68-69, Nat Inst Gen Med Sci fel, 69-70; from asst prof to assoc prof, 70-82, chmn, 76-78, assoc dean, Col Arts & Sci, 85-87, PROF BIOL, LEHIGH UNIV, 82- *Concurrent Pos:* Fogarty int fel, Autonomous Univ Madrid, 78-79. *Mem:* Am Soc Microbiol; AAAS; Sigma Xi. *Res:* Characterization of chromosome organization of bacteria; acquisitive evolution; degradation of xenobiotics. *Mailing Add:* Dept Biol Lehigh Univ Bethlehem PA 18015-3189

KRAY, LOUIS ROBERT, b San Bernardino, Calif, Oct 20, 38; m 58; c 5. PETROLEUM CHEMISTRY. *Educ:* Univ Calif, Riverside, BA, 61, PhD(org chem), 65. *Prof Exp:* Researcher, 65-67; RES CHEMIST ORG CHEM, CHEVRON RES CO, STAND OIL CO CALIF, RICHMOND, 67- *Mem:* Am Chem Soc. *Res:* Fuel additive synthesis and development. *Mailing Add:* 700 Bamboo Terr San Rafael CA 94903

KRAYBILL, EDWARD K(READY), b Lancaster, Pa, June 3, 17; m 39; c 1. ELECTRICAL ENGINEERING. *Educ:* Pa State Col, BS, 39, EE, 51; Univ Mich, MSE, 48, PhD, 66. *Prof Exp:* From instr to assoc prof elec eng, Duke Univ, 39-71, asst to dean, Col Eng, 53-62, asst dean, 62-66, assoc dean, 66-71; dir, Worthington Scranton Campus, Pa State Univ, 71-78, prof eng, 71-82, emer prof, 82-84; RETIRED. *Mem:* Am Soc Eng Educ; Illum Eng Soc; Inst Elec & Electronics Engrs; Nat Soc Prof Engrs; Am Asn Higher Educ. *Res:* Higher education. *Mailing Add:* 500 E Marylyn Ave No 265 State College PA 16801

KRAYBILL, HENRY LAWRENCE, b Washington, DC, Apr 13, 18; m 44; c 2. EXPERIMENTAL HIGH ENERGY PHYSICS. *Educ:* Univ Chicago, SB, 38, PhD(physics), 49. *Prof Exp:* From instr to prof physics, Yale Univ, 48-84; RETIRED. *Mem:* Am Phys Soc; Am Asn Physics Teachers. *Res:* High energy particles; bubble chamber analysis of hadron interactions. *Mailing Add:* 960 Benham Hamden CT 06514

KRAYBILL, HERMAN FINK, b Marietta, Pa, June 27, 14; c 3. BIOCHEMISTRY, TOXICOLOGY. *Educ:* Franklin & Marshall Col, BS, 36; Univ Md, MS, 38, PhD(biochem), 41. *Prof Exp:* Instr chem, Univ Md, 36-39; res chemist, Swift & Co, 41-43; res biochemist, Moorman Mfg Co, Ill, 46 & Nat Dairy Res Labs, Md, 46-48; res assoc, Nat Res Coun, DC, 48-49; res biochemist, Bur Animal Indust, USDA, 49-53; supvry biochemist & chief chem div, Army Med Nutrit Lab, 53-59; sr scientist, Curtiss Wright Corp, NJ, 59-60; sr biochemist & scientist adminstr, Div Radiation Health, Nat Cancer Inst, 60-63; chief pesticides prog, USPHS, 63-66; asst dir biol sci res, Bur Sci & asst dir sci coord, Bur Foods, Food & Drug Admin, 66-72; SCI COORDR ENVIRON CANCER, NAT CANCER INST, 72- *Concurrent Pos:* Lectr biochem, Univ Colo & Univ Denver, 55-59. *Honors & Awards:* Merit Award, NIH, 81. *Mem:* AAAS; Am Chem Soc; Am Inst Nutrit; Soc Toxicol; NY Acad Sci. *Res:* Food research; fat enzymes; animal and human nutrition; dairy products; meats and fishery products; allergy; cancer; toxicology of irradiated foods; pesticides. *Mailing Add:* 17708 Lafayette Dr Olney MD 20832

KRAYBILL, RICHARD R(EIST), b Dover, NH, July 31, 20; m 45; c 4. CHEMICAL ENGINEERING, POLYMER PROCESSING. *Educ:* Purdue Univ, BChE, 42; Univ Mich, MS, 43, PhD(chem eng), 53. *Prof Exp:* Asst res engr, Calif Res Corp, 44-46; from asst prof to assoc prof chem eng, Univ Rochester, 50-67; tech assoc develop, Mfg Tech Div, Eastman Kodak Co, Rochester, 67-83; RETIRED. *Concurrent Pos:* Sr lectr, Univ Rochester, 78-79. *Mem:* Am Chem Soc; Am Soc Eng Educ; fel Am Inst Chem Engrs; sr mem Soc Plastics Engrs; Soc Rheology. *Res:* Fluid flow; heat transfer; extrusion. *Mailing Add:* 1704 Laurie La Belleair FL 34616-1629

KRAYCHY, STEPHEN, b Redwater, Alta, Feb 18, 28; nat US; m 54; c 3. ORGANIC CHEMISTRY, NUCLEAR MEDICINE. *Educ:* Univ Alta, BSc, 50; Univ Wis, PhD(org chem), 55. *Prof Exp:* Asst mem, Sloan-Kettering Inst Cancer Res, 54-56; sr investr chem res, G D Searle & Co, 56-71, asst dir biochem res, 71-73, asst dir drug metab-radiochem, 73-75, mgr radiopharmaceut, Searle Labs, 75-78, SR RES SCIENTIST, G D SEARLE & CO, 78-; SEARLE FOOD RESOURCES, PARK FOREST; GROUP LEADER CHEMICAL DEVELOP, NUTRASWEET CO, UNIVERSITY PARK. *Mem:* Am Chem Soc. *Res:* Synthesis of steroids; steroid metabolism; microbiological transformations of steroids; drug metabolism; radiochemicals; research and development of radiopharmaceuticals; anti-infective agents; peptide synthesis. *Mailing Add:* Zonagen Inc 2408 Timberloch Pl B-4 The Woodlands TX 77380

KRAYNAK, MATTHEW EDWARD, b Scranton, Pa, Dec 19, 27; m 68. NUTRITION. *Educ:* Scranton Univ, BS, 50; Univ Tenn, MS, 52, PhD(biochem), 56. *Prof Exp:* Instr chem, Univ Tenn, 53-55; asst prof biochem & vis chmn dept, Indonesia, 56-60; asst prof, Univ Tenn, 61-62; from asst prof to assoc prof, 62-69, PROF CHEM & NUTRIT, UNIV OKLA, 69- *Concurrent Pos:* Mem, Okla Nutrit Task Force. *Mem:* AAAS; Am Chem Soc; Am Dietetic Asn. *Res:* Nutritional availability of plant galactosides; biochemistry of galactosemia and lactose intolerance. *Mailing Add:* Dept Chem Univ Okla Main Campus Norman OK 73019

KRBECHEK, LEROY O, b Thief River Falls, Minn, May 21, 34; m 60; c 3. ORGANIC CHEMISTRY. *Educ:* Univ NDak, BS, 57; Univ Mich, MS & PhD, 61. *Prof Exp:* Chemist, Aerospace Corp, 61-64 & Int Minerals & Chem Corp, 64-69; sr res chemist, James Ford Bell Res Ctr, Gen Mills, Inc, 69-72, sr res chemist, Gen Mills Chem Inc, 72-80, SR RES CHEMIST, HENKEL INC, 80- *Mem:* Am Chem Soc. *Res:* Organic azides and synthesis. *Mailing Add:* Henkel Res Corp 2330 Circadian Way Santa Rosa CA 95407

KRC, JOHN, JR, b Chicago, Ill, May 17, 20; m 49; c 5. CHEMICAL MICROSCOPY, CRYSTALLOGRAPHY. *Educ:* Univ Chicago, BS, 43. *Prof Exp:* Chemist, E J Brach & Sons, 47, Swift & Co, 48 & Armour Res Found, Ill Inst Technol, 49-61; sr res pharmacist, Warner-Lambert/Parke-Davis Pharmaceut Res Div, 61-80, res assoc, 80-82; RETIRED. *Concurrent Pos:* Adj prof, Sch Pharm, Univ Mich, Ann Arbor, 70-79. *Mem:* Am Crystallog Asn. *Res:* Chemical, x-ray and optical crystallography; chemical microscopy; phase diagrams; thermal stability; crystallization kinetics; solvation; polymorphism; recrystallization; nucleation and crystal growth; physical properties of nonhomogeneous solids. *Mailing Add:* 1514 Chipmunk Lane Oviedo FL 33765

KREAM, BARBARA ELIZABETH, b New York, NY, Mar 11, 48; m 80; c 2. BIOCHEMISTRY, ENDOCRINOLOGY. *Educ:* Mt Holyoke Col, BA, 69; Yale Univ, PhD(molecular biophys, biochem), 74. *Prof Exp:* NIH fel, Dept Biochem, Univ Wis, 74-77; res assoc, 77-78, instr med & endocrinol, 78-79, asst prof, 79-85, ASSOC PROF, DEPT MED, DIV ENDOCRINOL & METAB, UNIV CONN HEALTH CTR, 85- *Concurrent Pos:* Res grants, Am Diabetes Asn, 78-79, Juvenile Diabetes Found, 79-81, Proctor & Gamble Co, 81 & NIH, 81- *Mem:* Sigma Xi; Endocrine Soc; AAAS; Am Soc Bone & Mineral Res; Am Soc Biol Chemists. *Res:* Bone and calcium metabolism; mechanism of action of hormones; effect of insulin on collagen synthesis in bone; hormonal regulation of bone collagen synthesis. *Mailing Add:* 45 Rockledge Dr Avon CT 06001

KREAM, JACOB, b New York, NY, Apr 16, 19; m 42; c 4. BIOCHEMISTRY, CLINICAL CHEMISTRY. *Educ:* City Col NY, BS, 42; Columbia Univ, PhD(biochem), 52; Nat Registry Clin Chem, dipl; Nat Acad Clin Biochem, dipl. *Prof Exp:* Asst, Rockefeller Inst Med Res, 43; chemist, Kellex Corp, 43-44 & Pyridium Corp, 44-46; asst biochem, Columbia Univ, 46-49; biochemist, Inst Cancer Res, 50-52 & US Vet Admin Hosp, NY, 52-53; res assoc biochem, Columbia Univ, 53-54; chief, Dept Clin Chem, Hosp for Joint Dis, 54-65; dir, Core Lab, Clin Res Ctr & sr investr, Steriod Inst, Montefiure Hosp & Med Ctr, 65-82; assoc prof lab med, Albert Einstein Col Med, 78-82; dir, Biochem Labs, Inst Chronobiol, NY Hosp-Cornell Med Ctr, Cornell Univ Med Col, 82-86, assoc prof biochem psychiat, 82-86; CONSULT CLIN CHEM & ENDOCRINOL/HORMONE ASSAY, 86- *Concurrent Pos:* Mem bd examr, Bur Labs, NY Dept Health; lectr, Hunter Col, 51-65; lectr-consult, US Naval Hosp, St Albans, NY, 58-62; chmn, NY Sect, Am Asn Clin Chem, 59-60 & 78-79, Nat Comt, 70-75, Nat Comt Radionuclides & Radioassay, 74-76; consult clin radioimmunoassay, Union Carbide Corp, 75-79. *Mem:* AAAS; Am Chem Soc; Am Asn Clin Chem; Sigma Xi; Endocrine Soc. *Res:* Enzymology; purine and pyrimidine metabolism; purine analogs and cancer; nucleic acid chemistry; polypeptide metabolism; clinical chemistry; automated clinical methods; steroid analysis; radioimmunoassay; competitive protein binding analysis; episodic secretion of pituitary hormones and corticosteroids; radioisotopes, endocrinology of cancer, hormonal studies of mental illness. *Mailing Add:* Three Milford Lane Glen Cove NY 11542

KREAR, HARRY ROBERT, b Pittsburgh, Pa, Apr 13, 22. ETHOLOGY, ECOLOGY. *Educ:* Pa State Univ, BSF, 49; Univ Wyo, MS, 53; Univ Colo, PhD(ecol, ethology), 65. *Prof Exp:* Biologist wildlife res, Mont Fish & Game Dept, 53-54; explor & res, Arctic Wildlife Range Exped, 56; instr biol, Univ Colo, 60-61; asst prof, Mankato State Col, 65-66; chmn div sci & math, US Int Univ, Colo Alpine Campus, 67-73; assoc prof biol sci, Mich Technol Univ, 73-84; RETIRED. *Concurrent Pos:* Vis lectr zool, NSF Insts, Univ Colo, 59-64. *Mem:* Animal Behav Soc; Ecol Soc Am; Am Soc Mammalogists. *Res:* Ecology of selected vertebrates of Ungava; reproduction of cow fur seals pribilots; Arctic wildlife range; ecology of muskrats; behavior and ecology of sea otters amchitka; ecology and ethology of pikas. *Mailing Add:* Moraine Rte 944 Rams Horn Rd Mary's Lake Estes Park CO 80517

KREBILL, RICHARD G, b Upland, Calif, Mar 9, 36; m 58; c 3. FOREST PATHOLOGY. *Educ:* Univ Calif, Berkeley, BS, 58; Univ Wis, PhD(plant path), 62. *Prof Exp:* Plant pathologist, 62-76, ASST DIR, ROCKY MOUNTAIN FOREST & RANGE EXP STA, US FOREST SERV, 79- *Mem:* Am Phytopath Soc; Soc Am Foresters. *Res:* Tree diseases; rust fungi; shrub diseases; research administration. *Mailing Add:* 6209 Woodland Ogden UT 84403

KREBS, EDWIN GERHARD, b Lansing, Iowa, June 6, 18; m 45; c 3. BIOCHEMISTRY. *Educ:* Univ Ill, AB, 40; Washington Univ, MD, 43. *Hon Degrees:* Dr, Univ Geneva, 79. *Prof Exp:* Intern & asst resident, Barnes Hosp, St Louis, 44-45; NIH res fel, Washington Univ, 46-48, from asst prof to prof biochem, 48-68; prof biochem & chmn dept, Sch Med, Univ Calif, Davis, 68-77; investr, Sch Med, Univ Wash, Seattle, 77-80, chmn dept 77- 84, sr investr, Howard Hughes Med Inst, 70-90, PROF PHARMACOL, SCH MED, UNIV WASH, SEATTLE, 77- *Concurrent Pos:* Guggenheim fel, 59, 66; Gairdner found award, Toronto, Ont, 78 & Passano found award, Baltimore, Md, 88. *Honors & Awards:* George W Thorn Award Sci Excellence, 83; Res Achievement Award, Am Heart Asn, 87; Life Sci Award, 3M, 89; Albert Lasker Basic Med Res Award, 89; Louisa Gross Horwitz Award, Columbia Univ, 89. *Mem:* Nat Acad Sci; Am Acad Arts & Sci; Am Soc Biochem & Molecular Biol (pres, 85-86). *Res:* Enzyme chemistry; regulation of metabolism; mechanism of action of hormones; protein phosphorylation. *Mailing Add:* Dept Pharmacol SL-15 Univ Wash Seattle WA 98195

KREBS, JAMES JOHN, b St Louis, Mo, Feb 28, 32; m 72; c 2. EXPERIMENTAL SOLID STATE PHYSICS. *Educ:* St Louis Univ, BS, 54, PhD(physics), 59. *Prof Exp:* PHYSICIST MAGNETIC RESONANCE, US NAVAL RES LAB, 58- *Concurrent Pos:* Nat Res Coun res assoc, 58-59; vis fel, Princeton Univ, 75-76. *Mem:* Fel Am Phys Soc; Sigma Xi. *Res:* Investigation of electron-nuclear interactions by means of magnetic double resonance; resonance and optical absorption in exchange coupled systems; deep impurity resonance in III-V semiconductors; properties of ultra-thin magnetic single crystals. *Mailing Add:* Code 6340 Naval Res Lab Washington DC 20375-5000

KREBS, JAMES N, b Sauk Center, Minn, Apr 20, 24. JET ENGINE DESIGN & DEVELOPMENT. *Educ:* Northwestern Univ, BS, 45. *Prof Exp:* Design devep mgr & market mgr, Gen Elec Co, 46-78, vpres, Mil Eng Progs, 78-82, LYNN Eng Opers, 82-84 & technol & mgt assessment, aircraft eng group, 84-85; RETIRED. *Mem:* Nat Acad Eng; Am Inst Aeronaut & Astronaut. *Mailing Add:* 84 Harbor Ave Marblehead MA 01945

KREBS, JULIA ELIZABETH, b Baton Rouge, La, Mar 29, 43; m 80. TEACHING, ORNITHOLOGY. *Educ:* Oberlin Col, AB, 65; Boston Col, MEd, 69; Univ Ga, MSc, 72, PhD(zool & ecol), 77. *Prof Exp:* PROF BIOL, FRANCIS MARION COL, 77- *Mem:* Ecol Soc Am; Asn Biol Lab Educ. *Res:* Nutrient cycling; effect of man on natural systems; bird populations. *Mailing Add:* Dept Biol Francis Marion Col Florence SC 29501

KREBS, ROBERT DIXON, b Gowanda, NY, Mar 12, 31; m 54; c 3. SOIL MECHANICS, FOUNDATION ENGINEERING. *Educ:* Rutgers Univ, BS, 52, PhD(soil sci), 56; Purdue Univ, MSE, 59. *Prof Exp:* From asst to assoc prof agron, Va Polytech Inst, 55-57; instr eng geol, Purdue Univ, 58-59; ASSOC PROF CIVIL ENG, VA POLYTECH INST & STATE UNIV, 59-, ASST HEAD DEPT, 70- *Concurrent Pos:* Assoc, Hwy Res Bd, Nat Acad Sci-Nat Res Coun. *Mem:* AAAS; Am Soc Civil Engrs; Am Soc Eng Educ. *Res:* Soil mechanics, physics, mineralogy and chemistry; soils and geologic engineering; soil genesis and classification; soil stabilization; soil behavior. *Mailing Add:* Dept Civil Eng Va Polytech Inst & State Univ Blacksburg VA 24061

KREBS, WILLIAM H, b Detroit, Mich, Apr 6, 38; m 83; c 2. INDUSTRIAL HYGIENE, TOXICOLOGY. *Educ:* Univ Mich, Ann Arbor, BS, 60, MPH, 63, MS, 65, PhD, 70; Am Bd Indust Hyg, cert. *Prof Exp:* Res asst indust health, Sch Pub Health, Univ Mich, Ann Arbor, 62; indust hygienist, Lumbermens Mutual Casualty Co, Chicago, Ill, 63-64; indust hygienist, Indust Hyg Dept, 70-77, mgr, 77-81, dir, Toxic Mat Control Activ, 81-90, DIR, INDUST HYG ACTIV, GEN MOTORS CORP, 90- *Concurrent Pos:* Dir, Am Indust Hyg Asn, 76-79; vpres, Am Indust Hyg Asn, 86-87, pres-elect, 87-88, pres, 88-89; pres, Mich Indust Hyg Soc, 80-81; vpres, Int Occup Hyg Asn, 90- *Mem:* AAAS; NY Acad Sci; Am Indust Hyg Asn; Am Acad Indust Hyg; Brit Occup Hyg Soc; Am Pub Health Asn. *Res:* Formation of ferruginous bodies. *Mailing Add:* GM Indust Hygiene 30500 Mound Rd Box 9055 Warren MI 48090-9055

KREDICH, NICHOLAS M, b Chicago, Ill, Sept 23, 35; m 57; c 3. INTERNAL MEDICINE, BIOCHEMISTRY. *Educ:* Duke Univ, BA, 57; Univ Mich, MA, 60, MD, 62. *Prof Exp:* Intern internal med, Duke Hosp, Durham, NC, 62-63, asst resident, 63-64; res assoc molecular biol, Nat Inst Arthritis & Metab Dis, 64-66; staff assoc, 66-68; from asst prof to assoc prof, 68-80, PROF INTERNAL MED, MED CTR, DUKE UNIV, 80- *Concurrent Pos:* Nat Inst Arthritis & Metab Dis res grant, 68-81; investr, Howard Hughes Med Inst, 73-89. *Res:* Regulation of metabolic pathways, including feedback inhibition and repression and induction of enzymes; bacterial and human genetics; sulfur metabolism in bacteria; adenosine deaminase deficiency; immunodeficiency disease; genetics; molecular biology. *Mailing Add:* Dept Med & Biochem Duke Med Ctr Box 3100 Durham NC 27710

KREEGER, RUSSELL LOWELL, b Amherst, Ohio, Jan 24, 46; m 73; c 2. CELLULOSICS, PERSONAL CARE POLYMERS. *Educ:* Kent State Univ, BS, 68; Ohio State Univ, Columbus, PhD(org chem), 76. *Prof Exp:* SR RES SCIENTIST, SPECIALTY CHEM DIV, UNION CARBIDE CORP, 76- *Mem:* Am Chem Soc; Sigma Xi. *Res:* Research and development; natural polymer process research; cellulosic derivatives; personal care polymers; water-soluble polymers. *Mailing Add:* Four Cornfield Terr Flemington NJ 08822

KREER, JOHN B(ELSHAW), b Brooklyn, NY, Sept 25, 27; m 57; c 2. ELECTRICAL ENGINEERING. *Educ:* Iowa State Col, BS, 51; Univ Ill, MS, 54, PhD(elec eng), 56. *Prof Exp:* From instr to asst prof elec eng, Univ Ill, 55-59; from assoc prof to prof, Univ WVa, 59-64; assoc prof, 64-68, chmn, dept elec eng & systs sci, 77-87, PROF ELEC ENG, MICH STATE UNIV, 68- *Mem:* Inst Elec & Electronics Engrs; Am Soc Eng Educ. *Mailing Add:* Dept Elec Eng Mich State Univ East Lansing MI 48824-1226

KREEVOY, MAURICE M, b Boston, Mass, Aug 28, 28; m 53; c 2. CHEMICAL KINETICS, MEMBRANE DYNAMICS. *Educ:* Univ Calif, Los Angeles, BS, 50; Mass Inst Technol, PhD, 54. *Prof Exp:* Res assoc chem, Pa State Univ, 53-55; NSF fel, Univ Utah, 55-56; from asst prof to assoc prof, 56-64, PROF CHEM, UNIV MINN, MINNEAPOLIS, 64- *Concurrent Pos:* Consult, Gen Mills Inc, 59-78, Ventron Corp, 75-79, Henckel Am Inc, 78-85, Honeywell Inc, 85-86, Medtronic Inc, 88-; Sloan Found fel, 60-64; NSF sr fel, Oxford Univ, 62-63; partic, US Acad Sci exchange prog with Coun of Acad Socialist Fed Repub of Yugoslavia, 69-70. *Mem:* Am Chem Soc; The Chem Soc; Croatian Chem Soc; Sigma Xi. *Res:* Physical and theoretical organic chemistry; chemical kinetics and dynamics in solution; isotope effects; dynamics of membrane transport. *Mailing Add:* Dept Chem Univ Minn Minneapolis MN 55455

KREFT, ANTHONY FRANK, III, b Detroit, Mich, May 28, 48; m 79. ORGANIC CHEMISTRY, MEDICINAL CHEMISTRY. *Educ:* Univ Mich, BS, 70; Columbia Univ, MPh, 73, PhD(org chem), 76. *Prof Exp:* Supvr, Wyeth Res Labs, 78-87; prin scientist, 88, SUPVR, MED CHEM, WYETH-AYERST LABS, 87- *Concurrent Pos:* Fel, Stanford Univ, 76-78. *Mem:* Am Chem Soc; Inflammation Res Asn. *Res:* Design and synthesis of drugs of medicinal interest. *Mailing Add:* Wyeth-Ayerst Labs CN 8000 Princeton NJ 08540

KREGLEWSKI, ALEXANDER, thermodynamics, petroleum chemistry; deceased, see previous edition for last biography

KREH, DONALD WILLARD, b Frederick, Md, Mar 17, 37; m 66; c 4. ORGANIC CHEMISTRY. *Educ:* Univ Richmond, BS, 59, MS, 61; Va Polytech Inst, PhD(org chem), 66. *Prof Exp:* Chemist, Great Lakes Res Corp, 66-67; chemist, 67-68, SR CHEMIST, TENN EASTMAN CO, 69- *Mem:* Am Chem Soc. *Res:* Reactions and synthesis of small ring sulfides, sulfoxides and sulfones; synthesis of photographic chemicals, antioxidants, stabilizers, and industrial chemical intermediates. *Mailing Add:* Bldg 54B Tenn Eastman Co Kingsport TN 37760

KREH, E(DWARD) J(OSEPH), JR, b Pittsburgh, Pa, Feb 26, 15; m 38; c 3. ENGINEERING. *Educ:* Carnegie Inst Technol, BS, 37. *Prof Exp:* Design engr, Westinghouse Elec Corp, 37-42; plant engr, Camillus Cutlery Co, 46-51; mgr equip develop, Westinghouse Elec Corp, 51-56, mem div mgr staff, Stress Corrosion & Hydraul Fields, Bettis Plant, 56-58, div apparatus engr, 58-59, mgr, Nuclear Core Dept, 59-61, mgr, Core Mat Dept, 61-65, mgr cent labs, 65-67, mgr opers, Bettis Plant, 67-72, Prod Assurance, Westinghouse Pressurized Water Reactors Div, 72-79, consult engr, 79-82; CONSULT ENGR, O'DONNELL & ASSOC, 82- *Mem:* Am Soc Mech Engrs; Nat Asn Corrosion Engrs; Am Inst Mgt. *Res:* Thermal, mechanical and electrical design; corrosion studies; development of fabrication processes for nuclear reactors; development and management of quality assurance systems to assure reactor safety and reliability. *Mailing Add:* 624 Trotwood Circle Pittsburgh PA 15241

KREH, RICHARD EDWARD, b Waterbury, Conn, Dec 22, 41; m 67; c 2. BIOLOGICAL SCIENCES. *Educ:* Univ Conn, BS, 69; Va Polytech Inst & State Univ, MS, 74. *Prof Exp:* RES ASSOC FOREST BIOL, VA POLYTECH INST & STATE UNIV, 69- *Mem:* Soc Am Foresters. *Res:* Silviculture research on site preparation; root growth analysis of forest tree nursery growth seedlings; nitrogen dynamics of mine spoil soils for forestry reclamation; hybrid performance of selected pine crosses. *Mailing Add:* Reynolds Homestead Res Ctr Critz VA 24082

KREIBICH, GERT, b Komotau, Czech, Nov 14, 39; Ger citizen; m 66; c 1. CELL BIOLOGY. *Educ:* Univ Heidelberg, dipl chem, 65, Dr rer nat, 68. *Prof Exp:* Fel chem carcinogenesis, Ger Cancer Res Ctr, Heidelberg, 70; res assoc cell biol, Rockefeller Univ, 70-72; from asst prof to assoc prof, 72-82, PROF CELL BIOL, MED CTR, NY UNIV, 82- *Concurrent Pos:* Res fel, Ger Res Soc, 70-72; NIH res career develop award, 77-82; mem cell biol study sect, NIH, 78-82, Irma Hirschl Award, 82-86. *Mem:* Ger Soc Biol Chem; NY Acad Sci; Am Soc Cell Biol; Am Soc Biol Chemists. *Res:* Structure and function of subcellular membranes in eukaryotic cells; function of membrane bound polysomes in membrane biogenesis. *Mailing Add:* Dept Cell Biol Sch Med NY Univ 550 First Ave New York NY 10016

KREIBICH, ROLAND, b Glasert, Bohemia, July 30, 22; m 58. ORGANIC CHEMISTRY. *Educ:* Univ Graz, Magister Pharmaciae, 49, PhD(chem), 51. *Prof Exp:* Res chemist, Can Westinghouse, 52-54 & Durez Plastics, Inc, 54-56; proj engr, Gen Elec, 57-58; mgr, Polymer Res Dept, Weyerhaeuser Co, 58-82; RETIRED. *Res:* Polymers; resins. *Mailing Add:* 4201 S 344 St Auburn WA 98001-9545

KREIDER, DONALD LESTER, b Lancaster, Pa, Dec 5, 31; m 52; c 3. MATHEMATICAL LOGIC. *Educ:* Lebanon Valley Col, BS, 53; Mass Inst Technol, PhD(math), 59. *Prof Exp:* Instr, Lebanon Valley Col, 52-53; asst, Mass Inst Technol, 53-55, instr math, 55-60; from asst prof to assoc prof, 60-68, PROF MATH, DARTMOUTH COL, 68- *Mem:* Am Math Soc; Asn Symbolic Logic; Math Asn Am. *Res:* Recursive functions; automata theory. *Mailing Add:* Box 352 Norwich VT 05055

KREIDER, EUNICE S, b Ohio, June 5, 41; m 63. INTERNATIONAL PRODUCT DEVELOPMENT. *Educ:* Goshen Col, BA, 63; Purdue Univ, PhD(chem), 67; Northwestern Univ, MM, 77. *Prof Exp:* From res investr to sr res investr, G D Searle & Co, 67-74, group leader, 74-76, dir, Prog Planning & Qual Assurance, 76-78, assoc dir, 78-80, dir proj mgt, Preclin Res & Develop, 80-81; DIR, INT PROD DEVELOP, ORTHO PHARMACEUT CORP, 81- *Mem:* Sigma Xi; Am Chem Soc; NY Acad Sci. *Res:* Heterocyclic medicinal chemicals; drug safety testing and GLP'S; project management. *Mailing Add:* Witikonerstrasse 248 CH 8053 Zurich Switzerland

KREIDER, HENRY ROYER, b Baltimore, Md, Dec 31, 11; m 36; c 2. PHYSICAL CHEMISTRY, ORGANIC CHEMISTRY. *Educ:* Univ Toledo, BA, 33; Ohio State Univ, MS, 35, PhD(chem), 36. *Prof Exp:* Chemist, Am Med Asn, Chicago, 36-42 & Mead Johnson & Co, Ind, 42-45; exec asst to vpres, William S Merrell Co, 45-50, assoc dir res, 50-56; dir res, Chesebrough-Ponds, Inc, 56-59; exec vpres, Viobin Corp, 59-60; dir res, Sherman Labs, 60-68; dir prod develop, Cooper Labs, Inc, 68-71; CONSULT DRUG COSMETIC INDUSTS, 71- *Mem:* AAAS; Am Chem Soc; Soc Cosmetic Chem; Am Pharmaceut Asn; Asn Res Dirs. *Res:* Development of foods, drugs & cosmetics; vitamins; proteins; fats; micronutrients. *Mailing Add:* 830 N Shore Dr NE Apt 14A St Petersburg FL 33701

KREIDER, JACK LEON, b Afton, Okla, Mar 12, 41; m 67; c 3. ANIMAL SCIENCE, REPRODUCTIVE PHYSIOLOGY. *Educ:* Okla State Univ, BS, 68; Univ Ky, MS, 70, PhD(animal sci), 71. *Prof Exp:* Asst prof animal sci, Univ Mo-Columbia, 72; from asst prof to assoc prof animal sci, La State Univ,

Baton Rouge, 73-79; assoc prof animal sci, Tex A&M Univ, 79-84; PROF ANIMAL SCI & RESIDENT DIR, DEAN LEE RES STA, ALEXANDRIA, LA, 85- *Concurrent Pos:* Mem, La Forage & Greenlands Coun, Am Forage & Greenlands Coun. *Mem:* Am Soc Animal Sci; Am Registry Prof Animal Scientists. *Res:* Reproductive physiology and endocrinology of the mare as related to improving efficiency of production of horses, particularly the perparturient period; reproductive physiology of the stallion, particularly semen physiology; administration of research in beef cattle and and service management as well as agronomy and weed control. *Mailing Add:* Dean Lee Res Sta Alexandria LA 71302-9608

KREIDER, JOHN WESLEY, b Philadelphia, Pa, Mar 24, 37; m 63; c 2. CANCER RESEARCH, TUMOR IMMUNOLOGY. *Educ:* La Salle Col, AB, 59; Univ Pa, MD, 63. *Prof Exp:* Mem fac path, Med Sch, Univ Pa, 67-68; MEM FAC PATH, HERSHEY MED CTR, 68- *Concurrent Pos:* Career develop award, USPHS, 69; mem, Path B Study Sect, NIH, 79-, Am Cancer Soc Study Sect, VA Study Sect. *Honors & Awards:* Borden Award, AOA. *Mem:* Am Asn Pathologists; Am Asn Cancer Res; Am Asn Immunologists. *Res:* Host regulation of tumor growth; neoplastic cell differentiation; papillomavirus transformation. *Mailing Add:* Dept Path Hershey Med Ctr Hershey PA 17033

KREIDER, KENNETH GRUBER, b Lancaster, Pa, May 21, 37; m 61; c 3. THIN FILMS, SENSORS. *Educ:* Mass Inst Technol, SB, 59, SM, 61, ScD, 63. *Prof Exp:* Res supvr, United Aircraft Res Labs, 65-73; DIV CHIEF, & SR SCIENTIST, NAT BUR STANDARDS, 75- *Concurrent Pos:* chmn, Inst Elec & Electronics Engrs Tech Comt Sensor Standards; judge, Am Paper Inst Energy & Environ Awards, 78-90; adv panel, UPA Ctr Chem Elec, 83-86, Mass Inst Technol, Electronic Package, 90- *Honors & Awards:* Bronze Medal, US Dept Com. *Mem:* Am Vacuum Soc; Am Soc Testing & Mat; Am Inst Mining, Metall & Petrol Engrs; Mat Res Soc; Am Soc Metals. *Res:* Thin Film Sensors; instrumentation for harsh environments; metal matrix composites; infrared thermography; sputtering of thin films. *Mailing Add:* Nat Inst Standards & Technol Bldg 221 Rm A303 Gaithersburg MD 20899

KREIDER, LEONARD CALE, b Sterling, Ohio, Feb 16, 10; m 33; c 3. CHEMISTRY. *Educ:* Goshen Col, AB, 31; Ohio State Univ, MSc, 33, PhD(org chem), 36. *Prof Exp:* Preparations asst, Ohio State Univ, 31-32, asst chem, 32-36; res chemist, Rockefeller Inst, 36-37; from asst prof to prof chem, Bethel Col Kans, 37-49, chmn, Div Natural Sci, 46-49; chemist tech serv, Res Ctr, B F Goodrich Co, 49-50, pioneering res, 50-56 & org res, 56-59, sr res chemist polymerization res, 59-75; RETIRED. *Mem:* Am Chem Soc; Sigma Xi. *Res:* Development of catalyst for manufacture of cis-1, 4-polyisoprene and for Hydrin rubbers; oligosaccharides; alkaline degradation of carbohydrates; galacturonic acid chemistry; pigment reinforcement in rubber; alkali metal catalyzed rubbers; polyester urethane rubbers; rubber hysteresis; aluminum alkyls; metalloorganic catalyzed polymerizations. *Mailing Add:* 1320 Greencroft Dr Goshen IN 46526-5135

KREIDL, NORBERT J(OACHIM), b Atzgersdorf, Austria, July 3, 04; US citizen; m 34; c 3. PHYSICS, GLASS SCIENCE. *Educ:* Univ Vienna, PhD(physics), 28. *Hon Degrees:* Dr,Alfred Univ, 71, Univ Vienna, 78 & Univ Jona, 84. *Prof Exp:* Fel glass sci, Kaiser Wilhelm Inst, 29; mgr res, Schreiber & Nephews, Rapotin, Czech, 29-39; asst prof glass sci, Pa State Col, 39-43; head, Glass Div, Bausch & Lomb, Inc, 43-55, dir mat res & develop, 55-64; prof ceramics, Rutgers Univ, 64-66; prof, 66-75, EMER PROF CERAMIC ENG, UNIV MO-ROLLA, 76- *Concurrent Pos:* Consult numerous industs, 39-43 & 64-; chmn, infrared transmitters comt, Mat Adv Bd, Nat Acad Sci-Nat Res Coun, 68 & mem, submergencies comt, 69; pres, Int Comn Glass, 69-72; mem, Glass Comn, Univ Space Res Assoc, 73-, Sci Coun, 80-; adj prof chem eng, Univ NMex, 78-; adj prof physics, Col Santa Fe, 80-, adj prof sci, 86-; adj prof mat sci, Univ Ariz, Tucson, 86- *Honors & Awards:* Toledo Award, Am Ceramic Soc, 67, Jepson Award, 69. *Mem:* Am Ceramic Soc; hon fel Brit Soc Glass Technol; hon mem Ger Glass Technol Soc; Sigma Xi; Emer mem Am Chem Soc. *Res:* Glass structure and properties; radiation effects. *Mailing Add:* 1433 Canyon Rd Santa Fe NM 87501

KREIDL, TOBIAS JOACHIM, b Rochester, NY, May 6, 54. DIGITAL IMAGE PROCESSING. *Educ:* Univ Vienna, Austria, PhD(astron), 79. *Prof Exp:* Res assoc, Ruhr Univ, Bochum, WGer, 79-80; ASTRONR, LOWELL OBSERV, 80- *Concurrent Pos:* Lectr comput sci, Northern Ariz Univ, 81. *Mem:* Am Astron Soc; Sigma Xi. *Res:* Digital image processing and image processing systems; photometry of peculiar A-type stars; computer analysis of astronomical data of various nature. *Mailing Add:* Lowell Observ Mars Hill Rd 1400 W Flagstaff AZ 86001

KREIDLER, ERIC RUSSELL, b Lock Haven, Pa, July 21, 39; m 68; c 2. PHASE EQUILIBRIA, LUMINESCENCE. *Educ:* Pa State Univ, BS, 61, MS, 63, PhD(ceramic sci), 67. *Prof Exp:* Res chemist, Gen Elec Co, 66-80; ASSOC PROF CERAMIC ENG, OHIO STATE UNIV, 80- *Concurrent Pos:* Consult, Gen Elec Co, 80-82; contrib ed, Communications of the Am Ceramic Soc, 81-; assoc ed, Phase Diagrams for Ceramists, 81- *Mem:* Fel Am Ceramic Soc; Electrochem Soc; Mat Res Soc. *Res:* Determination of phase diagrams; crystal chemistry and luminescence of inorganic materials; phosphors and luminescence; glass-metal composites; high temperature ceramic superconductors; electronic ceramics; ceramic processing. *Mailing Add:* Ohio State Univ 381 Watts Hall 2041 Col Rd Columbus OH 43210

KREIER, JULIUS PETER, b Philadelphia, Pa, Nov 30, 26; m 55; c 2. PROTOZOOLOGY, IMMUNOLOGY. *Educ:* Univ Pa, VMD, 53; Univ Ill, MS, 59, PhD, 62. *Prof Exp:* Vet, Agr Res Serv, USDA, 53-56; instr vet physiol, Univ Ill, 56-59, instr vet path & hyg, 59-61, USPHS fel, 61-62; from asst prof to assoc prof, Ohio State Univ, 62-72, prof microbiol, 72-89; RETIRED. *Concurrent Pos:* Vis prof, State Univ de Sao Paulo, Botucatu, Brazil, 81; lectr, China Nat Ctr Prev Med, Inst Parasitol, 85. *Honors & Awards:* Fulbright Award, Inst Hyg, Montevideo, 77. *Mem:* Am Soc Trop Med Hygiene; Soc Protozoologists; Am Soc Parasitol; Am Asn Immunol; Am Soc Microbiol. *Res:* Host-parasite interactions, primarily the blood inhabiting protozoa; malaria parasites, babesia and trypanosomes; procaryotic protists which parasitize the blood; Anaplasmataceae and Bartonellaceae; pathogenesis of the anaemic changes associated with infection and the mechanisms by which the host controls the parasite population; immunological and biochemical events associated with disease and recovery from disease; isolation and identification of parasites and parasite parts for use as antigens. *Mailing Add:* 2047 Iuka Ave Columbus OH 43201

KREIFELDT, JOHN GENE, b Manistee, Mich, Oct 7, 34; m 64; c 2. ENGINEERING DESIGN, BIOMEDICAL ENGINEERING. *Educ:* Univ Calif, Los Angeles, BS, 61; Mass Inst Tech, MS, 64; Case Western Reserve Univ, PhD(biomed & human factors eng), 69. *Prof Exp:* From asst prof to assoc prof, 69-80, PROF ENG DESIGN & HUMAN FACTORS ENG, TUFTS UNIV, 80- *Concurrent Pos:* Vpres, Appl Ergonomics Corp; USPHS grants, Tufts Univ & New Eng Med Ctr, 71-81; Nat Res Coun associateship, NASA-Ames Res Ctr, 73; Dept Health, Educ & Welfare grant; NASA grants. *Mem:* Human Factors Soc. *Res:* Electromyographic processing and control; man-machine system design; air traffic control studies; computers in automation of radiotherapy treatment; multidemensional scaling in design; mathematical models of human operators; consumer product design; safety design. *Mailing Add:* Dept Eng Design Tufts Univ Medford MA 02155

KREIGHBAUM, WILLIAM EUGENE, b Elkhart, Ind, June 17, 34; m 61; c 2. MEDICINAL CHEMISTRY, SCIENCE COMMUNICATIONS. *Educ:* Wabash Col, AB, 56; Ind Univ, PhD(org chem), 60. *Prof Exp:* Bristol Labs res fel org chem, Ind Univ, 60-61; sr res scientist, 61-67, group leader, 68-69, sr investr, 70-73, prin investr, Mead Johnson Res Ctr, Mead Johnson & Co, 74-86; SR INFO ANALYST, BRISTOL MYERS SQUIBB CO, WALLINGFORD, 87- *Mem:* Am Chem Soc. *Res:* Synthesis and pharmacological activity of organic sulfur compounds; heterocyclic compounds containing sulfur or nitrogen; chemistry of the sympathetic nervous system. *Mailing Add:* Sci Info Dept 809 Bristol Myers Squibb Co Wallingford CT 06492-7660

KREILICK, ROBERT W, b Kalamazoo, Mich, Jan 3, 38; m 59; c 2. PHYSICAL CHEMISTRY. *Educ:* Wash Univ, AB, 59, PhD(magnetic resonance), 64. *Prof Exp:* From asst prof to assoc prof, 64-71, PROF CHEM, UNIV ROCHESTER, 71- *Concurrent Pos:* Alfred P Sloan fel, 69-71; consult, NIH; founder Adaptable Lab Software, 83. *Mem:* Am Chem Soc. *Res:* Nuclear magnetic resonance and electron spin resonance; biophysical chemistry; study of biologically important metal complexes; computer software for chemistry. *Mailing Add:* Dept Chem Univ Rochester Rochester NY 14627

KREILING, DARYL, b Minatare, Nebr, May 18, 36; m 56; c 3. MATHEMATICS. *Educ:* Chadron State Col, BS, 61; Bowling Green State Univ, MA, 63; Univ Wyo, PhD(math), 69. *Prof Exp:* Instr math, Univ Wyo, 66-69; from asst prof to assoc prof math, Western Ill Univ, 73-80, asst dean, 74-80; dean arts & sci, 80-87, PROF MATH, UNIV TENN, 80- *Mem:* Am Math Soc; Math Asn Am. *Res:* Associative and non-associative rings; radicals of rings and ring-like structures. *Mailing Add:* Dept Math & Comput Sci Univ Tenn Martin TN 38238

KREILING, WILLIAM H(ERMAN), b Brooklyn, NY, Dec 20, 23; m 51; c 2. CHEMICAL ENGINEERING. *Educ:* Polytech Inst Brooklyn, BChE, 49. *Prof Exp:* Chem res engr pilot plant design, M W Kellogg Co Div, Pullman, Inc, 42-49; dept chemist paper coatings, Lowe Paper Co, 49-53, res chemist, 53-55, gen foreman coatings dept, 55-57, tech supt finishing mill, 57-58, qual control supvr, 58-59; chem engr coating develop, Keuffel & Esser Co, 59-64, mgr process eng, 64-68; sect leader, paper & coatings group, Corp Res Ctr, Int Paper Co, 68-70, sr res assoc paper develop, Corp Res & Develop Div, 70-78; sr develop engr, Specialty Papers & Packaging Div, Ludlow Corp, 78-80, sr develop engr, Laminating & Coatings Div, 80-82; res engr, Timex Corp, 82-84; sr proj mgr, Luminescent Syst Inc, 84-88; RETIRED. *Concurrent Pos:* Chmn, Testing Div, Tech Asn Pulp & Paper Indust, 72-74. *Mem:* Tech Asn Pulp & Paper Indust; fel Am Inst Chem; NY Acad Sci; Soc Info Display. *Res:* Protective, decorative, printing and photographic coatings and specialty papers; electroluminescent coatings & technology; patents in electrophotography and electroluminescent technology. *Mailing Add:* 53 Hadley Village Rd South Hadley MA 01075-2184

KREIMER, HERBERT FREDERICK, JR, b Cincinnati, Ohio, Feb 19, 36; m 61; c 2. MATHEMATICS. *Educ:* Yale Univ, BS, 58, PhD(math), 62. *Prof Exp:* From asst prof to assoc prof, 62-77, PROF MATH, FLA STATE UNIV, 77- *Concurrent Pos:* Univ Res Coun grant, 64 & 67; vis assoc prof, Northwestern Univ, 65-66; NSF grant, 65-66 & 68-71. *Mem:* Am Math Soc; Math Asn Am; Sigma Xi. *Res:* Ring theory and homological algebra. *Mailing Add:* Dept Math Fla State Univ Tallahassee FL 32306

KREIN, PHILIP THEODORE, b Orange, Calif, Apr 22, 56. POWER ELECTRONICS, ELECTROSTATICS. *Educ:* Lafayette Col, BS, 78, AB, 78; Univ Ill, Urbana, MS, 80, PhD(elec eng), 82. *Prof Exp:* Physicist, Tektronix, Inc, Beaverton, Ore, 84-87; vis asst prof, 82-84, ASST PROF ELEC ENG, UNIV ILL, URBANA, 87- *Concurrent Pos:* Vis researcher, Sundstrand Corp, Rockford, Ill, 83; res initiative award, NSF, 88. *Mem:* Inst Elec & Electronic Engrs; Electrostatics Soc Am; Soc Imaging Sci & Technol. *Res:* Large-signal and nonlinear control issues in power electronics; advanced switching power converters; electrohydrodynamics and other applications of electrostatics; electric machines and drive systems. *Mailing Add:* Dept Elec & Computer Eng Univ Ill 1406 W Green Urbana IL 61801

KREIPKE, MERRILL VINCENT, b Evansville, Ind, Feb 14, 16; m 37; c 2. GEOTECHNICAL ENGINEERING, CIVIL ENGINEERING. *Educ:* Purdue Univ, BS, 36. *Prof Exp:* Resident engr, Off City Engr, Evansville, Ind, 36-39; inspector, US Army Engrs Dist, Ky, 39-41, jr engr, 41-42, asst engr, 42-44, engr, 46-51, civil engr, 51-56 & Off Chief Engrs, US Dept Army, 56-61, engr, Off Chief Res & Develop, 61-69, chief geophys sci br, US Army Res Off, 69-74, chief mil res & develop team, Chief Engrs, Dept Army, 74-75;

CONSULT, 75- *Concurrent Pos:* Permanent secy, quadripartite standing working group ground mobility, Armies of US, UK, Can & Australia, 59-66; proj officer for US, NATO long-term sci study land-based mobility, 66-; US nat leader, NATO long-term sci study mobility interface, 69-; exec mem & US leader, subgroup T-ground mobility, Tech Coop Prog, US, UK, Can & Australia, 69-; US nat leader, NATO long-term sci study on Arctic opers, 71- *Honors & Awards:* Meritorious Civilian Serv Medal, Dept Army, 66. *Mem:* Am Soc Civil Engrs; Soc Am Mil Engrs; Nat Soc Prof Engrs; Int Soc Soil Mech & Found Eng; Int Soc Terrain-Vehicle Systs. *Res:* Soil mechanics; terrain-vehicle interaction; quantitative terrain evaluation; design and construction of earth and rockfill dams; rapid earthwork; soil stabilization; soil surfacings. *Mailing Add:* 3060 Hazelton St Falls Church VA 22044

KREIS, RONALD W, b Passaic, NJ, Oct 20, 42; m 69; c 2. PHYSICAL CHEMISTRY. *Educ:* Ursinus Col, BS, 64; Univ Del, PhD(chem), 69. *Prof Exp:* Res scientist chem, Uniroyal Inc, 69-77; vpres res, 77-90, DIR QUAL RHÔNE-POULENC, ALCOLAC INC, 90- *Mem:* Am Chem Soc. *Res:* Latex and colloid chemistry; adhesion of rubber; paper chemicals; surfactants. *Mailing Add:* Alcolac Inc 3440 Fairfield Rd Baltimore MD 21226

KREIS, WILLI, b Ebnat, Switz, Nov 3, 24; m 62; c 4. CLINICAL PHARMACOLOGY, CHEMOTHERAPY. *Educ:* Univ Zurich, MD, 54; Univ Basel, PhD(org chem), 57. *Prof Exp:* Res mem biochem pharmacol, Sandoz, Ltd, Basel, Switz, 58-61; res assoc, Sloan-Kettering Inst Cancer Res, 61-64, assoc, 64-69; asst prof, Sloan-Kettering Inst Cancer Res, 67-72, chmn biochem unit, 74-75, assoc prof, Sloan-Kettering Div, Grad Sch Med Sci, Cornell Univ, 72-, assoc mem, 69- *Concurrent Pos:* Damon Runyon grant, 73; Nat Cancer Inst grant, 75; asst attend clin pharmacologist, Dept Med, Mem Hosp, 75-; Am Cancer Soc grants, 76, 78, 79, 80 & 81; assoc prof pharmacol & therapeut, Sloan-Kettering Div Grad Sch Med Sci, Cornell Univ, 80-, res prof, 82-, adj staff mem Mem Sloan-Kettering Cancer Ctr, 82- *Mem:* Swiss Med Soc; Swiss Chem Soc; Am Asn Cancer Res; NY Acad Sci; Am Soc Biol Chemists; Am Soc Clin Oncol. *Res:* Biochemical pharmacology of anticancer drugs; experimental and clinical pharmacology of cancer; biochemistry of nucleic acids; phase I and II evaluation of new anticancer agents. *Mailing Add:* Dept Med N Shore Univ Hosp 300 Commun Dr Manhasset NY 11030

KREISER, RALPH RANK, b Lebanon, Pa, Oct 7, 41; m; c 2. INORGANIC CHEMISTRY, ANALYTICAL CHEMISTRY. *Educ:* Lebanon Valley Col, BSc, 63; Brown Univ, MSc, 66; Univ Conn, PhD(chem), 69. *Prof Exp:* Asst solid state chem, Philips Nature Sci Lab, Eindhoven, Holland, 69-70; inst chem, Univ Conn, 71-72; PROF CHEM & CHMN DEPT, COMMUNITY COL RI, 72- *Concurrent Pos:* Mem comn math & sci, State of RI, 84-85. *Mem:* Am Chem Soc; Sigma Xi. *Res:* Interface of science and the arts in the areas of conservation and restoration. *Mailing Add:* Dept Chem Community Col RI 400 East Ave Warwick RI 02886

KREISER, THOMAS H(ARRY), b Ono, Pa, Aug 12, 35; m 79; c 3. ANALYTICAL CHEMISTRY. *Educ:* Lebanon Valley Col, BS, 58; Univ Nebr, MS, 60, PhD(chem), 65. *Prof Exp:* Res biochemist, Miles Labs Inc, 62-86; SR SCIENTIST, ENVIRON TEST SYSTS, INC, 87- *Mem:* Sigma Xi. *Res:* Analytical test systems research and development. *Mailing Add:* 30182 Blue Spruce Dr Elkhart IN 46514-9723

KREISHMAN, GEORGE PAUL, b Nurnberg, Ger, Jan 28, 46; US citizen; m 72; c 2. BIOPHYSICAL CHEMISTRY. *Educ:* Univ Wis-Milwaukee, BS, 67; Calif Inst Technol, PhD(chem), 72. *Prof Exp:* Fel chem, Int Chem Nuclear Corp, 71-72; fel biophys, Univ Pittsburgh, 72-74; vis teaching asst chem, Mich Technol Univ, 74-75; from asst prof to assoc prof, 75-90, PROF CHEM, UNIV CINCINNATI, 90- *Mem:* Am Chem Soc; Biophys Soc; Res Soc Alcoholism. *Res:* Application of nuclear magnetic resonance spectroscopy and electrochemical techniques to the study of biologically important systems. *Mailing Add:* Dept Chem Univ Cincinnati Cincinnati OH 45221

KREISLE, LEONARDT F(ERDINAND), b Austin, Tex, Oct 22, 22. MECHANICAL ENGINEERING. *Educ:* Univ Tex, BS, 44, MS, 51; Cornell Univ, PhD(mach design, eng mech & servomech), 55. *Prof Exp:* Instr, Univ Tex, 42-43, instr eng drawing, 43-44; mech engr struct design, Robert E McKee, Gen Contractor, 44-45; instr mech eng, 45-49, asst prof, 49-53 & 55-56, assoc prof, 56-67, PROF MECH ENG, UNIV TEX, AUSTIN, 67-, COUNR, COL ENG, 66- *Concurrent Pos:* Res engr, Univ Tex, Austin, 51 & 52-53; vis instr, Cornell Univ, 53-55. *Mem:* AAAS; Am Soc Mech Engrs; Am Soc Eng Educ; Nat Soc Prof Engrs. *Res:* Transient analyses of linear and nonlinear dynamic systems; design of machine elements for dynamic and fatigue conditions; measurement engineering. *Mailing Add:* Dept Mech Eng Univ Tex Austin TX 78712

KREISLER, MICHAEL NORMAN, b Bronx, NY, Oct 30, 40; m 63; c 3. HIGH ENERGY PHYSICS. *Educ:* Princeton Univ, AB, 62; Stanford Univ, MS, 63, PhD(physics), 66. *Prof Exp:* From instr to asst prof physics, Joseph Henry Labs, Princeton Univ, 66-72; assoc prof, 72-76, grad dean res, 75-77, PROF PHYSICS, UNIV MASS, AMHERST, 76- *Concurrent Pos:* Sci Assoc, Europ Orgn Nuclear Res, Geneva, 78-79; consult, Lawrence Livermore Labs, Los Alamos Labs. *Mem:* Am Phys Soc; Sigma Xi. *Res:* Investigation of the strong interactions of neutrons, especially cross sections; study of the decays of multi-pionic resonances such as the eta meson; search for rare phenomena in weak interactions, kaon decays, beta decay of Lambda hyperon; search for tachyons; polarization in inclusive reactions; charm searches; macron accelerators; hadronic production of particles with strange, charm, and bottom quarks; new approaches to ultra high speed computation. *Mailing Add:* Dept Physics Univ Mass GRC Tower C Amherst MA 01003

KREISMAN, NORMAN RICHARD, b Chicago, Ill, June 26, 43; m 75; c 1. NEUROPHYSIOLOGY. *Educ:* Ariz State Univ, BA, 65; Univ Mich, MS, 68; Med Col Pa, PhD(physiol), 71. *Prof Exp:* Instr, 71-73, asst prof, 73-79, ASSOC PROF PHYSIOL, SCH MED, TULANE UNIV, 79- *Mem:* Int Soc Cerebral Blood Flow & Metab; Soc Neurosci; Am Heart Asn. *Res:* Electrophysiological and metabolic relationships in brain in physiological and pathophysiological states; epilepsy; hypoxia. *Mailing Add:* Dept Physiol Sch Med Tulane Univ 1430 Tulane Ave New Orleans LA 70112

KREITH, FRANK, b Vienna, Austria, Dec 15, 22; nat US; m 51; c 3. SOLAR ENGINEERING, THERMODYNAMICS. *Educ:* Univ Calif, BS, 45; Univ Calif, Los Angeles, MS, 49; Univ Paris, Dr Univ Paris(sci), 65. *Prof Exp:* Res engr, Jet Propulsion Lab, Calif Inst Technol, 45-49; asst prof mech eng, Univ Calif, 51-53; assoc prof, Lehigh Univ, 53-59; chief thermal conversion, Solar Energy Res Inst, 77-84; prof mech & chem eng, 59-78, fac res assoc, Inst Behav Sci, 71-77, EMER PROF CHEM ENG, UNIV COLO, BOULDER, 78-; PRES, KREITH ENG INC, 88- *Concurrent Pos:* Consult, Proj Squid, 50, Air Prod, Inc, 55-57, Metals Disintegrating Co, 57-69, Beech Aircraft Co, 59 & Nat Ctr Atmos Res, 67-69; mem staff, Nat Bur Standards, 61-63; Fulbright grants & vis lectr, France, Israel & Spain, 64-65; mem, Nat Adv Group Aeronaut Res & Develop-NATO, 64-65; fac res asst, Inst Arctic & Alpine Res, 65-71; NATO sr fel, 75; pres, Environ Consult Serv, 75-77; sr res fel, Solar Energy Res Inst, 84-87, consult engr, 87-88. *Honors & Awards:* Worcester Award, 80; Max Jacob Award, 85; Charles Freeley Abbott Award, 88. *Mem:* Fel Am Soc Mech Engrs; Am Inst Chem Engrs; Inst Soc Solar Energy. *Res:* Heat transfer; boundary layer theory; solar engineering; solar energy thermal conversion; heat transfer & energy conservation; design of energy conversion & cogeneration systems; author of books on heat transfer, renewable energy conversion & solar building design; technical editor. *Mailing Add:* 1485 Sierra Dr Boulder CO 80302

KREITH, KURT, b Vienna, Austria, May 3, 32; US citizen; m 57. MATHEMATICS. *Educ:* Univ Calif, Berkeley, AB, 53, MA, 57, PhD(math), 60. *Prof Exp:* Asst prof math, Univ Calif, Davis, 60-63; phys sci officer, US Arms Control & Disarmament Agency, 63-65; assoc prof, 65-69, PROF MATH, UNIV CALIF, DAVIS, 69- *Mem:* Am Math Soc; Math Asn Am. *Res:* Differential equations and differential operators in Hilbert space. *Mailing Add:* Dept Math Univ Calif Davis CA 95616

KREITZBERG, CARL WILLIAM, b Missoula, Mont, Mar 25, 37; m 58; c 4. METEOROLOGY. *Educ:* Univ Wash, BS, 59, PhD(meteorol), 63. *Prof Exp:* Res physicist, Meteorol Lab, Air Force Cambridge Res Labs, 63-67; asst prof meteorol, Pa State Univ, 67-69; assoc prof, 70-76, PROF PHYSICS & ATMOSPHERIC SCI, DREXEL UNIV, 77- *Mem:* AAAS; Am Meteorol Soc; Sigma Xi. *Res:* Atmospheric structure, dynamics and prediction on the mesoscale, especially scale interactions due to clouds and boundary layer processes as determined from numerical simulation and field experiments. *Mailing Add:* 2311 N Feathering Rd Media PA 19063

KREITZMAN, STEPHEN NEIL, b New York, NY, Sept 6, 39; m 62. BIOCHEMISTRY, NUTRITION. *Educ:* Hofstra Univ, BA, 61; Mass Inst Technol, PhD(nutrit biochem & metab), 69. *Prof Exp:* Assoc prof nutrit biochem, 69-77, ASSOC PROF ORAL BIOL, SCH DENT, EMORY UNIV, 69- *Mem:* Int Asn Dent Res. *Res:* Role of enzymes in the demineralization of the calcified tissues of bones and teeth; dental caries; relationship between nutrition and learning in school children. *Mailing Add:* Dept Oral Surg S Clin Bldg Emory Univ 1327 Clifton Rd Atlanta GA 30322

KREIZINGER, JEAN DOLLOFF, b Presque Isle, Maine, Oct 17, 31; div; c 3. GENETICS. *Educ:* Univ Maine, BS, 53; Cornell Univ, MS, 56, PhD(genetics), 58; Univ Conn, MBA, 81. *Prof Exp:* Asst prof biol, Danbury State Col, 65-67; NIH res fel human genetics, Univ Tex M D Anderson Hosp & Tumor Inst, 67-69, res assoc biol, Univ Tex, Houston, 69-70; assoc prof, 70-79, asst dean arts & sci, 82-84, PROF BIOL, WESTERN CONN STATE UNIV, 79- *Concurrent Pos:* Vis prof, Cornell Univ, 81 & 86. *Mem:* Sigma Xi; Crop Sci Soc Am; Agron Soc Am. *Res:* Chemical mutagenesis; plant cytogenetics. *Mailing Add:* Four Ferris Rd Newtown CT 06470

KREJCI, ROBERT HENRY, b Shenandoah, Iowa, Nov 15, 43; m 68; c 2. SOLID ROCKET MOTORS, BALLISTIC MISSILES. *Educ:* Iowa State Univ, BS, 67, ME, 71; Nat Defence Univ, dipl, 91. *Prof Exp:* Officer, USAF Space Div, 69-73, Ballistic Missiles Off, 75-78; res assoc, Lawrence Livermore Lab, 73-75; mgr advan technol, Wasatch Div, 78-84, space motor progs, Strategic Div, 84-86, MGR SPEC PROJS, STRATEGIC DIV, THIOKOL CORP, 86- *Concurrent Pos:* Aeronaut engr, USAF Reserve, Wright Lab, Propulsion & Power Div, 85- *Mem:* Am Inst Aeronaut & Astronaut. *Res:* Develop and produce test vehicles to advance rocket motor propulsion understanding, including ignition mechanisms and motor dynamics. *Mailing Add:* 885 N 300 E Brigham City UT 84302

KREJSA, RICHARD JOSEPH, b Cleveland, Ohio, Apr 4, 33; m 62; c 6. CONODONT PALEOBIOLOGY, CYCLOSTOME TOOTH DEVELOPMENT. *Educ:* Mich State Univ, BS, 54; Univ Calif, Los Angeles, MA, 58; Univ BC, PhD(zool), 65. *Prof Exp:* Asst cur fishes, Scripps Inst Oceanog, Univ Calif, 58-59; instr gen zool & biol, Western Wash State Col, 64-65; vis asst prof zool, Univ Hawaii, 65-66; Nat Inst Dent Res trainee comp calcification, Col Physicians & Surgeons, Columbia Univ, 66-68; from asst prof to assoc prof, 68-78, PROF BIOL SCI, CALIF POLYTECH STATE UNIV, 78-, PROF POLIT SCI, 88- *Concurrent Pos:* NSF stipend, Summer Inst Animal Behav, Utah State Univ, 65; elected mem, San Luis Obispo County Bd of Supvrs, 73-80, chmn of the bd, 75, co-chmn, 76; co-chmn & bd dirs, San Luis Obispo Environ Ctr & hon life mem, 80; vis scholar, Museum Comp Zool, Harvard Univ, Cambridge, MA, 82-83; hon life mem, Red Wind Found, 80. *Honors & Awards:* Frederick H Stoye Award, Am Soc Ichthyologists & Herpetologists, 63. *Mem:* Am Soc Zoologists; Am Soc Ichthyologists & Herpetologists; Europ Soc Comp Skin Biol; AAAS; Pander Soc; Am Fisheries Soc. *Res:* Comparative morphology and embryology of vertebrate skin, teeth and scales; origin of craniata; paleobiology of conodonta; effects of gold mining and other historic natural resource exploitations on California salmonid fisheries. *Mailing Add:* Biol Sci Dept Calif Polytech State Univ San Luis Obispo CA 93407

KREKELER, CARL HERMAN, b Levenworth, Kans, Jan 12, 20; m 44; c 2. ZOOLOGY, ENTOMOLOGY. *Educ:* Concordia Sem, BA, 41; Univ Chicago, PhD(zool), 55. *Prof Exp:* Instr biol, Bethany Col, 42-44; PROF BIOL, VALPARAISO UNIV, 47- *Mem:* Ecol Soc Am; Soc Study Evolution; Soc Syst Zoologists; Nat Speleol Soc. Res; Speciation pattern in cave beetles; systematic entomology. *Mailing Add:* Dept Biol Valpraiso Univ Valparaiso IN 46383

KREKORIAN, CHARLES O'NEIL, b Los Angeles, Calif, Apr 17, 41; m 67; c 2. ANIMAL BEHAVIOR. *Educ:* Calif State Col, Los Angeles, BA, 63, MA, 66; Univ Toronto, PhD(zool), 70. *Prof Exp:* Res assoc behav res, Am Inst Res, 64-66; from asst prof to assoc prof, 70-79, PROF ZOOL, SAN DIEGO STATE UNIV, 79- *Mem:* AAAS; Animal Behav Soc. *Res:* Ethology of fish and reptiles with emphasis on their agonistic and gamopractic behavior. *Mailing Add:* Biol Dept San Diego State Univ San Diego CA 92182

KRELL, ROBERT DONALD, b Toledo, Ohio, Dec 2, 43; m 66; c 2. IMMUNOPHARMACOLOGY. *Educ:* Univ Toledo, BS, 66; Ohio State Univ, PhD(pharmacol), 72. *Prof Exp:* Fel, Sch Hyg & Pub Health, Johns Hopkins Univ, 72-73; sr scientist pharmacol, SmithKline Corp, 73-81; SR MGR, PULMONARY PHARMACOL SECT, STUART PHARMACEUT, 81-; AT ICI AM INC. *Mem:* Soc Neurosci; AAAS; Am Acad Allergy; Am Soc Pharmacol & Exp Therapeut; Am Thoracic Soc; Sigma Xi. *Res:* Biochemical, pharmacological, physiological and immunological investigation into the mechanisms of asthma, immediate-type hypersensitivity reactions and chronic obstructive pulmonary diseases. *Mailing Add:* Dept Biomed Res ICI Am Inc Wilmington DE 19897

KREMBS, G(EORGE) M(ICHAEL), b Merrill, Wis, Sept 2, 34; m 57; c 5. COMPUTER SYSTEMS INTEGRATION. *Educ:* Notre Dame Univ, BS, 56; Stanford Univ, PhD, 59. *Prof Exp:* Elec engr, Ampex Corp, 57-58; group supvr solid state mat, Philco Res Labs, 59-61, sect mgr solid state mat, Ford-Philco Appl Res Labs, 61-64; staff engr, Systs Develop Div, 64-65, mgr adv graphic technol, 65-68, mgr adv display systs, 68-70, mgr, Eng Dept Adv Systs Develop, 70-77, mgr display prod technol, 77-78, mgr adv display prod, 78-79, mgr adv display technol, 79-87, TOTAL SYSTS STRATEGIST, IBM CORP, 87- *Concurrent Pos:* Guest lectr, IBM Systs Res Inst, New York, 80-82. *Mem:* Inst Elec & Electronics Engrs; Asn Comput Mach. *Res:* Computer displays; electronic scanning; graphic image processing; electro-optical systems; television engineering; broadband communications; electron tube devices; solid state device and materials technology; transistor circuit design; design and development of multi-vendor computer networks for IBM large systems used by engineers and scientists. *Mailing Add:* IBM Corp MS 282 PO Box 100 Kingston NY 12401

KREMENAK, CHARLES ROBERT, b Newell, Iowa, Apr 17, 31; m 54; c 4. MAXILLOFACIAL GROWTH & DEVELOPMENT. *Educ:* Univ Iowa, DDS, 55, MS, 61. *Prof Exp:* Instr pedodontics, 59-61, asst prof orthodont, 61-69, asst prof otolaryngol & maxillofacial surg, 66-69, assoc prof, 69-72, prof orthodont, otolaryngol head & neck surg, 72-84, PROG DIR, UNIV IOWA, 69-, PROF ORTHODONT, 72- *Concurrent Pos:* Prog writer & consult, Encycl Britannica Films, Inc, 61-62; Nat Inst Dent Res fel, Univ Iowa, 63-64, Nat Inst Dent Res investr cleft palate prog proj, 65-71, co-prin investr, 71-91; pres, Craniofacial Biol Group, Int Asn Dent Res, 87-88, Am Cleft Palate Asn, 87-88. *Mem:* Int Asn Dent Res; AAAS; Am Cleft Palate Asn. *Res:* Maxillofacial growth, especially elucidation of maxillofacial growth control systems; cleft palate habilitation with emphasis on prevention of postsurgical growth aberration; role of postsurgical wound contraction in midfacial growth and development. *Mailing Add:* Maxillofacial Growth Div Univ Iowa Dent Res Lab Iowa City IA 52242

KREMENTZ, EDWARD THOMAS, b Newark, NJ, Apr 30, 17; m 46; c 5. SURGICAL ONCOLOGY, CLINICAL RESEARCH. *Educ:* Wesleyan Univ, AB, 39; Univ Rochester, MD, 43; Am Bd Surg, dipl, 52. *Prof Exp:* Asst surg, Yale Univ, 43, 44-48, asst resident surg path, 45-46, instr surg, 48-50, Childs fel med res, 48-49; from instr to assoc prof surg, 50-61, dir, Cancer Res Clin, 62-75, chief, sect oncol, 78, cancer teaching coordr, Am Cancer, 53-82, prof clin oncol, 77-83, PROF SURG, TULANE UNIV, 61- *Concurrent Pos:* Fel, New Haven Hosp, 43, 44-45, asst resident, 46-48, assoc resident, 48-49, chief resident, 49-50; sr vis surgeon, Charity Hosp, La & consult surg, various hosps, 50-; surgeon, Touro Infirmary, 57-63, sr assoc, 63-; prof clin oncol, Am Cancer Soc, 77-83. *Honors & Awards:* Lucy Wortham James Clin Cancer Res Award, Soc Surg Oncol, 85; Res Center Award, Nat Cancer Inst, Nat Inst Health, 62-67. *Mem:* Am Cancer Soc; Soc Univ Surgeons; Soc Exp Biol & Med; Am Col Surgeons; Am Asn Cancer Res; Int Surg Asn; Soc Surgical Oncol; Am Surg Asn; Am Soc Clin Oncol; Am Med Asn; Soc Int Surg. *Res:* Cancer chemotherapy; experimental surgery; immunotherapy; author or coauthor of 268 publications. *Mailing Add:* Tulane Sch Med Surg 1430 Tulane Ave New Orleans LA 70112

KREMER, JAMES NEVIN, b Montclair, NJ, July 19, 45; m 69; c 2. BIOLOGICAL OCEANOGRAPHY, ECOLOGY. *Educ:* Princeton Univ, BA, 67; Univ RI, PhD(oceanog), 75. *Prof Exp:* Res asst marine ecol, Grad Sch Oceanog, Univ RI, 70-75; asst prof, 76-83, ASSOC PROF BIOL SCI, UNIV SOUTHERN CALIF, 83- *Concurrent Pos:* NATO fel, 75-76. *Mem:* Am Soc Limnol & Oceanog; Estuarine Res Fedn; Am Geophys Union; Sigma Xi. *Res:* Marine plankton ecology; systems ecology and computer simulation, especially physical processes and nutrient dynamics in planktonic systems. *Mailing Add:* 509 Camino de Encanto Redondo Beach CA 90277-6532

KREMER, PATRICIA MCCARTHY, b San Francisco, Calif, Apr 24, 47; m 69; c 2. ZOOPLANKTON ECOLOGY. *Educ:* Stanford Univ, BA, 69; Univ RI, PhD(oceanog), 76. *Prof Exp:* Res asst, Grad Sch Oceanog, Univ RI, 73-74; ASST RES PROF, UNIV SOUTHERN CALIF, 85- *Concurrent Pos:* Allen Hancock Found res fel, Univ Southern Calif, 76-; mem at large, Am Soc Limnol & Oceanog, 81-84. *Mem:* Am Soc Limnol & Oceanog; Am Geophys Union; Sigma Xi. *Res:* Zooplankton ecology, soft-bodied forms. *Mailing Add:* 509 Camino Dr Encanto Redondo Beach CA 90277

KREMER, RUSSELL EUGENE, b Milford, Nebr, May 10, 54; m 84; c 1. CRYSTAL GROWTH, SEMICONDUCTOR CHARACTERIZATION. *Educ:* Goshen Col, BA, 75; Purdue Univ, MS, 78, PhD(physics), 83. *Prof Exp:* From asst prof to assoc prof appl physics & elec eng, Ore Grad Ctr, 83-87; STAFF SCIENTIST, CRYSTAL SPECIALTIES INT, 87- *Concurrent Pos:* Consult, United Epitaxial Technol, 85-86; adj assoc prof, Ore Grad Ctr, 87-90, Colo Univ, Colo Springs, 90. *Mem:* Am Phys Soc; Mat Res Soc; Sigma Xi; Am Asn Crystal Growth. *Res:* Growth and characterization of compound semiconductor single crystal material; development of a process to grow semi-insulating GaAs using a vertical bridgman technique. *Mailing Add:* Crystal Specialties Int 2853 Janitell Rd Colorado Springs CO 80906

KREMERS, HOWARD EARL, b Urbana, Ill, Sept 21, 17; m 40; c 3. INDUSTRIAL CHEMISTRY. *Educ:* Western Reserve Univ, AB, 39; Syracuse Univ, MS, 41; Univ Ill, PhD(chem), 44. *Prof Exp:* Chemist, Lindsay Light & Chem Co, 44-46, dir res, 46-51, dir res, Lindsay Chem Co, 51-56, secy, 56-58, vpres mkt develop, Lindsay Chem Div, Am Potash & Chem Corp, 59-60, dist mgr, 60-63, mgr mkt develop & tech serv, 63-69; mgr mkt res & develop, 69-70, mgr mkt serv, 70-71, DIR MKT SERV, KERR-MCGEE CHEM CORP, 72- *Mem:* Am Chem Soc. *Res:* Rare earths; thorium. *Mailing Add:* 2020 S Monroe St, No 620 Denver CO 80210-3755

KREMKAU, FREDERICK WILLIAM, b Mechanicsburg, Pa, Apr 30, 40; m 67; c 1. BIOACOUSTICS. *Educ:* Cornell Univ, BEE, 63; Univ Rochester, MS, 69, PhD(elec eng), 72. *Prof Exp:* Teaching asst elec eng, Univ Rochester, 67-69, res asst, 69-72; assoc prof, Yale Univ, 81-85; instr med, 72-74, res asst prof, 74-80, DIR, CTR MED ULTRASOUND, BOWMAN GRAY SCH MED, 85- *Concurrent Pos:* Consult, NSF, 74-75. *Mem:* Sigma Xi; Inst Elec & Electronics Engrs; AAAS; NY Acad Sci; Am Inst Ultrasound Med. *Res:* Biological effects of ultrasound; acoustic properties of biological material. *Mailing Add:* Ctr Med Ultrasound Bowman Gray Sch Med Winston-Salem NC 27103

KREMP, GERHARD OTTO WILHELM, b Berlin, Ger, Nov 14, 13; nat US; m 40; c 3. GEOLOGY. *Educ:* Reichs Univ, Posen, Ger, Dr rer nat, 45. *Prof Exp:* Sci asst geol, Univ Gottingen, 45-47; geologist, Geol Surv, Nordhein-Westfalen, Ger, 48-54; sr res assoc geol, Pa State Univ, 55-59; geologist, US Geol Surv, 59-60; prof geosci, 60-79, EMER PROF, UNIV ARIZ, 80- *Concurrent Pos:* Palynological consult, Kremp Palynologic Data Retrieval Res Proj, Atlantic Richfield, Am Oil Co, Chevron, Exxon, Gulf, Mobil, Texaco, Union Oil Co & Geol Surv Can & Phillips Petrol Co, 68- *Honors & Awards:* Prof Gunnar Erdtman Int Medal Palynological Soc, India, 70- *Mem:* Ger Paleont Soc; Palynological Soc India; Am Asn Stratig Palynologists. *Res:* Palynology; paleobotany; coal geology; paleontology. *Mailing Add:* 101 N Avenida Carolina Univ Ariz Tucson AZ 85711

KREMPL, ERHARD, b Regensburg, Ger, Mar 5, 34; wid; c 2. MECHANICS. *Educ:* Munich Tech Univ, Dipl Ing, 56, Dr Ing(mech of mat), 62. *Prof Exp:* Res proj engr, Munich Tech Univ, 56-64; mech of mat engr, Gen Elec Co, NY, 64-68; assoc prof mech, 68-75, PROF MECH & DIR, MECH MAT LAB, RENSSELAER POLYTECH INST, 75-; HEAD, DEPT MECH ENG, AERONAUT ENG & MECH, 87- *Concurrent Pos:* Fulbright fel, Austria, 85. *Honors & Awards:* Nadai Award, ASME, Am Acad Mech. *Mem:* Fel Am Soc Mech Engrs; Am Soc Exp Stress Anal; Am Soc Testing & Mat; Soc Eng Sci; fel Am Acad Mech; fel Japan Soc Promotion Sci. *Res:* Mechanics of deformation and fracture behavior of metals and composites; creep, fatigue, fracture; applications to power plant such as steam and gas turbines and nuclear reactors; constitutive equation theory to describe time-dependent material behavior; inelastic analysis. *Mailing Add:* Dept Mech & Aeronaut Rensselaer Polytech Inst Troy NY 12180-3590

KREMSER, THURMAN RODNEY, b Temple, Pa, Aug 29, 32; m 88; c 4. PHYSICS. *Educ:* Lehigh Univ, BS, 54, MS, 56; Temple Univ, PhD(physics), 68. *Prof Exp:* PROF PHYSICS & CHMN DEPT, ALBRIGHT COL, 56- *Mem:* Am Asn Physics Teachers. *Mailing Add:* Dept Physics Albright Col PO Box 15234 Reading PA 19612-5234

KREMZNER, LEON T, b Poland, Sept 16, 24; US citizen; m 56; c 3. BIOCHEMISTRY. *Educ:* Seton Hall Univ, BS, 49; Rutgers Univ, MS, 52, PhD(biochem), 55. *Prof Exp:* Res chemist biochem, Gen Foods Corp, 49-51; asst, Bur Biol Res, Rutgers Univ, 52-55; proj leader, Res Ctr, Gen Foods Corp, 55-58; res assoc neurochem, Col Physicians & Surgeons, Columbia Univ, 59-63; neurochemist, Bur Res, NJ Neuropsychiat Inst, 63-67; from asst prof to assoc prof neurochem, Col Physicians & Surgeons, Columbia Univ, 67-87; RETIRED. *Mem:* NY Acad Sci; Am Soc Biol Chemists. *Res:* Enzyme chemistry; intermediate metabolism; cholinergic system; histamine and polyamine metabolism; neurochemistry. *Mailing Add:* Box 92 Canaan St Canaan NH 03741

KRENDEL, EZRA SIMON, b New York, NY, Mar 5, 25; wid; c 3. HUMAN FACTORS ENGINEERING. *Educ:* Brooklyn Col, BA, 45; Mass Inst Technol, ScM, 47; Harvard Univ, AM, 49. *Hon Degrees:* MA, Univ Pa, 71. *Prof Exp:* From res engr to mgr engr, Psychol Lab, Labs Res & Develop, Franklin Inst, 49-63, tech dir opers res, Res Labs, 63-66; dir, Mgt Sci Ctr, Univ Pa, 67-69, chmn bd adv, Ctr, 69-70, prof statist & opers res, 66-90, PROF SYSTS, SCH ENG & APPL SCI, UNIV PA, 83, EMER PROF STATIST OPERS & RES, 90- *Concurrent Pos:* Consult to indust, res, non-profit, local & fed govt orgn, 67-; NATO vis guest lectr in univs & res insts, Greece, Turkey, Eng, Italy, France & Ger, 68-71; prin scientist, Systs Technol, Inc, 87-88. *Honors & Awards:* Louis E Levy Gold Medal, Franklin Inst, 60. *Mem:* Fel AAAS; fel Inst Elec & Electronics Engrs; fel Am Psychol Asn; fel Human Factors Soc; Ergonomics Soc. *Res:* Human control dynamics, tracking, decision making, power output and human error; command control and man-machine systems design. *Mailing Add:* 211 Cornell Ave Swarthmore PA 19081-1933

KRENER, ARTHUR JAMES, b Brooklyn, NY, Oct 8, 42; m; c 3. APPLIED MATHEMATICS, SYSTEMS THEORY. *Educ:* Col of the Holy Cross, BS, 64; Univ Calif, Berkeley, MA, 67, PhD(math), 71. *Prof Exp:* Asst prof, 71-76, assoc prof, 76-80, PROF MATH, UNIV CALIF, DAVIS, 80- *Concurrent Pos:* Res fel eng & appl physics, Harvard Univ, 74-75; Fullbright Hays fel, Univ Rome, 79; vis sr res fel, Imperial Col, London, 80-81. *Mem:* Soc Indust & Appl Math; Am Math Soc; Inst Elec & Electronics Engrs; Sigma Xi. *Res:* Nonlinear systems theory; stochastic processes. *Mailing Add:* Dept Math Univ Calif Davis CA 95616

KRENITSKY, THOMAS ANTHONY, b Throop, Pa, Sept 13, 38. BIOCHEMISTRY. *Educ:* Scranton Univ, BS, 59; Cornell Univ, PhD(biochem), 63. *Prof Exp:* Fel biochem, Sloan-Kettering Inst Cancer Res, 63-64; res assoc, Yale Univ, 64-66; sr res biochemist, Wellcome Res Labs, Burroughs Wellcome Co, 66-68, head enzym, 68-83, head, Div Exp Ther, 83-89, VPRES RES, WELLCOME RES LABS, BURROUGHS WELLCOME CO, 89- *Concurrent Pos:* Adj assoc prof, Dept Biochem & Nutrit, Univ NC, Chapel Hill, 76-; dir, Wellcome Fund, 90- *Honors & Awards:* Aaron Bendich Award, Sloan-Kettering Cancer Inst, 87. *Mem:* Am Chem Soc; Am Soc Biol Chemists. *Res:* Specificities, mechanisms and phylogenetic relationships of the enzymes involved in purine and pyrimidine metabolism; purine and pyrimidine hydroxylating enzymes, ribosyltransferases, phosphoribosyltransferases, nucleoside and nucleotide kinases, and nucleotide interconverting enzymes; nucleoside antiviral agents. *Mailing Add:* Burroughs Wellcome Co 3030 Cornwallis Rd Research Triangle Park NC 27709

KRENKEL, PETER ASHTON, b San Francisco, Calif, Jan 3, 30; m 85; c 3. WATER QUALITY MANAGEMENT, THERMAL POLLUTION. *Educ:* Univ Calif, Berkeley, BS, 56, MS, 58, PhD(environ eng), 60. *Prof Exp:* Instr, Col Eng, Univ Calif, Berkeley, 58-60; chmn & prof, Dept Environ & Water Resources Eng, Vanderbilt Univ, 60-74; dir, Div Environ Planning, Tenn Valley Authority, 74-78; exec dir, Water Resources Ctr, Desert Res Inst, Reno, Nev, 78-82; dean, Col Eng, 82-87, PROF CIVIL ENG, UNIV NEV, RENO, 88- *Concurrent Pos:* Lectr, Am Inst Chem Engrs, 68-; chmn thermal pollution, Nat Water Comn, 72-74; consult, WHO, 68-, Environ Protection Agency, 82-84, Gen Motors Corp, Monsanto Res Corp, Mead Corp, Stouffer Chem Corp, Inland container, Olin Corp, Korean Adv Inst Sci Tech, Ministry Water & Power, Repub of China, 86. *Honors & Awards:* Rudolf Hering Award, Am Soc Civil Engrs, 63; Serv, Integrity, Responsibility Award, Asn Gen Contractors, 84; Eminent Speaker, Inst Engr, Australia, 86. *Mem:* Am Inst Chem Engrs; Am Soc Civil Engrs; Am Acad Environ Engrs; Am Water Works Asn; Int Asn Water Pollution Res; Water Pollution Control Fedn. *Res:* Water quality management; thermal pollution; gas absorption in water; turbulent diffusion and mixing analysis; mercury in the aquatic environment. *Mailing Add:* 3500 Cashill Blvd Reno NV 89509

KRENOS, JOHN ROBERT, b New Britain, Conn, Sept 4, 45. CHEMICAL PHYSICS. *Educ:* Univ Conn, BA, 67; Yale Univ, MS, 68, PhD(chem), 72. *Prof Exp:* Fel chem, Harvard Univ, 72-73; asst prof, 73-78, ASSOC PROF CHEM, RUTGERS UNIV, 78- *Mem:* Am Phys Soc; Am Chem Soc. *Res:* Energy transfer in hyperthermal collisions and collisions involving electronically excited reactants; molecular beam chemiluminescence; model calculations of chemical reactions. *Mailing Add:* Dept Chem Rutgers Univ New Brunswick NJ 08903

KRENZ, JERROLD H(ENRY), b Buffalo, NY, Apr 24, 34. ELECTRICAL ENGINEERING. *Educ:* Univ Buffalo, BS, 56; Stanford Univ, MS, 58, PhD(elec eng), 64. *Prof Exp:* Engr antennas, Lockheed Missile Systs, Lockheed Aircraft Corp, 56-57; engr microwave tubes, Gen Elec Microwave Lab, 58-61; asst prof elec eng, 63-77, dir eng honors prog, 69-73, assoc prof, 77-81, PROF ELEC ENG, UNIV COLO, BOULDER, 81- *Concurrent Pos:* Consult, Gen Elec Co, 62. *Mem:* Inst Elec & Electronics Engrs; Int Solar Energy Soc; AAAS; Sigma Xi. *Res:* Energy systems and policy; modeling; economic studies. *Mailing Add:* Dept Elec Eng Univ Colo Campus Box 425 Boulder CO 80309-0425

KRENZELOK, EDWARD PAUL, b Ladysmith, Wis, Mar 11, 47. TOXICOLOGY, PHARMACY. *Educ:* Univ Wis, BS, 71; Univ Minn, PhD(pharm), 74. *Prof Exp:* From asst prof to assoc prof pharm, Univ Minn, 74-83; PROF PHARM & MED, PITTSBURGH POISON CTR, 83- *Concurrent Pos:* Dir toxicol, Hennepin Poison Ctr, Hennepin County Med Ctr, Minneapolis, 76-83 & mem fac toxicol, Dept Emergency Med, 78-83; Minneapolis Community Health Serv grant poison prev prog children in day care ctrs, 77-82; consult, Emergency Med Serv Div, Minn Dept Health, Minneapolis, 78 & Emergency Med Serv Div, Metrop Coun, St Paul, 78-83; chmn, Dept Prof Educ, Nat Poison Ctr Network, Pittsburgh, 78-83. *Mem:* Am Asn Poison Ctr; Am Soc Hosp Pharm; Am Acad Clin Toxicol; Nat Poison Ctr Network. *Res:* Poison education and prevention for preschoolers; study toxicity of acetaminophen, corrosives and caustics; drugs of abuse; nonprescription. *Mailing Add:* Pittsburgh Poison Ctr 3705 Fifth Ave at DeSoto St Pittsburgh PA 15213

KREPINSKY, JIRI J, b Prague, Czech, July 15, 34; Can citizen; m 88; c 2. ORGANIC CHEMISTRY, MOLECULAR GENETICS. *Educ:* Charles Univ, Prague, MSc, 57, Dr rer nat, 66; Czech Acad Sci, PhD(chem), 61. *Prof Exp:* Res asst org chem, Inst Org Chem & Biochem, Czech Acad Sci, 57-61, res assoc natural prod, 61-66 & 67-68; vis scientist, Inst Org Chem, Univ Milan, 66-67; fel synthesis natural prod, Univ NB, 68-70, lectr, 70-72; dir, Chem Res Lab of Simes, Milan, 72-75; sr res scientist, Dept Med Genetics, Univ Toronto, 76-80, assoc prof, 80-88, sr staff scientist, Ludwig Inst Cancer Res, Toronto Br, 81-88, prof med biophysics, 88-90, PROF, DEPT MED GENETICS, UNIV TORONTO, 88- *Concurrent Pos:* Mem bot exped, Soviet Cent Asia, 61; consult, Ont Cancer Inst, 78-81. *Mem:* Am Chem Soc; Royal Soc Chem; Chem Soc Can; NY Acad Sci. *Res:* Determination of structures of biologically important compounds, particularly glycoproteins; roles of oligosaccharides moieties of glycoproteins and glycolipids development and malignancy, in particular colon cancer; organic chemistry and mass spectrometry of carbohydrates and glycopeptides; development of protocols for population cancer screening. *Mailing Add:* Dept Molecular Med Genetics Sci Bldg Univ Toronto Toronto ON M5S 1A8 Can

KREPS, DAVID PAUL, b Pottstown, Pa, Jan 13, 43; m 65; c 2. MICROBIOLOGY, IMMUNOLOGY. *Educ:* Manchester Col, BS, 64; Ohio State Univ, MS, 68; Chicago Med Sch, PhD(microbiol), 76. *Prof Exp:* Teaching asst microbiol, Ohio State Univ, 65-67; ASSOC PROF BIOL, MANCHESTER COL, 67- *Concurrent Pos:* Res asst microbiol, Chicago Med Sch, 74-75; Res Corp res grant, 78-79; dir, NSF Int Soc Educ Planners grant, Manchester Col, 78-81. *Mem:* Sigma Xi; Am Soc Microbiol. *Res:* Immunological responses to salmonella typhimurium vaccines and cell fractions in inbred and outbred mice. *Mailing Add:* 104 S Sycamore North Manchester IN 46962

KRESCH, ALAN J, b New York, NY, June 25, 31; m 75; c 1. PHYSICAL CHEMISTRY, DATA PROCESSING. *Educ:* Cornell Univ, AB, 52; Rutgers Univ, PhD(solution kinetics), 61. *Prof Exp:* Sr res chemist, Nat Cash Register Co, Ohio, 60-73, STAFF RES ASSOC, APPLETON PAPERS INC, 73- *Mem:* AAAS; Am Chem Soc; Sigma Xi. *Res:* Solution kinetics of inorganic polymers; reversible photochemical reactions in solution; color technology; laboratory computer. *Mailing Add:* 39 S Meadows Dr Appleton WI 54915

KRESGE, ALEXANDER JERRY, b Wilkes-Barre, Pa, July 17, 26; m 50, 63; c 3. PHYSICAL ORGANIC CHEMISTRY. *Educ:* Cornell Univ, BA, 49; Univ Ill, PhD(chem), 53. *Prof Exp:* Asst, Univ Ill, 49-51; Fulbright scholar, Univ Col, Univ London, 53-54; res assoc, Purdue Univ, 54-55 & Mass Inst Technol, 55-57; assoc chemist, Brookhaven Nat Lab, 57-60; from asst prof to prof chem, Ill Inst Technol, 60-74; chmn chem group, 74-78, PROF CHEM, SCARBOROUGH COL, UNIV TORONTO, 74- *Concurrent Pos:* Guggenheim fel, 64, Killam fel, 84-86 & Yamada fel, 85; NSF sr fel, 64-65; vis lectr, Bedford Col, London, 64-65; vis prof , Oxford Univ, 65, Univ Toronto, 70-71, Univ Mich, 79, Univ Lausanne, 81, Tech Univ Denmark, 82, Univ San Paulo, 84, Fed Univ Santa Catarina, 84 & Kyoto Univ, 85; guest of Inst, Mass Inst Technol, 65; vis scientist, Fritz Haber Inst, 81 & Univ Goteborg, 83; Mardi Gras Lectr, La State Univ, 81. *Honors & Awards:* Mobay Lectr, Univ NH, 82-; Morley Medal, 88. *Mem:* Am Chem Soc; fel Royal Soc Can; Chem Inst Can; Sigma Xi. *Res:* Reaction mechanisms; isotope effects; acid-base catalysis; kinetics. *Mailing Add:* Dept Chem Univ Toronto Toronto ON M5S 1A1 Can

KRESGE, EDWARD NATHAN, b Noxen, Pa, Aug 14, 35; m 63. POLYMER CHEMISTRY. *Educ:* Univ Tampa, BS, 57; Univ Fla, PhD(chem), 61. *Prof Exp:* Res chemist, 61-63, proj leader elastomers, 63-75; HEAD ELASTOMERS EXPLOR RES, ELASTOMERS TECHNOL DIV, EXXON CHEM CO, 75-, CHIEF POLYMER SCIENTIST, 78- *Honors & Awards:* Chmn Gordon Res Conf, Elastomers, 87. *Mem:* AAAS; Am Chem Soc. *Res:* Elastomer, morphology, polymer rheology and physics. *Mailing Add:* Exxon Chem Co Polymer Group PO Box 45 Linden NJ 07036

KRESH, J YASHA, b L'vov, Soviet Union, July 13, 48; US citizen. ARTIFICIAL INTERNAL ORGANS, MODELING & SIMULATION. *Educ:* NJ Inst Technol, BS, 71; Rutgers Univ, MSBME, 73, PhD(biomed eng & cardiovasc physics), 77. *Prof Exp:* Res intern biomed eng, Rutgers Univ, 71-75; res intern comp-med, Mt Sinai-Rutgers Health Care Comp Lab, 75-77; res assoc, Dept Surg, Newark Beth Israel Med Ctr, 76-79; res asst prof surg, Thomas Jefferson Univ, 79-83, res mem, Ischemia Shock Res Inst, 81-86; PROF & DIR RES, DEPT CARDIOTHORACIC SURG, HAHNEMANN UNIV, 86-, PROF MED & DIR CARDIOVASC BIOPHYS & COMPUT, 86- *Concurrent Pos:* Assoc prof surg, Thomas Jefferson Univ, 83-86, assoc prof pharmacol, 84-86; adj prof bioeng, Biomed Eng & Sci Inst, Drexel Univ, 84- *Mem:* Sr mem Inst Elec & Electronics Engrs; sr mem Biomed Eng Soc; Am Heart Asn; Cardiovasc Syst Dynamics Soc; fel Am Col Cardiol; fel Acad Surg Res; Sigma Xi. *Res:* Cardiovascular system dynamics; heart assist devices; computers and cardiology; patient monitoring systems; physiological and biophysical sensors; closed-loop physiological control; numerous scientific publications. *Mailing Add:* Dept Cardiothoracic Surg Hahnemann Univ Mail Stop 110 Broad & Vine Sts Philadelphia PA 19102-1192

KRESHECK, GORDON C, b North Tonawanda, NY, Sept 3, 33; m 61; c 3. PHYSICAL BIOCHEMISTRY. *Educ:* Ohio State Univ, BS, 55, MS, 59, PhD(dairy technol), 61. *Prof Exp:* Res asst biochem, Nobel Med Inst, Stockholm, Sweden, 62-63; vis scientist, Procter & Gamble Co, 63; NIH res fel chem, Cornell Univ, 63-65; asst prof, 65-68, assoc prof, 68-78, PROF CHEM, NORTHERN ILL UNIV, 78-, DIR, CTR BIOCHEM & BIOPHYS STUDIES, 75- *Concurrent Pos:* Assoc ed, Bull Thermodynamics & Thermochem, 71-76. *Mem:* Am Chem Soc; Biophys Soc; Am Soc Biol Chemists; Sigma Xi. *Res:* Protein chemistry; solution calorimetry; surfactants. *Mailing Add:* Dept Chem Northern Ill Univ De Kalb IL 60115

KRESHOVER, SEYMOUR J, b New York, NY, June 22, 12; m 46; c 4. PATHOLOGY. *Educ:* NY Univ, BA, 34, MD, 49; Univ Pa, DDS, 38; Yale Univ, PhD(clin med, path), 42; Am Bd Oral Med, dipl. *Hon Degrees:* DSc, State Univ NY Buffalo, 61, Univ Pa, 67 & Boston Univ, 69; DOdont, Gothenburg Univ, 73; DSc, Univ Mich, 75. *Prof Exp:* Asst instr, Sch Med, Univ Ill, 38-39; clin asst dent surg, Yale Univ, 42-43; teaching fel histoanat, NY Univ, 46-47, instr, 47; prof oral path & dir dent res, Grad & Postgrad Study, Med Col Va, 49-56; assoc dir, Nat Inst Dent Res, 46-66, dir, 66-75; vis prof oral biol, State Univ NY, Buffalo, 75-80; RETIRED. *Concurrent Pos:* Assoc trustee, Bd Med Educ & Res, Univ Pa, 56-66; chmn comn dent res, Int Dent Fedn, 61-67. *Honors & Awards:* Pierre Fouchard Medal, 72; Callahan Medal, 72. *Mem:* Am Dent Asn; Am Pub Health Asn; Am Acad Oral Path; Int Asn Dental Res (pres, 62). *Res:* Dental histology and embryology; dental pathology; prenatal factors in congenital defects. *Mailing Add:* 838 John Anderson Dr Ormond Beach FL 32074

KRESINA, THOMAS FRANCIS, b Baltimore, Md, June 18, 54; m 78; c 3. IMMUNE NETWORK INTERACTIONS, SOMATIC CELL HYBRIDIZATION. *Educ:* Cath Univ Am, BS, 75; Univ Ala, Birmingham, PhD(biochem), 79. *Prof Exp:* NIH asst fel biochem, Univ Ala, Birmingham, 79-80; res assoc immunol, Brandeis Univ, 80-81, NIH asst, 81-82; sr res assoc rheumatology, Case Western Reserve Univ, 82-83, asst prof rheumatology & path, Univ Hosp, 83-87, asst prof environ health, Med Sch, 85-87; ASSOC PROF MED, PROG GEOG MED, MIRIAM HOSP & BROWN UNIV INT HEALTH INST, 87- *Concurrent Pos:* Prin investr scientist, Orthop Res & Educ Found grant, 85-86, NIH RO-1 res grant, 85- *Mem:* Am Asn Immunologists; NY Acad Sci; Orthop Res Soc; AAAS. *Res:* Elucidation of immunoregulatory mechanisms which can have application in understanding human disease; generation and molecular characterization of T cells which can suppress the erythema and adema associated with arthritis; analysis of the immune network and its potential usage in vaccine formulation in schistosomiasis; analysis of granulomas inflammation and hepatic pathology in schistosomiasis. *Mailing Add:* Dept Med Miriam Hosp 164 Summit Ave Providence RI 02726

KRESPAN, CARL GEORGE, b Erie, Pa, Aug 10, 26; m 49; c 3. ORGANIC CHEMISTRY. *Educ:* Univ Rochester, BS, 48; Univ Minn, PhD(org chem), 52. *Prof Exp:* Res chemist, 52-60, res supvr org chem, 60-70, RES SCIENTIST, CENT RES DEPT, E I DU PONT DE NEMOURS & CO, 70- *Mem:* Am Chem Soc. *Res:* Organic fluorine chemistry; free radical, sulphur, cyanocarbon, macroheterocycle chemistry, and fluoropolymer. *Mailing Add:* Du Pont Exp Sta PO Box 80328 Wilmington DE 19880-0328

KRESS, BERNARD HIRAM, b New York, NY, Apr 18, 17; c 3. ORGANIC CHEMISTRY. *Educ:* City Col New York, BS, 38; Columbia Univ, MA, 40, PhD(org chem), 47. *Prof Exp:* Asst biochem, Col Physicians & Surgeons, Columbia Univ, 38-40; jr biochemist neuropsychiat res unit, US Vet Admin, Long Island, 41-42; res chemist, Fed Telecommun Labs, NJ, 42-46 & Celanese Corp Am, 47-48; sr res chemist & group leader, Plaskon Div, Libbey-Owens-Ford Glass Co, 49-53; dir org res, Quaker Chem Prod Co, 53-64, mgr polymer res & develop, Quaker Chem Corp, 64-70, sr scientist, 70-82; PRES, KRESS ASSOCS, 82- *Mem:* Am Chem Soc; Am Asn Textile Chem & Colorists; Am Tech Asn Pulp & Paper Indust; Am Soc Lubrication Eng; fel Am Inst Chem; Sigma Xi. *Res:* High polymers; textile, metal and paper chemicals. *Mailing Add:* 4018 Kottler Dr Lafayette Hill PA 19444

KRESS, DONNIE DUANE, b American Falls, Idaho, Mar 17, 42; m 70; c 2. GENETICS, ANIMAL BREEDING. *Educ:* Univ Idaho, BS, 64; Univ Wis, MS, 66, PhD(genetics & animal sci), 69. *Prof Exp:* NIH fel quant genetics, Univ Minn, 69-70; asst & assoc prof, 70-80, PROF GENETICS & ANIMAL BREEDING, MONT STATE UNIV, 80- *Concurrent Pos:* Guest partic, Cong Vet Med & Animal Prod, Buenos Aires, Argentina, 85; sabbatical leave, Univ Nebr, Lincoln, 87. *Mem:* Am Soc Animal Sci; Sigma Xi. *Res:* Quantitative genetics and animal breeding; selection, genetic by environment interaction; maternal ability of beef cattle of varying biological types and beef sire evaluation. *Mailing Add:* Dept Animal & Range Sci Mont State Univ Bozeman MT 59717

KRESS, LANCE WHITAKER, b Camp Lejeune, NC, Sept 2, 45; m 69; c 3. PHYTOPATHOLOGY. *Educ:* Pa State Univ, BS, 68, MS, 72; Va Polytech Inst & State Univ, PhD(plant path), 78. *Prof Exp:* Jr res aide, Pa State Univ, 72-73; res assoc, Va Polytech Inst & State Univ, 75-80; asst ecologist, Argonne Nat Lab, 80-84, ecologist, 84-86; RES PLANT PATHOLOGIST, FOREST SERV, USDA, 86- *Mem:* Am Phytopath Soc; Air Pollution Control Asn; Sigma Xi. *Res:* Evaluating the impacts of low concentrations of air pollutants and pollutant combinations on the growth and marketable yield of important agricultural field crops and forest tree species. *Mailing Add:* USDA Forest Serv PO Box 12254 Research Triangle Park NC 27709

KRESS, LAWRENCE FRANCIS, b Milwaukee, Wis, Oct 5, 36; div; c 4. BIOCHEMISTRY. *Educ:* Marquette Univ, BS, 59, MS, 61, PhD(physiol), 64. *Prof Exp:* NSF fel biochem, Med Sch, Dartmouth Univ, 64-66; Am Heart Asn adv res fel, 66-68, sr res scientist, 68-78, res scientist IV, 78-80, RES CANCER SCIENTIST V, ROSWELL PARK CANCER INST, 80- *Mem:* Am Soc Biochem & Molecular Biol; Int Soc Toxinol. *Res:* Enzymology; proteolytic enzymes and their inhibitors; interactions between snake venom proteinases and plasma proteinase inhibitors. *Mailing Add:* Roswell Park Cancer Inst 666 Elm St Buffalo NY 14263-0001

KRESS, THOMAS JOSEPH, b Indianapolis, Ind, Oct 31, 40; m 65; c 4. ORGANIC CHEMISTRY. *Educ:* Xavier Univ Ohio, BS, 62, MS, 64; Ohio Univ, PhD(org chem), 67. *Prof Exp:* Res assoc, Ohio Univ, 67-68; sr org chemist, 68-74, res scientist, 74-80, RES ASSOC, ELI LILLY & CO, 80- *Mem:* Int Soc Heterocycle Chem; Am Chem Soc; Royal Soc Chem; Sigma Xi. *Res:* The synthesis and reactions of nitrogen heterocycles. *Mailing Add:* 2048 Ridgemere Pl Greenwood IN 46143

KRESS, THOMAS SYLVESTER, b Kingsport, Tenn, Dec 5, 33; m 56; c 3. NUCLEAR ENGINEERING, AEROSOL SCIENCE. *Educ:* Univ Tenn, BS, 56, MS, 65, PhD(eng sci), 71. *Prof Exp:* Engr aircraft nuclear propulsion, Pratt & Whitney Aircraft, 56-59; engr reactor safety, 59-76, prog mgr & group leader advan reactor syst, Nuclear Div, 76-80, MGR, NUCLEAR RES COUN, UNION CARBIDE CORP, 80- *Mem:* AAAS; Am Soc Mech Engrs; Am Nuclear Soc; Nat Soc Prof Engrs; Nat Mgt Asn. *Res:* Thermal sciences, heat ransfer, fluid transfer, fluid mechanics and thermodynamics; nuclear safety; aerosol science. *Mailing Add:* Y-12 Area Bldg 9108 Mail Stop 8088 Oak Ridge Nat Lab Oak Ridge TN 37831

KRESSE, JEROME THOMAS, b Buffalo, NY, Dec 29, 31; m 62; c 5. ORGANIC CHEMISTRY. *Educ:* Mich State Univ, BS, 58; Univ Fla, PhD(org chem), 65. *Prof Exp:* Asst prof chem, Muskingum Col, 65-66; asst prof, 66-69, assoc prof chem & chmn dept chem & physics, 69-74, chmn, Div Math & Natural Sci, 74-78, PROF CHEM, D'YOUVILLE COL, 74- *Mem:* AAAS; Am Asn Univ Prof; Am Chem Soc. *Res:* Studies of factors influencing the stereochemistry of the Wittig reaction; synthesis of amino acid antagonists. *Mailing Add:* 292 Mill Rd West Seneca NY 14224

KRESSEL, HENRY, b Vienna, Austria, Jan 24, 34; US citizen; m 56; c 2. ELECTROOPTICS. *Educ:* Yeshiva Col, BA, 55; Harvard Univ, MS, 56; Univ Pa, MBA, 59, PhD(mat sci), 65. *Prof Exp:* Engr, Semiconductor Div, Radio Corp Am, 59-61, group head microwave device, 61-63, group device physics, Tech Progs Lab, 65-66, mem tech staff, RCA Labs, 67-69, group head, 69-77, lab dir, 77-79, staff vpres, 79-83; MANAGING DIR, WARBURG PINCUS & CO, 83- *Concurrent Pos:* Co-founder, J Lightwave Technol; consult, Dept Defense. *Honors & Awards:* Achievement Award, RCA Corp, 62, 68 & 69; Centennial Medal, Inst Elec & Electronics Engrs, 84, David Sarnoff Award, 85. *Mem:* Nat Acad Eng; fel Am Phys Soc; fel Inst Elec & Electronics Engrs; Inst Elec & Electronics Engrs Laser & Electrooptics Soc. *Res:* New semiconductor devices, particularly in area of microwaves and optical devices; lasers; properties of defects in semiconductors; author of 120 technical publications; awarded 33 US patents. *Mailing Add:* E M Warburg Pincus & Co 466 Lexington Ave New York NY 10017

KRESSEL, HERBERT YEHUDE, b Brooklyn, NY, Nov 20, 47; c 2. RADIOLOGY. *Educ:* Brandeis Univ, Waltham, Mass, BA, 68; Univ Southern Calif, MD, 72. *Prof Exp:* Clin instr radiol, Univ Calif, San Francisco, 76-77; from asst prof to assoc prof, 77-85, PROF RADIOL, UNIV PA, PHILADELPHIA, 85- *Concurrent Pos:* NIH fel radiol, Univ Calif, San Francisco, 76; ed, Magnetic Resonance Ann, 85-88 & Magnetic Resonance Quart, 88-; mem, Comn Magnetic Resonance, Am Col Radiol, 87-, chmn, Comt Magnetic Resonance Clin Appln, 87-; med dir, RI Magnetic Resonance Imaging Network, Providence, 88-; prog chair, Soc Magnetic Resonance Med, 90. *Mem:* Asn Univ Radiologists; Soc Gastrointestinal Radiologists; Am Col Radiol; Radiol Soc NAm; Soc Magnetic Resonance Med (pres elect, 89-90, pres, 90-91). *Res:* Improved magnetic resonance for the abdomen and pelvis. *Mailing Add:* Dept Radiol Hosp Univ Pa Philadelphia PA 19104

KRESTA, JIRI ERIK, b Kosice, Czech, Apr 19, 34. POLYMER CHEMISTRY, CHEMICAL ENGINEERING. *Educ:* Inst Chem Technol, Prague, MChE, 57; Tech Univ Prague, MS, 64; Czech Acad Sci, PhD(polymer sci), 67. *Prof Exp:* Res assoc, Res Inst Synthetic Rubber, Zlin, Czech, 57-62; res scientist, Res Inst Macromolecular Chem, Brno, 62-69; res assoc, Dept Chem, Wayne State Univ, 69-71; RES PROF POLYMER SCI, DEPT CHEM & CHEM ENG, POLYMER INST, UNIV DETROIT, 71- *Mem:* Am Chem Soc; Soc Plastics Engrs; Czech Chem Soc; NY Acad Sci. *Res:* Reaction kinetics and catalysis of polyreactions; characterization, flammability degradation and stabilization of polymers, morphological and viscoelastic studies of polymers; research in polyurethanes; cellular materials, polyolefins; elastomers; thermostable polymers; plastics failure. *Mailing Add:* Polymer Inst Univ Detroit 4001 W McNichols Detroit MI 48221

KRESTENSEN, ELROY R, b New York, NY, Sept 6, 21; m 48; c 1. ENTOMOLOGY. *Educ:* Univ Fla, BSA, 49, MS, 51; Univ Md, PhD(entom), 62. *Prof Exp:* Asst entom, Univ Fla, 49-51, interim instr, 51-52; entomologist, Fla Bd Health, 52-54; from instr to assoc prof entom, Sharpsburg Res Ctr, Univ Md, 55-84; consult, 84-86; RETIRED. *Mem:* Entom Soc Am. *Res:* Insect pests and control methods for fruit. *Mailing Add:* 1050 St Clair St Hagerstown MD 21742

KRETCHMAR, ARTHUR LOCKWOOD, biochemistry, for more information see previous edition

KRETCHMER, NORMAN, b New York, NY, Jan 20, 23; m 42; c 3. NUTRITIONAL SCIENCE, PEDIATRICS & OBSTETRICS. *Educ:* Cornell Univ, BS, 44; Univ Minn, MS, 45, PhD(physiol chem), 47; State Univ NY, MD, 52; Am Bd Pediat, dipl. *Prof Exp:* Asst physiol chem, Univ Minn, 44-45, jr scientist, 45-47; asst prof biochem & path, Col Med, Univ Vt, 47-48; res assoc path, Long Island Col Med, State Univ NY, 48-52; Commonwealth fel med & intern, Montefiore Hosp, 52-53; asst prof biochem & pediat, Med Col, Cornell Univ, 53-56, from asst prof to assoc prof pediat, 56-59; from prof to Harold K Faber prof, Sch Med, Stanford Univ, 59-74, exec head dept, 56-59; dir, Nat Inst Child Health & Human Develop, 74-81; chmn, Dept Nutrit Sci, 83-88, PROF NUTRIT, UNIV CALIF, BERKELEY, 81-; PROF OBSTET & PEDIAT, UNIV CALIF, SAN FRANCISCO, 81- *Concurrent Pos:* Asst resident, NY Hosp, 53-55, asst pediatrician outpatients, 53-54, pediatrician, 54-55, asst attend pediatrician, 55-59; Commonwealth Fund traveling fel, Univ Paris & St Mary's Hosp, London, 57 & Atomic Energy Lab, Saclay, Paris, 65-66; pediatrician-in-chief, Stanford Hosp, 59-69; spec consult, WHO, 67-69; mem bd sci coun, Nat Inst Child Health & Human Develop, 69-71; chmn prog human biol, Stanford Univ, 69-72; vis prof, Univ Lagos, 70; pres, Int Orgn Study Human Develop, 70-; Guggenheim fel, 73-74; consult lectr, Nat Naval Med Ctr, 74-81. *Honors & Awards:* Johnson Award, Am Acad Pediat, 58, Borden Award, 69. *Mem:* Inst Med-Nat Acad Sci; AAAS; Am Pediat Soc; Am Soc Biol Chemists; Am Soc Clin Nutrit. *Res:* Perinatology; human development; developmental biology; maternal and child health; biochemical development of the intestine; lactose intolerance; pyrimidine biosynthesis; enzymatic adaptations to nutrients; diabetes mellitus. *Mailing Add:* Dept Nutrit Sci 309 Morgan Hall Univ Calif Berkeley CA 94720

KRETCHMER, RICHARD ALLAN, b Tracy, Minn, Dec 12, 40; m 67; c 3. SYNTHETIC ORGANIC CHEMISTRY. *Educ:* Univ Minn, BChem, 62; Univ Wis, PhD(org chem), 66; Chicago-Kent Col Law, JD, 75. *Prof Exp:* USPHS fel chem, Columbia Univ, 66-68; from asst prof to assoc prof chem, Ill Inst Technol, 68-76; ASSOC GEN PATENT ATTY, PATENTS & LICENSING DEPT, AMOCO CORP, 76- *Concurrent Pos:* Law firm assoc, 75-76; adj prof law, Chicago Kent Col Law, 83-84 & 87-89. *Res:* Organic chemistry; structure and synthesis; chemistry of natural products; the organic chemistry of mercury. *Mailing Add:* 270 Walker Ave Clarendon Hills IL 60514

KRETSCH, MARY JOSEPHINE, US citizen. NUTRITIONAL STATUS ASSESSMENT, DIETARY ASSESSMENT. *Educ:* Univ Minn, BS, 69; Univ Calif, San Francisco, RD, 70; Univ Calif, Davis, PhD(nutrit sci), 75. *Prof Exp:* Teaching asst nutrit, Univ Calif, Berkeley, 71-73, postdoctoral fel human nutrit, 75-77; nutrit scientist, Dept Defense, Letterman Army Inst Res, 77-80; RES NUTRIT SCIENTIST, WESTERN HUMAN NUTRIT RES CTR, AGR RES SERV, USDA, 80- *Concurrent Pos:* Dir, Human Metab Res Unit, Western Human Nutrit Res Ctr, 80-83, res leader, Ctr, 83-86. *Mem:* Am Inst Nutrit; Am Soc Clin Nutrit; Am Dietetic Asn. *Res:* Nutritional status assessment with expertise in dietary assessment and biological markers of dietary exposure; vitamin B-6; water soluble vitamins; human metabolic research study techniques; nutrition surveys. *Mailing Add:* USDA Agr Res Serv Western Human Nutrit Res Ctr PO Box 29997 San Francisco CA 94129

KRETSCHMER, ALBERT EMIL, JR, b New York, NY, Nov 15, 25; m 49; c 3. TROPICAL AGROSTOLOGY. *Educ:* Univ Fla, BA, 49; Rutgers Univ, PhD(soil chem), 52. *Prof Exp:* Soil chemist, Everglades Exp Sta, 52-55, AGRONOMIST, IFAS AGR RES & EDUC CTR, UNIV FLA, 55- *Concurrent Pos:* Consult soil chemist, Univ Fla-AID Prog, Costa Rica, 58-60, chief, 69-70; pvt consult tropical pastures, overseas. *Mem:* Am Soc Agron; Sigma Xi. *Res:* Evaluation of tropical pasture legumes and grasses; micro and macro nutrient requirements of forages; management of grass-legume mixtures. *Mailing Add:* IFAS Agr Res & Educ Ctr PO Box 248 Univ Fla Ft Pierce FL 34954

KRETSINGER, ROBERT, b Denver, Colo, Mar 20, 37. MOLECULAR BIOLOGY, BIOPHYSICS. *Educ:* Univ Colo, AB, 58; Mass Inst Technol, PhD(biophys), 64. *Prof Exp:* Helen Hay Whitney Found fel, Med Res Coun Lab Molecular Biol, Cambridge Univ, Eng, 64-65; fel, Inst Molecular Biol, Geneva, Switz, 66-67; assoc prof, 67-75, chmn dept biol, 79-84, PROF BIOL, UNIV VA, 75- *Mem:* Am Crystallog Asn. *Res:* Protein structure determination by x-ray crystallography; function and evolution of calcium modulated proteins; role of calcium as cytosolic messenger. *Mailing Add:* Dept Biol Univ Va Gilmer Hall 270 A Charlottesville VA 22903

KRETZ, RALPH, Can citizen. PETROLOGY, GEOCHEMISTRY. *Educ:* Univ Chicago, PhD(geol), 58. *Prof Exp:* Geologist, Geol Surv Can, 58-61; sr lectr geol, Univ Queensland, 61-65; assoc prof, 67-71, PROF GEOL, UNIV OTTAWA, 71- *Mem:* Geochem Soc; Mineral Asn Can. *Res:* Chemical composition and texture of metamorphic rocks. *Mailing Add:* Dept Geol Univ Ottawa Ottawa ON K1N 6N5 Can

KRETZMER, ERNEST R(UDOLF), b Ger, Dec 24, 24; nat US; m 54, 83; c 2. ELECTRONICS. *Educ:* Worcester Polytech Inst, BS, 45; Mass Inst Technol, SM, 46, ScD(elec eng), 49. *Prof Exp:* Mem tech staff, Mass Inst Technol, 45-49, res assoc, 49; dept head, 65-70, DIR, BELL TEL LABS, INC, 70- *Mem:* Fel Inst Elec & Electronics Engrs. *Res:* Pulse modulation; phase measurement; redundancy in television; coded facsimile; transistor applications; electronic telephone system development; data communication. *Mailing Add:* 118 N Polk Dr Sarasota FL 34236

KREUTEL, RANDALL WILLIAM, JR, b Norwood, Mass, May 3, 34. ELECTRICAL ENGINEERING, ELECTROPHYSICS. *Educ:* Northeastern Univ, BS, 61, MS, 64; George Washington Univ, DSc(electrophys), 78. *Prof Exp:* Res eng antennas, Sylvania Electron Syst, 57-66; mem tech staff, Communications Satellite Corp, 66-68, mgr, Antennas Dept, 68-77, sr staff scientist res & develop, 77-79, dir optical commun, 79-81, dir, div develop eng, Comsat Labs, 81- 84; dir, System Planning Corp, 84-87; dir, Sci-Atlanta, 87-89; DIR, ELECTROMAGNETIC SCI, 89- *Mem:* Inst Elec & Electronics Engrs; Int Sci Radio Union; Am Inst Aeronaut & Astronaut; Sigma Xi. *Res:* Antennas, microwave circuits, fiber optics and communications; satellite communications; electromagnetics. *Mailing Add:* 8050 Willow Tree Way Alpharetta GA 30202

KREUTNER, WILLIAM, b Brooklyn, NY, Feb 20, 41; m 63; c 2. PHARMACOLOGY, BIOCHEMISTRY. *Educ:* Brooklyn Col, BS, 62; Univ Minn, PhD(pharmacol), 67. *Prof Exp:* sect leader, 78-88, ASSOC DIR, SCHERING CORP, 88- *Concurrent Pos:* Adj asst prof biochem, Fairleigh Dickinson Univ, 72-79. *Mem:* Am Acad Allergy & Immunol; NY Acad Sci; Am Soc Pharmacol Exp Ther; Am Thoracic Soc; Am Col Allergy & Immunol. *Res:* Prostaglandins; leukotrienes; cyclic nucleotides; neuronal pathways and neuropeptides; antihistamines. *Mailing Add:* Schering Corp 60 Orange St Bloomfield NJ 07003

KREUTZ-DELGADO, KENNETH KEITH, b Aguadilla, PR. ROBOTICS, MACHINE INTELLIGENCE. *Educ:* Univ Calif, San Diego, BA, 76, MS, 78, PhD(systs sci), 85. *Prof Exp:* Mem tech staff, Mach Intel Systs Group, NASA Jet Propulsion Lab, Calif Inst Technol, 85-89; ASST PROF ROBOTICS, DEPT APPL MECH & ENG SCI, UNIV CALIF, SAN DIEGO, 89- *Concurrent Pos:* Vis assoc mech eng, Calif Inst Technol, 89-90; NSF presidential young investr, 90; tech ed, Inst Elec & Electronics Engrs J Robotics & Automation, 91- *Mem:* AAAS; Inst Elec & Electronics Engrs Robotics & Automation Soc; Inst Elec & Electronics Engrs Computer Soc; Inst Elec & Electronics Engrs Systs Man & Cybernet Soc; Inst Elec & Electronics Engrs Controls Soc. *Res:* Sensor-based real-time robot planning and control; robotic manufacturing, servicing and assembly; kinematics and dynamics of multibody systems with time-varying interconnection topologies. *Mailing Add:* Dept Appl Mech & Eng Sci Univ Calif San Diego La Jolla CA 92093-0411

KREUTZER, RICHARD D, b Evergreen Park, Ill, June 23, 36; m; c 2. CYTOGENETICS, BIOCHEMISTRY OF LEISHMANIA. *Educ:* Univ Ill, BS, 63, MS, 65, PhD(zool), 68. *Prof Exp:* Instr zool, Univ Ill, Urbana, 67-69; from asst prof to assoc prof, 69-79, PROF BIOL, YOUNGSTOWN STATE UNIV, 79- *Concurrent Pos:* Chief vector, Biol Sect, Gorgas Mem Lab, 77-79. *Mem:* Am Soc Zoologists; Am Mosquito Control Asn; Genetics Soc Am; Entom Soc Am; Am Soc Trop Med Hyg. *Res:* Genetics; invertebrates; parasitology; entomology; cytogenetics and evolution of anophelines; isozyme studies on insects and protozoan parasites. *Mailing Add:* Dept Biol Youngstown State Univ Youngstown OH 44555

KREUTZER, WILLIAM ALEXANDER, b Gunnison, Colo, Apr 13, 08; m 39; c 3. PLANT PATHOLOGY. *Educ:* Colo Agr Col, BS, 30, MS, 32; Iowa State Col, PhD(plant path), 39. *Prof Exp:* From instr to asst prof bot, Colo Agr Col, 31-34; asst bot & plant path, Iowa State Col, 34-36; from asst prof to prof, Colo State Col, 36-46; plant pathologist, Agr Lab, Shell Develop Co, 46-62; prof, 62-74, EMER PROF BOT & PLANT PATH, COLO STATE UNIV, 74- *Mem:* Fel AAAS; Am Phytopath Soc; Mycol Soc; NY Acad Sci. *Res:* Soil fungicides and soil microecology. *Mailing Add:* 868 Gregory Rd Ft Collins CO 80524

KREUZ, JOHN ANTHONY, b Buffalo, NY, Sept 18, 33; m 57; c 6. ORGANIC POLYMER CHEMISTRY. *Educ:* St Bonaventure Univ, BS, 55; Univ Notre Dame, PhD(org chem), 60. *Prof Exp:* Res chemist, E I Du Pont de Nemours & Co, Inc, 59-64, staff scientist, 65-77, res assoc, 77-85, sr res assoc, 85-88, RES FEL, E I DUPONT DE NEMOURS & CO, INC, 88- *Mem:* Am Chem Soc. *Res:* Alkaline decomposition of aliphatic disulfides; addition and condensation polymerizations; polyimides and other high temperature polymers; polymer surface chemistry, polyimide and other adhesives. *Mailing Add:* 1614 McCoy Rd Columbus OH 43220

KREUZER, HAN JURGEN, b Lahnstein, Ger, Aug 9, 42; Can citizen; m 86; c 1. SURFACE SCIENCE. *Educ:* Univ Bonn, dipl, 66, Dr rer nat, 67. *Prof Exp:* Asst theoret nuclear physis, Inst Theoret Nuclear Physics, Univ Bonn, 67-69; postdoctoral fel theoret physics, Theoret Physics Inst, Univ Alta, 69-71, from asst prof to prof physics, 71-85; Killam res prof, 82-90, PROF PHYSICS, DALHOUSIE UNIV, 90- *Concurrent Pos:* Lady Davies prof, Technion, Haifa, Israel, 77; vis fel, Wolfson Col, Oxford, 87; fel, Max-Planck Soc, Ger, 87; external sci mem, Fritz-Haber Inst, Berlin, 88-; bd mem, Int Soc Theoret Chem Physics. *Mem:* Can Asn Physicists; Chem Inst Can; Int Soc Theoret Chem Physics. *Res:* Theoretical surface science; transport processes at surfaces and interfaces; kinetics of adsorption, desorption, diffusion and reactions at surfaces; physics and chemistry in high electric fields. *Mailing Add:* Dept Physics Dalhousie Univ Halifax NS B3H 3J5 Can

KREUZER, JAMES LEON, organic chemistry; deceased, see previous edition for last biography

KREUZER, LLOYD BARTON, b Los Angeles, Calif, Aug 26, 40. MICROCOMPUTERS, MICROCOMPUTER OPERATING SYSTEMS. *Educ:* Swarthmore Col, BA, 62; Princeton Univ, PhD(physics), 66. *Prof Exp:* Mem tech staff physics, Bell Tel Labs, NJ, 66-73; vpres, Diax Corp, Calif, 73-74; mem tech staff, Hewlett-Packard Lab, 74-78; vpres eng, 78-81, vpres adv develop, Dynabyte Inc, 81-82; pres, Menlo Corp, 83-86,; PRES, KREUZER SOFTWARE CORP, 86- *Mem:* Am Phys Soc; Inst Elec & Electronics Engrs. *Res:* Nonlinear optics; optical parametric effects; experimental gravitation; air pollution detection by IR laser; optoacoustic spectroscopy; microcomputers and microcomputer software. *Mailing Add:* Acuson 1220 Charleston Rd PO Box 7393 Mountain View CA 94039

KREVANS, JULIUS RICHARD, b New York, NY, May 1, 24; m; c 5. HEMATOLOGY. *Educ:* NY Univ, BS, 44, MD, 46; Am Bd Internal Med, dipl, 56. *Hon Degrees:* LLD, Rush Univ, 84. *Prof Exp:* Intern, Queens Gen Hosp, 46-47; resident path, Flushing Hosp, 47; fel hemat, Johns Hopkins Univ, 50-51, asst resident, 51-52, resident, 52-63, dir, Blood Bank, 53-62, from asst prof to prof med, Sch Med, 60-71, asst dean, 62-63, dean acad affairs, 68-71; prof med & dean Sch Med, 71-82, CHANCELLOR, UNIV CALIF, SAN FRANCISCO, 82- *Concurrent Pos:* Vis hematologist, Baltimore City Hosps, 53-63, physician-in-chief, 63-; asst prof, Johns Hopkins Univ, 55-60; chmn, Asn Am Med Cols, 80-81 & Comt Humanistic Qual Internist, Am Bd Internal Med, 83; mem, numerous comts, Nat Found, govt agencies & orgn; consult, Sch Med, Univ Wash, Univ Colo & Med Serv Found, 86. *Honors & Awards:* Abraham Flexner Award, 83; Convocation Medal, Am Col Cardiol, 84; Belkin Mem Lectr, Albert Einstein Col Med, 86. *Mem:* Inst Med-Nat Acad Sci; Am Soc Hemat; assoc Am Col Physicians; Am Fed Clin Res; Int Soc Hemat. *Res:* Hematology. *Mailing Add:* Chancellor Univ Calif San Francisco CA 94143

KREVSKY, SEYMOUR, b Elizabeth, NJ, July 2, 20; m 44; c 2. RADIATION PROTECTION, TACTICAL COMMUNICATIONS SATELLITES-SYSTEMS-HF ANTENNAS & PROPAGATION. *Educ:* Newark Col Eng, BS, 42, MS, 50. *Prof Exp:* Dep dir eng, US Army Commun Systs Agency, Ft Monmouth, NJ, 42-58 & 68-80; sr mem tech staff, RCA Astro Electronics, Princeton, NJ, 58-68; sr engr, PRC Inc, Eatontown, NJ, 80-84; mem tech staff, Mitre Inc, 84-85; prin engr, Analytics Inc, Tinton Falls, NJ, 85-89, & C31 Systs Inc, Eatontown, NJ, 89-90; PRIN STAFF MEM, BDM INT INC, EATONTOWN, NJ, 90- *Concurrent Pos:* Pres, Int Test & Eval Asn, Ft Monmouth, 86-; mem, Nat Defense Exec Reserve, Region II, Fed Emergency Mgt Agency, Washington, DC, 87-; dipl, Am Asn Environ Engrs. *Mem:* Fel AAAS; fel Radio Club Am; Armed Forces Commun & Electronics Asn; Inst Elec & Electronic Engrs, Eng Mgt Soc (vpres, 89-91); Int Test & Eval Asn. *Res:* Engineering management; participative management in corporate, middle management and working level group management arenas; communications engineering; tactical systems engineering; wire antennas; high frequency propagation. *Mailing Add:* 69 Judith Rd Little Silver NJ 07739

KREWER, SEMYON E, b Moscow, Russia, Mar 10, 15; nat US; m 39; c 1. PHYSICS. *Educ:* Tech Hochsch Berlin, dipl, 37. *Prof Exp:* Asst to Prof Fermi & Szilard, Columbia Univ, 38-40; dir res, Photovolt Corp, 40-65, vpres, 59-65; OWNER, KREWER RES LABS, 65- *Concurrent Pos:* Consult energy, US Senator Moynihan, NY, 77- *Mem:* Optical Soc Am; Arthritis Found; Am Congress Rehab Med; NY Acad Sci. *Res:* Design of scientific instruments for industry and medical research; pH meters and electrodes; supersensitive photometers; densitometers; colorimeters; fluorescence meters; rheumatoid hand gymnasium; design of simple test for tightness of intrinsic hand muscles. *Mailing Add:* c/o E B Grossmann Apt 7D 215 W 78th St New York NY 10024

KREY, LEWIS CHARLES, b New York, NY, Oct 1, 44; m 67; c 3. NEUROENDOCRINOLOGY. *Educ:* Brown Univ, AB, 66; Duke Univ, PhD(physiol), 71. *Prof Exp:* Res assoc & fel physiol, Univ Pittsburgh Sch Med, 71-73, asst prof, 73-75; asst prof neuroendocrinol, 75-81, ASSOC PROF NEUROENDOCRINOL, ROCKEFELLER UNIV, 81- *Concurrent Pos:* Alfred P Sloan Found fel, 78-80; Irma T Hirschl Found fel, 80-85; assoc ed, Endocrinol, 83-88; Reproductive Endocrinol Study Sect, NIH, 85-88. *Mem:* Endocrine Soc; Sigma Xi; Int Soc Neuroendocrinol. *Res:* Role of hypothalamic and hypophyseal steroid receptors in the neuroendocrine regulation of anterior pituitary gland function; in particular, the control of gonadotropin release in several mammalian species. *Mailing Add:* Rockefeller Univ 1230 York Ave New York NY 10021

KREY, PHILIP W, b Brooklyn, NY, June 18, 27; m 52; c 5. RADIOCHEMISTRY, ENVIRONMENTAL SCIENCE. *Educ:* St Francis Col, BS, 48; Duquesne Univ, MS, 50. *Prof Exp:* Chemist, Nuclear Defense Lab, 50-55, chief radiochem div, 55-57; mgr radiochem div, Isotopes, Inc, 57-64; dir radioactivity in surface air prog, 65-67, dir stratospheric radioactivity prog, Health & Safety Lab, 67-75, environ scientist, US Energy Res & Develop Admin & US Dept Energy, 75-80, dir, Anal Chem Div, 80-88 ACTG DEP LAB DIR, ENVIRON MEASUREMENTS LAB, US DEPT ENERGY, 88- *Concurrent Pos:* Mem task group on C-14 waste disposal, Nat Coun Radiation Protection & Measurements, 75- *Mem:* AAAS; NY Acad Sci. *Res:* Behavior and transport of artificial and natural radioactivity; trace metal and gaseous pollutants in the environment, including soil, troposphere and stratosphere from both local and global sources of contamination. *Mailing Add:* Environ Measurements Lab US Dept Energy 376 Hudson St New York NY 10014

KREY, PHOEBE REGINA, b Ambridge, Pa; m 60; c 3. RHEUMATOLOGY. *Educ:* Northeastern Univ, BS, 55; Boston Univ, MD, 60. *Prof Exp:* Intern, Newton-Wellesley Hosp, 60-61; clin fel rheumatology, Boston City Hosp & res fel, Univ Hosp, 63-69; instr med, Boston Univ Med Sch, 69-74, asst prof, 74-75; asst prof med & dir rheumatology, 75-77, ASSOC PROF MED, COL MED & DENT, NJ, 77- *Mem:* Am Rheumatism Asn; Reticuloendothelial Soc; Electron Micros Soc Am. *Res:* Rheumatoid arthritis, fine structure and culture of the synovial membrane; gout, systemic lupus erythematons; immune experimental arthritis in animals. *Mailing Add:* Div Rheumatol Univ Med & Dent NJ Med Sch 100 Bergen St Newark NJ 07103

KREYSA, FRANK JOSEPH, b Stankov, Czech, Apr 21, 19; nat US; m 50; c 4. ORGANIC CHEMISTRY, RESOURCE MANAGEMENT. *Educ:* Macalester Col, BA, 40; Columbia Univ, MA, 43, PhD(org chem), 48. *Prof Exp:* Anal chemist, Rockefeller Inst, 40-41; asst chem, Col Pharm, Columbia Univ, 43-44; from instr to assoc prof chem, St John's Univ NY, 46-55; from sr res chemist to asst to vpres Europ develop, W R Grace & Co, NY & Md, 55-61; from sr prof assoc to vpres, Smithsonian Inst Sci Info Exchange, DC, 61-73; chief, Sci Serv Div, Bur Alcohol, Tobacco & Firearms, Treas Dept, Washington, DC, 73-82; RETIRED. *Concurrent Pos:* Tech consult, Chemo Puro Mfg Co, NY & NJ, 51-55; chmn bd trustees, Am Soc Safety Res, 66-70; comnr, Sci Manpower Comn, 69-73. *Mem:* AAAS; Am Inst Chem; Am Chem Soc. *Res:* Research and development management; forensic science; instrumentation; analytical chemistry. *Mailing Add:* 1186 Willoughby Ct Frederick MD 21702

KREZANOSKI, JOSEPH Z, b Mundare, Alta, Apr 14, 27; nat US; m 49, 54; c 3. PHARMACEUTICAL CHEMISTRY. *Educ:* Univ Calif, BS, 51, MS, 53, PhD(pharmaceut chem), 56. *Prof Exp:* Asst pharm, Univ Calif, 51-56; asst prof, Med Col Va, 56-59; dir pharmaceut res & develop, Barnes-Hind Labs, Inc, 59-67; vpres & tech adv, Flow Pharmaceut, Inc, 67-77; dir res & develop, Cooper Labs, Inc, 77-; AT COOPER VISION INC. *Honors & Awards:* Borden Award, 51; Brunswick Award, 51. *Mem:* AAAS; Am Pharmaceut Asn; Am Chem Soc; Asn Am Acad Dermat; Am Mgt Asn; Sigma Xi. *Res:* Physical pharmacy; mechanism of drug action at the cellular level; pharmaceutical formulation. *Mailing Add:* 810 Amber Lane Los Altos CA 94024-4617

KREZDORN, ROY R, b Shreveport, La, Jan 30, 10; m 35; c 2. ELECTRICAL ENGINEERING. *Educ:* Texas A&M Univ, BS, 32, PhD(elec eng), 52; Texas Univ, Austin, MS, 51. *Prof Exp:* Prof elec eng, Univ Texas Austin, 41-78; CONSULT, 78- *Concurrent Pos:* Mgt adv, Lower Colo River Authority, 44-75; chief elec eng engr, Fargo Eng Co, 41-51; owner, Texas Eng Assoc, 51-81; asst dir eng res, Univ Texas, 70-78. *Mem:* Fel Inst Elec & Electronics Engrs. *Res:* Various subjects concerning power distribution and transmission. *Mailing Add:* 1501 Hillmont St Austin TX 78704

KREZOSKI, JOHN R, b Kalamazoo, Mich, Jan 15, 47; m 72; c 2. MARINE BIOGEOCHEMISTRY. *Educ:* Kalamazoo Col, BA, 69; Univ Mich, MS, 76, PhD(natural resources), 81. *Prof Exp:* Proj assoc, 82-84, asst scientist, 84-88, ASSOC SCIENTIST, CTR GREAT LAKES STUDIES, UNIV WIS-MILWAUKEE, 88- *Concurrent Pos:* Guest fac res partic, Argonne Nat Lab, 82-84; dir, Dept Environ Health & Safety, 88- *Mem:* AAAS; Am Soc Limnol & Oceanog; Int Asn Great Lakes Res; Int Asn Theoret & Appl Limnol; NAm Benthological Soc. *Res:* Multiple radiotracer techniques to study benthic community structure, biogenic nutrient regeneration from sediments, and burial and redistribution of hazardous substances by aquatic and marine invertebrates. *Mailing Add:* Ctr Great Lakes Studies Univ Wis-Milwaukee Milwaukee WI 53201

KRIBEL, ROBERT EDWARD, b Pittsburgh, Pa, Sept 17, 37; m 59; c 4. PLASMA PHYSICS, MAGNETOHYDRODYNAMICS. *Educ:* Univ Notre Dame, BS, 59; Univ Calif, San Diego, MS, 66, PhD(physics), 68. *Prof Exp:* Res asst plasma physics, Gulf Gen Atomic Inc, Gulf Oil Corp, 63-65 & Univ Calif, San Diego, 65-68; staff assoc, Gulf Gen Atomic Inc, Gulf Oil Corp, 68-69; asst prof physics, Drake Univ, 70-73; vis assoc prof elec eng, Cornell Univ, 73-74; assoc prof & head, dept physics, James Madison Univ, 74-78; prof & head dept physics, Auburn Univ 78-86, actg dean, Col Sci, 86-88; VPRES ACAD AFFAIRS, JACKSONVILLE STATE UNIV, 88- *Concurrent Pos:* Lectr, Univ San Diego, 67-68; consult, Cornell Univ, 70-71 & Los Alamos Nat Lab, 83-; Los Alamos Nat Lab, 83- *Mem:* Am Phys Soc; AAAS; Sigma Xi. *Res:* Plasma production; confinement and stability. *Mailing Add:* VPres Acad Affairs Jacksonville State Univ Jacksonville FL 36265

KRICHER, JOHN C, b Philadelphia, Pa, Feb 7, 44; m 68. ECOLOGY. *Educ:* Temple Univ, BA, 66; Rutgers Univ, NB, PhD(zool), 70. *Prof Exp:* From asst prof to assoc prof, 70-80, PROF BIOL, WHEATON COL, MASS, 80- *Concurrent Pos:* Cottrell sci grants, Res Corp, 74 & 75; Earthwatch grants, 81-83. *Mem:* Am Inst Biol Sci; Ecol Soc Am; Am Ornith Union; Cooper Ornith Soc; Sigma Xi; Asn Field Ornith (pres, 84-87); Soc Study Evol. *Res:* Bird species diversity in relation to secondary succession; species diversity of intertidal communities; tropical bird species diversity; ecology of migrant birds in the tropics; range expansions of North American birds. *Mailing Add:* Biol Dept Wheaton Col Norton MA 02766

KRICHEVSKY, MICAH I, b Chicago, Ill, May 4, 31; m 52; c 2. MICROBIOLOGY, BIOCHEMISTRY. *Educ:* Univ Conn, BA, 52; Univ Ill, MS, 55, PhD(dairy sci), 58. *Prof Exp:* Asst dairy sci, Univ Ill, 53-57, 58; biochemist, Nat Inst Allergy & Infectious Dis, 58-59; biochemist, Nat Heart Inst, 59-61; biochemist, 61-68, chief environ mechanisms sect, 68-74, CHIEF MICROBIAL SYSTEMATICS SECT, NAT INST DENT RES, 74- *Mem:* Am Soc Microbiol; Sigma Xi. *Res:* Biochemical differentiation in slime molds; metabolic pathways in bacteria; automation and computer technology in biomedical research. *Mailing Add:* Microbial Syst Sect Nat Inst Dent Res NIH Park Bldg Rm 451 Bethesda MD 20892

KRICK, IRVING PARKHURST, b San Francisco, Calif, Dec 20, 06. METEOROLOGY, PHYSICS. *Educ:* Univ Calif, Berkeley, BA, 28; Calif Inst Technol, MS, 33, PhD(meteorol), 34. *Prof Exp:* Weather forecasting, Western Air Express, 32-33; staff mem meteorol, Calif Inst Technol, 33-35, from asst prof to prof, 35-48, head dept, 42-48; PRES WEATHER FORECASTING & MODIFICATION, IRVING P KRICK ASSOC INC, 50- *Concurrent Pos:* Consult, Meteorol Dept, Am Air Lines Inc, 35-36; major chief, Long Range Res & Forecasting Sect, US Army Air Force, 42-44, mem, Sci Adv Group, 45-46; dep dir weather serv, Europ Theatre Oper, 44; chief, Weather Info Sect, SHAEF 45; consult, US Adv Comt Weather, 56-57; consult & official weather engr, VIII Olympic Winter Games, 58-60; consult, White House, 65- *Honors & Awards:* Croix de Guerre Avec Etoile Vermeil, France, 44. *Mem:* Am Water Works Asn; Am Geophys Union; assoc fel Am Inst Aeronaut & Astronaut; fel Royal Soc Arts; Royal Meteorol Soc. *Res:* Extensive original work in weather forecasting and weather modification. *Mailing Add:* 610 S Belardo Rd No 1000 Palm Springs CA 92262

KRICK, MERLYN STEWART, b Shillington, Pa, Jan 13, 38; m 68. NUCLEAR SCIENCE. *Educ:* Albright Col, BS, 59; Univ Pa, PhD(physics), 66. *Prof Exp:* Res assoc physics, Univ Rochester, 66-68; res appointee, Los Alamos Sci Lab, 68-70; from asst prof to assoc prof nuclear eng, Kans State Univ, 70-75; MEM STAFF NUCLEAR SAFEGUARDS, LOS ALAMOS NAT LAB, UNIV CALIF, 75- *Mem:* Am Phys Soc; Am Nuclear Soc; Inst Elec & Electronics Engrs. *Res:* Nuclear safeguards; nuclear instrumentation; delayed neutron physics. *Mailing Add:* Los Alamos Nat Lab MS-E540 PO Box 1663 Los Alamos NM 87545

KRIDEL, DONALD JOSEPH, b Rochester, NY, Apr 2, 16; m 45; c 6. CHEMICAL ENGINEERING. *Educ:* Univ Rochester, BS, 37; Mass Inst Technol, ScD, 40. *Prof Exp:* Staff engr, Eastman Kodak Co, 40-64, asst supt, 64-66, supt, 66-78; RETIRED. *Concurrent Pos:* Civilian employee, Corp Engrs, Manhattan Proj, 43-45. *Mem:* Am Chem Soc; Am Inst Chem Engrs. *Res:* Photographic chemicals. *Mailing Add:* 217 Ridgewood Dr Victoria TX 77904

KRIDER, EDMUND PHILIP, b Chicago, Ill, Mar 22, 40; div; c 2. ATMOSPHERIC ELECTRICITY, ATMOSPHERIC PHYSICS. *Educ:* Carleton Col, BA, 62; Univ Ariz, MS, 64, PhD(physics), 69. *Prof Exp:* Nat Acad Sci resident res assoc, Manned Spacecraft Ctr, NASA, 69-71; asst res prof, Inst Atmospheric Physics, 71-75, from asst prof to assoc prof, 73-80, PROF, DEPT ATMOSPHERIC SCI & INST ATMOSPHERIC PHYSICS, UNIV ARIZ, 80-, HEAD & DIR, 86- *Concurrent Pos:* Prin investr numerous res grants & contracts, 71-; adv, NASA, 76; mem, Lightning & Sferics Subcomn, Int Comn Atmospheric Elec, 76-; assoc ed, J Geophys Res, 77-79; co-chief ed, J Atmos Sci, 90- *Honors & Awards:* Outstanding Contributions to Advan Appl Meteorol, Am Metrol Soc, 85. *Mem:* Sigma Xi; fel Am Meteorol Soc; Am Geophys Union; Am Asn Physics Teachers. *Res:* Lightning and atmospheric electricity; cosmic ray physics. *Mailing Add:* Inst Atmospheric Physics Univ Ariz Tucson AZ 85721

KRIDER, JAKE LUTHER, b Lewistown, Ill, Dec 12, 13; m 36. ANIMAL NUTRITION. *Educ:* Univ Ill, BS, 39, MS, 41; Cornell Univ, PhD(animal husb), 42. *Prof Exp:* Asst animal husb, Univ Ill, 39-40 & Cornell Univ, 40-42; assoc, Univ Ill, 42-43, asst prof swine husb, 43-46, assoc prof, 46-47, prof animal sci, 47-50, dir feed res & nutrit, 50-51; vpres & dir feed sales, McMillen Feed Div, Cent Soya Co, Inc, 51-56, vpres & dir pub rels, 56-59, vpres personnel develop & pub rels, 59-63; prof, 63-79, EMER PROF ANIMAL SCI, PURDUE UNIV, 79- *Honors & Awards:* Res Award, Am Feed Mfrs Asn, 49; Res Award, Am Soc Animal Prod, 49; E G Cherbonnier Nat Award, Grain & Feed Dealers Nat Asn, 66; Animal Indust Award, Am Soc Animal Sci, 78. *Mem:* Hon fel Am Soc Animal Sci (vpres, 67-68, pres, 68-69); Poultry Sci Asn. *Res:* Nutritive requirements of the baby pig; value of pastures; causes of reproductive failures in sows; vitamin B-12 in baby pig nutrition; dose range antibiotic protocols; choline requirement and choline-methionine responses of young pigs. *Mailing Add:* 2201 Camelback Trace 22 West Lafayette IN 47906-1886

KRIEBEL, HOWARD BURTT, b Philadelphia, Pa, July 31, 21; m 49; c 1. FOREST GENETICS, MOLECULAR BIOLOGY CELL CULTURE. *Educ:* Haverford Col, BA, 46; Yale Univ, MF, 48, PhD, 56. *Prof Exp:* Instr forestry, Univ NH, 49-52; from instr to assoc prof forestry, Ohio Agr Res & Develop Ctr, 53-62, from asst prof to prof bot & plant path, Ohio State Univ, 55-69, prof forestry, Ohio Agr Res & Develop Ctr, 62-88, prof genetics, 69-88, EMER PROF, OHIO STATE UNIV, 88- *Concurrent Pos:* Vis scientist, Royal Col Forestry, Stockholm, 63; actg chmn, Dept Forestry, Ohio Agr Res & Develop Ctr, 66-69; Fulbright lectr, Univ Zagreb, 71-72; Fulbright distinguished scientist, Yugoslavia, 86; coordr & mem exec bd, Div 2, Int Union Forestry Res Orgn, 91-95. *Mem:* Fel AAAS; fel Soc Am Foresters; Am Soc Plant Physiologists; hon mem Asn Genetic Socs Yugoslavia. *Res:* Heritability studies; incompatibility systems; molecular biology of embryogenesis; developmental regulation of gene expression; hybrid and ecotype testing. *Mailing Add:* Ohio Agr Res & Develop Ctr Ohio State Univ Wooster OH 44691-4096

KRIEBEL, MAHLON E, b Garfield, Wash, Nov 18, 36; m 56, 80; c 3. PHYSIOLOGY, NEUROMUSCULAR TRANSMITTER. *Educ:* Wash State Univ, BS, 58; Univ Wash, MS, 64, PhD(zool), 67. *Prof Exp:* Fel, Albert Einstein Col Med, 67-69; PROF PHYSIOL, STATE UNIV NY UPSTATE MED CTR, 69- *Concurrent Pos:* Mem, Marine Biol Lab, Woods Hole; vis prof, Univ Konstanz, Max-Plank-Gottingen & Univ Calif, Irvine; Alexander von Humboldt sr scientist award. *Mem:* Am Soc Cell Biologists; Soc Neurosci. *Res:* Transmitter release at the n-m junction; physiology of tunicate heart; neurophysioloy of fish oculomotor neurons; degranulation of mast cells; squid chromatophore nerve-muscle studies. *Mailing Add:* Dept Physiol State Univ NY Health Sci Col Med 750 E Adams St Syracuse NY 13210

KRIEBEL, RICHARD MARVIN, b WReading, Pa, Apr 12, 47; m 66; c 5. NEUROANATOMY. *Educ:* Albright Col, BA, 69; Temple Univ, PhD(anat), 74. *Prof Exp:* Instr & asst prof anat, Med Col Va, Va Commonwealth Univ, 73-75; from asst prof anat to assoc prof anat & neurobiol, Col Med, Univ Vt, 80-87; ASSOC PROF ANAT, PHILADELPHIA COL OSTEOP MED, 87- *Mem:* Am Asn Anatomists; Soc Neurosci. *Res:* Neuroendocrine mechanisms synaptology of thalamic nuclei in mammals with specific interest in lateral geniculate; automatic control both central and peripheral cardiovascular system. *Mailing Add:* Dept Anat Philadelphia Col Osteop Med 4150 City Ave Philadelphia PA 19131

KRIEBLE, JAMES G(ERHARD), b NJ, Oct 23, 20; m 43; c 3. CHEMICAL ENGINEERING. *Educ:* Princeton Univ, BS, 42, PhD(chem eng), 49. *Prof Exp:* Res assoc chem process eng, Res Lab, Gen Elec Co, NY, 49-57, process engr, Refractory Metals Lab, 57-61, mgr powder prod eng, 61-68, MGR ENG, REFRACTORY METAL POWDER & GAS OPER, GEN ELEC CO, 68- *Concurrent Pos:* Fel, Textile Res Inst, 49. *Mem:* Am Chem Soc; Am Inst Chem Engrs; NY Acad Sci. *Res:* Development and economic evaluation of processes for refractory metals and gases used in lamps. *Mailing Add:* 3646 Tolland Rd Shaker Heights OH 44120

KRIEG, ARTHUR F, b East Orange, NJ, Oct 23, 30; m 56; c 3. PATHOLOGY. *Educ:* Yale Univ, AB, 52; Tufts Univ, MD, 56. *Prof Exp:* Rotating intern, Western Reserve Univ, 56-57, resident path, 57-60, resident, New Eng Deaconess Hosp, 63-64; asst prof, State Univ NY Upstate Med Ctr, 64-68; assoc prof, 68-71, PROF PATH & DIR CLIN LABS, HERSHEY MED CTR, PA STATE UNIV, 71- *Mem:* Fel Acad Clin Lab Physicians & Scientists; fel Am Soc Clin Pathologists; fel Col Am Pathologists. *Res:* Clinical pathology; clinical laboratory management, computer applications to medical care. *Mailing Add:* Milton S Hershey Med Ctr Pa State Univ Hershey PA 17033

KRIEG, DANIEL R, b Taylor, Tex, May 19, 43; m 65; c 2. PLANT PHYSIOLOGY, BIOCHEMISTRY. *Educ:* Tex A&M Univ, BS, 65, PhD(plant physiol), 70. *Prof Exp:* From asst prof to assoc prof, 70-77, PROF PLANT PHYSIOL, TEX TECH UNIV, 77- *Concurrent Pos:* Assoc ed, Agron J. *Mem:* Am Soc Plant Physiologists; Crop Sci Soc Am; Am Soc Agron. *Res:* Sorghum and cotton, physiological responses to environmental stress; environmental effects on biochemical changes in germinating cotton seeds; drought tolerance and photosynthetic activity of sorghum; environmental effects on seed development of grain sorghum. *Mailing Add:* Plant Physiol Lab Plant & Soil Sci Dept Tex Tech Univ Lubbock TX 79409

KRIEG, DAVID CHARLES, b Bradford, Pa, June 10, 36; m 58; c 3. ANIMAL BEHAVIOR, VERTEBRATE ZOOLOGY. *Educ:* Mansfield State Col, BS, 58; St Bonaventure Univ, MS, 61, PhD(biol), 64. *Prof Exp:* Instr biol high sch, NY, 59-62; asst prof zool, State Univ NY Col Cortland, 62-64; grad asst, St Bonaventure univ, 64-67; ASSOC PROF BIOL, STATE UNIV NY COL NEW PALTZ, 67- *Concurrent Pos:* Frank M Chapman grants, Am Mus Natural Hist, 66, 71, 72 & 73. *Honors & Awards:* Marcia Brady Tucker Award, Am Ornith Union, 66. *Mem:* AAAS; Am Ornith Union; Am Soc Zoologists; Animal Behav Soc; Am Soc Ichthyologists & Herpetologists. *Res:* Comparative behavior of genus Sialia; hybridization of bluebirds in great plains. *Mailing Add:* Dept Biol State Univ NY Col New Paltz NY 12561

KRIEG, NOEL ROGER, b Waterbury, Conn, Jan 11, 34. MICROBIOLOGY. *Educ:* Univ Conn, BA, 55, MS, 57; Univ Md, PhD(microbiol), 60. *Prof Exp:* Asst bact, Univ Conn, 55-57 & microbiol, Univ Md, 57-60; from asst prof to prof, 60-83, ALUMNI DISTINGUISHED PROF MICROBIOL, VA POLYTECH INST & STATE UNIV, 83- *Concurrent Pos:* Chmn, Acad Teaching Excellence, Va Polytech Inst & State Univ, 82-83; Bergey's Manual Trust, 76- *Mem:* Am Soc Microbiol; Soc Gen Microbiol. *Res:* Bacterial systematics; microaerophily. *Mailing Add:* Dept Biol Va Polytech Inst & State Univ Blacksburg VA 24061-0406

KRIEG, RICHARD EDWARD, JR, b New York, NY, Oct 16, 42; m 66; c 3. BACTERIOLOGY. *Educ:* Rutgers Univ, New Brunswick, BS, 64; Iowa State Univ, MS, 66, PhD(bact), 68. *Prof Exp:* Res microbiologist, US Air Force Sch Aerospace Med, 68-70, biomed analyst, Aerospace Med Div, 70-73; res microbiologist, Armed Forces Inst Path, 73-74, chief bact, 84-80; instr bact, Univ Md, 73-84, tech adv, 77-84; med inspector, Biomed Sci Br, HQ Air Force Inspection & Safety Ctr, Norton AFB, Calif, 84-86; assoc prof, Div Trop Pub Health, Dept Prev Med & Biomet, Uniformed Serv Univ Health Sci, Bethesda, Md, 86-91; US DIR, MINISTRY HEALTH CENT MED LAB, US EPIDEMIOL RES CTR, BELIZE CITY, BELIZE, 91- *Concurrent Pos:* Res microbiologist, Biosci Div, USAF Sch Aerospace Med, Brooiks AFB, Tex, 68-79; biomed analyst, Aerospace Med Div, 70-73; res microbiologist, Geog Path Div, Armed Forces Inst Path, Wash, DC, 73-74, chief, Bacteriol Br, Microbiol Div, 74-80; adj asst prof, Dept Preventative Med & Biomet, Uniformed Serv Univ Health Sci, 80-81; prog dir, Sch Med Technol, Malcolm Grow USAF Med Ctr, Andrews AFB, Md, 80-84, chief, microbiol Serv, Dept path, 80-84, asst clin lab mgr, 83-84; secy, Soc Armed Forces Med Lab Scientists, 74, exec secy, 77, bd dirs, 82-, pres-elect, 83, pres, 84; chmn, Bact Taxon Comt, Soc Armed Forces Med Lab Scientists, 74, mycobact comt, 74, prog comt, 80; fac adv, Uniformed Serv Univ Health Sci, 87- *Mem:* Am Soc Microbiol; AAAS; Sigma Xi; fel Am Acad Microbiol. *Res:* Pathogenicity of mycobacterial skin infections; epidemiology and etiology of Legionnaires' disease; numerical taxonomy; deoxyribonucleic acid base ration analysis; tropical infectious diseases. *Mailing Add:* 13014 Turkey Br Pkwy Univ Md Rockville MD 20853

KRIEG, WENDELL JORDAN, b Lincoln, Nebr, Apr 13, 06; m 52; c 2. NEUROANATOMY. *Educ:* Univ Nebr, BSc, 28; NY Univ, MS, 31, PhD(anat), 35. *Prof Exp:* Instr anat, Univ Nebr, 28-29; instr, Col Dent, NY Univ, 29-32, from instr to asst prof, Col Med, 32-44; assoc prof neurol, Inst Neurol, 44-46, prof neurol & dir inst, Med Sch, 46-48, prof anat, 48-74, EMER PROF ANAT, MED SCH, NORTHWESTERN UNIV, CHICAGO, 74- *Concurrent Pos:* Mem corp, Marine Biol Lab, Woods Hole. *Mem:* Am Neurol Asn; Am Asn Anatomists. *Res:* Originator of electroneuroprosthesis; structure and connections of cerebral cortex and diencephalon of rat, monkey and man; illustration of nervous system; design of stereotaxic machines; models, reconstructions, and illustrations of the brain. *Mailing Add:* 1236 Hinman Ave Evanston IL 60202

KRIEGE, OWEN HOBBS, b Toledo, Ohio, Nov 6, 29; m 52; c 3. APPLIED CHEMISTRY. *Educ:* Ohio State Univ, BSc, 51, MSc, 52, PhD(anal chem), 54. *Prof Exp:* Staff mem, Los Alamos Sci Lab, 54-60; sr chemist, Res & Develop Ctr, Westinghouse Elec Corp, 60-66; group leader anal chem, Adv Mat Res & Develop Lab, Pratt & Whitney Aircraft 66-67; group leader anal & struct chem, 67-68, res supvr, 68-70, tech supvr appl chem, 71-76, asst mgr, Mat Eng & Res Lab, 76-87; RETIRED. *Mem:* Am Chem Soc; Sigma Xi. *Res:* Analytical chemistry of refractory materials; phase separations in superalloys; trace analysis in complex alloys; atomic absorption; polymer chemistry; electroplating. *Mailing Add:* 370 Deer Pass Dr Sedona AZ 86336

KRIEGEL, MONROE W(ERNER), b Giddings, Tex, July 30, 12; m 42; c 1. CHEMICAL ENGINEERING. *Educ:* Univ Tex, BS, 34, MS, 36, PhD(chem eng), 39. *Prof Exp:* Instr chem, Univ Tex, 34-36, res assoc, bur indust chem, 37-39; assoc prof chem eng, Tex Col Arts & Indust, 39-40; sr geochemist, Carter Oil Co, 40-45, res engr & group supvr, 45-49, head prod & pipe line res, 49-58; dir tech placement & col rels, Jersey Prod Res Co, 58-64; prof chem eng, eng exten, 64-78, asst dir exten, 64-66, dir, 66-78, EMER PROF CHEM ENG, OKLA STATE UNIV, 78-; CONSULT CONTINUING ENG EDUC, 78- *Concurrent Pos:* Field test engr, oil & gas div, Tex RR Comn, 39. *Honors & Awards:* Distinguished Serv Award & Pioneer Award, Am Soc Eng Educ. *Mem:* Soc Petrol Engrs; Am Inst Chem Engrs; Am Soc Eng Educ. *Res:* Microgas analysis; geochemistry; corrosion in hydrogen sulphide; geochemical method of prospecting for petroleum; personnel selection; research management; hiring, training and placement of technical personnel; continuing engineering education; industry-university relations. *Mailing Add:* 2123 Countryside Dr Stillwater OK 74074

KRIEGER, ALLEN STEPHEN, b New York, NY, Feb 23, 41; m 66; c 2. SOLAR PHYSICS, X-RAY OPTICS. *Educ:* Mass Inst Technol, BS, 62, PhD(physics), 67. *Prof Exp:* Res assoc cosmic ray physics, Ctr Space Res, Mass Inst Technol, 67-68; sr scientist, Am Sci & Eng Inc, 68-71; staff scientist, 72-73, sr staff scientist, 74-77, dir solar res, 78-79, vpres space systs, 80-83, sr vpres space sci, 84-86; PRES, RADIATION SCI, INC, 86- *Mem:* Am Phys Soc; Am Astron Soc. *Res:* Solar physics; x-ray astronomy; x-ray optics. *Mailing Add:* Radiation Sci Bldg 200 One Kendall Sq Suite 2200 Cambridge MA 02139

KRIEGER, BARBARA BROCKETT, b Madison, Wis, Jan 27, 47. CHEMICAL ENGINEERING, CHEMICAL PHYSICS. *Educ:* Univ Wis-Madison, BS, 68; Wayne State Univ, MS, 72, PhD(chem eng), 75. *Prof Exp:* Res engr, Inst Francais du Petrole, 68-69; res technician auto emission control, Gen Motor Res Labs, 70-71; res asst, Wayne State Univ, 71-75; asst prof, 75-80, ASSOC PROF CHEM ENG, UNIV WASH, 80- *Concurrent Pos:* Consult, Rocket Res Corp, 77-; Hanford Energy Develop Lab, 78- & Nat Acad Adv Bd, Environ Protection Agency, 76- *Mem:* Am Inst Chem Engrs; Am Chem Soc; AAAS; Sigma Xi. *Res:* Chemical kinetics, chemical physics and transport related to chemical reaction engineering as applied to high temperature-high energy phenomena such as combustion, pyrolysis, laser and plasma processing, air pollution and atmospheric chemistry. *Mailing Add:* 2906 Fuhrman Ave E Seattle WA 98102

KRIEGER, CARL HENRY, b Milwaukee, Wis, Aug 14, 11; wid; c 3. BIOCHEMISTRY, RESEARCH ADMINISTRATION. *Educ:* Univ Wis, BS, 33, MS, 38, PhD(biochem), 40. *Prof Exp:* Res assoc, Alumni Res Found, Univ Wis, 34-40, lab mgr, 40-50; dir gen labs, 50-55; dir basic res, Campbell Soup Co, 55-57, dir basic res & prod develop, 57-60, pres, 60-76, vpres prod res, 61-76, pres, Campbell Inst Food Res, 66-76; dir, Technol Resources, Inc, 75-76; RETIRED. *Concurrent Pos:* Mem indust adv comt canned meat, Res & Develop Assoc Mil Food/Packaging Syst, 59-62, dir, 73-74; mem sci res comt, Nat Canners Asn, 60-74; mem comt res life sci, Nat Acad Sci-Nat Res Coun, 66-69, mem comt fruit & vegetable prod, Adv Bd Mil Personnel Supplies, 68-71, mem, Food Nutrit Bd, 69-72, mem, Bd Agr & Renewable Resources, 73-76; mem tech adv comt, Inst Human Nutrit, Columbia Univ, 66-72; mem sci & tech adv comt, AID, 67-72; mem, Monell Chem Senses Ctr, Nat Adv Coun, 68-76; chmn food & nutrit liaison comt, Nutrit Found Inc, 70-72; mem sci adv bd, Nat Ctr Toxicol Res, 72-76; trustee, Campbell Soup Co Res Inst, 60-76. *Mem:* Fel AAAS; Am Chem Soc; Am Inst Nutrit; Animal Nutrit Res Coun; fel Inst Food Technol. *Res:* Fats; dairy products; fermentation; proteins; carbohydrates; flavors; microbiology and nutrition as applied to foods; development of new food products in heat processed, frozen and dehydrated food areas. *Mailing Add:* 1245 E Murray Holladay Rd Salt Lake City UT 84117-4939

KRIEGER, EDUARDO MOACYR, medical physiology, for more information see previous edition

KRIEGER, GARY LAWRENCE, b New York, NY, May 2, 48; m 74; c 2. HEALTH PHYSICS, RADIATION PROTECTION PHYSICS. *Educ:* NY Inst Technol, BS, 71; Univ Kans, MS, 76. *Prof Exp:* Health physics supvr, Siemens Corp, 76-79; assoc scientist, Brookhaven Nat Lab, 79-82; lead sr engr, Impell Corp, 82-84; proj health physicist, KLM Eng, 84-85; EMERGENCY PLANNING SCIENTIST, LONG ISLAND LIGHTING CO, 85- *Mem:* Health Physics Soc; Am Nuclear Soc. *Res:* Dose assessment modeling technique for accidental radiological releases to the environment; impact of this exposure on man. *Mailing Add:* Long Island Lighting Co 131 Hoffman Lane Central Islip NY 11722

KRIEGER, HENRY ALAN, b Denver, Colo, May 7, 36; m 57; c 2. PROBABILITY & MATHEMATICAL STATISTICS. *Educ:* Rensselaer Polytech Inst, BAE, 57; Brown Univ, PhD(appl math), 64. *Prof Exp:* Bateman res fel math, Calif Inst Technol, 64-65; asst prof, 65-68; from asst prof to assoc prof math, 68-83, PROF MATH, HARVEY MUDD COL, 83- *Concurrent Pos:* Vis assoc prof statist, Israel Inst Technol, 74-75; vis prof statist, Hebrew Univ Jeusalem, 81, Australian Nat Univ, 82; vis res scientist, Commonwealth Sci Res Orgn, Div Math Statist, 82. *Mem:* Am Math Soc; Math Asn Am; Soc Indust Appl Math; Sigma Xi. *Res:* Probability theory, particularly limit theorems; measure theory. *Mailing Add:* Dept Math Harvey Mudd Col Claremont CA 91711

KRIEGER, HOWARD PAUL, b Brooklyn, NY, July 2, 18; m 53; c 2. NEUROLOGY. *Educ:* Harvard Univ, SB, 41; Long Island Col Med, MD, 44; Am Bd Psychiat & Neurol, dipl. *Prof Exp:* Intern, Long Island Col Med Hosp, 44-45, resident, Long Island Col Med, 46-47; resident, 49-51, fel Columbia Univ, 51; asst attend neurologist, 52-59, assoc attend neurologist, 59-62; chief div neurol, Beth Israel Med Ctr, 60-77; ATTEND NEUROLOGIST, MT SINAI HOSP, 62-, PROF NEUROL, MT SINAI SCH MED, 75-; ATTEND NEUROLOGIST, BETH ISRAEL MED CTR, 60- *Concurrent Pos:* Fel, NY Univ-Bellevue Med Ctr, 47-48; USPHS fel, 48-49, spec fel, 51; Nat Found Infantile Paralysis fel, 52, 54; consult neurologist, US Marine Hosp, 57-75; clin prof neurol, Mt Sinai Med Sch, 66-75. *Mem:* Am Psychol Asn; Am Neurol Asn; Am Acad Neurol; Am Fedn Clin Res; NY Acad Sci; Sigma Xi. *Res:* Neurophysiology; neuroendocrinology. *Mailing Add:* 1200 Fifth Ave Mt Sinai Med Sch New York NY 10029

KRIEGER, IRVIN MITCHELL, b Cleveland, Ohio, May 14, 23; m 65; c 1. PHYSICAL CHEMISTRY. *Educ:* Case Inst Technol, BS, 44, MS, 48; Cornell Univ, PhD(phys chem), 51. *Prof Exp:* Asst, Cornell Univ, 47-48; instr chem, Case Western Reserve Univ, 49-51, from asst prof to assoc prof phys chem, 51-68, prof phys chem & macromolecular sci, 68-88, dir, Ctr Adhesives, Sealants & Coatings, 82-88, EMER PROF CHEM, CASE WESTERN RESERVE UNIV, 88- *Concurrent Pos:* vis prof, Nat High Sch Chem, Mulhouse, 87-; prof invité, Ecole Nationale Superieure de Chimie de Mulhouse, 87; assoc dir res, Univ Louis Pasteur, Strasbourg. *Honors & Awards:* Bingham Medal, Soc Rheol, 89. *Mem:* Am Chem Soc; Soc Rheol (pres, 77-79); Am Inst Chem Eng; Sigma Xi. *Res:* Rheology and statistical mechanics of colloids and polymers. *Mailing Add:* 15691 Fenemore Rd East Cleveland OH 44112

KRIEGER, JEANNE KANN, b Hartford, Conn, Apr 16, 44; m 66; c 2. PHYSICAL ORGANIC CHEMISTRY. *Educ:* Bryn Mawr Col, BA, 66; Mass Inst Technol, PhD(org chem), 71; Boston Col, MBA, 84. *Prof Exp:* Res assoc chem, Mass Inst Technol, 71-72, instr, 72-75, lectr, 75-78; proj leader, 78-80, asst to dir mgr, New England Nuclear, 81-83; area supvr, 83-87, MGR RES PROD OPERS, DUPONT MED PRODS, 87- *Mem:* Am Chem Soc; AAAS; Am Nat Standard. *Res:* Radioactive waste disposal; scintillation techniques. *Mailing Add:* 44 Webster Rd Lexington MA 02173

KRIEGER, JOHN NEWTON, b Philadelphia, Pa, May 3, 48; m 72. UROLOGY, INFECTIOUS DISEASES. *Educ:* Princeton Univ, AB, 70; Cornell Univ Med Col, MD, 74. *Prof Exp:* Asst surgeon, gen surg, New York Hosp-Cornell Med Ctr, 74-76, urol, 76-79, surgeon, 70-80; instr urol, Univ Va, 80-82; from asst prof to assoc prof, 82-90, PROF UROL, UNIV WASH, 90- *Concurrent Pos:* Scholar, Am Urol Asn, 80-82; attend surgeon, Univ Va Hosp, 80-82, Univ Hosp, Seattle, 82-, Harborview Med Ctr, Seattle, 82-, Children's Orthop Hosp, 82-; consult urol, Vet Admin Hosp, Seattle, 82- *Honors & Awards:* Cornell Award Excellence Surg, 74. *Mem:* Am Fedn Clin Res; Am Venereal Dis Asn; Am Urol Asn; Am Soc Microbiol; Infectious Dis Soc Am; Sigma Xi. *Res:* Genitourinary tract infections; bacteriuria; sexually transmitted diseases. *Mailing Add:* Dept Urol RL-10 Univ Wash Seattle WA 98195

KRIEGER, JOSEPH BERNARD, b Brooklyn, NY, July 10, 37; m 64; c 1. THEORETICAL SOLID STATE PHYSICS, ATOMIC PHYSICS. *Educ:* Columbia Univ, AB, 59, PhD(physics), 65. *Prof Exp:* From asst prof to assoc prof physics, Polytech Inst Brooklyn, 65-72; assoc prof, 72-74, PROF PHYSICS, BROOKLYN COL, 74-; EXEC OFFICER, DOCTORAL PROG PHYSICS, CITY UNIV NEW YORK, 90- *Concurrent Pos:* Vis assoc prof, Brooklyn Col, 71-72; chmn dept, 76-80; acad assoc, Calif Inst Technol, 79. *Mem:* Fel Am Phys Soc; Sigma Xi. *Res:* Transport theory in solids; density functional theory. *Mailing Add:* Dept Physics Brooklyn Col Brooklyn NY 11210

KRIEGER, ROGER B, b Milwaukee, Wis, May 4, 41; m 68, 79. MECHANICAL ENGINEERING. *Educ:* Univ Wis, Madison, BS, 64, PhD(mech eng), 68. *Prof Exp:* Res engr, French Inst Petrol, 68-69; SR STAFF RES ENGR, ENGINE RES DEPT, GEN MOTORS RES LABS, 69- *Mem:* Soc Automotive Engrs; Am Soc Mech Engrs; Combustion Inst. *Res:* Combustion; pollutant formation and destruction during combustion; combustion modelling and internal combustion engine simulation. *Mailing Add:* 636 Lakeview Birmingham MI 48009

KRIEGER, STEPHAN JACQUES, b San Francisco, Calif, Aug 2, 37; m 58; c 4. THEORETICAL PHYSICS. *Educ:* Univ Calif, Berkeley, BS, 59, PhD(physics), 63. *Prof Exp:* Res physicist, Carnegie Inst Technol, 63-66; assoc prof physics, Univ Ill, Chicago Circle, 71-78, prof, 78-80; STAFF SCIENTIST, LAWRENCE LIVERMORE NAT LAB, 80- *Mem:* Am Phys Soc. *Res:* Nuclear structure; many body problem. *Mailing Add:* L-297 Lawrence Livermore Nat Lab PO Box 808 Livermore CA 94550

KRIEGH, JAMES DOUGLAS, b Dodge City, Kans, Dec 29, 28. CIVIL ENGINEERING. *Educ:* Univ Colo, BS, 55, MS, 58. *Prof Exp:* Asst, Cryogenics Lab, Nat Bur Standards, Colo, 53-54; asst, Eng Exp Sta, Univ Colo, 54-55, instr civil eng, 55-58; from asst prof to prof civil eng, Univ Ariz, 58-86, prof eng mech, 81-86; RETIRED. *Concurrent Pos:* NSF fac fel, Univ Colo, 63-64; comt chmn, Hwy Res Bd, Nat Acad Sci-Nat Res Coun, 64-70. *Mem:* Am Soc Civil Engrs; Am Concrete Inst. *Res:* Epoxy resins for concrete construction and structural adhesives. *Mailing Add:* 40 E Calle Concordia Oro Valley AZ 85737

KRIEGSMAN, HELEN, b Pittsburg, Kans, Feb 27, 24. MATHEMATICS. *Educ:* Kans State Teachers Col Pittsburg, BS, 44, MS, 47; Ohio State Univ, PhD(math educ), 64. *Prof Exp:* Teacher, High Sch, Kans, 44-47; from instr to assoc prof, 47-67, PROF MATH & CHMN DEPT, PITTSBURG STATE UNIV, 67- *Mem:* Math Asn Am; Am Math Soc; Nat Coun Teachers Math. *Res:* Curriculum and methods of teaching mathematics on the secondary school and college levels. *Mailing Add:* Pittsburg State Univ Wilkinson Alumni Ctr Pittsburg KS 66762

KRIEGSMANN, GREGORY A, b Chicago, Ill, Sept 20, 46; m 69; c 2. APPLIED MATHEMATICS. *Educ:* Marquette Univ, BS, 69; Univ Calif, Los Angeles, MS, 70, PhD(appl math), 74. *Prof Exp:* Instr math, Courant Inst, NY Univ, 74-76; mem tech staff, Hughes Aircraft Co, 76-77; asst prof math, Univ Nebr, 77-79, assoc prof, 79-80; from assoc prof to prof appl math, Northwestern Univ, 80-90; PROF MATH, NJ TECH INST, 90- *Mem:* Soc Indust Appl Math; Am Math Soc; Acoust Soc Am. *Res:* Numerical and asymstotic analysis of wave propagation; bifurcation problems in the physical sciences. *Mailing Add:* Dept Math NJ Tech Inst Technol Newark NJ 07102

KRIENKE, ORA KARL, JR, b Seattle, Wash, Jan 31, 31; m 60; c 2. ASTRONOMY. *Educ:* Seattle Pac Col, BA, 53, MA, 55; Univ Wash, MS, 59 & 69, PhD(astron), 73. *Prof Exp:* Instr math, 53-59, asst prof physics & math, 59-63, assoc prof physics, math & philos, 63-71, PROF PHYSICS, MATH & PHILOS, SEATTLE PAC UNIV, 71-, DEAN, SCH NAT & MATH SCI, 80- *Concurrent Pos:* Vis lectr, Univ Wash, 64 & 68. *Mem:* AAAS; Am Asn Physics Teachers; Am Astron Soc. *Res:* Structure of galaxies, especially irregular type II galaxies. *Mailing Add:* Dept Math & Sci Seattle Pac Univ 3307 Third Ave W Seattle WA 98119

KRIENS, RICHARD DUANE, b Belmond, Iowa, Oct 16, 32; m 67; c 3. ORGANIC CHEMISTRY. *Educ:* Iowa State Teachers Col, BA, 56; Iowa State Univ, PhD(org chem), 63. *Prof Exp:* Teacher, High Sch, Iowa, 57-58; asst prof chem, Iowa Wesleyan Col, 63-65; assoc prof, 65-73, PROF CHEM, ASHLAND UNIV, 73- *Mem:* Am Inst Chem; Royal Soc Chem; Am Chem Soc. *Res:* Free radical organic chemistry; reaction mechanisms in organic chemistry. *Mailing Add:* Dept Chem Ashland Univ Ashland OH 44805

KRIER, CAROL ALNOTH, b Bismarck, NDak, July 22, 28; m 57; c 2. MATERIAL SCIENCE ENGINEERING, PRODUCTION ENGINEERING. *Educ:* St Martin's Col, BS, 50; Univ Pittsburgh, PhD(chem), 55. *Prof Exp:* Lab asst fuel oils, State Labs Dept, NDak, 45-46; asst instr chem, Pvt Sch, Wash, 48-50; asst, Univ Pittsburgh, 50-51, asst phys chem, 51-55; prin chemist, proj leader & sr scientist, Battelle Mem Inst, 55-62; metals res specialist, Boeing Co, 62-67, supvr mat stress & environ simulation, 67-70, mgr advan develop & spec studies, 70-71, eng mgr, Boeing Aerospace, 71-83, mat mfg tech, 84-86, MGR, PROD ENG, BOEING AEROSPACE & ELECTRONICS, 87- *Concurrent Pos:* Consult, Defense Metals Info Ctr, 59-62; mat adv bd mem, Nat Acad Sci, 61-69, Aerospace Industs Asn, 86-89. *Honors & Awards:* NASA-Apollo Achievement Award, 70; Apollo/Saturn V Roll of Honor, 71; Apollo 11 Manned Flight Awareness Award, 71. *Mem:* Am Chem Soc; Am Soc Metals; Sigma Xi. *Res:* Thermodynamics; calorimetry; cryogenics; theory of metals and alloys; extractive metallurgy; refractory, structural and electronic materials; high temperature coatings for metals; oxidation of metals and alloys; alkali metals; platinum-group metals; high temperature corrosion; physical metallurgy; vapor deposition; manufacturing processes; automation. *Mailing Add:* 4520 133 Rd Ave SE Bellevue WA 98006

KRIESBERG, JEFFREY IRA, b Far Rockaway, NY, July 7, 49; c 3. EXPERIMENTAL PATHOLOGY. *Educ:* State Univ NY, Albany, BS, 71; Univ Md, PhD(exp path), 75. *Prof Exp:* Res assoc, Dept Path, Univ Ala Med Ctr, 75-76; res fel, Dept Path, Med Sch, Harvard Univ, 76-77; from instr to asst prof, 80-89; from asst prof to assoc prof path, 80-89, assoc prof med, 83-89, PROF MED & PATH, HEALTH SCI CTR, UNIV TEX, SAN ANTONIO, 89-; CAREER SCIENTIST, VET ADMIN, 89- *Concurrent Pos:* Lectr, var univs, socs & hosps, 78-91; asst biologist, Dept Med, Mass Gen Hosp, 79-80; consult path, Sch Med, Yale Univ, Kidney Dis Inst, NY Dept Health, 79 & Cystic Fibrosis Core Ctr, Case Western Univ, 86-91; prin investr, NIH, 81-95; mem, Spec Study Sect, NIH, 84 & 87, Spec Planning Comt, Nat Inst Arthritis & Diabetes & Digestive Kidney Dis, 85, Path A Study Sect, 85-88, Coun Kidney Cardiovasc Dis, Am Heart Asn, 86, Prog Comt, Am Soc Nephrology, 88-89; prog chmn, Tissue Cell Cult, Am Soc Nephrology, 87; Nat Kidney Found fel, 87-88 & 89-90. *Mem:* AAAS; Am Soc Nephrology; Am Asn Pathologists; Am Soc Cell Biol; Am Diabetes Asn; Tissue Cult Asn; Int Soc Nephrology; Am Heart Asn. *Res:* Author of more than 50 technical publications. *Mailing Add:* Dept Path Health Sci Ctr Univ Tex 7703 Floyd Curl Dr San Antonio TX 78284

KRIESEL, DOUGLAS CLARE, b Owatonna, Minn, July 22, 37; m 59; c 3. MEDICINAL CHEMISTRY. *Educ:* Univ Minn, BS, 60, PhD(med chem), 65. *Prof Exp:* Prof med chem, Sch Pharm, Southwestern State Col, Okla, 65-72; RES SCIENTIST PHARMACEUT PRODS DIV, ABBOTT LABS, 76- *Concurrent Pos:* Res grants, Mead Johnson Labs, 66-67 & Okla Heart Asn, 67-68. *Mem:* Am Chem Soc; Am Pharmaceut Asn; Acad Pharmaceut Sci; Sigma Xi. *Res:* Synthesis formulation and development of pharmaceutical dosage forms of analgesics, antineoplastics, antivirals, and cardiovascular agents. *Mailing Add:* 56 Old W High St East Hampton CT 06424

KRIGBAUM, WILLIAM RICHARD, b Ill, Sept 29, 22; m 46; c 3. PHYSICAL CHEMISTRY. *Educ:* Millikin Univ, BS, 44; Univ Ill, MS, 48, PhD(chem), 49. *Hon Degrees:* DSc, Millikin Univ, 66. *Prof Exp:* Nat Res Coun fel, Cornell Univ, 49-50, res assoc & instr, 50-52; from instr to prof, 52-69, chmn dept, 76-79, JAMES B DUKE PROF CHEM, DUKE UNIV, 69- *Concurrent Pos:* Sloan res fel, 56-60; NSF sr fel, 59-60; mem adv panel chem, NSF. *Mem:* Am Chem Soc; Am Phys Soc; Am Crystallog Asn. *Res:* Wide and low angle x-ray diffraction; physical chemical studies of polymers in solution and in bulk state; physical chemistry of macromolecules. *Mailing Add:* 2504 Wilson St Durham NC 27705

KRIGSVOLD, DALE THOMAS, b Grant Co, Minn, June 21, 37; m 82. PLANT PATHOLOGY, PLANT ECOLOGY. *Educ:* Old Dominion Univ, BS, 73; Va Polytech Inst & State Univ, PhD(plant path), 79. *Prof Exp:* Plant pathologist, Trop res Div, United Fruit Co, 79-; POST-HARVEST HANDLING ADV, CHEMONICS INT. *Mem:* Am Phytopath Soc; Sigma Xi. *Res:* Etiology and control of post-harvest diseases on commercially produced tropical fruits and vegetables, primarily bananas; ecology of soil-borne plant pathogens, primarily fungal spore germination as affected by soil fungistasis and the host plant. *Mailing Add:* Chemonics Int 2000 M St NW Suite 200 Washington DC 20036

KRIKORIAN, ABRAHAM D, b Worcester, Mass, May 5, 37. PLANT PHYSIOLOGY, BIOCHEMISTRY. *Educ:* Mass Col Pharm, BS, 59; Cornell Univ, PhD(plant physiol), 65. *Prof Exp:* From teaching asst to teaching assoc plant physiol, Cornell Univ, 60-64, from instr to asst prof, 63-66; from asst prof to assoc prof biol sci, 66-81, assoc prof biochem, 81-88, PROF BIOCHEM, STATE UNIV NY, STONY BROOK, 88- *Concurrent Pos:* Ed, Ann Bot, 76-82; plant sci book rev consult, Quart Rev Biol, 79-; mem comt space res, Int Coun, Sci Union, 84-; gov bd, Am Soc Gravitational & Space Biol, 85-87. *Honors & Awards:* Cosmos Achievement Award, NASA, 75 & 81. *Mem:* AAAS; Am Soc Pharmacog; Int Soc Plant Morphologists; Int Asn Plant Tissue Cult; Soc Develop Biol; Tissue Cult Asn; Soc Econ Bot (vpres, 81-82, pres, 82-83); Bot Soc Am; Am Soc Plant Physiologists; Scand Soc Plant Physiol; Am Soc Gravitational & Space Biol (pres, 87-88); Int Palm Soc; Plant growth Regulator Soc Am. *Res:* Physiological and morphological aspects of growth and development in flowering plants; morphogenesis and biochemical differentiation; nitrogen metabolism; production of secondary products and expression of biochemical potentialities by cells and tissues grown in culture; clonal stability; totipotency of higher plant cells in terms of morphogenesis and biochemical competence. *Mailing Add:* Dept Biochem & Cell Biol State Univ NY Stony Brook NY 11794-5215

KRIKORIAN, ESTHER, b Chelsea, Mass. SOLID STATE PHYSICS, MATERIALS SCIENCE. *Educ:* Columbia Univ, BS, 50, PhD(phys chem), 57. *Prof Exp:* Lab asst phys chem, Columbia Univ, 49-50; fel, Brookhaven Nat Lab, 56-58; instr, Hunter Col, 60-61; staff scientist physics, 61-66, SR STAFF SCIENTIST PHYSICS, POMONA DIV, GEN DYNAMICS CORP, 66- *Mem:* Am Phys Soc; Am Vacuum Soc. *Res:* Crystal physics; infrared anisotropy in single crystals; a-decay in polonium-210; nucleation and growth of thin films of semiconductors, metals and dielectrics; epitaxial growth; organic semiconductors; semiconductor devices including electronic and electrooptical; integrated optics. *Mailing Add:* 1262 Deerfield Circle Upland CA 91786

KRIKORIAN, JOHN SARKIS, JR, b Providence, RI, Sept 18, 41. ELECTRICAL ENGINEERING, APPLIED MATHEMATICS. *Educ:* Univ RI, BS, 63; Syracuse Univ, MS, 67, PhD(elec eng), 68. *Prof Exp:* Res specialist elec eng, Elec Boat Div, Gen Dynamics Corp, Groton, Conn, 68-73; asst prof, Univ RI, 73-80; mem tech staff, Mitre Corp, 83-84; VPRES, SARKIS CORP, 81- *Concurrent Pos:* Lectr elec eng, Univ Conn, 69; adj prof, Brown Univ, 80-81. *Mem:* Sigma Xi. *Res:* Electrical engineering and related interdisciplinary activities. *Mailing Add:* Five Thayer Pl Warwick RI 02888

KRIKORIAN, OSCAR HAROLD, b Fresno, Calif, Nov 22, 30; m 53; c 2. HIGH TEMPERATURE CHEMISTRY. *Educ:* Fresno State Col, BS, 52; Univ Calif, PhD(chem), 55. *Prof Exp:* RES CHEMIST, LAWRENCE LIVERMORE NAT LAB, 55- *Mem:* Am Chem Soc. *Res:* High temperature chemistry; thermodynamic properties of gaseous species that exist at high temperatures; molten metal containment studies, estimation of heat capacities and thermal expansivities of refractory materials. *Mailing Add:* Lawrence Livermore Nat Lab PO Box 808 L-369 Livermore CA 94551

KRIKOS, GEORGE ALEXANDER, b Old Phaleron, Greece, Sept 17, 22; US citizen; m 49; c 3. PATHOLOGY. *Educ:* Univ Pa, DDS, 49; Univ Rochester, PhD(path), 59. *Hon Degrees:* Dr, Univ Athens, Greece, 81. *Prof Exp:* NIH res fel path, Univ Rochester, 54-58; from asst prof to prof, Sch Dent Med, Univ Pa, 58-68, chmn path, 64-68; prof pathobiol, 68-75, chmn dept, 68-73, assoc dean oral biol affairs, 75-76, prof 75-86, CLIN PROF ORAL BIOL, SCH DENT, UNIV COLO, 86- *Concurrent Pos:* Assoc prof oral path, Grad Sch Arts & Sci, Div Grad Educ, Sch Med, Univ Pa, 62-68; vis prof, Sch Dent, Univ Athens, Greece, 80-81. *Mem:* Am Asn Pathologists; Int Asn Dent Res; Sigma Xi. *Res:* Connective tissue research. *Mailing Add:* Dept Diag & Biol Sci Univ Colo Health Sci Ctr Sch Dent C285 4200 E Ninth Ave Denver CO 80262

KRILL, ARTHUR MELVIN, b Burlington, Colo, Oct 17, 21; m 44; c 3. MECHANICAL ENGINEERING, ENGINEERING GENERAL. *Educ:* Univ Colo, BS, 43, MS, 51; Indust Col Armed Forces, Dipl, 52. *Prof Exp:* Prod engr, Pratt-Whitney Aircraft Div, United Aircraft Corp, 42, exp test engr, 43-47; from instr to assoc prof mech engr, Col Eng, Univ Denver, 47-62, coord coop plan, 48-56, head admin eng, 51-52, proj supvr, 51-56; dir opers anal unit, Denver Res Inst, 55-62, head mech div, 56-62; pres, Falcon Res & Develop, 62-70; pres, Ken R White Co, 63-76; chmn, 76-78, PRES, ARTHUR M KRILL CONSULTS, 79-; CHMN, OGDEN DEVELOP CORN, 70- *Concurrent Pos:* Consult, Bond Eng Co, 50-52; mem, Colo State Air Pollution Variance Bd; mgr corp planning, Stearns Roger Corp, 82-83; mem, Metrop Air Qual Coun. *Mem:* Fel AAAS; Am Soc Mech Engrs; Am Soc Eng Educ; Nat Soc Prof Engrs; Am Inst Consult Engrs. *Res:* Theoretical and applied mechanics; operations research; behavioral sciences; magnetohydrodynamics. *Mailing Add:* 450 Westwood Dr Denver CO 80206

KRIM, MATHILDE, b Como, Italy, July 9, 26; US citizen; m 58; c 1. CYTOGENETICS, VIROLOGY. *Educ:* Geneva Univ, BS, 48, PhD(cytogenetics), 53. *Prof Exp:* Jr scientist & res assoc cancer res, Weizmann Inst, 53-59; res assoc virol, Div Virus Res, Med Col, Cornell Univ, 59-62; assoc, 62-75, ASSOC & MEM, SLOAN-KETTERING INST CANCER RES, 75-; AT AM FOUND AIDS RES, 88- *Concurrent Pos:* Mem, President's Comt Ment Retardation, 66-69, Nat Endowment for Humanities, 69-73, Comt of 100 for Nat Health Ins, 69-; consult spec virus cancer prog & mem adv comt, Nat Colorectal Cancer Prog, Nat Cancer Inst, 71; trustee, Rockefeller Found, 71-; co-chmn, Nat Comt to Save our Schs of Health, 71-; mem bd trustees, Nat Biomed Res Found; mem bd dirs, Inst Soc, Ethics & Life Sci; pres, Comn Study Ethnical Probs in Med, Biomed & Behav Res, 80-; secy, Adv Comt Health Protection & Dis Prev, Dept Health, Educ & Welfare, 69-70; mem panel consult conquest of cancer, Comt Labor & Pub Welfare, US Senate, 71. *Mem:* AAAS; Am Cancer Soc; Am Asn Ment Deficiency. *Res:* Structure of chromosomes; prenatal determination of sex; aberrations in human sexual development; cell biology and mechanisms of oncogenic transformation; interferon research. *Mailing Add:* Am Found AIDS Res 1515 Broadway Suite 3601 New York NY 10036

KRIMIGIS, STAMATIOS MIKE, b Chios, Greece, Sept 10, 38; US citizen; c 2. SPACE PHYSICS. *Educ:* Univ Minn, BPhys, 61; Univ Iowa, MS, 63, PhD(physics), 65. *Prof Exp:* Res assoc space physics, Univ Iowa, 65-66, asst prof physics, 66-68; from sr staff scientist to supvr space physics, Johns Hopkins Univ, 68-74, head space physics & instrumentation group, 74-81, chief scientist, 80-90, DEPT HEAD, SPACE DEPT, APPL PHYSICS LAB, JOHNS HOPKINS UNIV, 91- *Concurrent Pos:* Co-investr, Mariner IV & Injun IV, 63, Orbiting Geophys Observ-4, Explorer 33 & 35 & Injun V, 65, Mariner V Venus, 66, prin investr, Interplanetary Monitoring Platform on 7 & 8, 67; mem var NASA adv comts on space invests; prin investr, Multiple-Charged Energetic Trapped Nuclei in Radiation Belt, NSF, 71-74, Voyager Low Energy Charged Particle Exp, 72, Light Ion Release Exp, NASA, 73, Active Magnetospheric Particle Tracer Explorers, 77, Studies of Solar & Magnetospheric Particles, NSF, 77-83; assoc ed, J Geophys Res, Space Physics, 75-77; co-prin investr, Galileo Mission, Energetic Particle Detector Exp, 77 & co-investr, Int Solar-Polar Mission, LAN Exp, 77; mem, Space Sci Bd, Nat Acad Sci, 83-86, chmn, Comt Solar & Space Physics, 83-86, NASA Space & Earth Sci Adv Comt, 87-90; prin investr, Imaging Neutral Particle Detector, NASA Innovative Res Prog, 85-89; mem, Space Sci Working Group, Asn Am Univs; prin investr,Cassini mission to Saturn, 90- *Honors & Awards:* Except Sci Achievement Medal, NASA, 81 & 86. *Mem:* AAAS; fel Am Geophys Union; fel Am Phys Soc. *Res:* Space plasma physics; solar and heliospheric physics; geomagnetically trapped radiation; planetary magnetospheres; cosmic rays; particle instrumentation; over 240 publications in journals and books. *Mailing Add:* Appl Physics Lab Johns Hopkins Univ Laurel MD 20723-6099

KRIMM, SAMUEL, b Morristown, NJ, Oct 19, 25; m 49; c 2. BIOPHYSICS, POLYMER PHYSICS. *Educ:* Polytech Inst Brooklyn, BS, 47; Princeton Univ, MA, 49, PhD(phys chem), 50. *Prof Exp:* Fel, 50-52, from instr to assoc prof, 52-63, assoc dean res, Col Lit Sci & Arts, 72-75, chmn biophysics res div, 76-86, PROF PHYSICS, UNIV MICH, ANN ARBOR, 63-, DIR PROG PROTEIN STRUCT & DESIGN, 85- *Concurrent Pos:* Consult, Monsanto Co & Allied-Signal Corp; NSF sr fel, 62-63; chmn, Gordon Res Conf, 68; vis prof, Weizmann Inst, 70, Univ Mainz, 83 & Univ Paris, 91; sr fel, Univ Mich, 71-76; mem, Nat Bur Standards Polymers Div Eval Panel, Nat Acad Sci/Nat Res Coun, 73-76, chmn, 75-76; chmn biopolymers subgroup, Biophys Soc, 74-75; vchmn, Div Biol Physics, Am Phys Soc, 78, chmn, 79; mem, Mat Res Adv Comt, NSF, 81-86, chmn, 84-85, mem, Coun Mat Sci, DOE, 86 -; invited prof, French Govt, 91. *Honors & Awards:* High Polymer Physics Prize, Am Phys Soc, 77; Alexander von Humboldt Prize, 83. *Mem:* Fel AAAS; Am Chem Soc; Biophys Soc; fel Am Phys Soc; Am Crystallog Asn. *Res:* Infrared and Raman spectroscopy; x-ray diffraction; high polymers; protein structure; membrane structure. *Mailing Add:* Dept Physics Univ Mich Ann Arbor MI 48109

KRIMMEL, C PETER, b Erie, Pa, June 23, 17. MEDICINAL CHEMISTRY. *Educ:* Pa State Univ, BS, 39, Northwestern Univ, MS, 41; Pa State Univ, PhD(Chem), 45. *Prof Exp:* Instr chem, Pa State Univ, 44-45; res chemist, G D Searle & Co, 46-78; RETIRED. *Concurrent Pos:* Vchmn, Conf Med Chem, Gordon Res Confs, 64, chmn, 65, monitor Confs, 79-88; consult, 79- *Mem:* AAAS; Am Chem Soc. *Res:* Organic chemistry applied to pharmaceuticals; spasmolytics; diuretics; cardioactive drugs; anti-atherosclerotic drugs; antivirals and antibiotics; chemistry of sugars, kojic acid, adamantane, squaric acid, polyaromatic substituted aliphatic carboxylic acids; receptors; continuing education management and design. *Mailing Add:* 723 C Shoreline Rd Barrington IL 60010

KRIMMER, EDWARD CHARLES, b Youngstown, Ohio, Dec 31, 33; m 58; c 3. PHARMACOLOGY, PSYCHOPHARMACOLOGY. *Educ:* Univ Pittsburgh, BS, 68, PhD(pharmacol), 74. *Prof Exp:* Fel pharmacol, Univ Pittsburgh, 74-76, asst prof, 76-81, ASSOC PROF PHARMACOL, UNIV

PITTSBURGH, 81- *Concurrent Pos:* Consult, ICI US, 74-75; fel, Nat Inst Drug Abuse grant, 74-76; co-investr, NIMH grant, 76-; investr, Nat Inst Drug Abuse grant, 79-83; co-investr, Nat Inst Alcohol Abuse. *Mem:* AAAS; Behav Pharmacol Soc; Am Soc Pharmacol & Exp Therapeut; Soc Stimulus Properties Drugs (secy-treas, 78-80); Soc Neurosci; Res Soc Alcoholism; Sigma Xi. *Res:* Investigate the stimulus properties of various sedatives, axiolytics, narcotics and cannabinoids and the pharmacological antagonism or enhancement of these perceived effects. *Mailing Add:* 318 Elmwood Dr Kent OH 44240

KRINER, WILLIAM ARTHUR, b Pottsville, Pa, Feb 8, 31; m 57; c 2. INORGANIC CHEMISTRY. *Educ:* West Chester State Col, BS, 53; Univ Pa, PhD(inorg chem), 59. *Prof Exp:* Res chemist, Rohm and Haas Co, Pa, 59-61; lectr chem, Univ Pa, 62-65; asst prof, 65-70, ASSOC PROF CHEM, ST JOSEPHS UNIV, PA, 70- *Mem:* Am Chem Soc; Sigma Xi. *Res:* Small ring heterocyclics of Group IV preparation and reactivity; organometallic polymers; synthesis of novel organosilicon compounds; silicon hydride chemistry; boron cyanides and siloxy aluminum compounds. *Mailing Add:* Dept Chem St Josephs Univ Philadelphia PA 19131

KRING, JAMES BURTON, b Monett, Mo, May 25, 21; m 47; c 5. ENTOMOLOGY. *Educ:* Rockhurst Col, BS, 47; Kans State Col, MS, 48, PhD(entom), 52. *Prof Exp:* Asst Instr, Kans State Col, 50; asst biol, Rockhurst Col, 50; from asst entomologist to entomologist, Conn Agr Exp Sta, 51-77; head dept, dean & dir, Col Food & Nat Resource, Agr Exp Sta & Coop Ext Serv, 79-81; prof entom, Univ Mass, 77-82; ADJ PROF ENTOM, UNIV FLA, 82- *Concurrent Pos:* Mem grad fac, Univ Conn, 59-78. *Honors & Awards:* L O Howard Award, 82. *Mem:* Entom Soc Am (pres, 79); AAAS; Royal Entom Soc London; Entom Soc Can. *Res:* Ecology and systematics of Aphididae; control pests of vegetables and ornamentals; behavior of vectors of plant diseases. *Mailing Add:* Dept Math & Sci Roanne State Comm Col Patton Lane Harriman TN 37748

KRINITZSKY, E(LLIS) L(OUIS), b Norfolk, Va, July 1, 24; m 52. ENGINEERING GEOLOGY. *Educ:* Va Polytech Inst, BS, 45; Univ NC, MS, 47; La State Univ, PhD(geol), 50. *Prof Exp:* Asst prof geol, Southwestern La Inst, 46-47; geologist, Army Corps Engrs, 48-53; sr geologist, Creole Petrol Corp, 53-61; vis prof, Univ Houston, 62-63; CHIEF GEOL RES, WATERWAYS EXP STA, ARMY CORPS ENGRS, 63- *Honors & Awards:* Richard H Jahns Distinguished Lectr Eng Geol, 91. *Mem:* Fel Geol Soc Am; Earthquake Eng Res Inst; Am Soc Civil Engrs; Int Soc Rock Mech; Seismol Soc Am; Asn Eng Geologists. *Res:* Engineering geology; earthquake hazards, x-radiography. *Mailing Add:* Waterways Exp Sta Corps Engrs PO Box 631 Vicksburg MS 39180

KRINSKY, HERMAN Y, b Hudson, NY, Aug 6, 24; m 48; c 2. CHEMICAL ENGINEERING, THERMODYNAMICS. *Educ:* Univ Del, BChE, 48; Columbia Univ, MS, 51. *Prof Exp:* From instr to assoc prof, 51-66, chmn dept, 64-74, PROF CHEM ENG, PRATT INST, 66- *Mem:* Am Chem Soc; Am Inst Chem Eng; Sigma Xi. *Res:* Thermodynamics of irreversible processes, particularly as applied to transport. *Mailing Add:* 327 Oakford St West Hempstead NY 11552

KRINSKY, NORMAN IRVING, b Iron River, Mich, June 29, 28; m 60; c 2. BIOCHEMISTRY. *Educ:* Univ Southern Calif, BA, 48, MS, 50, PhD(biochem), 53. *Prof Exp:* USPHS fel, Harvard Univ, 53-55, Nat Coun to Combat Blindness fel, 55-56, instr biol, 56-59, lectr, 59-60, from asst prof pharmacol to prof biochem, 60-70, prof biochem & pharmacol, 70-87, PROF BIOCHEM, SCH MED, TUFTS UNIV, 87- *Concurrent Pos:* Vis prof, Univ Calif, Berkeley, 73; res assoc, Boston Vet Admin Med Ctr, 81-82. *Honors & Awards:* Lotte Arnrich lectr, Iowa State Univ, 90. *Mem:* Am Chem Soc; Biophys Soc; NY Acad Sci; Am Soc Photobiol (secy-treas, 75-81, pres, 82-83); Am Soc Biochem & Molecular Biol. *Res:* Function and metabolism of carotenoids; photosensitization; mechanisms of membrane damage. *Mailing Add:* Dept Biochem Tufts Univ Sch Med Boston MA 02111-1837

KRINSKY, SAMUEL, b Brooklyn, NY, Jan 14, 45; m 72; c 2. ACCELERATOR PHYSICS. *Educ:* Mass Inst Technol, BS, 66; Yale Univ, PhD(physics), 71. *Prof Exp:* Res assoc physics, Inst Theoret Physics, State Univ NY Stony Brook, 71-73; asst physicist, 73-75, assoc physicist, 75-78, physicist, 78-85, SR PHYSICIST, BROOKHAVEN NAT LAB, 85- *Mem:* Am Phys Soc. *Res:* Particle beam dynamics in storage rings, undulators and wigglers as sources of synchrotron radiation; free electron lasers. *Mailing Add:* NSLS Brookhaven Nat Lab Bldg 725B Upton NY 11973

KRINSKY, WILLIAM LEWIS, b Brooklyn, NY, Jan 10, 47; m 70; c 2. MEDICAL FORENSIC ENTOMOLOGY, BIOSYSTEMATICS. *Educ:* Yale Univ, AB, 67, MD, 74; Cornell Univ, PhD(entom), 74. *Prof Exp:* Res assoc med entom, Rocky Mountain Lab, Nat Inst Allergy & Infectious Dis, 74-77; from asst prof to assoc prof epidemiol, Sect Med Entom, dept epidemiol & public health, 77-87, ASSOC CLIN PROF EPIDEMIOL, SCH MED, YALE UNIV, 87- *Concurrent Pos:* Fac affil entomol, Peabody Mus Natural Hist, Yale Univ, 80- *Mem:* Entom Soc Am; Coleopterist Soc; Am Entom Soc. *Res:* Forensic entomology; medical entomology, acarology and parasitology; Coleoptera systematics. *Mailing Add:* Sect Med Entom Sch Med Yale Univ PO Box 3333 New Haven CT 06510

KRINSLEY, DANIEL B, b New York, NY, Jun 22, 23. NATURAL HAZARDS. *Educ:* Brooklyn Col, AB, 44, Brown Univ, MSc, 49, Univ Md, PhD(geomorphol), 70. *Prof Exp:* Chief environ impact anal prog, US Geol Surv, Washington, DC, 49-80; CONSULT GEOLOGIST, 80- *Mem:* fel Geol Soc Am; Asn Am Geography. *Mailing Add:* 2475 Virginia Ave NW Washington DC 20037

KRINSLEY, DAVID, b Chicago, Ill, Jan 9, 27; m 58; c 3. SEDIMENTOLOGY. *Educ:* Univ Chicago, PhB, 48, SB & SM, 50, PhD(geol), 56. *Prof Exp:* Asst geol, Univ Ill, 54-55, instr, 55-56; micropaleontologist & geochemist oceanog & geochem, Lamont Geol Observ, Columbia Univ, 56-57; from instr to prof geol, Queens Col, NY, 57-76, chmn dept geol & geog, 62-65, assoc dean fac, 66-70, actg dean fac, 70; chmn dept, 76-82, PROF GEOL, ARIZ STATE UNIV, 76- *Concurrent Pos:* Grants, Am Philos Soc, Petrol Res Fund, Am Chem Soc, NASA, NSF, NATO, Dept Energy; overseas fel, Churchill Col, Cambridge Univ, 70-71. *Mem:* Fel AAAS; Soc Econ Paleont & Mineral; Geol Soc Am; Sigma Xi. *Res:* Backscattered electron microscopy of fine grained sedimentary rocks. *Mailing Add:* Dept Geol Ariz State Univ Tempe AZ 85287-1404

KRIPALANI, KISHIN J, b Karachi, W Pakistan, Oct 3, 37; m 66; c 2. DRUG METABOLISM, BIOPHARMACEUTICS. *Educ:* Univ Bombay, BSc Hons, 57, BSc(tech), 59; Univ Calif, PhD(pharmaceut chem), 66. *Prof Exp:* Staff scientist, Worcester Found Exp Biol, Shrewsbury, Mass, 67-68, NIH fel steroid biochem, 68-69; RES GROUP LEADER DRUG METAB, SQUIBB INST MED RES, E R SQUIBB & SONS, INC, 69- *Mem:* Am Chem Soc; Am Soc Pharmacol Exp Therapeut; Int Soc Study Xenogiotics; Am Asn Pharmaceut Sci. *Res:* Drug metabolism; drug-protein interactions; biotransformations, and biopharmaceutics of drugs in animal species and humans, mechanism of drug-induced drug-enzyme interactions. *Mailing Add:* Bristol Myers-Squibb PO Box 4000 Princeton NJ 08543

KRIPKE, BERNARD ROBERT, b Washington, DC, Aug 25, 39; m 79; c 1. VISION, LEARNING DISABILITIES. *Educ:* Harvard Col, AB, 59; Harvard Univ, AM, 60, PhD(math), 64. *Prof Exp:* Staff mem math, Mass Inst Technol, 62-63; asst prof, Univ Tex, Austin, 63-64 & Univ Calif, Berkeley, 64-69; vis lectr vision, Hadassah Hosp, Hebrew Univ, 69-70; vis fel biophysics, Ohio State Univ, 70-72; res instr, Univ Utah Sch Med, 72-76, asst prof physiol, 76-82; vpres, Future Software Inc, 82-83; partner, WKCG Software Develop Corp, 83-86; sr design consult, Gen Data Syst, 88-91; MGR PPC ADV SYST CORP, ELECTRONIC DATA SYST, 91- *Concurrent Pos:* Consult, Utah State Budget Off, 80-81 & Utah State Div Data Processing, 81-82. *Mem:* AAAS; Soc Neurosci. *Res:* Effects of visual deprivation on cat striate cortex; hereditary learning disability; analytic functions of several complex variables; approximation in banach spaces. *Mailing Add:* 1530 N Key Blvd 706 Arlington VA 22209

KRIPKE, DANIEL FREDERICK, b Washington, DC, Oct 12, 41; c 2. SLEEP DISORDERS, BIOLOGICAL RHYTHMS. *Educ:* Harvard Col, BA, 61; Col Physicians & Surgeons, Columbia Univ, MD, 65. *Prof Exp:* Intern, Bronx Municipal Hosp Ctr, 65-66; resident psychiat, Albert Einstein Col Med, 68-71; from asst prof to assoc prof, 71-82, PROF, DEPT PSYCHIAT, UNIV CALIF, SAN DIEGO, 82- *Concurrent Pos:* Attend physician, Dept Psychiat, Univ Hosp, Univ Calif, San Diego, 71-; staff psychiatrist, San Diego Vet Admin Med Ctr & dir, Sleep Disorders Clin, 72-, clin investr biol rhythms, 72-76, dir psychiat, Emergency Evaluation & Crisis Serv, 76-77; ed, Sleep Res, 79-80; consult, US Surgeon Gen Proj Sleep, 80. *Mem:* AAAS; Sleep Res Soc; Int Soc Chronobiol; Soc Psychophysiol Res; Am Psychiat Soc. *Mailing Add:* Dept Psychiat 116A Vet Admin Med Ctr 3350 La Jolla Village Dr San Diego CA 92161

KRIPKE, MARGARET LOUISE (COOK), b Concord, Calif, July 21, 43; m 75; c 1. CANCER. *Educ:* Univ Calif, Berkeley, AB, 65, MA, 67, PhD(immunol), 70. *Prof Exp:* Teaching asst immunol, Univ Calif, Berkeley, 65-66; res assoc, Ohio State Univ, Columbus, 70-72; res assoc, Sch Med, Univ Louisville, 72; asst prof path, Col Med, Univ Utah, 72-75; sr prin scientist cancer, Frederick Cancer Res Ctr, 75-79, dir, Cancer Biol Prog, 79-83; PROF & CHMN DEPT IMMUNOL, M D ANDERSON CANCER CTR, UNIV TEX, 83- *Concurrent Pos:* Chancellors distinguished lectr, Univ Calif, Berkeley, 80. *Honors & Awards:* Edna Roe Mem lectr, Int Cong Photo Biol, 80; Lila Gruber Hon Award in Cancer Res. *Mem:* Am Asn Cancer Res; Am Soc Photobiol; Transplantation Soc; AAAS; Soc Investigative Dermat; Am Asn Immunol. *Res:* Mechanisms of immunologic responses to tumors; relationship between the immune system and carcinogenesis; nature and significance of tumor antigens using the system of experimental ultraviolet carcinogenesis; effects of ultraviolet radiation on immunologic processes. *Mailing Add:* Dept Immunol M D Anderson Cancer Ctr Univ Tex 1515 Holcombe Blvd Box 178 Houston TX 77030

KRIPPAEHNE, WILLIAM W, surgery; deceased, see previous edition for last biography

KRIPPNER, STANLEY CURTIS, b Edgerton, Wis, Oct 4, 32; m 66; c 2. CROSS-CULTURAL STUDIES, PSYCHOLOGY OF CONSCIOUSNESS. *Educ:* Univ Wis, Madison, BS, 54; Northwestern Univ, MA, 57, PhD(educ psychol), 61. *Prof Exp:* Dir, Child Study Ctr, Kent State Univ, Ohio, 61-64 & Dream Lab, Maimonides Med Ctr, Brooklyn, 64-73; PROF PSYCHOL, SAYBROOK INST GRAD SCH, SAN FRANCISCO, 72- *Concurrent Pos:* Lectr, Acad Pedag Sci, Moscow, 71 & Acad Scis, Beijing, China, 81; vis prof, Univ PR, Santurce, 72, Sonoma State Univ, 72-73, Univ Life Scis, Bogota, 74, Inst Psychodrama, Caracas, 75 & West Ga Col, Carrollton, Ga, 76. *Honors & Awards:* Cert Recognition, US Dept Health & Human Serv, 76. *Mem:* Fel Am Psychol Asn; Asn Humanistic Psychol (pres 74-75); Parapsychol Asn (pres, 83); fel Am Soc Clin Hypn; fel Soc Sci Study Sex; Am Soc Psychical Res; fel Am Psychol Soc. *Res:* Anomalous effects in dreams; indigenous healing systems; creative problem-solving in altered states of consciousness; the effects of "personal myths" on cognition and behavior; hypnotherapy and learning. *Mailing Add:* Saybrook Inst 1550 Sutter St San Francisco CA 94109

KRISCH, ALAN DAVID, b Philadelphia, Pa, Apr 19, 39; m 61; c 1. HIGH ENERGY PHYSICS. *Educ:* Univ Pa, BA, 60; Cornell Univ, PhD(physics), 64. *Prof Exp:* Instr physics, Cornell Univ, 64; from asst prof to assoc prof, 64-68, PROF PHYSICS, UNIV MICH, ANN ARBOR, 68- *Concurrent Pos:* Guggenheim fel, 71; trustee, Argonne Univ Assoc, 72-73 & 80-83, chmn, Argonne ZGS Users Group, 74-76 & 78-79; vis prof, Niels Bohr Inst, Copenhagen, 75-76; chmn, Int Comt for High Energy Spin Physics Symposia, 77-92 & Conf on Intersect between Particle & Nuclear Physics, 83-86. *Mem:* AAAS; fel Am Phys Soc. *Res:* Experiments on high energy elastic and inelastic scattering of strongly interacting particles; experiments on spin dependence of strong interactions; phenomenology of strong interactions; acceleration of polarized beams. *Mailing Add:* Randall Lab Physics Univ Mich Ann Arbor MI 48109

KRISCH, JEAN PECK, b Washington, DC, May 23, 39; m 61; c 1. HIGH ENERGY THEORETICAL PHYSICS, ASTROPHYSICS. *Educ:* Univ Md, BS, 60; Cornell Univ, MS, 62, PhD(physics), 65. *Prof Exp:* Teaching asst physics, Cornell Univ, 60-65; res assoc, 65-75, LECTR PHYSICS, UNIV MICH, 76- *Mem:* Am Asn Physics Teachers. *Res:* High energy particle physics; general relativity. *Mailing Add:* Randall Lab Univ Mich Ann Arbor MI 48104

KRISCH, ROBERT EARLE, b Philadelphia, Pa, Jan 29, 37; m 70; c 1. BIOPHYSICS. *Educ:* Univ Pa, BA, 56, MS, 62, PhD(physics), 64; Temple Univ, MD, 60. *Prof Exp:* Instr physics, Univ Pa, 64-65; asst biophysicist, Argonne Nat Lab, 65-72, biophysicist, Div Biol & Med Res, 72-77; spec fel, Dept Radiation Ther, Harvard Med Sch, 77-80; ASSOC PROF, DEPT RADIATION ONCOL, SCH MED, UNIV PA, 80- *Mem:* Radiation Res Soc; Am Soc Ther Radiol & Oncol. *Res:* Radiation biology. *Mailing Add:* Dept Radiation Oncol Univ Pa Philadelphia PA 19104

KRISCIUNAS, KEVIN L, b Chicago, Ill, Sept 12, 53. REAL-TIME DATA ACQUISITION & ANALYSIS, ASTRONOMICAL PHOTOMETRY. *Educ:* Univ Ill, Urbana-Champaign, BS, 74; Univ Chicago, MA, 76. *Prof Exp:* Programmer & onboard operator, Kuiper Airborne Observ, Ames Res Ctr, NASA, 77-82; PROGRAMMER & PUB RELS PERSON, JOINT ASTRON CTR, 82- *Concurrent Pos:* Astron lectr, WValley Col, Saratoga, Calif, 78-82. *Mem:* Int Astron Union; Am Astron Soc; Astron Soc Pac. *Res:* Photometry of variable stars; infrared photometry and spectroscopy; astronomical site evaluation; history of astronomy. *Mailing Add:* Joint Astron Ctr 665 Komohana St Hilo HI 96720

KRISE, GEORGE MARTIN, b San Antonio, Tex, May 12, 19; m 43; c 2. PHYSIOLOGY. *Educ:* Univ Tex, BS, 46, MA, 48, PhD(zool), 52. *Prof Exp:* From instr to asst prof biol, St Edward's Univ, 49-51; res scientist physiol, Univ Tex, 52-58,; admin officer, Dept Biol, 69-74, prof, 58-82, EMER PROF PHYSIOL, TEX A&M UNIV, 82- *Res:* Microbial physiology; effects of ionizing radiations on various species. *Mailing Add:* 2301 Hillside Dr Bryan TX 77802

KRISHAN, AWTAR, b Srinagar, India, Oct 11, 37; US citizen; m 55; c 3. CANCER RESEARCH & CHEMOTHERAPY. *Educ:* Panjab Univ, India, PhD(zool), 63; Univ Western Ont, PhD(anat), 64. *Prof Exp:* Res prof cytogenetics, Univ Minn, St Paul, 65-66; cytologist, Children's Cancer Res Found, Boston, 66-71, chief cancer res, Div Exp Path & Lab Cytokinetics, 72-77; assoc prof, 77-79, PROF ONCOL, MED SCH, UNIV MIAMI, 79- *Concurrent Pos:* Chief, Div Cytokinetic, Comp Career Ctr, State Fla, 77- & sci dir, 81- *Honors & Awards:* Collip Medal, Univ Western Ont, 65. *Mem:* Am Asn Cancer Res; Cell Kinetics Soc Am; Electron Micros Soc (pres, 75-76). *Res:* Tumor cell kinetics; effect of cancer chemotheraphy on tumor growth; use of laser flow cytometry for monitoring drug uptake. *Mailing Add:* Med Sch Univ Miami R-71 PO Box 016960 Miami FL 33101

KRISHEN, ANOOP, b Ludhiana, India, Aug 7, 27; m 57; c 2. ANALYTICAL CHEMISTRY. *Educ:* Univ Panjab, India, BSc, 48, MSc, 49; Univ Pittsburgh, PhD, 57. *Prof Exp:* Lectr chem, Govt Col, Ludhiana, India, 49-50; res asst anal chem, Nat Phys Lab, India, 50-52; sr res chemist, B F Goodrich Co Res Ctr, 57-62; chief chemist, Synthetics & Chem Ltd, India, 62-63; sr res chemist, 63-83, SECT HEAD, RES DIV, GOODYEAR TIRE & RUBBER CO, 83- *Mem:* Am Chem Soc. *Res:* Pyrolysis-gas chromatography; high speed liquid chromatography; instrumental analysis; gas chromatography; laboratory robotics; laboratory information management. *Mailing Add:* Goodyear Tire & Rubber Co Res Div 142 Goodyear Blvd Akron OH 44305-0001

KRISHEN, KUMAR, b Srinagar, India, June 22, 39; US citizen; m 61; c 3. ELECTRONICS, REMOTE SENSING. *Educ:* Univ Jammu & Kashmir, BA, 59; Univ Calcutta, BTech, 62, MTech, 63; Kans State Univ, MS, 66, PhD(electronics), 68. *Prof Exp:* Res fel, Univ Calcutta, 64-65; res asst & instr elec eng, Kans State Univ, 65-68, asst prof, 68-69; staff scientist & engr earth observ, Lockheed Electronics Co, 69-76; proj mgr earth resources microwave prog, Johnson Space Ctr, NASA, 76-78, mgr advan microwave prog, 78-82, mgr advan progs, Tracking & Commun Div, 82-88, asst to dir tech & advan progs mission support, 88-90, CHIEF TECHNOLOGIST, NIO, JOHNSON SPACE CTR, NASA, 90- *Concurrent Pos:* Consult applns investr, NASA, 69-76, proj leader, Skylab Microwave Sensors Eval Team, 73-75, mem, NASA active microwave workshops, 74-77, mem agr panel earth resources, 76-77, chmn water resources panel, Microwave Remote Sensing Symp, 77; reviewer, Radio Sci, 75-; mem, Synthetic Aperture Radar Team, NASA, 76-80 & Coun Sci & Technol, 88-; lectr, Univ Houston, 77-79; adj prof, Rice Univ, 90- *Mem:* Sr mem Inst Elec & Electronics Engrs; Am Soc Eng Educ; Am Inst Aeronaut & Astronaut; Radio Physics & Electronics Asn; Sigma Xi. *Res:* Applications of microwaves to the field of remote sensing of earth resources and ocean/weather phenomena and human health; developing specifications for space borne systems for earth resources, ocean and weather sensing and robotic vision. *Mailing Add:* NASA Johnson Space Ctr Code IA Houston TX 77058

KRISHER, LAWRENCE CHARLES, b Rochester, NY, Aug 21, 33. ATOMIC & MOLECULAR PHYSICS. *Educ:* Syracuse Univ, AB, 55; Harvard Univ, AM, 57, PhD(chem), 59. *Prof Exp:* NSF fel physics, Columbia Univ, 59-61, asst prof, 61-63; from asst prof to assoc prof, 63-75, PROF PHYSICS, INST PHYS SCI & TECHNOL, UNIV MD, COLLEGE PARK, 75- *Concurrent Pos:* Owner, Stockworks Co. *Res:* Microwave spectroscopy; molecular dynamics; molecular structure. *Mailing Add:* Inst Phys Sci & Technol Univ Md College Park MD 20742

KRISHNA, C R, b Bangalore, India, May 31, 39; m 74; c 3. ENGINEERING, COMBUSTION. *Educ:* Indian Inst Sci, ME, 61; State Univ NY, Stony Brook, PhD(eng), 74. *Prof Exp:* Engr, Hindustan Aeronaut Ltd, India, 61-69; res assoc, 74-76, MECH ENGR, BROOKHAVEN NAT LAB, 76- *Mem:* Combustion Inst; Am Soc Mech Engrs. *Res:* Fluidized beds; coal-slurries and liquid fuels. *Mailing Add:* Dept Appl Sci Brookhaven Nat Lab Upton NY 11973

KRISHNA, GOLLAPUDI GOPAL, b Anakapally, India, Aug 1, 52; US citizen; m 83. NEPHROLOGY, HYPERTENSION. *Educ:* Andhra Univ, MBBS, 74. *Prof Exp:* Asst prof med, Sch Med, Univ Calif, Los Angeles, 81-83; dir dialysis, Temple Univ Hosp, Philadelphia, 84-85, attend physician, 83-90; from asst prof to assoc prof med, Sch Med, Temple Univ, Philadelphia, 83-90; ASSOC PROF MED, UNIV PA, 90- *Concurrent Pos:* Staff physician, Ctr Health Sci, Univ Calif, Los Angeles, 81-83 & Wadsworth Vet Admin Med Ctr, Los Angeles, 82-83. *Mem:* Am Soc Nephrol; Int Soc Nephrol; Am Fedn Clin Res. *Res:* Role of potassium in the pathogenesis of hypertension; protein intake and kidney function; mechanism of progression of renal disease. *Mailing Add:* Renal Electrolyte Sect Univ Pa 422 Curie Blvd Philadelphia PA 19104-6144

KRISHNA, J HARI, b Madras, India, May 13, 48; m. RESEARCH PLANNING & SUPERVISION, WATER CONSERVATION. *Educ:* Osmania Univ, BS, 67; Kans State Univ, MS, 71; Utah State Univ, PhD(irrig eng & hydrol), 79. *Prof Exp:* scientist soil & water eng, Int Crops Res Inst Semi-Arid Tropics, 72-81; consult, Food & Agr Orgn, UN, 81-82; asst prof water resources, Utah State Univ, 82-84; res scientist hydrol, Tex A&M Univ, 84-88; DIR WATER RESOURCES RES CTR, UNIV VI, 88- *Concurrent Pos:* Vpres, Am Inst Hydrol, Tex sect, 87-88; mem, tech adv group, Consortium of Caribbean Univs, 89-, bd dirs, Int Rainwater Catchment Systs Asn, 89- & Am Soc Agr Engrs Tech Comt, 88-; chairperson & co-ed, Surface Water Tech Comm Int Symp Trop Hydrol, 89-90. *Honors & Awards:* Cert Outstanding Serv, Am Water Resources Asn, 90. *Mem:* Am Water Resources Asn; Am Inst Hydrol; Am Soc Agr Engrs; Am Water Works Asn; Int Water Resources Asn; World Asn Soil & Water Conserv. *Res:* Conduct and supervise research in the areas of water quality, hydrology and water conservation; author and co-author of approximately 40 technical/scientific papers and publications. *Mailing Add:* Water Res Ctr Univ VI St Thomas VI 00802

KRISHNA, KUMAR, b Dehradun, India, June 21, 30; US citizen; m 60. ZOOLOGY, ENTOMOLOGY. *Educ:* Agra Univ, BS, 50; Univ Lucknow, MS, 52; Univ Chicago, PhD(zool), 61. *Prof Exp:* Res asst, Forest Res Inst, India, 52-54; teaching asst biol, Univ Ill at Chicago Circle, 58-60; res assoc, Univ Chicago, 60-62; from instr to assoc prof, 62-74, PROF BIOL, CITY COL NEW YORK, 74- *Concurrent Pos:* Res assoc, Am Mus Natural Hist, 62-; NSF res grant, 62- *Mem:* Am Soc Zoologists; Soc Syst Zoologists; Int Union Study Soc Insects. *Res:* Taxonomy, ecology, zoogeography and evolution of termites; general evolutionary theory. *Mailing Add:* Dept Biol City Col New York 139th St at Convent Ave New York NY 10031

KRISHNA, N RAMA, b 45. BIOPHYSICAL CHEMISTRY. *Educ:* India Inst Technol, Kampur, PhD(physics), 72. *Prof Exp:* ASSOC PROF BIOCHEM & DIR, NUCLEAR MAGNETIC RESONANCE FACIL CANCER CTR, UNIV ALA, BIRMINGHAM, 76- *Concurrent Pos:* Leukemia Soc of Am Scholar, 82-87. *Mem:* AAAS; Biophys Soc; Soc Biol Chemists. *Res:* Nuclear magnetic resonance; biomolecular conformations. *Mailing Add:* Dept Biochem Univ Ala Birmingham AL 35294

KRISHNAIAH, PARUCHURI RAMA, statistics; deceased, see previous edition for last biography

KRISHNAMOORTHY, GOVINDARAJALU, b Tanjore, India, Jan 1, 31; m 56; c 1. STRUCTURAL MECHANICS, CIVIL ENGINEERING. *Educ:* Col Eng, Guindy, India, BSCE, 52; Ill Inst Technol, MSCE, 60, PhD(struct), 65. *Prof Exp:* Jr engr, Madras Hwy Dept, India, 52-57; from instr to asst prof civil eng, Ill Inst Technol, 61-68; from asst prof to assoc prof, 68-74, PROF CIVIL ENG, SCH ENG, SAN DIEGO STATE UNIV, 74- *Concurrent Pos:* Consult, IIT Res Inst, 67-, Rohr Corp, 69-70 & SAI, La Jolla. *Mem:* Am Soc Civil Engrs; Am Soc Eng Educ. *Res:* Buckling of shells; computer applications in structures; analysis and design of ocean structures; reinforced concrete masonry structures; dynamic and thermal response of mountings on main cooling pipe of nuclear reactors; computer graphics; computer aided design. *Mailing Add:* Dept Civil Eng San Diego State Univ San Diego CA 92182

KRISHNAMURTHY, LAKSHMINARAYANAN, b Kumbakonam, Madras, India, Oct 23, 41; US citizen; m. COMBUSTION THEORY, COMPUTATIONAL FLUID DYNAMICS. *Educ:* Univ Madras, BEng, 62; Indian Inst Sci, MEng, 64; Univ Calif, San Diego, PhD(eng sci), 72. *Prof Exp:* Scientist power eng, Cent Mech Eng Res Inst, India, 65-67; res engr, Univ Calif, San Diego, 72-73; staff scientist propulsion, Duvvuri Res Assocs, Chula Vista, Calif, 73-74; res staff mem aerospace eng, Princeton Univ, 75-76; res assoc propulsion, Purdue Univ, 77-78; res engr, 78-82, SR RES ENG FLUID MECH, UNIV DAYTON RES INST, 82- *Concurrent Pos:* Sr res fel, Cent Mech Eng Res Inst, Durgapur, India, 64-65; res fel, Univ Calif, San Diego, 67-68, res assist, 68-71; reviewer, Appl Mech Revs, 73-84 & Am Inst Aeronaut & Astronaut J, 82-; prin investr, Res Inst, Univ Dayton, 78-, instr, Sch Eng, 81-; mem propellants & combustion tech comt, Am Inst Aeronaut & Astronaut, 84-86. *Mem:* Am Inst Aeronaut & Astronaut; Am Soc Mech Engrs; Am Acad Mech; Combustion Inst; Sigma Xi; Soc Indust & Appl Math; Planetary Soc; Union Concerned Scientists. *Res:* Fluid mechanics and combustion; analytical and computational fluid dynamics of nonreacting and reacting gas flows. *Mailing Add:* Res Inst Univ Dayton KL 461 Dayton OH 45469-0001

KRISHNAMURTHY, RAMANATHAPUR GUNDACHAR, b Mysore, India, May 8, 31; m 60; c 3. FOOD CHEMISTRY, BIOCHEMISTRY. *Educ:* Univ Mysore, BS, 51; Rutgers Univ, MS, 64, PhD(food sci), 65. *Prof Exp:* Lab asst metall, Indian Inst Sci, India, 51-54; sci asst food technol, Cent Food Technol Res Inst, Mysore, 54-61; res asst food sci, 61-63, res fel, 63-65, asst res prof, Rutgers Univ, 65-66; res chemist, Best Foods Div, Corn Prod Co, 66-67; from group leader, to sr group leader edible oil prod, 67-87, TECHNOL MGR, RES & DEVELOP DIV, KRAFT, INC , 87- *Mem:* Am Chem Soc; Am Oil Chemists Soc; Inst Food Technol. *Res:* Autoxidation and thermal oxidation of fats and oils; investigation of flavors and flavor percursors in foods; chemistry and technology of oils and fats and products derived from them. *Mailing Add:* Kraft Inc 801 Waukegan Rd Glenview IL 60025

KRISHNAMURTHY, SUBRAMANIAN, statistics, condensed matter, for more information see previous edition

KRISHNAMURTHY, SUNDARAM, b Coimbatore, Madras, India, Nov 26, 44; m 75. ORGANIC CHEMISTRY. *Educ:* Univ Madras, BSc, 64, MSc, 66; Purdue Univ, PhD(chem), 71. *Prof Exp:* Sr res assoc chem, Purdue Univ, West Lafayette, 71-80; MEM RES STAFF, RES LAB, EASTMAN KODAK CO, 80- *Mem:* Am Chem Soc. *Res:* Synthesis and application of trialkylborohydrides in stereospecific and reguospecific organic synthesis; selective reductions; organometallics in organic synthesis. *Mailing Add:* Eastman Kodak Co Res Lab Bldg 82 Kodak Park Rochester NY 14650

KRISHNAMURTI, CUDDALORE RAJAGOPAL, b Cuddalore, India, Apr 16, 29; m 53; c 2. ANIMAL PHYSIOLOGY, ANIMAL BIOCHEMISTRY. *Educ:* Univ Madras, BVSc, 51, MVSc, 61; Univ Alta, PhD(animal nutrit), 66. *Prof Exp:* Asst lectr physiol, parasitol & bact, Madras Vet Col, 55-59; bacteriologist, Inst Vet Prev Med, Ranipet, 62; res assoc animal nutrit, Univ Alta, 66-67; assoc prof, 67-75, PROF & ACT HEAD, DEPT ANIMAL SCI, UNIV BC, 75- *Mem:* Agr Inst Can; Nutrit Soc Can; Can Soc Animal Sci. *Res:* Biochemical investigations on rumen microorganisms; fetal physiology; digestion, absorption and metabolism of nutrients by the ruminant animal, especially metabolic disorders. *Mailing Add:* Dept Animal Sci Univ BC 2075 Wesbrook Pl Vancouver BC V6T 1W5 Can

KRISHNAMURTI, PULLABHOTLA V, b Gudivada, India, Mar 1, 23; m 49; c 5. VETERINARY MICROBIOLOGY. *Educ:* Univ Madras, BVSc, 49, DVP, 58; Univ Wis-Madison, MS, 61; Tex A&M Univ, PhD(vet microbiol), 67. *Prof Exp:* State vet, Andhra Vet Serv, India, 48-54; instr vet sci & exten vet, Exten Training Ctr, 55-56; asst lectr, Andhra Vet Col, 57-58; res asst, Univ Wis-Madison, 59-63; researcher poultry dis, Hy-line Poultry Farms, Iowa, 64-65; res asst vet microbiol, Tex A&M Univ, 65-66; from asst prof to assoc prof, sch vet med, Tuskegee Inst, 66-69; assoc prof microbiol, 69-74, PROF BIOL, SAVANNAH STATE COL, 74- *Mem:* Am Vet Med Asn; Am Soc Parasitol; Poultry Sci Asn; Am Asn Avian Path. *Res:* Parasites and parasitism; plasmodium in Wisconsin chickens; cultivation of Histomonas meleagridis free of bacteria and its demonstration in tissues and cell cultures using fluorescent labeled antibody techniques; therapeutic agents in canine distemper. *Mailing Add:* Dept Biol Savannah State Col State Col Branch Savannah GA 31404

KRISHNAMURTI, RUBY EBISUZAKI, b Haney, BC, Oct 23, 34; m 60. PHYSICS, FLUID MECHANICS. *Educ:* Univ Western Ont, BSc, 57; Univ Chicago, MS, 60; Univ Calif, Los Angeles, PhD(physics), 67. *Prof Exp:* Res assoc fluid mech, Stanford Univ, 67; assoc prof oceanog, 71-75, SR RES ASSOC, GEOPHYS FLUID DYNAMICS INST, FLA STATE UNIV, 67-, PROF OCEANOG, 75- *Concurrent Pos:* Asst prof oceanog, Fla State Univ, 68-71. *Mem:* Am Phys Soc; Am Meteorol Soc; Am Geophys Union. *Res:* Geophysical fluid dynamics, particularly theoretical and experimental studies of convection and ocean circulation modelling. *Mailing Add:* Dept Oceanog Fla State Univ Tallahassee FL 32306

KRISHNAMURTI, TIRUVALAM N, b Madras, India, Jan 10, 32; nat US. METEOROLOGY, ATMOSPHERIC SCIENCE. *Educ:* Univ Delhi, BS, 51; Andhra Univ, India, MS, 53; Univ Chicago, PhD(meteorol), 59. *Prof Exp:* Prof meteorol, Univ Calif, Los Angeles, 60-67; assoc prof, 67-70, PROF METEOROL, FLA STATE UNIV, 70- *Concurrent Pos:* Mem, Global Atmospheric Res Prog working group on struct of the trop atmosphere, 69-71; mem, Adv Panel to Nat Oceanic & Atmospheric Admin on Nuclear Metal Conf, 71-73; consult synoptic subprog, GATE, Nat Acad Sci, 73-74; external examr, Univ Nairobi, Kenya, 73-76 & McGill Univ PhD students, 74-77; consult, MONEX Comt, World Meteorol Orgn, 75-76; vis lectr, Ctr Theoret Physics, Trieste, Italy, 75; mem, US Global Atmospheric Res Prog, 75-; assoc ed, J Atmospheric Sci, 75-; chmn, US MONEX Panel, Nat Acad Sci, 76-; mem, Working Group on Numerical Experimentation Global Atmospheric Res Prog, Joint Organizing Comt; mem, Comt Atmospheric Sci, Nat Acad Sci, 79- *Honors & Awards:* Half-Century Award, Am Meteorol Soc, 74, Rossby Award, 85; Creativity Award, NSF, 82. *Mem:* Fel Am Meteorol Soc; Am Geophys Union; fel Royal Meteorol Soc; Meteorol Soc Japan. *Res:* Dynamic and synoptic meteorology, including diagnostic and prognostic studies of tropical and mid-latitude systems using real input data together with analyses of tropical weather systems using satellite and aircraft information in sparse conventional data areas. *Mailing Add:* Dept Meteorol Fla State Univ Tallahassee FL 32306

KRISHNAN, ENGIL KOLAJ, b Bangalore, India, 42; US Citizen. RADIATION ONCOLOGY, TUMOR IMMUNOLOGY. *Educ:* Univ Mysore, BSc(physics) & (math), 64, Univ Mo, MS, 70, Cetec Univ, WI, MD, 82. *Prof Exp:* Supvr electronics, Elec Radar Develop Eng, 65-67; teaching assoc, 71-75, instr oncol, 75-82, ASST PROF SURG & DIR RES, UNIV KANS MED CTR, 82- *Concurrent Pos:* Consult, VA Med Ctr, 82- *Mem:* Am Asn Cancer Res; Am Asn Physicians Med; Am Asn Pediat Oncol; Soc Clin Trials. *Res:* Published approximately fifty journal articles in the areas of biophysics, immunology and oncology. *Mailing Add:* Dept Radiation Therap Surg & Res Univ Kansas Med Ctr Col Health Sci Rainbow Blvd & 39th Kansas City KS 66103

KRISHNAN, GOPAL, b Kancheepuram, India, Mar 15, 35; US citizen; m 67; c 2. CANCER IMMUNOLOGY, TRANSPLANTATION IMMUNOLOGY. *Educ:* Annamala; Univ, India, MA, 55, MSc, 56; Univ Madras, PhD(chem), 65. *Prof Exp:* res assoc immunol, Univ Pa, 67-74; asst prof biochem & res instr med, Jefferson Med Col, Philadelphia, 74-76; res assoc, Pa Col Podiatric Med, 76-77; immunologist, St Vincent Med Ctr, 77-82; DIR, IMMUNOL RES SECT, MERCY CATH MED CTR, 82-; DIR, TISSUE-TYPING LAB, OUR LADY OF LOURDES MED CTR, 84- *Concurrent Pos:* Lectr, Univ Madras Cols, India, 56-61. *Mem:* Am Asn Med Lab Immunologists; Am Soc Transplant Physicians; Am Asn Immunologists; NY Acad Sci. *Res:* Human cancer; transplantation immunology; transplant rejections and factors affecting the allograft survival. *Mailing Add:* Our Lady of Lourdes Med Ctr 1565 Haddon Ave Camden NJ 08103

KRISHNAN, K RANGA RAMA, b Madras, India, April 22, 56; m 87. AFFECTIVE DISORDERS, CHRONIC PAIN. *Educ:* Univ Madras, MBBS, 78. *Prof Exp:* Sr house officer, dept emergency med, Queen Elizabeth Hosp, Univ WI, Barbados, 80-81; fel neurobiol, 82-84, asst prof, div biol psychiat, 85-90, MED DIR, AFFECTIVE DISORDERS UNIT, DUKE UNIV MED CTR, DURHAM, NC, 85-, ASSOC PROF, 90- *Concurrent Pos:* Resident psychiat, John Umstead prog, 81-83, chief resident, Duke Umstead prog, 82-83; Rafaelsen fel, Cong Int Col Neuropsychopharmacologicum. *Honors & Awards:* Laughlin Award, Am Col Psychiat, 84. *Mem:* Int Asn Study of Pain; Soc Biol Psychiat; AAAS; NY Acad Sci; Int Soc Psychoneuroendocrinol. *Res:* Understanding the physiology & pathophysiology of the hypothalamo pituitary adrenal function in affective disorders; drug trials for the treatment of affective disorders; brain imaging, Magnetic Resonance Imaging and PET in affective disorders. *Mailing Add:* Duke Univ Med Ctr PO Box 3215 Durham NC 27710

KRISHNAN, KAMALA SIVASUBRAMANIAM, b Tiruchirappalli, India, Nov 12, 37; US citizen; m 58; c 3. OPTICAL PHYSICS, SOLID STATE SCIENCE. *Educ:* Univ Madras, MA, 57; Indian Inst Sci, Bangalore, DIISc, 60; Univ Fla, PhD(physics), 66. *Prof Exp:* Res scientist solid state physics, Res Div, Am Standard Inc, 65-68; sr res physicist, Stanford Res Inst, 68-77; mgr electro-optics, Systs Control Technol, Inc, 77-82; mgr optical signal processing, Litton Appl Technol, 82-85; SR STAFF ENGR, LOCKHEED MISSILES & SPACE CO, 85- *Concurrent Pos:* Vis prof mat sci, San Jose State Univ, 74-84. *Mem:* Am Phys Soc; Optical Soc Am; Sigma Xi; Soc Photo-Optical Instrumentation Engrs. *Res:* Optical and infrared systems; laser physics and applications; nonlinear optics; applied statistics; remote sensing; oceanography. *Mailing Add:* 180 Walter Hays Dr Palo Alto CA 94303

KRISHNAN, VENKATANAMA, b Madras, India, Oct 20, 29. ELECTRICAL & SYSTEMS ENGINEERING. *Educ:* Univ Madras, BSc, 48; Banares Hindu Univ, BSc, 53; Princeton Univ, MSE, 59; Univ Pa, PhD(elec eng), 63. *Prof Exp:* Instr chem, Loyola Col, Madras Univ, 48-49; sr res asst elec eng, Indian Inst Sci Bangalore, 53-56; instr, Princeton Univ, 57-58; asst prof, Villanova Univ, 58-61; assoc, Moore Sch Elec Eng, Pa, 61-64; asst prof, Polytech Inst Brooklyn, 64-66, assoc prof, 66-76; MEM FAC, DEPT ELEC ENG, SAN FRANCISCO STATE UNIV, 76- *Mem:* Inst Elec & Electronics Engrs. *Res:* Inertial navigation systems; control theory; electronic circuits and solid state physics. *Mailing Add:* Div Eng San Francisco State Univ 1600 Holloway Ave San Francisco CA 94132

KRISHNAN, VIAKALATHUR SANKRITHI, mathematics, for more information see previous edition

KRISHNAPPA, GOVINDAPPA, b Mysore, India, June 2, 36; Can citizen; m 64; c 2. SOUND INTENSITY MEASUREMENTS, AEROACOUSTICS. *Educ:* Univ Mysore, India, BEng, 58; Indian Inst Sci, MSc, 64; Univ Waterloo, Can, PhD(aeroacoust), 67. *Prof Exp:* Asst res officer aeroacoust, 67-73, assoc res officer aeroacoust & mach noise, 73-81, SR RES OFFICER MACHINERY NOISE, NAT RES COUN CAN, 81-, GROUP LEADER NOISE & VIBRATION, 85- *Concurrent Pos:* Mem & secy, Assoc Comt Mach Noise, Can Nat Comt, 81-; mem working group sound intensity, Int Standards Orgn & Int Elctrotech Comn, 84- *Mem:* Am Nat Standards Inst; fel Acoust Soc Am; Inst Noise Control Eng; Int Standards Orgn. *Res:* Machinery noise and vibration; developing techniques and methods of noise and vibration analysis and reduction; sound intensity measurements; structural noise reduction; groundborne vibrations; wind turbine noise. *Mailing Add:* Mach & Engine Tech Prog M-7 Nat Res Coun Can Ottawa ON K1A 0R6 Can

KRISHNAPPAN, BOMMANNA GOUNDER, b Madras, India, Jan 15, 43; Can citizen; m 72; c 2. HYDRAULICS, FLUID MECHANICS. *Educ:* Madras Univ, BE, 66; Univ Calgary, MSc, 68; Queen's Univ, Ont, PhD(civil eng), 72. *Prof Exp:* Res scientist, Can Ctr Inland Waters, 72-77; flow syst engr, Ont Hydro, 77-78; RES SCIENTIST HYDRAUL, NAT WATER RES INST, CAN CTR INLAND WATERS, ONT, 78- *Concurrent Pos:* Asst, Nat Res Coun Can, 66-72. *Mem:* Int Asn Hydraul Res. *Res:* Sediment transport in open channel flows; dispersion of mass in open channels; mathematical models for river morphology; thermal models. *Mailing Add:* 204 Lynbrook Dr Hamilton ON L9C 2L3 Can

KRISHNAPRASAD, PERINKULAM S, b Bombay, India, May 15, 49; m. SYSTEM THEORY. *Educ:* Indian Inst Technol, BTech, 72; Syracuse Univ, MS, 73; Harvard Univ, PhD(eng), 77. *Prof Exp:* Asst prof systs eng, Case Inst Technol, 77-80; from asst prof to assoc prof, 80-87, PROF ELEC ENG, UNIV MD, 87- *Mem:* Fel Inst Elec & ElectronicS Engrs; Am Math Soc; Am Inst Aeronaut & Astronaut. *Res:* System theory and applications to modeling; geometric methods applied to problems in systems; control theory and nonlinear mechanics; robotics; real-time control. *Mailing Add:* Dept Elec Eng Univ Md College Park MD 20742

KRISHNASWAMY, S V, b Villianallur, India, April 15, 40; US citizen; m 71; c 2. THIN FILM DEPOSITION, MICROWAVE ACOUSTICS. *Educ:* N Wadia Col, BSc, 59; Univ Poona, MSc, 61; Indian Inst Technol, MTech, 66; Pa State Univ, PhD(physics), 74. *Prof Exp:* Res assoc, Pa State Univ, 74-76, asst prof physics, 76-78, sr res assoc, 78-81; FEL SCIENTIST, WESTINGHOUSE RES & DEVELOP CTR, 81- *Concurrent Pos:* Sr scientist, Armament Res & Develop, Kirkee, India, 62-64; res assoc, Indian Inst Technol, Bombay, 64-68; mem prog comt, Thin Film Div, Am Vacuum Soc, 88-, Ultrasonics Div, Inst Elec & Electronics Engrs, 88- *Mem:* Am Phys Soc; Am Vacuum Soc; Indian Vacuum Soc. *Res:* Preparation and characterization of thin films of wide range of materials, using ion beam deposition, magnetron sputtering and other novel techniques; microwave acoustics and magnetics; high temperature super conducting applications. *Mailing Add:* Westinghouse Res & Develop Ctr MS 501-2C13 1310 Beulah Rd Pittsburgh PA 15235

KRISHTALKA, LEONARD, b Montreal, Can, Jan 30, 46; m 86; c 1. MAMMALIAN PALEONTOLOGY. *Educ:* Univ Alta, BSc, 69, MSc, 71; Tex Tech Univ, Lubbock, PhD(biol & vert paleont), 75. *Prof Exp:* Fel, Carnegie Mus Natural Hist, 75-76, res fel, 76-77, asst cur, 77-80, assoc cur, 80-87, CUR, ASST DIR SCI, CARNEGIE MUS NATURAL HIST, 87- *Concurrent Pos:* Adj lectr, Univ Pittsburgh, 76-77, adj asst prof, 77-80, adj assoc prof, 80-; ed, Sci Publs, 86- *Mem:* Soc Vert Paleont; AAAS; Paleont Soc. *Res:* Origin, evolution, relationships, paleocology and systematics of early Tertiary and Mesozoic mammals, especially primates, artiodactyls, insectivores and multituberculates; African Neogene hominids and microfaunal paleontology. *Mailing Add:* Sect Vert Fossils Carnegie Mus Carnegie Mus Natural Hist 4400 Forbes Ave Pittsburgh PA 15213

KRISS, JOSEPH P, b Philadelphia, Pa, May 15, 19; m 48; c 3. INTERNAL MEDICINE, NUCLEAR MEDICINE. *Educ:* Pa State Col, BS, 39; Yale Univ, MD, 43. *Prof Exp:* Asst med, Sch Med, Yale Univ, 43-44, instr, 44-45; res fel metab, Sch Med, Univ Wash, 46-48; asst med, 48-50, clin instr, 50-52, asst clin prof, 52-56, assoc clin prof, 56-57, assoc prof, 57-62, PROF MED & RADIOL, SCH MED, STANFORD UNIV, 62- *Concurrent Pos:* Intern, New Haven Hosp, Conn, 43-44, asst resident, 44, assoc resident, 44-45, resident, 45; consult, Vet Admin Hosp, San Francisco, 49-59 & Palo Alto, 59-; USPHS spec fel biochem, Stanford Univ, 66-67; Kaiser fel, Ctr Advan Study Behav Sci, 79-80. *Honors & Awards:* Henry J Kaiser Award, 75. *Mem:* Endocrine Soc; Am Soc Nuclear Med; Am Fedn Clin Res; Am Thyroid Asn; Asn Am Physicians. *Res:* Nuclear medicine; endocrinology; thyroid disorders; nuclear cardiology. *Mailing Add:* Dept Radiol Stanford Univ Sch Med Palo Alto CA 94305

KRISS, MICHAEL ALLEN, b San Diego, Calif, Dec 14, 40; m 63; c 3. PHYSICS, COMPUTER SCIENCES. *Educ:* Univ Calif, Los Angeles, AB, 62, MS, 64, PhD(physics), 69. *Prof Exp:* res assoc, Color Photog Div, Eastman Kodak, 69-79, mgr, Physics & Comput Sci Dept, Kodak, Japan, 85-88, RES ASSOC PHYSICS DIV, RES LAB, EASTMAN KODAK, 79-, MGR, EXTERNAL RES PROG, ELECTRONIC IMAGING RES LABS, 88- *Concurrent Pos:* Lectr, Univ Col, Univ Rochester, 76- *Mem:* Soc Photog Scientist & Engrs; Sigma Xi; Soc Motion Picture & TV; NY Acad Sci. *Res:* Photographic sciences; photographic research with emphasis on the mechanisms of color reproduction and image structure photographic film systems; development of methods to measure and evaluate the color reproduction and image structure of photographic and non-photographic systems; image processing by use of computers; electronic imaging systems. *Mailing Add:* Eastman Kodak Co Bldg 80 Kodak Park Rochester NY 14650-2039

KRISST, RAYMOND JOHN, b Lithuania, USSR, May 29, 37; m 62; c 3. INDUSTRIAL & MANUFACTURING ENGINEERING, OPERATIONS RESEARCH. *Educ:* Univ Conn, AB, 58; Mich State Univ, PhD(physics), 65. *Prof Exp:* Res asst physics, Mich State Univ, 62-65; res fel, Harvard Univ & res affil, Mass Inst Technol, 65-70; vis scientist, German Electron-Synchrotron Inst, 70-74; prin physicist, Nuclear Power Dept, Combustion Eng Inc, 74- 85; DIR ENG APPL CTR, UNIV HARTFORD, 85- *Concurrent Pos:* Instr, Nuclear Weapons Effects, US Army, a59 & nuclear physics, Mich State Univ, 60-63; adj fac atomic physics, Harvard Univ, 67-70; adj prof nuclear eng & numerical math, Univ Conn, 83-84, adj prof mech eng & elec eng, Univ Hartford, 85-; consult, Hartford Area Indust, 84-; bd adv, Corp Strategies Mag, 86- *Mem:* Am Mgt Asn; Am Phys Soc; Am Nuclear Soc; Am Asn Physics Teachers; Sigma Xi. *Res:* Nuclear, high-energy and accelerator physics; nuclear engineering; polarized electron beams; nuclear instrumentation. *Mailing Add:* Eng Appl Ctr Univ Hartford United Technol Hall Hartford CT 06117

KRISTA, LAVERNE MATHEW, b Webster, SDak, Dec 24, 31; m 64; c 3. VETERINARY ANATOMY. *Educ:* SDak State Univ, BS, 58, MS, 60; Univ Minn, PhD(poultry sci), 66, DVM, 69. *Prof Exp:* Res asst, SDak State Univ, 58-60; res asst poultry nutrit & phys, Univ Minn, 60-69; asst prof, 69-73, assoc prof, 73-81, PROF VET ANAT & HISTOL, AUBURN UNIV, 81- *Mem:* Sigma Xi; Am Asn Vet Anat; World Poultry Sci; Am Vet Med Asn; Sigma Xi. *Res:* Nutrition; atherosclerosis; cardiovascular physiology. *Mailing Add:* Vet Anat-Hist Vet Med Auburn Univ Auburn AL 36830

KRISTAL, MARK BENNETT, b New York, NY, Apr 19, 44; m 67; c 1. BIOPSYCHOLOGY, BEHAVIOR-ETHOLOGY. *Educ:* Rutgers Univ, BA, 65; Kans State Univ, MS, 70, PhD(psychol), 71. *Prof Exp:* Trainee behav genetics & neuroendocrinol, Jackson Lab, 71-73; from asst prof to assoc prof, 73-91, PROF PSYCHOL, STATE UNIV NY, BUFFALO, 91- *Concurrent Pos:* Dir, Biopsychol Prog, Psychol Dept, State Univ NY, Buffalo, 78-86; interim assoc dean, SUNY Buffalo Sch Health Related Professions, 86-88; NSF, NIMH, Nat Inst Drug Abuse grants; assoc dean, Suny Buffalo Fac Soc Sci, 89- *Mem:* AAAS; Int Soc Develop Psychobiol; Soc Neurosci; Animal Behav Soc. *Res:* Neural, endocrine, and genetic bases of maternal, ingestive and sexual behaviors; functions of the hypothalamus; limbic-hypothalamic function interactions; opiates and analgesia; endocrinology; pharmacology. *Mailing Add:* State Univ NY Dept Psychol Buffalo NY 14260

KRISTIAN, JEROME, b Milwaukee, Wis, June 5, 34; m 55; c 1. PHYSICS, ASTRONOMY. *Educ:* Univ Chicago, MS, 56, PhD(physics), 62. *Prof Exp:* Mem staff, Argonne Nat Lab, 57-59; vis lectr physics & math, Univ Tex, 62-64; asst prof astron, Univ Wis, 64-67; MEM STAFF, MT WILSON & LAS CAMPANAS OBSERV, CARNEGIE INST WASH, 67- *Mem:* Am Phys Soc; Am Astron Soc; Int Astron Union. *Res:* Cosmology; extra-galactic astronomy. *Mailing Add:* 813 Santa Barbara St Pasadena CA 91101

KRISTIANSEN, MAGNE, b Elverum, Norway, Apr 14, 32; nat US; m 57; c 2. PULSED POWER TECHNOLOGY. *Educ:* Univ Tex, BS, 61, PhD(elec eng), 67. *Prof Exp:* Res engr, Univ Tex, 63-66; from asst prof to prof elec eng, 66-77, P W HORN PROF ELEC ENG, 77-, TEX TECH UNIV, C B THORNTON PROF ELEC ENG, 90- *Concurrent Pos:* NSF grants, 67-86 & US Atomic Energy Comn grant, 67-72; US Army & Air Force grants, 69-; consult to various industs; vis staff mem, Los Alamos Nat Lab & Lawrence Livermore Nat Lab; contractor, Sandia Labs, Lawrence Livermore Nat Lab, Los Alamos Nat Lab, Defense Nuclear Agency, Strategic Defense Initiative Off & US Navy, 77-91; sci adv bd, US Air Force,81-85; assoc ed, Trans Plasma Sci, Inst Elec & Electronics Engrs, 79- *Honors & Awards:* Nuclear & Plasma Sci Soc Merit Award, Inst Elec & Electronic Engrs, 91; Peter Haas Award, 87. *Mem:* AAAS; fel Inst Elec & Electronic Engrs; fel Am Phys Soc; Am Soc Eng Educ; Nat Soc Prof Eng. *Res:* Plasma dynamics, pulsed power technology and physical electronics; high power switching and radio frequency wave propagation and technology; high power microwaves; author of 200 publications, co-author of one book, editor of series of books. *Mailing Add:* Dept Elec Eng MS 3102 Tex Tech Univ Lubbock TX 79409-4439

KRISTMANSON, DANIEL D, b Vancouver, BC, Oct 10, 29; m 56; c 2. CHEMICAL ENGINEERING. *Educ:* Univ BC, BASc, 53; Univ London, PhD(chem eng), 60. *Prof Exp:* Develop engr, Consolidated Mining & Smelting Co Can, Ltd, 53-56, fertilizer mfg, 60-62; from asst prof to assoc prof, 62-68, PROF CHEM ENG, UNIV NB, 68- *Concurrent Pos:* Sr chem engr, Monenco Consult, 84-85. *Mem:* Chem Inst Can; Air Pollution Control Asn. *Res:* Mixing in reactors and natural streams; atomization of liquids; aerial spraying. *Mailing Add:* Dept Chem Eng Univ NB Col Hill Box 4400 Fredericton NB E3B 5A3 Can

KRISTOFFERSEN, THORVALD, b Denmark, May 6, 19; nat US; m 48; c 1. FOOD SCIENCE. *Educ:* Royal Vet & Agr Col, Denmark, BS, 44; Iowa State Univ, MS, 48, PhD(dairy bact), 54. *Prof Exp:* Asst milk & milk prod, Govt Res Inst Denmark, 44-46; lab asst cheese, Iowa State Univ, 46-54; res assoc sanitizers, Ohio State Univ, 55-56, from asst prof to prof dairy technol, 56-84, chmn dept, 72-84; RETIRED. *Honors & Awards:* Pfizer-Paul Lewis Award Cheese Res, 65. *Mem:* Fel AAAS; Am Soc Microbiol; Am Dairy Sci Asn; Inst Food Technol. *Res:* Mechanism of flavor development in cheese; enzyme system of milk, its function and purpose; butter and its physical structure; dairy sanitizers; analysis and evaluation of dairy products. *Mailing Add:* 1433 Clubview Blvd S Worthington OH 43085

KRISTOL, DAVID SOL, b Brooklyn, NY, June 4, 38; m 75; c 2. BIOENGINEERING. *Educ:* Brooklyn Col, BS, 58; New York Univ, PhD(chem), 69. *Prof Exp:* Res asst, Jewish Hosp Brooklyn, 58-62; from asst prof to assoc prof, 66-81, PROF CHEM, NJ INST TECHNOL, 81-, DIR, BIOMED ENG PROG, 82- *Mem:* Am Chem Soc; Sigma Xi; AAAS; NY Acad Sci. *Res:* Effect of structure upon reactivity of organic molecules; reaction of amines with chlorocarbons; preparation of novel dentin bonding agents; effect of hormones upon leukocytes and the microcirculation. *Mailing Add:* Chem Div NJ Inst Technol 323 King Blvd Newark NJ 07102

KRITCHEVSKY, DAVID, b Kharkov, Russia, Jan 25, 20; nat US; m 47; c 3. LIPID METABOLISM, NUTRITION. *Educ:* Univ Chicago, BS, 39, MS, 42; Northwestern Univ, PhD(chem), 48. *Prof Exp:* Jr chemist, Ninol Labs, 41-42, chemist, 42-46; asst & quiz instr, Northwestern Univ, 46-48; Am Cancer Soc fel, Swiss Fed Inst Technol, 48-50; mem staff, Radiation Lab, Univ Calif, 50-52; res chemist, Lederle Labs Div, Am Cyanamid Co, 52-57; asst prof, Sch Med, 57-65, prof, Sch Vet Med, 65-67, Wistar prof biochem, 67-72, chmn grad group molecular biol, 71-84, MEM, WISTAR INST, UNIV PA, 57-, PROF BIOCHEM IN SURG, SCH VET MED, 72-, ASSOC DIR, 75- *Honors & Awards:* Borden Award, Am Inst Nutrit, 74; Am Col Nutrit Award, 78. *Mem:* AAAS; Soc Exp Biol & Med; Am Soc Biol Chem; Am Chem Soc; Sigma Xi. *Res:* Synthesis and metabolism of compounds labeled with isotopic carbon and hydrogen; experimental atherosclerosis; organic synthesis; steroids; biology of deuterium oxide; lipid metabolism; nutrition and cancer. *Mailing Add:* Wistar Inst 36th & Spruce Sts Philadelphia PA 19104-4268

KRITIKOS, HARALAMBOS N, b Tripolis, Greece, Mar 8, 33; US citizen; m 64. ELECTRICAL ENGINEERING. *Educ:* Worcester Polytech Inst, BS, 54, MS, 56; Univ Pa, PhD(elec eng), 61. *Prof Exp:* Res asst elec eng, Worcester Polytech Inst, 54-56; from asst instr to assoc prof, 56-76, PROF ELEC ENG, MOORE SCH ELEC ENG, UNIV PA, 76- *Concurrent Pos:* Res fel, Calif Inst Technol, 66; exec ed, Inst Elec & Electronics Engrs Trans of Geosci Electronics, 76-80. *Mem:* Inst Elec & Electronics Engrs. *Res:* Diffraction theory; antennas; propagation; microwave hazards; remote sensing; electromagnetic field theory. *Mailing Add:* Moore Sch Elec Eng Univ Pa 200 S 33rd St Philadelphia PA 19104

KRITSKY, GENE RALPH, b Minot, NDak, June 26, 53. ETHNOENTOMOLOGY, CICADA EVOLUTION. *Educ:* Indiana Univ, AB, 74; Univ Ill, MS, 76, PhD(entom), 77. *Prof Exp:* From asst prof to assoc prof biol, Tri-State Univ, 77-83; assoc prof, 83-87, CHAIRPERSON BIOL, COL MT ST JOSEPH, 85-, PROF, 87- *Concurrent Pos:* Entomologist, Tri-State Agri-Res, 79-83; adj curator, Cincinnati Mus Natural Hist, 90-; Fulbright Award, Coun Int Exchange Scholars, 81-82. *Mem:* Entom Soc Am; AAAS; Coleopterists Soc; Am Entom Soc. *Res:* Insect evolution; insect systematics; history of biology especially Darwin and entomology. *Mailing Add:* Dept Biol Col Mt St Joseph Mt St Joseph OH 45051

KRITZ, ARNOLD H, b Providence, RI, Jan 6, 35; m 57; c 3. PLASMA PHYSICS. *Educ:* Brown Univ, ScB, 56; Yale Univ, MS, 57, PhD(physics), 61. *Prof Exp:* Res asst, Yale Univ, 57-61; sr physicist, Space Sci Lab, Gen Dynamics/Astronaut, 61-63, staff scientist, 63-65; sr staff physicist, Aeronaut Res Assocs Princeton, Inc, 65-69; from asst prof to prof, 69-76, chmn dept physics & astron, 71-77, PROF PHYSICS, HUNTER COL, 76- *Concurrent Pos:* Lectr, New Haven Col, 59-60 & Southern Conn State Col, 60-61; asst prof, San Diego State Col, 63-64; consult, Oak Ridge Nat Lab, 74-; sci adj, Ctr Plasma Physics Res, Ecole Polytech Lausanne, 75-76 & 82-83; vis res fel, Princeton Univ, 77-; vis scientist, Lawrence Livermore Lab, 86- *Mem:* AAAS; Am Phys Soc; Am Asn Physics Teachers; Sigma Xi. *Res:* Nonequilibrium statistical mechanics; kinetic description of plasma; microwave interactions with inhomogeneous plasma; radio frequency heating of toroidal plasmas. *Mailing Add:* 695 Park Ave Hunter Col City Univ New York NY 10021

KRITZ, J(ACOB), b Brooklyn, NY, Dec 12, 18; m 49; c 2. ELECTRONICS, ELECTROACOUSTICS. *Educ:* City Col New York, BEE, 39; Polytech Inst Brooklyn, MEE, 49. *Prof Exp:* Engr signal corps, US War Dept, 40-44; MacKay Radio & Tel, 45-48; sr res engr, Radio Receptor Co, 48-49; W L Maxson Corp, 49-56; head res labs, Arma Div, Am Bosch Arma Corp, 56-61; pres, Janus Prod Inc, 61-66; eng mgr sonar systs, Marquardt Corp, 66-67; chief electroacoustic technol, Kollsman Instrument Corp, 67-69; Eng dept head, systs mgr div, Sperry Corp, 69-86, res dept head, Sperry Marine Systs, 72-86; CONSULT, 86- *Concurrent Pos:* Adj prof grad sch, Polytech Inst Brooklyn, 57-60. *Honors & Awards:* Dr Samuel N Burka Award, Am Inst Navig, 69. *Mem:* Sr mem Inst Elec & Electronics Engrs; Acoust Soc Am. *Res:* Navigation aids, especially poppler sonar, radio-radar, inertial, ultrasonic instrumentation including flowmeters and densitometers. *Mailing Add:* 692 Sherman Ct Westbury NY 11590

KRITZMAN, JULIUS, b Lawrence, Mass, Sept 15, 24; m 50; c 2. MEDICINE, HEMATOLOGY. *Educ:* Harvard Univ, AB, 47; Boston Univ, MD, 51; Am Bd Internal Med, dipl, 58, dipl hemat, 74, dipl med oncol, 75, dipl advan achievement internal med, 87. *Prof Exp:* Res fel hemat, New Eng Ctr Hosp, 53-54; clin instr med, Sch Med, 54-58, sr instr, 59-62, asst clin prof, 62-67, asst prof, 67-75, ASSOC CLIN PROF MED, SCH MED, TUFTS UNIV, 75-; STAFF INTERNAL MED, NEW ENG MED CTR HOSPS, 67- *Concurrent Pos:* Physician, Med Clin, Boston Dispensary, 58-; res assoc, Arthur G Rotch Lab, 60-; attend physician, Boston Vet Admin Hosp, 60-; asst vis physician, Beth Israel Hosp, 61-67, assoc in med, 67-; instr, Harvard Med Sch, 61-70; mass lectureship, Am Col Physicians, 90. *Mem:* Am Soc Hemat; Am Col Physicians. *Res:* Synthesis and function of antibodies, especially the use of in vitro systems for study of antibody synthesis; medical oncology. *Mailing Add:* 171 Harrison Ave Boston MA 02111

KRITZ-SILVERSTEIN, DONNA, b Brooklyn, NY, Aug 22, 57; m 86; c 1. WOMENS HEALTH, STATISTICS. *Educ:* Brooklyn Col, BS, 79; NY Univ, MA, 81, PhD(social psychol), 84. *Prof Exp:* Data processing supvr, Components & Effective Fertil Regulation Study, Downstate Med Ctr, State Univ NY, 82-84, asst prof, 84-86; res assoc, 86-89, ASST ADJ PROF EPIDEMIOL, UNIV CALIF, SAN DIEGO, 89- *Concurrent Pos:* Adj prof personality psychol, San Diego State Univ, 86-87. *Mem:* Am Psychol Asn; Western Psychol Asn. *Res:* Women's health, specifically the long term effects of pregnancy on the risk for diabetes, hypertension, gall bladder and other diseases; menopause; menstrual symptoms; employment and health. *Mailing Add:* Dept Community & Family Med Univ Calif San Diego 9500 Gilman Dr 0620 La Jolla CA 92093-0620

KRIVAK, THOMAS GERALD, b Johnstown, Pa, Aug 21, 40; m 63; c 3. INORGANIC CHEMISTRY. *Educ:* Univ Pittsburgh, BS, 62, MEd, 64; Univ Notre Dame, PhD(inorg chem), 69. *Prof Exp:* Res asst chem, Radiation Labs, Mellon Inst, 62-64; SR RES ASSOC, INDUST CHEM DIV, PPG INDUSTS, 69- *Mem:* Am Chem Soc; Sigma Xi. *Res:* Silica pigments for paint, paper, plastics and rubber applications. *Mailing Add:* Two Highview Circle Irwin PA 15642

KRIVANEK, NEIL DOUGLAS, b Milwaukee, Wis, June 11, 44; m 69. TOXICOLOGY. *Educ:* Univ Wis, BS, 66; Wayne State Univ, MS, 68, PhD(physiol), 72. *Prof Exp:* Instr toxicol, Dept Occup & Environ Health, Col Med, Wayne State Univ, 72-74; res toxicologist, 74, consult, 86, STAFF TOXICOLOGIST, HASKELL LAB TOXICOL & INDUST MED, E I DU PONT DE NEMOURS & CO, INC, 75- *Mem:* Am Chem Soc; Am Indust Hyg Asn; Am Bd Indust Hyg Asn; Sigma Xi; Am Bd Toxicol; Soc Toxicol. *Res:* Biochemical mechanisms of industrial toxicology; methods development for measuring toxic effects. *Mailing Add:* Haskell Lab E I du Pont de Nemours & Co Inc Newark DE 19711

KRIVANEK, ONDREJ LADISLAV, b Prague, Czech, Aug 1, 50; Brit citizen. ELECTRON MICROSCOPY, ELECTRON SPECTROSCOPY. *Educ:* Univ Leeds, BSc, 71; Univ Cambridge, PhD(physics), 76. *Prof Exp:* Res fel physics, Cavendish Lab, Cambridge, 75-76; res consult electron micros, Bell Lab, 76-77; asst res engr mat sci, Univ Calif, Berkeley, 77-80; asst prof solid state sci, 81-85, ADJ ASSOC PROF PHYSICS, ARIZ STATE UNIV, 85-; DIR RES, GATAN, INC, 85- *Mem:* Am Phys Soc; Electron Micros Soc Am; Brit Inst Physics. *Res:* High resolution and analytical electron microscopy; electron optics; electron energy loss spectroscopy; UHV electron microscopy. *Mailing Add:* Gatan Inc 6678 Owens Dr Pleasanton CA 94566

KRIVI, GWEN GRABOWSKI, b Huntington, NY, Feb 11, 50; c 1. GROWTH FACTORS, INFLAMMATION. *Educ:* Bucknell Univ, BA, 72; Mass Inst Technol, PhD(biochem), 78; Wash Univ, MBA, 90. *Prof Exp:* Sr res chemist, Dept Molecular Biol, Monsanto Co, 80-82, res specialist, 82-84, res group leader I, Dept Biol Sci, 84-85, res group leader II, 85-86, sci fel, 86-89, SR FEL, DEPT BIOL SCI, MONSANTO CO, 89- *Concurrent Pos:* Mem, Comt Biotechnol, Agr Res Serv, USDA, 86-; adj prof, Wash Univ, 91- *Mem:* AAAS; Am Chem Soc; Endocrine Soc. *Res:* Function relationships of growth proteins and their receptors; role of arachidonic acid metabolites in inflammation; animal cell engineering; immunoassay; cell biology; molecular genetics technologies in support of human pharmaceutical product development. *Mailing Add:* Monsanto Co 700 Chesterfield Village Pkwy Chesterfield MO 63198

KRIVIS, ALAN FREDERICK, b New York, NY, Sept 27, 31; m 56; c 3. ANALYTICAL CHEMISTRY. *Educ:* Columbia Univ, AB, 52, MA, 54; Univ Mich, MS, 56, PhD(electrochem), 58. *Prof Exp:* Bio-analyst, Sloan-Kettering Inst, 52-53; mem pigment synthesis staff, Res Div, Interchem Corp, 53; res assoc phys & anal chem, Upjohn Co, 57-61; group supvr, Chem Div Res, Olin Mathieson Chem Corp, 61-66; ASSOC PROF CHEM, UNIV AKRON, 66- *Mem:* AAAS; fel Am Inst Chem; Electrochem Soc; Am Chem Soc; Am Microchem Soc. *Res:* Analytical chemistry including chromatography and electro-chemistry of inorganic and organic systems; biologically active materials; water and air analyses. *Mailing Add:* 4235 W Streetsboro Rd Richfield OH 44286-0423

KRIVIT, WILLIAM, b Jersey City, NJ, Nov 28, 25; m 51; c 4. PEDIATRICS, HEMATOLOGY. *Educ:* Duke Univ, 42-44; Tulane Univ, MD, 48; Am Bd Pediat, dipl, 53, dipl pediat hemat, 75. *Prof Exp:* Intern, Charity Hosp, New Orleans, 48-49; resident pediat, Col Med, Univ Utah, 49-50, chief resident, 50-51; PROF PEDIAT, MED SCH, UNIV MINN, MINNEAPOLIS, 51-, HEAD DEPT, 79- *Mem:* Soc Pediat Res; Am Soc Hemat; Soc Exp Biol & Med; NY Acad Sci. *Mailing Add:* 1252 Ingerson Rd St Paul MN 55112

KRIVOY, WILLIAM AARON, b Newark, NJ, Jan 2, 28. PHARMACOLOGY. *Educ:* Georgetown Univ, BS, 48; George Washington Univ, MS, 49, PhD(pharmacol), 53. *Prof Exp:* Pharmacologist, Chem Corps Med Labs, Army Chem Ctr, Md, 50-54; USPHS res fel pharmacol, Sch Med, Univ Pa, 54-55 & Sch Med, Univ Edinburgh, 55-57; instr, Sch Med, Tulane Univ, 57-59; from asst prof to assoc prof, Col Med, Baylor Univ, 59-63; pharmacologist, Addiction Res Ctr, Nat Inst Drug Abuse, 68- 83; RETIRED. *Mem:* Am Soc Pharmacol & Exp Therapeut; Am Col Neuropsychopharmacol; Am Soc Clin Pharmacol & Therapeut; Biophys Soc. *Res:* Neuropharmacology and neurophysiology; autonomic agents and psychochemicals; neuroendocrinology; neurosciences; physiology; psychopharmacology. *Mailing Add:* 2525 Centerville Rd Dallas TX 75228

KRIZ, GEORGE JAMES, b Brainard, Nebr, Sept 20, 36; m 60; c 3. AGRICULTURAL ENGINEERING, SOIL SCIENCE. *Educ:* Iowa State Univ, BSAE, 60, MSAE, 62; Univ Calif, Davis, PhD(eng), 65. *Prof Exp:* Teaching asst agr eng, Iowa State Univ, 60-62; asst engr, Univ Calif, Davis, 64-65, lectr groundwater hydrol, 65; from asst prof to assoc prof biol & agr eng, NC State Univ, 65-73, assoc head dept, 69-73; PROF & DIR RES, NC AGR EXP STA, 73- *Concurrent Pos:* Consult, Int Basic Econ Corp & Indian Inst Technol, 67; educ & res dir, Am Soc Agr Engrs, 83-85. *Mem:* Fel Am Soc Agr Engrs. *Res:* Animal waste management; saturated flow in porous media, especially soil water relationships. *Mailing Add:* Box 7601 NC State Univ Raleigh NC 27695-7601

KRIZ, GEORGE STANLEY, b Santa Cruz, Calif, Oct 20, 39; m 89; c 3. PHYSICAL ORGANIC CHEMISTRY. *Educ:* Univ Calif, Berkeley, BS, 61; Ind Univ, PhD(org chem), 66. *Prof Exp:* Foreign asst chem, Univ Montpellier, 65-66; vis res assoc, Ohio State Univ, 66-67; asst prof, 67-72, exec asst to dean arts & sci, 75-77, assoc prof, 77-79, PROF CHEM, WESTERN WASH UNIV, 79- *Concurrent Pos:* Vis prof, Indiana Univ, 84-85. *Mem:* AAAS; Am Chem Soc; Royal Soc Chem. *Res:* Deuterium kinetic isotope effects; mechanisms of organic reactions; nuclear magnetic resonance spectroscopy; linear free energy relationships. *Mailing Add:* Dept Chem Western Wash Univ Bellingham WA 98225-9058

KRIZAN, JOHN ERNEST, theoretical physics; deceased, see previous edition for last biography

KRIZEK, DONALD THOMAS, b Cleveland, Ohio, June 25, 35; m 62; c 3. ENVIRONMENTAL PHYSIOLOGY, STRESS PHYSIOLOGY. *Educ:* Western Reserve Univ, BA, 57; Univ Chicago, MS, 58, PhD(bot), 64. *Prof Exp:* Res & develop officer, Arctic, Desert, Tropic Info Ctr, Res Studies Inst, Air Univ, 58-62; instr bot & biol, Univ Chicago, 64-66; res plant physiologist, Phyto-Eng Lab, USDA, 66-72, res plant physiologist, Plant Stress Lab, 72-89, RES PLANT PHYSIOLOGIST, CLIMATE STRESS LAB, BELTSVILLE AGR RES CTR, AGR RES SERV, USDA, 89-; ADJ ASSOC PROF, DEPT HORT, UNIV MD, 86- *Concurrent Pos:* Instr, Montgomery Ctr, Univ Ala, 59-61; mem, Am Soc Hort Sci Working Group Controlled Environ, 69-, Am Soc Agr Eng Plant & Animal Physiol Adv Group, 66-; USDA rep, White House Task Force Inadvertent Modification Stratosphere, 75-78; consult, NASA, 79-, ARS rep; NCR-101 comt Growth Chamber Use, 75-; chmn Inadvertent Modification of Stratosphere subcomt, Biol & Climate Effects Res, 75-78; chmn, Environ Protection Agency Interagency Task Group, Biol (non human) Effects, 76-78; mem Plant Phys Adv Group, Am Inst Biol Sci/ NASA, 79-; chmn, Am Inst Biol Sci, Kennedy Space Ctr Biomass prod tech panel, 87-; mem NASA Controlled Ecol Life Support Systs Discipline Working Group, 89-; chmn Comn Int de l'Eclairage Tech Comt Action Spectra in Plants, 88- *Mem:* Am Soc Plant Physiol; fel Am Soc Hort Sci; Brit Soc Exp Biol; Japanese Soc Plant Physiol; Scand Soc Plant Physiol; Am Inst Biol Sci; Am Soc Photobiol; AAAS; Int Soc Hort Sci. *Res:* Plant growth and development; senescence of vascular plants; photoperiodism and photomorphogenesis; plant growth regulators; plant stress; controlled environments; environmental physiology; carbon dioxide enrichment; plant growth chambers; ultra-violet radiation effects; water stress. *Mailing Add:* Climate Stress Lab USDA Rm 206 B-001 BARC-W Beltsville MD 20705-2350

KRIZEK, RAYMOND JOHN, b Baltimore, Md, June 5, 32; m 64; c 2. GEOTECHNICAL ENGINEERING. *Educ:* Johns Hopkins Univ, BE, 54; Univ Md, MS, 61; Northwestern Univ, PhD(soil mech), 63. *Prof Exp:* Instr civil eng, Univ Md, 57-61; lectr soil mech, Cath Univ, 61; from asst prof to assoc prof, 63-70, PROF CIVIL ENG, NORTHWESTERN UNIV, 70-, CHMN DEPT, 80- *Concurrent Pos:* Consult, 63- *Honors & Awards:* C A Hogentogler Award, Am Soc Testing & Mat, 70; Walter L Huber Res Prize, Am Soc Civil Engrs, 71. *Mem:* Am Soc Civil Engrs. *Res:* Soil-structure interaction of buried conduits; use of dredgings for landfill; relationship between soil fabric and its engineering properties; constitutive relations for soils; flow through porous media; disposal of solid waste materials; soil stabilization by grouting. *Mailing Add:* Dept Civil Eng Northwestern Univ Evanston IL 60208

KRIZEK, THOMAS JOSEPH, b Milwaukee, Wis, Dec 1, 32; m 59; c 3. PLASTIC SURGERY, RECONSTRUCTIVE SURGERY. *Educ:* Marquette Univ, BS, 54, MD, 57. *Hon Degrees:* MA, Yale Univ, 74. *Prof Exp:* Asst prof plastic surg, Johns Hopkins Univ & Univ Md, 66-68; assoc prof, Sch Med, Yale Univ, 68-73, prof plastic surg, 73-78, assoc dean grad & continuing educ, 75-77; prof surg, Col Physicians & Surgeons, Columbia Univ & Chief, Div Plastic & Reconstructive Surg, Columbia-Presby Med Ctr, 78-81; prof surg, Univ Southern Calif & chief, Div Plastic Surg, Los Angeles County, Univ

Southern Calif, Med Ctr, 81-84; prof surg & chief Plastic & Reconstructive Surg, 84-87, CHMN, DEPT SURG, UNIV CHICAGO, 87- *Mem:* Am Asn Plastic Surgeons; Asn Hand Surg (pres, 80-81); Am Asn Surg of Trauma; Am Burn Asn; Am Col Surgeons. *Res:* Surgical infection, particularly as related to burns and other trauma; surgery and epidemiology of head and neck cancer; aging. *Mailing Add:* 5841 S Maryland Ave Chicago IL 60637

KRNJEVIC, KRESIMIR, b Zagreb, Yugoslavia, Sept 7, 27; m 54; c 2. NEUROPHYSIOLOGY. *Educ:* Univ Edinburgh, MB, ChB, 49, BSc, 51, PhD(physiol), 53. *Prof Exp:* Beit Mem res fel, Univ Edinburgh, 51-54; res assoc & actg asst prof, Univ Wash, 54-56; from prin sci officer to sr prin sci officer, ARC Inst Animal Physiol, Eng, 59-65; Drake prof & chmn dept physiol, 78-87, HEAD DEPT ANESTHESIA RES, McGILL UNIV, 65- *Concurrent Pos:* Prof physiol, McGill Univ, 64-; officer, Order Can, 87; mem coun, Int Union Physiol Sci. *Honors & Awards:* Forbes lectr, 78; Gairdner Award, 84. *Mem:* Can Physiol Soc; Int Soc Neurochem; Soc Neurosci; Royal Soc Can; Int Union Physiol Sci. *Res:* Central synaptic mechanisms and disruptive effects of anolia and anesthesia. *Mailing Add:* Dept Anesthesia Res McGill Univ McIntyre Bldg 3655 Drummond St Montreal PQ H3G 1Y6 Can

KROC, ROBERT LOUIS, b Chicago, Ill, June 19, 07; m 34; c 2. PHYSIOLOGY. *Educ:* Oberlin Col, BA, 29, MA, 31; Univ Wis, PhD(zool, physiol), 33. *Hon Degrees:* DSc, Oberlin Col, 79. *Prof Exp:* Asst, Univ Wis, 30-33; from instr to asst prof zool, Ind Univ, 33-44; biologist, Res Div, Maltine Co, 44-47, dir physiol res, Chilcott Labs Div, 47-51, Warner-Chilcott Labs Div, Warner-Hudnut, Inc, 51-58 & Warner-Lambert Res Inst, 58-69; pres, The Kroc Found, 69-85; RETIRED. *Concurrent Pos:* mem bd dirs, Med Found Buffalo Res Inst, 85- & Sansum Med Res Found, Santa Barbara, 86- *Honors & Awards:* Addison B Scoville Award, Am Diabetes Asn, 84. *Mem:* Endocrine Soc; Am Physiol Soc; Am Thyroid Asn; NY Acad Sci; AAAS. *Res:* Endocrinology, especially thyroid and reproductive physiology; prothrombin time reagent simplastin. *Mailing Add:* 4737 Sierra Madre Rd Santa Barbara CA 93110

KROCHMAL, ARNOLD, b New York, NY, Jan 30, 19; m 56, 70; c 3. ECONOMIC BOTANY. *Educ:* NC State Col, BS, 42; Cornell Univ, MS, 51, PhD(econ bot), 52. *Prof Exp:* Instr hort, NMex Col Agr & Mech Arts, 47-49; Fulbright prof hort, Am Farm Sch, Greece, 52-53; instr agr, State Univ NY Agr & Tech Inst Delhi, 53-54; chmn, Sci Div, Morris Brown Col, 54-55; assoc prof hort, Ariz State Univ, 55-57; chief res adv, Univ Wyo Team, Afghanistan, 57-59; assoc prof agr, Western Carolina Col, 59-60; chmn dept hort, Pan Am Sch Agr, Honduras, 60-61; res botanist & asst in charge crops res div, VI, USDA, 61-66, proj leader & res botanist, Northeastern Forest Exp Sta, Berea, Ky, 66-71, prin econ botanist, Southeastern Forest Exp Sta, 71-82, prin econ botanist, Inst Trop Forestry, PR, 82-83; RETIRED. *Concurrent Pos:* Consult trop fruit prod, Surinam Govt, 64; consult, Montserrat, 64-66, Thai Tapioca Asn, Thailand, 64 & Agr Develop Corp, Jamaican Govt, 65; adj prof bot, NC State Univ, 71-77; adj prof biol, Univ NC, Asheville, 75-83; collabr, Nat Acad Sci, 75; working party leader, Int Union Forestry Res Orgns structure of trop rain forests, 77-; adj prof forestry, NC State Univ, 78-; mem, Nat Acad Sci, Romania, 79 & 80; chmn, Div Sci, World Univ, PR, 82-83; hon res assoc, Inst Econ Bot, NY Bot Garden, 82-; adj res prof biol, Atlanta Univ, 83-84; multi-nat agr team, OAS, Dominican Repub, 84; sr res fel, Dept Trop Studies, Agr Univ Wageninsen, Neth; guest lectr, Ctr Alternate Energy, Nurop, Denmark, 87 & Main Bot Garden, Moscow, 80; Alfred G Burnam Donor Fund res grants, 87, 88 & 90; assoc ed, Bot, Int J Pharmacog, 90- *Mem:* Caribbean Food Crops Soc (secy, treas, 63-64); Sigma Xi; Volunteers in Tech Assistance. *Res:* Production and associated problems of fruit, vegetable and industrial crops in tropics and arid regions; culture of wild plants for drug, dye and food sources; wildlife feed and nitrogen fixation; endangered, rare and threatened plants; natural areas; author of four books. *Mailing Add:* 119 Bell Rd Asheville NC 28805

KROCHMAL, JEROME J(ACOB), b New York, NY, Dec 17, 30; m 52; c 3. ORGANIZATION DEVELOPMENT, ORGANIZATIONAL EFFECTIVENESS. *Educ:* Ga Inst Technol, BCerE, 52. *Prof Exp:* Staff mem ceramics, Battelle Mem Inst, 52; staff mem ceramics & graphites, US Air Force Propulsion Lab, 54-57; staff mem, US Air Force Mat Lab, US Air Force, 57-60, sr proj officer, 60-69, sr plans analyst, 71-75, sr mgt analyst, Aeronaut Labs, 75-88; CONSULT, 75- *Concurrent Pos:* Staff consult, Mat Adv Bd, Nat Acad Sci-Nat Res Coun, 60-61; Stanford-Sloan fel, 69-70. *Mem:* Orgn Develop Network; Orgn Develop Inst; Asn Psychol Type; Acad Mgt; Am Soc Training & Develop. *Res:* Research and development organization effectiveness; instrumentation and simulations; synergistic problem solving in advanced technology; fostering creativity and innovation. *Mailing Add:* 232 Spalding Trail Atlanta GA 30328-1071

KROCHTA, WILLIAM G, b Piney Fork, Ohio, Sept 24, 30; m 54; c 3. ANALYTICAL CHEMISTRY. *Educ:* Mt Union Col, BS, 52; Purdue Univ, MS, 54, PhD(chem), 57. *Prof Exp:* Sr res chemist, Columbia-Southern Chem Corp, 56-59, supvr, 59-62; SR SUPVR, CHEM DIV, PPG INDUSTS, BARBERTON, 62- *Mem:* Am Chem Soc; Am Indust Hyg Asn. *Res:* Absorption spectroscopy; gas chromatography. *Mailing Add:* 237 Tanglewood Trail Wadsworth OH 44281

KROCK, HANS-JÜRGEN, b Cracow, Poland, Aug 30, 42; US citizen. ENVIRONMENTAL WATER QUALITY, OCEAN ENGINEERING. *Educ:* Ariz State Univ, BS, 65, MS, 67; Univ Calif, Berkeley, PhD, 72. *Prof Exp:* Pub health engr, Dept Health, State of Ariz, 65-66; res engr tertiary treat, Los Angeles County Sanit Dists, Calif, 67-68; assoc engr hydraul, Alameda County Flood Dist, Calif, 69; res assoc, Sanit Eng Res Lab, Univ Calif, Berkeley, 70-72; sr environ engr, M&E Pac, Inc, Honolulu, Hawaii, 72-80; ASSOC PROF OCEAN ENG & DIR, J K K LOOK LAB, UNIV HAWAII, MANOA, 80-; PRES, OCEAN ENG & ENERGY SYST INT, INC, HONOLULU, HAWAII, 88- *Concurrent Pos:* Consult engr, 80-; prin investr, Open Cycle Ocean Thermal Energy Conversion, 82-; mem nat comt on Ocean Energy, Am Soc Civil Engrs, 87- *Mem:* Fel Am Inst Chemists; Am Soc Civil Engrs; Water Pollution Control Fedn; Marine Technol Soc; Am Soc Limnol & Oceanog; AAAS; Am Chem Soc. *Res:* Physical, chemical and biological interactions in the oceans and estuaries; gas exchange characteristics in sea water related to ocean thermal energy conversion; water quality standards. *Mailing Add:* 3786 Pukalani Pl Honolulu HI 96816

KROCK, LARRY PAUL, STRESS PHYSIOLOGY. *Educ:* San Fernando Valley State Col, BA, 72; Calif State Univ, MA, 74; Tex A&M Univ, PhD(physiol), 85. *Prof Exp:* Res asst, Dept Phys Educ, Calif State Univ, 73-74, lectr, 75-77, dir, Sports Med Facil & Undergrad Curric Sports Med, 75-82, asst prof, 79-82; res asst, Dept Physiol & Appl Physics Lab, Col Vet Med, Tex A&M Univ, 83-84, asst, Dept Health & Phys Educ, 82-85; res physiologist environ physiol function, 85-87, RES PHYSIOLOGIST ACCELERATION EFFECTS FUNCTION, CREW TECHNOL DIV, SCH AEROSPACE MED, USAF, 87- *Concurrent Pos:* Res asst, Dept Med, Baylor Col Med, 84-85; adj prof, Univ Tex, San Antonio, 90- & Tex Lutheran Col, 91; mem, Coun Circulation, Am Heart Asn, Basic Res Rev Bd, Naval Res & Develop Command, 90 & Sci Prog Rev Comt, Aerospace Med Asn, 91. *Mem:* Aerospace Med Asn; Aerospace Physiologists Soc; AAAS; Am Col Sports Med; Am Heart Asn; Am Physiol Soc; Inst Elec & Electronics Engrs; Sigma Xi; NY Acad Sci. *Res:* Fatigue and recovery of skeletal muscle to high and low acceleration stress; response of the cardiovascular and musculoskeletal systems to extended duration microgravity stress; autonomic neural influence on cardiovascular function during and immediately following acceleration stress; dynamic response of the heart to increased acceleration stress; dynamic ventricular interaction during stress; stress physiology. *Mailing Add:* Flight Motion Effects Br Acceleration Crew Technol Div USAF Armstrong Lab Brooks AFB TX 78235-5000

KROEGER, DONALD CHARLES, b Boise, Idaho, Sept 18, 25; m 48; c 2. PHARMACOLOGY. *Educ:* Ore State Col, BS, 47; Purdue Univ, MS, 49, PhD(pharmacol), 51. *Prof Exp:* From asst prof to assoc prof pharmacol, Univ Houston, 51-56; from asst prof to assoc prof, 56-64, PROF PHARMACOL, UNIV TEX DENT BR, HOUSTON, 64-, CHMN DEPT, 68- *Mem:* AAAS; Am Pharmaceut Asn; Soc Exp Biol & Med; Am Soc Pharmacol & Exp Therapeut; Int Asn Dent Res. *Res:* Neuropharmacology; electrophysiological stimulation and recording of autonomic neural centers in the brain and relationship of chemical changes in tissues to brain stimulation. *Mailing Add:* 9325 Lugary Houston TX 77074

KROEGER, PETER G, b Swinoujscie, Poland, Apr 26, 30; US citizen; m 58; c 2. FLUID DYNAMICS, HEAT TRANSFER. *Educ:* Inst Technol, Aachen, WGer, ME, 56; Case Western Reserve Univ, PhD(mech eng), 72. *Prof Exp:* Res engr, Gen Elec Co, 62-67 & 72-74, Kennecott Copper Co, 67-72; RES ENGR, BROOKHAVEN NAT LAB, 74- *Mem:* Am Soc Mech Engrs; Sigma Xi; Am Nuclear Soc. *Res:* Analysis of thermal and fluid dynamics systems and processes; nuclear reactor thermohydraulics; phase change processes such as freezing and melting; dynamics of propulsion plants. *Mailing Add:* Seven Thornton Common Yaphank NY 11980

KROEGER, RICHARD ALAN, b Waterloo, Iowa, May 28, 55. GAMMA-RAY OBSERVATIONS, GAMMA-RAY DETECTOR DEVELOPMENT. *Educ:* Univ Northern Iowa, BA, 77; Univ Chicago, MS, 79, PhD(physics), 85. *Prof Exp:* Res assoc astrophys, Enrico Fermi Inst, 85; vis scientist astrophys, Univ Space Res Asn, 85-87; staff engr, Hughes Aircraft Corp, 87-89; ASTROPHYSICIST, NAVAL RES LAB, 89- *Mem:* Am Inst Physics; Am Astron Soc. *Res:* High energy astrophysics; gamma ray observations; measurements of cosmic rays. *Mailing Add:* Code 4150 Naval Res Lab 4555 Overlook Ave SW Washington DC 20375

KROEKER, RICHARD MARK, b Bakersfield, Calif, Sept 7, 52; m 78; c 2. ELECTRON TUNNELING. *Educ:* Washington Univ, St Louis, BA, 74; Univ Calif, Santa Barbara, PhD(physics), 79. *Prof Exp:* Fel, Physics Dept, Univ Calif, Santa Barbara, 79-80; fel, 80-81, SR ENGR, GEN PROD DIV, IBM CORP, SAN JOSE, 81- *Mem:* Am Phys Soc. *Res:* Physical chemistry of surfaces, including monolayer spectroscopy, radiation chemistry of thin films and mechanisms of heterogeneous catalysis; tribology and wear. *Mailing Add:* 700 Spring Hill Dr Morgan Hill CA 95037

KROEMER, HERBERT, b Weimar, Ger, Aug 25, 28; m 50; c 5. SEMICONDUCTORS. *Educ:* Univ Gottingen, dipl, 51, Dr rer nat, 52. *Hon Degrees:* Dr, Tech Univ Aachen, Germany, 84. *Prof Exp:* Res scientist, Ger Post Off Lab, 52-54 & labs, Radio Corp Am, 54-57; res group leader, Ger Philips Lab, 57-59; sr scientist, Varian Assocs, 59-66; head New Phenomena Sect, Semiconductor Res & Develop Lab, Fairchild, 66-68; prof elec eng, Univ Colo, Boulder, 68-76; PROF ELEC ENG, UNIV CALIF, SANTA BARBARA, 76- *Concurrent Pos:* Nat lectr, Inst Elec & Electronics Engrs Device Soc. *Honors & Awards:* J Ebers Award, Inst Elec & Electronics Engrs, 73; Sr Res Award, Am Soc Eng Educ, 82; Heinrich Walker Medal, 82; Jack A Morton Award, Inst Elec & Electronics Engrs, 86. *Mem:* Fel Am Phys Soc; fel Inst Elec & Electronics Engrs; Electrochem Soc. *Res:* Semiconductor physics and exploratory research on new device principles; physics and technology of semiconductor materials and devices; heterojunctions; molecular beam epitaxy. *Mailing Add:* Dept Elec & Comput Eng Univ Calif Santa Barbara CA 93106

KROENBERG, BERNDT, b Riga, Latvia, Oct 31, 36; m 59; c 1. FOOD SCIENCE, NUTRITION. *Educ:* Inst Divi Thomae, MS, 61, PhD(biochem), 63. *Prof Exp:* Instr Ger, Our Lady Cincinnati Col, 61-63; instr, Xavier Univ, Ohio, 62-63; group leader leaf chem, Brown & Williamson Tobacco Corp, 63-66; mgr emulsion develop, Celanese Coatings Co, 66-69; dir prod develop res, 69-76, dir licensing, Ross Labs Div, Abbott Labs, Columbus, 76-84; STAFF VPRES TECHNOL & DEVELOP,BAUSCH & LOMB INC, 84- *Mem:* AAAS; Am Chem Soc; Licencing Exec Soc. *Res:* Isolation and identification of natural products of animal and plant origin; amino acids, alkaloids, sterols; medicinal use of natural products; infant and geriatric foods; pediatric and obstetrics and gynecology drugs; diagnostic kits; enteral feeding pumps; contact lenses, ophthalmic drugs. *Mailing Add:* Bausch & Lomb Inc One Lincoln First Sq Rochester NY 14601

KROENERT, JOHN THEODORE, b Arkansas City, Kans, Nov 28, 21; m 47; c 2. ELECTRICAL ENGINEERING, COMMUNICATIONS. *Educ:* Purdue Univ, BSEE, 43. *Prof Exp:* Res engr sonar, US Navy Underwater Sound Lab, 43-51, br head servodyn group, 51-55; sr engr sonar develop, Ultrasonic Corp, 55-56; engr supvr airborne sonar, Light Mil Electronic Equip, Gen Elec Co, 56-58; sr engr, Raytheon Corp, 58-61, prin engr, Submarine Signal Div, 61-64, consult engr, 64-87; RETIRED. *Mem:* Sr mem Inst Elec & Electronics Engrs; fel Acoust Soc Am; assoc Sigma Xi. *Res:* Sonar signal processing; detection, classification, location and underwater telephone. *Mailing Add:* 349 New Meadow Rd Barrington RI 02806

KROENING, JOHN LEO, b Princeton, Minn, Aug 18, 34; m 56; c 4. PHYSICS. *Educ:* Univ Minn, Minneapolis, BS, 56, MS, 59, PhD(physics), 62. *Prof Exp:* Res assoc physics, Univ Minn, Minneapolis, 62-65, asst prof, 65-68, ASSOC PROF PHYSICS, UNIV MINN, DULUTH, 68- *Mem:* Am Geophys Union. *Res:* Atmospheric physics and electricity; small ion content; atmospheric ozone distribution; chemiluminescent detection; atmospheric aerosol-effect on ozone and small ion content; stratosphere-troposphere transport. *Mailing Add:* Sch Physics Univ Minn Duluth MN 55812

KROENKE, LOREN WILLIAM, b Milwaukee, Wis, July 2, 38. MARINE GEOLOGY. *Educ:* Univ Wis-Madison, BS, 60; Univ Hawaii, MS, 68, PhD(geol), 72. *Prof Exp:* Proj asst marine geol & geophys, Geophys & Polar Res Ctr, Univ Wis, 61-63; asst geophys, Hawaii Inst Geophys, Univ Hawaii, 63-66, jr geophysicist, 67-72, asst geophysicist, 72-75, assoc geophysicist marine geol & geophys, 75-86, GEOPHYSICIST, HAWAII INST GEOPHYS, UNIV HAWAII, 86- *Concurrent Pos:* Tech adv, Comt Coord, Joint Prospecting Mineral Resources SPac Offshore Areas, Econ Comn Asia & Far East, UN, 72-73; sci adv, US-Japan Coop Prog Marine Sea-Bottom Surv Panel, 73-74; marine geologist & regional adv, UN Develop Prog, 74-76; mem, Adv Panel Ocean Margin, Joint Oceanog Inst Deep Earth Sampling-Int Prog Ocean Drilling, 74- *Mem:* Am Geophys Union; Seismol Soc Am; Geol Soc Am; AAAS; Sigma Xi. *Res:* Marine geology and geodynamics of the Pacific Ocean Basin with particular reference to the formation and deformation of oceanic crust and continental margins in the southwest Pacific. *Mailing Add:* Hawaii Inst Geophys Univ Hawaii Sch Ocean & Earth Sci & Technol Honolulu HI 96822

KROENKE, WILLIAM JOSEPH, b Cleveland, Ohio, Aug 16, 34; m 61; c 2. MATERIALS SCIENCE, TECHNOLOGY TRANSFER. *Educ:* Case Inst Technol, BS, 56, PhD(inorg chem), 63. *Prof Exp:* Chemist, Nat Carbon Res Lab, Union Carbide Corp, 56-61; res chemist, Res Ctr, 63-65, sr res chemist, 65-70, res assoc, 70-74, sr res assoc, 74-80, RES & DEVELOP FEL & MGR, B F GOODRICH CO, 80- *Mem:* Am Chem Soc. *Res:* Inorganic and organometallic chemistry; goechemistry; smoke retarders; synthesis; coordination numbers; molecular and crystal structure; property-structure relationships; solid state chemistry; phase relationships; fire retardants; catalysis; polymers; high-temperature chemistry; carbon/carbon composites. *Mailing Add:* 8485 Sunnydale Dr Brecksville OH 44141

KROES, ROGER L, b Racine, Wis, Dec 3, 35; m 64; c 2. SOLID STATE PHYSICS. *Educ:* Marquette Univ, BS, 57; Univ Mo-Columbia, PhD(physics), 68. *Prof Exp:* PHYSICIST, MARSHALL SPACE FLIGHT CTR, NASA, 68- *Mem:* Am Phys Soc. *Res:* Color centers in alkaline earth oxides; optical properties of solids; crystal characterization. *Mailing Add:* 902 Coronado Ave Huntsville AL 35802

KROGDAHL, WASLEY SVEN, b Springfield, Ill, Jan 17, 19; m 42; c 3. COSMOLOGY. *Educ:* Univ Chicago, BS, 39, PhD(astron), 42. *Prof Exp:* Jr physicist, Naval Ord Lab, 42-43; instr physics, Army Specialized Training Prog, Ripon Col, 43-44; assoc prof math & astron, Univ SC, 44-45; instr, Yerkes Observ, 45-46; asst prof astron & astrophys, Dearborn Observ, Northwestern Univ, Ill, 46-58; assoc prof math & astron, 58-64, prof math & astron, 64-65, prof physics & astron, 65-86, EMER PROF, UNIV KY, 86- *Mem:* Int Astron Union. *Res:* Theoretical astrophysics; relativistic cosmology. *Mailing Add:* 3493 Castleton Way N Lexington KY 40517

KROGER, F(ERDINAND) A(NNE), b Amsterdam, Netherlands, Sept 11, 15; div; c 2. PHYSICAL CHEMISTRY, THERMODYNAMICS. *Educ:* Univ Amsterdam, BSc, 34, Drs, 37, PhD(phys chem), 40. *Prof Exp:* Res worker, Philips Res Labs, Netherlands, 38-40, res group leader, 40-58; sci adv, Mullard Res Labs, Eng, 58-64; prof mat sci & chem, 64-85, David Packard prof elec eng, 72-85, EMER PROF MAT SCI, UNIV SOUTHERN CALIF, 85- *Concurrent Pos:* Corresp, Royal Dutch Acad Sci, 78. *Honors & Awards:* Chaudron Gold Medal, Fr Soc High Temperatures & Refractors. *Res:* Solid state luminescence; compound semiconductors; imperfection chemistry. *Mailing Add:* 4604 Newport Ave San Diego CA 92107

KROGER, HANNS H, b Hamburg, Ger, Sept 25, 26; US citizen; m 59; c 2. INORGANIC CHEMISTRY, ELECTROCHEMISTRY. *Educ:* Univ Hamburg, Cand, 53, Dipl, 56, Dr rer nat (chem, mineral), 58. *Prof Exp:* Sci asst chem, Univ Hamburg, 56-59; chemist, Accumulatorenfabrik AG, Ger, 59-61; electrochemist, NY, 62-65, ELECTROCHEMIST, BATTERY BUS SECT, GEN ELEC CO, FLA, 65- *Mem:* Am Chem Soc. *Res:* Battery technology; analytical chemistry. *Mailing Add:* 3841 Second Ave Gainesville FL 32601

KROGER, HARRY, b Brooklyn, NY, Aug 13, 36; m 58; c 3. SOLID STATE ELECTRONICS, EXPERIMENTAL SOLID STATE PHYSICS. *Educ:* Univ Rochester, BS, 57; Cornell Univ, PhD(physics), 62. *Prof Exp:* Mem res staff, 62-68, group leader microwave semiconductors, 68-69, mgr semiconductor device dept, 69-75, mgr advan device dept, Sperry Res Ctr, 75-; AT MCC ECHELON, AUSTIN. *Mem:* Inst Elec & Electronics Engrs; Am Phys Soc. *Res:* Soft x-ray spectroscopy; semiconductor memories; conduction through thin insulators; microwave semiconductor devices; Josephson devices. *Mailing Add:* MCC 12100 Technology Blvd Austin TX 78727

KRGER, HELMUT KARL, b Kassel, W Ger, Mar 12, 49. COMPUTATIONAL PHYSICS, LATTICE GAUZE THEORY. *Educ:* Univ Mainz, W Ger, dipl, 73; Univ Bonn, PhD(physics), 77. *Prof Exp:* Res assoc, Univ Giessen, W Ger, 77-79 & 80; res vis, Oak Ridge Nat Lab, 79-80; res staff, KWU Siemens Co, Erlanger, W Ger, 81-82; res assoc, 82-84, PROF RES, UNIV LAVAL, QUEBEC, 84- *Concurrent Pos:* Fel Ger Acad Exchange Serv, N Atlantic Treaty Orgn, 79-80; fel Natural Sci & Eng Res Coun Can, 83- *Mem:* Am Phys Soc; Can Phys Soc; Inst Particle Physics. *Res:* Nuclear physics; qual control data; gauge theories. *Mailing Add:* Dept Physics Univ Laval Quebec PQ G1K 7P4 Can

KROGER, LARRY A, b Hastings, Nebr, Dec 6, 43; m 68; c 2. NUCLEAR PHYSICS, MEDICAL PHYSICS. *Educ:* Hastings Col, BA, 66; Univ Wyo, PhD(physics), 72. *Prof Exp:* Res assoc physics & NSF-Nat Res Coun fel, Nat Reactor Testing Sta, 71-73; res assoc, Univ Pa, 73-75; sr physicist, Emergency Care Res Inst, 75-78; HEALTH PHYSICIST, MED SYST DIV, GEN ELEC CO, 78- *Concurrent Pos:* Lectr energy. *Mem:* Am Inst Physics; AAAS; Am Asn Physicist Med; Sigma Xi; Am Phys Soc. *Res:* Medical applications of x and gamma rays and performance of radiology systems; study of radioactivity and nuclear decay schemes; radiation safety. *Mailing Add:* 1146 School Dr E Waukesha WI 53186

KROGER, MANFRED, b Bad Oeynhausen, Ger, May 19, 33; Can & US citizen; m 62; c 3. FOOD SCIENCE. *Educ:* Univ Man, BSA, 61; Pa State Univ, MS, 63, PhD(food chem). 66. *Prof Exp:* Asst dairy sci, 61-63, from instr to assoc prof, 66-78, PROF FOOD SCI & SCI TECHNOL & SOC, PA STATE UNIV, UNIVERSITY PARK, 78- *Concurrent Pos:* Tech ed dairy & other food jour; consult food indust; assoc ed, J Food Sci, sr assoc ed, Nutrit Forum. *Mem:* AAAS; Am Chem Soc; Am Dairy Sci; Inst Food Technologists; World Future Soc; Ger Dairy Sci Asn. *Res:* Food flavor chemistry; pesticide residue analysis; chemistry of food contaminants; environmental quality; food safety; food laws and regulations; instrumental analysis of fat and protein in foods; milk processing; dairy products manufacture; yogurt quality. *Mailing Add:* 104 Borland Lab Pa State Univ Univ Park PA 16802

KROGER, MARLIN G(LENN), electronics engineering, operations analysis, for more information see previous edition

KROGH, LESTER CHRISTENSEN, b Ruskin, Nebr, Aug 22, 25; m 46; c 2. ORGANIC CHEMISTRY. *Educ:* Univ Nebr, BS, 45, MS, 48; Univ Minn, PhD(chem), 52. *Hon Degrees:* DSc, Univ Minn, 90. *Prof Exp:* Asst, Univ Nebr, 46-48; asst, Univ Minn, 48-51; sr chemist, Cent Res Lab, Minn Mining & Mfg Co & mgr res group, Abr Lab, 55-59, mgr res & develop group, 59-60, asst tech dir, 60-62, tech dir, 62-64, dir, Chem Res Lab, Cent Res Lab, 64-65, corp tech planning & coordr, 65-69, gen mgr, New Bus Ventures Div, 69-70, exec dir, Cent Res Labs, 70-73, vpres, Com Chem Div, 73-81, vpres, res & develop, Indust & Consumer Sector, 81-82, vpres, res & develop, Minn Mining & Mfg Co, 82-88; sr vpres, res & develop, 3M, 88-90; RETIRED. *Concurrent Pos:* Mem, bd regents, Milwaukee Sch Eng Space Appln Bd, Comt Indust Appln Microgravity Environ, Nat Res Coun, adv panel, Off Tech Assessment, Cong US, Materials & Indust Appln Panel, Strategic Defense Orgn, Dept Defense, Nat Acad Eng; mem bd dirs, Indust Res Inst, Inc; trustee, Univ Nebr Found; Minn Mining & Mfr Ind Res Inst Inc; chmn, Minn High Technol Coun, 90-91. *Honors & Awards:* Maurice Holland Award, Indust Res Inst, 89; E B Barnes Award, Am Chem Soc, 91. *Mem:* Nat Acad Eng; fel AAAS; Am Chem Soc; Am Inst Chem Eng. *Res:* Preparation and reaction of polymers; fluorocarbons and the Michael reactions; coated abrasives; analysis of research projects; technology transfer; innovation in a large corporation. *Mailing Add:* 1390 W Skillman Ave St Paul MN 55113

KROGH, THOMAS EDVARD, b Peterborough, Ont, Jan 12, 36; m 61; c 4. GEOCHEMISTRY. *Educ:* Queen's Univ, Ont, BSc, 59, MSc, 61; Mass Inst Technol, PhD(isotope geol), 64. *Prof Exp:* mem sci staff, Carnegie Inst Geophys Lab, 66-75; DIR GEOCHRONOLOGY LAB, GEOL DEPT, ROYAL ONT MUS, 75- *Concurrent Pos:* Fel, Carnegie Inst Dept Terrestrial Magnetism, 64-66. *Honors & Awards:* Logan Medal, Geol Asn Can, 89. *Mem:* Geol Asn Can; fel Am Geophys Union. *Res:* Isotope geology; use of isotopic variation in nature as natural tracers in geological processes; geochronology and genesis of rock systems; uranium-lead dating of zircon and low level lead isotopic analyses. *Mailing Add:* Dept Geol 100 Queen's Park Toronto ON M5S 2C6 Can

KROGMANN, DAVID WILLIAM, b Washington, DC, Oct 21, 32; m 58; c 3. BIOLOGY. *Educ:* Cath Univ Am, AB, 53; Johns Hopkins Univ, PhD(biochem), 57. *Prof Exp:* Fel biochem, Johns Hopkins Univ, 57-58; res assoc, Univ Chicago, 58-60; from asst prof to prof, Wayne State Univ, 60-67; PROF BIOCHEM, PURDUE UNIV, WEST LAFAYETTE, 67- *Mem:* Am Soc Biol Chemists; Am Soc Plant Physiol. *Res:* Biological chemistry; biochemistry of electron transport and phosphate metabolism in photosynthesis; understanding of the photosynthetic mechanism in green plants by studying the structures of individual proteins and the interaction of these proteins in the photosynthetic membrane. *Mailing Add:* Dept Biochem Purdue Univ West Lafayette IN 47907

KROGSTAD, BLANCHARD ORLANDO, b Winger, Minn, Oct 6, 21; m 46; c 3. INSECT ECOLOGY, INVERTEBRATE ECOLOGY. *Educ:* Bemidji State Col, BA, 46; Univ Minn, MA, 48, PhD, 51. *Prof Exp:* Asst prof biol, St Olaf Col, 51-54; from asst prof to assoc prof, 54-63, PROF BIOL, UNIV MINN, DULUTH, 63-, HEAD DEPT, 78- *Concurrent Pos:* Mem staff, Rockefeller Found, Chapingo, Mex, 63-64; researcher, Mexican Inst Coffee, Xalapa, 70-71. *Mem:* Ecol Soc Am; Entom Soc Am; Sigma Xi; Am Inst Biol Sci. *Res:* Ecology of insects and other invertebrates. *Mailing Add:* R 1 Winger MN 56592

KROGSTAD, DONALD JOHN, b New York, NY, Feb 18, 43; m 65; c 2. MEDICINE, BIOLOGY. *Educ:* Bowdoin Col, AB, 65; Harvard Med Sch, MD, 69. *Prof Exp:* Intern, mass Gen Hosp, Boston, 69-70, asst resident, 70-71, sr resident, 75-76, clin & res fel med, 76-78; epidemic intel serv officer,

Parasitic Dis Div & Parasitic Dis Drug Serv, Ctr Dis Control, Ga, 71-73; lectr physiol & med, Med Asst Training Sch, Lilongue, Malawi, 73-75; ASST PROF MED & PATH, SCH MED, WASH UNIV, 78- *Concurrent Pos:* Physician, Lilongue Gen Hosp, 73-75; dir, Microbiol Lab, Barnes Hosp, St Louis, 78-; consult, Jewish Hosp, St Louis, 78- *Mem:* Am Soc Microbiol; Am Soc Trop Med & Hyg; Am Col Physicians; Am Asn Pathologists; Am Col Epidemiology. *Res:* Mechanisms of drug resistance; pathogenicity; epidemiology of nosocomial infection. *Mailing Add:* Micro Lab Barnes Hosp St Louis MO 63110

KROH, GLENN CLINTON, b Philadelphia, Pa, Dec 20, 41; m 67; c 1. PLANT ECOLOGY. *Educ:* Pa State Univ, BS, 66, MS, 70; Mich State Univ, PhD(plant ecol), 75. *Prof Exp:* ASST PROF BIOL, TEX CHRISTIAN UNIV, 75- *Mem:* Ecol Soc Am; Am Inst Biol Sci; AAAS. *Res:* The effects of intra and interspecific competition on the strategies of annual herbs with regard to how they partition available resources into roots, shoots and reproductive tissue. *Mailing Add:* Dept Biol Tex Christian Univ Ft Worth TX 76129

KROHMER, JACK STEWART, b Cleveland, Ohio, Nov 7, 21; m 46; c 3. DIAGNOSTIC & THERAPEUTIC RADIOLOGICAL PHYSICS. *Educ:* Western Reserve Univ, BS, 43, MA, 47; Univ Tex, PhD(biophysics), 61. *Prof Exp:* Sec assoc radiol physics, Atomic Energy Med Res Proj, Western Reserve Univ, 47-57; prof, Univ Tex, Southwestern Med Sch, 57-63; res prof, Roswell Park Mem Inst, 63-66; assoc, Geisinger Med Ctr, 66-72 & Radiol Assocs Erie, 72-79; prof radiol & radiation oncology, Wayne State Univ, Sch Med, 79-84; PRES, J S KROHMER RADIOL PHYSICS CONSULT, 84- *Concurrent Pos:* Chief physicist, Univ Hosp Cleveland, 47-57 & Parkland Mem Hosp, Dallas, Tex, 57-63; Physicist, Radiation Ctr, Fort Worth, Tex, 58-63; prof biophysics, State Univ NY, Buffalo, 63-66 & physics, Bucknell Univ, Lewisburg, Pa, 66-72; pres, Erie Clinic Inc, Pa, 73-76; dir, Div Radiol Physics, Wayne State Univ, 79-84; assoc mem staff, Mt Carmel Mercy Hosp & Detroit Receiving Hosp, 80-84, Harper-Grace Hosp, 82-84; bd trustees, Am Bd Radiol, 81-93; mem, Tex Radiation Adv bd, 86-, chmn, 89-, Georgetown Tex Hosp, Georgetown, Tex Health Care Syst bd, 87, chmn, 89-; consult, US Vet Admin Hosp, Temple, Tex, 84-90. *Honors & Awards:* William D Coolidee Award, Am Asn Physicists Med, 85; Marie Curie Award, Health Physics Soc, 85; Robert J Shalek Award, Am Asn Physicists Med, 88. *Mem:* Am Asn Physicists Med (pres, 74-75); Am Bd Radiol; Radiol Soc NAm (vpres, 79); Am Roetgen Ray Soc; Health Physics Soc; Sigma Xi. *Res:* Biological effects of radiation; radiation imaging; uses of radioactive materials; radiation therapy and radiation protection. *Mailing Add:* 117 Highview Rd Georgetown TX 78628

KROHN, ALBERTINE, chemistry; deceased, see previous edition for last biography

KROHN, BURTON JAY, b St Louis, Mo, Feb 25, 41; m 67; c 2. MOLECULAR PHYSICS. *Educ:* Vanderbilt Univ, BA, 64; Ohio State Univ, MS, 66, PhD(physics), 71. *Prof Exp:* Fel chem physics, Battelle Mem Inst, 71-72; res assoc infrared spectros, Fla State Univ, 73-74; STAFF MEMBER MOLECULAR PHYSICS, THEORET DIV, LOS ALAMOS NAT LAB, 74- *Mem:* Am Phys Soc. *Res:* Quantum-mechanical theory; modeling and computations of energies and properties of vibrating, rotating polyatomic molecules; analysis of positions of absorption lines and band- and line-intensities in high-resolution infrared spectra. *Mailing Add:* Los Alamos Nat Lab Group T-12 Mail Stop 569 Los Alamos NM 87545

KROHN, JOHN LESLIE, b Clarksville, Ark, Dec 7, 58; m; c 2. INSTRUMENTAL NEUTRON ACTIVATION ANALYSIS, NEUTRON SPECTRAL CHARACTERIZATION. *Educ:* Univ Ark, BS, 81, MS, 83. *Prof Exp:* Grad asst basic nuclear eng, Mech Eng Dept, Univ Ark, 81-83; eng res assoc, Nuclear Sci Ctr, Tex Eng Exp Sta, 84-86; mgr reactor opers, Tex A&M Univ, 86-88, asst dir, Nuclear Sci Ctr, 88-91. *Concurrent Pos:* Rad safety officer, AAE/BCS Traders, Inc, 86- *Mem:* Am Nuclear Soc; Am Soc Testing & Mat. *Res:* Instrumental neutron activation analysis; prompt gamma neutron activation analysis; characterization of neutron energy spectra and damage to electronic components by fast neutrons. *Mailing Add:* Dept Eng Ark Tech Univ Russellville AR 72801

KROHN, KENNETH ALBERT, b Stevens Point, Wis, June 19, 45; m 68; c 1. NUCLEAR CHEMISTRY. *Educ:* Andrews Univ, BA, 66; Univ Calif, Davis, PhD(chem), 71. *Prof Exp:* Instr radiation sci, Washington Univ, 71-73; from asst prof to assoc prof radiol, Sch Med, Univ Calif, Davis, 73-81, assoc dir, Crocker Nuclear Lab, 78-80; assoc prof, 81-85, PROF RADIOL & RADIATION ONCOL, SCH MED, UNIV WASH, SEATTLE, 85- *Concurrent Pos:* Mem comt radiopharmaceut & radioassay, Am Col Nuclear Physicians, 75-78. *Mem:* Am Chem Soc; Soc Nuclear Med; Radiation Res Soc; AAAS; Sigma Xi. *Res:* Application of isotopes to biological problems; development of new cyclotron produced radiopharmaceuticals for diagnostic procedures for cancer and heart disease; halogen and technetium chemistry and chemical effects of nuclear transformations. *Mailing Add:* Dept Radiol-Nuclear Med RC-05 Univ Wash Hosp Seattle WA 98195

KROL, ARTHUR J, b Chicago, Ill, Dec 3, 25. DENTISTRY. *Educ:* Loyola Univ, Ill, DDS, 47, BS, 48. *Prof Exp:* Instr full dentures, Sch Dent, Northwestern Univ, 53-54; prof prosthodont & chmn dept, Sch Dent, Loyola Univ, Ill, 54-64; prof prosthodont, Sch Dent, Univ of the Pac, 64-76; chief, Dent Serv, Vet Admin Med Ctr, San Francisco, 74-85; prof prosthodont, sch dent, Univ Calif, San Francisco, 74-85; RETIRED. *Concurrent Pos:* Consult, Vet Admin Hosps, Hines & West Side, Chicago, Ill, 57-61, Dwight, Ill, 61-64 & Palo Alto, Calif, 64-; consult, Vet Admin Hosp, Livermore, Calif, 67- & Letterman Army Hosp, San Francisco, 69- *Mem:* Am Dent Asn; Am Prosthodont Soc; Int Asn Dent Res; Int Col Appl Nutrit; Am Col Dent. *Res:* Prosthodontics; factors involved in denture retention, especially the application of the principles of hydraulics in establishing peripheral seal; new removable partial denture designs for esthetics and plaque reduction. *Mailing Add:* 765 14th Ave San Francisco CA 94118

KROL, GEORGE J, b Wilno, Poland, June 6, 36; US citizen; m 62; c 3. PHYSICAL CHEMISTRY, ANALYTICAL CHEMISTRY. *Educ:* Univ Rochester, BS, 58; Rutgers Univ, PhD(phys anal chem), 68. *Prof Exp:* Res technician, Med Sch, Univ Rochester, 58-59; chemist, Hoffmann-La Roche Inc, 60-62; res asst phys chem, Rutgers Univ, 63-67; sr res chemist, 67-75, sect head, Ayerst Labs, Inc, 75-78; supvr, 78-84, STAFF SCIENTIST, MILES PHARMACEUT INC, 84- *Mem:* Am Chem Soc; Am Asn Pharmaceut Scientists. *Res:* Biochemistry; radiochemistry and kinetics; analytical method development, especially chromatography; drug metabolism and pharmacokinetic analysis. *Mailing Add:* Prod Develop Lab Miles Pharmaceut Inc West Haven CT 06516

KROL, JOSEPH, b Warsaw, Poland, Jan 14, 11; US citizen; m 52. MECHANICAL & INDUSTRIAL ENGINEERING. *Educ:* Warsaw Tech Univ, MS, 37; Univ London, PhD(mech eng), 47. *Prof Exp:* Chief tech off, Boryszew Co, Ltd, Poland, 37-39; asst works engr, George Clark Co, Ltd, Eng, 40-41; tech off ammunition prod, Brit Ministry Supply, 41-45; res scientist, Univ London, 46-47; consult engr, Howard Smith Paper Mills Co, Ltd, Can, 48-51; assoc prof mech eng, Univ Man, 51-56; prof indust eng, 56-80, prof systs eng, 69-80, EMER PROF INDUST & SYSTS ENG, GA INST TECHNOL, 80- *Honors & Awards:* Stephenson Prize, Brit Inst Mech Eng, 51. *Mem:* AAAS; Am Soc Mech Eng; Sigma Xi; fel Inst Mech Eng; Inst Mgt Sci; Am Asn Univ Prof. *Res:* Economic analysis and computer simulation of complex systems. *Mailing Add:* 311 Tenth St NW Atlanta GA 30318

KROLIK, JULIAN H, b Detroit, Mich, Apr 4, 50; m 83; c 2. THEORETICAL ASTROPHYSICS. *Educ:* Mass Inst Technol, SB, 71; Univ Calif, Berkeley, PhD(physics), 77. *Prof Exp:* Mem, Inst Advan Study, Mass Inst Technol, 77-79, scientist, Ctr Theoret Physics & Space Res, 79-81; res assoc, Smithsonian Ctr Astrophys, Harvard Univ, 81-84; asst prof, 84-86, ASSOC PROF, PHYSICS & ASTRON, JOHNS HOPKINS UNIV, 86- *Concurrent Pos:* Lectr, dept astron, Harvard Univ, 81-84; vis prof, Col France, 84. *Mem:* Am Astron Soc. *Mailing Add:* Dept Physics & Astron Johns Hopkins Univ Baltimore MD 21218

KROLL, BERNARD HILTON, b Brooklyn, NY, Sept 8, 22; m 47; c 3. STATISTICS, SYSTEMS SCIENCE. *Educ:* Brooklyn Col, BA, 47. *Prof Exp:* Air transp economist, Civil Aeronaut Bd, 48-51; anal statistician, Nat Inst Ment Health, 51-58; supvry statistician, Nat Inst Neurol Dis & Blindness, 58-67, supvry systs analyst, Nat Inst Neurol & Commun Disorders & Stroke, 67-77, assoc chief of biomet & epidemiol, 77-81; RETIRED. *Concurrent Pos:* Consult, NIH, 80-81. *Mem:* Fel AAAS; Asn Systs Mgt; fel Am Pub Health Asn; fel Royal Soc Health. *Res:* Design of management and administrative systems involving the use of computers; management of epidemiologic and statistical research in medical and related fields; epidemiology, manpower and information systems. *Mailing Add:* 3507 Farthing Dr Wheaton MD 20906

KROLL, CHARLES L(OUIS), b Boston, Mass, Sept 15, 23; m 46, 56; c 6. CHEMICAL ENGINEERING. *Educ:* Yale Univ, BE, 44; Mass Inst Technol, SM, 49, ScD(chem eng), 51. *Prof Exp:* Field engr, E I du Pont de Nemours & Co, 46-47; res assoc chem eng, Mass Inst Technol, 48-50; res chemist, Celanese Corp Am, 51-52; head, eng sect microbiol pilot plant, Squibb Inst Med Res, Div Olin Mathieson Chem Corp, 52-55, tech data mgr, 55-64, assoc dir drug regulatory affairs, 64-72, dir regulatory opers, 72-87; RETIRED. *Concurrent Pos:* Regulatory affairs consult. *Mem:* Am Chem Soc; Am Inst Chem Engrs. *Res:* Heat transfer to air at high temperatures and low Reynolds numbers; theoretical supersonic performance of a ducted rocket; industrial fermentations. *Mailing Add:* PO Box 9486 Noank CT 06340

KROLL, EMANUEL, b New York, NY, Feb 26, 19; m 43; c 2. PHARMACEUTICALS, COSMETICS. *Educ:* City Col New York, BS, 42; Canisius Col, MS, 46. *Prof Exp:* Chemist, Chem Construct Co, 42-43 & Fedders Mfg Co, 43-44; res chemist, Manhattan Proj, Linde Air Prod Co, 44-46; vpres res, Chem Res Assocs, 46-51; dir res, Heparin, Inc, 51-52; tech dir pharmaceut divs, Int Latex Corp, 52-69; dir res & develop, Standard Chem Prod, Inc, Henkel Co, 69-74; DIR RES SERV, INT PLAYTEX, INC, 74- *Mem:* AAAS; Am Chem Soc; Soc Cosmetic Chem; Am Pharmaceut Asn. *Res:* Drug and cosmetic product development; protein, amino acid and enzyme research; organic synthesis; antiseptics; iodine compounds; ion exchange; chemical specialty product development; quality control; regulatory agency liaison administration. *Mailing Add:* Playtex Family Prod Inc PO Box 728 Paramus NJ 07652

KROLL, HARRY, b Chicago, Ill, Nov 28, 14; m 45; c 3. ELECTROLESS GOLD DEPOSITION, BRIGHT TIN LEAD ELECTROPLATING. *Educ:* Univ Ill, BS, 38, Univ Chicago, PhD(org chem), 42. *Prof Exp:* Chemist, Alrose chem co, Cranston, RI, 42-49; prin investr, Am Cancer Soc, Yale Univ, 51-58; res dir, Geigy Chem Corp, 51-58; prin investr, US Atomic Energy Comt, 58-64; res dir & vpres, Philip A Hunt Chem Corp, 64-82; SR SCIENTIST, TECHNIC INC, 83- *Concurrent Pos:* Adj prof, Univ RI, 74-75. *Mem:* Am Chem Soc. *Res:* Chelating afents, photo resist and metal deposition. *Mailing Add:* 615 Middle Rd E Greenwich RI 02818

KROLL, JOHN ERNEST, b Los Angeles, Calif, Aug 15, 40; m 73; c 3. FLUID DYNAMICS, PHYSICAL OCEANOGRAPHY. *Educ:* Univ Calif, Los Angeles, BS, 63, MS, 66; Yale Univ, PhD(eng & appl sci), 73. *Prof Exp:* Res fel phys oceanog, Nova Oceanog Lab, 72-75; instr, Mass Inst Technol, 75-76; asst prof, 76-80, ASSOC PROF APPL MATH, OLD DOMINION UNIV, 81- *Concurrent Pos:* Sr Res Assoc, Nat Res Coun Dept Oceanog, Naval Postgrad Sch, Monterey, Calif, 85-86. *Mem:* Soc Indust & Appl Math; Am Geophys Union. *Res:* Theoretical investigation of the generation and propagation of inertial oscillations and the stability of fronts and eddies. *Mailing Add:* Dept Math Old Dominion Univ Norfolk VA 23529

KROLL, MARTIN HARRIS, b Washington, DC, June 19, 52; m 75; c 3. CLINICAL CHEMISTRY, CLINICAL BIOCHEMISTRY. *Educ:* Univ Md, Col Park, BS, 74; Univ Sch Med, MD, 78. *Prof Exp:* Resident, path, Univ Md Hosp, 78-82; fel, clin chem, 82-84, MED STAFF, CLIN CHEM SERV,

CLIN PATH DEP, NIH, 84- *Concurrent Pos:* Fel, Nat Sci Found Summer Res, 73. *Honors & Awards:* Young Investr, Acad Clin Lab Physicians & Scientists, 83; Past Chmn Award, Capital sect, Am Asn Clin Chem, 90. *Mem:* Am Asn Clin Chem; Am Chem Soc; Acad Clin Lab Physicians & Scientists. *Res:* Assessment of interferences in clinical chemistry; standardization of cholesterol determinations; biological roles of magnesium; dynamics of biological systems. *Mailing Add:* NIH Bldg 10 Rm 2C-407 Bethesda MD 20892

KROLL, NORMAN MYLES, b Tulsa, Okla, Apr 6, 22; m 45; c 4. MATHEMATICAL PHYSICS. *Educ:* Columbia Univ, AB, 42, AM, 43, PhD(physics), 48. *Prof Exp:* Asst physics, Columbia Univ, 42-44, mem sci staff, radiation lab, 43-62, from asst prof to prof physics, 49-62; chmn dept, 63-65 & 83-88, prof, 62-91, EMER PROF PHYSICS, UNIV CALIF, SAN DIEGO, 91- *Concurrent Pos:* Nat Res Coun fel, Inst Advan Study, 48-49; Guggenheim fel, Rome, 55-56; mem staff, Jason Div, Inst Defense Anal, 60-81; NSF sr fel, Europ Orgn Nuclear Res, 65-66. *Mem:* Nat Acad Sci; fel Am Phys Soc; Sigma Xi. *Res:* Nonlinear optics; microwave and nuclear physics; magnetron design; quantized field theories; elementary particle theory; quantum field theory; free electron lasers; excelerator physics. *Mailing Add:* MS 0319 9500 Gilman Dr La Jolla CA 92093

KROLL, ROBERT J, b Cincinnati, Ohio, May 1, 28; m 54; c 5. AEROSPACE & STRUCTURAL ENGINEERING. *Educ:* Univ Cincinnati, BS, 49, MS, 56; Mich State Univ, PhD(appl mech), 62. *Prof Exp:* Engr reinforced concrete, Pollak Steel Co, 49-51; sr engr stress & design, Gen Elec Co, 52-57; from asst prof aeronaut eng to prof aerospace eng, Univ Cincinnati, 57-68, actg head dept, 70-71 & 78-79, Bradley Jones Prof aerospace eng, 75-82; RETIRED. *Concurrent Pos:* Consult, Gen Elec Corp, 57-, Univ Dayton Res Div, 62-63, Aeronca Mfg Corp, 64-, Rockwell Standard Corp, 64-, Cincinnati Shaper Co, 65- & Cincinnati Industs, 67-; NSF sci fac fel, 61-62; proj adv student design terms, 81-82 & 82-83. *Honors & Awards:* Weston Award, 49. *Mem:* Assoc fel Am Inst Aeronaut & Astronaut; Soc Exp Mech; Am Soc Eng Educ; Sigma Xi. *Res:* Thermal stresses and stability; energy methods applied to lightweight structures; structural testing using strain gages and photostress; experimental wave propagation. *Mailing Add:* 2579 Beechmar Dr Cincinnati OH 45230-1204

KROM, MELVEN R, b Hospers, Iowa, Oct 4, 31; m 53. MATHEMATICS. *Educ:* Univ Iowa, BA, 54, MS, 57; Univ Mich, PhD(math), 63. *Prof Exp:* Asst prof, 63-69, assoc prof, 69-76, PROF MATH, UNIV CALIF, DAVIS, 76- *Mem:* Asn Symbolic Logic; Math Asn Am; Am Math Soc. *Res:* Mathematical logic. *Mailing Add:* Dept Math Univ Calif Davis CA 95616

KROMAN, RONALD AVRON, b Minneapolis, Minn, Mar 30, 27; m 62; c 3. GENETICS. *Educ:* Univ Minn, PhD(zool), 57. *Prof Exp:* From asst prof to assoc prof, 59-69, PROF BIOL, CALIF STATE UNIV, LONG BEACH, 69- *Concurrent Pos:* Consult revision jr high & high sch math curricula, Sch Math Study Group & Calif State Dept Educ; statist consult, Arthritis Found, Mem Hosp, Long Beach, Calif; adj prof, Chapman Col, 75- *Mem:* Sigma Xi. *Res:* Soc Study Evolution; Am Genetic Asn; Genetics Soc Am; Am Inst Biol Sci; Sigma Xi. Res: Drosophila genetics; tumors; melanin metabolism; eye pigmentation; symbiotic associations in the Acarina. *Mailing Add:* Dept Biol Calif State Univ Long Beach CA 90840

KROMANN, PAUL ROGER, b Racine, Wis, Nov, 15, 29; m 60. PHYSICAL CHEMISTRY. *Educ:* Hope Col, AB, 52; Univ Calif, PhD, 57. *Prof Exp:* Chemist org synthesis, Hope Col, 52; asst chem, Univ Calif, 52-55; chemist, Plastics Fundamental Res Lab, Dow Chem USA, Mich, 57-71; ASST PROF CHEM, FT VALLEY STATE COL, 71- *Mem:* Am Chem Soc. *Res:* Radiation chemistry; photochemistry; fluorescence lifetimes. *Mailing Add:* 348 Clairmont Warner Robins GA 31088

KROMANN, RODNEY P, b Stockton, Calif, Sept 3, 31; m 79; c 4. ANIMAL NUTRITION, BIOENERGETICS. *Educ:* Calif State Polytech Univ, BS, 58; Univ Calif, Davis, MS, 60, PhD(nutrit), 66. *Prof Exp:* Asst prof nutrit, NMex State Univ, 63-68; nutritionist, Shell Develop Co, 68-70; asst prof nutrit, Wash State Univ, 71-73, assoc prof, 73-78; PRES, NUTRIT INT, 78- *Concurrent Pos:* Nutrit consult, 65-68. *Mem:* AAAS; Am Soc Animal Sci; Am Inst Nutrit; Am Dairy Sci Asn. *Res:* Animal bioenergetics and nutrient and energy metabolism; mathematical modeling of protein and fat synthesis and of heat loss for the determination of maintenance by a radiometer. *Mailing Add:* Nutrit Int PO Box 190 Ephrata WA 98823

KROMBEIN, KARL VON VORSE, b Buffalo, NY, May 26, 12; m 42; c 3. ENTOMOLOGY. *Educ:* Cornell Univ, BS, 34, AM, 35, PhD(entom), 60; Univ Peradeniya, Sri Lanka, PhD(zool), 80. *Prof Exp:* Asst, NY State Exp Sta, 36-38; entomologist, Niagara Sprayer & Chem Co, NY, 39-40; assoc entomologist, Bur Entom & Plant Quarantine, USDA, 41-42, entomologist & invests leader, 46-65; chmn dept entom, 65-71, sr entomologist, 71-80, SR SCIENTIST, SMITHSONIAN INST, 80- *Concurrent Pos:* Civilian consult, Surgeon Gen, US Air Force, 72-79, emer consult, 79- *Mem:* Fel AAAS; Am Entom Soc; hon mem Egyptian Entom Soc; Entom Soc Am. *Res:* Systematics, bionomics, ecology and behavior of Hymenoptera Aculeata. *Mailing Add:* Dept Entom NHB Stop 105 Smithsonian Inst Washington DC 20560

KROMER, LAWRENCE FREDERICK, b Sandusky, Ohio, Sept 1, 50. NEUROBIOLOGY, NEURAL REGENERATION. *Educ:* Univ Chicago, BA, 72, PhD(anat), 77. *Prof Exp:* Fel, Dept Histol, Univ Lung, Sweden, 77-79; asst res neuroscientist, Univ Calif, San Diego, 79-81; ASST PROF, DEPT ANAT & NEUROBIOL, UNIV VT, 81-; AT GEORGETOWN UNIV, 85- *Concurrent Pos:* Prin investr, pvt found res grants, NIH, 79-; fel, A P Sloan Found, 80-86. *Mem:* Soc Neurosci; Int Soc Develop Neurosci; Am Asn Anatomists; AAAS. *Res:* Development and regeneration in the mammalian central nervous system by utilizing and intracephalic implantation technique which allows the transplantation of embryonic and neonatal neural tissue into the brain of neonatal and adult rodents. *Mailing Add:* Dept Anat & Cell Biol Georgetown Univ Med Dent Bldg 3900 Reservoir Rd NW Washington DC 20007

KROMHOUT, ROBERT ANDREW, b Elgin, Ill, Oct 23, 23; m 50; c 3. CHEMICAL PHYSICS, COGNITIVE SCIENCES. *Educ:* Kans State Univ, BS, 47; Univ Ill, Urbana-Champaign, MS, 48, PhD(physics), 52. *Prof Exp:* Asst prof physics, Univ Ill, 52-56; from asst prof to assoc prof, 56-63, PROF PHYSICS, FLA STATE UNIV, 63- *Concurrent Pos:* Dir, Inst Cognitive Sci, 83-; head dept, Fla State Univ, 60-62. *Mem:* Am Phys Soc; Am Asn Physics Teachers; Sigma Xi; Cognitive Sci Soc. *Res:* Phase transitions; statistical mechanics; intermolecular forces; Van der Waal's interactions, phase transitions, statistical mechanisms, cognition and pedagogy. *Mailing Add:* Dept Physics Fla State Univ Tallahassee FL 32306-3016

KROMMINGA, ALBION JEROME, b Mille Lacs Co, Minn, June 20, 33; div; c 2. THEORETICAL PHYSICS. *Educ:* St Cloud State Col, BS, 55; Univ Minn, PhD(physics), 61. *Prof Exp:* Asst prof physics, Iowa State Univ, 61-63 & Idaho State Univ, 63-65; assoc prof, 65-69, PROF PHYSICS, CALVIN COL, 69- *Mem:* Am Phys Soc; Sigma Xi; Am Sci Affil. *Res:* Theory of electromagnetic, atomic, and nuclear reactions. *Mailing Add:* Dept Physics Calvin Col 3201 Burton St SE Grand Rapids MI 49546

KRON, GERALD EDWARD, b Milwaukee, Wis, Apr 6, 13; m 46; c 5. ASTRONOMY. *Educ:* Univ Wis, BS, 33, MS, 34; Univ Calif, PhD(astrophys), 38. *Prof Exp:* Jr astronr, Lick Observ, Univ Calif, 38-42, asst astronr, 42-47, assoc astronr, 47-52, astronr, 52-65; dir, US Naval Observ, Flagstaff Sta, 65-73; sr res fel, Australian Nat Univ, 74-76; RETIRED. *Concurrent Pos:* Res assoc, Mass Inst Technol, 40-41 & Calif Inst Technol, 42-45; physicist, US Naval Ord Test Sta, Calif, 45; astron res, Pinecrest Observ, Flagstaff, Ariz, 76-85, Pinecrest Observ, Hawaii, 85- *Mem:* AAAS; Am Astron Soc; Int Astron Union; Royal Astron Soc. *Res:* Electronic camera research and development; investigation of integrated properties of globular clusters; distribution of interstellar reddening. *Mailing Add:* 2929 Poni Moi Rd Honolulu HI 96815

KRONAUER, RICHARD ERNEST, b Paterson, NJ, Aug 5, 25; m 48; c 3. MECHANICAL ENGINEERING, BIOMEDICAL PHYSICS. *Educ:* Stevens Inst Technol, MechEng, 47; Harvard Univ, SM, 48, PhD, 51. *Prof Exp:* From instr to assoc prof, 51-64, PROF MECH ENG, HARVARD UNIV, 64- *Concurrent Pos:* Consult, Pratt & Whitney Aircraft Div, United Aircraft Corp, 51- 58, Flow Corp, 53-67, Baldwin-Lima-Hamilton Corp, 56-61, Arthur D Little Co, 60-67 & Campbell-Kronauer Assoc, 80-; NSF fel, 64 & 71; NIH int fel, 78. *Mem:* Am Soc Mech Engrs; AAAS; Sleep Res Soc; Asn Res Vision & Ophthal; Sigma Xi. *Res:* Fluid dynamics and turbulence; nonlinear oscillations; visual system information processing; human circadian oscillators. *Mailing Add:* 14 Chauncy St Unit 7 Cambridge MA 02138

KRONBERG, PHILIPP PAUL, b Toronto, Ont, Sept 16, 39; m 63; c 3. RADIO ASTRONOMY, ASTROPHYSICS. *Educ:* Queen's Univ Kingston, BSc, 61, MSc, 63; Univ Manchester, PhD(physics), 67. *Prof Exp:* Lectr physics, Univ Manchester & Jodrell Bank, 66-68; from asst prof to assoc prof, 68-78, PROF ASTRON, UNIV TORONTO, 78- *Concurrent Pos:* Mem, Assoc Comt Astron, Nat Res Coun Can, 71-74, ed, 74; mem grant selection comt, Univ Astron & Space Res Can, 74-78; Alexander von Humboldt sr res fel, Max-Planck-Inst fur Radioastronomie, Germany, 75-77; chmn, Assoc Univ Inc; vis comt, US Nat Radio Astron Observ, 82; mem, Univ Toronto Res Bd, 78-82; mem & chmn, VLA Adv Comt, 78-82; sr von Humboldt fel, 75; Guggenheim fel, 85; chmn, Nat Res Coun Millimetre Telescope Steering Comt, 83- *Mem:* Am Astron Soc; Int Astron Union; Can Astron Soc; Sigma Xi. *Res:* Measurement and analysis of magnetic fields in space; structure of our galaxy; extragalactic astrophysics; radioastronomy; high-energy processes in space. *Mailing Add:* Dept Astron Univ Toronto 60 Saint George St Toronto ON M5S 1A7 CAN

KRONBERGER, KARLHEINZ, b Vienna, Austria, Jan 24, 40. POLYMER CHEMISTRY, ORGANIC CHEMISTRY. *Educ:* Vienna Tech Univ, Dipl Ing, 64; Univ Nebr, PhD(org chem), 67. *Prof Exp:* Res assoc, Mass Inst Technol, 67-68; sr res chemist, 68-78, RES SECT MGR, RES DIV, ROHM AND HAAS CO, 78- *Mem:* Am Chem Soc. *Res:* Coatings and polymer research; materials for electronics industry. *Mailing Add:* Rohm & Haas Res Labs Spring House PA 19477

KRONE, LAWRENCE JAMES, b Chicago, Ill, Sept 8, 40; m 68; c 2. VECTOR CONTROL, SOLID WASTE MANAGEMENT. *Educ:* North Park Col, BS, 63; Yale Univ, MPH, 67; Univ Ill, PhD(entom), 71. *Prof Exp:* Asst prof environ health, Dept Health, Ore State Univ, 72-75; proj dir, Nat Environ Health Asn, 75-77, exec dir environ health, 77-82; CONSULT, PROF EXAM SERV, 78- *Concurrent Pos:* Pub health sanitarian, Benton County Health Dept, Ore, 73-74 & Nat Park Serv, 75; honorarium fac, Univ Denver & Univ Colo, 81. *Res:* Environmental health science, educational and professional roles in environmental health; entomology. *Mailing Add:* 514 Woodsedge Rd Dover DE 19901

KRONE, LESTER H(ERMAN), JR, b St Louis, Mo, Oct 8, 31; m 53; c 3. MANAGEMENT SCIENCE. *Educ:* Wash Univ, BS, 52, DSc(chem eng), 55; Univ Ill, MS, 53. *Prof Exp:* Group leader process develop dept, Uranium Div, Mallinckrodt Chem Works, 55-56; appl sci rep, Int Bus Mach Corp, 56-57, mgr appl sci, Mo, 57-58; mgr oper anal, Monsanto Co, 58-68; independent consult, 68-73; MGR PROD PLANNING & MKT, MCDONNELL-DOUGLAS AUTOMATION CO, 73- *Concurrent Pos:* Lectr, Wash Univ; assoc prof, Southern Ill Univ. *Mem:* Asn Comput Mach; Am Inst Chem Engrs; Opers Res Soc Am; Inst Mgt Sci. *Res:* Applied mathematics; computer techniques; operations research; mathematical programming, statistics and simulation. *Mailing Add:* 749 Chatelet Woods St Louis MO 63135

KRONE, RALPH WERNER, b Berlin, Ger, May 18, 19; US citizen; m 42; c 3. NUCLEAR PHYSICS. *Educ:* Antioch Col, BS, 42; Univ Ill, MS, 43; Johns Hopkins Univ, PhD, 49. *Prof Exp:* From asst prof to assoc prof, Univ Kans, 48-61, actg chmn dept, 65-66, prof physics, 61-82, prof physics & astron, 82; RETIRED. *Mem:* Fel Am Phys Soc. *Res:* Nuclear structure; spectroscopy of light and medium light nuclei. *Mailing Add:* Box 21 Washington NH 03280

KRONE, RAY B, b Long Beach, Calif, June 7, 22; m 46; c 2. HYDRAULIC & SANITARY ENGINEERING. *Educ:* Univ Calif, BS, 50, MS, 58, PhD(sanit eng), 62. *Prof Exp:* From asst to assoc res engr, Univ Calif, Berkeley, 50-64, lectr, 62-64; chmn dept civil eng, 68-72, assoc dean res, Col Eng, 72-88, from assoc prof to prof, 64-88, EMER PROF CIVIL ENG, UNIV CALIF, DAVIS, 88- *Concurrent Pos:* Consult, US Army Corps Engrs, 60-, Calif Atty Gen, 69- & engr firms. *Mem:* AAAS; Am Geophys Union; Water Pollution Control Fedn; Am Soc Civil Engrs; Estuarine Res Fedn. *Res:* Soil science; sediment transport in estuaries, particle transport in surface and ground waters; properties of water; soil water relations. *Mailing Add:* Dept Civil Eng Univ Calif Davis CA 95616

KRONENBERG, RICHARD SAMUEL, b Chicago, Ill, Aug 7, 38; m 63; c 3. MEDICINE, PHYSIOLOGY. *Educ:* Northwestern Univ, BA, 60, MD, 63. *Prof Exp:* USPHS res fel, Cardiovasc Res Inst, Med Ctr, Univ Calif, San Francisco, 68-70; asst prof med, Pulmonary Div, Univ Minn Hosp, 70-74, assoc prof med, 74-78, prof med & head, 78-84; PROF & CHMN DEPT MED, UNIV TEX HEALTH CTR, TYLER, 84- *Concurrent Pos:* Consult, Vet Admin Hosp, Minneapolis, 71-84. *Mem:* AAAS; Am Physiol Soc; fel Am Col Physicians; Am Fedn Clin Res; Am Thoracic Soc. *Res:* Repiratory physiology; pulmonary disease. *Mailing Add:* Dept Med Univ Tex Health Ctr PO Box 2003 Tyler TX 75710

KRONENBERG, STANLEY, b Krosno, Poland, May 3, 27; US citizen; m 53; c 2. PHYSICS. *Educ:* Univ Vienna, PhD(physics, math, chem), 52. *Prof Exp:* Physicist, 53-88; RES PHYSICIST, US ARMY COMMUN COMMAND, EW/RSTA CTR, 70- *Concurrent Pos:* Army liaison rep, Nat Res Coun-Nat Acad Sci, 61-; consult, Fed Emergency Mgt Agency & Dept Energy. *Mem:* Am Phys Soc. *Res:* Nuclear radiation detection; radiology. *Mailing Add:* Hollow Rd Shillman NJ 08558

KRONENTHAL, RICHARD LEONARD, b New York, NY, Oct 6, 28; m 49; c 2. POLYMER CHEMISTRY, SURGICAL DEVICES. *Educ:* Brooklyn Col, BS, 51; Polytech Inst New York, PhD(chem), 55. *Prof Exp:* Sr proj chemist org chem, Colgate Palmolive Co, 54-57; mgr dept org & polymer chem, Ethicon, Inc, Johnson & Johnson, 57-68, assoc dir res, 68-72, dir res, 72-84, dir res & develop, 84-89; PRES, KRONENTHAL ASSOCS, INC, 89- *Concurrent Pos:* Instr, Polytech Inst New York, 58-64. *Mem:* AAAS; Am Chem Soc; Am Soc Artificial Internal Organs; Am Inst Chem; Royal Soc Chem; Sigma Xi. *Res:* Chemistry of proteins; surgical devices; biomaterials; biodegradable polymers. *Mailing Add:* 33 Garwood Rd Fair Lawn NJ 07410

KRONENWETT, FREDERICK RUDOLPH, b Newark, NJ, July 29, 23; m 50; c 5. BACTERIOLOGY. *Educ:* Upsala Col, BSc, 48; Rutgers Univ, MS, 50, PhD(dairy bact), 54. *Prof Exp:* Dir qual control, Hohneker Dairy Co, 54-57; fel, Rutgers Univ, 58; from instr to prof biol sci, Fairleigh Dickinson Univ, 58-89; DIR, AM BIOL CONTROL LABS, 57- *Concurrent Pos:* Lectr, Upsala Col, 58-59; chmn Adv Panel Biol Indicators, US Pharmacopeia, mem, Steril Comt, US Pharmacopeia XVIII; consult, Bergen Pines Hosp, 59; dir bioanal lab, NJ Bd Med Examr, 59; expert adv, Int Atomic Energy Agency, Vienna, 72; specialist microbiol, Am Soc Clin Pathologists, 65; fel, Upsala Col, 72- *Mem:* Am Soc Microbiol; Am Pub Health Asn; Inst Food Technologists; NY Acad Sci; Sigma Xi. *Res:* Identification of thermophilic bacteria; thermal death time studies on Brucella abortus; food poisoning; general microbiology; sterilization and disinfection; irradiation microbiology and dosimetry; medical device development; water microbiology. *Mailing Add:* PO Box 505 Tenafly NJ 07670

KRONER, KLAUS E(RLENDUR), b Gottingen, Ger, July 19, 26; nat US; m 59; c 2. INDUSTRIAL ENGINEERING, ENGINEERING GRAPHICS. *Educ:* Col Wooster, BA, 49; NY Univ, BEE, 57; Am Int Col, MBA, 62. *Prof Exp:* Instr eng drawing, NY Univ, 50 & 51-55; instr eng graphics, Univ Maine, 55-57; from asst prof to assoc prof basic eng, Univ Mass Amherst, 57-69, assoc prof indust eng & opers res, 69-88, asst head, 81-86, EMER PROF INDUST ENG & OPERS RES, UNIV MASS, AMHERST, 88- *Concurrent Pos:* Res scientist, indust develop div, Ga Inst Technol, 71. *Honors & Awards:* Distinguished Serv Award, Graphics Div, Am Soc Eng Educ, 85. *Mem:* Am Soc Eng Educ. *Res:* Computer graphics; engineering economy; plant location and layout; industrial development. *Mailing Add:* PO Box 354 Leverett MA 01054

KRONFELD, DAVID SCHULTZ, b Auckland, NZ, Nov 5, 28; m 57; c 2. NUTRITION, VETERINARY PHYSIOLOGY. *Educ:* Univ Queensland, BVSc, 52, BSc, 54, MVSc, 57, DSc(biochem), 72; Univ Calif, Davis, PhD(physiol), 59; Am Col Vet Internal Med, dipl, 73. *Hon Degrees:* MA, Univ Pa, 71. *Prof Exp:* Demonstr physiol, Univ Queensland, 53-54, lectr vet physiol, 54-57; lectr vet med, Univ Calif, Davis, 58-59, asst prof physiol, 59-60; from asst prof to assoc prof pharmacol, 60-67, PROF NUTRIT, SCH VET MED, UNIV PA, 67-, CHIEF SECT NUTRIT, 81- *Mem:* Am Dairy Sci Asn; Am Physiol Soc; Am Vet Med Asn; Am Inst Nutrit. *Res:* Nutrition and high performance of exercise, growth, pregnancy and lactation; metabolic disorders, ketosis, hypercholesterolemia, hypoglycemia, hypocalcemia, hypomagnesemia; tracer methodology, kinetic analysis, regulatory models; preventive medicine, health economics. *Mailing Add:* Dept Animal Sci Va Tech Blacksburg VA 24061-0306

KRONGELB, SOL, b Jersey City, NJ, Aug 15, 32; m 53; c 3. PHYSICS. *Educ:* NY Univ, BS, 53; Mass Inst Technol, MS, 55, PhD, 58. *Prof Exp:* Asst physics, Res Lab Electronics, Mass Inst Technol, 54-58; assoc, 58-60, MEM RES STAFF, IBM, 60-, MGR, DEVICE FABRICATION & MAGNETICS, 86- *Mem:* Am Vacuum Soc; Inst Elec & Electronics Engrs. *Res:* Microwave spectroscopy; paramagnetic resonance; parametric devices; semiconductor technology; thin film adhesion; deposition and properties of magnetic and glass thin films; device fabrication technology. *Mailing Add:* Greenlawn Rd Katonah NY 10536

KRONK, HUDSON V, b Port Jervis, NY, Oct 6, 38; m 59; c 3. MATHEMATICS. *Educ:* Rensselaer Polytech Inst, BS, 59; Mich State Univ, MS, 60, PhD(math), 64. *Prof Exp:* Lectr math, Kalamazoo Col, 63; asst prof, 64-68, ASSOC PROF MATH, STATE UNIV NY, BINGHAMTON, 68- *Mem:* Am Math Soc; Math Asn Am. *Res:* Graph theory. *Mailing Add:* Dept Math State Univ NY Binghamton NY 13901

KRONMAL, RICHARD AARON, b Los Angeles, Calif, May 3, 39; m 60; c 3. BIOSTATISTICS. *Educ:* Univ Calif, Los Angeles, AB, 61, PhD(biostatist), 64. *Prof Exp:* From instr to assoc prof, 64-75, chmn biomath group, 73-85, PROF BIOSTATIST, SCH PUB HEALTH, UNIV WASH, 75- *Concurrent Pos:* Career develop award, 68-73; mem, Renal Cardiovasc Ad Comt, Food & Drug Admin; mem, Epidemiol Study Sect, NIH. *Mem:* Fel Am Statist Asn; Soc Study Human Biol; AAAS; Biomet Soc. *Res:* Mathematical statistics; statistical computing; public health and epidemiology. *Mailing Add:* Dept Biostatist Sch Pub Health Univ Wash Seattle WA 98195

KRONMAN, JOSEPH HENRY, b New York, NY, Apr 4, 31; m 61; c 3. ANATOMY, DENTISTRY. *Educ:* NY Univ, BS, 52, DDS, 55; Columbia Univ, cert orthodont, 59; Med Col Va, PhD(anat), 62. *Prof Exp:* Instr anat, Med Col Va, 61-62; assoc prof orthodont, 62-68, asst to dean, Grad Sch Arts & Sci, 64-69, dir postdoctoral studies, Sch Dent Med, 64-70, dir growth & develop div, 68-69, PROF ORTHODONT, SCH DENT MED, TUFTS UNIV, 68- *Mem:* Am Asn Orthodont; Am Asn Anatomists; Int Asn Dent Res; Int Soc Craniofacial Biol. *Res:* Growth and development; caries; histochemistry; endocrinology; physiology of salivary glands. *Mailing Add:* Sch Dent Med Tufts Univ Boston MA 02111

KRONMAN, MARTIN JESSE, b New York, NY, Sept 30, 27; m 65; c 3. PHYSICAL CHEMISTRY. *Educ:* Rutgers Univ, BS, 50; Temple Univ, PhD, 55. *Prof Exp:* Nat Heart Inst fel, Purdue Univ, 55-56; phys chemist, Eastern Regional Res Lab, USDA, 56-61; head biochem lab, US Army Natick Labs, 61-68; PROF BIOCHEM, STATE UNIV NY UPSTATE MED CTR, 68- *Honors & Awards:* Chem of Milk Award, Am Chem Soc, 68. *Mem:* AAAS; Am Chem Soc; Am Soc Biol Chemists. *Res:* Physico-chemical properties of proteins and nucleic acids; theory and technique of light scattering; optical rotation dispersion, absorption and emission spectra as applied to biological macromolecules; protein denaturation in enzyme action. *Mailing Add:* 100 Dewitt Rd Syracuse NY 13214

KRONMILLER, C W, ENGINEERING. *Prof Exp:* MEM STAFF, GEN ELEC CO. *Mailing Add:* General Electric Co 1635 Broadway Ft Wayne IN 46816

KRONSTAD, WARREN ERVIND, b Bellingham, Wash, Mar 3, 32; m 52; c 4. GENETICS, AGRONOMY. *Educ:* Wash State Univ, BS, 57, MS, 59; Ore State Univ, PhD(crop sci, genetics), 63. *Prof Exp:* Sr exp aide genetics, Wash State Univ, 59; from instr to assoc prof genetics & agron, 59-72, distinguished prof plant genetics, 89, PROF PLANT BREEDING & AGRON CROP SCI, ORE STATE UNIV, 72- *Concurrent Pos:* Consult, US AID, Turkey, 67-, Washington, DC, 69, Ecuador, Korea, & People's Repub China. *Honors & Awards:* Alexander von Humboldt Award, 81; Crop Sci Award, Crop Sci Soc Am, 83. *Mem:* Fel Am Soc Agron; fel Crop Sci Soc Am; Sigma Xi. *Res:* Cereal improvement; environment-genotype interaction; use of biometrical models to partition genetic variation; disease resistance; influence of chelating agents on genetic recombination; aluminum tolerance in plants. *Mailing Add:* Dept Crop Sci Ore State Univ Corvallis OR 97331

KRONSTEIN, KARL MARTIN, b Heidelberg, Ger, Feb 9, 28; US citizen; m 60; c 3. ALGEBRA. *Educ:* Georgetown Univ, BS, 51; Harvard Univ, AM, 52, PhD(math), 64. *Prof Exp:* Instr math, Reed Col, 57-58; from instr to asst prof, 58-69, ASSOC PROF MATH, UNIV NOTRE DAME, 69- *Concurrent Pos:* NATO res fel, Univ Frankfurt, 64-65. *Mem:* Am Math Soc; Math Asn Am; Sigma Xi. *Res:* Finite group theory, particularly Schur index. *Mailing Add:* Dept Math Univ Notre Dame Notre Dame IN 46556

KRONSTEIN, MAX, b Basel, Switz, Oct 7, 95; nat US; m 48; c 2. POLYMER CHEMISTRY. *Educ:* Univ Leipzig, PhD(chem), 22. *Prof Exp:* Chemist in charge prod varnishes & insulating mat, Elec Varnish Div, Brown Boveri Co, Ger, 22-38; develop chemist, William Zinsser & Co, NY, 39-43; chief chemist, Res Div, Titeflex, Inc, NJ, 43-46; from res assoc to sr res scientist & proj dir, Res Div, Sch Eng & Sci, NY Univ, 46-73; RES ASSOC CHEM, MANHATTAN COL, 73- *Mem:* Am Chem Soc; fel Am Inst Chem; Fedn Socs Paint Technol; Sigma Xi. *Res:* Formation of three-dimensional units in polymerization of organic material and metal-organics; formation of metal-organic complexes and basic principles of film-formation; mechanism of underwater coatings; surface reactions on galvanized steel; application of infrared spectroscopy in structural functions. *Mailing Add:* Dept Chem Manhattan Col Riverdale NY 10471

KRONTIRIS-LITOWITZ, JOHANNA KAYE, b Wheeling, WVa, June 7, 52; m 77; c 3. INVERTEBRATE NEUROBIOLOGY, CENTRAL REGULATION OF CARDIOVASCULAR FUNCTION. *Educ:* Case Western Reserve Univ, BS, 74, MS, 77; Cleveland State Univ, PhD(regulatory biol), 83. *Prof Exp:* Postdoctoral fel, Baylor Col Med, 83-85 & Univ Tex Med Ctr, Houston, 85-87; ASST PROF, YOUNGSTOWN STATE UNIV, 88- *Mem:* Sigma Xi; Soc Neurosci; Asn Women Sci; Women Neurosci. *Res:* Regulation of cardiovascular function and stress-evoked humoral responses in molluses especially Aplysia Californica. *Mailing Add:* Dept Biol Sci Youngstown State Univ 410 Wick Ave Youngstown OH 44555

KRONZON, ITZHAK, b Haifa, Israel, Sept 14, 39; US citizen; m 60; c 3. ECHOCARDIOGRAPHY, DOPPLER. *Educ:* Hebrew Univ, MD, 64; Am Col Physicians, dipl, 79 & 81. *Prof Exp:* Assoc prof, 78-83, PROF CLIN MED, NY UNIV, 83- *Concurrent Pos:* Vis prof, Tel Aviv Univ, 87-, Hebrew Univ, Jerusalem, 89- *Mem:* AMA; Am Heart Asn; Am Soc Echocardiography; fel Am Col Physicians; fel Am Col Cardiol; fel Am Col Chest Physicians. *Res:* Clinical applications of computorized technologies in the non-invasive diagnosis and evaluation of heart diseases; hemodynamics; angiography; cardiac imaging; non-invasive cardiology. *Mailing Add:* 560 First Ave New York NY 10016

KROODSMA, DONALD EUGENE, b Zeeland, Mich, July 7, 46; m 68; c 3. ANIMAL COMMUNICAION, BEHAVIORAL ECOLOGY. *Educ:* Hope Col, BA, 68; Ore State Univ, PhD(zool), 72. *Prof Exp:* Fel, animal behav, Rockefeller Univ, 72-74; asst prof, 74-80; assoc prof, 80-87, PROF ZOOL, UNIV MASS, AMHERST, 87- *Concurrent Pos:* Mem adv panel, psychobiol, NSF, 80, 85-88 & prin investr, grants, 76- *Mem:* Am Ornithologists Union; Animal Behav Soc; Cooper Ornith Soc. *Res:* Diversity of vocal behaviors among birds; development, evolution, and functions of these diverse vocal communication systems. *Mailing Add:* Dept Zool Univ Mass Amherst MA 01003

KROODSMA, ROGER LEE, b Zeeland, Mich, Jan 23, 44; div; c 3. WILDLIFE ECOLOGY, ORNITHOLOGY. *Educ:* Hope Col, BA, 66; ND State Univ, MS, 68, PhD(zool), 70. *Prof Exp:* Asst prof biol, Union Univ, Jackson, Tenn, 70-73; res assoc ecol, Univ Ga, Athens, 73-74; mgr environ impacts prog, 77-79, MEM RES STAFF, ENVIRON SCI DIV, OAK RIDGE NAT LAB, 74- *Mem:* Wildlife Soc; Am Ornithologists Union. *Res:* Community ecology of birds in man-affected habitats, such as transmission line rights-of-way and pine plantations; edge effect. *Mailing Add:* Environ Sci Div Oak Ridge Nat Lab PO Box 2008 Oak Ridge TN 37831

KROOK, LENNART PER, b Eksh-rad, Sweden, Aug 28, 24; m 58; c 2. VETERINARY PATHOLOGY. *Educ:* Royal Vet Col Sweden, DVM, 53, PhD(vet path), 57. *Prof Exp:* From asst prof to assoc prof vet path, Royal Vet Col Sweden, 51-57; assoc prof, Res Inst Nat Defense, Sundbyberg, 58; assoc prof, Sch Vet Med, Kans State Univ, 58-59; assoc prof, 59-65, PROF VET PATH, NY STATE COL VET MED, CORNELL UNIV, 65-, ASSOC DEAN POSTDOCTORAL EDUC, 81- *Mem:* Am Inst Nutrit; Int Acad Path; Sigma Xi. *Res:* Nutritional pathology. *Mailing Add:* 113 Pine Tree Rd Ithaca NY 14850

KROON, JAMES LEE, b Grand Rapids, Mich, Apr 24, 26; m 52; c 3. SCIENCE EDUCATION. *Educ:* Calvin Col, AB, 48; Purdue Univ, MS, 51, PhD(chem), 54. *Prof Exp:* Asst, Purdue Univ, 48-51; chemist, Dow Chem Co, 53-58, proj leader, 58-69; assoc prof, 69-79, PROF CHEM, BETHEL COL, 79- *Mem:* Am Chem Soc; Sigma Xi; Am Sci Affiliation. *Res:* Polarography; electrochemistry; aerosols. *Mailing Add:* Dept Natural Sci Bethel Col Mishawaka IN 46545

KROON, PAULUS ARIE, b Rotterdam, Neth, June 1, 45; NZ citizen; m 68; c 2. PHYSICAL BIOCHEMISTRY. *Educ:* Univ Auckland, BS, 67, MS, 68; Calif Inst Technol, PhD(chem), 75. *Prof Exp:* Res assoc, Life Sci Dept, Univ Pittsburgh, 75-77; SR RES FEL, MERCK SHARP & DOHME RES LABS, 77- *Concurrent Pos:* Res scientist, Biosci Div, Jet Propulsion Lab, 70-73. *Mem:* Am Soc Biochem & Molecular Biol; Am Heart Asn. *Res:* Structure, function and metabolism of plasma lipoproteins; molecular biology; cholesterol metabolism. *Mailing Add:* 101 North Ave Fanwood NJ 07023

KROON, R(EINOUT) P(IETER), b Hoorn, Holland, Aug 4, 07; nat US; c 3. MECHANICAL ENGINEERING. *Educ:* Polytech Acad, Zurich, MS, 30. *Prof Exp:* Develop engr, Westinghouse Elec Corp, 31-37, mgr, Exp Div & Steam Div, 37-45, eng mgr, Aviation Gas Turbine Div, 45-54; dir res, 54-57, chief engr adv, Design & Develop Dept, 57-58, adv systs planning, 58-60; prof, 60-75, EMER PROF MECH ENG, UNIV PA, 75- *Honors & Awards:* Spirit of St Louis Award, 50; Longstreth Medal, 63. *Mem:* Hon mem Am Soc Mech Engrs; assoc fel Am Inst Aeronaut & Astronaut. *Res:* Mechanics; turbomachinery; biomechanics. *Mailing Add:* Kendal at Longwood No 171 Kennett Square PA 19348

KROONTJE, WYBE, b Rotterdam, Neth, Aug 2, 22; nat US; m 48; c 1. AGRONOMY. *Educ:* Cornell Univ, BS, 51; Univ Nebr, MS, 53, PhD(soils), 56. *Prof Exp:* Asst, Univ Nebr, 51-56; assoc prof, 56-74, PROF AGRON, VA POLYTECH INST & STATE UNIV, 74- *Concurrent Pos:* Prof & chmn dept soil sci, Ataturk Univ, Turkey, 63-65. *Mem:* Am Soc Agron; Soil Sci Soc Am; Int Soil Sci Soc; Turkish Soil Sci Soc. *Res:* Soil fertility and management of tobacco, soybeans and field crops; transformation and uptake of nitrogen; development of agronomic and higher education. *Mailing Add:* Five Dogwood Circle Blacksburg VA 24060

KROP, STEPHEN, b New York, NY, Sept 24, 11; m 34; c 4. PHARMACOLOGY, TOXICOLOGY. *Educ:* George Washington Univ, BS, 39; Georgetown Univ, MS, 40; Cornell Univ, PhD(pharmacol), 42. *Prof Exp:* Asst pharmacol, Med Col, Cornell Univ, 39-42, instr, 42-44; from instr to asst prof, Sch Med, Yale Univ, 44-46; chief pharmacol sect, Med Div, Army Chem Ctr, Md, 46-48, dep chief physiol div, Chem Corps Med Labs, 52-57; assoc mem, Squibb Inst Med Res, 48-49; dir pharmacol div, Warner Inst Therapeut Res, Div Warner-Hudnut, Inc, 49-51; res assoc & asst dir chem-biol coord ctr, Nat Res Coun, 51-52; chief pharmacol dept & pres, Ethicon Res Found Div, Johnson & Johnson Co, NJ, 57-63; chief drug pharmacol br, US Food & Drug Admin, 63-79; CONSULT, 79- *Concurrent Pos:* Prof lectr, Med Sch, Georgetown Univ, 51-58 & 63-67; spec lectr, Med Sch, Univ Md, 52-57. *Mem:* Soc Toxicol; Am Soc Pharmacol & Exp Therapeut; Soc Exp Biol & Med; Am Physiol Soc; Harvey Soc. *Res:* Pharmacology of central nervous system; circulation; nerve-muscle; smooth muscle; local anesthetics; methods; respiration; toxicology; chemical warfare agents. *Mailing Add:* 7908 Birnam Wood Dr McLean VA 22102

KROPF, ALLEN, b Queens, NY, Oct 3, 29; m 50; c 3. BIOPHYSICAL CHEMISTRY. *Educ:* Queens Col, NY, BS, 51; Univ Utah, PhD(chem), 54. *Hon Degrees:* MA, Amherst Col, 69. *Prof Exp:* Chemist, Appl Physics Lab, Johns Hopkins Univ, 54-56; Am Cancer Soc res fel chem vision, Harvard Univ, 56-58; from instr to assoc prof chem, 58-68, PROF CHEM, AMHERST COL, 68-, JULIAN H GIBBS PROF NAT SCI, 85- *Concurrent Pos:* NSF sci fac fel, Univ Calif, Berkeley, 62-63; NIH spec res fel, Weizmann Inst Sci, 68-69; mem Visual Sci Sect, NIH, 73-77; vis prof biophys, Kyoto Univ, 75-76; vis prof phys chem, Hebrew Univ, 76; assoc chmn, Gordon Conf Visual Transduction, 78, 80; vis scholar biol, Harvard Univ, 82-83; vis scientist membrane sci, Weizmann Inst Sci, 86-87. *Mem:* Biophys Soc. *Res:* Photochemistry of visual pigments; preparation and properties of visual figments; chemistry of olfaction. *Mailing Add:* Dept Chem Amherst Col Amherst MA 01002

KROPF, DONALD HARRIS, b Watertown, Wis, Mar 8, 31; m 62; c 3. MEAT SCIENCE, ANIMAL HUSBANDRY. *Educ:* Univ Wis, BS, 52, PhD(animal husb), 57; Univ Fla, MS, 53. *Prof Exp:* Res asst meat sci, Univ Fla, 52-53 & Univ Wis, 53-56; asst prof animal husb, Clemson Col, 58-62; assoc prof, 62-72, PROF ANIMAL SCI & MEAT RES SCI, KANS STATE UNIV, 72- *Mem:* AAAS; Am Soc Animal Sci; Am Meat Sci Asn; Inst Food Technologists. *Res:* Meat color; effect of freezing system and rate packaging, display temperature and lighting on color; muscle histochemistry; carcass and live animal evaluation; processing and quality control. *Mailing Add:* Dept Animal Sci & Indust Weber Hall Kans State Univ Manhattan KS 66506-0201

KROPP, JAMES EDWARD, b Chicago, Ill, July 25, 39; m 61; c 4. ORGANIC CHEMISTRY. *Educ:* Wabash Col, BA, 61; Univ Colo, PhD(org chem), 65. *Prof Exp:* Res chemist, Film Dept, E I du Pont de Nemours, Inc, 65-67; SR RES SPECIALIST, MINN MINING & MFG CO, 67- *Mem:* Am Chem Soc. *Res:* Organic synthesis, mechanisms, stereochemistry; polymer synthesis, mechanisms, morphology, solvent effects; rheology; adhesion; new product development, marketing and production. *Mailing Add:* 3M Bldg 207-1W-08 Maplewood MN 55144

KROPP, JOHN LEO, b Salem, Ore, June 26, 34; m 59; c 4. PHYSICAL CHEMISTRY, SPECTROSCOPY. *Educ:* Univ Santa Clara, BS, 56; Univ Notre Dame, PhD(phys chem), 61. *Prof Exp:* Res assoc radiation lab, Univ Notre Dame, 61-62; mem tech staff, 62-80, sr proj engr, 80-84, PROJ MGR, TRW SPACE & TECHNOL GROUP, 84- *Mem:* AAAS; Sigma Xi; Am Inst Aeronaut & Astronaut. *Res:* Photochemistry of large organic molecules, especially fluorescence and phosphorescence of aromatic hydrocarbons; microgravity research experiment development; development of instrument systems especially space-oriented instruments. *Mailing Add:* 315 Via San Sebastian Redondo Beach CA 90277

KROPP, PAUL JOSEPH, b Springfield, Ohio, June 29, 35; m 63; c 2. ORGANIC CHEMISTRY. *Educ:* Univ Notre Dame, BS, 57; Univ Wis, PhD(org chem), 62. *Prof Exp:* Res chemist, Procter & Gamble Co, 61-70; PROF CHEM, UNIV NC, CHAPEL HILL, 70-, CHMN, CURRIC APPL SCI, 84- *Concurrent Pos:* Alfred P Sloan Found res fel, 72; vis prof, Univ Calif, Los Angeles, 77-78, Univ Bordeaux, 82 & Duke Univ, 89-90. *Mem:* Am Chem Soc. *Res:* Photochemistry; surface-mediated reactions; organic synthesis. *Mailing Add:* Dept Chem Univ NC Chapel Hill NC 27599-3290

KROPP, WILLIAM A, physics, for more information see previous edition

KROPP, WILLIAM RUDOLPH, JR, b Chicago, Ill, Nov 10, 36; m 63; c 2. COSMIC RAY PHYSICS, ELEMENTARY PARTICLE PHYSICS. *Educ:* DePaul Univ, BS, 58; Case Inst Technol, PhD(physics), 64. *Prof Exp:* Res assoc, Case Inst Technol, 64-66; asst res physicist, 66-67, asst prof physics, 67-73, assoc res physicist 73-80, RES PHYSICIST, UNIV CALIF, IRVINE, 80- *Honors & Awards:* Rossi Prize, Am Astrophys Soc, 89. *Mem:* Am Phys Soc. *Res:* Low background detection systems; neutrino interactions; cosmic rays. *Mailing Add:* Dept Physics Univ Calif Irvine CA 92717

KROSCHEWSKY, JULIUS RICHARD, b Taylor, Tex, Dec 14, 24; m 46; c 3. BOTANY, PHYTOCHEMISTRY. *Educ:* Univ Tex, BA, 47, MA, 49, PhD(bot), 67. *Prof Exp:* Instr biol, Lee Col, Tex, 49-52; Odessa Col, 54-59 & high sch, Calif, 60-61; assoc prof, St Edward's Univ, 61-67; prof biol, Bloomsburg State Col, 67-87; ASST DIR, EDWARD'S AQUIFER RES & DATA CTR, SOUTHWESTERN UNIV, 88- *Mem:* AAAS; Phytochem Soc NAm. *Res:* Determination of structures and taxonomic significances of flavonoid compounds of plants; plant tissue culture studies. *Mailing Add:* 2935 Philo Dr San Marcus TX 78666

KROSS, ROBERT DAVID, b Brooklyn, NY, Apr 25, 31; m 52; c 2. POLYMER CHEMISTRY, ANALYTICAL CHEMISTRY. *Educ:* Brooklyn Col, BS, 52; Iowa State Col, PhD(phys chem), 56. *Prof Exp:* Anal group leader, Rayonier, Inc, 57-59; chief chemist, Food & Drug Res Labs, 59-66; res dir, Foster D Snell, 66-69; dir, Kross Ref Lab, 69-80; vpres & dir res, Hydro Optics, Inc, 73-80; vpres, Mkt-Tech Indust Ltd, 81-83; Pres, KROSS-LINK LABS, 81-; VPRES DIR RES, ALCIDE CORP, 84- *Res:* Consulting and analysis in nutrition, pharmaceuticals, cosmetics, plastics and polymers; life sciences; biochemistry; infrared spectroscopy; physical analytical chemistry; federal regulations pertaining to chemically-oriented products; environmental sciences; microbiocidal systems for pharmaceutical, food and environmental applications. *Mailing Add:* Kross-Link Labs PO Box 374 Bellmore NY 11710

KROTHE, NOEL C, b Shickshinny, Pa, May 22, 38. GEOLOGY. *Educ:* Bloomsburg State Univ, BS, 61; Ind Univ, MAT, 69; Penn State Univ, MS, 73, PhD(geol), 76. *Prof Exp:* Teacher, sci, Tarrytown, NY & Baltimore, Md, 61-68; asst prof, 76-81, ASSOC PROF GEOL, IND UNIV, 81- *Mem:* Fel Geol Soc Am; Int Asn Hydrologists. *Res:* Chemical aspects of hydrology; isotopes in ground water. *Mailing Add:* Dept Geol Rm 425 Ind Univ 1005 E Tenth St Bloomington IN 47401

KROTKOV, ROBERT VLADIMIR, b Toronto, Ont, July 17, 29; m 58; c 4. ATOMIC PHYSICS. *Educ:* Queen's Univ, Ont, BA, 51, MA, 52; Princeton Univ, PhD, 58. *Prof Exp:* Instr physics, Palmer Lab, Princeton Univ, 56-58, res asst, 58-60; from instr to asst prof, Yale Univ, 60-66; assoc prof, 66-73, PROF PHYSICS, UNIV MASS, AMHERST, 73- *Mem:* Am Phys Soc; Sigma Xi. *Res:* Relativity and gravitation. *Mailing Add:* Dept Physics Univ Mass Hasbrouck Lab Amherst MA 01003

KROUSE, HOWARD ROY, b Norfolk Co, Ont, Jan 8, 35; m 58; c 2. PHYSICS, CHEMISTRY. *Educ:* McMaster Univ, BSc, 56, PhD(physics), 60. *Prof Exp:* From asst prof to prof physics, Univ Alta, 60-71, asst chmn dept, 70-71; PROF PHYSICS & CHMN DEPT, UNIV CALGARY, 71- *Concurrent Pos:* NATO fel & res assoc, Univ Calif, San Diego, 66-67; exchange scientist, USSR, 69; vis scientist, Japan, 71; res grants, Nat Res Coun Can, Defence Res Bd Can & Geol Surv Can. *Mem:* AAAS; Am

Geophys Union; Am Phys Soc; Geochem Soc; Can Asn Physicists; Sigma Xi. *Res:* Isotope fractionation studies in physical, geological, chemical, biological and environmental processes; mass spectrometry. *Mailing Add:* Dept Physics Univ Calgary Calgary AB T2N 1N4 Can

KROUSKOP, THOMAS ALAN, b Washington, DC, July 11, 45; m 68; c 3. BIOENGINEERING, BIOMATERIALS. *Educ:* Carnegie Inst Technol, BS, 67; Carnegie-Mellon Univ, MS, 69, PhD(civil eng, biotechnol), 71. *Prof Exp:* Design engr, Gen Analytics Inc, 68; from asst to assoc prof bioeng, Tex A&M Univ, 71-78; assoc prof, 78-87, PROF, BAYLOR COL MED, 87- *Concurrent Pos:* Consult, Sampson Corp, 70-71; consult, Tex Heart Inst, 75- *Honors & Awards:* BCM Res Achievement Award, 87. *Mem:* Am Soc Testing & Mat; Rehab Eng Soc NAm; Am Cong Rehab Med. *Res:* Design of prosthetic appliances; materials for use as medical implants; effects of mechanical stress on soft tissue metabolism; development of assistive devices for the physically handicapped. *Mailing Add:* Inst Rehab & Res 1333 Moursund Ave Houston TX 77030

KROW, GRANT REESE, b Reading, Pa, June 30, 41; m 70; c 2. ORGANIC CHEMISTRY. *Educ:* Albright Col, BS, 63; Princeton Univ, MFA, 64, PhD(chem), 67; Temple Univ, JD, 78. *Prof Exp:* Res assoc chem, Ohio State Univ, 67-69; form asst prof to assoc prof, 69-80, PROF CHEM, TEMPLE UNIV, 80- *Concurrent Pos:* Sci & legal consult, 78- *Mem:* Am Chem Soc; Sigma Xi. *Res:* Stereochemistry; chemistry of heterocycles; synthesis of anticancer natural products. *Mailing Add:* Dept Chem Temple Univ Philadelphia PA 19122

KROWN, SUSAN E, b Bronx, NY, Sept 8, 46; div; c 1. MEDICAL ONCOLOGY, AIDS RESEARCH. *Educ:* Barnard Col, AB, 67; State Univ NY Downstate Med Ctr, MD, 71. *Prof Exp:* Intern internal med, Mt Sinai Hosp, NY, 71-72, resident, 72-74; fel med oncol & clin immunol, 74-77, res assoc immunobiol, Sloan Kettering Inst Cancer Res, 77-84, ASSOC MEM MED, MEM SLOAN KETTERING CANCER CTR, 84-; ASSOC PROF MED, CORNELL UNIV MED COL, 83- *Concurrent Pos:* Clin asst & asst attend physician med, Mem Hosp, 77-82, assoc attend, 82-; mem oncol drugs adv comt, US Food & Drug Admin, 86-90, consult, 90-; vchair oncol comt, AIDS Clin Trials Group, Nat Inst Allergy & Infectious Dis, NIH, 87-89, chair, 90-; adv bd, Cancer Treat Reports, 84-87; bd dirs, Soc Biol Ther, 86-89; mem ed Bd, J Interferon Res, J Aquired Immune Deficiency Symdrone; mem, Int Soc Interferon Res; chair, AIDS Task Force, mem Sloan Kettering Cancer Ctr, 89-; mem, AIDS Clin Drug Discovery Comt, NIH, 90- *Mem:* Am Asn Cancer Res; Am Soc Clin Oncol; AAAS; Soc Biol Ther; Int Soc Interferon Res. *Res:* Clinical investigations of biological agents and antivirals in the treatment of cancer and AIDS; mechanisms of action of biological response modifiers; identification of immunological prognostic parameters in cancer and AIDS. *Mailing Add:* Mem Hosp 1275 York Ave New York NY 10021

KRUBINER, ALAN MARTIN, b New York, NY Apr 6, 41; m 62; c 2. ORGANIC CHEMISTRY. *Educ:* Queens Col, BS, 61; Univ Calif, Berkeley, PhD(org chem), 65; Seton Hall Univ, JD, 74. *Prof Exp:* Sr chemist, Hoffman-La Roche, Inc, 64-70, patent chemist, 70-74; patent atty, 74-77, supvry patent atty, 77-79, asst patent coun, 79-81, dir patent law & licensing, 81-83, VPRES, INTELLECTUAL PROPERTY LAW, SYNTEX CORP, 83- *Mem:* Am Chem Soc. *Res:* Proprietary information, primarily in health care area. *Mailing Add:* Syntex Corp Patent Dept 3401 Hillview Ave Box 10850 Palo Alto CA 94304

KRUCKEBERG, ARTHUR RICE, b Los Angeles, Calif, Mar 21, 20; m 42, 53; c 5. BOTANY. *Educ:* Occidental Col, AB, 41; Univ Calif, Berkeley, PhD(bot), 51. *Prof Exp:* Asst biol, Occidental Col, 39-41; field asst, Carnegie Inst, 41; asst biol, Stanford Univ, 41-42; asst bot, Univ Calif, Berkeley, 46-51; from instr to assoc prof, 51-63, chmn dept 71-77, prof, 64-88, EMER PROF, BOT, UNIV WASH, 88- *Concurrent Pos:* Instr, Occidental Col, 46. *Mem:* Am Soc Plant Taxon. *Res:* Experimental plant taxonomy; edaphic ecology of serpentine soils. *Mailing Add:* Dept Bot KB-15 Univ Wash Seattle WA 98195

KRUCZYNSKI, WILLIAM LEONARD, b Buffalo, NY, July 18, 43; m 66; c 2. RESOURCE MANAGEMENT. *Educ:* Canisius Col, BS, 65; Univ NC, Chapel Hill, PhD(marine biol), 71. *Prof Exp:* Asst prof biol, Hartwick Col, 70-74; asst prof res, Fla A&M Univ, 74-79; life scientist, 79-85, chief, wetlands protection sect, Region IV, 85-87, WETLANDS SPEC, US ENVIRON PROTECTION AGENCY, GULFBREEZE, FLA, 87- *Concurrent Pos:* Consult, Environ Anal, Inc, 72-75, Conservation Consult, Inc, 77 & Tex A&M Univ & Tex Instruments, Inc, 78-79; vis prof, Col Ctr Finger Lakes, 73; res assoc, Fla State Univ, 74-77; vpres, Environ Systs Serv, Inc, 74-76. *Honors & Awards:* Bronze Medal, US Environ Protection Agency. *Mem:* Am Soc Zoologists; AAAS; Ecol Soc Am; Estuarine Res Fedn. *Res:* Ecology of a crab symbiotic with mollusk hosts; systematics of marine isopod crustaceans; ecology of saline marshes. *Mailing Add:* 3680 Monteigne Dr Pensacola FL 32504

KRUEGEL, ALICE VIRGINIA, b Louisville, Ky, May 29, 39; m 65; c 2. ORGANIC CHEMISTRY, FORENSIC SCIENCE. *Educ:* Spalding Col, Louisville, BA, 61; Univ Ky, PhD(org chem), 72. *Prof Exp:* Res chemist, Spec Testing & Res Lab, 70-78, supvry chemist forensic drug chem, Western Field Lab, 78-88, PROG MGR, OFF FORENSIC SCI, DRUG ENFORCEMENT ADMIN, 88- *Mem:* Am Chem Soc; Am Acad Forensic Sci. *Res:* Forensic drug analysis; identification of impurities in illictly manufactured drugs. *Mailing Add:* Off Forensic Sci Drug Enforcement Admin Washington DC 20537

KRUEGER, ARLIN JAMES, b Lamberton, Minn, Oct 22, 33; m 78; c 3. STRATOSPHERIC OZONE, REMOTE SENSING. *Educ:* Univ Minn, BA, 55; Colo State Univ, PhD(atmospheric sci), 84. *Prof Exp:* Physicist, Naval Weapons Ctr, China Lake, Calif, 59-69; AEROSPACE TECHNOLOGIST & ASTROPHYS, GODDARD SPACE FLIGHT CTR, NASA, GREENBELT, MD, 69- *Concurrent Pos:* Prin invstr, ROCOZ rocket ozonesonde, 60-84; sr scientist, Nimbus 7 Total Ozone Mapping Spectrometer, TOMS, 73-; instrument scientist, Meteor 3/TOMS, 88-; prin investr, ADEOS/TOMS, 88-; instr scientist, Earth Probe/TOMS, 89- *Honors & Awards:* Except Achievement Award, Goddard Space Flight Ctr, 86. *Mem:* Am Geophys Union; Am Meteorol Soc; Sigma Xi; AAAS; Explorers Club. *Res:* Atmospheric chemistry; development of satellite and rocket instruments to measure ozone and sulfur dioxide; analysis of ozone variability and volcanic clouds; modelling of stratospheric chemistry; Mars atmosphere; remote sensing. *Mailing Add:* Goddard Space Flight Ctr NASA Code 916 Greenbelt MD 20771

KRUEGER, CHARLES ROBERT, b Milwaukee, Wis, June 21, 38; m 65; c 2. FORAGE CROP MANAGEMENT, FORAGE CROP UTILIZATION. *Educ:* Univ Wis-Madison, BS, 60, MS, 63, PhD(agron), 67. *Prof Exp:* Asst prof extension, Dept Agron, Univ Wis-Madison, 67-70; assoc prof res & teaching, Dept Plant Sci, SDak State Univ, 70-73, prof & head admin, res & teaching, 73-78; ASST DIR ADMIN, OHIO AGR RES & DEVELOP CTR, OHIO STATE UNIV, WOOSTER, 78- *Mem:* Am Soc Agron; Crop Sci Soc Am; Coun Agr Sci & Technol; Am Dairy Sci Soc; Am Soc Animal Sci. *Res:* Interdisciplinary research on forage crop production, management and utilization; determination of forage nutritive value by in vitro and in vivo methods; pasture and forage systems for beef cattle; establishment and use of warm-season perennial grasses; interseeding of native rangeland to improve productivity. *Mailing Add:* Assoc Dean Res Pa State Univ 229 Agr Admin Bldg University Park PA 16802

KRUEGER, DAVID ALLEN, b Sidney, Mont, Aug 21, 39; m 61; c 4. THEORETICAL PHYSICS. *Educ:* Mont State Univ, BS, 61; Univ Wash, PhD(physics), 67. *Prof Exp:* Wis Alumni Res Found res assoc physics, Univ Wis, Madison, 67-69; from asst prof to assoc prof, 69-80, PROF PHYSICS, COLO STATE UNIV, 80- *Concurrent Pos:* Vis prof, Watson Res Lab, Int Bus Mach Corp, Yorktown Heights, NY, 75 & Dept Energy, Energy Technol Ctr, Bartlesville, Okla, 80; mem tech staff, Sandia Lab, Albuquerque, 78-79; vis scientist, Marathon Oil Co Res Ctr, Littleton, Colo, 85-86. *Mem:* AAAS; Am Phys Soc; Am Asn Phys Teachers; Soc Petrol Engrs. *Res:* Theoretical many-body problem; dynamics of second order phase transitions; system size effects on phase transitions; fluids (stability of flow in porous media); thermal physics. *Mailing Add:* Dept Physics Colo State Univ Ft Collins CO 80523

KRUEGER, EUGENE REX, b Grand Island, Nebr, Mar 30, 35; m 57; c 3. MATHEMATICS, COMPUTER SCIENCES. *Educ:* Rensselaer Polytech Inst, BS, 57, MS, 60, PhD(appl math), 62. *Prof Exp:* Physicist, Res Ctr, Int Bus Mach Corp, 57-58; vis fel, Math Res Ctr, Univ Wis, 62-63; asst prof math, Univ Colo, Boulder, 63-68, from asst prof to prof math & comput sci, 68-74, dir, Comput Ctr, 67-74; vchancellor educ syst, Ore State Syst Higher Educ, 74-82; gen mgr, Control Data Corp, 82-85, vpres, 85-89; prof computer sci, Ore State Univ, 74-82; EXEC DIR, WILLIAM C NORRIS INST, 89- *Mem:* AAAS; Sigma Xi. *Res:* Fluid mechanics with emphasis on hydrodynamic stability; mathematical methods of physics; numerical analysis; interactive computer graphics. *Mailing Add:* 5601 Dewey Hill Rd Edina MN 55435

KRUEGER, GEORGE CORWIN, b Seattle, Wash, Nov 29, 22. OPTICS. *Educ:* Reed Col, AB, 45; Brown Univ, PhD(physics), 51. *Prof Exp:* From instr to assoc prof, 50-62, PROF PHYSICS, UNIV MAINE, ORONO, 62- *Concurrent Pos:* Air Force Weapons Lab, NMex, 65-66. *Mem:* Am Phys Soc; Optical Soc Am. *Res:* Atmospheric turbulence and heat transfer; physical optics; electromagnetic theory. *Mailing Add:* Dept Physics Univ Maine Orono ME 04473

KRUEGER, GERHARD R F, b Berlin, Ger, Nov 21, 36; m 60; c 3. ANATOMIC PATHOLOGY, LABORATORY IMMUNOLOGY. *Educ:* Intern med rotating, Hosps Free Univ, 62-64. *Prof Exp:* resident path, City Hosp Spandau, Berlin, 64-65; res pathologist, Nat Cancer Inst, NIH, 65-67, pathologist, Lab Path, 68-72; resident path, Dept Path, Free Univ, Berlin, 67-68; asst prof path, Dept Path, Univ Cologne, Ger, 72-74, prof path & head immunopath, 74-91; PROF PATH, DEPT PATH LAB MED, UNIV TEX, HOUSTON, 91- *Concurrent Pos:* Chmn, Int Inst Immunopath, Inc, Cologne-Washington, 89-; vis prof, Dept Path Lab Med, Univ Tex, Houston, 90-91. *Mem:* Am Asn Pathologists; Am Soc Microbiol; Hematopath Soc; Int Asn Res Leukemia Asn Dis. *Res:* Relationships between reactivated herpesvirus infections to immune deficiency, autoimmunity and lymphoproliferative disorders; influence of virus infection on cell membrane function. *Mailing Add:* Dept Path Lab Med Health Sci Ctr Univ Tex 6431 Fannin St Houston TX 77030

KRUEGER, JACK N, b St Paul, Minn, Aug 29, 22; m 58; c 2. ELECTRICAL & AGRICULTURAL ENGINEERING. *Educ:* Univ Minn, BEE, 44, MS, 49. *Prof Exp:* Instr elec eng, Univ Minn, 44-49; res prof agr eng, Univ Ky, 49-51; consult engr, 51-55; assoc prof elec eng, Va Polytech Inst, 55-56; head, Electromech Develop Dept, Pillsbury Mills, Inc, 56-58; assoc prof elec eng, SDak State Univ, 58-59; assoc prof, 59-69, PROF ELEC ENG, UNIV NDAK, 69- *Mem:* Inst Elec & Electronics Engrs. *Res:* Electrical power and machinery; electrical instrumentation; biomedical and industrial electronics; environmental conditioning of production, processing and storage areas for agricultural products; forensic engineering; solar and wind energy systems. *Mailing Add:* PO Box 7165 Grand Forks ND 58202

KRUEGER, JAMES ELWOOD, b Marinette, Wis, Apr 2, 26; m 53; c 3. ORGANIC CHEMISTRY. *Educ:* Univ Wis, BS, 49; Mass Inst Technol, PhD(org chem), 54. *Prof Exp:* Res fel chem, Harvard Med Sch, 53-55; chemist, Dow Chem Co, 55-61; chemist, Lederle Labs Div, Am Cyanamid Co, 61-79, qual auditor, qual mgt, 79-90; VPRES, QUAL ASSURANCE, IND MGT, 90- *Concurrent Pos:* Cert qual auditor, Am Soc Qual Control. *Mem:* AAAS; Am Chem Soc; Sigma Xi; Acad Pharmaceut Sci; Am Soc Qual Control. *Res:* Pharmaceutical science; research quality assurance. *Mailing Add:* Six Lucille Blvd New City NY 10956

KRUEGER, JAMES HARRY, b Milwaukee, Wis, May 18, 36; m 59; c 3. INORGANIC CHEMISTRY. *Educ:* Univ Wis-Madison, BS, 58; Univ Calif, Berkeley, PhD(chem), 61. *Prof Exp:* From asst prof to assoc prof, 61-76, PROF CHEM, ORE STATE UNIV, 76- *Mem:* Am Chem Soc. *Res:* Kinetics and mechanisms of inorganic reactions; synthesis and characterization of transition metal complexes containing thiolate amino acid ligands. *Mailing Add:* Dept Chem Ore State Univ Corvallis OR 97331-4003

KRUEGER, KEATHA KATHRINE, b Faulk Co, SDak, Nov 15, 21. BIOCHEMISTRY. *Educ:* Univ SDak, BA, 43; Univ Wis, MS, 45, PhD(biochem), 48. *Prof Exp:* Assoc biochem, Med Sch, Univ SDak, 48-49, asst prof, 49-56; from asst ed to assoc ed, Chem Abstracts Serv, 56-61, head dept biochem, 61-63; sci commun officer, Nat Inst Arthritis & Metab Dis, 63-74, diabetes prog dir, 74-77, diabetes ctr prog dir, Nat Inst Arthritis, Diabetes, Digestive & Kidney Dis, 77-83; RETIRED. *Concurrent Pos:* Exec secy, Nat Comn Diabetes, 75-76. *Honors & Awards:* Am Diabetes Asn Best Medal, 85. *Mem:* AAAS; Am Chem Soc; Am Diabetes Asn; Am Med Writers Asn. *Res:* Diabetes program administration; metabolic diseases. *Mailing Add:* 2506 13th Ave SE Aberdeen SD 57401

KRUEGER, PAUL A, b Madison, SDak, Feb 4, 06; m 32; c 5. ORGANIC CHEMISTRY. *Educ:* Sioux Falls Col, AB, 27; Northwestern Univ, MS, 29; Pa State Col, PhD(org chem), 32. *Prof Exp:* Jr patent exam, US Patent Off, 27-28; res chemist, Mallinckrodt Chem Works, 32-42, asst tech dir, 42-46 & prod develop, 46-53, mgr, Develop Dept, 54-56, dir mkt res indust chem, 56-60 & corporate develop, Indust Chem Div, 60-63, asst to pres, 64-71, asst secy, 66-71; DIR CORP DEVELOP, NEWHARD, COOK & CO, INC, 73- *Concurrent Pos:* Stock broker, Stix Friedman & Co, Inc, 72. *Mem:* Am Chem Soc; Com Develop Asn; Chem Mkt Res Asn. *Mailing Add:* 5052 Westminster Pl St Louis MO 63108

KRUEGER, PAUL CARLTON, b Louisville, Ky, June 1, 36; m 65; c 2. INORGANIC CHEMISTRY, PHYSICAL CHEMISTRY. *Educ:* Marquette Univ, BS, 58; Case Inst Technol, PhD(chem), 63. *Prof Exp:* Res chemist, Air Reduction Co, 63-65, sr res chemist, Cent Res Labs, 65-78, supvr, Indust Health Serv & Safety, 78-80, mgr environ affairs, 80-84, mat selection & performance, 84-87, MGR ENVIRON AFFAIRS, AIRCO, INC, 87- *Mem:* Am Chem Soc; Am Soc Metals. *Res:* Chelate, silicon and slag chemistry; coordination compounds; inorganic and high temperature polymers; ferroalloy production; cutting of metals; corrosion of stainless steel alloys; development of stainless steel alloys; industrial hygiene sampling and interpretation; failure analysis of equipment and compressed gas cylinders. *Mailing Add:* BOC Group Inc Group Tech Ctr 100 Mountain Ave Murray Hill NJ 07974

KRUEGER, PETER GEORGE, b Lodz, Poland, May 20, 40; US citizen; div; c 3. PHYSICS, BIOPHYSICS. *Educ:* Loma Linda Univ, La Sierra Campus, BA, 62; Univ Calif, Riverside, MA, 66, PhD(physics), 70. *Prof Exp:* Physicist, Naval Weapons Ctr, 62-80; Naval Air Systs Command, Washington, DC, 80-86; STAFF SCIENTIST, SECY DEFENSE, WASHINGTON, DC, 86- *Mem:* Am Phys Soc. *Res:* Technical management of military equipment exports; defense cooperation progams in support of national security and foreign policy. *Mailing Add:* 10704 Hunters Valley Rd Vienna VA 22181

KRUEGER, PETER J, b Altona, Man, Nov 11, 34; m 59; c 3. ORGANIC CHEMISTRY, SPECTROSCOPY. *Educ:* Univ Man, BSc, 55, MSc, 56; Oxford Univ, DPhil(infrared spectros), 58. *Prof Exp:* Fel org spectrochem sect, Div Pure Chem, Nat Res Coun Can, Ont, 58-59; from asst prof to assoc prof chem, Univ Alta, 59-66; head dept, Univ Calgary, 66-70, vdean fac arts & sci, 70-72, mem bd govs, 70-73, acad vpres & provost, 76-90, PROF CHEM, UNIV CALGARY, 66- *Concurrent Pos:* Vis scientist, Nat Res Coun Can, Ottawa, 66-67. *Honors & Awards:* Coblentz Award Spectros, Coblentz Soc, 67; Gerhard Herzberg Award, Spetros Soc Can, 73. *Mem:* Am Chem Soc; Chem Inst Can; Spectros Soc Can; Royal Soc Chem; Coblentz Soc; Soc Appl Spectros; AAAS. *Res:* University governance and administration; organic spectrochemistry; infrared and Raman spectra of organic compounds; normal coordinate analysis of vibrational spectra; molecular structure determination. *Mailing Add:* Dept Chem Univ Calgary 2500 University Dr NW Calgary AB T2N 1N4 Can

KRUEGER, ROBERT A, b Oak Park, Ill, Dec 29, 35; m 59; c 3. ORGANIC CHEMISTRY. *Educ:* Knox Col, AB, 57; Kans State Univ, PhD(org chem), 65. *Prof Exp:* Chemist, Visking Div, Union Carbide Corp, 57-58 & 60; res chemist, 65-69, sect leader, 69-72, tech adminr, 72-73, facility mgr, 73-75, vpres res & develop eng, 77-78, vpres additives & specialty polymers & develop function, 77-79, sr vpres elastomers, latex & specialty chem, 79-80, sr vpres & gen mgr, Polyvinyl Chloride Div, 80-81, sr vpres staff support serv, B F Goodrich Chem Co, 81-85; GEN MGR, ICI RESINS US, 85- *Concurrent Pos:* Consult, mgmt, res & develop. *Mem:* Am Chem Soc; Indust Res Inst. *Res:* Synthesis and reactions of carbenes; olefin synthesis; antioxidants; polyvinyl chloride stabilization; organometallic chemistry. *Mailing Add:* 160 Carlton Lane North Andover MA 01845

KRUEGER, ROBERT CARL, b Philadelphia, Pa, Oct 11, 20; m 47; c 2. BIOCHEMISTRY. *Educ:* Univ Pa, BS, 42; Columbia Univ, PhD(biochem), 48. *Prof Exp:* Res assoc immunochem, Col Physicians & Surgeons, Columbia Univ, 47-50; asst prof, 50-56, ASSOC PROF BIOL CHEM, COL MED, UNIV CINCINNATI, 56- *Concurrent Pos:* Res assoc, Brookhaven Nat Lab, 59-60; vis scientist, Univ Brussels, 66-67; NIH spec fel, 66-67. *Mem:* Am Soc Biol Chemists. *Res:* RNA synthesis; deoxyribonucleoprotein structure and function. *Mailing Add:* Dept Molecular Genet Univ Cincinnati Col Med 231 Bethesda Ave Cincinnati OH 45267

KRUEGER, ROBERT GEORGE, b Duluth, Minn, Apr 22, 38; m 60; c 2. IMMUNOLOGY, VIROLOGY. *Educ:* Col St Thomas, BS, 60; Univ Detroit, MS, 62; Univ Chicago, PhD(microbiol), 66. *Prof Exp:* Asst prof immunol, New York Med Col, 66-67; asst prof microbiol, Univ Wash, 67-71; assoc prof, Mayo Grad Sch Med, Univ Minn & Mayo Med Sch, 71-75; dir, Lab

Molecular Oncol, Christ Hosp Inst Med Res, Cincinnati, 75-81, dir, Div Tumor Biol, 78-81; prof exp med & assoc prof microbiol, Col Med, Univ Cincinnati, 75-81; regional med assoc, Smith Kline & French Labs, 81-89; CONSULT, 89- *Concurrent Pos:* USPHS grants, 67-81; Am Cancer Soc grant, 69-71; consult microbiol & immunol, Mayo Clin, 71-75. *Mem:* AAAS; Am Asn Immunologists; Soc Exp Biol & Med; Am Soc Microbiol; Am Diabetes Asn. *Res:* Genetic regulation of immunoglobulin synthesis; mechanisms of neoplastic transformation by oncogenic viruses; protein synthesis in differentiated mammalian cells; structure of oncogenic viruses. *Mailing Add:* 9623 Timbermill Ct Cincinnati OH 45231

KRUEGER, ROBERT HAROLD, b Sioux City, Iowa, Jan 25, 26; m 56; c 3. PHYSICAL CHEMISTRY. *Educ:* Morningside Col, BS, 50; Northwestern Univ, MS, 55; Loyola Univ, PhD(chem), 67. *Prof Exp:* Chemist, Cent Com Co, 51-54 & Diversey Corp, Ill, 54-56; res chemist, Roy C Ingersoll Res Ctr, Borg-Warner Corp, 56-62, sr res chemist, 62-70, group leader 70-78, amer phys chem, 78-89; SCIENTIST, BIRL LAB, NORTHWESTERN UNIV 90- *Mem:* Am Chem Soc; Nat Asn Corrosion Engrs; Soc Tribologists & Lubrication Engrs. *Res:* Development of materials for bearings and seals; cement additives; new fluids for refrigeration; corrosion, friction and wear of materials; development of corrosion and scale inhibitors; plasma treatment of metals; protective coatings; sensors. *Mailing Add:* 1125 Skylark Dr Palatine IL 60067

KRUEGER, ROBERT JOHN, b Milwaukee, Wis, Apr 15, 48; m 71; c 2. PHARMACOGNOSY, PHARMACEUTICAL CHEMISTRY. *Educ:* Univ Conn, BS, 71; Univ Iowa, PhD(pharm), 75. *Prof Exp:* from asst prof to assox prof, 75-86, PROF PHARMACOG, FERRIS STATE UNIV, 86- *Concurrent Pos:* NDEA fel, 71-74; Eli Lilly & Co fel, 74-75, PMA; vis scientist, Abbott, 83; chmn, Am Chem Soc Western MI Region, 85; pres, Sigma Xi, Grand Valley St Univ, 86-87. *Mem:* AAAS; Am Soc Pharmacog; Am Chem Soc; Phytochem Soc NAm; Am Asn Col Pharm; Sigma Xi. *Res:* Plant tissue culture; isolation and identification of natural products; xenobiotic metabolism by plant cell cultures; fungal ellicitor stimulation of plant cell culture. *Mailing Add:* 701 Magnolia Big Rapids MI 49307

KRUEGER, ROBERT WILLIAM, b Philadelphia, Pa, Nov 16, 16; m 41; c 2. PHYSICS. *Educ:* Univ Calif, Los Angeles, AB, 37, MA, 38, PhD(physics), 42. *Prof Exp:* Res physicist, Douglas Aircraft Co, 42-46; asst chief, Missiles Div, Rand Corp, 47-53; pres, Planning Res Corp, 54-73; pres, PROF SERV INT, 73- *Concurrent Pos:* Pres-founder, Nat Coun Prof Serv Firms, 71-74, dir, 71- *Mem:* Am Phys Soc; Opers Res Soc Am. *Res:* Operations research; systems engineering; aerodynamics; propulsion; quantum mechanics; spectroscopy. *Mailing Add:* 1016 Moraga Dr Los Angeles CA 90049

KRUEGER, ROLAND FREDERICK, b Fond du Lac, Wis, Oct 18, 18; m 43; c 3. WELLBORE MECHANICS, OIL RECOVERY. *Educ:* Ripon Col, BA, 39; Univ Ill, MA, 41. *Prof Exp:* Lab instr, Univ Ill, 39-41; physicist, Tenn Eastman Corp, 41-42; physicist, Holston Ord Works, 42; physicist, radiation lab, Univ Calif, 43, tech supvr, Manhattan Proj, Oak Ridge Nat Labs, 43-44, asst dept syst, 44-46; physicist & res engr, Douglas Aircraft Co, Calif, 46-48; Physicist & res engr, 48-51, sect leader, Prod Res Dept, 51-55, sr sect leader, 55-62, supvr, 62-81, mgr, 81-85, staff consult to pres sci & technol div, Union Oil Co, 85-86; PRIN, KGK PETROL CONSULTS, 86- *Concurrent Pos:* Distinguished lectr, Soc Petrol Engrs, 75-76, sr tech ed, J Petrol Technol, 79-82 & mem bd dirs, 83-86. *Honors & Awards:* Distinguished Serv Award, Soc Petrol Engrs, 82; Citation Serv, Am Petrol Inst, 83. *Mem:* Soc Petrol Engrs; Am Petrol Inst; AAAS; Asn Advan Eng. *Res:* Isotope separation; mass spectroscopy; combustion; spectrophotometry; thermodynamics; flow of fluids in porous media; production mechanics; formation damage in oil and gas wells; oil recovery processes. *Mailing Add:* 561 Peralta Hills Dr Anaheim CA 92807

KRUEGER, RUSSELL FRANCIS, microbiology, for more information see previous edition

KRUEGER, WALTER L(AWRENCE), plastics engineering, for more information see previous edition

KRUEGER, WILLIAM ARTHUR, b Milford, Iowa, Mar 24, 41. AGRONOMY. *Educ:* Univ Minn, BS, 63; Univ Ill, PhD(agron), 68. *Prof Exp:* Asst prof agron, 68-74, ASSOC PROF PLANT & SOIL SCI, UNIV TENN, KNOXVILLE, 74- *Mem:* Plant Growth Regulator Soc; Weed Sci Soc Am. *Res:* The effects that herbicides exert on the physiology of plants, particularly mode of action and basis of selectivity. *Mailing Add:* 337 Plant Sci Bldg Dept Plant Sci Univ Tenn Knoxville TN 37996

KRUEGER, WILLIAM CLEMENT, b Medford, Ore, Aug 19, 42; m 65; c 2. RANGE SCIENCE. *Educ:* St Mary's Col, Calif, BS, 64; Humboldt State Col, MS, 67; Utah State Univ, PhD(range sci), 70. *Prof Exp:* Res technician range sci, Intermountain Forest & Range Exp Sta, 67-70; asst prof, Humboldt State Col, 71-75, leader rangeland resources prog, 75-80, HEAD, DEPT RANGELAND RESOURCES, ORE STATE UNIV, 81- *Concurrent Pos:* Head, Dept Range Sci, Colo State Univ, 80-81. *Mem:* Soc Range Mgt. *Res:* Range restoration; interaction of livestock grazing, wildlife and timber production through integration of management systems. *Mailing Add:* Rangeland Resources Dept Ore State Univ Corvallis OR 97331

KRUEGER, WILLIAM E, b St Louis, Mo, June 26, 40. ORGANIC CHEMISTRY. *Educ:* Univ Notre Dame, BS, 62; Univ NH, PhD(org chem), 67. *Prof Exp:* Res fel org chem, Univ Ill, Chicago Circle, 66; from asst prof to assoc prof, 66-85, PROF CHEM, STATE UNIV NY COL PLATTSBURGH, 85- *Mem:* Am Chem Soc. *Res:* Partially reduced pyridines; nitrenes; phosphorus additions. *Mailing Add:* Dept Chem State Univ NY Col Plattsburgh Plattsburgh NY 12901

KRUEGER, WILLIE FREDERICK, b Riesel, Tex, Oct 12, 21; m 46; c 2. POULTRY HUSBANDRY, GENETICS. *Educ:* Tex A&M Univ, BS, 43, MS, 49; Univ Mo, PhD(genetics, animal breeding), 52. *Prof Exp:* Teacher & prin pub sch, Tex, 41-42; flock supvr, Tex Poultry Improv Asn, 42-43; instr poultry husb, Miss State Univ, 46-47; asst, Agr & Mech Col Tex, 47-49; asst, Univ Mo, 49-50, from instr to asst prof, 50-53; from asst prof to assoc prof, 53-59, PROF POULTRY SCI, TEX A&M UNIV, 59-, HEAD DEPT, 72- *Mem:* Poultry Sci Asn; Am Genetic Asn; Sigma Xi. *Res:* Application of the principles of population genetics to poultry; embryology; embryology and incubation of chicken and turkey eggs; environmental factors influencing chickens and turkeys. *Mailing Add:* Dept Poultry Sci Tex A&M Univ College Station TX 77843

KRUER, WILLIAM LEO, b Louisville, Ky, Apr 20, 42; m 65; c 3. PLASMA PHYSICS. *Educ:* Univ Louisville, BS, 64, MS, 65; Princeton Univ, MA, 67, PhD(astron), 69. *Prof Exp:* Res assoc, Plasma Physics Lab, Princeton Univ, 69-70, mem res staff, 70-72; GROUP LEADER LASER PLASMA THEORY & SIMULATION, LAWRENCE LIVERMORE LAB, 72- *Concurrent Pos:* Lectr, Univ Calif, Davis/Livermore, 76-; affil mem Ctr Plas Physics & Fusion Eng, Univ Calif, Los Angeles, 76- *Mem:* Fel Am Phys Soc. *Res:* Plasma theory; computer simulation of plasmas; nonlinear plasma waves; plasma heating; laser fusion. *Mailing Add:* 4055 Suffolk Way Pleasanton CA 94566

KRUESI, WILLIAM R, b Seattle, Wash, Mar 4, 21. PROTECTIVE EQUIPMENT. *Educ:* Union Col, BSEE, 42. *Prof Exp:* CONSULT, GEN ELEC CO, 42- *Mailing Add:* 34-30 Gulf Shore Blvd N Apt 27 Naples FL 33940

KRUG, EDWARD CHARLES, b New Brunswick, NJ, Aug 24, 47. ENVIRONMENTAL GEOCHEMISTRY. *Educ:* Rutgers Univ, BSc, 75, MSc, 78, PhD(environ geochem), 81. *Prof Exp:* asst soil scientist, Conn Agr Exp Sta, 80-85; ASSOC PROF SCIENTIST, ILL STATE WATER SURV, 85- *Mem:* Soil Sci Soc Am; Sigma Xi; Am Geophys Union; Int Humic Substances Soc; Soc Environ Geochem & Health; AAAS; Int Soc Soil Sci. *Res:* Reactions, weathering, and biogeochemical cycling of elements and materials in soil, water, and lake sediment. *Mailing Add:* Ill State Water Surv 2204 Griffith Dr Champaign IL 61820-7407

KRUG, JOHN CHRISTIAN, b Toronto, Can, July 11, 38; m 74; c 2. MYCOLOGY, LICHENOLOGY. *Educ:* Univ Toronto, BSc, 63, MA, 64, PhD(mycol), 70. *Prof Exp:* Fel mycol, Inst of Bot, Swiss Fed Inst Technol, 70-72; fel, Univ Waterloo, 73; res assoc & fel, Univ Toronto, 73-74; cur asst crytogamic bot, Royal Ont Mus, 74-75; res assoc mycol, 75-77, asst cur, 81-82, LECTR, UNIV TORONTO, 77-, CUR, 82- *Concurrent Pos:* Collabr, Excerpta Botanica Sectio A: Taxonomica et Chorologica, 75-; res fel, Royal Ontario Mus, 78- *Mem:* Can Bot Asn; Brit Mycol Soc; Mycol Soc Am; Am Bryolog & Lichenolog Soc; Brit Lichen Soc; Int Asn Lichenology; Asn for Taxon Study Flora of Trop Africa. *Res:* Systematic and phytogeographical studies of Coprophilous Ascomycetes; Ascomycetes of over-wintering twigs; phytogeographical studies of Candian arctic lichens; systematic studies of Ascomycetes from tropical soils. *Mailing Add:* Dept Bot Univ Toronto 25 Willcocks St Toronto ON M3B 1E7 Can

KRUG, SAMUEL EDWARD, b Chicago, Ill, Nov 15, 43; m 68; c 4. PSYCHOLOGICAL & EDUCATIONAL MEASUREMENT, PERSONALITY ASSESSMENT. *Educ:* Col Holy Cross, AB, 65; Univ Ill Urbana-Champaign, MA, 68, PhD(psychol), 71. *Prof Exp:* Managing dir, Inst Personality & Ability Testing, 71-82; PRES, METRITECH INC, 82- *Concurrent Pos:* Adj prof, Univ Ill, Urbana-Champaign, 87- *Mem:* Fel Am Psychol Asn; fel Soc Personality Assessment. *Res:* Applied psychological and educational measurement, including test development and computer-based interpretation of test results. *Mailing Add:* MetriTech Inc 111 N Market St Champaign IL 61820

KRUGER, CHARLES HERMAN, JR, b Oklahoma City, Okla, Oct 4, 34; m 77; c 4. PHYSICAL GAS DYNAMICS, PARTIALLY IONIZED PLASMAS. *Educ:* Mass Inst Technol, SB, 56, PhD(mech eng), 60; Univ London, dipl, Imp Col, 57. *Prof Exp:* From instr to asst prof mech eng, Mass Inst Technol, 59-60; res scientist, Res Labs, Lockheed Missiles & Space Co, Calif, 60-62; from asst prof to assoc prof, 62-70, chmn, 82-88, PROF MECH ENG, STANFORD UNIV, 70-, SR ASSOC DEAN, 88- *Concurrent Pos:* Sr fel, Nat Sci Found, 68-69; vis prof, Harvard Univ, 68-69 & Princeton Univ, 78-79; mem hearing bd, Bay Area Air Pollution Control Dist, 70-83; chmn, Steering Comt, Eng Aspects of Magnetohydrodyn, 78-81; vis scientist, Norwegian Inst Technol, 79; mem, Environ Studies Bd, Nat Acad Sci, 80-83, adv coun, MAE Dept, Princeton Univ, 81- *Honors & Awards:* Fluid & Plasmadynamics Award & Medal, Am Inst Aeronaut & Astronaut, 79. *Mem:* Am Phys Soc; Am Soc Mech Engrs; Am Inst Aeronaut & Astronaut; Combustion Inst. *Res:* Physical gas dynamics; partially ionized plasmas; magnetohydrodynamics; combustion; air pollution; plasma chemistry. *Mailing Add:* Terman Engr 214 Stanford Univ Stanford CA 94305

KRUGER, FRED W, b Chicago, Ill, Dec 17, 21; m 47; c 3. MECHANICAL ENGINEERING. *Educ:* Univ Purdue, BS, 43, BS, 47; Univ Notre Dame, MS, 54. *Prof Exp:* Instr mech eng, 47-55, chmn dept, 55-65, dean col eng, 65-72, vpres, 74-87, EMER PROF, VALPARAISO UNIV, 87- *Concurrent Pos:* Consult, McDonnell Aircraft Co, Caterpillar Tractor Co, Argonne Nat Lab, Northern Ill Gas Co & Ind State Bd Registr Prof Eng; mem city coun, City of Valparaiso, 72- & City Planning Comn, 76-85. *Mem:* Am Soc Mech Engrs; Am Soc Eng Educ. *Res:* Heat power systems. *Mailing Add:* Gellerson Eng Bldg Valparaiso Univ Valparaiso IN 46383

KRUGER, FREDRICK CHRISTIAN, b St Paul, Minn, Apr 1, 12; m 36; c 2. ECONOMIC GEOLOGY. *Educ:* Univ Minn, BS, 35, MS, 36; Harvard Univ, PhD, 41. *Prof Exp:* Asst, Univ Minn, 35-36; instr geol, Dartmouth Col, 36-38; asst, Harvard Univ, 38-41; from asst geologist to asst chief geologist, Cerro de Pasco Copper Corp, 41-49; lectr geol, Northwestern Univ, 49; from assoc prof to prof, Univ Tenn, 49-52; asst chief geologist, Reynolds Metals Co, 52-57; chief geologist to vpres mining & explor div, Int Minerals & Chem Corp, 57-66; head dept, 66-74, prof mineral eng, 66-78, Donald Steel chair econ geol, 71-78, assoc dean, 72-78, EMER PROF, STANFORD UNIV, 77-; CONSULT, 77- *Concurrent Pos:* Asst, Radcliffe Col, 38-41; Krumb lectr, Am Inst Mining, Metall & Petrol Eng, 68; lectr geol, Univ Texas, 78. *Honors & Awards:* Hardinge Award, Am Inst Mining, Metall & Petrol Eng, 72. *Mem:* Fel Am Geol Soc; Soc Econ Geol; Am Inst Mining, Metall & Petrol Eng; Can Inst Mining & Metall; Peruvian Geol Soc. *Res:* Mining geology; administration. *Mailing Add:* 145 Wildwood Way Woodside CA 94062

KRUGER, JAMES EDWARD, b Winnipeg, Man, Oct 14, 38; m 60; c 1. ENZYMOLOGY, CEREAL CHEMISTRY. *Educ:* Univ Man, BSc, 60, MSc, 63; Univ Sask, PhD(phys org), 65. *Prof Exp:* RES SCIENTIST CHEM CEREAL, GRAIN RES LAB, 66- *Mem:* Am Asn Cereal Chemists; Chem Inst Can; Am Asn Plant Physiologists. *Res:* Wheat systems such as amylase, proteases, polyphenol, and oxidases; high performance liquid chromatography of sugars and proteins; automation of analytical techniques in cereal chemistry; pre-harvest sprouting problems. *Mailing Add:* 1425-303 Main St Winnipeg MB R3C 3G8 Can

KRUGER, LAWRENCE, b New Brunswick, NJ, Aug 15, 29; m 61; c 2. NEUROANATOMY, NEUROPHYSIOLOGY. *Educ:* Wagner Col, BS, 49; Yale Univ, PhD(physiol), 54. *Prof Exp:* Asst physiol, Yale Univ, 50-53, asst neurophysiol, Inst Living, Conn, 53-54; USPHS fel physiol, Johns Hopkins Univ, 55-58; Nat Res Coun fel, Col France, 58; Nat Res Coun fel anat, Oxford Univ, 58-59; USPHS sr res fel anat, 59-60, from asst prof to assoc prof, 60-66, PROF ANAT, UNIV CALIF, LOS ANGELES, 66- *Concurrent Pos:* Lederle med fac award, 61-64; mem, Int Brain Res Orgn, UNESCO; Wellcome vis prof, 81. *Mem:* AAAS; Am Asn Anatomists; Am Physiol Soc; Soc Neurosci. *Res:* Cutaneous receptors and their central nervous system representation; organization of the visual system in vertebrates with particular reference to the midbrain; thalamo-cortical relations; electron microscopy of neural degeneration; pain. *Mailing Add:* Dept Anat 73-235 Health Sci Ctr Univ Calif Los Angeles CA 90024

KRUGER, OWEN L, b Oak Park, Ill, Dec 1, 32; m 54; c 4. CERAMICS, METALLURGY. *Prof Exp:* Metall engr, Ill Inst Technol, 54; metallurgist, Continental Foundry & Mach Co, 54-55; assoc metallurgist, Argonne Nat Lab, 57-69; assoc chief, Plutonium Technol & Mat, Thermodyn Div, Columbus Div, Battelle Mem Inst, 69-77; SR ENGR & COST ANALYST, EXXON NUCLEAR CO, INC, 77- *Mem:* Fel Am Ceramic Soc; Am Nuclear Soc. *Res:* Light water reactor fuel fabrication; plutonium ceramics and metallurgy; fast reactor fuel development. *Mailing Add:* 2360 Harris Ave Richland WA 99352

KRUGER, PAUL, b Jersey City, NJ, June 7, 25; div; c 3. NUCLEAR CIVIL ENGINEERING. *Educ:* Mass Inst Technol, BS, 50; Univ Chicago, MS, 52, PhD(nuclear chem), 54. *Prof Exp:* Asst instr nuclear & phys chem, Univ Chicago, 50-53; res physicist, Res Labs Div, Gen Motors Corp, 53-54; vpres & head, Dept Phys Sci, Nuclear Sci & Eng Corp, Pa, 54-60; mgr nuclear proj, Hazelton-Nuclear Sci Corp, 60-62; PROF NUCLEAR CIVIL ENG, STANFORD UNIV, 62- *Mem:* Fel Am Nuclear Soc; Am Soc Civil Engrs. *Res:* Nuclear methods in civil engineering and environmental sciences; environmental radioactivity; nuclear explosion engineering; geothermal engineering. *Mailing Add:* Dept Civil Eng Stanford Univ Stanford CA 94305

KRUGER, RICHARD PAUL, b Chicago, Ill, July 27, 44; m 69; c 2. ELECTRICAL ENGINEERING. *Educ:* Purdue Univ, BS, 67; Univ Mo, MS, 68, PhD(elec eng), 71; Univ NMex, MBA, 79. *Prof Exp:* Asst prof elec eng & radiol, Univ Southern Calif, 71-75; asst vpres, Sci Int Applns Inc, 82-89; staff mem elec eng, 75-82, GROUP LEADER ELEC ENG, LOS ALAMOS NAT LAB, UNIV CALIF, 89- *Concurrent Pos:* Consult, Rockwell Corp, 73, Aerospace Corp, 75 & Univ Southern Calif, 76-77. *Mem:* Sr mem Inst Elec & Electronics Engrs; Sigma Xi. *Res:* Computer image processing applied to biomedical and industrial images; industrial automation; pattern recognition; computed tomography. *Mailing Add:* Los Alamos Nat Lab Univ Calif MS-J488 Los Alamos NM 87545

KRUGER, ROBERT A(LAN), mechanical engineering, applied physics, for more information see previous edition

KRUGH, THOMAS RICHARD, b Pittsburgh, Pa, May 3, 43; m 70; c 1. BIOPHYSICAL CHEMISTRY. *Educ:* Univ Pittsburgh, BS, 65; Pa State Univ, PhD(phys chem), 69. *Prof Exp:* NIH fel, Stanford Univ, 69-70; from asst prof to assoc prof, 70-78, PROF CHEM, UNIV ROCHESTER, 78- *Mem:* AAAS; Biophys Soc; Am Chem Soc. *Res:* Biophysical chemistry; drug-nucleic acid complexes; carcinogen-nucleic acid complexes; structures of nucleic acids; nuclear magnetic resonance. *Mailing Add:* Dept Chem Univ Rochester Rochester NY 14627

KRUGLAK, HAYM, b Ukraine, Mar 24, 09; m 41; c 2. PHYSICS, SCIENCE EDUCATION. *Educ:* Univ Wis, BA, 34, MA, 36; Univ Minn, PhD, 51. *Prof Exp:* Instr pub sch, Wis, 36-38 & Milwaukee Voc Jr Col, 38-42, supvr radio training, 42-44; vis asst prof physics, Princeton Univ, 44-46; instr, Univ Minn, 46-51, asst prof physics, astron & gen studies, 51-54; from assoc prof to prof, 54-77, EMER PROF PHYSICS, WESTERN MICH UNIV, 78- *Concurrent Pos:* Vis scholar, Univ Ariz, 82-89. *Mem:* Am Asn Physics Teachers; Am Asn Univ Prof; fel AAAS; Am Asn Physics Teachers. *Res:* Performance tests in laboratory instruction; design of laboratory and demonstration apparatus; physics, mathematics and astronomy education. *Mailing Add:* Dept Physics Western Mich Univ Kalamazoo MI 49008-5151

KRUGMAN, SAUL, b New York, NY, Apr 7, 11; m 40; c 2. MEDICINE. *Educ:* Med Col Va, MD, 39. *Hon Degrees:* DSc, Ohio State Univ, 79. *Prof Exp:* NIH res fel, 48-50; assoc prof, 54-60, chmn dept, 60-74, PROF PEDIAT, SCH MED, NY UNIV, 60- *Concurrent Pos:* mem comn viral infections, US Armed Forces Epidemiol Bd, 59-73; dir pediat serv, Bellevue

& Univ Hosps, 60-74; mem nat adv coun, Nat Inst Allergy & Infectious Dis, 65-69, chmn infectious dis adv comt, mem comt viral hepatitis, Nat Res Coun; mem, WHO Expert Panel Virus Dis. *Honors & Awards:* James D Bruce Mem Award, 72; Howland Award, 81; Lasker Award, 83; Laudsteiner Award, 85; William Beaumont Prize, 88. *Mem:* Nat Acad Sci; Am Pediat Soc; Soc Pediat Res; Am Epidemiol Soc; Am Acad Pediat. *Res:* Infectious diseases of children with special emphasis on viral infections. *Mailing Add:* Dept Pediat NY Univ Sch Med New York NY 10016

KRUGMAN, STANLEY LIEBERT, b St Louis, Mo, June 8, 32; m 58; c 2. FOREST GENETICS, FOREST PHYSIOLOGY. *Educ:* Univ Mo, BS, 54; Univ Calif, MS, 56, PhD(plant physiol), 61. *Prof Exp:* Res asst forestry, Univ Calif, 58-60, res specialist, 60-62; plant physiologist forest genetics, Pac Southwest Forest Range & Exp Sta, 62-66, proj leader genetics res proj, Inst Forest Genetics, 66-71, chief br genetic & related res, 71-74, prin res forest geneticist, 74-81, DIR, TIMBER MGT RES, US FOREST SERV, 81- *Concurrent Pos:* Res assoc, Agr Exp Sta, Univ Calif, 62-71; biotechnol comt, USDA, 74-87, Nat Genetics Resources Bd, 78-86. *Honors & Awards:* William Schlich Memorial Medal, Soc Am Foresters, 90. *Mem:* Fel Soc Am Foresters; Soc Plant Physiol; fel AAAS. *Res:* Pigments, hormones and reproductive physiology as related to trees; forest and tree improvement; international research. *Mailing Add:* 6515 Dryden Dr McLean VA 22101

KRUH, DANIEL, b Brooklyn, NY, May 22, 34; m 61; c 1. PRODUCT DEVELOPMENT, NEW TECHNOLOGY. *Educ:* WVa Wesleyan Col, BS, 55; Rensselaer Polytech Inst, PhD(org chem), 63. *Prof Exp:* Res chemist, Hercules Powder Co, 63-66; specialist, Insulating Mat Dept, Gen Elec Co, 66-70, mgr wire enamel develop, 70-72; res assoc, Johnson & Johnson Dental Prods Co, 72-75, sr res scientist, 75-76, supvr polymer tech serv, 76-83; sr chemist, Electro-Science Labs Inc, 83-84; proj mgr, Int Hydron Corp, 84-87; sr res assoc, 87-88, mgr new technol, 88-89, MGR, LICENSING & ACQUISITIONS, BLOCK DRUG CO, INC, 89- *Concurrent Pos:* Elected mem bd dirs, Med Plastics Div, 87-90. *Honors & Awards:* Inventors Award, Gen Elec. *Mem:* Am Chem Soc; Soc Plastics Eng; Adhesion Soc; Am Inst Chem; Controlled Release Soc; Int Asn Dental Res; Licensing Execs Soc; Soc Cosmetic Chemists. *Res:* Antiradiation drugs; phthalocyanine pigments; polyimides; chemical and light cured dental materials; high performance polymers in coatings and composites; silicone gas permeable and tinted soft contact lenses; conductive polymer thick films; consumer drug products. *Mailing Add:* Eight Braddock St East Brunswick NJ 08816

KRUH, ROBERT FRANK, b St Louis, Mo, June 15, 25; m 48; c 2. PHYSICAL CHEMISTRY. *Educ:* Washington Univ, AB, 48, PhD(chem), 51. *Prof Exp:* Asst prof chem, DePauw Univ, 51-52; from asst prof to prof chem & dean col arts & sci, Univ Ark, 52-67; PROF CHEM & DEAN GRAD SCH, KANS STATE UNIV, 67- *Concurrent Pos:* Vis prof, Wash Univ, 60-61; mem coun res policy and grad educ, Nat Asn State Univs & Land Grant Cols, 68-72; mem policy comt, Coun Grad Schs, US, 69-73, chmn, Bd Dirs, 78-79; mem bd trustees, Argonne Univ Asn, 70-77; pres, Kans State Univ Res Found, 70-; mem, Grad Record Exam Bd, 77-, chmn, 80-81. *Mem:* Sigma Xi; AAAS; Am Chem Soc; fel Am Inst Chemists; Am Phys Soc. *Res:* Crystallography; x-ray diffraction; structure of liquids. *Mailing Add:* 2155 Blue Hills Rd Manhattan KS 66502

KRUIDENIER, FRANCIS JEREMIAH, helminthology; deceased, see previous edition for last biography

KRUISBEEK, ADA M, b Rotterdam, Netherlands, Oct 2, 38. DEVELOPMENTAL BIOLOGY. *Educ:* Leiden Univ, BS, 73, PhD(immunol), 78. *Prof Exp:* Vis scientist, 79-82, cancer expert, 82-86, SR INVESTR, NIH, 86- *Mem:* Am Asn Immunologists; Brit Soc Immunologists; Europ Soc Geront. *Mailing Add:* Biol Response Modifiers Prog Bldg 10 Rm 12N226 NCI NIH Bethesda MD 20892

KRUKOWSKI, MARILYN, b New York, NY, May 3, 32; m 55; c 1. PHYSIOLOGY, DEVELOPMENTAL BIOLOGY. *Educ:* Brooklyn Col, BA, 54; NY Univ, MS, 62, PhD(biol), 65. *Prof Exp:* Res asst endocrinol exp cardiovasc dis, NY Med Col, Flower & Fifth Ave Hosps, 56-62, res assoc, 62-66, asst prof pharmacol & coinvestr endocrinol and exp cardiovasc dis, 66-69, instr pharmacol, 64-66; from res asst prof to asst prof, 69-75, assoc prof, 75-87, PROF BIOL, WASH UNIV, 87- *Mem:* AAAS; Am Soc Cell Biol; Am Soc Bone & Mineral Res; Sigma Xi. *Res:* Bone development; origin, induction and cytodifferentiation of bone-resorbing cells. *Mailing Add:* 24 Washington Terr St Louis MO 63112

KRULL, IRA STANLEY, b New York, NY, Oct 21, 40; m 73; c 1. ANALYTICAL CHEMISTRY. *Educ:* City Col New York, BS, 62; NY Univ, MS, 66, PhD(chem), 68. *Prof Exp:* Weizmann fel chem, Weizmann Inst Sci, 70-73; asst scientist, Boyce Thompson Inst, 73-76; sr scientist chem, Thermo Electron Corp, 77-79; SR SCIENTIST, NORTHEASTERN UNIV, 79-, ASSOC PROF CHEM & FAC FEL, 84- *Concurrent Pos:* Union Carbide fel, 68-70. *Mem:* Am Chem Soc; The Chem Soc; Soc Electroanal Chem; Asn Off Anal Chemists. *Res:* Synthetic organic chemistry; mechanisms of chemical carcinogenesis; trace organic analytical chemistry; analytical method development; instrumentation research and development; drug analysis. *Mailing Add:* Northeastern Univ 360 Huntington Ave Boston MA 02115

KRULL, JOHN NORMAN, b Albany, NY, July 31, 39; m 63; c 2. ECOLOGY, WILDLIFE BIOLOGY. *Educ:* State Univ NY Col Forestry, Syracuse Univ, BS, 61, MS, 63, PhD(wetland ecol), 67. *Prof Exp:* Asst prof Southern Ill Univ, 67-71; assoc prof, 71-75, PROF WILDLIFE BIOL & CONSERV, CENT MICH UNIV, 75- *Mem:* Soc Am Foresters; Conserv Educ Asn; Soil & Water Conserv Soc; Wildlife Soc. *Res:* Wetland ecology, especially green-tree reservoir ecology, waterfowl use and production from various wetland habitats, and aquatic plant-invertebrate associations; wildlife management investigations; conservation. *Mailing Add:* Dept Biol Cent Mich Univ Mt Pleasant MI 48859

KRULWICH, TERRY ANN, b New York, NY, Apr 7, 43; m 73; c 3. BIOCHEMISTRY & BIOENERGETICS, MICROBIAL PHYSIOLOGY. *Educ:* Goucher Col, BA, 64; Univ Wis-Madison, MS, 66, PhD(bact), 68. *Hon Degrees:* DSc, Goucher Col, 87. *Prof Exp:* NIH trainee bact, Univ Wis, 68; postdoctoral fel molecular biol, Albert Einstein Col Med, 68-70; from asst prof to assoc prof, 70-81, PROF BIOCHEM & DEAN GRAD SCH, MT SINAI SCH MED, 81- *Concurrent Pos:* NSF predoctoral fel, 64-68, NSF postdoctoral fel, 68-70; NIH, Res Career Develop Award, 75-80; NIH, mem Cellular Molecular Basic Dis Comt, 78-82, mem, Microbiol Physiol Genetics Study Sect, 83-87. *Mem:* AAAS; Am Soc Microbiol; Am Chem Soc; Am Soc Biol Chemists; NY Acad Sci; Biophys Soc; Sigma Xi. *Res:* Microbial bioenergetics; alkalophilic and acidophilic bacteria; protonophore resistance. *Mailing Add:* Dept Biochem Mt Sinai Sch Med New York NY 10029

KRUM, ALVIN A, b Fresno, Calif, May 14, 28; m 54; c 2. PHYSIOLOGY. *Educ:* Univ Calif, AB, 50, PhD(physiol), 57. *Prof Exp:* Res physiologist, Univ Calif, 57-61; from asst prof to assoc prof, 61-73, PROF PHYSIOL, MED CTR, UNIV ARK, LITTLE ROCK, 73- *Concurrent Pos:* Fel, Steroid Training Prog, Univ Utah, 58-59; Lederle Med Fac Award, 64-67. *Mem:* AAAS; Am Physiol Soc; Endocrine Soc; Soc Exp Biol & Med; Sigma Xi. *Res:* Endocrinology; steroid biosynthesis; metabolism. *Mailing Add:* Dept Physiol Univ Ark Med Ctr Little Rock AR 72205

KRUM, JACK KERN, b Kansas City, Mo, Mar 17, 22; m 46; c 4. FOOD TECHNOLOGY. *Educ:* Hope Col, AB, 44; Mich State Univ, MS, 48; Univ Mass, PhD(food tech), 49. *Prof Exp:* Assoc prof food technol, Univ Tenn, 49-50; food technologist in charge prod control labs, Oscar Mayer & Co, 50-52; res chemist, Nat Biscuit Co, 52-56; asst tech dir, Sterwin Chem Inc, 56-61, tech dir, 61-69; asst res dir, R T French Co, 69-70, dir res & develop, 71-72, vpres & dir res & develop, 72-73; tech dir, ITT Paniplus Co, 73-79; PRES, TECHNIQUES, INC, 80- *Concurrent Pos:* Mem food additive comt, Flavoring Extract Mfrs Asn US, 61-72, chmn, 67-71; mem indust comt, Food Protection Comn, Nat Acad Sci, 64-66, White House Conf Food Nutrit & Health, 69 & tech comt, Grocery Mfrs Am, 70-73. *Honors & Awards:* Fel, Inst Food Technol, 83. *Mem:* Am Asn Cereal Chem; Inst Food Technol. *Res:* Food additives; new product development; fermentation; nutrition; food processing; dry mixes; food additives; dehydration; food flavors; stabilizers, emulsifiers and enrichment; research administration. *Mailing Add:* Techniques Inc Rte 2 Box 84 Paola KS 66071

KRUMBEIN, AARON DAVIS, b New York, NY, Apr 6, 21; m 50; c 3. REACTOR PHYSICS, PLASMA PHYSICS. *Educ:* Brooklyn Col, AB, 41; NY Univ, PhD(physics), 51. *Prof Exp:* Lab asst physics, Bartol Res Found, 41-42; asst, NY Univ, 42-46, asst, Cosmic Ray Proj, 46-47, consult physicist, Upper Atmosphere Res, Res Div, 47-49; sr scientist, Solid State Physics, US Naval Ord Lab, 55; asst prof physics, Univ Md, 50-56, adv scientist, Develop Div, United Nuclear Corp, 56-71; sr scientist, 71-86, CONSULT, SOREQ NUCLEAR RES CTR, YAVNE, ISRAEL, 86- *Concurrent Pos:* Vis asst prof, Yeshiva Univ, 61-70; vis sr res assoc, Plasma Physics Group, Univ Md, 78-79, vis prof, Plasma Lab, 85-86. *Mem:* Israel Phys Soc (treas, 75-78); Am Nuclear Soc; Israel Nuclear Soc; Am Phys Soc; Sigma Xi. *Res:* Gaseous electronics; nuclear detectors; nuclear reactor and reactor shielding physics; intermetallic semiconductors; Geiger-Müller counters; radiation transport; space physics; laser produced plasma; thermonuclear reactors. *Mailing Add:* Nine Lutzki St Rehovot 76251 Israel

KRUMBEIN, SIMEON JOSEPH, b Brooklyn, NY; m 57; c 3. ELECTROCHEMISTRY, SURFACE PHYSICS. *Educ:* Brooklyn Col, BS, 56; NY Univ, PhD(phys chem), 61. *Prof Exp:* Chemist, TRG, Inc, 56-57; phys chemist, Fuel Cell Lab, Gen Elec Co, 61-64; sr scientist, Res Div, Burndy corp, 64-69; asst prof, Stern Col, Yeshiva Univ, 69-75, assoc prof chem, 75-79; SR ENG SCIENTIST, RES DIV, AMP, INC, 77- *Concurrent Pos:* Consult, US Army Electronic Components Lab, 70; coadj asst prof chem, Univ Coll, Rutgers Univ, 72-74; consult, Res Div, AMP, Inc, 73-76. *Mem:* Am Chem Soc; Electrochem Soc; fel Am Inst Chem; Am Soc Testing & Mat; Sigma Xi. *Res:* Experimental methods in fuel cell research; electrical contact phenomena; metallic corrosion; experimental electrode kinetics; contact materials; environmental testing; electromigration; plating porosity. *Mailing Add:* 3106 Labyrinth Rd Baltimore MD 21208

KRUMDIECK, CARLOS L, b Lima, Peru, Nov 11, 32;; c 2. BIOCHEMISTRY, NUTRITION. *Educ:* San Marcos Univ, Lima, BMed & MD, 58; Tulane Univ La, PhD(biochem), 64. *Prof Exp:* Resident med, Hosp 2nd of May, Lima, Peru, 58-59; biochemist, Hektoen Inst Med Res, Chicago, Ill, 59-60; Rockefeller Found fel biochem, Tulane Univ La, 60-62 & 64, USPHS fel, 62-63; prof biochem, Univ Cayetano Heredia, Peru, 65-70; asst prof biochem & med, 67-70, assoc prof biochem, 70-81, assoc dir nutrit prog, 76-85, PROF BIOCHEM, UNIV ALA, BIRMINGHAM, 77-, ASSOC PROF PEDIAT, 76- *Honors & Awards:* Bordeu Award in Nutrit, 85. *Mem:* Am Soc Biol Chemists; Sigma Xi; Am Inst Nutrit; Am Soc Clin Nutrit; NY Acad Sci; AAAS; Physicians Social Responsibility. *Res:* Biochemistry of folic acid polyglutamates; biochemical assessment of nutrient status; pathogenesis of atherosclerosis; nutrition and cancer. *Mailing Add:* 1919 Seventh Ave S Univ Med Ala Birmingham AL 35294

KRUMHANSL, JAMES ARTHUR, b Cleveland, Ohio, Aug 2, 19; m 44; c 3. CONDENSED MATTER PHYSICS, MATERIALS SCIENCE. *Educ:* Univ Dayton, BS, 39; Case Inst Technol, MS, 40; Cornell Univ, PhD(physics), 43. *Hon Degrees:* DSc, Case Western Reserve, 80. *Prof Exp:* Instr physics, Cornell Univ, 43-44; physicist, Stromberg-Carlson Co, 44-46; from asst prof to assoc prof physics & appl math, Brown Univ, 46-48; from asst prof to assoc prof physics, Cornell Univ, 48-55; from asst dir res to assoc dir res, Nat Carbon Co, 55-58; dir lab atomic & solid state physics, 60-64, Horace White prof physics, 81-90, PROF PHYSICS, CORNELL UNIV, 59-, EMER HORACE WHITE PROF PHYSICS, 90- *Concurrent Pos:* Consult to var industs, 46-; mem adv comts, AEC, Dept Defense & Nat Acad Sci, 56-; ed, J Appl Physics, 57-60; Guggenheim fel, 59-60; assoc ed, Solid State Commun, 63- & Rev Mod Physics, 68-73; NSF sr fel, Oxford Univ, 66-67; mem gov bd,

Am Inst Physics, 73-; ed, Phys Rev Lett, 74-; asst dir, math phys sci & eng, NSF, 77-79; adj prof physics, Univ Pa, 80-82; fel, Los Alamos Nat Lab; dir, Allied Corp, 80-87; adj prof, Univ Mass, Amherst, 91; vis fel, All Souls, Oxford, 77, Royal Soc London, 83, Ganville & Caius, Cambridge, 83, Yamada Found Japan, 82 & 89. *Honors & Awards:* Fulbright lectr, 76. *Mem:* Fel Am Phys Soc (pres, 89); fel AAAS; Sigma Xi. *Res:* Theoretical physics; materials science; applied mathematics; biophysics; research administration. *Mailing Add:* 515 Station Rd Amherst MA 01002

KRUMM, CHARLES FERDINAND, b Macomb, Ill, Aug 3, 41; m 67. ELECTRICAL ENGINEERING. *Educ:* Univ Mich, BSE, 63, MSE, 65, PhD(elec eng), 70. *Prof Exp:* Res asst, Electron Physics Lab, Univ Mich, 65-69; sr res scientist, Res Div, Raytheon Co, 69-76; mem tech staff, 76-77, sect head, 77-78, asst dept mgr, 78-80, dept mgr, 80-86, MGR, MICOELECTRONICS LAB, HUGHES RES LAB, 86- *Mem:* Inst Elec & Electronics Engrs. *Res:* Design, fabrication and testing of Gallium Arsenide & Silicon integrated circuits including ion implantation; electron beam and optical lithography; high resolution dry processing and molecular beam epitaxy. *Mailing Add:* Aircraft Co Radar Sys Group/Bldg R01 MS 11A05/ PO Box 92426 Los Angeles CA 90009

KRUMMEL, WILLIAM MARION, b New York, NY, Aug 15, 28; m 62; c 3. BIOMEDICAL ENGINEERING, ELECTRICAL ENGINEERING. *Educ:* City Col New York, BEE, 49; Columbia Univ, MS, 63; NY Univ, PhD(biomed eng), 67. *Prof Exp:* Engr, Control Div, Gen Elec Co, 49-54, systs engr, 56-62; asst prof elec eng, Univ Conn, 67-69 & NY Univ, 69-71, adj assoc prof, 71-73; assoc prof math, Bronx Community Col, City Univ New York, 71-73; assoc dean, Westchester Community Col, 73-80; pres, Norwalk State Tech Col, 80-86; PRES, BRIDGEPORT ENG INST, 86- *Concurrent Pos:* Chair, New England Sect, Soc Mfg Engrs. *Mem:* AAAS; Inst Elec & Electronics Engrs; NY Acad Sci; Sigma Xi; Am Soc Eng Educ. *Res:* Application of mathematical and engineering techniques to experimental and theoretical biomedical problems; application of electrical engineering techniques to industrial process control. *Mailing Add:* Bridgeport Eng Inst Fairfield CT 06430

KRUMMENACHER, DANIEL, b Geneva, Switz, Mar 14, 25; m 51. GEOLOGY, CHEMISTRY. *Educ:* Univ Geneva, dipl chem eng, 52, PhD(geol), 59. *Prof Exp:* Res scientist, Univ Geneva, 59-62, assoc prof geochem, 62-68; from asst prof to prof geol, San Diego State Univ, 68-88; RETIRED. *Concurrent Pos:* Researcher mass spectros, Univ Calif, Berkeley, 60-61, res assoc, 67-68. *Mem:* Geol Soc Am. *Res:* Isotope geology; geochronometry. *Mailing Add:* Box 168 San Diego State Univ Rancho Sante Fe CA 92067

KRUMREI, W(ILLIAM) C(LARENCE), b Cleveland, Ohio, Mar 31, 24; m 52; c 2. PROCESS DEVELOPMENT, PRODUCT DEVELOPMENT. *Educ:* Case Inst Technol, BS, 50; Mass Inst Technol, MS, 51. *Prof Exp:* Develop engr, Process Develop Dept, Procter & Gamble Co, 51-54, tech brand mgr, Soap Prod Res Dept, 54-55, sect head, 55-58, assoc dir, Foods Div, Prod Develop Dept, 58-62, assoc dir, Household Soap Prod Develop Div, 62-63, dir prod develop, Household Soap Div, 63-70, dir tech govt rels, 70-72, sr dir, Corp Res & Develop Dept, 72-82; RETIRED. *Concurrent Pos:* Mem adv comt, NIH, 77-78. *Mem:* AAAS; Am Mgt Asn; Sigma Xi; NY Acad Sci; Am Indust Health Counc. *Res:* Oil, food and detergent chemistry and processing. *Mailing Add:* Six N Calibogue Cay Rd Hilton Head Island SC 29928

KRUPA, SAGAR, b Madras, India, Oct 11, 40; US citizen; m 67. PHYTOPATHOLOGY, AIR POLLUTION. *Educ:* Andhra Univ, BSc, 59; Univ Wis, MS, 68; Univ Uppsala, PhD(plant physiol), 71. *Prof Exp:* Swed Nat Sci Res Coun res fel physiol bot, Inst Physiol Bot, Univ Uppsala, 72; res fel soil sci, Dept Microbiol, 72-73, res fel plant path, 73-74, from asst prof to assoc prof, 74-85, PROF PLANT PATH & PLANT PHYSIOL, DEPT PLANT PATH, UNIV MINN, ST PAUL, 85-, ENVIRON PATHOLOGIST, 74- *Concurrent Pos:* Docent, Inst Physiol Bot, Univ Uppsala, 73- *Mem:* Am Phytopath Soc; Air Pollution Control Asn. *Res:* Atmospheric chemistry and effects of air pollution on vegetation. *Mailing Add:* Dept Plant Path Univ Minn St Paul MN 55108

KRUPKA, LAWRENCE RONALD, b New York, NY, Mar 7, 33; m 58. PLANT PATHOLOGY. *Educ:* Cornell Univ, BS, 54; Univ Del, MS, 56; La State Univ, PhD, 59. *Prof Exp:* Asst plant path, Univ Del, 54-56 & La State Univ, 56-58; asst pathologist, Univ Nebr, 59-60; biologist, Rohm and Haas Chem Co, Pa, 61-63; asst prof biol, Philadelphia Col Pharm, 63-65; from asst prof to assoc prof, 65-68, PROF NATURAL SCI, MICH STATE UNIV, 68-, CHMN, DEPT GREAT ISSUES & INTERDEPT COURSES, 73- *Concurrent Pos:* USPHS fel, 59-60; co-dir, sci, technol & human values, NSF. *Mem:* Mycol Soc Am; Am Phytopath Soc; Soc Indust Microbiol; Sigma Xi. *Res:* Host-parasite relationships of plant pathogens, especially rusts; respiration studies; enzyme and organic acid metabolism; drug toxicity; effects of science and technology upon society. *Mailing Add:* Dept Natural Sci Mich State Univ East Lansing MI 48823

KRUPKA, MILTON CLIFFORD, b New York, NY, Jan 1, 24; m 54; c 3. HIGH TEMPERATURE PHYSICAL CHEMISTRY, SYSTEMS ENGINEERING MATERIALS. *Educ:* City Univ New York, BA, 44; Univ NMex, MS, 58, PhD(phys chem), 62. *Prof Exp:* Jr chemist, SAM Labs, Columbia Univ, 44-45; asst chemist, Los Alamos Nat Lab, Univ Calif, 45-46, lab supvr explosives, 46-50, proj engr weapons develop, 50-59, mat scientist high temp chem, 59-71, staff mem & asst mgr high temp chem, 71-75, mem staff & prin investr, technol & environ assessment, New Energy Technol & Conversion Systs, 76-78, syst analyst & prin investr, 78-89; RETIRED. *Mem:* Fel Am Inst Chemists; Am Chem Soc; Sigma Xi; Sci Res Soc. *Res:* Elucidation of high temperature thermodynamics and properties of various refractory materials; ultrahigh pressure and temperature research; preparation of new materials and new superconductors; corrosion research in natural silicate melts; technology assessment; systems analysis for advanced energy systems. *Mailing Add:* 6401 Turnberry Lane NE Albuquerque NM 87111-5860

KRUPKA, RICHARD M(ORLEY), b Winnipeg, Man, Jan 17, 32; m 58; c 2. BIOCHEMISTRY. *Educ:* Univ Sask, BA, 53, MA, 55; McGill Univ, PhD(plant biochem), 57. *Prof Exp:* Fel enzyme kinetics, Univ Ottawa, 58-60; RES SCIENTIST, CAN DEPT AGR, 60- *Mem:* Can Fedn Biol Socs; Fedn Am Soc Exp Biol. *Res:* Transport kinetics; enzyme kinetics. *Mailing Add:* Res Ctr Can Agr 1400 Western Rd London ON N6G 2V4 Can

KRUPKE, WILLIAM F, b Springfield, Mass, Jan 30, 37; m 61. SOLID STATE PHYSICS. *Educ:* Rensselaer Polytech Inst, BS, 58; Univ Calif, Los Angeles, MA, 60, PhD(spectros), 66. *Prof Exp:* Mem tech staff laser res, Minneapolis Regulator Co, 61-62; mem tech staff solid state res, Aerospace Corp, 62-66; staff physicist, Hughes Aircraft Co, Culver City, 66-74; MEM STAFF, LAWRENCE LIVERMORE NAT LAB, 74- *Mem:* Am Phys Soc. *Res:* Spectroscopic research on solid and gaseous systems as related to the dynamics of lasers; electronic structure of rare earth elements in solids; transition intensities and energy transfer processes. *Mailing Add:* Lawrence Livermore Lab L-488 PO Box 5508 L-0275354 Livermore CA 94551

KRUPP, DAVID ALAN, b Blue Island, Ill, Feb 6, 53; m 86; c 2. BIOCHEMISTRY OF CORAL MUCUS, CALORIMETRY OF TISSUES. *Educ:* Univ Calif, Los Angeles, BA, 76; Univ Hawaii, PhD(zool), 82. *Prof Exp:* INSTR BIOL & MARINE SCI, WINDWARD COMMUNITY COL, 84-, FAC COORDR, MARINE OPTION PROG, 84- *Concurrent Pos:* Prin investr, Summer Prog Enhancement Basic Educ Oceanog, 85-90; co-prin investr, Marine Option Prog Sea Grant Projs, 85-91; sci fac coordr, Math-Sci Dept, Windward Community Col, 85-89, dept chairperson, 89-91; vis researcher, Marine Lab, Univ Guam, 89; lectr, summer advan res & training prog, Hawaii Inst Marine Biol, 91- *Mem:* AAAS; Am Soc Zoologists. *Res:* Aspects of the biology of reef corals: reproduction, production of mucus and its properties, diffusion barriers and carbon limitation, pollution effects on coral reefs; biochemistry of sea cucumber body walls. *Mailing Add:* Windward Community Col 45-720 Keaahala Rd Kaneohe HI 96744

KRUPP, EDWIN CHARLES, b Chicago, Ill, Nov 18, 44; m 68; c 1. ASTRONOMY. *Educ:* Pomona Col, BA, 66; Univ Calif, Los Angeles, MA, 68, PhD(astron), 72. *Prof Exp:* Cur, 72-74, actg dir, 74-76, DIR ASTRON, GRIFFITH OBSERV, CITY LOS ANGELES, 76- *Concurrent Pos:* Inst, El Camino Col, 69-74 & Univ Southern Calif, 74-75; consult, Los Angeles County Supt Sch, Community Col Consortium, 74-82; lectr & course coord, Univ Calif Exten, 75- *Honors & Awards:* Klumpke-Roberts Award for Contrib to Pub Understanding of Astron, Astron Soc of the Pac, 89. *Mem:* Am Astron Soc; Sigma Xi; fel Explorer's Club. *Res:* Archaeoastronomy. *Mailing Add:* Griffith Observ 2800 E Observatory Rd Los Angeles CA 90027

KRUPP, IRIS M, b New Orleans, La, May 1, 28; m. PARASITOLOGY, ALLERGY. *Educ:* La State Univ, BS; Tulane Univ, MS, 55, PhD(parasitol), 58, MD, 71. *Prof Exp:* Res asst, Sch Med, Tulane Univ, 49-53; instr vet parasitol, Mich State Col, 53-54; asst, 54-58, from instr to asst prof trop med & pub health, 59-66, assoc prof trop med & pub health, 66-81, CLIN ASSOC PROF MED, MED SCH, TULANE UNIV, 71- *Concurrent Pos:* Observer res technol, London Sch Hyg & Trop Med, 59-60; consult, USPHS grant, 60-65; intern, USPHS Hosp, 71-72; res fel, Vet Admin Hosp, New Orleans, 72-73; mem adv bd parasitic dis & consult to surgeon gen, Dept Army, 73-75; resident dermat, Charity Hosp, New Orleans, 73-76; pvt pract dermat, 76- *Mem:* Am Soc Dermat Surg; Int Soc Dermat Surg; Am Acad Dermat. *Res:* Immunology of parasitic infections; schizophrenia; dermatology. *Mailing Add:* 4357 Folse Dr Metairie LA 70002

KRUPP, MARCUS ABRAHAM, b El Paso, Tex, Feb 12, 13; m 41; c 4. MEDICINE, METABOLISM. *Educ:* Stanford Univ, AB, 34, MD, 39; Am Bd Int Med, dipl, 47. *Prof Exp:* Dir clin path, Vet Admin Hosp, San Francisco, 46-50; dir, Palo Alto Med Clin Lab, 50-80, dir, 50-86, EMER DIR, PALO ALTO MED RES FOUND, 86- *Concurrent Pos:* Asst clin prof, Sch Med, Stanford Univ, 46-56, assoc clin prof, 56-65, clin prof, 65- *Honors & Awards:* Albion Walter Hewlezz, 87. *Mem:* Am Col Physicians; Am Fedn Clin Res; AAAS. *Res:* Renal physiology; water and electrolyte metabolism; author of several medical textbooks. *Mailing Add:* 860 Bryant St Palo Alto CA 94301

KRUPP, PATRICIA POWERS, b New York, NY. GROSS ANATOMY, EXPERIMENTAL MORPHOLOGY. *Educ:* Beaver Col, BA, 64; Hahnemann Med Col, PhD(anat), 70. *Prof Exp:* Asst instr nursing anat & physiol, Pa State Univ, Ogontz Campus, 63-66; sr instr gross anat, Hahnemann Med Col, 70-71, asst prof, 71-78; ASSOC PROF ANAT, COL MED, UNIV VT, 78- *Mem:* Am Asn Anatomists; AAAS; NY Acad Sci; Am Fedn Clin Res; Am Women Sci. *Res:* Experimental modification of thyroid gland structure and function; effects of diet, drugs, spontaneous hypertension and chronic stimulation. *Mailing Add:* Dept Anat & Neurobiol Col Med Univ Vt Given Bldg Burlington VT 05405

KRUSBERG, LORIN RONALD, b July 18, 32; m; c 3. PLANT PATHOLOGY. *Educ:* Univ Del, BS, 54; NC State Col, MS, 56, PhD(plant path), 59. *Prof Exp:* From asst prof to assoc prof, 60-70, PROF PLANT PATH, UNIV MD, COLLEGE PARK, 70- *Concurrent Pos:* Fel, NSF, Rothamsted Exp Sta, Eng, 59-60. *Mem:* Am Phytopath Soc; fel Soc Nematol; Sigma Xi. *Res:* Biology and control of nematodes parasitic on corn, soybeans and vegetables. *Mailing Add:* Bot Dept Univ Md College Park MD 20742-5815

KRUSCHWITZ, WALTER HILLIS, b Edgerton, Ohio, July 20, 20; m 47; c 2. PHYSICS. *Educ:* Taylor Univ, AB, 42; Vanderbilt Univ, MA, 48; Univ Mich, PhD(higher ed), 61. *Prof Exp:* Assoc prof physics, Cumberland Univ, 48-50; asst prof physics & math, Union Univ, Tenn, 51-60, prof & head physics dept, 61-63; prof physics, Mobile Col, 63-67; assoc prof physics & educ, 67-69, assoc prof phys sci & physics, 69-72, ASSOC PROF PHYSICS, UNIV SFLA, 73- *Mem:* Am Asn Physics Teachers. *Res:* Measurement of the velocity of a gas immediately before combustion; undergraduate college physics research and its sponsorship; science education. *Mailing Add:* Dept Physics Univ SFla 4202 Fowler Ave Tampa FL 33620

KRUSE, ARTHUR HERMAN, b Easton, Kans, Feb 5, 28; m 54; c 2. MATHEMATICS. *Educ:* Univ Kans, BA, 49, MA, 51; Univ Chicago, PhD(math), 56. *Prof Exp:* Res assoc math, Univ Kans, 54-60, from instr to asst prof, 54-60; RES PROF MATH SCI, NMEX STATE UNIV, 60- *Mem:* Am Math Soc; Math Asn Am. *Res:* Topology; axiomatic set theory; analysis. *Mailing Add:* Res Ctr Arts Sci NMex State Univ PO Box RC Las Cruces NM 88003

KRUSE, CARL WILLIAM, b Aline, Okla, June 2, 27; m 49; c 4. COAL DESULFURIZATION, COAL ANALYSES. *Educ:* Bethany-Nazarene Col, AB, 50; Univ Kans, MS, 52; Univ Ill, PhD(chem), 58. *Prof Exp:* Res chemist, Phillips Petrol Co, 52-54 & Ill State Geol Surv, 54-56; group leader org chem, Phillips Petrol Co, 57-64, sect mgr, 64-68; assoc prof, Mid-Am Nazarene Col, 68-74, prof chem, 74-78, chmn div environ serv, 68-78; chemist, 78-80, head minerals eng sect, 80-84, SR SCIENTIST, ILL STATE GEOL SURV, 84- *Concurrent Pos:* Consult, Midwest Res Inst, 68-74 & Oak Ridge Nat Labs, Energy Res & Develop Admin, 75. *Mem:* Am Chem Soc. *Res:* Aromatic and aliphatic halides; alkylation of aromatics; acetylenic and olefinic compounds; organoaluminum compounds; energy problems; low temperature carbonization of coal; beneficiation of coal and coal char. *Mailing Add:* Ill State Geol Surv 615 E Peabody Dr Champaign IL 61820-6964

KRUSE, CONRAD EDWARD, b Philadelphia, Pa, Sept 14, 23; m 49; c 2. BACTERIOLOGY. *Educ:* Philadelphia Col Pharm, BSc, 49, DSc(bact), 53; Univ Wis, MSc, 51. *Prof Exp:* Instr bact, Philadelphia Col Pharm, 51-56; res assoc, can div, Crown Cork & Seal Co, Pa, 56-57 & Col Dept, Lea & Febiger, 57-60; asst prof biol sci, Drexel Inst, 60-67; assoc prof biol, Ursinus Col, 67-89; PROF ENVIRON ENG, TEMPLE UNIV, 90- *Concurrent Pos:* Dept supvr bact, Children's Hosp, Philadelphia, 53-56; instr, Misericordia Hosp, Philadelphia, 55-56. *Mem:* AAAS; Am Soc Microbiol; Am Pharmaceut Asn; Inst Food Technologists. *Res:* Industrial and medical microbiology; biochemistry; food technology; development of pharmaceutical and food products; development of biochemical fuel cell. *Mailing Add:* Dept Environ Eng CECSA Temple Univ 12th & Norris Sts Philadelphia PA 19122

KRUSE, FERDINAND HOBERT, b Council Bluffs, Iowa, July 9, 25; m 47; c 3. PHYSICAL CHEMISTRY. *Educ:* Iowa State Col, BSc, 45; Univ NMex, MSc, 51; Univ Calif, Los Angeles, PhD(phys chem), 56. *Prof Exp:* Chemist, Gen Elec Co, Mass, 45-46; instr inorg chem, Iowa State Col, 46-48; res chemist phys chem, Esso Res & Eng Co, 56-57; mem staff, Los Alamos Sci Lab, 57-67; SR CHEMIST CORP TECH, TECH CTR, OWENS-ILL, INC, 68- *Concurrent Pos:* Fulbright lectr, Univ Alexandria, 65-66. *Mem:* Am Crystallog Asn. *Res:* X-ray crystallography; single crystal structure analysis; optical properties of crystalline compounds; inorganic chemistry of actinide elements; chemistry of tellurium and selenium; digital computer programming and applications; analytical chemistry; x-ray fluorescence spectroscopy. *Mailing Add:* Owens Ill Ntc-B One Seagate Toledo OH 43666

KRUSE, JURGEN M, b Berlin, Ger, Apr 27, 27; nat US; m 86; c 3. CHEMISTRY. *Educ:* Harvard Univ, AB, 49; Purdue Univ, MS, 50, PhD(chem), 52. *Prof Exp:* Res chemist & group leader, E I du Pont de Nemours & Co, Inc, 52-64, res assoc, 63-67; mgr anal lab, Itek Corp, 67-75; specialist, US Postal Serv, 75-76; dir advan develop, Coulter Systs Corp, 76-80; Consult, Bruning, 80-89; CONSULT, XAAR, 90- *Mem:* Am Chem Soc. *Res:* Analytical chemistry; catalysis; reprographics. *Mailing Add:* Five Lochbourne Dr Clinton CT 06413-1411

KRUSE, KIPP COLBY, b Norfolk, Nebr, Nov 21, 49; m 75; c 2. BEHAVIORAL ECOLOGY. *Educ:* Wayne State Col, BSE, 71; Univ SDak, MA, 73; Univ Nebr, PhD(Ecol), 78. *Prof Exp:* Instr biol, Univ Nebr, 78-79; from asst prof to assoc prof, 79-87, PROF ZOOL, EASTERN ILL UNIV, 87- *Concurrent Pos:* Frederic Clements Ecol scholar, 78- *Mem:* Sigma Xi; Soc Study Amphibians & Reptiles; Herpetologists League. *Res:* Aspects of sexual selection in frogs, toads and waterbugs. *Mailing Add:* Dept Zool Eastern Ill Univ Charleston IL 61920

KRUSE, LAWRENCE IVAN, b Springfield, Vt, Dec 21, 58; m 80. PHARMACEUTICAL CHEMISTRY. *Educ:* Mass Inst Technol, BS, 76, PhD(chem), 79. *Prof Exp:* assoc sr investr med chem, Smith Kline & French Labs, 79-89; VPRES CHEM RES & DEVELOP, STERLING RES GROUP, 89- *Mem:* Am Chem Soc. *Res:* Design and synthesis of selective enzyme inhibitors; synthesis of unnatural amino acids and antitumor-antimetabolic agents of potential therapeutic ability; stereoelectronic effects in organic chemistry; indole and other heterocyclic chemistry. *Mailing Add:* Sterling Res Group Nine Great Valley Pkwy Malvern PA 19355

KRUSE, OLAN ERNEST, b Coupland, Tex, Sept 6, 21; m 42; c 2. PHYSICS. *Educ:* Tex Col Arts & Indust, BS, 42; Univ Tex, MA, 49, PhD, 51. *Prof Exp:* Jr radio engr, Signal Corps Labs, Camp Evans, NJ, 42-43; from asst prof to prof physics & chmn dept, Stephen F Austin State Col, 51-56; prof & chmn dept, 56-87, PROF PHYSICS, TEX A&I UNIV, 87- *Concurrent Pos:* Instr, Naval Officers Pre-Radar Sch, Harvard Univ, 43-44 & Bowdoin Col, 44-45. *Mem:* Am Asn Physics Teachers. *Res:* Electronic circuitry; mechanics; electricity and magnetism; multiple scattering of charged particles by thin foils. *Mailing Add:* Dept Physics Tex A&I Univ Kingsville TX 78363

KRUSE, PAUL WALTERS, JR, b Hibbing, Minn, Nov 24, 27; m 54; c 9. SOLID STATE PHYSICS, ELECTROOPTICS. *Educ:* Univ Notre Dame, BS, 51, MS, 52, PhD(physics), 54. *Prof Exp:* Physicist, Farnsworth Electronics Co, 54-56; sr res scientist, Res Ctr, Minn-Honeywell Regulator Co, 56-59, prin res scientist, 59-60, staff scientist, 60-69, sr staff scientist, 70-77, prin staff scientist, 78-80, prin res fel, Corp Technol Ctr, 80-83, CHIEF RES FEL, SENSOR & SYST DEVELOP CTR, HONEYWELL, INC, 83- *Concurrent Pos:* Mem, US Army Sci Adv Panel, 65-77, ground warfare panel, President's Sci Adv Comt, 70-72, comt mat for electromagnetic radiation detection devices, Nat Acad Sci, 71-73 & Army Countermine Adv Comt, 71-74; mem planning comt, Third Int Photoconductivity Conf, 69-71; chmn, US Army ERADCOM Technol Panel, 75-76; mem, US Army Near-Millimeter Wave Technol Base Develop Study, 76-77; mem, Army Sci Bd, 78-82, 85-90; Mem comt phys sci, adv bd military personnel supplies, Nat Res Coun, Nat Acad Sci, 69-71, NATO rev panel Pres's Sci Adv Comt & Vietnam Panel, 71-72 & US/USSR Tech Balance Assessment Study, US Naval Res Adv Comt, 79-80; mem, Comt Biol & Chem Sensor Technol, Nat Res Coun, Nat Acad Sci, 83-84; mem, Sci & Technol Operating Comt, Am Electronics Asn, 90- *Honors & Awards:* Recipient, H W Sweatt Award, Honeywell, Inc, 66; Alan Gordon Mem Award, Int Soc Optical Eng, 81. *Mem:* Fel Am Phys Soc; assoc fel Am Inst Aeronaut & Astronaut; fel Optical Soc Am; sr mem Inst Elec & Electronics Engrs. *Res:* Electrooptical physics; nonlinear optics; infrared detectors; crystal growth; solid state devices; lasers; photoeffects in high temperature superconductors. *Mailing Add:* Sensor & Syst Develop Ctr Honeywell Inc 10701 Lyndale Ave Bloomington MN 55420

KRUSE, ROBERT LEROY, b Jacksonville, Fla, Jan 7, 41. DATA STRUCTURES, ALGORITHM ANALYSIS. *Educ:* Pomona Col, BA, 60; Calif Inst Technol, MS, 62, PhD(math), 64. *Prof Exp:* Staff mem, Sandia Lab, 64-70; assoc prof math, Emory Univ, 73-76; chmn dept, 76-79, PROF MATH & COMPUT SCI, ST MARY'S UNIV, HALIFAX, 79- *Concurrent Pos:* Fulbright-Hays grant & vis reader, Univ Canterbury, NZ, 70-72; vis prof, Univ Alberta, 82; Erskine fel, Univ Canterbury, New Zealand, 90. *Mem:* Am Math Soc; Asn Comput Mach. *Res:* Data structures; applications of computers in abstract algebra; finite rings. *Mailing Add:* Dept Math & Comput Sci St Mary's Univ Halifax NS B3H 3C3 Can

KRUSE, ROBERT LOUIS, b Fairmont, Minn, Aug 23, 38; m 64; c 2. CHEMICAL ENGINEERING, POLYMER CHEMISTRY. *Educ:* Univ Minn, Minneapolis, BS, 60; Univ Ill, Urbana, MS, 62; Columbia Univ, EngSciD(chem eng), 67. *Prof Exp:* Chem engr, Esso Res & Eng Co, NJ, 61-64; chem engr, 67-70, res group leaser, 70-78, sci fel 78-85, SR SCI FEL, MONSANTO CO, 85- *Concurrent Pos:* Lectr polymer processing, Univ Mass, 67-68; lectr math, Western New Eng Col, 68-; lectr polymer characterization, St Joseph Col, Conn, 70-71. *Mem:* Am Chem Soc; Soc Rheol. *Res:* Polymerization kinetics; rheology and surface properties of polymers; molecular characterization; continuum properties; composites. *Mailing Add:* 444 Michael Sears Rd Belchertown MA 01007

KRUSE, ULRICH ERNST, b Berlin, Ger, May 22, 29; nat US. PHYSICS. *Educ:* Harvard Univ, PhD(physics), 54. *Prof Exp:* From instr to asst prof physics, Univ Chicago, 54-59; from asst prof to prof physics, 59-, EMER PROF PHYSICS, UNIV ILL, URBANA. *Mem:* Fel Am Phys Soc. *Res:* Experimental nuclear physics. *Mailing Add:* 429 Loomis Lab 1110 W Green St Urbana IL 61801

KRUSE, WALTER M, b Heide, Ger, Oct 6, 28; m 60; c 2. INORGANIC CHEMISTRY, PHYSICAL CHEMISTRY. *Educ:* Univ Cologne, PhD, 58. *Prof Exp:* Fel, Univ Chicago, 58-60; asst kinetics, Max Planck Inst Phys Chem, 60-64; res chemist, Hercules Res Ctr, Del, 64-71, res chemist, Atlas Chem Industs, Inc, 71-74, res chemist, ICI United States, Inc, 74-79, sr res chemist, ICI Americas, Inc, 79-88; RETIRED. *Concurrent Pos:* Res info scientist, 88-91. *Mem:* Am Chem Soc; Sigma Xi. *Res:* Preparative inorganic chemistry; complex chemistry; oxidation of olefins; hydrogenation of carbohydrates; homogeneous catalysis. *Mailing Add:* One Woodbury Ct Wilmington DE 19805

KRUSEN, EDWARD MONTGOMERY, b Philadelphia, Pa, Feb 7, 20; m 48; c 2. MEDICINE. *Educ:* Univ Pa, BA, 41, MD, 44; Univ Minn, MS, 50. *Prof Exp:* Med dir phys ther sch, Baylor Univ Med Ctr, 51-71, prof phys ther, Univ, 58-71, chief Dept Phys Med & Rehab, 50-85; RETIRED. *Concurrent Pos:* From asst prof to assoc prof phys med & rehab, Univ Tex Health Sci Ctr, 51-61, chmn dept, 51-55, clin prof phys med & rehab, 61-; area consult, Vet Admin, 54-; chmn med adv bd, United Cerebral Palsy, Tex, 55-73; consult, Elizabeth Kenney Found, 55-57 & Am Rehab Found, 57-; chmn, Coun Med Dirs Phys Ther Schs, 60; mem, Am Bd Phys Med & Rehab, 65-77, Gov Coun Develop Disabilities, 71 & Residency Review Comt, 76-82. *Honors & Awards:* Tex Award Planning Voc Rehab, 68; Distinguished Clinician, Am Acad Phys Med & Rehab, 85. *Mem:* AMA; Am Rheumatism Asn; Am Cong Rehab Med; Am Acad Phys Med & Rehab; Sigma Xi. *Res:* Cervical syndrome; backache; hemiplegia; exercise; rehabilitation of the elderly; electromyography. *Mailing Add:* 3500 Colgate Dallas TX 75225

KRUSHENSKY, RICHARD D, b Ferndale, Mich, June 3, 32; m 60; c 1. GEOLOGY, VOLCANOLOGY. *Educ:* Wayne State Univ, BS, 55, MS, 57; Ohio State Univ, PhD(geol), 60. *Prof Exp:* Geologist, Orinoco Mining Co Div, US Steel Corp, 60-62; geologist, Br Spec Projs, 62-63, Br Mil Geol, 63-64 & Br Int Geol, 64-70, geologist, Br Eastern Environ Geol & dep chief, Off Environ Geol, 74-79, geologist, Br Eastern Regional Geol, 79-83, GEOLOGIST & DEP CHIEF LATIN AM, OFF INT GEOL, US GEOL SURV, 83-, ASSOC INT CHIEF, US GEOL SURV, 88- *Mem:* Geol Soc Am. *Res:* Mineral exploration and training of local personnel in Turkey; study of Irazu Volcano, Costa Rica and other active and inactive volcanoes in Central America; volcanic petrology and petrography; volcanology; mineral exploration; regional geology of volcanogenic and intrusive rocks in Southwestern Puerto Rico. *Mailing Add:* Off Int Geol Mail Stop 917 Nat Ctr US Geol Surv Reston VA 22092

KRUSIC, PAUL JOSEPH, b Trieste, Italy, Nov 28, 34; m 56; c 3. ELECTRON SPIN RESONANCE, FREE RADICAL CHEMISTRY. *Educ:* Wesleyan Univ, BA, 59; Univ Calif, Berkeley, PhD(chem), 66. *Prof Exp:* Chemist, Cent Res Lab, Gen Elec Co, NY, 59-61; RES CHEMIST, CENT RES DEPT, E I DU PONT DE NEMOURS & CO, INC, 66- *Concurrent Pos:* Vis scholar, Centre d'Etudes Nucléaires de Grenoble, France, 81-82; vis prof, Ecole Normale Supérieure, Paris, 87. *Mem:* Am Chem Soc. *Res:* Microwave spectroscopy; electron spin resonance spectroscopy; free radical chemistry; organometallic reaction mechanisms; homogeneous catalysis; organometallic photochemistry; autoxidation of hydrocarbons; molecular ferromagnets. *Mailing Add:* Cent Res Dept E I du Pont de Nemours & Co Inc Wilmington DE 19880-0328

KRUSKAL, JOSEPH BERNARD, b New York, NY, Jan 29, 28; m 53; c 2. MATHEMATICS. *Educ:* Univ Chicago, PhB & BS, 48, MS, 49; Princeton Univ, PhD, 54. *Prof Exp:* Instr math, Princeton Univ, 55; res instr, Univ Wis, 56-58; asst prof, Univ Mich, 58-59; MEM TECH STAFF, BELL LABS, 59- *Concurrent Pos:* Vis prof, Yale Univ, 66-67, Columbia Univ, 76, Rutgers Univ, 77; assoc ed jour, Soc Indust & Appl Math, 67-70. *Mem:* Am Math Soc; Classification Soc NAm (pres, 74-77); fel AAAS; fel Am Statist Asn; Psychometric Soc (pres, 74-75). *Res:* Statistics, psychometrics and statistical linguistics. *Mailing Add:* 42 Oakland Rd Maplewood NJ 07040

KRUSKAL, MARTIN DAVID, b New York, NY, Sept 28, 25; m 50; c 3. MATHEMATICAL PHYSICS, APPLIED MATHEMATICS. *Educ:* Univ Chicago, BS, 45; NY Univ, MS, 48, PhD(math), 52. *Prof Exp:* Asst & instr math, NY Univ, 46-51; res scientist, Plasma Physics Lab, Princeton Univ, 51-59, sr res assoc & lectr astron, 59-61, prof astrophys sci, 61-89, dir, Prog Appl Math, 68-86, prof, 79-, emer prof math, 79-89; DAVID AILBERT PROF MATH, RUTGERS UNIV, NEW BRUNSWICK, 89- *Concurrent Pos:* Consult, Los Alamos Sci Lab, 53-59, radiation lab, Univ Calif, 54-57, Oak Ridge Nat Lab, 55-58, 63-, Radio Corp Am, 60-62 & IBM Corp, 63-; assoc head theoret div, Plasma Physics Lab, Princeton Univ, 56-64; NSF sr fel, Max Planck Inst Physics & Astrophysics, Germany, 59-60; vis prof, Weizman Inst Sci, Israel, 73-74; fel, Japanese Soc Prom Sci, Nagoya Univ, 79-80; mem bd dir, Soc Indust & Appl Math, 85-; co dir, Inst Termenoild Physics, Santa Barbara, 85. *Honors & Awards:* Fulbright lectr, Grenoble, France, 59 & 78; Gibbs lectr, Am Math Soc, 79; Dannie-Helmman Prize Math Physics, 83; Potts Gold Medal, Franklin Inst, 86; Appl Math & Numerical Anal, Nat Acad Sci, 89. *Mem:* Nat Acad Sci; Am Math Soc; Math Asn Am; Am Phys Soc; Soc Indust & Appl Math; Am Acad Arts & Sci. *Res:* Plasma physics; general mathematics; asymptotic phenomena; logic; magnetohydrodynamics; controlled fusion; relativity; minimal surfaces. *Mailing Add:* Dept Math Hill Centre Rutgers Univ Busch Campus New Brunswick NJ 08903

KRUSKAL, WILLIAM HENRY, b New York, NY, Oct 10, 19; m 42; c 3. STATISTICS. *Educ:* Harvard Univ, SB, 40, MS, 41; Columbia Univ, PhD(math statist), 55. *Prof Exp:* Mathematician, US Naval Proving Ground, 41-46; vpres, Kruskal & Kruskal, Inc, 46-48; lectr math, Columbia Univ, 49-50; from instr to prof, Univ Chicago, 50-73, chmn dept, 66-73, Ernest Dewitt Burton distinguished serv prof statist & col, 73- 89, dean, Div Social Sci, 74-84, EMER ERNEST DEWITT BURTON DISTINGUISHED SERV PROF STATIST & COL, UNIV CHICAGO, 90- *Concurrent Pos:* Vis asst prof, Univ Calif, Berkeley, 55-56; ed, Annals Math Statist, Inst Math Statist, 58-61; chmn math sci panel & mem cent planning comt, Behav & Soc Sci Surv Comt, 66-70; mem adv comt, Encycl Britannica, 66-76, adv comt probs census enumeration, Nat Res Coun, 69-71 & President's Comn Fed Statist, 70-71; fel, Ctr Advan Study in Behav Sci, 70-71; NSF sr fel, 70-71; trustee, Nat Opinion Res Ctr, 70-; dir, Social Sci Res Coun, 75-78; chmn comt Nat Statist, Nat Acad Sci-Nat Res Coun, 71-78; mem adv coun, NSF, 77-; mem adv comt, Fed Statist Syst Reorgn, 78-80; John Simon Guggenheim Mem Found fel, 79-80, mem educ adv board, 80-88. *Honors & Awards:* Ronald A Fisher Mem lectr, Comt Pres Statist Socs, 78; Samuel S Wilks Mem Medal, 78. *Mem:* Fel AAAS; fel Inst Math Statist (pres, 70-71); fel Am Statist Asn (vpres, 72-74, pres, 82); Int Statist Inst; Math Asn Am; fel Am Acad Arts & Sci; hon fel Royal Statist Soc. *Res:* Theoretical statistics, especially nonparametric analysis and analysis of variance; public policy aspects of statistics. *Mailing Add:* Dept Statist Univ Chicago Chicago IL 60637

KRUTAK, JAMES JOHN, SR, b Atlanta, Ga, Apr 5, 42; m 60, 80; c 5. CHEMISTRY. *Educ:* La State Univ, BS, 64; Univ NC, PhD(org chem), 68. *Prof Exp:* Sr res chemist, 69-75, RES ASSOC, TENN EASTMAN CO, EASTMAN KODAK CO, 75- *Concurrent Pos:* Coordr, Photog Chem Res Lab, Eastman Kodak, 73-83, asst to chem res dir, 84-85; lab head, Colorants Res Lab, 86-; mem sci adv comt, Cosmetic, Toiletry & Fragrance Asn. *Mem:* Am Chem Soc; Int Soc Heterocyclic Chem; Soc Cosmetic Chemists; Am Asn Textile Chemists & Colorists; Tech Asn Pulp & Paper Indust; Soc Plastics Engrs. *Res:* Synthesis of natural products; development of novel indole synthesis; study of mechanisms of indole forming reactions using stable and radioactive isotope tracer methodology; cycloaddition reactions; chemical uses of lasers; photographic dye and developer synthesis and related technology; dye design and synthesis for coloration of plastics, fibers, inks, coatings, personal and household care products, safe coloration and UV-screen systems for cosmetics, medical products, construction products and household products; synthetic paper compositions and advanced materials of construction from recyclable polymers. *Mailing Add:* Res Labs Eastman Chem Co PO Box 1972 Kingsport TN 37662

KRUTCHEN, CHARLES M(ARION), b Gadsen, Ala, Sept 7, 34; m 59; c 4. CHEMICAL ENGINEERING. *Educ:* Vanderbilt Univ, BE, 56; Cornell Univ, PhD(chem eng), 64. *Prof Exp:* Chem engr, Res Ctr, Hercules Powder Co, 56-58, res engr, Fiber Dept, 62-64; sr res engr, Res Div, W R Grace & Co, 64-66; staff chem engr, Gen Elec Res & Develop Ctr, 66-76; eng assoc, Res & Develop Lab, Edison, NJ, 76-77, res supvr, Plastics Div, 77-90, SR RES ASSOC, MOBIL CHEM, CANANDAIGUA, NY, 90- *Mem:* Am Inst Chem Engrs; Am Chem Soc; AAAS; Soc Plastics Engrs. *Res:* Chemical process studies; polymer fiber, film and foam technology; polymer processing; morphology-physical property relationships in polymers; polymer recycling; granted 25 patents. *Mailing Add:* Plastics Div 100 North St Canandaigua NY 14424

KRUTCHKOFF, DAVID JAMES, b Eureka, Calif, June 7, 38; m 66; c 3. ORAL PATHOLOGY, DENTISTRY. *Educ:* Univ Calif, Berkeley, AB, 60; Wash Univ, DDS, 64; Univ Mich, MS, 70. *Prof Exp:* Res asst path, Wash Univ, 67-68; Nat Inst Dent Res fel & instr oral path, Univ Mich, 68-70; asst prof oral path, Univ Louisville, 70-73; PROF ORAL DIAGNOSIS-PATH, UNIV CONN, 73- *Concurrent Pos:* Vis prof, Tohoku Univ, Sendai, Japan, 87. *Mem:* Int Asn Dent Res; Am Acad Oral Path. *Res:* Dental caries; clinical studies; dental enamel; infrared internal reflection spectroscopy; oral cancer; lichenoid dysplasia; forensic consultation. *Mailing Add:* Dept Anat Path Univ Conn Health Ctr Farmington CT 06030

KRUTCHKOFF, RICHARD GERALD, b Brooklyn, NY, Dec 23, 33; m 60; c 3. APPLIED STATISTICS. *Educ:* Columbia Univ, AB, 56, MA, 58, PhD(math statist), 64. *Prof Exp:* Instr physics, Wilkes Col, 58-60; lectr, Queens Col, NY, 60-64; from asst prof to assoc prof, 64-68, PROF STATIST, VA POLYTECH INST & STATE UNIV, 68- *Concurrent Pos:* Ed, J Statist Comput & Simulation, 72- *Mem:* AAAS; fel Am Statist Asn; Inst Math Statist; Biomet Soc; Water Pollution Control Fedn. *Res:* Statistical inference; empirical Bayes decision theory; water pollution statistics. *Mailing Add:* Dept Statist Va Polytech Inst & State Univ Blacksburg VA 24061

KRUTTER, HARRY, b Boston, Mass, Mar 17, 11; m 35; c 1. PHYSICS. *Educ:* Mass Inst Technol, 32, SM, 33, PhD(physics), 35. *Prof Exp:* Teaching fel physics, Mass Inst Technol, 33-35; instr, Purdue Univ, 35-36; asst prof petrol eng, Pa State Col, 36-42; res assoc & staff mem, Mass Inst Technol, 42-45; chief scientist, Field Sta, Naval Res Lab, 46-49, tech dir, Aero Electronics & Elec Lab, 49-56, chief scientist, 56-67, tech dir, 67-73, consult, Naval Air Develop Ctr, 73-85; CONSULT, DIGITAL SYST GROUP, 86- *Concurrent Pos:* Lectr, Wharton Sch, Univ Pa, 73-79. *Honors & Awards:* Distinguished Civilian Serv Award, USN, 56; Distinguished Civilian Serv Award, US Dept Defense, 57. *Mem:* Am Phys Soc; fel Inst Elec & Electronics Engrs; assoc fel Am Inst Aeronaut & Astronaut. *Res:* X-ray crystal structure; theory of metals; flow of gases through porous media; antennas; radar; electronics. *Mailing Add:* 310 S Easton Rd B 408 Glenside PA 19038

KRUTZSCH, HENRY C, b Fairbanks, Alaska, Feb 7, 42; m 74. PROTEIN BIOCHEMISTRY. *Educ:* Univ Calif, Riverside, BA, 64; Univ Iowa, PhD(org chem), 68. *Prof Exp:* Res chemist, Pioneering Res Div, Exp Sta, E I du Pont de Nemours & Co, 68-73; sr staff fel, Lab Chem, Nat Heart, Lung & Blood Inst, NIH, 73-77, Lab Immunogenetics, Nat Inst Allergy & Infectious Dis, 77-79, expert, 79-82, Lab Exp Carcinogenesis, Nat Cancer Inst, 82-86, spec expert, Diabetes Br, Nat Inst Diabetes & Digestive & Kidney Dis, 86-87, CANCER EXPERT, LAB PATH, NAT CANCER INST, NIH, 87- *Mem:* AAAS; Am Soc Biochem & Molecular Biol; Protein Soc; Am Chem Soc; Sigma Xi. *Res:* Protein biochemistry. *Mailing Add:* Path Lab NIH Bldg 10 Rm 2A23 Bethesda MD 20892

KRUTZSCH, PHILIP HENRY, b St Louis, Mo, July 12, 19; m 40; c 2. ANATOMY, ZOOLOGY. *Educ:* San Diego State Col, BA, 43; Univ Calif, MA, 48; Univ Kans, PhD(zool), 53. *Prof Exp:* Asst, Univ Calif, 47-48; asst instr, Univ Kans, 48-52, asst, 49; instr anat, Sch Med, Univ Pittsburgh, 53-54; asst prof, Univ Tex Health Sci Ctr, Dallas, 55-56; assoc prof, Sch Med, Univ Pittsburgh, 57-64; head dept, 64-73, PROF ANAT, COL MED, UNIV ARIZ, 64- *Mem:* Am Soc Mammal; Am Asn Anatomists; Am Soc Zoologists; Soc Study Reproduction. *Res:* Physiology of reproduction; studies of brown adipose tissue and prolonged sperm longevity. *Mailing Add:* Dept Anat Univ Ariz Col Med 1501 N Campbell Tucson AZ 85724

KRUUS, JAAN, b Kuimetsa, Estonia, July 23, 36; Can citizen; m 62; c 4. ELECTRICAL ENGINEERING, INSTRUMENTATION. *Educ:* Univ Toronto, BASc, 59; Univ Ill, MS, 61, PhD(elec eng), 63. *Prof Exp:* Engr, Spruce Falls Power & Paper Co, Ont, 59-60; asst prof elec eng, Queen's Univ, Ont, 63-65; assoc prof, Univ Ottawa, 65-69; head, Instrumentation Sect, Hydrol Sci Div, Inland Waters Br, 69-74, coordr remote sensing, Off Sci Adv, 74-78, planning analyst, Atmospheric Environ Serv, 78-80, DIR, DATA ACQUISITION SERV BR, ENVIRON CAN, 80- *Concurrent Pos:* Vpres, Comm Instruments & Methods Observation, World Meteorol Orgn, 81- *Mem:* Inst Elec & Electronics Engrs; Sigma Xi. *Mailing Add:* 15 Michigan Dr Willowdale ON M2M 3H9 Can

KRUUS, PEETER, b Tallinn, Estonia, July 8, 39; Can citizen; m 63; c 4. PHYSICAL CHEMISTRY. *Educ:* Univ Toronto, BSc, 61, PhD(phys chem), 65; Tech Univ Denmark, Lic Techn, 63. *Prof Exp:* From asst prof to assoc prof, 65-77, PROF CHEM, CARLETON UNIV, 77- *Concurrent Pos:* Sci adv, Sci Coun Can, 69-70; res scientist, Cominco Ltd, 79-80; adv, Int Orgn Consumers Unions, 87-88. *Mem:* Chem Inst Can. *Res:* Structure and dynamics in liquids and solutions; cavitation-induced chemical reactions; supercritical extraction. *Mailing Add:* Dept Chem Carleton Univ Colonial By Dr Ottawa ON K1S 5B6 Can

KRUUV, JACK, b Tartu, Estonia, June 1, 38; Can citizen; m 60; c 3. CRYOBIOLOGY, RADIOBIOLOGY. *Educ:* Univ Waterloo, BASc, 62, MSc, 63; Univ Western Ont, PhD(biophys), 66. *Prof Exp:* Ont Cancer Found fel & lectr, Victoria Hosp, 66; fel, Argonne Nat Lab, 66-67; from asst prof to assoc prof physics, 67-77, PROF PHYSICS & BIOL, UNIV WATERLOO, 77- *Concurrent Pos:* Vis prof biophysics, Pa State Univ, 74-75. *Mem:* Radiation Res Soc; Cryobiol Soc. *Res:* Cancer research; radiation biophysics; research with synchronized tissue culture cells; effects of low oxygen atmospheres on cells; radiotherapy of cancer cells; repair of radiation and freeze-thaw damage; cryobiology; multi-cellular tissue culture systems; hypothermia; hyperthermia; mechanisms of freeze-thaw damage. *Mailing Add:* Dept Physics Univ Waterloo Waterloo ON N2L 3G1 Can

KRYDER, MARK HOWARD, b Portland, Ore, Oct 7, 43; m 65; c 2. MAGNETIC RECORDING, MAGNETO-OPTIC RECORDING. *Educ:* Stanford Univ, BS, 65; Calif Inst Technol, MS, 66, PhD(elec eng, physics), 70. *Prof Exp:* NSF res fel, Calif Inst Technol, 69-71; scientist solid state physics, Univ Regensburg, 71-73; res staff mem, IBM Res Ctr, 73-75, mgr explor bubble devices, 75-78; assoc prof, 78-80, dir Magnetics Technol Ctr, 82-90, PROF ELEC & COMP ENG, CARNEGIE-MELLON UNIV, 80-, DIR DATA STORAGE SYSTS CTR, 90- *Concurrent Pos:* Consult, IBM, Gen Elec, Nat Semiconductor Corp, Motorola, & Alcoa; distinguished lectr, Inst Elec & Electronics Engrs, 85. *Mem:* Am Phys Soc; Inst Elec & Electronics Engrs. *Res:* Applied magnetics; magneto-optical, and magnetic recording devices including materials, fabrication, device design and use in systems. *Mailing Add:* Dept Elec & Comput Eng Carnegie-Mellon Univ Pittsburgh PA 15213

KRYGER, ROY GEORGE, b Brooklyn, NJ, June 7, 36; m 58; c 2. PHYSICAL ORGANIC CHEMISTRY. *Educ:* Atlantic Union Col, AB, 57; Stevens Inst Technol, MS, 66; Boston Univ, PhD(org chem), 73. *Prof Exp:* Chemist qual control, Lederle Lab Div, Am Cyanamid Co, 57-58; chemist food analyst, US Army Med Res & Nutrit Lab, 59-60; anal methods develop & antibiotic res & develop chemist, Lederle Lab Div, Am Cyanamid Co, 60-66; prof chem, Atlantic Union Col, 81-84, chmn dept, 66-84; PROF CHEM, LOMA LINDA UNIV, RIVERSIDE, CALIF, 84- *Mem:* AAAS; Nat Sci Teachers Asn. *Res:* Relative and absolute reactivities of free radicals and relation of structure of free radicals to their reactivity. *Mailing Add:* Dept Chem Loma Linda Univ Riverside CA 92515

KRYNICKI, VICTOR EDWARD, neurophysiology, for more information see previous edition

KRYNITSKY, JOHN ALEXANDER, b Far Rockaway Beach, NY, June 15, 18; m 49; c 1. PETROLEUM FUELS, CHEMISTRY. *Educ:* Univ Md, BS, 39; Univ NC, PhD(org chem), 43. *Prof Exp:* Asst chem, Univ NC, 39-43; chemist, Naval Res Lab, DC, 43-44 & 45-64; staff asst, Off Dir Defense Res & Eng, 64-67; dir tech opers, Defense Fuel Supply Ctr, 67-81; CONSULT, FUELS & PETROL PROD, 81- *Mem:* Am Chem Soc; Am Soc Testing & Mat. *Res:* Fuels; lubricants; petroleum products; organic materials; synthetic organic chemistry; detection and identification of organic substances; preparations and properties of some highly chlorinated hydrocarbons; hydrazine; aircraft and rocket fuels. *Mailing Add:* 4904 Cumberland Ave Chevy Chase MD 20815

KRYSAN, JAMES LOUIS, b Calmar, Iowa, Mar 12, 34; m 60; c 3. INSECT PHYSIOLOGY. *Educ:* Iowa State Teachers Col, BA, 61; Univ Ill, MS, 64, PhD(entom), 65. *Prof Exp:* Asst prof biol, St Mary's Col, Minn, 65-68; entomologist, Northern Grain Insects Res Lab, Sci & Educ Admin-Agr Res, USDA, 68-84; RES LEADER, RES ENTOMOLOGIST, YAKIMA RES LAB, AGR RES SERV, USDA, 84- *Mem:* Entom Soc Am; AAAS; Am Soc Nat; Coleopterists Soc. *Res:* Management of insect pests of pear; biosystematics of Diabrotica; seasonality of insects. *Mailing Add:* Yakima Res Lab 3706 W Nob Hill Blvd Yakima WA 98902

KRYSIAK, JOSEPH EDWARD, b Cleveland, Ohio, Jan 13, 37. TRIBOLOGY, AIRCRAFT GROUND EFFECTS SYSTEMS. *Educ:* Univ Dayton, BME, 64, MME, 73. *Prof Exp:* Mech engr, Wright-Patterson AFB, 65-73, Reliance Elec Co, 73-75 & Indust Applications Int, 76-79; MECH ENGR, LEWIS RES CTR, NASA, 79- *Mem:* NY Acad Sci; Am Soc Metals Int; Math Asn Am. *Res:* High speed, high temperature bearings; braking systems; gearing systems; aircraft ground effects systems; high speed rotational systems; tribology of friction couples; ceramic components; wind trubine generating systems; six US patents. *Mailing Add:* PO Box 676 Willoughby OH 44094

KRYTER, KARL DAVID, b Indianapolis, Ind, Oct 13, 14; m 46; c 3. PSYCHOPHYSIOLOGY, PSYCHOACOUSTICS. *Educ:* Butler Univ, AB, 39; Univ Rochester, PhD(psychol), 43. *Prof Exp:* Fel psycho-acoust, Harvard Univ, 42-46; asst prof psychol, Univ Wash, St Louis, 46-48; dir human factors oper res lab, US Air Force, 48-52 & oper appln lab, Cambridge Res Ctr, 52-57; head psychoacoust dept, Bolt Beranek & Newman, Inc, 57-65; STAFF SCIENTIST, SRI INT, 65- *Concurrent Pos:* Pres, Acousis Co; mem comt hearing, bioacoust & biomech, Nat Acad Sci-Nat Res Coun, 57-, chmn exec coun, 61-64; mem adv panel psychol & soc sci, Off Asst Secy Defense, 58-63; Int Orgn Standardization & Int Electrotechnol Comn, 62-; adv, President's Comt Sci & Technol. *Honors & Awards:* Distinguished Award Sci, Am Speech & Hearing Asn; Franklin V Taylor Award in Eng Psychol. *Mem:* Fel AAAS; Soc Eng Psychol (pres, 66); Am Psychol Asn; Acoust Soc Am (pres, 71); Human Factors Soc. *Res:* Audition; psychoacoustics; speech communication; electrophysiology. *Mailing Add:* PO Box 1017 Borrego Springs CA 92004

KRYWOLAP, GEORGE NICHOLAS, b Ukraine, Jan 4, 36; US citizen; m 57; c 1. MICROBIAL ECOLOGY. *Educ:* Drexel Univ, BS, 60; Pa State Univ, MS, 62, PhD(microbiol), 64. *Prof Exp:* From asst prof microbiol to assoc prof, Schs Pharm & Dent, 64-77, PROF MICROBIOL, UNIV MD, 77- *Concurrent Pos:* Vis prof, Col Dent, Univ Fla, 88-89. *Mem:* AAAS; Am Soc Microbiol; Int Asn Dent Res; Brit Soc Gen Microbiol. *Res:* Antibiotics; production of antibiotics by mycorrhizal fungi; microbial ecology of the oral cavity; microbiology of the periodontum. *Mailing Add:* Dept Microbiol Sch Dent Univ Md 666 W Baltimore St Baltimore MD 21201

KRZANOWSKI, JOSEPH JOHN, JR, b Hartford, Conn, Feb 4, 40; m 63; c 2. PHARMACOLOGY, PHYSIOLOGY. *Educ:* Univ Conn, BS, 62; Univ Tenn, MS, 65, PhD(pharmacol, physiol), 68; Barry Col, MA, 87. *Prof Exp:* Fel pharmacol, Med Sch, Wash Univ, 68-71; from asst prof to assoc prof, 71-83, PROF PHARMACOL, COL MED, UNIV SFLA, 83-, VCHMN DEPT, 81- *Concurrent Pos:* NIH pulmonary young investr award, 74-76; co-chmn, Am Acad Allergy & Immunol Role & Care Animals Res, 88- *Mem:* Am Soc Pharmacol & Exp Therapeut; NY Acad Sci; AAAS; Am Acad Allergy & Immunol; Am Col Clin Pharmacol. *Res:* Autonomic pharmacology; smooth muscle pharmacology; pulmonary and cardiac cyclic nucleotides; prostaglandins; pulmonary and immunopharmacol; Red Tide Toxin. *Mailing Add:* Dept Pharmacol & Therapeut Col Med Univ SFla Tampa FL 33612

KRZEMINSKI, LEO FRANCIS, metabolism, analytical biochemistry, for more information see previous edition

KRZEMINSKI, STEPHEN F, b Philadelphia, Pa, Dec 26, 43; m 66; c 3. PHYSICAL CHEMISTRY, ANALYTICAL CHEMISTRY. *Educ:* La Salle Col, BA, 65; Univ Pittsburgh, PhD(phys chem), 69. *Prof Exp:* Sr res chemist, Bristol Res Labs, 69-75, proj leader, Res Div, 75-77, mgr, Govt Regulatory Rels, 77-80, CORP MGR, GOVT RELS & PROD STANDARDS, AGR CHEM-NORTH AM, ROHM AND HAAS CO, 80- *Mem:* Am Chem Soc. *Res:* Transition metal chemistry; coordination compounds; Mossbauer spectroscopy; pesticide residues; metabolism and environmental fate of pesticides; drug delivery systems; pharmacokinetics. *Mailing Add:* Rohm and Haas Co Independence Mall West Philadelphia PA 19105

KRZYZANOWSKI, PAUL, b New York, NY, Feb 7, 64. COMPILER DESIGN. *Educ:* NY Univ, BS, 85; Cooper Union, BE, 85; Columbia Univ, MS, 87. *Prof Exp:* Mem tech staff, UNIX Systs Lab, 85-89, MEM TECH STAFF, COMPUT SYSTS RES, AT&T BELL LABS, 89- *Mem:* Inst Elec & Electronics Engrs. *Mailing Add:* AT&T Bell Labs 600 Mountain Ave Murray Hill NJ 07974-2070

KSHIRSAGAR, ANANT MADHAV, b Satara, India, Aug 16, 31; m 57; c 2. STATISTICS. *Educ:* Univ Bombay, MSc, 51; Univ Manchester, PhD(statist), 61; Univ Manchester, DSc, 76. *Hon Degrees:* DSC, Univ Manchester, 76. *Prof Exp:* Lectr statist, Univ Bombay, 51-63; sr sci officer, Defence Sci Lab, India, 63-68; asso prof statist, Southern Methodist Univ, 68-71; prof statist, Tex A&M Univ, 71-77; PROF BIOSTATIST, UNIV MICH, 77- *Mem:* Fel Am Statist Asn; fel Inst Math Statist; Int Inst Statist. *Res:* Design of experiments; multivariant and discriminant analysis; renewal theory, especially Markovian renewal theory. *Mailing Add:* Dept Biostatist Univ Mich Ann Arbor MI 48109-2029

KSIENSKI, A(HARON), b Warsaw, Poland, June 23, 24; nat US; m 54; c 2. ELECTRICAL ENGINEERING. *Educ:* Univ Southern Calif, MS, 52, PhD(elec eng), 58. *Prof Exp:* Consult engr, W L Schott Co, Calif, 52-53; staff engr, Wiancko Eng Co, 53-57; staff engr, Antenna Res Dept, Hughes Aircraft Co, 58-60, sr staff engr, 60, head res staff, 60-67; prof elec eng & tech dir commun systs, Electrosci Lab, 67-87, EMER PROF ELEC ENG, OHIO STATE UNIV, 87- *Concurrent Pos:* Lectr, Univ Southern Calif, 54-57; assoc ed, Antennas, Inst Elec & Electronics Engrs, 70-72. *Honors & Awards:* Lord Brabazon Award, Inst Radio & Electronic Engrs, London, 67 & 76. *Mem:* Fel Inst Elec & Electronics Engrs; Int Union Radio Sci. *Res:* Antennas and antenna systems; information theory; data processing; radar detection and identification; target identification; signal processing arrays communication. *Mailing Add:* Dept Elec Eng Ohio State Univ 2015 Neil Ave Columbus OH 43210

KSIR, CHARLES JOSEPH, b Albuquerque, NMex, May 19, 45; m 67; c 1. PSYCHOPHARMACOLOGY. *Educ:* Univ Tex, Austin, BA, 67; Ind Univ, Bloomington, PhD(psychol), 71. *Prof Exp:* Fel Neurobiol, Worcester Found Exp Biol, 71-72; asst prof, 72-76, assoc prof, 76-80, PROF PSYCHOL, UNIV WYO, 80- *Concurrent Pos:* Vis scientist, Salk Inst Biol Study, 81, Univ Cambridge, Eng, 83. *Mem:* Behav Pharmacol Soc; Psychonomic Soc; AAAS; Sigma Xi; Soc Neurosci; Am Psychol Asn. *Res:* Behavioral pharmacology; neuropharmacology. *Mailing Add:* Dept Psychol Box 3415 Univ Wyo Laramie WY 82071

KSYCKI, MARY JOECILE, b Du Bois, Ill, May 26, 13. RADIATION CHEMISTRY. *Educ:* St Louis Univ, BSc, 36, MS, 40, PhD(chem), 42. *Hon Degrees:* MI, St Mary's Col, 85. *Prof Exp:* Instr chem, Le Clerc Col, 42-43; prof & head dept, 43-49; prof, Webster Col, 49-50; head dept sci pvt sch, Ill, 50-54; prof & head dept, Notre Dame Col, Mo, 54-77; prof chem & chmn sci, St Mary's Col, Orchard Lake, Mich, 77-91; RETIRED. *Concurrent Pos:* Assoc radiation proj, Univ Notre Dame, 58; fel, Univ Okla, 61 & 63, Univ Ill, 62 & Kans State Univ, 64; NIH res award, 61-64; NSF res grant, 63-65; Japan Soc Promotion Sci res grant, Kyoto Univ, 71-72. *Honors & Awards:* Sci Prof of the Year Award, Nat Sci Teachers Asn, 60. *Mem:* Am Chem Soc; Nat Sci Teachers Asn. *Res:* Electrodeposition of molybdenum; alloxan diabetes; electrodeposition potentials of copper, nickel and cobalt; radiation chemistry; effects of drugs on iodine metabolism; protective power of nucleotides against gamma ray inactivation of ribonuclease. *Mailing Add:* Dept Chem St Mary's Col Orchard Lake MI 48324

KU, ALBERT B, b Changsha, China, June 15, 33; m 65; c 2. ENGINEERING MECHANICS, APPLIED MATHEMATICS. *Educ:* Univ Taiwan, BSCE, 56; Va Polytech Inst, MSCE, 61; Ohio State Univ, PhD(soil mech), 65. *Prof Exp:* Jr engr, Mil Construct Comt, 58-59; engr, Kai-Nan Eng Corp, 59-60; from asst prof to assoc prof, 64-76, chmn dept civil eng, 82-85, PROF ENG MECH, UNIV DETROIT, MERCY, 76- *Concurrent Pos:* Consult, Chrysler Corp, 66-, Burrough's Corp, 70- & Gen Motors, 72- *Mem:* Am Acad Mech. *Res:* Rheological properties of bituminous concrete; linear viscoelasticity; nonlinear mechanics; static and dynamic stability; finite element analysis; artificial intelligence; computer software. *Mailing Add:* Col Eng & Sci Univ Detroit Mercy Detroit MI 48221

KU, ANTHONY CHIA-HUNG, heat transfer, numerical analysis, for more information see previous edition

KU, AUDREY YEH, b Taiwan. ORGANIC CHEMISTRY, POLYMER CHEMISTRY. *Educ:* Providence Col, Taiwan, BS, 71; Eastern Ill Univ, MS, 73; State Univ NY, Buffalo, PhD(chem), 77. *Prof Exp:* Teaching & res asst chem, Eastern Ill Univ, 72-73 & State Univ NY, Buffalo, 73-77; fel chem, Ohio State Univ, 77-78; res chemist, Union Carbide Corp, 78-80; SR RES CHEMIST, MONSANTO CO, 80- *Mem:* Am Chem Soc; Sigma Xi. *Mailing Add:* 800 N Linderbergh Blvd U 2 I St Louis MO 63166

KU, BERNARD SIU-MAN, b Hong Kong. NEXT GENERATION SWITCHING TECHNOLOGIES, ADVANCED SOFTWARE ENGINEERING RESEARCH & DEVELOPMENT. *Educ:* Univ Hong Kong, BS, 81; Univ Tex, MBA, 83; Univ NTex, MS, 85; Southern Methodist Univ, PhD(telecommun), 91. *Prof Exp:* Chairperson, Computer Sci Dept, Univ Mary Hardin-Baylor, Belton, Tex, 85-88; computer mgr, Southern Methodist Univ, Dallas, Tex, 88-89; MEM TECH STAFF TELECOMMUN RES & DEVELOP ENG, ADV SWITCHING LAB, NEC AM, INC, IRVING, TEX, 89- *Concurrent Pos:* Teaching fel & res assistantship, Southern Methodist Univ, Dallas, Tex, 88; vis prof computer sci, Richland Col, 89. *Mem:* Int Commun Asn; Inst Elec & Electronics Engrs; Asn Comput Mach. *Res:* Advanced software technology for telecommunication switching software development, e.g. software reuse, case tools and integrated software engineering environment; telecommunication industry trends and technology assessment. *Mailing Add:* 3608 Legendary Lane Plano TX 75023

KU, CHIA-SOON, b Nanking, China, Apr 15, 46; m 84; c 1. TRIBOLOGY, APPLIED CHEMISTRY. *Educ:* Nat Taiwan Univ, BS, 68; Worcester Polytech Inst, MS, 71; Pa State Univ, PhD(chem eng), 77. *Prof Exp:* Asst prof chem eng, Lafayette Col, 77-78; sr chem engr, Appl Sci Div, Versar Inc, 78-79; SR RES ENGR, NAT BUR STANDARDS, 79- *Honors & Awards:* Bronze Medal Award, Dept Com, 87. *Mem:* Am Inst Chem Engrs; Am Chem Soc; Am Soc Automotive Engrs; Am Soc Tribologist & Lubricant Engrs. *Res:* Tribiology; oxidation bench test development for lubricating oils; oil degradation mechanisms; additive response studies for lubricants; friction and wear mechanism and bench test developments. *Mailing Add:* 14109 Saddle River Dr Gaithersburg MD 20878

KU, DAVID NELSON, b St Louis, Mo, Mar 15, 56; m 80; c 3. BIOFLUID DYNAMICS, VASCULAR SURGERY. *Educ:* Harvard Univ, BA, 78; Ga Inst Technol, MS, 82, PhD(aerospace eng & biofluid dynamics), 83; Emory Univ, MD, 84. *Prof Exp:* Surg resident, Univ Chicago, 84-85; dir, Vascular Lab, Hyde Park Community Hosp, 85-86; asst prof, 86-90, ASSOC PROF BIOENG, GA INST TECHNOL, 90-; ASST PROF & DIR VASCULAR LAB, EMORY UNIV, 86- *Concurrent Pos:* Fulbright Gastprofessor, Munich, Ger, 85; fel cardiovascular path, Univ Chicago, 85-86; NSF presidential young investrs award, 87; mem, Coun Arteriosclerosis, Am Heart Asn; Y C Fung young investr award, Am Soc Mech Engrs, 89. *Mem:* Am Soc Mech Engrs; Am Heart Asn; AMA; Am Col Angiol. *Res:* Biofluid dynamics research on the development, diagnosis, and treatment of arterial disease; mechanisms of atherogenesis and the thrombosis of athersclerotic arteries. *Mailing Add:* Sch Mech Eng Ga Inst Technol Atlanta GA 30332-0405

KU, EDMOND CHIU-CHOON, b Canton, China, Aug 11, 32; US citizen; m 59; c 2. BIOCHEMISTRY. *Educ:* Taiwan Prov Col, BS, 56; Va Polytech Inst, PhD(biochem), 62. *Prof Exp:* Res scientist biochem, Parke, Davis & Co, 63-67; NIH spec res fel, Cornell Univ, 67-69; sr staff scientist, 70-79, mgr, 80-85, SR RES FEL, CIBA-GEIGY CORP, 85- *Mem:* Am Chem Soc; Sigma Xi; NY Acad Sci; Inflammable Res Asn. *Res:* Enzyme kinetics and its application to the study of drug action at molecular level; regulatory mechanism involved in the biosynthesis and degradation of prostaglandins at subcellular level; cholesterol and triglyceride biosynthesis regulation. *Mailing Add:* Biochem Subdiv Ciba-Geigy Corp Summit NJ 07901

KU, HAN SAN, b Hsin-Chu, Taiwan, Nov 20, 35; m 65; c 2. PLANT PHYSIOLOGY, BIOCHEMISTRY. *Educ:* Nat Taiwan Univ, BS, 58; Osaka Univ, MS, 63; Univ Calif, Davis, PhD(plant physiol), 68. *Prof Exp:* Lectr agr, Taipei Agr Prof Sch, 57-58; asst plant physiol, Nat Taiwan Univ, 60-61; res asst, Univ Calif, Davis, 64-68, NIH fel ethylene biosynthesis, 68; NSF fel ethylene physiol, Purdue Univ, 68-70; res biologist, Allied Chem Res Lab, 70-71, biologist & res leader plant sci, 72-73; plant biochemist, Mich State Univ, 73-74; SR RES ASSOC BIOCHEM, T R EVANS RES CTR, 74- *Concurrent Pos:* Mgr, Biol Eval SDS Biotech; mgr, Agr Chem Biol Evaluation; dir & sr scientist, Environ Sci Ricenca, Inc; vpres, ISK Interprise. *Honors & Awards:* Japanese Food Sci Soc Award, 66; Campbell Award, Am Inst Biol Sci, 69. *Mem:* AAAS; Am Soc Plant Physiologists; Am Soc Agron; Am Soc Hort; Soil Sci Soc Am; Am Chem Soc; Japanese Silver Chem Soc. *Res:* Ethylene biogenesis and action in plant tissue; plant growth regulator; photorespiration; nitrogen fixation; crop production pesticide; pesticide metabolism; animal, plant metabotsim; Environ Fate EPA; allochemical mode of pesticide. *Mailing Add:* 6338 Coleridge Painesville OH 44077

KU, HARRY HSIEN HSIANG, b Peking, China, Mar 3, 18; US citizen; m 42, 81; c 1. MATHEMATICAL STATISTICS, ENGINEERING. *Educ:* Purdue Univ, BS, 40, MSE, 41; George Washington Univ, MS, 60, PhD, 68. *Prof Exp:* Civil engr, M W Kellogg Co, 42-43; math statistician, Ctr Appl Math, Nat Eng Lab, Nat Bur Standards, 59-79, chief statist eng div, 78-85; RETIRED. *Concurrent Pos:* Consult, Ctr Measurement Standards, Hsinchu, Taiwan; vis scientist, Cent Bur Nuclear Measurements, Geel, Belgium. *Mem:* AAAS; fel Am Statist Asn; Int Statist Inst. *Res:* Statistical analysis of measurement data in engineering and physical sciences; propagation of error, precision and accuracy; application of information theory in analysis of multi-dimensional contingency tables and Markov chains. *Mailing Add:* 9608 Glencrest Lane Kensington MD 20895

KU, HSU-TUNG, b Formosa, Oct 24, 33; m 64. TOPOLOGY. *Educ:* Taiwan Prov Norm Univ, BSc, 61; Tulane Univ, MSc, 64, PhD(math), 67. *Prof Exp:* Mem math, Inst Adv Study, 67-68; asst prof, 68-73, assoc prof, 73-79, PROF MATH, UNIV MASS, AMHERST, 79- *Concurrent Pos:* Vis mem math, Inst Adv Study, 77. *Mem:* Am Math Soc. *Res:* Transformation groups; algebraic topology. *Mailing Add:* Dept Math Univ Mass Amherst MA 01003

KU, MEI-CHIN HSIAO, b Formosa, Nov 1, 37; m 64. TOPOLOGY, GEOMETRY. *Educ:* Taiwan Norm Univ, BSc, 61; Syracuse Univ, MS, 64; Tulane Univ, La, PhD(math), 67. *Prof Exp:* Mathematician, Inst Advan Study, 67-68; from asst prof to assoc prof, 70-82, PROF MATH, UNIV MASS, AMHERST, 82- *Concurrent Pos:* Vis mem, Inst Advan Study, 77. *Mem:* Math Asn Am. *Res:* Transformation groups; geometry; PDE. *Mailing Add:* Dept Math & Statist Univ Mass Amherst MA 01003

KU, PEH SUN, b Shangtung, China, Aug 23, 22; US citizen; m 57; c 2. ENVIRONMENTAL PHYSICS. *Educ:* Nat Cent Univ, BS, 47; Univ Rochester, MS, 55; Yale Univ, DEng(chem eng), 60. *Prof Exp:* Chem engr, Chinese Petrol Corp, Taiwan, 47-54; res asst, Carnegie Inst Technol, 54-55; res engr, Boeing Co, 60-65; theoret physicist, Reentry Systs Div, Gen Elec Co, 65-69, staff scientist, 69-70; chem engr air pollution control, 70-73, sr engr, 73-78, SR ENVIRON ENGR, ENVIRON AFFAIRS, CONSOL EDISON CO, NY, 78- *Mem:* AAAS; Am Chem Soc; Am Phys Soc; Am Inst Aeronaut & Astronaut; Am Geophys Union. *Res:* Acoustics; high temperature thermodynamics and transport properties of matter; physical properties of matter under high pressures; chemical kinetics; control and dispersion of air pollutants in the atmosphere and their removal from industrial processes; water pollution and hazerdous wastes; novel methods of energy conversion. *Mailing Add:* 244 Old State Rd Berwyn PA 19312

KU, ROBERT TIEN-HUNG, b Shanghai, China, Jan 19, 47; US citizen; m 71; c 2. FIBER OPTICS, LASER SPECTROSCOPY. *Educ:* Univ Ill, Urbana, BS, 67, MS, 68, PhD(elec eng), 73. *Prof Exp:* Res asst, Gaseous Electronics Lab, Univ Ill, 67-73; mem tech staff, Optics Div, Lincoln Lab, Mass Inst Technol, 73-81; mem tech staff, Laser Develop Dept, 81-88, SUPVR, LIGHTWAVE DEVICE PACKAGING DEPT, AT&T BELL LABS, 88- *Concurrent Pos:* Fel, NSF. *Mem:* Am Phys Soc; Inst Elec & Electronics Engrs; Optical Soc Am; Sigma Xi. *Res:* Laser and optical components for fiber communication systems; laser spectroscopy studies of pollutant gases; laser diagnostics of plasma and chemically excited media; laser radars; high power lasers. *Mailing Add:* PO Box 13396 Reading PA 19612

KU, TEH-LUNG, b Shanghai, China, Aug 30, 37; m 70; c 3. GEOCHEMISTRY, GEOCHRONOLOGY. *Educ:* Nat Taiwan Univ, BS, 59; Columbia Univ, PhD(geochem), 66. *Prof Exp:* Post-doctoral, Lamont Geol Observ, Columbia Univ, 66-67; asst scientist, Woods Hole Oceanog Inst, 67-69; assoc prof, 69-75, PROF GEOL SCI, UNIV SOUTHERN CALIF, 75- *Concurrent Pos:* Guggenheim fel, 83-; Fulbright Sr Scholar, 83- *Mem:* AAAS; Am Geophys Union; Geol Soc Am; Geochem Soc. *Res:* Isotope geochemistry; chemical oceanography; geochronology; Pleistocene geology; climatology; hydrogeochemistry. *Mailing Add:* Dept Geol Sci Univ Southern Calif Los Angeles CA 90089-0740

KU, TIMOTHY TAO, b Chaochow, China, Mar 26, 26; m 50; c 2. FORESTRY. *Educ:* Nanking Univ, BS, 48; Mich State Univ, MF, 50, PhD(forest ecol, silvicult), 54. *Prof Exp:* Forester, T S Coile, Inc, Forest Land Consults, 56-58; from asst prof to assoc prof, 59-63, PROF FORESTRY, UNIV ARK, MONTICELLO, 63- *Concurrent Pos:* Vis prof forestry, Nat Taiwan Univ, Taipei, 83; fel, Soc Am Foresters, 85. *Mem:* Soc Am Foresters; Soil Sci Soc Am. *Res:* Silviculture; forest soils and ecology; biomass production and nutrient cycling in forest stands, site evaluation and classification, and applied silviculture. *Mailing Add:* Dept Forest Resources Univ Ark Monticello AR 71655

KU, VICTOR CHIA-TAI, b Nanking, China, July 15, 42; US citizen; m 72; c 2. ELECTRICAL ENGINEERING, COMPUTER SCIENCE. *Educ:* Cheng Kung Univ, Taiwan, BS, 66; Rutgers Univ, MS, 69; Univ Pittsburgh, PhD(elec eng), 73. *Prof Exp:* Sr engr, Stromberg Carlson Corp, Gen Dynamic Corp, 73-74; mem tech staff, RCA Labs, 75-76; mem tech staff elec eng, Bell Labs, 76-79; eng mgr, Wang Labs, Inc, 79, GEN MGR, WANG COMPUT LTD, TAIWAN, 80- *Mem:* Inst Elec & Electronics Engrs; Sigma Xi; Chinese Inst Engrs. *Res:* Design automation especially logic and fault simulation, testing, placement and routing; interactive graphic computer systems; computer manufacturing, research and development; ideographic computer design. *Mailing Add:* Taicom Systs Ltd 296 Jen AI Rd Sect 4 Taipei Taiwan

KU, VICTORIA FENG, b Peking, China, Mar 14, 30; US citizen; m 50; c 2. ORGANIC CHEMISTRY. *Educ:* Barat Col, BS, 50; Univ Ark, MS, 64, PhD(chem), 76. *Prof Exp:* Chemist, Mich State Health Dept Lab, 52-56 & Hercules Powder Co, 57-59; asst prof, 69-75, ASSOC PROF CHEM, UNIV ARK, MONTICELLO, 76- *Mem:* Am Chem Soc. *Mailing Add:* Univ Ark Dept Natural Sci Monticello AR 71655

KU, Y(U) H(SIU), b Wusih, China, Dec 24, 02; m 29; c 7. ELECTRICAL ENGINEERING, SYSTEMS ENGINEERING. *Educ:* Mass Inst Technol, SB, 25, SM, 26, ScD, 28. *Hon Degrees:* MA & LLD, Univ Pa, 72. *Prof Exp:* Prof elec eng & head dept, Chekiang Univ, 29-30; dean eng, Cent Univ, China, 31-32; pres, 44-45, prof elec eng, 47-49; dean eng, Tsing Hua Univ, 32-37; vminister, Ministry Ed, 38-44; educ comnr, Shanghai Munic Govt, 45-47; pres, Chengchi Univ, 47-49; vis prof, 52-54, prof, 54-71, EMER PROF ELEC & SYSTS ENG, UNIV PA, 72- *Concurrent Pos:* Dir, Aeronaut Res Inst, China, 34-37 & Electronics Res Inst, 35-37; head, Chinese Educ Mission to India, 43; chief Chinese deleg, Int Tech Cong, Paris, 46; mem gen assembly, Int Union Theoret & Appl Mech, 46-92; vis prof elec eng, Mass inst Technol, 50-52; consult, Gen Elec Co, NY, 51-55, Univac, 52-53, Radio Corp Am, 59-60; hon prof, Shanghai Jiao-Tong Univ, 79-, Xian, Southwestern, Northern Jiao-Tong Univ, 85-, Northeast Univ of Sci & Eng & Northwest Telecommunication Univ, 86-, Southeast Univ, Nanjing, 88-; hon adv, Huazhong Univ of Sci & Eng, 87- *Honors & Awards:* Gold Medal, Ministry Educ, 60; Lamme Medal, Inst Elec & Electronic Engrs, 72; Gold Medal, Chinese Inst Elec Engrs, 72; Pro Mundi Beneficio Medal, Brazilian Acad Humanities, 75. *Mem:* Am Soc Eng Educ; fel Inst Elec & Electronics Engrs; Brit Inst Elec Engrs; Chinese Inst Eng (vpres, 45-46); Chinese Inst Elec Engr (pres, 40-41). *Res:* Analysis and control of linear and nonlinear systems; electric energy conversion; transient circuit analysis. *Mailing Add:* Dept Systs Univ Pa Philadelphia PA 19104

KUAN, SHIA SHIONG, b Canton, China, Oct 18, 33; US citizen; m 73; c 1. DEVELOPMENT OF BIOSENSORS. *Educ:* Nat Chung Hsing Univ, BS, 53; WVa Univ, MS, 65, PhD(biochem), 68. *Prof Exp:* Dir, Nanchow Sugar Factory, 58-63; res asst, WVa Univ, 63-68; res fel, La State Univ, New Orleans, 68-70; sr res assoc, Univ New Orleans, 71-80; DIR, NATURAL TOXINS RES CTR, FOOD & DRUG ADMIN, 80- *Concurrent Pos:* Consult, Taiwan Sugar Res Inst, 68- & W China Univ Med Sci, 85-; adj prof chem, Univ New Orleans, 81- *Mem:* Am Chem Soc; Sigma Xi; NY Acad Sci; AAAS; Asn Anal Chemists; Inst Food Technologists. *Res:* Induction, isolation, purification and immobilization of enzymes and the application of immobilized enzymes to agricultural, chemical and clinical analyses; the isolation of naturally occuring toxins and the development of fast, simple and inexpensive procedures for the determination of these toxins. *Mailing Add:* Mycotoxin Res Ctr Food & Drug Admin 4298 Elysian Fields Ave New Orleans LA 70122

KUAN, TEH S, b Maymyo, Burma; US citizen; m. POLYMER CHEMISTRY. *Educ:* Univ BC, BSc, 65; Univ Southern Calif, PhD(chem), 69. *Prof Exp:* Fel chem physics, Univ Southern Calif, 69; res assoc spectros, Univ Calif, Los Angeles, 69-71; sr chemist photo chem, Eastman Kodak Co, 71-80; sr scientist, Atlantic Richfield Co, 80-85; staff scientist, Lockheed

Aeronaut Syst Co, 85-91, SR STAFF SCIENTIST, LOCKHEED MISSILES & SPACE CO, 91- *Mem:* Am Ceramic Soc; Mat Res Soc; Am Chem Soc. *Res:* Conductive polymers; ceramics; materials research; thin film sputtering technology; far-infrared technology; coating technology, ceramics-injection molding. *Mailing Add:* Lockheed PO Box 3213 Thousand Oaks CA 91360

KUAN, TUNG-SHENG, b Taiwan, Dec 13, 47; US citizen; m 77; c 2. ELECTRON MICROSCOPY, ELECTRONIC MATERIALS. *Educ:* Nat Taiwan Univ, BS, 70; Cornell Univ, MS, 73, PhD(mat sci), 77. *Prof Exp:* RES STAFF MEM, IBM THOMAS J WATSON RES CTR, 77- *Concurrent Pos:* Mem, Metall Soc, Am Inst Mech Engrs. *Mem:* Am Phys Soc; Mat Res Soc; Am Inst Mech Engrs; Electron Micros Soc Am. *Res:* Structural, mechanical and electrical properties of electronic materials; electron microscopy of thin film materials and interfaces. *Mailing Add:* Thomas J Watson Res Ctr IBM PO Box 218 Yorktown Heights NY 10598

KUBAS, GREGORY JOSEPH, b Cleveland, Ohio, Mar 12, 45; m 73; c 2. TRANSITION-METAL & SMALL MOLECULE CHEMISTRY. *Educ:* Case Inst Technol, BS, 66; Northwestern Univ, PhD(inorg chem), 70. *Prof Exp:* Fel chem, Princeton Univ, 71-72; mem staff chem, 72-88, LAB FEL, LOS ALAMOS NAT LAB, 88- *Mem:* Am Chem Soc. *Res:* Coordination chemistry of dihydrogen and sulfur dioxide, structure and reactivity of transition metal SO2 complexes; synthesis, characterization and structure of organometallic small molecule complexes of molybdenum and tungsten. *Mailing Add:* 345 Valle Del Sol Los Alamos NM 87544-3563

KUBENA, LEON FRANKLIN, b Caldwell, Tex, July 6, 40; m 68; c 2. NUTRITION, TOXICOLOGY. *Educ:* Tex A&M Univ, BS, 65, PhD(poultry sci), 70. *Prof Exp:* Res nutritionist, SCent Poultry Res Lab, Animal Sci Div, 70-75, RES ANIMAL SCIENTIST BIOCHEM & NUTRIT, FOOD ANIMAL PROTECTION RES LAB, AGR RES SERV, USDA, COLLEGE STATION, TEX, 76- *Concurrent Pos:* Mem fac, Tex A&M Univ, 80- *Mem:* Poultry Sci Asn; Asn Off Anal Chemists; AAAS; World Poultry Sci Asn; US Animal Health Assoc; Coun Agr Sci & Technol. *Res:* Interrelationships of environment and nutrition; toxicology of environmental toxicants in poultry with special emphasis on interaction of these toxicants. *Mailing Add:* 2010 Langford St College Station TX 77840

KUBERSKY, EDWARD SIDNEY, b Brooklyn, NY, Feb 25, 47; m 69; c 2. LIMNOLOGY, ECOLOGY. *Educ:* Brooklyn Col, CUNY, BS, 67; Ind Univ, MA, 68, PhD(zool), 73. *Prof Exp:* from instr to assoc prof biol, Upsala Col, 72-89, chmn dept, 79-84; CHMN & ASSOC PROF BIOL, ST FRANCIS COL, 89- *Concurrent Pos:* Proj dir, NSF grant, 72-74 & 76-77; Danforth assoc, 80-86; consult-Lake Restoration & Mgt. *Mem:* Int Asn Theoret & Appl Limnol; Freshwater Biol Asn; Am Soc Limnol & Oceanog; NAm Lake Mgt Soc. *Res:* Ecology and taxonomy of cladocera; lake restoration. *Mailing Add:* Dept Biol St Francis Col 180 Remsen St Brooklyn NY 11201

KUBES, GEORGE JIRI, b Prague, Czech, Feb 14, 34; Can citizen; m 59; c 2. PULP CHEMISTRY. *Educ:* Tech Univ, Prague, MSc, 58; Tech Univ, Bratislava, PhD(pulp chem). *Prof Exp:* Supvr res group, NBohemian Pulp & Paper Mill, Czech, 58-62, mgr res pulping, papermaking & pollution, 62-67; head pulping group, Pulp & Paper Res Inst Czech, 68; sr res chemist, CIP Res Ltd, Can, 69-72; res scientist pulping, 72-77, head Chem Pulping & Bleaching Sect, 77-85, SR SCIENTIST, PULP & PAPER RES INST CAN, 85-; AUXILIARY PROF, CHEM ENG DEPT, McGILL UNIV, 85- *Concurrent Pos:* Indust consult, Chem Eng Designing Inst, Czech, 58-68; lectr, Chem Eng Fac, Tech Univ, 62-67. *Mem:* Can Pulp & Paper Asn; Tech Asn Pulp & Paper Indust; Chem Inst Can. *Res:* Pulping with an effort to develop a sulphur-free pulping and chlorine-free bleaching process; consultant services to pulp and paper manufacturing industry all over the world; improvements in existing processes, pulping of tropical wood species and annual plants; viscosity of pulp; reaction kinetics in cellulose degradation during alkaline deliquification which resulted in the development of G-factor; differential thermal analysis of black liquor quality; reaction engineering studies and chemical kinetics, also fundamental studies in fluidized bed technology. *Mailing Add:* Chem Eng Dept McGill Univ 3420 University St Pulp & Paper Res Inst Montreal PQ H3A 2A7 Can

KUBICA, GEORGE P, b Little Falls, NY, June 18, 29; m 53; c 2. MEDICAL BACTERIOLOGY, BIOSAFETY. *Educ:* Cornell Univ, BA, 51; Univ Mich, MA, 52; Univ Wis, PhD(med bact), 55. *Prof Exp:* Actg chief, Tuberc Unit, Commun Dis Ctr, USPHS, 55-62, chief, 62-69; mem & head mycobact sect, Trudeau Inst, Inc, 69-74; microbiologist consult, 74-86, CHIEF, MYCOBACT CTR DIS CONTROL, 87- *Concurrent Pos:* Instr, Sch Med, Emory Univ, 59-69; assoc prof, Univ NC, 64-69; chmn, Assembly on Microbiol & Immunol, Am Thoracic Soc, 73-74; secy, Comt on Bacteriol, Int Union Against Tuberc, 76-82; adv/lectr, Int Tuberc Course, Japan, 84-; chmn, Mycobact Div, 85-86. *Honors & Awards:* Commendation Medal, Surg Gen, USPHS, 68. *Mem:* Am Thoracic Soc; Am Soc Microbiol; Int Union Against Tuberc; fel Am Acad Microbiol; Am Biol Safety Asn. *Res:* Tuberculosis. *Mailing Add:* 2383 Welton Pl Atlanta GA 30338

KUBICEK, JOHN D, b Owatonna, Minn, Feb 1, 43. MATHEMATICS. *Educ:* St Johns Univ, BA, 65, MS, 67; Univ Mo, PhD(math), 75. *Prof Exp:* PROF MATH, SOUTHWEST MO STATE UNIV, 81-, ACTG DIR, MATH CTR, 88- *Mem:* Am Math Soc. *Res:* Developmental education in colleges. *Mailing Add:* Dept Math Southwest Mo State Univ Springfield MO 65802

KUBIK, PETER W, b Penzberg, Ger, Dec 31, 49. NUCLEAR RESEARCH. *Educ:* Tech Univ Munich, Ger, MSc, 78, PhD(physics), 83. *Prof Exp:* Res fel, Inst Nuclear Physics, Tech Univ Munich, Ger, 78-83; Res assoc, 83-87, SR RES ASSOC, NUCLEAR STRUCT RES LAB, UNIV ROCHESTER, 87- *Mem:* Am Phys Soc. *Res:* Measurements of long-lived radioisotopes in natural samples like meteorites, glacial ice, ground water, ocean sediments, bones, etc, using accelerated mass spectrometry. *Mailing Add:* Nuclear Struct Res Lab Univ Rochester 271 E River Rd Rochester NY 14627

KUBIK, ROBERT N, b Honolulu, Hawaii, Nov 17, 31; m 55; c 4. COMPUTER SCIENCE, CONTROL SYSTEMS. *Educ:* Univ Calif, Berkeley, AB, 54. *Prof Exp:* Programmer, Atomic Energy Div, Babcock & Wilcox Co, 57-61, sr programmer, 61-62, prog supvr, 62-63, chief comput serv, 63-66; acct rep, IBM Corp, 67; prin engr, Nuclear Power Generating Dept, Babcock & Wilcox Co, 68-69, chief instrument develop, Res & Develop Div, 69-71, mgr process control, Lynchburg Res Ctr, 71-72, mgr indust systs, 72-76, mgr advan control & exp physics lab, Lynchburg Res Ctr, 76-79. *Mem:* Inst Elec & Electronics Engrs; Am Soc Nondestructive Testing; Am Nuclear Soc. *Res:* methodology; nondestructive examination. *Mailing Add:* PO Box 503 Point Reyes Station CA 94956

KUBIN, ROSA, b St Poelten, Austria, Dec 15, 06; nat US; m 31. BIOCHEMISTRY, VETERINARY PATHOLOGY. *Educ:* St Poeltner Obergym, Austria, BS, 25; Univ Vienna, MS, 29, PhD(org chem), 31. *Prof Exp:* Asst, Austrian Chem Works, 31-35 & Syngala, Inc, Austria, 35-38; AMA fel, Med Sch, Univ Ore, 38, Lilly fel, 39-40, asst, 40-41; asst prof biochem & clin path, Med Sch, Middlesex Univ, 41-47; asst prof chem, Univ Mass, 47-49; assoc prof biochem, New Eng Col Pharm, 50-51; consult vet pathologist, 51-90; RETIRED. *Concurrent Pos:* Lectr, Wellesley Col, 55-57 & Concord Acad, 57-61; teacher advan chem & molecular biol, Waltham High Sch, 61-73. *Mem:* Fel AAAS; Am Chem Soc; NY Acad Sci. *Res:* Medical biochemistry; bleaching of textiles; hormone extraction; laboratory methods applied in veterinary clinical pathology; diseases in veterinary medicine. *Mailing Add:* 865 Central Ave N Hill Needham MA 02192-1338

KUBINSKI, HENRY A, b Warsaw, Poland, Jan 15, 33; m 59; c 3. ONCOLOGY, MOLECULAR BIOLOGY. *Educ:* Univ Warsaw, MD, 55. *Prof Exp:* Res assoc microbiol, Sch Med, Univ Warsaw, 54-55, instr, 56-60; instr, Found Res Poliomyelitis & Multiple Sclerosis, Hamburg, Ger, 61-64; res assoc, 64, from asst prof to assoc prof, 65-74, PROF NEUROSURG, UNIV WIS-MADISON, 75- *Mem:* Am Asn Cancer Res; Am Soc Biol Chemists; Biophys Soc; NY Acad Sci; Am Soc Neurochem; Soc Neurosci. *Res:* Neurochemistry and oncology of the central nervous system; cellular regulatory mechanisms; chemical carcinogenesis. *Mailing Add:* 4605 Med Sci Ctr 1300 University Ave Madison WI 53706

KUBIS, JOSEPH J(OHN), b New York, NY, Apr 15, 38; m 68, 84; c 1. NONLINEAR MECHANICS, COMPUTATIONAL FLUID DYNAMICS. *Educ:* Mass Inst Technol, SB, 59; Princeton Univ, AM, 61, PhD(physics), 64. *Hon Degrees:* MA, Cambridge Univ, 67. *Prof Exp:* Asst prof physics, Tex A&M Univ, 64-67; sr res physicist, Cavendish Lab & fel, Clare Hall, Cambridge Univ, 67-69; asst prof physics, Mich State Univ, 69-71; mem tech staff, Theory & Comput Div, KMS Fusion, Inc, 72-82; prog specialist, Software Serv Corp, 83-85; SR RES ENGR, FORD MOTOR CO, 85- *Concurrent Pos:* Consult, Brookhaven Nat Lab, 67 & Los Alamos Sci Lab, 68 & 71. *Mem:* Am Phys Soc; Asn Comput Mach; Inst Elec & Electronics Engrs Computer Soc; Sigma Xi; Soc Indust & Appl Math. *Res:* Nonlinear crash mechanics; finite element methods; computational fluid dynamics; optical ray tracing; numerical analysis; scientific computing; software engineering. *Mailing Add:* 3489 Oak Dr Ypsilanti MI 48197-3747

KUBISEN, STEVEN JOSEPH, JR, b Iowa City, Iowa, June 21, 52; m 77. PHYSICAL ORGANIC CHEMISTRY, ORGANIC CHEMISTRY. *Educ:* Cornell Univ, AB, 74; Harvard Univ, MA, 75, PhD(org chem), 78. *Prof Exp:* res chemist, 78-81, proj scientist, 81-84, group leader, Union Carbide Corp, 84-86; polymer lab dir, Akeo Coatings Am, 86-87; TECHNOL MGR, GE ELECTROMAT DEPT, 87- *Mem:* Am Chem Soc; Sigma Xi; Inst Paper Chem. *Res:* Phosphate ester hydrolysis; process chemistry; epoxidation chemistry; natural oils chemistry; acrylic, urethane technology; radiation cure; epoxy resins. *Mailing Add:* 879 Severn Dr Coshocton OH 43812

KUBITSCHEK, HERBERT ERNEST, biophysics; deceased, see previous edition for last biography

KUBITZ, WILLIAM JOHN, b Freeport, Ill, Dec 27, 38; m 60; c 2. HARDWARE SYSTEMS, COMPUTER SCIENCE. *Educ:* Univ Ill, Urbana-Champaign, BS, 61, MS, 62, PhD(elec eng), 68. *Prof Exp:* Develop engr, Gen Elec, 62-64; teaching asst elec eng, 64-65, res asst comput sci, 65-68, res asst prof, 68-69, from asst prof to assoc prof, 70-85, PROF COMPUT SCI & ASSOC HEAD DEPT, UNIV ILL, 85- *Mem:* Inst Elec & Electronics Engrs; AAAS; Sigma Xi; Asn Comput Mach; Soc Info Display. *Res:* Automated chip and modul4e layout of digital circuits based on automatic generation from a topological data structure with size, shape and timing constraints; object oriented graphics for a networked workstation environment with application to user interfaces, design systems and visualization; computeer graphics. *Mailing Add:* Dept Comput Sci 252 DCL Univ Ill 1304 W Springfield Ave Urbana IL 61801

KUBLER, DONALD GENE, b Easton, Md, Apr 4, 23; m 47; c 4. ORGANIC CHEMISTRY. *Educ:* Univ SC, BS, 47; Univ Md, PhD(chem), 52. *Prof Exp:* Instr chem, Univ SC, 47-48; chemist, Develop Dept, Union Carbon Chem Co, WVa, 52-58; asst prof chem, Univ SC, 58-59 & Hampden-Sydney Col, 59-61; from assoc prof to prof, 61-85, chmn dept, 67-72, EMER PROF CHEM, FURMAN UNIV, 85- *Concurrent Pos:* NSF fac fel sci, Clemson Univ, 75; vis prof, Univ Sterling, 82; vis prof, Grinnell Col, 86-87. *Mem:* Am Chem Soc; Sigma Xi. *Res:* Structure and mechanism for acetal and carbohydrate hydrolysis; forensic chemistry. *Mailing Add:* 136 Conder Dr Marietta SC 29661

KUBLER, HANS, b Ger, Sept 11, 22; m 60, 84. FOREST PRODUCTS. *Educ:* Univ Hamburg, dipl wood tech, 50, Dr rer nat(wood sci), 57. *Prof Exp:* Res asst wood tech, Nat Res Ctr Wood Prod, Ger, 50-54, proj leader wood sci, 59-66; PROF WOOD SCI, UNIV WIS-MADISON, 67- *Concurrent Pos:* Proj leader, Ger Res Asn, 57-58; Fulbright fel, US Forest Prod Lab, Madison, Wis, 58-59; exchange scientist, Acad Wood Tech, Leningrad, USSR, 62; int ed, Forest Prod Res Soc, 67-74. *Mem:* Forest Prod Res Soc; Soc Wood Sci & Technol. *Res:* Wood at low temperatures; growth stresses in trees; drying of wood; cracks in stems of trees; self heating of wood and other organic materials; wood as building material; ignition and fire performance of wood; spiral grain in trees. *Mailing Add:* Forestry Dept Univ Wis 1630 Linden Dr Madison WI 53706

KUBO, RALPH TERUO, b Hilo, Hawaii, Mar 28, 42; m 67; c 2. MOLECULAR IMMUNOLOGY, IMMUNOCHEMISTRY. *Educ:* Univ Calif, Los Angeles, BA, 65; Univ Hawaii, MS, 67, PhD(microbiol), 70. *Prof Exp:* Asst prof microbiol, Univ Hawaii, Honolulu, 70-71; sr fel, Nat Jewish Ctr Immunol Respiratory Med, 71-73, mem immunol, 73-89; asoc prof, 80-89, PROF DEPT MICROBIOL, IMMUNOL & MED, MED SCH, UNIV COLO, 89-; SR FAC MEM, NAT JEWISH CTR IMMUNOL, 89- *Concurrent Pos:* Asst prof, Med Sch, Univ Colo, 75-80; assoc ed, J Immunol & Develop Comp Immunol, 78-82; Japan Soc for the Promotion of Sci, Fels, Univ Tokyo, 87; mem study sect, NIH, 89-93. *Mem:* Am Soc Microbiol; Sigma Xi; Am Asn Immunologists; NY Acad Sci; Am Soc Cell Biol. *Res:* Regulation of expression of mixed isotype class II MHC antigens; characterization of the structure and function of T-cell antigen recognition molecules. *Mailing Add:* Dept Med Nat Jewish Ctr Immunol 1400 Jackson St Denver CO 80206-2762

KUBOTA, MITSURU, b Eleele, Hawaii, Sept 25, 32; m 56; c 2. INORGANIC CHEMISTRY. *Educ:* Univ Hawaii, BA, 54; Univ Ill, MS, 58, PhD, 60. *Prof Exp:* From instr to assoc prof, 59-71, PROF CHEM, HARVEY MUDD COL, 71- *Concurrent Pos:* NSF fel, Univ NC, Chapel Hill, 66-67; Fulbright-Hays advan res fel, Univ Sussex, Eng, 73-74; NIH spec fel, Calif Inst Technol, 74-75; consult, Chevron Res, 81-82; NSF prof develop award, Univ Calif, Berkeley, 81-82. *Mem:* AAAS; Am Chem Soc; The Chem Soc. *Res:* Organometallic chemistry; homogeneous catalysis; inorganic synthesis. *Mailing Add:* Dept Chem Harvey Mudd Col Claremont CA 91711

KUBOTA, TOSHI, b Westmoreland, Calif, Feb 25, 26; m 52; c 3. AERONAUTICS. *Educ:* Univ Tokyo, BEng, 47; Calif Inst Technol, MS, 52, PhD(aeronaut), 57. *Prof Exp:* Asst, 52-57, res fel, 57-59, from asst prof to assoc prof, 59-71, PROF AERONAUT, CALIF INST TECHNOL, 71- *Concurrent Pos:* Consult, AER, Inc, 57-59, Calif Div, Lockheed Aircraft Corp, 59-61, NESCO, 61-62, NAm Aviation, 62-69, TRW Systs Group, 68- & Aerospace, 69- *Mem:* Am Inst Aeronaut & Astronaut; Phys Soc Japan; Sigma Xi. *Res:* Hypersonic aerodynamics and heat transfer. *Mailing Add:* 140 Lowell Ave Sierra Madre CA 91024

KUBU, EDWARD THOMAS, b New York, NY, Nov 19, 26; m 51. PHYSICAL CHEMISTRY, POLYMER PHYSICS. *Educ:* NY Univ, BA, 49; Princeton Univ, MA, 51, PhD(chem), 52. *Prof Exp:* Sect leader, Textile Physics Sect, Res Ctr, B F Goodrich Co, 52-59; supvr characterization res, Cent Res Lab, Allied Corp, 59-61, asst dir lab res, 61-62, dir lab res, 62-63, dir res & develop, Fibers Div, asst to pres, 68-70, tech dir, Fibers Div, 70-73, dir tech opers, 73-77, mgr Govt & Indust Liaison, 77-86; PVT CONSULT, 86- *Concurrent Pos:* Bd trustees exec comt, Textile Res Inst, 71-78; mem adv bd, Textile Res Inst Regulatory Tech Info Ctr. *Mem:* Am Chem Soc; Sigma Xi. *Res:* Physical, chemical and mechanical properties of high polymers; manufacture and use of synthetic fibers; impact of government regulations on synthetic fiber, plastics and chemical manufacture and use. *Mailing Add:* 4720 Southmoor Rd Richmond VA 23234

KUBY, STEPHEN A, biochemistry; deceased, see previous edition for last biography

KUBYNSKI, JADWIGA, b Kluczkowice, Poland, Feb 3, 48; Polish & Can citizen; m 72; c 2. SOLID STATE PHYSICS, ELECTROMAGNETISM. *Educ:* Tagiellonion Univ, MSc, 71; A Mickiewicz Univ, PhD(physics), 80. *Prof Exp:* Res fel physics, Dept Physics, A Mickiewicz, 72-80, res assoc, 81-87; postdoctoral res fel physics, 87-90, RES ASSOC PHYSICS, DEPT PHYSICS & ASTRON, UNIV CALGARY, 91- *Concurrent Pos:* Consult, SPEC Instruments Ltd, Calgary, 88- & ESSO, Calgary, 90- *Mem:* Can Asn Physicists. *Res:* Magnetic resonance spectroscopy; dynamic in-situ CW-EPR studies of coal; spin-labelled oil and erythrocyte membrane spectroscopy; dehydratation mechanisms in clay minerals; isolator-metal phase transitions; magnetic transitions. *Mailing Add:* Dept Physics & Astron Univ Calgary Calgary AB T2N 1N4 Can

KUC, JOSEPH, b New York, NY, Nov 24, 29; m 54; c 3. PLANT BIOCHEMISTRY, PLANT PATHOLOGY. *Educ:* Purdue Univ, BS, 51, MS, 53, PhD(biochem), 55. *Prof Exp:* Asst biochem, Purdue Univ, 51-54, from asst prof to prof, 55-74; PROF PLANT PATH, UNIV KY, 74- *Concurrent Pos:* Fulbright fel, 60 & 66; fel, Brazilian Coffee Inst, 69 & 71; Alexander von Humboldt Found res prize. *Honors & Awards:* Campbell Award, Am Phytopath Soc. *Mem:* Am Chem Soc; fel Am Phytopath Soc; Am Soc Plant Physiol; Phytochem Soc; fel Am Inst Chemists; Sigma Xi. *Res:* Biochemistry of disease resistance in plants; synthesis of natural products; plant immunization. *Mailing Add:* Dept Plant Path Univ Ky Lexington KY 40546-0091

KUCERA, CLARE H, b Laurel, Mont, Sept 20, 25; m 50, 79; c 5. OILWELL CHEMICALS, ORGANIC & INORGANIC CHEMISTRY. *Educ:* Mont State Col, BS, 51; Purdue Univ, MS, 54, PhD(chem), 56. *Prof Exp:* From chemist to res assoc, Dowell Div, Dow Chem Co, 56-84; consult, Oilwell Chem, 84-86; RES ASSOC, BASF CORP, 86- *Concurrent Pos:* Instr, Gen Chem, Tulsa Jr Col. *Mem:* Am Chem Soc; Sigma Xi. *Res:* Organic synthesis; stereochemistry of the 2-ethyl-cyclohexanols; organic acid corrosion inhibitors, especially acetylenic compounds; synthesis of acetylenic compounds; development of oilwell fracturing fluids; rheology of fracturing fluids; synthesis of orthophosphate esters; development of oilwell cements; synthesis of cement additives; synthesis and development of drilling fluids. *Mailing Add:* 11229 S 106 East Ave Bixby OK 74008

KUCERA, LOUIS STEPHEN, b New Prague, Minn, June 23, 35; m 59; c 4. VIROLOGY, AIDS. *Educ:* St John's Univ, Minn, BA, 57; Creighton Univ, MS, 59; Univ Mo, PhD(microbiol), 64. *Prof Exp:* Res bacteriologist, Radioisotope Serv, Vet Admin Hosp, Omaha, Nebr, 60; res asst virol, Sect Microbiol, Mayo Clin, 64-65, res assoc, 65-66, res assoc, Virol Lab, St Jude Res Hosp, Memphis, Tenn, 66-68, staff mem, 68-70; from asst prof to assoc prof, 70-80, PROF MICROBIOL, BOWMAN GRAY SCH MED, WAKE FOREST UNIV, 80- *Concurrent Pos:* Reviewer, Human Cell Biol Prog, NSF; cancer res, proceedings, Nat Acad Sci; site visit reviewer, NC Biotechnol Ctr; prin investr & vis scientist, German Cancer Res Ctr, Heidelberg, 86. *Mem:* AAAS; Am Soc Microbiol; fel Am Acad Microbiol; Sigma Xi; Am Soc Virol; Int Soc Antiviral Res; Am Asn Cancer Res. *Res:* Herpes simplex virus DNA sequences; prostaglandin synthesis and tumorigenesis; consequences of herpes simplex virus-alveolar macrophage interactions; tumor promoters; chemical carcinogens and herpes viruses; human immunodeficiency virus and herpes virus interactions; ether lipids and chemotherapy of HIV infections. *Mailing Add:* Dept Microbiol & Immunol Bowman Gray Sch Med Winston-Salem NC 27103-2797

KUCERA, THOMAS J, b Oak Park, Ill, Feb 22, 25; m 64; c 3. ORGANIC CHEMISTRY. *Educ:* Loyola Univ, Ill, BS, 45; Ill Inst Technol, MS, 52; Purdue Univ, PhD, 53. *Prof Exp:* Res chemist, Miner Labs, Mid-West Div, Arthur D Little, Inc, 45-50; Fulbright scholar, Univ Auckland, 53-54; res fel, Purdue Univ, 54-55; asst to pres, Mid-West Labs, 55-56; mgr chem res, Charles Bruning Co, Inc, 56-61; consult, 61-64; vpres res & eng, Apeco Corp, 64-81; CONSULT, 82- *Concurrent Pos:* Fulbright scholar, New Zealand. *Mem:* AAAS; Am Chem Soc; The Chem Soc; Soc Photog Sci & Eng. *Res:* Photoreproduction; organic photoreactions; inorganic photoconductors; electrostatics; organic mechanisms. *Mailing Add:* 9310 Hamlin Ave Evanston IL 60203-1302

KUCESKI, VINCENT PAUL, b Superior, Wis, Apr 1, 20; m 44; c 2. ORGANIC CHEMISTRY. *Educ:* Univ Wis, BS, 42, MS, 48, PhD(chem), 50. *Prof Exp:* Res chemist, Southern Cotton Oil Co, 50-52; sr chemist, 52-59, dir res, 59-71, vpres res, 71-74, VPRES RES & DEVELOP, C P HALL CO ILL, 74- *Mem:* Am Chem Soc; Am Inst Chemists; Am Oil Chemists Soc. *Res:* Oxidations of organic compounds; oils and fats; analytical organic chemistry. *Mailing Add:* 293 W Elmwood Chicago IL 60411

KUCHAR, NORMAN RUSSELL, b Cleveland, Ohio, June 22, 39; m 67; c 2. MATERIALS PROCESSING. *Educ:* Case Inst Technol, BS, 61, MS, 65; Case Western Reserve Univ, PhD(eng), 67. *Prof Exp:* Fluid dynamicist fluid physics, Space Sci Lab, Gen Elec Co, 67-69, group leader biofluid mech, Environ Sci Lab, 69-72, mech engr, 72-80, mgr process physics prog, 80-82, mgr, Process Technol Br, 83-90, MGR, PROCESS PHYSICS LAB, GEN ELEC CO, 90- *Mem:* Am Soc Mech Engrs; Am Inst Aeronaut & Astronaut; Am Soc Mat; Laser Inst Am. *Res:* Intelligent processing of materials; process modeling; computer-aided engineering; process sensors; laser technology. *Mailing Add:* Gen Elec Co Corp Res & Develop PO Box 8 Schenectady NY 12301

KUCHAREK, THOMAS ALBERT, b Cleveland, Ohio, Nov 16, 39; m 63; c 2. PLANT PATHOLOGY. *Educ:* Kent State Univ, BS, 62; Univ Minn, MS, 65, PhD(plant path), 69. *Prof Exp:* Asst plant path, Univ Minn, 62-65; instr, Okla State Univ, 65-68 & Univ Minn, 69; asst prof & asst exten plant pathologist, 70-74, assoc prof, 75-80, PROF PLANT PATH & EXTEN PLANT PATHOLOGIST, UNIV FLA, 80- *Mem:* Am Phytopath Soc; Sigma Xi. *Res:* Diagnosis and control of diseases on field crops and vegetables. *Mailing Add:* Dept Plant Path Univ Fla Gainesville FL 32611

KUCHEL, OTTO GEORGE, b Spis Stara Ves, Czech, June 22, 24; Can citizen; m 53; c 3. NEPHROLOGY. *Educ:* Charles Univ, Prague, MD, 50, PhD(endocrinol), 56, ScD(nephrology), 65. *Prof Exp:* Instr int med, Safarik Univ, Kosice, 56; asst prof, III Dept Med, Charles Univ, 57-65; instr, Vanderbilt Univ, 66; prof, III Dept Med, Charles Univ, Prague, 67-68; PROF NEPHROLOGY, CLIN RES INST, UNIV MONTREAL, 68-, DIR LAB SYMPATHETIC NERV SYST, INST, 75- *Concurrent Pos:* Mem serv nephrology, Hotel-Dieu Hosp & Univ Montreal, 68; mem hypertension task force, NIH, 75-79; mem, Coun High Blood Pressure Res, Cleveland. *Honors & Awards:* Res Award, Asn French Speaking Physicians Can, 72. *Mem:* Endocrine Soc; Royal Soc Med. *Res:* Clinical nephrology and endocrinology related to research of mechanisms of hypertension, particularly the role of the sympathetic nervous system, adrenals and the kidney. *Mailing Add:* Clin Res Inst Montreal 110 Pine Ave W Montreal PQ H2W 1R7 Can

KUCHERLAPATI, RAJU SURYANARAYANA, b Kakinada, India, Jan 18, 43; m; c 1. HUMAN GENETICS. *Educ:* Andhra Univ, India, BSc, 60, MSc, 62; Univ Ill, Urbana, PhD(genetics), 72. *Prof Exp:* Res fel human genetics, Yale Univ, 72-75; asst prof human genetics, Princeton Univ, 75-82; prof genetics, Univ Ill, 82-88; PROF GENETICS, ALBERT EINSTEIN COL MED, 89-, SAUL & LOLA KRAMER PROF, 89- *Concurrent Pos:* Damon Runyon Cancer Fund res fel, Yale Univ, 73-74, NIH fel, 74-75; mem, Mammalian Genetics Study Sect, NIH, 85-89. *Mem:* AAAS; Genetics Soc Am; Am Soc Microbiol. *Res:* Human gene mapping; study of regulation of gene action in human cells; gene transfer, gene therapy; homologous recombination. *Mailing Add:* Dept Molecular Genetics Albert Einstein Col Med Bronx NY 10461

KUCHINSKAS, EDWARD JOSEPH, b Maspeth, NY, Feb 11, 27; Wid; c 1. BIOCHEMISTRY. *Educ:* Queen's Col, NY, BS, 49; Cornell Univ, PhD, 54. *Prof Exp:* Instr biochem, Med Col, Cornell Univ, 54-56; from asst prof to assoc prof, 56-67, from asst dean to assoc dean sch grad studies, 67-73, PROF BIOCHEM, STATE UNIV NY DOWNSTATE MED CTR, 67- *Mem:* AAAS; Am Chem Soc; Soc Exp Biol Med; Harvey Soc; Am Soc Biol Chemists; NY Acad Sci. *Res:* Metabolic effects of cysteine analogues, especially vitamin requirements and enzyme activation; catalase; semisynthetic penicillins; S-methyl group oxidations; peroxidative mechanisms; metabolism of penicillamine. *Mailing Add:* Dept Biochem Box 8 State Univ NY Downstate Med Ctr Brooklyn NY 11203

KUCHLER, ROBERT JOSEPH, b Pittsburgh, Pa, Mar 28, 28; m 58; c 3. MICROBIOLOGY. *Educ:* Univ Pittsburgh, BS, 50, MS, 52; Univ Mich, PhD, 58. *Prof Exp:* Asst bact, WVa Univ, 52-54 & Univ Mich, 54-58; bacteriologist & head dept, William Singer Res Lab, Allegheny Gen Hosp, 58-62; from asst prof to prof bact, 62-75, dir coord grad prog microbiol, 78-81, PROF MICROBIOL, RUTGERS UNIV, 75- *Mem:* Am Soc Microbiol; Tissue Cult Asn; Am Soc Cell Biol; Sigma Xi. *Res:* Development of metazoan cell populations in tissue culture with emphasis on their permeability to small molecular species and on the organization of macromolecules within these cells; viral nucleic acids. *Mailing Add:* Dept Microbiol Rutgers Univ New Brunswick NJ 08903-1059

KUCHNIR, FRANCA TABLIABUE, b Russe, Bulgaria, July 18, 35; US citizen; m 60; c 2. MEDICAL PHYSICS, RADIOLOGICAL SCIENCE. *Educ:* Univ San Paulo, BS, 58; Univ Ill, MS, 62, PhD(physics), 65. *Prof Exp:* Res asst physics, Univ Ill, 60-65; fel, Argonne Nat Lab, 66-68; asst prof, Univ Ill, 69-70; asst physicist, Argonne Nat Lab, 70-71; trainee, 71-73, asst prof, 73-74, dir, sect med physics, 80- 84, ASSOC PROF MED PHYSICS, UNIV CHICAGO, 74- *Concurrent Pos:* Prog dir, Nat Res Serv Awards, Nat Cancer Inst, 79-88. *Mem:* Am Phys Soc; Am Asn Physicists Med; Radiol Soc NAm; Am Col Radiol; Am Soc Ther Radiol Oncol. *Res:* Experimental nuclear physics; radiation physics and dosimetry specifically related to neutrons as applied to radiobiology, radiation therapy and diagnosis. *Mailing Add:* Dept Radiation Oncol Univ Chicago 5841 S Maryland Ave Box 442 Chicago IL 60637

KUCHNIR, MOYSES, b Sao Paulo, Brazil, May 18, 36; US citizen; m 60; c 2. LOW TEMPERATURE PHYSICS, SUPERCONDUCTIVITY APPLICATIONS. *Educ:* Univ Sao Paulo, BS, 57; Univ Ill, Urbana, MS, 62, PhD(physics), 66. *Prof Exp:* Mem staff solid state physics, Argonne Nat Lab, 66-68, asst physicist, 68-73; proj assoc cryogenics, Univ Wis-Madison, 73; prof physics, Univ Estadual Campinas, Brazil, 74; PHYSICIST & APPL SCIENTIST, FERMI NAT ACCELERATOR LAB, 74- *Mem:* AAAS; Am Phys Soc. *Res:* Properties of quantum fluids; magnetic properties of superconductors; cryogenic and superconducting equipment and techniques; superconducting magnets for accelerators; properties of materials at low temperatures. *Mailing Add:* Fermilab MS 316 PO Box 500 Batavia IL 60510

KUCK, DAVID JEROME, b Muskegon, Mich, Oct 3, 37; m 77; c 1. COMPUTER SCIENCE. *Educ:* Univ Mich, Ann Arbor, BS, 59; Northwestern Univ, MS, 60, PhD(eng), 63. *Prof Exp:* Ford fel & asst prof elec eng, Mass Inst Technol, 63-65; from asst to assoc prof, 65-72, PROF COMPUT SCI, UNIV ILL, URBANA-CHAMPAIGN, 72-, DIR, CTR SUPERCOMPUT RES & DEVELOP, 84- *Concurrent Pos:* Nat Sci Found res grant, 70-; consult, Burroughs Corp, 72- & Los Alamos Nat Lab, 78-; assoc ed, Inst Elec & Electronics Engrs Trans Comput, 73-75, Asn Comput Mach Database Systs, 77-, Int J Comput & Info Sci, 77-, J Asn Comput Mach, 80- & J Digital Systs, 80-; pres, Kuck & Assoc, Inc; mem tech adv bd, Sequent Comput Systs, Portland, Ore, 85-, Sci Comput Systs, San Diego, Calif, 85-, Dana Group, Sunnyvale, Calif, 86-; mem comput sci & technol bd, Nat Res Coun, Washington, DC 86-; bd dirs, Supercomput Systs, Inc, 88- *Honors & Awards:* Emanuel R Piore Award, Inst Elec & Electronics Engrs, 87. *Mem:* Nat Acad Eng; Inst Elec & Electronics Engrs. *Res:* Parallel, pipeline and multiprocessor computation methods; interconnection networks; memory hierarchies; compilation of ordinary programs for such machines. *Mailing Add:* Ctr Supercomput Res & Develop 104 S Wright St Urbana IL 61801

KUCK, JAMES CHESTER, b New Orleans, La, Dec 24, 12; m 69. AGRICULTURAL BIOCHEMISTRY, ORGANIC CHEMISTRY. *Educ:* La State Univ, Baton Rouge, BS, 38. *Prof Exp:* Prof, New Orleans Pub Sch Syst, 38-41; prof math & sci, Rugby Mil Acad, New Orleans, 41-42; asst chemist, La State Bd Health, 42; sr res chemist, Celotex Corp, Marrero, La, 42-49; RES CHEMIST, SOUTHERN REGIONAL RES CTR, SCI & EDUC ADMIN, AGR RES, USDA, NEW ORLEANS, 53- *Mem:* Am Oil Chemists Soc; Sigma Xi. *Res:* Chemical and physical properties of oil-bearing seed, especially cottonseed, peanuts, soybean, sunflower, rape and properties relating to improvements in present industrial processing and development of higher quality and utility of these products. *Mailing Add:* 1373 Madrid St New Orleans LA 70122

KUCK, JOHN FREDERICK READ, JR, b Savannah, Ga, Jan 27, 18; m 49; c 5. BIOCHEMISTRY. *Educ:* Va Polytech Inst, BS, 39, MS, 40; Univ NC, PhD(biochem), 51. *Prof Exp:* Chemist, Nat Adv Comn Aeronaut, 40-46; prof chem, St Procopius Col, 50-51; res assoc surg, Col Med, Wayne State Univ, 51-56; res assoc, Kresge Eye Inst, 57-63; from asst prof to prof ophthal, Sch Med, Emory Univ, 63-88, asst prof biochem, 63-88; VIS PROF, GA TECH, 88- *Concurrent Pos:* Adj prof, Sch Chem, Ga Inst Technol, 87-88. *Mem:* AAAS; Am Chem Soc; Asn Res Vision & Ophthal; Sigma Xi. *Res:* Lens metabolism; diabetic and radiation cataracts; anti-cataractogenic drugs; Raman spectroscopy of lens. *Mailing Add:* Sch Chem & Biochem Ga Tech Atlanta GA 30332-0400

KUCK, JULIUS ANSON, b Willimantic, Conn, Jan 6, 07; m 32, 55; c 2. ORGANIC CHEMISTRY, HIGH PRESSURE LIQUID CHROMATOGRAPHY. *Educ:* Hamilton Col, AB, 28; Cornell Univ, PhD(org chem), 32. *Prof Exp:* Asst chem, Cornell Univ, 28-32; fel, City Col New York, 34-35, tutor, 35-36, from instr to assoc prof, 37-64; vis prof & actg chmn dept, Inter-Am Univ, PR, 65-66; from asst prof to assoc prof chem, Stamford Br, Univ Conn, 66-77; RES ASSOC DEPT CHEM, BANNOW SCI CTR, FAIRFIELD UNIV, 77- *Concurrent Pos:* Consult org microanalysis, Am Cyanamid Co, 43-64. *Mem:* Am Chem Soc; AAAS; Am Microchem Soc; Sigma Xi. *Res:* Synthetic organic chemistry; aliphatic boric acids; cardiac lactones; microanalysis; analytical methods and development; saxitoxin in paralytic shellfish poison; HPLC for natural product analysis; determination of alpha-tocopherol, retinol, and betacarotene in human serum. *Mailing Add:* 57 Mimosa Dr Cos Cob CT 06807

KUCZENSKI, RONALD THOMAS, b Detroit, Mich, July 27, 44. PSYCHOPHARMACOLOGY. *Educ:* Univ Notre Dame, BS, 66; Mich State Univ, PhD(biochem), 70. *Prof Exp:* Res psychobiologist psychiat, Univ Calif, San Diego, 70-73, asst prof, 73-74; ASST PROF PHARMACOL, VANDERBILT UNIV, 74-, ASST PROF BIOCHEM, 80- *Res:* Regulation of biochemical events of central nervous system synaptic transmission and relationship to effects of pharmacological manipulations on behavioral parameters. *Mailing Add:* Dept Psych UCSD M-003 La Jolla CA 92093

KUCZKOWSKI, JOSEPH EDWARD, b Buffalo, NY, Nov 18, 39; m 65; c 4. ALGEBRA, MATHEMATICS EDUCATION. *Educ:* Canisius Col, BS, 61; Purdue Univ, MS, 63, PhD(math), 68. *Prof Exp:* From instr to asst prof math, Purdue Univ, 66-71; from assoc prof to prof math, 71-87, asst dean, 84-87, ASSOC DEAN, SCH SCI, IND UNIV, PURDUE, 87- *Mem:* Math Asn Am; Nat Coun Teachers Math. *Res:* Subsemigroups of groups, with emphasis on nilpotent groups; semigroups satisfying certain non-tautological laws. *Mailing Add:* Dept Math Sci Ind Univ-Purdue Univ PO Box 647 Indianapolis IN 46223

KUCZKOWSKI, ROBERT LOUIS, b Buffalo, NY, Aug 2, 38; m 62; c 3. PHYSICAL INORGANIC CHEMISTRY. *Educ:* Canisius Col, BS, 60; Harvard Univ, MA, 62, PhD(chem), 64. *Prof Exp:* Nat Acad Sci res fel chem, Nat Bur Standards, 64-66; from asst prof to assoc prof, 66-74, PROF CHEM, UNIV MICH, ANN ARBOR, 74- *Mem:* Am Chem Soc. *Res:* Microwave spectroscopy of inorganic compounds. *Mailing Add:* Dept f Chem Univ Mich Ann Arbor MI 48109

KUCZMARSKI, EDWARD R, b Cleveland, Ohio, Sept 4, 49. CELL BIOLOGY. *Educ:* Hiram Col, BS, 71; Yale Univ, PhD(cell biol), 77. *Prof Exp:* Postdoctoral, Stanford Univ, 77-82; asst prof cell biol, Northwestern Univ Med Sch, 82-89; ASSOC PROF PHYSIOL, CHICAGO MED SCH, 89- *Mem:* AAAS; Am Soc Cell Biol; Am Physiol Soc; Am Soc Biochem & Molecular Biol. *Res:* Mechanism and regulation of cell motility; cytoskeleton and signal transduction. *Mailing Add:* Dept Physiol Chicago Med Sch 3333 Green Bay Rd North Chicago IL 60064

KUCZYNSKI, EUGENE RAYMOND, reliability & maintainability engineering systems, mathematics statistics, for more information see previous edition

KUCZYNSKI, GEORGE CZESLAW, b Krakow, Poland, Apr 20, 14; nat US; c 3. MATERIALS SCIENCE. *Educ:* Cracow Tech Univ, MA, 36; Univ South Wales, BSc, 42; Mass Inst Technol, ScD(metall), 46. *Prof Exp:* Teacher, pub sch, Poland, 36-39; teacher, Army Personnel Training Sch, Scotland, 42-43; asst, State Col Wash, 43-44; consult engr, Baldwin Locomotive Co, Pa, 45-46; sr engr, Sylvania Elec Prods, Inc, NY, 46-49; consult engr, Colombia, SAm, 49-50; aeronaut scientist, Nat Adv Comt Aeronaut, Ohio, 50-51; assoc prof, 51-55, prof metall, 55-80, PROF, DEPT METALL ENG & MAT SCI, UNIV NOTRE DAME, 80- *Concurrent Pos:* Lectr, Polytech Inst Brooklyn, 47-49. *Mem:* Am Phys Soc; fel Am Ceramic Soc. *Res:* Physics of solids; order-disorder transformations; diffusion in solids; sintering and powder technology; ceramics; semi-conductors; color centers. *Mailing Add:* Dept Metall Univ Notre Dame Notre Dame IN 46556

KUDENOV, JERRY DAVID, b Lynwood, Calif, Dec 19, 46; m 69; c 1. INVERTEBRATE ZOOLOGY, POLLUTION BIOLOGY. *Educ:* Univ Calif, San Diego, BA, 68; Univ of the Pac, MSc, 70; Univ Ariz, PhD(zool), 74. *Prof Exp:* RES SCIENTIST ZOOL & POLLUTION BIOL, MARINE POLLUTION STUDIES GROUP, FISHERIES & WILDLIFE DIV, AUSTRALIA, 74- *Mem:* Am Soc Zoologists; Sigma Xi; AAAS; Australian Marine Sci Asn; NZ Marine Sci Asn. *Res:* Feeding biology and population dynamics of spionid polychaetes in stressed or polluted habitats; taxonomy, systematics and zoogeography of polychaetous annelids. *Mailing Add:* Dept Biol Univ Alaska 3221 Providence Dr Anchorage AK 99508

KUDER, JAMES EDGAR, b Madang, New Guinea, Dec 28, 39; US citizen; m 62; c 2. ORGANIC CHEMISTRY. *Educ:* Capital Univ, BS, 62; Ohio Univ, PhD(org chem), 68. *Prof Exp:* Chemist water anal, US Geol Surv, Ohio, 62-63; res fel, Rensselaer Polytech Inst, 68-69; scientist, Res Labs, Xerox Corp, 69-77; RES ASSOC, CELANESE RES CO, 77- *Mem:* Am Chem Soc; The Chem Soc; Sigma Xi; Soc Photo-Optical Instrumentation Engrs. *Res:* Electronic structure and properties of organic dyes; photochemical rearrangements; quantum chemistry; reactions and spectroscopic studies of heterocyclic compounds; organic electrochemistry; optical recording materials. *Mailing Add:* Hoechst Celanese Corp 86 Morris Ave Summit NJ 07901

KUDER, ROBERT CLARENCE, b North Baltimore, Ohio, Dec 31, 18; m 42; c 7. PLASTICS CHEMISTRY. *Educ:* Ohio State Univ, AB, 39; Northwestern Univ, PhD(org chem), 42. *Prof Exp:* Jr chemist, Ethyl Gasoline Corp, 39; res chemist, Stand Oil Co, Ind, 42-46; asst prof chem, Univ Dayton, 46-48; sr res chemist, Barrett Div, Allied Chem Corp, 48-52, asst supvr res, 52-57, supvr polymer res, 57; tech dir, Bemis Bros Bag Co, 57-58; dir res & develop, Mol-Rez Div, Am Petrochem Corp, 58-63; res assoc, Gen Mills, Inc, 63-68; tech dir resins, Whittaker Corp, 68-77, tech dir, Minneapolis Coatings & Chem Div, 77-80; res chemist, Precision Cosmet Co, 80-83; RETIRED. *Concurrent Pos:* Consult, 84- *Mem:* Am Chem Soc. *Res:* Polyesters; polyurethanes. *Mailing Add:* 222 W Eagle Lake Dr Maple Grove MN 55369

KUDLICH, ROBERT A, b Pottsville, Pa, Jan 16, 26; m 51; c 1. ELECTRICAL ENGINEERING, COMPUTER SCIENCE. *Educ:* Lafayette Col, BSEE, 50; Harvard Univ, MS, 51; Univ Ill, PhD(elec eng), 54. *Prof Exp:* Mem tech staff comput systs & logical design, Bell Tel Lab, Inc, 54-59; dir comput res & develop, AC Spark Plug Div, Gen Motors Corp, 59-63, prog dir, AC Electronics Div, Wis, 63-69; mgr, Air Traffic Control Systs, Raytheon Co, Wayland, 69-74, prod assurance mgr, 74-80, mgr, Radar Systs Lab, Equip Div, 80-84; PROG MGR, RAMP RADARS, 84- *Concurrent Pos:* Consult, Defense Sci Bd Task Force Electronic Test Equip, 75-77. *Mem:* Inst Elec & Electronics Engrs; Am Fedn Info Processing Soc (vpres, 72-73); Sigma Xi. *Res:* Systems engineering; radar systems development. *Mailing Add:* 119 Bay State Rd Weston MA 02193

KUDMAN, IRWIN, b Brooklyn, NY, Feb 20, 36; m 59; c 2. METALLURGY, PHYSICS. *Educ:* NY Univ, BMetEng, 58, MS, 60. *Prof Exp:* Assoc engr, Radio Corp Am, 60-62, mem tech staff mats, Res Lab, RCA Corp, 62-72; vpres, Princeton Infrared Equip, Inc, 72-76; PRES, INFRARED ASN, INC, 76- *Res:* Optical, electrical and thermal properties of semiconductor materials and alloy systems. *Mailing Add:* Infrared Assocs Inc 1000 Rt 130 Cranbury NJ 08512-9523

KUDO, AKIRA, b Japan, Apr 6, 39; m 74; c 2. ENVIRONMENTAL SCIENCES, WATER CHEMISTRY. *Educ:* Kyoto Univ, Japan, BSc, 63, MSc, 65; Univ Tex, Austin, PhD(environ health eng), 69; Kyoto Univ, Japan, DEng, 79. *Prof Exp:* Sr res officer biol, Nat Res Coun Can, 71-90; SR RES OFFICER, INST ENVIRON CHEM, 90- *Concurrent Pos:* Vis prof, Univ Ottawa, Can, 75-; vis scientist, Japan, 76, 85 & France, 81, 83 & 85; assoc ed, J Environ Conserv Eng, 80- *Mem:* Int Asn Water Pollution Res; Am Soc Civil Eng; Am Water Pollution Control Fedn. *Res:* Distribution, transport, transformation, and transfer of heavy metal pollutants, including radioactive materials (plutonium etc), in the aquatic systems such as rivers, lakes, and estuaries; anaerobic treatment of wastewater. *Mailing Add:* Inst Environ Chem 100 Sussex Dr Ottawa ON K1A 0R6 Can

KUDO, ALBERT MASAKIYO, b New Westminster, BC, May 30, 37; m 62; c 2. PETROLOGY, VOLCANOLOGY. *Educ:* Univ Toronto, BS, 60; McMaster Univ, MS, 62; Univ Calif, San Diego, PhD(earth sci), 67. *Prof Exp:* Lab instr geol, McMaster Univ, 60-62; res asst geochem, Univ Calif, San Diego, 62-66; from asst prof to assoc prof, 66-85, PROF GEOL, UNIV NMEX, 85- *Concurrent Pos:* Petrologist-geochemist, Leg 45, Deep Sea Drilling Proj, 79; fel, Japan Soc Prom Sci, Tohuku Univ, Japan, 82; consult, Morrison-Knudsen Engrs, 88. *Mem:* Am Geophys Union; Geochem Soc; Geol Soc Am; fel Mineral Soc Am. *Res:* Igneous petrology; geochemistry. *Mailing Add:* Dept Geol Univ NMex Albuquerque NM 87131-1116

KUDZIN, STANLEY FRANCIS, b Jersey City, NJ, Mar 1, 26; m 50; c 3. ORGANIC CHEMISTRY. *Educ:* Fordham Univ, BS, 47, MS, 49, PhD(chem), 51. *Prof Exp:* Res & tech serv chemist, E I du Pont de Nemours & Co, 51-56; tech supvr, Ciba Co, Inc, 56-60; assoc prof org chem, Clemson Col, 60-61; ed, Acad Press, Inc, 61-62; assoc prof, 62-70, PROF CHEM, STATE UNIV NY COL NEW PALTZ, 70- *Concurrent Pos:* Consult, Acad Press, Inc. *Mem:* AAAS; Am Chem Soc. *Res:* Chemical education and literature. *Mailing Add:* 35 Ferris Lane Poughkeepsie NY 12601

KUEBLER, JOHN RALPH, JR, b Indianapolis, Ind, Oct 22, 24; m 57; c 2. INDUSTRIAL CHEMISTRY. *Educ:* Univ Wis, BS, 48; Univ Ill, MS, 49, PhD, 51. *Prof Exp:* Inorg res chemist, Mallinckrodt Chem Works, Mallinckrodt Chem Works, 51-85, qual control mgr, Calsicat Div, 70-81; RETIRED. *Mem:* Am Chem Soc; Am Soc Qual Control. *Res:* Inorganic stereochemistry; inorganic analytical methods development. *Mailing Add:* 3536 Breezeway Dr Erie PA 16506

KUEBLER, ROY RAYMOND, JR, biostatistics; deceased, see previous edition for last biography

KUEBLER, WILLIAM FRANK, JR, b Kansas City, Mo, Feb 21, 16; m 45; c 2. CHEMISTRY. *Educ:* Univ Kans, BA, 38. *Prof Exp:* Pharmaceut chemist, Peerless Serum Co, Kans, 38-40; pharmaceut tablet maker, Geo A Breon & Co, Mo, 40-41; chief control chemist, Thompson-Hayward Chem Co, 41-42, res chemist, 46-51; res chemist, Jensen-Salsbery Labs Div, Richardson-Merrell, Inc, 51-54, group leader, 54-57, dir pharmaceut res & develop, 57-59, res adminr, 59-61, asst to res dir proj planning & anal, 61-65; asst to tech dir, Marion Labs, Inc, 65-70 & Knoll Pharmaceut Co; exec admin asst to vpres-sci dir, Knoll Pharmaceut Co, 70-71, dir regulatory affairs, 71-72; exec vpres, Bavley-Kuebler Assocs, 72-74; tech dir, Bio-Eval, Inc, 74-79; CONSULT, 79- *Mem:* Am Chem Soc; Biomet Soc; Am Statist Asn; fel Am Inst Chem. *Res:* Clinical investigations; protocols; biostatistics; data interpretation. *Mailing Add:* 534 Fairmount Ave Chatham NJ 07928

KUECKER, JOHN FRANK, b Webster, SDak, Mar 21, 32; m 58; c 4. PHYSICAL CHEMISTRY. *Educ:* Northern State Col, BS, 54; SDak Sch Mines & Technol, BS, 58; Univ Nebr, MS, 63, PhD(chem), 65. *Prof Exp:* Teacher high sch, SDak, 54-55 & 56-57; asst prof chem, Doane Col, 63-65; from asst prof to assoc prof, 65-71, head dept, 69-72, PROF CHEM, KEARNEY STATE COL, 71- *Mem:* AAAS; Am Chem Soc; Nat Sci Teachers Asn. *Res:* Viscosity of aqueous salt solutions; ultracentrifugation of inorganic polymer solutions. *Mailing Add:* Dept Chem Kearney State Col Kearney NE 68849

KUEHL, GUENTER HINRICH, b Geesthacht, Ger, Jan 2, 28; m 57; c 3. CATALYSTS, ZEOLITES. *Educ:* Brunswick Tech Univ, dipl, 55, Dr rer nat(chem), 57. *Prof Exp:* Res chemist, Kali-Chemie AG, Ger, 60-61; sr res chemist, Cent Res Div, Socony Mobil Oil Co, 62-69; sr res chemist, Process Res & Develop Serv Div, 69-75, assoc, 75-83, RES ASSOC, PAULSBORO RES LAB, CATALYST RES & DEVELOP SECT, MOBIL RES & DEVELOP CORP, 83- *Concurrent Pos:* USPHS fel, Ind Univ, 57-59; counr, Am Chem Soc, 85-; mem, Synthesis Comn, Internal Geolite Asn, 90- *Mem:* Soc Ger Chem; Am Chem Soc; Int Zeolite Asn; Catalysis Soc. *Res:* Crystallization, modification, characterization, and chemistry of zeolites; catalyst research and development for application in petroleum and petrochemical industry; preparation and investigation of hydrogenphosphato-carbonato-apatites; phosphate complexes; preparation and properties of organo-metallic acetylene compounds. *Mailing Add:* Process Res & Tech Serv Div Mobil Res & Develop Corp Paulsboro NJ 08066

KUEHL, HANS H(ENRY), b Detroit, Mich, Mar 16, 33; m 65; c 2. ELECTRICAL ENGINEERING. *Educ:* Princeton Univ, BSEE, 55; Calif Inst Technol, MS, 56, PhD(elec eng), 59. *Prof Exp:* Mem tech staff, Hughes Aircraft Co, 58-59; res scientist plasmas, Eng Ctr, 59-60, from asst prof to assoc prof, 60-72, PROF ELEC ENG, UNIV SOUTHERN CALIF, 72-, CHMN, ELEC ENG & ELECTROPHYSICS, 87- *Concurrent Pos:* Res fel, Calif Inst Technol, 59-60. *Mem:* Am Phys Soc; fel Inst Elec & Electronics Engrs; Int Sci Radio Union. *Res:* Plasma physics; electromagnetic theory; antennas. *Mailing Add:* Dept Elec Eng PHE 604 Univ Southern Calif Los Angeles CA 90089-0271

KUEHL, LEROY ROBERT, b Ketchikan, Alaska, Aug 15, 31; m 59; c 3. BIOCHEMISTRY. *Educ:* Iowa State Univ, BS, 53; Ore State Univ, MS, 55; Univ Calif, Berkeley, PhD(comp biochem), 61. *Prof Exp:* NIH fel, Max Planck Inst Biol, Tübingen, Ger, 62-65; from instr to assoc prof, 65-80, PROF BIOCHEM, UNIV UTAH, 80- *Mem:* Fedn Am Soc Exp Biol; Am Chem Soc; AAAS. *Res:* Biochemistry of the cell nucleus; chromosomal proteins. *Mailing Add:* Dept Biochem Sch Med Univ Utah Salt Lake City UT 84112

KUEHLER, JACK D, ELECTRON OPTICS. *Educ:* Santa Clara Univ, BS & MS. *Prof Exp:* Assoc engr, San Jose, Calif Res Lab, IBM Corp, 58-67, dir, Raleigh Commun Lab, NC, 67-70, dir, San Jose & Menlo Park Develop Labs, 70-72, vpres, Gen Prod Div, 72-74, vpres develop, 74-77, pres, Syst Prod Div, 78-80, IBM vpres & pres, Gen Technol Div, 80-81, sr vpres, 82, IBM vchmn bd, 88-89, PRES & MEM EXEC COMT, IBM CORP, 89- *Concurrent Pos:* Asst group exec, systs develop, Data Processing Prod Group, 77-78, Infor Systs & Tech Group exec, 81, mem corp mgmt bd & US mfg, 85, mem bd dir, 86, exec vpres, 87; mem bd dir, Olin Corp & Nat Asn Mfrs. *Mem:* Nat Acad Eng; sr mem Inst Elec & Electronics Engrs. *Mailing Add:* IBM Corp Old Orchard Rd Armonk NY 10504

KUEHN, GLENN DEAN, b Terry, Mont, Apr 13, 42; m 65; c 1. BIOCHEMISTRY, MOLECULAR BIOLOGY. *Educ:* Concordia Col, BA, 64; Wash State Univ, PhD(biochem), 68. *Prof Exp:* NIH fel, Univ Calif, Los Angeles, 68-70; from asst prof to assoc prof, 70-75, PROF CHEM, NMEX STATE UNIV, 80- *Concurrent Pos:* Am Cancer Soc res support award, 71-76 & 81-83; NSF res support award, 73-79; NIH res support award, 74-77, 77-80, 80-85, 85-89, 90-95, USDA res support award, 86-91, USGS res support, 87-92; Roche Found Fel, Univ Berne, Switz, 79. *Mem:* AAAS; Am Chem Soc; Fedn Am Soc Exp Biol; Am Soc Microbiol; Soc Adv Chicanos & Nat Am Sci. *Res:* Biosynthesis and regulatory functions of polyamines; carbon dioxide fixation in autotrophs; regulatory enzymology; biochemical mechanisms of drought and heat tolerance in plants. *Mailing Add:* Grad Prog Molecular Biol Box 3MLS NMex State Univ Las Cruces NM 88003-0001

KUEHN, HAROLD HERMAN, b Orange, NJ, June 17, 27; m 55; c 2. MICROBIOLOGY, MYCOLOGY. *Educ:* Univ Mont, AB, 50; Univ Ill, MS, 52, PhD(bot), 54. *Prof Exp:* Asst prof biol, NMex Highlands Univ, 54-55; sr res mycologist, Grain Processing Corp, Iowa, 58-61; res mycologist, Campbell Soup Co, NJ, 61; res microbiologist, Rohm and Haas Co, Pa, 61-70; PROF BIOL, MERCER COUNTY COMMUNITY COL, 70- *Mem:* Bot Soc Am; Mycol Soc Am. *Res:* Industrial mycology; Gymnoascaceae; soil fungi; water molds; mucorales; fungal enzymes. *Mailing Add:* Dept Sci Mercer County Community Col PO Box B Trenton NJ 08690

KUEHN, JEROME H, b Minneapolis, Minn, July 20, 20; m 45; c 3. FISH BIOLOGY, FISHERIES ADMINISTRATION. *Educ:* Univ Minn, BS, 42, MS, 49. *Prof Exp:* Aquatic biologist aide, 46, aquatic biologist, 47-52, asst wildlife projs coordr, 52-56, supvr, Survs & Inventories Unit, 56-66, natural resource planning dir, 66-79, CHIEF FISHERIES, STATE NATURAL RESOURCES DEPT, MINN, 79- *Mem:* Am Fisheries Soc. *Res:* Techniques of fisheries survey procedures; development of fisheries management investigations; fish toxicants; natural resource planning; environmental impact review; water resources planning. *Mailing Add:* 120 Park Ave St Paul MN 55115

KUEHN, LORNE ALLAN, b Sault Saint Marie, Ont, Jan 28, 43; c 2. BIOPHYSICS. *Educ:* Univ Alta, BSc, 63; York Univ, PhD(physics), 68. *Prof Exp:* Defence serv sci officer physics, Defence Res Estab Toronto, 66-71; dir, biosci div, Defence & Civil Inst Environ Med, 71-82; dir, res & develop, human performance, 82-86, DIR, SCI & TECH INTEL, DEPT NAT DEFENSE, 86- *Concurrent Pos:* Mem, proj rev group develop pneumatic decompression comput, Dept Indust, Trade & Com, 71-72; tech secy, Adv Panel Arctic Med & Climatic Physiol, Defence Res Bd, 71-75; Can observer, Human Biol Working Party, Sci Comt for Anarctic Res, 75-; chmn, NAm Affairs Comt, Under Sea Med Soc, 80-81; SAS exp mgr, Can Astronaut Prog Off, 83-86. *Mem:* Arctic Inst NAm; Can Physiol Soc; Can Asn Physicists; assoc fel Aerospace Med Asn; Undersea Med Soc; assoc fel Aerospace Med Soc. *Res:* Development of decompression theories and operational computers; development of measuring techniques in environmental stress, particularly in extreme heat and extreme cold; diagnosis and treatment of accidental and occupational hypothermia; defense against biological and chemical warfare. *Mailing Add:* Sci Tech Intel Dept Nat Defense 101 Col-By Dr Ottawa ON K1A 0K2 Can

KUEHNE, DONALD LEROY, b Oak Park, Ill, Jan 24, 52. CHEMICAL ENGINEERING. *Educ:* Cornell Univ, BS, 73; Calif Inst Technol, MS, 75, PhD(chem eng), 79. *Prof Exp:* res engr, Chevron Oil Field Res Co, Chevron Corp, 78-88; SR RES ENGR, CHEVRON RES & TECHNOL COM, RICHMOND, CALIF, 88- *Mem:* Am Inst Chem Engrs; Soc Petrol Engrs. *Res:* Development and reservoir applications of enhanced oil recovery chemicals; technology for improving sweep efficiency in oil production. *Mailing Add:* 2309 N Maplewood St Orange CA 92665-3516

KUEHNE, MARTIN ERIC, b Floral Park, NY, May 29, 31; m 53; c 1. ORGANIC CHEMISTRY. *Educ:* Columbia Univ, AB, 52, PhD(chem), 56; Harvard Univ, MA, 53. *Prof Exp:* Sr chemist, Chem Res Dept, Ciba Pharmaceut Prod, Inc, 55-61; from asst prof to assoc prof & Sloan fel, 61-68, chmn dept, 76-78, PROF CHEM, UNIV VT, 68- *Mem:* Am Chem Soc. *Res:* Synthetic and degrative problems in natural products; general organic chemistry; medicinal chemistry. *Mailing Add:* Redstone Park 169 S Cove Rd Burlington VT 05401

KUEHNE, ROBERT ANDREW, b Austin, Tex, June 22, 27; m 55; c 4. ZOOLOGY. *Educ:* Southern Methodist Univ, BS, 49, MS, 50; Univ Mich, PhD(zool), 58. *Prof Exp:* Aquatic biologist, State Game & Fish Comn, Tex, 50-53; instr zool, Univ Mich, 57-58; from instr to asst prof, 58-66, ASSOC PROF ZOOL, UNIV KY, 66- *Concurrent Pos:* Sci fac fel, Freshwater Biol Asn, Eng, 65-66; vis prof, Hebrew Univ Jerusalem, 71. *Mem:* AAAS; Am Fisheries Soc; Am Soc Ichthyologists & Herpetologists; Ecol Soc Am. *Res:* Ecology, behavior and taxonomy of freshwater fishes; stream ecology. *Mailing Add:* Dept Biol Univ Ky Lexington KY 40506

KUEHNER, CALVIN CHARLES, b Put-in-Bay, Ohio, Dec 12, 22; m 47; c 3. MEDICAL MICROBIOLOGY, PUBLIC HEALTH. *Educ:* Ohio State Univ, BS, 49, MS, 50, PhD(mycol), 53. *Prof Exp:* Zymologist, Fermentation Div, Northern Regional Res Lab, Peoria, Ill, 51-54; asst prof microbiol, Univ Detroit, 54-58; assoc prof biol, Univ Windsor, 58-67; prof & chmn dept, St Dominic Col, 67-69; prof microbiol, Moraine Valley Community Col, 69-75; dir, Basic Sci Div, 75-89, PROF MICROBIOL & PUB HEALTH, NAT COL CHIROPRACTIC, 75- *Concurrent Pos:* Consult microbiol, Palos Med Labs, Palos Heights, Ill, 74-76; assoc ed, J Manipulative & Physiol Therapeut, 78- *Mem:* Am Inst Biol Sci; Am Pub Health Asn. *Res:* Teaching of general and medical micrbiology; community health problems. *Mailing Add:* Nat Col Chiropractic 200 E Roosevelt Rd Lombard IL 60148

KUEHNER, JOHN ALAN, b Lennoxville, Que, Oct 8, 31; div; c 3. NUCLEAR PHYSICS. *Educ:* Bishop's Univ, BSc, 51; Queen's Univ, Ont, MA, 54; Univ Liverpool, PhD(physics), 56. *Prof Exp:* Res officer, Chalk River Nuclear Labs, 56-66; PROF PHYSICS, MCMASTER UNIV, 66- *Mem:* Am Phys Soc; Can Asn Physicists; Royal Soc Can. *Res:* Nuclear structure studies using reactions induced with accelerated ion beams. *Mailing Add:* Dept Physics McMaster Univ Hamilton ON L8S 4K1 Can

KUEHNERT, CHARLES CARROLL, b Springdale, Ark, Nov 21, 30; m 68. PLANT MORPHOGENESIS. *Educ:* Mankato State Col, BA, 53; Purdue Univ, MS, 55, PhD(bot), 59. *Prof Exp:* Asst bot, Purdue Univ, 53-58, res asst, 59; Nat Res Coun Can fel & res assoc biol, Univ Sask, 60-62; res assoc, Brookhaven Nat Lab, 62-64; from asst prof to assoc prof bot, 64-71, ASSOC PROF BIOL, SYRACUSE UNIV, 71- *Concurrent Pos:* Res collabr, Brookhaven Nat Lab, 64-67, vis prof, 65-66, asst botanist, 70-71; adv except undergrad, NSF undergrad res partic prog, Syracuse Univ, 66-67 & adv except sec sch students, NSF Pre-Col Studies Ctr, 68 & 71; sci consult, L W Singer Publ Co, 67; organizing co-chmn, Int Conf Dynamics Meristem Cell Pop, Univ Rochester, 71. *Mem:* Bot Soc Am. *Res:* Dynamics of meristem cell populations; cell biology; cell population kinetics; developmental and experimental morphology. *Mailing Add:* Dept Biol Syracuse Univ 130 College Pl Syracuse NY 13244

KUEKER, DAVID WILLIAM, b Denver, Colo, Dec 14, 43; div. LOGIC. *Educ:* Univ Calif, Los Angeles, BA, 64, MA, 66, PhD(math), 67. *Prof Exp:* Actg asst prof math, Univ Calif, Los Angeles, 67-68; Hildebrandt res instr, Univ Mich, Ann Arbor, 68-70, asst prof, 70-73; asst prof, 73-76, assoc prof math, 76-84, PROF MATH, UNIV MD, COLLEGE PARK, 84- *Mem:* Am Math Soc; Asn Symbolic Logic. *Res:* Mathematical logic, especially model theory for finitary, infinitary and other non-classical languages. *Mailing Add:* Dept Math Univ Md College Park MD 20742

KUELLMER, FREDERICK JOHN, b Chicago, Ill, Mar 28, 24; m 48; c 4. COAL GEOCHEMISTRY, PETROLOGY. *Educ:* Univ Chicago, SB, 48, SM, 49, PhD(geol), 52. *Prof Exp:* Geologist, NMex Bur Mines & Mineral Resources, 52-64; prof geol & head dept, Univ Ill, Chicago Circle, 64-66; sr geologist, 66-74, vpres acad affairs, 66-76, actg chmn, Geosci Dept, 80-81, grad dean, 84-85, PROF GEOL, NMEX INST MINING & TECHNOL, 66-; VPRES, RES & ECON DEVELOP, 86- *Concurrent Pos:* Prof geol, Swiss Fed Inst Technol, 56-64, NSF res fel, 59-60; chmn Coal Div, Geol Soc Am, 87-88. *Mem:* Fel Geol Soc Am; fel Mineral Soc Am; Geochem Soc; Am Geophys Union; Sigma Xi. *Res:* Coal geochemistry and petrology; trace elements in coal; high pressure mineral equilibria; tertiary igneous rocks and porphysics. *Mailing Add:* NMex Tech Socorro NM 87801

KUEMMEL, DONALD FRANCIS, b Milwaukee, Wis, Dec 27, 27; m 49; c 3. ANALYTICAL CHEMISTRY. *Educ:* Marquette Univ, BS, 50, MS, 52; Purdue Univ, PhD(anal chem), 56. *Prof Exp:* Chemist, Allis-Chalmers Mfg Co, Wis, 51-53; res chemist, Procter & Gamble Co, 55-88; RETIRED. *Mem:* Am Chem Soc; Am Oil Chem Soc. *Res:* Chromatography; separations. *Mailing Add:* 3367 Nandale Dr Cincinnati OH 45239

KUEMMERLE, NANCY BENTON STEVENS, b Marshall, Minn, Mar 6, 48; m 71. MOLECULAR GENETICS, DNA REPAIR MECHANISMS. *Educ:* Univ Iowa, BS, 70; Univ Tex, MS, 75; Univ Tenn, PhD, 87. *Prof Exp:* Microbiol technician, Ill State Univ, 72-73; microbiol technologist, Med Ctr, Peoria Sch Med, Univ Ill, 73-74; sr res asst, Univ Tenn-ERDA Comp Animal Res Lab, Oak Ridge, 74-76; res assoc, biol div, Oak Ridge Nat Lab, 76-82; res assoc fel, Dept Molecular Biol, Princeton Univ, 87-90; RES ASSOC, DEPT BIOL, TUFTS UNIV, MEDFORD, MASS, 90- *Mem:* Am Chem Soc; Am Soc Microbiol; AAAS; Asn Women Sci. *Res:* Molecular genetics and biochemical dissection of mouse spermatogenesis; investigation of expression and the effects of genetic regions on fertility and sterility by cloning into yeast artificial chromosome-vectors and characterization by pulsed-field gel electrophoresis. *Mailing Add:* One Stonehill Dr Apt 2JF Stoneham MA 02180

KUENHOLD, KENNETH ALAN, b Cleveland, Ohio. GENERAL PHYSICS, ENGINEERING PHYSICS. *Educ:* Cornell Univ, BEP, 64; Ohio State Univ, PhD(physics), 73. *Prof Exp:* Asst prof, 73-76, ASSOC PROF PHYSICS, UNIV TULSA, 76-, CHMN ENG PHYSICS, 77- *Concurrent Pos:* Adj res partic, Oak Ridge Assoc Univs, 73-78. *Mem:* Am Phys Soc; Am Asn Physics Teachers; Sigma Xi. *Res:* Two-phase oil and gas instrumentation and measurement. *Mailing Add:* Dept Physics NCj 245 Univ Tulsa 600 S College Tulsa OK 74104

KUENZEL, WAYNE JOHN, b Philadelphia, Pa, Jan 22, 42; c 2. POULTRY PHYSIOLOGY, ORNITHOLOGY. *Educ:* Bucknell Univ, BS, 64, MS, 66; Univ Ga, PhD(zool), 69. *Prof Exp:* NIH fel neurophysiol, Cornell Univ, 71-73, res assoc, 73-74; asst prof poultry sci, 74-78, assoc prof physiol, 78-84, PROF PHYSIOL, UNIV MD, COLLEGE PARK, 84- *Concurrent Pos:* Sabbatical leave, Scotland, 81; Fulbright-Hays sr res fel award, Gt Brit. *Mem:* AAAS; World's Poultry Sci Asn; Poultry Sci Asn; Am Ornithologists Union; Am Soc Zoologists; Soc Neurosci. *Res:* Avian physiology; regulation of food and water intake; neuroanatomy; neurobiology; neural control of precocious puberty; neurosciences. *Mailing Add:* 6829 Pineway Hyattsville MD 20792

KUENZI, NORBERT JAMES, b Beaver Dam, Wis, Aug 5, 35; m 60; c 5. MATHEMATICS. *Educ:* Wis State Univ-Eau Claire, BS, 59; Univ Ill, Urbana, MA, 63; Univ Iowa, PhD(statist), 69. *Prof Exp:* Teacher high sch, Wis, 59-62; asst prof, 64-66 & 69-70, assoc prof, 70-80, PROF MATH, UNIV WIS-OSHKOSH, 80-, CHAIRPERSON DEPT, 76- *Mem:* Inst Math Statist; Am Statist Asn; Math Asn Am. *Res:* Probability theory; mathematical statistics. *Mailing Add:* Dept Math Univ Wis-Oshkosh 800 Algoma Blvd Oshkosh WI 54901

KUENZI, W DAVID, geology; deceased, see previous edition for last biography

KUENZLER, EDWARD JULIAN, b West Palm Beach, Fla, Nov 11, 29; m 65; c 2. WETLAND ECOLOGY, ENVIRONMENTAL NUTRIENT CYCLING. *Educ:* Univ Fla, BS, 51; Univ Ga, MS, 53, PhD(ecol), 59. *Prof Exp:* From res asst to res assoc marine biol, Woods Hole Oceanog Inst, 59-64, assoc scientist, 64-65; assoc prof environ sci & eng, 65-70, prof bot, 69-75, PROF ENVIRON SCI & ENG, UNIV NC, CHAPEL HILL, 70-, PROF MARINE SCI, 68- *Concurrent Pos:* Prog dir biol oceanog, NSF, Washington, DC, 71-72; prog area dir, Environ Chem & Biol, 80-83, Aquatic & Atmospheric Sci, 90-; dep chmn, Environ Sci Eng, 84-87; chmn, Univ NC marine sci curric, 68-73, prof ecol, 71- *Mem:* Estuarine Res Fedn; Ecol Soc Am; Am Soc Limnol & Oceanog; Soc Wetland Scientists. *Res:* Ornithology; spider populations; energy and nutrient flow through marine animal populations; nutrition of marine and freshwater phytoplankton; ecology of estuaries; aquatic and wetland ecology. *Mailing Add:* Dept Environ Sci & Eng Univ NC CB No 7400 Chapel Hill NC 27599-7400

KUEPER, THEODORE VINCENT, b Dubuque, Iowa, Aug 13, 41; m 63; c 3. EXPERIMENTAL STATISTICS. *Educ:* Iowa State Univ, BS, 63. *Prof Exp:* Scientist biochem, Beatrice Foods Co, 63, statistician exp statist, 63-67, head statist, 67-70, res mgr sci serv, 70-74, res mgr new prod develop, 74-78, dir formulated food res, 78-81, dir res & develop qual assurance grocery foods, 81-83, dir res & develop qual assurance cheese, 83-84, vpres qual assurance cheese, 84-87; consult & high sch chem teacher, 87-89; PRES, WIS DATA LAB, LTD, 89- *Res:* New food products, both consumer and industrial; statistical analysis, experimental design and computer applications in regard to food research and development; prediction of food sensitivities for patients of health practitioners. *Mailing Add:* 415 Cymric Ct Wales WI 53183

KUESEL, THOMAS ROBERT, b Richmond Hill, NY, July 30, 26; m 59; c 2. BRIDGES, TUNNELS. *Educ:* Yale Univ, BEng, 46, MEng, 47. *Prof Exp:* Mem staff, Parsons, Brinckerhoff, Quade & Douglas, NY, 47-63, proj mgr, San Francisco, 67-68, partner & sr vpres, 68-83, dir, 68-90, chmn bd, 83-91, EMER CHMN BD, PARSONS, BRINCKERHOFF, QUADE & DOUGLAS, NY, 91-; CONSULT ENGR, 91- *Concurrent Pos:* Asst mgr eng, Parsons Brinckerhoff-Tudor-Bechtel, San Francisco, 63-67; vchmn, OECD Tunneling Conf, Washington, DC, 70; mem, US Nat Comt Tunneling Technol, 72-74; chmn, Geotech Bd, Nat Res Coun, 87- *Honors & Awards:* Clemens Herschel Award, Boston Soc Civil Eng, 87; Ernest E Howard Award, Am Soc Civil Eng, 88. *Mem:* Nat Acad Eng; fel Am Soc Civil Eng; fel Am Consult Engrs Coun; Int Asn Bridge, Struct Eng; Brit Tunneling Soc. *Res:* Structural engineering; designer of 120 bridges and 135 tunnels world wide. *Mailing Add:* Five Wood Lane Charlottesville VA 22901

KUETHER, CARL ALBERT, b Ripon, Wis, Oct 15, 15; m 39; c 2. BIOCHEMISTRY. *Educ:* Miami Univ, AB, 36; Wayne State Univ, MS, 40; George Washington Univ, PhD(biochem), 43. *Prof Exp:* Asst chem, Miami Univ, 33-36 & Oberlin Col, 36-38; from asst to instr biochem, George Washington Univ, 40-43; from instr to sr instr, Western Reserve Univ, 43-46; asst prof, Univ Wash, 46-51; biochemist, Res Labs, Eli Lilly & Co, 51-60; assoc prof chem, Youngstown Univ, 60-62; assoc prog dir metab biol, NSF, 62-65; PROG ADMINR, NAT INST GEN MED SCI, 65- *Mem:* AAAS; Am Chem Soc. *Res:* Biosciences research administration. *Mailing Add:* Pharmacol Sci Prog Nat Inst Gen Med Sci Bethesda MD 20892

KUETTNER, KLAUS E, b Bunzlau, Ger, June 25, 33; m 75. BIOCHEMISTRY. *Educ:* Univ Freiburg, MS, 58; Univ Berne, PhD(pharmaceut chem), 61. *Prof Exp:* Res assoc biochem, Ciba Pharmaceut Co, Switz, 61-62; fel, Div Biol & Med Res, Argonne Nat Lab, 62-64; from instr to asst prof biol chem, Univ Ill Col Med, 64-65; res assoc, dept orthop surg, Presbyterian-St Luke's Hosp, 64-66, assoc biochemist, 66-79, asst prof, dept biochem, Rush Med Col, 71-72, assoc prof, dept orthop surg & biochem, Rush Col Health Sci & Rush Med Col, 72-77, PROF, DEPT BIOCHEM & ORTHOP SURG, RUSH MED COL, RUSH PRESBYTERIAN-ST LUKE'S CTR, SR BIOCHEMIST, DEPT ORTHOP SURG, 79-, SCI STAFF, 66-, CHMN, DEPT BIOCHEM, 80-; PROF, COOK COUNTY GRAD SCH MED, 77- *Concurrent Pos:* Ed adv bd, J Orthop Res, 81-; adv comt mem on res, Am Acad of Orthopaedic Surgeons, 86; organizer & chmn, Int Workshop Conf on Articular Cartilage Biochem, Wesbaden, Fed Repub of Ger, 87. *Honors & Awards:* Kappa Delta Award, Am Acad Orthop Surgeons & Orthop Res Soc; Carol Nachman, Mainz, Fed Repub Ger, 87. *Mem:* Soc Complex Carbohydrates; Orthop Res Soc; Int Asn Dent Res; Am Soc Biol Chemists; Am Soc Cell Biol; Am Chem Soc; Am Soc Biol Chemist; Am Rheumatism Asn; Am Soc Bone & Mineral Res; Am Asn Advan Sci; Soc Biol Chem; NY Acad Sci; Sigma Xi. *Res:* Biochemistry of connective tissue; biochemical changes during cartilage calcification, vascular and tumor invasion; bone formation and development. *Mailing Add:* Dept Biochem Rush Presbyterian St Luke Med Ctr 1753 W Congress Pkwy Chicago IL 60612

KUFF, EDWARD LOUIS, b Baltimore, Md, June 1, 24; m 47; c 1. MOLECULAR BIOLOGY, RETROVIRUSES. *Educ:* Johns Hopkins Univ, AB, 43, MD, 47; Washington Univ, PhD(cytol), 52. *Prof Exp:* Intern med, Barnes Hosp, Washington Univ, 47-48, instr anat, Sch Med, 48-52; head tumor-host rels sect, Lab Biochem, 65-68, actg chief, Lab Biochem, 69-73, MED OFFICER, USPHS, NAT CANCER INST, 52-, HEAD BIOSYNTHESIS SECT, 68-, DEP CHIEF, LAB BIOCHEM, 81- *Concurrent Pos:* Vis scientist, Virol Sect, Weizmann Inst Sci, Rehovot, Israel, 74-75. *Mem:* Am Soc Biol Chemists; Am Asn Cancer Res. *Res:* Biochemical basis of cell structure; protein and nucleic acid biosynthesis; molecular biology of endogenous retroviruses. *Mailing Add:* Lab Biochem Nat Cancer Inst NIH Bethesda MD 20892

KUFFNER, ROY JOSEPH, surface chemistry; deceased, see previous edition for last biography

KUFTINEC, MLADEN M, b Zagreb, Yugoslavia, Apr 18, 43; US citizen; div; c 1. ORTHODONTICS, NUTRITION. *Educ:* Univ Sarajevo, Yugoslavia, DStom, 65; Harvard Sch Dent Med, cert orthod, 68, DMD, 72. *Hon Degrees:* ScD, Mass Inst Technol, 71. *Prof Exp:* Instr nutrit & health, Cambridge Ctr Adult Educ, 71; res assoc & instr, Mass Inst Technol, 71-72; staff assoc orthod, Forsyth Dent Ctr, Boston, 72; assoc prof, Va Commonwealth Univ, 72-76; prof & chmn orthod, Sch Dent, Univ Louisville, 76-87, chmn growth & spec care, 87-90; PROF & CHMN, ORTHOD, NY UNIV, 90- *Concurrent Pos:* Consult, Cranio Facial Anamalies Team, Louisville, 76- & Nat Bd Dent Examr, 77-80; reviewer, J Dent Educ, 76-81, J Dent Res, 77-80 & J Am Orthod, 79- *Mem:* Am Asn Orthod; Am Dent Asn; Int Asn Dent Res; Nutrit Today Soc. *Mailing Add:* One Washington Sq Village Apt O-10 New York NY 10012

KUGEL, HENRY W, b 1940; US citizen. NUCLEAR FUSION RESEARCH. *Educ:* Canisius Col, BS, 62; Univ Notre Dame, PhD(physics), 67. *Prof Exp:* Res assoc nuclear physics, Univ Notre Dame, 67; res assoc, Univ Wis, 68-70; res fel nuclear & atomic physics, Rutgers Univ & Bell Lab, 70-72; asst prof atomic physics, Rutgers Univ, 72-78; PRIN RES PHYSICIST, PRINCETON PLASMA PHYSICS LAB, 78- *Mem:* Am Phys Soc; AAAS; Sigma Xi. *Res:* Neutral beam operations; PBX-M tokamak operation; neutral beam diagnostics; radiological studies; plasma surface interactions. *Mailing Add:* Princeton Plasma Physics Lab PO Box 451 Princeton NJ 08544

KUGEL, ROBERT BENJAMIN, b Chicago, Ill, May 2, 23; m 50; c 4. PEDIATRICS, ACADEMIC ADMINISTRATION. *Educ:* Univ Mich, AB, 45, MD, 46. *Hon Degrees:* Brown Univ, MS, 64. *Prof Exp:* Intern, Univ Hosp, Univ Mich, 47, resident pediat, 48-50; Commonwealth Fund fel, Child Study Ctr, Yale Univ, 50-52; instr, Univ, 52-53; res assoc, Sch Hyg & Pub Health & asst prof pediat, Johns Hopkins Univ, 55-56; assassoc prof, Col Med, Univ Iowa & dir child develop clin, Univ Hosp, 56-63; from prof med sci to prof child health, Brown Univ, 63-66; found prof pediat, 66-69, chmn dept pediat, 66-69; prof pediat & dean Col Med, Univ Nebr, Omaha, 69-74; prof pediat & vpres health sci, Univ NMex Health Sci ctr, 74-76; exec vchancellor, Univ Kans Med Ctr, 76-77; vpres community health plan, Georgetown Univ, Washington, DC, 77-80; MED DIR, CARDINAL COOKE HEALTH CARE CTR, NEW YORK, NY, 81- *Concurrent Pos:* Dir sch health, Baltimore Health Dept, Md, 55-56; mem, President's Comn on Ment Retardation, 66-70; consult, State Hosp & Sch, Woodward, Iowa; Ment Retardation Br, Div Hosp & Med Facilities, USPHS, Health Res Facilities Br, NIH & US Children's Bur; chief admin officer, Bernalillo County Med Ctr, Albuquerque, NMex, 74-76. *Honors & Awards:* Mildred Thomson Award, 73. *Mem:* Soc Res Child Develop; fel Am Asn Ment Deficiency; fel Am Acad Pediat; Am Pediat Soc; Am Col Physicians; AMA. *Res:* Child development; medical ecology; mental retardation. *Mailing Add:* 130th Station Hosp EFMS APO NY 09102

KUGELMAN, IRWIN JAY, b Brooklyn, NY, Feb 15, 37; m 58; c 4. ENVIRONMENTAL ENGINEERING. *Educ:* Cooper Union, BCE, 58; Mass Inst Technol, SM, 60, ScD(civil eng),63. *Prof Exp:* Asst prof civil eng, NY Univ, 62-65; res scientist process res, Am-Standard Corp, 65-70; res sanit engr, Phys & Chem Res Sect, Advan Waste Treatment Prog, 70-74, chief pilot & field eval, Munic Environ Res Lab, US Environ Protection Agency, 74-81; dir, Ctr Marine & Environ Syst, 81-88, CHMN, CIVIL ENG DEPT, LEHIGH UNIV, 85- *Concurrent Pos:* Sci Adv Comt, Univ Ill, 80-82 & 86-90 & Notre Dame, 80-82; Peer Review Panel, US Environ Protection Agency, 80-; Water Pollution Mgt Comt, Am Soc Civil Engrs, 85; New Jersey Govnrs Sci Adv Comt, 85-; consult, Nat Acad Sci, Potomic Estuary Study, 87-88; vice chair, Water Pollution Control Fedn Res Comt, 88-; dept chair coun, Am Soc Civil Engrs, 90- *Honors & Awards:* Bronze Medal, US Environ Protection Agency, 81. *Mem:* Sigma Xi; Am Soc Civil Engrs; Am Waterworks Asn; Water Pollution Control Fedn; Air Pollution Control Asn; Int Asn Water Pollution Res; Am Chem Soc. *Res:* Treatment of water and wastes; hazardous waste management. *Mailing Add:* Civil Eng Dept Fritz Lab Lehigh Univ Bethleham PA 18105

KUGLER, GEORGE CHARLES, b Philadelphia, Pa, Oct 26, 41. ANALYTICAL CHEMISTRY. *Educ:* La Salle Univ, BA, 63; Univ Pa, PhD(chem), 67; Fla Atlantic Univ, MBA, 78. *Prof Exp:* Lab head, ESB-Exide, 69-80, dir mkt, 80-84; prin scientist, Leeds & Northrup, 84-85; res dir, Integrated Ionics, 85-86; DIR RES & ENG, MINE SAFETY APPLIANCES CO, 87- *Mem:* Electrochem Soc; Am Chem Soc; Instrument Soc Am. *Mailing Add:* 216 Woodcroft Rd Baden PA 15005

KUGLER, LAWRENCE DEAN, b Orange, Calif, Feb 18, 41; m 62; c 2. GENERAL MATHEMATICS. *Educ:* Calif Inst Technol, BS, 62; Univ Calif, Los Angeles, MA, 65, PhD(math), 66. *Prof Exp:* PROF MATH, UNIV MICH-FLINT, 66- *Concurrent Pos:* NSF grants, 67-71. *Mem:* Am Math Soc; Math Asn Am. *Res:* Application of nonstandard analysis to the theory of almost periodic functions; division algebra. *Mailing Add:* Dept Math Univ Mich Flint MI 48502

KUH, ERNEST SHIU-JEN, b Peking, China, Oct 2, 28; nat US; m 57; c 2. ELECTRICAL ENGINEERING. *Educ:* Univ Mich, BS, 49; Mass Inst Technol, SM, 50; Stanford Univ, PhD, 52. *Prof Exp:* Mem tech staff, Bell Tel Labs, NJ, 52-56; assoc prof, Univ Calif, Berkeley, 56-62, Miller res prof, 65-66, chmn, Dept Elec Eng & Computer Sci, 68-72, dean Col Eng, PROF ELEC ENG, UNIV CALIF, BERKELEY, 62-, WILLIAM S FLOYD PROF ENG, 90- *Concurrent Pos:* Consult, Res Lab, Int Bus Mach Corp, 57-62; NSF sr fel, 62-63; mem adv panel elec sci, NSF, 76-77, vis comt, Gen Motors Inst, Sci Adv Bd, Mills Col & Peer Rev Panel, Nat Bur Standards; Alexander von Humboldt Award, 77; mem adv panel eng NSF, 79-; hon prof, Shanghai Jiao Tong Univ, 79, Tsinghua Univ, 85 & Tianiin Univ, 85; Brit Sci & Eng fel, 81; mem vis comt, Elec Eng & Comput Sci Dept, Mass Inst Technol; adv coun, Elec Eng Dept, Princeton Univ; bd counr, Sch Eng, USC. *Honors & Awards:* Guillemin-Cauer Award, Inst Elec & Electronics Engrs, 73; Lamme Award, Am Soc Eng Educ, 81; Educ Medal, Inst Elec & Electronics Engrs, 81, Centennial Medal, 83; Circuits & Systs Soc Award, 88. *Mem:* Nat Acad Eng; fel Inst Elec & Electronics Engrs; Acad Sinica; fel AAAS. *Res:* Network system theory and computer-aided design in microelectronics. *Mailing Add:* Dept Elec Eng & Comput Sci Univ Calif Berkeley CA 94720

KUHAJEK, EUGENE JAMES, b Chicago, Ill, Mar 4, 34; m 60, 70; c 2. CRYSTALLIZATION, ION EXCHANGE. *Educ:* Loyola Univ, Chicago, BS, 55; Univ Minn, PhD(inorg chem), 62. *Prof Exp:* Res chemist, Morton Thiokol, 62-78, MGR RES & DEVELOP, MORTON SALT DIV, MORTON INT, 78- *Mem:* Am Chem Soc; Am Soc Animal Sci; Am Water Works Asn; Int Desalination Asn. *Res:* Product development, product and process improvements relate to sodium chloride and potassium chloride. *Mailing Add:* Morton Salt 1275 Lake Ave Woodstock IL 60098

KUHAR, MICHAEL JOSEPH, b Scranton, Pa, Mar 10, 44; m 69; c 2. NEUROPHARMACOLOGY, NEUROSCIENCE. *Educ:* Univ Scranton, BS, 65; Johns Hopkins Univ, PhD(biophys, pharmacol), 70. *Prof Exp:* Fel psychiat, Sch Med, Yale Univ, 70-72; asst prof, 72-76, assoc prof, 76-81, PROF NEUROSCI, PHARMACOL & PSYCHIAT, SCH MED, JOHNS HOPKINS UNIV, 81-; CHIEF NEUROSCI BR, NAT INST DRUG ABUSE ADDICTION RES CTR, 85- *Honors & Awards:* Daniel H Efron Award, 81; Mathilde Solowey Award, 85. *Mem:* AAAS; Soc Neurosci; Am Soc Neurochem; Am Soc Pharmacol & Exp Therapeut; Am Col Neuropsychopharmacol. *Res:* Interaction of drugs with central nervous system neurotransmitters. *Mailing Add:* Neurosci Br Nat Inst Drug Abuse Addiction Res Ctr PO Box 5180 Baltimore MD 21224

KUHI, LEONARD VELLO, b Hamilton, Ont, Oct 22, 36; nat US; m 60; c 2. ASTROPHYSICS. *Educ:* Univ Toronto, BASc, 58; Univ Calif, Berkeley, PhD(astron), 63. *Prof Exp:* Carnegie fel astron, Mt Wilson & Palomar Observs, 63-65; from asst prof to assoc prof astron Univ Calif, Berkeley, 65-74, chmn dept, 75-76, dean phys sci, 76-82, prof astron, 74-89, dean Col Letters & Sci, 82-89; SR VPRES ACAD AFFAIRS & PROVOST, UNIV MINN, 89- *Concurrent Pos:* Foreign prof, Col de France, Paris, 72-73; vis prof, Joint Inst Lab Astrophys, Boulder, 69, Inst d'Astrophysique, Paris, 72-73 & Univ Heidelberg, Landessternwarte, 78; Alexander von Humboldt US sr scientist fel, 80-81. *Mem:* Fel AAAS; Astron Soc Pac (pres, 78-80); Int Astron Union; Am Astron Soc; Royal Astron Soc Can; Sigma Xi. *Res:* Pre-main sequence stellar evolution; extended stellar atmospheres and mass flow problems. *Mailing Add:* Acad Affairs 213 Morrill Hall Univ Minn 100 Church St SE Minneapolis MN 55455-0110

KUHL, DAVID EDMUND, b St Louis, Mo, Oct 27, 29; m 54; c 1. NUCLEAR MEDICINE, RADIOLOGY. *Educ:* Temple Univ, AB, 51; Univ Pa, MD, 55. *Prof Exp:* Asst instr radiol, Sch Med, Univ Pa, 58-61, from instr to prof, 61-76, chief, Nuclear Med Div, Univ Pa Hosp, 63-76, prof eng, Moore Sch Elec Eng, 74-76, vchmn, Dept Radiol, Hosp, 75-76; chief, Div Nuclear Med, Dept Radiol Sci, Sch Med, Univ Calif, Los Angeles, 76-84, assoc dir, Lab Biomed & Environ Sci & chief, Lab Nuclear Med, 76-84, prof radiol sci, 76-86, vchmn dept, 77-86; PROF INTERNAL MED & RADIOL, MED SCH, UNIV MICH, 86-, CHIEF, DIV NUCLEAR MED, 86- *Concurrent Pos:* Mem, Adv Comt, med uses isotopes, US Atomic Energy Comn, Nat Res Coun, 67-79, Comt Radiol, Nat Acad Sci, 67-71 & Radiation Study Sect, NIH, 68-73; chmn, Diagnostic Radiol Comt, Nat Cancer Inst, NIH, 73-77; fel, Coun Circulation, Am Heart Asn, 78; bd trustees, James T Case Radiol Found, 82-86; consult, Outstanding Investr Prog, Nat Cancer Inst, 84-; bd sci councillors, Brain Res Inst, Univ Chicago, 86-; dir, Positron Emission Tomography Ctr, Univ Mich, 86-; Javits neurosci investr award, NIH, 89. *Honors & Awards:* Nuclear Med Pioneer Citation, Soc Nuclear Med, 76; numerous hon lect awards, 77-; Ernst Jung Prize Med, Ernst Jung Found, Ger, 81; William C Menninger Mem Award, Am Col Physicians, 89. *Mem:* Inst Med-Nat Acad Sci; fel Am Col Radiol; Asn Univ Radiol; Int Soc Cerebral Blood Flow & Metabol; Sigma Xi; Soc Nuclear Med; Am Neurol Asn; fel Am Col Nuclear Physicians; Asn Am Physicians. *Res:* Radionuclide imaging. *Mailing Add:* 3955 Waldenwood Lane Ann Arbor MI 48105

KUHL, FRANK PETER, JR, b New York, NY, Oct 28, 35; m 64; c 5. ELECTRICAL ENGINEERING. *Educ:* Columbia Univ, BSEE, 57, MSEE, 58; Yale Univ, MEng, 61, DEng, 63. *Prof Exp:* Engr, Sperry Gyroscope Co, NY, 62-63; sr engr, Missile Systs Div, Digital Systs Dept, Raytheon Co, 63-65; asst prof elec eng, Union Col, NY, 65-67 & US Naval Acad, 68-73; eng specialist, Singer-Kearfott, 74-76; sr engr, Avionics Div, ITT, 76-78; PROJ LEADER, US ARMY ARMAMENT RES & DEVELOP COMMAND, 78- *Concurrent Pos:* Adj prof, Fla Inst Technol, 82-84. *Honors & Awards:* Region I Award, Inst Elec & Electronics Engrs. *Mem:* Sr mem Inst Elec & Electronics Engrs. *Res:* Pattern recognition by use of computers, specifically in the areas of handprinted letters and numbers; polarized radar backscatter of solid objects in free space; video images of airplanes. *Mailing Add:* 64 E Shawnee Trail Wharton NJ 07885

KUHLERS, DARYL LYNN, b Mason City, Iowa, Nov 12, 45; m 76; c 3. ANIMAL BREEDING. *Educ:* Iowa State Univ, BS, 67; Univ Wis, MS, 70, PhD(animal sci & genetics), 73. *Prof Exp:* Asst prof animal sci, Iowa State Univ, 74-78; assoc prof, 78-84, PROF ANIMAL & DAIRY SCI, AUBURN UNIV, 84- *Concurrent Pos:* Res assoc, Univ Wis, 73-74; consult, US Feed Grains Coun, 77, Am Soybean Asn, 82, 85 & 87 Thailand, 86. *Mem:* Am Soc Animal Sci; Coun Agr Tech. *Res:* Swine breeding; genetics; selection and swine production. *Mailing Add:* Animal & Dairy Sci Bldg Auburn Univ Main Campus Auburn AL 36849-4201

KUHLMAN, ELMER GEORGE, b Beaver Dam, Wis, Dec 15, 34; m 61; c 3. PLANT PATHOLOGY. *Educ:* Univ Wis, BS, 56; Ore State Univ, PhD(plant path), 61. *Prof Exp:* Plant pathologist, Southeastern Forest & Range Exp Sta, 61-68, prin plant pathologist, 68-71, supvry plant pathologist, 71-73, PRIN PLANT PATHOLOGIST, FORESTRY SCI LAB, SOUTHEASTERN FOREST & RANGE EXP STA, US FOREST SERV, 73- *Concurrent Pos:* Adj prof plant path, NC State Univ, 75-83; assoc ed, Plant Dis, 78-81; adjunct prof plant path, Univ Ga, 84-; assoc ed, Phytopath, 83-85; assoc ed, Southern J Applied Forestry, 87-90. *Mem:* Mycol Soc Am; Am Phytopath Soc; Soc Am Foresters. *Res:* Ecological studies of soil organisms; epidemiology of pitch canker disease; effect of environment and mycoparasites on sporulation by Cronartium fusiforme; hyperparasites and hypovirulence; taxonomy of Gibberella fujikuroi and Mortierella; resistance to fusiform rust and variation in veridence. *Mailing Add:* Forestry Sci Lab Carlton St Athens GA 30602

KUHLMAN, JOHN MICHAEL, b Akron, Ohio, June 1, 48; m 70; c 4. MECHANICAL ENGINEERING. *Educ:* Case Western Reserve Univ, BS, 70, MS, 73, PhD(eng), 75. *Prof Exp:* From asst prof to assoc prof mech eng, Dept Mech Eng & Mech, Old Dominion Univ, 74-85; assoc prof, 85-87, PROF MECH & AEROSPACE ENG, WVA UNIV, MORGANTOWN, 87- *Concurrent Pos:* NSF grant, 78-80, prin investr, Langley Res Ctr, NASA grants; prin investr, Naval Surface Weapons Ctr, 75-77, 78-83, & 85-90, co-prin investr, 81-84; prin investr, Air Force Wright Aeronaut Lab, 89-91. *Mem:* Am Soc Mech Engrs; Am Soc Eng Educ; assoc fel Am Inst Aeronaut & Astronaut; Sigma Xi. *Res:* Experimental and theoretical fluid mechanics and aerodynamics. *Mailing Add:* Mech & Aerospace Eng Dept WVa Univ Eng Sci Bldg Morgantown WV 26506-6101

KUHLMAN, GEORGE EDWARD, b Bronxville, NY, Apr 7, 42. ORGANIC CHEMISTRY, PHYSICAL CHEMISTRY. *Educ:* City Col New York, BS, 64; Syracuse Univ, MS, 65, PhD(org chem), 68. *Prof Exp:* Proj chemist, 68-71, res chemist, 71-78, staff res chemist, 78-83, SR RES CHEMIST, AMOCO CHEM CORP, 83- *Mem:* AAAS; Am Chem Soc. *Res:* Sulfur chemistry; aromatic acids; hydrocarbon oxidation. *Mailing Add:* Amoco Chem Corp PO Box 400 Naperville IL 60566

KUHLMANN, KARL FREDERICK, b Ogden, Utah, Feb 3, 37; m 77; c 2. BIOPHYSICAL CHEMISTRY, ANALYTICAL DATA SYSTEMS. *Educ:* Johns Hopkins Univ, BA, 59; Univ Utah, PhD(chem), 63. *Prof Exp:* Res fel chem, Harvard Univ, 62-64 & Int Bus Mach fel, Comput Ctr, 63-64; Alumni Res Found fel, Univ Wis, 64-65; asst prof, Dartmouth Col, 65-71; assoc prof, Stanford Univ, 71-73; phys chemist, Life Sci Div, Stanford Res Inst, 73-81; SR DEVELOP ENGR, NELSON ANALYSTS INC, 81- *Mem:* Am Phys Soc. *Res:* Magnetic resonance; structure and relaxation in liquids; drug design; binding of drugs to macromolecules. *Mailing Add:* 1115 Hermosa Way Menlo Park CA 94025

KUHLMANN-WILSDORF, DORIS, b Bremen, Ger, Feb 15, 22; US citizen; m 50. TRIBOLOGY, SLIDING ELECTRICAL CONTACTS. *Educ:* Gottingen Univ, BS, 44, MS, 46, PhD(mat sci), 47. *Hon Degrees:* DSc, Univ Witwatersrand, Johannesburg, SAfrica, 54. *Prof Exp:* Postgrad fel mat sci, Univ Göttingen, Ger, 47-48; postgrad fel physics, Bristol Univ, Eng, 49-50; lectr physics, Univ Witwatersrand, SAfrica, 50-56; from assoc prof to prof metall, Univ Pa, Philadelphia, 57-63; prof eng physics 63-66, UNIV PROF APPL SCI PHYSICS & MAT SCI, UNIV VA, CHARLOTTESVILLE, 66- *Concurrent Pos:* Vis prof physics, Pretoria Univ, SAfrica, 82- *Honors & Awards:* Medal for Excellence in Res, Am Soc Eng Educ, 65 & 66; Heyn Medal, Ger Soc Mat Sci, 88; Achievement Award, Soc Women Engrs, 89; Ragnar Helm Sci Achievement Award, Inst Elec & Electronics Engrs, 91. *Mem:* Fel Am Phys Soc; fel Am Soc Metals; Am Women Engrs; Am Inst Mech Engrs; Am Asn Univ Professors. *Res:* Publications in the areas of crystal defect theory; theory of workhardening and plastic deformation of metals, tribology, electrical sliding contacts. *Mailing Add:* 304 Dept Physics Univ Va Charlottesville VA 22901

KUHLTHAU, A(LDEN) R(OBERT), b New Brunswick, NJ, Apr 29, 21; m 43; c 3. TRANSPORTATION, SYSTEMS ENGINEERING. *Educ:* Wake Forest Col, BS, 42; Univ Va, MS, 44, PhD(physics), 48. *Prof Exp:* Asst physics, Wake Forest Col, 41-42 & Off Sci Res & Develop & Naval Bur Ord contracts, 42-48; asst prof, Univ NH, 48-51; asst dir, Ord Res Lab, Univ Va, 51-54, dir, Res Lab Eng Sci, 54-67, assoc dean, Sch Eng & Appl Sci, 59-67, assoc provost for res, 67-71, pres, Univ Space Res Asn, 69-75, prof aerospace eng, 59-77, prof transp, Dept Civil Eng, 77-86; RETIRED. *Mem:* Am Inst Aeronaut & Astronaut; Transp Res Bd. *Res:* Rarefied gas dynamics; human factors in transportation; air transportation systems. *Mailing Add:* 1817 Meadowbrook Heights Rd Charlottesville VA 22901

KUHN, CEDRIC W, b Milroy, Ind, Dec 23, 30; m 56; c 2. PLANT PATHOLOGY. *Educ:* Purdue Univ, BS, 56, MS, 58, PhD(plant path), 60. *Prof Exp:* Grad asst, Purdue Univ, 56-60; from asst plant pathologist to assoc plant pathologist, 60-65, head dept plant path, 66-68, assoc prof, 68-70, PROF PLANT PATH, UNIV GA, 70- *Mem:* Fel Am Phytopath Soc; Am Soc Virol. *Res:* Plant virus research. *Mailing Add:* Dept Plant Path Univ Ga Athens GA 30602

KUHN, CHARLES, III, b Cambridge, Mass, May 18, 33; m 59; c 3. PATHOLOGY. *Educ:* Harvard Univ, AB, 55; Washington Univ, MD, 59. *Prof Exp:* Intern path, Barnes Hosp, St Louis, Mo, 59-60, from asst resident to resident, 60-62; Am Cancer Soc fel, 62-63, from instr to assoc prof, 65-76, prof path, Sch Med, Wash Univ, 76-87; PROF PATH, SCH MED, BROWN UNIV, 87- *Concurrent Pos:* Vis prof biochem, Univ Manchester, UK, 80-81. *Mem:* Am Soc Cell Biologists; Am Thoracic Soc; Int Acad Path; Am Assoc Path. *Res:* Pulmonary ultrastructure; pulmonary connective tissue; experimental emphysema. *Mailing Add:* Dept Path Mem Hosp 111 Brewster St Pawtucket RI 02860

KUHN, DAISY ANGELIKA, b Heidelberg, Germany, Aug 3, 30. BACTERIOLOGY. *Educ:* Univ Calif, AB, 52; Univ Calif, PhD(microbiol), 60. *Prof Exp:* Asst bact, Univ Calif, 57-59; from instr to assoc prof microbiol, 59-71, PROF BIOL, CALIF STATE UNIV, NORTHRIDGE, 71- *Mem:* AAAS; Am Soc Microbiol; Brit Soc Gen Microbiol; Can Soc Microbiologists. *Res:* Systematics of bacteria; microbial ecology. *Mailing Add:* Dept Biol Calif State Univ Northridge CA 91330

KUHN, DAVID TRUMAN, b Tucson, Ariz, Apr 4, 40; m 68; c 2. GENETICS. *Educ:* Colo State Col, BA, 63; Univ Utah, MS, 65; Ariz State Univ, PhD(zool), 68. *Prof Exp:* Asst prof biol, Creighton Univ, 68-70; asst prof, 70-72, assoc prof, 72-79, PROF BIOL, UNIV CENT FLA, 79- *Concurrent Pos:* Sabbatical, Univ Geneva, Switz, 78; Sabbatical, Ariz State Univ, Tempe, 84. *Mem:* AAAS; Genetics Soc Am; Sigma Xi; Am Genetic Asn. *Res:* Developmental and molecular genetics of Drosophila. *Mailing Add:* Dept Biol Sci Univ Cent Fla Orlando FL 32816

KUHN, HANS HEINRICH, b Uzwil, Switz, Jan 12, 24; m 54; c 2. ORGANIC CHEMISTRY. *Educ:* Swiss Fed Inst Technol, ChemEng, 49. *Hon Degrees:* DSc, Swiss Fed Inst Technol, 54. *Prof Exp:* Res chemist, Dewey & Almy Div, W R Grace & Co, 57-60; group leader textile chem, 60-61, sect leader, 61-65, mgr polymer res, 65-79, SEN SCI, MILLIKEN RES CO, 79- *Concurrent Pos:* Asst to Prof H Hopff, Swiss Fed Inst Technol, 53-57; hon consul Switz for SC & NC. *Mem:* Am Chem Soc; Swiss Chem Soc; Swiss Chem Asn. *Res:* Chemistry of epoxy steroids, aliphatic and aromatic epoxides; polymer chemistry, specifically oriented toward textile applications including electroactive polymers. *Mailing Add:* 176 W Park Dr Spartanburg SC 29301

KUHN, HAROLD WILLIAM, b Santa Monica, Calif, July 29, 25; m 49; c 3. MATHEMATICS. *Educ:* Calif Inst Technol, BS, 47; Princeton Univ, MA, 48, PhD(math), 50. *Prof Exp:* Fine instr math, Princeton Univ, 49-50; Fulbright res scholar, dept sci, Univ Paris, 50-51; lectr, Princeton Univ, 51-52; from asst prof to assoc prof, Bryn Mawr Col, 52-59; assoc prof, 59-63, PROF MATH & ECON, PRINCETON UNIV, 63- *Concurrent Pos:* Exec secy, div math, Nat Acad Sci-Nat Res Coun, 57-58 & 59-61; NSF fel & vis mem, London Sch Econ, 58-59 & 71-72; sr consult, Mathematica, Inc, 61-; mem adv comt, Army Res Off, 62-65 & div math, Nat Res Coun, 63-65 & 69-71; NSF fel, Univ Rome, 65-66. *Mem:* Am Math Soc; Soc Indust & Appl Math (pres, 53-54); fel Economet Soc; Math Asn Am. *Res:* Mathematical economics; mathematical programming; combinatorial problems. *Mailing Add:* Dept Math Fine Hall Box 37 Princeton Univ Washington Rd Princeton NJ 08544

KUHN, HOWARD A, b Pittsburgh, Pa, Dec 6, 40; m 62; c 4. MANUFACTURING ENGINEERING, METALLURGICAL ENGINEERING. *Educ:* Carnegie-Mellon Univ, BS, 62, MS, 63, PhD(mech eng), 66. *Prof Exp:* Instr mech eng, Carnegie-Mellon Univ, 65-66; from asst prof to assoc prof metall eng, Drexel Univ, 66-74; from assoc prof to prof, 75-89, ADJ PROF METALL ENG & MECH ENG, UNIV PITTSBURGH, 89-; TECH VPRES, METALWORKING TECHNOL INC, 89- *Concurrent Pos:* Consult, Appl Res Lab, US Steel Corp, 63-65, RCA Defense Electronic Prod, 66 & Metals Lab, TRW, Inc, 67. *Mem:* Fel Am Soc Metals; Am Soc Mech Engrs; Am Powder Metall Inst; Soc Mfg Engrs. *Res:* Powder metallurgy; metal flow analysis; process design; mechanical and metallurgical analysis of net-shape forming processes (powder consolidation, precision forging), fracture during forming, expert systems for metalworking processes. *Mailing Add:* 5408 Peach Dr Gibsonia PA 15044

KUHN, JANICE OSETH, b Canacao, Philippines, June 29, 40; US citizen;; m; c 3. TOXICOLOGY. *Educ:* Univ RI, BS, 62, PhD(biochem), 70. *Prof Exp:* Asst prof biol sci, Va Polytech Inst & State Univ, 69-72; res assoc biochem, Univ SFla, 72-73; from asst prof to assoc prof chem, 73-83; res asst prof, Dept Med, Baylor Col Med, 83-88; toxicologist, 88-90, SR TOXICOLOGIST, STILLMEADOW INC, 90- *Mem:* Am Chem Soc; AAAS; NY Acad Sci; Sigma Xi; Soc Toxicol. *Res:* Role of intracellular calcium in cell injury and metabolic regulation; mechanisms of chemical injury. *Mailing Add:* 2118 S Fountain Valley Dr Missouri City TX 77459-1899

KUHN, KLAUS, b Breslau, May 1, 27; Ger citizen; m 56; c 3. CELL MATRIX. *Educ:* Univ Munich, dipl, 50 & 52; Max Planck Inst EiweiB-u Lederforschung, PhD(biochem), 55. *Prof Exp:* Fel Deutsche Forschungsgemeinschaft res, Inst Technol, Darmstadt, 56-58, asst prof, 58-60; assoc prof res, Univ Heidelberg, 60-66; dir, Dept Connective Tissue Res & exec dir, Max Planck Inst Biochem, 77-80; Fogarty scholar, Fogarty Intern Ctr, NIH, Bethesda, 81-82; chmn, Biol Med Sect, Sci Coun, 87-91, SCI MEM, MAX PLANCK INST, GESELLSCHAFT, 66- *Concurrent Pos:* Mem sci adv bd, Inst Arteriosclerosis Res, Münster, 81-91, Int Inst Cellular & Molecular Path, Brussels, 84-91, Inst Environ Res, Neuherberg nr Munich, 91- *Mem:* Hon mem Am Soc Biol Chemists; Europ Molecular Biol Orgn. *Res:* Structure and function of extracellular matrix problems; cell matrix interaction; molecular biology of collagen genes and other extracellular matrix constituents. *Mailing Add:* Max-Planck Inst Biochem Martinsried nr Munich D-8033 Germany

KUHN, LESLIE A, b South Falls, NY, May 10, 24; m 50; c 2. CARDIOLOGY. *Educ:* State Univ NY Downstate Med Ctr, MD, 48. *Prof Exp:* Assoc prof med, 66-75, assoc attend cardiologist, Hosp, 66-75, CLIN PROF MED, MT SINAI SCH MED, 75-, ATTEND CARDIOLOGIST, MT SINAI HOSP, NY, 75-, DIR, CORONARY CARE UNIT, 70- *Concurrent Pos:* Prin investr, Nat Heart Inst res grant, 60; fel coun clin cardiol, Am Heart Asn, 65-; consult cardiologist, US Vet Admin Hosp, Bronx, 69-; sr asst ed, J Am Col Cardiol; Consult, Coronary Care Unit, Mt Sinai Hosp, New York, 81- *Mem:* Fel Am Col Cardiol; fel Am Col Physicians; Am Fedn Clin Res; Am Soc Artificial Internal Organs; Am Col Chest Physicians. *Res:* Hemodynamic and cardiac metabolic effects of pharmacological agents and methods of mechanical circulatory support in experimental and clinical acute myocardial infarction with shock. *Mailing Add:* Cardiol Div Mt Sinai Sch Med 1050 Fifth Ave New York NY 10028

KUHN, MARTIN CLIFFORD, b Tucson, Ariz, Apr 4, 40; m 63; c 3. METALLURGY. *Educ:* Colo Sch Mines, Met Eng, 63, MS, 67, PhD(metall), 68. *Prof Exp:* Sr res engr, Anaconda Co, 68-72, supvr mineral processing, 72-74, mgr process technol, 74-75; proj mgr, Hazen Res, 75-76; mgr, Tech Develop Ctr, Mineral Sci Div, UOP Inc, 76-79; VPRES & GEN MGR, MINERALS SEPARATION CORP, 79- *Concurrent Pos:* Vpres new ventures, Mountain States Mineral Enterprises Inc. *Mem:* Am Inst Mining Metall & Petrol Engrs; Soc Mining Engrs; Mining & Metallurgical Soc Am; Am Mining Congress. *Res:* Mineral processing; froth flotation; hydrometallurgy; extractive metallurgy; heavy media and ultrasonics. *Mailing Add:* 5760 W Placita Del Risco Tucson AZ 85745

KUHN, MATTHEW, b Sacalaz, Rumania, Mar 19, 36; Can citizen; c 2. MICROELECTRONICS, TELECOMMUNICATIONS. *Educ:* Queen's Univ, Ont, BSc, 62; Univ Waterloo, MASc, 63, PhD(elec eng), 67. *Hon Degrees:* DEng, Univ Waterloo, Ont, Can, 85. *Prof Exp:* Advan Res Proj Agency fel, Div Eng, Brown Univ, 67-68; mem staff, device res & develop, Bell Tell Labs Inc, 68-70, supvr electroluminescence device develop, 70-73, res mgr, Elec Mat & Process Dept, Bell Northern, 73-76, mgr, Advan Technol Lab, 76-79, dir, Tech Dept, 79-84, ASST VPRES RESOURCE DEVELOP UNIV REL, 84- *Concurrent Pos:* Pres, Electron Devices Soc, Inst Elec & Electronics Engr, 80-81; mem, Nat Sci & Eng Res Coun Can, (Commun & Comput), 81-84; mem, Queen's Univ Adv Coun on Eng, 80-86; bd vis, Duke Fac Eng. *Mem:* Fel Inst Elec & Electronics Engrs; AAAS. *Res:* Solid state device physics; electroluminescence; semiconductor-insulator interface physics; solid state display development; silicon integrated circuit research; optoelectronics, fiber optics and telecommunications systems applications; management and administration of advanced technology research laboratory & telecommunications systems research programs. *Mailing Add:* Microelectronics Ctr N Carolina PO Box 122889 Research Triangle Park NC 27709

KUHN, NICHOLAS J, b Feb 15, 55. MATHEMATICS. *Educ:* Princeton Univ, AB, 76; Univ Chicago, MS, 77, PhD(math). 80. *Prof Exp:* Actg asst prof, Dept Math, Univ Wash, Seattle, 80-82; vis fel, Dept Math, 82-83, asst prof, Princeton Univ, 83-86; ASSOC PROF DEPT MATH, UNIV VA, 86- *Concurrent Pos:* Fel, Am Math Asn, 82-83; vis scholar, Northwestern Univ, Evanston, IL, 83; Sloan Found Fel, 85-87; vis fel, Cambridge Univ, Cambridge, Eng, 86-87. *Mem:* Am Math Soc; Math Asn Am. *Mailing Add:* Dept Math Astron Bldg Univ Va Charlottesville VA 22903-3199

KUHN, PETER MOUAT, b Janesville, Wis, Feb 2, 20; m 41. METEOROLOGY, ATMOSPHERIC CHEMISTRY & PHYSICS. *Educ:* Univ Wis, BS, 51, MS, 52, PhD(meteorol), 62. *Prof Exp:* Res meteorologist, US Weather Bur, Washington, DC, 52-54; res assoc meteorol, Univ Wis, 54-56; res meteorologist, US Weather Bur, Univ Wis, 56-67; proj leader, Radiation Group, Atmospheric Phys & Chem Lab, Nat Oceanic & Atmospheric Admin Res Labs, 67-70, chief Thermal Modification Br, 70-80; SR RES SCIENTIST, NORTHROP SERVS, INC, AMES RES CTR, NASA, CALIF, 80- *Concurrent Pos:* Staff meteorologist, WKOW-TV, 54-56; consult, atmospheric remote sensing, Boulder, Colo, 85- *Mem:* AAAS; fel Explorers Club; fel Optical Soc Am; Sigma Xi; Am Meteorol Soc; Am Inst Aeronaut & Astronaut. *Res:* Experimental meteorology, especially infrared radiation measurements surface through 30 kilometers. *Mailing Add:* 1780 Skyline Dr Stoughton WI 53589

KUHN, RAYMOND EUGENE, b Biloxi, Miss, Sept 6, 42; m 64. IMMUNOBIOLOGY, PARASITOLOGY. *Educ:* Carson-Newman Col, BS, 65; Univ Tenn, PhD(zool), 68. *Prof Exp:* Assoc prof, 68-80, PROF BIOL, WAKE FOREST UNIV, 80- *Mem:* Am Soc Parasitologists; Am Asn Immunol; Am Soc Trop Med & Hyg. *Res:* Immunology of parasitic diseases. *Mailing Add:* Dept Biol Wake Forest Univ PO Box 7325 Winston-Salem NC 27109

KUHN, THOMAS S, b Cincinnati, Ohio, July 18, 22; div; c 3. HISTORY OF PHYSICS, CONCEPTUAL CHANGE. *Educ:* Harvard Univ, SB, 43, AM, 46, PhD(physics), 49. *Hon Degrees:* LLD, Univ Notre Dame, 73; DHL, Bucknell Univ, 79. *Prof Exp:* Asst prof gen educ & hist sci, Harvard Univ, 51-56; prof hist sci, Univ Calif, Berkeley, 56-64; prof hist sci, Princeton Univ, 64-68; M Taylor Pine prof hist sci, 68-79, prof philos & hist sci, 79-83, LAURANCE S ROCKEFELLER PROF PHILOS, MASS INST TECHNOL, 83- *Concurrent Pos:* Lectr, Lowell Inst, 51; proj dir, Sources Hist Quantum Physics, 61-64; mem bd dir, Social Sci Res Coun, 64-66, Inst Advan Study, Princeton, 72-79 & Assembly Behav & Social Sci, Nat Acad Sci, Nat Res Coun, 80-83. *Honors & Awards:* Howard T Behrman Award, Princeton Univ, 77; George Sarton Medal, Hist Sci Soc, 82. *Mem:* Nat Acad Sci; Am Philos Soc; Am Acad Arts & Sci; Leopoldina Acad; Am Philos Asn. *Res:* Reconstruction of out-of-date scientific ideas; description and abstract analysis of the way language and ideas change in scientific development. *Mailing Add:* Dept Ling & Philos Mass Inst Technol 20D-213 Cambridge MA 02139

KUHN, TRUMAN HOWARD, geology; deceased, see previous edition for last biography

KUHN, WILLIAM E(RIK), b Toronto, Ont, Feb 27, 22; nat US; m 54; c 2. MATERIALS SCIENCE, POWDER METALLURGY. *Educ:* Univ Toronto, BASc, 44. *Prof Exp:* Res fel eng & metall, Ont Res Found, Can, 44-48; metall, Univ Notre Dame, 48-49; sr res engr, Metall Dept, Titanium Alloy Mfg Co, 49-52; sr engr, Carborundum Co, 52-58, supvr & mgr metall dept, 58-63; head metall dept, Spindletop Res, Inc, Ky, 63-68, mgr mat & processing div, 68-70; res assoc prof mat sci, Univ Cincinnati, 70-89; PRES, ZESTOTHERM INC, 84- *Concurrent Pos:* Pres, Dymatron Inc, 72-; dir, Hemotec Inc, 75-89; prin investr, NSF grant, 72-77 & NIH grant, 75-77. *Honors & Awards:* Leonard Medal, Can Inst Mining & Metall, 51. *Mem:* Sigma Xi; Am Soc Metals; Brit Inst Metals; Am Powder Metall Inst; Res & Develop Assocs Mil Food & Packaging. *Res:* Materials technology including ferrous and nonferrous metals; arc and plasma technology re arc melting, furnacing and synthesis of compounds; powder metallurgy; fibers and ultrafine particles; ceramics; bio-materials; biomaterials and powder metallurgy; mechanical alloying; comminution; electrochemical heaters for heating military rations; developed and manufactured flameless ration heaters for heating US military rations. *Mailing Add:* 2085 Fallon Rd Lexington KY 40504

KUHN, WILLIAM FREDERICK, b Kittanning, Pa, Apr 1, 30; m 53; c 4. ANALYTICAL CHEMISTRY, SPECTROCHEMISTRY. *Educ:* St Vincent Col, BS, 57; Univ Richmond, MS, 62. *Prof Exp:* Chemist, Philip Morris Inc, 57-59, res chemist, 59-62, group leader tech info, 62-64, sr scientist mass spectros, 64-69, facil leader Instrument Sect, 69-72, proj leader smoke condensation, 72-74, mgr biochem res, 74-81, mgr anal res, 81-84, dir appl res, dir res & develop support, 87-91, ASST TO VPRES RES & DEVELOP, PHILIP MORRIS USA, 91- *Honors & Awards:* William J Poehlman Award, Soc Appl Spectros, 75; St Vincent Gold Medal in chem. *Mem:* Soc Appl Spectros; AAAS; Am Chem Soc; Am Soc Mass Spectrometry; Sigma Xi; NY Acad Sci. *Res:* Spectroscopic methods; chromatographic techniques; computer applications; technical information; environmental pollution; tobacco and smoke composition; ionization phenomena; entomology. *Mailing Add:* Philip Morris USA PO Box 26583 Richmond VA 23261

KUHN, WILLIAM LLOYD, b Grafton, WVa, Dec 23, 25; m 51; c 2. PHARMACOLOGY. *Educ:* WVa Wesleyan Col, BS, 49; Univ Cincinnati, PhD(pharmacol), 57. *Prof Exp:* Jr biochemist, William S Merrell Co, 51-54; asst pharmacol, Univ Cincinnati, 54-55; pharmacologist, 56-57, dept head pharmacol, 57-62, assoc dir biol sci, 62-64, dir biol sci, 64-69, dir res planning & coord, 69-73, VPRES RES PLANNING & COORD, MERRELL-NAT LABS, 73- *Res:* Pharmacology and physiology of neuromuscular junction; central nervous system; heart and circulation; chemical structure; pharmacological activity relationships. *Mailing Add:* 318 Whitthorne Dr Cincinnati OH 45215

KUHN, WILLIAM R, b Columbus, Ohio, May 7, 38; m 57; c 3. PLANETARY ATMOSPHERES, CLIMATOLOGY. *Educ:* Capital Univ, BS, 61; Univ Colo, PhD(astro-geophys), 66. *Prof Exp:* Fel astro-geophys, Univ Colo, 66-67; asst prof, 67-71, assoc prof atmosphere & ocean sci, 72-77, PROF ATMOSPHERIC OCEANIC SCI, UNIV MICH, ANN ARBOR, 77-, DEPT CHMN, 80- *Mem:* Am Geophys Union; Am Astron Soc; AAAS. *Res:* Radiation and photochemical studies applicable to planetary atmospheres and the prebiotic earth atmosphere; climatology; photochemistry of outer planets. *Mailing Add:* Dept Atmospheric Sci Univ Mich Main Campus Ann Arbor MI 48109-2143

KUHNEN, SYBIL MARIE, b Haledon, NJ, Sept 12, 17. BOTANY. *Educ:* Montclair State Col, BA, 41; Columbia Univ, MA, 46; NY Univ, PhD(sci educ), 60. *Prof Exp:* Pub sch teacher, NJ, 41-43; asst bot, Columbia Univ, 43-46; from instr to assoc prof bact & bot, Montclair State Col, 46-66, chmn dept, 69-76, prof biol, 66-87; CONSULT, 55- *Mem:* AAAS; Bot Soc Am; Nat Sci Teachers Asn. *Res:* Plant ecology. *Mailing Add:* Five Charles St Clifton NJ 07013

KUHNERT, BETTY R, b New York, NY Dec 16, 44. OBSTETRICAL ANESTHESIA. *Educ:* Kent State Univ, PhD(biol), 72. *Prof Exp:* Asst prof, Case Western Reserve Univ, 75-84, assoc prof reproductive biol, 84-; ASSOC DIR, CLIN RES & DEVELOP, WYETH-AYERS RES. *Mem:* Fedn Am Socs Exp Biol; Am Soc Pharmacol & Exp Therapeut; Perinatal Res Soc; Sigma Xi; Soc Obstet Anesthia & Perinatology; Am Soc Clin Pharm & Therapeut. *Mailing Add:* Clin Commun Wyeth-Ayerst Res PO Box 8299 Philadelphia PA 19101

KUHNLEIN, HARRIET V, b Sadsburyville, Pa, Aug 14, 39; m; c 3. ECOLOGY, STUDY OF DIETS. *Educ:* Pa State Univ, BS, 61; Ore State Univ, MS, 69; Univ Calif, Berkeley, PhD(nutrit), 76. *Prof Exp:* Res fel, Int Ctr Med Res & Training, Calif, Colombia & Tulane Univ, 67; res asst, Inst Molecular Biol, Univ Ore, 69-72; NIH trainee nutrit sci, Univ Calif, Berkeley, 73-76; assoc prof nutrit, Sch Family & Nutrit Sci, Univ BC, 76-85; PROF & DIR, SCH DIETETICS & HUMAN NUTRIT, MACDONALD CAMPUS, MCGILL UNIV, 85- *Concurrent Pos:* Can rep, comt nutrit anthrop, Int Union Nutrit Sci, 83-; mem adv bd, Herb Res Found, 83-; mem, Nutrit Working Group, Int Strategic Issues for Health Prom, Med Serv Health & Welfare Can, 84; consult, PCB contamination inuit diet, Health & Welfare Can, 86; assoc ed, Ecol Food & Nutrit, 89- *Mem:* Am Dietetic Asn; Can Dietetic Asn; Soc Ethnobiol; Can Soc Nutrit Sci; Am Inst Nutrit. *Res:* Cultural and ecological determinants of diets and human nutritional status; nutrient levels in foods of indigenous people; Nuxalk of British Columbia, Baffin Inuit and other indigenous people. *Mailing Add:* Sch Dietetics & Human Nutrit Macdonald Campus McGill Univ Ste Anne de Bellevue PQ H9X 1C0 Can

KUHNLEY, LYLE CARLTON, b Buffalo, Minn, Dec 23, 25; m 53; c 4. MICROBIOLOGY. *Educ:* Univ Minn, BA, 49; Univ Tex, MA, 55, PhD(bact), 61. *Prof Exp:* Bacteriologist, Ariz State Dept Health, 49-50 & 53-55; from assoc to emer prof biol, Tex Tech Univ, 59-88; RETIRED. *Mem:* AAAS; Am Soc Microbiol. *Res:* Inducible enzyme formation; rumen microbiology; ecology of coliphage; geomicrobiological prospecting; resistance mechanisms. *Mailing Add:* PO Box 408 Monroe OR 97456-0496

KUHNS, ELLEN SWOMLEY, b Chester, Pa, Feb 6, 19; m 74; c 1. PHYSICS, RESEARCH ADMINISTRATION. *Educ:* Coe Col, BA, 41; Johns Hopkins Univ, PhD(physics), 46. *Prof Exp:* Instr astron, Teachers Col, Johns Hopkins Univ, 41-45; instr physics, Conn Col, 45-46; asst prof, NJ Col for Women, Rutgers Univ, 46-51; physicist, Naval Electronics Lab Ctr, 51-66, head, Optical Physics Div, 66-68, res mgr, 68-70, planning office, 70-74, tech prog mgt off, 74-77; dep independent res & independent exp develop dir, Naval Ocean Systs Ctr, 77-82; RETIRED. *Mem:* Fel Am Phys Soc. *Res:* Ultrasonics; underwater acoustics. *Mailing Add:* 875 Albion St San Diego CA 92106

KUHNS, JOHN FARRELL, b Albuquerque, NMex, Mar 2, 47; m 67; c 1. WATER CHEMISTRY, ICHTHYOLOGY. *Educ:* Univ Mo, Kansas City, BS, 69. *Prof Exp:* Co-owner, Fish Ltd, 67-69, Piscean Fantasy, 69; partner & pres, Mid-Continent Fish Ltd, 70; vpres & pres, Montserrat Educ & Sci Co, 70-73; res dir, Gen Drug & Chem Corp, 74-82; PRES, CORP SECY & RES DIR, AQUASCI RES GROUP INC, 82-; CO-OWNER, THE WRITTEN WORD, 79- *Concurrent Pos:* Founder, chmn bd & pres, Friends of the Aquarium, Inc, 77-; ed, J Aquaricult & Aquatic Sci, 79-, ed & compiler, Codex Fishery Chem, 84- & ed, Drum & Croaker, 89-; vis prof, Tex A&M Col Sta, Aqua Med Prog, 82-83; asst syst oper, Aquaria-Fish Forum, 84-; reviewer, AAAS Books & Films; mgr, Aquatic Data Ctr, 89-; pres, EECHO Systs, 89- *Mem:* AAAS; Am Chem Soc; Am Soc Ichthyologist & Herpetologist; Am Fisheries Soc; World Aquaculture Soc; Am Cichlid Asn; Int Asn Aquatic Animal Med. *Res:* Development of products for aquaculture, aquariculture and sport fisheries designed to control water quality in closed systems. *Mailing Add:* 7601 E Forest Lakes Dr NW Parkville MO 64152

KUHNS, WILLIAM JOSEPH, b Allentown, Pa, Sept 2, 18; m 61; c 7. CHEMISTRY. *Educ:* Muhlenberg Col, BS, 40; Lehigh Univ, MS, 42; Johns Hopkins Univ, MD, 48. *Prof Exp:* Chemist, Lederle Labs Div, Am Cyanamid Corp, 42-44; intern med, Salt Lake County Gen Hosp, 49-50; fel microbiol, Col Med, NY Univ, 50-51; vis investr & asst physician, Hosp, Rockefeller Inst, 51-54; assoc prof path, Sch Med, Univ Pittsburgh, 54-59; assoc prof path, Sch Med, NY Univ, 60-77; dir transfusion serv, NC Mem Hosp, 77-81, prof, 77-81, RES PROF PATH, UNIV NC, 81-; VIS INVESTR, HOSP SICK CHILDREN, TORONTO, 84- *Concurrent Pos:* Biochem res, Lister Inst, Univ London, 74-75; fel biochem, Res Inst, Hosp Sick Children, Toronto, 84-86; summer res, Marine Biol Lab, Woods Hole, MA, 88-90. *Mem:* Am Soc Clin Invest; Soc Exp Biol & Med; Am Asn Immunologists; Am Asn Pathologists; fel AAAS; fel Am Acad Microbiol; Am Soc Cell Biol; Soc Develop Biol; Am Soc Hematol. *Res:* Glycosyltransferases of O-linked glycans, sulfotransferases; blood groups and their precursors on cultured cells; blood groups on cells on culture; blood groups in infrahuman species; blood groups and antibodies in transplantation and cancer. *Mailing Add:* Hosp Sick Children 555 University Ave Toronto ON M5G 1X8

KUHR, RONALD JOHN, b Appleton, Wis, Dec 29, 39; m 61; c 3. AGRICULTURAL CHEMISTRY, INSECT TOXICOLOGY. *Educ:* Univ Wis, BS, 63; Univ Calif, Berkeley, PhD(agr chem), 66. *Prof Exp:* NIH fel, Pest Infestation Lab, Slough, Eng, 66-68; from asst prof to assoc prof insect toxicol, NY State Agr Exp Sta, 73-77; prof entom, assoc dir res & assoc dir, Agr Exp Sta, Cornell Univ, 77-80; prof & head, Dept Entom, 80-86, ASSOC DEAN & DIR RES, COL AGR LIFE SCI, NC STATE UNIV, 87- *Mem:* Am Chem Soc; Entom Soc Am. *Res:* Metabolism of carbamate insecticide chemicals in plants and insects; environmental degradation of pesticides. *Mailing Add:* NC State Univ Box 7643 Raleigh NC 27695-7643

KUHRT, WESLEY A, management; deceased, see previous edition for last biography

KUIDA, HIROSHI, b Ogden, Utah, Oct 23, 25; m 51; c 4. INTERNAL MEDICINE, PHYSIOLOGY. *Educ:* Univ Utah, BS, 49, MD, 51. *Prof Exp:* Intern med, Salt Lake County Gen Hosp & Univ Utah, 51-52, asst resident, 52-53; fel cardiol, Harvard Med Sch & Peter Bent Brigham Hosp, 53-54, res fel, 54-56; res fel physiol, Univ Minn, 56-57; chief resident med, 57-58, instr, 58-61, asst res prof, 61-64, assoc prof, 64-69, assoc prof physiol, 65-69, PROF MED & PHYSIOL, COL MED, UNIV UTAH, 69-, CHIEF, DIV CARDIOL, 80- *Concurrent Pos:* USPHS fel, 53-55, res career develop award, Am Heart Asn res fel, 55-57. *Mem:* Am Physiol Soc; Am Fedn Clin Res. *Res:* Pulmonary vascular hemodynamics; hemodynamics of endotoxin shock; pathophysiology of pulmonary hypertensive heart disease in cattle. *Mailing Add:* Dept Physiol Univ Utah Sch Med 50 N Medical Dr Salt Lake City UT 84132

KUIJT, JOB, b Velsen, Holland, May 25, 30; nat Can; m 59. PLANT ANATOMY, PLANT MORPHOLOGY. *Educ:* Univ BC, BA, 54; Univ Calif, MA, 55, PhD(anat), 58. *Prof Exp:* From instr to asst prof biol & bot, Univ BC, 59-68; assoc prof, 68-70, PROF BIOL, UNIV LETHBRIDGE, 70- *Concurrent Pos:* Adj prof, Univ Victoria, 89- *Mem:* Bot Soc Am. *Res:* Structure and taxonomy of parasitic angiosperms; systematics; floristics of Southern Alberta. *Mailing Add:* Dept Biol Univ Victoria Victoria BC V8W 2Y2 Can

KUIKEN, KENNETH (ALFRED), b Chicago, Ill, Oct 14, 18; m 44; c 2. BIOCHEMISTRY. *Educ:* Geneva Col, BS, 39; Univ Pittsburgh, PhD(biochem), 43. *Prof Exp:* Assoc nutritionist & assoc prof biochem & nutrit, Exp Sta, Agr & Mech Col Tex, 43-50; mem staff, Cellulose & Specialties Tech Div, 50-74, SR RES CHEMIST, BUCKEYE CELLULOSE CORP, 74- *Mem:* AAAS; Tech Asn Pulp & Paper Indust; Am Chem Soc; Soc Exp Biol & Med; fel Am Oil Chemists' Soc; Sigma Xi. *Res:* Microbiological methods of amino acid analysis; cottonseed processing; nutritional requirements of laboratory and farm animals; manufacture and application of wood and cotton cellulose; analytical methods for cellulose. *Mailing Add:* 4796 Gwynne Rd Memphis TN 38117

KUIPER, LOGAN KEITH, b Oskaloosa, Iowa, Sept 12, 40; m 72; c 3. GROUNDWATER HYDROLOGY, FLUIDS. *Educ:* Univ Iowa, BA, 62, MS, 65, PhD(physics), 69; Calif Inst Technol, MS, 63. *Prof Exp:* Res physics, SDak Sch Mines, 70; res geologist, Iowa Geol Surv, 72-78; HYDROLOGIST, US GEOL SURV, 79- *Mem:* Am Geophys Union; Soc Indust & Appl Math. *Res:* Groundwater hydrology and particularly the mathematical modelling; applied mathematics. *Mailing Add:* Los Alamos Nat Lab MS 0443 Los Alamos NM 87545

KUIPER, THOMAS BERNARDUS HENRICUS, b Amersfoort, Neth, July 14, 45; Can citizen; m 70. RADIOASTRONOMY. *Educ:* Loyola Col, Montreal, BSc, 66; Univ Md, PhD(astron), 73. *Prof Exp:* Sr scientist, 75-77, MEM TECH STAFF ASTRON, JET PROPULSION LAB, CALIF INST TECHNOL, 77- *Concurrent Pos:* Resident res assoc, US Nat Res Coun & Jet Propulsion Lab, 73-75. *Mem:* Am Astron Soc; Can Astron Soc; Int Astron Union; AAAS; Int Union Radio Sci. *Res:* Spectroscopy observations with emphasis on instrumentation and techniques; very large baseline interferometer; solar physics; radio search for extraterrestrial intelligence and evolution of civilization in space. *Mailing Add:* Jet Propulsion Lab 169-506 Calif Inst Technol 4800 Oak Grove Dr Pasadena CA 91109

KUIPER-GOODMAN, TINE, b Leeuwarden, Netherlands, Sept 11, 37; m 61; c 1. ELECTRON MICROSCOPY, TOXICOLOGY. *Educ:* McMaster Univ, BSc, 61, MSc, 63; Nat Res Coun Can & Ontario fels & PhD(histol, embryol), Univ Ottawa, 67. *Prof Exp:* RES SCIENTIST CELL TOXICOL & TOXICOLOGIST, HEALTH PROTECTION BR, BUR CHEM SAFETY, HEALTH & WELFARE, CAN, 66- *Mem:* Electron Micros Soc Am; Can Soc Cell Biol; Micros Soc Can; Can Asn Pathologists; Am Soc Toxicol. *Res:* Effect of exogenous substances that may be present in food on cell organelles of animal tissues; risk assessment of mycotoxins and natural toxicants present in food; development of quantitative morphological methods. *Mailing Add:* Nat Health & Welfare Bur Chem Safety Health Protect Bldg Rm 2-2 Tunney's Pasture ON K1A 0L2 Can

KUIPERS, BENJAMIN JACK, b Grand Rapids, Mich, Apr 7, 49; m 75; c 3. INTELLIGENT SYSTEMS. *Educ:* Swarthmore Col, BA, 70; Mass Inst Technol, PhD(math), 77. *Prof Exp:* Systs Programmer, Psychol Dept, Harvard Univ, 70-72; res assoc, Div Study Res Educ, Mass Inst Technol, 77-78; asst prof comput sci, Dept Math, Tufts Univ, 78-84; res assoc, lab computer sci, Mass Inst Technol, 84-85; ASSOC PROF, DEPT COMPUTER SCI, UNIV TEX, AUSTIN, 85- *Concurrent Pos:* Vis scientist, Lab Computer Sci, Mass Inst Technol, 80-84; asst prof med, Tufts Univ, 83-85, Univ Tex Health Sci Ctr, San Antonio, 88-90; consult, MCC, 85-87, USCG, 86-87, CISE, 86-88, Peat Marwick Found, 89-90. *Mem:* NY Acad Sci; Sigma Xi; Asn Comput Mach; Cognitive Sci Soc; Am Asn Artificial Intelligence; Comput Prof Social Responsibility; fel Soc Values Higher Educ. *Res:* Artificial intelligence; qualitative simulation and modeling of physical systems; artificial intelligence in medicine; knowledge representation; spatial learning, exploration and problem solving; robotics and intelligent control. *Mailing Add:* Comput Sci Dept Univ Tex Austin TX 78712

KUIPERS, JACK, b Grand Rapids, Mich, Mar 27, 21; m 48; c 5. MATHEMATICS. *Educ:* Calvin Col, AB, 43; Univ Mich, BSEE, 43, MSE, 59, Info & ContE, 66. *Prof Exp:* Asst to dir res, Elec Sorting Mach Co, 46-50; proj engr, Lear, Inc, 50-53; chief engr, R C Allen Bus Mach, Inc, 53-54; sr proj engr, Lear, Inc, 54-59; sr physicist, Cleveland Pneumatic Industs, 59-62; lectr aerospace eng, Inst Sci & Technol, Univ Mich, 62-65, assoc res engr, Univ, 65-67; PROF MATH, CALVIN COL, 67- *Concurrent Pos:* Consult, Precision Prod Dept, Nortronics Div, Northrop Corp, 67-, Precision Prod Dept & Avionics Div, Lear Jet Industs, 67-, Polhemus Navigation Sci, Inc & Advan Technol Systs, Div Austin Co, Cleveland, Ohio, 75- *Mem:* Inst Elec & Electronics Engrs; Math Asn Am; Sigma Xi. *Res:* Automatic control; analog-digital computer simulation; special purpose computer design; navigation and guidance control and instrumentation; coordinate converters for gyroscope inertial reference systems; mathematical models and optimization. *Mailing Add:* Dept Math Calvin Col 3201 Burton St Grand Rapids MI 49546

KUIST, CHARLES HOWARD, physical chemistry, polymer chemistry, for more information see previous edition

KUITERT, LOUIS CORNELIUS, b Spring Lake, Mich, Aug 20, 12; m 46; c 3. ENTOMOLOGY. *Educ:* Kalamazoo Col, BA, 39; Univ Kans, MA, 40, PhD(entom), 47. *Prof Exp:* Asst entomologist, State Entom Comn, Kans, 40-41; high sch instr, 41-42; asst instr biol, Univ Kans, 46-47; asst prof entom, Kans State Col, 47-48; asst entomologist, Agr Exp Sta, Univ Fla, 48-52, assoc entomologist, 52-55, entomologist, 55-61, head dept, 61-66, prof entom, 66-76; RETIRED. *Mem:* Entom Soc Am. *Res:* Control of insect and arachnid pests of woody ornamentals, pastures, tobacco; taxonomy of western hemisphere water scorpions; biology and control of tobacco insects. *Mailing Add:* 2325 NW 38th Dr Gainesville FL 32605-2663

KUIVILA, HENRY GABRIEL, b Fairport Harbor, Ohio, Sept 17, 17; m 43; c 3. ORGANIC CHEMISTRY. *Educ:* Ohio State Univ, BSc, 42, MA, 44; Harvard Univ, PhD(chem), 47. *Prof Exp:* Jr chemist, Manhattan Proj, Monsanto Chem Co, Ohio, 44-46; from asst prof to prof chem, Univ NH, 48-64; chmn dept, 64-69, 82-85, PROF CHEM, STATE UNIV NY, ALBANY, 64- *Concurrent Pos:* NSF sr fel & Guggenheim fel, 59; vis prof, Japan Soc Promotion of Sci, 73. *Mem:* Fel AAAS; Am Chem Soc; Royal Soc Chem. *Res:* Organic reaction mechanisms; organometallic chemistry. *Mailing Add:* Dept Chem State Univ NY Albany NY 12222

KUJATH, MAREK RYSZARD, b Poznan, Poland, Apr 25, 50; Can citizen. DYNAMICS OF MACHINES, MACHINE DESIGN. *Educ:* Warsaw Tech Univ, MASc, 74; Polish Acad Sci, PhD(mech eng), 80. *Prof Exp:* Proj mgr, Cent Inst Indust Safety, Poland, 78-81; postdoctoral mech eng, Norweg Inst Technol, 81-82; ASSOC PROF MECH ENG, TECH UNIV NS, 82- *Concurrent Pos:* Vis prof, Can Space Agency, 90-91. *Mem:* Am Soc Mech

Engrs. *Res:* Machine dynamics; machine design; time varying systems; robotics, rotors and vibration; signal processing; space structures; machine condition monitoring. *Mailing Add:* Dept Mech Eng Tech Univ NS PO Box 1000 Halifax NS B3J 2X4 Can

KUKAL, GERALD COURTNEY, b St Louis, Mo, Oct 1, 43; m 63; c 4. LOG ANALYSIS, PETROLEUM GEOLOGY. *Educ:* Southwest Mo Univ, BS, 67; Purdue Univ, MS, 73. *Prof Exp:* Teacher geol, Riverview Gardens Sch Dist, 67-70; teaching asst & instr geol, Purdue Univ, 70-73; field engr, Dresser Atlas, 73-77; SR GEOLOGIST & FORMATION EVALUATION SPECIALIST, CER CORP, 77- *Mem:* Soc Prof Well Log Analysts; Am Asn Petrol Geologists. *Res:* Development of log interpretation systems for low-permeability gas reservoirs. *Mailing Add:* 5010 Reno Ct Las Vegas NV 89119

KUKES, SIMON G, b Moscow, USSR; US citizen; m 80; c 2. CATALYSIS IN PETROLEUM. *Educ:* Moscow Inst Chem Technol, BS & MS, 70; USSR Acad Sci, Moscow, PhD(chem), 73. *Prof Exp:* Res chemist, Metalo-Org Inst, USSR Acad Sci, 73-77; postdoctoral org & phys chem, Rice Univ, 71-82; chemist, Phillips Petrol, 79-82, sr chemist, 82-84, res assoc catalysis petrol, 84-87; RES ASSOC CATALYSIS PETROL, AMOCO OIL CO, 87- *Concurrent Pos:* Alt counr, Pet Div, Am Chem Soc, 89- *Res:* Petroleum; catalysis; kinetics; author of over 30 publications; awarded 110 US patents. *Mailing Add:* 569 Braemar Ave Naperville IL 60563

KUKI, ATSUO, b Chicago, Ill; m; c 1. QUANTUM BIOPHYSICS, MULTICOMPONENT ELECTRON TRANSFER SYSTEMS. *Educ:* Yale Univ, BS, 78; Stanford Univ, PhD(phys chem), 85. *Prof Exp:* NIH-NRSA postdoctoral fel, Univ Ill, 85-86; ASST PROF CHEM, CORNELL UNIV, 86- *Concurrent Pos:* NSF presidential young investr, 89; Camille & Henry Dreyfus Found teacher-scholar, 89. *Mem:* Am Phys Soc; Am Chem Soc; Biophys Soc. *Res:* Chemical physics of electron transfer reactions; electronically active peptides of de novo design; quantum theory of electronic interactions; biophysics. *Mailing Add:* Dept Chem Cornell Univ Ithaca NY 14853-1301

KUKIN, IRA, b New York, NY, Apr 4, 24; m 54; c 3. ENVIRONMENTAL CHEMISTRY. *Educ:* City Col New York, BS; Harvard Univ, MA, 50, PhD(inorg chem), 51. *Hon Degrees:* Yeshiva Univ, 86. *Prof Exp:* Instr chem, Sampson Col, 46-48; group leader, Gulf Res & Develop Co, 51-57; res dir & scientist, res Sonneborn Chem & Ref Corp, Div Witco Corp, 57-63; PRES & FOUNDER, APOLLO CHEM CORP, 63- *Concurrent Pos:* Chmn, Apollo Technol Int. *Mem:* Am Chem Soc. *Res:* Energy conservation; pollution control; consultant with government agencies on air pollution. *Mailing Add:* 45 Edgemont Rd West Orange NJ 07052

KUKKONEN, CARL ALLAN, b Duluth, Minn, Jan 25, 45; m 68; c 2. SPACE MICROELECTRONICS, TECHNOLOGICAL ASSESSMENT. *Educ:* Univ Calif, Davis, BS, 68; Cornell Univ, MS, 71, PhD(physics), 75. *Prof Exp:* Res assoc physics, Purdue Univ, 75-77; res staff, Ford Motor Co, 77-84; DIR, CTR SPACE MICROELECTRONICS TECHNOL & MGR SUPERCOMPUT, JET PROPULSION LAB, CALIF INST TECHNOL, 84- *Mem:* Am Phys Soc. *Res:* Theory of electrons in metals; direct injection diesel engines; design and development of small high speed direct injection diesel engines for passenger cars; technological assessment of hydrogen as an alternative automotive fuel; concurrent computing; neural networks; solid state devices; photonics; custom microcircuits; supercomputing and computer sciences; solid state and theoretical physics; electrical engineering. *Mailing Add:* 5467 La Forest Dr La Canada CA 91011

KUKLA, MICHAEL JOSEPH, b Frankfort, Ger, Sept 23, 47; US citizen; m 69; c 3. PHARMACEUTICAL CHEMISTRY. *Educ:* Kalamazoo Col, BA, 69; Ohio State Univ, PhD(org chem), 74. *Prof Exp:* res investr chem, G D Searle & Co, 74-78; SR SCIENTIST, MCNEIL PHARMACEUT, 78- *Mem:* Am Chem Soc. *Res:* Synthesis of heterocyclic ring systems which may alter functions in the central nervous system. *Mailing Add:* 1551 Oak Hollow Dr Maple Glen PA 19002-2834

KUKOLICH, STEPHEN GEORGE, b Appleton, Wis, Feb 3, 40; m; c 3. PHYSICAL CHEMISTRY, STRUCTURAL CHEMISTRY. *Educ:* Mass Inst Technol, BS, 62, DSc(physics), 66. *Prof Exp:* Instr physics, Mass Inst Technol, 66-68; asst prof chem, Univ Ill, 68-69 & Mass Inst Technol, 69-74; assoc prof, 74-79, PROF CHEM, UNIV ARIZ, 79- *Concurrent Pos:* NSF res grants, 70-79 & 83-86, Am Chem Soc, 77-81 & 83-86. *Mem:* Am Phys Soc; Sigma Xi; Am Chem Soc. *Res:* Structures of weakly bound complexes, high resolution microwave spectroscopy; molecular Zeeman effect; molecular beam maser spectroscopy; electric and magnetic interactions in molecules; electron paramagnetic resonance spectroscopy of biological molecules; molecular relaxation studies. *Mailing Add:* Dept Chem Univ Ariz Tucson AZ 85721

KUKSIS, ARNIS, b Valka, Latvia, Dec 3, 27; nat Can; m 53; c 4. BIOCHEMISTRY. *Educ:* Iowa State Col, BS, 51, MS, 53; Queen's Univ, Ont, PhD(biochem), 56. *Prof Exp:* Res fel org chem, Royal Mil Col, Ont, 56-58. Prof Exp: Res assoc biochem, Queen's Univ, Ont, 58-59, asst prof, 60-65; asst prof, Banting & Best Dept Med Res, 65-68, assoc prof, 68-74, PROF, DEPT BIOCHEM & BANTING & BEST DEPT MED RES, C H BEST INST, UNIV TORONTO, 74- *Concurrent Pos:* Career investr, Med Res Coun Can, 60- & dir, Regional Gas Chromatography/Mass Spectrometry Lab, 72-; fel coun arteriosclerosis, Am Heart Asn; vis prof, Japanese Soc Prom Sci, 81. *Mem:* Am Oil Chem Soc; Can Biochem Soc; Am Soc Biol Chemists; Am Inst Nutrit; fel Royal Soc Can. *Res:* Composition of food fats; mechanics of lipid digestion and absorption; metabolism of triglycerides and phospholipids, sterols and bile acids; chromatographic separations and mass spectrometry of lipids. *Mailing Add:* Banting & Best Dept Med Res C H Best Inst Univ Toronto Toronto ON M5G 1L6 Can

KULA, ERIC BERTIL, b New York, NY, July 4, 29; m 51; c 2. PHYSICAL METALLURGY. *Educ:* Mass Inst Technol, BS, 48, MS, 52, ScD(metall), 54. *Prof Exp:* Metallurgist, Domnarvet's Steelworks, Sweden, 48-49; asst, Royal Inst Technol, Sweden, 49-50; asst, Mass Inst Technol, 50-54; SUPVRY METALLURGIST, US ARMY MAT TECHNOL LAB, WATERTOWN, MASS, 56-, BRANCH CHIEF, 73-, DIV DIR, 88- *Mem:* Am Soc Metals; Am Inst Mining, Metall & Petrol Engrs. *Res:* Mechanical behavior of metals; high strength steels; failure analysis. *Mailing Add:* 23 Mason St Lexington MA 02173

KULACKI, FRANCIS ALFRED, b Baltimore, Md, May 21, 42; m; c 2. MECHANICAL ENGINEERING, HEAT TRANSFER. *Educ:* Ill Inst Technol, BSME, 63, MSGE, 66; Univ Minn, PhD(mech eng), 71. *Prof Exp:* From asst prof to assoc prof mech eng, Ohio State Univ, 71-80; prof & chmn, dept mech & aerospace eng, Univ Del, Newark, 80-85; DEAN, COL ENG, COLO STATE UNIV, 86- *Concurrent Pos:* Consult to indust, govt labs & litigation. *Mem:* Sigma Xi; fel Am Soc Mech Engrs; Am Soc Eng Educ; fel AAAS. *Res:* Heat and mass transfer; convective heat transfer; hydrodynamic stability; electrofluid mechanics; nuclear waste disposal. *Mailing Add:* Col Eng Colo State Univ Ft Collins CO 80523

KULAK, GEOFFREY LUTHER, b Edmonton, Alta, Nov 26, 36; m 58; c 2. CIVIL ENGINEERING. *Educ:* Univ Alta, BSc, 58; Univ Ill, Urbana, MS, 61; Lehigh Univ, PhD(civil eng), 67. *Prof Exp:* Design engr, Bridge Br, Prov of Alta Dept Hwy, 58-60; instr civil eng, Univ Alta, 61-62; asst prof, NS Tech Col, 62-64; res asst, Lehigh Univ, 64-67; assoc prof, NS Tech Col, 67-70; PROF CIVIL ENG, UNIV ALTA, 70- *Concurrent Pos:* Mem, Res Coun Struct Conn, 67-; invited prof, Swiss Fed Inst Technol, 84. *Honors & Awards:* Moiseff Award, Am Soc Civil Engrs , 85. *Mem:* Am Soc Civil Engrs; Can Standards Asn; Can Soc Civil Eng. *Res:* Strength and behavior of steel structures; strength of high-strength bolts and welds; behavior of welded and bolted connections; fatigue strength of steel structures. *Mailing Add:* Dept Civil Eng Univ Alta Edmonton AB T6G 2G7 Can

KULAKOWSKI, ELLIOTT C, b Feb 18, 51; c 3. SCIENCE ADMINISTRATION. *Educ:* Fairfield Univ, BS, 72; Long Island Univ, MS, 75; Lehigh Univ, PhD(biochem), 80. *Prof Exp:* Med technologist, Cent Gen Hosp, Plainview, NY, 73; res & teaching asst, Mich State Univ, 74; clin chemist, Upjohn Co, 74-77; res assoc & instr chem, Lehigh Univ, 77-80; staff fel, hypertension endocrine br, NIH, 80-83, prog officer biochem, cardiac functions br, 83-84, sci prog adminr res grants, Cardiac Dis Br, Div Heart & Vascular Dis, Nat Heart Lung & Blood Inst, 84-88, sci prog adminr Ischemic Heart Dis, Specialized Ctr Res, 85-86, sr scientific adv, Nat Heart, Lung & Blood Inst, 86-89; ASSOC PROF BIOCHEM, SCH MED, TEMPLE UNIV, 89-, ASSOC VPROVOST HEALTH SCI RES DEVELOP, 89- *Mem:* Am Chem Soc; Am Soc Pharmacol & Therapeut. *Res:* Cardiovascular pharmacology and metabolism; clinical biochemistry. *Mailing Add:* Sch Med Temple Univ Old Med Sch Bldg Rm 100 3400 N Broad St Philadelphia PA 19140

KULANDER, BYRON RODNEY, b Huntington, WVa, Aug 27, 37; m 68; c 1. ROCK FRACTURES, FORELAND FOLD BELTS. *Educ:* Kent State Univ, BS, 62; WVa Univ, MS, 64, PhD(geol). 68. *Prof Exp:* From asst prof to assoc prof geol, Dept Geol, Alfred Univ, 66-79; PROF GEOL, DEPT GEOL SCI, WRIGHT STATE UNIV, 79-, CHMN GEOL, 89- *Concurrent Pos:* Coop res geologist, WVa Geol Surv, 69-; consult, Dept Energy, 76-78; Mound Labs, 81-82; Terra Tek, 83-85; BDM Corp, 85-88; Amoco Corp, 86-89; Oryx Energy Co, 89-90; vis prof, Dept Geol, WVa, Univ, 78. *Mem:* Am Asn Petrol Geologists; fel Geol Soc Am. *Res:* Rock fractures-fractured reservoirs, including application of fractography to fractured core and outcrop rocks; structural geology of foreland fold belts; application of geophysics to detached sedimentary rocks and basement structures. *Mailing Add:* Dept Geol Sci Wright State Univ Dayton OH 45435

KULANDER, KENNETH CHARLES, b St Paul, Minn, Nov 26, 43; m. MOLECULAR COLLISIONS, MULTIPHOTON PROCESSES. *Educ:* Cornell Univ, BS, 65; Univ Minn, PhD(phys chem), 72. *Prof Exp:* Postdoctoral, Chem Dept, Univ Minn, 72-75; sr res assoc, Daresbury Lab, Sci Res Coun, Warrington, Eng, 75-78; staff scientist, Laser Prog, Lawrence Livermore Nat Lab, 78-82, chemist, Chem Dept, 82-85, physicist, Physics Dept, 85-86, GROUP LEADER, THEORET ATOMIC & MOLECULAR PHYSICS GROUP, LAWRENCE LIVERMORE NAT LAB, 86- *Concurrent Pos:* Vis scientist, Max Planck Inst Quantum Optics, Garching, Ger, 82-83; chartered physicist, Inst Physics. *Mem:* Fel Am Phys Soc; Optical Soc Am; Inst Physics. *Res:* Atomic and molecular collision theory; multiphoton processes; laser interactions with atoms and molecules; development of computational methods for quantum dynamics. *Mailing Add:* Lawrence Livermore Nat Lab PO Box 808 Livermore CA 94550

KULCINSKI, GERALD LA VERN, b LaCrosse, Wis, Oct 27, 39; m 61; c 3. NUCLEAR ENGINEERING. *Educ:* Univ Wis, BS, 61, MS, 62, PhD(nuclear eng), 65. *Prof Exp:* Asst scientist nuclear rockets, Los Alamos Sci Lab, 63; sr res scientist & group leader radiation damage reactor mats, Battelle Northwest Lab, 65-72; adj prof nuclear eng, Ctr Grad Study, Richland, 68-71; PROF NUCLEAR ENG, UNIV WIS, 72-, DIR FUSION TECHNOL INST, 74- *Concurrent Pos:* Assoc ed, Nuclear Engr & Design, 83-; scientific adv bd, Karlsruhe Nuclear Lab, Karlsruhe, FRG; Grainger chair nuclear engrs, 84- *Honors & Awards:* Curtis McGraw Res Award, Am Soc Eng Educ, 78. *Mem:* Fel Am Nuclear Soc. *Res:* Fission reactors; fusion reactor design; materials; radiation damage; environmental effects; nuclear power. *Mailing Add:* Nuclear Eng/443 Engr Res Univ Wis Madison WI 53706

KULCZYCKI, ANTHONY, JR, b Easton, Pa, Dec 17, 44; m 69; c 2. IMMUNOCHEMISTRY, ALLERGY. *Educ:* Princeton Univ, BA, 66; Harvard Univ, MD, 70. *Prof Exp:* Intern med, Buffalo Gen Hosp, & E J Meyer Hosp, 70-71; med resident, State Univ NY, Buffalo, 71-72; res assoc, Nat Inst Arthritis, Metab & Digestive Dis, NIH, Bethesda, 72-74; NIH teaching res fel, Sch Med, Wash Univ, 74-76; from instr to asst prof med, 76-82, asst prof microbiol & immunol, 80-85, ASSOC PROF MED, DEPT MED, DIV ALLERGY & IMMUNOL, SCH MED, WASH UNIV, 82-, ASSOC PROF MICROBIOL & IMMUNOL, 85- *Concurrent Pos:* Assoc investr, Howard Hughes Med Inst, Wash Univ, 77-84; asst physician, dept med, Barnes Hosp, St Louis, 77- *Honors & Awards:* J D Lane Award, USPHS, 74. *Mem:* Am Asn Immunologists; fel Am Acad Allergy & Immunol; Am Soc Clin Invest; Sigma Xi; Collegium Internationale Allergologicum. *Res:* Receptors for antibodies (IgE and Fc receptors), their gene structures, functions and potential involvement in immunological diseases; immune complexes; infantile colic; allergic reactions to aspartame (NutraSweet). *Mailing Add:* Dept Int Med Wash Univ Sch Med 660 S Euclid Ave St Louis MO 63110

KULCZYCKI, LUCAS LUKE, b Jurjampol, Poland, Aug 19, 11; US citizen; c 2. PEDIATRICS. *Educ:* Univ Lwow, BSc, 34, DVM, 36; Univ Edinburgh, MB BCh, 44, MD, 46; Univ London, dipl pub health, 48; Royal Col Physicians & Surgeons Can, cert pediat, 58. *Prof Exp:* Resident physician med & surg, Raigmore Hosp, Dept Health, Scotland, 46-47; asst physician, London Exec Coun, Eng, 47-50; med dir pub health, Local Health Unit, Dept Health, Winnipeg, Can, 51-53; residential training pediat, Children's Hosp Med Ctr, Boston, Mass, 53-55, asst physician, 55-62; dir, Cystic Firbrosis Ctr, Children's Hosp Nat Med Ctr, 62-77; prof clin pediat, 73-78, PROF PEDIAT, MED SCH, GEORGETOWN UNIV, 78-; DIR, CYSTIC FIBROSIS CTR, GEORGETOWN UNIV HOSP, 77- *Concurrent Pos:* Res fel, Children's Hosp Med Ctr, Boston, 55-61; instr, Harvard Med Sch, 56-62; clin dir, Wrentham State Sch, 57-58; consult pediatrician, Dept Health Maine & Maine Med Ctr, Portland, 58-68; clin assoc prof pediat, Georgetown Univ, 62-67, assoc prof, 67-72; consult pediatrician & co-worker, Children's Hosp, Boston, Mass, 62-68; guest worker, NIH, 62-68; consult pediatrician, Children's Convalescent Hosp, Washington, DC, 63-75. *Honors & Awards:* Physician's Recognition Award, AMA, 69, 81 & 84. *Mem:* Fel Am Acad Pediat; fel Am Col Chest Physicians; AMA; fel NY Acad Sci; hon mem Polish Pediat Soc; fel Royal Soc Health; Lung & Thoracic Soc; Brit Med Asn; Can Med Asn; fel Am Pub Health Asn; fel Am Lung Asn. *Res:* Cystic fibrosis in caucasians and negroes; cyctic fibrosis, tuberculosis and allergy; upper respiratory tract in cystic fibrosis; hearing and cystic fibrosis; bronchoscopy and bronchial lavage in cystic fibrosis; impact of cystic fibrosis on the patient and his parents; patient home care; mucus retention and over-inflation as a basic pulmonary complication in cystic fibrosis; use and abuse of antibiotics in management of cystic fibrosis. *Mailing Add:* Childrens Hosp Nat Med Ctr Washington DC 20009

KULEVSKY, NORMAN, b New York, NY, July 28, 35; m 61; c 3. PHYSICAL CHEMISTRY. *Educ:* Brooklyn Col, BS, 56; Univ Mich, MS, 58, PhD, 63. *Prof Exp:* From asst prof to assoc prof, 62-74, PROF CHEM, UNIV NDAK, 74- *Mem:* Am Chem Soc. *Res:* Molecular and charge transfer complexes; hydrogen bonding studies. *Mailing Add:* Dept Chem Univ NDak Box 7185 Grand Forks ND 58202

KULFINSKI, FRANK BENJAMIN, b New Brunswick, NJ, May 30, 30; m 56; c 3. ECOLOGY, ENVIRONMENTAL STUDIES. *Educ:* Rutgers Univ, BS, 52; Univ Mass, MS, 54; Iowa State Col, PhD(ecol), 57. *Prof Exp:* Instr bot, Iowa State Col, 56-57; asst prof biol sci, Western Ill Univ, 57-60; assoc prof biol, Ill Wesleyan Univ, 60-69; assoc prof, 69-77, PROF BIOL SCI & COORDR ENVIRON STUDIES MS PROG, SOUTHERN ILL UNIV, EDWARDSVILLE, 77- *Concurrent Pos:* Vis assoc prof zool, Southern Ill Univ, Carbondale, 68; instr, Civil Serv Comn Workshop Environ Impact Statements, 73-77; consult biol portions environ impact statements, eng firms, 71-81. *Res:* Cytology; phycology; pathology; ecology. *Mailing Add:* Dept Biol Sci Southern Ill Univ Box 1651 Edwardsville IL 62026

KULGEIN, NORMAN GERALD, b Bridgeport, Conn, Mar 6, 34; m 60; c 2. ENGINEERING, APPLIED PHYSICS. *Educ:* Mass Inst Technol, BS, 55, MS, 56; Harvard Univ, PhD(eng, appl physics), 60. *Prof Exp:* Res scientist, Aerospace Sci Lab, 60-67, STAFF SCIENTIST & MGR AEROPHYS GROUP, LOCKHEED PALO ALTO RES LAB, 67- *Concurrent Pos:* Lectr, Univ Santa Clara, 63-; Stanford Univ, 69-70. *Mem:* Am Inst Aeronaut & Astronaut; Combustion Inst. *Res:* High temperature viscous flows; radiation gas dynamics; hydrodynamics; reentry vehicle hardening technology; infrared systems analysis. *Mailing Add:* LMSC Dept 9710 Bldg 201 3251 Hanover St Palo Alto CA 94304

KULHAWY, FRED HOWARD, b Topeka, Kans, Sept 8, 43; m 66. GEOTECHNICAL ENGINEERING. *Educ:* Newark Col Eng, BSCE, 64, MSCE, 66; Univ Calif, Berkeley, PhD(civil eng), 69. *Prof Exp:* Asst inst civil eng, Newark Col Eng, 64-66; soils engr, Storch Engrs, 66; res asst & jr res specialist, Univ Calif, Berkeley, 66-69; assoc, Raamot Assoc PC, 69-71; from asst prof to assoc prof, Syracuse Univ, 69-76; assoc prof, 76-81, PROF CIVIL ENG, CORNELL UNIV, 81- *Concurrent Pos:* Numerous consults to govt agencies, indust firms, eng & archit consults & attys, 69-; vis prof, Univ Cambridge, Univ Sydney, Univ Hawaii, 85-86; Fulbright Scholar, 85; hon tech affil, Int Asn of Found Drilling Contractors. *Honors & Awards:* Edmund Friedman Young Eng Award, Am Soc Civil Eng, 74; Walter L Huber Res Prize, Am Soc Civil Eng, 82; Cross-Can lectr, Can Geotech Soc, 88. *Mem:* Fel Am Soc Civil Eng; fel Geol Soc Am; Int Soc Rock Mech; Int Soc Soil Mech & Found Eng; Int Asn Eng Geol; Int Asn of Found Drilling Contractors. *Res:* Numerical methods applications in geotechnical engineering; soil and rock stress-strain-strength behavior; model and full-scale behavior of geotechnical structures; foundation engineering. *Mailing Add:* Sch Civil & Environ Eng Cornell Univ, Hollister Hall Ithaca NY 14853-3501

KULIER, CHARLES PETER, b Chicago, Ill, Aug 11, 35; m 59; c 2. SYNTHETIC ORGANIC CHEMISTRY, PHARMACEUTICAL CHEMISTRY. *Educ:* Ill Wesleyan Univ, BS, 57; Univ Kans, PhD(org chem), 62. *Prof Exp:* Asst chem, Ill Wesleyan Univ, 56-57 & Univ Kans, 57-61; res assoc & fel org chem, Johns Hopkins Univ, 62-63; from assoc res chemist to res chemist, 63-70, sr res chemist, 70-72, sr scientist, 72-75, res assoc, 76-89, INFO SERV, PARKE DAVIS DIV, WARNER-LAMBERT CO, 90- *Mem:* Am Chem Soc. *Res:* Organic synthesis of heterocyclic compounds and natural products; use of newer reaction methods for preparation of organic compounds of potential medicinal use; process research; information science and systems. *Mailing Add:* 1181 Oak Hampton Rd Holland MI 49424-2663

KULIK, MARTIN MICHAEL, b Brooklyn, NY, Apr 20, 32; m 62; c 3. PLANT PATHOLOGY. *Educ:* Cornell Univ, BS, 54; La State Univ, MS, 56, PhD(plant path), 59. *Prof Exp:* Plant pathologist, Seed Br, USDA, 61-63, res platn pathologist, Seed Qual Lab, 63-72, Seed Res Lab, 72-85, Germplasm Qual & Enhancement Lab, 86-90, PLANT PATHOLOGIST, SOYBEAN & ALFALFA RES LAB, AGR RES SERV, USDA, 90- *Mem:* Am Soc Agron; Am Phytopath Soc; Mycol Soc Am; Can Phytopath Soc; Am Inst Biol Sci. *Res:* Seedborne fungi, diseases of seeds and forages. *Mailing Add:* 5100 Moorland Lane Bethesda MD 20814

KULIKOWSKI, CASIMIR A, b Hertford, Eng, May 4, 44. ARTIFICIAL INTELLIGENCE. *Educ:* Univ Hawaii, PhD(elec eng), 70. *Prof Exp:* From asst prof to assoc prof, 70-77, PROF & CHMN COMPUT SCI DEPT, RUTGERS UNIV, 84- *Concurrent Pos:* Dir, lab sci dept, Rutgers Univ, 85- *Mem:* Inst Med-Nat Acad Sci; fel Am Acad Med Informatics; AAAS. *Mailing Add:* Dept Comp Sci Rutgers Univ New Brunswick NJ 08903

KULKA, JOHANNES PETER, b Vienna, Austria, Feb 7, 21; nat US. PATHOLOGY, PSYCHIATRY. *Educ:* Cornell Univ, AB, 41; Johns Hopkins Univ, MD, 44; Am Bd Path, dipl. *Prof Exp:* Intern path, Strong Mem Hosp, NY, 44-45; asst res, Mass Gen Hosp, Boston, 45-47; instr anat, Harvard Med Sch, 47-49, instr path, 49-52, assoc, 52-58, clin assoc, 58-61, from asst clin prof to assoc clin prof, 61-70; resident psychiat, McLean Hosp, 70-73; child psychiat trainee, South Shore Ment Health Ctr, Quincy, Mass, 73-74; gen physician health serv & clin instr med, Tufts Univ, 75-79; RETIRED. *Concurrent Pos:* Pathologist, Lovett Mem, 47-52; assoc path, Peter Bent Brigham Hosp, 55-58, asst med, 58-61, assoc staff, 61-70; pathologist, Robert B Brigham Hosp, Boston, 55-58, clin & res assoc, 58-61, pathologist, 61-70; clin fel psychiat, Harvard Med Sch, 70-73 & McLean Hosp, 73-74; mem courtesy staff, Lawrence Mem Hosp, Medford, Mass, 75-79. *Mem:* Am Asn Path & Bact; Am Soc Exp Path; Soc Exp Biol & Med. *Res:* Pathology of rheumatic diseases, cold injury and microcirculatory disorder; psychosomatic disorders. *Mailing Add:* RFD 1 Box 82 Lincoln MA 01773

KULKARNI, ANAND K, b Gokak, Karnatak, India, Oct 18, 46; m 71; c 1. ELECTRONIC MATERIALS, ELECTRONIC DEVICES. *Educ:* Karnatak Univ, Dharwad, India, BS, 67, MS, 70; Iowa State Univ, Ames, MS, 75; Univ Nebr-Lincoln, PhD(eng), 79. *Prof Exp:* Asst engr, Pvt Electronics Co, 71; jr sci asst, govt orgn, 71-72; res & teaching asst physics, Iowa State Univ, 73-75; res asst elec eng, Univ Nebr-Lincoln, 76-78; vis asst prof electronics, 78-80, asst prof, 80-85, ASSOC PROF ELECTRONICS, MICH TECHNOL UNIV, 85- *Concurrent Pos:* Prin investr, NSF, 83-86; co-investr, Int Bus Mach, 85-86, Ramtron Corp, 87-88, NSF, 89-92; proj dir, Mich State, 90-91. *Honors & Awards:* Ralph R Teetor Educ Award, Soc Automotive Engrs, Inc, 86. *Mem:* Inst Elec & Electronics Engrs; Am Vacuum Soc; Int Soc Hybrid Microelectronics. *Res:* Ohmic contacts, refractory metal and refractory silicide schottky contacts to gallium arsenide; ferroelectric thin film memory devices; analysis of materials and processes to improve solder joints and metal/ceramic contacts. *Mailing Add:* 105 W Houghton Ave Houghton MI 49931

KULKARNI, ANANT SADASHIV, b Kolhapur, India, July 31, 34; m 60; c 2. CLINICAL PHARMACOLOGY, IMMUNOLOGY. *Educ:* Podar Med Col, GFAM (MD), 58; Univ Minn, PhD(pharmacol), 66. *Prof Exp:* Intern med, Sisters Hosp, Buffalo, NY, 59-60; surg resident, St Anthony's Hosp St Louis, 60-61; res asst pharmacol, Univ Wis, 61-62; res asst pharmacol, Univ Minn, 62-65; sr scientist, Mead Johnson Res Ctr, 65-67; res pharmacologist, Dow Human Health Res Lab, 67-71, clin monitor, Med Dept, Dow Chem Co, 71-73; sect head neuropsychiat & assoc dir clin res, Abbott Lab, 73-75; assoc dir, 75-77, DIR CLIN RES, GEN MED & NEUROPSYCHIAT, G D SEARLE & CO, 78- *Concurrent Pos:* Vis lectr, Med Ctr, Ind Univ, Indianapolis, 68-71, clin asst prof, 71-73. *Mem:* Acad Psychosom Med; Am Soc Pharmacol & Exp Therapeut; Sigma Xi; Am Psychol Asn; Am Pharmaceut Asn. *Res:* CNS pharmacology; clinical psychopharmacology; animal behavior; rheumatology; drug behavior interactions. *Mailing Add:* G D Searle & Co Med Res Box 5110 Chicago IL 60680

KULKARNI, BIDY, b Maharashtra, India, Apr 18, 30; m 57; c 2. MATERNAL & CHILD HEALTH, ABNORMAL PREGNANCY. *Educ:* Univ Poona, India, MS, 56, PhD, 62. *Prof Exp:* Jr sci asst biochem & steroid chem, Nat Chem Lab, Poona, India, 52-56, sr sci asst steroid chem, 56-61; fel steroid biochem, Clark Univ, 61-64; fel org chem, Nat Res Coun Can, 64-66; sr sci officer biochem, Nat Chem Lab, Poona, 66-67; sect chief, Dept Endocrinol, Div Clin Sci, Southwest Found Res & Educ, 67-70; asst prof obstet & gynec, Pritzker Sch Med, Univ Chicago, 70-73; assoc prof obstet & gynec & dir reproductive endocrinol, Stritch Sch Med, Loyola Univ Chicago, 73-79; DIR, REPRODUCTIVE ENDOCRINOL LABS, DEPT OBSTET & GYNEC, COOK COUNTY HOSP, CHICAGO, 79-; ASSOC PROF OBSTET & GYNEC, CHICAGO MED SCH, 80- *Concurrent Pos:* Dir labs, Sect Gynecic Endocrinol, Michael Reese Hosp & Med Ctr, 70-73; dir perinatal ctr labs, Forster G McGaw Hosp, Maywood, 73-77; consult, Gottlieb Mem Hosp, 77-81; sci officer, Cook County Hosp, 79-; mem, Obstet & Gynec Delegation, USSR & Scandanavia, 87. *Honors & Awards:* Outstanding New Citizen of Year, 73. *Mem:* AAAS; Endocrine Soc; Soc Study Reproduction; Asn Clin Scientist; Nat Acad Clin Biochem; Chicago Gynec Soc; Am Fertil Soc; Int Fertil Soc. *Res:* Natural and contraceptive steroid hormone metabolism in man and nonhuman primates; methods in hormone assays involving competitive protein binding and radioimmunoassays; clinical endocrinology fetoplacental function population control research; steroid biochemistry; abnormal pregnancy; contraception; clinical chemistry. *Mailing Add:* Cook County Hosp M3322 1825 W Harrison Chicago IL 60612

KULKARNI, PADMAKAR VENKATRAO, b Inamhongal, India, Nov 1, 42; m 68; c 3. NUCLEAR CHEMISTRY, RADIOPHARMACEUTICALS. *Educ:* Janata Col, BS, 63; Rensselaer Polytech Inst, MS, 72, PhD(chem), 73. *Prof Exp:* Sci officer trainee radiochem, Bhabha Atomic Res Ctr, Bombay, 63-64, sci officer, 64-68; radiopharmaceut specialist, Cambridge Nuclear Radiopharm Corp, Mass, 72-73; isotope chemist, Abbott Diag Div, Abbot, Ill, 73-76; asst prof, 76-84, ASSOC PROF, UNIV TEX SOUTHWESTERN MED SCH, 84- *Concurrent Pos:* Radiopharmaceut scientist, Parkland Mem Hosp, Dallas, Tex, 76- *Mem:* Am Chem Soc; Soc Nuclear Med; AAAS. *Res:* Development of radioisotope labeled compounds as radiopharmaceuticals for diagnostic purposes; development of radioimmunoassay systesm; diagnostic nuclear cardiology; radioisotope tracer techniques in health sciences; contrast agents for magnetic resonance imaging. *Mailing Add:* Radiol Imaging Ctr 5323 Harry Hines Blvd Dallas TX 75235-9058

KULKARNI, PRASAD SHRIKRISHNA, b Karad, India, May 22, 43; US citizen. PHARMACOLOGY, OPHTHALMOLOGY. *Educ:* Downstate Med Sch, State Univ NY, New York, MS, 71, PhD, 74. *Prof Exp:* Fel, Washington Univ, St Louis, Mo, 74-76; teaching fel, dept ophthal, Columbia Univ, 76-78, res assoc, 78-80, asst prof, 80-87; ASSOC PROF, DEPT OPHTHAL, UNIV LOUISVILLE, 87- *Mem:* Am Soc Pharmacol & Exp Therapeut; Int Soc Eye Res; Asn Res Vision & Ophthal; Brit Pharmacol Soc; Inflam Res Asn. *Res:* Role of prostaglandins, leukotrienes and other arachidonic acid metabolites in ocular inflammation; mechanism of steroidal and nonsteroidal anti-inflammatory agents in ocular inflammation. *Mailing Add:* Ky Lions Eye Res Inst Univ Louisville Louisville KY 40202

KULKARNY, VIJAY ANAND, b Karwar, India, May 3, 47. FLUID DYNAMICS, APPLIED PHYSICS. *Educ:* Indian Inst Technol, Bombay, BTech, 69; Calif Inst Technol, MS, 70, PhD(aeronaut), 75. *Prof Exp:* From res fel to sr res fel aeronaut, Calif Inst Technol, 75-78; MEM TECH STAFF, ENG SCI LAB, TRW DEFENSE & SPACE SYSTS GROUP, 78- *Concurrent Pos:* Instr aeronaut, Calif Inst Technol, 76-77; consult, TRW Defense & Space Systs Group, 78. *Mem:* Am Phys Soc; Sigma Xi. *Res:* Gas dynamics; acoustics; shock waves and associated linear and nonlinear wave phenomena in multidimensions and inhomogeneous media; dynamics of vortex interactions and vorticity dominated flows; flow and acoustics of high energy pulsed gas lasers. *Mailing Add:* 30427 Via Rivera Rancho Palos Verdes CA 90274

KULKE, BERNHARD, b Freiburg, Germany, Nov 29, 32; US citizen; m 62; c 2. ELECTRON BEAMS, DIAGNOSTICS. *Educ:* Univ Colo, BS, 55; Stanford Univ, MS, 60, PhD(microwave electronics), 65. *Prof Exp:* Mem tech staff, Bell Tel Labs Inc, 55-58; asst prof elec eng, Syracuse Univ, 65-67; physicist, Electronics Res Ctr, NASA, 67-70; chief radar beacon sect, Dept Transp, Cambridge Univ, 70-74; group leader Accelerator Technol, 74-84, PROJ ENGR, BEAM RES PROG, LAWRENCE LIVERMORE NAT LAB, 84- *Mem:* Inst Elec & Electronics Engrs; Am Phys Soc. *Res:* Design and construction of state-of-the-art, high current, induction linear accelerator; X-band, cyclotron resonance oscillator; expermental, 100 kilowatt, traveling wave klystron; electron beam diagnostics, antenna design, radar systems analysis and solid state device characterization; magnet design. *Mailing Add:* Lawrence Livermore Univ Calif PO Box 808 Livermore CA 94550

KULKOSKY, PAUL JOSEPH, b Newark, NJ, Mar 3, 49; m 78. PHYSIOLOGICAL PSYCHOLOGY. *Educ:* Columbia Col, BA, 71, MA, 72; Univ Wash, PhD(psychol), 75. *Prof Exp:* Staff fel, Nat Inst Alcohol Abuse & Alcoholism, 76-80; from res assoc to instr, Cornell Univ Med Col, 80-82; from asst prof to assoc prof, 82-89, PROF PSYCHOL, UNIV SOUTHERN COLO, 89-, CHAIR, DEPT PSYCHOL, 88- *Concurrent Pos:* Affil prof psychol, Am Univ, 77-80; bd dir, 85-88, bd adv, Pueblo Zool Soc, 88-; prin investr, Nat Inst Health grant, 84-; Consortium Aquariums, Univs & Zoos, 89-; vchair, Psychol Sci Sect, Southwestern & Rocky Mountain Div, AAAS, 90-; coun, Undergrad Psychol Progs, 90-; regional liason, Rocky Mountain Area, 90- *Mem:* Psychonomic Soc; Soc Neurosci; Int Soc Biomed Res Alcoholism; Sigma Xi; NY Acad Sci; Soc Ingestive Behav; Am Psychol Soc; AAAS; Int Brain Res Orgn. *Res:* Regulatory behaviors in mammals, including the learned and physiological controls of ingestive behaviors. *Mailing Add:* Dept Psychol Univ Southern Colo Pueblo CO 81001

KULL, FREDERICK CHARLES, SR, b Newark, NJ, Apr 10, 19; m 43; c 5. BACTERIOLOGY. *Educ:* Villanova Univ, BS, 41; Ind Univ, MA, 49; Univ Mich, PhD(bact), 52; Am Bd Med Microbiol, dipl. *Prof Exp:* Chemist, Sherwin-Williams Co, NJ, 41-43; asst bact, Ind Univ, 47-49; from asst to instr, Univ Mich, 49-51; sr bacteriologist, Ciba-Geigy Pharmaceut Co, 52-58, dir virol, 58-59, dir bact, 59-61, dir sci info ctr, 61-68; adminr, 68-77, dir admin, res, develop & med, Burroughs Wellcome Co, 77-85; CONSULT, 85- *Concurrent Pos:* Instr, Rutgers Univ, 55-58; adj prof, Sch Pharm, Univ NC, 72-; consult, Nat Serv Exec Corp, 85-90, Glaxo Inc, 86-90; vpres, Exec Serv Corp Carolinas, 90- *Mem:* Am Acad Microbiol; Am Soc Microbiol; Sigma Xi. *Res:* Medical information; documentation; biological sciences; virology; enzymology. *Mailing Add:* 3804 St Marks Rd Durham NC 27707

KULL, FREDRICK J, b Marion, Ohio, Mar 9, 35; m 66; c 2. BIOCHEMISTRY. *Educ:* Kent State Univ, BS, 60; Ohio State Univ, MSc, 62; Brandeis Univ, PhD(biochem), 67. *Prof Exp:* Assoc biochemist, Biol Div, Oak Ridge Nat Lab, 62-63, investr, 67-69; asst prof, 69-75, ASSOC PROF BIOL, STATE UNIV NY, BINGHAMTON, 75- *Mem:* AAAS; Am Chem Soc; Sigma Xi; NY Acad Sci; Am Soc Biol Chemists. *Res:* Mechanism of yeast phosphoglucose isomerase; effects of hydrocortisone on enzyme induction and RNA synthesis; heterologous aminoacylation between Neurospora crassa and Escherichia coli; mammalian peptidyl-oligonucleotidyl complexes; mammalian aminoacylation reactions; mammalian alkaline ribonuclease and ribonuclease inhibitor. *Mailing Add:* Dept Biol Sci State Univ NY PO Box 6000 Binghamton NY 13902-6000

KULL, LORENZ ANTHONY, b Chicago, Ill, Dec 25, 37; m 85; c 2. NUCLEAR PHYSICS. *Educ:* Ill Inst Technol, BSc, 63; Mich State Univ, PhD(physics), 67. *Prof Exp:* Physicist, Gulf Gen Atomic, Inc, 67-69; physicist, 69-75, vpres & mgr appl sci & technol group, 75-79, exec vpres, 79-88, PRES & CHIEF OPER OFFICER, SCI APPLICATIONS INT CORP, 88- *Concurrent Pos:* Mem, Air Force Studies Bd, 84-90. *Mem:* Am Phys Soc; Am Nuclear Soc; Inst Elec & Electronics Engrs. *Res:* Development of nuclear materials assay instrumentation; experimental studies of direct particle transfer reactions with light nuclei; experimental studies of photoneutron cross-sections, threshold photoneutrons and photo fission; modeling and analysis of nuclear fuel cycle systems; design of military electronics, components and systems. *Mailing Add:* Sci Applns Int Corp 10260 Campus Point Dr San Diego CA 92121

KULLA, JEAN B, b Washington, DC, Sept 9, 49. PETROPHYSICS, ORGANIC GEOCHEMISTRY. *Educ:* Univ Md, BS, 72; Univ Champaign, Ill, MS, 75, PhD(geol), 79. *Prof Exp:* Res geologist, 78-79, sr res geologist, 80-81, RES SPECIALIST, EXXON PROD RES CO, 82- *Mem:* Am Asn Petrol Geologists; Sigma Xi; Grad Women Sci; Geochem Soc. *Res:* Physics and chemistry of shales and how these properties change with increases in temperature and pressure as the shale undergoes burial. *Mailing Add:* 10735 Boardwalk St Houston TX 77042

KULLBACK, JOSEPH HENRY, b Washington, DC, July 16, 33; m 60; c 3. MATHEMATICAL STATISTICS. *Educ:* George Washington Univ, BA, 55; Stanford Univ, MS, 57, PhD(math statist), 60. *Prof Exp:* Mathematician, Stanford Res Inst, 60-67; math statistician, US Naval Res Lab, 67-81; CHIEF SCIENTIST, GRUMMAN DATA SYSTS INC, 81- *Concurrent Pos:* Prof lectr, George Washington Univ, 77- *Mem:* Armed Forces Commun Electronics Asn. *Res:* Operations research; simulation techniques. *Mailing Add:* Grumman Data Systs 6862 Elm St McLean VA 22101

KULLBACK, SOLOMON, b Brooklyn, NY, Apr 3, 07; m 30, 74; c 2. MATHEMATICAL STATISTICS, APPLIED STATISTICS. *Educ:* City Col New York, BS, 27; Columbia Univ, MA, 29; George Washington Univ, PhD(math), 34. *Prof Exp:* Prof, 38-75, EMER PROF STATIST, GEORGE WASHINGTON UNIV, 75- *Mem:* Inst Math Statist; Am Statist Asn; Royal Statist Soc. *Res:* Information theory; analysis of count or categorical data. *Mailing Add:* Dept Statist George Washington Univ Washington DC 20052

KULLBERG, RUSSELL GORDON, b Flint, Mich, Aug 4, 22; m 45; c 2. BOTANY. *Educ:* Univ Mich, BS, 49, MS, 50; Mich State Univ, PhD(bot), 66. *Prof Exp:* Assoc prof, 65-77, PROF BIOL, SOUTHEAST MO STATE UNIV, 77- *Mem:* AAAS; Phycol Soc Am; Ecol Soc Am; Am Soc Limnol & Oceanog; Am Micros Soc. *Res:* Phycology; ecology of hot spring algae. *Mailing Add:* 1824 Westridge Cape Girardeau MO 63701

KULLER, ROBERT G, b Baltimore, Md, Nov 29, 26; m 59; c 5. MATHEMATICS. *Educ:* Swarthmore Col, AB, 48; Univ Mich, MS, 49, PhD(math), 55. *Prof Exp:* Instr math, Dartmouth Col, 53-55; from instr to asst prof, Wayne State Univ, 55-59; asst prof, Dartmouth Col, 59-60, 61-62 & Univ Colo, 62-65; assoc prof, Wayne State Univ, 65-68; ASSOC PROF MATH, NORTHERN ILL UNIV, 68- *Concurrent Pos:* Vis lectr, Nat Taiwan Univ, 60-61. *Mem:* Am Math Soc; Math Asn Am; Soc Indust & Appl Math. *Res:* Functional analysis; computers in undergraduate mathematics curriculum. *Mailing Add:* Dept Math Scis N Ill Univ Dekalb IL 60115

KULLERUD, GUNNAR, b Odda, Norway, Nov 12, 21; m 47; c 5. GEOCHEMISTRY, MINERALOGY-PETROLOGY. *Educ:* Tech Univ Norway, MSc, 46, PhD, 48; Univ Oslo, DSc, 54. *Hon Degrees:* Dr Techn, Tech Univ Norway, 82. *Prof Exp:* Instr, Univ Chicago, 48-49, res assoc, 49-52; res assoc, Univ Oslo, 52-54; sr staff mem, Geophys Lab, Carnegie Inst, 54-71; head dept, 70-76, PROF GEOSCI, PURDUE UNIV, WEST LAFAYETTE, 70- *Concurrent Pos:* Adj prof, Lehigh Univ, 62-71; vis prof, Univ Heidelberg, 64-70; hon collabr, Div Meteorites, Smithsonian Inst, 64-71; mem comt on chem of solar syst, Space Sci Bd, 64-72; consult prof, Tex Tech Univ, 68-71; co-ed Mineralium Deposita, 64-85 & Chem Geol, 66-86. *Honors & Awards:* A H Dumont Award, Belg Geol Soc, 65. *Mem:* Geochem Soc; Soc Econ Geologists; fel Am Mineral Soc; fel Geol Soc Am; Norweg Acad Sci; Royal Acad Arts & Sci Norway. *Res:* Sulfide phase equilibria; geochemistry of ore deposits, meteorites; mineralogy-geochemistry of coal; mineral separation. *Mailing Add:* Dept Earth Sci Purdue Univ Main Campus West Lafayette IN 47907

KULLGREN, THOMAS EDWARD, b Grand Rapids, Mich, Apr 10, 41; div; c 3. MECHANICAL ENGINEERING. *Educ:* US Air Force Acad, BS, 64; Stanford Univ, MS, 72; Colo State Univ, PhD(mech eng), 76. *Prof Exp:* From asst prof to assoc prof eng mech, US Air Force Acad, 76-82, prof & actg head, 82-84; DEAN & PROF MECH ENG, SAGINAW VALLEY STATE UNIV, 84- *Concurrent Pos:* Vis scientist, Air Force Mats Lab, 77, prin investr, Wind Energy Res Proj, 77-84; dir, Ctr Appl Technol Res, Saginaw Valley State Univ, 85 & Bus & Indust Develop Inst, 88-90. *Mem:* Am Soc Mech Engrs; Am Soc Engr Educ; Am Wind Energy Asn. *Res:* Numerical solution of three-dimensional problems in fracture mechanics; international wind energy resource assessments and feasibility studies; technology transfer and economic development. *Mailing Add:* PO Box 444 Midland MI 48640

KULLMAN, DAVID ELMER, b Kenosha, Wis, May 27, 40; m 65; c 2. MATHEMATICS EDUCATION. *Educ:* Northwestern Univ, BA, 62; Cornell Univ, MA, 63; Univ Kans, PhD(math), 69. *Prof Exp:* High sch teacher, Ill, 63-65; from asst prof to assoc prof, 69-81, PROF MATH, MIAMI UNIV, 81- *Mem:* Nat Coun Teachers Math; Sigma Xi; Math Asn Am; Can Soc Hist & Philos Math. *Res:* Geometry point-set topology; history of mathematics; problem solving and applications in school mathematics. *Mailing Add:* Dept Math & Statist Miami Univ Oxford OH 45056

KULLNIG, RUDOLPH K, b Kirchberg, Lower Austria, Oct 2, 18; US citizen; m 54. PHYSICAL ORGANIC CHEMISTRY. *Educ:* Univ Ottawa, Can, PhD(chem), 58. *Prof Exp:* Chemist, Bell-Craig Ltd, Ont, 52-55; asst res chemist, Sterling Res Group, 58-64, res chemist & group leader, 64-68, sect head, 68-90; ADJ PROF, RENSSELAER POLYTECH INST, TROY, NY, 83-, SR FEL, 90- *Mem:* Am Chem Soc; Am Crystallog Asn. *Res:* Nuclear magnetic resonance spectroscopy; indoles and other heterocyclic compounds; single cryst x-ray diffraction. *Mailing Add:* Sterling Res Group Rensselaer NY 12144

KULM, LAVERNE DUANE, b Mobridge, SDak, Feb 17, 36; m 62. GEOLOGICAL OCEANOGRAPHY. *Educ:* Monmouth Col, BA, 59; Ore State Univ, PhD(oceanog), 65. *Prof Exp:* From asst prof to assoc prof, 64-74, PROF OCEANOG, ORE STATE UNIV, 74- *Concurrent Pos:* Fel, Marathon Oil Co, 71. *Mem:* AAAS; Soc Econ Paleontologists & Mineralogists; fel Geol Soc Am; Am Geophys Union. *Res:* Continental margin structure, tectonics, sedimentation; deep-sea sedimentation. *Mailing Add:* Col Oceanog Ore State Univ Oceanog Admin Bldg 104 Corvallis OR 97331-5503

KULMAN, HERBERT MARVIN, b Sayre, Pa, June 12, 29; div; c 2. ENTOMOLOGY. *Educ:* Pa State Univ, BS, 52; Duke Univ, MF, 55; Univ Minn, PhD, 60. *Prof Exp:* Entomologist, Southeastern Forest Exp Sta, USDA, 56-57; asst prof entom, WVa Univ, 59-62; from asst to assoc prof forest entom, Va Polytech Inst, 62-69; assoc prof, 69-72, PROF ENTOM, UNIV MINN, ST PAUL, 72- *Concurrent Pos:* Latin Inst Forestry, Merida Vez, 70; NSF res group, Korea & Taiwan. *Mem:* Entom Soc Am. *Res:* Forest entomology, especially biological control and damage evaluation. *Mailing Add:* Dept Entom Univ Minn St Paul MN 55108

KULP, BERNARD ANDREW, b Columbus, Ohio, Aug 3, 23; m 52; c 9. PHYSICS. *Educ:* Univ Minn, BEE, 46; Ohio State Univ, MS, 47, PhD(physics), 55. *Prof Exp:* Asst, Ohio State Univ, 46-47; res metallurgist, Carnegie-Ill Steel Corp, 47-48; res engr, Battelle Mem Inst, 48-55; physicist, Linde Co, 55-58; physicist, Aerospace Res Lab, Wright-Patterson AFB, 58-69; chief scientist, Air Force Armament Lab, Eglin AFB, 69-75; CHIEF SCIENTIST & DIR LABS, AIR FORCE SYSTS COMMAND, 75- *Mem:* Fel AAAS; Am Phys Soc; Am Defense Preparedness Asn. *Res:* Solid state physics; radiation damage; electrooptics; ordnance engineering. *Mailing Add:* 13418 Queens Lane Ft Washington MD 20744

KULP, STUART S, b Pennsburg, Pa, July 19, 25; m 50; c 3. ORGANIC CHEMISTRY. *Educ:* Gettysburg Col, BA, 50; Lehigh Univ, MS, 53, PhD(chem), 57. *Prof Exp:* Instr gen & anal chem, Lehigh Univ, 55-57; PROF CHEM & HEAD DEPT, MORAVIAN COL, 57- *Honors & Awards:* E Emmet Reid Award, MARM, Am Chem Soc, 75. *Mem:* Am Chem Soc; Sigma Xi; Am Inst Chemists. *Res:* Synthesis and study of tautomerism of 2-cyanocycloalkanones, especially five and six numbered rings. *Mailing Add:* 365 Valley Park S Bethlehem PA 18018

KULSRUD, RUSSELL MARION, b Lindsborg, Kans, Apr 10, 28; m 55; c 3. PLASMA PHYSICS. *Educ:* Univ Md, BA, 49; Univ Chicago, MS, 52, PhD(physics), 54. *Hon Degrees:* MS, Yale Univ, 66. *Prof Exp:* Mem staff physics, Proj Matterhorn, Princeton Univ, 54-59, sr res assoc, 59-64, head theoret sect, Plasma Physics Lab, 64-66; prof appl sci & astron, Yale Univ, 66-67; PROF ASTROPHYS SCI, PRINCETON UNIV, 67- *Concurrent Pos:* Consult, Oak Ridge Nat Lab, 55, RCA Corp, 60 & Gen Atomic Div, Gen Dynamics Corp, 60. *Mem:* Fel Am Phys Soc; Int Astron Union; Am Astron Soc. *Res:* Plasma physics with application to controlled fusion reactor research; astrophysics. *Mailing Add:* Plasma Physics Lab Princeton Univ PO Box 451 Princeton NJ 08544

KULWICH, ROMAN, b New York, NY, Oct 18, 25; m 48; c 3. BIOCHEMISTRY. *Educ:* Univ Fla, BS, 49, PhD(animal nutrit), 51. *Prof Exp:* Animal nutritionist & animal husbandman, Animal Husb Res Div, Agr Res Serv, USDA, 51-57, biochemist & supvry chemist, Mkt Qual Res Div, Agr Mkt Serv, Plant Indust Sta, Beltsville, 57-62; grants assoc, Div Res Grants, NIH, 62-63; scientist adminr, Nat Inst Child Health & Human Develop, 63-64; endocrinol prog dir extramural prog, Nat Inst Arthritis & Metab Diseases, 64-69; health scientist adminr, Nat Ctr Health Serv Res & Develop, 69-71; asst for rev & eval, 71-73, asst dir extramural progs, Nat Inst Allergy & Infectious Dis, 73-78; CONSULT, 78- *Mem:* AAAS; Am Inst Nutrit. *Res:* Nutritional and biochemical research on trace mineral and sulfur metabolism in laboratory and farm animals involving the use of radioactive tracers; body composition research involving 4-pi low level gamma ray measurements; biomedical science administration. *Mailing Add:* 9504 SE 107th Pl Belleview FL 32620

KULWICKI, BERNARD MICHAEL, b Detroit, Mich, July 3, 35. CHEMICAL ENGINEERING, MATERIALS SCIENCE. *Educ:* Univ Detroit, BChE, 58; Univ Mich, MSE, 60, PhD(chem eng), 63. *Prof Exp:* Res fel, Inst Solid State Physics, Czech Acad Sci, 63-64; proj engr semiconductor mat res, 64-69, sect leader, 69-70, br mgr active mat develop, Mat & Elec Prod Group, Mat & Controls Div, 71-79, SR MEM TECH STAFF, ADVAN DEVELOP, TEX INSTRUMENTS INC, 80- *Concurrent Pos:* Assoc ed, J Am Ceramic Soc, 90- *Mem:* AAAS; Am Ceramic Soc; Electrochem Soc; Am Chem Soc; Mat Res Soc; Soc Advan Mat & Process Eng. *Res:* Phase equilibria; thermodynamic and electrical properties of semiconducting materials; ferroelectric materials; thermistors; ceramic varistors. *Mailing Add:* Tex Instruments Inc 34 Forest St M/S 10-13 Attleboro MA 02703

KUMAGAI, LINDY FUMIO, b Rock Springs, Wyo, Aug 5, 27; m 52; c 3. MEDICINE. *Educ:* Univ Utah, BA, 49, MS, 50, MD, 54. *Prof Exp:* Asst anat, Sch Med, Univ Utah, 49-54; med intern, Mass Mem Hosp, 54-55; USPHS res fel, Thorndyke Mem Lab, Boston City Hosp, Harvard Med Sch, 55-57; asst resident med, Univ Hosp, Utah, 57-58, from instr to assoc prof, Col Med, 58-69, asst dean, 68-69; assoc prof, 69-71, PROF INTERNAL MED, SCH MED, UNIV CALIF, DAVIS, 71-, HEAD ENDOCRINE SECT, 69-

Concurrent Pos: Clin investr, Vet Admin Hosp, Salt Lake City, Utah, 58-61; chief radioisotope serv, 61-69; assoc ed, Endocrinology, 73-77; mem, Endocrine Study Sect, Dept Health & Human Sci, NIH, 80-; mem, Calif Bd Med Quality Assurance, 80- *Mem:* Endocrine Soc. *Res:* Metabolism of adrenocortical and thyroidal hormones. *Mailing Add:* Dept Med-Endocrinol Univ Calif Med Ctr 4301 X St Sacramento CA 95817

KUMAI, MOTOI, b Nagano, Japan, Mar 22, 20; m 48; c 2. CLOUD PHYSICS, EARTH SCIENCES. *Educ:* Sci Univ Tokyo, BS, 41; Hokkaido Univ, PhD(physics), 57. *Prof Exp:* Res assoc physics, Hokkaido Univ, 42-55, lectr, 55-58; res assoc cloud physics, Univ Chicago, 58-61; res physicist atmospheric sci, US Army Cold Regions Res & Eng Lab, 61-90; RES PHYSICIST, SCI & TECHNOL CORP, 90- *Mem:* Am Meteorol Soc; Int Glaciol Soc; Sigma Xi; Phys Soc Japan; Meteorol Soc Japan; Clay Minerals Soc. *Res:* Physics of atmosphere, research on snow crystal nuclei, and ice fog nuclei; electron diffraction of ice and aerosols, and attenuation of infrared radiation; scanning electron microscopy and acid snow and rain. *Mailing Add:* 18 Marlyn Dr Burnt Hills NY 12027-9737

KUMAMOTO, JUNJI, b Sacramento, Calif, May 9, 24; m 50; c 4. PHYSICAL ORGANIC CHEMISTRY. *Educ:* Univ Calif, Los Angeles, BS, 50; Univ Chicago, PhD(phys org chem), 53. *Prof Exp:* NSF grant, Harvard Univ, 53-55; chemist, Shell Develop Co, 55-60 & res lab, IBM Corp, 60-66; LECTR & CHEMIST, UNIV CALIF, RIVERSIDE, 66- *Mem:* Am Chem Soc; NY Acad Sci; AAAS; Sigma Xi. *Res:* Reaction mechanisms of phosphate ester hydrolysis; free radical-metal ion reactions; relationship between structure and spectra; thermodynamic basis for temperature breaks in arrhenius plots. *Mailing Add:* Dept Bot & Plant Sci Univ Calif Riverside CA 92521

KUMAR, AJIT, b Bihar, India, Mar 2, 40; m 71; c 1. CELL BIOLOGY, MOLECULAR GENETICS. *Educ:* Bihar Univ, India, BSc, 58, MSc, 60; Univ Chicago, PhD(biol), 68. *Prof Exp:* Fel biochem, Albert Einstein Col Med, 68-71; res assoc, Harvard Med Sch, 71-75, asst prof microbiol & molecular genetics, 77; tutor biochem & molecular biol, Harvard Univ, 77-79; ASSOC PROF BIOCHEM, SCH MED, GEORGE WASHINGTON UNIV, 79- *Concurrent Pos:* Vis fel microbiol, Uppsala Univ, Sweden, 71; vis scientist biochem, Cambridge Univ, 77; assoc prof genetics, George Washington Univ, 80-; guest res scientist, Lab Biochem, Nat Cancer Inst, NIH, 80- *Mem:* Am Soc Cell Biol; Am Soc Biol Chemists. *Res:* RNA protein complexes and their role in eukaryotic gene expression. *Mailing Add:* Dept Biochem Sch Med George Washington Univ 2300 Eye St NW Washington DC 20037

KUMAR, ALOK, b Meerut, India, Sept 22, 51; m 80. COMPUTER AIDED DESIGN. *Educ:* Indian Inst Technol, Kanpur, India, BTech, 72, MTech, 76; Univ Houston, PhD(mech eng), 80. *Prof Exp:* Asst prof mech eng, Univ Wis-Platteville, 79-81; ASST PROF MECH ENG, UNIV DEL, NEWARK, 81- *Concurrent Pos:* Res & teaching asst, Indian Inst Technol, Kanpur, India, 73-76; res & teaching fel, Univ Houston, 76-79. *Mem:* Am Soc Mech Engrs; Soc Mfg Engrs; Am Soc Eng Educ. *Res:* Computer-aided design; kinematics; robotics and mechanical manipulator characterization; design and control; mathematical modeling of manufacturing processes; biomechanics. *Mailing Add:* Dept Elec Eng WVa Univ PO Box 6101 Morgantown WV 26506

KUMAR, CIDAMBI KRISHNA, b Madras, India, Sept 24, 37; m 68; c 2. ASTROPHYSICS, ATOMIC PHYSICS. *Educ:* Andhra Univ, India, BSc, 57; Univ Wis, MS, 65; Univ Mich, PhD(astron), 69. *Prof Exp:* Jr sci officer nuclear physics, AEC, India, 58-63; Carnegie fel atomic physics & astrophys, Carnegie Inst Washington Dept Terrestrial Magnetism, 70-72; asst prof, 72-77, ASSOC PROF PHYSICS & ASTRON, HOWARD UNIV, 77- *Concurrent Pos:* Res assoc, Carnegie Inst Washington Dept Terrestrial Magnetism, 73- *Mem:* Am Astron Soc; Sigma Xi. *Res:* Spectrophotometry of galaxies and comets; beam-foil spectroscopy; radio astronomy. *Mailing Add:* 4281 Bright Bay Way Ellicott City MD 21043

KUMAR, DEVENDRA, b Delhi, Sept 14, 44; Indian citizen; m 69; c 1. LASER SPECTROSCOPY, OPTOGALVANIC & PHOTOACOUSTIC SPECTROSCOPY. *Educ:* Univ Delhi, India, BSc, 63, MSc, 65, PhD(physics), 76. *Prof Exp:* Asst lectr, 65-68, lectr, physics, Kirori Mal Col, Univ Delhi, 68-78; res assoc, 78-85, ASST PROF CHEM, LA STATE UNIV, 86- *Concurrent Pos:* Asst prof, Dept Physics, Southern Univ, Baton Rouge, 86-87. *Mem:* Inst Elec & Elecronic Engr; Optical Soc Am; Am Chem Soc. *Res:* Laser optogalvonic spectroscopy in dc and rf discharges; study of highly excited Rydberg states of atoms and simple molecules; laser photoacoustic spectroscopy. *Mailing Add:* Dept Chem La State Univ Baton Rouge LA 70803

KUMAR, GANESH N, b Madras, India, Oct 4, 48; m 75; c 2. POLYMER SCIENCE, CHEMICAL ENGINEERING. *Educ:* Univ Madras, BTech, 70; Clarkson Col Technol, MS, 72; Case Western Reserve Univ, PhD(polymer sci), 75; MBA, 88. *Prof Exp:* Assoc scientist, Xerox Corp, 74-78; sr res scientist, Johnson & Johnson Dent Prod Co, 78-80, mgr polymer res, 80-81,; dir polymer sci, Vistakon Inc, Jacksonville, Fla, 83-85, dir polymer sci & quality assurance, 85-87, VPRES RES, VISTAKON INC, JACKSONVILLE, FLA, 87- *Concurrent Pos:* Res assistantship, Clarkson Univ, 70-71; res fel CWRU, 71-74. *Honors & Awards:* Gold Medal, Univ Madras, 70. *Mem:* Am Chem Soc; Am Phys Soc; NAm Thermal Analysis Soc; Sigma Xi; AAAS. *Res:* Polymer structure property relationships; polymer mechanical and rheological properties; polymer blends and composites. *Mailing Add:* Vistakon Inc PO Box 10157 Jacksonville FL 32247

KUMAR, K S P, US citizen. ELECTRICAL ENGINEERING. *Educ:* Purdue Univ, MS, 61, PhD(elec eng), 64. *Prof Exp:* PROF ELEC ENG, UNIV MINN, 64- *Mem:* Inst Elec & Electronics Engrs. *Res:* Adaptive control of processes; stochastic filtering algorithms; robotic control. *Mailing Add:* 1155 W Sandhurst Dr Roseville MN 55113

KUMAR, K SHARVAN, b Hyderabad, India, Oct 12, 56; m 85. HIGH TEMPERATURE STRUCTURAL MATERIALS & INTERMETALLICS, AEROSPACE LIGHTWEIGHT ALLOY DEVELOPMENT. *Educ:* Drexel Univ, Philadelphia, MS, 81, PhD(mat eng), 84. *Prof Exp:* Postdoctoral fel mat, Drexel Univ, 84-85; scientist, 85-89, SR SCIENTIST, MARTIN MARIETTA LABS, 89- *Mem:* Mat Res Soc; Metals Soc. *Res:* Lightweight materials for aerospace applications, particularly hot structures including jet engine components, missile fins, leading edges of wings and nose cones for hypersonic transportation; fundamental mechanisms controlling the microstructure and mechanical properties of such materials; author of 50 technical publications. *Mailing Add:* Martin Marietta Labs 1450 S Rolling Rd Baltimore MD 21227

KUMAR, KAPLESH, b Lucknow, India, Nov 9, 47; m 74; c 2. MATERIALS SCIENCE. *Educ:* Indian Inst Technol, Kanpur, BTech, 69, Stevens Inst Technol, MS, 71; Mass Inst Technol, ScD, 75. *Prof Exp:* Staff scientist, 75-80, chief, Mat Develop Sect, 80-88, CHIEF, MAT SCI & TECHNOL SECT, CHARLES STARK DRAPER LAB, INC, 88- *Honors & Awards:* Invention Disclosure Award, NASA. *Res:* Samarium-transition-metal permanent magnets; metal matrix composites; ion implantation; chemical vapor deposition; dimensional stability; flex lead corrosion; printed circuit board adhesion degradation; manganese-zinc ferrites; rotating electrical contacts; high Tc superconductors; cold fusion. *Mailing Add:* 25 Redwing Rd Wellesley MA 02181

KUMAR, MADHURENDU B, b Khagaria, India, Jan 4, 42; m; c 2. HYDROLOGY & WATER RESOURCES. *Educ:* Indian Sch Mines, Ranchi Univ, BS, 61, MS, 62; La State Univ, PhD(geol), 72. *Prof Exp:* Instr, Indian Sch Mines, Minerol & Petrol, 62-63; sci officer, Atomic Mines & Minerals Div, Dept Atomic Energy, Govt India, 63; sr exec oil geologist, Oil India, Ltd, 63-69; instr geol & geography, City Univ New York, Hunter Col, 74-77; res assoc geol, Inst Environ Studies, 75, sr investr geohydrol, 78-82, sr res assoc, La State Univ, 78-82; GEOLOGIST SUPVR, OIL & GAS DIV, OFF CONSERV, LA DEPT NATURAL RESOURCES, 82- *Concurrent Pos:* Consult, 72-; vis prof, La State Univ, 77-78, Univ Southwest La, 79, Southern Univ, 78-90. *Honors & Awards:* Sir Henry Hayden Medal, Mining, Metal & Geol Inst India, 78. *Mem:* Am Asn Petrol Geologists; Am Inst Prof Geologists; fel Geol Soc Am. *Res:* Subsurface of petroleum geology; salt domes; mine hydrology; petroleum resource conservation. *Mailing Add:* 7744 Wimbledon Ave Baton Rouge LA 70810

KUMAR, MAHESH C, b Montgomery, WPakistan, Sept 21, 35; m; c 4. VETERINARY MICROBIOLOGY, PUBLIC HEALTH. *Educ:* Univ Bihar, BVSc & AH, 58; Univ Minn, Minneapolis, MS, 64, PhD(vet microbiol), 67. *Prof Exp:* Vet asst surg, Animal Husb Dept, Bihar, India, 58-61; res fel, 67, res assoc vet microbiol, Univ Minn, St Paul, 67-76; dir vet servs, Mile High Turkey Hatchery, Longmont, Colo, 76-84; dir vet servs, Mill Farms Co, Paynesville, Minn, 84-85; dir vet serv, Koronis Mills, 86-89; DIR VET SERV, E B OLSON FARMS, ATWATER, MINN, 89- *Mem:* Am Vet Med Asn; Am Asn Avian Path; Poultry Sci Asn; Am Soc Microbiol; Conf Res Workers Animal Diseases. *Res:* Poultry diseases; mycoplasma and salmonella infections in turkeys; elimination of salmonella from turkeys and their environment; prevention and treatment of diseases in turkeys. *Mailing Add:* E B Olson Farms Div Jennie-O Foods Box 439 Atwater MN 56209

KUMAR, PANGANAMALA RAMANA, b Nagpur, India, April 21, 52; m 82; c 2. STOCHASTIC SYSTEMS, MANUFACTURING SYSTEMS. *Educ:* Indian Inst Technl, BTech, 73; Wash Univ, MS, 75, DSc, 77. *Prof Exp:* From asst prof to assoc prof math, Univ Md, Baltimore County, 77-84; assoc prof, 85-87, PROF ELEC & COMPUT ENG, UNIV ILL, URBANA-CHAMPAIGN, 87- *Honors & Awards:* Donald P Eckman Award, Am Automotive Control Coun, 85. *Mem:* Fel Inst Elec & Electronics Engrs. *Res:* Systems theory & its applications; adaptive control; stochastic systems; manufacturing systems; optimization; game theory; neural networks; communication networks. *Mailing Add:* Univ Ill CSL 1101 W Springfield Ave Urbana IL 61801

KUMAR, PRADEEP, b Allahabad, India, Jan 1, 49; m 87; c 1. NONLINEAR PHENOMENA, SUPERCONDUCTIVITY. *Educ:* Univ Lucknow, India, BSc, 66; Indian Inst Technol, Kanpur, MSc, 68; Univ Calif, San Diego, PhD(physics), 73. *Prof Exp:* Res assoc physics, Univ Wis-Milwaukee, 73-75; res assoc, Univ Southern Calif, 75-77, asst prof, 77-78; asst prof, 79-83, ASSOC PROF PHYSICS, UNIV FLA, 83- *Concurrent Pos:* Nordita prof, Helsinki Univ Technol, Finland, 78-79; guest prof, Nordita, Copenhagen, Denmark, 78-79. *Mem:* Am Phys Soc; NY Acad Sci; AAAS. *Res:* Theoretical ultra low temperature physics; nonlinear phenomena; solitons; magnetism; superconductivity. *Mailing Add:* Phys Dept Univ Fla Gainesville FL 32611

KUMAR, ROMESH, b Rajpura, India, Oct 18, 44; m 76; c 2. ENERGY SYSTEMS DESIGN & ANALYSIS, FUEL CELLS. *Educ:* Panjab Univ, BSc, 65; Univ Calif, Berkeley, MS, 68, PhD(chem eng), 72. *Prof Exp:* Appointee, 72-74, asst chem engr, 74-76, CHEM ENGR, ARGONNE NAT LAB, 76- *Mem:* Am Inst Chem Engrs. *Res:* Food processing; fast breeder reactor safety analysis; transformation and transport of environmental pollutants; fuel cells; Tex typesetting system and development and training; computer graphics and other software; awarded five patents and 100 research and development awards. *Mailing Add:* Chem Eng Div Argonne Nat Lab Argonne IL 60439-4837

KUMAR, S, b Ernakulam, India, July 31, 59. SOLID STATE PHYSICS. *Educ:* Indian Inst Technol, MS, 80; Pa State Univ, PhD(physics), 86. *Prof Exp:* Grad asst, Pa State Univ, 80-86; postdoctoral researcher, Ohio State Univ, 86-89; asst prof, Northeast Mo State Univ, 89-90; RES PHYSICIST, QUANTUM MAGNETICS INC, 90- *Mem:* Am Phys Soc. *Res:* Liquid helium and the design and construction of cryogenic apparatus; design, construction and testing of SQUID-based magnetometers. *Mailing Add:* Quantum Magnetics 11578 Sorrento Valley Rd No 30 San Diego CA 92121

KUMAR, S ANAND, b Bangalore, India, Mar 12, 36; m 63. BIOCHEMISTRY. *Educ:* Univ Mysore, BSc, 54; Univ Poona, MSc, 57; Indian Inst Sci, PhD(biochem), 63. *Prof Exp:* Fel biochem, Sch Med, Tufts Univ, Boston, Mass, 63-65; res assoc biochem, Scripps Clin & Res Found, La Jolla, Calif, 65-66; lectr biochem, Indian Inst Sci, Bangalore, 66-71; guest scientist biochem, Roche Inst Molecular Biol, Nutley, NJ, 71-72; vis asst prof biol, Hunter Col, City Univ NY, 72-77; RES SCIENTIST CLIN SCI, WADSWORTH CTR LABS & RES, NY STATE DEPT HEALTH, 78- *Concurrent Pos:* Adj prof, State Univ NY, 80-; mem gov bd, Astra Res & Develop Ctr, Bangalore, India, 84-, consult, Fine Chem Div, Astra-IDL, Ltd, 85- *Mem:* Sigma Xi; Am Soc Biol Chemists; NY Acad Sci. *Res:* Structure and mode action of enzymes; genetic transcriptions; bacterial RNA polymerase and its site-specific inhibitors; steroid hormone receptors-DNA interactions; steroid hormone controlled gene expression. *Mailing Add:* Astra Res Ctr PO Box 359 Malleswaram Bangalore 560003 India

KUMAR, SHIV SHARAN, b Bannu, India, Mar 15, 39; m 64; c 3. ASTRONOMY. *Educ:* Univ Mich, PhD(astron), 62. *Prof Exp:* Asst astron, Univ Observ, Univ Mich, 57-60; astrophysicist, Smithsonian Astrophys Observ, 60-61; res assoc, Goddard Inst Space Studies, 62-63; staff mem, Phys Res Lab, India, 63-65; asst prof, 65-68, ASSOC PROF ASTRON, UNIV VA, 68- *Mem:* Fel AAAS; Int Astron Union; Am Astron Soc; fel Royal Astron Soc. *Res:* Stellar atmospheres; stellar structure and evolution; origin of the solar system; celestial mechanics. *Mailing Add:* Dept Math-Astron Rm 314 Univ Va PO Box 3818 Univ Sta Charlottesville VA 22903

KUMAR, SHRAWAN, b Allahabad, India, July 1, 39; m; c 2. BIOMECHANICS, ERGONOMICS. *Educ:* Univ Allahabad, BSc, 59, MSc, 62; Univ Surrey, PhD(physiol), 71. *Prof Exp:* Lectr zool, Univ Allahabad, 62-66; pool officer orthop, All-India Inst Med Sci, 73-74; res assoc rehab med, Univ Toronto, 74-77; from asst prof to assoc prof, 77-81, PROF PHYS THER, UNIV ALTA, 82- *Concurrent Pos:* Fel eng, Univ Dublin, 71-73; asst prof biomed eng, Univ Alta, 77-80. *Mem:* Am Soc Biomech; Orthop Res Soc; Human Factors Soc; Human Factors Soc Can; Ergonomics Soc; Am Ind Hygien Asn. *Res:* Work physiology; occupational biomechanics; tissue biomechanics. *Mailing Add:* Dept Phys Ther Univ Alta Edmonton AB T6G 2G4 Can

KUMAR, SOMA, b Lucknow, India, May 16, 24; m 55; c 3. BIOCHEMISTRY. *Educ:* Univ Lucknow, BSc, 44, MSc, 45; Univ Md, PhD, 53. *Prof Exp:* Res assoc, Univ Md, 53-54; lectr biochem, Univ Lucknow, 54-56; asst prof, All-India Inst Med Sci, New Delhi, 56-58; from asst prof to assoc prof, 58-72, PROF CHEM, GEORGETOWN UNIV, 72- *Mem:* AAAS; Am Chem Soc; Am Soc Biol Chemists; Sigma Xi. *Res:* Biosynthesis of fatty acids; phospholipase A2. *Mailing Add:* Dept Chem Georgetown Univ Washington DC 20057-0001

KUMAR, SUDHIR, b Saharanpur, India, Oct 31, 33; m; c 3. RAILROAD ENGINEERING. *Educ:* Agra Univ, BSc, 50, MSc, 52; Indian Inst Sci, Bangalore, AIISc, 55; Pa State Univ, PhD(eng mech), 58. *Prof Exp:* Demonstr physics, Bareilly Col, Agra Univ, 50-51; res asst eng mech, Pa State Univ, 55-57, res assoc, 57-58, asst prof, 58-59; asst solid mech br, Eng Sci Div, Off Ord Res, US Army, 58-59, chief, 59-62, assoc dir, Eng Sci Div, US Army Res Off-Durham, 62-71; chmn, 71-78, PROF MECH & AEROSPACE ENG, ILL INST TECHNOL, 71-, DIR, RAILROAD ENG LAB, 78- *Concurrent Pos:* Vis lectr, Duke Univ, 58-62, vis assoc prof, 62-71 & NC State Univ, 68-71; vis lectr, Acad Railway Sci, People's Repub China, 83-84; mem, Fac Adv Comn, Ill Bd Higher Educ, 85-87. *Honors & Awards:* Octave Chanute Medal, Western Soc of Engrs, 86. *Mem:* Am Inst Aeronaut & Astronaut; Indian Soc Theoret & Appl Mech; Assoc Railway Engrs Am; Am Acad Mech; Am Soc Mech Engrs. *Res:* Wheel and rail interaction; railroad engineering; materials; helicopters. *Mailing Add:* Dept Mech & Aerospace Eng Ill Inst Technol Chicago IL 60616

KUMAR, SUDHIR, b Anjhi, India, Sept 16, 42; m 68; c 2. BIOCHEMISTRY, NEUROCHEMISTRY. *Educ:* Univ Rajasthan, India, BS, 59, MS, 61; Univ Lucknow, India, PhD(biochem), 66. *Prof Exp:* Res assoc pharmacol, Baylor Col Med, 67-68; sr res scientist, NY State Res Inst Neurochem, Columbia Univ, 68-69; chief biochemist pediat, Methodist Hosp, Brooklyn, 69-73; res biochemist hematol, Vet Admin Hosp, Brooklyn, 73-75; dir perinatal res & lab, Christ Hosp, Oak Lawn, Ill, 75-81, dir pediat res, 78-81; from asst prof to assoc prof, 76-86, PROF BIOCHEM & NEUROL SCI, MED SCH, RUSH UNIV, 86-; DIR, CLIN REGION LAB, HAZEL CREST, ILL, 88- *Concurrent Pos:* Int Brain Res Org res fel award, UNESCO & Govt France, 71; Dreyfus Med Found fel, 71 & 73; consult scientist med res, Vet Admin Hosp, Hines, Ill, 76-79; dir & pres, clin diagnostics, Hazel Crest, Oak Forest, Ill, 82-88; consult, Govt India & Unesco Prog, 87, 89 & 91. *Mem:* Fel NY Acad Sci; Am Soc Biol Chemists; Am Inst Nutrit; Am Soc Neurochem; Soc Exp Biol & Med; Am Soc Clin Nutrit; Soc Pediat Res; Am Soc Microbiol; Int Soc Neurochem; fel Nat Acad Clin Biochemists; fel Am Inst Chemists. *Res:* Study of vitamin B12 metabolism; effect of malnutrition on brain development and its correlation to mental retardation; changes in levels of nucleic acids and enzymes of purine catabolism; changes in amino acid levels; developing brain and metabolic disorders in newborn; perinatal medicine and development of screening tests for metabolic disorders in newborn; development of serum-free culture media and its use in growing neural and amniotic fluid cells; development of rapid tests for use in clinical laboratory; alzheimer's disease; effects of drugs in controlling syndromes associated with HIV (AIDS etc). *Mailing Add:* 18901 Springfield Flossmoor IL 60422

KUMAR, SURIENDER, b Panjab, India, Dec 5, 38; m 65; c 4. ORGANIC CHEMISTRY, BIOCHEMISTRY. *Educ:* Univ Delhi, BSc, 58, MSc, 60; Boston Univ, PhD(org chem), 67. *Prof Exp:* Lectr chem, Deshbandhu Col, Delhi, 60-62; fel, Cornell Univ, 66-68; fel, Univ Wis-Madison, 68; res chemist, Vet Admin Hosp, 68-70; asst prof biochem, 71-75, assoc prof, 75-81, PROF BIOCHEM, COL MED & DENT NJ, 81- *Mem:* AAAS; Am Chem Soc; fel Am Inst Chemists; NY Acad Sci; Am Soc Biol Chemists; Am Asn Cancer Res. *Res:* Cancer research; proteolytic enzymes; all transformation. *Mailing Add:* Dept Biochem Univ Md-NJ Med Sch Newark NJ 07103

KUMAR, VIJAY, b Punjab, India, Apr 15, 45; m 73. LABORATORY MEDICINE, IMMUNOLOGY. *Educ:* Panjab Univ, BS, 66, MS, 68; State Univ NY Buffalo, PhD(biochem), 73; Am Bd Med Microbiol, dipl; Am Bd Med Lab Immunol, dipl. *Prof Exp:* Teaching asst biochem, State Univ NY Buffalo, 69-73; fel, E J Meyer Mem Hosp, 73-74, fel immunol, Erie County Lab, E J Meyer Mem Hosp, 74-76; asst prof, Dept Microbiol, 80-86, ASST DIR, IF TESTING SERV, BUFFALO, 76-; ASSOC RES PROF, STATE UNIV NY, BUFFALO, 87-, PRES, IMMUNOL DIAG, 87- *Concurrent Pos:* From clin instr to clin asst prof, Dept Microbiol, State Univ NY Buffalo, 74-76. *Mem:* Am Soc Microbiol; Soc Investigative Dermat. *Res:* Isolation of proteins, enzymes, autoimmunity, and immunochemistry. *Mailing Add:* 219 Sherman Hall State Univ NY Buffalo NY 14214

KUMAR, VINAY, b Montgomery, India, Dec 24, 44; m 72; c 2. CANCER, IMMUNOLOGY. *Educ:* Poona Univ, BSc, 62; Punjab Univ, MBBS, 67; All India Inst Med Sci, MD(path), 72. *Prof Exp:* Tutor path, All India Inst Med Sci, 69-72; from instr to asst prof, Boston Univ, 72-78, assoc prof path & microbiol, Sch Med, 78-82; assoc prof, 82-83, PROF PATH, SOUTHWESTERN MED CTR, UNIV TEX, 83-, CHARLES T ASHWORTH PROF, 86- *Concurrent Pos:* Med Found Inc fel, Boston, 74-76; prin investr, Nat Cancer Inst, 77-; Am Cancer Soc res scholar award, 78; mem, Immunobiological Study Sect, NIH, 84-87. *Mem:* Am Asn Immunol; Sci Res Soc NAm. *Res:* Tumor immunology; virus induced cancer; bonemarrow transplantation. *Mailing Add:* Southwestern Med Sch Univ Tex 5323 Harry Hines Blvd F2-406 Dallas TX 75235

KUMAR, VIPIN, b Muzaffarnagar, Uttar Prad, Oct 21, 56; m 82; c 2. PARALLEL COMPUTING, ARTIFICIAL INTELLIGENCE. *Educ:* Univ Roorkee, Uttar Prad, India, BE, 77; Philips Int Inst, Eindhoven, Neth, ME, 79; Univ Md, College Park, PhD(computer sci), 82. *Prof Exp:* Asst prof, Dept Computer Sci, Univ Tex, Austin, 83-89; ASSOC PROF, DEPT COMPUTER SCI, UNIV MINN, 89- *Mem:* Inst Elec & Electronics Engrs; Asn Comput Mach; Am Asn Artificial Intel. *Res:* Algorithms for solving various scientific and artificial intelligence problems on massively parallel computers. *Mailing Add:* 5-215 EE/C Sci Bldg 200 Union St SE Minneapolis MN 55455

KUMARAN, A KRISHNA, b Govada, India, July 17, 32; US citizen; m 56; c 1. DEVELOPMENTAL BIOLOGY. *Educ:* Univ Madras, BSc, 50, MSc, 55, PhD(zool), 59. *Prof Exp:* Demonstr zool, Sri Venkateswara Univ, India, 55-57, lectr, 57-62; NIH trainee, Western Reserve Univ, 62-65; reader, Osmania Univ, India, 65-68; sr res assoc biol, Case Western Reserve Univ, 68-69; assoc prof, 69-73, PROF BIOL, MARQUETTE UNIV, 73- *Concurrent Pos:* NSF res grants, 70-; NIH res grant, 75-; vis scientist, Czech Acad Sci, 78; vis scientist biol, Harvard Univ, 83; USDA res grant, 86-; vis lectr, Univ Guam, 88. *Mem:* Fel AAAS; Entom Soc Am; Am Soc Cell Biol; Am Soc Zoologists; Int Soc Develop Biol; Am Soc Develop Biol. *Res:* Molecular entomology; role of hormones in control of insect development; hormonal control of specific gene expression during post embryonic development in insects. *Mailing Add:* Dept Biol Marquette Univ Milwaukee WI 53233

KUMARAN, MAVINKAL K, b Cannanore, Kerala, India, June 1, 46; m 71; c 2. CHEMISTRY. *Educ:* Univ Kerala, India, BS, 65, MS, 67; Univ London, PhD (chem thermodynamics), 76. *Prof Exp:* Lectr phys chem, Sree Narayana Col, Ind, 67-80; res fel thermodynamics, Massey Univ, NZ, 80-; res assoc, 81-84, res officer, 84-89, SR RES OFFICER, NAT RES COUN, CAN, 90- *Concurrent Pos:* Res fel, Calcut Univ, India, 72-73; commonwealth scholar, Univ London, 73-76. *Res:* Experimental and theoretical investigations on properties of liquids and liquid mixtures, especially in the critical region; experimental and theoretical investigation on heat and moisture transport properties of insulating materials. *Mailing Add:* Inst Res Construct Nat Res Coun Can Ottawa ON K1A 0R6 Can

KUMARI, DURGA, b New Delhi, India, Apr 27, 51. ORGANIC CHEMISTRY, ANALYTICAL CHEMISTRY. *Educ:* Meerut Univ, India, BS, 72, MS, 74; Delhi Univ, India, PhD(chem), 80. *Prof Exp:* Scientist chem, Indian Inst Technol, 79-81; res assoc chem, Univ Minn, Bemidji State Univ, Fla A&M Univ & Okla State Univ, 81-87; RESEARCHER CHEM, MAT DIV, FED HWY ADMIN, 87- *Concurrent Pos:* Fulbright scholar, Fulbright Off, Wash, DC, 77-78; postdoctoral fel chem, Okla State Univ, Fla A&M Univ, Bemidji State Univ, Univ Minn, 81-87. *Mem:* Am Chem Soc; Am Soc Mass Spectrometry; AAAS; Indian Women Sci Asn. *Res:* Isolation, identification and synthesis of novel compounds using organic chemistry; material science chemistry; author of various publications. *Mailing Add:* Fed Hwy Admin HNR-30 6300 Georgetown Pike McLean VA 22101

KUMAROO, KUZIYILETHU KRISHNAN, b Kerala, India, Apr 6, 31; m 67; c 2. BIOCHEMISTRY. *Educ:* Kerala Univ, India, BSc, 55; Univ NC, PhD(biochem), 69. *Prof Exp:* Chemist, Capsulation Serv, India, 55-56; clin chemist, Grant Med Col, Bombay, 56-57; petrol chemist, Kuwait Oil Co, Arabia, 57-63; res asst biochem, Univ NC, Chapel Hill, 63-68; NIH trainee, Univ Mich Med Ctr, 68-71; asst prof biochem, Univ NC, Chapel Hill, 71-79; RES BIOCHEMIST, DEPT HYPERBARIC MED, NAVAL MED RES INST, BETHESDA, MD, 79- *Mem:* Sigma Xi; Am Soc Biochem & Molecular Biol; Oxygen Soc; Nat Asn Retarded Citizens; Undersea Soc Int; Planetary Soc Int. *Res:* Biochemistry of circulating blood factors and cells that regulate pulmonary, cardiovascular and central nervous system functions in normal and pathological conditions; cerebral ischemia and thrombosis; biochemistry of decompression sickness; regulatory role of proteases; basic chromosomal proteins in differentiating cells; biochemistry of oxygen/hydrogen gases in diving. *Mailing Add:* Dept Hyperbaric Med Naval Med Res Inst Bethesda MD 20889

KUMBAR, MAHADEVAPPA M, b Tallur, India, Nov 15, 39; m; c 2. PHYSICAL CHEMISTRY, BIOPHYSICS. *Educ:* Karnatak Univ, India, BSc, 61, MSc, 63; Adelphi Univ, PhD(phys chem), 69. *Prof Exp:* Lectr, Parle Col, India, 63-65; res assoc biophys, 69-71, adj asst prof, 71-79, ADJ ASSOC PROF BIOPHYS, ADELPHI UNIV, 79-; CHIEF, INFO MGT SYSTS, PILGRIM PSYCHOL CTR, 87- *Concurrent Pos:* Adj fac, Nassau Community Col, 71- *Mem:* Am Chem Soc. *Res:* Statistical mechanics of macromolecules; dynamic and mechanical properties, conformational changes and conformational studies. *Mailing Add:* Dept Chem Adelphi Univ Garden City NY 11530

KUMBARACI-JONES, NURAN MELEK, b Instanbul, Turkey, Apr 3, 44; m 81; c 2. TEACHING, NEUROSCIENCES. *Educ:* Robert Col, Istanbul, Turkey, BS, 66; Columbia Univ, MS, 73, MA, 75, MPhil, 76, PhD(physiol), 77. *Prof Exp:* Chemist res & develop, Eczacibasi Pharmaceut Co, 66-71; NIH fel, dept physiol, Col Physicians & Surgeons, Columbia Univ, 77-79; asst prof, 79-84, ASSOC PROF CHEM, STEVENS INST TECHNOL, 84- *Mem:* Am Physiol Soc; Soc Neurosci; NY Acad Sci; Sigma Xi. *Res:* Effects of chemicals on nerve and muscle tissues; electrophysiological properties of excitable cells; determination of acetylcholinestenase activity; isolation and identification of bioactive compounds. *Mailing Add:* Dept Chem & Chem Eng Stevens Inst Technol Hoboken NJ 07030

KUMINS, CHARLES ARTHUR, b New York, NY, Jan 13, 15; m 40; c 3. SURFACE CHEMISTRY. *Educ:* City Col New York, BS, 36; Polytech Inst Brooklyn, MS, 41. *Prof Exp:* Jr chemist, Titanium Div, Nat Lead Co, NJ, 37-41; res chemist & group leader, Wyandotte Chem Co, Mich, 41-42; sr res chemist, Res Labs, Interchem Corp, 42-50, head dept inorg & phys chem, 50-54, dir dept, 54-59, dir textile chem, 59-63, asst dir, 63-67; dir res lab, Charles Bruning Co, 67-71; dir res & develop, Graphics Develop Lab, Addressograph-Multigraph Co, 71-73; dir res & develop, Tremco Corp, 73-80; vpres sci & technol, Sherwin Williams Co, 80-81; CONSULT, 82- *Concurrent Pos:* Secy & trustee, Paint Res Inst, 75- *Honors & Awards:* Roon Awards, Fed Socs Coatings Technol, 65 & 76; Matticello lectr, 79; Sci Achievement Award, Fed Soc Coatings Technol. *Mem:* AAAS; Am Chem Soc; fel Am Inst Chemists. *Res:* Rheology and particle size; colloids and surface chemistry; high temperature reactions; inorganic pigments; metallo-organics; iron compounds; alumina silicates; physical chemistry of polymers; transport phenomena in polymers; photochemistry; imaging systems; electrophotography; sealants and adhesives. *Mailing Add:* RD 4 Box 439 Easton MD 21601

KUMKUMIAN, CHARLES SIMON, b Meriden, Conn, June 17, 20; m 61; c 4. MEDICINAL CHEMISTRY. *Educ:* Temple Univ, BS, 44, MS, 51; Univ Md, PhD(med chem), 62. *Prof Exp:* Instr, Temple Univ, 47-51; instr chem, Univ Md, 57-60; asst ed, Chem Abstr Serv, 62-63; chemist, Bur Med, 64-68, supvry chemist, Bur Drugs, 68-72, ASST DIR CHEM, OFF DRUG RES & REVIEW, US FOOD & DRUG ADMIN, 72- *Mem:* Am Chem Soc. *Res:* Synthesis and biological activity of steroids; structure activity relationships; analytical chemistry. *Mailing Add:* 5919 Holland Rd Rockville MD 20851

KUMLER, MARION LAWRENCE, plant ecology, environmental studies; deceased, see previous edition for last biography

KUMLER, PHILIP L, b Columbus, Ohio, May 23, 41; div; c 1. ORGANIC CHEMISTRY, POLYMER CHEMISTRY. *Educ:* Miami Univ, BA, 62; Univ Rochester, PhD(chem), 67. *Prof Exp:* Jr chemist indust res, Procter & Gamble Co, 62, summer staff, Indust Soap & Chem Prod, 63 & 64; fac asst chem, Wabash Col, 64; NATO fel, Univ Copenhagen, 67-68; fel & res assoc, Univ Chicago, 68-69, NIH fel, 69-70, fel & res assoc, 70; from asst prof to assoc prof chem, Saginaw Valley State Col, 70-76; assoc prof, 76-80, PROF CHEM, STATE UNIV NY, FREDONIA, 80- *Concurrent Pos:* Vis res assoc macromolecular sci, Case Western Reserve Univ, 81; vis prof, Mich Molecular Inst, 87, 88, 89 & 90, State Univ NY, Buffalo, 89-90. *Mem:* AAAS; Am Chem Soc; Sigma Xi; NY Acad Sci; NAm Thermal Anal Soc. *Res:* polymer chemistry and physics; surface analysis of polymers. *Mailing Add:* Dept Chem State Univ NY Fredonia NY 14063

KUMLI, KARL F, b Denver, Colo, Oct 9, 27; m 53; c 2. ORGANIC CHEMISTRY. *Educ:* Kans State Teachers Col, AB, 55; Univ Kans, PhD(chem), 59. *Prof Exp:* Res chemist, Celanese Chem Corp, 59-61; res scientist, Weyerhaeuser Co, 61-64; from asst prof to assoc prof, 64-71, PROF CHEM, CALIF STATE UNIV, CHICO, 71-, CHMN DEPT, 75- *Mem:* Am Chem Soc. *Res:* Stereochemistry and mechanisms of reactions of the phosphorus atom; synthesis and characterization of polyoxymethylene, phenol-formaldehyde, epoxy. *Mailing Add:* 1340 Manchester Rd Chico CA 95926

KUMMER, JOSEPH T, b Baltimore, Md, Oct 21, 19; m 47; c 4. SOLID STATE CHEMISTRY. *Educ:* Johns Hopkins Univ, BE, 41, PhD(chem eng), 45. *Prof Exp:* Fel catalysis, Mellon Inst, 45-51; assoc scientist, Dow Chem Co, 51-60; sr staff scientist, Sci Lab, Ford Motor Co, 60-84; RETIRED. *Honors & Awards:* Mobay Award & Thomas Midgley Award, Am Chem Soc, 81. *Mem:* Nat Acad Engrs; Am Chem Soc. *Res:* Catalysis; electrochemistry; plant process design; heat engines. *Mailing Add:* 3904 Golfside Ypsilanti MI 48197

KUMMER, MARTIN, b Glarus, Switz, June 11, 36; m 67. MATHEMATICAL PHYSICS, APPLIED MATHEMATICS. *Educ:* Swiss Fed Inst Technol, predipl math, 57, dipl math phyiscs, 59, PhD(math physics), 62. *Prof Exp:* Asst theoret physics, Swiss Fed Inst Technol, 60-64; NSF res assoc, Univ Mich, 64-66; from asst prof to assoc prof, 66-75, PROF MATH, UNIV TOLEDO, 75- *Concurrent Pos:* Sabbatical leave, Courant Inst, NY Univ, 78; res grant, Swiss Fed Inst Technol, 81, Naval Res Lab, Summer, 87 & 88. *Mem:* Am Phys Soc; Math Asn Am; Int Asn Math Physicists; Am Math Soc. *Res:* Statistical mechanics; application of group theory to physical problems; differential equations, in particular their application to mechanics; hamiltonian, celestial and quantum mechanics. *Mailing Add:* 3411 Queenswood Blvd Toledo OH 43606

KUMMER, W(OLFGANG) H(ELMUT), b Stuttgart, Ger, Oct 10, 25; nat US; m 56; c 4. ELECTRICAL ENGINEERING. *Educ:* Univ Calif, BS, 46, MS, 47, PhD(elec eng), 54. *Prof Exp:* Asst elec eng, Univ Calif, 46-50, lectr, 50-53; mem tech staff, Bell Tel Labs, 53-59; head res sect, Antenna Dept, 59-66, sr scientist, Antenna Dept, 66-76, CHIEF SCIENTIST, RADAR

MICROWAVE LAB, HUGHES AIRCRAFT CO, CULVER CITY, 76- *Concurrent Pos:* Mem comns B & F, Int Sci Radio Union. *Mem:* Fel Inst Elec & Electronics Engrs (pres, Antennas & Propagation Soc, 74). *Res:* Electromagnetic theory; signal processing and electronically scanned antennas and tropospheric propagation beyond the horizon; microwave field; slot radiators in wave-guides. *Mailing Add:* W H Kummer Inc 1310 Sunset Ave Santa Monica CA 90405

KUMMEROW, FRED AUGUST, b Berlin, Ger, Oct 4, 14; nat US; m 42; c 3. FOOD SCIENCE. *Educ:* Univ Wis, BS, 39, MS, 41, PhD(biochem), 43. *Prof Exp:* Assoc nutritionist, Clemson Col, 43-45; assoc prof chem, Kans State Col, 45-50; assoc prof food chem, 50-59, PROF FOOD CHEM, UNIV ILL, URBANA, 59- *Concurrent Pos:* Mem, Assocs Food & Container Inst, Chicago; mem comn arteriosclerosis, Am Heart Asn. *Mem:* AAAS; Am Chem Soc; Sigma Xi; FASEB; Am Soc Microbiol. *Res:* Nutrition; biochemistry; fat and oil chemistry. *Mailing Add:* 205 Burnsides Res Lab Univ Ill 1208 W Pennsylvania Ave Urbana IL 61801

KUMMLER, RALPH H, b Jersey City, NJ, Nov 1, 40; m 62; c 3. CHEMICAL ENGINEERING, ENVIRONMENTAL ENGINEERING. *Educ:* Rensselaer Polytech Inst, BS, 62; Johns Hopkins Univ, PhD(chem eng), 66. *Prof Exp:* Res chemist, Gen Elec Space Sci Lab, 65-70; assoc prof, 70-74, CHMN & PROF CHEM ENG, WAYNE STATE UNIV, 74- *Concurrent Pos:* Consult, Gen Elec Space Sci Lab, 70-74, Phys Dynamics Inc, 70-78, Urban Sci Appl, 77-, KMS Fusion, 77, Urban Consult Inc, 87- & Limnotech, 87-88; mem sci adv bd, Environ Protection Agency, 76-78. *Mem:* Am Inst Chem Engrs; Am Chem Soc; Air & Waste Mgt Asn; Am Inst Chem. *Res:* Chemical kinetics; environmental chemistry and transport including computer simulation of natural and polluted air and aquatic environments; chemiluminescence and hydrocarbon reactivity; hazardous waste management. *Mailing Add:* Dept Chem Eng Wayne State Univ Detroit MI 48202

KUMOSINSKI, THOMAS FRANCIS, b Philadelphia, Pa, Apr 19, 41. PHYSICAL CHEMISTRY. *Educ:* Drexel Univ, BSc, 64; Temple Univ, PhD(phys chem), 73. *Prof Exp:* PHYS CHEMIST BIOCHEM, EASTERN REGIONAL LAB, USDA, 61- *Mem:* Am Chem Soc; Sigma Xi. *Res:* Physical chemistry of biological macromolecules; theoretical and experimental quantum chemistry; small angle x-ray and light scattering; general spectroscopy of small molecular weight biological systems, porphyrins, flavins, etc. *Mailing Add:* USDA East Reg Res Ctr 600 E Mermaid Lane Philadelphia PA 19118

KUMP, LEE ROBERT, b Minneapolis, Minn, Apr 3, 59; m 84; c 2. SEDIMENTARY GEOCHEMISTRY, ATMOSPHERIC & OCEANIC EVOLUTION. *Educ:* Univ Chicago, AB, 81; Univ SFla, PhD(marine sci), 86. *Prof Exp:* Geologist, US Geol Surv, 81-82; res asst, Univ SFla, 81-86; asst prof, 86-91, ASSOC PROF GEOSCI, PA STATE UNIV, 91- *Concurrent Pos:* Secy gen working group, Int Asn Geochem & Cosnochem. *Mem:* Geol Soc Am; Am Geophys Union; Geochem Soc; Clay Minerals Soc; Am Chem Soc; Int Asn Geochem & Cosnochem. *Res:* Charges in oceanic and atmospheric composition, and climate, through earth history; numerical models of biogeochemical cycles; clay diagenesis; marine chemistry of trace metals; neutron activation analysis of iron isotopes; chemical weathering. *Mailing Add:* Dept Geosci Pa State 210 Deike Bldg University Park PA 16802

KUMPEL, PAUL GREMMINGER, b Riverside, NJ, Sept 5, 35; div; c 2. TOPOLOGY. *Educ:* Trenton State Col, BS, 56; Brown Univ, PhD(math), 64. *Prof Exp:* Instr math, Lafayette Col, 56-59; asst, Brown Univ, 59-63, instr, 63-64; from asst prof to assoc prof math, 64-84, dir teacher prep, Div Math Sci, 71-74, PROF MATH, STATE UNIV NY, STONY BROOK, 84- *Concurrent Pos:* Sci Res Coun sr vis fel, Univ Hull, 71, dir undergrad prog math, 76-82; vis scholar, 78 & assoc chair, Wesleyan Univ, 87. *Mem:* Math Asn Am. *Res:* Topology of H-spaces. *Mailing Add:* Dept Math State Univ NY Stony Brook NY 11794

KUN, ERNEST, biochemistry, pharmacology, for more information see previous edition

KUN, KENNETH ALLAN, b Brooklyn, NY, July 14, 30; m 55; c 2. TECHNICAL PLANNING & IMPLEMENTATION. *Educ:* Brooklyn Col, BS, 52; Polytech Inst Brooklyn, MS, 55; Yale Univ, MS, 59, PhD(chem), 61. *Prof Exp:* Chemist, US Elec Mfg Co, 52-53, Warner-Chilcott Res Labs, 53-55 & Am Cyanamid Co, 55-57; asst redox polymers, Yale Univ, 57-60; sr res chemist, 60-66, Far East regional mgr indust chem, Foreign Opers Div, Tokyo, 66-72, sales/mkt coordr, Latin Am Opers, Int Div, 72-76, mem staff corp hq, Rohm & Haas Co, 76-78; dir mkt, Specialty Chem Div, Church & Dwight Co Inc, 78-79; dir specialty chem res, Calgon Corp, Div Merck & Co Inc, 79-81; chief tech officer & vpres, Polychrome Corp, Div Dainippon Ink & Chem, 81-83; PRES & CHIEF EXEC OFFICER, SYRACUSE RES CORP, 83- *Concurrent Pos:* Bd dirs, Am Asn Lab Accreditation; grad asst, NIH. *Mem:* AAAS; Am Chem Soc; Sigma Xi; Am Inst Chemists; NY Acad Sci. *Res:* Synthesis of monomers and polymers; application of polymers for surface coatings, fibers, redox polymers and ion-exchange resins; structures of porous solids; pharmaceuticals; batteries and dry cells; environmental chemistry; acid deposition; purification systems. *Mailing Add:* 5200 Silverfox Dr Merrill Lane Jamesville NY 13078

KUNA, SAMUEL, b Velke Levare, Czech, May 7, 12; nat US; m 36; c 2. PHARMACOLOGY. *Educ:* NY Univ, BA, 43, PhD(biol), 56; Temple Univ, MA, 49. *Prof Exp:* Mem, Merck Inst Therapeut Res, 34-43, res asst to dir, 43-50, res assoc & head pharmacol res unit, 50-57; head pharmacol dept, Bristol-Myers Prod Div, Bristol-Myers Co, 57-62, asst dir res & develop, 63-67; dir pharmacol & toxicol, Calgon Consumer Prod Co Div, Merck & Co, 67-74, dir biol res, 74-80; prof & dir Toxicol Prog, Grad Sch, 80-84, EMER PROF, RUTGERS UNIV, 84- *Concurrent Pos:* Head biol control dept, Merck Inst Therapeut Res, 47-50, 53-56; instr, Temple Univ, 47-50. *Mem:* Am Soc Pharmacol & Exp Therapeut; Soc Toxicol; NY Acad Sci. *Res:* Action of chemical agents on interchange of tissue fluids; analgesics; physiology of the stomach; psychopharmacology; product development. *Mailing Add:* 746 Hyslip Ave Westfield NJ 07090

KUNASZ, IHOR ANDREW, b Montlucon, France, Sept 24, 39; US citizen; m 65; c 2. ECONOMIC GEOLOGY. *Educ:* Case Western Reserve Univ, BA, 63; Pa State Univ, MS, 68, PhD(geol), 70. *Prof Exp:* Staff geologist, Foote Mineral Co, 70-72, chief geologist, 72-88; gen mgr, Cyprus Minera Chile, 88-90; CONSULT, 90- *Concurrent Pos:* Chmn, Indust Minerals Div, Soc Mining Eng, Am Inst Mining Eng, 87, bd mem, 89- *Mem:* Soc Mining Eng, Am Inst Mining Eng; Geol Soc Am; Geochem Soc; Shevchenko Sci Soc. *Res:* Economic and exploration geology associated with industrial minerals, especially saline deposits and lithium deposits of the world, precious metals and non-ferrous metals development. *Mailing Add:* Foote Mineral Co Rt 100 Exton PA 19341

KUNAU, ROBERT, JR, EXPERIMENTAL BIOLOGY. *Prof Exp:* PVT PRACT NEPHROLOGY, 89- *Mailing Add:* 7777 Forest Lane Suite B-245 Dallas TX 75230

KUNCE, HENRY WARREN, b St Louis, Mo, Apr 18, 25; m 48; c 5. COMPUTER SIMULATION, HUMAN-SOCIAL RELATIONS. *Educ:* Wash Univ, BA, 46; McCormick Theol Sem, Chicago, BD & MDiv, 49; Univ Miami, MSIE, 71, PhD(statist), 79. *Prof Exp:* Clergyman, Ohio & Mo, 49-61; planner-analyst parish develop, Bd Missions, United Presby Church, Mo & Fla, 61-70; systs analyst, Clin Campesina, Homestead, Fla, 70-71; res scientist sociocybernetics, Univ Miami, 71-72; CHIEF MGR SYST ENG, MGT INFO SYST & OPERS RES, METROP DADE COUNTY GOVT, FLA, 72- *Concurrent Pos:* Adj math fac, Univ Miami, Coral Gables, Fla, 80- *Mem:* Opers Res Soc Am; Am Inst Indust Engrs; Inst Mgt Sci; Soc Comput Simulation; AAAS; Sigma Xi. *Res:* Sociocybernetics, the application of cybernetics to the dynamics of human interaction and social structures, using systems analysis and computer simulation of stochastic and deterministic mathematical models, a discrete system, finite state automata, and a continuous model analyzed by phase space analysis; applicaitons made in management; counseling education; personnel problems. *Mailing Add:* 5025 SW 74th Terr Miami FL 33143

KUNDEL, HAROLD LOUIS, b New York, NY, Aug 15, 33. RADIOLOGY. *Educ:* Columbia Univ, AB, 55, MD, 59; Temple Univ, MS, 63. *Hon Degrees:* MA, Univ Pa, 80. *Prof Exp:* Intern med, Mary Imogene Bassett Hosp, Cooperstown, NY, 59-60; resident radiol, Temple Univ Hosp, 60-63; resident radiobiol, Sch Aerospace Med, 63-64; James Picker Found advan acad fel physiol, Sch Med, Temple Univ, 64-66, attend radiologist, Univ Hosp, 66-80, prof radiol, 68-80; WILSON PROF RES RADIOL, UNIV PA, 80- *Concurrent Pos:* attend radiologist, Hosp Univ Pa, 80- *Honors & Awards:* Mem Award, Asn Univ Radiologists, 63, Stauffer Award, 82. *Mem:* AAAS; Am Col Radiol; Asn Univ Radiologists. *Res:* Image information processing and analysis; visual perception; diagnostic decision making. *Mailing Add:* Dept Radiol Hosp Univ Pa Philadelphia PA 19104

KUNDERT, ESAYAS G, b Ruti, Switz, May 7, 18; US citizen; m 54; c 3. ALGEBRA. *Educ:* Swiss Fed Inst Technol, dipl, 45, Dr Math, 50. *Prof Exp:* Asst prof math, Univ Tenn, 50-51; from asst prof to prof, La State Univ, 51-62; prof, 62-82. EMER PROF MATH, UNIV MASS, AMHERST, 82- *Concurrent Pos:* US Army grant, 52-53. *Res:* Algebraic geometry and algebraic topology; elementary number theory. *Mailing Add:* Dept Math Univ Mass Amherst MA 01002

KUNDIG, FREDERICKA DODYK, b Nashville, Tenn, Apr 22, 24; c 1. CELL PHYSIOLOGY, VERTEBRATE PHYSIOLOGY. *Educ:* Univ Rochester, PhD(pharmacol), 61. *Prof Exp:* Res assoc biochem, Rackham Arthritis Res Unit, Univ Mich, Ann Arbor, 61-65 & Johns Hopkins Univ, 65-67; from asst prof to assoc prof, 67-73, PROF VERT ANAT, PHYSIOL & CELL BIOL, TOWSON STATE UNIV, 73- *Res:* Cell surface phenomena; hormones and enzyme activity. *Mailing Add:* Dept Biol Towson State Univ Baltimore MD 21204

KUNDIG, WERNER, BIOLOGY. *Prof Exp:* EMER PROF BIOL, HOOD COL, FREDERICK, MD, 88- *Mailing Add:* Star Rte Box 67 Eastsound WA 98245

KUNDSIN, RUTH BLUMFELD, b New York, NY, July 30, 16; c 2. MEDICAL MICROBIOLOGY. *Educ:* Hunter Col, BA, 36; Boston Univ, MA, 49; Harvard Univ, ScD(microbiol), 58. *Hon Degrees:* ScD, Lowell Tech Inst, 75. *Prof Exp:* Res bacteriologist, Sch Pub Health, Harvard Univ, 36-37 & Sch Med, Univ Pa, 37-38; res bacteriologist, Peter Bent Brigham Hosp, 51-58, asst surg, 58-64, mem assoc staff, 64-70, res assoc, Harvard Med Sch, 61-76, ASSOC PROF MICROBIOL & MOLECULAR GENETICS, HARVARD MED SCH, 76-, HOSP EPIDEMIOLOGIST, BRIGHAM & WOMEN'S HOSP, BOSTON, MASS, 70-; PRES, KUNDSIN LAB INC, 81- *Mem:* AAAS; Am Soc Microbiol; NY Acad Sci; Am Venereal Disease Asn. *Res:* Dynamics of disinfection as applied to environmental bacteriology; sanitary bacteriology; maintenance of standards for a hygienic environment; epidemiology of staphylococcal disease; skin disinfection; mycoplasmas and reproductive failure in humans. *Mailing Add:* 721 Huntington Ave Boston MA 02115

KUNDT, JOHN FRED, b Denver, Colo, Dec 21, 26; m 48; c 2. DENDROLOGY, FORESTRY EDUCATION. *Educ:* WVa Univ, BS, 52; NC State Univ, PhD(forestry), 72. *Prof Exp:* Mgt chief, Forest Mgt Serv, Div Forestry, Va, 52-57; forest supvr prod, Union Camp Paper Corp, 57-64; fel dendrol, NC State Univ, 64-69; asst prof bot, genetics & plant taxon, State Univ NY, 69-73; assoc prof & exten forestry specialist, Univ Md, 74-88; RETIRED. *Mem:* Soc Am Foresters. *Res:* Effects of adding composted sewage sludge to newly planted Pinus Taeda and Pinus Virginina; developing pine hybrids between Pinus Taeda and Pinus Rigida by selecting superior parental phenotypes, and producing experimental seed orchard; developing a Pinus Virginina seed orchard to produce seed for Christmas tree production; development of Paulownia tomentosa in plantations, initial establishment and cold hardiness development. *Mailing Add:* 103 Fox Run Laurel DE 19956

KUNDU, MUKUL RANJAN, b Calcutta, India, Feb 10, 30; m 58; c 3. RADIOPHYSICS, ELECTRONICS. *Educ:* Univ Calcutta, BSc, 49, MSc, 51. *Hon Degrees:* DSc, Univ Paris, 57. *Prof Exp:* Asst, Coun Sci & Indust Res, Univ Calcutta, 52-54; French Govt scholar radio astron, Ecole Normale Superieure & Meudon Observ, 54-56; asst, Nat Ctr Sci Res, Ministry Educ, France & Meudon Observ, 56-58; sr res fel, Nat Phys Lab, India, 58-59; res assoc solar radio astron, Observ, Univ Mich, 59-62; assoc prof astron, Cornell Univ, 62-65 & Tata Inst Fundamental Res, Bombay, 65-68; actg dir astron, 78-79, dir astron, 80-85, PROF PHYSICS & ASTRON, UNIV MD, COLLEGE PARK, 68- *Concurrent Pos:* Sr res assoc, Nat Acad Sci-Nat Res Coun, 67, 74-75 & 86-87. *Honors & Awards:* Sr US Scientist Award, Alexander von Humboldt Found, 78. *Mem:* Int Astron Union; sr mem Inst Elec & Electronic Engrs; Am Astron Soc; assoc Brit Inst Radio Eng; fel Royal Astron Soc; fel Am Phys Soc; Am Geophys Union. *Res:* Solar, stellar and galactic radio astronomy. *Mailing Add:* Astron Prog Space Sci Bldg Univ Md College Park MD 20742

KUNDU, SAMAR K, EXPERIMENTAL BIOLOGY. *Prof Exp:* SR SCIENTIST, DIAG DEPT, ABBOTT LABS, 83- *Mailing Add:* Dept 90J Abbott Labs Bldg AP20 Abbott Park IL 60064

KUNDUR, PRABHA SHANKAR, b Bangalore, India, March 18, 39; Can citizen; m 63; c 1. ELECTRIC POWER TRANSMISSION, POWER SYSTEM STABILITY & CONTROL. *Educ:* Mysore Univ, BE, 59; Indian Inst Sci, ME, 61; Univ Toronto, MASc, 65, PhD(elec eng), 67. *Prof Exp:* Reader elec eng, Bangalore Univ, India, 67-69; MGR, ANALYTICAL METHODS & SPECIALIZED STUDIES DEPT, ONT HYDRO, CAN, 69- *Concurrent Pos:* Asst prof elec eng, Mysore Univ, India, 61-63; adj prof power syst stability & control, Univ Toronto, 69- *Mem:* Fel Inst Elec & Electronics Engrs. *Res:* Development and application of improved methods of analysis and control of power system dynamic performance. *Mailing Add:* Syst Planning Div Ont Hydro Toronto ON M5G 1X6 Can

KUNELIUS, HEIKKI TAPANI, b Konginkangas, Finland, Mar 21, 40; Can citizen; m 71; c 2. PLANT SCIENCE, FORAGE MANAGEMENT. *Educ:* Univ Helsinki, BSA & MSc, 66; Univ Man, PhD(plant sci), 70. *Prof Exp:* Res asst plant path, Agr Res Ctr, 66; RES SCIENTIST PLANT SCI, AGR CAN, RES BR, 70-, HEAD, FORAGE-BEEF SECT, 85- *Concurrent Pos:* Fel plant sci, Univ Man, 70; mem expert comt, Forage Breeding, Can, 79-; study leave, NZ-Australia, 79-80. *Mem:* Can Soc Agron; Am Soc Agron; Agr Inst Can; Finnish Asn Agr Grad; Swed Seed Asn. *Res:* Physiology and management of forage grasses and legumes; pasture management; minimum tillage for pasture renovation. *Mailing Add:* Agr Can Res Sta PO Box 1210 Charlottetown PE C1A 7M8 Can

KUNESH, CHARLES JOSEPH, b Greensburg, Pa, Oct 18, 48; m 74; c 2. FINE PARTICLE TECHNOLOGY, PAPER SCIENCE. *Educ:* Carnegie-Mellon Univ, BS, 70; Univ Pittsburgh, PhD(phys chem), 73. *Prof Exp:* Res chemist, Pfizer Inc, 73-77, sr res chemist, 77-80, sr res scientist, 80-82, res mgr, 82-88, DIR RES, PFIZER INC, 88- *Mem:* Am Chem Soc; Tech Asn Pulp & Paper Indust; Am Soc Metals Int. *Res:* Synthesis and crystal engineering of fine particle inorganic materials, especially precipitated calcium carbonate; application of inorganic fine particles in filled and-or coated paper and in polymer composites. *Mailing Add:* Pfizer Inc Nine Highland Ave Bethlehem PA 18017

KUNESH, JERRY PAUL, b Kewaunee, Wis, Jan 19, 38; m 61; c 3. CLINICAL PHARMACOLOGY, MEDICINE. *Educ:* Iowa State Univ, DVM, 61, MS, 66, PhD(physiol), 69. *Prof Exp:* Instr vet med & surg, 61-62, NIH fel physiol, 64-65, asst prof physiol & pharmacol, 65-70, assoc prof pharmacol & med, 70-75, PROF VET MED & SURG, IOWA STATE UNIV, 75- *Mem:* Am Asn Swine Practitioners; Am Asn Bovine Practitioners; Am Vet Med Asn. *Res:* Porcine hemorrhagic syndromes and clinical evaluation of antimicrobial agents as well as their mechanisms of action. *Mailing Add:* Dept Vet Clin Sci Iowa State Univ 1720 Vet Med Col Ames IA 50011

KUNG, CHING, b Kwang Tung, China, Apr 28, 39; US citizen; m; c 3. GENETICS, NEUROBIOLOGY. *Educ:* Chung Chi Col, Chinese Univ Hong Kong, dipl, 63; Univ Pa, PhD(biol), 68. *Prof Exp:* Fel genetics, Ind Univ, Bloomington, 68-70; fel electrophysiol, Univ Calif, Los Angeles, 70-71; from asst prof to assoc prof molecular biol, Univ Calif, Santa Barbara, 71-74; assoc prof, 74-77, Hilldale prof, 90, PROF MOLECULAR BIOL & GENETICS, UNIV WIS-MADISON, 77- *Mem:* AAAS; Genetic Soc Am; Am Soc Cell Biol. *Res:* Genetic dissection of sensory transductions in microbes; ion channels of paramecium, yeast, and e coli. *Mailing Add:* Lab Molecular Biol Univ Wis Madison WI 53706

KUNG, ERNEST CHEN-TSUN, b Ping-tung, Taiwan, China, Jan 1, 31; m 59; c 3. METEOROLOGY. *Educ:* Nat Univ Taiwan, BS, 53; Univ Ariz, MS, 59; Univ Wis, PhD(meteorol), 63. *Prof Exp:* Specialist agron, Taiwan Prov Govt, 54-58; proj assoc meteorol, Univ Wis, 63; res meteorologist, Geophys Fluid Dynamics Lab, Environ Sci Serv Admin, 63-67; assoc prof, 67-70, PROF ATMOSPHERIC SCI, UNIV MO, COLUMBIA, 70- *Concurrent Pos:* NSF grants, 67-69, 70-75, 76-88; pres grad fac senate, Univ Mo-Columbia, 77-78; Nat Oceanic & Atmospheric Admin grant, 83-88. *Honors & Awards:* Res Award, Sigma Xi, 83. *Mem:* Fel Am Meteorol Soc; fel Royal Meteorol Soc; Meteorol Soc Japan; Sigma Xi. *Res:* Atmospheric general circulation; long-range forecasting; dynamic climatology. *Mailing Add:* Dept Atmospheric Sci Univ Mo Columbia MO 65211

KUNG, HAROLD HING CHUEN, b Hong Kong, Oct 12, 49; m 71. CATALYSIS, KINETICS. *Educ:* Univ Wis-Madison, BS, 71; Northwestern Univ, MS, 72, PhD(chem), 74. *Prof Exp:* Res chemist, E I du Pont de Nemours & Co, Inc, 74-76; from asst prof to assoc prof, 76-84, PROF CHEM ENG, NORTHWESTERN UNIV, 84-, CHMN, 86- *Honors & Awards:* P H Emmett Award, Catalysis Soc, 91. *Mem:* Am Chem Soc; Am Inst Chem Engrs. *Res:* Surface chemistry and physics; catalysis; chemical reaction engineering. *Mailing Add:* Dept Chem Eng Northwestern Univ Evanston IL 60208

KUNG, HSIANG-FU, b Chungking, China, Sept 4, 42; US citizen; m; c 2. BIOCHEMICAL PHYSIOLOGY. *Educ:* Nat Chung-hsing Univ, BS, 63; Vanderbilt Univ, PhD, 66. *Prof Exp:* Fel, Lab Biochem, Sect Enzymes, NIH, 69-70; sci asst, Max-Planck Inst Biol, 70-71; res fel, Dept Biochem, Roche Inst Molecular Biol, 71-73, asst mem, 73-80; res fel, Dept Molecular Genetics, Res Div, Hoffman-La Roche Inc, 80-81, sr res fel, 81-82, res leader, Dept Molecular Oncol, 82-86; CHIEF, LAB BIOCHEM PHYSIOL, NAT CANCER INST, DCT, BIOL RESPONSE MODIFIERS PROG, NIH, 86- *Concurrent Pos:* Vis asst prof, Dept Biochem, Col Med & Dent NJ, 71-73 & 73-76; adj prof, Dept Zool & Physiol, Rutgers State Univ, 76-78. *Mem:* Fedn Am Socs Exp Biol; AAAS. *Res:* Biochemical physiology. *Mailing Add:* Nat Cancer Inst Lab Biochem Physiol Frederick Cancer Res Facil Bldg 560 Frederick MD 21701-1013

KUNG, HSIANG-TSUNG, b Shanghai, China, Nov 9, 45; m 70; c 2. COMPUTER ARCHITECTURE, PARALLEL COMPUTATION. *Educ:* Nat Tsing Hua Univ, BS, 68; Univ NMex, MA, 70; Carnegie-Mellon Univ, PhD(math), 74. *Prof Exp:* Res assoc, 73-74, asst prof, 74-78, assoc prof, 78-82, PROF COMPUT SCI, CARNEGIE-MELLON UNIV, 82- *Concurrent Pos:* Archit consult, ESL, Inc, 82; Guggenheim fel, 83-84. *Mem:* Asn Comput Mach; Inst Elec & Electronics Engrs. *Res:* Computer algorithms; computational complexity; parallel computation; multiprocessors; very large scale integration; numerical analysis; computer architectures; supercomputers. *Mailing Add:* Sch Comput Sci Carnegie-Mellon Univ 5000 Forbes Ave Pittsburgh PA 15213

KUNG, PATRICK C, EXPERIMENTAL BIOLOGY. *Prof Exp:* VCHMN BD & CHIEF SCI OFFICER, T CELL SCI INC, 89- *Mailing Add:* T Cell Sci Inc 840 Memorial Dr Cambridge MA 02139

KUNG, SHAIN-DOW, b Lini, China, Mar 14, 35; US citizen; m 64; c 3. MOLECULAR BIOLOGY. *Educ:* Chung Hsing Univ, Taiwan, BSc, 58; Univ Guelph, MSc, 65; Univ Toronto, PhD(bot), 68. *Prof Exp:* Instr bot, Chung Hsing Univ, Taiwan, 58-62; res assoc biochem, Univ Toronto, 68-71 & biol, Univ Calif, Los Angeles, 71-74; from asst prof to assoc prof, 74-82, actg chmn, 82-84, prof biol, 82-86, assoc dean, Univ Md, Baltimore County, 85-87; actg dir, Ctr Agr Biotechnol, 86-88, PROF BOT, UNIV MD, COLLEGE PARK, 86-, DIR, 88- *Concurrent Pos:* Fulbright Award, 83. *Honors & Awards:* Philip Morris Award, 79. *Mem:* AAAS; Am Soc Plant Physiologists. *Res:* Biochemistry and genetics of chloroplast protein; properties, function and evolution of chloroplast DNA; molecular biology of genetic tumors. *Mailing Add:* Dept Bot Univ Md College Park MD 20742

KUNHARDT, ERICH ENRIQUE, b Montecristy, Dominican Rep, May 31, 49; m 76. ELECTROKINETICS, COMPUTATIONAL PHYSICS. *Educ:* New York Univ, BS, 69, MS, 72; Polytechnic Univ, PhD(electroph), 76. *Prof Exp:* From asst prof to prof elec eng & physics, Tex Tech Univ, 76-85; PROF ELECTROPHYSICS, POLYTECH UNIV, 85-, DIR, WEBER RES INST, 86- *Concurrent Pos:* Consult, Los Alamos Nat Lab, 79-87, GTE Labs, 82-85 & Lawrence Livermore Labs, 88-; adv bd, Transport Theory & Statist J, 82- *Honors & Awards:* Halliburton Found Award, 83. *Mem:* Am Phys Soc; Sigma Xi; Inst Elec & Electronics Engrs. *Res:* Experimental, theoretical and computational investigations of the non-equilibrium behavior of electron assemblies in matter under the influence of place-time varying electric and magnetic fields. *Mailing Add:* Weber Res Inst Polytech Univ Rte 110 Farmingdale NY 11735

KUNIN, ARTHUR SAUL, b Brooklyn, NY, Aug 11, 25; m 59; c 4. MEDICINE, PHYSIOLOGICAL CHEMISTRY. *Educ:* Columbia Univ, BA, 48; Univ Vt, MD, 52. *Prof Exp:* Intern med, Peter Bent Brigham Hosp, Boston, Mass, 52-54, jr asst resident, 53-54, sr asst resident, 56-57; NIH fel, Med Sch, Boston Univ, 54-56; from instr to assoc prof, 57-82, PROF, COL MED, UNIV VT, 82- *Concurrent Pos:* Lederle Med fac award, Col Med, Univ Vt, 65-68; NIH spec fel, Mass Gen Hosp & Harvard Med Sch, 62-64; Am Col Physicians Willard Thompson traveling scholar, Univ Col Hosp Med Sch, London, 64; vis prof, Inst Chem Med, Univ Bern, 70-71; fel Dept Physiol, Harvard Med Sch, 78-79. *Mem:* Am Col Physicians; Am Fedn Clin Res. *Res:* Renal physiology; mitochondrial metabolism; nutrition in renal disease; nephrology; diseases of metabolism; metabolic bone diseases and the intermediary metabolism of epiphyseal cartilage. *Mailing Add:* Col Med Unit Vt Burlington VT 05405

KUNIN, CALVIN MURRY, b Burlington, Vt, May 3, 29; m 52; c 3. INTERNAL MEDICINE, INFECTIOUS DISEASE. *Educ:* Columbia Univ, AB, 49; Cornell Univ, MD, 53. *Prof Exp:* Intern med, New York Hosp, 53-54; sr asst surg, USPHS, 54-56; asst resident, Peter Bent Brigham Hosp, Boston, 56-57; res fel, Harvard Med Sch, 57-59; from asst prof to assoc prof med & prev med, Sch Med, Univ Va, 59-70; prof med & assoc chmn dept, Univ Wis-Madison, 70-79; chief med, Vet Admin Hosp, 70-70; prof & chmn dept, 79-84, POMERENE PROF MED, COL MED, OHIO STATE UNIV, 83- *Mem:* Am Fedn Clin Res; Am Asn Immunologists; Am Assoc Physicians; Am Soc Clin Invest; Soc Exp Biol & Med; fel Infectious Dis Soc Am (past pres). *Res:* Epidemiology; antibiotic therapy; urinary tract infections. *Mailing Add:* 410 W Tenth Ave Dept Med Ohio State Univ Columbus OH 43210

KUNIN, ROBERT, b West New York, NJ, July 16, 18; m 42; c 2. PHYSICAL CHEMISTRY. *Educ:* Rutgers Univ, BS, 39, PhD(colloidal chem), 42. *Prof Exp:* Assoc chemist, Tenn Valley Authority, 42-44; sr chemist, Manhattan Proj, Columbia Univ, 44-45; Gulf fel, Mellon Inst, 45-46; lab head chg res & develop ion exchange resins, 46-59, res assoc, 59-70, sr staff assoc, Rohm & Haas Co, 70-76; CONSULT, 76- *Concurrent Pos:* Lectr, Univ Pa & Am Univ. *Honors & Awards:* Franklin Inst Gold Medal, 66. *Mem:* AAAS; Am Chem Soc; Am Inst Chem Engrs; Electrochem Soc; Israel Chem Soc; Sigma Xi. *Res:* Desalination; adsorption; liquid extraction; theory and application of ion exchange; inorganic chemistry of phosphates, uranium fluorides; analytical chemistry of inorganic constituents; ion exchange in silicates; electrochemistry of membrane processes; catalysis; water treatment and purification. *Mailing Add:* 860 Lower Ferry Rd No 2J Trenton NJ 08628

KUNISHI, HARRY MIKIO, b Honolulu, Hawaii, Aug 30, 32; m 59. SOIL CHEMISTRY. *Educ:* Univ Hawaii, BS, 55, MS, 56; Univ Wash, BS, 58; Univ Wis, PhD(soils), 63. *Prof Exp:* SOIL SCIENTIST, AGR RES SERV, USDA, 62- *Mem:* Int Soc Soil Sci; Am Soc Agron; Am Chem Soc; Sigma Xi. *Res:* Chemistry and mineralogy of potassium in soils; adsorption and movement of radionuclides in soils; phosphate reactions in field and streams; rates of phosphate supplied to plants by acid soils of southeastern United States; phosphorus management under no-tillage; model of phosphorus transport from agricultural fields; soil-water-plant-phosphorus reactions in fresh water and brackish water systems. *Mailing Add:* 4400 Sellman Rd Beltsville MD 20705

KUNISI, VENKATASUBBAN S, b Kottayam, India; m 74; c 1. ORGANIC CHEMISTRY. *Educ:* Univ Madras, India, BSc, MSc; Unliv Kans, PhD(chem), 75. *Prof Exp:* Fel, Emory Univ, 74-76; lectr, Tex A&M Univ, 76-79; assoc pharmacol, Univ Fla, 79-80; asst prof, 81-84, ASSOC PROF CHEM, UNIV NFLA, 84- *Mem:* Am Chem Soc. *Res:* Bio-organic mechanisms; chemical and enzyme catalysis; organic reaction mechanisms; solution kinetics; istope effectss. *Mailing Add:* Dept Natural Sci Bldg 3 Rm 2211 Univ NFla Jacksonville FL 32216

KUNKA, ROBERT LEONARD, b Chicago, Ill, Aug 30, 47; m 78; c 1. PHARMACOKINETICS, BIOPHARMACEUTICS. *Educ:* Univ Ill, BS, 70; Univ NC, Chapel Hill, PhD(pharmaceut), 77. *Prof Exp:* asst prof pharmaceut, Univ Pittsburgh, 78-84; assoc dir, Corp Med & Sci Affairs, 84-89; DIR, CLIN TRIALS, CHUGAI-UPJOHN INC, 89- *Mem:* Am Pharmaceut Asn; Acad Pharmaceut Sci; Am Asn Col Pharm. *Res:* Protein binding of drugs; effect of disease states on pharmacokinetics and biopharmaceutics; computer modeling of drugs; bioavailability testing; secretion of drugs in body fluids. *Mailing Add:* 406 W Gulf Rd Libertyville IL 60048

KUNKEE, RALPH EDWARD, b San Fernando, Calif, July 30, 27. BIOCHEMISTRY, ENOLOGY. *Educ:* Univ Calif, AB, 50, PhD(biochem), 55. *Prof Exp:* Asst biochemist, Univ Calif, 50-53; res biochemist, E I du Pont de Nemours & Co, 55-60; asst res biochemist, 60-63; PROF ENOL, UNIV CALIF, DAVIS, 63- *Mem:* AAAS; Am Soc Microbiol; Am Soc Enol (secy-treas, 81-83). *Res:* Intermediary metabolism and control; fermentation; microbiology. *Mailing Add:* Dept Viticult & Enol Univ Calif Davis CA 95616-8749

KUNKEL, HARRIOTT ORREN, b Olney, Tex, July 3, 22; m 60; c 2. NUTRITION, PHILOSOPHY OF AGRICULTURAL SCIENCE. *Educ:* Tex A&M Univ, BS, 43, MS, 48; Cornell Univ, PhD(biochem), 50. *Prof Exp:* Instr biochem, Univ Wis, 50-51; from asst prof to assoc prof animal sci & biochem, 51-57, assoc dir exp sta, 62-68, dean col agr & actg dir, Tex Agr Exp Sta, 68-72, dean agr, 72-88, PROF ANIMAL SCI BIOCHEM & BIOPHYS, TEX A&M UNIV, 57- *Mem:* Am Chem Soc; Am Soc Animal Sci; Am Soc Biochem & Molecular Biol; Am Inst Nutrit; Soc Exp Biol & Med. *Res:* Science administration; philosophy of agricultural science. *Mailing Add:* Dept Animal Sci Tex A&M Univ College Station TX 77843-2471

KUNKEL, JOSEPH GEORGE, b Oceanside, NY, Aug 17, 42; m 64; c 2. DEVELOPMENTAL BIOLOGY, INSECT PHYSIOLOGY. *Educ:* Columbia Col, AB, 64; Case Western Reserve Univ, PhD(biol), 68. *Prof Exp:* Trainee biomet, Case Western Reserve Univ, 68; NIH trainee develop biol, Yale Univ, 68-70; from asst prof to assoc prof, 70-85, PROF ZOOL, UNIV MASS, 85- *Concurrent Pos:* Instr biol, Yale Univ, 69-70; prin investr, NIH, "Cockroack Develop", NSF, "Role Oligosaccharides in Vitellogenesis", USDA, "Gypsy Moth pop biol, Storage Proteins as indices of nutritive quality"; vis scholar, dept biochem, Univ Calif, Berkeley, 77-78; adj prof molecular & cellular biol prog, Univ Mass, 84-, entom, 85-; vis scholar, Zool Inst, Univ Bern, Switz, 85-86. *Mem:* AAAS; Am Soc Zoologists. *Res:* Insect physiology and development; chemistry and function of vitellogenin; evolution; biometry; effect of oligosaccharides on proteins; role of ions in early development of oocytes. *Mailing Add:* Dept Zool Univ Mass Amherst MA 01003

KUNKEL, LOUIS M, b New York, NY, Oct 13, 49. MEDICINE. *Educ:* Gettysburg Col, BA, 71; Johns Hopkins Univ, PhD, 78. *Prof Exp:* Fel, Univ Calif, San Francisco, 78-80; res fel med, Children's Hosp Med Ctr, Boston, 80; res fel pediat, 80-82, from instr to assoc prof, 82-90, PROF PEDIAT & GENETICS, HARVARD MED SCH, 90-; CHIEF, DIV GENETICS, CHILDREN'S HOSP, BOSTON, 89- *Concurrent Pos:* George Meany postdoctoral fel, Muscular Dystrophy Asn, 80-82; lectr neurobiol, Harvard Med Sch, 83-89, tutor human genetics, 87-89; assoc investr, Howard Hughes Med Inst, 87-90, investr, 90- *Honors & Awards:* Duchenne-Erb-Preis, Ger Muscular Dystrophy Asn, 86; George Cotzias Mem Lectr, Am Acad Neurol, 88; Royal Soc Wellcome Found Prize, Eng, 88; Warren Alpert Found Prize, 88; Passano Found Young Scientist Award, 89; Nat Med Res Award, Nat Health Coun, 89; Pruzansky Lectr, March of Dimes Birth Defects Found, 89; Gairdner Found Int Award, 89; E Mead Johnson Award, 91; Silvio O Conte Decade of the Brain Award, 91. *Mem:* Nat Acad Sci; Muscular Dystrophy Asn (vpres, 87-89). *Res:* Linkage of human genetic diseases with DNA markers; molecular genetics of Duchenne muscular dystrophy; differential gene expression during development; structural organization of mammalian DNA. *Mailing Add:* Children's Hosp 300 Longwood Ave Boston MA 02115

KUNKEL, WILLIAM ECKART, b Berlin, Ger, Mar 25, 36; nat US; m 69; c 2. ASTRONOMY. *Educ:* Univ Calif, Berkeley, BA, 59; Univ Tex, Austin, PhD(astron), 67. *Prof Exp:* Jr astronr, Interam Observ Cerro Tololo, 67-70, assoc astronr, 70-77; mem staff, Brazilian Nat Observ, 77-79; astronomer, Max Planck Inst Astron, Ger, 80-83; AT LAS CAMPANAS OBSERV. *Mem:* Am Astron Soc; Int Astron Union. *Res:* Solar neighborhood flare stars; dwarf galaxies, photometry. *Mailing Add:* Las Campanas Observ Casilla 607 La Serena Chile

KUNKEL, WULF BERNARD, b Eichenau, Ger, Feb 6, 23; nat US; m 47; c 2. PLASMA PHYSICS. *Educ:* Univ Calif, BA, 48, PhD(physics), 51. *Prof Exp:* Asst res engr aerodyn, 51-54, lectr, 53-67, assoc res eng, 54-55, physicist, Lawrence Berkeley Lab, 56-70, PROF PHYSICS, UNIV CALIF, BERKELEY, 67-, GROUP LEADER MAGNETIC FUSION ENERGY RES PROJ, LAWRENCE BERKELEY LAB, 70- *Concurrent Pos:* Guggenheim fel, 55-56 & 72-73; consult, Aerospace Corp, 61-71; ed, Plasma Physics, 70-80; Alexander von Humboldt award, 80. *Mem:* Fel Am Phys Soc; Sigma Xi. *Res:* Physics of ionized gases; magnetohydrodynamics; controlled-fusion research. *Mailing Add:* Lawrence Berkeley Lab Univ Calif Berkeley CA 94720

KUNKLE, DONALD EDWARD, b New Kensington, Pa, Mar 9, 28; m 50; c 3. PHYSICS, MATHEMATICS. *Educ:* Lafayette Col, BS, 50. *Prof Exp:* Res physicist spectros, 50-55, group leader nondestructive testing, 55-62, process control, Alcoa Res Lab, 62-70; staff physicist process technol, Kaiser Aluminum & Chem Corp, 70-80, mgr prod anal & test syst, qual control, 80-86; NUCLEAR ENGR, DEPT NAVY, 87- *Mem:* Am Soc Non-destructive Testing; Anal Chem Appl Spectros; Am Soc Testing & Mat. *Res:* Applied emission spectroscopy; eddy current testing; new principles of radiation thickness gauging for non-ferrous rolling mills; closed loop process control systems; quality control in metals industry; nuclear instrumentation for submarine program. *Mailing Add:* 1200 Mira Mar Ave No 1419 Medford OR 97504-8562

KUNKLE, GEORGE ROBERT, b Elyria, Ohio, Mar 27, 34; m 58; c 4. ENVIRONMENTAL GEOLOGY, HYDROLOGY. *Educ:* Iowa State Univ, BS, 56; Univ Mich, MS, 58, PhD(geol), 61. *Prof Exp:* Geologist, Res Coun Alta, Can, 60-62 & US Geol Surv, 62-66; asst prof geol, Univ Toledo, 66-71; pres & environ consult, Earthview, Inc, 71-77; sr scientist, Jones & Henry Eng, LTD, 77-80; assoc & mgr, Neyer, Tiseo & Hindo, Ltd, 80-86; PRES & PRIN HYDROGEOLOGIST, G R KUNKLE & ASSOC, INC, 86- *Mem:* Nat Water Well Asn; Am Inst Prof Geologists. *Res:* Groundwater resources and environmental geology; influence of land use and natural processes on the quality and quantity of ground and surface waters. *Mailing Add:* 3650 Brigton Rd Howell MI 48843

KUNOS, GEORGE, b Budapest, Hungary, May 14, 42; Can citizen; m 67; c 2. MOLECULAR PHARMACOLOGY, RECEPTOR RESEARCH. *Educ:* Budapest Med Sch, MD, 66; McGill Univ, Montreal, PhD(pharmacol), 73. *Prof Exp:* Asst prof pharmacol, McGill Univ, Montreal, 74-79, from assoc prof to prof pharmacol & med, 79-88, LAB CHIEF PHYSIOL & PHARMACOL, NAT INST ALCOHOL ABUSE & ALCOHOLISM, BETHESDA, 87- *Mem:* Am Soc Pharmacol & Exp Therapeut; Am Soc Biochem & Moleuclar Biol; Int Soc Hypertension; Soc Exp Biol & Med. *Res:* Pharmacology, molecular biology and physiological regulation of drug and hormone receptors; neural mechanisms of blood pressure regulation; author of over 100 publications and four books. *Mailing Add:* 6606 Marywood Rd Bethesda MD 20817

KUNOV, HANS, b Copenhagen, Denmark, Mar 14, 38; Can citizen; m 64, 77; c 2. PHYSIOLOGICAL ACOUSTICS, ACOUSTICAL COMMUNICATION PROCESSES. *Educ:* Tech Univ Denmark, MASc, 63, PhD(elec eng), 66. *Prof Exp:* Postdoctoral fel biomed eng, Tech Univ Denmark, 66-67; from asst prof to assoc prof, 67-82, PROF BIOMED ENG, UNIV TORONTO, 82-, DIR, INST BIOMED ENG, 89- *Concurrent Pos:* Pres, Div 934533 Ont Inc, Artel Eng, 75-; dir, Elec Eng Consociates, 90-92; mem, Grant Selection Comt Elec Eng, Natural Sci & Eng Res Coun Can, 90-; assoc ed, Inst Elec & Electronics Engrs Trans Biomed Eng, 91- *Mem:* Inst Elec & Electronics Engrs; Acoust Soc Am; AAAS; Can Med & Biol Eng Soc; Sigma Xi; Instrument Soc Am. *Res:* Acoustics and hearing; acousto-mechanical models of the head; hearing assistive devices; signal processing by the ear; signal processing for amelioration of hearing deficit. *Mailing Add:* Inst Biomed Eng Univ Toronto Toronto ON M5S 1A4 Can

KUNSELMAN, A(RTHUR), RAYMOND, b Witchita Falls, Tex, Feb 22, 42. NUCLEAR PHYSICS, PARTICLE PHYSICS. *Educ:* Univ Calif, Berkeley, BA, 64, MA, 65, PhD(physics), 69. *Prof Exp:* Physicist, Lawrence Berkeley Lab, 69; FAC MEM PHYSICS, UNIV WYO, 69- *Concurrent Pos:* Consult, Rutherford Lab, Eng, 75. *Mem:* Am Asn Physics Teachers; Sigma Xi. *Res:* Muonic and hadronic atoms; leptonic conservations. *Mailing Add:* Dept Physics Univ Wyo Box 3905 Laramie WY 82071

KUNTZ, GARLAND PARKE PAUL, b Fort Worth, Tex. CORROSION, COMPUTER ASSISTED INSTRUCTION. *Educ:* Fla State Univ, BS, 66; Case Western Reserve Univ, MS, 69, PhD(chem), 72. *Prof Exp:* Fel, Dept Chem, Case Western Reserve Univ, 72-73; res assoc, Univ Alta, 73-75; ASSOC PROF PHYSICS & CHEM, CONCORDIA COL, EDMONTON, 75- *Mem:* Nat Asn Corrosion Engrs. *Res:* Computer assisted instruction. *Mailing Add:* Dept Sci Concordia Col 7128 Ada Blvd Edmonton AB T5B 4E4 Can

KUNTZ, IRVING, b New York, NY, Feb 16, 25; m 47, 77; c 2. POLYMER CHEMISTRY, ORGANIC CHEMISTRY. *Educ:* City Col New York, BS, 48; Polytech Inst Brooklyn, MS, 50, PhD(chem), 55. *Prof Exp:* Res chemist, Sprague Elec Co, Mass, 50-53; from sr chemist to sect head, Esso Res & Eng Co, 55-63, res assoc, 63-68, SR RES ASSOC, EXXON CHEM CO, 68- *Res:* Organic reaction mechanisms; polymer chemistry; ionic polymerizations; kinetics. *Mailing Add:* 725 Haven Pl Linden NJ 07036

KUNTZ, IRWIN DOUGLAS, JR, b Nashville, Tenn, Aug 31, 39; m 61; c 3. PHYSICAL CHEMISTRY. *Educ:* Princeton Univ, AB, 61; Univ Calif, Berkeley, PhD(chem), 65. *Prof Exp:* Asst prof chem, Princeton Univ, 65-71; assoc prof, 71-76, PROF CHEM, UNIV CALIF, SAN FRANCISCO, 77- *Mem:* Fel AAAS; Sigma Xi; Am Chem Soc. *Res:* Physical chemistry of liquid state; hydration of macromolecules; spectroscopic studies of biological materials and fast reactions in biological systems; design of ligands. *Mailing Add:* Dept Pharmaceut Chem Univ Calif San Francisco CA 94143-0446

KUNTZ, MEL ANTON, b Minneapolis, Minn, July 4, 39; m 67; c 2. GEOLOGY, PETROLOGY. *Educ:* Carleton Col, BA, 61; Northwestern Univ, MS, 64; Stanford Univ, PhD(geol), 68. *Prof Exp:* Asst prof geol, Amherst Col, 68-74; GEOLOGIST, US GEOL SURV, 74- *Mem:* Mineral Soc Am; Geol Soc Am. *Res:* Petrogenesis of epizonal and catazonal plutons; application of experimental studies to natural igneous and metamorphic rocks; geology of Colorado; petrogenesis of basalts; basalts of Snake River Plain, Idaho; geology of intermountain western United States; geology of the Idaho batholith. *Mailing Add:* US Geol Surv Mail Stop 913 Box 25046 Denver Fed Ctr Denver CO 80225

KUNTZ, RICHARD A, b Lakewood, NJ, Sept 7, 39; m 60; c 2. MATHEMATICS. *Educ:* Monmouth Col, BS, 64; Univ Md, MA, 67, PhD(math), 69. *Prof Exp:* Teaching asst math, Univ Md, 64-68; from asst prof to assoc prof, Monmouth Col, NJ, 68-76, chmn dept, 74-76, dean grad sch, 76-80, vpres admin, 80-87, PROF MATH, MONMOUTH COL, NJ, 76-, VPRES & DEAN, SCH INFO SCI, 87- *Mem:* Am Math Soc; Math Asn Am. *Res:* Abstract algebra; ideal theory in commutative rings. *Mailing Add:* Monmouth Col West Long Branch NJ 07764

KUNTZ, ROBERT ELROY, b Lawton, Okla, Feb 23, 16; m 38; c 3. PARASITOLOGY, HELMINTHOLOGY. *Educ:* Univ Okla, BA, 39, MS, 40; Univ Mich, PhD(zool), 47; Am Bd Med Microbiol, dipl. *Prof Exp:* Asst zool, Univ Okla, 38-40; teaching fel, Univ Mich, 40-43; mem staff, Naval Med Sch, Bethesda, Md, 43, head epidemiol teams, SPac, 43-45, res parasitologist, Naval Med Res Inst, 45-48, head parasitol dept, Naval Med Res Unit 3, Cairo, 48-53, instr, Naval Med Sch, 53-57, head parasitol dept, Naval Med Res Unit 2, Taipei, 57-62, res parasitologist & head tech serv dept, Naval Med Res Inst, 62-64; head parasitol dept, Southwest Found Res & Educ, 64-84; CONSULT, 85- *Concurrent Pos:* Exam parasitol, Fac Med, Ain Shams Univ, Cairo, 50-52; res prof Univ Md & adj prof, microbiol, Univ Tex Med Sch, San Antonio; consult, Parasitic Div, WHO. *Mem:* Am Soc Trop Med & Hyg; Am Soc Parasitologists; Am Micros Soc; Int Primatol Soc; SW Asn Parasitol. *Res:* Biology of schistosomes and other helminths; survey-type investigations on parasites of man and lower vertebrates; epidemiology of helminth diseases and zoogeography of parasites of vertebrates, especially the parasites of primates. *Mailing Add:* 14794 Cadillac Dr San Antonio TX 78248-1008

KUNTZ, ROBERT ROY, b Barry, Ill, Apr 10, 37; m 59; c 2. PHYSICAL CHEMISTRY. *Educ:* Culver-Stockton Col, BA, 59; Carnegie Inst Technol, MS, 62, PhD(chem), 63. *Prof Exp:* From asst prof to assoc prof, 62-71, chmn dept, 78-79, assoc chair, 79-81, PROF CHEM, UNIV MO, COLUMBIA, 71- *Concurrent Pos:* Assoc prog dir, NSF, 73-74. *Mem:* Am Soc Photobiol; Sigma Xi; Am Chem Soc; Am Phys Soc; Inter Am Photochem Soc. *Res:* Photolysis and radiolysis of organic compounds, free radical kinetics, radiation protection; flash photolysis; photobiology; photocatalysis. *Mailing Add:* Dept Chem Univ Mo Columbia MO 65211

KUNTZMAN, RONALD GROVER, b New York, NY, Sept 17, 33; m 55; c 2. BIOCHEMISTRY, PHARMACOLOGY. *Educ:* Brooklyn Col, BS, 55; George Washington Univ, MS, 57, PhD(biochem), 62. *Prof Exp:* Chemist, Nat Inst Health; sr biochemist, Wellcome Res Labs, Burroughs & Co, 62-70, dep head biochem pharmacol dept, 67-70; assoc dir dept biochem & drug metab, Hoffmann-La Roche, Inc, 70-72, assoc dir biol res, 72-73, asst vpres & dir therapeuts res, 73-81; vpres & dir Pharmaceut Res & Develop, Hoffmann-La Roche Inc, 80-84; VPRES RES & DEVELOP, HOFFMANN-LA ROCHE INC, NUTLEY, NJ, 84- *Concurrent Pos:* Res & develop steering comt, chmn subcomt on Adv Comt Systs, Comn Drugs for Rare Dis, Pharmaceut Mfrs Asn, 85-; chmn Drug Metab Div, Am Soc Pharmacol & Exp Therapeut; res adv coun, Nat Orgn Rare Dis; mem Adv Bd, Univ Pa Natural Sci Asn; mem Adv Bd, Univ Ariz Col Pharm Nat Adv Bd; mem, Am Soc Pharmacol & Exp Therapeut Coun; corp mem, Muscular Dystrophy Asn. *Honors & Awards:* John Jacob Abel Award, Am Soc Pharmacol & Exp Therapeut, 69. *Mem:* Am Soc Pharmacol & Exp Therapeut (secy-treas, 81-83); Am Soc Biol Chemists; Sigma Xi; Am Col Neuropsychopharmacol; Soc Toxicol; AAAS. *Res:* Biochemical effects and metabolism of drugs and steroid hormones, induced enzyme syntheses; syntheses metabolism and storage of biogenic amines; preclinical development of new drugs; pharmacokinetics and efficacy studies on new therapeutics. *Mailing Add:* Hoffmann-La Roche Inc 340 Kingsland St Nutley NJ 07110

KUNZ, ALBERT BARRY, b Philadelphia, Pa, Oct 2, 40; m 64; c 1. CHEMICAL PHYSICS, SOLID STATE SCIENCE. *Educ:* Muhlenberg Col, BS, 62; Lehigh Univ, MS, 64, PhD(physics), 66. *Prof Exp:* Res assoc physics, Lehigh Univ, 66-69; res asst prof, 69-71, from asst prof to assoc prof, 71-76, PROF PHYSICS, UNIV ILL, URBANA, 76- *Concurrent Pos:* Consult, US Air Force Aerospace Res Lab, 71 & E I du Pont de Nemours & Co, Inc, 73-79; adj prof physics, Mich Technol Univ, 81- *Mem:* Am Phys Soc; Sigma Xi. *Res:* Solid state, atomic and molecular theory; band theory of solids, solid state spectroscopy; spectra of ions, atoms and molecules; theory of ground state properties of polyatomic systems; theory of catalysis is being developed. *Mailing Add:* Dept Physics Mich Tech Univ Houghton MI 49931

KUNZ, ALBERT L, b Bloomington, Ind, Oct 3, 33; m 57; c 5. PHYSIOLOGY. *Educ:* Ind Univ, AB, 56, MD, 59; Ohio State Univ, MS, 65. *Prof Exp:* Instr, 62-63, asst prof, 65-69, assoc prof, 69-76, PROF PHYSIOL, OHIO STATE UNIV, 76- *Concurrent Pos:* Nat Heart Inst fel, 63-65; Alexander Von Humboldt fel, 74-75. *Honors & Awards:* Perkin's Award, Am Physiol Soc, 74. *Mem:* Am Physiol Soc; Am Heart Asn; Am Asn Univ Professors. *Res:* Respiratory control; anomalous viscosity of blood. *Mailing Add:* Dept Physiol Ohio State Univ 333 W Tenth Ave Columbus OH 43210

KUNZ, BERNARD ALEXANDER, b Montreal, Que, Mar 5, 52; m 80; c 1. DNA REPAIR, MUTAGENESIS. *Educ:* McGill Univ, Que, BSc, 74; Brock Univ, Ont, MSc, 76; York Univ, PhD(molecular genetics), 81. *Prof Exp:* Fogarty fel, Nat Inst Environ Health Sci, 81-83; res assoc, York Univ, 83-84, asst prof, 84-86; asst prof, 86-89, ASSOC PROF, UNIV MAN, 89- *Mem:* Genetics Soc Can; Environ Mutagen Soc. *Res:* Analysis of molecular mechanisms of mutation using genetic and recombinant DNA technique; influence of DNA repair and DNA precursor metabolism on genetic stability; isolation and characterization of plant DNA repair genes. *Mailing Add:* Microbiol Dept Univ Man Winnipeg MB R3T 2N2 Can

KUNZ, HAROLD RUSSELL, b Troy, NY, Oct 3, 31; m 56; c 2. THERMODYNAMICS, HEAT TRANSFER. *Educ:* Rensselaer Polytech Inst, BME, 53, MS, 58, PhD(heat transfer), 66. *Prof Exp:* Jr anal engr heat transfer & fluid mech res, Pratt & Whitney Aircraft Div, United Technologies Corp, 53-54, anal engr, 54-57, sr anal engr, 57-60, asst proj engr heat transfer res, 60-63, proj engr heat transfer & fuel cell res, 63-68, sr proj engr, 68-74, sr proj engr, Power Systs Div, 75, sr proj engr fuel cell res, Power Systs Div, 75-85; SR PROJ ENGR FUEL CELL RES, INT FUEL CELLS, 85- *Concurrent Pos:* Adj asst prof, Hartford Grad Ctr, 66-70, adj assoc prof, 70- *Mem:* Electrochem Soc; Am Soc Mech Engrs. *Res:* Thermodynamics; single-phase and two-phase fluid mechanics and heat transfer; electrochemistry; electrocatalysis. *Mailing Add:* Int Fuel Cells 195 Governors Hwy PO Box 739 South Windsor CT 06074-2419

KUNZ, HEINZ W, b Zurich, Switz, Feb 2, 38; Swiss & US citizen. PATHOLOGY. *Educ:* Univ Pittsburgh, PhD(immunogenetics), 78. *Prof Exp:* Res asst, Dept Path, Harvard Med Sch, Boston, 62-65; sr res asst, 65-69, tech assoc, 69-70, assoc, 70-71; clin asst staff, Presby-Univ Hosp, 71-84; from res asst prof to assoc prof path, Univ Pittsburgh, 76-90, asst dir, Div Exp Path, 87-90, assoc dir, 87-88, DIR GRAD PROG, UNIV PITTSBURGH, 88-, PROF PATH, 90- *Concurrent Pos:* Assoc staff, Presby-Univ Hosp, 84-; mem, Pittsburgh Cancer Inst, 86-; consult, Nat Res Coun, 90- *Mem:* AAAS; Am Chem Soc; Genetics Soc Am; Transplantation Soc; Am Asn Pathologists; Am Asn Immunologists; Int Soc Immunol & Reprod; Am Soc Immunol & Reprod; Am Asn Cancer Res. *Res:* Experimental pathology. *Mailing Add:* Dept Path Sch Med Univ Pittsburgh Pittsburgh PA 15261

KUNZ, KAISER SCHOEN, b New Middletown, Ind, Oct 16, 15; m 44; c 3. PHYSICS. *Educ:* Univ Ind, AB, 36; Univ Cincinnati, AM, 37, PhD(theoret physics), 39. *Prof Exp:* Instr math, Univ Cincinnati, 39-42; instr electronics, Cruft Lab, Harvard Univ, 42-45, res assoc, 45-46, res fel, 46-47, lectr appl math, comput lab, 47-49; assoc prof elec eng, Case Inst Technol, 49-51; from res physicist to head interpretation res dept, Schlumberger Well Surv Corp, 51-60; res prof physics & elec eng, 60-76, RES PROF PHYSICS, NMEX STATE UNIV, 76- *Mem:* Fel AAAS; Am Phys Soc; Inst Elec & Electronics Engrs; Am Math Soc; Am Asn Physics Teachers; Sigma Xi. *Res:* Propagation of electromagnetic waves in dynamic media; quantum electronics and lasers; electrodynamics; field theory; numerical analysis. *Mailing Add:* 2047 Crescent Dr Las Cruces NM 88005

KUNZ, SIDNEY EDMUND, b Fredericksburg, Tex, Dec 24, 35; m 60; c 3. ENTOMOLOGY, ECOLOGY. *Educ:* Tex A&M Univ, BS, 58, MS, 62; Okla State Univ, PhD(entom), 67. *Prof Exp:* Surv entomologist, Okla State Univ, 61-64, exten entomologist, 64-67; res entomologist, Agr Res Serv, USDA, Kerrville, 67-69, res entomologist, Col Sta, 69-77, res leader & res entomologist, sci & educ, 77-86, LAB DIR, AGR RES SERV, USDA, KERRVILLE, 86- *Concurrent Pos:* Entom consult, Food & Agr Orgn UN Develop Prog, Mauritius, 73-74 & USAID, Tanzania, IAEA, Somalia, 82. *Mem:* Entom Soc Am; Am Registry Prof Entomologists; Sigma Xi. *Res:* Biology, ecology and area integrated pest management control of biting flies of cattle, horn flies and stable flies. *Mailing Add:* Sci & Educ Agr Res Serv USDA PO Box 232 Kerrville TX 78028

KUNZ, THOMAS HENRY, b Kansas City, Mo, June 11, 38; m 62; c 2. ANIMAL PHYSIOLOGY, BEHAVIOR-ETHOLOGY. *Educ:* Cent Mo State Univ, BS, 61, MS, 62; Drake Univ, MA, 68; Univ Kans, PhD(ecol), 71. *Prof Exp:* Instr biol, Shawnee Mission Schs, 62-67; res fel, Kans Nat Hist Surv, 67-70; teaching fel, Univ Kans, 70-71; from asst prof to assoc prof, Boston Univ, 71-84, dir grad studies, 78-81, assoc chmn, 81-85, chmn, 85-90, DIR GRAD PROGS, ECOL & BEHAV EVOLUTION, BOSTON UNIV, 83-, PROF BIOL, 84- *Concurrent Pos:* Prin investr, NSF, 73-, Nat Geog Soc, 84-85; assoc ed, Am Midland Naturalist, 78-80; res assoc, Carnegie Mus Natural Hist, 78-; Orgn Am States grant, 84-85. *Honors & Awards:* Gerritt S Miller Award, 84. *Mem:* AAAS; Am Soc Mammalogists; Ecol Soc Am; Am Soc Naturalists; Soc Study Evolution. *Res:* Behavioral and physiological ecology of bats, with emphasis on social behavior, energetics, reproductive biology and feeding ecology of New World temperate and tropical species. *Mailing Add:* Dept Biol Boston Univ Boston MA 02215

KUNZ, WALTER ERNEST, b Chattanooga, Tenn, Apr 17, 18; m 43, 70; c 5. RADIATION PHYSICS. *Educ:* Davidson Col, BS, 40; Univ Tenn, PhD(physics), 54. *Prof Exp:* Physicist, Oak Ridge Nat Lab, 52-54, US Naval Res Lab, 54-58 & AEC, 58-62; STAFF MEM, LOS ALAMOS NAT LAB, 62- *Honors & Awards:* I R 100 Award, 83. *Mem:* Am Phys Soc. *Res:* Nuclear reaction at low energy; gamma ray measurements; extreme ultraviolet; solar x-rays; research and development on safeguards for fissionable materials; development of transuranic waste assay systems. *Mailing Add:* 319 Cordova Lane Santa Fe NM 87501

KUNZE, ADOLF WILHELM GERHARD, b Philadelphia, Pa, Aug 23, 36; m 67; c 2. GEOPHYSICS. *Educ:* Pa State Univ, BS, 63, PhD(geophys), 73. *Prof Exp:* Nat Res Coun res assoc lunar geophys, Johnson Space Ctr, NASA, Houston, 73-74; from asst prof to assoc prof, 74-85, PROF GEOL, UNIV AKRON, 85- *Concurrent Pos:* Vis prof, Inst Geophys, Kiel, Fed Repub Ger, 82, 90; Fulbright sr prof teaching/res award, 90. *Mem:* Am Geophys Union; Soc Explor Geophysicists; Asn Eng Geologists. *Res:* Engineering geophysics: shallow subsurface investigations using gravity/magnetic, electrical resistivity and seismic refraction methods; planetology, particularly interpretation of planetary gravity anomalies. *Mailing Add:* 1585 Ruth Ave Cuyahoga Falls OH 44221

KUNZE, DIANA LEE, b Winthrop, Mass, Dec 19, 39. MEDICAL PHYSIOLOGY. *Educ:* Stetson Univ, BS, 61; Emory Univ, MS, 66; Univ Utah, PhD(physiol), 70. *Prof Exp:* Researcher neurophysiol, Nat Ctr Sci Res, France, 71-72; res assoc cardiovasc physiol, Univ Utah, 72-73; asst prof cardiovasc physiol, 73-78, ASSOC PROF PHYSIOL & BIOPHYS, UNIV TEX MED BR GALVESTON, 78- *Res:* Studies of control mechanisms of cardiac rhythm by neural input and by local factors. *Mailing Add:* Dept Molecular Physiol & Biophys Baylor Col Med One Baylor Plaza Houston TX 77030

KUNZE, GEORGE WILLIAM, b Warda, Tex, Sept 16, 22; m 48; c 2. SOIL MINERALOGY. *Educ:* Tex A&M Univ, BS, 47, MS, 50; Pa State Univ, PhD(soil mineral), 52. *Prof Exp:* From asst prof to assoc prof, Tex A&M Univ, T52-60, from assoc dean to dean grad col, 67-84, prof soils & crop sci, 60-84; RETIRED. *Concurrent Pos:* Consult ed, Soil Sci, 58-; grad prog consult, Bangladesh Agr Univ, 70 & Grad Sch Agr Sci, Castelar, Arg, 72; mem, Fed Adv Comt for Affirmative Action in Employment Pract in Inst of Higher Educ, Adv to Secy of Labor & Secy of HEW, 74-76; vpres, Conf Southern Grad Schs, 79-80, pres, 80-81. *Mem:* Fel AAAS; fel Am Soc Agron; Soil Sci Soc Am; Clay Minerals Soc; fel Mineral Soc Am. *Res:* Soil chemistry. *Mailing Add:* PO Box 107 Warda TX 78960

KUNZE, JAY FREDERICK, b Pittsburgh, Pa, Feb 24, 33; m 56; c 3. PHYSICS, MEDICAL PHYSICS. *Educ:* Carnegie Inst Technol, BS, 54, MS, 55, PhD(nuclear physics), 59. *Prof Exp:* Proj physicist & asst, Carnegie Inst Technol, 54-58; physicist, Idaho Test Sta, Gen Elec Co, 58-65, mgr nuclear technol, 65-69; mgr oper & anal, Aerojet Nuclear Corp, 69-70, mgr reactor technol, LPT, 70-74; mgr geothermal & adv technol, EG&G Idaho, Inc, 74-78; vpres & gen mgr, Energy Serv Inc, 78-83; CHMN, NUCLEAR ENG, UNIV MO, 83- *Concurrent Pos:* Site leader, Air Force res solar eclipse expeds, 54-55; affil prof, Univ Idaho, 59-; assoc prof, Univ Utah, 69- *Mem:* Am Nuclear Soc; Nat Soc Prof Engrs. *Res:* Geothermal energy; experimental reactor physics and reactor analysis; energy engineering and conservation; astronomy. *Mailing Add:* 2201 Oak Cliff Dr Columbia MO 65203

KUNZE, OTTO ROBERT, b Warda, Tex, May 27, 25; m 51; c 4. ENGINEERING, AGRICULTURE. *Educ:* Tex A&M Univ, BS, 50; Iowa State Univ, MS, 51; Mich State Univ, PhD(agr eng), 64. *Prof Exp:* Agr & indust engr, Cent Power & Light Co, 51-56; assoc prof elec power & processing, Tex A&M Univ, 57-61 & 64-69, prof, 69-90; RETIRED. *Concurrent Pos:* Consult, post-harvest rice processing, India, 75 & 85; mem, Tex Air Control Bd, 78-90. *Mem:* Fel Am Soc Agr Engrs; Am Asn Cereal Chemists; AAAS; Nat Soc Prof Engrs; Sigma Xi. *Res:* Electric power and processing in agriculture; physical properties of agricultural products; hygroscopicity of rice and its effects on the grain; moisture absorption in low-moisture rough rice and rapid moisture removal in grains. *Mailing Add:* 1002 Milner College Station TX 77840

KUNZE, RAY A, b Des Moines, Iowa, Mar 7, 28; m 51; c 5. MATHEMATICS. *Educ:* Univ Chicago, BS, 50, MS, 51, PhD(math), 57. *Prof Exp:* Asst prof math, Brandeis Univ, 60-62; from assoc prof to prof, Wash Univ, 63-69; chmn dept, 69-74, PROF MATH, UNIV CALIF, IRVINE, 69- *Concurrent Pos:* Consult, Inst Defense Anal, 54-, Prentice Hall & McGraw Hill, 61- *Mem:* Am Math Soc. *Res:* Harmonic analysis; representations of Lie Groups. *Mailing Add:* Boyd Grad Studies Bldg Univ Ga Athens GA 30602

KUNZE, RAYMOND J, b La Grange, Tex, Oct 25, 28; m 51; c 2. SOIL PHYSICS. *Educ:* Tex A&M Univ, BS, 51, MS, 56; Iowa State Univ, PhD(soil physics), 60. *Prof Exp:* Asst, Tex A&M Univ, 54-56; from asst to res assoc, Iowa State Univ, 56-60, NSF res fel & asst prof soil physics, 60-62; soil scientist, 62-65, RES SOIL SCIENTIST, SCI & EDUC ADMIN-AGR RES, USDA, 65-; PROF SOIL SCI, MICH STATE UNIV, 70-, ASSOC CHMN, 80- *Concurrent Pos:* Assoc prof soils, Mich State Univ, 65-70; vis prof, Purdue Univ, 74. *Mem:* Am Soc Agron; Soil Sci Soc Am. *Res:* Measurement of unsaturated flow of moisture in soils; predictions, by computer techniques and analysis, of water movement and distribution in the profile based on measured characteristics of the soil; expertise in operating double-gamma beam for simultaneous, nondestructive, two-component soil and water analysis. *Mailing Add:* 101 Soil Sci Corp/Soil Sci Mich State Univ East Lansing MI 48824

KUNZLE, HANS PETER, b Kreuzlingen, Switz, Sept 1, 40; m 68; c 4. MATHEMATICAL PHYSICS. *Educ:* Swiss Fed Inst Technol, dipl, 64; Univ London, PhD(relativity), 67. *Prof Exp:* Res asst, Kings Col, Univ London, 67-68; lectr & asst res mathematician, Univ Calif, Berkeley, 68-70; asst prof, 70-73, assoc prof, 73-80, PROF MATH, UNIV ALTA, 80- *Concurrent Pos:* Vis sci, Centre Phys Theor Nat Inst Sci Res, Marseille, France, 75-76, Max Planck Inst Astrophy, Garching, 82-83. *Mem:* Am Math Soc; Am Phys Soc. *Res:* Mathematical problems in general relativity; applications of differential geometry to physics, especially relativistic mechanics and field theories. *Mailing Add:* Dept Math Univ Alta Edmonton AB T6G 2G1 Can

KUNZLER, JOHN EUGENE, b Willard, Utah, Apr 25, 23; m 50; c 4. SOLID STATE DEVICES, SUPERCONDUCTIVITY. *Educ:* Univ Utah, BS, 45; Univ Calif, PhD(phys chem), 50. *Prof Exp:* Asst, Purdue Univ, 45-46; from asst to res assoc, Univ Calif, 46-52; mem tech staff, AT&T Bell Labs, 52-61, head metal physics res dept, 61-69, dir, Electronic Mat & Device Lab, 69-79, dir, Electronic Mat, Processes & Devices Lab, 79-85, dir, Future Device Studies Ctr, 85-86; RETIRED. *Honors & Awards:* John Price Wetherill Award, Franklin Inst, 64; Int Prize New Mat, Am Phys Soc, 79; Kamerlingh Onnes Medal, Neth Asn Refrig, 79. *Mem:* Nat Acad Eng; Am Chem Soc; fel Am Phys Soc. *Res:* Electrical, thermal and magnetic properties of solids at low temperatures; Fermi surface; galvanomagnetic and magnetothermal effects; high purity metals; high-field superconductivity; superconducting magnets; low temperature heat capacity and related thermal effects. *Mailing Add:* Rte 2 PO Box 130 Port Murray NJ 07865

KUO, ALBERT YI-SHUONG, b Tayuan, Taiwan, Nov 4, 39; m 65; c 2. PHYSICAL OCEANOGRAPHY, HYDRODYNAMICS. *Educ:* Nat Taiwan Univ, BS, 62; Univ Iowa, MS, 65; Johns Hopkins Univ, PhD(fluid mech), 70. *Prof Exp:* Jr instr fluid mech, Johns Hopkins Univ, 67-69, res assoc, 70; assoc marine scientist, 70-78, sr marine scientist & head hydraulics sect, 78-81, PROF, COL WILLIAM & MARY, 81-; ASST DIR, VA INST MARINE SCI, 91- *Concurrent Pos:* From asst prof to assoc prof, Univ Va & Col William & Mary, 70-80; vis prof, Nat Taiwan Univ, 77-78. *Mem:* Am Soc Civil Engrs; Estuarine Res Fedn; Int Asn Hydraul Res. *Res:* Turbulence, diffusion, dispersion; estuarine mathematical model; estuarine hydrodynamics; sediment transport; coastal circulation. *Mailing Add:* Div Phys Oceanog & Environ Eng Va Inst Marine Sci Gloucester Point VA 23062

KUO, BENJAMIN CHUNG-I, b China, Oct 5, 30; m 54; c 1. ELECTRICAL ENGINEERING. *Educ:* Univ NH, BS, 54; Univ Ill, MS, 56, PhD(elec eng), 58. *Prof Exp:* Plant engr, Laible Mfg Co, 53-54; asst, 54-57, from asst prof to assoc prof, 58-66, PROF ELEC ENG, UNIV ILL, URBANA, 66- *Mem:* Inst Elec & Electronics Engrs. *Res:* Feedback control systems; sampled-data systems. *Mailing Add:* 3206 Valley Brook Dr Champaign IL 61821

KUO, CHAN-HWA, b Shanghai, China, Oct 7, 31; US citizen; m 57; c 3. ORGANIC CHEMISTRY. *Educ:* Hartwick Col, BS, 57; Rensselaer Polytech Inst, MS, 58; Polytech Inst Brooklyn, PhD(org chem), 75. *Prof Exp:* Res chemist, 58-74, SR RES CHEMIST, MERCK & CO, INC, 74- *Mem:* Sigma Xi. *Res:* Synthesis of griseofulvin, fluoro- and polychlorogriseofulvin, estrone, prostaglandin E1; synthesis and conformational analysis of pantetheine analogs; synthesis and relative configurational studies of the chiral lactone derived from thermozymocidin (myriocin). *Mailing Add:* 105 E Nassau Ave South Plainfield NJ 07080

KUO, CHAO-YING, b Taiwan, Apr 27, 40; m; c 2. PEDIATRICS. *Educ:* Nat Taiwan Normal Univ, BS, 64; Ind State Univ, MA, 70; Univ Iowa, PhD(microbiol), 74. *Prof Exp:* Teaching asst, Lab Invert, Zool & Human Physiol, Taiwan Normal Univ, 65-67, Gen Bot, Human Anat & Physiol, Ind State Univ, 68-70; asst immunol, Univ Iowa, 70-73, asst res scientist, Div Allergy & Immunol, Dept Med, 74-77, instr, 80-81; from instr to asst prof, Dept Med, Microbiol & Immunol, Univ Tenn, 80-84, res assoc, Dent Res Ctr, 88 & Dept Pediat, 89-90, ASST PROF, DEPT PEDIAT, CTR HEALTH SCI, UNIV TENN, 91- *Concurrent Pos:* Res assoc, Lady Davis Inst Med Res, Jewish Gen Hosp, Montreal, 73-74; assoc mem, Barbara Kopp Geriat Res Ctr, Auburn, NY, 77-80; grants, Leukemia Res Found, Inc, 78-79, Nat Cancer Inst, NIH, 79-83, Am Cancer Soc, 81-82 & LeBonheur Diabetes Res Fund, 91-92; res microbiologist, Res Serv, Vet Admin Med Ctr, Memphis, Tenn, 81-84; pvt enterprise, 84-88. *Mem:* Am Asn Immunologists. *Res:* Allergy & tumor immunology; monoclonal antibody production; islet transplantation and islet organoid development; aging and cancer; author of numerous scientific publications. *Mailing Add:* Dept Pediat Col Med Univ Tenn Rm B310 956 Court Ave Memphis TN 38163

KUO, CHARLES C Y, b Hubei, China. MATERIALS & PROCESSES, MICROELECTRONICS. *Educ:* Col Ord Eng, China, BS, 52; Lehigh Univ, Bethlehem, Pa, MS, 57, PhD(chem), 61. *Prof Exp:* Asst prof chem eng, Col Ord Eng, Taiwan, China, 52-57; res fel chem eng, Lehigh Univ, Bethlehem, Pa, 57-61; mem staff, res & develop mat, Bell Tel Lab, AT&T, Pa, 61-67; eng specialist, res & develop mat, Res & Develop Lab, GTE, NY, 67-71; dept head, res & develop mat, Engelhard Industs, NJ, 71-77; TECH DIR RES & DEVELOP MAT, CTS CORP, ELKHART, IND, 77- *Concurrent Pos:* Mem prog comt, Int Conf Electronic Components & Mat, 89-92. *Honors & Awards:* Tech Achievement Award, Int Soc Hybrid Microelectronics, 87. *Mem:* Am Chem Soc; Am Ceramic Soc; Am Soc Metals. *Res:* Materials and processes of hybrid microelectronics; thick and thin films; resistors; conductors; dielectrics; reliability; interconnection; packaging; author of more than 80 publications; granted 30 patents. *Mailing Add:* CTS Corp 905 West Blvd N Elkhart IN 46514

KUO, CHENG-YIH, b Tainan, Taiwan, Apr 2, 42; US citizen; m 78; c 1. POLYMER SCIENCE. *Educ:* Nat Taiwan Univ, BS, 66; Univ Akron, MS, 69, PhD(polymer sci), 73. *Prof Exp:* Fel, Inst Polymer Sci, Univ Akron, 73-75; sr chemist, 75-80, assoc scientist, 80-82, SCIENTIST MAT SCI, RES CTR, GLIDDEN CO, 82- *Mem:* Am Chem Soc. *Res:* Characterization of polymers and analysis of organic coatings. *Mailing Add:* Glidden Co PO Box 8827 Strongsville OH 44136

KUO, CHIANG-HAI, b Tainan, Taiwan, Feb 10, 36; m 59; c 2. CHEMICAL & PETROLEUM ENGINEERING. *Educ:* Nat Taiwan Univ, BS, 57; Univ Houston, MS, 61, PhD(chem eng), 64. *Prof Exp:* Teaching asst, Nat Taiwan Univ, 57-59; engr, Shell Develop Co, 62-64, res engr, 64-70; assoc prof, 70-77, PROF CHEM ENG, MISS STATE UNIV, 77- *Honors & Awards:* Award, Am Inst Chem Engrs, 64; Bronze Medal Award, US Environ Protection Agency, 75. *Mem:* Am Inst Chem Engrs; Am Inst Mining Metall & Petrol Engrs. *Res:* Mass transfer and chemical reactions; reaction kinetics; flow through porous media; heat transfer; petroleum recovery processes; air and water pollution control. *Mailing Add:* Dept Chem Eng Miss State Univ Mississippi State MS 39762

KUO, CHING-MING, b Taipei, Taiwan, Sept 23, 35; m 62; c 3. REPRODUCTIVE PHYSIOLOGY, ICHTHYOLOGY. *Educ:* Nat Taiwan Univ, BSc, 58; Scripps Inst Oceanog, Univ Calif, San Diego, PhD(marine biol), 70. *Prof Exp:* Lab instr zool, Nat Taiwan Univ, 60-64; res asst ichthyol, Scripps Inst Oceanog, 64-70; res assoc physiol, Oceanic Inst, 70-72, head aquaculture, 73-80, sr scientist physiol, 72-80; sr scientist physiol, Int Ctr Living Aquatic Resources Mgt, 80-85; vis prof, 85-87, DIR, INST FISHERIES SCI, NAT TAIWAN UNIV, 85-, PROF, 87- *Concurrent Pos:* Adv/consult serv, reproductive physiol, artificial propagation & penaeid shrimp culture technol. *Mem:* Am Soc Zoologists; Sigma Xi. *Res:* Process and mechanism involved in gonadal maturation and ovulation; reproductive physiology of Penaeid Shrimps; artificial propagation and genetic improvement of cultured fish; induced sex reversal and breeding of groupers. *Mailing Add:* Inst Fisheries Sci Nat Taiwan Univ Taipei 10746 Taiwan

KUO, CHO-CHOU, b Taiwan, Sept 12, 34; m 64; c 1. MEDICAL MICROBIOLOGY. *Educ:* Nat Taiwan Univ, MD, 60; Univ Wash, PhD(prev med), 70. *Prof Exp:* Fel, 67-71, asst prof, 71-76, assoc prof, 76-80, PROF PATHOBIOL, UNIV WASH, 80- *Mem:* Am Pub Health Asn; Am Col Prev Med; Am Soc Microbiol; Am Asn Immunologists. *Res:* Microbiology and immunology of the Chlamydia Trachomatis organisms which cause eye and genital infection, development of diagnostic methods and prevention of the disease. *Mailing Add:* Dept Path Univ Wash Seattle WA 98195

KUO, CHUNG-MING, b Chang-Hwa, Taiwan, China, Aug 6, 35; US citizen; m 66; c 2. CELLULOSE, PULP & PAPER SCIENCES. *Educ:* Chung-Shing Univ, Taiwan, BS, 58; Syracuse Univ, MS, 64, PhD(org chem), 69. *Prof Exp:* Res chemist, 68-71, sr res chemist, 71-81, RES ASSOC, EASTMAN CHEM CO, EASTMAN KODAK CO, 81- *Mem:* Am Chem Soc. *Res:* Chemistry and new and improved methods for the preparation of cellulose and its derivatives; modification of cellulose and cellulose derivatives for fibers, films and plastics end uses; new products based on cellulose and the related carbohydrate materials. *Mailing Add:* Eastman Chem Co Res Labs Eastman Kodak Co Kingsport TN 37662

KUO, ERIC YUNG-HUEI, b Chiayi, Taiwan, Aug 8, 34; US citizen; m 68; c 2. ONCOLOGY, VETERINARY MEDICINE. *Educ:* Nat Taiwan Univ, BS, 60; Univ Ill, Urbana, MS, 66, PhD(vet med sci), 70. *Prof Exp:* Asst vet parasitol, Dept Vet Med, Nat Taiwan Univ, 61-63; res asst vet physiol, Col Vet Med, Univ Ill, Urbana, 63-69; res assoc endocrinol, Dept Physiol, Sch Med, Boston Univ, 69-71; sr investr vet endocrinol, 71-74, prin investr oncol, Mason Res Inst, 75-76; vet med officer, USDA, 76-80. *Concurrent Pos:* Vet, Southeast Vet Hosp, Taiwan, 61-63; lectr, Grad Div, Anna Maria Col, Mass, 76- *Mem:* AAAS; Endocrine Soc; Am Vet Med Asn. China. *Res:* The roles of infection, infection hormonal imbalance, radiation and immunosuppression in mammary oncogenesis; the responses of hosts and tumors to surgery, radiation and chemotherapy. *Mailing Add:* 2455 Sedgefield Dr Chapel Hill NC 27514

KUO, FRANKLIN F(A-KUN), b China, Apr 22, 34; m 58; c 2. ELECTRICAL ENGINEERING. *Educ:* Univ Ill, BS, 55, MS, 56, PhD(elec eng), 58. *Prof Exp:* Asst prof elec eng, Polytech Inst Brooklyn, 58-60; mem tech staff, Bell Tel Labs, Inc, 60-66; prof elec eng, Univ Hawaii, 66-82; EXEC DIR, SRI INT, MENLO PARK, CALIF, 82- *Concurrent Pos:* Mem Cosine comt, Nat Acad Eng, 65-72; liaison scientist, US Off Naval Res, London, 71-72; consult ed, Prentice-Hall, Inc; dir info systs, Off Secy Defense, 76-77; coun mem, Asn Comput Mach. *Mem:* Fel Inst Elec & Electronics Engrs; Asn Comput Mach. *Res:* Digital computers; information transmission; computer networks; data communications. *Mailing Add:* Sri Int 333 Ravenswood Ave EL-290 Menlo Park CA 94025

KUO, HARNG-SHEN, b Hangchow, China, June 9, 35; m 67; c 3. CHEMISTRY, BIOCHEMISTRY. *Educ:* Cheng Kung Univ Taiwan, BS, 59; La State Univ, New Orleans, MS, 66; Pa State Univ, PhD(chem), 70. *Prof Exp:* Analyst cement, Taiwan Chi Hsin Co, 61; engr, Taiwan Fertilizer Co, 61-64; chemist & fel, Lawrence Berkeley Lab, 70-71; anal res chemist, 71-72, radiation safety officer, 72-75, SR STAFF SCIENTIST, CUTTER LABS, 75- *Mem:* Am Chem Soc. *Mailing Add:* Cutter Labs PO Box 1986 Fourth & Parker Sts Berkeley CA 94701

KUO, HSIAO-LAN, b Mancheng, China, Jan 7, 15; m 49; c 3. DYNAMIC METEOROLOGY, FLUID DYNAMICS. *Educ:* Tsing Hua Univ, BS, 37; Univ Chicago, PhD(meteorol), 48. *Prof Exp:* From res assoc meteorol to res meteorologist, Mass Inst Technol, 49-57; vis assoc prof meteorol, Univ Chicago, 57-58; supvr res meteorol & hurricane res proj, Mass Inst Technol, 58-62; PROF METEOROL, UNIV CHICAGO, 62- *Res:* Dynamics of planetary atmospheres and atmospheric vortices; general circulation; atmospheric radiation; high atmosphere; climate change. *Mailing Add:* Dept Geophys Sci Rm 449 Univ Chicago 5734 S Ellis Ave Chicago IL 60637

KUO, HUI-HSIUNG, b Ta-chia, Taichung, Taiwan, Oct 21, 41; US citizen; m 69; c 2. STOCHASTIC DIFFERENTIAL EQUATIONS, BROWNIAN FUNCTIONALS. *Educ:* Taiwan Univ, BA, 65; Cornell Univ, MA, 68, PhD(math), 70. *Prof Exp:* Asst prof, res & teaching, Univ Va, 71-75; vis asst prof res & teaching, State Univ NY, Buffalo, 75-76; assoc prof res & teaching, Wayne State Univ, 76-77; assoc prof, 77-82, PROF RES & TEACHING, LA STATE UNIV, 82- *Concurrent Pos:* Vis mem res, Courant Inst, NY Univ, 70-71, Ctr Stochastic Processes, Univ NC, 84; prin investr, NSF, 72-77, 78-84, 85-87 & 90-92; vis prof, Nagoya Univ, 84 & Univ Bielefeld, 86 & 88; res fel, Japan Soc Prom Sci, 84; mem, Comt Summer Inst, Am Math Soc, 83-86. *Mem:* Am Math Soc; Math Soc Japan; Inst Math Statist. *Res:* Stochastic differential equations, infinite dimensional stochastic analysis, theory of generalized Brownian functionals, probability and harmonic analysis on infinite dimensional spaces. *Mailing Add:* Dept Math La State Univ Baton Rouge LA 70803

KUO, JOHN TSUNG-FEN, b Hangchow, China, Apr 1, 22; m 57; c 3. GEOPHYSICS. *Educ:* Univ Redlands, BS, 52; Calif Inst Technol, MS, 54; Stanford Univ, PhD(geophys), 58. *Hon Degrees:* ScD, Univ Redlands, 78. *Prof Exp:* From instr to asst prof geol & geophys, San Jose State Col, 56-60; res scientist, Lamont Geol Observ, 60-64, from assoc prof to prof mining (geophysics), 64-82, Vinton prof mining (geophysics), 83-85, EWING & WORZEL PROF GEOPHYS, COLUMBIA UNIV, 85- *Concurrent Pos:* Res assoc, Stanford Univ, 58-60; NSF sr fel, Cambridge Univ, 70-71; consult; vis prof, Univ Texas, Austin, 77, Cornell Univ, 78, Technische Universität Clausthal, WGer, 87; distinguished US sr scientist, Alexander von Humboldt Award, 86-87. *Mem:* Seismol Soc Am; Am Geophys Union; Soc Explor Geophys; fel Geol Soc Am; fel Royal Astron Soc. *Res:* Acoustic, elastical, EM wave scattering and diffractions; geophysical exploration; solid earth and ocean dynamics. *Mailing Add:* Columbia Univ New York NY 10027

KUO, JYH-FA, b Taiwan, China, May 19, 33; m 65; c 2. BIOCHEMISTRY, PHARMACOLOGY. *Educ:* Nat Taiwan Univ, BS, 57; SDak State Univ, MS, 61; Univ Ill, Urbana, PhD(biochem), 64. *Hon Degrees:* MD, Linköping Univ, Sweden, 80. *Prof Exp:* Res biochemist, Lederle Labs, Am Cyanamid Co, 64-68; from asst prof to assoc prof pharmacol, Sch Med, Yale Univ, 68-72; assoc prof, 72-76, PROF PHARMACOL, SCH MED, EMORY UNIV, 76-, PROF BIOCHEM, 85- *Concurrent Pos:* Vis prof, Swedish Med Res Coun, Linköping Univ, Sweden, 70, Peking Univ, China, 83 & Max-Planck Inst Biophys Chem, Ger, 89; Res Career Develop Award, NIH, 71-75; Merit Award, NIH, 86-96. *Mem:* AAAS; Am Soc Biochem & Molecular Biol; Am Soc Pharmacol & Exp Therapeut. *Res:* Action of hormones; role of cyclic adenosine monophosphate, cyclic guanosine monophosphate, cyclic cytidine monophosphate, calcium and protein kinases in cellular function and metabolism; signal transduction; cancer and cardiac pathophysiology. *Mailing Add:* Dept Pharmacol Emory Univ Sch Med Atlanta GA 30322

KUO, LAWRENCE C, b Hong Kong; US citizen. MOLECULAR ENZYMOLOGY, RECOMBINANT DNA. *Educ:* Univ Chicago, PhD(biophys), 81. *Prof Exp:* Res fel chem, Harvard Univ, 81-85; ASST PROF CHEM, BOSTON UNIV, 85- *Concurrent Pos:* Res fel, Jane Coffin Childs Fund Med Res, 81-84; Pew scholar; NIH res career develop award. *Honors & Awards:* Harold Lamport Award, Biophys Soc. *Mem:* Am Chem Soc; AAAS. *Res:* Biologically related organic and inorganic chemistry; use of chemical methods and approaches to the solution of enzyme actions; the roles of metal ions in metalloenzymes; protein isomerization. *Mailing Add:* Dept Chem Boston Univ 590 Commonwealth Ave Boston MA 02215

KUO, MINGSHANG, b Kaohsiung, Taiwan, Oct, 11, 49; m 74; c 2. SPECTROSCOPY, CHROMATOGRAPHY. *Educ:* Nat Tsing-Hua Univ, Taiwan, BS, 71; Mich State Univ, PhD(chem), 79. *Prof Exp:* SCIENTIST ANALYTICAL CHEMIST NATURAL PROD, UPJOHN CO, 79- *Mem:* Am Chem Soc; Am Soc Pharmacog. *Res:* Isolation and identification of fermentation products; natural products chemistry. *Mailing Add:* Upjohn Co 7700 Portage Rd Kalamazoo MI 49001

KUO, PAO-KUANG, b Hopei, China, Feb 23, 35; m 61; c 2. THEORETICAL PHYSICS. *Educ:* Nat Taiwan Univ, BSc, 57; Univ Minn, PhD(physics), 64. *Prof Exp:* Instr physics, Cornell Univ, 64-66; from res assoc to instr, Mass Inst Technol, 66-68; vis lectr, Johns Hopkins Univ, 68-69; asst prof, 69-71, ASSOC PROF PHYSICS, WAYNE STATE UNIV, 71- *Mem:* Am Phys Soc. *Res:* Quantum electrodynamics; theory of elementary particles and coherence phenomena. *Mailing Add:* Dept Physics Wayne State Univ 5950 Case Ave Detroit MI 48202

KUO, PETER TE, b Fukien, China, Mar 21, 16; m 49; c 2. INTERNAL MEDICINE. *Educ:* St John's Univ, China, MD, 39; Univ Pa, MMSc, 49, DSc(med), 50. *Prof Exp:* From asst to asst prof med, Med Sch, St John's Univ, China, 40-46; from instr to prof, Sch Med, Univ Pa, 50-73, sr staff mem, Robinette Found Cardiovasc Res, Hosp Univ Pa, 52-73; prof med & dir cardiovasc div, Rutgers Med Sch, Col Med & Dent NJ, 73-82, John G Detwiler Prof Cardiol, 82-87; CONSULT MED & CARDIOL MED CTR, MUHLENBERG HOSP, ST PETER'S MED CTR, 73-; DIR, HYPERLIPIDEMIA PROG, HOUSTON VET ADMIN MED CTR, 87- *Concurrent Pos:* Consult cardiol & probs lipid metab; estab investr, Am Heart Asn, 55-60, Arteriosclerosis Coun, 58; USPHS career develop award, 61-66; hon prof, Second Med Col, Shanghai, China, 81; dir, Atherosclerosis Res, Univ Med & Dent, Robert Wood Johnson Med Sch, 82-87; clin prof med, Baylor Col Med, 87-88, prof med, 88- *Honors & Awards:* Sci Award, Am Chinese Asn. *Mem:* Fel Gerontol Soc; fel Am Col Physicians; fel Am Col Cardiol; Am Soc Clin Nutrit; Am Nutrit Inst; AAAS; Am Fedn Clin Res; Am Med Asn; fel Am Col Angiology; fel Am Col Chest Physicians. *Res:* Blood and tissue lipids and their relationship to the problem of arteriosclerosis; circulatory hemodynamics. *Mailing Add:* Prof Dept Med Baylor Col Med Houston TX 77030

KUO, SCOT CHARLES, b July 4, 61. CELL BIOLOGY & PHYSIOLOGY. *Educ:* Harvard Univ, BA, 82; Univ Calif, Berkeley, DPhil, 88. *Prof Exp:* Tech, Dept Chem, Harvard Univ, 81, res, Dept Biochem, 79-82; mem res proj, Dept Biochem, Univ Calif, Berkeley, 82-88; RES FEL, DEPT CELL BIOL, DUKE UNIV, 88- *Concurrent Pos:* Fel, Jane Coffin Childs Mem Fund Med Res, 89- *Mem:* Am Soc Cell Biol; Biophys Soc. *Res:* Fluorescence studies of the anion transporter, Band-3, from human erythrocytes; elucidating the mechanism of information processing in bacterial chemotaxis; interactions between cytoskeleton and membrane components; biophysics of microtubule-dependent motility. *Mailing Add:* Dept Cell Biol Box 3709 Med Ctr Duke Univ St Louis MO 63110

KUO, SHAN SUN, b Nanking, China, Nov 22, 22; m 58; c 1. APPLIED MATHEMATICS, COMPUTER SCIENCE. *Educ:* Nat Chung Cheng Univ, China, BEng, 44; Ohio State Univ, MSc, 48; Harvard Univ, MEng, 54; Yale Univ, DEng, 58. *Prof Exp:* Instr, Nat Chung Cheng Univ, China, 44-46; lectr, Formosa Inst Technol, 46-47; struct engr, Ohio State Univ, 48-52; engr, Carew Steel Prod Corp, 52-53; engr, Fay Spofford & Thorndike, 54-55; from asst prof to assoc prof civil eng, Tufts Univ, 58-64, dir comput ctr, 61-64; prof math, 64-77, PROF COMPUT SCI, UNIV NH, 77-, DIR COMPUT CTR, 64- *Mem:* Am Math Soc; Asn Comput Mach; Am Soc Civil Engrs; Am Soc Mech Engrs; Am Soc Eng Educ. *Res:* Numerical analysis; computer applications. *Mailing Add:* Dept Comp Sci Univ NH Durham NH 03824

KUO, SHIOU, b Ping-Tung, Taiwan, Oct 8, 43; m 69; c 3. SOIL CHEMISTRY, PHYSICAL CHEMISTRY. *Educ:* Chung-Hsing Univ, Taiwan, BS, 66; Utah State Univ, MS, 70; Univ Maine, PhD(soil chem), 73. *Prof Exp:* Res assoc soils, Iowa State Univ, 74-75; res assoc agron, Univ Calif, Davis, 75-78; ASST SOIL SCIENTIST, WESTERN WASH RES & EXTEN CTR, WASH STATE UNIV, 78- *Mem:* Am Soc Agron; Soil Sci Soc Am; Chinese Agr Asn; Sigma Xi. *Res:* Nitrogen transformations in soils and their relation to the nitrogen uptake by plant; cations and anions reactions with soil colloidal particals and the plant growth. *Mailing Add:* 2505 Manorwood Dr Puyallup WA 98501

KUO, THOMAS TZU SZU, b Peiping, China, July 31, 32; m 62; c 2. THEORETICAL PHYSICS. *Educ:* Naval Col Eng, Taiwan, BS, 54; Tsing Hua Univ, Taiwan, MS, 59; Univ Pittsburgh, PhD(physics), 64. *Prof Exp:* From instr to asst prof physics, Princeton Univ, 64-68; vis scientist, Argonne Nat Lab, 68 & 69; assoc prof, 68-72, PROF PHYSICS, STATE UNIV NY, STONY BROOK, 72- *Concurrent Pos:* Nordita guest prof physics, Univ Oslo, 74-75, 78 & 83; vis prof, Julich Nuclear Res Ctr, WGer, 79; hon prof, Inst High Energy Physics, China, Jilin Univ & Fudan Univ, 81. *Honors & Awards:* Humboldt Award Sr Am Scientists, 77. *Mem:* Fel Am Phys Soc. *Res:* Theoretical nuclear physics; nuclear structure and the free nucleon nucleon interaction; nuclear matter phase transitions; finite temperature; many body problems. *Mailing Add:* Dept Physics State Univ NY Stony Brook NY 11794

KUO, TZEE-KE, b Peking, China, Apr 13, 37; m 61. HIGH ENERGY PHYSICS. *Educ:* Nat Taiwan Univ, BS, 57; Univ Chicago, MS, 60; Cornell Univ, PhD(physics), 63. *Prof Exp:* Res assoc physics, Brookhaven Nat Lab, 63-65; asst prof, 65-68, assoc prof, 68-77, PROF PHYSICS, PURDUE UNIV, WEST LAFAYETTE, 77- *Mem:* Am Phys Soc. *Res:* Elementary particle physics. *Mailing Add:* Dept Physics Purdue Univ West Lafayette IN 47907

KUO, YEN-LONG, b Taipei, Taiwan, Nov 18, 36; m 66; c 1. ELECTRICAL ENGINEERING. *Educ:* Taipei Inst Technol, Taiwan, Dipl elec eng, 57; Okla State Univ, MS, 61; Univ Calif, Berkeley, PhD(elec eng), 66. *Prof Exp:* Actg asst prof elec eng, Univ Calif, Berkeley, 66; asst prof, Purdue Univ, 66-70; MEM TECH STAFF ELEC ENG, BELL TEL LABS INC, 70- *Mem:* Inst Elec & Electronics Engrs. *Res:* Computer-aided circuit analysis and synthesis; nonlinear distortion analysis; system theory. *Mailing Add:* AT&T Bell Tel Labs Inc 1600 Osgood St North Andover MA 01845

KUO, YING L, b Taipei, Taiwan, Aug 19, 58; m. TELECOMMUNICATION FIELD DESIGN, COMPUTER QUALITY CONTROL PLANNING. *Educ:* Van-Nam Inst Technol, BS, 81; Calif Century Univ, BS, 88, MS, 89; Century Univ NMex, PhD(computer eng), 91. *Prof Exp:* Chief engr, Res & Develop Dept, Microtel Inc, 82-84; supvr, Res & Develop Dept, All Best Inc, 84-86; mgr, Res & Develop Dept, Telemate Tech, Inc, 86-88; ENG STAFF, RES & DEVELOP DEPT, FOUNTAIN TECH INC, 89- *Res:* Microcomputer and microprocessor; hardware design and software design. *Mailing Add:* Fountain Tech Inc 49 Almond Dr Somerset NJ 08873

KUPCHELLA, CHARLES E, b Nanty Glo, Pa, July 7, 42; m 63; c 3. CANCER BIOLOGY, ENVIRONMENTAL EDUCATION. *Educ:* Ind Univ Pa, BSEd, 64; St Bonaventure, PhD(physiol), 68. *Prof Exp:* From asst prof to assoc prof, Bellarmine Col, 68-73; assoc dir, Cancer Res Ctr & assoc prof oncol, Sch Med, Univ Louisville, 73-79; chmn biol, Murray State Univ, 79-85; DEAN, OGDEN COL SCI TECHNOL & HEALTH, WESTERN KY UNIV, 85- *Concurrent Pos:* Secy-treas, Ky Sci & Technol Coun, 88-; chmn, Finance Comt, Am Asn Cancer Educ, 90-; mem, bd trustees, Nature Conservancy, Ky & bd dirs, Ky Ctr Pub Issues; consult environ progs var insts. *Mem:* AAAS; Sigma Xi; NAm Asn Environ Educ; Am Asn Cancer Educ. *Res:* Biology of cancer; biology of the glycosaminoglycans-involvement in wound repair; diseases of the skin; environmental science. *Mailing Add:* Ogden Col Sci Technol & Health TCCW 105 Western Ky Univ Bowling Green KY 42101

KUPCHIK, EUGENE JOHN, b Wallington, NJ, Aug 26, 29; m 65. ORGANOMETALLIC CHEMISTRY. *Educ:* Rutgers Univ, BS, 51, PhD(org chem), 59. *Prof Exp:* Res chemist, Union Carbide Plastics Co, 54-55; Alfred P Sloan res fel, 56; teaching fel, Du Pont, 57; instr org chem, Rutgers Univ, 58-60; from asst prof to assoc prof, 60-68, PROF ORG CHEM, ST JOHN'S UNIV, NY, 68- *Mem:* Am Chem Soc. *Res:* Organometallic chemistry; organotin compounds; biological properties of organometallic compounds; chemical graph theory; quantitative structure: activity relationships in chemistry, biology and pharmacy. *Mailing Add:* Dept Chem St John's Univ Jamaica NY 11439

KUPCHIK, HERBERT Z, b Brooklyn, NY, Dec 6, 40; m 64; c 2. CANCER. *Educ:* Bethany Col, BS, 62; Wayne State Univ, MS, 65, PhD(biochem), 67. *Prof Exp:* Asst chem, Wayne State Univ, 62-63, res asst biochem, 64-67; assoc med, Harvard Med Sch, 69-71, assoc biol chem, 71-72, prin res assoc biochem, 72-78; asst prof, 76-82, ASSOC PROF MICROBIOL, SCH MED, BOSTON UNIV, 82- *Concurrent Pos:* Instr biochem & org chem, Marygrove Col, 64-65; NIH fel enzym, Cancer Res Inst, New Eng Deaconess Hosp, Boston, 67-69; res fel enzym, Harvard Med Sch, 68-69; clin assoc, Thorndike Mem Lab, 69-73; res assoc, Mallory Gastroenterol Lab, Boston City Hosp, 69-74, sr res assoc, 74-80; res assoc, Sch Med, Boston Univ, 71-76, mem staff, Hubert H Humphrey Cancer Res Ctr, 80-; mem spec sci staff, Boston City Hosp & Mallory Inst Path, 78- *Mem:* Am Asn Pathologists; Am Soc Microbiol; Am Asn Cancer Res; Am Fedn Clin Res; NY Acad Sci. *Res:* Properties of invasive human tumors; in-vitro screening and evaluation of immunotherapeutic agents; development of monoclonal antibodies to human tumors; in-vitro transformation of human colonic adenomas to carcinomas. *Mailing Add:* Dept Microbiol Boston Univ Sch Med 80 E Concord St Boston MA 02118

KUPEL, RICHARD E, b Peoria, Ill, Nov 8, 20; m 46; c 1. ORGANIC CHEMISTRY, INORGANIC CHEMISTRY. *Educ:* Monmouth Col, Ill, BS, 48. *Prof Exp:* Chemist, Gen Elec Co, Wash, 48-50, supvr mass spectrometry lab, 50-52, chemist, Ohio, 52-54, unit leader instrumental anal lab, 54-61; asst chief lab phys & chem anal br, Nat Inst Occup Safety & Health, USPHS, 61-73, hazard eval coordr, 73-80; CONSULT, SKC INC, 80- *Concurrent Pos:* Chmn subcomt seven, Intersoc Comt Manual Methods Ambient Air Sampling & Anal. *Mem:* Am Chem Soc; Am Conf Govt Indust Hyg; Am Indust Hyg Asn; Soc Appl Spectros (pres elect, 65). *Res:* Quantitative analytical methods for analysis of trace elements in biological tissues and environmental samples using emission spectrographic, mass spectrometric, x-ray diffraction, gas chromatographic and spectrophotometric procedures and instrumentation; charcoal tube for sampling organic vapors; K-2 spot test of asbestos. *Mailing Add:* SKC Inc 3935 Freeman Ave Hamilton OH 45015

KUPER, J B HORNER, b New York, NY, Nov 5, 09; m 37; c 1. PHYSICS, ELECTRONICS. *Educ:* Williams Col, AB, 30; Princeton Univ, PhD(physics), 38. *Prof Exp:* From physicist to asst physicist, Wash Biophys Inst, 37-40; assoc physicist, NIH, 40-41; from mem staff to assoc group leader, Radiation Lab, Mass Inst Technol, 41-46; from sr engr to head dept, Fed Telecommun Labs, NY, 46; head electronics Div, 47-48, chmn instrumentation & health physics dept, 48-70, chmn environ sci study group, 70-72, asst to dir, 72-74, CONSULT, BROOKHAVEN NAT LAB, 75- *Concurrent Pos:* Ed, Rev Sci Instruments, Am Inst Physics, 54-79. *Mem:* Fel Am Phys Soc; Health Physics Soc; fel Inst Elec & Electronics Engrs. *Res:* General instrumentation; spectrophotometers; Geiger counters and other equipment for radioactive research; microwave plumbing; general electronics; health physics; electronic instrumentation and editorial work. *Mailing Add:* Brookhaven Nat Lab Upton NY 11973

KUPERMAN, ALBERT SANFORD, b New York, NY, Aug 1, 31; m 56; c 2. PHARMACOLOGY, EDUCATIONAL ADMINISTRATION. *Educ:* NY Univ, AB, 52; Cornell Univ, PhD(pharmacol), 57. *Prof Exp:* Res fel pharmacol, Med Col, Cornell Univ, 57-58, instr, 58-59; asst prof, Col Med, NY Univ, 59-61; asst prof, Med Col, Cornell Univ, 61-65; assoc prof, Hunter Col, 65-68, prof biol sci, 68; Rockefeller Found vis prof & actg chmn dept pharmacol, Fac Med Sci, Mahidol Univ, Thailand, 68-75; ASSOC DEAN EDUC AFFAIRS, ALBERT EINSTEIN COL MED, 75-, ASSOC PROF MOLECULAR PHARMACOL, 89- *Concurrent Pos:* USPHS fel, 57-59. *Mem:* Am Soc Pharmacol & Exp Therapeut; fel Am Col Clin Pharmacol. *Res:* General pharmacology; physiology and pharmacology of excitable cells. *Mailing Add:* Off Educ Albert Einstein Col Med Bronx NY 10461

KUPFER, CARL, b New York, NY, Feb 9, 28. OPHTHALMOLOGY. *Educ:* Yale Univ, AB, 48; Johns Hopkins Univ, MD, 52. *Hon Degrees:* DSc, Univ Pa, 82. *Prof Exp:* Intern & asst resident, Wilmer Eye Inst, Johns Hopkins Hosp, 52-53, lab asst biostatist, Med Sch, Johns Hopkins Univ, 53-54 & 57-58; from instr to asst prof ophthal, Harvard Med Sch, 60-66; prof & chmn dept, Sch Med & res affil, Primate Ctr, Univ Wash, 66-70; DIR, NAT EYE INST, 70- *Concurrent Pos:* Res fel ophthal, Wilmer Eye Inst, 57-58; res fel, Harvard Med Sch, 58-60; prog dir ophthal training grant, Mass Eye & Ear Infirmary, 62-66; mem vision res training comt, NIH, 63-64, mem neurol prog proj B, 67-69; mem adv comt basic & clinical res, Nat Soc Prev Blindness, 69-; clin assoc prof, Howard Univ, 70-; mem sci adv panel, Res to Prevent Blindness, Inc, 71-75; mem sci adv comt, Fight for Sight, 71-; chmn proj & priorities comt, Int Agency Prevention Blindness, 75-; mem bd dirs, Helen Keller Int Inc, 75- *Honors & Awards:* Pisart Vision Award. *Mem:* Inst Med-Nat Acad Sci; Am Physiol Soc; Asn Res Vision & Ophthal; Am Acad Ophthal; Am Ophthal Soc; Pan Am Ophthal Soc. *Res:* Intraocular pressure and neurophysiology; glaucoma; neuroophthalmology. *Mailing Add:* Nat Eye Inst Bldg 31 Rm 6A03 Bethesda MD 20892

KUPFER, DAVID, b Warsaw, Poland, Nov 27, 28; US citizen; m 61; c 3. BIOCHEMICAL PHARMACOLOGY, DRUG METABOLISM. *Educ:* Univ Calif, Los Angeles, BA, 52, PhD(biochem), 58. *Prof Exp:* Scientist, Worcester Found Exp Biol, 58-60; intermediate scientist & fel, Weizmann Inst Sci, 61-62; res scientist, Lederle Labs, Am Cyanamid Co, 62-71; SR SCIENTIST, WORCESTER FOUND EXP BIOL, 71- *Mem:* Am Chem Soc; Am Soc Biol Chemists; Soc Pharmacol & Exp Therapeut. *Res:* Drug-drug interactions; prostaglandin metabolism; hepatic monoxygenases; hormonal activity of environmental pollutants. *Mailing Add:* Worcester Found Exp Biol 222 Maple Ave Shrewsbury MA 01545

KUPFER, DAVID J, PSYCHIATRY. *Prof Exp:* PROF & CHMN, DEPT PSYCHIAT, UNIV PITTSBURGH; DIR RES, WESTERN PSYCHIAT INST & CLIN. *Mem:* Inst Med-Nat Acad Sci. *Mailing Add:* Western Psychiat Inst & Clin 3811 O'Hara St Pittsburgh PA 15261

KUPFER, DONALD HARRY, b Los Angeles, Calif, Oct 4, 18; m 52; c 2. STRUCTURAL GEOLOGY. *Educ:* Calif Inst Technol, BS, 40; Univ Calif, Los Angeles, AM, 42; Yale Univ, MS, 51, PhD(geol), 51. *Prof Exp:* Geologist, Gladding McBean & Co, 41-42 & US Geol Surv, 42-55; from asst prof to assoc prof, 55-66, prof, 66-80, EMER PROF GEOL, LA STATE UNIV, BATON ROUGE, 81- *Concurrent Pos:* Indust mineral consult, 58-; NSF sr fel, NZ, 62-63; Cent Treaty Orgn minerals mapping consult, Turkey, 66 & Pakistan, 67; fel, Salt Domes, Spain & Ger, 69, Can, Mexico, Israel, 79; pres, Geol Res Indust Minerals Corp, 77- *Honors & Awards:* A I Levorsen Award, 75. *Mem:* Fel AAAS; Am Asn Petrol Geologists. *Res:* Earthquakes; faults; salt domes; nonmetal mining; computer tectonics; areal geology; Gulf Coast geology; geopressures; energy resources. *Mailing Add:* 210 W Circle Dr Canon City CO 81212

KUPFER, JOHN CARLTON, b Los Angeles, CA, Feb, 12, 55; m 87. NUCLEAR EFFECTS ENGINEER, CRYOGENICS. *Educ:* Rice Univ, BA, 77; Univ Ariz, MS, 81 & PhD(physics), 85. *Prof Exp:* MEM TECH STAFF, ROCKWELL INT, 85- *Mem:* Am Phys Soc; Mat Res Soc. *Res:* Analysis of nuclear effects on military missile systems, particularly in regards to material and structural design choices for management of deposited nuclear energy; survivability and vulnerability analysis and assessments. *Mailing Add:* Rockwell Int Autonetics 3370 Miraloma Ave MS 031-HB13 D683 Anaheim CA 92803-3170

KUPFER, SHERMAN, b Jersey City, NJ, Apr 28, 26; m 51; c 3. INTERNAL MEDICINE, PHYSIOLOGY. *Educ:* Cornell Univ, MD, 48. *Prof Exp:* Res fel physiol, Sch Med, Western Reserve Univ, 49-50; res fel, Med Col, Cornell Univ, 50-51, from instr to asst prof physiol, 55-66; assoc prof med, 66-72, from assoc dean to sr assoc dean, 68-80, dep dean, 80-85, PROF MED MT SINAI SCH MED, 72-, ASSOC PROF PHYSIOL, 68- *Concurrent Pos:* Asst to dir med res, Mt Sinai Hosp, 56-58, res assoc, 58-60, asst attend physician, 60-65, dir clin res ctr, 63-, assoc attend physician, 65-72, attend physician, 72- *Mem:* Am Physiol Soc; Am Fedn Clin Res; ASAIO; Am Heart Asn; Harvey Soc; Am Soc Nephrology. *Res:* Renal and cardiovascular physiology. *Mailing Add:* Dept Med Mt Sinai Sch Med One E 100th St New York NY 10029-6574

KUPFERBERG, HARVEY J, b New York, NY, Jan 4, 33; m 62; c 2. PHARMACOLOGY. *Educ:* Univ Calif, Los Angeles, BS, 55; Univ Southern Calif, PharmD, 59; Univ Calif, San Francisco, PhD(pharmacol), 62. *Prof Exp:* USPHS fel pharmacol, Univ Calif, San Francisco, 60-62 & 62-63; staff fel, Nat Heart Inst, 63-65; from instr to asst prof, Univ Minn, Minneapolis, 65-71; PHARMACOLOGIST, EPILEPSY BR, NEUROL DIS PROG, NAT INST NEUROL & COMM DIS & STROKE, 71- *Concurrent Pos:* USPHS res grant, 66-69. *Mem:* AAAS; Am Pharmaceut Asn; Acad Pharmaceut Sci; Am Soc Pharmacol & Exp Therapeut; Soc Toxicol. *Res:* Pharmacodynamics; metabolism of drugs; mechanism of action of anticonvulsant drugs. *Mailing Add:* Epilepsy Br NIH Fed Bldg Rm 118 Bethesda MD 20892

KUPFERBERG, LENN C, b Flushing, NY, July 27, 51; m 76; c 2. ELECTROACTIVE POLYMERS, MATERIALS ANALYSIS. *Educ:* Trinity Col, Conn, BS, 73; Univ Rochester, NY, MA, 75, PhD(physics), 79. *Prof Exp:* Assoc fel physics, Mass Inst Technol, 78-80; asst prof physics, Worcester Polytech Inst, 80-84; sr res scientist, 84-87, SR DEVELOP SCIENTIST & MGR, MAT ANALYSIS LAB, RES DIV, RAYTHEON CO, 87- *Concurrent Pos:* Vis scientist physics, Mass Inst Technol, 80-84. *Mem:* Am Phys Soc; AAAS; Inst Elec & Electronics Engrs; Sigma Xi; Mat Res Soc. *Res:* Phase transitions and critical phenomina; magnetism and magnetic material; experimental physics; picosecond lasers and spectroscopy; dielectrics and piezoelectrics; polymer physics; materials analysis. *Mailing Add:* Raytheon Co Res Div 131 Spring St Lexington MA 02173

KUPFERMAN, ALLAN, b New York, NY, Aug 5, 35; m 59; c 2. PHARMACOLOGY, OPHTHALMOLOGY. *Educ:* Univ Bridgeport, BA, 59; Clark Univ, AM, 61; Univ Vt, PhD(pharmacol), 66. *Prof Exp:* Asst prof pharmacol, 69-78, ASSOC PROF PHARMACOL & OPHTHAL, SCH MED, BOSTON UNIV, 78- *Res:* Pharmacokinetics of topically applied steroids in the eye. *Mailing Add:* Dept Pharmacol-Therapeut Boston Univ Sch Med 80 E Concord St Boston MA 02118

KUPFERMAN, STUART L, b New York, NY, June 30, 37; m 66; c 2. METROLOGY. *Educ:* Polytech Inst Brooklyn, BS, 59; Harvard Univ, AM, 64, PhD(physics), 67. *Prof Exp:* Res assoc phys oceanog, Univ RI, 68-70; asst prof phys oceanog, Univ Del, 70-78; VIS INVESTR, WOODS HOLE OCEANOG INST, 78- *Concurrent Pos:* Grants, Univ Del Res Found, 71-72 & NSF, 71-78. *Mem:* AAAS; Am Geophys Union; Am Phys Soc; Am Meteorol Soc. *Res:* Automation of high accuracy calibration systems. *Mailing Add:* Div 7342 Sandia Nat lab PO Box 5800 Albuquerque NM 87185

KUPFERMANN, IRVING, b New York, NY, Jan 26, 38; m 65; c 2. NEUROPSYCHOLOGY. *Educ:* Univ Fla, BS, 59; Univ Chicago, PhD(biopsychol), 64. *Prof Exp:* Res fel, Harvard Med Sch, 65-66; from instr to assoc prof, NY Univ Med Sch, 66-74; assoc prof med psycol, 74-79, PROF PHYSIOL & PSYCHIAT, COL PHYSICIANS & SURGEONS, COLUMBIA UNIV, 79-; ASSOC RES SCIENTIST, NY STATE PSYCHIAT INST, 73- *Concurrent Pos:* Mem, NIMH Neuropsychol Study Sect, 75-; assoc ed, J Neurosci, Neurosci Letts & Brain Behav Sci. *Honors & Awards:* Richard Temple Award, Univ Chicago, 65; Res scientist Develop Award, NIMH, 69, Merit Award, 90. *Mem:* Soc Neurosci. *Res:* Invertebrate behavior and learning; neural mechanisms of learning and motivation; Aplysia; feeding behavior in Aplysia. *Mailing Add:* Columbia Univ 722 W 168th St New York NY 10032

KUPIECKI, FLOYD PETER, b Bronson, Mich, May 1, 26; m 50; c 2. BIOCHEMISTRY. *Educ:* Western Mich Univ, BS, 50; Univ Notre Dame, PhD(chem), 53. *Prof Exp:* Res chemist, Mich Chem Corp, 53-55; res assoc org chem & biochem, Univ Pa, 55-56; from res assoc to instr biochem, Univ Mich, 56-59; Fulbright fel, Biochem Inst, Helsinki, Finland, 59-60; RES SCIENTIST, UPJOHN CO, 60- *Mem:* Am Soc Biol Chemists; Am Chem Soc; Sigma Xi. *Res:* Diabetes research; lipid metabolism and adipose tissue enzymes; metabolism in islets of diabetic animals. *Mailing Add:* 5409 Circlewood Dr Kalamazoo MI 49001

KUPKE, DONALD WALTER, b Omaha, Nebr, Mar 16, 22; m 49; c 5. BIOCHEMISTRY. *Educ:* Valparaiso Univ, AB, 47; Stanford Univ, MS, 49, PhD(chem), 52. *Prof Exp:* Nat Res Coun fel med sci, Carlsberg Lab, Denmark, 52-53 & Uppsala Univ, Sweden, 53-54; USPHS fel, Carlsberg Lab & Stanford Univ, 55; mem staff, Carnegie Inst, Stanford Univ, 55-56; from asst prof to assoc prof, 57-66, chmn dept, 64-66, PROF BIOCHEM, SCH MED, UNIV VA, 66- *Concurrent Pos:* Vis prof, Otago Univ, NZ, 84. *Mem:* AAAS; Am Soc Biol Chemists; Am Chem Soc; Biophys Soc; Am Soc Plant Physiol. *Res:* Protein and virus biophysical chemistry; chloroplast proteins; magnetic balancing methods; density, viscosity and osmotic pressure; volume change on metal-ion coordination to biological molecules; hydration changes of DNA. *Mailing Add:* Dept Biochem Univ Va Sch Med Charlottesville VA 22908

KUPPENHEIMER, JOHN D, JR, b Orange, NJ, Sept 15, 41; div. OPTICS. *Educ:* Lafayette Col, BS, 63; Boston Univ, MA, 65; Worcester Polytech Inst, PhD(physics), 69. *Prof Exp:* Fel physics, Worcester Polytech Inst, 69-70, asst prof, 70-71; scientist, Diffraction Ltd, Inc, 71-72; asst mgr, Diffraction Ltd, Div, 72-73, dir Optical Metrology Lab, 73-79, sr prin physicist, 79-84, ENG FEL, SANDERS ASSOCS, 84-; PROF PHYSICS, TUFTS UNIV, 84- *Concurrent Pos:* Adj prof physics, Univ Lowell, 75-87. *Honors & Awards:* Tech Achievement Award, Sanders Assoc, 84; Chmn's Award, Sanders Asn, 86; Robert E Gross Award, Lockheed Corp, 87. *Mem:* Optical Soc Am; Sigma Xi. *Res:* Quantum optics; photon count statistics; lasers; optical constants of semiconductors; atmospheric optics; optical guidance; optical counter measures; development of IR lasers; development of IR countermeasures; applications of non-imaging optics to laser pumping. *Mailing Add:* 100 Brookfield Rd Tewksbury MA 01876

KUPPERIAN, JAMES EDWARD, JR, space science, astronomy; deceased, see previous edition for last biography

KUPPERMAN, HERBERT SPENCER, b Newark, NJ, Apr 12, 15; m 42, 76; c 3. ENDOCRINOLOGY. *Educ:* Univ Wis, BA, 36, MA, 37, PhD(endocrinol), 40; Med Col Ger, MD, 45. *Prof Exp:* Nat Res Coun fel exp med, Sch Med, Univ Ga, 42-46, sr res fel endocrinol, 46-47, res assoc, 47; res assoc therapeut, 47-53, ASSOC PROF MED, COL MED, NY UNIV, 53-; DIR, ROCHE CLIN LABS, 75- *Mem:* Endocrine Soc; fel AMA; fel Am Col Physicians; fel Am Col Obstet & Gynec; fel Acad Psychosom Med; Sigma Xi; fel Am Col Pharm; fel AAAS; fel NY Acad Sci. *Res:* Physiology of reproduction; antihormones; pharmacology; endocrines and cardiovascular drugs; clinical and assay endocrinology. *Mailing Add:* 650 First Ave New York NY 10019

KUPPERMAN, MORTON, b New York, NY, Mar 19, 18; m 46. PROBABILITY. *Educ:* City Col New York, BS, 38; George Washington Univ, MA, 50, PhD(math statist), 57. *Prof Exp:* Statistician, Gen Staff, US War Dept, 40-41 & Europ Cent Inland Transp Orgn, France, 46; statistician, Med Statist Div, Off Army Surgeon Gen, 47-55; mathematician, Nat Security Agency, 55-73; sr lectr math statist, Univ Leicester, Eng, 73-78; RETIRED. *Concurrent Pos:* Prof lectr, George Washington Univ, 57-73. *Mem:* Inst Math Statist; Royal Statist Soc; Math Asn Am; Am Statist Asn; Sigma Xi. *Res:* Distribution theory; application of information theory to multivariate analysis and statistical inference; counterexamples in probability and statistics. *Mailing Add:* 5904 Mt Eagle Dr Apt 214 Alexandria VA 22303

KUPPERMAN, ROBERT HARRIS, b New York, NY, May 12, 35; m 67; c 1. APPLIED MATHEMATICS, OPERATIONS RESEARCH. *Educ:* NY Univ, BA, 56, PhD(appl math), 62. *Prof Exp:* Instr math, NY Univ, Pratt Inst & Hunter Col, 56-60; sr engr, Jet Propulsion Lab, Calif Inst Technol, 60-62; exec adv opers res, Douglas Aircraft Co, Inc, 62-64; mem sr staff, Inst Defense Anal, 64-67; asst dir, Natural Resource Anal Ctr, Exec Off of Pres, 67-70, dep asst dir, President's Off Emergency Preparedness, 70-71, asst dir, 71-73; chief scientist, US Arms Control & Disarmament Agency, 73-79; EXEC DIR SCI TECHNOL & SR ADV, CTR STRATEGIC & INT STUDIES, GEORGETOWN UNIV, 79- *Concurrent Pos:* Prin engr, Repub Aviation Corp, 59-60; consult, US Civil Serv Comn, 65 & US Army Security Agency, Army Intel & Army Electronic Warfare Bd, 66; lectr, Univ Md, 65, vis prof govt & polit, 74-76; expert consult, Exec Off President, 67-68; dep exec dir, President's Property Rev Bd, 70-73; mem, Army Sci Bd, 79-, Coun Foreign Relations, 84-; pres, Kuppeman Assocs, Inc, 79-; sr lab fel, Los Alamos Nat Lab, 80- *Honors & Awards:* Outstanding Serv Awards, Exec Off President, 68-71, Order of Paul Revere Patriot, 70; Presidential Citations, 71-73,. *Mem:* Fel NY Acad Sci; fel Opers Res Soc; Soc Indust & Appl Math; Int Inst Strategic Studies; Mil Opers Res Soc. *Res:* Strategic analysis and arms race stability; conversational computer systems and crisis management; conventional arms transfers; terrorism. *Mailing Add:* 2832 Ellicott St NW Washington DC 20008

KUPPERMANN, ARON, b Sao Paulo, Brazil, May 6, 26; nat US; m 51; c 4. CHEMICAL PHYSICS. *Educ:* Univ Sao Paulo, Brazil, ChemE, 48, CE, 53; Univ Notre Dame, PhD(phys chem), 56. *Prof Exp:* Asst prof phys chem, Cath Univ Sao Paulo, 49-50 & chem, Inst Aeronaut Technol, 50-51; head anodizing sect, Ajax Indust & Trade Co, 52; res assoc phys chem, Radiation Proj, Univ Notre Dame, 53-55; from instr to assoc prof, Univ Ill, 55-63; PROF CHEM PHYSICS, CALIF INST TECHNOL, 63- *Concurrent Pos:* Resident res assoc, Argonne Nat Lab, 57; res assoc, Inst Atomic Energy, Sao Paulo, 59-60; Reilly lectr, Univ Notre Dame, 65; NSF fel, 68-69; Guggenheim fel, 76-77; consult, Jet Propulsion Lab, 65-69, TRW Systs Group, 70-77, World Bank, 83-; chmn joint chem study group, Nat Acad Sci-Nat Res Coun, Brazil, 73-76. *Honors & Awards:* Venable lectr, Univ NC, 67; Werner lectr, Univ Kans, 68. *Mem:* Fel Am Inst Chem; fel Am Phys Soc; Am Chem Soc; AAAS. *Res:* Experimental and theoretical chemical dynamics; collisions in crossed molecular beams; laser spectroscopy and photochemistry; radiation chemistry; low energy electron impact phenomena, experiment and theory; variable angle photoelectron spectroscopy. *Mailing Add:* Dept Chem 127-72 Calif Inst Technol Pasadena CA 91104

KUPPERS, JAMES RICHARD, b Newland, Ind, Aug 4, 20; m 44; c 4. PHYSICAL CHEMISTRY. *Educ:* Univ Fla, BS, 43, PhD(chem), 57; La State Univ, MS, 47. *Prof Exp:* Food technologist, United Fruit Co, 47-49, assoc biochemist, 49-54; res chemist textile fibers dept, E I du Pont de Nemours & Co, 57-60; assoc prof chem, Pfeiffer Col, 60-64; from assoc prof to prof chem, Univ NC, Charlotte, 65-86; RETIRED. *Mem:* Am Chem Soc. *Res:* Solution thermodynamics. *Mailing Add:* Dept Chem Univ NC Charlotte NC 28223

KUPSCH, WALTER OSCAR, b Amsterdam, Neth, Mar 2, 19; nat Can; m 45; c 3. GEOLOGY. *Educ:* Univ Amsterdam, BSc, 43; Univ Mich, MS, 48, PhD(geol), 50. *Prof Exp:* Dir, Inst North Studies, 65-72, dir, Churchill River Study, 73-76, prof 50-86, EMER PROF, UNIV SASK, 86- *Concurrent Pos:* Prin geologist, Geol Surv, Sask, 50-56, consult, 56-; ed, Musk-Ox; mem, Sci Coun Can, 76-82; mem, NWT Sci Adv Bd, 76-82, petrol adv, 80-83, mem North Dev Adv Coun, 85-88. *Honors & Awards:* fel Royal Soc Can. *Mem:* Fel Arctic Inst NAm; fel Geol Asn Can; fel Geol Soc Am; Am Asn Petrol Geol. *Res:* Stratigraphy; geomorphology; glacial geology. *Mailing Add:* Dept Geol Sci Univ Sask Saskatoon SK S7N 0W0 Can

KUPSTAS, EDWARD EUGENE, b Eynon, Pa, Aug 1, 21; m 57; c 5. ORGANIC CHEMISTRY. *Educ:* Fordham Col, BS, 51, MS, 53, PhD(chem), 58. *Prof Exp:* Res chemist, Textile Fibers Dept, E I du Pont de Nemours & Co, Inc, 55-88; RETIRED. *Mem:* Am Chem Soc. *Res:* Structure and synthesis of ichtiamin; dyes; polymers; polyesters. *Mailing Add:* 1614 Hardee Rd Kingston NC 28501

KURACHI, KOTOKU, b Amagi City, Japan, Nov 16, 41; m; c 2. HUMAN GENETICS. *Educ:* Kyushu Univ, Japan, BS, 65, MS, 67, PhD, 70. *Prof Exp:* Res assoc, Dept Biochem, Kyushu Univ, Japan, 70; sr fel, Dept Biochem, Univ Wash, Seattle, 70-72 & 74-75, Dept Biol Struct, 72-74, from sr res assoc to res assoc prof biochem, 75-86; assoc prof, 86-90, PROF, DEPT HUMAN GENETICS & CELLULAR & MOLECULAR BIOL PROG, UNIV MICH,

90- *Concurrent Pos:* Vis lectr, Ctr Biochem & Biophys Sci & Med, Harvard Med Sch, 83-86; travel award, Am Soc Biochem & Molecular Biol, 88; mem, Res Peer Rev Comt, Am Heart Asn, Mich, 89-; consult. *Mem:* Am Chem Soc; Am Soc Biol Chem & Molecular Biol; AAAS; Am Soc Human Genetics; Am Soc Hemat. *Res:* Human genetics; author of numerous scientific publications. *Mailing Add:* Dept Human Genetics Med Sch Univ Mich 3712 Med Sci II Bldg Ann Arbor MI 48109-0618

KURAJIAN, GEORGE MASROB, b Highland Park, Mich, Oct 28, 26; m 55; c 3. MECHANICAL DESIGN, SOLID MECHANICS. *Educ:* Univ Detroit, BME, 48, ME, 63; Univ Mich, MSE, 53. *Prof Exp:* From instr to asst prof eng mech, Univ Detroit, 48-64; from asst prof to assoc prof, 64-72, PROF MECH ENG & ENG MECH, UNIV MICH, DEARBORN, 72-, CHMN DEPT MECH ENG, 75- *Concurrent Pos:* Consult, indust & govt agencies, 54- *Mem:* Am Soc Eng Educ; fel, Am Soc Mech Engrs; Soc Exp Stress Anal; Indust Math Soc; Int Asn Vehicle Design; Int Asn Struct Mech Reactor Technol. *Res:* Design and stress analysis of structural shells; space frames; amphibious vehicles; chemical machinery; automotive components; automotive dynamometers and test cells; physical testing laboratory projects; mechanical design; finite element; solid mechanics; theories of failure; fatigue. *Mailing Add:* Dept Mech Eng Sch Eng Univ Mich 4901 Evergreen Rd Dearborn MI 48128-1491

KURAMITSU, HOWARD KIKUO, b Los Angeles, Calif, Oct 18, 36; m 70; c 2. BIOCHEMISTRY. *Educ:* Univ Calif, Los Angeles, BS, 57, PhD(biol chem), 62. *Prof Exp:* Jr res biochemist, Sch Med, Univ Calif, Los Angeles, 61-62; res fel bact, Harvard Med Sch, 62-63; res assoc microbiol, Sch Med, Univ Southern Calif, 63-67; from asst prof to assoc prof, 67-79, PROF MICROBIOL, MED SCH, NORTHWESTERN UNIV, CHICAGO, 79- *Concurrent Pos:* NIH Oral Biol & Med Study Sect, 84-89. *Mem:* AAAS; Am Soc Biol Chemists; Am Soc Microbiol; Int Asn Dent Res. *Res:* Regulation of carbohydrate metabolism in oral microorganisms; isolation and characterization of genes involved in pathogenic proper ties of oral microorganisms. *Mailing Add:* Dept Pediat Dent & Microbiol Univ Tex Health Sci Ctr 7703 Floyd Curl Dr San Antonio TX 78284

KURATA, F(RED), chemical engineering; deceased, see previous edition for last biography

KURATA, MAMORU, b Nagoya, Japan, Apr 27, 36. SEMICONDUCTOR DEVICE MODELING. *Educ:* Yokohama Nat Univ, Bachelor, 61; Univ Tokyo, Dr(elec eng), 73. *Prof Exp:* Guest researcher semiconductors, Tech Univ Aachen, Ger, 64- 66; engr, Illum Div, Toshiba Corp, 61-64, researcher, Res & Develop Ctr, 67-82, sr researcher, 82-86, CHIEF RES SCIENTIST, RES & DEVELOP CTR, TOSHIBA CORP, 86- *Mem:* Fel Inst Elec & Electronics Engrs. *Res:* Semiconductor device modeling with its application to high power devices; gate turn-off thyristors; high speed devices such as heterojunction bipolar transistors; author of several books. *Mailing Add:* Soshigaya 3-17-4-209 Setagaya Tokyo 157 Japan

KURATH, DIETER, b Evanston, Ill, Oct 17, 21; m 45; c 4. THEORETICAL NUCLEAR PHYSICS. *Educ:* Brown Univ, AB, 42; Univ Chicago, PhD(physics), 51. *Prof Exp:* Asst, Univ Chicago, 47-51; assoc physicist, 51-60, SR PHYSICIST, ARGONNE NAT LAB, 60- *Concurrent Pos:* Guggenheim fel, 57-58; vis prof, Univ Wash, 61-62 & State Univ NY Stony Brook, 69-70; sr vis fel, Nuclear Physics Lab, Oxford Univ, 73-74, Univ Melbourne, 88. *Mem:* Am Phys Soc. *Res:* Shell model of nuclear structure. *Mailing Add:* Argonne Nat Lab Argonne IL 60439

KURATH, PAUL, b St Gallen, Switz, June 18, 24; nat US; div; c 5. ORGANIC CHEMISTRY. *Educ:* Swiss Fed Inst Technol, dipl, 48, DSc, 51. *Prof Exp:* Res assoc & fel pharmaceut chem, Univ Kans, 51-53 & org chem, Univ Rochester, 54-58; res chemist, Res Div, Abbott Labs, 58-88; RETIRED. *Mem:* Am Chem Soc; The Chem Soc; Swiss Chem Soc; Sigma Xi. *Res:* Organic synthesis; natural products; steroids; antibiotics; peptides. *Mailing Add:* 2801 Gradville Ct Apt 109 Waukegan IL 60085

KURATH, SHELDON FRANK, b Moscow, Idaho, Mar 29, 28; m 54; c 3. POLYMER CHEMISTRY, RHEOLOGY. *Educ:* Univ Wis, BS, 50, MS, 51, PhD(chem), 54. *Prof Exp:* Res aide, Inst Paper Chem, Lawrence Univ, 53-65; assoc prof phys chem, 65-69, PROF CHEM, UNIV WIS-OSHKOSH, 69- *Mem:* Am Chem Soc; Soc Rheol; Am Inst Chem Eng; Tech Asn Pulp & Paper Indust. *Res:* Non-Newtonian flow of polymers and pigment suspensions; polymer viscoelasticity; colloid chemistry. *Mailing Add:* 2413 S Greenview Appleton WI 54915

KURCHACOVA, ELVA S, b Oriente, Cuba, Aug 5, 21; m 44; c 2. ORGANIC CHEMISTRY. *Educ:* Univ Havana, DSc(physics, chem), 45. *Prof Exp:* Dir res, Linner Labs, Cuba, 45-61; assoc res chemist, Miles Labs, Inc, 61-78, sr assoc res scientist, 78-89; RETIRED. *Concurrent Pos:* Pres & dir, Yelene Prod, 53-61. *Mem:* AAAS; Am Chem Soc; NY Acad Sci. *Res:* Pharmaceuticals; organic synthesis; development of medicinal drugs. *Mailing Add:* 3355 Jaywood Terr No J 112 Boca Raton FL 33431-6579

KURCZEWSKI, FRANK E, b Erie, Pa, May 24, 36; m 59; c 4. ENTOMOLOGY. *Educ:* Allegheny Col, BS, 58; Cornell Univ, MS, 62, PhD(insect taxon), 64. *Prof Exp:* Res assoc entom, Univ Kans, 64-66, vis asst prof, 66; from asst prof to prof entom, 66-77, PROF ENVIRON & FOREST BIOL, STATE UNIV NY COL ENVIRON SCI & FORESTRY, 77- *Concurrent Pos:* NSF fel, 64-65; NIH fel, 65-66. *Res:* Comparative behavior and systematics of digger wasps; insect behavior. *Mailing Add:* Dept Environ Sci State Univ NY Col Environ Sci Syracuse NY 13210

KURCZYNSKI, THADDEUS WALTER, b Hamtramck, Mich, Oct 31, 40; m 63, 79; c 2. HUMAN GENETICS, NEUROLOGY. *Educ:* Univ Mich, BS, 62, MS, 63; Case Western Reserve Univ, PhD(human genetics), 69, MD, 70; Am Bd Psychiat & Neurol, with spec competence in child neurol; Am Bd Med Genetics, cert clin genetics & clin biochem genetics. *Prof Exp:* From intern

to resident neurol, Univ Mich Hosps, 70-73; resident pediat, Children's Hosp Mich, 73-74; fel pediat neurol, Albert Einstein Col Med, 74-76; asst prof, Dept Pediat, Div Pediat Neurol, Dept Med, Div Neurol & Human Genetics & Genetics Ctr, Case Western Reserve Univ, 76-81; ASSOC PROF, DEPT PEDIAT & NEUROL & DIR, GENETICS CTR NORTHWEST OHIO, MED COL OHIO, 81- *Mem:* Am Soc Human Genetics; Am Acad Neurol; Child Neurol Soc. *Res:* Medical genetics; pediatric neurology. *Mailing Add:* Dept Pediat Med Col Ohio C S 10008 Toledo OH 43699

KUREY, THOMAS JOHN, b Boston, Mass, Feb 21, 37. NUCLEAR PHYSICS, REACTOR PHYSICS. *Educ:* Boston Col, BS, 58; Pa State Univ, MS, 61, PhD(physics), 63. *Prof Exp:* Physicist, Knolls Atomic Power Lab, Gen Elec Co, 64-80; AT MED SYST, GEN ELEC CO, MILWAUKEE. *Mem:* Am Phys Soc; Am Nuclear Soc. *Res:* Beta and gamma spectroscopy; applications of solid state nuclear detectors; electron spin resonance study of decay of unstable free radicals in gamma irradiated solids; reactor physics analytical methods; critical experiments. *Mailing Add:* 2130 La Rochelle Ct Brookfield WI 53005

KURFESS, JAMES DANIEL, b Perrysburg, Ohio, Nov 8, 40; div; c 2. ASTROPHYSICS. *Educ:* Case Inst Technol, BS, 62, MS, 63, PhD(physics), 67. *Prof Exp:* Res assoc space sci, Rice Univ, 67-69; ASTROPHYSICIST, E O HULBURT CTR SPACE RES, US NAVAL RES LAB, 69- *Concurrent Pos:* Mem data base group study uses sci balloons, Nat Acad Sci, 75; mem, Sci Adv Panel Long Duration Balloon Develop Prog; prin investr, Oriented Scintillation Spectros Exp, Gamma Ray Observ, NASA; mem, Comt Space Astron & Astrophys, Nat Acad Sci, 83-87; secy-treas, Div Astrophysics, Am Phys Soc, 80-84, vchairperson, 90. *Mem:* Am Phys Soc; Am Astron Soc. *Res:* Hard x-ray and gamma-ray observations of solar and extra-solar sources using balloons and satellite instrumentation; development of long duration balloon-borne capabilities. *Mailing Add:* Naval Res Lab Code 4150 Washington DC 20375-5000

KURIAKOSE, AREEKATTUTHAZHAYIL, b Palai, India, Aug 20, 33; Canadian citizen; m 60; c 3. MATERIAL SCIENCE, CERAMICS PROCESSING & CHARACTERIZATION. *Educ:* Univ Madras, India, BSc, 53, MA, 55, PhD(chem), 61. *Prof Exp:* Lectr chem, St Thomas Col, Palai, India, 55-56; res engr, Norton Res Corp Can Ltd, 66-69, sr res engr, 69-75, supvr mat res, 75-81; RES SCIENTIST, DEPT ENERGY, MINES & RESOURCES, GOVT CAN, 81- *Concurrent Pos:* Ed-in-chief, J Can Ceramic Soc, 87-89. *Mem:* Can Ceramic Soc; Am Ceramic Soc; Sigma Xi. *Res:* High temperature chemistry; abrasive materials; ceramics microstructure and properties; solid electrolytes and energy storage and generating systems; toughened ceramics; silicon carbide. *Mailing Add:* Can Dept Energy Mines & Resources 405 Rochester St Ottawa ON K1A 0G1 Can

KURIGER, WILLIAM LOUIS, b Waterloo, Iowa, Aug 7, 33; m 56; c 7. ELECTRICAL ENGINEERING. *Educ:* Univ Iowa, BSEE, 58; Iowa State Univ, ME, 63, PhD(elec eng), 66. *Prof Exp:* Engr, Collins Radio Co, 58-64; From asst prof, to assoc prof, 66-80, PROF ELEC, ENG, UNIV OKLA, 80- *Mem:* Inst Elec & Electronics Engrs; Optical Soc Am; Sigma Xi. *Res:* Laser applications; electronics. *Mailing Add:* 912 Schulze Dr Norman OK 73071

KURIHARA, NORMAN HIROMU, b Oxnard, Calif, Mar 23, 38; m 65. ORGANIC CHEMISTRY. *Educ:* Univ Calif, Santa Barbara, BA, 61; Univ Calif, Davis, PhD(org chem), 65. *Prof Exp:* Res fel, Univ Calif, 65-66; RES SPECIALIST, AGR ORG DEPT, DOW CHEM CO, 66- *Mem:* Am Chem Soc; Sigma Xi. *Res:* Agricultural and pesticide chemistry. *Mailing Add:* Dow Chem Co 2800 Mitchell Dr Walnut Creek CA 94598

KURIHARA, YOSHIO, b Korea, Oct 24, 30; Japan citizen; m 60; c 2. METEOROLOGY. *Educ:* Univ Tokyo, BA, 53, PhD(geophys), 62. *Prof Exp:* Tech officer, Japan Meteor Agency, 53-59; res officer, Meteor Res Inst, 59-63; res meteorologist, Geophys Fluid Dynamics Lab, US Weather Bur, 63-65; res officer, Meteor Res Inst, 65-67; res meteorologist, Environ Sci Serv Admin, 67-70, RES METEOROLOGIST, GEOPHYS FLUID DYNAMICS LAB, NAT OCEANIC & ATMOSPHERIC ADMIN, 70- *Concurrent Pos:* Vis lectr, Princeton Univ, 71-86. *Honors & Awards:* Meteorol Soc Japan Award, 75; B Miller Award, Am Meteorol Soc, 84. *Mem:* fel Am Meteorol Soc; Am Geophys Union; Meteorol Soc Japan. *Res:* Numerical analysis of meteorological data; construction of statistical-dynamical model of the atmosphere; simulation of the hurricane. *Mailing Add:* Geophys Fluid Dynamics Lab Princeton Univ PO Box 308 Princeton NJ 08542

KURIS, ARMAND MICHAEL, b New York, NY, May 16, 42. PARASITOLOGY, MARINE ECOLOGY. *Educ:* Tulane Univ, BS, 63; Univ Calif, Berkeley, MA, 66, PhD(zool), 71. *Prof Exp:* Asst prof zool, Univ Fla, 73-74; asst prof zool & marine sci, Univ NC, Chapel Hill, 74-75; ASSOC PROF BIOL SCI, UNIV CALIF, SANTA BARBARA, 75- *Concurrent Pos:* NIH fel, G W Hooper Found, Univ Calif, San Francisco, 71-72; NIH fel, Dept Zool, Univ Mich, Ann Arbor, 72-73; actg asst prof, Bodega Marine Lab, Univ Calif, Bodega Bay, 73, 74 & 75; prin investr, Marine Sci Inst, Univ Calif, Santa Barbara, 78- *Mem:* Am Soc Ichthyol & Herpetol; AAAS; Ecol Soc Am; Soc Protozool; Am Soc Parasitol; Crustacean Soc. *Res:* Parasite ecology; biological control; crustacean biology; molting physiology; nemertean biology; competition; parasitic castration; shrimp taxonomy; limb regeneration; population biology; prawn aquaculture. *Mailing Add:* Dept Biol Sci Univ Calif Santa Barbara CA 93106

KURITZKES, ALEXANDER MARK, b Leipzig, Ger, May 3, 24; nat US; wid; c 2. ORGANIC CHEMISTRY. *Educ:* Univ Calif, BA, 48; Univ Basel, PhD(chem), 59. *Prof Exp:* Res chemist, R J Strasenburgh Co, NY, 49-52; RES CHEMIST, MATTIN LABS, MEARL CORP, 59- *Mem:* AAAS; Am Chem Soc; Sigma Xi. *Res:* Isolation and determination of structures of natural products; organic analytical chemistry; spectroscopy. *Mailing Add:* Henry L Mattin Labs Mearl Corp Ossining NY 10562

KURIYAMA, KINYA, b Kyoto, Japan, July 11, 32; m 59; c 2. DRUG RECEPTOR, GAMA-AMINOBUTYRIC ACID. *Educ:* Kyoto Prefectural Univ Med, MD, 57, PhD(pharmacol), 63. *Prof Exp:* Res assoc pharmacol, Johns Hopkins Univ Sch Med, 63-64; sr res scientist, City Hope Nat Med Ctr, 64-67; assoc prof pharmacol, Loma Linda Univ Sch Med, 67-69; prof neurochem, State Univ NY Sch Med, 70-71; from instr to asst prof, 58-63, PROF PHARMACOL & CHMN DEPT, KYOTO PREFECTURAL UNIV MED, 71-, DIR, GRAD SCH, 79- *Concurrent Pos:* Mem coun, Japanese Pharmacol Soc, 78-; Japanese Soc Neuropsychopharmacol, 87-90 & Asian W Pac Pharmacol Soc, 88-; ed, Neurochem Int, 82-; assoc ed, Alcohol & Alcoholism, 85- *Honors & Awards:* Sci Award, Japanese Med Asn, 82. *Mem:* Int Soc Neurochem; Int Soc Biomed Res Alcoholism (pres, 91-); Am Soc Pharmacol & Exp Therapeut. *Res:* Neurochemical and pharmacological studies on amino acid neurotransmitters, drug receptors, drug dependence and signal transductions in exitable cells. *Mailing Add:* 69-1 Iwagakakiuchi-cho Kamigamo Kita-Ku Kyoto 603 Japan

KURIYAMA, MASAO, b Tokyo, Japan, Oct 29, 31; m 58; c 1. X-RAY PHYSICS, MATERIALS SCIENCE. *Educ:* Tokyo Metrop Univ, BS, 53; Univ Tokyo, MS, 55, DSc(physics), 58. *Prof Exp:* Res assoc, Tokyo Metrop Univ, 58-59; res assoc x-ray physics, Inst Solid State Physics, 59-62; sr scientist, Westinghouse Elec Corp, 62-66; assoc prof physics, Univ Tokyo, 66-67; physicist, Inst Mat Sci & Eng, Nat Bur Standards, 67-80, supvry physicist, 80-90; CONSULT, 90- *Concurrent Pos:* Vis prof, Nat Lab High Energy Physics, Japan, 86. *Honors & Awards:* Silver Medal, US Dept Com, 74, IR-100, 79. *Mem:* Am Crystallog Asn; Am Phys Soc; Phys Soc Japan. *Res:* Magnetism; x-ray dynamical diffraction; crystal perfection; crystal growth; x-ray inelastic scattering; synchrotron radiation topography; x-ray microscopy; x-ray nondestructive evaluation; x-ray tomographic imaging. *Mailing Add:* 20337 Bickleton Pl Gaithersburg MD 20879

KURKJIAN, CHARLES R(OBERT), b Wanamassa, NJ, Dec 7, 29; m 55; c 3. CERAMICS. *Educ:* Rutgers Univ, BS, 52; Mass Inst Technol, ScD(ceramics), 55. *Prof Exp:* Res assoc glass, Mass Inst Technol, 55-57; fel, Univ Sheffield, England, 57-59; MEM TECH STAFF INORG CHEM, BELL TEL LABS, 59- *Mem:* Fel Am Ceramic Soc; fel Brit Soc Glass Technol. *Res:* Glass; ceramics; general high temperature inorganic chemistry. *Mailing Add:* 82 Harrison Brook Dr Basking Ridge NJ 07920

KURKOV, VICTOR PETER, b Zrenjanin, Yugoslavia, Mar 29, 36; US citizen; m 57; c 2. ORGANIC CHEMISTRY, CATALYSTS. *Educ:* NY Univ, BChE, 63; Columbia Univ, MA, 65, PhD(org chem), 67. *Prof Exp:* Res asst biochem, Col Med, NY Univ, 58-63; res chemist, 67-74, sr res chemist, chem res dept, 74-83, SR RES ASSOC, CHEVRON RES CO, 83- *Mem:* Am Chem Soc. *Res:* Free radical reactions; oxidation; homogeneous catalysis; new petrochemical processes; polymer chemistry. *Mailing Add:* Chevron Res & Technol Co 100 Chevron Way Richmond CA 94802

KURLAND, ALBERT A, b Wilkesbarre, Pa, June 29, 14; m 41; c 2. PSYCHIATRY. *Educ:* Univ Md, BS, 36, MD, 40. *Prof Exp:* Staff psychiatrist, Spring Grove State Hosp, State of Md, 49-53, dir med res, 53-60, dir res, Dept Ment Hyg, 60-69; dir, Md Psychiat Res Ctr, 69-77; RES PROF PSYCHIAT, SCH MED, UNIV MD, 79- *Mem:* AMA; Am Psychiat Asn. *Res:* Chlorpromazine in the treatment of schizophrenia; clinical reaction and tolerance to lysergic acid diethylamine tartrate in chronic schizophrenia; the drug placebo and its psychodynamic and conditional reflex action; comparative effectiveness of eight phenothiazines; author of over 185 publications in clinical Psychopharmacology. *Mailing Add:* Taylor Manor Hosp 6317 Park Heights NE Baltimore MD 21215

KURLAND, JEFFREY ARNOLD, b New York, NY, Nov 19, 43; m 67; c 2. SOCIOBIOLOGY, PRIMATOLOGY. *Educ:* Cornell Univ, BA, 67; Harvard Univ, PhD(anthrop), 76. *Prof Exp:* Res assoc primatol, Primate Res Inst, Kyoto Univ, 72-73; instr anthrop, Harvard Univ, 74-75; asst prof, 75-84, ASSOC PROF ANTHROP, PA STATE UNIV, UNIV PARK, 84- *Mem:* AAAS; Animal Behav Soc; Int Primatol Soc; Soc Study Evolution; Am Asn Phys Anthropologists; Sigma Xi. *Res:* Primate sociobiology and behavioral ecology; crab-eating, rhesus, Japanese and barbary macaques; human sociobiology. *Mailing Add:* Dept Anthrop Pa State Univ Main Campus University Park PA 16802

KURLAND, JONATHAN JOSHUA, b Boston, Mass, Jan 11, 39; m 64; c 2. PHYSICAL ORGANIC CHEMISTRY. *Educ:* Univ Pa, BA, 60; Harvard Univ, MA, 67, PhD(chem), 68. *Prof Exp:* Res assoc chem, Columbia Univ, 67-68; chemist, 68-75, proj scientist, 75-84, RES SCIENTIST, UNION CARBIDE CHEM & PLASTICS CO INC, 84- *Mem:* Am Chem Soc. *Res:* Oxidation and free-radical chemistry; process research and development acetyls and acrylics. *Mailing Add:* 1617 Kirklee Rd Charleston WV 25314-2426

KURLAND, LEONARD T, b Baltimore, Md, Dec 24, 21; m 42; c 5. MEDICINE, EPIDEMIOLOGY. *Educ:* Johns Hopkins Univ, BA, 42, DrPH, 51; Univ Md, MD, 45; Harvard Univ, MPH, 48. *Prof Exp:* Intern, Univ Md Hosp, 45-46; asst resident, Glenn Dale Tuberculosis Sanatorium, 46; asst dir, Div Tuberculosis Control & Sanatoria, State of Mass, 46-47; epidemiologist, NIMH, 48-54; chief, Epidemiol Br, Nat Inst Neurol Dis & Blindness, NIH, 55-64, consult, Collab & Field Prog, 65-86; chmn, Dept Med Statist, Epidemiol & Population Genetics, Mayo Clin & Mayo Found, 64-86, prof epidemiol, Grad Sch Med, 65-91; chief epidemiol br, Nat Inst Neurol Dis & Blindness, 55-64; prof epidemiol, 64-86, SR CONSULT, MAYO GRAD SCH MED, CHMN, DEPT MED STATIST & EPIDEMIOL, MAYO CLIN, 87-, SR CONSULT, DEPT HEALTH SCI RES, SECT CLIN EPIDEMIOL, MAYO CLIN & MAYO FOUND, 87- *Concurrent Pos:* Res assoc, Dept Epidemiol, Johns Hopkins Univ, 50-51, assoc epidemiol, Sch Hygiene & Pub Health, 74-81, sr assoc, Dept Epidemiol, 82-88; Res assoc neurol, biometry & med statist, Mayo Clin, 53-55, mem res comt, 66-70, mem computer comt, 72-74, mem clin coord comt, 73-75, mem subcomt cancer prog eval, 73-74; fel neuropath, Armed Forces Inst Path, 55-56; vis lectr, Dept

Neurol, Med Col SC, 56; prof lectr neurol, Georgetown Univ, 57; clin asst prof neurol, Howard Univ, 59-62, prof, 62-64; adj fac mem, San Diego State Univ, 85; assoc ed, Am J Epidemiol, 72-; contrib ed, Am J Indust Med, 80-89; co-ed, Neuroepidemiol, 81-85. *Honors & Awards:* Golden Sci Hope Chest Award, Nat Multiple Sclerosis Soc, 66; Allan A Bailey Lectr, Univ Sask, 81; Sigma Xi; assoc mem Am Acad Neurol; Int Soc Geog Path; Int Epidemiol Asn; Soc Epidemiol Res; Asn Teachers Prev Med; Am Col Prev Med; fel Soc Adv Med Systs; Int Soc Pharmacoepidemiol. *Res:* Human ecology; medical record systems; geographic pathology; human genetics as applied to neurology; epidemiology of chronic disease; author of three books and 421 technical papers. *Mailing Add:* 1165 Plummer Circle Rochester MN 55901

KURLAND, ROBERT JOHN, b Denver, Colo, Apr 2, 30; m 64; c 5. NUCLEAR MAGNETIC RESONANCE. *Educ:* Calif Inst Technol, BS, 51; Harvard Univ, MA, 53, PhD(chem physics), 55. *Prof Exp:* Res assoc, Nat Bur Standards-Nat Res Coun, 56-58; from instr to assoc prof chem, Carnegie Mellon Univ, 58-68; assoc prof chem, State Univ NY Buffalo, 68-85; SR RESEARCHER, DEPT SPEC IMAGING-RADIOL, GEISINGER MED CTR, 85- *Mem:* Am Asn Phys Med; Soc Magnetic Resonance Imaging. *Res:* Magnetic imaging and spectroscopy. *Mailing Add:* Dept Spec Imaging Geisinger 29-00 Danville PA 17821

KURMES, ERNEST A, b Brooklyn, NY, Jan 19, 31; m 56; c 2. FORESTRY. *Educ:* Lehigh Univ, BA, 53; Yale Univ, MS, 57, MF, 58, PhD(forest ecol), 61. *Prof Exp:* Asst prof forestry, Southern Ill Univ, Carbondale, 61-67; assoc prof, 67-80, PROF FORESTRY, NORTHERN ARIZ UNIV, 80- *Mem:* Fel AAAS; Soc Am Foresters; Ecol Soc Am; Sigma Xi. *Res:* Forest ecology; regeneration of forest tree species. *Mailing Add:* Sch Forestry Fac Box 4098 Northern Ariz Univ Flagstaff AZ 86011

KURNICK, ALLEN ABRAHAM, b Kaunas, Lithuania, Mar 15, 21; nat US; m 42; c 2. CHEMISTRY, NUTRITION. *Educ:* Calif State Polytech Univ, BS, 53; Agr & Mech Univ, Tex, MS, 55, PhD(biochem, nutrit), 57. *Prof Exp:* Asst biochem & nutrit, Agr & Mech Univ, Tex, 53-57; asst prof poultry sci, Univ Ariz, 57-59, prof & head dept, 59-62; assoc mgr tech & res serv, 62-66, dir tech serv, 66-73, dir, Dept Agr & Animal Health, 73-75, gen mgr, 75-79, VPRES-DIR RES & DEVELOP, ROCHE CHEM DIV, HOFFMANN-LA ROCHE INC, 80- *Mem:* AAAS; Am Chem Soc; Poultry Sci Asn; Am Inst Nutrit. *Res:* Unidentified factors required for reproduction; enzyme systems in the developing chick embryo; metabolism and nutrition of mineral elements; vitamins in animal nutrition; biochemistry. *Mailing Add:* 2431 Unicornio St Rancho La Costa CA 92009-4931

KURNICK, JOHN EDMUND, b New York, NY, Feb 9, 42; m 69; c 2. HEMATOLOGY, ONCOLOGY. *Educ:* Harvard Univ, BA, 62; Univ Chicago, MD, 66. *Prof Exp:* Intern, Univ Wash Hosps, 66-67; resident med, Stanford Univ Hosps, 67-68; fel hemat, 68-70, asst prof med, Univ Colo Med Ctr & chief, Hemat Serv, Denver Vet Admin Hosp, 73-78; ASSOC CLIN PROF MED (HEMAT/ONCOL), UNIV CALIF, IRVINE, 79- *Mem:* Am Col Physicians; Am Fedn Clin Res; Am Soc Clin Oncol; Am Soc Hemat; Int Soc Exp Hemat. *Res:* Hematopoietic cellular differentiation and control of granulopoiesis; erythropoiesis in anemias of chronic diseases and uremia; chemotherapy of malignant disorders. *Mailing Add:* 11411 Brookshie Suite 103 Downey CA 90241

KURNICK, NATHANIEL BERTRAND, b Brooklyn, NY, Nov 8, 17; m 40; c 3. BIOCHEMISTRY, MEDICINE. *Educ:* Harvard Univ, BA, 36, MD, 40; Am Bd Internal Med, dipl, 51, cert oncol, 73, cert hemat, 74. *Prof Exp:* Workman fel med & biochem, Mass Gen Hosp, Harvard Univ, 40-41; intern, Mt Sinai Hosp, NY, 41-42, resident med, 46-47; Nat Res Coun & Am Cancer Soc res fel biochem & cytochem, Rockefeller Inst, 47-48 & Karolinska Nobel Inst, Stockholm, 48-49; asst prof med & dir lab cell res, Med Sch, Tulane Univ, 49-54; assoc clin prof med, Univ Calif, Los Angeles, 54-65, assoc internist, 59-65; assoc prof med in residence & assoc internist, 65-68, chmn div med, 66-71, CLIN PROF MED, UNIV CALIF, IRVINE, 68- *Concurrent Pos:* Vis physician, Charity Hosp, New Orleans, La, 49-54 & Touro Infirmary, 52-54; consult, Charity Hosp, Pineville, La, 49-54; mem staff, Vet Admin Hosp, Long Beach, Calif, 54-59, consult, 59-; vis physician, Harbor Gen Hosp, Torrance Calif, 54-59, consult, 59-66; vis physician, Los Angeles County Hosp, 65-68; chmn dept med, Long Beach Community Hosp, 66-67; staff mem var hosps; dir oncol-hemat lab, Long Beach Community Hosp, 81- *Mem:* Histochem Soc; Am Soc Hemat; Soc Exp Biol & Med; Int Soc Hemat; Int Soc Exp Hemat. *Res:* Nucleic acids; chemistry and metabolism; nucleolytic enzymes; cytochemistry; hematology; oncology; radiation biology. *Mailing Add:* Long Beach Comm Hosp Long Beach CA 90804

KURNOW, ERNEST, b New York, NY, Oct 21, 12; m 38; c 3. STATISTICS. *Educ:* City Col New York, BS, 32, MS, 33; NY Univ, PhD(econ), 51. *Prof Exp:* From instr to prof econ, Schs Bus, NY Univ, 48-62, chmn quant anal area, 62-76, chmn doctoral prog, Grad Sch Bus Admin, 76-85, prof statist, 62-86, EMER PROF STATIST, SCHS BUS, NY UNIV, 86- *Concurrent Pos:* Lincoln Found grant, 58-61; study dir, Tri-State Transp Comt, 64-66, Finance Mass Transit, 71-72 & Gov Spec Comn, 71-72; Fulbright grant, Athens, Greece, 66-67; consult, Tri-State Regional Planning Comn, 73-75 & New York Temp Comn City Finances, 75-76; dir, Careers Bus Prog, 78-86. *Mem:* Fel Am Statist Asn; Int Statist Inst; Inst Mgt Sci; Am Econ Asn; Am Inst Decision Sci. *Res:* Applications of statistics in fields of transportation and state and local government; design of sampling studies. *Mailing Add:* Three Washington Sq Village Apt 17I New York NY 10012

KUROBANE, ITSUO, b Tochigi, Japan, Dec 23, 44; m 73; c 1. APPLIED MICROBIOLOGY, PLANT BIOTECHNOLOGY. *Educ:* Univ Tokyo, MS, 72, PhD(agr sci), 75. *Prof Exp:* assoc researcher biol, Dalhousie Univ, 75-79; res assoc, Univ Wis-Madison, 79-80; res assoc, Med Sch, Northwestern Univ, 80-82; sr res scientist, Hoechst Japan Ltd, 82-88; proj mgr, Rhone-Poulenc Agrochimie, Lyon, France, 88-90; GEN MGR, TSUKUBA-AKENO RES CTR, RHONE-POULENC AGRO, JAPAN, 90- *Concurrent Pos:* Guest scientist, Atlantic Regional Lab, Nat Res Coun Can, 75-79. *Mem:* Am Chem Soc; Can Inst Chem; Japan Soc Agr Chem; Japan Asn Bioindust. *Res:* Biosynthesis and biological activity of secondary metabolites; structure-activity interactions of biologically active compounds; establishment of new screening systems for agrochemicals; biotechnology with microorganisms and plants. *Mailing Add:* Tsukuba-Akeno Res Ctr Rhone-Poulenc Agro 1500-3 Kitahara Mukoueno Akeno Ibaraki 300-45 Japan

KURODA, PAUL KAZUO, b Fukuoka, Japan, Apr 1, 17; nat US; m 53; c 3. CHEMISTRY. *Educ:* Univ Tokyo, BS, 39, ScD(inorg chem), 44. *Prof Exp:* Asst prof chem, Univ Tokyo, 44-49; fel, Univ Minn, 49-52; from asst prof to prof, 52-81, DISTINGUISHED PROF CHEM, UNIV ARK, FAYETTEVILLE, 81- *Concurrent Pos:* Assoc chemist, Argonne Nat Lab, 57-58. *Honors & Awards:* Nuclear Appln Award, Am Chem Soc, 78. *Mem:* AAAS; Am Phys Soc; Geochem Soc; Am Chem Soc; Am Geophys Union; Sigma Xi. *Res:* Natural radioactivity; nuclear and radiochemistry; cosmochemistry; geochemistry; spontaneous fission; low-level counting; radioactive fallout. *Mailing Add:* Rm 8 Chem Bldg Dept Chem Univ Ark Fayetteville AR 72703

KUROHARA, SAMUEL S, b Hilo, Hawaii, Apr 21, 31; m 56; c 3. RADIOBIOLOGY, RADIOTHERAPY. *Educ:* Wash Univ, BA, 53, MD, 57; Univ Rochester, PhD(radiobiol), 64. *Prof Exp:* Intern gen med, Jewish Hosp, St Louis, 57-58; res radiol, Strong Mem Hosp, 58-61, asst radiotherapist, 61-64; radiotherapist, US Naval Hosp, San Diego, 64-66; assoc dir radiother & assoc chief cancer res, Roswell Park Mem Inst, NY, 66-68; asst dir radiother, Med Ctr, 68-74, prof radiol, 68-74, CLIN PROF RADIOL, SCH MED, UNIV SOUTHERN CALIF, 75-; ASSOC RADIOTHERAPIST, WHITTIER ONCOL MED CLIN, 75- *Concurrent Pos:* Instr radiol, Sch Med, Univ Rochester, 61-64; clin consult, Roswell Park Mem Inst, NY, 68-; specialist physician, Los Angeles County Univ Southern Calif Med Ctr, 68-; consult, Tech Serv Corp, 60-71, Good Samaritan Hosp, Los Angeles, 69-72, Whittier Oncol Med Clin, 71- & Alpha Omega Serv, 73- *Mem:* Radiol Soc NAm; AMA; Am Soc Therapeut Radiol; Radiation Res Soc; Sigma Xi. *Res:* Computer applications in the study of medical and biological data; computer application to automated system in radiotherapy; effects of ionizing radiation on normal tissues; mechanisms of tumor control with radiation. *Mailing Add:* 825 Oak Knoll Circle Pasadena CA 91106

KUROKAWA, KANEYUKI, b Tokyo, Japan, Aug 14, 28; m 57; c 2. MICROWAVES, OPTICAL COMMUNICATIONS. *Educ:* Univ Tokyo, Bachelor Eng, 51, Dr Eng, 58. *Prof Exp:* Asst prof microwave eng, Univ Tokyo, 57-63; mem tech staff, Bell Labs, 63-64, supvr, 64-75; dep dir, 75-79, dir, 79-85, MANAGING DIR, FUJITSU LABS, 85- *Concurrent Pos:* Vis prof, Inst Indust Sci, Univ Tokyo, 86-89. *Mem:* Fel Inst Elec & Electronics Engrs; Asn Comput Mach. *Res:* Microwave circuit theory; parametric amplifier; balanced transistor amplifier; solid state oscillators theory; solid state switches optical fiber communication; analysis of head crash of hard discs; technical management. *Mailing Add:* Fujitsu Labs Ltd 1015 Kamiodanaka Nakaharaku Kawasaki 211 Japan

KUROKI, GARY W, EXPERIMENTAL BIOLOGY. *Educ:* Ft Lewis Col, BS, 79; Univ Iowa, MS, 82, PhD(bot), 85. *Prof Exp:* Res fel, Dept Bot, Univ Iowa, 80-85; postdoctoral scholar, Dept Biochem & Biophys, Univ Calif, Davis, 86-88, Dept Bot & Plant Sci, Riverside, 88-90; RES SCIENTIST, DNA PLANT TECHNOL CORP, 90- *Mem:* Am Soc Biochem & Molecular Biol; Am Soc Plant Physiologists. *Res:* Elucidation of the catabolic pathway for (R)-amygdalin in mature Prunus serotina seeds; analysis of lipid metabolism in plants; analysis of the kinetic characteristics of chorismate mutase 1 purified from Solanum tuberosum tubers; author of numerous scientific publications. *Mailing Add:* DNA Plant Technol Corp 6701 San Pablo Ave Oakland CA 94608-1239

KUROSAKA, MITSURU, b Shenyang, China, Mar 26, 35; US citizen; m 63; c 3. MECHANICAL ENGINEERING, APPLIED MATHEMATICS. *Educ:* Univ Tokyo, BS, 59, MS, 61; Calif Inst Technol, PhD(mech eng), 68. *Prof Exp:* Design engr, Hitachi Ltd, 61-63; grad res & teaching asst, Calif Inst Technol, 63-67; eng specialist, AiResearch Mfg Co, 67-69; fluid mech engr, Gen Elec Res & Develop Ctr, 69-77; assoc prof, 77-79, prof mech & aerospace eng, Univ Tenn Space Inst, 79- 87, PROF AERONAUT & ASTRONAUT, UNIV WASH, 87- *Concurrent Pos:* Consult, Gen Elec Co, ARO, Inc & AiResearch Mfg Co & Calspan & Pratt & Whitney Can; vis prof, Mass Inst Technol, 84-85. *Honors & Awards:* Gen H H (Hap) Arnold Award, Am Inst Aeronaut & Astronaut, 83. *Mem:* Assoc fel Am Inst Aeronaut & Astronaut; Am Soc Mech Engrs; Sigma Xi. *Res:* Aerothermodynamics of gas turbines; aeroacoustics; unsteady flow, aeroelasticity; thermodynamics and heat transfer; fluid dynamics. *Mailing Add:* Dept Aeronaut & Astronaut Univ Wash Seattle WA 98195

KUROSE, GEORGE, b Eatonville, Wash, June 13, 24; m 56; c 3. CHEMICAL ENGINEERING. *Educ:* Columbia Univ, BS, 49, MS, 50. *Prof Exp:* Chem engr, 50-55, res chem engr, 55-62, sr res chem engr, 62-77, GROUP LEADER, AM CYANAMID CO, 77- *Mem:* Am Chem Soc; Am Inst Chem Eng. *Res:* Process development; process design; process and economic evaluation; synthetic fiber process development. *Mailing Add:* Ten Wayfaring Rd Norwalk CT 06851

KUROSKI-DE BOLD, MERCEDES LINA, b Cordoba, Argentina, Sept 23, 42; Can citizen; m 68; c 5. IMMUNOCYTOCHEMISTRY. *Educ:* Nat Univ, Cordoba, Argentina, BSc, 68; Queen's Univ, Can, MSc, 72, PhD(path), 74. *Prof Exp:* Res fel path, Queen's Univ, Kingston, 68-74, instr, 72-74, res assoc, 77-85, asst prof, 85-86; ASST PROF PATH, UNIV OTTAWA, & SCIENTIST, HEART INST, 86- *Concurrent Pos:* Bd dirs, Child Life & Play Ottawa Liaison; bd dirs, Can-Arg Inst. *Mem:* Soc Exp Biol & Med; Int Soc Heart Res; Can Cardiovasc Soc. *Res:* Correlation between the structure and function of mammalian and non-mammalian cells; sequence cardionatrins; functional morphology of mammalian atrial cardiocytes. *Mailing Add:* Ottawa Civic Hosp Univ Ottawa Heart Inst 1053 Carling Ave Ottawa ON K1Y 4E9 Can

KUROSKY, ALEXANDER, b Windsor, Ont, Can, Sept 12, 38; US citizen; m 63; c 3. PROTEIN STRUCTURE, PLASMA PROTEINS. *Educ:* Univ BC, BSc, 65; Univ Toronto, MSc, 69, PhD, 72. *Prof Exp:* Res technician, Dept Agr, Harrow, Ont & Vancouver, BC, Can, 59-64; res & develop chemist, Can Breweries, Ltd, Toronto, 65-67; Prov Ont fel, 68-71; from asst prof to assoc prof, 75-82, PROF HUMAN BIOL CHEM & GENETICS, UNIV TEX MED BR, GALVESTON, 82- *Concurrent Pos:* NIH & NSF res grants, Burkitt Found, 76- *Mem:* Am Soc Biol Chemists; Am Chem Soc; Can Biochem Soc; AAAS; Sigma Xi; Am Soc Human Genetics. *Res:* Biochemistry; structure, function and genetics. *Mailing Add:* Dept Human Biol Chem & Genetics Univ Tex Med Br Galveston TX 77550

KUROWSKI, GARY JOHN, b Fargo, NDak, Mar 22, 31; m 63; c 3. NUMERICAL ANALYSIS. *Educ:* Univ Minn, BS, 53, MS, 54; Carnegie Inst Technol, PhD(math), 59. *Prof Exp:* Res assoc math, Off Ord Res, Duke Univ, 59-63; from asst prof to assoc prof math, 63-72, dir comput ctr, 69-71, PROF MATH, UNIV CALIF, DAVIS, 72- *Mem:* Am Math Soc; Asn Comput Mach; Soc Indust & Appl Math. *Res:* Applied mathematics, especially discrete and semi-discrete analogues of the classic fields of analysis and their application to numerical analysis. *Mailing Add:* Dept Math Univ Calif Davis CA 95616

KUROYANAGI, NORIYOSHI, b Tokyo, Japan, Feb 7, 30; m 59; c 3. DIGITAL TELECOMMUNICATIONS SYSTEMS, HIGH SPEED DIGITAL TRANSMISSIONS. *Educ:* Tokyo Inst Technol, Bachelor, 54, DrEng(computer), 62. *Prof Exp:* Mem tech staff, Nippon Tel & Tel Pub Corp, 54-64, staff engr, Elec Commun Labs, 64-71, dept head, 71-74, sr staff engr, 74-77, dep dir, 78-81, dir, 81-86; PROF TELECOMMUN, TOKYO ENG UNIV, 86- *Honors & Awards:* Donald W McLellan Meritorious Serv Award, Inst Elec & Electronics Engrs, Commun Soc, 87. *Mem:* Inst Elec & Electronics Engrs Commun Soc. *Res:* High speed computer-arithmetic circuits; high speed digital transmission systems; synchronization techniques; enhanced digital communication networks. *Mailing Add:* Tokyo Eng Univ 1404-1 Katakura Hachiohji Tokyo 192 Japan

KURSHAN, JEROME, b Brooklyn, NY, Mar 10, 19; m 46; c 2. MATHEMATICS. *Educ:* Columbia Univ, AB, 39; Cornell Univ, PhD(physics), 43. *Prof Exp:* Asst physics, Columbia Univ, 39 & Cornell Univ, 39-43; res physicist, RCA Labs, 43-55, mgr employ & training 55-59, mgr, Res Serv Lab, 59-66, mgr mkt, 66-73, mgr admin serv, 73-83, mgr admin proj, 83-87; RETIRED. *Concurrent Pos:* Instr, Rutgers Univ, 44. *Mem:* Am Phys Soc; Inst Elec & Electronics Engrs. *Res:* Ion sources; gated amplifiers; frequency modulated magnetrons; automatic frequency control oscillators; transistors; semiconductors physics; materials analysis; computer applications; research management; research administration. *Mailing Add:* 73 Random Rd Princeton NJ 08540

KURSS, HERBERT, b Brooklyn, NY, Mar 30, 24; m 63. MATHEMATICS. *Educ:* Cooper Union, BEE, 43; Polytech Inst Brooklyn, MEE, 52; NY Univ, PhD, 57. *Prof Exp:* Tech writer radar, Techlit Consult, Inc, 47; instr elec eng, US Merchant Marine Acad, 47-48; res assoc appl math, Microwave Res Inst, Polytech Inst Brooklyn, 48-54; res asst math, NY Univ, 54-57; res assoc appl math, Microwave Res Inst, Polytech Inst Brooklyn, 57-59, res asst prof, 59-62; assoc prof, 62-69, prof, 69-90, EMER PROF MATH, ADELPHI UNIV, 90- *Mem:* Am Math Soc; Math Asn Am; Sigma Xi. *Res:* Problems associated with ordinary differential equations and with electromagnetic theory. *Mailing Add:* Dept Math Adelphi Univ Garden City NY 11530

KURSTEDT, HAROLD ALBERT, JR, b Columbus, Ohio, Sept 15, 39; m 61; c 3. NUCLEAR ENGINEERING. *Educ:* Va Mil Inst, BS, 61; Univ Ill, Urbana, MS, 63, PhD(nuclear eng), 68. *Prof Exp:* Instr mech eng, Va Mil Inst, 61-62; res & develop coordr, Ballistic Res Labs, Aberdeen Proving Ground, Md, 66-68; asst prof nuclear eng, Col Eng, Ohio State Univ, 68-70; prog mgr, Fed Systs Div, Indust Nucleonics Corp, 70-76; ASSOC PROF MECH & NUCLEAR ENG, VA POLYTECH INST & STATE UNIV, 76- *Mem:* Am Soc Civil Eng; Am Soc Eng Educ; Am Nuclear Soc; Sigma Xi. *Res:* Nuclear reactor kinetics and heat transfer, particularly experimental and analytical techniques in pulsed thermal and fast reactors; nuclear instrumentation and control; nondestructive inspection. *Mailing Add:* Rte 1 Box 102 Newport VA 24128

KURSUNOGLU, BEHRAM N, b Bayburt, Turkey, Mar 14, 22; m 52; c 3. THEORETICAL PHYSICS. *Educ:* Univ Edinburgh, BSc, 49; Cambridge Univ, PhD, 52. *Hon Degrees:* DSc, Fla Inst Technol, 82. *Prof Exp:* Res assoc, Cornell Univ, 52-54; sr fel, Yale Univ, 55; vis prof physics, Miami Univ, 54-55; dean fac nuclear sci & technol, Mid E Tech Univ, Ankara, Turkey, 56-58; PROF PHYSICS, UNIV MIAMI, 58-, DIR, CTR THEORET STUDIES, 65- *Concurrent Pos:* Adv, Turkish Gen Staff Atomic Matters, 56-58; mem, Turkish Atomic Energy Comn, 56-58; Turkish mem sci comt, NATO, 58; consult, Brit Atomic Energy Res Estab, 61 & Max Planck Inst Physics & Astrophys, 61; consult, Oak Ridge Nat Lab, 62-64, chmn, Annual High Energy Physics Conf, Fla, 64-83; chmn, Annual Int Sci Forum, Energy, 77-; pres & chmn bd, Global Foundation Inc, 78- *Honors & Awards:* Presidential Sci Prize, 72; Sci Award, Asn of Turkish Am Scientists of USA, 88. *Mem:* Fel Am Phys Soc; Am Asn Physics Teachers; Turkish Am Sci Asn. *Res:* Theoretical high energy, relativity and plasma physics; arms control; energy. *Mailing Add:* Ctr Theoret Studies Univ Miami PO Box 249055 Coral Gables FL 33124-9055

KURT, CARL EDWARD, b Muskogee, Okla, June 3, 43; m 69; c 1. CIVIL ENGINEERING, STRUCTURAL ENGINEERING. *Educ:* Okla State Univ, BS, 65, MS, 66, PhD(civil eng), 69. *Prof Exp:* Instr civil eng, Okla State Univ, 69; sr engr strength, McDonnell Douglas Astronaut Corp, 69-74; from asst prof to assoc prof, Dept Civil Eng, Auburn UNiv, 74-82; PROF, DEPT CIVIL ENG, UNIV KANS, 82- *Concurrent Pos:* Pres, EnGraph. *Mem:* Am Soc Civil Engrs; Nat Water Well Asn; Am Soc Testing & Mat; Am Inst Steel Construct. *Res:* Behavior of structural materials and structural steel design; structural analysis and stability; engineering properties of thermoplastic water well casings; hydraulic analysis and model studies; computer aided design. *Mailing Add:* Dept Civil Eng Univ Kans Lawrence KS 66045

KURTA, ALLEN, b Detroit, Mich, Sept 6, 52; m 80; c 1. BATS, ENDANGERED SPECIES. *Educ:* Mich State Univ, BS, 75, MS, 80; Boston Univ, PhD(biol), 86. *Prof Exp:* Asst prof biol, Nazareth Col, 85-86; res assoc biol, Boston Univ, 86-88; ASST PROF BIOL, EASTERN MICH UNIV, 88- *Concurrent Pos:* Consult, 80- *Mem:* Am Soc Mammalogists; Am Soc Naturalists; Am Soc Zoologists. *Res:* Physiology, ecology and natural history of bats and other mammals. *Mailing Add:* Dept Biol Eastern Mich Univ Ypsilanti MI 48197

KURTENBACH, AELRED J(OSEPH), b Dimock, SDak, Jan 3, 34; m 60; c 5. ELECTRICAL ENGINEERING. *Educ:* SDak Sch Mines & Technol, BS, 61; Univ Nebr, MS, 62; Purdue Univ, PhD(elec eng), 68. *Prof Exp:* Instr elec eng, SDak State Univ, 62-65, asst prof, 65-69, assoc prof, 69-72; PRES, DAKTRONICS, INC, BROOKINGS, SDAK, 69- *Concurrent Pos:* Instr elec eng, Purdue Univ, 65-66. *Mem:* Inst Elec & Electronics Engrs. *Res:* Biomedical telemetry; pulse-code modulation telemetry. *Mailing Add:* Daktronics Inc PO Box 128 Brookings SD 57006

KURTTI, TIMOTHY JOHN, b Minneapolis, Minn, Mar 8, 42; m 85. INSECT PATHOLOGY, INSECT PHYSIOLOGY. *Educ:* Univ Minn, BA, 65, PhD(entom), 74. *Prof Exp:* From jr scientist to asst scientist insect microbiol, 66-69, res asst, 69-70, res fel, 70-73, res assoc insect microbiol, Dept Entom, Fisheries & Wildlife, Univ Minn, St Paul, 73-77; scientist, Int Lab for Res on Animal Dis, Nairobi, Kenya, 77-80; asst res prof microbiol, Waksman Inst Microbiol, Rutgers Univ, 80-85; asst prof entom, 86-89, ASSOC PROF ENTOM, UNIV MINN, 89- *Mem:* Soc Invert Path; Entom Soc Am; Sigma Xi; Am Soc Parasitol; Soc Protozool. *Res:* Insect microbiology; bovine and tick tissue culture; development physiology of insects; insect nutrition; biological calorimetry; theileriosis; intracellular parasitism; microbial control; Lyme disease. *Mailing Add:* Dept Entomol Univ Minn St Paul MN 55108

KURTZ, A PETER, b Staten Island, NY, June 12, 42; m 64; c 2. PESTICIDE CHEMISTRY, STRUCTURE-ACTIVITY RELATIONSHIPS. *Educ:* Fordham Univ, BS, 63; Columbia Univ, MA, 64, PhD(org chem), 68. *Prof Exp:* Res chemist, Letterman Army Inst Res, 69-71; res chemist, Tech Ctr, Rhone-Poulenc Agr Co, 71-78, group leader pesticide chem, 78-85, sr group leader pesticide chem, 85- 88, MGR SCI INFO MGMT & SYSTS, TECH CTR, RHONE-POULENC AGR CO, 88- *Mem:* Am Chem Soc. *Res:* Development of quantitative structure; activity relationships and application to the design of selectively toxic pesticide chemicals; computerized chemical/biological information storage and analysis systems. *Mailing Add:* Tech Ctr Rhone-Poulenc Agr Co PO Box 12014 Research Triangle Park NC 27709

KURTZ, ANTHONY DAVID, b New York, NY, May 3, 29; m 55, 85; c 2. PHYSICAL METALLURGY. *Educ:* Mass Inst Technol, SB, 51, SM, 52, ScD(phys metall), 55. *Prof Exp:* Res asst, Mass Inst Technol, 51-54, staff mem, Lincoln Lab, 54-55; sr engr, Transistor Prod, Inc Div, Clevelite Corp, 56; supvr appl res, Semiconductor Div, Minneapolis-Honeywell Regulator Co, 56-59; dir res & develop, 59-66, PRES, KULITE SEMICONDUCTOR PROD, 66- *Honors & Awards:* Si Fluor Technol Award, Instrument Soc Am, 78. *Mem:* Am Soc Metals; Am Phys Soc; Inst Elec & Electronics Engrs. *Res:* Solid state physics and tranducer design; semiconductor devices and materials; diffusion in solids; imperfections in metals and semiconductors; experimental mechanics; stress analysis. *Mailing Add:* Kulite One Willow Tree Rd Leonia NJ 07605

KURTZ, CLARK N, b Stillwater, Minn, Nov 24, 37; m 65; c 2. OPTICS. *Educ:* SDak Sch Mines & Technol, BS, 59; Univ Ill, MS, 63; Univ Rochester, PhD(elec eng), 67. *Prof Exp:* Design engr electronics, Eastman Kodak Co, 59-62, sr res physicist, 66-72, res assoc optics, 72-79, lab head res labs, 79-84, dir Optical Rec Prod, Photo Div, 84-85, Adv Dev Mgr Mass Mem, 85-88, sr tech asst to dir res, 88-89, DIR INFO & COMPUTER TECH DIV, RES, EASTMAN KODAK CO, 89- *Concurrent Pos:* Adv bd, Nat Ctr Supercomputing, Univ Ill, Cornell Theory Ctr. *Honors & Awards:* Charles Ives Award, Soc Photog Scientists & Engrs, 72. *Mem:* Optical Soc Am; Am Inst Physics. *Res:* Physics of forming images on paper in copying; theory of light propagation in waveguides; design of optical screens and diffusers; optical disk; light waves. *Mailing Add:* 4355 W Lake Rd Canandaigua NY 14424

KURTZ, DAVID ALLAN, b Evanston, Ill, Jan 31, 32. PESTICIDE CHEMISTRY, CHEMOMETRICS. *Educ:* Knox Col, AB, 54; Pa State Univ, MS, 58, PhD(org chem), 60. *Prof Exp:* Instr gen chem, Pa State Univ, 59-60; sr chemist, HRB-Singer, Inc, 60-62; res assoc appl chem, Mat Res Lab, 62-66, asst prof pesticides anal, 67-71, ANALYTICAL CHEMIST, PESTICIDES RES LAB, PA STATE UNIV, UNIVERSITY PARK, 71- *Concurrent Pos:* Co-leader, Coop Regional Proj NE-115, Pa Agr Exp Sta, 78-83. *Mem:* Am Chem Soc; Chemometrics Soc; Soc Environ Toxicol & Chem; Asn Off Anal Chemists; Int Union Pure & Appl Chem; Int Asn Great Lakes Res. *Res:* Methods of analysis for pesticides and herbicides; chemometric methods (statistical calibration methods in trace residue analysis for pesticides and environmental compounds); pesticide analysis in marine and marine atmosphere environments; long range transport of pesticides. *Mailing Add:* 118 E S Hills Ave State College PA 16801

KURTZ, DAVID WILLIAMS, b Altoona, Pa, July 27, 42; m 72; c 3. PHOTOCHEMISTRY, EDUCATION. *Educ:* Houghton Col, NY, BS, 64; Syracuse Univ, PhD(chem), 71. *Prof Exp:* Res assoc photochem, Univ Wis, 71-73; from asst prof to assoc prof, 73-83, PROF ORG CHEM, OHIO NORTHERN UNIV, 84- *Concurrent Pos:* Petrol Res Found fel, Iowa State Univ, 83. *Mem:* Am Chem Soc; Sigma Xi. *Res:* Torsional processes in photochemical transformations and other syntheitic applications in organic chemistry. *Mailing Add:* Dept Chem Ohio Northern Univ Ada OH 45810

KURTZ, EDWIN BERNARD, JR, b Wichita, Kans, Aug 11, 26; m 52; c 2. SCIENCE EDUCATION. *Educ:* Univ Ariz, BS, 48, MS, 49; Calif Inst Technol, PhD, 52. *Prof Exp:* Instr bot, Univ Ariz, 47, from asst prof to prof, 51-68, actg head dept, 54-55; prof biol & head dept, Kans State Teachers Col, 68-72; prof life sci & chmn dept, 72-89, EMER PROF, UNIV TEX PERMIAN BASIN, 89- *Concurrent Pos:* Asst dir educ, AAAS, Washington, DC, 65-67. *Mem:* AAAS; Am Inst Biol Sci; Nat Sci Teachers Asn. *Res:* Science education. *Mailing Add:* 1620 N Kutch Dr Univ Tex Permian Basin Flagstaff AZ 86001

KURTZ, GEORGE WILBUR, b Harrisburg, Pa, Dec 8, 28; m 53; c 3. ANALYTICAL CHEMISTRY. *Educ:* Pa State Univ, BS, 50, MS, 52, PhD(dairy sci), 54. *Prof Exp:* Res chemist, Swift & Co, 54; head flavor & phys chem lab, Armed Forces Qm Food & Container Inst, Chicago, 56-62; chemist, Dalare Assocs, 62-69; mgr qual control, Monroe, NC, 69-79, plant mgr, 79-89, CONSULT, R P SCHERER, CLEARWATER, 89- *Mem:* Am Chem Soc; Inst Food Technol. *Res:* Analytical chemistry; food technology. *Mailing Add:* 2389 Wind Gap Pl Clearwater FL 34625

KURTZ, HAROLD JOHN, b Brookings, SDak, Feb 18, 31; m 53; c 3. VETERINARY PATHOLOGY. *Educ:* SDak State Univ, BS & MS, 54; Univ Minn, DVM, 58, PhD(vet path), 66. *Prof Exp:* Instr vet surg, 60-61 & vet med, 61-62, fel vet path, 62-66, from asst prof to assoc prof, 66-74, PROF VET PATH, COL VET MED, UNIV MINN, ST PAUL, 74- *Mem:* Am Vet Med Asn; Am Col Vet Path; Int Cad Path. *Res:* Dissecting aortic rupture in turkeys; neuropathology and pathology of animal diseases; comparative pathology; edema disease of swine. *Mailing Add:* 236-E Vet Diag Lab Univ Minn Col Vet Med St Paul MN 55108

KURTZ, LAWRENCE ALFRED, b Providence, RI, Dec 29, 40. NUMERICAL ANALYSIS. *Educ:* Univ RI, BS, 62; Univ Conn, MS, 65; Univ Tenn, Knoxville, PhD(math), 69. *Prof Exp:* Engr, Eastman Kodak Co, 62-63; teaching asst math, Univ Conn, 63-65; teaching asst, Univ Tenn, Knoxville, 65, instr, 70; statist comput analyst, 71, asst prof, Hollins Col, 70-76; ASSOC PROF MATH, UNIV MONTEVALLO, 77- *Mem:* Am Math Soc; Soc Indust & Appl Math. *Res:* Computational fluid dynamics; numerical analysis of partial differential equations; mathematics education. *Mailing Add:* Dept Math & Phys Montevallo AL 35115

KURTZ, LESTER TOUBY, b Howard Co, Ind, Nov 7, 14; m 40; c 2. AGRONOMY. *Educ:* Purdue Univ, BS, 38; Univ Ill, PhD(agron), 43. *Prof Exp:* Asst soil chemist, Dept Agron, Univ Ill, Urbana, 38-43, assoc, 43-44, instr US army spec training prog, 43-45, from asst prof to assoc prof, 44-50, prof soil fertil & fertilizers, Exp Sta, 50-82; RETIRED. *Concurrent Pos:* Guggenheim fel, soils lab, Agr Res Serv, USDA, 53-54; Fulbright fel, Waite Inst, SAustralia, 61-62 & soil & fertilizer br, Tenn Valley Authority, Ala, 68-69; Lady Davis fel, Soils & Fertilizer Div, Israel Inst Tech, Haifa, Israel. *Mem:* AAAS; fel Am Soc Agron; Am Chem Soc; Soil Sci Soc Am. *Res:* Soil chemistry and fertility; analytical chemistry; phosphate fixation in Illinois soils; fate of fertilizer nitrogen in soils as indicated by nitrogen 15; role of fertilizer in water pollution. *Mailing Add:* Dept Agron Turner Hall Univ Ill 1102 S Goodwin Urbana IL 61801

KURTZ, MARGOT, b WGer, Aug 30, 41; m 65; c 1. MEDICAL EDUCATION. *Educ:* Mich State Univ, BA, 72, MA, 73, PhD, 76. *Prof Exp:* Fac behav sci, Lansing Community Col, 74-76; coordr, 76-77, co-dir preceptor prog, 77-, co-dir jr partnership progs, 80-, asst prof, 85-87, ASSOC PROF, DEPT FAMILY MED, MICH STATE UNIV, 87- *Res:* Educational and psychological aspects of clinical training; behavioral aspects of physician-patient relationship; processes in medical interviewing; psychological aspects of student personal development and relationship patterns while in medical school. *Mailing Add:* Dept Family Med Mich State Univ East Lansing MI 48824

KURTZ, MARK EDWARD, b Trenton, Mo, Nov 8, 46; m 52, 71; c 2. WEED SCIENCE, CROP ROTATION & FIBER RESEARCH. *Educ:* Mo Valley Col, BS, 69; Miss State Univ, MS, 77, PhD(weed sci), 80. *Prof Exp:* ASSOC PLANT PHYSIOLOGIST, RES DEPT, DELTA BR, MISS AGR & FORESTRY EXP STA, STONEVILLE, MS, 80- *Concurrent Pos:* Adj assoc prof, Dept Plant Path & Weed Sci, Miss State Univ, 81- *Mem:* Weed Sci Soc Am; Int Kenaf Asn. *Res:* Soybean and rice cotton weed control, utilizing existing techniques and improvising new ideas to solve unanswered problems as they arise in varied cropping systems; evaluating herbicides in a soybean-rice rotation; herbicide tolerance in Kenaf (Hibiscus Cannabinus). *Mailing Add:* Box 197 Stoneville MS 38776

KURTZ, MYRA BERMAN, b New York, NY, July 20, 45; m 70; c 1. MICROBIOLOGY, GENETICS. *Educ:* Goucher Col, AB, 66; Harvard Univ, PhD(microbiol), 71. *Prof Exp:* Res assoc, State Univ NY, Albany, 71-72; assoc prof microbiol, Fed Univ Sao Carlos, Brazil, 72-74; res assoc, Rutgers Univ, 75-76, asst res prof, Waksman Microbiol, 76-82; sr res scientist, Squibb Inst, Princeton, NJ, 82-87; sr res fel, 88-89, DIR, MERCK, SHARP & DOHME RES LABS, RAHWAY, NJ, 89- *Concurrent Pos:* Ed exp mycol. *Mem:* Am Soc Microbiol; AAAS; Sigma Xi. *Res:* Molecular genetics of the dimorphic human pathogenic fungus; Candida albicans; systemic fungal disease. *Mailing Add:* 16 Redwood Rd Martinsville NJ 08836

KURTZ, PETER, JR, b Chicago, Ill, May 12, 27; m 50. CERAMIC ENGINEERING, THERMODYNAMICS. *Educ:* Univ Mo, Rolla, BS, 52, MS, 53; Univ Calif, Los Angeles, PhD(eng), 64. *Prof Exp:* Grad res engr, Univ Calif, Los Angeles, 53-55, jr res engr, 55-57, asst res engr, 57-58, assoc eng, 56-62, actg asst prof, 63-64, asst prof, 64-68; prof eng, 68-80, chmn, Div Sci, 71-80, PROF PHYSICS, BIOLA COL, 80- *Mem:* Am Asn Physics Teachers; Sigma Xi. *Res:* Mechanical properties of ceramic materials; thermodynamic properties of multicomponent polyphase systems. *Mailing Add:* Dept Math & Sci Biola Univ 13800 Biola Ave La Mirada CA 90639

KURTZ, RICHARD LEIGH, b Dubuque, Iowa, Oct 19, 56; m 87; c 1. SURFACE SCIENCE, SYNCHROTRON RADIATION. *Educ:* Brandeis Univ, BA, 78; Yale Univ, MS, 79, PhD(appl Physics), 83. *Prof Exp:* Nat Res Coun postdoctoral assoc, Nat Bur Standards, 83-85; res physicist, Nat Inst Standards & Technol, 85-91; ASSOC PROF PHYSICS, LA STATE UNIV, 90- *Mem:* Am Phys Soc; Am Vacuum Soc; Sigma Xi; AAAS. *Res:* Synchrotron radiation applications in surface science; photoelectron spectroscopy; surface electronic structure; molecular adsorption; stimulated desorption; transition-metal oxides; high temperature superconductors; high Tc thin films. *Mailing Add:* Dept Physics & Astron La State Univ 202 Nicholson Hall Baton Rouge LA 70803-4001

KURTZ, RICHARD ROBERT, b Moose Jaw, Sask, Mar 20, 45; m 70; c 2. SYNTHETIC ORGANIC CHEMISTRY. *Educ:* Univ Calgary, BSc, 67; Mass Inst Technol, PhD(org chem), 71. *Prof Exp:* Assoc org chem, John C Sheehan Inst Res, 71-73; researcher, 73-80, res head, 80-83, res mgr, 83-84, ASSOC DIR, UPJOHN CO, 84- *Mem:* Am Chem Soc; AAAS; NY Acad Sci. *Res:* Pharmaceutical research and development; synthesis of heterocycles. *Mailing Add:* Chem Process Res & Develop Upjohn Co Unit 1500 Kalamazoo MI 49001

KURTZ, STANLEY MORTON, b Philadelphia, Pa, May 11, 26; m 57; c 2. PATHOLOGY. *Educ:* George Washington Univ, BS, 49, MS, 50, PhD(anat), 53; Univ Ala, MD, 58. *Prof Exp:* Instr anat, Bowman Gray Sch Med, Wake Forest Col, 52-54; instr anat, Med Sch, Univ Ala, 54-58; sr res fel path, Univ Pittsburgh, 58-61; assoc prof, Med Ctr, Duke Univ, 61-65; dir dept toxicol, Res Labs, Parke Davis & Co, 65-76; prof path, Med Univ SC, Charleston & staff pathologist, Charleston Vet Admin Hosp, 76-85; PROF PATH, SOUTHWESTERN MED CTR, DALLAS & STAFF PATHOLOGIST, DALLAS VET ADMIN HOSP, 85- *Concurrent Pos:* Consult, Vet Admin Hosps, 61-65, Nat Acad Sci-Nat Res Coun, 72- & Nat Inst Drug Abuse, 74- *Mem:* AAAS; Am Soc Path & Bact; Am Soc Exp Path; Soc Toxicol. *Res:* Cytology; electron microscopy; development, structure and pathology of mammalian renal glomerulus. *Mailing Add:* Lab Serv Vet Admin Med Ctr 4500 S Lancaster Rd Dallas TX 75216

KURTZ, STEVEN ROSS, b Washington, DC, Oct 3, 53; m 78. ELECTRONIC MATERIALS, PHOTODETECTOR DEVELOPMENT. *Educ:* Bucknell Univ, BS, 75; Univ Ill, Urbana-Champaign, MS, 77, PhD(physics), 80. *Prof Exp:* MEM TECH STAFF, SANDIA NAT LAB, 80- *Mem:* Am Phys Soc. *Res:* Electronic materials, semiconductors and insulators; photoconductivity and transport in layered semiconductors and disordered materials; optical and electron paramagnetic resonance spectroscopies; photodetector development, novel infrared materials and photodetectors. *Mailing Add:* Div 1163 Sandia Nat Lab Albuquerque NM 87185

KURTZ, STEWART K, b Bryn Mawr, Pa, June 9, 31; m 51; c 4. ELECTROOPTICS. *Educ:* Ohio State Univ, BSc, 56, MSc, 57, PhD(physics), 60. *Prof Exp:* Staff scientist, Bell Tel Labs, Inc, 60-69; group dir explor res, Philips Labs Div, NAm Philips Corp, 69-80; WITH CLAIROL APPLIANCES, 80- *Mem:* Am Phys Soc; sr mem Inst Elec & Electronics Engrs; NY Acad Sci. *Res:* High resolution infrared molecular spectra; paramagnetic resonance, cross-relaxation; ferroelectrics, primarily optical and electrooptical properties; nonlinear optical materials; powder survey methods; second harmonic coefficients; Raman scattering; crystal growth; electro-crystallization; phase transitions; crystal chemistry. *Mailing Add:* Mat Res Lab Pa State Univ University Park PA 16802

KURTZ, THOMAS EUGENE, b Oak Park, Ill, Feb 22, 28; m 53, 74; c 3. MATH STATISTICS, COMPUTER SYSTEMS. *Educ:* Knox Col, BA, 50; Princeton Univ, PhD(math), 56. *Hon Degrees:* DSc, Knox Col, 85. *Prof Exp:* From instr to assoc prof math, 56-66, dir, Comput Ctr, 59-75, dir, Off Acad Comput, 75-78, vchmn, Prog Comput Info Sci, 80-88, PROF MATH, DARTMOUTH COL, 66-; VCHMN, TRUE PASIC INC, 83- *Concurrent Pos:* Consult, Vet Admin, White River Junction, Vt; mem Pierce panel, President's Sci Adv Coun Comput in Higher Educ, 65-67; chmn coun, EDUCOM, 73-74; chmn bd, NERComP, Inc, 74- *Honors & Awards:* Pioneer Award, Am Fed Info Processing Socs, 74. *Mem:* Am Statist Asn; Asn Comput Mach; Inst Elec & Electronics Engrs. *Res:* Computer languages; computer systems and their applications; computer use in education; statistics applications. *Mailing Add:* PO Box 962 Hanover NH 03755

KURTZ, THOMAS GORDON, b Kansas City, Mo, July 14, 41; m 63; c 2. STOCHASTIC PROCESSES. *Educ:* Univ Mo-Columbia, BA, 63; Stanford Univ, MS, 65, PhD(math), 67. *Prof Exp:* Vis lectr math, Univ Wis-Madison, 67-69, from asst prof to assoc prof, 69-75, dept chair, 85-88, PROF MATH, UNIV WIS-MADISON, 75-, DIR, CTR MATH SCI, 90- *Mem:* Math Asn Am; Soc Indust & Appl Math; Am Math Soc; fel Inst Math Statist; Opers Res Soc Am. *Res:* Probability theory and stochastic processes; Markov processes; operator semigroups, particularly the relationship between Markov processes and operator semigroups; approximation for stochastic process; filtering; stochastic control. *Mailing Add:* Dept Math Univ Wis Madison WI 53706

KURTZ, VINCENT E, b Duluth, Minn, Apr 12, 26; m 53; c 5. PALEONTOLOGY, STRATIGRAPHY. *Educ:* Univ Minn, BA, 46, MS, 49; Univ Okla, PhD(geol), 60. *Prof Exp:* Geologist, Aurora Gasoline Co, Colo, 52-55, dist geologist, Kans, 56-58; consult geologist, Mich, 60-65; from asst prof to assoc prof earth sci, 65-74, PROF GEOL, SOUTHWEST MO STATE COL, 74- *Mem:* Paleont Soc. *Res:* Late Cambrian and early Ordovician stratigraphy; paleontology and paleoecology of trilobites, inarticulate brachiopods and conodonts. *Mailing Add:* Dept Geog & Geol SW Mo State Univ 901 S National Springfield MO 65804

KURTZ, WILLIAM BOYCE, b Austin, Tex, July 15, 41; m 65; c 2. FOREST RESOURCE ECONOMICS. *Educ:* NMex State Univ, BS, 63, MS, 66; Univ Ariz, PhD(natural resource econ), 71. *Prof Exp:* Range conservationist, Soil Conserv Serv, USDA, 63-64; res assoc watershed mgt, Univ Ariz, 68-69; proj economist, Daniel Mann Johnson & Mendenhall, 69; sr economist, Voorhies, Trindle & Nelson, Orange County, 69-70; asst prof natural resources mgt, Calif Polytech State Univ, 70-75; assoc prof, 75-80, PROF FORESTRY, UNIV MO-COLUMBIA, 80- *Mem:* Sigma Xi; Soc Range Mgt; Soil Conserv Soc Am. *Res:* Multiple use economics of forest resources. *Mailing Add:* Sch Natural Resources Univ Mo Agr Bldg Rm 1 30 Columbia MO 65211

KURTZE, DOUGLAS ALAN, b Mt Vernon, NY, Oct 15, 54. PATTERN FORMATION, STATISTICAL PHYSICS. *Educ:* Lehigh Univ, BA & BS, 74; Cornell Univ, MS, 78, PhD(physics), 80. *Prof Exp:* Res physicist, Carnegie-Mellon Univ, 79-82; asst prof physics, Clarkson Univ, 82-90; PROF PHYSICS, NDAK STATE UNIV, 90- *Concurrent Pos:* Resident vis, AT&T Bell Lab, 83-84; res physicist, Inst Theoret Physics, 87. *Mem:* Am Phys Soc; Soc Indust & Appl Math; Am Asn Crystal Growth. *Res:* Instabilities and pattern formation in solidifying and crystallizing systems; analytical and numerical methods for moving boundary problems; zeros of partition functions. *Mailing Add:* Dept Physics NDak State Univ Fargo ND 58105-5566

KURTZKE, JOHN F, b Brooklyn, NY, Sept 14, 26; m; m 50; c 7. NEUROLOGY, EPIDEMIOLOGY. *Educ:* St John's Univ, NY, BS, 48; Cornell Univ, MD, 52; Am Bd Psychiat & Neurol, dipl neurol, 58. *Prof Exp:* Intern, Kings County Hosp, 52-53; resident, Vet Admin Hosp, Bronx, 53-56; instr neurol, Jefferson Med Col, 58-61, asst neurologist, Hosp & Clin, 58-63, assoc clin neurol, Col, 61-63, asst prof, 63; clin assoc prof, 63-65, assoc prof, 65-68, PROF NEUROL & COMMUNITY MED, SCH MED, GEORGETOWN UNIV, 68-, VCHMN, DEPT NEUROL, 76- *Concurrent Pos:* Rear adm, Med Corps, USNR, 44-86; chief neurol serv, Vet Admin Hosp, Coatesville, Pa, 56-63, assoc chief staff res, 57-62; examr, Am Bd Psychiat & Neurol, 61-; consult neurol, Naval Hosp, Bethesda, 66-; chief neurol serv, Vet Admin Ctr, Washington, DC, 63-; mem exec comt coop study of adrenocorticotropic hormone in multiple sclerosis, Nat Inst Neurol Dis & Blindness, 64-71; Vet Admin rep, Neurol Study Sect, Div Res Grant, 64-72; mem, World Fedn Neurol, comn geog neurol, epidemiol & statist, 64-76, comn Multiple Sclerosis, 67-, comn neuroepidemiol, 77-, mem med adv bd, Nat Multiple Sclerosis Soc, 66-; consult to Surgeon Gen, Dept Navy, 70-; mem int med adv bd, Int Fedn Multiple Sclerosis Soc, 72-; consult ad hoc comt spinal cord injury, Nat Inst Neurol Dis & Stroke, 73-76; mem epilepsy adv comt, NIH, 74-77; mem, task force on neurol serv, Joint Comn Neurol, 71-75 & work group on epidemiol, Nat Multiple Sclerosis Soc, 73; chmn work group on epidemiol, biostatist & pop genetics, Comn for Control of Huntington's Dis, 76-77; liaison officer, USN Med Sch, 79-86; established investr, Nat Mult Sclerosis Soc, 87; mem working group on design clin studies, Nat Multiple Sclerosis Soc, 76-84; mem, Comt Nat Needs for Neurol, Am Acad Neurol, 79-85; mem, Manpower Comt, Soc Medal Consult Armed Forces, 84-; mem, Inst Clin Rev Sub panet, Nat Inst Neurol Dis & Stroke, 89- *Honors & Awards:* Zimmerman Lectr, Stanford Univ, 80; Hope Chest Award, Nat Mult Sclerosis Soc, 82; Gold Vicennial Medal, Georgetown Univ, 82; Tarbox Lectr, Tex Tech Univ Health Sci, 89; Geigy Lectr in Multiple Sclerosis, Univ Western Ont, 89. *Mem:* Fel Am Col Epidemiol; Am Neurol Asn; fel Am Col Physicians; fel Am Acad Neurol; Am Epidemiol Soc; fel Am Col Prev Med; hon mem Danish Neurol Soc; hon foreign mem French Neurol Soc. *Res:* Neuroepidemiology. *Mailing Add:* 7509 Salem Rd Falls Church VA 22043-3209

KURTZMAN, CLETUS PAUL, b Mansfield, Ohio, July 19, 38; m 62; c 3. MYCOLOGY. *Educ:* Ohio Univ, BS, 60; Purdue Univ, MS, 62; WVa Univ, PhD(mycol), 67. *Prof Exp:* Microbiologist, 67-70, zymologist, 70-81, RES LEADER & HEAD AGR RES SERV CULTURE COLLECTION, NAT CTR AGR UTILITY RES, USDA, 81- *Concurrent Pos:* Adj prof mycol, Ill State Univ, 81-; US re, Int Comn Yeasts, 80-, Int Mycol Asn, 83- *Honors & Awards:* J Roger Porter Award, 90. *Mem:* Mycol Soc Am; Sigma Xi; Am Soc Microbiol; AAAS; US Fedn Cult Collections (vpres, 77-78, pres, 78-80); Soc Gen Microbiol. *Res:* Yeast taxonomy; genetic, molecular and physiological aspects of interspecific and intergeneric relationships. *Mailing Add:* USDA Nat Ctr Agr Utility Res 1815 N University St Peoria IL 61604

KURTZMAN, RALPH HAROLD, JR, b Minneapolis, Minn, Feb 21, 33; m 55; c 2. BIOCHEMISTRY, MYCOLOGY. *Educ:* Univ Minn, BS, 55; Univ Wis, MS, 58, PhD(plant path biochem), 59. *Prof Exp:* Asst prof plant path, Univ RI, 59-62; asst prof biol, Univ Minn, 62-65; BIOCHEMIST, WESTERN REGIONAL RES LAB, USDA, 65- *Concurrent Pos:* NASA contract res with A H Brown & A O Dahl, Univ Minn, 63; lectr, Pakistan & Thailand, 74, Pakistan & India, 78 & Univ Helsinki, 80; guest scientist, VTT Tech Res Ctr Finland, 80; consult mushroom growing, 81. *Mem:* Int Mushroom Soc Tropics; Mushroom Growers Asn Gt Brit; Mycol Soc Am; Am Mushroom Inst. *Res:* Physiology of plant diseases, particularly Dutch elm disease; alkaloid metabolism of fungi; fungal decomposition of cellulose and lignin; mushroom production from wastes; bureaucratic vasilation; mushroom physiology. *Mailing Add:* Western Regional Res Lab USDA 800 Buchanan St Albany CA 94710

KURUCZ, ROBERT LOUIS, b Columbus, Miss, Sept 7, 44. ASTROPHYSICS. *Educ:* Harvard Col, AB, 66, PhD(astron), 73. *Prof Exp:* Res fel, Harvard Col Observ, 73-74; PHYSICIST, SMITHSONIAN ASTROPHYS OBSERV, 74- *Mem:* Am Astron Soc; Int Astron Union. *Res:* Stellar atmospheres; solar physics; radiative transfer; atomic and molecular physics. *Mailing Add:* Smithsonian Astrophys Observ 60 Garden St Cambridge MA 02138

KURUP, VISWANATH PARAMESWAR, b Thattayil, India, Jan 20, 36; US citizen; m 62; c 3. MEDICAL MYCOLOGY, IMMUNOLOGY. *Educ:* Univ Poona, BS, 57, MS, 59; Univ Delhi, PhD(med mycol), 67. *Prof Exp:* Fel med mycol, Ohio State Univ, 68-70; microbiologist, St Anthony Hosp, Columbus, Ohio, 70-73; asst prof, 73-77, assoc prof, 77-86, PROF MED, MED COL WIS, 86-; MICROBIOLOGIST MED MYCOL, VET ADMIN MED CTR, 73- *Concurrent Pos:* Vis prof, PR, 81-; Fulbright fel, Finland, 84. *Mem:* Am Soc Microbiol; Mycol Soc Am; Int Soc Human & Animal Mycol; Am Acad Allergy & Clin Immunol; Med Mycol Soc Am; fel Am Acad Microbiol. *Res:* Isolation and purification of antigens and allergens associated with HP from pathogenic fungi; characterization of the antigens and development of immunological tests for the early diagnosis of hypersensitivity lung disease and immune regulation in HP. *Mailing Add:* Vet Admin Res Serv 151 5000 W National Ave Milwaukee WI 53295

KURYLA, WILLIAM C, b Cuyahoga Falls, Ohio, Sept 3, 34; m 57; c 2. ORGANIC CHEMISTRY. *Educ:* Kent State Univ, BSc, 56; Univ Minn, MSc, 58, PhD(org chem), 60. *Prof Exp:* Microanalyst, Univ Minn, 56-59; res & develop chemist, Chem Div, 60-69, res scientist, Tech Ctr, 69-71, group leader, 71-73, mgr recruiting & univ rels, 73-77, technol mgr, Occup Health, Res & Develop Dept, Tech Ctr, 77-80, sr group leader, indust hyg & environ anal, 78-80, corp mgr appl toxicology serv, 80-84, corp mgr into resources & technol, 84-85, dir prod safety CIPS group, 85-86, corp mgr prod & distrib risk, 86-87, ASST DIR PROD SAFETY, UNION CARBIDE CORP, 80- *Concurrent Pos:* Adj prof, WVa State Col. *Mem:* Am Chem Soc; Am Inst Chem; Am Ind Hyg Asn; Sigma Xi; NY Acad Sci; Am Col Toxicol. *Res:* Polyurethane chemistry and technology; ketene acetals and indole chemistry; radiochemical studies with polymers; textile and fiber chemicals; flame retardants; health effects of chemicals. *Mailing Add:* Four Peaceful Dr New Fairfield CT 06812

KURYLO, MICHAEL JOHN, III, b Meriden, Conn, July 20, 45; m 66; c 4. CHEMICAL KINETICS OF FREE RADICALS. *Educ:* Boston Col, BS, 66; Catholic Univ Am, PhD(phys chem), 69. *Prof Exp:* Nat Res Coun res assoc, 69-71, RES CHEMIST, NAT BUR STANDARDS, 71- *Concurrent Pos:* Mem, Panel Lab Measurement & Data Eval, NASA, 78-; sci asst to dir, Nat Measurement Lab, Nat Bur Standards, 79-80. *Honors & Awards:* Bronze Medal, US Dept Com, 83. *Mem:* Am Chem Soc; Am Phys Soc; Interam Photochem Soc. *Res:* Rates and mechanisms of gas phase reaction of importance to atmospheric chemistry and combustion processes; effects of internal reactant energy on reaction dynamics; temperature and pressure effects in free radical reactions. *Mailing Add:* Ctr Chem Tech Nat Inst Standards & Technol Gaithersburg MD 20899

KURZ, JAMES ECKHARDT, b Louisville, Ky, Oct 8, 34; m 63; c 2. PHYSICAL CHEMISTRY, POLYMER CHEMISTRY. *Educ:* Centre Col, AB, 56; Duke Univ, MA, 58, PhD(phys chem), 61. *Prof Exp:* Sr res chemist, Monsanto Co, 61-66, sr res specialist, 66-75, sr group leader, 75-82, mgr res, 82-84, mgr res & develop, 84-85; PRIN TECH SPECIALIST, MCDONNELL-DOUGLAS CORP, 86- *Mem:* Am Chem Soc; Soc Advan Mat & Process Eng. *Res:* Characterization of polymers by dilute solution methods; column fractionation of polymers and gel permeation chromatography; physical, mechanical and thermal characterization of polymers and polymer structure; membrane structure and use in industrial processes; structure, property and applications of polymers; structural composites; processing science of aerospace materials. *Mailing Add:* 14317 Aitken Hill Chesterfield MO 63017

KURZ, JOSEPH LOUIS, b St Louis, Mo, Dec 13, 33. PHYSICAL ORGANIC CHEMISTRY. *Educ:* Wash Univ, AB, 55, PhD(chem), 58. *Prof Exp:* Res fel, Harvard Univ, 58-60; res chemist, Cent Basic Res Lab, Esso Res & Eng Co, 60-64; from asst prof to assoc prof, 64-73, PROF CHEM, WASH UNIV, 73- *Concurrent Pos:* Vis prof, Wash Univ, 63-64. *Mem:* Am Chem Soc; AAAS. *Res:* Mechanisms, kinetics and thermodynamics of reactions in solution; mechanisms of homogeneous catalysis; transition state structure; kinetic and equilibrium isotope effects. *Mailing Add:* RR No 1 Box 179E Lonedell MO 63060

KURZ, MICHAEL E, b Detroit, Mich, Mar 5, 41; m 64; c 4. ORGANIC CHEMISTRY. *Educ:* St Mary's Col, BA, 63; Case Western Reserve Univ, PhD(chem), 67. *Prof Exp:* Instr & res assoc chem, Columbia Univ, 67-68; from asst prof to assoc prof, 68-76, actg chmn dept, 74-75, PROF CHEM, ILL STATE UNIV, 76-, CHMN DEPT, 87- *Concurrent Pos:* Petrol res fund res grant, 69-72 & 85-87, NSF grant, 85, 89; Sabbatical Res, La State Univ, 76-77, Univ Ill, 88. *Mem:* Am Chem Soc; Int Assoc Arson Investr. *Res:* Free radical aromatic substitution utilizing oxidative and photolytic methods of radical generation; oxidative aromatic substitutions; trace residue analysis; fire debris. *Mailing Add:* Dept Chem Ill State Univ Normal IL 61761

KURZ, RICHARD J, b Springfield, Ill, Oct 10, 35; m 58. SPACE RESEARCH, INSTRUMENTATION. *Educ:* Univ Ill, BS, 57, MS, 58; Univ Calif, Berkeley, PhD(physics), 63. *Prof Exp:* Asst physics, Lawrence Radiation Lab, Univ Calif, 59-62, physicist, 62-64; NSF fel, LePrince-Ringuet Lab, Ecole Polytech, Paris, 64-65; physicist, Lawrence Radiation Lab, Univ Calif, Berkeley, 65-66; Nat Acad Sci res assoc, Goddard Space Flight Ctr, NASA, Md, 66-67, chief physics br, Planetary & Earth Sci Div, Manned Spacecraft Ctr, Houston, 67- 72; ASST MGR, ADVAN SYSTS DEPT, TRW SYSTS GROUP, 72- *Mem:* Am Phys Soc. *Res:* Space research instrumentation development. *Mailing Add:* 92 Headland Dr Rancho Palos Verdes CA 90274

KURZ, RICHARD KARL, b New York, NY, Feb 4, 36; m 60; c 5. PHOTOGRAPHIC CHEMISTRY. *Educ:* St John Fisher Col, NY, BS, 57; Univ Ill, Champaign-Urbana, PhD(org chem), 61. *Prof Exp:* Res Chemist, Res Labs, Eastman Kodak Co, 61-69, res assoc, 69-74, lab head photog chem, 74-81, sr lab head, 81-85, prod develop mgr, Graphic Imaging Syst Div, 85-89, SR TECH STAFF, GRAPHIC IMAGING SYSTS DIV, EASTMAN KODAK CO, 89- *Mem:* Am Chem Soc; Soc Photog Scientists & Engrs. *Res:* Design and development of advanced photographic materials for use in radiography, graphic arts, micrographics and instrumentation recording applications; environmental conformance technology. *Mailing Add:* 40 True Hickory Dr Rochester NY 14615

KURZ, WOLFGANG GEBHARD WALTER, b Innsbruck, Austria, June 9, 33; m 63; c 3. MICROBIOLOGY, PLANT PHYSIOLOGY. *Educ:* Univ Vienna, PhD(microbiol, biochem), 58. *Prof Exp:* Res asst microbiol & biochem, Royal Inst Technol, Sweden, 55-63; fel, Prairie Regional Lab, Nat Res Coun Can, 63-65; res scientist, Tech Res Coun Sweden, 65-67; assoc res officer microbiol & fermentation technol, Nat Res Coun Can, 67-73, sr res officer Microbiol & Fermentation Technol, Prairie Regional Lab, 73-87, head, Biotechnol Sect, 81-86, head plant prod technol, 83-86, INT PROJ COORDR, PLANT BIOTECHNOL INST, NAT RES COUN CAN, 85-, HEAD BIOTECHNOL DEPT, 89- *Honors & Awards:* Can Soc Microbiologists Award, 79. *Mem:* Fel Chem Inst Can; Can Soc Microbiologists; Int Asn Plant Tissue Cult. *Res:* Biological dinitrogen fixation; continuous cultivation of microbes and plant cells; fermentation biology; enzymology; microbial physiology; process development; apparatus design; biosynthesis of secondary metabolites by microbes and plant cells. *Mailing Add:* Plant Biotechnol Inst Nat Res Coun Can Saskatoon SK S7N 0W9 Can

KURZE, THEODORE, b Brooklyn, NY, May 18, 22; div; c 4. NEUROSURGERY. *Educ:* Wash Col, BS, 43; Long Island Col Med, MD, 47; Am Bd Neurol Surg, dipl, 56. *Prof Exp:* Intern, St Monica's Hosp, Phoenix, 47-48; resident neurosurg, Vet Admin Wadsworth Hosp, Los Angeles, 48-49; resident neurosurg, Los Angeles County Gen Hosp, 51-54; instr neurol surg, Med Ctr, Univ Calif, Los Angeles, 55-56; from instr to assoc prof, 56-57, PROF NEUROL SURG, SCH MED, UNIV SOUTHERN CALIF, 67-, CHMN DEPT, 63-; DIR NEUROL SURG, LOS ANGELES COUNTY-UNIV SOUTHERN CALIF MED CTR, 67- *Concurrent Pos:* Asst, Sch Med, Univ Southern Calif, 53-54; pvt pract, Calif, 56-61; head physician, Los Angeles County Gen Hosp, 61-65, chief neurol surg, 65-67; mem, Am Bd Neurol Surg, 68-74. *Mem:* Am Col Surgeons; Soc Neurol Surgeons; Am Asn Neurol Surgeons; Am Acad Neurol Surgeons; Cong Neurol Surg; Sigma Xi. *Res:* Microtecniques in neurological surgery; humoral response to cerebral injury; acoustic tumor surgery; biomedical ethics. *Mailing Add:* 9402 Presby Univ Hosp Pittsburgh PA 15213

KURZHALS, PETER R(ALPH), b Berlin, Ger, Aug 20, 37; div. AEROSPACE ENGINEERING, INFORMATION SYSTEMS. *Educ:* Va Polytech Univ BS, 60, MS, 62, PhD(aerospace eng), 66; Harvard Univ, PMD, 73. *Prof Exp:* Aerospace engr, Space Sta Off, Langley Res Ctr, NASA, 60-62, head stability & control sect, 62-70, controls & performance br, 70, stability & control br, 70-71, chief guidance & control br, Hq Washington, DC, 71-73, dir, Guidance, Control & Info Systs Div, 74-76, dir, Electronics Div, 76-78, dir, Space Systs Div, 78-79, ASST DIR RES & TECHNOL, NASA GODDARD SPACE FLIGHT CTR, 79- *Concurrent Pos:* tech consult, Radio Tech Comn for Aeronaut, 74-79; mem guidance & control panel, Wash Opers Res & Mgt Sci Coun, 81- *Honors & Awards:* Inventions & Contrib Award, Langley Res Ctr, NASA, 65 & 67. *Mem:* Assoc fel Am Inst Aeronaut & Astronaut; Aerospace Group Res & Develop. *Res:* Aircraft and spacecraft guidance and control, sensing and data acquisition, data processing and transfer, mission operations, systems analysis, electronics, vehicle design. *Mailing Add:* Code 500 NASA Goddard Space Flight Ctr Greenbelt MD 20771

KURZROCK, RAZELLE, b Toronto, Ont, Sept 29, 54; m 85. LEUKEMIA, MEDICINE. *Educ:* Univ Toronto, BS, 73, MD, 78. *Prof Exp:* ASSOC PROF MED ONCOL & HEMAT, Univ Tex M D ANDERSON CANCER CTR, 89- *Mem:* AMA; AAAS; Am Soc Hemat; Am Soc Clin Oncol; Am Asn Cancer Res. *Res:* Elucidation of the molecular genetic mechanisms responsible for leukemia and the treatment of this disorder. *Mailing Add:* Univ TX M D Anderson Cancer Ctr 1515 Holcombe Blvd Houston TX 77030

KURZWEG, FRANK TURNER, b Plaquemine, La, Aug 7, 17; m 56; c 2. SURGERY. *Educ:* Harvard Univ, SB, 38; Harvard Med Sch, MD, 42; Univ Minn, MS, 47. *Prof Exp:* Instr surg, Med Sch, Tulane Univ, 49-56; from assoc prof to prof, Med Sch, Univ Miami, 56-68; prof, Surg Div, Sch Med, La State Univ, Shreveport, 68-80, head dept & div, 68-76; RETIRED. *Mem:* Am Col Surgeons. *Res:* General, thoracic and vascular surgery. *Mailing Add:* 6632 Sugar Creek Rd S Mobile AL 36609

KURZWEG, ULRICH H(ERMANN), b Jena, Ger, Sept 16, 36; US citizen; m 63; c 1. FLUID MECHANICS, APPLIED MATHEMATICS. *Educ:* Univ Md, BS, 58; Princeton Univ, MA, 59, PhD(physics), 61. *Prof Exp:* Fulbright res grant appl math, Univ Freiburg, 61-62; res scientist physics, United Technol Res Labs, Conn, 62-64, sr theoret physicist, 64-68; assoc prof, 68-76, PROF ENG SCI, UNIV FLA, 76- *Concurrent Pos:* Adj asst & assoc prof, Hartford Grad Ctr, Rensselaer Polytech Inst, 63-68. *Mem:* AAAS; Am Phys Soc; Sigma Xi; NY Acad Sci. *Res:* Hydrodynamic and hydromagnetic stability of rotating flows; thermal instability of electrically conducting fluids; molecular spectroscopy; two-phase magnetohydrodynamic flows; numerical solutions of partial differential equations; optics of solar concentrators; heat exchange and gas separation by high frequency oscillations; fluid mechanics of high frequency pulmonary ventilation. *Mailing Add:* Dept Aerospace Eng Mechanics & Eng Sci Univ Fla Gainesville FL 32611

KUSALIK, PETER GERARD, b Taber, Alta, Can, July 17, 59; m 84; c 2. COMPUTER SIMULATION, LIQUID STATE THEORY. *Educ:* Univ Lethbridge, BSc, 81; Univ BC, MSc, 84, PhD(chem), 87. *Prof Exp:* Vis fel, Res Sch Chem, Australian Nat Univ, 87-89; ASST PROF, DEPT CHEM, DALHOUSIE UNIV, 89- *Concurrent Pos:* NSERC postdoctoral fel, Australian Nat Univ, 87-89, postdoctoral fel, Res Sch Chem, 89; NSERC univ res fel, Dept Chem, Dalhousie Univ, 89- *Mem:* Chem Inst Can. *Res:* Computer simulation and theoretical studies of polar solvents and electrolyte solutions; non-equilibrium molecular dynamics techniques; applied field simulations; dynamics of polar liquids and solutions; solvation. *Mailing Add:* Dept Chem Dalhousie Univ Halifax NS B3H 4J3 Can

KUSANO, KIYOSHI, b Nagasaki, Japan, Feb 1, 33; m 61. NEUROPHYSIOLOGY. *Educ:* Kumamoto Univ, BSc, 56; Kyushu Univ, DSc, 60. *Prof Exp:* Instr physiol, Med Sch, Kumamoto Univ, 56-59 & Tokyo Med & Dent Univ, 59-60; jr res zoologist, Univ Calif, Los Angeles, 60-61; res assoc neurol, Columbia Univ, 61-63; asst prof physiol, Tokyo Med & Dent Univ, 63-65; from asst prof to assoc prof psychiat, Med Sch, Ind Univ, 65-70; assoc prof, 70-73, PROF BIOL, ILL INST TECHNOL, 73- *Concurrent Pos:* Corp mem, Marine Biol Lab, 67; USPHS res grant, Med Sch, Ind Univ, 67-69, NSF grant, 68-70; USPHS res grant, Ill Inst Technol, 71-; mem, Physiol Study Sect, NIH, 79- *Mem:* Am Physiol Soc; Soc Gen Physiol; Soc Neurosci; Biophys Soc. *Res:* Synaptology; comparative neurophysiology. *Mailing Add:* Dept Biol Ill Inst Technol 3101 S Dearborn St Chicago IL 60616

KUSCH, POLYKARP, b Germany, Jan 26, 11; US citizen; m 35, 60; c 5. PHYSICS. *Educ:* Case Inst Technol, BS, 31; Univ Ill, MS, 33, PhD(physics), 36. *Hon Degrees:* DSc, Case Inst Technol, 56, Ohio State Univ, 59, Univ Ill & Colby Col, 61, Gustavus Adolphus Col, 63, Yeshiva Univ, 76, Incarnate Word Col, 80 & Columbia Univ, 83. *Prof Exp:* Asst physics, Univ Ill, 31-36 & Univ Minn, 36-37; instr, Columbia Univ, 37-41; develop engr, Westinghouse Elec & Mfg Co, 41-42; res assoc, Div War Res, Columbia Univ, 42-44; mem tech staff, Bell Tel Labs, Inc, 44-46; from assoc prof to prof physics, Columbia Univ, 46-72, chmn dept, 49-52 & 60-63, exec dir radiation lab, 52-60, vpres & dean fac, 69-70, exec vpres & provost, 70-71; prof, 72-74, Eugene McDermott prof, Syst Chair, 74-80, regental prof, 80-82, EMER REGENTAL PROF PHYSICS, UNIV TEX, DALLAS, 82- *Concurrent Pos:* Consult, IBM Corp, 52-57; fel, Ctr Advan Study in Behav Sci, Stanford Univ, Calif, 64-65. *Honors & Awards:* Nobel Prize in Physics, 55. *Mem:* Nat Acad Sci; Am Acad Arts & Sci; Am Phys Soc; Am Philos Soc; Am Asn Physics Teachers; fel AAAS; Am Asn Univ Profs; Sigma Xi. *Res:* Atomic and molecular beams; application of technique to spectroscopy at electronically generated frequencies and problems in chemical physics; optical molecular spectroscopy. *Mailing Add:* Univ Tex Dallas PO Box 830688 Richardson TX 75083

KUSCHNER, MARVIN, b New York, NY, Aug 13, 19; m 48; c 1. PATHOLOGY. *Educ:* NY Univ, AB, 39, MD, 43. *Prof Exp:* Asst path, Col Med, NY Univ, 47-49; from instr to assoc prof, Col Physicians & Surgeons, Columbia Univ, 49-55; prof, Col Med, NY Univ, 55-70, dir path, Univ Hosp, 68-70; prof path, Health Sci Ctr, 70-87, dean sch med, 72-87, DISTINGUISHED SERV PROF PATH, STATE UNIV NY, STONY BROOK, 87- *Concurrent Pos:* Asst pathologist, Bellevue Hosp, 49-54, actg dir path, 54-55, dir, 55-70; res prof path & environ med, NY Univ, Col Med, 70-; trustee, Assoc Univs, Inc, 80-86; chmn, Environ Health Sci Rev Comt, Nat Inst Environ Health Sci, NIH, 81-; mem, Environ Health Comt, Sci Adv Bd, US Environ Protection Agency, 83-85. *Mem:* Am Asn Path & Bact; Am Soc Exp Pathologists; Am Soc Clin Pathologists; Am Asn Cancer Res; Int Acad Path. *Res:* Pathology of cardiopulmonary diseases. *Mailing Add:* 64 E Gate Dr Huntington NY 11743

KUSERK, FRANK THOMAS, b Philadelphia, Pa, Mar 26, 51; m 75; c 2. MICROBIAL ECOLOGY, ENVIRONMENTAL MICROBIOLOGY. *Educ:* Univ Notre Dame, BS, 73; Univ Del, Newark, PhD(biol), 78. *Prof Exp:* Asst prof, 77-84, ASSOC PROF BIOL, MORAVIAN COL, BETHLEHEM, PA, 84-, DIR, ACAD COMPUT, 88- *Concurrent Pos:* Res fel, Marine Biol Lab, Woods Hole, Mass, 79; res assoc, Acad Natural Sci Philadelphia, 81-82; consult ecol, ITT Res Inst, Chicago, Ill, 84- *Mem:* Am Inst Biol Sci; Ecol Soc Am; Sigma Xi; Nat Cent Sci Educ. *Res:* Microbial ecology; population dynamics and community structure of a group of soil amoebae, the dictyostelid cellular slime molds; uptake and utilization of dissolved organic carbon by streambed micro-organisms. *Mailing Add:* Dept Biol Moravian Col Bethlehem PA 18018

KUSHICK, JOSEPH N, b New York, NY, July 18, 48; m 70; c 2. THEORETICAL BIOPHYSICAL CHEMISTRY. *Educ:* Columbia Col, AB, 69; Columbia Univ, PhD(chem phys), 75. *Hon Degrees:* Am, Amherst Col, 88. *Prof Exp:* Res assoc chem, Univ Chicago, 74-76; from asst prof to assoc prof, 76-88, PROF CHEM, AMHERST COL, 88- *Concurrent Pos:* Fel NSF, 76-78; vis scholar, Harvard Univ, 79-80; Camille and Henry Dreyfus teacher scholar grant, 80; adj assoc prof physiol & biophys, Mt Sinai Sch Med, NY, 86-88, adj prof, 88- *Honors & Awards:* L P Hammett Award, Columbia Univ, 73. *Mem:* Biophys Soc; Am Chem Soc. *Res:* Computer simulation of biological molecules; statistical mechanics. *Mailing Add:* Dept Chem Amherst Col Amherst MA 01002

KUSHIDA, TOSHIMOTO, b Tokyo, Japan, Feb 13, 20; m 46; c 3. SOLID STATE PHYSICS. *Educ:* Hiroshima Univ, BSc, 44, ScD(physics), 56; Harvard Univ, MSc, 56. *Prof Exp:* Asst physics, Hiroshima Univ, 44-48, from instr to prof, 48-61; RES SCIENTIST, SCI LAB, FORD MOTOR CO, 61- *Concurrent Pos:* Res fel, Harvard Univ, 56-58. *Mem:* Fel Am Phys Soc; Phys Soc Japan; Inst Elec & Electronics Engrs; Sigma Xi. *Res:* Nuclear magnetic resonance; high pressure physics; the puli susceptibility of alkali metals as a function of pressure. *Mailing Add:* 22836 Nona Dearborn MI 48124

KUSHINSKY, STANLEY, b Brooklyn, NY, Sept 20, 30; div. BIOANALYTICS, DRUG METABOLISM. *Educ:* City Col New York, BS, 51; Columbia Univ, MA, 52; Univ Boston, PhD(chem), 55; Nat Registry Clin Chem, dipl, 68. *Prof Exp:* Res asst, Worcester Found Exp Biol, 52-55; steroid chemist, Dept Surg, Sch Med, Univ Southern Calif, 55-57, res assoc, 57-58, adj asst prof surg & biochem, 58-62, asst prof, 62-64, asst prof biochem, 64-65; from assoc res biochemist to res biochemist, Dept Obstet & Gynec, Sch Med, Univ Calif, Los Angeles, 65-70; dir biochem res, Rees-Stealy Clin Res Found, 70-79; prin scientist, Syntex Res, 79-81, sr scientist, 82-84, dept head, anal & metab chem, 81-86, asst dir, Inst Pharmacol & Metabol, 84, dir bioanal chem & metabol, 84-86, DIR BIOANALYSIS & METABOL RES, SYNTEX RES, 86- *Concurrent Pos:* USPHS res career develop award, 65-70; adj prof chem, San Diego State Univ, 73-76. *Mem:* Fel AAAS; Am Chem Soc; Am Asn Clin Chem; Nat Acad Clin Biochem; Endocrine Soc; Am Asn Pharmaceut Scientists. *Res:* Synthesis, isolation and metabolism of steroid hormones; enzyme kinetics; betaglucuronidase; gas chromatographic and radioimmunologic determination of steroids in blood and tissues; high performance liquid chromatography; drug metabolism; bioavailability; pharmacokinetics. *Mailing Add:* Syntex Res 3401 Hillview Ave Palo Alto CA 94304

KUSHMERICK, MARTIN JOSEPH, b Pa, May 21, 37; m 62; c 4. RADIOLOGY, BIOPHYSICS. *Educ:* Univ Scranton, BS, 58; Univ Pa, MD, 63, PhD(molecular biol), 66. *Prof Exp:* Asst prof biochem, Univ Pa, 66-67; staff assoc, Lab Phys Biol, Nat Inst Arthritis & Metab Dis, NIH, 67-69; hon res assoc & Brit-Am exchange fel, Am Heart Asn, Univ Col, Univ London, 69-70; asst prof, 70-76, ASSOC PROF PHYSIOL, HARVARD MED SCH, 76-, ASSOC PROF BIOL, HARVARD UNIV, 78- *Concurrent Pos:* NIH Res Career Develop Awards, 76-81. *Mem:* Am Physiol Soc; AAAS; Biochem Soc Eng; Biophys Soc; Soc Gen Physiologists; Am Soc Biol Chem; Soc Mag Res Med. *Res:* Muscle physiology; energetics, metabolism and their control; mechanism and control of contraction; nuclear magnetic resonance. *Mailing Add:* Imaging Res Lab SB-05 Univ Wash Seattle WA 98195

KUSHNARYOV, VLADIMIR MICHAEL, b Odessa, USSR, Jan 2, 31; m 54; c 1. ELECTRON MICROSCOPY. *Educ:* 1st Moscow Med Inst, MD, 54; Acad Med Sci, USSR, PhD(microbiol), 61, DSc, 69. *Prof Exp:* Chief & lab prof, Moscow Inst Vaccines & Serim, 75-77; from asst prof to assoc prof med microbiol, 78-80, DIR, ELECTRON MICROS INSTNL FAC, MED COL WIS, 80-, PROF MICROBIOL, 89- *Concurrent Pos:* Lectr, Moscow Postgrad Med Sch, 60-77. *Mem:* AAAS; Am Soc Microbiol; NY Acad Sci; Int Soc Interferon Res; Electron Micros Soc Am. *Res:* Interaction of biologically active ligands-diptheria toxin, staphycoccal toxic shock syndrome toxin, interferons with mammalian cells; analysis of internalization of ligands by cells, employing quantitative immunocytochemistry, electron microscopy and biochemical techniques. *Mailing Add:* 805 E Henry Clay 106 Milwaukee WI 53217

KUSHNER, ARTHUR S, b New York, NY, May 9, 40; m 64; c 2. ORGANIC CHEMISTRY, BIOLOGICAL INSECT CONTROLS. *Educ:* Univ Evansville, BA, 62; Pa State Univ, PhD(org chem), 66; Cleveland State Univ, MBA, 78. *Prof Exp:* Fel chem, Univ Chicago, 66-68; asst prof, Cleveland State Univ, 68-74; supvr film chem sect, Photohorizons Div, Horizons Res, Inc, Cleveland, 74-75, mgr tech support serv, 75-76; group leader cooling water prod, Mogul Div, Dexter Corp, Chagrin Falls, 76-78, prod mgr, 78-79, sales mgr, 79-80; prod develop mgr, Woodhill Permatex Div, Loctite Corp, Cleveland, 80-81; mkt mgr, Chromatix, 81-84; vpres marketing & sales, 84-88, DIR OPERS, BIOSYS INC, 88- *Concurrent Pos:* Dir, Kushner Electroplating Sch, 78- *Mem:* Am Soc Metals; Am Electroplaters & Surface Finishers Soc; Am Chem Soc; Sigma Xi. *Res:* Synthesis and reactions of bridged polycyclic systems; corrosion inhibition; treatment of cooling water; preparation of new corrosion inhibition materials; photochemistry of free-radical film systems; corrosion inhibition; metal finishing and electroplating; development of biological insect controls. *Mailing Add:* 732 Glencoe Ct Sunnyvale CA 94087

KUSHNER, DONN JEAN, b Lake Charles, La, Mar 29, 27; m 49; c 3. MICROBIAL BIOCHEMISTRY, MICROBIAL PHYSIOLOGY. *Educ:* Harvard Univ, SB, 48; McGill Univ, MSc, 50, PhD(biochem), 52. *Prof Exp:* Asst, Res Inst, Mont Gen Hosp, 52-53; Nat Found Infantile Paralysis fel, 53-54; res officer bact physiol & genetics, Forest Insect Lab, 54-61; assoc res officer, Nat Res Coun Can, 61-65; from assoc prof to prof, 65-89, EMER PROF BIOL, UNIV OTTAWA, 89-; PROF MICROBIOL, UNIV TORONTO, 89- *Concurrent Pos:* With Nat Inst Med Res, London, Eng, 58-59, Inst Pasteur, Paris, 72, MacDonald Col, 80 & Cornell Univ, 81-; co-ed, Can J Microbiol, 83-, Archives of Microbiol & Inst Jacques Monod, Paris, 86-87; vis prof microbiol & environ studies, Univ Toronto, 88- *Honors & Awards:* Ottawa Biol & Biochem Soc Award, 86. *Mem:* Am Soc Biol Chem; Am Soc Microbiol; Can Soc Microbiol (pres, 80-81). *Res:* Physiology of halophilic and psychrophilic bacteria; action of microorganisms in natural environments on polymers and heavy metals; bacterial drug resistance. *Mailing Add:* Dept Microbiol Univ Toronto Toronto ON M5S 1A8 Can

KUSHNER, HAROLD J(OSEPH), b New York, NY, July 29, 33; m 60. MATHEMATICS, OPERATIONS RESEARCH. *Educ:* City Col New York, BSc, 55; Univ Wis, MSc, 56, PhD(elec eng), 58. *Prof Exp:* Staff mem, Lincoln Lab, Mass Inst Technol, 58-63 & Res Inst Advan Studies, Martin-Marietta Corp, Md, 63-64; PROF APPL MATH & ENG, BROWN UNIV, 64-, CHMN, DIV APPL MATH, 88- *Mem:* Inst Math Statist; Soc Indust & Appl Math; Opers Res Soc Am; Inst Elec & Electronics Engrs. *Res:* Theoretical study of automatic control and communication systems, especially when random phenomenon are of some significance; applied probability; stochastic systems theory. *Mailing Add:* Dept Appl Math Brown Univ Providence RI 02912

KUSHNER, HARVEY, b Philadelphia, Pa, Nov 2, 50; m 73; c 4. MATHEMATICAL BIOMEDICAL STATISTICS. *Educ:* Temple Univ, AB, 72, MA, 74, PhD(math), 78. *Prof Exp:* PROF MATH & BIOSTATIST & CHMN, DEPT BIOMET, HAHNEMANN UNIV, 78- *Concurrent Pos:* Adj assoc prof, Dept Math, Temple Univ, 79- & Grad Sch, Med Col Pa, 80-81 & 85-86; treas, Biomed Comput Res Inst. *Mem:* Am Math Soc; Am Statist Asn. *Res:* Time series analysis; categorical data analysis. *Mailing Add:* 9743 Redd Rambler Rd Philadelphia PA 19115

KUSHNER, HARVEY D(AVID), b New York, NY, Dec 28, 30; m 51; c 3. OPERATIONS RESEARCH. *Educ:* Johns Hopkins Univ, BE, 51. *Prof Exp:* Engr, Mach Evals Group, Bur Ships, US Navy Dept, 51 & Performance & Sci Sect, 52-53; mem tech staff, Cent Res Lab, Melpar, Inc Div, Westinghouse Air Brake Co, 53-54 & Flight Simulator Dept, 54-55; res engr, Reliance Group Inc, 55-57, group leader, 57-59, prog dir, 59-61, vpres & dir, Eastern Div, 61-62, vpres & dir, Phys Systs Div, 63-64, dir, Govt & Indust Systs Div, 64-68, exec vpres, 62-69, vpres, 71-77; pres, Disclosure, Inc, 72-77; pres, ORI Inc, 69-83, chmn & chief exec officer, 83-85, CHMN & CHIEF EXEC

OFFICER, THE ORI GROUP, INC, 85- *Concurrent Pos:* Consult, Appl Physics Lab, Johns Hopkins Univ, 57-58 & Nat Acad Sci Comt Undersea Warfare, 63-64. *Mem:* Opers Res Soc Am; Inst Mgt Sci; Am Inst Aeronaut & Astronaut; sr mem Inst Elec & Electronics Engrs; Am Soc Mech Engrs; fel NY Acad Sci. *Res:* Systems analysis; operations research. *Mailing Add:* Kushner Mgt Planning Corp 9607 Kingston Rd Kensington MD 20895

KUSHNER, IRVING, b New York, NY, Jan 16, 29; m 55; c 3. RHEUMATOLOGY, MEDICINE. *Educ:* Columbia Univ, BA, 50; Wash Univ, MD, 54. *Prof Exp:* Intern med, New Haven Hosp, 54-55; asst resident, 2 & 4 med serv, Boston City Hosp, 57-58; demonstr med, Case Western Reserve Univ, 58-59, instr, 60-61, sr instr, 61-64, from asst prof to assoc prof, 64-73; prof, WVa Univ, 73-74; PROF MED, SCH MED, CASE WESTERN RESERVE UNIV, 74- *Concurrent Pos:* USPHS res fel, 58-59; Helen Hay Whitney Found res fel, 59-62; foreign fel, Inst Sci Res Cancer, France, 62-63; sr int fel, Fogarty Ctr, NIH, 76; med dir, Metro Health Ctr Rehab, 85- *Mem:* Am Col Physicians; Am Rheumatism Asn; Am Fedn Clin Res; Soc Exp Biol & Med; NY Acad Sci. *Res:* Acute phase reaction; C-reactive protein. *Mailing Add:* Dept Med Metro Health Med Ctr Cleveland OH 44109

KUSHNER, LAWRENCE MAURICE, b New York, NY, Sept 20, 24; m 72; c 2. MATERIALS SCIENCE, TOXIC & HAZARDOUS SUBSTANCE REGULATION. *Educ:* Queens Col, BS, 45; Princeton Univ, AM, 47, PhD, 49. *Prof Exp:* Teaching asst, Princeton Univ, 47-48; staff mem, Nat Bur Standards, 48-56, chief, Metal Physics sect, 56-61 & Metall Div, 61-66, dep dir, Inst Appl Tech, 66-68, dir, Inst for Appl Technol, 68-69, dep dir, Bur, 69-73, actg dir, 72-73; comnr, Consumer Prod Safety Comn, 73-77; coordr policy develop, Nat Bur Standards, 77-80; sr staff scientist, Mitre Corp, 80-85, consult scientist, 85-89. *Concurrent Pos:* Lectr chem, Am Univ, 52-60; mem ad hoc int group metal physics, Org Econ Coop & Develop, 61; fel, Sci & Technol Fel Prog, Dept Com, 64-65; mem, Md Gov Sci Adv Coun, 72-75; adj prof eng & pub policy, Carnegie-Mellon Univ, 81- *Honors & Awards:* Students Medal, Am Inst Chemists, 45; Gold Medal, Dept Com, 68; Meritorious Serv Award, Am Nat Students Inst, 73. *Mem:* Fel AAAS; hon mem Am Soc Testing & Mat; Am Chem Soc; Am Phys Soc; Sigma Xi (pres, 76). *Res:* Physical chemistry of surface active agents; relationship between physical properties of materials and their molecular and crystal structures. *Mailing Add:* 9528 Briar Glenn Way Gaithersburg MD 20879

KUSHNER, SAMUEL, b Auburn, NY, Apr 25, 15; m 37; c 4. PHARMACEUTICAL CHEMISTRY. *Educ:* Univ Mich, BS, 39, MS, 40, PhD(chem), 42. *Prof Exp:* Res chemist, Am Cyanamid Co, 42-44, group leader, 44-55, unit leader, 55-56, head dept med chem of infectious dis, 56-71, head chem of infectious res dis sect, Lederle Labs, Am Cyanamid Co, 71-81; RETIRED. *Mem:* AAAS; Am Chem Soc; NY Acad Sci. *Res:* Structure of penicillin; chemotherapy related to tropical diseases; tuberculosis and virus; structure and synthesis of antibiotics; antineoplastics. *Mailing Add:* 138 Highview Ave Nanuet NY 10954

KUSHNER, SIDNEY RALPH, b New York, NY, Dec 14, 43; m; c 2. MOLECULAR GENETICS, ENZYMOLOGY. *Educ:* Oberlin Col, BA, 65; Brandeis Univ, PhD(biochem), 70. *Prof Exp:* NIH fel molecular biol, Univ Calif, Berkeley, 70-71; NIH fel biochem, Med Sch, Stanford Univ, 71-73, from asst prof to assoc prof biochem, 73-80, assoc prof, 80-82, PROF GENETICS, UNIV GA, 82-, HEAD DEPT, 87- *Concurrent Pos:* NIH res career develop award, 75; assoc ed, Gene, 80-; Microbiol Genetics Study Sect, NIH, 81-85; bd dirs, Am Type Cult Collection, 88- *Mem:* Am Soc Microbiol; AAAS; Am Soc Biol Chemists; Genetics Soc Am. *Res:* Genetic control and enzymology of recombination and DNA repair; functional expression of eukaryotic DNA in prokaryotes; maintenance of eukaryotic DNA in prokaryotes; analysis of messenger RNA degradation. *Mailing Add:* Dept Genetics Univ Ga Athens GA 30602

KUSHNICK, THEODORE, b Brooklyn, NY, Mar 29, 25; m 49; c 3. PEDIATRICS, MEDICAL GENETICS. *Educ:* Ohio State Univ, BS, 44, MS, 47; Harvard Med Sch, MD, 51. *Prof Exp:* Intern med, Boston City Hosp, 51-52; resident pediat, Boston Children's Med Ctr, 52-53, 54-55; clin res asst, Boston Children's Cancer Res Found, 55-56; pvt pract, NJ, 56-59; clin instr pediat, Col Med & Dent NJ, Newark, 58-59, from asst prof to prof, 59-79, dir, div human genetics, 61-79; PROF PEDIAT & DIR MED GENETICS, SCH MED, E CAROLINA UNIV, 79- *Concurrent Pos:* Jr physician, Wrentham State Sch, Mass, 53; resident psychiat, Boston State Hosp, 53-54; clin instr, Harvard Med Sch, 55-56; genetic consult, Nat Found March of Dimes Spec Birth Defects Treatment Ctr, Babies Hosp, Newark; Mead-Johnson res grant, 60-62; NIH res grant, 61-63; Nat Found-March of Dimes Med Serv Prog Grants, 74-75 & 76-77; Sci Adv Comm, Nat Tay-Sachs & Allied Dis Asn, 74- *Mem:* AAAS; Am Asn Ment Deficiency; fel Am Acad Pediat; Am Soc Human Genetics; NY Acad Sci. *Res:* Clinical pediatrics; clinical genetics mental retardation; cytogenetics; immunology. *Mailing Add:* Dept Pediat Sch Med E Carolina Univ Greenville NC 27858-4354

KUSHWAHA, RAMPRATAP S, b India, July 11, 43. LIPID METABOLISM. *Educ:* Wash State Univ, PhD(nutrit), 73. *Prof Exp:* SCIENTIST, SOUTHWEST FOUND BIOMED RES, 82- *Concurrent Pos:* Adj assoc prof, Dept Path, Univ Tex Health Sci Ctr, 82- *Mem:* Am Heart Asn; Am Inst Nutrit; Med & Health Sci; Biol Sci. *Res:* Metabolic and molecular basis of genetic dyslipoproteinemias in pedigreed baboons; metabolic and molecular mechanisms by which dietary factors such as cholestrol and sex steroid hormones modulate lipoprotein metabolism in normal and dyslipoproteinemic subjects. *Mailing Add:* Dept Physiol & Med Southwest Found Biomed Res San Antonio TX 78284

KUSIAK, ANDREW, b Kozia Wola, Poland, June 14, 49; Can citizen; m 74; c 3. ENGINEERING DESIGN, MANUFACTURING & ARTIFICIAL INTELLIGENCE CONCURRENT ENGINEERING. *Educ:* Warsaw Tech Univ, BS, 72, MS, 74; Polish Acad Sci, PhD(oper res), 79. *Prof Exp:* Proj mgr, Dept Automation, Inst Mgt & Org, 79-81; asst prof automation, Tech Univ, Nova Scotia, 82-85; assoc prof automation, Univ Man, Can, 85-88; PROF & CHAIR ARTIFICIAL INTEL, UNIV IOWA, 88- *Concurrent Pos:* Ed, Artificial Intel in Indust, 86-, Appl Artificial Intel, 88-; res award, Univ Man, 86; chmn, Int Conf Adv Prod, 87, Int Conf Artificial Intel, 90; vis prof, Inst Adv Studies, Vienna, Austria, 91. *Mem:* Sr mem Soc Mfg Engrs; Oper Res Soc Am; sr mem Am Asn Artificial Intel; Inst Indust Engrs; Int Fedn Automation & Control; Int Fedn Info Processing. *Res:* Knowledge-based systems for design of products and manufacturing systems; group technology; design of facilities; process planning; concurrent engineering; design automation; design methodologies. *Mailing Add:* Dept Indust Eng Univ Iowa Iowa City IA 52242

KUSIC, GEORGE LARRY, JR, b Aliquippa, Pa, Aug 26, 35; m 69; c 1. CONTROL ENGINEERING, COMPUTER SCIENCE. *Educ:* Carnegie Inst Technol, BSEE, 57, MSEE, 66, PhDEE, 68. *Prof Exp:* Res engr, Sikorsky Aircraft Co, 57-59; elec develop engr, TRW Corp, 59-63; asst prof, 67-77, ASSOC PROF ELEC ENG & GRAD PROG COORDR, UNIV PITTSBURGH, 77- *Concurrent Pos:* NASA-Am Soc Eng Educ fac res fel, 69; sr Fulbright-Hays lect grant, Univ Belgrade, 70-71; consult, NSF-Agency Int Develop India Prog, 68, IBM Data Processing Div, 68 & Westinghouse Res Lab, 69-70. *Mailing Add:* Elec Eng & Benedum 443 Univ Pittsburgh 4200 Fifth Ave Pittsburgh PA 15260

KUSIK, CHARLES LEMBIT, b New York, NY, Apr 24, 34. WASTE MINIMIZATION, RECYCLING. *Educ:* Mass Inst Technol, BS, 56; NY Univ, DSc(chem eng), 61. *Prof Exp:* Scientist opers res, Mass Inst Technol, 61-62; scientist gas dynamics, Avco Corp, 63-64; MNR metals & energy mgt, 64-88, DIR TECHNOL & PROD DEVELOP, ARTHUR D LITTLE, INC, 89- *Mem:* Am Inst Chem Engrs; Am Chem Soc; Am Inst Mining, Metall & Petrol Engrs. *Res:* Energy assessments; recycling; process development; economics; commercial feasibility studies. *Mailing Add:* Arthur D Little Inc 20 Acorn Park Cambridge MA 02138

KUSKA, HENRY (ANTON), b Chicago, Ill, July 28, 37; m 64; c 3. PHYSICAL CHEMISTRY. *Educ:* Cornell Col, BA, 59; Mich State Univ, PhD(phys chem), 65. *Prof Exp:* Res assoc & res fel phys chem, Mich State Univ, 64-65; ASSOC PROF PHYS CHEM, UNIV AKRON, 65- *Mem:* Am Chem Soc. *Res:* Spectroscopy; nuclear magnetic resonance; electron spin resonance; electron-nuclear double resonance; infrared, visible ultraviolet. *Mailing Add:* Dept Chem Univ Akron Akron OH 44325

KUSKO, ALEXANDER, b New York, NY, Apr 4, 21; m 41; c 2. ELECTRICAL ENGINEERING. *Educ:* Purdue Univ, BS, 42; Mass Inst Technol, SM, 44, ScD, 51. *Prof Exp:* Asst, 42-44, from instr to assoc prof, 46-58, LECTR ELEC ENG, MASS INST TECHNOL, 58-; DIV DIR, FAILURE ANALYSIS ASSOC, 88- *Concurrent Pos:* Pres, Alexander Kusko Inc, 56-88. *Mem:* Inst Elec & Electronics Engrs. *Res:* Energy conversion and control. *Mailing Add:* Failure Analysis Assoc 115 Flanders Rd Westborough MA 01581

KUSLAN, LOUIS ISAAC, b New Haven, Conn, Feb 14, 22; m 47; c 2. HISTORY OF SCIENCE. *Educ:* Univ Conn, BS, 43; Yale Univ, MA, 49, PhD(sci educ), 54. *Prof Exp:* Instr high schs, Conn, 43-46; asst chem, Univ Conn, 46-47, instr, Waterbur Br, 47-49; from instr to asst prof sci, Southern Conn State Univ, 50-56, assoc prof chem, 56-60, prof chem & chmn dept sci, 60-66, dean arts & sci, 66-78, prof chem, 78-88; RETIRED. *Concurrent Pos:* Fel chem, Yale Univ, 58-59, hist sci, 62-63. *Mem:* Am Chem Soc; Hist Sci Soc. *Res:* History of analytical and American chemistry; elementary science education; nineteenth century American chemistry. *Mailing Add:* 90 Robin Wood Rd Hampden CT 06157

KUSNETZ, HOWARD L, b New York, NY, Nov 2, 29; m 50; c 2. INDUSTRIAL HYGIENE, INDUSTRIAL SAFETY. *Educ:* New York Univ, BA, 49; Columbia Univ, MSc, 50; Univ Cincinnati, BSc, 62. *Prof Exp:* Various positions, US Pub Health Serv, 51-64, dir eng, 64-71; mgr, indust hygiene, 71-75, MGR, SAFETY & INDUST HYGIENE, SHELL OIL CO, 75- *Concurrent Pos:* Vis prof, Sch Med, Hebrew Univ, Jerusalem, 67-; adj prof, Sch Pub Health, Univ Tex, 72-; chmn, Tex Occup Safety Bd, 81-85,. *Mem:* Am Bd Indust Hygiene, (secy, 81-83); Am Acad Indust Hygiene; fel AAAS; Am Indust Hygiene Asn (vpres, 83-84, pres-elect, 84-85, pres, 85-86). *Res:* Methods of prevention and control of chemical and physical factors in work environments which may lead to ilness or injury. *Mailing Add:* Shell Oil Co PO Box 4320 Houston TX 77210-4320

KUSPIRA, J, b Yorkton, Sask, Nov 20, 28; m 58; c 4. CYTOGENETICS. *Educ:* Univ Sask, BSc, 51, MSc, 52; Univ Alta, PhD(genetics), 55. *Prof Exp:* Asst cytogeneticist, 55-57, assoc res prof cytogenetics, 58-62, assoc prof, 62-70, PROF GENETICS, UNIV ALTA, 70-, ASSOC DEAN SCI, 72- *Mem:* Am Genetic Asn; Genetics Soc Can. *Res:* Cytogenetic analysis of tetraploid and hexaploid wheats. *Mailing Add:* Dept Genetics Univ Alta Edmonton AB T6G 2M7 Can

KUSSE, BRUCE RAYMOND, b Rochester, NY, Aug 10, 38. PLASMA PHYSICS. *Educ:* Mass Inst Technol, SB, 60, SM, 64, PhD(elec eng), 69. *Prof Exp:* Sr scientist, Eastern Sci & Technol Div, EG&G, 69-70; res assoc plasma physics, Res Lab Electronics, Mass Inst Technol, 70; res assoc, Lab Plasma Studies, 70-71, asst prof, 71-76, ASSOC PROF PLASMA PHYSICS, CORNELL UNIV, 76- *Mem:* Sigma Xi; Am Phys Soc. *Res:* Plasma physics-experimental studies of intense, relativistic beam-plasma interactions, particularly in toroidalgeometry. *Mailing Add:* 144 N Sunset Dr Ithaca NY 14850

KUSSMAUL, KEITH, b Sterling, Ill, Apr 9, 39; m 65; c 4. DESIGN & ANALYSIS OF EXPERIMENTS, STATISTICAL PROCESS CONTROL. *Educ:* Univ Mich, BS, 60, MS, 61; NC State Univ, PhD(statist), 66. *Prof Exp:* Mathematician, Int Bus Mach Corp, 61-62; STATISTICIAN, WESTINGHOUSE ELEC CORP, 66- *Concurrent Pos:* Lectr indust eng, Univ Pittsburgh, 67-70, vis lect prog, Soc Indust Appl Math, 77-83. *Mem:* Am Statist Asn; Biomet Soc; Am Soc Qual Control. *Res:* Design and analysis of industrial experiments; statistical methods; general linear hypothesis; general statistical consulting. *Mailing Add:* Math Dept Westinghouse Sci & Technol Ctr Pittsburgh PA 15235

KUST, ROGER NAYLAND, b Berwyn, Ill, Apr 20, 35; m 57; c 2. INORGANIC CHEMISTRY, PHYSICAL CHEMISTRY. *Educ:* Purdue Univ, BS, 57; Iowa State Univ, PhD(fused salts), 63. *Prof Exp:* Asst prof inorg chem, Tex A&M Univ, 64-65 & Univ Utah, 65-71; sr inorg chemist, Ledgemont Lab, Kennecott Copper Corp, 71-77, group leader chem, 77-78, sect head chem, 78-79; sr staff engr, Exxon Minerals Co, 79-80, sect head, 80-81, mgr, Minerals Processing Res Div, 81-87; MGR METALS RECOVERY, TETRA TECHNOLOGIES, INC, 87- *Mem:* AAAS; Am Chem Soc; Electrochem Soc; Am Acad Arts & Sci; Am Inst Chemists; Metall Soc. *Res:* Acid-base reactions in fused salts; electrochemical investigations in nonaqueous media with emphasis on fused salts; chemistry of metallurgical processes. *Mailing Add:* 8408 Crescent Wood Lane Spring TX 77379

KUSTIN, KENNETH, b Bronx, NY, Jan 6, 34; m 56; c 3. INORGANIC CHEMISTRY, PHYSICAL CHEMISTRY. *Educ:* Queens Col, NY, BSc, 55; Univ Minn, Minneapolis, PhD(inorg chem), 59. *Prof Exp:* USPHS fel, Max Planck Inst Phys Chem, 59-61; from asst prof to assoc prof, 61-72, chmn dept, 74-77, PROF CHEM, BRANDEIS UNIV, 72- *Concurrent Pos:* Vis prof, Dept Pharmacol, Harvard Med Sch, 77-78; Fulbright lectr, 78; prog dir, NSF, 85-86. *Mem:* Am Chem Soc. *Res:* Inorganic biochemistry; oscillating reactions; fast reactions. *Mailing Add:* Dept Chem Brandeis Univ Box 9110 Waltham MA 02254-9110

KUSTOM, ROBERT L, b Chicago, Ill, July 11, 34; c 3. ION ACCELERATION & FOCUSING, POWER ELECTRONIC NETWORKS. *Educ:* Ill Inst Technol, BSEE, 56, MSEE, 58; Univ Wis-Madison, PhD(elec eng), 69. *Prof Exp:* Elec engr particle detect develop, Accelerator Res Facil Div, Argonne Nat Lab, 58-69, elec engr radio frequency separators & microwave discharge chambers, 69-71, group leader zero gradient synchrotron operations, 71-73, assoc div dir, Plasma Support Syst, Tokamaks & Accelerator Exp Area, 73-78, mgr accelerator syst intense pulsed neutron source, 78-79, div dir accelerator res & develop, 79-81, assoc proj dir electron accelerator, 81-83, sr elec engr, Seven Giga Electronvolt Storage Ring-Advan Photon Source Proj, 83-88, interim dir, Advan Photon Source Accelerator Syst Div, 89-90, SR ELEC ENGR, ADVAN PHOTON SOURCE, GROUP LEADER, RF GROUP, ARGONNE NAT LAB, 90- *Concurrent Pos:* Vis scientist, Rutherford High Energy Lab, Didcot, UK, 70-71; Tokamak Fusion Test Reactor eng rev comt, Princeton Plasma Physics Lab, 75-77; vis prof, Elec & Comput Eng Dept, Univ Wis-Madison, 78-79 & 80-81, consult, Superconductive Energy Storage Group, 81, adj prof, Elec & Computer Eng Dept, 83- *Mem:* Inst Elec & Electronics Engrs; Sigma Xi. *Res:* Development of ion acceleration, focussing, and detection techniques, and the electrodynamic interactions between ions and electromagnetic fields; theoretical and experimental development of superconductive energy storage and transfer techniques using power electronic circuits and electronic circuits and electrodynamic devices. *Mailing Add:* Argonne Nat Lab 9700 S Cass Ave Argonne IL 60439

KUSTU, SYDNEY GOVONS, b Baltimore, Md, Mar 18, 43. BACTERIOLOGY, GENETICS. *Educ:* Harvard Univ, BA, 63; Univ Calif, Davis, PhD(biochem), 70. *Prof Exp:* Asst prof, 74-80, ASSOC PROF BACT, UNIV CALIF, DAVIS, 80- *Mem:* AAAS; Am Soc Microbiol; Am Soc Biol Chemists; Sigma Xi. *Res:* Regulation of bacterial nitrogen metabolism. *Mailing Add:* Dept Microbiol & Immunol Univ Calif Berkeley CA 94720

KUSUDA, TAMAMI, b Seattle, Wash, June 24, 25; m 55; c 3. MECHANICAL ENGINEERING. *Educ:* Univ Tokyo, BS, 47; Univ Wash, Seattle, MS, 52; Univ Minn, PhD(mech eng), 55. *Prof Exp:* Staff engr, Worthington Corp, 55-62; mech engr, Nat Bur Standards, 62-70, asst chief environ eng sect, 70-74, chief thermal eng sect, 74-78, chief thermal anal prog, 78-83, chief bldg physics div, 83-86; CONSULT, US DEPT COM, OFF JAPANESE & TECH LIT, 87- *Concurrent Pos:* Prof lectr, George Wash Univ, 87- *Honors & Awards:* Wolverline Award, Am Soc Heating, Refrig & Air-Conditioning Engrs, 57; Crosby-Field Award, Am Soc Heating, Refrig & Air Conditioning Engrs,; Gold Medal, US Dept Com, 80. *Mem:* Am Soc Mech Engrs; fel Am Soc Heating, Refrig & Air-Conditioning Engrs; hon mem Automated Procedure Eng Consult; AAAS. *Res:* Heat transfer and thermodynamics related to environmental science, such as air conditioning, heating, ventilating, refrigeration, ground heat exchange and psychrometrics; energy conservation; indoor air quality. *Mailing Add:* Dept Com Nat Inst Standards & Technol Bldg 226 Rm B307 Gaithersburg MD 20899

KUSWA, GLENN WESLEY, b Milwaukee, Wis, Dec 11, 40; wid; c 1. EXPERIMENTAL PHYSICS, TECHNOLOGY TRANSFER. *Educ:* Univ Wis-Madison, BS, 62, MS, 64, PhD(physics), 70. *Prof Exp:* Physicist, Sandia Labs, 70-74; physicist, Laser & Isotope Separation Technol Off, US Energy Res & Develop Admin, 74-76; mgr, Particle Beam Fusion Res Dept, Sandia Labs, 76-81; tech adv asst secy defense progs, US Dept Energy, 82-83; mgr future options planning, 84-85, MGR TECHNOL TRANSFER DEPT, SANDIA LABS, 86-, SUPVR, PLASMA DIAG, 88- *Mem:* Am Phys Soc; AAAS; Sigma Xi. *Res:* Plasma guns; measurement of distribution functions; holographic interferometry; production of dense plasmas; electrical breakdown in vacuum; interaction of electron and ion beams with matter; inertially driven fusion technology using lasers, ions, electron beams; large scale research management; defense science technology transfer and licensing. *Mailing Add:* 1115 San Rafael Ave NE Albuquerque NM 87112-1130

KUSY, ROBERT PETER, b Worcester, Mass, Oct 19, 47; m 69; c 2. DENTAL RESEARCH, MEDICAL RESEARCH. *Educ:* Worcester Polytech Inst, BS, 69; Drexel Univ, MS, 71, PhD(mat eng), 73. *Prof Exp:* Res asst mats, Dept Metall Eng, Drexel Univ, 69-72; res assoc, Med Sch, Univ NC, 72-74, asst prof oral biol, 74-79, assoc prof orthod & Dent Res Ctr, Dent Sch, 79-89, assoc prof biomed eng, Med Sch, 85-89, PROF DEPT ORTHOD, DENT RES CTR & BIOMED ENG, 89-, ADJ PROF APPL SCI CURRIC, SCH MED UNIV NC, 90- *Concurrent Pos:* Co-investr, Duke-NC Eng Res Ctr, 88- *Mem:* Am Soc Metals; Am Chem Soc; Soc Plastics Engr; NAm Thermal Anal Soc; Int Asn Dent Res; Int Metallog Soc; Microbeam Anal Soc; Soc Biomat. *Res:* Properties of dental and medical materials; fractography and fracture work energy of polymers; fabrication of high strength/high modulus fibers; laser scattering experiments; thermal analysis and radiation properties of polymers; biosensors; ion implantation of dental and medical materials; failure analysis. *Mailing Add:* Bldg No 210H Univ NC Chapel Hill NC 27599-7455

KUSZAK, JEROME R, b May 26, 51. PATHOLOGY. *Educ:* Wayne State Univ, BS, 72, MS, 76, PhD(anat), 80. *Prof Exp:* Instr anat, cell biol & path, Rush Med Col, 80-83, asst prof path, Conjoint Appointment Anat, 83-87; DIR ELECTRON MICROS, DEPT PATH, RUSH-PRESBY, ST LUKE'S MED CTR, 83- *Concurrent Pos:* New investr award, Nat Eye Inst, NIH, 83-86; assoc prof path, Conjoint Appointment Anat, Rush Med Col, 87- , assoc prof ophthal, 90- *Honors & Awards:* Alcon Res Inst Award, 90. *Mem:* Inst Soc Eye Res; Asn Res Vision & Ophthal; Am Soc Cell Biol. *Res:* Pathology. *Mailing Add:* Dept Path Rush-Presby St Luke Med Ctr 1653 W Congress Pkwy Chicago IL 60612

KUTAL, CHARLES RONALD, b Chicago, Ill, Aug 9, 44; m 73. PHOTOCHEMISTRY. *Educ:* Knox Col, Ill, AB, 65; Univ Ill, Urbana-Champaign, MS, 68, PhD(chem), 70. *Prof Exp:* Res assoc chem, Univ Southern Calif, 70-72; from instr to assoc prof, 73-85, PROF CHEM, UNIV GA, 85- *Concurrent Pos:* Res fel, Nat Res Coun-Nat Acad Sci, 72-73; vis scientist, IBM Res Labs, 86. *Honors & Awards:* Res Award, Sigma Xi, 79. *Mem:* Am Chem Soc; AAAS; Sigma Xi. *Res:* Photochemical and photophysical investigations of transition metal and organometallic complexes, development of solar energy storage systems, photolithography. *Mailing Add:* Dept Chem Univ Ga Athens GA 30601

KUTAS, MARTA, b Hungary, Sept 2, 49; US citizen. PSYCHOLOGY, PSYCHOPHYSIOLOGY. *Educ:* Oberlin Col, BA, 71; Univ Ill, Urbana-Champaign, MA, 74, PhD(biol psychol), 77. *Prof Exp:* Vis res assoc, Dept Psychol, Univ Ill, 77-78; res neuroscientist, 78-80, asst res neuroscientist, 80-84, ASSOC RES NEUROSCIENTIST, UNIV CALIF, SAN DIEGO, 85- *Honors & Awards:* Early Career Contribution Psychol Award, Am Psychiat Asn. *Mem:* Soc Psychophyisol Res; Int Neuropsychol Soc; Women in Neurosci. *Res:* Brain function, including recording and interpreting pattern of brain waves (event related potentials) from the scalp as humans try to comprehend the oral, written or pictorial world. *Mailing Add:* Dept Neurosci M-008 Univ Calif Sch Med La Jolla CA 92093

KUTCHAI, HOWARD C, b Detroit, Mich, Feb 21, 42; m; c 1. PHYSIOLOGY, BIOCHEMISTRY. *Educ:* Univ Mich, BS, 63; Univ Calif, San Francisco, PhD(physiol), 67. *Prof Exp:* NIH trainee, Univ Mich; fel, Univ Olso , 69-70 & Johns Hopkins Univ, 70-72; from asst prof to assoc prof, 72-81, PROF PHYSIOL, SCH MED, UNIV VA, 81- *Concurrent Pos:* Assoc ed, Biophys J, 88- *Mem:* AAAS; Am Physiol Soc; Biophys Soc; Soc Gen Physiol. *Res:* Function of the calcium-ATP ase of sarcoplasmic reticulum, diffusion boundary layers; oxygen transport in red blood cells; influence of membrane lipids on transport processes; biophysics. *Mailing Add:* Dept Physiol Univ Va Charlottesville VA 22908

KUTCHER, STANLEY PAUL, b Toronto, Can, Dec 16, 51; m; c 3. ADOLESCENT AFFECTIVE DISORDERS, PSYCHOPHARMACOLOGY. *Educ:* McMaster Univ, BA, 74, MA, 75, MD, 79; Royal Col Physicians & Surgeons, FRCP, 83. *Prof Exp:* Assoc prof psychiat & phys rehab med, Univ Toronto, 86-91, asst prof, Sch Grad Studies, 87-88; head, Div Adolescent Psychiat, 86-91, DIR ADOLESCENT PSYCHIAT SERVS, SUNNYBROOK MED CTR. *Concurrent Pos:* Vis clin scientist psychiat, Med Res Coun Gt Brit, 83-84. *Mem:* Am Psychiat Asn; Can Phys Asn; Can Col Neuropharmacol. *Res:* Psychopharmacology of adolescent affective disorders. *Mailing Add:* Dept Psychiat Univ Toronto Sunnybrook Med Ctr 2075 Bayview Ave Toronto ON M4N 3M5 Can

KUTCHES, ALEXANDER JOSEPH, animal nutrition; deceased, see previous edition for last biography

KUTIK, LEON, b New York, NY, Mar 6, 27; m 63; c 3. ORGANIC CHEMISTRY, TECHNICAL MANAGEMENT. *Educ:* City Col New York, BS, 49; Univ Chicago, MBA, 73. *Prof Exp:* Chemist, Clover Leaf Paint & Varnish Corp, 51-55; chemist, Cent Res Labs, Interchem Corp, 55-57, sr chemist, 57-59, group leader, 59, asst dept dir appl res finishes & adhesives, 59-63, prog mgr, 63-67, mgr graphic sci, 67-68; tech dir chem coatings, De Soto Inc, 68-70, mgr resin develop, Chem Coatings Div, 70-73, mgr indust res, 73-77, tech mgr construct coatings, 77-80, mgr mfg & tech serv, Sherwin Williams Corp, 80-83; MGR, COM APPLN LAB, W R GRACE & CO, DAVISON DIV TECH CTR, 83- *Honors & Awards:* Roy H Kienle Award, 65. *Mem:* Am Chem Soc; Soc Paint Technol; Nat Asn Corrosion Engrs. *Res:* Organic coatings for metal, paper, fiberboard, plywood, wood and plastics; adhesives for packaging and structural applications; methods of application for industrial coatings; powder coatings; matting agents; corrosion inhibiting pigments. *Mailing Add:* 704 St Paul Ave Reisterstown MD 21136

KUTILEK, MICHAEL JOSEPH, b Baltimore, Md, July 1, 43; m 68; c 1. WILDLIFE ECOLOGY. *Educ:* San Diego State Univ, BS, 66, MS, 68; Mich State Univ, PhD(fisheries & wildlife), 75. *Prof Exp:* Res technician biol, Calif Dept Fish & Game, 65; wildlife biologist, Kenya Nat Parks, US Peace Corps, 69-71; asst prof, 75-80, ASSOC PROF BIOL, SAN JOSE STATE UNIV, 80- *Mem:* Wildlife Soc. *Res:* Foraging strategies of herbivorous large mammals of Africa particularly, grazing and browsing ungulates; ecological and evolutionary aspects of plant-herbivore interactions. *Mailing Add:* Dept Biol Sci San Jose State Univ San Jose CA 95192

KUTKUHN, JOSEPH HENRY, b Weehawken, NJ, Mar 28, 27; m 53; c 5. ECOLOGY, FISHERIES. *Educ:* Colo State Univ, BS, 53; Iowa State Univ, MS, 54, PhD(fishery mgt), 56. *Prof Exp:* Fishery biologist, Dept Fish & Game, Calif, 56-58; asst lab dir, Tex, 58-65, NC, 65-69, Mich, 72-75, lab dir Mich, 76-82, ASSOC DIR, FISHERY RESOURCES, US FISH & WILDLIFE SERV, DEPT INTERIOR, WASHINGTON, DC, 83-

Concurrent Pos: Consult, UN Develop Prog, Food & Agr Orgn, Fishery Res & Develop Proj, Lima, Peru, 70-71. *Mem:* Am Fisheries Soc; Am Inst Fishery Res Biol; Sigma Xi. *Res:* Dynamics of exploited fish and shellfish resources. *Mailing Add:* Rte 4 Box 4512A Grayling MI 49738

KUTLER, BENTON, b Council Bluffs, Iowa, May 21, 20; m 44; c 5. DENTISTRY. *Educ:* State Univ Iowa, BA, 42; Creighton Univ, DDS, 45. *Prof Exp:* Instr prev med, Col Med, Univ Nebr Med Ctr, 45-65, clin asst prof oral surg & gen dent, 65-91, clin asst prof community dent & preceptor, Col Dent, 79-91; RETIRED. *Concurrent Pos:* Preceptor trainer, Creighton Univ Sch Dent, 71-78, vis lectr, 72-80. *Mem:* Sigma Xi; fel AAAS; fel Int Col Dentists; fel Am Col Dentists; fel Acad Gen Dent; Am Dent Asn; Am Soc Dent for Children; Int Asn Dent Res; Am Asn Endodontists; Am Acad Implant Dent; Int Cong Oral Implantologists. *Res:* Fluoride applications for adult caries; caries research methods. *Mailing Add:* 9909 Essex Dr Omaha NE 68114

KUTNER, ABRAHAM, b Lynn, Mass, Mar 28, 19; m 47; c 3. ORGANIC POLYMER CHEMISTRY, PHOTOCHEMISTRY. *Educ:* Ohio State Univ, PhD(org chem), 50. *Prof Exp:* Res chemist, Schering Corp, 46-47; res chemist, Hercules, Inc, 50-73, sr res chemist, 73-80, res scientist, 80-85; RETIRED. *Concurrent Pos:* Vol teaching prog sci. *Mem:* Am Chem Soc. *Res:* Organic synthesis; polymers; polymer additives; stabilization; polymer reactions; photochemistry applications. *Mailing Add:* 12971 Bucklard Ct West Palm Beach FL 33414-6229

KUTNER, LEON JAY, b Camden, NJ, Mar 25, 28; c 2. MEDICAL MICROBIOLOGY. *Educ:* Temple Univ, AB, 49; Pa State Univ, MS, 50, PhD(bact), 53; Univ Pittsburgh, MD, 63. *Prof Exp:* Asst, Pa State Univ, 49-53; res assoc virol, Sloan-Kettering Inst Cancer Res, 56-59; intern, Second Med Div, Bellevue Hosp, New York, 63-64; asst prof microbiol in surg, Med Col, Cornell Univ, 64-73; asst prof path in residence, 73-77, ASSOC CLIN PROF PATH, UNIV CALIF, SAN DIEGO, 77-; CHIEF MICROBIOL LAB, LAB SERV, VET ADMIN HOSP, SAN DIEGO, 73- *Concurrent Pos:* Assoc scientist, Hosp Spec Surg, New York, 64-73. *Mem:* Am Soc Microbiol; NY Acad Sci. *Res:* Resistance to infectious disease. *Mailing Add:* Lab Serv Vet Admin Med Ctr 3350 La Jolla Village Dr San Diego CA 92161

KUTNER, MICHAEL HENRY, b Hartford, Conn, Sept 24, 37; m 66; c 2. LINEAR MODELS, VARIANCE COMPONENTS. *Educ:* Cent Conn State Col, BS, 60; Va Polytechnic Inst, MS, 62; Tex A&M Univ, PhD(statist), 71. *Prof Exp:* Asst prof math & statist, Col William & Mary, 62-67; asst prof statist, Tex A&M Univ, 70-71; from asst prof to assoc prof, 71-80, PROF BIOSTATIST, EMORY UNIV, 81-, ASSOC DEAN ACAD AFFAIRS, 90- *Concurrent Pos:* Lectr, NASA, Langley Air Force Base, 63-67, Ctr Dis Control, 80-; bd mem, Am Statist Asn, 81-83, pres, Atlantic Chap, 84-86, vchair publ comt, 88-90; regional adv bd, Biometric Soc, 81-83; assoc ed, Am Statistician, 86-88; dir biostatistics, Emory Univ, 87-; prog chair, Summer Res Conf, 89; consult, Norwich Eaton Pharmaceut, 89-91. *Honors & Awards:* H O Hartley Award, Former Students Tex A&M Univ, 84. *Mem:* Inst Math Statist; fel Am Statist Asn; Biomet Soc. *Res:* Repeated measures on analysis of variance; coauthor of two textbooks. *Mailing Add:* Div Biostatist Emory Univ 1599 Clifton Rd Atlanta GA 30329

KUTNEY, JAMES PETER, b Lamont, Alta, May 2, 32; m 53; c 3. ORGANIC CHEMISTRY. *Educ:* Univ Alta, BSc, 54; Univ Wis, MSc, 56; Wayne State Univ, PhD(org chem), 58. *Prof Exp:* Res fel org chem, Syntex Res Labs, Mex, 58-59; from instr to assoc prof, 59-66, PROF CHEM, UNIV BC, 66- *Concurrent Pos:* NATO scholar, Bonn, WGer, 65; consult, MacMillan Bloedel, 68-78; vis prof, Japan Soc Prom Sci, 75; bd dir, Canadian Patents & Develop Ltd, 76-81; mem, Sci Coun BC, 78-81; mem, adv panel Biotechnol, NSERC, 82-84. *Honors & Awards:* Merck, Sharp & Dohme Award, Chem Inst Can, 68. *Mem:* Am Chem Soc; The Chem Soc; fel Chem Inst Can; Swiss Chem Soc. *Res:* Chemistry, biosynthesis and biodegradation of natural products and related biologically active compounds, particularly synthesis, isolation and structure elucidation of alkaloids, steroids and terpenes; biotechnology, plant cell cultures; microbial transformations. *Mailing Add:* Dept Chem Univ BC 2036 Main Mall Vancouver BC V6T 1Y6 Can

KUTSCHA, NORMAN PAUL, b Irvington, NJ, Sept 24, 37; m 62; c 2. FOREST PRODUCTS. *Educ:* State Univ NY Col Forestry, Syracuse Univ, BS, 59, PhD(wood prod eng), 67; Univ Wis-Madison, MS, 61. *Prof Exp:* Forest prod technologist, US Forest Prod Lab, 59-62; asst prof wood prod eng, State Univ NY Col Forestry, Syracuse Univ, 67-68; from asst prof to assoc prof wood technol, Sch Forest Resources, Univ Maine, Orono, 68-77; SCIENTIST, WEYERHAEUSER CO, 77- *Concurrent Pos:* Partic, McIntire-Stennis Res Proj, Maine Agr Exp Sta, USDA, 69- *Mem:* Soc Wood Sci & Technol; Electron Micros Soc Am; Int Asn Wood Anat; Forest Prod Res Soc; Sigma Xi. *Res:* Light and electron microscopic studies of wood as a developing tissue in the growing tree and as a raw material for various products. *Mailing Add:* 4235 S 324 Pl Auburn WA 98001

KUTSHER, GEORGE SAMUEL, b Reading, Pa, July 16, 21; wid; c 5. ANALYTICAL CHEMISTRY. *Educ:* Albright Col, BS, 47; Lehigh Univ, MS, 49. *Prof Exp:* Head res anal dept, Nitrogen Div, Allied Chem & Dye Corp, 49-58, supvr opers eng, Allied Chem Corp, 58-60, supvry res chemist, 60-69, sr engr, Gas Purification Dept, 69-81, sr process engr, Selexol Dept, 81-82; mgr eng serv, Norton Co, 82-83, Tech consult Selexol, 83-87; RETIRED. *Mem:* Am Chem Soc. *Res:* Gas purification; selexol gas purification process. *Mailing Add:* 76 Woodland Rd Dover NJ 07801

KUTSKY, ROMAN JOSEPH, b Allentown, Pa, May 13, 22; div; c 3. BIOLOGY, CHEMISTRY. *Educ:* Princeton Univ, AB, 44; Univ Calif, MA, 49, PhD(zool), 53. *Prof Exp:* Asst physics, Princeton Univ, 44-46; asst zool, Univ Calif, 46-49, asst specialist plant path, 49-51, res fel biochem, Donner Lab, 53-57; res biochemist, Vet Admin Hosp, 57-67; assoc prof biol, Tex Woman's Univ, 67-73; prof life sci, Bishop Col, 73-77; chemist, Army Med Ctr, El Paso, 77-78; supvr & staff chemist, Bonneville Power Admin, 78-86; RETIRED. *Concurrent Pos:* Consult, Microchem Specialties Co, 59; NASA res grant, 73-76. *Mem:* AAAS; Am Chem Soc; Tissue Cult Asn; NY Acad Sci. *Res:* Biochemical extractions of biologically active materials; cellular biochemistry and physiology; physical biochemistry; tissue culture growth and form; vitamins and hormones; continuous flow preparative electrophoresis; effects of antioxidants; hormones and vitamins in tissue culture. *Mailing Add:* 5719 NE Hazel Dell Dr No D Vancouver WA 98663

KUTTAB, SIMON HANNA, b Jerusalem, Palestine, Apr 17, 46; US citizen; m 78; c 2. MEDICINAL CHEMISTRY. *Educ:* Am Univ Beirut, Lebanon, BSc, 68; Univ Kans, PhD(med chem), 74. *Prof Exp:* Asst res pharmacologist med chem, Univ Calif, Davis, 74-75 & Univ Calif, San Francisco, 75-76; asst prof med chem, Northeastern Univ, 76-81; ASSOC PROF CHEM, BIR-ZEIT UNIV, WEST BANK, ISRAEL, 81- *Mem:* Am Chem Soc; AAAS; Sigma Xi; Acad Am Pharmaceut Asn. *Res:* Design and synthesis of compounds of biological interest; breakdown of xenobiotics and specific absorption rate correlations using such advanced analytical techniques as gas chromatography, high performance liquid chromatography and gas chromatography-mass spectrometry. *Mailing Add:* Dept Chem Bir-Zeit Univ Box 14 Bir-Zeit West Bank Israel

KUTTEH, WILLIAM HANNA, b Statesville, NC, Mar 18, 54; m 88. REPRODUCTIVE ENDOCRINOLOGY & IMMUNOLOGY. *Educ:* Wake Forest Univ, BA, 75; Univ Ala, Birmingham, PhD(immunol), 81; Bowman Gray Sch Med, MD, 85. *Prof Exp:* Res asst immunol, Duke Univ Med Ctr, 75-78; predoctoral fel microbiol, Univ Ala, Birmingham, 78-81, res instr obstet & gynec, 85-89; postdoctoral fel biochem, Bowman Gray Sch Med, 81-85; instr fel reproduction endocrinol, 89-91, ASST PROF OBSTET & GYNEC & DIR REPRODUCTIVE IMMUNOL, UNIV TEX SOUTHWESTERN MED CTR, 91- *Concurrent Pos:* Res award, Sigma Xi, 80 & 84; chief resident obstet & gynec, Univ Ala, Birmingham, 89, consult, dept microbiol, 89-; distinguished res award, Am Fertil Soc, 91. *Mem:* Am Asn Immunologists; Am Fertil Soc; Am Col Obstet & Gynec; Soc Mucosal Immunol; NY Acad Sci; Am Asn Gynec Laparoscopists. *Res:* Secretory immune system of the female reproductive tract; immune response to human ovarian cancer; recurrent pregnancy loss; cholesterol metabolism and steroid production. *Mailing Add:* Dept Obstet & Gynec Univ Tex Southwestern Med Ctr 5323 Harry Hines Blvd Dallas TX 75235-9032

KUTTER, ELIZABETH MARTIN, b Chicago, Ill, Aug 11, 39; c 2. MOLECULAR BIOLOGY. *Educ:* Univ Wash, BS, 62; Univ Rochester, PhD(biophys), 68. *Prof Exp:* Res assoc biol, Univ Va, 69-72; MEM FAC BIOPHYS, EVERGREEN STATE COL, 72- *Concurrent Pos:* Res grants, NIH, 73-76, NSF, 70-72 & 76-, mem, NIH Dir Adv Comt on Recombinant DNA, 75-79; mem, NSF Adv Comt Ethics & Values in Sci & Technol, 78-80; vis scientist, Dept Biochem, Univ Calif, San Francisco, 78-79; teacher ethics & molecular biol, AAAS, Chataqua, 77-80; mem bd dir, John Bastyr Col, 79-; Nat Acad Sci exchange mem USSR, 90. *Mem:* Biochem Soc; AAAS; Am Soc Microbiol; Genetics Soc. *Res:* Biochemical developments during bacteriophage T4 infection of Escherichia coli, especially regulation of transcription and events altering host metabolism; roles of 5-hydroxymethylcytosine in T4 DNA. *Mailing Add:* Dept Molecular Biol Evergreen State Col Olympia WA 98505

KUTTLER, JAMES ROBERT, b Burlington, Iowa, Aug 8, 41; m 63; c 3. DIFFERENTIAL EQUATIONS, EIGENVALUES. *Educ:* Rice Univ, BA, 62, Univ Md, MA, 64, PhD(math), 67. *Prof Exp:* MATHEMATICIAN, APPLIED PHYSICS LAB, JOHNS HOPKINS UNIV, 63- *Concurrent Pos:* Lectr, math, Johns Hopkins Univ, GWC Whiting Sch Eng, Continuing Prof Progs, 81- *Res:* Differential equations, electromagnetics, eigenvalues. *Mailing Add:* Appl Physics Lab Johns Hopkins Univ Laurel MD 20723-6099

KUTZ, FREDERICK WINFIELD, b Wilmington, Del, Sept 29, 39; m 63; c 2. ECOLOGY, MEDICAL ENTOMOLOGY. *Educ:* Univ Del, BS, 62, MS, 64; Purdue Univ, PhD(entom), 72. *Prof Exp:* Res fel & assoc entom, Univ Del, 62-64; res asst & instr, Purdue Univ, 66-69; entomologist & parasitologist, Insect Control & Res Inc, 69-72; ECOLOGIST, US ENVIRON PROTECTION AGENCY, 72- *Concurrent Pos:* Mem sci adv panel, Onchocerciasis Control Prog, WHO, 74-80, monitoring panel, Fed Working Group Pest Mgr, 75-77 & subcomt, Comt Environ Carcinogens, Nat Cancer Inst, 76-; adj prof, Univ Miami Sch Med, 80- *Mem:* Entom Soc Am; Am Soc Trop Med & Hyg; Am Mosquito Control Asn; Sigma Xi. *Res:* Arthropod-insect pest management, particularly of medical significance and chemical and biological monitoring in humans and environmental components; environmental processes and effects of chemicals. *Mailing Add:* 4967 Moonfall Way Columbia MD 21044

KUTZKO, PHILIP C, b Brooklyn, NY, Nov 24, 46; m 67; c 1. NUMBER THEORY. *Educ:* City Col New York, BA, 67; Univ Wis, MA, 68, PhD(math), 72. *Prof Exp:* Instr math, Univ Wis, Green Bay, 68-69, instr, Rock County Ctr, 69-72; instr, Princeton Univ, 72-74; asst prof, 74-77, assoc prof, 77-80, PROF MATH, UNIV IOWA, 80- *Mem:* Am Math Soc. *Res:* Representation theory of p-adic linear groups and applications to non-abelian classfield theory. *Mailing Add:* Div Math Sci Univ Iowa Iowa City IA 52242

KUTZMAN, RAYMOND STANLEY, b St Cloud, Minn, April 16, 49. TOXICOLOGY. *Educ:* St Cloud State Col, BA, 71; Univ Notre Dame, MS, 74; NC State Univ, PhD(zool), 77. *Prof Exp:* Res assoc, Chem Dept, 77-79, from asst scientist to assoc scientist, Med Dept, Brookhaven Nat Lab, 79-85; chem mgr, Nat Toxicol Prog, Nat Inst Environ Health Sci, 85. *Concurrent Pos:* Dir, Toxic Hazard Res Unit, Armstrong Aerospace Res Lab, USAF. *Mem:* AAAS; Soc Toxicol; Int Soc Study Xenobiotics; Soc Risk Anal; Am Soc Pharmacol & Exp Therapeut. *Res:* Biodistribution of xenobiotic agents after inhalation exposure; risk assessment; genetic disposition as an underlying factor in biochemical and physiological aspects of toxicity. *Mailing Add:* Mitre Corp HSD/VAU Bldg 624 Brooks AFB TX 78235-5000

KUTZSCHER, EDGAR WALTER, b Leipzig, Ger, Mar 21, 06; nat US; m 45; c 2. PHYSICS. *Educ:* Univ Berlin, PhD(physics), 33, Dr phil habil(appl physics), 36. *Hon Degrees:* DrEng, Hannover Univ, 63. *Prof Exp:* Asst physics, Univ Berlin, 30-33 & Inst Technol, Berlin, 33; physicist, Dept Defense, Ger, 34-37; dir res, Electroacoust Co, Kiel, Ger, 37-45 & univ exten, Flensburg, Ger, 46-47; physicist infrared, US Navy, 47-51 & solid state physics, Santa Barbara Res Ctr, Calif, 51-53; head dept radiation sensors & technol, Lockheed Aircraft Corp, 54-72; CONSULT PHYSICIST, 72- *Concurrent Pos:* Asst prof, Inst Technol, Berlin, 37-45. *Honors & Awards:* Todt Prize, 44. *Mem:* Optical Soc Am. *Res:* Infrared physics and detectors. *Mailing Add:* 15450 Briarwood Dr Sherman Oaks CA 91403

KUWAHARA, STEVEN SADAO, b Lahaina, Hawaii, July 20, 40; m 73; c 2. BIOCHEMISTRY HEMOSTASIS, ANALYTICAL BIOCHEMISTRY. *Educ:* Cornell Univ, BS, 62; Univ Wis, MS, 65, PhD(biochem), 67. *Prof Exp:* Res assoc biochem, Univ Wash, 66-67; asst prof chem, Calif State Col, Long Beach, 67-71; asst res biologist, Dept Develop & Cell Biol, Univ Calif, Irvine, 71-73; biochemist & unit chief, Bur Dis Control & Lab Serv, Mich Dept Pub Health, 73-76, sect chief prod anal, 76-78, sect chief biochem & bioassay, 78-82; mgr test technol, 82-87, mgr, QA Labs, MGR TEST TECHNOL, HYLAND THERAPEUT, LOS ANGELES, 90- *Concurrent Pos:* Spec res fel, NIH, 71-73; adj res assoc, Dept Med, Mich State Univ, 81-82; adv coun, Dept Chem & Biochem, Calif State Univ Long Beach, 91- *Mem:* NY Acad Sci; Am Chem Soc; Soc Exp Biol & Med; Am Fedn Clin Res; Am Soc Microbiol; AAAS. *Res:* Biochemistry of blood coagulation factors; biochemistry of plasma proteins; analytical biochemistry. *Mailing Add:* Hyland Div Baxter Healthcare 1720 Flower St Duarte CA 91010

KUWANA, THEODORE, b Idaho Falls, Idaho, Aug 3, 31. ANALYTICAL CHEMISTRY, ELECTROCHEMISTRY. *Educ:* Antioch Col, BS, 54; Cornell Univ, MS, 56; Univ Kans, PhD(anal chem), 59. *Prof Exp:* Res chemist, Aerojet-Gen Corp Div, Gen Tire & Rubber Co, 59; fel, Calif Inst Technol, 59-60; asst prof anal chem, Univ Calif, Riverside, 60-66; from assoc prof to prof, Case Western Reserve Univ, 66-71; PROF CHEM, OHIO STATE UNIV, 71-; AT CTR BIOANALYTICAL RES, KANS UNIV, LAWRENCE. *Concurrent Pos:* Chmn, Gordon Res Conf Anal Chem, 64. *Mem:* AAAS; Am Chem Soc; Royal Soc Chem. *Res:* Organic electrode processes; photoelectrochemistry and electroluminescence. *Mailing Add:* Ctr Bioanalytical Res 2095 Constant Ave Lawrence KS 66046

KUYATT, CHRIS E(RNIE EARL), b Grand Island, Nebr, Nov 30, 30; m 49; c 4. ELECTRON OPTICS, ATOMIC PHYSICS. *Educ:* Univ Nebr, BS, 52, MS, 53, PhD(physics), 60. *Prof Exp:* Res asst, dept physics, 53-59, res assoc & instr atomic physics, Univ Nebr, 59-60; physicist, Nat Bur Standards, 60-69, actg chief electron physics sect, 69-70, chief electron & optical physics sect, 70-73, chief surface & electron physics sect, 73-78, chief, Radiation Physics Div, 78-79, dir, Ctr Radiation Res, 79-91; COORDR, RADIATION MEASUREMENT SERVS, NAT INST STANDARDS & TECHNOL, 91- *Concurrent Pos:* Interagency Radiation Res Comt, 80-84; com sci & technol fel, Nat Sci Found, 83-84; mem, Nat Inst Standards & Technol Liaison with Nat Coun Radiation Projections & Measurement, 80-, chmn, Radiation Safety Comt, 79-; mem, Interagency Steering Comt Acad Res Facil, 84. *Honors & Awards:* US Dept Com Silver Medal, 64. *Mem:* Fel Am Phys Soc; AAAS; Sigma Xi. *Res:* Electron scattering; polarized electrons; electron monochromators; electron energy analyzers; electron optics. *Mailing Add:* Rm C229 Bldg 245 Nat Inst Standards & Technol Gaithersburg MD 20899

KUYPER, LEE FREDERICK, b Mitchell, SDak, Feb 28, 49; c 1. ORGANIC CHEMISTRY, MEDICINAL CHEMISTRY. *Educ:* Ouachita Univ, BS, 71; Univ Ark, PhD(org chem), 77. *Prof Exp:* Res assoc, Univ NC, 76-77; SR SCIENTIST, BURROUGHS WELLCOME CO, 77- *Mem:* Am Chem Soc. *Res:* Molecular modeling; drug design and synthesis. *Mailing Add:* Burroughs Wellcome Co 3030 Cornwallis Rd Research Triangle Park NC 27709

KUZEL, NORBERT R, b Angus, Minn, May 23, 23; m 49. ANALYTICAL CHEMISTRY, INSTRUMENTATION. *Educ:* NDak State Univ, BS, 48, MS, 49. *Prof Exp:* Anal chemist, Eli Lilly & Co, 49-59, dept head anal develop, 59-63, sr anal chemist, 63-67, res scientist, 68-73, res assoc, 73-84; RETIRED. *Mem:* Am Chem Soc; Instrument Soc Am. *Res:* Development of analytical methods; residue analysis; laboratory and process automation and computerization. *Mailing Add:* 4611 Berkshire Lane Indianapolis IN 46226

KUZMA, JAN WALDEMAR, b Warsaw, Poland, Apr 24, 36; US citizen; m 63; c 3. BIOSTATISTICS. *Educ:* Andrews Univ, BA, 59; Columbia Univ, MS, 61; Univ Mich, PhD(biostatist), 63. *Prof Exp:* Lectr biostatist & dir clin trials unit, Univ Calif, Los Angeles, 63-67; chmn dept biostatist, 67-73, PROF BIOSTATIST & CHMN DEPT BIOSTATIST & EPIDEMIOL, SCH PUB HEALTH, LOMA LINDA UNIV, 73- *Concurrent Pos:* Consult biostatistician, Loma Linda Univ, 64-67. *Mem:* Am Statist Soc; Biomet Soc; Am Pub Health Asn; AAAS. *Res:* Lifestyle and longevity; health care costs; general statistical methodology. *Mailing Add:* 1280 E San Bernardino Ave Redlands CA 92374

KUZMA, JOSEPH FRANCIS, b Austria, Mar 14, 15; US citizen; m 41; c 7. PATHOLOGY. *Educ:* Univ Ill, BS, 37, MD, 40; Marquette Univ, MS, 42. *Prof Exp:* From instr to prof, Marquette Univ, 46-74, dir dept, 53-69; prof path, 74-80, CLIN PROF PATH, MED COL WIS, 80- *Concurrent Pos:* Dir lab, Milwaukee Hosp, 43-47 & Milwaukee County Hosp, 47-54 & 64-69. *Mem:* Am Soc Clin Path; fel AMA; Am Asn Path & Bact; Am Col Physicians; Col Am Path; Sigma Xi. *Res:* Mammary tumors and diseases; sulfonamide reactions; experimental arthritis; kidney diseases; radioactive strontium; bone cancer. *Mailing Add:* 1115 Honey Creek Pkwy Wauwatosa WI 53213

KUZMAK, JOSEPH MILTON, b Man, Can, Mar 7, 22; m 42; c 3. PHYSICAL CHEMISTRY, PAPER COATINGS. *Educ:* Univ Man, BSc, 49, MSc, 50; McGill Univ, PhD(phys chem), 53. *Prof Exp:* Res officer, Nat Res Coun Can, 53-57; res chemist, Am Viscose Corp, Pa, 57-67; sr res chemist, St Regis Paper Co, 67-84; sr scientist, Champion Int Corp, 84-86; RETIRED. *Mem:* Am Chem Soc; Tech Asn Pulp & Paper Indust. *Res:* Mechanism of moisture movement in porous materials; chemical modification of regenerated cellulose; chemical modification of pulp and paper; paper coatings and coating process. *Mailing Add:* 24 Collingswood Rd New City NY 10956

KUZMANOVIC, B(OGDAN) O(GNJAN), b Belgrade, Yugoslavia, July 16, 14; m 50; c 1. CIVIL ENGINEERING. *Educ:* Univ Belgrade, Dipl Eng, 37; Serbian Acad Sci, Dr Tech Sc, 56. *Prof Exp:* Asst designer bridges, Ministry Transp, Belgrade, 38-41, sr designer, 45-53; from asst prof to assoc prof struct, Univ Sarajevo, 53-58; Brit Coun & Gilchrist Ed Trust res fel, Sheffield Sci Sch, Yale, 58-59; prof, Univ Khartoum, 59-60, dean fac eng & head dept civil eng, 60-63, prof & head dept, 63-65; prof struct, Univ Kans, 65- 81; SR STRUCT ENGR, BEISWENGER, HOCH & ASSOCS, MIAMI BEACH, FLA, 81- *Concurrent Pos:* Off Civil Defense res grant, 67. *Mem:* Fel Am Soc Civil Engrs; sr mem Int Asn Bridge & Struct Engrs. *Res:* Theory of elasticity and plasticity; plastic analysis and design of steel and concrete structures; metal fatigue. *Mailing Add:* Beiswenger Hoch & Assocs 1190 NE 163rd St N Miami Beach FL 33160

KUZNESOF, PAUL MARTIN, b Bronx, NY, Aug 13, 41; m 82; c 1. REGULATORY FOOD CHEMISTRY, INORGANIC CHEMISTRY. *Educ:* Brown Univ, ScB, 63; Northwestern Univ, PhD, 67. *Prof Exp:* Fel inorg mat res div, Lawrence Radiation Lab, Univ Calif, 67-69; asst prof chem, San Francisco State Col, 69-70; prof chem, Univ Campinas, Brazil, 70-75; mem staff, chem div, Naval Res Lab, 78-79; assoc prof, Agnes Scott Col, Decatur, Ga, 79-83; SUPVRY CHEMIST, FOOD & COLOR ADDITIVES REV SECT, DIV FOOD CHEM & TECHNOL, FOOD & DRUG ADMIN, 84- *Concurrent Pos:* Grant, FAPESP Res Found, Sao Paulo, 71; lectr, Univ Mich, 75-76; vis assoc prof, Trinity Col, Hartford, Conn, 76-78; grants, Res Corp, 77 & NSF, 81; vis scholar, Northwestern Univ, 82. *Mem:* Am Chem Soc; Sigma Xi; AAAS. *Res:* Synthesis and electronic properties of boron-nitrogen compounds; hydrides of the lighter main group elements; electron donor-acceptor interactions; electroactive polymers. *Mailing Add:* Div Food Chem & Technol HFF-415 Food & Drug Admin 200 C St SW Washington DC 20204

KVAAS, T(HORVALD) ARTHUR, b Des Moines, Iowa, Jan 8, 19; m 42; c 2. PHYSICS, ENGINEERING. *Educ:* Univ Calif, Los Angeles, BA, 40, MA, 42. *Prof Exp:* Phys sci res engr, Res Lab, Douglas Aircraft Co, 42-46, phys scientist, proj Rand, 46-48; proj engr, Rand Corp, 48-52; sect chief missiles adv design, Douglas Aircraft Co, 52-57; mgr, Synthesis Sect, Tech Mil Planning Oper, Gen Elec Co, 57-60, prof staff, 60-62, tech anal & appln oper, 62-63, mgr tech environ studies, tempo, ctr advan studies, 63-70; pres, ADCON Corp, 70-74; vpres & opers mgr, Moseley Assocs, 74-76, consult, 77; opers mgr, Cetec Broadcast Corp, 78-; CONSULT ACOUSTICS, 80- *Concurrent Pos:* Mem comt, Am Standards Asn Comt Acoust Terminology, 46-48; Am Rocket Soc rep, Cong Int Astronaut Fedn, Amsterdam, 58. *Mem:* Assoc fel Am Inst Aeronaut & Astronaut; Acoustical Soc Am. *Res:* Technological and environmental forecasting and planning with particular emphasis on future technologies, technical resources and their application to future human needs; corporate long range strategic business planning. *Mailing Add:* 933 Roble Lane Santa Barbara CA 93103

KVALNES-KRICK, KALLA L, b Sept 15, 60; m; c 2. BIOCHEMISTRY. *Educ:* Univ Cincinnati, BS, 82; Hahnemann Univ PhD(biochem), 87. *Prof Exp:* Postdoctoral fel, Dept Biochem, Albert Einstein Col Med, 87-90; POSTDOCTORAL FEL, DEPT BIOCHEM, UNIV NC, 90- *Concurrent Pos:* NIH grants, 88-91 & 89-90. *Mem:* Am Chem Soc; AAAS; Am Soc Biochem & Molecular Biol. *Res:* Protein, native and mutant, purification; chemical modification of proteins. *Mailing Add:* Dept Biochem & Biophys Univ NC 430 Fac Lab Off Bldg CB No 7260 Chapel Hill NC 27599-7260

KVALSETH, TARALD ODDVAR, b Brunkeberg, Norway, Nov 7, 38; m 64; c 3. STATISTICS, MATHEMATICAL MODELING. *Educ:* Univ Durham, Eng, BSc, 63; Univ Calif, Berkeley, MS, 66, PhD(indust eng), 71. *Prof Exp:* Asst prof indust eng, Ga Inst Technol, 71-74; sr lectr, Norweg Inst Technol, 74-79; assoc prof mech eng, 79-82, PROF MECH ENG, UNIV MINN, 82-, HEAD INDUST ENG, 83- *Concurrent Pos:* Co-prin investr, Nat Inst Occup Safety & Health, 87- *Mem:* Int Ergonomics Asn (vpres, 82-85); AAAS; Sigma Xi; Human Factors Soc; Ergonomics Soc; Inst Indust Engrs. *Res:* Human factors engineering with emphasis on human performance measures, quantitative models, statistical methods, industrial ergonomics and safety. *Mailing Add:* 108 Turnpike Rd Golden Valley MN 55416

KVAM, DONALD CLARENCE, b Escanaba, Mich, Oct 20, 32; m 54; c 3. PHARMACOLOGY. *Educ:* Ferris State Col, BS, 54; Univ Wis, PhD(pharmacol), 60. *Prof Exp:* Sr pharmacologist, Mead Johnson & Co, 60-63, group leader pharmacol, 63-64; supvr biol res, 64-67, mgr biol res, 67-71, mgr pharmacol, 71-78, mgr clin pharmacol, 78-82, ASSOC DIR CLIN PHARMACOL, 3M PHARMACEUTICALS, 82- *Concurrent Pos:* Lectr, Col Med Sci, Univ Minn. *Mem:* AAAS; NY Acad Sci; Am Soc Pharmacol & Exp Therapeut; Am Asn Pharm Scientists; Am Col Clin Pharmacol; Am Soc Clin Pharmacol & Therapeut. *Res:* Clinical pharmacology, phase 1 and pharmacokinetics studies. *Mailing Add:* 3M Pharmaceuticals Bldg 270-3A-01 3M Ctr St Paul MN 55144-1000

KVEGLIS, ALBERT ANDREW, b Brooklyn, NY, Feb 10, 34; m 61; c 3. POLYMER CHEMISTRY, ORGANIC CHEMISTRY. *Educ:* Queens Col, NY, BS, 56; Stevens Inst Technol, MS, 65. *Prof Exp:* Sr res chemist polymers, Plastics Div, Allied Chem Corp, 56-71; res chemist polymers, Trimflex Div, Teleflex Corp, 71; methods develop, Biomed Sci, Inc, 71-72; group leader, Polymers & Vehicles, Inmont Corp, 72-82; MGR, INK VEHICLES, SUNCHEMICAL CORP, 82- *Mem:* Am Chem Soc. *Res:* Polymer synthesis and characterization, development of resins and vehicles for inks and coatings, synthesis of flame retardant monomers and additives for plastics, modification of polymers. *Mailing Add:* Six Buckingham Circle Pine Brook NJ 07058-9712

KVENVOLDEN, KEITH ARTHUR, b Cheyenne, Wyo, July 16, 30; m 59; c 2. ORGANIC GEOCHEMISTRY. *Educ:* Colo Sch Mines, GpE, 52; Stanford Univ, MS, 58, PhD(geol), 61. *Prof Exp:* Jr geologist, Socony Mobil Oil Co, Venezuela, 52-54, sr res technologist petrol geochem, Mobil Field Res Lab, Tex, 61-66; res scientist, Ames Res Ctr, NASA, Calif, 66-71, chief, Chem Evol Br, 71-74, chief, Planetary Biol Div, 74-75; GEOLOGIST, US GEOL SURV, 75- *Concurrent Pos:* Consult assoc prof geol, Stanford Univ, 67-73, consult prof, 73-; chmn, Jodies Adv Panel Organic Geochem, 74-80; mem, US Nat Comt Geochem, 80-83, chmn, 84-86; chmn, Gordon Res Conf on Organic Geochem, 84; mem, US Sci Adv Comt, Ocean Drilling Prog, 85-86. *Honors & Awards:* Meritorious Serv Award, US Dept Interior. *Mem:* Am Asn Petrol Geol; fel Geol Soc Am; Geochem Soc; Am Geophys Union; fel AAAS; fel Explorers Club. *Res:* Organic geochemistry of modern and ancient sediments; petroleum geochemistry; environmental geochemistry; organic chemistry of meteorites; origin and evolution of life; geochemistry of amino acids; geochemistry of hydrocarbon gases and gas hydrates. *Mailing Add:* 2433 Emerson St Palo Alto CA 94301

KVIST, TAGE NIELSEN, b Copenhagen, Denmark, Jan 17, 42; US citizen; m 65; c 3. DEVELOPMENTAL ANOMALIES, TERATOLOGY. *Educ:* Univ BC, BS, 66, MS, 69; Univ Pa, Philadelphia, PhD(biol), 73. *Prof Exp:* Teaching asst develop, Univ BC, Vancouver, 66-67; teaching fel biol, Univ Pa, Philadelphia, 69-72, res assoc develop, 73-76; lectr comp embryol, Rosemont Col, Pa, 72; chief neurosurg res, congenital anoms, Joseph Stokes Jr Res Inst, 73-76; from asst prof to assoc prof 76-87, ASST DEAN BASIC SCI, PHILADELPHIA COL OSTEOP MED, 86-, PROF & CHMN ANAT, 87- *Concurrent Pos:* Sci res adv, Pa Gov Conf Handicapped Individuals, 76; consult, NIH Sci Rev Group, 79-; rev bd human res, Philadelphia Col Obstet Med, 81-; guest lectr, Sch Nursing, Univ Pa, 82-87; Mem, Nat Comt Res Neurol Commun Dis, Spina Bifida Asn Am, 85-; reviewer, March of Dimes Birth Defects Found Grant; mem, Inst Self Study Task Force & Exec Fac, Philadelphia Col Obsteop Med, 85-; chmn Animal Care & Utilization, Philadelphia Col Osteop Med, 86-, curriculum 86-, computer asst instruction, 86-, Strategic Planning Task Force, 87, dir Sch Allied Health, 86- rep, Health Sci Libraries Consortium, 87- *Honors & Awards:* Lindback Found Award, Christian R & Mary F Lindback Found, 85. *Mem:* Soc Develop Biol; Teratol Soc; Am Asn Anatomists; Spina Bifida Asn Am; Am Asn Clin Anatomists; Sigma Xi; Humanity Gifts Registry Pa Anatomists (bd mem, 86-). *Res:* Birth defects involving the central nervous system; Spina Bifida Cystica; Anencephalus; Hydrocephalus; connective tissue macromolecule formation in developing embryos and in rheumatoid arthritis. *Mailing Add:* Dept Anat Philadelphia Col Osteop Med 4150 City Ave Philadelphia PA 19131

KWAAN, HAU CHEONG, b Hong Kong, Sept 30, 31; US citizen; m 58; c 2. INTERNAL MEDICINE, HEMATOLOGY. *Educ:* Univ Hong Kong, MB & BS, 52, MD, 58; FRCP(E), 67; Am Bd Internal Med, cert internal med, 69, cert hemat, 74, cert med oncol, 79. *Prof Exp:* House physician, Univ Med Unit, Queen Mary Hosp, 52-53; sr clin asst med, Univ Hong Kong, 53-55, asst lectr, 56-59, lectr, 59-61; sr investr physiol, James F Mitchell Found, DC, 62-65; assoc prof, 66-72, PROF MED, MED SCH, NORTHWESTERN UNIV CHICAGO, 72- *Concurrent Pos:* China Med Bd NY fel pharmacol, 58-59; vis res fel, Col Physicians & Surgeons, Columbia Univ, 58-59; clin asst prof, Sch Med, Georgetown Univ, 64-65; mem coun thrombosis, Am Heart Asn, 64-; chief hemat-oncol sect, Vet Admin Lakeside Med Ctr, Chicago, 67-; attend physician, Northwestern Mem Hosp, Chicago, 69-; sr Fulbright travel scholar, 74. *Mem:* Fel Am Col Physicians; AMA; Am Physiol Soc; Am Soc Hemat; Am Fedn Clin Res. *Res:* Blood coagulation; fibrinolysis; thrombosis. *Mailing Add:* Dept Med Northwestern Univ Med Sch 33 E Huron St Northwestern Univ Med Sch Chicago IL 60611

KWAK, JAN C T, b Schagen, Neth, May 6, 42; m 65; c 3. POLYMER SCIENCE, COLLOID SCIENCE. *Educ:* Univ Amsterdam, MSc, 64, PhD, 67. *Prof Exp:* Res assoc molten salts, Neth Orgn Advan Pure Res, 64-68; res chemist, Sea Water Conversion Lab, Univ Calif, Berkeley, 68-70; from asst prof to assoc prof, 70-83, PROF CHEM, DALHOUSIE UNIV, 83-, CHAIR, DEPT CHEM, 86- *Mem:* Am Chem Soc; Chem Inst Can. *Res:* Polymer solutions; surfactants; surfactant nuclear magnetic resonance; biophysical chemistry; flocculation studies; coal beneficiation; colloid chemistry. *Mailing Add:* Dept Chem Dalhousie Univ Halifax NS B3H 4H6 Can

KWAK, NOWHAN, b Seoul, Korea, Sept 16, 28; US citizen; m 58; c 2. HIGH ENERGY PHYSICS. *Educ:* Seoul Nat Univ, BS, 52; Emory Univ, MS, 56; Univ Rochester, MA, 59; Tufts Univ, PhD(physics), 62. *Prof Exp:* Res assoc high energy physics, Tufts Univ, 62-65; from asst prof to assoc prof, 64-78, PROF PHYSICS & ASTRON, UNIV KANS, 78- *Concurrent Pos:* Vis scientist, Deutsches Elektronen-Synchrotron, WGer, 80-81, Cern, 74-76; sr fel, Austrian Acad Sci, 73-74. *Mem:* Am Phys Soc. *Res:* Experimental high energy physics. *Mailing Add:* Dept Physics Univ Kans Lawrence KS 66045

KWAK, YUN SIK, b Taegu, Korea, Aug 21, 37. EXPERIMENTAL BIOLOGY. *Educ:* Kyungpook Nat Univ, MD, 61; Union Univ, Albany, PhD(molecular biol & path), 72; Am Bd Path, cert anal & clin, 78, cert chempath, 80. *Prof Exp:* Teaching fel, Dept Biochem, Sch Med, Kyungpook Univ, 64-66, from instr to asst prof, 66-69; from res instr to asst prof, Dept Path, Albany Med Col, 69-78; assoc prof, Dept Path, Sch Med, Wright & State Univ, 79-81; ASST PROF, DEPT PATH, SCH MED, CASE WESTERN RESERVE UNIV, 81- *Concurrent Pos:* Chief resident, Dept Path, Albany Med Ctr Hosp, 73-74; staff pathologist-in-chg clin chem, Vet Admin Hosp, Albany, NY, 75-77, mem numerous comts, 77-, chief clin path sect, Lab Serv, Cleveland, 77-79, chief lab serv, Dayton, 79-81, Cleveland, 81-, spec asst to med ctr dir, I/C Med Info Mgt Sect, Vet Admin Med Ctr, 89- *Mem:* Am Asn Clin Chem; fel Am Soc Clin Pathologists; Am Asn Pathologists; NY Acad Sci; fel Nat Acad Clin Biochem; fel Col Am Pathologists; AMA. *Res:* Biochemical aspects of atherogenesis; effective utilization of laboratory information in clinical medicine and laboratory management; author of numerous scientific publications. *Mailing Add:* Vet Admin Med Ctr 10701 East Blvd Cleveland OH 44106

KWAN, JOHN YING-KUEN, b Hong Kong, Apr 5, 47; m 73; c 2. ASTROPHYSICS. *Educ:* Utah State Univ, BS, 69; Calif Inst Technol, PhD(physics), 72. *Prof Exp:* Res fel astrophys, Calif Inst Technol, 73 & Inst Advan Study, 73-74; asst prof astrophys, State Univ NY Stony Brook, 75-76; mem tech staff, Bell Labs, 76-80; assoc prof, 81-87, PROF ASTROPHYS, UNIV MASS, 87- *Mem:* Am Astron Soc; Inst Elec & Electronics Engrs. *Res:* Theoretical studies of astrophysical masers, interstellar molecular clouds, quasars, young stellar objects. *Mailing Add:* Dept Physics & Astron Univ Mass Amherst MA 01003

KWAN, KING CHIU, b Hong Kong, Jan 14, 36. DRUG METABOLISM, PHARMACOKINETICS. *Educ:* Univ Mich, BS, 56, MS, 58, PhD(pharmaceut chem), 62. *Prof Exp:* Res chemist, R P Scherer Corp, 62-63; lectr pharm, Univ Mich, 63-64; res assoc, 64-66, unit head, 66-69, pharmacokinetic specialist, 69-70, sr res fel, 70-76, sr invest, 76-79, sr dir, biopharmaceut, 79-81, EXEC DIR DRUG METAB, MERCK SHARP & DOHME RES LABS, 81- *Concurrent Pos:* Mem, Pharmacol Study Sect, NIH, 80-83. *Mem:* Am Pharmaceut Asn; Am Asn Pharmaceut Scientists; NY Acad Sci; Am Soc Pharmacol & Exp Therapeuts; Int Soc Study Xenobiotics; Sigma Xi. *Res:* Pharmaceutical research and development; drug metabolism; pharmacokinetics; biopharmaceutics. *Mailing Add:* Merck Sharp & Dohme Res Labs West Point PA 19486

KWAN, PAUL WING-LING, b Hong Kong, Nov 7, 42; US citizen; m 77. CANCER, CELL BIOLOGY. *Educ:* Univ Md, BS, 66; Clark Univ, MA, 71, PhD(biol), 75. *Prof Exp:* Teaching asst biol, Clark Univ, 67-73; res assoc, 75-85, RES ASSOC PROF, DEPT PATH, SCH MED, TUFTS UNIV, 85- *Res:* Application of immunohistochemistry to the study of tumor biology and clinical diagnosis; pathogenesis of benign prostatic hyperplasia and prostate cancer in animal models. *Mailing Add:* Dept Path 136 Harrison Ave Boston MA 02111

KWAN-GETT, CLIFFORD STANLEY, b Emmaville, NSW, Oct 14, 34; m 61; c 2. SURGERY. *Educ:* Univ Sydney, BSc, 54, BE, 56, MD, 63. *Prof Exp:* Engr, Australian Postmaster Gen Dept, 60-61; resident med off, Lanceston Gen Hosp, Tasmania, 64-66; asst res prof, 68-70, ASSOC RES PROF SURG, UNIV UTAH, 70- *Concurrent Pos:* Fel med, Cleveland Clin Found, 66-67; consult, Aerojet Gen Corp, Calif, 66-68; cardiol, thoracic & vascular surgeon. *Mem:* Am Soc Artificial Internal Organs; Biomed Eng Soc; AMA. *Res:* Developing total replacement artificial hearts to replace the irreparable human heart; use of artificial heart assist devices; development and use of artificial kidneys, especially for home use by patients. *Mailing Add:* Western Cardiovasc Assoc 1055 E 3900 S Salt Lake City UT 84112

KWART, HAROLD, physical chemistry, organic chemistry; deceased, see previous edition for last biography

KWARTLER, CHARLES EDWARD, b Stanislau, Austria, Oct 5, 11; US citizen; m 41; c 3. ORGANIC CHEMISTRY. *Educ:* NY Univ, BS, 32, PhD(org chem), 36. *Prof Exp:* Microchem technician, NY Univ, 34-36, asst instr, 36-38; res chemist, Winthrop Chem Co, 39-43, dir pilot lab, 43-45, head process develop lab, Winthrop-Stearns, Inc, 45-51, chief chemist, Winthrop Prod, Inc, 47-51; dir res & develop, Gamma Chem Co, 51-55, vpres, 55-56, exec vpres, 57-66; consult, Ashland Chem Co, 76-78; CONSULT, SOUTHLAND CORP, 78- *Concurrent Pos:* Mgr process develop, Ashland Chem Co, 69-71, asst to pres & tech coordr, 71-76; chmn bd trustees, Warren County Community Col Comn, 81- & Warren County Community Col Found, 84-; mem bd dir, Hackettstown Community Hosp, 86- *Mem:* Chemists' Club; Sigma Xi; Nat Hon Soc Scientists; Nat Hon Soc Chemists; emer mem Am Chem Soc. *Res:* Anesthetics; antimalarials; antiseptics; analgesics; antispasmodics; sulfanilamides; quaternary ammonium compounds; arsenicals; radiopaques; diuretics; synthetic sex hormones; organic antimony compounds; synthetic detergents; 8-hydroxyquinoline; synthetic herbicides and pecticides. *Mailing Add:* 102 Petersburg Rd Hackettstown NJ 07840

KWATNY, EUGENE MICHAEL, b Philadelphia, Pa, Oct 25, 43; m 66; c 2. BIOMEDICAL ENGINEERING. *Educ:* Drexel Univ, BS, 66, MS, 68, PhD(biomed eng), 71. *Prof Exp:* Biomed engr, Aerospace Crew Equipment Dept, US Naval Air Develop Ctr, 66-71; PRIN INVESTR VISUAL SYSTS & DIR COMPUT & INFO SCI, KRUSEN CTR RES & ENG & ASST PROF REHAB MED, SCH MED, TEMPLE UNIV, 71- *Concurrent Pos:* Adj asst prof visual sci & biomed eng, Pa Col Optom, 74- *Mem:* Inst Elec & Electronics Engrs. *Res:* Sensory aids for rehabilitation; bioelectric signal processing; computers in medicine and biology. *Mailing Add:* Dept Comput Sci 38-24 Temple Univ Broad & Montgomery Sts Philadelphia PA 19122

KWATRA, SUBHASH CHANDER, b India, Nov 12, 41; m 66; c 2. DIGITAL SATELLITE COMMUNICATIONS. *Educ:* Birla Inst Technol, BE, 62, MS, 70; Univ South Fla, PhD(elec eng), 75. *Prof Exp:* Lectr eng, Birla Inst Technol & Sci, 65-70; from asst prof to prof, 77-86, PROF ENG, UNIV TOLEDO, 86- *Concurrent Pos:* Prin investr, Lewis Res Ctr, NASA, 79- *Mem:* Inst Elec & Electronics Engrs. *Res:* Digital signal processing; satellite communications. *Mailing Add:* Col Eng Dept Elec Eng Univ Toledo 2801 W Bancroft Toledo OH 43606

KWEI, TI-KANG, b Shanghai, China, Mar 19, 29; US citizen; m 54; c 3. POLYMER CHEMISTRY, PHYSICAL CHEMISTRY. *Educ:* Chiao Tung Univ, BS, 49; Univ Toronto, MASc, 54; Polytech Inst Brooklyn, PhD(chem), 58. *Prof Exp:* Polymer chemist, Stand Oil Co, Ind, 58-59; polymer chemist, Interchem Corp, 59-61, sr chemist, 61-63, group leader polymer chem, 63-65; mem tech staff, Bell Labs, 65-81; vpres res, Indust Technol Res Inst, Taiwan, 81-84; MEM TECH STAFF, POLYTECH UNIV, 84- *Honors & Awards:* Achievement Award, Chinese Inst Eng, USA; Res Award, Sigma Xi. *Mem:* Am Chem Soc. *Res:* Thermodynamics of polymer mixtures; viscoelasticity and surface chemistry of polymers; transport phenomena in polymers. *Mailing Add:* Polytech Univ Brooklyn NY 11201

KWENTUS, GERALD K(ENNETH), b St Louis, Mo, Jan 10, 37. CHEMICAL ENGINEERING. *Educ:* Wash Univ, BS, 60; Mass Inst Technol, PhD(chem eng), 67. *Prof Exp:* Sr res engr, Org Div, Monsanto Co, 66-70, res specialist, 70-75, sr res specialist, 75-80, sr res group leader, 80-82, MGR, RES & DEVELOP, MONSANTO CHEM INTERMEDIATES CO, 82- *Mem:* Am Inst Chem Engrs. *Res:* Preparative chromatography; fractional distillation; chemical kinetics; heat transfer. *Mailing Add:* 9526 Pine Spray Ct St Louis MO 63126

KWIATEK, JACK, b Kansas City, Mo, Feb 9, 24; m 48; c 3. INDUSTRIAL ORGANIC CHEMISTRY, POLYMER CHEMISTRY. *Educ:* Univ Ill, BS, 44; Cornell Univ, PhD(chem), 50. *Prof Exp:* Org res chemist, M W Kellogg Co, 50-54; res assoc, Gen Elec Co, 54-58; sr res assoc, Nat Distillers & Chem Corp, 58-88; RES SCIENTIST, USI DIV, QUANTUM CHEM CORP, 88- *Concurrent Pos:* Adj asst prof, Eve Col, Univ Cincinnati, 61-65; sr fel, Weizmann Inst Sci, 68-70. *Honors & Awards:* Chemist of the Year, Am Chem Soc, 74. *Mem:* Am Chem Soc; Catalysis Soc. *Res:* Homogeneous and heterogeneous catalysis; hydrogenation; syngas reactions; oxidation; carbonylation; coordination compounds; organometallics; free radical reactions; organophosphorus compounds; conductive polymers; biodegradable polymers. *Mailing Add:* 3135 N Farmcrest Dr Cincinnati OH 45213

KWIRAM, ALVIN L, b Man, Can, Apr 28, 37; m 64; c 2. PHYSICAL CHEMISTRY, CHEMICAL PHYSICS. *Educ:* Walla Walla Col, BS(chem) & BA(physics), 58; Calif Inst Technol, PhD(chem), 63. *Prof Exp:* Instr & res assoc chem, Calif Inst Technol, 62-63; res assoc physics, Stanford Univ, 63-64; instr chem, Harvard, 64-67, lectr, 67-70; assoc prof chem, Univ Wash, 70-75, chmn dept, 77-87, vice provost, 87-88, sr vice provost, 88-90, PROF CHEM, UNIV WASH, 75-, VICE PROVOST RES, 90- *Concurrent Pos:* Woodrow-Wilson fel, 58; Alfred D Sloan fel, 68-70; Mem, exec comt & secy-treas, Div Phys Chem, Am Chem Soc, 76-86, counr, 86-, founding comt & bd dirs, Coun Chem Res, 80-84, chmn, 82-83; mem, Comt Sci, Am Chem Soc, 88-; chair-elect, Chem Div, AAAS, 91- *Honors & Awards:* Eastman-Kodak Sci Award, 62; Univ-Indust Rel Award, Coun Chem Res, 86. *Mem:* Am Chem Soc; fel Am Phys Soc; Sigma Xi; fel AAAS. *Res:* Magnetic resonance in solids and molecular crystals; electron-nuclear double resonance; optical detection of magnetic resonance; structure and dynamics in ground and excited states of molecules. *Mailing Add:* Off Res Admin 312 AH-20 Univ Wash Seattle WA 98195

KWITEROVICH, PETER O, JR, b Danville, Pa, June 24, 40; m 65; c 3. LIPOPROTEIN METABOLISM. *Educ:* Holy Cross Col, AB, 62; Dartmouth Med Sch, BMS, 64; Johns Hopkins Univ, MD, 66. *Prof Exp:* Intern pediat, Childrens Hosp Med Ctr, 66-67, staff assoc lipoprotein res, Molecular Dis Br, NIH 67-70; resident pediat, Johns Hopkins Hosp, 70-72; from asst prof to assoc prof, 72-84, PROF PEDIAT & MED, JOHNS HOPKINS UNIV MED SCH, 84- *Concurrent Pos:* Chief lipid res, Atherosclerosis Unit, Johns Hopkins Univ, 76; prin investr lipid res clin, Johns Hopkins Univ, 71-; chmn, Dietary Intervention study in children, 87; mem nutrit study sect, NIH, 87- *Honors & Awards:* Blakeslee Award, Am Heart Asn. *Mem:* Soc Pediat Res; Am Soc Clin Invest. *Mailing Add:* CMSC 601 Johns Hopkins Hosp 600 N Wolfe St Baltimore MD 21205

KWITOWSKI, PAUL THOMAS, b Buffalo, NY, Nov 14, 39; m 63; c 4. INORGANIC CHEMISTRY. *Educ:* Canisius Col, BS, 61; Univ Wis, PhD(inorg chem), 67. *Prof Exp:* Res chemist, Airco-Speer Res Labs, NY, 66-69; assoc prof, 69-80, PROF CHEM, NIAGARA COMMUNITY COL, 80-, CHMN, DEPT PHYS SCI, 71- *Mem:* Am Chem Soc. *Res:* Gas chromatography; catalysis of organic reactions; halocarbon and organometallic chemistry; spectroscopy; chemistry of refractive compounds; chemical vapor deposition. *Mailing Add:* Dept Eng Math & Phys Sci Niagara County Community Col 3111 Saunders Sanborn NY 14132

KWITTER, KAREN BETH, b Brooklyn, NY, Mar 20, 51; m 79; c 2. GASEOUS NEBULAE, EVOLUTION OF LOW-MASS STARS. *Educ:* Wellesley Col, BA, 72; Univ Calif, Los Angeles, MA, 74, PhD(astron), 79. *Prof Exp:* Asst prof, 79-86, ASSOC PROF ASTRON, 86-, CHMN, ASTRON DEPT, WILLIAMS COL, 88- *Concurrent Pos:* Harlow Shapley vis lectr, Am Astron Soc, 81-; vis asst prof astron, Univ Ill, 83-84. *Mem:* Am Astron Soc; Sigma Xi; Int Astron Union; AAAS. *Res:* Gaseous nebulae, their chemical compositions and physical conditions, including nebulae around Wolf-Rayet stars and planetary nebulae; stellar evolution from planetary nebulae nucleus to sub dwarf to white dwarf. *Mailing Add:* Dept Astron Williams Col Williamstown MA 01267-2693

KWOCK, LESTER, b San Francisco, Calif, June 21, 42; m 68; c 1. BIOCHEMISTRY, RADIATION BIOCHEMISTRY. *Educ:* San Jose State Univ, BS, 65; San Diego State Univ, MS, 68; Univ Calif, Santa Barbara, PhD(chem), 73. *Prof Exp:* Instr radiation biol, 73-76, ASST PROF, SCH MED, TUFTS UNIV, 76-; AT DEPT RADIOL, UNIV NC, CHAPEL HILL. *Concurrent Pos:* NIH fel, Sch Med, Tufts Univ & Tufts-New England Med Ctr, 74-76; Nat Cancer Inst grant, Tufts-New England Med Ctr, 76- *Mem:* Am Chem Soc; Radiation Res Soc; Sigma Xi; AAAS. *Res:* Membrane transport; effects of ionizing radiation on biological systems; radioprotectors and radiosensitizers for normal and neoplastic cells. *Mailing Add:* Campus Box 7510 Old Clinic Bldg Univ NC Sch Med Chapel Hill NC 27599

KWOK, CLYDE CHI KAI, b Shanghai, China, May 26, 37; m 62; c 1. MECHANICAL ENGINEERING. *Educ:* McGill Univ, BEng, 61, MEng, 62, PhD, 67. *Prof Exp:* Res asst mech eng, McGill Univ, 61-64; prin scientist, Aviation Elec Ltd, 64-69; assoc prof mech eng, 69-77, PROF ENG, SIR GEORGE WILLIAMS CAMPUS, CONCORDIA UNIV, 77- *Mem:* Am Soc Mech Engrs; Am Inst Aeronaut & Astronaut. *Res:* Research and development of basic fluidic devices particularly the design and analysis of vortex type devices; fluid control elements and systems. *Mailing Add:* Dept Mech Engr Concordia Univ Sir G Williams 1455 De Maisonneuve Montreal PQ H3G 1M8 Can

KWOK, HOI S, b Hong Kong, China, Mar 1, 51; US citizen; m 78; c 3. LASER-MATTER INTERACTION. *Educ:* Northwestern Univ, BS, 73; Harvard Univ, MS, 75, PhD(physics), 78. *Prof Exp:* Teaching fel physics, Harvard Univ, 77-78; res fel chem, Lawrence Berkeley Lab, Univ Calif, 78-80; from asst prof to assoc prof, 80-85, PROF ENG, STATE UNIV NY BUFFALO, 85- *Concurrent Pos:* Prin investr grants, NSF, US Dept Energy, 81-; sci adv, Photochem Res Assocs, 82- & Excel Technologies, 87- *Honors & Awards:* Presidential Young Investr Award, 84. *Mem:* Am Phys Soc; sr mem Inst Elec & Electronics Engrs; Am Chem Soc; Optical Soc Am. *Res:* Ultrafast laser spectroscopy of semiconductors and superconductors; application of lasers to chemical and material processing. *Mailing Add:* Dept Elec & Comput Eng Bonner Hall State Univ NY Buffalo NY 14260

KWOK, MUNSON ARTHUR, b San Francisco, Calif, Apr 28, 41; m 77. GAS LASERS, LASER APPLICATIONS. *Educ:* Stanford Univ, BS, 62, MS, 63, PhD(aeronaut & astronaut), 67. *Prof Exp:* NSF fel, Stanford Univ, 67-68; mem tech staff, 68-76, staff scientist, 76-77, res scientist, 77-84, mgr, 84-87, DEPT HEAD, AEROSPACE CORP, 87- *Mem:* Am Phys Soc; Soc Photo-Optical Instrumentation Engrs; Am Inst Aeronaut & Astronaut; Am Soc Mech Engrs; Optical Soc Am. *Res:* Fluid mechanics; thermophysics; gas lasers; chemical lasers; kinetic theory; plasmas; high temperature gasdynamics; optical element metrology. *Mailing Add:* Aerospace Corp PO Box 92957 MS M5/753 Los Angeles CA 90009

KWOK, SUN, b Hong Kong, Sept 15, 49; Can citizen; m 73; c 2. ASTRONOMY. *Educ:* McMaster Univ, BSc, 70; Univ Minn, Minneapolis, MS, 72, PhD(physics), 74. *Prof Exp:* Fel astron, Dept of Physics, Univ BC, 74-76; asst prof, Dept of Physics, Univ Minn, Duluth, 76-77; res assoc, Ctr Res Exp Space Sci, York Univ, 77-78; res assoc astron, Herzberg Inst Astrophysics, 78-83; from asst prof to assoc prof, 83-88, PROF ASTRON, UNIV CALGARY, 88- *Concurrent Pos:* Course dir Atkinson Col, York Univ, 77-78; vis fel, Joint Inst Lab Astrophys, Univ Colo, 89-90. *Mem:* Am Astron Soc; Can Astron Soc; Int Astron Union. *Res:* Stellar mass loss; the late stages of stellar evolution; planetary nebulae; infrared astronomy. *Mailing Add:* Dept Physics & Astron Univ Calgary Calgary AB T2N 1N4 Can

KWOK, THOMAS YU-KIU, b Hong Kong; m; c 2. ELECTRONICS MATERIALS, VLSI TECHNOLOGY. *Educ:* Mass Inst Technol, BS, 76, MS, 78, PhD(mat sci), 82. *Prof Exp:* RES STAFF MEM, IBM RES DIV, THOMAS J WATSON RES CTR, 82- *Mem:* Am Phys Soc; Inst Elec & Electronics Engrs; Mat Res Soc. *Res:* Grain boundary structure and diffusion; microstructure of metallic thin films and submicron metal lines; electromigration, mechanical properties and reliability of multilevel interconnection; very-large-scale integration metallization; molecular dynamics and computer simulations. *Mailing Add:* IBM Res Div Thomas J Watson Res Ctr PO Box 218 Yorktown Heights NY 10598

KWOK, WO KONG, b Hong Kong, Jan 13, 36; m 63. ORGANIC CHEMISTRY, POLYMER CHEMISTRY. *Educ:* Nat Taiwan Univ, BS, 58; E Tenn State Univ, MA, 63; Ill Inst Technol, PhD(phys & org chem), 67. *Prof Exp:* Chemist, S China Bleaching & Dyeing Factory, 58-61; res chemist, Exp Sta, 66-76, sr res chemist, Kinston, 76-81, SR RES CHEMIST, CHESTNUT RUN, E I DU PONT DE NEMOURS & CO, INC, 81-, RES ASSOC, 84- *Concurrent Pos:* Ill Inst Technol Res Inst fel, 65-67. *Mem:* Am Chem Soc. *Res:* Elimination reaction kinetics and mechanism; polymer degradation mechanism; nonwoven technology. *Mailing Add:* 11 McCormick Dr W Riding Hockessin DE 19707

KWOLEK, STEPHANIE LOUISE, b New Kensington, Pa, July 31, 23. POLYMER CHEMISTRY & PROCESSING. *Educ:* Carnegie Inst Technol, BS, 46. *Hon Degrees:* DSc, Worcester Polytechnic Inst, 81. *Prof Exp:* Chemist, Fibers Dept, Pioneering Res Lab, Exp Sta, E I Du Pont de Nemours & Co, Inc, 46-59, from res chemist to sr res chemist, 59-74, res assoc, 74-86; RETIRED. *Concurrent Pos:* Consult. *Honors & Awards:* Publ Award, Am Chem Soc, 59; Howard N Potts Medal, Franklin Inst, 76; Mat Achievement Citation, Am Soc Metals, 78; Chem Pioneer Award, Am Inst Chemists, 80; Creative Invention Award, Am Chem Soc, 80; Eng/Technol Award, Soc Plastics Engrs, 85; Harold DeWitt Smith Mem Award, Am Soc Testing & Mat, 88. *Mem:* Am Chem Soc; Sigma Xi; Am Inst Chem; Int Union Pure & Appl Chem. *Res:* Condensation polymers; high temperature polymers; low temperature interfacial and solution polymerizations; high tenacity and high modulus fibers and films; liquid crystalline polymers, solutions and melts. *Mailing Add:* 312 Spalding Rd Wilmington DE 19803

KWON, TAI HYUNG, b Yechon, Korea, Sept 15, 32; m 69; c 1. SOLID STATE PHYSICS. *Educ:* Univ Ga, BS, 63, MS, 65, PhD(physics), 67. *Prof Exp:* Res fel, Ga Inst Technol, 67-69; from asst prof to assoc prof, 69-90, PROF PHYSICS, UNIV MONTEVALLO, 90- *Concurrent Pos:* Frederick Gardner Cottrell Res Corp grant, 71- *Mem:* Am Phys Soc. *Res:* Statistical physics; spin dynamics; neutron scattering; Heisenberg system; magnetism; lattice dynamics; anharmonicity. *Mailing Add:* Dept Physics Univ Montevallo Montevallo AL 35115

KWON-CHUNG, KYUNG JOO, b Seoul, Korea, Oct 5, 33; m 57; c 3. MEDICAL MYCOLOGY. *Educ:* Ewha Womans Univ, Korea, BS, 56, MS, 58; Univ Wis, MS, 63, PhD(bact), 65. *Prof Exp:* Instr microbiol, Ewha Womans Univ, 59-61; res asst bact, Univ Wis, 61-65; RES MICROBIOLOGIST, NIH, 68- *Concurrent Pos:* Fel, Univ Wis, 65; vis fel med mycol, NIH, 66-68. *Honors & Awards:* Director's Award, NIH, 77; Award, Int Soc Human & Animal Mycol, 82. *Mem:* Mycol Soc Am; Am Soc Microbiol; Med Mycol Soc of the Americas; Int Soc Human & Animal Mycol. *Res:* Morphology, genetics and pathogenicity of fungi. *Mailing Add:* Rm 11N104 Bldg 10 NIH Bethesda MD 20014

KWONG, JOSEPH N(ENG) S(HUN), b Chung Won, China, Oct 28, 16; nat US; m 42; c 3. CHEMICAL ENGINEERING. *Educ:* Stanford Univ, BA, 37; Univ Mich, MS, 39; Univ Minn, PhD(chem eng), 42. *Prof Exp:* Chem engr, Minn Mining & Mfg Co, 42-44; chemist, Shell Develop Co, Calif, 44-47, chemist & res engr, 47-51; sr chemist, Minn Mining & Mfg Co, 51-60, res specialist, 60-80; RETIRED. *Mem:* Am Chem Soc; Am Inst Chem Engrs. *Res:* Process and resin development, design and evaluation; thermodynamics; polyethylene terephthalate polymer and film technology. *Mailing Add:* 1399 N Hamline Ave St Paul MN 55108

KWONG, MAN KAM, b Canton, China, Feb 2, 47; m 70; c 2. MATHEMATICS. *Educ:* Univ Hong Kong, BSc, 68; Univ Chicago, MSc, 70, PhD(math), 73. *Prof Exp:* Lectr math, Hong Kong Baptist Col, 73-75, Hong Kong Polytech, 75-77; from asst prof to assoc prof, 77-83, PROF MATH, NORTHERN ILL UNIV, 83- *Concurrent Pos:* Resident scientist, Argonne Nat Lab, 82-83, spec term appointment, 85- *Mem:* Sigma Xi; Am Math Soc. *Res:* Ordinary differential equations; functional analysis; inequalities. *Mailing Add:* Math & Comp Sci Div Argonne Nat Lab Argonne IL 60439

KWONG, YUI-HOI HARRIS, b Hong Kong, Oct 10, 57; m 85. GRAPH THEORY, INTEGER SEQUENCES. *Educ:* Univ Mich, BS, 80, MS, 81; Univ Pa, PhD(math), 87. *Prof Exp:* ASST PROF MATH & COMP SCI, STATE UNIV NY COL, FREDONIA, 87- *Mem:* Am Math Soc; Math Asn Am; Sigma Xi. *Res:* Graph labeling; gextremal graph theory; divisibility; congruences of integer sequences. *Mailing Add:* Dept Math & Comp Sci State Univ NY Col Fredonia Fredonia NY 14063

KWUN, KYUNG WHAN, b Seoul, Korea, Mar 7, 29; m 57; c 3. MATHEMATICS. *Educ:* Seoul Nat Univ, BS, 52; Univ Mich, MS, 54, PhD(math), 58. *Prof Exp:* Instr math, Univ Mich, 57-58; res assoc, Tulane Univ, 58-59; vis assoc prof, Seoul Nat Univ, 59-60, from instr to asst prof, 60-62; vis lectr, Fla State Univ, 62-64, assoc prof, 64-65; assoc prof, 65-66, PROF MATH, MICH STATE UNIV, 66- *Concurrent Pos:* Vis lectr, Univ Wis, 61-62; mem, Inst Advan Study, 64-65. *Mem:* Am Math Soc. *Res:* Topology, particularly theory of manifolds. *Mailing Add:* Dept Math Mich State Univ East Lansing MI 48824

KYAME, GEORGE JOHN, textile physics, textile engineering; deceased, see previous edition for last biography

KYAME, JOSEPH JOHN, b New Orleans, La, Mar 12, 24; wid; c 1. MATHEMATICAL PHYSICS. *Educ:* Tulane Univ, BS, 44, MS, 45; Mass Inst Technol, PhD(physics), 48. *Prof Exp:* Asst & instr, 44-45, asst prof, 48-58, ASSOC PROF PHYSICS, TULANE UNIV, 58- *Mem:* Am Phys Soc; Sigma Xi. *Res:* Electromagnetic theory; piezoelectricity; thermodynamics. *Mailing Add:* 32 Warbler St New Orleans LA 70124

KYANKA, GEORGE HARRY, b Syracuse, NY, July 17, 41; m 66; c 2. MECHANICAL ENGINEERING, WOOD SCIENCE. *Educ:* Syracuse Univ, BS, 62, MS, 66, PhD(mech eng), 75. *Prof Exp:* Res engr gas turbines, Caterpillar Tractor Co, 62-64; res asst aero eng, Syracuse Univ, 64-66; asst prof mech tech, Onondaga Col, 67-68; from asst prof to assoc prof, 68-80, PROF WOOD ENG, COL ENVIRON SCI & FORESTRY, STATE UNIV NY, 80-, CHMN DEPT, 85- *Concurrent Pos:* NSF res fel, Syracuse Univ, 67; proj dir, NSF, 70-73, 76- & Weyerhaeuser Corp, 78-; adj prof, Onondaga Col Archit, 70-; consult engr, 70-; dir, Educ Opportunity Prog in Forestry, 73-76. *Mem:* Am Soc Mech Engrs; Am Soc Testing & Mat; Soc Exp Stress Anal; Forest Prod Res Soc; Am Acad Mech; Soc Exp Mech. *Res:* Mechanical properties of wood and wood products; testing and design of wood products; professional responsibility and products liability in product design; wood in architecture and art. *Mailing Add:* Dept Wood Prod Eng State Univ NY Syracuse NY 13031

KYBA, EVAN PETER, b Canora, Sask, June 27, 40; m 62; c 1. ORGANIC CHEMISTRY, SYNTHETIC MEDICINAL CHEMISTRY. *Educ:* Univ Sask, BA, 62; Univ Ala, PhD(org chem), 71. *Prof Exp:* Teacher chem, Regina Col Inst, Sask, 62-65; Nat Res Coun Can fel, Univ Calif, Los Angeles, 71-72; from asst prof to assoc prof, 72-85, prof chem, Univ Tex, Austin, 85-87; dir, 88-90, SR DIR MED CHEM, ALCON LAB, INC, 90- *Mem:* Am Chem Soc; The Chem Soc. *Res:* Reactive intermediates; organophosphorus chemistry; stereochemistry; synthesis of unusual small heterocycles and multiheteromacrocycles; homogeneous catalysis; organometallic chemistry; medicinal chemistry. *Mailing Add:* Alcon Labs Inc Conner Res Ctr 6201 S Freeway Ft Worth TX 76134

KYBETT, BRIAN DAVID, b Oxford, Eng, May 10, 38; m 63; c 1. PHYSICAL CHEMISTRY. *Educ:* Univ Wales, BSc, 60, PhD(chem), 63. *Prof Exp:* Res assoc chem, Rice Univ, 63-65; from asst prof to assoc prof, 65-81, PROF CHEM & DIR, ENERGY RES UNIT, UNIV REGINA, 81- *Mem:* Chem Inst Can; Royal Soc Chem; Sigma Xi; Am Chem Soc; Can Inst Mining. *Res:* Lattice energies; thermochemistry; reactivity of coal. *Mailing Add:* Energy Res Unit Univ Regina Regina SK S4S 0A2 Can

KYBURG, HENRY, b New York, NY, Oct 9, 28; m 60; c 8. UNCERTAIN INFERENCE INDUCTIVE LOGIC. *Educ:* Yale Univ, BA, 48; Columbia Univ, MA, 52, PhD(philos), 55. *Prof Exp:* Asst prof math, Wesleyan Univ, 58-61; res assoc, Rockefeller Univ, 61-62; PROF PHILOS, UNIV ROCHESTER, 63-, PROF COMPUTER SCI, 86- *Mem:* Fel AAAS; Am Philos Asn; Am Math Soc; Am Asn Artificial Intel; Asn Comput Mach. *Res:* Uncertain inference; inductive logic; representation of uncertainty; decision under uncertainty. *Mailing Add:* Univ Rochester Rochester NY 14627

KYCIA, THADDEUS F, b Montreal, Que, Aug 10, 33; m 57; c 1. HIGH ENERGY PHYSICS. *Educ:* McGill Univ, BS, 54, MS, 55; Univ Calif, Berkeley, PhD(high energy physics), 59. *Prof Exp:* Res asst high energy physics, 59-61, from asst physicist to assoc physicist, 61-66, physicist, 66-72, SR PHYSICIST, BROOKHAVEN NAT LAB, 72- *Mem:* Fel Am Phys Soc. *Res:* Development of Cerenkov detectors; measurement of total cross sections; search for resonances; measurement of magnetic moment of hyperons; study of rare K meson decay. *Mailing Add:* Brookhaven Nat Lab Upton NY 11973

KYDD, DAVID MITCHELL, b Jersey City, NJ, May 17, 03; m 29; c 1. MEDICINE, METABOLISM. *Educ:* Princeton Univ, BS, 24; Harvard Univ, MD, 28. *Prof Exp:* Sterling res fel, Yale Univ, 29, Sax res fel, 29-30, from instr to asst prof med, Sch Med, 30-34; asst prof, Albany Med Col, 36-47; assoc prof, Sch Med, Yale Univ, 47-52; from assoc prof to prof, 52-69, EMER PROF MED, STATE UNIV NY DOWNSTATE MED CTR, 70- *Concurrent Pos:* Assoc, Bassett Hosp, NY, 34-47. *Mem:* Am Soc Clin Invest; Harvey Soc; Soc Exp Biol & Med; Am Inst Nutrit. *Res:* Electrolyte disturbances; thyroid diseases. *Mailing Add:* 48 Grove St Cooperstown NY 13326-1427

KYDD, GEORGE HERMAN, cardiopulmonary physiology; deceased, see previous edition for last biography

KYDD, PAUL HARRIMAN, b New Haven, Conn, Nov 25, 30; m 56; c 2. PHYSICAL CHEMISTRY, CHEMICAL ENGINEERING. *Educ:* Princeton Univ, AB, 52; Harvard Univ, MA, 53, PhD(phys chem), 56. *Prof Exp:* Fel phys chem, Harvard Univ, 56-57; phys chemist, Gen Elec Res Lab, 57-66; lectr, Harvard Univ, 59-60; mgr chem processes, Gen Elec Res & Develop Ctr, 66-75; vpres technol, Hydrocarbon Res, Inc, 75-83; PRES, PARTNERSHIPS LTD INC, 83- *Mem:* AAAS; Sigma Xi; Am Inst Chem Eng. *Res:* Coal liquifaction, gasification; petroleum production and refining; gas turbines, power generation; renewable resources, chemical intermediates; microcomputer applications. *Mailing Add:* BOC Group 100 Mountain Ave Murray Hill NJ 07974

KYDES, ANDY STEVE, b Spilia, Greece, Jan 21, 45; US citizen; c 2. NUMERICAL ANALYSIS, ENERGY SYSTEMS ANALYSIS. *Educ:* Harvard Univ, AB, 68; State Univ NY, Stony Brook, MS, 73, PhD(numerical anal), 74. *Prof Exp:* Instr math & physics, Milton Acad, Mass, 68-71; asst prof math, State Univ NY, Stony Brook, 74-76; ENERGY SYST ANALYST, DEPT ENERGY & ENVIRON, BROOKHAVEN NAT LAB, 76- *Mem:* Oper Res Soc Am; Inst Mgt Sci. *Res:* Energy systems analysis; multi-criteria analysis; optimization. *Mailing Add:* 3551 N Nottingham St Arlington VA 22207

KYDONIEFS, ANASTASIOS D, b Athens, Greece, Mar 6, 28; m 53; c 2. APPLIED MATHEMATICS, CONTINUUM MECHANICS. *Educ:* Univ Nottingham, MSc, 65, PhD(theoret mech), 67. *Prof Exp:* Res fel theoret mech, Univ Nottingham, 67-69, sr res asst, 67-68; asst prof math, 68-73, ASSOC PROF MATH, LEHIGH UNIV, 73- *Res:* Finite elasticity and its applications to biomechanics. *Mailing Add:* Dept Math & Physics Aristotle Univ Thessalonike Sch Technol Thessaloniki 54006 Greece

KYHL, ROBERT LOUIS, b Omaha, Nebr, July 27, 17; m 43; c 1. ELECTRICAL ENGINEERING. *Educ:* Univ Chicago, SB, 37; Mass Inst Technol, PhD(physics), 47. *Prof Exp:* Asst physics, Univ Chicago, 40-41; res assoc radiation lab, Mass Inst Technol, 41-45, insulation lab, 45-47 & electronics res lab, 47-48; res assoc, Hansen Lab, Stanford Univ, 48-54 & res lab, Gen Elec Co, 54-56; from assoc prof to prof, 56-83, EMER PROF ELEC ENG, MASS INST TECHNOL, 83- *Honors & Awards:* Baker Award, Inst Radio Eng, 58. *Mem:* Am Phys Soc; Inst Elec & Electronics Engrs. *Res:* Microwave spectroscopy and power tubes; electron accelerators; solid state masers. *Mailing Add:* Dept Elec Eng Mass Inst Technol 77 Massachusetts Ave Cambridge MA 02139

KYHOS, DONALD WILLIAM, b Los Angeles, Calif, Apr 10, 29; m 61; c 3. PLANT TAXONOMY, PLANT CYTOGENETICS. *Educ:* Whittier Col, AB, 51, MS, 56; Univ Calif, Los Angeles, PhD(bot), 64. *Prof Exp:* NIH fel biol, Stanford Univ, 64-65; from asst prof to assoc prof, 65-74, PROF BOT, UNIV CALIF, DAVIS, 74- *Concurrent Pos:* Australian Res Grant Comt fel, 72. *Mem:* Am Soc Plant Taxon; Bot Soc Am; Soc Study Evolution; Sigma Xi. *Res:* Plant systematics and evolutionary cytogenetics. *Mailing Add:* Dept Bot Univ Calif Davis CA 95616

KYKER, GARY STEPHEN, inorganic chemistry, polymer chemistry, for more information see previous edition

KYLE, BENJAMIN G(AYLE), b Atlanta, Ga, Dec 4, 27; m 52; c 4. CHEMICAL ENGINEERING. *Educ:* Ga Inst Technol, BChE, 50; Univ Fla, MSE, 55, PhD(chem eng), 58. *Prof Exp:* From asst prof to assoc prof, 58-64, PROF CHEM ENG, KANS STATE UNIV, 64- *Mem:* Am Chem Soc; Am Inst Chem Engrs. *Res:* Thermodynamics; mass transfer. *Mailing Add:* Dept Chem Eng Kans State Univ Manhattan KS 66506

KYLE, HERBERT LEE, b Monmouth, Ill, June 28, 30; m 60; c 1. EARTH RADIATION BUDGET, SOLAR PHYSICS. *Educ:* Univ Ariz, BS, 54; Univ NC, MS, 59, PhD(atomic physics), 64. *Prof Exp:* SPACE PHYSICIST, GODDARD SPACE FLIGHT CTR, NASA, 59- *Mem:* Am Phys Soc; Am Geophys Union. *Res:* Remote sensing of the earth's radiation budget; remote sensing for the cloud properties and cloud cover; atmospheric radiative transfer theory and numerical analysis; measurement of the solar constant and it's variations; application of computers to scientific problems; creation of satellite climate data sets with emphasis on long-term sensor calibration and numerical analysis. *Mailing Add:* Code 636 Goddard Space Flight Ctr Greenbelt MD 20771

KYLE, MARTIN LAWRENCE, b Akron, Ohio, Jan 2, 35; m 57; c 3. CHEMICAL ENGINEERING. *Educ:* Univ Notre Dame, BS, 56; Purdue Univ, MS, 61; Univ Chicago, MBA, 71. *Prof Exp:* Chem engr, E I DuPont de Nemours & Co, 56-57; chem engr, 60-78, ASST LAB DIR, ARGONNE NAT LAB, 78- *Mem:* Am Chem Soc. *Res:* Battery development; solar energy; coal technology. *Mailing Add:* Argonne Nat Lab 9700 S Cass Ave Argonne IL 60439-6324

KYLE, PHILIP R, b Wellington, NZ, Dec 3, 47. VOLCANOLOGY. *Educ:* Victoria Univ, PhD(geol), 76. *Prof Exp:* PROF GEOCHEM, NMEX INST MINING TECHNOL, 81- *Mem:* Fel Geol Soc Am. *Mailing Add:* Dept Geosci NMex Inst Mining Technol Socorro NM 87801

KYLE, ROBERT ARTHUR, b Bottineau, NDak, March 17, 28; m 54; c 4. HEMATOLOGY, MONOCLONAL GAMMOPATHICS. *Educ:* Univ NDak, BS, 48; Northwestern Univ Med Sch, MD, 52; Univ Minn, MS, 58. *Prof Exp:* Clin asst hemat, Tufts Univ Sch Med, 60-61; postdoctoral fel hemat, Nat Cancer Inst, 60-61; CONSULT MED, MAYO MED SCH, 61- *Concurrent Pos:* William H Donner prof med & Lab Med, Mayo Med Sch, 81-87; prin invest, Acute Leukemia Group B, 71-72; chmn, myeloma comn, Eastern Coopr Oncol Group, 84-; consult, PDQ Info Bank, Nat Cancer Inst, 84- *Mem:* Central Soc Clin Res; Am Assoc Cancer Res; Am Soc Hematol; fel Am Col Physicians; Am Soc Clin Oncol. *Res:* Monoclonal gammopathies; includes multiple myeloma, amyloidosis, macroglobulinemia, and related plasma cell proliferative process. *Mailing Add:* 200 First St SW Rochester MN 55905

KYLE, THOMAS GAIL, b Crawford, Okla, Sept 12, 36; m 58; c 2. INFRARED STUDIES, METEOROLOGICAL TRACERS. *Educ:* Univ Okla, BS, 60, MS, 62; Univ Denver, PhD(physics), 65. *Prof Exp:* Res officer, Commonwealth Sci & Indust Res Orgn, Australia, 65-66; res physicist, Univ Denver, 66-71; scientist, Nat Ctr Atmospheric Res, 71-76; STAFF MEM, LOS ALAMOS NAT LAB, 76- *Concurrent Pos:* Vis prof, Clemson Univ, 81-82, Denver Univ, 82-83, Oklahoma State Univ, 83-84. *Mem:* Optical Soc Am; Sigma Xi. *Res:* Computational physics; theoretical and experimental studies of infrared spectra; laser studies; studies of the composition and radiative properties of the atmosphere; weather modification; artificial intelligence; atmospheric modeling; image processing. *Mailing Add:* Woodstock Hill Sandridge St Albans Hertsfordshire AL49 HQ England

KYLE, WENDELL H(ENRY), b Louisville, Ohio, June 8, 20; m 45; c 5. GENETICS. *Educ:* Iowa State Univ, BS, 43; Univ Wis, PhD(genetics), 49. *Prof Exp:* Asst genetics, Univ Wis, 45-48, proj assoc, 48-49; animal geneticist, US Sheep Exp Sta, Idaho, 49-57; asst nat coord, Regional Poultry Breeding Proj, Purdue Univ, 57-58, assoc prof univ & leader, Pioneering Res Lab Pop Genetics, USDA, 58-66; scientist adminstr, Div Res Grants, NIH, 66-78; RETIRED. *Mem:* AAAS; Am Inst Biol Sci; Biomet Soc; Genetics Soc Am; Sigma Xi. *Res:* Population genetics of Tribolium, mice, poultry and livestock; biometry. *Mailing Add:* 3011 McComas Ave Kensington MD 20895

KYLSTRA, JOHANNES ARNOLD, b Manado, Neth EIndies, Nov 30, 25; m 56; c 2. MEDICINE, PHYSIOLOGY. *Educ:* Univ Leiden, MD, 58. *Prof Exp:* prof pulmonary & allergy, Duke Univ, 72-89; RETIRED. *Honors & Awards:* Lockheed Award, Marine Technol Soc, 70. *Mem:* AAAS; Am Physiol Soc; Undersea Med Soc (pres, 73-74). *Res:* Liquid breathing and lung lavage. *Mailing Add:* 4415 Malvern Rd Durham NC 27707

KYNCL, J JAROSLAV, b Prague, Czech, Aug 16, 36; US citizen; m 61; c 2. HYPERTENSION, ADRENERGIC RECEPTORS. *Educ:* Masaryk Univ, Czech, MS, 59; Komensky Univ, Czech, PhD(pharmacol), 63; Czech Acad Sci, Prague, ScC, 67. *Prof Exp:* Sr res scientist, Res Inst Pharm & Biochem, Prague, Czech, 63-68; A V Humboldt fel cardiovasc res, Univ Heidelberg, Ger, 68-72; res fel, Cleveland Clin Res Div, 70-72; E VOLWILER RES FEL, ABBOTT LABS RES & DEVELOP, 72- *Concurrent Pos:* Mem, Coun High Blood Pressure, Am Heart Asn. *Honors & Awards:* Pfizer Lectr, Clin Res Inst Montreal, 80. *Mem:* Fel Am Heart Asn; Am Endocrine Soc; Am Soc Pharmacol & Exp Therapeut; Am Soc Hypertension; Int Soc Hypertension. *Res:* Pathophysiology of cardiovascular and endocrine disorders; novel concepts and specific agents useful in therapy; author of numerous scientific publications. *Mailing Add:* 964 Lake Rd Lake Forest IL 60045

KYRALA, ALI, b New York, NY, Dec 14, 21; m 66; c 3. MATHEMATICAL PHYSICS, ASTROGEOPHYSICS. *Educ:* Mass Inst Technol, BSc, 47; Stanford Univ, MSc, 48; Harvard Univ, SM, 57; Vienna Tech Univ, DSc(physics), 60. *Prof Exp:* Instr math, Univ Santa Clara, 47-48 & Univ Mass, 51-53; mathematician & physicist, Lessells & Assocs, Mass, 53-58; math physicist, Goodyear Aerospace Corp, 58-60; staff scientist, semiconductor div, Motorola, Inc, 60-62; Fulbright prof & award math & physics, Univ Alexandria, 63-64; PROF PHYSICS, ARIZ STATE UNIV, 64- *Concurrent Pos:* Vis prof math, Am Univ Beirut & Univ Libanaise, 68-70 & Univ Petrol & Minerals, Dhahran/Saudi Arabia, 75-77; theory of lightning, NASA-Am Soc Eng Educ, 81, 82, 84, 85. *Mem:* Am Phys Soc; Am Math Soc; Europ Phys Soc; Brit Interplanetary Soc; Am Astron Asn; Inst Elec & Electronics Engrs. *Res:* Theory of metal fatigue; theory of lightning; partial differential and integral equations; microwaves; function theory; theoretical solid state physics; electro biophysics; sub-quantum theory; astrogeophysics; relativity; statistical physics; plasma physics; planetary sciences. *Mailing Add:* Dept Physics Ariz State Univ Tempe AZ 85287

KYRALA, GEORGE AMINE, b Bhamdoun, Lebanon, Apr 20, 46; US citizen; m 73; c 2. OPTICS, LASER FUSION & INTERACTIONS. *Educ:* Am Univ Beirut, BS, 67; Yale Univ, MPh, 69, PhD(physics), 74. *Prof Exp:* Fel physics, Joint Inst Lab Astrophysics, Univ Colo, 74-76; res assoc optics & physics, Optical Sci Ctr & Dept Physics, Univ Ariz, 76-78, res fel lasers & spectros, Dept Physics, 78-79; MEM STAFF, PHYSICS DIV, LOS ALAMOS NAT LAB, 79- *Concurrent Pos:* Lectr, Dept Physics, Univ Colo, 75; vis fac & consult, Al-Hazen Res Ctr, Baghdad, 75; Rockefeller fel, Am Univ Beirut, Lebanon, 66-67; Gibbs fel, Yale Univ, 67-68. *Mem:* Am Inst Phys; Arab Phys Soc; Int Optical Eng Soc. *Res:* Charge transfer in atomic collision; electron scattering from excited atoms; laser construction and use in ultra high resolution spectroscopy; laser fusion experiments and optics; plasma xray spectroscopy; ultrafast diagnostics. *Mailing Add:* MS E-526 Los Alamos Nat Lab Los Alamos NM 87545

KYRIAKIS, JOHN M, MEDICINE. *Prof Exp:* RES FEL, MASS GEN HOSP, 87-; INSTR MED, HARVARD MED SCH, 90- *Mailing Add:* Dept Med Mass Gen Hosp MGH E Bldg 149 13th St Charlestown MA 02129

KYRIAKOPOULOS, NICHOLAS, b Atalanti, Greece, Nov 14, 37; m 67; c 2. SYSTEMS THEORY & CONTROLS, DIGITAL SIGNAL PROCESSING. *Educ:* George Washington Univ, BEE, 60, MS, 63, DSc, 68. *Prof Exp:* Electronic engr, Harry Diamond Labs, Dept Army, 60-62; aerospace engr, Goddard Space Flight Ctr, NASA, 62-64; instr elec eng, George Washington Univ, 64-66, from asst prof to assoc prof, elec eng & comput sci, 66-79, PROF ENG, GEORGE WASHINGTON UNIV, 80- *Concurrent Pos:* NASA-Am Soc Eng Educ fac fel, Goddard Space Flight Ctr, NASA, 67-68; consult, Nat Biomed Res Found, 66-67; Howard Res Corp, 67 & RCA Serv Co, 67-; summer fac fel, 74-76; vis prof, Nat Tech Univ, Athens, 72-73; NIK Assocs, 77-; sr scientist, US Arms Control & Disarmament Agency, 79-; consult, Int Atomic Energy Agency, 82- *Honors & Awards:* Sigma Xi. *Mem:* Inst Elec & Electronics Engrs. *Res:* Computer-aided analysis and design; performance evaluation of space communication systems; digital signal processing; monitoring and data collection systems; process automation and control; applications of technology to arms control; security systems. *Mailing Add:* Dept Elec Eng George Washington Univ 2121 Eye St NW Washington DC 20052

KYRIAZIS, ANDREAS P, b Aigion, Greece, Jan 19, 32; m 65; c 1. PATHOLOGY. *Educ:* Nat Univ Athens, MD, 57; Univ Thessaloniki, DrSc(path), 62; Jefferson Med Col, PhD(path), 68. *Prof Exp:* Resident path, Univ Thessaloniki, 60-64; attend pathologist, Piraeus Gen Hosp, Greece, 64; resident path, Jersey City Med Ctr, 65; res assoc, Univ Chicago, 68-70, asst prof path, 70-75; assoc prof path, Univ Cincinnati, 78-82; PROF PATH, UNIV MED & DENT NJ, 82- *Concurrent Pos:* Vis scientist, Argonne Cancer Res Hosp, Univ Chicago, 68-; vis prof, dept gen & tumor immunol, Hebrew Univ Hadassah Med Sch, Jerusalem, Isreal. *Mem:* Reticuloendothelial Soc; Am Asn Pathologists; NY Acad Sci; Int Acad Path; Am Asn Cancer Res; AAAS; Tissue Cult Asn. *Res:* Tissue culture; tumor biology; experimental tumor chemotherapy. *Mailing Add:* Dept Path Univ Med & Dent NJ NJ Med Sch 100 Bergen St Newark NJ 07103

KYSER, DAVID SHELDON, b Houston, Tex, Aug 29, 36; m 59; c 3. EXPERIMENTAL SOLID STATE PHYSICS. *Educ:* Univ Tex, BS, 58 & 60, MA, 63, PhD(physics), 65. *Prof Exp:* Res assoc chem phys, Inst Study Metals, Univ Chicago, 64-66; PHYSICIST, NAVAL WEAPONS CTR, 66- *Mem:* Am Phys Soc; Sigma Xi. *Res:* Modulation of optical properties of semiconductors. *Mailing Add:* 1727 W Langley Ridgecrest CA 93555

KYTE, JACK ERNST, b Pasadena, Calif, May 21, 47. BIOCHEMISTRY. *Educ:* Carleton Col, BA, 67; Harvard Univ, PhD(biochem), 72. *Prof Exp:* Damon Runyon Fund fel biochem, 72-74; ASST PROF BIOCHEM, UNIV CALIF, SAN DIEGO, 74- *Res:* Molecular structure of proteins which catalyze the transport of matter across biological membranes. *Mailing Add:* Dept Chem D-006 Univ Calif at San Diego La Jolla CA 92093

KYTHE, PREM KISHORE, b India, Jan 29, 30; US citizen; m 55; c 2. WAVE STRUCTURE, GEOMETRIC & COEFFICIENT PROBLEMS. *Educ:* Agra Univ, BSc, 50, MSc, 55; Aligarh Muslim Univ, PhD(math), 61. *Prof Exp:* Lectr math, Aligarh Muslim Univ, India, 58-60; from lectr to asst prof, Indian Inst Technol, Bombay, India, 60-67; from vis assoc to assoc prof, 67-74, PROF MATH, UNIV NEW ORLEANS, LA, 74- *Concurrent Pos:* Fel, UNESCO, 63; consult, Inst Human Learning, Univ Calif, Berkeley, 64; vis prof, math dept, Imperial Col, London, 73 & comput sci dept, Univ Ill, Urbana-Champaign, 86. *Mem:* Am Acad Mech. *Res:* Univalent functions; boundary-value problems in continuum mechanics; laplace transforms; wave theory; wave structure in unsteady free convection flows in a rotating medium; geometric and coefficient problems in some subclasses of univalent functions. *Mailing Add:* Dept Math Univ New Orleans-Lake Front New Orleans LA 70148

KYUNG, JAI HO, b Seoul, Korea, Dec 26, 47; m 73. ORGANIC CHEMISTRY. *Educ:* Seoul Nat Univ, BS, 69; Brown Univ, PhD(org chem), 75. *Prof Exp:* Res chemist, 74-76, SR RES CHEMIST, ASHLAND CHEM CO, 76- *Mem:* Am Chem Soc. *Res:* Synthesis of organic ligands for recovery of metals for mining industry and commercial development of solvent extraction of hydrometallurgy; homogeneous and heterogeneous catalysis of petrochemicals and industrial intermediate chemicals. *Mailing Add:* Ashland Chem Co PO Box 2219 Columbus OH 43216

L

LA, SUNG YUN, b Seoul, Korea, Sept 24, 36. PHYSICS. *Educ:* WVa Wesleyan Col, BS, 59; Univ Conn, MS, 62, PhD(physics), 64. *Prof Exp:* Res assoc appl physics, mat res lab, Pa State Univ, 65-68; assoc prof, 68-78, PROF PHYSICS & EARTH SCI, WILLIAM PATERSON COL NJ, 78- *Mem:* Am Phys Soc. *Res:* Theoretical studies of ionic crystals; defects investigated by electron spin resonance technique, cohesive energy and compressibility. *Mailing Add:* Dept Physics William Paterson Col NJ Wayne NJ 07470

LAAKSO, JOHN WILLIAM, b Minn, Jan 28, 15; m 41; c 4. BIOCHEMISTRY. *Educ:* Winona State Col, BS, 38; Mont State Col, MS, 49; Univ Minn, PhD(biochem), 56. *Prof Exp:* Teacher high schs, Minn, 38-42; instr math, Mont State Col, 46-47; from instr to assoc prof chem, St Cloud State Univ, 48-63, chmn dept, 66-73, prof chem, 63-80; RETIRED. *Mem:* Am Chem Soc; Sigma Xi. *Res:* Synthesis and biological assay of orotic acid analogs. *Mailing Add:* 3496 1060th St South Haven IN 55382

LAALE, HANS W, b Copenhagen, Denmark, Apr 20, 35; Can citizen. EXPERIMENTAL EMBRYOLOGY, TERATOLOGY. *Educ:* Bob Jones Univ, BSc, 59; Univ Western Ont, MSc, 61; Univ Toronto, PhD(zool), 66. *Prof Exp:* Asst lectr biol, Hong Kong Baptist Col, 61-63; lectr, Chinese Univ Hong Kong, 66-67; from asst prof to assoc prof, 67-85, PROF ZOOL, UNIV

MAN, 85- *Concurrent Pos:* Vis prof, Tunghai Univ, Taiwan, 73- & Univ BC, 80- *Mem:* Can Soc Zoologists; Sigma Xi. *Res:* Teleost embryology and teratology; in vitro fish embryo culture: differentiation and organogenesis. *Mailing Add:* Dept Zool Univ Man Winnipeg MB R3T 2N2 Can

LAALI, KHOSROW, b Tehran, Iran, July 5, 51. ORGANIC CHEMISTRY. *Educ:* Univ Tehran, BS, 73; Univ Manchester, PhD(chem), 77. *Prof Exp:* Res fel, Sci Res Coun, King's Col, Univ London, 77-79; Nat Ctr Sci Res fel, Univ Strasbourg, 79-80; sci co-worker-demonstr, Univ Amsterdam, 80-81; res assoc, Swiss Fed Inst Technol, 81-82; res scientist, Hydrocarbon Res Inst, Univ Southern Calif, 82-85; ASST PROF CHEM, KENT STATE UNIV, 85- *Mem:* Am Chem Soc; Royal Soc Chem; Sigma Xi; Swiss Chem Soc. *Res:* Physical organic, organometalic, fluorine and hydrocarbon chemistry; reactive intermediates; superacid chemistry; Friedel-crafts, NMR and modern mass spectrometry. *Mailing Add:* Dept Chem Kent State Univ Kent OH 44242

LAANE, JAAN, b Paide, Estonia, June 20, 42; US citizen; m 66; c 2. PHYSICAL CHEMISTRY, SPECTROSCOPY. *Educ:* Univ Ill, BS, Urbana, 64; Mass Inst Technol, PhD(chem), 67. *Prof Exp:* Asst prof chem, Tufts Univ, 67-68; chmn, Div Phys & Nuclear Chem 77-87; from asst prof to assoc prof, 68-76, PROF CHEM, TEX A&M UNIV, 76-, SR POLICY ADV, KORIYAMA, JAPAN, 90- *Concurrent Pos:* Vis scientist, Los Alamos Sci Lab, NMex, 64 & 67-68; vis prof, Univ Bayreuth, WGer, 79-80, 81, 83; dir, Inst Pac Asia, 87-90. *Honors & Awards:* Kendall Award, 64; Kodak Award, 67; Alexander von Humboldt US Sr Scientist Award, 79. *Mem:* Am Chem Soc; Am Phys Soc; Soc Appl Spectros; Coblentz Soc (treas, 85-); fel Am Inst Chem. *Res:* Far-infrared spectroscopy of small ring compounds; potential energy functions; organometallic syntheses; infrared and raman spectroscopy; nitrogen-oxygen chemistry. *Mailing Add:* Dept Chem Tex A&M Univ College Station TX 77843

LAASPERE, THOMAS, b T :nnasilma, Estonia, Mar 17, 27; US citizen; m 55; c 3. RADIOPHYSICS. *Educ:* Univ Vt, BS, 56; Cornell Univ, MS, 58, PhD(commun eng), 60. *Prof Exp:* Res assoc radiophys, Cornell Univ, 60-61; from asst prof to assoc prof, 61-70, PROF ENG, THAYER SCH ENG, DARTMOUTH COL, 70- *Concurrent Pos:* Mem comn IV, US Nat Comt, Int Sci Radio Union, 64- *Mem:* Inst Elec & Electronics Engrs. *Res:* Scattering of radio waves in the troposphere and ionosphere; whistlers and other audio-frequency electromagnetic waves; space research; electric rates, load management. *Mailing Add:* Dept Eng Scis Dartmouth Col Thayer Sch Hanover NH 03755

LAATSCH, RICHARD G, b Fairmont, Minn, July 14, 31; m 66; c 5. MATHEMATICAL ANALYSIS. *Educ:* Cent Mo State Col, BS, 53; Univ Mo, MA, 57; Okla State Univ, PhD(math), 62. *Prof Exp:* Instr math, Univ Tulsa, 57-60; asst prof, Okla State Univ, 62; from asst prof to assoc prof, 62-70, PROF MATH & ASSOC CHMN DEPT, MIAMI UNIV, 70- *Mem:* Math Asn Am. *Res:* Subaddive functions; topological vector spaces and cones of functions. *Mailing Add:* Dept Math Miami Univ 123 Bachelor Hall Oxford OH 45056

LABANA, SANTOKH SINGH, b Maritanda, India, Nov 15, 36; m 64. ORGANIC CHEMISTRY, POLYMER SCIENCE. *Educ:* Univ Panjab, India, BSc, 57, MSc, 59; Cornell Univ, PhD(org chem), 63. *Prof Exp:* Lectr chem, G H G Col, Sadhar, India, 58-60; res chemist, Univ Calif, Berkeley, 63-64; scientist, Xerox Corp, 64-67; prin scientist assoc, 67-70, staff scientist, 70-72, MGR POLYMER SCI DEPT, FORD MOTOR CO, 72- *Concurrent Pos:* Chmn coatings & films, Gordon Res Conf, 80 & Org Coatings & Plastics Chem Div, Am Chem Soc, 81; mem adv bd, J Coatings & Technol. *Mem:* Am Chem Soc. *Res:* Synthetic organic chemistry; polymer syntheses; mechanism of organic reactions; physical and thermal properties of polymers with special reference to network polymers; radiation induced polymerizations, coating and composites. *Mailing Add:* Box 2037 Dearborn MI 48123

LABANAUSKAS, CHARLES K, b Upyna, Lithuania, Jan 3, 23; nat US. PLANT PHYSIOLOGY. *Educ:* Hohenheim Agr Univ, dipl, 47; Univ Ill, MS, 53, PhD, 54. *Prof Exp:* From asst horticulturist to assoc horticulturist, 55-68, HORTICULTURIST, CITRUS RES CTR, UNIV CALIF, RIVERSIDE, 68, PROF HORT SCI, COL NATURAL & AGR SCI, 68- *Concurrent Pos:* Lectr, Univ Calif, Riverside, 65-68. *Mem:* Am Soc Hort Sci. *Res:* Mineral metabolism in plants. *Mailing Add:* Dept Bot & Plant Sci Univ Calif Riverside CA 92521

LABANICK, GEORGE MICHAEL, b Passaic, NJ, Sept 27, 50; m 79; c 2. ZOOLOGY, HERPETOLOGY. *Educ:* Col William & Mary, BS, 72; Ind State Univ, MA, 74; Southern Ill Univ, PhD(zool), 78. *Prof Exp:* Asst prof biol, Emory & Henry Col, 78-79; from asst prof to assoc prof, 79-89, PROF BIOL, UNIV SC, SPARTANBURG, 89-, DIV CHAIR, 90- *Mem:* Sigma Xi; Am Soc Ichthyologists & Herpetologists; Herpetologists' League; Soc Study Amphibians & Reptiles; Soc Study Evolution. *Res:* Mimicry and other defense mechanisms; salamander ecology and systematics. *Mailing Add:* Div Natural Sci & Eng Univ SC Spartanburg SC 29303

LABAR, MARTIN, b Radisson, Wis, May 15, 38. POPULATION BIOLOGY, BIOETHICS. *Educ:* Wis State Univ, Superior, BA, 58; Univ Wis, MS, 63, PhD(genetics, zool), 65. *Prof Exp:* Assoc prof, 64-66, PROF SCI, CENT WESLEYAN COL, 66-, CHMN DIV SCI, 64- *Mem:* Am Sci Affiliation; Ecol Soc Am; Soc Study Evolution. *Res:* Use of computer simulation in teaching population biology; bioethics. *Mailing Add:* Div Sci Cent Wesleyan Col Central SC 29630

LABARGE, ROBERT GORDON, b Buffalo, NY, July 11, 40. CHEMISTRY. *Educ:* Univ Rochester, BS, 62; Carnegie-Mellon Univ, PhD(chem), 66; Cent Mich Univ, MBA, 78. *Prof Exp:* Res chemist, Consumer Prod Dept, 66-73, dir acad educ, 73-75, res specialist, Designed Prod Dept, 75-78, mgr, New Prod Develop, 78-83, MGR, OPPORTUNITIES IDENTIFICATION, DOW CHEM CO, 83- *Concurrent Pos:* Adj prof chem, Cent Mich Univ, 74,

econ & mgt, Delta Col, 78-82; adj prof, Northwood Inst, 82. *Mem:* Am Chem Soc; Inst Food Technologists; Am Oil Chemists Soc; Sigma Xi. *Res:* New product exploration and development. *Mailing Add:* 2713 Mt Vernon Midland MI 48640

LABARRE, ANTHONY E, JR, b New Orleans, La, July 18, 22; m 43, 77; c 2. MATHEMATICS. *Educ:* Tulane Univ, BE, 43, MS, 47; Univ Okla, PhD(math), 57. *Prof Exp:* Instr math, Tulane Univ, 46-48; asst prof, Univ Idaho, 48-50; instr, Univ Okla, 50-54 & Univ Wyo, 54-56; from asst prof to assoc prof, Univ Idaho, 56-61; chmn dept, Calif State Univ, Fresno, 61-66, prof math, 61-90; RETIRED. *Mem:* Math Asn Am; Am Math Asn Two Year Col. *Res:* Functional analysis, differential geometry. *Mailing Add:* Dept Math Calif State Univ Fresno CA 93740

LABARTHE, DARWIN RAYMOND, b Berkeley, Calif, Aug 5, 39. EPIDEMIOLOGY. *Educ:* Princeton Univ, AB, 61; Columbia Univ, MD, 65; Univ Calif, Berkeley, MPH, 67, PhD(epidemiol), 75. *Prof Exp:* Epidemiologist, Com Corps, Heart Dis & Stroke Control Prog, San Francisco, 67-69; dep chief & sr epidemiologist, Epidemiol Field & Training Sta, USPHS Heart Dis & Stroke Control Prog, San Francisco, 69-70; from assoc res epidemiologist to assoc prof epidemiol, Sch Pub Health, Univ Tex Health Sci Ctr, Houston, 70-73; consult epidemiol, dept med statist & epidemiol, Mayo Clin & Mayo Found, 74-77; PROF EPIDEMIOL, SCH PUB HEALTH, UNIV TEX, 77- *Concurrent Pos:* Dep dir, Coord Ctr, Hypertension Detection & Followup Prog, Nat Heart & Lung Inst, 71-73; consult, Task Force Automated Blood Pressure Devices, Nat Heart & Lung Inst, 73-74; consult, coord ctr, Hypertension Detection & Followup Prog, Nat Heart & Lung Inst, 74-; co-investr & co-dir, Study Incidence & Natural Hist Genital Tract Anomalies & Cancer in Offspring exposed in Utero to synthetic Estrogens, Nat Cancer Inst, 74-; chmn & dir, US Seminar in Cardiovasc Epidemiol, Am Heart Asn, 75-; dir design & anal, Baylor Col Med, 77-; dep dir, Beta-Blocker Heart Attack Trial Coord Ctr. *Mem:* Fel Am Heart Asn; fel Am Col Prev Med; Soc Epidemiol Res (pres, 72-73); Am Pub Health Asn; Int Soc Cardiol. *Res:* Epidemiology and prevention, especially of cardiovascular and other chronic conditions; drugs; intra-individual variability of blood pressure and other personal characteristics. *Mailing Add:* Box 20186 Astrodome Sta Houston TX 77025

LABATE, SAMUEL, b Easton, Pa, Dec 19, 18; m 49; c 2. ACOUSTICS. *Educ:* Lafayette Col, AB, 40; Mass Inst Technol, MS, 48. *Prof Exp:* Asst instr math, Univ Pa, 40-41; engr, E I du Pont de Nemours & Co, 41-42; consult engr, Bolt & Beranek, 48-49, consult engr, 49-53, exec vpres, 53-69, pres, 69-76, dir, Bolt Beranek & Newman, Inc, 53-83, chmn bd, 76-83; RETIRED. *Mem:* Fel Acoust Soc Am; Acad Appl Sci. *Res:* Engineering; applied architectural and physical acoustics. *Mailing Add:* PO Box 724 Cataumet MA 02534

LABAVITCH, JOHN MARCUS, b Covington, Ky, Oct 15, 43. PLANT PHYSIOLOGY, BIOCHEMISTRY. *Educ:* Wabash Col, AB, 65; Stanford Univ, PhD(plant physiol), 73. *Prof Exp:* Instr biol, Wabash Col, 65-67; NIH fel biochem, Univ Colo, 72-76; asst pomologist, 76-80, ASSOC POMOLOGIST, UNIV CALIF, DAVIS, 80-, LECTR POMOL, 76- *Mem:* Am Soc Plant Physiologists; AAAS. *Res:* Cell wall metabolism of fruit. *Mailing Add:* Dept Pomol Univ Calif Davis CA 95616

LABBE, ROBERT FERDINAND, b Portland, Ore, Nov 12, 22; m 55; c 3. CLINICAL CHEMISTRY, NUTRITION. *Educ:* Univ Portland, BS, 47; Ore State Col, MS, 49, PhD(biochem), 51. *Prof Exp:* AEC fel med sci, Col Physicians & Surgeons, Columbia Univ, 51-53; res instr, Med Sch, Univ Ore, 53-55, res asst prof, 55-57; res asst prof pediat & lectr biochem, 57-60, res assoc prof pediat, 60-68, prof pediat, 68-74, PROF LAB MED, MED SCH, UNIV WASH, 74-, HEAD, CLIN CHEM DIV, 80- *Concurrent Pos:* Vis asst prof, Inst Enzyme Res, Univ Wis, 56-57; vis researcher, Commonwealth Sci & Indust Res Orgn, Australia, 65; NIH spec fel, 65 & career develop award, 66-70. *Honors & Awards:* Ames Award, Am Asn Clin Chemists. *Mem:* Fel AAAS; Am Chem Soc; Am Soc Biol Chemists; Acad Clin Lab Physicians & Scientists; Am Asn Clin Chemists; Am Soc Clin Nutrit. *Res:* Heme biosynthesis; iron metabolism; related metabolic diseases; clinical nutrition; ascorbic acid metabolism, low power lasers. *Mailing Add:* Harborview Med Ctr Univ Wash Seattle WA 98104

LABBE, RONALD GILBERT, b Berlin, NH, July 16, 46; m 78; c 2. MICROBIOLOGY. *Educ:* Univ NH, BA, 68; Univ Wis, MS, 70, PhD(bact), 76. *Prof Exp:* Res assoc microbiol, Food Res Inst, Univ Wis, 76; from asst prof to assoc prof, 76-87, PROF FOOD MICROBIOL, DEPT FOOD SCI, UNIV MASS, 88- *Mem:* Sigma Xi; Int Asn Milk, Food & Environ Sanitarians; Am Soc Microbiol; Inst Food Technologists. *Res:* Clostridium perfringens food poisoning; germination and sporulation of bacterial spores. *Mailing Add:* Dept Food Sci Univ Mass Amherst MA 01003

LABELLA, FRANK SEBASTIAN, b Middletown, Conn, Sept 23, 31; m 52; c 3. PHARMACOLOGY. *Educ:* Wesleyan Univ, BA, 52, MA, 54; Emory Univ, PhD(basic health sci), 57. *Prof Exp:* Asst biol, Wesleyan Univ, 52-54; asst physiol, Emory Univ, 54-55, asst histol, 55-57, instr physiol, 57-58; lectr pharmacol, 58-60, from asst prof to assoc prof, 60-67, PROF PHARMACOL & THERAPEUT, FAC MED, UNIV MAN, 67- *Concurrent Pos:* Am Heart Asn fel, Emory Univ, 57-58; Can Rheumatism & Arthritis Soc res fel, Univ Man, 58-61; estab investr, Am Heart Asn, 61-66, mem coun arteriosclerosis; career investr, Med Res Coun Can, 66-; mem, Manitoba Environ Res Comt & Int Narcotics Res Conf. *Honors & Awards:* John J Abel Award, Am Soc Pharmacol & Exp Therapeut, 67; E W R Steacie Prize in Natural Sci, Pharmacol Soc Can, 69; Upjohn Award, 82. *Mem:* Can Biochem Soc; Endocrine Soc; AAAS; Can Asn Gerontol. *Res:* Cellular pharmacology and biochemistry; neurochemistry; aging; endocrine pharmacology; neuroendocrinology. *Mailing Add:* Dept Pharmacol & Therapeut Univ Man Fac Med Winnipeg MB R3E 0W3 Can

LABELLE, EDWARD FRANCIS, b Worcester, Mass, Aug 11, 48; m; c 2. EXPERIMENTAL BIOLOGY. *Educ:* Col Holy Cross, AB & MS, 70; Univ Mich, Ann Arbor, PhD(biochem), 74. *Prof Exp:* Grad teaching asst, Dept Biol Chem, Univ Mich, 70-74; postdoctoral fel, Sect Biochem, Cornell Univ, 74-76; asst prof chem, Western Ill Univ, 76-78; from asst prof to assoc prof, Div Biochem, Med Br, Univ Tex, Galveston, 78-87; RES SCIENTIST, GRAD HOSP, BOCKUS RES INST, PHILADELPHIA, PA, 87- *Concurrent Pos:* Prin investr, Western Ill Univ, NIH, Am Diabetes Asn, Muscular Dystrophy Asn & Am Heart Asn, 77-89; mem, Cancer Ctr Rev Comt, Med Br, Univ Tex, 78-86, Biochem Grad Student Adv Comt, 81-87 & Biohazards Comt, 83-84; co-investr, NIH, 82-91 & NSF, 84-85; adj assoc prof, Dept Physiol, Univ Pa, Philadelphia, 88- *Mem:* Am Chem Soc; AAAS; Am Soc Biol Chemists; Soc Gen Physiologists. *Mailing Add:* Grad Hosp Bockus Res Inst 415 S 19th St Philadelphia PA 19146

LABEN, ROBERT COCHRANE, b Darien Center, NY, Nov 16, 20; m 46; c 4. GENETICS, ANIMAL HUSBANDRY. *Educ:* Cornell Univ, BS, 42; Okla Agr & Mech Col, MS, 46; Univ Mo, PhD(animal breeding), 50. *Prof Exp:* Asst animal husb, Okla Agr & Mech Col, 46-47; asst dairy husb, Univ Mo, 47-49, asst instr, 49-50; instr animal husb & jr animal husbandman, 50-52, asst prof & asst animal husbandman, 52-58, assoc prof & assoc animal husbandman, 58-64, prof, animal husbandman & dir comput ctr, 64-69, dept vchmn, 78-82, prof animal sci & geneticist, exp sta, 69-86, EMER PROF ANIMAL SCI, EXP STA, UNIV CALIF, DAVIS, 86- *Mem:* Am Soc Animal Sci; Biomet Soc; Am Dairy Sci Asn; Am Genetic Asn; Sigma Xi. *Res:* Breeding and genetics of farm livestock. *Mailing Add:* 502 Oak Ave Davis CA 95616

LABER, LARRY JACKSON, b Lincoln, Vt, July 9, 37; m 63. PLANT PHYSIOLOGY. *Educ:* Univ Vt, BS, 59, MS, 61; Univ Chicago, PhD(bot), 67. *Prof Exp:* Asst prof, Pa State Univ, 65-66; NSF trainee, Univ Ga, 67-69; NIH trainee, Brandeis Univ, 69-70; asst prof, 70-76, ASSOC PROF BOT, UNIV MAINE, ORONO, 77- *Mem:* AAAS; Am Soc Plant Physiol; Japanese Soc Plant Physiol. *Res:* Choroplast development; photophosphorylation; carbon dioxide fixation. *Mailing Add:* 15 Back River Rd Apt 4 Dover NH 03820

LABERGE, GENE L, b Ladysmith, Wis, Mar 15, 32; m 62; c 2. GEOLOGY. *Educ:* Univ Wis, BS, 58, MS, 59, PhD(geol), 63. *Prof Exp:* Sponsored res officer, Commonwealth Sci & Indust Res Orgn, Melbourne, Australia, 63-64; Nat Res Coun Can fel, Geol Surv Can, 64-65; from asst prof to assoc prof, 65-74, PROF GEOL, UNIV WIS-OSHKOSH, 74- *Concurrent Pos:* Mem staff, Wis Geol & Natural Hist Surv, 72- *Mem:* AAAS; Geol Soc Am; Soc Econ Geologists. *Res:* Origin of Precambrian iron formations; Precambrian geology and mineral deposits of Wisconsin. *Mailing Add:* Dept Geol Univ Wis Oshkosh 800 Algoma Blvd Oshkosh WI 54901

LABERGE, WALLACE E, b Grafton, NDak, Feb 7, 27; m 58; c 3. ENTOMOLOGY. *Educ:* Univ NDak, BSc, 49, MS, 51; Univ Kans, PhD(entom), 55. *Prof Exp:* Asst cur, Snow Entom Mus & instr entom, Univ Kans, 54-55, asst prof, 55-56; asst prof zool, Iowa State Univ, 56-59; assoc prof entom, Univ Nebr, 59-65; assoc taxonomist, 65-67, TAXONOMIST, ILL NATURAL HIST SURV, 67-; PROF ENTOM, UNIV ILL, URBANA, 70- *Mem:* Entom Soc Am; Soc Study Evolution; Soc Syst Zool; Am Entom Soc; Sigma Xi. *Res:* Systematics of Hymenoptera, Apoidea and Braconidae. *Mailing Add:* 607 E Peabody Dr Champaign IL 61820

LABERGE, WALTER B, b Chicago, Ill, Mar 29, 24; m 82; c 5. ASTROPHYSICS. *Educ:* Univ Notre Dame, BS in Naval Sci, 44, BS, 47, PhD(physics), 50. *Prof Exp:* Vpres, Defense Div, Philco-Ford, 57-70; tech dir, Naval Weapons Ctr, China Lake, Calif, 70-73; asst secy Air Force, Res & develop, 73-75; asst secy gen, NATO, Brussels, 75-76; under secy Army, 76-79; prin dep to Dr William Perry, under Secy Defense, 79-81; exec asst to pres, Lockheed Corp, 81-82, vpres, planning & technol, 82-84, vpres & gen mgr, Res & Develop Div, 84-86, vpres corp develop, 86-88; CHAIR, ACQUISITION MGT POLICY, DEFENSE SYSTS MGT COL, FT BELVOIR, VA, 89- *Concurrent Pos:* Mem, Bd Army Sci & Technol, Nat Acad Eng, 82-88, Army Sci Bd, 83-89; vis scholar, Stanford Univ, 89- *Mem:* Nat Acad Eng. *Res:* One of the principal inventors of the Sidewinder air to air missile. *Mailing Add:* 910 Via Palo Alto Aptos CA 95003

LABES, MORTIMER MILTON, b Newton, Mass, Sept 9, 29; m 53, 72; c 6. CHEMICAL PHYSICS. *Educ:* Harvard Univ, AB, 50; Mass Inst Technol, PhD, 54. *Prof Exp:* Asst, Mass Inst Technol, 51-54, res chemist, Sprague Elec Co, 54-57; sr res chemist, Franklin Inst, 57-59, sr staff chemist, 59-60, lab mgr, 60-61, tech dir chem div, 61-66; prof chem, Drexel Inst, 66-70; PROF CHEM, TEMPLE UNIV, 70- *Mem:* Am Chem Soc; Am Phys Soc; Sigma Xi. *Res:* Chemistry and physics of organic solid state; molecular complexes; liquid crystals; electronic properties of polymers; synthesis and properties of carbon fibers and carbon composites. *Mailing Add:* Dept Chem Temple Univ Philadelphia PA 19122

LABIANCA, DOMINICK A, b Brooklyn, NY, Feb 4, 43; m 73; c 1. ORGANIC CHEMISTRY, POLYMER CHEMISTRY. *Educ:* Polytech Inst Brooklyn, BS, 65; Univ Mich, PhD(chem), 69. *Prof Exp:* NSF fel org photochem, Calif Inst Technol, 69-70; res chemist, Res & Develop, Bound Brook Tech Ctr, Union Carbide Corp, 70-72; asst prof to assoc prof, New Sch Lib Arts, 72-80, assoc prof, 80-83, PROF, DEPT CHEM, BROOKLYN COL, CITY UNIV NY, 83- *Concurrent Pos:* Consult, expert witness in driving while intoxicated & related cases, 85- *Mem:* NY Acad Sci; Sigma Xi; Am Chem Soc; Nat Sci Teachers Asn. *Res:* Chemical education; chemistry/humanities integration, curriculum development; chemistry of breath-alchol testing. *Mailing Add:* 189 Ribbon St Franklin Square NY 11010

LABIANCA, FRANK MICHAEL, b Brooklyn, NY, Aug 17, 39; m 70; c 2. UNDERWATER ACOUSTICS, ACOUSTIC SIGNAL PROCESSING. *Educ:* Polytech Inst Brooklyn, BEE, 61, MS, 63, PhD(elec eng), 67. *Prof Exp:* Instr elec eng, Polytech Inst Brooklyn, 61-67; mem tech staff, 67-85, TECH SUPVR, AT&T BELL LABS, 85- *Concurrent Pos:* Prin investr, Off Naval Res & other govt agencies. *Mem:* Inst Elec & Electronics Engrs; Acoust Soc Am; Sigma Xi. *Res:* Propagation in surface ducts and underwater channel, scattering of sound from the ocean surface, radiation from cavitating propellers and the origins of ambient noise; adaptive array processing for underwater acoustic detection of signals in noise; speech recognition; digital filter design; LMS techniques in algorithm design. *Mailing Add:* AT&T Bell Labs Whippany Rd Whippany NJ 07981

LABINGER, JAY ALAN, b Los Angeles, Calif, July 6, 47; m 70; c 1. INORGANIC CHEMISTRY, ORGANOMETALLIC CHEMISTRY. *Educ:* Harvey Mudd Col, BS, 68; Harvard Univ, PhD(chem), 74. *Prof Exp:* Res assoc chem, Princeton Univ, 73-74, instr, 74-75; asst prof chem, Univ Notre Dame, 75-81; sr res chemist, Occidental Res Corp, 81-83; res adv, ARCO, 83-86; MEM PROF STAFF & ADMINR, BECKMAN INST, CALIF INST TECHNOL, 86- *Concurrent Pos:* Assoc ed, Chem Revs, 79-81. *Mem:* Am Chem Soc. *Res:* Synthetic and mechanistic organo-transition metal chemistry; activation of small molecules by transition metal complexes; homogeneous and heterogeneous catalysis; alkane activation. *Mailing Add:* Calif Inst Technol 139-74 Pasadena CA 91125

LABISKY, RONALD FRANK, b Aberdeen, SDak, Jan 16, 34; m 58; c 2. WILDLIFE BIOLOGY, FISHERIES SCIENCE. *Educ:* SDak State Univ, BS, 55; Univ Wis, MS, 56, PhD(wildlife ecol-zool), 68. *Prof Exp:* Field asst game bird res, Ill State Natural Hist Surv, 56-57, from asst proj leader to proj leader, 57-59, from asst wildlife specialist to assoc wildlife specialist, 59-72, wildlife specialist, 72-76; actg dir & asst dir, Univ Fla, 76-78; prof & wildlife coordr, Sch Forest Resources & Conserv, 78-84, chair, 84-87, PROF, DEPT WILDLIFE & RANGE SCI, UNIV FLA, 84- *Concurrent Pos:* Chmn, Nat Fish & Wildlife Resources Res Coun, 78-; chmn, Fish & Wildlife Comn, Nat Asn State Univ & Land Grant Col, 84-; assoc ed, J Wildlife Mgt, 83-86, adv bd, Critical Rev in Natural Resources Mgt, 87-88. *Honors & Awards:* Grad Fac Adv Award, Am Fisheries Soc, 78; Special Recognition Serv Award, Wildlife Soc, 88. *Mem:* Am Fisheries Soc; Wildlife Soc; Am Ornith Union; Wilson Ornith Soc; Am Soc Mammal. *Res:* Ecology and physiology of gallinaceous game birds, doves and waterfowl; population ecology, social biology and spatial distribution of pheasants and white-tailed deer; ecological, ethological, physiological and nutritive factors influencing distribution and abundance of terrestrial and aquatic wildlife; biology and management of spiny lobsters and deep-water reef fishes; ecology and management of nongame and endangered wildlife, particularly red-cockaded woodpeckers. *Mailing Add:* Dept Wildlife & Range Sci Univ Fla Gainesville FL 32611-0304

LABODA, HENRY M, b Kingston, Pa, Dec 9, 50; m. EXPERIMENTAL BIOLOGY. *Educ:* Wilkes Col, BS, 72; Temple Univ, MS, 76; Hahnemann Univ, PhD(biol chem), 81. *Prof Exp:* Med technologist, Robert Packer Mem Hosp, Sayre, Pa, 72-74; biochemist, EM Sci, Gibbstown, NJ, 81-82; clin chemist, Cooper Hosp, Camden, NJ, 82-83; NIH postdoctoral trainee, Dept Physiol & Biochem, Med Col Pa, Philadelphia, 83-86; fel, Biophysics Inst, Boston Univ Med Ctr, 86-88; SR SCIENTIST, SCHERING-PLOUGH HEALTHCARE PROD, MEMPHIS, TENN, 88- *Concurrent Pos:* Mem, Prog Comt, Am Chem Soc, 90-92. *Mem:* Am Soc Biochem & Molecular Biol; Am Chem Soc. *Res:* Method development for enzyme assays; preparative isolation and purification of enzymes and apolipoproteins; chromatographic separation methods; spectrophotometric methods; enzyme-linked immunosorbent assay, antibody production and purification; surface chemistry techniques including lipid monolayers, force-area isotherms and enzymatic hydrolysis. *Mailing Add:* Skin Biol Res Schering-Plough HealthCare Prod 3030 Jackson Ave Memphis TN 38151

LA BONTE, ANTON EDWARD, b Minneapolis, Minn, May 6, 35; m 59. MACHINE INTELLIGENCE, TECHNOLOGY TRANSFER. *Educ:* Univ Minn, Minneapolis, BS, 57, MSEE, 60, PhD(elec eng), 66. *Prof Exp:* Instr elec eng, Univ Minn, Minneapolis, 59-62, res fel micromagnetics, 62-63, instr elec eng, 63-65; sr scientist, Control Data Corp, 66-69; mgr systs anal, Aerospace, Navigation & Space Systs, 69-75, sr tech consult, corp res & eng, 75-89; CONSULT, 89- *Mem:* Inst Elec & Electronics Engrs; Cognitive Sci Soc; Am Asn Artificial Intel. *Res:* Machine intelligence with emphasis on knowledge-based information and decision systems; applications of computers to digital image analysis; technology transfer strategies and techniques. *Mailing Add:* 4729 30th Ave S Minneapolis MN 55406

LABORDE, ALICE L, b Jan 8, 47. CHEMISTRY, BIOLOGY. *Educ:* La Col, BS, 70; Univ Southwestern La, MS, 72; Univ Tex, Austin, PhD(microbiol), 79. *Prof Exp:* Sr res scientist, Infectious Dis Res, 79-86, SR RES SCIENTIST, CHEM & BIOL SCREENING, UPJOHN CO, 86- *Mem:* Am Soc Microbiol; Soc Indust Microbiol; Am Soc Biochem & Molecular Biol. *Res:* Development of assays to detect novel therapeutic agents which may be present in fermentation cultures of actinomycetes, and in extracts of plants, algae or in marine microorganisms; elucidating the role of interleukin-1 in bone remodelling using osteoblastic cells as a model system. *Mailing Add:* Chem & Biol Screening Upjohn Co 301 Henrietta St Kalamazoo MI 49001

LABOSKY, PETER, JR, b Manville, NJ, Jan 9, 37; m 67; c 2. WOOD CHEMISTRY, PULP & PAPER. *Educ:* Rutgers Univ, BS, 63; Va Polytechnic Inst, MS, 67, PhD(wood technol), 70. *Prof Exp:* Res engr, Westvaco Corp, 70-74; assoc exten, Clemson Univ, 74-79; assoc pulp & paper, 79-85, PROF WOOD CHEM, PA STATE UNIV, 85- *Mem:* Tech Asn Pulp & Paper Indust; Forest Prod Res Soc; Soc Wood Sci & Technol. *Res:* Relationship of fiber properties to paper properties; biopulping; kraft pulping of hardwoods; bark chemistry; author of many publications. *Mailing Add:* 309 FRL Pa State Univ University Park PA 16802

LABOUNTY, JAMES FRANCIS, SR, b Minneapolis, Minn, Dec 14, 42; m 69; c 2. LIMNOLOGY, LAKE MANAGEMENT. *Educ:* Univ Nev, Las Vegas, BS & BA, 67, MS, 68; Ariz State Univ, Tempe, PhD(zool), 74. *Prof Exp:* Fish & wildlife biologist, US Bur Reclamation, Boulder City, Nev, 69-72, environ specialist, Phoenix, Ariz, 72-74, res biologist, Eng & Res Ctr, Denver, 74-80, tech specialist, 80-84, HEAD, ENVIRON SCI SECT, ENG & RES

CTR, US BUR RECLAMATION, DENVER, 84- *Concurrent Pos:* Prin investr, US Bur Reclamation, Denver, 74-84; lectr, Univ Colo, Denver, 77-78; bd dir, NAm Lake Mgt Soc. *Mem:* NAm Lake Mgt Soc; Am Fisheries Soc; Southwestern Asn Naturalists; Desert Fishes Coun; Am Soc Ichthyologists & Herpetologists. *Res:* Performing applied investigations on the ecology of lakes, reservoir and streams and in particular the nature of eutrophication research program for the US Bureau of Reclamation. *Mailing Add:* 13222 W LaSalle Circle Lakewood CO 80228-4932

LABOV, JAY BRIAN, b Philadelphia, Pa, Sept 19, 50; m 75; c 2. PRE-COLLEGE SCIENCE EDUCATION. *Educ:* Univ Miami, Fla, BS, 72; Univ RI, MS, 74, PhD(biol sci), 79. *Prof Exp:* Instr biol, Wash & Lee Univ, 78-79; asst prof, 79-84, ASSOC PROF BIOL, COLBY COL, 84- *Concurrent Pos:* Vis investr, Jackson Lab, 81; vis scientist, Monell Chem Senses Ctr, 85-86; fel, W K Kellogg Found Nat Fel Prog, 88-91. *Mem:* Animal Behav Soc; AAAS; Am Soc Zoologists; Am Soc Mammalogists; Nat Sci Teachers Asn; Sigma Xi. *Res:* Behavioral and physiological aspects of mammalian reproduction; pre-college science education. *Mailing Add:* Dept Biology Colby Col Waterville ME 04901-9989

LABOWS, JOHN NORBERT, JR, b Wilkes-Barre, Pa, June 27, 41; m 64; c 4. ORGANIC CHEMISTRY. *Educ:* Lafayette Col, BS, 63; Cornell Univ, PhD(org chem), 67. *Prof Exp:* Asst prof org chem, Wilkes Col, 67-70, assoc prof chem, 70-78; asst mem, Monell Chem Senses Ctr, 78-80, assoc mem, 80-85; res assoc, 85-87, sr res assoc, 87-90, ASSOC RES FEL, COLGATE PALMOLIVE CO, 90- *Concurrent Pos:* Nat Cancer Inst fel, Fels Res Inst, Temple Univ, 70-71. *Mem:* Am Chem Soc; Sigma Xi; Asn Chemoreception Sci. *Res:* Gas chromatography mass spectrometry analysis; physical chemistry of flavors and fragrances; chemical communication; odor/fragrance analysis. *Mailing Add:* Colgate-Palmolive Res Ctr 909 River Rd Piscataway NJ 08855-1343

LABRECQUE, DOUGLAS R, EXPERIMENTAL BIOLOGY. *Prof Exp:* Dir liver serv, 79, PROF INTERNAL MED, UNIV IOWA, 87-; CHIEF GI, IOWA CITY VET ADMIN HOSP, 82- *Mailing Add:* Dept Internal Med Liver Serv Univ Iowa Hosps & Clins Iowa City IA 52242

LABREE, THEODORE ROBERT, b Lafayette, Ind, June 25, 31; m 51; c 2. BACTERIOLOGY, FOOD TECHNOLOGY. *Educ:* Purdue Univ, BS, 58, MS, 60. *Prof Exp:* Res asst food technol, Purdue Univ, 58-59; assoc bacteriologist, Mead Johnson & Co, 59-62, scientist bact, 62-65, mgr med admin, 65-73; tech dir regulatory affairs, Riviana Foods, Inc, 73-86; coordr & mgr qual assurance, Trbesweet Co Inc, 86-89; TECH DIR, CORP RES & DEVELOP QUAL ASSURANCE, ERLY JUICE INC, 89- *Mem:* Am Soc Microbiol; Inst Food Technologists. *Res:* Active oxygen method; spore destruction of food spoilage organisms; new methods development; microbiology; government regulations, processing, packaging, labeling; good manufacturing practices regulations; low acid; sanitation; food plant; warehouse evaluation; pesticides; fumigation. *Mailing Add:* 9839 Canoga Ln Houston TX 77080

LABRIE, DAVID ANDRE, b Baltimore, Md, Mar 23, 37; m 62; c 2. MICROBIOLOGY, BIOCHEMICAL GENETICS. *Educ:* Bethany Col, WVa, AB, 61; NMex Highlands Univ, MS, 65; NC State Univ, PhD(microbiol), 68. *Prof Exp:* NIH fel, Univ Tex, Austin, 68-70; from asst prof to assoc prof, 70-81, PROF BIOL, WTEX STATE UNIV, 81-, DEPT HEAD, 86- *Mem:* AAAS; Sigma Xi. *Res:* Microbiology genetics of antibiotic resistance and ultraviolet light. *Mailing Add:* Dept Biol & Geosci WTex State Univ Canyon TX 79016

LABRIE, FERNAND, b June 28, 37; Can citizen; m 63; c 4. ENDOCRINOLOGY, BIOCHEMISTRY. *Educ:* Laval Univ, BA, 57, MD, 62, PhD(endocrinol), 67; FRCP(C), 73. *Prof Exp:* From asst prof to assoc prof physiol, 66-73, HEAD LAB MOLECULAR ENDOCRINOL, HOSP CTR, LAVAL UNIV, 69-, PROF PHYSIOL, 73- *Concurrent Pos:* Med Res Coun Can fels, Laval Univ, 63-66, Univ Cambridge, 66-67, Univ Sussex, 67-68 & centennial fel, Lab Molecular Biol, Univ Cambridge, 68-69; Med Res Coun Can scholar, Laval Univ, 69-; Med Res Coun Can assoc, Laval Univ, 73-; dir molecular endocrinol, Med Res Coun Group, 73- *Mem:* Am Soc Biol Chemists; Am Physiol Soc; Endocrine Soc; Can Physiol Soc; Can Biochem Soc. *Res:* Mechanism of action of hypothalamic regulatory hormones in the anterior pituitary gland, mammalian messenger RNA; hormone dependent breast cancer; reproductive physiology and biochemistry; hormones and brain. *Mailing Add:* Ctr Hops de Univ Lavel 2705 Blvd Laurier Ste Foy Quebec PQ G1V 4G2 Can

LABUDDE, ROBERT ARTHUR, b Flint, Mich, May 28, 47; m 69; c 2. LASER ANNEALING, DIGITAL CODING. *Educ:* Univ Mich, Ann Arbor, BS, 69; Univ Wis-Madison, PhD(chem), 73. *Prof Exp:* Res asst chem, Univ Wis, 68-73, lect comput sci, 73, asst scientist, math res ctr, 73-74; instr appl math, Mass Inst Technol, 74-75; asst prof math & comput sci, Old Dominion Univ, Norfolk, Va, 76-79; exec vpres, Labudde Eng Corp, 83-86; PRES, LEAST COST FORMULATIONS, LTD, 79- *Concurrent Pos:* Consult, Gen Systs Div, IBM, 75-78, Res Ctr, Allied Tech Corp, 77-78, Digital Design Labs, 79, Burroughs Corp, 80-82 & Optical Coating Labs, Inc, 82-83; secy, ERB Leasing Co, Inc, 83-86; mem bd dirs, Tech Express Inc, 90. *Mem:* Soc Indust & Appl Math; Asn Comput Mach; Am Soc Testing Mats; Am Soc Qual Control; Inst Indust Engrs; Inst Food Technologists; Sigma Xi. *Res:* Theoretical modeling of laser-based optical disk memory systems, including thermal, optical and thin-film properties; noise sources and kinetics of environmental degraddation; digital coding and communication. *Mailing Add:* 824 Timberlake Dr Virginia Beach VA 23464-3239

LABUTE, JOHN PAUL, b Tecumseh, Ont, Feb 26, 38; m 61; c 3. MATHEMATICS. *Educ:* Univ Windsor, BSc, 60; Harvard Univ, MA, 61, PhD(math), 65. *Prof Exp:* Nat Res Coun Can res fel, Col France, 65-67; asst prof, 67-70, ASSOC PROF MATH, MCGILL UNIV, 70- *Mem:* Can Math Cong; Am Math Soc. *Res:* Algebra and number theory. *Mailing Add:* Dept Math McGill Univ 805 Sherbrooke St W W Montreal PQ H3A 2K6 Can

LABUZA, THEODORE PETER, b Perth Amboy, NJ, Nov 10, 40; m 85; c 2. FOOD SCIENCE, PHYSICAL CHEMISTRY. *Educ:* Mass Inst Technol, SB, 62, PhD(food sci), 65. *Prof Exp:* From instr to assoc prof food eng, Mass Inst Technol, 65-71; assoc prof, 71-72, PROF FOOD TECHNOL, UNIV MINN, ST PAUL, 72- *Concurrent Pos:* Food processing consult. *Honors & Awards:* Samuel Cate Precott Res Award, Inst Food Technologists, 72, Cruess Award, 73, Babcock Hart Award, 88; Howard Lectureship, Univ Ill, 87. *Mem:* Fel Inst Food Technologists (pres, 88-89); Am Inst Chem Eng; Am Chem Soc; Am Asn Cereal Chemists; Asn Food & Drug Officials; Sigma Xi. *Res:* Physical chemical factors involved in water in foods; reaction kinetics and prediction of food storage life; stability of intermediate moisture foods; nutrient degradation in processing; kinetics of microbial growth and death; plant tissue culture; edible films. *Mailing Add:* Dept Food Sci & Nutrit Univ Minn St Paul MN 55108

LACASCE, ELROY OSBORNE, JR, b Fryeburg, Maine, Jan 17, 23. ACOUSTICS. *Educ:* Bowdoin Col, AB, 43; Harvard Univ, AM, 51; Brown Univ, PhD(physics), 55. *Prof Exp:* Instr physics, Bowdoin Col, 43 & 47-49, instr math, 51; physicist, Naval Res Lab, 44; foreign serv officer, US Dept State, 45-46; teacher, High Sch, 46-47; asst, Brown Univ, 51-54; from instr to assoc prof physics, 54-69, chmn dept, 77-88, PROF PHYSICS, BOWDOIN COL, 69- *Concurrent Pos:* Res assoc, Yale Univ, 60-61; NSF fac fel, 60-61; vis investr, Woods Hole Oceanog Inst, 68-69, guest investr, 75-76 & 82-83. *Mem:* Acoust Soc Am. *Res:* Ultrasonics and underwater sound. *Mailing Add:* Dept Physics Bowdoin Col Brunswick ME 04011

LACEFIELD, GARRY DALE, b McHenry, Ky, Aug 22, 45; m 67; c 2. FORAGE PRODUCTION & MANAGEMENT. *Educ:* Western Ky Univ, BS, 70, MS, 71; Univ Mo, PhD(agron & physiol), 74. *Prof Exp:* Lab instr & instr plant sci, Western Ky Univ, 69-71; teaching asst, Univ Mo, 71-74; from asst exten prof to assoc exten prof, 74-82, EXTEN FORAGE SPECIALIST, 74- & EXTEN PROF, UNIV KY, 82- *Honors & Awards:* Exten Specialist of the Year, 83. *Mem:* Am Soc Agron; Am Forage & Grassland Coun. *Res:* Development and implementation of improved practices in forage establishment, production and utilization. *Mailing Add:* Res & Educ Ctr PO Box 469 Princeton KY 42445

LACELLE, PAUL (LOUIS), b Syracuse, NY, July 4, 29; m 53; c 4. HEMATOLOGY, PHYSIOLOGY. *Educ:* Houghton Col, BA, 51; Univ Rochester, MD, 59. *Prof Exp:* Intern & resident, Strong Mem Hosp, Univ Rochester, 59-62; from sr instr to assoc prof, 67-74, chmn dept, 77, PROF BIOPHYS, SCH MED, UNIV ROCHESTER, 74- *Concurrent Pos:* USAEC res fel, 63-65; USPHS spec fel, Univ Saarland, 65-66; Buswell fel, Sch Med, Univ Rochester, 66-67; NIH res grant, 70- *Honors & Awards:* Sr Humboldt Award. *Mem:* Biophys Soc; Am Phys Soc; Am Soc Hemat; Am Fedn Clin Res; Biorheology, Europ Microcirculation. *Res:* Biophysical properties of blood cells; microcirculation. *Mailing Add:* Box BPH 601 Elmwood Ave Rochester NY 14642-8408

LACEWELL, RONALD DALE, b Plainview, Tex, Apr 15, 40; m 62; c 3. RESOURCE ECONOMICS, PRODUCTION ECONOMICS. *Educ:* Tex Tech Univ, BS, 63, MS, 67; Okla State Univ, PhD(agr econ), 70. *Prof Exp:* Statistician, Bur Census, US Dept Commerce, 63-64; instr agr econ, Tex Tech Univ, 65-66; economist, Econ Res Serv, US Dept Agr, 67-70; from asst prof to assoc prof, 70-78, PROF AGR ECON, TEX A&M UNIV, 78- *Concurrent Pos:* Consult, Govt, Legal & Corp. *Mem:* Am Agr Econ Asn. *Res:* Economics of water resources emphasizing agriculture; alternative energy sources and impacts of energy price adjustments; economics and environmental impacts of integrated pest management systems used for crop production. *Mailing Add:* Dept Agr Econ Tex A&M Univ College Station TX 77843

LACEY, BEATRICE CATES, b New York, NY, July 22, 19; m 38; c 2. PSYCHOPHYSIOLOGY. *Educ:* Cornell Univ, AB, 40; Antioch Col, MA, 58. *Prof Exp:* From instr to prof psychol, Antioch Col, 56-77, adj prof, 77-82; mem staff, Fel Res Inst, Wright State Univ, 53-82, sr investr, 66-72, sr scientist, 72-82, fel prof, 77-82, actg sci dir, 79-82, clin prof psychiat, 82-89, EMER FELS PROF, WRIGHT STATE UNIV, 89- *Concurrent Pos:* Co-prin investr, USPHS grant, 60-82; assoc ed, Psychophysiol, 75-78. *Honors & Awards:* Distinguished Sci Contrib Award, Am Psychol Asn, 76; Psychol Sci Gold Medal Award, Am Psychol Found, 85. *Mem:* Fel Soc Exp Psychologists; fel Acad Behav Med Res; Soc Psychophysiol Res (pres, 78-79); Soc Neurosci. *Res:* Psychophysiology of the autonomic nervous system. *Mailing Add:* 1425 Meadow Lane Yellow Springs OH 45387-1221

LACEY, ELIZABETH PATTERSON, b Cleveland, Ohio. PLANT ECOLOGY. *Educ:* Univ Colo, BA, 69; Univ Mich, MS, 74, PhD(bot), 78. *Prof Exp:* ASST PROF BOT, UNIV NC, GREENSBORO, 78- *Mem:* Bot Soc Am; Brit Ecol Soc; Ecol Soc Am; Soc Study Evolution; Sigma Xi. *Res:* Plant population biology; evolution of life history patterns. *Mailing Add:* Dept Biol Univ NC Greensboro NC 27412

LACEY, HOWARD ELTON, b Leaky, Tex, Feb 9, 37; m 58; c 4. MATHEMATICS. *Educ:* Abilene Christian Col, BA, 59, MA, 61; NMex State Univ, PhD(math), 63. *Prof Exp:* Asst prof math, Abilene Christian Col, 63-64 & Univ Tex, Austin, 64-67; res assoc, NASA Manned Spacecraft Ctr, 67-68; from assoc prof to prof math, Univ Tex, Austin, 68-80, mem grad fac, 68-80, vchmn dept, 75-77; dept head math, 80-91, ASSOC DEAN, COL SCI, UNIV TEX A&M, 91- *Concurrent Pos:* Res assoc, Inst Math, Polish Acad Sci, Warsaw, 72-73. *Mem:* Am Math Soc; Math Asn Am. *Res:* Functional analysis; classical Banach spaces. *Mailing Add:* Dept Math Tex A&M Univ College Station TX 77843

LACEY, JOHN I, b Chicago, Ill, Apr 11, 15; m 38; c 2. PSYCHOPHYSIOLOGY, NEUROPHYSIOLOGY. *Educ:* Cornell Univ, BA, 37, PhD(psychol), 41. *Prof Exp:* Instr psychol, Queens Col, NY, 41-42; res assoc, Antioch Col, 46-73, from assoc prof to prof psychophysiol & chmn dept, 48-77; sr scientist, Fels Res Inst, 73-77; fels prof & chmn dept, 77-82, FELS EMER PROF PSYCHIAT, SCH MED, WRIGHT STATE UNIV, 82-

Concurrent Pos: William James fel, Am Physchol Soc, 89; lectr, Ohio State Univ, 50 & sch med, Univ Louisville, 55; fel, Commonwealth Fund, 57-59; mem ment health, behav sci & exp psychol study sects, USPHS, 56-60, res career develop comt, 64-65; mem adv panel life sci facilities, NSF, 60-61; mem coun, Am Psychol Asn, 64-68, 70-73, & 78-79, pres, div 6, 69-70, bd dirs, 74-77; mem clin prog-proj rev comt, NIMH, 66-71, chmn, 70-71; mem bd sci counrl, Nat Inst Aging, 77-80; prof, Antioch Col, 78-82; chmn, sect J, AAAS, 85-86. *Honors & Awards:* Distinguished Contrib Award, Soc Psychophysiol Res, 70; Distinguished Sci Contrib Award, Am Psychol Asn, 76; Sci Gold Medal Award, Am Psychol Found, 85. *Mem:* Nat Acad Sci; fel Soc Exp Psychologists; Soc Neurosci; fel Acad Behav Med; fel Am Psychol Asn; Int Brain Res Orgn. *Res:* Psychophysiology of the autonomic nervous system and psychosomatic medicine; brain physiology and behavior. *Mailing Add:* 1425 Meadow Lane Yellow Springs OH 45387

LACEY, RICHARD FREDERICK, b Vallejo, Calif, May 29, 31; m 71. MAGNETISM. *Educ:* Mass Inst Technol, SB, 52, PhD(physics), 59. *Prof Exp:* Sr engr, Sylvania Lighting Prod Co, 59-62; sr scientist, Am Sci & Eng Co, 62-63; sr physicist, Varian Assocs, 63-67; physicist, 67-69, STAFF SCIENTIST, HEWLETT-PACKARD LABS, 69- *Mem:* Inst Elec & Electronics Engrs; Am Phys Soc. *Res:* Atomic structure; radio-frequency spectroscopy; physical and quantum electronics. *Mailing Add:* Hewlett-Packard Labs 1651 Page Mill Rd Palo Alto CA 94304

LACH, JOHN LOUIS, physical pharmacy; deceased, see previous edition for last biography

LACH, JOSEPH T, b Chicago, Ill, May 12, 34; m 65. PHYSICS. *Educ:* Univ Chicago, AB, 53, MS, 56; Univ Calif, Berkeley, PhD(physics), 63. *Prof Exp:* Res assoc physics, Yale Univ, 63-65, asst prof, 66-69; chmn, Fermi Lab Physics Dept, 74-75, MEM STAFF, FERMI NAT ACCELERATOR LAB, 69-; RES AFILIARE, ILL STATE GEOL SURVEY, 87- *Concurrent Pos:* Joint res prog, Leningrad Nuclear Physics Inst (USSR). *Mem:* Am Phys Soc. *Res:* Elementary particle physics; physics electronic data processing. *Mailing Add:* 28 W 364 Indian Knoll Trail West Chicago IL 60185

LACHAINE, ANDRE RAYMOND JOSEPH, b Ottawa, Ont, Sept 22, 45; m 68; c 3. OPTICS. *Educ:* Univ Ottawa, BSc Hons, 67, MSc, 70, PhD(physics), 76. *Prof Exp:* Instr physics, Univ NB, 76; from asst prof to assoc prof, 76-90, PROF PHYSICS, ROYAL MIL COL CAN, 90- *Concurrent Pos:* Investr contract, 77- *Res:* Photoacoustics and photothermal physics. *Mailing Add:* Dept Physics Royal Mil Col Kingston ON K7K 5L0 Can

LACHANCE, DENIS, b Quebec, Que, Feb 2, 39; m 64; c 3. FOREST PATHOLOGY. *Educ:* Laval Univ, BSc, 62; Univ Wis-Madison, PhD(phytopath), 66. *Prof Exp:* res scientist forest path, Laurentian Forest Res Ctr, 66-79, HEAD, FOREST INSECT & DIS SURV SECT, CAN FORESTRY SERV, 79-, RES PROJ LEADER, ENVIRON STRESS ON FORESTS, 89- *Mem:* Can Phytopath Soc (secy-treas, 71-73); Can Inst Forestry; Int Soc Plant Path. *Res:* Decay of conifers; root diseases. *Mailing Add:* Laurentian Forestry Ctr Forestry Can Ste-Foy PQ G1V 4C7 Can

LA CHANCE, LEO EMERY, b Brunswick, Maine, Mar, 1, 31; m 55; c 3. GENETICS. *Educ:* Univ Maine, AB, 53; NC State Col, MS, 55, PhD(genetics), 58. *Prof Exp:* Res assoc biol, Brookhaven Nat Lab, 58-60; insect geneticist, 60-63, proj leader, Insect Genetics & Radiation Biol Sect, Metab & Radiation Res Lab, Entom Res Div, Agr Res Serv, USDA, 63-69; sci officer & head, Insect Eradication & Pest Control Sect, Joint Food & Agr Orgn-Int Atomic Energy Agency, Austria, 69-71; proj leader, Insect Genetics & Radiation Biol Sect, Agr Res Serv, USDA, 71-77, dir, 77-82, supvr insect geneticist & nat tech adv, Metab & Radiation Res Lab, 82-86; DEP DIR, JOINT FOOD & AGR ORGN INT ATOMIC ENERGY AGENCY, AUSTRIA, 86- *Mem:* AAAS; Genetics Soc Am; Radiation Res Soc; Entom Soc Am; Fedn Am Scientists; Am Inst Biol Sciences. *Res:* Insect genetics and radiation biology; population genetics of screwworms, mechanism of hybrid sterility in Heliothis species; insect cytology and cytogenetic effects of radiation; factors influencing chromosome aberrations and dominant lethal mutations induced by radiation and chemicals; insect reproduction. *Mailing Add:* Int Atomic Energy Agency Wagramerstrasse Five PO Box 200 A-1400 Vienna Austria

LACHANCE, MURDOCK HENRY, b Detroit, Mich, Dec 12, 20; m 43; c 1. ELECTRO OPTICS. *Educ:* Mich Technol Univ, BS, 42, MS, 47. *Prof Exp:* Prin phys metallurgist, Battelle Mem Inst, 47-57; sr res metallurgist, Whirlpool Res Labs, 57-62; sr scientist, Xerox Electro-Optical Systs, 62-83; CONSULT, 83- *Concurrent Pos:* Com pilot & flight instr, Purdue Aeronaut Corp, 42-44; naval aviator, US Naval Reserves, 44-55. *Mem:* Sigma Xi. *Res:* Corrosion of stainless steels; development of refractory alloys, controlled porosity tungsten for ion propulsion and semiconductors for Peltier cooling; failure mechanisms of electro-optical devices; author and patentee in foregoing areas. *Mailing Add:* 260 S Chester Ave Pasadena CA 91106

LACHANCE, PAUL ALBERT, b St Johnsbury, Vt, June 5, 33; m 55; c 4. NUTRITION, FOOD SCIENCE. *Educ:* St Michael's Col, Vt, BSc, 55; Univ Ottawa, PhD(biol, nutrit), 60. *Hon Degrees:* DSc, St Michael's Col, 82. *Prof Exp:* Res biologist, Aerospace Med Res Lab, Wright-Patterson AFB, Ohio, 60-63; coord flight food & nutrit, NASA Manned Spacecraft Ctr, 63-67; assoc prof food sci, 67-72, dir sch feeding effectiveness res proj, 69-72, PROF NUTRIT & FOOD SCI, RUTGERS UNIV, 72-, DIR, GRAD PROG FOOD SCI, 87- *Concurrent Pos:* Lectr, Univ Dayton, 63; Chair, Univ Senate & Fac Rep Bd Gov, Rutgers Univ, 90-, actg chair, Dept Food Sci, 90- *Honors & Awards:* Gemini Prog Achievement Award, NASA, 66. *Mem:* Fel Inst Food Technologists; Am Soc Clin Nutrit; Am Pub Health Asn; NY Acad Sci; Am Dietetic Asn; fel Am Col Nutrit. *Res:* Aerospace food and nutrition; nutritional toxicology; nutritional aspects of food processing; micronutrient nutrification. *Mailing Add:* 34 Taylor Rd RD 4 Princeton NJ 08540-9521

LACHAPELLE, RENE CHARLES, b Joliette, Que, Jan 28, 30; US citizen; m 59; c 3. MEDICAL MICROBIOLOGY. *Educ:* Seminaire de Joliette, BA, 50; Univ Montreal, BSc, 53; Syracuse Univ, MS, 57, PhD(microbiol), 62. *Prof Exp:* Lab admin dir clin path, Syracuse Mem Hosp, 62-66; assoc prof biol, Univ Dayton, 66-74; dir, Sch Allied Health, 77 & 82, CHAIRPERSON, DEPT MED TECHNOL, UNIV VT, 74- *Mem:* Sigma Xi; AAAS; Am Soc Med Tech; Am Soc Microbiol; Can Soc Microbiol. *Res:* Morphogenesis and serological properties of Candida albicans; monomine oxidase and serotonin in germfree animals; skin bacteria in long-term space flights; educational aspects of medical technology; coagglutination of streptococcal groups. *Mailing Add:* 302 Rowell Bldg Univ Vt Burlington VT 05401

LACHENBRUCH, ARTHUR HEROLD, b New Rochelle, NY, Dec 7, 25; m 50; c 3. GEOPHYSICS. *Educ:* Johns Hopkins Univ, BA, 50; Harvard Univ, MA, 54, PhD(geophys), 58. *Prof Exp:* GEOPHYSICIST, US GEOL SURV, 51- *Concurrent Pos:* Vis prof, Dartmouth Col, 63. *Honors & Awards:* Kirk Bryan Award, Geol Soc Am, 63; Distinguished Serv Award, Dept Interior, 78; Walter H Bucher Medal, Am Geophys Union, 89. *Mem:* Nat Acad Sci; fel AAAS; fel Am Geophys Union; fel Royal Astron Asn; fel Arctic Inst NAm. *Res:* Solid earth geophysics; terrestrial heat flow; tectonophysics; permafrost. *Mailing Add:* Br Tectonophysics US Geol Surv 345 Middlefield Rd Menlo Park CA 94025

LACHENBRUCH, PETER ANTHONY, b Los Angeles, Calif, Feb 5, 37; m 62. BIOSTATISTICS. *Educ:* Univ Calif, Los Angeles, BA, 58, PhD(biostatist), 65; Lehigh Univ, MS, 61. *Prof Exp:* Asst math, Lehigh Univ, 58-59; programmer, Douglas Aircraft Co, 59-60; sr opers res analyst, Syst Develop Corp, 60-61; res scientist, Am Inst Res, 61-62; USPHS fel biostatist, Univ Calif, Los Angeles, 62-65; from asst prof to prof biostatist, 65-76; Univ NC, Chapel Hill, 75-76; prof prev med, Univ Iowa, 76; AT SCH PUB HEALTH, UNIV CALIF, LOS ANGELES. *Honors & Awards:* Mortimer Spiegelman Gold Medal Award, Am Pub Health Asn, 71; fel, Am Statist Asn, 79. *Mem:* AAAS; fel Am Statist Asn; Biomet Soc; Am Pub Health Asn; Royal Statist Soc; Sigma Xi. *Res:* Discriminant analysis; statistical epidimology; computer analysis of data; survival analysis. *Mailing Add:* Sch Pub Health Univ Calif Los Angeles Los Angeles CA 90024

LACHER, ROBERT CHRISTOPHER, b Atlanta, Ga, Oct 14, 40. TOPOLOGY, APPLIED MATHEMATICS. *Educ:* Univ Ga, BS, 62, MA, 64, PhD(math), 66. *Prof Exp:* Asst prof math, Univ Calif, Los Angeles, 66-67; vis mem math, Inst Advan Study, 67-68; from asst prof to assoc prof, 68-75, PROF MATH, FLA STATE UNIV, 75- *Concurrent Pos:* Alfred P Sloan fel, 70-72; mem, Inst Advan Study, 72; NSF res grants, 72- *Mem:* AAAS; Sigma Xi; Am Math Soc. *Res:* Geometric topology; cell-like mappings and generalized manifolds; embedding problems; catastrophe theory. *Mailing Add:* Dept Comput Sci Fla State Univ Tallahassee FL 32306

LACHER, THOMAS EDWARD, JR, b Pittsburgh, Pa, Aug 9, 49; m 78; c 2. TROPICAL ECOLOGY, CONSERVATION BIOLOGY. *Educ:* Univ Pittsburgh, BS, 72, PhD(biol sci), 80. *Prof Exp:* Teaching asst biol, Univ Pittsburgh, 72-79; asst prof zool, Univ Brazil, 79-81; from asst prof to assoc prof environ studies, Western Wash Univ, 81-89; DIR TROP ECOL, CLEMSON UNIV, 89- *Concurrent Pos:* Vis prof, Univ Fed Minas Gerais, 86; consult, World Wildlife Fund, 86, Empresa Brazil Pesquisa Agropecvaria, 88, Kellogg Found, 91; mem, Species Survival Comm IUCN, 89-, Working Group Comt Nat Inst Environ, 90- *Mem:* AAAS; Am Soc Mammalogists; Ecol Soc Am; Asn Trop Biol; Am Asn Naturalists; Soc Conserv Biol. *Mailing Add:* Archbold Trop Res Ctr Clemson Univ Clemson SC 29634-1019

LACHIN, JOHN MARION, III, b New Orleans, La, July 4, 42. CLINICAL TRIALS. *Educ:* Tulane Univ, BS, 65; Univ Pittsburgh, ScD(biostatist), 72. *Prof Exp:* Epidemiologist & dir, Div Prog Info & Eval, Va, 72-73; from asst res prof to res prof statist, George Washington Univ, 73-84, asst dir, Biostatist Ctr, 80-85, co-dir, 85-88, PROF STATIST, GEORGE WASHINGTON UNIV, 84-, DIR, BIOSTATIST CTR, 88- *Concurrent Pos:* Adj asst prof biometry, Va Commonwealth Univ, 72-73; mem, Serv Res & Epidemiol Studies Rev Comt, NIMH, 77-80, Gastrointestinal Drugs Adv Comt, Food & Drug Admin, 78-82, Data Monitoring Comt, NIH Nat Coop Dialysis Study, 79-81, Opers Comt, Vet Admin Coop Study on Hypertension, 81-86; dir, Biostatist Coord Ctr, Nat Coop Gallstone Study, 78-83, Lupus Nephritis Collab Study, 81-86 & Diabetes Control & Complications Trial, 82-; mem policy adv bd, Hypertension Prevention Trial, NIH, 82-86, chmn treatment effects monitoring & adv comt, Glaucoma Laser Trial, 83. *Mem:* Biomet Soc; Am Statist Asn; Soc Epidemiol Res; Soc Clin Trials; Inst Math Stat; Int Soc Clin Biostatist. *Mailing Add:* Biostatist Ctr George Washington Univ 6110 Executive Blvd Rockville MD 20852

LACHMAN, IRWIN MORRIS, b New York, NY, Aug 2, 30; m 59; c 2. CERAMICS ENGINEERING. *Educ:* Rutgers Univ, BSc, 52; Ohio State Univ, MSc, 53, PhD(ceramic eng), 55. *Prof Exp:* Sr scientist ceramics, Thermo Mat, Inc, 57-58; staff mem, Sandia Corp, 58-60; RES ASSOC CERAMICS, CORNING GLASS WORKS, 60- *Mem:* AAAS; fel Am Ceramic Soc; Brit Ceramic Soc; Soc Automotive Engrs. *Res:* Mechanical and thermal properties of ceramics. *Mailing Add:* Corning Inc Sullivan Park Lab Corning NY 14831

LACHMAN, LAWRENCE B, b Denver, Colo, Nov 13, 47. CELLULAR BIOLOGY. *Educ:* Univ Colo, BA, 69; Boston Univ, PhD(biochem), 73. *Prof Exp:* Res assoc pharmacol, Sch Med, Yale Univ, 73-76; res assoc, Dept Microbiol & Immunol, Div Immunol, Duke Univ Med Ctr, 76-78, med res assoc, 78-79, med res asst prof, 79-82; sr staff scientist, Immunex Corp, Seattle, Wash, 82-83; assoc prof, 83-88, PROF, DEPT CELL BIOL, M D ANDERSON CANCER CTR, UNIV TEX, HOUSTON, 88- *Concurrent Pos:* Mem, Duke Comprehensive Cancer Ctr, 81; ed-in-chief, Lymphokine Res, 82-; mem, Biol Response Modifiers Prog, Decision Network Comt, 82-83; NIH grants, 84-87, 86-89, 88-89 & 87-92; mem, AIDS & Related Res Rev Group, NIH-Dept Health & Human Serv, 90- *Mem:* Am Asn Immunol. *Res:* Lymphokines and cytokines; interleukin-1; author or co-author of numerous publications; recipient of one patent. *Mailing Add:* Dept Cell Biol Univ Tex M D Anderson Cancer Ctr 1515 Holcombe Blvd Houston TX 77030

LACHMAN, LEON, b Bronx, NY, Jan 29, 29; m 51; c 2. PHARMACY. *Educ:* Columbia Univ, BSc, 51, MSc, 53; Univ Wis, PhD, 56. *Hon Degrees:* Dr, Columbia Univ, 76. *Prof Exp:* Asst dir pharm, Res & Develop Div, Ciba Pharmaceut Co, NJ, 56-68, dir, 68-69; vpres develop & control, Du Pont Pharmaceut, 69-79; sr vpres sci & technol, United Lab Inc, 79-81; PRES LACHMAN CONS SERVS INC, WESTBURY, NY, 81- *Concurrent Pos:* Vis scientist, Am Asn Cols Pharm; mem bd trustees, Col Pharm, Columbia Univ, 74-78; vis prof, Rutgers Univ Col Pharm, 83- *Honors & Awards:* Indust Pharmaceut Technol Award, Acad Pharmaceut Sci, 70. *Mem:* Fel Acad Pharmaceut Sci; Parenteral Drug Asn (pres 81-); Am Chem Soc; Am Asn Pharmaceut Scientists. *Res:* Process and equipment design; research and development of pharmaceutical dosage forms; analytical research; quality control practices; medical research; regulatory affairs. *Mailing Add:* Lachman Cons Servs Inc 1600 Stewart Ave Suite 604 Westbury NY 11590-6611

LACHNER, ERNEST ALBERT, b New Castle, Pa, Apr 3, 15; m 39; c 4. SYSTEMATIC ICHTHYOLOGY. *Educ:* Pa State Teachers Col, Slippery Rock, BS, 37; Cornell Univ, PhD(ichthyol), 46. *Prof Exp:* Teacher high sch, Pa, 37-39; asst zool, Cornell Univ, 40-42; assoc prof fishery biol, Pa State Col, 47-49; assoc cur, Nat Mus Natural Hist, 49-65, cur 65-66, supvr & cur, Div Fishes, 66-82; RETIRED. *Concurrent Pos:* Guggenheim fel, 56 & 59. *Mem:* AAAS; Am Soc Ichthyol & Herpet (vpres, 54, pres-elect, 66, pres, 67); Am Fisheries Soc; Am Soc Limnol & Oceanog. *Res:* Systematics and morphology of marine and fresh water fishes; ecology; life history of fishes. *Mailing Add:* Div Fishes US Nat Mus Washington DC 20560

LACHS, GERARD, b Essen, Ger, Aug 2, 34; US citizen; m 57; c 2. ELECTRONICS. *Educ:* NY Univ, BS, 56; Univ Rochester, MS, 61; Syracuse Univ, PhD(elec eng), 64. *Prof Exp:* Asst engr, Sperry Gyroscope Co, 57-58; sr res staff mem commun, Gen Dynamics/Electronics, 58-61; instr elec eng, Syracuse Univ, 61-64; from asst prof to prof elec eng, Pa State Univ, 64-84; PROF ELEC ENG, UNIV SFLA, 84- *Mem:* Inst Elec & Electronics Engrs; Acoust Soc Am. *Res:* Coherent fiber optic communication systems; digital communication systems; coding of orthogonal wave shapes; audition and psycho-acoustics; bioengineering. *Mailing Add:* Dept Elec Eng Univ SFla Tampa FL 33620

LACK, LEON, b New York, NY, Jan 7, 22; m 48; c 5. BIOCHEMISTRY. *Educ:* Brooklyn Col, AB, 43; Mich State Univ, MS, 48; Columbia Univ, PhD(biochem), 53. *Prof Exp:* Fel, Duke Univ, 53-55; from instr to asst prof pharmacol, Sch Med, Johns Hopkins Univ, 55-64; from asst prof to assoc prof, 65-71, PROF PHARMACOL, MED CTR, DUKE UNIV, 71- *Mem:* Am Soc Biol Chemists; Am Soc Pharmacol & Exp Therapeuts. *Res:* Metabolism of aromatic substances; intestinal active transport; pharamacology of androgen related disorders. *Mailing Add:* Lab Molecular Pharmacol Box 3185 Duke Univ Med Ctr Durham NC 27706

LACKEY, CAROLYN JEAN, b Shelby, NC, Nov 24, 48. COMMUNITY NUTRITION. *Educ:* Univ NC, Greensboro, BSHE, 71; Univ Tenn, Knoxville, MS, 73, PhD(food sci), 74. *Prof Exp:* Asst prof foods & nutrit, Purdue Univ, 74-76; from asst prof to assoc prof community nutrit, Mich State Univ, 76-83; EXTEN PROF, FOOD & NUTRIT, NC AGR EXTEN SERV, NC STATE UNIV, 83- *Concurrent Pos:* Proj dir nutrit educ grant, Mich Dept Educ, 78. *Mem:* Soc Nutrit Educ; Inst Food Technologists. *Res:* Investigation of determinants of food behavior and food behavior modification; development, implementation and evaluation of food and nutrition education materials. *Mailing Add:* NC State Univ Box 7605 Raleigh NC 27695-7605

LACKEY, HOMER BAIRD, b Freewater, Ore, Nov 23, 20; m 42; c 3. APPLIED CHEMISTRY. *Educ:* Ore State Univ, BS, 47, MS, 48. *Prof Exp:* Asst, Ore State Univ, 47-48; res chemist, Cent Res Dept, 48-55, supvr prod res, Chem Prod Div, 55-68, mgr prod res, 68-80, MGR REGULATORY AFFAIRS, CHEM PROD DIV, CROWN ZELLERBACH CORP, 80- *Concurrent Pos:* Consult, chem concrete, 82- *Mem:* Am Chem Soc; Am Soc Test & Mat; Am Concrete Inst. *Res:* Forest byproduct utilization. *Mailing Add:* 804 NW 19th Ave Camas WA 98607-9316

LACKEY, JAMES ALDEN, b Glens Falls, NY, Nov 25, 38; div; c 2. MAMMALOGY. *Educ:* Cornell Univ, BS, 61; Calif State Univ, San Diego, MA, 67; Univ Mich, PhD(zool), 73. *Prof Exp:* ASST PROF ZOOL, NY STATE UNIV COL, OSWEGO, 73- *Mem:* Am Soc Mammalogists; Sigma Xi; AAAS. *Res:* Reproduction, growth, development and population ecology of mammals. *Mailing Add:* Biol Dept State Univ NY Oswego NY 13126

LACKEY, LAURENCE, US citizen. GEOMORPHOLOGY, ENGINEERING GEOLOGY. *Educ:* Principia Col, BS, 69; Univ Mich, PhD(geol), 74. *Prof Exp:* Asst prof geol, Mich State Univ, 74-75; asst prof geol, Memphis State Univ, 75-78; asst prof, 79-80, ASSOC PROF & CHMN GEOL, PRINCIPIA COL, 81- *Concurrent Pos:* Dir, Tenn Earthquake Info Ctr, 77-78; consult eng geol, 75- & Off Energy Info Validation, US Dept Energy, 78; prin investr, US Nuclear Regulatory Comn Contract, 77-78; mem proposal rev panel, Inst Sci Equip Prof, NSF, 81. *Mem:* AAAS; Geol Soc Am; Am Quaternary Asn; Nat Asn Geol Teachers. *Res:* Computer applications in geology and quaternary geology. *Mailing Add:* 3940 Regency Dr Colorado Springs CO 80906

LACKEY, ROBERT T, b Kamloops, BC, May 18, 44; m 67; c 2. FISHERIES & WILDLIFE MANAGEMENT. *Educ:* Humboldt State Univ, BS, 67; Univ Maine, Orono, MS, 68; Colo State Univ, PhD(fisheries & wildlife), 71. *Prof Exp:* Asst prof, Va Polytech Inst & State Univ, 71-74, sect leader, Fisheries Sci, 71-72 & 75-77, assoc prof fisheries, 74-79; group leader, Nat Water Res Analysis Group, 79-81; sr ecologist, 81-85, assoc br chief, 85-87, BR CHIEF, ENVIRON RES LAB, CORVALLIS, 87-; ASSOC PROF, ORE STATE UNIV, 82- *Concurrent Pos:* Res grants, Off Econ Opportunity & Celanese Corp, Off Water Res, 72-77, US Nat Marine Fisheries Serv, 72-78, US Dept Agr, 73-78 & US Forest Serv, 75-78; consult, US Fish & Wildlife Serv, 74-76, Brandermill Corp, 74-75 & US Army Corps Engrs, 75-76; fish & wildlife

adminr, US Fish & Wildlife Serv, 76-77; vis prof, George Mason Univ, 76-77, Univ Mich, 78. *Mem:* Inst Fishery Res Biologists; Am Fisheries Soc. *Res:* Effects of acid rain, global climate change, and strato spheric ozone depletion on aquatic and terrestrial resources; fisheries management, including structure and management of aquatic renewable natural resources; systems analysis; environmental assessment. *Mailing Add:* Environ Protection Agency 200 SW 35th St Corvallis OR 97333

LACKEY, WALTER JACKSON, b Shelby, NC, Feb 6, 40; m 61; c 2. CERAMIC & METALLURGICAL ENGINEERING. *Educ:* NC State Univ, BS(metall eng) & BS(ceramic eng), 61, MS, 63, PhD(ceramic eng), 70. *Prof Exp:* Res scientist, Battelle-Northwest Lab, 63-65; mat engr, Douglas Aircraft Corp, 65-66; res asst electronic ceramics, NC State Univ, 66-69; group leader, Metals & Ceramics Div, Oak Ridge Nat Lab, 69-84; PRIN RES SCIENTIST, GA INST TECHNOL, 86- *Mem:* Am Ceramic Soc; Am Soc Metals; fel Am Ceramic Soc. *Res:* Fabrication, characterization and testing of nuclear fuels and waste forms; mechanisms and measurement of electrical conduction in ceramic insulators; ceramic coatings and composites; ceramic superconductors. *Mailing Add:* 3129 Wendwood Dr Marietta GA 30062

LACKMAN, DAVID BUELL, bacteriology, public health; deceased, see previous edition for last biography

LACKNER, HENRIETTE, b Vienna, Austria, Feb 27, 22; US citizen; m 49; c 3. HEMATOLOGY. *Educ:* Univ Leeds, MB & ChB, 45, MD, 48. *Prof Exp:* Jr lectr med, Univ Cape Town, 55-62; res assoc, 63-65, instr med, 65-67, from asst prof clin med to asst prof med, 67-75, ASSOC PROF CLIN MED, SCH MED, NY UNIV, 75- *Concurrent Pos:* Res asst, Groote Schuur Hosp, Cape Town, SAfrica, 55-62, asst physician, Arthritis Clin, 55-56, physician-in-chg & asst physician med outpatient clin, 56-62; res scientist, Am Nat Red Cross, 63-; clin asst vis physician, Bellevue Hosp, New York, 65-75, assoc vis physician, 75-; asst, Univ Hosp, 66-75, assoc med, 75- *Mem:* Soc Study Blood; Am Soc Hemat; Med Soc Ny. *Res:* Blood coagulation disorders and pathological fibrinolysis. *Mailing Add:* Dept Med NY Univ Med Ctr New York NY 10016-6451

LACKO, ANDRAS GYORGY, b Budapest, Hungary, Nov 10, 36; Can citizen; m 64; c 4. BIOCHEMISTRY, MICROBIOLOGY. *Educ:* Univ BC, BSA, 61, MSc, 63; Univ Wash, PhD(biochem), 68. *Prof Exp:* Res asst biochem, Univ Wash, 63-68; asst mem, Albert Einstein Med Ctr, 69-71; mem staff, Med Sch, Temple Univ, 71-72; asst prof med, 72-75; assoc prof, 75-83, PROF BIOCHEM, TEX COL OSTEOP MED, 83- *Concurrent Pos:* NIH fel, Albert Einstein Col Med, 68-69. *Mem:* Am Soc Biochem & Molecular Biol; Am Heart Asn. *Res:* Structure and function of enzymes and lipoproteins. *Mailing Add:* Dept Biochem Tex Col Osteop Med Ft Worth TX 76107

LACKS, SANFORD, b New York, NY, Jan 28, 34; m 59; c 3. GENETICS, MOLECULAR BIOLOGY. *Educ:* Union Univ, NY, BS, 55; Rockefeller Univ, PhD, 60. *Prof Exp:* Instr biol, Harvard Univ, 60-61; from asst geneticist to geneticist, 61-82, SR GENETICIST, BROOKHAVEN NAT LAB, 82- *Mem:* Am Soc Microbiol; Am Soc Biol Chem; Genetics Soc Am. *Res:* Bacterial transformation; DNA repair; DNA methylation; restriction enzymes; folate biosynthesis. *Mailing Add:* Biol Dept Brookhaven Nat Lab Upton NY 11973

LACKSONEN, JAMES W(ALTER), b Ashtabula, Ohio, Oct 17, 36; m 57; c 2. CHEMICAL ENGINEERING. *Educ:* Ohio State Univ, BChE & MSc, 59, PhD(chem eng), 64. *Prof Exp:* Res engr, Battelle Mem Inst, 60-65; proj engr, Pittsburgh Plate Glass Co, 65-66; sr develop engr chem-plastics div, Gen Tire & Rubber Co, Ohio, 66-67; asst prof, 67-80, ASSOC PROF CHEM ENG, UNIV TOLEDO, 80-, ASST DEAN, COL ENG, 71- *Mem:* AAAS; Am Inst Chem Engrs; Am Chem Soc; Electrochem Soc. *Res:* Kinetics and surface chemistry processes; fuel cells; transport of gases in microporous media; reactor design; reinforced plastics; foam; mass transfer. *Mailing Add:* Dept Chem Eng Univ Toledo 2801 W Bancroft Toledo OH 43606

LA CLAIRE, JOHN WILLARD, II, b Utica, NY, July 1, 51. PHYCOLOGY, MARINE BOTANY. *Educ:* Cornell Univ, BS, 73; Univ SFla, MA, 75; Univ Calif, Berkeley, PhD(bot), 79. *Prof Exp:* Asst prof, 79-85, ASSOC PROF BOT, DEPT BOT, UNIV TEX, 85- *Concurrent Pos:* Prin investr cell biol sect, NSF, 81-; prin investr plant growth develop, USDA, 87-; vis assoc prof, Dept Cell Biol, Stanford Univ, 89. *Mem:* AAAS; Am Soc Cell Biol; Bot Soc Am; Brit Phycol Soc; Int Phycol Soc; Phycol Soc Am. *Res:* Cell biology and developmental biology of algae; cell motility phenomena; cellular wound healing, mitosis, cytokinesis and cytoplasmic streaming; cytoskeleton of plant cells. *Mailing Add:* Dept Bot Univ Tex Austin TX 78713-7640

LACOSS, RICHARD THADDEE, b Gardner, Mass, Aug 19, 37; m 84. SIGNAL PROCESSING, ESTIMATION. *Educ:* Columbia Univ, AB, 59, BS, 60; Univ Calif, Berkeley, MS, 62, PhD(elec eng, info & control theory), 65. *Prof Exp:* Mem staff, 65-69, GROUP LEADER, LINCOLN LAB, MASS INST TECHNOL, 69- *Mem:* Inst Elec & Electronics Engrs; Sigma Xi; Asn Comput Mach. *Res:* Signal processing; underground nuclear test detection; system engineering; intelligent systems; atmospheric acoustics; distributed algorithms. *Mailing Add:* No 8 Chauncy Lane Cambridge MA 02138

LACOSTE, RENE JOHN, b New York, NY, Feb 19, 27. ANALYTICAL CHEMISTRY, PESTICIDE CHEMISTRY. *Educ:* Rensselaer Polytech Inst, BS, 50; Univ Chicago, MS, 53. *Prof Exp:* Chemist, Am Dent Asn, 50-53; chemist, Rohm & Haas Co, 53-68, sr chemist, 68-69, int registr agr & sanit chem, 69-75, regional regulatory mgr, 75-80, foreign regulatory mgr agr chem, 80-90; RETIRED. *Concurrent Pos:* Agr indust rep UN, Codex Comt Pesticide Residues, 70-90. *Mem:* AAAS; Am Chem Soc; Am Inst Chem; NY Acad Sci; Sigma Xi. *Res:* Physical and chemical methods of analysis; electrochemical analysis; separations of organic mixtures; developed methods for analysis of chemicals in products of industrial production; national and international industry and government groups concerned with drafting and adopting regulations regarding production and use of agricultural pesticide chemicals. *Mailing Add:* Park Dr Manor Lincoln Dr & Harvey St Philadelphia PA 19144-4306

LACOUNT, ROBERT BRUCE, b Martinsburg, WVa, Sept 16, 35; m 64; c 2. ORGANIC CHEMISTRY. *Educ:* Shepherd Col, BS, 57; Univ Pittsburgh, MLitt, 62, PhD(org chem), 65. *Prof Exp:* Res assoc fundamental org chem, Mellon Inst, 58-65; from asst prof chem to prof & chmn dept chem, 65-71, PROF & CHMN DEPT CHEM & PHYSICS, WAYNESBURG COL, 71- *Concurrent Pos:* Res grant, Petrol Res Fund, 65-67; res chemist, US Bur Mines, 70-75 & Energy Res & Develop Admin, 75-77 & Dept Energy, 77- *Mem:* Am Chem Soc. *Res:* Synthetic organic chemistry; organic sulfur chemistry; production of low-sulfur fuels from coal. *Mailing Add:* Waynesburg Col Waynesburg PA 15370

LACOUTURE, PETER GEORGE, b Worcester, Mass, Oct 26, 51; m 78; c 3. PHARMACOLOGY, PHYSIOLOGY. *Educ:* Col Holy Cross, AB, 73; Mass Col Pharm & Allied Health Sci, MS, 81, PhD(pharmacol & physiol), 86. *Prof Exp:* Sr res asst physiol, Harvard Sch Pub Health, 80-85; res affil toxicol, Div Clin Pharmacol & Toxicol, Children's Hosp, Boston, 85-87; res fel path, Harvard Med Sch & develop officer, Clin Labs, Children's Hosp Boston, Mass, 87; sr clin scientist, 88-90, ASST DIR, WYETH-AYERST RES, 90- *Concurrent Pos:* Consult, Mass Poison Control Syst, 77-87 & Drug Epidemiol Unit, Boston Univ Med Ctr, 80-87; instr pediat, Harvard Med Sch, 88; asst med, Dept Med, Children's Hosp Boston, 88-89; examr, Am Bd Appl Toxicol, 89. *Mem:* Am Acad Clin Toxicol; Am Soc Pharmacol & Exp Therapeut; Am Soc Clin Pharmacol & Therapeut. *Res:* Rheumatology, anti-inflammatory drugs; immunology, anti-allergy and pulmonary drugs; general toxicology; pharmacoepidemiology; regulatory affairs; author of over 100 technical publications. *Mailing Add:* 226 Baldwin Dr West Chester PA 19380

LACROIX, GUY, b Que, Apr 10, 30; m 60; c 2. MARINE ECOLOGY. *Educ:* Laval Univ, BA, 52, LPh, 53; Univ Montreal, BSc, 57, MSc, 59; Laval Univ, DSc, 68. *Prof Exp:* Zooplanktonologist, Grande Riviere Marine Biol Sta, Que, 58-68; from asst prof to assoc prof, 68-74, dir, Grad Studies Biol, 83-85, PROF BIOL OCEANOG, DEPT BIOL, LAVAL UNIV, 74-, ASSOC DIR, SCH GRAD STUDIES, 85- *Concurrent Pos:* Exec secy, Interuniv Group Oceanog Res, Que, 70-77; mem bd, Laval Univ, 74-77 & Sci Comt Oceanic Res, Can Nat Comt, 74-78; ed, Can Naturalist, 79-86. *Mem:* Marine Biol Asn UK; Am Soc Limnol & Oceanog; Plankton Soc Japan. *Res:* Zooplankton; invertebrate zoology; primary production; marine invertebrates. *Mailing Add:* Dept Biol Laval Univ Ste-Foy Quebec PQ G1K 7P4 Can

LACROIX, JOSEPH DONALD, b Windsor, Ont, Apr 7, 25; nat US; m 51; c 3. BOTANY. *Educ:* Univ Western Ont, BA, 47; Univ Detroit, MS, 50; Purdue Univ, PhD(bot), 53. *Prof Exp:* Res asst mycol, Parke, Davis & Co, 51; from instr to assoc prof, 53-74, PROF BIOL, UNIV DETROIT, 74- *Concurrent Pos:* Kellogg fel, Univ Mich, 71-72. *Mem:* Bot Soc Am; Am Inst Biol Sci. *Res:* Scanning electron microscopy and electron probe analysis of silicification patterns in plant species; effects of gravity on plant tissue; radiosensitivity of higher plants. *Mailing Add:* Univ Detroit 4001 W McNichols Rd Detroit MI 48221

LACROIX, LUCIEN JOSEPH, b St Louis, Sask, May 14, 29; m 52; c 5. PLANT PHYSIOLOGY, HORTICULTURE. *Educ:* Univ Sask, BSA, 57, MSc, 58; Iowa State Univ, PhD(plant physiol), 61. *Prof Exp:* Technician field husb, Univ Sask, 54-57; res assoc plant sci, 61-63, from asst to assoc prof, 63-72, PROF PLANT SCI, UNIV MAN, 72- *Mem:* Am Soc Plant Physiol; Can Soc Plant Physiol; Sigma Xi. *Res:* Potato tuberization; greenhouse research; winter hardiness. *Mailing Add:* Dept Plant Sci Univ Man Winnipeg MB R3T 2N2 Can

LACROIX, NORBERT HECTOR JOSEPH, b Sarsfield, Ont, Can, Oct, 26, 40; m 65; c 3. MATHEMATICS, BIOLOGICAL SCIENCES. *Educ:* Univ Ottawa, BSc, 62; Univ Notre Dame, PhD(math), 66. *Prof Exp:* Instr math, Univ Notre Dame, 62-66; from asst prof to assoc prof, 66-77, chmn dept, 70-77, PROF MATH, LAVAL UNIV, 77- *Mem:* Can Math Soc; Asn Canadienne-Francaise Advan Sci. *Res:* Organisational principles in developmental and structural biology; mathematical models. *Mailing Add:* Dept Math Laval Univ Quebec PQ G1K 7P4 Can

LACY, ANN MATTHEWS, b Boston, Mass, May 29, 32. MICROBIAL GENETICS. *Educ:* Wellesley Col, BA, 53; Yale Univ, MS, 56, PhD(microbiol), 59. *Prof Exp:* Asst dept genetics, Carnegie Inst, 53-54; instr genetics, 59-61, asst prof, 61-67, assoc prof, 67-73, chmn dept biol sci, 69-72, PROF BIOL SCI, GOUCHER COL, 73- *Concurrent Pos:* Res fel, Glasgow Univ, 68-69. *Mem:* AAAS; Genetics Soc Am; Bot Soc Am; Am Inst Biol Sci; Sigma Xi. *Res:* Gene structure and function, and gene regulation in Neurospora crassa. *Mailing Add:* Dept Biol Sci Goucher Col Towson MD 21204

LACY, GEORGE HOLCOMBE, b Washington, DC, Nov 13, 43; m 64; c 2. PHYTOPATHOLOGY, BACTERIAL GENETICS. *Educ:* Calif State Univ, Long Beach, BS, 66, MS, 71; Univ Calif, Riverside, PhD(phytopath), 75. *Prof Exp:* Lab technician qual control, Am Chem & Plastics Co, Stauffer Chem Co, Calif, 64-65; biol sci instr, US Peace Corps, Corozal Town, Belize, 66-68; scientist II soil microbiol, Jet Propulsion Lab, Calif Inst Technol, 69-71; res assoc plant path, Univ Wis-Madison, 75-77; asst plant pathologist, Conn Agr Exp Sta, 77-80; from asst prof to assoc prof, 80-88, PROF PLANT PATH, VA POLYTECH INST & STATE UNIV, 88- *Concurrent Pos:* NIH grant, 75-76, NSF grant, Univ Wis-Madison, 76-77, 87-, USDA grants, 83-89, Environ Protection Agency, NSF grants, 87-91. *Mem:* Am Phytopath Soc; Am Microbiol Soc. *Res:* Molecular basis for plant pathogenesis and biological control of plant disease and biological control of plant disease and ecological impact of release of genetically engineered microorganisms into the environment. *Mailing Add:* Plant Molecular Biol Va Polytech Inst & State Univ Blacksburg VA 24061-0330

LACY, JULIA CAROLINE, b Detroit, Mich, July 10, 46. ENVIRONMENTAL BIOLOGY. *Educ:* Univ Mich, BA, 68, MS, 72. *Prof Exp:* Chief ecologist environ biol, Stearns Roger, Inc, 72-74; SR SCIENTIST BIOL ENVIRON BIOL, RADIAN CORP, 74- *Mem:* Ecol Soc Am; Am Inst Biol Sci; World Future Soc; Sigma Xi. *Res:* Policy analysis technology assessment with emphasis on energy resource use, particularly Texas lignite and analysis of policy complexes affecting resource development and commercialization of new technologies. *Mailing Add:* 4110 Iollewild Rd Austin TX 78731

LACY, LEWIS L, b Bluefield, WVa, Mar 25, 41; m 64; c 2. MICROGRAVITY SCIENCES, MATERIAL ANALYSIS & PHYSICS FOR PETROLEUM PRODUCTION. *Educ:* Va Polytech Inst & State Univ, BS, 63, MS, 65; Univ Tenn, Knoxville, PhD(physics), 71. *Prof Exp:* Res assoc solid state physics, Los Alamos Sci Lab, 64; exp physicist, Nuclear & Plasma Physics Div, Space Sci Lab, Marshall Space Flight Ctr, NASA, 65-68, mat & appl physics scientist, Space Sci Lab, 68-77, br chief, Solid State & Solidification Br, 77-81, sr res specialist, 81-84; SR RES STAFF, EXXON PROD RES CO, 84- *Concurrent Pos:* Comt advance in hydraulic fracturing, Soc Petrol Engrs, 86-88. *Honors & Awards:* Marshall Space Flight Ctr Dirs Commendation Award, NASA, 71 & NASA Manned Flight Awareness Award, 74; NASA Group Achievement Award, Johnson Space Flight Ctr, 76; NASA New Technol Award, 80, 84. *Mem:* Am Phys Soc; Soc Petrol Engrs. *Res:* Gleeble welding simulation experiments and fracture toughness of offshore platform steels; hydraulic fracture geometry and orientation, triaxial borehole seismic, tiltmeter arrays and fracture mapping; solidification and crystal growth, containerless supercooling and low-gravity solidification, low temperature and superconducting materials. *Mailing Add:* Six Postvine Ct Spring TX 77381

LACY, MELVYN LEROY, b Henry, Nebr, Oct 24, 31; m 54; c 2. PLANT PATHOLOGY, EPIDEMIOLOGY. *Educ:* Univ Wyo, BS, 59, MS, 61; Ore State Univ, PhD(plant path), 64. *Prof Exp:* From asst prof to assoc prof, 65-78, PROF PLANT PATH, MICH STATE UNIV, 78- *Honors & Awards:* Distinguished Serv Award, Am Phytopath Soc. *Mem:* Am Phytopath Soc; Can Phytopath Soc; Am Potato Asn. *Res:* Soil-borne fungus diseases, epidemiology and control; pesticides for disease control; epidemiology, disease forecasting and disease management. *Mailing Add:* Dept Bot & Plant Path Mich State Univ East Lansing MI 48823

LACY, PAUL ESTON, b Trinway, Ohio, Feb 7, 24; m 45; c 2. PATHOLOGY. *Educ:* Ohio State Univ, BA, 45, MSc & MD, 48; Univ Minn, PhD(path), 55. *Prof Exp:* Asst instr anat, Ohio State Univ, 44-48; intern, White Cross Hosp, Columbus, Ohio, 48-49; Nat Cancer Inst fel, Med Sch, Wash Univ, 55-56; from instr to assoc prof, 56-61, asst dean, 59-61, Mallinckrodt prof path & chmn dept, 61-85, ROBERT L KROC PROF PATH, MED SCH, WASH UNIV, 85- *Concurrent Pos:* Mem path B study sect, NIH, 61-66, chmn, 66-67; mem adv comt res personnel, Am Cancer Soc, 66-70; mem basic sci adv comt, Nat Cystic Fibrosis Res Found, 67-69; assoc ed, Diabetes, 73-; mem nat comn diabetes, NIH, 74-75 & mem nat adv environ health sci coun, 74-77; mem, Nat Diabetes Adv Bd, NIH, 85, 86; path, Barnes & Allied Hosps & St Louis Hosp, 85- *Honors & Awards:* Banting Mem Lectr, Brit Diabetic Ans, 63, Banting Award, 70; Elliot Proctor Joslin Mem Lectr, 66; Rollin Rurner Woodyatt Mem Lectr, 70; Richard M Jaffe Lectr, 69; Maude Abbolt Lectr, Int Acad Path, 86; H P Smith Mem Award Lectr, Am Soc Clin Pathologists, 86. *Mem:* Inst Med Nat Acad Sci; Am Asn Anat; Am Soc Ecp Path; Am Asn Path & Bact; Am Diabetes Asn; assoc mem Royal Soc Med; Am Soc Cell Biol; Am Soc Clin Pathologists; Int Acad Path; Int Diabetes Fedn; fel AAAS. *Res:* Endocrine pathology; experimental diabetes; published numerous articles in various journals. *Mailing Add:* Dept Path Wash Univ Med Sch St Louis MO 63110

LACY, PETER D(EMPSEY), b Jacksonville, Fla, Dec 6, 20; m 50, 60, 79; c 2. ELECTRONICS ENGINEERING. *Educ:* Univ Fla, BS, 42; Stanford Univ, MS, 47, PhD, 52. *Prof Exp:* Instr, Univ Fla, 42; asst, Stanford Univ, 46-49; consult, Varian Assocs, 49; engr, Hewlett-Packard Co, 50-60; chmn bd, Wiltron Co, 60-89; RETIRED. *Mem:* Fel Inst Elec & Electronics Engrs; Sigma Xi. *Res:* Electron devices; microwave systems; automated measurement systems. *Mailing Add:* 27635 Red Rock Rd Los Altos CA 94022

LACY, W(ILLARD) C(ARLETON), b Waterville, Ohio, July 17, 16; m 40; c 6. GEOLOGICAL ENGINEERING. *Educ:* DePauw Univ, AB, 38; Univ Ill, MS, 40; Harvard Univ, PhD(geol), 50. *Prof Exp:* Geologist, Titanium Alloy Mfg Co, 42-43; petrologist, Cerro de Pasco Corp, 46-50, from asst chief to chief geologist, 50-55; prof geol, Univ Ariz, 55-64, prof mining & geol eng & head dept, 64-71; found chair geol, 72-81, EMER PROF GEOL, JAMES COOK UNIV, NORTH QUEENSLAND, 81- *Concurrent Pos:* Vis lectr, Harvard Univ, 53; prin, Lacy & Assocs, Consults. *Honors & Awards:* Henry Krumb lectr, 85; Fulbright lectr, Univ Queensland, 67; Ben Dickerson Award, 91. *Mem:* fel Geol Soc Am; Soc Econ Geol; Am Inst Mining, Metall & Petrol Engrs; Int Soc Rock Mech; hon fel Australasian Inst Mining & Metall. *Res:* Mining geology; localization of ore deposits; ground stabilization. *Mailing Add:* PO Box 1422 Green Valley AZ 85622

LACY, W(ILLIAM) J(OHN), b Meriden, Conn, May 26, 28; m 50; c 3. CHEMICAL ENGINEERING. *Educ:* Univ Conn, BS, 50, New York Univ, 51, Orins, 57, PhD. *Hon Degrees:* DSc, Paul Sabatier Univ, France, 83. *Prof Exp:* Asst chemist & res assoc, NY Univ, 50-51; chemist & sr proj engr, Eng Res & Develop Labs, Va, 51-58; chief test sta, Oak Ridge Nat Lab, Tenn, 58-59; chief radiochemist, Off Civil Defense & Mobilization, 59-62, asst dir Post Attack Res Div, Off Civil Defense, Washington, DC, 62-67; chief indust pollution control res & develop, Fed Water Pollution Control Admin, Environ Protection Agency, 67-71, dir, Appl Sci & Technol Div, 71-74, prin eng sci adv, 74-79, dir, Water, Waste & Hazardous Mat Res, 79-83; PRES, LACY & CO, ENVIRON & INDUST CONSULTS, 83- *Concurrent Pos:* Partic sanit eng conf, Atomic Energy Comn, 52, 54 & 56, chmn adv comt spec weapons, 56-58; mem, Nat Adv Bd Water Decontamination, 54-56; lectr, numerous US univs; dep dir, US Deleg to USSR, 75, 76 & 78; head, US Deleg UN Environ Prog, Paris, 75 & 78; cochmn, Third Int Conf, Sorrento, Italy, 76; deleg, Tokyo Conf, 77; rep Environ Protection Agency, World Cong Berlin, 77; scientific dir, US Deleg to India, 78, Egypt, 79 & Italy, 81; co-chair III, IV,

V & VI, Int Conf Chem, Protect Environ, Poland, 81 & 89, France, 83, Belgium, 85, Italy, 87 & Hong Kong, 90; chmn (ad hoc) EPA Hazardous Waste Res Lab Comt in Waste Minimization; vpres Int & Executive Comts Chem Protection Environ. *Honors & Awards:* Leonard Gloub Chem Award; Gold Medal, Am Water Works Asn, 60; Bronze Medal, Environ Protection Agency, 83; US Distinguished Serv Medal, 84; Lublin Polytech Univ Medal, 88; Madam Curie Medal, Poland, 89. *Mem:* Am Soc Testing & Mat; Am Chem Soc; Sigma Xi; fel Am Inst Chem Engrs; Fac Inst Regulatory Sci; Int Ozone Asn. *Res:* Industrial waste water treatment; radioactive water decontamination; reactor waste disposal problems; hazardous waste monitoring and disposal; author of 184 publications. *Mailing Add:* 9114 Cherry Tree Dr Alexandria VA 22309

LACY, WILLIAM WHITE, b Atlanta, Ga, Sept 18, 23; m 53; c 5. MEDICINE. *Educ:* Davidson Col, BS, 47; Harvard Med Sch, MD, 51. *Prof Exp:* Intern med, Sch Med, Johns Hopkins Univ, 51-52; asst resident, Duke Univ Hosp, 52-53 & Vanderbilt Univ Hosp, 53-54; from instr to asst prof, 54-73, ASSOC PROF MED, SCH MED, VANDERBILT UNIV, 73- *Concurrent Pos:* Am Heart Asn res fel, 57-61, estab investr, 61- *Mem:* Sigma Xi. *Res:* Metabolism in patients with cardiovascular and renal disease. *Mailing Add:* Dept Med Vanderbilt Univ Sch Med Nashville TN 37232

LAD, PRAMOD MADHUSUDAN, b Bombay, India, Dec 25, 48; nat US; m 78; c 1. RECEPTOR PHARMACOLOGY, MEMBRANE BIOPHYSICS & BIOCHEMISTRY. *Educ:* London Univ, BSc, 70; Cornell Univ, PhD(chem), 74. *Prof Exp:* DIR RES, KAISER REGIONAL RES LAB, KAISER FOUND INST, LOS ANGELES, 81- *Concurrent Pos:* Vis assoc res fac, Calif Inst Technol, Pasadena, 81-; adj assoc prof toxicol, Univ Southern Calif, 86- *Mem:* Am Soc Pharmacol & Exp Therapeut; Biophys Soc; Endocrine Soc; AAAS; Am Chem Soc; Reticuloendothelial Soc. *Res:* Major pathways of receptor mediated signaling in cells of the immune response; cells involved in inflammatory reactions such as neutrophils, lymphocytes, and mast cells; platelet-leukocyte interactions and their role in thrombus formation. *Mailing Add:* 6748 N La Eresa Dr San Gabriel CA 91775-1108

LAD, ROBERT AUGUSTIN, b Chicago, Ill, May 8, 19; m 44; c 9. CHEMISTRY. *Educ:* Univ Chicago, SB, 39, SM, 41, PhD(inorg chem), 46. *Prof Exp:* Asst, Nat Defense Res Comt, Univ Chicago, 42-46; aeronaut res scientist, Nat Adv Comt Aeronaut, Lewis Res Ctr, NASA, 46-59, chief, Mat Sci Br, 59-78; RETIRED. *Concurrent Pos:* Mem solid state sci panel, Nat Acad Sci-Nat Res Coun, 63-78. *Mem:* AAAS; Am Phys Soc; fel Am Inst Chem. *Res:* Physics and chemistry of surfaces; radiation chemistry; materials science. *Mailing Add:* 3114 W 159th St Cleveland OH 44111

LADA, ARNOLD, b New York, NY, May 26, 26; m 47; c 3. BIOCHEMISTRY, ORGANIC CHEMISTRY. *Educ:* Brooklyn Col, BS, 47; Georgetown Univ, MS, 51, PhD, 53. *Prof Exp:* Chemist, Glyco Prod Co, 47-48; biochemist, NIH, 49-50; res chemist, Food & Drug Admin, 50-54; biochemist, Toni Co Div, Gillette Co, 54-57; chief tech servs, Onyx Oil & Chem Co, Kewanee Industs Inc, 57-60, chief tech servs, Onyx Chem Corp, 60-65, gen mgr, Onyx Chem Co Div, 65-83, pres, 74-83; RETIRED. *Mem:* AAAS; Am Chem Soc; Soc Cosmetic Chem; Chem Specialties Mfrs Asn; Cosmetics Toiletries Fragrances Asn. *Res:* Surface active agents; industrial applications; antimicrobial agents. *Mailing Add:* Six Harbor Way Monmouth Beach NJ 07750

LADA, CHARLES JOSEPH, b Webster, Mass, Mar 18, 49; m 84; c 2. ASTRONOMY. *Educ:* Boston Univ, BA, 71; Harvard Univ, AM, 72, PhD(astron), 75. *Prof Exp:* Fel, Ctr Astrophysics, Harvard Col Observ & Smithsonian Astrophys Observ, 75-77; res fel, Harvard Univ, 77-78; Bart Bok fel astron, Steward Observ, Univ Ariz, 78-80, from asst prof to assoc prof, 80-90; SR ASTROPHYSICIST, SMITHSONIAN ASTROPHYS OBSERV, 90- *Concurrent Pos:* Alfred P Sloan Found fel, 81-84; vis assoc prof, Univ Calif, Berkeley, 88-89. *Mem:* Am Astron Soc; Int Astron Union. *Res:* Formation of stars; interstellar gas dynamics; structure and evolution of interstellar molecular clouds; structure and evolution of our galaxy. *Mailing Add:* Smithsonian Astrophys Observ 60 Garden St Cambridge MA 02138

LADANYI, BRANKA MARIA, b Zagreb, Yugoslavia, Sept 7, 47; Can citizen; m 74. THEORETICAL CHEMISTRY, PHYSICAL CHEMISTRY. *Educ:* McGill Univ, BSc, 69; Yale Univ, MPhil, 71, PhD(chem), 73. *Prof Exp:* Vis asst prof chem, Univ Ill, Urbana, 74; res assoc chem, Yale Univ, 74-79; from asst prof to assoc prof chem, 79-87, PROF CHEM, COLO STATE UNIV, 87- *Concurrent Pos:* Camille & Henry Dreyfus Teacher-Scholar, 83-87. *Honors & Awards:* Alfred P Sloan fel, 82-85. *Mem:* Am Chem Soc; Am Phys Soc; AAAS. *Res:* Statistical mechanics of fluids; structure of molecular liquids; propagation and scattering of light in fluids; statistical mechanics of polymer solutions; solvent effects on chemical reactions. *Mailing Add:* Dept Chem Colo State Univ Ft Collins CO 80523

LADANYI, BRANKO, b Zagreb, Yugoslavia, Dec 14, 22; Can citizen; m 46; c 3. GEOTECHNICAL ENGINEERING. *Educ:* Univ Zagreb, BEng, 47, Univ Louvain, PhD(civil eng), 59. *Prof Exp:* Design engr found & hydraul struct, Dept Transport, Zagreb, 47-52; asst prof soil mech & found eng, Univ Zagreb, 52-58; res engr soil mech, Belgian Geotech Inst, Ghent, 58-62; from assoc prof to prof geotech eng, Laval Univ, Que, 62-67; prof geotech eng, Dept Civil Eng, 77-82, PROF CIVIL ENG, ECOLE POLYTECH, UNIV MONTREAL, 67- *Concurrent Pos:* Dir, Northern Eng Centre, 72-; fel, Can Soc Civil Engrs. *Honors & Awards:* R F Legget Geotech Award, 81; E E De Beer Geotech Award, 87. *Mem:* Fel Eng Inst Can; Can Inst Mining & Metall; fel Am Soc Civil Engrs; fel Royal Soc Can; Tunnelling Asn Can (pres, 82-84); Can Rock Mech Asn (pres, 84-87); fel Can Acad Eng. *Res:* Soil and rock mechanics; permafrost engineering; mechanics of permafrost and ice; problems of foundation engineering, tunnelling and Arctic offshore construction. *Mailing Add:* Dept Civil Eng Ecole Polytech Univ Montreal Box 6079 Sta A Montreal PQ H3C 3A7 Can

LADAS, GERASIMOS, b Lixuri, Greece, Apr 25, 37; US citizen; m 65; c 2. DIFFERENTIAL EQUATIONS. *Educ:* Nat Univ Athens, BS, 61; NY Univ, 66, MS,66, PhD(math), 68. *Prof Exp:* Fel, NY Univ, 64-68; asst prof math, Fairfield Univ, 68-69; from asst prof to assoc prof, 69-75, chmn dept, 72-78, PROF MATH, UNIV RI, 75- *Mem:* Am Math Soc. *Res:* Ordinary, functional and abstract differential equations. *Mailing Add:* Math Dept Tyler Hall Univ RI Kingston RI 02881

LADD, CHARLES CUSHING, III, b Brooklyn, NY, Nov 23, 32; m 54; c 4. CIVIL ENGINEERING, GEOTECHNICAL ENGINEERING. *Educ:* Bowdoin Col, AB, 55; Mass Inst Technol, SB, 55, SM, 57, ScD(soil eng), 61. *Prof Exp:* From instr to assoc prof, 57-70, PROF CIVIL ENG, MASS INST TECHNOL, 70- *Concurrent Pos:* Vis consult, Haley & Aldrich, Inc, 67-68; vis sr scientist, Norwegian Geotech Inst, 83. *Honors & Awards:* Croes Medal, 73; Norman Medal, 76; Terzaghi Lectr, 86; Hogentogler Award, 90. *Mem:* Nat Acad Eng; Am Soc Civil Engrs; Am Soc Testing & Mat; Am Soc Eng Educ; Nat Soc Prof Engrs. *Res:* Engineering properties of soils, soft ground and offshore construction as applied to civil engineering projects. *Mailing Add:* Rm 1-348 Dept Civil Eng Mass Inst Technol Cambridge MA 02139

LADD, CONRAD MERVYN, b Lakewood, Ohio, Dec 16, 26; m 47; c 4. POWER PLANT PROJECT MANAGEMENT. *Educ:* Univ Mich, BS, 49. *Prof Exp:* Student engr, Duquesne Light Co, 49-51; intermediate engr, Westinghouse Atomic Power Div, 51-52; sr engr, Atomic Power Develop Asn/Commonwealth Asn, 52-59; prod mgr, Brush Beryllium Corp, 59-63; mkt mgr, Atomics Int, Div Rockwell Int, 63-74; asst mgr mkt, Stone & Webster Engrs, 74-76; mgr bus develop & proj exec, Stearns Roger, Inc, 76-85; CHMN & CHIEF EXEC OFFICER, SR MGT CONSULT, INC, 85- *Concurrent Pos:* Mem exec comt, Power Div, Am Soc Mech Engrs, 78-83, chmn, 81-82, chmn, Task Force Clean Coal Technol, 85-86, co-chmn, Task Force Acid Rain, 85-86, chmn energy comt, 87-90. *Mem:* Fel Am Soc Mech Engrs. *Res:* Engineering, development and testing of basic nuclear fabrication (fuel), equipment and systems; two infield patents. *Mailing Add:* Sr Mgt Consults Inc 7077 S Madison Way Littleton CO 80122

LADD, JOHN HERBERT, b Kewanee, Ill, Sept 6, 18; m 39; c 3. PHYSICAL CHEMISTRY, PHYSICS. *Educ:* Univ Ill, BS, 40, MS, 42, PhD(phys chem), 47. *Prof Exp:* Asst phys chem, Univ Ill, 40-42; sr chemist, Eastman Kodak Co, 47-48, develop engr, 48, sr develop engr, 48-53, tech assoc, 53-57, sr develop proj engr, 57-69, sr staff mem, Kodak Res Labs, 69-81; RETIRED. *Concurrent Pos:* Instr, Univ Rochester, 49-51. *Mem:* Am Phys Soc. *Res:* Electron emission from metals; color photographic printers; color films in television; television test charts and standards; analog computers; digital equipment instrumentation, optical testing; digital hardware; photosensors; laser applications. *Mailing Add:* Seven Barnswallow Dr Pittsford NY 14534-2315

LADD, KAYE VICTORIA, b Seattle, Wash, Aug 26, 41. INORGANIC CHEMISTRY, PHYSICAL CHEMISTRY. *Educ:* Reed Col, BA, 63; Brandeis Univ, MA, 65, PhD(inorg chem), 74. *Prof Exp:* Staff scientist chem biol, Tyco Labs Inc, 65-68; assoc prof chem, Suffolk Univ, 68-75; MEM FAC CHEM, EVERGREEN STATE COL, 75- *Concurrent Pos:* Consult, New Eng Aquarium, 70-75 & Corff & Shapiro, 77. *Mem:* Am Chem Soc; AAAS; Sci Inst Publ Info. *Res:* Environmental inorganic research, especially the transport of trace metals in metabolic process within and between organisms. *Mailing Add:* 1704 24th Ave NW Olympia WA 98374

LADD, SHELDON LANE, b Merced, Calif, Sept 21, 41; m 62; c 2. GENETICS, AGRONOMY. *Educ:* Calif State Univ, Fresno, BS, 63; Univ Calif, Davis, PhD(genetics), 66. *Prof Exp:* Captain asst chief forensic toxicol, US Air Force Sch Aerospace Med, 66-71; plant breeder sugarcane genetics, Dept Genetics & Path, Hawaiian Sugar Planters' Asn, 71-76; assoc prof, 76-80, PROF AGRON, COLO STATE UNIV, 80- *Concurrent Pos:* Dir, Asn Off Seed Certifying Agencies, exec dir, Colo Seed Growers Asn & head seed cert, State Colo, 76- *Mem:* Crop Sci Soc Am; Am Soc Agron. *Res:* Plant breeding of agronomic species; cell and tissue culture of agronomic and revegetation species; seed quality and vigor. *Mailing Add:* Dept Agron Colo State Univ Ft Collins CO 80523

LADD, THYRIL LEONE, JR, b Albany, NY, Oct 10, 31; m 56; c 3. ENTOMOLOGY. *Educ:* State Univ NY, Albany, AB, 56, MA, 58; Cornell Univ, PhD(entom), 63. *Prof Exp:* Res entomologist, 62-69, RES LEADER, HORT INSECTS RES LAB, AGR RES SERV, USDA, 69- *Concurrent Pos:* Adj prof entom, Ohio State Univ, 71- *Mem:* AAAS; Entom Soc Am; Am Entom Soc; Sigma Xi; Coun Agr Sci & Technol. *Res:* Effects of radiation and chemicals on insect reproduction; integrated insect control; insect responses to attractants and repellents; effects of insect feeding on plant yields; improved procedures for applications of insecticides. *Mailing Add:* Japanese Beetle Res Lab USDA Ohio Agr Res & Develop Ctr Wooster OH 44691

LADDA, ROGER LOUIS, b Highland, Ill, Oct 28, 36; m 57; c 5. PEDIATRICS, CLINICAL GENETICS. *Educ:* Wesleyan Univ, BA, 58; Sch Med, Univ Chicago, MD, 63. *Prof Exp:* Res & clin fel, Children's Serv & Genetics Unit, Mass Gen Hosp, Sch Med, Harvard Univ, 72-74; from asst prof to assoc prof, 74-83, PROF PEDIAT, COL MED, PA STATE UNIV, 83-, CHIEF, DIV GENETICS, 74- *Concurrent Pos:* App mem, sci adv bd, Geront Res Ctr, Nat Inst Aging, 82-84. *Mem:* Am Pediat Soc; Soc Pediat Res; AAAS; Sigma Xi. *Res:* Growth regulating factors and cell division. *Mailing Add:* Dept Pediat Milton S Hershey Med Ctr PO Box 850 Hershey PA 17033

LADDE, GANGARAM SHIVLINGAPPA, b Jalkot, India, Mar 9, 40; US citizen; m 65; c 3. DIFFERENTIAL EQUATIONS, APPLIED SYSTEMS ANALYSIS. *Educ:* Marathwada Univ, BSc, 63, MSc, 65; Univ RI, PhD(math), 72. *Prof Exp:* Teaching asst, Univ RI, 67-71, instr, 71-73; from asst prof to assoc prof, State Univ NY, Potsdam, 73-80; PROF MATH, UNIV TEX, ARLINGTON, 80- *Concurrent Pos:* Fel, Univ Santa Clara, Calif, 74, res assoc, 81 & 87; grant-in-aid, Res Found, State Univ NY, Albany, 78-79; vis prof math, Univ Rome, Italy, 78 & Univ Tex, Arlington, 79-80; ed,

Stochastic Anal & Appln. *Mem:* Am Math Soc; Indian Math Soc; Sigma Xi; Soc Indust & Appl Math; Inst Elec & Electronics Engrs; NY Acad Sci. *Res:* Biomathematics; competitive analysis; differential games; deterministic analysis; mathematical modeling in biological, medical, physical, and social sciences; nonlinear boundary value problems; oscillation theory; stability theory; stochastic andlysis; systems analysis; filtering and control theory. *Mailing Add:* PO Box 19408 Arlington TX 76019

LADDU, ATUL R, b Aug 23, 40; US citizen. EXPERIMENTAL BIOLOGY. *Educ:* MB & BS, 62, MD, 67. *Prof Exp:* Instr pharmacol, Maulana Azad Med Col, New Delhi, 63-68; postdoctoral fel, Dept Pharmacol, Med Col Wis, 68-71, from 1st instr to asst prof, 71-73; group leader, Cardiovasc Div, Lederle Labs, Div Am Cyanamid Co, 73-75; sr clin invest assoc, Ciba-Geigy Corp, 75-76; asst med dir, Ives Labs, Inc, Div of Am Home Prod Inc, 76, assoc med dir, 76-78, proj leader, 78-82; dir clin res, DuPont Critical Care, 82-88; DIR CLIN RES, CARDIOVASC & NEUROSCI, ABBOTT LABS, 88- *Concurrent Pos:* Assoc ed, J Clin Pharmacol, Int J Clin Pharmacol. *Mem:* Fel Am Col Cardiol; fel Am Soc Clin Pharmacol & Therapeut; fel Am Col Clin Pharmacol; Am Soc Hypertension; Am Fedn Clin Res; NY Acad Sci; Am Soc Pharmacol & Exp Therapeut; Soc Exp Biol & Med; Sigma Xi; Int Study Group Res Cardiac Metab. *Res:* Conducting clinical trials with cardiovascular products and neuropharmacological agents; author or co-author of several publications. *Mailing Add:* Abbott Labs Abbott Park D42B Bldg AP9A Abbott Park IL 60064-3500

LADE, ROBERT WALTER, b Fond du Lac, Wis, Apr 3, 35; m 56; c 4. SOLID STATE ELECTRONICS. *Educ:* Marquette Univ, BEE, 58, MS, 61; Carnegie Inst Technol, PhD(elec eng), 62. *Prof Exp:* Instr, Marquette Univ, 59-61 & Carnegie Inst Technol, 61-62; assoc prof, NC State Univ, 63-67; prof elec eng & chmn dept, Marquette Univ, 67-77; mem staff, RWL Eng, 77-84; res mgr, Eaton Corp, 84-91; CONSULT, 91- *Concurrent Pos:* Engr, AC Spark Plug, Wis, 59-60. *Mem:* Inst Elec & Electronics Engrs. *Res:* Electrical properties of free and passivated semiconductor surfaces under the influence of high energy radiation fields; computer-aided circuit design and power semiconductor device design. *Mailing Add:* PO Box 2831 Ft Myers FL 33902

LA DELFE, PETER CARL, b Woburn, Mass, Feb 19, 43; m 78. PHYSICS, OPTICS. *Educ:* Clarkson Col Technol, BS, 68, MS, 71. *Prof Exp:* Sr physicist optical films, Spectrum Systs Div, Barnes Eng Co, 71-74; consult, Private Pract, 74-77; staff mem optical films, Los Alamos Nat Lab, 78-79, sect leader, Coating Sect, 79-82, staff mem, 82-85, proj leader, Phoenix Beam Transport, 85-88, MGR, AURORA OPTICS, LOS ALAMOS NAT LAB, 88- *Mem:* Am Vacuum Soc; Optical Soc Am; Int Soc Optical Eng. *Res:* Design and development of optical interference filters including research in the materials science of producing optical films. *Mailing Add:* 600 Los Pueblos Los Alamos NM 87544

LADEN, KARL, b Brooklyn, NY, Aug 10, 32; m 56; c 5. BIOCHEMISTRY. *Educ:* Univ Akron, BS, 54; Northwestern Univ, PhD(chem), 57. *Prof Exp:* Asst, Northwestern Univ, 54-57; res chemist, William Wrigley Jr Co, 57-58; res biochemist, Toni Co Div, 59-60, res supvr, 60-62, mgr biol res, 62-64, asst lab dir, Gillette Med Res Inst, 64-67, mgr biomed sci dept, Gillette Res Inst, 67-68, vpres biomed sci, Gillette Res Inst, Gillette Co, 68-76, pres, 71-76; vpres res & develop, Carter Prod Div, Carter Wallace, Inc, 76-86; INDEPENDENT CONSULT, TECHNOL TRANSFER, NEW PROD DEVELOP, 87- *Concurrent Pos:* Consult, Indust Bio-Test Labs, Inc, 55-57; lectr, Northwestern Univ, 58-64 & Am Univ, 64-65; ed, J Soc Cosmetic Chem, 67-71. *Mem:* Am Chem Soc; Soc Cosmetic Chem (pres, 77). *Res:* Biochemistry and physiology of skin and hair. *Mailing Add:* PO Box 398 Cranbury NJ 08512

LADENHEIM, HARRY, b Vienna, Austria, Oct 17, 32; US citizen; m 55; c 2. REACTOR ENGINEERING, PROCESS ENGINEERING. *Educ:* City Col New York, BS, 54; Polytech Inst Brooklyn, PhD(org polymer chem), 58. *Prof Exp:* Fel chem, Ill Inst Technol, 58-59; res chemist, Esso Res & Eng Co, 59-63; res chemist, Houdry Process & Chem Co, 63-70, sr res chemist, 70-76, sr develop engr, 76-79, prin res engr, Air Prods & Chem Inc, Marcus Hook, Pa, 79-81, process engr, 82-87, sr process engr, Air Prods & Chem Inc, Paulsboro, NJ, 82-87; CONSULT, 87- *Mem:* Am Chem Soc. *Res:* Mechanisms in physical organic chemistry; exploratory and process study in petroleum technology; use of polymers as enzyme models; catalysis; catalytic chemistry. *Mailing Add:* Tech Dept Air Prods & Chem Inc Billingsport Rd Paulsboro NJ 08066

LADENSON, JACK HERMAN, b Philadelphia, Pa, Apr 8, 42; m 68; c 2. CLINICAL CHEMISTRY. *Educ:* Pa State Univ, BS, 64; Univ Md, PhD(anal chem), 71. *Prof Exp:* NSF fel, 66-70; fel clin chem, Hartford Hosp, 70-72; asst dir clin chem, Barnes Hosp, 72-76, co-dir, 76-79; from asst prof to assoc prof, 72-84, PROF PATH & MED, SCH MED, WASH UNIV, 84-, HEAD CLIN CHEM & COMPUT, 80-; DIR CLIN CHEM, BARNES HOSP, 80- *Concurrent Pos:* Med bd dirs, Am Bd Clin Chem, 79-85 & Am Asn Clin Chem, 81-83 & 85-87. *Mem:* Am Asn Clin Chem (pres, 86); Acad Clin Lab Physicians & Scientists. *Res:* Activity measurements in biological fluids; binding of ions of biological importance; measurement of isoenzymes utilizing monoclonal antibodies. *Mailing Add:* Div Lab Med Sch Med Wash Univ Barnes Hosp St Louis MO 63110

LADERMAN, A(RNOLD) J(OSEPH), b Pittsburgh, Pa, Apr 27, 30; m 59; c 2. FLUID MECHANICS, HEAT TRANSFER. *Educ:* Univ Calif, Berkeley, BS, 51, MS, 57, PhD(mech eng), 60. *Prof Exp:* Res engr mech eng, Mech Equip Unit, Boeing Co, 51-55; assoc res engr, Propulsion Dynamics Lab, Univ Calif, Berkeley, 57-65; supvr, Exp Fluid Mech Sect, 66-71, prin scientist, Fluid Mech Dept, 65-79, PRIN SCIENTIST, MECH ENG DEPT, AERONUTRONIC DIV, FORD AEROSPACE CORP, 79- *Concurrent Pos:* Lectr, Univ Calif, Berkeley, 60-61, 64; consult, Repub Aviation Co, NY, 62, Jet Propulsion Lab, Calif, 63-64 & Sandia Corp, 63-65. *Mem:* Am Inst Aeronaut & Astronaut; Combustion Inst; Sigma Xi. *Res:* Non steady gas dynamics of reactive media; shock and detonation wave phenomena; two phase flow; high power lasers; transition and turbulence in compressible boundary layers. *Mailing Add:* 2745 Temple Hills Dr Laguna Beach CA 92651

LADERMAN, JULIAN DAVID, b New York, NY, Oct 15, 48. MATHEMATICS, COMPUTER SCIENCE. *Educ:* NY Univ, BA, 70, MS, 72, PhD(comput sci), 76. *Prof Exp:* From instr to asst prof math, 73-83, ASSOC PROF MATH & COMPUT SCI, LEHMAN COL, 84- *Concurrent Pos:* Consult, Systs Revisited, 78- *Mem:* Am Math Soc; Math Asn Am; Asn Comput Mach. *Res:* Computational complexity; mathematical programming; statistics; game theory; probability; operations research; programming languages; numerical analysis. *Mailing Add:* 2600 Netherland Ave Apt 824 Bronx NY 10463

LADINSKY, HERBERT, b New York, NY, July 22, 35; m 67. PHARMACOLOGY, BIOCHEMISTRY. *Educ:* City Col New York, BS, 58; State Univ NY, PhD(pharmacol), 66. *Prof Exp:* Fel pharmacol, Royal Caroline Inst, Stockholm, 66-67 & Mario Negri Inst, 67-69; NIH fel, 66-68; lab chief, Mario Negri Inst, 69-84; HEAD DEPT BIOCHEM, INST DE ANGELI, SPA, MILAN, 85- *Concurrent Pos:* Consult, Multinat Pharmaceut Co, 74-84; vis scientist, Weizman Inst, Rehovot, Israel, 79. *Mem:* Ital Pharmacol Soc; Int Soc Neurochem; Am Soc Neurosci. *Res:* Biochemistry, neurochemistry, and neuropharmacology of the central and peripheral cholinergic and serotonergic systems; location and function of subtypes of muscarinic and serotoninergic receptors; brain lesions and drug effects; neuronal circuits containing cholinergic neurons; nerve and receptor plasticity. *Mailing Add:* Dept Biochem & Molecular Pharmacol Inst De Angeli Boehringer-Ingelheim Italia Milan 20139 Italy

LADINSKY, JUDITH L, b Los Angeles, Calif, June 16, 38; m 61; c 2. ENDOCRINOLOGY, PUBLIC HEALTH. *Educ:* Univ Mich, BS, 61; Univ Wis-Madison, MS, 64, PhD(reprod physiol), 68. *Prof Exp:* Med technologist, Clin Labs, St Mary's Hosp, Mich, 55-56; res asst, Dept Neuropath, Univ Mich, 56-58, Dept Anat, 58-60 & Dept Surg, 60-61; proj assoc, Dept Gynec-Obstet, 61-68, from instr to asst prof prev med, 68-74, ASSOC PROF PREV MED, SCH MED, UNIV WIS-MADISON, 75-, DIR OFF INT HEALTH, 85- *Concurrent Pos:* Consult, Ministries of Health, Southeast Asia. *Mem:* Am Soc Cell Biol; NY Acad Sci; Tissue Cult Asn; Am Pub Health Asn; Asn Teachers Prev Med; Nat Coun Int Health; Nat Asn Public Health Policy (secy, 84-86); Sigma Xi. *Res:* Cell kinetics of normal and neoplastic tissues; automated methods of cancer detection; endocrinology of tumors; community medicine; neonatology; health care delivery; international health. *Mailing Add:* Dept Prev Med 101 Bradley Hosp Med Sch Univ Wis 1300 Univ Ave Madison WI 53706

LADISCH, MICHAEL R, b Upper Darby, Pa, Jan 15, 50; m 75; c 2. BIOCHEMICAL ENGINEERING. *Educ:* Drexel Univ, BS, 73; Purdue Univ, MS, 74, PhD(chem eng), 77. *Prof Exp:* Res eng biochem, 77-78, from asst prof to assoc prof, 78-81, PROF FOOD, AGR & CHEM ENG, PURDUE UNIV, 85-, GROUP LEADER RES & PROCESS ENG, LAB RENEWABLE RESOURCES, 78- *Honors & Awards:* Peterson Award, Am Chem Soc, 77; US Presidential Young Investr Award, 84; James M Van Laren Serv Award, Am Chem Soc, 90. *Mem:* Am Chem Soc; Am Inst Chem Engrs; Am Soc Automotive Engrs. *Res:* Cellulose conversion; bioseparations; enzyme and chemical kinetics; fermentation with strict araerobes. *Mailing Add:* Lab Renewable Resources Eng Potter Ctr Purdue Univ West Lafayette IN 47906

LADISCH, STEPHAN, b Garmisch-Partenkirchen, WGermany, July 18, 47; US citizen; m 74; c 2. TUMORGENICITY. *Educ:* Univ Pa, Philadelphia, BS, 69, MD, 73; Am Bd Pediat, cert hemat-oncol, 78. *Prof Exp:* Intern & resident pediat, Children's Hosp Med Ctr, Boston, Mass, 73-75; clin assoc, pediat oncol br, Nat Cancer Inst, Bethesda, Md, 75-77, investr, 77-78; from asst prof to assoc prof, 78-86, PROF PEDIAT, DIV HEMAT-ONCOL, SCH MED, UNIV CALIF, LOS ANGELES, 86-, SR MEM, HUMAN IMMUNOBIOL GROUP, 78- *Concurrent Pos:* NSF res grants, 68-69 & 72; NIH res grants, 80-84 & 83-, Career Develop Award, 82-87; Nat Found-March of Dimes res grant, 80-84; vis scientist, Lausanne Br, Ludwig Inst for Cancer Res, Epalinges, Switz, 81-82, Inst Pasteur, Paris, 86-87; scholar, Leukemia Soc Am, 82-87. *Mem:* Am Soc Hemat; AAAS; Soc Pediat Res; Am Asn Immunologists; Am Fedn Clin Res. *Res:* Modulation of the immune response by gangliosides, membrane glycolipids shed by tumor cells and role of this process in tumorgenicity. *Mailing Add:* Dept Pediat Sch Med Univ Calif Los Angeles CA 90024-1752

LADISH, JOSEPH STANLEY, b Worcester, Mass, Aug 9, 43; m 64; c 2. ATOMIC & MOLECULAR PHYSICS. *Educ:* Mass Inst Tech, BS, 65; Yale Univ, MS, 66, MPhil, 67 & PhD(physics), 74. *Prof Exp:* RES STAFF, LOS ALAMOS NAT LAB, 74- *Concurrent Pos:* Adj fac, Univ NMex, Los Alamos, 83- *Mem:* Am Phys Soc. *Res:* Production of polarized electrons; carbon dioxide laser fusion research; detection by Cerenkav radiation; optical transition radiation. *Mailing Add:* 327 Rover Blvd Los Alamos NM 87544

LADMAN, AARON J(ULIUS), b Jamaica, NY, July 3, 25; m 48, 82; c 3. ANATOMY. *Educ:* NY Univ, AB, 47; Ind Univ, PhD(anat), 52. *Prof Exp:* Res fel anat, Harvard Med Sch, 52-55, assoc anat, 55-61; assoc prof, Med Units, Univ Tenn, 61-64; prof anat & chmn dept, Univ NMex, 64-81; dean, Sch Allied Health, 81-86, PROF ANAT, HAHNEMANN UNIV, PHILADELPHIA, 81- *Concurrent Pos:* USPHS career develop award, 62-64; mem res career award comt, Nat Inst Gen Med Sci, 67-71; ed, Anat Record, 68- *Mem:* Am Soc Cell Biol; Coun Biol Ed; Histochem Soc; Am Asn Anatomists (2nd vpres, 80-81 & 1st vpres, 81-82); fel AAAS. *Res:* Cytochemistry; electron microscopy; endocrinology; experimental cytology; retina lung; tumor biology; scientific writing. *Mailing Add:* Hahnemann Univ MS 408 Broad & Vine Sts Philadelphia PA 19102-1192

LADNER, DAVID WILLIAM, b Meadville, Pa, June 22, 47; m 73; c 2. QUANTITATIVE STRUCTURE-ACTIVITY RELATIONSHIPS. *Educ:* Pa State Univ, BS, 69; Univ Ga, PhD(org chem), 74. *Prof Exp:* Res assoc org systhesis, Dept Chem, Univ Wash, 74-75; res fel org synthesis, Syntex, SA, Mex, DF, 75-76; sr scientist, N L Industs, 76-77; res chemist herbicide synthesis, Agr Res Div, 77-80, sr res chemist herbicide synthesis, Agr Res

Div, 80-83, GROUP LEADER HERBICIDE SYNTHESIS, AGR RES DIV, AM CYANAMID, 83- *Mem:* Am Chem Soc. *Res:* Design and synthesis of herbicides and plant regulants, particularly those affecting enzymes in amino acid biosynthetic pathways; quantitative structure-activity relationships: techniques to model activity, translocation and uptake of pesticides. *Mailing Add:* PO Box 400 Princeton NJ 08543-0400

LADNER, SIDNEY JULES, b Houston, Tex, Mar 12, 36; m 59; c 2. PHYSICAL CHEMISTRY. *Educ:* Univ Houston, BS, 59, PhD(phys chem), 65. *Prof Exp:* Fel chem, Univ NMex, 65-66; chemist, Shell Develop Co, Tex, 66-67; asst prof, 67-69, ASSOC PROF CHEM, HOUSTON BAPTIST UNIV, 69- *Mem:* Am Chem Soc. *Res:* Molecular spectroscopy; decay processes and the decay kinetics of molecules in excited electronic energy states; chemical education. *Mailing Add:* Dept Chem Houston Baptist Univ 7502 Fondren Rd Houston TX 77074

LADO, FRED, b La Coruna, Spain, June 5, 38; US citizen; m 60; c 3. PHYSICS. *Educ:* Univ Fla, BS, 60, PhD(physics), 64. *Prof Exp:* Fel physics, Univ Fla, 64-65; staff mem, Los Alamos Sci Lab, 65-68; from asst prof to assoc prof, 68-85, PROF PHYSICS, NC STATE UNIV, 85- *Concurrent Pos:* Fulbright sr lectr, Spain, 71-72; guest scientist, Int Ctr Theoret Physics, Trieste, Italy, 87. *Mem:* Am Phys Soc. *Res:* Statistical mechanics; equilibrium and non-equilibrium theory of liquids; many-body problem. *Mailing Add:* Dept Physics NC State Univ Raleigh NC 27695-8202

LADSON, THOMAS ALVIN, b Hyattsville, Md, Sept 29, 17; m 48; c 2. VETERINARY MEDICINE. *Educ:* Univ Pa, VMD, 39. *Prof Exp:* Vet, 39-60; field rep, Livestock Sanit Serv, Md State Bd Agr, 60-62, dir, 62-73, head vet sci dept, Univ Md, College Park, 62-73; actg dir, Div Animal Indust, Md Dept Agr, 73-77, chief, 77-80; RETIRED. *Mem:* Am Vet Med Asn; US Animal Health Asn. *Res:* Large animal medicine; regulatory veterinary medicine. *Mailing Add:* Rte 1 Box 89B White Point Leonardtown MD 20650

LA DU, BERT NICHOLS, JR, b Lansing, Mich, Nov 13, 20; m 47; c 4. BIOCHEMICAL PHARMACOLOGY. *Educ:* Mich State Col, BS, 43; Univ Mich, MD, 45; Univ Calif, PhD(biochem), 52. *Prof Exp:* Intern, Rochester Gen Hosp, NY, 45-46; asst biochem, Mich State Col, 46-47 & Univ Calif, 47-50; from sr asst surgeon to med dir, NIH, 50-63; prof pharmacol & chmn dept, Med Sch, NY Univ, 63-74; chmn dept, 74-80, prof, 74-88, EMER PROF PHARMACOL, UNIV MICH MED SCH, ANN ARBOR, 88-, DIR RES, DEPT ANESTHESIOL, 91- *Concurrent Pos:* Res assoc, Goldwater Mem Hosp Res Serv, NY Univ, 50-54, instr, Bellevue Med Ctr, 51-54. *Mem:* Am Soc Biol Chemists; Am Chem Soc; Am Soc Pharmacol & Exp Therapeut (pres, 78-79); Am Soc Human Genetics; NY Acad Sci (pres, 70). *Res:* Drug metabolism; metabolism of tyrosine; inborn errors of metabolism; pharmacogenetics. *Mailing Add:* Dept Anesthesiol R4038 Kresge II Univ Mich Med Sch Ann Arbor MI 48109-0572

LADUKE, JOHN CARL, b Jackson, Mich, Nov 21, 50; m 73; c 1. SYSTEMATIC BIOLOGY. *Educ:* Tex Tech Univ, BS, 73, MS, 75; Ohio State Univ, PhD(bot), 80. *Prof Exp:* asst prof, 80-85, ASSOC PROF BIOL, UNIV NDAK, 85- *Honors & Awards:* Ralph E Alston Award, Bot Soc Am, 79. *Mem:* Int Asn Plant Taxon; Am Soc Plant Taxonomists; Bot Soc Am; Soc Syst Zool; Sigma Xi. *Res:* Plant systematics; chemosystematics; sphaeralcea (Malvacae) including gathering morphological, cytological and flavonoid chemical data. *Mailing Add:* PO Box 8238 Grand Forks ND 58202

LADWIG, HAROLD ALLEN, b Manilla, Iowa, May 11, 22; m 46; c 2. NEUROLOGY. *Educ:* Univ Iowa, MD, 47, BA, 52; Am Bd Psychiat & Neurol, dipl. *Prof Exp:* Clin instr neurol, Univ Minn, 50-53; from instr to asst prof, 54-56, ASSOC PROF, NEUROL & PSYCHIAT, SCH MED, CREIGHTON UNIV, 66-; AT DEPT NEUROL, UNIV NEBR MED CTR. *Concurrent Pos:* Dir electroencephalog lab, Creighton Mem St Joseph's Hosp, 54-, asst dir rehab ctr, 54-58, assoc dir, 58-64; attend physician, Vet Admin Hosp, 54-59, consult physician, 59-63; mem med staff, Nebr Children's Ther Ctr, 56-64; dir, Electroencephalog Lab, Children's Mem Hosp, 63-70 & Archbishop Bergan Mercy Hosp, 64- *Mem:* Am Electroencephalog Soc; AMA; Am Col Physicians; Am Cong Rehab Med; Am Acad Neurol. *Res:* Diagnostic neurology and rehabilitation of neurological patients; care of the aged; electroencephalography. *Mailing Add:* 1600 Canal Dr Wilson NC 27893

LAEMLE, LOIS K, b New York, NY, May 26, 41. DEVELOPMENTAL NEUROBIOLOGY, VISUAL SYSTEMS. *Educ:* City Univ New York, BS, 62; Columbia Univ, PhD(anat), 68. *Prof Exp:* Fel anat, Albert Einstein Col Med, 68; res fel neurosci, Rose F Kennedy Ctr Ment Retardation, 69-72; asst prof, 72-77, ASSOC PROF NEUROANAT & HISTOL, UNIV MED & DENT NJ, 77- *Concurrent Pos:* Prin investr, NIH res grant, 73-81 & 82-86; vis prof, Dept Ophthal, NY Med Sch, 83; assoc ed, Am J Anat, 89- *Mem:* Am Asn Anatomists; Soc Neurosci; Sigma Xi; NY Acad Sci; Cajal Club. *Res:* Morphology and development of the central nervous system; fiber connections, neuronal maturation, and neurotransmitters in the visual system; mechanisms of arcadian timekeeping. *Mailing Add:* Dept Anat Univ Med & Dent NJ Newark NJ 07103

LAEMMLE, JOSEPH THOMAS, b Louisville, Ky, Feb 7, 41; m 65; c 3. ORGANIC CHEMISTRY. *Educ:* Bellarmine Col, BA, 64; Ga Inst Technol, MS, 68, PhD(org chem), 71; Ga State Univ, MBA, 76. *Prof Exp:* From asst res chemist to asst, Ga Inst Technol, 67-73; asst prof chem, Kennesaw Col, 73-77; sr scientist, Alcoa Labs, 77-80, staff scientist, 80-81, tech supvr, 81, sr tech supvr, 81-86, DIV MGR, SURFACE TECH DIV, ALCOA LABS, 86- *Mem:* Am Chem Soc; Soc Tribologist & Lubrication Engrs; Sigma Xi. *Res:* Determination of the structure of organometallic compounds; lubricant testing and development; descriptions of organometallic reaction mechanisms and sterechemistry of additions with Ketones; metal working lubricants including formation, handling, reclamation and disposal. *Mailing Add:* Surface Tech Div Alcoa Tech Ctr Alcoa Lab Alcoa Ctr PA 15069

LAEMMLEN, FRANKLIN, b Reedley, Calif, Mar 8, 38; m 61; c 2. PLANT PATHOLOGY, ENTOMOLOGY. *Educ:* Univ Calif, Davis, BS, 60, PhD(plant path), 70; Purdue Univ, West Lafayette, MS, 67. *Prof Exp:* Res asst plant path, Univ Calif, Davis, 66-70; asst prof, Univ Hawaii, 70-72; from asst prof to assoc prof plant path, Mich State Univ, 76-80; FARM ADV, IMP CO, UNIV CALIF, 80- *Concurrent Pos:* Assoc ed, Plant Dis Reporter, 74-77; vis colleague, Dept Plant Path, Univ Calif, Berkeley, 78-79. *Mem:* Sigma Xi; Am Phytopath Soc. *Res:* Extension plant pathology; ornamental plant diseases; plant disease diagnostic laboratory; field and vegetable crop diseases. *Mailing Add:* Univ Calif Coop Exten 1050 E Holton Rd Holtville CA 92250

LAERM, JOSHUA, b Waynesboro, Pa, Sept 27, 42; m 81. FUNCTIONAL MORPHOLOGY. *Educ:* Pa State Univ, BA, 65; Univ Ill, MS, 72, PhD(zool), 76. *Prof Exp:* Asst prof, 76-81, ASSOC PROF ZOOL, UNIV GA, 81-, DIR, MUS NATURAL HIST, 78- *Mem:* AAAS; Soc Vert Paleont; Am Soc Mammalogists; Soc Study of Evolution; Am Soc Zoologists. *Res:* Evolution and functional morphology of vertebral column in fossil fishes; mammalian systematics; vertebrate natural history. *Mailing Add:* Dept Zool Univ Ga Athens GA 30602

LAESSIG, RONALD HAROLD, b Marshfield, Wis, Apr 4, 40; m 66; c 1. CLINICAL CHEMISTRY, PUBLIC HEALTH. *Educ:* Wis State Univ-Stevens Point, BS, 62; Univ Wis-Madison, PhD(anal chem), 65. *Prof Exp:* Fel, Princeton Univ, 65-66 & Ctr Dis Control, Atlanta, 66; asst dir, State Lab Hyg, 70-79, chief chem, 66-79; assoc prof, 71-76, PROF PREV MED & PATH, MED CTR, UNIV WIS-MADISON, 76-; DIR, STATE LAB HYG, 79- *Concurrent Pos:* Mem, Inst Bd Anal Chem, 71-; chmn diag prod comt, Food & Drug Admin, 72-74; pres, Nat Comt Clin Lab Standards, 80-82; mem bd, Am Asn Clin Chem, 85-88. *Honors & Awards:* DIFCO Award, Am Pub Health Asn, 74; Outstanding Contrib through Serv to Profession of Clin Chem Award, Am Asn Clin Chem, 90, Natelson Award for Advan Clin Chem, 90. *Mem:* Am Asn Clin Chem; Am Chem Soc; Am Pub Health Asn. *Res:* Automation; newborn screening; computerization of laboratory operation; public health laboratory applications of test procedures. *Mailing Add:* State Lab Hyg 465 Henry Mall Univ Wis Madison WI 53706

LAETSCH, THEODORE WILLIS, b St Louis, Mo, Jan 7, 40; m 61. MATHEMATICAL ANALYSIS. *Educ:* Washington Univ, St Louis, BS, 61; Mass Inst Technol, SM, 62; Calif Inst Technol, PhD(appl math), 68. *Prof Exp:* Asst prof physics, Col of Idaho, 62-65; asst prof math, Ill State Univ, 68-70; ASSOC PROF MATH, UNIV ARIZ, 71- & HEAD MATH DEPT, 78. *Concurrent Pos:* Nat Acad Sci-Nat Res Coun resident res associateship, Wright-Patterson AFB, Ohio, 70-71. *Mem:* Am Math Soc; Soc Indust & Appl Math. *Res:* Functional analysis in partially ordered spaces; boundary value problems for ordinary and partial differential equations. *Mailing Add:* Dept Math Univ Ariz Tucson AZ 85721

LAETSCH, WATSON MCMILLAN, b Bellingham, Wash, Jan 19, 33; m 58; c 2. BOTANY. *Educ:* Wabash Col, AB, 55; Stanford Univ, PhD(biol), 61. *Hon Degrees:* DSc, Wabash Col, 85. *Prof Exp:* Fulbright fel, Univ Delhi, India, 56-57; asst prof biol, State Univ NY, Stony Brook, 61-63; from asst prof to assoc prof bot, 63-71, assoc dir Lawrence Hall Sci, 69-72, dir Univ Bot Garden, 69-73, dir Lawrence Hall Sci, 72-80, PROF BOT, UNIV CALIF, BERKELEY, 71-, VCHANCELLOR UNDERGRAD AFFAIRS, 80- *Concurrent Pos:* Fulbright fel, Univ Delhi, India, 56-57; NSF sr fel, Univ Col, London, 68-69; pres, Asn Sci-Tech Ctrs, 77-78; mem, Indo-US subcomn Educ & Cult, 83- *Mem:* AAAS; Bot Soc Am; Am Soc Plant Physiol; Soc Exp Biol & Med. *Res:* Plant development; structure and function of the photosynthetic apparatus; science education. *Mailing Add:* Dept Bot Univ Calif Berkeley CA 94704

LAEVASTU, TAIVO, b Vihula, Estonia, Feb 26, 23; m 49; c 3. OCEANOGRAPHY, METEOROLOGY. *Educ:* Gothenburg & Lund, Fil Kand, 51; Univ Wash, Seattle, MS, 54; Univ Helsinki, PhD(oceanog), 61. *Prof Exp:* Fisheries officer, Swedish Migratory Fish Comt, 51-53; res assoc oceanog, Univ Wash, Seattle, 54-55; fisheries oceanogr, Food & Agr Orgn, UN, 55-62; assoc prof oceanog, Univ Hawaii, 62-64; res oceanogr, US Fleet Numerical Weather Facil, 64-71, chief oceanog div, Environ Prediction Res Facil, Naval Postgrad Sch, 71-76; ECOSYST MODELING EXPERT, NAT OCEANIC & ATMOSPHERIC ADMIN, NAT MARINE FISHERIES SERV, 76- *Concurrent Pos:* UNESCO lectr, Bombay, India, 59; mem panel disposal radioactive waste into sea & fresh water, Int Atomic Energy Agency, 59-62; mem working group fisheries prob comn maritime meteorol, World Meteorol Orgn, 60-62; NSF & US Navy res grants, 62-64; mem, World Meteor Orgn/UNESCO Panel on Oceanic Water Balance, 72- & Sea Use Coun, Sci-Tech Bd, 71-76. *Honors & Awards:* Mil Oceanog Award, 69. *Mem:* Am Geophys Union; Can Meteorol & Oceanog Soc; Am Inst Fishery Res Biol. *Res:* Fisheries hydrography and oceanography; marine chemistry; sea-air interactions; oceanographic forecasting; numerical modeling in oceanography and meteorology; marine ecosystem modeling. *Mailing Add:* Nat Oceanic & Atmospheric Admin 7600 Sand Point Way NE Seattle WA 98115-0070

LA FEHR, THOMAS ROBERT, b Los Angeles, Calif, Feb 6, 34; m 57; c 5. GEOPHYSICS. *Educ:* Univ Calif, Berkeley, AB, 58; Colo Sch Mines, MSc, 62; Stanford Univ, PhD(geophys), 64. *Prof Exp:* Geophysicist, US Geol Surv, 62-64; geophysicist, Geophys Assocs, Int, 64-66; vpres tech develop, GAI-GMX Inc, 66-67, dir tech develop, GAI-GMX Div, EG&G Inc, 67-69; assoc prof, 69-75, ADJ PROF GEOPHYS, COLO SCH MINES, 75-; PRES, EDCON, 75- *Concurrent Pos:* Lectr, Stanford Univ, 64; consult, GAI-GMS Div, EG&G, Inc, 69-70 & Explor Data Consult, 70- *Mem:* Hon mem Soc Explor Geophys; Am Asn Petrol Geol; Am Geophys Union. *Res:* Gravity and magnetic exploration; potential field theory; integrated seismic, gravity, well log data; borehole gravity. *Mailing Add:* 2303 Table Heights Dr Golden CO 80401

LAFERRIERE, ARTHUR L, b Willimantic, Conn, Dec 3, 33; m 55; c 3. ORGANIC CHEMISTRY, INORGANIC CHEMISTRY. *Educ:* Brown Univ, BS, 55; Rutgers Univ, MS, 58; Univ RI, PhD(chem), 60. *Prof Exp:* Res chemist, Minerals & Chem Corp, 56-58 & Am Cyanamid Co, 60-62; PROF CHEM, RI COL, 62- *Mem:* Am Chem Soc. *Res:* Inorganic solution chemistry; organic redox mechanisms. *Mailing Add:* Dept Chem RI Col 600 Mt Pleasant Ave Providence RI 02908

LAFEVER, HOWARD N, b Hagerstown, Ind, May 13, 38; m 58; c 2. PLANT BREEDING, GENETICS. *Educ:* Purdue Univ, BS, 59, MS, 61, PhD(plant breeding & genetics), 63. *Prof Exp:* Instr bot, Wis State Univ, LaCrosse, 63; asst prof genetics, Purdue Univ, 63; res geneticist, Boll Weevil Res Lab, USDA, 63-65; asst prof, 65-71, prof genetics & plant breeding, 71-77, PROF AGRON, OHIO AGR RES & DEVELOP CTR, 77- *Honors & Awards:* Crops & Soils Award, Am Soc Agron, 77. *Mem:* Am Soc Agron; Crop Sci Soc Am. *Res:* Breeding new wheat varieties for distribution and production in midwest; genetic studies of wheat. *Mailing Add:* Ohio State Univ Ohio Agr Res & Develop Ctr Wooster OH 44691

LAFFERTY, JAMES FRANCIS, b Pampa, Tex, Dec 23, 27; m 56; c 3. BIOMEDICAL ENGINEERING, MECHANICAL ENGINEERING. *Educ:* Univ Ky, BS, 55; Univ Southern Calif, MS, 57; Univ Mich, MS, 66, PhD(nuclear eng), 67. *Prof Exp:* Asst, Wenner-Gren Res Lab, Univ Ky, 54-55; mem tech staff, Hughes Aircraft Co, 55-57; asst prof nuclear eng, Werner-Gren Res Lab, Univ Ky, 57-62, assoc prof mech eng, 62-73, actg lab dir, 67-73, prof mech eng & lab dir, 73-85, prof & dir, Biomed Eng Ctr, 85-90, EMER PROF & DIR BIOMED ENG CTR, UNIV KY, 90-; BIOMED ENG CONSULT, 90- *Concurrent Pos:* NSF fac fels, Univ Mich, 61-62 & 65-66. *Mem:* Orthop Res Soc; Am Soc Mech Eng; Soc Automotive Eng. *Res:* Biomechanics of the skeletal and cardiovascular systems; response of biosystems to impact, vibration, acceleration. *Mailing Add:* Wenner-Gren Res Lab Univ Ky Lexington KY 40506-0070

LAFFERTY, JAMES M(ARTIN), b Battle Creek, Mich, Apr 27, 16; m 42; c 4. PHYSICAL ELECTRONICS, POWER ELECTRONICS. *Educ:* Univ Mich, BSE, 39, MS, 40, PhD(elec eng), 46. *Prof Exp:* Mem staff, Eastman Kodak Co, NY, 39; radio proximity fuse res, Carnegie Inst Wash, 41; res assoc, Gen Elec Co, 42-56, mgr plasma & vacuum physics br, 56-68, mgr gen physics lab, 68-72, mgr physics & elec eng lab, 72-74, mgr physics & electronic eng lab, 74-75, mgr electronic power conditioning & control lab, 75-78, mgr power electronics lab, Res & Develop Ctr, 78-81; CONSULT, 81- *Concurrent Pos:* Group leader, People to People Citizen Embassador Prog, 84- 88. *Honors & Awards:* Naval Ord Develop Award, 46; Lamme Medalist, Inst Elec & Electronics Engrs, 79. *Mem:* Nat Acad Eng; fel AAAS; fel Am Phys Soc; fel Inst Elec & Electronics Engrs; hon mem Am Vacuum Soc, past pres; Int Union Vacuum Sci Tech Applns, (pres, 81-83). *Res:* Electrometer and microwave tubes; electron guns; lanthanum boride cathodes; color television picture tubes; gas discharge tubes; hot-cathode magnetron ionization gauge; triggered vacuum gap; vacuum switch; electric vehicles. *Mailing Add:* 1202 Hedgewood Lane Schenectady NY 12309

LAFFERTY, WALTER J, b Wilmington, Del, Feb 10, 34; m 58; c 6. PHYSICAL CHEMISTRY. *Educ:* Univ Del, BS, 56; Mass Inst Technol, PhD(phys chem), 61. *Prof Exp:* Res assoc, Johns Hopkins Univ, 61-62; CHEMIST, NAT INST STANDARDS & TECHNOL, 62- *Concurrent Pos:* Leverhulme vis fel, Univ Reading, England, 70-71; vis prof, Univ Paris (VI), 87. *Honors & Awards:* Silver Medal, Dept Com. *Mem:* Am Phys Soc. *Res:* Infrared and microwave spectroscopy. *Mailing Add:* Molecular Physics Div Nat Inst Standards & Technol Gaithersburg MD 20899

LAFFIN, ROBERT JAMES, b New Haven, Conn, Apr 16, 27; m 51; c 4. MICROBIOLOGY. *Educ:* Yale Univ, BS, 49, PhD(microbiol), 55; Am Bd Med Lab Immunol, dipl. *Prof Exp:* Instr microbiol, Womans Col, Univ NC, 53-55; from instr to assoc prof, Creighton Univ, 55-62; instr obstet & gynec, med sch, Tufts Univ, 62-64; asst prof, 64-67, assoc prof, 67-76, PROF MICROBIOL, ALBANY MED COL, 76- *Concurrent Pos:* Immunologist, St Margaret's Hosp, 62-64. *Mem:* Sigma Xi; Am Soc Microbiol; Am Asn Immunologist; Clin Immunol Soc; Asn Med Lab Immunologist. *Res:* Study of Host immune responses to tumor-specific transplantation antigens; study of protein-protein interactions on metal-coated slides. *Mailing Add:* Dept Microbiol Albany Med Col Union Univ Albany NY 12208

LAFFLER, THOMAS G, b Detroit, Mich, May 10, 46; m 68; c 3. GENETICS. *Educ:* Mass Inst Technol, BS, 68; Univ Wash, PhD(genetics), 74. *Prof Exp:* Fel oncol, McArdle Lab Cancer Res, 74-78, res assoc, 78-80; ASST PROF MICROBIOL, MED & DENT SCHS, NORTHWESTERN UNIV, 80- *Mem:* Genetics Soc Am; Sigma Xi. *Res:* Study of molecular basis of cell cycle control using Physarum polycephalum as a lower enkorycote model; cell cycle mutants and the regulation of tubulin biosyntghesis and DNA replication. *Mailing Add:* 1117 N Harvey Oak Park IL 60302

LAFFOON, JOHN, b Albia, Iowa, June 20, 55. ELECTRON MICROSCOPY. *Educ:* Univ Iowa, BS, 79. *Prof Exp:* electron microscopy, Univ Iowa Dent Col, 79-88; HEAD, ULTRA MICROTOMY DENT RES, 79- *Res:* Electron microsopy in dental research for the past 9 years; tem and sem and micro probe, soft and hard tissue. *Mailing Add:* Electron Probe Microanalysis Facility Dows Inst Dent Res Univ Iowa Col Dent N441 Dental Sci Bldg Iowa City IA 52242

LAFLAMME, GASTON, b St Prosper, Beauce, Que, July 15, 45; m 80; c 1. EPIDEMIOLOGY, MYCOLOGY. *Educ:* Laval Univ, Que, BScApp, 68, MSc, 71; Swiss Fed Inst Technol, Zurich, DScApp(forest path & mycol), 75. *Prof Exp:* Res scientist forest path, Can Forestry Serv, Nfld, 75-78; forest pathologist, Dept Land & Forest, Que, 78-80; RES SCIENTIST, CAN FORESTRY SERV, STE-FOY, QUE, 80- *Mem:* Can Phytopath Soc; Poplar Coun Can; Can Forestry Inst. *Res:* Decay of living trees; taxonomy of fungi; scleroderris canker and other tree diseases; endophytic fungi. *Mailing Add:* PO Box 3800 3418 Sarnah St Ste Foy PQ G1H 2K4 Can

LA FLEUR, JAMES KEMBLE, b Los Angeles, Calif, Apr 23, 30; m 64; c 3. MECHANICAL ENGINEERING. *Educ:* Calif Inst Technol, BSME, 52, Pepperdine, MBA, 80. *Prof Exp:* Engr, AiResearch Mfg Co, 52-56; pres Kemsco Inc, 56-57, Dynamic Res Inc, 57-60 & LaFleur Corp, 60-65, chmn bd, 65-66; pres, Indust Cryogenics Inc, 66-75; chmn, pres & chief exec officer, GTI Corp, 75-90; RETIRED. *Mem:* Am Soc Mech Engrs; Cryogenic Soc Am; Int Solar Energy Soc; Int Asn Hydrogen Energy; Am Wind Energy Asn. *Res:* Applying knowledge gained in the development of normal turbo-machinery to the field of low temperature to develop new low temperature processes. *Mailing Add:* 4337 Talofa Ave Toluca Lake CA 91602

LAFLEUR, KERMIT STILLMAN, b Waterville, Maine, Feb 14, 15; m 39; c 1. TEXTILE CHEMISTRY, SOIL SCIENCE. *Educ:* Colby Col, BA, 37; Clemson Univ, MS, 64, PhD, 66. *Prof Exp:* Asst chemist, Wyandotte Worsted Co, Maine, 37-40, chief chemist, 40-46; chief chemist, Deering Milliken Maine Mills, 47-52; res chemist, Excelsior Mills, 52-56, tech supt, 56-58; group leader wool res, Deering Milliken Res Corp, 59-62, consult chemist, 62-66, sr scientist, 66-67; assoc prof soil chem, Clemson Univ, 67-75, prof, 75-80; RETIRED. *Mem:* AAAS; Am Soc Agron; Soil Sci Soc Am. *Res:* Wool chemistry; soil chemistry. *Mailing Add:* 206 Hunter Ave Clemson SC 29631

LAFLEUR, LOUIS DWYNN, b Elton, La, Dec 28, 40; m 64; c 4. ACOUSTICS, ULTRASONICS. *Educ:* Univ Southwestern La, BS, 62; Univ Houston, PhD(physics), 69. *Prof Exp:* Res scientist assoc, Defense Res Lab, Univ Tex, Austin, 64-65; aerospace technologist, Manned Spacecraft Ctr, NASA, 65-66; asst prof physics, Drury Col, 69-70; asst prof, 70-74, ASSOC PROF PHYSICS, UNIV SOUTHWESTERN LA, 74- *Mem:* Am Phys Soc; Am Asn Physics Teachers; Acoust Soc Am. *Res:* Acoustics and ultrasonics; electroacoustics; mossbauer spectroscopy. *Mailing Add:* Dept Physics Univ Southwestern La Box 44210 Lafayette LA 70504

LAFLEUR, ROBERT GEORGE, b Albany, NY, Mar 31, 29; m 50; c 3. HYDROLOGY & WATER RESOURCES. *Educ:* Univ Rochester, AB, 50; Rensselaer Polytech Inst, MS, 53, PhD, 61. *Prof Exp:* Instr geol, 52-55, from asst prof to assoc prof, 55-82, PROF, GLACIAL GEOL & WATER RESOURCES, RENSSELAER POLYTECH INST, 82- *Concurrent Pos:* Consult, NY State Educ Dept & US Geol Surv. *Mem:* Fel Geol Soc Am; Nat Asn Geol Teachers; Am Geophys Union; Arctic Inst NAm; Soc Econ Paleontologists & Mineralogists; Sigma Xi; Am Inst Prof Geologist; Am Inst Hydrol. *Res:* Glacial geology; hydrogeology. *Mailing Add:* Dept Geol Rensselaer Polytech Inst Troy NY 12181

LAFON, EARL EDWARD, b Oklahoma City, Okla, Apr 24, 40; m 61. SOLID STATE PHYSICS. *Educ:* Univ Okla, BS, 62, MS, 64, PhD(physics), 67. *Prof Exp:* Adj asst prof physics & res assoc, Res Inst, Univ Okla, 67, NSF fel, 67-68; asst prof physics, 68-71, assoc prof, 71-74, PROF PHYSICS, OKLA STATE UNIV, 74- *Mem:* Am Phys Soc; Am Math Soc. *Res:* Electronic structure and band theory. *Mailing Add:* Dept Physics Okla State Univ Stillwater OK 74075

LAFON, GUY MICHEL, b Bordeaux, France, June 5, 43. GEOCHEMISTRY, PHYSICAL CHEMISTRY. *Educ:* Paris Sch Mines, Civil Ing Mines, 64; Univ Alta, MSc, 65; Northwestern Univ, Ill, PhD(geol), 69. *Prof Exp:* Res Found fel, State Univ NY Binghamton, 69-70, asst prof geol, 70-72; asst prof geol, Johns Hopkins Univ, 72-79; SR RES GEOLOGIST RES ASSOC, EXXON PROD RES CO, 79- *Mem:* Sigma Xi; AAAS; Geochem Soc; Soc Econ Paleont & Mineral; Geol Soc Am. *Res:* Geochemistry of natural water systems; thermodynamic properties of brines and minerals; equilibrium models; experimental study of mineral-fluid reactions, hydrothermal simulation of geological processes. *Mailing Add:* PO Box 2189 Houston TX 77252-2189

LAFON, STEPHEN WOODROW, b Owasso, Mich, Aug 3, 53; m 74; c 2. ANTIMICROBIAL THERAPY. *Educ:* Olivet Nazarene Col, BA, 75; WVa Univ, MSc, 78. *Prof Exp:* Res asst, Dept Exp Ther, Burroughs Wellcome Co, 77-78, res scientist I, 78-81, res scientist II, 81-86, prog coordr, Dept Proj Coord, 86-87, clin res assoc III, Dept Infectious Dis, 87-89, CLIN RES SCIENTIST I, DEPT INFECTIOUS DIS, BURROUGHS WELLCOME CO, 89- *Mem:* Assoc mem Am Soc Biol Chemists. *Res:* Metabolism and pharmacokinetics of numerous potential therapeutic agents; purine metabolism and nucleic acid synthesis in mammalian cells and protozoa; DNA damage assays and repair research; early HIV diseases and opportunist infections; author of numerous publications. *Mailing Add:* Dept Antimicrobial Ther Burroughs Wellcome Co 3030 Cornwallis Rd Research Triangle Park NC 27709

LAFOND, ANDRE, b Montreal, Que, July 1, 20; m 46; c 3. FORESTRY. *Educ:* Jean-de-Brebeuf Col, BA, 42; Laval Univ, BA, 45, BASc, 46; Univ Wis, PhD, 51. *Prof Exp:* Forester, Que Forest Serv, 46-51; pres Res Found, 75, PROF FOREST ECOL & PHYSIOL, LAVAL UNIV, 51-, DEAN FAC FORESTRY, 71- *Concurrent Pos:* Consult, Que Northshore Paper Co, World Bank & Can Int Develop Agency. *Mem:* Can Soc Soil Sci; Can Soc Plant Physiol; French-Can Asn Advan Sci; Can Inst Forestry; Int Soc Soil Sci. *Res:* Forest ecology, particularly soil vegetation relationships; forest physiology, particularly mineral nutrition of trees and fertilization; forest management, particularly site classification. *Mailing Add:* 2071 Marie-Victorin St Nicolas PQ G0S 2Z0 Can

LA FOND, EUGENE CECIL, b Bridgport, Wash, Dec 4, 09; m 35; c 2. OCEANOGRAPHY. *Educ:* San Diego State Col, AB, 32; Andhra Univ, India, DSc, 56. *Prof Exp:* Asst, Scripps Inst, Univ Calif, 33-40, oceanogr, 40-47; prof oceanog, Andhra Univ, India, 52-53 & 55-56; specialist oceanog, US State Dept, 56-57; sr scientist, Atomic Submarine US Ship Skate, North Pole, 58; marine biologist, Scirpps Inst & Int Coop Admin, 60-61; chief scientist, US Prog Biol, Int Indian Ocean Exped, Woods Hole Oceanog Inst, 62-63; dep dir off oceanog & dep secy, Int Oceanog Comn, UNESCO, Paris, France, 63-64; supvry res oceanogr, Navy Electronics Lab, Naval Undersea Res & Develop Ctr, 64-68, sr scientist & consult oceanog, 68-73; secy gen,

70-87, PRES, COMN OCEANOG COOP WITH DEVELOP COUNTRIES, INT ASN PHYS SCI OCEAN, 83- *Concurrent Pos:* Gen mgr, La Fond Oceanic Consults, 58- *Honors & Awards:* Ocean Sci Award, Am Geophys Union, 82; 100 Years of Int Geophys Medal, Soviet Acad Sci, 83; Distinguished Serv Award, Int Asn Phys Sci Ocean, 87. *Mem:* Marine Technol Soc; Soc Limnol & Oceanog (vpres, 54-55); Am Geophys Union; Int Asn Phys Sci Ocean (secy-gen, 70-87); Maratime Res Soc. *Res:* Physical oceanography. *Mailing Add:* LaFond Oceanic Consult Box 7325 San Diego CA 92107

LAFONTAINE, EDWARD, b Yakima, Wash, Jan 2, 52. AIR & WATER POLLUTION CONTROL, THERMAL PROCESS DESIGN. *Prof Exp:* Pres, T&C Res, 78-80; tech dir, Cameron-Yakima Inc, 81-87; vpres, Atesia USA, Inc, 87-88; PRES/OWNER, TELTECH CO, 88- *Concurrent Pos:* Consult, Ralston Brokers Int, 88-, Pac Aqua-Tech, Ltd, 90-; tech dir, Intercon Pac Inc, 90- *Mem:* Int Carbon Soc; Nat Pollution Control Fedn; Nat Air Pollution Control Asn; Am Water Works Asn. *Res:* Chemically treated activated carbons for special applications; manufacture of special purpose activated carbons; conversion of organic waste from land fills into activated carbon; process development for the disposal of waste tires. *Mailing Add:* PO Box 2784 Yakima WA 98907

LAFONTAINE, JEAN-GABRIEL, b Sherbrooke, Que, Aug 4, 28; m 52; c 3. CELL BIOLOGY, ELECTRON MICROSCOPY. *Educ:* Laval Univ, Lic es Sci, 50; Univ Wis, MS, 52, PhD(zool), 54. *Prof Exp:* Res asst cytol, Sloan Kettering Inst, 54-56, Rockefeller Inst, 56-58 & Montreal Cancer Inst, 58-60; asst prof path, Med Sch, 60-64, assoc prof biol, 64-68, PROF BIOL, SCI FAC, LAVAL UNIV, 68- *Concurrent Pos:* Damon Runyon fel, 54-56. *Honors & Awards:* Quebec Asn Advan Sci Prize, 82. *Mem:* Am Soc Cell Biol; Can Soc Cell Biol; Royal Soc Can, 84; NY Acad Sci, 85. *Res:* Cytochemistry and ultrastructure of the cell nucleus. *Mailing Add:* Dept Biol Fac Sci Laval Univ Quebec PQ G1K 7P4 Can

LAFORNARA, JOSEPH PHILIP, b Buffalo, NY, Dec 5, 42; wid; c 2. CHEMISTRY, ENVIRONMENTAL SCIENCES. *Educ:* Canisius Col, BS, 64; Univ Fla, PhD(inorg chem), 70. *Prof Exp:* Res chemist, Edison Water Qual Lab, Fed Water Qual Admin, Dept Interior, 70-71; res chemist, Nat Environ Res Ctr, 71-75, Oil & Hazardous Mat Spills Br, Indust Environ Res Lab, 75-78, phys scientist, Environ Response Team, 78-87, CHIEF ENVIRON RESPONSE TEAM, US ENVIRON PROTECTION AGENCY, 87- *Concurrent Pos:* Tech adv, Hazardous Mat Adv Comt, Nat Res Coun-Nat Acad Sci, 71-; mem, Task Force for Nitrosamine Control & Task Force for Kepone Control, US Environ Protection Agency, 75-; chmn, Hazardous Mat Div, Am Soc Testing & Mat Comt, No F-20, Spill Control Syst, 78- *Mem:* Am Soc Testing & Mat; Am Chem Soc; Water Pollution Control Fedn. *Res:* Application of chemical technology to control of spills of hazardous materials; chemical analysis of inorganic and organic water and air pollutants; ultimate disposal of chemical wastes. *Mailing Add:* 894 Joan Ct North Brunswick NJ 08902

LAFOUNTAIN, JAMES ROBERT, JR, b Richmond, Va, Jan 8, 44; m 70; c 2. CELL BIOLOGY. *Educ:* Princeton Univ, AB, 66; State Univ NY, Albany, PhD(biol sci), 70. *Prof Exp:* Fel, Eidgenossische Technische Hochschule, Switz, 71-72; asst prof, 72-77, assoc prof biol sci, 77-86, PROF BIOL SCI, SUNY, 86- *Mem:* Am Soc Cell Biol; Electron Micros Soc Am. *Res:* Physiology of cell division and cell motility. *Mailing Add:* Dept Biol Sci State Univ NY 657 Cooke Hall Buffalo NY 14260

LAFOUNTAIN, LESTER JAMES, JR, b Marinette, Wis, Sept 27, 42; m 64; c 1. GEOLOGY. *Educ:* Univ Wis, BS & MS, 64; Univ Colo, PhD(geol), 73. *Prof Exp:* Field geologist, US Steel Corp, 64, party chief geol, 65; geologist, Texaco Inc, 66; res assoc rock mech, Dept Geol, Univ NC, 71-74; asst proj geologist, D'Appolonia Geophys Corp, 75-76, proj geologist, 76-77, chief geologist int oper, 78-79, gen mgr corp, 80-84; vpres, Technos, 84-85; asst prog mgr, Battelle Off Nuclear Waste Isolation, 85-88; VPRES, CONVERSE ENVIRON, CONSULTS CALIF, 88- *Mem:* Geol Soc Am; Am Geophys Union; AAAS; Sigma Xi; Am Soc Civil Engrs; Nat Water Well Asn; Soc Explor Geophysicists. *Res:* The mechanisms and physical aspects of rock dilation, stick slip and earthquake precursors; the tectonics of the mid-continent and its relationship to seismicity. *Mailing Add:* 3393 E Foothill Blvd No B Pasadena CA 91107-3112

LAFRAMBOISE, JAMES GERALD, b Windsor, Ont, July 26, 38; m 62; c 2. PLASMA PHYSICS. *Educ:* Univ Windsor, BSc, 57; Univ Toronto, BASc, 59, MA, 60, PhD(aerospace studies), 66. *Prof Exp:* Asst prof math, Univ Windsor, 65-67; asst prof physics, 67-71, assoc prof, 71-77, PROF PHYSICS, YORK UNIV, 77- *Concurrent Pos:* Mem prog team, WISP Shuttle Exp; assoc ed, J Geophys Res, Space Physics, 83-85. *Mem:* Can Asn Physicists; Am Geophys Union; Planetary Soc. *Res:* High-voltage electrical charging of spacecraft; electrode devices for plasma diagnostics; high-voltage antennas in space plasmas; spacecraft-plasma interactions. *Mailing Add:* Dept Physics York Univ 4700 Keele St Toronto ON M3J 1P6 Can

LAFRAMBOISE, MARC ALEXANDER, b Windsor, Ont, May 18, 15; m 49; c 2. MATHEMATICS. *Educ:* Univ Ottawa, BA, 42; Univ Mich, MA, 46, MSc, 49. *Prof Exp:* Prin & teacher pub & separate schs, Ont, 34-42; asst prof math, Assumption Col, 42-50; from asst prof to assoc prof math, Univ Detroit, 53-76, adj prof, 76-80; RETIRED. *Concurrent Pos:* Dean eve div, Assumption Col, 46-49. *Res:* Mathematics education. *Mailing Add:* 1477 Dufferin Pl Windsor ON N8X 3K3 Can

LAFRANCHI, EDWARD ALVIN, b Petaluma, Calif, July 23, 28; m 54; c 3. ELECTRICAL ENGINEERING. *Educ:* Univ Santa Clara, BS, 50. *Prof Exp:* Opers engr, Lawrence Livermore Lab, 53-56, design engr, 56-58, group leader, 58-66, div leader, 66-73, dept head, dept electronics eng, 73-86, DEP ASSOC DIR ENG, LAWRENCE LIVERMORE LAB, 86- *Mem:* Inst Elec & Electronics Engrs. *Res:* Computer science and engineering; engineering management. *Mailing Add:* Dept Electronics Eng Lawrence Livermore Lab Livermore CA 94550

LAFRENZ, DAVID E, b Waterloo, Iowa, Aug 13, 47; m; c 2. MICROBIOLOGY. *Educ:* Ariz State Univ, BS, 71; Univ Iowa, PhD, 78. *Prof Exp:* Postdoctoral fel, Howard Hughes Med Inst, Stanford Univ Sch Med, 78-81, postdoctoral res affil, 81-83; res health sci specialist, Vet Admin Med Ctr, Iowa City, 83-88, dir, Animal Care Facil, 83-88; assoc res scientist, Dept Internal Med, 87-88, adj asst prof, 87-88, ASST PROF & ASSOC DIR, CELL & IMMUNOBIOL CORE FACIL & DIR FLOW CYTOMETRY FACIL, DEPT MICROBIOL, SCH MED, UNIV MO, COLUMBIA, 88-; RES MICROBIOLOGIST, H S TRUMAN MEM VET ADMIN HOSP, COLUMBIA, 88- *Concurrent Pos:* Mem, Res & Develop Comt, Vet Admin Med Ctr, Iowa City, Iowa, 87-88. *Mem:* Am Asn Immunologists; AAAS; Am Soc Microbiol; Soc Anal Cytol. *Res:* Cellular immunology; immunologic memory; molecular immunology; molecular biology. *Mailing Add:* H S Truman Mem Vet Admin Hosp 800 Hospital Dr Columbia MO 65201

LAFUSE, HARRY G, b Liberty, Ind, Jan 22, 30; m 54; c 1. ELECTRICAL ENGINEERING. *Educ:* Purdue Univ, BSEE, 57; Univ Ill, MSEE, 58, PhD, 62. *Prof Exp:* Instr elec eng, Univ Ill, 61-62; asst prof elec eng, Univ Notre Dame, 62-65, assoc prof, 65-81; elec engr, Phillips Eng Co, 81-83; STAFF ENGR, BENDIX GUID SYSTS DIV, 83- *Concurrent Pos:* Consult, Bendix Corp, 62-69 & McCarthy & Assocs, 70-83. *Mem:* Sigma Xi; Nat Soc Prof Engrs. *Res:* Electromagnetic field theory; network analysis and synthesis; high frequency transmission systems; microwave theory; guidance electronics systems. *Mailing Add:* 1611 Tudor Lane South Bend IN 46614

LAFUZE, JOAN ESTERLINE, b Indianapolis, Ind. PEDIATRICS. *Educ:* Ind Univ, AB, 59; Ball State Univ, MS, 75, Ind Univ, PhD(physiol), 81. *Prof Exp:* Lab tech & asst supvr, Gen Lab, Methodist Hosp, 59-60; student technologist, Sch Med Technol, St Vincent Hosp, 60-61; med technologist, Richmond Med Lab, 63-65; teaching supvr, In-Serv Lab, Educ Prog, Reid Mem Hosp, 65-68; educ coordr, Ind Voc Tech Col, 68-71; res & teaching asst, Dept Physiol & Biophys, Sch Med, Ind Univ, 75-81; res technologist pediat hemat-oncol, James Whitcomb Riley Hosp Children, Ind, 81-85; RES ASSOC PEDIAT HEMAT-ONCOL, SCH MED, IND UNIV, 85-; ASST PROF BIOL, IND UNIV E, 87- *Concurrent Pos:* Asst prof, Dept Physiol & Biophys, Sch Med, Ind Univ, 84-87. *Mem:* Am Soc Clin Pathologists; Am Physiol Soc; Tissue Cult Asn; AAAS. *Res:* Construction of a Molt-3 cDNA library; construction and characterization of a subtracted T-cell ALL cDNA library; construction of a subtracted T-cell ALL probe; sequencing; effect of neutrophil activation on respiration, blood pressure and absolute granulocyte count of rabbits and cats; use of antioxidants to attenuate the in vivo and in vitro effects of chemotactic agents; adherence of neutrophils to culture vascular endothelium; transendothelial migration of activated neutrophils. *Mailing Add:* Dept Pediat Riley Hosp Children 702 Barnhill Dr Indianapolis IN 46202

LAGAKOS, STEPHEN WILLIAM, b Philadelphia, Pa, June 18, 46; m 68; c 2. BIOSTATISTICS. *Educ:* Carnegie-Mellon Univ, BS, 68; George Washington Univ, MPhil & PhD(math & statist), 72; Harvard Univ, AM, 86. *Prof Exp:* Math statistician, Naval Ord Sta, 68-70; statistician biostatist, Statist Lab, State Univ NY Buffalo, 72-80, asst prof statist sci, 73-80; assoc prof, Harvard Sch Pub Health, 80-85, PROF, HARVARD SCH PUB HEALTH, 86- *Concurrent Pos:* Coord statistician, Working Party Ther Lung Cancer, 72-; protocol statistician, Eastern Coop Oncol Group, 72- *Honors & Awards:* Spiegelman Gold Medal, 86. *Mem:* Biomet Soc; Royal Statist Soc; Inst Math Statist; Int Asn Study Lung Cancer; Am Statist Asn. *Res:* The planning, design and analysis of clinical trials with particular emphasis on survival-type data. *Mailing Add:* Dept Biostatist Harvard Sch Pub Health 677 Huntington Ave Boston MA 02115

LAGALLY, MAX GUNTER, b Darmstadt, Ger, May 23, 42; US citizen; m 69; c 3. MATERIALS SCIENCE, SURFACE PHYSICS. *Educ:* Pa State Univ, BS, 63; Univ Wis, MS, 65, PhD(physics), 68. *Prof Exp:* Vis fel physics, Fritz Haber Inst, Max Planck Soc, 68-69; instr physics & res assoc surface physics, 70-71, from asst prof to assoc prof 71-77, PROF MAT SCI, UNIV WIS-MADISON, 77-, DIR THIN-FILM DEPOSITION & APPLS CTR, 85- *Concurrent Pos:* Sloan Found fel, 73-77; vis scientist surface physics, Sandia Nat Lab, 74; H I Romnes fel, 76-80; John Bascom prof surface sci & technol, Univ Wis-Madison, 86-; Gordon Godfrey vis prof physics, Univ New S Wales, Sydney, Australia, 87; Humboldt sr res fel, Jülich, Ger, 92. *Mem:* Fel Am Phys Soc; Mat Res Soc; Am Vacuum Soc; Am Chem Soc; Am Soc Metals Int; fel Australian Inst Physics. *Res:* Crystallographic and electronic properties of surfaces, thin films, and interfaces; surface disorder; multilayer thin films for x-ray optics; diffraction; scanning tunneling microscopy. *Mailing Add:* Dept Mat Sci & Eng & Dept Physics Univ Wis Madison WI 53706

LA GANGA, THOMAS S, b Caldwell, NJ, July 23, 27; m 75; c 5. ENDOCRINOLOGY, CLINICAL CHEMISTRY. *Educ:* Drew Univ, AB, 51; Rutgers Univ, MS, 66, PhD(animal sci), 67. *Prof Exp:* Clin chemist, Princeton Hosp, NJ, 67-69; asst dir endocrinol, 69-79, TECH COORDR, BIO-SCI LABS, 79- *Mem:* AAAS. *Res:* Improved methodology in chemical and biological hormone assays; endocrine physiology of the mammary gland; experimental hypertension; automated analytical systems; thyroid cancers and autoantibodies. *Mailing Add:* 10954 Belmar Ave Northridge CA 91326

LAGANIS, DENO, b Detroit, Mich, July 17, 19; m 48; c 3. POLYMER CHEMISTRY, ORGANIC CHEMISTRY. *Educ:* Wayne State Univ, BS, 41. *Prof Exp:* Develop chemist rubber, US Rubber Co, 41-43; develop chemist plastics, Ford Motor Co, 43-45; group leader res & develop resins, Reichhold Chem Inc, 45-54; mgr plastics, Chem Compounding Co, 54-55; GROUP LEADER RES & DEVELOP RESINS, SCHENECTADY CHEM INC, 55- *Mem:* Am Chem Soc; Soc Mfg Engrs. *Res:* Polymer chemistry related to electrical insulation resins for surface coatings; unsaturated polyesters for various uses; polyurethanes for coatings and foams; epoxies as coatings and potting compounds; leveling resins for floor polishes. *Mailing Add:* 2331 Algonquin Rd Schenectady NY 12309

LAGANIS, EVAN DEAN, b Detroit, Mich, June 6, 53; m 77; c 2. INFRARED SYNTHESIS, ORGANOSILICONE CHEMISTRY. *Educ:* State Univ NY, Geneseo, BA, 75; Dartmouth Col, PhD(org chem), 80. *Prof Exp:* Fel organometallic & cyclophane chem, Univ Ore, 79-81; res chemist organofluorine, organosilicon & polyacetylene chem, Cent Res & Develop Dept, 81-84, RES CHEMIST, OPTICAL DISK MEDIA, PHOTOSYSTS & ELECTRONIC PROD DEPT, E I DU PONT DE NEMOURS & CO, INC, 85- *Mem:* Am Chem Soc. *Res:* Preparation of infrared dyes for use as the active layer in Optical Disk Media; organosilicon reagents for organic synthesis. *Mailing Add:* 204 W Crest Rd Wilmington DE 19803

LAGARIAS, JEFFREY CLARK, b Pittsburgh, Pa, Nov 16, 49. MATHEMATICS. *Educ:* Mass Inst Technol, SB & SM, 72, PhD(math), 74. *Prof Exp:* MEM TECH STAFF, BELL TELEPHONE LABS, 74- *Concurrent Pos:* Vis asst prof, Univ Md, 78-79; vis assoc prof comput sci, Rutgers Univ, 84. *Honors & Awards:* Lester Ford Award, Math Asn Am, 87. *Mem:* Am Math Soc; Math Asn Am; Soc Indust & Appl Math; Inst Elec & Electronics Engrs; Asn Comput Mach. *Res:* Computational complexity theory; cryptography; number theory; discrete mathematics. *Mailing Add:* AT&T Bell Labs Rm 2C-373 Murray Hill NJ 07974

LAGARIAS, JOHN S(AMUEL), b Rochester, NY, July 4, 21; m 47; c 3. PHYSICS, ELECTRONICS. *Educ:* Rensselaer Polytech Inst, BS, 48. *Prof Exp:* Engr, Res Dept, Westinghouse Elec Corp, 48-51; physicist, Koppers Co, Inc, Pa, 51-53, mgr, Precipitation Br, 53-55, mgr, Metal Prod Res, 56-61, mgr, Physics & Phys Chem Lab, 61-63; mgr, Res & Develop, Am Instrument Co, 63-65; vpres, Resources Res, Inc, Va, 65-66, exec vpres, 66-67, pres, 67-71; dir environ qual, Kaiser Engrs Inc, 71-84; PRES, LAGARIAS ASSOCS, INC, 84-; BD MEM, CALIF AIR RESOURCES BD, 85- *Concurrent Pos:* Conf chmn, 2nd Int Clean Air Cong, 70; 1st Int Conf Electro Precipitation, 81. *Mem:* Fel, hon mem Air Pollution Control Asn (pres, 68-69); Am Phys Soc; sr mem Inst Elec & Electronics Engrs; Am Acad Environ Engrs; fel Int Soc Electrostatic Precipitation. *Res:* Industrial gas cleaning equipment including electrostatic precipitators, bag filters, and scrubbers; environmental controls. *Mailing Add:* Lagarias Assocs 276 Donald Dr Moraga CA 94556

LAGE, GARY LEE, b Hinsdale, Ill, Nov 11, 41; m 64; c 2. PHARMACOLOGY, TOXICOLOGY. *Educ:* Drake Univ, BS, 63; Univ Iowa, MS, 65, PhD(pharmacol), 67; Am Bd Toxicol, dipl, 80. *Prof Exp:* From asst prof to assoc prof pharmacol, Sch Pharm, Univ Kans, 67-73; assoc prof pharm, Univ Wis-Madison, 73-78; prof toxicol & dir toxicol progs, 78-84, CHMN DEPT PHARMACOL & TOXICOL, PHILADELPHIA COL PHARM & SCI, 84- *Concurrent Pos:* USPHS res career develop award, 75-80. *Mem:* Soc Toxicol (treas, 85-87); Am Pharmaceut Asn; AAAS; Am Soc Pharmacol & Exp Therapeut; Am Asn Col Pharm. *Res:* Study of drug distribution and metabolism in relation to drug toxicity, distribution and/or metabolism, especially cardiac glycosides; hepatotocity mechanisms. *Mailing Add:* Philadelphia Col Pharm & Sci 43rd St & Kingsessing Mall Philadelphia PA 19104

LAGE, JANICE M, b Exeter, Calif, July 5, 51. PATHOLOGY. *Educ:* Calif State Univ, Fresno, BS, 73; Wash Univ, Mo, MD, 80; Am Bd Path, cert, 85. *Prof Exp:* Instr path, Sch Med, Stanford Univ, 80; instr path, Sch Med, Wash Univ, 81, instr obstet/gynec, 82, asst path, 83; instr, 84-87, ASST PROF, HARVARD MED SCH, 87-, PATHOLOGIST, BRIGHAM & WOMEN'S HOSP, 87- *Concurrent Pos:* Resident path, Wash Univ, 81-82, obstet & gynec, 82-83, fel, Dept Path, 83-84; assoc pathologist, Brigham & Women's Hosp, Boston, Mass, 84-87; NIH grant, 90. *Mem:* US Acad Path; Can Acad Path; Am Asn Pathologists; Int Soc Gynec Pathologists; Soc Pediat Pathologists. *Res:* Gestational trophoblastic diseases; perinatal and obstetric pathology, with emphasis on congenital malformations; application of flow cytometry to surgical pathology of obstetric and gynecologic tumors. *Mailing Add:* Dept Path Brigham & Women's Hosp 75 Francis St Boston MA 02115

LAGERGREN, CARL ROBERT, b St Paul, Minn, Nov 21, 22; m 47; c 3. PHYSICS. *Educ:* State Col, Wash, BS, 44, MS, 49; Univ Minn, PhD(physics), 55. *Prof Exp:* Sr physicist, Hanford Atomic Prod Oper, Gen Elec Co, 55-65; mgr mass spectrometry, 65-68, RES ASSOC, RADIOL SCI DEPT, PAC NORTHWEST LABS, BATTELLE MEM INST, 68- *Mem:* Am Phys Soc; Sigma Xi. *Res:* Mass spectrometry; electron impact phenomena; isotopic abundances; surface ionization; ion optics. *Mailing Add:* 2110 Howell Ave Richland WA 99352

LAGERSTEDT, HARRY BERT, b Glen Ridge, NJ, Aug 2, 25; m 52; c 5. PLANT PHYSIOLOGY, HORTICULTURE. *Educ:* Ore State Univ, BS, 54, MS, 57; Tex A&M Univ, PhD(plant physiol), 65. *Prof Exp:* From instr to asst prof hort, Ore State Univ, 57-67, assoc prof, 67-; res horticulturist, Agr Res Serv, 67-; AT NORTHWEST GERMPLASM REPOSITORY. *Mem:* Am Soc Plant Physiol; Am Soc Hort Sci; Sigma Xi. *Res:* Plant growth regulators; nut crops. *Mailing Add:* 1700 NW Kings Blvd Corvallis OR 97330

LAGERSTROM, JOHN E(MIL), b Galesburg, Ill, Dec 12, 22; m 47; c 3. ELECTRICAL ENGINEERING. *Educ:* Iowa State Univ, BS, 44, MS, 51, PhD(elec eng), 58. *Prof Exp:* Instr elec eng, Iowa State Univ, 46-49, from asst prof to prof, 49-66, res asst prof, A-C network anal, 50-57, asst to dean eng, 57-58, asst dean, 58-63, assoc dean, 63-66, actg dir tech inst, 60-62, actg head dept archit & archit eng, 62-64; dean eng, SDak State Univ, 66-71; chmn, Dept Elec Eng, 71-76, DIR ENG EXTEN, UNIV NEBR, LINCOLN, 76- *Concurrent Pos:* Consult, Winpower Mfg Co, Iowa, 56-59; adv to rector, Nat Univ Eng, Peru, Ford Found-Iowa State Univ contract, 64-66. *Mem:* Nat Soc Prof Eng; Inst Elec & Electronics Engrs. *Res:* Electrical machines; power transmission and distribution; power system protection; system stability; economic power system loading. *Mailing Add:* 2301 Jameson S Lincoln NE 68512

LAGHARI, JAVAID ROSOOLBUX, b Hyderabad, Pakistan, June 25, 50; nat US; m 82; c 1. ELECTRICAL ENGINEERING. *Educ:* Sind Univ, Pakistan, BE, 71; Middle East Tech Univ, Turkey, MS, 75; State Univ NY, Buffalo, PhD(elec eng), 80. *Prof Exp:* Asst eng, Indus Grindery, Pakistan, 71-72; asst exec engr, Airports Develop Agency, Pakistan, 75-76; asst prof, 80-88, ASSOC PROF ELEC ENG, SUNY BUFFALO, 89- *Concurrent Pos:* Prin investr, Air Force Off Sci Res, 83-90, Off Naval Res, 86-87; secy, Radiation Soc Comt, Inst Elec & Electronic Engrs, 86-, mem, Comt Man & Radiation; mem, Tech Prog Comt, Int Symp Elec Insulation, 88-90, chmn, 92; mem, Tech Prog Comt, Int High Voltage Symp, 89. *Mem:* Inst Elec & Electronics Engrs. *Res:* High voltage and pulsedpower; electrical insulation and dielectrics, as applicable to space power technology, including energy storage and transport devices; high speed diagnostics; Supervision and guidance under graduate and engineering studies. *Mailing Add:* 191 Sprucewood Terr Williamsville NY 14221

LAGLER, KARL FRANK, fisheries, zoology; deceased, see previous edition for last biography

LAGO, JAMES, b New York, NY, Nov 7, 21; m 68. PROCESS RESEARCH & DEVELOPMENT, FERMENTATION & ISOLATION. *Educ:* Polytech Inst Brooklyn, BChE, 44; Mass Inst Technol, MS, 47. *Prof Exp:* Asst, Manhattan Proj, 44-46; jr engr, Merck & Co Inc, Rahway, 47-51, group leader chem eng, 51-57, sect mgr, 57-64, mgr, 64-69, dir chem eng res & develop, 69-79, vpres process res & develop, 79-85; CONSULT, 86- *Mem:* Nat Acad Eng; Am Chem Soc; Am Inst Chem Engrs; AAAS. *Res:* Development of processes for the preparation of medicinals; design and startup of manufacturing facilities for the processes developed. *Mailing Add:* PO Box 1699 White Salmon WA 98672-1699

LAGO, PAUL KEITH, b Worthington, Minn, June 24, 47; m 69; c 1. ENTOMOLOGY. *Educ:* Bemidji State Col, BA, 69, MA, 71; NDak State Univ, PhD(entom), 77. *Prof Exp:* PROF BIOL, UNIV MISS, 76- *Mem:* Coleopterists Soc; Entom Soc Am; Am Entom Soc; NAm Benthological Soc; Sigma Xi. *Res:* Insect taxonomy, principally coleoptera and aquatic insects; insect ecology. *Mailing Add:* Dept Biol Univ Miss University MS 38677

LAGOWSKI, JEANNE MUND, b St Louis, Mo, Nov 17, 29; m 54. ORGANIC CHEMISTRY. *Educ:* Bradley Univ, BS, 51, MS, 52; Univ Mich, PhD(org chem), 57. *Prof Exp:* Instr anal chem, Bradley Univ, 51-52; res chemist, Mich State Univ, 56-57; res fel phys org chem, Cambridge Univ, 57-59; assoc res scientist biochem genetics, 59-63, res scientist, 63-73, lectr zool, 73-74, asst dean, Div Gen & Comp Studies, 72-78, assoc prof zool, 74-81, asst dean, 78-81, ASSOC DEAN, COL NATURAL SCI & PROF ZOOL, UNIV TEX, AUSTIN, 81- *Concurrent Pos:* Res career develop award, NIH, 64-69; assoc, Danforth Found, 77- *Mem:* Am Chem Soc; Int Soc Heterocyclic Chemists. *Res:* Chemistry of nitrogen heterocycles; biochemical genetics. *Mailing Add:* Dept Zool Univ Tex Austin TX 78712

LAGOWSKI, JOSEPH JOHN, b Chicago, Ill, June 8, 30; m 54. INORGANIC CHEMISTRY. *Educ:* Univ Ill, BS, 52; Univ Mich, MS, 54; Mich State Univ, PhD(inorg chem), 57; Cambridge Univ, PhD(inorg chem), 59. *Prof Exp:* From asst prof to assoc prof, 59-67, PROF CHEM, UNIV TEX, AUSTIN, 67- *Honors & Awards:* Piper Prof Award, Nat Chem Mfg Asn, 81; Chem Educ Award, Am Chem Soc, 89. *Mem:* Am Chem Soc; The Chem Soc. *Res:* Liquid ammonia solutions; organometallic compounds; borazines and derivatives; electrochemistry; development of computer-based teaching methods; non-aqueous solution chemistry; metal atom reactions. *Mailing Add:* Dept Chem Univ Tex Austin TX 78712

LAGRAFF, JOHN ERWIN, b Schenectady, NY, July 24, 40; m 62; c 2. AERODYNAMICS, GAS TURBINE HEAT TRANSFER. *Educ:* Mass Inst Technol, BS, 62; Oxford Univ, DPhil(eng sci), 70. *Prof Exp:* Assoc scientist, Res & Advan Develop Div, Avco Corp, 62-66; PROF FLUIDS-AERODYN, DEPT MECH & AERODYN ENG, 70-, DIR AEROSPACE ENG PROG, SYRACUSE UNIV, 86- *Concurrent Pos:* Vis prof, Oxford Univ, 83. *Honors & Awards:* Ralph Teetor Award, Soc Automotive Engrs, 72; Nat Fac Adv Award, Am Inst Aeronaut & Astronaut, 90. *Mem:* Assoc fel Am Inst Aeronaut & Astronaut; Am Soc Mech Engrs; Am Asn Univ Prof; Am Asn Eng Educ. *Res:* Unsteady aerodynamics; heat transfer associated with gas turbines. *Mailing Add:* 144 Westminster Ave Syracuse NY 13210

LAGRANGE, WILLIAM SOMERS, b Ames, Iowa, Apr 23, 31; m 54; c 3. FOOD MICROBIOLOGY, DAIRY BACTERIOLOGY. *Educ:* Iowa State Univ, BS, 53, PhD(dairy bact), 59. *Prof Exp:* Exten technologist dairy mfg, Univ Ky, 59-62; EXTEN FOOD TECHNOLOGIST, IOWA STATE UNIV, 62- *Mem:* Int Asn Milk, Food & Environ Sanit; Inst Food Technol; Am Soc Microbiol; Am Dairy Sci Asn. *Res:* Dairy manufacturing quality control; dairy and foods microbiology; foods processing and control. *Mailing Add:* Dept Food Sci Iowa State Univ Ames IA 50011

LAGREGA, MICHAEL DENNY, b Yonkers, NY, July 19, 44; m 70; c 2. HAZARDOUS WASTE MANAGEMENT. *Educ:* Manhattan Col, BE, 66; Syracuse Univ, MS, 71, PhD(environ eng), 72. *Prof Exp:* Proj mgr, O'Brien & Gere Engrs, Inc, NY, 66-72; asst prof environ eng, Drexel Univ, 72-74; asst prof, 74-78, assoc prof, 78-86, PROF CIVIL ENG, BUCKNELL UNIV, 86- *Concurrent Pos:* Staff consult, Buchart-Horn Engrs & Planners, 74-81; prin consult, Roy F Weston, Inc, 81-84; dir, Hazardous Planning, Pa Dept Environ Resources, 84-85; consult, Hazardous Waste, Environ Resources Mgt, Inc. *Mem:* Water Pollution Control Fedn; Int Asn Water Pollution Res & Control; Asn Environ Eng Prof; fel Am Soc Civil Engrs; Am Acad Environ Engrs. *Res:* Management of hazardous wastes; physical-chemical processes for water pollution control. *Mailing Add:* Dept Civil Eng Bucknell Univ Lewisburg PA 17837

LAGRONE, ALFRED H(ALL), electrical engineering, for more information see previous edition

LAGUNOFF, DAVID, b New York, NY, Mar 14, 32; m 58; c 3. PATHOLOGY. *Educ:* Univ Chicago, MD, 57. *Prof Exp:* Asst microbiol, Univ Miami, 51-53; intern, San Francisco Hosp, Calif, 57-58; from instr to prof path, Univ Wash, 60-79; PROF & CHMN PATH, ST LOUIS UNIV, 79- *Concurrent Pos:* Nat Heart Inst fel path, Univ Wash, 58-59, USPHS trainee, 59-60; Nat Heart Inst spec fel physiol, Carlsberg Lab, Denmark, 62-64; Nat Cancer Inst spec fel path, Sir William Dunn Sch Exp Path, Oxford Univ, 69-70. *Mem:* Am Asn Path; Am Soc Cell Biol. *Res:* Mast cell structure and function; cell secretion; inflammation; pulmonary edema. *Mailing Add:* Dept Path Sch Med St Louis Univ 1402 S Grand Blvd St Louis MO 63104

LAGUNOWICH, LAURA ANDREWS, b Nov 29, 60; m; c 1. ANATOMY. *Educ:* Dickinson Col, BSc, 83; Thomas Jefferson Univ, PhD(path & cell biol), 87. *Prof Exp:* Postdoctoral fel, Dept Anat & Develop Biol, Thomas Jefferson Univ, 87-90; RES ASST PROF, NEUROTOXICOL LABS, DEPT PHARMACOL & TOXICOL, COL PHARM, RUTGERS UNIV, 90- *Concurrent Pos:* NIH training grant develop biol & teratology, 87; Nat Res Serv award, Nat Eye Inst, 87-90; instr, Dept Anat, Pa Sch Podiat Med, 88-89; Johnson & Johnson scholar develop neurotoxicol, Environ & Occup Health Sci Inst, 91-96, assoc mem, 91- *Mem:* Int Soc Differentiation; Soc Develop Biol; Soc Neurosci. *Res:* Genetic control of disease; clinical aspects of neoplasia; environmental and nutritional disorders; teratology; translational and posttranslational control of proteins; hemostasis and red blood cells; cardiac function and blood pressure. *Mailing Add:* Dept Pharmacol & Toxicol Rutgers Univ Col Pharm Piscataway NJ 08854-0789

LAGUROS, JOAKIM GEORGE, b Istanbul, Turkey, Feb 4, 24; US citizen; m 57; c 1. SOIL MECHANICS, HIGHWAY ENGINEERING. *Educ:* Robert Col, Istanbul, BS, 46; Iowa State Univ, MS, 55, PhD(soil mech), 62. *Prof Exp:* Asst engr, Naval Shipyard, Turkey, 48-51; instr civil eng, Robert Col, 51-54, asst prof, 56-59; res asst soils, Exp Sta, Iowa State Univ, 54-56, 59-62; asst prof soil mech, Univ Ohio, 62-63; assoc prof, 63-69, prof soils & hwys, 69-80, PROF CIVIL ENG & ENVIRON SCI, UNIV OKLA, 80- *Concurrent Pos:* Consult, Netherlands Harbor Works Co, Turkey, 52, Robert Col, 58 & McFadzen, Everly & Assocs, 61; mem physicochem phenomena soils comt, Hwy Res Bd, Nat Acad Sci-Nat Res Coun, 64-67. *Mem:* Am Soc Civil Engrs; Am Soc Eng Educ; Clay Minerals Soc. *Res:* Behavior of soils under load application; improvement of soil properties by admixtures; quality control of materials. *Mailing Add:* Dept Civil Eng Univ Okla Main Campus Norman OK 73019

LAHA, RADHA GOVINDA, b Calcutta, India, Oct 1, 30; US citizen. ANALYTICAL MATHEMATICS, PURE MATHEMATICS. *Educ:* Univ Calcutta, BSc, 49, MSc, 51, PhD(math), 57. *Prof Exp:* Mem staff math, Res & Training Sch, Indian Statist Inst, Calcutta, 52-57, lectr, 57; Smith-Mundt-Fulbright fel, Cath Univ Am, 57-58, res asst prof, 58-60; reader, Div Theoret Res & Training, Indian Statist Inst, 60-61; vis res fel, Inst Statist, Univ Paris & Swiss Fed Inst Technol, 61-62; from asst prof to prof, Cath Univ Am, 62-72; PROF MATH, BOWLING GREEN STATE UNIV, 72- *Concurrent Pos:* Vis res fel, Mass Inst Technol, 68-69 & Inst Advan Study, Canberra, Australia, 80; vis mem, Inst Advan Study, Princeton, 74. *Mem:* Fel Inst Math Statist; Int Statist Inst; Am Math Soc. *Res:* Analytical and abstract probability; harmonic analysis and representation theory of groups; application of probability and analysis to number theory. *Mailing Add:* Dept Math Bowling Green State Univ Bowling Green OH 43403

LAHAIE, IVAN JOSEPH, b Bay City, Mich, May 21, 54. ELECTROMAGNETIC SCATTERING & IMAGING. *Educ:* Mich State Univ, BS, 76; Univ Mich, MS, 77, PhD(elec eng), 81. *Prof Exp:* Res asst, Radiation Lab, Univ Mich, 76-80; RES ENGR, ENVIRON RES INST MICH, 80- *Mem:* Inst Elec & Electronics Engrs; Sigma Xi; Optical Soc Am. *Res:* Electromagnetic and optical imaging systems and radar systems; scattering, inverse scattering, and coherence for imaging systems. *Mailing Add:* Environ Res Inst Mich PO Box 8618 Ann Arbor MI 48107

LAHAM, QUENTIN NADIME, b Oshkosh, Wis, Feb 18, 27; m 50; c 3. HISTOLOGY, EMBRYOLOGY. *Educ:* Ripon Col, BA, 49; Marquette Univ, MS, 51; Univ Ottawa, PhD, 59. *Prof Exp:* Lectr gen biol, Univ Ottawa, 51-54, from asst prof to assoc prof, 54-58, prof histol & embryol, 68-76, chmn biol, 69-81, prof biol, 77-81; RETIRED. *Concurrent Pos:* Nuffield fel, 60-61. *Mem:* Teratology Soc; Soc Develop Biol; Can Soc Zoologists; Can Soc Cell Biol. *Res:* Ontogeny of enzyme systems during embryogenesis; influence of heavy metals on developing embryos. *Mailing Add:* 282 St Catherine Dr Daly City CA 94015

LAHAM, SOUHEIL, b Port-au-Prince, Haiti, Apr 17, 26; m 55; c 2. INHALATION TOXICOLOGY, CANCER. *Prof Exp:* Res assoc, Univ Paris, 54-56; guest scientist, Can Dept Health & Welfare, 56-58, head biochem sect, Environ Health Directorate, 58-62, chief environ toxicol prog, Occup Health Div, 62-71, sr res scientist & consult, 62-79, HEAD INHALATION TOXICOL UNIT, ENVIRON HEALTH DIRECTORATE, CAN DEPT HEALTH & WELFARE, 79-, HEAD, OCCUP TOXICOL RES SECT. *Concurrent Pos:* Guest scientist, Nat Res Coun Can, 56-58; vis prof, Univ Ottawa, 71-, Univ Quebec, 73-, Ohio State Univ, Columbus, 74-, Carleton Univ, 77- & Univ Calif, Berkeley, 81- *Mem:* Am Indust Hyg Asn; Pharmacol Soc Can; Europ Soc Toxicol; Soc Toxicol; AAAS. *Res:* Inhalation toxicity and metabolism of toxic and carcinogenic substances; environmental and occupational cancer; chemical carcinogenesis; neurotoxicology; peripheral neuropathy induced by industrial chemicals; inhalation toxicity of indoor air pollutants. *Mailing Add:* 249 Latchford Rd Ottawa ON K1Z 5W3 Can

LAHEY, M EUGENE, b Ft Worth, Tex, Dec 28, 17; m 42; c 6. PEDIATRICS. *Educ:* Univ Tex, BA, 39; St Louis Univ, MD, 43. *Prof Exp:* Nat Res Coun fel med sci, Univ Utah, 49-51, asst prof pediat, 51-52; from asst prof to assoc prof, Univ Cincinnati, 52-58; head dept, 58-74, prof, 58-83, EMER PROF PEDIAT, UNIV UTAH, 83- *Concurrent Pos:* Mem med adv bd, Leukemia Soc, 58- & hemat training grant comt, Nat Inst Arthritis & Metab Dis, 59-63; mem, Scope Panel, US Pharmacopeia, 60-; mem residency rev comt, AMA, 61-65, pres, 65-; res dir, Children's Hosp, East Bay, 64-65; vis prof, Children's Hosp, Honolulu. *Mem:* Am Soc Hemat; Am Pediat Soc; Soc Pediat Res. *Res:* Pediatric hematology. *Mailing Add:* Dept Pediat Univ Utah Med Ctr Salt Lake City UT 84132

LAHEY, RICHARD THOMAS, JR, b St Petersburg, Fla, Feb 20, 39; m 61; c 3. HEAT TRANSFER, FLUID MECHANICS. *Educ:* US Merchant Marine Acad, BS, 61; Rensselaer Polytech Inst, MS, 64; Columbia Univ, ME, 66; Stanford Univ, PhD(mech eng), 71. *Prof Exp:* Engr, Knolls Atomic Power Lab, 61-64; res assoc, Columbia Univ, 64-66; mgr core & safety develop, Nuclear Energy Div, Gen Elec, 66-75; chmn dept Nuclear Eng, 75-87, PROF NUCLEAR ENG & ENG PHYSICS, RENSSELAER POLYTECH INST, 87- *Concurrent Pos:* Mem, Sci Adv Comt, EG&G Idaho Inc, 76-; mem Advan Code Rev Group & LOFT Rev Group, US Nuclear Regulatory Comn, 76-; comnr, Eng Manpower Comn, 81-; pres, R T Lahey Inc, 81-83; Fulbright fel, Magdalen Col, Oxford Univ; adj prof, Univ Pisa, Italy & Claude Bernard Univ, France. *Honors & Awards:* Glen Murphy Award, Am Soc Eng Educ; Tech Achievement Award, Am Nuclear Soc. *Mem:* Fel Am Soc Mech Engrs; fel Am Nuclear Soc; Sigma Xi; NY Acad Sci; Am Asn Eng Educ. *Res:* Two-phase flow and boiling heat transfer technology; nuclear reactor thermal-hydraulics and safety. *Mailing Add:* Dept Nuclear Eng & Eng Physics Rensselaer Polytech Inst Troy NY 12181

LAHIRI, SUKHAMAY, b Calcutta, India, Apr 1, 33; m 65. PHYSIOLOGY. *Educ:* Univ Calcutta, BSc, 51, MSc, 53, DPhil(physiol), 56; Oxford Univ, DPhil(physiol), 59. *Prof Exp:* Govt of WBengal scholar, Oxford Univ, 56-59; asst prof physiol, Presidency Col, Univ Calcutta, 59-65, hon lectr, Univ, 60-65; vis fel & asst prof, State Univ NY Downstate Med Ctr, 65-67; sr res assoc, Cardiovasc Inst, Michael Reese Hosp & Med Ctr, Chicago, Ill, 67-69; assoc prof environ physiol, 69-73, ASSOC PROF PHYSIOL, UNIV PA, 73- *Honors & Awards:* Premchand-Roychand Gold Medal, Univ Calcutta, 62. *Mem:* NY Acad Sci; Am Physiol Soc. *Res:* High altitude physiology; regulation and adaptation; gas exchange; chemoreceptors. *Mailing Add:* Dept Physiol Univ Philadelphia 8400 RB Philadelphia PA 19104-5085

LAHIRI, SYAMAL KUMAR, b Rangoon, Burma, Jan 1, 40; m 70; c 3. MATERIALS SCIENCE, ADVANCED CERAMICS. *Educ:* Univ Calcutta, BE, 61; Univ Notre Dame, MS, 64; Northwestern Univ, PhD(mat sci), 69. *Prof Exp:* Sr sci asst, Defence Metall Res Lab, Govt of India, 61-62; res staff mem, T J Watson Res Ctr, IBM Corp, 68-84; Nat Phys Lab & Cent Electronics Inst, New Delhi, India, 84-86; T J Watson Res Ctr, IBM Corp, 87; CENT GLASS RES INST, CALCUTTA, INDIA. *Concurrent Pos:* Vis scientist, Nat Phys Lab & Indian Inst Technol, New Delhi, India, 78-79; adj prof, IIT, Uharagpur, India, 88-89. *Honors & Awards:* Outstanding Invention Award, IBM Corp, 76. *Mem:* Am Phys Soc; Inst Elec & Electronics Engrs; fel Inst Engrs India. *Res:* Thin film properties; fabrication of thin film devices; physical metallurgy; Josephson tunneling devices; microelectronic packaging; mechanical properties of materials; electronic ceramics. *Mailing Add:* Cent Glass & Ceramic Res Inst Jadavpur Calcutta 700032 India

LAHITA, ROBERT GEORGE, b Elizabeth, NJ, Dec 30, 45; m 71; c 2. IMMUNOLOGY, RHEUMATOLOGY. *Educ:* St Peter's Col, BS, 67; Thomas Jefferson Univ, MD, 73, PhD(microbiol), 73. *Prof Exp:* From asst prof to assoc prof immunol, Rockefeller Univ, 80-87; assoc prof med & pharmacol, Cornell Med Col, 80-90; ASSOC PROF, COLUMBIA, 91- *Concurrent Pos:* Lectr, Mt Sinai Med Ctr, 81-; consult, Medcom, 81-; mem exec bd, NY Arthritis Found, 82-, NY Serv Life Eval Found, Serv Life Eval Found Am, 83-; physician, Rockefeller Hosp, 83-; attend physician, Hosp Joint Dis, NY; clin scholar, Rockefeller Univ; attend phys, Hosp Spec Surg, NY; chief Rheumatol, St Luke's Roosevelt Hosp, NY; chmn bd, Lupus Found Am; bd gov, NY Arthritis Found. *Mem:* Am Rheumatism Asn; Am Soc Microbiol; NY Acad Sci; Harvey Soc; AAAS. *Res:* Effect of sex steroids on immune response; disease systemic lupus erythematosus. *Mailing Add:* Columbia Univ 432 W 58th St New York NY 10019

LAHOTI, GOVERDHAN DAS, b Jaipur, India, May 4, 48; m 75; c 3. MATERIALS SCIENCE ENGINEERING. *Educ:* Univ Burdwan, BEng, 69, Univ Calif, Berkeley, MS, 70, PhD(mech eng), 73. *Prof Exp:* Teaching asst mech eng, Univ Calif, Berkeley, 70-72, res asst, 72-73; res scientist, Batelle Mem Inst, Columbus, Ohio, 74-81; RES SCIENTIST, TIMKEN RES, CANTON, OHIO, 82- *Concurrent Pos:* Consult mech engr, 73-74. *Honors & Awards:* Gold Medal, Univ Burdwan, 70. *Mem:* Soc Mfg Engrs; Am Soc Metals; Am Soc Mech Engrs. *Res:* Development and optimization of metalworking processes, such as forging, rolling, and extrusion; computer aided modeling of metalworking techniques. *Mailing Add:* 3977 Bramshaw Rd NW Canton OH 44718

LAHR, CHARLES DWIGHT, b Philadelphia, Pa, Feb 6, 45; m 69, 86; c 4. MATHEMATICAL ANALYSIS. *Educ:* Temple Univ, BA, 66; Syracuse Univ, MA, 68, PhD(math), 71. *Prof Exp:* Mathematician, Bell Labs, 71-73; vis asst prof math, Savannah State Col, 73-74 & Amherst Col, 74-75; from asst prof to assoc prof math, Dartmouth Col, 75-84, assoc dean fac sci & dean grad studies, 81-84, prof math & comput sci & dean fac, 84-89, PROF MATH & COMPUT SCI, DARTMOUTH COL, 84- *Mem:* Am Math Soc; Sigma Xi; Math Asn Am; AAAS. *Res:* Banach algebras, particularly convolution algebras in harmonic analysis. *Mailing Add:* Dept Math & Comput Sci Dartmouth Col Hanover NH 03755

LAHR, GILBERT M, b Detroit, Mich, Sept 18, 22. METALLURGICAL ENGINEERING. *Educ:* Gen Motors Inst, BSIndustE, 46. *Prof Exp:* Metallurgist, Detroit Diesel Eng Div, Gen Motors, 47-62, asst chief metallurgist, 62-77, chief metallurgist in charge mat eng, Mat Qual Control, Failure Anal & Metall Processing Diesel Engines, 77-85; RETIRED. *Mem:* Fel Am Soc Metals. *Res:* High strength cold-worked steel; development of materials and processes for manufacturing of cylinder liners; holder of two patents. *Mailing Add:* 45152 Byrne Ct Northville MI 48167

LAHTI, LESLIE ERWIN, b Floodwood, Minn, July 27, 32; m 56; c 3. CHEMICAL ENGINEERING. *Educ:* Tri State Col, BS, 54; Mich State Univ, MS, 58; Carnegie Inst Technol, PhD(chem eng), 64. *Prof Exp:* Glass technologist, Corning Glass Works, 55-57; develop engr, Ren Plastics, 57; assoc prof, Tri State Col, 58-60; asst prof chem eng, Purdue Univ, 63-67; from assoc prof to prof chem eng & chmn dept, 67-80, DEAN ENG, UNIV TOLEDO, 80- *Concurrent Pos:* Consult, Am Oil, Great Lakes Chem, 67-69; Inland Chem Co, 70-; Stubbs, Overbeck, 79-81. *Honors & Awards:* Shreve Prize, 67. *Mem:* Am Inst Chem Engrs; Am Chem Soc; Am Soc Eng Educ; Nat Soc Prof Engrs. *Res:* Fundamentals of nucleation and crystallization from solutions; polymerization processes. *Mailing Add:* Dept Chem Eng Univ Toledo Toledo OH 43606

LAHTI, ROBERT A, b Kingsford, Mich; m 85; c 3. CENTRAL NERVOUS SYSTEM DRUG RESEARCH, SCHIZOPHRENIA RESEARCH. *Educ:* Western Mich Univ, BS, 60; Univ NC, PhD(biochem), 68. *Prof Exp:* Teacher chem, East Grand Rapids, Mich, 60-61 & Kalamazoo, Mich, 61-62; chemist, 62-64, SCIENTIST, UPJOHN CO, 67- *Concurrent Pos:* Vis scientist, Univ Uppsala, Uppsala, Sweden, 78-79; guest scientist, Lab Cell Biol, NIMH, Bethesda, Md, 88-89. *Mem:* Soc Neurosci; Fedn Am Soc Exp Biol; Am Soc Pharmacol & Exp Therapeut. *Res:* Central nervous system drug discovery development and mechanism of action in schizophrenia, anxiety, depression and analgesia using pharmacological, biochemical, endocrine and receptor binding techniques. *Mailing Add:* Cent Nervous Syst Res Upjohn Labs Kalamazoo MI 49001

LAI, CHII-MING, b Taiwan, Repub China, May 25, 35; m; c 2. METABOLISM. *Educ:* Kaohsiung Med Col, BS, 65; Univ Ga, MS, 71; State Univ NY, Buffalo, PhD(pharmaceut), 77. *Prof Exp:* Pharmacist, Develop Dept, Taiwan Tanabe Pharmaceut Co, 66-69; teaching asst pharm, Sch Pharm, Univ Ga, 69-71; res asst pharmaceut, Sch Pharm, State Univ NY, Buffalo, 71-77; res investr, Am Critical Care, McGaw Park, Ill, 77-78, sr res investr, 78-81, res fel, 81-86; res fel, Du Pont Critical Care, Newark, Del, 86-89; sr res assoc, Du Pont Pharmaceut, 89-90, SR RES ASSOC, STINE-HASKELL RES CTR, DU PONT MERCK PHARMACEUT CO, 91- *Mem:* Am Pharmaceut Asn; Am Asn Pharmaceut Scientists; Am Soc Pharmacol & Exp Therapeut; NY Acad Sci. *Res:* Biopharmaceutics, pharmacokinetics, drug metabolism and detoxication; animal screening and screening techniques; linear and nonlinear model fitting and stimulation; statistical methods; bioanalytical methods development; isotope tracer techniques; control theory and biological feedback mechanisms; physical pharmacy and bio-organic chemistry. *Mailing Add:* Dept Metabolics & Pharmacokinetics Du Pont Pharmaceut Elkton Rd PO Box 30 Newark DE 19714

LAI, CHING-SAN, b Taiwan, Nov 27, 46; m 71; c 2. CELL BIOPHYSICS, MEMBRANE BIOPHYSICS. *Educ:* Nat Taiwan Norm Univ, BS, 70; Univ Hawaii, PhD(biophysics), 78. *Prof Exp:* Fel biophysics, Univ Hawaii, 78; res assoc, 79-80, asst prof biophysics, 81-84, ASSOC PROF, MED COL WIS, 85- *Concurrent Pos:* NIH grantee, 82- *Mem:* Biophys Soc; AAAS. *Res:* Molecular dynamics of cell adhesive glycoproteins; development and applications of electron spin resonance spectroscopy to biomedical systems. *Mailing Add:* Nat Biomed Electron Spin Resonance Ctr Med Col Wis 8701 Watertown Plank Rd Milwaukee WI 53226

LAI, CHINTU (VINCENT C), b Changhua, Formosa, Aug 5, 30; m 63; c 2. COMPUTATIONAL HYDRAULICS, HYDROMECHANICS. *Educ:* Taiwan Univ, BS, 54; Univ Iowa, MS, 57; Univ Mich, PhD(civil eng), 62. *Prof Exp:* Res hydraul engr, Washington, DC, 61-63, Ore, 63-65 & Arlington, Va, 65-73, RES HYDROLOGIST, WATER RESOURCES DIV, US GEOL SURV, RESTON, 73- *Mem:* Am Soc Civil Engrs; Asn Comput Mach; Int Asn Hydraul Res; Am Geophys Union; Sigma Xi. *Res:* Computational hydraulics-surface water problems; transient flows in closed and open conduits; numerical modelling and computer simulation of unsteady flows in rivers, estuaries, embayments, closed conduits and other areas in hydromechanics and hydrologic process. *Mailing Add:* 6814 Glenmont St Falls Church VA 22042

LAI, CHUN-YEN, ENZYMOLOGY, BACTERIAL TOXIN. *Educ:* Univ Ill, PhD(biochem), 61. *Prof Exp:* RES INVESTR, ROCHE RES CTR, 73-; ADJ PROF BIOCHEM, MED COL, CORNELL UNIV, 79-; RES CHIEF, HOFFMANN-LA ROCHE, INC. *Mailing Add:* H M Jackson Found Res Labs 1500 E Gude Dr Rockville MD 20850

LAI, DAVID CHIN, b Beijing, China, Nov 11, 31; US citizen; m 63; c 2. ELECTRICAL ENGINEERING. *Educ:* Nat Taiwan Univ, BSEE, 54; Johns Hopkins Univ, DEng, 60. *Prof Exp:* Asst prof eng, Brown Univ, 60-62; assoc prof elec eng, Northeastern Univ, 62-65; assoc prof, 65-71, prof elec eng, Univ Vt, 71-85. *Concurrent Pos:* Vis prof elec eng, Stanford Univ, 71-75; sr staff mem, Martin Marietta Orlando Aerospace, 85- *Mem:* Inst Elec & Electronics Engrs. *Res:* Signal processing; radar signals; pattern recognition; automatic target recognition; multi-sensor fusion. *Mailing Add:* 175 Spring Chase Circle Altamonte Springs FL 32714

LAI, DAVID YING-LUN, b Canton, China, Aug 1, 47; US citizen; m 84; c 2. STRUCTURAL-ACTIVITY RELATIONSHIPS, RISK ASSESSMENT. *Educ:* Chinese Univ, Hong Kong, BSc, 70; Med Col, Ga, PhD(biochem), 75. *Prof Exp:* Instr biol & biochem, Dept Biol, Chinese Univ, Hong Kong, 70-72; instr, Dept Med, Tulane Univ Med Ctr, 77-79; sr toxicologist consult, Sci Applications Int Corp, 79-87; TOXICOLOGIST, US ENVIRON PROTECTION AGENCY, 87- *Mem:* Soc Toxicol; Am Col Toxicol; Am Asn Cancer Res; Am Soc Pharmacol & Exp Therapeut; Soc Risk Anal; Europ Asn Cancer Res. *Res:* Development and evaluation of hazard and risk assessment of toxic substances. *Mailing Add:* US Environ Protection Agency TS-796 401 M St SW Washington DC 20460

LAI, ELAINE Y, b British Hong Kong, Nov 11, 49; m 81. CELL BIOLOGY. *Educ:* Iowa State Univ, BS, 73; Brandeis Univ, PhD(biol), 78. *Prof Exp:* SR RES ASSOC BIOL, BRANDEIS UNIV, 82- *Concurrent Pos:* Vis lectr cell-free translation, Univ NC, Chapel Hill, 79; organizer EMBO course, Max Planck Inst, WGer, 86; Max-Planck fel, Munich, 86. *Honors & Awards:* Estherlee Runoto Gilbert Merit Award, 77. *Mem:* Am Soc Cell Biol. *Res:* Regulation of eukaryotic gene expression during cell differentiation; emphasis on dissection of the coordinate expression of flagellar calmodulin and tubulin genes in the amebo-flagellate, Naegleria gruberi. *Mailing Add:* Dept Biol Brandeis Univ Waltham MA 02254-9110

LAI, FONG M, b Taiwan, Aug 17, 42; m 69; c 2. CARDIOVASCULAR PHARMACOLOGY. *Educ:* Taipei Med Col, BS, 66; Taiwan Univ, MS, 69; Med Col Va, PhD(pharmacol), 74. *Prof Exp:* Res fel, Roche Inst Molecular Biol, 74-76; GROUP LEADER & PRIN PHARMACOLOGIST, LEDERLE LABS, AM CYANAMID CO, 76- *Mem:* Am Soc Pharmacol & Exp Therapeut. *Res:* Mechanisms of the hypertension and the cerebral vasculative pharmacology. *Mailing Add:* Lederle Labs Am Cyanamid Co Bldg 56D Rm 263 Pearl River NY 10965

LAI, JAI-LUE, b Taipei, Taiwan, Dec 9, 40; US citizen; m 68; c 2. ACOUSTICS, STRUCTURAL DYNAMICS. *Educ:* Nat Taiwan Univ, BS, 62; Polytech Inst Brooklyn, MSE, 66; Princeton Univ, PhD(mech eng), 69. *Prof Exp:* Engr satellite struct, RCA Corp, 67; assoc res & develop fel, B F Goodrich Co, 68-89; DIR, GENCORP AUTOMOTIVE, 87- *Mem:* Am Inst Aeronaut & Astronaut; Am Soc Mech Engrs; Soc Advan Mat Process Eng; Soc Petrol Engrs; Soc Automotive Engrs. *Res:* Application of new material composite structure as new products or components. *Mailing Add:* 4421 Conestoga Terr Akron OH 44321

LAI, JUEY HONG, b Taipei, Taiwan, Dec 4, 36; US citizen; m 68; c 2. PHYSICAL CHEMISTRY, POLYMER CHEMISTRY. *Educ:* Nat Taiwan Univ, BS, 59; Univ Wash, MS, 63, PhD(phys chem), 69. *Prof Exp:* Res specialist polymer, Univ Minn, 69-73; from prin res scientist to sr prin res scientist polymer mat, 73-83, STAFF SCIENTIST POLYMER MAT, CHEM SENSORS, HONEYWELL SENSORS & SIGNAL PROCESSING LABS, 83- *Honors & Awards:* H W Sweatt Award, Honeywell Inc, 80. *Mem:* Am Chem Soc; Sigma Xi; Am Inst Chemists. *Res:* Polymer materials for electronics; electron resists for electron beam microfabrication, membrane technology for gas removal, solid state chemistry and chemical sensors. *Mailing Add:* 3025 Carlsbad Ct Burnsville MN 55337

LAI, KAI SUN, b Hong Kong, China; US citizen. ENGINEERING. *Educ:* Pa State Univ, BSc, 59. *Prof Exp:* Engr, Aerojet Gen Corp, 59-62; sr analyst, Atlantic Res Corp, 62-68; mgr, 68-80, sr scientist, 80-82, ENG SPECIALIST, TELEDYNE MCCORMICK SELPH, 82- *Res:* Combustion process and thermochemical analysis of solid fuels and additives, including boranes; propulsion for aerospace applications and use of explosives and pyrotechnics for safety applications. *Mailing Add:* 855 W Eighth St Gilroy CA 95020

LAI, KUO-YANN, b Miao-Li, Taiwan, Sept 13, 46; m 72; c 3. PHYSICAL CHEMISTRY, SURFACE & COLLOID SCIENCE. *Educ:* Cheng Kung Univ, Taiwan, BS, 69; Univ Tex, El Paso, MS, 74; Clarkson Col Technol, PhD(chem), 77. *Prof Exp:* Res chemist, Colgate-Palmolive Co, 77-80, sr res chemist, 80-83, res assoc, 83, sect head, 83-86, sr sect head chem res, 86-87, MGR, ORAL PROD DEVELOP, COLGATE-PALMOLIVE CO, 87- *Concurrent Pos:* Robert A Welch fel, 72-74; NSF fel, 74-77. *Honors & Awards:* Pres Award for Tech Excellence, Colgate-Palmolive, 85. *Mem:* Am Chem Soc; Am Oil Chemists Soc. *Res:* Adhesional wetting; scavenging of aerosols; surfactants and detergents; oral hygiene prods. *Mailing Add:* Technol Ctr Colgate-Palmolive Co 909 River Rd Piscataway NJ 08855-1343

LAI, MICHAEL MING-CHIAO, b Tainan, Taiwan, Sept 8, 42; m 71; c 2. VIROLOGY, MOLECULAR BIOLOGY. *Educ:* Nat Taiwan Univ Col Med, MD, 68; Univ Calif, Berkeley, PhD(molecular biol), 73. *Prof Exp:* Med officer, Chinese Marine Corps, 68-69; postgrad molecular biologist, Univ Calif, Berkeley, 73; from asst prof to assoc prof microbiol, 73-83, PROF MICROBIOL & NEUROL, SCH MED, UNIV SOUTHERN CALIF, 83-; INVESTR, HOWARD HUGHES MED INST, 90- *Concurrent Pos:* Prin investr grants, Nat Cancer Inst & Am Cancer Soc, 73-85, NIH, 75-, Nat Sci Found, 79-87 & Nat Multiple Sclerosis Soc, 82- *Mem:* Am Soc Microbiol; Am Soc Virol; Fedn Am Soc Exp Biol; AAAS. *Res:* Molecular biology of hepatitis viruses and coronaviruses; mechanism of viral pathogenesis. *Mailing Add:* Dept Microbiol Sch Med Univ Southern Calif 2011 Zonal Ave Los Angeles CA 90033

LAI, PATRICK KINGLUN, b Hong Kong, Oct 10, 44; Australian citizen; c 2. IMMUNOVIROLOGY, IMMUNOPATHOLOGY. *Educ:* Univ Western Australia, PhD(microbiol), 78. *Prof Exp:* Res fel biol, Univ Ottawa, Can, 78-79 & immunol, Univ Col London, UK, 79-82; sr res officer immunol, Royal Postgrad Med Sch, UK, 82-84; from instr to asst prof immunol, Univ Nebr Med Ctr, 84-87; asst mem, 87-90, MEM VIROL, TAMPA BAY RES INST, 90- *Concurrent Pos:* Vis scholar, Int Agency Res Cancer, France, 75; WHO fel, Rush-Presbyterian St Luke Med Ctr, Chicago, 76-77; res fel, Imp Cancer Res Funds, UK, 79-82; Europ Molecular Biol Orgn fel, Univ Zurich, 81; vis prof, Alta Heritage Fund, Univ Alta, 82. *Mem:* Am Asn Immunologists; AAAS; NY Acad Sci. *Res:* Interaction between viruses and cellular components that give diseases; how cellular factors, including immunomodulators, regulate virus replication and pathogenesis. *Mailing Add:* Tampa Bay Res Inst 10900 Roosevelt Blvd St Petersburg FL 33716

LAI, POR-HSIUNG, PROTEIN. *Prof Exp:* FOUNDER, PROTEIN INST INC, 90- *Mailing Add:* Protein Inst Inc PO Box 550 Broomall PA 19008-0550

LAI, RALPH WEI-MEEN, b Tou-Lu, Taiwan, Dec 17, 36; US citizen; m 66; c 2. SURFACE CHEMISTRY, MINERAL SCIENCE & ENGINEERING. *Educ:* Cheng Kung Univ, Taiwan, BS, 59; SDak Sch Mines & Technol, MS, 64; Univ Calif, Berkeley, PhD(mat sci & eng), 70. *Prof Exp:* Res scientist mat res, Cyprus Mines Corp, 69-72; mineral processing scientist process develop, Anglo-Am Clays Corp, 73-74; sr proj engr metall eng, Kennecott Develop Ctr, Kennecott Copper Corp, 74-85; SCIENTIST, US DEPT ENERGY, 85- *Concurrent Pos:* Pres, Western Prospect Co, 78- *Mem:* Am Inst Mining, Metall & Petrol Engrs; Japan Inst Mining & Metall; Clay Minerals Soc. *Res:* Surface chemistry of oxide minerals; coal preparation and utilization. *Mailing Add:* 2305 Hidden Timber Dr Pittsburgh PA 15241

LAI, SAN-CHENG, b Taiwan, China, Dec 8, 40; m 68. CHEMICAL ENGINEERING. *Educ:* Taiwan Univ, BS, 63; Univ Mo-Rolla, MS, 66, PhD(chem eng), 68. *Prof Exp:* Sr res engr, Am Potash & Chem Corp, 68-77; MEM TECH STAFF, ATOMICS INT, 77- *Mem:* Am Inst Chem Eng; Electrochem Soc; Am Chem Soc. *Res:* Electrodeposition of manganese dioxide; alkaline battery and magnesium can battery development; sodium chlorate and perchlorate process development; manganese metal process improvement; fluidized bed cell development. *Mailing Add:* 1478 Kingston Circle Westlake Village CA 91362

LAI, SHU TIM, b Hong Kong, May 23, 38; US citizen; m 72. SPACECRAFT INTERACTIONS, SPACE PHYSICS. *Educ:* Brandeis Univ, MA, 67, PhD(physics), 71. *Prof Exp:* Mem res staff, Lincoln Lab, Mass Inst Technol, 78-79; sr mem res staff, Boston Col, 79-80; RES PHYSICIST, USAF GEOPHYS LAB, 81- *Mem:* Am Geophys Union; Am Phys Soc; Am Asn Physics Teachers; Inst Elec & Electronics Engrs; Am Inst Aeronaut & Astronaut. *Res:* Space plasma physics, spacecraft charging; electron, ion and neutral beams emitted from spacecrafts; atmospheric physics; spacecraft interactions with space environment; digital signal processing. *Mailing Add:* PO Box 273 Burlington MA 01803

LAI, TZE LEUNG, b Hong Kong, June 28, 45; m 75; c 2. MATHEMATICS, STATISTICS. *Educ:* Univ Hong Kong, BA, 67; Columbia Univ, MA, 70, PhD(statist), 71. *Prof Exp:* From asst prof to prof statist, Columbia Univ, 71-87; PROF STATIST, STANFORD UNIV, 87- *Concurrent Pos:* Vis assoc prof math, Univ Ill, Urbana-Champaign, 75-76; vis prof statist, Stanford Univ, 78-79; John Simon Guggenheim fel, 83-84; vis prof, Math Sci Res Inst, Berkeley, 83. *Honors & Awards:* Comt of Presidents Statist Soc Award, 83. *Mem:* Fel Am Statist Asn; fel Inst Math Statist; Sigma Xi; AAAS; NY Acad Sci; Int Statist Inst; Biometric Soc; Drug Info Asn. *Res:* Sequential methods in statistics; statistical quality control and clinical trials; time series analysis; limit theorems in probability; renewal theory and random walks; martingales and potential theory; system identification and control; cardiorespiratory physiology; medical informatics. *Mailing Add:* Dept Statist Stanford Univ Stanford CA 94305

LAI, W(EI) MICHAEL, b Amoy, China, Nov 29, 31; US citizen; m 63; c 2. MECHANICAL ENGINEERING. *Educ:* Nat Taiwan Univ, BS, 53; Univ Mich, Ann Arbor, MS, 59, PhD(eng mech), 62. *Prof Exp:* From asst prof to prof mech, Rensselaer Polytech Inst, 61-89; PROF MECH ENG & ORTHOP BIOENG, COLUMBIA UNIV, 87- *Honors & Awards:* Melville Medalist, Am Soc Mech Engrs. *Mem:* Am Math Soc; fel Am Soc Mech Engrs; Am Soc Biomech; AAAS; Ortho Res Soc. *Res:* Hydrodynamic stability; continuum mechanics; biomechanics. *Mailing Add:* 14 Bedford Dr Latham NY 12110

LAI, YING-SAN, b Taiwan, China, Sept 9, 37; US citizen; m 66; c 3. VALVE DESIGN & DEVELOPMENT, VALVE APPLICATIONS. *Educ:* Nat Taiwan Univ, BS, 60; Univ Iowa, MS, 63; Northwestern Univ, PhD(mech eng), 73. *Prof Exp:* Design engr, CBI Industs, 63-69, stress analyst, 72-73; chief engr, Valve Div, Dresser Industs, 73-83, eng dir, Dresser Dewrance Ltd, UK, 83-84, dir eng, Valve Div, 84-90, VPRES ENG, DRESSER INDUSTS, 90- *Concurrent Pos:* Mem, Indust Technol Adv Coun, Northwestern State Univ La; mem, Am Petrol Inst. *Mem:* Am Soc Mech Engrs; Nat Mgt Asn. *Res:* Pressure relief valves and line valves for industrial applications. *Mailing Add:* 4806 Westgarden Blvd Alexandria LA 71303

LAI, YUAN-ZONG, b Taiwan, Repub of China, Mar 11, 41; m 68; c 2. WOOD CHEMISTRY. *Educ:* Nat Taiwan Univ, BS, 63; Univ Wash, MS, 66 & 67, PhD(wood chem), 68. *Prof Exp:* From res asst to res assoc wood chem, Col Forest Resources, Univ Wash, 64-70; sr res assoc wood chem, Univ Mont, 70-75; asst prof wood chem, dept forestry, Mich Technol Univ, 75-77, ASSOC PROF FORESTRY, 77-; AT EMPIRE STATE PAPER RES INST, STATE UNIV NY, SYRACUSE. *Mem:* Tech Asn Pulp & Paper Indust; Am Chem Soc; Sigma Xi. *Res:* Lignin, cellulose and extractive chemistry; thermal properties of wood components. *Mailing Add:* Empire State Paper Res Inst State Univ NY Syracuse NY 13210

LAI, YU-CHIN, b Feb 2, 49; m 79; c 2. POLYMER CHEMISTRY. *Educ:* Nat Tsing Hua Univ, Taiwan, BS, 71; Carnegie-Mellon Univ, MS, 75; Univ Fla, PhD(chem), 80. *Prof Exp:* Res assoc, Univ Mass, 80-81; res chemist polymer chem, Corp Res & Develop, Allied corp, 81-86; sr polymer chemist, 86-90, SR SCIENTIST, BAUSCH & LOMB, 90- *Mem:* Am Chem Soc. *Res:* Synthesis of organic compounds: monomers and polymers; kinetics and mechanism of polymerization; structure-properties relationships in polymers. *Mailing Add:* Contact Lens Div Res & Develop Bausch & Lomb Inc 1400 N Goodman St Rochester NY 14692-0450

LAIBLE, JON MORSE, b Bloomington, Ill, July 25, 37; m 59; c 4. ALGEBRA. *Educ:* Univ Ill, Urbana, BS, 59, PhD(math), 67; Univ Minn, Minneapolis, MA, 61. *Prof Exp:* Asst prof math, Western Ill Univ, 61-64; from asst prof to assoc prof, 64-79, PROF MATH, EASTERN ILL UNIV, 79-, DEAN, COL LIB ARTS & SCI, 81- *Mem:* Math Asn Am; Sigma Xi. *Mailing Add:* Dean Col Lib Arts & Sci Eastern Ill Univ Charleston IL 61920

LAIBLE, ROY C, b Boston, Mass, June 16, 24. POLYMER PHYSICS. *Educ:* Northeastern Univ, BS, 45; Boston Univ, MA, 48; Mass Inst Technol, PhD, 70. *Prof Exp:* Res assoc polymerization, Univ RI, 50-52; org chemist, Cent Intel Agency, 52-53; org chemist, US Army Natick Labs, 53-58, phys sci adminr, 58-63, physics scientist, 63-70, chief, Textile Res Sect, 70-76, chief, Polymers & Org Mat Br, 76-; RETIRED. *Concurrent Pos:* Secy of Army res & study fel viscoelastic properties polymers, Sweden & Scotland, 62-63. *Mem:* Am Chem Soc. *Res:* Allyl polymerization; viscoelastic properties of fibrous and non-fibrous polymers; ballistic properties of polymers. *Mailing Add:* 101 Overbrook Dr Wellesley MA 02181

LAIBOWITZ, ROBERT (BENJAMIN), b Yonkers, NY, Mar 24, 37; m 58; c 3. APPLIED PHYSICS. *Educ:* Columbia Col, BA, 59; Columbia Univ, BS, 60, MS, 63; Cornell Univ, PhD(appl physics), 67. *Prof Exp:* RES STAFF MEM, IBM RES CTR, 66- *Mem:* Am Phys Soc; fel Am Vacuum Soc. *Res:* Electrical and optical properties of materials; superconductivity. *Mailing Add:* T J Watson Res Ctr PO Box 218 Yorktown Heights NY 10598

LAIBSON, PETER R, b New York, NY, Dec 11, 33; m 63; c 1. OPHTHALMOLOGY. *Educ:* Univ Vt, BA, 55; State Univ NY Downstate Med Ctr, MD, 59; Am Bd Ophthal, dipl, 65. *Prof Exp:* NIH fel corneal dis, Retina Found & Mass Eye & Ear Infirmary, 64-65; ASSOC PROF OPHTHAL, SCH MED, TEMPLE UNIV, 66-; PROF OPHTHAL, SCH MED, THOMAS JEFFERSON UNIV, 73- *Concurrent Pos:* Attend surgeon & dir cornea serv, Wills Eye Hosp; consult lectr, US Naval Hosp, Philadelphia, 68-; mem ophthal staff, Lankenau Hosp, Philadelphia. *Mem:* Asn Res Ophthal; Am Acad Ophthal & Otolaryngol; AMA; Am Ophthal Soc. *Res:* Corneal diseases and surgery of the cornea, particularly viral external diseases, herpes simplex and adenoviruses. *Mailing Add:* Ninth & Walnut Philadelphia PA 19107

LAIDLAW, HARRY HYDE, JR, b Houston, Tex, Apr 12, 07; m 46; c 1. APICULTURE. *Educ:* La State Univ, BS, 33, MS, 34; Univ Wis, PhD(entom, genetics), 39. *Prof Exp:* From minor sci helper to agent, USDA, 29-34 & 35-39; asst zool & entom, La State Univ, 33-34, asst exp sta, 34-35; prof biol sci, Oakland City Col, 39-41; apiarist, State Dept Agr & Indust, Ala, 41-42; entomologist hqs, 1st Army, New York, 46-47; asst prof entom & asst apiculturist, 47-53, assoc prof entom & assoc apiculturist, 53-59, prof entom & apiculturist, 59-74, prof genetics, 71-74, assoc dean col agr, 60-64, EMER PROF ENTOM, EXP STA, UNIV CALIF, DAVIS, 74- *Concurrent Pos:* Wis Alumni Res Found asst, Univ Wis, 37-39; NIH grant, Univ Calif, Davis, 63-66, NSF grant, 66-73. *Honors & Awards:* C W Woodworth Award, Entom Soc Am, 81; Gold Merit Award, Int Fedn Beekeepers Asn, 86. *Mem:* Fel AAAS; Genetics Soc Am; Am Soc Zool; Entom Soc Am; Am Soc Nat; Int Bee Res Asn. *Res:* Genetics, breeding and anatomy of the honeybee; factors influencing the development of queen bees; artificial insemination of queen bees. *Mailing Add:* 761 Sycamore Lane Davis CA 95616

LAIDLAW, JOHN COLEMAN, b Toronto, Ont, Feb 28, 21; m 57; c 2. ENDOCRINOLOGY. *Educ:* Univ Toronto, BA, 42, MD, 44, MA, 47; Univ London, PhD(biochem), 50; FRCP(C), 55. *Prof Exp:* Jr intern, Toronto Gen Hosp, 44; demonstr biochem, Univ Toronto, 46-47; lectr biochem, Univ London, 47-50; sr intern med, Toronto Gen Hosp, 50-51; res fel, Harvard Med Sch, 51-53, instr, 53-54; assoc, 54-56, from asst prof to prof med, Univ Toronto, 56-75, dir inst med sci, 67-75; prof med & chmn dept, McMaster Univ, 75-81, DEAN, FAC HEALTH SCI, 81-; AT MED RES COUN, OTTAWA, ONT. *Concurrent Pos:* Asst, Peter Bent Brigham Hosp, Boston, 51-53, jr assoc, 53-54; physician, Toronto Gen Hosp, 54-59, sr physician, 59-75. *Mem:* Endocrine Soc; Am Soc Clin Invest; Res; Can Soc Clin Invest (pres, 62); Can Soc Endocrinol & Metab. *Mailing Add:* Dept Med McMaster Univ Fac Health Sci 1200 Main Street W Hamilton ON L8N 3Z5 Can

LAIDLAW, WILLIAM GEORGE, b Wingham, Ont, Mar 13, 36; m 61; c 2. THEORETICAL CHEMISTRY. *Educ:* Univ Western Ont, BSc, 59; Calif Inst Technol, MSc, 61; Univ Alta, PhD(theoret chem), 63. *Prof Exp:* NATO fel, Math Inst, Univ Oxford, 64-65; from asst prof to assoc prof, 65-73, PROF CHEM, UNIV CALGARY, 73- *Concurrent Pos:* Vis prof, Univ Waterloo, Univ Brussels, Univ Hawaii & Univ Strasbourg. *Mem:* Chem Inst Can. *Res:* Hydrodynamics; flow of fluids in porous media; fluid systems near instabilities; molecular orbital calculations. *Mailing Add:* Dept Chem Univ Calgary Calgary AB T2N 1N4 Can

LAIDLER, KEITH JAMES, b Liverpool, Eng, Jan 3, 16; m 43; c 3. PHYSICAL CHEMISTRY. *Educ:* Oxford Univ, BA, 37, MA, 55, DSc, 56; Princeton Univ, PhD(phys chem), 40. *Prof Exp:* Res chemist, Nat Res Coun Can, 40-42; sci officer, Can Armaments Res & Develop Estab, 42-44, chief sci officer & supt phys & math wing, 44-46; from asst prof to assoc prof chem, Cath Univ Am, 46-55; chmn dept, 61-66, vdean fac pure & appl sci, 62-66, prof, 55-81, EMER PROF CHEM, UNIV OTTAWA, 81- *Concurrent Pos:* Commonwealth vis prof, Sussex Univ, 66-67. *Honors & Awards:* Medal, Chem Inst Can, 71; Queen's Jubilee Medal, 77; Centenary Medal, Royal Soc Can, 82, Henry Marshall Tory Medal, 87. *Mem:* Royal Soc Chem; fel Royal Soc Can; fel Chem Inst Can. *Res:* Chemical kinetics of gas reactions; surface, solution and enzyme reactions; photochemistry; history of physical chemistry. *Mailing Add:* Dept Chem Univ Ottawa Ottawa ON K1N 9B4 Can

LAI-FOOK, JOAN ELSA I-LING, b Port of Spain, Trinidad, Aug 3, 37. ZOOLOGY. *Educ:* Univ Col WI, BSc, 61; Western Reserve Univ, PhD(biol), 66. *Prof Exp:* Asst prof, 66-73, ASSOC PROF ZOOL, UNIV TORONTO, 73- *Mem:* Am Soc Cell Biol. *Res:* Fine structure of insect development and physiology. *Mailing Add:* Dept Zool Univ Toronto Toronto ON M5S 1A1 Can

LAI-FOOK, STEPHEN J, b Trinidad & Tobago, Aug 28, 40; US citizen; m; c 2. BIOMEDICAL ENGINEERING. *Educ:* Loughborough Univ, Eng, BTech, 64; Southhampton Univ, Eng, MScEng, 66; Univ Wash, Seattle, PhD(mech eng), 72. *Prof Exp:* Res engr, Boeing Co, 66-69; fel, Dept Aerospace Eng & Mech, Univ Minn, 73-74; instr biophysics, Mayo Med Sch,

75-78, assoc prof physiol & med, 78-81, assoc prof physiol & biophysics, 81; assoc mem, Cardiovasc Res Inst, 84-87; PROF BIOMED ENG, WENNER GREN RES LAB, UNIV KY, 87-, PROF PHYSIOL & BIOPHYS, 91- Concurrent Pos: NIH young invest award, 76, res career develop award, 80; assoc consult thoracic dis, Mayo Clin, 77-80, consult, 80-81; adj asst prof physiol & med, Univ Calif, San Francisco, 81, instr orgn physiol, Pulmonary Physiol Lab, 82-85, lectr, Sch Pediat, 83, 84 & 86, adj assoc prof physiol, 84-87; mem, Respiratory & Appl Study Sect, NIH, 90- Mem: Am Physiol Soc; Am Soc Mech Engrs; Biomed Eng Soc; Microcirculation Soc; Am Thoracis Soc; Am Heart Asn; Am Acad Mech. Res: Pulmonary mechanics; mechanical properties of the lung mechanics of lung interstitium and the pleural space in relation to liquid and solute exchange; mathematical modeling of physiological systems. Mailing Add: Ctr Biomed Eng Univ Ky Wenner Gren Res Lab Lexington KY 40506

LAIKEN, NORA DAWN, b Chicago, Ill, June 28, 46; m 67; c 2. MEDICAL PHYSIOLOGY, MEDICAL EDUCATION. Educ: Univ Chicago, BS, 67; Rockefeller Univ, PhD(life sci), 70. Prof Exp: USPHS fel & res assoc phys biochem, Inst Molecular Biol, Univ Ore, 70-71; USPHS fel & res assoc phys biochem, 71-72, sci curric adv physics, biol & chem, Adaptive Learning Prog, 72-73, lectr med, 74-76, asst prof, 76-80, asst adj prof med, 80-83, DIR TUTORIAL PROG, UNIV CALIF, SAN DIEGO, 73-, LECTR MED, 83-, ASST DEAN CURRIC & STUDENT AFFAIRS, 83- Concurrent Pos: Mem test adv Asn Am Med Col, 74-78; mem, Undergrad Teaching Proj Comt, Am Gastroenterol Asn, 85- Res: Development of innovative instructional materials and methods in the basic medical sciences, particularly in medical physiology and pharmacology; applications of computers to biomedical problems. Mailing Add: 9500 Gilman Dr 0606 Univ Calif San Diego La Jolla CA 92093-0606

LAINE, RICHARD MASON, b San Fernando, Calif, Oct 31, 47. ORGANOMETALLIC CHEMISTRY. Educ: Calif State Univ, Northridge, BS, 69; Univ Southern Calif, PhD(chem), 73. Prof Exp: Fel chem, Univ Del, 73-74 & Dept Chem & Dept Chem & Nuclear Eng, Univ Calif, Santa Barbara, 74-76; fel chem, Stanford Res Inst, 76-77, PHYS INORG CHEMIST, SRI INT, 77- Concurrent Pos: Prin investr, NSF Chem Eng Grant, 78-79; proj leader, NIH Grant, 78-81. Mem: Am Chem Soc; Catalysis Soc; Sigma Xi. Res: Homogeneous catalysis of the water-gas shift reaction and the catalysis of related reactions wherein water serves as a source of hydrogen. Mailing Add: SRI Int 333 Ravenswood Ave Menlo Park CA 94025

LAINE, ROGER ALLAN, b Cloquet, Minn, Jan 28, 41; c 2. BIOCHEMISTRY. Educ: Univ Minn, Minneapolis, BA, 64; Rice Univ, Houston, PhD(biochem), 70. Prof Exp: Fel biochem, Mich State Univ, East Lansing, 70-72; fel pathobiol, Univ Wash, 72-74; from asst prof to assoc prof biochem, Col Med, Univ Ky, 75-88; chmn, Dept Biochem, La State Univ, Baton Rouge, 88-90; CHIEF SCIENTIST, GLYCOMED, INC, 88-, PROF, DEPT BIOCHEM & CHEM. Mem: Am Chem Soc; Soc Complex Carbohydrates; Am Soc Mass Spectrometry; Am Soc Biol Chemists. Res: Biochemistry of cell membrane components; gas-liquid-chromatography and mass spectrometry in carbohydrate analysis. Mailing Add: Dept Biochem La State Univ Baton Rouge LA 70803

LAING, CHARLES CORBETT, b Brooklyn, NY, Dec 24, 25; m 59; c 3. ECOLOGY. Educ: Univ Chicago, PhB, 50, PhD, 54. Prof Exp: Instr bot, Univ Tenn, 54-56; asst prof, Univ Wyo, 56-59 & Univ Nebr, Lincoln, 59-66; mem fac biol, 66-68, ASSOC PROF BIOL, OHIO NORTHERN UNIV, 68- Mem: Ecol Soc Am; Brit Ecol Soc; Sigma Xi. Res: Ecology of sand dunes; population ecology of dune grasses; grasslands. Mailing Add: Dept Biol Ohio Northern Univ 500 S Main Ada OH 45810

LAING, FREDERICK M, plant physiology; deceased, see previous edition for last biography

LAING, JOHN E, b Ottawa, Ont, Oct 17, 39; m 64; c 2. ENTOMOLOGY, ECOLOGY. Educ: Carleton Univ, BSc, 64; Univ Calif, Berkeley, PhD(entom), 68. Prof Exp: Asst res entomologist & lectr, Div Biol Control, Univ Calif, Berkeley, 68-73; asst prof, 73-78, ASSOC PROF ENVIRON BIOL, UNIV GUELPH, 78- Mem: Entom Soc Can (secy, 78-81); Entom Soc Am; Ecol Soc Am; Int Asn Ecol; Int Orgn Biol Control; Sigma Xi. Res: Ecology of tetranychid mites; populations dynamics of arthropods; ecology and control of orchard pests; biological control of insect pests and weeds. Mailing Add: 28 Karen Dr Guelph ON N1G 2N9 Can

LAING, PATRICK GOWANS, b Barnes, Eng, Nov 8, 23; US citizen; m 56; c 4. ORTHOPEDIC SURGERY. Educ: Univ Southampton, MB & BS, 40; FRCS, 48; FRCS(C), 54, Am Bd Orthop Surg, dipl, 60. Prof Exp: House surgeon, Kings Col Hosp, London, Eng, 45-46; registr orthop surg, Royal Hampshire County Hosp, Winchester, 46-47, gen & orthop surg, Queen Mary's Hosp, Sidcup, 48, orthop surg, Lewisham Hosp, London, 48-50 & Pembury Hosp, Kent, 50-52; sr registr, Bradford Hosp, Yorkshire, 52-54; chief resident surg, Vet Hosp, St John, NB, 54-55; assoc prof orthop surg, 56-63, CLIN PROF ORTHOP SURG, UNIV PITTSBURGH, 63-; CHIEF SERV, VET ADMIN HOSP, 56- Concurrent Pos: Fel cerebral palsy, Univ Pittsburgh, 55-56. Mem: Orthop Res Soc; Am Orthop Asn; Am Soc Testing & Mat; NY Acad Sci; Brit Orthop Asn. Res: Blood supply and the dynamics of circulation in bones and joints; metallurgy and engineering in orthopedics; radioisotopes in clinical orthopedics. Mailing Add: Aiken Med Bldg 532 S Aiken Ave Suite 408 Pittsburgh PA 15232

LAING, RONALD ALBERT, b Seattle, Wash, Dec 9, 33. BIOPHYSICS. Educ: Reed Col, BA, 56; Rice Univ, MA, 58, PhD(low temperature physics), 60. Prof Exp: Asst prof physics, Tulane Univ, 60-68; sr scientist, Space Sci Inc, 68-70; vis scientist, Univ Tokyo, 69-70; ASSOC PROF OPHTHAL, MED SCH, BOSTON UNIV, 70- Concurrent Pos: NSF sci fac fel, Harvard Univ, 65-66; NIH fel, Mass Inst Technol, 66-67; vis lectr, Univ Mass, Boston, 67-68; consult, Space Sci Inc, 67-68. Mem: Biophys Soc; Asn Res in Vision & Ophthalmol; Optical Soc Am; AAAS; Sigma Xi. Res: Ophthalmic biophysics; bioengineering. Mailing Add: 1024 Massachusetts Ave Lexington MA 02173

LAIPIS, PHILIP JAMES, b Charleston, SC, Apr 20, 44; m 70; c 2. MOLECULAR BIOLOGY, GENETICS. Educ: Calif Inst Technol, BS, 66; Stanford Univ, PhD(genetics), 72. Prof Exp: Nat Cancer Inst fel, Princeton Univ, 72-74; from asst prof to assoc prof, 74-86, PROF BIOCHEM, UNIV FLA, 86- Concurrent Pos: Vis scholar biochem, Harvard Univ, 81-82; vis affil, Whitehead Inst, Mass Inst Technol, 87-88. Mem: AAAS; Am Soc Microbiol; Sigma Xi; Am Soc Biol Chem; Am Soc Cell Biol. Res: Gene organization and variation in mammalian mitochondrial DNA; mechanisms of maternal inheritance, mitochondrial amplification and embryonic distribution of mitochondria on mammals; gene organization and variation in mammalian carbonic anhydrase genes; site directed mutation of human carbonic anhydrase isozymes and expression in bacterial systems. Mailing Add: Box J-245 JHM Health Ctr Univ Fla Gainesville FL 32610

LAIR, ALAN VAN, b Anna, Tex, May 2, 48; m 78. MATHEMATICS. Educ: NTex State Univ, BA, 70; Tex Tech Univ, MS, 72, PhD(math), 76. Prof Exp: from asst prof to assoc prof math, Univ SDak, 76-82; asst prof, 82-87, ASSOC PROF MATH, AIR FORCE INST TECHNOL, 87- Mem: Am Math Soc; Soc Indust & Appl Math. Res: Parabolic and elliptic partial differential equations. Mailing Add: 116 Holmes Dr Fairborn OH 45324

LAIRD, CAMPBELL, b Ardrishaig, Scotland, June 17, 36; m 64; c 3. PHYSICAL METALLURGY. Educ: Cambridge Univ, BA, 59, MA, 63, PhD(metall), 63. Prof Exp: Res fel metall, Christ's Col, Cambridge Univ, 61-65; prin scientist, Sci Lab, Ford Motor Co, 63-68; prof metall, 68-80, PROF MAT SCI & ENG, UNIV PA, 80- Concurrent Pos: Battelle vis prof, Ohio State Univ, 68- Mem: Am Inst Mining, Metall & Petrol Engrs; Am Soc Testing & Mat; Electron Micros Soc Am; Royal Inst Gt Brit; Brit Inst Metals; fel Am Soc Metals. Res: Fracture of materials, especially by fatigue; super-conductivity; diffusional phase transformations; electron microscopy; cyclic stress-strain response of materials. Mailing Add: 951 Weadley Rd Radnor PA 19087

LAIRD, CHARLES DAVID, b Portland, Ore, May 12, 39; m 61; c 4. CELL BIOLOGY, HUMAN GENETICS. Educ: Univ Ore, BA, 61; Stanford Univ, PhD(genetics), 66. Prof Exp: NIH fel genetics, Univ Wash, 67-68; asst prof zool, Univ Tex, Austin, 68-71; assoc prof zool & adj assoc prof genetics, 71-75, PROF ZOOL & ADJ PROF GENETICS, UNIV WASH, 75-, RES AFFIL, CHILD DEVELOP & MENTAL RETARDATION CTR, 89- Concurrent Pos: Distinguished vis lectr, Univ Tex, Austin, 77; vis scholar, Cambridge Univ, 78-79; mem, Fred Hutchinson Cancer Res Ctr, 90- Honors & Awards: Wassenberg Mem lectr, San Diego State Univ, 90. Mem: AAAS; Am Soc Human Genetics; Genetic Soc Am. Res: Encoding the three dimensional structure of chromosomes; mechanisms of transcription control; chromosome structure and function; human genetics. Mailing Add: Dept Zool Univ Wash NJ 15 Seattle WA 98195

LAIRD, CHRISTOPHER ELI, b Anniston, Ala, Nov 29, 42; m 66; c 2. NUCLEAR PHYSICS. Educ: Univ Ala, BS, 63, MS, 63, PhD(physics), 70. Prof Exp: From asst prof to assoc prof, 67-75, PROF PHYSICS, EASTERN KY UNIV, 75- Concurrent Pos: Fac res mem, Vanderbilt Univ, 69 & Argonne Nat Lab, 77-78; vis res prof, Univ Ky, 79-81 & 86. Mem: Am Phys Soc; Sigma Xi; Sci Res Soc; Sigma Xi. Res: Theoretical and experimental nuclear physics with primary emphasis in beta decay; atomic effects during beta decay; proton induced nuclear reactions; experimental measurement of proton-induced reaction cross-sections analysis of this data using various nuclear models. Mailing Add: Dept Physics Moore 351 Eastern Ky Univ Richmond KY 40475

LAIRD, CLEVE WATROUS, b Montclair, NJ, Mar 29, 38; m 65; c 2. PHYSIOLOGY. Educ: Gettysburg Col, BA, 61; Univ Nebr, MS, 63; Rutgers Univ, PhD(endocrinol & reproductive physiol), 67. Prof Exp: Biol sect head, Hycel Inc, 69-72; res assoc, Biores Inc, 72-74; sr scientist, Block Eng, 74-77, sr scientist & group leader, Union Carbide, med prod, 77-80, dir, Lab Serv Clin, Chem & Physiol, Remote Imaging Syst, 81-86; PRES & CONSULT, DRIAL CONSULT, 86- Concurrent Pos: Cyto chem consult, Coulter Electronics; staff, NATO mammalian genetics course, 68, adj assoc prof, Baylor Col Med, 69-72, adj asst prof, Boston Univ, 72-77, instr, Harvard Med Sch, 74-80. Mem: Am Physiol Asn; Am Heart Asn; Am Asn Clinical Chem; Am Soc Vet Clin Pathologists; Soc Reproductive Physiol; Am Soc Anatomists. Res: Clinical chemistry. Mailing Add: 3216 Sheri Dr Simi Valley CA 93063

LAIRD, DONALD T(HOMAS), b Sykesville, Pa, Dec 12, 26; m 48; c 4. COMPUTER SCIENCE. Educ: Pa State Univ, BS, 46, PhD(physics), 55; Cornell Univ, MS, 51. Prof Exp: Asst physics, Cornell Univ, 46-49; res assoc, Ord Res Lab, 49-55, asst prof elec eng, 55-58, eng res, 58-61, assoc prof, 61-64, ASSOC PROF COMPUT SCI, PA STATE UNIV, 64-, DIR COMPUT CTR, 58- Concurrent Pos: Prog dir comput sci, NSF, 63-64. Mem: Soc Indust & Appl Math; Asn Comput Mach. Res: Computer programming systems including supervisors and language processors. Mailing Add: 445 Waupelani Dr K11 State College PA 16801-4473

LAIRD, HUGH EDWARD, II, b Phoenix, Ariz, Mar 30, 39; m 61; c 2. NEUROPHARMACOLOGY, NEUROCHEMISTRY. Educ: Univ Ariz, BS, 62, PhD(pharmacol), 74. Prof Exp: Res assoc, 69-70, teaching assoc, 71-73, instr, 73-74, asst prof, 74-80, asst provost grad studies, Grad Col, 81-82, ASSOC PROF PHARMACOL & TOXICOL, COL PHARM UNIV ARIZ, 80- Concurrent Pos: Teaching asst pharmeceut, Col Pharm, Univ Ariz, 68-70; consult, Ariz Poison Control Info Ctr, 74-79; prin investr, Nat Inst Neurol & Commun Dis & Stroke, 78-82. Mem: Acad Pharmaceut Sci; AAAS; Am Epilepsy; NY Acad Sci; Soc Neurosci; Sigma Xi; Am Soc Pharmacol & Exp Therepeut. Res: Receptor coupling and the molecular events regulating cell growth. Mailing Add: 4745 E Towner Tucson AZ 85712

LAIRD, WILSON MORROW, b Erie, Pa, Mar 4, 15; m 38, 90; c 4. GEOLOGY. *Educ:* Muskingum Col, BA, 36; Univ NC, MA, 38; Univ Cincinnati, PhD(geol), 42. *Hon Degrees:* DSc, Muskingum Col, 64, Univ NDak, 85. *Prof Exp:* Asst geol, Univ NC, 36-38 & Univ Cincinnati, 38-40; from asst prof to prof, 40-69, EMER PROF GEOL, UNIV NDAK, 69- *Concurrent Pos:* State geologist, NDak, 41-69, state geologist emer, 69-; geologist, US Geol Surv, 44-48; consult geologist, 47-48; dir off oil & gas, Dept Interior, Washington, DC, 69-71; dir comt explor, Am Petrol Inst, 71-79; consult, 79- *Honors & Awards:* Am Asn Petrol Geol Pres Award, 48. *Mem:* Fel Geol Soc Am; Am Asn Petrol Geol; hon mem Asn Am State Geol (vpres, 48, pres, 50); AAAS. *Res:* Stratigraphy of the upper Devonian and lower Mississippian of southwestern Pennsylvania and the northern Rockies; paleontology of brachiopods; physiography and glacial geology; oil conservation; petroleum and ground water geology; geomorphology. *Mailing Add:* 101 Spanish Oak Lane Kerrville TX 78028

LAITIN, HOWARD, b Brooklyn, NY, Nov 18, 31; m 61; c 3. SYSTEMS ANALYSIS & DESIGN. *Educ:* Brooklyn Col, BA, 52; Harvard Univ, MA, 53, PhD(statist, pub health & econ), 56. *Prof Exp:* Med economist, Hosp Coun Greater New York, 54-56; dir, Michael Saphier & Assoc, 56; proj dir, Army Med Serv, 57-59; sr economist, Rand Corp, 59-62; mgr prog anal, 62-82, CHIEF SCIENTIST, HUGHES AIRCRAFT CO, 82- *Concurrent Pos:* Clin assoc prof pub health, Univ Calif, Los Angeles, 59-73; adj prof, Sch Eng, Univ Southern Calif, 66-; adv to various orgn & govt agencies. *Res:* Technical analysis; military affairs; public health; solid and hazardous waste management; air pollution; transportation; safety; economic studies. *Mailing Add:* 4916 White Ct Torrance CA 90503

LAITINEN, HERBERT AUGUST, b Ottertail Co, Minn, Jan 17, 15; m 40; c 3. ANALYTICAL CHEMISTRY. *Educ:* Univ Minn, BCh, 36, PhD(phys chem), 40. *Prof Exp:* Asst phys chem, Univ Minn, 36-39; from instr to prof, 40-74, head div anal chem, 53-67, EMER PROF CHEM, UNIV ILL, URBANA, 74-; GRAD RES PROF, UNIV FLA, 74- *Concurrent Pos:* Guggenheim fel, 53, 62; ed, Anal Chem, 66-79; Nat Acad Sci exchange visitor, Yugoslavia, 69; vis prof, Seoul Nat Univ, Korea, 71-80. *Honors & Awards:* Fisher Award, 61; Synthetic Org Chem Mfrs Asn Award Environ Chem, 75; Gadolin Medal, 84. *Mem:* AAAS; Electrochem Soc; Am Chem Soc; hon mem Japan Soc Anal Chem; hon fel Royal Soc Chem; hon mem Finnish Acad Arts & Sci. *Res:* Electrochemistry; polarography; amperometric titrations; diffusion; polarization of microelectrodes; fused salts; environmental science; surface chemistry. *Mailing Add:* Dept Chem Univ Fla Gainesville FL 32611

LAITY, DAVID SANFORD, b Mt Kisco, NY, Nov 20, 26; m 50; c 2. CHEMICAL ENGINEERING. *Educ:* Haverford Col, BA, 49; Mass Inst Technol, MS, 50; NY Univ, ScD(chem eng), 56. *Prof Exp:* Plant process engr & supvr, Eng Serv Div, E I du Pont de Nemours & Co, 50-59; supv res engr, Process Design Div, 59-66, staff econ analyst, Comptrollers Anal Div, 67-69, asst proj mgr, Belg Refinery, 69-71, staff planner, Chevron Oil Europe, 71-73, mgr Process Design Div, Chevron Corp, 73-86, VPRES PROCESS RES DEPT, CHEVRON RES CO, 86- *Concurrent Pos:* Chmn, Adv Coun, Sch Chem Eng, Cornell Univ & Worcester Polytech Inst. *Mem:* Am Inst Chem Eng; Sigma Xi. *Res:* Process engineering management in chemical and petroleum industries. *Mailing Add:* 96 Silverwood Dr Lafayette CA 94549

LAITY, JOHN LAWRENCE, b Helena, Mont, Feb 23, 42; m 64; c 2. INDUSTRIAL CHEMISTRY, PETROLEUM CHEMISTRY. *Educ:* Stanford Univ, BS, 64; Univ Wash, PhD(chem), 68. *Prof Exp:* CHEMIST & SUPVR, SHELL OIL CO, 68- *Mem:* Am Chem Soc. *Res:* Photochemical smog; automotive and engine research; combustion; gasoline and oil additives; compositions of fuels and solvents; exhaust emissions; catalysts; atmospheric reactions; air and water pollution; polymer chemistry. *Mailing Add:* 14207 W Hersdale Houston TX 77077

LAITY, RICHARD WARREN, b Mt Kisco, NY, Sept 16, 28; m 51; c 5. ELECTROCHEMISTRY, PHYSICAL CHEMISTRY. *Educ:* Haverford Col, AB, 50, MS, 51; Iowa State Univ, PhD(phys chem), 55. *Prof Exp:* From instr to asst prof chem, Princeton Univ, 55-65; PROF CHEM, RUTGERS UNIV, NEW BRUNSWICK, 65- *Concurrent Pos:* Consult, Monsanto Res Corp, 59-69 & Standard Oil Co (Ohio), 60-68; AEC res contract, 60-71; Frontiers in Chem lectr, Cleveland, Ohio, 62; consult ed, Prentice-Hall, Inc, 62-68; chmn, Gordon Res Conf Molten Salts, 67-69; NSF res grant, 72. *Mem:* Am Chem Soc; Electrochem Soc; Sigma Xi. *Res:* Properties of molten salts; electrochemistry; irreversible thermodynamics; transport properties of extremely concentrated aqueous electrolytes. *Mailing Add:* Sch Chem Rutgers Univ Busch Campus New Brunswick NJ 08903

LAJTAI, EMERY ZOLTAN, b Hungary, Oct 28, 34; Can citizen; m 59. GEOLOGY. *Educ:* Univ Toronto, BASc, 50, MASc, 61, PhD(Pleistocene geol), 66. *Prof Exp:* Soils engr, Subway Construct Br, Toronto Transit Comn, 61-63; eng geologist, H G Acres & Co Ltd, Ont, 63-65; vis lectr eng geol, 65-67, asst prof, 67-70, assoc prof eng geol & rock mech, 70-77, PROF GEOL, UNIV NB, 77- *Concurrent Pos:* Nat Res Coun Can res grants, 66-68, 71-74 & 74-77; Govt Can, Geol Surv grants, 67-68 & 71-72. *Mem:* Can Geotech Soc; Can Rock Mech Group. *Res:* Brittle fracture of rocks under compressive loading with application in structural and engineering geology. *Mailing Add:* Dept Geol Univ Man Winnipeg MB R3T 2N2 Can

LAJTHA, ABEL, b Budapest, Hungary, Sept 22, 22; nat US; m 53; c 2. BIOCHEMISTRY. *Educ:* Eotvos Lorand Univ, Budapest, PhD(chem), 45. *Hon Degrees:* Dr, Univ Padua, Italy. *Prof Exp:* Asst prof biochem, Eotvos Lorand Univ, 45-47; asst prof, Inst Muscle Res, Mass, 49-50; sr res scientist, NY State Psychiat Inst, 50-57, assoc res scientist, 57-62; prin res scientist, 62-66, DIR, NY STATE RES INST NEUROCHEM, 66-; PROF EXP PSYCHIAT, SCH MED, NY UNIV, 71- *Concurrent Pos:* Fel, Zool Sta, Italy, 47-48; res fel, Royal Inst Gt Brit, 48-49; asst prof, Col Physicians & Surgeons, Columbia Univ, 56-69; pres, Am Soc Neurochem & Int Soc Neurochem. *Mem:* Int Brain Res Orgn; Am Soc Biol Chemists; Am Acad Neurol; Am Col Neuropsychopharmacol; Int Soc Neurochem; Am Chem Soc. *Res:* Neurochemistry; amino acid and protein metabolism of the brain and the brain barrier system. *Mailing Add:* Dept Psychiat NY Univ Sch Med 550 First Ave New York NY 10016

LAKATOS, ANDRAS IMRE, b Budapest, Hungary, Aug 23, 37; US citizen; m 72; c 1. THIN FILM DEVICES, LIQUID CRYSTAL DISPLAYS. *Educ:* Alfred Univ, BS, 62, MS, 63; Cornell Univ, PhD(appl physics), 67. *Prof Exp:* Scientist photoelec properties displays, 66-78, MGR THIN FILM DEVICE AREA, WEBSTER RES CTR, XEROX, INC, 78- *Mem:* Sr mem Inst Elec & Electronics Engrs; fel Soc Info Display; Am Phys Soc; Soc Photog Sci & Eng. *Res:* Development of thin film transistors for the addressing of one and two dimensional marking or display arrays. *Mailing Add:* Xerox Corp 0114-44d 800 Philips Rd Webster NY 14580

LAKATTA, EDWARD G, b Scranton, Pa, May 10, 44. CARDIOLOGY, MUSCLE PHYSIOLOGY. *Educ:* Georgetown Univ, MD, 70. *Prof Exp:* DIR, GERONTOL RES CTR, NAT INST AGING, NIH, 76- *Mailing Add:* Nat Inst Aging NIH Gerontol Res Ctr 4940 Eastern Ave Baltimore MD 21224

LAKE, CHARLES RAYMOND, b Nashville, Tenn, July 6, 43; m 67; c 2. PSYCHOPHARMACOLOGY. *Educ:* Tulane Univ, BS, 65, MS, 66; Duke Univ, PhD(physiol & pharmacol), 71, MD, 72. *Prof Exp:* Resident psychiat, Duke Univ Med Ctr, 72-74; res assoc, 74-75, clin assoc, Lab Clin Sci, 75-77, attending physician, Sect Exp Therapeut, NIMH, 78-80; PROF PHARMACOL & PROF PSYCHIAT, UNIFORMED SERV UNIV HEALTH SCI, 79- *Concurrent Pos:* Psychiat consult, Nat Naval Med Ctr, Bethesda, Md, 80- *Mem:* Am Soc Pharmacol & Exp Therapeut; Soc Biol Psychiat; Am Soc Neurochem; Am Col Neuropsychopharmacol; Int Soc Hypertension. *Res:* Biogenic amine metabolism as related to neuropsychiatric disease and bloodpressure regulation; endorphins and neuropsychiatric disorders; endogenous opioid and catecholamine interrelationships. *Mailing Add:* Univ Health Sci 4301 Jones Bridge Rd Bethesda MD 20814

LAKE, GEORGE RUSSELL, b Washington, DC, June 12, 53; m; c 2. DYNAMICS OF GALAXIES, COSMOLOGY. *Educ:* Haverford Col, BA, 75; Princeton Univ, MA, 77, PhD(physics), 80. *Prof Exp:* Fel, Univ Calif, Berkeley, 79-80 & Churchill Col, Cambridge, 80-81; mem tech staff, Bell Labs, 81-86; ASSOC PROF, UNIV WASH, 86- *Honors & Awards:* Dudley Prize. *Mem:* Am Astron Soc. *Res:* Dynamics of galaxies, galaxy formation and cosmology; low mass stars; search for extraterrestrial intelligence. *Mailing Add:* Dept Astron Univ Wash FM 20 Seattle WA 98195

LAKE, JAMES ALBERT, b Nebr; m 67; c 2. MOLECULAR & CELL BIOLOGY. *Educ:* Univ Colo, BA, 63; Univ Wis, Madison, PhD(physics), 67. *Prof Exp:* Fel physics, Univ Wis, 67; NIH fel molecular biol, Mass Inst Technol, 67-68 & Children's Cancer Res Found, Mass, 68-70; res fel, Harvard Univ, 69-70; asst prof cell biol, Rockefeller Univ, 70-72; from asst prof to assoc prof cell biol, Med Sch, NY Univ, 72-76; PROF MOLECULAR BIOL, UNIV CALIF, LOS ANGELES, 76- *Honors & Awards:* Irma T Hirschl Found Award, 74; Burton Award, Electron Micros Soc Am, 75. *Mem:* AAAS; Biophys Soc; Electron Micros Soc Am; Cell Biol Soc; Am Soc X-ray Crystallog. *Res:* Molecular structure of biological molecules; ribosome function and structure; protein synthesis; molecular evolution. *Mailing Add:* Molecular Biol Inst Univ Calif 405 Hilgard Ave Los Angeles CA 90024

LAKE, LORRAINE FRANCES, b St Louis, Mo, Feb 12, 18. ANATOMY, REHABILITATION MEDICINE. *Educ:* Wash Univ, BS, 50, MA, 54, PhD(anat), 62. *Prof Exp:* Instr phys ther, 49-54, instr anat & phys ther, 54-58, dir phys ther, 59-60, asst dir, 60-67, assoc dir, Irene Walter Johnson Inst Rehab, 67-79, asst prof, 68-80 EMER ASST PROF PHYS THER, WASH UNIV, 80- *Concurrent Pos:* Assoc dir phys ther curric & chg clin training, Sch Med, Wash Univ, 58-63; Woodcock Mem lectr, Univ Calif, 59; consult, Surgeon Gen, US Air Force, 65-67 & birth defects treatment ctr, Nat Found, 65-69. *Res:* Normal and abnormal neuromuscular function; electromyocardiographic investigations of normal human movement; human teratology. *Mailing Add:* 6220 Loughborough Ave St Louis MO 63109-3635

LAKE, ROBERT D, b Lansing, Mich, Sept 7, 30; m 64; c 2. POLYMER CHEMISTRY. *Educ:* Mich State Univ, BS, 52; Ind Univ, PhD(org chem), 56. *Prof Exp:* Am Petrol Inst fel, Northwestern Univ, 56-57; fel chem res, Mellon Inst, 57-60; scientist, Koppers Co, Inc, 60-66, group mgr explor res, 66-72, sr scientist, Res Dept, 72-77; proj tech coordr, Reichhold Chem, 77-80, develop group mgr, 80-89, mgr, flame retardant resins, 89-90; CONSULT, 90- *Mem:* Soc Plastics Engrs. *Res:* Synthesis and properties of vinyl and condensation polymers; preparation and properties of unsaturated polyester resins; smoke and flammability behavior of polymers; development of thermoset polyester molding compounds; flame retardant and corrosion resistant polyester and vinylester resins. *Mailing Add:* 215 Thornwood Ct Coraopolis PA 15108

LAKE, ROBIN BENJAMIN, b Warren, Ohio, Sept 8, 38; m 63; c 3. BIOMETRICS, BIOMEDICAL ENGINEERING. *Educ:* Rensselaer Polytech Inst, BEE, 60; Case Western Univ, PhD(biomed eng), 69; Harvard Univ, AM, 64. *Prof Exp:* Researcher cell biol & motor neuron degeneration, Harvard Univ, 62-65; researcher systs theory, Systs Res Ctr & Cybernet Systs Group, Case Western Reserve Univ, 65-69, res assoc biomed eng, 69, sr instr, Schs Med & Eng, 69-70, sr instr biomet, Sch Med, 70-72, dir, Biomet Comput Lab, 70-76, ASST PROF BIOMET & BIOMED ENG, SCH MED, 72-, DIR COMPUT APPLN TRAINING PROG, 73-; SOHIO RES & DEVELOP. *Concurrent Pos:* Consult comput mfr, 70-; prin scientist, Monolithic Systs Corp, 76- *Mem:* AAAS; Inst Elec & Electronics Engrs; Asn Comput Mach; Biomed Eng Soc. *Res:* Software systems; mitigation of environmental hazards; applications of AI; health care quality assurance. *Mailing Add:* BP Am Res & Develop 4440 Warrensville Ctr Rd Cleveland OH 44128-2837

LAKEIN, RICHARD BRUCE, b Baltimore, Md, Mar 5, 41; m 64; c 2. COMPUTER NETWORKS. *Educ:* Yale Univ, BA, 62; Univ Md, MA, 64, PhD(math), 67. *Prof Exp:* Lectr math, Univ Md, 67-68; asst prof math, State Univ NY, Buffalo, 68-74; mathematician, 75-82, COMPUTER SYSTS ANALYST, NAT SECURITY AGENCY, 82- *Mem:* Am Math Soc; Asn Comput Mach. *Res:* High speed computer networks; number theory. *Mailing Add:* 8711 Bunnell Dr Potomac MD 20854

LAKES, RODERIC STEPHEN, b New York, NY, Aug 10, 48; m 71. LASERS & OPTICS, MECHANICAL ENGINEERING. *Educ:* Rensselaer Polytech Inst, BS, 69, PhD(physics, biophys), 75. *Prof Exp:* Res assoc appl sci, Yale Univ, 75-77; asst prof physics, Tuskegee Inst, 77-78; vis asst prof biomed eng, Rensselaer Polytech Inst, 78; from asst prof to assoc prof, 78-86, PROF BIOMED ENG, UNIV IOWA, 86- *Concurrent Pos:* NIH fel, Yale Univ, 75-77; prin investr proj in bone biomech, NIH, 79-82; vis prof, Dept Mat, Queen Mary Col, Univ London, 84; prin investr, Study Viscoelastic Elastomers, 86-88. *Honors & Awards:* Burlington Northern Found Award, 87. *Mem:* Am Phys Soc; Orthop Res Soc; Sigma Xi; Am Soc Mech Engrs; AAAS. *Res:* Novel structured materials; bone biomechanics and bioelectricity; properties of piezoelectric solids and composite materials; applied optics; holographic interferometry; four patents. *Mailing Add:* Biomed Eng Dept Univ Iowa Iowa City IA 52242

LAKEY, WILLIAM HALL, b Medicine Hat, Alta, Nov 12, 27; m 57; c 4. GENITO-URINARY SURGERY. *Educ:* Univ Alta, BSc, 49, MD, 53; FRCPS(C), 60. *Prof Exp:* PROF SURG, FAC MED, UNIV ALTA, 60-, DIR DIV UROL, UNIV HOSP, EDMONTON, 71- *Honors & Awards:* Surg Res Medal, Royal Col Physicians & Surgeons Can, 56. *Mem:* Fel Am Col Surg; Am Urol Asn; Can Urol Asn; Can Acad Genito-Urinary Surg; Am Asn Genito-Urinary Surg. *Res:* Renal transplantation; renal hypertension and use of diagnostic tests; kidney preservation. *Mailing Add:* Dept Surg-Urol Univ Alta Fac Med Edmonton AB T6G 2G3 Can

LAKHTAKIA, AKHLESH, b Lucknow, India, July 1, 57; m 82; c 1. T-MATRIX WAVE SCATTERING, ELECTRODYNAMICS OF CHIRAL MEDIA. *Educ:* Banaras Hindu Univ, India, BTech, 79; Univ Utah, MS, 81 & PhD(elec eng), 83. *Prof Exp:* Res asst, Univ Utah, 79-83; postdoctoral scholar, 83-84, ASST PROF, PA STATE UNIV, 84- *Concurrent Pos:* Am ed, Speculations in Sci & Technol, 90- *Mem:* Sigma Xi; Optical Soc Am. *Res:* Author and co-author of over 150 papers and conference publications on wave-material interaction, electromagnetics, chiral media, fractals, chaos, bioelectromagnetics. *Mailing Add:* 227 Hammond Bldg University Park PA 16802

LAKIN, JAMES D, b Harvey, Ill. ALLERGIES. *Educ:* Northwestern Univ, BSc, 68, PhD, 68, MD, 69. *Prof Exp:* Instr, Dept Microbiol, Northwestern Univ Med Sch, 66-68; Passavant Hosp Sch Nursing, 66-68 & Wesley Mem Hosp Sch Nursing, 66-68; intern, Dept Internal Med, Univ Mich Med Ctr, 70, resident, 71-72, fel, Allergy Sect, 73; lt commander, Med Corps, Naval Med Res Inst, USNR, 74-76, staff investr & lab dir, Allergy Res Lab, Nat Naval Med Ctr, 74, dir allergy-res training, 74-76; asst prof, Georgetown Univ Med Ctr, 75-76; clin asst prof, 76-82, clin assoc prof, 82-88, CLIN PROF INTERNAL MED & PEDIAT, UNIV OKLA HEALTH SCI CTR, OKLAHOMA CITY, 88-, CHMN, DEPT ALLERGY, 89-; PVT PRACT, OKLA ALLERGY CLIN, 75- *Concurrent Pos:* Dep dir, Clin Immunol Fiv, Naval Med Res Inst, Nat Naval Med Ctr, 74-75; dir, Adolescent Allergy Clin, Children's Mem Hosp, 76-, Frontiers of Sci, 79-81, Okla Allergy Clin, 80-, Okla Med Res Found, 82-; mem, Res Comt, Presby Hosp, Oklahoma City, 77; mem, Comt Allergy & Clin Immunol, Am Col Chest Physicians, 77; diving med officer, Nat Oceanic & Atmospheric Admin, 88; staff mem, Baptist Med Ctr, St Anthony Hosp, Children's Mem Hosp, Univ Hosp, Presby Hosp, Mercy Mem Hosp & Deaconess Hosp, Oklahoma City, Okla; spec consult, Vet Admin Hosp, Oklahoma City, Okla. *Mem:* Am Asn Immunologists; fel Am Col Physicians; fel Am Col Chest Physicians; AMA; Am Soc Internal Med; Fedn Am Socs Exp Biol; fel Am Acad Allergy & Immunol; Undersea Med Soc. *Res:* Classification of hypersensitivity reactions; immune response and hypersensitivity reactions. *Mailing Add:* Okla Allergy Clin Okla Univ Health Sci Ctr PO Box 20827 Oklahoma City OK 73126

LAKKARAJU, H S, b Bapatla, India, Sept 20, 46; m 69; c 2. SURFACE SCIENCE, NONLINEAR OPTICS. *Educ:* Andhra Univ, India, BSc, 65, MSc, 67; Fairleigh Dickinson Univ, MS, 73; State Univ NY, Buffalo, PhD(physics), 79. *Prof Exp:* Sr sci asst, Radiosci Div, Nat Phys Lab, 70-71; PROF PHYSICS, SAN JOSE STATE UNIV, 81- *Concurrent Pos:* Vis asst prof, Tex A&M Univ, 79-81. *Mem:* Am Phys Soc; Optical Soc Am; Int Soc Optical Eng. *Res:* Laser spectroscopy and nonlinear optics, and the applications of these in the surface science; condensed matter physics and biophysics. *Mailing Add:* Physics Dept San Jose State Univ San Jose CA 95192

LAKOSKI, JOAN MARIE, b Poughkeepsie, NY, Mar 28, 53; m 78; c 2. NEUROPHARMACOLOGY, NEUROENDOCRINOLOGY. *Educ:* Mount Holyoke Col, AB, 75; Univ Iowa, PhD(pharmacol), 81. *Prof Exp:* Biologist, NIH, 75-77; postdoctural neuropharmacol, Yale Univ Sch Med, 81-84; ASST PROF NEUROPHARMACOL, NEUROENDOCRINOL & NEUROCHEM, UNIV TEX MED BR, GALVESTON, 84-; ADJ MEM, MARINE BIOL INST, 88- *Concurrent Pos:* Mem, Initial Rev Group Pharmacol Spec Rev Comt, Nat Inst Drug Abuse, 87; res career develop award, Nat Inst Aging, 89-94; mem, Teaching & Eval Mat Subcomt, Am Soc Pharmacol & Exp Therapeut, 89-91; assoc ed, Molecular & Cellular Neurosci, 89-; mem, Coord Ctr Aging Adv Comt, 90-; chmn, Comt Res, Univ Tex Med Br, 90-; chair, Preprof Subcomt on Educ, Am Soc Pharmacol & Exp Therapeut, 91-93; co-chmn, Symp Serotonin & Drugs of Abuse, 91. *Mem:* Soc Neurosci; Endocrine Soc; Sigma Xi; Serotonin Club. *Res:* Neuropharmacologic basis of age-related changes in central nervous system function that mediate reproductive aging; role of sertonin neuronal systems in drug-induced effects of cocaine, steroids, histamine and aging. *Mailing Add:* Dept Pharmacol J-31 Univ Tex Med Br Galveston TX 77550

LAKOWICZ, JOSEPH RAYMOND, b Philadelphia, Pa, Mar 15, 48; c 2. BIOPHYSICS, BIOCHEMISTRY. *Educ:* La Salle Col, BA, 70; Univ Ill, Urbana, MS, 72, PhD(biochem), 73. *Prof Exp:* NATO fel biochem, Oxford Univ, 73-74; asst prof biochem, univ Minn, 75-80; assoc prof, 80-84, PROF BIOCHEM, SCH MED, UNIV MD, 84- *Concurrent Pos:* Estab investr, Am Heart Asn, 77. *Mem:* AAAS; Am Chem Soc; Biophys Soc; Am Soc Photobiol; Sigma Xi; Am Soc Biol Chemists; Optical Soc Am; Am Phys Soc; Protein Soc. *Res:* Fluorescence spectroscopy; membrane transport of chlorinated hydrocarbons and carcinogens; rapid relaxation phenomena in biopolymers; frequency-domain fluorometry; energy transfer; molecular dynamics; time-domain fluorometry. *Mailing Add:* Biochem Dept Univ Md 660 W Redwood St Baltimore MD 21201

LAKRITZ, JULIAN, b Antwerp, Belg, Feb 13, 30; US citizen; c 2. ORGANIC CHEMISTRY. *Educ:* NY Univ, BA, 52; Univ Mich, MS, 54, PhD(org chem), 60. *Prof Exp:* Res chemist, Esso Res & Eng Co, 58-68; dir res & develop, Am Permac Inc, Garden City, NY, 68-75; tech dir, 75-90, VPRES SALES & MKT, ANSCOTT-SIGNAL CHEM CO, 75- *Mem:* Am Chem Soc; Am Asn Textile Chem & Colorists. *Res:* Chemistry and technology for solvent processing of textiles. *Mailing Add:* Two Livingston Ave Edison NJ 08820

LAKS, DAVID BEJNESH, b Brooklyn, NY, Mar 31, 62; m 84; c 1. DEFECT CALCULATIONS IN SOLIDS. *Educ:* Columbia Univ, BS, 83, MS, 84, PhD(solid state physics), 90. *Prof Exp:* POSTDOCTORAL, SOLAR ENERGY RES INST, 90- *Mem:* Am Phys Soc; Mat Res Soc. *Res:* Theoretical investigations of defects in zinc-selenide and doping problems in wide band-gap semiconductors; semiconductor interface structures; Auger recombination in semiconductors. *Mailing Add:* Solar Energy Res Inst 1617 Cole Blvd Golden CO 80401

LAKS, PETER EDWARD, b Brisbane, Australia, Jul 15, 53; Can Citizen; m 73; c 3. WOOD PRESERVATION. *Educ:* Simon Fraser Univ, BSc, 76, MSc, 79; Univ British Columbia, PhD(wood sci), 84. *Prof Exp:* SR RES SCI, MICH TECHNOL UNIV, 85- *Concurrent Pos:* Adj prof, 85- *Mem:* Am Chem Soc; Am Soc Pharmacog. *Mailing Add:* Inst Wood Res Mich Technol Univ Houghton MI 49931

LAKSHMAN, M RAJ, b Calcutta, India, Aug 3, 38; US citizen; m 66; c 2. LIPIDS, ALCOHOLISM. *Educ:* Poona Univ, BS, 58, MS, 59; Inst Sci, PhD(biochem), 66. *Prof Exp:* NRC fel, Dept Nat Health & Welfare, Can, 66-67; sr res adv, Rockefeller Found, Bangkok, 67-71; proj assoc, Vet Admin Med Ctr, Wis, 71-74; vis scientist, NIMH, 74-78; res chemist, Nat Inst Alcohol Abuse & Alcoholism, 78-80; CHIEF LIPID RES, VET ADMIN MED CTR, WASHINGTON, DC, 80-; ASSOC RES PROF, GEORGE WASHINGTON UNIV, 83- *Concurrent Pos:* Res prof, Am Univ, 76; mem, Exp Adv Group Nutrit, Food & Drug Admin, 83. *Mem:* Am Soc Biol Chem; Res Soc Alcoholism; Am Inst Nutrit. *Res:* Lipid and lipoprotein regulation, alcoholic hyperlipidemia hormonal control, regulation of alcohol dehydrogenase, mechanism of absorption and action of tetrachloro-dibenzo p-dioxin. *Mailing Add:* Lipid Res Lab Vet Admin Med Ctr 50 Irving St Washington DC 20422

LAKSHMANAN, FLORENCE LAZICKI, biochemistry, for more information see previous edition

LAKSHMANAN, P R, b Jamshedpur, Bihar, India, Apr 28, 39. ORGANIC POLYMER CHEMISTRY. *Educ:* Univ Calcutta, BS, 58; Univ Bombay, BS, 61; NDak State Univ, MS, 65, PhD(polymers & coating), 66. *Prof Exp:* Sect supvr, 66-80, dir new prods res & develop, 80-85, RES CHEMIST PLASTICS & SR RES CHEMIST COATINGS & ADHESIVES, GULF OIL CHEM CO, 66-; TECH DIR, BAYCHEM INT, INC, 83- *Mem:* Am Chem Soc; Oil & Color Chemists Asn; Soc Plastic Engrs; Am Inst Chemists. *Res:* Relationship between structure and performance of adhesives and coatings; mechanism of adhesion, polymer blends and alloys. *Mailing Add:* 15311 Ripplestream St Houston TX 77068

LAKSHMANAN, VAIKUNTAM IYER, b Pazhaya Kayal, Madras, India, July 10, 40; m 68. METALLURGICAL CHEMISTRY, HYDROMETALLURGY. *Educ:* Univ Bombay, BSc, 61, MSc, 63, PhD(chem), 68. *Prof Exp:* Chief chemist, H&R Johnson India (PVT) Ltd, 68-69; res fel minerals eng, Univ Birmingham, 69-72; lectr, 72-75; fel metall chem sect, Canmet, Dept Energy, Mines & Resources, Ottawa, 75-76; assoc scientist extractive metall sect, Noranda Res Ctr, Montreal, 76-77; res chemist, Eldorado Nuclear Ltd, 77-81; MGR MINERAL RESOURCES, ORTECH INT, 81- *Honors & Awards:* Bosworth Smith Inst Award, Brit Inst Mining & Metall, 74. *Mem:* Metall Soc; Royal Inst Chem; Brit Inst Mining & Metall; Soc Chem Indust; Can Inst Mining & Metall; Am Soc Minning Eng. *Res:* Solution chemistry; solution treatment precipitation; solvent extraction; ion exchange; treatment of effluents; radiotracer studies; precious and rad metals recovery. *Mailing Add:* Ortech Int 2395 Speakman Dr Mississauga ON L5K 1B3 Can

LAKSHMIKANTHAM, VANGIPURAM, b Hyderabad, India, Aug 8, 26; m 42; c 3. MATHEMATICS. *Educ:* Osmania Univ, India, PhD(math), 59. *Prof Exp:* Res assoc math, Univ Calif, Los Angeles, 60-61; assoc prof, Univ Alta, 63-64; prof & chmn dept, Marathwada Univ, India, 64-66; prof & chmn dept, Univ RI, 66-73; prof math & chmn dept, 73-86, ASHBEL SMITH PROF, UNIV TEX, ARLINGTON, 87- *Concurrent Pos:* Ed, Nonlinear Anal; assoc ed, Jour Math Anal & Applns, Applicable Anal, Appl Math & Comput & Jour Math & Phys Sci; vis mem, Math Res Ctr, Univ Wis-Madison, 61-62, Res Inst Advan Study, 62-63. *Honors & Awards:* Distinguished Res Award, 81. *Mem:* Am Math Soc; Indian Math Soc; Indian Nat Acad Sci; Soc Indust & Appl Math. *Res:* Differential inequalities; theory and applications, including stability theory by Liapunov's second method; nonlinear analysis. *Mailing Add:* Dept Appl Math Fla Inst Technol 150 W University Blvd Melbourne FL 32901

LAKSHMINARAYAN, S, b Madras, India, July 2, 43; US citizen. RESPIRATORY DISEASES. *Educ:* India Inst Med Sci, New Delhi, MBBS, 64; Royal Col Physicians, London, MRCP, 69, FRCP, 83. *Prof Exp:* Fel, Pulmonary Div, Westminster Hosp, London, 69-70, Brompton Hosp, 70-71 & Univ Colo Med Ctr, 71-72; from instr to asst prof med, Pulmonary Div, Univ Colo Med Ctr, 72-75; assoc prof, 75-82, PROF MED, SCH MED, UNIV WASH, 82-; MED DIR RESPIRATORY THER UNIT & PULMONARY FUNCTION LAB & CHIEF PULMONARY CRITICAL CARE MED SECT, VET ADMIN MED CTR, SEATTLE, WASH, 75- *Concurrent Pos:* Vis prof, St Johns Med Col, Bangalore, India, 85-86, Royal Postgrad Med Sch & Hammersmith Hosp, London, 86; chmn, Intensive Care Comt, Vet Admin Med Ctr, Seattle, 88- *Mem:* Am Col Chest Physicians; Am Thoracic Soc; Am Fedn Clin Res; Am Heart Asn; Am Physiol Soc; NY Acad Sci. *Res:* Pulmonary medicine; respiratory therapy. *Mailing Add:* Dept Respiratory Dis Vet Admin Med Ctr 1660 S Columbian Way Seattle WA 98106

LAKSHMINARAYANA, B, b Shimoga, India, Feb 15, 35; m 65; c 2. AEROSPACE & MECHANICAL ENGINEERING. *Educ:* Univ Mysore, BE, 58; Univ Liverpool, PhD(mech eng), 63, DEng, 82. *Prof Exp:* Asst engr, Kolar Gold Fields, India, 58-60; from asst prof to assoc prof, 63-74, dir computational fluid dynamic studies, 80-86, EVAN PUGH PROF AEROSPACE ENG, PA STATE UNIV, 86- *Concurrent Pos:* Consult, Pratt & Whitney Aircraft, 72-, Garret Turbine Engine, 80-, Teledyne CAE, 83-, Buffalo Fone Co, 84; Distinguished Alumni Prof Aerospace Eng, Pa State Univ, 85. *Honors & Awards:* Henry R Worthington Prize, 77; Outstanding Res Award, 77, 82; Premier Res Award, 83; Pandrey Lit Award, Am Inst Aeronaut & Astronaut; Freeman Scholar Award, Am Soc Mech Engrs, 90. *Mem:* Sigma Xi; Am Soc Eng Educ; AAAS; fel Am Soc Mech Engrs; fel Am Inst Aeronaut & Astronaut. *Res:* Three dimensional inviscid and viscid flow through rotor; rotor wake flow; rotor end wall flows; unsteady flow; transonic flow and acoustics of turbomachinery; aircraft and space propulsion; fluid mechanics; computational fluid dynamics; turbomachinery three dimensional flow field measurement and computation; automotive torque converter flow field. *Mailing Add:* Dept Aerospace Eng Pa State Univ University Park PA 16802

LAKSHMINARAYANA, J S S, b Penumantra, India, Sept 22, 31; m 60; c 2. PHYCOLOGY, WATER POLLUTION. *Educ:* Andhra Univ, India, BSc, 52; Banaras Hindu Univ, MSc, 54, PhD(bot), 60. *Prof Exp:* Lectr bot, Banaras Hindu Univ, 55; scientist & head biol, Nat Environ Eng Res Inst, India, 59-70; assoc prof biol, 70-80, PROF, DEPT BIOL, UNIV MONCTON, 80- *Concurrent Pos:* Fr Govt fel, ASTEF, Paris, 65-66; lectr, Visvesvaraya Regional Col Eng, India, 66-69; fel, Mem Univ Nfld, 70. *Mem:* Fel Linnean Soc UK; fel Marine Biol Asn, India. *Res:* Algology, limnology and oceanography in relation to pollution; coastal zone management; primary productivity in relation to fishery development. *Mailing Add:* Dept Biol Univ Moncton Moncton NB E1A 3E9 Can

LAKSHMINARAYANAN, KRISHNAIYER, b Bikshandarkoil, India, July 5, 24; m 60; c 2. BIOCHEMISTRY, INDUSTRIAL MICROBIOLOGY. *Educ:* Univ Madras, BSc, 45, MSc, 50, PhD(biochem), 55. *Prof Exp:* Jr chemist, King Inst Prev Med, India, 45-47, biochemist, Stanley Hosp, Madras, 50-51; asst prof microbiol, Birla Col, Pilani, 52; Imp Chem Industs fel, Nat Inst Sci India, 55-56; Nat Res Coun Can fel, Univ Manitoba, 56-58; Sci & Indust Res, Govt India, 59-60; plant biochemist, Cent Bot Lab, Allahabad, 60-61; res scientist indust microbiol, John Labatt Ltd, 61-62, sr res scientist, 62-63, proj leader, 63-67, sr indust enzymologist, Dawe's Fermentation Prod, Inc, 67-69, dir fermentation develop, 69-71, res & develop, 71; mgr process develop, Searle Biochemics, Div G D Searle & Co, 71-75; PRES, BIO-TECH INC, BENSENVILLE, ILL, 75- *Concurrent Pos:* Hon lectr, Univ Western Ont, 64- *Mem:* Fel Chem Inst Can; fel Royal Inst Chemists. *Res:* Microbial enzymology; plant biochemistry; toxicology; immunology; chromatography; microtechniques; industrial fermentations; enzyme production; immobilization. *Mailing Add:* 1310 N Belmont Ave Arlington Heights IL 60004

LAKSO, ALAN NEIL, b Auburn, Calif, Jan 3, 48. POMOLOGY, PLANT PHYSIOLOGY. *Educ:* Univ Calif, Davis, BS, 70, PhD(plant physiol), 73. *Prof Exp:* Asst prof pomol, 73-80, assoc prof hort sci dept, 80-86, PROF, NY STATE AGR EXP STA, CORNELL UNIV, 86- *Honors & Awards:* Gourley Award, Am Soc Hort Sci, 80. *Mem:* Int Soc Hort Sci; Am Soc Hort Sci; Am Soc Enol & Viticult; Sigma Xi. *Res:* Environmental physiology and the physiological bases of yield and quality of apples and grapes. *Mailing Add:* Dept Hort Sci NY State Agr Exp Sta Geneva NY 14456

LAL, DEVENDRA, b Banaras, India, Feb 14, 29; m 55. NUCLEAR PHYSICS, GEOCHEMISTRY. *Educ:* Banaras Hindu Univ, BSc, 47, MSc, 49; Univ Bombay, PhD(physics), 58. *Hon Degrees:* Hon DSc, Banaras Hindu Univ, Varanasi, 81. *Prof Exp:* From res fel to sr prof, Tata Inst Fundamental Res, Bombay, India, 49-72; dir, 72-83, sr prof, 83-89, FEL, PHYS RES LAB, AHMEDABAD, INDIA, 89-; PROF NUCLEAR GEOPHYS, SCRIPPS INST OCEANOG, UNIV CALIF, SAN DIEGO, 67- *Concurrent Pos:* Mem sci adv comt to Cabinet, Gov't India, 81-82; mem sci comt Indo-US joint comn sci & technol; founding fel, Third World Acad Sci, Trieste, Italy. *Honors & Awards:* Krishnan Medal, Indian Geophys Union, 65; S S Bhatnagar Award, 67; Krishnan Medal Lectr, Indian Nat Sci Acad, 81; Hawaharlal Nehru Award for Phys Sci. *Mem:* Foreign assoc Nat Acad Sci; fel Indian Acad Sci; fel Indian Nat Sci Acad; assoc mem Royal Astron Soc. *Res:* Cosmic rays; astrophysics; meteoritics; oceanography; meteorology; hydrology; geophysics. *Mailing Add:* GRD 0220 Scripps Inst Oceanog La Jolla CA 92093-0220

LAL, HARBANS, PHARMACOLOGY. *Educ:* Univ Kans, MS, 58; Univ Chicago, PhD(pharmacol), 62. *Prof Exp:* Res fel, Univ Chicago, 58-61; res pharmacologist, IIT Res Inst, 61-65; res assoc neurol & psychiat, Med Sch Northwestern Univ, 62-65; assoc prof pharmacol & toxicol, Univ Kans, 65-67; res & develop scientist, Janssen Pharmaceut, 73-74; from assoc prof to prof pharmacol & toxicol, Univ RI, 67-80, prof psychol, 70-80; res assoc, RI Inst Ment Health, 69-78, clin psychopharmacologist, 78-80; dir psychopharmacol, Inst Behav Med, 77-79; PROF & CHMN DEPT PHARMACOL, TEX COL OSTEOP MED, UNIV NTEX, 80-, PROF BIOL SCI, 80- *Concurrent Pos:* Adj prof chem & behav, Tex Christian Univ, 80-; teaching asst & lectr, Univ Chicago; co-instr, Chicago Med Sch, Chicago Col Osteop Med, Univ Kans, Univ RI, Brown Univ Med Sch, Tex Col Osteop Med & Univ North Tex; nat grant rev panels ad hoc appts, NSF, Nat Inst Drug Abuse Study Sect, Harry Frank Guggenheim Found, Human Embryol & Develop Study Sect; consult, Boehinger Pharmaceut Co, Burroughs Welcome Pharmaceut Co, Hoechst-Roussel Pharmaceut Co, Upjohn Pharmaceut Co, Ciba-Geigy Pharmaceut Co, McNeill Pharmaceut Co, Ortho Pharmaceut Co & Sterling-Winthrop Pharmaceut Co; grants, var corp & orgn. *Mem:* Am Col Neuropsychopharmacol; Asn Med Sch Pharmacol; Am Soc Pharmacol & Exp Therapeut; Soc Neurosci; Soc Toxicol; Behav Pharmacol Soc; fel Am Col Clin Pharmacol; Soc Biol Psychiat; Am Psychol Asn; Fedn Am Socs Exp Biol. *Res:* Pharmacology; toxicology; psychopharmacology. *Mailing Add:* Dept Pharmacol Tex Col Osteop Med 3500 Camp Bowie Blvd Ft Worth TX 76017-2690

LAL, JOGINDER, b Amritsar, India, July 2, 23; nat US; m 51; c 2. POLYMER CHEMISTRY. *Educ:* Punjab Univ, India, BSc Hons, 44, MSc, 46; Polytech Inst Brooklyn, MS, 49, PhD, 51. *Prof Exp:* Prof chem, Jain Col, Ambala, India, 45-47 & Hindu Col Amritsar, 51-52; head polymer res, H D Justi & Son, Inc, Pa, 52-56; res scientist, Goodyear Tire & Rubber Co, 56-67, sect head, 67-75, mgr polymer res, 75-82, sr res & develop assoc, 83-85; ADJ PROF, INST BIOMEN ENG, UNIV AKRON, 87-; CONSULT, LAL ASSOCS, 87- *Concurrent Pos:* Mem adv bd, J Polymer Sci, 67-90, mem adv comt, Chem Technol Prog, Univ Akron; counr & mem several nat comts, Am Chem Soc; prog chmn, Akron Polymer Lect Group 60-61, chmn, 61-62; cong sci counr, 74-86; chmn awards comt, Polymer Div, Am Chem Soc, 79-82, alternate counr, Rubber Div, 85-87; vchmn, Gordon Res Conf Elastomers, 82, chmn, 83; co-ed & assoc ed of three books on polymer chem; frequent lectr, nat & int meetings & indust, univ & govt res labs. *Honors & Awards:* Gold Medal, Hindu Col Amritsar, 41; Akron Summit Polymer Conf Award, 76; IR-100 Award, 79; First Serv Award, Polymer Div, Am Chem Soc, 83; Melvin Mooney Distinguished Technol Award, Rubber Div, Am Chem Soc, 89. *Mem:* Am Chem Soc. *Res:* Dental prosthetic materials; stereoregular polymers; mechanism of polymerization catalysts; chemical reactions of polymers; block copolymers; coatings; relationship between structure and properties of polymers; vulcanization; monomer synthesis; inventor Hexsyn rubber; reinforcement; author of over 100 US patents and journal publications. *Mailing Add:* Lal Assocs 855 Shullo Dr Akron OH 44313-5852

LAL, MANOHAR, b Lakki Marwat, India, Apr 11, 34; m 63; c 3. ENGINEERING SYSTEM MODELS, MATHEMATICAL PHYSICS. *Educ:* Allahabad Univ India, BSc, 55; Indian Inst Sci, DIISc, 58; Univ Ill, Urbana, MS, 61, PhD(elec eng), 63. *Prof Exp:* Lectr electronics & commun, Univ Roorkee, India, 58-60, asst prof to prof, 63-74; prof elec eng, Wichita State Univ, 74-78; sr res scientist math physics, 78-80, sr staff scientist, 80-82, RES ASSOC PROD RES, RES CTR, AMOCO PROD CO, 82- *Concurrent Pos:* Khosla res award, Univ Roorkee, India, 71; vis res prof, Coord Sci Lab, Univ Ill & Elec Eng & Comput Sci Dept, Univ Santa Clara, 75; adj prof, Univ Tulsa, 81. *Mem:* Inst Elec & Electronics Engrs. *Res:* Engineering systems modeling and control, fluid mechanics, drilling and production research in petroleum, fluid flow and wave propagtion in earth models and digital signal processing. *Mailing Add:* Amoco Prod Co Res Ctr PO Box 591 Tulsa OK 74102

LAL, MOHAN, b Dharmkot, Punjab, May 8, 32; Can citizen; m 64. MATHEMATICS. *Educ:* D M Col, Punjab, India, BA, 52; Aligarh Muslim Univ, MSc, 55; Univ BC, PhD(nuclear physics), 62. *Prof Exp:* Lectr physics, D A V Col, Punjab, India, 55-57; res asst, Univ BC, 57-61; res assoc, Univ Alta, 62-63; asst prof math & physics, Mt Allison Univ, 63-64; from asst prof to assoc prof math, 64-75, PROF MATH, MEM UNIV NFLD, 75- *Concurrent Pos:* Comput specialist, Fed & Prov Land Inventory Studies, Dept Mines & Natural Resources, Can, 67. *Mem:* Can Math Cong. *Res:* Numerical analysis; applied mathematics and elementary number theory. *Mailing Add:* Dept Math Mem Univ Nfld St John's NF A1C 5S7 Can

LAL, RATTAN, b Karyal, Punjab, Sept 5, 44; m 71; c 4. SOIL PHYSICS, TROPICAL SOILS. *Educ:* Punjab Agr Univ, Ludhiana, India, BSc, 63; Indian Agr Res Inst, New Delhi, MSc, 65; Ohio State Univ, PhD(soil physics), 68. *Prof Exp:* Sr res fel soil physics, Univ Sydney, Australia, 68-69; soil physicist soil physics, Int Inst Trop Agr, Ibadan, Nigeria, 70-87; assoc prof, 87-89, PROF SOIL PHYSICIS, OHIO STATE UNIV, COLUMBUS, OHIO, 89- *Concurrent Pos:* Vpres, Int Comt Continental Erosion, IAHS, UK, 82-87 & World Asn Soil & Water Conserv, Ankeny, Iowa, 83-87, pres, 88-91; chmn, Working Group ISSS, Soil Erosion Res Methods, 83-88; bd mem, Int Soil Tillage Res Orgn, Holland, 84-88, pres, 88-91; coordr, Upland Prod Systs, Int Inst Trop Agr, Ibadan, Nigeria, 84-87; bd mem, Orgn Trop Studies, 89- *Honors & Awards:* Int Soil Sci Award, Soil Sci Soc Am, 88; Distinguished Scientist Award, Asn Sci Indian Origin, 90. *Mem:* World Asn Soil & Water Conserv (pres, 88-91); Int Soil Tillage Res Orgn (pres, 88-91); fel Soil Sci Soc Am; fel Am Soc Agron; Int Soc Soil Sci; Soil & Water Conserv Soc. *Res:* Management of soil and water resources with particular relevance to the tropics and sub-tropics; processes of soil degradation under intensive management including accelerated erosion, compaction, anaerobiosis, transport of sediment-related pollutants, emission of radiatively active gases from soils and deterioration of soil structure. *Mailing Add:* Dept Agron Ohio State Univ 2021 Coffey Rd Columbus OH 43210

LAL, RAVINDRA BEHARI, b Agra, India, Oct 5, 35; m 62; c 1. PHYSICS. *Educ:* Agra Univ, BSc, 55, MSc, 58, PhD(physics), 63. *Prof Exp:* Lectr physics, REI Col, Agra Univ, 58-59 & Delhi Polytech, 63-64; Nat Acad Sci-Nat Res Coun resident res assoc, Marshall Space Flight Ctr, NASA, 64-67; asst prof, Indian Inst Technol, Delhi, 68-70; sr res assoc, Univ Ala, Huntsville, 71-73; asst prof physics, Paine Col, 73-75; assoc prof, 75-79, PROF

PHYSICS, ALA A&M UNIV, 79- *Honors & Awards:* New Technol Invention Award, 81, 83. *Mem:* Am Phys Soc; Sigma Xi; Am Asn Crystal Growth. *Res:* solid state physics; crystal growth and characterization of materials; magnetic and electrical properties of II-VI and III-V compounds; infra-red detector materials; manufacturing in space; selected by NASA as an investigator for an experiment on International Microgravity Lab (1ML-1) scheduled for 1990. *Mailing Add:* Dept Physics Ala A&M Univ PO Box 71 Normal AL 35762

LAL, SAMARTHJI, b London, Eng, Mar 23, 38; Can citizen; m 74; c 1. NEUROPSYCHIATRY. *Educ:* Univ London, MB, BS, 62; McGill Univ, dipl psychiat, 67; FRCP(C), 70; Am Bd Psychiat & Neurol, 78. *Prof Exp:* Med Res Coun Can res fel psychiat, 67-71; chief consultation serv, Montreal Gen Hosp, 71-75, assoc psychiatrist, 74-78; from asst prof to assoc prof, 73-83, PROF PSYCHIAT, MCGILL UNIV, 83-; DIR CLIN & BASIC RES PSYCHIAT, MONTREAL GEN HOSP, 75-, SR PSYCHIATRIST, 78- *Concurrent Pos:* Consult psychiatrist, Queen Mary Vet Hosp, 71-78; staff psychiatrist, Montreal Gen Hosp, 71-, consult, Psychiat Consultation Serv, 75-; staff psychiatrist, Douglas Hosp, 76-, bd dirs, Res Ctr, 80-; examr, Royal Col Phys Surg, Can, 81-82; bd of dirs, Douglas Hosp Res Ctr, 80- *Honors & Awards:* Hein Lehmann Award, 86. *Mem:* Can Soc Clin Invest; Can Psychiat Asn; fel Am Psychiat Asn; fel Can Col Neuropsychopharmacol; Soc Biol Psychiat; Int Soc Psychoneuroendrinol. *Res:* Monoaminergic mechanisms in anterior pituitary secretion and in neurological and psychiatric disorders. *Mailing Add:* Dept Psychol McGill Univ Fac Med 1033 Pine Ave W Montreal PQ H3A 1A1 Can

LALA, PEEYUSH KANTI, b Chittagong, Brit India, Nov 1, 34; m 62; c 2. CANCER, IMMUNOLOGY. *Educ:* Univ Calcutta, MB, BS, 57, PhD(med biophysics), 62. *Prof Exp:* Demonstr path, Calcutta Med Col, 59-60; demonstr path & hemat, NRS Med Col, 61-62; res assoc biol & med res, Argonne Nat Lab, 63-64; res scientist, Radiobiol Lab, Univ Calif, San Francisco, 64-66; res assoc biol & health physics, Chalk River Nuclear Labs, Atomic Energy Can Ltd, 67-68; from asst prof to assoc prof, 68-77, prof anat, McGill Univ, 77-83; PROF & CHAIR ANAT, UNIV WESTERN ONT, 83- *Concurrent Pos:* Fulbright travel scholar, 62; res dir, Med Res Coun Can grant, 68- & Nat Cancer Inst Can grant, 69-; USPHS grant, 75-; vis prof, Walter & Eliza Hall Inst Med Res, Melbourne Univ, 77-78; counr, Int Soc Reprod Immunol, 86; vpres, Can Asn Anat, 89-90, pres, 91-92; JCB grant award, Can Asn Anat, 90. *Mem:* Am Asn Immunol; Am Asn Cancer Res; Int Soc Exp Hemat; Int Soc Reprod Immunol; Am Asn Anat; Am Soc Reprod Immunol (vpres, 85); Reticuloendothelial Soc. *Res:* Studies of cell population kinetics during normal hematopoiesis, leukemias and cancer; host-tumor cell interactions in vivo; imūnobiology of tumor-hwst and fetomaternal relationship; author of more than 100 research publications and 11 book chapters on hematology, immulogy, cancer and reproduction; discoverer of a new mode of cancer therapy currently applied to human trial; reproduction. *Mailing Add:* Dept Anat Univ Western Ont London ON N6A 5C1 Can

LALANCETTE, JEAN-MARC, b Drummondville, Que, Apr 21, 34; m 58; c 3. INORGANIC CHEMISTRY, ENVIRONMENTAL CHEMISTRY. *Educ:* Univ Montreal, BSc, 57, MSc, 58, PhD(chem), 61. *Prof Exp:* From asst prof to prof chem, Univ Sherbrooke, 60-80; vpres, res & develop, Net Asbestos Soc, 80-85; PRES & CONSULT CHEMIST ENVIRON & HIGH TEMPERATURE CHEM, INOTEL INC, 85- *Honors & Awards:* Manning Award, 85; Award, Can-French Asn Advan Sci, 85. *Mem:* Chem Inst Can. *Res:* Organometallic chemistry; chemistry of graphite intercalates, both catalytic and synthetic properties; photochemical reactions; use of natural materials for protection of environment; peat moss. *Mailing Add:* Inotel Inc 470 Duvernay Sherbrooke PQ J1L 1J4 Can

LALANCETTE, ROGER A, b Springfield, Mass, July 30, 39; m 67, 78; c 2. ANALYTICAL CHEMISTRY, STRUCTURAL CHEMISTRY. *Educ:* Am Int Col, BA, 61; Fordham Univ, PhD(anal chem), 67. *Prof Exp:* Res fel, Brookhaven Nat Lab, 66-67; res chemist photopolymerization, Photo Prod Dept, E I du Pont de Nemours & Co, Inc, NJ, 67-69; asst prof anal chem, 69-76, assoc prof, 76-87, PROF CHEM, RUTGERS UNIV, NEWARK, 87- *Mem:* Am Chem Asn; Am Mat Soc. *Res:* Preparation and structural studies; x-ray powder and single crystal analysis; GC/MS of pesticides in soil and water. *Mailing Add:* Dept Chem Rutgers Univ Newark NJ 07102

LALAS, DEMETRIUS P, b Athens, Greece, Sept 28, 42; m 67; c 2. DYNAMIC METEOROLOGY, ENVIRONMENTAL FLUID DYNAMICS. *Educ:* Hamilton Col, AB, 62; Cornell Univ, MAeroE, 65, PhD(Aerospace), 68. *Prof Exp:* Asst prof, Dept Eng Mech & Dept Mech Eng, 68-73, assoc prof, 73-79, PROF, DEPT MECH ENG, WAYNE STATE UNIV, 79- *Concurrent Pos:* Vis fel, Coop Inst Res Environ Eng, Univ Colo, 73-74, consult, 74-75; assoc prof, Dept Meteorol, Univ Athens, Greece, 76-77, prof & chmn dept & dir, Meteorol Inst, 79-83; pres, Ctr Renewable Energy Sources, Athens, Greece. *Mem:* Am Meteorol Soc; Am Geophys Union; Greek Meteorol Soc; Am Soc Mech Eng. *Res:* Dynamics of micro and mesoscale wave dynamics, their excitation, stability and properties; physics and dynamics of two phase flows in the atmosphere and the laboratory; solar and wind energy; air pollution modelling; computational fluid mechanics; lubrication theory. *Mailing Add:* Dept Mech Eng 307/667 Merrek Wayne State Univ 5950 Cass Ave Detroit MI 48202

LALCHANDANI, ATAM PRAKASH, b India, Oct 20, 43; US citizen; div; c 1. OPERATIONS RESEARCH, PLANNING. *Educ:* Indian Inst Technol, Bombay, BTech, 63; Cornell Univ, MS, 66, PhD(oper res), 67. *Prof Exp:* Sr oper res analyst, Procter & Gamble, 67-69; dir appl syst, Optimum Systs Inc, Santa Clara, Calif, 69-73; dir indust serv, Control Anal Corp, 73-77; dir planning & anal, Nat Semiconductor Corp, 77-81; treas, 81-83, VPRES, FINANCE & ADMIN, NAT ADV SYSTS, 83- *Concurrent Pos:* Vis lectr, Grad Sch Bus, Univ Santa Clara, 69-77, Univ Cincinnati, 69 & Xavier Univ, Ohio, 69. *Mem:* Opers Res Soc Am; Inst Mgt Sci; Financial Exec Inst. *Res:* Finance. *Mailing Add:* 262 Angela Dr Los Altos CA 94022

LALEZARI, PARVIZ, b Hamadan, Iran, Aug 17, 31; m 58; c 2. MEDICINE, PHYSIOLOGY. *Educ:* Univ Teheran, MD, 54. *Prof Exp:* from asst prof to assoc prof med, 67-79, PROF MED, ALBERT EINSTEIN COL MED, 79-; DIR IMMUNOHEMAT & BLOOD BANK, MONTEFIORE HOSP & MED CTR, 60- *Concurrent Pos:* City New York Res Coun res grant, 60-64; NIH res grant, 65- *Mem:* Am Soc Hemat; Am Soc Clin Invest; Am Asn Immunol. *Res:* Leukocyte immunology; red cell immunology and autoimmune diseases. *Mailing Add:* Montefiore Hosp & Med Ctr 111 E 210th St Bronx NY 10467

LALIBERTE, GARLAND E, b Walkerburn, Man, Dec 28, 36; m 59; c 2. AGRICULTURAL ENGINEERING. *Educ:* Univ Sask, BEng, 56, MSc, 61; Colo State Univ, PhD(agr eng), 66. *Prof Exp:* Engr, Can Dept Agr, 56-61, res scientist, 61-67; assoc prof, 67-69, prof agr eng & head dept, Univ Man, 69-86; PROF AGR ENG, UNIV MAN, 86-, DEAN ENG, 89- *Concurrent Pos:* Agr Inst Can fel, 83; Can Soc Agr Eng fel, 84. *Honors & Awards:* Maple Leaf Award, Can Soc Agr Eng, 81; Agr Eng of Year Award, Am Soc Agr Eng, 90. *Mem:* Am Soc Agr Engrs; Can Soc Agr Eng (pres, 78-79); Sigma Xi; fel Agr Inst Can. *Res:* Drainage engineering; irrigation engineering; soil and water conservation. *Mailing Add:* Fac Eng Univ Man Winnipeg MB R3T 2N2 Can

LALIBERTE, LAURENT HECTOR, b Ottawa, Ont, Can, Nov 7, 43; m 66; c 2. ELECTROCHEMISTRY, CORROSION. *Educ:* Univ Ottawa, BSc, 66, PhD(chem), 69. *Prof Exp:* Fel chem, Univ Ottawa, 69-71; scientist corrosion, Pulp & Paper Res Inst Can, 71-78; sr res assoc, 78-79, group mgr, 79-87, MGR TECH SERV, INT PAPER, PINEVILLE MILL, 87- *Honors & Awards:* Weldon Medal, Can Pulp & Paper Asn. *Mem:* Nat Asn Corrosion Engrs; Tech Asn Pulp & Paper Indust. *Res:* Corrosion of materials used in pulp and paper industry process equipment. *Mailing Add:* Int Paper PO Box 2448 Mobile AL 36652

LALICH, JOSEPH JOHN, b Slunj, Yugoslavia, Nov 23, 09; nat US; m 41. PATHOLOGY. *Educ:* Univ Wis, BS, 33, MS, 36, MD, 37. *Prof Exp:* Fel exp med, Univ Kans, 38-42; from instr to assoc prof, 46-56, PROF PATH, MED SCH, UNIV WIS-MADISON, 56- *Mem:* AAAS; fel Soc Exp Biol & Med. *Res:* Hemorrhagic and traumatic shock; hemostasis; coagulation; hemoglobinuric nephrosis; experimental lathyrism; myocardial necrosis after allylamine ingestion; monocrotaline induced cor pulmonale. *Mailing Add:* 6306 Mound Dr Middleton WI 53562

LALL, ABNER BISHAMBER, b Bareilly, UP, India, Jan 28, 33; m 68; c 1. VISUAL PHYSIOLOGY, SENSORY SYSTEMS. *Educ:* Univ Delhi, BSc, 54; Boston Univ, STB, 59; Syracuse Univ, MS, 62; Univ Md, PhD(zool), 71. *Prof Exp:* Investr, Eye Res Found Bethesda, 69-72; asst prof neurophysiol & comp physiol, City Col NY, 72-74; fel neurophysiol, Johns Hopkins Univ Sch Med, 74-75; sr assoc, Howard Univ Col Med, 76-79; asst prof, Skidmore Col, 77-78; asst prof, Univ Miami, 82-83; assoc res scientist, Dept Biol, McCollum Pratt Inst, 79-82 & 83-85, res scientist, Dept Biophys, John Hopkins Univ, 85-88; ASSOC PROF ZOOL, HOWARD UNIV, 88- *Concurrent Pos:* Asst prof, Univ Miami, 82-83; Grass Found fel, neurophysiology, Marine Biol Lab, Woods Hole, Mass, 70. *Mem:* Asn Res Vision & Opthal; Am Soc Zoologists; AAAS; Soc Neurosci. *Res:* Sensory mechanisms of vision and neural mechanisms of instinctive visual behavior among anthropods, amphibians and fishes. *Mailing Add:* Thomas C Jenkins Dept Biophys John Hopkins Univ Baltimore MD 21218

LALL, B KENT, b Sargodha, India, Feb 4, 39; US citizen; m 70; c 1. CIVIL ENGINEERING, TRANSPORTATION ENGINEERING. *Educ:* Panjab Univ, India, BSc, 61; Univ Roorkee, ME, 64; Univ Birmingham, PhD(transp), 69. *Prof Exp:* Teaching fel hwy, Univ Roorkee, 61-64; asst prof & lectr civil eng, Indian Inst Technol, Delhi, 64-75; assoc prof, Univ Manitoba, 75-77; assoc prof, 77-84, PROF CIVIL ENG, PORTLAND STATE UNIV, 84- *Concurrent Pos:* Mem, Pub Transport Comt, Urban Transp Div, Am Soc Civil Engr, 79-, mem, Control Group, 83-, vchmn, 86-88, chmn, 88-90, Geometric Design Hwys Comt, 80-, mem, Control Group, 84-, mem, Transp Educ Control Group, 81-86, secy, Exec Comt, Urban Transp Div, 90-; vis prof, Univ Adelaide, S Australia, 85; consult, Nat Rds Bd, Ministry Works, NZ, 86; mem, Unsig Intersect Comt, Transp Res Bd, 90- *Mem:* Fel Am Soc Civil Engrs; Sigma Xi. *Res:* Urban transportation; transportation planning and systems; pavement design; highway and traffic engineering; highway materials and construction; highway capacity and geometric design. *Mailing Add:* Dept Civil Eng Portland State Univ Box 751 Portland OR 97207-0751

LALL, PRITHVI C, b Panjab, India, Sept 20, 31. NUCLEAR PHYSICS, ACOUSTICS. *Educ:* Panjab Univ, MS, 54; Oregon State Univ, PhD(physics), 62; George Washington Univ, JD, 69. *Prof Exp:* Asst prof, Howard Univ, 62-71; ASSOC COMNR, DEPT NAVY, NAVAL UNDERWATER SYST CTR, 71- *Mem:* Am Phys Soc; Sigma Xi. *Mailing Add:* Off Patent Counsel Naval Underwater Syst Ctr Bldg 142 Newport RI 02841

LALL, SANTOSH PRAKASH, b Motihari, Bihar, India, Sept 8, 44; Can citizen; m 74; c 4. NUTRITIONAL BIOCHEMISTRY. *Educ:* Allahabad Univ, BSc, 64; Univ Guelph, MSc, 68, PhD(nutrit), 73. *Prof Exp:* Res asst animal nutrit, Allahabad Agr Inst, 64-65; RES SCIENTIST FISH NUTRIT, HALIFAX LAB, 74- *Concurrent Pos:* Res asst, Nutrit Dept, Univ Guelph, 68, fel, 73. *Mem:* Can Soc Nutrit Sci; Aquacult Asn Can; NY Acad Sci. *Res:* Nutrient requirements of salmonid fishes in fresh water and sea water. *Mailing Add:* Dept Fisheries & Oceans PO Box 550 Halifax NS B3J 2S7 Can

LALLEY, PETER AUSTIN, human genetics, biochemical genetics, for more information see previous edition

LALLEY, PETER MICHAEL, b Scranton, Pa, Jan 21, 40; m 63; c 4. NEUROPHYSIOLOGY, NEUROPHARMACOLOGY. *Educ:* Philadelphia Col Pharm & Sci, BSc, 63, MSc, 65, PhD(pharmacol), 70. *Prof Exp:* Fel neuropharmacol, Sch Med, Univ Pittsburgh, 70-73, lectr neurosci pharmacol, 72-73, asst prof, 73-74; asst prof pharmacol, Col Med, Univ Fla, 74-76; asst prof, 76-80, ASSOC PROF PHYSIOL, SCH MED, UNIV WIS-

MADISON, 80- *Concurrent Pos:* Consult, US Pharmacopoeia & Dispensing Info; vis prof physiol, Univ Heidelberg, Fed Repub Germany, 85, 86, 87, Univ Göttingen, Fed Repub Ger, 88. *Mem:* Soc Neurosci; Sigma Xi; Am Physiol Soc; AAAS; Am Pharmaceut Asn; Am Heart Asn. *Res:* Identifying the neurotransmitters in the brainstem and spinal cord which control respiration and blood pressure, and determining the conditions under which they are operative. *Mailing Add:* Dept Physiol Ctr Health Sci Univ Wis 1300 Univ Madison WI 53706

LALLI, CAROL MARIE, b Toledo, Ohio, Dec 5, 38. MARINE BIOLOGY. *Educ:* Bowling Green State Univ, BS & BEd, 60, MA, 62; Univ Wash, PhD(zool), 67. *Prof Exp:* Lectr zool, McGill Univ, 68-69, from asst prof to assoc prof marine sci, 69-79; RES ASSOC, UNIV BC, 80- *Mem:* Marine Biol Asn. *Res:* Ecological studies of planktonic and benthonic gastropod molluscs. *Mailing Add:* Zool Dept Univ BC Vancouver BC V6T 2A9 Can

LALLY, PHILIP M(ARSHALL), b New York, NY, Sept 30, 25; m 47; c 3. ELECTRICAL ENGINEERING. *Educ:* Mass Inst Technol, SB, 48, SM, 49. *Prof Exp:* Asst elec eng, Mass Inst Technol, 48-49; engr, Electron Tube Dept, Sperry Gyroscope Co, 49-54, sr engr, 54-55, eng sect head, 55-57, eng supvr res & develop, Electronic Tube Div, Sperry-Rand Corp, 57-59, eng dept head, 59-60, asst prod eng supt, 60, prod eng mgr, 60-64, mgr res & advan devices, 64-68; mgr eng, Low Power Prod Line, 68-85, adv develop mgr, 85-90, MGR, ELEC ENG, TELEDYNE MEC, 90- *Concurrent Pos:* Lectr, Adelphi Col, 55-56 & Univ Fla, 58-59. *Mem:* Inst Elec & Electronics Engrs. *Res:* Microwave vacuum tubes, especially traveling wave tubes and klystrons. *Mailing Add:* Dept 252-1 Teledyne Mec PO Box 10007 Palo Alto CA 94303

LALLY, VINCENT EDWARD, b Brookline, Mass, Oct 13, 22; m 53; c 3. METEOROLOGY, ELECTRONICS. *Educ:* Univ Chicago, BS, 44; Mass Inst Technol, BS, 48, MS, 49. *Prof Exp:* Engr, Bendix-Friez, Md, 49-51; chief meteorol instrument sect, Air Force Cambridge Res Labs, 51-58; res mgr, Tele-Dynamics Div, Am Bosch Arma Corp, 58-61; PROG HEAD, NAT CTR ATMOSPHERIC RES, 61-, COMT ON SPACE RES, 65- *Honors & Awards:* Cleveland Abbe Award, Am Met Soc, 90. *Mem:* AAAS; fel Am Meteorol Soc; Sigma Xi. *Res:* Meteorological instruments and measurement systems. *Mailing Add:* 4330 Comanche Dr Boulder CO 80303

LALONDE, ROBERT THOMAS, b Bemidji, Minn, May 7, 31; m 57; c 7. TOXICOLOGY, GEOCHEMISTRY. *Educ:* St John's Univ, Minn, BA, 53; Univ Colo, PhD, 57. *Prof Exp:* Sr res engr chem, Jet Propulsion Lab, Calif Inst Technol, 57-58; res assoc, Univ Ill, 58-59; from asst prof to assoc prof, 59-68, PROF CHEM, STATE UNIV NY, 68- *Concurrent Pos:* NIH fel, 65-66, Fed Rep Ger Exchange, 80. *Mem:* Am Chem Soc; Am Soc Pharmacol; Environ Mutagen Soc; Sigma Xi. *Res:* Chemistry of natural products; chemistry of alkaloids, terpenoids, steroids and fatty acid derivatives; synthesis and structure-activity relations of halogen containing mutagens; geogenesis of organo-sulfur compounds. *Mailing Add:* Dept Chem State Univ NY Col Environ Sci & Forestry Syracuse NY 13210

LALOR, WILLIAM FRANCIS, b Dublin, Ireland, Sept 30, 35; US citizen; m 61; c 4. COTTON PRODUCTION & PROCESSING, TROPICAL FOOD CROPS. *Educ:* Univ Col, Dublin, B Agr Sc, 58; Mich State Univ, MS, 62; Iowa State Univ, PhD(agr eng), 68. *Prof Exp:* Lectr agr, Univ Col, Dublin, 62-65; prof agr eng, Auburn Univ, 68-71; scientist agr eng, Int Inst Tropical Agr, 71-73; dir, Agr Res, 73-90, VPRES, AGR, COTTON INC, 91- *Concurrent Pos:* Prof engr, dept consumer-affairs, Calif. *Mem:* Am Soc Agr Engrs. *Res:* Development of production and processing systems for cotton and cottonseed. *Mailing Add:* Cotton Inc PO Box 30067 Raleigh NC 27622

LAM, CHAN F, b Kwantung, China, Oct 23, 43; m 67; c 2. ELECTRICAL ENGINEERING. *Educ:* Calif Polytech State Univ, BS, 65; Clemson Univ, MS, 67, PhD(elec & comp eng), 70. *Prof Exp:* Res asst, Grad Inst Technol, Univ Ark, 65-66; res asst comp anal, Clemson Univ, 66-70; dir, Opers & Chief Prog, 71-72, from asst prof to assoc prof biomed eng, 70-80, dir time share & hybrid comput syst, 75-80, dir, Biomed Comput Ctr, 80-85, PROF BIOMED, MED UNIV SC, 80-, DIR, BIOMED IMAGE & SIGNAL PROCESSING LAB, 87- *Concurrent Pos:* Spec Study Sect, NIH, 79, Biomed Res Technol Review comt, 86-90; vis res prof, dept elec eng, Cheng Kung Univ, Tainan, Taiwan, 85. *Mem:* Sigma Xi; Inst Elec & Electronics Engrs; Pattern Recognition Soc; Soc Math Biol. *Res:* Modeling of enzyme kinetic reaction mechanisms; spinal cord injury detection; biomedical signal and image processing. *Mailing Add:* Dept Biomet Med Univ SC Charleston SC 29425

LAM, DANIEL J, b Hong Kong, Dec 30, 30; m 59; c 3. PHYSICAL METALLURGY & CHEMISTRY. *Educ:* Rensselaer Polytech Inst, BMetE, 56, MMetE, 58, PhD(phys metall), 60. *Prof Exp:* Res assoc metall, Rensselaer Polytech Inst, 56-58, instr, 58-60; asst metallurgist, 60-66, assoc metallurgist, 66-72, metallurgist, 72-74, SR SCIENTIST, ARGONNE NAT LAB, 74-, GROUP LEADER, 78- *Mem:* AAAS; Am Phys Soc; Am Inst Mining, Metall & Petrol Engrs. *Res:* Electronic structure and related physical and chemical properties of actinide metals, alloys and compounds; electronic structure and related physical properties of multicomponent oxides. *Mailing Add:* Mat Sci Div Argonne Nat Lab 9700 S Cass Ave Argonne IL 60439

LAM, FUK LUEN, b Hong Kong, Nov 7, 37; US citizen; m 68; c 2. CHEMISTRY. *Educ:* Univ SC, PhD(org chem), 66. *Prof Exp:* Fel chem, Mass Inst Technol, 66-67; Brandeis Univ, 68-69; res assoc, 70-75, ASSOC CHEM ONCOGENESIS, SLOAN-KETTERING INST, 75- *Mem:* Am Chem Soc. *Res:* Photochemistry and chemical reactions. *Mailing Add:* 92 Westminster Rd Chatham NJ 07928

LAM, GABRIEL KIT YING, b Hong Kong, Jan 1, 47; m 74; c 3. RADIATION BIOPHYSICS, CANCER RADIOTHERAPY. *Educ:* Univ Hong Kong, BSc, 70; Univ Western Ont, MSc, 71; Univ Toronto, PhD(biophysics), 74. *Prof Exp:* STAFF BIOPHYSICIST, BC CANCER RES CTR, 76- *Concurrent Pos:* Hon asst prof, Univ BC, 81- *Mem:* Radiation Res Soc; Am Asn Physicists Med. *Res:* Biophysical studies in the use of meson radiation for cancer radiotherapy; theoretical studies of radiation action. *Mailing Add:* Batho Biomed Facil TRIUMF Univ BC 4004 Wesbrook Mall Vancouver BC V6T 2A3 Can

LAM, GILBERT NIM-CAR, b Shanghai, China, Nov 10, 51. PHARMACOKINETICS, DRUG METABOLISM. *Educ:* State Univ NY Buffalo, BS, 76; Univ Ill, PhD(pharm), 81. *Prof Exp:* RES BIOCHEMIST, E I DUPONT DE NEMOURS & CO, INC, 81- *Mem:* Am Pharmaceut Asn. *Res:* Pharmacokinetics; biopharmaceutics; drug metabolism and analytical methodology of pharmaceuticals. *Mailing Add:* 20 Coach Hill Ct Newark DE 19711

LAM, HARRY CHI-SING, b Hong Kong, Nov 10, 36. THEORETICAL HIGH ENERGY PHYSICS. *Educ:* McGill Univ, BSc, 58; Mass Inst Technol, PhD(physics), 63. *Prof Exp:* Res assoc physics, Univ Md, 63-65; from asst prof to assoc prof, 65-75, chmn deprt, 76-80, PROF PHYSICS, MCGILL UNIV, 75- *Concurrent Pos:* Asst ed, Can J Physics, 73- *Mem:* Am Phys Soc; Can Asn Physicists. *Res:* Quantum field theory; particle theory. *Mailing Add:* Rutherford Physics Bldg McGill Univ 3600 University St Montreal PQ H3A 2T8 Can

LAM, JOHN LING-YEE, b Hong Kong, May 28, 40; US citizen. CLASSICAL & QUANTUM ELECTRODYNAMICS. *Educ:* Rice Univ, BA, 62; Calif Inst Technol, PhD(physics), 67. *Prof Exp:* Res fel, Calif Inst Technol, 66-68, Univ Miami, 68-69, Nat Res Coun Can, 69-71 & Max Planck Inst Physics & Astrophysics, Munich, 71-73; sr res physicist, Dikewood Corp, 74-81; SR RES PHYSICIST, NORTHROP CORP, 81- *Mem:* Am Phys Soc. *Res:* Interaction between radiation and matter in both the classical and quantum regimes, and in both the microscopic and macroscopic aspects. *Mailing Add:* 4821 Kent-Des Moines Rd Apt 303 Kent WA 98032

LAM, KAI SHUE, b Hong Kong, Feb 22, 49. PHYSICS, CHEMICAL PHYSICS. *Educ:* Univ Calif, Berkeley, AB, 70; Mass Inst Technol, PhD(physics), 76. *Prof Exp:* Fel & instr chem physics, 76-80, SR RES ASSOC CHEM PHYSICS, DEPT CHEM, UNIV ROCHESTER, 80- *Mem:* Am Phys Soc; Sigma Xi. *Res:* Atomic and molecular collision physics; atom-surface collisions; interaction of collision systems with laser radiation; spectral line broadening. *Mailing Add:* Dept Chem Univ Rochester Rochester NY 14627

LAM, KIN LEUNG, m; c 1. COMPUTATIONAL FLUID MECHANICS, NUCLEAR REACTOR SAFETY. *Educ:* Univ Calif, Berkeley, BS, 81; Univ Calif, Santa Barbara, PhD(chem eng), 89. *Prof Exp:* STAFF MEM, LOS ALAMOS NAT LAB, 89- *Mem:* Am Inst Chem Engrs; Soc Indust & Appl Math; Am Soc Mech Engrs; Am Phys Soc; Am Inst Aeronaut & Astronaut. *Res:* Computational fluid dynamics with applications in nuclear and chemical plant safety problems involving turbulence, multiphase flow and heat transfer phenomena. *Mailing Add:* Los Alamos Nat Lab MS K559 Los Alamos NM 87545

LAM, KUI CHUEN, b Hong Kong, Sept 22, 43; m 74; c 1. MATHEMATICAL PROGRAMMING & ACCURACY ANALYSIS, ASTRONAUTICAL GUIDANCE. *Educ:* Univ Hong Kong, BSc gen hons, 67, BSc spec hons, 68; Univ Ore, MS, 71, PhD(physics), 74. *Prof Exp:* Res assoc physics, Univ Ga, 74-75; asst prof physics, Pahlaui Univ, Shiraz, Iran, 75-76; lab supvr foreign languages, Western Carolina Univ, 76; res specialist atmospheric physics, Cloud Physics Dept, Univ Mo, Rolla, 76-79; res assoc radio astron, Mass Inst Technol, 79-80; TECH STAFF ASTRONAUT GUID, C S DRAPER LABS, INC, 80- *Mem:* Am Phys Soc. *Res:* Numerical analysis; mathematical physics; control and decision astronautical guidance; particle theory. *Mailing Add:* Mass Inst Technol BR PO Box 172 Cambridge MA 02139

LAM, KWOK-WAI, b Kowloon, Hong Kong, Sept 21, 35; m 61; c 2. BIOCHEMISTRY. *Educ:* ETex Baptist Col, BS, 57; Univ Pittsburgh, PhD(biochem), 63. *Prof Exp:* Nat Inst Child Health & Human Develop fel enzymol & geront, 63-65, assoc, 65-66; assoc enzymol, Retina Found, Boston, 66-73; res assoc prof biochem, 73-81, res prof ophthal, Albany Med Col, 81; PROF OPHTHAL, UNIV TEX HEALTH SCI CTR. *Concurrent Pos:* NIH career develop award, 67; asst prof biochem, Sch Med, Boston Univ, 70-73. *Mem:* Nat Acad Clin Biochem; Am Chem Soc; Asn Res Vision & Ophthal; Fedn Am Socs Exp Biol; Nat Registry Clin Chem. *Res:* Mechanism of oxidative phosphorylation; clinical enzymology. *Mailing Add:* Dept Ophthal Univ Tex Health Sci Ctr 7703 Floyd Curl Dr San Antonio TX 78284

LAM, LEO KONGSUI, b Hong Kong, Sept 12, 46. ATOMIC PHYSICS, CHEMICAL PHYSICS. *Educ:* Univ Hong Kong, BSc, 69; Columbia Univ, MA, 70, PhD(physics), 75. *Prof Exp:* Res assoc physics, Joint Inst Lab Astrophys, Univ Col, 75-77; res asst prof physics, Univ Mo-Rolla, 77-79; mem fac, Univ Southern Calif, 79-81; mem tech staff, 81-88, SR MEM TECH STAFF, GUIDANCE & CONTROL SYSTS DIV, LITTON INDUSTS INC, 88- *Concurrent Pos:* Guest worker physics, Boulder Labs, Nat Bur Standards, 75-77. *Mem:* Am Phys Soc; Sigma Xi; Inst Elec & Electronics Engrs. *Res:* Atomic, molecular and chemical physics; optical double resonance spectroscopy; laser spectroscopy; atom-molecule kinetics; low temperature plasma. *Mailing Add:* Litton Industs Inc 5500 Canoga Ave Mail Sta 12 Woodland Hills CA 91367

LAM, NGHI QUOC, b Vietnam, Oct 4, 45; US citizen; m 69; c 3. METAL PHYSICS & RADIATION EFFECTS. *Educ:* Laval Univ, BS, 68; McMaster Univ, PhD(mat sci), 71. *Prof Exp:* Fel metal physics, 71-74, asst scientist, 74-77, scientist, 77-88, SR SCIENTIST, ARGONNE NAT LAB, 88- *Concurrent Pos:* Adj prof, Div Med Physics & Bioeng, Chicago Med Sch, 86-88; vis scientist, Ctr Nuclear Studies, Saclay, France, 79-80 & 82. *Honors & Awards:* Material Sci Res Award, Dept Energy, 84. *Mem:* Am Phys Soc; Mat Res Soc. *Res:* Radiation effects; atomic defects; diffusion; segregation; phase transformation; sputtering; ion implantation; electron microscopy; computer modeling and simulations. *Mailing Add:* Mat Sci Div Argonne Nat Lab Argonne IL 60439

LAM, SAU-HAI, b Macao, Dec 18, 30; m 59. AERONAUTICAL ENGINEERING. *Educ:* Rensselaer Polytech Inst, BAeroEng, 54; Princeton Univ, PhD(aeronaut eng), 58. *Prof Exp:* Asst, Princeton Univ, 56-58, res assoc, 58-59; asst prof, Cornell Univ, 59-60; from asst prof to assoc prof aeronaut eng, Princeton Univ, 60-67, chmn, Eng Physics Prog, 72-81, assoc dean eng, 80-81, co-chmn, prog appl & computational math, 83-86, chmn, dept mech & aerospace eng, 83-89, PROF AERONAUT ENG, PRINCETON UNIV, 67-, EDWIN WILSEY '04 CHAIRED PROF, 73- *Concurrent Pos:* Sr NSF fel, 66-67. *Mem:* Am Inst Aeronaut & Astronaut; Am Phys Soc; Am Soc Mech Engrs. *Res:* Theoretical gas dynamics; boundary layer theory; ionized gas flows. *Mailing Add:* Dept Mech & Aerospace Eng Princeton Univ D214 Eng Quadrangle Princeton NJ 08544

LAM, SHEUNG TSING, b Hong Kong, Dec 11, 34; Can citizen. APPLIED NUCLEAR PHYSICS. *Educ:* Univ Hong Kong, BSc, 59; Univ Ottawa, MSc, 62; Univ Alta, PhD(physics), 67. *Prof Exp:* Demonstr physics, Univ Hong Kong, 59-60; Can Nat Coun fel & res assoc nuclear physics, Univ Toronto, 67-70; asst prof nuclear physics, Univ Va, 70-72; staff physicist & safety officer, Nuclear Res Ctr, 72-86, AMS PROJ COORDR, UNIV ALTA, 87- *Concurrent Pos:* Attached staff mem, Chalk River Nuclear Labs, Atomic Energy Can Ltd, 67-70; Frederick Gardner Cottrell Res Corp grant, 71-72. *Mem:* Am Phys Soc. *Res:* Nuclear structure studies using electrostatic accelerators and fast neutron induced fission studies; neutron-neucleus scattering and analysis using optical potentials; study of 3-body interacton using n-D breakup reaction; trace element analysis using proton-induced X-ray emission; trace isotope analysis using accelerator mass spectrometry. *Mailing Add:* Nuclear Res Ctr Univ Alta Edmonton AB T6G 2N5 Can

LAM, STANLEY K, b Hong Kong; US citizen. SEPARATION SCIENCE, CLINICAL CHEMISTRY. *Educ:* Calif State Univ, BA, 74; State Univ NY, Buffalo, PhD(chem), 80. *Prof Exp:* ASSOC PROF LAB MED, ALBERT EINSTEIN COL MED, 78- *Mem:* Am Chem Soc; Am Asn Clin Chem. *Res:* Chromatographic methods for the monitoring of therapeutic agents and development of chromatographic techniques. *Mailing Add:* Albert Einstein Col Med 1300 Morris Park Ave Bronx NY 10461

LAM, TENNY N(ICOLAS), b Hong Kong, Nov 28, 40; m 66; c 1. TRANSPORTATION ENGINEERING, OPERATIONS RESEARCH. *Educ:* Univ Calif, Berkeley, BS, 63, MEng, 64, DEng(transp sci), 67. *Prof Exp:* Asst prof civil eng, Univ Mo-Columbia, 66-68; sr res engr, Dept Theoret Physics, Gen Motors Res Labs, 68-74; from assoc prof to prof civil eng, Univ Calif, Davis, 74-86; RETIRED. *Concurrent Pos:* Assoc ed, Transp Sci, 74-77 & 80-86. *Mem:* Opers Res Soc Am; Am Soc Civil Engrs. *Res:* Traffic flow theory; transportation systems planning and analysis. *Mailing Add:* 3100 Shelter Cove Univ Calif Davis CA 95616

LAM, TSIT-YUEN, b Hong Kong, Feb 6, 42; m 70; c 4. ALGEBRA. *Educ:* Hong Kong Univ, BA, 63; Columbia Univ, PhD(math), 67. *Prof Exp:* Fel math, Univ Ill, Urbana, 67; instr, Univ Chicago, 67-68; lectr, Univ Calif, 68-69, from asst prof to assoc prof, 69-77, vchmn dept, 75 & 80-81, Miller prof, 78-79, PROF MATH, UNIV CALIF, BERKELEY, 76- *Concurrent Pos:* Alfred P Sloan Found fel, 72-74; Guggenheim fel, 81-82. *Honors & Awards:* Steele Prize, Am Math Soc, 83. *Mem:* Am Math Soc. *Res:* Finite groups and group representation theory; quadratic forms; ring theory. *Mailing Add:* Dept Math Univ Calif Berkeley CA 94720

LAM, VINH-TE, b Saigon, SVietnam, Dec 12, 39; Can citizen; m 80. PHYSICAL CHEMISTRY. *Educ:* Univ Montreal, BSc, 62, PhD(phys chem), 67. *Prof Exp:* Prof org chem, Col St Laurent, 66-67; fel, Nat Res Coun Can, 67-69; lectr phys chem & Nat Res Coun Can grant, Univ Sherbrooke, 69-72; PROF CHEM, COL BOIS-DE-BOULOGNE, 72- *Mem:* Am Chem Soc; Chem Inst Can. *Res:* Thermodynamics; thermochemistry; static and dynamic microcalorimetry; critical phenomena; surface and polymer chemistry; molecular interactions; structure of liquids and solutions. *Mailing Add:* 6728 Chateaubriand Montreal PQ H2S 2N8 Can

LAM, YIU-KUEN TONY, b Hong Kong, June 5, 47; US citizen; m 77; c 3. ORGANIC CHEMISTRY. *Educ:* Chinese Univ, Hong Kong, BSc, 71; Univ NB, PhD(org chem), 74. *Prof Exp:* Res assoc, Univ Tex, Austin, 75-77; asst prof, Univ Alta, 77-79; sr res chemist, 79-83, RES FEL, MERCK & CO, INC, 84- *Mem:* AAAS; Am Chem Soc; Am Soc Pharmacog. *Res:* Discovery and chemistry of novel biologically interesting principles from microbial, herbal and animal sources. *Mailing Add:* 25 Hamilton Lane N Plainsboro NJ 08536

LAMANNA, CARL, microbiology, toxicology, for more information see previous edition

LAMANNA, JOSEPH CHARLES, b Bronxville, NY, July 12, 49; m 71; c 3. NEUROPHYSIOLOGY, OPTICAL INSTRUMENTATION. *Educ:* Georgetown Univ, BS, 71; Duke Univ, PhD(physiol), 75. *Prof Exp:* NIH fel & res assoc physiol, Duke Univ Med Ctr, 75-77; from asst prof to assoc prof, dept neurol & physiol-biophys, Med Sch, Univ Miami, 77-90; PROF, DEPT NEUROL, CASE WESTERN RESERVE UNIV, 90- *Mem:* Am Physiol Soc; Optical Soc Am; Int Soc Oxygen Transp Tissues; Soc Neurosci; Biomed Eng Soc. *Res:* Determining the role of oxygen and oxidative energy metabolism in the function of the central nervous system in mammals, utilizing optical monitoring techniques. *Mailing Add:* Dept Neurol Wearn S-52 Case Western Reserve Univ Hosps Abington Rd Cleveland OH 44106

LA MANTIA, CHARLES R, b New York, NY, June 12, 39; m 61; c 2. CHEMICAL ENGINEERING. *Educ:* Columbia Univ, BA, 60, BS, 61, MS, 63, ScD(chem eng), 65. *Prof Exp:* Res & develop proj off, Defense Atomic Support Agency, 65-67; vpres chem & metall eng, Arthur D Little Inc, Cambridge, 67-81; pres, Koch Process Systs Inc, Westborough, Mass, 81-86; PRES, ARTHUR D LITTLE, INC, CAMBRIDGE, 86- *Concurrent Pos:* Mem staff, Charles F Bonilla & Assocs, 65. *Mem:* Am Inst Chem Engrs. *Res:* Chemical process design, analysis and development; air pollution control; energy technology; cryogenic technology. *Mailing Add:* Three Goodwin Rd Lexington MA 02173

LA MAR, GERD NEUSTADTER, b Brasov, Romania, Dec 21, 37; US citizen; m 64; c 2. STRUCTURAL CHEMISTRY. *Educ:* Lehigh Univ, BS, 60; Princeton Univ, PhD(chem), 64. *Prof Exp:* NSF fel, 64-66; NATO fel, 66-67; res chemist, Shell Develop Co, 67-70; from asst prof to assoc prof, 71-74, PROF CHEM, UNIV CALIF, DAVIS, 74- *Concurrent Pos:* Fel, Alfred P Sloan Found, 72, John Simon Guggenheim Mem Found, 75. *Mem:* Am Chem Soc. *Res:* The use of magnetic resonance spectroscopy as a tool for elucidating structure-function relationships in metallo-enzymes and their model complexes. *Mailing Add:* Dept Chem Univ Calif Davis CA 95616

LAMARCA, MICHAEL JAMES, b Jamestown, NY, June 4, 31; m 54; c 3. DEVELOPMENTAL BIOLOGY. *Educ:* State Univ NY Albany, AB, 53; Cornell Univ, PhD(zool), 61. *Prof Exp:* Instr zool, Rutgers Univ, 61-63, asst prof, 63-65; from asst prof to assoc prof biol, 65-76, chmn dept, 70-74, PROF BIOL, LAWRENCE UNIV, 76- *Concurrent Pos:* NSF res grant, 63-65; resident dir, Assoc Cols Midwest Argonne Semester Prog, Argonne Nat Lab, 68-69; NSF sci fac fel biol sci, Purdue Univ, 71-72; vis lectr biol chem, Harvard Med Sch, 77-78. *Res:* RNA and protein synthesis in echinoderm, amphibian, and mammalian development. *Mailing Add:* Dept Biol Lawrence Univ Main Campus Box 599 Appleton WI 54912

LAMARCHE, FRANÇOIS, b Montreal, Que, Jan 2, 60; m 86; c 2. SURFACE ACTIVITY OF PROTEINS, PROTEIN STRUCTURE. *Educ:* Univ Que Trois-Rivil0res, BSc, 82, PhD(biophys), 88. *Prof Exp:* RES SCIENTIST PROTEIN STRUCT, FOOD RES DEVELOP CTR, AGR CAN, 88- *Mem:* Biophys Soc; Protein Soc. *Res:* Study of the behavior of proteins at air-water and oil-water interface; importance of structural factors of protein on their surface properties. *Mailing Add:* Food Res & Develop Ctr Agr Can St-Hyacinthe PQ J2S 8E3 Can7

LAMARCHE, J L GILLES, b Montreal, Que, May 31, 27. PHYSICS. *Educ:* Univ Montreal, BSc, 50; Univ BC, MA, 53, PhD(physics), 57. *Prof Exp:* From asst prof to assoc prof, 57-70, PROF PHYSICS, UNIV OTTAWA, 70- *Mem:* Am Phys Soc; Asn Canadienne Francaise pour D l'Avancement des Sciences; Can Asn Physicists. *Res:* Low temperature physics; semimagnetic semiconductor magnetism; nuclear magnetism. *Mailing Add:* Dept Physics Univ Ottawa Ottawa ON K1N 6N5 Can

LAMARCHE, PAUL H, b Boston, Mass, Sept 5, 29; m 52; c 5. GENETICS, PEDIATRICS. *Educ:* Boston Col, BS, 56; Boston Univ, MD, 60; Mass Inst Technol, ScM, 74. *Hon Degrees:* MA, Brown Univ. *Prof Exp:* Res assoc path & dir genetics lab, RI Hosp, 63-75, med dir, Birth Defects Ctr, 65-75, med dir child develop ctr, 66-75, assoc physician-in-chief pediat, 69-75; chief pediat & genetics, 74-85, MED DIR, EASTERN MAINE MED CTR, 85-; PROF GENETICS, UNIV MAINE, ORONO, 74- *Concurrent Pos:* Asst pediatrician, Providence Lying-In Hosp, 63-74, consult, 66-74; prin investr Nat Cancer Inst grant, 64-69; prof pediat, Sch Med, Tufts Univ, 81; consult, Child Study Ctr, Brown Univ, 67-74. *Mem:* AAAS; Genetics Soc Am; Tissue Cult Asn. *Res:* Genetics and cytogenetics of teratogenesis and oncogenesis; electron microscopy of fine structure of somatic cellular phenotypes normal and abnormal in the human. *Mailing Add:* 489 State St Bangor ME 04401

LAMARCHE, PAUL HENRY, JR, b Norwood, Mass, April 21, 53. PLASMA-MATERIALS INTERACTIONS, ULTRA-HIGH VACUUM SCIENCE. *Educ:* Boston Col, BS, 75; Yale Univ, MS, 76, PhD(physics), 81. *Prof Exp:* Res physicist, Exxon Prod Res Co, 81-82; res assoc, Univ Chicago, 82-84; HEAD VACUUM SYSTS GROUP, PRINCETON UNIV, 84- *Concurrent Pos:* Consult, 84-; chmn, Plasma Sci & Technol Div, Am Vacuum Soc, 89. *Mem:* AAAS; Am Phys Soc; Am Vacuum Soc. *Res:* The interaction of energetic plasma ions with solids as embodied in fusion research devices; interfacial processes at vacuum-wall boundary; consulting work on vacuum vessel and system design. *Mailing Add:* Princeton Univ PO Box 451 Princeton NJ 08543

LAMASTRO, ROBERT ANTHONY, b New York, NY, Sept 11, 56; m 81; c 1. GLASS TECHNOLOGY. *Educ:* Rutgers Univ, BA & BS, 79, MS, 81, PhD(ceramic eng), 82. *Prof Exp:* Glass technologist, 82-84, MGR GLASS RES & DEVELOP, WHEATON INDUSTS, 84-; ADJ PROF, CUMBERLAND CO COL, 89- *Mem:* Am Ceramic Soc; Soc Glass Technol; Parenteral Drug Asn; Am Chem Soc; AAAS. *Res:* Development of specialty glass formulations for the pharmaceutical and cosmetic packaging industries. *Mailing Add:* Wheaton Industs 1101 Wheaton Ave Millville NJ 08332

LAMATTINA, JOHN LAWRENCE, b Brooklyn, NY, Jan 22, 50; m 71; c 3. HETEROCYCLIC CHEMISTRY, MEDICINAL CHEMISTRY. *Educ:* Boston Col, BS, 71; Univ NH, PhD(chem), 75. *Prof Exp:* RES SCIENTIST MED CHEM, PFIZER INC, 77-, DIR MED CHEM, 87- *Concurrent Pos:* NIH fellow, Princeton Univ, 75-77. *Mem:* Am Chem Soc. *Res:* Design and synthesis of compounds which possess intriguing biological properties. *Mailing Add:* Pfizer Inc Eastern Point Rd Groton CT 06340

LAMAZE, GEORGE PAUL, b Algiers, Algeria, Jan 15, 45; US citizen; m 65; c 2. NEUTRON DEPTH PROFILING, EXPERIMENTAL NUCLEAR PHYSICS. *Educ:* Fla State Univ, BA, 65; Duke Univ, PhD(physics), 72. *Prof Exp:* Physicist neutron standards, Nat Bur Standards, 72-89; PHYSICIST, INORG ANALYSIS RES DIV, NAT INST STANDARDS & TECHNOL, 89- *Concurrent Pos:* Sci asst to Rep George Brown, Calif, 78-79; secy comt on Nuclear Technol & Applns, Am Soc Testing & Mat, vchair, subcomt on Nuclear Radiation Metrol. *Mem:* Am Phys Soc; Am Soc Testing & Mat. *Res:* Measurement of neutron depth profiling; radioactivity measurements; cold neutron fluence rates. *Mailing Add:* Nat Inst Standards & Technol Bldg 235 Gaithersburg MD 20899

LAMB, DAVID E(RNEST), b Pampa, Tex, Apr 6, 32; m 56; c 5. CHEMICAL ENGINEERING, COMPUTER SCIENCE. *Educ:* Yale Univ, BE, 53; Princeton Univ, MS, 54, PhD(chem eng), 62. *Prof Exp:* Instr chem eng, Princeton Univ, 56; res engr, Sun Oil Co, Pa, 57; asst prof chem eng, 58-63, assoc prof & dir comput ctr, 63-65, PROF CHEM ENG, STATIST &

COMPUT SCI & CHMN DEPT STATIST & COMPUT SCI, UNIV DEL, 65- *Concurrent Pos:* Consult, Sun Oil Co, 57-, Ethyl Corp, 60 & Prentice-Hall, Inc, 60-; mem theory comt, Am Automatic Control Coun, 61-; mem steering comt, Simulation Coun, 65-; mem adv comt, Off Comput Activities, NSF, 68-71. *Mem:* Asn Comput Mach; Simulation Coun. *Res:* Continuous and discrete system simulation; computer graphics. *Mailing Add:* 708 Nottingham Rd Newark DE 19711

LAMB, DENNIS, b Chicago, Ill, Feb 3, 41; m 81; c 1. CLOUD PHYSICS. *Educ:* Kalamazoo Col, BA, 63; Univ Wash, PhD(atmospheric sci), 70. *Prof Exp:* Gen physicist data assessment, Naval Weapons Ctr, China Lake, Calif, 63-65; NATO res assoc meteorol, Univ Frankfurt, 71-72; from asst to res prof, Atmospheric Sci Ctr, Desert Res Inst, Univ Nev, Reno, 72-86; ASSOC PROF METEROL DEPT, PA STATE UNIV, UNIVERSITY PARK, 86- *Mem:* Am Meteorol Soc; Sigma Xi; Am Geophys Union. *Res:* Nucleation and growth of solids from the liquid and vapor phases; formation of cloud nuclei; cloud physics/weather modification; atmospheric chemistry. *Mailing Add:* Meteorol Dept Penn State Univ 503 Walker Bldg University Park PA 16802

LAMB, DONALD JOSEPH, b Pittsburgh, Pa, Oct 29, 31; m 56; c 2. PHARMACY. *Educ:* Ohio State Univ, BSc, 54, MSc, 55, PhD(pharm), 60. *Prof Exp:* Res assoc pharmaceut res & develop, 60-65, res head, 65-70, res mgr pharmaceut res, 70-83, dir proj support, 83-88, DIR PROJ MGT, UPJOHN CO, 88- *Mem:* Am Pharmaceut Asn; Am Acad Pharmaceut Sci; Am Chem Soc; Sigma Xi. *Res:* Design and evaluation of drug dosage forms, including design and evaluation of drugs to fit specific dosage forms. *Mailing Add:* 5128 Allardowne St Kalamazoo MI 49001

LAMB, DONALD QUINCY, JR, b Manhattan, Kans, June 30, 45; m 78; c 1. ASTROPHYSICS. *Educ:* Rice Univ, BA, 67; Univ Liverpool, MSc, 69; Univ Rochester, PhD(physics), 74. *Prof Exp:* Res asst prof physics, Univ Ill, 73-75, from asst prof to prof, 75-80; physicist, Smithsonian Ctr Astrophys, Harvard Univ, 80-85; PROF ASTRON & ASTROPHYS, UNIV CHICAGO, 85-, DEPT CHMN, 88- *Concurrent Pos:* Marshall scholar, 67-69; vis assoc prof physics, Mass Inst Technol, 78-79; vis scientist, Smithsonian Ctr Astrophys, Harvard Univ, 79-80; lectr astron, Harvard Univ, 80-85; trustee, Aspen Ctr Physics, 81-87, sect, 85-86, adv bd, 87-; vis prof physics, Inst Theoret Physics, Univ Calif, Santa Barbara, 87. *Mem:* fel Am Phys Soc; Am Astron Soc; fel Royal Astron Soc; Brit Inst Physics; Europ Phys Soc. *Res:* Evolution and structure of white dwarfs and neutron stars; physics of compact x-ray and gamma-ray sources, supernovae; properties of matter at high densities. *Mailing Add:* Dept Astron & Astrophys Univ Chicago 5640 S Ellis Ave Chicago IL 60637

LAMB, DONALD R(OY), b Yuma, Colo, May 6, 23; m 43; c 3. CIVIL ENGINEERING. *Educ:* Hastings Col, BA, 47; Univ Wyo, BS, 51, MS, 53, CE, 58; Purdue Univ, PhD, 62. *Prof Exp:* Supt high schs, Nebr, 46-47, coach, 47-49; from supply instr to assoc prof civil eng, 51-70, PROF CIVIL ENG & HEAD DEPT, UNIV WYO, 70- *Mem:* Am Soc Eng Educ; Am Soc Civil Engrs; Nat Soc Prof Engrs; Sigma Xi. *Res:* Use of radioisotopes in the study of portland cement, asphalt concrete and soils; portland cement concrete and associated aggregates; transportation; recreational engineering; engineering geology. *Mailing Add:* 1354 Indian Hills Dr Laramie WY 82070

LAMB, FREDERICK KEITHLEY, b Manhattan, Kans, June 30, 45; m 71; c 2. ASTROPHYSICS. *Educ:* Calif Inst Technol, BS, 67; Oxford Univ, DPhil(theoret physics), 70. *Prof Exp:* Instr & res assoc, 70-72, from asst prof to assoc prof, 72-78, PROF PHYSICS, UNIV ILL, URBANA, 78- *Concurrent Pos:* Fel physics, Magdalen Col, Oxford Univ, 70-72; assoc, Ctr Advan Study, Univ Ill, Urbana, 73-74; res fel, Alfred P Sloan Found, 74-78; vis fel, Inst of Astronomy, Cambridge, UK, 75-76; fel commorer, Churchill Col, Cambridge, UK, 75-76; vis assoc, Caltech, 77-78; fel John Simon Guggenheim Found, 85-86; vis scholar, Stanford Univ Ctr Space Sci & Astrophysics; sci fel, Ctr Int Security & Arms Control, Stanford Uni, 85-86. *Mem:* Fel Am Phys Soc; Am Astron Soc; fel Royal Astron Soc; Int Astronomical Union. *Res:* White dwarfs, neutron stars, and black holes; plasma theory and applications to pulsars and cosmic X-ray sources; the interaction of radiation with matter; arms control and international security. *Mailing Add:* Dept Physics Univ Ill 1110 W Green St Urbana IL 61801

LAMB, GEORGE ALEXANDER, b Glens Falls, NY, Sept 25, 34; m 56; c 3. PEDIATRICS, INFECTIOUS DISEASES. *Educ:* Swarthmore Col, BS, 55; State Univ NY Upstate Med Ctr, MD, 59. *Prof Exp:* Intern pediat, State Univ NY Upstate Med Ctr, 59-60, resident, 60-62, from asst prof to assoc prof, 64-72; assoc prof prev & social med, Harvard Med Sch, 72-79; PROF PEDIAT, BOSTON UNIV SCH MED, 79- *Concurrent Pos:* Fel infectious dis, 64- *Res:* Infectious diseases of children, especially the epidemiology of respiratory illnesses; community child health. *Mailing Add:* Boston Dept Health & Hosps 818 Harrison Ave Boston MA 02118

LAMB, GEORGE LAWRENCE, JR, b Norwood, Mass, Apr 28, 31; m 59; c 4. PHYSICS. *Educ:* Boston Col, BS, 53, MS, 54; Mass Inst Technol, PhD(physics), 58. *Prof Exp:* Staff mem, Los Alamos Sci Lab, 58-63; physicist, United Aircraft Res Labs, Conn, 63-74; FAC MEM & PROF MATH & PROF OPTICAL SCI, UNIV ARIZ, 74- *Mem:* Am Phys Soc; Sigma Xi; Acoust Soc Am. *Res:* Nonlinear waves and solitons; acoustic wave propagation. *Mailing Add:* 2942 Ave Del Conquistador RR 2 Univ Ariz Tucson AZ 85749

LAMB, GEORGE MARION, b Little Rock, Ark, Dec 23, 28; m 53; c 2. MICROPALEONTOLOGY, STRATIGRAPHY. *Educ:* Emory Univ, BA, 50, MS, 54; Univ Colo, Boulder, PhD(geol), 64. *Prof Exp:* Geologist, Standard Oil Calif, Inc, 55-61; prof geol, 64-89, ASST TO PRES, UNIV SALA, 89- *Mem:* Am Asn Petrol Geol; Geol Soc Am; Am Inst Prof Geologists. *Res:* Ecology and paleoecology of Foraminifera; biostratigraphic relationships; groundwater and environmental geology; beach erosion and development. *Mailing Add:* Off of the Pres Univ SAla 307 Unit Blvd Mobile AL 36688

LAMB, H RICHARD, b Philadelphia, Pa, Sept 18, 29; m 69; c 3. PSYCHIATRY. *Educ:* Univ Pa, BA, 50; Yale Univ, MD, 54. *Prof Exp:* Chief rehab serv, San Mateo County Ment Health Serv, 60-76; assoc prof, 76-80, PROF PSYCHIAT, SCH MED, UNIV SOUTHERN CALIF, 80- *Concurrent Pos:* Consult, NIMH, 75-; ed-in-chief, New Directions Ment Health Serv J, 78-; mem comt rehab, Am Psychiat Asn, 78-84, Comt Chronically Mentally Ill, 86-, chmn, Comt Coord Hosp & Community Psychiat Serv, 83-; mem Group Advan Psychiat. *Mem:* Fel Am Psychiat Asn; fel Am Col Psychiatrists. *Res:* Social and community psychiatry and community mental health with a major focus on the long-term severely disabled psychiatric patient in the community. *Mailing Add:* Dept Psychiat Univ Southern Calif Sch Med 1934 Hosp Pl Los Angeles CA 90033

LAMB, J(AMIE) PARKER, JR, b Boligee, Ala, Sept 21, 33; m 55; c 2. MECHANICAL & AEROSPACE ENGINEERING. *Educ:* Auburn Univ, BS, 54; Univ Ill, MS, 58, PhD(mech eng), 61. *Prof Exp:* Proj engr, Flight Control Lab, Wright Air Develop Ctr, Ohio, 55-57; asst prof eng mech, NC State Univ, Raleigh, 61-63; from asst prof to prof mech eng, 63-81, chmn dept, 70-76, assoc dean, Col Eng, 76-81, chmn aerospace eng, 81-88, ERNEST COCKRELL JR MEM PROF, UNIV TEX, AUSTIN, 81- *Concurrent Pos:* Consult, ARO, Inc, Tenn, 63-65, Tracor, Inc, Tex, 65-67, NASA, Ala, 69-70, Vought Aerospace Corp, Tex, 69-70 & Mobil Oil Corp, Tex, 77-78; assoc tech ed, J Fluids Eng, 76-79; gen chmn, Tenth US Nat Cong, Appl Mech, 86. *Honors & Awards:* Founders Award, Am Soc Mech Engrs, 75; Centennial Award, Am Soc Mech Engrs, 80; Joe J King Prof Eng Award, 84. *Mem:* Fel Am Soc Mech Engrs; assoc fel Am Inst Aeronaut & Astronaut; Am Soc Eng Educ; Nat Soc Prof Engrs. *Res:* Heat transfer and fluid mechanics in separated flow regions; compressible turbulent boundary layers; heat, mass and momentum transfers in free turbulent jets; energy conversion processes for low temperature sources. *Mailing Add:* Dept Mech Eng Univ Tex Austin TX 78712

LAMB, JAMES C(HRISTIAN), III, b Warsaw, Va, Aug 20, 24. SCIENCE EDUCATION, RESEARCH ADMINISTRATION. *Educ:* Va Mil Inst, BS, 47; Mass Inst Technol, SM, 48, SE, 52, ScD(sanit eng), 53; Environ Engrs Intersoc Bd, dipl. *Prof Exp:* Instr civil eng, Va Mil Inst, 48-50; asst sanit eng, Mass Inst Technol, 51-53, res assoc, 53-55; sanit engr, Am Cyanamid Co, 55-59; assoc prof, 59-65, PROF SANIT ENG, UNIV NC, CHAPEL HILL, 65- *Concurrent Pos:* Consult engr, 48-50, 52-55, 59-; lectr, Washington & Lee Univ, 49-50 & Exten Div, State Dept Educ, Mass, 51-52; adj prof, Newark Col Eng, 56-59; Judge, US Nuclear Regulatory Comn, 74- *Mem:* Am Soc Civil Engrs; Am Water Works Asn; Water Pollution Control Fedn. *Res:* Industrial wastes; sewage treatment; water supply; saline water conversion; corrosion; stream pollution; refuse disposal; steam pollution, regulatory controls and standards; civil engineering. *Mailing Add:* Dept Environ Sci & Eng Univ NC Chapel Hill NC 27514

LAMB, JAMES FRANCIS, b Denton, Tex, Oct 3, 37; m 58; c 2. NUCLEAR MEDICINE, NUCLEAR CHEMISTRY. *Educ:* Univ NTex, BS, 60, MS, 61; Univ Calif, Berkeley, PhD(chem), 69. *Prof Exp:* Res chemist, Lawrence Radiation Lab, Univ Calif, 69-70; prin radiochemist, Medi-Physics, Inc, 70-74, assoc dir, 74-81, dir res & develop, 81-85; PRES, IMAGENTS INC, 85-; PRES, IMP INC, 90- *Concurrent Pos:* Reviewer, J Nuclear Med, 78-; fac, Int Symp, Washington, DC, 83, High Country Nuclear Med Conf, Vail, Colo, 83-90, Diag Nuclear Med, Univ Calif-San Francisco, 85, Lawrence Berkeley Labs, 88; invited speaker, Conf Appln Accelerators Res & Indust, Univ NTex, 88; organizing comt, Int Symp Radiopharmaceut Synthesis, Qual Assurance & Regulatory Control, Am Chem Soc, 90- *Mem:* Am Chem Soc; Soc Nuclear Med; AAAS; Sigma Xi. *Res:* Nuclear chemistry in nuclear medical and radiopharmaceutical applications. *Mailing Add:* IMP Inc 8050 El Rio Houston TX 77054

LAMB, JAMES L, b Los Angeles, Calif, Jan 17, 25; m 45; c 2. MICROPALEONTOLOGY. *Educ:* Univ Southern Calif, BS, 53. *Prof Exp:* Paleontologist, Richfield Oil Corp, 53-57 & Creole Petrol Corp, 57-64; paleontologist, Exxon Prod Res Co, 64-81; CONSULT, 81- *Mem:* Soc Econ Paleont & Mineral; Am Asn Petrol Geologists; Venezuelan Asn Geol, Mining & Petrol. *Res:* Historical geology and paleontology; geologic distribution of planktonic foraminifera; tertiary microfossils; Pleistocene epoch. *Mailing Add:* 1358 Lawnridge St Medford OR 97504

LAMB, MINA MARIE WOLF, b Sagerton, Tex, Aug 14, 10; m 41; c 1. NUTRITION. *Educ:* Tex Tech Col, BA, 32, MS, 37; Columbia Univ, PhD(nutrit, chem), 42. *Prof Exp:* Teacher, elem & high sch, 33-35; teacher & res worker food & nutrit, 35-37; from lab asst to prof, 40-69, Margaret W Weeks distinguished prof, 69-75, head dept food & nutrit, 55-69, lectr & adv foreign students, 60-71, EMER PROF FOOD & NUTRIT, TEX TECH UNIV, 75- *Honors & Awards:* Piper Award, 65; Medallion Award, from Am Dietetic Assoc, 86; Serv Award, 87. *Mem:* AAAS; Am Dietetic Asn; Am Home Econ Asn. *Res:* Basal metabolism of college girls and children of various ages older than two years; needs of children and adults; dietary studies of children, college girls and families; animal feeding work with albino rats determining growth and reproduction responses to various diets and foods. *Mailing Add:* 6002 W 34th St Lubbock TX 79407-1241

LAMB, NEVEN P, b New York, NY, May 18, 32; m 57; c 2. BIOLOGICAL ANTHROPOLOGY. *Educ:* Pa State Univ, BA, 54; Univ Ariz, PhD(anthrop), 69. *Prof Exp:* Res asst morphogenetics, Jackson Mem Lab, Maine, 60-61; from asst prof to assoc prof anthrop, Portland State Univ, 65-73; ASSOC PROF ANTHROP, TEX TECH UNIV, 73- *Concurrent Pos:* Vis prof, Univ Ariz, 71-72; prin investr, NIH, 74-76; proj dir curric improvement, NSF, 78-81. *Mem:* AAAS; fel Am Asn Anthropologists; Am Asn Phys Anthropologists; Sigma Xi. *Res:* Human evolution; population biology; mating patterns and genetic systems; anthropometry of North American Indians; socio-cultural and biological aspects of mate selection among Papago Indians. *Mailing Add:* Dept Anthrop Tex Tech Univ Lubbock TX 79409

LAMB, RICHARD C, b Lexington, Ky, Sept 8, 33; m 59; c 4. GAMMA RAY ASTRONOMY, ELEMENTARY PARTICLE PHYSICS. *Educ:* Mass Inst Technol, BS, 55; Univ Ky, PhD(physics), 63. *Prof Exp:* Asst scientist, Argonne Nat Lab, 63-67; assoc prof physics, 67-72, PROF PHYSICS, IOWA STATE UNIV, 72- *Concurrent Pos:* Vis scientist, NASA-Goddard Space Flight Ctr, 75-76; prin investr prog observational gamma-ray astron, 80-; sr associateship, Jet Propulsion Lab, Nat Res Coun, 82-83. *Mem:* Am Phys Soc; Am Astron Soc. *Res:* Very high energy gamma ray astronomy using the atmospheric Cerenkov technique; identification of gamma ray sources. *Mailing Add:* Dept Physics Iowa State Univ Ames IA 50011

LAMB, ROBERT ANDREW, b London, Eng, Sept 26, 50; m 89. VIROLOGY, MOLECULAR BIOLOGY. *Educ:* Univ Birmingham, BSc, 71; Univ Cambridge, PhD(virol), 74. *Prof Exp:* Res assoc, Rockeveller Univ, 74-77, from asst prof to assoc prof virol, 77-82; JOHN EVANS PROF MOLECULAR & CELLULAR BIOL, NORTHWESTERN UNIV, 83-, INVESTR, HOWARD HUGHES MED INST, 91- *Concurrent Pos:* Fulbright-Hays travel award, 74-77; Irma T Hirschl career scientist award, 79-83; assoc ed, Virol, 80-, ed, J Virol, 87-; estab investr, Am Heart Asn, 82-87. *Honors & Awards:* Phoebe Weinstein Award for Negative Strand Virus Res, 80; Wallace P Rowe Award for Excellence in Virol Res, 90. *Mem:* Am Soc Cell Biol; Am Soc Microbiol; Am Soc Biochem & Molecular Biol; Soc Gen Microbiol; Am Soc Virol; AAAS. *Res:* Virology; replication of influenza virus and paramyxoviruses; cell biology of integral membrane proteins. *Mailing Add:* Howard Hughes Med Inst & Dept Biochem Molecular & Cell Biol Northwestern Univ 2153 Sheridan Rd Evanston IL 60208-3500

LAMB, ROBERT CARDON, b Logan, Utah, Jan 8, 33; m 53; c 5. DAIRY SCIENCE. *Educ:* Utah State Univ, BS, 56; Mich State Univ, MS, 59, PhD(dairy cattle breeding), 62. *Prof Exp:* Instr dairy sci, Mich State Univ, 58-60; asst prof, Utah State Univ, 61-64; res dairy husbandman, Agr Res Serv, USDA, 64-72, res leader, 72-90; HEAD ANIMAL DAIRY & VET SCI, UTAH STATE UNIV, 90- *Mem:* Am Dairy Sci Asn. *Res:* Use of incomplete records in dairy cattle selection; genetics by nutrition interactions; inheritance of abnormalities in livestock; feed utilization efficiency in dairy cattle; dairy herd management; exercise for dairy cows; integrated reproduction management; stress in dairy cattle; dairy cattle housing; use of BST in dairy cattle. *Mailing Add:* 1027 Thrushwood Dr Logan UT 84321

LAMB, ROBERT CHARLES, b Union Co, SC, Sept 28, 28; m 50; c 2. ORGANIC CHEMISTRY. *Educ:* Presby Col, SC, BS, 48; Univ Ga, MS, 55; Univ SC, PhD(chem), 58. *Prof Exp:* Instr chem, Presby Col, SC, 50-51; asst prof, Univ Ga, 58-66; assoc prof & chmn dept, Augusta Col, 66; chmn dept, 66-77, PROF CHEM, ECAROLINA UNIV, 66- *Mem:* Am Chem Soc; Sigma Xi. *Res:* Organic peroxides; free radicals in solution; chemical kinetics. *Mailing Add:* Dept Chem ECarolina Univ PO Box 2787 Greenville NC 27834

LAMB, ROBERT CONSAY, b Saskatoon, Sask, May 11, 19; nat US; m 49; c 3. POMOLOGY, GENETICS. *Educ:* Univ Sask, BSA, 41; Univ Minn, MS, 47, PhD(hort), 54. *Prof Exp:* Asst hort, Univ Minn, 46-48; from asst prof to prof, 48-88, EMER PROF POMOL, NY STATE COL AGR & LIFE SCI, 88- *Concurrent Pos:* Orgn Europ Econ Coop sr vis fel sci, John Innes Inst, Eng, 62; mem plant explor team, Int Bd Plant Genetic Resources, Nepal, 84- *Mem:* Am Soc Hort Sci; Int Soc Hort Sci; Can Soc Hort Sci; Am Pomol Soc (pres, 80-82). *Res:* Breeding disease resistant apple varieties; introduced Liberty and Freedom apples which are resistant to apple scab, mildew, cedar apple rust and fire blight. *Mailing Add:* NY State Agr Exp Sta Geneva NY 14456

LAMB, ROBERT EDWARD, b Sharon, Pa, July 12, 45; m 73; c 2. ANALYTICAL CHEMISTRY. *Educ:* St Louis Univ, AB, 69, BS, 70; Univ Ill, MS, 74, PhD(anal chem), 75. *Prof Exp:* Lectr anal chem, Sch Chem Sci, Univ Ill, 75; asst prof chem, Southern Methodist Univ, 75-78; from asst prof to assoc prof, 78-88, PROF CHEM, OHIO NORTHERN UNIV, 88- *Mem:* Am Chem Soc. *Res:* Pulse polarography and stripping analysis; liquid chromatography; analysis of trace metal complexes; environmental applications of analytical techniques. *Mailing Add:* Dept Chem Ohio Northern Univ Ada OH 45810

LAMB, SANDRA INA, b New York, NY, Apr 20, 31; m 50; c 4. ORGANIC CHEMISTRY, ENVIRONMENTAL CHEMISTRY. *Educ:* Univ Calif, Los Angeles, BS, 54, PhD(phys org chem), 59. *Prof Exp:* Instr chem, Santa Monica City Col, fall 59; asst prof, San Fernando Valley State Col, 60-61; instr, Exten Div, Univ Calif, 61-69; from asst prof to assoc prof chem, 69-76, Mt St Mary's Col, Calif, 71-76, chmn dept phys sci & math, 69-75; LECTR CHEM, UNIV CALIF, LOS ANGELES, 76- *Concurrent Pos:* Asst res pharmacologist, Med Sch, Univ Calif, 66-, lectr, 70. *Mem:* AAAS; Am Chem Soc; Sigma Xi. *Res:* Analytical applications of gas chromatography in chemistry and medicine with special interest in analysis of acetylcholine and various cholinergic agents; mechanism of action of muscarinic agents; analytical applications of gas chromatography and ion chromatography air pollution; synthesis of small ring compounds. *Mailing Add:* Dept Chem Univ Calif 3060 Young Han 405 Seven Gard Ave Los Angeles CA 90024

LAMB, WALTER ROBERT, b Weiser, Idaho, Sept 26, 22; m 46; c 1. PHYSICS. *Educ:* Univ Calif, AB, 48. *Prof Exp:* Physicist, US Naval Radiol Defense Lab, 48-59; solid state physicist, Res & Develop Dept, Raytheon Semiconductor Co, 59-63, Fairchild Semiconductor, 63-64 & Union Carbide Corp, 64-65; mgr advan processing, Stewart-Warner Microcircuits, 65-68; physicist, Fairchild Semiconductor Corp, 68-71; physicist, Raytheon Co, 71-77; ENG MGR, FAIRCHILD SEMICONDUCTOR CORP, 78- *Mem:* AAAS; Am Phys Soc. *Res:* Solid state, nuclear radiation, optical, luminescent, thermodynamic and gravitational phenomena. *Mailing Add:* 148 Jacinto Way Sunnyvale CA 94086

LAMB, WILLIAM BOLITHO, b Chicago, Ill, May 6, 37; m 61; c 2. CHEMICAL ENGINEERING. *Educ:* Princeton Univ, BSE, 58; Univ Del, MChE, 63, PhD(chem eng), 65. *Prof Exp:* Res engr, 65-68, res supvr, 68-69, group mgr, 69-75, tech supt, Film Dept, 75-78, planning mgr, 79-80, prod mgr, Polymer Prod Dept, 80-86, regional mgr, La, 86-89, western sales mgr, 89-90, MGR, FILM PRODUCTS, E I DU PONT DE NEMOURS & CO, INC, 90- *Mem:* Am Inst Chem Eng. *Res:* Mass transfer in gas-liquid systems; polymer rheology; extrusion and processing of polymers for packaging film applications; fluoropolymers. *Mailing Add:* Four Sorrel Dr Surrey Park Wilmington DE 19803

LAMB, WILLIS EUGENE, JR, b Los Angeles, Calif, July 12, 13; m 39. QUANTUM MECHANICS, ATOMIC PHYSICS. *Educ:* Univ Calif, BS, 34, PhD(physics), 38; Oxford Univ, MA, 56; Yale Univ, MA, 61. *Hon Degrees:* ScD, Univ Pa, 54; LHD, Yeshiva Univ, 65; ScD, Gustavus Adolphus Col, 75; ScD, Columbia Univ, 90. *Prof Exp:* Asst physics, Univ Calif, 34-35, 36-37; instr, Columbia Univ, 38-43, assoc, 43-45, from asst prof to prof physics, 45-52; prof, Stanford Univ, 51-56; fel, New Col & Wykeham prof, Oxford Univ, 56-62; Henry Ford II prof, Yale Univ, 62-72, Josiah Willard Gibbs prof, 72-74; prof physics & optical sci, 74-90, REGENTS PROF, ARIZ RES LABS, UNIV ARIZ, 90- *Concurrent Pos:* Mem staff, Radiation Lab, Columbia Univ, 43-52; Loeb lectr, Harvard Univ, 53-54; Guggenheim fel, 60-61; consult, Philips Labs, Inc, NASA, Bell Tel Labs & Perkin- Elmer Corp. *Honors & Awards:* Rumford Medal, Am Acad Arts & Sci, 53; Nobel Prize Physics, 55; Award, Res Corp, 55. *Mem:* Nat Acad Sci; fel Am Phys Soc; hon mem NY Acad Sci; hon fel Brit Inst Physics; hon fel Royal Soc Edinburgh. *Res:* Theoretical physics; atomic and nuclear structure; microwave spectroscopy; fine structure of hydrogen and helium; magnetron oscillators; statistical mechanics; masers and lasers; quantum theory of measurement. *Mailing Add:* Optical Sci Ctr Univ Ariz Tucson AZ 85721

LAMBA, RAM SARUP, b Calcutta, India, Dec 29, 41; US citizen; m 69; c 2. INORGANIC CHEMISTRY, ORGANIC CHEMISTRY. *Educ:* Delhi Univ, India, BSc, 62, MSc, 64; ETex State Univ, DEd(inorg chem, educ), 73. *Prof Exp:* Res asst, Indian Inst Petrol, Dehradun, India, 64-65; chemist & supt dyeing & finishing, Beaunit Corp of NC, Humacao, PR, 68-69; instr chem, Inter Am Univ PR, 69-70, asst prof & chmn dept, 70-71, chmn dept natural sci, 73-77, assoc prof chem, math & physics, 73-83, dean acad affairs, 77-82, prof, 83-87, DISTINGUISHED PROF CHEM, INTER AM UNIV PR, 87- *Mem:* Royal Inst Chem; Am Chem Soc; The Chem Soc; fel Inst Educ Leadership. *Res:* To develop innovative methods in the teaching of college chemistry and to integrate with biological sciences; synthesis and study of chromium (III), complexes; construction of low cost equipment in chemistry. *Mailing Add:* Dept Chem Inter Am Univ PO Box 1293 Hato Rey PR 00919

LAMBA, SURENDAR SINGH, b India, Mar 3, 34; m 67; c 1. PHARMACY, PHARMACOGNOSY. *Educ:* Agra Univ, BSc, 54; Univ Rajasthan, BPharm, 57; Panjab Univ, India, MPharm, 60; Univ Nebr, MS, 63; Univ Colo, PhD(pharmacog), 66. *Prof Exp:* Assoc prof, 66-67, PROF PHARMACOG, FLA A&M UNIV, 67- *Concurrent Pos:* Vis prof, Univ Panama, 77-78 & Univ Benin, Nigeria, 81-82. *Honors & Awards:* Lederle Fac Award, 75. *Mem:* Am Pharmaceut Asn; Acad Pharmaceut Sci; NY Acad Sci; Am Soc Pharmacog; Sigma Xi. *Res:* Tissue culture studies; effects of growth retardants on growth and alkaloid biosynthesis; phytochemistry; antisickling agents. *Mailing Add:* Dept Pharm-Pharmacog Fla A&M Univ PO Box 367 Tallahassee FL 32307

LAMBDIN, PARIS LEE, b St Charles, Va, Oct 13, 41; m 64; c 2. ENTOMOLOGY. *Educ:* Lincoln Mem Univ, BA, 64; Va Polytech Inst & State Univ, MS, 72, PhD(entom), 74. *Prof Exp:* Teacher biol, Bassett High Sch, 64-66; PROF ENTOM, DEPT ENTOM & PLANT PATHOL, UNIV TENN, 74- *Honors & Awards:* Sigma Xi Res Award. *Mem:* Entom Soc Am. *Res:* Systematics of species in the superfamily Coccoidea; biological control of vegetable insect pests. *Mailing Add:* Dept Entom & Plant Pathol Univ Tenn Knoxville TN 37901-1071

LAMBE, EDWARD DIXON, b Prince Rupert, BC, July 25, 24; m 50; c 4. PHYSICS. *Educ:* Univ BC, BASc, 48, MASc, 49; Princeton Univ, PhD(physics), 59. *Prof Exp:* Asst prof physics, Washington Univ, 56-61; assoc prof, 61-65, asst vchancellor, 66-70, dir, Instructional Resources Ctr, 67-74, PROF PHYSICS, STATE UNIV NY, STONY BROOK, 65- *Concurrent Pos:* Exec secy, Comn Col Physics, 62-64, secy, 64-68. *Mem:* Am Asn Physics Teachers. *Res:* Electron and nuclear magnetic resonance; beta and gamma ray polarization; learning processes in physics. *Mailing Add:* 7605 NW 41st St Coral Springs FL 33065

LAMBE, JOHN JOSEPH, b Cork, Ireland, Dec 1, 26; US citizen; m 50; c 2. SOLID STATE PHYSICS. *Educ:* Univ Mich, BSE, 48, MS, 50; Univ Md, PhD(physics), 54. *Prof Exp:* Eng physics, Airborne Instruments Lab, 48-51; physicist solid state physics, Naval Res Lab, 51-56; physicist microwave res, Univ Mich, 56-59; staff scientist solid state physics, Ford Motor Co, 59-79; consult, Jet Propulsion Lab, Pasadena, Calif, 79-86; CONSULT, 86- *Mem:* Fel Am Phys Soc; Am Phys Soc. *Res:* Solid state physics; magnetic resonance; luminescence; super conductivity; electron tunneling. *Mailing Add:* 205 224th Ave SE Redmond WA 98053

LAMBE, ROBERT CARL, b Minneapolis, Minn, Nov 25, 27; m 50; c 2. PLANT PATHOLOGY. *Educ:* Univ Southern Calif, AB, 52; Univ Calif, MS, 55; Ore State Col, PhD(plant path), 60. *Prof Exp:* Jr plant pathologist, Ore State Col, 58-60; plant pathologist, Area Exten, Tex A&M Univ, 60-63; exten plant pathologist, Iowa State Univ, 63-67; ASSOC PROF PLANT PATH, VA POLYTECH INST & STATE UNIV, 67- *Res:* Fungicides and extension plant pathology. *Mailing Add:* PO Box 1167 Sanibel FL 33957

LAMBE, T(HOMAS) WILLIAM, b Raleigh, NC, Nov 28, 20; m 47. GEOTECHNICAL ENGINEERING. *Educ:* NC State Univ, BS, 42; Mass Inst Technol, SM, 44, ScD(soil mech), 48. *Prof Exp:* Struct detailer, Am Bridge Co, Pa, 42; field engr, Olsen Consult Engrs, Edenton, NC, 42; instr civil eng, Univ NH, 42-43; field engr airbase construct, US Navy, Brunswick,

Maine, 43; struct & found engr, Univ Calif, San Francisco, 44; soil engr, Dames & Moore, San Francisco, 44-45; from instr to asst prof soil mech, 45-52, assoc prof & dir, Soil Stabilization Lab, 52-59, prof geotech eng & head, Geotech Div, 59-69, Edmund K Turner prof, 69-81, EMER EDMUND K TURNER PROF CIVIL ENG, MASS INST TECHNOL, 81- Concurrent Pos: Consult geotech eng, 45- Honors & Awards: Collingswood Prize, Am Soc Civil Engrs, 52, Arthur M Wellington Prize, 61 & 84, Norman Medal, 64, Terzaghi Lectr, 70 & Karl Terzaghi Award, 75; Desmond Fitzgerald Medal, Brit Soc Civil Engrs, 54 & 56; Rankine Lectr, Brit Inst Civil Engrs, 73; R P Davis Lectr, Univ WVa, 73; Terzaghi Mem Lectr, Istanbul, Turkey, 73; Moh Lectr, Taipei, Taiwan, 80, Indonesia & Singapore, 81; Ardaman Lectr, Univ Fla, 85; Shaw Lectr, NC State Univ, 85. Mem: Nat Acad Eng; hon mem Am Soc Civil Engrs; fel Brit Inst Civil Engrs; hon mem Venezuelan Soc Soil Mech & Found Eng; hon mem Southeast Asian Soc Geotech Eng. Res: Soil testing, stabilization and mineralogy; soil engineering; earth and rock dams. Mailing Add: 1641 Harbor Cay Lane Longboat Key FL 34228

LAMBE, THOMAS ANTHONY, b Victoria, BC, Dec 27, 30; m 64. OPERATIONS RESEARCH, ENGINEERING SCIENCE. Educ: Univ BC, BASc, 52; Stanford Univ, MSc, 58, PhD(eng sci), 68. Prof Exp: Engr, Can Westinghouse, 52-54; res engr, BC Res Coun, 54-57, proj leader opers res, 58-65; assoc prof indust eng, Univ Toronto, 68-74; ASSOC PROF, SCH PUB ADMIN, UNIV VICTORIA, 74- Mem: Can Oper Res Soc; Opers Res Soc Am. Res: Economic analysis of engineering systems, particularly the transportation and natural resource industries; decision theory and individual choice behavior. Mailing Add: Sch Pub Admin Univ Victoria Box 1700 Victoria BC V8W 2Y2 Can

LAMBEK, JOACHIM, b Leipzig, Ger, Dec 5, 22; nat Can; m 48; c 3. MATHEMATICS. Educ: McGill Univ, BSc, 46, MSc, 47, PhD, 51. Prof Exp: Assoc prof math, 54-63, PROF MATH, MCGILL UNIV, 63- Concurrent Pos: Mem, Inst Advan Study, 59-60. Mem: Am Math Soc; Math Asn Am; Can Math Cong; Sigma Xi. Res: Algebra. Mailing Add: Dept Math McGill Univ 805 Sherbrooke St W Montreal PQ H3A 2K6 Can

LAMBERG, STANLEY LAWRENCE, b Brooklyn, NY, Oct 2, 33; m 63; c 2. HEMATOLOGY, HISTOLOGY. Educ: Brooklyn Col, BS, 55; Oberlin Col, MA, 57; Tufts Univ, MS, 62; NY Univ, PhD(biol), 68. Prof Exp: Teaching asst biol, Oberlin Col, 55-57; chief technician biochem, Sch Med, Cornell Univ, 57-58; res fel, Sch Med, Tufts Univ, 58-61; Nat Inst Dent Res fel, Col Dent, NY Univ, 61-66; lectr biol, City Col New York, 66-67; asst prof biol, Conolly Col, Long Island Univ, 67-70; from asst prof to assoc prof, 70-75, PROF MED LAB TECHNOL, STATE UNIV NY COL TECHNOL, FARMINGDALE, 75- Concurrent Pos: Asst res scientist, Guggenheim Inst Dent Res, NY Univ, 68-69; adj asst prof, Conolly Col, Long Island Univ, 70-73, adj assoc prof, 73-75, adj prof, 75-78; adj instr, 81-86, adj asst prof, 86-87; adj assoc prof, Suffolk County Comt Col, 87- Honors & Awards: Founder's Day Award, NY Univ, 69. Mem: AAAS; NY Acad Sci; Sigma Xi; Nat Soc Histotechnol. Res: Mitochondrial phosphorylation reactions during embryonic development; effect of ultraviolet irradiation and various inhibitors and uncoupling reagents on mitochondrial phosphorylation reactions. Mailing Add: Dept Med Lab Technol State Univ NY Col Farmingdale NY 11735

LAMBERSON, HAROLD VINCENT, JR, b Albany, NY, July 29, 45. MEDICINE. Educ: Union Col, Schenectady, NY, BS, 67, MS, 69; Albany Med Col, MD & PhD, 75. Prof Exp: asst prof path, asst dir clin path & dir, Diagnostic Virol Lab & Microbiol Sect, Upstate Med Ctr, State Univ NY, 78-82, actg dir, Clin Immunol Sect, 80-81; DIR, AM RED CROSS BLOOD SERV, 82-; ASSOC PROF PATH, STATE UNIV NY, HSC SYRACUSE, 85- Mem: Am Soc Clin Pathologists; Am Soc Microbiol. Res: Rapid laboratory diagnosis of infectious diseases; transfusion related viral infections; effects of blood donation on donor immune function. Mailing Add: 216 Dewitt Rd Syracuse NY 13214

LAMBERSON, LEONARD ROY, b Stanwood, Mich, Nov 18, 37; m 75; c 3. INDUSTRIAL & MANUFACTURING ENGINEERING. Educ: Gen Motors Inst, BME, 61; NC State Univ, MS, 63; Tex A&M Univ, PhD(indust eng), 67. Prof Exp: Prod foreman, Chevrolet Div, Gen Motors Corp, 61-64; from asst prof to prof indust eng, Gen Motors Inst, 64-70, chmn dept, 69-70; asst prof, Tex A&M Univ, 65-68; assoc prof, Wayne State Univ, 70-79, prof indust eng, 79-89, chmn dept, 82-89; DEAN ENG, WESTERN MICH UNIV, KALAMAZOO, 89- Concurrent Pos: Reliability Engr, US Army Tank Auto Command, 77-78. Honors & Awards: Craig Award, Am Soc Qual Control, 78. Mem: Inst Indust Engrs; Am Soc Qual Control; Am Soc Eng Educ. Res: Development of techniques and procedures to improve the reliability of commercial products. Mailing Add: 8420 Valleywood Lane Portage MI 49002

LAMBERT, ALAN L, b New York, NY, Nov 28, 43. MATHEMATICS. Educ: Univ Miami, BS, 66, MS, 67; Univ Mich, PhD, 70. Prof Exp: PROF MATH, UNIV NC, CHARLOTTE, 83- Mem: Am Math Soc; Irish Math Soc. Res: Properties of composition operators. Mailing Add: Math Dept Univ NC Charlotte NC 28223

LAMBERT, BRIAN KERRY, b Spokane, Wash, Nov 21, 41; m 63; c 2. INDUSTRIAL ENGINEERING. Educ: Tex Tech Col, BS, 64, MS, 66, PhD(indust eng), 67. Prof Exp: Asst prof, 67-71, ASSOC PROF INDUST ENG, TEX TECH UNIV, 71- Mem: Soc Mfg Engrs; Inst Indust Engrs; Am Soc Eng Educ. Res: Manufacturing research and development, specifically machining operations research and systems analysis, specifically reliability. Mailing Add: Dept Indust Eng Wichita State Univ Wichita KS 67208

LAMBERT, CHARLES CALVIN, b Rockford, Ill, Apr 10, 35; m 65; c 2. DEVELOPMENTAL BIOLOGY, REPRODUCTIVE BIOLOGY. Educ: San Diego State Univ, BA, 64, MS, 66; Univ Wash, PhD(zool), 70. Prof Exp: NIH traineeship, Univ Wash, 70; from asst prof to assoc prof, 70-79, PROF ZOOL, CALIF STATE UNIV, FULLERTON, 79- Concurrent Pos: Vis investr, Friday Harbor Labs, 74-, Hopkins Marine Sta, 78, Bermuda Biol Sta, 80, Shimoda Marine Res Ctr, 80 & Kewalo Marine Lab, 85; vis prof, Friday Harbor Labs, 81 & Shimoda Marine Res Ctr (UNESCO/ICRO course), 82. Mem: Am Soc Zoologists; Soc Develop Biol; AAAS; Am Soc Cell Biol; Int Soc Develop Biol; Int Cell Res Orgn. Res: Development and physiology of marine invertebrates. Mailing Add: Dept Biol Calif State Univ Fullerton CA 92634

LAMBERT, DIANE, US citizen. MATHEMATICAL STATISTICS. Educ: Univ Rochester, PhD(statist), 79. Prof Exp: From asst prof to assoc prof statist, Carnegie-Mellon Univ, 80-86; MEM TECH STAFF, AT&T BELL LABS, 86- Concurrent Pos: Adv panel mem, Statist Income Div, US Internal Revenue Serv, 80-88; vis assoc prof statist, Univ Chicago, 84-86; assoc ed, J Am Statist Asn, 84-90; bd mem, Bd Math Sci, Nat Res Coun, 90- Mem: Inst Math Statist (exec secy, 90-); fel Am Statist Asn. Res: Developing, analyzing and applying innovative statistical models for nonstandard applications such as the risk of disclosure in publicly released databases and the probability of detecting low levels of environmental contaminants. Mailing Add: AT&T Bell Labs Rm 2C-256 600 Mountain Ave Murray Hill NJ 07974-2070

LAMBERT, EDWARD HOWARD, b Minneapolis, Minn, Aug 30, 15; m 40, 75. MEDICAL PHYSIOLOGY, NEUROMUSCULAR DISORDERS. Educ: Univ Ill, BS, 36, MS, 38, MD, 39, PhD(physiol), 44. Prof Exp: Instr med technol, Herzl Jr Col, 41-42; assoc med, Off Sci Res & Develop, Col Med, Univ Ill, 42-43; res asst, 43-45, from instr to prof physiol, 45-58, prof physiol, Mayo Grad Sch Med, Univ Minn, 58-73, prof physiol & neurol, Mayo Med Sch, 73-85; PROF NEUROL, UNIV MINN MED SCH, 85- Concurrent Pos: Consult, Mayo Clin, 45-85; mem pub adv groups, NIH; prin investr, NIH grant, 81- Honors & Awards: Presidential Cert Merit, 47; Tuttle Award, Aerospace Med Asn, 52. Mem: Am Acad Neurol; Soc Neurosci; Am Physiol Soc; Aerospace Med Asn; Am Asn Electromyography & Electrodiag (pres, 58); hon mem Am Neurol Asn. Res: Neurophysiology; neuromuscular disorders in man; electromyography; neuromuscular transmission. Mailing Add: 202 14th St NE Rochester MN 55904

LAMBERT, FRANCIS LINCOLN, b Staunton, Va, Oct 8, 23; m 67; c 2. PHYSIOLOGY. Educ: George Washington Univ, BS, 49, MS, 51; Harvard Univ, PhD(biol), 58. Prof Exp: Instr zool, George Washington Univ, 49-52; asst prof biol, 55-61, assoc prof physiol & biophys, 61-80, prof & chmn biol sci, 80-84, PROF PHYSIOL & BIOPHYS, UNION COL, NY, 85- Concurrent Pos: Jacques Loeb assoc marine biol, Rockefeller Inst, 60-61; educ consult, US Agency Int Develop, India, 65-68. Mem: AAAS; Am Soc Zool; Sigma Xi. Res: Invertebrate physiology; cellular neurophysiology. Mailing Add: 720 Riverside Ave Scotia NY 12302

LAMBERT, FRANK LEWIS, b Minneapolis, Minn, July 10, 18; m 43. ORGANIC CHEMISTRY. Educ: Harvard Univ, BA, 39; Univ Chicago, PhD(org chem), 42. Prof Exp: Res & develop chemist, Edwal Labs, Ill, 42-43, develop chemist, 43-44, head develop dept, 46-47; instr chem, Univ Calif, Los Angeles, 47-48; from asst prof to prof, 48-80, EMER PROF CHEM, OCCIDENTAL COL, 81-; SCI CONSULT, J PAUL GETTY MUS, 82- Concurrent Pos: NSF fac fel, 57-58, 70-71. Mem: Am Chem Soc. Res: Polarography of organic halogen compounds; halogenation of organic compounds. Mailing Add: 1105 Olancha Dr Los Angeles CA 90065

LAMBERT, GEORGE, b Etobicoke, Ont, Oct 8, 23; US citizen; m 48; c 3. VETERINARY MICROBIOLOGY. Educ: Univ Guelph, DVM, 47; Iowa State Univ, MS, 66. Prof Exp: Instr vet path, Ont Vet Col, Univ Guelph, 47-48; asst prof, WVa Univ, 48-50; coop agt, Univ Wis & USDA, 50-53; asst state vet epidemiol, Va Dept Agr, 53-57; res vet bact, Nat Animal Dis Lab, 57-65, res virol, 65-67; asst dir biol dept, Diamond Labs, Inc, 67-70; chief virol res lab, 70-75, asst dir, 75-80, assoc dir, 80-85, RES LEADER, IMMUNOL RES, NAT ANIMAL DIS CTR, 85- Mem: Am Vet Med Asn; US Animal Health Asn; Conf Res Workers Animal Dis. Res: Administration of animal disease research. Mailing Add: Rte 1 Boone IA 50036

LAMBERT, GLENN FREDERICK, b Columbus, Ohio, Nov 21, 18; m 45; c 2. BIOCHEMISTRY. Educ: DePauw Univ, AB, 40; Univ Ill, PhD(biochem), 44. Prof Exp: Spec res asst, Univ Ill, 45-46; res chemist, Abbott Labs, 46-60, sr res pharmacologist, 61-73, sr res chemist, 73-89, RETIRED. Concurrent Pos: Mem, Coun Arteriosclerosis & Coun Thrombosis, Am Heart Asn. Mem: Am Chem Soc; Sigma Xi. Res: Biochemistry and nutrition of amino acids; fat emulsions for intravenous therapy; atherosclerosis; thrombolytic drugs; high pressure liquid chromatography separation and analysis of peptides. Mailing Add: 318 Judge Ave Waukegan IL 60085

LAMBERT, HELEN HAYNES, b Baton Rouge, La, July 25, 39; div; c 2. ENDOCRINOLOGY. Educ: Wellesley Col, BA, 61; Univ NH, MS, 63, PhD(zool), 69. Prof Exp: Instr zool, Univ NH, 67-68; asst prof biol, Simmons Col, 69-70; asst prof, 70-75, ASSOC PROF BIOL, NORTHEASTERN UNIV, 75- Concurrent Pos: Mem, Sex Info & Educ Coun US. Mem: AAAS; Sigma Xi; Am Inst Biol Sci; Am Soc Zool. Res: Environmental factors affecting reproduction and sexual behavior; sex determination and development of sex differences. Mailing Add: Dept Biol Northeastern Univ Boston MA 02115

LAMBERT, HOWARD W, b Oakland, Calif, Aug 2, 37; m 57; c 3. TOPOLOGY. Educ: Univ Calif, Berkeley, BA, 60; Iowa State Univ, MS, 61; Univ Utah, PhD(math), 66. Prof Exp: Asst prof, 66-71, assoc prof, 71-78, prof math, Univ Iowa, 78-80; PROF MATH, WESTERN NMEX UNIV, 80- Mem: Am Math Soc; Math Asn Am; Am Asn Univ Professors; Soc Indust & Appl Math. Res: Upper semi-continuous decompositions of topological spaces, 3-manifolds. Mailing Add: Dept Math Sci Eastern NMex Univ Portales NM 88130

LAMBERT, JACK LEEPER, b Pittsburg, Kans, Mar 2, 18; m 43; c 4. ANALYTICAL CHEMISTRY, INORGANIC CHEMISTRY. *Educ:* Kans State Teachers Col, Pittsburg, BA & MS, 47; Okla State Univ, PhD(chem), 50. *Prof Exp:* Instr chem, Kans State Teachers Col, Pittsburg, 47-48; asst, Okla State Univ, 48-50; from instr to prof, 50-88, EMER PROF CHEM, KANS STATE UNIV, 88- *Concurrent Pos:* Assoc prog dir, NSF, Washington, DC, 65-66. *Mem:* Am Chem Soc; Sigma Xi; fel AAAS. *Res:* Methods research in analytical chemistry; reagents for trace analysis in air, water and blood; insoluble, demand-type disinfectants for water. *Mailing Add:* Dept Chem Kans State Univ Manhattan KS 66506

LAMBERT, JAMES LEBEAU, b Sanford, Fla, Feb 11, 34. ORGANIC CHEMISTRY, ANALYTICAL CHEMISTRY. *Educ:* Spring Hill Col, BS, 59; Johns Hopkins Univ, PhD(chem), 63. *Prof Exp:* From asst prof to assoc prof, prof chem, 68-79, acad dean 76-78, PROF CHEM, SPRING HILL COL, 79-, CHMN DEPT, 82- *Mem:* AAAS; Am Chem Soc; Sigma Xi. *Res:* Mechanisms of organic reactions; carbanions. *Mailing Add:* Dept Chem Spring Hill Col Mobile AL 36608-1791

LAMBERT, JAMES MORRISON, b Chicago, Ill, Feb 18, 28; m 53; c 3. NUCLEAR PHYSICS. *Educ:* Johns Hopkins Univ, BA, 55, PhD(physics), 61. *Prof Exp:* Instr physics, Johns Hopkins Univ, 60-61; asst prof, Univ Mich, 61-63; from asst prof to assoc prof, 64-74, PROF PHYSICS, GEORGETOWN UNIV, 74- *Concurrent Pos:* Res consult, Naval Res Lab, 66- *Mem:* Am Phys Soc; AAAS; Sigma Xi. *Res:* Experimental medium energy nuclear physics; nuclear reaction studies using particle accelerators; experimental surface physics. *Mailing Add:* Dept Physics Georgetown Univ Washington DC 20057

LAMBERT, JEAN WILLIAM, b Ewing, Nebr, June 10, 14; m 43; c 2. AGRONOMY. *Educ:* Univ Nebr, BS, 40; Ohio State Univ, MS, 42, PhD(agron), 45. *Prof Exp:* Instr agron, Ohio State Univ, 43-45; from asst prof to prof, 46-82, EMER PROF AGRON & PLANT GENETICS, UNIV MINN, ST PAUL, 82- *Concurrent Pos:* Consult, Am Soybean Asn, 63, Food & Agr Orgn, Hungary, 75, US Info Agency, Romania, 76, Agr Corp Am, USSR, 79 & Food & Agr Orgn, Poland, 80; res consult, Chilean Agr Prog, Rockefeller Found, 64; partic, vis scientist prog, Am Soc Agron; tech ed, Agron J, 71-73. *Honors & Awards:* Agron Achievement Award, Am Soc Agron, 83. *Mem:* Fel Am Soc Agron; life mem Am Soybean Asn; fel Crop Sci Soc Am; Sigma Xi. *Res:* Bromegrass cultural research; varietal improvement in barley and soybeans; barley and soybean genetics. *Mailing Add:* Dept Agron & Plant Genetics Univ Minn 1509 Gortner St Paul MN 55108

LAMBERT, JERRY ROY, b Benton, Ill, Sept 16, 36; div; c 3. AGRICULTURAL ENGINEERING, AGRICULTURAL INFORMATION SYSTEMS. *Educ:* Univ Fla, BAgrE, 58, MS, 62; NC State Col, PhD(agr eng), 64. *Prof Exp:* Design eng trainee, Soil Conserv Serv, USDA, 58-60; from asst prof to assoc prof, 64-72, PROF AGR ENG, CLEMSON UNIV, 72-, COMPUT COORDR, 85- *Mem:* Am Soc Eng Educ; Am Soc Agr Engrs. *Res:* Water relations of plants; evapotranspiration; water movement in soils; simulation of agricultural systems; microcomputer applications to agriculture information delivery systems. *Mailing Add:* Dept Agr Eng Clemson Univ Main Campus Clemson SC 29634

LAMBERT, JOHN B(OYD), b Billings, Mont, July 5, 29; m 53, 58; c 6. METALLURGY & PHYSICAL METALLURGICAL ENGINEERING, MATERIALS SCIENCE ENGINEERING. *Educ:* Princeton Univ, BS, 51; Univ Wis, PhD(chem eng), 56. *Prof Exp:* Res engr, Indust & Biochem Dept, E I du Pont de Nemours & Co, 56-63, sr res engr, Pigments Dept, Del, 63-68; mkt mgr, Fansteel Inc, 68-70, plant mgr, 70-71, tech mgr, 72-73, mgr mfg eng, VR/Wesson Div, 73-81, vpres & tech dir, 81-88, vpres & gen mgr, Metals Div, 88-90, VPRES & TECH DIR, METALS DIV, FANSTEEL INC, 90- *Honors & Awards:* Co-Recipient, Charles Hatchett Award, Inst Metals, 86. *Mem:* Am Chem Soc; Am Inst Chem Engrs; Electrochem Soc; Am Soc Metals; Sigma Xi; Soc Mfg Engrs. *Res:* Inorganic colloid chemistry; physical and powder metallurgy; surface chemistry, drying, machining and metal cutting; ceramic cutting tools; technical management; refractory metals, including tantalum, niobivim, and hard metals. *Mailing Add:* Fansteel Inc No One Tantalum Pl N Chicago IL 60064

LAMBERT, JOSEPH B, b Ft Sheridan, Ill, July 4, 40; m 67; c 3. ORGANIC CHEMISTRY. *Educ:* Yale Univ, BS, 62; Calif Inst Technol, PhD(org chem), 65. *Prof Exp:* From asst prof to assoc prof, 65-74, dir, Integrated Sci Prog, 82-85, chmn dept, 86-89, PROF CHEM, NORHTWESTERN UNIV, 74- *Concurrent Pos:* Alfred P Sloan Found fel, 68-70; Guggenheim fel, 73; Japan Soc Prom Sci fel, 78; vis scholar, Polish Acad Sci, 81 & Chinese Acad Sci, 88; Nat Acad Sci exchange fel, 85; vis assoc, Brit Mus Res Lab, 73; ed-in-chief, J Phys Org Chem, 86-; USAF Off Sci Res fel, 90. *Honors & Awards:* Eastman Kodak Award, 65; Nat Fresenius Award, 76; Norris Award for Teaching of Chem, Am Chem Soc, 87; Fryxell Award in Sci Archaeol, Soc Am Archaeol, 89. *Mem:* Fel AAAS; Sigma Xi; Am Chem Soc; Royal Soc Chem; fel Brit Interplanetary Soc; Soc Archeol Sci (pres int off, 86-87); Int Soc Magnetic Resonance. *Res:* Nuclear magnetic resonance spectroscopy, organic reaction mechanisms, organosilicon and organotin chemistry, applications of analytical chemistry to archaeology. *Mailing Add:* Dept Chem Northwestern Univ Evanston IL 60208-3113

LAMBERT, JOSEPH MICHAEL, b Philadelphia, Pa, Nov 19, 42; m 73; c 3. APPROXIMATION THEORY, NUMERICAL METHODS. *Educ:* Drexel Univ, BS, 65; Cornell Univ, MA, 67; Purdue Univ, PhD(math), 70. *Prof Exp:* from asst prof to assoc prof math, 70-81, asst dean, Col Sci, 79-82, actg head, 80-82, ASSOC PROF & DEPT HEAD COMPUT SCI, PA STATE UNIV, 82- *Concurrent Pos:* Vis assoc prof math, Univ Tenn, Knoxville, 77-78; vis assoc prof comput sci, Cornell Univ, Ithaca, 86-87. *Mem:* Am Math Soc; Asn Comput Mach; Inst Elec & Electronics Engrs. *Res:* Functional analysis; approximation theory; numerical analysis; operations research; software metrics. *Mailing Add:* Comput Sci Pa State Univ Whitmore Labs University Park PA 16802

LAMBERT, JOSEPH PARKER, b Bronte, Tex, Oct 6, 21; m 45; c 5. DENTISTRY. *Educ:* Baylor Univ, DDS, 52. *Prof Exp:* Instr, 52-56, PROF PROSTHETICS & CHMN DEPT, COL DENT, BAYLOR UNIV, 56- *Mem:* Am Dent Asn. *Mailing Add:* 3707 Gaston Suite 602 Dallas TX 75246

LAMBERT, LLOYD MILTON, JR, b Olympia, Wash, May 10, 29; m 52; c 3. SOLID STATE PHYSICS. *Educ:* US Naval Acad, BS, 52; Univ Calif, MS, 58, MA, 63, PhD(physics), 64. *Prof Exp:* Res engr, Sperry Gyroscope Co, 57; assoc elec eng, Univ Calif, 58; res engr & mgr phys electronics dept, Aeronutronic Div, Ford Motor Co, 58-63; staff scientist, Aerospace Corp, 63-65; from assoc prof to prof elec eng, Univ Vt, 65-77, chmn physics dept, 77-85, prof physics dept, 77-91; CONSULT, 91- *Concurrent Pos:* Res fel, Norges Teknisk-Naturvitenskaplige Forskningsrad, 71-72, sr fel, 77-78; Nat Acad Sci exchange, Rumania, 77 & USSR, 77. *Mem:* Am Phys Soc. *Res:* Solid state devices; magnetic storage devices; semiconductors; thin film devices; optical properties of solids; transport properties of semiconductors. *Mailing Add:* Dept Physics Univ Vt Burlington VT 05401

LAMBERT, MARY PULLIAM, b Birmingham, Ala, Apr 27, 44; m 67; c 3. BIOCHEMISTRY, NEUROBIOLOGY. *Educ:* Birmingham-Southern Col, BS, 66; Northwestern Univ, PhD(biochem), 71. *Prof Exp:* Instr biochem, 70-72, fel reproductive biol, Northwestern Univ, Ill, 81-82; NIH res fel, 82-83, RES ASSOC, NORTHWESTERN UNIV, 83- *Mem:* Sigma Xi; Soc Neurosci. *Res:* Neurotransmitter receptors; receptor biochemistry; receptor development; mechanism of receptor function. *Mailing Add:* Dept Neurobiol & Physiol Northwestern Univ Evanston IL 60208

LAMBERT, MAURICE C, b Roosevelt, Utah, Apr 14, 18; m 42; c 5. PHYSICAL CHEMISTRY. *Educ:* Brigham Young Univ, BS, 39, MA, 41. *Prof Exp:* Assoc chemist, Indust Lab, Mare Island Naval Shipyard, 41-46; chemist, Hanford Atomic Prod Oper, 48-64; sr chemist, Gen Elec Co, 64; sr res scientist, Battelle Northwest Labs, 65-70; sr res scientist, Westinghouse Hanford Co, 70-82; RETIRED. *Mem:* Am Chem Soc; Soc Appl Spectros. *Res:* X-ray spectrometry, absorptiometry and diffraction; atomic absorption and flame emission spectrometry; separations of trace elements; gas-solid reactions; properties of inorganic oxides; fused salt studies; surface analysis by electron spectroscopy; automation of analytical techniques. *Mailing Add:* 1617 Hains Ave Richland WA 99352

LAMBERT, PAUL WAYNE, b Ft Worth, Tex, Oct 27, 37; m 59; c 1. GEOMORPHOLOGY, PHOTOGRAPHY. *Educ:* Tex Tech Univ, BA, 59; Univ NMex, MS, 61, PhD(geol), 68. *Prof Exp:* Geologist, Texaco Inc, NMex, 61-62; asst prof geol, Cent Mo State Col, 65-68; assoc prof, WTex State Univ, 68-70; geologist, Dept Prehist, Nat Inst Anthrop & Hist, Mex, 72-73; geologist, US Geol Surv, 73-81; ASSOC PROF GEOL, WTEX STATE UNIV, 81- *Concurrent Pos:* Res grants, Geol Soc Am & Sigma Xi, 68-69 & NSF, 69-70. *Mem:* Geol Soc Am; Am Quaternary Asn; Soc Am Archaeol; Asn Am Geographers; Royal Geog Soc; Explorers Club. *Res:* photographic documentation of physical and cultural geographic features in western United States and Mexico. *Mailing Add:* Dept Biol & Geosci WTex State Univ Canyon TX 79016-0907

LAMBERT, REGINALD MAX, b Delta, Ohio, Feb 25, 26; m 52; c 3. BACTERIOLOGY, IMMUNOLOGY. *Educ:* Butler Univ, BA, 50; Univ Buffalo, MA, 52, PhD(bact, immunol), 55. *Prof Exp:* Asst bact & immunol, Sch Med, State Univ NY Buffalo, 51-55, instr, 55-57, assoc, 57-59, asst prof, 59-64; asst prof path, Col Med, Univ Fla, 64-67; assoc dir, Blood Group Res Unit, 55-64 & 67-76, ASSOC PROF MICROBIOL, SCH MED, STATE UNIV NY, BUFFALO, 67- *Concurrent Pos:* Consult, E J Meyer Mem Hosp, Buffalo, 58, 60-63 & 67- & Buffalo Gen Hosp, 63-64; dir blood bank, Shands Teaching Hosp, Univ Fla, 64-67; dir, Buffalo Regional Red Cross Bldg Prog, 73- *Mem:* AAAS; Am Soc Microbiol; Int Soc Blood Transfusion; Int Soc Hemat; Sigma Xi. *Res:* Blood groups; immunohematology; transfusion genetics. *Mailing Add:* 750 Stolle St Elma NY 14059

LAMBERT, RICHARD BOWLES, JR, b Clinton, Mass, Apr 20, 39; m 64. PHYSICAL OCEANOGRAPHY. *Educ:* Lehigh Univ, AB, 61; Brown Univ, ScM, 64, PhD(physics), 66. *Prof Exp:* Fulbright fel, aerodyn, Munich Tech, 66-67; from asst prof to assoc prof oceanog, Univ RI, 67-75; prog dir phys oceanog, NSF, 75-77; RES OCEANOGR, SCI APPLICATIONS, INC, 77-, MGR, OCEAN PHYSICS DIV, 79-, ASST VPRES, 80- *Mem:* AAAS; Am Geophys Union. *Res:* Hydrodynamic stability; oceanic turbulence; diffusion energy transfer; air-sea interaction. *Mailing Add:* 11312 Gainsborough Rd Potomac MD 20854

LAMBERT, RICHARD ST JOHN, b Trowbridge, Eng, Nov 11, 28; m 52; c 6. PETROLOGY, GEOCHEMISTRY. *Educ:* Univ Cambridge, BA, 52, PhD(petrol), 55, MA, 56; Oxford Univ, MA, 56. *Prof Exp:* Asst lectr geol, Univ Leeds, 55-56; lectr, Oxford Univ, 56-70; PROF GEOL & CHMN DEPT, UNIV ALTA, 70- *Concurrent Pos:* Vis prof, Univ Alta, 63-64; fels, Iffley Col, Oxford Univ, 65-66 & Wolfson Col, 66-70. *Mem:* Geol Soc London; Geol Asn Can; Am Geophys Union; Geochem Soc; Mineral Soc Gt Brit & Ireland. *Res:* Mineralogy, petrology, geochemistry and isotope geology of medium to high grade metamorphic rocks; theory of metamorphic processes; thermal history of the earth; geological time-scale. *Mailing Add:* Dept Geol Univ Alta Edmonton AB T6G 2E2 Can

LAMBERT, ROBERT F, b Warroad, Minn, Mar 14, 24; m 51; c 3. ELECTRICAL ENGINEERING, ACOUSTICS. *Educ:* Univ Minn, BEE, 48, MS, 49, PhD, 53. *Prof Exp:* Asst elec eng, 48-49, from instr to assoc prof, 49-59, assoc dean, Inst Technol, 67-68, PROF ELEC ENG, UNIV MINN, MINNEAPOLIS, 59- *Concurrent Pos:* Vis asst prof, Mass Inst Technol, 53-55; vis scientist, III Phys Inst, Univ Goettingen, Ger, 64 & NASA Langley Res Ctr, Hampton, Va, 79; acoust consult. *Honors & Awards:* John Johnson Mem Educ Award, Inst of Noise Control Engrs, 84. *Mem:* Am Soc Eng Educ; fel Acoust Soc Am; fel Inst Elec & Electronics Engrs; Sigma Xi; Inst Noise Control Engrs. *Res:* Signal analysis including random processes and noise; acoustics including flow ducts, wave filters, porous materials, wave propagation, noise control; communication technology including ink jet printing, ultrasonic scanning, speech and electro-acoustics; random processes. *Mailing Add:* Dept Elec Eng Inst Technol Univ Minn Minneapolis MN 55455

LAMBERT, ROBERT HENRY, b Bayshore, NY, Nov 3, 30; div; c 2. ATOMIC PHYSICS. *Educ:* St Lawrence Univ, BS, 52; Harvard Univ, MS, 54, PhD(physics), 63. *Prof Exp:* Instr physics, Univ NH, 55-57; asst, Harvard Univ, 57-60; from asst prof to assoc prof, 61-68, PROF PHYSICS, UNIV NH, 68- *Concurrent Pos:* Cent Univ res grants, Univ NH, 62-63, 65-66; NSF grant, 65-71. *Mem:* Am Phys Soc. *Res:* Measurement of hyperfine structure using optical pumping. *Mailing Add:* Dept Physics De Meritt Hall Univ NH Durham NH 03824

LAMBERT, ROBERT J, b Dubuque, Iowa, Dec 23, 21; m 42; c 2. MATHEMATICS. *Educ:* Drake Univ, BA, 43; Iowa State Univ, MS, 48, PhD(math), 51. *Prof Exp:* Instr math, Drake Univ, 43-44; instr, Iowa State Univ, 46-51; mathematician, Nat Security Agency, 51-53; from asst prof to assoc prof math, Univ, 53-64, sr mathematician, Ames Lab, 64-78, PROF MATH & COMPUT SCI, COMPUT CTR, IOWA STATE UNIV, 64-, ASSOC DIR, 78- *Concurrent Pos:* Consult, Nat Security Agency & Collins Radio Co. *Mem:* Am Math Soc; Math Asn Am; Soc Indust & Appl Math. *Res:* Matrix theory; finite fields; numerical analysis; partial differential equations; ordinary differential equations; development of numerical software packages for solving differential equations, linear systems and eigensystems. *Mailing Add:* 3301 Ross Rd Ames IA 50010

LAMBERT, ROBERT JOHN, b Faribault, Minn, Mar 14, 27. PLANT GENETICS. *Educ:* Univ Minn, BS, 52, MS, 58; Univ Ill, PhD(plant breeding, genetics), 63. *Prof Exp:* Res asst plant breeding & genetics, Univ Minn, 56-58; from res asst to res assoc, 58-64, from instr to assoc prof, 64-76, PROF PLANT BREEDING & GENETICS, UNIV ILL, URBANA, 76- *Concurrent Pos:* Supvr world collection of maize mutants, Maize Genetics Coop. *Mem:* AAAS; Crop Sci Soc Am; Genetics Soc Am; Am Asn Cereal Chemists. *Res:* Investigations of plant geometry of maize and breeding in high yield environment; selection and development of modified protein maize strains. *Mailing Add:* Dept Agron Univ Ill 1102 S Goodwin Ave Urbana IL 61801

LAMBERT, ROGER GAYLE, b Minneapolis, Minn, Jan 22, 30; m 56; c 3. PLANT PHYSIOLOGY. *Educ:* Univ Minn, BS, 53, PhD(plant physiol), 61. *Prof Exp:* Instr plant physiol, Univ Minn, 57-61; asst prof plant physiol & path, 61-64, from actg head to head dept biol, 63-66, assoc prof plant physiol, 64-68, PROF PLANT PHYSIOL, UNIV LOUISVILLE, 68- *Concurrent Pos:* Fel bot & plant path, Potato Virus Lab, Colo State Univ, 70-71. *Mem:* Am Soc Plant Physiol; Sigma Xi. *Res:* Plant competition and trophic structure of ecosystems. *Mailing Add:* Dept Biol Univ Louisville Louisville KY 40292

LAMBERT, ROGERS FRANKLIN, b Kamas, Utah, July 12, 29; m 51; c 4. ORGANIC CHEMISTRY. *Educ:* Brigham Young Univ, BS, 53; Purdue Univ, PhD(org chem), 58. *Prof Exp:* Chemist, US Bur Mines, 53; res chemist, Ethyl Corp, 58-61; res supvr, Thiokol Chem Corp, 61-65; PROF CHEM, RADFORD COL, 65- *Mem:* Am Chem Soc. *Res:* Polymers; chemical reductions; transition metal carbonyls; reactions of heterocyclics. *Mailing Add:* Dept Chem Radford Univ Radford VA 24142

LAMBERT, RONALD, photographic chemistry; deceased, see previous edition for last biography

LAMBERT, ROYCE LEONE, b Coatesville, Ind, Nov 3, 33; m 53; c 3. SOILS, AGRONOMY. *Educ:* Purdue Univ, Lafayette, BS, 64, MS, 66, PhD(soil physics), 70. *Prof Exp:* ASSOC PROF SOILS, CALIF POLYTECH STATE UNIV, SAN LUIS OBISPO, 69- *Concurrent Pos:* Soil conservationist, Nat Park Serv, 71. *Mem:* Am Soc Agron; Soil Sci Soc Am; Soil Conserv Soc Am. *Res:* Soil management. *Mailing Add:* Dept Soil Sci Calif Polytech State Univ San Luis Obispo CA 93407

LAMBERT, WALTER PAUL, b Glendale, WVa, Sept 25, 44; m 72. CIVIL ENGINEERING. *Educ:* Univ Cincinnati, BS, 67, MS, 69; Univ Tex, Austin, PhD(civil eng), 75. *Prof Exp:* Fel, Univ Tex, Austin, 72-75; res area mgr, Med Bioeng Res & Develop Lab, US Army, Ft Detrick, 75-77, environ eng staff officer, Hq, US Army Med Res & Develop Command, 77-81; mgr, Res & Develop, Roy F Weston, Inc, 81-; AT JAMES M MONTGOMERY INC. *Concurrent Pos:* Lectr, Hood Col, 79-81. *Honors & Awards:* Medal, Int Ozone Inst, 77. *Mem:* Am Water Works Asn; Water Pollution Control Fedn; Soc Am Military Engrs; Am Defense Preparedness Asn; Am Soc Testing & Mat. *Res:* Mathematical modeling of water resource systems; wastewater reuse; research management decision theory; human factors engineering; hazardous materials; decontamination of contaminated soils and sediments. *Mailing Add:* James M Montgomery Inc 250 Madison Ave Pasadena CA 91109-7009

LAMBERT, WILLIAM M, JR, b Wausau, Wis, Apr 6, 36. MATHEMATICS. *Educ:* Univ Wis, BA, 58; Univ Calif, Los Angeles, MA, 59, PhD(math), 65. *Prof Exp:* Teaching asst math, Univ Wis, 57-58; teaching asst, Univ Calif, Los Angeles, 59-60, res asst, 60-63; from asst prof to assoc prof, Loyola Univ, Calif, 63-69; assoc prof, Univ Detroit, 69-74; prof, Dept Math, Univ Costa Rica, 74-85; RETIRED. *Mem:* Math Asn Am; Asn Symbolic Logic. *Res:* Effective processes of general algebraic structures; metamathematics of algebra. *Mailing Add:* Apdo 111 Sabanilla 2070 Montes de Oca Costa Rica

LAMBERTI, JOSEPH W, b Toronto, Ont, Dec 20, 29; m 55; c 6. PSYCHIATRY. *Educ:* Univ Ottawa, MD, 54; Royal Col Physicians & Surgeons Can, cert psychiat, 61. *Prof Exp:* Asst psychiatrist, Winnipeg Psychiat Inst, 60-61; sr psychiatrist, 61-63; asst prof, 63-69, ASSOC PROF PSYCHIAT, MED CTR, UNIV MO, COLUMBIA, 69- *Concurrent Pos:* Dir

consult serv & lectr, Med Ctr, Univ Mo, 63-67, consult, Peace Corps, 64- & Univ Press, 65- *Mem:* Am Psychiat Asn; cor mem Can Psychiat Asn; Sigma Xi. *Res:* Treatment of common sexual disorders; study of antisocial behavior; study of affective disorders. *Mailing Add:* Dept Psychiat Univ Mo Med Ctr Columbia MO 65201

LAMBERTS, AUSTIN E, b East Saugatuck, Mich, Nov 30, 14; div; c 4. MARINE ZOOLOGY, NEUROSURGERY. *Educ:* Calvin Col, AB, 36; Univ Mich, Ann Arbor, MD, 41, MS, 50; Am Bd Neurosurg, dipl, 52; Univ Hawaii, PhD(marine zool), 73. *Prof Exp:* Resident & instr neurosurg, Univ Mich, 45-50; pvt pract neurosurg, St Mary's Hosp, Grand Rapids, 50-68; teaching asst marine zool, Univ Hawaii, 69-73; INDEPENDENT RES, REEF ECOL, 73- *Concurrent Pos:* Consult neurosurg, St Mary's Hosp, Grand Rapids, 50-76; vol physician displaced dependents camp site 8, Africa & Asia; res grant, Nat Geog Soc, 74 & 78, Nat Sci Asn, 78. *Mem:* Fel Explorers Club; Am Asn Neurosurgeons; Cong Neurosurgeons; Am Med Asn. *Res:* Study of natural life cycles of reef corals and unexplained coral kills; coral growth using the dye alizarin; effects of pesticides on coral growth; collecting and identification of modern Pacific reef corals. *Mailing Add:* 1520 Leffingwell NE Grand Rapids MI 49505

LAMBERTS, BURTON LEE, b Fremont, Mich, Oct 24, 19; m 60; c 2. BIOCHEMISTRY. *Educ:* Calvin Col, BS, 49; Mich State Univ, PhD(chem), 58. *Prof Exp:* Chemist, Northern Regional Res Lab, Ill, 51-54; asst chem, Mich State Univ, 55-58, instr, 58-60; chief biochemist, Dent Res Facil, Naval Dent Res Inst, 60-88; RETIRED. *Mem:* Am Chem Soc; Int Asn Dent Res. *Res:* Dental caries; salivary gland secretions; relationship of oral microbial products to periodontal disease. *Mailing Add:* Biochem Div Naval Dent Res Inst USN Base Great Lakes IL 60088

LAMBERTS, ROBERT L, b Fremont, Mich, Sept 8, 26; m 51; c 6. PHOTOGRAPHIC OPTICS, PHYSICAL OPTICS. *Educ:* Calvin Col, AB, 49; Univ Mich, MS, 51; Univ Rochester, PhD(optics), 69. *Prof Exp:* Res assoc, 51-80, sr res assoc, Kodak Res Labs, Eastman Kodak Co, 80-83; teaching, Roberts-Wesleyan Col, 83-84; teaching, Daystar Univ Col, Nairobi, Kenya, 84-87; TEACHING, NAZARETH COL ROCHESTER, 83- *Concurrent Pos:* Partner, Sine Patterns, Penfield, NY. *Mem:* Fel Optical Soc Am; Soc Photog Sci & Eng; Optical Soc Am. *Res:* Image structure of optical systems and photographic materials; physical optics. *Mailing Add:* 236 Henderson Dr Penfield NY 14526

LAMBERTSEN, CHRISTIAN JAMES, b Westfield, NJ, May 15, 17; m 44; c 4. PHARMACOLOGY. *Educ:* Rutgers Univ, BS, 39; Univ Pa, MD, 43. *Prof Exp:* Intern, Hosp Univ Pa, 43; from instr to assoc prof pharmacol, Univ Pa, 46-53, prof exp therapeut, 62-87, dir, Inst Environ Med, 70-87, PROF PHARMACOL, SCH MED, UNIV PA, 53-, PROF MED, UNIV HOSP, 78-, FOUND DIR, INST ENVIRON MED, 87- *Concurrent Pos:* Markle scholar, 48-53; assoc med, Univ Hosp, Univ Pa, 48-77; mem panel shipboard & submarine med, Off Secy Defense Res Develop Bd, 50-53; vis res assoc prof, Univ Col, London, 51-52; mem comt undersea warfare & comt naval med res, Nat Res Coun, 53- & panel underwater swimmers, 53-56; basic sci secy, Nat Bd Med Exam, 54-, mem pharmacol comt, 54-55; consult, US Army Chem Ctr, 55-59; consult & lectr, Off Surgeon Gen, US Navy, 57-60; consult neuropharmacol, Del State Hosp, 58-61; consult, Sci Adv Bd, US Air Force, 59-61, mem adv panel med sci, Off Secy Defense; chmn, Man in Space Comt, Space Sci Bd, Nat Acad Sci, 60-62, consult, 62-; mem, US Oceanogr Adv Bd, 70-; chmn comt manned undersea activity, Off Secy Navy; mem comt undersea physiol & med, Nat Res Coun, 72- & comt hyperbaric oxygenation. *Honors & Awards:* Ocean Sci & Eng Award, Marine Technol Soc, 72; Environ Sci Award, NY Acad Sci, 74; Environ Sci Award, Aerospace Med Asn, 79. *Mem:* Nat Acad Eng; Fel Am Soc Clin Pharmacol & Therapeut; Am Physiol Soc (pres, 54-55); Am Soc Clin Invest; Am Soc Pharmacol & Exp Therapeut. *Res:* Respiratory physiology and pharmacology; aerospace and diving medicine; breathing apparatus for underwater swimmers. *Mailing Add:* Inst Environ Med Univ Pa Med Ctr One John Morgan Bldg Philadelphia PA 19104-6068

LAMBERTSEN, ELEANOR C, HEALTH SERVICES DELIVERY. *Prof Exp:* NURSING & HEALTH SERV CONSULT, 88- *Mem:* Inst Med-Nat Acad Sci. *Mailing Add:* 510 E 77th St New York NY 10021

LAMBERTSEN, RICHARD H, b Philadelphia, Pa, Feb 11, 53. PATHOBIOLOGY, PATHOPHYSIOLOGY. *Educ:* Univ Pa, BA, 75, VMD, 79, PhD(comp med & exp hemat), 80. *Prof Exp:* Postdoctoral scholar, Woods Hole Oceanog Inst, 81-82; asst prof physiol, Col Vet Med, Univ Fla, 82-88, dir, Inst Biomed Aquatic Studies, 84-88; RES ASSOC, ECOSYSTS TECHNOL TRANSFER, INC, 88- *Concurrent Pos:* Vis scientist, Inst Exp Path, Univ Iceland, 81-88; guest investr, Coastal Res Ctr, Woods Hole Oceanog Inst, 82, 83, 86 & 91; mem sci comn, Int Whaling Comn, 86; consult, Div Res Resources, NIH, 87; me, Working Group on Assessment of Risk Associated with Maritime Shipment Dangerous Prod, NATO/CCMS, 87; adv ecotoxicol, Ctr Doc & Res Marine Pollution, France, 87-89; fel, Comt Challenges of Modern Soc, NATO, 87-89. *Mem:* Int Asn Aquatic Animal Sci; Am Soc Anatomists; Soc Marine Mammal; Am Soc Zoologists; AAAS; Oceanog Soc. *Res:* Investigations of fundamental and societal implications of the maximization of kinetic energy through processes of organic and technologic evolution; spatiokinetic organization of hematopoietic microenvironments; baleen whale feeding mechanics and biological momentum transfer processes; large whale pathobiology and pathophysiology; marine ecotoxicology and conservation; maintenance of freedom of scientific inquiry; marine policy. *Mailing Add:* Coastal Res Ctr Woods Hole Oceanog Inst Woods Hole MA 02543

LAMBERTSON, GLEN ROYAL, b Paonia, Colo, Jan 14, 26; m 50; c 3. ACCELERATION SCIENCE, PARTICLE DYNAMICS. *Educ:* Univ Colo, BS, 48; Univ Calif, Berkeley, MA, 51. *Prof Exp:* Res physicist, Lawrence Radiation Lab, Berkeley, 51-54 & 55-63 & Brookhaven Nat Lab, 54-55; res physicist, 64-71, STAFF SR SCIENTIST, LAWRENCE BERKELEY LAB,

71-, GROUP LEADER, 73- *Concurrent Pos:* Vis scientist, Europ Orgn Nuclear Res, Geneva, 63-64. *Mem:* Am Phys Soc; AAAS. *Res:* Analysis, design specification, & development of particle acceleration components; items that interact electromagnetically with the beam. *Mailing Add:* MS 47-112 Lawrence Berkeley Lab Berkeley CA 94720

LAMBETH, DAVID N, b Carthage, Mo, Mar 18, 47; m 69; c 1. MAGNETISM. *Educ:* Univ Mo-Columbia, BS, 69; Mass Inst Technol, PhD(physics), 73. *Prof Exp:* SR RES PHYSICIST, EASTMAN KODAK CO, 73- *Mem:* Inst Elec & Electronics Engrs; Magnetics Soc; Am Phys Soc. *Res:* Magnetism and magneto-optics of thin film materials. *Mailing Add:* 620 Lake Rd Webster NY 14580

LAMBETH, DAVID ODUS, b Carthage, Mo, June 16, 41; m 62; c 2. BIOCHEMISTRY, ORGANIC CHEMISTRY. *Educ:* Univ Mo-Columbia, BSEd, 62; Purdue Univ, MS, 67; Univ Wis-Madison, PhD(biochem), 71. *Prof Exp:* Instr chem, Columbia Pub Schs, 62-67; NIH fel biochem, Univ Mich, 71-73; asst prof chem, Univ SFla, 73-77; asst prof, 77-79, ASSOC PROF BIOCHEM, SCH MED, UNIV NDAK, 79- *Mem:* Sigma Xi; Am Chem Soc; Am Soc Biol Chemists; Sigma Xi. *Res:* Chemistry and biochemistry of sulfides and thiols; enzymology; metabolism of iron in biological systems. *Mailing Add:* 1909 20th Ave S Grand Forks ND 58201

LAMBETH, VICTOR NEAL, b Sarcoxie, Mo, July 5, 20; m 46; c 2. VEGETABLE CROPS, TOMATO BREEDING. *Educ:* Univ Mo, BS, 42, MA, 48, PhD(hort), 50. *Prof Exp:* Asst instr, 39-42, from asst prof to prof, 59-90, EMER PROF HORT, UNIV MO-COLUMBIA, 90- *Concurrent Pos:* NSF vis prof, Thailand, 81, res, 82-84; consult, Liberia, 87. *Mem:* AAAS; Am Soc Plant Physiol; fel Am Soc Hort Sci; Int Soc Hort Sci. *Res:* Soil fertility and plant nutrition; raw product quality; tomato breeding; water relationships; post harvest physiology; international agriculture. *Mailing Add:* 1-40 New Agr Bldg Univ Mo Columbia MO 65211

LAMBIRD, PERRY ALBERT, b Reno, Nev, Feb 7, 39,; m 60; c 4. PATHOLOGY. *Educ:* Stanford Univ, BA, 58; Johns Hopkins Univ, MD, 62; Oklahoma City Univ, MBA, 73. *Prof Exp:* Fel internal med, Johns Hopkins Univ, 62-63, fel path, 65-69; consult, Health Serv Admin, USPHS, 63-65; PATHOLOGIST, MED ARTS LAB, 69-, CLIN PROF PATH & ORTHOPED SURG, 70- *Concurrent Pos:* Consult pathologist, Off Med Examr, 69-80; adj asst prof, Oklahoma City Univ, 73-82; proj pathologist, Am Cancer Soc-Nat Cancer Inst BCDP, 74-81; proprietor, Lambird Mgt Consult Serv, 74-; consult, Okla Breast Cancer Control Network, 75-80; reviewer, Southern Med J, J AMA & Diag Cytopath, 75-; mem bd dirs, Am Path Found, 78-84, pres, 83-84; deleg, AMA, 80-, mem, Coun Med Serv, 85-; regent, Uniformed Serv, Univ Health Sci, 83-88; mem, Task Force Entitlements & Human Asst, US House Rep, 83-; pres, Independent Path Inst, Inc, 84-; bd govs, Col Am Pathologists, 84- *Honors & Awards:* CCE Award, Am Soc Clin Path, 83-91; Distinguished Serv Medal, Uniformed Servs, Univ Health Sci, 88; Outstanding Pathologist Award, Am Path Found, 84; Distinguished Practr, Nat Academies Practice, 90. *Mem:* Am Med Asn; Am Path Found (pres, 82-83); Arthur Purdy Stout Soc Surg Pathologists; fel Col Am Pathologists; Am Soc Cytol; fel Am Soc Clin Pathologists. *Res:* Breast cancer pathology and control; health care financing and administration; medical systems management. *Mailing Add:* Med Arts Lab 100 Pasteur 1111 N Lee St Olkahoma City OK 73103

LAMBOOY, JOHN PETER, b Kalamazoo, Mich, Dec 6, 14; m 42; c 4. BIOCHEMISTRY. *Educ:* Kalamazoo Col, AB, 37, MS, 38; Univ Ill, MA, 39; Univ Rochester, PhD(physiol chem), 42. *Prof Exp:* From instr to assoc prof physiol, Univ Rochester, 46-63; prof chem pharmacol & sect head biochem pharmacol, Eppley Inst Cancer Res, Col Med, Univ Nebr, 63-68, prof biochem, 64-69; assoc dean grad sch, Baltimore Campuses, 69-71, prof biol chem, Sch Med, 69-74, dean grad studies & res, 71-74, prof biochem & chmn dept, 74-85, EMER PROF BIOCHEM, SCH DENT, UNIV MD, BALTIMORE, 85- *Mem:* Fel AAAS; Am Chem Soc; Am Soc Biol Chem; Sigma Xi. *Res:* Synthesis and biological acitivity of vitamin analogs, amino acid analogs, anesthetics, sympathomimetic amines, bacteriostatic agents, carcinolytic agents and carcinogenic agents. *Mailing Add:* 904 Huntsman Rd Baltimore MD 21204

LAMBORG, MARVIN, biochemistry, for more information see previous edition

LAMBORN, BJORN N A, b Stockholm, Sweden, Apr 2, 37. PLASMA PHYSICS. *Educ:* Univ Calif, Berkeley, AB, 58, MA, 60; Univ Fla, PhD(physics, math), 62. *Prof Exp:* Instr physics, Univ Miami, 63; res physicist, Inst Plasmaphysik, GmbH, Munich, Ger, 63-65; from asst prof to assoc prof physics, 65-75, chmn dept, 70-73, PROF PHYSICS, FLA ATLANTIC UNIV, 75-, CHMN DEPT, 79- *Concurrent Pos:* Bd trustees, Thiouracil. *Mem:* Sigma Xi; Am Phys Soc. *Res:* Theoretical plasma physics; wave interaction in relativistic plasmas; nonadiabatic particle motion; diffusion; nonlinear wave coupling. *Mailing Add:* Dept Physics Fla Atlantic Univ Boca Raton FL 33431

LAMBRAKIS, KONSTANTINE CHRISTOS, b Piraeus, Greece, Jan 30, 36; US citizen. AEROSPACE ENGINEERING. *Educ:* Univ Bridgeport, BSEE, 62, MSME, 65; Rensselaer Polytech Inst, PhD(aerospace eng), 71. *Prof Exp:* Develop elec engr, Skinner Pridision Industs, 61-64; sr mech eng, MB Electronics, 64-66; asst prof thermodynamics, 66-69, assoc prof gas dynamics, 69-72, chmn mech eng dept, 74-76, PROF ENG & GAS DYNAMICS, UNIV NEW HAVEN, 72-, DEAN ENG, 76- *Concurrent Pos:* Consult, United Nuclear Corp, 84-85, Times Fiber Commun Inc, 86-87, Textron Inc, 86-87. *Mem:* Am Soc Mech Eng; AAAS; Am Soc Elec Eng; Am Soc Aeronaut Eng. *Res:* Compressible fluid flow and thermal sciences; new computational techniques for solving non-linear partial differential equations occurring in thermal-fluid sciences and field theory. *Mailing Add:* Dept Mech Eng Univ New Haven 300 Orange Ave West Haven CT 06516

LAMBRECHT, RICHARD MERLE, b Salem, Ore, Apr 8, 43; div; c 3. PHYSICAL CHEMISTRY, NUCLEAR MEDICINE. *Educ:* Ore State Univ, BS, 65; Univ Nebr, PhD(phys chem), 69. *Prof Exp:* Res assoc chem, 69-70, assoc, 70-74, CHEMIST, BROOKHAVEN NAT LAB, 74- *Concurrent Pos:* Consult, Capintec, Inc, 75-; ed, J Radioanalytical Chem, 78-83. *Mem:* Soc Nuclear Med; Europ Soc Nuclear Med; Am Chem Soc. *Res:* Radiopharmaceutical chemistry and nuclear medicine with emphasis on accelerator production and use of short-lived radionuclides; positron emission tomography; exotic atoms; chemical effects of nuclear transformation. *Mailing Add:* ANSTO Bio Med & Health Prog Menai 2234 Australia

LAMBREMONT, EDWARD NELSON, b New Orleans, La, July 29, 28; m 81, 90; c 4. ENTOMOLOGY, NUCLEAR SCIENCE. *Educ:* Tulane Univ, BS, 49, MS, 51; Ohio State Univ, PhD(entom), 58. *Prof Exp:* Asst zool, Tulane Univ, 48-51; asst entom, Ohio State Univ, 54-56; entomologist, Insect Physiol, Entom Res Div, Agr Res Serv, USDA, La, 58-66; assoc prof nuclear sci, 66-74, PROF NUCLEAR SCI & DIR NUCLEAR SCI CTR, LA STATE UNIV, BATON ROUGE, 74- *Concurrent Pos:* Vis scientist, Oak Ridge Assoc Univs, Med & Health Sci Div, 77-, bd dirs, 79-84; consult nuclear sci & technol, pub info & radiation safety, energy issues; southeast regional dir, bd dir, Sigma Xi, 83-90; vis scientist, Int Atomic Energy Agency, Vienna, 88- *Mem:* AAAS; Entom Soc Am; Am Oil Chem Soc; Sigma Xi. *Res:* Physiology and biochemistry of insects, especially lipid metabolism, synthesis and utilization of fatty acids, phospholipids and glycerolipids and lipid enzyme systems; radiotracer and nuclear science methodology as applied to biological problems; insect radiation biology and physiology of tumorous tissues. *Mailing Add:* Nuclear Sci Ctr La State Univ Baton Rouge LA 70803-5820

LAMBROPOULOS, PETER POULOS, b Tripolis, Greece, Oct 5, 35; US citizen. THEORETICAL PHYSICS. *Educ:* Athens Tech Univ, dipl, 58; Univ Mich, MSE, 62, MS, 63, PhD(nuclear sci), 65. *Prof Exp:* Engr, Orgn Telecommun, Greece, 59-60; sr physicist, Bendix Res Lab, Mich, 65-67; asst physicist, Argonne Nat Lab, Ill, 67-72; vis fel, Joint Inst Lab Astrophys, 72-73; from asst to assoc prof, Tex A&M Univ, 73-75; assoc prof, 75-79, co-chmn dept, 81-83, PROF PHYSICS, UNIV SOUTHERN CALIF, 79- *Mem:* AAAS; Am Phys Soc; Fedn Am Scientists. *Res:* Atomic physics; interaction of radiation with matter; quantum optics; strong electromagnetic fields. *Mailing Add:* Dept Physics Univ Southern Calif University Park Los Angeles CA 90089-0484

LAMBSON, ROGER O, b Provo, Utah, Feb 5, 39; m 59; c 3. ANATOMY. *Educ:* Univ Mont, BA, 61; Tulane Univ, PhD(anat), 65. *Prof Exp:* From instr to prof anat, Col Med, Univ Ky, 65-84, assoc dean student affairs & dir admis, 71-75, assoc dean acad affairs, 75-76, assoc dean basic sci, 76-84; vchancellor, health policy & prog develop, 84-88, PROF ANAT, UNIV KANS, 84-, VCHANCELLOR ADMIN, 89- *Mem:* AAAS; Am Asn Anatomists; Am Asn Med Clins. *Mailing Add:* VChancellor Admin Univ Kansas Med Ctr Kansas City KS 66103

LAMBUTH, ALAN LETCHER, b Seattle, Wash, Jan 5, 23; m 44; c 4. ADHESIVE TECHNOLOGY, WOOD UTILIZATION. *Educ:* Univ Wash, BS, 47. *Prof Exp:* Res chemist, Western Div, Monsanto Co, 47-57, res group leader, Plastics Div, 58-68; asst mgr mfg tech serv, 69-70, mgr prod develop, 71-76, mgr res & develop, 77-81, MGR PROD & PROCESS DEVELOP, TIMBER & WOOD PROD DIV, BOISE CASCADE CORP, 82- *Concurrent Pos:* Reviewer of publ & res proposals, NSF, Forest Prod Res Soc & USDA forest prod labs & exp stas, 75-; mem, US-Can Binational Comt Plywood, 85 & Adv Comt USDA Southern Forest Exp Sta; chmn, Prod/Standards Comt, Am Plywood Asn, Treated Wood Prod Comt, Nat Forest Prod Asn; chmn, PS-1 Standing Comt Plywood, US Dept Com & Adv Bd, Univ Calif Forest Prod Lab. *Honors & Awards:* Monsanto Award, 69; Borden Award, Forest Prod Res Soc, 84. *Mem:* Forest Prod Res Soc (pres, 82-83); Am Chem Soc; Am Inst Timber Construct; Am Soc Testing & Mat; Int Union Forestry Res Orgn; Nat Forest Prod Asn; Am Plywood Asn; Western Wood Prod Asn. *Res:* Adhesive technology and innovation; wood resource utilization; wood product development; chemical utilization of biomass; wood engineering. *Mailing Add:* 7240 Cascade Dr Boise ID 83704

LAMDEN, MERTON PHILIP, b Boston, Mass, Sept 7, 19; m 42; c 2. BIOCHEMISTRY. *Educ:* Univ Mass, BS, 41; Mass Inst Technol, PhD(food technol), 47. *Prof Exp:* Asst, Mass Inst Technol, 41-42, mem res staff, Food Technol Labs, 43-44; from asst prof to prof, 47-85, EMER PROF BIOCHEM, COL MED, UNIV VT, 85- *Concurrent Pos:* Commonwealth Fund fel & NSF-Orgn Europ Econ Coop sr vis fel, Univ Col, Univ London, 61-62; vis res biochem, Dept Food Sci & Technol, Univ Calif, Davis, 75. *Mem:* AAAS; Am Chem Soc; Am Inst Nutrit; Brit Biochem Soc. *Res:* Vitamin content and retention in foods; nutritional status of humans; biochemical studies on ascorbic and oxalic acids; role of ascorbic acid in metabolism. *Mailing Add:* 17 Wildwood Dr Burlington VT 05401

LAMDIN, EZRA, b Cleveland, Ohio, Nov 25, 23; m 69; c 3. CLINICAL TRIALS, HYPERTENSION. *Educ:* Harvard Univ, AB, 47, MD, 51; Am Bd Internal Med, dipl, 58. *Prof Exp:* Intern med, Harvard Med Serv, Boston City Hosp, 51-52, asst resident, 52-53; USPHS fel, Sch Med, Yale Univ, 53-55; asst, Sch Med, Boston Univ, 55-56, instr, 56-58; clin instr, Sch Med, Tufts Univ, 58-60; res fel biochem, Brandeis Univ, 61-62; asst prof med & physiol, Col Med, Univ Cincinnati, 62-67; assoc prof med, Sch Med, Univ Pittsburgh, 67-69; asst med dir, Am Heart Asn, 69-73, dir div sci affairs, 73-75; asst med dir, Ayerst Labs, 75-77, assoc med dir, 77-79; dir, med affairs, ICI Americas, Inc, 79-88; CONSULT PHARMACEUT INDUST, 88- *Concurrent Pos:* Teaching fel med, Harvard Med Sch, 52-53; resident, Vet Admin Hosp, Boston, 55-56, staff physician, 56-60, clin investr, 58-60; clin & res fel med, Mass Gen Hosp, Boston, 58-59; res collabr, Med Res Dept, Brookhaven Nat Lab, 63-69; chief med serv, Vet Admin Hosp, Pittsburgh, 67-69; adj assoc prof physiol, Mt Sinai Sch Med, 69-79. *Mem:* AAAS; Am Fedn Clin Res; AAAS; Am Heart Asn; Am Soc Clin Pharmacol & Therapeut; Sigma Xi; affil Royal Soc Med. *Res:* Renal physiology and disease; biochemical aspects of membrane phenomena and active transport; intermediary metabolism; diabetes and obesity; mechanism of hormone action; clinical pharmacology; mechanism and treatment of hypertension. *Mailing Add:* 817 Cherry Hill Rd Princeton NJ 08540

LAME, EDWIN LEVER, b Evanston, Ill, Feb 23, 04; m 40; c 2. INTERNAL MEDICINE. *Educ:* Mass Inst Technol, BS, 26; Univ Pa, MD, 33; Am Bd Internal Med, dipl, 45; Am Bd Radiol, dipl, 45. *Prof Exp:* Asst med, Sch Med, Univ Pa, 36-42, asst radiol, 42-47; dir radiol, Presby Hosp, 47-66; chief radiol, Vet Admin Hosp, Coatesville, 66-77; RETIRED. *Concurrent Pos:* Fel internal med & chest dis, physician & asst to dir, Dept Res Respiratory Dis, Germantown Hosp, 36-42; asst physician, Pa Hosp, 37-43; asst pediatrist & chief chest clin, Children's Hosp, 38-45; fel radiol, Hosp Univ Pa, 42-45; chief radiol, Jeanes Hosp, 45-48, consult, 48-60; assoc prof clin radiol, Sch Med & asst prof radiol, Grad Sch Med, Univ Pa, 57-66. *Mem:* Radiol Soc NAm; Am Roentgen Ray Soc; AMA; fel Am Col Physicians; fel Am Col Radiol. *Res:* Pulmonary radiologic interpretation; pelvic and vertebral osteomyelitis arising from the urinary tract; cholecystitis; clinical and radiologic criteria; gastrointestinal barium; protection and dose reduction in diagnostic radiology; radiologic signs of preclinical heart failure. *Mailing Add:* 1400 Waverly Rd Gladwyne PA 19035

LAMEIRO, GERARD FRANCIS, b Paterson, NJ, Oct 3, 49. TIME BASED MANAGEMENT, PROCESSED BASE MANAGEMENT. *Educ:* Colo State Univ, BS, 71, MS, 73, PhD(mech eng), 77. *Prof Exp:* NSF fel solar energy, 77, Solar Energy Appln Lab, Colo Energy Res Inst fel, Colo State Univ, 74-77, sr scientist, Solar Energy Res Inst, 77-78; asst prof mgt sci & info systs, Colo State Univ, 78-82; pres, Successful Automated Off Systs, Inc, 82-84; prod mgr, Hewlett-Packard Co, 84-88; columnist, HP Chronicle, 88; INDEPENDENT MGT STRATEGIST, COMPUTER CORP, 88- *Concurrent Pos:* Instr solar energy, Univ Colo, 76-78; consult, Solar Energy Res Inst, 78; consult, Comput Networking, 88-89. *Honors & Awards:* Nat Distinguished Service Award, Asn Energy Engrs, 81. *Mem:* Asn Energy Engrs (pres-elect, 79, pres, 80, Nat Bd Dirs, 80- 81); Asn Comput Mach; Am Soc Qual Control; Inst Elec & Electronics Engrs Computer Soc; Inst Indust Eng. *Res:* Time based and process based management; computer science techniques applied to the solution of real world problems; national policy work; simulation models. *Mailing Add:* 3313 Downing Ct Ft Collins CO 80526-2315

LAMELAS, FRANCISCO JAVIER, b Havana, Cuba, Feb 6, 59; US citizen. SYNCHROTRON X-RAY SCATTERING, MOLECULAR BEAM EPITAXY GROWTH & SURFACE SCIENCE. *Educ:* Univ Wis-Milwaukee, BS, 80, Madison, MS, 82; Univ Mich, Ann Arbor, MS, 87, PhD(physics), 90. *Prof Exp:* Sr assoc engr, Int Bus Mach, East Fishkill, NY, 82-84; MEM TECH STAFF, AT&T BELL LABS, MURRAY HILL, NJ, 90- *Mem:* Am Phys Soc. *Res:* Synchrotron-based x-ray scattering studies of thin film and surface structures; molecular beam epitaxy growth of metallic superlattices; semiconductor growth processes. *Mailing Add:* AT&T Bell Labs Rm 7C-216 600 Mountain Ave Murray Hill NJ 07974

LAMENSDORF, DAVID, b NY, Nov 22, 37. ELECTRICAL ENGINEERING. *Educ:* Cornell Univ, BEE, 60; Harvard Univ, SM, 61, PhD(appl physics), 67. *Prof Exp:* MEM TECH STAFF, SPERRY RES CTR, 67- *Concurrent Pos:* Res assoc, Univ Col London, 72-73. *Mem:* AAAS; sr mem Inst Elec & Electronics Engrs; Sigma Xi. *Res:* Electromagnetic theory; transient analysis of antennas; microwave antennas, networks and electronics. *Mailing Add:* Mitre Corp MS N230 Burlington Rd Bedford MA 01730

LA MERS, THOMAS HERBERT, b New York, NY, Apr 23, 45; m 69. ELECTRO CHEMICAL SENSORS, BIOSENSORS. *Educ:* Antioch Col, BSES, 68. *Prof Exp:* Machine designer, Vernay Labs, 68-69; aircraft designer, SR Corp, 69; prod res & develop, 69-89, DESIGN CONSULT, MECH DEVICES, YELLOW SPRINGS INSTRUMENT CO INC, 89- *Concurrent Pos:* Consult, Ventura Labeling Co, 84-85. *Res:* Integrating science and technology to accomplish new objectives in engineering design. *Mailing Add:* 777 Dayton St Yellow Springs OH 45387

LAMEY, HOWARD ARTHUR, b Bloomington, Ind, Dec 20, 29; m 56; c 5. EXTENSION TEACHING, ROW CROP DISEASES. *Educ:* Ohio Wesleyan Univ, BA, 51; Univ Wis-Madison, PhD(plant path), 54. *Prof Exp:* Res asst, Dept Plant Path, Univ Wis, 51-54, proj assoc, 54-55, 57-58; asst plant pathologist, US Army, Md, 56-57; plant pathologist, USDA, Camaguey, Cuba, 58-60, sr res plant pathologist, Baton Rouge, La, 60-69; plant pathologist, Int Inst Trop Agr, Ibadan, Nigeria, 69-71, coordr plant protection prog & chmn res comt, 70-71; proj mgr, Food & Agr Orgn, Suweon, Korea, 71-75, consult, Cent Am & Mex, 76; EXTEN PLANT PATHOLOGIST & PROF, EXTEN SERV, NDAK STATE UNIV, 77- *Concurrent Pos:* Rice virologist, Food & Agr Orgn, UN, Bangkok, Thailand, 66-67, proj mgr, 67-68, consult, 70. *Mem:* Fel AAAS; Am Phytopath Soc; Am Soc Sugar Beet Technologists. *Res:* Diseases of sugarbeets, sunflower, dry edible beans, and soybeans; fungicides, foliar and seed treatment, epidemiology. *Mailing Add:* Dept Plant Path NDak State Univ Box 5012 Fargo ND 58105

LAMEY, STEVEN CHARLES, b Lock Haven, Pa, Mar 5, 44; m 70. ANALYTICAL CHEMISTRY, ORGANIC CHEMISTRY. *Educ:* Lock Haven State Col, BA, 68; WVa Univ, MS, 73, PhD(anal chem), 75. *Prof Exp:* Res chemist organic synthesis, Am Aniline Corp, 68-70; RES CHEMIST ENERGY RES, MORGANTOWN ENERGY TECHNOL CTR, 75- *Mem:* Am Chem Soc; AAAS; Coblentz Soc; Sigma Xi. *Res:* Characterization of coal tars; organic composition of shale; instrument development for coal characterization; alkali corrosion; characterization of coal combustion products; mild gasification of coal. *Mailing Add:* Morgantown Energy Technol Ctr Collins Ferry Rd Morgantown WV 26505

LAMIE, EDWARD LOUIS, b Kingsley, Mich, Aug 27, 41; m 60; c 7. COMPUTER SCIENCE. *Educ:* San Diego State Univ, AB, 69; Univ Southern Calif, MS, 71; Mich State Univ, PhD(comput sci), 74. *Prof Exp:* Mem tech staff, Rockwell Int, 69-71; assoc prof comput sci, Cent Mich Univ, 71-82, prof & chmn dept, 82-; AT DEPT COMPUT SCI, CALIF STATE UNIV, STANISLAUS. *Mem:* Asn Comput Mach; Inst Elec & Electronics Engrs. *Res:* Database systems; discrete simulation; artificial intelligence. *Mailing Add:* Dept Comput Sci Calif State Univ Stanislaus 801 W Monte Vista Ave Turlock CA 95380

LAMM, A UNO, transductor; deceased, see previous edition for last biography

LAMM, FOSTER PHILIP, b Whittier, Calif, April 18, 50; m 77; c 2. POLYMER CHEMISTRY. *Educ:* Univ Calif, San Diego, BA, 72; Wesleyan Univ, PhD(chem), 79. *Prof Exp:* RES SCIENTIST, UNITED TECHNOLOGIES RES CTR, 79- *Mem:* Am Chem Soc; Soc Mfg Engrs. *Res:* Chemistry and processing of high performance adhesives; new materials and processing techniques for electrical insulating; organic materials failure analysis. *Mailing Add:* 56 Clinton Dr South Windsor CT 06074

LAMM, MICHAEL EMANUEL, b Brooklyn, NY, May 19, 34; m 61; c 2. IMMUNOLOGY, PATHOLOGY. *Educ:* Univ Rochester, MD, 59; Western Reserve Univ, MS, 62; Am Bd Path, dipl, 65. *Prof Exp:* From intern to resident path, Univ Hosps, Cleveland, Ohio, 59-62; res assoc chem, NIH, 62-64; from asst prof to prof path, Sch Med, NY Univ, 64-81; PROF & CHMN PATH, CASE WESTERN RESERVE UNIV, 81- *Mem:* NY Acad Sci; Am Soc Biol Chemists; Soc Exp Biol & Med; Am Asn Immunol; Am Asn Pathologists. *Res:* Mucosal immunity; immunopathology. *Mailing Add:* Inst Path Case Western Reserve Univ 2085 Adelbert Rd Cleveland OH 44106

LAMM, WARREN DENNIS, b West Reading, Pa, Jan 27, 47; m 71; c 1. BEEF CATTLE MANAGEMENT. *Educ:* Delaware Valley Col Sci & Agr, BS, 69; Iowa State Univ, MS, 72; Univ Nebr, PhD(ruminant nutrit), 76. *Prof Exp:* Exten agt, Exten Serv, Colo State Univ, 72-73; asst prof beef cattle nutrit & eval, Va Polytech Inst & State Univ, 76-81; EXTEN BEEF SPECIALIST, DEPT ANIMAL SCI, COLO STATE UNIV, 81- *Mem:* Am Soc Animal Sci; Am Forage & Grassland Coun; Coun Agr Sci & Technol. *Res:* Beef nutrition and management; energetic efficiency; protein utilization; use of underutilized feedstuffs and animal waste refeeding. *Mailing Add:* Colo State Univ 203 Administration Ft Collins CO 80523

LAMMERS, WIM, b Rome, Italy, Oct 12, 47; Dutch citizen. PHYSIOLOGY. *Educ:* Univ Amsterdam, MD, 81; Univ Limburg, Neth, PhD(physiol), 87. *Prof Exp:* Asst prof physiol, Fac Med, Univ Limburg, Neth, 75-88; asst prof, 88-89, ASSOC PROF PHYSIOL, FAC MED & HEALTH SCI, UNITED ARAB EMIRATES UNIV, 89-, ASST DEAN, RES & GRAD STUDIES, 90- *Mem:* Am Physiol Soc; NY Acad Sci; Europ Soc Cardiol. *Res:* Normal and abnormal electrical propagation in cardiac and smooth muscle. *Mailing Add:* Dept Physiol Fac Med & Health Sci PO Box 17666 Al Ain United Arab Emirates

LAMMIE, PATRICK J, PARASITOLOGY. *Prof Exp:* RES BIOLOGIST, CTR DIS CONTROL, 89- *Mailing Add:* Parasitol Br Ctr Dis Control MS-F-13 1600 Clifton Rd Atlanta GA 30333

LAMMI-KEEFE, CAROL J, b Acushnet, Mass, June 11, 47. NUTRITION. *Educ:* Univ Minn, PhD(nutrit), 80. *Prof Exp:* ASST PROF NUTRIT SCI, UNIV CONN. *Mailing Add:* Dept Nutrit Sci U-17 Univ Conn Storrs CT 06268

LAMOLA, ANGELO ANTHONY, b Newark, NJ, Aug 12, 40; m 63; c 2. PHOTOBIOLOGY, PHOTOCHEMISTRY. *Educ:* Mass Inst Technol, BS, 61; Calif Inst Technol, PhD(chem), 65. *Prof Exp:* Asst prof chem, Univ Notre Dame, 64-66; mem res staff, Bell Labs, 66-80, head, Molecular Biophys Res Dept, 80-85; dir chem res, Polaroid Corp, 85-88; VPRES RES & DEVELOP, SHIPLEY CO, 88- *Concurrent Pos:* Mem photobiol comt, Nat Res Coun-Nat Acad Sci; ed, Molecular Photochem; adj prof, Dept Dermat, Columbia Col Physicians & Surgeons, 75-85; mem study sect biophys & biophys chem, NIH, 78-83; sci adv, Advan Magnetics Corp, 86- *Honors & Awards:* Baekeland Award, Am Chem Soc, 77; Welch Found lectr, 78; Snider Found lectr, Univ Toronto, 81. *Mem:* AAAS; Am Chem Soc; Am Soc Photobiol (pres, 76-77); Int Soc Optical Eng. *Res:* Photochemistry; imaging chemistry; medical applications of fluorescence. *Mailing Add:* Shipley Co 2300 Washington St Newton MA 02162-1469

LAMON, EDDIE WILLIAM, b Yuba City, Calif, Aug 30, 39; c 3. CANCER, IMMUNOLOGY. *Educ:* Univ NAla, BS, 61; Med Col Ala, MD, 69; Karolinska Inst, Sweden, DSc, 74. *Prof Exp:* Asst biologist, Southern Res Inst, 64-65; from intern surg to resident gen surg, Univ Ala Sch Med, 69-71; from asst prof to assoc prof, 74-79, PROF SURG & MICROBIOL, UNIV ALA, BIRMINGHAM, 79-, SR SCIENTIST, CANCER RES & TRAINING CTR, 79-; CHIEF TUMOR IMMUNOL RES, BIRMINGHAM VET ADMIN MED CTR, 74-, CHMN RES & DEVELOP COMT, 84-, STAFF PHYSICIAN, SURG SERV, 84- *Concurrent Pos:* Guest investr, Karolinska Inst, Sweden, 71-74; mem, Comt Cancer Immunodiagnosis, Nat Cancer Inst, Div Cancer Biol & Diag, 76-78; res career develop award, Nat Cancer Inst, 75-80, cancer spec prog advisory comt, 77, construct study sect, Nat Cancer Inst, 82; assoc ed, Jour Immunol, 76-80; radiation study sect, NIH, 83-87; fac scholar, Josiah Macy Jr Found, 80-81. *Mem:* Am Asn Univ Professors; Am Asn Immunologists; AAAS; Sigma Xi; NY Acad Sci; Int Asn Comp Res Leukemia & Related Diseases; Am Asn Path; Am Asn Cancer Res; Soc Exp Biol & Med; Am Soc Microbiologists; Transplantation Soc. *Res:* Studies of the immune response to virus induced tumors with emphasis on the interactions of antibodies and lymphocytes in tumor cell destruction and idiolypic regulation of immunity; role of retinoids in tumor immunity. *Mailing Add:* Dept Surg Univ Sta Univ Ala Birmingham Birmingham AL 35294

LAMONDE, ANDRE M, b St Lambert, Que, Oct 5, 36; m 66. PHARMACY. *Educ:* Univ Montreal, BPharm, 61; Purdue Univ, MSc, 63, PhD(indust pharm), 65. *Prof Exp:* Asst prof pharm, Univ Montreal, 65-68; regulatory affairs coordr, Med Div, 68-70, qual control mgr, 70-72, QUAL CONTROL DIR, SYNTEX LTD, 72- *Res:* Basic pharmaceutics and pharmaceutical analysis. *Mailing Add:* 151 Anselme Lavigne Ddes O Roxboro PQ H9A 1P4 Can

LAMONDIA, JAMES A, b Springfield, Mass, Dec 3, 57. NEMATOLOGY, SOIL MICROBIOLOGY. *Educ:* Fitchburg State Col, 79; Cornell Univ, MS, 82, PhD(plant path), 84. *Prof Exp:* Res assoc plant path, Cornell Univ, 84-86; asst scientist res, 86-89, ASSOC SCIENTIST RES, DEPT PLANT PATH, CONN AGR EXP STA, 89- *Concurrent Pos:* Assoc ed, J Nematol, 88-91.

Mem: Am Phytopath Soc; Soc Nematologists. *Res:* Ecology and integrated management of soilborne plant pathogens, especially plant parasitic nematodes and fungi involved in complex diseases. *Mailing Add:* Dept Plant Path & Ecol Valley Lab Conn Agr Exp Sta PO Box 248 Windsor CT 06095

LAMONT, GARY BYRON, b St Paul, Minn, Feb 14, 39; m 66; c 3. CONTROL ENGINEERING, COMPUTER SCIENCE. *Educ:* Univ Minn, BPhysics, 61, MSEE, 67, PhD(control sci), 70. *Prof Exp:* Develop engr, Honeywell Inc, 61-65, systs analyst, 65-67; from asst prof to assoc prof, 70-80, PROF ELEC ENG, AIR FORCE INST TECHNOL, 80- *Mem:* Inst Elec & Electronics Engrs; Am Soc Eng Educ; Asm Comput Mach. *Res:* Control estimation theory; applications of small computers; computer structures; intelligent systems; operating systems. *Mailing Add:* Air Force Inst Technol Wright-Patterson AFB AFT/ENG Dayton OH 45433

LAMONT, JOHN THOMAS, b Lockport, NY, Oct 2, 38; m 64; c 3. MEDICINE, GASTROENTEROLOGY. *Educ:* Canisius Col, BS, 60; Univ Rochester, MD, 65. *Prof Exp:* Fel, Mass Gen Hosp, 71-73; instr, Harvard Med Sch, 73-75, asst prof med, 75-80; assoc prof & chief, 80-85, PROF GASTROENTEROL, UNIV HOSP, BOSTON UNIV, 85- *Concurrent Pos:* Consult gastroenterologist, Peter Bent Brigham Hosp, Boston Hosp Women & W Roxbury Vet Admin Hosp, 75-80 & Univ & Boston City Hosp, 80-; NIH career investr award, 75; res grants, Am Cancer Soc & Nat Found for Ileitis Colitis, 77-78. *Mem:* Am Soc Clin Res; Am Gastroenterol Asn; Am Soc Study Liver Dis. *Res:* Structure and function of colonic glycoproteins; biochemistry of intestinal tract in health and disease; colon cancer; gallstones, ulcer; bacterial toxins. *Mailing Add:* 720 Harrison Ave Boston MA 02118

LAMONT, JOHN W(ILLIAM), b Cape Girardeau, Mo, Mar 7, 42; m 68; c 2. ELECTRICAL ENGINEERING, COMPUTER SCIENCE. *Educ:* Univ Mo-Rolla, BSEE, 64; Univ Mo-Columbia, MSEE, 66, PhD, 70. *Prof Exp:* Instr elec eng, Univ Mo-Columbia, 66-70; asst prof, Univ Southern Calif, 70-73; asst prof, dept elec eng, Univ Tex, Austin, 73-77; proj mgr, Elec Power Res Inst, 77-87; AT IOWA STATE UNIV, 87- *Mem:* Nat Soc Prof Engrs; Inst Elec & Electronics Engrs; Sigma Xi. *Res:* Application of computers to power systems. *Mailing Add:* 1005 Idaho Ave Ames IA 50010

LAMONT, PATRICK, b Dublin, Ireland, Aug 29, 36. INTELLIGENT SYSTEMS, SOFTWARE SYSTEMS. *Educ:* Glasgow Univ, BSc, 58, PhD(math), 62. *Prof Exp:* Asst lectr math, Royal Col Sci & Technol, Scotland, 61-62; Dept Sci & Indust Res traveling fel, State Univ Utrecht & Univs Gottingen & Munich, 62-64; lectr pure math, Univ Birmingham, 64-70; assoc prof math, St Mary's Col, Ind, 70-74; chmn, Deep Springs Col, California, 74-76; asst prof math, Monmouth Col, Ill, 76-79; asst prof quant info sci, 79-83, ASSOC PROF COMPUT SCI, WESTERN ILL UNIV, 83- *Concurrent Pos:* Asst prof, Univ Notre Dame, 66-68; assoc prof math & comput sci, St Johns Univ, 88-89. *Mem:* Am Math Soc; London Math Soc; Asn Comput Mach; Inst Elec & Electronics Engrs Comput Soc. *Res:* Arithmetic theory of nonassociative algebras and intelligent software systems; cryptology. *Mailing Add:* Comput Sci Dept Western Ill Univ Macomb IL 61455

LAMONT, SUSAN JOY, b Hammond, Ind, Dec 26, 53; m 74; c 1. IMMUNOGENETICS, DISEASE RESISTANCE. *Educ:* Trinity Christian Col, BA, 75; Univ Ill Med Ctr, PhD(anat), 80. *Prof Exp:* Postdoctoral fel, Univ Mass, 80-83; asst prof, 83-87, ASSOC PROF, ANIMAL SCI, IOWA STATE UNIV, 87- *Concurrent Pos:* Vis scientist, Spelderholt, Neth, 90. *Mem:* Conf Res Workers Animal Dis; Am Asn Immunologists; Poultry Sci Asn; World Poultry Sci Asn; Int Soc Animal Genetics. *Res:* Structure and function of the chicken major histocompatibility complex; genetic resistance to disease in poultry; biomedical disease models in avian species. *Mailing Add:* Dept Animal Sci Iowa State Univ 201 Kildee Hall Ames IA 50011

LA MONTAGNE, JOHN RING, b Mexico City, Mex, Jan 1, 43; US citizen; m 68. MICROBIOLOGY. *Educ:* Univ Tex, BA, 65, MA, 67; Tulane Univ, PhD(microbiol), 71. *Prof Exp:* Teaching asst, Dept Microbiol, Univ Tex, 65-67; fel, Dept Microbiol & Immunol, Sch Med, Tulane Univ, 67-70, teaching asst, 70-71; res assoc, Dept Microbiol, Sch Med, Univ Pittsburgh, 71-74, instr, 74-76; influenza prog officer, Nat Inst Allergy and Infectious Dis, NIH, 76-84, viral vaccines prog officer, 83-86, dir, AIDS Prog, 86-87, actg dir, Microbiol & Infectious Dis Prog, 87-88, DIR, DIV MICROBIOL & INFECTIOUS DIS, NAT INST ALLERGY & INFECTIOUS DIS, NIH, BETHESDA, 88- *Concurrent Pos:* Mem, Task Force Reye's Syndrome, Pub Health Serv, 81- *Mem:* Am Soc Microbiol; AAAS; Am Soc Virol; Infectious Dis Soc Am; Am Soc Trop Med & Hyg. *Res:* Microbiology. *Mailing Add:* Div Microbiol & Infectious Dis Nat Inst Allergy & Infectious Dis Bldg 31 Rm 7A52 Bethesda MD 20892

LAMOREAUX, PHILIP ELMER, b Chardon, Ohio, May 12, 20; m 43; c 3. GEOLOGY, HYDROLOGY. *Educ:* Denison Univ, BA, 43, DSc; Univ Ala, MA, 49. *Prof Exp:* Jr geologist, US Geol Surv, 43-45, asst geologist, 45-47, dist geologist, 47-48, div hydrologist, 58-59, chief groundwater br, 59-61; state geologist & oil & gas supvr, Geol Surv Ala, 61-77; lectr geol, Univ Ala, Tuscaloosa, 48-59, from assoc prof to prof, 61-78, dir, Environ Inst Waste Mgt Studies, 67-78; RETIRED. *Concurrent Pos:* Consult, Egypt, 53, 59, 61, 63-64, 65, 70, 80, Thailand, 54, 61, 76, Philippines, 61, Surinam, 63, Mauritania, Africa, Senegal & Colombia, 64; chmn, bd trustees, Geol Soc Am; mem, bd trustee, Denison Univ, 87. *Honors & Awards:* Ian Campbell Medal, Am Geol Inst, 90; Commander's Medal, US Corps Eng, 90. *Mem:* Nat Acad Eng; Int Union Geol & Geophys; Geol Soc Am; Am Asn Petrol Geologists; Int Asn Hydrogeologists; Am Inst Prof Geologists; Soc Econ Geol; Am Inst Hydrol. *Res:* Groundwater geology; fluoride in groundwater; stratigraphy of Gulf Coastal Plain; hydrogeology of Karst areas. *Mailing Add:* PO Box 2310 Tuscaloosa AL 35403

LAMORTE, MICHAEL FRANCIS, b Altoona, Pa, Feb 20, 26; m 57; c 3. SOLID STATE ELECTRONICS, ELECTRICAL ENGINEERING. *Educ:* Va Polytech Inst & State Univ, BS, 50; Polytech Inst New York, MEE, 51. *Prof Exp:* Mgr solid state, RCA Corp, 59-67; pres solid state, Laser Diode Labs, 67-69; pres comput, Mathatronics Corp, 70-71; gen mgr med, Diamondhead Corp, 71-72; pres electronics, Princeton Synergestek Prod, 72-76; SR ENGR SOLID STATE, RES TRIANGLE INST, 76- *Concurrent Pos:* Fel engr, Westinghouse Elec Corp, 55-59; solid state physicist, Fort Monmouth, 52-55. *Honors & Awards:* Eng Achievement Award, RCA Corp, 60. *Mem:* Inst Elec & Electronics Engrs; Am Inst Physics; AAAS. *Res:* Semiconductor device physics; properties of materials; device design; device and material technology; special research interest in application of III-IV materials to electrooptical, microwave and logic devices. *Mailing Add:* Appl Eng Software Inc Box 222 Timberly Dr Rte 7 Durham NC 27707

LAMOTTE, CAROLE CHOATE, b Washington, DC, May 15, 47; m 70; c 2. IMMUNOCYTOCHEMISTRY, ELECTRON MICROSCOPY. *Educ:* Univ Okla, BS, 67; Georgetown Univ, MS, 69; Johns Hopkins Univ, PhD(physiol), 72. *Prof Exp:* NIH fel anat, Sch Med, Johns Hopkins Univ, 72-74, instr anat, 74-75, asst prof anat, 75-77; res assoc neurosurg, 77-78, asst prof neuroanat & neurosurg, 78-83, ASSOC PROF NEUROSURG, SCH MED, YALE UNIV, 83- *Concurrent Pos:* Prin investr, NIH Grant, 78-95; regular mem, NIH Neurol Sci & Study Sect, 85-88. *Honors & Awards:* Jacob Javitz Neurosci Investr Award, 88-95. *Mem:* Am Asn Anatomists; Soc Neurosci; Am Pain Soc. *Res:* Anatomy and physiology of pain and temperature sensation; sprouting and reorganization in injured spinal cord. *Mailing Add:* Dept Neurobiol-Anat Yale Univ Sch Med 333 Cedar St New Haven CT 06510

LAMOTTE, CLIFFORD ELTON, b Alpine, Tex, June 24, 30; m 74; c 2. PLANT PHYSIOLOGY. *Educ:* Tex A&M Univ, BS, 53; Univ Wis, PhD(bot), 60. *Prof Exp:* Res assoc biol, Princeton Univ, 60-61; from instr to asst prof, Boston Univ, 61-66; assoc prof bot & plant path, 66-80, PROF BOT, IOWA STATE UNIV, 80- *Concurrent Pos:* Fac leave Univ Col Wales, UK, 74-75. *Mem:* Bot Soc Am; AAAS; Am Soc Plant Physiol. *Res:* Hormonal regulation of development and orientation in plants; plant morphogenesis using tissue culture methods; growth and development in plants. *Mailing Add:* Dept Bot Iowa State Univ 353 Bessey Ames IA 50011

LAMOTTE, LOUIS COSSITT, JR, b Clinton, SC, Jan 21, 28; m 48; c 5. MICROBIOLOGY, EPIDEMIOLOGY. *Educ:* Duke Univ, AB, 48; Univ NC, MSPH, 51; Johns Hopkins Univ, ScD(virol, entom), 58. *Prof Exp:* Bacteriologist, State Bd Health, NC, 48-51; virologist, Chem Corps, US Dept Army, 51-58; chief virus invest unit, Dis Ecol Sect, Tech Br, Nat Commun Dis Ctr, 58-65, asst chief, Dis Ecol Sect, 65-66, chief community studies, Pesticides Prog, 66-69, dept chief, 69-70, chief, Microbiol Br, 70-72, dir, Licensure & Proficiency Testing Div, 72-82, dir, Div Tech Eval & Assistance, Ctr Dis Control, 82-86; RETIRED. *Concurrent Pos:* Mem grad fac, Colo State Univ, 59-66; adj prof, Ga State Univ, 71- *Honors & Awards:* Pub Health Serv Super Serv Award, 1981. *Mem:* AAAS; Sigma Xi; Am Soc Trop Med & Hyg; Am Pub Health Asn; Am Soc Microbiol. *Res:* Epidemiology of arthropod-borne viruses; virology, bacteriology, parasitology and epidemiology of infectious diseases. *Mailing Add:* Ctr Dis Control 1600 Clifton Rd NE Bldg 3 Rm 43 Atlanta GA 30333

LAMOTTE, ROBERT HILL, b Washington, DC, Nov 4, 40; m 70. NEUROSCIENCES. *Educ:* Trinity Col, BS, 63; Kans State Univ, PhD(psychol), 68. *Prof Exp:* instr, 70-73, asst prof neurophysiol, Sch Med & asst prof psychol, Johns Hopkins Univ, 73-77; ASSOC PROF ANESTHESIOL & PHYSIOL, MED SCH, YALE UNIV, 77- *Concurrent Pos:* Fel neurophysiol, Johns Hopkins Univ, 73-77. *Mem:* Soc Neurosci; AAAS. *Res:* Neurophysiology and psychophysics of somesthesis. *Mailing Add:* Dept Anesthesiol Yale Univ Sch Med 333 Cedar St New Haven CT 06510

LAMOUREUX, CHARLES HARRINGTON, b West Greenwich, RI, Sept 14, 33; m 54; c 2. BOTANY. *Educ:* Univ RI, BS, 53; Univ Hawaii, MS, 55; Univ Calif, Davis, PhD(bot), 61. *Prof Exp:* Asst bot, Univ Hawaii, 53-55; jr plant pathologist, Calif State Dept Agr, 55; asst bot, Univ Calif, Davis, 55-59; from asst prof to assoc prof, 59-71, PROF BOT, 71-, ASSOC DEAN ACAD AFFAIRS, COLS ARTS & SCI, UNIV HAWAII, 85- *Concurrent Pos:* Res assoc, Bernice P Bishop Mus, 62-; guest scientist, Nat Biol Inst Indonesia, 72-73, 79-80. *Mem:* Sigma Xi; Bot Soc Am. *Res:* Plant morphology; phenology; island biology; pteridology. *Mailing Add:* Dept Botany Univ Hawaii 3190 Maile Way Honolulu HI 96822

LAMOUREUX, GERALD LEE, b Bottineau, NDak, Apr 13, 39; m 69. BIOCHEMISTRY. *Educ:* Minot State Col, BS, 61; NDak State Univ, PhD(chem), 66. *Prof Exp:* RES CHEMIST, METAB & RADIATION RES LAB, AGR RES, USDA, ARS, 66- *Concurrent Pos:* Adj prof, NDak State Univ, 74- *Mem:* Am Chem Soc; Sigma Xi. *Res:* Elucidation of metabolic pathways utilized by plants and animals in the metabolism of herbicides, insecticides and other exenobiotics; glutathione-S-transferase mediated reactions; isolation and identification of natural products. *Mailing Add:* USDA Biosci Res Lab State Univ Sta Fargo ND 58105

LAMOUREUX, GILLES, b Marieville, Que, Mar 2, 34; c 5. MEDICINE, IMMUNOLOGY. *Educ:* St Mary's Col, BA, 56; Univ Montreal, MD, 61, MSc, 63, PhD(immunol), 67. *Prof Exp:* HEAD IMMUNODIAG LAB, INST ARMAND FRAPPIER, 67-; ASST PROF IMMUNOL, UNIV MONTREAL, 68- *Concurrent Pos:* Fel, Walter & Eliza Hall Inst Med Res, Melbourne, Australia; Multiple Sclerosis Soc Can grant. *Mem:* Can Soc Immunol; Fr Soc Immunol; Transplantation Soc. *Res:* Immune studies in multiple sclerosis and experimental allergic encephalomyelitis; mechanism of action of lymphocytes in specific cell-mediated immunity; stimulation and immunosuppression; practical aspect in autoimmune diseases, transplantation and cancer; genetic control of natural resistance in multiple scperosis and normal population. *Mailing Add:* 3293 de Musset Mascouche PQ H7W 4Z3 Can

LAMOYI, EDMUNDO, b Oaxaca, Mex, May 6, 52; m 90; c 1. IMMUNOCHEMISTRY, IMMUNOPARASITOLOGY. *Educ:* Nat Univ Mex, BS, 75; Brandeis Univ, PhD(biol), 81. *Prof Exp:* Postdoctoral res assoc, Brandeis Univ, 81-83; vis assoc, Nat Inst Allergy & Infectious Dis, NIH, 83-86; SR INVESTR, BIOMED RES INST, NAT UNIV MEX, 86- *Concurrent Pos:* Nat researcher, Nat Syst Res, Pub Educ Secretariat, Mex, 86; lectr, Sci & Humanities Col, Nat Univ Mex, 87- *Mem:* Am Asn Immunologists. *Res:* Studies of the human immune response to the parasite entamoeba histolytica; analysis of parasite antigens with monoclonal antibodies. *Mailing Add:* Dept Immunol Inst Invest Biomed UNAM APDO 70-228 Mexico City DF 04510 Mexico

LAMP, GEORGE EMMETT, JR, industrial engineering; deceased, see previous edition for last biography

LAMP, HERBERT F, b Davenport, Iowa, Aug 6, 19; m 47; c 6. PLANT ECOLOGY. *Educ:* Chicago Teacher Col, BEd, 41; Univ Chicago, SM, 47, PhD(bot), 51. *Prof Exp:* Instr bot, Fla State Univ, 47-50; teacher biol, Ill Teachers Col Chicago-South, 50-59, from assoc prof to prof, 59-64, chmn dept natural sci, 56-64; prof biol, 64-66, prof biol sci & chmn dept, 66-83, assoc dean, Arts & Sci, 83-84, EMER PROF BIOL SCI, NORTHEASTERN ILL UNIV, 85- *Concurrent Pos:* Res assoc, Univ Chicago, 52-56. *Mem:* Bot Soc Am; Ecol Soc Am; Sigma Xi. *Res:* Physiological ecology of range grasses; bromus inermis leyss; prairie ecology; commmunity ecology. *Mailing Add:* 180 Linden Ave Elmhurst IL 60126

LAMP, WILLIAM OWEN, b Omaha, Nebr, June 19, 51; m 73; c 2. POPULATION ECOLOGY, BIOLOGICAL CONTROL. *Educ:* Ohio State Univ, MS, 76; Univ Nebr, BS, 72, PhD(entom), 80. *Prof Exp:* Res assoc, Ill Natural Hist Surv, 80-85; ASST PROF ENTOM, UNIV MD, 85- *Mem:* Entom Soc Am; Weed Sci Soc Am. *Res:* Ecological interactions in agroecosystems, especially mutiple pest interactions, and their economic and environmental impact for corp protection. *Mailing Add:* Dept Entom Univ Md College Park MD 20742

LAMPE, FREDERICK WALTER, b Chicago, Ill, Jan 5, 27; m 49; c 5. PHYSICAL CHEMISTRY. *Educ:* Mich State Col, BS, 50; Columbia Univ, AM, 51, PhD(chem), 53. *Prof Exp:* Asst chem, Columbia Univ, 50-53; res chemist, Humble Oil & Refining Co, 53-56, sr res chemist, 56-60, res specialist, 60; assoc prof chem, 60-65, head dept, 83-88, PROF CHEM, PA STATE UNIV, 65- *Concurrent Pos:* Consult, Socony Mobil Oil Co, 61-69, Sci Res Instruments Corp, 67-77 & W H Johnston Labs, Inc, 60-67; NSF sr fel & guest prof, Univ Freiburg, 66-67. *Honors & Awards:* US Sr Scientist Award, Alexander von Humboldt Found, 73-74 & 84. *Mem:* AAAS; Am Chem Soc; fel Am Phys Soc; Royal Soc Chem. *Res:* Photochemistry; radiation chemistry; reactions of free radicals and of gaseous ions; mass spectrometry. *Mailing Add:* 542 Ridge Ave State Col PA 16803

LAMPE, KENNETH FRANCIS, b Dubuque, Iowa, Dec 3, 28; m 49; c 1. PHARMACOLOGY, SYNTHETIC ORGANIC & NATURAL PRODUCTS CHEMISTRY. *Educ:* Univ Iowa, BA, 49, MS, 51, PhD(pharm), 53. *Prof Exp:* Instr chem, Mont State Col, 49-50; instr pharm, Univ Iowa, 53; from asst prof to assoc prof pharmacol, Sch Med, Univ Miami, 54-67, prof pharmacol & anesthesiol, 67-81; sr scientist, 80-83, CHMN DRUG DIV, AMA, 83- *Concurrent Pos:* Fel, Yale Univ, 53-54. *Mem:* Am Chem Soc; Am Acad Clin Toxicol. *Res:* Clinical toxicology of plants and mushrooms. *Mailing Add:* 260 E Chestnut Apt 3807 Chicago IL 60611

LAMPE, MARTIN, b Brooklyn, NY, Apr 29, 42; m 64; c 2. PLASMA PHYSICS, CHARGED PARTICLE BEAMS. *Educ:* Harvard Univ, AB, 62; Univ Calif, Berkeley, MA, 63, PhD(physics), 67. *Prof Exp:* Res assoc, NY Univ, 67-69; res physicist, 69-75, SUPVR RES PHYSICIST, NAVAL RES LAB, 75- *Honors & Awards:* E O Hulburt Award, Naval Res Lab, 85. *Mem:* Am Phys Soc; Inst Elec & Electronics Engrs. *Res:* Theory of plasma instabilities, non-linear theory and propagation of relativistic electron beams. *Mailing Add:* Naval Res Lab Code 4792 Washington DC 20375-5000

LAMPEN, J OLIVER, b Holland, Mich, Feb 26, 18; m 44; c 3. MICROBIOLOGY, MOLECULAR BIOLOGY. *Educ:* Hope Col, AB, 39; Univ Wis, MS, 41, PhD(biochem), 43. *Hon Degrees:* LHD, Hope Col, 74. *Prof Exp:* Biochemist, Am Cyanamid Co, 43-46; res assoc, Med Sch, Washington Univ, 46-47, instr biochem, 47-48, asst prof biol chem, 48-49; assoc prof microbiol, Sch Med, Western Reserve Univ, 49-53; dir div biochem res, Squibb Inst Med Res, Olin Mathieson Chem Corp, 53-58; dir, Waksman Inst, 58-80, prof, 80-88, EMER PROF MICROBIOL, RUTGERS UNIV, 88- *Honors & Awards:* Lilly Award, 52. *Mem:* AAAS; Am Soc Microbiol; Am Soc Biol Chemists; Am Acad Microbiol; Brit Biochem Soc; Soc Gen Microbiol. *Res:* Secretion of enzymes by microorganisms: site of exoenzyme formation, mechanism of secretion, control of synthesis; antibiotics and cell membrane. *Mailing Add:* Waksman Inst Rutgers Univ Piscataway NJ 08855-0759

LAMPERT, CARL MATTHEW, b Portland, Ore, Feb 20, 52; m 75. ALTERNATIVE ENERGY ENGINEERING, OPTICAL SWITCHING TECHNOLOGY. *Educ:* Univ Calif, Berkeley, BS(electronic eng) & BS(mat sci), 74, MS 77, PhD(mat sci), 79. *Prof Exp:* Electronics technician, Contra Costa Col, San Pablo, 70- 72, computer programmer, US Forestry Serv, 72-74; res asst, 74-79, STAFF SCIENTIST, LAWRENCE BERKELEY LAB, BERKELEY, CA, 79- *Concurrent Pos:* Conf chmn, Int Optical Eng Soc, 82-, proj leader, Int Energy Agency, Paris, France, 84-,; ed-in-chief, Solar Energy Mat J, 82-; lectr, Inst Theoret Physics, Trieste, Italy, 85-; consult, Lyon & Lyon, Los Angeles, 85, UN Develop Pros, New York, 85-, Optical Coating Labs Inc, Santa Rosa, 88-, Innotech, Trumbull, Conn, 86- *Honors & Awards:* Fulbright Scholar, 83. *Mem:* Fel Int Optical Eng Soc; Inst Elec & Electronics Engrs; Am Vacuum Soc; Int Solar Energy Soc; Sigma Xi. *Res:* Research and development of new optical materials and coatings for buildings, automotive and aerospace glazing applications; engineering and materials of energy conversion components; large scale optical switching materials; author of over 60 papers, lectured in 13 countries. *Mailing Add:* Lawrence Berkeley Lab Univ Calif One Cyclotron Rd MS 62-203 Berkeley CA 94720

LAMPERT, SEYMOUR, b Brooklyn, NY, Mar 5, 20; m 48; c 3. SOLAR ENERGY & ALTERNATE ENERGY APPLICATIONS. *Educ:* Ga Inst Technol, BS, 43; Calif Inst Technol, MS, 47, AE, 48, PhD(aeronaut eng & math), 54. *Prof Exp:* Instr math, Ga Sch Technol, 43-44; res scientist, Ames Lab, Nat Adv Comt Aeronaut, 44-51; res engr, Jet Propulsion Lab, Calif Inst Technol, 51-54; asst prof eng, Univ Southern Calif, 54-55; chief engr, Odin Assocs, 55-56; mgr appl mech, Ford Aeronotronic, 56-62; dir advan systs, North Am Aviation Space & Info Systs, 62-67; vpres, Syst Assoc Inc, 67-71; PROF ENG, UNIV SOUTHERN CALIF, 75-; VPRES & MEM, BD DIRS, DAVATO CORP, 80- *Concurrent Pos:* Consult, Jet Propulsion Lab, Calif Inst Technol, 67-68; sci adv, Dept Defense, 68-69; ed-in-chief, J Solar Sci, 81-82. *Res:* Structures for spacecraft; spacecraft design; methodology for developing area transportation; fluid mechanics; wing theory; solar energy. *Mailing Add:* PO Box 4719 Irvine CA 92716-4719

LAMPERTI, ALBERT A, b Bronx, NY, Oct 24, 47; m 72; c 2. RADIATION BIOLOGY. *Educ:* Manhattan Col, BS, 69; Univ Cincinnati, PhD(anat), 73. *Prof Exp:* asst prof anat, Col Med, Univ Cincinnati, 73-80; ASSOC PROF, DEPT ANAT, SCH MED, TEMPLE UNIV, 80- *Mem:* Am Asn Anatomists; Radiation Res Soc. *Res:* Reproductive neuroendocrinology; radiation biology. *Mailing Add:* Dept Anat Temple Univ Sch Med 3420 N Broad St Philadelphia PA 19140

LAMPERTI, JOHN WILLIAMS, b Montclair, NJ, Dec 20, 32; m 57; c 4. MATHEMATICS. *Educ:* Haverford Col, BS, 53; Calif Inst Technol, PhD(math), 57. *Prof Exp:* From instr to asst prof math, Stanford Univ, 57-61; vis asst prof, Dartmouth Col, 61-62; res assoc, Rockefeller Inst, 62-63; assoc prof, 63-68, PROF MATH, DARTMOUTH COL, 68- *Concurrent Pos:* Sci exchange visitor, USSR, 70; vis prof, Aarhus Univ, 72-73, Nat Atonomous Univ Nicaragua, 90; consult, Am Friends Serv Comt, 80, 85. *Mem:* Fedn Am Scientists; fel Inst Math Statist; Union Concerned Scientists. *Res:* Probability theory, particularly properties of stochastic processes. *Mailing Add:* Dept Math Dartmouth Col Hanover NH 03755

LAMPI, RAUNO ANDREW, b Gardner, Mass, Aug 12, 29; m 51; c 4. FOOD TECHNOLOGY, CHEMICAL ENGINEERING. *Educ:* Univ Mass, BS, 51, MS, 55, PhD(food technol), 57. *Prof Exp:* Res instr food technol, Univ Mass, 53-57; tech dir, New Eng Appl Prod Co, 59-62; mgr, Food Technol Sect, Cent Eng Labs, FMC Corp, 62-66; packaging technologist, Container Div, Natick Labs, US Army, 66-67, res phys scientist, Packaging Div, 67-69, chief systs develop br, Packaging Div, 69-76, chief, Food Equip Div, 76-88, Chief, Adv Equip Br, Technol Acquisition Div, Natick Res Develop & Eng Ctr, 88-89; CONSULT, 90- *Concurrent Pos:* Asst mgr indust instrumentation, Food Div, Foxboro Co, 57-59. *Honors & Awards:* Rohland Isker Award, Res & Develop Assocs, 69; Indust Achievement Award, Inst Food Technol, 78. *Mem:* Inst Food Technol. *Res:* Development of continuous applesauce and juice processes; stability characteristics of freeze dried foods; thermal processing of foods in flexible packages and flat metal containers; development of food service systems and equipment. *Mailing Add:* 20 Wheeler Rd Westboro MA 01581

LAMPKY, JAMES ROBERT, b Battle Creek, Mich, June 19, 27; m 50, 71; c 6. MICROBIOLOGY. *Educ:* Eastern Mich Univ, BS, 59; Univ Mo, MA, 61, PhD(microbiol), 66. *Prof Exp:* From instr to asst prof bact, Wis State Univ, 63-66; asst prof, 66-70, assoc prof, 70-76, PROF BACT, CENT MICH UNIV, 76- *Mem:* Am Soc Microbiol; Soc Indust Microbiologists; Mycol Soc Am. *Res:* Cellulolytic fruiting myxobacteria of the genus Polyangium with emphasis on morphology and ultrastructure. *Mailing Add:* Dept Biol Cent Mich Univ Mt Pleasant MI 48859

LAMPMAN, GARY MARSHALL, b South Gate, Calif, Oct 8, 37; m 71; c 2. ORGANIC CHEMISTRY. *Educ:* Univ Calif, Los Angeles, BS, 59; Univ Wash, PhD(chem), 64. *Prof Exp:* From asst prof to assoc prof, 64-73, PROF CHEM, WESTERN WASH UNIV, 73- *Mem:* Am Chem Soc. *Res:* Conformational analysis in small ring compounds; synthesis and reactions of strained compounds; organometallic chemistry. *Mailing Add:* Dept Chem Western Wash Univ Bellingham WA 98225

LAMPORT, DEREK THOMAS ANTHONY, b Brighton, Eng, Dec 1, 33; m 63; c 5. BIOCHEMISTRY. *Educ:* Univ Cambridge, BA, 58, PhD(biochem), 63. *Prof Exp:* Staff scientist, Res Inst Advan Studies, Martin Marietta Corp, Md, 61-64; from asst prof to assoc prof, 64-74, PROF BIOCHEM, DOE PLANT RES LAB, MICH STATE UNIV, 74- *Mem:* Am Chem Soc. *Res:* Plant cell wall proteins; role of hydroxyproline-rich glycoproteins, notably extensin, in plant growth. *Mailing Add:* DOE Plant Res Lab Mich State Univ East Lansing MI 48824

LAMPORT, LESLIE B, b New York, NY, Feb 7, 41. CONCURRENT & DISTRIBUTED COMPUTER SCIENCE. *Educ:* Mass Inst Technol, BS, 60; Brandeis Univ, MA, 63, PhD(math), 72. *Prof Exp:* Prof computer sci, Malboro Col, 65-69, Mass Computer Assocs, 70-77 & SRI Int, 77-85; SR CONSULT ENGR, DIGITAL EQUIPMENT CORP, 85- *Mem:* Nat Acad Eng. *Mailing Add:* Digital Equip Corp 1130 Lytton Ave Palo Alto CA 94301

LAMPPA, GAYLE K, GENETICS, BIOLOGY. *Educ:* Reed Col, BA, 73; Univ Wash, PhD(bot), 80. *Prof Exp:* Postdoctoral fel, Lab Plant Molecular Biol, Rockefeller Univ, 81-84; ASST PROF, DEPT MOLECULAR GENETICS & CELL BIOL, UNIV CHICAGO, 85- *Concurrent Pos:* Damon Runyon-Walter Winchell Res Fund postdoctoral fel, 82-84, NIH postdoctoral fel, 82-85; Zeisler fac award, 85; Andrew Mellon fel, 85; ad hoc reviewer, Dept Energy & NSF; chmn, Univ Biosafety Comt, Univ Chicago, 86-87, mem, Ctr Photochem & Photobiol & Comt Genetics; mem study sect, Prog Plant Growth & Develop, Competitive Res Grants Off, USDA, 88, 89 & 90. *Mem:* Sigma Xi; Am Soc Plant Physiol; Int Soc Plant Molecular Biol; Am Soc Cell Biol. *Mailing Add:* Dept Molecular Genetics & Cell Biol Univ Chicago 920 E 58th St Chicago IL 60637

LAMPSON, BUTLER WRIGHT, b Washington, DC, Dec 23, 43; m 67; c 2. COMPUTER SCIENCE. *Educ:* Harvard Univ, AB, 64; Univ Calif, Berkeley, PhD(comput sci), 67. *Hon Degrees:* Dsc, 86, Swiss Fed Inst Technol, Zurich, Switz, DSc, 86. *Prof Exp:* From asst prof to assoc prof comput sci, Univ Calif, Berkeley, 67-71; res fel, Xerox Palo Alto Res Ctr, 71-80, sr res fel, 80-; CORP CONSULT ENGR, DIGITAL EQUIP CORP, CAMBRIDGE, MASS. *Concurrent Pos:* Dir syst develop, Berkeley Comput Corp, 69-71. *Honors & Awards:* Software Systs Award, Asn Comput Mach, 85. *Mem:* Nat Acad Eng; Asn Comput Mach. *Res:* Programming languages and operating systems. *Mailing Add:* Cambridge Res Lab Digital Equip Corp One Kendall Sq Bldg 700 Cambridge MA 02139

LAMPSON, FRANCIS KEITH, b Minneapolis, Minn, Aug 7, 24; m 45; c 4. METALLURGICAL ENGINEERING, MATERIALS ENGINEERING. *Educ:* Univ Ill, BS, 49. *Prof Exp:* Jr metallurgist, NEPA Div, Fairchild Eng & Air Corp, 49-51; exp metallurgist, Allison Div, Gen Motors Corp, 51-54; group leader, Mat & Processing, Marquardt Co, Van Nuys, Calif, 54-57; Pacific Coast area tech rep, Allegheny-Ludlum Steel Corp, Los Angeles, 57-65; DIR MAT ENGRS, MARQUARDT CO, DIV CCI CORP, 65- *Concurrent Pos:* Pres, F K Lampson Assocs, 74- *Mem:* Am Soc Metals; Soc Aeroscpace Mat Process Engrs; Am Soc Testing Mats; Am Inst Mining Engrs. *Res:* Propulsion technology materials; ferrous and super alloy materials; refractory materials and related disilicide coatings; biomedical materials. *Mailing Add:* 10000 Aldea Ave N Northridge CA 91325

LAMPSON, GEORGE PETER, b Colman, SDak, June 12, 19; m 48; c 3. BIOCHEMISTRY. *Educ:* SDak State Univ, BS, 42; Univ Wis, MS, 50. *Prof Exp:* Res assoc biochem, Ortho Pharmaceut Corp, 50-59; sr res biochemist, Dept Virus & Cell Biol, Merck Inst Therapeut Res, 59-85; RETIRED. *Mem:* NY Acad Sci; Am Chem Soc; Am Soc Biol Chemists. *Res:* Purification and characterization of chicken embryo interferon as a low molecular protein; synthetic and natural double-stranded RNA as inducers of inderferon and host resistance; virus chemistry. *Mailing Add:* 2012 Keystone Dr Hatfield PA 19440

LAMPSON, LOIS ALTERMAN, IMMUNE RESPONSES. *Educ:* Univ Calif, Berkeley, PhD(immunol), 76. *Prof Exp:* ASST PROF ANAT, SCH MED, UNIV PA, 79- *Res:* Role of major histocompatibility complex in neural tissue; immune response to neural tumor. *Mailing Add:* Dept Neurol Harvard Med Sch Ctr Neurol Dis Biosci Res Bldg Brigham & Women's Hosp Boston MA 02115

LAMPTON, MICHAEL LOGAN, b Williamsport, Pa, Mar 1, 41. X-RAY ASTRONOMY. *Educ:* Calif Inst Technol, BS, 62; Univ Calif, Berkeley, PhD(physics), 67. *Prof Exp:* ASST RES PHYSICIST, SPACE SCI LAB, UNIV CALIF, BERKELEY, 67- *Concurrent Pos:* NSF fel, Univ Calif, Berkeley, 68-69. *Mem:* Am Geophys Union; Am Astron Soc. *Res:* Ultraviolet astronomy. *Mailing Add:* Space Sci Lab Univ Calif Berkeley CA 94720

LAMSTER, HAL B, b New York, NY, July 12, 41; m 67; c 1. GENERAL COMPUTER SCIENCE, OPERATIONS RESEARCH. *Educ:* NY Univ, BS, 63, MS, 70. *Prof Exp:* Cosult, Gen Anal, 68-69; pres, Time Sharing Sci Inc, 70-71; EXEC VPRES, TELMAR GROUP INC, 71- *Concurrent Pos:* Chair comt, Asn Comput Mach, 86- *Mem:* Inst Elec & Electronics Engrs; Asn Comput Mach. *Res:* Microcomputing application; data communications; man-machine interactions. *Mailing Add:* 399 E 72nd St New York NY 10021

LAMSTER, IRA BARRY, b New York, NY, Mar 6, 50; m 71; c 2. DENTISTRY, BIOCHEMISTRY. *Educ:* Queens Col, NY, BA, 71; Univ Chicago, SM, 72; State Univ NY, Stony Brook, DDS, 77; Harvard Univ, MMSc, 80; Am Bd Oral Med, cert, 84. *Prof Exp:* Res fel, Lab for Surg Res, Med Sch, Harvard Univ & teaching fel periodontol, Sch Dent Med, 77-80; from asst prof to assoc prof peridont & oral med, Fairleigh Univ, 80-88; ASSOC PROF DENT & DIR PERIODONT, COLUMBIA UNIV, 88- *Concurrent Pos:* Prin investr, res contracts & grants, Lever Res Inc, Warner-Lambert Co, Johnson & Johnson & Block Drug Co, 80-, NIH dent student res training award, Nat Inst Dent Res, 82 & NIH grants, 82-86 & 85-; Salisbury fel. *Mem:* Sigma Xi; Am Dent Asn; Am Acad Periodontol; AAAS; Am Acad Oral Med. *Res:* Host response in human periodontal disease; diagnostic techniques for periodontal disease; application of a biochemical profile of gingival crevicular fluid. *Mailing Add:* Columbia Univ Sch Dent & Oral Surg 630 W 168 St New York NY 10032

LAMUTH, HENRY LEWIS, b Painesville, Ohio, Apr 15, 42; m 69; c 2. APPLIED PHYSICS, ELECTRICAL ENGINEERING. *Educ:* Ohio State Univ, BS, 66, PhD(physics), 70. *Prof Exp:* Assoc res physicist optics, Willow Run Labs, Inst Sci & Technol, Univ Mich, 70-72; mem tech staff electro-optics, Orlando Div, Martin Marietta Aerospace, 72-74; mgr, Sensors, Electronics & Controls Group, Columbus Labs, Battelle Mem Inst, 74-; AT SPERRY CORP. *Mem:* AAAS; Am Phys Soc; Sigma Xi; Inst Elec & Electronics Engrs. *Res:* Electro-optics, infrared, radar sensors and sensing; missile systems; atmospheric transmission; fiber and integrated optics; communication systems and techniques with optical radiation; electronic control systems; analog-digital electronic systems. *Mailing Add:* 15 Kimber Ct East Northport NY 11731

LAMY, PETER PAUL, b Breslau, Ger, Dec 14, 25; US citizen; m 51; c 3. GERIATRICS & GERONTOLOGY, CLINICAL PHARMACOLOGY. *Educ:* Philadelphia Col Pharm & Sci, BSc, 56, MSc, 58, PhD(biopharmaceut), 64. *Hon Degrees:* DSc, Albany Col Pharm, 88. *Prof Exp:* Instr pharm, Philadelphia Col Pharm & Sci, 56-63; assoc prof, 67-72, asst dean, geriat, 88, PROF PHARM, SCH PHARM, UNIV MD, BALTIMORE, 73-, DIR, CTR STUDY PHARM & THERAPEUT ELDERLY, 78- *Concurrent Pos:* Asst to dir pharm, Jefferson Med Col, 59-62; instr, Woman's Hosp, Philadelphia, 60-62; lectr, Sch Nursing, Cath Univ Am; consult, USPHS Hosp, 66-72, Levindale Hebrew Geriat Ctr & Hosp, Baltimore, 72-, John L Deaton Med Ctr, Baltimore, 73-, Vet Admin Hosps, Baltimore & Washington, DC; Am Asn Cols Pharm vis scientist, 68-72; consult geriat & geront; vchmn, task force on aging Univ Md at Baltimore; ed, Contemp Pharm Pract, 78-80; mem, Nat Aging Res Planning Panel, 81-; dir, Instnl Pharm Progs, 68-, Ctr Study Ther Elderly, 80- & Univ Md, Baltimore Ctr Aging, 84; chmn, White House Conf Aging, 80; prof epidemiol & prev med, Sch Med, Univ Md, 85-, prof family med, 87- *Mem:* Fel AAAS; fel Am Col Clin Pharmacol; Am Soc Clin Pharmacol Therapeut; Sigma Xi; fel Am Gerontol Soc; fel Am Geriat Soc; Am Soc Consult Pharmacists; Am Soc Hosp Pharmacists. *Res:* Drug transport mechanisms in vivo and in vitro; drug interactions; drug equivalencies and efficiencies. *Mailing Add:* Sch Pharm Univ Md 20 N Pine St Baltimore MD 21201

LAN, CHUAN-TAU EDWARD, b Taiwan, China, Apr 21, 35; US citizen; m 61; c 3. ENGINEERING MECHANICS, CIVIL ENGINEERING. *Educ:* Nat Taiwan Univ, BS, 58; Univ Minn, MS, 63; NY Univ, PhD(aeronaut), 68. *Prof Exp:* Asst civil engr hydraul, Bd Water Supply, NY, 63-65; assoc res scientist aeronaut, Aerospace Labs, NY Univ, 68; from asst prof to assoc prof, 68-78, PROF AERONAUT, UNIV KANS, 78- *Concurrent Pos:* Consult, Vigyan Res Assoc, Inc, 86-; reviewer, J Aircraft & Am Inst Aeronaut & Astronaut J, 75-; consult, Aeronaut Res Lab, Taiwan, 77-; prin investr, NASA Langley Res Ctr, 73- *Honors & Awards:* Cert of Recognition, NASA Langley Res Ctr, 78, 80, 82 & 86. *Mem:* Am Inst Aeronaut & Astronaut. *Res:* Steady and unsteady aerodynamics. *Mailing Add:* Dept Aerospace Eng Univ Kans Lawrence KS 66045

LAN, MING-JYE, b Kaohsiung, Taiwan, Repub China, Nov 3, 53; m 80; c 1. POLYMER SYNTHESIS, POLYMER COMPOSITE. *Educ:* Nat Tsing-Hua Univ, Taiwan, BS, 76; Univ Mich, MS, 81, PhD(chem & macromolecular sci & eng), 85. *Prof Exp:* Teaching asst org chem, Univ Mich, 79-84; res chemist, 85-87, SR RES CHEMIST, ALLIED-SIGNAL INC, 87- *Mem:* Am Chem Soc; Soc Advan Mat & Process Eng. *Res:* Bipolar membranes for water splitting; synthesis and characterization of biopolymers as anti-viral agents; synthesis of monomers and polymers; thermoset polymers for composites. *Mailing Add:* 543 Bovidae Crest Naperville IL 60565

LAN, SHIH-JUNG, b Kwangtung, China, Sept 15, 38; m 67; c 2. DRUG METABOLISM. *Educ:* Univ Tunghai, BS, 60; Okla State Univ, MS, 64; Univ Minn, PhD(biochem), 68. *Prof Exp:* Fel physiol chem, Univ Wis, 67-69; res investr drug metab, 69-73, sr res investr, 73-74, RES GROUP LEADER, SQUIBB INST MED RES, 75- *Mem:* Am Chem Soc; Am Soc Pharmacol & Exp Therapeut. *Res:* Microsomal drug metabolism enzyme systems; mechanism of drug action. *Mailing Add:* Drug Metab Dept Squibb Inst Med Res PO Box 4000 Princeton NJ 08540

LANA, EDWARD PETER, b Duluth, Minn, Oct 17, 14; m 42; c 3. HORTICULTURE. *Educ:* Univ Minn, BS, 42, MS, 43, PhD, 48. *Prof Exp:* Canning crops res, Fairmont Canning Co, Minn, 43-47; asst prof hort, Iowa State Univ, 47-56; prof & chmn dept, 56-81, EMER PROF HORT, NDAK STATE UNIV, 81- *Mem:* Am Soc Hort Sci; Sigma Xi. *Res:* Crop breeding; cultural studies. *Mailing Add:* Dept Hort NDak State Univ Fargo ND 58102

LANAM, RICHARD DELBERT, JR, b Denver, Col, May 31, 43; m 65; c 3. METALLURGY OF PLATINUM GROUP METALS. *Educ:* Northwestern Univ, BS, 66; Drexel Univ, MS, 69, PhD(mat eng), 72. *Prof Exp:* Eng Specialist, Olin Corp, 70-74; Sr metall eng, Pfizer Inc, 74-79; metall section head, 79-82, tech oper mgr, 82-88, TECH/ENG MGR, ENGELHARD CORP, 88- *Mem:* Am Soc Metals; Am Inst Metall Eng. *Res:* General refining and use of platinum group metals; sputter-coated ruthenium for electrical contact applications; author of ten publications and holder of nine US patents. *Mailing Add:* 655 Fourth Ave Westfield NJ 07090

LANCASTER, CLEO, b Rocky Mount, NC, Dec 10, 48. ANIMAL PHYSIOLOGY. *Educ:* Elizabeth City State Univ, NC, BS, 71; Western Mich Univ, MS, 79. *Prof Exp:* Res asst, Brookhaven Nat Lab, 71; SR RES BIOLOGIST, UPJOHN CO, 71- *Mem:* AAAS; NY Acad Sci. *Res:* Experimental gastroenterology; development of experimental models of ulcers, gastric lesions, pancreatitis and surgical methods to study gastric secretion; helped develop the cytoprotection concept, antisecretory property of prostaglandins and the antipancreatitis effect of opioid agonists. *Mailing Add:* Safety Pharmacol Upjohn Co Kalamazoo MI 49001

LANCASTER, GEORGE MAURICE, b Penrith, Eng, July 18, 34; m 64; c 2. DIFFERENTIAL GEOMETRY. *Educ:* Univ Liverpool, BSc, 56; Univ Sask, PhD(math), 67. *Prof Exp:* Res analyst, Weapons Res Div, A V Roe & Co Ltd, Woodford, Eng, 56-58; opers res analyst, Northern Elec Co Ltd, Montreal, 58-60; lectr math, Royal Roads Mil Col, 60-64; asst prof, Univ Sask, 67-70; assoc prof, 70-80, head dept, 78-87, PROF MATH, 80, DEAN SCI & ENG, ROYAL RDS MIL COL, 87- *Concurrent Pos:* Nat Res Coun Can grants, Univ Sask, 68-7O; spec lectr, Univ Victoria, 71. *Mem:* Am Math Soc; Can Math Soc; Tensor Soc. *Res:* Differential geometry; imbedding of Riemannian manifolds. *Mailing Add:* Dept Math Royal Rds Mil Col Victoria BC V0S 1B0 Can

LANCASTER, JACK R, JR, b Memphis, Tenn, Aug 27, 48. BIOENERGETICS, ELECTRON TRANSFER. *Educ:* Univ Tenn, Martin, BS, 70, Ctr Health Sci, PhD(biochem), 74. *Prof Exp:* Res assoc biochem, Cornell Univ, 74-76 & Duke Univ, 76-80; asst prof, 80-83, ASSOC PROF BIOCHEM, UTAH STATE UNIV, 83- *Concurrent Pos:* NSF trainee, 70-73; estab investr, Am Heart Asn, 83-88; mem, coun basic sci, Am Heart Asn. *Mem:* Am Chem Soc; Am Soc Biol Chemists; Biophys Soc; Sigma Xi; Am Heart Asn. *Res:* Basic biochemical mechanisms of (1) immune cytotoxicity and (2) energy utilization in the methanogenic bacteria. *Mailing Add:* Dept Chem & Biochem Utah State Univ Logan UT 84322

LANCASTER, JAMES D, b Randolph, Miss, June 11, 19; m 42; c 3. AGRONOMY. *Educ:* Miss State Col, BS, 47, MS, 48; Univ Wis, PhD, 54. *Prof Exp:* Asst agronomist, Exp Sta & asst prof agron, 51-57, agronomist, Exp Sta & prof agron, Miss State Univ, 57-85; RETIRED. *Mem:* Am Soc Agron; Soil Sci Soc Am; AAAS. *Res:* Soil fertility and testing; fertilizer evaluation; crop fertilization. *Mailing Add:* Rte 1 Box 444 Starkville MS 39759

LANCASTER, JESSIE LEONARD, JR, b Horatio, Ark, Jan 26, 23; m 46; c 4. ENTOMOLOGY. *Educ:* Univ Ark, BSA, 47; Cornell Univ, PhD(econ entom), 51. *Prof Exp:* Asst, Cornell Univ, 47-51; from asst prof to assoc prof entom, 51-60; PROF ENTOM, UNIV ARK, FAYETTEVILLE, 60- *Concurrent Pos:* NIH spec res fel, Rocky Mountain Lab, 63-64. *Mem:* Entom Soc Am. *Res:* Medical veterinary entomology and mosquito control. *Mailing Add:* Rte 9 Box 262 Fayetteville AR 72703

LANCASTER, JOHN, b Bolton, Miss, Aug 30, 37; m 64. MICROBIAL GENETICS. *Educ:* Miss State Univ, BS, 59, MS, 61; Univ Tex, PhD(microbiol), 64. *Prof Exp:* NIH trainee, Univ Tex, 63-64; from asst prof to assoc prof, 68-83, assoc dean, 84-88, PROF MICROBIOL, UNIV OKLA, 83- *Concurrent Pos:* NIH grant, 66; dir scholar-leadership enrichment prog, Univ Okla, 87-, dir lab animal resources, 88- *Mem:* AAAS; Am Soc Microbiol; Hist Sci Soc; Am Asn Lab Animal Sci. *Res:* Mechanism of conjugation in Escherichia coli; animal behavior in relations to animal rights; biological perspectives on environmental ethics. *Mailing Add:* Univ Okla 770 Van Vleet Oval Rm 135 Norman OK 73019-0245

LANCASTER, MALCOLM, b Amarillo, Tex, July 28, 31; m 59; c 4. CARDIOLOGY, GERIATRICS. *Educ:* Univ Tex Southwestern Med Sch, MD, 56; Univ Colo, Denver, MS, 60. *Prof Exp:* Chief med serv, 48th Tactical Hosp, Royal Air Force, Lakenheath, Eng, 60-63; chief cardiopulmonary serv & chmn dept med, US Air Force Hosp, Wright-Patterson AFB, 65-66; chief internal med br, Sch Aerospace Med, Brooks AFB, Tex, 66-73, chief clin sci div, 72-78; chief med serv & clin dir, San Antonio State Chest Hosp, 78-81; PRES, SYSTEMICS, INC, SAN ANTONIO, TEX, 81-; ASSOC PROF CLIN MED, DEPT FAMILY PRACT, UNIV TEX HEALTH SCI CTR, SAN ANTONIO, 85- *Honors & Awards:* Casimir Funk Award, Asn Mil Surgeons US, 71; USAF Res & Develop Award, 71; John Jeffries Award, Am Inst Aeronaut & Astronaut, 74; Arnold D Tuttle Award, Aerospace Med Asn, 75. *Mem:* Fel Aerospace Med Asn; fel Am Col Cardiol; fel Am Col Physicians; Am Geriat Soc; fel Am Col Prev Med. *Res:* Medical aspects of aerospace operations; cardiovascular disease epidemiology; computers and electrocardiography; computer aided design; geriatrics aerospace medicine. *Mailing Add:* 101 Hibiscus San Antonio TX 78213

LANCASTER, OTIS EWING, b Pleasant Hill, Mo, Jan 28, 09; m 42; c 5. AERONAUTICAL ENGINEERING, MATHEMATICS. *Educ:* Cent Mo State Teachers Col, 29; Univ Mo, MA, 34; Harvard Univ, PhD(math), 37; Calif Inst Technol, AeroE, 45. *Prof Exp:* Teacher, Oak Grove High Sch, Mo, 29-30 & Independence Jr High Sch, Mo, 30-33; instr & tutor math, Harvard Univ, 36-37; from instr to asst prof, Univ Md, 37-42; head appl math br, Res Div, Bur Aeronaut, Navy Dept, Washington, DC, 46-54, asst dir, 54; mem planning staff, Internal Revenue Serv, 54-55; dir statist & econ staff, Bur Finance, Post Off Dept, 55-57; George Westinghouse prof eng educ, 57-75, assoc dean eng, 67-75, EMER WESTINGHOUSE PROF & ASSOC DEAN, PA STATE UNIV, UNIVERSITY PARK, 75- *Concurrent Pos:* Consult, NASA-Univ rels; chief math & statist, Interstate Com Comn, 76-80; statist consult, 81- *Honors & Awards:* Distinguished Citation, Am Soc Eng Educ. *Mem:* Am Soc Eng Educ (pres, 77-78); assoc fel Am Inst Aeronaut & Astronaut. *Res:* Engineering education; aircraft propulsion; statistics; improvement of teaching. *Mailing Add:* 268 Ellen Ave State College PA 16801

LANCE, EUGENE MITCHELL, b Brooklyn, NY, Aug 2, 33; m; c 4. SURGERY. *Educ:* Cornell Univ, BA, 54, MD, 58; Am Bd Orthop Surg, dipl, 67; London Univ, PhD, 68. *Prof Exp:* Asst instr surg, Med Col Cornell Univ, 59-69, instr, 67-69, asst prof orthop, 69-71; head, Div Surg Sci Clin Res Ctr, Northwick Park Hosp, London, 71-74; res assoc, Cancer Ctr Hawaii, 74-79; PROF ORTHOP SURG, JOHN A BURNS SCH MED, UNIV HAWAII, 75-; VPRES, ORTHOP ASSOC INC, 76-; DIR, IMMUNOL RES LAB, SHRINERS HOSP, HONOLULU UNIT, 84- *Concurrent Pos:* Resident surgeon, NY Hosp, 59-61; resident surgeon orthop, Hosp Spec Surg, 61-64, fel, 64-66; dir, Trauma Serv Bellevue Hosp, Cornell Div, 64-66; vis scientist, Nat Inst Med Res, 66-68; asst attend surgeon, Hosp Spec Surg & NY Hosp, 68-70, assoc attend surgeon, 70-71; consult orthop surgeon, Northwick Park Hosp, London, 71-74; attend surgeon, Queens Med Ctr, St Francis Hosp, Shriners Hosp Crippled Children & Kapiolani Women's & Children's Med Ctr, 74-; orthop surgeon, Arthritis Ctr Hawaii, 74-81. *Honors & Awards:* William Mecklenberg Polk Award; Lewis Clark Wagner Award. *Mem:* Fel Am Acad Orthop Surgeons; Am Soc Human Genetics; Am Asn Immunologists; AAAS; Am Asn Bone & Joint Surgeons. *Res:* Orthopaedics; immunology; surgery. *Mailing Add:* Orthop Assoc Hawaii Inc 1380 Lusitana St Suite 608 Honolulu HI 96813

LANCE, GEORGE M(ILWARD), b Youngstown, Ohio, Dec 4, 28; m 64; c 5. MECHANICAL & ELECTRICAL ENGINEERING. *Educ:* Case Inst Technol, BS, 52, MS, 54. *Prof Exp:* Instr eng, Case Inst Technol, 52-54; res engr, TRW, Inc, 54-56; lectr mech eng, Univ Wash, St Louis, 56-60; sr systs engr, Moog Servocontrols, Inc, 60-61; from asst prof to prof, Univ Iowa, 61-70, assoc dean undergrad progs & student affairs, 74-79, chmn eng prog, 74-85, prof mech eng, 70-91, EMER PROF MECH ENG, UNIV IOWA, 91- *Mem:* Am Soc Mech Engrs; Am Soc Eng Educ. *Res:* Theory of automatic control; hydraulic servosystems; system design. *Mailing Add:* Col Eng Univ Iowa Iowa City IA 52240

LANCE, JOHN FRANKLIN, b Vaughn, NMex, May 21, 16; m 42; c 2. GEOLOGY. *Educ:* Col Mines & Metal, Univ Tex, BA, 37; Calif Inst Technol, MS, 46, PhD(paleont, petrog), 49. *Prof Exp:* Asst prof geol, Whittier Col, 48-50; from asst prof to prof paleont, Univ Ariz, 50-67; staff assoc, Div Inst Prog, 63-65 & 67-70, exec asst div environ sci, 71-75, PROG DIR GEOL, NSF, 75- *Concurrent Pos:* Geologist, US Geol Surv, 52-63. *Mem:* Geol Soc Am; Paleont Soc; Soc Vert Paleontologists; Soc Study Evolution. *Res:* Pliocene and Pleistocene mammalian fossils. *Mailing Add:* 2221 E Copper St Tucson AZ 85719

LANCE, R(ICHARD) H, b Geneva, Ill, Nov 29, 31; m 53; c 3. MECHANICS. *Educ:* Univ Ill, Urbana, BSME, 54; Ill Inst Technol, MSME, 57; Brown Univ, PhD(solid mech), 62. *Prof Exp:* Test engr, Minneapolis Honeywell Regulator Co, 54; mech engr, Ingersoll Milling Mach Co, Ill, 57-58; res assoc, Brown Univ, 62; from asst prof to assoc prof theoret & appl mech, 62-81, assoc dean eng, 74-86, PROF THEORET & APPL MECH, CORNELL UNIV, 81- *Concurrent Pos:* Lectr, Int Bus Mach Corp, NY, 66; sr scientist, Hughes Aircraft Co, 80-81, 86-87. *Mem:* Am Soc Mech Engrs; AAAS; Am Acad Mech. *Res:* Mechanical behavior of solids; engineering structural mechanics; plasticity; numerical methods in engineering. *Mailing Add:* Col Eng Cornell Univ Thurston Hall Ithaca NY 14853

LANCE, VALENTINE A, b London, Eng, Feb 14, 40; wid; c 2. REPRODUCTIVE PHYSIOLOGY, COMPARATIVE ENDOCRINOLOGY. *Educ:* Long Island Univ, BS, 66; Col William & Mary, MA, 68; Univ Hong Kong, PhD(zool), 74. *Prof Exp:* Demonstr zool, Univ Hong Kong, 68-74; res assoc endocrinol, Boston Univ, 74-78; ASST PROF PHYSIOL, LA STATE UNIV, 78-; ENDOCRINOLOGIST, SAN DIEGO ZOO, 87- *Concurrent Pos:* Rev Gen & Comp Endocrinol; rev J of Exp Zool; rev peptides; rev herpetologica; rev J of Wildlife Dis; grant rev, NSF; asst prof, Sch Med, Tulane Univ, 82-87. *Mem:* Endocrine Soc; Soc Study Reproduction; Am Soc Zoologists; Soc Study Amphibians & Reptiles; Herpetologists League; Sigma Xi; AAAS. *Res:* Evolution of the endocrine system; evolution of pituitary control of gonadal steroidogenesis; role of hypothalamic hormones and related reptiles in non-mammalian vertebrates; hormonal control of seasonal reproduction; physiology and endocrinology of non-mammalian vertebrates; endocrinology of pregnancy, pancreatic hormones and peptides in non-mammalian and mammalian vertebrates. *Mailing Add:* Res Dept San Diego Zoo PO Box 551 San Diego CA 92112

LANCET, MICHAEL SAVAGE, b Detroit, Mich, Dec 11, 44; m 69; c 1. PHYSICAL CHEMISTRY, CHEMICAL ENGINEERING. *Educ:* Rose Hulman Inst Technol, BS, 66; Univ Chicago, MS, 71, PhD(nuclear chem), 72. *Prof Exp:* Res assoc nuclear cosmo chem, Carnegie-Mellon Univ, 72-74; RES SCIENTIST COAL CONVERSION RES, CONSOL COAL CO, 74- *Concurrent Pos:* Fel, Carnegie-Mellon Univ, 72-74. *Mem:* AAAS; Am Chem Soc; Sigma Xi. *Res:* Conversion of coal to substitute natural gas and synthetic liquids; utilization of all other forms of energy. *Mailing Add:* Consol Coal Co Res & Develop 4000 Brownsville Rd Library PA 15129

LANCHANTIN, GERARD FRANCIS, b Detroit, Mich, Mar 27, 29; m 55; c 5. BIOCHEMISTRY. *Educ:* Seton Hall Univ, BS, 50; Univ Wyo, MS, 51; Univ Southern Calif, PhD(biochem), 54. *Prof Exp:* Res assoc, Sch Med, Univ Southern Calif, 54-55, from asst prof to assoc prof biochem, 57-69; dir biochem, St Joseph Med Ctr, 74-85; RETIRED. *Concurrent Pos:* Consult, Los Angeles County Gen Hosp, 57-; adj prof biochem, Sch Med, Univ Southern Calif, 69- *Honors & Awards:* Chaney Award Clin Chem, 81. *Mem:* Fel AAAS; Am Asn Clin Chem; Am Chem Soc; Am Soc Hemat; Am Soc Biol Chemists. *Res:* Clinical biochemistry; blood coagulation. *Mailing Add:* 65 S Lasenda Dr South Laguna CA 92677-3311

LANCIANI, CARMINE ANDREW, b Leominster, Mass, May 16, 41; m 64; c 2. ECOLOGY. *Educ:* Cornell Univ, BS, 63, PhD, 68. *Prof Exp:* Interim asst prof zool, 68-70, asst prof zool & biol sci, 70-73, assoc prof zool, 73-80, PROF ZOOL, UNIV FLA, 80- *Mem:* Ecol Soc Am; Entom Soc Am; Am Soc Limnol & Oceanog; Soc Study Evolution. *Res:* Population ecology of aquatic organisms, particularly life cycles, growth, reproduction and competition of parasitic water mites; effect of parasitism on host ecology. *Mailing Add:* Dept Zool Univ Fla 416 Bar Bldg Gainesville FL 32611

LANCMAN, HENRY, b Warsaw, Poland, Mar 19, 32; US citizen; m 63; c 2. NUCLEAR SPECTROSCOPY, PHOTONUCLEAR REACTIONS. *Educ:* Moscow Univ, MS, 58; Inst Nuclear Res, Polish Acad Sci, PhD(physics), 66. *Prof Exp:* Asst res physics, Inst Nuclear Res, Warsaw, Poland, 58-61, sr asst res, 61-65, head, Lab Gamma Ray Spectroscopy, 65-66; res assoc, Columbia Univ, 67-68; instr, 68-70, from asst prof to assoc prof, 70-78, PROF PHYSICS, CITY UNIV NY, BROOKLYN COL, 78- *Concurrent Pos:* Consult, Am Inst Physics, 68-71; vis prof physics, Rijks Univ, Utrecht, Holland, 74-75; res grants, PSC-BHE, 76-88, NSF, 77-78 & US Dept Energy, 79-85. *Mem:* Am Phys Soc; Sigma Xi. *Res:* Higher order effects in beta decay; lifetimes of nuclear states; nuclear resonance flourescence and absorbtion of gamma rays; photofission. *Mailing Add:* Dept Physics Brooklyn Col Bedford Ave & Ave H Brooklyn NY 11210

LAND, CECIL E(LVIN), b Lebanon, Mo, Jan 8, 26; m 47; c 2. ELECTRONICS ENGINEERING, SOLID STATE PHYSICS. *Educ:* Okla State Univ, BS, 49. *Hon Degrees:* DSc, Okla Christian Col, 78. *Prof Exp:* Prof engr, Electronics Div, Westinghouse Elec Corp, Md, 49-56; staff mem, 56-83, DISTINGUISHED MEM TECH STAFF, SANDIA NAT LABS, 83- *Concurrent Pos:* Chmn, Inst Elec & Electronics Engrs UFFC-S Ferroelectrics Comt, 78- *Honors & Awards:* Nat Soc Prof Engrs Award, 73; Frances Rice Darne Mem Award, Soc Info Display, 76; Recognition Award, Inst Elec & Electronics Engrs UFFC-S, 86, Achievement Award, 90. *Mem:* Fel Soc Info Display; fel Inst Elec & Electronics Engrs; Am Phys Soc; fel Am Ceramics Soc; Optical Soc Am. *Res:* Ferroelectric ceramic electrooptic and piezoelectric materials and devices; ferroelectric thin films and devices. *Mailing Add:* Div 1112 Sandia Nat Labs Albuquerque NM 87185-5800

LAND, CHARLES EVEN, b San Francisco, Calif, July 13, 37; m 60; c 2. STATISTICS. *Educ:* Univ Ore, BA, 59; Univ Chicago, MA, 64, PhD(statist), 68. *Prof Exp:* Res assoc statist, Atomic Bomb Casualty Comn, Nat Acad Sci, 66-68; asst prof, Ore State Univ, 68-73; res assoc statist, Atomic Bomb Casualty Comn & Radiation Effects Res Found, 73-75; expert math statistician, Biometry Br, 75-77, health statistician, Environ Epidemiol Br, 77-84, HEALTH STATISTICIAN, RADIATION EPIDEMIOL BR, NAT CANCER INST, NIH, 84- *Concurrent Pos:* Mem, Nat Coun Radiation Protections Measurements; mem comt 1 on risk assessment, Int Comn Radiol Protection. *Mem:* Radiation Res Soc; Inst Math Statist; fel Am Statist Asn;

Biomet Soc; AAAS; Am Epidemiol Soc. *Res:* Risk analysis; mathematical statistics; inference problems associated with transformations of data; radiation carcinogenesis in human populations; epidemiology; biometry. *Mailing Add:* Radiation Epidemiol Br Nat Cancer Exec Plaza N Rm 408 Bethesda MD 20892

LAND, DAVID J(OHN), b Boston, Mass, Feb 15, 39. ATOMIC & MOLECULAR PHYSICS, MATERIALS ANALYSIS. *Educ:* Boston Col, BS, 59; Brown Univ, PhD(physics), 66. *Prof Exp:* Res asst physics, Brown Univ, 59-66; Nat Acad Sci res assoc, 66-68, RES PHYSICIST, NAVAL SURFACE WEAPONS CTR, 68- *Mem:* Am Phys Soc; Optical Soc Am. *Res:* Atomic collision physics. *Mailing Add:* Naval Surface Warfare Ctr R 41 Silver Spring MD 20903-5000

LAND, EDWIN HERBERT, physics; deceased, see previous edition for last biography

LAND, GEOFFREY ALLISON, b Jeannette, Pa, July 9, 42; m 66; c 4. MEDICAL MYCOLOGY, MEDICAL MICROBIOLOGY. *Educ:* Univ Tex, Arlington, BSc, 68; Tex Christian Univ, MSc, 70; Tulane Univ, PhD(med mycol), 73. *Prof Exp:* Inhalation therapist, Baylor Univ Med Ctr, Dallas, 66-68; teaching asst microbiol, Tex Christian Univ, 68-70; chmn & dir med mycol, Wadley Inst Molecular Med, 74-79; dir mycol & assoc dir microbiol, Univ Cinn Med Ctr, 79-81; DIR MICROBIOL & IMMUNOL, METHODIST HOSP, 81- *Concurrent Pos:* Vis asst prof, NC Cent Univ, 73-74; instr & lectr, Tex Soc Clin Microbiologists, 74- & NTex Soc Med Technologists, 75-; vis scientist, Dept Virol, Cent Pub Health Lab State Serum Inst, Helsinki, 81; adj prof biol & chem, NTex State Univ, 75-; assoc ed, J Oncol & Hematol, 75-79; adj assoc prof biol, Tex Christian Univ, 81-; consult, Clin Microbiol Labs, Univ Tex Health Sci Ctr, Dallas, 81- *Mem:* Am Soc Microbiol; Med Mycol Soc Am; Int Soc Human & Animal Mycol. *Res:* The molecular basis and early diagnosis of fungal infections in the compromised host; production of the antiviral interferon and its possible chemotherapeutic role for treating viral diseases. *Mailing Add:* Dept Mycol Wadley Inst Molecular Med Dallas TX 75235

LAND, LYNTON S, b Baltimore, Md, Dec 30, 40. GEOLOGY, GEOCHEMISTRY. *Educ:* Johns Hopkins Univ, AB, 62, MA, 63; Lehigh Univ, PhD(geol), 66. *Prof Exp:* Res fel geol, Calif Inst Technol, 66-67; asst prof, 67-77, PROF GEOL, UNIV TEX, AUSTIN, 77- *Mem:* Soc Econ Paleontologists & Mineralogists; Int Asn Sedimentol. *Res:* Sedimentology; carbonate sedimentation; diagenesis; sedimentary geochemistry; stable isotope geochemistry. *Mailing Add:* Dept Geol Univ Tex Austin TX 78712

LAND, MING HUEY, b Hsinchu, Taiwan, July 10, 40; US citizen; m 70; c 2. CAD APPLICATIONS. *Educ:* Taiwan Normal Univ, BS, 63; Northern Ill Univ, MS, 68; Utah State Univ, EdD, 70. *Prof Exp:* Asst prof technol, Eastern Ill Univ, 70-71; prof indust educ, Miami Univ, 71-83; chair technol, 83-89, DEAN FINE & APPL ARTS, APPALACHIAN STATE UNIV, 89- *Concurrent Pos:* Fulbright lectr, Chungnam Nat Univ Korea, 80-81; vis prof, Northeast Univ Technol, China, 86. *Honors & Awards:* Spec Recognition Award, Int Technol Educ Asn, 90. *Mem:* Am Soc Eng Educ; Am Voc Asn; Int Technol Educ Asn; Nat Asn Indust Technol. *Res:* Theories of engineering graphics; descriptive geometry; applications in computer graphics. *Mailing Add:* 115 University Circle Boone NC 28607

LAND, PETER L, b Leasburg, Mo, Nov 20, 29; m 66; c 3. SOLID STATE PHYSICS, CERAMICS. *Educ:* Univ Mo, BS, 58, MS, 60, PhD(physics), 64. *Prof Exp:* Res scientist, Metall & Ceramics Lab, Aerospace Res Labs, 64-75, RES SCIENTIST, HARDENED MAT BR, AIR FORCE WRIGHT AERONAUT LABS, WRIGHT-PATTERSON AFB, OHIO, 75- *Mem:* Sigma Xi. *Res:* New or improved nonlinear optical materials; optical properties of solids; laser effects on materials. *Mailing Add:* 502 Land Dr Dayton OH 45440

LAND, ROBERT H, b Portland, Maine, Sept 17, 24. NUCLEAR PHYSICS, SCIENTIFIC & ENGINEERING COMPUTER APPLICATIONS. *Educ:* Univ Maine, BS, 49; Mass Inst Technol, PhD(nuclear physics), 57. *Prof Exp:* Asst physicist, 56-60, assoc physicist, 60-85, PHYSICIST, AGRONNE NAT LAB, 85- *Mem:* Am Phys Soc; Asn Comput Mach; Sigma Xi. *Res:* Photoproduction of charged pi-mesons from deuterium; reactor neutron diffusion theory; molecular physics and solid state physics calculations using digital computers. *Mailing Add:* Argonne Nat Lab B205 9700 S Cass Ave Argonne IL 60439

LAND, WILLIAM EVERETT, b Baltimore, Md, Aug 23, 08; m 42. PHYSICAL CHEMISTRY. *Educ:* Johns Hopkins Univ, BS, 28, PhD(phys chem), 33. *Prof Exp:* Chemist, Devoe & Raynolds Co, 28-29; instr chem, Emory Univ, 33-37; asst dir res, Glidden Co, 37-42; dep head high explosives sect, Bur Ord, 42-51, head high explosives res & develop, 51-60, div engr, Mines & Explosives Div, Bur Naval Weapons, 60-65, asst dir, Mine Warfare Proj Off, 65-66, div engr, Mine Warfare Div, Naval Ord Systs Command, 66-68, CONSULT NAVAL ORD SYSTS COMMAND, BUR NAVAL WEAPONS, NAVY DEPT, 68- *Concurrent Pos:* Mem ammunition & high explosives panel, Res & Develop Bd, 48-51, chmn, 51-53; US leader explosives panel, Tripartite Tech Coop Prog, 61-65. *Mem:* Am Chem Soc. *Res:* Adsorption; catalysis; titanium pigments; microscopy; chemical engineering; ordnance engineering. *Mailing Add:* 9200 Beech Hill Dr Bethesda MD 20817

LANDAHL, HERBERT DANIEL, b Fancheng, China, Apr 23, 13; US citizen; m 40; c 3. MATHEMATICAL BIOLOGY. *Educ:* St Olaf Col, AB, 34; Univ Chicago, SM, 36, PhD(math biophys), 41. *Prof Exp:* Asst, Psychomet Lab, Univ Chicago, 37-38, asst math biophys, Dept Physiol, 39-42, res assoc, 42-45, asst prof, 45-48, from assoc prof to prof math biol, 48-58, prof biophys, 64-68, secy comt math biol, 48-64, actg chmn, 64-67; prof, 68-80, EMER PROF BIOPHYS & BIOMATH, UNIV CALIF, SAN FRANCISCO, 80- *Concurrent Pos:* Res career award, NIH, 62-68; chief ed,

Bull Math Biol, 73-81. *Honors & Awards:* Career Achievement Award, Soc Toxicol, 87. *Mem:* Biomet Soc; Biophys Soc; Soc Math Biol (vpres, 72-82, pres, 82-); Biomed Eng Soc. *Res:* Mathematical biophysics of cell division, nerve excitation and central nervous system; removal of aerosols and vapors by the human respiratory tract; biological effects of radiation; population interaction; biological periodicities; insulin production and release mechanisms. *Mailing Add:* 472 Lansdale Ave San Francisco CA 94127

LANDAU, BARBARA RUTH, b Pierre, SDak, Apr 28, 23. PHYSIOLOGY. *Educ:* Univ Wis, BS, 45, MS, 49, PhD, 56. *Prof Exp:* Instr phys educ, Rockford Col, 45-47; instr physiol, Mt Holyoke Col, 49-51; instr, St Louis Univ, 56-59; from instr to asst prof, Univ Wis, 59-62; asst prof zool, Univ Idaho, 62-64; from asst prof to assoc prof, 64-82, EMER PROF PHYSIOL, BIOPHYS & BIOL STRUCT, UNIV WASH, 83- *Res:* Neural aspects of temperature regulation, hibernation, cell activity at reduced temperature; author of textbooks of anatomy and physiology. *Mailing Add:* Dept Physiol & Biophys Univ Wash Seattle WA 98195

LANDAU, BERNARD ROBERT, b Newark, NJ, June 24, 26; m 56; c 3. MEDICINE, BIOCHEMISTRY. *Educ:* Mass Inst Technol, SB, 47; Harvard Univ, MA, 49, PhD(chem), 50; Harvard Med Sch, MD, 54. *Prof Exp:* Med house officer, 54-55, sr res physician, Peter Bent Brigham Hosp, Boston, Mass, 58-59; asst prof biochem & from asst prof to assoc prof med, Case Western Reserve Univ, 59-67; dir dept biochem, Merck Inst Therapeut Res, 67-69; prof pharmacol, 70-78, PROF MED, CASE WESTERN RESERVE UNIV, 69-, PROF BIOCHEM, 79- *Concurrent Pos:* Clin assoc, Nat Cancer Inst, 55-57; USPHS res fel biochem, Harvard Med Sch, 57-58, tutor, 57-59; estab investr, Am Heart Asn, 59-64. *Mem:* Endocrine Soc; Soc Biol Chem; Am Physiol Soc; Asn Am Physicians; Am Diabetes Asn. *Res:* Carbohydrate metabolism; endocrinology; diabetes mellitus. *Mailing Add:* Dept Med Case Western Reserve Univ 2074 Abington Rd Cleveland OH 44106

LANDAU, BURTON JOSEPH, b Boston, Mass, May 6, 33; m 57; c 2. MICROBIOLOGY, VIROLOGY. *Educ:* Boston Univ, AB, 54; Univ NH, MS, 57; Univ Mich, PhD(microbiol), 67. *Prof Exp:* Sr res virologist, Merck Inst Therapeut Res, Merck, Inc, 64-65; res assoc microbiol, Univ Mich, 65-67; sr instr, 67-69, asst prof, 69-74, ASSOC PROF MICROBIOL, 74-, ASST DEAN CURRIC, SCH MED, HAHNEMANN UNIV, 87- *Honors & Awards:* Lindback Award. *Mem:* Am Soc Microbiol; Soc Gen Microbiol; Tissue Culture Asn; Am Soc Virol. *Res:* Use of differentiating cell cultures to grow viruses with restricted host ranges; virus-host cell interactions leading to virus induced transformation; coxsackievirus virus infections. *Mailing Add:* Dept Microbiol & Immunol Hahnemann Univ MS 410 Philadelphia PA 19102

LANDAU, DAVID PAUL, b St Louis, Mo, June 22, 41; m 66; c 2. MAGNETISM, STATISTICAL MECHANICS. *Educ:* Princeton Univ, BA, 63; Yale Univ, MS, 65, PhD(physics), 67. *Prof Exp:* Asst res physics, Nat Ctr Sci Res, Grenoble, France, 67-68; lectr eng & appl sci, Yale Univ, 68-69; from asst prof to prof, 69-84, RES PROF PHYSICS, UNIV GA, 84- *Concurrent Pos:* Guest scientist, KFA Jülich, WGer, 74; Alexander von Humboldt fel, Univ Saarland, 75; Sr US scientist Humboldt fel, Univ Mainz, 88. *Honors & Awards:* Jesse Beams Medal, 87. *Mem:* Fel Am Phys Soc; Sigma Xi. *Res:* Critical phenomena associated with phase transitions; properties of magnetic solids. *Mailing Add:* Ctr Simulational Physics Univ Ga Athens GA 30602

LANDAU, EDWARD FREDERICK, organic chemistry, science education; deceased, see previous edition for last biography

LANDAU, EMANUEL, b New York, NY, Nov 28, 19; m 48; c 1. EPIDEMIOLOGY, BIOSTATISTICS. *Educ:* City Col New York, BA, 39; Am Univ, PhD(econ), 66. *Prof Exp:* Bus economist, Econ Date Anal Br, Off Price Admin, 41-42 & 46-47; chief, Family Statist Sect, Bur Census, 48-56; mem staff, Calif State Dept Pub Health, 57-59; chief, Biomet Sect, Div Air Pollution, USPHS, 59-62; head, Lab & Clin Trials Sect, Nat Cancer Inst, 62-64; statist adv, Nat Air Pollution Control Admin, 65-69; epidemiologist, Adminr Res & Develop, Environ Health Serv, 69-71; epidemiologist, Adminr Res & Monitoring, Environ Protection Agency, 71; chief, Epidemiol Studies Br, Bur Radiol Health, Food & Drug Admin, USPHS, 71-75; PROJ DIR & STAFF EPIDEMIOLOGIST, AM PUB HEALTH ASN, 75- *Concurrent Pos:* Mem, Career Serv Bd Math & Statist, Dept Health, Educ & Welfare, 65-69; mem, Comt Long-term Training Outside Serv, USPHS, 66-68; adv air qual criteria, WHO, Switz, 67; mem, comt study lung cancer among uranium miners, USPHS, 67; adv air qual criteria, Karolinska Inst, Sweden, 68; Nat Air Control Admin tech liaison rep, Adv Comt Toxicol, Nat Acad Sci, 68-69; adv, Dept Transp, 72-74; assoc ed, J Air Pollution Control Asn, 72 & J Clin Data & Anal, 74-; consult, Bur Radiol Health, Food & Drug Admin, 75-83; chmn comt Statist & Environ, Am Statist Asn, 84-85. *Honors & Awards:* Superior Serv Award, Dept Health, Educ & Welfare, 63. *Mem:* Fel Am Pub Health Asn; fel Royal Soc Health; Air Pollution Control Asn; Am Statist Asn; Soc Occup & Environ Health. *Res:* Problems of environmental health; public health statistics; chronic disease epidemiology. *Mailing Add:* Am Pub Health Asn 1015 15th St NW Washington DC 20005

LANDAU, JOSEPH VICTOR, b New York, NY, Jan 9, 28; m 50; c 3. MOLECULAR BIOLOGY. *Educ:* City Col New York, BS, 47; NY Univ, MSc, 49, PhD, 53. *Prof Exp:* USPHS asst, NY Univ, 49-51; USPHS res asst, Naples Zool Sta, Italy, 52; Runyon Cancer Res fel, NY Univ, 52-55; instr physiol, Russell Sage Col, 56-57; res assoc oncol, Albany Med Col, Union, 57-66; PROF BIOL & HEAD ACCELERATED BIOMED PROG, RENSSELAER POLYTECH INST, 67-, CHMN DEPT BIOL, 72- *Concurrent Pos:* Chief biol sect, Basic Sci Res Lab, Vet Admin Hosp, Albany, 57-66; adj assoc prof, Rensselaer Polytech Inst, 64-67. *Mem:* Am Soc Cell Biol; Biophys Soc; Am Inst Biol Sci; Am Soc Microbiol; Sigma Xi. *Res:* Protein and nucleic acid synthesis; barobiology; contractility. *Mailing Add:* 94 Meadowland St Delmar NY 12054

LANDAU, JOSEPH WHITE, b Buffalo, NY, May 23, 30; m 85; c 5. MEDICINE, DERMATOLOGY. *Educ:* Cornell Univ, BA, 51, MD, 55; Am Bd Pediat, dipl, 62; Am Bd Dermat, dipl, 65, cert dermatopath, 75. *Prof Exp:* Intern, Gen Hosp, Buffalo, 55-56; resident pediat, Children's Hosp, Buffalo, 56, Children's Hosp, Boston, Mass, 59-60 & Med Ctr, Univ Calif, Los Angeles, 60-61; asst res dermatologist, 64, from asst prof to assoc prof med & dermat, 64-74, attend physician, Student Health Serv, 65-86, ASSOC CLIN PROF MED DERMAT, MED CTR, UNIV CALIF, LOS ANGELES, 74- *Concurrent Pos:* USPHS fel hemat, Children's Hosp, Los Angeles, 61-62 & fel mycol, Med Ctr, Univ Calif, Los Angeles, 62-63; attend physician, Wadsworth Vet Admin Hosp, 66- *Mem:* Am Acad Dermat; Soc Invest Dermat. *Res:* Host-parasite relationships in mycology; genodermatoses. *Mailing Add:* Suite 106 Med Arts Bldg 2200 Santa Monica Blvd Santa Monica CA 90404

LANDAU, MATTHEW PAUL, b New York, NY, Dec 16, 49; m 85; c 2. AQUACULTURE, CRUSTACEAN PHYSIOLOGY. *Educ:* St John's Univ, BS, 72; Long Island Univ, MS, 76; Fla Inst Technol, PhD(oceanog), 83. *Prof Exp:* Res fel biol, NY Ocean Sci Lab, 74-75; technician biochem, USDA, Gainesville, Fla, 75-78 & biol, Univ WFla, 78-79; postdoctoral aquacult, Harbor Br Ocenog Inst, 83-84 & biol, Univ Conn, 85-88; res scientist aquacult, Oceanic Inst, 84-85; asst prof, 88-90, ASSOC PROF MARINE SCI, STOCKTON STATE COL, 90- *Mem:* World Aquacult Soc; Am Soc Zoologists; Crustacean Soc; Am Fisheries Soc. *Res:* General aquaculture systems, especially as they relate to crustaceans; reproductive endocrinology of crustaceans, in particular concerning the mandibular organ. *Mailing Add:* Dept Marine Sci Stockton State Col Pomona NJ 08240

LANDAU, RALPH, b Philadelphia, Pa, May 19, 16; m 40; c 1. CHEMICAL ENGINEERING. *Educ:* Univ Pa, BS, 37; Mass Inst Technol, ScD(chem eng), 41. *Prof Exp:* Asst chem eng, Mass Inst Technol, 38-41; process develop engr, M W Kellogg Co, 41-43; head, Chem Dept, Kellex Corp, 43-45; process develop engr, M W Kellogg Co, 46; exec vpres, Sci Design Co, Inc, 46-63; pres, Halcon Int, Inc, 63-75, chmn, 75-81, chmn, Halcon SD Group, Inc, 81-82; CONSULT PROF ECON, STANFORD UNIV, 82-; OWNER, LISTOWEL INC, 82- *Concurrent Pos:* Dir, Aluminum Co Am, 77-; adj prof mgt, tech & soc, Univ Pa, 77- *Honors & Awards:* Petrochem & Petrol Div Award, Am Inst Chem Engrs, 73; Chem Indust Medal, Soc Chem Indust, 73; Winthrop-Sears Medal, Chem Indust Asn, 77; Award, Newcomen Soc NAm, 78; Award, Asn Consult Chemists & Chem Engrs, 78; Perkin Medal, 81. *Mem:* Nat Acad Eng (vpres, 81-); fel Am Inst Chem Engrs; fel NY Acad Sci; Am Chem Soc; Dirs Indust Res. *Res:* Commercial and technical research, development and manufacture in chemical process industries; international operations of chemical industry; economics of technology and public policy implications. *Mailing Add:* Listowel Inc Two Park Ave New York NY 10016-5601

LANDAU, RICHARD LOUIS, b St Louis, Mo, Aug 8, 16; m 43; c 3. ENDOCRINOLOGY. *Educ:* Wash Univ, BS & MD, 40. *Prof Exp:* From asst prof to prof, 48-88, EMER PROF MED, UNIV CHICAGO, 88- *Concurrent Pos:* Ed, Perspectives in Biol & Med, 73- *Mem:* Am Soc Clin Invest; Endocrine Soc; AMA; Sigma Xi. *Res:* Hormonal regulation of growth processes; reproductive endocrinology; metabolic influence of progesterone; effect of steroid hormones on electrolyte metabolism. *Mailing Add:* Dept Med Univ Chicago Box 435 Chicago IL 60637

LANDAU, WILLIAM, b Jersey City, NJ, July 3, 27; m 63; c 2. MICROBIOLOGY. *Educ:* Univ Conn, BA, 49; Yale Univ, MS, 51; Univ Pa, PhD(pub health, prev med), 58. *Prof Exp:* Res assoc biochem, Roswell Park Mem Inst, 58-61; assoc dir microbiol dept, Presby-St Luke's Hosp, 61-74; asst prof, 62-74, ASSOC PROF BACT, RUSH MED CTR, 74-; ASSOC SCIENTIST, PRESBY-ST LUKE'S HOSP, CHICAGO, 74- *Mem:* Am Soc Microbiol; NY Acad Sci. *Res:* Clinical bacteriology. *Mailing Add:* 8345 Kenton Ave Skokie IL 60076

LANDAU, WILLIAM M, b St Louis, Mo, Oct 10, 24; m 47; c 4. NEUROLOGY, NEUROPHYSIOLOGY. *Educ:* Washington Univ, MD, 47. *Prof Exp:* From instr to assoc prof, 52-63, PROF NEUROL, SCH MED, WASH UNIV, 63-, HEAD DEPT, 70-, CO-HEAD, DEPT NEUROL & NEUROSURG, 75- *Concurrent Pos:* Sr asst surgeon & neurophysiologist, NIMH & Nat Inst Neurol Dis & Blindness, 52-54; vis prof, Univ Munich, 63; pres, Am Bd Psychiat & Neurol, 75; chmn, Nat Comt Res Neurol & Commun Disorders, 80- *Mem:* Am Physiol Soc; Am EEG Soc; Am Neurol Asn (pres, 77); Asn Univ Profs Neurol (pres, 78); Am Acad Neurol. *Res:* Sensory and motor systems. *Mailing Add:* Dept Neurol & Neurosurg Wash Univ Med 660 S Euclid Ave St Louis MO 63110

LANDAUER, MICHAEL ROBERT, b New York, NY, Sept 24, 46. ANIMAL BEHAVIOR. *Educ:* Rutgers Univ, BS, 68; Univ Ill, Urbana, MS, 70, PhD(biopsychol), 75. *Prof Exp:* Res assoc psychol, Beaver Col, 74-76; vis asst prof biol, Barnard Col, Columbia Univ, 76-79; fel toxicol & pharmacol, Med Col, Va Commonwealth Univ, 79-82, res assoc pharmacol, Med Col Va, 82-84; RES TOXICOLOGIST, ARMED FORCES RADIOBIOL RES INST, 84-, PROJ MGR BEHAV SCI DEPT, 90- *Concurrent Pos:* Lectr, Philadelphia Zoo, 76; grant, NIMH, 79; vis res scientist, US Army Chem Res & Develop Ctr, 82-84; grant recipient, Vet Admin & Dept Defense, 91. *Mem:* Radiation Res Soc; Animal Behavior Soc; Am Soc Zoologists; Europ Soc Radiation Biol; Am Col Toxicologists; Asn Govt Toxicologists; Behav Toxicol Soc; Int Neurotoxicol Soc. *Res:* Behavioral effects of ionizing radiation and chemical radiation protectors; neurotoxicology; psychopharmacology. *Mailing Add:* Behav Sci Dept Armed Forces Radiobiol Res Inst Bethesda MD 20889-5145

LANDAUER, ROLF WILLIAM, b Stuttgart, Ger, Feb 4, 27; nat US; m 50; c 3. SOLID STATE PHYSICS, COMPUTER TECHNOLOGY. *Educ:* Harvard Univ, SB, 45, AM, 47, PhD(physics), 50. *Prof Exp:* Physicist, Lewis Lab, Nat Adv Comt Aeronaut, 50-52; physicist, 52-61, dir phys sci, 61-66, asst dir res, 66-69, IBM FEL, THOMAS J WATSON RES CTR, IBM CORP, 69- *Mem:* Nat Acad Sci; Nat Acad Eng; fel Inst Elec & Electronics Engrs; fel Am Phys Soc; fel AAAS. *Mailing Add:* Thomas J Watson Res Ctr IBM Corp PO Box 218 Yorktown Heights NY 10598

LANDAW, STEPHEN ARTHUR, b Paterson, NJ, June 20, 36; c 2. INTERNAL MEDICINE, HEMATOLOGY. *Educ:* Univ Wis-Madison, BS, 57; George Washington Univ, MD, 59; Univ Calif, Berkeley, PhD(med physics), 69; Am Bd Internal Med, dipl, 72, cert hemat, 72, cert med oncol, 75; Am Bd Nuclear Med, dipl, 72. *Prof Exp:* Intern, Mt Sinai Hosp, NY, 59-60, asst resident internal med, 60-61; Nat Heart Inst fel med physics, Donner Lab, Univ Calif, Berkeley, 63-70; asst physician, Donner Lab, Univ Calif, Berkeley, 70-73, lectr med physics, Univ, 70-72; assoc prof med & radiol, 73-78, PROF MED, STATE UNIV NY HEALTH SCI CTR, 78- *Concurrent Pos:* Attend staff physician, Alameda County Hosp, Oakland, Calif, 69-73, chief isotope lab, 71-73; Nat Heart & Lung Inst career develop award, 70-73; assoc chief staff res, Vet Admin Hosp, Syracuse, NY, 73-; mem attend staff med, Vet Admin Hosp, Univ Hosp & Crouse-Irving Mem Hosp, 73-; NIH fel med & hemat, Med Col Va, 62-63. *Mem:* Fel Am Col Physicians; Am Soc Hemat; Am Fedn Clin Res; Soc Nuclear Med; Soc Exp Biol & Med; Soc Pediat Res. *Res:* Bilirubin kinetics; quantitative red blood cell kinetics; polycythemic disorders; carbon monoxide kinetics. *Mailing Add:* Dept Med State Univ NY Health Sci Ctr 750 E Adams St Syracuse NY 13210

LANDAY, ALAN LEE, b Pittsburgh, Pa, Feb 13, 55. MICROBIOLOGY. *Educ:* Pa State Univ, BS, 76; Univ Pittsburgh, PhD, 81. *Prof Exp:* Postdoctoral, Cellular Immunobiol Unit, Univ Ala Sch Med, Birmingham, 81-83; asst prof immunol/microbiol, 83-88, asst prof path, 85-88, DIR CLIN IMMUNOL LAB & FLOW CYTOMETRY LAB, OFF CONSOL SERVS, RUSH-PRESBY-ST LUKE'S MED MED CTR, CHICAGO, 83-, ASSOC PROF IMMUNOL/MICROBIOL, PATH & MED, 88- *Concurrent Pos:* Grants, Leukemia Res Found, 85-86, Rush Univ, 85-86, Rush Med Ctr, 85-86, Am Cancer Soc, 85-86, NIH, 86-89 & 87-92, Loyd Frye Found, 88-91, Coulter Immunol, 90-91, AMAC Corp, 91-92; mem, Standards Subcomt Flow Cytometry, Immunol Res Comt, Flow Cytometry Adv Comt, NIH; chmn, Nat Comt Clin Lab & Qual Control & Standards Comt, Soc Anal Cytol; adv, Freedon Proj, Col Am Pathologists Diag & NASA Space Sta. *Mem:* AAAS; NY Acad Sci; Int Soc Anal Cytol; Am Soc Histocompatability & Immunogenetics; Am Soc Clin Path; Am Fedn Clin Res; Am Asn Immunologists; Clin Immunol Soc; Am Soc Hemat; Am Asn Pathologists. *Res:* Effects of Vitamin C on growth characteristics of cultured cells; effects of radiation, chemotherapy & immunotherapy on a transplantable rat gliosarcoma tumor model; biology; pathology; numerous technical publications. *Mailing Add:* Dept Immunol & Microbiol Rush-Presby-St Luke's Med Ctr 1753 W Congress Pkwy Chicago IL 60612

LANDBORG, RICHARD JOHN, b Manchester, Iowa, May 13, 33; m 55; c 4. CHEMISTRY, SCIENCE EDUCATION. *Educ:* Luther Col, Iowa, BA, 55; Univ Iowa, MS, 57, PhD(chem), 59. *Prof Exp:* Part-time instr chem, Cornell Col, 57-58; asst prof, 59-63, chmn dept, 65-67, ASSOC PROF CHEM, AUGUSTANA COL, SDAK, 63- *Concurrent Pos:* Fulbright exchange prof, Univ Santa Maria Antigua, Panama, 67. *Mem:* AAAS; Am Chem Soc; Sigma Xi. *Res:* Chemistry of diazomethane particularly the addition cyclization reactions with activated olefinic systems. *Mailing Add:* 1109 W 37th St Sioux Falls SD 57105

LANDE, ALEXANDER, b Hilversum, Neth, Jan 5, 36; US citizen. THEORETICAL NUCLEAR PHYSICS. *Educ:* Cornell Univ, BA, 57; Mass Inst Technol, PhD(theoret physics), 64. *Prof Exp:* Instr, Palmer Phys Lab, Princeton Univ, 63-66; NSF fel, Niels Bohr Inst, 66-68, asst prof, 68-70; vis assoc prof, Nordic Inst Theoret Atomic Physics, 70-72; assoc prof, 72-80, PROF PHYSICS, INST THEORET PHYSICS, STATE UNIV GRONINGEN, 80- *Concurrent Pos:* Chmn, Inst Theoret Physics, 76-83. *Mem:* Am Phys Soc; Europ Phys Soc; Neth Phys Soc. *Res:* Theoretical nuclear structure. *Mailing Add:* Inst Theoret Physics Groningen Univ PO Box 800-WSN Groningen 9700 AV Netherlands

LANDE, KENNETH, b Vienna, Austria, June 5, 32; nat US; m; c 3. ASTROPHYSICS, ELEMENTARY PARTICLE PHYSICS. *Educ:* Columbia Univ, AB, 53, AM, 55, PhD(physics), 58. *Prof Exp:* Asst physics, Columbia Univ, 54-57; from instr to assoc prof, 59-74, PROF PHYSICS, UNIV PA, 74- *Concurrent Pos:* Actg chmn, Astron & Astrophys, 84-; assoc ed astrophys, Phys Rev Letters, 87. *Mem:* Am Phys Soc; Sigma Xi; Am Astron Soc. *Res:* Neutrino physics; cosmic rays. *Mailing Add:* Dept Physics Univ Pa Philadelphia PA 19174

LANDE, RUSSELL SCOTT, b Jackson, Miss, Aug 10, 51. POPULATION GENETICS, EVOLUTION. *Educ:* Univ Calif, Irvine, BS, 72; Harvard Univ, PhD(biol), 76. *Prof Exp:* ASST PROF BIOPHYS & THEORET BIOL, UNIV CHICAGO, 78- *Mem:* Fel genetics, Univ Wis-Madison, 76-78. *Res:* Population genetics and evolution, especially of quantitative characters and chromosomal rearrangements. *Mailing Add:* Dept Ecol Univ Chicago 940 E 57th St Chicago IL 60637

LANDE, SAUL, b Philadelphia, Pa, Aug 7, 30; m 54; c 4. BIOCHEMISTRY, ORGANIC CHEMISTRY. *Educ:* Ursinus Col, BS, 48; Univ Pittsburgh, PhD(biochem), 60. *Prof Exp:* Sr res chemist, Squibb Res Inst, 61-63; assoc prof biochem in med, 63-76, ASSOC CLIN PROF DERMAT, SCH MED, YALE UNIV, 77- *Mem:* Am Chem Soc. *Res:* Chemistry of biologically active peptides. *Mailing Add:* 15 Spector Rd Woodbridge CT 06525

LANDE, SHELDON SIDNEY, b Chicago, Ill, July 16, 41; m 64; c 2. ENVIRONMENTAL CHEMISTRY. *Educ:* Ill Inst Technol, BS, 62; Mich State Univ, PhD(chem), 66. *Prof Exp:* Multiple fel petrol, Mellon Inst, 68-70; res chemist, Gulf Res & Develop Co, 70-71; res assoc water chem, Grad Sch Pub Health, Univ Pittsburgh, 71-72; pub health adminr, Allegheny County Health Dept, Pa, 72-75; res assoc, Syracuse Univ Res Corp, 75-79; ENVIRON SPECIALIST, 3M CO, 79- *Mem:* Air & Waste Mgt Asn; Am Chem Soc; Soc Environ Toxicol & Chem; Soc Risk Analysis. *Res:* Fate of organic substances in soil and water; analysis of organic chemicals in the environment; chemical risk analysis. *Mailing Add:* Environ Lab 3M Co PO Box 33331 St Paul MN 55133

LANDECKER, PETER BRUCE, b New York, NY, Oct 1, 42. SPACECRAFT INSTRUMENTATION. *Educ:* Columbia Univ, BA, 63; Cornell Univ, PhD(exp physics), 68. *Prof Exp:* Instr physics, Cornell Univ, 67-68; asst res physicist, Univ Calif, Irvine, 68-70; res assoc, Columbia Univ, 70-74; mem tech staff, Aerospace Corp, 74-82; LAB SCIENTIST, SPACE & COMMUN GROUP, HUGHES AIRCRAFT CO, 82- *Concurrent Pos:* Instr, El Camino Col, 77 & 86-88; consult, Columbia Univ, 74-75, Aerospace Corp, 82-83, Sumware Corp, 86; prin investr, Solar X-ray satellite Payload, 74-82. *Honors & Awards:* Hughes Aircraft Co Inventor Awards, 83, 88, 89, 90, 91. *Mem:* Int Astron Union; Am Phys Soc. *Res:* Instruments on remote sensing spacecraft; x-ray astronomy; cosmic ray physics; solar physics; spacecraft and photography; star sensing attitudes determination; patent #4,679,753 plus 67 publications. *Mailing Add:* Hughes Space & Commun Group S41-B326 PO Box 92919 Airport Sta Los Angeles CA 90009

LANDEFELD, THOMAS DALE, b Columbus, Ohio, Mar 24, 47; div; c 2. REPRODUCTIVE ENDOCRINOLOGY, BIOCHEMISTRY. *Educ:* Marietta Col, AB, 69; Univ Wis-Madison, BS & PhD(reproductive endocrinol), 73. *Prof Exp:* Fel endocrinol div, Med Col, Cornell Univ, 73-74; fel obstet & gynec dept, Sch Med, Wash Univ, 74-76; from res assoc to sr res assoc, 76-78, asst res scientist, dept path, 78-82, asst prof, 82-87, ASSOC PROF, DEPT PHARMACOL, UNIV MICH, ANN ARBOR, 87- *Concurrent Pos:* Prin investr, NIH res grant, Univ Mich, 78-90, co-investr, 79-90 & 83-90, asst dean res & grad studies, 89- *Mem:* Endocrine Soc; Soc Study Reproduction; Sigma Xi; NY Acad Sci; Am Soc Biol Chemists. *Res:* Pituitary gonadotropins; isolation, purification, and biochemical characterization; mechanisms and control of biosynthesis; mRNA purification and translation; gene expression and regulation; recombinant DNA and cloning. *Mailing Add:* Dept Pharmacol Univ Mich Ann Arbor MI 48109

LANDEL, ROBERT FRANKLIN, b Pendleton, NY, Oct 10, 25; m 53; c 6. PHYSICAL CHEMISTRY, RHEOLOGY. *Educ:* Univ Buffalo, BA, 50, MA, 51; Univ Wis, PhD(phys chem), 54. *Prof Exp:* Res assoc, Univ Wis, 54-55; sr res engr, 55-59, chief solid propellant chem sect, 59-61, chief polymer res sect, 61-75, mgr propulsion & mat res sect, 75-76, mgr energy & mat res sect, 76-79 & 80-82, div technologist, control & energy conversion div, 79-80, mgr mat res & biotechnol sect, 82-83, dept mgr appl mech technol sect, 83-85, SR RES SCI, JET PROPULSION LAB, CALIF INST TECHNOL, 81- *Concurrent Pos:* Sr res fel, Calif Inst Technol, 65-69; sr Fulbright fel, Italy, 71-72; sr fel, Ctr Res Macromolecules, France, 72; vis prof, Swiss Fed Tech Inst, Lausanne, Switz; res affil, Rancho Los Amigos Hosp, Downey, Calif, 76-; consult, Sandia Corp, 83. *Honors & Awards:* Except Sci Achievement Award, NASA, 76, Except Serv Award, 88; Humboldt Prize, Ger, 90. *Mem:* Am Phys Soc; Am Chem Soc; Soc Rheology (pres, 85-87). *Res:* Mechanical properties and failure of high polymers; polymer solutions and slurries. *Mailing Add:* 1027 Sunmore Lane Altadena CA 91001

LANDER, ARTHUR DOUGLAS, b Brooklyn, NY, Sept 12, 58. COGNITIVE SCIENCE. *Educ:* Yale Univ, BS, 79; Univ Calif, San Francisco, PhD(neurosci), 85. *Prof Exp:* Assoc, Howard Hughes Med Inst, Ctr Neurobiol & Behav, Col Physicians & Surgeons, Columbia Univ, 85-87; Edward J Poitras asst prof human biol & exp med, 88-91, ASST PROF, DEPT BRAIN & COGNITIVE SCI & DEPT BIOL, MASS INST TECHNOL, 87- *Concurrent Pos:* David & Lucile Packard fel sci & eng, 88-93. *Mem:* Soc Neurosci; Soc Develop Biol; NY Acad Sci; Am Soc Cell Biol. *Res:* Molecular mechanisms of axon outgrowth and guidance; cellular responses to extracellular matrix; biologic functions of proteoglycans. *Mailing Add:* Dept Brain & Cognitive Sci Mass Inst Technol 77 Massachusetts Ave Cambridge MA 02139

LANDER, HORACE N(ORMAN), b Cambridge, Mass, May 28, 23; m 43; c 5. METALLURGY. *Educ:* Mass Inst Technol, BS, 51, ScD(metall), 55. *Prof Exp:* Group leader metall, Metal Hydrides, Inc, 54-55; supvr process metall, Jones & Laughlin Steel Corp, 56-60; asst dir res & develop, Youngstown Sheet & Tube Co, 60-62, dir, 62-70; vpres res & develop, Molybdenum & Specialty Metals Div, 70-75, SR VPRES RES & DEVELOP, AM METAL CLIMAX, INC, GREENWICH, 75- *Honors & Awards:* Hunt Award, Am Inst Mining, Metall & Petrol Engrs, 59. *Mem:* Am Soc Metals; Am Inst Mining, Metall & Petrol Engrs; Am Iron & Steel Inst. *Res:* Physical chemistry of metals; industrial research administration; coordination of process and product development programs with management and market objectives. *Mailing Add:* 34 Bayview Dr Swampscott MA 01907-2627

LANDER, JAMES FRENCH, b Bristol, Va, Aug 24, 31; m 60; c 3. GEOPHYSICS, TSUNAMIS. *Educ:* Pa State Univ, BS, 58; Am Univ, MS, 62, MA, 68. *Prof Exp:* Geophysicist, US Coast & Geol Surv, Nat Oceanic & Atmospheric Admin, 58-62, chief seismol invests sect, 62-63, chief seismol invests br, Environ Res Labs, 63-73, chief, Nat Earthquake Info Ctr, 66-73; dep dir, Nat Geophys & Solar-Terrestrial Data Ctr, Nat Oceanic & Atmospheric Admin, 73-88; RES ASST, UNIV COLO, 88- *Concurrent Pos:* At Exec Off of President, Off Emergency Preparedness, 70-71; dir, World Data Ctr-A Solid Earth Geophys, 73-83. *Mem:* AAAS; Seismol Soc Am; Am Geophys Union; Sigma Xi. *Res:* Tsunamis seismicity; earthquake intensity; earthquake engineering; strong motion seismology; diaster studies; natural hazard risks; digital data bases. *Mailing Add:* Univ Colo CIRES Campus Box 449 Boulder CO 80309

LANDER, PHILIP HOWARD, b Montreal, Que, Sept 17, 41; m 67; c 3. DIAGNOSTIC RADIOLOGY, SPINAL IMAGING & DIAGNOSIS. *Educ:* McGill Univ, Can, BSc, 64, MD, 66. *Prof Exp:* Asst prof 77-87, ASSOC PROF DIAG RADIOL, McGILL UNIV, 87-; SR RADIOLOGIST, SIR MORTIMER B DAVIS GEN HOSP, 72- *Mem:* Radiol Soc NAm; NY Acad Sci; Can Med Asn; Am Roentgen Ray Soc. *Res:* Correlation of Paget's disease, bone with radiographic pathology, histopathology and clinical findings involving the weight bearing joints and the spine; diagnostic imaging of painful intervertebral segments of the cervical and lumbar spine; correlation to provocative and analgesic clinical tests; investigation of electrocoagulation of dorsal ramii of painful vertebral segments. *Mailing Add:* Jewish Gen Hosp 3755 Cote St Catherine Rd Montreal PQ H3T 1E2 Can

LANDER, RICHARD LEON, b Oakland, Calif, Apr 23, 28; Div; c 3. PHYSICS. *Educ:* Univ Calif, Berkeley, BA, 50, PhD(physics), 58; Ohio State Univ, MA, 51. *Prof Exp:* Staff physicist, Lawrence Radiation Lab, Univ Calif, 58-60; res specialist nuclear physics, Boeing Co, 60-61; assoc res physicist, Univ Calif, San Diego, 61-66; assoc prof physics, 66-70, assoc dean res, Grad Div, 70-73, PROF PHYSICS, UNIV CALIF, DAVIS, 70- *Concurrent Pos:* Vis scientist, Europ Orgn Nuclear Res, Switz, 66-67. *Mem:* AAAS; Am Phys Soc. *Res:* Experimental elementary particle physics. *Mailing Add:* Dept Physics Univ Calif Davis CA 95616

LANDERL, HAROLD PAUL, b Pittsburgh, Pa, Apr 26, 22; m 44; c 3. ORGANIC CHEMISTRY. *Educ:* Carnegie Inst Technol, BS, 43, MS, 47, DSc(chem), 48. *Prof Exp:* Asst, Nat Defense Res Comt, Calif Inst Technol, 44-46; res chemist, Jackson Lab, E I Du Pont de Nemours & Co, Inc, 48-54, supvr res & develop, Tech Lab, 54-60, head textile dye appln div, 60-70, asst dir dyes & chem tech lab, 70-72, tech mgr dyes & chem, Belg, 72-76, tech mgr dyes, Chem, Dyes & Pigments Dept, 76-80, staff consult, Employee Rels Dept, 80-82; RETIRED. *Mem:* Am Chem Soc; Am Asn Textile Chemists & Colorists. *Res:* Application of dyes to fibers. *Mailing Add:* 1503 Fresno Rd Wilmington DE 19803

LANDERS, EARL JAMES, b Greybull, Wyo, Dec 17, 21; m 51; c 2. INVERTEBRATE ZOOLOGY. *Educ:* Univ Wyo, AB, 50, MS, 52; NY Univ, PhD(zool), 58. *Prof Exp:* Instr zool, Univ Wyo, 55-56; from asst prof to assoc prof biol sci, Tex Western Col, 56-60, actg chmn dept, 59-60; assoc prof zool, 60-70, PROF ZOOL, ARIZ STATE UNIV, 70- *Mem:* AAAS; Am Soc Parasitol. *Res:* Parasitic protozoa life cycles; transmembrane electrolyte transport. *Mailing Add:* 3034 S Country Club Way Tempe AZ 85282

LANDERS, JAMES WALTER, b Norfolk, Nebr, Oct 19, 27; m 52; c 3. PATHOLOGY. *Educ:* Univ Nebr, MD, 53. *Prof Exp:* Assoc prof, 62-80, CLIN PROF PATH, SCH MED, WAYNE STATE UNIV, 80- *Concurrent Pos:* Assoc pathologist, William Beaumont Hosp, Royal Oak, Mich, 64-68 & St John Hosp, Detroit, 68- *Mem:* Am Soc Clin Path; Col Am Path; Int Acad Path; Sigma Xi. *Res:* Neuropathology. *Mailing Add:* 1507 Sunningdale Grosse Pointe MI 48236

LANDERS, JOHN HERBERT, JR, b Stockton, Mo, Jan 24, 21; m 43; c 3. ANIMAL NUTRITION. *Educ:* Univ Mo, BS, 42, MS, 50; Kans State Univ, PhD(animal nutrit), 66. *Prof Exp:* County agent agr, Univ Mo, 45-49, instr animal sci, 49-50; from asst prof to prof & exten animal scientist, 50-77, EMER PROF, ORE STATE UNIV, 77- *Mem:* Am Romney Sheep Breeders Asn (secy); Am Soc Animal Sci. *Res:* Counseling and advising livestock growers in more efficient production of meat and fiber. *Mailing Add:* 29515 NE Weslinn Dr Corvallis OR 97333

LANDERS, KENNETH EARL, plant physiology; deceased, see previous edition for last biography

LANDERS, ROGER Q, JR, b Menard, Tex, July 23, 32; m 54; c 2. PLANT ECOLOGY, RANGE MANAGEMENT. *Educ:* Tex A&M Univ, BS, 54, MS, 55; Univ Calif, Berkeley, PhD(bot), 62. *Prof Exp:* From asst prof to assoc prof plant ecol, Iowa State Univ, 62-71, prof, 71-79; EXTEN RANGE SPECIALIST, TEX A&M UNIV SYST, 79- *Mem:* Ecol Soc Am; Soil Conserv Soc Am; Soc Range Mgt. *Res:* Grasslands; management of grazing land by chemical, mechanical and biological methods; prescribed burning to control undesirable brush and cactus and enhance productivity of desirable forage species for livestock and wildlife. *Mailing Add:* Res & Exten Ctr Tex A&M Univ Syst 7887 N Hwy 87 San Angelo TX 76901

LANDES, CHESTER GREY, b Lowell, Ind, Dec 22, 03; m 34; c 3. CELLULOSE & PAPER CHEMISTRY. *Educ:* Ohio State Univ, BS, 26. *Prof Exp:* Chem engr, Mead Corp, Tenn & Ohio, 26-34; supt, Fitchburg Paper Co, Mass, 34-36; sr group leader chem, Am Cyanamid Co, 36-58; tech dir, Wica Chem, Inc, 59-60; assoc prof, 60-72, EMER ASSOC PROF WOOD & PAPER SCI, NC STATE UNIV, 72- *Concurrent Pos:* Paper chem consult, 60-87. *Honors & Awards:* Coating & Graphic Arts Div Award, Tech Asn Pulp & Paper Indust. *Mem:* Am Chem Soc; Tech Asn Pulp & Paper Indust. *Res:* Paper sizing materials; wet strength and dry strength resins for paper; paper coating raw materials and additives; retention and drainage aids for paper making; pulp and paper chemicals and resins in general. *Mailing Add:* 215 Aldersgate Circle Asheville NC 28803

LANDES, HUGH S(TEVENSON), b Waynesboro, Va, July 4, 24; m 46; c 1. PHYSICS, ELECTRICAL ENGINEERING. *Educ:* Univ Va, BEE, 53, PhD(physics), 56. *Prof Exp:* Asst prof physics, Univ SC, 56-57; asst prof elec eng, 57-60, ASSOC PROF ELEC ENG, UNIV VA, 60- *Mem:* Nat Soc Prof Engrs; Am Asn Physics Teachers; Inst Elec & Electronics Engrs. *Res:* Electric circuit theory; electromagnetic field theory; microwave devices; ferrite phenomena. *Mailing Add:* Dept Elec Eng Univ Va Thornton Hall Charlottesville VA 22903

LANDES, JOHN D, b Sellersville, Pa, June 28, 42; m 64; c 4. FRACTURE MECHANICS. *Educ:* Lehigh Univ, BS, 64, MS, 65, PhD(mech), 70. *Prof Exp:* Res assoc, Pratt & Whitney Aircraft, 65-66; grad asst mech, Lehigh Univ, 66-70; sr engr, Westinghouse Elec Co, 70-76, fel engr, 76-78, adv engr, 78-85, mgr, Am Welding Inst, 85-87; PROF ESM, UNIV TENN, 87- *Honors & Awards:* Irwin Medal, Am Soc Testing & Mat, 80, Award of Merit, 89. *Mem:* Am Soc Testing & Mat; Am Welding Soc; Soc Eng Sci. *Res:* Research in fracture and fatigue, fracture of ductile materials; metals and polymers including testing standards and methods of applications. *Mailing Add:* 1921 Nolina Rd Knoxville TN 37922

LANDESBERG, JOSEPH MARVIN, b New York, NY, Apr 21, 39; m 64; c 2. ORGANIC CHEMISTRY. *Educ:* Rutgers Univ, BS, 60; Harvard Univ, MA, 62, PhD(chem), 65. *Prof Exp:* NIH res fel, Columbia Univ, 64-66; asst prof, Univ, 66-70, assoc prof, 70-75, PROF CHEM, GRAD SCH ARTS & SCI, ADELPHI UNIV, 75- *Mem:* AAAS; Am Chem Soc; Royal Soc Chem; Am Asn Univ Prof. *Res:* Heterocyclic chemistry; synthetic applications of organometallic compounds; synthesis of strained, small-membered rings. *Mailing Add:* Dept Chem Adelphi Univ Garden City NY 11530

LANDESMAN, EDWARD MILTON, b Brooklyn, NY, Mar 19, 38. MATHEMATICS. *Educ:* Univ Calif, Los Angeles, BA, 60, MA, 61, PhD(math), 65. *Prof Exp:* Asst prof in residence math, Univ Calif, Los Angeles, 65-66; asst prof, Univ Calif, Santa Cruz, 66-68; asst prof, Univ Calif, Los Angeles, 68-69; asst prof, 69-71, assoc prof, 71-80, PROF MATH, CROWN COL, UNIV CALIF, SANTA CRUZ, 80- *Concurrent Pos:* Air Force Off Sci Res grant, Univ Calif, Santa Cruz, 70-71. *Mem:* AAAS; Am Math Soc; Math Asn Am. *Res:* Partial differential equations; combinatorial theory; calculus. *Mailing Add:* Bd Studies in Math Univ Calif Santa Cruz CA 95064

LANDESMAN, HERBERT, b Newark, NJ, Apr 22, 27; m 53; c 2. INORGANIC CHEMISTRY. *Educ:* Harvard Univ, BS, 48; Purdue Univ, PhD(chem), 51. *Prof Exp:* Res chemist, Naval Ord Test Sta, 51-52, Olin Mathieson Chem Corp, 52-59 & Nat Eng Sci Co, 59-66; chem consult, West Precipitation Group, Joy Mfg Co, 66-68; vpres, Environ Resources, Inc, 68-69; PROF CHEM, LOS ANGELES SOUTHWEST COL, 69- *Mem:* Air Pollution Control Asn; Am Chem Soc. *Res:* Organosilicon chemistry; chemistry of boron hydrides; fire extinguishants; fluorocarbons; hazards analysis; air and water pollution. *Mailing Add:* Dept Life & Phys Sci Los Angeles Southwest Col 1600 W Imperial Hwy Los Angeles CA 90047

LANDESMAN, RICHARD, b Brooklyn, NY, Jan 30, 40. DEVELOPMENTAL BIOLOGY. *Educ:* NY Univ, BA, 61, MS, 63; Univ BC, PhD(zool), 66. *Prof Exp:* NIH fel biol, Mass Inst Technol, 66-69; ASSOC PROF ZOOL, UNIV VT, 69- *Concurrent Pos:* NIH & Nat Inst Dent Res, 84-85. *Mem:* Soc Develop Biol. *Res:* Cellular and molecular basis of limb regeneration; fracture healing in the newt; differentiation and morphogenesis of the regenerating newt limb-role of hormones and secondary growth factors in regeneration; osteoinductive potential of demineralized bone matrix and bone proteins. *Mailing Add:* Dept Zool Univ Vt Burlington VT 05405-0086

LANDGRAF, RONALD WILLIAM, b Freeport, Ill, Mar 7, 39; m 62; c 2. FATIGUE, FRACTURE. *Educ:* Carnegie Inst Technol, BS, 61; Univ Ill, Urbana, MS, 66, PhD(theoret & appl mech), 68. *Prof Exp:* Mat engr, Micro Switch Div, Honeywell, Inc, 61-65; res assoc theoret & appl mech, Univ Ill, Urbana, 66-68; res scientist, Sci Res Staff, 68-77, mem eng & res staff, 77-79, STAFF SCIENTIST, ENG & RES STAFF, FORD MOTOR CO, 79- *Concurrent Pos:* Assoc ed, Fatigue of Eng, Mat & Struct, 79- *Mem:* Am Soc Metals; Am Inst Mining, Metall & Petrol Engrs; Am Soc Testing & Mat; Soc Automotive Engrs; Inst Elec & Electronics Engrs. *Res:* Cyclic deformation and fracture behavior of metals and alloys; influence of metallurgical structure on fatigue crack initiation and propagation; development of fatigue design procedures. *Mailing Add:* 1501 Nelson Blacksburg VA 24060

LANDGRAF, WILLIAM CHARLES, b Elizabeth, NJ, Jan 10, 28; m 53; c 3. PHARMACEUTICAL CHEMISTRY, SOFTWARE SYSTEMS. *Educ:* Seton Hall Univ, BS, 50; Stanford Univ, PhD(chem), 59; Univ Santa Clara, MBA, 75. *Prof Exp:* Sr scientist chem, Lockheed Res Labs, Lockheed Missile Systs Div, 58-61; proj leader & lab supt, Ampex Corp, 61-63; mgr & res scientist, Varian Assocs, Calif, 63-70; MGR, SYNTEX LABS, 70- *Mem:* Am Chem Soc; Am Pharmaceut Asn. *Res:* Physical, biophysical and organic chemistry; kinetics; computer assisted experimentation; analytical chemistry; physical organic chemistry. *Mailing Add:* 762 Stone Lane Palo Alto CA 94303

LANDGREBE, ALBERT R, b New Rochelle, NY, Mar 4, 33; m 58; c 2. CHEMISTRY. *Educ:* Fordham Univ, BS, 57; Univ Md, PhD(chem), 64. *Prof Exp:* Inorg chemist, USDA, 60-63; radiochemist, Nat Bur Standards, Md, 63-68; chemist & chmn comt sci & tech symposia, AEC, 68-75; BR CHIEF CHEM STORAGE, ENERGY RES & DEVELOP ADMIN, 75- *Mem:* AAAS; Soc Nuclear Med; Sigma Xi; Am Chem Soc. *Res:* Use of radioisotopes in analytical and inorganic chemistry; radio chromatographic methods; substoichiometric radioisotopic dilution analysis; removal of radioisotopes from milk; activation analysis; trace and micro analysis. *Mailing Add:* Almar Res Lab 3201 Dunnington Rd Beltsville MD 20705

LANDGREBE, DAVID ALLEN, b Huntingburg, Ind, Apr 12, 34; m 59; c 3. ELECTRICAL ENGINEERING. *Educ:* Purdue Univ, BSEE, 56, MSEE, 58, PhD(elec eng), 62. *Prof Exp:* From asst prof to assoc prof, 62-70, dir, Lab Applications Remote Sensing, 69-81, assoc dean eng & dir, Eng Exp Sta, 81-84, PROF ELEC ENG, PURDUE UNIV, 70- *Concurrent Pos:* Consult, Earlham Col, 63 & Douglas Aircraft Co, 64-70; mem tech staff, Bell Tel Labs, Murray Hill, NJ, 56; electronics engr, Interstate Electronics Corp, Anaheim, Calif, 58-59; res scientist, Douglas Aircraft Co, Newport Beach, Calif, 62; consult, Earlham Col, 63 & Douglas Aircraft Co, 64-70. *Honors & Awards:* Except Sci Achievement Medal, NASA, 73; Geosci & Remote Sensing Soc Except Serv Award, Inst Elec & Electronics Engrs, Edinburgh, Scotland, 88. *Mem:* Fel Inst Elec & Electronics Engrs; Am Soc Eng Educ; AAAS. *Res:* Representation and analysis of signals; data processing. *Mailing Add:* Prof Elec Eng Purdue Univ West Lafayette IN 47907

LANDGREBE, JOHN A, b San Francisco, Calif, May 6, 37; m 61; c 2. ORGANIC CHEMISTRY. *Educ:* Univ Calif, Berkeley, BS, 59; Univ Ill, Urbana, PhD(org chem), 62. *Prof Exp:* From asst prof to assoc prof, 62-71, assoc chmn, 67-70, chmn, 70-80, PROF CHEM, UNIV KANS, 71- *Mem:* Am Chem Soc; Royal Soc. *Res:* Organic reaction mechanisms; small ring compounds; carbene intermediates; reactions of carbonyl ylides. *Mailing Add:* Dept Chem Univ Kans Lawrence KS 66044

LANDGREN, CRAIG RANDALL, b St Paul, Minn, Dec 20, 47; m 83. PLANT TISSUE CULTURE, PLANT DEVELOPMENT. *Educ:* Albion Col, BA, 69; Harvard Univ, MA, 70, PhD(biol), 74. *Prof Exp:* Asst prof, George Mason Univ, 74-77; vis asst prof, Univ Ore, 76-77; from asst prof to assoc prof biol, Middlebury Col, 77-89, chmn dept, 82-88, dir, Northern Studies, 84-87, chmn, Nat Sci Div, 85-88, dir, Sci Ens Prog, 87-90, dir, Freshman Writing Prog, 90-91, PROF BIOL, MIDDLEBURY COL, 89- *Concurrent Pos:* Res grant, George Mason Found, 75-76; res assoc, Univ Ore, 78-81; vis scientist, US-USSR Nat Acad Sci Exchange Prog, 80. *Mem:* Sigma Xi. *Res:* Studies in the culture and differentiation of isolated plant cells and plant cell protoplasts; genetic engineering through organelle transplantation and cell fusion; plant stress physiology; over wintering of plants; ethics in science. *Mailing Add:* Dept Biol Middlebury Col Middlebury VT 05753

LANDGREN, JOHN JEFFREY, b St Paul, Minn, Nov 16, 47; m 77; c 3. MATHEMATICS, FUNCTIONAL ANALYSIS. *Educ:* Univ Minn, BS, 69, MS, 71, PhD(math), 76. *Prof Exp:* Vis asst prof math, Ga Inst Technol, 76-78; asst prof math, Univ Tenn, 78-80; SR RES SCIENTIST, GA INST TECHNOL, 80- *Mem:* Sigma Xi; Soc Indust & Appl Math. *Res:* Electronic warfare; simulation of radar/jamming systems; applied mathematics; radar countermeasures and radar signal processing; adaptive antenna arrays. *Mailing Add:* Syst Eng Lab Ga Technol Res Inst Atlanta GA 30332

LANDING, BENJAMIN HARRISON, b Buffalo, NY, Sept 11, 20; m; c 4. PEDIATRIC PATHOLOGY, GENETICS. *Educ:* Harvard Univ, AB, 42; Harvard Med Sch, MD, 45. *Prof Exp:* Intern, Children's Hosp, Boston, 45-46; res pathologist, Children's Med Ctr, Boston, 48-50, asst pathologist, 50-52, assoc pathologist, 52-53; from asst to instr & assoc path, Harvard Med Sch, 48-53; from asst prof to assoc prof path & pediat, Col Med, Univ Cincinnati, 53-61; pathologist in chief & dir labs, 61-88, Winzer prof path, 76-88, RES PATHOLOGIST, CHILDREN'S HOSP, LOS ANGELES, 88-; PROF PATH & PEDIAT, UNIV SOUTHERN CALIF, 61- *Concurrent Pos:* Res pathologist, Free Hosp for Women & Boston Lying-In-Hosp, 49; dir pathologist, Children's Hosp & Res Found, Cincinnati, 53-61. *Honors & Awards:* Farber Mem Lect, Soc Pediat Path, 84. *Mem:* Histochem Soc; Endocrine Soc; Am Asn Path (asst secy, 53-57); Int Acad Path; Soc Pediat Path (pres, 72-73). *Res:* Histochemistry of metabolic diseases; pediatric pathology; morphometry of malformation syndromes; enteric nervous system involvement in neurologic diseases; neurosciences. *Mailing Add:* Children's Hosp 4650 Sunset Blvd Los Angeles CA 90027

LANDIS, ABRAHAM L, b New York, NY, May 25, 28; m 57; c 2. CHEMISTRY. *Educ:* City Col NY, BS, 51; Univ Kans, PhD(chem), 55. *Prof Exp:* Asst chem, Univ Kans, 51-55; aeronaut res scientist, Nat Adv Comt Aeronaut, 55-56; sr res chemist, Atomics Int Div, NAm Rockwell, Inc, 56-61; sr staff chemist, Hughes Aircraft Co, El Segundo, Calif, 61-80, sr scientist, 80-86; SR STAFF SCIENTIST, LOCKHEED AERONAUT SYST CO, CALIF, 86- *Mem:* Am Chem Soc; Sigma Xi; Am Inst Chem; Am Ceramic Soc; Soc Advan Mat & Process Eng. *Res:* High temperature polymers; polymer chemistry; vacuum technology; organic synthesis; organometallic polymers; aerospace materials. *Mailing Add:* 10935 Canby Ave Northridge CA 91326

LANDIS, ARTHUR MELVIN, b Lancaster, Pa, Jan 21, 44; m 68; c 2. HETEROPOLY COMPLEXES, SEPARATIONS SCIENCE. *Educ:* Elizabethtown Col, BS, 66; Ohio Univ, MS, 70; Georgetown Univ, PhD(chem), 77. *Prof Exp:* Head teaching fel chem, Georgetown Univ, Washington, DC, 70-75, vis prof, 73-75; vis prof chem, Dickinson Col, Carlisle, Pa, 75-77; sr res chemist, UOP, Inc, Signal-Allied Co, Des Plaines, Ill, 77-82; asst prof chem, Col Our Lady of The Elms, Chicopee, Kans, 83-87; asst prof, 87-91, ASSOC PROF CHEM, EMPORIA STATE UNIV, EMPORIA, KS, 91- *Concurrent Pos:* Vis prof chem, Georgetown Univ, 83; lectr chem, Western New Eng Col, 84-; consult, Springfiled Wire, 86-89; mem, Gordon Res Conf, 80-82. *Mem:* Am Chem Soc; Sigma Xi. *Res:* Heteropolies and polyoxometallates; preparation and properties of organo-heteropoly polymers; configurations and conformations of species; separation theory of solutes via interaction with inorganic solids, for example, preparative liquid chromatography. *Mailing Add:* 1008 Burns St Emporia KS 66801-6314

LANDIS, DENNIS MICHAEL DOYLE, b Boston, Mass, Aug 12, 45. NEUROLOGY, NEUROSCIENCE. *Educ:* Harvard Col, AB, 67; Harvard Med Sch, MD, 71; Am Bd Internal Med, dipl, 75; Am Bd Psychiat & Neurol, dipl, 79. *Prof Exp:* From instr to asst prof neurol, Harvard Med Sch, 78-83, asst prof neurol-neurosci, 83-85; assoc prof neurol, develop genetics & anat, 85-88, assoc prof neurol & Ctr Neurosci, 88-90, PROF NEUROL & NEUROSCI, SCH MED, CASE WESTERN RESERVE UNIV, 90-; ASSOC NEUROLOGIST, UNIV HOSPS CLEVELAND, 85- *Concurrent Pos:* Instr neurobiol, Marine Biol Labs, Woods Hole, Mass, 77-; teacher investr award, Nat Inst Neurol & Communicative Dis & Stroke, 78, Javits Neurosci investr award, 89; asst neurol, Mass Gen Hosp, 78-79, attend physician, Neurol Serv & Neurol Consult Serv & dir, Muscular Dystrophy Asn Clin, 78-85, asst neurologist, 79-85; assoc neuropathologist, Eunice Kennedy Shriver Ctr Ment Retardation, Waltham, Mass, 78-85; lectr, Mass Gen Hosp, Marine Biol Lab & Case Western Reserve Univ, 78-; attend physician, Neurol Serv, Vet Admin Hosp, Cleveland, 85- & Neurol & Consult Serv, Univ Hosps Cleveland, 85-; dir, Lab Neurocytol, 85- *Mem:* Soc Neurosci; Am Acad Neurol; Am Soc Cell Biol; Am Neurol Asn; Soc Exp Neuropath. *Res:* Structure and function at synaptic junctions in the central nervous system; membrane and cytoplasmic structure in astrocytes; astrocyte function during development, in the adult, and in the response to injury. *Mailing Add:* Dept Neurol & Neurosci Sch Med E-604 Case Western Reserve Univ 2109 Adelbert Rd Cleveland OH 44106-4901

LANDIS, E K, b Pulaski, Va, June 17, 30; m 65; c 1. CHEMICAL ENGINEERING. *Educ:* Va Polytech Inst, BS, 54; Univ Va, MChE, 55; Carnegie Inst Technol, PhD(chem eng), 59. *Prof Exp:* From asst prof to prof chem eng, Univ Ala, Tuscaloosa, 59-88; RETIRED. *Concurrent Pos:* Consult, US Army Missile Command, 61- & US Bur Mines, 62-; Ford Found prof, Carnegie Inst Technol, 64. *Mem:* Am Inst Chem Engrs; Am Soc Eng Educ. *Res:* Combustion instability; thermodynamics of solution; mass transfer in fixed beds; mass and energy transfer across living cell walls. *Mailing Add:* 21 High Forest Tuscaloosa AL 35406

LANDIS, EDWARD EVERETT, b Marion, Kans, June 15, 07; m 34; c 2. PSYCHIATRY. *Educ:* NCent Col, Ill, BA, 28; Northwestern Univ, MD, 34; Am Bd Psychiat & Neurol, dipl, 46. *Prof Exp:* Nat Comt Ment Hyg fel child psychiat, Louisville Ment Hyg Clin, 37-38; instr, 38-39, from asst prof to prof, 41-75, actg chmn dept, 73, vchmn dept, 74-75, PROF EMER PSYCHIAT, SCH MED, UNIV LOUISVILLE, 75- *Concurrent Pos:* Rockefeller fel neurol & res, Johnson Found, Univ Pa, 39-40; founder & dir dept electroencephalog, Sch Med, Univ Louisville & dept psychiat, Louisville & Jefferson County Children's Home, 40-47, mem gov bd, 51-; clin dir psychiat, Louisville Gen Hosp, 42-45, med dir outpatient clin, 45-49; asst dir psychiat, Louisville Ment Hyg Clin, 42-51 & Norton Mem Infirmary, 45-; med dir psychiat, Norton Psychiat Clin, 49-75; psychiat consult, Louisville Vet Admin, Kosair Crippled Children's, Jewish & Our Lady of Peace Hosps. *Mem:* Fel Am Col Physician; fel Am Psychiat Asn; Am Acad Child Psychiat; AMA. *Res:* Grantham type pre-frontal lobotomy; D-Lysergic acid diethylamide; vitamin B-complex in alcohol addiction. *Mailing Add:* 13908 US 60 Anchorage KY 40245

LANDIS, FRED, b Munich, Ger, Mar 21, 23; nat US; wid; c 3. MECHANICAL ENGINEERING. *Educ:* McGill Univ, BE, 45; Mass Inst Technol, SM, 49, ScD, Mass Inst Technol, 50. *Prof Exp:* Design engr, Can Vickers, Ltd, 45-47; asst, Mass Inst Technol, 48-50; asst prof mech eng, Stanford Univ, 50-52; thermodyn res engr, Northrop Aircraft, Inc, 52-53; from asst prof to prof mech eng, NY Univ, 61-73, chmn dept, 63-73; prof & dean intercampus progs, Polytech Univ,NY, 73-74; dean, Col Eng & Appl Sci, 74-83, PROF MECH ENG, UNIV WIS-MILWAUKEE, 84- *Concurrent Pos:* Mem, NY State Comn on Primary & Secondary Educ, 71; staff consult, Pratt & Whitney Aircraft, 57-88; bd gov, Am Soc Mech Engrs, 89- *Honors & Awards:* Centennial Medallion, Am Soc Mech Engrs, 80. *Mem:* Hon mem Am Soc Mech Engrs (vpres, 85-89); assoc fel Am Inst Aeronaut & Astronaut; fel Am Soc Eng Educ. *Res:* Thermodynamics; fluid mechanics; heat transfer; manpower economics. *Mailing Add:* Col Eng & Appl Sci Univ Wis PO Box 784 Milwaukee WI 53201

LANDIS, JAMES NOBLE, mechanical engineering, steam power plants; deceased, see previous edition for last biography

LANDIS, JOHN W, b Kutztown, Pa, Oct 10, 17; m 41; c 2. NUCLEAR POWER SAFETY. *Educ:* Lafayette Col, BS, 39. *Hon Degrees:* ScD, Lafayette Col. 60. *Prof Exp:* Res Engr, Eastman Kodak Co, Rochester, NY, 39-43; officer, US Navy, 43-46, consult, Bur Ordnance, 46-50, head sci & eng test develop, Educ Testing Serv, 48-50; proj and reactor engr, Atomic Energy Comn, 50-53; dir customer rel, atomic energy div, Babcock & Wilcox Co, 53-55, mgr, opers, Lynchburg, Va, 55-62 & atomic energy div, 62-65, gen mgr, opers, Wash, 65-68; group vpres, Gulf Gen Atomic Co, 68-70, pres, 70-75; SR VPRES & DIR, STONE & WEBSTER ENG CO, 75- *Concurrent Pos:* Chmn adv domt on isotopes radiation development & four other adv comts, US Atomic Comn, 57-70; vchmn mgt comt, Nat Environ Studies Prof, 74-; mem, fusion adv panel, US House Rep, 79-; dir, Cent Fidelity Banks, 79-; charter mem, magnetic fusion adv comn, US Dept Energy, 82-84, vchmn, energy res adv bd, 84-; chmn, Comt Protection Environ, US Energy Asn & Energy Res Adv Bd; NAm regional coordr, World Energy Coun; mem, Nat Energy Outlook Comt & Secy Energy Adv Bd, Dept Energy. *Honors & Awards:* George Washington Kidd Award, 84. *Mem:* Nat Acad Eng; fel Am Nuclear Soc (pres, 71-72); fel Am Soc Mech Engrs; Am Nat Standards Inst (pres, 74-77); Am Soc Macro-Eng (pres, 85-); Fusion Power Assocs; Sigma Xi. *Res:* Color sensitometry; guided missiles; radiation detection; advanced nuclear reactors; environmental protection; alternative energy systems; published numerous articles in various journals. *Mailing Add:* Stone & Webster Eng Corp 245 Summer St PO Box 2325 Boston MA 02107

LANDIS, PHILLIP SHERWOOD, b York, Pa, July 29, 22; m 44; c 2. ORGANIC CHEMISTRY. *Educ:* Franklin & Marshall Col, BS, 43; Univ Ky, MS, 47; Northwestern Univ, PhD(chem), 58. *Prof Exp:* Chemist, Cities Serv Refining Corp, 43-45; res chemist, Mobil Oil Corp, 47-63, res assoc, 63-66, sr res assoc, 66-69, mgr prod res group, Mobil Res & Develop Corp, 69-83; ADJ PROF, GLASSBORO STATE COL, 81- *Concurrent Pos:* Consult, Mobil Res & Develop, 83-85, Int Lubricants Inc, 86. *Honors & Awards:* Outstanding Sci Award, Am Chem Soc, 86. *Mem:* Am Chem Soc. *Res:* Mechanisms and kinetics of organic reactions; pyrolysis of organic compounds; organo-sulfur compounds; petrochemicals; radical reactions; chemistry of jojoba oil. *Mailing Add:* Chem Dept Glassboro State Col Glassboro NJ 08028

LANDIS, STORY CLELAND, b New York, NY, May 14, 45; m 69. NEUROBIOLOGY. *Educ:* Wellesley Col, BA, 67; Harvard Univ, MA, 70, PhD(biol), 73. *Prof Exp:* NIH fel neuropath, 73-75, RES FEL NEUROBIOL, HARVARD MED SCH, 75-, INSTR, 80-; ASSOC PROF, DEPT PHARMACOL, CASE WESTERN RES UNIV,- *Mem:* Am Asn Anatomists; Soc Neurosci; Am Soc Cell Biol. *Res:* Developmental neurobiology; cell biology. *Mailing Add:* Ctr Neurosci Case Western Reserve Univ Cleveland OH 44106

LANDIS, VINCENT J, b Minneapolis, Minn, Oct 27, 28; m 50; c 6. INORGANIC CHEMISTRY. *Educ:* Wash State Univ, BS, 50; Univ Minn, PhD(inorg chem), 57. *Prof Exp:* From instr to assoc prof, 54-65, PROF CHEM, SAN DIEGO STATE UNIV, 65- *Concurrent Pos:* Richland fac fel, Univ Wash, 64-65. *Mem:* Am Chem Soc. *Res:* Metal coordination compounds; radiochemistry. *Mailing Add:* Dept Chem San Diego State Univ San Diego CA 92182

LANDIS, WAYNE G, b Washington, DC, Jan 20, 52. AQUATIC TOXICOLOGY, METABOLISM OF XENOBIOTICS. *Educ:* Wake Forest Univ, BA, 74; Ind Univ, MA, 78, PhD(zool), 79. *Prof Exp:* Assoc instr biol, dept biol, Ind Univ, 74-79; environ & health scientist, environ & health studies group, Franklin Res Ctr, 79-82; RES BIOLOGIST, TOXICOL DIV, CHEM RES & DEVELOP CTR, 82- *Mem:* Genetics Soc Am; Soc Protozoologists; Am Soc Testing & Mat; Soc Study Evolution; AAAS; Sigma Xi; Soc Environ Toxicol & Chem. *Res:* Toxicity, fate and impact on community structure of environmental toxicants on aquatic systems; characterization of the DFPases in Tetrahymena thermophila; ecology and evolution of paramecium; structure-activity derivations for toxicologic endpoints. *Mailing Add:* 4158 Ridgewood Ave Bellingham WA 98226

LANDISS, DANIEL JAY, b Alton, Ill, June 11, 43. ANALOG ELECTRONICS, COMPUTERS IN EDUCATION. *Educ:* Wash Univ, St Louis, BS, 64, MS, 66. *Prof Exp:* Res engr, Mallinckrodt Inst Radiol, 66-67; dir, Eng Tech Serv, Wash Univ, 67-69 & Biomed Eng Lab, Lewis-Howe Co, 69-72; chief engr, Artronix, Inc, 72-73; PROF TECHNOL, ST LOUIS COMMUNITY COL, 74- *Concurrent Pos:* Pres & owner, Quest Instruments Ltd, 80-; assoc, Senne, Kelsey & Assocs, 85-; vis prof, Czech Tech Univ, Prague, 88-89. *Res:* Biomedical ultrasound diagnostics. *Mailing Add:* 5053 Westminster Pl St Louis MO 63108

LANDMAN, ALFRED, b Vienna, Austria, June 29, 33; US citizen; m 67; c 2. ATOMIC PHYSICS. *Educ:* Univ Pa, AB, 54; Columbia Univ, PhD(physics), 63. *Prof Exp:* Lectr physics, Brooklyn Col, 59-62; res assoc, Columbia Univ, 63; res assoc chem physics, Inst Study Metals, Univ Chicago, 63-64; res engr physics, Gen Tel & Electronics Labs, 65-66, adv res engr, 66-67; physicist, NASA Electronics Res Ctr, 67-70; GEN ENGR, TRANSP SYSTS CTR, US DEPT TRANSP, 70- *Mem:* Am Phys Soc. *Res:* Spectroscopy; optical pumping; gas laser molecular stark effect; synthetic fuels. *Mailing Add:* 29 Tyler Rd Lexington MA 02173

LANDMAN, DONALD ALAN, b New York, NY, Apr 23, 38; m 70; c 2. THEORETICAL & COMPUTATIONAL PLASMA, ATOMIC PHYSICS. *Educ:* Columbia Univ, AB, 59, MA, 61, PhD(physics), 65. *Prof Exp:* Asst prof physics, NY Univ, Bronx, 65-69; res scientist, Cornell Aeronaut Lab, Buffalo, 70 & Advan Res Instrument Systs Inc, 71; from assoc astron to astron, Inst Astron, Univ Haw, 72-85; RES PHYSICIST, MISSION RES CORP, 85- *Mem:* NY Acad Sci; Am Phys Soc; Int Astron Union. *Res:* Theoretical and computational plasma and atomic physics applied to atmospheric problems; research in solar physics. *Mailing Add:* Mission Res Corp 735 State St PO Drawer 719 Santa Barbara CA 93102

LANDMAN, OTTO ERNEST, b Mannheim, Ger, Feb 15, 25; nat US; m 48; c 3. MICROBIAL GENETICS. *Educ:* Queens Col, BS, 47; Yale Univ, MS, 48, PhD(microbiol), 51. *Prof Exp:* USPHS fel, Calif Inst Technol, 51-52; res assoc bact, Univ Ill, 53-56; chief microbial genetics br, US Army Biol Labs, Ft Detrick, 56-61, sr investr, 61-63; assoc prof biol, Georgetown Univ, 63-66, prof, 66-90; RETIRED. *Concurrent Pos:* NIH spec fel, Ctr Molecular Genetics, Nat Ctr Sci Res, Gif-Sur-Yvette, France, 68-69; vis investr, Nat Inst Med Res, Mill Hill, London, 75-76. *Mem:* AAAS; Am Soc Microbiol; Genetics Soc Am. *Res:* Protoplasts and L forms of bacteria; cell division; wall biosynthesis and transformation in bacteria; phage infection of protoplasts; gene expression in dicaryotic bacterial system; inheritance of acquired characteristics. *Mailing Add:* Dept Biol Georgetown Univ Washington DC 20057

LANDMANN, WENDELL AUGUST, b Waterloo, Ill, Dec 29, 19; m 44; c 2. BIOCHEMISTRY. *Educ:* Univ Ill, BS, 41; Purdue Univ, MS, 44, PhD(biochem), 51. *Prof Exp:* Chemist, US Naval Res Lab, 43-46; asst chem, Purdue Univ, 46-51; res chemist, Armour & Co, 51-55; assoc biochem, Argonne Nat Lab, 55-57; chief div anal & phys chem, Am Meat Inst Found, Ill, 57-64; prof animal sci, biochem & nutrit, Tex A&M Univ 64-70, King Ranch chair, 64-72, prof biochem & biophys, 70-85, prof animal sci & food sci, 78-85; RETIRED. *Mem:* AAAS; Am Chem Soc; Inst Food Technol; Am Meat Sci Asn; Am Inst Nutrit. *Res:* Lysosomal enzymes; human nutrition; protein chemistry; chemistry of muscle tissues; nutrient bioavailability. *Mailing Add:* 1602 Dominik Dr College Station TX 77843

LANDMESSER, LYNN THERESE, b Santa Ana, Calif, Nov 30, 43. PHYSIOLOGY, NEUROBIOLOGY. *Educ:* Univ Calif, Los Angeles, BA, 65, PhD(neurophysiol), 69. *Prof Exp:* NIH fel, Physiol Dept, Col Med, Univ Utah, 69-71, Dept Regulatory Biol, Univ Conn, 71-72; from asst prof to prof biol, Dept Biol, Yale Univ, 72-83; prof, Physiol Sect, Biol Sci Group, 83-85, PROF, DEPT PHYSIOL & NEUROBIOL, UNIV CONN, STORRS, 85- *Concurrent Pos:* NIH grant, 72-85, NSF grant, 88-91; assoc ed, Develop Biol, 77-81, J Neurosci, 81-85; mem, Sci Adv Comt, Nat Spinal Cord Injury Found, 79-87; mem, NIH Study Sect Neurol B, 82-85, Social Issues Comt, Soc Neurosci, 82-85; Grass Found vis scientist, Anat Dept, Emory Univ, Ga, 84; Jacob Javits investr award, 85-92; Arturo Rosenblueth distinguished prof, Ctr Advan Studies, Mexico City, 87; dir, NIH Postdoctoral Training Grant Neurosci, 88-; Wiersma vis prof neurosci, Calif Inst Technol, 89; mem sci adv bd, Nat Inst Child Health & Develop, NIH, 89- *Honors & Awards:* Yntema Mem Lectr, State Univ NY, Syracuse, 81; Twelfth Ann Trotter Lectr, Dept Anat & Neurobiol, Washington Univ, 87. *Mem:* Soc Neurosci; Am Physiol Soc; Soc Develop Biol (pres, 88-89); fel AAAS. *Res:* Neurophysiology. *Mailing Add:* Dept Physiol & Neurobiol Univ Conn Box U-42 75 N Eagleville Rd Storrs CT 06269-3042

LANDO, BARBARA ANN, b Elizabeth, NJ, Dec 7, 40; m 65. ALGEBRA. *Educ:* Georgian Court Col, BA, 62; Rutgers Univ, New Brunswick, MS, 64, PhD(math), 69. *Prof Exp:* Instr math, Douglass Col, Rutgers Univ, New Brunswick, 69; from asst prof to assoc prof, 79-82, PROF MATH, UNIV ALASKA, FAIRBANKS, 82- *Mem:* Am Math Soc; Math Asn Am; Inst Elec & Electronics Engrs; Asn Comput Mach. *Res:* Differential algebra; formal language theory. *Mailing Add:* Dept Math Univ Alaska Fairbanks AK 99775

LANDO, JEROME B, b Brooklyn, NY, May 23, 32; m 62; c 3. POLYMER SCIENCE. *Educ:* Cornell Univ, BA, 53; Polytech Inst Brooklyn, PhD(chem), 63. *Prof Exp:* Fel, Polytech Inst Brooklyn, 63; res chemist, Camille Dreyfus Lab, Res Triangle Inst, 63-65; asst prof polymer sci & eng, 65-68, assoc prof, 68-74, chmn dept, 78-85, PROF MACROMOLECULAR SCI, CASE WESTERN RESERVE UNIV, 74- *Concurrent Pos:* Humboldt Found Sr Am Scientist Award, 74; vis prof, Univ Mainz, 74; mem adv bd, J Molecular Electronics, Mat Lett & Polymers Advan Technol. *Mem:* Am Chem Soc; Am

Crystallog Asn; Am Phys Soc; Sigma Xi. *Res:* Polymer physical chemistry; solid state reactions, especially polymerization reactions and polymer crystal structure; synthesis of stereoregular polymers, pyroelectric and piezoelectric polymers; electronic properties of polymers and thin film. *Mailing Add:* Dept Macromolecular Sci Case Western Reserve Univ Cleveland OH 44106

LANDOLFI, NICHOLAS F, b Ashtabula, Ohio, Sept 2, 55; m 88. MOLECULAR IMMUNOLOGY. *Educ:* Ohio State Univ, BS, 77; Miami Univ, MS, 80; Univ Tex, PhD(immunol), 84. *Prof Exp:* Res fel, Univ Tex Southwestern Med Sch, 85-88 & Leukemia Soc Am, 87-88; STAFF SCIENTIST, PROTEIN DESIGN LABS, 88- *Mem:* Am Asn Immunologists; AAAS; Sigma Xi. *Res:* Structure and function analysis of molecules involved in immune response; differential control of genes in the cells of the immune system. *Mailing Add:* Protein Design Labs 2375 Garcia Ave Mountain View CA 94043

LANDOLL, LEO MICHAEL, b Cleveland, Ohio, Oct 11, 50; m 71; c 2. POLYMER CHEMISTRY. *Educ:* Kent State Univ, BA, 70; Univ Del, MBA, 82; Univ Akron, PhD(polymer sci), 75. *Prof Exp:* Res chemist, 74-79, sr res chemist polymer synthesis, 79-82, res scientist, 82-87, PROJ LEADER, HERCULES INC, 87- *Mem:* Am Chem Soc. *Res:* Polymer synthesis and processing related to thermoplastic and thermoset systems, fibers, films, property correlations; synthesis and structure property relations of natural and synthetic water soluble polymers; steric stabilization of particles in suspension. *Mailing Add:* Hercules Res Ctr Hercules Inc Wilmington DE 19894

LANDOLPH, JOSEPH RICHARD, II, b Upper Darby, Pa, Nov 9, 48; m 80; c 2. BIOCHEMISTRY. *Educ:* Drexel Univ, BS, 71; Univ Calif, Berkeley, PhD(chem), 76. *Prof Exp:* Res asst chem, Dept Chem, Univ Calif, Berkeley, 71-76; res fel chem carcinogenesis, Comprehensive Career Ctr, 77-80, asst prof res path, 80-82, asst prof, microbiol & path, 82-87, ASSOC PROF MICROBIOL, PATH TOXICOL, SCH MED, UNIV SOUTHERN CALIF, 87- *Concurrent Pos:* Prin investr, Nat Cancer Inst, Nat Inst Environ Health Sci & Environ Protection Agency grants, 83-; ad hoc reviewer, Nat Inst Environ Health Sci & Environ & Dept Energy, 85; mem, grant rev bd, Environ Protection Agency, 85-88; consult, Am Petrol Inst, 85-; Howard Hughes fels, Nat Res Coun, 88. *Mem:* Am Soc Biochem & Molecular Biol; Am Asn Cancer Res; Am Chem Soc; Environ Mutagen Soc; Am Soc Cell Biol; Soc Toxicologists. *Res:* Mechanisms of chemical carcinogenesis studied in cultured mammalian cells; molecular biology of oncogene activation caused by chemical carcinogens; human carcinogenesis. *Mailing Add:* Norris Cancer Hosp & Res Inst Rm 517 Univ Southern Calif Sch Med 1441 Eastlake Ave Los Angeles CA 90033

LANDOLT, ARLO UDELL, b Highland, Ill, Sept 29, 35; m 66; c 5. ASTRONOMY. *Educ:* Miami Univ, BA, 55; Ind Univ, MA, 60, PhD(astron), 62. *Prof Exp:* Scientist aurora & airglow, US Int Geophys Year Comt, 56-58; from asst prof to assoc prof, 62-68, PROF PHYSICS & ASTRON, LA STATE UNIV, BATON ROUGE, 68-, DIR OBSERV, 70- *Concurrent Pos:* Mem first wintering-over party, Int Geophys Year Amundson-Scott S Pole Sta, Antarctica, 57; Grad Res Coun res grants, La State Univ, Baton Rouge, 64-76; res grants, NSF, 64-77 & Res Corp, 65 & NASA, 65-67; secy, sect D Astron, AAAS, 70-78, US Nat Comt, Int Astron Union, 80-89; Air Force Off Sci Res, grants, 77-87; prog dir, NSF, Washington, DC, 75-76; guest investr, Dyer Observ, Vanderbilt Univ, Goethe Link Observ, Ind Univ, Kitt Peak Nat Observ, Cerro Tololo Inter-Am Observ & Las Campanas Observ, La Serena, Chile; Space Telescope Sci Inst grant, 85-90; pres fac senate, La State Univ, Baton Rouge, 79-80, actg chmn, dept physics & astron, 72 & 73. *Mem:* Fel AAAS; Int Astron Union; Am Astron Soc (secy, 80-89); Royal Astron Soc Eng; Am Polar Soc; Sigma Xi; Explorer's Club. *Res:* Photographic and photoelectric narrow band phoeoelectric and spectroscopic investigations of star clusters, variable stars, and eclipsing binaries; standard photometric systems; variable stars; galactic structure. *Mailing Add:* Dept Physics & Astron La State Univ Baton Rouge LA 70803

LANDOLT, JACK PETER, b Zurich, Switz, Mar 17, 34; Can citizen; m 64; c 5. SPATIAL DISORIENTATION, BIODYNAMICS. *Educ:* Univ Ottawa, BASc, 59, MSc, 62; Iowa State Univ Sci & Technol, PhD(elec eng), 68. *Prof Exp:* Defence scientist oper res, Can Army Oper Res Estab, 61-65; defence scientist vestibular physiol, 68-75, group head motion sickness biodynamics, 75-76, sect head disorientation biodynamics, 76-80, sect head biophys, 80-86, SECT HEAD, AEROSPACE PHYSIOL, DEFENCE & CIVIL INST ENVIRON MED, 86- *Concurrent Pos:* Mem, Can Adv Comt, Int Standardization Org, 77-89; Aerospace Med Panel, Adv Group, Aerospace Res & Develop, NATO, 78- *Mem:* Soc Neurosci; Barany Soc. *Res:* Neurobiology of peripheral vestibular apparatus; space adaptation syndrome and circular vection; vestibular-visual interactions; motion cues in simulation sickness; impact protection of the human; author or coauthor of over 85 publications; high altitude and high acceleration physiology. *Mailing Add:* Defence & Civil Inst Environ Med 1133 Sheppard Ave W Downsview ON M3M 3B9 Can

LANDOLT, MARSHA LAMERLE, b Houston, Tex, Jan 19, 48. FISH PATHOLOGY, TOXICOLOGY. *Educ:* Baylor Univ, BS, 69; Univ Okla, MS, 70; George Washington Univ, PhD(path), 76. *Prof Exp:* Asst prof, 75-79, asst dir, 80-83, assoc prof, 79-86, ASSOC DEAN, 83-, PROF FISHERIES, SCH FISHERIES, UNIV WASH, 86- *Concurrent Pos:* Histopathologist, Eastern Fish Dis Lab, US Dept Interior, Leetown, WVa, 70-74; path clerk, Dept Animal Health, Nat Zool Park, Smithsonian Inst, Washington, DC, 74-75. *Res:* Development of in vitro and in vivo test systems for use in genotoxic research; fish pathology. *Mailing Add:* Sch Fisheries Univ Wash Seattle WA 98195

LANDOLT, PAUL ALBERT, b Shubert, Nebr, July 10, 12; m 35; c 1. PHYSIOLOGY. *Educ:* Nebr State Teachers Col, Peru, BA, 33; Univ Nebr, MS, 51, PhD(zool, physiol), 60. *Prof Exp:* Teacher, High Schs, Nebr, 36-42; field dir mil welfare, Am Red Cross, Mariannas Islands, 42-46; instr biol sci,

Scottsbluff Jr Col, Nebr, 46-53; from instr to prof 53-77, EMER PROF PHYSIOL, UNIV NEBR, LINCOLN, 77- *Concurrent Pos:* Instr, Southeast Community Col, Lincoln, 77-88. *Mem:* AAAS; Am Soc Cell Biol. *Res:* Vertebrate physiology; tissue culture; problems related to effects of air pollutants on lung tissue. *Mailing Add:* 431 Glenhaven Dr Lincoln NE 68505

LANDOLT, ROBERT GEORGE, b Houston, Tex, Apr 4, 39; m 62; c 3. ORGANIC CHEMISTRY. *Educ:* Austin Col, BA, 61; Univ Tex, PhD(org chem), 65. *Prof Exp:* Res assoc org chem, Univ Ill, 65-67; from asst prof to assoc prof org chem, Muskingum Col, 67-80, chmn dept, 71-74; sr scientist, Radian Corp, 80-81; chmn, chem dept, 81-85, PROF, TEX WESLEYAN UNIV, 81-, CHMN, CHEM DEPT, 90- *Concurrent Pos:* Resident consult, Columbus Labs, Battelle Mem Inst, 74-75; consult & contractor, US Navy, 82-84; Cong fel, Am Chem Soc, 86-87. *Mem:* AAAS; Am Asn Univ Prof; Am Chem Soc. *Res:* Abnormal claisen rearrangement and reactions in aprotic polar solvents; nitroso aromatic compounds; oxidation of coal and coal model compounds; computer assisted information retrieval; origin of organic pollutants in water; hypochlorite catalyzed reactions. *Mailing Add:* Dept Chem Tex Wesleyan Univ Ft Worth TX 76105

LANDOLT, ROBERT RAYMOND, b Sherman, Tex, May 11, 37; m 70; c 3. HEALTH PHYSICS. *Educ:* Austin Col, BA, 59; Univ Kans, MS, 61; Purdue Univ, PhD(bionucleonics), 68. *Prof Exp:* Reactor health physicist, Phillips Petrol Co, Idaho, 61-64; from instr to assoc prof, 64-81, PROF HEALTH SCI, PURDUE UNIV, WEST LAFAYETTE, IND, 81- *Mem:* Health Physics Soc. *Res:* Isotopic neutron sources; low-level radioactive waste management. *Mailing Add:* Sch Health Sci Purdue Univ West Lafayette IN 47906

LANDON, ERWIN JACOB, b Cleveland, Ohio, Jan 22, 25; m 65. BIOCHEMISTRY. *Educ:* Univ Chicago, BS, 45, MD, 48; Univ Calif, PhD(biochem), 53. *Prof Exp:* Intern, Harper Hosp, Detroit, Mich, 48-49; asst prof, 59-67, ASSOC PROF PHARMACOL, SCH MED, VANDERBILT UNIV, 67- *Concurrent Pos:* Sr res fel pharmacol, Sch Med, Yale Univ, 57-59. *Mem:* Am Chem Soc; Am Soc Pharmacol & Exp Therapeut; Sigma Xi. *Res:* Biochemistry of renal transport; cell calcium regulation. *Mailing Add:* Dept Pharmacol Vanderbilt Univ Nashville TN 37232

LANDON, JOHN CAMPBELL, b Hornell, NY, Jan 3, 37; m 58; c 4. VIROLOGY, CANCER. *Educ:* Alfred Univ, AB, 59; George Washington Univ, MS, 62, PhD(biol), 67. *Prof Exp:* Biologist, Nat Cancer Inst, 60-65; head virol, Litton Bionetics, Inc, 65-68, dir dept virol & cell biol, 68-71, dir spec prog develop, 71-72, dir sci, Frederick Cancer Res Ctr, 72-75; pres, Mason Res Inst, 75-82; PRES, BIOQUAL INC, 82-; PRES, SEMA INC, 85- *Concurrent Pos:* Pres & chief exec officer, Diagnow Corp, 86- *Mem:* AAAS; Tissue Cult Asn; Am Soc Cell Biol; NY Acad Sci; Am Soc Microbiol. *Res:* Viral oncology; tissue culture; general human and simian virology; cell biology; environmental biology. *Mailing Add:* 8213 Raymond Ln Potomac MD 20854

LANDON, ROBERT E, b Chicago, Ill, June 1, 05; m 35. GEOLOGY. *Educ:* Univ Chicago, SB, 26, PhD(geol), 29. *Prof Exp:* Instr, YMCA Col, 28-29; geologist, Anaconda Copper Mining Co, 29-30; asst geologist, US Geol Surv, 30-31; instr geol, Colo Col, 31-33; consult mining geologist, 33-39; sr mining securities analyst, US Securities & Exchange Comn, 40-45; geologist, Mobil Oil Corp, NY, 45-70; geol consult, 70-81; RETIRED. *Mem:* Fel Geol Soc Am; Am Asn Petrol Geol. *Mailing Add:* 3460 S Race St Englewood CO 80110

LANDON, SHAYNE J, b Alexandria Bay, NY, July 21, 54; m 79. ORGANOSILICON & ORGANOMETALLIC CHEMISTRY. *Educ:* State Univ NY, Plattsburgh, BA, 76, MA, 79; Univ Del, PhD, 84. *Prof Exp:* Res assoc, Pa State Univ, 84-85; sr chemist, Spec Chems Div, 85-90, PROJ SCIENTIST, UNION CARBIDE CORP, TARRYTOWN, 90- *Mem:* Am Chem Soc; Sigma Xi; NY Acad Sci. *Res:* Organosilicon chemistry; silicone surfactants; urethanes; spectroscopy; reaction mechanisms and kinetics. *Mailing Add:* Union Carbide Corp Old Saw Mill River Rd Tarrytown NY 10591

LANDOR, JOHN HENRY, b Canton, Ohio, Sept 30, 27; m 53; c 6. SURGERY. *Educ:* Univ Chicago, PhB, 48, MD, 53. *Prof Exp:* Instr surg, Sch Med, Univ Chicago, 58; from instr to prof, Sch Med, Univ Mo-Columbia, 59-69; prof, Col Med, Univ Fla, 69-72; PROF & CHIEF GEN SURG, COL MED & DENT NJ, RUTGERS MED SCH, 72-; CHIEF, DEPT SURG, RARITAN VALLEY HOSP, GREEN BROOK, 73- *Concurrent Pos:* Commonwealth Found fel, Royal Postgrad Med Sch, London, 66-67. *Mem:* Am Col Surgeons; Soc Univ Surgeons; Am Gastroenterol Asn; Soc Surg Alimentary Tract; Int Soc Surgeons. *Res:* Physiology of the stomach. *Mailing Add:* Brooklyn Vet Admin Med Ctr 800 Poly Pl Brooklyn NY 11209

LANDOWNE, DAVID, b Chicago, Ill, Dec 26, 42; m 66; c 2. PHYSIOLOGY, BIOPHYSICS. *Educ:* Mass Inst Technol, BS, 63; Harvard Univ, PhD(physiol), 68. *Prof Exp:* Res assoc pharmacol, Sch Med, Yale Univ, 68-72; asst prof physiol, 72-75, ASSOC PROF PHYSIOL, SCH MED, UNIV MIAMI, 75- *Concurrent Pos:* Grass Found fel, 70; NSF fel, Univ London, 70-71. *Mem:* Biophys Soc; Soc Gen Physiol. *Res:* Excitable membranes; ion movements and optical measurements of the movement of excitable molecules. *Mailing Add:* Dept Physiol & Biophys Univ Miami Sch Med PO Box 016430 Miami FL 33101

LANDOWNE, MILTON, b New York, NY, Nov 19, 12; m 41; c 5. INTERNAL MEDICINE, CIRCULATORY PHYSIOLOGY. *Educ:* City Col New York, BS, 32; Harvard Univ, MD, 36. *Prof Exp:* Intern, Mt Sinai Hosp, New York, 36-39; Libman fel, Michael Reese Hosp, Chicago, 39-41; instr, Univ Chicago, 41-46, asst prof med, 46-48; chief cardiovasc res unit, Vet Admin Hosp, 48-49; assoc chief geront sect, Nat Heart Inst, 49-57; med dir, Levindale Hebrew Home & Infirmary, Baltimore, Md, 57-65; dir med lab, US Army Res Inst Environ Med, 65-76, med adv, 76-85; RETIRED. *Concurrent*

Pos: Asst prof, Johns Hopkins Univ, 55-65; head div cardiol & chronic dis, Sinai Hosp, Baltimore, 58-65; asst clin prof, Harvard Univ, 65-74. *Mem:* AAAS; Am Soc Clin Invest; Am Physiol Soc; Soc Exp Biol & Med; Am Heart Asn. *Res:* Disorders of the circulation; biology aging; clinical medicine; physiology of blood and circulation; metabolic and renal diseases; environmental medicine. *Mailing Add:* 67 Woodchester Dr Weston MA 02193

LANDRETH, KENNETH S, b Galax, Va, Aug 22, 47; m 76; c 2. ANATOMY. *Educ:* Univ Wash, Seattle, PhD(biol struct), 80. *Prof Exp:* Sr res scientist, Okla Med Res Found, 82-85; assoc prof, microbiol, 85-91, PROF MICROBIOL & IMMUNOL, SCH MED, WVA UNIV, 91- *Concurrent Pos:* Res fel, Sloan-Kettering Cancer Ctr; mem, Mary Babb Randolph Cancer Ctr, WVa, 90- *Mem:* Am Asn Immunologists. *Res:* Regulation of B lymphocyte production in the bone marrow; lemopoiesis. *Mailing Add:* Sch Med WVa Morgantown WV 26506

LANDRETH, RONALD RAY, b Mattoon, Ill, June 15, 49; m 71; c 3. ENVIRONMENTAL CHEMISTRY. *Educ:* Northwestern Univ, BA, 71; Pa State Univ, PhD(chem), 75. *Prof Exp:* Res fel chem, Pa State Univ, 72-75; from res engr to sr res engr, Inland Steel Co, 75-85, group leader environ res, 79-85, sr environ consult, 85-86, scientist, 86-89, PROJ MGR BOILER LIMESTONE INJECTION, INLAND STEEL CO, 84-; PROG MGR, INLAND STEEL INDUSTS, 89- *Mem:* Air Pollution Control Asn; Am Chem Soc; Am Iron & Steel Inst; Water Pollution Control Fedn. *Res:* Characterization and control of air, water and solid pollutants in an industrial area; development of businesses based upon these and other opportunities. *Mailing Add:* New Ventures Dept Inland Steel 3001 E Columbus Dr E Chicago IN 46312

LANDRIGAN, PHILIP J, b Boston, Mass, Jun 14, 42; m 76; c 3. ENVIRONMENTAL EPIDEMIOLOGY, OCCUPATIONAL MEDICINE. *Educ:* Harvard Med Sch, MD, 67; Univ London MSc, 77. *Prof Exp:* Chief environ hazards activ, CTR Disease Control, 70-73; dir, surveillance hazard & eval, Occup epidemiol, Nat Inst Occup Safety & Health, 79-85; PROF ENVIRON MED, MT SINAI SCH MED, 85- *Concurrent Pos:* Chair comt environ hazard, Am Acad Pediat, 83-87; vchair, bd environ sci & toxicol, Nat Acad Sci, 81-86. *Mem:* NY Acad Sci; Am Epidemiol Soc; Royal Soc Med; Am Acad Pediat; Am Pub Health Asn; Am Occup Med Asn. *Res:* Occupational and environmental epidemiology; Lema poisoning; occupational respiratory diseases; reproductive dysfunction and neurotoxicology. *Mailing Add:* Dept Commun Med Box 1058 Mt Sinai Sch Med New York NY 10029

LANDROCK, ARTHUR HAROLD, b New York, NY, May 19, 19; m 42; c 5. PLASTICS, MATERIALS ENGINEERING. *Educ:* Queens Col, BS, 41; Boston Univ, AM, 50. *Prof Exp:* Chemist packaging res, Paper Containers Div, Continental Can Co, 45-47; res asst & tech asst food packaging, Mass Inst Technol, 47-53; sr food packaging scientist, Film Div, Olin Mathieson Chem Corp, 53-55, sr res chemist, viscose res, Gen Res Orgn, 55-56; prod standards mgr, M&M'S Candies, Div Food Mfrs, 56-60; MAT ENGR, DEPT DEFENSE PLASTICS TECH EVAL CTR, ARMAMENT RES, DEVELOP & ENG CTR, US ARMY, 61- *Concurrent Pos:* US deleg & mem tech adv group plastics, Int Orgn Standardization, 77-; mem, Comt Terminology, Am Soc Testing & Mat, 83-89, ed reviewer, Standards in Plastics & Adhesives; liaison rep, comt combustion toxicity, Nat Mat Adv Bd, 84-; mem ed bd, J Testing & Eval, 88-, J Plastics & Elastomers, 78- *Honors & Awards:* Ed Excellence Award, Am Soc Testing & Mat, 77, D-20 Award Excellence, 88, Outstanding Achievement Award, 89, Comt Terminology Mem Award, 90. *Mem:* Am Soc Testing & Mat; Int Orgn Standardization; Soc Advan Mat & Process Eng; Sigma Xi; Soc Plastics Engrs; Am Soc Mat Int. *Res:* Plastics and adhesives; flammability of plastics and adhesive bonding; terminology; standardization. *Mailing Add:* Plastics Tech Eval Ctr Bldg 355-N US Army ARDEC Picatinny Arsenal NJ 07806-5000

LANDRUM, BILLY FRANK, b Atlanta, Ga, June 7, 20; m 48; c 3. ORGANIC CHEMISTRY, POLYMER CHEMISTRY. *Educ:* Emory Univ, AB, 47, MS, 49, PhD(chem), 50. *Prof Exp:* Res chemist polymer chem, M W Kellogg Co, 50-53, res supvr pilot plant, 53-57; head polymer sect, Minn Mining & Mfg Co, 57-62; mgr advan projs, FMC Corp, NJ, 62-66; staff scientist, Whittaker Corp, 66-67; staff scientist, Com Develop Dept, Ciba-Geigy Corp, 67-74, mgr, Mkt Develop, Plastics & Additives Div, 74-78, tech serv mgr, comput mat dept, 81-85; RETIRED. *Mem:* Am Chem Soc; Sigma Xi; Soc Advan Mat & Processing Eng. *Res:* Organo-metallic reactions; polymers; fluorocarbons; urethanes; coal and coke; activated carbon; composite materials. *Mailing Add:* 761 Isabella Way Fairfield CA 94533

LANDRUM, LESLIE ROGER, b St Louis, Mo, Dec 1, 46; m 73. SYSTEMATICS, PHYTOGEOGRAPHY. *Educ:* NY State Col Forestry, BS, 69; Univ Mich, MS, 75, PhD(bot), 80. *Prof Exp:* Peace Corp vol, Sch Forestry, Univ Chile, Santiago, 69-73; teaching asst, Div Biol Sci, Univ Mich, Ann Arbor, 73-80; B A Krukoff res assoc, NY Bot Garden, Bronx, 80-83; Tilton fel, Calif Acad Sci, San Francisco, 83-86; vis lectr, San Francisco State Univ, 85 & Univ Calif, Berkeley, 86; RES SCIENTIST & HERBARIUM CUR, DEPT BOT, ARIZ STATE UNIV, TEMPE, 86- *Concurrent Pos:* Grants, numerous insts & schs, 77-90; comn mem, Orgn Flora Neotropica, 86-; co-ed, Vascular Plants Ariz Proj, 87- & J Ariz-Nev Acad Sci, 91- *Mem:* Am Soc Plant Taxonomists; Int Asn Plant Taxon; Soc Econ Bot. *Res:* Systematics of American Myrtaceae; Flora and phytogeography of temperate South America, especially Chile; analysis of phylogenetic patterns; phytogeography of Southern South America; flora of Arizona; numerical phylogenetic analysis; author of numerous publications. *Mailing Add:* Dept Bot Ariz State Univ Tempe AZ 85287-1601

LANDRUM, RALPH AVERY, JR, b Memphis, Tenn, Oct 2, 26; m 49; c 3. GEOPHYSICS. *Educ:* Rice Inst, BS, 49; Univ Tulsa, MS, 64. *Prof Exp:* Asst seismic observer, Amerada Petrol Corp, 49-51; res seismic observer, Stanolind Oil & Gas Co, 51-56; res engr, Pan Am Petrol Corp, 56-63, staff res engr, 63-67, staff res scientist, 67-71; res assoc, Res Ctr, Amoco Prod Co, 71-74; SR RES GEOPHYSICIST, WESTERN GEOPHYS CO AM, 74- *Mem:* Am Soc Explor Geophys; Inst Elec & Electronics Engrs; Europ Asn Explor Geophys. *Res:* Exploration geophysics; design of seismic instrumentation; mathematics of seismic data processing. *Mailing Add:* 1707 Valley Vista Dr Houston TX 77077

LANDRY, FERNAND, b Levis, PQ, Can, Jan 13, 30; m 55; c 4. EXERCISE PHYSIOLOGY. *Educ:* Univ Ottawa, BSc, 54; Univ Ill, Urbana, MS, 55, PhD(phys educ exercise physiol), 68. *Hon Degrees:* Dr, Univ Ottawa, Can, 90. *Prof Exp:* From teacher-researcher, phys educ to dept head, Univ Ottawa, 55-68; dept head phys activ sci, Laval Univ, 68-81, head phys activ sci lab, 81-84. *Honors & Awards:* Medal, Que Govt, 74, Int Olympic Acad, 81. *Mem:* Can Asn Sport Sci (pres, 71-72); Int Coun Sport Sci & Phys Educ (vpres, 72-); fel Am Col Cardiol; fel Am Col Sports Med. *Res:* Short-term and chronic effects of physical activity and sports; use of physical activity in the prevention of and/or rehabilitation from generative diseases; causes of variations in suseptibility sensitivity to training stimuli; effects of training in patients with prosthetic aortic valves; exercise hypertension. *Mailing Add:* Dept Phys Activ Sci Laval Univ Quebec PQ G1K 7P4 Can

LANDRY, MICHAEL RAYMOND, b Berlin, NH, Apr 16, 48; m 72; c 2. MARINE ZOOPLANKTON, PROTOZOAN ECOLOGY. *Educ:* Univ Calif, Santa Barbara, BA, 70; Univ Wash, PhD(oceanog), 76, MBA(fin, oper mgmt), 86. *Prof Exp:* Res biologist, Scripps Inst Oceanog, Univ Calif, San Diego, 76-78; res asst prof, Univ Wash, 78-83, res assoc prof biol oceanog, Sch Oceanog, 83-87; assoc prof, 87-89, PROF OCEANOG, DEPT OCEANOG, UNIV HAWAII, 89-, ASSOC CHMN, 89- *Concurrent Pos:* Coordr, Northwest Regional Oceanog Prog, US Dept Energy, 82-85. *Mem:* Am Soc Limnol & Oceanog; Western Soc Naturalists; Soc Protozoologists; Intern Soc Copepodiologists; AAAS. *Res:* Feeding behavior and population dynamics of marine zooplankton; marine microbial ecology; marine ecosystem research and modeling; carbon and nitrogen cycling. *Mailing Add:* Dept Oceanog & Hawaii Inst Geophys Univ Hawaii-Manoa Honolulu HI 96822

LANDRY, RICHARD GEORGES, b Manchester, NH, Nov 7, 42; m 66; c 3. APPLIED STATISTICS. *Educ:* Oblate Col, BA, 64; Boston Col, MEd, 67, PhD(res & statist), 70. *Prof Exp:* Asst prof, 69-73, assoc prof measurement & statist, 70-80, PROF, CTR TEACHING & LEARNING, UNIV NDAK, 80- *Concurrent Pos:* Eval auditor, numerous ESEA Title III Projs, 70-; eval consult, Grand Rapids Sch Dist, Minn, 73-; res coordr, Nat Inst Educ Proj, Univ N Dak, 73- *Mem:* Am Asn Univ Prof; Am Educ Res Asn; Am Statist Asn; Nat Coun Measurement Educ. *Res:* Applied educational statistics; educational measurement and evaluation in affective domain; foreign language learning and creativity. *Mailing Add:* Measure & Statist Univ ND Box 8158 Grand Forks ND 58202

LANDRY, STUART OMER, JR, b New Orleans, La, Sept 30, 24; m 50; c 2. ZOOLOGY. *Educ:* Harvard Univ, BS, 49; Univ Calif, PhD(zool), 54. *Prof Exp:* Curatorial asst, Mus Vert Zool, Calif, 50-52; asst zool, Univ Calif, 52-53, assoc, 53-54; from instr to asst prof anat, Univ Mo, 54-59; assoc prof biol & chmn dept, La State Univ, 59-63; actg dean grad sch, 66-68, PROF BIOL, STATE UNIV NY BINGHAMTON, 63- *Mem:* AAAS; Soc Syst Zool; Am Soc Mammal; Am Asn Anat; Am Soc Zoologists. *Res:* Comparative anatomy and classification of mammals; functional anatomy of mammals. *Mailing Add:* Dept Biol State Univ NY Binghamton NY 13901

LANDS, WILLIAM EDWARD MITCHELL, b Chillicothe, Mo, July 22, 30; c 4. BIOCHEMISTRY. *Educ:* Univ Mich, Ann Arbor, BS, 51; Univ Ill, PhD(biol chem), 54. *Prof Exp:* NSF fel, Calif Inst Technol, 54-55; from instr to prof chem, Univ Mich, Ann Arbor, 55-80; head biochem, Univ Ill Med Ctr, 80-85, prof, 85-91; DIR, DIV BASIC RES, NAT INST ALCOHOL ABUSE & ALCOHOLISM, 91- *Concurrent Pos:* Chmn subcomt biochem nomenclature, Nat Acad Sci, 62-64; Danforth Assoc, 66-; ed, Can J, 72-78; ed, Biochem & Biophys Acta, J Lipid Res, 78- *Honors & Awards:* Gold Medal Bond Award, Am Oil Chemists Soc, 65; Glycerine Res Award, 69; Verhagen Lectr, 79-81; Pfizer Biomed Res Award, 85. *Mem:* AAAS; Am Chem Soc; Am Soc Biol Chemists; Am Oil Chemists Soc; Sigma Xi; Am Inst Nutrit. *Res:* Metabolism of glycerides and long-chain aliphatic acids and aldehydes; formation of membranes and regulation of membrane function; prostaglandin biochemistry and control of its biosynthesis. *Mailing Add:* Nat Inst Abuse & Alcoholism 5600 Fishers Lane Rm 16C06 Rockville MD 20857

LANDSBAUM, ELLIS M(ERLE), b Chicago, Ill, Feb 28, 25; m 52; c 2. CHEMICAL ENGINEERING. *Educ:* Ill Inst Technol, BSc, 49; Northwestern Univ, MSc, 53, PhD(chem eng), 55; Univ Calif, Los Angeles, cert nuclear tech, 67, cert bus mgt, 72. *Prof Exp:* Jr engr, Socony Oil Co, 49-51; res group supvr, Jet Propulsion Lab, Calif Inst Technol, 55-61; sect mgr, 61-84, STAFF ENGR, PROPULSION DEPT, AEROSPACE CORP, 84- *Mem:* Am Inst Aeronaut & Astronaut; Combustion Inst. *Res:* Solid propellant rockets; combustion; nozzles; system analysis. *Mailing Add:* 518 N Alta Dr Beverly Hills CA 90210

LANDSBERG, ARNE, b Des Moines, Iowa, June 10, 33; c 2. CHEMICAL ENGINEERING. *Educ:* Univ Colo, BS, 55; Ore State Univ, PhD(chem eng), 64. *Prof Exp:* Chem engr, US Bur Mines, 61-64; peace corps vol for prof, Chem Eng Prog, Costa Rica, 64-66; CHEM RES ENGR, US BUR MINES, ALBANY, 66- *Mem:* Sigma Xi. *Res:* Chemical kinetics of gas-solid reactions; vapor-solid equilibrium. *Mailing Add:* 1430 NW Greenwood Pl Corvallis OR 97330

LANDSBERG, JOHANNA D (JOAN), b Medford, Ore, July 15, 40. ANALYTICAL CHEMISTRY, RESEARCH ADMINISTRATION. *Educ:* Ore State Univ, BS, 62, MS, 64. *Prof Exp:* Res asst, dept food sci & technol, Ore State Univ, 76-77; res chemist, Bend Res, Inc, 77-78; res chemist, 79-89, PROF LEADER, PAC NORTHWEST RES STA, SILVICULT LAB, FOREST SERV, USDA, BEND, ORE, 89- *Concurrent Pos:* Prin investr, US-Spain Res Proj, 86-89. *Mem:* Am Chem Soc; Sigma Xi; AAAS; Soc Am

Foresters. *Res:* Effects of prescribed fire on soil, forest floor and foliar nutrients in Central Oregon ponderosa pine and mixed conifer lands. *Mailing Add:* Silvicult Lab USDA Forest Serv 1027 NW Trenton Ave Bend OR 97701

LANDSBERG, LEWIS, b New York, NY, Nov 23, 38. METABOLISM. *Educ:* Williams Col, AB, 60; Yale Univ, MD, 64. *Prof Exp:* From instr to asst prof med, Sch Med, Yale Univ, 69-72; from asst prof to prof med, Harvard Med Sch, 72-90; IRVING S CUTTER PROF & CHMN, DEPT MED, MED SCH, NORTHWESTERN UNIV, 90-, DIR, CTR ENDOCRINOL, METAB & NUTRIT, 90- *Concurrent Pos:* Assoc physician, Yale-New Haven Hosp, 69-71, Beth Israel Hosp, 74-79; attend physician, West Haven Vet Admin Hosp, 70-72, Yale-New Haven Hosp, 71-72; assisting physician, Boston City Hosp, 72-73, assoc vis physician, 73-74; physician, Beth Israel Hosp, 79-88, sr physician, 88-90; physician-in-chief, Dept Med, Northwestern Mem Hosp, 90- *Mem:* Am Fedn Clin Res; Endocrine Soc; NY Acad Sci; Am Heart Asn; Am Soc Pharmacol & Exp Therapeut; fel Am Col Physicians; AAAS; Am Physiol Soc; Am Soc Clin Invest; Asn Am Physicians. *Res:* Catecholamines and the sympathoadrenal system; nutrition and the sympathetic nervous system; obesity and hypertension; numerous publications. *Mailing Add:* Div Endocrinol & Metab Beth Israel Hosp 330 Brookline Ave L-423 Boston MA 02215

LANDSBERGER, FRANK ROBBERT, b Amsterdam, Netherlands, Aug 10, 43; US citizen; div; c 1. PHYSICAL BIOCHEMISTRY, VIROLOGY. *Educ:* Cornell Univ, BA, 64; Brown Univ, PhD(physics), 70. *Prof Exp:* Res asst physics, Brown Univ, 64-69; res fel biochem, Div Endocrinol, Sloan-Kettering Inst Cancer Res, 69-71; asst prof chem, Ind Univ, Bloomington, 71-74; asst prof, 74-80, assoc prof & Andrew W Mellon Found fel, 80-87, ADJ ASSOC PROF, ROCKEFELLER UNIV, 87- *Mem:* AAAS; Biophys Soc; fel NY Acad Sci; Am Soc Microbiol; Am Chem Soc; Sigma Xi. *Res:* Use of physical biochemical studies of the structure and function of biological and model membranes with emphasis on enveloped viruses and parasites and their interaction with cell surfaces. *Mailing Add:* 441 E 89th St New York NY 10128

LANDSBURG, ALEXANDER CHARLES, b Saginaw, Mich, Dec 23, 42; m 67; c 4. COMMERCIAL SHIP DESIGN, SHIP MANEUVERABILITY. *Educ:* Univ Mich, BS, 66, MS, 69; Harvard Bus Sch, PMD, 79. *Prof Exp:* Trainee, Maritime Admin, 66-67, naval archit, Off Ship Construct, 67-74, mgr, ship design develop, 74-76, chief, Environ Activ, Off Shipbldg Costs, 74-76, mgr, computer-aided ship design, 76-78, computer-aided cost anal, Off Shipbldg Costs, 78-85, liaison, Off Advan Ship Opers, 85-87, naval architect, Off Ship Construct, 87-88, PROG MGR, SHIP PERFORMANCE & SAFETY, OFF TECHNOL ASSESSMENT, MARITIME ADMIN, 88- *Mem:* Soc Naval Architects & Marine Engrs; Am Soc Naval Engrs; Human Factors Soc. *Res:* Innovative ship design and operations; computer-aided ship design and operations; ship maneuverability design; information and training for shipboard personnel; ship design and operations for military sealift; cost analysis. *Mailing Add:* 307 Williamsburg Dr Silver Spring MD 20901

LANDSHOFF, ROLF, b Berlin, Ger, Nov 30, 11; nat US; m 41; c 4. MATHEMATICAL PHYSICS. *Educ:* Berlin Tech Inst, DrIng, 36; Univ Minn, PhD(theoret physics), 38. *Prof Exp:* Asst physics, Univ Minn, 36-40; prof, Col St Thomas, 40-44; scientist, Los Alamos Sci Lab, 44-56; sr mem, Lockheed Palo Alto Res Lab, 56-76; CONSULT, 77- *Concurrent Pos:* Vis lectr, Weizmann Inst, 63-64. *Mem:* Fel Am Phys Soc. *Res:* Atomic physics; hydrodynamics; statistical mechanics. *Mailing Add:* 525 E Crescent Dr Palo Alto CA 94301

LANDSMAN, DAVID, b Cape Town, S Africa, Oct 5, 53; m 82; c 1. MOLECULAR BIOLOGY, GENETICS. *Educ:* Univ Cape Town, S Africa, BSc, 76, PhD(biochem), 84. *Prof Exp:* Vis fel, 84-87, VIS ASSOC, LAB MOLECULAR CARCINOGENESIS, NAT CANCER INST, NIH, 87- *Mem:* Am Soc Biochem & Molecular Biol; Am Soc Cell Biol. *Res:* Structure and function of interphase chromatin and nuclei; molecular and cellular interactions controlling the regulation of gene expression. *Mailing Add:* Nat Ctr Biotechnol Info Nat Libr Med Bldg 38A Rm 8N807 Bethesda MD 20892

LANDSMAN, DOUGLAS ANDERSON, b Dundee, Scotland, May 31, 29; div; c 4. PHYSICAL CHEMISTRY. *Educ:* Univ St Andrews, BSc, 49, Hons, 50, PhD(thermodynamics), 57. *Prof Exp:* Sr sci officer, Chem Div, Atomic Weapon Res Estab, UK Atomic Energy Authority, Eng, 53-57; Nat Res Coun Can fel, 57-58; Harwell sr fel, Chem Div, Atomic Energy Res Estab, UK Atomic Energy Authority, Eng, 58-60, prin sci officer, 60-65; res supvr, Mat Eng Res Lab, Pratt & Whitney Aircraft Div, Middletown, 65-72; SR PROJ ENGR, INT FUEL CELLS, 72- *Mem:* Fel Royal Soc Chem; Am Chem Soc. *Res:* Thermodynamics of ionization in aqueous solutions; chemistry of the hydrogen isotopes; isotope separation; gas chromatography; chemonuclear reactors; energy conversion; fuel cells. *Mailing Add:* 221 Sisson Ave No 305 Hartford CT 06105-3112

LANDSTREET, JOHN DARLINGTON, b Philadelphia, Pa, Mar 13, 40; m 86; c 2. STELLAR MAGNETISM, STELLAR ATMOSPHERES. *Educ:* Reed Col, BA, 62; Columbia Univ, MA, 63, PhD(physics), 66. *Prof Exp:* Instr physics, Mt Holyoke Col, 65-66, asst prof, 66-67; res assoc astron, Columbia Univ, 67-70, asst prof, 70; from asst prof to assoc prof, 70-76, PROF ASTRON, UNIV WESTERN ONT, 76- *Concurrent Pos:* Mem, Grant Selection Comt for Space & Astron, Natural Sci & Eng Res Coun Can, 80-83, chmn, 82-83; mem sci adv comt, Can Ctr Space Sci, 81-83; mem sci adv coun, Can-France-Hawaii Telescope Corp, 80-83, vchmn, 81, chmn, 82-83; vis scientist, Inst Theoret Astrophysics, Univ Heidelberg, 84-85; mem, Coun Admin, Astron Observatory, Mont Megantic, 85-; distinguished res prof, Univ Western Ont, 87-88; mem bd dirs, Telescope Corp, Can, France, Hawaii, 91- *Mem:* Am Astron Soc; Can Astron Soc; Royal Astron Soc; Int Astron Union. *Res:* Observation of circular and linear polarization in stars and extragalactic objects, especially white dwarfs; observation and interpretation of stellar magnetism and of stellar spectra. *Mailing Add:* Dept Astron Univ Western Ont London ON N6A 3K7 Can

LANDSTROM, D(ONALD) KARL, b Portland, Ore, Oct 12, 37. SOLAR PHYSICS, HVAC. *Educ:* Mass Inst Technol, BS, 59. *Prof Exp:* Supvr electron micros, Goodyear Atomic Corp, 59-63; res engr mat environ, NAm Aviation, 63-65; prin res scientist solar mat anal, 65-80, PROJ MGR ENERGY & THERMAL TECHNOL, COLUMBUS LABS, BATTELLE MEM INST, 80- *Honors & Awards:* granted 4 patents. *Mem:* AAAS; Int Solar Energy Soc; Am Soc Heating Refrigerating & Air Conditioning Engrs. *Res:* Solar energy; environmental impact analysis; thermal analysis; materials research; physical and chemical analysis; electron microscopy; electron probe analysis; nuclear waste; energy and environmental systems; agricultural controlled environment system; heat pumps and alternative energy systems; gas fired heat pumps; HVAC systems. *Mailing Add:* Battelle Mem Inst 505 King Ave Columbus OH 43201

LANDT, JAMES FREDERICK, biology; deceased, see previous edition for last biography

LANDUCCI, LAWRENCE L, b St Paul, Minn, May 20, 39; m 69; c 3. ORGANIC CHEMISTRY, NUCLEAR MAGNETIC RESONANCE. *Educ:* Univ Minn, BS, 62, PhD(org chem), 67. *Prof Exp:* RES CHEMIST & NUCLEAR MAGNETIC RESONANCE SPECTROSCOPIST, US FOREST PROD LAB, 67- *Concurrent Pos:* Organizer, First Int Workshop Nuclear Magnetic Resonance & Wood Sci, Vancouver, BC, 85. *Mem:* Am Chem Soc. *Res:* Lignin and lignin model compound chemistry; methods of lignin degradation; mechanism of anthraquinone pulping; nuclear magnetic resonance characterization of lignin and reaction products; application of state-of-the-art nuclear magnetic resonance methods to characterization of wood components. *Mailing Add:* 3198 Shady Oak Lane Verona WI 53592

LANDWEBER, LAWRENCE H, b New York, NY, Nov 29, 42; m 66. COMPUTER SCIENCE. *Educ:* Brooklyn Col, BS, 63; Purdue Univ, MS & PhD(computer sci), 66. *Prof Exp:* From asst prof to assoc prof, 67-77, PROF COMPUTER SCI & CHMN DEPT, UNIV WIS-MADISON, 77- *Mem:* Am Math Soc; Asn Comput Mach; Sigma Xi. *Res:* Theoretical computer science; computer networks; computer conferencing and mail. *Mailing Add:* Univ Wis 1210 W Dayton St Madison WI 53706

LANDWEBER, LOUIS, b New York, NY, Jan 8, 12; m 35; c 2. SHIP HYDRODYNAMICS, INTEGRAL EQUATIONS. *Educ:* City Col New York, BS, 32; George Washington Univ, MA, 35; Univ Md, PhD(physics), 51. *Prof Exp:* Head, hydrodynamics div, David W Taylor Naval Ship Res & Develop Ctr, 32-54; PROF FLUID MECH, INST HYDRAUL RES, UNIV IOWA, 54- *Concurrent Pos:* Prin Investr, Off Naval Res, 54- & Mobil Oil Co, 85- *Honors & Awards:* Ward Medal Math & Beldon Prize Physics, City Col New York, 32; Davidson Medal, Soc Naval Architects & Marine Engrs, 78; David Taylor Lectr, David Taylor Naval Ship Res & Develop Ctr, 78; Weinblum Mem Lectr, German Inst Ship Construct & Soc Naval Architects & Marine Engrs, 81. *Mem:* Nat Acad Eng; Soc Naval Architects & Marine Engrs; fel Am Acad Mech; corresp mem Maritime & Aeronaut Tech Asn. *Res:* Analytical, numerical & experimental research about the flow of one or more bodies or ships moving through a fluid, their added masses and the forces and moments acting upon them. *Mailing Add:* Inst Hydraul Res Univ Iowa Iowa City IA 52242

LANDWEBER, PETER STEVEN, b Washington, DC, Aug 17, 40; m 64; c 2. MATHEMATICS. *Educ:* Univ Iowa, BA, 60; Harvard Univ, MA, 61, PhD(math), 65. *Prof Exp:* Asst prof math, Univ Va, 65-68; asst prof, Yale Univ, 68-70; assoc prof, 70-74, PROF MATH, RUTGERS UNIV, NEW BRUNSWICK, 74- *Concurrent Pos:* Mem sch math, Inst Advan Study, 67-68 & 86-87; NATO fel, Univ Cambridge, 74-75; NSF grad fel commemorative lectr, 89; chmn, Russian Translations Comt, Am Math Soc, 89-91. *Mem:* Am Math Soc; Math Asn Am. *Res:* Cobordism theory of differential manifolds. *Mailing Add:* Dept Math Rutgers Univ New Brunswick NJ 08903

LANDWEHR, JAMES M, b Philadelphia, Penn, Jan 12, 45; m 67; c 3. STATISTICAL APPLICATIONS & COLLABORATIONS, RESEARCH ON APPLIED STATISTICS METHODOLOGIES. *Educ:* Yale Univ, BA, 66; Univ Chicago, PhD(statist), 72. *Prof Exp:* Lectr & asst prof statist, Univ Mich, Ann Arbor, 70-73; TECH STAFF & SUPVR, AT&T BELL LABS, MURRAY HILL, 73- *Concurrent Pos:* Panel mem, Nat Acad Sci, Nat Res Coun, 84-88; co-prin investr, Quant Lit Proj, Am Statist Asn, Nat Coun Teachers Math, 84-87. *Mem:* Fel AAAS; fel Am Statist Asn; Inst Math Statist; Math Asn Am; Classification Soc. *Res:* Statistical experimentation and data analysis in manufacturing; categorical data analysis and logistic regression; graphical methods; precollege statistics education; software metrics; statistical applications. *Mailing Add:* AT&T Bell Labs Rm 2C-257 600 Mountain Ave Murray Hill NJ 07974-2070

LANDY, ARTHUR, b Philadelphia, Pa, Mar 17, 39; m 65; c 2. BIOCHEMICAL GENETICS, GENE REGULATION. *Educ:* Amherst Col, BA, 61; Univ Ill, PhD(microbiol & biochem), 65. *Prof Exp:* Res fel biochem genetics, Med Res Coun Lab Molecular Biol, Cambridge, Eng, 66-68; asst prof, 68-74, assoc prof, 75-77, PROF MED SCI, BROWN UNIV, 78- *Concurrent Pos:* NATO fel, 66-67; fel, Am Cancer Soc, 68 ; fac res assoc, 75-81, mem adv bd, 78-; assoc ed, Cell, 79-; mem recombinant DNA adv comt, NIH, 81-; chmn nucleic acids, Gordon Res Conf, 83; mem, Microbiol Genetics Study Sect, NIH, 87- *Mem:* Am Soc Microbiol; Am Soc Biol Chemists. *Res:* Gene structure and regulation in prokaryotes and eukaryotes; organization of eukaryote genes; mechanisms of site-specific recombination. *Mailing Add:* Div Biol & Med Brown Univ Box G Providence RI 02912

LANDY, MAURICE, IMMUNOLOGY. *Prof Exp:* ADV, SALK INST, 83- *Mailing Add:* 7550 Eads Ave No 405 La Jolla CA 92037

LANDZBERG, ABRAHAM H(AROLD), b New York, NY, Sept 10, 29; m 55; c 3. SEMICONDUCTOR PROCESS ENGINEERING. *Educ:* NY Univ, BSME, 51; Princeton Univ, MSE, 53. *Prof Exp:* Develop engr, Gen Elec Co, 52-59; dept mgr appl mech, Res Div, IBM, 59-65, dept mgr integrated circuits, Components Div, 65-70, advan mfg mgr, 71-76, sr eng mgr, semiconductor develop eng, 76-85, sr mgr, Packaging Develop Eng, 85-90, CONSULT, 90- Mem: Inst Elec & Electronics Engrs. Res: Physics of failure of integrated circuits; mechanical properties of materials; applied mechanics, application to electronic computer components; turbines; electrical machinery; manufacturing systems and processes for integrated circuits; electronic component packaging development; microelectronics reliability. Mailing Add: 685 Fieldstone Ct Yorktown Heights NY 10598

LANE, ALEXANDER Z, b Detroit, Mich, July 22, 29; m 56. BIOCHEMISTRY, MEDICINE. Educ: Univ Detroit, BS, 50; Wayne State Univ, PhD(biochem), 54, MD, 58. Prof Exp: Intern med, Bon Secours Hosp, Grosse Pointe, Mich, 58-59; clin investr, Parke Davis & Co, Mich, 60-62, dir clin pharmacol, 62-66; dir med res, Bristol Labs, 66-70, vpres & med dir, 70-77; sr vpres res opers, 76-77, PRES, PHARMACEUT RES DIV, SHERING-PLOUGH CORP, 77- Concurrent Pos: Nat Cancer Inst fel occup med, Wayne State Univ, 59-60; lectr, Univ Mich, 65-66. Mem: Fel Am Soc Clin Pharmacol & Therapeut; Am Chem Soc; Am Soc Microbiol. Res: Correlation of animal and clinical pharmacological data; experimental design of clinical studies; analysis, performance and interpretation of clinical chemical tests; antibiotics; cancer chemotherapy; narcotic antagonists. Mailing Add: Galloping Hill Rd Kenilworth NJ 07033

LANE, ALFRED GLEN, b Stoutland, Mo, Aug 21, 32; m 57; c 2. ANIMAL NUTRITION. Educ: Univ Mo, BS, 59, MS, 60, PhD(animal nutrit), 65. Prof Exp: Instr voc agr, Parkersburg Community Sch, Iowa, 60-63; asst dairy husb, Univ Mo-Columbia, 63-65, asst prof, 65-70; mgr diary res, Allied Mills, Inc, 70-77; pvt consult, 77-82; DAIRY SPECIALIST, TEX A&M UNIV, 82- Mem: Am Dairy Sci Asn; Am Soc Animal Sci. Res: Ruminant nutrition; physiology. Mailing Add: Rte 2 Box 1 Stephenville TX 76401

LANE, ARDELLE CATHERINE, b Port Angeles, Wash, Mar 8, 22; m 67. PHYSIOLOGY. Educ: Seattle Pac Col, BS, 44; Northwestern Univ, MS, 47; Univ Ill, PhD(physiol), 54. Prof Exp: From instr to prof, 52-86, EMER PROF PHYSIOL, DENT SCH, NORTHWESTERN UNIV, 86- Mem: Am Physiol Soc. Res: Controlling mechanisms of gastric secretion. Mailing Add: 8803 Parkway Dr Highland IN 43622

LANE, BENJAMIN CLAY, b Raleigh, NC, Feb 8, 52. PHARMACOLOGY. Educ: Wash State Univ, BS, 74, MS, 76, PhD(bact), 80. Prof Exp: Res scholar, Div Clin Immunol & Rheumatic Dis, Med Ctr, Univ Southern Calif, 79-81; res assoc, Div Rheumatic Dis, Med Ctr, Duke Univ, 81-85; res pharmacologist, 85-88, sr res investr, 88-89, SR RES SCIENTIST, IMMUNOSCI PROG, ABBOTT LABS, ILL, 89- Concurrent Pos: Lab instr, Dept Bact & Pub Health, Wash State Univ, 75-78; NIH fel, Med Ctr, Duke Univ, 81-82 & Arthritis Found res fel, 83-84. Mailing Add: Abbott Labs D-47L AP9 Abbott Park IL 60064-3500

LANE, BENNIE RAY, b Deming, NMex, July 2, 35; m 56; c 4. MATHEMATICS. Educ: Colo State Col, BA, 56, MA, 57; George Peabody Col, PhD(math), 62. Prof Exp: Asst prof math, Univ Chattanooga, 59-61; instr appl math, Vanderbilt Univ, 61-62; asst prof math, Colo State Col, 62-63; from asst prof to assoc prof, George Peabody Col, 63-66; chmn dept, 66-78, PROF MATH, EASTERN KY UNIV, 66- Mem: Am Math Asn. Res: Mathematics education; teaching mathematics by television; abstract algebra; programmed instruction. Mailing Add: Dept Math Eastern Ky Univ Richmond KY 40475

LANE, BERNARD OWEN, b Greensboro, NC, Oct 5, 25; m 51. INVERTEBRATE PALEONTOLOGY. Educ: Univ NC, BS, 50; Brown Univ, MSc, 55; Univ Southern Calif, PhD(geol), 60. Prof Exp: Lectr geol, Univ Nev, 59-60 & 61-62; from asst prof to prof geol, Calif State Polytech Univ, Pomona, 62-83. Concurrent Pos: Consult, Earth Sci Curriculum Proj, 65-66. Mem: Paleont Soc; Nat Asn Geol Teachers. Res: Paleontology and stratigraphy of the early Pennsylvanian in the cordilleran region of North America. Mailing Add: Dept Physics & Earth Sci Calif State Polytech Univ Pomona CA 91768

LANE, BERNARD PAUL, b Brooklyn, NY, June 27, 38; m 62; c 3. PATHOLOGY. Educ: Brown Univ, AB, 59; NY Univ, MD, 63. Prof Exp: NIH trainee exp path, Sch Med, NY Univ, 65-66, from asst prof to assoc prof, 66-71; assoc prof, 71-76, PROF PATH, HEALTH SCI CTR, STATE UNIV NY STONY BROOK, 76- Concurrent Pos: Attend pathologist, Bellevue & NY Univ Hosps, 69-71; attend pathologist, Vet Admin Hosp, Northport, NY, 71- & Stony Brook Univ Hosp, 79-; vis scientist, Armed Forces Inst Path, 71; chief cell injury labs, Armed Forces Inst Path. Mem: Am Soc Cell Biol; Int Acad Path; Am Asn Path; Am Soc Clin Path; Am Asn Cancer Res. Res: Experimental pathology; electron microscopy; cellular injury; chemical carcinogenesis. Mailing Add: Dept Path State Univ NY Health Sci Ctr Stony Brook NY 11790

LANE, BYRON GEORGE, b Toronto, Ont, May 16, 33; m 61. BIOCHEMISTRY. Educ: Univ Toronto, BA, 56, PhD(biochem), 59. Prof Exp: Jr res asst biochem, Med Ctr, Univ Calif, San Francisco, 59-60; res assoc, Rockefeller Inst, 60-61; asst prof, Univ Alta, 61-63, assoc prof, 64-68; PROF BIOCHEM, UNIV TORONTO, 68- Honors & Awards: Ayerst Award, Can Biochem Soc, 71. Mem: Am Soc Biol Chemists. Res: Biochemical investigations of germin, the marker protein for onset of growth in germinating wheat embryos. Mailing Add: Dept Biochem Univ Toronto Toronto ON M5S 1A8 Can

LANE, CARL LEATON, b Raleigh, NC, Feb 11, 28; m 52; c 1. FOREST SOILS. Educ: NC State Univ, BS, 52, MS, 61; Purdue Univ, PhD(forest soil microbiol), 68. Prof Exp: Forest mgr, State Hosp Butner, NC, 52-59; from asst prof to prof forestry, Clemson Univ, 60-90, coordr, Res & Grad Progs, Dept Forestry, 85-90; RETIRED. Mem: Soc Am Foresters. Res: Forest soils microbiology; forest soil tree disease relationships; nitrogen fixation; effluent utilization. Mailing Add: 150 Folgers St Clemson SC 29631

LANE, CHARLES A, b Wichita, Kans, Nov 18, 32. ORGANIC CHEMISTRY. Educ: Univ Okla, BS, 54; Yale Univ, MS, 59; Univ Calif, PhD(chem), 63. Prof Exp: Org chemist, Lederle Labs, Am Cyanamid Co, 56-58; asst prof org chem, Univ Nigeria, 61-63; asst prof, Univ Calif, 63-64; asst prof, 64-73, ASSOC PROF ORG CHEM, UNIV TENN, KNOXVILLE, 73- Mem: AAAS; Am Chem Soc. Res: Theoretical organic chemistry. Mailing Add: Dept Chem Univ Tenn Knoxville TN 37996

LANE, CHARLES EDWARD, zoology; deceased, see previous edition for last biography

LANE, DONALD WILSON, b Fayetteville, Tenn, June 23, 34; m 60; c 3. PETROLEUM GEOLOGY. Educ: Dartmouth Col, BA, 56; Univ Ill, MS, 58; Rice Univ, PhD(geol), 61. Prof Exp: Geologist, Tenneco Oil Co, 61-70; regional geologist, Royal Resources Corp, 70; staff geologist, Geol Surv Wyo, 70-73; mgr exp geol & Rocky Mountain area, Mich & Wis Pipeline Co, 73-76; consult geologist, 76-84 & 87-88; sr geologist, Britoil Ventures, 85-86; GEOLOGIST, ERM-SOUTHWEST, 88- Mem: Am Asn Petrol Geol. Res: Lower Paleozoic stratigraphy and hydrocarbon potential of the northeastern United States; Wyoming stratigraphy and stratigraphic resources; Lower Cretaceous stratigraphy of northwestern Colorado; eocene and cretaceous stratigraphy of Texas Gulf Coast; Pennsylvania stratigraphy of North Texas. Mailing Add: 12214 Mossycup Houston TX 77024

LANE, EDWIN DAVID, b Vancouver, BC, May 9, 34; m 58; c 2. FISH CULTURE, FISH ECOLOGY. Educ: Univ BC, BSc, 59, MSc, 62; Univ Tex, PhD(zool), 66. Prof Exp: Staff mem, Fisheries Invest Off, NZ Marine Dept, 62-63; res scientist, Res Br, Ont Dept Land & Forest, 66-68; assoc prof zool, Calif State Univ, Long Beach, 68-74, res grant, 68-69 & 71-74; head coop res, Can Wildlife Serv, Can Dept Environ, 74-79; COORDR, FISH CULTURE DEPT, MALASPINA COL, NANAIMO, BC, 79- Concurrent Pos: Fisheries expert, Food & Agr Orgn, 67- Mem: Can Soc Zoologists; NZ Limnol Soc; Am Fish Soc. Res: Salmonid culture especially feeding; ecology of fishes, especially in streams, estuaries and coastal bay systems; environmental impact of development, especially in the North. Mailing Add: Fish Cult Dept Malaspina Col 900 Fifth St Nanaimo BC V9R 5S5 Can

LANE, ERIC TRENT, b Baton Rouge, La, Aug 30, 38; m; c 2. MICROCOMPUTER ANIMATION GRAPHICS. Educ: La State Univ, BS, 60; Rice Univ, MA, 63, PhD(physics), 67. Prof Exp: Vis lectr physics, La State Univ, New Orleans, 63-65; assoc prof, 67-77, PROF PHYSICS, UNIV TENN, CHATTANOOGA, 77- Concurrent Pos: NSF grant, microcomput course develop. Honors & Awards: Proj Seriphim Grand Prize Software, 87. Mem: Am Phys Soc; Am Asn Physics Teachers; Sigma Xi. Res: Microcomputer animation graphics for science and physics education; research applied to improvement of teaching and human relationships. Mailing Add: Dept Physics Univ Tenn Chattanooga TN 37403-2598

LANE, ERNEST PAUL, b Greene Co, Tenn, Nov 14, 33; m 61; c 2. TOPOLOGY. Educ: Berea Col, BA, 55; Univ Tenn, MA, 57; Purdue Univ, PhD(math), 65. Prof Exp: Programmer, Army Ballistic Missile Agency, Ala, 57-58; instr math, Berea Col, 58-60; asst prof, Va Polytech Inst, 65-70; assoc prof, 70-75, PROF MATH, APPALACHIAN STATE UNIV, 75- Mem: Math Asn Am; Am Math Soc. Res: Abstract spaces; metrization; real-valued functions on abstract spaces. Mailing Add: Dept Math Appalachian State Univ Boone NC 28608

LANE, FORREST EUGENE, b Enola, Ark, June 24, 34; m 54; c 5. PLANT PHYSIOLOGY, PLANT BIOCHEMISTRY. Educ: Univ Ark, BA, 56, MEd, 59, MS, 63; Univ Okla, PhD(plant physiol), 65. Prof Exp: Teacher, Hall High Sch, Ark, 57-58; instr biol, Univ Ark, 58-63; asst prof biol, Kans State Col Pittsburg, 65-67; asst prof, Dept Bot & Bact, 67-69, ASSOC PROF BOT & BACT, UNIV ARK, FAYETTEVILLE, 69- Concurrent Pos: NSF fel, 63-65. Mem: Sigma Xi; Am Soc Plant Physiol; Scand Soc Plant Physiol; Bot Soc Am; Phytochem Soc NAm. Res: Dormancy in plant structures such as seeds, fruits, tubers and buds; relationship between dormancy and plant phenolics; enzymes associated with hormone control and plant growth. Mailing Add: 5450 Huntsville Rd Univ Ark Fayetteville AR 72701

LANE, FRANK, b New York, NY, Dec 21, 23; m 45; c 1. AEROSPACE ENGINEERING, APPLIED MECHANICS. Educ: Univ Mich, BS, 44; Stevens Inst Technol, MS, 48; NY Univ, ScD(aeronaut eng), 54. Prof Exp: Asst & assoc prof underwater ord res, Ord Res Lab, Pa State Univ, 48-53; res assoc aerospace eng, NY Univ, 53-54, res assoc prof, 54-56; sr scientist aerospace eng & appl mech, Gen Appl Sci Labs, 56-60, sci supvr, 60-65, tech asst to pres, 65-67, chief scientist, 67-71, vpres, 69-71; CHIEF SCIENTIST & VPRES, KLD ASSOCS INC, 71- Concurrent Pos: Mem hydroballistics adv comt, Bur Ord, 50-51; adj assoc prof, NY Univ, 56-65; mem panel aeroelasticity & struct dynamics, Bur Naval Weapons Adv Comt Aeroballistics, 63-65. Mem: Am Inst Aeronaut & Astronaut. Res: Aeroelasticity; structural dynamics; aerodynamics and hydrodynamics; wave propagation; aerodynamic noise. Mailing Add: KLD Assocs Inc 300 Broadway Huntington Station NY 11746

LANE, GARY (THOMAS), b Center, Ky, Nov 8, 41; m 63; c 3. ANIMAL NUTRITION, BIOCHEMISTRY. Educ: Berea Col, BS, 63; Purdue Univ, West Lafayette, MS, 65, PhD(animal nutrit), 68. Prof Exp: Res asst animal nutrit, Purdue Univ, West Lafayette, 63-67; asst prof, 67-73, assoc prof animal nutrit, Tex A&M Univ, 73-77; assoc exten prof dairy, Univ Ky, 77-86; DIR TECH SERV, BERKMANN MILLS, 86- Mem: Am Dairy Sci Asn; Am Soc Animal Sci. Res: Ration additives for ruminants; ration and its relation to milk composition and yield; mechanisms of milk synthesis; chemical preservation of high-moisture grain. Mailing Add: Berkmann Mills 901 W Walnut Danville KY 40422

LANE, GEORGE ASHEL, b Norman, Okla, May 9, 30; m 52. APPLIED CHEMISTRY, THERMODYNAMICS & MATERIAL PROPERITIES. *Educ:* Grinnell Col, AB, 52; Northwestern Univ, PhD(phys chem), 55. *Prof Exp:* Asst chem, Grinnell Col, 51-52; asst, Northwestern Univ, 52-55; spec projs chemist, 55-56, staff asst, 56-58, chemist, 58-63, proj leader, 63-66, sr res chemist, 66-69, res specialist, 69-73, sr res specialist, 73-80, RES ASSOC, DOW CHEM USA, 80- *Honors & Awards:* IR-100 Award, 80. *Mem:* Am Chem Soc; Sigma Xi. *Res:* Solar energy; energy storage; oxygen isotope effects; trout ecology; auto crash protection; rocket propellant testing and evaluation; pyrotechnics; ceramics. *Mailing Add:* 3802 Wintergreen Dr Midland MI 48640

LANE, GEORGE H, b Milford, NH, Feb 19, 24; m 48; c 2. PHYSICS. *Educ:* Amherst Col, BA, 47; Yale Univ, MS, 49; Univ Conn, PhD(physics), 61. *Prof Exp:* Asst instr physics, Univ Conn, Hartford Br, 49-51; from instr to asst prof, Franklin & Marshall Col, 54-60; from asst prof to assoc prof, 60-66, dir grad studies, 68-75, 79-80, head dept, 62-79, PROF PHYSICS, NORWICH UNIV, 66- *Mem:* AAAS; Am Asn Physics Teachers; Sigma Xi; Astr Soc Pac. *Res:* Atomic collisions; mass spectrometry. *Mailing Add:* Dept Physics Norwich Univ Northfield VT 05663

LANE, H CLIFFORD, b Detroit, Mich, June 15, 50. EXPERIMENTAL BIOLOGY. *Educ:* Univ Mich, BS, 72, MD, 76; Am Bd Internal Med, cert, 79; Am Bd Infectious Dis, cert, 84; Am Bd Allergy & Immunol, cert, 86. *Prof Exp:* Intern internal med, Univ Hosp, Ann Arbor, Mich, 76-77, resident, 77-79; clin assoc, Lab Immunoregulation, 79-82, sr investr, 82-89, DEP CLIN DIR, NAT INST ALLERGY & INFECTIOUS DIS, NIH, BETHESDA, MD, 85-, CHIEF, CLIN & MOLECULAR RETROVIROL SECT, LAB IMMUNOREGULATION, 89-, CLIN DIR, INST, 91- *Concurrent Pos:* Assoc prof lectr med, George Washington Univ, 85-; mem, AIDS Clin Drug Develop Comt, Nat Inst Allergy & Infectious Dis, NIH, 87- & adv bd, Food & Drug Admin, 89- *Mem:* Am Fedn Clin Res; Am Asn Immunologists; Am Col Physicians; fel Infectious Dis Soc Am. *Res:* Mechanisms of activation, proliferation and differentiation of human lymphoid cells in the normal immune response; clinical, pathophysiologic, immunologic, molecular biologic, virologic and therapeutic aspects of disease states including the vasculitis syndromes, Sjogren's syndrome, and AIDS. *Mailing Add:* Nat Inst Allergy & Infectious Dis-NIH 9000 Rockville Pike Bldg 10 Rm 11B13 Bethesda MD 20892

LANE, HAROLD RICHARD, b Danville, Ill, Mar 7, 42; m 68. PALEONTOLOGY, STRATIGRAPHY. *Educ:* Univ Ill, Urbana, BS, 64; Univ Iowa, MS, 66, PhD(geol), 69. *Prof Exp:* SR RES SCIENTIST, RES CTR, AMOCO PROD CO, 68- *Mem:* Brit Palaeont Asn; Int Palaeont Asn; Soc Econ Paleont & Mineral. *Res:* The evolution, biostratigraphy and systematic paleontology of the microfossils, conodonts, especially in Devonian through Middle Pennsylvanian strata of North America. *Mailing Add:* 6542 Auden Houston TX 77005

LANE, HELEN W, BIOMEDICINE. *Educ:* Univ Calif, Berkeley, BS, 68; Univ Wis-Madison, MS, 71; Univ Fla, PhD(animal nutrit), 78. *Prof Exp:* Asst gastroenterol, Dept Internal Med, Univ Wis-Madison, 70-72; instr, Dept Food Sci, Univ Fla, 72-74, asst gastroenterol, Dept Internal Med, 74-75, grad teaching asst, Prog Clin & Commun Dietetics, 75-77; asst prof, Prog Nutrit & Dietetics, Univ Tex Health Sci Ctr, Houston, 77-82, assoc prof, Prog Nutrit & Dietetics & Grad Sch Biomed Sci, 82-84; prof, Dept Nutrit & Foods, Auburn Univ, 84-89; DIR NUTRIT BIOCHEM, JOHNSON SPACE CTR, HOUSTON, 89-, DIR CLIN LABS, 91- *Concurrent Pos:* Adj appt, Dept Chem, Houston Baptist Univ, 78-84, adj prof, Dept Prev Med, Univ Tex Med Br, Galveston, 89-; grant reviewer, USDA, 81, 83-85, 87 & 89, Nat Cancer Inst, NIH, 86, Am Inst Cancer Res Study Sect, 86-91 & Am Dietetic Asn, 89. *Mem:* Am Asn Cancer Res; Am Dietetic Asn; Am Home Econ Asn; Am Inst Nutrit; Am Soc Clin Nutrit; Inst Food Technologists; Sigma Xi. *Res:* Nutritional sciences; preventive medicine; dietetics. *Mailing Add:* Biomed Opers & Res Br SD4 Johnson Space Ctr Houston TX 77058

LANE, JAMES DALE, b Las Cruces, NMex, Aug 28, 37; m 58; c 1. VERTEBRATE ZOOLOGY. *Educ:* NMex State Univ, BS, 59, MS, 62; Univ Ariz, PhD(zool), 65. *Prof Exp:* Asst biol, NMex State Univ, 59-62; asst zool, Univ Ariz, 62-65, asst geochronology, 65; asst prof biol, 65-70, PROF ZOOL, MCNEESE STATE UNIV, 70- *Mem:* AAAS; Soc Syst Zool; Am Soc Mammal; Soc Vert Paleont; Am Inst Biol Sci. *Res:* Ecology and systematics of various mammalians taxons, especially rodents. *Mailing Add:* Dept Biol McNeese State Univ Lake Charles LA 70609

LANE, JOHN D, PHARMACOLOGY. *Prof Exp:* PROF PHARMACOL, TEX COL OSTEOP MED, 87- *Mailing Add:* Tex Col Osteop Med Ft Worth TX 76107

LANE, JOSEPH ROBERT, b Chicago, Ill, Mar 3, 17; m 49. PHYSICAL METALLURGY, METALLURGICAL ENGINEERING. *Educ:* Univ Ill, BS, 43; Mass Inst Technol, ScD(metall), 50. *Prof Exp:* Metallurgist, Univ Chicago, Metall Lab, 43-45; res asst, Mass Inst Technol, 45-50; br head, Naval Res Lab, 50-55; senior staff metallurgist, Nat Mat Adv Bd, Nat Acad Sci, 55-89; RETIRED. *Mem:* Fel Am Soc Metals; Soc Mfg Engrs; Am Inst Mining, Metall & Petrol Engrs; Soc Advan Mat & Process Eng. *Res:* Superalloys, refractory metals and various aerospace structural materials. *Mailing Add:* 7211 Rebecca Dr Alexandria VA 22307

LANE, KEITH ALDRICH, b Gridley, Kans, Nov 11, 21; m 45; c 2. ANALYTICAL CHEMISTRY. *Educ:* Oglethorpe Univ, AB, 42, MA, 43. *Prof Exp:* Chemist, Mutual Chem Co Am, 43-51, group leader analytical res, 51-58; chemist, Solvay Process Div, Allied Chem Corp, 58-64, group leader analytical res, 64-70, environ chemist, Indust Chems Div, 70-83; RETIRED. *Mem:* Am Chem Soc. *Res:* All phases of chromium chemistry; organic and inorganic analytical method development; environmental studies and pollution control. *Mailing Add:* 122 Royal Rd Liverpool NY 13088

LANE, LEONARD JAMES, b Tucson, Ariz, Apr 25, 45; m 64; c 2. HYDROLOGY. *Educ:* Univ Ariz, BS, 70, MS, 72; Colo State Univ, PhD(civil eng), 75. *Prof Exp:* Hydrologist, Agr Res Serv, USDA, 70-81; staff mem, Univ Calif, 81-84; HYDROLOGIST, AGR RES SERV, USDA, 84- *Concurrent Pos:* Fac affil civil eng, Colo State Univ, 73-74; adj assoc prof renewable natural resouces, Univ Ariz, 82- *Honors & Awards:* Superior Serv Award, USDA, 81; Arthur S Flemming Award, 83. *Mem:* Am Geophys Union; Am Soc Civil Engrs; Am Water Resources Asn; Brit Geomorphol Res Group; Am Soc Agr Engrs; Sigma Xi. *Res:* Hydrology of semiarid regions; runoff and sediment simulation models incorporating geomorphic features, land use and management; improved erosion prediction technology; climatic fluctuations and change. *Mailing Add:* US Dept Agr Agr Res Serv 2000 E Allen Rd Tucson AZ 85719

LANE, LESLIE CARL, b Stamford, Conn, Apr 5, 42; m 66, 77; c 2. PLANT VIROLOGY. *Educ:* Univ Wis, BS, 65, PhD(biochem), 71. *Prof Exp:* Fel virol, John Innes Inst, Norwich, Eng, 71-73; fel virol, 73-75, asst prof, 75-81, ASSOC PROF PLANT PATH, UNIV NEBR-LINCOLN, 81- *Mem:* Am Phytopath Soc; Electrophoresis Soc; Am Soc Virol. *Res:* Structure and replication of plant viruses; virus directed protein and nucleic acid synthesis; virus-host interactions; gel electrophoretic separations; fluorescence detection methods. *Mailing Add:* Dept Plant Path Univ Nebr Lincoln NE 68583

LANE, LOIS KAY, PROTEIN CHEMISTRY. *Educ:* Dartmouth Col, PhD(biol), 73. *Prof Exp:* ASSOC PROF, DEPT PHARMACOL & CELL BIOPHYS, COL MED, UNIV CINCINNATI, 77- *Mailing Add:* Dept Pharmacol & Cell Biophys Col Med Univ Cincinnati 231 Bethesda Ave Cincinnati OH 45267

LANE, MALCOLM DANIEL, b Chicago, Ill, Aug 10, 30; m 51; c 2. CELL BIOLOGY. *Educ:* Iowa State Univ, BS, 51, MS, 53; Univ Ill, PhD, 56. *Prof Exp:* Res asst, Iowa State Univ, 51-53; from assoc prof biochem & nutrit to prof, Va Polytech Inst, 56-64; from assoc prof biochem to prof Sch Med, NY Univ, 64-70; prof physiol chem, 70-78, DELAMAR PROF BIOL CHEM & CHMN DEPT, JOHNS HOPKINS UNIV, 78- *Concurrent Pos:* Sr fel, Max Planck Inst Cell Chem, 62-63; mem Biochem Study Sect, NIH, 70-74. *Honors & Awards:* Mead Johnson Award, Am Inst Nutrit, 66. *Mem:* Nat Acad Sci; Am Chem Soc; Am Soc Biochem & Molecular Biol; Am Inst Nutrit; AAAS; Am Soc Cell Biol; Harvey Soc; Am Acad Arts & Sci. *Res:* Enzymology regulation of enzyme activity; enzymatic carboxylation; fatty acid and carbohydrate metabolism; cholesterol biosynthesis; lipogenic differentiation and gene expression; very low density lipoprotein synthesis and secretion; hormone action. *Mailing Add:* Dept Physiol Chem Johns Hopkins Univ Sch Med Baltimore MD 21205

LANE, MEREDITH ANNE, b Mesa, Ariz, Aug 4, 51; div. SYSTEMATICS, EVOLUTION. *Educ:* Ariz State Univ, BS, 74, MS, 76; Univ Tex, PhD(bot), 80. *Prof Exp:* assoc prof bot, Dept EPO Biol, Univ Colo, Boulder, 80-89; DIR MCGREGOR HERBARIUM & ASSOC PROF BOT, UNIV KANS, 89- *Concurrent Pos:* Vis asst prof, Dept Bot, Univ Tex, 82; chmn, Systs Sect, Bot Soc Am, 84-86; actg cur, Rocky Mountain Herbarium, Univ Wyo, 85-86; consult ed, plant taxon, Encyc Sci & Eng, McGraw-Hill; adj asst prof bot, Dept Bot, Univ Wyo, Laramie, 86-89; mem coun, Int Orgn Plant Biosysts, 90-92. *Honors & Awards:* Cooley Award, Am Soc Plant Taxonomists, 82. *Mem:* Int Asn Plant Taxon; Am Soc Plant Taxonomists (secy, 87-89); Bot Soc Am; Int Orgn Plant Biosysts. *Res:* Angiosperm systematics, specifically of southwestern North American and Mexican Compositae (Astereae), using cytotaxonomic, ecogeographic, scanning electron microscopic, biogeographic and cladistic techniques; pollination biology of Compositae, especially micromorphological pollinator cues; forensic botany. *Mailing Add:* McGregor Herbarium Univ Kansas Lawrence KS 66047

LANE, MONTAGUE, b New York, NY, Aug 28, 29; m 57; c 2. CLINICAL PHARMACOLOGY, CLINICAL ONCOLOGY. *Educ:* NY Univ, BA, 47; Chicago Med Sch, MB, 52, MD, 53; Georgetown Univ, MS 57. *Prof Exp:* Res assoc radiobiol, Cancer Res Lab, City New York, 47-48; res assoc oncol, Chicago Med Sch, 50-52; intern med, Jewish Hosp Brooklyn, NY, 52-53, asst res, 53-54; clin assoc pharmacol, Nat Cancer Inst, 54-56; asst resident med, USPHS, Rochester, 56-57; investr clin pharmacol, Nat Cancer Inst, 57-60; from asst prof to assoc prof, 60-67, PROF PHARMACOL & MED, BAYLOR COL MED, 67-, HEAD, DIV CLIN ONCOL, 69- *Concurrent Pos:* Instr, Sch Med, George Washington Univ, 57-60; consult, Vet Admin Hosp, 63-; mem pharmacol & therapeut study sec, Nat Cancer Inst, 66-69, consult chemother study sect, 69-72, mem nat new agents & spec Krebiozen rev comts, 72-73, mem cancer clin invests rev comt, 73-79, mem subcomt med oncol, Am Bd Internal Med, 73-80, consult, Food & Drug Admin, 86-; mem subcomt med oncol, Am Bd Internal Med. *Mem:* Am Soc Hemat; Am Soc Pharmacol & Exp Therapeut; Soc Exp Biol & Med; Am Asn Cancer Res; Am Soc Clin Pharmacol & Therapeut (pres, 71-72); Sigma Xi; Am Soc Clin Oncol. *Res:* Cancer chemotherapy alkylating agents; antimetabolites; riboflavin deficiency; drug screening; ferro-kinetics; biological response modifiers. *Mailing Add:* 1514 Bissonnet Houston TX 77005

LANE, NANCY JANE, b Halifax, NS, Nov 23, 36; m 69; c 2. DEVELOPMENTAL CELL BIOLOGY, NEUROBIOLOGY. *Educ:* Dalhousie Univ, BSc, 58, MSc, 60; Oxford Univ, DPhil(cytol), 63; Cambridge Univ, PhD, 68. *Hon Degrees:* ScD, Cambridge Univ, 81; LLD, Dalhousie Univ, 85. *Prof Exp:* Res asst prof path, Albert Einstein Col Med, 64-65; res staff biologist, Yale Univ, 65-68; sr prin sci officer & head electron micros, Agr Res Coun Res Unit Insect Neurophysiol & Pharmacol, Dept Zool, 68-90, WELLCOME FEL, CAMBRIDGE UNIV, 91- *Concurrent Pos:* Res fel, Girton Col, Cambridge Univ, 68-70, off fel & lectr cell biol, 70-, grad tutor, 75- *Mem:* AAAS; Am Soc Cell Biol; Brit Soc Cell Biol (secy, 82-90); Soc Exp Biol; fel Royal Micros Soc; fel Zool Soc. *Res:* Freeze-fracture and tracer analysis of invertebrate central nervous systems; immunocytochemical studies on neuropile and neurosecretory systems in invertebrates cells; accessibility of nervous systems to tracer molecules; development and turnover of junctions in arthropod tissues; structure and biochemistry of tight and septate junctions as well as arthropod gap junctions. *Mailing Add:* Dept Zool Downing St Cambridge CB2 3EJ England

LANE, NEAL F, b Oklahoma City, Okla, Aug 22, 38; m 60; c 2. ATOMIC PHYSICS. *Educ:* Univ Okla, BSc, 60, MS, 62, PhD(physics), 64. *Prof Exp:* NSF res fel physics, Queen's Univ, Belfast, 64-65; vis fel, Joint Inst Lab Astrophys, Univ Colo, 65-66; from asst prof to assoc prof, physics, Rice Univ, 66-71, chmn dept, 77-82, prof space physics & astron, 77-84, PROF PHYSICS, RICE UNIV, 72-, PROVOST, 86- *Concurrent Pos:* Alfred P Sloan res fel, 67-73; vis fel, Joint Inst Lab Astrophys, Univ Colo, 75-76; dir, Physics Div, NSF, 79-80; distinguished vis scientist, Univ Ky, 80; chancellor, Univ Colo, Colo Springs, 84-85. *Honors & Awards:* Distinguished Karcher Lectr, Univ Okla, 83. *Mem:* Fel Am Phys Soc; fel AAAS; Am Asn Phys Teachers; Sigma Xi. *Res:* Theoretical studies of collision processes involving electrons, atoms and molecules. *Mailing Add:* 2219 Shakespeare St Houston TX 77030-1112

LANE, NORMAN GARY, b French Lick, Ind, Feb 19, 30; m 58; c 3. PALEONTOLOGY. *Educ:* Oberlin Col, AB, 52; Univ Kans, MS, 54, PhD(geol), 58. *Prof Exp:* From asst prof to prof geol, Univ Calif, Los Angeles, 58-73; chmn dept, 84-87, PROF PALEONT, IND UNIV, BLOOMINGTON, 73- *Concurrent Pos:* Fulbright scholar, Univ Tasmania, 55-56; Fulbright prof, Trinity Col, Dublin, 71-72; res assoc paleont, Smithsonian Inst, 71- *Mem:* Paleont Soc (pres, 88); Soc Econ Paleontologists & Mineralogists; Soc Vert Paleont; Paleont Asn. *Res:* Functional morphology and community relations of fossil crinoids. *Mailing Add:* Dept Geol Ind Univ Bloomington IN 47405

LANE, ORRIS JOHN, JR, b Sigourney, Iowa, Apr 21, 32; m. ROAD PAVEMENT. *Educ:* Iowa State Univ, BS, 57. *Prof Exp:* Asst dist mat engr, Iowa Hwy Comn, 57-62, spec invest eng, 62-63, Portland cement concrete engr, 63-73; dist mat engr, 73-83, Portland cement concrete engr, 83-87, TESTING ENGR, IOWA DEPT TRANSP, 87- *Concurrent Pos:* Mem, C-13 Tech Comn Concrete Pipe, Am Soc Testing & Mat, 67-73 & 83-, Standards Comt Direct Design Buried Concrete Pipe, Am Soc Civil Engrs, 90- *Mem:* Nat Soc Prof Engrs. *Res:* A principle in development of the Iowa System for concrete bridge floor repair and rehabilitation; a principle in development of the Iowa Fast Track system for concrete pavement which permits early return of the new pavement to service. *Mailing Add:* 1111 Garfield Ames IA 50010

LANE, RAYMOND OSCAR, b Asbury Park, NJ, Sept 25, 24; m 49; c 3. NUCLEAR PHYSICS. *Educ:* Iowa State Univ, PhD(physics), 53. *Prof Exp:* Res asst, Inst Atomic Res, Iowa State Col, 49-53; assoc physicist, Argonne Nat Lab, 53-66; prof physics, 66-74, DISTINGUISHED PROF PHYSICS, OHIO UNIV, 74- *Mem:* Am Phys Soc. *Res:* Penetration of electrons in matter; beta ray spectroscopy; neutron scattering; neutron polarization; nuclear structure. *Mailing Add:* Ten Canterbury Dr Athens OH 45701

LANE, RICHARD DURELLE, b Detroit, Mich, May 14, 53; m 75; c 3. MONOCLONAL ANTIBODY TECHNOLOGY, ENZYMOLOGY. *Educ:* Bowling Green State Univ, BS, 75; Med Col Va, PhD(anat), 80. *Prof Exp:* NIH res technician, 75-76, instr, 80-83, asst prof, 84-87, ASSOC PROF, DEPT ANAT, MED COL OHIO, 87- *Mem:* Sigma Xi; AAAS; Am Asn Immunologists; Am Asn Anatomists; Reticuloendothelial Soc. *Res:* Investigating neoplastic cell produced factors which influence tumoricidal macrophages; production of monoclonal antibodies to these tumor produced factors; immunochemical investigation of cardiac muscle cell enzymes; role of calpains and protein Kinase C in monocyte development. *Mailing Add:* Dept Anat Med Col Ohio CS No 10008 Toledo OH 43699

LANE, RICHARD L, b Franklinville, NY, Mar 11, 35; m 58; c 4. CERAMICS, ELECTRONIC MATERIALS. *Educ:* State Univ NY, Alfred, BS, 57, PhD(ceramic sci), 62. *Prof Exp:* Res scientist, Appl Res Lab, Xerox Corp, 62-65, Fundamental Res Lab, 65-68; proj leader, Hamco Mach & Electronics Corp, 68-70, mgr res & develop, 70-74, dir eng, Hamco Div, 74-78; dir technol ctr, Kaytex Corp, 78-87; PROF MICROELECTRONIC ENG, ROCHESTER INST TECHNOL, NY, 87- *Honors & Awards:* Tech Innovation Awards, NASA. *Mem:* Am Ceramic Soc; Am Inst Ceramic Engrs; Electrochem Soc; Am Asn Crystal Growth. *Res:* Physics of glass, electrical and optical properties; chemistry of glass, surface properties and reactions, high temperature reactions and crystal-glass interactions; chemical and physical properties of crystalline ceramic materials; semiconductor materials processing; crystal growth; abrasive machining. *Mailing Add:* 1350 Penfield Ctr Rd Penfield NY 14526

LANE, RICHARD NEIL, b Richmond, Va, Sept 1, 44. MATHEMATICS, ELECTRICAL ENGINEERING. *Educ:* Calif Inst Technol, BS, 65, PhD(math), 68. *Prof Exp:* Mem prof staff math, Gen Elec Co, 68-69; sr scientist, Systs Applns, Inc, 69-70, dir commun studies, 70-75; CONSULT, 75- *Mem:* Am Math Soc; Inst Elec & Electronics Engrs. *Res:* Systems analysis and modeling of cost and performance of communications systems, especially mobile radio and common carrier systems. *Mailing Add:* Lane Westly & Assoc 2813 Easton Dr Hillsborough CA 94010

LANE, ROBERT HAROLD, b Tampa, Fla, Sept 4, 44. BIOINORGANIC CHEMISTRY, INORGANIC CHEMISTRY. *Educ:* Univ NC, BS, 66; Univ Fla, PhD(chem), 71. *Prof Exp:* Assoc phys biochem, Ore Grad Ctr, 71-72, sr res fel, 73; instr, Univ Ga, 73-75, asst prof chem, 75-80; SR RES CHEMIST, OCCIDENTAL RES CORP, 80- *Concurrent Pos:* NIH fel, Ore Grad Ctr, 73. *Mem:* Am Chem Soc. *Res:* Theoretical and experimental aspects of transition metal electron transfer in multinuclear systems; highly dispersed metal species on ordered supports. *Mailing Add:* 2522 Curlew Circle Anchorage AK 99515

LANE, ROBERT KENNETH, b Brandon, Man, Can, Feb 7, 37; m 61; c 1. ENVIRONMENTAL SCIENCE, RESOURCE MANAGEMENT. *Educ:* Brandon Col, BSc, 57; Ore State Univ, MS, 62, PhD(oceanog), 65. *Prof Exp:* Forecaster, Meteorol Serv Can, 57-58; oceanogr, Fisheries Res Bd Can, 59-61; instr oceanog, Ore State Univ, 63-66; res scientist, Environ Can, 66-75, sci adv, 75-79; dir water proj, Can West Found, 79-81; DIR, ENVIRON PROTECTION SERV, ENVIRON CAN, 82- *Concurrent Pos:* Chmn, Int Joint Comn Upper Great Lakes Water Qual Bd, 72-75 & Poplar River Bd, 78-80; prin investr, Landsat & Skylab, NASA. *Honors & Awards:* Centennial Medal, Govt of Can, 67, Merit Award, 77. *Mem:* Am Soc Limnol & Oceanog; Am Geophys Union; Royal Meteorol Soc; Int Asn Gt Lakes Res (pres, 74-75); Am Meteorol Soc; Sigma Xi. *Res:* Physical oceanography and limnology, especially heat and radiation exchange; remote sensing; meteorology. *Mailing Add:* Environ Protection Serv Twin Atria-2 2nd Floor Environ Can 4999 - 98th Ave Edmonton AB T6B 2X3 Can

LANE, ROBERT SIDNEY, b Worcester, Mass, Mar 7, 44; m 68; c 2. MEDICAL ENTOMOLOGY, MICROBIOLOGY. *Educ:* Univ Calif, Berkeley, BA, 66, PhD(entom), 74; San Francisco State Col, MA, 69. *Prof Exp:* Asst pub health biologist, Vector Biol & Control Sect, State Dept Health, Calif, 74-77, assoc pub health biologist, 77-79; assoc specialist, 80-84, ASSOC PROF, DEPT ENTOM SCI, UNIV CALIF, BERKELEY, 84- *Concurrent Pos:* Lectr, Biol Dept, San Francisco State Univ, 79-80; fel, Calif Acad Sci. *Mem:* Entom Soc Am; Wildlife Dis Asn; Am Soc Trop Med & Hyg; Soc Vector Ecologists; Sigma Xi. *Res:* Biosystematics of Tabanidae (Diptera); ecology and epidemiology of tick-borne disease, particularly Lyme disease and other spirochetoses. *Mailing Add:* Dept Entom Sci Univ Calif 201 Wellman Hall Berkeley CA 94720

LANE, ROGER LEE, b Mt Carmel, Ill, July 4, 45; m 68; c 3. ZOOLOGY, EMBRYOLOGY. *Educ:* Univ Nebr, BS, 68, MS, 71, PhD(zool), 74. *Prof Exp:* Vis lectr zool, John F Kennedy Col, 73-75; asst prof, 75-80, ASSOC PROF BIOL, KENT STATE UNIV, 80- *Mem:* Crustacean Soc; Am Malacol Union; Am Micros Soc; Am Soc Zoologists; Nat Asn Biol Teachers. *Res:* Histology and histochemistry of terrestrial isopod crustaceans and terrestrial pulmonate gastropods. *Mailing Add:* Kent State Univ Ashtabula OH 44004

LANE, STEPHEN MARK, b Scott Air Force Base, Ill, Nov 22, 48. PHYSICS. *Educ:* San Jose State Univ, BA, 71; Univ Calif, Davis, MS, 73, PhD(appl sci), 79. *Prof Exp:* PHYSICIST, LASER FUSION, LAWRENCE LIVERMORE LAB, 78- *Mem:* Am Phys Soc. *Res:* Atomic and nuclear spectroscopy as applied to laser fusion research. *Mailing Add:* Lawrence Livermore Lab PO Box 508 Mail Code L-473 Livermore CA 94550

LANE, WALLACE, b Chicago, Ill, Aug 31, 11; wid; c 2. PREVENTIVE MEDICINE, CULTURAL ANTHROPOLOGY. *Educ:* Univ Kans, AB, 33, MA, 35, MD, 39; Johns Hopkins Univ, MPH, 51. *Prof Exp:* Staff physician student health serv, Univ Kans, 46-48, clin assoc prof med & microbiol, 52-56; dir Bi-County Health Dept, Kans, 48-52; dir, Div Adult Health, Seattle-King County Health Dept, 56-58; chief div, Wash State Dept Pub Health, 58-68, dir, 68-73; RETIRED. *Concurrent Pos:* Dir, Health Dept, Kansas City, Kans, 52-56; lectr, Sch Med, Univ Kans, 55-56; clin assoc prof prev med, Univ Wash, 57-73. *Mem:* AAAS; AMA; Am Pub Health Asn; Sigma Xi. *Res:* Role of behavioral science in medicine and public health. *Mailing Add:* 1817 Governor Stevens Ave Olympia WA 98501-3711

LANE, WILLIAM JAMES, b Zanesville, Ohio, Dec 5, 25; m 50; c 3. ANALYTICAL CHEMISTRY. *Educ:* Denison Univ, BA, 47; Miami Univ, MS, 53; Iowa State Univ, PhD(chem), 57. *Prof Exp:* Asst chemist, AEC, Mound Lab, Monsanto Chem Co, 48-51; res asst anal chem, Ames Lab, Iowa State Univ, 52-57; analytical chemist, Columbia-Southern Chem Corp, Pittsburgh Plate Glass Co, 57-60, analytical group supvr, 60-64, res analytical chemist, Chem Div, 64-69; res coordr analytical chem, Universal Oil Prod Co, 69-74, group leader analytical chem res & develop, 74-78, lab supvr, spectroscopy & analytical res, 78-81, LAB SUPVR, ANALYTICAL RES & SCHEDULING, UOP, INC, 81- *Mem:* Am Chem Soc. *Res:* General wet chemical analysis; polarography; chromatography; spectrophotometry. *Mailing Add:* 444 W Norman Ct Des Plaines IL 60016

LANEWALA, MOHAMMED A, b Dohad, India; m 68; c 2. CHEMICAL ENGINEERING. *Educ:* Univ Calcutta, BSc, 56, MSc, 59; Univ Toronto, MASc, 61; NY Univ, PhD(chem eng), 67. *Prof Exp:* Develop engr, Molecular Sieve Dept, Linde Div, Union Carbide Corp, 64-69, proj leader, Eastview, 69-77, comput tech coordr, 77-84, consult, 84-88; CONSULT, UOP, 88- *Mem:* Am Soc Testing & Mat; Am Mensa. *Res:* Catalysis; adsorption; petroleum processes; computer technology; laboratory automation. *Mailing Add:* 5909 Ole Mill Rd Mobile AL 36609

LANEY, BILLIE EUGENE, b Joplin, Mo, Aug 27, 35; m 69; c 5. THEORETICAL PHYSICS. *Educ:* Univ Tulsa, BS, 60; NMex State Univ, MS, 62, PhD(physics), 64. *Prof Exp:* Sr res physicist oil explor, Schlumberger Res & Develop Labs, 64-66; sr res physicist space physics, NAm Rockwell, Inc, 66-68; chief scientist nuclear weapons, Braddock, Dunn & MacDonald, Inc, 68-71; prin res physicist, Cornell Aeronaut Labs, 71-73; vpres opers comput aided design, R & M Systs, Inc, 73-76; PRES ENERGY CONVERSION, EX-CAL, INC, 76- *Concurrent Pos:* Consult, Cities Serv Res & Develop Labs, 60-62, Calspan Corp, 73-, Sci Appl, Inc, 74-, EG&G, Inc, 74- & BDM Corp, 75-; tech referee, Am Phys Soc, 76- *Mem:* Am Phys Soc; Am Inst Aeronaut & Astronaut; Am Soc Testing & Mat. *Res:* Product oriented research in energy conversion mechanisms, electromagnetic interactions, thermodynamic heat transfer, seismic wave propagation, radiation transfer and gas dynamics, for application to the design of biomedical instrumentation, micrographics, and underground nuclear test equipment. *Mailing Add:* 5800 Truches NE No 120 Albuquerque NM 87109

LANFORD, OSCAR E, III, b New York, NY, Jan 6, 40; m 61; c 1. MATHEMATICAL PHYSICS, ANALYSIS & FUNCTIONAL ANALYSIS. *Educ:* Wesleyan Univ, BA, 60; Princeton Univ, MA, 62, PhD(physics), 66. *Hon Degrees:* ScD, Wesleyan Univ, 90. *Prof Exp:* Instr math, Princeton Univ, 65-66; asst prof, Univ Calif, Berkeley, 66-67; vis prof physics, Inst Advan Sci Studies, 67-68; from asst prof to prof math, Univ Calif, Berkeley, 75-87; prof physics, Inst des Hantes Etudes Scientifiques, Bures-sur-Yvette, France, 82-87; PROF MATH, SWISS FED INST TECHNOL, ZURICH, 87- *Concurrent Pos:* Alfred Sloan Found res fel, 69-71; mem, Inst Advan Studies, 70; exchange prof, Univ Aix Marseille, 71. *Honors & Awards:*

Award in Appl Math & Numerical Analysis, Nat Acad Sci, 86. *Mem:* Am Math Soc. *Res:* Mathematical physics, especially statistical mechanics and quantum field theory; Mathematical physics, especially statistical mechanics and dynamical systems theory. *Mailing Add:* D-Math ETH - Zentrum Zurich 8092 Switzerland

LANFORD, ROBERT ELDON, b Ft Worth, Tex, Apr 18, 51. IMMUNOLOGY, VIROLOGY. *Educ:* Univ Tex, BS, 74; Baylor Col Med, PhD(virol), 79. *Prof Exp:* Asst prof, Dept Virol & Epidemiol, Baylor Col Med, Houston, Tex, 82-84; asst scientist, 84-85, assoc scientist, 86-89, SCIENTIST, DEPT VIROL & IMMUNOL, SOUTHWEST FOUND BIOMED RES, SAN ANTONIO, TEX, 90- *Concurrent Pos:* Adj asst prof, Dept Microbiol, Univ Tex Health Sci Ctr, San Antonio, 84-85, adj assoc prof, 86-; co-ed, Viral Immunol. *Mem:* Am Soc Microbiol; Am Soc Virol; Am Soc Cell Biol. *Res:* Hepatitis B virus. *Mailing Add:* Dept Virol & Immunol Southwest Found Biomed Res PO Box 28147 San Antonio TX 78228-0147

LANFORD, WILLIAM ARMISTEAD, b Albany, NY, Nov 15, 44; m 66; c 3. MATERIALS PHYSICS, APPLIED PHYSICS. *Educ:* Univ Rochester, BS, 66, PhD(physics), 72. *Prof Exp:* Res assoc, Mich State Univ, 71-72, asst prof, 72-73; from asst prof to assoc prof physics, Yale Univ, 73-79; assoc prof, 79-83, PROF PHYSICS, STATE UNIV NY, ALBANY, 83-, DIR, ACCELERATOR LAB, 88- *Concurrent Pos:* Consult, Exxon, 79-, Sotheby's, 79, IBM, 80-, Nat Semiconductor, 84 & Bell Labs, 85-; assoc ed, Appl Nuclear Sci, 79-, ed, Radiation Effects, 83-89; fel, Alfred P Sloan Found, 79-83; mem, Int Adv Comt, Ion Beam Anal Conf, 81-; vchmn, NY State sect, Am Phys Soc, 85-87, chmn, 87-89, coun rep, 89-91; chmn, Physics Dept, State Univ NY, Albany, 90- *Mem:* Am Ceramic Soc; Am Phys Soc; Mat Res Soc; Hist Metall Soc; Böhmische Phys Soc; Sigma Xi; AAAS. *Res:* Electronic materials; thin films; hydrogen in solids; effects of sea-level cosmic rays on microelectronics; storage of ultracold neutrons; nuclear reaction analysis of hydrogen in metals; nuclear structure; beta-decay and solar neutrinos; archaeometry; glass surfaces; reaction between water and glass. *Mailing Add:* Dept Physics State Univ NY Albany NY 12222

LANG, ANTON, b Petersburg, Russia, Jan 18, 13; nat US; m 46; c 3. DEVELOPMENTAL PLANT BIOLOGY. *Educ:* Univ Berlin, Dr Nat Sci, 39. *Hon Degrees:* LLD, Univ Glasgow, UK, 81. *Prof Exp:* Sci asst plant physiol, Max-Planck Inst Biol, Ger, 39-49; res assoc genetics, McGill Univ, 49; vis prof genetics & agron, Agr & Mech Col Tex, 50; from res fel to sr res fel plant physiol, Calif Inst Technol, 50-52; from asst prof to assoc prof bot, Univ Calif, Los Angeles, 52-59; prof biol, Calif Inst Technol, 59-65; dir, US Dept Energy Plant Res Lab, 65-78, prof, 78-83, EMER PROF, US DEPT ENERGY PLANT RES LAB & BOT & PLANT PATH, MICH STATE UNIV, 83- *Concurrent Pos:* Lady Davis Found Can fel, 49; NSF sr fel, Max Planck Inst Biol, Ger & Hebrew Univ, Israel, 58-59; consult develop biol prog, NSF, 60-64, consult, adv comt biol & med sci, 68-70; partic sci exchange prog, Nat Acad Sci-Acad Sci, USSR, 63, 68, 75-76 & 87; trustee, Argonne Univs Asn, 65-69; chmn comt effects on herbicides in Vietnam, Nat Acad Sci-Nat Res Coun, 72-74; hon vpres, XII Int Bot Cong, Leningrad, 75 & XIV Int Bot Cong, WBerlin, 88; mem, Comt USSR & Eastern Europe, Nat Acad Sci, 64-67 & 77-78; mem, Pres Comt, Nat Medal Sci, 76-79; vis prof, Dept Bot & Plant Sci, Univ Calif Riverside, 84. *Honors & Awards:* Sr Scientist Award, Sigma Xi & Mich State Univ, 69; Stephen Hales Price & Charles Reid Barnes Life Mem Award, Am Soc Plant Physiologists, 76; Silver Medal, Ma Hort Soc, 81. *Mem:* Nat Acad Sci; fel AAAS; Am Acad Arts & Sci; Leopoldina Acad Naturalists; Am Soc Plant Physiol (vpres, 63, pres, 71); hon mem Ger Bot Soc; Soc Develop Biol (pres, 68-69). *Res:* Physiology of flowering; plant hormone physiology. *Mailing Add:* MSU-DOE Plant Res Lab Mich State Univ East Lansing MI 48824-1312

LANG, BRUCE Z, b St Joseph, Mo, May 31, 37; m 59; c 2. PARASITOLOGY, IMMUNOLOGY. *Educ:* Chico State Col, BS, 60; Univ NC, Chapel Hill, MSPH, 61, PhD(parasitol), 66. *Prof Exp:* Vis asst prof zool, Univ Okla, 66-67; from asst prof to assoc prof, 67-74, chmn dept, 78-80, PROF BIOL, EASTERN WASH UNIV, 74- *Concurrent Pos:* NIH fel zool, Univ Okla, 66-67; NSF grants, 69-72; affil fac mem, Univ Idaho, 75-; USDA grants, 77-83; mem grad fac, Wash State Univ, 77- *Mem:* AAAS; Am Soc Parasitol; Am Soc Zoologists; Am Soc Trop Med & Hyg; Am Inst Biol Sci. *Res:* Host-parasite relationships; ecology of parasitism; ecology of fresh-water gastropod molluscs. *Mailing Add:* Dept Biol Eastern Washington Univ Cheney WA 99004

LANG, C MAX, b Paris, Ill, Dec 29, 37; m 65; c 3. LABORATORY ANIMAL MEDICINE. *Educ:* Univ Ill, BS, 59, DVM, 61. *Prof Exp:* CHMN, DEPT COMP MED, COL MED, MILTON S HERSHEY MED CTR, PA STATE UNIV, 66-, ASST DEAN CONTINUING EDUC, 83-,. *Concurrent Pos:* George T Harrel prof & dir, Cancer Res Ctr, Pa State Univ, 84-87. *Honors & Awards:* Res Award, Am Asn Lab Animal Sci, 79 & 80; Charles River Prize, 87. *Mem:* Am Col Lab Animal Med (secy-treas, 81-90, pres-elect, 90-). *Res:* Environmental factors that can influence interpretation of research data. *Mailing Add:* Milton S Hershey Med Ctr Pa State Univ PO Box 850 Hershey PA 17033

LANG, CALVIN ALLEN, b Portland, Ore, June 13, 25; m 49; c 4. GERONTOLOGY, NUTRITION. *Educ:* Princeton Univ, AB, 47; Johns Hopkins Univ, ScD(biochem), 54. *Prof Exp:* Res collabr, Brookhaven Nat Labs, 49-51; asst scientist insect biochem, Conn Agr Exp Sta, 54-56; res assoc, Sch Hyg & Pub Health, Johns Hopkins Univ, 56-59; from asst prof to assoc prof, 59-72, dir biomed aging res prog, 71-74, PROF BIOCHEM, SCH MED, UNIV LOUISVILLE, 72-, DIR, LOUISVILLE LONGITUDINAL LONGEVITY PROG, 83- *Concurrent Pos:* Fel, Sch Hyg & Pub Health & McCollum-Pratt Inst, Johns Hopkins Univ, 56-59; NIH fel, 57-59 & res career develop award, 67-72; Nat Sigma Xi lectureship, 90. *Honors & Awards:* Tanner Lectr, Inst Food Tech, 73. *Mem:* Am Soc Biol Chem; Am Inst Clin Nutrit; Soc Exp Biol & Med; fel Geront Soc (vpres, 71); Sigma Xi. *Res:* Biochemistry of growth and aging; insect and nutritional biochemistry; glutathione detoxification and aging. *Mailing Add:* Dept Biochem MDR 412 Univ Louisville Sch Med Louisville KY 40292

LANG, CHARLES H, b Pittsburgh, Pa, July 7, 54; m 81; c 3. INSULIN RESISTANCE, TRACER METHODOLOGY. *Educ:* Westminster Col, BS, 76; Hahnemann Med Col, MS, 79, PhD(physiol), 81. *Prof Exp:* Fel metab, 81-84, res asst prof metab & shock, 84-86, asst prof metab & shock, 86-89, ASSOC PROF METAB & SHOCK, LA STATE UNIV MED CTR, 89- *Mem:* Shock Soc; Am Physiol Soc; Am Diabetes Asn; Internal Endotoxin Soc; Res Soc Alcoholism; Surg Infection Soc. *Res:* The role of glucose counterregulatory hormones in mediating insulin action in various pathophysiological conditions, including sepsis, diabetes, chronic alcoholism and burn; the influence of various cytokines on carbohydrate homeostasis and their putative role in endotoxemia and sepsis. *Mailing Add:* Dept Physiol La State Univ Med Ctr 1901 Perdido St New Orleans LA 70112

LANG, CONRAD MARVIN, b Chicago, Ill, July 1, 39; m 61; c 3. PHYSICAL CHEMISTRY. *Educ:* Elmhurst Col, BS, 61; Univ Wis-Madison, MS, 64; Univ Wyo, PhD(chem), 70. *Prof Exp:* Teaching asst chem, Univ Wis-Madison, 61-63, res asst, 63-64; from instr to assoc prof, 64-78, PROF CHEM, UNIV WIS-STEVENS POINT, 78- *Concurrent Pos:* Consult, Crowns, Merklin, Midthun & Hill, Attorneys at Law, 70-; NSF res grant, Univ Wis-Stevens Point, 71-73; W B King vis prof, Iowa State Univ, 76-77; bd dirs, Am Chem Soc, 89- *Honors & Awards:* Outstanding Contribs to Chem Award, Cent Wis Sect Am Chem Soc, 83. *Mem:* Am Chem Soc. *Res:* Application of electron spin resonance to molecular structure and macromolecular aspects of binary fluid mixtures; semiempirical quantum chemical calculations on systems of biological interest; physiochemical aspects of vision; chemical philately; chemical demonstrations; history of science. *Mailing Add:* Dept Chem Univ Wis Stevens Point WI 54481

LANG, DAVID (VERN), b Willmar, Minn, July 11, 43; m 68; c 3. SEMICONDUCTORS, PHYSICS. *Educ:* Concordia Col, Moorhead, Minn, BA, 65; Univ Wis-Madison, PhD(physics), 69. *Prof Exp:* Res assoc physics, Univ Ill, Urbana, 69-70, res asst prof, 70-72; mem tech staff, 72-81, head, Semiconductor Electronics Res Dept, Bell Labs, 81-87; DIR, SOLID-STATE ELECTRONICS RES LAB, AT&T BELL LABS, 87- *Honors & Awards:* Morris E Leeds Award, Inst Elec & Electronics Engrs, 88. *Mem:* Fel Am Phys Soc; sr mem Inst Elec & Electronics Engrs. *Res:* Capacitance spectroscopy (DLTS); defects in III-V semiconductors; recombination enhanced solid state defect reactions; radiation damage in semiconductors; gap states in amorphous semiconductors. *Mailing Add:* Bell Labs 600 Mountain Ave Murray Hill NJ 07974

LANG, DENNIS ROBERT, cell physiology, for more information see previous edition

LANG, DIMITRIJ ADOLF, b Berlin, Ger, Aug 30, 26; m 59; c 2. BIOPHYSICS, MOLECULAR BIOLOGY. *Educ:* Univ Frankfurt, MS, 53, PhD(biophys), 59. *Prof Exp:* Res asst biophys, Max Planck Inst Biophys, 53-58 & Hyg Inst, Univ Frankfurt, 58-65; from asst prof to assoc prof biol, 65-72, PROF BIOL, UNIV TEX, DALLAS, 72- *Concurrent Pos:* NIH res career develop awards, 67-71 & 72-76. *Mem:* Electron Micros Soc Am; Biophys Soc. *Res:* Electronics; physics of ionizing radiations; high-output x-ray machines; standard dosimetry of x-rays; electron microscopy of bacteria, viruses and nucleic acids; physical chemistry of nucleic acids. *Mailing Add:* 822 Saint Lukes Richardson TX 75080

LANG, ENID ASHER, b Los Angeles, Calif, Aug 28, 44. PSYCHIATRY. *Educ:* Radcliffe Col, AB, 66; Univ Southern Calif, MD, 70; Columbia Univ, MS, 74. *Prof Exp:* Med intern, Beth Israel Hosp, New York, 71-72; resident psychiat, Columbia Psychiat Inst, 72-74; res fel, Columbia Univ Health Serv, 74-75; fac mem psychiat, Sch Med, NY Univ, 75-80; FAC MEM PSYCHIAT, MT SINAI SCH MED, NY, 81- *Concurrent Pos:* Dir, Group Psychiat & Training Psychiat Residents, Bellevue Hosp, Sch Med, NY Univ, 75-, dir, Socialization Prog Psychiat Outpatients, 75-; process groups psychiat residents training, Mt Sinai Med Ctr. *Mem:* Am Psychiat Asn; NY Acad Sci; Am Women's Med Asn. *Res:* A longitudinal comparative study of treatment-outcome of discharged psychiatric outpatients who receive group therapy with medication versus individual therapy with medication only. *Mailing Add:* Ten Innes Rd Scarsdale NY 10583

LANG, ERICH KARL, b Vienna, Austria, Dec 7, 29; US citizen; m 56; c 2. RADIOLOGY. *Educ:* Columbia Univ, MS, 51; Univ Vienna, MD, 53. *Prof Exp:* Assoc radiol, Johns Hopkins Hosp & Univ, 56-59, assoc radiologist, 59-61; radiologist, Methodist Hosp, Indianapolis, Ind, 61-67; prof radiol & chmn dept, Sch Med, La State Univ, Shreveport, 67-76; PROF RADIOL & CHMN DEPT, SCH MED, LA STATE UNIV, NEW ORLEANS, 76-; PROF RADIOL, SCH MED, TULANE UNIV, 76- *Mem:* AMA; Radiol Soc NAm; Soc Nuclear Med; Am Col Radiol; Am Col Chest Physicians. *Res:* Diagnostic, vascular roentgenographic examinations; diagnostic roentgenographic evaluation of tumors and tumor diagnosis. *Mailing Add:* Dept Radiol La State Univ Med Ctr New Orleans LA 70110

LANG, FRANK ALEXANDER, b Olympia, Wash, May 14, 37; m 59; c 2. SYSTEMATIC BOTANY. *Educ:* Ore State Univ, BS, 59; Univ Wash, MS, 61; Univ BC, PhD(bot), 65. *Prof Exp:* Asst prof biol, Whitman Col, 65-66; from asst prof to assoc prof, 66-77, chmn dept, 76-80, PROF BIOL, SOUTHERN ORE STATE COL, 77-, CHMN DEPT, 90- *Concurrent Pos:* Vis scholar, Harvard Univ, Herbaria, 81-82. *Mem:* Am Soc Plant Taxon; Int Asn Plant Taxon. *Res:* Biosystematics and cytotaxonomy of vascular plants; flora of the Siskiyou Mountains. *Mailing Add:* Dept Biol Southern Ore State Col Ashland OR 97520

LANG, FRANK THEODORE, b New York, NY, Jan 25, 38; m 63; c 3. PHYSICAL CHEMISTRY. *Educ:* St Francis Col, NY, BS, 59; Rensselaer Polytech Inst, PhD(phys chem), 64. *Prof Exp:* Res assoc photochem, Univ Sheffield, 64-65; res assoc-instr energy transfer, Univ NC, Chapel Hill, 65-67; assoc prof phys chem, 67-80, PROF CHEM, FAIRLEIGH DICKINSON UNIV, FLORHAM-MADISON, 80-, CHMN, DEPT CHEM, 70-73 & 89- *Mem:* Am Chem Soc. *Res:* Charge transfer complexes; flash photolysis; low temperature photochemistry; energy transfer processes; luminescene studies; oscillating reactions. *Mailing Add:* Dept Chem Fairleigh Dickinson Univ Madison NJ 07940

LANG, GEORGE E, JR, b Chicago, Ill, June 29, 42; m 68; c 2. TOPOLOGY. *Educ:* Loyola Univ Chicago, BS, 64; Univ Dayton, MS, 66; Purdue Univ, PhD(math), 70. *Prof Exp:* Teaching asst, Univ Dayton, 64-66 & Purdue Univ, 67-69; ASSOC PROF MATH & COMPUTER SCI, FAIRFIELD UNIV, 70- *Mem:* Am Math Soc; Math Asn Am; Am Asn Univ Profs; Asn Comput Mach. *Res:* Homotopy theory; subgroups of homotopy groups; direct limits of CW complexes with an eye to group theoretic applications. *Mailing Add:* Dept Math Fairfield Univ Fairfield CT 06430

LANG, GERALD EDWARD, b Chicago, Ill, Mar 1, 45; m 73; c 1. PLANT ECOLOGY. *Educ:* Western Ill Univ, BS, 67; Univ Wyo, MS, 69; Rutgers Univ, PhD(bot), 73. *Prof Exp:* Res instr terrestrial ecol, Dartmouth Col, 73-76; from asst prof to assoc prof, 76-84, PROF BIOL & ASST DEAN, WVA UNIV, 84- *Concurrent Pos:* NSF res grants, 74-76, 76-79, 77, 79-82, 83-85 & 84; Army Corps Engrs res contract, 79-82, Environ Protection Agency res contract, 82-85; NASA res grant, 85-88. *Mem:* AAAS; Brit Ecol Soc; Ecol Soc Am. *Res:* Vegetation patterns and biogeochemical processes in high-elevation wetland ecosystems in the Appalachian Mountains; decomposition rates for leaf and wood litter; concomitant elemental mineralization patterns in forest ecosystems. *Mailing Add:* Dept Biol WVa Univ Morgantown WV 26506-6057

LANG, GERHARD HERBERT, b Neunkirchen, WGermany, Jan 6, 27; Can citizen; m 59; c 2. MEDICAL MICROBIOLOGY, VETERINARY MEDICINE. *Educ:* Univ Lyons, France, DVM, 55; Pasteur Inst, Paris, cert bact, 55; Univ Toronto, MVSc, 62. *Prof Exp:* Assoc vet, Beauquesne, France, 55-57; asst prof, 61-74, ASSOC PROF VET VIROL, ONT VET COL, UNIV GUELPH, 74- *Concurrent Pos:* Animal health expert virol, Food & Agr Orgn, Rome, 71-73. *Mem:* Am Asn Avian Pathologists; World Vet Poultry Asn; Can Asn Microbiologists; Am Soc Microbiologists; Can Vet Med Asn. *Res:* Animal and human virus infections, their diagnosis and control; medical and veterinary microbiology. *Mailing Add:* Dept Vet Microbiol & Immunol Univ Guelph Guelph ON N1G 2W1 Can

LANG, GERHARD PAUL, b Omaha, Nebr, Feb 20, 17; m 44; c 4. INORGANIC CHEMISTRY. *Educ:* Valparaiso Univ, BA, 43; Wash Univ, MA, 55. *Prof Exp:* Chemist, Visking Corp, 44-45; chemist, Uranium Div, Mallinckrodt Chem Works, 46-64; res chemist, Emerson Elec Mfg Co, 64-66; SR ENGR, MCDONNELL DOUGLAS CORP, 66- *Mem:* Am Chem Soc; Sigma Xi. *Res:* Uranium chemistry; liquid-liquid extraction; radiochemistry; aerospace chemistry. *Mailing Add:* 7430 Hiawatha Ave Richmond Heights MO 63117

LANG, HARRY GEORGE, b Pittsburgh, Pa, June 2, 47; m 73. PHYSICS. *Educ:* Bethany Col, WVa, BS, 69; Rochester Inst Technol, MS, 74; Univ Rochester, EdD, 79. *Prof Exp:* Asst prof, 70-80, prof physics, Rochester Inst Technol, Nat Tech Inst Deaf, 80-84, coordr, Off Fac Develop, 84-90, PROF, DEPT EDUC RES & DEVELOP, ROCHESTER INST TECHNOL, NAT TECH INST DEAF, 90- *Concurrent Pos:* Consult, Proj Handicapped Sci, AAAS, 78-, Res Better Sch, Inc & Am Printing House for Blind; vis prof, Univ Rochester, 81-; ed, Testing Phys Handicapped Students in Sci. *Mem:* Nat Sci Teachers Asn; AAAS; Am Educ Res Asn; Nat Asn Res Sci Teaching; Asn Educ Teachers Sci. *Res:* Test measurement theory; criterion referenced measurement; science curriculum research for handicapped students; digital computer analysis and synthesis of speech; historical contributions of deaf persons to science; research on effective teaching. *Mailing Add:* Nat Tech Inst for the Deaf One Lomb Mem Dr Rochester NY 14623

LANG, HELGA M (SISTER THERESE), b York, Pa, Dec 2, 28. BIOCHEMISTRY. *Educ:* Duquesne Univ, BS, 50; Univ Iowa, MS, 52; Marquette Univ, PhD(biochem), 71. *Prof Exp:* Res asst, Inst Cancer Res, Philadelphia, 52-54 & McCollum Pratt Inst, Johns Hopkins Univ, 54-56; instr biol, Nazareth Col Rochester, 56-59; instr chem, St Agnes High Sch, 59-67; from asst prof to assoc prof 71-83, CHMN CHEM, NAZARETH COL, ROCHESTER, 77-, PROF CHEM, 83- *Concurrent Pos:* Fel, Sch Med, Univ Rochester, 71-72, Cornell Univ, 73 & Max Planck Inst WGer, 75 & 76-77 & 81, Physiol Chem Inst, Univ Düsseldorf, WGer, 87. *Mem:* Am Chem Soc; AAAS; Nat Sci Teachers Asn; Sigma Xi. *Res:* Enzyme kinetic studies of pepsin; primary structure studies of collagen. *Mailing Add:* Nazareth Col Rochester 4245 E Ave Rochester NY 14618-3990

LANG, JAMES FREDERICK, b Dayton, Ohio, Mar 19, 31; m 58; c 2. DRUG METABOLISM. *Educ:* Univ Cincinnati, BS, 58, MS, 70. *Prof Exp:* Res asst toxicol & drug metab, Christ Hosp Inst Med Res, Subsid Elizabeth Gamble Deaconess Home Asn, 58-63; from res asst biochem to biochemist, Merrell Dow Pharmaceuticals, Inc, 63-72, sect head drug metab, 72-82, sr res biochemist, Merrell Dow Res Inst, 82-85, group leader, Drug Metab Clin Res Support, 85-90. *Mem:* Am Chem Soc. *Res:* Isolation and identification of drug metabolites, pharmacokinetics; development and application of analytical methods for trace analysis of drug residues in biological media. *Mailing Add:* 2894 Pineridge Ave Cincinnati OH 45208-2818

LANG, JOHN CALVIN, JR, b Montclair NJ, May 6, 42; m 66; c 1. PHYSICAL CHEMISTRY. *Educ:* Wesleyan Univ, BA, 64; Cornell Univ, MS, 68, PhD(chem), 72. *Prof Exp:* Fels, Cornell Univ, 72 & 73-75, Univ Reading, Eng, 72-73; res scientist phys chem, Procter & Gamble Co, 75-84; res assoc, Arco Oil & Gas Co, 84-86; ALCON LABS, 86-; ADJ ASSOC PROF CHEM, UNIV TEX ARLINGTON, 86- *Concurrent Pos:* Adv, Ctr Surface Sci & Eng, Univ Fl. *Mem:* Am Phys Soc; Sigma Xi; Am Chem Soc; Controlled Release Soc. *Res:* Phase equilibria; phase transitions; critical phenomena; thermodynamics of aqueous solutions; magnetic resonance; light scattering; surfactant, polymer and colloid physical chemistry; drug delivery, especially ophthalmologic; drug assessment. *Mailing Add:* 2106 Riverforest Dr Arlington TX 76017

LANG, JOSEPH EDWARD, b Covington, Ky, Aug 10, 42. THEORETICAL PHYSICS. *Educ:* Thomas More Col, AB, 64; Univ Ill, MS, 65, PhD(physics), 70. *Prof Exp:* NSF fel, Lawrence Radiation Lab, Univ Calif, Berkeley, 70-71; asst prof physics, 71-80, ASSOC PROF PHYSICS, THOMAS MORE COL, 80- *Concurrent Pos:* Vis scientist, Air Force Mat Lab, Wright-Patterson AFB, 78-79. *Mem:* Am Phys Soc; Am Asn Physics Teachers, Sigma Xi. *Res:* Elementary particle physics; magnetic anisotropy; computers in education. *Mailing Add:* Wright Bros Br PO Box 301 Dayton OH 45409

LANG, KENNETH LYLE, b Cuba City, Wis, Apr 12, 36; m 61; c 1. FRESHWATER ECOLOGY. *Educ:* Iowa State Col, BS, 59; Univ Iowa, MS, 66, PhD(zool), 70. *Prof Exp:* Assoc prof biol, 70-77, ASSOC PROF ZOOL, HUMBOLDT STATE UNIV, 77- *Mem:* Am Soc Zool; Am Soc Limnol & Oceanog. *Res:* Freshwater zooplankton populations; dispersion patterns and species diversity of benthic and planktonic Cladoceran assemblages. *Mailing Add:* Dept Biol Humboldt State Univ Arcata CA 95521

LANG, LAWRENCE GEORGE, b Pittsburgh, Pa, Mar 25, 31; m 53; c 2. SOLID STATE PHYSICS. *Educ:* Carnegie Inst Technol, BS, 52, MS, 53, PhD(physics), 57. *Prof Exp:* Res physicist, Carnegie Inst Technol, 56-57; eng specialist, Philco Corp, 57-58; res physicist, Carnegie Inst Technol, 58-60, asst prof physics, 60-64; res physicist, Atomic Energy Res Estab, Eng, 63-65; assoc prof physics, Carnegie Inst Technol, 64-66; res physicist, Atomic Energy Res Estab, Eng, 66-73; PROF PHYSICS, PA STATE UNIV, 73- *Concurrent Pos:* Nat Acad Sci-Nat Res Coun fel, 63-64. *Mem:* Am Phys Soc. *Res:* Angular correlation of radiation from positron annihilation in solids; Mossbauer effect in compounds; paramagnetic and diamagnetic salts; biological macromolecules. *Mailing Add:* Dept Physics Pa State Univ University Park PA 16802

LANG, MARTIN, ENVIRONMENTAL ENGINEERING. *Prof Exp:* PVT CONSULT ENGR, 89- *Mem:* Nat Acad Eng. *Mailing Add:* 11 A Pine Dr N Roslyn NY 11576

LANG, MARTIN T, b Yokohama, Japan, May 7, 36; US citizen; m 65; c 3. COMPUTER IN MATHEMATICS EDUCATION. *Educ:* N Cent Col, BA, 59; Univ Kans, MA, 63; Univ Tex-Austin, PhD(math educ), 73. *Prof Exp:* From lectr to asst prof math, San Diego State Univ, 64-69; from asst prof to assoc prof, 69-78, PROF MATH, CALIF POLYTECH STATE UNIV, 78- *Concurrent Pos:* Asst instr math, Univ Kans, 63-65; lectr, Grossmont Community Col, 68-69. *Mem:* Math Asn Am; Nat Coun Teachers Math; Am Asn Univ Profs; Nat Educ Asn. *Res:* Use of computers in math education at the college level; computer augmented instruction. *Mailing Add:* 1444 Tanglewood Ct San Luis Obispo CA 93401

LANG, NEIL CHARLES, b Montreal, Que, Jan 24, 48; US citizen. CHEMICAL PHYSICS. *Educ:* McGill Univ, BSc, 68; Mass Inst Technol, PhD(phys chem), 74. *Prof Exp:* CHEMIST, LASER PROG, LAWRENCE LIVERMORE LAB, 76- *Concurrent Pos:* Fel, Nat Res Coun Can, 74-76. *Mem:* Am Phys Soc; Sigma Xi. *Res:* Chemical physics, reaction dynamics and energy transfer processes in gas phase collisions; applications in laser technology and isotope enrichment. *Mailing Add:* 670 Vernon St No 404 Oakland CA 94610

LANG, NORMA JEAN, b Memphis, Tenn, July 25, 31. PHYCOLOGY. *Educ:* Ohio State Univ, BS, 52, MA, 58; Ind Univ, PhD(bot), 62. *Prof Exp:* NIH res fel algae, Univ Tex, 62-63; from asst prof to assoc prof bot, 63-74, PROF BOT, UNIV CALIF, DAVIS, 74- *Concurrent Pos:* NSF res grant, 65-67; Guggenheim fel, Westfield Col, London, 69-70. *Honors & Awards:* Darbaker Award in phycol, 69. *Mem:* Phycol Soc Am (treas, 71-73, vpres, 74, pres, 75); Brit Phycol Soc; Electron Micros Soc Am; Int Phycol Soc; Bot Soc Am. *Res:* Electron microscopic studies of development in unicellular green algae, colonial green algae and cyanobacteria, especially differentiation of heterocysts; desiccation survival and development of colonial form. *Mailing Add:* Dept Bot Univ Calif Davis CA 95616

LANG, NORMA M, b Wausau, Wis, Dec 27, 39; c 2. NURSING & PUBLIC POLICY. *Educ:* Alverno Col, BSN, 61; Marquette Univ, MSN, 63, PhD(educ admin), 74. *Prof Exp:* Staff Nurse & asst instr, St Joseph's Hosp, 61-62, Instr & Coordr Med-Surg Nursing, St Mary's Sch Nursing, 64-65; instr & coordr med-surg nursing, St Marys sch nursing, 64-65; intr & asst prof, 65-68, from asst prof to assoc prof, 68-75, prof & proj dir res develop grant, 77-79, PROF & DEAN, SCH NURSING, UNIV WIS, MILWAUKEE, 80- *Concurrent Pos:* Medicus Corp, 86; nursing coordr, Wis Regional Med Prog, 68-73; res assoc, Sch Nursing, Univ Wis, Milwaukee, 77, ctr scientist, Urban Res Ctr, 77-79. *Mem:* Inst Med-Nat Acad Sci; fel Am Acad Nursing; Am Nurses' Asn; Am Heart Asn; Am Asn Univ Profs; Am Pub Health Asn. *Res:* Nursing; nursing and policy; education administration; numerous articles and publications. *Mailing Add:* Univ Wis Sch Nursing PO Box 413 Milwaukee WI 53201

LANG, NORTON DAVID, b Chicago, Ill, July 5, 40; m 69; c 2. SOLID STATE & SURFACE PHYSICS. *Educ:* Harvard Univ, AB, 62, AM, 65, PhD(physics), 68. *Prof Exp:* Asst res physicist, Univ Calif, San Diego, 67-69; STAFF MEM, IBM CORP, 69- *Concurrent Pos:* Assoc ed, Phys Review Letters, 80-83; chmn of the fel comt, Div of Condensed Matter Physics, Am Phys Soc, 85-87, chmn, Davisson-Germer prize comt, 90. *Honors & Awards:* Davisson-Germer Prize, Am Phys Soc, 77. *Mem:* Fel NY Acad Sci; fel Am Phys Soc; Europ Phys Soc. *Res:* Solid state physics. *Mailing Add:* IBM Res Ctr Yorktown Heights NY 10598

LANG, PETER MICHAEL, b Vienna, Austria, Sept 3, 30; US citizen; m 74. CHEMICAL & NUCLEAR ENGINEERING. *Educ:* Mass Inst Technol, SB, 51, SM, 52, ScD(chem eng), 55. *Prof Exp:* Res asst chem eng, Mass Inst Technol, 52-53 & 54-55; chem engr, Shell Develop Co, 55-56; nuclear engr, ACF Industs, 56-59 & Allis-Chalmers Mfg Co, 59-66; staff consult, NUS Corp, 66-73; div mgr, Aerojet Nuclear Co, 73-77; PROG MGR, DEPT ENERGY, 77- *Concurrent Pos:* Lectr, Univ Md, 60; vis scientist, Swiss Fed

Inst Reactor Res, 62-64. *Mem:* Am Nuclear Soc. *Res:* Development of nuclear power plant improvements, design, economics, and safety of nuclear plants; nuclear fuel improvements; ultra-safe reactor designs. *Mailing Add:* US Dept Energy Washington DC 20545

LANG, PHILIP CHARLES, b Jamestown, NY, Nov 16, 34; m 55; c 3. ORGANIC CHEMISTRY. *Educ:* Allegheny Col, BS, 57; Ohio Univ, MS, 59; Rensselaer Polytech Inst, PhD(org chem), 66. *Prof Exp:* Res chemist, Diamond Alkali Co, 59-62; assoc res chemist, Sterling Winthrop Res Inst, 62-67; res chemist, GAF Corp, 67-73; res chemist, Toms River Chem Corp, 73-80; SR RES CHEMIST, CIBA-GEIGY CORP, 80- *Mem:* Am Chem Soc. *Res:* Synthetic medicinal chemistry; heterocyclic and acteylene compounds; aromatics and synthetic dyes. *Mailing Add:* Ciba-Geigy Pharm Div 556 Morris Ave Summit NJ 07901

LANG, RAYMOND W, b Syracuse, NY, Aug 1, 30; m 53; c 5. MICROBIOLOGY. *Educ:* LeMoyne Col, NY, BS, 52; Mich State Univ, MS, 57, PhD(microbiol), 59. *Prof Exp:* Asst prof microbiol, St John's Univ, NY, 59-62; fel bact & immunol, Sch Med, State Univ NY Buffalo, 62-63; from instr to asst prof bact & immunol, 63-68; assoc prof, 68-72, PROF MED MICROBIOL, COL MED, OHIO STATE UNIV, 72- *Concurrent Pos:* Consult urol res sect, Millard Fillmore Hosp, 64-66; consult training prog, Nat Inst Dent Res, 72-75; consult, Ohio Dept Health, 83- *Mem:* AAAS; Am Soc Microbiol; Am Asn Immunol. *Res:* Immunochemistry of tissue antigens; autoimmunity. *Mailing Add:* Dept Med Microbiol Immunol Ohio State Univ Col Med Columbus OH 43210

LANG, ROBERT LEE, agronomy; deceased, see previous edition for last biography

LANG, ROBERT PHILLIP, b Chicago, Ill, June 15, 32. PHYSICAL CHEMISTRY. *Educ:* Univ Ill, BS, 55; Univ Chicago, MS, 60, PhD(phys chem), 62. *Prof Exp:* From instr to assoc prof 62-74, chmn dept, 71-88, PROF CHEM, QUINCY COL, 74- *Concurrent Pos:* NSF res grant, 65-68 & 70-72. *Mem:* AAAS; Am Chem Soc; Am Asn Physics Teachers. *Res:* Thermodynamic and electronic spectral characteristics of molecular complexes of iodine; amino acid sequences in enzymes. *Mailing Add:* Dept Chem Quincy Col Quincy IL 62301

LANG, ROGER H, b New York, NY, July 8, 40. COMMUNICATION THEORY. *Educ:* Polytech Inst Brooklyn, BSEE, 62, MSEE, 64, PhD(electrophys), 68. *Prof Exp:* PROF ENG APPL SCI, GEORGE WASHINGTON UNIV, 70-, CHMN DEPT, 84- *Concurrent Pos:* Fel Nat Res Consult. *Mem:* Fel Inst Elec & Electronics Engrs. *Mailing Add:* George Washington Univ 725 23rd St Washington DC 20052

LANG, SERGE, b Paris, France, May 19, 27. NUMBER THEORY, ALGEBRAIC GEOMETRY. *Educ:* Calif Tech Univ, BS, 46; Princeton Univ, PhD(math), 51. *Prof Exp:* Instr math, Princeton Univ, 51-52 & Univ Chicago, 53-55; fel, Inst Advan Studies, 52-53; from asst prof to prof, Columbia Univ, 55-71; PROF MATH, YALE UNIV, 72- *Concurrent Pos:* Vis scholar, Inst Advan Study, 52-53; Fulbright fel, 56-57; Humboldt award, 84. *Honors & Awards:* Cole Prize, Am Math Soc, 59; Prix Carriere, French Acad Sci, 67. *Mem:* Nat Acad Sci; Am Math Soc. *Mailing Add:* Dept Math Yale Univ 12 Hillhouse Ave New Haven CT 06520

LANG, SOLOMON MAX, hydrology, for more information see previous edition

LANG, STANLEY ALBERT, JR, b Cleveland, Ohio, Mar 30, 44. ORGANIC CHEMISTRY. *Educ:* John Carroll Univ, BS, 66; Brown Univ, PhD(org chem), 70. *Prof Exp:* Res fel, Ohio State Univ, 70, Nat Cancer Inst fel, 71; res chemist, Lederle Labs, 72-74, group leader, Info Dis Ther Sect, 74-77, group leader, 77-80, HEAD, CHEM DEPT, INFECTIOUS DIS THER SECT, MED RES DIV, AM CYANAMID INC, 80- *Mem:* Am Chem Soc. *Res:* Synthetic organic chemistry; medicinal drugs; antibiotics, anticancer agents; immunoregulants. *Mailing Add:* Seven Colony Dr Blauvelt NY 10913-1319

LANG, THOMAS G(LENN), b San Jose, Calif, July 28, 28; m 62; c 2. MECHANICAL ENGINEERING. *Educ:* Calif Inst Technol, BS, 48 & 50; Univ Southern Calif, MS, 53; Pa State Univ, PhD(aerospace eng), 68. *Prof Exp:* Operator, Southern Calif Coop Wind Tunnel, 48-49; stress analyst, NAm Aviation, Inc, 50-51; designer, US Naval Ord Test Sta, 51-52, hydrodynamicist, 52-58, head oceanic res group, 58-61, head hydrodyn res group, 61-66; US Naval Ord Test Sta scholar, Pa State Univ, 66-68; tech consult, Ocean Technol Dept, Naval Undersea Warfare Ctr, 68; head advan design, Systs Anal Group, Naval Undersea Ctr & Naval Ocean Systs Ctr, 68-70, head advan concepts group, 70-73, head, Advan Concepts Div, 73-78; RETIRED. *Concurrent Pos:* Consult, 78-79; pres Semi-Submerged Ship Co, 79- *Mem:* Am Inst Aeronaut & Astronaut; Marine Technol Soc; Soc Naval Architects & Marine Engrs. *Res:* Hydrodynamics, especially stability and control, propulsion, boundary layer control, vented hydrofoils, sea animal hydrodynamics, polymer additives for drag reduction; semisubmerged ship design. *Mailing Add:* 417 Loma Larga Dr Solana Beach CA 92075

LANG, WILLIAM HARRY, b Etna, Pa, Mar 29, 18; m 46; c 2. PETROLEUM CHEMISTRY, FUEL SCIENCE. *Educ:* Grove City Col, BS, 40. *Prof Exp:* Sr res chemist, Cent Res Div, Mobil Res & Develop Corp, 40-81; RETIRED. *Mem:* Am Chem Soc; Catalysis Soc. *Res:* Development of new and alternate fuels and energy sources to use as a petroleum substitute; co-inventor of methanol to gasoline and methanol to olefins processes. *Mailing Add:* Box 186 RR 2 Johnny Brook Rd Richmond VT 05477

LANG, WILLIAM WARNER, b Boston, Mass, Aug 9, 26; m 54; c 1. ACOUSTICAL & NOISE CONTROL ENGINEERING. *Educ:* Iowa State Univ, BS, 46, PhD(physics), 58; Mass Inst Technol, SM, 49. *Prof Exp:* Acoust engr, Bolt, Beranek & Newman, Inc, 49-51; instr physics, US Naval Post-Grad Sch, 51-55; spec engr, E I du Pont de Nemours & Co, Inc, 55-57; adv physicist, IBM Corp, 58-64, sr physicist & mgr acoust lab, 64-75, prog mgr acoust technol, 75-90, SR TECH STAFF MEM, IBM CORP, 91- *Concurrent Pos:* Mem eval panel, Mech Div, Nat Bur Standards, 74-75, chmn, 75-76; adj prof physics, Vassar Col, 79- *Honors & Awards:* Achievement Award, Inst Elec & Electronics Engrs Group Audio & Electroacoustics, 72, Centennial Medal, 84; Silver Medal, Acoust Soc Am, 84. *Mem:* Nat Acad Eng; AAAS; Inst Noise Control Eng (pres, 78, 88); fel Acoust Soc Am; fel Inst Elec & Electronics Engrs; fel Audio Eng Soc. *Res:* Acoustics; effects and control of noise; theory and design of acoustical materials. *Mailing Add:* 29 Hornbeck Ridge Poughkeepsie NY 12603

LANGACKER, PAUL GEORGE, b Evanston, Ill, July 14, 46; m 83. THEORETICAL PHYSICS, ELEMENTARY PARTICLE PHYSICS. *Educ:* Mass Inst Technol, BS, 68; Univ Calif, Berkeley, MA, 69, PhD(physics), 72. *Hon Degrees:* MS, Univ Pa, 81. *Prof Exp:* Res assoc, Rockefeller Univ, 72-74; res assoc, 74-75, from asst prof to assoc prof, 75-85, PROF PHYSICS, UNIV PA, 85- *Concurrent Pos:* Alexander von Humboldt award, 87-88. *Mem:* Fel Am Phys Soc; AAAS. *Res:* Theoretical elementary particle physics. *Mailing Add:* Dept Physics Univ Pa Philadelphia PA 19104-6396

LANGAGER, BRUCE ALLEN, b Willmar, Minn, Jan 17, 42; m 64; c 2. POLYMER CHEMISTRY. *Educ:* Augsburg Col, BA, 64; Univ Minn, Minneapolis, PhD(org chem), 68. *Prof Exp:* Sr res chemist, 68-74, res specialist, 74-78, TECH SUPVR, 3M CO, 78- *Mem:* Sigma Xi. *Res:* Life sciences; surface chemistry. *Mailing Add:* 3104 13th Terr NW New Brighton St Paul MN 55112

LANGAN, THOMAS AUGUSTINE, b Providence, RI, July 25, 30; m 60; c 2. BIOCHEMISTRY. *Educ:* Fordham Univ, BS, 52; Johns Hopkins Univ, PhD(biochem), 59. *Prof Exp:* Mem res staff, Med Nobel Inst, Stockholm, Sweden, 59-60; guest investr biochem, Rockefeller Inst, 60-61; mem res staff, Wenner-Gren Inst, Stockholm, Sweden, 61-62; res assoc biochem, Rockefeller Inst, 62-65; staff scientist, C F Kettering Res Lab, 65-67, investr, 67-70, sr investr, 70-71; assoc prof, 71-83, PROF PHARMACOL, MED SCH, UNIV COLO, DENVER, 83- *Concurrent Pos:* NSF fel, 59-62; from asst prof to assoc prof, Antioch Col, 67-71. *Mem:* Am Soc Biol Chemists; Am Soc Cell Biol; Am Soc Pharmacol & Exp Therapeut. *Res:* Metabolism and function of histones and nuclear phosphoproteins; control of histone phosphorylation by cyclic adenosine monophosphate and cell growth; regulation of nucleic acid synthesis in eukaryotes; effects of histone phosphorylation on chromotin structure; role of cdc2 protein kinase substrates in control of cell cycle progression. *Mailing Add:* Dept Pharmacol Univ Colo Med Ctr 4200 E Ninth Ave Denver CO 80262

LANGAN, WILLIAM BERNARD, b Wayne Co, Pa, Oct 31, 13; m 57; c 2. PHYSIOLOGY. *Educ:* Univ Scranton, BS, 36; Columbia Univ, MA, 37 & 44; Fordham Univ, PhD(zool), 42. *Prof Exp:* Asst sci, Teachers Col, Columbia Univ, 36-37; asst biol, Fordham Univ, 40-42; instr physiol, New York Med Col, 42-47, assoc, 47-50, asst prof prof, 50-60, asst prof pharmacol, 58-60; sect head physiol, Food & Drug Res Labs, Inc, 60-61; NIH spec fel, State Univ NY Downstate Med Ctr, 61-62, asst prof, 62-63; assoc prof biol, 63-80, ASST TO DIR RES, VILLANOVA UNIV, 81- *Concurrent Pos:* Lectr, Hunter Col, 60-62; consult, Food & Drug Res Labs, Inc, 61-63; training prog partic, Int Lab Genetics & Biophys, Italy, 65; vis prof, Dept Pharmacol, Thomas Jefferson Med Col, Philadelphia, 80- *Mem:* Fel AAAS; Harvey Soc; Endocrine Soc; Am Soc Zool; NY Acad Sci; Sigma Xi. *Res:* Steroids and cardiac electrophysiology; steroids and ovulation in sub-mammalian species; mechanism of hormone action at the cellular level. *Mailing Add:* Five Langan Ave Hawley PA 18428

LANGDELL, ROBERT DANA, b Pomona, Calif, Mar 14, 24; m 48; c 2. PATHOLOGY. *Educ:* George Washington Univ, MD, 48. *Prof Exp:* Intern, Henry Ford Hosp, Detroit, Mich, 48-49; fel path, Sch Med, Univ NC, Chapel Hill, 49-51, from instr to assoc prof, 51-64, PROF PATH, SCH MED UNIV NC, CHAPEL HILL, 64- *Concurrent Pos:* USPHS sr res fel, 56-62 & career develop award, 62-67; mem hemat study sect, USPHS, 67-70; ed-in-chief, Transfusion, 72-82; pres, Am Asn Blood Banks, 72-73; mem panel rev blood & blood derivatives, Food & Drug Admin, 75-80. *Mem:* AMA; Am Soc Clin Path; Col Am Path. *Res:* Hematologic pathology; physiology of blood coagulation; hemorrhagic disorders. *Mailing Add:* Dept Path Univ NC Sch Med Chapel Hill NC 27514

LANGDON, ALLAN BRUCE, b Edmonton, Alta, Dec 14, 41; m 66; c 3. PLASMA THEORY, COMPUTATIONAL PHYSICS. *Educ:* Univ Man, BSc, 63; Princeton Univ, PhD(astrophys), 69. *Prof Exp:* Actg asst prof elec eng, Univ Calif, Berkeley, 67-69; STAFF PHYSICIST, LAWRENCE LIVERMORE LAB, 70- *Concurrent Pos:* Lectr elec eng, Univ Calif, Berkeley, 69-73 & 81-82; affil mem, Ctr Plasma Physics & Fusion Eng, Univ Calif, Los Angeles, 77-78. *Mem:* Fel Am Phys Soc; Sigma Xi; Can Asn Physicists; Asn Comput Mach. *Res:* Plasma theory; computational physics; computer simulation of plasmas; numerical analysis. *Mailing Add:* L-472 Lawrence Livermore Lab Box 808 Livermore CA 94550

LANGDON, EDWARD ALLEN, b Los Angeles, Calif, Feb 9, 22. MEDICINE, RADIOLOGY. *Educ:* Western Reserve Univ, BS, 42; Univ Mich, MD, 45. *Prof Exp:* From asst prof to assoc prof, 59-70, prof radiol, Chief Radiother Div & asst dean student affairs, 70-78, PROF RADIOL ONCOL, VCHMN, DEPT RADIOL ONCOL & ASST DEAN STUDENT AFFAIRS, SCH MED, UNIV CALIF, LOS ANGELES, 70- *Mem:* AMA; Am Col Radiol; Radiol Soc NAm; Soc Nuclear Med; Asn Univ Radiol. *Res:* Radiation therapy. *Mailing Add:* Dept Radiol Radio Ther Div Univ Calif Med Ctr 405 Hilgard Ave Los Angeles CA 90024

LANGDON, GLEN GEORGE, JR, b Morristown, NJ, June 30, 36; m 63; c 1. COMPUTER ENGINEERING, COMPUTER SCIENCE. *Educ:* Wash State Univ, BSEE, 57; Univ Pittsburgh, MSEE, 63; Syracuse Univ, PhD(elec eng), 68. *Prof Exp:* Engr, Westinghouse Elec Corp, 60-63; engr, 63-73, res staff computer sci, IBM Corp, 74-87; PROF COMPUTER ENG, UNIV CALIF, SANTA CRUZ, 87- *Concurrent Pos:* Lectr, Syracuse Univ, NY, 68-69, Univ Santa Clara, 75-78, 80 & Stanford Univ, 84-85; vis prof, Univ Sao Paulo, Brazil, 71-72; mem gov bd, Comput Soc, Inst Elec & Electronics Engrs, 84-85. *Mem:* Fel Inst Elec & Electronics Engrs; Asn Comput Mach. *Res:* Computer and logic design; image compression and processing. *Mailing Add:* 220 Horizon Way Aptos CA 95003

LANGDON, HERBERT LINCOLN, b Malone, NY, July 7, 35; m 72. GROSS ANATOMY, DEVELOPMENTAL ANATOMY. *Educ:* St Lawrence Univ, BS, 57; Univ Mo, MA, 63; Univ Miami, PhD(biol struct), 72. *Prof Exp:* Asst prof biol, Miami-Dade Community Col, 65-68; asst prof anat, 72-78, ASSOC PROF ANAT, SCH DENT MED, UNIV PITTSBURGH, 78- *Concurrent Pos:* Res consortium, Cleft Palate Ctr, Univ Pittsburgh, 76. *Mem:* Am Cleft Palate Asn; Sigma Xi; Am Asn Anatomists; Soc Craniofacial Genetics. *Res:* Normal and abnormal morphology and development of human tongue and velopharyngeal mechanism; craniofacial development and growth. *Mailing Add:* Univ Pittsburgh 620 Salk Hall Pittsburgh PA 15261

LANGDON, KENNETH R, b Cache, Okla, Aug 20, 28; m 61; c 3. PLANT TAXONOMY, ENDANGERED SPECIES. *Educ:* Okla State Univ, BS, 58, MS, 60; Univ Fla, PhD(plant path, nematol, bot), 63. *Prof Exp:* Nematologist & botanist, div Plant Indust, Fla Dept Agr & Consumer Serv, 63-91; RETIRED. *Mem:* Soc Europ Nematologist; Int Asn Plant Taxon; Am Soc Plant Taxonomists. *Res:* Plant systematics. *Mailing Add:* 2216 NW 49th Terr Gainesville FL 32605

LANGDON, ROBERT GODWIN, b Dallas, Tex, Jan 18, 23; m 45; c 4. BIOCHEMISTRY. *Educ:* Univ Chicago, MD, 45, PhD(biochem), 53. *Prof Exp:* From instr to prof physiol chem, Sch Med, Johns Hopkins Univ, 53-67; prof biochem & chmn dept, Col Med, Univ Fla, 67-69; chmn dept, Univ Va, 77-82, prof biochem, 69-; RETIRED. *Concurrent Pos:* USPHS fel, Univ Chicago, 51-53; Lederle award, 54-57. *Mem:* Am Soc Biol Chemists; Am Chem Soc. *Res:* Membrane biochemistry; glucose transport; mechanism of hormone action. *Mailing Add:* Buck Creek Farm PO Box 464 Nemo TX 76070

LANGDON, WILLIAM KEITH, b Hubbardston, Mich, May 8, 16; m 40; c 4. ORGANIC CHEMISTRY. *Educ:* Mich State Univ, BS, 38, MS, 40. *Prof Exp:* Res chemist, Chrysler Corp, 39-43; res chemist, Wyandotte Chem Corp, 43-69, res supvr org chem, BASF Wyandotte Corp, 69-81; RETIRED. *Mem:* Am Chem Soc; Sigma Xi. *Res:* New organic chemical synthesis including heterocylic nitrogen compounds, surfactants, acetals, epoxides and plastic intermediates. *Mailing Add:* 8091 O'Donnell Grosse Ile MI 48138

LANGDON, WILLIAM MONDENG, b LaGrange, Ill, Aug 18, 14; m 42, 71; c 4. CHEMICAL ENGINEERING. *Educ:* Univ Ill, BS, 35, MS, 39, PhD(chem eng), 41. *Prof Exp:* Testing engr & metallurgist, Int Harvester Co, Ill, 35-37; asst chem & chem eng, Univ Ill, 37-41, asst prof chem eng, 43-46; res chem engr, Standard Oil Develop Co, NJ, 41-42; sr mat engr, War Prod Bd, Wash, 43-45; assoc prof chem eng, Rensselaer Polytech Inst, 46-48; assoc dir res, Graham Crowley & Assocs, Inc, 48-55; sr engr, Ill Res Inst, 55-69, prof, 69-79, emer prof chem eng, Ill Inst Technol, 79-84; RETIRED. *Mem:* Am Chem Soc; Am Inst Chem Engrs. *Res:* Column operation; vapor-liquid equilibrium; azeo and extractive distillation; synthetic organic chemical manufacturing; environmental control and design. *Mailing Add:* 1570 W Camino Urbano Green Valley AZ 85614

LANGE, BARRY CLIFFORD, b Philadelphia, Pa, June 14, 52; m 74; c 2. PROCESS RESEARCH, FORMULATION CHEMISTRY. *Educ:* Stevens Inst Technol, BS, 74; Pa State Univ, PhD(chem), 79. *Prof Exp:* Process chemist, 80-84, agr chem discovery chemist, 84-87, RES MGR, ROHM & HAAS CO, 87- *Mem:* Am Chem Soc. *Res:* Research management, design of biologically active compounds, chemical process research, kinetics and mechanisms of organic reactions; synthesis of penicillin and cephalosporin analogs and other antibiotics; physical organic chemistry including enolate alkylation; polymer chemistry; agricultural and food chemistry; biochemistry. *Mailing Add:* 1031 Barley Way Lansdale PA 19446-3200

LANGE, BRUCE AINSWORTH, b Springfield, Mass, Aug 3, 48; m 77; c 2. CHEMISTRY OF SURFACES, CEMENT CHEMISTRY. *Educ:* Lowell Technol Inst, BS, 70; Univ NH, PhD(chem), 74. *Prof Exp:* Fel chem, Northwestern Univ, 74-75, Univ Cincinnati, 75-76; res chemist, Nat Inst Occup Safety & Health, Cincinnati, Ohio, 76-79; group leader chem, W R Grace & Co, Cambridge, Mass, 79-83; group leader, 83-86, BR MGR, ANALYTICAL CHEM, ATOMIC ENERGY CAN, PINAWA, MAN, 86- *Mem:* Am Ceramic Soc. *Res:* X-ray crystallography; inorganic synthesis; radiopharmaceuticals; development of analytical methods for hazardous materials; surface chemistry; cement chemistry; comminution of minerals; process design; computer modelling; analytical chemistry. *Mailing Add:* Whiteshell Labs Pinawa MB R0E 1L0 Can

LANGE, CHARLES FORD, b Chicago, Ill, Feb 16, 29; m 53; c 3. BIOCHEMISTRY, IMMUNOLOGY. *Educ:* Roosevelt Univ, BS, 51, MS, 53; Univ Ill, PhD(biochem), 59; Am Bd Med Lab Immunol, dipl, 81. *Prof Exp:* Res assoc biochem, Univ Ill, 60-61; res assoc, Hektoen Inst Med Res, Cook County Hosp, 61-63, head phys chem, 63-69; assoc prof microbiol, 70-75, PROF MICROBIOL, STRITCH SCH MED, LOYOLA UNIV, 75- *Concurrent Pos:* Consult immunologist, Hines Vet Admin Hosp, 75- *Mem:* Am Chem Soc; Am Soc Biochem & Molecular Biol; Transplantation Soc; Am Soc Microbiol; Am Asn Immunol; AAAS. *Res:* Urinary glycoproteins; immunochemistry of streptococcal related glomerulonephritis; streptococcal M-proteins; transplantation antigens; immunology of aging; autoimmune diseases; monoclonal antitissue antibodies. *Mailing Add:* Dept Microbiol Stritch Sch Med Loyola Univ Maywood IL 60153

LANGE, CHARLES GENE, b Chattanooga, Tenn, Mar 30, 42; m 63; c 3. APPLIED MATHEMATICS. *Educ:* Tri-State Col, BS, 63; Case Inst Technol, MS, 65; Mass Inst Technol, PhD(appl math), 68. *Prof Exp:* Asst prof, 68-75, ASSOC PROF MATH, UNIV CALIF, LOS ANGELES, 75- *Mem:* Soc Indust & Appl Math; Am Math Soc. *Res:* Nonlinear random processes; singular perturbation techniques; nonlinear stability theory; elasticity; wave propagation. *Mailing Add:* Dept Math Univ Calif 405 Hilgard Ave Los Angeles CA 90024

LANGE, CHRISTOPHER STEPHEN, b Chicago, Ill, Feb 11, 40; m 64, 73; c 2. RADIATION BIOPHYSICS, GENOME STRUCTURE. *Educ:* Mass Inst Technol, Cambridge, BS, 61; Oxford Univ, Eng, DPhil, 68. *Prof Exp:* MRC res asst radiobiol, Churchill Hosp, Headington, Oxford, 61-62; NHS res officer radiobiol, Christie Hosp & Holt Radium Inst, Manchester, Eng, 62-68, NHS sr res officer, 68-69; asst prof radiol, radiobiol, biophys, Sch Med & Dent, Univ Rochester, 69-80; PROF RADIATION ONCOL & DIR RADIATION RES, STATE UNIV NY, HEALTH SCI CTR BROOKLYN, 80- *Concurrent Pos:* Prin investr, grants USDOE, NSF, NCI, NIGMS, Mathers Found, etc, 72-; vis radiobiologist, Am Inst Biol Sci, 73-79; vis prof chem, Univ Calif San Diego, 75-76; presidential distinguished lectr, Univ Hirosaki, Japan, 79; guest scientist, Brookhaven Nat Lab, 83-; tumor biol comt mem, Radiation Ther Oncol Group, Nat Proj for NIH/NCI, 88-; scholar adv comt, Kosciuszko Found, 89-; res career develop award, USDHEW/NCI, 72-77. *Mem:* Radiation Res Soc; Biophys Soc; NY Acad Sci; Sigma Xi; AAAS. *Res:* Molecular and cellular bases of cellular and organisimal radiation effects and aging; DNA damage and repair; DNA structure in mammalian chromosomes; viscoelastometry and other hydrodynamic behavior of DNA; differentiation control (polarity) in tissues; assays for improved cancer therapy. *Mailing Add:* State Univ NY Health Sci Ctr 450 Clarkson Ave Box 1212 Brooklyn NY 11203

LANGE, EUGENE ALBERT, b Stevens Pt, Wis, Oct 22, 23; m 51. STRUCTURAL INTEGRITY TECHNOLOGY, FRACTURE MECHANICS. *Educ:* Univ Wis, BS, 45, MS, 51. *Prof Exp:* Res metal, Univ Wis, 51-53; res engr, Gray Iron Res Inst, 53-56; surv res metal, Naval Res Lab, 56-80; consult, Eugene A Lange, 80-90; RETIRED. *Concurrent Pos:* Lectr, Union Col, Schenectady, NY, 70-88. *Honors & Awards:* Res Pub Award, Naval Res Lab, 68. *Mem:* Am Soc Metals. *Res:* Casting technology; non-magnetic steels; fracture mechanics; dynamic fracture toughness. *Mailing Add:* 5101 River Rd Apt 606 Bethesda MD 20816

LANGE, GAIL LAURA, b Chicago, Ill, June 28, 46. ALGEBRA. *Educ:* Univ Md, BS, 67; Univ NH, MS, 69, PhD(math), 75. *Prof Exp:* Instr math, Univ Maine, Farmington, 72-75, asst prof, 75-80; CONSULT, 80- *Mem:* Am Math Soc; Math Asn Am. *Res:* Investigation of which finite p-groups can be the Frattini subgroup of finite p-groups. *Mailing Add:* Shaw Hill Rd RFD 1 Box 1203 Farmington ME 04938

LANGE, GORDON DAVID, b Douglas, Ariz, Jan 15, 36; c 3. NEUROPHYSIOLOGY, BIOLOGICAL OCEANOGRAPHY. *Educ:* Calif Inst Technol, BS, 58; Rockefeller Univ, PhD(biophys), 65. *Prof Exp:* Res assoc biophys, Rockefeller Univ, 65-66; asst res neuroscientist, 66-68, asst prof, 68-74, assoc prof neurosci, Univ Calif, San Diego, 74-; LAB NEUROPHYSIOL, NAT INST NEUROL & COMMUN DISORDERS & STROKE, NIH. *Mem:* Soc Neurosci; NY Acad Sci; Sigma Xi. *Res:* Neurophysiology of sensory systems; studies of the dynamics of interactions among nerve cells; mathematical and computer models of interactions of organisms. *Mailing Add:* Nat Inst Neurol Disorders & Stroke Div Intramural Res Bldg 36 Rm 2A03 Bethesda MD 20892

LANGE, GORDON LLOYD, b Edmonton, Alta, Mar 1, 37; m 64; c 2. TERPENOID SYNTHESIS, PHOTOCHEMICAL ASYMMETRIC INDUCTION. *Educ:* Univ Alta, BSc, 59; Univ Calif, Berkeley, PhD(org chem), 63. *Prof Exp:* Res chemist, Procter & Gamble Co, 62-65; lectr org chem, Univ Western Ont, 65-67; from asst prof to assoc prof chem, 67-84, PROF CHEM & BIOCHEM, UNIV GUELPH, 85- *Honors & Awards:* Union Carbide Award for Chem Educ, 86. *Mem:* Am Chem Soc; Chem Inst Can. *Res:* Synthesis of natural products and compounds with potential biological activity; organic photochemistry; structural elucidation of natural products. *Mailing Add:* Dept Chem & Biochem Univ Guelph Guelph ON N1G 2W1 Can

LANGE, IAN M, b New York, NY, Nov 11, 40. GEOLOGY, GEOCHEMISTRY. *Educ:* Dartmouth Col, BA, 62, MA, 64; Univ Wash, PhD(geol), 68. *Prof Exp:* Asst prof geol, Fresno State Col, 68-73; assoc prof, 73-77, PROF GEOL, UNIV MONT, 77-, CHMN DEPT. *Mem:* Geochem Soc; Geol Soc Am. *Res:* Isotope geology, economic geology. *Mailing Add:* Dept Geol Univ Mont Missoula MT 59801

LANGE, JAMES NEIL, JR, b Bridgeport, Conn, May 4, 38; m 58; c 2. PHYSICS. *Educ:* Pa State Univ, PhD(physics), 64. *Prof Exp:* From asst prof to assoc prof, 65-71, PROF PHYSICS, OKLA STATE UNIV, 71-, REGENTS PROF PHYSICS, 81-; VPRES, INTERCOMP INC, STILLWATER, 85- *Concurrent Pos:* Vis prof, Univ Nottingham, 76, Nat Univ Mex, 78; sr vis fel, Gt Brit, 76. *Mem:* Am Phys Soc; Am Geophys Union. *Res:* Acoustics; geophysics. *Mailing Add:* Intercomp Inc Stillwater OK 74075

LANGE, KENNETH L, b Angola, Ind, June 16, 46; m 70; c 2. GENE MAPPING, COMPUTATIONAL PROBABILITY & STATISTICS. *Educ:* Mich State Univ, BS, 67; Mass Inst Technol, MS, 68, PhD(math), 71. *Prof Exp:* Asst prof math, Univ NH, 71-72; NIH postdoctoral fel biomath, Univ Calif, Los Angeles, 72-74, from asst prof to assoc prof, 74-83, PROF BIOMATH, UNIV CALIF, LOS ANGELES, 83-, DEPT CHAIR, 85- *Concurrent Pos:* NIH res career develop award, Univ Calif, Los Angeles, 79-84; vis prof statist, Mass Inst Technol, 83-84 & Harvard Univ, 90-91; mem, Joint Comt Math Life Sci, Am Math Soc & Soc Indust & Appl Math, 84-89. *Mem:* Soc Indust & Appl Math; Am Soc Human Genetics; Am Statist Asn. *Res:* Biomathematical modeling in genetics, medical imaging, demography, and physiology; applied stochastic processes and computational statistics. *Mailing Add:* Dept Biomath Univ Calif 10833 La Conte Ave Los Angeles CA 90024

LANGE, KLAUS ROBERT, b Berlin, Germany, Jan 15, 30; US citizen; m 51; c 2. PHYSICAL CHEMISTRY, SURFACE CHEMISTRY. *Educ:* Univ Pa, AB, 52; Univ Del, MS, 54, PhD(phys chem), 56. *Prof Exp:* Res chemist, Atlantic Refining Co, 55-59; sr res chemist, Philadelphia Quartz Co, 59-67, res assoc, 67-69; lab mgr, Betz Lab Inc, 69-74; tech dir, Quaker Chem Corp, 75-86, sr tech adv, 86-90; CONSULT, 90- *Concurrent Pos:* Mem bd dir, Chem Data Systs, 70-75. *Mem:* Am Chem Soc; Tech Asn Pulp & Paper Inst; Am Inst Chem; Am Soc Metals. *Res:* Physical adsorption; heterogeneous catalysis; surface chemistry of silica and related solids; silicate solutions, fundamental properties; colloidal suspensions; detergency; pollution control; polymer applications; lignin and paper chemistry. *Mailing Add:* 805 Lombard St Philadelphia PA 19147

LANGE, KURT, internal medicine, nephrology; deceased, see previous edition for last biography

LANGE, LEO JEROME, b New Rockford, NDak, Aug 29, 28; m 55; c 4. MATHEMATICS. *Educ:* Regis Col, Colo, BS, 52; Univ Colo, MA, 56, PhD(math), 60. *Prof Exp:* Instr math, Univ Colo, 52-56, asst, 58-60; mathematician, Boulder Labs, Nat Bur Standards, 56-60; from asst prof to assoc prof, 60-83, PROF MATH, UNIV MO-COLUMBIA, 83-, DEPT CHAIR, 88- *Mem:* Math Asn Am; Am Math Soc. *Res:* Continued fractions; complex analysis; approximations & expansions. *Mailing Add:* Dept Math Univ Mo Columbia MO 65211

LANGE, LESTER HENRY, b Concordia, Mo, Jan 2, 24; m 47, 62; c 5. MATHEMATICS. *Educ:* Valparaiso Univ, AB, 48; Stanford Univ, MS, 50; Univ Notre Dame, PhD, 60. *Prof Exp:* Instr math, Valparaiso Univ, 50-53, asst prof, 54-56; instr, Univ Notre Dame, 56-57 & 59-60; from asst prof to prof, San Jose State Univ, 60-70, actg head dept, 61-62, chmn dept, 62-70, dean sch sci, 70-88, EMER DEAN & EMER PROF MATH, SAN JOSE STATE UNIV, 88-; SPEC ASST TO DEAN, MOSS LANDING MARINE LAB, CALIF, 88- *Honors & Awards:* L R Ford Sr Award, Math Asn Am, 72. *Mem:* AAAS; Math Asn Am; London Math Soc; Nat Coun Teachers Math. *Res:* Complex variable; topology. *Mailing Add:* Spec Asst to Dean Moss Landing Marine Lab Moss Landing CA 95039

LANGE, ROBERT CARL, b Stoneham, Mass, Aug 26, 35; m 59, 82; c 2. MAGNETIC RESONANCE IMAGING, RADIOLOGICAL PHYSICS. *Educ:* Northeastern Univ, BS, 57; Mass Inst Technol, PhD(chem), 62. *Prof Exp:* Group leader physics, Monsanto Res Corp, 62-69; asst prof, 69-76, ASSOC PROF RADIOL PHYSICS, SCH MED, YALE UNIV, 76- *Concurrent Pos:* Consult, RJ Schulz Assocs, 73-; tech dir magnetic resonance imaging, Yale New Haven Hosp, 69- *Mem:* AAAS; Am Chem Soc; Am Phys Soc; Soc Nuclear Med; Sigma Xi. *Res:* Magnetic resonance imaging; computer applications to medicine. *Mailing Add:* Dept Radiol MR Sect Sch Med Yale Univ New Haven CT 06510

LANGE, ROBERT DALE, b Redwood Falls, Minn, Jan 24, 20; m 44; c 2. HEMATOLOGY. *Educ:* Macalester Col, AB, 41; Wash Univ, MD, 44. *Prof Exp:* Dir, St Louis Regional Blood Ctr, 48-51; instr med, Washington Univ, 51-53; instr clin med, Univ Minn, 53-54; asst prof med, Washington Univ, 56-61; chief physician, Vet Admin Hosp, 61-62; assoc prof med, Med Col, Univ Ga, 62-65; dir, 77-81, prof, 78-85, chmn dept, 78-81, RES PROF, MEM RES CTR, UNIV TENN, KNOXVILLE, 65-, EMER PROF & CHMN, DEPT MED BIOL, 85- *Concurrent Pos:* Consult, Milledgeville State Hosp & Vet Admin Hosp, Augusta, Ga; hematologist, Atomic Bomb Casualty Comn, 51-53. *Mem:* Am Soc Hemat; Am Fedn Clin Res; fel Am Col Physicians; fel Int Soc Hemat. *Res:* Internal medicine. *Mailing Add:* Univ Tenn Mem Res Ctr 1924 Alcoa Hwy Knoxville TN 37920

LANGE, ROBERT ECHLIN, JR, b Janesville, Wis; m 70. WILDLIFE DISEASES, WILDLIFE DISEASE PREVENTION. *Educ:* Colo State Univ, BS, 68, 70, MS, 73, DVM, 74. *Prof Exp:* Wildlife vet wildlife dis, NMex Dept Game & Fish, 74-80; adj prof wildlife dis, NMex State Univ, 78-80; field diagnostician, Nat Wildlife Health Lab, 80-88; DEP ASST REGIONAL DIR, FED AID, FISH & WILDLIFE, 88- *Concurrent Pos:* Wildlife dis consult, Colo Wild Animal Dis Ctr, 72-74; adj prof biol dept, NMex Highlands Univ, 75-77. *Mem:* Wild Animal Dis Asn; Am Vet Med Asn; Wildlife Soc; Am Asn Wildlife Vet (secy tres, 81-). *Res:* Investigation into the game management implications of wildlife diseases in elk, Rocky Mountain bighorn sheep and other mammals; wildlife disease management in migratory waterfowl; mammals and wildlife disease prevention. *Mailing Add:* Fish & Wildlife Div Ft Snelling Fed Bldg Twin Cities MN 55111

LANGE, ROLF, b E Wuppertal, Ger, Feb 9, 32; US citizen. ATMOSPHERIC SCIENCE, PHYSICS. *Educ:* Univ Calif, Los Angeles, BA, 62; Univ Calif, Davis, MS, 72, PhD, 85. *Prof Exp:* Physicist numerical modeling, Lawrence Livermore Lab, 62-68; physicist rock mech, Physics Int, 68-71; PHYSICIST ATMOSPHERIC SCI, LAWRENCE LIVERMORE NAT LAB, 71- *Res:* Numerical modeling of atmospheric transport and turbulent diffusion; study of air pollution; development of numerical air pollution models for the atmospheric boundary layer. *Mailing Add:* Lawrence Livermore Nat Lab PO Box 808 Livermore CA 94550

LANGE, WILLIAM JAMES, b Sandusky, Ohio, Jan 20, 30; m 51; c 4. SURFACE PHYSICS. *Educ:* Oberlin Col, AB, 51; Mass Inst Technol, PhD(physics), 56. *Prof Exp:* Asst phys electronics, Mass Inst Technol, 51-56; physicist, Res Labs, Westinghouse Elec Corp, 56-64, mgr vacuum physics, 64-; RETIRED. *Mem:* Am Phys Soc; Am Vacuum Soc. *Res:* Ultrahigh vacuum; interaction of gases with surfaces. *Mailing Add:* 3917 Hickory Hill Rd Murrysville PA 15668

LANGE, WINTHROP EVERETT, b Appleton, Wis, Sept 22, 25; m 48; c 2. PHARMACY. *Educ:* Univ Wis, BS, 52, MS, 53, PhD(pharm, chem), 55. *Prof Exp:* Asst pharm, Univ Wis, 52-54; asst prof, SDak State Col, 55-58; from asst prof to assoc prof, 58-66, prof & chmn dept, Mass Col Pharm, 66-68; dir labs, 68-74, VPRES, INT DIR TECH SERV, PURDUE FREDERICK CO, 74- *Concurrent Pos:* Adj prof, A&M Schwartz Col Pharm & Health Sci, 78-82. *Mem:* Am Chem Soc; Am Pharmaceut Asn; Soc Cosmetic Chem; Acad Pharmaceut Sci; Int Fedn Socs Cosmetic Chem (pres, 76-77). *Res:* Synthesis of metal chelates as pro-drugs; pharmaceutical analysis. *Mailing Add:* Purdue Frederick Co 100 Connecticut Ave Norwalk CT 06856-3590

LANGE, YVONNE, b Durban, SAfrica, Apr 5, 41; m, 82. MEMBRANE BIOGENESIS, CHOLESTEROL MOVEMENT IN CELLS. *Educ:* London Univ, BSc, 62; Oxford Univ, DPhil(theoret physics), 66. *Prof Exp:* Instr biophysics, Harvard Med Sch, 70-72, lectr, 73-75; res fel biophysics, Sch Med, Boston Univ, 75-76, from asst prof to assoc prof biochem, 76-81; assoc prof, 81-86, PROF BIOCHEM & PATH, RUSH MED COL, 84- *Mem:* Am Soc Biol Chemists; Am Soc Cell Biol. *Res:* Intracellular movement of newly synthesized cholesterol in cultured cells with the objective of elucidating mechanisms by which eukaryotic cells regulate their membrane cholesterol content. *Mailing Add:* Dept Biochem Rush Univ 1653 W Congress Pkwy Chicago IL 60612

LANGEBARTEL, RAY GARTNER, b Quincy, Ill, Apr 27, 21; m 45; c 4. MATHEMATICS, ASTRONOMY. *Educ:* Univ Ill, AB, 42, AM, 43, PhD(math), 48. *Prof Exp:* Asst math, 46-48, from instr to assoc prof, 48-70, PROF MATH, UNIV ILL, URBANA, 70- *Concurrent Pos:* Vis res assoc, Stockholm Observ, 50-51. *Res:* Function theory; stellar dynamics. *Mailing Add:* Dept Math Univ Ill 302 Altgeld Hall Urbana IL 61801

LANGEL, ROBERT ALLAN, b Pittsburgh, Pa, May 25, 37; m 59; c 3. GEOPHYSICS. *Educ:* Wheaton Col, AB, 59; Univ Md, College Park, MS, 71, PhD(physics), 73. *Prof Exp:* Physicist, Optics Br, US Naval Res Lab, 59-62; physicist, Commun Br, 62-63, Fields & Plasmas Br, 63-74, GEOPHYSICIST GEOMAGNETISM, GEOPHYS BR, GODDARD SPACE FLIGHT CTR, NASA, 74- *Concurrent Pos:* Proj scientist for Magsat Spacecraft; vis scholar, Bullard Labs, Cambridge Univ, 83-84; ed, J Geophys Res, 85. *Honors & Awards:* Except Performance Award, NASA, 81, Except Sci Achievement Medal, 82. *Mem:* Am Geophys Union; Int Asn Geomagnetism & Aeronomy. *Res:* Utilization of surface and near-earth satellite magnetic field measurements to study lithospheric magnetic anomalies, upper mantle conductivity and core-mantle processes; derivation of geomagnetic field models. *Mailing Add:* Code 921 Goddard Space Flight Ctr Greenbelt MD 20771

LANGELAND, KAARE, b Saltdal, Norway, Nov 3, 16. DENTAL MATERIALS, EXPERIMENTAL PATHOLOGY. *Educ:* Vet Col Norway, grad, 38; Norweg State Dent Sch, DDS, 42; Univ Oslo, PhD, 57. *Prof Exp:* asst, Dent Diag Dept, Gaustad Hosp, 42-52; res assoc, Norweg Inst Dent Res, 52-63; assoc prof oral histol & chmn dept, univ & proj dir, Minn Mining & Mfg, State Univ NY Buffalo, 63, prof oral biol, 64-69; prof gen dent, 69-71, PROF ENDODONT & CHMN DEPT, SCH DENT MED, UNIV CONN, 71- *Concurrent Pos:* USPHS grants, Res Found, State Univ NY Buffalo, 63; 3M grant, 63-69; USPHS grants, 63-68, 69-70 & 77-79; Serco grant, 66-67; Univ Conn Res Found grants, 70-73 & 76-77; Off Naval Res Contract, 71-77; teacher, Norweg State Dent Sch, 48-49; ed, Scand Dent J, 57-63; vis lectr, Boston Univ; vchmn comn dent res, Int Dent Fedn, chmn working group biol testing dent mat; mem coun standardization dent mats & devices, Am Mat Standardization Inst; Norweg state rep, Inter-Nordic Comt Planning Nordic Bur Standards Dept Mat, 61-62; mem working group dent terminology, Int Orgn Standardization, 65. *Honors & Awards:* Badge of Honor in Silver & Prize, Norweg Dent Asn, 59. *Mem:* Norweg Dent Asn; hon mem Dent Asn SAfrica; hon mem Dent Asn South Rhodesia; hon mem SAfrican Prosthodont Soc; corresp mem Finnish Dent Soc. *Res:* Experimental pathology regarding biomaterials; evaluation of the biologic properties of methods; devices, and materials used in dentistry before they are released for general use. *Mailing Add:* 340 Westmont St West Hartford CT 06117

LANGELAND, WILLIAM ENBERG, organic chemistry, for more information see previous edition

LANGENAU, EDWARD E, JR, b Brooklyn, NY, Oct 28, 46; m 69; c 2. WILDLIFE BIOLOGY, PSYCHOLOGY. *Educ:* Rensselaer Polytech Inst, BS, 68; Mich State Univ, MS, 73, PhD(wildlife mgt), 76; Mich State Univ, MPA, 81. *Prof Exp:* WILDLIFE RES BIOLOGIST, STATE MICH, 74- *Mem:* Soc Am Foresters; Wildlife Soc. *Res:* Public behavior; white-tailed deer behavior; forest recreation; attitude toward clearcutting; hunter behavior; natural resource policy analysis; public administration. *Mailing Add:* 736 Hickory Ln Williamston MI 48895

LANGENBERG, DONALD NEWTON, b Devils Lake, NDak, Mar 17, 32; m 53; c 4. SOLID STATE PHYSICS. *Educ:* Iowa State Univ, BS, 53; Univ Calif, Los Angeles, MS, 55, Berkeley, PhD(physics), 59. *Hon Degrees:* MA, Univ Pa, 71. *Prof Exp:* Actg instr physics, Univ Calif, Berkeley, 58-59; NSF fel, 59-60; from asst prof to assoc prof physics, Univ Pa, 60-67, dir lab res struct matter, 72-74, vprovost grad studies & res, 74-79, prof physics, 67-, prof elec eng & sci, 76-; bd dirs, Univ Ill, Chicago, 90; CHANCELLOR, UNIV MD, 91- *Concurrent Pos:* Sloan Found fel, 62-64; Guggenheim Found fel, 66-67; assoc prof, Advan Normal Sch, Univ Paris, 66-67; distinguished vis scientist, Mich State Univ, 69; mem, Nat Acad Sci-Nat Acad Eng-Nat Res Coun Panel Adv to Cryogenics Div, Nat Bur Standards, 69-70, chmn, 70-75; mem, Comn I, Int Union Radio Sci, 69-; vis prof, Calif Inst Technol, 71; guest researcher, Cent Inst Low Temperature Study, Bayer Acad Sci & Tech Univ Munich, 74; mem, Adv Comt Res, NSF, 74-77, Coun Govt Relations & chmn, Adv Coun, 77-80, dep dir, NSF, 80-; trustee, Assoc Univs, Inc, 75-80; mem, Nat Comn Res, 78-80. *Honors & Awards:* John Price Wetherill Medal, Franklin Inst, 75. *Mem:* Fel AAAS; fel Am Phys Soc. *Res:* Cyclotron resonance and Fermi surface studies in metals and semiconductors; tunneling and Josephson effects in superconductors; precision measurement and fundamental physical constants; low temperature physics; nonequilibrium phenomena in superconductors. *Mailing Add:* Hidden Waters 3112 Old Court Rd Baltimore MD 21208

LANGENBERG, F(REDERICK) C(HARLES), b New York, NY, July 1, 27; m 53; c 2. METALLURGICAL ENGINEERING. *Educ:* Lehigh Univ, BS, 50, MS, 51; Pa State Univ, PhD(metall eng), 55. *Prof Exp:* Engr, US Steel Corp, 51-53; vis fel, Mass Inst Technol, 55-56; from supvr pyrometall to vpres, Tech, Crucible Steel Co, 56-68, vpres, Res, Develop & Eng, 68; pres, Trent Tube Div, Colt Industs, Wis, 68-70; pres, Jessop Steel Co, 70-75; pres, Am Iron & Steel Inst, 75-79; pres, 79-82, pres & chief exec officer, 82-83, CHMN & CHIEF EXEC OFFICER, INTERLAKE, INC, 83- *Honors & Awards:* Pittsburgh Nite Lectr, Am Soc Metals, 70, Andrew Carnegie Lectr, 76. *Mem:* Fel Am Soc Metals; Am Inst Mining, Metall & Petrol Eng; Am Iron & Steel Inst; Sigma Xi. *Res:* Process research and development in the smelting, melting, casting and processing of ferrous base materials; management of research and development. *Mailing Add:* Interlake Corp 505 Warrenville Rd Lisle IL 60532-4387

LANGENBERG, PATRICIA WARRINGTON, b Des Moines, Iowa, Sept 10, 31; m 53; c 4. STATISTICS. *Educ:* Iowa State Univ, BS, 53; Temple Univ MA, 75, PhD(math), 78. *Prof Exp:* From instr to asst prof math, LaSalle Col, 75-80; asst prof statist, Temple Univ, 80-84; assoc prof biomet, Sch Pub Health, Univ Ill, 84-90; ASSOC PROF BIOSTATS, UNIV MD, BALTIMORE, 90- *Mem:* Am Statist Asn; Biomet Soc; Asn Women Math; Am Asn Univ Profs; Inst Math Statist; Caucus for Women Statist. *Res:* Clinical trials; biostatistics; mathematical statistics. *Mailing Add:* Dept Epidemiol & Prev Med Univ Md Baltimore 660 W Redwood St Baltimore MD 21201

LANGENBERG, WILLEM G, b Djombang, Indonesia, Apr 16, 28; US citizen; m 55; c 3. PLANT PATHOLOGY, PLANT VIROLOGY. *Educ:* Calif State Col Long Beach, BS, 63; Univ Calif, Berkeley, PhD(plant path), 67. *Prof Exp:* Assoc prof plant path, 67-80, PROF, LIFE SCI DEPT, UNIV NEBR-LINCOLN, 80-; RES PLANT PATHOLOGIST, AGR RES SERV, USDA, 67- *Mem:* Am Phytopath Soc. *Res:* Study of plant-virus-vector relationships with labeled antibodies or viruses; light and electron microscopy radioautography. *Mailing Add:* Dept Plant Pathol Univ Nebr Lincoln NE 68583

LANGENHEIM, JEAN HARMON, b Homer, La, Sept 5, 25; div. PLANT BIOCHEMICAL ECOLOGY & EVOLUTION. *Educ:* Univ Tulsa, BS, 46; Univ Minn, MS, 49, PhD(bot, geol), 53. *Prof Exp:* Investr, Nat Geol Serv, Colombia, 53; res assoc, Univ Calif, Berkeley, 54-59; teaching assoc, Univ Ill, 59-62; Asn Univ Women fel, Harvard Univ, 62-63; res assoc, Bot Mus, 63-66; from asst prof to assoc prof biol, 66-73, chmn, Biol Dept, 74-76, PROF BIOL, UNIV CALIF, SANTA CRUZ, 73- *Concurrent Pos:* Mem teaching staff & bd trustees, Rocky Mountain Biol Lab, 54-65; lectr, Mills Col, 55-56; from instr to asst prof, San Francisco Col Women, 56-59; scholar, Radcliffe Inst Independent Study, 63-64; mem exec comt, Org Trop Studies, 72-77, acad vpres, 75-77; vis prof biol, Harvard Univ, 74; mem, NSF comn floral inventory Amazon, 75-87; mem, ecol adv comm, Environ Protection Agency, 77-81; mem, US Nat Comt, Int Union Biol Sci, 89-94. *Mem:* Int Soc Chem Ecol (vpres, 85); Bot Soc Am; Ecol Soc Am (vpres, 80-85, pres, 85-); Soc Study Evolution; Asn Trop Biol (pres, 85); Soc Econ Biol. *Res:* Paleoecological studies of amber; evolutionary and physio-ecological studies of tropical resin-producing and other terpene-producing plants; concepts of ecology; plants and human affairs. *Mailing Add:* Dept Biol Univ Calif Santa Cruz CA 95064

LANGENHEIM, RALPH LOUIS, JR, b Cincinnati, Ohio, May 26, 22; m 46, 62, 70; c 2. PALEONTOLOGY, STRATIGRAPHY. *Educ:* Univ Tulsa, BS, 43; Univ Colo, MS, 47; Univ Minn, PhD(geol), 51. *Prof Exp:* Asst prof geol, Coe Col, 50-52; asst prof paleont, Univ Calif, 52-59; from asst prof to assoc prof, 59-67, PROF GEOL, UNIV ILL, URBANA, 67-, CUR PALEONT, MUS NATURAL HIST, 88- *Concurrent Pos:* Foreign expert, Inst Geol Nac, Colombia, 53; adv, Geol Surv Can, 57 & Cent Geol Surv, Repub China, 81; foreign assoc, Geol Surv Iran, 73; partner, Lanman Assocs, Consult Geologists, 74- *Mem:* Geol Soc Am; Paleont Soc (secy, 62-70); Am Asn Petrol Geol; Soc Econ Paleont & Mineral; Soc Geol Suisse; Geol Soc London. *Res:* Invertebrate paleontology and stratigraphy; Paleozoic of western and central North America; Tertiary of southern Mexico; Permian and Carboniferous of Iran; petroleum and energy geology; approximately 125 publications. *Mailing Add:* Dept Geol 245 NHB Univ Ill 1301 W Green St Urbana IL 61803

LANGENHOP, CARL ERIC, mathematics, for more information see previous edition

LANGER, ARTHUR M, b New York, NY, Feb 18, 36; m 62, 67; c 4. MINERALOGY, ENVIRONMENTAL SCIENCES. *Educ:* Hunter Col, BA, 56; Columbia Univ, MA, 62, PhD(mineral), 65. *Prof Exp:* Res assoc mineral, Columbia, 62-65; res assoc environ sci, 65-66, asst prof mineral, 66-68, ASSOC PROF MINERAL, MT SINAI SCH MED, 68-, ASSOC PROF POLYPEPTIDE, MEMBRANE RES, 85-; DIR, ENVIRON SCI LAB INST APPL SCI, BROOKLYN COL, CITY UNIV NEW YORK, 85- *Concurrent Pos:* Adj assoc prof, Queens Col, NY, 68-70; consult, NIH, Bethesda Md, 74, Int Agencies Res on Cancer, Lyon WHO, 74, Inst Public Health, Norway, 77, Ministry Mines, Johannesburg, SAfrica, 77, Int Metalworkers Fedn, Geneva, 80, Ctr Dis Control, Atlanta, 84, Nat Acad Sci, 84, Consumer Prod Safety Comn, Washington, DC, 86. *Mem:* AAAS; Geol Soc Am; Electron Probe Analytical Soc Am; Geochem Soc; Mineral Soc Am. *Res:* Metamorphic and igneous petrology; clay mineralogy; secondary mineralization; instrumentation; microparticulate identification, analysis and interaction in the human environment. *Mailing Add:* Environ Sci Mt Sinai Sch Med New York NY 10029

LANGER, DIETRICH WILHELM, b Berlin, Ger, Aug 13, 30; US citizen; m 61; c 4. SOLID STATE PHYSICS. *Educ:* Tech Univ Berlin, MS, 57, PhD(physics), 60. *Prof Exp:* From res physicist to sr res physicist, Aerospace Res Labs, Wright-Patterson Air Force Base, 57-65, group leader, electronic properties, Semiconductors Group, 65-87; PROF ELEC ENG, UNIV PITTSBURGH, 87- *Concurrent Pos:* Eve lectr, Univ Dayton, 59-60; grant, Ecole Normale Superieure, Paris, 64-65; adv, Max Planck Inst Solid State Study, 72-73; fel Indust Col Armed Forces, 75-76. *Honors & Awards:* Alexander von Humboldt Award, 72. *Mem:* Fel Am Phys Soc; sr mem Inst Elec & Electronics Engrs. *Res:* Optical, electronic and electrooptical properties of semiconductors and devices; materials research; research and development administration. *Mailing Add:* Elec Eng Dept Univ Pittsburgh 348 BEH Pittsburgh PA 15261

LANGER, GLENN A, b Nyack, NY, May 5, 28; m 54; c 1. MEDICINE. *Educ:* Colgate Univ, BA, 50; Columbia Univ, MD, 54. *Prof Exp:* Intern, Mass Gen Hosp, 54-55; asst res, Columbia-Presby Med Ctr, 57-58; sr resident, Mass Gen Hosp, 59-60; clin instr, Los Angeles County Cardiovasc Res Lab & Med Ctr, Univ Calif, Los Angeles, 60-62; asst prof med, Columbia Univ, 63-66; assoc prof, 66-69, assoc dir, Cardiovasc Res Lab, 66-87, vchmn, dept physiol, 67-87, PROF MED & PHYSIOL, MED CTR, UNIV CALIF, LOS ANGELES, 69-, CASTERA PROF CARDIOL, 78-, DIR, CARDIOVASC RES LAB & ASSOC DEAN RES, 87- *Concurrent Pos:* Chmn exec comt, Basic Sci Coun & bd dirs, Am Heart Asn, 76-78; Griffith vis prof cardiol, 79; Macy fac scholar, 79-80. *Mem:* Am Heart Asn; Am Physiol Soc; Am Soc Clin Invest; Am Asn Physicians; Soc Gen Physiologists. *Res:* Myocardial physiology and metabolism. *Mailing Add:* Dept Med & Physiol Univ Calif Med Ctr 10833 LeConte Ave Los Angeles CA 90024

LANGER, HORST GÜNTER, b Breslau, Ger, Dec 29, 27; US citizen; m 55; c 2. INORGANIC CHEMISTRY. *Educ:* Brunswick Tech Univ, Dipl, 54, Dr rer nat(chem), 56. *Prof Exp:* Asst inorg anal chem, Brunswick Tech Univ, 51-56; res assoc inorg chem, Ind Univ, 56-58; res chemist, 58-64, sr res chemist, 64-68, ASSOC SCIENTIST, DOW CHEM USA, 68- *Mem:* Am Chem Soc; Am Soc Mass Spectrometry; Int Confedn Thermal Analysis; fel NAm Thermal Analytical Soc; Ger Chem Soc. *Res:* Analytical, dental and organometallic chemistry; mass spectrometry; thermal analysis; fire retardants; catalysts. *Mailing Add:* 28 Joyce Rd Wayland MA 01778-4516

LANGER, JAMES STEPHEN, b Pittsburgh, Pa, Sept 21, 34; m 58; c 3. STATISTICAL MECHANICS. *Educ:* Carnegie Inst Technol, BS, 55; Univ Birmingham, PhD, 58. *Prof Exp:* Instr physics, Carnegie-Mellon Univ, 58-64, from asst prof to prof, 64-82, assoc dean, Mellon Inst Sci, 71-74; PROF PHYSICS, UNIV CALIF, SANTA BARBARA, 82-, DIR, INST THEORET PHYSICS, 89- *Concurrent Pos:* Vis assoc prof, Cornell Univ, 66-67; Guggenheim fel, Harvard Univ, 74-75. *Mem:* Nat Acad Sci; fel Am Phys Soc; Am Acad Arts & Sci; fel AAAS. *Res:* Theoretical solid state physics; kinetics of phase transformations. *Mailing Add:* Inst Theoret Physics Univ Calif Santa Barbara CA 93106

LANGER, LAWRENCE MARVIN, b New York, NY, Dec 22, 13; m 36. NUCLEAR PHYSICS. *Educ:* NY Univ, BS, 34, MS, 35, PhD(physics), 38. *Hon Degrees:* DSc, Ind Univ, 88. *Prof Exp:* Asst physics, NY Univ, 34-38; from instr to prof, 38-59, actg chmn dept, 61-62 & 65-66, chmn, 66-73, EMER PROF PHYSICS, IND UNIV, BLOOMINGTON, 79- *Concurrent Pos:* Res assoc, Radiation Lab, Mass Inst Technol, 41-42; alternate group leader, Los Alamos Atomic Bomb Proj, 43-45; sci consult, US War Dept, 45; expert consult, AEC, 48-74, US Energy Res & Develop Admin, 74-; consult & observer, Greenhouse Atomic Bomb Tests, Marshall Islands, 51; adv consult, Nat Res Coun, 57-60; dir, Off Naval Res & NSF res nuclear spectros, 63-; adv consult, Nuclear Data Proj, Nat Acad Sci-Nat Res Coun. *Honors & Awards:* Samuel F B Morse Medal. *Mem:* AAAS; fel Am Phys Soc. *Res:* Nuclear physics; artificial and natural radioactivity; beta ray spectra; neutron scattering; D-D reaction; beta and gamma coincidence measurements; Cockroft-Walton accelerator; Geiger Counter; cyclotron; counting equipment; microwave radar; underwater sound; ballistics; nuclear spectroscopy and nuclear weapons; double beta decay; mass of the neutrino; shapes of the allowed and forbidden beta spectra. *Mailing Add:* Dept Physics Ind Univ Bloomington IN 47401

LANGER, R M, b New York, NY, Dec 1, 99. OPTICS, BIOPHYSICS. *Educ:* Calif Inst Technol, PhD(physics), 26. *Prof Exp:* Sr physicist, Tech Opers, 50-52; mgr, Bege Co, 52-70; RETIRED. *Mem:* Am Phys Soc; AAAS. *Mailing Add:* 46 Park St Arlington MA 02174

LANGER, ROBERT SAMUEL, b Aug 29, 48; US citizen; m 88; c 1. BIOENGINEERING, BIOMEDICAL SCIENCE. *Educ:* Cornell Univ, BS, 70; Mass Inst Technol, ScD, 74. *Prof Exp:* From asst prof to assoc prof, Dept Nutrit & Food Sci, 78-85, prof, Dept Appl Biol Sci, 85-88, GERMESHAUSEN PROF, DEPT CHEM ENG, MASS INST TECHNOL, 88-; RES ASSOC SURG, BOSTON CHILDREN'S HOSP, 74- *Concurrent Pos:* Vis asst prof nutrit & food sci, Mass Inst Technol, 77-78; consult, numerous co; ed, Biomaterials, 83-; pres, Controlled Release Soc, 91-92. *Honors & Awards:* Creative Polymer Chem Award, Am Chem Soc; Food, Pharmaceut & Bioeng Award, Am Inst Chem Engrs, Prof Progress Award; Founders Award, Controlled Release Soc; Clemson Award, Soc Biomat. *Mem:* Nat Acad Sci-Inst Med; Am Inst Chem Engrs; Sigma Xi; Biomed Engr Soc; Controlled Release Soc; Am Soc Artificial Internal Organs; Am Chem Soc; Soc Biomat. *Res:* Polymer drug delivery systems; tumor neovascularization; application of enzymes in medicine; biomaterials. *Mailing Add:* Dept Chem Eng Mass Inst Technol 77 Mass Ave Cambridge MA 02139

LANGER, SIDNEY, b New York, NY, Dec 15, 25; c 1. INORGANIC CHEMISTRY. *Educ:* NY Univ, AB, 49; Ill Inst Technol, PhD(chem), 55. *Prof Exp:* Chemist, Oak Ridge Nat Lab, 54-60; mem res staff, Gen Atomic Div, Gen Dynamics Corp, 60-69, group leader, Nuclear Fuels Group, Res & Develop Div, Gulf Gen Atomic, 69-71, mgr gas cooled fast breeder reactor fuels, 77-81, GROUP LEADER FUELS & MAT DEVELOP, GAS-COOLED FAST REACTOR PROJECT, GEN ATOMIC CO, 71-, MEM SR RES STAFF, 69-; SR SCIENTIST, SCI APPL INT, 89- *Concurrent Pos:* Prin prog specialist, TMI-2 Accident Eval Prog, EG&G Idaho, Inc, 85-89. *Mem:* AAAS; Sigma Xi; Am Chem Soc; fel Am Nuclear Soc. *Res:* Nuclear reactor chemistry; physical chemistry and thermodynamics of high temperature systems; phase equilibria; fission product behavior in fuels; fission product release; fuel processing and reprocessing; nuclear reactor safety. *Mailing Add:* Sci Appln Int Corp 10210 Campus Point Dr San Diego CA 92121-1522

LANGER, WILLIAM DAVID, b New York, NY, Sept 28, 42. ASTROPHYSICS, PHYSICS. *Educ:* NY Univ, BS, 64; Yale Univ, MS, 65, PhD(physics), 68. *Prof Exp:* Assoc fel astrophys, Goddard Inst Space Studies, NASA, 68-70; NSF-NATO fel physics, Niels Bohr Inst, Copenhagen, Denmark, 70-71; res asst prof astrophys, NY Univ, 71-76; asst prof, Univ Pa, 76-78; assoc prof astron, Univ Mass, 78-80; PRIN RES PHYSICIST, PLASMA PHYSICS LAB, PRINCETON UNIV, 80- *Concurrent Pos:* Res assoc astrophys, Goddard Inst Space Studies, NASA, 74-76; consult radio astron, Bell Tel Labs, 76-81. *Mem:* AAAS; Am Astron Soc; Am Phys Soc; Sigma Xi. *Res:* astrophysics; plasma physics. *Mailing Add:* Princeton Univ PO Box 451 Princeton NJ 08543

LANGERMAN, NEAL RICHARD, b Philadelphia, Pa, Mar 11, 43. CHEMICAL & ENVIRONMENTAL SAFETY. *Educ:* Franklin & Marshall Col, AB, 65; Northwestern Univ, PhD(chem), 69. *Prof Exp:* NIH fel chem, Yale Univ, 69-70; asst prof biochem, Sch Med, Tufts Univ, 70-75; asst prof chem, 75-83, ASSOC PROF, UTAH STATE UNIV, 83-; CHEM SAFETY ASN, SAN DIEGO, CALIF. *Mem:* Am Soc Biol Chemists; Biophys Soc; Calorimetry Soc; Undersea Med Soc; Am Chem Soc. *Res:* Thermodynamic studies of protein reactions, especially flavin-flavoprotein interactions; microcalorimetry; fluorescence spectroscopy; analytical ultracentrifugation; chemical safety; management of hazardous waste; chemical resource recovery. *Mailing Add:* Chem Safety Asn 9163 Chesapeake Dr San Diego CA 92123-1002

LANGFELDER, LEONARD JAY, b Lynbrook, NY, Feb 5, 33; m 55; c 3. GEOTECHNICAL & COASTAL ENGINEERING. *Educ:* Univ Fla, BSCE, 59, MSE, 60; Univ Ill, PhD(civil eng), 64. *Prof Exp:* Res assoc coastal eng, Univ Fla, 60-62; vis lectr civil eng, Univ Ill, 64; from asst prof to prof civil eng, NC State Univ, 69-78, prof marine sci & eng & head dept, 78-80, dir ctr marine, Earth, Atmospheric Sci, 69-78, prof & head dept marine, earth & sci, 80- 82, 83-85; asst secy, Nat Resources, NC, 82-83; vpres & mgr dir, Harbor Br, Oceanog Inst, Inc, 86-88; CONSULT ENGR, 88- *Concurrent Pos:* Mem soils, geol & found cmt, Hwy Res Bd, Nat Acad Sci-Nat Res Coun, 65- *Mem:* Am Soc Civil Engrs. *Res:* Soil properties, principally shear strength and compaction properties of cohesive soils; improved foundation engineering principles; coastal processes. *Mailing Add:* 1970 Compass Cove Dr Vero Beach FL 32963

LANGFITT, THOMAS WILLIAM, b Clarksburg, WVa, Apr 20, 27; m 53; c 3. NEUROSURGERY. *Educ:* Princeton Univ, AB, 49; Johns Hopkins Univ, MD, 53. *Prof Exp:* Intern gen surg, Johns Hopkins Univ Hosp, 53-54; asst resident gen surg, Johns Hopkins Univ Hosp, 54-55, from asst resident to chief resident neurosurg, 57-61; assoc, Univ Pa, 61-63, from asst prof to assoc prof, 63-68, Charles Frazier prof neurosurg, Med Sch, 68-87, head dept, Univ Hosp, 68-87, vpres health affairs, Univ Pa, 74-87; PRES & CHIEF EXEC OFFICER, GLEN MEADE TRUST CO, 87- *Concurrent Pos:* Res fel, Johns Hopkins Univ Hosp, 57-60; contractor, US Army Chem Corp, 57-; head sect neurosurg, Pa Hosp, 61-68. *Mem:* Inst Med-Nat Acad Sci; Cong Neurol Surg; Am Col Surgeons; Am Asn Neurol Surg; AMA. *Res:* Pathophysiology and metabolism in acute brain injuries. *Mailing Add:* Glenmead Corp 229 S 18th St Philadelphia PA 19103

LANGFORD, ARTHUR NICOL, b Ingersoll, Ont, July 30, 10; m 38, 76; c 3. GENERAL BOTANY, SCIENCE EDUCATION. *Educ:* Queen's Univ, Ont, BA, 31; Univ Toronto, MA, 33, PhD(plant path, genetics), 36. *Prof Exp:* Asst bot, Univ Toronto, 33-36, bact, 37; lectr, Bishops Univ, 38-45, from asst prof to prof biol, 46-78, emer prof, 78-81; RETIRED. *Concurrent Pos:* Head dept biol, Bishops Univ, 51-71, chmn dept, 71-75; co-operant, Ctr for Info on Am, Swaziland, S Africa, 76- *Res:* Plant ecology of forests and peat bogs; gradient analysis. *Mailing Add:* 30 Salibury Levis PQ G6V 6G6 Can

LANGFORD, COOPER HAROLD, III, b Ann Arbor, Mich, Oct 14, 34; m 59. PHYSICAL INORGANIC CHEMISTRY. *Educ:* Harvard Univ, AB, 56; Northwestern Univ, PhD(chem), 59. *Prof Exp:* NSF fel, Univ Col, London, 59-60; instr chem, Amherst Col, 60-61, from asst prof to assoc prof, 61-67; from assoc prof Chem to prof, Carleton Univ, 67-80; prof chem & chmn dept, 80-87, assoc vrector res, 87-90, DIR PHYS & MATH SCI, NAT SCI & ENG RES COUN CAN, CONCORDIA UNIV, 90- *Concurrent Pos:* Vis asst prof, Columbia Univ, 64; Alfred P Sloan Found fel, 68-70; consult, Inland Waters Directorate, Can, 73; consult, Nat Health & Welfare Can, 77; mem, Chem Grants Comt, Nat Res Coun Can, 75-78, chmn, 77-78; chem chmn, Coun Can Univ, 81-83; mem, strategic grants comt, open sect, Nat Sci Eng Res Coun Can, 85- *Mem:* Fel AAAS; Am Chem Soc; Royal Soc Chem; fel Chem Inst Can. *Res:* Applications to energy problems; inorganic photochemistry; kinetics in analysis; solution physical chemistry. *Mailing Add:* Dept Chem Concordia Univ Montreal PQ H3G 1M8 Can

LANGFORD, DAVID, b New York, NY, May 6, 34; div; c 2. MECHANICAL ENGINEERING, NUCLEAR ENGINEERING. *Educ:* NY Univ, BS, 56, MS, 57; Ill Inst Technol, MS, 59; Rensselaer Polytech Inst, DEngSc, 65. *Prof Exp:* From asst physicist to assoc physicist, IIT Res Inst, 57-59; from sr anal engr to asst proj engr, Pratt & Whitney Aircraft, United Aircraft Corp, 59-66; from asst prof to assoc prof mech eng, Drexel Univ, 66-72; regional radiation noise rep, US Environ Protection Agency, 72-82; engr, US Nuclear Reg Comn, 82-91; RETIRED. *Mem:* AAAS; Am Phys Soc; Am Nuclear Soc; Am Soc Mech Engrs; NY Acad Sci. *Res:* Psychology and education of engineers; nuclear engineering; plasma physics and magnetohydrodynamics; superconductivity; conduction heat transfer; fluid flow; environmental science. *Mailing Add:* 1404 Regal Rd Clearwater FL 34616-2321

LANGFORD, EDGAR VERDEN, b Parry Sound, Ont, Can, Apr 27, 21; m 44; c 4. VETERINARY BACTERIOLOGY. *Educ:* Mem, Royal Col Vet Surgeons, Univ Toronto, DVM, VS, 49, DVPH, 50. *Prof Exp:* Vet surgeon diag & res, Can Dept Agr, 50-56; animal pathologist, BC Dept of Agr, 56-67; res sci diag, Can Dept Agr, 67-80; RETIRED. *Mem:* Am Soc Microbiol; Can Soc Microbiologists; Can Vet Med Asn; Int Soc Mycoplasmologists; Brit Vet Asn. *Res:* Mycoplasma relationship to disease in the ruminant; incidence of infection in the systems of the ruminant; serological response of the host; development of diagnostic serological methods; pathogenicity of Mycoplasma. *Mailing Add:* PO Box 240 Monarch AB T0L 1M0 Can

LANGFORD, ERIC SIDDON, b New York, NY, May 23, 38; m 59; c 1. MATHEMATICS. *Educ:* Mass Inst Technol, SB, 59; Rutgers Univ, MS, 60, PhD(math), 63. *Prof Exp:* Res specialist, Autonetics Div, NAm Rockwell, 63-64; asst prof math, Naval Postgrad Sch, 64-69; assoc prof math, Univ Maine, Orono, 69-77, prof, 77-82; PROF MATH, CALIF STATE UNIV, CHICO, 82- *Concurrent Pos:* Vis assoc, Daniel H Wagner, Assocs, 66; collab ed, Am Math Monthly, 69-71, assoc ed, 71-75; vis assoc prof math, Calif Inst Technol, 72-73; vis scholar, Univ Calif, Berkeley, 76; vis distinguished prof math, Calif Polytech State Univ, San Luis Obispo, 77-78. *Honors & Awards:* L R Ford Award, Math Asn Am, 71. *Res:* Geometrical aspects of Banach spaces; Riesz spaces. *Mailing Add:* Dept Math Calif State Univ Chico CA 95929-0525

LANGFORD, FLORENCE, b Celina, Tex, Sept 20, 12. NUTRITION. *Educ:* Tex Woman's Univ, BS, 32, MA, 38; Iowa State Univ, PhD(nutrit), 60. *Prof Exp:* High sch teacher, Tex, 32-34; teacher home econ, Kilgore Jr Col, 35-37; instr nutrit & home mgt, Univ Tenn, 38-41; from asst prof to assoc prof, 42-61, actg dean, 70-72, PROF FOOD & NUTRIT, TEX WOMAN'S UNIV, 62- *Mem:* Am Chem Soc; Am Dietetic Asn; Am Home Econ Asn; Am Pub Health Asn; Int Fedn Home Econ; Sigma Xi. *Res:* Food and nutrition; human nutrition; energy metabolism. *Mailing Add:* Tex Woman's Univ Sta Box 23835 Denton TX 76204

LANGFORD, FRED F, b Toronto, Ont, Dec 19, 29; m 53; c 3. ECONOMIC GEOLOGY. *Educ:* Univ Toronto, BA, 53; Queen's Univ, Ont, MA, 55; Princeton Univ, PhD(geol), 60. *Prof Exp:* Geologist, Imp Oil Co, 53-54; geologist, Ont Dept Mines, 54-57; assoc prof geol, Univ Kans, 58-62; assoc prof, 62-71, PROF GEOL SCI, UNIV SASK, 71- *Mem:* Mineral Asn Can; Geol Asn Can. *Res:* Economic geology; potash geology; uranium deposits. *Mailing Add:* Dept Geol Sci Univ Sask Saskatoon SK S7N 0W0 Can

LANGFORD, GEORGE, b Chicago, Ill, Dec 26, 36; m 68; c 2. PHYSICAL & MECHANICAL METALLURGY. *Educ:* Mass Inst Technol, SB, 59, ScD(metall), 66. *Prof Exp:* Instr metall, Mass Inst Technol, 60-62; scientist, Edgar C Bain Lab Fundamental Res, US Steel Corp, Pa, 66-71; res specialist, Monsanto, Chemstrand Res Ctr, Inc, Durham, 72-75; ASSOC PROF MAT ENG, DREXEL UNIV, 75- *Mem:* AAAS; Am Soc Metals; Am Inst Mining, Metall & Petrol Engrs. *Res:* Optical and electron metallography of heavily deformed metals and theory of their strain hardening; wire drawing; steel casting by diffusion solidification and liquid infiltration. *Mailing Add:* 32 Bodine Rd Berwyn PA 19312

LANGFORD, GEORGE MALCOLM, b Halif, NC, Aug 26, 44; m 68; c 3. CELL BIOLOGY, MICROTUBULES. *Educ:* Fayetteville State Univ, BS, 66; Ill Inst Technol, MS, 69, PhD(cell biol), 71. *Prof Exp:* Fel biophys cytol, Univ Pa, 71-73; asst prof cell biol, Univ Mass, 73-76, Col Med, Howard Univ, 77-79; assoc prof cell biol, 79-88, PROF, SCH MED, UNIV NC, CHAPEL HILL, 88- *Concurrent Pos:* NIH fel, 71-73; Macy fel, Marine Biol Lab, 76 & 77, Steps fel, 78; consult, NIH, 81-, Marine Biol Lab, 82-; NSF Adv Comt Cell Molecular Biol, 84-; trustee, Marine Biol Lab, 84-; Cell Biol Prog Dir, NSF, 88-89. *Mem:* AAAS; Am Soc Cell Biol; Sigma Xi. *Res:* Vesicle transport on axoplasmic microtubules and the structure and function of microtubules; wave propagation in the microtubular axostyle; properties of axonal and dendritic microtubules; axonal transport. *Mailing Add:* Dept Physiol CB No 7545 Sch Med Univ NC Chapel Hill NC 27599

LANGFORD, HERBERT GAINES, internal medicine, physiology; deceased, see previous edition for last biography

LANGFORD, PAUL BROOKS, b Lockesburg, Ark, Aug 11, 30; m 59; c 1. PHYSICAL ORGANIC CHEMISTRY. *Educ:* Okla State Univ, BS, 52, MS, 54; Ga Inst Technol, PhD(chem), 62. *Prof Exp:* Instr chem, Ga Inst Technol, 56-62; from asst prof to assoc prof, 62-70, PROF CHEM, DAVID LIPSCOMB COL, 70-, CHMN DEPT, 80- *Mem:* Am Chem Soc. *Res:* Rates and mechanisms of reactions of organic halogen compounds; charge transfer complex compounds. *Mailing Add:* Dept Chem David Lipscomb Col 3901-4001 Granny White Pike Nashville TN 37204

LANGFORD, ROBERT BRUCE, b San Francisco, Calif, Mar 7, 19; m 57. ORGANIC CHEMISTRY, CHEMICAL EDUCATION. *Educ:* Univ Calif, Los Angeles, BS, 48; Univ Southern Calif, MS, 63, PhD, 72. *Prof Exp:* Chemist petrol anal, Southern Pac Co, 49-54; res chemist pesticides, Stauffer Chem Co, 54-58; lab mgr org synthesis, Cyclo Chem Corp, 58-61; teacher chem, Los Angeles City Schs, 61-64; from instr to prof, 64-86, chmn dept, 68-74, EMER PROF CHEM, EAST LOS ANGELES COL, 86- *Mem:* Am Chem Soc; Sigma Xi. *Res:* Organic synthesis; sulfur compounds; photochemistry. *Mailing Add:* 644 Haverkamp Dr Glendale CA 91206

LANGFORD, RUSSELL HAL, b North Platte, Nebr, Nov 14, 25; m 46; c 4. HYDROLOGY, ENVIRONMENTAL CHEMISTRY. *Educ:* Univ Nebr, BSc, 49. *Prof Exp:* Hydrologist, US Geol Surv, 49-66, asst chief, Off Water Data Coord, 66-68, chief, Off Water Data Coord, 68-80, assoc chief hydrologist, 80-85; RETIRED. *Concurrent Pos:* Mem, Int Souris-Red River Eng Bd, Int Joint Comn US & Can; alt chmn, US Nat Comt Sci Hydrol; US mem comn hydrol, World Meteorol Org, Intergovt Coun, Int Hydrolog Prog, UNESCO. *Mem:* Am Chem Soc; Am Water Works Asn; Am Geophys Union; Water Pollution Control Fedn; AAAS. *Res:* Water chemistry; geochemistry. *Mailing Add:* 1002 Country Club Dr Vienna VA 22180

LANGFORD, WILLIAM FINLAY, b Thunder Bay, Ont, Sept 11, 43; m 70; c 2. DIFFERENTIAL EQUATIONS, BIFURCATION THEORY. *Educ:* Queens Univ, Can, BSc, 66; Calif Inst Technol, PhD(appl math), 71. *Prof Exp:* From asst prof to assoc prof math, McGill Univ, 70-82; assoc prof, 82-87, PROF MATH, UNIV GUELPH, CAN, 88-, CHAIR, DEPT MATH & STATIST, 90- *Concurrent Pos:* Res visitor, Univ Nice, France, 79-80; adj prof app math, Univ Waterloo, 83-; vis prof math, Univ Houston, 85-87; mem bd dirs, Can Math Soc, 85-; vis prof, Tianjin Univ China, 89- *Mem:* Am Math Soc; Soc Indust & Appl Math; Can Math Soc; Can Appl Math Soc. *Res:* Theory of bifurcation for nonlinear differential equations; effects of symmetry; numerical algorithms for bifurcation problems and applications in science and engineering. *Mailing Add:* Dept Math & Statist Univ Guelph Guelph ON N1G 2W1 Can

LANGHAAR, HENRY LOUIS, b Bristol, Conn, Oct 14, 09; m 37; c 3. APPLIED MECHANICS. *Educ:* Lehigh Univ, BS, 31, MS, 33, PhD(math), 40. *Prof Exp:* Test engr, Ingersoll Rand Co, NJ, 33-36; seismographer, Carter Oil Co, 36-37; instr math, Purdue Univ, 40-41; struct engr, Consol-Vultee Aircraft Corp, 41-47; from assoc prof to prof, 47-78, EMER PROF THEORET & APPL MECH, UNIV ILL, URBANA, 78- *Honors & Awards:* Von Karman Medal, Am Soc Civil Engrs. *Mem:* AAAS; fel Am Soc Mech Engrs; Sigma Xi. *Res:* Stress analysis; elasticity and buckling theories; theory of plates and shells; dimensional analysis and model theory. *Mailing Add:* Lake Rd Box 530 Corydon IN 47112

LANGHAM, ROBERT FRED, b Grand Ledge, Mich, Jan 31, 12; m 37; c 5. PATHOLOGY. *Educ:* Calvin Col, AB, 35; Mich State Univ, MS, 37, DVM, 42, PhD, 50. *Prof Exp:* From instr to assoc prof, 38-51, actg chmn, Dept Path, Col Vet Med, 73-75, PROF VET PATH, MICH STATE UNIV, 51- *Mem:* Am Vet Med Asn; Am Col Vet Path; Conf Res Workers Animal Dis; Int Acad Path. *Res:* Leptospirosis; neoplasms and joint disease in animals; co-author and author. *Mailing Add:* 330 Shoesmith Haslett MI 48840

LANGHANS, ROBERT W, b Flushing, NY, Dec 29, 29; m 52; c 3. FLORICULTURE. *Educ:* Rutgers Univ, BS, 52; Cornell Univ, MS, 54, PhD(floricult), 56. *Prof Exp:* From asst prof to assoc prof, 56-68, PROF FLORICULT, CORNELL UNIV, 68- *Honors & Awards:* Blauvelt Award, 55; Kenneth Post Award, Am Soc Hort Sci, 65. *Mem:* Am Soc Hort Sci; Am Soc Plant Physiol; Int Soc Hort Sci. *Res:* Effects of photoperiod and temperature on growth and flowering. *Mailing Add:* Dept Hort Cornell Univ Col Agr 15-D Plant Sci Bldg Ithaca NY 14850

LANGHEINRICH, ARMIN P(AUL), b Planitz, Germany, Sept 1, 26; US citizen; m 49; c 3. CHEMISTRY, FUEL ENGINEERING. *Educ:* Univ Utah, BS, 58, MA, 62. *Prof Exp:* Asst chemist, Utah Cooper Div, 59, jr scientist, Res Ctr, Western Mining Div, 59-62, from asst scientist to scientist, 62-71, sr scientist, Res Dept, Metal Mining Div, 71-79, MGR, TECH & ADMIN SERV, KENNECOTT MINERALS CO, 79- *Concurrent Pos:* Spec instr chem, Salt Lake Ctr Continuing Educ, Brigham Young Univ, 63-77. *Mem:* Am Chem Soc; Am Soc Testing & Mat (secy, 77-82). *Res:* Application of x-ray fluorescence techniques to laboratory and on-stream analyses; energy dispersion x-ray analysis with conventional and radioisotope excitation; optical emission spectroscopy applied to geochemical samples and to refined copper; development of high purity copper standards; environmental analysis. *Mailing Add:* 230 M St Salt Lake City UT 84103

LANGHOFF, CHARLES ANDERSON, b New Orleans, La, June 27, 47; m 73; c 3. CHEMICAL PHYSICS. *Educ:* Tulane Univ, BS, 69; Calif Inst Technol, PhD(chem), 74. *Prof Exp:* Fel chem, IBM Res Lab, San Jose, 73-75; asst prof chem, Ill Inst Technol, 75-81; DOW CHEM, 81- *Concurrent Pos:* Res Corp grant, 77-78. *Mem:* Am Chem Soc; Am Phys Soc. *Res:* Picosecond spectroscopy; dynamic of liquids; theory of radiationless transitions; nonlinear optics. *Mailing Add:* Dow Chem Co 1776 Bldg Midland MI 48674

LANGHOFF, PETER WOLFGANG, b New York, NY, Jan 19, 37; m 62; c 3. COMPUTATIONS, APPLIED SCIENCE. *Educ:* Univ Hofstra, BS, 58; State Univ NY Buffalo, PhD(physics), 65. *Prof Exp:* Physicist, Cornell Aeronaut Labs, Inc, Cornell Univ, 62-65; fel, Harvard Univ, 67-69; from asst prof to assoc prof, 69-77, PROF CHEM, IND UNIV, BLOOMINGTON, 77- *Concurrent Pos:* Vis prof, Univ Colo, Boulder, 76 & Univ Paris, Orsay, 81, vis fel, Joint Inst Lab Astrophy, 76, Nat Res Coun fel, 78-79; prof chem & fac assoc, Supercomput Res Inst, Fla State Univ, 85-86. *Mem:* Am Phys Soc; Am Chem Soc. *Res:* Atomic and molecular physics; interaction of radiation and matter; atomic and molecular structure; molecular photoionization; supercomputer computations; advanced solar energy technology; hydrogen production. *Mailing Add:* Dept Chem Ind Univ Bloomington IN 47405

LANGILLE, ALAN RALPH, b Amherst, NS, Apr 2, 38; m 67; c 1. CROP PHYSIOLOGY. *Educ:* McGill Univ, BS, 60; Univ Vt, MS, 62; Pa State Univ, PhD(agron), 67. *Prof Exp:* From asst prof to assoc prof agron, 73-79, PROF AGRON & BOT, UNIV MAINE, ORONO, 79- *Concurrent Pos:* Vis prof, dept plant path, Kan State Univ, 79-80. *Mem:* Am Soc Agron; Am Soc Hort Sci; Potato Asn Am. *Res:* Hormonal control of tuber initiation and subsequent growth in the potato; growth regulator physiology; salt tolerance in conifers; protoplast regeneration. *Mailing Add:* Dept Plant Soil & Environ Sci Univ Maine Orono ME 04469

LANGILLE, BRIAN LOWELL, b Victoria, BC, July 26, 47. CARDIOVASCULAR CELL BIOLOGY. *Educ:* Univ BC, BSc, 69, MSc, 70, PhD(zool), 75. *Prof Exp:* Asst prof biophys, Univ Western Ont, 77-79, asst prof physiol, 79-85; ASSOC PROF PATH, UNIV TORONTO, 85- *Concurrent Pos:* Sr res fel, Heart & Stroke Found, Ont, 78-79, career investr, 89-; mem, Coun Atherosclerosis, Am Heart Asn; counr, Biophys Soc Can, 91- *Mem:* Am Asn Pathologists; AAAS; Biophys Soc Can; Can Atherosclerosis Soc; Am Heart Asn; NY Acad Sci. *Mailing Add:* Max Bell Res Ctr Toronto Hosp 200 Elizabeth St Toronto ON M5G 2C4 Can

LANGILLE, ROBERT C(ARDEN), b Yarmouth, NS, Apr 22, 15; m 52; c 3. ELECTRONICS. *Educ:* St Francis Xavier Univ, Can, BSc, 37; Dalhousie Univ, MSc, 40; Univ Toronto, PhD(phys chem), 44. *Prof Exp:* Res officer, Can Army Oper Res Group, 44-46; from sci officer to supt electronics lab, 47-69, DIR GEN, COMMUN RES CTR, DEPT COMMUN, TELECOMMUN ESTAB, DEFENCE RES BD CAN, 69- *Mem:* Int Union Radio Sci; Inst Elec & Electronics Engrs; Arctic Inst NAm; Royal Meteorol Soc; Can Asn Physicists. *Res:* Solid state physics; general electronics; space research; tropospheric radio propagation. *Mailing Add:* 1875 Ferncroft Crescent Ottawa ON K1H 7B5 Can

LANGLAND, BARBARA JUNE, magnetic recording, equaltization, for more information see previous edition

LANGLAND, OLAF ELMER, b Madrid, Iowa, May 30, 25; wid; c 4. RADIOLOGY, DENTISTRY. *Educ:* Univ Iowa, DDS, 51, MS, 61; Am Bd Oral & Maxillo-Facial Radiol, dipl, 81; Am Bd Oral Med, dipl, 84. *Prof Exp:* From instr to assoc prof oral diag & radiol, Col Dent, Univ Iowa, 59-69, head dept, 64-69; prof oral diag-med-radiol & head dept, Sch Dent, La State Univ, New Orleans, 69-75; PROF & HEAD, DIV OF DENT RADIOL, DENT DIAG SCI, SCH DENT, UNIV TEX HEALTH SCI CTR, SAN ANTONIO, 75-, PROF, DEPT RADIOL, SCH MED, 75- *Concurrent Pos:* USPHS grant, Col Dent, Univ Iowa, 64-66; consult, Wilford Hall US Air Force Hosp, Lackland, Tex, 68-77; mem subcomt proposed dent x-ray mach, Am Nat Standards Inst, 68-74; staff dentist, Charity Hosp, New Orleans, 69-75; vis prof, Univ Fed Alagoas, Maceio, Brazil, 73; chmn, oral diag/med sect, Am Asn Dent Schs, 73 & dent radiol sect, 84; mem Nat Bd Test Const Comt Oral Path & Dent Radiol, Am Dent Asn, 79-83. *Honors & Awards:* Merit Award, Coun Int Rels, Am Dent Asn, 75. *Mem:* Fel Am Col Dent; Am Acad Dent Radiol (pres, 84); Orgn Teachers Oral Diag (pres, 73); Am Asn Dent Schs; Int Asn Oral & Maxillo-Facial Radiol. *Res:* Panoramic radiography; educational research in dentistry; clinical research in oral manifestations of systemic disease; application of modern intensifying screens in diagnostic radiology. *Mailing Add:* Sch Dent Univ Tex Health Sci Ctr San Antonio TX 78284

LANGLANDS, ROBERT P, b New Westminster, BC, Oct 6, 36; m 56; c 4. MATHEMATICS. *Educ:* Univ BC, BA, 57, MA, 58; Yale Univ, PhD(math), 60. *Hon Degrees:* DSc, Univ BC, McMaster Univ, City Univ New York Grad Ctr, 85, Univ Waterloo, 88 & Université de Paris VII, 89. *Prof Exp:* Instr math, Princeton Univ, 60-61, lectr, 61-62, from asst prof to assoc prof, 62-67; prof math, Yale Univ, 67-72; PROF MATH, INST ADVAN STUDY, 72- *Concurrent Pos:* Mem, Inst Advan Study, 62-63; Miller fel, Univ Calif, Berkeley, 64-65; Sloan fel, 64-66. *Honors & Awards:* Wilbur Cross Medal, Yale Univ, 75; Cole Prize, Am Math Soc, 82; Common Wealth Award, 84; Nat Acad Sci Award in Math, 88. *Mem:* Royal Soc Can; Royal Soc London; Am Math Soc; Can Math Soc; AAAS; Sigma Xi. *Res:* Group representations; automorphic forms. *Mailing Add:* Inst Advan Study Princeton NJ 08540

LANGLEBEN, MANUEL PHILLIP, b Poland, Apr 9, 24; nat Can; m 48; c 3. GLACIOLOGY, MICROMETEOROLOGY. *Educ:* McGill Univ, BSc, 49, MSc, 50, PhD(physics), 53. *Prof Exp:* Res atmospheric physics, 53-57, lectr, 57-59, from asst prof to assoc prof, 59-69, dir, McGill Ctr Northern Studies & Res, 77-80, PROF PHYSICS, MCGILL UNIV, 69-, ASST CHMN, 84- *Mem:* Royal Meteorol Soc; Glaciol Soc; Am Geophys Union; Sigma Xi; fel Royal Soc Can. *Res:* Physics of ice; sea ice; ice drift. *Mailing Add:* Rutherford Physics Bldg 3600 University St Montreal PQ H3A 2T8 Can

LANGLEY, ALBERT E, b July 2, 43. PHARMACOLOGY, TOXICOLOGY. *Educ:* Waynesburg Col, BS, 67; Ohio State Univ, Columbus, PhD(pharmacol), 74. *Prof Exp:* Postdoctoral fel, Dept Pharmacol, Med Ctr, Univ Colo, Denver, 74-76; scientist, Cardiovasc Sect, Warner-Lambert Res Inst, Morris Plains, NJ & Ann Arbor, Mich, 76-77; from asst prof to prof, Dept Pharmacol & Toxicol, Wright State Univ, 77-90, vchair dept, 80-82, chair dept, 82-85, ASSOC DEAN ACAD AFFAIRS, SCH MED, WRIGHT STATE UNIV, DAYTON, OHIO, 90- *Concurrent Pos:* Course dir med pharmacol, Wright State Univ, 78-83, mem, Lab Animal Utilization Comt, 82-87 & chmn, Radiation Safety Comt, 84-; res grants, var insts & asns, 79-; invited lectr, var asns, univs & insts, 82-; consult, Eurand Am, Inc, 84- *Mem:* Am Soc Pharmacol & Exp Therapeut; Asn Med Sch Pharmacol; Soc Toxicol. *Res:* Autonomic and ocular drugs; antihypertensives, antilipidemics, and antiarrythmics; diuretics; cardiotonics; histamine and antihistamines; vasoactive peptides; drugs used to treat migraine; alcohols; antianginals; respiratory drugs; thrombolytics. *Mailing Add:* Sch Med Wright State Univ PO Box 927 Dayton OH 45401-0927

LANGLEY, G R, b Sydney, NS, Oct 6, 31; c 3. INTERNAL MEDICINE, HEMATOLOGY. *Educ:* Mt Allison Univ, BA, 52; Dalhousie Univ, MD, 57; FRCP(C), 61. *Hon Degrees:* FRCP(E), 83. *Prof Exp:* Asst resident internal med, Victoria Gen Hosp & Toronto Gen Hosp, 57-60; j Arthur Haatz fel hemat, Univ Melbourne, 60-61; Med Res Coun Can fel, Sch Med & Dent, Univ Rochester, 61-62 & Dalhousie Univ, 63; lectr internal med, 63-64, from asst prof to assoc prof med, 64-68, head dept, 74-82, PROF MED, DALHOUSIE UNIV, 68- *Concurrent Pos:* John & Mary R Markle scholar med, 63-68; consult, Armed Forces Hosp, Halifax, 64-; head dept med, Camp Hill Hosp, 69-74; head dept med, 74-82, sr physician, Victoria Gen Hosp, 82-; Wightman vis prof, Royal Col Physicians & Surgeons, 90. *Honors & Awards:* Queens Jubilee Medal, 77; Medal, Nat Can Inst, 85. *Mem:* Fel Am Col Physicians; Can Soc Clin Invest; Can Med Asn; Am Soc Hemat; Royal Col Physicians & Surgeons Can. *Res:* Hematological oncology; quantitative analysis of bioethical issues. *Mailing Add:* Dept Med Victoria Gen Hosp Halifax NS B3H 2Y9 Can

LANGLEY, KENNETH HALL, b Ft Collins, Colo, Sept 1, 35; m 59; c 2. BIOPHYSICS. *Educ:* Mass Inst Technol, SB, 58; Univ Calif, Berkeley, PhD(physics), 66. *Prof Exp:* Actg asst prof & res assoc physics, Univ Calif, Berkeley, 66; asst prof, 66-73, assoc prof physics, 73-81, PROF PHYSICS, UNIV MASS, AMHERST, 81- *Concurrent Pos:* Cofounder, Langley-Ford

Instruments. *Mem:* Am Phys Soc; AAAS. *Res:* Experimental dynamic light scattering from polymers and biological macromolecules; dynamic nuclear orientation; light scattering from critical point fluids; biological systems. *Mailing Add:* Dept Physics & Astron Univ Mass Amherst MA 01003

LANGLEY, MAURICE N(ATHAN), b Dorchester, Nebr, July 6, 13; m 39; c 3. AGRICULTURAL ENGINEERING. *Educ:* Colo Agr & Mech Col, BS, 39. *Prof Exp:* Soil surveyor, Agr Exp Sta, Colo State Col, 39-40; jr soil surveyor, Soil Conserv Serv, USDA, Tex, 40-42, asst soil technologist & chief land classification div, 46-48, land use specialist & head land use & settlement div, 49-56, agr engr & chief opers div, 56-59, hydraul engr, Land & Water Br, 59-60, chief, Irrig Br, 60-62, asst chief div irrig & land use, 62-64, chief water & land, 64-73; vpres, Bookman-Edmonston Eng, Inc, 73-88; SR CONSULT, 88- *Concurrent Pos:* Pres, US Comt Irrig, Drainage & Flood Control, 70, 71 & 72; US nat chmn, US Comt Irrig & Drainage, 70-72 & Int Comn Irrig & Drainage, vpres, 74, 75 & 76; bd dirs, Am Water Found, 84-88. *Honors & Awards:* Meritorious Serv Award, US Bur Reclamation, 56 & 69; Outstanding Serv Award, USBR, 47; Meritorious Serv Award, USBR, 56 & 69; Int Water for Peace Conf Commendation, 67; Distinguished Serv Award, Dept of Interior, 68; Outstanding Serv & Leadership Award, US Comt on Irrig & Drainage, 84; Distinguished Serv Award, Nat Water Resources Asn, 89. *Mem:* Am Soc Agr Engrs; Soil Sci Soc Am; Am Soc Agron; Int Water Resources Asn; Am Water Works Asn; Int Comn Irrig & Drainage (pres, 73-75); Am Water Found; Int Eng Comt Am Consult Engrs Coun; Nat Water Resources Asn; Am Registry Cert Prof Agronomist & Soil Scientist. *Res:* Land and water resources development and administration. *Mailing Add:* 6825 Algonquin Ave Bethesda MD 20817-4813

LANGLEY, NEAL ROGER, b Sumas, Wash, July 27, 39. POLYMER CHEMISTRY. *Educ:* Univ Wash, BS, 61; Univ Wis, PhD(chem), 68. *Prof Exp:* Sr res scientist, Pac Northwest Lab, Battelle Mem Inst, 67-68; chemist, 68-80, ASSOC RES SCIENTIST, DOW CORNING CORP, 80- *Mem:* Am Chem Soc. *Res:* Structure and viscoelastic properties of cross-linked polymers; structure and thermo-mechanical properties of ceramic fibers. *Mailing Add:* Dow Corning Corp PO Box 994 C 043 D1 Midland MI 48686-0994

LANGLEY, ROBERT ARCHIE, b Athens, Ga, Oct 21, 37; div; c 3. VACUUM TECHNOLOGY, MATERIAL SCIENCE. *Educ:* Ga Inst Technol, BS, 59, MS, 60, PhD(physics), 63. *Prof Exp:* Asst physics, Ga Inst Technol, 59-63; physicist, Air Force Cambridge Res Labs, 63-65 & Oak Ridge Nat Lab, 66-68; staff mem, Sandia Corp, 68-78; prog coordr, Oak Ridge Nat Lab, 78-80; prog coordr, Int Atomic Energy Agency, 80-81; STAFF SCIENTIST, OAK RIDGE NAT LAB, 81- *Mem:* Am Phys Soc; Am Nuclear Soc; Am Vacuum Soc; Health Physics Soc. *Res:* Plasma-wall interactions in controlled fusion devices; ion implantation; hydrogen and helium migration in metals; surface physics; ion backscattering; nuclear microanalysis. *Mailing Add:* Rte 4 Box 188L Kingston TN 37763

LANGLEY, ROBERT CHARLES, b NJ, Apr 11, 25; m 54; c 1. ORGANIC CHEMISTRY. *Educ:* St Peters Col, BS, 49. *Prof Exp:* Chemist, E I du Pont de Nemours & Co, 50-54; res dir, Hanovia Div, 54-62, SECT HEAD, RES & DEVELOP DIV, ENGELHARD INDUSTS, INC, 62- *Mem:* Am Chem Soc; Am Ceramic Soc. *Res:* Organic compounds of metals; thin films; gas purification. *Mailing Add:* 214 Old Forge Rd Millington NJ 07946

LANGLOIS, BRUCE EDWARD, b Berlin, NH, Sept 16, 37; m 60; c 2. FOOD MICROBIOLOGY. *Educ:* Univ NH, BS, 59; Purdue Univ, PhD(dairy microbiol), 62. *Prof Exp:* Asst prof dairy, Purdue Univ, 62-64; from asst prof to assoc prof dairy sci, 64-74, PROF ANIMAL SCI, UNIV KY, 74- *Mem:* Am Soc Microbiol; Inst Food Technologists; Int Asn Milk, Food & Environ Sanit; Sigma Xi. *Res:* Staphyloccocal mastitis; transferable drug resistance in farm animals; microflora of dairy and meat products. *Mailing Add:* Univ Ky 204 WP Garrigus Bldg Univ Ky Lexington KY 40546-0215

LANGLOIS, GORDON ELLERBY, b Burley, Idaho, Aug 30, 18; m 44; c 3. PHYSICAL CHEMISTRY. *Educ:* Northwestern Univ, BS, 42; Univ Calif, PhD(phys chem), 52. *Prof Exp:* Sr res chemist, Calif Res Corp, Chevron Res Co, 43-69, sr res chemist, 69-73, mgr, Synthetic Fuels Div, 73-78. *Mem:* Am Chem Soc. *Res:* Catalytic reactions of hydrocarbons, as polymerization, alkylation and isomerization; synthetic fuels technology. *Mailing Add:* 15 Doral Dr Moraga CA 94556

LANGLOIS, WILLIAM EDWIN, b Providence, RI, Oct 23, 33; m 54; c 2. HYDRODYNAMICS. *Educ:* Univ Notre Dame, ScB, 53; Brown Univ, PhD(appl math), 57. *Prof Exp:* Res engr, Polychem Dept, E I du Pont de Nemours & Co, 56-59; STAFF MEM, SAN JOSE RES LAB, IBM CORP, 59- *Concurrent Pos:* Adj lectr, Univ Del, 56-58 & Univ Santa Clara, 61-69, 85; vis prof, Univ Notre Dame, 70-71. *Res:* Theory of viscous and viscoelastic fluid flow. *Mailing Add:* IBM Res Almaden Res Ctr 650 Harry Road San Jose CA 95120-6099

LANGLYKKE, ASGER FUNDER, b Pleasant Prairie, Wis, July 17, 09; m 39; c 4. MICROBIOLOGY. *Educ:* Univ Wis, BS, 31, MS, 34, PhD(biochem), 36. *Hon Degrees:* ScD, Trinity Col, 65. *Prof Exp:* Foreman, Procter & Gamble Co, Ill, 31-32; asst, Univ Wis, 32-36, Dow fel, 36-37; res chemist, Hiram Walker & Sons, Inc, 37-40; supt butyl alcohol plant, Cent Lafayette, PR, 40-43; div head, Northern Regional Res Lab, USDA, Ill, 43 & 45-47; chief pilot plant div & chief tech officer, Chem Warfare Serv, Md & Ind, 43-45; dir, Microbiol Develop, E R Squibb & Sons, 47-49, dir res & develop labs, 49-64, vpres, 64-68; exec dir, Am Acad Microbiol & Am Soc Microbiol, 68-74; proj mgr, Frederick Cancer Res Ctr, 77-79; RETIRED. *Concurrent Pos:* Consult, Res & Develop Bd, US Army, 45-53 & Chem Corp, 53-66; off dir, Defense Res & Eng, 60-63; mem, Comt Agr Sci, USDA, 64-68; Adv Bd Res & Grad Educ, Rutgers Univ & Sci Adv Comt, Rutgers Inst Microbiol, 64-68; vpres, Appl Chem Div Comt, Int Union Pure & Appl Chem, 77-81; adj prof, Rutgers Univ, 68-74; exec staff scientist, Genex Corp, 79-85. *Honors & Awards:* Barnett Cohen Award, Am Soc Microbiol, 78; James M Van Lanean Award,

Am Chem Soc, 83. *Mem:* Fel AAAS; fel NY Acad Sci; emer Mem Biochem Soc UK; Am Inst Chem Engrs; Am Soc Microbiol. *Res:* Fermentation processes for antibiotics steroids, industrial and pharmaceutical chemicals; biochemical engineering. *Mailing Add:* 240 Dill Ave Frederick MD 21701

LANGMUIR, ALEXANDER D, b Santa Monica, Calif, Sept 12, 10. EPIDEMIOLOGY. *Educ:* Cornell Univ, MD, 35; Johns Hopkins Univ, 40. *Prof Exp:* Itern, Boston City Hosp, 35-37; epidemiologist, Comn Acute Respiratory Dis, 42-46; assoc prof epidemiol, Johns Hopkins Univ, 46-49; chief epidemiologist, Communicable Dis Ctr, USPHS, 49-70; clin prof prev med & Community Health, Emory Univ, 49-70; vis prof epidemiol, Harvard Univ, 70-77; RETIRED. *Mem:* Inst Med-Nat Acad Sci; Int Epidemiol Asn; Infectious Dis Soc Am; Am Pub Health Asn; Am Epidemiol Soc; Am Soc Clin Invest; Royal Soc Med. *Mailing Add:* RFD Box 451 Chilmark MA 02535

LANGMUIR, DAVID BULKELEY, b Los Angeles, Calif, Dec 14, 08; m 42; c 3. PHYSICS. *Educ:* Yale Univ, BS, 31; Mass Inst Technol, ScD(physics), 35. *Prof Exp:* Res physicist, Radio Corp Am Mfg Co, NJ, 35-41; liaison officer, Off Sci Res & Develop, Washington, DC & London, 41-45; secy, Guided Missiles Comt, Joint Chiefs of Staff, Washington, DC, 45-46; dir planning div, Res & Develop Bd, 46-48; exec officer, Prog Coun, AEC, 48-50, liaison officer, Chalk River, 50-54; mem tech staff, Ramo-Wooldridge Corp, 54-56, dir res lab, Ramo-Wooldridge Div, Thompson Ramo Wooldridge Corp, 56-60 & TRW Space Technol Labs, 60-65, dir, Phys Res Ctr, TRW Systs, Calif, 65-70, dir res planning, 70-73; RES CONSULT, 73- *Concurrent Pos:* Prof lectr, George Washington Univ, 48-50; mem sci adv group, Off Aerospace Res, US Air Force, 63-70, chmn, 69-70; chmn, Joint Workshop Indust Innovation, Nat Acad Sci, Taiwan, 75. *Mem:* Fel Am Phys Soc; fel Inst Elec & Electronics Engrs; Am Inst Aeronaut & Astronaut; Sigma Xi. *Res:* Thermionics; television light valves; high frequency tubes; diffusion in metals; ionic propulsion. *Mailing Add:* 350 21st St Santa Monica CA 90402

LANGMUIR, DONALD, b Nashua, NH, Apr 5, 34; c 2. GEOCHEMISTRY. *Educ:* Harvard Univ, AB, 56, MA, 61, PhD(geol sci), 65. *Prof Exp:* Geochemist, Water Resources Div, US Geol Surv, 64-66; lectr water resources, Rutgers Univ, 66-67; from asst prof to prof geochem, Pa State Univ, University Park, 67-78; PROF GEOCHEM, DEPT CHEM & GEOCHEM, COLO SCH MINES, 78- *Concurrent Pos:* Adj prof geochem, Desert Res Inst, Univ Nev, Reno, 74-75; assoc ed, Geochim Cosmochim Acta, 75-80; pres, Hydrochem Systs Corp, 79-; vis prof inorg chem, Univ Sidney, Australia, 80; dir, Earth Search, Inc, 81-85; mem, President's Nuclear Waste Tech Rev Bd, 89; pres, Colo Mountain Club, 90. *Mem:* Fel AAAS; Am Chem Soc; Am Geophys Union; fel Mineral Soc Am; Geochem Soc; Sigma Xi; Soc Environ Geochem & Health. *Res:* Geochemistry of subsurface waters; thermodynamic properties of minerals and dissolved species in water; adsorption of dissolved inorganic species on geological materials; geochemistry of exploration for ore deposits, of solution mining, and of groundwater pollution and restoration. *Mailing Add:* Dept Chem & Geochem Colo Sch Mines Golden CO 80401

LANGMUIR, MARGARET ELIZABETH LANG, b Chicago, Ill, Nov 11, 35; m 62; c 2. PHYSICAL CHEMISTRY, PHOTOCHEMISTRY. *Educ:* Culver-Stockton Col, BA, 56; Purdue Univ, PhD(chem), 63. *Prof Exp:* Instr anal & inorg chem, Wellesley Col, 60-63; phys chemist, Pioneering Res Div, US Army Natrick Labs, 63-69; consult, 69-74; res assoc, Northeastern Univ, 76-78; sr scientist, EIC Corp, 78-81; SR SCIENTIST, COVALENT ASSOC, INC, 82- *Mem:* AAAS; Am Chem Soc; Electrochem Soc; Sigma Xi. *Res:* Organic photochemistry; acidity functions; flash photolysis; fast reaction mechanisms; excited state proton transfer; fluorescence, phosphorescence and charge transfer spectra; electrochemistry; implantable electrodes; isomerization; semiconductor electrochemistry; photoelectrochemical cells; solar cells; polymer modified electrodes, ion selective electrodes, ion selective fluorophores; electrochemistry. *Mailing Add:* Nine Bent Brook Rd Sudbury MA 01776

LANGMUIR, ROBERT VOSE, b White Plains, NY, Dec 20, 12; m 39; c 2. PHYSICS. *Educ:* Harvard Univ, AB, 35; Calif Inst Technol, PhD(physics), 43. *Prof Exp:* Physicist, Consol Eng Corp, Calif, 39-42 & Gen Elec Co, 42-48; sr res fel, 48-50, asst prof physics, 50-52, assoc prof elec eng, 52-57, prof, 57-80, EMER PROF ELEC ENG, CALIF INST TECHNOL, 80- *Concurrent Pos:* Consult, TRW, Inc, 57- *Mem:* Am Phys Soc; Inst Elec & Electronics Engrs. *Res:* Mass spectroscopy; synchrotrons; secondary emission cathode in magnetrons; starting of synchrotrons. *Mailing Add:* 3383 N Lake Ave Altadena CA 91001

LANGNER, GERALD CONRAD, b Austin, Minn, Feb 13, 44; m 73; c 1. PHYSICS. *Educ:* St John's Univ, BA, 66; NDak State Univ, MA, 68. *Prof Exp:* Physicist control syst eng, US Navy, 68-69; scientist II physics, Albuquerque Div, EG&G Inc, 69-73; MEM STAFF NONDESTRUCTIVE TESTING, LOS ALAMOS NAT LAB, 74- *Mem:* Am Soc Nondestructive Testing; Am Phys Soc. *Mailing Add:* Los Alamos Nat Lab 90 Mimbres Dr Los Alamos NM 87544

LANGNER, RALPH ROLLAND, environmental chemistry; deceased, see previous edition for last biography

LANGNER, RONALD O, b Chicago, Ill, May 10, 40; m 63; c 2. BIOCHEMICAL PHARMACOLOGY. *Educ:* Blackburn Col, BA, 62; Univ RI, MS, 66, PhD(pharmacol), 69. *Prof Exp:* ASSOC PROF PHARMACOL, UNIV CONN, 69- *Mem:* Am Biol Chem. *Res:* Metabolism of collagen and its relationship to experimental atherosclerosis. *Mailing Add:* Dept Pharmacol Univ Conn Storrs CT 06268

LANGNER, THOMAS S, psychiatry, epidemiology, for more information see previous edition

LANGONE, JOHN JOSEPH, b Cambridge, Mass, Aug 20, 44; m 67; c 2. BIO-ORGANIC CHEMISTRY. *Educ:* Boston Col, BS, 66; Boston Univ, PhD(org chem), 72. *Prof Exp:* Fel org chem, Boston Univ, 71-72; sr res assoc biochem, Brandeis Univ, 72-75; STAFF FEL IMMUNOCHEM, NAT CANCER INST, 75- *Mem:* Am Chem Soc; Am Soc Biol Chemists. *Res:* Immunopharmacology and immunochemistry of biologically active compounds; cancer immunochemistry. *Mailing Add:* 12709 Twinbrook Pkwy Rockville MD 20852

LANGRANA, NOSHIR A, b Bombay, India, Oct 1, 46; US citizen; m 72; c 2. COMPUTER AIDED DESIGN, BIOMECHANICS. *Educ:* Univ Bombay, India, BE, 68; Cornell Univ, MS, 71, PhD(mech eng), 75. *Prof Exp:* Asst prof, 76-82, ASSOC PROF MECH DESIGN, COL ENG, RUTGERS UNIV, 82- *Concurrent Pos:* Adj assoc prof surg, NJ Med Sch, 82- *Honors & Awards:* Ralph R Teetor Award, Soc Automotive Engrs, 77. *Mem:* Am Soc Mech Engrs; Orthopaedic Res Soc. *Res:* Relationship of computer-aided technology in the investigation of musculoskeletal problems; designing and developing new materials, diagnostic procedures and devices for medical use. *Mailing Add:* Mech Engr Rutgers Univ Busch Campus New Brunswick NJ 08903

LANGRETH, DAVID CHAPMAN, b Greenwich, Conn, May 22, 37; m 66; c 2. THEORETICAL CONDENSED MATTER PHYSICS, THEORETICAL SURFACE PHYSICS. *Educ:* Yale Univ, BS, 59; Univ Ill, MS, 61, PhD(physics), 64. *Prof Exp:* Res assoc physics, Univ Chicago, 64-65 & Cornell Univ, 65-67; from asst prof to prof, 67-80, assoc chmn dept & dir grad progs, 70-73, PROF II PHYSICS, RUTGERS UNIV, NEW BRUNSWICK, 80- *Concurrent Pos:* Rutgers Res Coun fel, Univ Calif, San Diego, 73-74; guest prof, Nordita, Copenhagen, 75-76; prin investr, NSF grants, 69-; coordr surface prog, Inst Theoret Physics, Santa Barbara, 83; mem, Davisson-Germer Prize Comt, 84-, chmn, 88; vis prof, Chalmers Univ, Göteborg, Sweden, 88. *Mem:* AAAS; Am Phys Soc. *Res:* Theoretical solid state physics, specializing in the many body problem; theoretical surface physics. *Mailing Add:* Dept Physics & Astron Rutgers Univ PO Box 849 Piscataway NJ 08855-0849

LANGRIDGE, ROBERT, b Essex, Eng, Oct 26, 33; m 60; c 2. MOLECULAR BIOLOGY. *Educ:* Univ London, BSc, 54, PhD(crystallog), 57. *Prof Exp:* Res fel biophys, Yale Univ, 57-59; res assoc biophys, Mass Inst Technol, 59-61 & Children's Hosp, Med Ctr, Boston Univ, 61-66; res assoc, Harvard Univ, 63-64, lectr, 64-66; prof biophys & info sci, Univ Chicago, 66-68; prof biochem, Princeton Univ, 68-76; PROF BIOCHEM & BIOPHYS, SCH MED, UNIV CALIF, 77-, PROF, DEPT PHARMACEUT CHEM, 77-, DIR, COMPUTER GRAPHICS LAB, 77- *Mem:* Inst Med-Nat Acad Sci; Biophys Soc; Am Crystallog Asn; Am Chem Soc; AAAS. *Res:* X-ray diffraction and physical-chemical studies of the structures of biological macromolecules, particularly nucleic acids, nucleoproteins, viruses and ribosomes; applications of high speed digital computers. *Mailing Add:* Dept Pharmaceut Chem Univ Calif San Francisco CA 94143-0446

LANGRIDGE, WILLIAM HENRY RUSSELL, b New York, NY, Jan 30, 38; m 60; c 2. VIROLOGY, DEVELOPMENTAL BIOLOGY. *Educ:* Univ Ill, Urbana, BS, 62, MS, 64; Univ Mass, PhD(biochem), 74. *Prof Exp:* NIH fel virol, Boyce Thompson Inst, Ithaca, NY, 74-87; PROF, DEPT PLANT SCI, UNIV ALTA, 87- *Mem:* AAAS; Soc Invert Path; Am Soc Microbiol; Sigma Xi. *Res:* Metabolism of baculoviruses and vertebrate and insect poxviruses; the mechanism of infection and the structure of the virus genome and virus protein; plant molecular biology. *Mailing Add:* Dept Plant Sci 6-30 Med Sci Bldg Univ Alta Sch Med Edmonton AB T6G 2H7 Can

LANGSAM, MICHAEL, b Brooklyn, NY, Nov 4, 38; m 62; c 3. POLYMER CHEMISTRY. *Educ:* Rensselaer Polytech Inst, BS, 59; Polytech Inst Brooklyn, PhD(polymer sci), 64. *Prof Exp:* Res chemist polymer, BF Goodrich Lab, Brecksville, Ohio, 64-67; sr res chemist, 67-75, MGR POLYMER PROCESS DEVELOP, AIR PROD & CHEM, INC, 75- *Mem:* Sigma Xi; Am Chem Soc. *Res:* Polymer synthesis, rheology, polymer kinetics, and particle morphology. *Mailing Add:* 1114 N 26 St Allentown PA 18104

LANGSDORF, ALEXANDER, JR, b St Louis, Mo, May 30, 12; m 41; c 2. NUCLEAR PHYSICS. *Educ:* Washington Univ, AB, 32; Mass Inst Technol, PhD(physics), 37. *Prof Exp:* Nat Res Found fel, Univ Calif, 38; instr physics, Washington Univ, 39-43; physicist, Metall Lab, Argonne Nat Lab, 43-45, sr physicist & successor, 45-77; RETIRED. *Concurrent Pos:* Asst physicist, Mallinckrodt Inst Radiol, 39-43; vis scientist, Atomic Energy Res Estab, UK, 59-60; assoc ed, Appl Physics Lett, 79-; consult, Argonne Nat Lab, 77- *Mem:* Fel Am Phys Soc; Fedn Am Scientist. *Res:* Neutron cross-section and polarization measurements using neutron chain reactor and electrostatic accelerator; cyclotron, electrostatic and dynamitron construction, design and operation; theory of design of experiments; corona studies; diffusion cloud chamber. *Mailing Add:* 645 S Meacham Rd Schaumburg IL 60193

LANGSDORF, WILLIAM PHILIP, b Cambridge, Ohio, Apr 6, 19; m 41; c 1. PHYSICAL ORGANIC CHEMISTRY. *Educ:* Ohio State Univ, BSc, 41; Mass Inst Technol, PhD(chem), 49. *Prof Exp:* Res chemist, Indust & Biochem Dept, E I du Pont de Nemours & Co, Inc, 49-62, res assoc, 62-79; RETIRED. *Mem:* Sigma Xi. *Res:* Organic reactions; mechanism of organic reactions; kinetics. *Mailing Add:* 2407 Rambler Rd Graylyn Crest Wilmington DE 19810

LANGSETH, MARCUS G, b Lebanon, Tenn, Nov 24, 32; m 63. GEOPHYSICS, OCEANOGRAPHY. *Educ:* Waynesburg Col, BS, 54; Columbia Univ, PhD(geol), 64. *Prof Exp:* SR RES ASSOC GEOPHYS, LAMONT-DOHERTY GEOL OBSERV, COLUMBIA UNIV, 58- *Concurrent Pos:* Adj prof, Columbia Univ. *Mem:* AAAS; Am Geophys Union; Geol Soc Am; Sigma Xi. *Res:* Terrestrial heat flow; lunar heat flow; oceanographic instrumentation; submarine geology. *Mailing Add:* Lamont-Doherty Geol Observ Palisades NY 10964

LANGSETH, ROLLIN EDWARD, b St Paul, Minn, Apr 13, 40; m 66; c 2. ELECTRICAL ENGINEERING, MATHEMATICS. *Educ:* Univ Minn, Minneapolis, BS, 62, MSEE, 65, PhD(elec eng), 68. *Prof Exp:* mem tech staff, Bell Tel Labs, 68-80; DIST MGR, AM TEL & TEL CO, 81- *Mem:* Inst Elec & Electronics Engrs. *Res:* Communication in multipath propagation media; diversity systems; effects of noise and interference; data networks; satellite systems. *Mailing Add:* AT&T Bell Labs Crawfords Corner Rd Rm 3M 526 Holmdel NJ 07733

LANGSJOEN, ARNE NELS, b Dalton, Minn, Apr 6, 19; m 43; c 3. ORGANIC CHEMISTRY. *Educ:* Gustavus Adolphus Col, BA, 42; Univ Iowa, MS, 43, PhD, 49. *Prof Exp:* From asst prof to assoc prof, Gustavus Adolphus Col, 48-56, chmn dept, 56-66, prof chem, 56-86; RETIRED. *Concurrent Pos:* NSF res fel, Uppsala & Royal Inst Technol, Sweden, 58-59; vis prof chem, Tughai Univ, Taichung, Taiwan, 69-70. *Mem:* Am Chem Soc. *Res:* Biological chemistry. *Mailing Add:* Dept Chem Gustavus Adolphus Col St Peter MN 56082

LANGSJOEN, PER HARALD, b Fergus Falls, Minn, Aug 9, 21; m 45; c 5. MEDICINE, CARDIOLOGY. *Educ:* Gustavus Adolphus Col, AB, 43; Univ Minn, Minneapolis, MB, 50, MD, 51. *Prof Exp:* Intern & residency internal med, Letterman Army Hosp, San Francisco, Calif, 50-54; staff physician, Coco Solo Hosp, Cristobal, CZ, 54-56; resident cardiol, Fitzsimmons Army Hosp, Denver, Colo, 56-58; chief cardiovasc serv, Wm Beaumont Gen Hosp, El Paso, Tex, 58-60; chief cardiovasc serv, Scott & White Clin, 60-75; CLIN PROF, COL MED, TEX A&M UNIV, 75- *Concurrent Pos:* Fel coun clin cardiol, Am Heart Asn, 65-; consult, US Army Hosp, Ft Hood, Tex, 61-75 & Vet Admin Hosp, Temple, 61-75. *Mem:* Fel Am Col Physicians; fel Am Col Cardiol. *Res:* Rheologic aspects of the circulatory system in health and in disease states. *Mailing Add:* Scott & White Clin Temple TX 76501

LANGSLEY, DONALD GENE, b Topeka, Kans, Oct 5, 25; m 55; c 3. PSYCHIATRY, PSYCHOANALYSIS. *Educ:* State Univ NY Albany, AB, 49; Univ Rochester, MD, 53. *Prof Exp:* Intern, USPHS Hosp, San Francisco, 53-54; resident psychiat, Langley Porter Clin, Sch Med, Univ Calif, San Francisco, 54-59, NIMH career teacher award psychiat, 59-61; from asst to assoc prof psychiat, Sch Med, Univ Colo, 61-68; prof psychiat & chmn dept, Sch Med, Univ Calif, Davis, 68-76; prof psychiat & chmn dept, Sch Med, Univ Cincinnati, 76-81; EXEC VPRES, AM BD MED SPECIALTIES, 81-; PROF PSYCHIAT, NORTHWESTERN UNIV MED SCH, 82- *Concurrent Pos:* Dir inpatient serv, Colo Psychiat Hosp, Denver, 61-68; dir ment health serv, Sacramento Med Ctr, Univ Calif, Davis, 68-73; mem psychiat training comt, NIMH, 71-75, mem psychiat test comt, Nat Bd Med Examr, 72-76; chief staff, Sacramento Med Ctr, 74-75; chmn dept defense select comt psychiat care eval, NIMH, 75-78; dir, Am Bd Psychiat & Neurol, 75-80. *Honors & Awards:* Hofheimer Award, Am Psychiat Asn, 71. *Mem:* Am Psychiat Asn (vpres, 77-79, pres, 80-81); Am Psychoanalysis Asn; AMA. *Res:* Family crisis therapy; evaluation of therapy, medical education and psychiatric education. *Mailing Add:* One Rotary Ctr Ste 805 Evanston IL 60201

LANGSTON, CHARLES ADAM, b 1949. SEISMOLOGY. *Educ:* Case Western Reserve Univ, BS, 72; Calif Inst Technol, MS, 74, PhD(geophys), 76. *Prof Exp:* Res asst geophys, Calif Inst Technol, 72-74, Louis D Beaumont fel, 74-75, res asst, 75-76, res fel, 76-77; from asst prof to assoc prof, 77-86, PROF GEOPHYS, PA STATE UNIV, 86- *Concurrent Pos:* Assoc ed, J Geophys Res, 80-82; dir seismic observ, Pa State Univ, 85-; bd dirs, Seismol Soc Am, 88-91; mem comt seismol, Nat Res Coun, 90- *Mem:* Am Geophys Union; Seismol Soc Am (pres, 90-91); AAAS. *Res:* Theoretical and observation seismology; wave propagation in elastic media; seismic source parameter estimation; crustal and upper mantle structure. *Mailing Add:* Dept Geosci Pa State Univ 440 Deike Bldg University Park PA 16802

LANGSTON, CLARENCE WALTER, b Gainesville, Tex, July 4, 24; m 48; c 2. MICROBIOLOGY. *Educ:* Southern Methodist Univ, BS, 49; North Texas State Univ, MA, 51; Univ Wis, PhD(bact), 55. *Prof Exp:* Bacteriologist, Kraft Foods Co, 50-51; asst, Univ Wis, 51-54; bacteriologist, Dairy Cattle Res Br, Agr Res Serv, USDA, 54-62; chief virus & rickettsial div, Directorate Biol Opers, Pine Bluff Arsenal, US Army, Ark, 62-66; head bact sect, Midwest Res Inst, 66-71; PRES, LANGSTON LABS, INC, KANS & PR, 71- *Mem:* Am Acad Microbiol; Am Soc Microbiol; Soc Indust Microbiol. *Res:* Dairy and food bacteriology; microbiology and chemistry of fermentations; physiology of bacteria; taxonomy and nomenclature of bacteria; research development and production of viruses and rickettsiae. *Mailing Add:* 4921 W 96th St Overland Park KS 66207

LANGSTON, DAVE THOMAS, b Chickasha, Okla, Apr 19, 45; m 65; c 4. ENTOMOLOGY. *Educ:* Southwestern Okla State Univ, BS, 67; Okla State Univ, MS, 70; Univ Ariz, PhD(entom), 74. *Prof Exp:* Res asst, Okla State Univ, 67-70; res asst, 70-74, ASSOC PROF ENTOM, UNIV ARIZ, 80-, EXTEN SPECIALIST ENTOM, 74- *Mem:* Entom Soc Am. *Res:* Insect management as it influences production agriculture, specifically those insects which are economically important in the Southwestern United States. *Mailing Add:* 5902 S College Ave Temple AZ 85283

LANGSTON, GLEN IRVIN, b Marion, Ohio, Nov 22, 56; m 87; c 1. RADIO ASTRONOMY. *Educ:* Mass Inst Technol, BS, 81, PhD(physics), 87. *Prof Exp:* Scientist, Max Planck Inst, 87-88 & Naval Res Lab, 88-89; SCIENTIST, NAT RADIO ASTRON OBSERV, 89- *Res:* Gravitational lensing. *Mailing Add:* Nat Radio Astron Observ Edgemont Rd Charlottesville VA 22903

LANGSTON, HIRAM THOMAS, b Rio de Janeiro, Brazil, Jan 12, 12; US citizen; m 41; c 3. THORACIC SURGERY. *Educ:* Univ Louisville, AB, 30, MD, 34; Univ Mich, MS, 41; Am Bd Surg, dipl, 42; Am Bd Thoracic Surg, dipl, 48. *Prof Exp:* Intern, Garfield Mem Hosp, Washington, DC, 34-35, resident path, 35-37; from asst resident to resident surg, Univ Mich Hosp, 37-39, resident thoracic surg, 39-40, instr, Univ, 40-41; assoc surg, Med Sch, Northwestern Univ, Ill, 41-42, asst prof, 46-48; assoc prof, Col Med, Wayne

State Univ, 48-52; prof surg, 52-78, clinical prof, 78-80, EMER PROF, COL MED, UNIV ILL, 80- *Concurrent Pos:* Chief surgeon, Chicago State Tuberc Sanitarium, State Dept Pub Health, 52-71; mem staff, Grant & St Joseph's Hosps, 52- *Mem:* Am Thoracic Soc; Am Asn Thoracic Surg (secy, 56-61, vpres, 68-69, pres, 69-70); fel Am Col Surg; Soc Thoracic Surgeons; Am Surg Asn. *Res:* Surgery for diseases of the chest; tuberculosis and cancer of the lung. *Mailing Add:* 2800 N Sheridan Rd Chicago IL 60657

LANGSTON, JAMES HORACE, b Garrison, Tex, Oct 8, 17; m 84. ORGANIC CHEMISTRY. *Educ:* Stephen F Austin State Col, BA, 37; Univ NC, MA, 39, PhD(org chem), 41. *Prof Exp:* Asst chem, Univ NC, 37-40; res chemist, Columbia Chem Div, Pittsburgh Plate Glass Co, 41-46; from assoc prof to prof textile chem & dyeing, Clemson Col, 46-58; prof chem, Samford Univ 58-84, head dept, 58-78, chmn div natural sci & math, 68-72; RETIRED. *Concurrent Pos:* Fulbright lectr, Cent Univ & Nat Polytech Sch, Ecuador, 59-60, Fulbright lectr & res consult, Nat Univ Honduras, 67-68. *Mem:* Am Chem Soc; fel Am Inst Chemists. *Res:* Drugs; polymers; plastics; plasticizers; catalysis; fibers; textile finishing materials; sulfone formation. *Mailing Add:* 516 One Mill Pl Augusta GA 30909-3785

LANGSTON, JIMMY BYRD, b Shelby, Miss, Apr 10, 27; m 72; c 2. PHYSIOLOGY, PHARMACOLOGY. *Educ:* Delta State Col, BS, 49; Univ Miss, Jackson, MS, 58, PhD(physiol), 59. *Prof Exp:* From instr to asst prof physiol, Med Ctr, Univ Miss, Jackson, 60-64; sect head toxicol, Alcon Labs, Inc, 64-65, dir urol, 65-72; res scientist, ALZA Res, 73-77; DIR TOXICOL, COOPER LABS, INC, 77- *Concurrent Pos:* NIH fel, Univ Miss 59-60; spec NIH fel, Univ Gothenburg, 62-63; NIH develop award, Med Ctr, Univ Miss, 63-64. *Mem:* Am Physiol Soc. *Res:* Renal physiology; cardiovascular physiology; ocular pharmacology; pharmaceutical toxicology. *Mailing Add:* 11345 Moonsail Helotes TX 78023

LANGSTON, WANN, JR, b Oklahoma City, Okla, July 10, 21; m 46; c 2. VERTEBRATE PALEONTOLOGY. *Educ:* Univ Okla, BS, 43, MS, 47; Univ Calif, PhD(paleont), 51. *Prof Exp:* Instr geol, Tex Tech Col, 46-48; preparator, Mus Paleont, Univ Calif, 49-54, lectr, 51-52; vert paleontologist, Nat Mus Can, 54-62; res scientist, Tex Mem Mus, 62-, dir, Vertebrate Palenot Lab, 69-; prof dept geol sci, Univ Tex, Austin, 75-, Yaeger prof, 83-; VPRES LAB, BALCONES RES CTR, AUSTIN. *Concurrent Pos:* Res assoc, Cleveland Mus Natural Hist, 74- *Mem:* Geol Soc Am; Soc Vert Paleont (vpres, 73-74, pres, 74-75); Am Soc Icthyol & Herpet; Am Asn Petrol Geologists; Sigma Xi. *Res:* Fossil amphibians and reptiles; stucture and relationships of extinct crocodylia and large pterosaurs; cretaceous non-mammalian tetrapods in Texas. *Mailing Add:* Balcones Res Ctr 10100 Burnet Rd Austin TX 78758

LANGSTROTH, GEORGE FORBES OTTY, b Montreal, Que, July 13, 36; m 60; c 2. PHYSICS. *Educ:* Univ Alta, BSc, 57; Dalhousie Univ, MSc, 59; Univ London, PhD(physics), 62. *Prof Exp:* Res assoc physics, 62-63, from asst prof to assoc prof, 67-69, asst dean grad studies, 67-68, actg dean, 68-69, dean, 69-72, PROF PHYSICS, DALHOUSIE UNIV, 69- *Concurrent Pos:* Mem, Defence Res Bd Can, 71-77. *Mem:* Can Asn Physicists. *Res:* Microwave breakdown in gases; ions in afterglows; positron annihilation; optical properties of metals. *Mailing Add:* Dept Physics Dalhousie Univ Halifax NS B3H 4H6 Can

LANGVARDT, PATRICK WILLIAM, b Dodge City, Kans, Mar 20, 50. MASS SPECTROMETRY, CHROMATOGRAPHY. *Educ:* Kans State Teachers Col, BS, 72; Purdue Univ, MS, 74. *Prof Exp:* RES LEADER ANALYTICAL CHEM, DOW CHEM CO, 74- *Mem:* Am Soc Mass Spectrometry; Sigma Xi; Am Chem Soc. *Res:* Analytical chemistry in support of toxicology studies; metabolite identification; analytical toxicology; trace determinations in biological matrices; automated sample preparation and analysis. *Mailing Add:* 1501 Montague St Midland MI 48640

LANGWAY, CHESTER CHARLES, JR, b Worcester, Mass, Aug 15, 29; m 59; c 4. GEOLOGY. *Educ:* Boston Univ, AB, 55, MA, 56; Univ Mich, PhD(geol, glaciol), 65. *Prof Exp:* Res geologist, US Army Snow, Ice & Permafrost Res Estab, 56-59; res assoc properties of snow & ice, Res Inst, Univ Mich, 59-61; res glaciologist, US Army Cold Regions Res & Eng Lab, 61-65, chief snow & ice br, 66-77; PROF GEOL SCI & CHMN DEPT, STATE UNIV NY BUFFALO, 77- *Concurrent Pos:* Mem panel glaciol, Comt Polar Res, Nat Acad Sci, 69-75, secy, Int Comt Ice Core Studies; chmn dept geol sci, State Univ NY Buffalo, 75- *Mem:* AAAS; fel Geol Soc Am; Am Geophys Union; Am Polar Soc; fel Arctic Inst NAm; Sigma Xi. *Res:* Basic and applied research related to the properties of snow and ice, including field and laboratory techniques of analyzing shallow and deep ice cores for stratigraphy and age dating; isotopic and ionic constituents, terrestrial and extraterrestrial inclusions. *Mailing Add:* Dept Geol Sci 4240 Ridge Lea Rd Rm 9 Buffalo NY 14260

LANGWEILER, MARC, b Astoria, NY, Jan 27, 52. PATHOLOGY. *Educ:* Cornell Univ, BS, 72, DVM, 75, MS, 80, PhD, 83. *Prof Exp:* Vet clinician, Flushing Vet Hosp, NY, 75-76; Henry Bergh Mem Hosp, New York, 76-77; grad res asst, NY State Col Vet Med, Cornell Univ, Ithaca, 77-79, grad vet asst, 79-82; postdoctoral res fel, NIH, Dept Microbiol/Immunol, Duke Univ Med Ctr, 82-84, clin immunol fel, 83-84; res assoc, Dept Path, State Univ NY Med Ctr, 84-85; res tech, Dept Radiation Biol/Oncol, Sch Med, E Carolina Univ, 86-88; INSTR, DARTMOUTH MED SCH & TECH SPECIALIST, MARY HITCHCOCK MEM HOSP, DEPT PATH, DARTMOUTH-HITCHCOCK MED CTR, 88- *Mem:* Am Asn Immunologists; Am Asn Vet Immunologists; Am Vet Med Asn; Am Asn Med Lab Immunologists; Int Soc Analytical Cytol. *Res:* Microbiology; immunology; veterinary medicine. *Mailing Add:* Dept Path Dartmouth-Hitchcock Med Ctr Hanover NH 03756

LANGWIG, JOHN EDWARD, b Albany, NY, Mar 5, 24; m 46; c 1. FOREST PRODUCTS, WOOD SCIENCE. *Educ:* Univ Mich, Ann Arbor, BS, 48; State Univ NY Col Forestry, Syracuse Univ, MS, 68, PhD(wood sci), 71. *Prof Exp:* Instr wood prod eng, State Univ NY Col Forestry, Syracuse Univ, 69-70;

from asst prof to prof, 71-86, chmn, dept forestry, 75-81, exten prof, 81-86, EMER PROF WOOD SCI, OKLA STATE UNIV, 86- *Mem:* Soc Wood Sci & Technol; Forest Prod Res Soc; Soc Am Foresters; Sigma Xi; Tech Asn Pulp & Paper Indust. *Res:* Neutron activation analysis of trace elements in wood and effects on physical properties; physical properties of wood-polymer composites. *Mailing Add:* Dept Forestry Okla State Univ Stillwater OK 74078-0491

LANGWORTHY, HAROLD FREDERICK, b White Plains, NY, Aug 1, 40; m 65; c 3. MATHEMATICS, OPTICS. *Educ:* Rensselaer Polytech Inst, BS, 62; Univ Minn, PhD(math), 70. *Prof Exp:* Res assoc, Res Labs, Eastman Kodak Co, Rochester, 67-79, lab head, 79-81, asst dir, 81-83, dir physics div, Kodak Res Labs, 83-86, dir, Res Labs, CISG, 86-89, DIR RES IISG, EASTMAN KODAK CO, ROCHESTER, 89- *Mem:* Optical Soc Am; AAAS. *Res:* Mathematical optics; rheology. *Mailing Add:* 732 Hightower Way Webster NY 14580

LANGWORTHY, JAMES BRIAN, b Billings, Mont, Feb 18, 34; m 65; c 2. NUCLEAR PHYSICS, MATHEMATICAL PHYSICS. *Educ:* Univ Colo, BS, 56; Univ Md, MS, 66. *Prof Exp:* Physicist math physics, Radiation Div, 59-69, res physicist particle transport, Theory Br, Nuclear Physics Div, 69-76, RES PHYSICIST RADIATION DAMAGE, RADIATION SURVIVABILITY BR, CONDENSED MATTER & RADIATION SCI DIV, NAVAL RES LAB, 76- *Concurrent Pos:* Pres, Youth Resources Ctr, Inc, 71-87. *Mem:* Am Phys Soc; AAAS; Sigma Xi. *Res:* Radiation damage, hardening and shielding; energetic particle transport by Monte Carlo computer codes; microdosimetry in single event upset of memory cells; chord distributions. *Mailing Add:* Code 4613 Naval Res Lab Washington DC 20375-5000

LANGWORTHY, THOMAS ALLAN, b Oak Park, Ill, Aug 7, 43; div; c 2. MICROBIAL PHYSIOLOGY. *Educ:* Grinnell Col, AB, 65; Univ Kans, PhD(microbiol), 71. *Prof Exp:* Res assoc, 71-72, from asst prof to assoc prof, 72-82, PROF MICROBIOL, UNIV SD, 82- *Honors & Awards:* Alexander von Humboldt US Sr Scientist Prize, 84. *Mem:* Am Soc Microbiol; AAAS; Am Acad Microbiol; Int Orgn Mycoplasmology. *Res:* Structure and function of the membranes and cell surfaces of bacteria from extreme environments, mycoplasmas, archaebacteria and cellular evolution. *Mailing Add:* Dept Microbiol Sch Med Univ SD Vermillion SD 57069

LANGWORTHY, WILLIAM CLAYTON, b Watertown, NY, Sept 3, 36; m 58; c 2. ORGANIC CHEMISTRY, ENVIRONMENTAL CHEMISTRY. *Educ:* Tufts Univ, BSChem, 58; Univ Calif, Berkeley, PhD(org chem), 62. *Prof Exp:* NIH fel chem, Mass Inst Technol, 61-62; asst prof, Alaska Methodist Univ, 62-65; from asst prof to prof, Calif State Col, Fullerton, 65-73; prof chem & head dept, 73-76, dean sch sci & math, Calif Polytech State Univ, San Luis Obispo, 76-83; VPRES ACAD AFFAIRS, FT LEWIS COL, DURANGO, CO, 83- *Concurrent Pos:* Assoc dean, Sch Lett, Arts & Sci, Calif State Col, Fullerton, 70-73, dir environ studies prog, 70-72. *Mem:* AAAS; Am Chem Soc; Sigma Xi. *Res:* Physical organic chemistry; organic reactions in liquid ammonia; environmental chemistry, especially analysis and effects of trace pollutants. *Mailing Add:* Ft Lewis Col Durango CO 81301

LANHAM, RICHARD HENRY, JR, b Shelbyville, Ill; m 59; c 2. EDUCATIONAL ADMINISTRATION. *Educ:* Ohio Col Podiatric Med, DPM, 58; Univ Louisville, MEd, 80. *Prof Exp:* Pvt pract, Clarksville, Ind, 59-85; PROF PODIATRIC MED, VPRES & DEAN ACAD AFFAIRS, CALIF COL PODIATRIC MED, 85- *Concurrent Pos:* Clin investr, Sutter Biomedical, 79-85 & Dow Corning Corp, 79-82; adj clinician, Ohio Col Podiatric Med, 81- *Mem:* Am Podiatric Med Asn (pres, 85-86); Am Col Foot Surgeons (Am Col Foot Orthopedists (pres, 70-71); Am Pub Health Asn; Am Bd Podiatric Orthopedics; Am Bd Podiatric Surg. *Res:* Clinical investigation of three designs for foot implants; author of articles on foot surgery, drug use surveys and case reports. *Mailing Add:* 53 Everson San Francisco CA 94131

LANHAM, URLESS NORTON, b Grainfield, Kans, Oct 17, 18; m 45; c 3. ENTOMOLOGY. *Educ:* Univ Colo, BA, 40; Univ Calif, Berkeley, PhD(entom), 48. *Prof Exp:* Asst zool, Univ Calif, Los Angeles, 40-41, biol, Scripps Inst, 41-42, entom, 47-48; from instr to asst prof zool, Univ Mich, 48-56, res asst & consult, NSF Proj Insect Ecol, 56-61; vis cur entom, 61-62, assoc cur, 62-73, CUR ENTOM & PROF NATURAL HIST, UNIV COLO MUS, BOULDER, 73- *Concurrent Pos:* Consult, Biol Sci Curric Study, Univ Colo, Boulder, 63-66, lectr, Inst Develop Biol, 66-67, asst prof biophys, 68-71; adv ed, Columbia Univ Press, 64-72; assoc prof, Div Natural Sci, Monteith Col, Wayne State Univ, 59-61; consult, Smithsonian Inst, 67. *Mem:* Kans Entom Soc; Pac Coast Entom Soc. *Res:* Faunistics and systematics of Apoidea, especially of the genus Andrena. *Mailing Add:* Univ Colo Mus Boulder CO 80309

LANIER, GERALD NORMAN, forest entomology; deceased, see previous edition for last biography

LANIER, LEWIS L, b Memphis, Tenn, July 1, 53. EXPERIMENTAL BIOLOGY. *Educ:* Va Polytech Inst & State Univ, BS, 75; Univ NC, PhD(microbiol & immunol), 78. *Prof Exp:* Fel, Dept Path, Damon Runyon-Walter Winchell Cancer Fund, Sch Med, Univ NMex, 78-81, res asst prof, 81; sr res scientist, Becton Dickinson Monoclonal Ctr, 81-88, assoc res dir, 88-90; RES ASSOC, CANCER RES INST, SCH MED, UNIV CALIF-SAN FRANCISCO, 87-; SR SCI STAFF, DEPT IMMUNOL, DNAX RES INST MOLECULAR & CELLULAR BIOL, INC, 91- *Concurrent Pos:* Transmitting ed, Int Immunol, 88-; mem, Inst Health Reviewers Reserve, NIH, 89-93; assoc ed, J Immunol, 88- *Mem:* Am Asn Immunologists; Soc Anal Cytol; Clin Immunol Soc; Sigma Xi. *Res:* Experimental biology. *Mailing Add:* DNAX Res Inst Molecular & Cell Biol 901 California Ave Palo Alto CA 94304-1104

LANIER, ROBERT GEORGE, b Chicago, Ill, Oct 27, 40; div; c 2. NUCLEAR PHYSICS. *Educ:* Lewis Col, BS, 62; Fla State Univ, PhD(nuclear chem), 68. *Prof Exp:* US AEC fel, Chem Div, Oak Ridge Nat Lab, 67-69; fel, 69-71, STAFF MEM, NUCLEAR CHEM DIV, LAWRENCE LIVERMORE NAT LAB, UNIV CALIF, 71- *Mem:* AAAS; Am Phys Soc. *Res:* Experimental low energy nuclear structure studies; charged-particle reaction spectroscopy, cross section measurements and in-beam gamma ray spectroscopy. *Mailing Add:* Nuclear Chem Div L 232 Univ Calif Lawrence Livermore Nat Lab PO Box 808 Livermore CA 94550

LANING, J HALCOMBE, b Kansas City, Mo, Feb 14, 20; m 43; c 4. MANUFACTURING AUTOMATION. *Educ:* Mass Inst Technol, BA, 40, PhD(appl math), 47. *Prof Exp:* Sr staff mem, Instrumentation Lab, Mass Inst Technol, 45-73; head, mfg & comput dept, C S Draper Lab, Inc, 73-82, sr fel automation, 82-85, head, Automation Technol Dept, 85-89; RETIRED. *Mem:* Nat Acad Eng; Asn Comput Mach; Am Math Soc; Am Inst Aeronaut & Astronaut; Soc Indust & Appl Math; Inst Elec & Electronics Engrs. *Mailing Add:* 130 Temple St West Newton MA 02165

LANING, STEPHEN HENRY, b Albany, NY, Oct 18, 18; m 46; c 2. ANALYTICAL CHEMISTRY. *Educ:* Union Col, NY, BS, 41; Rutgers Univ, PhD(phys chem), 47. *Prof Exp:* Asst, Rutgers Univ, 41-44, instr, 45-46; res chemist, Chem Div, Pittsburg Plate Glass Co, 47-61, supvr res & tech serv, 61-67, SUPVR RES & TECH SERV, PPG INDUSTS, INC, 67- *Honors & Awards:* Hon Award, Pittsburg Plate Glass Co, 62. *Mem:* Am Chem Soc; AAAS; Soc Appl Spectros. *Res:* Development of methods for x-ray analysis of glass, silica and titania pigments, minerals, cements and other types of materials. *Mailing Add:* 1219 Greenvale Ave PO Box 31 Akron OH 44313

LANKFORD, CHARLES ELY, bacteriology; deceased, see previous edition for last biography

LANKFORD, J(OHN) L(EWELLYN), b Hampton, Va, Sept 13, 20; m 45; c 2. PROPULSION, HYPERBALLISTICS. *Educ:* Va Polytech Inst, BS, 42. *Prof Exp:* Aeronaut res scientist, Nat Adv Comt Aeronaut, 45-53; res engr, Cent Res Lab, Melpar, 53-54; head, dept gas dynamics, Exp Inc, 54-58; aeronaut res engr, Naval Ord Lab, 58-62; chief adv studies lunar logistics flight systs, NASA, 62-64; actg chief, Missile Dynamics Div, Naval Ord Lab, 64-65, prin investr & mgr rain erosion & hypersonic reentry mat, Naval Surface Weapons Ctr, White Oak, Md, 65-70, consult & prog mgr, US Navy, 70-75; RETIRED. *Concurrent Pos:* Consult, Bur Weapons, US Navy, 64-75; consult, energy & heat transfer, 76-; assoc staff & consult energy prog, Univ Md, 79-80; lectr, Montgomery Col, 78-79. *Mem:* Assoc fel Am Inst Aeronaut & Astronaut; AAAS. *Res:* Supersonics; hypersonics; propulsion aerodynamics; inlets; aeroballistics. *Mailing Add:* 1717 Marymont Rd Silver Spring MD 20906

LANKFORD, WILLIAM FLEET, b Charlottesville, Va, Jan 9, 38. NUCLEAR & SOLID STATE PHYSICS. *Educ:* Univ Va, BA, 60; Univ SC, MS, 64, PhD(physics), 69. *Prof Exp:* Instr physics, Univ NC, Greensboro, 62-63; fel, Col William & Mary, 69; from asst prof to assoc prof, 59-78, PROF PHYSICS, GEORGE MASON UNIV, 78- *Mem:* Am Inst Physics; Am Phys Soc. *Res:* Experimental solid state research. *Mailing Add:* Dept Physics George Mason Univ 4400 Univ Dr Fairfax VA 22030

LANKS, KARL WILLIAM, b Philadelphia, Pa, Nov 1, 42. PATHOLOGY, MOLECULAR BIOLOGY. *Educ:* Pa State Univ, BS, 63; Temple Univ, MD, 67; Columbia Univ, PhD(path), 71. *Prof Exp:* Intern, Columbia-Presby Hosp, 67-68; instr path, Columbia Univ, 71-72; from asst prof to assoc prof, 74-88, PROF PATH, STATE UNIV NY DOWNSTATE MED CTR, 88- *Concurrent Pos:* NIH res fel, Dept Chem, Harvard Univ, 71-72. *Mem:* Am Soc Biol Chemists; Am Soc Exp Path; Am Soc Cell Biol. *Res:* Structure and metabolism of messenger ribonucleic acids; mechanism of cell attachment; regulation of protein and RNA messenger synthesis in cultured cells; structure and function of heat shock proteins. *Mailing Add:* 450 Clarkson Ave Brooklyn NY 11203

LANMAN, ROBERT CHARLES, b Bemidji, Minn, Oct 2, 30; m 57; c 4. BIOCHEMICAL PHARMACOLOGY, TOXICOLOGY. *Educ:* Univ Minn, BS, 56, PhD(pharmacol), 67. *Prof Exp:* Teaching asst, Col Pharm, Univ Minn, 58-59; pharmacologist, Sect Biochem Drug Action, Lab Chem Pharmacol, Nat Heart Inst, 62-66; assoc prof, 66-81, PROF PHARMACOL & MED, UNIV MO-KANSAS CITY, 81-, CHMN, DIV PHARMACOL, 87- *Concurrent Pos:* Consult, Marion Labs. *Mem:* NY Acad Sci; Am Asn Cols Pharm; Am Soc Pharmacol & Exp Therapeut; Fedn Am Socs Exp Biol; Am Asn Pharmaceut Scientists; Am Pharmaceut Asn. *Res:* Passage of drugs across body membranes; mechanism and kinetics of drug absorption, distribution, metabolism and excretion. *Mailing Add:* 8202 W 72nd St Overland Park KS 66204-1725

LANN, JOSEPH SIDNEY, b Washington, DC, Sept 16, 17; m 45; c 3. ORGANIC CHEMISTRY. *Educ:* Univ Md, BS, 37, PhD(org chem), 41. *Prof Exp:* Res chemist, Jackson Lab, E I du Pont de Nemours & Co, Inc, 46-54, dir, Freon Prod Lab, 54-64, asst mgr new prod & mkt develop, Freon Prod Div, 64-65, asst dist mgr, 65-68, mgr develop prod, Freon Prod Div, 68-80. *Mem:* Am Chem Soc. *Res:* Surface active agents and neoprene; organic compounds; fluorinated hydrocarbons. *Mailing Add:* 608 Haverhill Rd Sharpley Wilmington DE 19803

LANNER, RONALD MARTIN, b Brooklyn, NY, Nov 12, 30; m 57; c 2. FOREST GENETICS, TREE PHYSIOLOGY. *Educ:* Syracuse Univ, BS, 52, MF, 58; Univ Minn, Minneapolis, PhD(forestry), 68. *Prof Exp:* Res forester, Pac Southwest Forest & Range Exp Sta, US Forest Serv, 58-64; PROF FOREST GENETICS & DENDROL, UTAH STATE UNIV, 67- *Concurrent Pos:* Consult, Forestry Proj, Food & Agr Orgn, UN, Taiwan, 69; res fel, Univ Fla, 73-74; vis prof, Univ Wash, 82-83. *Mem:* Soc Am Foresters. *Res:* Morphogenesis and growth of woody plants; evolution and ecology of pines with bird-dispersed seeds; author of books about trees. *Mailing Add:* Dept Forest Res Utah State Univ Logan UT 84322

LANNERS, H NORBERT, b Volkmarsen, Ger, June 23, 43; m 73. ELECTRON MICROSCOPY. *Educ:* Univ Tubingen, Ger, Dr rer nat, 73. *Prof Exp:* Fel, Rockefeller Univ, 73-76; res assoc, Cornell Univ Med Col, 76-80; ASST PROF, ROCKEFELLER UNIV, 80- *Concurrent Pos:* Vis scientist, Agr Exp Sta, Univ Puerto Rico, 76; adj res assoc, Rockefeller Univ, 76-79. *Mem:* Soc Protozoologists; Electron Microsc Soc Am. *Res:* Cultivation of human malaria parasites. *Mailing Add:* Dept Pub Health Tulane Univ New Orleans LA 70118

LANNERT, KENT PHILIP, b Belleville, Ill, Nov 29, 44; m 70; c 2. INDUSTRIAL ORGANIC CHEMISTRY. *Educ:* Southern Ill Univ, Carbondale, BA, 66; Vanderbilt Univ, PhD(chem), 69. *Prof Exp:* Res chemist, 69-72, res specialist, 72-76, sr res specialist, 76-82, group leader, 82-83, sr group leader, 83-86, MGR TECHNOL, MONSANTO CO, 86- *Mem:* Am Chem Soc. *Res:* Organic synthesis; chelation; synthesis of chelants and other detergent related chemicals; phosphates processing. *Mailing Add:* Monsanto Co 800 N Lindbergh Blvd St Louis MO 63166

LANNI, FREDERICK, BIOLOGY. *Prof Exp:* SR RES SCIENTIST, BIOL DEPT, CARNEGIE-MELLON UNIV, 90- *Mailing Add:* Biol Sci Dept Carnegie-Mellon Univ 4400 Fifth Ave Pittsburgh PA 15213

LANNIN, JEFFREY S, b New York, NY, Aug 21, 40; m 71; c 1. MATERIALS SCIENCE ENGINEERING. *Educ:* Purdue Univ, BS, 62; Univ Ill, Urbana, MS, 64; Stanford Univ, PhD(solid state physics), 71. *Prof Exp:* Physicist thin films, Fairchild Semiconductor Res Lab, 66-67 & semiconductor physics, Lockheed Palo Alto Res Labs, 67-68; staff physicist raman scattering, Max Planck Inst Solid State Res, 71-74; vis scientist optical properties, Argonne Nat Lab, 74-75; vis asst prof mat sci, Univ Del, 75-76; asst prof, 76-81, assoc prof, 81-86, PROF PHYSICS, PA STATE UNIV, 86- *Concurrent Pos:* Consult, Lockheed Palo Alto Res Lab, 68-71. *Mem:* Am Phys Soc; Sigma Xi; Am Vacuum Soc. *Res:* Raman and neutron scattering in ordered and disordered solids and liquids; amorphous materials; thin film physics; cluster vibrational and electronic properties. *Mailing Add:* Dept Physics Pa State Univ University Park PA 16802

LANNING, DAVID D(AYTON), b Baker, Ore, Mar 30, 28; m 50; c 3. NUCLEAR ENGINEERING. *Educ:* Univ Ore, BA, 51; Mass Inst Technol, PhD(nuclear eng), 63. *Prof Exp:* Physicist, Hanford Atomic Prod Oper, Gen Elec Co, 51-57; res assoc & reactor supt, Mass Inst Technol, 57-62, from asst prof to assoc prof nuclear eng, 62-65, asst dir res reactor, 62-65; unit mgr reactor physics, Battelle-Northwest, 65-66, sect mgr, 66-69; PROF NUCLEAR ENG, MASS INST TECHNOL, 69- *Concurrent Pos:* Consult, Stone & Webster Eng Corp, 77-78, Boston Edison Co; mem, Monticello Nuclear Power Reactor Safety Audit Comt, Nat Res Coun, NSF. *Mem:* Am Nuclear Soc. *Res:* Nuclear engineering education; design, safety, control and operation of nuclear reactor systems. *Mailing Add:* Dept Nuclear Eng Mass Inst Technol Cambridge MA 02139

LANNING, FRANCIS CHOWING, b Denver, Colo, Jan 5, 08; m 34; c 2. CHEMISTRY. *Educ:* Univ Denver, BS, 30, MS, 31; Univ Minn, PhD(phys chem), 36. *Prof Exp:* Asst, Univ Denver, 30-31; anal chemist, Minn State Hwy Dept, 36-42; from instr to prof, 42-78, EMER PROF CHEM, KANS STATE UNIV, 78- *Mem:* Am Chem Soc; Sigma Xi; Am Soc Hort Sci. *Res:* Organosilicon compounds; silicon and other minerals in plants; chemical composition of limestones. *Mailing Add:* Dept Chem Kans State Univ Manhattan KS 66504

LANNING, WILLIAM CLARENCE, b Boicourt, Kans, Dec 9, 13; m 36; c 3. PHYSICAL CHEMISTRY. *Educ:* Sterling Col, AB, 34; Univ Kans, AM, 36, PhD(chem), 38. *Prof Exp:* Asst instr chem, Univ Kans, 35-38; res chemist & sect mgr, Naval Res Lab, 38-45; res chemist, Phillips Petrol Co, 46-73, sect mgr, 57-73; res chemist, Bartlesville Energy Res Ctr, ERDA, 74-80; RETIRED. *Mem:* Am Chem Soc; Sigma Xi. *Res:* Non-aqueous solutions; inorganic preparations; catalytic hydrocarbon and petroleum processes; fundamentals of crude oil production; refining of synthetic crude oils. *Mailing Add:* 1530 Pecan Pl Bartlesville OK 74003

L'ANNUNZIATA, MICHAEL FRANK, b Springfield, Mass, Oct 14, 43; m 73; c 3. EDUCATION FELLOWSHIPS MANAGEMENT. *Educ:* St Edward's Univ, BS, 65; Univ Ariz, MS, 67, PhD(agr chem & soils), 70. *Hon Degrees:* Hon Teaching Dipl, Central Univ Ecuador, 78. *Prof Exp:* Res chemist herbicides, Amchem Prods, Inc, 71-72; res assoc, Univ Ariz, 72-73; prof & res investr, Univ Chapingo, Mex, 73-75; res investr, Nat Inst Nuclear Energy, Mex, 75-77; assoc officer, Int Atomic Energy Agency, 77-80, second officer, 80-83, first officer, 83-86, SR OFFICER, HEAD FEL & TRAINING SECT, INT ATOMIC ENERGY AGENCY, VIENNA, 86- *Concurrent Pos:* Consult health veg, Caborca, Mex, 72; consult, Govt Nicaragua, 78, Costa Rica, 78 & 80, Guatemala, 79, Columbia, 79, Arg, 80, Panama, 80, Spain, 80, Uruguay, 80, US Dept State, 85 & 89, USSR State Comm on Utilization Atomic Energy, 80 & 85, Fed Rep Ger, 87, France, 87 & 88, Poland, 88 & 90, People's Repub China, 88, Thailand, 88, Vietnam, 88, Czech, 89, Sweden, 89 & 90, Ger Dem Repub, 89, Mex, 89, Hungary, 90, Israel, 90, Belg, 90 & Italy, 90; consult & lectr, Atomic Sch Trop Agr, Cardenas, Mex, 73 & Atomic Energy Comn, Quito, Ecuador, 78; vis lectr, Timiryazev Agr Acad, Moscow, Inst Nuclear Appln Vet Sci, Turkey & Univ Guanajuato, Mex, 81; mem bd gov, Uppsala Univ, Int Sci Progs, Uppsala, Sweden, 88-91. *Res:* Nuclear techniques in the agricultural and biological sciences; detection and measurement of radionuclide tracers; liquid scintillation and Cherenkov counting of radioisotope tracers; use of isotopes in the elucidation of biochemical pathways and mechanisms. *Mailing Add:* PO Box 181214 Coronado CA 92178-1214

LANNUTTI, JOSEPH EDWARD, b Cedar Hollow, Pa, May 4, 26; m 54; c 3. EXPERIMENTAL HIGH ENERGY PHYSICS. *Educ:* Pa State Univ, BS, 50; Univ Pa, MS, 53; Univ Calif, PhD(physics), 57. *Prof Exp:* Admin asst, Pa RR, 43-47; asst, Univ Pa, 51-52; physicist, NAm Aviation Inc, 52-53; asst, Univ Calif, 54-57; from asst prof to assoc prof, 57-64, PROF PHYSICS, FLA

STATE UNIV, 64- *Concurrent Pos:* Physicist, Lawrence Radiation Lab, Univ Calif, 59-60; consult, Oak Ridge Nat Lab. *Mem:* Am Phys Soc. *Res:* Physics of elementary particles. *Mailing Add:* Dept Physics Fla State Univ Tallahassee FL 32306

LANOU, ROBERT EUGENE, JR, b Burlington, Vt, Feb 13, 28; m 60; c 4. EXPERIMENTAL PHYSICS. *Educ:* Worcester Polytech Univ, BS, 52; Yale Univ, PhD(physics), 57. *Prof Exp:* Physicist, Lawrence Radiation Lab, Univ Calif, 57-59; mem fac, 59-66, PROF PHYSICS, BROWN UNIV, 66-, CHMN DEPT, 85- *Concurrent Pos:* Chmn, High Energy Discussion Group, Brookhaven Nat Lab, 81-83; consult, Brookhaven Nat Lab & US Dept Energy; mem, High Energy Physics Adv Panel Comt, Future High Energy Comput, Solar Neutrino Res; mem, exec comt, Div Particles & Fields, Am Phys Soc. *Mem:* AAAS; fel Am Phys Soc. *Res:* Elementary particle physics. *Mailing Add:* Dept Physics Brown Univ Providence RI 02912

LA NOUE, KATHRYN F, b Camden, NJ, Dec 21, 34; m 76; c 4. BIOCHEMISTRY, CELL PHYSIOLOGY. *Educ:* Bryn Mawr Col, AB, 56; Yale Univ, PhD(biochem), 60. *Prof Exp:* Res chemist, US Army Surg Res Unit, 61-67; NIH fel, Johnson Res Found, Sch Med, Univ Pa, 68-70, res assoc, 70, asst prof, 71-74; assoc prof, 74-81, PROF PHYSIOL, MILTON S HERSHEY MED CTR, PA STATE UNIV, 81- *Concurrent Pos:* Dr W D Stroud estab investr, Am Heart Asn, 71-76. *Mem:* AAAS; Am Chem Soc; Am Soc Biol Sci; Biophys Soc; Am Physiol Soc. *Res:* Control of mitochondrial metabolism; membrane transport mechanisms; mechanism of transmembrane signalling. *Mailing Add:* Dept Physiol Milton S Hershey Med Ctr Hershey PA 17033

LANOUX, SIGRED BOYD, b New Orleans, La, Nov 1, 31; m 54; c 2. INORGANIC CHEMISTRY. *Educ:* Southwestern La Univ, BS, 57; Tulane Univ, PhD(inorg chem), 62. *Prof Exp:* Res assoc, Univ Ill, Urbana, 61-62; res chemist, Textile Fibers Dept, E I du Pont de Nemours & Co, Inc, 62-66; from asst prof to assoc prof, 66-74, head dept chem, 72-86, PROF CHEM, UNIV SOUTHWESTERN LA, 74-, DEAN SCI, 86- *Mem:* Am Chem Soc. *Res:* Phosphazene chemistry. *Mailing Add:* 104 Ridgewood Lafayette LA 70506-3222

LANPHERE, MARVIN ALDER, b Spokane, Wash, Sept 29, 33; m 61; c 3. GEOLOGY, GEOCHEMISTRY. *Educ:* Mont Sch Mines, BS, 55; Calif Inst Technol, MS, 56, PhD(geol), 62. *Prof Exp:* Geologist, US Geol Surv, Calif, 62-67, dep asst chief geologist, Washington, DC, 67-69, RES GEOLOGIST, US GEOL SURV, CALIF, 69- *Concurrent Pos:* Vis prof, Stanford Univ, 72; vis fel, Australian Nat Univ, 75-76. *Honors & Awards:* Meritorious Serv Award, Interior Dept. *Mem:* Geol Soc Am; Am Geophys Union. *Res:* Geochronology, application of techniques of radioactive age determination of rocks and minerals to geological problems; isotope tracer studies of geological processes. *Mailing Add:* US Geol Surv 345 Middlefield Rd Menlo Park CA 94025

LANPHIER, EDWARD HOWELL, b Madison, Wis, May 29, 22; m 78. MEDICINE, ENVIRONMENTAL PHYSIOLOGY. *Educ:* Univ Wis, BS, 46; Univ Ill, MS & MD, 49; MDiv, Nashotah House, 76. *Prof Exp:* Am Col Physicians res fel physiol, Grad Sch Med, Univ Pa, 50-51; asst med officer & physiologist, Exp Diving Unit, US Navy, 52-58, diving med officer, Eniwetok Proving Ground, 58, med officer, Underwater Demolition Team, Norfolk, Va, 58-59; from asst prof to assoc prof physiol, Sch Med, State Univ NY, Buffalo, 59-73; SR SCIENTIST PREV MED, UNIV WIS-MADISON, 76-, ASST DIR BIOMED RES, THE BIOTRON, 78- *Honors & Awards:* Behnke Award, Undersea & Hyperbaric Med Soc, 77. *Mem:* Am Physiol Soc; Undersea & Hyperbaric Med Soc. *Res:* Respiratory physiology; submarine and diving medicine; physiological problems of immersion and exposure to increased pressure; hyperbaric medicine. *Mailing Add:* Biotron 2115 Observatory Dr Madison WI 53706-1087

LANSBURY, PETER THOMAS, b Vienna, Austria, Feb 24, 33; US citizen; m 57; c 3. ORGANIC CHEMISTRY. *Educ:* Pa State Univ, BS, 53; Northwestern Univ, PhD(chem), 56. *Prof Exp:* Res scientist chem, E I du Pont de Nemours & Co, Inc, 56-58; lectr org chem, Univ Del, 58-59; from asst prof to assoc prof, 59-65, PROF CHEM, STATE UNIV NY BUFFALO, 65- *Concurrent Pos:* Prin investr, NSF res grants, 60-; vis prof chem, Univ Ill, Urbana, 63-64; fel, Alfred P Sloan Found, 63-67; consult, Hooker Chem Corp, 65-74; res award, Ciba-Geigy Corp, 73-74. *Mem:* Am Chem Soc; Sigma Xi. *Mailing Add:* Dept Chem State Univ NY 357 Acheson Hall Buffalo NY 14214

LANSDELL, HERBERT CHARLES, b Montreal, Que, Dec 22, 22; US citizen; div; c 2. NEUROPSYCHOLOGY, PHYSIOLOGICAL PSYCHOLOGY. *Educ:* Sir George Williams Eve Col, BSc, 44; McGill Univ, PhD(psychol), 50. *Prof Exp:* Asst prof psychol, McGill Univ, 49-50; defense res sci officer, Defense Res Med Labs, 50-54; asst prof psychol, Univ Buffalo, 54-58; res supvr, 58-70, HEALTH SCI ADMINR, NAT INST NEUROL DIS & STROKE, 70- *Mem:* Fel Am Psychol Asn; fel AAAS; Soc Neurosci; Psychomet Soc; Int Brain Res Orgn. *Res:* Statistical analysis of psychological test results obtained from neurological patients to investigate brain function, especially hemispheric and sex differences; test construction. *Mailing Add:* Fed Bldg Rm 916 Nat Insts Health Bethesda MD 20892

LANSDOWN, A(LLEN) M(AURICE), b Winnipeg, Manitoba, Sept 7, 39; m 61; c 3. CIVIL, STRUCTURAL & TRANSPORTATION ENGINEERING. *Educ:* Univ Manitoba, BSc, 61; Southampton Univ, PhD(struct), 65. *Prof Exp:* Nat Res Coun Can fel, 64-65; asst prof, 65-67, head dept, 67-73, assoc prof, 67-69, PROF CIVIL ENG, UNIV MAN, 69- *Concurrent Pos:* Mem, Defence Res Bd, 73-78, adv Dept of Defence, 78-; provost, Univ Col, 78-81. *Mem:* Eng Inst Can; Can Soc for Climat Energy; Can Geotechnical Soc. *Res:* Housing in small remote northern communities: self-help, sweat-equity and house performance; transportation in developing regions; enhancement of self reliance, labour intensive methods. *Mailing Add:* Civil Eng Dept Fac Eng Univ Man Winnipeg MB R3T 2N2 Can

LANSFORD, EDWIN MYERS, JR, b Houston, Tex, June 26, 23; m 50; c 3. BIOCHEMISTRY, MICROBIOLOGY. *Educ:* Univ Calif, Los Angeles, BA, 46; Rice Univ, BA, 48; Univ Tex, MA, 51, PhD(biochem), 51. *Prof Exp:* Fel, Univ Ill, 51-52; res scientist, Clayton Found Biochem Inst, Univ Tex, 53-67; PROF BIOCHEM, SOUTHWESTERN UNIV, GEORGETOWN, TEX, 67- *Mem:* Am Chem Soc; Sigma Xi; NY Acad Sci. *Res:* Microbial intermediary metabolism; amino acid activating enzymes; metabolic effects of alcohol; single carbon unit metabolism and its control. *Mailing Add:* Rte 7 Box 9 Leander TX 78641

LANSING, ALLAN M, b St Catherines, Ont, Sept 12, 29; m 51; c 3. CARDIOVASCULAR SURGERY, ORGAN TRANSPLANTATION. *Educ:* Univ Western Ont, MD, 53, PhD(physiol), 57; FRCS(C), 59. *Hon Degrees:* LHD, Bellarmine Col, Louisville, Ky, 85; DSc, Transylvania Univ, Lexington, Ky, 85. *Prof Exp:* Nat Res Coun Can scholar, Univ Western Ont, 55-57; asst prof surg & physiol, Fac Med, Univ Western Ont, 61-63; assoc prof, Sch Med, Univ Louisville, 63-69, chief sect cardiovasc surg, 69-74, prof surg, 69-80, prog surg thoracic & cardiovasc, 80-84; DIR CARDIOVASC SURG, HUMANA HEART INST INT, LOUISVILLE, KY, 83- *Concurrent Pos:* Markle scholar med sci, 61-; bd mem, Nat Kidney Found, Louisville, Western Ky, Bellarmine Col, Louisville, Ky & Transylvania Univ, Lexington, Ky, 85- *Honors & Awards:* Hon G Ferguson Trophy, Western Ont Fac Med, 53. *Mem:* Fel Am Col Surg; fel Am Col Cardiol; Soc Univ Surg; Royal Col Surg Can; Warren H Cole Soc. *Res:* Cardiovascular physiology and shock; pulmonary atelectasis; renal transplantation; open heart surgery; chemical research in heart replacement, including transplantation and the mechanical heart and the major fields at present; promotion and improvement of college education. *Mailing Add:* 2266 Med Arts Bldg Louisville KY 40217-1412

LANSING, NEAL F(ISK), JR, energy, environmental science; deceased, see previous edition for last biography

LANSINGER, JOHN MARCUS, b July 20, 32; US citizen; m 53; c 2. GEOPHYSICS, IONOSPHERIC PHYSICS. *Educ:* Lewis & Clark Col, BS, 54; Univ Alaska, MS, 56. *Prof Exp:* Instr geophys, Univ Alaska, 56-57; sr engr, Philco Corp, 57-59; staff assoc, Boeing Sci Res Labs, Northwest Environ Technol Labs, Inc, 59-69, vpres, 69-76; STAFF MEM, PHYS DYNAMICS INC, 76- *Mem:* Am Geophys Union; Inst Elec & Electronics Engrs; Air Pollution Control Asn; Am Phys Soc. *Res:* Environmental sciences; ionospheric research; atmospheric propagation at optical wavelengths. *Mailing Add:* 9301 26th Pl NW Seattle WA 98117

LANSKI, CHARLES PHILIP, b Chicago, Ill, Oct 19, 43. MATHEMATICS. *Educ:* Univ Chicago, SB, 65, SM, 66, PhD(math), 69. *Prof Exp:* From asst prof to assoc prof, 69-81, PROF MATH, UNIV SOUTHERN CALIF, 82- *Concurrent Pos:* NSF grant, 71-78. *Mem:* Am Math Soc; Math Asn Am. *Res:* Noncommutative ring theory. *Mailing Add:* Dept Math Univ Southern Calif Los Angeles CA 90007

LANSON, HERMAN JAY, b Utica, NY, Feb 22, 13; m 35; c 3. ORGANIC CHEMISTRY. *Educ:* Syracuse Univ, BS, 34, MS, 36; Polytech Inst New York, PhD(org chem), 45. *Prof Exp:* Org res chemist, Nuodex Prod Co, Inc, NJ, 36-40; res chemist, HD Roosen Co, NY, 40-43; chief chemist, Crown Oil Co Prod Corp, 43-45; supvr, Vehicle Res & Prod, Grand Rapids Varnish Corp, 45-50; supvr, resin & plasticizer develop, Chem Mat Dept, Gen Elec Co, 50-57; vpres & tech dir, US Vehicle & Chem Co, 57-61; vpres & res dir, Lanson Chem Corp, 61-70, pres & res dir, Washburn Lanson Corp, 70-75; pres & res dir, Lanchem Corp, 75-84; TECH CONSULT, 85- *Concurrent Pos:* Lectr, Washington Univ, Roosevelt Univ, Chicago, Univ Houston, Univ Mo & St Louis Univ. *Honors & Awards:* St Louis Gateway Award. *Mem:* Am Oil Chem Soc; Soc Plastics Engrs; fel Am Inst Chemists; Am Chem Soc. *Res:* Synthetic resins; drying oils; protective coatings; electrical insulation materials; development of water-soluble polymers, synthetic latexes, synthetic resins for electrical insulation and coatings. *Mailing Add:* 12020 Gardengate Dr St Louis MO 63146

LANTER, ROBERT JACKSON, b Middletown, Ohio, Nov 9, 14; m 47; c 2. PHYSICS, NUCLEAR WEAPON ENGINEERING. *Educ:* Univ Utah, BA, 42; Univ NMex, MS, 57. *Prof Exp:* Alt group leader physics, 46-48, STAFF MEM PHYSICS & ENG, LOS ALAMOS NAT LAB, UNIV CALIF, 49- *Mem:* Am Phys Soc; AAAS. *Res:* Neutron production and detection; developed 1-meter diameter liquid scintillation counter for detection of bursts of fewer than 100 neutrons; helped develop D-T pulsed neutron sources; silver counter for detecting and counting deutron-deutron reaction and deutrium-tritium reaction neutrons. *Mailing Add:* 2438 Club Rd Los Alamos NM 87544

LANTERMAN, ELMA, b Elkhart, Ill, Jan 25, 17. ANALYTICAL CHEMISTRY. *Educ:* Univ Ill, BS, 40; Ind Univ, MA, 48, PhD(analytical chem), 51. *Prof Exp:* Chem technician, Mayo Clin, 41-42; chemist, Devoe & Raynolds Co, 42-43 & Metal & Thermit Corp, 43-46; asst, Univ Ind, 46-51; asst prof physics, NC State Col, 51-53; supvr, Metal Groups, Indianapolis Naval Ord Plant, 53-55; sr res chemist, Am Can Co, 55-59; sr res chemist, Borg-Warner Corp, 59-61, scientist, 61-63, mgr analytical chem, 63-74, staff scientist, Res Ctr, 74-82; RETIRED. *Concurrent Pos:* Comnr, Environ Comn, Des Plaines, Ill, 74-79. *Mem:* Am Chem Soc; Electron Micros Soc Am; Am Crystallog Asn; Soc Appl Spectros (treas, 70-74); Sigma Xi. *Res:* X-ray diffraction and spectroscopy; electron microscopy; technology forecasting. *Mailing Add:* 1124 E Vilas Marshfield WI 54449

LANTERO, ORESTE JOHN, JR, b Chicago, Ill, Aug 26, 42; m 67; c 2. BIOCHEMISTRY, ORGANIC CHEMISTRY. *Educ:* Purdue Univ, BS, 64; NDak State Univ, PhD(biochem), 71. *Prof Exp:* Res fel, Merrell Nat Lab, 71-73; RES SCIENTIST BIOCHEM, MILES, INC, 73- *Mem:* AAAS. *Res:* Isolation and characterization of enzymes; preparation and characterization of immobilized enzymes. *Mailing Add:* 59731 Ridgewood Dr Goshen IN 46526

LANTHIER, ANDRE, medicine, biochemistry; deceased, see previous edition for last biography

LANTOS, P(ETER) R(ICHARD), b Budapest, Hungary, July 18, 24; nat US; m 47; c 4. CHEMICAL ENGINEERING. *Educ:* Cornell Univ, BChE, 45, PhD(chem eng), 50. *Prof Exp:* Develop chemist, Gen Elec Co, 46-47; res engr, E I du Pont de Nemours & Co, Inc, 50-55, res supvr, 55-60; mgr appln & prod develop, Celanese Plastics Co, 61-63, mgr res & develop, 64-69; dir develop, Sun Chem Corp, 69-70, vpres res & develop, 70-75; gen mgr, Plastics Div, Rhodia Inc, 75; dir res & develop, Arco Polymers, Inc, 76-77, vpres, 78-79; PRES, TARGET GROUP, INC, 80- *Concurrent Pos:* Chmn, Chem Mkts & Econ, Div Am Chem Soc; chmn, Res Mgt Group, Philadelphia. *Mem:* Asn Consult Chemists & Chem Engrs; Am Chem Soc; Am Inst Chem Engrs; Plastic Inst Am; Soc Plastics Engrs. *Res:* Polymers; plastics; fibers. *Mailing Add:* 1000 Harston Lane Target Group Inc Philadelphia PA 19118

LANTZ, THOMAS LEE, b Clarksburg, WV, July 12, 36; m 61; c 2. POWER HYDRAULICS, LUBRICATION OF INDUSTRIAL MACHINERY. *Educ:* WVa Univ, BS, 59; Univ Pittsburgh, MA, 67. *Prof Exp:* MTCE engr, 59-67, PLANT HYDRAUL & LUBRICATION ENGR, WHEELING-PITTSBURGH STEEL, 67- *Concurrent Pos:* Dir, Am Soc Lubrication Engrs, 74-84; comt mem, Asn Iron & Steel Engrs, 75-; instr tech math & physics, Belmont Tech Community Col, 77-87; instr hydraul, drafting, eng, WVa Northern Community Col, 87- *Mem:* Fel Am Soc Lubrication Engrs (vpres, 74-84). *Res:* Power hydraulics; lubrication of industrial machinery; several technical articles. *Mailing Add:* 124 E Cardinal St Wheeling WV 26003

LANYI, JANOS K, b Budapest, Hungary, June 5, 37; US citizen; m 88; c 3. BIOCHEMISTRY. *Educ:* Stanford Univ, BS, 59; Harvard Univ, MA, 61, PhD(biochem), 63. *Prof Exp:* NIH fel genetics, sch med, Stanford Univ, 63-65; Nat Acad Sci-Nat Res Coun res assoc biochem, 65-66; res scientist, Planetary Biol Div, Ames Res Ctr, NASA, 66-80; PROF PHYSIOL & BIOPHYS, UNIV CALIF, IRVINE, 80- *Honors & Awards:* Except Sci Achievement Medal, NASA, 77; H Julian Allen Award, 78; Alexander von Humbolt Prize, 79. *Mem:* Biophys Soc; Am Soc Biol Chem. *Res:* Structure and function of enzymes and membranes in halophilic microorganisms; bacteriorhodopsin, halorhodopsin; energetics and mechanism of proton and chloride transport. *Mailing Add:* Dept Physiol & Biophys Univ Calif Irvine CA 92717

LANYON, HUBERT PETER DAVID, b Halesowen, Eng, June 25, 36; m 79; c 5. SOLID STATE PHYSICS. *Educ:* Cambridge Univ, BA, 58, MA, 62; Leicester Univ, PhD(physics), 61. *Prof Exp:* Res demonstr physics, Leicester Univ, 58-61; res assoc elec eng, Univ Ill, Urbana, 61-63; mem tech staff, RCA Labs, 63-66; assoc prof elec eng, Carnegie Inst Technol, 66-67; assoc prof, 67-77, PROF ELEC ENG, WORCESTER POLYTECH INST, 77- *Mem:* Inst Elec & Electronics Engrs; Sigma Xi. *Res:* Physics of solid state devices; device modelling; integrated circuits; solar cell physics; instrumentation. *Mailing Add:* Elec Eng Dept Worcester Polytech Inst 100 Institute Rd Worcester MA 01609

LANYON, WESLEY EDWIN, b Norwalk, Conn, June 10, 26; m 51; c 2. ORNITHOLOGY. *Educ:* Cornell Univ, AB, 50; Univ Wis, MS, 51, PhD(zool), 55þ. *Prof Exp:* Instr zool, Univ Ariz, 55-56; asst prof, Miami Univ, 56-57; asst cur, Am Mus Natural Hist, 57-63, assoc cur, 63-67, cur, 67-86; RETIRED. *Concurrent Pos:* Res dir, Kalbfleisch Field Res Sta, Am Mus Natural Hist, Huntington, 58-80; adj prof, City Univ New York, 68- *Honors & Awards:* Brewster Award, Am Ornith Union, 68. *Mem:* Am Ornith Union (pres, 76-78); Cooper Ornith Soc; Wilson Ornith Soc; Soc Syst Zool; Soc Study Evolution. *Res:* Application of compararive behavior and ecology of closely related populations of birds to avian systematics, especially vocalizations in avian taxonomy; use of morphology in the systematics of higher categories of birds. *Mailing Add:* 138 Prince St New York NY 10012

LANZA, GIOVANNI, b Trieste, Italy, Aug 5, 26; m 50; c 4. PHYSICS. *Educ:* Univ Trieste, PhD, 50. *Prof Exp:* Asst prof quantum theory, Univ Trieste, 50-52; prof, Univ Cagliari, Univ Sardinia & Univ Padua, 52-54; vis physicist nuclear physics, Mass Inst Technol, 54-55; res fel, Harvard Univ, 55-58; assoc prof, 58-60, PROF PHYSICS, NORTHEASTERN UNIV, 60- *Concurrent Pos:* Fulbright & Smith Mundt scholars, 54; consult, Lab Electronics, Inc, 58- & Saunders Assocs, 63- *Mem:* Ital Phys Soc; Sigma Xi. *Res:* Magnetohydrodynamics; plasma physics; energy conversion techniques; nuclear physics; physics of upper atmosphere. *Mailing Add:* 1994 Mass Ave Lexington MA 02173

LANZA, GUY ROBERT, b Englewood, NJ, Jan 27, 39; m 68; c 2. AQUATIC ECOLOGY. *Educ:* Fairleigh Dickinson Univ, BS, 61; Univ Ky, MS, 69; Va Polytech Inst & State Univ, PhD(zool), 72. *Prof Exp:* Res biologist, Merck Inst Therapeut Res, 63-69; aquatic ecologist, Smithsonian Inst, 71-73 & NY Univ, 73-75; AQUATIC ECOLOGIST, UNIV TEX, DALLAS, 75-, ASSOC PROF ENVIRON SCI, 75- *Concurrent Pos:* Consult ecologist, Int Ctr Med Res & Training, Malaysia, 72-73; asst dir, Aquatic ecol prog, NY Univ Med Ctr, 73-75; subcomt partic, Nat Comn Water Qual, 75. *Mem:* AAAS; Water Pollution Control Fedn. *Res:* Structure and function of aquatic ecosystems; pollution ecology and the environmental physiology and energetics of aquatic organisms. *Mailing Add:* Environ Sci Prog Univ Tex Dallas Box 688 Richardson TX 75080

LANZA, RICHARD CHARLES, b New York, NY, Apr 28, 39; m 63; c 1. PHYSICS. *Educ:* Princeton Univ, AB, 59; Univ Pa, MS, 61, PhD(physics), 66. *Prof Exp:* Res assoc physics, 66-68, asst prof, 68-74, mem res staff, 74-83, PRIN SCIENTIST, MASS INST TECHNOL, 83- *Concurrent Pos:* Assoc radiol, Harvard Med Sch & Peter Bent Brigham Hosp, 74-; res fel, Mass Gen Hosp, 75-76. *Mem:* AAAS; Inst Elec & Electronics Engrs; Am Phys Soc. *Res:* Experimental particle physics, nuclear and electronic instrumentation, medical instrumentation, especially in radiology and nuclear medicine; imaging for nondestructive testing and evaluation. *Mailing Add:* Dept Nuclear Eng Rm 24-036 Mass Inst Technol Cambridge MA 02139

LANZA-JACOBY, SUSAN, m 78; c 1. NUTRITION, INFECTION. *Educ:* Hunter Col, BS, 65; Columbia Univ, MS, 68; Rutgers Col, PhD(nutrit biochem), 79. *Prof Exp:* ASSOC PROF, THOMAS JEFFERSON UNIV, 79- *Concurrent Pos:* Lectr, Hunter Col, 68-72; instr, Philadelphia Gen Hosp Sch Nursing, 72-77. *Mem:* Am Inst Nutrit; Am Dietetic Asn; Shock Soc; Am Soc Parenteral Nutrit. *Res:* Lipid metabolism in infection, tumor growth, and parenteral feeding. *Mailing Add:* Dept Surg Thomas Jefferson Univ 1025 Walnut St Philadelphia PA 19107

LANZANO, BERNADINE CLARE, b Stanberry, Mo, Oct 29, 33; m 57. COMPUTER SCIENCE, MATHEMATICS. *Educ:* Benedictine Col, BS, 55. *Prof Exp:* Mathematician, Lockheed Corp, 56-57; SR STAFF ENGR COMPUTER SCI, TRW SYSTS GROUP, 57- *Mem:* Math Asn Am; Soc Indust & Appl Math. *Res:* Design and development of computer software in the areas of trajectory analysis, optimization, targeting, and in the fields of information processing of financial and scientific data; management of database management systems applications software development, installation, and operational support. *Mailing Add:* 8614 Pappas Way Annandale VA 22003

LANZANO, PAOLO, b Cairo, Egypt, Nov 29, 23; nat US; m 57. APPLIED MATHEMATICS, SPACE PHYSICS. *Educ:* Univ Rome, BS, 43, PhD(math), 47. *Prof Exp:* Asst prof math, Univ Rome, 46-49 & St Louis Univ, 50-56; design specialist, Douglas Aircraft Co, Calif, 56-58; mem tech staff, Space Tech Labs, Calif, 58-60; res scientist, Nortronics Div, Northrop Corp, 60-61; prin scientist, Space & Info Systs Div, NAm Aviation, 61-71; assoc prof math, Nicholls State Univ, 71-72; head math res ctr, 72-76, SR RES SCIENTIST, NAVAL RES LAB, 76- *Concurrent Pos:* Fel, Inst Advan Studies, Univ Rome, 46-48. *Mem:* Am Math Soc; Am Geophys Union Soc; Soc Indust & Appl Math; assoc fel Am Inst Aeronaut & Astronaut. *Res:* Celestial mechanics; theory of relativity; Riemannian geometry; space physics; geodesy; geophysics. *Mailing Add:* 8614 Pappas Way Annandale VA 22003

LANZEROTTI, LOUIS JOHN, b Carlinville, Ill, Apr 16, 38; m 65; c 2. GEOPHYSICS, SPACE PHYSICS. *Educ:* Univ Ill, BS, 60; Harvard Univ, AM, 63, PhD(physics), 65. *Prof Exp:* Fel, 65-67, MEM TECH STAFF, BELL LABS, 67- *Concurrent Pos:* Mem, Space Sci Adv Comt, NASA, 75-79, Space & Earth Sci Adv Comt, 84-88, Adv Coun, 84-; mem, Space Sci Bd, Nat Acad Sci, 79-83, 88-, Polar Res Bd, 82-90; adj prof, Univ Fla, 78-; regents lectr, Univ Calif, Los Angeles, 87. *Honors & Awards:* NASA Distinguished Pub Serv Medal; Antarctica Geog Feature named in honor, Mt Lanzeratti. *Mem:* Nat Acad Eng; fel Am Phys Soc; fel Am Geophys Union; Inst Elec & Electronics Engrs; Soc Terrestial Magnetism & Elec Japan; assoc fel Am Inst Aeronaut & Astronaut; fel AAAS; Int Acad Astronaut. *Res:* Particles and fields in planetary magnetospheres; solar cosmic ray composition and propagation; ionosphere-magnetosphere coupling; planetary magnetospheres; geomagnetic depth sounding; impacts of space effects on technologies. *Mailing Add:* AT&T Bell Labs Murray Hill NJ 07974

LANZEROTTI, MARY YVONNE DEWOLF, b Phoenix, Ariz, Nov 7, 38; m 65; c 2. PHYSICAL CHEMISTRY. *Educ:* Univ Calif, Berkeley, BS, 60; Harvard Univ, PhD(phys chem), 65. *Prof Exp:* Chemist, US Naval Ord Test Sta, 60; asst, Harvard Univ, 60-64; res chemist, Mithras Inc, Mass, 64-65; RES PHYS SCIENTIST, E & W DIV, US ARMY ARDEC, 65- *Mem:* AAAS; Am Chem Soc; Am Phys Soc; Am Soc Mech Engrs; Am Soc Lubrication Engrs; Am Defense Preparedness Asn. *Res:* Mechanical behavior of materials under high power spectral analysis of fracture surface topography acceleration. *Mailing Add:* E & W Div Bldg 3022 US Army ARDEC Picatinny Arsenal NJ 07806-5000

LANZKOWSKY, PHILIP, b Cape Town, SAfrica, Mar 17, 32; m 55; c 5. PEDIATRICS, HEMATOLOGY. *Educ:* Univ Cape Town, MB ChB, 54, MD, 59; Royal Col Physicians & Surgeons, dipl child health, 60; Am Bd Pediat, dipl, 66, cert pediat hematol-oncol, 75; FRCP(E), 73. *Prof Exp:* From intern to sr intern, Groote Schuur Hosp, Univ Cape Town, 55-56; gen pract, 56-57; from registr to sr registr, Red Cross War Mem Children's Hosp, 57-60, consult pediatrician & pediat hematologist, 63-65; asst prof pediat, New York Hosp-Cornell Med Ctr, 65-67, assoc prof, 67-70; PROF PEDIAT, STATE UNIV NY STONY BROOK, 70- *Concurrent Pos:* Dr C L Herman res grants, 58 & 64; Cecil John Adams mem traveling fel & Hill-Pattison-Struthers bursary, 60; Benger Labs traveling grant, 61; registr pediat unit, St Mary's Hosp Med Sch, Univ London, 61; clin & res fel pediat hemat, Duke Univ, 61-62; res fel, Col Med, Univ Utah, 62-63; lectr, Univ Cape Town, 63-65; dir pediat hemat, New York Hosp-Cornell Med Ctr, 65-70; pediatrician-in-chief, chmn pediat & chief pediat hemat, Long Island Jewish-Hillside Med Ctr, 70-; pediatrician-in-chief, Queens Hosp Ctr, 70-; mem pediat adv comt NY City Dept Health, 70-73. *Honors & Awards:* Joseph Arenow Prize, 59. *Mem:* Am Soc Hemat; Am Acad Pediat; Am Soc Clin Oncol; Am Asn Cancer Res; Am Pediat Soc. *Res:* Nutritional anemias in children, especially iron, folate and protein deficiency; pediatric oncology. *Mailing Add:* L I Jewish Med Ctr New Hyde Park NY 11040

LANZKRON, R(OLF) W(OLFGANG), b Hamburg, Ger, Dec 9, 29; nat US; m 61; c 3. ELECTRICAL ENGINEERING. *Educ:* Milwaukee Sch Eng, BS, 54; Univ Wis, MS, 55, PhD, 56. *Prof Exp:* Asst, Univ Wis, 55-56; asst res & develop, Univac Div, Sperry Rand Corp, 56-57; design engr, Martin Co, 57-62; chief flight projs, Div, Apollo, NASA, 62-68; prog mgr, Fed Aviation Admin Display Systs, 68-73, prog mgr Air Force AN-TPN/19 Prog, 73-78, opers mgr graphic oper, 78-81, graphic systs mgr, Raytheon Co, 81-83; DEP DIR, AIR TRAFFIC CONTROL, 83- *Honors & Awards:* Outstanding Achievement Award, NASA, 64. *Mem:* Am Inst Aeronaut & Astronaut; Sigma Xi; Math Asn Am; Inst Elec & Electronics Engrs; Nat Soc Prof Engrs. *Res:* Guidance and control of ballistic missiles and space vehicles; digital and analog computer design and usage and ground systems engineering; flight vehicle integration; design of ground support equipment and automated test equipment; advanced technology of display system and computers. *Mailing Add:* Eight Hickory Dr Medfield MA 02052

LANZL, LAWRENCE HERMAN, b Chicago, Ill, Apr 8, 21; m 47; c 2. PHYSICS, MEDICAL PHYSICS. *Educ:* Northwestern Univ, BS, 43; Univ Ill, MS, 47; PhD(physics), 51; Am Bd Health Physics, dipl, Am Bd Radiol, dipl, Am Bd Med Physics, dipl, 90. *Prof Exp:* Asst, Dearborn Observ, Ill, 41-42; asst, Northwestern Univ, 42-43; asst, Univ Chicago, 44; jr scientist, Los Alamos Sci Lab, 44-45; asst, Univ Ill, 46-50; assoc physicist, Argonne Nat Lab, 51; sr physicist, 51-55, from asst prof to assoc prof, Dept Radiol, Sch Med, 55-68, prof med physics, div biol sci, 68-80, EMER PROF, DIV BIOL SCI, PRITZKER SCH MED & FRANKLIN MCLEAN MEM RES INST, UNIV CHICAGO, 80-; PROF MED PHYSICS & CHMN DEPT MED PHYSICS, COL HEALTH SCI, RUSH UNIV, RUSH PRESBY ST LUKE'S MED CTR, CHICAGO, 80- *Concurrent Pos:* At Argonne Cancer Res Hosp, 51-; first officer, Int Atomic Energy Comn, Vienna, Austria, 67-68; bd mem, Am Int Sch, Vienna, 67-68; mem Nat Coun Radiation Protection, 67 & US Nat Comt Med Physics, 70-; chmn Radiation Protection Adv Coun, State of Ill, 71-; mem tech adv panel, Los Alamos Meson Physics Facil, Los Alamos, NMex & steering comt, radiol physics ctr, Univ Tex M D Anderson Hosp & Tumor Inst; consult, Int Atomic Energy Agency, Vienna, Hines Vet Admin Hosp, Ill, NIH & WHO, Geneva; consult radiation dosimetry prob, India, 73, Turkey, 75, Saudia Arabia, 75, Iran, 76, Israel, 77, Ghana, 78, Thailand, 79 & Nigeria, 81; chmn, Comn Accreditation Educ Progs Med Ph. *Honors & Awards:* Coolidge Award, Am Asn Physicists Med, 78; Landauer Award, 89. *Mem:* AAAS; Asn Med Physicists India; Brit Hosp Physicists' Asn; fel Am Asn Physicists Med (pres, 67-68); Radiol Soc NAm; fel Health Physics Soc; Sigma Xi; fel Am Col Radiol; Int Orgn Med Physics (pres, 85-88); Int Union Phys & Eng Sci Med (pres, 88-91). *Res:* Accelerators; radiation as applied to medicine; radiation physics; high energy x-rays and electrons; isotopes in medicine; nuclear reactors; health physics; radiation dosimetry. *Mailing Add:* 5750 S Kenwood Ave Chicago IL 60637

LANZONI, VINCENT, b Kingston, Mass, Feb 23, 28; m 60; c 3. PHARMACOLOGY, CLINICAL MEDICINE. *Educ:* Tufts Univ, PhD(pharmacol), 53; Boston Univ, MD, 60. *Prof Exp:* Instr pharmacol, Sch Med, Tufts Univ, 53-54; from intern to resident, Boston City Hosp, 60-63; asst prof pharmacol & instr med, 63-66, assoc prof pharmacol & med, 66-73, assoc dean sch med, 69-75, prof pharmacol, Sch Med, Boston Univ, 73-75; DEAN & PROF MED, NJ MED SCH, 75- *Concurrent Pos:* Res fel, NIH, 53-54; fel med, Boston City Hosp. *Res:* Cardiovascular pharmacology. *Mailing Add:* One Fairview Rd Millburn NJ 07041

LAO, BINNEG YANBING, b Szechwan, China, Feb 25, 45; US citizen; m 70; c 2. ELECTRONICS ENGINEERING, APPLIED PHYSICS. *Educ:* Univ Calif, Los Angeles, BS, 67; Princeton Univ, MA, 69, PhD(physics), 71. *Prof Exp:* NSF fel solid state physics, Ctr Theoret Physics, Univ Md, 71-73; sr res physicist instrumentation, Eastern Div Res Labs, Dow Chem Co, 73-76; prin physicist appl physics, Bendix Res Labs, Bendix Corp, 76-80; mgr, Microelectronics Magnavox Res Oper, Magnavox Adv Prod & Syst Co, 80-86; consult, 86-88; VPRES, SIERRA MONOLITHICS, INC, 88- *Mem:* Am Phys Soc; Sigma Xi; Inst Elec & Electronics Engrs. *Res:* Surface acoustic wave devices and gallium arsenide integrated circuits; analytical instrument development; optical and dielectrical properties of matter; nonlinear and excitonic effects in semiconductors; sensors and solid state devices. *Mailing Add:* 3449 Coolheights Rancho Palos Verdes CA 90274

LAO, CHANG SHENG, b Shanghai City, China, Dec 10, 35; US citizen; m 66; c 3. CLINICAL TRIALS REVIEW, IN-VITRO DIAGNOSTIC TESTS ANALYSIS. *Educ:* Nat Taiwan Univ, BA, 60; Univ Mass, MS, 66; Yale Univ, PhD(biostatist), 73. *Prof Exp:* Statistician, E I du Pont de Nemours & Co, Inc, 66-68; med res scientist, Pa Dept Health, 73-74; epidemiologist, US Environ Protection Agency, 80; math statistician, Ctr Drugs Eval & Res, Food & Drug Admin, 74-80 & Ctr Radiol Health Devices, 80-86, supvry math statistician, Ctr Drugs Eval & Res, 86-87, MATH STATISTICIAN, CTR RADIOL HEALTH DEVICES, FOOD & DRUG ADMIN, 87- *Concurrent Pos:* Statist reviewer, Asn Off Analytical Chemists, 77-; lectr med statist, Howard Univ Sch Med, 83. *Mem:* Am Statist Asn; Biomet Soc; Asn Off Analytical Chemists. *Res:* Statistical reviews of medical devices clinical trials; analyze the statistical and epidemiological data related to public health; design and analyze laboratory data on heart valve study; ethylene oxide study; in-vitro diagnostic tests. *Mailing Add:* 15429 Narcissus Way Rockville MD 20853

LAO, YAN-JEONG, b Nanking, China, Feb 5, 36; US citizen; m 65; c 2. CHEMICAL ENGINEERING, ENVIRONMENTAL HEALTH. *Educ:* Nat Taiwan Univ, BS, 58; Univ Mich, MS, 62; PhD(chem eng), 69. *Prof Exp:* Res engr chem eng, E I du Pont de Nemours & Co, Inc, 69-71; sr engr, Monsanto Co, 72-73; from asst prof to assoc prof, 73-81, PROF ENVIRON HEALTH, ECAROLINA UNIV, 81-, CHMN DEPT, 85- *Mem:* Nat Environ Health Asn; Am Indust Hyg Asn; Sigma Xi. *Res:* Monitoring and analyzing environmental pollutants, the study of their effects and health related problems. *Mailing Add:* Dept Environ Health ECarolina Univ Greenville NC 27858

LAPALME, DONALD WILLIAM, b Woonsocket, RI, July 27, 37; m 61; c 4. PHOTOGRAPHIC ENGINEERING, PHYSICAL CHEMISTRY. *Educ:* St John's Univ NY, BS, 59, MS, 61, PhD(chem), 68. *Prof Exp:* Sr res chemist photo eng, Photo Div, 67-72, sr prod engr, 72-74, prod mgr reprographics, 74-76, tech dir process eng, Res & Develop, 76-78, plant mgr prod, Photo Div, 78-81, VPRES OPERS, GAF BLDG MATS CORP, 81- *Mem:* Soc Photographic Scientists & Engrs (pres, 73-74); Am Chem Soc. *Res:* Photographic science. *Mailing Add:* 53 Pike Dr Apt 1A Wayne NJ 07470-2464

LAPERRIERE, JACQUELINE DOYLE, b Northampton, Mass, Dec 31, 42; div; c 2. AQUATIC FISHERIES. *Educ:* Univ Mass, BS, 64; Iowa State Univ, MS, 71, PhD(water resources), 81. *Prof Exp:* Water resource technician, 71-72, res biologist, 72-73, aquatic biologist, 73-74, instr water resources, 74-79, asst prof, Inst Water Resources, 79-85, asst prof fisheries, 80-85, ASSOC PROF FISHERIES & WATER RESOURCES, COL NATURAL SCI & INST WATER RESOURCES, UNIV ALASKA, FAIRBANKS, 85-, ASST LEADER, ALASKA COOP FISHERY RES UNIT, 80- *Concurrent Pos:* Andrew W Mellon Found travel grant, 80. *Mem:* Am Fisheries Soc; Am Soc Limnol & Oceanog; Int Asn Limnol; NAm Benthol Soc; NAm Lake Mgt Soc; Sigma Xi. *Res:* Stream ecology, primary and secondary production, water quality and effects of development; limnology of subarctic lakes, thermal regime, chemical cycling and production; toxicity testing of Alaskan fishes. *Mailing Add:* PO Box 81547 College AK 99708

LAPETINA, EDUARDO G, PLATELET AGGREGATION, PHOSPHOLIPIDS. *Educ:* Univ Buenos Aires, Argentina, PhD(biochem), 67. *Prof Exp:* GROUP LEADER MOLECULAR BIOL, WELLCOME RES LABS, BURROUGHS WELLCOME CO, 76- *Res:* Thrombosis. *Mailing Add:* Wellcome Res Labs Burroughs Wellcome Co 3030 Cornwallis Rd Research Triangle Park NC 27709

LAPEYRE, GERALD J, b Riverton, Wyo, Jan 3, 34; m 60; c 3. SOLID STATE PHYSICS, SURFACE PHYSICS. *Educ:* Univ Notre Dame, BS, 56; Univ Mo, MA, 58, PhD(physics), 62. *Prof Exp:* From asst prof, to assoc prof physics, 62-74, PROF PHYSICS, MONT STATE UNIV, 74- *Mem:* Fel Am Phys Soc; Am Asn Physics Teachers; Am Vacuum Soc. *Res:* Solid state physics surface science with emphasis on synchrotron photoemission and electronic structure. *Mailing Add:* Dept Physics Mont State Univ Bozeman MT 59715

LAPEYRE, JEAN-NUMA, b Los Angeles, Calif, Oct 17, 45. MOLECULAR PATHOLOGY. *Educ:* Univ Calif, Los Angeles, BS, 67, MS, 69; Univ Southern Calif, PhD(molecular biol), 75. *Prof Exp:* Lectr cell biol & dent biochem, Univ Southern Calif, 73-75; Fogarty Int fel, Swiss NSF, Univ Geneva, 75-77; proj investr, Exp Path Sect, M D Anderson Cancer Ctr, Univ Tex, 77-78, res assoc, 78-79, from instr to asst prof & asst biochemist, 79-85, ASSOC BIOCHEMIST & ASSOC PROF BIOCHEM, EXP PATH SECT, M D ANDERSON CANCER CTR, UNIV TEX, 85- *Concurrent Pos:* Lectr chem embryol, Molecular Biol Grad Sch, Univ Geneva, 76; vis exchange scientist, Nat Ctr Sci Res, Inst Molecular Biophys, Orleans, France, 82; mem, NIH Study Sect, Clin Sci IV, 85-89, NSF Rev Bd, 90-; consult, DNA Sci, Inc, Houston & Cytol Technol, Inc. *Mem:* Am Soc Biochem & Molecular Biol; AAAS; Biophys Soc; Am Asn Cancer Res. *Res:* DNA chemical synthetic methods and DNA sequencing; DNA structure and conformation; author of numerous scientific publications. *Mailing Add:* Dept Molecular Path M D Anderson Cancer Ctr 1515 Holcombe Blvd Box 89 Houston TX 77030

LAPHAM, LOWELL WINSHIP, b New Hampton, Iowa, Mar 20, 22; div; c 4. NEUROPATHOLOGY. *Educ:* Oberlin Col, BA, 43; Harvard Med Sch, MD, 48. *Prof Exp:* Sr instr path, Western Reserve Univ, 55-57, from asst prof to assoc prof, 57-64; assoc prof path, 64-69, PROF NEUROPATH, MED CTR, UNIV ROCHESTER, 69- *Concurrent Pos:* Nat Multiple Sclerosis Soc fel cytochem, Case Western Reserve Univ, 56-58; consult neuropath, Univ Rochester Affil Hosps. *Mem:* Am Asn Neuropath; Sigma Xi. *Res:* Studies of developmental diseases of nervous system; brain tumors; nature and function of glia; effects of environmental substances on nervous system. *Mailing Add:* 1380 Elmwood Ave No 4 Rochester NY 14620-3150

LAPICKI, GREGORY, b Warsaw, Poland; US citizen. INNER SHELL IONIZATION. *Educ:* Warsaw Univ, MS, 67; NY Univ, PhD(physics), 75. *Prof Exp:* Post doc physics, NY Univ, 75-76, res scientist, 77-78; vis asst prof physics, Tex A&M Univ, 79-80; asst prof physics, Northwestern State Univ La, 80-81; assoc prof, 81-87, PROF PHYSICS, E CAROLINA UNIV, 88- *Concurrent Pos:* Prin investr, Nat Bur Standards, 82-84; participant, Oak Ridge Nat Lab, 81-87; panel reviewer, Off Naval Technol Postdoc Fel, 87. *Mem:* Am Phys Soc; Sigma Xi. *Res:* Penetration of charged particles in matter; development of theories of inner-shell direct ionization and electron capture; study of asymmetric and symmetric ion-atom collisions; published over 50 refereed articles. *Mailing Add:* Dept Physics E Carolina Univ Greenville NC 27858

LAPIDES, JACK, b Rochester, NY, Nov 27, 14; m 48, 75. UROLOGY. *Educ:* Univ Mich, BA, 36, MA, 38, MD, 41. *Prof Exp:* USPHS fel, Nat Cancer Inst, 48-50; from instr to prof surg, Med Sch, Univ Mich, Ann Arbor, 50-84, head, Sect Urol, 68-83, emer prof surg, 85-86; RETIRED. *Concurrent Pos:* Rockefeller res assoc, 36-38; assoc physician surg, Wayne County Gen Hosp, 50-81, chief sect urol, 58-83 & Vet Admin Hosp, Ann Arbor, 54-83; mem, Comt Genito-Urinary Systs, Nat Acad Sci-Nat Res Coun, 66-70. *Mem:* AMA; Am Urol Asn; fel Am Col Surg; Am Asn Genito-Urinary Surg. *Res:* Renal, ureteral and bladder physiology; fluid and electrolyte balance; urinary incontinence; urinary infection. *Mailing Add:* 2805 N Wagner Rd Ann Arbor MI 48103

LAPIDUS, ARNOLD, b Brooklyn, NY, Nov 6, 33; m 52. MATHEMATICS. *Educ:* Brooklyn Col, BS, 56; NY Univ, MS, 60, PhD(math), 67. *Prof Exp:* Asst math, Courant Inst, NY Univ, 58-60, asst res scientist, AEC comput facility, 61-65, assoc res scientist, 61-68; math analyst, Comput Applns Inc-NASA, 68-69, sci prog mgr, 69-71; asst prof math, 71-76, ASSOC PROF QUANT ANALYSIS, FAIRLEIGH DICKINSON UNIV, 77- *Mem:* Soc Indust & Appl Math; Math Asn Am; Am Math Soc. *Res:* Partial and ordinary differential equations; Monte Carlo methods; scientific programming; artificial intelligence; tedious algebra by computer; fluid dynamics by computer; shock calculations; numerical methods and analysis; linear programming. *Mailing Add:* 160 Rockwood Pl Englewood NJ 07631

LAPIDUS, HERBERT, b New York, NY, Aug 10, 31; m 52; c 2. PHARMACY, PHARMACOLOGY. *Educ:* Columbia Univ, BS, 53, MS, 55; Rutgers Univ, PhD(pharm), 67. *Prof Exp:* Proj leader, Julius Schmid Co, 57-60; proj leader pharm, Bristol-Myers Co, 60-63, group leader, 63-67, dept head, 67-70; tech dir, 70-77, VPRES RES & DEVELOP, COMBE INC, 77- *Mem:* Am Chem Soc; Sigma Xi; Soc Cosmetic Chemists; NY Acad Sci; Am Soc Clin Pharmacol & Therapeut. *Res:* Development of pharmaceutical dosage forms, especially sustained release medication, biopharmaceutics and percutaneous absorption. *Mailing Add:* Combe Inc 1101 Westchester Ave White Plains NY 10604

LA PIDUS, JULES BENJAMIN, b Chicago, Ill, May 1, 31; m 54, 70; c 4. MEDICINAL CHEMISTRY. *Educ:* Univ Ill, BS, 54; Univ Wis, MS, 57, PhD(pharmaceut chem), 58. *Prof Exp:* From asst prof to assoc prof, 58-67, assoc dean res, Grad Sch, 72-74, vprovost res & dean, 74-84, PROF MED CHEM, OHIO STATE UNIV, 67-; PRES, COUN GRAD SCHS, US, 84- *Concurrent Pos:* Consult, Pharmacol & Toxicol Training Grants Comt, Nat Inst Gen Med Sci, NIH, 65-67, Prog Comt, 71-75; Pres Coun Grad Sch US, 84- *Mem:* Am Chem Soc; fel AAAS. *Res:* Structure-action relationships; autonomic pharmacology. *Mailing Add:* Coun Grad Schs One Dupont Circle NW Suite 430 Washington DC 20036

LAPIDUS, MICHEL LAURENT, b Casablanca, Morocco, July 4, 56; m 80; c 1. FUNCTIONAL ANALYSIS, MATHEMATICAL PHYSICS. *Educ:* Univ Pierre & Marie Curie, Paris VI, MS, 77, DEA, 78, Dr(math), 80, Doctoral d'Etatès Sci Math, 86. *Prof Exp:* Res assoc, math, Rectorat de Paris, Inst Pure Math, Univ Paris VI, 78-80; Georges Lurcy fel math, Univ Calif, Berkeley, 79-80; asst prof math, Univ Southern Calif, Los Angeles, 80-85; vis asst prof, Univ Iowa, Iowa City, 85-86; assoc prof math, Univ Ga, Athens, 86-90; PROF MATH, UNIV CALIF, RIVERSIDE, 90- *Concurrent Pos:* Award, Fac Res & Innovation Fund, Univ Southern Calif, 84; mem, Math Sci Res Inst, Berkeley, 84-85; vis prof, Yale Univ, 90-91; creative res medal, Univ Ga, Athens, 89-90. *Mem:* NY Acad Sci; Math Asn Am; French Math Soc; AAAS; Int Asn Math Physicists; Am Asn Univ Prof; Am Math Soc; Am Phys Soc. *Res:* Mathematical research in linear and nonlinear functional analysis in partial differential equations, and in mathematical physics; study of the Trotter-Lie formula and modification of the Feynman integral; Feynman path integrals in quantum mechanics; Feynman's operational calculus for noncommuting operators; eigenvalues and eigenfunctions of elliptic boundary value problems with indefinite weights; spectral and fractal geometry; vibrations of fractal drums. *Mailing Add:* Dept Math Univ Calif Riverside CA 92521-0135

LAPIDUS, MILTON, b New York, NY, May 8, 22; m 58; c 2. BIOCHEMISTRY. *Educ:* Univ Wis, BS, 48, MS, 53, PhD(biochem), 56. *Prof Exp:* Res chemist, Abbott Labs, 48-51; sr fel, Eastern Regional Res Lab, USDA, 56-59; SR BIOCHEMIST, WYETH LABS, INC, 59- *Mem:* AAAS; Am Chem Soc. *Res:* Isolation of vitamin B12b; microbiological transformation of steroids; chromatographic purification of viruses; prostaglandin biosynthesis and isolation; synthesis of penicillins and sweeteners; synthesis of peptides, complement inhibitors. *Mailing Add:* 412 Yorkshire Way Rosemont PA 19010-1119

LAPIERRE, WALTER A(LFRED), electrical & mechanical engineering; deceased, see previous edition for last biography

LAPIERRE, YVON DENIS, b Bonnyville, Alta, Oct 19, 36; m 60; c 3. PHARMACOLOGY. *Educ:* Ottawa Univ, BA, 57, MD, 61; Univ Montreal, MSc, 70; FRCP(C), 72. *Prof Exp:* Lectr, 70-73, from asst prof to assoc prof, 73-81, PROF PSYCHIAT & PHARMACOL, UNIV OTTAWA, 81-, CHMN, PSYCHIAT DEPT, 86- *Concurrent Pos:* Sci dir psychiat, Pierre Janet Hosp, Que, 70-76; dir psychopharmacol, Ottawa Gen Hosp, 76-79; chmn, Expert Standing Comt Psychotrop Drugs, 82; dir res & psychiat, Royal Ottawa Hosp, 79-86, dir outpatient clin, 80-85, psychiatrist in chief, 86- *Honors & Awards:* Tait-MacKenzie Medal, Ottawa Acad Med, 80; Medal of hon, Can Col Neuropschopharmacol, 88. *Mem:* Can Col Neuropsychopharmacol; fel Royal Col Physicians & Surgeons Can; Soc Biol Psychiat; Col Int Neuropsychopharmacol; fel Am Psychiat Asn; fel Am Col Psychiat. *Res:* Drugs used in the treatment of psychiatric disorders plus biochemical clinical and electrophysiological research in the underlying biological factors contributing to mental illness. *Mailing Add:* Royal Ottawa Hosp 1145 Carling Ave Ottawa ON K1Z 7K4 Can

LAPIETRA, JOSEPH RICHARD, b New York, NY, July 20, 32; m. PHYSICAL CHEMISTRY, THERMODYNAMICS. *Educ:* Marist Col, BA, 54; Cath Univ Am, PhD(chem), 61. *Prof Exp:* Teacher high sch, NY, 54-56; instr chem, Cath Univ Am, 60-61; from asst prof to assoc prof, 64-77, acad dean, 69-75, PROF CHEM, MARIST COL, 77- *Mem:* Am Chem Soc. *Res:* Chemistry of transition metal complexes; thermochemistry; history of science; electrochemistry. *Mailing Add:* Marist Col North Rd Poughkeepsie NY 12601

LAPIN, A I E, b Montreal, Que, May 13, 38; m 64; c 3. ELECTRICAL ENGINEERING. *Educ:* McGill Univ, BEng, 60; Univ Sheffield, PhD(elec eng), 63. *Prof Exp:* Mem tech staff, Bell Tel Labs Inc, 63-70; ENGR STAFF SPECIALIST, GEN DYNAMICS CORP, 70- *Mem:* Inst Elec & Electronics Engrs. *Res:* Microwave diode and transistor circuitry; microwave systems for tactical missiles. *Mailing Add:* Gen Dynamics Corp PO Box 2507 Pomona CA 91769

LAPIN, ABRAHAM, b Cairo, Egypt, Sept 30, 23; US citizen; wid; c 2. CHEMICAL ENGINEERING, MATHEMATICS. *Educ:* Univ Mich, BScE(chem eng) & BScE(math), 49; Polytech Inst Brooklyn, MSc, 55; Lehigh Univ, PhD(chem eng), 63. *Prof Exp:* Asst port engr, Am Israeli Shipping Co, Inc, 49-51; sales mgr, Dapor Trading Co, Inc, 51-52; mat engr, US Corps Eng, 53; chem engr, Mineral Beneficiation Lab, Columbia Univ, 54; chem engr, Air Prod & Chem, Inc, 55-57, proj engr, 57-59, group leader, 59-63, sect mgr cryogenic eng res & develop, 63-75; RETIRED. *Concurrent Pos:* Lectr, Pa State Univ, 61 & Lehigh Univ, 64. *Mem:* Am Chem Soc; Am Inst Chem Engrs; fel Am Inst Chemists; Am Soc Heating, Refrig & Air-Conditioning Eng; Am Soc Testing & Mat. *Res:* Cryogenic engineering; low temperature separation; distillation; heat transfer; insulation; fluid flow. *Mailing Add:* 845 Palmer Ave The Pavillion Mamaroneck NY 10543

LAPIN, DAVID MARVIN, b New York, Apr 12, 39; m 67; c 2. BIOLOGY. *Educ:* NY Univ, BA, 60, MS, 63, PhD(biol), 68. *Prof Exp:* From instr to assoc prof biol, 66-76, PROF BIOL SCI, CHMN DEPT, FAIRLEIGH DICKINSON UNIV, 76- *Concurrent Pos:* Grants in aid, Fairleigh Dickinson Univ, 68-72. *Mem:* AAAS; Am Soc Hemat. *Res:* Kinetics of hematopoiesis; humoral regulation of hematopoiesis. *Mailing Add:* 1124 E 31st St Brooklyn NY 11236

LAPIN, EVELYN P, b Montreal, Que, Aug 29, 33; m; c 3. NEUROCHEMISTRY, ENZYMOLOGY. *Educ:* McGill Univ, BSc, 54, PhD(biochem), 57. *Prof Exp:* Am Cancer Soc fel, Albert Einstein Col Med, 57-59; instr math & chem, Herzliah Acad, 62-65; lectr biochem, McGill Univ, 65-66; NIH fel, Mt Sinai Sch Med, 70-73; instr neurochem, 74-81, RES ASST PROF, MT SINAI SCH MED, 81- *Concurrent Pos:* Lectr, Queen's Col, NY, 73-; vis asst prof, Stern Col Women, Yeshiva Univ, NY; vis assoc prof, Columbia Univ. *Mem:* Brit Biochem Soc; Am Soc Neurochem; Can Fedn Univ Women; Int Soc Neurochem; NY Acad Sci. *Res:* Subcellular compartmentalization of respiratory activity and energy metabolism; protein and lipid chemistry of nervous system, localization, separation and identification and function; biochemical mechanisms of hydrocarbon neurotoxins in central and distal axonopathy; brain catecholamines, dopamine, sub P, CCK in selected areas in animal Parkinsonian model. *Mailing Add:* 142-05 Roosevelt Apt 325 Flushing NY 11354

LAPKIN, MILTON, b New York, NY, July 29, 29; m 64; c 3. POLYMER CHEMISTRY. *Educ:* Polytech Inst New York, BS, 51, PhD(org chem), 55. *Prof Exp:* Group leader, Olin Corp, 55-59, sect leader, 59-60, dir res, 69-73; dir res, Beatrice Foods, Polyvinyl Chem Indust, 73-86; DIR RES, ICI RESINS, USA, 86- *Mem:* Am Chem Soc; AAAS; Am Inst Chem; Fedn Soc Coating Technol. *Res:* Free radical polymerization; polyvinyl chloride; acrylic and methacyclic polymers; epoxies; propylene oxide; ethylene oxide; polyutheranes; suspension polymerization; emulsion polymerization. *Mailing Add:* 194 A Greenwood Ave Beverly Farms MA 01915-5429

LAPLANCHE, LAURINE A, b New York, NY, July 4, 38; m. PHYSICAL CHEMISTRY, PHARMACEUTICAL CHEMISTRY. *Educ:* Univ Md, BS, 59; Mich State Univ, PhD(chem), 63. *Prof Exp:* Asst prof physics, WVa State Col, 64-65; from asst prof to assoc prof, 65-91, PROF CHEM, NORTHERN ILL UNIV, 91- *Concurrent Pos:* Vis prof pharmaceut chem, Univ Calif, San Francisco, 84-85. *Mem:* Am Chem Soc; Sigma Xi. *Res:* Nuclear magnetic resonance; two dimensional nuclear magnetic resonance; energy barriers to internal rotation; molecular structure of biological molecules; proton exchange; hydrogen bonding; lanthanide shift reagents and conformational analysis. *Mailing Add:* Dept Chem Northern Ill Univ De Kalb IL 60115

LAPLAZA, MIGUEL LUIS, b Zaragoza, Spain, Mar 20, 38; m 69; c 4. MATHEMATICS. *Educ:* Univ Barcelona, MD, 60; Univ Madrid, PhD(math), 65. *Prof Exp:* Instr math, Univ Barcelona, 60-61; from asst prof to assoc prof, Univ Madrid, 60-66; assoc prof math, 67-76, PROF MATH, UNIV PR, MAYAGUEZ, 76- *Mem:* Am Math Soc; Math Asn Am. *Res:* Category theory. *Mailing Add:* 14-M-7-B Terrace Mayaguez PR 00708

LAPOINTE, JACQUES, b Montreal, Que, Nov 22, 42; m 67; c 1. PROTEIN BIOSYNTHESIS, MOLECULAR GENETICS. *Educ:* Univ Montreal, BSc, 64, MSc, 66; Yale Univ, PhD(molecular biophys), 72. *Prof Exp:* Adj prof, 73-77, assoc prof, 77-81, PROF BIOCHEM, UNIV LAVAL, 81- *Concurrent Pos:* Vis assoc prof molecular biophys & biochem, Yale Univ, 80-81. *Mem:* Am Soc Microbiol; Am Soc Biochem & Molecular Biol; Can Biochem Soc. *Res:* Regulation of the expression of genes encoding aminoacyl-t RNA synthetases in gram-negative and gram-positive bacteria; structure-function studies of bacterial aminoacyl-t RNA systhetases. *Mailing Add:* Biochem Sci Dept Univ Laval Quebec City PQ G1K 7P4

LA POINTE, JOSEPH L, b Harvey, Ill, Sept 7, 34; m 54, 66; c 3. ZOOLOGY. *Educ:* Portland State Col, BA, 60; Univ Calif, Berkeley, PhD(zool), 66. *Prof Exp:* Assoc instr zool, Univ Calif, Berkeley, 64-66; Nat Inst Child Health & Human Develop fel, 66-68; asst prof, 68-71, assoc prof biol, 71-80, PROF ENDOCRINOL, NMEX STATE UNIV, 80- *Concurrent Pos:* NIH res grant, 72-73. *Mem:* Am Soc Ichthyol & Herpet; Am Soc Zool; Brit Soc Endocrinol; Europ Soc Comp Endocrinol. *Res:* Effect of perietal eye on circadian rhythms in lizards; thermoregulation in antusiid lizards; ultrastructure of reptilian pituitary; physiology of neurohypophysial hormones in lower vertebrates; fat mobilization in lizards. *Mailing Add:* Dept Biol NMex State Univ Las Cruces NM 88003

LAPOINTE, LEONARD LYELL, b Iron Mountain, Mich, June 28, 39; m 63; c 2. SPEECH PATHOLOGY. *Educ:* Mich State Univ, BA, 61; Univ Colo, MA, 66, PhD(speech path), 69. *Prof Exp:* Dir speech path commun dis, Bd Educ, Menasha, Wis, 61-64; speech pathologist, Gen Rose Mem Hosp, 66; asst prof phonetics, Univ Colo, Denver, 68-69; coordr & instr audiol & speech path, Vet Admin Med Ctr, Gainesville, 69-84, res investr speech sci, 71-84; CHMN, DEPT SPEECH & HEARING SCI, ARIZ STATE UNIV, TEMPE, 84- *Concurrent Pos:* Fel neurogenic commun dis, Vet Admin Hosp, Denver, 68-69; adj prof commun dis, Univ Fla, 69-, mem res fac neuroling, Ctr Neurol-Behav Ling Res, 74-; consult, Vet Admin Med Ctr, Phoenix & Vet Admin Outpatient Clin, Los Angeles. *Honors & Awards:* Award, Sci Exhib, XV World Cong Logopedics & Phoniatrics, 71. *Mem:* Acad Aphasia; fel Am Speech-Lang-Hearing Asn; Int Asn Logopedics & Phoniatrics; Int Neuropsychol Soc; Am Cleft Palate Asn. *Res:* Development of measurement strategies of human oral sensation-perception; oral physiology and neurolinguistics; diagnosis and treatment strategies in aphasia and related neurogenic communication impairments; developing reading tests for aphasia. *Mailing Add:* Chmn Dept Speech & Hearing Sci Ariz State Univ Tempe AZ 85287-0102

LAPONSKY, ALFRED BAER, b Cleveland, Ohio, Nov 24, 21; m 57; c 2. PHYSICAL ELECTRONICS. *Educ:* Lehigh Univ, BS, 43, MS, 47, PhD(physics), 51. *Prof Exp:* Instr physics, Lehigh Univ, 47-51; physicist electron physics, Res Lab, Gen Elec Co, 51-63; assoc prof elec eng, Univ Minn, Minneapolis, 63-66; FEL ENGR, INDUST & GOVT TUBE DIV, WESTINGHOUSE ELEC CORP, 66- *Mem:* Am Phys Soc; Sigma Xi. *Res:* Electron physics; electro-optics; image sensing and display techniques. *Mailing Add:* 176 Greenridge Dr Horseheads NY 14845

LAPORTE, LÉO FRÉDÉRIC, b Englewood, NJ, July 30, 33; m 56, 85; c 3. GEOLOGY. *Educ:* Columbia Univ, AB, 56, PhD(geol), 60. *Prof Exp:* From instr to prof geol, Brown Univ, 59-71; PROF EARTH SCI, UNIV CALIF, SANTA CRUZ, 71- *Concurrent Pos:* Co ed, Palaios, 86-89. *Mem:* Soc Vert Paleont; Geol Soc Am; Soc Sedimentary Geol; Hist Sci Soc; Hist Earth Sci Soc; AAAS. *Res:* Paleoecology and environmental stratigraphy; history and evolution of life; history of paleontology. *Mailing Add:* Earth Sci Univ Calif Santa Cruz CA 95064

LAPORTE, RONALD E, b Buffalo, NY, May 29, 49; m 71. EPIDEMIOLOGY. *Educ:* Univ Buffalo, BA, 71; Univ Pittsburgh, MS, 73, PhD(psychol), 76. *Prof Exp:* EPIDEMIOLOGIST, DEPT EPIDEMIOL, UNIV PITTSBURGH, 78- *Concurrent Pos:* Fel epidemiol, Univ Pittsburgh, 76- *Mem:* AAAS; Soc Epidemiol Res; Am Psychol Asn. *Res:* Chronic disease epidemiology; investigating possible protective factors of coronary heart disease and diabetes epidemeology. *Mailing Add:* Sch Pub Health 111 Ph Bldg Univ Pittsburgh Pittsburgh PA 15260

LAPOSA, JOSEPH DAVID, b St Louis, Mo, July 21, 38; m 68; c 3. PHYSICAL CHEMISTRY. *Educ:* St Louis Univ, BS, 60; Univ Chicago, MS, 62; Loyola Univ, Ill, PhD(chem), 65. *Prof Exp:* NIH fel chem, Cornell Univ, 65-67; from asst prof to assoc prof, 67-86, PROF CHEM, MCMASTER UNIV, 86- *Res:* Molecular luminescence. *Mailing Add:* Dept Chem McMaster Univ 1280 Main St W Hamilton ON L8S 4L8 Can

LAPOSATA, MICHAEL, b Johnstown, Pa, Apr 22, 52. EXPERIMENTAL BIOLOGY. *Educ:* Bucknell Univ, BS, 74; Johns Hopkins Univ, MD, 81, PhD(cellular & molecular biol), 82; Am Bd Path, dipl, 89. *Prof Exp:* Postdoctoral res fel, Div Hemat-Oncol, Dept Med, Sch Med Wash Univ, St Louis, Mo, 81-82, resident, Div Lab Med, Depts Path & Med, 83-84, chief resident, 84-85; asst dir, Hemostasis Lab, Hosp Univ Pa, 85-86, co-dir, 86-89; asst prof path & lab med, Med Sch Univ Pa, 85-89; DIR CLIN LABS & CHIEF, DIV CLIN LABS, MASS GEN HOSP, 89-; ASSOC PROF PATH, HARVARD MED SCH, 89- *Concurrent Pos:* Sheryl N Hirsch Award, Lupus Found, 87-88; assoc physician Lab Med, Dept Med & assoc pathologist, Dept Path, Mass Gen Hosp, 89-; mem, Coun Thrombosis, Am Heart Asn. *Mem:* Am Asn Clin Res; Am Asn Pathologists; Acad Clin Lab Physicians & Scientists; Am Heart Asn; NY Acad Sci; Am Soc Clin Pathologists. *Res:* Cellular and molecular biology; pathology; hematology, coagulation and blood transfusion. *Mailing Add:* Dir Clin Labs Mass Gen Hosp 249 Gray Bldg Boston MA 02114

LAPOSTOLLE, PIERRE MARCEL, b Vanves, France, May 29, 22; m 47; c 5. DYNAMICS OF PARTICLES IN ELECTROMAGNETIC FIELDS, SPACE CHARGE PHENOMENA. *Educ:* Paris Univ, PhD(traveling wave tube theory), 47. *Prof Exp:* Engr, Nat Ctr Telecommun, Paris, 45-54, sci dir, 72-78; physicist, Europ Orgn Nuclear Res, Geneva, 54-71; sci adv, Ganil Caen, France, 78-85; CONSULT, 85- *Concurrent Pos:* Consult, Europ Orgn Nuclear Res & Los Alamos Nat Lab, 85- *Mem:* Fel Inst Elec & Electronics Engrs. *Res:* Theory of the interaction between an electromagnetic wave and a beam of particles; amplification in a traveling wave tube; acceleration of particles in a linear accelerator. *Mailing Add:* Three Rue Victor Daix Neuilly Sur Seine 92200 France

LAPP, H(ERBERT) M(ELBOURNE), b Alameda, Sask, Feb 2, 22; m 50; c 3. AGRICULTURAL ENGINEERING. *Educ:* Univ Sask, BE, 49; Univ Minn, MS, 62. *Prof Exp:* Water develop, Prairie Farm Rehab Admin, Dominion Govt Can, 49-51; exten engr, Man Dept Agr, 51-53; from asst prof to assoc prof, Univ Man, 53-70, head dept, 58-67, prof agr eng, 70-; RETIRED. *Concurrent Pos:* Colombo plan adv, Khon Kaen Univ, Thailand, 65-67; local prog head develop grad studies in agr eng for Latin Am region, Nat Agrarian Univ, Peru, 67-70; consult, Can Int Develop Agency, Nigeria, 73 & 76, F F Sloney Co, Consult Engrs, Vancouver, Honduras, 74, Int Bank Reconstruct & Develop, Philippines, 76, Pakistan, 77, Brazil Agr Res Dept, 81, US Agency for Int Develop, Peru, 82, Can Int Develop Agency, 82, Int Develop Res Ctr, India, 83 & Lavalin-Crippen Int, Honduras, 83. *Mem:* Fel Can Soc Agr Eng (pres, 75-76); Am Soc Agr Engrs; Agr Inst Can. *Res:* Farm structure; soil and water. *Mailing Add:* 592 Borebank St Winnipeg MB R3N 1E9 Can

LAPP, M(ARSHALL), b Buffalo, NY, Aug 20, 32; m 58, 80; c 2. LASER GAS DIAGNOSTICS, LIGHT SCATTERING. *Educ:* Cornell Univ, BEngPhys, 55; Calif Inst Technol, PhD(eng sci), 60. *Prof Exp:* Physicist, Corp Res & Develop, Gen Elec Co, 60-80, actg mgr, Combustion Inst, 81-; SANDIA NAT LAB. *Concurrent Pos:* Sci Res coun sr vis fel, Sch Physics, Univ Newcastle, 68-69. *Mem:* AAAS; fel Am Phys Soc; fel Optical Soc Am; Am Inst Aeronaut & Astronaut; fel Brit Inst Physics. *Res:* Optical diagnostics of flames; laser Raman spectroscopy; radiative properties of metal vapors and gases; optical diagnostics of gases and surfaces; atomic and molecular physics; physics of fluids. *Mailing Add:* Sandia Nat Labs Org 8300 A Livermore CA 94551-0969

LAPP, MARTIN STANLEY, b Toronto, Ont. NUMERICAL TAXONOMY, COMPUTER SIMULATION. *Educ:* York Univ, BSc, 69, MSc, 72; Univ Alta, PhD(plant path), 77. *Prof Exp:* Res assoc, 77-80, RES OFFICER NUMERICAL TAXON, NAT RES COUN CAN, 80- *Mem:* Can Phytopath Soc; Can Bot Asn. *Res:* Numerical taxonomy of conifers using the leaf-oil terpene patterns; preinfection phase of the life cycle of leaf pathogens. *Mailing Add:* 763 Wilkinson Way Saskatoon SK S7N 3E4 Can

LAPP, N LEROY, MEDICINE. *Prof Exp:* PROF MED, WVA SCH MED, 78- *Mailing Add:* Med Dept WVa Univ Med Ctr Morgantown WV 26506

LAPP, NEIL ARDEN, b Bloomington, Ill, Dec 7, 42; m 67; c 2. PLANT PATHOLOGY, NEMATOLOGY. *Educ:* Goshen Col, BA, 64; WVa Univ, MS, 67; NC State Univ, PhD(plant path), 70. *Prof Exp:* plant pathologist, Plant Protection Div, NC Dept Agr, 71-81; FIELD DEVELOP FEL, MERCK & CO, 81- *Concurrent Pos:* Adj asst prof plant path, NC State Univ, 71-77, adj assoc prof, 77- *Mem:* Am Phytopath Soc; Soc Nematologists. *Res:* Plant disease and nematode survey and detection; chemical control. *Mailing Add:* 7208 Madiera Ct Raleigh NC 27615

LAPP, P(HILIP) A(LEXANDER), b Toronto, Ont, May 12, 28; m 52; c 3. REMOTE SENSING. *Educ:* Univ Toronto, BASc, 50; Mass Inst Technol, SM, 51, ScD(instrumentation), 54. *Prof Exp:* Instr aeronaut eng, Mass Inst Technol, 52-53, res assoc, 53-54; systs eng, De Havilland Aircraft Can Ltd, 54-56, proj engr, Guided Missile Div, 55-60, chief engr, Spec Prod & Appl Res Div, 60-65, dir tech opers, 65-68; sr vpres, Spar Aerospace Prod Ltd, 68-69; PRES, PHILIP A LAPP LTD, 69- *Concurrent Pos:* Chmn working group sensors, Can Ctr Remote Sensing, 70-76; mem, Can Accreditation Bd, 71-75; mem, Bd Govs, York Univ, 80-; fel, Ryerson Polytech Inst, 85. *Mem:* Am Inst Aeronaut & Astronaut; Inst Elec & Electronics Engrs; Can Res Mgt Asn. *Res:* Dynamics of vehicles; guidance and control of missiles and aircraft; military and industrial instrumentation and automatic control; educational planning and research on public policy; remote sensing applied to resource management and environmental monitoring. *Mailing Add:* Philip A Lapp Ltd 14A Hazelton Toronto ON M5R 2E2 Can

LAPP, THOMAS WILLIAM, b Joliet, Ill, Oct 6, 37; m 61. ENVIRONMENTAL CHEMISTRY, RISK ASSESSMENT. *Educ:* Coe Col, BA, 59; Kans State Univ, MS, 61, PhD(inorg chem), 63. *Prof Exp:* Fel radiation chem, NAm Aviation Sci Ctr, 63-64; asst prof nuclear chem, Univ WVa, 64-66; assoc chemist, Midwest Res Inst, Mo, 66-69; mem staff, Univ Mo-Kansas City, 70-74; assoc chemist, Midwest Res Inst, 74-77, sr chemist, 77-87, PRIN ENVIRON SCIENTIST, 87- *Res:* Production and utilization of industrial chemicals suspected of possessing toxic properties; environmental transport and fale; risk assessment; hazardous waste incineration studies. *Mailing Add:* Midwest Res Inst 401 Harrison Oak Blvd Suite 350 Cary NC 27513

LAPP, WAYNE STANLEY, b Stevensville, Ont, Can, Nov 11, 36; m 64; c 4. TRANSPLANTATION BIOLOGY. *Educ:* Univ Toronto, BSA, 62, MSA, 64; McGill Univ, Montreal, PhD(transplantation physiol), 67. *Prof Exp:* Fel transplantation immunol, Karolinska Inst, Stokholm, 67-68; from asst prof to assoc prof, 68-82, PROF PHYSIOL IMMUNOL, DEPT PHYSIOL, MCGILL UNIV, 82- *Concurrent Pos:* Guest prof immunol, German Cancer Rec Ctr, Heidelberg, 77-78; assoc mem, dept med, McGill Univ, Montreal, 80-, prof, Centre Clin Immunol Transplantation Biol; counr, Can Soc Immunol. *Mem:* Can Physiol Soc; Can Soc Immunol; Am Asn Immunol; Transplantation Soc; NY Acad Sci. *Res:* Graft-versus-host induced immunosuppression of T and B lymphocyte functions; effect of the GVH reaction on T and B cell ontogency and thymus function. *Mailing Add:* Dept Physiol McGill Univ 3655 Drummond St Montreal PQ H3G 1Y6 Can

LAPPAS, LEWIS CHRISTOPHER, b Lynn, Mass, May 14, 21; m 49; c 3. PHARMACEUTICAL CHEMISTRY. *Educ:* Mass Col Pharm, BS, 43, MS, 48; Purdue Univ, PhD(pharmaceut chem), 51. *Prof Exp:* Res scientist, Eli Lilly & Co, 51-84; CONSULT, 84- *Mem:* Am Chem Soc; Am Pharmaceut Asn. *Res:* Drug encapsulation processes; basic gelatin research as applied to capsular forms; study pf filmogens as drug release mechanisms; stabilization of drugs and drug forms; pharmaceutical aspects of drug absorption; investigation of new antimicrobials in drug and cosmetic formulations. *Mailing Add:* 12240 Brompton Rd Carmel IN 46032

LAPPE, RODNEY WILSON, b Breese, Ill, Sept 12, 54; c 2. CARDIOVASCULAR. *Educ:* Blackburn Col, BA, 76; Ind Univ, PhD(pharmacol), 80. *Prof Exp:* Postdoctoral pharmacol, Univ Iowa, 80-82; res scientist, Hypertension Sect, Wyeth Labs, Inc, 82-85, mgr, 85-87, mgr, Vascular Dis Sect, Wyeth-Ayerst Res, 87-88; res fel, Hypertension Sect, Rorer Cent Res, 88-90; DIR CARDIOVASC RES, CIBA-GEIGY PHARMACEUT DIV, 90- *Concurrent Pos:* Res fel, Iowa Cardiovasc Ctr Inst, 82; chmn, Cardiovasc Subcomt, 87 & 88; mem, Animal Use Comt, Wyeth-Ayerst Res, 88. *Mem:* Am Soc Hypertension; Am Soc Pharmacol & Exp Therapeut; Inter-Am Soc Chemother; Coun High Blood Pressure. *Res:* Cardiovascular system. *Mailing Add:* Cardiovasc Res Ciba-Geigy Pharmaceut Div 556 Morris Ave Summit NJ 07901

LAPPIN, GERALD R, b Caro, Mich, Apr 14, 19; m 45; c 2. CHEMISTRY. *Educ:* Alma Col, BS, 41; Northwestern Univ, PhD(org chem), 46. *Prof Exp:* Asst, Northwestern Univ, 41-43, interim instr chem, 44, asst, 44-46; asst prof, Antioch Col, 46-49 & Univ Ariz, 49-51; sr res chemist, 51-69, RES ASSOC, TENN EASTMAN CO DIV, EASTMAN KODAK CO, 69- *Concurrent Pos:* Consult, Vernay Labs, Ohio, 46-49. *Mem:* Am Chem Soc. *Res:* Additives for foods; plastics and petroleum products; chemistry of polyesters; technology forecasting as applied to research and development planning. *Mailing Add:* 4047 Skyland Dr Kingsport TN 37664

LAPPLE, CHARLES E, b New York, NY, Feb 11, 16; m 41; c 2. CHEMICAL ENGINEERING, FLUID & PARTICLE MECHANICS. *Educ:* Columbia Univ, BS, 36, ChE, 37. *Prof Exp:* Chem engr, E I du Pont de Nemours & Co, 37-45, process engr, 45-46, chem engr, 46-48, res proj eng, 48-50; assoc prof chem eng, Ohio State Univ, 50-55; sr scientist, SRI Int, 55-79; CONSULT, 79- *Concurrent Pos:* Lectr, Columbia Univ, 41-48 & Univ Del, 48-50; indust consult, 50-; consult, Atomic Energy Comn, 50-58, USPHS, 61-66 & US Dept Interior, 66- *Honors & Awards:* Colburn Award, Am Inst Chem Engrs, 46. *Mem:* Am Chem Soc; Am Inst Chem Engrs. *Res:* Fluid mechanics; heat transfer; dust and mist collection; particle dynamics; aerosols; atmospheric pollution abatement; fine particle technology. *Mailing Add:* 260 Calle Linda Fallbrook CA 92028-9425

LAPPLE, WALTER C(HRISTIAN), b Mt Vernon, NY, Apr 8, 21; m 45, 90; c 2. CHEMICAL ENGINEERING. *Educ:* Cooper Union, BS, 43; Univ Kans, MS, 57. *Prof Exp:* Indust engr, E I du Pont de Nemours & Co, 43-46; res chem engr, Varcum Chem Corp, 46-47; chem engr, Dorr Co, 47-51; sr chem engr, Midwest Res Inst, 51-52, head process sect, Chem Eng Dept, 52-54, asst mgr, 54-57; sr chem engr, Oliver Iron Mining Div, US Steel Corp, 57-59 & FMC Corp, 60-65; sr chem engr, Babcock & Wilcox Co, 65-73, res specialist, 73-83; RETIRED. *Res:* Chemical processing; metallurgical operations; industrial waste abatement; fluidized bed technology, fundamentals, drying and high temperature reactions. *Mailing Add:* 1407 NE 11th Terr Kansas City MO 64155-1481

LAPPORTE, SEYMOUR JEROME, b Chicago, Ill, Mar 26, 30; m 64; c 2. ORGANIC CHEMISTRY, PETROLEUM CHEMISTRY. *Educ:* Univ Chicago, MS, 53; Univ Calif, Los Angeles, PhD(chem), 57. *Prof Exp:* Asst, Univ Calif, Los Angeles, 53-56; sr res assoc, Chevron Res Co, 56-74, mgr, Pioneering Div, 74-86, sr res scientist, Pioneering Div, Chevron Res Co, 80-86; PROG OFFICER, NSF, 87- *Concurrent Pos:* Teacher, Exten, Univ Calif, 60-; vis scholar, Stanford Univ, 68-69. *Mem:* Am Chem Soc; Royal Soc Chem. *Res:* Organic reaction mechanisms; organometallics; oxidation; transition metal chemistry; homogenous catalysis; ultraviolet stabilization. *Mailing Add:* NSF 1800 G St NW Rm 340 Washington DC 20550

LAPRADE, MARY HODGE, b Oakland, Calif, Feb 6, 29; m 58; c 2. ZOOLOGY. *Educ:* Wilson Col, AB, 51; Radcliffe Col, AM, 52, PhD(biol), 58. *Prof Exp:* Instr biol, Simmons Col, 52-55; instr zool, 58-60 & 64-65, LECTR BIOL SCI, SMITH COL, 65-, DIR, CLARK SCI CTR, 73- *Concurrent Pos:* Instr, NSF In Serv Inst High Sch Biol Teachers, 65-66. *Mem:* Sigma Xi. *Res:* Growth and regeneration, particularly in crustaceans; fine structure of endocrine organs in crustaceans. *Mailing Add:* 15 Dragon Circle Easthampton MA 01027

LAPSLEY, ALWYN COWLES, b Albemarle Co, Va, Mar 12, 20. PHYSICS. *Educ:* Univ Va, BEE, 41, MS, 44, PhD(physics), 47. *Prof Exp:* Instr physics, Univ Va, 41-43, asst, Manhattan Proj & Navy Fire Control, 43-46; sr physicist, Photo Prod Dept, E I du Pont de Nemours & Co, 47-51, res engr, Atomic Energy Div, 51-60; lectr nuclear eng, Reactor Facil, Sch Eng & Appl Sci, Univ Va, 60-77, sr scientist, 60-85, res assoc prof, 77-85; RETIRED. *Mem:* Am Phys Soc; Am Nuclear Soc. *Res:* Mechanics; physical optics; Kerr effect with high frequency fields; nuclear physics; ion chambers; reactor kinetics; isotope separation. *Mailing Add:* 1609 Inglewood Dr Charlottesville VA 22901

LAPUCK, JACK LESTER, b Jamaica Plain, Mass, Aug 28, 24; m 48; c 3. FOOD CHEMISTRY, BACTERIOLOGY. *Educ:* Northeastern Univ, BS, 46; Univ Mass, MS, 49; Calvin Coolidge Col, Dsc, 60. *Prof Exp:* Food sanitarian, Montgomery County Health Dept, Md, 50-51; food chemist, Food & Drug Res Labs, NY, 51; chemist, Waltham Labs, Inc, 51-55, lab dir & vpres, 55-66; OWNER & MGR, LAPUCK LABS, 66- *Concurrent Pos:* Instr, Univ Exten, Mass Dept Educ, 55-; instr, Boston State Col. *Mem:* AAAS; Am Chem Soc; Am Soc Microbiol; Inst Food Technologists; NY Acad Sci; Nat Environ Health Asn. *Res:* Food technology; microbiology; analytical chemistry. *Mailing Add:* 50 Hunt St Watertown MA 02172

LAQUATRA, IDAMARIE, NUTRITION. *Educ:* Pa State Univ, BS, 75, MS, 79, PhD(appl nutrit), 83. *Prof Exp:* Clin dietitian, Custom Mgt Corp, 76-77; instr & consult, Pa State Univ, University Park, 78-82; pvt nutrit consult, State Col, Pa, 80-82; consult, Clin Nutrit Staff, Vet Admin Med Ctr, Bronx, NY, 83-84; nutritionist, Heinz USA, 84-89, MGR NUTRIT SERV, WEIGHT WATCHERS DIV, HEINZ USA, PITTSBURGH, PA, 90- *Concurrent Pos:* Grad asst, Pa State Univ, 77-81, lab asst, 78 & 79; postdoctoral fel, Prev Cardiol Prog, Univ Med & Dent NJ, 82-84; adj fac, Home Econ Dept, Montclair State Col, NJ, 83-84; mem, Col Bd Adv, Col Home Econ, Ohio State Univ, 86-89, adv group, Plan V Dietetics Prog, Pa State Univ, 88- & Coun Res, Am Dietetic Asn, 89- *Mailing Add:* Weight Watchers Tech Div Nutrit Serv Heinz USA PO Box 57 Pittsburgh PA 15230

LAQUER, HENRY L, b Frankfurt-am-Main, Ger, Nov 28, 19; nat US; m 47; c 4. CRYOGENICS, SUPERCONDUCTIVITY. *Educ:* Temple Univ, AB, 43; Princeton Univ, MA, 45, PhD(phys chem), 47. *Prof Exp:* Res chemist, Ladox Labs, Pa, 46; mem staff, Los Alamos Sci Lab, Univ Calif, 47-77; consult, 77-83; PRIN, CRYOPOWER ASSOCS, 84- *Concurrent Pos:* Adj prof, Los Alamos Residence Ctr, Univ NMex, 70-73. *Mem:* Am Phys Soc. *Res:* High magnetic fields; applied superconductivity; dielectric studies; elastic properties of metals; cryogenics; high temperature superconductivity materials and applications. *Mailing Add:* Rte 1 Box 445 Espanola NM 87532

LARABELL, CAROLYN A, b Detroit, Mich, Dec 16, 47. BIOCHEMISTRY, BIOPHYSICS. *Educ:* Mich State Univ, BA, 70; Ariz State Univ, BS, 81, PhD(zool), 88. *Prof Exp:* Supvr, Dept Neurophysiol, Good Samaritan Hosp, Phoenix, 71-75; teaching asst, Dept Zool, Ariz State Univ, 83, res assoc, 83-88; postdoctoral scholar, Dept Biochem & Biophys & Dept Zool, Davis, 88-90, MGR, WEST COAST FACIL INTERMEDIATE VOLTAGE ELECTRON MICROS, LAWRENCE BERKELEY LAB, UNIV CALIF, BERKELEY. *Concurrent Pos:* NIH reprod biol training grant, Univ Calif, Davis, 89. *Mem:* Am Soc Cell Biol; Soc Develop Biol; AAAS; Electron Micros Soc Am. *Res:* Modification of the egg extracellular matrix at fertilization; mechanisms of egg activation, including the role of inositol lipid hydrolysis and generation and propagation of the calcium wave; egg cytoskeleton and its modification during fertilization and development. *Mailing Add:* 160A Donner Lab Lawrence Berkeley Lab Univ Calif Berkeley CA 94720

LARA-BRAUD, CAROLYN WEATHERSBEE, b Waco, Tex, Jan 4, 40; m 70. BIOCHEMISTRY. *Educ:* Univ Tex, Austin, BA, 62, PhD(chem), 69. *Prof Exp:* Res assoc biochem, Clayton Found Biochem Inst & lectr home econ, Nutrit Div, Univ Tex, Austin, 71-73; asst res scientist biochem 73-75, asst prof, 75-80, ASSOC PROF HOME ECON, UNIV IOWA, 80- *Mem:* AAAS; Am Chem Soc; Sigma Xi, Inst Food Technol; Am Home Econ Asn. *Res:* Intermediary metabolism; regulation of inducible enzyme systems. *Mailing Add:* Dept Home Econ 38MH Univ Iowa Iowa City IA 52242-1371

LARACH, SIMON, b Brooklyn, NY, Apr 21, 22; m 48; c 2. PHYSICAL INORGANIC CHEMISTRY, SOLID STATE CHEMISTRY. *Educ:* City Col New York, BS, 43; Princeton Univ, MA, 51, PhD(chem), 54. *Prof Exp:* Res chemist, Third Res Div, Goldwater Hosp, Col Med, NY Univ, 43 & 46; res chemist luminescence & solid state, David Sarnoff Res Ctr, Radio Corp Am, 46-59, head photoelectronic, magnetic & dielec res, 59-61, assoc lab dir, 61-67, overseas fel, 69, fel, 67-87; PRES, DEVTECH INC, 87- *Concurrent Pos:* Vis fel, Princeton Univ, 67-68; Indust Res Inst-Am Chem Soc liaison lectr, 68; vis prof, Hebrew Univ Jerusalem, 69-70, adj prof, 71-73; UN consult, UNESCO Div Tech Educ & Res, 71; prin lectr, NATO Adv Study Inst, Norway, 72; vis prof, Swiss Fed Inst Technol, 72 & Princeton Univ, 73; div ed electronics, J Electrochem Soc; res prof radiol, Hahnemann Med Col & Hosp, 74-; adj prof radiol, Col Physicians & Surgeons, Columbia Univ, 78-86; invited lectr, People's Repub China, 87 & Taiwan, 87. *Mem:* Am Chem Soc; fel Am Phys Soc; fel Am Inst Chem; Electrochem Soc. *Res:* Synthesis and properties of electronically-active solids; medical ultrasound. *Mailing Add:* One Windsor Rd Great Neck NY 11021-3920

LARAGH, JOHN HENRY, b Yonkers, NY, Nov 18, 24; m 74; c 3. PHYSIOLOGY, MEDICINE. *Educ:* Cornell Univ, MD, 48. *Prof Exp:* Intern med, Presby Hosp, New York, 48-49, asst resident, 49-50, asst, 50-55, instr, 55-57, assoc, 57-59, from asst prof to prof clin med, Col Physicians & Surgeons, Columbia Univ, 67-75; HILDA ALTSCHUL MASTER PROF MED, MED COL, CORNELL UNIV, 75-, DIR CARDIOVASC CTR, NY HOSP-CORNELL MED CTR, 75- *Concurrent Pos:* Nat Heart Inst trainee, 50-51; asst physician, Presby Hosp, New York, 50-54, from asst attend physician to assoc attend physician, 54-69, attend physician, 69-75, vchmn in chg med affairs, Bd Trustees, 74; NY Heart Asn res fel, 51-52; mem med adv bd, Coun High Blood Pressure Res, Am Heart Asn, 61, chmn, 68-72; consult cardiovasc study sect, USPHS, 64-68 & heart prog proj A, 67-72; dir, Hypertension Ctr & Nephrol Div, Columbia-Presby Med Ctr, 71-75; mem policy adv bd, Hypertension Detection & Follow-Up Prog, Nat Heart & Lung Inst, 71-, mem bd sci coun, 74-; mem adv bd, Am Soc Contemporary Med & Surg, 74. *Honors & Awards:* Stouffer Prize Med Res, 69. *Mem:* Am Soc Clin Invest; fel Am Col Physicians; Am Soc Nephrol; assoc Harvey Soc; Asn Am Physicians. *Res:* Cardiovascular and renal diseases; endocrinology. *Mailing Add:* NY Hosp 525 E 68th St New York NY 10021

LARAMORE, GEORGE ERNEST, b Ottawa, Ill, Nov 5, 43; m 64; c 1. PHYSICS, MEDICINE. *Educ:* Purdue Univ, BS, 65; Univ Ill, Urbana, MS, 66, PhD(physics), 69; Univ Miami, MD, 76. *Prof Exp:* NSF fel, Univ Ill, Urbana, 69-70, res assoc physics, 70-71; res physicist, Sandia Labs, 71-75; ASST PROF, DEPT RADIATION ONCOL, UNIV WASH, 78- *Mem:* Am Phys Soc; Am Vacuum Soc; AMA. *Res:* Theory of low-energy electron diffraction; fast neutron radiotherapy for human malignancies; interaction of fast electrons with solids. *Mailing Add:* Dept Radiation Oncol Univ Wash Med Ctr 1959 NE Pacific St MS RC-08 Seattle WA 98195

L'ARCHEVÊQUE, REAL VIATEUR, b Montreal, Que. NUCLEAR ENGINEERING, ELECTRONICS. *Educ:* Polytech Inst, Montreal, BSc, 60; Univ London, PhD(electronics), 64. *Prof Exp:* Head, Electronics Br, Atomic Energy of Can Ltd, 65-77; asst to pres, 77-80, PRES, CANATOM, 80- *Res:* Hybrid microelectronics; thick film microcircuits; nuclear instruments development and design; on-line computer data acquisition systems development. *Mailing Add:* SNC Inc Two Place Felix Martin Montreal PQ H2Z 1Z3 Can

LARCOM, LYNDON LYLE, b Olean, NY, Apr 11, 40. BIOPHYSICS, PHOTOBIOLOGY. *Educ:* Carnegie-Mellon Univ, BS, 62; Univ Pittsburgh, MS, 65, PhD(biophysics), 68. *Prof Exp:* NIH fel chem, Univ Pittsburgh, 68-70, res assoc, 70-72; CONSULT, 72- *Mem:* Biophys Soc; Am Soc Photobiol; Am Soc Microbiol; Am Chem Soc; Sigma Xi. *Res:* Mechanisms of DNA damage and repair; the biophysical properties of nucleic acids; molecular quantum mechanics; DNA-protein interactions; virus structure; mechanisms of carcinogenesis. *Mailing Add:* Dept Physics Clemson Univ Clemson SC 29634-1911

LARD, EDWIN WEBSTER, b Ala, July 17, 21; m 45; c 4. ANALYTICAL CHEMISTRY. *Educ:* Ark State Col, BS, 49; Memphis State Univ, MA, 61. *Prof Exp:* Chemist, Ethyl Corp, 49-52 & Chemstrand Corp, 52-54; sr chemist, Nitrogen Prod Div, W R Grace & Co, Tenn, 54-62, res supvr, Res Div, Clarksville, Md, 62-74; chem engr, Naval Sea Systs Command, Washington, DC, 74-84; RETIRED. *Mem:* Am Chem Soc. *Res:* Trace gas analysis with infrared; separation and determination of argo, oxygen and nitrogen by chromatography; trace analysis of acetylene, methane, carbon monoxide and carbon dioxide; synthesis of aryl dimethyl sulfonium chloride compounds; chemical warfare agents; unsaturates in auto emissions; water energy conservation on naval ships; issued 23 patents and author of 17 publications. *Mailing Add:* 12703 Beaverdale Lane Bowie MD 20715

LARDNER, JAMES F, b Davenport, Iowa, May 24, 24. COMPUTER DESIGN, SCIENCE ADMINISTRATION. *Educ:* Cornell Univ, BME, 45. *Prof Exp:* Eng & mfg mgr, Deere & Co, 46-56, mgr, Overseas Mfg Group, 56-62, managing dir, John Deer Iberica SA, 62-67, asst gen mgr, Des Moines Works, 67-68, Harvester Works, 68-69, mgr, Corp Plant & Prod Eng Dept, 69-70, corp dir mfg eng, 70-80, vpres mfg develop, 80-82, vpres, Govt Prod & Component Sales, 82-85, vpres, Component Group, 85-90; dir, Am Standard, 84-90; DIR, POTASH CORP, SASK, 89- *Concurrent Pos:* Mem, Comt Computer-Aided Mfg, Nat Res Coun, 80-81, Panel Computers in Design & Mfg, 83, Comt Indust/Acad Coop in Mfg, 84- 85, Mfg Studies Bd, 84-, Comn Eng & Tech Systs, 84-86, Cross- Disciplinary Eng Res Comt, 85-87, Comt Labor Mkt Adjustments, 87, Comt Defense Mfg Strategy, Panel Pvt Contractors, 90-, chmn, Mfg Studies Bd, 90-; dir, Computer Aided Mfg Int Inc, 81-84; mem, Panel Eng Res Ctrs, Nat Acad Eng, 84, Foundations Mfg Comt, 89-; mem, Panel Mfg Eng, Bd of Assessment Nat Inst Standards & Technol, 86-91. *Mem:* Nat Acad Eng. *Mailing Add:* 2752 Nichols Lane Davenport IA 52803

LARDNER, ROBIN WILLMOTT, b Leicester, Eng, Feb 9, 38; m 58, 79; c 4. APPLIED MATHEMATICS, MECHANICS. *Educ:* Cambridge Univ, BA, 59, PhD(appl math), 63. *Hon Degrees:* ScD, Cambridge Univ, 87. *Prof Exp:* Res assoc physics, Columbia Univ, 61-63; NATO fel appl math & theoret physics, Peterhouse Col, Cambridge Univ, 63-65; lectr math & physics, Univ E Anglia, 65-67; assoc prof, 67-70, chmn dept, 71-73, PROF MATH, SIMON FRASER UNIV, 70- *Concurrent Pos:* Prof appl math, Univ Petrol & Minerals, Dhahran, Saudi Arabia, 82-87. *Res:* Solid mechanics, particularly dislocation theory and fracture; nonlinear vibrations of continuous media; asymptotic solutions of nonlinear partial differential equations; numerical hydrodynamical models; numerical solution of partial differential equations. *Mailing Add:* Dept Math Simon Fraser Univ Burnaby BC V5A 1S6 Can

LARDNER, THOMAS JOSEPH, b New York, NY, July 19, 38; m 64; c 3. ENGINEERING MECHANICS. *Educ:* Polytech Inst Brooklyn, BAeroE, 58, MS, 59, PhD(appl mech), 61. *Prof Exp:* Res assoc appl mech, Polytech Inst Brooklyn, 59-61; res engr, Jet Propulsion Lab, Calif Inst Technol, 62-63; instr math, Mass Inst Technol, 63-67, asst prof appl math, 67-70, assoc prof mech eng, 70-73; prof theoret & appl mech, Univ Ill, Urbana, 73-78; PROF CIVIL ENG, UNIV MASS, 78- *Concurrent Pos:* Fulbright lectr, Univ Nepal, 65-66; consult, Polytech Inst Brooklyn & Jet Propulsion Lab, Calif Inst Technol. *Mem:* Fel Am Soc Mech Engrs; Soc Indust & Appl Math. *Res:* Applied mathematics and mechanics; applied solid mechanics. *Mailing Add:* 175 Amity St Amherst MA 01002

LARDY, HENRY ARNOLD, b Roslyn, SDak, Aug 19, 17; m 43; c 4. BIOCHEMISTRY. *Educ:* SDak State Univ, BS, 39; Univ Wis, MS, 41, PhD(biochem), 43. *Hon Degrees:* DSc, SDak State Univ, 78. *Prof Exp:* Fel, Nat Res Coun, Banting Inst, Univ Toronto, 44-45; from asst prof to prof, 45-50, VILAS PROF BIOL SCI, UNIV WIS-MADISON, 66-, CHMN, RES DEPT, ENZYME INST, 50- *Honors & Awards:* Neuberg Medal, 56; Lewis Award, Am Chem Soc, 49; Wolf Found Award Agr, 81; Nat Award Agr Excellence, Agr Mkt Asn, 82; Carl Hartman Award, Soc for Study of Reproduction, 84; Amory Prize, Am Acad Arts & Sci, 84; William Rose Award Biochem, Am Soc Biol Chemists, 88. *Mem:* Nat Acad Sci; Am Philos Soc; Am Chem Soc; Am Soc Biol Chem (pres, 64); Am Acad Arts & Sci. *Res:* Enzymes; intermediary metabolism; hormones. *Mailing Add:* Enzyme Inst Univ Wis-Madison Madison WI 53705

LARDY, LAWRENCE JAMES, b Sentinel Butte, NDak, Aug 23, 34; m 56; c 2. NUMERICAL ANALYSIS. *Educ:* NDak State Col, Dickinson, BS, 57; Univ NDak, MS, 59; Univ Minn, PhD(math), 64. *Prof Exp:* Instr math, Univ NDak, 59-60 & Univ Minn, 62-64; from asst prof to assoc prof, 64-74, dept chmn, 82-88, PROF MATH, SYRACUSE UNIV, 74- *Concurrent Pos:* Res fel, Yale Univ, 67-68; vis assoc prof, Univ Md, 73-74. *Mem:* Soc Indust & Appl Math; Math Asn Am; Am Math Soc; Sigma Xi. *Res:* Functional analysis. *Mailing Add:* 4838 Westfield Dr Manlius NY 13104

LAREW, H(IRAM) GORDON, b Independence, WVa, June 5, 22; m 46; c 3. CIVIL ENGINEERING, SOIL MECHANICS. *Educ:* Univ WVa, BS, 44; Purdue Univ, MS, 51, PhD, 60. *Prof Exp:* Jr engr, NY Cent Syst, 46; instr civil eng, Purdue Univ, 47-56; assoc prof, 56-64, PROF CIVIL ENG, UNIV VA, 64- *Concurrent Pos:* Consult, 54- *Mem:* Fel Am Soc Civil Engrs; Am Soc Eng Educ; Am Arbitration Asn. *Res:* Utilization of solid wastes; the effects of repeated loads upon soils; soil strength; earth dams; blast damage. *Mailing Add:* Dept Civil Eng Thronton Hall Univ Va Charlottesville VA 22903

LARGE, ALFRED MCKEE, b Listowel, Ont, Mar 7, 12; nat US; m 43; c 1. SURGERY. *Educ:* Univ Toronto, BA, 33, MD, 36. *Prof Exp:* Instr surg, Sch Med, Washington Univ, 43-44; from asst prof to assoc prof clin surg, Col Med, Wayne State Univ, 46-80; RETIRED. *Concurrent Pos:* Attend surgeon, St John Hosp, Ben Secours Hosp & Cottage Hosp; pvt pract. *Mem:* AMA. *Res:* Clinical surgery. *Mailing Add:* 72 Trollemy Grosse Pointe MI 48230

LARGE, RICHARD L, b Rochester, Ind, June 9, 40; m 62; c 3. SOIL SCIENCE, AGRONOMY. *Educ:* Purdue Univ, BSc, 62; Okla State Univ, MSc, 66; Ohio State Univ, PhD(soil sci), 69. *Prof Exp:* Agronomist, US Testing Co, Inc, 69-71; VPRES, A&L AGR LABS, INC, 71- *Mem:* Soil Sci Soc; Am Soc Agron; Coun Agr Sci & Technol. *Res:* Soil fertility; crop nutrition; land application of sludge. *Mailing Add:* 3610 Huckleberry St Memphis TN 38116

LARGENT, DAVID LEE, b San Francisco, Calif, Oct 30, 37; m 70. MYCOLOGY. *Educ:* San Francisco State Col, BA, 60, MA, 63; Univ Wash, PhD(bot), 68. *Prof Exp:* Instr bot, Foothills Jr Col, 63; instr bot & biol, Phoenix Jr Col, 63-64; asst prof bot, 68-74, assoc prof, 74-77, PROF BOT, HUMBOLDT STATE COL, 77- *Mem:* Am Soc Plant Taxon; Mycol Soc Am; Am Bryol & Lichenological Soc; Sigma Xi. *Res:* Taxonomy and ecology of the Rhodophylloid fungi on the Pacific coastal states of America; cryptogamic botany. *Mailing Add:* 141 Carter Lane Eureka CA 95501

LARGENT, MAX DALE, b Winchester, Va, Feb 28, 23; m 54; c 2. DENTISTRY. *Educ:* Med Col Va, DDS, 50. *Prof Exp:* From instr to prof, pedodontics, Med Col Va, 52-72; prof & assoc dean, Baylor Col Dent, 72-90; RETIRED. *Concurrent Pos:* Chmn dept pedodontics, Med Col Va, 69-72, dir postgrad pedodontics, 57-69. *Mem:* Am Dent Asn; Am Asn Dent Schs; fel Am Col Dent. *Mailing Add:* 9222 Loma Vista Dallas TX 75243

LARGIS, ELWOOD EUGENE, BROWN ADIPOSE TISSUE METABOLISM, HYPOGLYCEMIC DRUGS. *Educ:* Univ NDak, PhD(biochem), 70. *Prof Exp:* RES PHARMACOLOGIST, DEPT METAB DIS, LEDERLE LABS, 73- *Mailing Add:* Dept Metab Dis Am Cyanamid Med Res Div N Middletown Rd Pearl River NY 10965

LARGMAN, COREY, m; c 2. BIOCHEMISTRY. *Educ:* Reed Col, BA, 66; Mass Inst Technol, PhD(org chem), 70. *Prof Exp:* Postdoctoral fel biochem, Univ Calif, Berkeley, 71-74; asst adj prof, 78-82, assoc res prof, 84-89, RES PROF INTERNAL MED & BIOL CHEM, SCH MED, UNIV CALIF, DAVIS, 90-; ASSOC CAREER SCIENTIST, VET ADMIN, 85- *Concurrent Pos:* Asst dir, Enzym Res Lab, Martinez Vet Med Ctr, Calif, 74-82, actg assoc chief staff res, 89, dir, Core Biochem Lab, 82-; vis scientist, Dept Biochem, Univ Calif, San Francisco, 83-84; res prog specialist basic sci, Vet Admin, Washington, DC, 87-90; merit rev bd, 90-; Dept Nutrit Res Training Ctr, Univ Calif, Davis, 90- *Mem:* Am Soc Biol Chemists; AAAS. *Res:* Expression of homeotic genes during differentiation; structure and function of proteolytic enzymes; role of homeobox genes in hematopoiesis; elastase structure and function. *Mailing Add:* Vet Admin Med Ctr 150 Muir Rd Res Rm 108 Martinez CA 94553

LARGMAN, THEODORE, b Philadelphia, Pa, Nov 16, 23; m 59; c 4. ORGANIC CHEMISTRY. *Educ:* Temple Univ, AB, 48; Ind Univ, PhD(org chem), 52. *Prof Exp:* Sr res chemist, Nitrogen Div, Allied Signal Corp, 62-66, scientist, Cent Res Labs, 66-68, res group leader, Corp Chem Res Lab, 51-81, res assoc, 81-88, sr res assoc, 88-89; RETIRED. *Concurrent Pos:* Consult, Triad Enterprises, 90- *Res:* Organic synthesis; fine chemicals; agricultural pesticides; flame retardant chemicals and polymers; uranium extraction; polymer adhesion; bioresorbable polymers; fiber research. *Mailing Add:* Seven Upper Field Rd Morristown NJ 07960

LARI, ROBERT JOSEPH, b Aurora, Ill, 31; m 56; c 4. PHYSICS. *Educ:* St Procopius Col, BS, 53; Univ Notre Dame, MS, 55. *Prof Exp:* Instr physics, St Procopius Col, 57-61; PHYSICIST, ARGONNE NAT LAB, 61- *Mem:* Am Asn Physics Teachers; Asn Comput Mach. *Res:* Particle accelerator magnet design, other electromagnetic devices design. *Mailing Add:* 800 W Marywood Aurora IL 60504

LARIMER, FRANK WILLIAM, b Mt Pleasant, Mich, Feb 26, 48; m 72. PROTEIN ENGINEERING, MOLECULAR GENETICS. *Educ:* Albion Col, BA, 71; Fla State Univ, MS, 73, PhD(genetics), 75. *Prof Exp:* Res scientist, Chem Mutagenesis Prog, Biol Div, 76-84, SR RES SCIENTIST, PROTEIN ENG PROG, BIOL DIV, OAK RIDGE NAT LAB, 84- *Concurrent Pos:* Adj prof, Oak Ridge Grad Sch Biomed Sci, Univ Tenn, 77-; adj prof, biotech concentration, Bio Consortium, 85-, Cell, Molecular & Develop Biol Prog, Univ Tenn, Knoxville, 89- *Mem:* Am Soc Microbiol; Am Soc Biochem & Molecular Biol; Genetics Soc Am; Int Soc Plant Molecular Biol; Protein Soc. *Res:* Protein engineering: enzyme structure/function analysis, modeling of binding sites and catalytic complexes, subunit interactions and assembly; molecular genetics of mutagenesis and DNA repair. *Mailing Add:* Biol Div Oak Ridge Nat Lab PO Box 2009 Oak Ridge TN 37831

LARIMER, JAMES LYNN, b Washington Co, Tenn, Jan 7, 32; div; c 2. NEUROBIOLOGY. *Educ:* ETenn State Univ, BS, 53; Univ Va, MA, 54; Duke Univ, PhD, 59. *Prof Exp:* From asst prof to assoc prof, 59-68, actg chmn dept, 73-74, PROF ZOOL, UNIV TEX, AUSTIN, 68- *Concurrent Pos:* Guggenheim fel, 67-68; mem physiol study sect, NIH, 72-86; mem marine sci panel, NSF, 86. *Honors & Awards:* Javits Neurosci Res Award, 88-95. *Mem:* Soc Neurosci; Am Physiol Soc; Sigma Xi; fel AAAS. *Res:* Comparative physiology; behavior and neurophysiology of invertebrates. *Mailing Add:* Dept Zool Univ Tex Austin TX 78712

LARIMER, JOHN WILLIAM, b Pittsburgh, Pa, Sept 4, 39; m 65; c 2. GEOCHEMISTRY. *Educ:* Lehigh Univ, BA, 62, MS, 63, PhD(geol), 66. *Prof Exp:* NASA-AEC res assoc geochem, Enrico Fermi Inst, Univ Chicago, 66-69; from asst prof to assoc prof, 74-77, PROF GEOL, ARIZ STATE UNIV, 77- *Concurrent Pos:* NATO fel, Max Planck Inst, 75-76; vis prof, Calif Inst Technol, 79; prin investr, Lunar & Planetary Sci Prog, NASA, 72- *Honors & Awards:* Nininger Award, 66. *Mem:* AAAS; Geochem Soc; Am Geophys Union; Meteoritical Soc; Sigma Xi. *Res:* Cosmochemistry; mineralogy and composition of meteorites. *Mailing Add:* Dept Geol Ariz State Univ Tempe AZ 85287

LARIMORE, RICHARD WELDON, b Rogers, Ark, Feb 10, 23; m 47; c 3. FISH BIOLOGY. *Educ:* Univ Ark, BS, 46; Univ Ill, MS, 47; Univ Mich, PhD(zool), 50. *Prof Exp:* Asst aquatic biol, Univ Ill, Urbana, 46-54, assoc, 54-58, prof zool, 70-76, prof environ eng, 69-88, AQUATIC BIOLOGIST, ILL STATE NATURAL HIST SURV, 58- *Concurrent Pos:* Fishery expert, Food & Agr Orgn, 63-64 & 72-73; sr lectr, Fulbright Comn, 80, 83, 90. *Honors & Awards:* Fisheries Pub Award, Wildlife Soc, 57; Am Fisheries Soc Award, 60. *Mem:* Am Soc Ichthyol & Herpet; Am Fisheries Soc; Am Inst Fishery Res Biol. *Res:* Ecology of stream and reservoir fishes; dynamics of cooling lakes; utilization of tropical aquatic resources. *Mailing Add:* 277 Nat Res Bldg Univ Ill Champaign IL 61820

LARIS, PHILIP CHARLES, b Perth Amboy, NJ, Sept 5, 31; m 56; c 4. PHYSIOLOGY. *Educ:* Rutgers Univ, BS, 52; Princeton Univ, MA, 54, PhD(physiol), 56. *Prof Exp:* Instr biol, Univ Calif, 56-58, from asst prof to assoc prof, 58-65; assoc prof, Franklin & Marshall Col, 65-66; assoc prof, 66-75, PROF BIOL, UNIV CALIF, SANTA BARBARA, 75- *Mem:* Am Physiol Soc; Soc Gen Physiol. *Res:* Cell permeability membrane potentials; amino acid and ion transport. *Mailing Add:* Dept Biol Sci Univ Calif Santa Barbara CA 93106

LARK, CYNTHIA ANN, b Shawnee, Okla, Dec 31, 28; m 51; c 4. MICROBIOLOGY. *Educ:* Mt Holyoke Col, BA, 50; St Louis Univ, PhD, 62. *Prof Exp:* Lab technician, Carnegie Inst, 50-51 & Sloan-Kettering Inst Cancer Res, 51-53; res asst, Univ Geneva, 55-56 & St Louis Univ, 59-62; NIH fel, Washington Univ, 62-63; asst prof microbiol, Kans State Univ, 63-70; assoc res prof biochem, 70-72, ASSOC PROF BIOL, UNIV UTAH, 72- *Res:* Microbial genetics; molecular biol; DNA reproducing bacteria. *Mailing Add:* Dept Biol Univ Utah 201 S Biol Bldg Salt Lake City UT 84112

LARK, KARL GORDON, b Lafayette, Ind, Dec 13, 30; m 51; c 4. MOLECULAR BIOLOGY, GENETICS. *Educ:* Univ Chicago, PhB, 49; NY Univ, PhD(microbiol), 53. *Prof Exp:* Am Cancer Soc res fel, Statenserum Inst, Denmark, 53-55; Nat Found res fel, Biophys Lab, Univ Geneva, 55-56; instr microbiol, Sch Med, St Louis Univ, 56-57, sr instr, 57-58, from asst prof to assoc prof, 58-63; prof, Kans State Univ, 63-70; chmn dept, 70-77, PROF BIOL, UNIV UTAH, 77- *Concurrent Pos:* Nat Inst Gen Med Sci career develop award, 63-70; mem, Genetics Panel, NSF, 66-69; consult, Eli Lilly & Co, 68-74; mem Genetics Study Sect, NIH, 71-75; ad hoc mem, Nat Inst Gen Med Sci Coun, 78-79, mem, 79- *Mem:* Biophys Soc; Am Soc Cell Biol; Am Soc Biol Chem; Am Soc Microbiol. *Res:* Cell growth and division; DNA replication and segregation in bacteria and eucaryotes; plant genetics and tissue culture. *Mailing Add:* Dept Biol Univ Utah 201 Biol Sci Bldg Salt Lake City UT 84112

LARK, NEIL LAVERN, b Baker, Ore, Sept 10, 34; m 58; c 2. NUCLEAR PHYSICS. *Educ:* Chico State Col, AB, 55; Cornell Univ, PhD(phys chem), 60. *Prof Exp:* Res asst, Los Alamos Sci Lab, 55-56; res asst, Brookhaven Nat Lab, 57, jr res assoc, 58-59, res assoc, 60-61; NATO fel, Inst Nuclear Res, Amsterdam, Netherlands, 61-62, res assoc, 62; from asst prof to prof natural sci, Raymond Col, 62-75, PROF PHYSICS, UNIV PAC, 75- *Concurrent Pos:* Res asst, Univ NMex, 54; consult & res collabr, Los Alamos Sci Lab, 69; Ford Found fel & Fulbright grantee, Niels Bohr Inst, Copenhagen, Denmark, 67-68; NSF lectr, Tex A&M Univ, 70; res collabr, Brookhaven Nat Lab, 71-72, vis physicist, 75; vis fel, Australian Nat Univ, 76; vis astronr, Univ Hawaii, 86, 87. *Honors & Awards:* Sigma Xi. *Mem:* Am Phys Soc; Am Asn Physics Teachers; Am Astron Soc; Am Asn Univ Prof. *Res:* Nuclear spectroscopy; teaching of physics and astronomy. *Mailing Add:* Dept Physics Univ Pac Stockton CA 95211-0197

LARKE, R(OBERT) P(ETER) BRYCE, b Blairmore, Alta, Nov 14, 36; m 60; c 3. VIROLOGY, INFECTIOUS DISEASES. *Educ:* Queen's Univ, Ont, MD & CM, 60; Univ Toronto, DClSc, 66. *Prof Exp:* Intern, St Michael's Hosp, Toronto, Ont, 60-61; resident pediat, Hosp for Sick Children, Toronto, Ont, 61-62, res fel virol, Res Inst, 62-66; instr prev med, Sch Med, Case Western Reserve Univ, 66-68, sr instr, 68; asst prof pediat, McMaster Univ, 69-72, asst prof path, 71-72, assoc prof pediat & path, 72-75; assoc prof pediat, Univ Alta, 75-76, clin virologist, Prov Lab Pub Health, 75-85, hon prof med, Div Infectious Dis, 77-86; assoc med dir, Can Red Cross Blood Transfusion Serv, Edmonton, Alta, 85-88; PROF PEDIAT, UNIV ALTA, 76-, PROF MED MICROBIOL & INFECTIOUS DIS, 86-; DIR, PROV AIDS PROG, ALTA HEALTH, 88- *Concurrent Pos:* Fel Microbiol, Sch Hyg, Univ Toronto, 63-66; Med Res Coun Can res fel, 64-66; dir virol lab, St Joseph's Hosp, Hamilton, Ont, 71-75; vis scientist, Viral Oncol Unit, Pasteur Inst, Paris, France, 86-87; dep med dir, Can Red Cross Blood Transfusion Serv, 88- *Honors & Awards:* Parkin Prize, Royal Col Physicians, Edinburgh, 66. *Mem:* Am Soc Microbiol; Am Soc Virol; Am Pediat Soc; Infectious Dis Soc Am; Can Pediat Soc; Can Soc Clin Invest. *Res:* Clinical virology and infectious diseases of children and adults; mechanisms of host resistance to viral infections; interaction between viruses and blood platelets; epidemiology of viral infections; immunization against hepatitis B. *Mailing Add:* Dept Pediat Univ Alta Edmonton AB T6G 2R7 Can

LARKIN, DAVID, b London, Eng, Oct 6, 41. ELECTROCHEMISTRY. *Educ:* Loughborough Univ Technol, BTech, 65, PhD(electrochem), 68; Royal Inst Chem, ARIC, 70. *Prof Exp:* Asst prof chem, Fla Technol Univ, 72-73; asst prof, 73-80, ASSOC PROF CHEM, TOWSON STATE UNIV, 80- *Mem:* Am Chem Soc; Royal Soc Chem; Sigma Xi. *Res:* Study of electro kinetics at solid metal electrodes. *Mailing Add:* Dept Chem Towson State Univ Towson MD 21204

LARKIN, EDWARD CHARLES, b Waltham, Mass, Aug 7, 37; m 65; c 2. HEMATOLOGY. *Educ:* Harvard Col, AB, 59; Yale Univ Sch Med, MD, 63. *Prof Exp:* Asst prof med, Univ Tex Med Br, 70-74, assoc prof, 74-75; chief hemat oncol, Marinez Va Hosp, 75-84; from assoc prof to prof med, 75-83, PROF MED & PATH, UNIV CALIF, DAVIS, 83- *Mem:* Am Soc Hemat; Int Soc Hemat; Am Col Physicians; Western Soc Clin Res. *Mailing Add:* Vet Admin Univ Calif Med Hosp 150 Muir Rd Martinez CA 94553

LARKIN, EDWARD P, b Watertown, Mass, Sept 27, 20; m 49; c 3. FOOD MICROBIOLOGY, VIROLOGY. *Educ:* Mass State Col, BS, 46; Univ Mass, MS, 48, PhD(bact), 54. *Hon Degrees:* Hon BS, Mass State Col, 43. *Prof Exp:* Instr bact, Univ Mass, 48-54, asst prof, 54-61; dir, Med Sch Study, CIC, Purdue Univ, 60-61; adminr & assoc mem virol, Inst Med Res, NJ, 61-68; chief biophys lab, New Bolton Ctr, Sch Vet Med, Univ Pa, 68-70; chief virol br, Ctr Food Safety & Appl Nutrit, Food & Drug Admin, 70-87; RETIRED. *Concurrent Pos:* Vis assoc, Univ Pa, 61-65; Leukemia Soc scholar, 65-70. *Mem:* Am Soc Microbiol; Sigma Xi; Int Asn Comp Res on Leukemia & Related Dis (secy, 69-71). *Res:* Food virology; physical characterization of viruses; kinetic studies of physical and chemical inactivation of viruses. *Mailing Add:* 400 Stanton Ave Terrace Park OH 45174

LARKIN, FRANCES ANN, nutrition, for more information see previous edition

LARKIN, JEANNE HOLDEN, developmental biology; deceased, see previous edition for last biography

LARKIN, JOHN MICHAEL, b York, Nebr, Aug 11, 37; m 65; c 3. ORGANIC CHEMISTRY. *Educ:* Univ Nebr, Lincoln, BS, 59; Univ Colo, Boulder, MS, 63, PhD(org chem), 65. *Prof Exp:* Chemist, Qual Water Br, US Geol Surv, Nebr, 59-60; from chemist to sr chemist, Texaco Inc, Beacon, 65-70, res chemist, 70-74, sr res chemist, 74-76; sr proj chemist, Jefferson Chem Co, Austin, Tex, 76-87; SUPVR RES SECT, TEXACO CHEM CO, AUSTIN, TEX, 87- *Mem:* Am Chem Soc. *Res:* Organic synthesis; structure elucidation; reaction mechanisms; organic nitrogen compounds; free radical rearrangements; steroidal heterocycles; air pollution; lubricant additive synthesis; chemicals from synthesis gas. *Mailing Add:* Texaco Chem Co PO Box 15730 Austin TX 78761

LARKIN, JOHN MONTAGUE, b Philadelphia, Pa, Apr 7, 36; m 62; c 2. MICROBIAL PHYSIOLOGY, MICROBIAL ECOLOGY. *Educ:* Ariz State Univ, BS, 61, MS, 63; Wash State Univ, PhD(microbiol), 67. *Prof Exp:* Asst microbiol, Ariz State Univ, 61-63 & Wash State Univ, 63-67; from asst prof to assoc prof, 67-81, PROF MICROBIOL, LA STATE UNIV, BATON ROUGE, 81- *Mem:* AAAS; Am Soc Microbiol; Am Soc Limnol Oceanog. *Res:* bacterial taxonomy; biology of gliding bacteria; filamentous sulfur bacteria. *Mailing Add:* Dept Microbiol La State Univ Baton Rouge LA 70803

LARKIN, K(ENNETH) T(RENT), b Fowler, Colo, Nov 11, 20; m 46; c 2. ELECTRICAL ENGINEERING. *Educ:* Southern Methodist Univ, BS, 43. *Prof Exp:* Jr engr, Lone Star Gas Co, 39-42; instr radio eng, Southern Methodist Univ, 42-43; scientist, US Naval Res Lab, DC, 43-45; mgr receiver & indicators br, Waltham Labs, Raytheon Mfg Co, 46-55, dept mgr radar & radar develop, Santa Barbara Labs, 55-56; mgr telecommun dept, Lockheed Missiles & Space Co, 56-57, mgr electronics div, 57-58, assoc dir electronics res & develop, 58-62, assoc dir eng electronics, 62-63, asst dir eng, Res & Develop Div, 63, dir, 63-65, dir info systs, 65-71; pres, Technicon Med Info Systs Corp, 71-; PALO ALTO MGT GROUP. *Mem:* Sr mem Inst Elec & Electronics Engrs. *Res:* Computer systems, especially as applied to health care. *Mailing Add:* 215 Golden Hills Dr Portola Valley CA 94025

LARKIN, LAWRENCE A(LBERT), b Kansas City, Mo, Jan 5, 37; m 60. FINITE ELEMENT ANALYSIS, ELASTO-PLASTIC FLOW. *Educ:* Univ Kans, BA, 59, MS, 60, PhD(civil eng), 64. *Prof Exp:* Asst prof civil eng, State Univ NY Buffalo, 64-69; res scientist, 70-74, CONSULT ENGR, DATA SYSTS DIV, A O SMITH CORP, 74- *Mem:* Am Soc Civil Engrs. *Res:* Development of finite element computer codes; finite element computation of elasto-plastic response of vehicle structures during crash conditions. *Mailing Add:* 7728 W Coventry Dr Franklin WI 53132

LARKIN, LYNN HAYDOCK, b Highland Co, Ohio, Jan 29, 34; div; c 2. ANATOMY, REPRODUCTIVE BIOLOGY. *Educ:* Otterbein Col, BS, 56; Univ Colo, PhD(anat), 67. *Prof Exp:* Instr anat, Sch Med, Univ Colo, 66-67, res assoc molecular, cellular & develop biol, 67-68; from asst prof to assoc prof, 68-79, PROF ANAT, COL MED, UNIV FLA, 79- *Mem:* Sigma Xi; Am Asn Anatomists; Soc Study Reproduction; Am Asn Clin Anatomists. *Res:* Role of relaxin in pregnancy and parturition. *Mailing Add:* Dept Anat & Cell Biol Box J235 JHMHC Col Med Univ Fla Gainesville FL 32610

LARKIN, PETER ANTHONY, b Auckland, NZ, Dec 11, 24; m 48; c 5. FISHERIES. *Educ:* Univ Sask, BA, 45, MA, 46; Oxford Univ, DPhil, 48. *Hon Degrees:* LLD, Univ Sask, 89. *Prof Exp:* Chief fisheries biologist, Game Comn, BC, 48-55; dir inst fisheries & prof zool, Univ BC, 55-63; dir biol sta, Fisheries Res Bd Can, BC, 63-66; dir, Inst Fisheries, Univ BC, 66-69, prof zool, 66-75, actg head dept, 69-70, prof, Inst Animal Resource Ecol, 69-75, head zool dept, 72-75, dean fac grad studies, 74-84, assoc vpres res, 81-84, vpres res, 84-89; RETIRED. *Concurrent Pos:* Nuffield Found fel, 61-62; mem, Sci Coun Can, 71-; mem Can nat comt, Spec Comt on Probs of Environ, 71-; mem, Fisheries Res Bd Can, 72-; bd dirs, BC Packers, Ltd, 80-, Sci Coun BC, 84-88, Int Develop Res Ctr, 84-, Nat Sci Eng Res Coun Can & BC Coun Fedn, 86- *Honors & Awards:* Can Centennial Medal, 67; Queen's Jubilee Medal, 77; Fry Medal, Can Soc Zool, 78; Award of Excellence, Am Fisheries Soc, 83; Achievement Award, Am Inst Fishery Res Biologists, 86. *Mem:* Am Fisheries Soc; Can Nature Fedn; Can Soc Environ Biologists; Can Soc Zoologists (pres, 72). *Res:* Population studies in fisheries biology. *Mailing Add:* Inst Animal Resource Ecol Univ BC Vancouver BC V6T 1W5 Can

LARKIN, ROBERT HAYDEN, b New York, NY, Mar 26, 46; m 67; c 4. ANALYTICAL CHEMISTRY. *Educ:* Providence Col, BS, 68; Univ Mass, PhD(phys chem), 72. *Prof Exp:* Res assoc chem, Mass Inst Technol, 72-73; res chemist, Rohm & Haas Co, 73-79, sect mgr analytical res, 79-84, sect mgr environ sci, 84-86, DIR REGULATORY AFFAIRS, ROHM & HAAS CO, 86- *Mem:* Am Chem Soc; Nat Agr Chem Asn. *Res:* Determination of the environmental fate and metabolism of agricultural pesticides. *Mailing Add:* 207 Hemlock Circle North Wales PA 19454

LARKIN, WILLIAM (JOSEPH), b Morristown, NJ, Aug 18, 18; m 49; c 1. NUCLEAR ENGINEERING, ENGINEERING MANAGEMENT. *Educ:* Mass Inst Technol, BS & MS, 49; Univ Tenn, MS, 56. *Prof Exp:* Design draftsman, Ford Instrument Co, NY, 42; mech engr, Los Alamos Sci Lab, 44-46; asst dir eng sch, Mass Inst Technol, 49-50; engr, US AEC, Dept Energy, 50-55, chief, Reactor Projs Br, 55-66, actg dir, Reactor Div, 66-69, dir, Off Nuclear & Criticality Safety, 69-70, dir off safety, 70-73, chief Uranium Enrichment Expansion Proj Br, Res Develop Admin, Res Develop Admin, 73-76, dir, Planning, Anal, Control & Reports Div, 77-78; consult engr, 78-88; RETIRED. *Res:* Uranium enrichment plant engineering; construction project planning and management control; operational safety. *Mailing Add:* 112 Blue Ridge Ct Oak Ridge TN 37830

LARKIN, WILLIAM ALBERT, b Boston, Mass, Dec 17, 26; m 53; c 7. FLAME RETARDANTS. *Educ:* Boston Col, BS, 49. *Prof Exp:* Dist mgr, McKesson & Robbins Inc, 53-57; dist mgr, Atochem NAm, 57-61, mgr tech serv, 61-68, mgr prod develop, 68-73, tech dir, 73-80, plant gen mgr, 80-81, DIR TECHNOL, ATOCHEM NAM, 82- *Mem:* Soc Plastics Engrs; Licensing Exec Soc. *Res:* Polymer stabilizers; flame retardants; modifiers; glass coatings for strengthening and electroconductivity; organometallic, organotin chemistry and applications; holder of 25 US patents. *Mailing Add:* 38 Sylvania Ave Avon By The Sea NJ 07717

LARKINS, BRIAN ALLEN, b Bellville, Kans, Aug 12, 46; m 69; c 2. PLANT PHYSIOLOGY, PLANT BIOCHEMISTRY. *Educ:* Univ Nebr, BSEd, 69, PhD(bot), 74. *Prof Exp:* Res assoc biochem genetics, 75-76, from asst prof to assoc prof, 76-83, PROF GENETICS, PURDUE UNIV, 84- *Concurrent Pos:* Head, plant sci, Univ Ariz; Houde distinguished prof, Purdue Univ, 86. *Honors & Awards:* Charles Albert Schull Award, Am Soc Plant Physiologists, 83. *Mem:* Am Soc Plant Physiologists; AAAS; Sigma Xi; Am Soc Biochem & Molecular Biol; Soc Develop Biol. *Res:* Protein and nucleic acid biosynthesis; seed storage protein metabolism; regulation of gene activity during seed formation. *Mailing Add:* Dept Plant Sci Univ Ariz Tucson AZ 85721

LARKINS, HERBERT ANTHONY, fishery biology, population dynamics, for more information see previous edition

LARKINS, THOMAS HASSELL, JR, b Dickson, Tenn, Mar 1, 39; m 60; c 3. INORGANIC CHEMISTRY. *Educ:* Austin Peay State Col, BS, 60; Vanderbilt Univ, MA, 62, PhD(chem), 63. *Prof Exp:* Sr res chemist, Tenn Eastman Co, 63-80, RES ASSOC, EASTMAN CHEM DIV, EASTMAN KODAK CO, 80- *Mem:* Am Chem Soc. *Res:* Coordination chemistry; catalysis; coal gasification; chemicals from coal. *Mailing Add:* 4408 Beechcliff Dr Kingsport TN 37664

LARKY, ARTHUR I(RVING), b Bound Brook, NJ, Feb 27, 31; m 81; c 6. ELECTRICAL ENGINEERING, COMPUTER SCIENCE. *Educ:* Lehigh Univ, BS, 52; Princeton Univ, MS, 53; Stanford Univ, PhD(elec eng), 57. *Prof Exp:* Res asst prof elec eng, 56-60, assoc prof, 60-64, PROF ELEC & COMPUTER ENG, LEHIGH UNIV, 64- *Concurrent Pos:* Consult, Bell Tel Labs, 61- *Mem:* Inst Elec & Electronics Engrs. *Res:* Computer design; automated testing. *Mailing Add:* CSEE Dept 215 Packard Bldg 19 Bethlehem PA 18015

LARMIE, WALTER ESMOND, b Smithfield, RI, Sept 6, 20; m 43; c 3. FLORICULTURE. *Educ:* Univ RI, BS, 49, MS, 54. *Prof Exp:* Res asst prof to prof hort, Univ RI, 49-83, chmn dept plant & soil sci, 72-83; RETIRED. *Mem:* Am Soc Hort Sci; Am Hort Soc; Soc Am Florists; NCTRH. *Res:* Storage of flowers and plants; weed control; growth regulators; propagation and culture of poinsettias; floral design. *Mailing Add:* 65 Blackbird Rd West Kingston RI 02892

LARMORE, LAWRENCE LOUIS, b Washington, DC, Nov 23, 41; m 64; c 2. MATHEMATICS. *Educ:* Tulane Univ, BS, 61; Northwestern Univ, PhD(math), 65. *Prof Exp:* Asst prof math, Univ Ill, Chicago Circle, 65-68 & Occidental Col, 68-70; assoc prof, 70-77, PROF MATH, CALIF STATE COL, DOMINGUEZ HILLS, 77- *Mem:* Am Math Soc. *Res:* Algebraic topology; obstruction theory; classification of liftings, embeddings and immersions; twisted extraordinary cohomology. *Mailing Add:* Dept Computer Sci Univ Calif Riverside CA 92521

LARMORE, LEWIS, b Anderson, Ind, July 29, 15; m 39; c 2. SOLAR PHYSICS, MILITARY INFRARED APPLICATIONS. *Educ:* Ind Univ, AB, 37, MA, 38, MA, 39; Univ Calif, Los Angeles, PhD(physics), 52. *Prof Exp:* Instr physics, Univ Utah, 42-43; asst prof physics, Ariz State Univ, 47-49; physicist, Rand Corp, 51-54; head, aerophys Dept, Lockheed Missile & Space Co, Lockheed Corp, 54-56, cor res adv, Lockheed Aircraft Corp, 56-59, chief scientist, Lockheed Calif Co, 59-64; vpres & res dir, Douglas Aircraft Corp, 64-70; chief scientist, Astronaut Co, McDonnell-Douglas Corp, 70-72; dir technol, Off Naval Res, 72-77, sci dir, Western Regional Off, 77-81, physicist, 81-86; RETIRED. *Concurrent Pos:* Vis asst prof, Univ Calif, Los Angeles, 50 & 51; trustee, W Coast Univ, 69-89. *Honors & Awards:* Goddard Award, Int Soc Optical Eng, 74. *Mem:* Fel Am Acad Arts & Sci; fel Optical Soc Am; fel Int Soc Optical Eng (vpres, 77, pres, 85); fel AAAS; fel Am Astronaut Soc (pres, 66-68); Int Acad Astronaut. *Res:* Solar physics including the motions, distribution and effects of solar flares and prominences; acoustical holography including acoustical imaging and techniques; atmospheric transmission of infrared radiation; infrared characteristics of the solar flash spectrum. *Mailing Add:* 1245 Calle Estrella San Dimas CA 91773-4081

LARNER, JOSEPH, b Brest-Litovsk, Poland, Jan 9, 21; nat US; m 47; c 3. BIOCHEMISTRY, PHARMACOLOGY. *Educ:* Univ Mich, BA, 42; Columbia Univ, MD, 45; Univ Ill, MS, 49; Wash Univ, PhD(biochem), 51. *Hon Degrees:* Dr, Univ Barcelona, 83- *Prof Exp:* Instr biochem, Wash Univ, 51-53; asst prof, Noyes Lab Chem, Univ Ill, 53-57; from assoc prof to prof pharmacol, Sch Med, Western Reserve Univ, 57-65; Hill prof metab enzym, Col Med Sci, Univ Minn, Minneapolis, 65-69; PROF PHARMACOL & CHMN DEPT, SCH MED, UNIV VA, 69- *Concurrent Pos:* Travel award, Int Cong Biochem, 55; NIH res career award, 63-64; Commonwealth Fund fel, Lab Molecular Biol, Cambridge Univ, 63-64; mem metab study sect, NIH, 62-66, mem training comt, Nat Inst Arthritis & Metab Dis, 66; mem subcomt enzymes, Nat Res Coun-Nat Acad Sci, 64-; mem rev bd, Am Cancer Soc, 70-74. *Honors & Awards:* David Rumbaugh Award, Juv Diabetes Found, 80; Diaz Christobal Award, Int Diabetes Fedn, 82; Banting Lect Can Diabetes Asn, Univ Toronto, 81; Diaz Christobal Award, Int Diabetes Fedn, 82; Res Award, Japan Soc Starch Res, 85; Res Award, Asn Am Med Col, 87; Banting Medal, Am Diabetes Asn, 87. *Mem:* Inst Med; Am Soc Biol Chemists; fel Royal Col Med; Am Soc Pharmacol & Exp Therapeut; hon mem Japan Biochem Soc; Am Chem Soc; hon mem Can Soc Endocrinol. *Res:* Enzymatic aspects of intermediary carbohydrate metabolism, genetic and hormonal control. *Mailing Add:* Dept Pharmacol Univ Va Sch Med Charlottesville VA 22903

LARNER, KENNETH LEE, b Chicago, Ill, Nov 1, 38; m 76; c 2. EXPLORATION GEOPHYSICS. *Educ:* Colo Sch Mines, GpE, 60; Mass Inst Technol, PhD(geophys), 70. *Prof Exp:* Res scientist image enhancement, EG&G, Inc, 67-69; sr res geophysicist, Western Geophys Co, 70-74, mgr res & develop explor geophys, 75-80, vpres & develop, 80-88; GREEN CHAIR PROF GEOPHYS, COLO SCH MINES, GOLDEN, COLO, 88- *Concurrent Pos:* Lectr geophys, Univ Houston, 73-74. *Mem:* Soc Explor Geophysicists; Seismol Soc Am; Europ Asn Explor Geophysicists; Sigma Xi. *Res:* Seismic signal enhancement and wave propagation; estimation of geophysical parameters from seismic measurements. *Mailing Add:* 8132 Citation Trail Evergreen CO 80439

LARNEY, VIOLET HACHMEISTER, b Chicago, Ill, May 19, 20; m 50. MATHEMATICS. *Educ:* Ill State Univ, BEd, 41; Univ Ill, AM, 42; Univ Wis, PhD(math), 50. *Prof Exp:* Teacher high sch, Ill, 42-44; asst math, Univ Wis, 44-48, instr, Exten Div, 48-50; asst prof, Kans State Univ, 50-52; from assoc prof to prof math, State Univ NY, Albany, 52-82, emer prof, 82; RETIRED. *Mem:* Math Asn Am; Am Math Soc. *Res:* Abstract algebra. *Mailing Add:* 573 Leisure World Mesa AZ 85206-3129

LARNTZ, KINLEY, b Coshocton, Ohio, Oct 2, 45; m 65; c 3. APPLIED STATISTICS, MATHEMATICAL STATISTICS. *Educ:* Dartmouth Col, AB, 67; Univ Chicago, PhD(statist), 71. *Prof Exp:* PROF APPL STATIST, UNIV MINN, ST PAUL, 71- *Mem:* Am Statist Asn; Royal Statist Soc; Biomet Soc; Inst Math Statist; Sigma Xi; AAAS. *Res:* Analysis of qualitative data; comparison of small sample distributions for chi-square goodness-of-fit statistics; data analysis applied statistical methods. *Mailing Add:* Dept Appl Statist Univ Minn St Paul MN 55108

LAROCCA, ANTHONY JOSEPH, b New Orleans, La, May 15, 23; m 54; c 6. PHYSICS, ELECTRICAL ENGINEERING. *Educ:* Tulane Univ, BS in EE, 49; Univ Mich, MS, 52. *Prof Exp:* Res asst physics, Tulane Univ, 49-51; res asst, Inst Sci & Technol, Univ Mich, Ann Arbor, 56-58, res assoc, 58-60, from assoc res physicist to res physicist, 60-73; RES PHYSICIST, ENVIRON RES INST MICH, 73- *Concurrent Pos:* Adj prof, Univ Mich, 73- *Mem:* AAAS; Optical Soc Am. *Res:* Infrared and optical technology; radiation phenomena; propagation and attenuation; techniques of measurement of radiation; design and use of electrooptical devices; standards of radiation; study of techniques of remote sensing of environment. *Mailing Add:* 2600 Englave Ann Arbor MI 48103

LA ROCCA, JOSEPH PAUL, b La Junta, Colo, July 5, 20; m 47; c 4. MEDICINAL CHEMISTRY. *Educ:* Univ Colo, BS, 42; Univ NC, MS, 44; Univ Md, PhD(pharmaceut chem), 48. *Prof Exp:* Res chemist, Naval Res Lab, 47-49; from assoc prof to prof pharm, Univ Ga, 49-88, head, dept med chem, 70-88; RES & DEVELOP, BECTON-DICKINSON LABWARE, 88- *Mem:* Am Chem Soc; Am Pharmaceut Asn. *Res:* Synthetic sedative-hypnotics; anticonvulsant compounds; chemotherapy of cancer. *Mailing Add:* Res & Develop Becton-Dickinson Labware Two Bridgewater Lane Lincoln Park NJ 07035

LAROCHELLE, JACQUES, b Quebec, Can, Sept 4, 46; m 72; c 2. ANIMAL PHYSIOLOGY, CELL PHYSIOLOGY. *Educ:* Laval Univ, BA, 66, BSc, 71. *Hon Degrees:* DSc, Laval Univ, 76. *Prof Exp:* Lectr physiol, Laval Univ, 74-75; Que Ministry Educ fel & vis scholar zool, Duke Univ, 76; asst prof, 77-80, ASSOC PROF PHYSIOL, LAVAL UNIV, 80- *Mem:* Can Soc Zoologists. *Res:* Temperature regulation and locomotion in birds and mammals. *Mailing Add:* Dept Biol Laval Univ Quebec PQ G1K 7P4 Can

LA ROCHELLE, JOHN HART, b Longmeadow, Mass, Aug 17, 24; m 48; c 4. PHYSICAL CHEMISTRY. *Educ:* Univ Mass, BS, 48; Northeastern Univ, MS, 50; Univ Mich, PhD(chem), 55. *Prof Exp:* Res chemist, Shell Chem Co, 55-60, sr res chemist, 60-68, res supvr, 68-72, res supvr, Shell Develop Co, 72-74, staff chemist, Geismar Plant, 74-83; RETIRED. *Mem:* AAAS; Am Inst Chemists; Sigma Xi. *Res:* Dielectric polarization of gases. *Mailing Add:* 2757 Woodland Ridge Baton Rouge LA 70816

LAROCK, BRUCE E, b Berkeley, Calif, Dec 24, 40; m 68; c 2. CIVIL ENGINEERING, HYDRODYNAMICS. *Educ:* Stanford Univ, BS, 62, MS, 63, PhD(civil eng), 66. *Prof Exp:* From asst prof to assoc prof, 66-79, PROF CIVIL ENG, UNIV CALIF, DAVIS, 79- *Concurrent Pos:* Sr US scientist award, Alexander von Humboldt-Stiftung, 86-87. *Mem:* Am Soc Civil Engrs. *Res:* Hydraulics and fluid mechanics; finite element methods. *Mailing Add:* Dept Civil Eng Univ Calif Davis CA 95616

LAROCK, RICHARD CRAIG, b Berkeley, Calif, Nov 16, 44. ORGANIC CHEMISTRY. *Educ:* Univ Calif, Davis, BS, 67; Purdue Univ, PhD(chem), 72. *Prof Exp:* NSF fel chem, Harvard Univ, 71-72; from instr to assoc prof, 72-85, PROF CHEM, IOWA STATE UNIV, 85- *Concurrent Pos:* Du Pont Young fac scholar, Du Pont Chem Co, 75-76; A P Sloan fel, 77-79. *Mem:* Am Chem Soc; Royal Soc Chem; Sigma Xi. *Res:* Synthesis of biologically active compounds; new synthetic methods; organometallic and hetrocyclic chemistry. *Mailing Add:* Dept Chem 228 Gilman Hall Iowa State Univ Ames IA 50011

LA ROCQUE, JOSEPH ALFRED AURELE, geology; deceased, see previous edition for last biography

LAROS, GERALD SNYDER, II, b Los Angeles, Calif, July 19, 30; m 58; c 3. ORTHOPEDIC SURGERY. *Educ:* Northwestern Univ, BS, 52, MD, 55; Univ Iowa, MS, 70. *Prof Exp:* Orthop resident, Vet Admin Hosp, Hines, Ill, 55-60 & Shriners Hosp, Honolulu, Hawaii, 60-61; pvt pract, 61-68; NIH fel, Univ Iowa, 68-70, asst prof orthop surg, 70-71; assoc prof, Univ Ark, 71-73; prof surg & chmn sect orthop surg, Univ Chicago, 73-83; PROF & CHMN, DEPT ORTHOP SURG, TEX TECH UNIV, 84- *Concurrent Pos:* Mem & secy-treas, Am Bd Orthop Surg, 81-85, pres, 86-87. *Mem:* Am Acad Orthop Surg; Orthop Res Soc; Am Col Surgeons; Am Orthop Asn. *Res:* Electromicroscopy of bone tendon and ligaments. *Mailing Add:* Dept Orthop Surg Tex Tech Univ Health Sci Ctr Lubbock TX 79430

LAROSE, ROGER, pharmacy, for more information see previous edition

LAROSSA, ROBERT ALAN, b New York, NY, Jan 29, 51; m; c 2. AGRICULTURE PRODUCTION. *Educ:* Johns Hopkins Univ, BA, 73; Yale Univ, MPhil, 76, PhD(molecular biophys & biochem), 77. *Prof Exp:* Postdoctoral fel, Dept Molecular Biophys & Biochem, Yale Univ, 77; postdoctoral fel, Dept Biochem, Stanford Univ, 77-80, res associate, 80; prin investr, Cent Res & Develop Dept, 80-90, SR RES BIOLOGIST, AGR PRODS DEPT, E I DU PONT DE NEMOURS & CO, INC, 90- *Concurrent Pos:* Teaching asst, Yale Univ, 73-74; Am Chem Soc postdoctoral fel, 77-79, NSF postdoctoral fel, 79-80; invited lectr, numerous insts & univs, 79-; lectr molecular genetics, DuPont Continuing Educ Prog, 81; adj asst prof, Sch Life & Health Sci, Univ Del, 83-86; spec reviewer, Microbial Genetics & Physiol Study Sect II, NIH, 89. *Mem:* Am Soc Microbiol; Genetics Soc Am; Am Soc Biochem & Molecular Biol. *Mailing Add:* Stine-Haskell Res Ctr E I du Pont de Nemours & Co Inc Elkton Rd PO Box 30 Newark DE 19714

LAROW, EDWARD J, b Albany, NY, Dec 22, 37; m 63; c 4. AQUATIC ECOLOGY, INVERTEBRATE ZOOLOGY. *Educ:* Siena Col, BS, 60; Kans State Univ, MS, 65; Rutgers Univ, PhD(zool), 68. *Prof Exp:* From asst prof to assoc prof, 68-74, chmn dept, 71-80, PROF BIOL, SIENA COL, NY, 74- *Concurrent Pos:* Nat Res Coun Int Biol Prog grant, 70-74; vis lectr, State Univ NY Albany, 70-75 & Col Environ Sci & Forestry, 76-85. *Mem:* Ecol Soc Am; Am Soc Limnol & Oceanog; Int Soc Limnol. *Res:* Biological rhythms and their role in the vertical migration of zooplankton; secondary production of zooplankton; effect of acid precipitation on zooplankton populations. *Mailing Add:* Dept Biol Siena Col Loudonville NY 12211

LARRABEE, ALLAN ROGER, b Flushing, NY, Feb 24, 35; m 60; c 3. BIOCHEMISTRY. *Educ:* Bucknell Univ, BS, 57; Mass Inst Technol, PhD, 62; Ore Inst Sci & Technol, MS, 86. *Prof Exp:* Staff fel biosynthesis of fatty acids, NIH, 64-66; asst prof chem, Univ Ore, 66-72; from assoc prof to prof chem, Memphis State Univ, 72-83; SYSTS PROGRAMMER, BOEING COMPUT SERVS, BELLEVUE, 86- *Concurrent Pos:* NIH res grant, 67-70; NSF res grant, 70-75; consult 81- 86. *Mem:* Am Soc Biol Chemists. *Res:* Role of vitamin B-12, folic acid and pantothenate as coenzymes; biosynthesis of fatty acids; multienzyme complexes; protein turnover; parallel computers. *Mailing Add:* Boeing Computer Servs PO Box 24346 MS-7L-48 Seattle WA 98124-0346

LARRABEE, MARTIN GLOVER, b Boston, Mass, Jan 25, 10; m 32, 44; c 2. NEUROCHEMISTRY, NEUROPHYSIOLOGY. *Educ:* Harvard Univ, AB, 32; Univ Pa, PhD(biophys), 37. *Hon Degrees:* MD, Univ Lausanne, 74. *Prof Exp:* Asst, Univ Pa, 34-35, fel med physics, 37-40; asst prof physiol, Med Col, Cornell Univ, 40-41; fel, Johnson Found, Univ Pa, 41-42, assoc, 42-43, from asst prof to assoc prof biophys, 43-48; assoc prof, 49-63, PROF BIOPHYS, JOHNS HOPKINS UNIV, 63- *Mem:* Nat Acad Sci; Am Physiol Soc; Am Soc Neurochem; Int Soc Neurochem; Soc Neurosci (treas, 72-75). *Res:* Metabolism in relation to physiological function and embryological development in sympathetic ganglia. *Mailing Add:* Dept Biophys Johns Hopkins Univ Baltimore MD 21218

LARRABEE, R(OBERT) D(EAN), b Flushing, NY, Nov 29, 31; m 53; c 2. OPTICAL METROLOGY, SEM METROLOGY. *Educ:* Bucknell Univ, BS & MS, 53; Mass Inst Technol, SM, 55, ScD, 57; Rider Col, MBA, 76. *Prof Exp:* Res engr, David Sarnoff Res Ctr, RCA Corp, 57-71, Sr engr, Advan Technol Lab, 72-76; physicist, 76-81, GROUP LEADER, NAT INST STANDARDS & TECHNOL, 81- *Concurrent Pos:* Mem adj fac, Univ Md, 81-82. *Mem:* AAAS; sr mem Inst Elec & Electronics Engrs; Am Phys Soc; Sigma Xi. *Res:* Semiconductor materials and devices; solid-state plasma physics; semiconductor microwave oscillators and amplifiers; infrared physics and detectors; characterization of semiconductor materials; optical and scanning electron microscope submicrometer metrology. *Mailing Add:* 18801 Woodway Dr Derwood MD 20855

LARRABEE, RICHARD BRIAN, b Sacramento, Calif, Apr 29, 40; div; c 2. POLYMER CHEMISTRY. *Educ:* Univ Santa Clara, BS, 61; Univ Chicago, MS & PhD(chem), 67. *Prof Exp:* NIH postdoctoral fel, Ohio State Univ, 66-67; RES STAFF, IBM CORP, 67- *Mem:* AAAS; Am Chem Soc. *Res:* Materials science; magnetic disk technology. *Mailing Add:* 6340 Menlo Dr San Jose CA 95120

LARRICK, JAMES WILLIAM, b Englewood Colo, Jan 4, 50; m 89. HUMAN MONOCLONAL ANTIBODIES, CANCER THERAPEUTICS. *Educ:* Colo Col, BA, 72; Duke Univ, PhD(immunol), 79, MD, 80. *Prof Exp:* Intern, Stanford Med Ctr, Palo Alto, 81; scientist & proj leader, Cetus Immune Res Labs, 82-84, sr scientist, 85, dir, 85-88, dir explor res, 90; FOUNDER, PANORAMA RES, 88- *Concurrent Pos:* Thomas J Watson fel. *Mem:* Am Asn Immunologists; Am Fedn Clin Res; AAAS. *Res:* Discovery and characterization of novel biopharmaceuticals for therapy and diagnosis of diseases; human monoclonal antibodies, cytokines. *Mailing Add:* Genelabs Inc 505 Penobscot Dr Redwood City CA 94063

LARROWE, BOYD T, b Merriam, Kans, May 6, 23; m 59; c 4. ELECTRICAL ENGINEERING. *Educ:* Univ Kans, BS, 50; Univ Ill, MS, 51. *Prof Exp:* Mem staff digital computer design, Univ Ill, 50-51; res assoc, Univ Mich, 52-57; sr engr, Strand Eng Co, 57-62; chief engr, Computer Displays, Burroughs Corp Labs, Mich, 62-64; res engr, Inst Sci Technol, Univ Mich, 64-73; res engr, Environ Res Inst Mich, 73-89; RETIRED. *Mem:* Inst Elec & Electronics Engrs. *Res:* Digital computer design; automatic radar image interpretation; display techniques; real time electronic processor design for synthetic aperture radar. *Mailing Add:* 403 Seneca St Tecumseh MI 49286-1022

LARROWE, VERNON L, b Galax, Va, Feb 21, 21; m 66; c 1. ELECTRICAL ENGINEERING. *Educ:* Univ Kans, BS, 50; Univ Ill, MS, 51; Univ Mich, PhD(elec eng), 64. *Prof Exp:* Asst elec eng, Univ Ill, 50-51; res assoc, Inst Sci & Technol, Univ Mich, Ann Arbor, 51-53, assoc res engr, 53-57, head analog comput lab, 53-65, res engr, 57-65; res engr, Infrared & Optics Lab, Willow Run Labs, 65-73; res engr, 73-83, sr res engr, 83-86, EMER MEM TECH STAFF, ENVIRON RES INST MICH, 89- *Mem:* Inst Elec & Electronics Engrs; Soc Computer Simulation; Soc Photo & Instrumentation Engrs; AAAS; NY Acad Sci. *Res:* Application of electronic analog computers; data processing; information theory; remote sensing; automatic pattern recognition; high density digital recording; servo controls. *Mailing Add:* Environ Res Inst Mich Box 8618 Ann Arbor MI 48107

LARRY, JOHN ROBERT, b Mt Clare, WVa, Nov 13, 39; m 63; c 3. PHYSICAL CHEMISTRY. *Educ:* WVa Univ, BS, 61; Ohio State Univ, PhD(chem), 66. *Prof Exp:* Res chemist, 66-74, sr res chemist, 74-76, res supvr, 77-84, RES MGR, ELECTRONICS DEPT, RES & DEVELOP DIV, E I DU PONT DE NEMOURS & CO, INC, WILMINGTON, DEL, 84- *Honors & Awards:* Tech Achievement Award, Int Soc Hybrid Microelectronics, 82. *Mem:* Am Chem Soc; Int Soc Hybrid Microelectronics. *Res:* Charge transfer and molecular complexes; ultracentrifugation; emulsion polymerization; colloid chemistry; rheology; solid state conductors; solid state resistors; multilayer capacitors. *Mailing Add:* DuPont Co Exp Sta PO Box 80334 Wilmington DE 19880-0334

LARSEN, ARNOLD LEWIS, b Audubon, Iowa, Sept 7, 27; m 50; c 2. BOTANY, AGRONOMY. *Educ:* Univ Iowa, BA, 50; Iowa State Univ, MS, 61, PhD(econ bot), 63. *Prof Exp:* Farmer, Audubon County, Iowa, 53-57; from assoc to res asst bot, Seed Lab, Iowa State Univ, 57-63; botanist, Field Crops & Animal Prod Seed Res Lab, Agr Res Serv, USDA, 63-70; RES ASSOC PROF SEED TECHNOL, COLO STATE UNIV, 70-, DIR, COLO SEED LAB, 70- *Honors & Awards:* Award of Merit, Asn Off Seed Analysts. *Mem:* Am Soc Agron; Crop Sci Soc Am; Asn Off Seed Analysts; Coun Agr Sci & Technol; Int Seed Testing Asn; hon mem Soc Com Seed Technologists. *Res:* Developmental procedures for determining quality of seeds. *Mailing Add:* E-10 Plant Sci Colo State Univ Ft Collins CO 80523

LARSEN, AUBREY ARNOLD, b Rockford, Ill, Sept 27, 19; m 43; c 5. MEDICINAL CHEMISTRY. *Educ:* Antioch Col, BS, 43; Mich State Col, MS, 44; Cornell Univ, PhD(org chem), 46. *Prof Exp:* Mem staff, Sterling-Winthrop Res Inst, 46-60; asst dir org chem, Mead Johnson & Co, Ind, 60-63, dir chem res, 63-67, vpres phys sci, 67-70; vpres & sci dir, Bristol-Myers Int, 70-75; vpres res & develop, Mead Johnson & Co, 75-82; RETIRED. *Mem:* AAAS; NY Acad Sci; Am Chem Soc; Sigma Xi. *Res:* Pharmaceutical and nutritional research and development. *Mailing Add:* 2920 Cypress Ct Evansville IN 47711

LARSEN, AUSTIN ELLIS, b Provo, Utah, Nov 1, 23; m 45; c 4. VETERINARY MEDICINE, MICROBIOLOGY. *Educ:* Wash State Univ, BS, 48, DVM, 49; Univ Utah, MS, 56, PhD(microbiol), 69. *Prof Exp:* Vet pvt pract, Utah, 49-68; clin instr microbiol, Col Med, Univ Utah, 61-68, asst prof, 68-80, dir Vivarium, 68-87, assoc prof cellular, viral & molecular virol, 80-87; RETIRED. *Concurrent Pos:* Vet & dir res lab, Fur Breeders Agr Coop Lab, 52-68, consult vet, 68-; consult, Schering Corp, 63-; non-med med asst res, Vet Admin, 69-; mem, Adv Comt Fur Farmers Res Inst; mem, Health Task Force of Utah. *Mem:* Am Soc Microbiol; Am Soc Exp Path; Am Soc Lab Animal Pract; Sigma Xi. *Res:* Slow virus research. *Mailing Add:* 1825 S 2300 E Salt Lake City UT 84108

LARSEN, BARBARA SELIGER, b Englewood, NJ, Oct 22, 56; m 80; c 2. MASS SPECTROMETRY. *Educ:* Santa Clara Univ, BS, 78; Univ Del, PhD(phys chem), 83. *Prof Exp:* Res assoc, Int Diagnostics Tech, 78-79; NSF postdoctoral fel, Johns Hopkins Univ, 83-84; RES SCIENTIST, CENT RES, DUPONT CO, 84- *Mem:* Am Soc Mass Spectrometry; Am Chem Soc; Int Protein Soc. *Res:* Mass spectrometry as applied to the area of life sciences including sequence of proteins, identification of metabolites; new ionization techniques to expand the application of mass spectrometry. *Mailing Add:* Exp Sta Dupont Co PO Box 80228 Wilmington DE 19880-0228

LARSEN, CHARLES MCLOUD, b Staten Island, NY, Dec 6, 24; m 48; c 4. MATHEMATICS EDUCATION. *Educ:* Cornell Univ, AB, 45, AM, 50; Stanford Univ, PhD(educ), 60. *Prof Exp:* From instr to prof, 54-88, EMER PROF MATH, SAN JOSE STATE UNIV, 88- *Concurrent Pos:* Vis prof, Beijing Agr Univ, China, 85 & 86-87. *Mem:* Math Asn Am; Nat Coun Teachers Math. *Res:* History of mathematics; mathematics education. *Mailing Add:* 1675 Ellis Hollow Rd Ithaca NY 14850

LARSEN, DAVID M, b Hawthorne, NJ, Mar 8, 36; m 58; c 3. THEORETICAL PHYSICS. *Educ:* Mass Inst Technol, SB, 57, PhD(physics), 62. *Prof Exp:* Nat Res Coun-Nat Bur Standards fel physics, Nat Bur Standards, Washington, DC, 62-64; staff physicist, Lincoln Lab, 64-76, STAFF PHYSICIST, FRANCIS BITTER NAT MAGNET LAB, MASS INST TECHNOL, 76- *Mem:* Fel Am Phys Soc. *Res:* Impurities in semiconductors; polaron theory. *Mailing Add:* Six Fessenden Way Lexington MA 02173

LARSEN, DAVID W, b Chicago, Ill, Feb 21, 36; m 63. PHYSICAL CHEMISTRY. *Educ:* Dana Col, BA, 58; Northwestern Univ, PhD(phys chem), 63. *Prof Exp:* Res assoc nuclear magnetic resonance spectros, Washington Univ, 63-64; asst prof, 64-66, assoc prof chem, 66-80, PROF, UNIV MO-ST LOUIS, 80- *Mem:* Am Chem Soc. *Res:* Nuclear magnetic resonance spectroscopy; exchange reactions of Lewis acids and bases in non-aqueous media; ionic interactions in aqueous media. *Mailing Add:* Dept Chem 8001 Nat Bridge Rd Univ Mo-St Louis St Louis MO 63121

LARSEN, DON HYRUM, b Provo, Utah, Sept 22, 17; m 41; c 3. INDUSTRIAL MICROBIAL, MEDICAL MICROBIOLOGY. *Educ:* Brigham Young Univ, BS, 40; Univ Nebr, MA, 42; Univ Utah, PhD, 50. *Prof Exp:* Bacteriologist, Commercial Solvents Co, Inc, 43-46; instr bact, Univ Utah, 47-50; asst prof, Univ Nebr, 50-52; from asst prof to assoc prof, 52-60, chmn dept, 55-60 & 65-72, PROF BACT, BRIGHAM YOUNG UNIV, 60- *Concurrent Pos:* Res assoc, Naval Biol Lab, Calif, 60-61; consult, Vitamins Inc, 74- *Mem:* AAAS; Am Soc Microbiol; Soc Indust Microbiol; Soc Appl Microbiol. *Res:* Microbial physiology; pathogenic microbiology; ergosterol production by yeasts; single cell protein from waste materials. *Mailing Add:* 140 E 2000 S Orem UT 84057

LARSEN, EARL GEORGE, biochemistry, for more information see previous edition

LARSEN, EDWARD WILLIAM, b Flushing, NY, Nov 12, 44; m 74. APPLIED MATHEMATICS. *Educ:* Rensselaer Polytech Inst, BS, 66, PhD(math), 71. *Prof Exp:* Asst prof math, NY Univ, 71-76; assoc prof, Univ Del, 76-77; MEM STAFF MATH, LOS ALAMOS NAT LAB, 77- *Concurrent Pos:* Ed, Transport Theory & Statist Physics, 75- & J Appl Math, 76- *Mem:* Soc Indust & Appl Math; Am Nuclear Soc. *Res:* Asymptotic expansions; spectral theory; numerical analysis; transport theory. *Mailing Add:* Dept Nuclear Eng Univ Mich Ann Arbor MI 48109

LARSEN, EDWIN MERRITT, b Milwaukee, Wis, July 12, 15; m 46; c 3. INORGANIC CHEMISTRY. *Educ:* Univ Wis, BS, 37; Ohio State Univ, PhD(chem), 42. *Prof Exp:* Chemist, Rohm and Haas, Pa, 37-38; asst chem, Ohio State Univ, 38-42; instr, Univ Wis, 42-43; group leader, Manhattan Dist Proj, Monsanto Chem Co, Ohio, 43-46; from asst prof to assoc prof, 46-58, prof chem, Univ Wis-Madison, 58-86, assoc chmn dept, 77-86; RETIRED. *Concurrent Pos:* Vis prof, Univ Fla, 58; Fulbright lectr, Inst Inorg Chem, Vienna Tech Inst, 66-67; Wis Fusion Tech Inst, 77- *Mem:* AAAS; Am Chem Soc; Am Nuclear Soc. *Res:* Chemistry of the transitional elements; reduced states; solid state chemical and physical properties; chemistry of nuclear fusion. *Mailing Add:* Dept Chem Univ Wis Madison WI 53705

LARSEN, ELLEN WYNNE, b Paterson, NJ, Apr 28, 42; US & Can citizen; c 1. DEVELOPMENTAL GENETICS. *Educ:* Univ Mich, BSc, 63, MSc, 67, PhD(zool), 69. *Prof Exp:* Fel cytogenetics, York Univ, 69-70; fel genetics, Simon Fraser Univ, 70-74; asst prof, 74-79, ASSOC PROF GENETICS, DEPT ZOOL, UNIV TORONTO, 79- *Mem:* AAAS; Genetics Soc Am; Genetics Soc Can. *Res:* Genes and morphogenesis. *Mailing Add:* Dept Zool Univ Toronto Toronto ON M5S 1A1 Can

LARSEN, ERIC RUSSELL, b Port Angeles, Wash, July 7, 28; m 51; c 4. ORGANIC CHEMISTRY. *Educ:* Univ Wash, BS, 50; Univ Colo, PhD(org chem), 54. *Prof Exp:* Res chemist, Chem Eng Lab, 56-59, proj leader, 59-62, sr res chemist, 62-64, group leader, Halogens Res Lab, 64-68, assoc res scientist, 68-74, RES SCIENTIST, HALOGENS RES LAB, DOW CHEM CO, 74- *Mem:* AAAS; Am Chem Soc; Sigma Xi. *Res:* Primary research in bromine and fluorine chemistry; flammability of organic compounds, inluding polymers and in the mechanism of flame suppression; fire research, especially flame retardancy in plastics. *Mailing Add:* 314 W Meadowbrook Dr Midland MI 48640

LARSEN, FENTON E, b Preston, Idaho, Mar 22, 34; m 54; c 4. HORTICULTURE, POMOLOGY. *Educ:* Utah State Univ, BS, 56; Mich State Univ, PhD(hort), 59. *Prof Exp:* From asst horticulturist to assoc horticulturist, 59-73, PROF & HORTICULTURIST, WASH STATE UNIV, 73- *Concurrent Pos:* Consult, Columbia, 75, Venezuela, 77, Costa Rica, 79, 87, 88, Indonesia, 82, Ecuador, 85 & Jordan, 88; vis prof, Univ Jordan, Amman, 79. *Mem:* Am Soc Hort Sci; Am Pomol Soc; Int Plant Propagators Soc; Sigma Xi. *Res:* Pomology; propagation; rootstocks, growth regulators; leaf abscission and branching of nursery stock. *Mailing Add:* Dept Hort Wash State Univ Pullman WA 99164-6414

LARSEN, FREDERICK DUANE, b St Johnsbury, Vt, Mar 20, 30; m 52; c 4. GEOMORPHOLOGY. *Educ:* Middlebury Col, BA, 52; Boston Univ, MA, 60; Univ Mass, PhD(geol), 72. *Prof Exp:* From instr to asst prof geol, 57-89, PROF, NORWICH UNIV, 80- *Concurrent Pos:* US Geol Surv, 68-82, VT & NH Geol Surv, 83- *Mem:* Geol Soc Am; SEPM; NAGT. *Res:* Glacial geology if the Connecticut Valley of Mass, Vermont and New Hampshire; glacial geology of central Vermont. *Mailing Add:* Dept Earth Sci Norwich Univ Mil Col Northfield VT 05663

LARSEN, HAROLD CECIL, b Granite, Utah, June 15, 18; m 56; c 3. AERODYNAMICS, ASTRODYNAMICS. *Educ:* Univ Utah, BSc, 41; Calif Inst Technol, MSc, 46, AE, 55. *Prof Exp:* Prod engr, Lockheed Aircraft Corp, Calif, 41; from asst prof to assoc prof aerodyn, 46-56, PROF AERODYN & HEAD DEPT, USAF INST TECHNOL, 56- *Mem:* Am Inst Aeronaut & Astronaut; Sigma Xi. *Res:* Wind energy conversion; vortex theory of the cyclogiro and application to vertical axis wind turbine; Giro mill and Madaras rotor. *Mailing Add:* 2829 Rugby Rd Dayton OH 45406

LARSEN, HARRY STITES, b Pittsburgh, Pa, Aug 12, 27; m 56; c 3. FORESTRY. *Educ:* Rutgers Univ, BS, 50; Mich State Univ, MS, 53; Duke Univ, PhD, 63. *Prof Exp:* Forester, Southern Timber Mgt Serv, 53-56; asst prof, 59-71, assoc prof silvicult, Dept Forestry, 71-80, ASSOC PROF FORESTRY, AUBURN UNIV, 80- *Mem:* Ecol Soc Am; Soc Am Foresters. *Res:* Tree physiology; nursery seedling quality. *Mailing Add:* Sch Forestry Auburn Univ Auburn AL 36849

LARSEN, HOWARD JAMES, b Duluth, Minn, Jan 21, 25; m 46; c 2. DAIRY NUTRITION. *Educ:* Univ Wis, BS, 50; Iowa State Col, MS, 52, PhD(dairy husb), 53. *Prof Exp:* Assoc & instr dairy husb, Iowa State Col, 54-55; from asst prof to assoc prof, 55-57, PROF DAIRY SCI, MARSHFIELD EXP STA, UNIV WIS, 75- *Concurrent Pos:* Consult, US Feed Grains Coun, 81. *Mem:* Fel Am Inst Nutrit; Coun Agr Sci & Technol; Am Soc Animal Sci; Am Dairy Sci Asn; Sigma Xi. *Res:* Forage utilization by dairy cattle; forage and concentrate preservation and utilization by dairy cattle; environmental studies with ruminants. *Mailing Add:* Ashland Res Sta Ashland WI 54806

LARSEN, HOWLAND AIKENS, b Seattle, Wash, June 29, 28; m 62; c 6. CHEMICAL ENGINEERING. *Educ:* Mass Inst Technol, SB, 50; Univ Ill, MS, 51, PhD(chem eng), 57. *Prof Exp:* Jr engr, Shell Develop Co, 51-53; res engr, 57-64, sr res engr, 64-66, res supvr, 67-68, sr supvr, Fluorocarbons Div, 69-80, RES ASSOC, PLASTICS PROD DEPT, E I DU PONT DE NEMOURS & CO, INC, 80- *Mem:* Am Chem Soc. *Res:* Thermoplastics; irreversible chemical effects of high pressure and shear. *Mailing Add:* E I du Pont de Nemours & Co Inc PO Box 1217 Parkersburg WV 26102

LARSEN, JAMES ARTHUR, b Rhinelander, Wis, Mar 14, 21; m 68; c 4. PLANT ECOLOGY. *Educ:* Univ Wis-Madison, BS, 46, MS, 56, PhD(ecol), 68. *Prof Exp:* Sci ed, Univ Wis-Madison, 56-82; RETIRED. *Concurrent Pos:* Res northern boreal plant ecol & environ studies, 58-88. *Mem:* Ecol Soc Am; Bot Soc Am; Arctic Inst NAm; Nat Asn Sci Writers; Am Meteorol Soc. *Res:* Arctic and boreal forest ecology and bioclimatology; ecologico-economic studies of northern lands; arctic botany; problems of scientific communication and of public information on science; environmental and conservation writer. *Mailing Add:* Box 1496 1700 Larsen Dr Rhinelander WI 54501

LARSEN, JAMES BOUTON, b Detroit, Mich, July 28, 41; m 64; c 2. COMBUSTION TOXICOLOGY, NATURAL TOXINS. *Educ:* Kalamazoo Col, BA, 63; Univ Miami, MS, 66, PhD(marine biol), 68. *Prof Exp:* Fel biochem, Colo State Univ, 67-68; asst prof biol, Hamline Univ, 68-73; asst prof, 73-76, ASSOC PROF BIOL, UNIV SOUTHERN MISS, 76- *Concurrent Pos:* Investr, Marine Biol Lab, Woods Hole, 82; res prof, Sch Pharm, Univ Conn, 87. *Mem:* AAAS; Sigma Xi; NY Acad Sci; Am Soc Zoologists; Int Soc Toxinology. *Res:* Physiology, toxicology and pharmacology of natural toxins; physiological effects of carbon monoxide; combustion toxicology of polymers. *Mailing Add:* Dept Biol Sci Univ Southern Miss Southern Sta Box 9236 Hattiesburg MS 39406-9236

LARSEN, JAMES VICTOR, b Salt Lake City, Utah, June 16, 42; m 66. CHEMICAL & MATERIALS ENGINEERING. *Educ:* Univ Utah, BS, 67; Univ Md, College Park, MS, 69, PhD(chem eng), 71. *Prof Exp:* Chem engr, US Naval Ord Lab, 67-71; sales mgr, Eimco Div, 71-73, gen mgr, Eimcomet Plastics, 73-76, gen mgr extractor opers, Eimco PMD, 76-77, PRES, MOLDED PROD DIV, ENVIROTECH CORP, 77-, PRES, ENG DEPT. *Mem:* Am Inst Chem Engrs; Am Inst Mining Engrs. *Res:* Nonmetallic materials; carbon fiber composites; plastics. *Mailing Add:* 2906 Kennedy Dr Salt Lake City UT 84108

LARSEN, JOHN HERBERT, JR, b Tacoma, Wash, July 20, 29; m 51. VERTEBRATE ZOOLOGY. *Educ:* Univ Wash, BA, 55, MS, 58, PhD(zool), 63. *Prof Exp:* Instr embryol, Univ Wash, 60; instr biol, Univ Puget Sound, 60-61; cur & instr zool, Univ Wash, 61-62, NIH res fel, 63-64, USPHS sr fel electron micros, 64-65; from asst prof to assoc prof, 65-75, chmn dept, 78-84, PROF ZOOL, WASH STATE UNIV, 75-, DIR, ELECTRON MICROS CTR, 83- *Mem:* Am Inst Biol Sci; Am Soc Zoologists; Sigma Xi; Soc Syst Zool; Am Soc Ichthyologists & Herpetologists. *Res:* Evolution and functional morphology of feeding systems in amphibians; implications of neoteny to urodele evolution; mechanisms of ovulation in lower vertebrates. *Mailing Add:* Electron Micros Ctr Wash State Univ Pullman WA 99164-4210

LARSEN, JOHN W, b Hartford, Conn, Oct 30, 40; c 3. ORGANIC CHEMISTRY, PHYSICAL CHEMISTRY. *Educ:* Tufts Univ, BS, 62; Purdue Univ, PhD(chem), 67. *Prof Exp:* Res fel chem, Univ Pittsburgh, 66-68; from asst prof to assoc prof chem, Univ Tenn, Knoxville, 68-78, prof, 78-; chem div, Oak Ridge Nat Lab, 76-; DEPT CHEM, LEHIGH UNIV. *Concurrent Pos:* Exxon Res & Engr Co. *Honors & Awards:* Storch Award, Am Chem Soc. *Mem:* AAAS; Am Chem Soc; Sigma Xi. *Res:* Coal chemistry; thermodynamics of organic intermediates; chemistry in strong acid solutions and molten salts; solvent-solute interactions. *Mailing Add:* Dept Chem Lehigh Univ Bethlehem PA 18015

LARSEN, JON THORSTEN, thermonuclear fusion, x-ray lasers, for more information see previous edition

LARSEN, JOSEPH REUBEN, insect physiology; deceased, see previous edition for last biography

LARSEN, KENNETH MARTIN, b Ogden, Utah, June 26, 27; m 55; c 9. APPLIED MATHEMATICS. *Educ:* Univ Utah, BA, 50; Brigham Young Univ, MA, 56; Univ Calif, Los Angeles, PhD(math), 64. *Prof Exp:* Teaching asst math, Brigham Young Univ, 54-55 & Univ Calif, Los Angeles, 56-60; from asst prof to prof, 60-89, EMER PROF MATH, BRIGHAM YOUNG UNIV, 89- *Mem:* Sigma Xi; Math Asn Am; Soc Indust & Appl Math. *Res:* Numerical analysis; ordinary and partial differential equations; plasma confinement and stability. *Mailing Add:* 292 TMCB Dept Math Brigham Young Univ Provo UT 84602

LARSEN, LAWRENCE HAROLD, b Staten Island, NY, July 22, 39; m 69; c 1. OCEANOGRAPHY, HYDRODYNAMICS. *Educ:* Stevens Inst Technol, BS, 61; Johns Hopkins Univ, PhD(hydrodyn), 65. *Prof Exp:* NSF fel meteorol, Univ Oslo, 65-66; vis res asst prof, Johns Hopkins Univ, 66-67; res asst prof, 67-71, RES ASSOC PROF OCEANOG, UNIV WASH, 71- *Concurrent Pos:* Prog dir phys & chem oceanog, NSF, 72-73. *Mem:* Am Meteorol Soc; Am Geophys Union. *Res:* Physical oceanography; wave motion; estuaries; sediment dynamics; electroencephalogram; Alzheimer's diagnostics; computers. *Mailing Add:* Dept Psychiat Behav Sci RP-10 Univ Wash Seattle WA 98195

LARSEN, LELAND MALVERN, b Blair, Nebr, Aug 20, 15; m 40; c 3. ALGEBRA. *Educ:* Dana Col, BA, 41; Univ Nebr, MA, 48, PhD, 67. *Prof Exp:* Pub sch teacher, Nebr, 36-41, supt schs, 41-43; instr, Univ Nebr, 43-44; assoc prof, 48-67, head dept, 67-80, PROF MATH, KEARNEY STATE COL, 67- *Mem:* Math Asn Am; Nat Coun Teachers Math. *Res:* Theory of fields; various algorithms. *Mailing Add:* 3217 First Ave Kearney State Col 25th St & Ninth Ave Kearney NE 68847

LARSEN, LLOYD DON, b Terre Haute, Ind, Sept 13, 44; m 69; c 6. FOOD MICROBIOLOGY. *Educ:* Brigham Young Univ, BS, 69, MS, 74; Univ Minn, PhD(food microbiol), 79. *Prof Exp:* Sr scientist, Carnation Res Lab, 79-84; MICROBIOLOGIST, DUGWAY, UTAH, 84- *Mem:* Am Soc Microbiol; Inst Food Technol. *Res:* Plasmids in industrial microorganisms including group N streptococci; improving fermentations via biotechnology; development of substitute dairy products. *Mailing Add:* 115 E 2075 S Orem UT 84058-8174

LARSEN, LYNN ALVIN, b Grand Forks, NDak, Aug 2, 43; m 66; c 2. NUTRITION. *Educ:* Univ NDak, BS, 65; Univ Wash, PhD(inorg chem), 71. *Prof Exp:* Res assoc, Div Nephrology, Dept Med, Univ Wash, 71-77; chemist, GRAS Rev Br, Div Food & Color Additives, Ctr Food Safety & Appl Nutrit, Food & Drug Admin, 77-78, consumer safety officer, 78-81, assoc dir prog develop, Div Nutrit, 81-90, sci policy analyst, 90-91, DIR, EXEC OPERS STAFF, CTR FOOD SAFETY & APPL NUTRIT, FOOD & DRUG ADMIN, 91- *Concurrent Pos:* Wash State Heart Asn fel, Div Nephrology, Dept Med, Univ Wash, 72-74; instr, Prog Educ Gifted, Prince William County, Va, 79-83. *Mem:* Am Chem Soc; AAAS. *Mailing Add:* Exec Opers Staff Ctr Food Safety & Appl Nutrit Food & Drug Admin HFF-6 200 C St SW Washington DC 20204

LARSEN, MARILYN ANKENEY, b Chicago, Ill, Dec 13, 40; m; c 2. PHYSICAL CHEMISTRY, INORGANIC CHEMISTRY. *Educ:* Univ Wis-Madison, BS, 62; Univ Wash, PhD(phys chem), 67. *Prof Exp:* From asst prof to assoc prof, 67-82, chmn dept, 73-82, asst to provost, 82-86, dean, sch sci & prof studies, 86-89 PROF CHEM, MONMOUTH COL, 81- *Concurrent Pos:* Counr, Am Chem Soc, 82-88. *Honors & Awards:* Virgil F Payne Award, Am Chem Soc. *Mem:* Am Chem Soc; Nat Sci Teachers Asn. *Res:* infrared spectroscopy. *Mailing Add:* Dept Chem Monmouth Col West Long Branch NJ 07764

LARSEN, MARLIN LEE, b Grand Island, Nebr, Nov 22, 42; m 64; c 3. CLINICAL CHEMISTRY. *Educ:* Kearney State Col, BS, 64; Wash State Univ, PhD(chem), 68. *Prof Exp:* dir res & develop, ICN Med Labs, Inc, 68-79; CONSULT, 79-; PRES, HARVARD LAB & XRAY INC, ROSEBURG, ORE, 81- *Mem:* Am Asn Clin Chemists; Am Asn Bioanalysts; Am Chem Soc. *Res:* Automation and methodology research of medical laboratory procedures. *Mailing Add:* 160 Rivershore Dr Roseburg OR 97470-9425

LARSEN, MAX DEAN, b Pratt, Kans, Jan 23, 41; m 62. ALGEBRA. *Educ:* Kans State Teachers Col, BA, 61; Univ Kans, MA, 63, PhD(math), 66. *Prof Exp:* From asst prof to assoc prof, 66-73, dean, Col Arts & Sci, 74-82, PROF MATH, UNIV NEBR-LINCOLN, 73- *Concurrent Pos:* NSF res grant, 68-70. *Mem:* Am Math Soc; Math Asn Am; Nat Coun Teachers Math; Sigma Xi. *Res:* Extension of integral domain concepts to general commutative rings, particularly valuation theory and Prüfer rings; module theory over commutative rings. *Mailing Add:* 641 Haverford Circle Lincoln NE 68510

LARSEN, MICHAEL JOHN, b London, Eng, Apr 27, 38; m 70; c 2. MYCOLOGY, FOREST PATHOLOGY. *Educ:* Syracuse Univ, BSc, 60; State Univ NY, MSc, 63, PhD(mycol pathol), 67. *Prof Exp:* Res scientist, Can Forestry Serv, 66-70; RES SCIENTIST, US FOREST SERV, USDA, 70- *Concurrent Pos:* Adj assoc prof, Univ Wis-Madison, 71- & Mich Technol Univ, 78- *Mem:* Mycol Soc Am; Am Phytopath Soc; Int Asn Plant Taxon & Nomeclature; Int Mycol Soc. *Res:* Speciation, taxonomy, physiology of North American wood inhabiting fungi and their ecological roles in forest ecosystems. *Mailing Add:* Ctr Forest Mycol Res Forest Serv USDA Pinchot Dr Madison WI 53705

LARSEN, MIGUEL FOLKMAR, b Caracas, Venezuela, June 2, 53; US citizen; m 78. MESOSCALE & RADAR METEOROLOGY. *Educ:* Univ Rochester, BS, 71; Cornell Univ, MS, 77, PhD(meteorol), 79. *Prof Exp:* Res assoc ionospheric physics, Cornell Univ, 79-84; ASST PROF PHYSICS, CLEMSON UNIV, 84- *Concurrent Pos:* Mem, sci adv comt, Arecibo Observ, 85-87; prin investr, Air Force Off Sci Res grant, 85-88 & NASA grant, 86-88. *Mem:* Am Geophys Union; Am Meteorol Soc. *Res:* Mesoscale meteorological research using radar wind profilers; studies of neutral and ion interactions in the auroral zone thermosphere using sounding rockets and radars. *Mailing Add:* 102 Blue Ridge Dr Clemson SC 29631

LARSEN, PAUL M, BOTANY. *Prof Exp:* RES ASSOC BOT, UNIV MD, 87- *Mailing Add:* Bot Dept Univ Md College Park MD 20742

LARSEN, PETER FOSTER, b Mt Kisco, NY. BIOLOGICAL OCEANOGRAPHY. *Educ:* Univ Conn, BA, 67, MS, 69; Col William & Mary, PhD(marine sci), 74. *Prof Exp:* Lectr chem, Norwalk Community Col, 68-70; res asst marine biol, Va Inst Marine Sci, 70-72, asst marine scientist ecol, 72-73; state oceanogr, Maine Dept Marine Resources, 73-76; res scientist, 76-77, SR RES SCIENTIST, BIGELOW LAB OCEAN SCI, 77-; PRIN INVESTR, KENNEBEC AREA RES ENDOWMENT, 91- *Concurrent Pos:* Pres, Coastal Sci, 73-; consult, Res Inst Gulf Maine, 73- & Bigelow Lab Ocean Sci, 75-76; gov bd, Estuarine Res Fedn, 80-84 & secy, 81-83. *Mem:* New Eng Estuarine Res Soc (secy-treas, 80-82, pres, 82-84, past pres, 84-86); Estuarine & Coastal Sci Asn; Estuarine Res Fedn. *Res:* The documentation of benthic community structure and function in the estuarine and marine environments of the Gulf of Maine region; environmental quality of the Kennebec River estuary and its influences on the coastal ocean. *Mailing Add:* Bigelow Lab Ocean Sci West Boothbay Harbor ME 04575

LARSEN, PHILIP O, b Audubon Co, Iowa, Dec 1, 40; m 61; c 3. PLANT PATHOLOGY. *Educ:* Iowa State Univ, BS, 63; Univ Ariz, MS, 67, PhD(plant path), 69. *Prof Exp:* Asst prof plant path, Ohio State Univ, 69-73, assoc prof, 73-76, from assoc prof to prof bot, 76-85; PROF & DEPT HEAD, DEPT PLANT PATH, UNIV MINN, 85- *Mem:* Am Phytopath Soc. *Res:* Diseases of turf grasses. *Mailing Add:* 495 Borlang Hall Univ Minn 1991 Upper Buford Circle St Paul MN 55108

LARSEN, RALPH IRVING, b Corvallis, Ore, Nov 26, 28; wid; c 4. ENVIRONMENTAL ENGINEERING. *Educ:* Ore State Univ, BS, 50; Harvard Univ, MS, 55, PhD(air pollution, indust hyg), 57. *Prof Exp:* Sanit engr, Div Water Pollution Control, USPHS, 50-54, chief tech serv, State & Community Serv Sect, Nat Air Pollution Control Admin, 57-61, chief biomet sect, field studies br, 63-65, asst chief br, 65-67, res engr, Off Criteria & Standards, 67-71; environ res engr, Environ Opers Br, Meteorol & Assessment Div, Atmospheric Sci Res Lab, 71-88, ENVIRON ENGR, EXPOSURE ASSESSMENT RES DIV, ATMOSPHERIC RES & EXPOSURE ASSESSMENT LAB, US ENVIRON PROTECTION AGENCY, 88- *Concurrent Pos:* Lectr, adj fac, Inst Air Pollution Training, Environ Protection Agency, 69- *Honors & Awards:* Commendation Medal, USPHS, 79. *Mem:* Air & Waste Mgt Asn; Sigma Xi. *Res:* Studies on the concentration, effects and control of air pollution; mathematical modeling; computer analyses; statistical analyses. *Mailing Add:* Environ Res Ctr Environ Protection Agency MD-56 Research Triangle Park NC 27711

LARSEN, ROBERT PAUL, b Vineyard, Utah, Dec 1, 26; m 48; c 4. HORTICULTURE. *Educ:* Utah State Univ, BS, 50; Kans State Univ, MS, 51; Mich State Univ, PhD(hort), 55. *Prof Exp:* From asst prof to prof hort, Mich State Univ, 55-68; prof hort & supt, Tree Fruit Res Ctr, Wash State Univ, 68-82; VPRES UNIV EXTEN, UTAH STATE UNIV, 82- *Mem:* Fel Am Soc Hort Sci (pres, 75-76); Int Soc Hort Sci; fel AAAS. *Res:* Physiology, nutrition and management of tree fruit crops. *Mailing Add:* 1175 Cedar Heights Logan UT 84321

LARSEN, ROBERT PETER, b Kalamazoo, Mich, Jan 1, 21; m 48; c 2. PHYSICAL CHEMISTRY. *Educ:* Kalamazoo Col, AB, 42; Brown Univ, PhD(chem), 48. *Prof Exp:* Instr phys chem, Brown Univ, 44-46; instr anal chem, Amherst Col, 47-48; asst prof phys chem, Ohio Wesleyan Univ, 48-51; assoc chemist & anal group leader, 51-73, GROUP LEADER, METAB & ENVIRON BEHAVIOR PLUTONIUM, RADIOL & ENVIRON RES DIV, ARGONNE NAT LAB, 73-, TRANSP TECHNOL ANALYST. *Mem:* Am Chem Soc. *Res:* Inorganic analytical chemistry in reactor fuel process development; uranium and plutonium analysis and the measurement of fission yields; metabolism of plutonium and other actinide elements in mammals, the behavior of these elements in the environment, and their determination in materials of biological and environmental origin. *Mailing Add:* 5333 N Sheridan Rd Apt 10-I Chicago IL 60640-2562

LARSEN, RONALD JOHN, b Chicago, Ill, Jan 1, 37; m 62; c 1. MATHEMATICAL ANALYSIS. *Educ:* Mich State Univ, BS, 57, MS, 59; Stanford Univ, PhD(math), 64. *Prof Exp:* Instr math, Yale Univ, 63-65; asst prof, Cowell Col, Univ Calif, Santa Cruz, 65-70; assoc prof math, Wesleyan Univ, 70-75; vis assoc prof, State Univ NY Binghamton, 75-76 & State Univ NY Albany, 76-77; adj assoc prof, Clarkson Col Technol, 77-78. *Concurrent Pos:* Fulbright-Hays advan res grant, Univ Oslo, 68-69, vis asst, 73-74; Fulbright-Hays travel award to Norway, 73-74. *Mem:* Am Math Soc; Math Asn Am; Norweg Math Soc. *Mailing Add:* RD 4 Canton NY 13617

LARSEN, RUSSELL D, b Muskegon, Mich, June 6, 36; m 58; c 2. CHEMICAL PHYSICS, STATISTICS. *Educ:* Kalamazoo Col, BA, 57; Kent State Univ, PhD(chem), 64. *Prof Exp:* Teaching asst chem, Univ Cincinnati, 58-60; asst instr, Kent State Univ, 64; res assoc, Princeton Univ, 64-65; Robert A Welch fel, Rice Univ, 65-66; asst prof, Ill Inst Technol, 66-72; asst prof chem, Tex A&M Univ, 72-76; actg assoc prof, dept chem, Univ Nev, Reno, 76-77; assoc prof, dept chem, Univ Mich, 77-83; DEPT CHEM, TEX TECH UNIV, LUBBOCK, 83- *Mem:* Am Chem Soc; Am Phys Soc; Am Statist Asn; Sigma Xi; AAAS. *Res:* Chemical and biomedical signal processing; spectral analysis; Walsh functions; spline representations; zero-based signal representations; chemical education, chaos. *Mailing Add:* Dept Epidemiol Univ Pittsburgh Grad Sch Pub Health Pittsburgh PA 15261

LARSEN, SIGURD YVES, b Brussels, Belg, Aug 14, 33; US citizen. THEORETICAL PHYSICS. *Educ:* Columbia Univ, AB, 54, MA, 56, PhD(physics), 62. *Prof Exp:* Asst physics, Columbia Univ, 54-57 & 60-62; consult, Nat Bur Standards, 62; Nat Acad Sci-Nat Res Coun assoc, 62-63, physicist, Washington, DC, 63-68; assoc prof, 68-75, PROF PHYSICS & CHMN DEPT, TEMPLE UNIV, 75- *Concurrent Pos:* Consult, Los Alamos Sci Lab, 64, Lawrence Radiation Lab, 67-72 & Nat Bur Standards, 68-71; mem panel quantum fluids, Int Union Pure & Appl Chem, 66-; vis prof, Mex Inst Petrol, 71-72 & Nat Univ Mex, 72. *Mem:* Am Phys Soc; Ital Phys Soc; Sigma Xi. *Res:* Statistical physics; quantum theory; numerical analysis. *Mailing Add:* Dept Physics Temple Univ Philadelphia PA 19122

LARSEN, STEVEN H, b Bringham City, Utah, Aug 28, 44; m 68; c 4. CHLAMYDIA TRACHOMATIS, CLONING. *Educ:* Utah State Univ, BA, 68, MS, 70; Univ Wis, PhD(biochem), 74. *Prof Exp:* Teaching asst chem & physics, Utah State Univ, 66-70; trainee biochem, Univ Wis, 70-74, fel genetics, 74-75; fel microbiol, John Hopkins Univ Sch Med, 75-79, instr, 77; ASST PROF MICROBIOL, IND UNIV SCH MED, 79- *Concurrent Pos:* Consult, Human Amylase Cloning Group, Dept Genetics, Ind Univ, 80-; Eli Lilly Young Scientist Res Awards, 83-85. *Mem:* Am Soc Microbiol; AAAS. *Res:* Development of eukaryotic cloning vectors; chlamydia trachomatics genetics. *Mailing Add:* Dept Microbiol & Immunol Ind Univ Sch Med 625 Barnhill Dr Indianapolis IN 46223

LARSEN, TED LEROY, b Jerome, Idaho, Mar 18, 35; m 57; c 2. PHYSICS, SEMICONDUCTORS. *Educ:* Univ Calif, Berkeley, BS, 61, MS, 62; Stanford Univ, PhD(mat sci), 70. *Prof Exp:* Mem tech staff, Hewlett-Packard Co, 62-65, res & develop group leader, 68-71, res & develop sect mgr semiconductors, 71-79; DIR RES & DEVELOP, ENG OPTOELECTRONICS DIV, GEN INSTRUMENT CORP, 79- *Mem:* Electrochem Soc; Inst Elec & Electronics Engrs. *Res:* III-V compound materials and devices for optoelectronic and microwave applications, including single crystal and epitaxial growth, crystalline defects, impurity diffusion and minority carrier recombination processes; III-IV and silicon device and product development including led lamps and displays and optocouplers. *Mailing Add:* 1099 Los Robles Palo Alto CA 94306

LARSEN, WILLIAM L(AWRENCE), b Crookston, Minn, July 16, 26; m 54; c 2. CORROSION, FAILURE ANALYSIS. *Educ:* Marquette Univ, BME, 48; Ohio State Univ, MS, 50, PhD(metall eng), 56. *Prof Exp:* Res assoc metall eng, Ohio State Univ, 51-56; res metallurgist, E I du Pont de Nemours & Co, Inc, 56-58; asst prof mech eng & chem, 58-62, from assoc metallurgist to metallurgist, Ames Lab, 58-69, assoc prof 69-73, PROF METALL, IOWA STATE UNIV, 73- *Concurrent Pos:* Consult mat design, mat failure & educ progs for superior students. *Mem:* Am Soc Metals Int; Nat Col Hons Coun; Nat Asn Corrosion Engrs; Am Soc Eng Educ; Am Soc Testing & Mat. *Res:* Metallurgical engineering design and failure analysis; expert witness-product liability. *Mailing Add:* Dept Mat Sci & Eng Iowa State Univ Ames IA 50011

LARSEN-BASSE, JORN, b Maribo, Denmark, Oct 14, 34; US citizen; m 59; c 1. MATERIALS SCIENCE, CORROSION. *Educ:* Tech Univ Denmark, MS, 58, PhD(metall), 61. *Prof Exp:* Actg asst prof metall, Tech Univ Denmark, 59-61; researcher, Soderfors Bruk, Sweden, 61-62; res assoc mat sci, Stanford Univ, 63; asst prof, San Jose State Col, 63-64; from asst prof to prof mech eng, Univ Hawaii, 64-86, chmn dept, 76-80 & 82-85; PROF MECH ENG, GA INST TECHNOL, 86- *Concurrent Pos:* Ford Found resident indust, 68-69; hon vis prof, Univ New South Wales, 71 & Commonwealth Sci & Indust Res Orgn, Melbourne, 72; prog dir, NSF, 88- *Mem:* Am Inst Mining, Metall & Petrol Engrs; Nat Asn Corrosion Engrs; Am Soc Mech Engrs; Sigma Xi; Soc Tribologists & Lubrication Engrs. *Res:* Abrasion resistance of metals; corrosion in marine and volcanic environments; ocean thermal energy conversion. *Mailing Add:* Tribology Prog NSF 1800 G St NW No 1108 Washington DC 20550

LARSH, HOWARD WILLIAM, b East St Louis, Ill, May 29, 14; m 38; c 1. MEDICAL MYCOLOGY. *Educ:* McKendree Col, BA, 36; Univ Ill, MS, 38, PhD, 41. *Prof Exp:* Asst bot, Univ Ill, 38-41; instr, Dept Bot & Bact, Univ Okla, 41-42; plant pathologist, Bur Plant Indust, Soils & Agr Eng, USDA, 43-45; assoc prof bact, med, mycol & plant path, 45-48, chmn, dept plant sci, 45-62 & dept microbiol & bot, 66-76, PROF PLANT SCI, UNIV OKLA, 48-, RES PROF, 62- *Concurrent Pos:* Spec consult, McKnight State Tuberc Hosp & Nat Communicable Disease Ctr, USPHS, 50-; med mycologist consult & co-dir lab, Mo State Chest Hosp, Mt Vernon, 55-, dir labs, 77; consult, Manned Spacecraft Ctr, NASA; res reviewer immunol & infectious diseases comt, Vet Admin Hosp, Oklahoma City, 72; cert bioanal lab dir, Am Bd

Bioanal, 74-; ed rev bd, J Clin Microbiol, 74-; ed-in-chief, Sabouraudia, 75-79; consult med mycol. *Mem:* Fel AAAS; fel Am Pub Health Asn; fel Am Acad Microbiol; Bot Soc Am; Mycol Soc Am; Sigma Xi. *Res:* Medical mycosis; systemic mycoses; histoplasmosis; cryptococcosis. *Mailing Add:* 611 Broad Lane Norman OK 73069

LARSON, ALLAN BENNETT, b Chicago, Ill, Feb 9, 43; m 71; c 2. INDUSTRIAL PHARMACY, COSMETIC CHEMISTRY. *Educ:* Drake Univ, BS, 66; Univ Wis, MS, 69; Purdue Univ, PhD(phys & indust pharm), 72. *Prof Exp:* Sr res pharmacist, Dorsey Labs Div, Sandoz-Wander Inc, 72-76; sr pharmaceut scientist, Richardson-Merrell, Inc, 76-77, group leader, Vick Divs Res & Develop, 77-80, mgr tech serv, Vicks Personal Care Div, 80-84; dir, process res & develop, Ayerst Labs, 84-87, DIR PROCESS RES & DEVELOP, WYETH-AYERST RES, AM HOME PROD, 87- *Mem:* Am Pharmaceut Asn; Acad Pharmaceut Sci; Soc Cosmetic Chemists; Am Asn Pharmaceut Scientists; Drug Info Asn. *Res:* Pharmaceutical and skin care product development and stability, preformulation, uniformity of mixing, scale-up technology, process improvements, process validation. *Mailing Add:* Rd 2 Box 112 C Peru NY 12972-9714

LARSON, ANDREW HESSLER, b Peru, Ill, Sept 14, 31; m 55; c 3. EXTRACTIVE CHEMICAL METALLURGY. *Educ:* Mo Sch Mines, BS, 53, MS, 54; Univ Mo, PhD(metall), 59. *Prof Exp:* From asst prof to assoc prof metall eng, Mo Sch Mines, 59-63; from assoc prof to prof, Colo Sch Mines, 63-68; sr res metallurgist, Bunker Hill Co, 68-69, mgr res & develop, 69-78; dir prod develop & eng, Metals Div, Gould Inc, 78-84; dir metals eng, GNB Inc, 84-88; METALL CONSULT, 88- *Concurrent Pos:* Ford Found Prog prod specialist, Tex Div, Dow Chem Co, 66-67. *Mem:* Am Soc Metals; Am Inst Mining, Metall & Petrol Engrs; Sigma Xi; Can Inst Mining & Metall. *Res:* High temperature thermodynamic and kinetic studies of metallurgical systems involving the extraction and refining of metals. *Mailing Add:* 4116 Strawberry Lane Eagan MN 55123-1421

LARSON, ARVID GUNNAR, b Chicago, Ill, July 26, 37; m 89; c 1. SIGNAL PROCESSING, COMPUTER ARCHITECTURE. *Educ:* Ill Inst Tech, BS, 59; Stanford Univ, MS, 66, PhD(elec eng), 73. *Prof Exp:* Res engr, Stanford Res Inst, 64-74; mgr, appl res group, Planning Res Corp, 74-78; proj mgr, Syst Planning Corp, 78-80; div mgr, Advan Res & Appln Corp, 80-85; vpres, Analytic Disciplines Inc, 85-86; prin, Booz, Allen & Hamilton Inc, 86-89; sr vpres, Syntek Eng & Computer Systs, 89-90; RES PROF, GEORGE MASON UNIV, 91- *Concurrent Pos:* Bd dir, Res Inst Info Sci & Eng; chmn, Prof Activ Coun, Am Asn Eng Socs; vchmn, US Activ Bd; chmn, Technol Policy Coun & US Technol Policy Conf, Inst Elec & Electronics Engrs. *Honors & Awards:* Centennial Medal, Inst Elec & Electronics Engrs. *Mem:* Sigma Xi; fel Inst Elec & Electronics Engrs. *Res:* Computer systems and network architecture, high speed signal processing; software engineering; systems science. *Mailing Add:* 6921 Espey Lane McLean VA 22101-5455

LARSON, BENNETT CHARLES, b Buffalo, NDak, Oct 9, 41; m 69; c 2. X-RAY DIFFRACTION, CRYSTAL DEFECTS. *Educ:* Concordia Col, BA, 63; Univ NDak, MS, 65; Univ Mo, PhD(physics), 70. *Prof Exp:* Physicist, 69-79, group leader, 80-90, SECT HEAD, SOLID STATE DIV, OAK RIDGE NAT LAB, 90- *Honors & Awards:* Bertram E Warren Diffraction Physics Award, Am Crystallog Asn, 85. *Mem:* Am Phys Soc; Am Crystallog Asn; Mat Res Soc. *Res:* X-ray diffraction study of intrinsic and induced defects in crystalline solids using diffuse scattering; x-ray diffraction study of pulsed-laser annealing. *Mailing Add:* Solid State Div PO Box 2008 Oak Ridge Nat Lab Oak Ridge TN 37831-6030

LARSON, BRUCE LINDER, b Minneapolis, Minn, June 24, 27; m 54; c 3. BIOCHEMISTRY, NUTRITION. *Educ:* Univ Minn, BS, 48, PhD(biochem), 51. *Prof Exp:* Asst, Univ Minn, 48-51; from instr to prof biol chem, Univ Ill, Urbana, 51-90, mem nutrit sci fac, 72-90, emer prof biol chem, 90-; RETIRED. *Concurrent Pos:* Prin investr, NSF, 59-74, 79-82, NIH, 72-75, Nat Dairy Coun, 76-79. *Honors & Awards:* Am Chem Soc Award, 66; Fulbright lectr, Arg, 65. *Mem:* AAAS; Am Chem Soc; Am Soc Biol Chemists; Am Dairy Sci Asn. *Res:* Lactation and mammary gland metabolism in the formation of mammary secretions, including origin and transport of precursors, synthesis and cellular ejection of products and biological significance of colostrum and milk; lactation. *Mailing Add:* Dept Animal Sci 315 Animal Sci Lab Univ Ill 1207 W Gregory Dr Urbana IL 61801

LARSON, CHARLES CONRAD, b Pettibone, NDak, Nov 17, 14; m 58; c 2. FORESTRY ADMINISTRATION. *Educ:* Univ Minn, BS, 40; Univ Vt, MS, 43; Inst Pub Admin, NY, cert, 50; State Univ NY Col Forestry, Syracuse Univ, PhD(forestry admin), 52. *Prof Exp:* Summer res asst, Lake States Forest Exp Sta, 40-41; asst forest supvr, Crossett Lumber Co, Ark, 44; res & state exten forester, Univ Vt, 44-47; res assoc, Inst Pub Admin, NY, 48-49; forest econmist & res assoc forestry, 50-58, from assoc prof to prof world forestry, 59-83, dir int forestry, 64-79, dean, Sch Environ & Resource Mgt, 71-79, EMER PROF FORESTRY, COL ENVIRON SCI & FORESTRY, STATE UNIV NY, 83- *Concurrent Pos:* Ford Found oversease res fel, 54-; consult, Syracuse Univ Res Inst, 57-59; vis prof & proj leader, AID-State Univ NY Res Found Assistance Contract, Col Forestry, Univ Philippines, 59-62; mem bd dirs, Orgn Trop Studies, 71-; chmn educ comn, Int Union Socs Foresters, 71-79. *Mem:* Soc Am Foresters; Int Soc Trop Foresters (pres, 79-81). *Res:* Forestry administration, policy and economics; tropical vegetation; world forestry development with emphasis on the tropics; range management. *Mailing Add:* 200 Ridgecrest Rd Syracuse NY 13214

LARSON, CHARLES FRED, b Gary, Ind, Nov 27, 36; m 59; c 2. MECHANICAL ENGINEERING. *Educ:* Purdue Univ, BS, 58; Fairleigh Dickinson Univ, Rutherford, MBA, 73. *Prof Exp:* Proj engr, Combustion Eng, Inc, 58-60; asst dir, Welding Res Coun, 60-75; EXEC DIR, INDUST RES INST, INC, 75- *Concurrent Pos:* Secy, Indust Res Inst Res Corp, 75- *Mem:* Fel AAAS; Am Soc Mech Engrs; Soc Res Adminr. *Res:* Pressure vessels; fatigue of welded structures; fracture toughness of metals; non-destructive examination; heavy-section steels; research management; research and technical management. *Mailing Add:* 1550 M St NW Washington DC 20005

LARSON, CLARENCE EDWARD, b Cloquet, Minn, Sept 20, 09; m 34, 57; c 3. PHYSICAL CHEMISTRY. *Educ:* Univ Minn, BS, 32; Univ Calif, PhD(chem), 36. *Prof Exp:* Instr chem, Univ Calif, 32-36; res assoc, Mt Zion Res Found, Calif, 36-37; assoc prof chem, Col of the Pac, 37-39, prof & head dept, 39-40; chief anal sect, Radiation Lab, Univ Calif, 40-43; head tech staff, Electromagnetic Plant, Carbide & Carbon Chem Co Div, Union Carbide Corp, 43-46, dir res & develop, 46-49, supt, 49-50, dir, Oak Ridge Nat Lab, 50-55, vpres chg res, Nat Carbon Co Div, 55-59, assoc mgr res admin, 59-61, vpres, Union Carbide Nuclear Co Div, 61-64, pres, Nuclear Div, Union Carbide Nuclear Co, 64-69; comnr, Energy Res & Develop Admin, Washington, DC, 69-74; RETIRED. *Concurrent Pos:* Am Nuclear Soc fel, 62; mem bd dirs, Oak Ridge Inst Nuclear Studies, 63-65; energy consult, Systs Control, Inc, 64-74; comnr, US Atomic Energy Comn, 69-74, energy consult, 74-83; deleg, UN Conf Peaceful Uses of Atomic Energy. *Honors & Awards:* Soc Advan Mgt Award, 65. *Mem:* Nat Acad Eng; AAAS; Am Chem Soc; fel Am Inst Chemists; Am Nuclear Soc; Pioneers Sci & Technol Hist Asn (pres, 83-). *Res:* Inorganic chemistry of biological systems; separation methods for isotopes; radiochemistry; analytical methods; uranium chemistry; colloids; chemical separation methods. *Mailing Add:* 6514 Bradley Blvd Bethesda MD 20817

LARSON, CURTIS L(UVERNE), b Cottonwood, Minn, Oct 10, 20; m 44; c 3. AGRICULTURAL & CIVIL ENGINEERING. *Educ:* Univ Minn, BAgrE, 43, MS, 49; Stanford Univ, PhD(civil eng), 65. *Prof Exp:* From instr to assoc prof agr eng, 48-65, PROF AGR ENG, INST AGR, UNIV MINN, ST PAUL, 65- *Concurrent Pos:* Water mgt consult, Colombia, 72-73, Nicaragua, 75, Chile, 76 & Panama, 78. *Mem:* Am Soc Agr Engrs; Am Geophys Union; Am Water Resources Asn; Soil Conserv Soc Am. *Res:* Surface water hydrology; watershed modeling; erosion control; water resources. *Mailing Add:* 2232 Ferris Lane St Paul MN 55113

LARSON, D WAYNE, b Prince Albert, Sask, Dec 2, 38; m 62; c 4. PHYSICAL CHEMISTRY. *Educ:* Univ Sask, BA, 61, MA, 62; Univ Toronto, PhD(inorg chem), 65. *Prof Exp:* Mem res staff, Can Forces Inst Aviation Med, 65-68; asst prof, 68-71, ASSOC PROF CHEM, UNIV REGINA, 71 - *Concurrent Pos:* Asst to dean, Fac Grad Studies & Res, Univ Regina, 73-79. *Mem:* Chem Inst Can; Can Nuclear Soc. *Res:* Science education, attitudes toward science; technology assessment, ethics, public policy; social issues, perceptions of risk, energy resources; uranium development in Saskatchewan. *Mailing Add:* Dept Chem Univ Regina Regina SK S4S 0A2 Can

LARSON, DANIEL JOHN, b Minneapolis, Minn, Nov 8, 44; m 66; c 2. LASER SPECTROSCOPY, NEGATIVE IONS. *Educ:* St Olaf Col, Northfield, Minn, BA, 66; Harvard Univ, Cambridge, Mass, MA, 67, PhD(physics), 71. *Prof Exp:* From asst prof to assoc prof, physics, Harvard Univ, 70-78; assoc prof, 78-87, PROF PHYSICS, UNIV VA, 87- *Concurrent Pos:* Vis scientist, Nat Bur Standards, 85-86; vis prof, Chalmers Univ, Gothenburg, Swed, 86. *Mem:* Am Phys Soc; Optical Soc Am. *Res:* High resolution optical and microwave spectroscopy of atoms and ions, especially negative ions, using tunable lasers and atom and ion storage and beam techniques; atomic structure and interactions in strong fields. *Mailing Add:* Dept Physics Univ Va McCormick Rd Charlottesville VA 22901

LARSON, DONALD ALFRED, b Chicago, Ill, Sept 15, 30; m 53; c 3. BOTANY. *Educ:* Wheaton Col, Ill, BS, 53; Univ Ill, MS, 55, PhD(bot), 59. *Prof Exp:* Assoc prof bot, Univ Tex, Austin, 59-69, prof & dir health prof, 69-73; assoc vpres health sci, 73-81, PROF BIOL HEALTH SCI, STATE UNIV NY BUFFALO, 73- *Mem:* Bot Soc Am; Am Soc Cell Biologists. *Res:* Electron microscopy; cytology; palynology; taxonomic uses; paleobotany. *Mailing Add:* 202A Hockstetter Hall N Campus State Univ NY Buffalo NY 14260

LARSON, DONALD CLAYTON, b Wadena, Minn, Jan 29, 34; m 60; c 3. SOLID STATE PHYSICS. *Educ:* Univ Wash, BS, 56; Harvard Univ, SM, 57, PhD(appl physics), 62. *Prof Exp:* Asst prof physics, Univ Va, 62-67; assoc prof, 67-83, PROF PHYSICS, DREXEL UNIV, 83- *Mem:* Am Phys Soc; Int Solar Energy Soc; Am Soc Testing & Mat. *Res:* Electrical properties of metallic, organic and amorphous semiconducting films; biomechanics; solar energy; insulation studies; integrated optics. *Mailing Add:* Dept Physics Drexel Univ Philadelphia PA 19104

LARSON, DONALD W, b Avoca Minn, Aug 7, 40; m 62; c 3. AGRICULTURAL ECONOMICS. *Educ:* SDak State Univ, BS, 62; Mich State Univ, MS, 64, PhD(agr econ), 68. *Prof Exp:* Asst prof agr econ, Mich State Univ, 68-70; from asst prof to assoc prof, 70-82, PROF AGR ECON, OHIO STATE UNIV, 82- *Concurrent Pos:* Mkt consult, 73, Loan consult, US AID, 74, 85, 86; consult, mkt & price policy, World Bank, 78, 80, 82, 84, 86. *Mem:* Am Agr Econ Asn; Int Asn Agr Economists; Brazilian Agr Econ Asn. *Res:* Grain marketing and transportation systems, marketing and market policy in developing countries; rural financial markets in developing countries. *Mailing Add:* Ohio State Univ Dept Agr Econ Columbus OH 43210

LARSON, EDWARD WILLIAM, JR, b New Haven, Conn, Apr 17, 23; m 52; c 2. STRUCTURAL ENGINEERING. *Educ:* Ind Technol Col, BSCE, 43; Northwestern Univ, MS, 48, PhD(civil eng), 53. *Prof Exp:* Jr engr, Bur Ships, US Dept Navy, 43-44, physicist, David Taylor Model Basin, 44-46; res assoc struct, Northwestern Univ, 49-52, instr struct eng, 52-53; engr, Hughes Tool Co, Calif, 53-55; res engr, Lockheed Aircraft Corp, 55-57; supv stress anal, Rocketdyne Div, Rockwell Int, 57-62, group leader dynamic sci, 62-63, sect chief & mgr turbomach, 63-69, mgr tech specialties, 69-75, assoc chief engr, 75-79, dir, design technol, 79-86; RETIRED. *Concurrent Pos:* Lectr, Univ Calif, Los Angeles, 53-64, Calif State Univ, 86- *Mem:* Soc Exp Stress Analysis; Am Inst Aeronaut & Astronaut; Sigma Xi; Am Soc Eng Educ. *Res:* Failure investigation of engine components under actual operating conditions and structural behavior of liquid rocket engines to transient, steady state and flow-induced vibrations. *Mailing Add:* 18621 Ringling Tarzana CA 91356

LARSON, EDWIN E, b Los Angeles, Calif, Jan 5, 31. GEOPHYSICS, GEOLOGY. *Educ:* Univ Calif, Los Angeles, BA, 54, MA, 58; Univ Colo, PhD(geol), 65. *Prof Exp:* Explor geologist, Humble Oil & Refining Co, 57-60; NSF fel, 65-66; from asst prof to assoc prof, 66-75, PROF GEOL SCI, UNIV COLO, BOULDER, 75- *Mem:* AAAS; Geol Soc Am; Am Geophys Union; Sigma Xi. *Res:* Investigation of rock magnetic properties; paleomagnetism and its application to the solution of geological problems; lunar magnetism. *Mailing Add:* Dept Geol Univ Colo Campus Box 250 Boulder CO 80309-0250

LARSON, ELAINE LUCILLE, b Apr 27, 43. NURSING. *Educ:* Univ Wash, Seattle, BS, 65, MS, 69, PhD(epidemiol), 81. *Prof Exp:* Staff nurse, Univ Wash Hosp, Seattle, 65-66, clin specialist, cardiovasc nursing, 66-67, hosp epidemiologist, 67-70, nurse coordr, staff develop, 76-77, nursing res coordr, 77-81, assoc dir nursing, 81-83; clin nurse, Robert Wood Johnson Sch Nursing, Univ Pa, 83-85; M ADELAIDE NUTTING CHAIR CLIN NURSING, SCH NURSING, JOHNS HOPKINS UNIV, 85-, PROF & DIR, CTR NURSING RES, 90- *Concurrent Pos:* Instr, cardiopulmonary resuscitation, Regional Med Prog & Wash State Red Cross, 66-67; asst prog dir, Sch Nursing, Univ Wash, 69-70, clin asst prof, Dept Physiol Nursing, 78-81; consult, St Elizabeth's Hosp, 71, Health Sci Learning Resource Ctr, Univ Wash, 78, Wash State Dept Social & Health Serv & Ctrs Dis Control, 80, Philadelphia Naval Hosp, 84, Purdue Frederick Co, Norwalk, 85, Keck Found & Columbia Univ, NY, 86; biomed Res Support grant, Robert Wood Johnson Found clin nurse fel, Ctr for Dis Control grant; nurse practr coronary care, Kaiser-Permanente Sunnyside Med Ctr, Portland, 75-76; ed, riv mag, 83-87, Annals Int Med, 88, Sour Am Med Asn, 83-; mem, Cert Bd Infection Control, 83-86, Res Comt, Asn Practr Infection, 86-88, Task Force Severity Illness Adjusters, Soc Hosp Epidemiologists Am, 87-88, Comt Study Resources Clin Invest, Inst Med, 88, Gov Coun & Priorities & Planning Comt, Am Acad Nursing, 88-89, Adv Bd Health Sci Policy, Inst Med, 89-92, Gov Coun Prog Planning Comt, 90-92 & Planning Panel Health Effects Stress & Emotions, 91. *Mem:* Inst Med-Nat Acad Sci; Am Nurses Asn; Am Pub Health Asn; Asn Practitioners in Infection Control; Soc Hosp Epidemiol Am; Soc Res in Nursing Educ Forum; Am Soc Microbiol; fel Am Acad Nursing; Sigma Xi. *Res:* Author of over 99 articles, journals and books. *Mailing Add:* Johns Hopkins Sch Nursing 460-E Houck Bldg 600 N Wolfe St Baltimore MD 21218

LARSON, ERIC GEORGE, b Chicago, Ill, Apr 25, 57; m 85; c 1. PHOTOCHEMISTRY. *Educ:* Purdue Univ, BS, 80; Northwestern Univ, MS, 81, PhD(org chem), 85. *Prof Exp:* Sr chemist, Corp Res Lab, 84-86 & Indust Abiasives Div, 86-88, RES SPECIALIST, INDUST ABRASIVES DIV, 3M CO, 88- *Mem:* Am Chem Soc. *Res:* High performance polymers; structural adhesives; rapid cure resins; radiation curable resins; photochemistry; new polymer synthesis. *Mailing Add:* 251-1A-03 3M Ctr St Paul CT 06340

LARSON, EVERETT GERALD, b Logan, Utah, Nov 12, 35; m 63; c 4. SOLID STATE PHYSICS, MOLECULAR PHYSICS. *Educ:* Mass Inst Technol, SB, 57, SM, 59, PhD(electron correlation), 64. *Prof Exp:* Asst, Mass Inst Technol & mem staff, Lincoln Labs, 57-64; asst prof physics, Brigham Young Univ, 64-68; vis asst prof chem, Univ Ga, 68-69; assoc prof, 69-75, PROF PHYSICS, BRIGHAM YOUNG UNIV, 75- *Mem:* Am Phys Soc. *Res:* Atomic molecular and solid state theory; correlation effects; cooperative phenomena; theory of irreversible processes. *Mailing Add:* Brigham Young Univ 291 Eyring Sci Provo UT 84602

LARSON, FRANK CLARK, b Columbus, Nebr, Jan 17, 20; m 48; c 3. MEDICINE. *Educ:* Nebr State Teachers Col, AB, 41; Univ Nebr, MD, 44; Am Bd Internal Med, dipl. *Prof Exp:* Instr, 50-51, asst clin prof, 51-56, from asst prof to assoc prof, 56-63, PROF MED & PATH, UNIV WIS-MADISON, 63-, DIR CLIN LABS, HOSP, 58- *Concurrent Pos:* Asst chief med & tuberc serv, Vet Admin Hosp, Madison, 51-56, chief invest med serv, 52-56. *Mem:* Endocrine Soc; Am Col Physicians; Cent Soc Clin Res; Sigma Xi. *Res:* Thyroid metabolism. *Mailing Add:* B4/251 CSC University Hosps Madison WI 53792

LARSON, FREDERIC ROGER, b Los Angeles, Calif, Mar 26, 42; m 74; c 1. FOREST MANAGEMENT, SYSTEMS ANALYSIS. *Educ:* Northern Ariz Univ, BSF, 66, MS, 68; Colo State Univ, PhD(forestry), 75. *Prof Exp:* Forester, Kaibab & Nez Perce Nat Forests, 65-67, res forester, Rocky Mountain Exp Sta, 67-78, RES FORESTER, PAC NORTHWEST EXP STA, US FOREST SERV, 79- *Concurrent Pos:* Prof, Northern Ariz Univ Forestry Sch, 71-72, adj prof, 75-78. *Mem:* Soc Am Foresters; Am Soc Photogram; Soc Range Mgt; Wildlife Soc; Sigma Xi. *Res:* Quantifying and simulating growth and management of southwestern coniferous forests; computer simulation models ecosystems; forest multi-resource inventories. *Mailing Add:* 201 E 9th Ave Suite 303 Anchorage AK 99501

LARSON, GARY EUGENE, b Jersey Shore, Pa, Aug 10, 36; m 60; c 2. BOTANY, BIOPHYSICS. *Educ:* State Univ NY Albany, BS, 58, MS, 60; Rutgers Univ, PhD(bot), 64. *Prof Exp:* Asst bot, Rutgers Univ, 61-62, instr biol, Douglass Col, 62-64; asst prof, 64-66, assoc prof, 66-81, chmn dept, 68-81, PROF BOT, BETHANY COL, WVA, 81-, DIR HEALTH SCI PROG, 81- *Concurrent Pos:* WVa Heart Asn grant, 67-68; teacher & dir, Col Educ Prog, WVa Penitentiary, 68-78; Peace Corps sci curric adv, Gambian Govt, 70-71; dir, Nat Defense Educ Act Title I Proj, 72-73; exec dir, Brooke-Hancock Comprehensive Health Planning Asn, 74-75; vis Fulbright Lectr, Dept Biol, Univ Ile, Nigeria, 77-78; AAAS rep to West African Sci Asn meeting, Lome, Togo, 79. *Mem:* Fel AAAS; Sigma Xi; Am Inst Biol Sci; Audubon Soc. *Res:* Bioelectric potentials surrounding the roots of plants; computer aided instruction; audio-tutorial education. *Mailing Add:* Dept Biol Bethany Col Bethany WV 26032

LARSON, GEORGE H(ERBERT), b Lindsborg, Kans, Jan 28, 15; m 41; c 1. AGRICULTURAL ENGINEERING. *Educ:* Kans State Univ, BS, 39, MS, 40; Mich State Univ, PhD, 55. *Prof Exp:* Asst, Kans State Univ, 39-40; asst instr agr eng, Univ Wis, 40-42; instr, Panhandle Agr & Mech Col, 42; jr instr, US Dept Navy, 42-43; assoc prof agr eng, 46-50, head dept, 56-70, prof agr eng, 50-84, EMER PROF AGR ENG, KANS STATE UNIV, 84- *Concurrent Pos:* Prof & proj leader agr eng, Nebr Mission, USAID, Bogota, Colombia, 70-72; prof agr eng & head dept agron with Kans State Univ Proj, USAID, Ahmadu Bello Univ, Nigeria, 72-74; consult with Kans State Univ Proj, USAID, Central Luzon State Univ, Philippines, 78-80. *Mem:* Am Soc Agr Engrs; Nat Soc Prof Engrs. *Res:* Utilizing liquefied petroleum gas for weed control by flaming; power and machinery, utilizing liquefied petroleum gas in tractors; operating costs of field machinery; agricultural mechanization in developing countries. *Mailing Add:* 419 Oakdale Dr Manhattan KS 66502

LARSON, GERALD LOUIS, b Tacoma, Wash, Jan 14, 42; m 66; c 2. ORGANIC CHEMISTRY, SYNTHETIC INORGANIC & ORGANOMETALLIC CHEMISTRY. *Educ:* Pac Lutheran Univ, BSc, 64; Univ Calif, Davis, PhD(chem), 68. *Prof Exp:* Fel chem, Wash State Univ, 68-69; fel, Mass Inst Technol, 69-70; from asst prof to prof chem, Univ PR, Rio Piedras, 70-86; mgr res prods, Huls Am Petrarch Systs, 86-88, mgr res chem, 88-90; HEAD, SILICON APPLICATIONS, HULS AM, 90- *Concurrent Pos:* Vis prof, Polytech Inst Mex, 82, Bari Univ, Italy, 83, Wurzburg Univ, 86 & La State Univ, 84-85; chmn, PR Sect, Am Chem Soc, 76. *Honors & Awards:* Igarravides Award, Am Chem Soc, 87. *Mem:* Am Chem Soc; Sigma Xi. *Res:* Organosilicon chemistry with an emphasis on the synthetic applications of organosilanes; new routes to organosilicon compounds, new applications of organosilicones in industry. *Mailing Add:* Huls Am Inc Bartram Rd Bristol PA 19007

LARSON, GUSTAV OLOF, b Cedar City, Utah, Dec 24, 26; m 55; c 4. ORGANIC CHEMISTRY. *Educ:* Univ Utah, BS, 48, MA, 51; Wash State Univ, PhD, 59. *Prof Exp:* Asst prof chem, Utah State Univ, 57-62; assoc prof & head dept, Westminster Col, Utah, 62-65; NSF fac fel, Univ Colo, 65-66; assoc prof, 66-71, PROF CHEM, FERRIS STATE UNIV, 71- *Mem:* AAAS; Am Chem Soc. *Res:* Organic reaction mechanisms; stereochemistry; teaching aids, including patents on paper steromodels and pK-pH calculator. *Mailing Add:* Dept Chem Ferris State Col Big Rapids MI 49307

LARSON, HAROLD JOSEPH, b Eagle Grove, Iowa, Nov 16, 34; m 62; c 4. MATHEMATICAL STATISTICS. *Educ:* Iowa State Univ, BS, 56, MS, 57, PhD(math, statist), 60. *Prof Exp:* Instr statist, Iowa State Univ, 59-60; math statistician, Stanford Res Inst, 60-62; prof opers anal, 62-80, PROF OPERS RES & STATIST, NAVAL POSTGRAD SCH, 80- *Concurrent Pos:* Consult, Autonetics Div, NAm Aviation, Inc, 63-64, Data Dynamics, Inc, 65-67 & Field Res Corp, 65-69; Fulbright prof, Univ Sao Paulo, 70-71. *Mem:* Am Statist Asn. *Res:* Probability theory; general statistical methods. *Mailing Add:* Dept Opers Analysis Naval Postgrad Sch Monterey CA 93940

LARSON, HAROLD OLAF, b Port Wing, Wis, May 27, 21. ORGANIC CHEMISTRY. *Educ:* Univ Wis, BS, 43; Purdue Univ, MS, 47; Harvard Univ, PhD, 50. *Prof Exp:* Navigator, Pan Am Airways, 44-45; chemist, Hercules Powder Co, 50-54; res fel chem, Harvard Univ, 54-55; asst prof, Univ WVa, 55-57; res fel, Purdue Univ, 57-58; from asst prof to assoc prof, 58-72, PROF CHEM, UNIV HAWAII, 72- *Mem:* Am Chem Soc. *Mailing Add:* Dept Chem Univ Hawaii Honolulu HI 96822

LARSON, HAROLD PHILLIP, b Hartford, Conn, July 13, 38; m 60; c 2. ASTRONOMY. *Educ:* Bates Col, BS, 60; Purdue Univ, MS, 63, PhD(physics), 67. *Prof Exp:* Res assoc physics, Purdue Univ, 67-68; fel, Aime-Cotton Lab, Nat Ctr Sci Res, France, 68-69; from asst prof to assoc prof astron, 71-76, RES PROF, LUNAR & PLANETARY LAB, UNIV ARIZ, 83- *Mem:* Am Astron Soc. *Res:* Infrared astronomy of planetary atmospheres and surfaces. *Mailing Add:* 2373 Miraval Segundo Tucson AZ 85718

LARSON, HARRY THOMAS, b Berkeley, Calif, Oct 16, 21; div; c 3. COMPUTER SECURITY, LARGE SCALE COMMAND & CONTROL SYSTEM ENGINEERING. *Educ:* Univ Calif, Berkeley, BS, 47; Univ Calif, Los Angeles, MS, 54. *Prof Exp:* Computer engr, Nat Bur Standards Inst Numerical Annl, 49-51; syst engr, Hughes Aircraft Co, 51-54; mgr bus appl dept, Ramo Woolbridge Corp, 54-56; prog engr, aeronaut div, Philco-Ford Corp, 56-68; asst dir, software & info systs div, TRW Systs, 68-69; dir planning, CalComp, 69-74; sr scientist, Hughes Aircraft, 78-87; PRES, LARBRIDGE ENTERPRISES, 87- *Concurrent Pos:* Chmn, Nat Prof Group Electronic Comput, Inst Radio Engrs, 53-54, guest ed, 60-61; lectr, computer design, Univ Southern Calif Grad Sch, 54-55; mem, bd gov, Am Fed Info Processing Soc, 56-59; chmn, tech prog, Joint Comput Conf, 67; mem, Army Sci Bd, 88- *Honors & Awards:* Centennial Med, Inst Elec & Electronics Engrs, 84. *Mem:* Fel Inst Elec & Electronics Engrs; Soc Info Displays. *Res:* Social implications of science and technology; public understanding of technology and science; architecture of large scale computerized systems. *Mailing Add:* 236 Calle Aragon No A Laguna Hills CA 92653-3433

LARSON, HUGO R, physical metallurgy, melting practice; deceased, see previous edition for last biography

LARSON, INGEMAR W, b Clarissa, Minn, Dec 4, 28; m 62. ZOOLOGY, PARASITOLOGY. *Educ:* Concordia Col, Moorhead, Minn, AB, 51; Kans State Univ, MS, 57, PhD(parasitol), 64. *Prof Exp:* Instr biol, Concordia Col, 51-52; asst zool, Kans State Univ, 55-57, from instr to asst zool, 57-63; asst parasitol, Biol Sta, Univ Mich, 57; res assoc, Ore State Univ, 63-66; from asst prof to prof biol, Augustana Col, Ill, 63-83, chmn dept biol, 77-83; CONSULT, 83- *Mem:* Am Soc Parasitologists; Am Micros Soc; Sigma Xi. *Res:* Parasitic protozoa; floodplain insects. *Mailing Add:* Dept Biol Augustana Col Rock Island IL 61201

LARSON, JAMES D, b Kansas City, Mo, Feb 16, 35. BEAM TRANSPORT ANALYSIS, ACCELERATOR TECHNOLOGY. *Educ:* Mass Inst Technol, BS, 57; Calif Inst Technol, MS, 59, PhD(physics), 65. *Prof Exp:* RETIRED. *Mem:* Am Phys Soc. *Mailing Add:* 10011 E 35th St Terr Independence MO 64052

LARSON, JAY REINHOLD, b Urbana, Ill, Dec 6, 32; m 58; c 4. MECHANICAL & NUCLEAR ENGINEERING. *Educ:* Univ Ill, Urbana, BS, 55; Univ Wash, MS, 60; Purdue Univ, PhD(mech eng), 64. *Prof Exp:* Engr, Gen Elec Co, 60-61; assoc scientist, Aerojet Nuclear Co, 64-77; SCIENTIST & ENG SUPVR, EG&G IDAHO INC, 77- *Concurrent Pos:* Affil prof, Univ Idaho. *Mem:* Am Soc Mech Engrs; Nat Soc Prof Engrs. *Res:* Heat transfer; nuclear reactor safety research. *Mailing Add:* 1033 E 25 St Idaho Falls ID 83401

LARSON, JERRY KING, b Willmar, Minn, May 15, 41; m 64; c 2. CLINICAL PHARMACOLOGY. *Educ:* Macalester Col, BA, 63; Mass Inst Technol, PhD(org chem), 67. *Prof Exp:* Res chemist, Chas Pfizer & Co, Inc, 67-70; asst to dir clin res, 70-76, MGR CLIN SCI SERV, PFIZER CENT RES, PFIZER INC, 76- *Mem:* Am Chem Soc. *Res:* Clinical trials with new potential drug candidates, particularly the administration and monitoring of such trials. *Mailing Add:* Pfizer Inc Eastern Point Rd Groton CT 06340

LARSON, JOHN GRANT, b Galesburg, Ill, Aug 17, 33; m 59; c 2. CATALYSIS, SURFACE SCIENCE. *Educ:* Bradley Univ, BS, 55; Univ Ill, Urbana, PhD(phys chem), 62. *Prof Exp:* Proj officer, Mat Lab, Wright Field, US Air Force, 55-57; fel, Mellon Inst, 61-64; group leader, Gulf Res & Develop Co, 64-66, supvr chem physics sect, 66-69, dir, 70-73; head phys chem dept, 73-85, HEAD PHYSICS DEPT, GEN MOTORS RES LABS, 85- *Mem:* Catalysis Soc; Am Chem Soc; Sigma Xi; Am Phys Soc. *Res:* Catalysis, both oxide and supported metals especially for emission control; combustion chemistry; surface chemistry; magnetic materials; metal physics; layered materials. *Mailing Add:* Physics Dept Gen Motors Res Labs 30500 Mound Rd Warren MI 48090-9055

LARSON, JOSEPH STANLEY, b Stoneham, Mass, June 23, 33; m 58; c 2. WETLAND ECOLOGY. *Educ:* Univ Mass, BS, 56, MS, 58; Va Polytech Inst, PhD(zool), 66. *Prof Exp:* Exec secy, Wildlife Conserv Inc, Mass, 58-59; state ornithologist & asst to dir, Mass Div Fisheries & Game, 59-60; head conserv educ div, Natural Resources Inst, Univ Md, 60-62, res asst prof wildlife, 65-67; adj asst prof wildlife biol & asst unit leader, Mass Coop Wildlife Res Unit, 67-69, assoc prof, 69-77, chmn dept, 80-83, PROF, DEPT FORESTRY & WILDLIFE MGT, UNIV MASS, AMHERST, 77-, DIR, ENVIRON INST, 83- *Concurrent Pos:* Consult wetland ecol to var pub, pvt & int agencies & foreign govts; exec chmn, Nat Wetlands Tech Coun, 77-; deleg, Ramsar Conv Wetlands Int Importance, 87 & 90; chmn, US Nat Ramsar Comt, 89-; adv wetlands, Int Union Conserv Nature & Natural Resources, Switz. *Honors & Awards:* Conserv Award for Except Pub Serv in the Cause of Conserv, Chevron, 90. *Mem:* AAAS; Wildlife Soc; Ecol Soc Am; Am Soc Mammal. *Res:* Wetland ecology and management; beaver behavior. *Mailing Add:* 27 Arnold Rd Pelham MA 01002-9757

LARSON, KENNETH ALLEN, b Havre, Mont, July 6, 35; m 61; c 4. IMMUNOLOGY, VETERINARY MEDICINE. *Educ:* Wash State Univ, DVM, 61, MS, 65, PhD(immunol), 66. *Prof Exp:* Vet pvt pract, Mont, 61-63; asst prof vet med, Colo State Univ, 66-69, assoc prof vet med & microbiol, 69-76; PRES, ELARS BIORES LABS INC, 76- *Mem:* AAAS; Am Soc Microbiol; Am Vet Med Asn. *Res:* Immunology of tumors in animals. *Mailing Add:* 1305 Teakwood Dr Ft Collins CO 80521

LARSON, KENNETH BLAINE, b Mesa, Ariz, Jan 16, 30; m 58; c 4. BIOPHYSICS, BIOMATHEMATICS. *Educ:* Colo Sch Mines, MetE, 51; Mass Inst Technol, SM, 54 & 58, PhD(metall & physics), 64. *Prof Exp:* Instr & res asst metall, Mass Inst Technol, 51-53; staff mem, Engr Res & Develop Labs, US Army, Ft Belvoir, Va, 54 & White Sands Proving Ground, NMex, 54-55; res asst metall, Mass Inst Technol, 58-64; staff mem, Gen Atomic Div, Gen Dynamics Corp, Calif, 64-65; res staff, Cent Res Dept, Monsanto Co, 65-70; INSTR BIOMED ENG & RES ASSOC, BIOMED COMPUT LAB, SCH MED, WASHINGTON UNIV, 70- *Concurrent Pos:* Consult, NIH, 78- *Mem:* AAAS. *Res:* Ion-exchange resins; diffusion in liquid metals; measurement of thermal diffusivity and heat capacity of thin films; experimental design and statistics; structure of liquids; semiconductor crystal growth; mathematical modeling in physiology; tracer kinetics; biomedical engineering; physics of cancer radiotherapy. *Mailing Add:* Sch Med Wash Univ 700 Euclid Ave St Louis MO 63110

LARSON, KENNETH CURTIS, b Madison, Wis, July 7, 40; m 64; c 2. PAPER CHEMISTRY, WOOD CHEMISTRY. *Educ:* Calif Inst Technol, BS, 62; Lawrence Univ, MS, 64, PhD(wood chem), 67. *Prof Exp:* Sci specialist res & develop, 67-74, sci assoc, 74-77, proj head, 77-80, Mgr Res & Develop, 80-81, DIR HOUSEHOLD CONVENIENCE PROD DEVELOP, SCOTT PAPER CO, PHILADELPHIA, 81- *Mem:* Tech Asn Pulp & Paper Indust. *Res:* Colloid and surface chemistry of wood pulp fibers; product and process development. *Mailing Add:* Scott Paper Co Scott Plaza 3 Philadelphia PA 19113

LARSON, LARRY LEE, b Horton, Kans, Nov 18, 39; m 63; c 2. REPRODUCTIVE PHYSIOLOGY. *Educ:* Kans State Univ, BS, 62, MS, 65, PhD(animal breeding), 68. *Prof Exp:* NIH fel, Cornell Univ, 68-70, asst prof reproductive physiol, 70-72; asst prof, 72-77, ASSOC PROF REPRODUCTIVE PHYSIOL, UNIV NEBR-LINCOLN, 77- *Mem:* Soc Study Reproduction; Brit Soc Study Fertil; Am Soc Animal Sci; Am Dairy Sci Asn. *Res:* Estrus control and determination; conception failures; management factors to improve reproductive performance. *Mailing Add:* Dept Animal Sci Univ Nebr Lincoln NE 68583-0908

LARSON, LAURENCE ARTHUR, b Cleveland, Ohio, Mar 17, 30; m 56; c 2. PLANT PHYSIOLOGY, BIOLOGY EDUCATION. *Educ:* Ohio Univ, BS, 56; Univ Tenn, MS, 59; Purdue Univ, PhD(amino acid metab), 63. *Prof Exp:* Phys oceanogr, US Navy Hydrographic Off, 56-57; instr bot, Univ Tenn, 59; from asst prof to assoc prof, 63-75, PROF BOT, OHIO UNIV, 75- *Mem:* Am Inst Biol Sci; Asn Am Biol Teachers; Inst Religion Age Sci. *Res:* Germination physiology. *Mailing Add:* Dept Bot Ohio Univ Col Arts & Sci Athens OH 45701

LARSON, LAWRENCE T, b Waukegan, Ill, Dec 3, 30; m 57; c 3. ECONOMIC GEOLOGY, MINERALOGY. *Educ:* Univ Ill, BS, 57; Univ Wis, MS, 59, PhD(geol), 62. *Prof Exp:* From asst prof to prof geol, Univ Tenn, Knoxville, 61-75; prof geol & chmn dept, 75-90, PROF ECON GEOL, MACKAY SCH MINES, UNIV NEV, RENO, 90- *Concurrent Pos:* Consult, Oak Ridge Nat Lab, 63-69 & mining firms, 68-, United Nations Develop Prog, WHO & US Dept Energy; partner, Appl Explor Concepts, 72-79; Fulbright lectr, Turkey, 85-86, UN Develop Prog, 90. *Mem:* Geol Soc Am; Am Inst Mining, Metall & Petrol Engrs; Soc Econ Geologists. *Res:* Manganese mineralogy and ore deposits; ore microscopy; uranium mineralization in Great Basin; geologic thermometry; ceramic and high alumina clay deposits; applied geochemistry and gold exploration; ore deposit geology of Turkey. *Mailing Add:* Dept Geol Sci Univ Nev Mackay Sch Mines Reno NV 89507

LARSON, LEE EDWARD, b Bristol, Conn, Dec 2, 37; m 63; c 2. PHYSICS. *Educ:* Univ NH, PhD(ionospheric physics), 67. *Prof Exp:* Instr physics, Allegheny Col, 61-63; from asst prof to assoc prof, 66-78, PROF PHYSICS, DENISON UNIV, 78- *Honors & Awards:* Distinguished Serv Citation, Am Asn Physics Teachers. *Mem:* Am Asn Physics Teachers; Am Geophys Union. *Res:* High resolution molecular spectroscopy. *Mailing Add:* Dept Physics Denison Univ Granville OH 43023

LARSON, LESTER LEROY, b Amherst Junction, Wis, Feb 12, 23; m 51; c 3. VETERINARY MEDICINE, DISEASES OF REPRODUCTION. *Educ:* Univ Minn, BS, 50, DVM & MS, 53, PhD(vet med), 57; Am Col Theriogenologists, dipl, 70. *Prof Exp:* Vet, 53-54; from instr to asst prof vet med, Univ Minn, 54-58; assoc vet, Am Breeders Serv, 58-73, vet & head vet dept, 74-85; RETIRED. *Concurrent Pos:* Chmn comt exam, Am Col Theriogenologists, 73; guest lectr genital path in the bull, Univ Venezuela, 78; pvt vet pract, 85- *Honors & Awards:* M S D Agvet Dairy Prev Med Award, Am Asn Bovine Practitioners, 85. *Mem:* Am Vet Med Asn; US Animal Health Asn. *Res:* Neuroanatomical and neurophysiological aspects of reproductive process-male; surgical technique for anesthesia of penis-bull; control of venereal diseases of cattle through artificial insemination; actuarial studies of bull under conditions of an artificial insemination center. *Mailing Add:* 1183 Cty T Amherst Junction WI 55407-9216

LARSON, LESTER MIKKEL, b Devils Lake, NDak, Aug 2, 18; m 43; c 3. CHEMISTRY. *Educ:* Lawrence Col, BA, 40; Univ Wis, MS, 50, PhD(biochem), 51. *Prof Exp:* Chemist, Abbott Labs, 41-48; instr biochem, Univ Wis, 51-52; biochemist, E I Du Pont de Nemours & Co, Inc, 52-64; assoc prof, 65-69, prof chem, Del State Col, 69-; RETIRED. *Concurrent Pos:* Pres, Larson Corp. *Mem:* Am Chem Soc; Am Inst Chem; Sigma Xi. *Res:* Arsenicals; antibiotics; polymers; separations; biomedical materials; orthotics. *Mailing Add:* 14110 White Rock Dr Sun City West AZ 85375-5640

LARSON, M(ILTON) B(YRD), b Portland, Ore, July 3, 27; m 50; c 4. MECHANICAL ENGINEERING. *Educ:* Ore State Univ, BS, 50, MS, 55; Yale Univ, MEng, 51; Stanford Univ, PhD(mech eng), 61. *Prof Exp:* From instr to assoc prof mech eng, Ore State Univ, 52-64; dean eng, Univ NDak, 64-68; PROF MECH ENG, ORE STATE UNIV, 69- *Concurrent Pos:* Ford Found resident, Washington Works, E I Du Pont de Nemours & Co, Inc, 68-69. *Mem:* Am Soc Mech Engrs; Am Soc Eng Educ. *Res:* Heat transfer. *Mailing Add:* Dept Mech Eng Ore State Univ Corvallis OR 97331

LARSON, MAURICE A(LLEN), b Iowa, July 19, 27; m 53; c 3. CHEMICAL ENGINEERING. *Educ:* Iowa State Univ, BS, 51, PhD(chem eng), 58. *Prof Exp:* Chem engr res & develop, Dow Corning Corp, 51-52, chem engr prod, 53-54; res asst, 54-55, instr & res assoc, 55-58, from asst prof to prof, 58-77, chmn dept, 78-83, DISTINGUISHED PROF CHEM ENG, IOWA STATE UNIV, 77- *Concurrent Pos:* NSF sci fac fel, Stanford Univ, 65-66; Shell vis prof, Univ Col, Univ London, 71-72; vis prof, Univ Queensland, 81 & Inst Sci & Technol, Univ Manchester, 84-85. *Honors & Awards:* Wester Fund Award, Am Soc Engr Educ, 70. *Mem:* Am Chem Soc; Am Soc Eng Educ; fel Am Inst Chem Engrs. *Res:* Process dynamics and control; crystallization; analysis of particulate system; fertilizer technology. *Mailing Add:* Dept Chem Eng Iowa State Univ Ames IA 50011

LARSON, MERLYN MILFRED, b Story City, Iowa, Sept 11, 28; m 54; c 6. FOREST PHYSIOLOGY, TROPICAL FORESTRY. *Educ:* Colo State Univ, BS, 54; Univ Wash, MF, 58, PhD, 62. *Prof Exp:* Jr forester, US Forest Serv, 55, res forester, Rocky Mountain Forest & Range Exp Sta, 55-64, forest physiologist, 64-66; assoc prof silvicult, 66-70, PROF FORESTRY, OHIO AGR RES & DEVELOP CTR, 70-; PROF NATURAL RESOURCES, OHIO STATE UNIV, 70- *Concurrent Pos:* Asst, Univ Wash, 58-59; consult, Cath Univ, Dominican Repub, 83, G B Plant Univ Agr Technol, India, 88; vis prof, Trop Agr Res & Training Ctr, Costa Rica, 84. *Mem:* Soc Am Foresters; Int Soc Trop Foresters. *Res:* Forest regeneration; tree physiology; forest nurseries. *Mailing Add:* Div Forestry OARDC Ohio State Univ Wooster OH 44691

LARSON, NANCY MARIE, b Dickinson, ND, Sept 30, 46; m 68; c 1. NEUTRON PHYSICS, DATA ANALYSIS & EVALUATION. *Educ:* Mich State Univ, BS, 67, MS, 69, PhD(theoret physics), 72. *Prof Exp:* SR RES STAFF, COMPUTER APPLN DEPT, OAK RIDGE NAT LAB, MARTIN MARIETTA ENERGY SYSTS, 72- *Mem:* Am Phys Soc. *Res:* Generalized least square (Bayes method); Reich-Moore R-matrix theory; analysis of neutron scattering data; data reduction and uncertainty propagation. *Mailing Add:* Oak Ridge Nat Lab Bldg 6010 PO Box 2008 Oak Ridge TN 37831-6356

LARSON, OMER R, b Roseau, Minn, Dec 1, 31; m 60; c 4. ENTOMOLOGY. *Educ:* Univ NDak, BA, 54; Univ Minn, MS, 60, PhD(fish parasites), 63. *Prof Exp:* Instr biol, Minot State Col, 63-64; from asst prof to assoc prof, Univ NDak, 64-76, actg chmn dept, 66-67, assoc dean arts & sci, 69-70 & 75-76, chmn dept, 78-80, PROF BIOL, UNIV NDAK, 76- *Concurrent Pos:* Vis prof, dept biol, USAF Acad, 90-91. *Mem:* Am Soc Parasitologists; Wildlife Dis Asn; Sigma Xi; Soc Vector Ecologists; Am Entom Soc. *Res:* Helminth life cycles; parasites and diseases of fish; biogeography and cold tolerance of fleas. *Mailing Add:* Biol Dept Univ NDak Grand Forks ND 58202

LARSON, PAUL STANLEY, b Cannon Falls, Minn, June 9, 07; m 36; c 1. PHARMACOLOGY. *Educ:* Univ Calif, AB, 30, PhD(physiol), 34. *Prof Exp:* Instr physiol, Sch Med, Georgetown Univ, 34-39; assoc physiol & pharmacol, Med Col Va, 40; lectr pharmacol, Col Med, Wayne Univ, 40-41; res assoc pharmacol, Med Col Va, 41-46, from assoc prof to prof, 46-63, chmn dept, 55-72, Haag prof, 63-75; CONSULT, 75- *Mem:* Am Physiol Soc; Am Soc Pharmacol & Exp Therapeut; Am Chem Soc; Soc Toxicol (pres, 63-64); NY Acad Sci. *Res:* Protein and purine metabolism; water balance; potassium metabolism; anti-spasmodics; biological actions of nicotine; chemical nature of irritants; toxicology. *Mailing Add:* Dept Pharmacol Med Col Va Richmond VA 23298

LARSON, PHILIP RODNEY, b North Branch, Minn, Nov 26, 23; m 48; c 2. FOREST PHYSIOLOGY, PLANT ANATOMY. *Educ:* Univ Minn, BS, 49, MS, 52; Yale Univ, PhD(forestry), 57. *Prof Exp:* Res forester, Fla, 52-54, plant physiologist, Lake States Forest Exp Sta, 56-62, LEADER PHYSIOL WOOD FORMATION, PIONEERING RES UNIT, N CENT FOREST EXP STA, US FOREST SERV, 62- *Honors & Awards:* Distinguished Serv Award, USDA, 75; Barrington Moore Biol Res Award, Soc Am Foresters, 75; New York Bot Garden Award, Bot Soc Am, 77. *Mem:* Am Soc Plant Physiologists; Soc Am Foresters; fel Int Acad Wood Sci; Bot Soc Am; Int Asn Wood Anatomists. *Res:* Wood formation; vascular anatomy; physiology of growth and development. *Mailing Add:* 3012 S Rifle Rd Rhinelander WI 54501

LARSON, RACHEL HARRIS (MRS JOHN WATSON HENRY), b Wake Forest, NC, Aug 1, 13; m 56, 71; c 3. DENTAL RESEARCH. *Educ:* Appalachian State Univ, BS, 40; Georgetown Univ, MS, 49, PhD(biochem), 58. *Prof Exp:* Teacher, High Sch, Ga, 40-41 & NC, 41-42; chemist, Lab Indust Hyg Res, NIH, 42-48, res chemist, Nat Inst Dent Res, 48-76, chief, Prev Methods Develop Sect, Nat Caries Prog, 72-76; consult, Vet Admin, Washington, DC, 77-85; RETIRED. *Concurrent Pos:* Vis scientist, Royal Dent Col, Denmark, 70-71. *Mem:* Am Inst Nutrit; Int Asn Dent Res. *Res:* In vivo studies of caries inhibitory effects of trace elements in the rat. *Mailing Add:* 9301 Fernwood Rd Bethesda MD 20817

LARSON, REGINALD EINAR, b Milo, Maine, July 27, 34; m 66; c 1. AIR MASS TRACERS, NUCLEAR SPECTROSCOPY. *Educ:* Univ Maine, BS, 55; Univ Md, MS, 66. *Prof Exp:* RES PHYSICIST, NAVAL RES LAB, 59-; NUCLEAR PHYSICS CONSULT, CONSOL CONTROLS CORP, 85- *Mem:* Am Phys Soc; Am Meteorol Soc; Am Geophys Union; Sigma Xi. *Res:* Nuclear physics; atmospheric physics research; author or coauthor of over 80 publications. *Mailing Add:* 1106 Montezuma Ft Washington MD 20744

LARSON, RICHARD ALLEN, b Minot, NDak, July 9, 41; m 63; c 2. ENVIRONMENTAL ORGANIC CHEMISTRY. *Educ:* Univ Minn, Minneapolis, BA, 63; Univ Ill, Urbana, PhD(org chem), 68. *Prof Exp:* USPHS fels, Univ Liverpool, 68-69 & Cambridge Univ, 69-70; res assoc bot, Univ Tex, Austin, 70-72; asst cur, Stroud Water Res Ctr, Acad Natural Sci Philadelphia, 72-79; PROF, INST ENVIRON STUDIES, UNIV ILL, URBANA, 79- *Concurrent Pos:* Spec lectr, Univ Pa, 73-74; NRC sr res assoc, EPA, Athens, Ga, 85-86. *Mem:* Phytochem Soc NAm; Am Chem Soc; Am Soc Photobiol; Sigma Xi. *Res:* Natural products; phytochemistry; aquatic organic chemistry; photobiology. *Mailing Add:* 1101 W Peabody Dr Urbana IL 61801

LARSON, RICHARD BONDO, b Toronto, Ont, Jan 15, 41; Can citizen. STAR FORMATION, GALACTIC EVOLUTION. *Educ:* Univ Toronto, BSc, 62, MA, 63; Calif Inst Technol, PhD(astron), 68. *Prof Exp:* From asst prof to assoc prof, 68-75, dir undergrad studies, 71-81, chmn dept, 81-87, PROF ASTRON, YALE UNIV, 75- *Concurrent Pos:* Tinsley vis prof, Univ Tex, 90. *Mem:* Int Astron Union; Am Astron Soc; Royal Astron Soc Can; Royal Astron Soc; Sigma Xi; Astron Soc Pac. *Res:* Theoretical studies of star formation and early stellar evolution; stellar dynamics, formation and evolution of galaxies. *Mailing Add:* Dept Astron Yale Univ Box 6666 New Haven CT 06511

LARSON, RICHARD GUSTAVUS, b Pittsburgh, Pa, May 16, 40; div; c 3. COMPUTER MATHEMATICS. *Educ:* Univ Pa, AB, 61; Univ Chicago, MS, 62, PhD(math), 65. *Prof Exp:* Instr math, Mass Inst Technol, 65-67; from asst prof to assoc prof, 67-77, PROF MATH, UNIV ILL AT CHICAGO, 77- *Honors & Awards:* Fulbright Lectr, Philippines, 78. *Mem:* Am Math Soc; Asn Comput Mach; Math Asn Am; Soc Indust & Appl Math. *Res:* Algebraic and arithmetic structure of Hopf algebras; symbolic algorithms; computational problems of algebra; theoretical computer science. *Mailing Add:* Dept Math MC 249 Univ Ill at Chicago Circle Chicago IL 60680

LARSON, RICHARD I, b Chicago, Ill, Dec 29, 37; m 61; c 3. CHEMICAL ENGINEERING. *Educ:* Northwestern Univ, BSChE, 60; Cornell Univ, MChE, 63; Rensselaer Polytech Inst, PhD(chem eng), 69. *Prof Exp:* Consult, Atomic Energy Comn, 66-68; Engr, 63-66, proj engr, Knolls Atomic Power Lab, Schenectady, 68-77, PRIN ENGR, NUCLEAR FUEL DEPT, GEN ELEC CO, 75- *Mem:* Am Inst Chem Engrs; Am Chem Soc; Am Nuclear Soc. *Res:* Nuclear fuel processing, turbulent flow-heat and mass transfer; colloidal and surface chemistry; numerical solution; ordinary and partial differential equations; system analysis; fluidization production equipment. *Mailing Add:* 417 Pettigrew Wilmington NC 28401

LARSON, ROBERT ELOF, b Spokane, Wash, Oct 9, 32; m 57; c 3. PHARMACOLOGY, TOXICOLOGY. *Educ:* Wash State Univ, BS, & BPharm, 57, MS, 62; Univ Iowa, PhD(pharmacol), 64. *Prof Exp:* Pharmacist, Manito Pharm, Wash, 57-59; staff fel toxicol, Nat Cancer Inst, 64-65; from asst prof to assoc prof, 65-76, chmn dept, 70-76, PROF PHARMACOL & TOXICOL, SCH PHARM, ORE STATE UNIV, 76- *Mem:* Soc Toxicol; Am Soc Pharmacol & Exp Therapeut; Soc Exp Biol Med; Sigma Xi. *Res:* Hepatotoxicity and nephrotoxicity of halogenated hydrocarbons; toxicity of nitrosoureas. *Mailing Add:* Dept Pharmacol & Toxicol Ore State Univ Sch Pharm Corvallis OR 97331

LARSON, ROGER, b Stratford, Iowa, Jan 12, 43. MARINE GEOPHYSICS, TECTONICS. *Educ:* Iowa State Univ, BS, 65; Univ Calif, San Diego, PhD(oceanog), 70. *Prof Exp:* Sr res assoc, Lamont-Doherty Geol Observ, Columbia Univ, 76- 80; PROF MARINE GEOPHYS, UNIV RI, 80- *Mem:* Fel Geol Soc Am; Geophys Union. *Mailing Add:* Grad Sch Oceanog Univ RI Narragansett RI 02882

LARSON, ROLAND EDWIN, b Ft Lewis, Wash, Oct 31, 41; m 60; c 2. MATHEMATICS. *Educ:* Lewis & Clark Col, BS, 66; Univ Colo, Boulder, MA, 68, PhD(math), 70. *Prof Exp:* From asst prof to assoc prof, 70-83, PROF MATH, BEHREND COL, PA STATE UNIV, 83- *Concurrent Pos:* Mem. Nat Coun Teachers Math. *Mem:* Am Math Soc; Math Asn Am; Nat Coun Teachers Math; Am Math Asn Two Year Col. *Res:* Author of numerous publications. *Mailing Add:* Dept Math Behrend Col Pa State Univ Erie PA 16563

LARSON, ROY AXEL, b Cloquet, Minn, Feb 5, 31; m 53; c 4. HORTICULTURE. *Educ:* Univ Minn, BS, 53, MS, 57; Cornell Univ, PhD(floricult), 61. *Prof Exp:* Assoc prof, 61-69, PROF HORT, NC STATE UNIV, 69- *Mem:* Am Soc Hort Sci. *Res:* Floriculture, particularly investigations on effects of environment and regulators on flowering and plant growth. *Mailing Add:* NC State Univ Box 7609 Hort Raleigh NC 27605

LARSON, RUBY ILA, b Hatfield, Sask, Can, May 30, 14. CYTOGENETICS. *Educ:* Univ Sask, BS, 42 & 43, MA, 45; Univ Mo, PhD(genetics), 52. *Prof Exp:* Cytogeneticist, Cereal Div, Dom Exp Sta, Sask, 45-48; cytogeneticist, Sci Serv Lab, Can Dept Agr, 48-59, cytogeneticist, Res Sta, 59-79; RETIRED. *Res:* Wheat cytogenetics. *Mailing Add:* 2503 12 Ave S Lethbridge AB T1K 0P4 Can

LARSON, RUSSELL EDWARD, b Minneapolis, Minn, Jan 2, 17; m 39; c 3. AGRICULTURAL EDUCATION, ACADEMIC ADMINISTRATION. *Educ:* Univ Minn, BS, 39, MS, 40, PhD(genetics, plant breeding), 42. *Hon Degrees:* DSc, Del Valley Col Sci & Agr, 66. *Prof Exp:* Asst hort, Univ Minn, 39-41; asst res prof agron & asst agronomist, Exp Sta, RI State Col, 41-44; asst prof veg gardening, 44-45, assoc prof plant breeding, 45-47, head dept hort, 52-61, dir agr & home econ exten, 61-63, dean & dir, Col Agr, 63-74, provost, 72-77, prof, 47-77, EMER PROVOST, EMER DEAN & EMER PROF HORT, PA STATE UNIV, UNIVERSITY PARK, 77-; CONSULT AGR, 77- *Concurrent Pos:* Sci aide, Mex Agr Prog, Rockefeller Found, 60; chmn comn educ agr & natural resources, Nat Acad Sci-Nat Res Coun, 66- *Honors & Awards:* Vaughan Award, Am Soc Hort Sci, 48. *Mem:* AAAS; Am Soc Hort Sci; Genetics Soc Am; Am Genetics Asn. *Res:* Administration of agricultural research development and education. *Mailing Add:* 608 Elmwood St State Col PA 16801

LARSON, RUSSELL L, b Bridgewater, SDak, Dec 9, 28; wid; c 1. BIOCHEMISTRY. *Educ:* SDak State Univ, BS, 57, MS, 59; Univ Ill, PhD(biochem), 62. *Prof Exp:* Fel microbiol, Ore State Univ, 62-64; RES CHEMIST, USDA, UNIV MO-COLUMBIA, 65- *Mem:* Am Chem Soc; Phytochem Soc NAm; Am Soc Plant Physiol. *Res:* Biochemistry of disease resistance in plants; alteration in metabolic processes in maize as a result of alteration in the genetic systems in maize. *Mailing Add:* Univ Mo 2517 Highland Dr Columbia MO 65203

LARSON, SANFORD J, b Chicago, Ill, Apr 9, 29; m 57; c 3. NEUROANATOMY, NEUROSURGERY. *Educ:* Wheaton Col, Ill, BA, 50; Northwestern Univ, MD, 54, PhD(anat), 62. *Prof Exp:* Resident neurosurg, Northwestern Univ, 55-57 & 59-61; USPHS res fel, 61-62; dir neurosurg educ, Cook County Hosp, Chicago, 62-63; assoc prof, 63-68, PROF NEUROSURG, MED COL WIS, 68-, CHMN DEPT, 63- *Concurrent Pos:* Chief neurosurg, Vet Admin Hosp, Wis, Milwaukee County Gen Hosp & Froedtert Mem Lutheran Hosp; consult, Columbia & Milwaukee Children's Hosps, Wis & Shriners Hosps Crippled Children, Chicago, Ill. *Mem:* Soc Univ Surg; Am Asn Neurol Surg; Am Col Surg; Soc Neurol Surgeons. *Res:* Neurophysiology; neurological surgery. *Mailing Add:* Dept Neurosurg Med Col Wisc 8700 W Wisconsin Ave Milwaukee WI 53226

LARSON, THOMAS D, b Sept 28, 28; c 3. TRANSPORTATION MANAGEMENT. *Educ:* Pa State Univ, BS, 52, MS, 59, PhD(civil eng), 62. *Prof Exp:* Prof, Col Bus Admin & Col Eng, Pa State Univ, 62-89; ADMINR, FED HWY ADMIN, 89- *Concurrent Pos:* Secy transp, Commonwealth Pa, 79-87; chair, Nat Gov's Task Force on New Fed Transp Legis. *Honors & Awards:* Secy's Gold Medal Award for Outstanding Achievement, Secy Transp, 90. *Mem:* Nat Acad Eng; fel Nat Soc Civil Engrs; fel Nat Acad Pub Admin. *Mailing Add:* Fed Hwy Admin 400 Seventh St SW Rm 4218 Washington DC 20590

LARSON, THOMAS E, b Waupaca, Wis, Apr 13, 26; m 49; c 3. PHYSICAL CHEMISTRY, ORGANIC CHEMISTRY. *Educ:* Lewis & Clark Col, BA, 50; Johns Hopkins Univ, MA, 51, PhD(phys chem), 56. *Prof Exp:* Jr instr chem, Johns Hopkins Univ, 50-53, res asst, Inst Coop Res, 53-56; staff mem, 56-75, alt group leader WX-2, 75-84, GROUP LEADER EXPLOSIVE TECHNOL, LOS ALAMOS NAT LAB, 84- *Mem:* Am Chem Soc. *Res:* Kinetics of exchange reactions in boron hydrides; sensitivity of explosives to various stimuli; radioactive materials. *Mailing Add:* Los Alamos Nat Lab Box 1663 Los Alamos NM 87544

LARSON, VAUGHN LEROY, b Mondovi, Wis, Feb 21, 31; m 61; c 1. VETERINARY MEDICINE. *Educ:* Univ Minn, BS, 58, DVM, 60, PhD(vet med), 68. *Prof Exp:* Veterinarian, 60-61; fel bovine leukemia, Univ Minn, 61-68; med scientist animal res, Brookhaven Nat Lab, 68; from asst prof to assoc prof, 68-74, PROF VET MED, UNIV MINN, ST PAUL, 74- *Mem:* Am Vet Med Asn; Am Asn Equine Practitioners; Int Leukemia Res Asn; Sigma Xi. *Res:* Transmission and pathogenesis studies on bovine leukemia; clinical, pathological and therapeutic studies on chronic obstructive pulmonary disease in the equine. *Mailing Add:* Vet Teaching Hosp/225F Univ Minn St Paul MN 55108

LARSON, VERNON C, b Stambough, Mich, Apr 8, 23; m 46; c 3. AGRICULTURE, ACADEMIC ADMINISTRATION. *Educ:* Mich State Univ, BS, 47, MS, 50, EdD(educ admin), 54. *Prof Exp:* From asst prof to assoc prof dairy & asst to dean, Mich State Univ, 47-59; prof agr & asst dean, Am Univ, Beirut, 59-62; prof dairy agr & dir int agr prog, Kans State Univ, 62-65; prof agr sci & dean agr, Ahmadu Bello Univ, Nigeria & chief party, AID & Kans State Univ Team in Nigeria, 66-68; dir int agr progs, Kans State Univ, 68-70; chief party, AID & Kans State Univ Team in India, 70-72; dir int agr progs, 72-87, ASST PROVOST & DIR INT PROG, KANS STATE UNIV, 87- *Concurrent Pos:* AID consult, Jordan, 60, Cyprus & Sudan, 61, Kenya, 68, Colombia, 69, Sierre Leon, 75, Philippines, 77, Paraguay, 78 & Peru, 78, Morocco, 85, Nepal, 85, Botswana, 87, BArazil 89. *Mem:* AAAS; Int Asn Agr Students; Nat Asn State Univs & Land-Grant Cols. *Res:* Agricultural administration. *Mailing Add:* 1951 Bluestem Terr Manhattan KS 66502

LARSON, VINCENT H(ENNIX), b Clinton, Minn, Feb 23, 17; m 55. MECHANICAL ENGINEERING. *Educ:* Univ Minn, BS, 51, ME, 64, PhD(mech eng), 65. *Prof Exp:* Proj engr, Research, Inc, Minn, 54-58, chief anal engr, 58-63; assoc prof mech eng & astronaut sci, Northwestern Univ, 65-72; prof mech eng, 72-85, chmn dept, 72-78, EMER PROF MECH ENG, CLEVELAND STATE UNIV, 85- *Mem:* Am Inst Aeronaut & Astronaut; Am Soc Mech Engrs. *Res:* Control system theory and practice; systems engineering; conception; analysis, evaluation, synthesis and optimization of mechanical and electromechanical systems; analysis of robotic mechanisms. *Mailing Add:* 790 N Arm Dr Orono MN 55364

LARSON, VIVIAN M, b Erie, NDak, Oct 3, 31. VIROLOGY. *Educ:* NDak State Col, BS, 53; Univ Mich, MPH, 58, PhD(virol), 63. *Prof Exp:* Bacteriologist, Detroit Dept Health Labs, Mich, 53-59; from res fel to sr res fel virol, Merck Sharp & Dohme Res Labs, 63-71, dir, NIH Virus Lab, 71-80, dir, 80-90; RETIRED. *Mem:* Am Soc Microbiol; Am Acad Microbiol; AAAS. *Res:* Cell biology; immunology; viral; vaccines. *Mailing Add:* 362 Park Dr Harleysville PA 19438

LARSON, WILBUR JOHN, b Rockford, Ill, Nov 19, 21; m 45; c 3. ANALYTICAL CHEMISTRY. *Educ:* Augustana Col, Ill, BA, 46; Univ Wis, MS, 48, PhD, 51. *Prof Exp:* Chemist, Mallinckrodt Chem Works, 51-57, head analytical develop lab, 57-63, asst dir qual control, 64-72; RES ASSOC, MALLINCKRODT INC, 72- *Mem:* Am Chem Soc; Sigma Xi. *Res:* Trace analysis; quality control; electronic grade chemicals; reagent chemicals. *Mailing Add:* 1364 Stein Ave Ferguson MO 63135-1709

LARSON, WILBUR S, b Downing, Wis, Jan 28, 23; m 53; c 1. INORGANIC CHEMISTRY, ANALYTICAL CHEMISTRY. *Educ:* Wis State Univ, River Falls, BS, 45; Univ Wyo, MS, 58, PhD(inorg chem), 64. *Prof Exp:* Teacher high schs, Wis, 44-55; asst gen chem, Univ Wyo, 55-63; assoc prof, 63-76, PROF CHEM, UNIV WIS-OSHKOSH, 76- *Mem:* Am Chem Soc; Sigma Xi. *Res:* Colorimetric sulfide, sulfite and thiosulfate analysis; stability of antimony addition compounds; equilibrium exchange mechanisms of metallic sulfides; mechanism studies of antimony pentachloride. *Mailing Add:* Dept Chem Univ Wis Oshkosh WI 54901

LARSON, WILLIAM EARL, b Creston, Nebr, Aug 7, 21; m 47; c 4. SOIL SCIENCE. *Educ:* Univ Nebr, BS, 44, MS, 46; Iowa State Univ, PhD, 49. *Hon Degrees:* DSc, Univ Nebr, 82. *Prof Exp:* Asst prof agron, Iowa State Univ, 49-50; soil scientist, USDA, Mont State Col, 51-54; from assoc prof to prof soils, Iowa State Univ, 54-67; prof, 67-89, EMER PROF SOIL SCI, UNIV MINN, ST PAUL, 89- *Concurrent Pos:* Soil scientist, USDA, Iowa State Univ, 54-67; vis prof, Univ Ill, 60 & Univ Minn, 63; Fulbright scholar, Australia, 65-66. *Honors & Awards:* Soil Sci Award, Am Soc Agron; Bennett Award, Soil Conserv Soc, 85; Black Award, 87. *Mem:* Fel AAAS; fel Am Soc Agron; fel Soil Sci Soc Am; Int Soc Soil Sci; fel Soil Conserv Soc Am. *Res:* Soil structure and mechanics; water infiltration; nutrient interrelations in plants; crop response to soil moisture levels and soil temperature; tillage requirements of crops; utilization of sewage wastes on land. *Mailing Add:* 435 A Borlaug Hall Univ Minn St Paul MN 55108

LARTER, EDWARD NATHAN, b Can, Feb 13, 23; m 45; c 3. GENTICS, PLANT BREEDING. *Educ:* Univ Alta, BSc, 51, MSc, 52; State Col Wash, PhD(genetics, plant breeding), 54. *Prof Exp:* Assoc prof genetics & plant breeding, Univ Sask, 54-69; ROSNER RES CHAIR PROF PLANT SCI & DIR TRITICALE RES PROG, UNIV MAN, 69- *Mem:* Genetics Soc Can; Can Soc Agron; Agr Inst Can. *Res:* Plant breeding and cytogenetics of barley, triticale and related species. *Mailing Add:* Dept Plant Sci Univ Man Winnipeg MB R3T 2N2 Can

LARTER, RAIMA, b Kingsville, Tex, May 1, 55; m 77; c 2. NONLINEAR DYNAMICS. *Educ:* Mont State Univ, BS, 76; Ind Univ, Bloomington, PhD(phys chem), 80. *Prof Exp:* Res assoc chem, Princeton Univ, 80-81; asst prof, 83-87, ASSOC PROF CHEM, IND UNIV-PURDUE UNIV, INDIANAPOLIS, 88- *Concurrent Pos:* Vis asst prof chem, Ind Univ-Purdue Univ, Indianapolis, 81-83; prin investr, Petrol Res Fund, Am Chem Soc & NSF, 87- *Mem:* Am Chem Soc; Am Phys Soc; Asn Women Sci (secy, 86-88). *Res:* Chemical oscillations, chaos, self-organization phenomena and biochemical and biophysical applications. *Mailing Add:* Chem Dept Ind Univ-Purdue Univ 1125 E 38th St Indianapolis IN 46205

LARTIGUE, DONALD JOSEPH, b Baton Rouge, La, Sept 7, 34; m 62; c 2. BIOCHEMISTRY, ENZYMOLOGY. *Educ:* La State Univ, BS, 57, MS, 59, PhD(biochem), 65. *Prof Exp:* Analyst, US Food & Drug Admin, 59-60; marine biologist, Marine Lab, Univ Miami, 60; biochemist, USPHS Hosp, Carville, La, 60-62; asst biochem, La State Univ, 62-63, assoc, 63-65; biochemist, R J Reynolds Tobacco Co, NC, 65-70; sr clin biochemist, J T Baker Chem Co, 70-71; sr res biochemist, Corning Glass Works, 71-74; clin chemist, Ochsner Clinic, 74-76; clin chemist, Southern Baptist Hosp, 76-87; assoc prof med technol, 87-90, ASST PROF PATH, LA STATE UNIV MED CTR, 90-; CLIN CHEMIST, VET ADMIN MED CTR, NEW ORLEANS, 90- *Concurrent Pos:* Vis prof, Wake Forest Univ, 68. *Mem:* Am Asn Clin Chemists. *Res:* Clinical chemistry; industrial enzymology; immobilized enzymes. *Mailing Add:* 8809 Southdown Lane River Rdg New Orleans LA 70123

LARUE, JAMES ARTHUR, b Rivesville, WVa, Apr 24, 23; m 47; c 2. MATHEMATICS. *Educ:* WVa Univ, AB, 48, MS, 49; Univ Pittsburgh, PhD(math), 61. *Prof Exp:* Asst prof math, Morris Harvey Col, 49-54; from asst prof to prof math, Fairmont State Col, 54-88, chmn dept, 65-88; RETIRED. *Res:* Mathematical analysis; divergent series. *Mailing Add:* 1114 S Park Dr Fairmont WV 26554

LA RUE, JERROLD A, b San Bernardino, Calif, June 22, 23; m 46; c 2. METEOROLOGY. *Educ:* Univ Calif, Los Angeles, BA, 48. *Prof Exp:* Gen meteorologist, US Weather Bur, Nat Weather Serv, 51-55, forecast meteorologist, 55-57, meteorologist proj analyst, 57-60, quant precipation meteorologist, 60-64, sect supvr meteorol, 64-68, br chief, 68-69, meteorologist in chg, Weather Serv Off, Washington, DC, 69-80; RETIRED. *Mem:* Am Meteorol Soc; Nat Weather Asn. *Res:* Objective methods adapted to operational meteorology. *Mailing Add:* 5550 Rio Vida Lane Sebastopole CA 95472

LARUE, ROBERT D(EAN), b Meridian, Idaho, June 4, 22; m 46; c 3. COMPUTER GRAPHICS, COMPUTER AIDED DESIGN. *Educ:* Univ Idaho, BS, 44 & 49, MS, 51. *Prof Exp:* Engr, Amalgamated Sugar Co, Idaho, 46-48; instr civil eng, Univ Idaho, 51-52; asst prof mech eng, SDak Sch Mines & Technol, 52-54 & Colo State Univ, 54-58, from assoc prof to prof, 58-65; assoc prof, 65-71, PROF ENG GRAPHICS, OHIO STATE UNIV, 71- *Honors & Awards:* Award, Am Soc Eng Educ, 56. *Mem:* Am Soc Eng Educ. *Res:* Improvement of educational methods and text material in area of engineering design. *Mailing Add:* Dept Eng Graphics Ohio State Univ 240 Hitchcock H Columbus OH 43210

LARUE, THOMAS A, b Winnipeg, Man, May 1, 35; div; c 2. BIOCHEMISTRY. *Educ:* Univ Man, BSc, 56, MSc, 58; Univ Iowa, PhD(biochem), 62. *Prof Exp:* From asst res officer to sr res officer, 62-78, Nat Res Coun Can; BIOCHEMIST, BOYCE THOMPSON INST PLANT RES, CORNELL UNIV, 78- *Honors & Awards:* Can Soc Microbiol Award, 79. *Mem:* Am Soc Plant Physiologists; Can Soc Microbiol; Sigma Xi. *Res:* nitrogen fixation. *Mailing Add:* Boyce Thompson Inst Tower Rd Ithaca NY 14853-1801

LA RUSSA, JOSEPH ANTHONY, b New York, NY, May 10, 25; m 46; c 3. ELECTRO-OPTICS, ENGINEERING PHYSICS. *Educ:* City Col NY, BME, 49; Columbia Univ, MS, 55. *Prof Exp:* Sr vpres & tech dir, Farrand Optical Co Inc, 52-88; PRES, SURG MICROSYST INC, 72- *Concurrent Pos:* Lectr flight simulation, Univ Dayton, 80; bd mem, Microsurg Res Found, 75- *Honors & Awards:* De Florez Award, Am Inst Aeronaut & Astronaut, 68. *Mem:* NY Acad Sci; Am Inst Aeronaut & Astronaut. *Res:* Development of many optical systems for spaceflight simulators including Mercury, Gemini, Apollo, LEM, T-27 and for major air force simulators; developed many surgical instruments for ophthalmological surgeons; research on implantable artificial kidney; implantable artificial heart; implantable heart valves; 31 patents in engineering technologies. *Mailing Add:* 451 Rutledge Dr Yorktown Heights NY 10598

LA RUSSO, NICHOLAS F, MEDICINE BIOLOGY. *Prof Exp:* PROF & CHMN, DIV GASTROENTEROL & DIR, CTR BASIC RES DIGESTIVE DIS, MAYO MED SCH CLIN & FOUND, 77- *Mailing Add:* Dept Med Mayo Med Sch Clin & Found Second St SW Rochester MN 55905

LASAGA, ANTONIO C, b Havana, Cuba, Dec 17, 49; US citizen; m 73; c 2. GEOCHEMISTRY, PHYSICAL CHEMISTRY. *Educ:* Princeton Univ, BA, 71; Harvard Univ, MS, 73, PhD(chem physics), 76. *Prof Exp:* Res asst chem, Harvard Univ, 73-76, lectr chem & geol, 76-77; from asst prof to prof geochem, Pa State Univ, 77-84; PROF GEOCHEM, YALE UNIV, 84- *Concurrent Pos:* NSF prin investr, 78-85; vis assoc prof, Yale Univ, 81; Guggenheim Fel, 88. *Honors & Awards:* F W Clarke Medal, Geochem Soc, 79; Award, Mineral Soc Am, 85. *Mem:* Geochem Soc; Am Geophys Union; fel Mineral Soc Am. *Res:* Kinetics and thermodynamics of geochemical processes, particularly modeling diagenetic reactions in the oceans; geochemical cycles; diffusion in silicates; non-equilibrium aspects of geothermometry; quantum mechanics of bonding in silicates; structure of silicate melts. *Mailing Add:* Dept Geol Yale Univ PO Box 6666 New Haven CT 06511

LASAGNA, LOUIS (CESARE), b New York, NY, Feb 22, 23; m 46; c 7. PHARMACOLOGY. *Educ:* Rutgers Univ, BS, 43; Columbia Univ, MD, 47. *Prof Exp:* Asst & instr pharmacol, Sch Med, Johns Hopkins Univ, 50-52, from asst prof to assoc prof med & from asst prof to assoc prof pharmacol & exp therapeut, 54-70; prof pharmacol, toxicol & med, Univ Rochester, 70-84; DEAN, SACKLER SCH GRAD BIOMED SCI & ACAD DEAN, MED SCH, TUFTS UNIV, 85- *Concurrent Pos:* Vis physician, Columbia Res Serv, Goldwater Mem Hosp, 51, 53 & 54; clin & res fel, Mass Gen Hosp, 52-54; lectr, Sch Med, Boston Univ, 52-54; res assoc, Harvard Univ, 53-54. *Mem:* Inst Med, Nat Acad Sci; Asn Am Physicians; Am Soc Pharmacol & Exp Therapeut; Am Soc Clin Invest; Am Fedn Clin Res. *Res:* Hypnotics; analgesics; psychological responses to drugs; placebos; clinical trials; prescribing patterns. *Mailing Add:* Sackler Sch 136 Harrison Ave Tufts Univ Boston MA 02111

LASALA, EDWARD FRANCIS, b Lynn, Mass, June 15, 28; m 53; c 5. PHARMACEUTICAL CHEMISTRY, ORGANIC CHEMISTRY. *Educ:* Mass Col Pharm, BS, 53, MS, 55, PhD(pharmaceut chem), 58. *Prof Exp:* From instr to asst prof, 58-67, assoc prof, 67-77, PROF CHEM & CHMN DEPT, MASS COL PHARM, 77- *Mem:* Am Chem Soc; Am Pharmaceut Asn. *Res:* Synthesis and biological studies of medicinal agents, chiefly analgesics and antiradiation agents. *Mailing Add:* Dept Chem Mass Col Pharm 179 Longwood Ave Boston MA 02115

LASATER, HERBERT ALAN, b Paris, Tenn, Sept 11, 31; m 59, 84; c 2. MATHEMATICAL STATISTICS, APPLIED STATISTICS. *Educ:* Univ Tenn, BS, 57, MS, 62; Rutgers Univ, PhD(statist), 69. *Prof Exp:* Instr statist, Univ Tenn, Knoxville, 57-58; assoc statistician, Nuclear Div, Union Carbide Corp, 58-62; asst prof, 62-65 & 68-71, ASSOC PROF STATIST, UNIV TENN, KNOXVILLE, 71-; EXEC VPRES, TENN ASSOCS, INC, 84- *Concurrent Pos:* Lectr statist, Univ Tenn, 59-62; ed, J Qual Technol, 71-74. *Mem:* Am Statist Asn; fel Am Soc Qual Control; Sigma Xi. *Res:* Statistical quality control techniques. *Mailing Add:* PO Box 710 Alcoa TN 37701

LASCA, NORMAN P, JR, b Detroit, Mich, Oct 20, 34; m 65; c 2. QUATERNARY GEOLOGY, GEOARCHAEOLOGY. *Educ:* Brown Univ, AB, 57; Univ Mich, MS, 61, PhD(geol), 65. *Prof Exp:* Res asst geol, Univ Mich, 60-61, teaching fel, 61-65; NATO res fel, Inst Geol, Univ Oslo, 65-66; asst prof , 66-71, assoc prof geol, 71-76, asst to chancelor, 75-77, assoc dean grad sch, 77-80, chmn dept, 80-81 & 84-88, actg vchancellor, 81-82, PROF GEOL SCI, UNIV WIS-MILWAUKEE, 76- *Concurrent Pos:* Assoc scientist, 71-76, sr scientist, Ctr Great Lakes Studies, Univ Wis-Milwaukee, 76-; fel acad admin, Am Coun Educ, 74-75. *Mem:* Fel Geol Soc Am; Am Asn Quaternary Res; Int Asn Quaternary Res; Glaciol Soc; Swedish Soc Anthrop & Geog; Sigma Xi. *Res:* Glacial geology and geomorphology of polar regions; river and lake ice formation and processes; Glacial-Pleistocene geology in Wisconsin, Geoarcheology in upper midwest. *Mailing Add:* Dept Geosci Univ Wis PO Box 413 Milwaukee WI 53201

LASCELLES, JUNE, b Sydney, Australia, Jan 23, 24. MICROBIOLOGY, BIOCHEMISTRY. *Educ:* Univ Sydney, BSc, 44, MSc, 47; Oxford Univ, DPhil(microbial biochem), 52. *Prof Exp:* Mem external sci staff, Med Res Coun, Eng, 53-60; lectr microbial biochem, Oxford Univ, 60-65; prof, 65-89, EMER PROF MICROBIOL, UNIV CALIF, LOS ANGELES, 89- *Concurrent Pos:* Rockefeller fel, 56-57; consult panel microbial chem, NIH, 73-77; ed, J Bact, 80-89. *Mem:* Am Soc Microbiol; Am Soc Biol Chemists; Brit Biochem Soc; Brit Soc Gen Microbiol. *Res:* Biochemistry of microorganisms; tetrapyrrole synthesis and regulation; bacterial photosynthesis. *Mailing Add:* Dept Microbiol Univ Calif 405 Hilgard Ave Los Angeles CA 90024

LASCHEVER, NORMAN LEWIS, b Hartford, Conn, July 27, 18; m 42; c 4. ELECTRICAL ENGINEERING. *Educ:* Mass Inst Technol, BS, 40; Northeastern Univ, MSEE, 70. *Prof Exp:* Sect chief airborne commun & navig hq, Air Tech Command, Wright-Patterson AFB, Ohio, 46-55; asst dir res eng, Lab Electronics, Mass, 55-62; mgr radar eng, RCA Corp, 62-71, chief engr, Aerospace Systs Div, 71-76, prin scientist, Automated Systs Div, 76-83; RETIRED. *Mem:* Sr mem Inst Elec & Electronics Engrs; Sigma Xi. *Res:* Instrumentation of aircraft dynamics; development of aircraft communication and navigation equipment, airborne Doppler navigators and aerospace radar and transponder equipment. *Mailing Add:* 21310 Millbrook Ct Boca Raton FL 33498

LASDAY, ALBERT HENRY, b Pittsburgh, Pa, Apr 8, 20; m 42; c 2. ENVIRONMENTAL MANAGEMENT. *Educ:* Univ Pittsburgh, BS, 41; Harvard Univ, MA, 47; Carnegie Inst Technol, DSc(physics), 51. *Prof Exp:* Engr, Carnegie Inst Technol, 47-50; physicist, Preston Labs, 51-55; dir res, Houze Glass Corp, 55-58; pres, A H Lasday Co, 58-63; proj mgr, Micarta Div, Westinghouse Elec Corp, SC, 63-65; group leader mat develop, Texaco Exp, Inc, 65-68; supvr res, Richmond Res Labs, 68-71, SR COORDR, ENVIRON CONSERV & TOXICOL, TEXACO INC, 71- *Concurrent Pos:* Consult, Res & Develop Bd, Off Secy Defense, 51. *Mem:* Am Phys Soc; Sigma Xi; AAAS; NY Acad Sci; Am Petrol Inst; Nat Ocean Industs Asn. *Res:* Air and water conservation; water pollution control for petroleum and petrochemical waste effluents; oil spill cleanup management. *Mailing Add:* 37 King Dr Poughkeepsie NY 12603

LASEK, RAYMOND J, b Chicago, Ill, Nov 25, 40; m 64; c 3. NEUROBIOLOGY. *Educ:* Utica Col, BA, 61; State Univ NY, PhD(anat), 67. *Prof Exp:* NIH res fel neuropath, McLean Hosp, Harvard Med Sch, 66-68 & neurobiol, Univ Calif, San Diego, 68-69; asst prof, 69-73, assoc prof, 74-78, PROF ANAT, CASE WESTERN RESERVE UNIV, 79- *Res:* Axonal transport; regulation of growth and differentiation in neurons; evolutionary neurobiology. *Mailing Add:* Med Ctr Case Western Reserve Unit 2119 Abington Rd Cleveland OH 44106

LASERSON, GREGORY, mechanical engineering; deceased, see previous edition for last biography

LASETER, JOHN LUTHER, biochemistry, for more information see previous edition

LA SEUR, NOEL EDWIN, b Stanhope, Iowa, June 25, 22; m 44; c 3. METEOROLOGY. *Educ:* Univ Chicago, SB, 47, SM, 49, PhD(meteorol), 53. *Prof Exp:* From asst to instr, Univ Chicago, 48-52; asst prof, 53-56, assoc prof, 56-58, PROF METEOROL, FLA STATE UNIV, 58- *Mem:* AAAS; Am Meteorol Soc; Am Geophys Union; Royal Meteorol Soc. *Res:* Synoptic meteorology of temperature and tropical latitudes. *Mailing Add:* Dept Meteorol Fla State Univ Tallahassee FL 32306

LASFARGUES, ETIENNE YVES, b Milhars, France, May 5, 16; nat US; m 50; c 2. MICROBIOLOGY, ONCOLOGY. *Educ:* Univ Paris, BS, 35, DVM, 41. *Prof Exp:* Roux Found res fel microbiol, Pasteur Inst, France, 42-44, asst virol, 44-47, head lab virol, 50-55; Am Cancer Soc res fel cytol, Inst Cancer Res, Pa, 47-50; assoc microbiol, Col Physicians & Surgeons, Columbia Univ, 55-59, asst prof, 59-66; assoc mem, Dept Cytol Biophys, Inst Med Res, 66-77, head, Dept Tumor Cell Biol, 77-82; RETIRED. *Concurrent Pos:* Mem, Breast Cancer Task Force Comt, Nat Cancer Inst, 79-82. *Honors & Awards:* Jensen Prize, French Acad Med, 46; Silver & Bronze Medal, Pasteur Inst, 69. *Mem:* Am Soc Cell Biologists; Tissue Cult Asn; Int Soc Cell Biologists; Am Asn Cancer Res. *Res:* Viral oncology; cell transformation; cytogenetics. *Mailing Add:* 706 Camden Ave Moorestown NJ 08057

LASH, JAMES (JAY) W, b Chicago, Ill, Oct 24, 29; m 55; c 1. DEVELOPMENTAL BIOLOGY, ANATOMY. *Educ:* Univ Chicago, PhD(zool), 54. *Hon Degrees:* MA, Univ Pa, 71. *Prof Exp:* Nat Cancer Inst, Univ Pa, 55-57, instr, 57-59, assoc, 59-61, from asst prof to assoc prof, 61-69, PROF ANAT, SCH MED, UNIV PA, 69- *Concurrent Pos:* Helen Hay Whitney Found fel, 58-61; estab investr, Helen Hay Whitney Found, 61-66; NIH career develop award, 66-70; mem, maternal & child health res comt, Nat Inst Child Health & Human Develop, 71-75, nat adv comt, 77-80, nat adv coun, 76-80 & bd sci counrs, 81-85; mem, sci adv comt, Ctr Oral Health Res, Univ Pa, 73-76; mem, nat adv comt, Primate Res Ctr, Davis, Calif, 73-80. *Honors & Awards:* Paulo Found Award, 69. *Mem:* Soc Cell Biol; Soc Develop Biol. *Res:* Developmental biology. *Mailing Add:* Dept Anat Univ Pa Sch Med Philadelphia PA 19104

LASHEEN, ALY M, b Cairo, Egypt, Dec 27, 19; nat US; m 54; c 3. PLANT PHYSIOLOGY, HORTICULTURE. *Educ:* Cairo Univ, BS, 42; Univ Calif, Los Angeles, 49; Agr & Mech Col Tex, PhD(plant physiol, hort), 54. *Prof Exp:* Asst hort, Agr & Mech Col Tex, 50-53, res assoc plant physiol, 54-55; jr plant pathologist, Wash State Univ, 55-57, asst prof hort, 57-61; assoc prof plant physiol, AID Contract-Univ Ky, Indonesia, 61-65, prof, 65-67, prof hort, 67-77; prof hort, Aid Contract, Univ Minn, 77-79; RETIRED. *Concurrent Pos:* Hort adv & chief party, Univ Minn proj, Morocco, 70-79. *Mem:* Am Soc Hort Sci; Am Soc Plant Physiol. *Res:* Chemical analysis of macro and micro elements in plants; biochemical analysis of sugars and amino, organic and nucleic acids in plants; dormancy in seeds; effects of additives on plants; cold hardiness in plants; nature of dwarfing in apples. *Mailing Add:* 3607 Salisbury Dr Lexington KY 40510

LASHEN, EDWARD S, b New York, NY, Aug 11, 34; m 57; c 3. MICROBIOLOGY. *Educ:* Brooklyn Col, BS, 56; Rutgers Univ, MS, 62, PhD(microbiol), 65. *Prof Exp:* Res asst blood res, Jewish Chronic Dis Hosp, Brooklyn, NY, 56-57; res asst hemat & pharmacol, Wallace Labs Div, Carter Prod, Inc, 57-61; sr microbiologist, 64-75, head proj leader biocides, 75-81, RES SECT MGR BIOCIDES, RES LABS, ROHM & HAAS CO, 81- *Mem:* Am Soc Testing & Mat; Am Soc Microbiol; Soc Indust Microbiol. *Res:* Microbial transformation of thiourea and substituted thioureas; biodegradation of surfactants by sewage sludge and river water microflora; broad spectrum anti-microbial agents for application as industrial biocides. *Mailing Add:* 95 Valley Dr Furlong PA 18925

LASHER, GORDON (JEWETT), b Denver, Colo, Feb 1, 26; m 53; c 3. ASTROPHYSICS. *Educ:* Rensselaer Polytech Inst, BS, 49; Cornell Univ, PhD(theoret physics), 54. *Prof Exp:* Staff physicist, Lawrence Radiation Lab, Univ Calif, 53-55; assoc physicist, res lab, IBM Corp, 55-56, proj physicist, 56-58, RES STAFF MEM, IBM-THOMAS J WATSON RES CTR, 58- *Concurrent Pos:* Vis prof, Cornell Univ, 69-70; vis scholar, Stanford Univ, 80. *Mem:* Fel Am Phys Soc; Int Astron Union. *Res:* Applied mathematics with computer applications; solid state physics; superconductivity; theory of liquid crystals; general relativity; cosmology; theory of supernovae; lattice gauge theory; silicon dioxide and its silicon interface. *Mailing Add:* IBM-Watson Res Ctr Yorktown Heights NY 10598

LASHEWYCZ-RUBYCZ, ROMANA A, b Newark, NJ, Mar 1, 52; m 81. SCIENCE EDUCATION. *Educ:* Seton Hall Univ, BS, 74; State Univ NY, Buffalo, PhD(inorg chem), 79. *Prof Exp:* Asst prof, 80-86, ASSOC PROF CHEM, HOBART & WILLIAM SMITH COLS, 86-, CHAIR, 89- *Concurrent Pos:* Vis asst prof, State Univ NY, Buffalo, 79-80, vis assoc prof, 91; vis asst prof, Univ Rochester, 84; mem exec comt, Geneva Chap, Sigma Xi, 87-89, deleg, 89- *Mem:* Am Chem Soc; Am Inst Physics; Am Crystallog Asn; Sigma Xi. *Res:* Investigation of group VIII transition metal complexes; preparation of mixed-metal ruthenium-rhenium tetranuclear clusters and study of their behavior with acetylenes; kinetics of substitution reactions involving metal ions in metalloporphyrins; photosubstitution behavior of carbonyl complexes. *Mailing Add:* Dept Chem Hobart & William Smith Cols Geneva NY 14456

LASHLEY, GERALD ERNEST, b Johnstown, Pa, Sept 26, 35; m 55; c 3. NUMERICAL ANALYSES. *Educ:* Eastern Nazarene Col, BS, 57; Boston Univ, AM, 61; Boston Univ, EdD(math educ), 69. *Prof Exp:* High sch teacher, Mass, 58-63; assoc prof math, Eastern Nazarene Col, 64-72, chmn dept, 70-72; chmn div natural sci, 72-76, dir, Acad Computer Ctr, 77-80, PROF COMPUTER SCI & MATH, MT VERNON NAZARENE COL, 80- *Concurrent Pos:* Dir & instr, NSF In-Serv Inst Secondary Teachers, 64-72. *Mem:* Asn Comput Mach; Math Asn Am. *Mailing Add:* Dept Math & Computer Sci Point Loma Nazarene Col 3900 Lomaland Dr San Diego CA 92106

LASHMET, PETER K(ERNS), b Ann Arbor, Mich, Aug 28, 29; m 56; c 5. CHEMICAL ENGINEERING. *Educ:* Univ Mich, BSE(chem eng) & BSE(math), 51, MSE, 52; Univ Del, PhD(chem eng), 62. *Prof Exp:* Process design engr, M W Kellogg Co, 52-53; from chem engr to sr chem engr, Air Prod & Chem, Inc, 58-60, mgr cryostat eng, 60-62, sr cryogenic specialist, 62-65; ASSOC PROF CHEM ENG, RENSSELAER POLYTECH INST, 65-, DEPT EXEC OFFICER, 77- *Mem:* Am Chem Soc; Am Inst Chem Engrs. *Res:* Catalytic properties of ion exchange resins; cryogenic refrigeration processes; miniature and compact heat exchangers; heat exchanger dynamics; computer simulation of chemical processes; chemical separation processes. *Mailing Add:* Dept Chem Eng Rennselaer Polytech Inst Troy NY 12181

LASHMORE, DAVID S, b Hempstead, NY, July 9, 46; m 70; c 3. MATERIALS SCIENCE, PHYSICAL METALLURGY. *Educ:* Univ Fla, BS, 69; Mich Technol Univ, MS, 70; Univ Va, PhD(mat sci), 77. *Prof Exp:* Res assoc metallurgist coatings, 77-79, GROUP LEADER ELECTRODEPOSITED COATINGS, NAT BUR STANDARDS, 79- *Concurrent Pos:* Consult, B-08 Comt Aluminum Coatings, Am Soc Testing & Mat, 78- *Mem:* Sigma Xi; Am Soc Metals; Am Electroplaters; Am Soc Testing & Mat; Am Soc Mech Engrs; Electrochem Soc. *Res:* Surface and interface properties; transmission electron microscopy; diffraction; electrocrystallization; anodizing. *Mailing Add:* Sect 561 Dept Com Washington DC 20234

LASHNER, BRET AUERBACH, b Philadelphia, Pa, July 27, 54; m 88. GASTROENTEROLOGY, INTERNAL MEDICINE. *Educ:* Haverford Col, AB, 76; NY Univ, MD, 80; Univ Ill, Chicago, MPH, 88. *Prof Exp:* ASST PROF MED & GASTROENTEROL, UNIV CHICAGO, 86- *Mem:* Am Gastroenterol. *Res:* Clinical epidemiology of gastrointestinal diseases, principally inflammatory bowel diseases and colon cancer. *Mailing Add:* Univ Chicago Med Ctr 5841 S Maryland Ave Box 400 Chicago IL 60637

LASHOF, JOYCE COHEN, b Philadelphia, Pa, Mar 27, 26; m 50; c 3. MEDICINE. *Educ:* Duke Univ, AB, 46. *Hon Degrees:* DSc, Med Col Pa, 83. *Prof Exp:* Intern, Bronx Hosp, New York, 50-51, asst resident med, 51-52; asst resident med, Montefiore Hosp, 52-53; Nat Found Infantile Paralysis fel, Yale Univ, 53-54; asst med & physician, Student Health Serv, Univ Chicago, 54-56, from instr to asst prof, Sch Med, 56-60; dir sect prev med & clin labs, Rush-Presby-St Luke's Med Ctr, 61-66, dir sect community med, 66-72, chmn dept prev med, 72-73; dir, Dept Pub Health, State of Ill, 73-77; dep asst secy health progs, Dept HEW, 77-78; asst dir, Off Technol Assessment, US Congress, 78-81; dean, Sch Pub Health, 82-91, PROF PUB HEALTH, UNIV CALIF, BERKELEY, 91- *Concurrent Pos:* Staff physician, Union Health Serv, Inc, 60-61; asst attend physician, Presby-St Luke's Hosp, 60-61, assoc attend physician, 61-; clin asst prof med, Sch Med, Univ Ill, 61-64, assoc prof prev med, 64-71; prof prev med, Rush Med Col, 71-77. *Mem:* Inst Med-Nat Acad Sci; fel Am Col Physicians; fel Am Pub Health Asn (pres, 91); Soc Med Adminr; fel Asn Teachers Prev Med; fel Am Col Prev Med. *Res:* Internal medicine; medical care; author of numerous articles and publications. *Mailing Add:* Sch Pub Health Univ Calif Berkeley CA 94720

LASHOF, RICHARD KENNETH, b Philadelphia, Pa, Nov 9, 22; m 50; c 3. MATHEMATICS. *Educ:* Univ Pa, BS, 43; Columbia Univ, PhD(math), 54. *Prof Exp:* From instr to prof math, 54-88, chmn dept, 67-70, EMER PROF MATH, UNIV CHICAGO, 88- *Concurrent Pos:* NSF fel, 60-61; mem-at-large, Nat Res Coun; vis prof, Univ Calif, Berkeley, 85-89. *Mem:* Am Math Soc; fel AAAS. *Res:* Algebraic topology; differential geometry. *Mailing Add:* 379 Gravatt Dr Berkeley CA 94705

LASHOF, THEODORE WILLIAM, b Philadelphia, Pa, June 27, 18; m 50; c 3. PHYSICS. *Educ:* Univ Pa, BS, 39, PhD(physics), 42. *Prof Exp:* Mem staff, Radiation Lab, Mass Inst Technol, 42-45; from instr to asst prof physics, Reed Col, 45-49; assoc prof, Mich Col Mining & Technol, 49-50; physicist, Mass Sect, Nat Inst Standards & Technol, 50-54, paper sect, 54-62, appl polymer standards sect, 62-64 & eval criteria sect, 64-67, actg chief paper standards sect, 67-70, chief performance criteria sect, 70-74, prog mgr lab performance, 74-78, res assoc, 78-81, GUEST WORKER, NAT BUR STANDARDS, 81- *Honors & Awards:* Testing Div Medal, Tech Asn Pulp & Paper Indust, 72; Award of Merit, Am Soc Testing & Mat, 75; US Dept of Com Silver Medal, 77; W J Youden Award in Interlaboratory Testing, 88. *Mem:* Fel Am Soc Testing & Mat; fel Tech Asn Pulp & Paper Indust. *Res:* Physical properties of paper; interlaboratory standardization; consumer product performance and safety; testing laboratory evaluation. *Mailing Add:* 10125 Ashburton Lane Bethesda MD 20817

LASHOMB, JAMES HAROLD, b Potsdam, NY, Oct 25, 42; m 69; c 2. ENTOMOLOGY. *Educ:* Cornell Univ, BS, 70; Univ Md, College Park, MS, 73, PhD(entom), 75. *Prof Exp:* Res assoc entom, Miss State Univ, 75-78; asst prof, 78-84, ASSOC PROF ENTOM, RUTGERS UNIV, 84- *Mem:* AAAS; Entom Soc Am; Ecol Soc Am; Int Orgn Biol Control Noxious Animals & Plants. *Res:* Sampling of insect populations and natural enemies; parasitic insect distribution within plants and parasitic insect biology. *Mailing Add:* Dept Entom PO Box 231 New Brunswick NJ 08903

LASIECKA, IRENA, b Warsaw, Poland, Feb 4, 48; m 71; c 1. APPLIED MATHEMATICS. *Educ:* Univ Warsaw, MS, 72, PhD(appl math), 75. *Prof Exp:* Asst prof appl math, Polish Acad Sci, Warsaw, 75-78; vis scholar control theory, Univ Calif, Los Angeles, 78-80; from assoc prof to prof differential equation, math dept, Univ Fla, Gainesville, 80-87; PROF CONTROL THEORY, UNIV VA, 87- *Concurrent Pos:* Assoc ed, J Appl Math & Optimization, 84-, Int J Math & Math Sci, 85-, Soc Indust & Appl Math J Control. *Mem:* Int Fedn Info Processes. *Res:* Applied mathematics; control theory and optimization; partial differential equations; numerical analysis; author of approximately 70 research papers published in major journals. *Mailing Add:* Dept Appl Math Univ Va Charlottesville VA 22903

LASINSKI, BARBARA FORMAN, b New York, NY, Nov 28, 41; m 71; c 1. PLASMA PHYSICS. *Educ:* Barnard Col, BA, 62; Univ Rochester, PhD(physics), 68. *Prof Exp:* Res assoc high energy physics, Enrico Fermi Inst, Univ Chicago, 68-71; PHYSICIST PLASMA PHYSICS, LAWRENCE LIVERMORE LAB, 72- *Mem:* Am Phys Soc. *Res:* Computational simulation of plasmas; laser-plasma interactions; computational physics. *Mailing Add:* L-472 Lawrence Livermore Lab PO Box 808 Livermore CA 94550

LASKA, EUGENE, b New York, NY, Mar 17, 38; m 59; c 3. MATHEMATICS, STATISTICS. *Educ:* City Col New York, BS, 59; NY Univ, MS, 61, PhD(math), 63. *Prof Exp:* Asst res scientist, Comput Lab, Res Div, NY Univ, 59-61, res assoc math, Courant Inst Math Sci, 61-62; systs engr, Int Bus Mach Corp, NJ, 62-63; RES PORF PSYCHIAT, NY UNIV MED CTR, 79-; dir info sci div, Nathan S Kline Inst Psychiat Res, Orangeburg, NY, 63-84, DIR, STATIST SCI & EPI DIV, 84-, WHO COLLABORATING CTR RES & TRAINING, MENT HEALTH PROG MGT, NATHAN S KLINE INST PSYCHIAT RES, ORANGEBURG, NY, 85- *Concurrent Pos:* NIMH grants, 67-78; consult comput in psychiat, USSR, Israel, Italy, Iran, Peru & Indonesia, China, Chile, Ecuador, Cuba, 68; mem comput & biomath sci study sect, NIH, 72-76; assoc comnr, NY State Dept Ment Hyg, 80. *Mem:* Fel AAAS; Inst Math Statist; fel Am Statist Asn; Am Soc Clin Pharmacol & Therapeut; Biomet Soc. *Res:* Mathematical statistics, including estimation theory; applied statistics, including biostatistics and clinical trial methodology; computers applications in medicine. *Mailing Add:* 34 Dante St Larchmont NY 10538

LASKAR, AMULYA LAL, physics; deceased, see previous edition for last biography

LASKAR, RENU CHAKRAVARTI, b Bhagalpur, India, Aug 8, 32; m 62; c 2. MATHEMATICS. *Educ:* Univ Bihar, BS, 54; Univ Bhagalpur, MS, 57; Univ Ill, Urbana, PhD(group theory), 62. *Prof Exp:* Lectr math, Ranchi Women's Col, India, 57-59; lectr, Indian Inst Technol, Kharagpur, 62-65; fel, Univ NC, Chapel Hill, 65-68; assoc prof, 68-76, PROF MATH, CLEMSON UNIV, 76- *Concurrent Pos:* Guest mathematician, Univ Paris, 75-76. *Mem:* Am Math Soc; Math Asn Am. *Res:* Group theory; combinatorial mathematics, especially graph theory. *Mailing Add:* Dept Math Clemson Univ 201 Sikes Hall Clemson SC 29631

LASKARIS, TRIFON EVANGELOS, b Cairo, Egypt, Jan 14, 44; US citizen; m 71; c 2. MECHANICAL & ELECTRICAL ENGINEERING. *Educ:* Nat Tech Univ Athens, BS, 66; Rensselaer Polytech Inst, MS, 71, PhD(mech eng), 74. *Prof Exp:* Mech engr advan eng, Large Steam-Turbine Generator Dept, 67-73, cryog br, Power Generation & Propulsion Lab, 73-77, MGR APPL SUPERCONDUCTIVITY PROG, ENGR SYSTS LAB, GEN ELEC CO, 77- *Honors & Awards:* Gold & Silver Patent Awards, Corp Res & Develop, Gen Elec Co; IR-100 Award, Indust Res Mag, 77. *Mem:* Greek Chamber Engrs. *Res:* Superconductivity; cryogenics; computational fluid dynamics; rotating machinery; intense magnetic fields. *Mailing Add:* Res & Develop Bldg K-1 Rm EP117 Gen Elec Schenectady NY 12301

LASKER, BARRY MICHAEL, b Hartford, Conn, Aug 12, 39; m 70; c 2. ASTROPHYSICS. *Educ:* Yale Univ, BS, 61; Princeton Univ, MA, 63, PhD(astrophys sci), 64. *Prof Exp:* NSF fel, Mt Wilson & Palomar Observ, 65-67; asst prof astron, Univ Mich, 67-69; staff astronr, Cerro Tololo Interam Observ, Chile, 69-81; MEM STAFF, SPACE TELESCOPE SCI INST, 81- *Mem:* Am Astron Soc; Int Astron Union. *Res:* Supernova remnants; the interstellar medium in galaxies; astronomical instrumentation; catalogs of stars and galaxies. *Mailing Add:* Space Telescope Sci Inst 3700 San Martin Dr Baltimore MD 21218

LASKER, GABRIEL (WARD), b York, Eng, Apr 29, 12; US citizen; m 49; c 3. BIOLOGICAL ANTHROPOLOGY, HUMAN ANATOMY. *Educ:* Univ Mich, AB, 34; Harvard Univ, MA, 41, PhD(phys anthrop), 45. *Prof Exp:* From instr to prof, 46-82, EMER PROF ANAT, SCH MED, WAYNE STATE UNIV, 82- *Concurrent Pos:* Viking Fund grant, Paracho, Michoacan, Mex, 48; ed, Human Biol, 53-87; mem staff, Dept Anthrop, Univ Wis, 54-55; Fulbright fel, Peru, 57-58; adj prof anthrop, Wayne State Univ, 82-; fel commoner, Churchill Col, Cambridge Univ, 83-84. *Mem:* Fel AAAS (vpres, 68); Am Asn Phys Anthrop (secy-treas, 46-50, vpres, 60-62, pres, 63-65); Am Asn Anat; fel Am Anthrop Asn; hon life mem Soc Study Human Biol; Human Biol Coun (pres, 83-84); Soc Mexicana de Antrop Biol. *Res:* Demographic aspects of human biology; human genetics; physical anthropology; physical characteristics of Chinese, Mexicans and Peruvians; population structure Britain; biological aspects of human migration. *Mailing Add:* Dept Anat Wayne State Univ Sch Med Detroit MI 48201

LASKER, GEORGE ERIC, b Prague, Czech; Can citizen. INFORMATION SCIENCES, PSYCHOLOGY. *Educ:* Prague Tech Univ, EC, 57; Charles Univ, Prague, DP, 61. *Prof Exp:* Asst prof math, Univ Sask, 65-66; assoc prof comput sci, Univ Man, 66-68; PROF COMPUTER SCI, UNIV WINDSOR, 68- *Concurrent Pos:* Nat Res Coun Can grant, Univ Man, 66-68; ed bd, Int J Gen Systs, 74-; vis scholar, Dept Comput & Commun Sci, Univ Mich, Ann Arbor, 74-; distinguished vis, Inst Elec & Electronics Engrs Comput Soc Prog, 79-82; pres, Int Cong Appl Systs Res & Cybernet, 80-; chmn Soc Gen Systs Res conf, 82-83; dir Int Inst Advan Studies Systs Res Cybernet, 83; dir Int Conf Systs Res, Informatics & Cybernet, Baden, WGer, 85. *Mem:* Can Comput Sci Asn; World Orgn Gen Systs & Cybernet; NY Acad Sci. *Res:* Diagnostic methodology; artificial intelligence; biomimetic engineering; expert systems; cerebrometic design of computers; simulation models; behavioral prediction; computer controlled conditioning; mathematical psychology; forecasting methodology; computer applications; psychotronics; psychocybernetics; quality of life; expert educational systems; systems models of brain; synergetics. *Mailing Add:* Sch Computer Sci Univ Windsor Windsor ON N9B 3P4 Can

LASKER, HOWARD ROBERT, b New York, NY, Feb 8, 52; m; c 1. MARINE BENTHIC ECOLOGY, ECOLOGY OF GORGONIAN CORALS. *Educ:* Univ Rochester, BS, 72, MS, 73; Univ Chicago, PhD(geophys sci), 78. *Prof Exp:* Fel, Rosenstiel Sch Marine & Atmospheric Sci, Univ Miami, 78-79; asst prof, 79-86, ASSOC PROF BIOL SCI, STATE UNIV NY, BUFFALO, 86- *Mem:* Ecol Soc Am; Am Soc Zool; Am Soc Naturalists; Int Soc Reef Studies. *Res:* Ecology of corals and coral reefs; physiological ecology of algal-coelenterate symbioses; population biology of benthic invertebrates; predator-prey interactions between grazers and gorgonian corals. *Mailing Add:* Dept Biol Sci State Univ NY Buffalo NY 14260

LASKER, REUBEN, comparative physiology; deceased, see previous edition for last biography

LASKER, SIGMUND E, b New York, NY, Sept 5, 23; m 65; c 1. PHYSICAL CHEMISTRY, BIOPHYSICS. *Educ:* Brooklyn Col, BS, 49; NY Univ, MS, 51; Stevens Inst Technol, PhD(phys chem), 65. *Prof Exp:* Res assoc surg, NY Med Col, 57-60, asst instr biochem, 58-60; res assoc phys chem, Stevens Inst Technol, 60-66; asst prof biophys, 66-72, asst prof pharmacol, 72-75, assoc prof, 75-82, PROF, MED (RES), NY MED COL, 82- *Concurrent Pos:* Adj assoc prof, Rockerfeller Univ, 68-80. *Mem:* Soc Magnetic Resonance Med; Biophys Soc; Am Chem Soc; NY Acad Sci; Am Heart Asn; Sigma Xi; NY Acad Med. *Res:* Magnetic resonance imaging; polyelectrolytes; experimental thermal injuries; anticoagulants. *Mailing Add:* Dept Med NY Med Col Valhalla NY 10595

LASKI, BERNARD, b Ont, Nov 11, 15; m 47; c 2. PEDIATRICS, HEMATOLOGY. *Educ:* Univ Toronto, MD, 39, FRCP(C), 49; Am Acad Pediat, FAAP, 49. *Prof Exp:* PROF PEDIAT, UNIV TORONTO, 77- *Concurrent Pos:* Chief Pediat, Mt Sinai Hosp, 54-; sr physician, Hosp for Sick Children, 49. *Mem:* Am Pediat Soc; Am Soc Pediat Res; Can Pediat Soc; Am Soc Hemat; Am Acad Pediat. *Mailing Add:* 99 Avenue Rd Toronto ON M5S 2R8 Can

LASKIN, ALLEN I, b Brooklyn, NY, Dec 7, 28; m 54, 73; c 2. BIOTECHNOLOGY. *Educ:* City Col New York, BS, 50; Univ Tex, MA, 52, PhD, 56. *Prof Exp:* Res scientist, Univ Tex, 53-55; sr res microbiologist, Squibb Inst Med Res, 55-64, res supvr microbiol, 64-67, asst dir microbiol, 67-69; res assoc, 69-74, sr res assoc, 74-84, head, Biosci Res, Exxon Res & Eng Co, 71-; assoc dir, NJ Ctr Adv Biotechnol & Med, 85-86; vpres res & develop, Matrix Res Labs, 86-88, pres, 88-89; vpres res & develop, Ethigen Corp, 86-89; RES FEL, CHARLES A DANA RES INST, DREW UNIV, 89- *Concurrent Pos:* Found microbiol lect, Am Soc Microbiol, 72. *Honors & Awards:* Found Microbiol Lectr, Am Soc Microbiol, 72; I M Lewis Award, 77; Selman A Waksman Hon Lectr, Theobald Smith Soc, 74. *Mem:* Soc Indust Microbiol (pres, 78-79); Am Soc Microbiol; Am Chem Soc; fel Am Acad Microbiol; fel NY Acad Sci. *Res:* Cell-free protein synthesis; mode of action of antibiotics; mechanisms of bacterial resistance to antibiotics; microbial transformations of steroids; petroleum microbiology; microbial enzymes; transformations and biodegradation of hydrocarbons and related compounds; lipid pharmacology; antiviral chemotherapy. *Mailing Add:* RD 2 Box 392T Somerset NJ 08873

LASKIN, DANIEL M, b New York, NY, Sept 3, 24; m 45; c 3. ORAL SURGERY, MAXILLOFACIAL SURGERY. *Educ:* Ind Univ, BS & DDS, 47; Univ Ill, MS, 51; Am Bd Oral & Maxillofacial Surg, dipl; FRCS; FRCPS(G). *Prof Exp:* Intern oral surg, Jersey City Med Ctr, 47-48; clin asst, Col Dent, Univ Ill, 49-50, res asst, 50-51, from instr to assoc prof, 51-60, prof oral & maxillofacial surg, 60-73, head dept, 73-83; PROF & CHMN DEPT ORAL & MAXILLOFACIAL SURG, MED COL VA, 84- *Concurrent Pos:* Resident, Cook County Hosp, 50-51, chmn dept oral surg, 67-77, dep chmn, 77-; attend oral surg, Hosps, 52-; clin prof surg, Univ Ill Hosp, 61-83, head dept dent, 76-83; ed, Am Asn Oral Maxillofacial & Surgeons Forum, J Oral & Maxillofacial Surg & Oral & Maxillofacial Implants; dir, Temporamandibular Joint & Facial Pain Res Ctr, 61-83; attend oral surgeon, Skokie Valley Community Hosp, 77-83, Swed Covenant Hosp, 79-83 & Edgewater Hosp, head dept, 70-83; consult oral surgeon, Ill Masonic Hosp, 65-83 & Bethany Methodist Hosp, 78-83; chmn & atten oral surgeon, Div Oral Surg, Cook County Hosp, 67-76, dep chmn, 76-83; attend oral & maxillofacial surgeon, Richmond Eye & Ear Hosp & Med Col Va Hosps, 84-; head, dept dent, Med Col Va Hosps, 86-; ed-in-chief, J Oral & Maxillofacial Surg, 72-; bd dirs, Am Pain Soc, 78-80. *Honors & Awards:* Res Recognition Award, Am Asn Oral & Maxillofacial Surgeons, 78; W Harry Archer Award, 81; Francis N Reichman Lect, 71; Simon P Hullihen Mem Award, WVA Univ, 76; Arnold K Maislen Mem Award, 77; William J Gies Oral Surg Award, 79; Cordwainer Lect, Univ London, 80; Heidbrink Award, Am Dent Soc Anesthesiol, 83; Rene LeFort Medal, Brazilian Col Oral & Maxillofacial Surg & Traumatology, 85; Distinguished Serv Award, Am Soc Oral Surgeons, 72; Arnold K Maislen Mem Award Oral Surg, 77; Hinman Medallion, 81; Edward C Hinds Lectr, 90. *Mem:* Sigma Xi; Int Asn Dent Res; Soc Exp Biol & Med; Am Soc Exp Path; Am Dent Asn; Am Asn Dent Res; fel AAAS; Am Asn Oral & Maxillofacial Surgeons (pres, 76-77); Int Asn Oral & Maxillofacial Surgeons (pres, 83-86, secy-gen, 89-). *Res:* Temporomandibular joint; metabolism of bone and cartilage; sutural growth; calcification and resorption of bone. *Mailing Add:* Dept Oral & Maxillofacial Surg Va Commonwealth Univ Box 566 MCV Sta Richmond VA 23298-0001

LASKOWSKI, EDWARD L, b Cleveland, Ohio, Mar 13, 43; m 72; c 2. COMPUTER INTEGRATED MANUFACTURING, SENSORS. *Educ:* Univ Detroit, BEE, 66; Ohio State Univ, MSEE, 68; Cleveland State Univ, DEng, 78. *Prof Exp:* Res engr, Gen Elec Co, 68-79; mgr advan systs, Bendix Corp, 79-85, dir res, 85-86; VPRES TECHNOL, ACME CLEVELAND CORP, 86- *Mem:* Inst Elec & Electronics Engrs; Soc Mfg Engrs; Nat Elec Mfrs Asn. *Res:* Intelligent sensors and systems; manufacturing computer systems; systems modeling; analysis of physical processes to be used as component of control system model; marketing new products. *Mailing Add:* 6154 Winchester Dr Seven Hills OH 44131

LASKOWSKI, LEONARD FRANCIS, JR, b Milwaukee, Wis, Nov 16, 19; m 46; c 3. MEDICAL MICROBIOLOGY, CLINICAL MICROBIOLOGY. *Educ:* Marquette Univ, BS, 41, MS, 48; St Louis Univ, PhD(med bact), 51; Am Bd Med Microbiol, dipl. *Prof Exp:* Instr bact, 46-48 & 51-53, sr instr, 53-54, asst prof microbiol, 54-57, from asst prof to assoc prof path, 57-69, PROF PATH, SCH MED, ST LOUIS UNIV, 69-, ASSOC PROF INTERNAL MED, 77- *Concurrent Pos:* China Med Bd fel, Latin Am, 57; fel trop med, La State Univ, 57; consult, St Mary's Group Hosp, 57-; attend, Vet Admin Hosp, health & tech training coordr, Latin Am Peace Corps Projs, 62-66; dir clin microbiol sect, St Louis Univ Hosps, 65; consult clin microbiol, John Cochran Vet Hosp, 66-, St Elizabeth's Hosp, Belleville, Ill, 68- & St Louis County Hosp, 69-; referee mycol, USPHS Commun Dis Ctr Prog for Testing Clin Diag Labs, 68-; consult, Jefferson Barracks, Vet Hosp, 72-, Mogul Diag Div, Mogul Corp, 72-79, St Francis Hosp, 74-77, St Elizabeth's Hosp, Granite City, Ill, 78- & St Louis County & City Med Examr, 78- *Mem:* NY Acad Sci; Am Soc Microbiol; AAAS; fel Am Acad Microbiol; Med Mycol Soc Americas. *Res:* Mechanism of intracellular parasitism; mechanism of action of therapeutic compounds. *Mailing Add:* 6229 Robertsville Rd Villa Ridge MO 63089

LASKOWSKI, MICHAEL, JR, b Warsaw, Poland, Mar 13, 30; nat US; m 57; c 2. BIOCHEMISTRY. *Educ:* Lawrence Col, BS, 50; Cornell Univ, PhD(phys chem), 54. *Prof Exp:* Asst chem, Cornell Univ, 50-52, USPHS fel, 52-56, instr, 56-57; from asst prof to assoc prof, 57-65, PROF CHEM, PURDUE UNIV, 65- *Concurrent Pos:* Chmn, Gordon Res Conf Physics & Phys Chem Biopolymers, 66, Proteolytic Enzymes & Inhibitors, 82; mem, Biophys & Biophys Chem Study Sect, NIH, 67-71; counr, Am Chem Soc, 84-87; vis prof, Yale Univ, 71, Alberta Univ, 81, Osaka Univ, 85, & Univ Calif San Francisco, 86. *Honors & Awards:* McCoy Award, Purdue Univ, 75; Alfred Jurzykowski Found Award, 77. *Mem:* AAAS; Am Chem Soc (treas, Div Biol Chem, 81-84); Am Soc Biol Chemists; Polish Inst Arts & Sci Am; AAAS; Biophys Soc; Protein Soc. *Res:* Protein chemistry; role of individual amino acid residues in proteinase inhibitor-proteinase interaction; evolution; proteolytic enzymes and their inhibitors. *Mailing Add:* Dept Chem Purdue Univ West Lafayette IN 47907-3699

LASKOWSKI, MICHAEL BERNARD, b Chicago, Ill, Apr 20, 43. EXPERIMENTAL BIOLOGY. *Educ:* Loyola Univ, BS, 66; Univ Okla, PhD(physiol), 70. *Prof Exp:* Muscular Dystrophy Asn postdoctoral fel, Dept Biol Sci, Northwestern Univ, 70-71, Dept Pharmacol, Vanderbilt Univ, 71-72, instr, 72-74, asst prof, 74-76; from asst prof to prof physiol, St Louis Univ Sch Med, 76-88, asst dean students, 85-88, actg chmn dept, 87-88; PROF PHYSIOL, DEPT BIOL SCI, UNIV IDAHO, 88-, DIR, WAMI PROG, 88- *Concurrent Pos:* Andrew Mellon career develop award, 74-76; vis scientist, Dept Physiol & Biophys, Wash Univ, 84-85; dir, Regional Med Educ Prog, Univ Idaho/Wash State Univ, 88. *Mem:* Sigma Xi; Am Physiol Soc; Am Acad Neurol; Soc Neurosci. *Mailing Add:* WAMI Med Prog Univ Idaho Moscow ID 83843

LASKY, JACK SAMUEL, b New York, NY, Mar 14, 30; m 52; c 2. ORGANIC CHEMISTRY, POLYMER CHEMISTRY. *Educ:* City Col New York, BS, 51; Univ Md, PhD(chem), 55. *Prof Exp:* Res chemist, US Rubber Co, 55-69; dir polymer res, 69-70, vpres res, 70-76, VPRES RES & ENG, OKONITE CO, 76- *Mem:* Am Chem Soc; Inst Elec & Electronics Engrs; Power Eng Soc. *Res:* Synthetic polymers; rubber; plastics; stereospecific polymerization; heterogeneous catalysis; kinetics of reactions; electrical insulating and covering materials. *Mailing Add:* 29 Newman Ave Verona NJ 07044

LASLETT, LAWRENCE JACKSON, b Boston, Mass, Jan 12, 13; m 39; c 3. PHYSICS. *Educ:* Calif Inst Technol, BS, 33; Univ Calif, PhD(physics), 37. *Prof Exp:* Rockefeller Found fel, Inst Theoret Physics, Copenhagen Univ, 37-38, Oersted fel, 38; instr physics, Univ Mich, 39 & Ind Univ, 39-42; mem staff, Radiation Lab, Mass Inst Technol, 41-45; from asst prof to prof physics, Iowa State Univ, 46-63; STAFF MEM, ACCELERATOR & FUSION RES DIV & ACCELERATOR THEORY GROUP, LAWRENCE BERKELEY LAB, UNIV CALIF, 63- *Concurrent Pos:* Head nuclear physics br & consult, Off Naval Res, DC, 52-53, sci liaison officer, London Br, 60-61; vis res prof, Univ Ill, 55-56 & Univ Wis, 56-57; head high energy physics br, US AEC, 61-63; lectr physics, Univ Calif, 65-67. *Mem:* AAAS; fel Am Phys Soc. *Res:* Accelerator theory and design; beta-ray spectra; electron and gamma-ray spectra; photonuclear research. *Mailing Add:* 726 Cragmont Ave Berkeley CA 94708

LASLEY, BETTY JEAN, b Winston-Salem, NC, July 10, 27. MICROBIOLOGY. *Educ:* Drew Univ, AB, 49; Rutgers Univ, MS, 53; NY Univ, PhD, 68. *Prof Exp:* Sr med technologist, Path Lab, NJ State Hosp, Greystone Park, 49-51; assoc investr chem microbiol, Warner Chilcott Res Labs Div, Warner-Lambert Co, 53-56; asst physiol, Med Col, NY Univ, 56; asst leukemia div, Sloan-Kettering Inst Cancer Res, 56-57; asst abstractor, Am Cyanamid Co, 57-58; asst metab & endocrinol, Med Ctr, 58-70, res assoc, 70-74, INSTR, MED CTR, NY HOSP, CORNELL UNIV, 74- *Mem:* NY Acad Sci. *Res:* Metabolism; immunobiology and endocrinology. *Mailing Add:* Payne Whitney Cornell Med Ctr NY Hosp 525 E 68th St New York NY 10021

LASLEY, BILL LEE, b Ottumwa, Iowa, June 4, 41. REPRODUCTIVE PHYSIOLOGY. *Educ:* Calif State Univ, Chico, BA, 63; Univ Calif, Davis, PhD(physiol), 72. *Prof Exp:* Teacher math & sci, Roseville Union High Sch, 64-67; fel reproductive biol, Rockefeller Found, 72-75; RES ENDOCRINOLOGIST, SAN DIEGO ZOOL GARDEN, 75- *Mem:* Endocrin Soc; Soc Gyn Invest; Soc Study Reproduction. *Res:* Reproductive endocrinology with emphasis on comparative studies. *Mailing Add:* Univ Calif Davis ITEH Old Davis Rd Davis CA 95616

LASLEY, JOHN FOSTER, b Liberal, Mo, Jan 26, 13; m 40; c 1. ANIMAL BREEDING. *Educ:* Univ Mo, BSA, 38, AM, 40, PhD(physiol reprod), 43. *Prof Exp:* Asst animal husb, Univ Mo, 38-43; agr exten agent & supvr agr & stock raising, US Indian Serv, 43-49; prof animal husb, Univ Mo-Columbia, 49-79; RETIRED. *Concurrent Pos:* Mem, Gov Sci Adv Comts, Mo. *Mem:* AAAS; Am Soc Animal Sci; Am Genetic Asn. *Res:* Physiology of spermatozoa; breeding problems of range cattle; staining method for differentiation of live and dead spermatozoa; improvement of beef cattle through selection and crossing. *Mailing Add:* 2207 Bushnell Dr Columbia MO 65201

LASLEY, STEPHEN MICHAEL, b Louisville, Ky, Feb 12, 50; m 74; c 2. NEUROTOXICOLOGY, NEUROCHEMISTRY. *Educ:* Univ Louisville, MEng, 73, PhD(neuropsychopharmacol), 79. *Prof Exp:* Fel neurotoxicol, Col Med Univ Cincinnati, 79-83, res assoc, 82-83; sci assoc pharmacol, Tex Col Osteop Med, Ft Worth, 83-85, res asst prof pharmacol, 85-86; ASST PROF, DEPT BASIC SCI, COL MED, UNIV ILL, PEORIA, 86- *Mem:* AAAS; Soc Neurosci; Soc Toxicol; Am Soc Pharmacol & Exp Therapeut. *Res:* Neurochemical mechanisms involved in epileptogenesis. *Mailing Add:* Dept Basic Sci Col Med Univ Ill PO Box 1649 Peoria IL 61656

LASS, NORMAN J(AY), b Brooklyn, NY, Sept 20, 43; m 67; c 2. SPEECH PATHOLOGY, AUDIOLOGY. *Educ:* Brooklyn Col, City Univ NY, BA, 65; Purdue Univ, MS, 66, PhD(speech & hearing sci), 68. *Prof Exp:* Res fel, Bur Child Res, Univ Kans Med Ctr, Kansas City, 68-69; from asst prof to assoc prof, 69-77, chmn dept, 74-83, PROF SPEECH PATH & AUDIOL, WVA UNIV, 77- *Concurrent Pos:* Consult, Audiol & Speech Path Serv, Vet Admin Med Ctr, Martinsburg, WVa, 75-83. *Mem:* Fel Am Speech-Lang Hearing Asn; Acoust Soc Am; Am Asn Phonetic Sci; Int Soc Phonetic Sci; Am Cleft Palate Asn. *Res:* Usefulness of time-altered speech; listener attitudes toward vocal characteristics and communicative disorders; hearing aid effect phenomenon; public and professional knowledge of and attitudes toward hearing loss and hearing aids. *Mailing Add:* Dept Speech Path & Audiol WVa Univ PO Box 6122 Morgantown WV 26506-6122

LASSEN, LAURENCE E, b Milwaukee, Wis, Dec 16, 32; m 59; c 2. FORESTRY. *Educ:* Iowa State Univ, BS, 54, MS, 58; Univ Mich, PhD(forestry), 67. *Prof Exp:* Res forest prod technologist, Forest Prod Lab, Wis, 58-67, proj leader, 67-71; staff asst res admin, 71-72, chief forest prod technol res, Washington, DC, 72-74, dep dir, 74-76, dir, Southern Forest Exp Sta, New Orleans, 76-83, DIR, INTERMOUNTAIN RES STA, US FOREST SERV, OGDEN, UT, 83- *Mem:* Soc Am Foresters; Sigma Xi; Soc Range Mgt. *Res:* Forestry research. *Mailing Add:* Intermountain Res Sta US Forest Serv 324 25th St Ogden UT 84401

LASSER, ELLIOTT CHARLES, b Buffalo, NY, Nov 30, 22; m 44; c 4. MEDICINE, RADIOLOGY. *Educ:* Harvard Univ, BS, 43; Univ Buffalo, MD, 46; Univ Minn, MS, 53. *Prof Exp:* Instr radiol, Grad Sch Med, Univ Minn, 52-53; assoc, Sch Med, Univ Buffalo, 53-54, asst prof, 54-56; prof & chmn dept, Sch Med, Univ Pittsburgh, 56-68; chmn dept, 68-77, PROF RADIOL, SCH MED, UNIV CALIF, SAN DIEGO, 68- *Concurrent Pos:* Consult, Vet Admin Hosp, Pittsburgh, Pa, 57-68. *Mem:* AAAS; Radiol Soc NAm; AMA; Am Col Radiol. *Res:* Radiology of the vascular system. *Mailing Add:* Univ Hosp San Diego Co 225 Dickinson San Diego CA 92103

LASSER, HOWARD GILBERT, b New York, NY, Nov 24, 26; m 50; c 3. CHEMICAL ENGINEERING, ELECTROCHEMISTRY. *Educ:* Lehigh Univ, BS, 50; Columbia Univ, ChemE, 51. *Hon Degrees:* Dr Ing, Darmstadt Polytech Inst, Ger, 56. *Prof Exp:* Chem engr indust gases, Eng Res & Develop Labs, Army Corps Engrs, 51-53, cryog, Bur Ships, Dept Navy, 53-55, metal prod, Gen Serv Admin, 55-56, metallic & org coatings, Eng Res & Develop Ctr, Army Corps Engrs, 56-68 & electrochem, Electronics Command, 68-73; mat eng consult coatings & electrochem, Naval Facil Eng Command, Dept Navy, 73-83; OWNER, MAT RES CONSULTS, SPRINGFIELD, VA, 83- *Concurrent Pos:* Chmn, Coatings Comt, Dept Defense, 74-82; consult organic & electrodeposited coatings & corrosion processes. *Honors & Awards:* Sci Res Award, Sigma Xi, 68. *Mem:* Fel Oil & Colour Chemists Asn; Am Electroplaters Soc; Am Inst Chem Engrs; fel Am Inst Chemists; fel AAAS. *Res:* Corrosion prevention through the use of metallic and organic coatings; development of anodic films on aluminum; transport theory as related to corrosion processes. *Mailing Add:* Mat Res Consults 5912 Camberly Ave Springfield VA 22150

LASSER, MARVIN ELLIOTT, solid state physics, for more information see previous edition

LASSETER, KENNETH CARLYLE, b Jacksonville, Fla, Aug 12, 42; m 63, 77; c 2. CLINICAL PHARMACOLOGY, INTERNAL MEDICINE. *Educ:* Stetson Univ, BS, 63; Univ Fla, MD, 67. *Prof Exp:* From intern to resident med, Univ Ky Hosp, 67-69, USPHS fel cardiol, 69-70; fel clin pharmacol, 70-71, asst prof, 71-74, assoc prof pharmacol & med, 74-81, CLIN ASSOC PROF PHARMACOL, UNIV MIAMI, 81-; VPRES & MED DIR, CLIN PHARMACOL ASSOC, INC, 81-; ADJ ASSOC PROF PHARMACOL, BARRY UNIV SCH MED, 87- *Concurrent Pos:* Attend physician, Jackson Mem Hosp, 71- *Honors & Awards:* Res Award, Interstate Postgrad Med Asn, 74. *Mem:* Am Soc Pharmacol & Exp Therapeut; Am Soc Clin Pharmacol & Therapeut; Am Col Physicians; Am Col Clin Pharmacol; Sigma Xi. *Res:* Cardiovascular pharmacology. *Mailing Add:* Clin Pharmacol Assoc Inc 2060 NW 22 Ave Miami FL 33142

LASSETTRE, EDWIN NICHOLS, chemical physics; deceased, see previous edition for last biography

LASSILA, KENNETH EINO, b Hancock, Mich, Apr 27, 34; m; c 3. HIGH ENERGY PHYSICS, THEORETICAL NUCLEAR PHYSICS. *Educ:* Univ Wyo, BS, 56; Yale Univ, MS, 59, PhD(theoret physics), 61. *Prof Exp:* Res assoc theoret physics, Case Inst Technol, 61-63; asst prof physics, Iowa State Univ, 63-64; sr res assoc, Res Inst Theoret Physics, Univ Helsinki, 64-66; res assoc, Stanford Univ, 66; from asst prof to assoc prof physics, 66-69, PROF PHYSICS, IOWA STATE UNIV, 69- *Concurrent Pos:* Fulbright res fel, Res Inst Theoret Physics, Univ Helsinki, 65; prof, Nordic Inst Theoret Atomic Physics, 72; Fulbright lectr, Univ Oulu, Finland & Univ Oslo, Norway, 73, Fermi Nat Accelerator Lab, 80-81 & 88; organizer, Fifth Int Workshop Weak Interations, 78; docent, Univ Helsinki, 73- *Mem:* AAAS; fel Am Phys Soc; Finnish Phys Soc. *Res:* Nucleon-nucleon interaction; radiation-field theory; elementary particle interactions. *Mailing Add:* Dept Physics Iowa State Univ Ames IA 50010

LASSITER, CHARLES ALBERT, b Murray, Ky, Feb 20, 27; m 46; c 2. ANIMAL SCIENCE. *Educ:* Univ Ky, BS, 49, MS, 50; Mich State Univ, PhD, 52. *Prof Exp:* From asst prof to assoc prof dairy sci, Univ Ky, 52-55; assoc prof, Mich State Univ, 56-59, prof dairy sci & chmn dept, 59-76; PROF ANIMAL SCI & HEAD DEPT, NC STATE UNIV, 76- *Mem:* Am Soc Animal Sci; Am Dairy Sci Asn. *Res:* Dairy cattle nutrition, especially pasture and forage studies dealing with more applied aspects; nutritional problems of young dairy calves. *Mailing Add:* Dept Animal Sci NC State Univ Box 7621 Raleigh NC 27695

LASSITER, RAY ROBERTS, b Hertford Co, NC, Apr 8, 37; m 60; c 2. ECOLOGY, ENVIRONMENTAL SCIENCE. *Educ:* NC State Col, BS, 59, MS, 62; NC State Univ, PhD(animal ecol), 71. *Prof Exp:* Instr biol, Campbell Col, NC, 62-63; asst statistician quant ecol, NC State Univ, 63-67; systs analyst environ statist, Pollution Surveillance Br, SE Water Lab, Environ Protection Agency, 67-70, biol statistician environ sci, Freshwater Ecosysts Br, 70-75; chief environ systs, Br Environ Sci, Athens Environ Res Lab, Environ Protection Agency, 75-81; resident res scientist, Inst Ecol, Univ Ga, 81-83; RES ECOLOGIST, ATHENS ENVIRON RES LAB, 83- *Concurrent Pos:* Mem, Inst Ecol, Univ Ga, 77-; res grant, Environ Protection Agency, 81. *Res:* Quantitative ecology; population and community dynamics; experimental ecology using laboratory ecosystems; mathematical modeling of ecosystem processes; fate of toxic chemicals in aquatic ecosystems; predicting effects of toxic chemicals on aquatic ecosystem populations and functions. *Mailing Add:* Inst Ecol Univ Ga Athens GA 30602

LASSITER, WILLIAM EDMUND, b Wilmington, NC, July 21, 27; m 56; c 4. PHYSIOLOGY, NEPHROLOGY. *Educ:* Harvard Univ, AB, 50, MD, 54. *Prof Exp:* Intern & asst resident med, Mass Gen Hosp, Boston, 54-56; sr asst resident, NC Mem Hosp, 56-57; Donner res fel, Mass Gen Hosp, Boston, 57-58; Life Ins Med Res Fund fel, Univ NC, Chapel Hill, 58-60; from instr to assoc prof, 60-70, PROF MED, SCH MED, UNIV NC, CHAPEL HILL, 70- *Concurrent Pos:* Estab investr, Am Heart Asn, 62-67, mem coun kidney in cardiovasc dis, 71-; vis investr, Physiol Inst, Berlin, 63-64; Markle scholar, 63-68; Nat Inst Arthritis & Metab Dis career develop award, 67-72; mem cardiovasc & pulmonary res A study sect, NIH, 69-73 & gen med B study sect, 78-82; sect ed, Renal & Electrolyte Physiol, Am J Physiol & J Appl Physiol, 74-76; co-dir NIH surv res needs nephrol & urol, 74-78. *Mem:* AAAS; Am Soc Clin Invest; Am Physiol Soc; fel Am Col Physicians; Am Soc Nephrology. *Res:* Micropuncture studies of mammalian kidney function; tubular transport processes in mammalian kidney; calcium and phosphorus metabolism. *Mailing Add:* Dept Med CB No 7005 Univ NC Sch Med Chapel Hill NC 27599-7005

LASSITER, WILLIAM STONE, b Spring Hope, NC, July 7, 39; m 71; c 2. ENGINEERING MANAGEMENT, NOISE & POLLUTION CONTROL. *Educ:* NC State Univ, BS, 61, PhD(mech eng), 71; Col William & Mary, MS, 68, MBA, 73. *Prof Exp:* Facil Engr, 63-65, mat engr, 65-70, pollution sensing res, 70-75, ENG MGR ENG ANALYSIS, LANGLEY RES CTR, NASA, 76- *Concurrent Pos:* Adj instr, George Washington Univ, Christopher Newport Col & Golden Gate Univ. *Mem:* Am Soc Mech Engrs. *Res:* Noise control; pollution sensing; thermal, structural and fluid analysis. *Mailing Add:* Langley Res Ctr NASA Mail Stop 431 Hampton VA 23665

LASSLO, ANDREW, b Mukacevo, Czech, Aug 24, 22; nat US; m 55; c 1. MEDICINAL CHEMISTRY. *Educ:* Univ Ill, MSc, 48, PhD(pharmaceut chem), 52, MSLS, 61. *Prof Exp:* Asst chem, Univ Ill, 47-51, Univ fel, 51-52; res chemist, Org Chem Div, Monsanto Chem Co, 52-54; asst prof med chem, dept pharmacol, Emory Univ, 54-60; prof med chem & chmn dept, 60-89, alumni distinguished serv prof & chmn dept, 89-90, EMER PROF MED CHEM, COL PHARM, HEALTH SCI CTR, UNIV TENN, MEMPHIS, 90- *Concurrent Pos:* 1st lieutenant to captain, Med Serv Corps, US Army Reserve, 53-62; prin investr, USPHS, 58-64, 66-72 & 82-89, Geschickter Fund Med Res, 59-65, Nat Sci Found, 64-66, US Army Med Res & Develop Command, 64-67 & Gustavus & Louise Pfeiffer Res Found, 81-87; consult, Gesckickter Fund Med Res, 61-62; dir, postgrad training prog for sci librn, Nat Libr Med, 66-72 & inorg med chem, US Food & Drug Admin, 71; chmn subcomt, pre and postdoctoral training, Am Soc Pharmacol & Exp Therapeut, 74-78; producer & moderator TV & radio series, Health Care Perspectives, 76-78; ed, Surface Chem & Dent Integuments, 73, Blood Platelet Function & Med Chem, 84. *Honors & Awards:* Commendation, Sigma Xi, 76. *Mem:* Fel AAAS; fel Am Inst Chemists; sr mem Am Chem Soc; Am Soc Pharmacol & Exp Therapeut; fel Am Asn Pharmaceut; fel Acad Pharmaceut Sci. *Res:* Synthesis and study of compounds with pharmacodynamic potentialities; exploration of relationships between the molecular constitution of synthetic entities and their biodynamic response; science information and library resources; seven US and nine foreign patents in field; authored numerous articles in leading scientific and professional journals. *Mailing Add:* 5479 Timmons Ave Memphis TN 38119-6932

LASSMAN, HOWARD B, m; c 3. PHARMACOLOGY. *Educ:* State Univ NY, Buffalo, BS, 63, MS, 67, PhD(biochem pharmacol), 72. *Prof Exp:* Asst cancer res scientist, Roswell Park Mem Inst, NY State Dept Health, Buffalo, 63-67, cancer res scientist, 67-69; sr res pharmacologist, Hoechst-Roussel Pharmaceuticals Inc, 72-75, res assoc, Preclin Pharmacol Sect & radiation safety officer, 75-77, asst dir clin res-internal med, 77-78, assoc dir, 78-79, actg dir clin pharmacol, 80-81, DIR CLIN PHARMACOL, HOECHST-ROUSSEL PHARMACEUTICALS, INC, NJ, 72- *Concurrent Pos:* Chmn, Res Sem Comt, Hoechst-Roussel Pharmaceuticals, Inc, 73; mem, Pharmaceut Found Adv Coun, Univ Tex, Austin, 84 & 89, organizing comt, Int Indust Pharm Conf, 85 & 90. *Mem:* Am Soc Clin Pharmacol & Therapeut; Am Col Clin Pharmacol; Am Asn Pharmaceut Sci; Drug Info Asn; Am Soc Pharmacol & Exp Therapeut; NY Acad Sci; Inflammation Res Asn; Int Inflammation Club; AAAS; Sigma Xi. *Res:* Clinical and clinicopharmacological investigation of new drugs for National Drug Association approval and studies of new indications for marketed drugs. *Mailing Add:* Hoechst-Roussel Pharmaceuticals Inc Rte 202-206 PO Box 2500 Somerville NJ 08876-1258

LAST, ARTHUR W(ILLIAM), metallurgical engineering; deceased, see previous edition for last biography

LAST, JEROLD ALAN, b New York, NY, June 5, 40; m 75; c 3. BIOCHEMISTRY, INHALATION TOXICOLOGY. *Educ:* Univ Wis, BS, 59, MS, 61; Ohio State Univ, PhD(biochem), 65. *Prof Exp:* Biochemist, Corn Prod Co, Ill, 61-62; sr res scientist, Squibb Inst Med Res, NJ, 67-69 & Rockefeller Univ, 69-70; asst managing ed, Proc Nat Acad Sci, USA, 70-71, managing ed, 71-73; consult, NIH, 71-73; res assoc, Dept Biochem & Molecular Biol, Harvard Univ, 73-76; from asst prof to assoc prof, 76-83, PROF, DEPT INTERNAL MED, PULMONARY DIV, MED CTR, UNIV CALIF, DAVIS, 83- *Concurrent Pos:* Am Cancer Soc fel biochem, Med Sch, NY Univ, 66-67; Fulbright prof, Montevideo, Uruguay, 83; vis prof, Bishop Col, Dallas, 82; exp lung res, 83, toxicol & appl pharmacol, 84, toxicol, 80 -84; vis prof Catholic Univ PR, Ponce, 85. *Honors & Awards:* Frank R Blood Award, Soc Toxicol, 80. *Mem:* Am Soc Biol Chem; Am Fedn Clin Res; Brit Biochem Soc; Soc Toxicol; Am Thoracic Soc. *Res:* Antibiotics biosynthesis and mechanism of action; lung disease; protein biosynthesis; nucleic acids; collagen biosynthesis; mucus glycoproteins; lung biochemistry; pulmonary fibrosis; health effects of air pollutants. *Mailing Add:* Dept Med Div Pulmonary Med Univ Calif Davis CA 95616

LAST, JOHN MURRAY, b Tailem Bend, Australia, Sept 22, 26; m 57; c 3. EPIDEMIOLOGY. *Educ:* Univ Adelaide, MB & BS, 49, MD, 68; Univ Sydney, DPH, 60. *Prof Exp:* Australian Postgrad Med Found fel, Social Med Res Unit, Med Res Coun UK, 61-62; lectr pub health, Univ Sydney, 62-63;

asst prof epidemiol, Univ Vt, 63-64; sr lectr social med, Univ Edinburgh, 65-69; PROF EPIDEMIOL & COMMUNITY MED & CHMN DEPT, UNIV OTTAWA, 70- *Concurrent Pos:* Mem, Nat Health Grant Rev Comt, 70-76 & Epidemiol Study Sect, NIH, 72-76; consult med educ, WHO, 72-73, 74, 76 & environ health, 84, 90, & 91; sci ed, Can J Pub Health, assoc ed, Am J Prev Med. *Honors & Awards:* Distinguished Serv Award, Am Col Prev Med, 84. *Mem:* Int Asn Epidemiol; fel Am Pub Health Asn; fel Royal Australasian Col Physicians; fel Am Col Prev Med (pres, 87-89); fel Fac Community Med, Eng. *Res:* Environmental epidemiology; international health; biomedical ethics (history/philosophy of medicine); author. *Mailing Add:* Facil Med Univ Ottawa Ottawa ON K1H 8M5 Can

LAST, ROBERT L, b New York, NY, Nov 10, 58. AMINO ACID BIOSYNTHESIS, UV-B RESPONSE IN PLANTS. *Educ:* Ohio Wesleyan Univ, BA, 80; Carnegie-Mellon Univ, PhD(biol sci), 86. *Prof Exp:* Postdoctoral fel, Whitehead Inst Biomed Res, 86-89; ASST SCIENTIST, BOYCE THOMPSON INST PLANT RES, CORNELL UNIV, 89- *Concurrent Pos:* Adj asst prof, Sect Genetics & Develop, Cornell Univ, 90-; NSF presidential young investr award, 90. *Mem:* AAAS; Am Soc Plant Physiologists; Genetics Soc Am; Sigma Xi. *Res:* Regulation of amino acid biosynthesis in flowering plants using genetics and molecular biology techniques. *Mailing Add:* Plant Molecular Biol Prog Boyce Thompson Inst Plant Res Ithaca NY 14853

LASTER, DANNY BRUCE, b Scotts Hill, Tenn, Nov 29, 42; m 60; c 2. ANIMAL SCIENCES. *Educ:* Univ Tenn, Knoxville, BS, 63; Univ Ky, Lexington, MS, 64; Okla State Univ, PhD(animal breeding), 70. *Prof Exp:* Res specialist, Univ Ky, Lexington, 65-68; asst prof endocrin, Iowa State Univ, 70-71; res leader, Reproduction Res Unit, Clay Ctr, Agr Res Serv, USDA, Nebr, 71-78, nat prog leader & assoc dep adminr, Beef & Sheep, 81-88; DIR, ROMAN L HRUSKA MEAT ANIMAL RES CTR, CLAY CTR, NEBR, 88- *Mem:* Am Soc Animal Sci. *Res:* Twinning, dystocia and embryonic mortality in beef cattle. *Mailing Add:* US Meat Animal Res Ctr Agr Res Serv USDA PO Box 166 Clay Center NE 68933-0166

LASTER, LEONARD, b New York, NY, Aug 25, 28; m 56; c 3. MEDICINE, SCIENCE POLICY. *Educ:* Harvard Univ, AB, 49, MD, 50; Am Bd Internal Med, dipl, 61; Am Bd Gastroenterol, dipl, 66. *Prof Exp:* From intern to resident med, Mass Gen Hosp, 50-53; vis investr purine metab, Pub Health Res Inst, New York, Inc, 53-54; sr clin investr, Nat Inst Arthritis & Metab Dis, 54-58; chief sect gastroenterol, Metab Dis Br, Nat Inst Arthritis & Metab Dis, 59-69; staff mem, Off Sci & Technol, Exec Off of President, 69-71, asst dir human resources, 71-74; vpres acad & clin affairs & dean, Col Med, State Univ NY Downstate Med Ctr, 74-78; PROF MED, SCH MED & PRES, ORE HEALTH SCI UNIV, 78- *Concurrent Pos:* Res fel gastroenterol, Mass Mem Hosps, 58-59; clin instr, Sch Med, George Washington Univ, 55-58, prof lectr, 66- *Mem:* Am Soc Biol Chemists; Am Fedn Clin Res; Am Gastroenterol Asn; Am Col Physicians; Am Soc Clin Invest. *Res:* Biochemical aspects of human disease; inborn errors of metabolism; disturbances of the gastrointestinal tract. *Mailing Add:* 47 Pine Arden Dr West Boylston MA 01583

LASTER, RICHARD, b Vienna, Austria, Nov 10, 23; nat US; m 48; c 2. CHEMICAL ENGINEERING. *Educ:* Polytech Inst Brooklyn, BChE, 43. *Prof Exp:* Asst lab dir eng res & develop, Cent Labs, Gen Foods Corp, 44-54, res mgr, Walter Baker Div, 54-58, mgr mfg & eng, Franklin Baker Div, 58-60, oper mgr, 60-62, oper mgr, Atlantic Gelatin Div, 62-64, opers mgr res & new prod develop, Jello Div, 64-67, dir corp qual assurance, 67, opers mgr, Maxwell House Coffee Div, 67-69, asst gen mgr corp, 69-71, vpres corp & pres, Maxwell House Coffee Div, 71-72, group vpres coffee & food serv, 72-74, exec vpres & dir, 74-82; chmn, 88, PRES, CHIEF EXEC OFFICER & DIR, DNA PLANT TECHNOL CORP, 82- *Concurrent Pos:* Dir, Bowater Inc; chmn, Purchase Col Found; trustee, Polytech Univ. *Honors & Awards:* Food & Bioeng Award, Am Inst Chem Engrs, 72. *Mem:* AAAS; Am Chem Soc; Am Inst Chem Engrs; NY Acad Sci; Inst Food Tech. *Res:* Chemical engineering research as applied to food processing; spray drying; atomization and leaching; chemistry of fats and oils; chocolate processing. *Mailing Add:* 23 Round Hill Rd Chappaqua NY 10514

LASTER, WILLIAM RUSSELL, JR, b Ala, Oct 20, 26; m 61; c 1. VETERINARY MEDICINE. *Educ:* Auburn Univ, DVM, 51. *Prof Exp:* Virologist, 51-56, head cancer screening sect, 56-66, HEAD, CANCER SCREENING DIV, SOUTHERN RES INST, 66- *Mem:* Am Asn Cancer Res. *Res:* Cancer chemotherapy. *Mailing Add:* Cancer Screening Div Southern Res Inst PO Box 55305 Birmingham AL 35255-5305

LASURE, LINDA LEE, b Bartlesville, Okla, Nov 23, 46. MICROBIAL GENETICS. *Educ:* St Cloud State Col, BS, 68; Syracuse Univ & State Univ NY, PhD(genetics), 73. *Prof Exp:* Res fel, New York Bot Garden, 72-74; res scientist microbiol, Miles Lab Inc, 74-79, sr res scientist, 79-80, supvr mutation & screening, 80-84, dir bioprod res, 84-89; sr dir, 89-90, VPRES, US LABS, PANLABS, INC, 90- *Mem:* Sigma Xi; AAAS; Genetics Soc Am; Am Soc Microbiol; Mycol Soc Am. *Res:* Studies of mutagenesis, inheritance, and physiological control of sexual reproduction in the lower fungi, fungal spore germination, and selection of strains of fungi with improved yields of enzymes and organic acids; enzyme synthesis and molecular biology of filamentous fungi. *Mailing Add:* Pan Labs 11804 Northcreek Pkwy So Bothell WA 98011-8805

LASWELL, TROY JAMES, b Ottawa, Ky, Nov 12, 20; m 43; c 1. GEOLOGY. *Educ:* Berea Col, AB, 42; Oberlin Col, AM, 48; Univ Mo, PhD(geol), 53. *Prof Exp:* Instr geol, Univ Mo, 48-53; from asst prof to assoc prof, Washington & Lee Univ, 53-57; from assoc prof to prof, La Polytech Inst, 57-62; head dept, 62-85, prof, 62-85, EMER PROF GEOL & GEOG, MISS STATE UNIV, 85- *Concurrent Pos:* Geologist, Mo Geol Surv, 52-53; consult geologist, Va Mins, Inc, 55-56, South River Mining Co, 57, Humble Oil & Refining Co, 55-58. *Mem:* Geol Soc Am; Am Asn Petrol Geologists; Am Asn Geol Teachers; Sigma Xi. *Res:* Stratigraphy; sedimentation. *Mailing Add:* PO Box 824 Mississippi State MS 39762

LASZEWSKI, RONALD M, b Chicago, Ill, June 22, 47. EXPERIMENTAL NUCLEAR PHYSICS, ACOUSTICS. *Educ:* Univ Ill, Urbana, BS, 69, MS, 72, PhD(physics), 75. *Prof Exp:* Res assoc physics, Argonne Nat Lab, 75-78; from res asst prof physics to sr res physicist, 78-90, PRIN RES PHYSICIST, UNIV ILL, URBANA, 90- *Mem:* Am Phys Soc. *Res:* Photonuclear physics and medium energy physics. *Mailing Add:* Nuclear Physics Lab Univ Ill 23 Stadium Dr Champaign IL 61820

LASZLO, CHARLES ANDREW, b July 8, 35; Can citizen; m; c 2. BIOMEDICAL ENGINEERING. *Educ:* McGill Univ, BEng, 61, MEng, 66, PhD(biomed eng), 68. *Prof Exp:* Design engr, Northern Elec Co & RCA Victor Co, Ltd, Montreal, Que, 61-62; biomed engr, OTL Res Labs, Royal Victoria Hosp, Montreal, Que, 62-68; assoc prof, Biomed Eng Unit & Dept Otolaryngol, McGill, 68-74, dir, Biomed Eng Unit, 70-74; assoc dir, Div Health Systs, 74-85, PROF, DEPT ELEC ENG, UNIV BC, 80-, DIR, CLIN ENG PROG, 80- *Concurrent Pos:* Mem, Adv Comt Med Devices, Bur Med Devices, Health Protection Br, Health & Welfare Can, 79-, Comt Hearing Aid Standards, 86-, Comt Telecommun Devices, Can Standards Asn, 88, Res Grant Eval Comt, Health Develop Fund Sci Coun, BC, 89-, Assoc Comt Res & Develop Rehab Disabled Persons, Nat Res Coun, 90-; assoc men, Dept Health Care & Epidemiol & Sch Audiol & Speech sci, Univ BC, 80-; consult, Royal Columbian Hosp, 83-, Royal Jubilee Hosp, Children's Hosp & St Paul Hosp, 84-, G F Strong Rehab Ctr, 86-; chmn, Bd Hearing Aid Dealers & Consult, Ministry Health, BC, 87- *Honors & Awards:* Award of Merit, Can Hard of Hearing Asn, 90. *Mem:* Sr mem Inst Elec & Electronics Engrs; sr mem Instrument Soc Am; fel Am Asn Med Systs & Info. *Res:* Over 120 publications; electrical engineering; biomedical engineering. *Mailing Add:* Elec Eng Dept Univ BC 2356 Main Mall Vancouver BC V6T 1Z4

LASZLO, TIBOR S, b Oraviczafalu, Hungary, Apr 25, 12; US citizen. INDUSTRIAL CHEMISTRY, MATERIALS SCIENCE ENGINEERING. *Educ:* Royal Hungarian Univ Technol & Econ Sci, DSc(chem eng), 35. *Prof Exp:* Res & develop engr oil industs, Hungary, 35-48; asst dir inorg high temperature chem, Fordham Univ, 48-51, process engr res dept, Calif Tex Oil Co, 51-53; dir high temperature lab, Fordham Univ, 53-57; proj specialist high temperature ceramics, Wright Aeronaut Div, Curtiss Wright Corp, 57-58; prin staff scientist, High Temperature Inorg Lab, Res & Adv Develop Div, Avco Corp, 58-65, sr consult scientist, 65-66; prin scientist, Res Ctr, Philip Morris, Inc, 66-77; CONSULT, 77- *Concurrent Pos:* Consult specialist, 53-57; NSF grant, 55-57; foreign ed, La Revue des Hautes Temperatures et des Refractaires, 65-; mem bd gov & symp dir, Int Microwave Power Inst. *Res:* Petroleum and vegetable oil processing; high temperature generation; solar furnaces; high refractory compounds; solar radiation simulation; temperature control coatings; thermal radiation measurements; microwave industrial technology; cellulose-water bond. *Mailing Add:* 4600 S Four Mile Dr 226 Arlington VA 22204

LATA, GENE FREDERICK, b New York, NY, May 17, 22; m 51; c 5. BIOCHEMISTRY. *Educ:* City Col New York, BS, 42; Univ Ill, MS, 48, PhD(biochem), 50. *Prof Exp:* Jr chemist, Gen Foods Corp, 42, chemist, 46; asst chem, Univ Ill, 47, asst biochem, 48-50; from instr to asst prof, 50-62, fel med, 65-66, ASSOC PROF BIOCHEM, UNIV IOWA, 62- *Concurrent Pos:* Vis lectr, Huntington Labs, Harvard Med Sch, 65-66; consult, Am Dent Asn; chmn, Iowa Sect, Am Chem Soc, 87. *Mem:* Fel AAAS; Am Chem Soc; Am Soc Biol Chemists; NY Acad Sci; Sigma Xi (pres Iowa Chap, 87). *Res:* Peroxisomal enzymes and their control; hormonal control of enzyme actions; steroid interactions and transport; aging and steroid hormone dynamics; angiogenesis factors. *Mailing Add:* Dept Biochem Univ Iowa Iowa City IA 52242

LATANISION, RONALD MICHAEL, b Richmondale, Pa, July 2, 42; m 64; c 2. CORROSION ENGINEERING, MATERIALS PROCESSING. *Educ:* Pa State Univ, BS, 64; Ohio State Univ, PhD(metall eng), 68. *Prof Exp:* Actg head, Mat Sci Group, Martin Marietta Labs, Baltimore, Md, 70-74; prof mat sci & eng, 75-83, SHELL PROF MAT SCI, MASS INST TECHNOL, 83-, DIR, H H UHLIG CORROSION LAB, 75-, DIR, MAT PROCESSING CTR, 84- *Concurrent Pos:* Sci adv, US House Rep Comt Sci & Technol, 82-83. *Honors & Awards:* Campbell Award, Nat Asn Corrosion Engrs; Krumb Lectr, 84; McFarland Award, Pa State Univ, 86; Humbolt Sr Sci Prize, 74-75. *Mem:* Nat Acad Eng; Nat Asn Corrosion Engrs; Am Inst Mining, Metall & Petrol Engrs; Electrochem Soc. *Res:* Corrosion of new materials (composites, magnetic alloys, etc); materials for construction of engineering systems. *Mailing Add:* Mat Processing Ctr Mass Inst Technol 77 Massachusetts Ave Rm 8-202 Cambridge MA 02139

LATCH, DANA MAY, b New York, NY, Aug 29, 43; c 1. ALGEBRA, TOPOLOGY. *Educ:* Harpur Col, BA, 65; Queens Col, NY, MA, 67; City Univ New York, PhD(math), 71. *Prof Exp:* Teaching asst math, Queens Col, NY, 65-66, lectr, 66-67; asst prof, Douglass Col, Rutgers Univ, 71-74 & Lawrence Univ, 74-76; asst prof, 76-79, ASSOC PROF MATH, NC STATE UNIV, 79- *Concurrent Pos:* NSF res grant, Rutgers Univ, 72-73 & NC State Univ, 78-80 & 81-83; Carnegie Found fac develop award, Lawrence Univ, 75; study grant, Ger Acad Exchange Serv, Univ Konstanz, Ger, 78; vis prof math, Universitat Konstanz, Ger, 79; vis scholar comput sci, Univ NC, Chapel Hill, 82; Alexander von Humboldt res fel, Munich, Fed Repub Ger, 83. *Mem:* Am Math Soc; Asn Women Math. *Res:* Algebraic topology of small categories; applications of category theory to the theory of program behaviors; applications of categorical methods to homotopy theory and the theory of localizations. *Mailing Add:* Dept Math NC State Univ Raleigh NC 27607

LATEEF, ABDUL BARI, b Faisalabad, Pakistan, Apr 4, 39; m 70. FORENSIC SCIENCE, CHEMISTRY. *Educ:* Punjab Univ, Pakistan, BS, 59, MS, 61; Univ Newcastle, PhD(chem), 66. *Prof Exp:* Nat Res Coun Can fel, Univ Calgary, 66-69; from instr to asst prof chem, 69-71, from asst prof to assoc prof, 71-81, PROF FORENSIC SCI, YOUNGSTOWN STATE UNIV, 81-; DIR, TRI-STATE LABS INC, 81- *Mem:* Am Acad Forensic Sci; Forensic Soc, London; Acad Criminal Justice; Sigma Xi. *Res:* Spectroscopic and chromatographic analytical techniques in forensic science; role of forensic science in criminal justice system. *Mailing Add:* Tri-State Labs 19 E Frong St Youngstown OH 44503

LATHAM, ALLEN, JR, b Norwich, Conn, May 23, 08; m 33; c 4. ENGINEERING, BIOENGINEERING. *Educ:* Mass Inst Technol, BS, 30. *Prof Exp:* Jr engr, E I du Pont de Nemours & Co, WVa, 30-35; chief engr, Polaroid Corp, Mass, 36-41; sr mech engr, Arthur D Little, Inc, 41-51, vpres, 51-59, sr vpres, 59-67; pres, 500 Inc, 67-68; pres, 72-76, chmn bd, 72-82, FOUNDER, HAEMONETICS CORP, 82- *Concurrent Pos:* Chmn bd, Cryogenic Technol, 68-72. *Honors & Awards:* Mortem Grove-Rasmussen Mem Award, Am Asn Blood Banks. *Mem:* Nat Acad Eng; Am Soc Mech Engrs; Am Inst Chem Engrs; Instrument Soc Am. *Res:* Cryogenics; blood processing. *Mailing Add:* Haemonetics Corp 400 Wood Rd Braintree MA 02184

LATHAM, ARCHIE J, b Blackfoot, Idaho, June 26, 26; m 63; c 3. PLANT PATHOLOGY, MYCOLOGY. *Educ:* Idaho State Col, BS, 56; Univ Idaho, MS, 59; Univ Ill, PhD(plant path), 61. *Prof Exp:* Ranger, Yellowstone Park, Wyo, 56; res asst plant path, Univ Idaho, 56-58 & Univ Ill, 58-61; res biologist, Gulf Res & Develop Co, Kans, 61-67; asst prof bot & plant path, Auburn Univ, 67-78, assoc prof, 78-85; CONSULT, 85- *Mem:* Am Phytopathathological Soc; Mycol Soc Am. *Res:* Fruit and nut diseases. *Mailing Add:* Dept Plant Path Auburn Univ Main Campus Auburn AL 36849

LATHAM, DAVID WINSLOW, b Boston, Mass, Mar 19, 40; m 60; c 5. ASTRONOMY. *Educ:* Mass Inst Technol, BS, 61; Harvard Univ, MA, 65, PhD(astron), 70. *Prof Exp:* ASTRONR, SMITHSONIAN ASTROPHYS OBSERV, 65-; LECTR ASTRON, HARVARD UNIV, 71- *Mem:* Am Astron Soc; Soc Photog Scientist & Engrs; Royal Astron Soc; Sigma Xi. *Res:* Stellar spectroscopy; stellar chemical abundances and nucleosynthesis; detection of light at low levels. *Mailing Add:* 107 Bristers Hill Rd Concord MA 01742

LATHAM, DEWITT ROBERT, b Chugwater, Wyo, Oct 26, 28; m 55; c 2. PETROLEUM CHEMISTRY. *Educ:* Univ Wyo, BS, 50. *Prof Exp:* Anal chemist, Rocky Mountain Arsenal, US Army Chem Corps, Colo, 52-54; chemist, Laramie Energy Res Ctr, US Bur Mines, 54-60, res chemist, 60-64, PROJ LEADER, LARAMIE ENERGY TECH CTR, DEPT ENERGY, 64- *Mem:* Am Chem Soc; Sigma Xi. *Res:* Nitrogen and oxygen compounds in petroleum; development of methods of analysis for petroleum shale, oil and coal liquids; separation and characterization of fossil fuel energy sources. *Mailing Add:* 1213 Custer St Laramie WY 82070

LATHAM, DON JAY, b Lewiston, Idaho, Dec 21, 38; m 60; c 3. ATMOSPHERIC SCIENCES. *Educ:* Pomona Col, BA, 60; NMex Inst Mining & Technol, MS, 64, PhD, 68. *Prof Exp:* Res asst, NMex Inst Mining & Technol, 61-67; sr res scientist, Rosenstiel Sch Marine & Atmospheric Sci, Univ Miami, 67-68, asst prof atmospheric sci, 68-72, assoc prof, 72-75; RES METEOROLOGIST & PHYSICIST, NORTHERN FOREST FIRE LAB, 75- *Concurrent Pos:* NSF grants, Univ Miami, 69-75; fac affil, Univ Mont, 75-; mem, Atmospheric Elec Comt, Am Meteorol Soc, 81-83. *Mem:* Am Meteorol Soc; Am Geophys Union. *Res:* Atmospheric electricity; radar meteorology; combustion physics; artificial intelligence. *Mailing Add:* Intermountain Fire Sci Lab Box 8089 Missoula MT 59807

LATHAM, GARY V, geophysics, for more information see previous edition

LATHAM, MICHAEL CHARLES, b Kilosa, Tanzania, May 6, 28; m 74; c 2. NUTRITION, TROPICAL PUBLIC HEALTH. *Educ:* Univ Dublin, BA, 49, MB, BCh & BAO, 52; Univ London, DTM&H, 58; Harvard Univ, MPH, 65. *Hon Degrees:* FFCM, Royal Col Physicians, London, 73. *Prof Exp:* House surgeon, High Wycombe Hosp, Buckinghamshire, Eng, 52-53; rotating physician, Methodist Hosp, Los Angeles, 53-54; sr house officer, NMiddlesex Hosp, London, 54-55; med officer, Tanzania Ministry of Health, 55-64, dir nutrit unit, 62-64; res assoc & asst prof nutrit, Harvard Univ, 64-68; PROF INT NUTRIT & DIR, PROG INT NUTRIT, DIV NUTRIT SCI, CORNELL UNIV, 68- *Concurrent Pos:* Vis exchange fel, Methodist Hosp, Los Angeles, 53-54; contrib ed, Nutrit Rev, 68-74; chmn panel, White House Conf Food, Nutrit & Health & vchmn panel, Follow-Up Conf, 69-71; consult, WHO, Manila, 70 & UN Food & Agr Orgn, 64 & Zambia, 70; mem, Nat Acad Sci-Nat Res Coun Int Nutrit Comt, 70-76; UNICEF consult, Thailand & Malaysia, 73; mem exec comt pest control, Nat Acad Sci, 73-78; mem expert adv panel nutrit, WHO, 74-; vis prof, Univ Nairobi, Kenya, 74-75; World Bank consult, Indonesia, 75; US AID consult, Guyana, 77 & 78, Tanzania, 79 & Philippines, 81; mem comt nutrit & infections, Nat Acad Sci, 78-81, Bd Int Health, Nat Acad Sci & Inst Med, 83-88; assoc scientist, Kenya Med Res Inst, Nairobi; adj prof, Inst Agronomique et Veterinaire Hassan II, Morocco, 86-; mem bd dir, Soc Nutrit Ed, 86-, ed bd, Nutrit-Int J Appl & Basic Nutrit Sci, 87, panel collaborators, J Acta Tropica, Switz, 87-, bd dirs, Soc Nutrit Educ, 86-89, Exec Comt, Soc Int Nutrit Res, 89- & World Alliance Breastfeeding Action, 91-; consult, Swedish Int Develop Agency, Tanzania, 86, Meals for Millions Found, Calif, 86 & UNESCO, Paris, 88; team leader, UNICEF-WHO Rev JNSP, Ethiopia, 87-88; vis scientist & chief res officer, Kenya Med Res Inst, Nairobi, Kenya, 89-90; external examr, Univ Nairobi, Kenya 89-90. *Honors & Awards:* Food Cycle Trophy Award, Ministry Health, Tanzania, 78. *Mem:* Am Inst Nutrit; Brit Nutrit Soc; Am Soc Clin Nutrit; fel Royal Soc Trop Med & Hyg; fel Am Pub Health Asn. *Res:* International nutrition problems; nutrition and health of low income populations; xerophthalmia; infant feeding practices; nutrition and intellectual development; protein-energy malnutrition of children; evaluation of applied nutrition programs; lactose intolerance; nutritional surveillance; nutrition and parasitic infections; author of co-author of over 300 publications including 6 books. *Mailing Add:* Div Nutrit Sci Cornell Univ Ithaca NY 14853

LATHAM, PATRICIA SUZANNE, b Annapolis, Md, Aug 22, 46. HEPATOLOGY. *Educ:* Simmons Col, BS, 68; Sch Med, Univ Southern Calif, MD, 72. *Prof Exp:* ASST PROF MED & PATH, UNIV MD HOSP & BALTIMORE VET ADMIN HOSP, 81- *Mem:* Am Asn Study Liver Dis; Am Gastroenterol Asn. *Res:* Investigation of liver diseases with emphasis on structure and function; hepatocytes; Kupffer cells; endotoxin; viral pathogenesis. *Mailing Add:* Univ Md Hosp 22 Greene St Baltimore MD 21201

LATHAM, ROSS, JR, b Chicago, Ill, Dec 18, 32; m 61; c 2. INORGANIC CHEMISTRY. *Educ:* Principia Col, BS, 55; Univ Ill, MS, 57, PhD(inorg chem), 61. *Prof Exp:* Instr chem, Lafayette Col, 59-60; chemist, Esso Res & Eng Co, NJ, 61-66; assoc prof, 66-72, PROF CHEM, ADRIAN COL, 72- *Res:* Titanium alkoxide-halide chemistry. *Mailing Add:* Dept Chem Adrian Col 110 S Madison St Adrian MI 49221-2575

LATHEM, WILLOUGHBY, b Atlanta, Ga, Oct 9, 23; m 51; c 5. MEDICINE. *Educ:* Emory Univ, BS, 44, MD, 46; Am Bd Internal Med, dipl, 54. *Prof Exp:* Asst med, Col Physicians & Surgeons, Columbia Univ, 52-53; asst clin prof, Yale Univ, 53-56; from asst prof to assoc prof, Sch Med, Univ Pittsburgh, 56-64; sci rep, Off Int Res, NIH, Eng, 62-66; dep dir biomed sci, Rockefeller Found, 66-72, assoc dir health sci, 72-78, regional officer, Asia, Bangkok & Thailand, 75-77, field staff, Salvador, Bahia & Brazil, 78-80; vpres & med dir, 80-85, VPRES SCI AFFAIRS, STERLING INT, STERLING DRUG, INC, 85- *Concurrent Pos:* Hon res assoc, Univ Col, Univ London, 62-65; vis prof, Mt Sinai Sch Med, 80. *Mem:* Am Soc Clin Invest; Harvey Soc; Am Fedn Clin Res. *Res:* International health. *Mailing Add:* Sterling Drug Inc 90 Park Ave New York NY 10016

LATHERS, CLAIRE M, b Brooklyn, NY. CARDIO-RENAL DRUG PRODUCTS, PHARMACOLOGY. *Educ:* Union Univ, BS, 69; State Unit NY, Buffalo, PhD(pharmacol), 73. *Prof Exp:* NIH postdoctoral fel, Med Col Pa, 73-75, from instr to assoc prof, Dept Pharmacol, 75-88; PHARMACOLOGIST, CARDIO-RENAL DRUG PROD, FOOD & DRUG ADMIN, 89- *Concurrent Pos:* Pharmacol lectr, Smith Kline & French, 75-83, Pa Col Optom, 76-77, Gwynedd Mercy Col, 78-89 & Hoeschst-Roussel Med Col Pa, 82-83; mem, Subcomt Women in Pharmacol, Pharmacol Soc, 84-92; vis prof, Dept Pharmacol & Toxicol, Philadelphia Col Pharm & Sci, 87, Dept Pharmacol, Schs Med & Dent, State Univ NY, Buffalo, 87-88; vis scientist sabbatical, Johnson Space Ctr, NASA, 88, vis scientist, 89- *Mem:* Aerospace Med Asn; Am Soc Pharmacol & Exp Therapeut; Soc Exp Biol & Med; Sigma Xi; NY Acad Sci; Am Fedn Clin Res; Int Study Group Res Cardiac Metab; Am Col Clin Pharmacol (treas, 90-). *Mailing Add:* Div Cardio-Renal Drug Prod Food & Drug Admin 5600 Fishers Lane Rockville MD 20857

LATHROP, ARTHUR LAVERN, b Kittitas, Wash, Nov 21, 18; m 46. PHYSICS. *Educ:* Wash State Univ, BS, 43, Univ Ill, MS, 46; Rice Univ, PhD(physics), 52. *Prof Exp:* Instr physics, Wash State Univ, 43-44; physicist, Ames Aeronaut Lab, Moffett Field, Calif, 44-45; instr physics, Univ Tulsa, 47-49; res engr, Boeing Airplane Co, Wash, 52-53; res asst physics, Inst Paper Chem, Lawrence, 53-58, res assoc, 58-65; from asst prof to prof physics, Western Ill Univ, 65-88; RETIRED. *Concurrent Pos:* Res Corp grant. *Mem:* Am Asn Physics Teachers. *Res:* Superconductivity; physical properties of paper; radiative transfer theory. *Mailing Add:* 1739 Center St Walla Walla WA 99362

LATHROP, EARL WESLEY, b Oakley, Kans, Mar 1, 24; m 49; c 1. SYSTEMATIC BOTANY. *Educ:* Walla Walla Col, BA, 50, MA, 52; Univ Kans, PhD(bot), 57. *Prof Exp:* Instr biol, Can Union Col, 52-54; assoc prof bot, La Sierra Col, 57-64; assoc prof biol, 64-80, PROF BIOL, LOMA LINDA UNIV, 80- *Mem:* Am Range Mgt Soc. *Res:* Floristics. *Mailing Add:* Dept Nat Sci Loma Linda Univ Loma Linda CA 92350

LATHROP, JAY WALLACE, b Bangor, Maine, Sept 6, 27; m 48, 85; c 6. SOLID STATE PHYSICS, ELECTRICAL ENGINEERING. *Educ:* Mass Inst Technol, BS, 48, MS, 49, PhD(physics), 52. *Prof Exp:* Electronic scientist, Nat Bur Standards, 52-58; mgr advan technol, Semiconductor Components Div, Tex Instruments Inc, 58-68; PROF ELEC ENG, CLEMSON UNIV, 68-, DIR, CTR SEMICONDUCTOR DEVICE RELIABILITY RES, 84- *Mem:* Fel Inst Elec & Electronics Engrs. *Res:* Microelectronics; integrated circuits; solar cells; semiconductor devices reliability. *Mailing Add:* Rte 1 Box 279 West Union SC 29696

LATHROP, KATHERINE AUSTIN, b Lawton, Okla, June 16, 15; m 38; c 5. NUCLEAR MEDICINE. *Educ:* Okla State Univ, BS, 36 & 39, MS, 39. *Prof Exp:* Asst, Univ Wyo, 42-44; jr chemist, Biomed Div, Manhattan Proj Metall Lab, 45-47; assoc biochemist, Argonne Nat Lab, Univ Chicago, 47-54, res assoc, Franklin Mclean Mem Res Inst, 54-85, asoc prof, 67-78, PROF RADIOL, UNIV CHICAGO, 78- *Concurrent Pos:* Mem comt nuclear med, Am Nat Standards Inst; adv comt radiopharmaceut, US Pharmacopeia; from instr to asst prof, Univ Chicago, 54-67; mem med internal radiation dose comt, Soc Nuclear Med, 67-, chmn, 77-85. *Honors & Awards:* Spec Pres Award, Soc Nuclear Med. *Mem:* AAAS; Radiation Res Soc; NY Acad Sci; Soc Nuclear Med; Sigma Xi. *Res:* Development of radionuclides for diagnostic and therapeutic purposes, including their production, purification and incorporation into various chemical forms; biological behavior and dosimetry in laboratory animals and humans. *Mailing Add:* 5514 S Woodlawn Ave Chicago IL 60637

LATHROP, KAYE DON, b Bryan, Ohio, Oct 8, 32; m 57; c 2. COMPUTER SCIENCE. *Educ:* US Mil Acad, BS, 55; Calif Inst Technol, MS, 59, PhD(mech eng, physics), 62. *Prof Exp:* Staff mem reactor math, Los Alamos Sci Lab, 62-67; staff mem & group leader reactor physics methods develop, Nuclear Analysis & Reactor Physics Dept, Gen Atomic Div, Gen Dynamics Corp, 67-68; T-1 alt group leader, 68-72, T-1 group leader, 72-75, T-div asst div leader, 73-75, assoc div leader nuclear safeguards, Reactor Safety & Technol Div, 75-77, alt div leader, Energy Div, 77-78, div leader, Computer Sci & Serv Div, 78-79, assoc dir eng sci, Los Alamos Nat Lab, 79-84; ASSOC LAB DIR & PROF, STANFORD LINEAR ACCELERATOR CTR, 84- *Concurrent Pos:* Vis prof, Univ NMex, 64-65, adj prof, 66-67; mem, US Energy Res & Develop Admin Adv Comt for Reactor Physics, 73-77; vis comt, Reactor Physics Div, Argonne Nat Lab, 78-83; mgt adv comt, Y-12 Div, Union Carbide Corp, 79-82; mem, Am Nuclear Soc deleg People's Repub China, 80; eng adv comt, Univ NMex, 80-84; eng nat adv comt, Univ Mich, 83-; steering comt, energy eng res prog, Joint Mass Inst Technol- Idaho Nat Eng Lab, 85-90; mem, Nat Acad Sci study, Mat Control & Acct Dept Energy

Nuclear Fuel Complex, 87-88; mem, energy res adv bd panel, access cand reactor technologies new prod reactor, 88; external adv comt, Nuclear Technol & Eng Div, Los Alamos Nat Lab, 88-; nat lab adv comt, Dept Energy Nat Energy Strategy, 89- *Honors & Awards:* E O Lawrence Mem Award, 76. *Mem:* Fel Am Nuclear Soc (treas, 77-79); Am Phys Soc; Nat Acad Eng. *Res:* Analytic and numerical solutions of equations of neutron and photon transport; reactor safety; computer systems and communications. *Mailing Add:* Stanford Linear Accelerator Ctr Bin 07 PO Box 4349 Stanford CA 94309

LATHROP, RICHARD C(HARLES), b Wauwatosa, Wis, Sept 6, 24; m 56, 83. ELECTRICAL & AERONAUTICAL ENGINEERING. *Educ:* Univ Wis, BS, 48, MS, 50, PhD(elec eng), 51. *Prof Exp:* Res asst, Univ Wis, 51; proj engr, Wright Air Develop Ctr, US Air Force, 51-55; staff mem Test Pilot Sch, 55-59, commandant, 59-61, assoc prof elec eng, US Air Force Acad, 61-65, prof & head dept, Pakistan Air Force Col Aeronaut Eng, 65-67, pilot, Fourth Air Commando Squadron, 67-68, exec officer, Cent Inertial Guide Test Facil, Air Force Missile Develop Ctr, Holloman AFB, NMex, 68-70, tech dir, Air Force Flight Test Ctr, 71-74; CONSULT ENGR, 74- *Mem:* Sr mem Inst Elec & Electronics Engrs; assoc fel Am Inst Aeronaut & Astronaut; sr mem Simulation Coun. *Res:* Analog and digital computers; automatic controls; aircraft flight testing; aircraft dynamics; microcomputers. *Mailing Add:* 39929 Dyott Way Palmdale CA 93551

LATHWELL, DOUGLAS J, b Mich, Mar 28, 22; m 48; c 3. SOIL SCIENCE. *Educ:* Mich State Univ, BS, 47; Ohio State Univ, PhD(soil sci), 50. *Prof Exp:* From asst prof to assoc prof, 50-61, PROF SOIL SCI, CORNELL UNIV, 61- *Concurrent Pos:* Fulbright res scholar, Neth, 64; vis prof, Univ Reading, Eng, 79. *Mem:* Soil Sci Soc Am; fel Am Soc Agron; fel AAAS; Int Soc Soil Sci. *Res:* Soil fertility; plant nutrition. *Mailing Add:* 236 Emerson Hall Agron Cornell Univ Ithaca NY 14853

LATIES, ALAN M, b Beverly, Mass, Feb 8, 31; c 2. OPHTHALMOLOGY. *Educ:* Harvard Col, AB, 54; Baylor Univ, MD, 59; Am Bd Ophthal, dipl, 65. *Prof Exp:* Intern, Mt Sinai Hosp, New York, 59-60; resident, Hosp Univ Pa, 60-63, instr, 63-64, assoc, 64-66, asst prof, 66-70, Given prof, 70-83, HAROLD G SCHEIE RES PROF OPHTHAL, MED SCH, UNIV PA, 83- *Concurrent Pos:* NIH trainee, 61-63, spec fel, 63-64; assoc ophthal, Children's Hosp Philadelphia, 63-; asst attend physician, Philadelphia Gen Hosp, 63-; attend ophthal, Vet Admin Hosp, 63-; mem vision res & training comt, Nat Eye Inst. *Honors & Awards:* Res to Prevent Blindness Professorship Award, 64; Friedenwald Award for Res in Ophthal, 72. *Mem:* AMA; Asn Res Vision & Ophthal; Histochem Soc; Am Asn Anat; Am Acad Ophthal & Otolaryngol. *Res:* Histochemistry; visual pathways; experimental myopia. *Mailing Add:* Scheie Eye Inst 51 N 39th St Philadelphia PA 19104

LATIES, GEORGE GLUSHANOK, b Sevastopol, Russia, Jan 17, 20; US citizen; m 47. PLANT PHYSIOLOGY. *Educ:* Cornell Univ, BS, 41; Univ Minn, MS, 42; Univ Calif, PhD(plant physiol), 47. *Prof Exp:* Asst, Div Plant Nutrit, Univ Calif, 42-43, sr asst, 43-47; res fel biol, Calif Inst Technol, 47-50, sr res fel, 50-52 & 55-58; asst prof bot, Univ Mich, 52-55; plant physiologist, Exp Sta & assoc prof hort sci, Univ, 59-63, PROF PLANT PHYSIOL, DEPT BIOL, UNIV CALIF, LOS ANGELES, 63- *Concurrent Pos:* Rockefeller Found fel, Sheffield & Cambridge Univs, 49-50; res botanist, Univ Mich, 52-55; mem physiol chem study sect, USPHS, 63-65; Guggenheim fel, Commonwealth Sci & Indust Res Orgn, Australia, 66-67. *Mem:* AAAS; Am Soc Plant Physiol (vpres, 64-65); Bot Soc Am; Scand Soc Plant Physiol. *Res:* Respiratory regulatory mechanisms and respiratory pathways in plant tissues; permeability and salt transport; biochemical aspects of growth and development. *Mailing Add:* Dept Biol Univ Calif 405 Hilgard Ave Los Angeles CA 90024

LATIES, VICTOR GREGORY, b Racine, Wis, Feb 2, 26; m 56; c 3. PSYCHOPHARMACOLOGY. *Educ:* Tufts Univ, AB, 49; Univ Rochester, PhD(psychol), 54. *Prof Exp:* Res assoc, Univ Rochester, 53-54; teaching intern, Brown Univ, 54-55; from instr to asst prof, Sch Med, Johns Hopkins Univ, 55-65; assoc prof radiation biol, biophys, pharmacol & psychol, 65-71, PROF TOXICOL, PHARMACOL & PSYCHOL, UNIV ROCHESTER, 71-, DIR, TOXICOL TRAINING PROG, 78- *Concurrent Pos:* Mem preclin psychopharmacol res rev comt, NIMH, 67-71; exec ed, J Exp Anal Behav, 66-, ed, 73-77; mem, Nat Res Coun Panel on Carbon Monoxide, Nat Acad Sci, 73-77, mem bd toxicol & environ health hazards, 77-80, mem toxicol info prog comt, 82-85; mem Sci Rev Panel Health Res, Environ Protection Agency, 81-89, Indoor Air Qual/Total Human Exposure Comt, 87-; mem bd sci affairs, Am Psychol Asn, 83-86; pres, Div Psychopharmacol, Environ Protection Agency, 68-69, Div Exp Analytical Behav, 78-82. *Mem:* Am Psychol Asn; Am Soc Pharmacol & Exp Therapeut; Behav Pharmacol Soc (pres, 66-68); Soc Toxicol; Asn Behav Analysis; Soc Exp Analytical Behav (secy-treas, 66-). *Res:* Behavioral pharmacology; experimental analysis of behavior; behavioral toxicology. *Mailing Add:* Environ Health Sci Ctr Univ Rochester Sch Med & Dent Rochester NY 14642

LATIMER, BRUCE MILLIKIN, b Hamilton, Ohio, Aug 23, 53; m 86. BIOMECHANICS OF LOCOMOTION, HOMINID PALEONTOLOGY. *Educ:* Univ Ariz, BA, 75; Case Western Reserve Univ, MA, 78; Kent State Univ, PhD(biomed sci), 88. *Prof Exp:* From asst cur to assoc, 85-88, CUR & HEAD, PHYS ANTHROP, CLEVELAND MUS NATURAL HIST, 88-; ASST PROF ANAT, CASE WESTERN RESERVE UNIV, 88- *Mem:* Am Asn Phys Anthropologists. *Res:* Pleistocene hominis evolution; comparative primate anatomy. *Mailing Add:* Cleveland Mus Natural Hist Wade Oval Univ Circle Cleveland OH 44106

LATIMER, CLINTON NARATH, b New York, NY, Aug 30, 24; m 56; c 3. NEUROPHYSIOLOGY. *Educ:* Columbia Univ, AB, 48; Syracuse Univ, PhD(physiol), 56. *Prof Exp:* Group leader neuropharmacol, Am Cyanamid Co, 58-76, study dir toxicol, Lederle Labs Div, 76-80; DIR DRUG SAFETY EVAL, DIR, RES DEVELOP PHARMACEUT DIV, PENNWALT CORP, 80- *Concurrent Pos:* Nat Inst Neurol Dis & Blindness res fel neurophysiology,

Univ Wash, 56-58. *Mem:* AAAS; Am Soc Pharmacol & Exp Therapeut; Soc Neurosci; Am Physiol Soc; Am Col Toxicol; Sigma Xi. *Res:* Neuropharmacology; function of the central nervous system as delineated by extra and intracellular recordings of neuronal activity under influence of drugs and in control states; psychopharmacology; safety evaluation of all classes of compounds; physiologic toxicology; long-term alteration of function by drugs; computer applications; research administration, toxicology, pathology, drug safety evaluation. *Mailing Add:* 2785 Rush-Mendon Rd Honeoye Falls NY 14472

LATIMER, HOWARD LEROY, b Seattle, Wash, July 18, 29; m 57; c 4. PLANT GENETICS, PLANT ECOLOGY. *Educ:* Wash State Univ, BS, 51, MS, 55; Claremont Cols, PhD(bot), 59. *Prof Exp:* From asst prof to assoc prof, 58-68, PROF BIOL, CALIF STATE UNIV, FRESNO, 68- *Mem:* Soc Study Evolution; Ecol Soc Am. *Res:* Reproductive ecology of plants. *Mailing Add:* Dept Biol Calif State Univ Fresno CA 93740

LATIMER, PAUL HENRY, b New Orleans, La, Nov 25, 25; m 52; c 3. BIOPHYSICS. *Educ:* Univ Ill, MS, 50, PhD(biophys), 56. *Prof Exp:* Instr physics, Col William & Mary, 50-51; asst bot, Univ Ill, 53-56; res fel plant biol, Carnegie Inst Technol, 56-57; asst prof physics, Vanderbilt Univ, 57-62; assoc prof, 62-71, PROF PHYSICS, AUBURN UNIV, 71- *Concurrent Pos:* Investr, Howard Hughes Med Inst, 57-62; consult, Southern Res Inst, Birmingham, 76; contractor, US Army, 77. *Mem:* Biophys Soc; Am Phys Soc; Optical Soc Am. *Res:* Light scattering; biological optics; fluorescence; photosynthesis. *Mailing Add:* 530 Forestdale Dr Auburn AL 36830

LATORELLA, A HENRY, b Winthrop, Mass, Mar 12, 40; m 64; c 2. GENETICS, ALGOLOGY. *Educ:* Boston Col, BS, 61, MS, 64; Univ Maine, Orono, PhD(zool, genetics), 71. *Prof Exp:* Asst prof biol, Salem State Col, 66-68; ASSOC PROF BIOL, STATE UNIV NY COL GENESEO, 70- *Mem:* Am Genetic Asn; Phycol Soc Am; AAAS; Genetics Soc Am; Sigma Xi. *Res:* Algal genetics and physiology; regulation of DNA replication and genetics and biochemistry of salinity adaptation by phytoflagellates. *Mailing Add:* Dept of Biol State Univ NY Col Geneseo NY 14454

LATORRE, DONALD RUTLEDGE, b Charleston, SC, May 4, 38; m 60; c 2. MATHEMATICS. *Educ:* Wofford Col, BS, 60; Univ Tenn, MA, 62, PhD(math), 64. *Prof Exp:* Asst prof math, Univ Tenn, 67; from asst prof to assoc prof, 67-76, PROF MATH, CLEMSON UNIV, 76- *Mem:* Am Math Soc; Math Asn Am. *Res:* Abstract algebra, especially semigroups. *Mailing Add:* Dept Math Sci Clemson Univ Clemson SC 29631

LATORRE, ROBERT GEORGE, b Toledo, Ohio, Jan 9, 49; m 80. SHIP HYDRODYNAMICS, NAVAL ARCHITECTURE. *Educ:* Univ Mich, BScE, 71, MS, 72; Univ Tokyo, MScE, 75, DEng, 79. *Prof Exp:* Asst prof naval archit & marine eng, Univ Mich, 79-83; res scientist, Bassin d Essais des Carenes, Paris, 83-84; assoc prof, 84-87, PROF & CHMN NAVAL ARCHIT & MARINE ENG, UNIV NEW ORLEANS, 89- *Concurrent Pos:* Res scientist, David W Taylor Navel Ship Res & Develop Ctr, 80 & 81; assoc prof, dept mech eng, Univ Tokyo, Japan, 86-87. *Honors & Awards:* Halburton Award Teaching & Res; Medal, Inst Indust Sci, Tokyo Univ. *Mem:* Am Soc Mech Engrs; Soc Naval Archit & Marine Engrs; Japan Soc Naval Archit; Royal Inst Naval Archit. *Res:* Development of the design of shallow water river pushboats and estimating methods for ship resistance; towed ship safety and towed vessel course stability; design and use of 125 by ft by 7 ft deep ship - offshore model testbasin in deep and shallow water; numerical hydrodynamics; ship design using computer aided engineering; cavitation noise and underwater acoustics. *Mailing Add:* Univ New Orleans 911 Eng Bldg New Orleans LA 70148

LATORRE, V(ICTOR) R(OBERT), b Brooklyn, NY, Nov 17, 31; m 57; c 2. ELECTRICAL ENGINEERING. *Educ:* Univ Ariz, BSEE, 56, MS, 57, PhD(elec eng), 60. *Prof Exp:* From instr to asst prof elec eng, Univ Ariz, 57-61; res specialist commun systs, Boeing Co, Wash, 61-64; asst prof elec eng, Univ Calif, Davis, 64-67, electronics engr, 67-72, assoc div head, Electronic Eng Dept, 72-76; RES ENGR, LAWRENCE LIVERMORE LAB, UNIV CALIF, LIVERMORE, 76- *Concurrent Pos:* Sr res engr, Frederick Res Corp, Ariz, 60-61; lectr, Seattle Univ, 62; consult, electromagnetic radiation effects on equipment & humans, microwave non-destructive anal of mat. *Mem:* Sr mem Inst Elec & Electronics Engrs; Am Soc Eng Educ; Sigma Xi; NY Acad Sci. *Res:* Biotelemetry systems; meteoric and tropospheric scatter propagation; design and statistical analysis of secure and jam-resistant communication systems; transportation systems research; microwave non-destructive analyses; electromagnetic interference. *Mailing Add:* Lawrence Livermore Lab L-156 Univ Calif Box 808 Livermore CA 94550

LATOUR, PIERRE RICHARD, b Buffalo, NY, Apr 15, 40; m 62; c 2. CHEMICAL ENGINEERING, AUTOMATIC CONTROL. *Educ:* Va Polytech Inst, BSChE, 62; Purdue Univ, MS, 64, PhD(chem eng), 66. *Prof Exp:* Coop student, E I du Pont de Nemours & Co, NC, 58-61; res engr, Math Group, Houston Res Lab, Shell Oil Co, Tex, 66, Comput Control Group, 66-67; mathematician, Theory & Anal Off, Manned Spacecraft Ctr, NASA, Houston, 67, tech asst to chief, Simulation Br, 67-68; br Shell Oil Co, NY, sr engr, head off, Mfg Technol Dept, 69; mgr process control eng, Davis Comput Systs, Inc, NY & Tex, 70 & Biles & Assocs, Inc, 71-77; dir & chmn bd, Setpoint Japan Inc, 84-90; CONSULT ENGR & VPRES, SETPOINT INC, TEX, 77- *Concurrent Pos:* Lectr, Purdue Univ, 66, Houston Res Lab, Shell Oil Co, 66, 67, Univ Houston, 68, Lehigh Univ, 69 & Univ Calif, Santa Barbara, 80, 81. *Mem:* Am Chem Soc; Am Inst Chem Engrs; Instrument Soc Am; Sigma Xi. *Res:* Automatic process control; applied mathematics; digital computation; crude oil distillation; computer control of petroleum refining; cracking; simulation; process dynamics; plant economics; optimization; process engineering; computer control justification; sales; business development; marketing. *Mailing Add:* 810 Herdsman Dr Houston TX 77079

LATOURETTE, HAROLD KENNETH, b Seattle, Wash, Apr 10, 24; m 44; c 4. ORGANIC CHEMISTRY. *Educ:* Whitman Col, AB, 47; Univ Wash, PhD(organic chem), 51. *Prof Exp:* Res assoc, Univ Wash, 48-51; res chemist, Westvaco Chem Div, FMC Corp, WVa, 51-54, dir pioneering res, Westvaco Chlor-Alkali Div, 54-56, supvr org res, Cent Res Labs, NJ, 56-57, mgr org res & develop, Becco Chem Div, NY, 57-58, mgr org chem res, Inorg Chem Div, NJ, 58-62, Europ tech dir, Chem Div & vpres, FMC Chem, SA, Switz, 62-65, mgr planning & eval, Cent Res Dept, NJ, 65-72, mgr eval, Chem Group, 72-76, environ mgr Toxic Substances, FMC Corp, Pa, 77-83; RETIRED. *Mem:* Sigma Xi; Am Chem Soc. *Res:* Organic reaction mechanisms; aromatic substitution; peroxides; epoxides; isocyanates; phosphorus and sulfur organics; halogenation; industrial processes. *Mailing Add:* 1526 Deception Rd Anacortes WA 98221

LATOURETTE, HOWARD BENNETT, b Detroit, Mich, Aug 26, 18; m 42; c 4. RADIOLOGY. *Educ:* Oberlin Col, AB, 40; Univ Mich, MD, 43. *Prof Exp:* From instr to assoc prof radiol, Univ Mich, 49-59; prof radiol, Col Med, Univ Iowa, 59-80, head radiation ther, 78-80. *Concurrent Pos:* Mem staff radiation ther sect, Univ Hosps, Univ Iowa. *Mem:* AAAS; Radiol Soc NAm. *Res:* Clinical use of radiation in treatment of cancer and resulting survival studies; tumor registry organization and function alteration of radiosensitivity of tumors. *Mailing Add:* 701 Oakknoll Dr Iowa City IA 52246

LA TOURRETTE, JAMES THOMAS, b Miami, Ariz, Dec 26, 31; m 55; c 4. PHYSICS. *Educ:* Calif Inst Technol, BS, 53; Harvard Univ, MA, 54, PhD(physics), 58. *Prof Exp:* Res fel physics, Harvard Univ, 57-58, lectr, 58-59; NSF fel, Univ Bonn, 59-60; physicist, Gen Elec Res Lab, 60-62; sr supvry scientist, TRG, Inc Div, Control Data Corp, 62-66, sect head gas laser, 66-67; PROF ELEC ENG & COMPUTER SCI, POLYTECH UNIV, 67-; ASSOC DIR, WEBER RES INST, 87- *Mem:* AAAS; Am Phys Soc; Inst Elec & Electronics Engrs. *Res:* Gas laser research and applications; saturated resonance spectroscopy and laser frequency stabilization. *Mailing Add:* Dept Elec Eng Polytech Univ 333 Jay St Brooklyn NY 11201

LATSCHAR, CARL ERNEST, b Newton, Kans, May 24, 19; m 41; c 5. PHYSICAL CHEMISTRY. *Educ:* Kans State Univ, BS, 41, MS, 47; Univ Wis, PhD(phys chem), 50. *Prof Exp:* Res chemist, 50-62, SR RES CHEMIST, E I DU PONT DE NEMOURS & CO, INC, 62- *Mem:* Am Chem Soc. *Res:* Textile fiber process and product development. *Mailing Add:* PO Box 1273 Salina KS 67402-1273

LATSHAW, DAVID RODNEY, b Allentown, Pa, Nov 4, 39; m 71; c 2. ANALYTICAL CHEMISTRY. *Educ:* Muhlenberg Col, BS, 61; Lehigh Univ, MS, 63, PhD(chem), 66. *Prof Exp:* Res asst, Lehigh Univ, 63-66; res chemist, 66-69, group leader, 69-80, MGR, ANALYTICAL SERV, ALLENTOWN LABS, AIR PROD & CHEM INC, 80- *Mem:* Am Chem Soc; Sigma Xi. *Res:* Infrared spectroscopy; atomic absorption. *Mailing Add:* 944 Belford Rd Allentown PA 18103

LATSHAW, J DAVID, b Reading, Pa. ANIMAL SCIENCE & NUTRITION. *Educ:* Pa State Univ, BS, 64; Wash State Univ, PhD(nutrit), 70. *Prof Exp:* From asst prof to assoc prof, 70-84, PROF POULTRY SCI, OHIO STATE UNIV, 84- *Mem:* Poultry Sci Asn; Am Inst Nutrit. *Res:* Nutritional needs of egg-type and meat-type chickens; nutrients such as selenium, amino acids, and energy; effects of feedstuff processing. *Mailing Add:* Poultry Sci Dept Ohio State Univ, 674 W Lane Ave Columbus OH 43210

LATTA, BRUCE MCKEE, physical organic chemistry; deceased, see previous edition for last biography

LATTA, BRYAN MICHAEL, b Oshawa, Ont, Oct 25, 46. SOLID STATE THIN FILM ELECTROCHROMICS, PARTICLE SOLID INTERACTION THEORY & MODELING. *Educ:* Queen's Univ, BSc, 70, MSc, 75, PhD(physics), 78. *Prof Exp:* Asst prof physics, Acadia Univ, 78-81; Nat Sci & Eng Res Coun Can Univ Res fel, Mem Univ Nfld, 81-84, asst prof physics, 85; ASSOC PROF PHYSICS, ACADIA UNIV, 85- *Concurrent Pos:* Prin investr, Thin Film Lab, Acadia Univ, 88- *Mem:* Can Asn Physicists; Am Phys Soc; Soc Photo Instrumentation Engrs; Int Soc Optical Eng. *Res:* Use detailed solid state charge distributions to evaluate the electronic and nuclear stopping power; Monte-Carlo and numerical modeling of particle-solid collision phenomena; multiple layer thin film coating and optical performance evaluation; Particle solid interaction theory and modeling. *Mailing Add:* Physics Dept Acadia Univ Wolfville NS B0P 1X0 Can

LATTA, GORDON, b Vancouver, BC, Mar 8, 23; m 45; c 3. MATHEMATICS. *Educ:* Calif Inst Technol, PhD(math), 51. *Prof Exp:* Lectr math, Univ BC, 47-48; asst, Calif Inst Technol, 48-51; Fine instr math & hydrodyn, Princeton Univ, 51-52; asst prof, Univ BC, 52-53; from asst prof to prof math, Stanford Univ, 53-67; prof math, Univ Va, 67-81; PROF MATH, NAVAL POSTGRAD SCH, 81- *Res:* Singular perturbation problems for differential equations; linear integral equations and Wiener-Hopf techniques. *Mailing Add:* Dept Math Naval Postgrad Sch Monterey CA 93940

LATTA, HARRISON, b Los Angeles, Calif, Apr 5, 18; m 41, 85; c 4. PATHOLOGY, BIOPHYSICS. *Educ:* Univ Calif, Los Angeles, AB, 40; Johns Hopkins Univ, MD, 43. *Prof Exp:* Intern, Church Home & Hosp, Md, 44; from asst resident to resident path, Johns Hopkins Hosp, 44-46, instr, Johns Hopkins Univ, 45-46; res assoc biol, Mass Inst Technol, 49-51; asst prof path, Sch Med, Case Western Reserve Univ, 51-54; assoc prof, 54-60, PROF PATH, SCH MED, UNIV CALIF, LOS ANGELES, 60- *Concurrent Pos:* Res fel, Children's Hosp, Boston & Harvard Med Sch, 48-49. *Mem:* AAAS; Electron Micros Soc Am; Am Soc Cell Biol; Am Soc Exp Path; Am Asn Path & Bact. *Res:* ultrastructure and diseases of the kidney. *Mailing Add:* Dept Path 1P-250 CHS Med Sch UCLA Los Angeles CA 90024-1732

LATTA, JOHN NEAL, b Ottumwa, Iowa, Apr 11, 44; m 66; c 3. OTHER COMPUTER SCIENCE, ELECTRICAL ENGINEERING. *Educ:* Brigham Young Univ, BES, 66; Univ Kans, MS, 69, PhD(elec eng), 71. *Prof Exp:* Mem tech staff holography, RCA Labs, NJ, 67; res asst optics, Ctr Res Eng Sci, Univ Kans, 67-68; mem tech staff holography, Bell Tel Labs, NJ, 69; res engr, Radar & Optics Lab, Univ Mich, Ann Arbor, 69-73; sr res engr, Environ Res Inst Mich, 73-77; sr staff scientist, Sci Applns, Inc, 77-83; pres, Adroit Systs, Inc, 83-89; PRES, FOURTH WAVE, INC, 89- *Mem:* Inst Elec & Electronic Engrs; Asn Comput Mach; Sigma Xi. *Res:* Multimedia computing; digital image processing; visual displays; system engineering; digital system architecture; software engineering; holography; optical design. *Mailing Add:* PO Box 6547 Alexandria VA 22306

LATTA, WILLIAM CARL, b Niagara Falls, NY, May 18, 25; m 50; c 5. FISH BIOLOGY. *Educ:* Cornell Univ, BS, 50; Univ Okla, MS, 52; Univ Mich, PhD(fishery biol), 57. *Prof Exp:* Aquatic biologist, State Conserv Dept, NY, 50; in-chg, Inst Fisheries Res, 66-76, FISHERY BIOLOGIST, INST FISHERIES RES, STATE DEPT NATURAL RESOURCES, MICH, 55-, CHIEF RES SECT, FISHERIES DIV, 76- *Concurrent Pos:* Adj prof fisheries & wildlife, Sch Natural Resources, Univ Mich, Ann Arbor, 73-; adj prof fisheries & wildlife, Dept Fisheries & Wildlife, Mich State Univ, East Lansing, 85- *Mem:* Am Fisheries Soc; Am Soc Ichthyol & Herpet; Wildlife Soc; Ecol Soc Am; Am Inst Fishery Res Biologists. *Res:* Fish population dynamics; management of freshwater fisheries. *Mailing Add:* Inst for Fisheries Res Univ Mus Annex Ann Arbor MI 48109

LATTER, ALBERT L, b Kokomo, Ind, Oct 17, 20; m 49; c 2. THEORETICAL PHYSICS. *Educ:* Univ Calif, Los Angeles, BA, 41, PhD(physics), 51. *Prof Exp:* Staff physicist radiation lab, Univ Calif, 41-46; staff physicist, Rand Corp, 52-60, head physics dept, 60-71; pres, R&D Assocs, 71-85, VPRES & MEM BD DIRS, LOGICON, 85- *Concurrent Pos:* Mem Air Force Sci Adv Bd, 56-72, chmn nuclear panel, 63-67; mem US sci deleg to test ban negotiations, Geneva, 59; mem adv group, Ballistic Systs Div, 61-67; mem sci adv group on effects, Defense Nuclear Agency, 64-; mem adv group, Space & Missile Systs Orgn, 66-; mem Defense Sci Bd, 67-70, chmn penetration panel, 66-68, mem vulnerability task force, 66-, chmn strategic task force, 69-72, mem net assessment task force, 71-72; mem sci adv group, Joint Strategic Target Planning Staff, 68-72. *Mem:* Am Phys Soc. *Res:* Nuclear physics; quantum mechanics; nuclear weapons design and effects; strategic weapons systems. *Mailing Add:* 4346 Redwood Ave 104 Marina del Rey CA 90292

LATTER, RICHARD, b Chicago, Ill, Feb 20, 23; m 62; c 4. THEORETICAL PHYSICS. *Educ:* Calif Inst Technol, BS, 42, PhD(theoret physics), 49. *Prof Exp:* Physicist, Rand Corp, 49-56, chief physics div, 56-60, mem res coun, 60-71; vpres, R&D Assocs, 71-86; CONSULT, 86- *Concurrent Pos:* Lectr, Calif Inst Technol, 55-57; adv, AEC, 58-71 & Dept Defense, 58- *Honors & Awards:* E O Lawrence Award. *Mem:* Am Phys Soc. *Res:* Solid state, atomic, and nuclear physics; hydrodynamics; high pressure physics. *Mailing Add:* 3419 Via Lido Suite 217 Newport Beach CA 92663

LATTERELL, JOSEPH J, b St Cloud, Minn, Nov 2, 32; m 64; c 4. ANALYTICAL CHEMISTRY. *Educ:* St John's Univ, Minn, BA, 59; Purdue Univ, MS, 62; Univ Colo, PhD(analytical chem), 64. *Prof Exp:* Instr analytical chem, Univ Wis, 64; asst prof, John Carroll Univ, 64-67; from asst prof to assoc prof, 67-78, PROF ANALYTICAL CHEM, UNIV MINN, MORRIS, 78- *Concurrent Pos:* Danforth assoc, 81-86. *Mem:* Am Chem Soc. *Res:* Heavy metal migration and plant uptake of phosphorous in soils amended with wastewater and sewage sludge; chemical analyses of the bottom deposits of Minnesota lakes. *Mailing Add:* Dept Chem Univ Minn Morris MN 56267-2134

LATTERELL, RICHARD L, b Paynesville, Minn, Mar 14, 28; m 51. GENETICS. *Educ:* Univ Minn, Duluth, BA, 50; Pa State Univ, MS, 55; Cornell Univ, PhD(genetics), 58. *Prof Exp:* Nat Cancer Inst fel genetics, Brookhaven Nat Lab, 58-60; geneticist, Div Radiation & Organisms, Smithsonian Inst, 60-63; geneticist, Union Carbide Res Inst, 63-68; assoc prof biol, 68-82, PROF BIOL, SHEPARD COL, 83- *Concurrent Pos:* Vis assoc prof agron, Colo State Univ, 79-80. *Mem:* Genetics Soc Am; Bot Soc Am; Sigma Xi; Am Inst Biol Sci. *Res:* Radiation genetics of maize; space biology; stress tolerance of higher plants; plant cytogenetics. *Mailing Add:* Dept Biol Shepherd Col Shepherdstown WV 25443

LATTES, RAFFAELE, b Italy, May 22, 10; nat US; m 36; c 2. MEDICINE. *Educ:* Univ Turin, MD, 33; Columbia Univ, MedSciD, 46. *Prof Exp:* Asst gen surg, Univ Turin, 34-38; instr path, Woman's Med Col Pa, 41-43; instr surg & surg path, 43-46, asst prof path, Postgrad Hosp, 46-48, asst prof, 48-51, prof surg path, 51-78, EMER PROF SURG PATH, COL PHYSICIANS & SURGEONS, COLUMBIA UNIV, 78-; SR CONSULT SURG PATH, NY PRESBY HOSP, 78- *Concurrent Pos:* Consult, Roswell Park Mem Inst, Knickerbocker Hosp, NY, Roosevelt Hosp, St Lukes Hosp, Hosp Joint Dis & Vet Admin Hosp. *Mem:* Harvey Soc; Am Asn Path & Bact; Am Asn Cancer Res; Col Am Path; Asn Am Med Cols; NY Acad Med. *Res:* Surgical pathology. *Mailing Add:* 630 W 168th St New York NY 10032

LATTIME, EDMUND CHARLES, b Newburyport, Mass, Jan 18, 51; m 74. IMMUNOLOGY. *Educ:* Gettysburg Col, BA, 73; Rutgers Univ, MS, & PhD(zool), 77. *Prof Exp:* Fel immunol, 77-79, res assoc, 79-84, INSTR IMMUNOL, GRAD SCH, CORNELL UNIV, SLOAN-KETTERING INST, 81- *Concurrent Pos:* Scholar, Leukemia Soc Am, 85; lab asst mem immunol, Sloan-Kettering Inst Cancer Res, 85-; mem, Oncol Rev Bd, US Vet Admin, 88- *Mem:* Am Asn Immunologists; Am Asn Cancer Res; Am Chem Soc. *Res:* Natural immunity with emphasis on mechanisms of lytic factor release and actions. *Mailing Add:* Mem Sloan Kettering Cancer Ctr 425 E 68th St New York NY 10021

LATTIMER, JAMES MICHAEL, b Marion, Ind, Apr 12, 50; c 3. ASTROPHYSICS, COSMOCHEMISTRY. *Educ:* Univ Notre Dame, BS, 72; Univ Tex, Austin, PhD(astron), 76. *Prof Exp:* Res assoc, Univ Chicago, 76; res assoc astron, Univ Ill, Urbana-Champaign, 76-79; from asst prof to assoc prof, 79-88, PROF ASTRON, STATE UNIV NY, STONY BROOK, 88- *Concurrent Pos:* Alfred P Sloan res fel, 82-84; Ernst F Fullam fel, 85-86; vis prof, Nordita, Copenhagen, Denmark, 85-86. *Mem:* Am Astron Soc; Int Astron Union; Am Phys Soc; Am Geophys Union. *Res:* Supernovae; neutron stars; equation of state at high densities and temperatures; meteoritics; formation of the solar system; grain formation in novae and supernovae; nuclear physics. *Mailing Add:* Dept Earth & Space Sci State Univ NY Stony Brook NY 11794

LATTIMER, ROBERT PHILLIPS, b Kansas City, Mo, Feb 2, 45; c 2. ANALYTICAL CHEMISTRY, POLYMER CHEMISTRY. *Educ:* Univ Mo-Columbia, BS, 67; Univ Kans, PhD(chem), 71. *Prof Exp:* Fel chem, Univ Mich, Ann Arbor, 72-74; SR RES & DEVELOP ASSOC ANALYTICAL CHEM, BF GOODRICH CO, 74- *Honors & Awards:* Sparks-Thomas Award. *Mem:* Am Chem Soc; Am Soc Mass Spectrometry; Sigma Xi. *Res:* Analysis of polymers; desorption ionization mass spectrometry; degradation of polymers. *Mailing Add:* B F Goodrich Res & Develop Ctr 9921 Brecksville Rd Brecksville OH 44141

LATTIN, DANNY LEE, b Smith Center, Kans, Jan 9, 42; m 61; c 2. MEDICINAL CHEMISTRY. *Educ:* Univ Kans, BS, 65; Univ Minn, Minneapolis, PhD(med chem), 70. *Prof Exp:* From asst prof to assoc prof med chem, 70-80, PROF MED CHEM, COL PHARM, UNIV ARK, MED SCI CAMPUS, LITTLE ROCK, 80- *Mem:* Am Chem Soc; Am Asn Cols Pharm; Am Asn Pharm Sci. *Res:* Synthesis of novel opiate antagonists; synthesis and evaluation of novel opiate receptor affinity labels; study of sterochemical properties of drug receptors. *Mailing Add:* Dept Med Chem Col Pharm Univ Ark Med Sci Campus Little Rock AR 72205

LATTIN, JOHN D, b Chicago, Ill, July 27, 27; m 53; c 3. ENTOMOLOGY. *Educ:* Iowa State Univ, BS, 50; Univ Kans, MA, 51; Univ Calif, Berkeley, PhD(entom), 64. *Prof Exp:* Aquatic entomologist, Dept Limnol, Acad Natural Hist, Philadelphia, 51; jr vector control specialist, Bur Vector Control, Calif Dept Pub Health, 54-55; asst entomologist, Agr Exp Sta, 55-61, from instr to assoc prof, 55-68, asst dean sci, 67-73, PROF ENTOM, ORE STATE UNIV, 68- *Concurrent Pos:* Cur, Syst Entom Lab, Ore State, 61-66 & 74-; NSF fac fel, Univ Wageningen, 65-66; consult, USDA, 81-82. *Honors & Awards:* Loyd Carter Award, Ore State Univ, 61. *Mem:* Entom Soc Am; Soc Syst Zool. *Res:* Systematics of the Pentatomoidea, Leptopododoidea and Miridae; origin, distribution and phylogeny of the Heteroptera; evolution and zoogeography of the insecta; aquatic entomology; scientific education; education of talented students; introduced insects; applied systematic entomology. *Mailing Add:* Dept Entom Ore State Univ Corvallis OR 97331

LATTMAN, EATON EDWARD, b Chicago, Ill, May 15, 40; m 66; c 1. MOLECULAR BIOPHYSICS. *Educ:* Harvard Col, BA, 62; Johns Hopkins Univ, 69. *Prof Exp:* NIH fel biophysics, Johns Hopkins Univ, 69-70, res scientist, 70-73; fel, Max Planck Inst Biochem, 74; NIH fel biophys, Brandeis Univ, 74-76; from asst prof to assoc prof biophys, 76-90, PROF BIOPHYS, SCH MED, JOHNS HOPKINS UNIV, 90- *Concurrent Pos:* Chmn, NIH Biomed Sci Study Sect. *Honors & Awards:* Res Career Develop Award, NIH. *Mem:* Am Phys Soc; Am Crystallog Asn. *Res:* Determination of the three-dimensional structure of biological macromolecules by x-ray diffraction; globins; site-directed mutants. *Mailing Add:* Dept Biophys & Biophys Chem Johns Hopkins Univ Med Sch Baltimore MD 21205-2185

LATTMAN, LAURENCE HAROLD, b New York, NY, Nov 30, 23; m 46; c 2. GEOLOGY. *Educ:* City Col New York, BChE, 48; Univ Cincinnati, MS, 51, PhD(geol), 53. *Prof Exp:* Instr geol, Univ Cincinnati, 51 & Univ Mich, 52-53; photogeologist, Gulf Oil Corp, 53-56, asst head photogeol sect, 56-57; from asst prof to prof geomorphol, Pa State Univ, 57-70; prof geol & head dept, Univ Cincinnati, 70-75; dean, Col Mines & Mineral Indust & Col Eng, Univ Utah, 75-83; PRES, NMEX INST MINING TECHNOL, 83- *Concurrent Pos:* Mem, Nat Res Coun, 59-62. *Honors & Awards:* Fulbright Lectr, Moscow State Univ, 75. *Mem:* Fel Geol Soc Am; Am Soc Photogram; Am Asn Petrol Geologists; Sigma Xi. *Res:* Remote sensing of environment; geomorphology; fracture analysis on aerial photographs. *Mailing Add:* NMex Inst Mining Technol Socorro NM 87801

LATTUADA, CHARLES P, b Danville, Ill, May 8, 33; m 55, 79; c 4. BACTERIOLOGY. *Educ:* Ind State Univ, BS, 56; Univ Wis, MS, 58, PhD(bact), 63. *Prof Exp:* Instr bact, Univ Wis, 58-59; consult microbiol, Res Prod Corp, Wis, 59-61; bacteriologist, Wis Alumni Res Found, 61-64; consult, A R Schmidt Co, Inc, 62-64, dir res, 64-69; dir germ-free res, ARS/Sprague-Dawley Div, 69-74, Mogul Corp, gen mgr microbiol lab, 68-74, vpres, Gibco Diag Div, 74-76; vpres, Harlan Industs, Inc, 77-81; exec vpres, 81-84, PRES, GRANITE DIAGNOSTICS,INC, 84-; DIR, BIO-TROL, INC, 77- *Concurrent Pos:* Vpres & mgr, A R S Serum Co, Inc, 62-68; adj asst prof dept biol, Elow Col, 83- *Mem:* Am Soc Microbiol; Am Asn Lab Animal Sci; Asn Microbiol Media Mfrs; Asn Gnotobiotics; NY Acad Sci. *Res:* Dairy and food industries; botulinum food poisoning; derivation of new strains of germ-free animals and their maintenance and associated technology; intestinal flora of laboratory animals; process, development and management of microbiological culture media manufacture; process, development and management of diagnostic kit development. *Mailing Add:* Granite Diag Inc PO Box 908 Burlington NC 27215

LATZ, ARJE, b Lithuania, July 1, 27; US citizen; m 60; c 1. PSYCHOPHARMACOLOGY. *Educ:* Univ Maine, BA, 58; Boston Univ, MA, 60, PhD(psychol), 63. *Prof Exp:* Res assoc, 63-67, asst prof, 67-71, ASSOC PROF PSYCHOPHARMACOL, SCH MED, BOSTON UNIV, 71- *Concurrent Pos:* NIH fel, 63-65; Boston Med Found fel, 65-67; consult ed, Psychopharmacologia, 69- *Res:* Effects of amphetamine on behavior differentially maintained by excitatory or inhibitory processes; the effects of morphine and morphine antagonists on the cortical EEG of free moving rats. *Mailing Add:* Dept Psychiat Boston Univ Sch Med 88 E Newton St F-2 Boston MA 02118

LATZ, HOWARD W, b Rochester, NY, Jan 9, 33; m 53; c 3. ANALYTICAL CHEMISTRY. *Educ:* Rochester Inst Technol, BS, 59; Univ Fla, MS, 61, PhD(chem), 63. *Prof Exp:* Res specialist, Union Carbide Corp, 63-66; assoc prof, 66-74, PROF ANALYTICAL CHEM, OHIO UNIV, 74- *Mem:* Am Chem Soc; Sigma Xi. *Res:* Luminescence methods of analysis; electrophoresis of organic ions; analytical applications of dye lasers. *Mailing Add:* Dept Chem Ohio Univ Athens OH 45701

LAU, BRAD W C, b Macaw, China, Mar 23, 50; US citizen. BIOCHEMISTRY. *Educ:* Univ NDak, BS, 72, MS, 75, PhD(biochem), 79. *Prof Exp:* Res assoc, Human Nutrit Res Ctr, USDA, 79-82; postdoctoral fel res, Biol Res Lab, Syracuse, Wis, 82-85, res & serv, Med Ctr, Wash Univ, 86-88; SR CHEMIST RES & DEVELOP, SIGMA DIAGNOSTICS, 88- *Concurrent Pos:* Young investr award, Am Clin Lab Physicians & Scientists, 86. *Mem:* Am Asn Clin Chem; Am Soc Molecular Biol & Biochem; Sigma Xi; Clin Ligand Assay Soc. *Res:* Developing test reagent kits for analyses of metabolites and other chemicals in human specimens including semen, plasma, urine, and cerebral spinal fluid to be used in the clinical chemistry laboratory in hospitals and reference laboratories. *Mailing Add:* Clin Res & Develop Sigma Diagnostics 545 S Ewing St Louis MO 63103

LAU, CATHERINE Y, b Hong Kong, Feb 11, 51; Can citizen; m 77; c 2. MOLECULAR BIOLOGY. *Educ:* Ind Univ, BS, 72; Yale Univ, MPhil, 74, PhD(biochem), 76. *Prof Exp:* Res fel immunol, Ont Cancer Inst & Princess Margaret Hosp, Toronto, 76-77; supvr immunopharmacol, 77-84, proj mgr immunopharmacol, 85-87, MGR BIOTECHNOLOGY RES, ORTHO PHARMACEUT, LTD, CAN, 87- *Concurrent Pos:* Asst prof, Dept Immunol, Univ Toronto, 84- *Mem:* Sigma Xi; Am Asn Immunologists; Int Soc Immunopharmacol. *Res:* Discover immunologically active molecules and develop them into human therapeutics; suppressor lymphokine; large scale production, purification and gene cloning. *Mailing Add:* Ortho Pharmaceut Ltd 19 Green Belt Dr Don Mills ON M3C 1L9 Can

LAU, CHEUK KUN, b Hong Kong, Sept 3, 51; Can citizen; m 84. LEUKOTRIENES, SYNTHETIC METHODS. *Educ:* McMaster Univ, BSc, 74; Univ BC, PhD(chem), 78. *Prof Exp:* Res fel chem, Wayne State Univ, 78-80; SR RES CHEMIST MED CHEM, MERCK FROSST CAN INC, 80- *Honors & Awards:* Boris Monsaroff Mem Medal, Chem Inst Can, 74. *Mem:* Am Chem Soc; Chem Inst Can. *Res:* Synthesis of leukotrienes and its related products of the lipoxygenase enzymes and the design of inhibitors of this enzyme. *Mailing Add:* 476 Cr Boyer Isle Bizard PQ H9C 9Z7 Can

LAU, FRANCIS YOU KING, b Honolulu, Hawaii, Jan 5, 24; m 48; c 4. MEDICINE. *Educ:* Loma Linda Univ, MD, 47. *Prof Exp:* From clin instr to clin asst prof med, Loma Linda Univ, 54-59; asst prof, Sch Med, Univ Calif, San Francisco, 59-60; assoc prof, 60-72, PROF MED, LOMA LINDA UNIV, 72-; CLIN PROF MED & RADIOL, SCH MED, UNIV SOUTHERN CALIF, 79- *Concurrent Pos:* Assoc prof med, Univ Southern Calif, 64-70, prof, 70-79; mem attend staff & chief, Adult Cardiovasc Catheterization Lab, Los Angeles County-Univ Southern Calif Med Ctr, 65-79, chief cardiol, 70-79; consult, Vet Admin & Glendale Hosps; fel coun clin cardiol, Am Heart Asn. *Mem:* Fel Am Col Cardiol; fel Am Col Physicians. *Res:* Cardiology; cardiac catheterization, arrhythmias and pacemakers; artificial heart-lung preparations; balloon valvuloplasty. *Mailing Add:* 1720 Brooklyn Ave No 1231 Los Angeles CA 90033

LAU, JARK CHONG, b Singapore, Oct 18, 35; US citizen; m 61; c 2. LASER APPLICATIONS. *Educ:* Calif Inst Technol, MS, 63, AE, 64; Univ Southampton, Eng, PhD(fluid mech), 71. *Prof Exp:* Assoc prof fluid & thermodyn, Univ Singapore, 71-75; tech consult fluid & acoust, Lockheed-Ga Co, 75-80; sr res fel, 80-90, PRIN RES FEL, KIMBERLY-CLARK CORP, 90- *Mem:* Assoc fel Am Inst Aeronaut & Astronaut; Laser Inst Am. *Res:* Characterize the structure of free shear turbulence; development of better techniques of forming fibrous webs; consultant service to many mills; development of a laser technology base. *Mailing Add:* Kimberly-Clark Corp 1400 Holcomb Bridge Rd Roswell GA 30076

LAU, JOHN H, b China, June 17, 46; US citizen; m 72; c 1. MECHANICAL ENGINEERING, ENGINEERING PHYSICS. *Educ:* Nat Taiwan Univ, BE, 70; Univ BC, MASc, 73; Univ Wis-Madison, MS, 74; Fairleigh Dickinson Univ, MS, 81; Univ Ill-Urbana, PhD(theoret & appl mech), 77. *Prof Exp:* Teaching asst mech, Univ Ill-Urbana, 74-77; engr struct, Exxon Prod & Res Co, 77-78; res assoc paper physics, Int Paper Co Res Ctr, 78-79; sr engr mech eng, Ebasco Serv Inc, 79-81 & nuclear eng, Bechtel Power Corp, 81-83; mem tech staff stress anal, Sandia Nat Labs, 83-84; MEM TECH STAFF ELECTRONICS & MECH ENG, HEWLETT PACKARD LABS, 84- *Concurrent Pos:* Session chmn, tech comt mem & workshop speaker, Inst Elec & Electronics Engrs & Am Soc Mech Eng Tech Conf; assoc tech ed, Inst Elec & Electronics Engrs, Trans Components, Hybrids & Mfg Technol, Am Soc Mech Engrs, Trans J Electronic Packaging. *Mem:* Sigma Xi; NY Acad Sci; AAAS; Inst Elec & Electronics Engrs. *Res:* Structural engineering; applied mechanics; nuclear engineering; operations research; materials science engineering; mechanical vibrations; electronics packaging and interconnection. *Mailing Add:* 961 Newell Rd Palo Alto CA 94303

LAU, JOSEPH T Y, b Hong Kong, Mar 21, 53; US citizen; m 88. GLYCOBIOLOGY. *Educ:* Univ Wash, Seattle, BS, 75; Purdue Univ, PhD(biochem), 81. *Prof Exp:* Postdoctoral fel molecular biol, Johns Hopkins Sch Med, 81-84, res assoc, 84-85; cancer res scientist III, 86-90, CANCER RES SCIENTIST IV, ROSWELL PARK CANCER INST, 90-; ASST RES PROF MOLECULAR BIOL, STATE UNIV NY, 86- *Mem:* Soc Complex Carbohydrates; Am Soc Biol Chemists; AAAS. *Res:* Function and regulation of glycoconjugates; molecular biology of glycosyltransferases; expression of cellular differentiation epitopes. *Mailing Add:* Dept Cell & Molecular Biol Roswell Park Cancer Inst Elm & Carlton Sts Buffalo NY 14263

LAU, KENNETH W, b Lamoure, NDak, Nov 27, 41; m 63; c 2. CHEMISTRY. *Educ:* Univ NDak, BS, 63, PhD(chem), 69; Univ Sask, Regina, MSc, 68. *Prof Exp:* Chemist, Rock Island Arsenal Lab, US Army, 63-64; res chemist, 69-73, develop supvr, Film Dept, Richmond, Va, 73-76, area supt-tech, Plastic Prod & Resins Dept, 76-79, mkt mgr, Fluoropolymers Div, 79-85, MKT MGR, SPECIALITY POLYMERS DIV, E I DU PONT DE NEMOURS & CO, INC, 85- *Mem:* Am Chem Soc; Sigma Xi. *Res:* Ylid chemistry. *Mailing Add:* 3213 Kammerer Dr Wilmington DE 19803

LAU, L(EUNG KU) STEPHEN, b Shanghai, China, Sept 9, 29; US citizen; m 59; c 3. HYDROLOGY, HYDRAULIC & SANITARY ENGINEERING. *Educ:* Univ Calif, Berkeley, BS, 53, MS, 55, PhD(hydraul & sanit eng), 59. *Prof Exp:* Res engr sea water intrusion, Univ Calif, Berkeley, 53-54 & ground water pollution, 55-56; from asst prof to assoc prof, 59-67, assoc dir water resources res ctr, 64-70, actg dir, 70, dir, 71-90, PROF CIVIL ENG, UNIV HAWAII, 67- *Concurrent Pos:* Ground water consult, Honolulu Bd Water Supply, 60-64; dir, US Pub Health Serv training grant, Univ Hawaii, 64-65; vis assoc prof, Univ Calif, Berkeley, 65-66; Fulbright vis prof, Univ Malaya, 73-74; adj res assoc, East-West Ctr, 80-81; consult, Univ Guam, 72 & 75, World Health Orgn, 74, AMAX, 78, World Bank, 78 & UN, 81. *Honors & Awards:* George Warren Fuller Award, Am Waterworks Asn, 89. *Mem:* Am Waterworks Asn; Am Geophys Union; Am Soc Civil Engrs; Water Pollution Control Fedn. *Res:* Water reuse; groundwater; water pollution assessment; hydrologic evaluation of water resources; flood computations; water resources research administration; water resources policies; Hydrology and water resources of the Hawaiian Islands; hyddrology instruction. *Mailing Add:* Water Resources Res Ctr Univ Hawaii Honolulu HI 96822

LAU, NGAR-CHEUNG, b Hong Kong, July 21, 53; US citizen; m 79; c 1. ATMOSPHERIC GENERAL CIRCULATION. *Educ:* Chinese Univ, Hong Kong, BSc, 74; Univ Wash, PhD(atmospheric sci), 78. *Prof Exp:* Res asst atmospheric sci, Univ Wash, 74-78; vis scientist, Geophys Fluid Dynamics Prog, 78-81, mem res staff, 81-84, RES METEOROLOGIST ATMOSPHERIC SCI, GEOPHYS FLUID DYNAMICS LAB, PRINCETON UNIV, NAT OCEANIC & ATMOSPHERIC ADMIN, US DEPT COM, 84- *Concurrent Pos:* Asst atmospheric sci, Nat Ctr Atmospheric Res, 76-77; lectr with rank of assoc prof, prof atmospheric & oceanic sci, Princeton Univ, 82-; mem, Coun Equatorial Pac Ocean Climate Studies, US Dept Com, 85-; US deleg, US/China Monsoon Workshop, 87, US/Japan Elnino- Southern Oscillation Workshop, 87; ed, Comt, Dynamics Atmospheres & Oceans, 88-; mem, Comt Climate Variations, Am Meterol Soc, 90- *Honors & Awards:* Clarence Leroy Meisinger Award, Am Meteorol Soc, 90. *Mem:* Fel Am Meteorol Soc; World Meteorol Orgn. *Res:* Data diagnostics of meteorological systems; diagnosis of large-scale circulation systems in observed and model-simulated atmospheres; processing of meteorological data sets for studying nature and causes of atmospheric variability on different time scales. *Mailing Add:* Geophys Fluid Dynamics Lab Nat Oceanic & Atmospheric Admin Princeton Univ PO Box 308 Princeton NJ 08542

LAU, PHILIP T S, b Kuala Lumpur, Malaysia, Feb 13, 35; US citizen; m 59; c 3. ORGANIC CHEMISTRY, PHOTOGRAPHIC SCIENCE. *Educ:* Alfred Univ, BS, 58; Syracuse Univ, PhD(org chem), 62. *Prof Exp:* Fel & res assoc chem, Univ Calif, Berkeley, 62-63; sr res chemist, 63-68, res assoc, 68-79, SR RES ASSOC, COLOR PHOTOG DIV, RES LABS, EASTMAN KODAK CO, 79- *Concurrent Pos:* Lectr Photog Chem; chmn, Eastman Kodak Res Coun, 89-91; mem, Eastman Kodak distinguished inventors gallery. *Mem:* Am Chem Soc. *Res:* New synthetic methodologies to novel heterocyclic and aromatic compounds; regioselective and regiospecific reactions, photographic developers, couplers, stabilizers, dyes, polymer synthesis and applications. *Mailing Add:* 345 St Andrews Dr Rochester NY 14626

LAU, ROLAND, b China, May 5, 43; US citizen; m 67; c 1. ORGANIC CHEMISTRY. *Educ:* Wayne State Univ, BA, 65; Purdue Univ, PhD(chem), 72. *Prof Exp:* Clin chemist, Henry Ford Hosp, 65; res chemist, Ash-Stevens, Inc, 65-66; fel, Syntex Res Div, Syntex Corp, 72-73; res investr org chem, E R Squibb, Inc, 73-77, sr res investr, 77-80; MGR PHARMACEUT CHEM, HARDWICKE CHEM CO, SUBSID ETHYL CORP, 80- *Mem:* Am Chem Soc. *Res:* Commercial developments of medicinal agents. *Mailing Add:* Interchem Corp 120 Rte 17 N Paramus NJ 07652

LAU, S S, b Chungking, China, July 31, 41; US citizen. ELECTRONICS ENGINEERING. *Educ:* Univ Calif, Berkeley, BS, 64, MS, 66, PhD, 69. *Prof Exp:* Mem staff, Bell Labs, 69-72; mem staff, Calif Inst Technol, 72-80; prof microelectronics, Univ Calif, San Diego, 80-; WONG'S ELECTRONICS CO LTD, HONG KONG. *Mem:* Bohemian Phys Soc; Inst Elec & Electronics Engrs. *Res:* Metal semiconductor interactions; ion-beam processes; Rutherford backscattering spectrometry; electronic materials. *Mailing Add:* Dept Elec Eng & Computer Sci Univ Calif San Diego R-007 La Jolla CA 92093

LAU, YIU-WA AUGUST, b Aug 22, 48; US citizen. MATHEMATICS. *Educ:* Univ Houston, BS & MS, 68, PhD(math), 71. *Prof Exp:* From asst prof to assoc prof math, NTex State Univ, 71-80; mem staff, Johnson Space Ctr, NASA, 80-81; SR RES SPECIALIST, EXXON PROD RES, 85- *Mem:* Am Math Soc. *Res:* Topological semigroups; topology. *Mailing Add:* Exxon Prod Res PO Box 2189 Houston TX 77001

LAUB, ALAN JOHN, b Edmonton, Alta, Aug 6, 48; US citizen. SCIENTIFIC COMPUTATION, NUMERICAL ANALYSIS. *Educ:* Univ BC, BSc, 69; Univ Minn, MS, 72, PhD(control sci), 74. *Prof Exp:* Asst prof systs eng, Case Western Reserve Univ, 74-75; asst prof elec eng, Univ Toronto, 75-77; res scientist control eng, Lab Info & Decision Systs, Mass Inst Technol, 77-79; assoc prof elec eng, Univ Southern Calif, 79-83; PROF ELEC ENG, UNIV CALIF, SANTA BARBARA, 83-, CHMN DEPT, 89- *Concurrent Pos:* Assoc ed, Inst Elec & Electronics Engrs Trans Automatic Control, 79-81, Int J Control, 80-87, J on Control & Optimization, Soc Indust & Appl Math, 85-89, Math of Control, Signals & Systs, 86- & Linear & Multilinear Algebra, 87-; dir, Am Automatic Control Coun, Inst Elec & Electronics Engrs, 90-91; pres, Control Systs Soc, 91- *Mem:* Fel Inst Elec & Electronics Engrs; Soc Indust & Appl Math; Asn Comput Mach. *Res:* Numerical analysis; mathematical software; scientific computation; computer-aided control system design; linear and large-scale control and filtering theory. *Mailing Add:* Elec & Computer Eng Dept Univ Calif Santa Barbara CA 93106

LAUB, RICHARD J, b San Francisco, Calif, Aug 4, 45. NON-ELECTROLYTE SOLUTIONS THEORIES. *Educ:* Regis Col, BS, 67; Univ Calif, Los Angeles, MS, 71; Univ Hawaii, PhD(chem), 74. *Prof Exp:* Res fel chem, UK Sci Res Coun, Univ Col Swansea, Wales, 74-78; asst prof chem, Ohio State Univ, 78-; DEPT CHEM, SAN DIEGO STATE UNIV. *Mem:* Am Chem Soc; Sigma Xi; Royal Soc Chem. *Res:* Separations science, analytical and physicochemical applications of gas and liquid chromatography; physical chemistry; mass spectrometry. *Mailing Add:* Dept Chem San Diego State Univ San Diego CA 92182-0328

LAUB, RICHARD STEVEN, b Brooklyn, NY, Nov, 15, 45; m 74; c 1. PALEONTOLOGY, SEDIMENTOLOGY. *Educ:* Queen's Col, NY, BA, 66; Cornell Univ, MS, 68; Univ Cincinnati, PhD(paleont), 76. *Prof Exp:* CUR GEOL, BUFFALO MUS SCI, 73- *Concurrent Pos:* Supvr, Morse Creek Fossil Salvage Proj, 76; adj fac, Empire State Col, 76-78; corresp, Fossil Cnidaria Newsletter, 76-; adj fac geol, State Univ NY Buffalo, 77-; mem, Distinguished Lectureship Comt, Buffalo Soc Natural Sci, 78- *Mem:* Paleont Soc; Paleont Asn; Am Asn Petrol Geologists; Sigma Xi; Paleont Res Inst. *Res:* Early Paleozoic corals, their systematics, morphology, ecology and distribution; axial torsion in rugose corals; systematics and biology of auloporid tabulate corals. *Mailing Add:* Geol Dept Buffalo Mus Sci Humboldt Pkwy Buffalo NY 14211

LAUBACH, GERALD D, b Bethlehem, Pa, Jan 21, 26; m 53; c 3. PHARMACEUTICAL CHEMISTRY. *Educ:* Univ Pa, AB, 47; Mass Inst Technol, PhD(chem), 50. *Hon Degrees:* Mt Sinai Sch Med, City Univ New York, DHLett, 88; DL, Conn Col, 86; Hofstra Univ, DSc, 79. *Prof Exp:* Lab scientist, Pfizer Inc, 50-58, mgr, Med Prods Res, 58-61, dir, Dept Med Chem, 61, group dir med res, 63, vpres, Med Prods Res, 64, pres, Pfizer Pharmaceut, 69-71, exec vpres, 71, pres, Pfizer, Inc, 72-91; RETIRED. *Concurrent Pos:* Ensign, US Naval Reserve, 44-46; dir, CIGNA Corp, Philadelphia, Millpore Corp, Mass; mem bd, Food & Drug Law Inst. *Honors & Awards:* Int Palladium Medal, Am Sect, Soc Indust Chem, France, 85. *Mem:* Nat Acad Eng; hon fel Am Inst Chemists; NY Acad Sci; Inst Med-Nat Acad Sci; AAAS; Soc Chem Indust; Am Chem Soc; Am Mgt Asn. *Res:* Industrial chemistry; corporate competition; pharmaceutical manufacturing. *Mailing Add:* 50 E 89th St New York NY 10128

LAUBER, JEAN KAUTZ, b Seattle, Wash, Aug 30, 26; m 56; c 3. ZOOLOGY, PHYSIOLOGY. *Educ:* Whitman Col, BA, 48; Washington Univ, MA, 51; Univ NMex, PhD(zool), 59. *Prof Exp:* Sr lab technician electron micros, Sch Med, Univ Wash, 51-53, res assoc, 53-56; instr zool, Univ Idaho, 58-60, asst prof, 60-62; asst prof, Wash State Univ, 62-65; from asst prof to assoc prof, 65-76, hon assoc prof ophthal, 75-76, PROF ZOOL, HON PROF OPHTHAL & ASSOC VPRES ACAD, UNIV ALTA, 76- *Concurrent Pos:* Assoc vpres, Acad Univ Alta, 76-80. *Mem:* AAAS; Am Soc Zool; Am Physiol Soc; Soc Exp Biol & Med; Electron Micros Soc Am. *Res:* Endocrine physiology; photobiology; ophthalmology; electron microscopy; embryology. *Mailing Add:* Dept Zool Univ Alta Edmonton AB T6G 2E2 Can

LAUBER, THORNTON STUART, b Cornwall, Ont, Jan 5, 24; US citizen; m 56; c 4. ELECTRIC POWER ENGINEERING. *Educ:* Cornell Univ, BSEE, 44; Ill Inst Technol, MSEE, 51; Univ Pa, PhD(elec eng), 64. *Prof Exp:* Engr, Commonwealth Serv, Inc, Mich, 46-50; develop engr, Large Power Transformer Dept, Gen Elec Co, 51-58, sr anal engr, Power Circuit Breaker Dept, 58-69; PROF ELEC POWER ENG, RENSSELAER POLYTECH INST, 69-, S B CRARY PROF ENG, 70- *Concurrent Pos:* Consult, Gen Elec Co & Hydro-Quebec; comt mem, Am Nat Standards Inst; mem, Int Conf Large High Voltage Elec Systs, 72. *Mem:* Sr mem Inst Elec & Electronics Engrs; Am Soc Mech Engrs; Sigma Xi. *Res:* Electric power engineering; power system transients; electromagnetic field theory. *Mailing Add:* 1005 Seminole Rd Scotia NY 12302

LAU-CAM, CESAR A, b Lima, Peru, Nov 24, 40; m 67; c 2. PHARMACOGNOSY, PHYTOCHEMISTRY. *Educ:* San Marcos Univ, Lima, BS, 63; Univ RI, MS, 66, PhD(pharmacog), 69. *Prof Exp:* Instr pharm bot, San Marcos Univ, Lima, 62-63; lab instr pharmacog, Univ RI, 67; from asst prof to assoc phytochem, 69-80, PROF PHARMACEUT SCI, DEPT PHARMACEUT SCI, COL PHARM & ALLIED HEALTH PROFESSIONS, ST JOHN'S UNIV, NY, 80- *Concurrent Pos:* Sci adv, US Food & Drug Admin, 79- *Mem:* AAAS; Am Soc Pharmacog; Am Pharmaceut Asn; Phytochem Soc NAm; NY Acad Sci; Acad Pharmaceut Sci. *Res:* Analytical methods applied to natural products; clinical chemistry; pharmacology natural compounds; toxicology; pharmaceutical analysis. *Mailing Add:* Dept Pharmaceut Sci St John's Univ Grand Central Utopia Jamaica NY 11439

LAUCHLE, GERALD CLYDE, b Williamsport, Pa, Sept 20, 45; div; c 2. FLUIDS ENGINEERING, ACOUSTICS. *Educ:* Pa State Univ, BS, 68, MS, 70, PhD(eng acoust), 74. *Prof Exp:* From res asst to sr res assoc, Pa State Univ, 68-85, prof acoust, 68-90, sr scientist, 85-90, MEM GRAD PROG ACOUST, APPL RES LAB, PA STATE UNIV, 90- *Mem:* Acoust Soc Am; Am Soc Mech Engrs; Inst Noise Control Eng. *Res:* Basic and applied research on the noise generated by fluid flow; general acoustics; hydrodynamic drag reduction. *Mailing Add:* Grad Prog Acoust Appl Res Lab PO Box 30 State College PA 16804

LAUCK, DAVID R, b Alton, Ill, June 6, 30; m 53; c 2. ENTOMOLOGY. *Educ:* Univ Ill, BS, 55, MS, 58, PhD(entom), 61. *Prof Exp:* Cur invert entom, Chicago Acad Sci, 59-61; from asst prof to assoc prof, 61-70, chmn div biol sci, 66-72, PROF ZOOL, HUMBOLDT STATE UNIV, 70- *Mem:* Entom Soc Am; Entom Soc Can. *Res:* Aquatic and forest entomology. *Mailing Add:* Div Biol Sci Humboldt State Univ Arcata CA 95521

LAUCK, FRANCIS W, b Madison, Wis, Oct 4, 18; m 40; c 4. MECHANICAL & CHEMICAL ENGINEERING. *Educ:* Univ Wis, BS, 40, MS, 52. *Prof Exp:* Engr, A O Smith Corp, Wis, 40-46, lab supvr electrodes, 46-49, lab supvr appliance, 50-53, plant engr, 53-54, proj engr, 54-59, res scientist, 59-62, res scientist govt contracts, 62-63; from asst prof to assoc prof thermodyn & eng analysis, Univ Nebr, Lincoln, 64-74; CHIEF ENGR, R J G ENG, INC, 74- *Mem:* Am Soc Mech Engrs. *Res:* Energy conversion; engineering analysis; mobile structures. *Mailing Add:* R J G Eng Inc PO Box 6426 Lincoln NE 68506

LAUD, PURUSHOTTAM WAMAN, b Bombay, India, Nov 25, 48. MATHEMATICAL STATISTICS, APPLIED PROBABILITY. *Educ:* Bombay Univ, BSc, 69; Lamar Univ, MS, 71; Univ Mo-Columbia, MA, 73, PhD(statist), 77. *Prof Exp:* Asst prof statist, 77-80, ASST PROF MATH, NORTHERN ILL UNIV, 80- *Mem:* Inst Math Statist. *Res:* Bayesian nonparametric inference; reliability theory; stochastic processes. *Mailing Add:* Dept Math Sci Northern Ill Univ De Kalb IL 60115

LAUDE, HORTON MEYER, b Beaumont, Tex, Feb 25, 15; m 46; c 2. AGRONOMY, BOTANY. *Educ:* Kans State Univ, BS, 37; Univ Chicago, PhD(bot), 41. *Prof Exp:* Agent, USDA, Univ Chicago, 40-41; asst prof agron & asst agronomist, 46-52, assoc prof & assoc agronomist, 52-58, prof & agronomist, 58-81, EMER PROF AGRON, EXP STA, UNIV CALIF, DAVIS, 81- *Mem:* Fel Am Soc Agron; Am Soc Plant Physiol; Soc Range Mgt; Crop Sci Soc Am; Sigma Xi. *Res:* Physiology and ecology of dry range plants, irrigated forages and cereals; plant growth substances; seed production; resistance to environmental stresses of heat, cold and drought. *Mailing Add:* 818 Oeste Dr Davis CA 95616

LAUDENSLAGER, JAMES BISHOP, b Harrisburg, Pa, June 8, 45. CHEMICAL PHYSICS. *Educ:* Temple Univ, AB, 67; Univ Calif, Santa Barbara, PhD(phys chem), 71. *Prof Exp:* Res assoc space sci, Jet Propulsion Lab, 71-73, sr res sci chem physics, 73-86; VPRES, LASER SYSTS & ADVAN INT SYSTS, IRVINE, CALIF, 86- *Mem:* Am Phys Soc; Optical Soc Am; Soc Photoelectronics & Optics. *Res:* Fundamental properties of charge transfer and metastable rare gas reactions in the gas phase for use in laser development and mass spectrometry. *Mailing Add:* Nine Parker Irvine CA 92718

LAUDENSLAGER, MARK LEROY, b Charlotte, NC, May 13, 47; c 2. PSYCHONEUROIMMUNOLOGY, PHYSIOLOGICAL PSYCHOLOGY. *Educ:* Univ NC, AB, 69; Univ Calif, Santa Barbara, PhD(psychol), 75. *Prof Exp:* Teaching asst introductory psychol, Univ Calif, Santa Barbara, 70-71, res asst neuropsychol, 71-72, teaching asst physiol psychol, 73, lectr, 74; NIMH fel, Scripps Inst Oceanog, 75-77; asst res psychologist, Univ Calif, Santa Barbara, 77-80; lectr, Denver Univ, 81-83; asst clin prof, 84-86, asst prof, 86-90, ASSOC PROF, HEALTH SCI CTR, UNIV COLO, 91-; ASST RES PROF, DENVER UNIV, 83- *Concurrent Pos:* NIMH fel, Health Sci Ctr, Univ Colo, 80. *Mem:* Int Soc Develop Psychobiol; AAAS; Am Physiol Soc; Soc Neurosci. *Res:* effects of stress and affective disorders on immunocompetence and health, role of emotions, social support and temperament on health. *Mailing Add:* Dept Psychiat Univ Colo Health Sci Ctr 700 Delaware St Denver CO 80204

LAUDER, JEAN MILES, b Haverhill, Mass, June 29, 45. CELL BIOLOGY, ANATOMY. *Educ:* Univ Maine, Orono, BA, 67; Purdue Univ, PhD(biol sci), 72. *Prof Exp:* NIMH staff fel, Lab Neuropharmacol, St Elizabeth's Hosp, Washington, DC, 72-74; res assoc, Lab Neuromorphol, Dept Behav Sci, Univ Conn, Storrs 74-75, asst prof-in-residence, Dept Behav Sci, 76-78; assoc prof, Dept Anat, 78-85, PROF, DEPT CELL BIOL & ANAT, SCH MED, UNIV NC, CHAPEL HILL, 85- *Concurrent Pos:* Mem bd dirs & asst treas, Inst Develop Neurosci & Aging, 83-; dir, Prog Develop Neurosci, Div Behav & Neural Sci, NSF, 84-85; vchmn, Gordon Res Conf Cent Nervous Syst-Neural Develop, 85-87, chmn, 87-89 & vchair, Gordon Conf Neural Develop, 87, chair, 89. *Mem:* Soc Neurosci; Women Neurosci; Int Soc Develop Neurosci (pres-elect, 86-88, pres, 88-90); AAAS; Am Asn Anatomists; Am Soc Cell Biol; Int Soc Psychoneuroendocrinol; Sigma Xi; Int Brain Res Orgn; NY Acad Sci. *Mailing Add:* Cell Biol & Anat Dept Univ NC Sch Med CB No 7090 Chapel Hill NC 27599-7090

LAUDERDALE, JAMES W, JR, b Washington, DC, Dec 21, 37; m 62; c 3. REPRODUCTIVE PHYSIOLOGY, ENDOCRINOLOGY. *Educ:* Auburn Univ, BS, 62; Univ Wis, MS, 64, PhD(reproductive physiol, endocrinol), 68. *Prof Exp:* SCIENTIST, UPJOHN CO, 67- *Honors & Awards:* Animal Physiol & Endocrinol Award, Am Soc Animal Sci, 86. *Mem:* Am Soc Animal Sci; Soc Study Reproduction. *Res:* Reproductive, growth and endocrinological function of large animals. *Mailing Add:* Upjohn Co 7000 Portage Rd Kalamazoo MI 49001

LAUDERDALE, ROBERT A(MIS), JR, b Harriman, Tenn, July 27, 22; m 48; c 1. SANITARY ENGINEERING. *Educ:* Univ Tenn, BS, 44, MS, 48; Mass Inst Technol, PhD(sanit eng), 58. *Prof Exp:* Jr chem engr, Tenn Valley Authority, 44-45; assoc health physicist, Oak Ridge Nat Lab, 48-52; res assoc sanit eng, Mass Inst Technol, 52-58, asst prof, 57-58; PROF CIVIL ENG, UNIV KY, 58-, DIR, WATER RESOURCES INST, 65- *Mem:* Am Chem Soc; Water Pollution Control Fedn. *Res:* Industrial and radioactive waste treatment. *Mailing Add:* Dept Civil Eng Univ Ky Lexington KY 40506

LAUDISE, ROBERT ALFRED, b Amsterdam, NY, Sept 2, 30; m 57; c 5. SOLID STATE CHEMISTRY. *Educ:* Union Univ, NY, BS, 52; Mass Inst Technol, PhD(inorg chem), 56. *Prof Exp:* Mem tech staff, 56-60, head, Crystal Chem Res Dept, 60-73, asst dir mat res, 73-76, dir mat res, 76-79, physics & inorg chem res, 80-86, mat chem res, 86-90, DIR MAT & PROCESSING RES, BELL LABS, 90- *Concurrent Pos:* Consult, President's Sci Adv Comt, 60-70; panel mem mat adv bd, Nat Acad Sci, 65-70; vis comt, Nat Inst Standards & Technol, 68-75, 86-, mat sci, Mass Inst Technol, 80-; ed, J Crystal Growth, 74-; adj prof mat sci, Mass Inst Technol, 87- *Honors & Awards:* Sawyer Award, Conf Frequency Control, 74; Int Award in Crystal Growth, Int Orgn Crystal Growth, 81; Mat Chem Prize, Am Chem Soc, 90. *Mem:* Nat Acad Eng; Nat Acad Sci; Int Orgn Crystal Growth (pres, 77-84); fel Mineral Soc Am; Am Ceramic Soc; sr mem Inst Elec & Electronics Engrs; fel AAAS; Am Asn Crystal Growth (pres, 68-75). *Res:* Materials research; crystal growth; hydrothermal chemistry; quartz; ferroelectrics; non-linear optical materials; magnetic materials; physical chemistry; superconductors. *Mailing Add:* 65 Lenape Lane Berkeley Heights NJ 07922

LAUDON, THOMAS S, b Sac City, Iowa, June 14, 32; m 56; c 5. SEDIMENTARY PETROLOGY, GRAVITY. *Educ:* Univ Wis, BSc, 55, MSc, 57, PhD(geol), 63. *Prof Exp:* Chmn dept geol, 69-72, PROF GEOL, UNIV WIS-OSHKOSH, 63- *Concurrent Pos:* NSF res grants, 62-67, 70-71, 84-85 & Nat Acad Sci res grants, 71-72; consult geol & geophys, 79- *Mem:* Sigma Xi; Geol Soc Am; Am Asn Petrol Geologists; Am Geophys Union; Nat Assn Geol Teachers. *Res:* Geologic exploration of Antarctica; gravity, tectonics and sedimentation in modern and ancient mobile belts. *Mailing Add:* Dept Geol Univ Wis Oshkosh WI 54901

LAUENROTH, WILLIAM KARL, b Carthage, Mo, July 31, 45. PLANT ECOLOGY. *Educ:* Humboldt State Col, BS, 68; NDak State Univ, MS, 70; Colo State Univ, PhD(plant ecol), 73. *Prof Exp:* Sr res ecologist, Nat Resource Ecol Lab, 75-81, PROF, RANGE SCI DEPT, COLO STATE UNIV, 81- *Honors & Awards:* Outstanding Achievement Award, Soc Range Mgt, 89. *Mem:* Ecol Soc Am; Soc Range Mgt; Bot Soc Am; AAAS; Int Soc Ecol Modelling. *Res:* Primary production and water relations of native plant communities, particularly temperate grasslands; ecosystem analysis of natural and agricultural ecosystems including simulation modelling. *Mailing Add:* Dept Range Sci Colo State Univ Ft Collins CO 80521

LAUER, B(YRON) E(LMER), b Glencoe, Okla, Apr 14, 07; m 31; c 1. SCIENCE EDUCATION, CHEMICAL ENGINEERING. *Educ:* Ore State Univ, BS, 27; Univ Minn, MS, 29, PhD(chem eng), 31. *Prof Exp:* Asst instr, Univ Minn, 27-31; res chem engr, Northern Paper Mills, 31-34 & Crown-Zellerbach Corp, 34-35; from asst prof to prof chem eng, NC State Univ, 35-42; prof, 46-75, head dept, 47-61, EMER PROF CHEM ENG, UNIV COLO, BOULDER, 75- *Concurrent Pos:* Consult, Nat Bur Standards, 61-64, PEC Corp, 64-65; Interam Transp, 68-71, Jet Propulsion Lab, 83- & Western Interstate Comn Higher Educ, 82-; ed, Televised Higher Educ, Assoc Western Univs, Inc, 75-82. *Mem:* Am Chem Soc; Am Soc Eng Educ; Tech Asn Pulp & Paper Indust; Am Inst Chem Engrs. *Res:* Pulp and paper; testing methods; heat transfer; oil shale retorting; information storage and retrieval. *Mailing Add:* 546 14th St Boulder CO 80302-7806

LAUER, DAVID ALLAN, b Creston, Iowa, Sept 25, 44; m 64; c 2. SOIL SCIENCE, AGRONOMY. *Educ:* Iowa State Univ, BS, 66; Colo State Univ, MS, 69, PhD(soil sci), 71. *Prof Exp:* Res assoc agron, Cornell Univ, 71-75; asst prof, Univ Ga, 75-76; RES SOIL SCIENTIST, AGR RES, USDA, 76- *Mem:* Am Soc Agron; Crop Sci Soc; Soil Sci Soc; AAAS; Sigma Xi. *Res:* Application of soil chemical and plant physiological principles to management of crop production systems with emphasis on plant nutrition. *Mailing Add:* Agr Res Serv USDA PO Box 30 Prosser WA 99350

LAUER, EUGENE JOHN, b Red Bluff, Calif, Apr 11, 20; m 48; c 5. PLASMA PHYSICS. *Educ:* Univ Calif, BS, 42, PhD(physics), 51. *Prof Exp:* Asst physics, 46-51, PHYSICIST, LAWRENCE LIVERMORE LAB, UNIV CALIF, 51- *Mem:* Am Phys Soc. *Res:* Discharge through gases; particle accelerator design; nuclear and plasma physics; controlled thermonuclear energy; relativistic beams. *Mailing Add:* 2221 Martin Ave Pleasanton CA 94566

LAUER, FLORIAN ISIDORE, b Richmond, Minn, Sept 13, 28; m 55; c 4. HORTICULTURE. *Educ:* Univ Minn, BS, 51, PhD, 57. *Prof Exp:* From asst prof to assoc prof, 59-66, PROF POTATO BREEDING, UNIV MINN, ST PAUL, 66- *Mem:* Am Soc Hort Sci; Potato Asn Am. *Res:* Potato breeding and genetics. *Mailing Add:* Dept Hort Inst Agr Univ Minn St Paul MN 55108

LAUER, GEORGE, b Vienna, Austria, Feb 18, 36; US citizen; wid; c 2. PHYSICAL CHEMISTRY. *Educ:* Univ Calif, Los Angeles, BS, 61; Calif Inst Technol, PhD(chem), 67. *Prof Exp:* Staff assoc, Sci Ctr, NAm Rockwell Corp, 62-66, mem tech staff, 66-71, group leader, Measurement Sci, 71-72, dir air qual monitoring res, 72-75, mgr environ res & technol, 76-78, dir, Environ Monitoring & Serv Ctr, Rockwell Int Sci Ctr, 78-84; pres, EMSI, Inc, 81-88; SR CONSULT, ATLANTIC RICHFIELD, INC, 88- *Mem:* AAAS; Am Chem Soc; Air Pollution Control Asn Am; Am Meteorol Soc. *Res:* Development of chronocoulometric techniques in electrochemistry; study of electroactive adsorbed species at electrodes; development of computer controlled instrumentation; instrumental methods in air pollution monitoring; air quality simulation modeling; environmental assessment. *Mailing Add:* 6009 Maury Ave Woodland Hills CA 91367

LAUER, GERALD J, b Montgomery City, Mo, Oct 18, 34; m 63. AQUATIC ECOLOGY, LIMNOLOGY. *Educ:* Quincy Col, BS, 56; Univ Wash, MS, 59, PhD(zool), 63. *Prof Exp:* Lab aide limnol, Quincy Col, 54-56; asst, Univ Wash, 56-59; teacher high sch, Mo, 59-60; staff biologist, USPHS, 60-62, prin biologist, Southeast Water Lab, 63-66, chief training br, 66; leader, Fisheries Coop Unit & assoc prof fisheries & limnol res, Ohio State Univ, 66-67; assoc cur limnol dept, Acad Natural Sci, Philadelphia, 67-69; asst dir, Lab Environ Studies, Med Ctr, NY Univ, 69-75; SR SCIENTIST & VPRES, EA SCI & TECHNOL, 75- *Concurrent Pos:* Adj assoc prof biol, NY Univ, 70- *Mem:* AAAS; Am Soc Limnol & Oceanog; Am Fisheries Soc; Am Littoral Soc; Ecol Soc Am. *Res:* Population dynamics and community diversity of aquatic organisms; effects of pollutants and other environmental stresses on aquatic life. *Mailing Add:* Three Washington Ctr Newburgh NY 12550

LAUER, JAMES LOTHAR, b Vienna, Austria, Aug 2, 20; nat US; m 55; c 2. MATERIALS SCIENCE. *Educ:* Temple Univ, AB, 42, MA, 44; Univ Pa, PhD(physics), 48. *Prof Exp:* Asst instr org chem, Temple Univ, 42-44; phys chemist, Sun Oil Co, 44-45, res physicist, 47-54, sr physicist, 54-58, res assoc,

58-62, res scientist, 62-77; PROF MECH ENG, 78-, DIR, INST WEAR CONTROL & TRIBOLOGY, RENSSELAER POLYTECH INST, 85- *Concurrent Pos:* Lectr, Univ Del, 52-56; asst prof, Univ Pa, 52-54; fel aerospace eng, Univ Calif, San Diego, 64-65; prin investr, Air Force Off Sci Res & NASA-Lewis Res Ctr, 73-85, Off Naval Res, 78-85, NSF res grant, 87- & US Army Res Off, 86-; lectr, Coblentz, 78; lectr friction & lubrication, Gordon Res Conf, 82. *Mem:* Am Chem Soc; Am Phys Soc; Soc Appl Spectros; Optical Soc Am; Spectros Soc Can. *Res:* Fourier spectroscopy; Raman and infrared spectroscopy; refraction and dispersion; theory of molecular structure; mathematical physics; x-ray spectra of polymers; combustion; shock waves in gases; adsorption and desorption; applications of molecular, mainly infrared emission, spectroscopy to problems of lubrication; optical methods of non-contacting surface analysis; electrostatics; boundary lubrication, studied by modern methods of surface analysis and applied to machinery at high operating temperatures. *Mailing Add:* Seven Northeast Lane Ballston Lake NY 12019

LAUER, ROBERT B, b York, Pa, Mar 9, 42; m 60; c 3. MATERIALS SCIENCE, SOLID STATE SCIENCE. *Educ:* Franklin & Marshall Col, AB, 64; Univ Del, MS, 66, PhD(physics), 70. *Prof Exp:* Sr physicist, Itek Corp, 69-74; mem tech staff, 74-81, ACTG TECH MGR, GTE LABS, 81- *Mem:* Am Phys Soc; Sigma Xi. *Res:* Luminescence and photoconductivity of semiconductors, semi-insulators and insulators; defect structure of solids; materials preparation; liquid phase epitaxial growth; semiconductor lasers; high radiance light emitting diodes; photodetectors. *Mailing Add:* GTE Labs 40 Sylvan Rd Waltham MA 02154

LAUERMAN, LLOYD HERMAN, JR, b Everett, Wash, Feb 5, 33; m 55; c 3. VETERINARY MICROBIOLOGY, VETERINARY IMMUNOLOGY. *Educ:* Wash State Univ, BA, 56, DVM, 58; Univ Wis, MS, 59, PhD, 68; Am Col Vet Microbiol, dipl, 74. *Prof Exp:* Res asst vet med, Univ Wis, 58-60, proj asst, 60-63; res dir vet biol, Biol Specialties Corp, 63-68; Colo State Univ-AID prof microbiol, Univ Nairobi, 68-72; assoc prof, Colo State Univ, 72-81, head bact sect, Diag Lab, Dept Path, 73-81; VET, DIAG LAB, MICRO SECT, STATE ALA, 81- *Concurrent Pos:* Vis assoc prof microbiol, Sch Dent, Univ Colo, Denver, 74-75. *Mem:* Am Soc Microbiologists; Am Asn Vet Lab Diagnosticians; Am Vet Med Asn; US Am Health Asn; Am Asn Vet Immunol. *Res:* Diagnosis and prevention of infectious diseases of animals; research and development of laboratory diagnostic techniques and animal vaccines. *Mailing Add:* Vet Diag Lab PO Box 2209 Auburn AL 36830

LAUF, PETER KURT, b Wuerzburg, Ger, Sept 25, 33; US citizen; c 3. PHYSIOLOGY, IMMUNOLOGY. *Educ:* Univ Freiburg, MD, 60. *Prof Exp:* Res assoc path, Inst Path, Univ Freiburg, 60-62; res fel biochem, Max Planck Inst Immunobiol, Freiburg, 62-64; res assoc path, Inst Path, Univ Marburg, Ger, 64; res assoc biochem, Child Res Ctr Mich, Detroit, 65-67; asst prof biochem, Wayne State Univ, 66-67; from asst prof to prof physiol, Duke Univ Med Ctr, 68-85, asst prof immunol, 70-80; PROF PHYSIOL & BIOPHYS & CHMN, SCH MED, WRIGHT STATE UNIV, DAYTON, OHIO, 85- *Concurrent Pos:* NIH res career develop award, 71-75; Golding distinguished prof, Wright State Univ. *Mem:* Soc Gen Physiol; Am Soc Hemat; Biophys Soc; Am Asn Immunol; Am Soc Physiol. *Res:* Membrane physiology; active and passive cation; modulation of membrane transport processes by immunological reactions; isolation and identification of membrane transport proteins and functionally associated surface antigens. *Mailing Add:* Dept Physiol & Biophys Wright State Univ Sch Med Dayton OH 45401-0927

LAUFER, ALLAN HENRY, b New York, NY, Mar 27, 36; m 59; c 2. PHOTOCHEMISTRY, CHEMICAL KINETICS. *Educ:* NY Univ, BA, 56; Lehigh Univ, MS, 58, PhD(phys chem), 62. *Prof Exp:* Res chemist, Gulf Res & Develop Co, 62-64; res chemist, Nat Bur Standards, 64-83; chemist & prog mgr, Chem Sci Div, Off Basic Energy Sci, 83-86, BR CHIEF, FUNDAMENTAL INTERACTIONS BR, US DEPT ENERGY, 86- *Mem:* AAAS; Am Chem Soc; Am Phys Soc; Sigma Xi; Inter-Am Photochem Soc. *Res:* Vacuum ultraviolet photochemistry; gas phase radical reactions and kinetics; chemistry of excited states. *Mailing Add:* Chem Sci Div Basic Energy Sci US Dept Energy Washington DC 20545

LAUFER, DANIEL A, b Affula, Israel, May 30, 38; US citizen; m 68; c 3. ORGANIC CHEMISTRY. *Educ:* Mass Inst Technol, BS, 59; Brandeis Univ, PhD(org chem), 64. *Prof Exp:* Res assoc org chem, Columbia Univ, 63-64; res fel biol chem, Harvard Med Sch, 64-66; from asst prof to assoc prof, 66-80 PROF CHEM, UNIV MASS, BOSTON, 80- *Concurrent Pos:* NIH trainee, 65-66, res grant, 67-70, 72-74 & 79-84. *Mem:* Am Chem Soc; AAAS. *Res:* Peptide synthesis; new reagents for organic synthesis macrocyclic complexations analytic applications of NMR. *Mailing Add:* Dept Chem Univ Mass Boston MA 02125

LAUFER, HANS, b Ger, Oct 18, 29; nat US; m 53; c 3. DEVELOPMENTAL BIOLOGY. *Educ:* City Col New York, BS, 52; Brooklyn Col, MA, 54; Cornell Univ, PhD(zool), 58. *Prof Exp:* Asst, Cornell Univ, 55-57; Nat Res Coun fel embryol, Carnegie Inst, 57-59; asst prof biol, Johns Hopkins Univ, 59-65; assoc prof biol, 65-72, PROF BIOL, UNIV CONN, 72- *Concurrent Pos:* Vis scholar, Case Western Reserve Univ, 62; Lalor fel, Marine Biol Lab, Woods Hole, Mass, 62-63, staff embryol course, 67-72, mem, Corp Marine Biol Lab, 62-, trustee, 78-82, mem exec comt, 79-80; assoc ed, J Exp Zool, 69-73, 89-; mem nat bd on grad educ, Conf Bd Assoc Res Couns, 71-75; vis prof, Karolinska Inst, Stockholm, 72 & Yale Univ, 80; NATO fel rev panel, NSF, 74 & 76; partic, Nat Acad Sci-Czech Acad Sci Exchange Prog, 74 & 77; chmn, Div Develop Biol, Am Soc Zoologists, 81-82; vis prof, Harvard Univ, 88-89. *Honors & Awards:* Rosenstiel vis scholar, Brandeis Univ, 73; hon prof, Charles Univ, Prague, 74; Lady Davis vis prof, Hebrew Univ, Jerusalem, 88; Sr Res Serv Award Harvard Univ, 89. *Mem:* Am Soc Zoologists; Soc Develop Biol; Am Soc Cell Biol; Tissue Cult Asn; fel AAAS. *Res:* Developmental physiology and biochemistry; molecular interactions in development; proteins and enzymes in ontogeny, regeneration and metamorphosis; chromosomal puffing in Diptera; gene action as related to development; hormone action during invertebrate development and reproduction. *Mailing Add:* Biol Sci Group U-42 Univ Conn Storrs CT 06268

LAUFER, IGOR, b Czech, Aug 8, 44; Can citizen; m 67; c 2. GASTROINTESTINAL RADIOLOGY, GALLSTONE DIAGNOSIS & TREATMENT. *Educ:* Univ Toronto, BSc, 66, MD, 67. *Prof Exp:* Intern, New Mt Sinai Hosp, Toronto, 67-68; med resident, Toronto Western Hosp, 68-69, asst resident,diag radiol, 69-71; chief resident, Beth Israel Hosp, Boston, 71-72; staff radiol, McMaster Univ Med Ctr, 72-76; assoc prof, 76-80, PROF, RADIOL, UNIV PA SCH MED, 80-; CHIEF GASTROINTESTINAL RADIOL, HOSP UNIV PA, 76- *Concurrent Pos:* Clin fel radiol, Harvard Med Sch, 69-72; co-dir gallstone-lithotripay unit, Hosp Univ Pa, 88-; consult diag radiol, NIH. *Honors & Awards:* Cannon Medal, Soc Gastrointestinal Radiologists, 89. *Mem:* Soc Gastrointestinal Radiologists (pres, 87); Am Roentgen Ray Soc; Am Col Radiol; Radiol Soc NAm; Am Gastroenterol Asn. *Res:* Early diagnosis of inflammatory and neoplastic disorders of the gastrointestinal tract; double contrast radiology; gallstone lithotripay. *Mailing Add:* Dept Radiol Hosp Univ Pa 3400 Spruce St Philadelphia PA 19104

LAUFER, ROBERT J, b Pittsburgh, Pa, May 10, 32; m 67; c 2. ORGANIC CHEMISTRY. *Educ:* Carnegie Inst Technol, BS, 53, MS, 56, PhD(org chem), 58. *Prof Exp:* Proj supvr res div, Consol Coal Co, 58-68; proj leader, Int Flavors & Fragrances Inc, 68-69, assoc dir fragrance res, 69-72, dir, 72-75; dir aromatic technol, Norda Inc, 75-77, corp dir res & develop, 77-78, vpres aromatics, 79-80, vpres & gen mgr, Orbis Prod Div, 80-81, vpres corp planning, 81-82; RETIRED. *Mem:* AAAS; Am Chem Soc. *Res:* Research management; aroma chemicals; terpenoids; fragrance applications research. *Mailing Add:* Six Greenhill Rd Colts Neck NJ 07722

LAUFERSWEILER, JOSEPH DANIEL, b Columbus, Ohio, Aug 13, 30; m 59; c 2. ECOLOGY, BOTANY. *Educ:* Univ Notre Dame, BS, 52; Ohio State Univ, MSc, 54, PhD(ecol), 60. *Prof Exp:* Instr bot & ecol, Ohio State Univ, 59-61; asst prof biol, Drake Univ, 61-63; asst prof, 63-77, ASSOC PROF BIOL, UNIV DAYTON, 77- *Concurrent Pos:* Bd trustees, Dayton Mus Natural Hist. *Mem:* Bot Soc Am; Ecol Soc Am; Am Inst Biol Sci; Nat Asn Biol Teachers. *Res:* Reproduction of plant communities; distribution of original vegetation and its influence on man. *Mailing Add:* Dept Biol Univ Dayton Dayton OH 45469-2320

LAUFF, GEORGE HOWARD, b Milan, Mich, Mar 23, 27. LIMNOLOGY, ZOOLOGY. *Educ:* Mich State Univ, BS, 49, MS, 51; Cornell Univ, PhD(limnol, zool), 53. *Prof Exp:* Fisheries res technician, Mich State Dept Conserv, 50; asst phycol, Point Barrow, Alaska, 51; biol asst, Cornell Univ, 51-52, zool, 52-53; instr, Univ Mich, 53-57, from asst prof to assoc prof, 57-62; res coordr, Sapelo Island Res Found, Ga, 62-64; dir, W K Kellogg Biol Sta, 64-90, PROF ZOOL, FISHERIES & WILDLIFE, MICH STATE UNIV, 64- *Concurrent Pos:* Res assoc, Great Lakes Res Inst, 54-59, Oak Ridge Inst Nuclear Studies & Oak Ridge Nat Lab, 60; assoc prof & dir marine inst, Univ Ga, 60-62. *Mem:* AAAS; Am Inst Biol Sci; Am Soc Limnol & Oceanog (treas, 58-61, secy, 58-64, 67-70, vpres, 71-72, pres, 72-73); Ecol Soc Am; Int Asn Theoret & Appl Limnol. *Mailing Add:* W K Kellogg Biol Sta Mich State Univ Hickory Corners MI 49060

LAUFFENBURGER, DOUGLAS ALAN, b Des Plaines, Ill, May 6, 53; m 79; c 2. MOLECULAR & CELLULAR BIOENGINEERING. *Educ:* Univ Ill, BS, 75; Univ Minn, PhD(chem eng), 79. *Prof Exp:* Asst prof, Univ Pa, 79-84, assoc prof, 84-87, prof chem eng, 87-90; PROF, UNIV ILL, 90- *Concurrent Pos:* Vis prof, Inst Appl Math Univ Heidelberg, 80; mem, grad group bioeng, Univ Penn, 79-90; mem, grad group cell biol, Univ Penn, 89-90; J S Guggenheim fel, 89. *Honors & Awards:* Alan P Colburn Award, Am Inst Chem Engrs, 88. *Mem:* Am Inst Chem Engrs; Biomed Eng Soc; Soc Math Biol; Am Soc Cell Biol. *Res:* Quantitative investigation and mathematical modeling of receptor-mediated cell phenomena, including growth, adhesion, migration, chemotaxis, endocytosis. *Mailing Add:* Dept Chem Eng Univ Ill Lab 297 Roge-Adams Urban IL 61801

LAUFFENBURGER, JAMES C, b Buffalo, NY, Aug 23, 38; m 61; c 5. SOLID STATE PHYSICS. *Educ:* Canisius Col, BS, 60; Univ Notre Dame, PhD(solid state physics), 65. *Prof Exp:* Asst prof, 64-72, ASSOC PROF PHYSICS, CANISIUS COL, 72- *Mem:* Sigma Xi; Am Asn Physics Teachers. *Res:* X-rays. *Mailing Add:* Dept Physics Canisius Col 2001 Main St Buffalo NY 14208

LAUFFER, DONALD EUGENE, b Lebanon, Pa, July 29, 40; m 64; c 3. PHYSICS. *Educ:* Ohio State Univ, BS, 64, PhD(physics), 68. *Prof Exp:* Sr res physicist, Phillips Petrol Co, 68-76, supvr geophys, 76-84, br mgr geophys, 84-86, explor & prod planning mgr, 86-88, SR RES ASSOC, PHILLIPS PETROL CO, 88- *Mem:* Am Phys Soc; Inst Elec & Electronics Engrs; Am Chem Soc. *Res:* Geophysics; wave propagation and signal processing. *Mailing Add:* 814 Revere Way E Bartlesville OK 74006

LAUFFER, MAX AUGUSTUS, JR, b Middletown, Pa, Sept 2, 14; m 36, 64; c 4. BIOPHYSICS. *Educ:* Pa State Univ, BS, 33, MS, 34; Univ Minn, PhD(biochem), 37. *Prof Exp:* Asst, Univ Minn, 35-36, instr biochem, 36-37; fel plant path, Rockefeller Inst, 37-38, asst, 38-41, assoc, 41-44; assoc res prof, 44-47, res prof physics & physiol chem, 47-49, chmn dept physics, 47-48, prof biophys, 49-63, head dept, 49-56 & chmn dept, 63-67, assoc dean res in natural sci, 53-54, dean div natural sci, 56-63, Andrew Mellon prof biophys, 63-84, chmn dept biophys & microbiol, 71-76, EMER PROF, UNIV PITTSBURGH, 84- *Concurrent Pos:* Spec lectr, Stanford Univ, 41; prin investr, Comt Med Res, 44-46; Priestley lectr, Pa State Univ, 46; Gehrmann lectr, Univ Ill, 51; vis prof, Theodor Kocher Inst, Bern Univ, 52, Max Planck Inst Virus Res, Tübingen, 65-66 & Univ Philippines, 67; mem Nat Res Coun comn macromolecules, 47-53, mem panel virol & immunol, sect biol, comt growth, 53-56; mem sci adv comt, Boyce Thompson Inst Plant Res Inc, 53-82; co-ed, Adv Virus Res, 53-85; mem prog proj comt, Nat Inst Gen Med Sci, 61-63, chmn, 62-63, mem adv coun, 63-67; mem sci adv bd, Delta Regional Primate Res Ctr, Tulane Univ, 64-67; ed, Biophys J, 69-73; consult to the provost, Univ Pittsburgh, 84-86; adj prof, Lebanon Valley Col, 88-90. *Honors & Awards:* Award, Eli Lilly & Co, 45; Pittsburgh Award, Am Chem Soc, 58.

Mem: Am Chem Soc; Am Soc Biol Chemists; Biophys Soc (pres elect, 60, pres, 61); Soc Exp Biol & Med; Fedn Am Sci. *Res:* Electrokinetics; ultracentrifugation; viscometry; biophysics of viruses; kinetics of virus disintegration; size and shape of macromolecules; polymerization of virus protein; hydration of proteins; entropy-driven processes in biology. *Mailing Add:* 190 Lauffer Rd Univ Pittsburgh Middleton PA 17057

LAUFMAN, HAROLD, b Milwaukee, Wis, Jan 6, 12; m 40; c 2. SURGERY. *Educ:* Univ Chicago, BS, 32, Rush Med Col, MD, 37; Northwestern Univ, MS, 46, PhD(surg), 48; Am Bd Surg, dipl. *Prof Exp:* From clin asst to prof surg, Med Sch, Northwestern Univ, 40-65; prof, 65-77, emer prof surg, Albert Einstein Col Med, 77; PROF LECTR SURG, MT SINAI SCH MED, 79-; PRES, HLA SYSTS, INC, 79- *Concurrent Pos:* Assoc attend surgeon, Cook County Hosp, 46-48; attend, Hines Vet Admin Hosp, 48-50; adj attend, Michael Reese Hosp, Chicago, 48-54; attend, Passavant Mem & Vet Admin Res Hosps, 54-65; James IV traveling prof, Israel, 62; dir, Inst Surg Studies, Montefiore Hosp, 65-81, emer dir, 81- *Mem:* Am Surg Asn; Soc Vascular Surg; fel Am Col Surg; Am Med Writers Asn (pres, 69); Asn Advan Med Instrumentation (pres, 74-75); Sigma Xi. *Res:* Surgical physiology, especially mesenteric and peripheral vascular diseases; surgical design and facilities engineering. *Mailing Add:* 31 E 72nd St New York NY 10021

LAUG, GEORGE MILTON, biology; deceased, see previous edition for last biography

LAUGHLIN, ALEXANDER WILLIAM, b Hot Springs, Ark, Nov 9, 36; m 69; c 3. GEOCHEMISTRY, ECONOMIC GEOLOGY. *Educ:* Mich Technol Univ, BSc, 58; Univ Ariz, MSc, 60, PhD(geol), 69. *Prof Exp:* Res assoc isotope geochem, Univ Ariz, 66-69; res assoc geol, Univ NMex, 69-70; from asst prof to assoc prof, Kent State Univ, 70-74; mem staff, 74-78, & 80-87, group leader, 78-80, DEP GROUP LEADER, LOS ALAMOS NAT LAB, 87- *Concurrent Pos:* Adj prof, Univ NMex, 75. *Mem:* Am Geophys Union; Int Asn Geochem & Cosmochem; Geol Soc Am. *Res:* Geochronology; trace element geochemistry; origin of ultramafic inclusions and basalts; economic geology; geothermal energy extraction from dry hot rock; geothermal exploration techniques, petrology of pre-Cambrian rocks; potassic mafic rocks. *Mailing Add:* EES-1 MS-D-462 Los Alamos Nat Lab Univ Calif Los Alamos NM 87545

LAUGHLIN, ALICE, b Malone, NY, Feb 19, 18. BIOCHEMISTRY, ANALYTICAL CHEMISTRY. *Educ:* St Joseph Col, Conn, BS, 49; Univ Vt, MS, 54; Columbia Univ, EdD(col sci teaching), 65. *Prof Exp:* Lab technician, Clin Lab, Staten Island Hosp, 49-50; teaching asst biochem, Univ Vt, 50-52; asst biochemist, Vt State Agr Exp Sta, 52-56; res chemist, Nat Biscuit Co, 56-57; res asst hemat & chemother, Columbia-Presby Med Ctr, Columbia Univ, 57-61; sci instr, Sch Nursing, St Michael Hosp, 61-62; from asst prof sci to assoc prof chem, 62-74, chmn dept, 69-70, PROF CHEM, JERSEY CITY STATE COL, 74- *Concurrent Pos:* Resource mem long range planning bd, Sch Nursing, St Francis Hosp, 63-; consult, NSF meeting on chem curric in jr cols, Rutgers Univ, 69. *Mem:* Am Chem Soc; Am Asn Higher Educ; fel Am Inst Chemists; NY Acad Sci; Am Microchem Soc; Am Asn Univ Women (pres, 87-89). *Res:* Human nutrition; science education. *Mailing Add:* 1225 76th St 8C North Bergen NJ 07047

LAUGHLIN, DAVID EUGENE, b Philadelphia, Pa, July 15, 47; m 71; c 4. PHYSICAL METALLURGY, ELECTRON MICROSCOPY. *Educ:* Drexel Univ, BSc, 69; Mass Inst Technol, PhD(metall), 73. *Prof Exp:* res assoc, Nat Bur Standards, 73-74; from asst prof to assoc prof, 74-82, PROF METALL, CARNEGIE-MELLON UNIV, 82- *Concurrent Pos:* Pres, Trinity Christian Sch Bd Dirs, 78-83; consult, 80-; assoc ed, Metall Trans, 82, ed. *Mem:* Fel Am Soc Metals; Am Inst Mining, Metall & Petrol Engrs; Am Soc Eng Educ; Am Sci Affil. *Res:* X-ray diffraction; innovative teaching; phase transformations; differential scanning calorimetry; electron microscopy; magnetic materials. *Mailing Add:* Dept Metall & Mat Sci Carnegie-Mellon Univ Pittsburgh PA 15213

LAUGHLIN, ETHELREDA R, b Cleveland, Ohio, Nov 13, 22; div; c 1. BIOCHEMISTRY, SCIENCE EDUCATION. *Educ:* Case Western Reserve Univ, AB, 42, MS, 44, PhD(sci educ), 62. *Prof Exp:* Instr chem, anat & physiol, St John Col, 49-51; teacher high sch, Ohio, 53-62; assoc prof sci educ & org chem, Ferris State Col, 62-63; instr chem, 63-65, prof chem, 68-85, HEAD, DEPT SCI, CUYAHOGA COMMUNITY COL, WESTERN CAMPUS, 65-, EMER PROF CHEM, 90- *Concurrent Pos:* Vis prof biochem, Case Western Reserve Univ, 70-71; sci educ res grants, NSF, 76-78. *Mem:* Nat Educ Asn; Am Chem Soc; Audubon Soc; Sigma Xi; Sierra Club. *Mailing Add:* 6486 State Rd 12 Concord Sq C12 Parma OH 44134

LAUGHLIN, JAMES STANLEY, b Guilford, Mo, Sept 23, 36; m 60; c 2. STATISTICS, OPERATIONS RESEARCH. *Educ:* Northwest Mo State Univ, BS, 58; Univ Northern Colo, MA, 61; Univ Denver, PhD(higher educ, math), 68. *Prof Exp:* Teacher chem & math, Grand Community Schs, Boxholm, Iowa, 58-59; teacher math, Denver Pub Schs, Colo, 59-67, res asst, 67-68; assoc prof math, Kans State Teachers Col, 68-75, dir instnl studies, 70-75; dir instnl res, 75-83, ASST PROF MATH, IDAHO STATE UNIV, 84- *Concurrent Pos:* Kans State Dept Educ grant, 69-70. *Mem:* Math Asn Am; Am Educ Res Asn; Nat Coun Teachers Math. *Res:* Teaching of mathematics. *Mailing Add:* Dept Math Idaho State Univ Pocatello ID 83209

LAUGHLIN, JOHN SETH, b Canton, Mo, Jan 26, 18; m 79; c 3. MEDICAL PHYSICS. *Educ:* Willamette Univ, AB, 40; Haverford Col, MS, 42; Univ Ill, PhD(physics), 47. *Hon Degrees:* DSc, Willamette Univ, 68. *Prof Exp:* Asst, Haverford Col, 40-42; asst, Univ Ill, 42-43; res assoc, Off Sci Res & Develop, 44-45, asst, 46; asst prof spec res, 47-48, asst prof radiol, Col Med, 48-51, assoc prof, 51-52; assoc prof biophys, Sloan-Kettering Div, Med Col, Cornell Univ, 52-55; vpres, 66-72, CHIEF, DIV BIOPHYS, SLOAN-KETTERING INST CANCER RES, 52-; PROF BIOPHYS, SLOAN-KETTERING DIV, MED COLL, CORNELL UNIV, 55- *Concurrent Pos:* Attend physicist, Mem Hosp, 52-, chmn, Dept Med Physics, 52-88. *Honors & Awards:* William

D Coolidge Award, Am Asn Physicists in Med, 74; John Wiley Jones Lectr, Rochester Inst Technol, 77; Aebersold Award, Soc Nuclear Med, 84; Janeway Lectr, Am Radium Soc, 86; Gold Medal, Am Col Radiol, 88. *Mem:* Radiol Soc NAm; Radiation Res Soc (pres, 70-71); Soc Nuclear Med; Am Asn Physicists in Med (pres, 64-65); Health Physics Soc (pres, 60-61). *Res:* Neutron-proton and neutron-deutron interaction; high pressure cloud chamber design; interaction of high energy electrons with nuclei; application of betatron to medical therapy; radiation dosimetry; high energy gamma ray scanning; development of digital and computer controlled scanning; isotope metabolic studies; 3-dimensional photon treatment planning. *Mailing Add:* Div Biophys 444 E 68th St Mem Sloan-Kettering Cancer Ctr New York NY 10021

LAUGHLIN, R(OBERT) G(ARDINER) W(ILLIS), b London, Eng, Oct 11, 42; m 67; c 3. CHEMICAL ENGINEERING. *Educ:* Univ Col, Univ London, BSc, 64, PhD(chem eng), 67. *Prof Exp:* MGR, BIOTECHNOL & CHEM ENG CTR, ENVIRON TECHNOL DIV, ORTECH INT, 67- *Concurrent Pos:* Chair, Ont Minister Environ Hazardous Waste Adv Comt. *Mem:* Can Soc Chem Eng; Asn Prof Engrs Ont. *Res:* Environmental engineering; industrial waste exchange; spontaneous combustion phenomena; industrial and municipal solid waste utilization; industrial waste treatment; biomass to energy conversion systems; waste treatment systems; waste reduction and minimization. *Mailing Add:* Biotechnol & Chem Eng Ctr Ortech Int Sheridan Park Mississauga ON L5K 1B3 Can

LAUGHLIN, ROBERT GENE, b Sullivan, Ind, Aug 9, 30; div; c 2. PHYSICAL ORGANIC CHEMISTRY. *Educ:* Purdue Univ, BS, 51; Cornell Univ, PhD(org chem), 55. *Prof Exp:* Fel org chem, Hickrill Res Labs, NY, 55-56; res chemist, 56-68, SECT HEAD, MIAMI VALLEY LABS, PROCTER & GAMBLE CO, 68- *Mem:* Royal Soc Chem; Am Chem Soc. *Res:* Synthesis, phase science, and colloid science of surfactant molecules; the correlation of surfactant molecular structure with aqueous phase equilibria, and other physical properties; synthesis of aliphatic organophosphorus compounds; chemistry of positive-halogen compounds; synthesis of organophosphorus and organosulfur compounds. *Mailing Add:* 11641 Bank Rd Cincinnati OH 45251

LAUGHLIN, WILLIAM SCEVA, b Canton, Mo, Aug 26, 19; m 44; c 2. PHYSICAL ANTHROPOLOGY. *Educ:* Willamette Univ, BA, 41; Haverford Col, MA, 42; Harvard Univ, AM, 48, PhD(anthrop), 49. *Hon Degrees:* DSc, Willamette Univ, 68. *Prof Exp:* Asst anthrop, Harvard Univ, 47-48; asst prof, Univ Ore, 49-53, assoc prof, 53-55; from assoc prof to prof, Univ Wis, 55-69, chmn dept, 60-62; prof biobehav sci, 69-85, PROF ECOL & EVOLUTIONARY BIOL, UNIV CONN, 85- *Concurrent Pos:* Ed, J Am Asn Phys Anthrop, 58-63; mem anthrop study sect, Nat Res Coun, 59-62; fel, Ctr Advan Study Behav Sci, Stanford Univ, 64-65. *Mem:* AAAS; fel Am Soc Human Genetics; Soc Am Archaeol; Am Asn Phys Anthrop; Am Anthrop Asn. *Res:* Population genetics; blood group genetics and skeletal history of human isolates, Indians and Eskimo-Aleut stock; skeletal analysis of Eskimos and Indians; peopling of New World from Siberia on Bering land bridge coast. *Mailing Add:* Dept Ecol & Evolutionary Biol PO Box U-154 3107 Horsebarn Hill Rd Storrs CT 06269-4154

LAUGHLIN, WINSTON MEANS, b Fountain, Minn, May 2, 17; m 47; c 4. SOIL SCIENCE. *Educ:* Univ Minn, BS, 41; Mich State Univ, MS, 47, PhD(soil sci), 49. *Prof Exp:* Soil surveyor, Univ Minn, 40-41; asst, Mich State Univ, 41-42; SOIL SCIENTIST, ALASKA AGR EXP STA, 49- *Mem:* AAAS; Am Soc Agron; Soil Sci Soc Am; Am Sci Affil; Int Soc Soil Sci. *Res:* Soil fertility, chemistry and classification with emphasis on Arctic conditions. *Mailing Add:* Agr & Forestry Exp Sta 533 E Fireweed Palmer AK 99645

LAUGHNAN, JOHN RAPHAEL, b Spring Green, Wis, Sept 27, 19; m 42; c 3. GENETICS. *Educ:* Univ Wis, BS, 42; Univ Mo, PhD(genetics), 46. *Prof Exp:* Asst prof biol, Princeton Univ, 47-48; asst prof bot, Univ Ill, 48-51, assoc prof, 51-54; prof field crops, Univ Mo, 54-55; prof bot & chmn dept, 55-59, head dept bot, 63-65, PROF BOT & PLANT GENETICS, UNIV ILL, URBANA, 59- *Concurrent Pos:* Gosney fel, Calif Inst Technol, 48; Guggenheim fel, 60-61. *Mem:* Genetics Soc Am; Bot Soc Am; Am Soc Naturalists. *Res:* Functional genetics; fine structure and recombination in maize and Drosophila; intrachromosomal recombination mechanisms; development of sweet corn carrying sh2 gene. *Mailing Add:* Dept Plant Biol Univ Ill 505 S Goodwin Ave Urbana IL 61801

LAUGHNER, WILLIAM JAMES, JR, b Wickliffe, Ohio, Oct 22, 43; m 68; c 3. GENETICS, MOLECULAR BIOLOGY. *Educ:* Thiel Col, BA, 65; Ind Univ, Bloomington, PhD(genetics & bot), 70. *Prof Exp:* PROF BIOL, HIRAM COL, 70- *Concurrent Pos:* Assoc, Dept Med Genetics, Med Sch, Ind Univ, 74-75; vis assoc prof, bot dept, Univ Fla, 82-83. *Mem:* Genetics Soc Am. *Res:* Heterosis in plants based on enzyme unit interactions. *Mailing Add:* Dept Biol Hiram Col Hiram OH 44234

LAUGHON, ROBERT BUSH, b Greensboro, NC, Apr 20, 34; m 57; c 2. HAZARDOUS WASTE DISPOSAL, ENVIRONMENTAL MANAGEMENT. *Educ:* Colo Col, BA, 60; Univ Colo, Boulder, MS, 63; Univ Ariz, PhD(geol), 70. *Prof Exp:* Geologist, US Geol Surv, 62-63 & Anaconda Co, 64-65; instr, Univ Ariz, 66-67; geologist, Manned Spacecraft Ctr, NASA, 67-76; proj mgr, Nuclear Div, Union Carbide Corp, 76-78; mgr, Geol Explor Dept, 78-81, chief geoscientist, Proj Mgt Div, 81-90, GEOTECHNOL MGR, MICH LLRW PROJ, BATTELLE MEM INST, 90- *Concurrent Pos:* Lectr, Moody Col, Tex A&M Univ, 74. *Res:* Waste management; mineralogy crystallography. *Mailing Add:* 657 Indian Mound Rd Columbus OH 43213

LAUGHTER, ARLINE H, IMMUNOLOGY. *Prof Exp:* SR RES ASSOC, DEPT IMMUNOL, VET ADMIN MED CTR, 68- *Mailing Add:* Vet Admin Med Ctr 2002 Holcombe Blvd Bldg 211 Rm 226 Houston TX 77030

LAUGHTON, PAUL MACDONELL, b Toronto, Ont, Sept 8, 23; m 46; c 4. ORGANIC CHEMISTRY. *Educ:* Univ Toronto, BA, 45; Dalhousie Univ, MSc, 47; Univ Wis, PhD(chem), 50. *Prof Exp:* Res assoc, Univ Wis, 50; Nat Res Coun Can fel, Dalhousie Univ, 50-51; from asst prof to assoc prof chem, 51-65, PROF CHEM, CARLETON UNIV, 65- *Concurrent Pos:* Vis prof, Stanford Univ, 62 & Kings Col, London, 72-73; Am Chem Soc-Petrol Res Fund fel, Univ Calif, Berkeley, 62-63. *Mem:* Am Chem Soc; fel Chem Inst Can; Royal Soc Chem; Sigma Xi. *Res:* Mechanism studies; isotope effects. *Mailing Add:* 928 Muskoka Ave Ottawa ON K2A 3H9 Can

LAUKHUF, WALDEN LOUIS SHELBURNE, b Maysville, Ky, July 25, 43; m 67; c 2. CHEMICAL ENGINEERING. *Educ:* Univ Louisville, BSChE, 66, MSChE, 67, PhD(chem eng), 69. *Prof Exp:* Engr, Humble Oil & Refining Co, 69; US Air Force, 69-73, proj engr, Air Force Rocket Propulsion Lab, 69-71, develop eng specialist, Air Force Mat Lab, 71-73; from asst prof to assoc prof, 73-86, PROF CHEM ENG, UNIV LOUISVILLE, 86- *Mem:* Am Inst Chem Engrs; Am Chem Soc; Am Soc Eng Educ. *Res:* Adsorption; process controls; mass transfer and thermodynamics; digital control. *Mailing Add:* 14104 Tree Crest Ct Louisville KY 40245

LAUKONIS, JOSEPH VAINYS, b Mich, Apr 1, 25; m 54. PHYSICS. *Educ:* Univ Detroit, BS, 51; Univ Cincinnati, PhD, 57. *Prof Exp:* SR RES PHYSICIST, GEN MOTORS CORP, 57- *Mem:* Am Phys Soc. *Res:* Iron whiskers; metal surfaces; high temperature oxidation; formability of sheet metals; electron microscopy; ultrahigh vacuums; oxidation-reduction catalysts. *Mailing Add:* 32405 Northampton Warren MI 48093

LAUL, JAGDISH CHANDER, b India, Sept 1, 39; m 70; c 3. GEOCHEMISTRY, COSMOCHEMISTRY. *Educ:* Punjab Univ, India, BS, 59; Purdue Univ, MS, & PhD(radio geochem), 69. *Prof Exp:* Res assoc, Enrico Fermi Inst, Univ Chicago, 69-71 & Radiation Ctr, Ore State Univ, 71-75; SR RES SCIENTIST CHEM, PHYS SCI DEPT, PAC NORTHWEST DIV, BATTELLE MEM INST, 72- *Honors & Awards:* Group Achievement Award, NASA, 73. *Mem:* Am Chem Soc; Geochem Soc; Meteoritical Soc. *Res:* Studies of trace elements and their implications in lunar, meteorite, terrestrial, environmental, nuclear waste and fossile fuel samples; development of radioanalytical methods and instrumentation in neutron activation area. *Mailing Add:* 607 Cherrywood Ln Richland WA 00352

LAULAINEN, NELS STEPHEN, b Longview, Wash, Oct 22, 41; m 85; c 5. ATMOSPHERIC SCIENCE. *Educ:* Univ Wash, BS, 63, MS, 65, PhD(physics), 68. *Prof Exp:* Res scientist physics, First Phys Inst, Heidelberg Univ, 68-70; res assoc med physics, Med Radiation Physics & Radiol, Univ Wash, 70-71, res assoc geophys & astron, Geophys Prog, 71-74; sr res scientist, Atmospheric Sci, 74-82, mgr atmospheric & precipitation chem sect, Earth Sci, 82-87, PROG MGR, ATMOSPHERIC SCI DEPT, BATTELLE-PAC NORTHWEST LABS, 87- *Concurrent Pos:* Res scientist, Fulbright Travel Stipend, 68-70; lectr physics, Univ Wash, 72-74; tech adv, US Environ Protection Agency, Washington, DC, 81-82; vis scientist, Fraunhofer Inst Atmospheric Environ Res, Garmisch-Partenkirchen, Ger, 89-90. *Mem:* Am Phys Soc; Sigma Xi; Am Geophys Union. *Res:* Atmospheric aerosol physics; solar radiation and its interaction with atmospheric constituents; atmospheric pollutant transformation and removal processes; climate effects of energy production; acid deposition field studies and modeling. *Mailing Add:* Battelle-Northwest Labs Earth Sci Richland WA 99352

LAUNDRE, JOHN WILLIAM, b Green Bay, Wis, Jan 30, 49; m 73. WILDLIFE BEHAVIOR, ECOLOGY. *Educ:* Univ Wis, Green Bay, BSc, 71; Northern Mich Univ, MA, 74; Idaho State Univ, PhD(ecol), 79. *Prof Exp:* Instr anat, Northeast Wis Tech & Voc Inst, 74-76; ASST PROF ECOL, SOUTHWEST STATE UNIV, 79- *Mem:* Wildlife Soc; Animal Behav Soc; Am Soc Mammalogist; Sigma Xi. *Res:* Ecology and behavior of mammals, including coyotes, deer, cats and small mammals. *Mailing Add:* 2865 Janet Pocatello ID 83201

LAUNER, PHILIP JULES, b Philadelphia, Pa, Nov 20, 22; m 47; c 3. ANALYTICAL CHEMISTRY. *Educ:* Drew Univ, AB, 43; Columbia Univ, MA, 47. *Prof Exp:* Teaching asst quant analysis, Columbia Univ, 47-48; proj chemist petrol chem, Res Dept, Standard Oil Co (Ind), 48-55; specialist spectros, Silicone Prod Dept, Gen Elec Co, 55-72; pres, 73-87, DIR RES, LAB FOR MATS INC, 87- *Mem:* Am Chem Soc; Soc Appl Spectros; fel Am Inst Chemists. *Res:* Infrared spectroscopy; silicone technology; identification of industrial materials. *Mailing Add:* PO Box 2551 Glenville NY 12325

LAUPUS, WILLIAM E, b Seymour, Ind, May 25, 21; m 48; c 4. MEDICINE. *Educ:* Yale Univ, BS, 43, MD, 45. *Prof Exp:* Instr pediat, Med Col, Cornell Univ, 50-52; from asst prof to prof, Med Col Ga, 59-63; prof pediat & chmn dept, Med Col Va, 63-75; dean, 75-88, EMER DEAN, SCH MED, E CAROLINA UNIV, 88- *Concurrent Pos:* Examr, Am Bd Pediat, 67-91, bd officer, 73-78, pres, 76-77; vchancellor, E Carolina Univ, 83-88, prof prev med, 88-91; pres, Am Bd Med Specialties, 86-87. *Mem:* Fel Am Acad Pediat; Am Fedn Clin Res; AMA; Am Pediat Soc. *Res:* Pediatric medicine. *Mailing Add:* Sch Med E Carolina Univ Greenville NC 27834

LAURA, PATRICIO ADOLFO ANTONIO, b Buenos Aires, Arg, June 13, 35; m 59; c 5. MECHANICAL VIBRATIONS, HEAT CONDUCTION. *Educ:* Univ Buenos Aires, Arg, CE, 59; Cath Univ, PhD(appl mech & math), 65. *Prof Exp:* Prof appl math & mech, Cath Univ, 60-70; DIR & RES SCIENTIST, INST APPL MECH, 75- *Concurrent Pos:* Consult, David Taylor Model Basin, US Navy, 65-70; prof, Nat Univ of the South, Arg, 70-; assoc ed, Ocean Eng, 80- *Honors & Awards:* First Nat Eng Prize of Arg, 85. *Mem:* Arg Nat Acad Sci; fel Am Acad Mech; fel Acoust Soc Am. *Res:* Analytical research on mechanical vibrations and heat conduction problems; acoustic emission method in monitoring the status of cable systems; mobility aids for the disabled; devices for nuclear reactor plants. *Mailing Add:* Horneros 160 Bahia Blanca 8000 Argentina

LAURANCE, NEAL L, b Winsted, Minn, Aug 19, 32; m 53; c 4. COMPUTER SCIENCE. *Educ:* Marquette Univ, BS, 54, MS, 55; Univ Ill, PhD(physics), 60. *Prof Exp:* Res scientist, Sci Lab, 60-73, mgr, Control Systs Dept, 73-79, mgr analytical sci, Eng & Res Staff, 79-85, SR STAFF SCIENTIST, FORD MOTOR CO, 85- *Mem:* AAAS; Am Phys Soc; Asn Comput Mach; sr mem Inst Elec & Electronics Engrs. *Res:* simulation systems; manufacturing automation, OSI computer networks; real-time systems. *Mailing Add:* 876 Heather Way Ann Arbor MI 48104

LAURENCE, ALFRED EDWARD, b Breslau, Ger, Dec 12, 10; nat US; m 49; c 3. INDUSTRIAL CHEMISTRY. *Educ:* LLD, Breslau, Ger, 33; dipl, Poitiers, 34, Caen, 36. *Prof Exp:* Chemist, serv lab, Am Corn Prod Co, India, 37-38; chem sales technologist, Shell Oil Co, 38-40; chemist & metallurgist, Indian Smelting & Ref Co, 40-41; chemist, res & develop dept, Atlantic Ref Co, Pa, 41-46; chemist & technologist, Shell Chem Corp, Shell Oil Co, Calif & NY, 46-52; export mgr, org chem, Propane Co, Eng, 52-54; mgr indust & chem prods, Europ Res Off, Minn Mining & Mfg Co, Ltd, 54-61; dir res, Int Develop & Invest Co, Ltd, Bahamas, 62-63; CONSULT, 68- *Concurrent Pos:* Consult, UN & UNESCO, 64; econ planning adv, Lewis Berger, Gt Brit Ltd, 65-66; researcher, Economist Intel Univ, London, 67-68; vis prof, Univ Utah, 69, 70. *Mem:* Am Chem Soc; Royal Chem Soc; fel Am Sociol Asn. *Res:* Organic chemistry; petrochemistry; analytical methods; corporation finance and economics; international and patent law; sociology of industry. *Mailing Add:* La Mer Seven Sisters Rd St Lawrence PO38 1UZ Isle of Wight England

LAURENCE, GEOFFREY CAMERON, b Quincy, Mass, Mar, 24, 43; m; c 2. FISH BIOLOGY, ECOLOGICAL MODELLING. *Educ:* Univ Maine, BA, 65; Fla State Univ, MS, 67; Cornell Univ, PhD(fishery sci), 71. *Prof Exp:* Asst biol, Fla State Univ, 65-67; res asst fishery sci, Cornell Univ, 67-70; fishery biologist, Bur Sport Fisheries & Wildlife, 70-72; SUPVRY FISHERY BIOLOGIST, NAT MARINE FISHERIES SERV, NAT OCEANIC ATMOSPHERIC ADMIN, 72- *Honors & Awards:* Gold Medal, US Dept Com, 81. *Mem:* Am Fisheries Soc; Am Inst Fishery Res Biologists; Am Soc Limnol & Oceanog; Sigma Xi. *Res:* Larval fish ecology and physiology as they pertain to growth and survival; ecological and mathematical modelling of larval survival in relation to stock-recruitment problems in fishery management; biological oceanography. *Mailing Add:* 1164 Slocum Rd Saunderstown RI 02874-1601

LAURENCE, JOHN A, b Berea, Ohio, Dec 16, 45; m 71; c 2. ENVIRONMENTAL STRESS. *Educ:* Pa State Univ, BS, 71; Univ Minn, MS, 73, PhD(plant path), 76. *Prof Exp:* Asst scientist, 77-82, assoc scientist, 82-88, SCIENTIST, BOYCE THOMPSON INST PLANT RES, 88- *Concurrent Pos:* Adj prof, Dept Plant Path, Cornell Univ, 83- *Res:* Assessment of effects of air pollutants on plant growth and yield. *Mailing Add:* Boyce Thompson Inst Cornell Univ Ithaca NY 14853-1801

LAURENCE, KENNETH ALLEN, b Cleveland, Ohio, Nov 4, 28; m 49. MICROBIOLOGY. *Educ:* Marietta Col, AB, 51; Univ Iowa, MS, 53, PhD, 56. *Prof Exp:* NIH fel immunol, Univ Iowa, 56-57; instr microbiol & immunol, Med Col, Cornell Univ, 57-59, asst prof, 59-60; asst med dir, Pop Coun Inc, 60-68, assoc dir, Biomed Div, 68-76; prof & head, Dept Biol Sci, Univ Idaho, 76-80, dir grants & contracts, 80-83; prin contact officer & alternate trustee, Consortium Int Develop, Tucson, Ariz, 80-83, dep exec dir, 83-85; RES DEVELOP COORDR, UNIV IDAHO, 85- *Concurrent Pos:* Ford Found consult physiol reprod, Egyptian Univs Prog, 66-67 & 70-; proj specialist, Ford Found, Cairo, Egypt, 73-74; consult, Egyptian Univ, Rockefeller Found, 80; Supreme Coun Univ Foreign Relations Unit, Cairo, Egypt, 84. *Mem:* AAAS; Am Soc Microbiol; Soc Study Reprod; Sigma Xi; Soc Res Adminr; Int Agr Res & Develop. *Res:* Physiology of reproduction; immunology; parasitology; medical bacteriology; immunologic studies of the reproductive processes. *Mailing Add:* Univ Res Off Univ Idaho Moscow ID 83843-4199

LAURENCE, ROBERT L(IONEL), b West Warwick, RI, July 13, 36; m 59; c 3. CHEMICAL ENGINEERING, POLYMER SCIENCE. *Educ:* Mass Inst Technol, BS, 57; Univ RI, MS, 60; Northwestern Univ, PhD(chem eng), 66. *Hon Degrees:* DSc, Inst Nat Polytechnique Toulouse, Univ Toulouse, France, 89. *Prof Exp:* Res engr, Elec Boat Div, Gen Dynamics Corp, Conn, 57-58; engr, Eng Res Lab, E I du Pont de Nemours & Co, Del, 60-61, engr, Elastomers Dept, Tex, 61-63; asst prof chem eng, Johns Hopkins Univ, 65-68; assoc prof, 68-73, head dept, 82-90, PROF CHEM ENG, POLYMER SCI & ENG, UNIV MASS, AMHERST, 73- *Concurrent Pos:* Vis prof, Imp Col Sci & Technol, 74-75, Universidad Nac del Sur, Bahia Blanca, Arg, 78 & Col de France, 82, Gen Elec Corp Res & Develop, 89-90, Ecole Nat Superscuse des Ingénieues, Génie Chemique, 90. *Mem:* Fel Am Inst Chem Engrs; Am Soc Eng Educ; Am Chem Soc. *Res:* Fluid mechanics-hydrodynamic stability; polymerization reaction engineering; diffusion in polymers; polymer processing. *Mailing Add:* Dept Chem Eng Univ Mass Amherst MA 01002

LAURENCOT, HENRY JULES, b Brooklyn, NY, Dec 14, 29; m 61; c 4. DRUG METABOLISM, ANIMAL SCIENCE. *Educ:* St Peter's Col, NJ, BS, 51; Fordham Univ, MS, 55, PhD(biol), 65. *Prof Exp:* Asst plant physiologist, Boyce Thompson Inst Plant Res, Inc, 57-66; RES INVESTR, HOFFMAN-LA ROCHE, INC, 66- *Mem:* Am Soc Pharmacol & Exp Therapeut; NY Acad Sci; Sigma Xi; AAAS. *Res:* In vivo and in vitro metabolic studies of radioactive experimental drugs. *Mailing Add:* Hoffmann-La Roche Inc Kingsland St Nutley NJ 07110

LAURENDEAU, NORMAND MAURICE, b Lewiston, Maine, Aug 16, 44; m 72; c 2. MECHANICAL ENGINEERING, PHYSICAL CHEMISTRY. *Educ:* Univ Notre Dame, BS, 66; Princeton Univ, MSE, 68; Univ Calif, Berkeley, PhD(mech eng), 72. *Prof Exp:* From asst prof to assoc prof, 72-82, PROF MECH ENG, PURDUE UNIV, 82- *Concurrent Pos:* Res engr, Arthur D Little, Inc, 80-81; dir, Coal Res Ctr, Purdue Univ, 81-84; consult, Sandia Nat Labs, 84- *Mem:* Am Soc Mech Engrs; Am Chem Soc; Combustion Inst; Am Soc Eng Educ; Optical Soc Am. *Res:* Combustion; chemical kinetics; coal combustion and gasification; combustion diagnostics; air pollution; heterogeneous char reactions; laser-induced fluorescence. *Mailing Add:* Sch Mech Eng Purdue Univ West Lafayette IN 47907

LAURENSON, ROBERT MARK, b Pittsburgh, Pa, Oct 25, 38; m 61; c 2. MECHANICAL ENGINEERING. *Educ:* Mo Sch Mines, BS, 61; Univ Mich, Ann Arbor, MSE, 62; Ga Inst Technol, PhD(mech eng), 69. *Prof Exp:* Dynamics engr, McDonnell Aircraft Co, 62-64, sr dynamics engr, McDonnell Douglas Corp, 68-71, group dynamics engr, 71-74, staff engr, 74-75, TECH SPECIALIST, MCDONNELL DOUGLAS CORP, 75- *Concurrent Pos:* Lectr, Dept Eng Mech, St Louis Univ, 69-71. *Honors & Awards:* Award, McDonnell Douglas Astronaut Co, McDonnell Douglas Corp, 71. *Mem:* Am Inst Aeronaut & Astronaut; Am Soc Mech Engrs. *Res:* Vibrations of structures and machine elements; response of elastic and flexible structures to transient loadings. *Mailing Add:* 349 Beaver Lake Dr St Louis MO 63141

LAURENT, PIERRE, b Thionville, France, July 25, 25; m 50; c 3. ELECTRON MICROSCOPY. *Educ:* Acad Lille, BSc, 44; Univ Nancy, MSc, 48; Univ Paris, Sorbonne, DSc, 58. *Prof Exp:* Teacher natural sci, Col d'Armenti07res, 48-49; asst prof, Univ Lille, 49-50; res fel, CNRS, Paris, 50-56, sr researcher, 57-68, res dir, Col France, 69-73, RES DIR, CNRS, STRASBOURG, 74- *Concurrent Pos:* Vis prof, Michael Rease Hosp, Chicago, 68, CVP Div Hosp, Univ Pa, Philadelphia, 76-80. *Mem:* Am Physiol Soc; Soc Exp Physiol; Am Soc Zoologists. *Res:* Structural and physiological approach of adaptative processes in vertebrates, more particularly fishes; osmoregulation, circulation and respiration; gill morphology and physiology, acid-base regulation and ion transport; immunocytochimie of carbonic anhydrase, cortisol, prolactine. *Mailing Add:* 18 Rue de la Scierie Ittenheim F 67117 France

LAURENT, ROGER, b Geneva, Switz, June 23, 38; m 62; c 3. GEOLOGY. *Educ:* Univ Geneva, Lic es sci, 62, Ing Geol, 64, Dr es Sci(geol & mineral sci), 67. *Prof Exp:* Asst mineral, Univ Geneva, 63-65, res asst geochronol, Sci Res Nat Corp Switz, 65-67; asst prof mineral & petrog, Middlebury Col, 67-71; adj prof, 71-73, assoc prof, 73-79, PROF PETROL, LAVAL UNIV, 79- *Concurrent Pos:* Vis prof, Univ Nancy, France, 80. *Mem:* Fel Geol Asn Can; Swiss Soc Mineral & Petrol; Swiss Geol Soc; fel Geol Soc Am. *Res:* Study of ophiolites, asbestos and chromite in the field and in the labs by petrography; petrology; geochronometric determinations and geochemical studies; study of platinum in ultramafic rocks. *Mailing Add:* Dept Geol Laval Univ Quebec PQ G1K 7P4 Can

LAURENT, SEBASTIAN MARC, b Wallace, La, Jan 21, 26; m 49; c 7. CATALYSIS, ZEOLITE SCIENCE. *Educ:* Loyola Univ, La, BS, 49. *Prof Exp:* Res physicist catalysis, Esso Res Labs, Humble Oil & Refining Co, 57-67; res physicist, 67-77, sr res physicist mat sci, 77-80, sr res physicist, 80-86, res & develop specialist, 86-89, RES & DEVELOP ADV, ZEOLITE APPL RES, ETHYL CORP, 90- *Mem:* Catalysis Soc; Int Zeolite Asn. *Res:* Application of zeolites to improve nutritional responses in animal diets; bone development, egg shell development and physiological responses under heat stress environments; scanning electron microscopy; x-ray crystallography; x-ray fluorescence spectroscopy; adsorption; ion exchange; animal science research. *Mailing Add:* Ethyl Corp PO Box 14799 Baton Rouge LA 70898

LAURENZI, BERNARD JOHN, b Philadelphia, Pa, Dec 23, 38. CHEMICAL PHYSICS. *Educ:* St Joseph's Col, Pa, BS, 60; Univ Pa, PhD(chem), 65. *Prof Exp:* Res chemist, Rohm & Haas Chem Co, 60-61; NSF fel, Pa State Univ, 65-66; asst prof chem, Univ Tenn, Knoxville, 66-68 & Bryn Mawr Col, 68-69; from asst prof to assoc prof, 69-83, PROF CHEM, STATE UNIV NY, ALBANY, 84- *Mem:* Am Chem Soc; Am Phys Soc. *Res:* Quantum chemistry; use of Green's functions in atomic and molecular calculations; properties of isoelectronic molecules. *Mailing Add:* 115A Willow St Guilderland NY 12084

LAURENZI, GUSTAVE, b Orange, NJ, July 19, 26; c 2. MEDICINE. *Educ:* NY Univ, BA, 49; Georgetown Univ, MD, 53; Am Bd Internal Med, dipl; Am Bd Pulmonary Dis, dipl. *Prof Exp:* Intern path, Mallory Inst Path, Boston City Hosp, 53-54; intern med, Yale Med Serv, Grace-New Haven Hosp, Conn, 54-55; asst resident, Columbia Med Serv, Bellevue Hosp, NY, 55-56; chief resident physician chest serv, Bellevue Med Ctr, 58-59; asst prof med & dir div respiratory dis, NJ Col Med & Dent, 60-63, assoc prof med, 63-68; assoc prof, 70-75, PROF MED, SCH MED, TUFTS UNIV, 75- *Concurrent Pos:* Res fel, Cardiopulmonary Lab, Columbia-Presby Med Ctr, 56-57; USPHS res fel, 57-58; Nat Found training fel, Am Trudeau Soc, 58-59; Channing res fel bact & immunol, Mallory Inst Path, Harvard Univ & res fel, Am Thoracic Soc, 59-60; Am Thoracic Soc Edward L Trudeau fel, 62-64; consult, Harvard Med Serv, Boston City Hosp, 59-60; chief med, St Vincent Hosp, Worcester, Mass, 68-; dir respiratory care serv, Newton-Wellesley Hosp. *Mem:* Fel Am Col Physicians; Am Thoracic Soc; Am Fedn Clin Res. *Res:* Chest disease; chronic bronchitis and pulmonary emphysema. *Mailing Add:* Newton-Wellesley Hosp 2000 Washington St Newton Lower Falls MA 02162

LAURIE, GORDON WILLIAM, b Hamilton, Ont, Dec 28, 53; m 81; c 1. CELL BIOLOGY, EXTRACELLULAR MATRIX. *Educ:* McMaster Univ, BSc, 76; McGill Univ, MSc, 79, PhD(anat), 82. *Prof Exp:* From vis fel to vis assoc, Nat Inst Health, 83-88; ASST PROF, DEPT ANAT & CELL BIOL, UNIV VA, 88- *Honors & Awards:* Ralph D Lillie Award, Am Histochem, Soc, 82. *Mem:* Am Asn Anatomists; Am Soc Cell Biol; Soc Develop Biol. *Res:* Molecular assembly and sythesis of basement membrane and its relation to cell receptors. *Mailing Add:* Dept Anat & Cell Biol Univ Va Charlottesville VA 22908

LAURIE, JOHN SEWALL, b Gloucester, Mass, May 30, 25. EXPERIMENTAL BIOLOGY. *Educ:* Ore State Univ, BS, 50; Johns Hopkins Univ, ScD, 56. *Prof Exp:* Res fel parasite physiol, Inst Parasitol, McGill Univ, 56-57; instr zool, Tulane Univ, 57-59; asst prof exp biol, Univ Utah, 59-62; mem staff water pollution study, USPHS, 62-63; assoc prof, 63-80, PROF BIOL, E CAROLINA UNIV, 80- *Mem:* Am Soc Parasitol; Wildlife Dis Asn; Am Soc Zoologists; Sigma Xi. *Res:* Physiology of parasites; physiology, ultrastructure and ecology of helminth parasites. *Mailing Add:* Dept Biol E Carolina Univ Greenville NC 27834

LAURIE, VICTOR WILLIAM, b Columbia, SC, June 1, 35; m 65; c 2. PHYSICAL CHEMISTRY. *Educ:* Univ SC, BS, 54; Harvard Univ, AM, 56, PhD(chem), 58. *Prof Exp:* Fel, Nat Res Coun-Nat Bur Standards, 57-59; NSF fel, Univ Calif, 59-60; asst prof chem, Stanford Univ, 60-66; assoc prof chem, Princeton Univ, 66-71, prof, 71-80; CONSULT, 80- *Concurrent Pos:* Alfred P Sloan fel, 63-67; John S Guggenheim fel, 70. *Mem:* Sigma Xi; AAAS; Am Chem Soc; Am Phys Soc. *Res:* Molecular spectroscopy and structure. *Mailing Add:* 109 Kingsway Commons Princeton NJ 08540

LAURIENTE, MIKE, b Trail, BC, June 26, 22; US citizen; m 56; c 1. METALLURGY, PHYSICS. *Educ:* Mich Technol Univ, BS, 43, MS, 47; Johns Hopkins Univ, DrEng, 55. *Prof Exp:* Metallurgist, Int Harvester Co, Ill, 47-49; res staff asst metal physics, Johns Hopkins Univ, 49-55; fel engr, Aerospace Div, Westinghouse Elec Corp, Md, 56-62, adv engr, 62-71; asst secy transp, Off Systs Eng, US Dept Transp, Washington, DC, 71-83; TECH MGR, GODDARD SPACE FLIGHT CTR, NASA, GREENBELT, MD, 83- *Concurrent Pos:* Consult, Ballistics Res Lab, Ord Dept, US Army, Aberdeen Proving Ground, 51-55. *Mem:* Am Phys Soc; fel Am Soc Metals; Sigma Xi. *Res:* Magnetic thin films; active and passive electronic devices; radiation damage; electrical insulation; magnetic anisotropic metals; technological advances having application to transportation safety and security; fracture mechanics; nondestructive testing; space shuttle environment. *Mailing Add:* 6608 White Gate Rd Clarksville Ridge Clarksville MD 21029

LAURIN, ANDRE FREDERIC, b Ste Anne de Bellevue, Que, Jan 18, 29; m; c 3. PETROLOGY, STRUCTURAL GEOLOGY. *Educ:* Univ Montreal, BSc, 51; McGill Univ, MSc, 54; Laval Univ, DSc(petrol), 57. *Prof Exp:* Geologist, Que Dept Natural Resources, 56-65, regional geologist, Grenville, 65-69, dir mineral deposits serv, 69-71, dir geol serv, 71-78, gen dir geol & mineral res, 78-80, asst dep minister, Que Dept Natural Resources, 80-88; PRES & CEO, SOQUEM, 88- *Concurrent Pos:* Lectr tour numerous univs, France & Ger, 68; chmn, Geol Div, Can Inst Mining & Metall, 75, Tech Comt, 78th Ann Meeting, 76, gen chmn, 84th Ann Meeting, 82 & vpres, dist No 2, 82-84; vchmn & chmn, Can Geosci Coun & Keynote speaker at the Coun Inst Mining & Metall in Singapore, 86. *Honors & Awards:* A O Dufresne Award, Can Inst Mining & Metall, 85; Distinguished Serv Award, Prospectors & Dev Asn, 87. *Mem:* Fel Geol Soc Am; Geol Asn Can; Can Inst Mining & Metall; Prospectors & Develop Asn; Asn des Pros du Québec. *Res:* Mapping and supervision of the mapping of the Grenville geological province, Quebec on a regional scale. *Mailing Add:* 2011 Chapdelaine St Ste Foy PQ G1V 1M4 Can

LAURIN, PUSHPAMALA, b Bangalore City, India; US citizen; m 64; c 2. ELECTROMAGNETISM. *Educ:* Gujarat Univ, India, BSc, 56; Karnatak Univ, India, MSc, 58; Univ Mich, Ann Arbor, MSE, 62, PhD(physics), 67. *Prof Exp:* Jr sci asst, Nat Sugar Inst, Kanpur, India, 58-59; sales engr, Toshniwal Bros, Bombay, 59-60; res asst meteorol, Univ Mich, Ann Arbor, 61-62, asst res physicist, Radiation Lab, 62-67; asst prof math, Eastern Mich Univ, 67-68; res scientist physics, McDonnell-Douglas Corp, Mo, 68-69; lectr elec eng & physics, Southern Ill Univ, 70-71; instr electronics & physics, Harper Col, 71-76; DIR INFO RESOURCES, GOULD INC, ROLLING MEADOWS, ILL, 76- *Concurrent Pos:* Vpres, Greame Publ Co, Wilborheim, Mass, 84-86 & AT&T Bell Labs, 86-89; sr resource mgr, Cellular Intrastruct Group, Motorola Inc, 89- *Mem:* Inst Elec & Electronics Engrs. *Res:* Electromagnetic interactions. *Mailing Add:* Three Lydia Cts Hawthorn Wood IL 60047

LAURITZEN, PETER O, b Valparaiso, Ind, Feb 14, 35; m 63; c 2. ELECTRICAL ENGINEERING. *Educ:* Calif Inst Technol, BS, 56; Stanford Univ, MS, 58, PhD(elec eng), 61. *Prof Exp:* Mem tech staff, Fairchild Semiconductor Div, 61-65; from asst prof to assoc prof, 65-73, PROF ELEC ENG, UNIV WASH, 73- *Concurrent Pos:* Adj prof social mgt technol, Univ Wash, 77-; eng mgr, Avtech Corp, 79-80. *Mem:* Inst Elec & Electronics Engrs; Am Soc Eng Educ; AAAS. *Res:* Semiconductor devices and instrumentation; power electronics. *Mailing Add:* Dept Elec Eng Ft-10 Univ Wash Seattle WA 98195

LAURMANN, JOHN ALFRED, b Cambridge, Eng, Aug 8, 26; nat US; m 57; c 2. GLOBAL CLIMATE CHANGE. *Educ:* Cambridge Univ, BA, 47, MA, 51; Cranfield Univ, MSc, 51; Univ Calif, PhD(eng sci), 58. *Prof Exp:* Asst res engr, Univ Calif, 53-58; head aerodyn res, Space Technol Labs, Inc, 58-59; aeronaut res scientist, NASA, 59-60; staff scientist, Lockheed Missile & Space Co, 60-63; tech staff mem, Gen Res Corp, 63-69; staff mem, Inst Defense Anal, 69-70; sr staff mem, Nat Acad Sci, Washington, DC, 70-74; consult scientist, 75-76; sr res assoc, Dept Mech Eng, Stanford Univ, 76-82; exec scientist, Gas Res Inst, 82-88; RES SCIENTIST, MARINE SCI INST, UNIV CALIF, SANTA BARBARA, 88- *Concurrent Pos:* Lectr, Univ Calif, 55-58, San Jose State Col, 60 & Calif Exten Div, 60-63; consult, Int Inst Appl Systs Anal, Nat Ctr Atmospheric Res, Rand Corp, Am Petrol Inst, Elec Power Res Inst, Gas Res Inst, Int Energy Agency, UN, US Environ Protection Agency, US Dept Energy, Nat Oceanic & Atmospheric Admin & Hawaii Natural Energy Res Inst; consult prof, Dept Civil Eng, Stanford Univ, 89- *Mem:* Am Geophys Union; Sigma Xi. *Res:* Greenhouse gas issues: energy, climate and policy; climatology. *Mailing Add:* 3372 Martin Rd Carmel CA 93923

LAURS, ROBERT MICHAEL, b Oregon City, Ore, Jan 27, 39; m 82; c 2. OCEANOGRAPHY, FISHERIES. *Educ:* Ore State Univ, BS, 61, MS, 63, PhD(oceanog), 67. *Prof Exp:* OCEANOGR, SOUTHWEST FISHERIES CTR, LA JOLLA LAB, NAT MARINE FISHERIES SERV, 67- *Concurrent Pos:* Sci adv, Am Fisherman's Res Found, 71-; sci subgroup, Intergovernment Oceanog Comn, Intergrated Global Ocean Sta Syst, 81- *Honors & Awards:* Silver Medal, Dept Com, 80. *Mem:* AAAS; Marine Biol Asn UK; Am Soc Limnol & Oceanog; Am Inst Fishery Res Biologists; Eastern Pac Oceanic Conf; Am Fisheries Soc. *Res:* Fishery forecasting; environmental conditions affecting the distribution and abundance of tunas; albacore tuna ecology; vertical distribution and migration of micronektonic organisms; satellite oceanography. *Mailing Add:* La Jolla Lab PO Box 271 Nat Marine Fisheries Serv La Jolla CA 92037

LAURSEN, E(MMETT) M(ORTON), b Fairmount, NDak, Jan 24, 19; m 51; c 3. HYDRAULIC ENGINEERING. *Educ:* Univ Minn, BCE, 41; Univ Iowa, PhD(mech, hydraul), 58. *Prof Exp:* Asst, St Anthony Falls Hydraul Lab, Minn, 41-42, asst scientist, 45; jr engr, Al Johnson Construct Co, 42-43; asst, Inst Hydraul Res, Univ Iowa, 45-47, res assoc, 47-48, res engr, 48-58; assoc prof civil eng, Mich State Univ, 58-62; head dept, 62-68, PROF CIVIL ENG, UNIV ARIZ, 62- *Concurrent Pos:* Consult, 85- *Honors & Awards:* Hilgard Prize, Am Soc Civil Engrs, 59, Res Prize, 61. *Mem:* Am Soc Civil Engrs; Am Geophys Union; Int Asn Hydraul Res. *Res:* Sediment transportation; fluid mechanics and its applications. *Mailing Add:* Dept Civil Eng Univ Ariz 926 W Comobabi Dr Tucson AZ 85704

LAURSEN, GARY A, b Seattle, Wash, Aug 13, 42; m 63; c 2. MYCOLOGY, ARCTIC ECOLOGY. *Educ:* Western Wash Univ, BA, 65; Univ Mont, MST, 70; Va Polytech Inst & State Univ, PhD(bot/mycol), 75. *Prof Exp:* Biol instr, Toppenish, Wash, Sch Dist, 65-71; res asst mycol, Va Polytech Inst & State Univ, 71-75, res assoc, 75-76; asst dir sci, Naval Arctic Res Lab, 76-80, PRIN INVESTR ANIMAL RES FACIL, UNIV ALASKA, 76-; proj officer cold prog, Off Naval Res, Arlington, 80-82; ADJ ASSOC PROF MYCOL, UNIV ALASKA, FAIRBANKS, 83- *Concurrent Pos:* Actg tech dir admin & sci, Naval Arctic Res Lab, Univ Alaska, 76-77 & Remote Sensing Prog, 77-78; prof officer cold prog, Off Naval Res, Arlington, 80-82. *Honors & Awards:* Sigma Xi Res Award, Va Polytech Inst & State Univ, 77. *Mem:* Mycol Soc Am; Sigma Xi. *Res:* Systematic and ecological treatments of Arctic, Alpine and Meratime tundra fleshy fungi, their role and significance within these environmentally harsh ecosystems. *Mailing Add:* 4830 Villanova Fairbanks AK 99709

LAURSEN, PAUL HERBERT, b Ord, Nebr, Mar 28, 29; m 59; c 2. GENERAL CHEMISTRY. *Educ:* Dana Col, BA, 54; Ore State Univ, PhD(org chem), 61. *Prof Exp:* From asst prof to assoc prof chem, 59-64, academic dean, 76-78, PROF CHEM, NEBR WESLEYAN UNIV, 64-, PROVOST, 78- *Concurrent Pos:* NSF sci fac fel, Univ Calif, Los Angeles, 67-68. *Mem:* AAAS; Am Chem Soc. *Res:* Synthesis of nitrogen heterocycles; identification of natural products. *Mailing Add:* 3148 N 75th Ct Lincoln NE 68507

LAURSEN, RICHARD ALLAN, b Normal, Ill, May 1, 38; m 71; c 2. BIO-ORGANIC CHEMISTRY. *Educ:* Univ Calif, Berkeley, BS, 61; Univ Ill, PhD(chem), 64. *Prof Exp:* NIH fel, Harvard Univ, 64-66; from asst prof to assoc prof chem, 66-76, PROF CHEM, BOSTON UNIV, 76- *Concurrent Pos:* NIH res career develop award, 69-74; guest scientist, Max Planck Inst Molecular Genetics, 71; Alfred P Sloan fel, 72-74; mem sci adv comt on clin invest, Am Cancer Soc, 75-79, NSF Biol Instrumentation Prog, 84-88; specialist protein chem, int adv panel, Chinese Prov Univ Develop, World Bank, Changsha, People's Repub China, 90. *Honors & Awards:* Pehr Edman Award, 88. *Mem:* Fel AAAS; Am Chem Soc; Am Soc Biochem & Molecular Biol. *Res:* Development of new methods for protein sequence analysis and peptide synthesis; structure, function and evolution of proteins; studies on brain myelin proteolipid and antifreeze proteins. *Mailing Add:* Dept Chem Boston Univ 590 Commonwealth Ave Boston MA 02215

LAUSCH, ROBERT NAGLE, b Chambersburg, Pa, Feb 22, 38; m 68; c 2. IMMUNOBIOLOGY. *Educ:* Muhlenberg Col, BS, 60; Pa State Univ, MS, 62; Univ Fla, PhD(microbiol), 66. *Prof Exp:* Fel virol, Baylor Col Med, 66-69; asst prof, 69-75, assoc prof microbiol, Col Med, Pa State Univ, 75-77; assoc prof, 77-85, PROF MICROBIOL & IMMUNOL, COL MED, UNIV SALA, 85- *Mem:* Am Soc Microbiol; Am Asn Cancer Res; Am Asn Immunologists; Fedn Am Socs Exp Biol. *Res:* Vitus immunology; study of the host immune response to membrane antigens found on the surface of virus infected cells and how such cells may escape immune destruction. *Mailing Add:* Dept Microbiol & Immunol Col Med Univ SAla Mobile AL 36688

LAUSH, GEORGE, b Barrackville, WVa, Sept 17, 21; m 56. MATHEMATICS. *Educ:* Univ Pittsburgh, BS, 43; Cornell Univ, PhD, 49. *Prof Exp:* Asst chem, Univ Pittsburgh, 43-44; res assoc, Manhattan Dist, Univ Rochester, 44-46; asst math, Cornell Univ, 46-49; from asst prof to assoc prof, 49-62, PROF MATH, UNIV PITTSBURGH, 62- *Mem:* Am Math Soc; Math Asn Am. *Res:* Infinite series; real functions; functional analysis. *Mailing Add:* 181 Pearce Mill Rd Pittsburgh PA 15090

LAUSHEY, LOUIS M(CNEAL), b Columbia, Pa, May 13, 17; m 49. CIVIL ENGINEERING. *Educ:* Pa State Univ, BS, 42; Carnegie Inst Technol, MS, 47, DSc, 51. *Prof Exp:* Struct draftsman, Am Bridge Co, NJ, 42; instr civil eng, Carnegie Inst Technol, 42-44, 46-48, from asst prof to assoc prof, 48-54; prof & head dept, Norwich Univ, 54-58; William Thoms Prof Civil Eng & Head Dept, 58-78, Geier Prof Eng Educ & Interim Dean, Col Eng, 83-86, EMER PROF & DEAN, UNIV CINCINNATI, 87- *Concurrent Pos:* Former partner, D'Appolonia, Laushey & Peck, Consult Engrs; consult to govt & industs. *Mem:* Am Soc Civil Engrs; Soc Am Mil Engrs; Am Soc Eng Educ; Am Geophys Union; Int Asn Hydraul Res. *Res:* Fluid mechanics; structural design and analysis. *Mailing Add:* Dept Civil & Environ Eng Univ Cincinnati Cincinnati OH 45221

LAUSHMAN, ROGER H, b Ida Grove, Iowa, May 3, 50; m 83; c 2. POPULATION GENETICS, CONSERVATION BIOLOGY. *Educ:* Univ Kans, BS, 79; Iowa State Univ, MS, 83; Univ Ga, PhD(bot), 88. *Prof Exp:* Postdoctoral fel, Friday Harbor Labs, Univ Wash, 88-89; VIS ASST PROF EVOLUTION & ECOL, OBERLIN COL, 89- *Mem:* Bot Soc Am; Ecol Soc Am; Soc Conserv Biol; Soc Study Evol; Am Soc Naturalists. *Res:* Population genetics of aquatic vascular plants, particularly those with water-pollination; conservation biology, particularly of threatened plant species and aquatic habitats. *Mailing Add:* Biol Dept Oberlin Col Oberlin OH 44074

LAUSON, HENRY DUMKE, b New Holstein, Wis, Aug 20, 12; m 77. PHYSIOLOGY. *Educ:* Univ Wis, BS, 36, PhD(physiol), 39, MD, 40. *Prof Exp:* Asst med, Univ Wis, 36-39; intern, Univ Kans Hosps, 40-41; asst resident med, Henry Ford Hosp, 41-42; Off Sci Res & Develop fel physiol & med, Col Med, NY Univ, 42-43, instr physiol, 43-46; assoc & assoc physician, Rockefeller Inst, 46-50; assoc prof physiol in pediat, Med Col, Cornell Univ, 50-55; chmn dept, 55-78, prof physiol, 55-78, EMER PROF PHYSIOL & BIOPHYSICS, ALBERT EINSTEIN COL MED, 78- *Concurrent Pos:* Assoc prof physiol, Med Col, Cornell Univ, 51-55; consult prog-proj comt, Nat Inst Arthritis & Metab Dis, 61-65. *Mem:* Am Physiol Soc; Am Soc Exp Biol & Med; Am Soc Clin Invest; Harvey Soc (secy, 52-55); Am Fed Clin Res; Am Soc Nephrol. *Res:* Pituitary-ovary interrelations; blood pressure in human right heart; renal physiology; nephrotic syndrome; metabolism of antidiuretic hormone. *Mailing Add:* 215 Willow Bend Kiel WI 53042

LAUTENBERGER, WILLIAM J, b Flushing, NY, Mar 11, 43; m 67; c 2. APPLIED STATISTICS, PHYSICAL CHEMISTRY. *Educ:* Muhlenberg Col, BS, 64; Univ Pa, PhD(phys chem), 67. *Prof Exp:* Res chemist, Dye Div, Org Chem & Res & Develop Dept, Jackson Labs, NJ, 67-71, res chemist, Org Chem Dept, Exp Sta, Wilmington, 71-74, statist prog consult, 74-78, develop specialist, indust hyg, Du Pont Fabrics & Finishes Dept, Appl Technol Div, 78-81, RES SUPVR, ELECTRONICS DEPT, E I DU PONT DE NEMOURS & CO, WILMINGTON, 81- *Mem:* Am Chem Soc; Sigma Xi. *Res:* Reaction mechanisms; photochemistry; heterogenous catalysis; mechanisms of dyeing; emulsion science; solid-solid adsorption; consulting in design and analysis of experiments; air sampling; gas diffusion mechanisms; gas adsorption, desorption phenomena, electronic materials, semiconductors and packaging materials PWB laminates, photo polymers. *Mailing Add:* 506 Ott Rd Bala Cynwyd PA 19004-2510

LAUTENS, MARK, b Hamilton, Ont, July 9, 59. ASYMMETRIC SYNTHESIS, ORGANOMETALLIC CATALYSIS. *Educ:* Univ Guelph, BSc, 81; Univ Wis-Madison, PhD(chem), 85. *Prof Exp:* Postdoctoral, Harvard Univ, 85-87; ASST PROF CHEM, UNIV TORONTO, 87- *Concurrent Pos:* Alfred P Sloan Found fel, 91. *Mem:* Am Chem Soc; Can Soc Chem. *Res:* Synthesis of biologically active or structurally novel compounds; metal catalyzed reactions for control of stereochemistry; novel polymers using metal catalysts. *Mailing Add:* Dept Chem Univ Toronto Toronto ON M5S 1A1

LAUTENSCHLAEGER, FRIEDRICH KARL, b Gefell, Ger, June 27, 34; m 60; c 2. ORGANIC CHEMISTRY. *Educ:* Univ Heidelberg, BA, 56; Univ Toronto, MA, 60. *Prof Exp:* Res asst org chem, Univ Toronto, 57-59; res chemist, NAm Res Centre, Dunlop Co, Ltd, 61-72, GROUP LEADER, DUNLOP RES CENTRE, 72- *Mem:* Chem Inst Can; Am Chem Soc. *Res:* Stereochemistry of organic compounds; synthesis of small ring compounds; organic sulfur chemistry; reactive intermediates; vulcanization chemistry and physics. *Mailing Add:* 2562 Cushing Rd Mississauga ON L5K 1X1 Can

LAUTENSCHLAGER, EUGENE PAUL, b Chicago, Ill, Apr 5, 37; m 61; c 1. BIOMATERIALS. *Educ:* Ill Inst Technol, BS, 58; Northwestern Univ, MS, 60, PhD(mat sci), 66. *Prof Exp:* Res metallurgist, Allis-Chalmers Mfg Co, 60-62; from asst prof to assoc prof, 66-74, PROF BIOL MAT, NORTHWESTERN UNIV, 74-, DIV DIR, 88- *Concurrent Pos:* NIH career develop award, 71-75; consult, Bioeng Comt, Am Acad Orthop Surgeons, 73-85; vis prof, Free Univ Berlin, 84. *Honors & Awards:* F4 Award of Merit, Am Soc Testing & Mat. *Mem:* Am Soc Testing & Mat; Am Inst Mining, Metall & Petrol Eng; Int Asn Dent Res; Am Soc Metals; Acad Dent Mat. *Res:* Biological and dental materials; medical implant materials; kinetics of cementing media for implant stabilization; computer assisted instruction. *Mailing Add:* Dept Biol Mat Northwestern Univ 311 E Chicago Chicago IL 60611

LAUTERBACH, GEORGE ERVIN, b Bushnell, Ill, June 13, 27; m 49; c 5. PHYSICAL CHEMISTRY. *Educ:* Monmouth Col, BS, 49; Bradley Univ, MS, 53; Purdue Univ, PhD(biochem), 58. *Prof Exp:* Chemist, Starch & Dextrose Div, Northern Regional Res Lab, 49-53; res org chemist, Paper Lab, Kimberly Clark Corp, 57-65 & Pioneering & Advan Develop Lab, 65; assoc prof chem & res assoc, Div Natural Mat & Systs, Inst Paper Chem, 65-78; MGR NEW PRODS RES & DEVELOP, MARINE COLLOIDS INC, 78- *Mem:* Am Chem Soc; Tech Asn Pulp & Paper Indust; Am Asn Cereal Chemists. *Res:* Polysaccharide chemistry; enzymic and chemical modification of starch; high temperature starch cooking; hemicelluloses; starch in paper coatings; top sizes and internal sizing of paper products. *Mailing Add:* Marine Colloids Inc PO Box 213 Thomaston ME 04861

LAUTERBACH, JOHN HARVEY, b Jersey City, NJ, Apr 2, 44; m 87. CARBOHYDRATE CHEMISTRY, TOBACCO CHEMISTRY. *Educ:* Worcester Polytech Inst, BS, 66; Ohio State Univ, MSc, 68, PhD(chem), 70. *Prof Exp:* Asst anal chem, Ohio State Univ, 66-67; chief asst, 67-68 & org chem, 68-69, res assoc, 70-; chemist anal chem, Union Carbide Chem & Plastics Co, 70-71; proj supvr, Cent Tech Eval, Nat Starch & Chem Corp, 71-73, mgr, 74-78; mgr chem, Pillsbury Co, 78-79; dir chem, Prof Serv Indust, Inc, 79-80; anal res div head, 80-87, mgr mat res, 87-89, MGR RES SERV, BROWN & WILLIAMSON TOBACCO CORP, 89- *Concurrent Pos:* Chmn, Environ Comn, Borough Raritan, NJ, 76-78. *Mem:* Am Chem Soc; fel Am Inst Chemists; Royal Soc Chem; Sigma Xi; Am Soc Testing & Mat; Am Mgt Asn. *Res:* Polysaccharide chemistry, particularly structure determinations and modifications; high performance liquid chromatography and nuclear magnetic resonance spectroscopy; new techniques for managing service departments in research and development organizations; tobacco chemistry; gas chromatography, liquid chromatography and mass spectrometry. *Mailing Add:* 1400 Willow Ave Louisville KY 40204-1438

LAUTERBACH, RICHARD THOMAS, b Rochester, Pa, Dec 29, 46; m 69; c 1. POLYMER CHEMISTRY, ORGANIC CHEMISTRY. *Educ:* Johns Hopkins Univ, BS, 68; Northwestern Univ, PhD(chem), 75. *Prof Exp:* Group leader water soluble polymers, Daubert Chem Co, 75-80, group leader protective coatings, 80-81; GROUP LEADER, OIL FIELD SERVS, RICHARDSON CO, 81- *Mem:* Am Chem Soc; Soc Cosmetic Chem. *Res:* Water soluble polymers for wastewater treatment and paper production additives; corrosion preventive coatings. *Mailing Add:* 632 S Tenth Ave La Grange IL 60525

LAUTERBUR, PAUL CHRISTIAN, b Sidney, Ohio, May 6, 29; m 58, 84; c 3. PHYSICAL CHEMISTRY, MEDICAL IMAGING. *Educ:* Case Inst Technol, BS, 51; Univ Pittsburgh, PhD(chem), 62. *Hon Degrees:* Dr, Univ Liege, Belgium, 84, Nicolaus Copernicus Med Acad, Poland, 88; DSc, Carnegie Mellon Univ, 87, Wesleyan Univ, 89, State Univ NY, 90. *Prof Exp:* Res asst, Mellon Inst, 51-52, res assoc, 52-53, jr fel, 53, fel, 55-63; from assoc prof to prof chem, State Univ NY Stony Brook, 63-83, res prof radiol, 78-85, leading prof chem, 83-84, univ prof, 84-85; prof, 85-90, DISTINGUISHED UNIV PROF, COL MED, UNIV ILL, CHICAGO, 90-; PROF, DEPT MED INFO SCI & DEPT CHEM, COL MED, UNIV ILL, URBANA-CHAMPAIGN, 85-, DIR, BIOMED MAGNETIC RESONANCE LAB, 85- *Concurrent Pos:* Mem, Sci Coun Galileo Galilei Found, Coun Int Soc Magnetic Resonance; chmn subcomt E-13.7 on nuclear magnetic resonance, Soc Testing & Mat, 60-62; Alfred P Sloan fel, 65-67; vis scholar, dept chem, Stanford Univ, 69-70; ed-in-chief, Magnetic Resonance Med, 82-83; adj prof, State Univ NY Stony Brook, 85-; prof, ctr advan study, Univ Ill, Urbana-Champaign, 87-, prof bioeng, Dept Elect & Comput Eng & prof biophys, Dept Physiol & Biophys, 88- *Honors & Awards:* Gold Medal, Soc Magnetic Resonance Med, 82; Biol Physics Prize, Am Phys Soc, 83; Jesse Beams Lectr, Univ Va, 83; Smith Kline & French Lectr, Univ Col London, 83; Howard N Potts Medal, Franklin Inst, 84; Albert Lasker Clin Res Award, 84; H H Iddles Lectr chem, Univ NH, 84; Kosar Mem Award, Soc Photog Scientists & Engrs, 85; Charles F Kettering Prize, Gen Motors Cancer Res Found, 85; Gairdner Found Int Award, 85; Distinguished Res Biomed Sci Award, Asn Am Med Col, 86; Roentgen Medal, 87; Medal Honor, Inst Elec & Electronics Engrs, 87; Nat Medal Sci, 87; Nat Medal Technol, 88. *Mem:* Nat Acad Sci; fel Am Phys Soc; Am Chem Soc; fel AAAS; Soc Magnetic Resonance Med (pres, 81-83); Sigma Xi; Radiol Soc NAm; assoc mem Inst Elec & Electronics Engrs; Biophys Soc; Soc Neurosci. *Res:* Nuclear magnetic resonance studies of structure and properties of molecules, crystals and biological systems; imaging by magnetic resonance zeugmatography, including biological and medical applications; Published numerous articles in various scientific journals. *Mailing Add:* Univ Ill 1307 W Park Urbana IL 61807

LAUTT, WILFRED WAYNE, b Lethbridge, Alta, Can, June 29, 46; m 68; c 2. LIVER FUNCTIONS, HEPATIC CIRCULATION. *Educ:* Univ Alta, BSc, 68; Univ Man, MSc, 70, PhD(pharmacol), 72. *Prof Exp:* Med Res Coun Can fel toxicol, Univ Montreal, 72-74; asst prof & res scholar, Can Liver Found, Univ Sask, 74-78, from assoc prof to prof physiol, 78-82; PROF PHARMACOL & THERAPEUT & SECT HEAD HEPATORENAL RES UNIT, UNIV MAN, 84- *Concurrent Pos:* J A F Stevenson vis prof, Can Physiol Soc, 80; mem coun, Can Physiol Soc, 82-85; sci officer grants comt, Can Heart Found, 82-88; pres, Can Asn Study of Liver, 87-88; mem med adv bd, Can Liver Found, 82- *Mem:* Microcirculatory Soc; Can Physiol Soc; Am Physiol Soc; Can Pharmacol Soc; Am Asn Study Liver Dis; Can Asn Study Liver Dis (vpres, 85, pres, 88); Am Soc Pharmacol & Exp Therapeut; Can Soc Clin Investr. *Res:* Peripheral vascular physiology and hepatic physiology, pharmacology and toxicology; vascular and metabolic consequences of autonomic nerve activity in the liver; local control of intestinal and hepatic blood flow. *Mailing Add:* Dept Pharmacol & Therapeut Univ Man Fac Med 770 Bannatyne Ave Winnipeg MB R3E 0W3 Can

LAUTZENHEISER, CLARENCE ERIC, b Lincoln, Nebr, May 21, 21; m 48; c 3. METALLURGY. *Educ:* Mass Inst Technol, BS, 52. *Prof Exp:* Res engr, Dow Chem Co, 52-53, maintenance engr, 52-60, maintenance specialist, 60-62; sr res engr, 62-67, mgr metall eng, 67-69, from asst dir to dir spec eng serv, 69-74, vpres, Qual Assurance Systs & Eng Div, 74-85, VPRES EMER, NDE SCI & TECHNOL DIV, SOUTHWEST RES INST, 85-; PRES, INT ENGRS, 86- *Concurrent Pos:* Consult reliability & qual assurance div, NASA, 65-67. *Mem:* Am Soc Metals; fel Am Soc Nondestructive Testing (pres, 75-76); Am Welding Soc; fel Am Soc Mech Engrs; Nat Asn Corrosion Engrs. *Res:* New methods of magnesium production; corrosion; failure analysis; welding; nondestructive inspection in petrochemical industry fabrication quality control and in-service inspection of nuclear reactor power systems. *Mailing Add:* Int Engrs PO Box 800 Medina TX 78055

LAUVER, DEAN C, b Warren, Pa, June 8, 20; m 46; c 3. AERONAUTICAL & ASTRONAUTICAL ENGINEERING. *Educ:* Pa State Univ, BS, 42. *Prof Exp:* Jr engr, Mesta Mach Co, Pa, 42; stress analyst, Air Frame Design Div, Bur Aeronaut, Dept Navy, 46-49, head struct design & anal unit, 49-57, tech asst & asst head contract struct design sect, 57-59, air craft & missile specialist, Chief of Naval Opers, 59-61; chief struct & mat br, Fed Aviation Agency, 61-62; tech dir air progs, Off Naval Res, 62-67, dep asst chief technol, 67-80; SR PROJ ENGR, AEROSTRUCTURES, INC, 84- *Concurrent Pos:* Mem Fed Aviation Agency-Air Force-NASA working groups supersonic transport res progs & Fed Aviation Agency rep on NASA screening comt on mat, 62; Off Naval Res mem Dept Defense res & eng panel supporting res & technol, Aeronaut & Astronaut Coord Bd, 66-; Navy liaison officer, Adv Group Aerospace Res & Develop, NATO, 76-80. *Mem:* Am Inst Aeronaut & Astronaut. *Res:* Naval vehicle and weapon research and technology, including sensors, electronics, acoustics and operational analysis. *Mailing Add:* 6538 Cedarwood Ct Falls Church VA 22041

LAUVER, MILTON RENICK, b Springfield, Ohio, Sept 14, 20; m 48; c 4. PLASMA PHYSICS. *Educ:* Wittenberg Col, AB, 42; Western Reserve Univ, MS, 44, PhD(phys chem), 48. *Prof Exp:* Res chemist, Westvaco Div, Food Mach & Chem Corp, 48-53; res chemist chromium chem div, Diamond Alkali Co, 53-58; res chemist, Lewis Res Ctr, NASA, 58-82; RETIRED. *Mem:* Am Chem Soc. *Res:* Gas dynamics; nuclear fusion plasmas; electrochemistry and phosphoric acid fuel cell technology. *Mailing Add:* 28385 Holly Dr North Olmsted OH 44070

LAUVER, RICHARD WILLIAM, b Monmouth, Ill, Mar 15, 43; m 72. PHYSICAL CHEMISTRY. *Educ:* Knox Col, AB, 65; Univ Ill, Urbana, PhD(chem), 70. *Prof Exp:* Nat Res Coun res assoc, 72-74, RES CHEMIST, LEWIS RES CTR, NASA, 74- *Mem:* Am Chem Soc; Soc Adv Mat & Process Eng; Soc Plastic Eng; Coblentz Soc. *Res:* Physical and chemical characterization of polymer materials. *Mailing Add:* 20090 Carolyn Ave Rocky River OH 44116

LAUX, DAVID CHARLES, b Sarver, Pa, Jan 1 45; m 70; c 1. IMMUNOLOGY, BACTERIOLOGY. *Educ:* Washington & Jefferson Col, BA, 66; Miami Univ, MS, 68; Univ Ariz, PhD(microbiol), 71. *Prof Exp:* Fel immunol, Dept Microbiol, Sch Med, Pa State Univ, 71-73; from asst prof to assoc prof, 73-78, PROF IMMUNOL, UNIV RI, 83-, CHMN, DEPT MICROBIOL, 88- *Concurrent Pos:* Res grant, Nat Cancer Inst, 74, 78 & Nat Inst Allergy & Infectious Dis, 80, 83, 86; vis scientist, Oxford Univ, 88. *Mem:* Am Asn Cancer Res; Am Soc Microbiol; Am Asn Immunol. *Res:* Investigation of factors responsible for tumor mediated suppression of cellular immune reactivity; molecular basis of large intestine colonization; T-lymphocyte mediated cytotoxicity. *Mailing Add:* Dept Microbiol 318 Morrill Hall Univ RI Kingston RI 02881

LAUZIER, RAYMOND B, b St Stephen, New Brunswick, Can, July 22, 48. JUVENILE SALMONID ECOLOGY. *Educ:* Univ Ottawa, BSc, 74; Simon Fraser, MSc, 80. *Prof Exp:* Biologist, Environcon Ltd, 80-83; RES BIOLOGIST, GOVT CAN DEPT FISHERIES & OCEANS, 83- *Concurrent Pos:* Consulting biologist, F F Slaney & Co, 77-80. *Res:* Juvenile chinook salmon ecology; investigations on over wintering habitat; migration timing; seasonal rearing densities; survival rates and impacts of linear developments on salmon rivers. *Mailing Add:* Dept Fisheries & Oceans 4160 Marine Dr West Vancouver BC V7V 1N6 Can

LAUZON, RODRIGUE VINCENT, b Ottawa, Ont, Oct 24, 37; US citizen; m 65; c 2. COLLOID SCIENCE, PAPER CHEMISTRY. *Educ:* Univ Toronto, BA, 60; Univ Conn, MS, 62; Clarkson Col Technol, PhD(colloid chem), 71. *Prof Exp:* Res specialist latex, Dow Chem Co, 70-74; res assoc colloid chem, Celanese Res Co, 74-75; dir, latex res & develop, Dart Industs, 75-78; sr res scientist colloid, NL Baroid, NL Industs, Inc, 78-82; RES SCIENTIST & GROUP LEADER, HERCULES INC, 83- *Concurrent Pos:* Chmn colloid & surface chem group, Southeastern Sect, Am Chem Soc, 79-83; vis scientist, Indust Res Inst, 82; colloid consult, 82-83. *Honors & Awards:* Meritorious Award for Eng Innovation; Petrol Engr Int, Offshore Technol Conf, 82. *Mem:* Am Chem Soc; Trade Asn Pulp & Paper Indust; Asn Consult Chemists & Chem Engrs; Mat Res Soc. *Res:* Inorganic colloids; clays, oxides, halides; polymer colloids: latexes, water-soluble polymers, natural polymers; rheology of disperse systems; drilling muds; foams and emulsions; coagulation; adsorption from solution; electrokinetics; paper and paper chemicals. *Mailing Add:* 63 Quail Hollow Dr Hockessin DE 19707

LAVAGNINO, EDWARD RALPH, b Fall River, Mass, Apr 3, 30; m 58; c 4. ORGANIC CHEMISTRY. *Educ:* Southeastern Mass Technol Inst, BS, 52; Univ Mass, MS, 55. *Prof Exp:* Assoc sr org chemist, 54-68, sr org chemist, 68-75, RES SCIENTIST, CHEM RES DIV, ELI LILLY & CO, 75- *Mem:* Am Chem Soc. *Res:* Preparative organic chemistry; hydrogenation and high pressure reactions. *Mailing Add:* Eli Lilly & Co Dept MC Bldg 88-2 Indianapolis IN 46285

LAVAIL, JENNIFER HART, b Evansville, Ind, Apr 2, 43; m 70; c 2. NEUROANATOMY, NEUROEMBRYOLOGY. *Educ:* Trinity Col, DC, BA, 65; Univ Wis, PhD(anat), 70. *Prof Exp:* From instr to asst prof neuropath, Harvard Med Sch, 73-76; assoc prof, 76-86, PROF ANAT, UNIV CALIF, SAN FRANCISCO, 83- *Concurrent Pos:* Vis fel anat, Sch Med, Washington Univ, 68-69, res fel, 69-70; Nat Inst Neurol Dis & Stroke res fel neuropath, Harvard Med Sch, 70-73, spec fel, 73-76; Alfred P Sloan fel, 76-79; mem neurol study sect, NIH, 82-86; bd sci counr, Nat Inst Neurol & Commun Disorders & Stroke, 86- *Mem:* Am Asn Anatomists; Soc Neurosci. *Res:* Development of the central nervous system; anterograde and retrograde axonal transport. *Mailing Add:* Dept Anat Univ Calif San Francisco CA 94143-0452

LAVAIL, MATTHEW MAURICE, b Abilene, Tex, Jan 7, 43; m 70. NEUROSCIENCES, CELL BIOLOGY. *Educ:* NTex State Univ, BA, 65; Univ Tex Med Br, PhD(anat), 69. *Prof Exp:* Res fel, Harvard Med Sch, 69-73, asst prof neuropath, 73-76; actg chmn, 81-82, ASSOC PROF ANAT, MED SCH, UNIV CALIF, SAN FRANCISCO, 76- *Concurrent Pos:* Nat Eye Inst fel neuropath, Harvard Med Sch, 70-73; res assoc neurosci, Children's Hosp Med Ctr, 73-76; Nat Eye Inst res career develop award, 74-79. *Honors & Awards:* Res Award, Sigma Xi, 70; Fight for Sight Citation, 75; Sundial Award, Retina Found, 76; Friedenwald Award, Asn Res Vision & Ophthal, 81. *Mem:* Asn Res Vision & Ophthal; Soc Cell Biol; AAAS; Am Asn Anat; Soc Neurosci. *Res:* Photoreceptor-pigment epithelial cell interactions; retinal development; inherited retinal degeneration; neuroembryology; retrograde axonal transport. *Mailing Add:* Dept Anat Univ Calif San Francisco CA 94143

LAVAL, WILLIAM NORRIS, b Seattle, Wash, Jan 27, 22; m 63; c 2. GEOLOGY. *Educ:* Univ Wash, BS, 43, MS, 48, PhD(geol), 56. *Prof Exp:* Field asst, US Geol Surv, 42, geologist, 43-45 & 48-49; geologist, Corps Eng, US Army, 49-51; resident geologist, Yale Dam, Ebasco Serv, Inc, Wash, 51-53; asst prof geol, Colo State Univ, 56-60; assoc prof geol & geol eng, SDak Sch Mines & Technol, 60-62; chmn div natural sci, 63-75, PROF GEOL & EARTH SCI, LEWIS-CLARK STATE COL, 63- *Concurrent Pos:* Consult, 54-56 & 62-63. *Mem:* Fel Geol Soc Am; Sigma Xi. *Res:* Stratigraphy, structure and petrology of Columbia Plateau; environmental geology. *Mailing Add:* Lewis-Clark State Col Lewiston ID 83501

LAVALLE, PLACIDO DOMINICK, b New York, NY, May 13, 37; div; c 3. GEOMORPHOLOGY. *Educ:* Columbia Univ, BA, 59; Univ Southern Ill, MA, 61; Univ Iowa, PhD(geog), 65. *Prof Exp:* Res asst geog, Univ Iowa, 62-63; asst prof, Univ Calif, Los Angeles, 64-67 & Univ Ill, Urbana, 67-69; ASSOC PROF GEOG, UNIV WINDSOR, 69- *Concurrent Pos:* Vis prof, Keele Univ, Eng, 76; consult, Parks, Can. *Mem:* AAAS; Asn Am Geogrs; Can Asn Geogr; Nat Speleol Soc. *Res:* Soil geography; quantative analysis of karst geomorphology in Kentucky and Puerto Rico; spatial patterns of soil toxin distribution in Lebec, California; dynamics of shoreline change at Point Pelee, Ontario, Canada. *Mailing Add:* Dept Geog Univ Windsor Windsor ON N9B 3P4 Can

LAVALLEE, ANDRE, b Joliette, Que, Can, Aug 31, 36; m 60; c 3. WHITE PINE BLISTER RUST. *Educ:* Montreal Univ, BA, 56; Laval Univ, BSc, 60, DSc, 69; McGill Univ, MSc, 63. *Prof Exp:* Biologist forest path, 60-69, res scientist, 70-75, sect head, Forest Inst Dis Surv, 75-78, prog mgr forest protection, 79-84, SCIENTIST FOREST PATH, CAN FORESTRY SERV, QUE, 85- *Mem:* Can Phytopath Soc; Can Inst Forestry. *Res:* Characterization of white pine plantation sites with respect to white pine blister rust and weevil susceptibility; elaborate and evaluate integrated forest pest control prescriptions within intensive forest management practices. *Mailing Add:* Laurentian Forestry Ctr 1055 rue du Peps Ste Foy PQ G1V 4C7 Can

LAVALLEE, DAVID KENNETH, b Malone, NY, Oct 1, 45; m 71; c 3. INORGANIC CHEMISTRY, BIOCHEMISTRY. *Educ:* St Bonaventure Univ, BS, 67; Univ Chicago, SM, 68, PhD(inorg chem), 71. *Prof Exp:* USPHS fel, Dept Anat, Univ Chicago, 71-72; asst prof chem, Colo State Univ, 72-78; assoc prof, 78-83, PROF CHEM, HUNTER COL, 84- *Concurrent Pos:* Vis scientist, Argonne Nat Lab, 71-72; mem grad fac biochem & chem, City Univ New York, 78-; res collabr, Brookhaven Nat Lab, 79-; Fulbright fel, 85-86. *Honors & Awards:* Catalyst Award, Chem Mfrs Asn, 86. *Mem:* Am Chem Soc; AAAS; Sigma Xi; NY Acad Sci; Soc Nuclear Med. *Res:* Synthesis, spectroscopy, reaction mechanisms and structural chemistry of metalloporphyrins; electron transfer reaction mechanisms. *Mailing Add:* Dept Chem Hunter Col CUNY 695 Park Ave New York NY 10021

LAVALLEE, LORRAINE DORIS, b Holyoke, Mass, May 31, 31. MATHEMATICS. *Educ:* Mt Holyoke Col, AB, 53; Univ Mass, Amherst, MA, 55; Univ Mich, PhD(math), 62. *Prof Exp:* From instr to asst prof, 59-70, assoc head dept math & statist, 71-72 & 77, ASSOC PROF MATH, UNIV MASS, AMHERST, 70- *Mem:* Am Math Soc; Asn Women in Math; Math Asn Am. *Res:* General topology. *Mailing Add:* Dept Math & Statist Lederle Grad Res Ctr Towers Univ Mass Amherst MA 01003

LAVANCHY, ANDRE C(HRISTIAN), b Switz, Nov 17, 22; nat US; m 53; c 3. MECHANICAL ENGINEERING. *Educ:* Swiss Fed Inst Technol, ME, 46. *Prof Exp:* Design & test engr, Brown & Boveri, Switz, 46-49; res engr, Sharples Res Labs, 50-55, chief anal engr, 56-65, MGR CENTRIFUGE DEVELOP, PENNWALT CORP, 65- *Concurrent Pos:* Design engr, Am Viscose Corp, 55. *Mem:* Am Soc Mech Engrs; Am Chem Soc. *Res:* Stress and vibration analysis; applied dynamics; powder technology; liquid-liquid and solid-liquid separation; fluid flow; aerodynamics; thermodynamics. *Mailing Add:* 675 Timber Lane Devon PA 19333

LAVANISH, JEROME MICHAEL, b Cleveland, Ohio, Mar 10, 40; m 65. ORGANIC CHEMISTRY. *Educ:* Case Inst Technol, BS, 62; Yale Univ, MS, 63, PhD(chem), 66. *Prof Exp:* Res chemist, PPG Industs, 66-81, mgr biochem synthesis, 81-88; GROUP LEADER, AGR RES DIV, AM CYANAMID CO, 88- *Mem:* Am Chem Soc; Sigma Xi. *Res:* Synthetic organic chemistry; synthesis and action of herbicides and plant growth regulators; design and action of agrochemicals. *Mailing Add:* Agr Res Div Am Cyanamid Co PO Box 400 Princeton NJ 08543-0400

LAVE, ROY E(LLIS), JR, b Homewood, Ill, Sept 23, 35; m 60; c 2. INDUSTRIAL ENGINEERING, OPERATIONS RESEARCH. *Educ:* Univ Mich, BS & MBA, 58, MS, 60; Stanford Univ, PhD(indust eng), 65. *Prof Exp:* Asst opers res, Univ Mich, 57-60; asst prof indust eng, Stanford Univ, 62-68, assoc prof & assoc chmn dept, 68-72; DIR & CHMN BD, SYSTAN, INC, 66- *Concurrent Pos:* Consult, Rand Corp, 61-65, Coun Int Prog Mgt, 63, Agency Int Develop, 65-67 & Ford Found, 66; assoc prof eng-econ systs & dir fed internships & interdisciplinary projs, Stanford Univ, 67-68; consult, Inter-Am Develop Bank, 68-72 & Unido, 68-70; dir, Phoenix Housing Develop Corp, 69-77 & Consumer Alliance, 71-73; coun mem, City Los Altos, 74-82, mayor, 76-78; comnr, Metrop Transp Comn, San Francisco Bay Area, 80-87; mem, Paratransit Comt, Transp Res Bd, 82-85. *Honors & Awards:* Outstanding Young Fac, Am Soc Eng Educ, 71. *Mem:* Opers Res Soc Am; Inst Mgt Sci; Inst Indust Engrs; Transp Res Bd. *Res:* Systems analysis in international development planning; managements control systems; national and urban transportation planning and policy analysis; decision budgeting; information systems design; project management. *Mailing Add:* Systan Inc PO Box U Los Altos CA 94022-4021

LAVELLE, ARTHUR, b Fargo, NDak, Nov 29, 21; m 47; c 1. NEUROCYTOLOGY. *Educ:* Univ Wash, BS, 46; Johns Hopkins Univ, MA, 48; Univ Pa, PhD(anat), 51. *Prof Exp:* Asst zool, Univ Wash, 44-46; jr instr biol, Johns Hopkins Univ, 46-48; asst instr anat, Sch Med, Univ Pa, 48-51; from instr to prof, 52-88, EMER PROF ANAT, UNIV ILL COL MED, 88- *Concurrent Pos:* USPHS fel, Univ Pa, 51-52; USPHS-NIH res grant, Univ Ill, 53-70; vis prof dept anat & brain res inst & Guggenheim fel, Univ Calif, Los Angeles, 68-69. *Mem:* AAAS; Am Asn Anat; Biol Stain Comn (pres, 81-86); Soc Develop Biol; Am Soc Cell Biol; Soc Neurosci. *Res:* Neurocytology; cytological development of nerve cells; experimental alteration of development of nerve cells. *Mailing Add:* 462 Highland Ave Elmhurst IL 60126

LAVELLE, FAITH WILSON, b St Johnsbury, Vt, Mar 14, 21; m 47; c 1. HISTOLOGY, NEUROEMBRYOLOGY. *Educ:* Mt Holyoke Col, BA, 43, MA, 45; Johns Hopkins Univ, PhD(biol), 49. *Prof Exp:* Lab instr zool, Mt Holyoke Col, 43-45; admin asst zool, Univ Pa, 48-51, instr anat, Med Sch, 51-52; lectr, Univ Ill Col Med, 52-53, instr, 53-55, res assoc anat, 55-70; from asst prof to prof anat, Stritch Sch Med, Loyola Univ, Chicago, 70-86, actg chmn dept, 84-85; RETIRED. *Concurrent Pos:* USPHS res grant, Univ Ill Col Med, 53-70; emer prof anat, Stritch Sch Med, Loyola Univ, Chicago, 87- *Mem:* Am Asn Anat; Soc Neurosci; Sigma Xi; AAAS. *Res:* Experimental alteration of development of nerve cells; proteins in neural development. *Mailing Add:* 462 Highland Ave Elmhurst IL 60126

LAVELLE, GEORGE CARTWRIGHT, virology, toxicology, for more information see previous edition

LAVELLE, JAMES W, limnology; deceased, see previous edition for last biography

LAVELLE, JOHN WILLIAM, b Sacramento, Calif, Apr 26, 43; m 71. GEOLOGICAL OCEANOGRAPHY, PHYSICAL OCEANOGRAPHY. *Educ:* Univ Calif, Berkeley, BA, 65; Univ Calif, San Diego, MS, 68, PhD(physics), 71. *Prof Exp:* Marine geophysicist, 72-73, geol oceanogr, Environ Res Labs, Miami, 73-77, OCEANOGR, PAC MARINE ENVIRON LAB, NAT OCEANOG & ATMOSPHERIC ADMIN, 77- *Mem:* Am Geophys Union. *Res:* Theoretical studies of centered on particle transport and deposition in marine environments. *Mailing Add:* Pac Marine Environ Lab 7600 Sand Point Way NE Seattle WA 98115

LAVENDEL, HENRY W, b Warsaw, Poland, Apr 23, 19; m 51; c 2. CHEMICAL ENGINEERING, METALLURGY. *Educ:* Univ MIlan, PhD(indust chem), 51. *Prof Exp:* Res scientist powder metall, Am Electro Metal Co, NY, 52-55; res assoc, Sintercast Corp, 55-57; assoc chem engr, Argonne Nat Lab, 57-60; sr staff scientist, Palo Alto Res Labs, Lockheed Missiles & Space Co, Lockheed Aircraft Co, 61-87; RETIRED. *Mem:* Sigma Xi; Am Inst Mining, Metall & Petrol Engrs; Am Soc Metals; Am Chem Soc. *Res:* High temperature materials. *Mailing Add:* 1511 Hamilton Ave Palo Alto CA 94303

LAVENDER, DENIS PETER, b Seattle, Wash, Oct 13, 26; m 57. PLANT PHYSIOLOGY. *Educ:* Univ Wash, BS, 49; Ore State Univ, MSc, 58, PhD, 62. *Prof Exp:* Res asst, Ore State Bd Forestry, 50-57, in chg forest physiol, Forest Res Ctr, 57-63; assoc prof, Ore Forest Res Lab, Ore State Univ, 63-70, prof forest physiol, Sch Forestry, 70-; AT UNIV BC, VANCOUVER, CAN. *Mem:* Sigma Xi. *Res:* Development of hardy coniferous seedlings; nutrition of second growth Douglas fir stands; reduction of the juvenile period of conifers; dormancy in Douglas fir seedlings and conifers; mineral nutrition and precocious flowering in conifers. *Mailing Add:* Univ BC 193-2357 Main Mall Vancouver BC V6T 1W5

LAVENDER, DEWITT EARL, b Jackson Co, Ga, Nov 9, 38; m 58; c 3. MATHEMATICS, STATISTICS. *Educ:* Univ Ga, BS, 62, MA, 63, PhD(math, statist), 66. *Prof Exp:* Asst prof math, 66-68, ASSOC PROF MATH, GA SOUTHERN COL, 68-, HEAD DEPT, 70- *Mem:* Am Math Soc; Math Asn Am; Inst Math Statist. *Res:* Mathematical statistics. *Mailing Add:* Dept Math Ga Southern Univ Statesboro GA 30460

LAVENDER, JOHN FRANCIS, b Nov 16, 29; US citizen; m 69; c 3. VIROLOGY, MICROBIOLOGY. *Educ:* Drake Univ, BA, 51; Univ Ill, Champaign, MS, 53; Univ Calif, Los Angeles, PhD(infectious dis), 62. *Prof Exp:* NIMH fel virol, Univ Calif, Los Angeles, 62-63; sr virologist, 64-72, res virologist, 72-85, RES SCIENTIST MOLECULAR BIOL, ELI LILY & CO, 85- *Mem:* Am Soc Microbiol; Sigma Xi; NY Acad Sci. *Res:* Psychological stress and viral disease resistance; drugs and the entry of viruses across the blood brain barrier; development of parainfluenza, rabies, canine distemper and measles vaccines; Herpes Simplex vaccines types 1 and 2; chemotherapy of virus diseases; molecular biology; cell culture for mass production of proteins. *Mailing Add:* 543 West Dr Woodruff Pl Indianapolis IN 46201

LAVER, MURRAY LANE, b Warkworth, Ont, Mar 7, 32; m 63; c 2. ORGANIC CHEMISTRY, WOOD CHEMISTRY. *Educ:* Ont Agr Col, BScA, 55; Ohio State Univ, PhD(org chem), 59. *Prof Exp:* Res chemist food sci, Westreco Co, 59-63; res chemist wood sci, Rayonier Can, Inc, 63; res scientist, Weyerhaeuser Co, 64-66, prof specialist res div, 66-68; res instr chem, Univ Wash, 68-69; ASSOC PROF FOREST PRODS CHEM, ORE STATE UNIV, 69- *Concurrent Pos:* Vis lectr biol chem, Harvard Univ, 77-78. *Mem:* Am Chem Soc; NY Acad Sci; Tech Asn Pulp & Paper Ind; AAAS; Am Asn Univ Prof. *Res:* Pulp and paper, carbohydrate, food and wood chemistry. *Mailing Add:* Col Forestry Ore State Univ Corvallis OR 97331

LAVERDIERE, MARC RICHARD, b Coaticook, Que, May 28, 46; m 69; c 3. SOIL CONSERVATION, SOIL CHEMISTRY. *Educ:* Laval Univ, BSc, 69, MSc, 71; Cornell Univ, PhD(agron), 76. *Prof Exp:* HEAD DEPT SOIL SCI & PROF SOIL CONSERV & LAND USE, LAVAL UNIV, 85- *Concurrent Pos:* Res Off Pedogenesis, 71-73; res scientist, Clay Miner, Can Dept Agr, 76-85; guest lectr, Fac Agr, Alimentation Dept Soils, Laval Univ, 77. *Mem:* Soil Conserv Soc; Am Soc Agron; Soil Sci Soc Am; Can Soil Sci Soc; Int Soc Soil Sci. *Res:* Soil conservation; control of soil degradation; surface composting. *Mailing Add:* 761 Des Vosges Ste Foy PQ G1X 2Y7 Can

LAVERNIA, ENRIQUE JOSE, b Havana, Cuba, July 30, 60; US citizen; m 86. SOLIDIFICATION PROCESSING, METAL MATRIX COMPOSITES. *Educ:* Brown Univ, BS, 82; Mass Inst Technol, MS, 84, PhD(mat eng), 86. *Prof Exp:* Res asst, Mass Inst Technol, 82-86, res assoc, 86-87; ASSOC PROF, DEPT MECH ENG, UNIV CALIF, IRVINE, 87- *Concurrent Pos:* Prin investr, Naval Air Develop Ctr, Air Force Off Sci Res, Army Res Off, NSF, Off Naval Res & var industs, 87-; NSF presidential young investr, 88-91; secy, Synthesis & Anal Mat Processing Comt, Metall Soc; fac career develop award, Univ Calif, Irvine, 89; presidential young investr, NSF, 89-94; young investr award, Off Naval Res, 90-93; Alumni Co Am fel, 91. *Mem:* Sigma Xi; Am Soc Metals; Mat Res Soc; Metall Soc; Am Powder Industs Fedn. *Res:* Structure and mechanical behavior of metals and alloys processed under rapid solidification conditions; spray atomization and deposition of metals and alloys; solidification processing of metal matrix composites; mathematical modeling of solidification. *Mailing Add:* Dept Mech & Aerospace Eng Univ Calif Irvine CA 92717

LAVERTY, JOHN JOSEPH, b Chicago, Ill, May 27, 38; m 67; c 2. ORGANIC CHEMISTRY. *Educ:* Eastern Ill Univ, BS, 64; Univ Ariz, MS, 66. *Prof Exp:* Res chemist, 66-71, assoc sr res chemist, 71-78, sr res chemist, 78-81, STAFF RES SCIENTIST, POLYMER DEPT, GEN MOTORS RES LABS, 81- *Mem:* Am Chem Soc. *Res:* Plastic engineering and polymer physics; structure-property relationship of block copolymers and polymer blends; durability of engineering plastics. *Mailing Add:* Gen Motors Res Labs 30500 Mound Rd Warren MI 48090-9055

LAVETT, DIANE KATHRYN JURICEK, b Ft Benning, Ga, Feb 29, 44; c 2. GENETICS, CELL BIOLOGY. *Educ:* Emory Univ, BS, 69, PhD(biol), 73; Southern Conn State Col, MS, 80. *Prof Exp:* Vis instr genetics, Emory Univ, 74-75; fel DNA biochem, Yale Univ, 75-77, res assoc somatic cell genetics, 77-80; asst prof genetics, Emory Univ, 80-83; ASSOC PROF GENETICS, STATE UNIV NY, CORTLAND, 86- *Concurrent Pos:* Am Cancer Soc Fel, Yale Univ, 76-77, fel NIH grant, 75-76; NSF grant, Emory Univ, 69-72; mem, Fertile & Gamete Physiol Training Prog, Marine Biol Lab, 74; expert witness, 80-; genetic coun, Scottish Rite Hosp, 83-86; vis scholar, Emory Univ, 91-92. *Mem:* Sigma Xi; Am Soc Cell Biol; Am Women Sci. *Res:* Cytogenetics; somatic cell genetics; chromosome structure; human genetics; virus-chromosome interactions; gene transfer; DNA biochemistry; gene regulation; regulation of the human globin complex. *Mailing Add:* Dept Biol State Univ NY Cortland NY 13045

LAVI, ABRAHIM, b Iran, Jan 12, 34; US citizen; m 59; c 2. ELECTRICAL ENGINEERING, COMPUTER SCIENCE. *Educ:* Purdue Univ, BScEE, 57; Carnegie Inst Technol, MS, 58, PhD(elec eng), 59. *Prof Exp:* Res asst, 57, from asst prof to assoc prof elec eng, 59-68, PROF ELEC ENG, CARNEGIE-MELLON UNIV, 68- *Concurrent Pos:* Consult, Graphic Arts Technol Found, Pa; mem staff, Dept Energy, Washington, DC, 76-78. *Mem:* Sr mem Inst Elec & Electronics Engrs; Marine Technol Soc; Int Solar Energy Soc. *Res:* System modelling and optimization; low temperature difference energy conversion; control and instrumentation. *Mailing Add:* Dept Elec Eng Carnegie Inst Technol Pittsburgh PA 15213

LAVIA, LYNN ALAN, b Des Moines, Iowa, Oct 30, 48; m 87; c 4. OBSTETRICS & GYNECOLOGY, CANCER. *Educ:* Briar Cliff Col, BS, 70; Drake Univ, MA, 76; Univ Nebr, PhD(med sci), 81. *Prof Exp:* Instr anat & physiol, Iowa Western Community Col, 76-78; grad fel internal med, Univ Nebr Med Ctr, 78-80; dir res, Obstet & Gynec Dept, Univ Kans Med Sch, Wichita, 81-89; asst prof anat & embryol, Biol Dept, Wichita State Univ, 88-90; SR SCIENTIST RES, IMAGE ANALYTICAL RES, 89- *Concurrent Pos:* Medic, US Army, 70-72; emergency med tech, Mercy Hosp Emergency Rm, 72-74; mem fac, morphometry course, US Armed Forces Inst Path, 89-; consult, Derby Coastal Oil Refinery, 90-; peer reviewer, Am Asn Physicians Assts, 91-92. *Mem:* Int Soc Stereology; Am Asn Physician Assts. *Res:* Currently involved in how estrogen-stimulated stroma-epithelial cell interaction in uterine endometrium induces hyperplasia and neoplasia. *Mailing Add:* 1124 N Crestway Wichita KS 67208

LA VIA, MARIANO FRANCIS, b Rome, Italy, Jan 29, 26; nat US; m 59; c 7. PATHOLOGY. *Educ:* Univ Messina, MD, 49. *Prof Exp:* Asst gen path, Univ Palermo, 50-52; asst path, Univ Chicago, 52-57, instr anat, 57-60; from asst prof to assoc prof, Sch Med, Univ Colo, Denver, 60-68; prof, Bowman Gray Sch Med, 68-71; prof path, Sch Med, Emory Univ, 71-; AT DEPT LAB MED, UNIV SC. *Mem:* Sigma Xi; Am Asn Immunol; Am Soc Exp Path; Am Soc Cell Biol; Soc Exp Biol & Med. *Res:* Cellular and biochemical mechanism of antibody production. *Mailing Add:* Dept Lab Med Univ SC 171 Ashley Ave Charleston SC 29425

LAVIGNE, ANDRE ANDRE, organic chemistry; deceased, see previous edition for last biography

LAVIGNE, DAVID M, b Watford, England, Mar 18, 46; Can citizen; m 68; c 3. BIOLOGY MANAGEMENT OF MARINE MAMMALS. *Educ:* Univ Western Ont, BSc, 68; Univ Guelph, MSc, 71, PhD(zool), 73. *Prof Exp:* from asst prof to assoc prof, 73-87, PROF ZOOL, UNIV GUELPH, 87- *Concurrent Pos:* Mem, Seal Specialists Group, Int Union Conserv Nature & Natural Resources, 77-; vis scientist, Brit Antarctic Surv, Cambridge, UK, 80-81; chmn, Sect CSZ, Wildlife Biologists, 85-86. *Res:* Population ecology of marine mammals; ecological energetics and life history traits of mammalian populations; management of marine mammal population. *Mailing Add:* Dept Zool Univ Guelph Guelph ON N1G 2W1 Can

LAVIGNE, MAURICE J, b Ottawa, Ont, July 12, 18. NUCLEAR MATERIALS. *Educ:* Univ Que, BA, 38, BSc, 43, PhD(sci), 54. *Prof Exp:* Head nuclear metall, Chalk River Sta, Dept Mines, 46-56; head metall & eng, Westinghouse Can, 56-62; mgr, Falconbridge Co, 62-73; res dir, Dept Mines, 73-82; RETIRED. *Mem:* Fel Am Soc Metals; Am Inst Mining Metall & Petrol Engrs. *Mailing Add:* Three Saguenay Dr Aylmer PQ J9J 1A3 Can

LAVIGNE, ROBERT JAMES, b Herkimer, NY, May 30, 30; m 76; c 4. ENTOMOLOGY. *Educ:* Am Int Col, BA, 52; Univ Mass, MS, 58, PhD(entom), 61. *Prof Exp:* Res instr entom, Univ Mass, 56-59; from asst prof to assoc prof, 59-71, PROF ENTOM, UNIV WYO, 71- *Concurrent Pos:* Chief party, Wyo Team, Somalia, 85-88. *Mem:* Entom Soc Am; Pan-Pac Entom Soc; Soc Range Mgt; Soc Am Acridiol; Australian Entom Soc. *Res:* Insect taxonomy, especially Diptera; insect behavior, especially robber flies, Asilidae, and horse flies, Tabanidae; environmental entomology; biocontrol of weeds; forest entomology; rangeland entomology. *Mailing Add:* Box 3354 University Sta Laramie WY 82071

LAVIK, PAUL SOPHUS, b Camrose, Alta, Feb 11, 15; US citizen; m 41; c 4. BIOCHEMISTRY. *Educ:* St Olaf Col, AB, 37; Univ Wis, MS, 41, PhD(biol chem), 43; Western Reserve Univ, MD, 59. *Prof Exp:* Instr biochem, Sch Med, La State Univ, 43; from instr to asst prof, Baylor Col Med, 43-47; asst prof biochem, 47-52, assoc prof, 52-70, ASSOC CLIN PROF RADIOL, SCH MED, CASE WESTERN RESERVE UNIV, 70- *Concurrent Pos:* Staff physician radiotherapy, Cleveland Clin, 74-81; emergency clin, 81-; pres, Radiation Therapy Consults, 82- *Mem:* Am Soc Biol Chemists; Radiation Res Soc; Am Asn Cancer Res; Am Soc Therapeut Radiologists; Am Col Radiol. *Res:* Nucleic acid metabolism; radiation biochemistry; radiation therapy. *Mailing Add:* 2202 Acacia Park Dr 2702 Cleveland OH 44124

LAVILLA, ROBERT E, b New York, NY, May 8, 26. PHYSICAL CHEMISTRY. *Educ:* Bethany Col, BS, 53; Cornell Univ, PhD(phys chem), 60. *Prof Exp:* X-RAY SPECTROS, NAT INST STANDARDS TECHNOL, 60- *Concurrent Pos:* Nat Res Coun-Nat Bur Standards fel, 60-61. *Mem:* Am Phys Soc. *Res:* Electron diffraction of solids and gases; optical properties of materials; x-ray absorption and emission; symchrotron radiation. *Mailing Add:* Nat Inst Standards Technol Gaithersburg MD 20899

LAVIN, EDWARD, b Springfield, Mass, Jan 6, 16; m 48; c 3. POLYMER CHEMISTRY. *Educ:* Univ Mass, BS, 36; Tufts Col, MS, 37. *Prof Exp:* Res chemist, Shawinigan Resins Corp, 38-53, group leader, 53-56, sect leader, 56-60, dir appln res, 60-63, res mgr, 63-65; mgr res, 65-76, mgt technol, Monsanto Co, 76-82; RETIRED. *Concurrent Pos:* Monsanto Acad Award, Harvard Univ, 49-50. *Mem:* AAAS; Am Chem Soc; Inst Elec & Electronics Engrs; Soc Aerospace Mat & Process Eng. *Res:* Vinyl high polymers; polyvinyl acetal resins; polyimides; adhesives; reprographic polymer coating. *Mailing Add:* 94 Wheelmeadow Dr Longmeadow Springfield MA 01106

LAVIN, J GERARD, b Manchester, Eng, Oct 7, 32; US citizen; m 62; c 2. MATERIALS SCIENCE ENGINEERING, ENGINEERING PHYSICS. *Educ:* Univ Mich, MScE, 59, ScD(chem eng), 63. *Prof Exp:* Res engr, Pioneering Res Lab, Du Pont Fibers, Inc, 63-68, supvr res & develop, Pioneering & Carothers Res Labs, 68-71 & Kingston Plant, NC, 71-73, sr supvr, Camden Plant, SC, 73-76, process supt, Cape Fear Plant, NC, 76-78, tech supt, Old Hickory Plant, Tenn, 78-81, mfg mgr, 82-83, res mgr, Eng Non-Woven Struct Div, 83-85, SR RES FEL, PIONEERING RES LAB, DU PONT FIBERS, INC, 85- *Mem:* Textile Inst; Sigma Xi; Am Carbon Soc; Mat Res Soc. *Res:* Man-made fibers; non-woven fabric forms; carbon fibers research and development cultures and their results; thermally conductive composites. *Mailing Add:* 15 Wellesley Rd Swarthmore PA 19081

LAVIN, PETER MASLAND, b Philadelphia, Pa, Apr 16, 35; m 58; c 2. GEOPHYSICS. *Educ:* Princeton Univ, BSE, 57; Pa State Univ, PhD(geophys), 62. *Prof Exp:* From instr to assoc prof, 60-83, PROF GEOPHYS, PA STATE UNIV, 83- *Mem:* Soc Explor Geophys; Am Geophys Union. *Res:* Exploration geophysics with emphasis on gravity and magnetic interpretation; time-series analysis; crustal structure and tectonic history; environmental and groundwater geophysics. *Mailing Add:* Dept Geosci 443 Deike Bldg Pa State Univ University Park PA 16802

LAVIN, PHILIP TODD, b Rochester, NY, Nov 21, 46; m 70; c 2. BIOSTATISTICS. *Educ:* Univ Rochester, AB, 68; Brown Univ, PhD(appl math), 72. *Prof Exp:* Res asst prof appl math, Brown Univ, 72-74; res asst prof biostatist, State Univ NY, Buffalo, 74-77; from asst prof to assoc prof biostatist, Sch Pub Health, 77-86, ASSOC PROF SURG, MED SCH, HARVARD UNIV, 87-; PRES, BOSTON BIOSTATIST, INC, 83- *Concurrent Pos:* Biostatistician, US deleg Japan, Sci Exchange Comt Gastric Oncol, Nat Cancer Inst, 75; prin investr, Gastrointestine Tumor Study Group, 75-79; Ditoh Study, 85-89; coord statistician, Eastern Coop Oncol Group, 76-78; biostatistician, Dana-Farber Cancer Inst, 77-84; pres, Consult Statist, Inc, 84-87; panel mem ophthalmic, dent & radiologic devices, Food & Drug Admin, 83-86. *Mem:* Am Statist Asn; Biomet Soc; Pattern Recognition Soc; Drug Info Asn; Soc Chem Trials; Regulatory Affairs Prof Soc. *Res:* Biometry; clinical trials; pattern analysis; experimental design; statistical computing; health care evaluation policy science; standardization; specialist in design and analysis of clinical trials for drugs, biologics, and medical devices; research interests in screening studies, longitudinal data bases, and natural history of disease. *Mailing Add:* Boston Biostatistics, Inc Three Cahill Park Dr Framingham MA 01701

LAVINE, ADRIENNE GAIL, b Norristown, Pa, Aug 14, 58; m 82; c 1. HEAT TRANSFER. *Educ:* Brown Univ, ScB, 79; Univ Calif, Berkeley, MS, 83, PhD(mech eng), 84. *Prof Exp:* ASST PROF MECH ENG, UNIV CALIF, LOS ANGELES, 84- *Concurrent Pos:* NSF presidential young investr award, 88. *Honors & Awards:* Taylor Medal, Int Inst Prod Engrs, 90. *Mem:* Am Soc Mech Engrs; Am Phys Soc. *Res:* Thermal aspects of manufacturing processes; coupled thermal and electrical behavior of superconducting electronic devices; natural and mixed convection heat transfer; numerical heat transfer. *Mailing Add:* Mech Aerospace & Eng Dept Univ Calif Los Angeles CA 90024-1597

LAVINE, JAMES PHILIP, b Syracuse, NY, Dec 3, 44; m 71; c 1. THEORETICAL & COMPUTATIONAL PHYSICS. *Educ:* Mass Inst Technol, BS, 66; Univ Md, PhD(physics), 71. *Prof Exp:* Res assoc physics, Univ Liege, 71-73; res asst prof, Laval Univ, 73-74; res assoc physics, Univ Rochester, 74-76; PHYSICIST, EASTMAN KODAK CO, 76- *Mem:* Am Phys Soc; Soc Indust & Appl Math; Mat Res Soc; Electron Micros Soc Am; Am Asn Adv Sci; Electrochem Soc. *Res:* Semiconductor device modeling and simulation of semiconductor device fabrication; optical properties of solids; transport in solids. *Mailing Add:* Res Labs Bldg 81 Eastman Kodak Co Rochester NY 14650-2008

LAVINE, LEROY S, b Jersey City, NJ, Oct 28, 18; m 46; c 2. ORTHOPEDIC SURGERY. *Educ:* NY Univ, AB, 40, MD, 43; Am Bd Orthop Surg, dipl, 55. *Prof Exp:* Res instr orthop surg, Col Med, Ind Univ, 51-52; from instr to prof, 65-80, EMER PROF ORTHOP SURG, COL MED, STATE UNIV NY DOWNSTATE MED CTR, 80-; LECTR, HARVARD MED SCH, 82- *Concurrent Pos:* Consult, Am Mus Natural Hist, 55- & Brooklyn Vet Admin Hosp, 65-; consult orthop surgeon, Long Island Jewish Hosp, 57-; adj prof biol, Grad Sch, NY Univ, 66-; clin prof, Med Sch, State Univ NY Stony Brook, 72-; vis orthop surgeon, Mass Gen Hosp, 82- *Mem:* Fel Am Col Surg; Am Asn Phys Anthrop; fel Am Acad Orthop Surg; fel NY Acad Sci; fel NY Acad Med. *Res:* Bone growth and metabolism and mechanisms of calcification; clinical research in bone healing and nerve root compression syndromes; physical properties of piezoelectricity in bone; electrical enhancement of bone growth. *Mailing Add:* Spaulding Rehab Hosp 125 Nashua St Boston MA 02114

LAVINE, RICHARD BENGT, b Philadelphia, Pa, June 27, 38; m 65; c 1. MATHEMATICAL ANALYSIS. *Educ:* Princeton Univ, AB, 61; Mass Inst Technol, PhD(math), 65. *Prof Exp:* Instr math, Aarhus Univ, 65-66; asst prof, Cornell Univ, 66-71; vis prof, Inst Theoret Physics, Univ Geneva, 71; mem staff, Inst Advan Study, 71-72; assoc prof, 72-76, PROF MATH, UNIV ROCHESTER, 76- *Mem:* Am Math Soc. *Res:* Mathematics of quantum mechanics; functional analysis. *Mailing Add:* Dept Math Univ Rochester Wilson Blvd Rochester NY 14627

LAVINE, ROBERT ALAN, b Chicago, Ill, Feb 18, 41. NEUROPHYSIOLOGY, PSYCHOLOGY. *Educ:* Univ Chicago, BS, 62, PhD(physiol), 69. *Prof Exp:* From instr to asst prof, 69-75, ASSOC PROF PHYSIOL & NEUROL, SCH MED, GEORGE WASHINGTON UNIV, 80- *Concurrent Pos:* NIH fel, George Washington Univ, 69-70; guest scientist, NIMH, 73-; int res fel rev, Fogarty & NIH, 83- *Mem:* Soc Neurosci; Soc Psychophysiol Res. *Res:* Human psychophysiology and experimental neuropsychology; clinical applications of averaged evoked potentials; clinical neurophysiology, computer applications in psychology. *Mailing Add:* Dept Physiol George Washington Univ Med Sch 2300 I St NW Washington DC 20037

LAVIOLETTE, FRANCIS A, plant pathology; deceased, see previous edition for last biography

LA VIOLETTE, PAUL ESTRONZA, b New York, NY, Apr 11, 30; m 75; c 2. SATELLITE OCEANOGRAPHY. *Educ:* Univ Ill, BS, 52. *Prof Exp:* Oceanogr descriptive oceanog, Naval Oceanogr Off, 60-76, oceanogr remote sensing, Naval Ocean Res & Develop Activity, 76- 90; OCEANOGR REMOTE SENSING, MISS STATE UNIV RES CTR, 90- *Concurrent Pos:* Prin investr, US & Mex Ocean Exp, 71, Sahara Upwelling Explor, US & Spain Ocean Exp, 73 & Grand Banks Explor, US & Can Ocean Exp, Little Window II, 79-80. *Mem:* Am Geophys Union; AAAS; Marine Technol Soc; Res Soc Am. *Res:* Basic and applied research in satellite oceanography. *Mailing Add:* Res Ctr Bldg 1103 Miss State Univ Stennis Space Ctr MS 39529-6000

LAVKULICH, LESLIE MICHAEL, b Coaldale, Alta, Apr 28, 39; m 62; c 2. SOIL SCIENCE. *Educ:* Univ Alta, BSc, 61, MSc, 63; Cornell Univ, PhD(soil sci), 67. *Prof Exp:* From asst prof to assoc prof, 66-75, head dept, 80-90, PROF SOIL SCI, UNIV BC, 75- *Mem:* Am Soc Agron; Can Agr Res Coun; Am Soc Soil Sci; Can Soc Soil Sci (pres, 80-81); Int Soc Soil Sci. *Res:* Soil genesis and classification; weathering of minerals; soil clay mineralogy; soil-plant relationships; mine waste characterization; resource allocation. *Mailing Add:* Dept Soil Sci Univ BC Vancouver BC V6T 2A2 Can

LAVOIE, ALVIN CHARLES, b Fall River, Mass, Jan 26, 56. EMULSION POLYMERIZATION, ACRYLATES. *Educ:* Southeastern Mass Univ, BS, 77; Univ Wis-Madison, PhD(org chem), 82. *Prof Exp:* SR SCIENTIST, ROHM AND HAAS CO, 81- *Mem:* Am Chem Soc; AAAS. *Res:* Utilization of vinyl sulfides as enolonium equivilants in organic synthesis; applications of organosulfer intermediates in organic synthesis; emulsion polymerization of acrylates. *Mailing Add:* 303 Abbey Lane Lansdale PA 19446

LAVOIE, EDMOND J, b New York, NY, Jan 11, 50; m 71. SYNTHETIC ORGANIC & NATURAL PRODUCTS CHEMISTRY. *Educ:* Fordham Univ, BS, 71; State Univ NY, Buffalo, PhD(med chem), 75. *Prof Exp:* Res assoc, 75-76, assoc, 76-79, HEAD SECT METAB BIOCHEM, DIV ENVIRON CARCINOGENESIS, AM HEALTH FOUND, 77-, ASSOC MEM, 80- *Concurrent Pos:* Asst res prof, Dept Urol, NY Med Col. *Mem:* Am Chem Soc; Am Asn Cancer Res; Environ Mutagen Soc. *Res:* Experimental and environmental carcinogenesis; environmental analysis; tobacco sciences. *Mailing Add:* Rutgers Univ Col Pharm PO Box 789 Piscataway NJ 08855

LAVOIE, MARCEL ELPHEGE, b Manchester, NH, July 16, 17; m 42; c 2. ZOOLOGY. *Educ:* St Anselm's Col, BA, 40; Univ NH, MS, 52; Syracuse Univ, PhD(zool), 56. *Prof Exp:* Instr chem, St Anselm's Col, 46-47; instr biol, Univ NH, 50-52; lectr zool, Syracuse Univ, 52-55; asst prof zool, Univ NH, 55-61, assoc prof, 61-84; RETIRED. *Res:* Mammalian anatomy and physiology. *Mailing Add:* 43 Madbury Rd Durham NH 03824

LAVOIE, RONALD LEONARD, b Manchester, NH, Apr 21, 33; m 59; c 2. METEOROLOGY. *Educ:* Univ NH, BA, 54; Fla State Univ, MS, 56; Pa State Univ, PhD, 68. *Prof Exp:* Chief observer, Mt Wash Observ, 57-59; asst prof meteorol, Univ Hawaii, 59-68; assoc prof, Pa State Univ, 68-72; assoc dir meteorol prog, NSF, 72-73; dir, Environ Modification Off, Nat Oceanic & Atmospheric Admin, 73-79, dir, Atmospheric Progs Off Res Develop, 79-82, Chief, Prog Res & Develop, Nat Weather Serv, 82-90, DIR, OFF METEOROL, NAT OCEANIC & ATMOSPHERIC ADMN, 90- *Concurrent Pos:* NSF sci fac fel, 63-64. *Mem:* AAAS; Am Meteorol Soc; Am Geophys Union. *Res:* Cloud physics and weather modification; numerical modeling on the mesoscale; tropical meteorology. *Mailing Add:* Nat Weather Serv 1325 East West Hwy Silver Spring MD 20910

LAVY, TERRY LEE, b Greenville, Ohio, Feb 9, 36; m 55; c 3. WEED SCIENCE, SOIL CHEMISTRY. *Educ:* Ohio State Univ, BS, 58, MS, 59; Purdue Univ, PhD(plant nutrit), 62. *Prof Exp:* Lab supvr soil chem classification, Ohio State Univ, 58-59; from asst prof to prof agron, Univ Nebr-Lincoln, 62-78; DIR, PESTICIDE RESIDUE LAB, UNIV ARK, 78 - *Mem:* Weed Sci Soc Am; Am Soc Agron; Am Chem Soc. *Res:* Factors affecting the mobility and degradation of pesticides in the soil profile; evaluating exposure of pesticide applicators; monitoring irrigation and domestic wells for pesticide contamination. *Mailing Add:* Dept Agron Altheimer Lab Univ Ark Fayetteville AR 72701

LAW, ALAN GREENWELL, b Seaham, Eng, Aug 21, 36; m 62; c 4. APPROXIMATION & COMPUTING, MEDICAL IMAGING. *Educ:* Univ BC, BA, 58, MA, 61; Ga Inst Technol, PhD(math), 68. *Prof Exp:* Instr math, Ga Inst Technol, 61-68; from asst prof to assoc prof math, 68-76, PROF COMPUTER SCI, UNIV REGINA, 76-, HEAD COMPUTER SCI, 90- *Concurrent Pos:* M A Ferst Res Award, Sigma Xi, 68; vis staff mem, Los Alamos Meson Physics Facil, 76-77; vis prof, Univ Col Swansea, Wales, 84, Univ Minn, Duluth, 85. *Mem:* Math Asn Am; Sigma Xi; Soc Indust & Appl Math; Can Info Processing Soc. *Res:* Numerical analysis and computing; modelling in sciences; digital image processing; large-scale computational problems; interdisciplinary analysis in natural and computational sciences. *Mailing Add:* Comp Sci Dept Univ Regina Regina SK S4S 0A2

LAW, ALBERT G(ILES), b Ottawa, Ill, July 1, 31; m 54; c 4. CIVIL ENGINEERING, HYDROLOGY. *Educ:* Univ Ill, Urbana, BS, 54; Univ Wis, MS, 60, PhD(civil eng), 65. *Prof Exp:* Design engr, Warzyn Eng Co, Wis, 56-58; instr civil eng, Univ Wis, 58-62; from asst prof to assoc prof civil eng, Clemson Univ, 62-77; MGR HYDROL SCI, ROCKWELL HANFORD OPERS, 77- *Concurrent Pos:* Vis assoc prof, Colo State Univ, 71-72; dir water resources engr, Clemson Univ & consult, US Geol Surv, 75-77. *Mem:* Nat Water Well Asn; Soc Mining Engrs; Am Soc Civil Engrs; Am Geophys Union; Am Water Resources Asn. *Res:* Ground water hydrology, ground water flow, radioactive waste disposal. *Mailing Add:* 417 Snyder Rd Richland WA 99352

LAW, AMY STAUBER, b Philadelphia, Pa, June 26, 38; div. CLINICAL BIOCHEMISTRY. *Educ:* Mt Holyoke Col, AB, 59; Univ Del, MS, 63, PhD(chem), 69. *Prof Exp:* Res chemist, AviSun Corp, 61-65; chief lab sect, Meat Inspection Div, Del State Bd Agr, 68-69; CLIN BIOCHEMIST, MED CTR DEL, 69- *Mem:* AAAS; Am Chem Soc; Am Asn Clin Chemists; Asn Women Sci. *Res:* Development of clinical methods; clinical applications of protein biochemistry; hemoglobinopathies; clinical toxicology. *Mailing Add:* Spec Chem-Christiana Hosp Med Ctr Del Newark DE 19718-6001

LAW, BRUCE MALCOLM, b Lower Hutt, NZ, June 23, 56; m 83; c 1. CONDENSED MATTER SURFACE PHYSICS, LIGHT SCATTERING. *Educ:* Victoria Univ, BS, 78, BS, 79, PhD(physics), 85. *Prof Exp:* RES ASSOC, UNIV MD, 85- *Mem:* Am Phys Soc. *Res:* Condensed matter surface physics/chemistry; application of light scattering techniques to both bulk & surface phases; non-equilibrium steady states. *Mailing Add:* Dept Physics Kans State Univ Cardwell Hall Manhattan KS 66506

LAW, CECIL E, b Vancouver, BC, Nov 27, 22; m 45; c 6. OPERATIONS RESEARCH. *Educ:* Univ BC, BA, 50. *Prof Exp:* Head animal field exp sect, Suffield Exp Sta, Defence Res Bd, 51, head arctic oper res sect, Defence Res Northern Lab, 51-54, head weapons effects & field trials sect, Can Army Oper Res Estab, 55-58, head oper gaming & tactics sect, 58-60; supvr opers res, Can Industs Ltd, 60-61, opers res mgr, 61-62; sr opers res analyst, Can Nat Rwy, 62-64, coordr opers anal, 64-66; prof oper res, Queen's Univ, 66-90, Prof computer sci, 69-83, exec dir, Can Inst Guided Transport, 71-83, dir computer lab, 84-88, dir, Inst Community & Occup Health, 85-88, EMER PROF OPERS RES SCH BUS, QUEEN'S UNIV, ONT, 90- *Concurrent Pos:* Lectr, Exten Dept, McGill Univ, 64-66; dir, Visway TPT Inc, 83-88. *Honors & Awards:* Coronation Medal, 53. *Mem:* Opers Res Soc Am; Am Inst Indust Eng; Can Opers Res Soc (vpres, 66, pres, 67). *Res:* Wildlife ecology and population dynamics, particularly Arctic; operations research, especially military and civil operational gaming and simulation; theoretical and applied critical path analysis and program evaluation and review technique; transportation research; micro-computer applications. *Mailing Add:* Sch Bus Queen's Univ Kingston ON K7L 3N6 Can

LAW, DAVID H, b Milwaukee, Wis, July 24, 27; m 49; c 5. INTERNAL MEDICINE, GASTROENTEROLOGY. *Educ:* Cornell Univ, AB, 50, MD, 54. *Prof Exp:* Intern med, NY Hosp, 54-55, asst resident, 55-57, asst physician to outpatients, 57-58, physician to out-patients & dir personnel health serv, 58-60; med dir out-patient dept & chief div gastroenterol, Vanderbilt Univ Hosp, 60-69; PROF MED, SCH MED, UNIV NMEX, 69-; CHIEF MED SERV, ALBUQUERQUE VET ADMIN HOSP, 69- *Concurrent Pos:* NIH fel, Nat Cancer Inst, 57-58; spec consult interdept comt nutrit for nat defense, NIH, 62-63; attend physician, Thayer Vet Admin Hosp, 62-69. *Mem:* Fel Am Col Physicians; Am Fedn Clin Res; Am Soc Clin Nutrit; Am Gastroenterol Asn; Am Inst Nutrit. *Res:* Inflammatory bowel disease; malabsorption; gastric secretion; medical care; nutrition; out-patient clinics; delivery of health care. *Mailing Add:* Dept Vet Affairs 810 Vermont Ave Washington DC 20420

LAW, DAVID MARTIN, b Bluefield, WVa. GROWTH & DEVELOPMENT, PHYTOHORMONES. *Educ:* Bluefield State Col, BS, 75; WVa Univ, MS, 78; Pa State Univ, PhD(plant physiol), 84. *Prof Exp:* Teaching asst, WVa Univ, 76-78; teaching asst, Pa State Univ, 78-80, res asst, 80-85; RES ASSOC, CORNELL UNIV, 85-, LECTR, 88- *Mem:* Am Soc Plant Physiologists; Am Inst Biol Sci; Sigma Xi; Scandinavian Soc Plant Physiol. *Res:* Regulation of indoleacetic acid biosynthesis; control of stem elongation; control of tomato fruit ripening; interaction of phytohormones; phytohormone analysis; prolongation of food storage life. *Mailing Add:* Dept Plant Biol Cornell Univ 259 Plant Sci Bldg Ithaca NY 14853-5908

LAW, ERIC W, b Taipei, Taiwan, July 14, 49; m; c 2. DIAGENESIS-METAMORPHISM, NEURAL NETWORK. *Educ:* Case Western Reserve Univ, PhD(geol), 83. *Prof Exp:* ASST PROF GEOL, MUSKINGUM COL, 84- *Res:* Petrology; geochronology; Taiwan; subduction; slate; K-Ar; computer; neural network; diagenesis; clay minerals; granite; metamorphism; sandstone. *Mailing Add:* Geol Dept Muskingum Col New Concord OH 43762

LAW, FRANCIS C P, b Hong Kong, Oct 12, 41; Can citizen; m 80; c 4. DRUG METABOLISM, TOXICOLOGY. *Educ:* Univ Alta, BS, 66, MS, 69; Univ Mich, PhD(drug metab), 72. *Prof Exp:* Vis fel, Nat Inst Environ Health Sci, 72-75; asst prof drug metab & toxicol, Col Pharm, Dalhousie Univ, 75-81, assoc prof, 81-82; assoc prof environ toxicol prog, Dept Biol Sci, 83-87, PROF, SIMON FRASER UNIV, 87- *Mem:* Pharmacol Soc Can; Soc Toxicol Can; Am Soc Pharmacol & Exp Therapeut. *Res:* Disposition, metabolism and toxicity of drugs; environmental pollutants and other chemicals in living organisms including man. *Mailing Add:* Dept Biol Sci Environ Toxicol Prog Simon Fraser Univ Burnaby BC V5A 1S6 Can

LAW, GEORGE ROBERT JOHN, b Vermilion, Alta, June 4, 28; US citizen; m 50; c 2. POULTRY GENETICS. *Educ:* Univ BC, BSA, 50; Wash State Univ, MS, 57; Univ Calif, Davis, PhD(genetics), 61. *Prof Exp:* Immuno-geneticist, Hy-Line Int, Pioneer Hi-Bred Int Inc, 61-72; assoc prof animal sci, Colo State Univ, 73-78, asst to dean, Col Agr Sci, 78-81, counr, Div Continuing Educ, 81-88, int counr, Col Agr Sci, 88-89; AGR PROJ MGR, SAN DIEGO STATE UNIV FOUND, 89- *Res:* Immuno-genetic studies of turkeys and chickens including blood type variation, serum protein, egg white protein and isozyme polymorphism; teaching, research and application of studies to breeding of poultry. *Mailing Add:* ENARP San Diego State Univ Found 6330 Alvarado Ct San Diego CA 92120

LAW, HSIANG-YI DAVID, b Hong Kong, Feb 12, 49; US citizen; m 73; c 2. OPTOELECTRONICS, SEMICONDUCTOR MATERIAL. *Educ:* Univ Wash, BSEE, 72; Cornell Univ, MSEE, 75, PhD(elec eng), 77. *Prof Exp:* Mem tech staff, Sci Ctr, Rockwell Int Corp, 77-80; mgr, Semiconductor Device Lab, Technol Res Ctr, TRW Inc, 80-84; VPRES TECHNOL, PCO, INC, 84- *Concurrent Pos:* Lectr optoelectronic class; technol eval of cos. *Mem:* Sr mem Inst Elec & Electronics Engrs; Am Phys Soc. *Res:* III-V alloys material study; avalanche photodiodes; double heterostructure lasers; integrated optoelectronic devices; ionization coefficients of III-V materials; ion implantation, anodic oxidation and other surface passivation methods. *Mailing Add:* 29776 Woodbrook Dr Agoura Hills CA 91301

LAW, JAMES PIERCE, JR, water pollution, for more information see previous edition

LAW, JIMMY, b Seremban, Malaysia, Sept 23, 42; m 69; c 2. THEORETICAL PHYSICS, COMPUTATIONAL PHYSICS. *Educ:* Univ London, BSc, 63, PhD(physics), 68. *Prof Exp:* Teaching fel physics, McMaster Univ, 66-69; from asst prof to assoc prof, 69-84, PROF PHYSICS, UNIV GUELPH, 84- *Mem:* Brit Inst Physics; Can Asn Physicists. *Res:* Theoretical calculations in nuclear and hypernuclear physics; inner shell vacancy creation mechanisms in atomic physics; Kaonic Hydrogen anomaly and Kaonic Atoms system; Simpson anomaly in the tritium beta decay spectrum; parallel computing of computers. *Mailing Add:* Dept Physics Univ Guelph Guelph ON N1G 2W1 Can

LAW, JOHN, b Cleveland, Ohio, Dec 8, 30; m 53; c 4. POWER SYSTEMS, ELECTRIC MACHINERY. *Educ:* Case Inst Technol, BS, 57; Univ Wis-Madison, MS, 60, PhD(elec eng), 62. *Prof Exp:* Instr, Univ Wis, 57-61; assoc prof elec eng, Mont State Univ, 62-63; sr elec engr, Carrier Corp, 63-65, chief engr, 65-74, sr staff engr, 74; assoc prof, 75-79, PROF ELEC ENGR, UNIV IDAHO, 79- *Concurrent Pos:* Vis assoc prof elec eng, Bogazici Univ, Turkey, 74-75; NSF fel, Elec Power Res Inst, 78; vis engr, Eng Soc Comn Energy, 79; vis prof elec eng, Wash State Univ, 80; planning engr, Idaho Power Co, 81; engr, Idaho Nat Eng Lab, 77; consult, Idaho Power Co, 76; mem, Fac Improvement Comt, NSF, 78. *Mem:* Inst Elec & Electronics Engrs; Nat Soc Prof Engrs. *Res:* Computer methods in power systems analysis; response of AC servomotors with nonsinusoinal and discontinuous impedance source voltage. *Mailing Add:* Elec Eng Dept Univ Idaho Buchanan Eng Lab Moscow ID 83843

LAW, JOHN HAROLD, b Cleveland, Ohio, Feb 27, 31; m 56. ENTOMOLOGY. *Educ:* Case Inst Technol, BS, 53; Univ Ill, PhD(chem), 57. *Prof Exp:* Res fel, Harvard Univ, 57-58; instr chem, Northwestern Univ, 58-59; from instr to asst prof, Harvard Univ, 59-65; prof biochem, Univ Chicago, 65-, prof chem, 67-; AT BIOCHEM DEPT, UNIV ARIZ. *Mem:* Am Chem Soc; Am Soc Biol Chemists; fel AAAS. *Res:* Insect biochemistry; lipid metabolism; protein chemistry. *Mailing Add:* Biochem Dept Biosci W Univ Ariz Tucson AZ 85721

LAW, LLOYD WILLIAM, b Ford City, Pa, Oct 28, 10; m 42; c 2. ONCOLOGY. *Educ:* Univ Ill, BS, 31; Harvard Univ, AM, 35, PhD(biol), 37. *Prof Exp:* Instr high sch, Ill, 31-33; asst, Harvard Univ, 36-37; res assoc, Stanford Univ, 37-38; Finney-Howell med res fel physiol genetics, Jackson Mem Lab, 38-41, Commonwealth Fund fel cancer res, 41-42; sci dir, Jackson Mem Lab, 46-47; sr geneticist, 47-54, mem sci directorate, 71-80, SCIENTIST DIR, NAT CANCER INST, 54-, CHIEF, LAB CELL BIOL, 71- *Concurrent Pos:* Harvard Univ Parker fel, Stanford Univ, 37-38; trustee, Jackson Mem Lab, 47-; mem study sect cancer chemother, NIH, 56-59, drug eval panel, Nat Serv Ctr, 56-59, sci adv bd, Roswell Park Mem Inst, 57-62, adv bd, Children's Cancer Found, expert adv panel cancer, WHO, 60-65 & adv sci bd, Hektoen Inst Chicago, 64-; mem, US Nat Comt, Int Union Against Cancer, 68-72 & 72-76, & Am Can Soc Rev Bd, 70-74. *Honors & Awards:* A F Rosenthal Award, AAAS, 58; G H A Clowes Award, Am Asn Cancer Res, 65; Meritorious Serv Award, USPHS, 65, Distinguished Serv Award, 69; Alexander Pascoli Prize, Univ Perugia, 69; G B Mider Lect Award, 70; hon mem Am Asn Cancer Res, 87; hon mem Europe Asn Cancer Res. *Mem:* Am Soc Exp Path; Am Asn Cancer Res (pres, 68-69); Transplantation Soc; Soc Exp Biol & Med; Soc Exp Leukemia; Am Asn Immunologists; Europ Asn Cancer Res. *Res:* Genetics; factors affecting development of leukemia and breast tumors; immunogenetics of the mouse; tumor immunology; chemotherapy of neoplasms. *Mailing Add:* Lab Cell Biol Nat Cancer Inst Bethesda MD 20892

LAW, MARGARET ELIZABETH, b Birmingham, Eng, May 6, 34; m 57. EXPERIMENTAL HIGH ENERGY PHYSICS. *Educ:* Univ Birmingham, Eng, Bsc, 55, PhD(high energy physics), 58; Boston Univ, MBA, 84. *Prof Exp:* Nat Res Coun fel nuclear physics, McMaster Univ, 58-60; res fel, Harvard Univ, 61-67, res assoc, 67-71, sr res assoc high energy physics, 71-78, lectr, 72-78, 83-85, registr, Fac Arts & Sci, 78-89, SR LECTR, HARVARD UNIV, 86-, DIR PHYSICS LAB, 89- *Mem:* Am Phys Soc; Brit Inst Physics. *Res:* Experimental research in strong interactions. *Mailing Add:* Physics Dept Harvard Univ Cambridge MA 02138

LAW, PAUL ARTHUR, b Lowell, Mass, Sept 19, 34; m 60, 81; c 4. CLINICAL CHEMISTRY. *Educ:* Lowell Technol Inst, BS, 56; Mich State Univ, PhD(org chem), 62. *Prof Exp:* Res chemist, Dow Corning Corp, 56-58; fel polypeptide synthesis, Fla State Univ, 62-63; sr res chemist, Eastman Kodak Co, 63-69, tech assoc, 69-78, gen supvr, 78-80, asst supt med prod, 80-86, GEN MGR MOTION PICTURE & TELEVISION PROD, EASTMAN KODAK CO, 86- *Mem:* Am Asn Clin Chemists. *Res:* Color photographic chemistry. *Mailing Add:* Eastman Kodak Co 343 State St Rochester NY 14650

LAW, PETER KOI, b Chengsha, China, Feb 25, 46; Chinese & Can citizen; c 2. NEUROMUSCULAR ELECTROPHYSIOLOGY. *Educ:* McGill Univ, BSc, 68; Univ Toronto, MSc, 69, PhD(neurophysiol), 72. *Prof Exp:* Fel, Med Ctr, McMaster Univ, 72-75; asst prof neurol, Sch Med, Vanderbilt Univ, 75-79; assoc prof, 79-88, PROF NEUROL, PHYSIOL & BIOPHYS, UNIV TENN, MEMPHIS, 88- *Concurrent Pos:* Sr investr, Jerry Lewis Neuromuscular Dis Res Ctr, Nashville, Tenn, 75-79; electromyography consult, Vanderbilt Univ Hosp, Nashville, 75-79, Baptist Hosp, 75-79 & St Thomas Hosp, 75-79; Muscular Dystrophy Asn Can fel, 72-75; Muscular Dystrophy Asn res grant, 75-88, Vanderbilt Univ Res Coun res grant, 75-79; NIH res grant, 78-79; NSF res grant, 79-82, NIH grants, 83-93. *Mem:* Asn Am Med Cols; AAAS; NY Acad Sci; Soc Neurosci; Can Soc Neurosci. *Res:* Developmental membrane biophysics; motor-unit electrophysiology; muscular dystrophy; myogenesis and muscle regeneration; genetic complementation and therapy; acupuncture analgesia; myogenic cell transplant treatment for muscle disease. *Mailing Add:* Dept Neurol Univ Tenn 956 Ct Ave No A218 Memphis TN 38163

LAW, WILLIAM BROUGH, b Elko, Nev, Oct 11, 32; m 56; c 3. PLASMA PHYSICS. *Educ:* Univ Nev, BSc, 54; Ohio State Univ, PhD(nuclear physics), 60. *Prof Exp:* Physicist, Armour Res Found, 60; staff mem, Sandia Lab, 60-65; asst prof, 65-68, ASSOC PROF PHYSICS, COLO SCH MINES, 68- *Mem:* Am Phys Soc. *Res:* Gamma ray spectroscopy; accelerator physics. *Mailing Add:* Dept Physics Colo Sch Mines Golden CO 80401

LAWFORD, GEORGE ROSS, b Toronto, Ont, Feb 27, 41; m 66; c 2. BIOCHEMISTRY, CELL BIOLOGY. *Educ:* Univ Toronto, BSc, 63, PhD(biochem), 66. *Prof Exp:* Can Med Res Coun fel, 66-68; asst prof biochem, McMaster Univ, 68-73; mem staff, Weston Res Ctr, 73-77, tech dir & gen mgr, 77-89; PRES, ORTECH INT, 89- *Concurrent Pos:* Mem, Nat Biotechnol Adv Comt Can. *Mem:* Can Res Mgt Asn; Can Inst Food Sci & Technol; Am Asn Cereal Chemists. *Res:* Functional significance of interactions between subcellular components; regulation of protein biosynthesis and the adenyl cyclase system; food chemistry; fermentation; research management. *Mailing Add:* Ortech Int 2395 Speakman Dr Mississauga ON L5K 1B3 Can

LAWING, WILLIAM DENNIS, b Charlotte, NC, Mar 29, 35; m 57; c 3. STATISTICS. *Educ:* NC State Col, BS, 57, MS, 59; Iowa State Univ, PhD(statist), 65. *Prof Exp:* Statistician, Res Triangle Inst, 65-69; ASSOC PROF INDUST & EXP STATIST, UNIV RI, 69- *Concurrent Pos:* Adj prof, Duke Univ, 66-67; vis lectr, Iowa State Univ, 67-68; adj assoc prof, NC State Univ, 68-69. *Mem:* Am Statist Asn. *Res:* Industrial applications of statistics; quality control; operations research; sequential analysis; decision theory; survey sampling. *Mailing Add:* Dept Comput Sci & Ext Statist Univ RI Kingston RI 02881

LAWLER, ADRIAN RUSSELL, b Etowah, Tenn, Nov 25, 40; div; c 2. AQUACULTURE, TOXICOLOGY. *Educ:* Univ Rochester, AB, 62; Col William & Mary, Va, MS, 64, PhD(marine biol), 71. *Prof Exp:* NSF summer student, Va Inst Marine Sci, 62, grad asst, 62-71; teaching fel, 71-73, assoc marine biologist, 73-75, co-chmn toxicol prog & head exp organism culture, 76-84, MARINE BIOLOGIST, GULF COAST RES LAB, 75-, AQUARIUM SUPVR, MARINE EDUC CTR, BILOXI, 84- *Concurrent Pos:* Consult, Nat Aquaculture Info Syst, 75- *Res:* Culture of marine and freshwater organisms, birds and mammals for experimentation; toxicity testing; external parasites of marine and freshwater fishes; larval fish development; display of organisms in Mississippi's marine science aquarium; disease diagnosis and control in captive organisms on public display; aquarium design and maintenance. *Mailing Add:* Aquarium Supvr Marine Educ Ctr & Aquarium 115 E Beach Blvd Biloxi MS 39530

LAWLER, EUGENE L(EIGHTON), b New York, NY, July 28, 33; m 79; c 2. APPLIED MATHEMATICS, COMPUTER SCIENCE. *Educ:* Fla State Univ, BS, 54; Harvard Univ, AM, 57, PhD(appl math), 63. *Prof Exp:* Sr elec engr, Sylvania Elec Prod Co, Mass, 59-62; from asst prof to prof elec eng, Univ Mich, 62-70; PROF COMPUT SCI, UNIV CALIF, BERKELEY, 71- *Mem:* Asn Comput Mach; Soc Indust & Appl Math; Inst Mgt Sci. *Res:* Combinatorial optimization; complexity of computations. *Mailing Add:* Dept Comput Sci Univ Calif 571 Evans Hall Berkeley CA 94720

LAWLER, GEORGE HERBERT, b Kingston, Ont, June 13, 23; m 47; c 5. FISHERIES. *Educ:* Queen's Univ Ont, BA, 46; Univ Western Ont, MSc, 48; Univ Toronto, PhD, 59. *Prof Exp:* Demonstr, Zool Lab, Queen's Univ Ont, 44-46 & Univ Western Ont, 46-48; assoc div ichthyol, Royal Ont Mus Zool, 48-50; asst scientist, Fisheries Res Bd, Man, 50-57, assoc scientist, Ont, 57-61, sr scientist, Man, 60-72, from asst dir to dir, Freshwater Inst, 72-74, gen dir, 74-86, co-chmn regulatory rev comm, Dept Fisheries & Oceans, Serv, 86-88; RETIRED. *Concurrent Pos:* Hon prof, Dept Zool, Univ Man, 68-77; consult, 89- *Mem:* Am Fisheries Soc; Am Inst Fishery Res Biol. *Res:* Population dynamics; parasitology and taxonomy of fishes. *Mailing Add:* Freshwater Inst 501 University Crescent Winnipeg MB R3T 2N6 Can

LAWLER, GREGORY FRANCIS, b Alexandria, Va, July 14, 55. PROBABILITY. *Educ:* Univ Va, BA, 76; Princeton Univ, MA, 77, PhD(math), 79. *Prof Exp:* Asst prof, 79-85, ASSOC PROF MATH, DUKE UNIV, 85- *Concurrent Pos:* Vis mem, Courant Inst Math Sci, 81-82 & 86-87; Sloan Res Fel, 86-88. *Mem:* Am Math Soc; Int Asn Math Physics; Int Math Statist. *Res:* Random walks; processes from mathematical physics such as self-avoiding walks and random environments. *Mailing Add:* Dept Math Duke Univ Durham NC 27706

LAWLER, JAMES E, b Jan 15, 46; m; c 1. PSYCHOLOGY. *Educ:* Cornell Col, BA, 67; Wake Forest Univ, MA, 70; Univ NC, Chapel Hill, PhD(physiol psychol), 73. *Prof Exp:* Grad asst, Wake Forest Univ, 67-68; NIH fel neurobiol, Sch Med, Univ NC, 70-72, res asst psychiat, 72-73; res assoc, Cardiovasc Labs, Sch Pub Health, Harvard Univ, 73-75; from asst prof to assoc prof, 75-85, PROF, DEPT PSYCHOL, UNIV TENN, KNOXVILLE, 85-, PROF, DEPT ANAT & NEUROBIOL, CTR HEALTH SCI, 87- *Concurrent Pos:* Mem, High Blood Pressure Res Coun, Am Heart Asn, 77- & Sudden Cardiac Death Rev Comt, Nat Heart, Lung & Blood Inst, NIH, 78; consult progs & planning, Am Psychol Asn, 80; vis assoc prof, Dept Med, Cardiovasc Res & Training Ctr, Sch Med, Univ Ala, 83-84; vis scientist, Minority Access Res Career, Fedn Am Socs Exp Biol, 83-; res incentive award, Sci Alliance, Univ Tenn, 85- & chancellor's fac res scholar award, 87. *Mem:* AAAS; Am Physiol Soc; Soc Behav Med; Soc Neurosci; Soc Psychophysiol Res. *Mailing Add:* Psychophysiol Lab Dept Psychol Univ Tenn Knoxville TN 37996

LAWLER, JAMES EDWARD, b St Louis, Mo, June 29, 51; m 73; c 2. SPECTROSCOPY. *Educ:* Univ Mo, BS, 73; Univ Wis-Madison, MS, 74, PhD(physics), 78. *Prof Exp:* Res assoc, Physics Dept,Stanford Univ, 78-80; from asst prof to assoc prof, 80-89, PROF PHYSICS, UNIV WIS, 89- *Mem:* Am Phys Soc; Optical Soc Am; Sigma Xi. *Res:* Physics of gas discharge plasmas and laser interactions with gas discharge plasmas; laser spectroscopy; nonlinear optics; laboratory astrophysics; atomic physics. *Mailing Add:* Dept Physics Univ Wis 1150 University Ave Madison WI 53706

LAWLER, JAMES HENRY LAWRENCE, b Detroit, Mich, Jan 31, 36; m 64; c 8. CHEMICAL ENGINEERING. *Educ:* Univ Louisville, BChE, 59, MEngr, 72; Brigham Young Univ, MS, 66; Univ Utah, ME & PhD(chem engr), 69. *Prof Exp:* Radio engr, Louisville Free Pub Libr, 52-54, Radio Sta WKLO, 57-58 & WKYW, 58-59; engr, Wright Patterson AFB, 59-62 & Boeing Co, 62-65; asst prof chem, Dixie Col, 69-73; head nuclear eng technol, Trident Technol Col, SC, 73-77; chmn chem technol, Univ Dayton, 77-80; staff engr, Rockwell Hanford Oper, 80-86; engr, Gen Dynamics Nat Aerospace Plane, 86-91; PHYSICS RES ENGR, SUPER CONDUCTING SUPER COLLIDER, 91- *Concurrent Pos:* Atomic Energy Comn fel, Kans State Univ, 72. *Mem:* Am Inst Chem Eng; Am Chem Soc; Inst Elec & Electronic Engrs. *Res:* Materials; cyclic history; socio-mathematics; quantization of space-time; unified theory of gravity, strong and weak nuclear forces, electromagnetic forces, proton theory. *Mailing Add:* 1912 Cedar Desoto TX 75115-7908

LAWLER, JOHN PATRICK, b Brooklyn, NY, Jan 30, 34; m 57; c 7. SANITARY ENGINEERING, MATHEMATICS. *Educ:* Manhattan Col, BCE, 55; NY Univ, MCE, 58; Univ Wis, PhD(sanit eng), 60. *Prof Exp:* Civil engr, F G Davidson, Inc, NY, 55-56; instr civil eng, Manhattan Col, 56-58; asst prof, Rutgers Univ, 60-65; partner, Quirk, Lawler & Matusky Engrs, 65-77; PARTNER, LAWLER, MATUSKY & SKELLY ENGRS, 77- *Concurrent Pos:* Consult, Humble Oil & Refining Co, NJ, 61-62; vis assoc prof, Manhattan Col, 63-, lectr summer inst stream anal, 64-66; assoc, Cosulich & Quirk, Water Resources Engrs, 64-65; lectr summer inst water resources, Clemson Univ, 65 & 66. *Mem:* Am Soc Civil Engrs; Water Pollution Control Fedn; Sigma Xi. *Res:* Mathematical analysis of the transport processes and reaction kinetics associated with stream and estuarine pollution; water resources systems; water and waste treatment operations. *Mailing Add:* Lawler Matusky & Skelly Engrs One Blue Hill Plaza Pearl River NY 10965

LAWLER, MARTIN TIMOTHY, b Rochester, Minn, Apr 6, 37; m 59, 81; c 3. FLUID MECHANICS, HEAT TRANSFER. *Educ:* Milwaukee Sch Eng, BS, 61; Case Inst Technol, MS, 65, PhD(eng), 67. *Prof Exp:* Instr mech eng & physics, Milwaukee Sch Eng, 59-61; res asst, Fluid, Thermal & Aerospace Sci, Case Inst Technol, 64-67; prin res scientist, Corp Res Ctr, Honeywell Inc, Minn, 67-70; vpres, Swenberg Eng Inc, 70-72; MANAGING PARTNER, LAWLER & ASSOCS, 72- *Concurrent Pos:* Nat Defense fel, Comt Acad Sci & Eng, Inst Technol, 61-64. *Mem:* Am Soc Mech Engrs; Nat Soc Prof Engrs; Am Soc Heating, Refrig & Air Conditioning Engrs; Sigma Xi; Am Asn Energy Engrs. *Res:* Fluid-particle and two-phase flows; heat transfer; solar systems, energy consumption and conservation. *Mailing Add:* 5491 Scioto Darby Rd Hilliard OH 43026

LAWLER, RONALD GEORGE, b Centralia, Wash, May 19, 38. PHYSICAL ORGANIC CHEMISTRY. *Educ:* Calif Inst Technol, BS, 60; Univ Calif, Berkeley, PhD(chem), 64. *Prof Exp:* Res assoc chem, Columbia Univ, 63-65; from asst prof to assoc prof, 65-73, PROF CHEM, BROWN UNIV, 73- *Concurrent Pos:* NSF fel, 63-64; Alfred P Sloan res fel, 70-71. *Mem:* Am Chem Soc. *Res:* Theoretical organic chemistry; electron and nuclear magnetic resonance; chemistry of free radicals and radical ions; radiation chemistry; in vivo nmr. *Mailing Add:* Dept Chem Brown Univ Providence RI 02912

LAWLESS, EDWARD WILLIAM, b Jacksonville, Ill, Apr 9, 31; m 59; c 6. TECHNICAL MANAGEMENT, ENVIRONMENTAL CHEMICALS ASSESSMENT. *Educ:* Ill Col, AB, 53; Univ Mo, PhD(phys chem), 60. *Hon Degrees:* DSc, Ill Col, 79. *Prof Exp:* Assoc chemist, Midwest Res Inst, 59-64, sr chemist, 64-66, prin chemist, 66-73, head, Technol Assessment Sect, 73-82, sr adv, Technol & Health Assessment, 82-87, HEAD, ENVIRON ASSESSMENT SECT, MIDWEST RES INST, 87- *Mem:* AAAS; Am Chem Soc; World Future Soc; Soc Risk Anal; Int Asn Impact Assessment; Air & Waste Mgt Asn. *Res:* Hazardous waste incineration and management; technology forecast, risk assessment and societal effects analysis; environmental chemistry and pollution control; evaluation of health and environmental hazards of industrial, consumer product and agricultural chemicals; chemistry of pesticides, fluorine, metal hydrides; correlations of chemical structures with properties; analysis of agricultural innovation; chemical kinetics. *Mailing Add:* Eng & Environ Technol Dept Midwest Res Inst 425 Volker Blvd Kansas City MO 64110

LAWLESS, GREGORY BENEDICT, b Covington, Va, Jan 5, 40; m 66; c 2. PHARMACEUTICAL CHEMISTRY. *Educ:* Fordham Univ, BS, 62; St Johns Univ NY, MS, 65; Temple Univ, PhD(pharmaceut chem), 69. *Prof Exp:* Tech rep, E I Du Pont de Nemours & Co, Inc, 69-71, regional sales mgr, 71-73, nat sales mgr, 73-74, mkt mgr, Instrument Prod, Sci & Process Div, 74-77, bus develop mgr, Riston Prod Div, 77-78, dir, Electronic Mat Div, 78-79, exec vpres, Endo Labs, Inc, 79-81, dir, Dept Plans Div, 81-88; PRES & CHIEF OPER OFFICER, CHIRON CORP, 88- *Mem:* Am Chem Soc. *Res:* Strategic planning for photo products department. *Mailing Add:* Chiron Corp 4560 Horton St Emeryville CA 94608

LAWLESS, JAMES GEORGE, b Brooklyn, NY, Aug 18, 42; m 66; c 1. ANALYTICAL CHEMISTRY. *Educ:* Lafayette Col, BS, 64; Purdue Univ, MS, 66; Kans State Univ, PhD(chem), 69. *Prof Exp:* Res scientist mass spectrometry, 69-, DIV CHIEF, EARTH SYST SCI DIV, AMES RES CTR, NASA. *Concurrent Pos:* Co-investr returned lunar samples, NASA, 70- *Mem:* Am Soc Mass Spectrometry; Geochem Soc; Meteoritic Soc. *Res:* Mass spectrometry of organic compounds; analysis of lunar samples and meteorites for carbon compounds. *Mailing Add:* MS 239-20 Ames Res Ctr Moffett Field CA 94035

LAWLESS, KENNETH ROBERT, b Key West, Fla, Aug 21, 22; m 52; c 4. MATERIALS SCIENCE. *Educ:* Lynchburg Col, BS, 46; Univ Va, PhD(chem), 51. *Prof Exp:* Fulbright fel, Univ Norway, 51-52; res assoc chem, Univ Va, 52-60, from asst prof to assoc prof, 60-68, prof mat sci, 68-80, chmn dept, 76-86; RETIRED. *Mem:* Electron Micros Soc Am; Am Crystallog Asn; Inst Mining, Metall & Petrol Engrs; Microbeam Anal Soc. *Res:* Chemistry and physics of solids and surfaces; x-ray diffraction; electron diffraction and electron microscopy. *Mailing Add:* Dept Math Sci Univ Va Thornton Hall Charlottesville VA 22903

LAWLESS, PHILIP AUSTIN, b Tulsa, Okla, June 7, 43; m 72; c 2. ENGINEERING PHYSICS. *Educ:* Rice Univ, BA, 65; Duke Univ, PhD(physics), 74. *Prof Exp:* PHYSICIST, RES TRIANGLE INST, 74- *Res:* Theoretical and experimental investigations of electrostatic precipitators for particulate control. *Mailing Add:* PO Box 12194 Res Triangle Park NC 27709

LAWLESS, WILLIAM N, b Denver, Colo, Sept 15, 36; m 57; c 3. SOLID STATE PHYSICS. *Educ:* Colo Sch Mines, EMet, 59; Rensselaer Polytech Inst, PhD(physics), 64. *Prof Exp:* Fel solid state physics, Swiss Fed Inst Technol, 64-66; sr res physicist, Res & Develop Labs, Corning Glass Works, 66-68, res assoc physics, 69-80; AT LAKE SHORE CRYOTRONICS, INC, 80- *Concurrent Pos:* Guest worker, Cryogenics Div, Nat Bur Stand, Boulder, 73-75. *Mem:* AAAS; Am Inst Physics; Cryogenic Soc Am; Am Phys Soc. *Res:* Ferroelectricity; doped alkali halides; glass-ceramic technology. *Mailing Add:* 921 Eastwind Dr Suite 110 Westerville OH 43081

LAWLEY, ALAN, b Birmingham, Eng, Aug 29, 33; m 60; c 3. METALLURGY, MATERIALS SCIENCE. *Educ:* Univ Birmingham, BSc, 55, PhD(phys metall), 58. *Prof Exp:* Res assoc metall, Univ Pa, 58-61; lab mgr phys metall, Res Labs, Franklin Inst, Pa, 61-66; assoc prof metall eng, 66-69, chmn, Mat Eng Dept, 69-79, PROF METALL ENG, DREXEL UNIV, 79- *Concurrent Pos:* Consult, Open Univ, UK, 75, Cabot Corp, 81. *Honors & Awards:* Krumb Lectr, Am Inst Mining Metall & Petrol Eng, 85. *Mem:* Am Inst Mining, Metall & Petrol Engrs (pres, 87); fel Am Soc Metals Int; Am Soc Eng Educ; Am Powder Metall Inst; Inst Metals; Electron Microscope Soc Am. *Res:* Powder metallurgy; composite materials; physical and mechanical metallurgy; failure analysis. *Mailing Add:* Dept Mat Eng Drexel Univ Philadelphia PA 19104

LAWLEY, THOMAS J, DERMATOLOGY. *Prof Exp:* PROF & CHMN DERMAT, EMORY UNIV, 89- *Mailing Add:* Dermat Dept Sch Med Emory Univ 201 Woodruff Mem Bldg Atlanta GA 30322

LAWMAN, MICHAEL JOHN PATRICK, b Sept 30, 49; Brit citizen; m. EXPERIMENTAL BIOLOGY. *Educ:* Guildford Co Tech Col, HNC, 72; Ewell Co Tech Col, LIBiol, 73, MIBiol, 75; Univ Surrey, PhD, 79. *Prof Exp:* Asst sci officer, Dept Exp Path, Animal Virus Res Inst, 69-72, sci officer, 72-76, higher sci officer, 76-79; NIH res fel, Dept Microbiol & Immunol, Univ Tenn, 79-80; asst prof virol-immunol, Dept Microbiol, Sch Vet Med, Auburn Univ, 80-81; asst prof immunol, Dept Prev Med, Col Vet Med, Univ Fla, 81-83, Dept Comp & Exp Path, 83-85, asst prof, Dept Immunol & Med Microbiol, Col Med, 82-85; res scientist & prog coordr immunol, Vet Infectious Dis Orgn, Univ Sask, 85-88; ASSOC PROF IMMUNOL, DEPT IMMUNOL & MED MICROBIOL, COL MED, UNIV FLA, 88-, ADJ ASSOC PROF & DIR RES, DEPT PEDIAT, DIV HEMAT-ONCOL, 89- *Concurrent Pos:* Prin investr, USDA, 81-84, 82-86 & 84-87, Natural Sci & Eng Res Coun Can, 86-89 & 87-90; NSF travel scholar, Ruminant Immunol Conf, Kenya, 84, Nat Acad Sci travel scholar, 85; consult, Upjohn, 84, Nat Acad Sci, 85 & USAID, 88-90; co-investr & consult porcine immunity to African swine fever virus, USAID, 87-90. *Mem:* Inst Biol; Conf Res Workers Animal Dis; Asn Am Immunologists; Am Soc Microbiol; Am Vet Immunologist Asn; NY Acad Sci. *Res:* Foot-and-mouth disease virus in British deer; epizootic haemorrhagic disease of deer virus in domestic farm animals and in British deer; in vitro characteristics of epizootic haemorrhagic disease of deer virus; pathogenesis of bluetongue virus in sheep; peste des petits ruminants; vertical transmission of bluetongue virus in sheep; electrophoretic studies on the double stranded RNA genome of orbviruses; stimulation and growth of macrophage of various animals species in in vitro culture. *Mailing Add:* Dept Immunol & Med Microbiol Univ Fla Col Med Box J-266-JHMHC Gainesville FL 32610

LAWRASON, F DOUGLAS, b St Paul, Minn, July 30, 19; m 44; c 3. INTERNAL MEDICINE. *Educ:* Univ Minn, BA, 41, MA & MD, 44. *Prof Exp:* Instr anat, Med Sch, Univ Minn, 41-43; from intern to resident, Sch Med, Yale Univ, 44-49, from instr to asst prof med, 49-50; prof assoc, Nat Res Coun, 50-53; asst prof & asst dean, Sch Med, Univ NC, 53-55; prof internal med, provost med affairs & dean med ctr, Univ Ark, 55-61; exec dir med res, Merck Sharp & Dohme Res Labs, Pa, 61-66, vpres med res, 66-69; prof internal med, Univ Tex Health Sci Ctr Dallas, 69-73, assoc dean acad affairs, 69-72, dean, 72-73; sr vpres, Sci Affairs & pres, Res Div, 73-80, SR VPRES SCI, SCHERING-PLOUGH CORP, 80- *Concurrent Pos:* James Hudson Brown res fel, Yale Univ, 48-49; mem hemat study sect, NIH, 51-53; comt blood & related probs, Nat Acad Sci, 53-57; inst grant comt, Am Cancer Soc, 55-60; training grant comt, Nat Res Coun, 59-64; mem bd dirs, Morristown Mem Hosp, 80- & NJ State Sci Adv Comt, 80- *Mem:* AAAS; Am Fedn Clin Res; Am Col Cardiologists; NY Acad Sci; Am Soc Internal Med. *Res:* Cancer and leukemia in inbred strains of mice; hematology; medical education and administration; research management. *Mailing Add:* Morristown Mem Hosp 53 Spring Valley Rd Morristown NJ 07920

LAWRENCE, ADDISON LEE, b Cape Girardeau, Mo, Dec 19, 35; m 91; c 8. SHRIMP CULTURE. *Educ:* Southeast Mo State Univ, BSc, 56; Univ Mo, MA, 58, PhD(physiol), 62. *Prof Exp:* Asst prof biol, Westminster Col, Fulton, Mo, 61-62; postdoctoral fel physiol, Stanford Univ, 62-64; prof, Univ Houston, 64-79; PROF NUTRIT & MARICULT, TEX A&M UNIV, 79- *Concurrent Pos:* Vis instr, Hopkins Marine Sta, Stanford Univ, 63; assoc dir res, Univ Houston, 75-77, dir, Marine Sci Prog, 77-78; maricult coordr, Tex A&M Sea Grant Prog, 79-86. *Mem:* Am Soc Zoologists; Soc Exp Biol Med; World Aquacult Soc; Nat Shellfish Soc; Western Soc Naturalists; Crustacean Soc. *Res:* Nutrition; physiology; biochemistry; shrimp mariculture; shrimp reproduction; shrimp raceway and pond production; shrimp larviculture; author of over 120 technical publications. *Mailing Add:* PO Drawer 1725 Port Aransas TX 78373

LAWRENCE, ALONZO WILLIAM, b Rahway, NJ, Apr 11, 37; m 60; c 3. ENVIRONMENTAL ENGINEERING. *Educ:* Rutgers Univ, BS, 59; Mass Inst Technol, MS, 60; Stanford Univ, PhD(civil eng), 67. *Prof Exp:* Asst prof civil eng, Drexel Inst Technol, 65-67; from asst prof to assoc prof environ eng, Cornell Univ, 67-76; vpres environ resources & occup health, 76-81, vpres, sci & technol, 81-84, VPRES & GEN MGR, CHEM SYSTS SECTOR, KOPPERS CO, PITTSBURGH, 84- *Honors & Awards:* Eng Sci Award, Am Asn Environ Eng Prof, 77. *Mem:* Am Soc Civil Engrs; Water Pollution Control Fedn. *Res:* Wastewater treatment technology; wastewater reclamation and reuse; biokinetics; solid wastes disposal; occupational safety and health; technological innovation; strategic management of technology. *Mailing Add:* 4299 Old New England Allison Park PA 15101

LAWRENCE, CHRISTOPHER WILLIAM, b London, Eng, Oct 2, 34; m 61; c 3. GENETICS, RADIOBIOLOGY. *Educ:* Univ Wales, BSc, 56; Univ Birmingham, PhD(genetics), 59. *Prof Exp:* Sci officer radiation biol, Wantage Labs, UK Atomic Energy Authority, 59-61, sr sci officer, 61-70; assoc prof, 70-82, PROF RADIATION BIOL, UNIV ROCHESTER, 82- *Concurrent Pos:* Vis asst prof radiation biol, Univ Rochester, 69. *Mem:* AAAS; Genetics Soc Am; Biophys Soc. *Res:* Radiation molecular genetics of Saccharomyces cerevisiae. *Mailing Add:* Dept Biophys Univ Rochester Med Sch 601 Elmwood Rochester NY 14642

LAWRENCE, DALE NOLAN, b Covington, Ky, Feb 24, 44; m 73; c 2. INTERNAL MEDICINE & INFECTIOUS DISEASES, GENETICS OF HUMAN IMMUNE RESPONSES. *Educ:* Duke Univ, MD, 69; Emory Univ, MPH, 89; Am Bd Internal Med, cert, 73; Am Bd Infectious Dis, cert, 75. *Prof Exp:* Intern, resident & fel internal med & infectious dis, Univ Tex Health Sci Ctr, San Antonio, 69-73; epidemiol intel serv officer, Field Serv Div, Ctr Dis Control, USPHS, Miami, Fla, 73-75, med officer immunol & parasitol, Clin Immunol Lab, Immunochem Br, Parasitol Div, Bur Labs, 75-77, chief immunogenetics, Clin Med Br, Divs Immunol & Host Factors, Ctr Infectious Dis, 79-82, task force mem & epidemiologist, AIDS Task Force & Div Host Factors, 82-89; sect chief clin develop, 89-91, SR SCI ADV COORDR FOR AIDS VACCINE, INT STUDIES, VACCINE RES & DEVELOP BR, BASIC RES & DEVELOP PROG, DIV AIDS, NAT INST ALLERGY & INFECTIOUS DIS, NIH, 91- *Concurrent Pos:* Resident gen prev med & epidemiol, Ctrs for Dis Control, 74-76, physician mem med adv bd, 79-81; med officer, Genetics Res Prog, NSF, Amazonas, Brazil, 76; vis scientist human leukocyte antigen genetics, Genetics Lab, Dept Biochem, Univ Oxford, UK, 78-79; ed consult, J AMA, J Nat Cancer Inst, J Infectious Dis, J AIDS, Sci & Am J Epidemiol, 79-; vis fel, Harvard Inst Health Res, 84-85; field investr team leader, Ctrs Dis Control-Mayo Clin Collab Reinvest: Swine Influenza Vaccine-Guillain Barre Syndrome, 85-87; consult, Global Prog on AIDS, WHO, 91-; prog officer AIDS panel, US-Japan Coop Med Sci Prog, 91- *Honors & Awards:* Spec Award, Nat Hemophilia Found, 90. *Mem:* Fel Infectious Dis Soc Am; fel Am Col Epidemiol; Asn Immunologists; fel Am Col Physicians; fel Am Col Prev Med; Am Soc Histocompatibility & Immunogenetics. *Res:* Early infectious disease epidemiologic investigations; studies of human leukocyte antigen genetics; vaccine immunology and development of AIDS vaccine. *Mailing Add:* NIH-NIAID-DAIDS 6003 Executive Blvd Rockville MD 20892

LAWRENCE, DAVID A, b Paterson, NJ, Jan 9, 45; m 67. IMMUNOLOGY. *Educ:* Rutgers Univ, BA, 66; Boston Col, MS, 68, PhD(biol), 71. *Prof Exp:* USPHS fel, Scripps Clin & Res Found, 71-74; from asst prof to assoc prof microbiol & immunol, 74-84, assoc prof med, 84-88, PROF MICROBIOL & IMMUNOL, ALBANY MED COL, 84-, PHARM & TOXICOL, 88-, PATH, 89- *Concurrent Pos:* NIH Toxicol Study Sect; Epa Health Res Rev Panel. *Mem:* NY Acad Sci; Am Soc Microbiol; Am Asn Immunologists; Soc Toxicol. *Res:* Cellular and subcellular events resulting from antigen activation and regulation of immune response; tumor immunology; immunotoxicology. *Mailing Add:* Dept Microbiol & Immunol Albany Med Col 43 New Scotland Ave Albany NY 12208

LAWRENCE, DAVID JOSEPH, b Johnson City, NY, June 15, 51. ELECTRONICS, SEMICONDUCTORS. *Educ:* Syracuse Univ, BS, 73; Cornell Univ, MS, 75, PhD(electrophys), 77. *Prof Exp:* Develop engr, Western Elec Co, 77-78; RES PHYSICIST SEMICONDUCTOR MAT, EASTMAN KODAK RES LABS, 78- *Mem:* Electrochem Soc; Inst Elec & Electronics Engrs. *Res:* III-V compounds; epitaxial growth; solid state light emitters and detectors. *Mailing Add:* 25 Woodmill Dr Kodak Park Bldg 81 Rochester NY 14626

LAWRENCE, DAVID REED, b Woodbury, NJ, Oct 11, 39; m 66. GEOLOGY, INVERTEBRATE PALEONTOLOGY. *Educ:* Johns Hopkins Univ, AB, 61; Princeton Univ, PhD(geol), 66. *Prof Exp:* Asst geol, Princeton Univ, 63-64, prof assoc, 66; asst prof, 66-69, ASSOC PROF GEOL & MARINE SCI, UNIV SC, 69- *Concurrent Pos:* NSF sci fac fel, Univ Tubingen, Ger, 71-72. *Mem:* Int Paleont Union; Geol Soc Am; Paleont Soc. *Res:* Evolutionary, ecologic and biogeographic aspects of fossil invertebrates; taphonomy; historiography of the earth sciences. *Mailing Add:* Dept Geol Univ SC Columbia SC 29208

LAWRENCE, DONALD BUERMANN, b Portland, Ore, Mar 8, 11; m 35. PLANT ECOLOGY, ETHNOBOTANY. *Educ:* Johns Hopkins Univ, PhD(plant physiol), 36. *Prof Exp:* Researcher Sigma Xi grant, 36-37; from instr to prof, 37-76, EMER PROF BOT, UNIV MINN, MINNEAPOLIS, 76- *Concurrent Pos:* Mem, Johns Hopkins Univ exped, Jamaica, 32, Am Geog Soc exped, Glacier Bay, Alaska, 41, 49, 50, 52, 55, S Chile & Arg, 59 & 67; spec consult, US Air Force, 48; dir terrestial ecosyst proj, Hill Found, 57-61; Fulbright res fel, NZ, 64-65; US del Int Sea Ice Conf, Reykjavik, Nat Acad Sci Nat Res Coun, 71; Glacier Bay Nat Park & Preserv study, Nat Park Serv, 67, 68, 72, 73, 82, 83, 85 & 88; Sigma Xi Nat Lectr Northeastern Tour, 56-57. *Honors & Awards:* Cert Commendation, Ecol Soc Am, 90. *Mem:* fel Am Geog Soc; Soc Econ Bot; Am Name Soc. *Res:* Vegetation development; physiographic ecology; ecological life histories of plants; causes of climatic change; glaciology; Asian plant names and place names in the New World; history of plant dispersal by man; economic botany. *Mailing Add:* 2420 34th Ave S Minneapolis MN 55406-1427

LAWRENCE, DONALD GILBERT, b Kingston, Ont, Jan 18, 32; m 56; c 3. NEUROLOGY, NEUROANATOMY. *Educ:* Bishop's Univ, BSc, 53; McGill Univ, MDCM, 57; Royal Col Physicians & Surgeons, FRCP(C), 74. *Prof Exp:* Res fel neuroanat, Western Reserve Univ, 65-66; Nat Multiple Sclerosis Soc fel neurophysiol, Univ Lab Physiol, Oxford Univ, 66-68; from asst prof to assoc prof neuroanat, Erasmus Univ, 68-72; head, lab neuroanat, Montreal Neuro Inst, 72-84; assoc prof neurol & neurosurg, 72-80, ASSOC PROF ANAT, MCGILL UNIV, 72-, PROF NEUROL & NEUROSURG, 80-, ASSOC DEAN, FAC MED, 84- *Concurrent Pos:* Asst physician, Montreal Gen Hosp, 72-77, assoc physician, 77- *Honors & Awards:* Osler Medal, Am Asn Hist Med, 58. *Mem:* Cajal Club; Am Asn Hist Med; Am Asn Anat; Soc Neurosci; Am Acad Neurol; Can Asn Neurosci. *Res:* Anatomical, behavioral and clinical investigations of motor pathways in the central nervous system; regeneration in the central nervous system. *Mailing Add:* Neurosci Unit Montreal Gen Hosp 1650 Cedar Ave Montreal PQ H3G 1A4 Can

LAWRENCE, FRANCIS JOSEPH, b Glen Arm, Md, May 12, 25; m 51; c 4. PLANT BREEDING. *Educ:* Univ Md, BS, 51, MS, 58, PhD(hort, bot), 65. *Prof Exp:* Asst hort, Univ Md, 53-62, from instr to asst prof, 62-65; RES HORTICULTURIST, CORVALLIS RES STA, USDA, 65- *Mem:* Am Soc Hort Sci; Am Pomol Soc. *Res:* Breeding of Fragaria and Rubus. *Mailing Add:* 1430 NW 27th Corvallis OR 97702

LAWRENCE, FRANKLIN ISAAC LATIMER, b Brooklyn, NY, July 14, 05; m 34; c 1. ORGANIC CHEMISTRY. *Educ:* NY Univ, BS, 28, MS, 30. *Prof Exp:* Jr engr, Atlantic Ref Co, Pa, 30-34, tech sales supvr, 34-42; develop engr, Permutit Co, NY, 42-44; res dir, Kendall Ref Div, Witco Chem Corp, 44-68, vpres res & develop, 68-70, CONSULT, 70- *Honors & Awards:* Cert Appreciation, Am Petrol Indust. *Mem:* Am Chem Soc; Soc Automotive Eng; fel Am Inst Chem. *Res:* Petroleum chemistry; ion exchange. *Mailing Add:* 23 Lake Hunter Dr Lakeland FL 33803

LAWRENCE, FREDERICK VAN BUREN, JR, b Hyannis, Mass, May 16, 38; m 62; c 3. METALLURGY, CIVIL ENGINEERING. *Educ:* Swarthmore Col, BS, 60; Mass Inst Technol, SM, 62, CE, 65, ScD(mat sci), 68. *Prof Exp:* PROF METALL & CIVIL ENG, UNIV ILL, URBANA, 68- *Concurrent Pos:* Ed, J Mat Civil Eng, Am Soc Civil Engrs. *Mem:* Am Welding Soc; Am Soc Metals; Am Inst Mining, Metall & Petrol Engrs; Sigma Xi; Am Soc Civil Engrs. *Res:* Fatigue strength of welded joints; microstructure of cementitious materials. *Mailing Add:* Dept Civil Eng 2129 CEB Univ Ill Urbana IL 61801

LAWRENCE, GENE ARTHUR, electrical engineering, for more information see previous edition

LAWRENCE, GEORGE EDWIN, b Berkeley, Calif, Mar 25, 20; m 43; c 4. ZOOLOGY. *Educ:* Univ Calif, AB, 46, MA, 49, PhD, 60. *Prof Exp:* Asst zool, Univ Calif, 46-47; from instr to assoc prof, 47-67, prof life sci, Bakersfield State Col, 67-, head dept, 70-87; RETIRED. *Mem:* Wilderness Soc; Cooper Ornith Soc; Ecol Soc Am; Nat Biol Teachers Asn. *Res:* Natural history of mammals and birds; ecology of chaparral association; temperature tolerance of small mammals. *Mailing Add:* Hwy Contract 8 Tehachapi CA 93561

LAWRENCE, GEORGE MELVIN, b Salt Lake City, Utah, Mar 26, 37; m 59; c 3. PHYSICS. *Educ:* Univ Utah, BS, 59; Calif Inst Technol, PhD(physics), 63. *Prof Exp:* Res assoc astrophys sci, Princeton Univ, 63-65, staff physicist, 65-67; res scientist, McDonnell Douglas Advan Res Lab, Calif, 67-70; vis fel, Joint Inst Lab Astrophys-Lab Atmospheric & Space Physics, 70-71, res assoc, 71-74, FEL, LAB ATMOSPHERIC & SPACE PHYSICS, UNIV COLO, BOULDER, 74- *Concurrent Pos:* Vis assoc, Calif Inst Technol, 68-70. *Mem:* Fel Am Phys Soc; Am Geophys Union. *Res:* Transition probabilities; cross sections; physical chemistry; detectors; space science. *Mailing Add:* Lab Atmospheric & Space Phys 55th Univ Colo Boulder CO 80309

LAWRENCE, HENRY SHERWOOD, b New York, NY, Sept 22, 16; m 43; c 3. IMMUNOLOGY, MEDICINE. *Educ:* NY Univ, AB, 38, MD, 43; Am Bd Internal Med, dipl; FRCPS(G), 77. *Prof Exp:* Intern, 3rd Med Div, Bellevue Hosp, NY, 43-44, from asst resident to chief resident, 46-48; asst, 47-49, from instr to prof, 49-79, dir, NY Univ Cancer Inst, 74-79, JEFFREY BERGSTEIN PROF MED, MED SCH, NY UNIV, 79-, HEAD, INFECTIOUS DIS & IMMUNOL UNIT, 59-, CO-DIR, NY UNIV-

BELLEVUE MED SERV, 64- *Concurrent Pos:* Wyckoff fel, NY Univ, 48-49; dir student health serv, NY Univ-Bellevue Med Serv, 50-57; Commonwealth Found fel, Univ Col, Univ London, 59; USPHS career develop award, 60-65; assoc mem streptococcal comt, Armed Forces Epidemiol Bd; consult, Allergy & Immunol Study Sect, USPHS, 60-65, chmn, 63-65, mem, comt cutaneous syst & comt tissue transplantation, Div Med Sci, Nat Res Coun-Nat Acad Sci, chmn, comt tissue transplantation, 63-65, mem, Nat Res Coun, 70; lectr, Harvey Soc, 73, counr, 74-77; consult & chmn, Allergy & Infectious Dis Panel, Health Res Coun City of New York, infectious dis prog comt res serv, Vet Admin & res comts, Arthritis Found, Am Cancer Soc & Am Thoracic Soc; dir, NY Univ Ctr AIDS Res, 89- *Honors & Awards:* Von Pirquet Gold Medal Award, Annual Forum Allergy, 72; Distinguished Contribs in Sci as Related to Med Award & Bronze Medal, Am Col Physicians, 73; NY Acad Med Sci Medal, 74; Bristol Award, Infectious Dis Soc Am, 74; Chapin Medal, 75; Lila Gruber Award for Cancer Res, Am Acad Dermat,75. *Mem:* Nat Acad Sci; Am Soc Clin Invest; Am Asn Immunologists; Harvey Soc (secy, 57-60); Asn Am Physicians; hon fel Am Acad Allergy; fel Am Col Physicians; Infectious Dis Soc; Soc Exp Biol & Med. *Res:* Infection and immunity; cellular immunology; T-cell hybridoma production; purification and characterization of inducer and suppressor factor in dialysates containing transfer factor. *Mailing Add:* Infectious Dis & Immunol Div NY Univ Med Ctr New York NY 10016

LAWRENCE, IRVIN E, JR, b Raleigh, NC, Apr 18, 26. EMBRYOLOGY, HISTOLOGY. *Educ:* Univ NC, AB, 50; Univ Wyo, MS, 55; Univ Kans, PhD(anat), 63. *Prof Exp:* Teacher high sch, NC, 51-54; instr biol, Louisburg Col, 55-57; asst prof zool, Univ Wyo, 60-64; assoc prof biol, 64-70, assoc prof, 70-78, PROF ANAT, ECAROLINA UNIV, 78- *Concurrent Pos:* Univ Res fel, Univ Wyo, 63-64; USPHS res grant, 64-65; NIH res grant, 75- *Mem:* Am Soc Zool; Soc Develop Biol; Sigma Xi; Pan-Am Asn Anat; Soc Study Reproduction. *Res:* Biogenic amines in development; epithelial-mesenchymal interactions in organogenesis of pancreatic islets and of ovary; ovarian nerves and reproductive function. *Mailing Add:* Dept Anat ECarolina Univ Sch Med Greenville NC 27834

LAWRENCE, JAMES FRANKLIN, b Okemah, Okla, Aug 20, 50. MATHEMATICS. *Educ:* Okla State Univ, BS, 72; Univ Wash, PhD(math), 75. *Prof Exp:* Instr math, Univ Tex, Austin, 75-77; res assoc, Nat Bur Standards, 77-79; ASST PROF, UNIV KY, 79- *Concurrent Pos:* Vis asst prof, Univ Mass, Boston, 80-81. *Mem:* Math Asn Am; Am Math Soc; Sigma Xi. *Res:* Field of combinatorics; study of oriented matroids. *Mailing Add:* Dept Math Sci George Mason Univ Fairfax VA 22030

LAWRENCE, JAMES HAROLD, JR, b Beatrice, Nebr, Feb 9, 32; m 55; c 2. MECHANICAL ENGINEERING. *Educ:* Tex Tech Col, BS, 56, MS, 60; Tex A&M Univ, PhD(mech eng), 65. *Prof Exp:* From instr to asst prof, 56-62, assoc prof, 64-71, chmn dept, 72-83, PROF MECH ENG, TEX TECH UNIV, 71- *Mem:* Am Soc Eng Educ; Am Soc Mech Engrs; Am Soc Heat, Refrig & Air-Conditioning Engrs. *Res:* Conduction; convection; radiation heat transfer; systems engineering. *Mailing Add:* Dept Mech Eng Tex Tech Univ Lubbock TX 79409

LAWRENCE, JAMES NEVILLE PEED, b Norfolk, Va, May 29, 29; m 48; c 1. HEALTH PHYSICS, PHYSICS. *Educ:* Johns Hopkins Univ, BA, 50; Vanderbilt Univ, MA, 58, PhD(physics), 68. *Prof Exp:* Res asst health physics, Los Alamos Nat Lab, Univ Calif, 51-54, mem staff, 54-68, assoc group leader, 68-80, asst group leader, 80-88, SR SCIENTIST, LOS ALAMOS NAT LAB, UNIV CALIF, 88- *Mem:* Health Physics Soc. *Res:* Theoretical treatment of nuclear fission, especially liquid drop applications; health physics, especially dosimetry, internal exposure calculations and radionuclide identification. *Mailing Add:* 206 El Conejo Los Alamos NM 87544

LAWRENCE, JAMES VANTINE, b Middletown, Ohio, July 2, 18; m 42; c 3. BACTERIOLOGY. *Educ:* Univ Ill, BS, 43; Ohio State Univ, MS, 48, PhD(bact), 50. *Prof Exp:* From asst prof to prof bact, Ohio Univ, 50-80, prof zool microbiol, 80-85; RETIRED. *Concurrent Pos:* Vis prof, WVa Univ, 51 & 52; clin lab dir, Athens State Hosp, 53, 54 & 56. *Mem:* Am Soc Microbiol; Am Pub Health Asn; Sigma Xi. *Res:* General and pathogenic bacteriology and parasitology. *Mailing Add:* Dept Zool & Microbiol Ohio Univ Athens OH 45701

LAWRENCE, JEANNE BENTLEY, b Sweetwater, Tex, Dec 10, 51; c 3. GENE MAPPING, IN SITU HYBRIDIZATION. *Educ:* Stepens Col, BA, 73; Rutgers Univ, MS, 75; Brown Univ, PhD(molecular & cell biol), 82. *Prof Exp:* Res assoc, Yale Univ, 75-77; postdoctoral fel, 82-85, instr cell biol, Dept Anat, 85-88, ASST PROF, DEPT CELL BIOL, UNIV MASS MED CTR, 88- *Concurrent Pos:* Res career & develop award, Nat Ctr Human Genome Res, NIH, 90. *Honors & Awards:* Jr Outstanding Cell Biologist Career Develop Award, Women Cell Biol, 89. *Mem:* AAAS; Am Soc Cell Biologists; Am Asn Human Genetics. *Res:* Analysis of genome and nuclear organization using fluorescence in situ hybridization; human gene mapping and the functional relationship of DNA/RNA to nuclear structure. *Mailing Add:* Cell Biol Dept Med Ctr Univ Mass 55 Lake Ave N Worcester MA 01655

LAWRENCE, JOHN, b UK, Mar 23, 43; Can citizen; m 65; c 2. ENVIRONMENTAL CHEMISTRY, ANALYTICAL CHEMISTRY. *Educ:* Bristol Univ, BSc, 64, PhD(chem), 67. *Prof Exp:* Res assoc electrochem, Univ Ottawa, 67-69 & Colo State Univ, 69-71; res chemist semiconductos, Bell-Northern Res, 71-73; res scientist environ toxic contaminants, 73-80, res mgr, environ anal methodology, 80-87, DIR RES & APPLN BR, NAT WATER RES INST, CAN, 87- *Mem:* Int Asn Great Lakes Res. *Res:* Water chemistry; water treatment; toxic contaminants; analytical methods; quality assurance; hydraulics; environmental research. *Mailing Add:* Nat Water Res Inst PO Box 5050 Burlington ON L7R 4A6 Can

LAWRENCE, JOHN HUNDALE, medicine; deceased, see previous edition for last biography

LAWRENCE, JOHN KEELER, b New York, NY, Oct 11, 40; m 67; c 1. GENERAL RELATIVITY, ASTROPHYSICS. *Educ:* Harvard Univ, AB, 62; Northeastern Univ, MS, 64, PhD(physics), 68. *Prof Exp:* Vis asst prof physics & astron, Univ Ga, 67-68; Univ asst physics, Univ Vienna, 68-71; vis res assoc, Northeastern Univ, 71-72; fel, Univ Windsor, 72-73; asst prof, 73-76, assoc prof physics & astron, 76-80, PROF PHYSICS & ASTRON, CALIF STATE UNIV, 80-, DEPT CHMN PHYSICS & ASTRON, 76- *Mem:* Am Phys Soc; AAAS; Austrian Phys Soc; Sigma Xi. *Res:* Deflection of null radiation by gravitational fields; cosmological models; cosmological coincidences; galactic structure; active galaxies and quasars. *Mailing Add:* Dept Physics & Astron Calif State Univ Northridge CA 91330

LAWRENCE, JOHN M, b Cape Girardeau, Mo, Oct 11, 37. PHYSIOLOGY. *Educ:* Southeast Mo State Col, BS, 58; Univ Mo, AM, 60; Stanford Univ, PhD(biol), 66. *Prof Exp:* Instr physiol, Stanford Univ, 64-65; from asst prof to assoc prof, 65-75, PROF DEPT BIOL, UNIV SFLA, 75- *Concurrent Pos:* Fel, Marine Biol Lab, Hebrew Univ Israel, 69-70. *Mem:* Am Soc Zool; Marine Biol Asn UK; Sigma Xi. *Res:* Nutritional and reproductive physiology of marine invertebrates. *Mailing Add:* Dept Biol Univ SFla Tampa FL 33620

LAWRENCE, JOHN MCCUNE, b Carmichaels, Pa, Feb 17, 16; m 38; c 3. BIOCHEMISTRY. *Educ:* Carnegie Inst Technol, BS, 37, MS, 39; Univ Pittsburgh, PhD(biochem), 43. *Prof Exp:* Res asst, Mellon Inst, 39-41; Nutrit Found fel, Dept Dairy Indust, Cornell Univ, 43-45, from instr to asst prof biochem, 45-48; assoc chemist, Wash State Univ, 48-58, agr chemist, 58-81; RETIRED. *Mem:* AAAS; Am Soc Plant Physiol; Sigma Xi. *Res:* Enzymes; proteins; amino acids; biochemistry of seed germination and nutritional quality of seeds. *Mailing Add:* Northeast 1225 Orchard Dr Pullman WA 99163

LAWRENCE, JOHN MEDLOCK, b Cedar Bluff, Ala, Sept 25, 19; m 47; c 3. FISHERIES. *Educ:* Ala Polytech Inst, BS, 41, MS, 43; Iowa State Univ, PhD, 56. *Prof Exp:* Asst fish culturist, 46-57, assoc fish culturist, 57-63, prof fisheries, Auburn Univ, 63-81; RETIRED. *Mem:* Am Fisheries Soc; Weed Sci Soc Am. *Res:* Methods for control of aquatic weeds and their effects upon fish production. *Mailing Add:* 1037 Terrace Acres Auburn AL 36830

LAWRENCE, JOSEPH D, JR, b Anderson, SC, Nov 6, 24; m 49; c 3. SYSTEMS DESIGN & SYSTEMS SCIENCE. *Educ:* Va Polytech Inst, BA, 51. *Prof Exp:* Engr/mgr, Tex Instruments, 66-70; mgr digital systs eng, Camco, 70-73; vpres, Datac Co, 73-77, pres, 77-78; mgr res & develop, Reliability Inc Tex, 78-82, vpres & dir technol, Reliability Inc, 82-87, pres, Reliability Japan Inc, 87- 90, VPRES, RELIABILITY INC, 87- *Mailing Add:* 115 Gershwin Houston TX 77079

LAWRENCE, KENNETH BRIDGE, b Susquehanna, Pa, June 18, 18; m 45; c 2. MECHANICAL ENGINEERING. *Educ:* Pa State Univ, BS, 39, MS, 48. *Prof Exp:* Asst engr, SKF Industs, Inc, 40-41 & 44-45; instr math, Pa State Univ, 46-47, from instr to assoc prof mech eng, 47-56; sr res engr, Res Div, Curtiss-Wright Corp, 56-59, proj engr, Curtiss Div, 59-63; head dept, 63-68, prof mech eng, Manhattan Col, 63-; RETIRED. *Mem:* Soc Automotive Engrs; Am Soc Eng Educ. *Res:* Lubrication, friction and wear; lightweight reinforced materials and construction; industrial products. *Mailing Add:* Fifth Ave N Matamoras PA 18336

LAWRENCE, KENT L(EE), b Beatrice, Nebr, Jan 23, 37; m 61; c 3. MECHANICAL ENGINEERING, ENGINEERING MECHANICS. *Educ:* Tex Tech Col, BS, 59, MS, 60; Ariz State Univ, PhD(eng mech), 65. *Prof Exp:* Instr mech, Univ Ill, 60-61; assoc prof mech eng, 61-62 & 64-77, PROF MECH ENG, UNIV TEX, ARLINGTON, 77- *Concurrent Pos:* Consult, Aerotechnol Dept, Bell Helicopter Co, 65- *Mem:* Am Inst Aeronaut & Astronaut; Am Soc Mech Engrs; Soc Exp Stress Anal; Am Helicopter Soc. *Res:* Vibrations; dynamics; structural mechanics; finite element methods. *Mailing Add:* 1309 Windmill Ct Arlington TX 76013

LAWRENCE, KURT C, b Decatur, Ga, July 9, 62; m 85; c 2. DIELECTRICS. *Educ:* Univ Ga, BS, 85, MS, 87. *Prof Exp:* AGR ENGR, AGR RES SERV, USDA, 85- *Mem:* Assoc mem Am Soc Agr Engrs. *Res:* Measurement of dielectric properties of agricultural products; determination of moisture content by dielectric properties measurements. *Mailing Add:* USDA Agr Res Serv Russell Res Ctr Box 5677 Athens GA 30613

LAWRENCE, MERLE, b Remsen, NY, Dec 26, 15; m 42; c 3. PHYSIOLOGY. *Educ:* Princeton Univ, AB, 38, MA, 40, PhD(psychol), 41. *Prof Exp:* Nat Res Coun fel, Johns Hopkins Univ, 41; from asst prof to assoc prof psychol, Princeton Univ, 46-52; assoc prof physiol acoust, Med Sch, Univ Mich, Ann Arbor, 52-57, prof otolaryngol & psychol, 57-85, prof physiol, 59-65, res assoc, Inst Indust Health, 52-85, dir Kresge Hearing Res Inst, 61-82, EMER PROF UNIV MICH, ANN ARBOR, 85- *Concurrent Pos:* Consult, Surgeon Gen Off, 53- & Secy Defense, 55-58; mem commun disorders res training comt, Nat Inst Neurol Dis & Stroke, 61-65 & commun sci study sect, NIH, 65-70, mem communicative disorders rev comt, Nat Inst Neurol & Communicative Disorders & Stroke, 72-76, Nat Adv Neurol & Commun Dis & Stroke Coun, 76-80. *Honors & Awards:* Gold Medal, Am Otol Soc; Award of Merit, Am Acad Opthalmol & Otolaryngol. *Mem:* Fel Acoust Soc Am; Am Laryngol, Rhinol & Otol Soc; Am Otol Soc; Col Oto-Rhino-Laryngol Amicitiae Sacrum; Asn Res Otolaryngol; Soc Univ Otolaryngologist; Am Acad Otolaryngol. *Res:* Physiology of hearing. *Mailing Add:* 1535 Shorelands Dr E Vero Beach FL 32963-2648

LAWRENCE, PAUL J, b Hazleton, Pa, Dec 18, 40; m 63; c 1. BIOCHEMISTRY. *Educ:* King's Col, Pa, BS, 62; Univ Wis, Madison, MS, 64, PhD(biochem), 67. *Prof Exp:* From asst prof to assoc prof biochem, Col Med, Univ Utah, 68-77; dir immunol res & develop, Smith Kline Instruments, 77-81, dir, diagnostics res & develop, 79-; PRES, LAWRENCE ASSAY, INC; PRES, LITMUS CONCEPTS INC. *Concurrent Pos:* Fel biochem, Univ Wis, 67-68; NIH fel, 67-69. *Mem:* AAAS; Am Chem Soc; Fedn Am Socs Exp Biol; Am Soc Microbiol; NY Acad Sci. *Res:* Mechanisms of drug action; development of diagnostic tests. *Mailing Add:* Litmus Concepts Inc 3485-A Kifer Rd Santa Clara CA 95051

LAWRENCE, PAULINE OLIVE, b Nov 10, 45; US citizen; m 76. INSECT ENDOCRINOLOGY. *Educ:* Univ WI, Jamaica, BSc Hons, 68; Univ Fla, Gainesville, MS, 72, PhD(entomol), 75. *Prof Exp:* Asst entomologist res, Ministry Agr, Jamaica, WI, 68-69; grad res fel, Dept Entomol, Univ Fla, Gainesville, 69-72, res asst, 72-75, from asst prof to assoc prof, 76-89, PROF RES & TEACHING, DEPT ZOOL, UNIV FLA, GAINESVILLE, 89- *Concurrent Pos:* Prin investr, regulatory biol, NSF, 81-84, 85-88 & 90-; vis assoc prof, Dept Entom, Cornell Univ, 84-85; mem & chmn, Nat Res Coun & NSF Minority Grad Fel Eval Panel Biol Sci, Biochem, Biophys & Biomed Sci, 87-89; assoc, Danforth Found; USDA Competitive Grants rev panel; competitive res grants, entom & nematol, USDA, 87-90; McKnight Found fel, 86-87. *Honors & Awards:* Career Advan Award Women, NSF, 88. *Mem:* AAAS; Am Soc Zoologists; Entom Soc Am; Sigma Xi; Tissue Cult Asn; Europ Soc Endocrinol. *Res:* Parasite biology and the influence of host hormones on parasite development; development of resistance by hosts to parasites and resulting feed-back on host and parasite populations; parasite ecology and behavior; the physiology of host-parasite interactions. *Mailing Add:* Dept Zool Univ Fla 223 Bartram Hall Gainesville FL 32611

LAWRENCE, PHILIP LINWOOD, b New Bedford, Mass, Mar 27, 23; m 48; c 2. GEOPHYSICS, OIL SEARCH. *Educ:* Colo Sch Mines, GeolE, 49, Southern Methodist Univ, MS, 60. *Prof Exp:* Res physicist, Magnolia Petrol Co, Tex, 49-62; staff adv, NY, 63-65, unit supvr, Geophys Serv Ctr, Tex, 66-68, corp geophysicist, 69-81, CONSULT, OIL EXPLORATION, MOBIL OIL CORP, 81- *Concurrent Pos:* Lectr elec eng, Southern Methodist Univ, 60-62. *Mem:* Sigma Xi. *Res:* Seismic, magnetic, gravity data gathering, processing and interpretation techniques and development for mineral exploration. *Mailing Add:* 467 Harvest Glen Richardson TX 75081

LAWRENCE, PHILIP SIGNOR, applied statistics, public health, for more information see previous edition

LAWRENCE, RAYMOND JEFFERY, b Cornwall, NY, Feb 25, 39; m 63; c 2. SHOCK WAVE PHYSICS. *Educ:* Lawrence Col, BA, 61; Univ NMex, MS, 70. *Prof Exp:* Res physicist shock wave physics, Air Force Weapons Lab, Kirtland AFB, NMex, 63-67; MEM TECH STAFF SHOCK WAVE PHYSICS, SANDIA LABS, 67- *Mem:* Am Phys Soc. *Res:* Shock wave physics; numerical wave propagation computer code development and application; constitutive model development; application of these fields to weapons development, energy research, weapon effects, and other dynamic phenomena. *Mailing Add:* 1308 Kirby St NE Albuquerque NM 87112

LAWRENCE, RICHARD AUBREY, b Logan, Utah, Feb 17, 47; m 70; c 5. BIOCHEMISTRY, NUTRITION. *Educ:* Brigham Young Univ, BS, 71; Univ Wis-Madison, PhD(biochem), 75. *Prof Exp:* Fel med, Univ Tex Health Sci Ctr, Dallas, 75-77, instr med, 77-78; instr, Sch Med, La State Univ, Shreveport, 78-79, asst prof med & biochem, 79-; RES ASST PROF, DEPT MED, UNIV TEX HEALTH SCI CTR. *Mem:* Am Inst Nutrit; Sigma Xi. *Res:* Selenium nutrition and metabolism; role of lipid peroxidation in tissue injury and disease and physiological mechanisms for protection against it; mechanism of lipid peroxidation; mechanisms of oxygen toxicity. *Mailing Add:* Dept Med Univ Tex Hlth Sci Ctr 7703 Floyd Curl Dr San Antonio TX 78284

LAWRENCE, ROBERT D, b Ithaca, NY, May 24, 43; m 66; c 2. GEOLOGY. *Educ:* Earlham Col, BA, 65; Stanford Univ, PhD(geol), 68. *Prof Exp:* Asst prof geol, Earlham Col, 68-70; asst prof, 70-77, ASSOC PROF GEOL, ORE STATE UNIV, 77- *Concurrent Pos:* Fulbright Prof, Peshawar Univ, Pakistan, 81-82 & 86-87. *Mem:* Geol Soc Am; Geologischen Vereinigung; Am Geophys Union; Khyben Geol Soc. *Res:* Major faults of western North America; tectonics of Pakistan; tectonic history of Pacific Northwest. *Mailing Add:* Dept Geosci Wilkinson Hall 104 Ore State Univ Corvallis OR 97331

LAWRENCE, ROBERT G, b Wilmington, NY, Feb 14, 21; m 46; c 2. ZOOLOGY. *Educ:* Eastern Nazarene Col, AB, 44; Boston Univ, MA, 46; Okla State Univ, PhD, 64. *Prof Exp:* Teacher, Henry Ford's Boys Sch, Mass, 45-46; prof biol & head dept, Bethany Nazarene Col, 47-68; assoc dean, Mid-Am Nazarene Col, 68-71, prof biol sci, 68-75, dir instnl res, 71-74, acad dean, 74-75; PROF BIOL, MT VERNON NAZARENE COL, OHIO, 75-, ACAD DEAN, 75-, VPRES ACAD AFFAIRS, 76- *Concurrent Pos:* Chmn div natural sci, Bethany Nazarene Col, 49-68; mem, Am Conf Acad Deans. *Mem:* Am Ornith Union; Nat Audubon Soc; Wilson Soc; Am Asn Higher Educ. *Res:* Ornithology; relation of weather factors to migration of water fowl. *Mailing Add:* Mt Vernon Nazarene Col 800 Martinsburg Rd Mt Vernon OH 43050

LAWRENCE, ROBERT MARSHALL, b Kennecott, Alaska, June 28, 23; m 50; c 9. ANESTHESIOLOGY. *Educ:* Univ Rochester, MD, 49. *Prof Exp:* Resident surg, Strong Mem Hosp, 49-54, resident anesthesiol, 54-56; from instr to assoc prof surg & med, 56-68, PROF ANESTHESIOL, SCH MED, UNIV ROCHESTER, 68- *Concurrent Pos:* Consult, Vet Admin Hosps, Batavia & Canandaigua, 56- *Mem:* Am Soc Anesthesiol; Am Asn Respiratory Ther. *Res:* Oxygen toxicity; respiratory therapy. *Mailing Add:* Univ Rochester Med Ctr Box 604 Rochester NY 14642

LAWRENCE, ROBERT SWAN, b Philadelphia, Pa, Feb 6, 38; m 60; c 5. INTERNAL MEDICINE, PREVENTIVE MEDICINE. *Educ:* Harvard Univ, AB, 60; Harvard Med Sch, MD, 64. *Prof Exp:* Intern & jr resident, Mass Gen Hosp, 64-66; asst surgeon epidemiol, Epidemic Intel Serv, Ctr Dis Control, USPHS, Atlanta, 66-69; sr resident, Mass Gen Hosp, 69-70; from asst prof med to assoc prof, Sch Med, Univ NC, 70-74; asst prof, 74-80, ASSOC PROF MED, HARVARD MED SCH, 80-, DIR, DIV PRIMARY CARE, 74- *Concurrent Pos:* Chief, Dept Med, Cambridge Hosp, 80-; chmn bd, prev med div, Inst Med-Nat Acad Sci, 84, chmn, US Prev Serv Task, Dept Health & Human Serv, 84-89; ed, Am J Prev Med, 90- *Honors & Awards:* Spec Recognition Award, Am Col Prev Med, 88. *Mem:* Inst Med-Nat Acad Sci; Am Col Physicians; Soc for Res & Educ Primary Care Med (pres, 78-79); Am Fedn for Clin Res; Soc Teachers Prev Med; Am Col Prev Med. *Res:* Health beliefs of patients; medical sociology; primary care education; manpower issues. *Mailing Add:* Cambridge Hosp 1493 Cambridge St Cambridge MA 02139

LAWRENCE, SIGMUND J(OSEPH), b Chicago, Ill, May 20, 18; div; c 6. CHEMICAL ENGINEERING. *Educ:* Ill Inst Technol, BS, 39; Univ Iowa, MS, 42, PhD(chem eng), 43. *Prof Exp:* Apprentice engr, Caterpillar Tractor Co, Ill, 39-40; res chem engr, Armour Res Found, Ill Inst Technol, 40-41; chem engr, Shell Develop Co, Calif, 43-46; chem engr, Gen Elec Co, Wash, 46-48, res assoc, Knolls Atomic Power Lab, 48-51, Res Lab, 51-52, process engr, Chem Div, Silicone Prods Dept, 52-57, chem engr, Gen Eng Lab, 57-59, appln engr, Process Comput Control, Syst Sales & Eng, 60-67, appln engr spec sensors, Instrument Dept, 67, mgr sensor progs, 67-69, consult engr indust process systs, Reentry & Environ Systs Div, 69-73; SUPVRY ENGR DESIGN CHEM PLANTS, CATALYTIC INC, PHILADELPHIA, 74- *Mem:* Am Chem Soc; Am Inst Chem Eng. *Res:* Process instrumentation, automation and control; computer monitoring and control; air, water and waste monitoring; pollution control; instrument development and design; process analysis and design. *Mailing Add:* 100 Linda Lane Media PA 19063

LAWRENCE, THOMAS, plant breeding, for more information see previous edition

LAWRENCE, VINNEDGE MOORE, b Bangor, Maine, Feb 19, 40; m 66; c 1. ENTOMOLOGY, ECOLOGY. *Educ:* Miami Univ Ohio, BS, 62, MA, 64; Purdue Univ, PhD(entom), 68. *Prof Exp:* Instr biol, Xavier Univ Ohio, 64-65; from asst prof to assoc prof, 68-86, PROF BIOL, WASHINGTON & JEFFERSON COL, 86- *Concurrent Pos:* Mem citizen's adv coun, Pa Dept Environ Resources, 71-81. *Mem:* AAAS; Am Inst Biol Sci; Sigma Xi; NAm Benthological Soc. *Res:* Population dynamics of Odonata naiads in farm-pond ecosystems; distribution of Odonata in Pennsylvania; microdistribution of capniid Plecoptera; breeding behavior of Henslow's sparrow. *Mailing Add:* Dept Biol Washington & Jefferson Col Washington PA 15301-4801

LAWRENCE, WALTER, JR, b Chicago, Ill, May 31, 25; m 47; c 4. SURGERY. *Educ:* Univ Chicago, PhB, 45, SB, 46, MD, 48. *Prof Exp:* Intern surg, Johns Hopkins Hosp, 48-49, asst resident & asst, Sch Med, Johns Hopkins Univ, 49-50, Halsted fel, Johns Hopkins Hosp, 50; resident, Mem Ctr Cancer & Allied Dis, 51-52 & 54-56; res fel exp surg, Mem Ctr Cancer & Allied Sci, 56; instr surg, Med Col, Cornell Univ, 57-58, asst prof, 58-63, clin assoc prof, 63-66; chmn div surg oncol, Med Col Va, 66-90, Am Cancer Soc prof clin oncol, 72-77, dir, Massey Cancer Ctr, 74-88, PROF SURG, MED COL VA, 66-, EMER DIR, 88- *Concurrent Pos:* Asst mem, Sloan-Kettering Inst Cancer Res, 57-60, assoc mem & assoc chief div exp surg, 60-66; clin asst attend surgeon, Mem Hosp, 57-59, asst attend surgeon, 59-62, assoc vis surgeon, 62-; asst vis surgeon, James Ewing Hosp, 57-62, assoc vis surgeon, 62-66; mem surg staff, NY Hosp, 57-66. *Honors & Awards:* Sloan Award Cancer Res, 64; Horsley Award, 73; Distinguished Serv Award, Univ Chicago, 76. *Mem:* Halsted Soc (pres, 75); fel Am Col Surg; Am Surg Asn; Soc Surg Oncol (pres, 80); Soc Univ Surgeons; Soc Head & Neck Surgeons. *Res:* Surgery, particularly cancer and cancer research. *Mailing Add:* Box 11 Dept Surg Med Col Va Richmond VA 23298

LAWRENCE, WALTER EDWARD, b Albany, NY, May 22, 42; m 69. THEORETICAL SOLID STATE PHYSICS. *Educ:* Carnegie Inst Technol, BS, 64; Cornell Univ, PhD(physics), 70. *Prof Exp:* Res assoc physics, Stanford Univ, 69-71; asst prof, 71-77, ASSOC PROF PHYSICS, DARTMOUTH COL, 77- *Mem:* Am Phys Soc. *Res:* Solid state theory, principally superconductivity; transport theory of metals. *Mailing Add:* Dept Physics Dartmouth Col Hanover NH 03755

LAWRENCE, WILLARD EARL, b Chassell, Mich, Apr 8, 17; m 43; c 5. STATISTICS. *Educ:* Marquette Univ, BS, 51, MS, 53; Univ Wis, MS, 62, PhD(statist), 64. *Prof Exp:* From instr to assoc prof, 53-69, asst chmn dept, 58-63, prof, 69-87, chmn dept math & statist, 73-79, EMER PROF MATH & STATIST, MARQUETTE UNIV, 87- *Concurrent Pos:* Consult, NSF, 66-69; statistician, Oak Ridge Nat Lab, 79-80. *Mem:* Math Asn Am. *Res:* Experimental design; response surface designs which minimize variance and bias errors; designs for mixtures; probability. *Mailing Add:* Dept Math Statist & Comput Sci Marquette Univ 540 N 15th St Milwaukee WI 53233

LAWRENCE, WILLIAM CHASE, b Cambridge, Mass, July 10, 34; m 55; c 3. VIROLOGY, MOLECULAR BIOLOGY. *Educ:* Univ Mass, BS, 55; Univ Pa, VMD, 59, PhD(microbiol), 66. *Prof Exp:* From asst prof to assoc prof, 65-82, PROF MICROBIOL, UNIV PA, 82- *Concurrent Pos:* USPHS res grant, 67-74; USDA Res Grant, 75- *Mem:* AAAS; Am Vet Med Asn; Am Soc Microbiol; NY Acad Sci; Sigma Xi. *Res:* Molecular virology. *Mailing Add:* Dept Pathobiol Sch Vet Med Univ of Pa 3800 Spruce St Philadelphia PA 19104

LAWRENCE, WILLIAM HOMER, b Magnet Cove, Ark, Mar 20, 28; div; c 4. TOXICOLOGY, PHARMACOLOGY. *Educ:* Col Ozarks, BS, 50; Univ Md, MS, 52, PhD(pharmacol), 55. *Prof Exp:* Instr materia medica, Sch Nursing, Univ Md, 51, asst pharmacol, Sch Pharm, 51-54; from asst prof to assoc prof pharmacol & physiol, Col Pharm, Univ Houston, 56-66; from assoc prof to prof toxicol, Univ Tenn Health Sci Ctr, 66-83, asst dir mat sci toxicol labs, 67-75, head animal toxicol sect, 67-85, assoc dir mat sci toxicol labs, 75-85, actg chmn dept drug & mat toxicol, 81-83, PROF MED CHEM, UNIV TENN HEALTH SCI CTR, 83-, VCHMN, 90- *Concurrent Pos:* Vis scientist, Univ Tex, 63, consult, Drug-Plastic Res & Toxicol Labs, 63-68, vis scientist, dept pharmacol & toxicol, Med Br at Galveston, 65, lectr, 64-67; vis scientist, dept pharmacol & toxicol, Univ Tex Med Br, Galveston, 65; consult, Vet Admin Hosp, Houston, 64-66 & Memphis, 67-72; mem, Gen Toxicity & Screening Task Force, Int Dent Fedn; adv, Neurostimulation Subcomt, Am Asn Med Implants, 74-78; mem ed bd, J Toxicol Environ Heath, 78-, J Pharmacol Sci, 82-88; dipl, Acad Toxicol Sci, 84-, fel, 89. *Mem:* Soc Toxicol; Am Pharmaceut Asn; Sigma Xi. *Res:* Toxicity of biomaterials and medical devices, especially dental materials, blood bags, intravenous administration tubings, extracorporeal devices, implantable devices and carcinogenic studies of plastics; in vivo activity and toxicity of some novel inhibitors of platelet aggregation. *Mailing Add:* Dept Med Chem Univ Tenn Health Sci Ctr Memphis TN 38163

LAWRENCE, WILLIAM MASON, b Brooktondale, NY, Oct 2, 18; m 42; c 2. FISHERIES, NATURAL RESOURCES. *Educ:* Cornell Univ, BS, 38, PhD(fishery biol), 41. *Prof Exp:* Biometrician bur game, NY State Conserv Dept, 41-42, sr aquatic biologist, 46-52, chief bur fish, 52-55, dir fish & game, 55-58, asst comnr, 58-64, dep comnr, NY State Dept Environ Conserv, 64-74; consult, 74-85; RETIRED. *Concurrent Pos:* US comnr, Great Lakes Fishery Comn; chmn, Atlantic States Marine Fisheries Comn, 65-88 & 71-73. *Honors & Awards:* Seth Gordon Award, Int Asn Fish & Wildlife, 76. *Mem:* Am Fisheries Soc (pres, 59); Int Asn Fish & Wildlife Agencies (pres, 69). *Res:* Fish and wildlife conservation; water resources. *Mailing Add:* 991 White Church Rd Brooktondale NY 14817

LAWRENZ, FRANCES PATRICIA, b Milwaukee, Wis, Nov 6, 47; c 1. EVALUATION, IN-SERVICE TEACHER TRAINING. *Educ:* Univ Minn, Minneapolis, BS, 68, MA, 71, PhD(educ, chem & math), 74. *Prof Exp:* Volunteer, Peace Corps, Ankara, Turkey, 69-70; asst prof chem, St Mary's Jr Col, 71-74; sci consult, Nat Assessment Ed Prog, 74-75; asst prof sci & phys sci educ, State Univ NY, Buffalo, 75-77; proj mgr, NSF, 77-78; consult adult educ, Govt Yukon, 78-79; evaluator health, Area Heatlh Educ Ctr, 79-80; ASST PROF SCI EDUC, ARIZ STATE UNIV, 81- *Concurrent Pos:* Surv researcher, Ariz Dept Energy, 82-83; eval consult, YMCA, 82-, Native Am Sci Educ Asn, 84-, Med Sch, Univ Ariz, 84-, Northern Ariz Univ, 85-; ed, Ariz Health Educ Newsletter & assoc ed, Sch Sci & Math, 83- *Mem:* Am Educ Res Asn; Nat Asn Res Sci Teaching; Asn Educ Teachers Sci; Sch Sci & Math Asn; Eval Network; Am Sch Health Asn. *Res:* Evaluation of in-service teacher training programs in physical science and health, improvement of the quality of training and determination of what variables affect outcomes. *Mailing Add:* Dept Circulation & Instr Col Educ Univ Minn 159 Pillsbury Dr SE Minneapolis MN 55455

LAWREY, JAMES DONALD, b Arlington, Va, Dec 15, 49. LICHENOLOGY. *Educ:* Wake Forest Univ, BS, 71; Univ SDak, MA, 73; Ohio State Univ, PhD(bot), 77. *Prof Exp:* ASSOC PROF BIOL, GEORGE MASON UNIV, 77- *Mem:* Am Bryol & Lichenological Soc; Ecol Soc Am; Brit Lichen Soc; Mycol Soc Am; Bot Soc Am. *Res:* Population and community ecology of lichens; ecological significance of lichen secondary compounds; use of lichens as biological indicators of atomospheric pollution. *Mailing Add:* Biol Dept George Mason Univ 4400 University Dr Fairfax VA 22030

LAWRIE, DUNCAN H, b Chicago, Ill, Apr 26, 43; m. COMPUTER SCIENCE, ELECTRICAL ENGINEERING. *Educ:* DePauw Univ, BA, 66; Purdue Univ, BSEE, 66; Univ Ill, MS, 69, PhD(comput sci), 73. *Prof Exp:* Sr res programmer, Univ Ill, Urbana-Champaign, 70-73, vis res asst prof, 73-74, from asst prof to assoc prof, 74-84, PROF COMPUTER SCI, UNIV ILL, URBANA-CHAMPAIGN, 84-, HEAD, DEPT COMPUTER SCI, 90- *Mem:* Fel Inst Elec & Electronics Engrs; Asn Comput Mach. *Res:* Computer system organization; especially very large systems; memory hierarchies. *Mailing Add:* Dept Comput Sci Univ Ill 1304 W Springfield Ave Urbana IL 61801

LAWROSKI, HARRY, b Dalton, Pa, Oct 10, 28; m 62. CHEMICAL & NUCLEAR ENGINEERING. *Educ:* Pa State Univ, BS, 50, MS, 56, PhD(chem eng), 59. *Prof Exp:* Asst, Pa State Univ, 50-56, instr petrol refining, 56-58; assoc chem engr, Idaho Div, Argonne Nat Lab, 58-63, tech mgr, Zero Power Plutonium Reactor, 63-68, supt EBR II Opers, 68-73; gen mgr environ, Nuclear Serv Corp, 73-76; asst gen mgr, Idaho Chem Progs, Allied Chem Corp, 76-79; CONSULT, 79- *Concurrent Pos:* Instr, Nat Reactor Testing Sta, Univ Idaho, 59-; chmn, Nuclear Div, Am Inst Chem Engrs, 67-68. *Mem:* Fel Am Nuclear Soc (treas, 71-77, vpres, 79-80, pres, 80-81); fel Am Inst Chem Engrs. *Res:* Reactor engineering; petroleum refining. *Mailing Add:* 2375 Belmont Ave Idaho Falls ID 83404

LAWROSKI, STEPHEN, b Scranton, Pa, Jan 17, 14; m 47; c 2. NUCLEAR ENGINEERING. *Educ:* Pa State Univ, BS, 34, MS, 39, PhD(chem eng), 43. *Prof Exp:* From asst to supvr res & develop, Petrol Refining Lab, Pa, 34-43; res chem engr, Standard Oil Develop Co, NJ, 43-44, asst sect chief, 46; group leader, Manhattan Proj, Math Lab, Univ Chicago, 44-46; adv prof trainee, Clinton Lab, Oak Ridge, Tenn, 46-47; dir, Chem Eng Div, Argonne Nat Lab, 47-63, assoc lab dir, 63-70, sr engr, 70-80; RETIRED. *Concurrent Pos:* Consult, US Army Chem Corps, 60-66; mem gen adv comt, US AEC, 64-70; mem adv comt reactor safeguards, US Nuclear Regulatory Comn, 74-81. *Honors & Awards:* Robert E Wilson Award, 71. *Mem:* Nat Acad Eng; AAAS; Am Chem Soc; fel Am Nuclear Soc; Sigma Xi; Am Inst Chem Engrs. *Res:* Petroleum technology; separations processes in atomic energy field; composition and utilization of several typical hydrocarbon naphthas; nuclear technology. *Mailing Add:* 1700 Robin Lane Apt 345 Lisle IL 60532

LAWS, EDWARD ALLEN, b Columbus, Ohio, Feb 4, 45. OCEANOGRAPHY. *Educ:* Harvard Univ, BA, 67, PhD(chem physics), 71. *Prof Exp:* Instr oceanog, Fla State Univ, 71-74; asst prof oceanog, 74-80, ASSOC PROF OCEANOG, UNIV HAWAII, 80- *Mem:* AAAS; Am Soc Limnol & Oceanog; Ecol Soc Am; Phycol Soc Am. *Res:* Metabolism of carbon and nitrogen by marine phytoplankton; importance of conditioning in regulating growth characteristics and metabolism; response of phytoplankton communities to nutrient enrichments. *Mailing Add:* Dept Oceanog Univ Hawaii at Manoa Honolulu HI 96822

LAWS, EDWARD RAYMOND, JR, b New York, NY, Apr 29, 38; m 62; c 4. NEUROSURGERY. *Educ:* Princeton Univ, AB, 59; Johns Hopkins Univ, MD, 63. *Prof Exp:* Intern, Johns Hopkins Hosp, 63-64; asst chief toxicol, Commun Dis Ctr, USPHS, Ga, 64-66; asst prof neurol surg & neurol surgeon, Johns Hopkins Hosp, 66-72; from assoc prof to prof neurol surg, Mayo Med Sch, Mayo Clin, 72-87, neurol surgeon, 72-87; PROF & CHMN, MED CTR, GEORGE WASHINGTON UNIV, 87- *Concurrent Pos:* USPHS res grants, 60-62; Henry Strong Denison fel, 62-63; fel surg, Johns Hopkins Univ, 63-64. *Mem:* Sigma Xi. *Res:* Neurooncology; pituitary surgery. *Mailing Add:* Dept Neurosurg George Washington Univ Med Ctr 2150 Pennsylvania Ave NW Washington DC 20037

LAWS, KENNETH LEE, b Pasadena, Calif, May 30, 35; m 65; c 2. METEOROLOGY, ELECTRONICS. *Educ:* Calif Inst Technol, BS, 56; Univ Pa, MS, 59; Bryn Mawr Col, PhD(physics), 62. *Prof Exp:* Instr physics, Hobart & William Smith Cols, 58-59; from asst prof to assoc prof, 62-78, from asst dean to assoc dean, 71-77, PROF PHYSICS, DICKINSON COL, 78- *Mem:* Am Asn Physics Teachers; Am Meteorol Soc; Sigma Xi. *Res:* Physics of dance. *Mailing Add:* Dickinson Col Carlisle PA 17013-2896

LAWS, LEONARD STEWART, b Pocasset, Okla, Dec 29, 17; m 43; c 4. MATHEMATICS. *Educ:* Willamette Univ, AB, 39; Stanford Univ, MA, 41; Mich State Univ, EdD, 53. *Prof Exp:* Asst math, Stanford Univ, 39-41; asst math & mech, Univ Minn, 41-42, from instr to asst prof, 42-52; dean & regstr, 53-55, chmn, Natural Sci Div, 55-74, PROF MATH, SOUTHWESTERN COL, KANS, 55- *Concurrent Pos:* Asst, Mich State Univ, 47 & 53; NSF fel, Stanford Univ, 61-62; mgt consult, 64- *Mem:* Am Soc Qual Control; Am Statist Asn. *Res:* Industrial reliability; design of experiments. *Mailing Add:* 311 Houston St Winfield KS 67156

LAWS, PRISCILLA WATSON, b New York, NY, Jan 18, 40; m 65. NUCLEAR PHYSICS. *Educ:* Reed Col, BA, 61; Bryn Mawr Col, MA, 63, PhD(physics), 66. *Prof Exp:* Asst physics, Bryn Mawr Col, 61-63; sr tech aide, Bell Labs, 62; from asst prof to assoc prof, 65-79, PROF PHYSICS, DICKSON COL, 79- *Concurrent Pos:* Mem med radiation adv comt, Bur Radiation Health, Food & Drug Admin, Dept Health, Educ & Welfare, 74-78; consult, Off Technol Assessment, US Cong. *Mem:* Am Asn Physics Teachers; Health Physics Soc. *Res:* Nuclear beta decay; environmental radiation; effects of medical x-rays, energy and environment; use of microcomputers in science education; research and development on microcomputer based laboratory materials for science education in K-Colleges; research on energy efficient control systems for residential use. *Mailing Add:* Dept Physics & Astron Dickinson Col Carlisle PA 17013

LAWSON, ANDREW COWPER, II, b Chicago, Ill, Oct 21, 46. PHYSICS. *Educ:* Pomona Col, BA, 67; Univ Calif, San Diego, MS, 69, PhD(physics), 72. *Prof Exp:* Asst res physicist, Inst Pure & Appl Physics, Univ Calif, San Diego, 72-77; asst prof physics, Pomona Col, 77-82; assoc prof, Mech Engr, Calif State Univ, Long Beach, 82-83; STAFF MEM, PHYS METALLURGY GROUP, LOS ALAMOS NAT LAB, 83- *Mem:* AAAS; Am Phys Soc; Am Asn Physics Teachers; Am Crystallog Asn. *Res:* Superconductivity in relation to crystal structure; behavior of the electrical resistance of metals; occurance of crystallographic transformations at low temperatures; applications of neutron scattering to materials science. *Mailing Add:* 300 Aragon Ave Los Alamos NM 87544

LAWSON, ANTON ERIC, b Lansing, Mich, Oct 24, 45; m 68; c 3. SCIENCE EDUCATION. *Educ:* Univ Ariz, BS, 67; Univ Ore, MS, 69; Univ Okla, PhD(sci educ), 73. *Prof Exp:* Teacher math & sci, Ralston Intermediate Sch, Belmont, Calif, 69-71; instr sci educ, Univ Okla, 72-73; res assoc biol educ, Purdue Univ, 73-74; res educr, Univ Calif, Berkeley, 74-77; asst prof, 77-80, ASSOC PROF SCI EDUC, ARIZ STATE UNIV, 80- *Honors & Awards:* Res in Sci Teaching Award, Nat Asn for Res in Sci Teaching, 76. *Mem:* Nat Asn Res Sci Teaching; Asn Educ Teachers Sci; Sch Sci & Math Asn; Nat Asn Biol Teaching; AAAS. *Res:* Development of formal reasoning; psychology of teaching science and mathematics. *Mailing Add:* 9227 S Juniper Tempe AZ 85284

LAWSON, BENJAMIN F, b Montgomery, Ala, May 29, 31; m 56; c 3. ORAL MEDICINE, PERIODONTOLOGY. *Educ:* Auburn Univ, BS, 53; Emory Univ, DDS, 61; Ind Univ, MSD, 63; Univ Ala, cert periodont, 68. *Prof Exp:* Asst prof oral diag & chmn dept, Sch Dent, Emory Univ, 63-66; asst prof, Sch Dent, Univ Ala, 66-68; from assoc prof to prof oral med & periodont, Sch Dent, Univ SC, 68-72, chmn dept, 68-70, chmn dept, Col Dent Med, 70-72, dean, Col Appl Health Sci, 72-90; RETIRED. *Concurrent Pos:* NIH training grant, 61-63; drug study grants, Eli Lilly & Co, 62, 65, 67 & 70; gen practr, 63-66; consult, Vet Admin, Birmingham & Charleston, Ft Benning, Ft Jackson & Charleston Co & Med Univ hosps, 64 & 67-69; Warner-Lambert drug study grants, 70; int consult health educ, Pakistan, 81 & Thailand, 83. *Mem:* Am Dent Asn; Am Soc Allied Health Professions. *Res:* Histology-histopathology of dental pulp, ceramic and titanium implants; vital pulpal therapy. *Mailing Add:* Col Allied Health Sci/Dean Allied Health Med Univ SC 171 Ashley Ave Charleston SC 29425-2701

LAWSON, CHARLES ALDEN, b Philadelphia, Pa, May 2, 51. ROCK MAGNETISM, EXPERIMENTAL PETROLOGY. *Educ:* Univ Calif, Santa Cruz, BS, 73; Princeton Univ, MA, 76, PhD(geol), 81. *Prof Exp:* Geologist, Bechtel Corp, 73-74; res fel, Geophys Lab, Carnegie Inst Washington, 77-79, Princeton Univ, 81-82 & NASA-Johnson Space Ctr, 82-84; geologist, 84-87, PROG OFFICER, US GEOL SURV, 87- *Concurrent Pos:* Rock magnetism working group comt, Int Asn Geomagnetism & Aeronomy, 83-; Tellers Comt, Mineral Soc Am, 85-86. *Mem:* Am Crystallographic Asn; Am Geophys Union; Electron Micros Soc Am; Int Asn Geomagnetism & Aeronomy; Mineral Soc Am; Sigma Xi. *Res:* Chemical and physical properties of iron-titanium oxides; correlation of microstructures and magnetic properties of ilmenite-hematite minerals; scanning and transmission electron microscopy of geologic materials. *Mailing Add:* Sci Attache Am Embassy Tel Aviv APO New York NY 09672-0001

LAWSON, CHARLES L, b Idaho, 31. MATHEMATICS. *Educ:* Univ Calif Berkeley, BS, 52; UCLA, PhD(math), 61. *Prof Exp:* SUPVR APPL MATH GROUP, JET PROPULSION LAB, CALIF TECH INST, 60- *Mem:* Asn Comput Mach; Soc Indust & Appl Math; Math Asn Am. *Res:* Algorithms in numerical linear algebra; development of mathematical software. *Mailing Add:* Caltech Jet Propulsion Lab Pasadena CA 91109

LAWSON, DANIEL DAVID, b Tucson, Ariz, Jan 13, 29; m 57; c 2. ORGANIC CHEMISTRY, POLYMER CHEMISTRY. *Educ:* Univ Southern Calif, BS, 57, MS, 59. *Prof Exp:* Biomed res fel, Charles Cook Hastings Found, 59-61; res polymer scientist, 61-71, MEM TECH STAFF,

POLYMER RES SECT, JET PROPULSION LAB, CALIF INST TECHNOL, 71- *Mem:* AAAS; Am Chem Soc; Royal Soc Chem; fel Am Acad Forensic Sci; Brit Soc Chem Indust. *Res:* Physical organic chemistry of polymers; synthesis of new biomaterials; use of thermoluminescence as applied to criminalistics; high energy batteries. *Mailing Add:* 5542 Halifax Rd Arcadia CA 91006

LAWSON, DAVID EDWARD, b Moncton, NB, Sept 17, 39; m 67. SEDIMENTOLOGY. *Educ:* Univ NB, BSc, 60, MSc, 62; Univ Reading, PhD(geol), 71. *Prof Exp:* Res geologist, Sedimentology Res Lab, Univ Reading, 62-66; lectr, 66-68; asst prof, 68-76, ASSOC PROF SEDIMENTOLOGY, UNIV WATERLOO, 76- *Mem:* Int Asn Sedimentol; Soc Econ Paleontologists & Mineralogists; fel Geol Asn Can. *Res:* Environmental fluvial sedimentology; nearshore sedimentation; primary sedimentary structures; volcanic sediments; continental shelf sedimentation; Torridonian sediments of northwest Scotland. *Mailing Add:* Dept Earth Sci Univ Waterloo Waterloo ON N2L 3G1 Can

LAWSON, DAVID FRANCIS, b Chicago, Ill, June 24, 45; m 85; c 3. POLYMER ORGANIC CHEMISTRY, FLAMMABILITY & SMOKE GENERATION OF POLYMERS. *Educ:* Lewis Univ, BA, 67; Iowa State Univ, PhD(org chem), 71. *Prof Exp:* Instr chem, Iowa State Univ, 68-69; res scientist, 70-75; sr res scientist, 75-81, ASSOC SCIENTIST, CENT RES LABS, BRIGESTONE/FIRESTONE, INC, 82- *Concurrent Pos:* Vis scholar, Ohio Acad Sci, 90- *Mem:* Am Chem Soc; Adhesion Soc; Sigma Xi. *Res:* Organic polymer chemistry; polymer synthesis; combustion, smoke, and flammability of polymers; synthetic and physical organic chemistry; elastomer synthesis and chemistry; engineering thermoplastics; adhesives; polymer surface chemistry and adhesion. *Mailing Add:* 11621 Garden Lane NW Uniontown OH 44685

LAWSON, DAVID MICHAEL, b Denver, Colo, Nov 13, 43; m 66; c 3. REPRODUCTIVE ENDOCRINOLOGY, LACTATION. *Educ:* Va Polytech Inst, BS, 65, MS, 67; Cornell Univ, PhD(physiol), 70. *Prof Exp:* NIH trainee physiol, Cornell Univ, 71; res assoc, 71-73, asst prof, 73-79, ASSOC PROF PHYSIOL, SCH MED, WAYNE STATE UNIV, 79- *Mem:* Endocrine Soc; Soc Exp Biol & Med. *Res:* Control of prolactin secretion; molecular nature of prolactin form synthesis through release; endocrine control of mammary gland function. *Mailing Add:* Dept Physiol 5374 Scott Hall Wayne State Univ Sch Med 540 E Canfield Detroit MI 48201

LAWSON, DEWEY TULL, b Kinston, NC, Feb 6, 44; m 66; c 2. ACOUSTICS. *Educ:* Harvard Univ, AB, 66; Duke Univ, PhD(physics), 72. *Prof Exp:* Res assoc physics, Lab Atomic & Solid State Physics, Cornell Univ, 72-74; asst prof physics, Duke Univ, 74-79; SR PHYSICIST, RES TRIANGLE INST, 79- *Concurrent Pos:* Consult archit & environ acoust & adj prof physics, Duke Univ, 80- *Mem:* Am Phys Soc; Am Asn Physics Teachers. *Res:* Low temperature physics, especially liquid and solid helium. *Mailing Add:* PO Box 12194 Res Triangle Inst Research Triangle Park NC 27709

LAWSON, EDWARD EARLE, b Winston-Salem, NC, Aug 6, 46; m 69; c 2. PEDIATRICS, NEONATOLOGY. *Educ:* Harvard Univ, BA, 68; Sch Med, Northwestern Univ, MD, 72. *Prof Exp:* Residency pediat, Children's Hosp, Boston, 72-75; fel neonatology, Harvard Med Sch, 75-78; instr pediat, 77-78; from asst prof to assoc prof, 78-86, PROF PEDIAT, UNIV NC, CHAPEL HILL, 87-, DIR, DIV NEONATAL/PERINATAL MED, 88- *Concurrent Pos:* E L Trudeau fel, Am Lung Asn, 78-81; prin investr, NIH res grants, 79-; attend pediatrician, NC Mem Hosp, 78-; res career develop award, NIH, 82-87; Alexander von Humboldt res fel, W Ger, 85-86. *Mem:* Am Thoracic Soc; Soc Pediat Res; Am Physiol Soc; Am Acad Pediat; Am Pediat Soc; Perinatal Res Soc. *Res:* Neural mechanisms of central respiratory control, particularly in newborns. *Mailing Add:* Dept Pediat Univ NC Chapel Hill NC 27599-7220

LAWSON, FRED AVERY, b Washington Co, Ark, Oct 25, 19. ENTOMOLOGY. *Educ:* Univ Ark, BSc, 43; Ohio State Univ, MS, 47, PhD(entom), 49. *Prof Exp:* Asst prof entom, Univ Tenn, 49-52; from asst prof to assoc prof, Kans State Univ, 52-60; assoc prof, Colo State Univ, 61-63; assoc prof, 63-70, PROF ENTOM, UNIV WYO, 70- *Mem:* Entom Soc Am; Electron Micros Soc Am; Coleopterists Soc. *Res:* Insects; electron microscopy; morphology, physiology, biology and taxonomy of Coleoptera. *Mailing Add:* Dept Entom Univ Wyo Box 3354 University Sta Laramie WY 82071

LAWSON, HERBERT BLAINE, JR, b Norristown, Pa, Jan 4, 42; m 64; c 2. MATHEMATICS. *Educ:* Brown Univ, ScB & AB, 64; Stanford Univ, PhD(math), 68. *Prof Exp:* Lectr math, Univ Calif, Berkeley, 68-70; vis prof, Inst Pure & Appl Math, Rio de Janeiro, Brazil, 70-71; from assoc prof to prof math, Univ Calif, Berkeley, 71-80; PROF MATH, STATE UNIV NY, STONY BROOK, 80- *Concurrent Pos:* Sloan Found fel, Univ Calif, Berkeley, 70-72; mem, Inst Advan Study, Princeton, 72-73 & IHES, Bures-sur-Yvette, France, 77-78; Guggenheim fel, 83, Sloan fel & JSPS fel, Kyoto Univ, Japan, 86; vis prof, Ecole Polytech, France, 83-84, Tata Inst Fundamental Res, Bombay, India, 87; mem, Coun Am Math Soc, 86-, Nat Comt Math. *Honors & Awards:* Steele Prize, Am Math Soc, 75. *Mem:* Am Math Soc. *Res:* Minimal surfaces; Riemannian geometry; foliations; several complex variables; mathematical physics; algebraic geometry. *Mailing Add:* Dept Math State Univ NY Stony Brook NY 11794

LAWSON, JAMES EVERETT, b Derby, Va, Jan 8, 33; m 60; c 2. ZOOLOGY. *Educ:* ETenn State Univ, BS, 58, MA, 59; Va Polytech Inst, PhD(zool), 67. *Prof Exp:* Instr biol, ETenn State Univ, 59-61; instr, Va Polytech Inst, 61-62; assoc prof, 64-76, PROF BIOL, E TENN STATE UNIV, 76- *Concurrent Pos:* Dir, Off Preprofessional Advert. *Res:* Ecology and systematics of pseudoscorpions. *Mailing Add:* Dept Health Sci ETenn State Univ Johnson City TN 37614

LAWSON, JAMES LLEWELLYN, physics; deceased, see previous edition for last biography

LAWSON, JAMES ROBERT, b Boston, Mass, May 18, 45; m 87. ANALOG CIRCUITS & SYSTEMS, SIGNAL PROCESSING. *Educ:* Univ Rochester, NY, BA, 70. *Prof Exp:* Engr electronic circuit design, Comtech Govt Systs Div, Hauppauge, NY, 83-84; prin engr electronic circuit design, Div ISC, Telephonics Corp, Huntington, NJ, 84-85; prin engr proj eng, Gull, Inc, Smithtown, NJ, 85; sr analog engr electronic circuit design, Chemco Technologies, Glen Cove, NY, 86-87; prin mem tech staff, 87-88, TECH DIR ELECTRONIC CIRCUIT DESIGN & PROJ ENG, ELECTRONIC ASSOCS, INC, WEST LONG BRANCH, NJ, 89- *Res:* Extremely high-accuracy computing grade linear and nonlinear analog circuits; analog computer systems; analog signal processing circuits; data-flow digital signal processing circuits; application-optimized custom window functions; high-end audio quality analog circuits. *Mailing Add:* 256N RD No 9 Rte 571 Jackson NJ 08527

LAWSON, JAMES W, ANTI-RHYTHMIC DRUGS. *Educ:* Univ Okla, PhD(pharmacol), 55. *Prof Exp:* PROF, DEPT MED CHEM, UNIV TENN, MEMPHIS, 64- *Mailing Add:* Ctr Health Sci Univ Tenn 800 Madison Ave Memphis TN 38163

LAWSON, JIMMIE BROWN, b Checotah, Okla, June 29, 34; m 54; c 2. PETROLEUM CHEMISTRY. *Educ:* Southeastern State Col, BS, 56; Tex Christian Univ, MA, 60; Rice Univ, PhD(chem), 64. *Prof Exp:* Chemist, Gen Dynamics Corp, 56-60; res chemist, Shell Develop Co, 64-72, sr chem engr, Shell Oil Co, 72-74, SR PETROL ENGR, SHELL DEVELOP CO, 74- *Mem:* Am Chem Soc; Soc Petrol Eng. *Res:* Properties of semiconductors; x-ray crystallography; corrosion and metal coatings; petroleum production; development of chemical systems for the tertiary recovery of petroleum; surface chemistry. *Mailing Add:* 6010 Yarwell Houston TX 77096

LAWSON, JIMMIE DON, b Waukegan, Ill, Dec 6, 42; m 64; c 2. TOPOLOGICAL ALGEBRA. *Educ:* Harding Col, BS, 64; Univ Tenn, PhD(math), 67. *Prof Exp:* Asst prof math, Univ Tenn, 67-68; from asst prof to assoc prof, 68-75, PROF MATH, LA STATE UNIV, BATON ROUGE, 76-, CHMN, 89- *Concurrent Pos:* Vis assoc prof, Univ Houston, 76; NSF grants, prin investr, 69-90; Alexander von Humboldt fel & hon Fullbright fel, Tech Univ, Darmstadt, WGer, 80-81; vis prof, Oxford Univ, 84; ed, Semigroup Forum, 88- *Mem:* Am Math Soc. *Res:* Topological algebra; algebraic topology and semigroups; topology; continuous lattices; lie semigroups; control theory; author of various articles; co-author of two research monographs; co-editor of one research monograph. *Mailing Add:* Dept Math La State Univ Baton Rouge LA 70803

LAWSON, JOEL S(MITH), b New York, NY, July 3, 24; m 46; c 4. PHYSICS, ELECTRONICS. *Educ:* Williams Col, BA, 47; Univ Ill, MS, 49, PhD(physics), 53. *Prof Exp:* Res assoc, Control Systs Lab, Univ Ill, 53-55, from res asst prof to res assoc prof, 55-57; sr staff mem, Sci Eng Inst, 58-65; spec asst electronics, Off Asst Secy Navy, 65-67; res & eng consult, Comdr-in-Chief Pac, 67-68; dir, Naval Labs, Washington, DC, 68-74; tech dir, Naval Electronics Systs, 74-81, chief scientist, 81-84; CONSULT, 85- *Mem:* Fel AAAS; Am Phys Soc; Inst Elec & Electronics Engrs. *Res:* Applications of digital data processing techniques to information handling; communications systems; radar techniques; development of advanced electronic devices for military applications; theory of military command control. *Mailing Add:* 4773 C Kahala Ave Honolulu HI 96816

LAWSON, JOHN D, b Eugene, Ore, Apr 19, 49. COMPUTER SCIENCE & EDUCATION. *Educ:* Univ Ore, BA, 82, MA, 83, PhD(computer sci & educ), 85. *Prof Exp:* Asst prof computer sci, Towson State Univ, 85-89; dir instrnl comput, 89-90, DIR, COMPUTER & INFO TECHNOL, LEWIS CLARK STATE COL, 90- *Concurrent Pos:* Chair, Spec Interest Group Computer Users & Educ, Asn Comput Mach. *Mem:* Asn Comput Mach. *Mailing Add:* Computer & Info Technol Dept Lewis Clark State Col Eighth & Sixth Sts Lewiston ID 83501-2698

LAWSON, JOHN DOUGLAS, b Meaford, Ont, Sept 2, 37; m 60; c 3. COMPUTER SCIENCE, NUMERICAL ANALYSIS. *Educ:* Univ Toronto, BASc, 59; Univ Waterloo, MSc, 60, PhD(appl math), 65. *Prof Exp:* Teaching fel math, 59-60, lectr, 60-64, from asst prof to assoc prof, 64-73, assoc dean math, 68-71, chmn dept comput sci, 74-78 & 83-84, prof comput sci, Univ Waterloo, 73-87, PRES, ALGOMA UNIV COL, 87- *Concurrent Pos:* Asst scientist, Med Div, Oak Ridge Inst Nuclear Studies, 64- 65; vis lectr, Univ Dundee, 71-72. *Res:* Numerical solution of ordinary differential equations; programming languages for scientific applications; approximation theory. *Mailing Add:* Pres Algoma Univ Col Sault Ste Marie ON P6A 2G4 Can

LAWSON, JOHN EDWARD, b Detroit, Mich, Feb 25, 31; m 57; c 2. MEDICINAL CHEMISTRY. *Educ:* Wayne State Univ, BS, 53; Vanderbilt Univ, MA, 56, PhD(org chem), 57. *Prof Exp:* Fel, Mass Inst Technol, 57-58; GROUP LEADER, MEAD JOHNSON & CO, 58- *Mem:* Am Chem Soc. *Res:* Pharmaceutical Chemistry; organic synthesis; heterocyclic chemistry; natural products. *Mailing Add:* Five Old Pasture Ct Wallingford CT 06492-2565

LAWSON, JOHN EDWIN, b Shoal Lake, Man, Nov 28, 33; m 58; c 1. ANIMAL BREEDING. *Educ:* Univ Man, BSAgr, 56, MSA, 63. *Prof Exp:* Res scientist beef cattle breeding, Res Br, Agr Can Res Sta, 57-88; RETIRED. *Concurrent Pos:* Head, Animal Sci Sect, Agr Can Res Sta. *Res:* Evaluation of cattle breeds and crosses for reproductive performance and efficiency of production in specific environments; investigation of genotype-environment interactions; direct and correlated response to single trait selection. *Mailing Add:* 2843 Lakeview Dr Lethbridge AB T1K 3G2 Can

LAWSON, JUAN (OTTO), b Bluefield, WVa, Apr 18, 39; m 63; c 2. PHYSICS. *Educ:* Va State Col, BS, 60; Howard Univ, MS, 62, PhD(physics), 66. *Prof Exp:* From asst prof to assoc prof, 67-74, asst dean, Grad Sch, 70-71; dean, Col Sci, 75-80, PROF PHYSICS, UNIV TEX, EL PASO, 74- *Mem:* Am Phys Soc; Sigma Xi. *Res:* Mathematical physics; solid state theory. *Mailing Add:* Dept Physics Univ Tex El Paso TX 79968

LAWSON, KENNETH DARE, b Clinchport, Va, May 12, 34; m 56; c 1. ELECTRON MICROSCOPY. *Educ:* East Tenn State Univ, BA, 59; Univ Fla, MS, 61, PhD(chem), 63. *Prof Exp:* SECT HEAD, MIAMI VALLEY LABS, PROCTER & GAMBLE CO, 63- *Mem:* AAAS; Am Chem Soc. *Res:* Nuclear magnetic resonance spectroscopy; structure of mesomorphic phases; biophysics. *Mailing Add:* 460 Whitestone Ct Cincinnati OH 45231-2716

LAWSON, KENT DELANCE, b Binghamton, NY, Feb 17, 21; m 42; c 3. THEORETICAL PHYSICS. *Educ:* Cornell Univ, BA, 43; Rensselaer Polytech Inst, MS, 51, PhD(physics), 56. *Prof Exp:* Instr physics, Rensselaer Polytech Inst, 46-52; mem fac, Bennington Col, 53-66; prof physics, 65-75, DISTINGUISHED TEACHING PROF, STATE UNIV NY COL, ONEONTA, 75- *Concurrent Pos:* Res & educ prof consult, 56-70; dir EDUX prog, 70- *Mem:* Am Phys Soc; Am Asn Physics Teachers. *Res:* Eduction theory: unified theory of physical reality, man and awareness; particle theory. *Mailing Add:* Dept Physics State Univ NY Col Oneonta NY 13820

LAWSON, LARRY DALE, b Elkhart, Ind, Dec 23, 46; m 73; c 6. LIPID BIOCHEMISTRY. *Educ:* Purdue Univ, BS, 68; Brigham Young Univ, MS, 73; Univ Ill, PhD(nutrit sci), 79. *Prof Exp:* Fel, Hormel Inst, Univ Minn, 78-83; res assoc, Univ NC, 83-85; res scientist, Murdock Healthcare, 86-90; DIR RES, PLANT BIOACTIVES RES INST, 90- *Mem:* Am Oil Chemists Soc; Sigma Xi; Am Chem Soc. *Res:* Lipid biochemistry, intestinal prostaglandins and organosulfur biochemistry of garlic. *Mailing Add:* Plant Bioactives Res Inst 35 W Center Oram UT 84058

LAWSON, MERLIN PAUL, b Jamestown, NY, Jan 12, 41; m 64; c 3. CLIMATOLOGY. *Educ:* State Univ NY Buffalo, BA, 63; Clark Univ, MA, 66, PhD(geog climat), 73. *Prof Exp:* Instr climat, Northeastern Univ, 67-68; from asst prof to assoc prof climat, 68-82, chmn dept geog, 80-87, PROF CLIMAT, UNIV NEBR-LINCOLN 82- *Concurrent Pos:* Asst vchancellor res & assoc dean grad studies, Univ Nebr- Lincoln, 87. *Mem:* Sigma Xi; Am Meteorol Soc; Asn Am Geogrs. *Res:* Historical climate of the Great American Desert; severe droughts since 1700 in the western United States; descriptive climatic change; dendroclimatology. *Mailing Add:* 301 Administration Univ Nebr Lincoln NE 68588-0434

LAWSON, MILDRED WIKER, b New London, Conn, Nov 10, 22; m 63. MATHEMATICS, COMPUTER SCIENCES. *Educ:* Univ Md, BS, 47, MA, 49. *Prof Exp:* Asst math, Univ Md, 47-49, instr, 49-50; cartog compilation aide, Corps Engrs, Army Map Serv, 50-51, mathematician, 51-55, proj leader math & comput prog, 55-57, asst chief prog, 57-58; from assoc mathematician to sr mathematician, appl physics lab, Johns Hopkins Univ, Laurel, 58-85; RETIRED. *Res:* Analysis and programming of computer solutions of problems arising in scientific projects; computer language training. *Mailing Add:* 5113 Durham Rd E Columbia MD 21044

LAWSON, NEAL D(EVERE), b 1916; m 41; c 2. CHEMICAL ENGINEERING. *Educ:* Pa State Col, BS, 38, MS, 42, PhD(chem eng), 46. *Prof Exp:* Asst petrol ref lab, Pa State Col, 38-42, instr, 42-47; group leader petrol chem lab, E I Du Pont de Nemours & Co, Inc, 47-50, head chem div, 50-56, additives mgr, Cent Region, 56-58, tech mgr, Ill, 58-56, develop specialist, 65-66, supvr new prod develop, 66-77, spec asst tech info res & develop, 77-80; RETIRED. *Mem:* Am Chem Soc; Soc Automotive Engrs; Am Soc Lubrication Engrs; Am Soc Testing & Mat. *Res:* Stability of petroleum products; additives for lubricating oils and gasoline; fuel oil stabilizers; additives and thickeners for greases. *Mailing Add:* 1005 Chickadee Lane Penn Wood S W Chester PA 19380

LAWSON, NORMAN C, b Glasgow, Scotland, Nov 3, 29; Can citizen; m 60; c 3. PLANT BREEDING. *Educ:* Glasgow Univ, BSc, 53; Univ Reading, dipl agr, 54; McGill Univ, MSc, 58, PhD, 61. *Prof Exp:* Res off, Exp Farm, Can Dept Agr, BC, 61-65, res scientist, Res Sta, Sask, 65-67; ASSOC PROF AGRON, MCGILL UNIV, 67-, DIR, DIPL PROG, 73- *Mem:* Am Soc Agron; Crop Sci Soc Am; Genetics Soc Can; Agr Inst Can; Can Soc Agron. *Res:* Genetics and breeding of forage and oil crop species. *Mailing Add:* Dept Plant Sci MacDonald Col 21111 Lakeshore Rd Ste Anne Bell PQ H9X 1C0 Can

LAWSON, ROBERT BARRETT, b Oakland, Calif, Aug 24, 11; m 39; c 2. PEDIATRICS. *Educ:* Harvard Univ, BA, 32, MD, 36; Am Bd Pediat, dipl, 41. *Prof Exp:* From asst prof to prof pediat & dir dept, Bowman Gray Sch Med, 40-54; prof & chmn dept, Sch Med, Univ Miami, 54-62, actg dean, Sch Med, 61-62; prof pediat, Sch Med, Northwestern Univ, 62-71, chmn dept, 62-70, vpres health sci, 70-71; chief staff, Variety Children's Hosp, 71-81; RETIRED. *Concurrent Pos:* Pro Sch Pub Health, Univ NC & assoc, Sch Med, Duke Univ, 40-42; pediat consult, State Bd Health, NC, 40-42; Nat Res Coun fel, Univ Calif & Yale Univ, 45; chief staff, Children's Mem Hosp, Chicago, Ill, 62-71; clin prof pediat, Sch Med, Univ Miami, 71- *Mem:* Soc Pediat Res; Am Pediat Soc; AMA; Am Acad Pediat. *Res:* Infectious disease. *Mailing Add:* 5450 Banyan Dr Miami FL 33156

LAWSON, ROBERT BERNARD, b New York, NY, June 20, 40; div; c 3. NEGOTIATIONS, PSYCHOLOGY. *Educ:* Monmouth Col, NJ, BA, 61; Univ Del, MA, 63, PhD(psychol), 65. *Prof Exp:* NASA fel, 62-64; from asst prof to assoc prof, 66-74, PROF PSYCHOL, UNIV VT, 74-, VPRES RES & GRAD DEAN, 78- *Concurrent Pos:* Consult, IBM, 74-77; mem bd govs, Univ Press of New Eng, 78-, chmn, 79-80; numerous res grants, NIH, NSF, NASA & USAID. *Mem:* AAAS; NY Acad Sci; Sigma Xi; Am Psychol Asn. *Res:* Human perception and cognition; stereoscopic vision; psychology of negotiation. *Mailing Add:* Dean Grad Studies Univ Vt Agr Col 85 S Prospect St Burlington VT 05405

LAWSON, ROBERT DAVIS, b Sydney, Australia, July 14, 26; nat US; m 50; c 3. THEORETICAL PHYSICS. *Educ:* Univ BC, BASc, 48, MASc, 49; Stanford Univ, PhD, 53. *Prof Exp:* Asst, Univ BC, 47-48; asst, Stanford Univ, 49-53; jr res physicist, Univ Calif, 53-57; res physicist, Enrico Fermi Inst Nuclear Studies, Ill, 57-59; assoc physicist, 59-66, SR PHYSICIST, ARGONNE NAT LAB, 66- *Concurrent Pos:* Vis physicist, UK Atomic Energy Authority, Harwell, Eng, 62-63; Weizmann sr fel, Weizmann Inst Sci, Israel, 67-68; vis prof, State Univ NY, Stony Brook, 72-73; Nordita fel, Niels Bohr Inst, Copenhagen, 76-77. *Mem:* Am Phys Soc. *Res:* Nuclear physics. *Mailing Add:* Appl Physics 316 Argonne Nat Lab Argonne IL 60439

LAWSON, WILLIAM, b New York, NY, Nov 23, 34; m 65; c 1. FACIAL PLASTIC SURGERY, MAXILLOFACIAL SURGERY. *Educ:* NY Univ, BA, 56, DDS, 61, MD, 65. *Prof Exp:* PROF OTOLARYNGOL, MT SINAI SCH MED, 82-, ATTEND SURGEON, MT SINAI HOSP, NY, 82- *Concurrent Pos:* chief head & neck surg, Bronx Vet Hosp, 75-; attend surgeon otolaryngol, Elmhurst Gen Hosp, 82- *Mem:* Am Acad Facial Plastic Reconstruct Surg (vpres, 87-89); Am Soc Head & Neck Surgeons; Am Soc Maxillofacial Surgeons; Am Laryngol Soc; Am Rhinologic, Laryngol, Otol Soc; Am Col Surgeons. *Res:* Melanocytic system of head and neck, discovery of melanocytes in larynx, nasal cavity, orbit and dental lamina; experimental proof of the neurocrine nature of glomus cells of the head and neck; paraganglyonic chemoreceptor system; discovery of paraganglia in laryna; author of sourcebook with 3000 references. *Mailing Add:* 19 E 98 St New York NY 10029

LAWSON, WILLIAM BURROWS, b Detroit, Mich, June 8, 29; div; c 1. BIOCHEMISTRY. *Educ:* Wayne State Univ, BS, 51; Univ Md, PhD(chem), 56. *Prof Exp:* Res assoc, Mass Inst Technol, 55-56 & 57-58, Nat Cancer Inst fel, 56-57; sr asst scientist, Nat Insts Health, Md, 58-60; Nat Cancer Inst fel, Max Planck Inst Biochem, 60-61; sr res scientist, 61-66, ASSOC RES SCIENTIST, WADSWORTH CTR LABS & RES, NY STATE DEPT HEALTH, 66- *Concurrent Pos:* Mem, Thrombosis Coun, Am Heart Asn. *Mem:* Am Chem Soc; Am Soc Biol Chem; Int Soc Thrombosis & Hemostasis; Am Heart Asn. *Res:* Amino acids; synthesis and degradation of peptides; chemical modification and inhibition of enzymes; blood coagulation. *Mailing Add:* Escuela De Quimica Unan-Leon Leon Nicaragua

LAWTON, ALEXANDER R, III, b Nov 8, 38. MEDICINE. *Educ:* Yale Univ, BA, 60; Vanderbilt Univ, MD, 64. *Prof Exp:* Intern, Vanderbilt Univ Hosp, 64-65, asst resident, 65-66; clin assoc, Lab Clin Invest, Nat Inst Allergy & Infectious Dis, 66-68, clin investr, 68-69; from asst prof to assoc prof, 71-76, PROF PEDIAT & MICROBIOL, SCH MED, UNIV ALA, BIRMINGHAM, 76-; AT SCH MED, VANDERBILT UNIV, NASHVILLE. *Concurrent Pos:* NIH spec fel immunol, Sch Med, Univ Ala, Birmingham, 69-71. *Mem:* Sigma Xi. *Res:* Pediatrics; microbiology; immunology. *Mailing Add:* Sch Med Vanderbilt Univ Med Ctr 10 Pediat D-3237 Nashville TN 37232

LAWTON, ALFRED HENRY, b Carson, Iowa, July 26, 16; m 40, 74; c 3. GERIATRICS. *Educ:* Simpson Col, AB, 37; Northwestern Univ, MS, 39, BM, 40, MD, 41, PhD(physiol), 43. *Hon Degrees:* ScD, Simpson Col, 58. *Prof Exp:* Intern, Passavant Mem Hosp, Chicago, 40-41; resident, Henry Ford Hosp, Detroit, 41-42; asst prof med, physiol & pharm, Sch Med, Univ Ark, 46-47; dean & prof physiol & pharm, Univ NDak, 47-48; chief res div, US Vet Admin, Washington, DC, 48-51; med res adv, US Dept Air Force, 51-55; asst dir prof serv res & educ & chief intermediate serv, US Vet Admin Ctr, Bay Pines, 55-62; dir study ctr, Nat Inst Child Health & Human Develop, 62-66; from asst to assoc dean acad affairs, Univ SFla, 66-70, actg vpres acad affairs, 70-73; exec dir tech adv comn aging res, US Dept Health, Educ & Welfare, 73-74; dir geriat res, Educ & Clin Ctr & assoc chief staff res & develop, Vet Admin Ctr, 75-78; clin prof community health & family prac, Univ Fla Sch Med & dir gerontol, 78-82, DIR GERONTOL EDUC, ADVENT CHRISTIAN VILLAGE, 82- *Concurrent Pos:* Asst clin prof, Sch Med, George Washington Univ, 48-55; liaison mem coun arthritis & metab dis, USPHS, 50-55; mem exec coun, Nat Res Coun, US Armed Forces Vision Comt & Nat Coun Aging; actg dean, Col Med, Univ SFla, 68-70. *Mem:* Am Physiol Soc; Am Soc Pharmacol & Exp Therapeut; Am Geriat Soc; Geront Soc; Am Pub Health Asn. *Res:* Aging; chronic diseases. *Mailing Add:* Advent Christian Village Dowling Park FL 32060

LAWTON, EMIL ABRAHAM, b Detroit, Mich, Oct 12, 22; c 3. SYNTHETIC CHEMISTRY, WATER CHEMISTRY. *Educ:* Wayne State Univ, AB, 46; Purdue Univ, PhD(inorg chem), 52. *Prof Exp:* Res chemist, Nat Bur Standards, 52-53; proj leader, Battelle Mem Inst, 53-57; prog mgr, Rocketdyne Div, Rockwell Int, 57-72; sect head, Wasatch Div, Thiokol Corp, 72-75; proj mgr, Neus, Inc, 75-76; vpres res & develop, Tech & Mgt Consults, 76-77; mgr advan prog, Shock Hydrodynamics Div, Whittaker Corp, 77-82; CONSULT, 89- *Concurrent Pos:* Mem tech staff, Jet Propulsion Lab, Calif Inst Technol, 83-89. *Honors & Awards:* Civilian Patriotic Award, US Army, 84. *Mem:* Am Chem Soc; AAAS. *Res:* Combustion; fuels and propellants; lubricants. *Mailing Add:* PO Box 5218 Sherman Oaks CA 91413

LAWTON, JOHN G, b Vienna, Austria, June 3, 23; c 7. ELECTRICAL ENGINEERING. *Educ:* City Col New York, BEE, 49; Mass Inst Technol, SM, 51; Cornell Univ, PhD, 60. *Prof Exp:* Jr engr electronics, Stanford Res Inst, 51; from asst engr to prin staff engr, Cornell Aeronaut Lab, Inc, 54-70, staff scientist, 70-78; pres, J & J Technologies, Inc, 78-83; PRES, LAWTRONICS, INC, 83- *Mem:* Inst Elec & Electronics Engrs. *Res:* Communications; modulation; information and communications theory; fire control; guidance equipment; computers; statistics. *Mailing Add:* 80353 Westfield Rd Seville OH 44273

LAWTON, RICHARD G, b Berkeley, Calif, Aug 29, 34; m 58; c 5. SYNTHETIC ORGANIC CHEMISTRY, BIO-ORGANIC CHEMISTRY. *Educ:* Univ Calif, Berkeley, BS, 56; Univ Wis, PhD(chem), 62. *Prof Exp:* Asst isolation & identification, Merck Sharp & Dohme Res Labs, 56-57; asst org chem, Univ Wis, 59-62; from instr to assoc prof, 62-70, PROF ORG CHEM, UNIV MICH, ANN ARBOR, 70- *Concurrent Pos:* Vis prof, Univ Wis, 70-71; consult, Colgate-Palmolive Res Labs, 70- & German Wool Res Inst, Aachen, WGer, 79-80; John S Guggenheim Found award, 79-80. *Mem:* AAAS; Am Chem Soc. *Res:* Synthetic organic chemistry including peptide chemistry, alkaloids, terpenes and polycyclic aromatic hydrocarbons. *Mailing Add:* 2888 Daleview Ann Arbor MI 48103

LAWTON, RICHARD L, b Council Bluffs, Iowa, June 18, 18; m 43; c 7. SURGERY. *Educ:* Univ Omaha, AB, 39; Univ Nebr, BS & MD, 43. *Prof Exp:* Intern, St Mary's & St Louis Hosps, Mo, 43-44; resident, US Vet Hosp, Omaha, Nebr, 46-50, asst chief surg, 50-51; pvt pract, 52-53; staff physician, Vet Admin Hosp, 53-58, chief cancer chemother, 58-76, asst chief surg, 63-76, dir renal dialysis, 64-76; chief, Div Transplantation, & actg chief oncol, Sch Med, Tex Tech Univ, 76-78, prof & vchmn, Dept Surg, 76-80, surg, 63-80, dir renal dialysis, 64-80; MEM STAFF, GEN SURG & SURG ONCOL, FRANKLIN GEN HOSP, 80- *Concurrent Pos:* Instr, Creighton Univ, 50-51; from clin asst prof to clin assoc prof, Univ Iowa, 54-64; assoc prof, 64-68, prof surg, 68-; res grant angiol, Univ Iowa, 64-65. *Mem:* Am Asn Cancer Res; Am Soc Artificial Internal Organs; fel Am Col Surg; Soc Exp Biol & Med; Am Soc Clin Oncol. *Res:* Renal dialysis; cancer chemotherapy; transplant and preservation. *Mailing Add:* RR 6 Box 111B Iowa City IA 52240

LAWTON, RICHARD WOODRUFF, b New York, NY, June 22, 20; m 46; c 2. PHYSIOLOGY, BIOPHYSICS. *Educ:* Dartmouth Col, AB, 42; Cornell Univ, MD, 44. *Prof Exp:* Fel physiol sci, Med Sch, Dartmouth Col, 46-48; from instr to asst prof physiol, Med Col, Cornell Univ, 48-54; assoc prof, Sch Med, Univ Pa, 54-58; mgr bioastronaut sect, Missile & Space Div, Pa, Gen Elec Co, 58-67, mgr bioastronaut sect, Res & Eng Space Systs Orgn, 67-69, Life Systs, 69-70, med res dir, Med Develop Oper, Chem & Med Div, 70-71, mgr ventures develop, Med Ventures Oper, Med Syst Bus Div, 72-74, consult med systs, Corp Res & Develop, 74-85; RETIRED. *Concurrent Pos:* Head physiol sect, Aviation Med Lab, Naval Air Develop Ctr, Pa, 54-58; adj assoc prof, Sch Med, Univ Pa, 58-70. *Mem:* Am Physiol Soc. *Res:* Public health and epidemiology; elasticity of body tissues; aerospace physiology; cardiovascular physiology. *Mailing Add:* 1340 Stanley Lane Schenectady NY 12309

LAWTON, ROBERT ARTHUR, b Sanford, Colo, July 2, 32; m 57; c 6. ELECTRONICS, ELECTROOPTICS. *Educ:* Brigham Young Univ, BS, 60; Univ Colo, MS, 68, PhD(elec eng), 72. *Prof Exp:* Physicist radio frequency standards, 60-70, pulse measurement, 70-74, PHYSICIST PULSED ELECTRO-OPTICS, NAT BUR STANDARDS, 74-, GROUP LEADER, ELECTROMAGNETIC WAVEFORM METROLOGY, 80- *Concurrent Pos:* Sci & Technol Fel energy & power, US House Rep, Dept Com, 78-79. *Honors & Awards:* Super Accomplishment, Nat Bur Standards, 66 & 75, Outstanding Performance, 77 & Spec Achievement, 78. *Mem:* Inst Elec & Electronics Engrs; Optical Soc Am; Sigma Xi. *Res:* Electrical and optical pulse generation and measurement including time domain autocorrelation and power spectral analysis; laser pulse demodulation; signal detection in noise; electromagnetic field effects in materials. *Mailing Add:* 2640 Cornell Cir Boulder CO 80303

LAWTON, STEPHEN LATHAM, b Milwaukee, Wis, Nov 12, 39; m 67; c 2. STRUCTURAL CHEMISTRY. *Educ:* Univ Wis, BS, 63; Iowa State Univ Sci & Technol, MS, 66. *Prof Exp:* Res chemist, Socony Mobil Oil Co, Inc, 66-69, sr res chemist, 69-75, assoc chemist, 75-78, RES ASSOC, MOBIL RES & DEVELOP CORP, 78- *Mem:* Am Chem Soc; Am Crystallog Asn. *Res:* X-ray crystallography; crystal and molecular structures of inorganic and organometallic compounds and zeolites; computer applications in chemistry. *Mailing Add:* Mobil Res & Develop Corp Res Dept Paulsboro NJ 08066

LAWTON, WILLIAM HARVEY, b Indianapolis, Ind, Nov 1, 37. APPLIED STATISTICS. *Educ:* Univ Calif, Berkeley, AB, 59, MA, 62, PhD(statist), 65. *Prof Exp:* Teaching asst statist, Univ Calif, Berkeley, 60-62, teaching assoc, 63-65; math analyst, 62-63, consult math & statist, 65-75, supvr appl math sect, 75-77, supvr corp com serv, 77-79, DIR MKT RES, ASIA, AFRICA, AUSTRALIA-ASIA REGION, EASTMAN KODAK CO, 79- *Concurrent Pos:* Lectr, Univ Rochester, 65-70; ed, Technometrics, 74-77. *Honors & Awards:* Shewell Award, Am Soc Qual Control, 71; Wixcuxon Award, Am Statist Asn, 71 & 75. *Mem:* Fel Am Statist Asn; fel Am Soc Qual Control; Sigma Xi; AAAS. *Res:* Probability; new tools for mathematical model building for problems in the physical sciences and management sciences. *Mailing Add:* 137 Glenview Lane Rochester NY 14609

LAWVERE, FRANCIS WILLIAM, b Muncie, Ind, Feb 9, 37; m 66; c 5. ALGEBRA. *Educ:* Ind Univ, BA, 60; Columbia Univ, MA & PhD(math), 63. *Prof Exp:* Syst analyst, Litton Industs, Inc, 62-63; asst prof math, Reed Col, 63-64; NATO fel, Swiss Fed Inst Technol, 64-65, res assoc, 65-66; asst prof, Univ Chicago, 66-67; assoc prof, Grad Ctr, City Univ New York, 67-68; Sloan fel, Swiss Fed Inst Technol, 68-69; res prof, Dalhousie Univ, 69-71; vis prof, Aarhus Univ, 71-72 & Nat Res Inst Italy, 72-74; PROF MATH, STATE UNIV NY BUFFALO, 74- *Mem:* Am Math Soc. *Res:* Foundations of category theory; categorical foundations of mathematics; algebraic theories and equational doctrines; axiomatic theory of topoi; closed categories and metric spaces; synthetic differential geometry; functorial thermodynamics and continuum mechanics. *Mailing Add:* Dept Math State Univ NY 106 Diefendorf Hall Buffalo NY 14214

LAWWILL, STANLEY JOSEPH, b London, Ohio, May 23, 16; m 40; c 5. MATHEMATICS. *Educ:* Univ Cincinnati, AB, 37, MA, 39, PhD(math), 41. *Prof Exp:* Instr math, Northwestern Univ, 41-44; mathematician appl math group, Columbia Univ, 44; gunnery analyst, 2nd & 20th air forces, US Air Force, 44-46, opers analyst, HQ Strategic Air Command, 46-48, dep chief opers anal off, 48-50, chief atomic capabilities div, 50-54, dep chief scientist, 54-58; tech dir sci anal off, Melpar, Inc Div, Westinghouse Air Brake Co, 58; pres, Anal Servs, Inc, 58-81; RETIRED. *Mem:* Opers Res Soc Am; Sigma Xi. *Res:* Orthogonal functions; overconvergence of approximations in terms of rational harmonic functions; operations analysis; weapon systems evaluation. *Mailing Add:* 6532 Copa Ct Falls Church VA 22044

LAWYER, ARTHUR L, EXPERIMENTAL BIOLOGY. *Educ:* Univ Calif, Davis, BS; Yale Univ, MPhil, 77, PhD(molecular biochem & biophys), 79. *Prof Exp:* Rockefeller Found fel, Calvin Lab, Lawrence Berkeley Labs, Univ Calif, 79-81; res biochemist, Biotechnol Group, Chevron Chem Co, 81-86, regulatory specialist-state liaison, 86-87, sr state regulatory specialist, Valent USA & Chevron Chem Co, 87-89; mgr, State Govt Affairs, Valent USA Corp,

89-90, sr proj mgr, 90; CHMN, CALIF-ENVIRON AGENCY TASK FORCE, WESTERN AGR CHEM ASN, 91-; DIR, STATE AFFAIRS DIV, TECHNOL SERV GROUP, 91- *Concurrent Pos:* Mem, State Affairs Comt, Nat Agr Chem Asn, 87-90; chmn, Proposition 65 Task Forces, Western Agr Chem Asn, 87-90, Govt Affairs Comt & Calif Rep Comt, 88-90, Environ Initiative Task Force, 89-90. *Res:* Thirteen publications covering enzymatic mechanisms, photosynthetic metabolism, plant biotechnology and prediction of groundwater contamination. *Mailing Add:* State Affairs Div Technol Serv Group Steuart Street Tower 27th Floor One Market Plaza San Francisco CA 94105

LAX, ANNELI, b Kattowitz, Ger, Feb 23, 22; nat US; m 48; c 2. MATHEMATICS EDUCATION. *Educ:* Adelphi Col, BS, 42; NY Univ, MA, 45, PhD, 55. *Prof Exp:* Asst aeronaut, Inst Mech Sci, NY Univ, 43-44, asst math, 45-55, asst res scientist, Wash Sq Col, 55-65, assoc prof math, 65-71, prof, 72-; MEM-AT-LARGE, COURANT INST. *Concurrent Pos:* Instr, Wash Sq Col, 45-55; ed, New Math Library, 60- *Honors & Awards:* Polya Award, 77. *Mem:* AAAS; Am Math Soc; Math Asn Am. *Res:* Mathematical analysis and exposition; partial differential equations. *Mailing Add:* Courant Inst 251 Mercer St Rm 710 New York NY 10012

LAX, BENJAMIN, b Miskolz, Hungary, Dec 29, 15; nat US; m 42; c 2. SOLID STATE PHYSICS, PLASMA PHYSICS. *Educ:* Cooper Union, BS, 41; Mass Inst Technol, PhD(physics), 49. *Hon Degrees:* DSc, Yeshiva Univ, 75. *Prof Exp:* Mech engr, US Eng Off, 41-42; radar officer, Radiation Lab, Mass Inst Technol, 44-46; mem staff, Air Force Cambridge Res Ctr, 46-51; mem staff, Lincoln Lab, 51-53, head, Ferrites Group, 53-55 & Solid State Group, 55-57, assoc head, Commun Div, 57-58, head, Solid State Div, 58-64, assoc dir, Lincoln Lab, 64-65, dir, 60-81, prof, 65-86, EMER PROF, MASS INST TECHNOL, 86-, EMER DIR & PHYSICIST, FRANCIS BITTER NAT MAGNETIC LAB, 81- *Concurrent Pos:* Assoc ed, J Appl Physics, 57-59 & Barnes Eng Co, 78-; mem, Coun Am Phys Soc, 63-67, Joint Coun Quantum Electronics, 64-81 & chmn, 66-68, Solid State Sci Panel, Nat Res Coun, 70-81, Assoc Prog Adv Comt & chmn, Physics Panel, 83-84; chmn, 10th Int Conf Physics Semiconductors, & Int Union Pure & Appl Physics Comn Quantum Electronics, 76-; mem bd dirs, Infrared Indust, 73-76 & Barnes Eng Co, 78-; consult, Raytheon Co, 76-, Gen Motors Res Lab, 83- & Amoco Res Ctr, Standard Oil Co, 85-, Lincoln Lab, Mass Inst Technol, 86-; Guggenheim fel, 81-82; adv comt, Am Friends of Jerusalem Col Technol, 82-83; adv bd, EPSCOR, Ky & adv comt, Weber Res Inst, 86- *Honors & Awards:* Buckley Prize, Am Phys Soc, 60; Gano Dunn Medal, Cooper Union Alumni Asn, 69. *Mem:* Nat Acad Sci; NY Acad Sci; Sigma Xi; fel Am Phys Soc; fel Am Acad Arts & Sci; fel Optical Soc Am; fel AAAS. *Res:* Nonlinear effects in solids and plasmas; interaction of submillimeter radiation with plasmas and solids; laser produced plasmas; semiconductors and ferrites; radar and millimeter waves. *Mailing Add:* Francis Bitter Nat Mag Lab Mass Inst Technol Bldg NW14-4104 Cambridge MA 02139

LAX, EDWARD, b Toronto, Ont, Aug 29, 31; US citizen; m 60; c 1. CRYOGENICS. *Educ:* Univ Calif, Los Angeles, AB, 52, MA, 59, PhD(physics), 60. *Prof Exp:* Sr physicist, Ultrasonic Systs, Inc, 60-61; mem tech staff, Aerospace Corp, 61-67; MEM TECH STAFF, AUTONETICS DIV, ROCKWELL INT CORP, 67- *Mem:* Am Phys Soc; Inst Elec & Electronics Engrs. *Res:* Electron-phonon effects in metals at low temperatures; acoustical-optical effects; cryogenic heat transfer and thermodynamics; hypersonics and delay lines; infra-red systems. *Mailing Add:* 5637 Wilhelmina Ave Woodland Hills CA 91367

LAX, LOUIS CARL, b Toronto, Ont, Apr 25, 30; m 58. PHYSIOLOGY, MEDICINE. *Educ:* Univ Toronto, BA, 52, MA, 53, MD, 57, PhD(physiol), 66. *Prof Exp:* Res assoc & fel, Banting & Best Dept Med Res, Univ Toronto, 52-61; clin fel, Clin Invest, Sunnybrook Hosp, Toronto, 56-61; physician & scientist, Med Res Ctr, Brookhaven Nat Lab, 61-64; vis scientist, 64-66; assoc prof surg & physiol, Sch Med, Univ Calif, Los Angeles, 64-69; asst med dir, Hosp Prod Div, Abbott Lab, Abbott Park, Ill, 69-71; assoc prof surg, Abraham Lincoln Sch Med, Univ Ill Med Ctr, 70-75; vpres & med dir, Telemed Cardioi-Pulmonary Systs Div, Becton, Dickinson & Co, Hoffman Estates, Ill, 71-81; assoc med dir, Cardiovasc-Renal Clinical Res, G D Searle, Skokie, Ill, 84-85; dir med affairs & services, Controlled Release Technol, 87-90; PRES & MED DIR, AMBULATORY MONITORING LABS, INC, WAUKEGAN, 81- *Concurrent Pos:* Res collabr dept med, Brookhaven Nat Lab, 64-71; consult, RAND Corp, Santa Monica, Calif, 67-69; Epoxylite Corp, S El Monte, Calif, 68-69 & Toxigenics Inc, Decatur, Ill, 80-84; vis prof, Bowman Gray Sch Med, Univ NC, 64-69; med dir, Telemed Corp, 71-73. *Mem:* AAAS; Biophys Soc; NY Acad Sci; Can Physiol Soc; Inst Elec & Electronic Engrs. *Res:* Ambulatory monitoring and analysis of physiological parameters; study of compartmental kinetics in living systems; influence of hormones and postsurgical trauma on fluid and electrolyte distribution in the body; mathematical simulation of living systems; computer processed 12-lead scaler electrocardiography; medical applications of computers; medical electronic devices; drug delivery systems; clinical investigation of cardiovascular drugs. *Mailing Add:* 2705 McAree Rd Waukegan IL 60087-3556

LAX, MELVIN, b New York, NY, Mar 8, 22; m 49; c 4. SOLID STATE PHYSICS, QUANTUM OPTICS. *Educ:* NY Univ, BA, 42; Mass Inst Technol, SM, 43, PhD(physics), 47. *Prof Exp:* Res physicist, Underwater Sound Lab, Mass Inst Technol, 42-45, res assoc physics, 47; from asst prof to prof, Syracuse Univ, 47-55; mem tech staff, Bell Tel Labs, 55-71, head theoret physics dept, 62-64; DISTINGUISHED PROF PHYSICS, CITY COL NEW YORK, 71- *Concurrent Pos:* Consult, Crystal Br, US Naval Res Lab, 51-55, US Army Res Off, 71-, Bell Labs, 72- & Los Alamos Nat Lab, 76-; lectr, Princeton Univ, 60, Oxford Univ, 61-62; mem basic res adv comt, Nat Acad Sci, 66-69. *Mem:* Nat Acad Sci; fel Am Phys Soc; AAAS; Optical Soc Am. *Res:* Meson creation and absorption; multiple scattering; phase transitions; optical and electrical properties of solids; impurity bands; classical and quantum relaxation and noise; group theory in solids; quantum communication theory; nonlinear optical properties in deforming solids; optics, lasers, continuum mechanics; two dimensional transport and tunneling in semiconductors. *Mailing Add:* Dept Physics City Col NY New York NY 10031

LAX, MELVIN DAVID, b Boston, Mass, Mar 20, 47. RANDOM DIFFERENTIAL EQUATIONS. *Educ:* Rensselaer Polytech Inst, BS, 69, MS, 71, PhD(math), 74. *Prof Exp:* Lectr math, Southern Ill Univ, Carbondale, 74-76; vis asst prof, Okla State Univ, 76-77; from asst prof to assoc prof, 77-86, PROF MATH, CALIF STATE UNIV, LONG BEACH, 86- *Concurrent Pos:* Mem, panel reviewers, Cent J Math & Related Regions, 77- *Mem:* Am Math Soc; Soc Indust & Appl Math. *Res:* Approximate solution of random differential equations and random integral equations. *Mailing Add:* Dept Math Calif State Univ Long Beach CA 90840

LAX, PETER DAVID, b Hungary, May 1, 26; nat US; m 48; c 2. PARTIAL DIFFERENTIAL EQUATIONS. *Educ:* NY Univ, AB, 47, PhD, 49. *Prof Exp:* From asst prof to assoc prof math, NY Univ, 49-63, asst to dir, Math Ctr, 59-63, dir, AEC Comput & Appl Math Ctr, 63-72, head, Dept Math & Comput Sci & dir, Courant Inst Math Sci, 72-80, DIR, COURANT MATH & COMPUT LAB, NY UNIV, 80- *Concurrent Pos:* Consult, Los Alamos Sci Lab, 50; mem, Nat Sci Bd, 80-86. *Honors & Awards:* Nat Medal of Sci; Wolf Prize. *Mem:* Nat Acad Sci; Am Acad Sci; NY Acad Sci; Math Asn Am; Am Math Soc (pres, 79-80); USSR Acad Sci. *Res:* Theory of partial differential equations; numerical analysis; scattering theory; functional analysis; fluid dynamics. *Mailing Add:* Courant Inst Math Sci NY Univ 251 Mercer St New York NY 10012

LAXER, CARY, b Brooklyn, NY, July 16, 55. COMPUTER AIDED INSTRUCTION, COMPUTER GRAPHICS. *Educ:* New York Univ, BA, 76; Duke Univ, PhD(biomed eng), 80. *Prof Exp:* Res asst prof comput sci, Duke Univ, 80-81; asst prof, 81-84, ASSOC PROF COMPUT SCI, ROSE-HULMAN INST TECHNOL, 84- *Mem:* Inst Elec & Electronics Engrs; Asn Comput Mach; Am Soc Eng Educ. *Res:* Computer analysis of cardiac electrical signals with relation to myocardial infarct geometry. *Mailing Add:* Rose-Hulman Inst Technol Campus Box 100 5500 Wabash Ave Terre Haute IN 47803-3999

LAXPATI, SHARAD R, b Bombay, India, July 16, 38; m 83; c 2. ANTENNAS, ELECTROMAGNETIC THEORY. *Educ:* Gujarat Univ, India, BE, 57; Univ Ill, MS, 61, PhD(elec eng), 65. *Prof Exp:* Jr scientific officer, Reactor Control Div, Atomic Energy Estab, India, 58-60; asst prof elec eng, Pa State Univ, 65-69; asst prof info eng, 69-73, ASSOC PROF ELEC ENG & COMPUT SCI, UNIV ILL, CHICAGO, 73- *Concurrent Pos:* Vis sr assoc, Syst Res Ltd, Richmond, UK, 76-77; dir, Matrix Publishers Priv LTD, India, 78-85; consult, Naval Res Lab, Washington, DC, 79-90, Locus, Inc, Alexandria, VA, 84; owner, LMS Engineering, Chicago, IL, 85-; consult, Symmetron Inc, Fairfax, Va, 87- *Mem:* Inst Elec & Electronics Engrs; Int Union Radio Sci; Inst Elec Engr, London. *Res:* Radiation and propagation of electromagnetic waves; applied mathematics; educational technology systems; optical communication. *Mailing Add:* Dept Elec Eng & Comput Sci m/c 154 PO Box 4348 Chicago IL 60680

LAY, DAVID CLARK, b Los Angeles, Calif, Mar 1, 41; m 70; c 3. OPERATOR THEORY, LINEAR ALGEBRA. *Educ:* Aurora Col, BA, 62; Univ Calif, Los Angeles, MA, 65, PhD(math), 66. *Prof Exp:* Teaching asst math, Univ Calif, Los Angeles, 63-64, asst prof, 66-71; assoc prof, 71-77, PROF MATH, UNIV MD, COLLEGE PARK, 77- *Concurrent Pos:* NSF res grants, Univ Md, College Park, 68-73, 76-77 & 90-91; res grant, Neth Orgn Advan Pure Res, 73; vis prof, Univ Amsterdam, 80; Air Force USR grants, 87-90. *Mem:* Am Math Soc; Math Asn Am; Sigma Xi; Soc Indust & Appld Math. *Res:* Functional analysis; spectral theory of linear operators; operator-valued analytic functions. *Mailing Add:* Dept Math Univ Md College Park MD 20742

LAY, DOUGLAS M, b Jackson, Miss, July 3, 36; m 61; c 2. ANATOMY, ZOOLOGY. *Educ:* Millsaps Col, BS, 58; La State Univ, MS, 61; Univ Chicago, PhD(anat), 68. *Prof Exp:* Instr anat, Univ Chicago, 68-69; asst prof zool & cur mammals, Univ Mich, 69-73; ASSOC PROF ANAT, UNIV NC, CHAPEL HILL, 73- *Mem:* Am Soc Mammal; Am Soc; Soc Study Evol; Soc Syst Zool; Soc Vert Palaeont. *Res:* The adaptive significance of specializations of mammals for life in deserts, particularly the structure and function of the ear in desert rodents; origin, evolution, functional anatomy, biology and systematics of rodents. *Mailing Add:* Dept Anat Div Health Sci Univ NC Chapel Hill NC 27514

LAY, JOACHIM E(LLERY), mechanical engineering, for more information see previous edition

LAY, JOHN CHARLES, b Ponca City, Okla, Mar 6, 48; wid; c 3. VETERINARY DIAGNOSTIC PATHOLOGY, PULMONARY PATHOLOGY. *Educ:* Univ Mo, Columbia, BS, 71, DVM, 75; Cornell Univ, PhD, 86. *Prof Exp:* Assoc scientist vet med, Inhalation Toxicol Res Inst, Lovelace Found for Med Educ & Res, 75-77; gen vet practr, Lakin, Kans, 77-78; fel trainee exp path, NY State Col Vet Med, Cornell Univ, 78-83; ASST PROF, DEPT VET PATH, TEX A&M UNIV, 83- *Mem:* Am Vet Med Asn. *Res:* Bovine Respiratory Disease; pulmonary inflammatory response and mechanisms of deep lung clearance and lung defense. *Mailing Add:* Ctr Environ & Lung Biol Univ NC Chapel Hill NC 27514

LAY, KENNETH W(ILBUR), b Ringgold Co, Iowa, Feb 4, 39; m 64; c 3. CERAMICS. *Educ:* Iowa State Univ, BS, 61; Northwestern Univ, PhD(mat sci), 66. *Prof Exp:* Res scientist, 65-73, commun & admin mgr mat sci & engr, 73-75, mgr ceramics processing, 75-87, MAT SCIENTIST, GEN ELEC RES & DEVELOP CTR, 87- *Mem:* Fel Am Ceramic Soc; Mat Res Soc. *Res:* Ceramics processing and properties; diffusion; nuclear fuels; non-stoichiometric compounds; oxide superconductors. *Mailing Add:* Gen Elec Res & Develop Ctr PO Box 8 Schenectady NY 12301

LAY, STEVEN R, b Los Angeles, Calif, Nov 28, 44; m 71; c 2. MATHEMATICS. *Educ:* Aurora Col, BA, 66; Univ Calif, Los Angeles, MA, 68, PhD(math), 71. *Prof Exp:* From asst prof to prof, 71-80, PROF MATH, AURORA UNIV, 80- *Mem:* Am Math Soc; Math Asn Am; Nat Coun Teachers Math. *Res:* Combinatorial geometry and convexity; the separation of convex sets; mathematics education. *Mailing Add:* Dept Math Aurora Univ Aurora IL 60506

LAY, THORNE, b Casper, Wyo, Apr 20, 56. SEISMOLOGY. *Educ:* Univ Rochester, BS, 78; Calif Inst Technol, MS, 80, PhD(geophys), 83. *Prof Exp:* Postdoctoral researcher, Calif Inst Technol, 83; from asst prof to assoc prof seismol, Univ Mich, 84-89, dir, Seismog Sta, 87-89; PROF SEISMOL, UNIV CALIF, SANTA CRUZ, 89-, DIR, INST TECTONICS, 90- *Concurrent Pos:* Sloan fel, Sloan Found, 85-87; Shell fac fel, Shell Found, 85-88; presidential young investr, NSF, 85-90; panelist, US Geol Surv Rev Panel, 88-90, Air Force Geophys Lab Panel, 88-92, Air Force Tech Appln Ctr Panel, 87-, Defense Advan Res Projs Agency Proposal Rev Panel, 90-93; ed, EOS, Am Geophys Union, 89-91; chmn, SEDI Comt, US Am Geophys Union, 90-92. *Honors & Awards:* Macelwane Medal, Am Geophys Union, 91. *Mem:* Seismol Soc Am; fel Am Geophys Union; fel Royal Astron Soc; AAAS; Soc Explor Geophysicists; Am Asn Prof Geologists. *Res:* Earthquake seismology; structure of the earths interior; author of over 85 papers in professional journals and books. *Mailing Add:* Inst Tectonics Univ Calif Santa Cruz Santa Cruz CA 95064

LAYBOURNE, PAUL C, b Akron, Ohio, Dec 21, 19; m 45; c 3. PSYCHIATRY. *Educ:* NY Med Col, MD, 44; Ind Univ, MS, 49. *Prof Exp:* Assoc, 49-51, from asst prof to assoc prof, 51-65, PROF PSYCHIAT, MED CTR, UNIV KANS, 65-, ASSOC PROF PEDIAT, 54-, PROF PSYCHIATRY & FAMILY PRACT, 80- *Concurrent Pos:* Consult, Spofford Home, Kansas City, Mo, 50-; dir, Atchison County Guid Clin, 51-59; supvr extern prog, State Hosps, 53-55; dir child guid clin, Mercy Hosp, 57-66. *Mem:* AMA; fel Am Psychiat Asn. *Res:* Child psychiatry. *Mailing Add:* Univ Kans Med Ctr Kansas City KS 66103

LAYCHOCK, SUZANNE GALE, b Brooklyn, NY, Apr 28, 49. PHOSPHOLIPIDS, CYCLIC NUCLEOTIDES. *Educ:* Brooklyn Col, BS, 71; City Univ NY, MA, 73; Med Col Va, PhD(pharmacol), 76. *Prof Exp:* Res assoc pharmacol, Vanderbilt Univ, 77-78; from asst prof to assoc prof pharmacol, Med Col Va, 78-89; PROF PHARMACOL, STATE UNIV NY, BUFFALO, 89- *Mem:* Endocrine Soc; Am Soc Pharmacol & Exp Therapeut; Am Diabetes Asn. *Res:* Investigation of the signal transduction mechanisms which regulate insulin secretion from islets of Langerhans; role of phospholipid turnover, cyclic nucleotides, eicosanoids and calcium in regulation of metabolism and insulin release in the beta cell. *Mailing Add:* Pharmacol & Therapeut Dept 102 Farber Hall Sch Med State Univ NY Buffalo NY 14214

LAYCOCK, MAURICE VIVIAN, b Liverpool, Eng, Sept 3, 38; Can citizen; m 66; c 2. NATURAL PRODUCTS, MARINE ALGAE. *Educ:* Liverpool Univ, BSc, 62, PhD(plant physiol), 65. *Prof Exp:* RES OFFICER BIOCHEM, NAT RES COUN CAN, 68- *Res:* Protein chemistry; biochemistry of nitrogen compounds in algae. *Mailing Add:* Nat Res Coun 1411 Oxford St Halifax NS B3H 3Z1 Can

LAYCOCK, WILLIAM ANTHONY, b Ft Collins, Colo, Mar 17, 30; m 55; c 2. PLANT ECOLOGY. *Educ:* Univ Wyo, BS, 52, MS, 53; Rutgers Univ, PhD(bot), 58. *Prof Exp:* Asst, Rutgers Univ, 55-58; range scientist, Intermountain Forest & Range Exp Sta, 58-74, asst dir, Rocky Mountain Forest Range Exp Sta, US Forest Serv, 74-76, range scientist, agr res serv, Colo State Univ, USDA, 76-85; HEAD DEPT RANGE MGT, UNIV WYO, 85- *Concurrent Pos:* Collabr range sci, Utah State Univ, 64-74; NZ Nat Res Adv Coun-NZ Forest Serv sr res fel, 69-70; coordr site atr western coniferous biome, US Int Biol Prog, 71-72; affil fac, Colo State Univ, 74- *Honors & Awards:* Outstanding Achievement Award, Soc Range Mgt, 85. *Mem:* Ecol Soc Am; fel Soc Range Mgt (pres, 88); Soil & Water Conserv Soc. *Res:* Ecology and management of rangelands; autecology of range species; snow management on rangelands. *Mailing Add:* Dept Range Mgt Box 3354 University Sta Laramie WY 82071

LAYDEN, GEORGE KAVANAUGH, b Greenport, NY, Apr 13, 29; m 60; c 3. MATERIALS SCIENCE. *Educ:* Lafayette Col, BS, 53; Pa State Univ, MS, 59, PhD(ceramic technol), 61. *Prof Exp:* mat scientist, Unitet Technol Res Ctr, 61-; RETIRED. *Mem:* Am Ceramic Soc; AAAS; Sigma Xi. *Res:* Fabrication and characterization of aerospace materials, including fibrilar carbon and graphite, structrual ceramics, nickel based superalloys and fiber and whisker reinforced glass/ceramic matrix composites. *Mailing Add:* 1071 Farmington Ave West Hartford CT 06107

LAYE, RONALD CURTIS, b New York, NY, Nov 27, 45; m 67; c 5. CLINICAL HEALTH PSYCHOLOGY, APPLIED PSYCHOPHYSIOLOGY. *Educ:* Clarkson Col Technol, BS, 67; Univ Alta, MSc, 73, PhD(psychol), 76. *Prof Exp:* Instr psychol & Clin psychologist, Univ Alta, 73-76; vis asst prof, State Univ NY, Oswego, 76-77; dept Head, 88-90, INSTR PSYCHOL, FRASER VALLEY COL, 77-; CONSULT, 82- *Concurrent Pos:* Clin psychologist & div, Chilliwack Mental Health Ctr, 79- & pvt pract, 82- *Mem:* Soc Behav Med; Asn Appl Psychophysiol & Biofeedback; Can Psychol Asn. *Res:* The use of psychophysiological profiling in assessment and treatment of stress disorders; biofeedback methodology; personality and health. *Mailing Add:* 1904 Arbutus St Vancouver BC V6J 3X7 Can

LAYER, ROBERT WESLEY, b Brooklyn, NY, Aug 11, 28; wid; c 3. ORGANIC CHEMISTRY. *Educ:* NY Univ, AB, 50; Univ Cincinnati, PhD, 55. *Prof Exp:* Control chemist, Naugatuck Chem Co, 50-52; sr tech mgr, B F Goodrich Co, 55-57, sr res chemist, 57-71, res assoc, 71-79, sr res assoc, 79-90, RES FEL, B F GOODRICH CO, 90- *Mem:* Am Chem Soc; Rubber & Polymer Divs, Akron Rubber Group. *Res:* Rubber chemicals; reactions of ozone; chemistry of p-phenylenediamines and of anils; antioxidants; accelerators. *Mailing Add:* B F Goodrich Co 9921 Brecksville Rd Brecksville OH 44141

LAYLOFF, THOMAS, b Granite City, Ill, Jan 29, 37; m; c 3. ANALYTICAL CHEMISTRY. *Educ:* Wash Univ, AB, 58, MA, 61; Univ Kans, PhD(anal chem), 64. *Prof Exp:* From asst prof to prof, 64-76, ADJ PROF CHEM, ST LOUIS UNIV, 76-; DIR, FDA DIV DRUG ANALYSIS, 76- *Concurrent*

Pos: Sci adv nat ctr drug anal, Food & Drug Admin, 67-76. *Mem:* AAAS; Sigma Xi; Am Chem Soc. *Res:* Solution behavior of reactive organic intermediates; biopotentiometry; electron transport; data acquisition and processing. *Mailing Add:* PO Box 207 Granite City IL 62040

LAYMAN, DON LEE, b Johnston, Pa, May 20, 38; m 64; c 3. ANATOMY. *Educ:* Juniata Col, BS, 61; Syracuse Univ, MS, 65; George Washington Univ, PhD(cell biol), 70. *Prof Exp:* Biologist, Nat Eye Inst & Nat Inst Dent Res, 68-70, res chemist, Lab Biochem, Nat Inst Dent Res, 70-71; NIH spec fel, Dept Path, Univ Wash, Seattle, 71-73; res asst prof, Dept Path, Baylor Col Med, 73-75; asst prof, Dept Med, Ore Health Sci Univ, 75-77 & Dept Anat, 77-84; ASSOC PROF, DEPT ANAT, MED CTR, LA STATE UNIV, 84- *Mem:* Sigma Xi; Am Soc Cell Biol; NY Acad Sci; AAAS; Tissue Cult Asn; Am Asn Dent Res. *Res:* Periodontitis. *Mailing Add:* Dept Anat La State Univ Med Ctr 1100 Florida Ave New Orleans LA 70119

LAYMAN, DONALD KEITH, b Kewanee, Ill, Feb 15, 50; m 73; c 2. HUMAN NUTRITION, EXERCISE PHYSIOLOGY. *Educ:* Ill State Univ, BS, 72, MS, 74; Univ Minn, PhD(nutrition), 78. *Prof Exp:* From asst prof to assoc prof, 78-89, PROF NUTRIT, UNIV ILL, 89-, CHAIR, DIV FOODS & NUTRIT, 90- *Concurrent Pos:* Consult, nutrition, Regional Office Educ, Ill, 79-80, Nat Aeronaut & Space Admin, 83, Shriner's Hosp Barred Children, 86; adj prof, Inst Agron & Vet, Rabat, Morroco, 86- *Honors & Awards:* Young Investr Award, NIH, 82; BioServ Award, Am Inst Nutrit, 86. *Mem:* Am Inst Nutrition; Inst Food & Technol; Sigma Xi. *Res:* protein energy requirements, regulation of protein turnover and energy metabolism on skeletal muscle; obesity, cellular growth and development of skeletal muscle, nutrition, exercise and cancer. *Mailing Add:* 274 Bevier Hall, Univ Ill 905 S Goodwin Urbana IL 61801

LAYMAN, WILBUR A, b Blair, Nebr, Jan 9, 29; m 53; c 5. ANALYTICAL CHEMISTRY, PHYSICAL CHEMISTRY. *Educ:* Dana Col, BS, 53; Univ Nebr, MS, 58; Mont State Univ, PhD(anal chem), 63. *Prof Exp:* Chemist, Harris Labs, 54-58; instr chem, Hastings Col, 58-60 & Dana Col, 62-63; asst prof, S Dak State Univ, 63-66; assoc prof, Adams State Col, 66-67; prof chem & chmn, phys sci dept, Easternmont Col, 67-82; RETIRED. *Mem:* Am Chem Soc. *Res:* Stability constants of metal complexes; polarized infrared spectroscopy of thin crystal films. *Mailing Add:* 619 Ave C Billings MT 59102

LAYMAN, WILLIAM ARTHUR, b West New York, NJ, Feb 8, 29; m 83; c 1. PSYCHIATRY. *Educ:* St Peter's Col, NJ, BS, 51; Georgetown Univ, MD, 55; Am Bd Psychiat & Neurol, dipl, 62. *Prof Exp:* Intern, Hackensack Hosp, NJ, 55-56; jr resident psychiat, Vet Admin Hosp, Lyons, 56-57; sr resident, Fairfield Hills Hosp, Newtown, Conn, 57-59; from instr to assoc prof, 59-74, clin prof, 74-77, PROF PSYCHIAT, UNIV MED & DENT NJ, 77- *Concurrent Pos:* Clin fel psychiat, Sch Med, Yale Univ, 58-59. *Mem:* Am Asn Univ Prof; Am Psychiat Asn; Am Acad Forensic Scis. *Res:* Nonverbal communication; psycotherapeutic technique. *Mailing Add:* Dept Psychiat NJ Med Sch Newark NJ 07103

LAYNE, CLYDE BROWNING, b El Paso, Tex, Feb 19, 47; m 69; c 1. APPLIED PHYSICS. *Educ:* Princeton Univ, AB, 69; Univ Calif, Davis, MS, 73, PhD(appl sci), 75. *Prof Exp:* Physicist, Div Laser Physics, Lawrence Livermore Lab, Univ Calif, 69-80; TECH STAFF, COMBUSTION PHYSICS DIV, SANDIA NAT LAB, LIVERMORE, CALIF, 80- *Mem:* Am Phys Soc; Optical Soc Am. *Res:* Relaxation and energy transfer in ions in solids; atomic vapor-laser isotope separation of uranium; lasers applied to combustion diagnostics. *Mailing Add:* Div 8151 Sandia Nat Labs Livermore CA 94550

LAYNE, DONALD SAINTEVAL, b Lime Ridge, Que, Apr 5, 31; m 59; c 3. BIOCHEMISTRY. *Educ:* McGill Univ, BSc, 53, MSc, 55, PhD, 57. *Prof Exp:* Fel biochem, Univ Edinburgh, 57-58; res assoc psychiat, Queen's Univ Ont, 58-59; from scientist to sr scientist, Worcester Found Exp Biol, 59-66; head physiol & endocrinol sect, Food & Drug Directorate, Can, 66-68; prof biochem, Univ Ottawa, 68-79; vpres, Connaught Lab, Ltd, 79-82; VPRES, TORONTO GEN HOSP, 82- *Mem:* AAAS; Endocrine Soc; Am Soc Biol Chemists; Royal Soc Can. *Res:* Biochemistry of estrogenic hormones. *Mailing Add:* Toronto Gen Hosp CCRW 1-800 101 College St Toronto ON M5G 1L7 Can

LAYNE, JAMES NATHANIEL, b Chicago, Ill, May 16, 26; m 50; c 5. VERTEBRATE BIOLOGY. *Educ:* Cornell Univ, BA, 50, PhD(zool), 54. *Prof Exp:* Asst vert zool, Cornell Univ, 50-54, asst prof zool, 63-67; asst prof zool, Southern Ill Univ, 54-55; from asst prof to assoc prof biol, Univ Fla, 55-63; res dir, Archbold Biol Sta & Archbold Cur Dept Mammal & mem bd dir, Archbold Expeds, Am Mus Natural Hist, 67-76, exec dir & mem bd dirs, 76-85, SR RES BIOLOGIST, ARCHBOLD BIOL STA, 85- *Concurrent Pos:* Vis scientist, Pvt Ecol Sect, Lab Perinatat Physiol, NINDB, 61-62; from asst cur to assoc cur biol sci, Fla State Mus, 55-63, res assoc, 63-65; adj prof, Univ S Fla, 68-; consult, WHO, 69; res assoc, Am Mus Natural Hist, 82- *Honors & Awards:* C Hart Merriam Award, Am Soc Mammalogists, 76. *Mem:* AAAS; Am Soc Zoologists; Am Soc Mammal (vpres, 65-70, pres, 70-72); Wildlife Soc; Ecol Soc Am; Orgn Biol Field Stas (vpres, 84-85, pres, 86-87). *Res:* Mammalian ecology, systematics, behavior and morphology; general vertebrate biology. *Mailing Add:* Archbold Biol Sta PO Box 2057 Lake Placid FL 33852

LAYNE, PORTER PRESTON, b Martin, Ky, Sept 20, 45; m 72; c 2. BIOCHEMISTRY. *Educ:* Univ Ky, BS, 66, PhD(biochem), 71; Boston Univ, MBA, 78. *Prof Exp:* Fel protein chem, Sch Med, Tufts Univ, 71-73, asst prof, 74-80; assoc investr develop, 80-81, asst dir, clin data sci, 82-83, asst dir, Preregis Affairs, 83-84, assoc dir clin info, 85, dir, 85-86, SR INVESTR DEVELOP, SMITH KLINE CORP, 82-, DIR REGULATORY AFFAIRS, 86- *Concurrent Pos:* Actg Chmn, Div Chem, Sch Med, Tufts Univ, 75-76. *Mem:* Sigma Xi; Drug Info Asn; Regulatory Affairs Prof Soc. *Res:* Extra chromosomal inheritance; bacterial physiology; enzyme mechanism (phosphoglucomutase); adenylate cyclase and tuftsin research. *Mailing Add:* RD Four Sylvan Dr Pottstown PA 19464

LAYNE, RICHARD C, b St Vincent, WI, Dec 14, 36; Can citizen; m 63; c 2. PLANT BREEDING. *Educ:* McGill Univ, BSc, 59; Univ Wis, MS, 60, PhD(agron, plant path), 63. *Prof Exp:* RES SCIENTIST, RES STA, AGR CAN, 63- *Concurrent Pos:* Head, Hort Sci Sect, Harrow Res Sta. *Honors & Awards:* Shepard Award, Am Pomol Soc, 67 & 83; Carrol R Miller Award, Am Soc Hort Sci, 77, 78, 82, 85. *Mem:* Can Soc Hort Sci (pres, 77); Am Soc Hort Sci; Am Pomol Soc (pres, 91-); Sigma Xi; Int Soc Hort Sci. *Res:* Breeding of cultivars and rootstocks of peach, nectarine and apricot; breeding for cold hardiness and disease resistance; environmental and genetic factors affecting cold hardiness; rootstock-scion physiology; peach orchard management systems. *Mailing Add:* Can Agr Res Sta Harrow ON N0R 1G0 Can

LAYNG, EDWIN TOWER, b Greenville, Pa, Jan 20, 09; m 46; c 2. CHEMISTRY. *Educ:* Allegheny Col, ScB, 30; NY Univ, PhD(chem), 33. *Prof Exp:* Asst, NY Univ, 30-33; res chemist, M W Kellogg Co, NJ, 34-42, assoc dir res, 43; dir res, Hydrocarbon Res, Inc, 43-44, asst to the pres, 44-46, vpres, 47-63, exec vpres, 64-72, pres, 72-74; vpres, Dynalectron Corp, 64-74; consult, 74-77; RETIRED. *Mem:* Am Chem Soc; Am Inst Chem Eng; Am Petrol Inst. *Res:* Reaction kinetics; catalysis in petroleum processes; synthetic fuels; coal gasification; coal liquefaction. *Mailing Add:* Eight Surrey Rd Summit NJ 07901

LAYSON, WILLIAM M(CINTYRE), b Lexington, Ky, Sept 24, 34; m 67; c 2. PHYSICS, MECHANICAL ENGINEERING. *Educ:* Mass Inst Technol, BS, 56, PhD(physics), 63. *Prof Exp:* Res physicist, Univ Calif, 62-63, lectr physics, 63-64; sr systs engr tech staff, Pan Am World Airways, 64-67; mem tech staff, Gen Res Corp, 67-69, mgr dept, 69-70; vpres & mgr, Wash Div, 70-76, VPRES CONTINUIUM MECH DIV, SCI APPLN INC, 76- *Mem:* Am Phys Soc; Am Inst Aeronaut & Astronaut; Inst Elec & Electronics Engrs. *Res:* Nuclear weapons effects; dust and debris clouds; radiation transport; ground coupling; fireball effects; radar systems; fluid mechanics; electromagnetic propagation; atmospheric physics; systems analysis. *Mailing Add:* 8301 Sumrind Dr McClean VA 22101

LAYTON, DAVID WARREN, b Woburn, Mass, Sept 19, 48; m 77; c 3. RISK ANALYSIS, WATER RESOURCES. *Educ:* Bridgewater State Col, BA, 70; Univ Ariz, PhD(water resources admin), 75. *Prof Exp:* ENVIRON SCIENTIST RISK ANALYSIS WATER RESOURCES & ENERGY, LAWRENCE LIVERMORE NAT LAB, UNIV CALIF, 75- *Mem:* Am Water Resources Asn; Soc Risk Anal. *Res:* Risk analysis; environmental studies; pollution control technologies; geothermal energy. *Mailing Add:* Environ Sci Div PO Box 5507 Livermore CA 94550

LAYTON, E MILLER, JR, physical chemistry, chemical physics; deceased, see previous edition for last biography

LAYTON, EDWIN THOMAS, JR, b Sept 13, 28; m 82; c 1. HISTORY OF TECHNOLOGY. *Educ:* Univ Calif, Los Angeles, BA, 50, MA, 53, PhD(hist), 56. *Prof Exp:* Instr hist, Univ Wis, 56-57, Ohio State Univ, 57-60; asst prof, Purdue Univ, 60-65; assoc prof hist sci & technol, Case Western Reserve Univ, 65-75; PROF HIST SCI & TECHNOL, UNIV MINN, 75- *Concurrent Pos:* Adv ed, Isis, 79-81, Business & Prof Ethics J, 81-; mem, Nuclear Manpower Studies Comn, Nat Acad Sci-Nat Res Coun, 81. *Honors & Awards:* Dexter Prize, Soc Hist Technol, 70, Leonardo da Vinci Medal, 90. *Mem:* Soc Hist Technol (pres, 85-86); Hist Sci Soc; fel AAAS; Soc Social Studies Sci. *Res:* Interaction of science and technology in nineteenth century America; nature and role engineering sciences; history of engineering. *Mailing Add:* Dept Mech Eng Univ Minn Minneapolis MN 55455

LAYTON, JACK MALCOLM, b Ossian, Iowa, Sept 27, 17; m 43; c 2. PATHOLOGY. *Educ:* Luther Col, AB, 39, DSc, 74; Univ Iowa, MD, 43; Am Bd Path, dipl, 50. *Prof Exp:* Intern, Univ Iowa Hosp, 43, asst path, 46-47, instr, 47-49, assoc, 49-50, from asst prof to prof, 50-67; prof path & head dept, Col Med, Univ Ariz, 67-88; RETIRED. *Concurrent Pos:* Actg dean col med & actg dir med ctr, Univ Ariz, 71-73; trustee, Am Bd Path, 74-; dir clin path, Ariz Health Servs, 67-88. *Mem:* Am Soc Clin Path (pres, 73); Col Am Path; Am Asn Path & Bact; Am Soc Exp Path; Int Acad Path (pres, 75-76); Sigma Xi. *Res:* Virology; host-parasite relationships in viral and rickettsial diseases; comparative pathology of inflammation; ultramicroscopic pathologic anatomy of infectious diseases; biological activities of teratomas; influenza and psittacosis-lymphogranuloma groups of viruses. *Mailing Add:* Dept Path Med Ctr Univ Ariz Tucson AZ 85724

LAYTON, RICHARD GARY, b Salt Lake City, Utah, Dec 24, 35 005 WH; m 63; c 3. PHYSICS, CLOUD PHYSICS. *Educ:* Univ Utah, BA, 60, MA, 62; Utah State Univ, PhD(physics), 65. *Prof Exp:* Asst physics, Univ Utah, 60-62; asst res physicist electro-dynamics Labs, Utah State Univ, 63-64; from asst prof to assoc prof physics, State Univ NY Col Fredonia, 65-69; assoc prof, 69-83, PROF PHYSICS, NORTHERN ARIZ UNIV, 83- *Concurrent Pos:* Res Assoc, Lowell Observ, 71 & 72; sci collabr, Grand Canyon Nat Park, 72-75; interim chair, physics dept, Northern Ariz Univ, 85-86, chair, 86- *Mem:* Am Asn Physics Teachers; Am Phys Soc; Sigma Xi; Int Asn Colloid & Interface Scientists. *Res:* Ice nucleation surfaces, ellipsometry; atmospheric optics; frost damage prevention for plants; science education. *Mailing Add:* Dept Physics Box 5763 Northern Ariz Univ Flagstaff AZ 86011

LAYTON, THOMAS WILLIAM, b Kaysville, Utah, Feb 24, 27; m 47; c 1. PHYSICS. *Educ:* Calif Inst Technol, BS, 51, PhD(physics, math), 57. *Prof Exp:* Res engr, Jet Propulsion Lab, Calif Inst Technol, 53-55; mem tech staff, Inertial Guid Dept, Thompson-Ramo-Wooldridge, Inc, 55-59, mgr, 59-64, SR STAFF ENGR, DEFENSE & SPACE SYSTS GROUP, TRW, INC, 64- *Mem:* Am Phys Soc; Am Inst Aeronaut & Astronaut. *Res:* Cosmic rays; navigation and guidance systems for ballistic missiles and space flight vehicles. *Mailing Add:* 4836 W Elmdale Dr Rolling Hills CA 90274

LAYTON, WILLIAM ISAAC, b Cameron, Mo, Sept 26, 13; m 41; c 2. GEOMETRY. *Educ:* Univ SC, BS, 34, MS, 35; George Peabody Col, PhD(math), 48. *Prof Exp:* Asst math, Univ SC, 34-35 & Univ Chicago, 36-37; instr, Amarillo Jr Col, 41-42, head dept, 42-46; prof math & head dept, Austin Peay State Univ, 46-48; assoc prof, Auburn Univ, 48-49; dean instruction, Frostburg State Univ, Md, 49-50; coordr, Data Processing Ctr, Stephen F Austin State Univ, 50-79; PROF MATH, NEWBERRY COL, 79-, CHMN DEPT MATH COMPUT SCI & PHYSICS, 82- *Concurrent Pos:* Instr eng sci & mgt war training, Tex Tech Univ, 42-45; NSF lectr, Tex Acad Sci, 60. *Mem:* Math Asn Am; Am Math Soc; Am Coun Teachers Math; Sigma Xi. *Res:* Higher geometry. *Mailing Add:* Dept Math & Comput Sci Physics Newberry Col 2100 College St Newberry SC 29108

LAYZER, ARTHUR JAMES, b Cleveland, Ohio, Aug 21, 27; m 64; c 1. THEORETICAL PHYSICS. *Educ:* Case Western Reserve Univ, BS, 50; Columbia Univ, PhD(physics), 60. *Prof Exp:* Res scientist, Courant Inst, NY Univ, 60-63; asst prof, 64-67, ASSOC PROF PHYSICS, STEVENS INST TECHNOL, 68-, ASSOC PROF ENG PHYSICS, 77- *Concurrent Pos:* Vis res scientist, Brookhaven Nat Lab, 66; resident visitor comput music, Acoust Div, Bell labs, 67-83; US Dept Educ grant deafness res & reading, 78-80. *Mem:* AAAS; Am Educ Res Asn. *Res:* Quantum mechanical theory of critical phenomena, especially superconductivity; sound-analogic text representations in deafness and language; score-mediated generation of music and text; test measures and analysis in reading and writing; language and art media applications of computer science. *Mailing Add:* Stevens Inst Technol Castle Point Sta Dept Phyics/Engr Physics Hoboken NJ 07030

LAYZER, DAVID, b Ohio, Dec 31, 25; m 49, 59; c 6. ASTRONOMY. *Educ:* Harvard Univ, AB, 47, PhD(theoret astrophys), 50. *Prof Exp:* Nat Res Coun res fel, 50-51; lectr astron, Univ Calif, Berkeley, 51-52; res assoc physics, Princeton Univ, 52-53; res assoc, 53-55, res fel & lectr, 55-60, prof astron, 60-80, DONALD H MENZEL PROF ASTROPHYS, HARVARD UNIV, 80- *Concurrent Pos:* Consult, Geophys Corp Am, 59-65. *Honors & Awards:* Bok Prize, 60. *Mem:* Am Acad Arts & Sci; Am Astron Soc; Int Astron Union; Royal Astron Soc. *Res:* Cosmology and cosmogony; theoretical astrophysics and atomic physics; ionospheric physics. *Mailing Add:* Dept Astron Harvard Col Observ Cambridge MA 02138

LAZAR, BENJAMIN EDWARD, b Vatra-Dornei, Romania, Feb 27, 30; US citizen; m 54; c 1. CIVIL ENGINEERING. *Educ:* Bucharest Polytech Inst, Dipl eng, 54; McGill Univ, grad dipl, 68; Sir George Williams Univ, DE, 70. *Prof Exp:* Engr, Precast Concrete Co, 54-61, Found Can Eng Corp, 61-64, Swan Wooster Eng Corp Ltd, 64-67 & Rust Eng, 67-68; res engr, Sir George Williams Univ, 68-70, asst prof civil eng, 70-71; asst prof, 71-80, ASSOC PROF CIVIL TECHNOL, UNIV HOUSTON, 80- *Concurrent Pos:* Can Steel Industs Construct Coun grant, Sir George Williams Univ, 70-71. *Mem:* Am Soc Civil Engrs. *Res:* Structural behavior and technology of civil engineering systems. *Mailing Add:* 5446 Cheena Houston TX 77096

LAZAR, NORMAN HENRY, b Brooklyn, NY, June 21, 29. PHYSICS. *Educ:* City Col New York, BS, 49; Ind Univ, MS, 51, PhD(physics), 53. *Prof Exp:* Physicist, Oak Ridge Nat Lab, Union Carbide Nuclear Co, 53-80; AT TRW DEFENSE & SPACE SYST, 80- *Mem:* Fel Am Phys Soc. *Res:* Controlled thermonuclear reactions; beta and gamma ray spectroscopy; plasma physics. *Mailing Add:* One Space Park Bldg R1 Rm 2120 TRW Defense & Space Syst Redondo Beach CA 90278

LAZARCHICK, JOHN, b Pottsville, Pa, Nov 1, 42; m 60; c 1. HEMATOLOGY & HEMOSTASIS. *Educ:* Lafayette Col, AB, 64; Thomas Jefferson Med Univ, MD, 68. *Prof Exp:* Asst prof hemat, Univ Conn Health Ctr, 77-79; asst prof, 79-82, ASSOC PROF HEMAT, MED UNIV SC, 82- *Mem:* Am Fedn Clin Res; AAAS; Asn Clin Scientists; World Hemophilia Fedn. *Res:* Role of protein kinase C in platelet function; release reaction and the pathophysiologic basis of fibronectin evaluation in preeclampsia. *Mailing Add:* Dept Lab Med Med Univ SC 171 Ashley Ave Charleston SC 29425

LAZARETH, OTTO WILLIAM, JR, b Brooklyn, NY, Sept 16, 38. SPACE NUCLEAR POWER, REACTOR PHYSICS. *Educ:* Wagner Col, BS, 61; Queens Col, MA, 68; City Univ NY, PhD(physics), 73. *Prof Exp:* Res asst physics, Queens Col, 65-67; physics assoc, 67-73, PHYSICIST, BROOKHAVEN NAT LAB, 74- *Mem:* Am Phys Soc; Am Nuclear Soc. *Res:* Nucleonics; radiation damage and effects in solids; modeling physical systems with computers. *Mailing Add:* Bldg 701 Brookhaven Nat Lab Upton NY 11973

LAZARIDIS, ANASTAS, b Istanbul, Turkey, Dec 8, 40; US citizen; m 66; c 2. ENERGY ENGINEERING, HEAT TRANSFER. *Educ:* Robert Col, Turkey, BS, 63; Columbia Univ, MS, 64, EngScD, 69. *Prof Exp:* Res engr, Exxon Res & Eng Co, 65, Heat & Mass Transfer Lab, Columbia Univ, 65-68, Textile Fibers Dept, E I du Pont de Nemours & Co Inc, 69-72; heat transfer specialist-process eng, Day & Zimmermann Inc, Philadelphia, 72-75; regional mkt mgr process control, Control Automation Technol Co, 75-76; pres, Helios Inc, Wilmington, Del, 76-83; asst prof, 83-89, ASSOC PROF, WIDENER UNIV, CHESTER, PA, 89- *Concurrent Pos:* Fulbright scholar award, Columbia Univ, 63-64; Boese scholar award, 64-65; lectr mech eng, Manhattan Col, 65-67 & Richmond Col, City Univ New York, 67-68; adj prof mech eng, Drexel Univ, 72-75; tech paper reviewer, Heat Transfer J & Solar Energy Eng J, Am Soc Mech Engrs, Int J Heat & Mass Transfer; fac res grants, Am Soc Eng Educ/Dept Energy, 83, US Army, 85 & USAF, 88-90. *Mem:* Am Soc Eng Educ; Am Soc Mech Eng. *Res:* Energy; heat transfer; thermodynamics of energetic materials; textile engineering; combustion and detonation; mathematical modeling. *Mailing Add:* Sch Eng Widener Univ Chester PA 19013

LAZARIDIS, CHRISTINA NICHOLSON, b New York, NY, Jan 12, 42; m 66; c 2. MATERIALS SCIENCE, ELECTRONIC MATERIALS. *Educ:* Mt Holyoke Col, BA, 62; Columbia Univ, MA, 63, PhD(chem), 66. *Prof Exp:* Res chemist, Colgate-Palmolive Co, 66-68; RES ASSOC, E I DU PONT DE NEMOURS & CO INC, 68- *Mem:* Am Chem Soc; Int Soc Hybrid Microelectronics. *Res:* Photosensitive systems, including conventional silver halide as well as novel photopolymeric materials; thick film materials for electronics, polymeric and cermet systems, membrane touch switch inks, under voltage curable products, conductors, dielectrics and registors for screen-printing applications and photoresists; polyimides and photodefinable polyimides. *Mailing Add:* Electronics Dept du Pont Co Exp Sta PO Box 80336 Wilmington DE 19880-0336

LAZARIDIS, NASSOS A(THANASIUS), b Athens, Greece, Oct 6, 43; m 73; c 2. MECHANICAL METALLURGY, PHYSICAL METALLURGY. *Educ:* Nat Tech Univ, Athens, BS & MS, 67; Univ Wis-Madison, MS, 70, PhD(metall eng), 71. *Prof Exp:* Prod engr metal working, Nat Can Corp, Greece, 73-74; lectr & proj assoc, Univ Wis-Madison, 74-75; sr res engr, Inland Steel Co, 75-79, spec consult, Sec Steel Refining & Continuous Casting, 79-80, supvr res engr, Res Lab, 80-85, sect mgr, Mat Eng, Prod Appl Res, 85-88, sect mgr flat prod, Qual Dept, 87-88; METALL TECHNOL MGR, I/N KOTE, JOINT VENTURE INLAND STEEL CO & NIPPON STEEL CORP, 89- *Concurrent Pos:* Proj assoc, Am Motors Corp, 74-75; adj prof, Purdue Univ, Calumet Campus, 78-86. *Mem:* Am Soc Metals Int; Am Soc Testing & Mat; Am Inst Mining, Metall & Petrol Engrs; Greek Tech Chamber Prof Engrs; Sigma Xi; Soc Automotive Engr. *Res:* Fracture toughness; fatigue; product development; formability, effect of metallurgy on machinability; hot deformation of steel and other alloys; materials engineering; hot drip and electrogalvanizing of sheet steel, product testing, corrosion resistance. *Mailing Add:* I-N Kote 30755 Edison Rd One Vision Dr New Carlisle IN 46552-9728

LAZARO, ERIC JOSEPH, b Muttra, India, Dec 28, 21; m 49; c 3. SURGERY. *Educ:* Univ Madras, MBBS, 46; Georgetown Univ, MS, 55; Am Bd Surg, dipl, 55; FRCS(C), 56. *Prof Exp:* Instr surg, Georgetown Univ, 54-57; assoc prof, All India Inst Med Sci, 58-60, prof thoracic surg, 60-61, prof surg, 61-62; assoc prof, 62, PROF SURG, COL MED & DENT NJ, 62- *Concurrent Pos:* Rockefeller Found fels, 57 & 62; Colombo Plan fel, 61. *Mem:* AAAS; fel Am Col Surg; Soc Surg Alimentary Tract; AMA. *Res:* Pathogenesis of pancreatitis; biologic effects of splenic extracts. *Mailing Add:* Dept Surg Univ Med Dent NJ NJ Med Sch Newark NJ 07103-2757

LAZAROFF, NORMAN, b Brooklyn, NY, Nov 24, 27; m 58; c 2. GEOCHEMICAL ACTIVITIES OF MICROORGANISMS. *Educ:* Syracuse Univ, AB, 50, MS, 52; Yale Univ, PhD(microbiol), 60. *Prof Exp:* Asst enzymol, Res Found, State Univ NY, 55; bacteriologist, Schwarz Labs, Inc, 55-56; proj leader, Evans Res & Develop Corp, 56-57, consult, 57-59; fel microbiol, Brandeis Univ, 60-61; microbiologist, BC Res Coun, 61-62; asst prof biol sci, Univ Southern Calif, 62-64; sr res scientist, Res Corp, Syracuse, 64-66; ASSOC PROF BIOL, STATE UNIV NY BINGHAMTON, 66- *Mem:* Am Soc Microbiol; Am Soc Photobiol; Phycol Soc Am; Brit Soc Gen Microbiol. *Res:* Chemolithotrophy; cyanobacterial photophysiology; environmental microbiology. *Mailing Add:* Dept Biol Sci State Univ NY Binghamton NY 13901

LAZAROW, PAUL B, b May 6, 45. CELL BIOLOGY, ANATOMY. *Educ:* Univ Chicago, AB, 67; Rockefeller Univ, PhD(biochem cytol), 72. *Prof Exp:* NIH int fel, Lab Molecular Embryol, Naples, Italy, 72-73; Damon Runyon fel, Dept Biol Sci, Stanford Univ, Palo Alto, Calif, 73-75; from asst prof to assoc prof, Rockefeller Univ, New York, NY, 75-89; PROF & CHMN, DEPT CELL BIOL & ANAT, MT SINAI SCH MED, NEW YORK, NY, 89- *Res:* Peroxisome biogenesis. *Mailing Add:* Cell Biol & Anat Dept Mt Sinai Sch Med Fifth Ave & 100th St New York NY 10029-6574

LAZARTE, JAIME ESTEBAN, b Lima, Peru, July 26, 43. HORTICULTURE, DEVELOPMENT PHYSIOLOGY & PROTEIN PRODUCTION. *Educ:* Agrarian Univ, BS & Ing Agr, 66; Rutgers Univ, MS, 70, PhD(hort), 76. *Prof Exp:* Lab supvr tissue cult, Rutgers Univ, 73; res assoc veg crops, Univ Fla, 76-78; asst prof hort, Tex A&M Univ, 78-82; pres, PRI, 83-88; res, Tex A&M Univ, 89-91; RES HARVARD MED SCH, 91- *Concurrent Pos:* Consult, Biotechnol; fulbright scholar, 90 & 91. *Mem:* Am Soc Hort Sci; Int Plant Propagators. *Res:* Tissue culture of horticultural crops; flower initiation and sex expression; sex modification; morphology and embryology of horticultural crops; asparagus officinalis L; proteins; plant tissue culture; biotechnology; molecular biology; protein production; cadmium. *Mailing Add:* Ctr Blood Res Harvard Med Sch 800 Huntington Ave Boston MA 02115

LAZARUS, ALLAN KENNETH, b Bangor, Maine, May 20, 31; m 57; c 3. CHEMISTRY. *Educ:* NY Univ, BA, 52, MS, 55, PhD(org chem), 57. *Prof Exp:* Chemist, Cities Serv Res & Develop Co, 57-59, Inorg Chem Div, FMC Corp, 59-65 & Esso Res & Eng Co, 65-66; group leader synthetic lubricants, Intermediates Div, Tenneco Chem, Inc, 66-71; ASST PROF CHEM, TRENTON STATE COL, 72 - *Res:* Stereochemistry; organic synthesis; product and process development; fuels; automatic transmission fluids; synthetic lubricants. *Mailing Add:* Dept Chem Trenton State Col CN 4700 Hillwood Lakes Trenton NJ 08650-4700

LAZARUS, DAVID, b Buffalo, NY, Sept 8, 21; m 43; c 4. SOLID STATE PHYSICS. *Educ:* Univ Chicago, PhD(physics), 49. *Prof Exp:* Instr electronics, Univ Chicago, 42-43; res assoc, Radio Res Lab, Harvard Univ, 43-45; asst physics, Univ Chicago, 46-49, instr, 49; from instr to prof, 49-87, EMER PROF PHYSICS, UNIV ILL, URBANA, 87- *Concurrent Pos:* Guggenheim fel, 68-69; vis prof, Univ Paris, 68-69, Harvard Univ & Mass Inst Technol, 78-79; chmn, Coun Mat Sci, US Dept Energy, 81-86; ed-in-chief, Am Phys Soc, 80-91; mem gov bd, Am Inst Physics, 81- *Mem:* Fel Am Phys Soc; fel Am Asn Physics Teachers; AAAS. *Res:* Defect and electronic properties of solids; high pressure physics. *Mailing Add:* 502 W Vermont Ave Urbana IL 61801

LAZARUS, GERALD SYLVAN, b New York, NY, Feb 16, 39; m 61; c 4. DERMATOLOGY. *Educ:* Colby Col, BS, 59; George Washington Univ, MD, 63. *Prof Exp:* Med intern, Univ Mich Med Ctr, 63-64, med resident, 64-65; clin assoc, Med Neurol Br, Nat Inst Neurol Dis & Blindness, 65-67; prin investr, Lab Histol & Path, Nat Inst Dent Res, 67-68; clin & res assoc, Dept Dermat, Harvard Med Sch, 68-70, chief resident dermat, 69-70; vis scientist, Strangeways Labs, Univ Cambridge, Eng, 70-72; assoc prof med & co-dir dermat training prog, Albert Einstein Col Med, 72-75; chmn, div dermat, Duke Univ Med Ctr, 75-78, prof med, 75-, J Lamar Calloway chair prof dermat, 78-; AT SCH MED, UNIV PA. *Concurrent Pos:* Carl Herzog fel, 70-72; res fel, Arthritis Found, 70-72, sr investr, 72-77; consult dermat, Addenbrookes Hosp, Cambridge, Eng, 70-72; vis fel, Clare Hall, Cambridge; head sect dermat, Dept Med, Montefiore Hosp, 72-75. *Mem:* Am Rheumatism Asn; Soc Invest Dermat; Am Fedn Clin Res; fel Am Col Physicians; Royal Soc Med. *Res:* Study of the role of proteinases in catabolic processes in skin and evaluation of the mechanisms by which these proteinases can instigate an inflammatory response. *Mailing Add:* Dermat 211 Clin Res Bldg Univ Pa 422 Curie Blvd Philadelphia PA 19104

LAZARUS, LAWRENCE H, PHYSIOLOGY, PHARMACOLOGY. *Educ:* Univ Calif, Los Angeles, PhD(cellular physiol), 66. *Prof Exp:* RES CHEMIST, NAT INST ENVIRON HEALTH SCI, NIH, 77-, HEAD, PEPTIDE NEUROCHEM SECT, 83- *Res:* Peptide biochemistry; peptide mood action. *Mailing Add:* Inst Environ Health Sci PO Box 12233 Research Triangle Park NC 27709

LAZARUS, MARC SAMUEL, b Brooklyn, NY, Sept 9, 46; c 1. PHYSICAL INORGANIC CHEMISTRY. *Educ:* City Univ New York, BS, 68; Princeton Univ, MA, 71, PhD(chem), 74. *Prof Exp:* Res assoc chem, Lawrence Berkeley Lab, Univ Calif, 73-74; from asst prof to assoc prof, 74-84, PROF CHEM, HERBERT H LEHMAN COL, CITY UNIV NEW YORK, 84- *Concurrent Pos:* Res collabr, Brookhaven Nat Lab, 74-84. *Mem:* Am Chem Soc; Am Phys Soc. *Res:* Applications of x-ray photoelectron spectroscopy to the study of transition metal compounds and alloys. *Mailing Add:* Dept Chem Herbert H Lehman Col Bronx NY 10468

LAZARUS, ROGER BEN, b New York, NY, June 3, 25; m 46; c 5. THEORETICAL PHYSICS. *Educ:* Harvard Univ, AB, 47, MA, 48, PhD(physics), 51. *Prof Exp:* Mem staff, 51-58, group leader, 58-68, div leader, 68-73, MEM STAFF, LOS ALAMOS SCI LAB, 73- *Mem:* NY Acad Sci; Am Phys Soc; Asn Comput Mach. *Res:* Simultaneous partial differential equations; computing machines. *Mailing Add:* 2539 35th St Los Alamos NM 87544

LAZAR-WESLEY, ELIANE M, b Strasbourg, France, Jan 24, 53; m. PHYSIOLOGY. *Educ:* Univ Louis Pasteur, France, BSc, 75, PhD, 82. *Prof Exp:* Researcher, Nat Ctr Sci Res, France, 79-82, sr researcher, 82-84 & Lab Molecular Oncol, Inst Sci Cancer Res, 87-89; vis fel, Lab Chemoprev, Nat Cancer Inst, NIH, Bethesda, 84-87, SR STAFF FEL, LAB PHYSIOL & PHARMACOL STUDIES, NAT INST ALCOHOL ABUSE & ALCOHOLISM, NIH, ROCKVILLE, MD, 89- *Mailing Add:* Lab Physiol & Pharmacol Studies Nat Inst Alcohol Abuse & Alcoholism 12501 Washington Ave Rockville MD 20852

LAZAY, PAUL DUANE, b Philadelphia, Pa, June 2, 39; m 61; c 1. SOLID STATE PHYSICS. *Educ:* Trinity Col, Conn, BS, 61; Mass Inst Technol, PhD(physics), 68. *Prof Exp:* Supvr, Optical Measurement & Process Automation, Bell Labs, 69-83; vpres eng, ITT-EOPD, Roanoke, Va, 83-86; vpres eng, 86-87, vpres mkt, pres & gen mgr, 87-88, PRES & CHIEF EXEC OFF, TELCO SYSTEMS CORP, 88- *Concurrent Pos:* Bd dirs, SpecTran Corp, Sturbridge, Mass, 87- *Mem:* Am Phys Soc; Optical Soc Am; Sigma Xi; IEEE. *Res:* Optical fiber research and optical communications. *Mailing Add:* Telco Systs Corp 63 Nahatan St Norwood MA 02062

LAZDA, VELTA ABULS, b Riga, Latvia, Dec 16, 39; US citizen; m 62. IMMUNOLOGY, MOLECULAR BIOLOGY. *Educ:* Purdue Univ, BS, 62; Northwestern Univ, PhD(microbiol), 67. *Prof Exp:* Res assoc, Immunol Div, Res Inst, Am Dent Asn Health Found, 69-77; ASST PROF, DEPT SURG, UNIV ILL MED CTR, 77- *Concurrent Pos:* Am Cancer Soc fel ribosome struct, Northwestern Univ, 67-69; USPHS career develop award. *Mem:* AAAS; Am Asn Immunologists; Am Soc Microbiol. *Mailing Add:* Histocompat Lab Reg Organ Bank Ill 800 S Wells Suite 190 Chicago IL 60607

LAZELL, JAMES DRAPER, b New York, NY, Sept 5, 39. POPULATION BIOLOGY, BIOGEOGRAPHY. *Educ:* Univ South, Sewanee, BA, 61; Univ Ill, MS, 63; Harvard Univ, MA, 66; Univ RI, PhD(biol), 70. *Prof Exp:* Head, Dept Sci, Palfrey Street Sch, 66-74; sci staff, Mass Audubon Soc, 67-76, sanctuary dir, 75-79; PRES, CONSERV AGENCY, 79- *Concurrent Pos:* Collabr, Nat Park Serv, 69-; prin investr, Earthwatch, 73-; assoc, Mus Comp Zool, 77-, Bishop Mus, 87-, Miss Mus Natural Sci, 90-; curatorial affil, Yale Peabody Mus, 82- *Mem:* Am Soc Ichthyologists & Herpetologists; Am Soc Mammalogists; Am Soc Zoologists; Soc Study Amphibians & Reptiles; fel Explorers Club. *Res:* Vertebrate systematics; endangered species conservation; ecology; demographics; distribution of rare and little-known animals; exploration as of 1990; author of 120 scientific papers and three books. *Mailing Add:* Six Swinburne St Jamestown RI 02835

LAZERSON, EARL EDWIN, b Detroit, Mich, Dec 10, 30; m 66; c 2. ALGEBRA, NUMBER THEORY. *Educ:* Wayne State Univ, BS, 53; Univ Mich, MA, 54, PhD, 82. *Prof Exp:* Actg chmn, Dept Math, 71-72, chmn, 72-73, dean, Sch Sci & Technol, 73, actg vpres & provost, 76-77, vpres & provost, 77-79, actg pres, 79-80, PROF MATH, SOUTHERN ILL UNIV, 73-, PRES, 80- *Concurrent Pos:* NSF res grant, 62-67. *Mem:* Am Math Soc; Math Asn Am; London Math Soc; Soc Math de France. *Mailing Add:* Off Pres Southern Ill Univ Edwardsville IL 62026

LAZO, JOHN STEPHEN, b Philadelphia, Pa, Dec 15, 48; m 74; c 1. ONCOLOGY, BIOCHEMICAL PHARMACOLOGY. *Educ:* Johns Hopkins Univ, BA, 71; Univ Mich, Ann Arbor, PhD(pharmacol), 76. *Prof Exp:* Lab asst, Dept Pharmacol, Thomas Jefferson Med Col, 68-69; lab asst, Dept Chem, Johns Hopkins Univ, 71; USPHS-NIH trainee, Dept Pharmacol, Univ Mich, Ann Arbor, 71-76; fel, 76-78, from asst prof to assoc prof pharmacol, Yale Univ, 78-87; PROF PHARMACOL & CHMN DEPT, SCH MED, UNIV PITTSBURGH, 87- *Concurrent Pos:* Vis scientist, Sloan-Kettering Inst Cancer Res, 85-86. *Mem:* Am Soc Pharmacol & Exp Therapeut; Tissue Culture Asn; Am Asn Cancer Res; NY Acad Sci; AAAS. *Res:* Pharmacology and toxicology of antitumor agents; use of cultured cells to study drug actions; action of drugs on lung tissue. *Mailing Add:* 5128 Pembroke Pl Pittsburgh PA 15232

LAZO-WASEM, EDGAR A, b Guatemala City, Guatemala, Jan 18, 26; nat US; m 50; c 3. PHARMACEUTICS. *Educ:* Carroll Col, Mont, BA, 49; Mont State Univ, MA, 51; Purdue Univ, PhD(endocrinol, pharmacol), 54. *Prof Exp:* Chief physiologist, Wilson Labs, 54-58, dir control, 58-63; dir sci serv, Strong Cobb Arner Inc, 63-64, vpres & sci dir, 65-67; vpres & tech dir, Wampole Labs, 67-74, vpres & tech dir, Wallace Labs, 74-77; vpres tech, 77-84, VPRES QUAL, BARNES-HIND INC, 85- *Mem:* AAAS; Endocrine Soc; Soc Exp Biol & Med; Am Physiol Soc. *Res:* Enzyme analysis; pharmaceutical development; clinical diagnostic reagents; ophthalmics. *Mailing Add:* 555 Middlefield Rd Mountainview CA 94043

LAZOWSKA, EDWARD DELANO, b Washington, DC, Aug 3, 50; m 78; c 2. COMPUTER SYSTEMS. *Educ:* Brown Univ, 72; Univ Toronto, MSc, 74, PhD, 77. *Prof Exp:* PROF COMPUT SCI, UNIV WASH, 77- *Concurrent Pos:* Vis scientist, DEC Systs Res Ctr, 84-85. *Res:* Computer systems; modelling & analysis; design & implementation; distributed parallel systems. *Mailing Add:* Comput Sci FR-35 Univ Wash Seattle WA 98195

LAZZARA, RALPH, b Tampa, Fla, Aug 14, 34; m 59; c 3. MEDICINE. *Educ:* Univ Chicago, BA, 55; Tulane Univ, MD, 59. *Prof Exp:* Instr med, Tulane Univ, 60-67; asst prof med, Sch Med, Univ Miami, 71-72, assoc prof, 72-77; chief sect cardiol, Vet Admin Hosp, 74-78; PROF MED, SCH MED, UNIV OKLA, 78- *Concurrent Pos:* Asst, Charity Hosp, New Orleans, 60-64; resident, Med Sch, Tulane Univ & Charity Hosp, 63-64; fel, Col Physicians & Surgeons, Columbia Univ, 64-65; staff mem & dir cardiovasc res lab, Ochsner Clin & Ochsner Found Hosp, New Orleans, 65-67; staff cardiologist & chief sect electrophysiol, Mt Sinai Hosp, Miami Beach, 70-72; dir coronary care unit, Vet Admin Hosp, Miami, 72-75; chief sect cardiol, Univ Okla Health Sci Ctr & Vet Admin Hosp, Okla City, 78- *Mem:* Am Physiol Soc; Soc Chem Invest. *Res:* Cardiac electrophysiology. *Mailing Add:* Univ Okla Health Sci Ctr PO Box 26901 Oklahoma City OK 73190

LAZZARI, EUGENE PAUL, biochemistry; deceased, see previous edition for last biography

LAZZARINI, ALBERT JOHN, b Lucca, Italy, Oct 11, 52; US citizen; m; c 4. SYSTEMS ENGINEERING, OPTICAL SYSTEMS DESIGN. *Educ:* Mass Inst Technol, SB, 74, PhD(nuclear physics), 78. *Prof Exp:* Res assoc Except Nuclear Physics, Univ Wash, 78-80, res asst prof physics 80-84; sr scientist, 84-88, PRIN INVESTR, KAMAN INST/KAMAN SCI CORP, 88- *Mem:* Am Soc Phys; Int Soc Optical Eng. *Res:* Overall responsibility for end-to-end system design and performance analysis of complex electro-optical systems operating in the visible and infrared; design of adaptive optics systems. *Mailing Add:* Kaman Sci Corp 1500 Garden of the Gods Rd Colorado Springs CO 80933-7463

LAZZARINI, ROBERT A, b New York, NY, Oct 14, 31. MOLECULAR GENETICS. *Educ:* Univ Calif, Los Angeles, BS, 55, PhD(biol chem), 60. *Prof Exp:* Postdoctoral fel, Johns Hopkins Univ, 60-63; staff fel, Lab Biochem, Nat Inst Dent Res, NIH, 63-64, Lab Molecular Biol, Nat Inst Neurol & Commun Dis & Stroke, 64-65, res scientist, 65-70, head, Sect Regulation Nucleic Acid Synthesis, 70-74, Sect Molecular Virol, 74-81, chief, Lab Molecular Genetics, 81-88; DIR, BROOKDALE CTR MOLECULAR BIOL, MT SINAI SCH MED, NY, 88- *Concurrent Pos:* Session chmn, Gordon Res Conf, 74 & 75; chmn, Conf Mech Viral Resistance, Mult Sclerosis Soc, Colo, 80; mem, Virol Study Sect, Div Res Grants, NIH, 80 & Neurol C Study Sect, 88-89; chmn, RNA Virus Div, Am Soc Microbiol, 81-83; mem, Adv Coun Basic Sci, Mult Sclerosis Soc, 81-87, Adv Coun Microbiol & Virol, Am Cancer Soc, 82-84 & Alzheimer Dis Adv Bd, Am Health Assistance Found, 89- *Honors & Awards:* Quastel Lectr Molecular Sci, McGill Univ, Can, 85; Pfizer Lectr, Univ Montreal, Can, 85; Meritorious Exec Rank Award Sr Serv, 86. *Mem:* Am Soc Biol Chemists; Am Soc Microbiol; Am Soc Neurochem; Am Soc Virol. *Mailing Add:* Mt Sinai Sch Med PO Box 1126 New York NY 10029

LE, CHÁP THAN, b Vietnam, Aug 1, 48; m 75. STATISTICS, BIOMETRICS. *Educ:* Calif State Univ, BA, BS & MA, 71; Univ NMex, PhD(statist), 78. *Prof Exp:* ASST PROF BIOMET, UNIV MINN, 78- *Mem:* Biomet Soc. *Res:* Reliability and life testing; nonparametric statistics; survey sampling; theory of survivorship; epidemiology. *Mailing Add:* Univ Minn Box 197 Mayo Minneapolis MN 55455

LE, HENG-CHUN, BIOCHEMISTRY. *Prof Exp:* PROF BIOCHEM, MT SINAI SCH MED, 71- *Mailing Add:* Biochem Dept Mt Sinai Sch Med Fifth Ave & 100th St Box 1020 New York NY 10029

LEA, ARDEN OTTERBEIN, b Cleveland, Ohio, Oct 19, 26; m 52; c 2. ENTOMOLOGY. *Educ:* Univ Rochester, BA, 48; Ohio State Univ, MSc, 50, PhD(entom), 57. *Prof Exp:* Res assoc insecticide testing, Ohio State Univ, 50-51, insect nutrit, 56-58; USPHS med entomologist, Onchocerciasis Proj, Pan Am Sanit Bur, Guatemala, 51-53; chief physiol sect, Entom Res Ctr, State Bd Health, Fla, 58-69; assoc prof entom, 69-74, PROF ENTOM, UNIV GA, 74- *Concurrent Pos:* USPHS spec res fel, Denmark, 60-61; mem trop med & parasitol study sect, NIH, 74-78; mem sci adv panel onchocerciasis, WHO. *Mem:* AAAS; Am Soc Tropical Med & Hyg; Am Mosquito Control Asn. *Res:* Endocrine physiology of Diptera; physiology and behavior of mosquitoes; peptide hormones. *Mailing Add:* Dept Entom Univ Ga Athens GA 30602

LEA, JAMES DIGHTON, b Monticello, Ill, Apr 9, 33; m 53; c 2. SYSTEMS SCIENCE. *Educ:* Tex Western Col, BA, 57; Univ Tex, MA, 60, PhD(physics), 63. *Prof Exp:* Sr res scientist physics, Esso Prod Res Co, Stand Oil NJ, 63-69, sr prof systs analyst, Humble Oil & Refining Co, 69-73, explor systs adv, Exxon Co USA, 73-75, RES ASSOC SYSTS, EXXON PROD RES CO, EXXON CORP, 75- *Mem:* Am Math Soc; Am Phys Soc; Soc Explor Geophysicists; Am Asn Petrol Geologists. *Res:* Application of computer science to geological problems. *Mailing Add:* 6230 Bayou Bridge Houston TX 77096

LEA, JAMES WESLEY, JR, b Lebanon, Tenn, Mar 17, 41; m 66; c 2. TOPOLOGY, ALGEBRA. *Educ:* Tenn Polytech Inst, BS, 63, MS, 65; La State Univ, PhD(math), 71. *Prof Exp:* Instr math, Univ Tenn, Martin, 65-66; instr, Tenn Technol Univ, 66-67; from asst prof to assoc prof, 71-81, PROF MATH, MIDDLE TENN STATE UNIV, 81- *Concurrent Pos:* Vis assoc prof math, Univ Tenn, Knoxville, 76-77. *Mem:* Am Math Soc; Math Asn Am; Sigma Xi. *Res:* Lattice theory. *Mailing Add:* Middle Tenn State Univ Murfreesboro TN 37132

LEA, MICHAEL ANTHONY, b Leeds, Eng, Dec 26, 39; m 61; c 2. BIOCHEMISTRY. *Educ:* Univ Birmingham, BSc, 61, PhD(biochem), 64. *Prof Exp:* Res assoc pharmacol, Sch Med, Ind Univ, 64-66, instr, 66-67; from asst prof to assoc prof, 67-78, PROF BIOCHEM, UNIV MED & DENT NJ, 78- *Mem:* AAAS; Am Chem Soc; Am Soc Cell Biol; Am Asn Cancer Res; Am Soc Biochem Molecular Biol. *Res:* Control of nucleic acid synthesis and growth in normal and neoplastic cells. *Mailing Add:* Dept Biochem Univ Med & Dent NJ Newark NJ 07103-2714

LEA, ROBERT MARTIN, b New York, NY, Nov 4, 31; m 53; c 1. PHYSICS. *Educ:* Union Col, NY, BS, 53; Yale Univ, PhD(physics), 57. *Prof Exp:* Chmn dept, 70-74 & 87-89, PROF PHYSICS, CITY COL NEW YORK, 57- *Concurrent Pos:* Vis physicist, Brookhaven Nat Lab, 59-70; prin investr, NSF res grants, City Col New York, 59-70, dir, NSF dept develop grant, 70-74, NRCSE, 81-88. *Mem:* Am Phys Soc; Am Asn Physics Teachers. *Res:* High energy experimental physics. *Mailing Add:* Dept Physics CUNY NY City Col Convent Ave at 138th New York NY 10031

LEA, SUSAN MAUREEN, b Cardiff, Wales, UK, July 10, 48; m 74; c 1. ASTROPHYSICS. *Educ:* Cambridge Univ, BA, 69, MA, 73; Univ Calif, Berkeley, PhD(astron), 74. *Prof Exp:* Res assoc astrophys, Ames Res Ctr, NASA, 74-76; res fel, Univ Md, College Park, 76-77; asst res astronomer, Univ Calif, Berkeley, 77-80; assoc prof, 80-84, PROF, DEPT PHYSICS & ASTRON, SAN FRANCISCO STATE UNIV, 84- *Mem:* Am Astron Soc; Royal Astron Soc; Int Astron Union; Am Phys Soc. *Res:* High energy astrophysics, especially x-ray and radio astronomy; numerical hydrodynamics; compact galactic x-ray sources, clusters of galaxies, intergalactic matter and cosmology. *Mailing Add:* Dept Physics & Astron San Francisco State Univ 1600 Holloway Ave San Francisco CA 94132

LEA, WAYNE ADAIR, b Helena, Mont, Jan 16, 40; c 7. COMPUTER RECOGNITION OF SPEECH. *Educ:* Mont State Col, BS, 62, MS, 64; Mass Inst Technol, SM & EE, 66, Purdue Univ, PhD, 72. *Prof Exp:* Res assoc, Electronics Res Lab, Mont State Col, 62-64; NSF fel, Electronics Res Lab, Mass Inst Technol, 64-66; proj leader, Electronics Res Ctr, NASA, 66-70; instr elec eng, Res Found, Purdue Univ, 70-72; prin investr, Defense Systs Div, Sperry Univac, 72-77; res scientist, Speech Commun Res Lab, 77-81; DIR, SPEECH SCI PUBL, 80-, DIR, SPEECH SCI INST, 85- *Concurrent Pos:* Adj assoc prof, Dept Linguistics, Univ Southern Calif, 78-83 & Dept Elect & Comput Eng, Univ Calif, Santa Barbara, 81-83; consult speech recognition, various co, 79-; chmn, Acad Forensic Appl Commun Sci, 78-82; sr vpres & dir, Ctr Speech Res & Educ, VCS Corp, 83-84; actg mgr, Speech Group, A I Lab, NYNEX S&T, 87. *Mem:* Sr mem Inst Elec & Electronics Engrs; Acoust Soc Am; Am Acad Forensic Sci; Am Asn Phonetic Sci; Sigma Xi. *Res:* Computer recognition of speech; intonation, linguistic stress and rhythm; acoustic phonetics, forensic phonetics. *Mailing Add:* 823 Jonathan Dr Apple Valley MN 55124-0428

LEABO, DICK ALBERT, b Walcott, Iowa, Oct 30, 21; m 55; c 1. APPLIED STATISTICS, ECONOMIC STATISTICS. *Educ:* Univ Iowa, BS, 49, MA, 50, PhD(statist, econ), 53. *Prof Exp:* Res asst econ, Bur Bus & Econ Res, Univ Iowa, 48-49, res assoc, 49-53, asst prof econ & asst dir, 53-56; asst prof econ & asst dir, Bur Econ & Bus Res, Mich State Univ, 56-57; from asst prof statist to assoc prof, 57-63, assoc dean, Grad Sch Bus, 62-65, dir PhD prog, 65-81, prof, 63-79, Fred M Taylor Distinguished Prof, 78-84, EMER PROF STATIST, UNIV MICH, ANN ARBOR, 84- *Concurrent Pos:* Consult, Brookings Inst, 57-59 & NCent Accrediting Asn, 74-84; exchange prof, Rotterdam Sch Econ, 65. *Mem:* Am Statist Asn; Am Econ Asn; Asn Bus Economists. *Res:* Regional economic research and the application of regression and correlation techniques. *Mailing Add:* 2115 Nature Cove Ct No 105 Ann Arbor MI 48104-4977

LEACH, BARRIE WILLIAM, b Winnipeg, Man, Nov 25, 45; m 66; c 2. APPLIED MATHEMATICS, ELECTRICAL ENGINEERING. *Educ:* Univ Man, BSc, 67, MSc, 68, PhD(elec eng), 72. *Prof Exp:* asst res officer, 72-80, ASSOC RES OFFICER, FLIGHT RES, NAT RES COUN CAN, 80- *Mem:* Inst Elec & Electronics Engrs. *Res:* Optimal parameter and state estimation; aeromagnetics; geophysical and anti-submarine warfare applications; digital filtering techniques multi-sensor navigation techniques. *Mailing Add:* Nat Res Coun Montreal Rd Ottawa ON K1A 0R6 Can

LEACH, BERTON JOE, b Tuscola, Ill, Mar 30, 32; m 55; c 2. ZOOLOGY, SCIENCE ADMINISTRATION. *Educ:* Washington Univ, AB, 57; Univ Mo, MA, 60, PhD(zool), 63. *Prof Exp:* Asst zool, Univ Mo, 58-60, instr, 60-62, USPHS res fel, 62-63; asst prof, George Washington Univ, 63-66; asst prog dir, Undergrad Student Prog, NSF, 66-67, asst prog dir col sci improv prog, 67-68; prof biol, Cent Methodist Col, 68-70, F H Dearing prof, 70-74, chmn dept biol & geol, 68-74; exec secy, Cardiovasc & Pulmonary Study Sect, Div Res Grants, NIH, 74-76; chief consult, Berton J Leach Assoc, Rockville, Md,

76-78; sr scientist, Capital Syst Group, 78-81; RETIRED. *Concurrent Pos:* NSF res grant, 63-65; assoc prof, George Washington Univ, 66; USPHS res evaluator, 66-67; vis scholar, Harvard Univ, 69; med technol educ adv, Jewish Hosp, St Louis, 72; comput data bases, US Govt Contracts Pvt Indust, 76-81; gen reader, Marine Biol Lab, 85-87; vol lab instr neuroanat, Georgetown Univ Med Sch, 88; adj assoc prof, Georgetown Univ Med Sch, 89-91. *Mem:* Sigma Xi; Am Soc Mammalogists. *Res:* Cancer construction data base; vertebrate biology computer courseware; mammalian neuroanatomy; mammalian behavior. *Mailing Add:* 12707 Weiss St Rockville MD 20853

LEACH, CHARLES MORLEY, b Sacramento, Calif, Oct 28, 24; m 49; c 3. PLANT PATHOLOGY, MYCOLOGY. *Educ:* Queen's Univ, Ireland, BS, 49, BAgr, 50; Ore State Univ, PhD(plant path), 56. *Prof Exp:* Instr bot, 51-57, from asst plant pathologist to assoc plant pathologist, 57-66, PROF PLANT PATH, ORE STATE UNIV, 66- *Concurrent Pos:* NSF fel, Univ Bristol, 62-63; NZ sr sci fel, 73-74 & Cambridge Univ, 84-85. *Mem:* Am Phytopath Soc; Mycol Soc Am; Brit Mycol Soc; Int Soc Plant Path; Can Plant Path Soc; Int Asn Aerobiology. *Res:* Biology of plant pathogenic fungi, especially reproduction and spore discharge; seed-borne diseases of agricultural crops; electrical nature of plant surfaces. *Mailing Add:* Dept Bot & Plant Path Ore State Univ Corvallis OR 97331

LEACH, EDDIE DILLON, microbiology, toxicology, for more information see previous edition

LEACH, ERNEST BRONSON, b Huchow, China, Dec 21, 24; US citizen; div. MATHEMATICS. *Educ:* Case Inst Technol, BS, 49; Mass Inst Technol, PhD(math), 53. *Prof Exp:* From instr to asst prof, 53-59, ASSOC PROF MATH, CASE WESTERN RESERVE UNIV, 59- *Concurrent Pos:* Partic, Indo-Am Prog, Indian Inst Technol, Kanpur, 63-64; mem staff, Northwestern Univ Proj, Univ Khartoum, 66-67; partic inelec proj, Inst Nat Elec & Electronics, Boumerdes, Algeria, 76-78. *Res:* Algebraic topology; functional analysis. *Mailing Add:* 6312 Eastondale Rd Mayfield Heights OH 44124

LEACH, FRANKLIN ROLLIN, b Gorman, Tex, Apr 2, 33; m 56, 70; c 6. BIOCHEMISTRY. *Educ:* Hardin-Simmons Univ, BA, 53; Univ Tex, PhD(chem), 57. *Prof Exp:* Res scientist I, Biochem Inst, 53-56; Nat Acad Sci fel med sci, Univ Calif, 57-59; res assoc, 59-60, from asst prof to assoc prof biochem, 60-68, chmn grad fac genetics, 76-78, PROF BIOCHEM, OKLA STATE UNIV, 68- *Concurrent Pos:* Soc Am Bacteriologists pres fel, Univ Ill, 60; NIH res career develop award, 62-72; res fel, Calif Inst Technol, 65-66. *Mem:* AAAS; Am Chem Soc; Am Soc Microbiol; Am Soc Biochem & Molecular Biol; Am Soc Photobiol; Coun Biol Ed. *Res:* Bioluminescence; analytical biochemistry; enzymology; environmental biochemistry. *Mailing Add:* Dept Biochem Okla State Univ Stillwater OK 74078-0454

LEACH, JAMES L(INDSAY), b Lawrenceville, Ill, Apr 9, 18; c 3. MECHANICAL ENGINEERING. *Educ:* Univ NMex, BS, 42; Univ Ill, MS, 50; Ill State Univ, PhD, 76. *Prof Exp:* Engr, Texas Co, Ill, 40-41, US Ord Res & Gage Dept, Washington, DC, 42, Pac Bridge Co, US Navy Cont Dock & Yards Comb & Control Eng, Southwest Pac, 43; from asst prof to assoc prof, 48-59, PROF MECH ENG, UNIV ILL, URBANA, 59- *Concurrent Pos:* Consult, USAF, 50-51, Aluminum Co Am, 52-53, Caterpillar Tractor Co, 55-56 & Universal Bleacher Co, 56-; guest prof, Indian Inst Tech, Kharagpur, Indian, 60-61; installation engr, Universal Bleacher Co, Kuwait, 62; guest prof, Sch Bengal Eng Col, Calcutta, 64, Inst Coimbatore, Indian, Pradesh Agr Univ, 67; adv to dean eng, Utter Pradesh Agr Univ, 69; consult, Acad Educ Develop, Tehran, Iran, 73-75. *Mem:* Am Foundrymen's Soc; Sigma Xi; Am Soc Mech Engrs; Am Soc Mil Engrs; Am Soc Petrol Engrs; Nat Soc Prof Engr; Fedn Am Scientists. *Res:* Carbon dioxide process for hardening molds and cores for casting metal; shell molding. *Mailing Add:* RR 2 Flat Rock IL 62427

LEACH, JAMES MOORE, organic chemistry; deceased, see previous edition for last biography

LEACH, JAMES WOODROW, thermodynamics, heat transfer, for more information see previous edition

LEACH, JOHN KLINE, b Buffalo, NY, July 11, 22; m 45; c 5. CARDIOLOGY, PHYSIOLOGY. *Educ:* Baldwin-Wallace Col, BS, 43; Albany Med Col, MD, 47; Am Bd Internal Med, dipl, 68; Am Bd Cardiovasc Dis, dipl, 69. *Prof Exp:* Instr med, Albany Med Col, 55-62, asst prof physiol, 62-63; assoc chief staff res, Vet Admin Ctr, Wadsworth, Kans, 63-64; from asst prof to assoc prof med, 66-76, from asst prof to prof med & physiol, 72-85, EMER PROF MED & PHARMOL, MED SCH, UNIV, NMEX, 86- *Concurrent Pos:* Clin asst & asst attend, Albany Hosp, 55-62; NIH res fel physiol, Albany Med Col, 61-63; res grant cardiol, Univ NMex, 69-72; lectr, Med Ctr, Univ Kans, 63-64; chief cardiol sect, Vet Admin Hosp, Albuquerque, 66-72 & 77-83, assoc chief staff res, 69-73; attend med, Bernalillo County Med Ctr, Albuquerque, NMex, 66-; consult med, Bataan Hosp, Albuquerque, 69-83; consult med & cardiol, St Joseph Hosp, Albuquerque, 69-83; consult cardiol, Presby Hosp, Albuquerque, 69-83; vis assoc prof, Med Ctr, Univ Calif, Los Angeles, 71-72. *Mem:* Fel Am Col Cardiol; Am Fedn Clin Res; Am Heart Asn; fel Am Col Physicians; Am Physiol Soc. *Res:* Cardiovascular research; cardiac muscle mechanics and hemodynamics. *Mailing Add:* Sch Med Dept Pharmacol Univ NMex Albuquerque NM 87131

LEACH, KAREN LYNN, US citizen; m; c 1. CELL BIOLOGY. *Educ:* Ohio Wesleyan Univ, BA, 77; Univ Mich, Ann Arbor, PhD(pharmacol), 81. *Prof Exp:* Postdoctoral fel, Lab Cellular Carcinogenesis & Tumor Prom, Nat Cancer Inst, 84; res scientist, 84-89, SR RES SCIENTIST, DEPT CELL BIOL, UPJOHN CO, KALAMAZOO, 90- *Concurrent Pos:* Organizer, Sixth Int Symp Cell Endocrinol, Lake Placid, NY, 90; assoc ed, J Immunol, 90-92; reviewer, Cancer Res, J Biol Chem, J Immunol & J Cell Biol; invited lectr, numerous univs. *Mem:* Soc Neurosci; Am Soc Cell Biol; Am Soc Biochem & Molecular Biol. *Mailing Add:* Dept Cell Biol Upjohn Co 301 Henrietta St Kalamazoo MI 49001

LEACH, LEONARD JOSEPH, b Rochester, NY, Aug 3, 24; m 53; c 3. TOXICOLOGY, ENVIRONMENTAL HEALTH. *Educ:* Brigham Young Univ, BS, 49. *Prof Exp:* Phys chemist, Army Chem Ctr, Md, 51-52; assoc indust hyg, Atomic Energy Proj, 52-55, unit chief toxicol, 55-57, instr indust hyg, 57-65, ASST PROF RADIATION BIOL & BIOPHYS, SCH MED & DENT, UNIV ROCHESTER, 65- *Concurrent Pos:* Speaker, Gordon Res Conf Toxicol & Safety Eval, NH, 60. *Mem:* Am Soc Toxicol; Am Acad Indust Hyg; Pan-Am Med Asn; Am Indust Hyg Asn; NY Acad Sci. *Res:* Inhalation toxicity of airborne agents related to air pollution and all aspects of environmental health. *Mailing Add:* 47 Pickdale Dr Rochester NY 14626

LEACH, ROBERT ELLIS, b Sanford, Maine, Nov 25, 31; m 55; c 6. MEDICINE, ORTHOPEDIC SURGERY. *Educ:* Princeton Univ, BA, 53; Columbia Univ, MD, 57. *Prof Exp:* Chmn orthop, Lahey Clin Found, 67-70; PROF ORTHOP SURG, MED SCH, BOSTON UNIV, 70- *Concurrent Pos:* Dir orthop serv, Boston City Hosp, 70-76; lectr, Tufts Univ, 70-; Am, Brit & Can traveling fel, 71; chmn, Sports Med Coun, US Olympic Comt. *Mem:* Am Orthop Asn; Am Orthop Soc Sports Med; Am Acad Orthop Surgeons. *Res:* Synthetic ligament reconstruction. *Mailing Add:* Dept Surg Boston Univ Sch Med 80 E Concord St Boston MA 02118

LEACH, ROLAND MELVILLE, JR, b Framingham, Mass, Aug 27, 32; m 54; c 3. TRACE MINERAL NUTRITION. *Educ:* Univ Maine, BS, 54; Purdue Univ, MS, 56; Cornell Univ, PhD(nutrit), 60. *Prof Exp:* Asst prof animal nutrit, Cornell Univ, 60-68, chemist, Plant Soil & Nutrit Lab, USDA, 59-68; assoc prof poultry sci, 68-73, PROF POULTRY SCI, PA STATE UNIV, UNIVERSITY PARK, 73- *Honors & Awards:* Am Feed Mfrs Asn Award, Poultry Sci Assoc, 80. *Mem:* Poultry Sci Asn; Am Inst Nutrit; AAAS. *Res:* Mineral nutrition of animals; role of trace elements in bone formation. *Mailing Add:* 205 Animal Industs Bldg Pa State Univ University Park PA 16802

LEACH, RONOLD, b Baltimore, Md, Feb 5, 44. SOFTWARE SYSTEMS. *Educ:* Univ Md, BS, 64, MA, 66; Johns Hopkins Univ, PhD(comput sci), 71, MS, 83. *Prof Exp:* PROF, HOWARD UNIV, 69- *Mem:* Asn Comput Mach; Comput Soc; Am Math Asn; Inst Elec & Electronics Engrs. *Mailing Add:* Ten E Lee St No 2101 Baltimore MD 21204

LEACH, WILLIAM MATTHEW, b Pine Mountain, Ky, June 26, 33; m 60; c 3. CELL BIOLOGY, RADIOBIOLOGY. *Educ:* Berea Col, BA, 56; Univ Tenn, MS, 62, PhD(zool), 65. *Prof Exp:* USAEC res assoc zool & entom, Inst Radiation Biol, Univ Tenn, 64-66; res biologist, Radiation Bio-Effects Prog, Nat Ctr Radiol Health, 66-67, chief radiation cytol lab, Div Biol Effects, Bur Radiol Health, 67-71, chief exp studies br, Bur Radiol Health, 71-84, dir Div Life Sci, 85-86, ASSOC DIR SCI, OFF SCI & TECHNOL, CTR DEVICES & RADIOL HEALTH, FOOD & DRUG ADMIN, 86- *Concurrent Pos:* Adj prof genetics, George Washington Univ, 72- *Mem:* AAAS; Am Soc Cell Biol; Am Genetic Asn; Sigma Xi. *Res:* Cell responses to radiation in relation to the cell cycle; cell synthetic activities during the cell cycle; behavior of particulates and molecules in cells; developmental biology; microwave radiation research. *Mailing Add:* 1805 Brisbane St Silver Spring MD 20902

LEACH-HUNTOON, CAROLYN S, b Leesville, La, Aug 25, 40; m 69; c 1. ENDOCRINOLOGY, PHYSIOLOGY. *Educ:* Northwestern State Col, La, BS, 62; Baylor Univ, MS, 66, PhD(physiol), 68. *Prof Exp:* Med technologist, Univ Tex M D Anderson Hosp & Tumor Inst, Houston, 62-64, spec med technologist, 64-68; head endocrine & biochem labs, 68-74, chief space metab & biochem br, 74-76, chief biomed labs br, 76-84, assoc dir, 84-87, DIR, SPACE & LIFE SCI, JOHNSON SPACE CTR, NASA, 87- *Concurrent Pos:* Consult, Proj Sea Lab, US Navy, 68-69, Proj Tektite I, US Navy-NASA-Gen Elec-Dept of Interior, 69 & Proj Tektite II, NASA-Dept of Interior, 70; Nat Res Coun-Nat Acad Sci res assoc, Manned Spacecraft Ctr, NASA, 68-70; adj instr physiol, Baylor Col Med, 68-70, adj asst prof, 70-; assoc investr, Inst Environ Med, Sch Med, Univ Pa, 70; mem res staff, Univ Tex Marine Biomed Inst. *Honors & Awards:* Louis H Bauer Founder's Award, Aerospace Med Asn; Paul Best Award for Res Aerospace Physiol; Outstanding Woman in Sci, Am Women in Sci, 84. *Mem:* Fel Aerospace Med Asn; Int Astronaut Fedn; Endocrine Soc; Am Physiol Soc; Am Soc Med Technol; Am Soc Clin Path; Am Inst Aeronaut & Astronaut. *Res:* The study of the physiological adaptation of man to changing environments, particularly the endocrine mechanisms involved in adaptation; aerospace medicine. *Mailing Add:* Dir Space & Life Sci Code SA NASA L B Johnson Space Ctr Houston TX 77058

LEACOCK, ROBERT A, b Detroit, Mich, Oct 3, 35; m 61; c 3. THEORETICAL PHYSICS, HIGH ENERGY PHYSICS. *Educ:* Univ Mich, BS, 57, MS, 60, PhD(physics), 63. *Prof Exp:* Instr physics, Univ Mich, 63-64; Am-Swiss Found Sci Exchange fel theoret physics, Europ Orgn Nuclear Res, 64-65; assoc physics, 65-67, from asst prof to assoc prof, & physicist, Ames Lab, 67-87, PROF, PHYSICS, IOWA STATE UNIV, 87- *Concurrent Pos:* Corning Glass Works Found fel, 62-63. *Mem:* Am Phys Soc. *Res:* Theoretical physics. *Mailing Add:* Dept Physics Iowa State Univ Ames IA 50011

LEACOCK, ROBERT JAY, b New York, NY, Mar 1, 39; m 64; c 2. ASTRONOMY. *Educ:* Univ Fla, BS, 60, MS, 62, PhD(astron), 71. *Prof Exp:* Instr physics, Pensacola Jr Col, 62-63; res asst astron, 63-71, asst prof, 71-76, ASSOC PROF PHYS SCI & ASTRON, UNIV FLA, 76-, ASSOC CHMN, 80- *Mem:* AAAS; Am Astron Soc. *Res:* Nonthermal radio observations of the major planets; optical variations of extragalactic radio sources. *Mailing Add:* Astron 211 Ssrb Univ Fla Gainesville FL 32611

LEADABRAND, RAY LAURENCE, b Pasadena, Calif, Oct 12, 27; m 55; c 1. ELECTRONICS ENGINEERING. *Educ:* San Jose State Col, BS, 50; Stanford Univ, MS, 53. *Prof Exp:* Field engr, Philco Corp, 50-52; asst, Stanford Univ, 52-55; res engr, SRI Int, 55-58, head propagation group, 59-60, mgr radio physics lab, 60-68, sr vpres, Eng Res Group, 80-85, EXEC DIR, ELECTRONICS & RADIO SCI DIV, SRI INTERNATIONAL, 68-; SR VPRES, SCI APPLNS INT CORP, 85- *Mem:* AAAS; fel Inst Elec & Electronics Engrs; Am Geophys Union; Int Union Radio Sci; Sigma Xi. *Res:* Ionospheric radio propagation; auroral radar; moon and satellite reflection and transmission; propagation studies related to nuclear explosions and missile flight; radio and radar astronomy research of solar systems. *Mailing Add:* 80 Joaquin Rd Portola Valley CA 94025

LEADBETTER, EDWARD RENTON, b Barnesboro, Pa, Jan 26, 34; m 56; c 4. MICROBIOLOGY. *Educ:* Franklin & Marshall Col, BS, 55; Univ Tex, PhD(bact), 59. *Hon Degrees:* MA, Amherst Col, 70. *Prof Exp:* Instr, Amherst Col, 59-61, from asst prof to assoc prof, 61-70, chmn dept, 67-71, prof biol, 70-78; prof biol sci group, 78-85, head dept, 78-83, head dept microbiol, 84-85, PROF, DEPT MOLECULAR & CELL BIOL, UNIV CONN, 85- *Concurrent Pos:* NSF fel, Hopkins Marine Sta, Pacific Grove, Calif, 62-63; NIH spec fel, Univ Mass, 66-67; vis prof, Hampshire Col, 71; instr, Marine Biol Lab, Woods Hole, 71-78, mem corp, 71-; NATO sr fel, Univ Seville, 72; Nat Acad Sci-Hungarian Acad Sci Exchange, Szeged, 85, 88; found microbiol lectr, Am Soc Microbiol, 89-90; prog div, Cellular Biochem, NSF, 90-91. *Mem:* AAAS; Am Soc Microbiol. *Res:* Microbial ecology, physiology, and biochemistry; amine metabolism; photosynthesis; myxobacteria; oral microbiology; hydrocarbon oxidation; ultrastructure; gliding motility, sulfonate formation and degradation. *Mailing Add:* Molecular & Cell Biol Dept Univ Conn U-131 Storrs CT 06269-2131

LEADER, GORDON ROBERT, b Milwaukee, Wis, Jan 27, 16; m 46; c 4. PHYSICAL CHEMISTRY. *Educ:* Univ Wis, BS, 37; Univ Minn, PhD(phys chem), 40. *Prof Exp:* Res chemist, Monsanto Chem Co, Mo, 40-42, Nat Defense Res Comt, Northwestern Univ, 42-43 & Manhattan Dist Proj, Univ Chicago, 43-47; asst prof chem, Univ Ky, 47-51; res chemist, Mallinckrodt Chem Works, Mo, 51-53 & Olin Mathieson Chem Corp, 53-58; sr res chemist, Thiokol Chem Corp, 58-64; sr res chemist, Pennwalt Corp, 64-86; RETIRED. *Mem:* Am Chem Soc. *Res:* Chemical process development; Raman and nuclear magnetic resonance spectroscopy; radiochemistry; conductance and dielectric constants of organic solutions. *Mailing Add:* 1661 Weedon Rd Wayne PA 19087

LEADER, JOHN CARL, b St Louis, Mo, Oct 25, 38. STATISTICAL OPTICS, COHERENCE THEORY. *Educ:* Rensselaer Polytech Inst, BS, 60, PhD(nuclear physics), 69. *Prof Exp:* Sr engr, McDonnel Aircraft Reconnaissance Lab, 69-74, sr scientist, 74-81, chief scientist radiation sci, 81-87, DIR RES, MCDONNELL DOUGLAS RES LABS, MCDONNELL DOUGLAS CORP, 87- *Mem:* Optical Soc Am; Soc Photo-Optical Instrumentation Engrs; Union Radio Sci Int. *Res:* Theoretical research on radiation scattering from rough surfaces; optical propagation through atmospheric turbulence; partial coherence theory; laser radar; charged particle beam propagation. *Mailing Add:* 1016 Julianna Dr Manchester MO 63011

LEADER, JOHN CARL, b St Louis, Mo, Oct 25, 38; m 61; c 2. STOCASTIC-OPTICS, LASER INTERACTIONS. *Educ:* Rensselaer Polytech Inst, BS, 60, PhD(nuclear physics), 69. *Prof Exp:* Sr engr electronics, McDonnell Aircraft Co, 69-74; sr scientist, 74-80, chief scientist, 81-87, DIR, MCDONNELL DOUGLAS RES LABS, 87. *Concurrent Pos:* Ed, Proc Soc Photo-Optical Instrumentation Engrs, 83. *Mem:* Optical Soc Am; Union Radio Sci Int; Soc Photo-Optical Instrumentation Engrs. *Res:* Theoretical investigations of laser scattering from rough surfaces, inhomogeneous volumes, and propagation through turbulent media; computational electromagnetics; optical physics; charged particle beams; combustion production of aerosols. *Mailing Add:* 1016 Julianna Manchester MO 63011

LEADER, ROBERT WARDELL, b Tacoma, Wash, Jan 16, 19; m 40, 69; c 3. COMPARATIVE PATHOLOGY. *Educ:* Wash State Univ, BS & DVM, 52, MS, 55. *Hon Degrees:* DMedSci, Univ Toledo, 76. *Prof Exp:* Instr vet path, Wash State Univ, 52-55; USPHS fel, Univ Calif, 55-56; asst prof vet path, Wash State Univ, 56-60; assoc prof, Rockefeller Univ, 65-71; prof animal path, Univ Conn, 71-75; PROF PATH & CHMN DEPT, MICH STATE UNIV, 75- *Concurrent Pos:* Mem path training comt, NIH, 66-71, mem virol study sect, 71-; mem bd div, Mark Morris Found, 71- *Mem:* Am Vet Med Asn; Am Soc Exp Path; Am Col Vet Path; NY Acad Sci; Int Acad Path; Sigma Xi. *Res:* Studies of model diseases in animals with objective of elucidating pathogenetic mechanisms of similar diseases in man; chronic degenerative and connective tissue diseases. *Mailing Add:* Dept Path A136 E FCC Mich State Univ East Lansing MI 48824

LEADER, SOLOMON, b Spring Lake, NJ, Nov 14, 25. MATHEMATICS. *Educ:* Rutgers Univ, BS, 49; Princeton Univ, MA, 51, PhD(math), 52. *Prof Exp:* From instr to assoc prof, 52-61, PROF MATH, RUTGERS UNIV, NEW BRUNSWICK, 61- *Mem:* Am Math Soc; Math Asn Am; Sigma Xi. *Res:* Functional analysis; general topology. *Mailing Add:* Dept Math Rutgers Univ New Brunswick NJ 08903

LEADERS, FLOYD EDWIN, JR, b Denison, Iowa, Dec 11, 31; m 75; c 1. PHARMACOLOGY, RESEARCH ADMINISTRATION. *Educ:* Drake Univ, BS, 55; Univ Iowa, MS, 60, PhD(pharmacol), 62. *Prof Exp:* From instr to asst prof pharmacol, Med Ctr, Univ Kans, 62-67; head pharmacol res, Alcon Labs, Tex, 67-72; dir res serv, Plough, Inc, Tenn, 72-73; dir res & develop labs, Pharmaceut Div, Pennwalt Corp, NY, 73-78; PRES, TECH EVAL & MGT SYSTS INC, 78- *Concurrent Pos:* NIH res grant, 62-67; consult, Midwest Res Inst, 65-67; adj asst prof, Univ Tex Southwestern Med Sch; adj prof, Purdue Univ, 78-83, lectr/course dir, Inst Appl Pharm Sci. *Mem:* AAAS; Am Soc Pharmacol & Exp Therapeut; Soc Exp Biol & Med; Drug Info Asn; NY Acad Sci; Licensing Execs Soc. *Res:* Managing all aspects of pharmaceutical drug research and development, both ethical and proprietary, including data management, regulatory compliance and regulatory submissions; drug-vehicle systems; physiology and pharmacology of the eye, cardiovascular system and autonomic nervous systems; use of computers in biomedical data management. *Mailing Add:* Tech Eval & Mgt Sys Inc 5151 Beltline Rd Ste 1110 Dallas TX 75240

LEADON, BERNARD M(ATTHEW), b Farmington, Minn, Nov 29, 17; m 46; c 10. FLUID MECHANICS. *Educ:* Col St Thomas, BS, 38; Univ Minn, MS, 42, PhD(fluid mech), 55. *Prof Exp:* Engr, Pac Gas & Elec Co, Calif, 41; asst aerodyn, Univ Minn, 42, instr, 42-43; aerodynamicist, Curtiss-Wright Corp, NY, 43-44, sr aerodynamicist, 44-45; head propulsion exp sect, Cornell Aeronaut Lab, 45-46; chief aerodynamicist, Rosemount Aeronaut Lab, 46-48, scientist, 48-57; sr staff scientist, Gen Off, Gen Dynamics/Convair, 57-64; PROF ENG SCI, UNIV FLA, 64- *Concurrent Pos:* Instr, Univ Buffalo, 44-45; lectr, Univ Minn, 46-57; consult, Minneapolis-Honeywell Regulator Corp, 56-57, Gen Dynamics/Convair, 64-65, US Air Force, 66, AMF Beaird, 69, Martinez & Costa & Assocs, 71 & 78 & Pratt & Whitney Aircraft Corp, 74 & 76-82; vis prof, San Diego State Col, 62-64; NATO fel, 72; chmn, Third US Nat Conf Wind Eng, 78. *Honors & Awards:* Raymond L Bisplinghoff Award, 85, 86. *Mem:* Am Phys Soc; Am Inst Aeronaut & Astronaut; Sigma Xi. *Res:* Heat transfer; aerodynamics. *Mailing Add:* 412 NE 13th Ave Gainesville FL 32601

LEAF, ALEXANDER, b Yokohama, Japan, Apr 10, 20; nat US; m 43; c 3. INTERNAL MEDICINE. *Educ:* Univ Wash, BS, 40; Univ Mich, MD, 43. *Hon Degrees:* AM, Harvard Univ, 61. *Prof Exp:* Instr med, Univ Mich, 47-49; assoc, 53-56, from asst prof to assoc prof, 56-65, Jackson prof clin med, 66-81, Ridley Watts prof prev med & chmn, Dept Prev Med & Clin Epidemiol, 80-90, JACKSON EMER PROF CLIN MED, SCH MED, HARVARD UNIV, 90- *Concurrent Pos:* From asst physician to assoc physician, Mass Gen Hosp, 53-62, physician, 62-, chief med serv, 66-81; John Simon Guggenheim Mem Found fel, Balliol Col, Oxford Univ, 71-72. *Mem:* Inst Med-Nat Acad Sci; Am Soc Clin Invest; Am Physiol Soc; Asn Am Physicians; Am Acad Arts & Sci. *Res:* Ion transport and membrane physiology; kidney physiology; nutrition. *Mailing Add:* Mass Gen Hosp Boston MA 02114

LEAF, BORIS, b Yokohama, Japan, Mar 4, 19; nat US; m 47; c 3. STATISTICAL MECHANICS. *Educ:* Univ Wash, BS, 39; Univ Ill, PhD(phys chem), 42. *Prof Exp:* Spec asst chem, Univ Ill, 42-43, instr phys chem, 43-44; assoc chemist, Metall Lab, Univ Chicago, 44-45; Jewett fel, Yale Univ, 45-46; assoc prof physics, Kans State Univ, 46-54, prof, 54-65; chmn dept, 65-82, PROF PHYSICS, STATE UNIV NY COL CORTLAND, 65- *Concurrent Pos:* Spec asst, Nat Defense Res Comt, Univ Ill, 45; res fel, Brussels, 58-60; prof, State Univ NY Binghamton, 67-71; fac fel & grants-in-aid, Res Found, State Univ NY, 70-72, scholar exchange prof, 74-; vis prof, Cornell Univ, 73-74; assoc ed, Am J Physics, 76-79; partic, State Univ NY-Moscow State Univ fac exchange, USSR, 81. *Mem:* Fel AAAS; fel Am Phys Soc; Am Asn Physics Teachers; NY Acad Sci. *Res:* Thermodynamic theory; transport processes; quantum theory. *Mailing Add:* 11039 39th Ave NE Seattle WA 98125

LEAHY, DENIS ALAN, Taber, Alberta, June 13, 52. X-RAY ASTRONOMY, SUPERNOVA REMNANTS. *Educ:* Univ Waterloo. BSc, 75; Univ BC, MSc, 76, PhD(physics), 80. *Prof Exp:* Res assoc, Marshall Space Flight Ctr, 80-82; sessional instr, 82-83, univ res fel physics, 83-88, ASST PROF, DEPT PHYSICS, UNIV CALGARY, 88- *Mem:* Am Astron Soc; Can Astron Soc; Sigma Xi. *Res:* X-ray astronomy, data analysis and interpretation; supernova remnants; x-ray binaries; symbiotic stars; pulsar magnetospheres. *Mailing Add:* Dept Physics Univ Calgary 2500 University Dr NW Calgary AB T2N 1N4 Can

LEAHY, RICHARD GORDON, b Buffalo, NY, Mar 6, 29; m 53; c 3. GEOCHEMISTRY. *Educ:* Yale Univ, BS, 52; Harvard Univ, AM, 54, PhD(geol), 57. *Prof Exp:* Asst geochem, Yale Univ, 52-53; asst to dir & res assoc geol, Woods Hole Oceanog Inst, 56-60; dir labs, Div Eng & Appl Physics, 60-68, asst to pres civic & govt rels, 70-71, ASSOC DEAN FAC ARTS & SCI, HARVARD UNIV, 68- *Concurrent Pos:* Mem US tech panel geochem, Int Geophys Year, 57-58; chmn, Consortium Sci Comput, 86-88; exec dir, New Eng Consortium Undergrad Sci Educ, 88- *Mem:* AAAS; Am Geophys Union; NY Acad Sci. *Res:* Geochemistry of heavy isotopes in sea water and marine sediments; chemical processes of submarine weathering; variation of carbon dioxide in the atmosphere and its relation to air mass properties. *Mailing Add:* Sci Ctr Harvard Univ Cambridge MA 02138

LEAHY, SISTER MARY GERALD, b San Francisco, Calif, Oct 11, 17. INSECT PHYSIOLOGY, ACAROLOGY. *Educ:* Univ Southern Calif, BA, 45; Cath Univ Am, MA, 47; Univ Notre Dame, PhD(biol), 62. *Prof Exp:* From asst prof to assoc prof, 47-70, chmn dept, 62-65, PROF BIOL, MT ST MARY'S COL, CALIF, 70- *Concurrent Pos:* NSF res grants, 62-64, 65-67 & 70-71; fel trop pub health, Harvard Univ, 66; WHO grant, Israel Inst Biol Res, Ness Ziona & Hebrew Univ Jerusalem, 68-69; sr res scientist, Nairobi, Kenya, 73-74; collab scientist, EAfrican Vet Res Orgn, Kenya, 73-74; prin investr, NIH grant, 74-77; vis scientist, Ga Southern Col, 74-75; exchange scientist tick pheromones & hormones, Poland, Czech & Russia, Nat Acad Sci. *Mem:* AAAS; Entom Soc Am; Am Inst Biol Sci. *Res:* Mosquitoes and ticks; pheromones and reproductive physiology. *Mailing Add:* Dept Biol Mt St Mary's Col 12001 Chalon Rd Los Angeles CA 90049

LEAIST, DEREK GORDON, b Akron, Ohio, Jan 5, 55; Can citizen; m 86. ELECTROCHEMISTRY, TRANSPORT PROPERTIES. *Educ:* Queen's Univ, Kingston, Ont, BSc, 77; Yale Univ, MSc, 78, PhD(phys chem), 80. *Prof Exp:* Res assoc chem, Nat Res Coun, Ottawa, Can, 81-82; from asst prof to assoc prof, 82-90, PROF, CHEM DEPT, UNIV WESTERN ONT, LONDON, CAN, 90- *Honors & Awards:* Lash Miller Award, Electrochem Soc, 89. *Res:* Theoretical and experimental studies of diffusion in liquids, with emphasis on coupled transport in multicomponent electrolyte mixtures; micelle solutions; thermal diffusion in liquids. *Mailing Add:* Dept Chem Univ Western Ont London ON N6A 5B7 Can

LEAK, JOHN CLAY, JR, b Washington, DC, Aug 31, 28; m 54; c 3. ORGANIC CHEMISTRY, RADIO PHARMACEUTICALS. *Educ:* Univ Vt, BS, 49; Univ Ill, PhD(chem), 54. *Prof Exp:* Asst, Univ Ill, 51-54, res assoc animal nutrit, 55-56; Fulbright scholar, Ger, 54-55; chemist, Isotopes Specialties Co, 56-59, dir carbon-14 dept, 59-60; res dir, Cyclo Chem Corp, 60-61; tech dir, ChemTrac Corp, 61-62, vpres, 62-65, oper mgr, Baird-Atomic, Inc, Mass, 62-65; mgr chem dept, Tracerlab, 65-67, sr staff chemist, 67-69; sr staff chemist, ICN Pharmaceut, Inc, 69-75, mgr prod opers, Life Group, 75-76; res chemist, 76-79, RES CHEMIST, FOOD & DRUG ADMIN, 79- *Res:* Mechanism of organic reactions; metabolic fate of labeled hydroxy-proline in rats; synthesis of labeled compounds; applications for stable and radioactive isotopes; manufacture and control of radiopharmaceuticals. *Mailing Add:* 1326 Bay Ave Annapolis MD 21403

LEAK, LEE VIRN, b Chesterfield, SC, July 22, 32; m 64; c 2. CELL BIOLOGY, ELECTRON MICROSCOPY. *Educ:* SC State Col, BS, 54; Mich State Univ, MS, 59, PhD(cell biol), 62. *Prof Exp:* Asst prof biol sci, Mich State Univ, 62; res fel electron micros, Mass Gen Hosp & Harvard Med Sch, 62-64; asst surg, Mass Gen Hosp, 64-65; asst biol, Harvard Med Sch, 65-68, from instr to asst prof anat, 67-71; chmn dept, 71-81, PROF ANAT, COL MED, HOWARD UNIV, 71-, PROF, GRAD SCH ART & SCI, 76-, RES PROF, 83- *Concurrent Pos:* USPHS res grant, 66-; consult, Shriners Burns Res Inst, 67-71; mem, Anat Sci Training Comt, 72-73; Am Heart Asn res grant, 67-70; mem, Div Biol & Agr, Nat Res Coun, 72-75; ed staff, Anat Rec; mem, Nat Bd Med Examr, 73-76; mem, Marine Biol Lab Corp, 73-; mem, Div Cancer Biol, Diag Bd, Nat Cancer Inst, 79; mem panel basic biomed sci, Nat Res Coun, 80; mem adv coun, Pulmonary Dis Br, Nat Heart, Lung & Blood Inst, 82-86; mem Res Comt, Am Heart Asn. *Honors & Awards:* Adelle Melbourne Holmes Mem Award, Am Heart Asn. *Mem:* Am Asn Anat; Am Soc Cell Biol; Am Soc Zool; Genetics Soc Am; Sigma Xi; Am Physiol Soc; Int Soc Lymphol; Tissue Cult Assoc. *Res:* Biology of the lymphatic vascular system and its role during the inflammatory response; pulmonary lymphatic drainage; ontogeny of the lymphatic system; pericarditis; chemotaxis; cell adhesion & migration. *Mailing Add:* Dept Anat Col Med Howard Univ 520 W St SW Washington DC 20059

LEAKE, DONALD L, b Cleveland, Okla, Nov 6, 31; m 64; c 3. BIOMATERIALS, BONE GRAFTING. *Educ:* Univ Southern Calif, AB, 53, MA, 57; Harvard Univ, DMD, 62; Stanford Univ, MD, 69. *Prof Exp:* NIH postdoctoral fel oral surg, Harvard Univ, 64-66, instr, 66-67; assoc prof oral & maxillofacial surg, 70-74, dir, Dent Res Inst, 82-86, PROF ORAL & MAXILLOFACIAL SURG, SCHS MED & DENT, UNIV CALIF, LOS ANGELES, 70-, CHIEF ORAL & MAXILLOFACIAL SURG, HARBOR-UCLA MED CTR, 70- *Concurrent Pos:* Foreign prof, Grad Sch, Asn Med Arg, 90; consult, NIH. *Honors & Awards:* First Prize, Plastic Surg Educ Found, 83. *Mem:* Biomed Eng Soc; Soc Biomat; AAAS; fel Am Col Surgeons; Sigma Xi. *Res:* Biomaterials for reconstructive surgery; biomechanics and pathophysiology of the temporomandibular joint; clinical research in oral and maxillofacial and reconstructive surgery; bone grafting. *Mailing Add:* Oral & Maxillofacial Surg The Harbor-UCLA Med Ctr 1000 W Carson St Torrance CA 90509

LEAKE, LOWELL, JR, b Denver, Colo, May 25, 28; m 59; c 2. MATHEMATICS EDUCATION. *Educ:* Tufts Col, AB, 50; Univ Wis, MS, 56, PhD(math, educ), 62. *Prof Exp:* Traffic chief, Northwestern Bell Tel Co, 50-54; high sch teacher, Ill, 56-58; from instr to assoc prof math, 60-74, PROF MATH, UNIV CINCINNATI, 74- *Mem:* Math Asn Am; Nat Coun Teachers Math; Am Asn Univ Prof. *Res:* Training of secondary and elementary mathematics teachers at undergraduate and graduate levels; learning of mathematics; piaget and probability; microcomputers in education. *Mailing Add:* Dept Math Sci/025 Old Chem 819C Univ Cincinnati Cincinnati OH 45221

LEAKE, PRESTON HILDEBRAND, b Proffit, Va, Aug 8, 29; m 54; c 2. ORGANIC CHEMISTRY. *Educ:* Univ Va, BS, 50; Duke Univ, MA, 53, PhD(chem), 54. *Prof Exp:* Res supvr org chem, Nitrogen Div, Allied Chem Corp, 54-60; asst res dir, Albemarle Paper Mfg Co, 60-65; asst to managing dir, Res & Develop Dept, Am Tobacco Co, 65-68, asst managing dir, 68-70, asst dir, 70-87, dir res & development dept, 87-89, VPRES RES, AM TOBACCO CO, 89- *Concurrent Pos:* Adj prof, Richmond Prof Inst, 63-64. *Honors & Awards:* Distinguished Serv Award, Va Sect, Am Chem Soc, 76. *Mem:* Am Chem Soc; Am Inst Chemists; Tech Asn Pulp & Paper Indust. *Res:* Polycyclic aromatic chemistry: Psychorr synthesis; amino acids and cyanuric acid derivatives; polyethylene; sizing; silica fume; specialty and filter papers; tobacco. *Mailing Add:* Am Tobacco Co R&D Dept PO Box 899 Hopewell VA 23860

LEAKE, WILLIAM WALTER, b Johnstown, Pa, Apr 24, 26; m 57; c 2. ORGANIC CHEMISTRY. *Educ:* Duquesne Univ, BS, 51; Duke Univ, MA, 53; Univ Pittsburgh, PhD(chem), 58. *Prof Exp:* Res chemist, Monsanto Chem Co, 57-60; from asst prof to assoc prof chem, 61-77, assoc dean, 68-80, DEAN ACAD ADMIN, MIL ADV & VET REP & DIR PROGS, WASHINGTON & JEFFERSON COL, 80- *Mem:* Am Chem Soc. *Res:* Instrumental methods of analysis. *Mailing Add:* Chief Acad Dean Washington & Jefferson Col Washington PA 15301

LEAKEY, JULIAN EDWIN ARUNDELL, b Settle, Yorkshire, UK, June 12, 51; m 79; c 1. BIOCHEMICAL PHARMACOLOGY, PROTEIN CHEMISTRY. *Educ:* Univ Dundee, UK, PhD, 76. *Prof Exp:* Fel toxicol, Inst Environ Health & Sci, Res Triangle Park, 77-78; res fel, Dept Biochem, Univ Dundee, 78-85; SR STAFF TOXICOL, DIV REP DEUTOXICOL, NAT BIOCHEM PHARMACOL CTR TOXICOL RES, 85-; ADJUNCT PROF, UNIV ARK, MED SCI, 87- *Concurrent Pos:* vis scientist, Develop Toxicol, Nat Ctr Toxicol Res, 83. *Mem:* Teratology Soc; Int Soc Xenobiotics. *Res:* Hormonal development, nutrition and environmental regulation of drug metabolizing enzymes in relation to toxicology, pharmacokinetics and carcinogenesis. *Mailing Add:* HFT 134 Nat Ctr Toxicol Res Jefferson AR 72079

LEAL, JOSEPH ROGERS, b New Bedford, Mass, Sept 14, 18; m 44; c 4. ORGANIC CHEMISTRY. *Educ:* Univ Mass, BS, 49; Ind Univ, PhD(chem), 53. *Prof Exp:* Res asst, Corn Prod Refining Co, 40-42; asst chemist, Revere Copper & Brass Co, 42-43 & 45-46; res chemist, Am Cyanamid Co, 52-57, tech rep govt res liaison, Washington, DC, 57-63, mgr contract rels, 63-67; sr staff assoc, Celanese Res Co, 67-83; PRES, CRESCENT CONSULTS, 83- *Mem:* AAAS; Am Chem Soc; NY Acad Sci; Am Inst Chemists. *Res:* High temperature resistant aromatic and heterocyclic polymers; nonflammable fibers; high strength, high modulus reinforcement materials. *Mailing Add:* 10 S Crescent Maplewood NJ 07040

LEAL, L GARY, b Bellingham, Wash, Mar 18, 43; m 65; c 3. FLUID MECHANICS, POLYMER PHYSICS. *Educ:* Univ Wash, BS, 65; Stanford Univ, MS, 67, PhD(chem eng), 69. *Prof Exp:* Nat Sci Found fel, Cambridge Univ, 69-70; from asst prof to prof, chem eng, Calif Inst Technol, 70-84, Chevron Prof, 84-89; PROF & CHAIR, CHEM & NUCLEAR ENG, UNIV CALIF SANTA BARBARA, CALIF, 90- *Concurrent Pos:* Petrol Res Fund grant, Calif Inst Technol, 70-; prin investr, Nat Sci Found, 71- & Off Naval Res, 75-; consult, Richards of Rockford, 75-79, Firestone Res, 81-87, Dynamics Technol Inc, 82- & Dowell-Schlumberger, 85-89; consult ed, J Am Inst Chem Engrs, 85-87; assoc ed, Int J Multiphase Flow, 85-; Guggenheim Fel, Cambridge Univ, 76-77. *Honors & Awards:* Allan Colburn Award, Am Inst Chem Engrs, 78; Allan Colburn Mem lectr, Univ Del, 78. *Mem:* Nat Acad Engrs; Soc Rheology; Brit Soc Rheology; fel Am Phys Soc; Am Inst Chem Engrs. *Res:* Fluid mechanics; suspension mechanics; rheology; mechanical and optical properties of polymeric liquids; multiphase flows; colloid physics. *Mailing Add:* 1560 Hillcrest Rd Santa Barbara CA 93103

LEAMON, TOM B, b Ossett, Yorkshire, UK, April 16, 40; US citizen; m 67; c 3. ERGONOMICS, HUMAN FACTORS & SAFETY. *Educ:* Univ Manchester, UK, BS, 61; Cranfield Inst Technol, UK, MS, 68, PhD(indust eng), 82; Univ Aston, UK, MS, 70. *Prof Exp:* Mgr ergonomics, Pilkington Glass, UK, 64-70; dir ergonomics, Ergo Lab, Cranfield Inst, 70-75; br head ergonomics, Nat Coal Bd, UK, 75-81; dir safety, Univ Ill, Chicago, 81-82; prof & chair indust eng, Northern Ill Univ, 82-87; prof & chair indust eng, Tex Tech Univ, 87-91; VPRES & DIR, RES CTR, LIBERTY MUTUAL INS CO, 91- *Concurrent Pos:* Chartered eng, Eng Coun Great Brit, 66; European eng, Europ Fed Europ Engrs; assoc reader ergonomics, Univ Loughborough, UK; indust secy, Ergonomics Soc, 69-72, conf secy, 71-73, gen secy, 73-79; chair, Tech Interest Group Indust Ergonomics, Human Factors Soc, 83-85; chair, Ergo Div, Inst Indust Engrs, 90-91; consult to var groups. *Honors & Awards:* Sir Ben Williams Silver Medal, Inst Prod Engrs, 69. *Mem:* Fel Human Factors Soc; fel Ergonomics Soc; fel Inst Prod Engrs; Inst Indust Eng; Am Soc Safety Eng. *Res:* Ergonomics and human factors considerations in industry and daily living, including slips and falls, computer application, workplace design, equipment design and the human and work performance of the disabled. *Mailing Add:* Liberty Mutual Ins Co 71 Frankland Rd Hopkinton MA 01748

LEAMY, HARRY JOHN, b Alton, Ill, Nov 15, 40; m; c 3. PHYSICAL METALLURGY, PHYSICS. *Educ:* Univ Mo-Rolla, BS, 63; Iowa State Univ, PhD(metall), 67. *Prof Exp:* Res fel metall, Max Planck Inst Metall Res, 67-69; mem tech staff, Mat Physics Res Group, 69-77, MEM TECH STAFF, ELECTRONIC MAT RES DEPT, BELL LABS, 77- *Concurrent Pos:* Vis scientist, Philips Res Labs, Eindhoven, The Neth, 76-77. *Mem:* Am Soc Metals; Am Inst Mining, Metall & Petrol Eng; Am Asn Crystal Growth; Am Phys Soc; Electron Micros Soc Am; Mat Res Soc. *Res:* Alloy properties and crystal growth; electron microscopy of crystal lattice defects; magnetic materials; metallic glasses; semiconductor defects; laser beam processing of semiconductor materials. *Mailing Add:* AT&T Bell Labs 3000 Skyline Dr Mesquite TX 75149

LEAMY, LARRY JACKSON, b Alton, Ill, Nov 15, 40; m 65; c 2. QUANTITATIVE GENETICS. *Educ:* Eastern Ill Univ, BS, 62; Univ Ill, Urbana, MS, 65, PhD(zool), 67. *Prof Exp:* Asst prof, 67-71, assoc prof, 71-76, PROF BIOL, CALIF STATE UNIV, LONG BEACH, 76-, CHMN DEPT, 78- *Concurrent Pos:* Calif State Univ Found new fac grant, Calif State Univ, Long Beach, 67-68, fac grant-in-aid, 69-70, 73, 78-81; NSF grant, 81-82; vis prof entomol & genetics, Univ Wis-Madison, fac grant-in-aid, 82-86. *Mem:* Genetics Soc Am; Am Genetic Asn; Soc Study Evolution; Behavior Genetics Soc; AAAS. *Res:* Quantitative genetics of mice. *Mailing Add:* Dept Biol Univ NC UNCC Sta Charlotte NC 28223

LEAN, DAVID ROBERT SAMUEL, b Peterborough, Ont, Oct 18, 37; div; c 3. BIOLOGY, CHEMISTRY. *Educ:* Univ Toronto, BASc, 62, PhD(zool), 73. *Prof Exp:* Engr chem, Union Carbide Corp, 63-67; lectr ecol, Univ Toronto, 71-73; RES SCIENTIST BIOL & LIMNOL, CAN DEPT ENVIRON, 72- *Mem:* Am Soc Limnol & Oceanog; Int Asn Theoret & Appl Limnol; Soc Prof Engrs. *Res:* Interrelationships of carbon, nitrogen, phosphorous and iron on algae growth and decomposition in lakes. *Mailing Add:* Nat Water Res Inst 867 Lake Shore Rd Box 5050 Burlington ON L7R 4A6 Can

LEAN, ERIC GUNG-HWA, b Fukien, China, Jan 1, 38; m 65; c 1. ACOUSTICS, OPTICS. *Educ:* Cheng Kung Univ, Taiwan, BS, 59; Univ Wash, MS, 63; Stanford Univ, PhD(elec eng), 67. *Prof Exp:* Res asst elec eng, Univ Wash, 61-63; res asst microwave acoust, Hanson Lab Physics, Stanford Univ, 63-67, res assoc microwave acoust & laser, 67; mem res staff nonlinear optics, T J Watson Res Ctr, 67-69, mgr acoust & optical physics, 69-71, mgr, Optical Solid State Technol, 71-79, mgr, Printer Technol, 79-81, MGR, OPTICAL & MAGNETIC STORAGE TECHNOL, T J WATSON RES CTR, IBM CORP, 81- *Mem:* Fel IEEE; Sigma Xi; Optical Soc Am. *Res:* Microwave acoustic waves in solids; nonlinear optics; optical signal processing devices; surface wave devices; laser applications; integrated optics; fiber optics; printers. *Mailing Add:* 10 Berrybrook Circle Chappaqua NY 10514

LEANDER, JOHN DAVID, b Mt Vernon, Wash, Apr 8, 44; m 65; c 3. PSYCHOPHARMACOLOGY. *Educ:* Pac Lutheran Univ, BA, 66; Western Wash State Col, MA, 67; Univ Fla, PhD(psychol), 71; Ind Univ, MBA, 85. *Prof Exp:* Fel neurobiol prog, Univ NC, Chapel Hill, 71-73, instr, 73-74, from asst prof to assoc prof pharmacol, 74-81; res scientist, 81-84, sr res scientist, 85-89, RES ADV, LILLY RES LAB, ELI LILLY & CO, 90- *Mem:* AAAS; Behavioral Pharmacol Soc; Am Soc Pharmacol & Exp Therapeut. *Res:* Behavioral pharmacology; effects of drugs on behavior and the interaction of drugs with ongoing behavior. *Mailing Add:* 8127 Groton Lane Indianapolis IN 46260

LEANING, WILLIAM HENRY DICKENS, b Whakatane, NZ, Feb 24, 34; m 56; c 4. TECHNICAL MANAGEMENT, PARASITOLOGY. *Educ:* Univ Sydney, BVSc, 56. *Prof Exp:* Vet gen pract, Nth Canterbury Vet Club, NZ & Putaruru Vet Club, NZ, 57-62; vet tech dir appl res parasitol, Merck Sharp & Dohme NZ Ltd, 62-69; from dir mkt develop large animal prod, 69-72, sr dir clin res animal sci res, 72-74, exec dir animal sci res develop res & admin, 75-81, EXEC DIR TECH SERV, MERCK SHARP & DOHME CO, INC, AGVET, 81- *Mem:* NZ Vet Asn; Am Asn Vet Parasitologist; World Asn Adv Vet Parasitol; Am Vet Med Asn; Am Asn Indust Vet. *Res:* Concepts of applied preventive medicine on whole herd/flock basis throughout productive life of animal/bird; primary areas helminthology, entomology; innovation in formulation and treatment application; applied parasitology and agri-economic benefits of year-round parasite control programs. *Mailing Add:* MSD-AGVET Div Merck & Co PO Box 2000 Rahway NJ 07065

LEAP, DARRELL IVAN, b Huntington, WVa, Oct 19, 37. WATER RESOURCES, AQUIFER ANALYSIS. *Educ:* Marshall Univ, BS, 60; Ind Univ, MA, 66; Pa State Univ, PhD(geol), 74. *Prof Exp:* Geologist, SDak State Geol Surv, 66-71; instr geol, Univ SDak, 66-69; hydrologist, US Geol Surv, 74-80; ASSOC PROF HYDROGEOL, PURDUE UNIV, 80- *Concurrent Pos:* Prin investr, Hydrol Nev Test Site, Nev Nuclear Waste-Storage Invest, US Dept Energy, 78-80, Ground Water Contamination Studies, Purdue Univ. *Honors & Awards:* Super Serv Award, USDA, 88. *Mem:* Am Geophys Union; AAAS; Sigma Xi; Geol Soc Am; Nat Asn Ground Water Scientists & Engrs. *Res:* Regional aquifer systems for recharge-discharge relationships and water resources; ground-water modeling; ground-water tracers; radioactive-waste disposal; flow in fractured rocks; flow in glaciated terranes; glacial geology; ground water contamination. *Mailing Add:* Dept Earth & Atmospheric Sci Purdue Univ West Lafayette IN 47907

LEAR, BERT, b Logan, Utah, June 10, 17; m 50; c 1. PLANT PATHOLOGY. *Educ:* Utah State Univ, BS, 41; Cornell Univ, PhD(plant path), 47. *Prof Exp:* Agt, Exp Sta, USDA, 41-43; Dow Chem Co fel & res assoc, Cornell Univ, 47-48, asst prof plant path, 48-52; nematologist, NMex Col, 52-53; from asst nematologist to nematologist, Univ Calif, Davis, 53-74, prof nematol, 63-85; RETIRED. *Mem:* Am Phytopath Soc; Soc Nematol; Orgn Trop Am Nematologists. *Res:* Soil treatment for control of nematodes; fate of chemicals in soils and plants when applied for control of nematodes; role of plant parasitic nematodes in diseases of plants. *Mailing Add:* 1036 The Fairway Santa Barbara CA 93108

LEAR, CLEMENT S C, orthodontics, for more information see previous edition

LEARN, ARTHUR JAY, b Lewistown, Mont, Mar 25, 33; m 59; c 2. SOLID STATE PHYSICS. *Educ:* Reed Col, BA, 54; Mass Inst Technol, PhD(physics), 58. *Prof Exp:* Mem tech staff, TRW Systs, Calif, 58-67; mem staff, Electronics Res Ctr, NASA, 67-70; sr mem res staff, Fairchild Camera & Instrument Corp, 70-76; prog mgr, Intel Corp, 76-81; eng mgr, Supertex, Inc, 81-83; Exec Mem Tech Staff, Anicon, Inc,; VPRES, CVD TECHNOL, SILICON VALLEY GROUP INC, 83- *Mem:* Am Phys Soc; Electrochem Soc; Am Vacuum Soc. *Res:* X-ray diffraction; electron microscopy; properties of thin films; thin film superconductor and semiconductor devices; metallization; chemical vapor deposition; thermal oxidation. *Mailing Add:* 10822 Wilkinson Ave Cupertino CA 95014

LEARNED, JOHN GREGORY, b Plattsburgh, NY, Apr 12, 40; m 63; c 2. NEUTRINO ASTRONOMY, VHE GAMMA ASTRONOMY. *Educ:* Columbia Col, NY, AB, 61; Univ Pa, Philadelphia, MS, 63; Univ Wash, Seattle, PhD(physics), 68. *Prof Exp:* PROF PHYSICS, DEPT PHYSICS & ASTRON, UNIV HAWAII. *Concurrent Pos:* Tech dir, Dumand Proj, Hawaii Dumand Ctr. *Res:* First high energy neutrino telescope for operation in deep ocean; particle astrophysics. *Mailing Add:* Dept Physics & Astron Univ Hawaii 2505 Correa Rd Honolulu HI 96822

LEARNED, ROBERT EUGENE, b Glendale, Calif, July 3, 28; m 56; c 2. ECONOMIC GEOLOGY, GEOCHEMISTRY. *Educ:* Occidental Col, AB, 55; Univ Calif, Los Angeles, MA, 62; Univ Calif, Riverside, PhD(geol), 66. *Prof Exp:* Geologist, Aerogeophys Co, Calif, 55-56; asst prof geol, Chapman Col, 65-67; GEOLOGIST, US GEOL SURV, 67- *Mem:* Geol Soc Am; Geochem Soc; Asn Explor Geochemists; Soc Econ Geologists; Asn Geoscientists Int Develop; Sigma Xi. *Res:* Geology and geochemistry of ore deposits; geochemical exploration methods. *Mailing Add:* Am Embassy La Paz Econ Sect APO Miami FL 34032

LEARSON, ROBERT JOSEPH, b Boston, Mass, May 6, 38; m 62; c 3. FISH PROCESSING TECHNOLOGY. *Educ:* Suffolk Univ, Boston, Mass, BS, 61. *Prof Exp:* Chemist, Werby Labs Inc, 61-62 & Bur Com Fisheries, 62-69; res chemist, 69-71, res food technologist, 71-78, dep lab dir, 78-82, actg dep ctr dir, 83, LAB DIR, NAT MARINE FISHERIES SERV, 83- *Concurrent Pos:* Tech adv, Nat Blue Crab Indust Asn, 75-, Mid Atlantic Fishery Develop Found, 80-, New Eng Fishery Develop Found, 80- & Refrig Res Found, 84-; lectr, Salem State Univ, 79-81; tech consult, Morocco Study Team, Nat Marine Fisheries Serv, 82, chmn, Qual Improv Task Force, 82-84; adj prof, Univ Mass, 84- *Honors & Awards:* Arthur S Fleming Award, 76. *Mem:* Inst Food Sci; Asn Off Anal Chem; Atlantic Fisheries Tech. *Res:* Processing and preservation of fishery products; determination of fish quality; new processing concepts; utilization of processing waste, irradiation of seafoods and chemical species identification. *Mailing Add:* Nat Marine Fisheries Serv Emerson Ave Gloucester MA 01930

LEARY, FRANCIS CHRISTIAN, b West Hartford, Conn, April 23, 49; m 76; c 3. ALGEBRA. *Educ:* Univ Conn, BA, 71; State Univ NY, Albany, MA, 74, PhD(math), 79. *Prof Exp:* Instr & asst prof, Skidmore Col, 79-80; asst prof math Transylvania Univ, 80-85; ASST PROF MATH, ST BONAVENTURE UNIV, 85- *Concurrent Pos:* Adj asst prof, Union Col, Schenectady, 75-76; lectr math, Waterbury State Tech Col & Post Jr Col, 73. *Mem:* Am Math Soc; assoc mem Sigma Xi; Math Asn Am. *Res:* Rings and modules; homological algebra. *Mailing Add:* 222 N Third St Olean NY 14760

LEARY, HARVEY LEE, JR, NUTRITION. *Educ:* NC State Univ, BS, 71, MS, 75, PhD(animal sci-immunol), 78. *Prof Exp:* Res asst, Dept Animal Sci, NC State Univ, 72-78; postdoctoral res assoc, Dept Dairy Sci, Univ Ill, Urbana-Champaign, 78-82 & Dept Biochem, Col Health Sci, Univ Kans, 82-84; sr res scientist, Immuno Biotech, Inc, Overland Park, Kans, 83-85; sr scientist, 85-88, SR RES SCIENTIST, MEAD JOHNSON NUTRIT GROUP, BRISTOL-MYERS SQUIBB CO, 88- *Mem:* Am Asn Immunologists; Am Soc Microbiol. *Res:* Radiometric and enzyme immunoassay development; Western blotting; gel precipitation analysis; preparation and characterization of radiotracers and antibody-enzyme conjugates; immunization of animals and purification of antibody; enzymatic digestion of immunoglobulias and purification of biologically active fragments; antigen localization methods with both light and electron microscopy; protein and virus purification and characterization. *Mailing Add:* Mead Johnson Nutrit Group Food & Nutrit Res Bristol-Myers Squibb Co 2400 W Lloyd Expressway Evansville IN 47721

LEARY, JAMES FRANCIS, b Portsmouth, NH, Apr 12, 48. PATHOLOGY. *Educ:* Mass Inst Technol, BS(aeronaut & astronaut) & BS(phil & hist), 70; Univ NH, MS, 74; Pa State Univ, PhD(biophys), 77. *Prof Exp:* Res fel biophys & instrumentation, Los Alamos Scientific Lab, 77-78; asst prof path, 81-85, ASSOC PROF PATH & LAB MED, MED SCH, UNIV ROCHESTER, 85- *Concurrent Pos:* Vis staff mem, Los Alamos Scientific Lab, 78-80. *Mem:* Soc Anal Cytol; AAAS; NY Acad Sci. *Res:* Development of new automated laser flow cytometric instrumentation and clinically useful diagnostic tests; molecular biology; cell differentiation; somatic cell genetics and developmental biology; transplantation. *Mailing Add:* Dept Path Box 626 Univ Rochester Med Ctr Rochester NY 14642

LEARY, JOHN DENNIS, b New Bedford, Mass, Dec 6, 34; m 57; c 4. PHYTOCHEMISTRY, NATURAL PRODUCTS. *Educ:* Mass Col Pharm, BS, 56, MS, 58; Univ Conn, PhD(pharmacog), 64. *Prof Exp:* Asst pharmacog, Univ Conn, 58-59; asst prof, Ore State Univ, 59-61; from asst prof to assoc prof phytochem, St John's Univ, NY, 63-68; assoc prof pharmacog & bot, 68-74, assoc prof biochem, 74-83, chmn, dept chem & physics, 84-87, PROF BIOCHEM, MASS COL PHARM, & ALLIED HEALTH SCI, 83- *Mem:* Am Asn Col Pharm; Am Soc Pharmacog. *Res:* Phytochemical studies, primarily Solanaceae and Meliaceae; chemotaxonomy and biogenesis; microbial transformation of organic compounds. *Mailing Add:* 179 Longwood Ave Boston MA 02115

LEARY, JOHN VINCENT, genetics; deceased, see previous edition for last biography

LEARY, RALPH JOHN, b Elizabeth, NJ, Nov 3, 29; m 52; c 6. ORGANIC CHEMISTRY. *Educ:* Seton Hall Univ, BS, 51; Univ Ill, PhD(chem), 57. *Prof Exp:* Jr chemist, Merck & Co, Inc, 51-54; group leader, Esso Res & Eng Co, 57-67, group leader & res assoc, 67-75, lab head, Exxon Chem Co USA, 75-77, res assoc, 78-80, sect head, Exxon Res & Eng Co, 78-87, sr res assoc, 80-87; RETIRED. *Mem:* Am Chem Soc; Sigma Xi. *Res:* Gas chromatography and automation of laboratory instruments; Am Soc Testing & Mat. *Mailing Add:* 211 Oak Lane Cranford NJ 07016

LEARY, ROLFE ALBERT, b Waterloo, Iowa, Mar 5, 38; m 67; c 2. FOREST MENSURATION. *Educ:* Iowa State Col, BS, 59; Purdue Univ, MS, 61, PhD(forest mgt), 68. *Prof Exp:* Vol forester, US Peace Corps, St Lucia, WIndies, 61-63; instr forest mensuration, Southern Ill Univ, 64-65; mensurationist, 68-72, PRIN MENSURATIONIST FORESTRY, US FOREST SERV, N CENT FOREST EXP STA, 72- *Mem:* AAAS; Ecol Soc Am; Sigma Xi; Coun Unified Res & Educ. *Res:* Boundary value problem method of calibrating forest growth models; generalized forest growth projection system; philosophy and methods of forest research; multiple-use decision-making. *Mailing Add:* 1992 Folwell Ave St Paul MN 55108

LEAS, J(OHN) W(ESLEY), b Delaware, Ohio, June 14, 16; m 43. ELECTRICAL ENGINEERING. *Educ:* Ohio State Univ, BS, 38. *Prof Exp:* Sales engr, Armstrong Cork Co, Pa, 38-41; electronic engr airborne radar, Airborne Instrument Lab, NY, 46-47; consult engr air navig, Air Transport Asn, DC, 47-49; electronic res scientist, Air Navig Develop Bd, Dept Commerce, DC, 49-51; chief engr, Electronic Data Processing Div, Radio Corp Am, 51-60, mgr data commun & custom proj dept, Electronic Data Processing Div, 60-63; gen mgr Valley Forge Div, Control Data Corp, 64-72; PRES, J W LEAS & ASSOCS, TRANSP CONSULT, 72- *Concurrent Pos:* Mem tech staff telecommun res estab, Ministry Aviation, Eng, 42-43; asst head eng, US Naval Res Lab, 43-46; tech adv, US State Dept, 46; consult, Civil Aeronaut Admin, 48-49 & Dept Defense, 58-63; mem adv group comput, Dept Defense, 55. *Mem:* Fel Inst Elec & Electronics Engrs. *Res:* Development, design and management of engineering digital computers; general management of data communications and custom projects in data processing field. *Mailing Add:* J W Leas & Assoc 910 Potts Lane Bryn Mawr PA 19010

LEASK, R(AYMOND) A(LEXANDER), b Edmonton, Alta, Jan 15, 19; m 42; c 4. PULP AND PAPER. *Educ:* Univ Alberta, BSc, 41, MSc, 47. *Prof Exp:* Chemist, Brit-Am Oil Co, 41; res chemist, Can Int Paper Co, 42-44; res asst, Res Coun Alberta, 45-47; chem engr, Powell River Co, 48-50; supvr pulping sect, Cent Res Div, Abitibi Power & Paper Co, 50-64; dir res, Bauer Bros Co, 64-71; process engr, Sandwell & Co, 71-73; asst res dir, Ont Paper Co, 73-83; RETIRED. *Concurrent Pos:* Consult, 83- *Mem:* Fel Tech Asn Pulp & Paper Indust; Can Pulp & Paper Asn; Asn Prof Engrs; Rotary Int. *Res:* Pulping methods on a pilot-plant scale and on a commercial scale; improving paper machine performance and improving newsprint quality. *Mailing Add:* Three Mayholme Ct St Catharines ON L2N 4C1 Can

LEATH, KENNETH T, b Providence, RI, Apr 29, 31; m 55; c 4. PLANT PATHOLOGY. *Educ:* Univ RI, BS, 59; Univ Minn, MS & PhD(phytopath), 66. *Prof Exp:* Res technician cereal rusts, Coop Rust Lab, Minn, 59-66, PLANT PATHOLOGIST, REGIONAL PASTURE RES LAB, USDA, 66- *Concurrent Pos:* Adj prof, Dept Plant Path, Pa State Univ, 66- *Honors & Awards:* Merit Cert, Am Forage & Grassland Coun. *Mem:* Am Phytopath Soc; Am Soc Agron; Am Forage & Grassland Coun; Int Soc Root Res. *Res:* Clover and alfalfa diseases; host-parasite interaction; biocontrol. *Mailing Add:* Regional Pasture Res Lab ARS-USDA University Park PA 16802

LEATH, PAUL LARRY, b Moberly, Mo, Jan 9, 41; m 62; c 2. SOLID STATE PHYSICS. *Educ:* Univ Mo-Columbia, BS, 61, MS, 63, PhD(physics), 66. *Prof Exp:* Res assoc theoret physics, Oxford Univ, 66-67; from asst prof to assoc prof, Rutgers Univ, New Brunswick, 67-78, assoc dept, 73-75, assoc provost, 78-87, PROF PHYSICS & PROVOST SCI, RUTGERS UNIV, NEW BRUNSWICK, 87- *Mem:* AAAS; Am Phys Soc; Sigma Xi; Brit Inst Physics; NY Acad Sci. *Res:* Theoretical solid state physics; inelastic neutron scattering; vibrational and electronic properties of alloys; anharmonic crystals; disordered and dilute magnets; percolation processes; breakdown phenomena. *Mailing Add:* Off Provost Rutgers Univ New Brunswick NJ 08903

LEATHEM, WILLIAM DOLARS, b Chicago, Ill, Jan 6, 31; m 52; c 3. MEDICAL NUTRITION, MEDICAL PARASITOLOGY. *Educ:* Univ Wis, BS, 61, MS, 63, PhD(zool), 65. *Prof Exp:* Asst zool, Univ Wis, 62; asst prof biol, Wis State Univ, Whitewater, 65-66; asst prof, Univ Wis, Waukesha Ctr, 66-69, NSF grants, 67-69; res assoc, Norwich Pharmacal Co, 69-74, asst dir clin nutrit, Eaton Labs, 74-76; med monitor nutrit, 76-78, MGR NUTRIT RES, ABBOTT LABS, 78- *Mem:* AAAS; Am Soc Parasitol; Soc Protozool; Am Soc Trop Med & Hyg; Am Soc Parenteral & Enteral Nutrit; Sigma Xi. *Res:* General parasitology and protozoology. *Mailing Add:* 541 Beck Rd Lindenhurst IL 60046

LEATHER, GERALD ROGER, b Smithsburg, Md, Oct 16, 37; m 63; c 3. PLANT PHYSIOLOGY. *Educ:* Shepherd Col, BS, 68; Hood Col, MA, 73; Va Polytech Inst & State Univ, PhD(plant physiol), 76. *Prof Exp:* Biologist, US Army, 70-74; biologist, 74-76, PLANT PHYSIOLOGIST, US DEPT AGR, 76- *Concurrent Pos:* Lectr, Hood Col, 76-; adj prof, Va Polytech Inst & State Univ, 80- *Mem:* Weed Sci Soc Am; Am Soc Plant Physiologists; Sigma Xi. *Res:* Secondary plant chemicals in the allelopathic interaction of weeds and crops; dormancy and germination mechanisms in weed seeds; maternal influence on the physiology and biochemistry of seed dormancy. *Mailing Add:* Weed Physiol USDA-ARS Bldg 1301 Ft Detrick Frederick MD 21701

LEATHERLAND, JOHN F, b Nottingham, Eng, July 5, 43; Can citizen; m 65; c 2. ENDOCRINOLOGY, PHYSIOLOGY. *Educ:* Sheffield, Eng, BSc, 64; Leeds, Eng, PhD(endocrinol), 67. *Prof Exp:* Postdoctoral fel endocrinol, Univ BC, 67-69, Univ Hull, UK, 69-71; from asst prof to assoc prof, 71-83, PROF PHYSIOL, DEPT ZOOL, UNIV GUELPH, CAN, 83- *Concurrent Pos:* Vis fel, Univ Bath, UK, 79; vis prof, Murdoch Univ, Australia, 80 & 87. *Honors & Awards:* Excellence in Res, Sigma Xi, 89. *Mem:* Soc Endocrinol; Can Soc Zool; Sigma Xi. *Res:* Endocrine control of metabolism, growth and reproduction of fish, particularly the role of the pituitary, thyroid and adrenal glands. *Mailing Add:* Dept Zool Univ Guelph Guelph ON N1G 2W1 Can

LEATHERMAN, ANNA D, b Centre Square, Pa, Jan 17, 09. PLANT ECOLOGY, BOTANY. *Educ:* Goshen Col, BS, 38, AB, 39; Cornell Univ, MA, 47; Univ Tenn, PhD, 55. *Prof Exp:* Teacher, Rural Sch, Va, 39-44 & Pub Sch, 44-47; from instr to prof biol, Upland Col, 47-64; asst, Univ Tenn, 52-53; prof, Bethel Col, 64-66; prof biol, 66-77, EMER PROF BOT, SPRING ARBOR COL, 77- *Mem:* Ecol Soc Am. *Res:* Ecological life history of Lonicera japonica Thunb. *Mailing Add:* 6650 W Butler Dr Apt 2 Glendale AZ 85302

LEATHERMAN, NELSON E(ARLE), b Grand Rapids, Mich, Mar 22, 39; m 68. BIOENGINEERING. *Educ:* Univ Mich, BSE, 62, MSE, 63, PhD(bioeng), 67. *Prof Exp:* Res asst physiol, Univ Mich, 67-68; res assoc, Ind Univ, Bloomington, 68-77; res assoc, Vt Lung Ctr, Vt, 77-; AT DEPT MED DUKE UNIV. *Mem:* Biomed Eng Soc. *Res:* Modeling of biological control systems; particularly identification of nonlinear systems. *Mailing Add:* Dept Med Box 3861 Med Ctr Duke Univ Durham NC 27706

LEATHERMAN, STEPHEN PARKER, b Charlotte, NC, Nov 6, 47; m 87. OCEANOGRAPHY, STRATIGRAPHY-SEDIMENTATION. *Educ:* NC State Univ, BS, 70; Univ Va, PhD(environ sci), 76. *Prof Exp:* Asst prof geol, Boston Univ, 75-77; dir res unit, Univ Mass, 77-81; from asst prof to assoc prof geog, Univ Md, 81-87, DIR, LAB COASTAL RES, UNIV MD, 87- *Concurrent Pos:* Mem Nat Acad Sci Comt Sea Level Impacts, 84-87; petrol geologist, Texaco Inc, Houston, Tex, 70-72; sci adv & team leader, Earthwatch Sci Expeds, Belmont, Mass, 76-81; mem Expert Panel Selection Global Coastal Biospheres, UNESCO, 80; tech consult, US Dept Interior Task Force Barrier Islands, 80-81; consult, Heritage Coast Proj, Wales, UK, 85-; dir bd, Climate Inst, Washington DC, 87-; expert testimony, US Senate, 86, 87 & 88. *Mem:* Soc Econ Paleontologists & Mineralogists; fel Geol Soc Am; corresp mem Int Geol Correlation Prog; AAAS. *Res:* Beaches and barrier islands; has authored/edited 8 books and over 50 refereed journal articles, including Science and Nature. *Mailing Add:* Dept Geog Univ Md Col Park MD 20742

LEATHERS, CHESTER RAY, b Claremont, Ill, May 15, 29; m 53; c 4. MYCOLOGY. *Educ:* Eastern Ill Univ, BS, 50; Univ Mich, MS, 51, PhD(bot), 55. *Prof Exp:* Res mycologist, Biol Warfare Labs, US Army, Md, 55-57; asst prof, 57-61, ASSOC PROF BOT, ARIZ STATE UNIV, 61-, EMER PROF MICROBIOL, 90- *Mem:* Fel AAAS; Mycol Soc Am; Am Phytopath Soc; Am Inst Biol Sci. *Res:* Mycology and plant pathology, particularly fleshy fungi, cereal and vegetable diseases; allergenic fungi; medical mycology. *Mailing Add:* Dept Bot & Microbiol Ariz State Univ Tempe AZ 85287-1601

LEATHERWOOD, JAMES M, b Waynesville, NC, Mar 22, 30; m 56; c 1. ANIMAL NUTRITION. *Educ:* Berea Col, BS, 52; NC State Univ, MS, 57, PhD(animal sci), 61. *Prof Exp:* From instr to assoc prof, 57-69, PROF ANIMAL SCI, NC STATE UNIV, 69- *Concurrent Pos:* Fel biochem, Duke Univ, 60-61. *Mem:* Am Soc Microbiol; Am Soc Animal Sci; Am Inst Nutrit. *Res:* Carbohydrate metabolism; enzymatic cellulose degradation; bacterial cell walls and microbiology of the rumen; animal efficiency and energetics. *Mailing Add:* Rte 1 Box 678B Candler NC 28715

LEATHRUM, JAMES FREDERICK, b Dover, Del, Dec 24, 37; m 60; c 3. COMPUTER SCIENCE, COMPUTER ENGINEERING. *Educ:* Univ Del, BChE, 59; Princeton Univ, MA, 61, PhD(chem eng), 63. *Prof Exp:* Proj scientist, Union Carbide Corp, 65-67; from asst prof to assoc prof comput sci, Univ Del, 67-80; PROF ELEC & COMPUT ENG, CLEMSON UNIV, 80- *Concurrent Pos:* Consult, US Army, 68-83 & Burroughs Corp, 69-73. *Mem:* Asn Comput Mach; Am Inst Chem Engrs; Sigma Xi; Inst Elec & Electronics Engrs. *Res:* Programming systems for real time and interactive computers; software engineering. *Mailing Add:* 103 Lancelt Dr Camelt Clemson SC 29631

LEAV, IRWIN, b Brooklyn, NY, July 4, 37; m 61; c 2. PATHOLOGY. *Educ:* Ohio State Univ, BA, 59, DVM, 65; Am Col Vet Pathologists, Dipl, 70. *Prof Exp:* Res fel path, Harvard Med Sch, 65-68, NIH spec fel, 68-70; assoc prof, 70-87, PROF PATH, SCH MED, VET MED, TUFTS UNIV, 87-; ASST PATH, HARVARD MED SCH, 68- *Concurrent Pos:* Res assoc, Steroid Biochem Lab, 69-; assoc dir path, Angell Mem Hosp, 74-76; consult, US Armed Forces Inst Environ Med, 74- & Angell Mem Hosp, 76-; grant, Nat Cancer Inst, 78-81; assoc dean basic sci, 78-83, assoc dean res, Sch Vet Med, Tufts Univ, 83-; Nat Cancer Inst, grant, 81- *Mem:* Int Acad Path; Am Col Vet Pathologists; Am Vet Med Asn. *Res:* Mechanisms of action of sex hormones on normal, hyperplastic and neoplastic male accessory sex organs. *Mailing Add:* Dept Path/Anat Tufts Univ Med Sch Dent & Vet Med 136 Harrison Ave Boston MA 02111

LEAVENS, PETER BACKUS, b Summit, NJ, June 20, 39; m 80; c 1. MINERALOGY. *Educ:* Yale Univ, BA, 61; Harvard Univ, MA, 64, PhD, 66. *Prof Exp:* Res assoc, Dept Mineral Sci, Smithsonian Inst, 65-67; from asst prof to assoc prof, 67-88, PROF GEOL, UNIV DEL, 88- *Concurrent Pos:* Res assoc, Smithsonian Inst, 67-; cur minerals, Univ Del, 78-; regist prof geologist, State Del, 87- *Mem:* Mineral Soc Am. *Res:* Crystal structure analysis; description of new mineral species; conditions of mineral occurrence and stability; mineralogy and geochemistry of pegmatites; carbonate metamorphism. *Mailing Add:* Dept Geol Univ Del Newark DE 19716

LEAVENWORTH, HOWARD W, JR, b Waterbury, Conn, June 3, 28; m 55; c 3. METALLURGY, RESEARCH ADMINISTRATION. *Educ:* Stevens Inst Technol, ME, 51; Yale Univ, MS, 53. *Prof Exp:* Metallurgist, Franklin Inst, 53-55; sr scientist, Pratt & Whitney Aircraft Co, 55-61 & Oak Ridge Nat Lab, 55-57; prog mgr solid state physics, Air Force Off Sci Res, 61-62; asst mgr, Am Mach & Foundry Co, 62-67; metallurgist, Bur Mines, Albany Res Ctr, 67-88, res supvr, 76-88; RETIRED. *Honors & Awards:* NASA Award, 67; President's Award, Am Soc Microbiol, 70 & 71; Meritorious Serv, 83. *Mem:* Fel AAAS; Sigma Xi; Am Inst Mining, Metall & Petrol Engrs. *Res:* Alloy development; corrosion; surface science; wear; extractive metallurgy; concrete; wire rope; powder metallurgy. *Mailing Add:* 20 Bayview Lane Port Townsend WA 98368

LEAVENWORTH, RICHARD S, b Oak Park, Ill, Sept 30, 30; m 55. INDUSTRIAL ENGINEERING & OPERATIONS RESEARCH, QUALITY CONTROL & CAPITAL BUDGETING. *Educ:* Stanford Univ, BSIE, 61, MSIE, 62, PhD(indust eng), 64. *Prof Exp:* Eng asst, Light Div, Dept Pub Utilities, Tacoma, Wash, 56-59; asst prof indust eng, Va Polytech Inst, 64-66; prof, 66-68, actg chmn dept, 79, EMER PROF,INDUSR & SYSTS ENG, UNIV FLA, 88- *Concurrent Pos:* Consult, Off Transp Res, US Dept Commerce, 65-66, mfg educ serv, Gen Elec Co, NY, 65-67 & Manhattan Industs, 70-73; ed, The Eng Economist, 76-80; adv, Eng Soc Comn Energy, 78-79; consult, Off Chief Economist, TVA, 84-88 & expert consult, Temple, Barker & Sloane, 84; consult, Total Qual Mgt, Naval Aviation Depot, Jacksonville, Fla, 88. *Mem:* Fel Inst Indust Engrs (vpres, 77-79 & 82-84); Am Soc Qual Control; Am Soc Eng Educ. *Res:* Engineering economics; statistical quality control; development of process control systems for Naval Aviation depots. *Mailing Add:* Dept Indust & Systs Eng 303 Weil Hall Univ Fla Gainesville FL 32611

LEAVIS, PAUL CLIFTON, b May 17, 44; US citizen. MUSCLE RESEARCH. *Educ:* Univ Notre Dame, BS, 66; Tufts Univ, PhD(physiol), 71. *Prof Exp:* Lab instr, Med Sch, Tufts Univ, 69-71; res fel, Dept Muscle Res, Boston Biomed Res Inst, 72-75, res assoc, 75-78, staff scientist, 78-86; asst prof physiol, Sch Dent Med & Sch Vet Med, Tufts Univ, 78-89; prin scientist, 86-88, SR SCIENTIST, DEPT MUSCLE RES, BOSTON BIOMED RES INST, 88-; ASSOC PROF PHYSIOL, SCHS DENT & VET MED, TUFTS UNIV, 89- *Concurrent Pos:* Res fel, Dept Neurol, Harvard Med Sch, 74-78, res assoc, Dept Neurol, 78; lectr, Tufts Univ, 74-; estab investr, Am Heart Asn, 78-83, mem, Molecular Aspects Excitable Tissues Res Study Comt, 86-89. *Mem:* AAAS; Biophys Soc; NY Acad Sci; Am Soc Biol Chemists. *Res:* Biochemistry of contractile and regulatory proteins in mammalian skeletal and cardiac muscle and non-muscle motile cells; structure and function of calcium binding proteins; effects of protein-protein interactions on the metal binding proteins of troponin C; use of rare earth ions as probes of metal binding sites; supramolecular structure of muscle protein complexes; molecular cytoskeleton structure and function. *Mailing Add:* Dept Muscle Res Boston Biomed Res Inst 20 Staniford St Boston MA 02114

LEAVITT, CHRISTOPHER PRATT, b Boston, Mass, Nov 20, 27; m 59; c 5. PHYSICS. *Educ:* Mass Inst Technol, BS, 48, PhD(physics), 52. *Prof Exp:* Res assoc physics, Brookhaven Nat Lab, NY, 52-54, assoc physicist, 54-56; from asst prof to assoc prof physics, 56-65, actg chmn dept physics & astron, 58-60, PROF PHYSICS, UNIV NMEX, 65- *Concurrent Pos:* Consult, Res Directorate, Phyiscs Div, Kirtland AFB, 56-60; directorate res & develop, Air Force Missile Develop Ctr, Holloman AFB, 56-60; mem particles & fields subcomt, NASA, 65-67; mem tech adv panel, Los Alamos Meson Physics Facility, 69-71, mem, Nuclear Physics Steering Comt, 70-71. *Mem:* Am Phys Soc. *Res:* Nuclear and high energy physics; cosmic rays; space physics. *Mailing Add:* Dept Physics & Astron Univ NMex Albuquerque NM 87131

LEAVITT, FRED W, b Elizabeth, NJ, Oct 16, 28. APPLIED MATHEMATICS. *Educ:* Newark Col Eng, BS, 50; Rensselaer Polytech Inst, MS, 55, PhD(chem eng), 57. *Prof Exp:* Chem engr, Biol Labs, US Army, Camp Detrick, 52-54; chem engr, 57-61, group leader, 61-63, SR ENGR, LINDE DIV, UNION CARBIDE CORP, 63- *Mem:* AAAS; Am Chem Soc. *Res:* Development of adsorptive separation processes; automatic data logging; development of computer systems for reducing, analyzing and correlating data and making design calculations; chemical kinetics and equilibria. *Mailing Add:* Basic Res Dept Bldg 10 PO Box 44 Tonawanda NY 14151-0044

LEAVITT, JOHN ADAMS, b Lewis, Colo, Dec 8, 32; m 55; c 5. ION BEAM ANALYSIS, ATOMIC PHYSICS. *Educ:* Univ Colo, BA, 54, Harvard Univ, MA, 56, PhD, 60. *Prof Exp:* From asst prof to assoc prof, 60-71, PROF PHYSICS, UNIV ARIZ, 71- *Mem:* Am Phys Soc. *Res:* Developing and using new techniques for analysizing thin films using ion beams from a 6 MV Van de Graaff accelerator. *Mailing Add:* Dept Physics Univ Ariz Tucson AZ 85721

LEAVITT, JULIAN JACOB, b Boston, Mass, Sept 4, 18; m 43; c 3. LICENSING, TECHNOLOGY ACQUISITION. *Educ:* Harvard Univ, AB, 39, AM, 40, PhD(org chem), 42. *Prof Exp:* Res chemist, Nat Defense Res Comt, Harvard Univ, 42 & Univ Pa, 42-44; res chemist, Calco Chem Div, 44-54, Res Div, 54-58, Org Chem Div, 58-64, mgr explor res, 64-69, tech dir, Decision Making Systs Dept, 69-70, asst to mgr com develop, 70-74, mgr licensing & technol, 74-77, dir licensing & technol acquisition chem, 77-81, dir licensing chemicals, Am Cyanamid Co, 81-85; VPRES, CHEMISTS GROUP, INC, 86- *Concurrent Pos:* Ed, Sect 41, Chem Abstr Serv, 61-; mem, Adv Bd Mil Personnel Supplies & Comt Textile Dyeing & Finishing, Nat Acad Sci-Nat Res Coun, 63-68. *Mem:* AAAS; Am Chem Soc; Licensing Exec Soc. *Res:* Dyes; applied photochemistry. *Mailing Add:* 227 Silver Hill Lane Stamford CT 06905-3122

LEAVITT, MARC LAURENCE, b St Louis, Mo, Apr 30, 47; m; c 1. NEUROSCIENCE. *Educ:* Southern Ill Univ, BA, 69, MA, 71; Univ Iowa, PhD(physiol & biophys), 75. *Prof Exp:* Postdoctoral fel, Dept Pharmacol & Med, Col Med, Univ Ky, Lexington, 75-78; asst prof, Biol Dept, Southwest Mo State Univ, Springfield, 78-82, assoc prof, Biomed Sci Dept, 82-86; SR SCIENTIST, NEUROSCI RES LAB, ALLEGHENY-SINGER RES INST, PITTSBURGH, 86-; ASST PROF PSYCHIAT, MED COL PA, ALLEGHENY CAMPUS, 88- *Concurrent Pos:* Kroc Found postdoctoral fel, 75-77; vis investr, Dept Anesthesia Res, Michael Reese Hosp & Med Ctr, Chicago, 80; sr res fel hypertension res, Allegheny-Singer Res Inst, 85-86; mem sci adv bd, Cryomed Sci, Inc, 90- *Mem:* Am Physiol Soc; Am Soc Hypertension; Soc Neurosci; NY Acad Sci; AAAS; Soc Cryobiol. *Res:* Psychopharmacology of behavior-aggression and delerium; techniques for profound hypothermic cardiac arrest in conjunction with blood substitution; pharmacological modulation of intracranial pressure. *Mailing Add:* Neurosci Res Lab Allegheny-Singer Res Inst 320 E North Ave Pittsburgh PA 15212-9986

LEAVITT, WENDELL WILLIAM, b Conway, NH, Jan 15, 38; m 59; c 4. ENDOCRINOLOGY, REPRODUCTIVE BIOLOGY. *Educ:* Dartmouth Col, AB, 59; Univ NH, MS, 61, PhD(zool), 63. *Prof Exp:* Res analyst zool, Agr Exp Sta, Univ NH, 60-63, res assoc endocrinol & instr zool, 63-64; asst prof biol, Univ Cincinnati, 64-67, from asst prof to prof physiol, Col Med, 67-77; sr scientist endocrinol, Worcester Found Exp Biol, 77-83; PROF BIOCHEM, HEALTH SCI CTR & PROF OBSTET & GYNEC, SCH MED, TEX TECH UNIV, 83- *Concurrent Pos:* Vis scientist, Univ Wis-Madison, 67-68 & Med Ctr, Vanderbilt Univ, 72-73; mem, regulatory biol panel, NSF, 77-79; prof obstet & gynec, Med Sch, Univ Mass, 78-83; adj prof, Boston Univ, 78-83; mem Pop Res Comn, Nat Inst Child Health & Human Develop, NIH, 83-87. *Mem:* Sigma Xi; Am Soc Zoologists; Endocrine Soc; Soc Study Reproduction; Am Physiol Soc; Am Soc Cell Biol; Am Soc Biol Chemists. *Res:* Mechanism of pituitary function in relation to gonadotrophin secretion; control of female reproductive processes; estrogens and pituitary function; neuroendocrinology and aging of the reproductive system; steroid hormone receptor systems; mechanism of steroid hormone action. *Mailing Add:* Dept Biol Sci Wichita State Univ Wichita KS 67208

LEAVITT, WILLIAM GRENFELL, b Omaha, Nebr, Mar 19, 16; wid; c 3. ALGEBRA. *Educ:* Univ Nebr, AB, 37, MA, 38; Univ Wis, PhD(math), 47. *Prof Exp:* Actg instr math, Univ Wis, 46; from instr to assoc prof, 47-56, chmn dept, 54-64, prof math, 56-86, EMER PROF MATH, UNIV NEBR-LINCOLN, 86- *Concurrent Pos:* NSF fel, 59-60; Univ Nebr Res Coun vis fel, Leeds, Eng, 73. *Mem:* Am Math Soc; Math Asn Am. *Res:* Ring theory; theory of modules; theory of radicals. *Mailing Add:* Dept Math Univ Nebr Lincoln NE 68588

LEAVY, PAUL MATTHEW, b Jackson, Mich, Apr 15, 23; m 44; c 2. REMOTE SENSING, SELF-REGULATING AUTONOMOUS VEHICLE OPERATION. *Educ:* US Naval Acad, BS, 44. *Prof Exp:* Mgr navy projs, Arma Div, Am Bosch Arma, 54-57; mgr adv progs, Lab Electronics, 57-59; pres, Trident Corp, 59-62; mgr, Radiation Div, Sanders Assocs, 62-72; vpres, ITT EOPD, 72-78; mgr mkt, Western Div, GTE EOO, 78-84; dir, Develop Progs, Teledyne CME, 84-90; PRES, PAUL LEAVY ASSOCS, 90- *Concurrent Pos:* Consult, Nat Acad Sci, ASW, 59-60; mem, Counter Terrorism, Nat Acad Eng, 72. *Mem:* Soc Photo-Optical Instrumentation Engrs; Am Inst Aeronaut & Astronaut; AAAS; Unmanned Vehicle Asn. *Res:* Original invention and development of various active OCM & IRCM systems; development of C W sensing lasar radar; development of chemistry/bio mass spectrometer; development of novel communication systems. *Mailing Add:* 8388 Riesling Way San Jose CA 95135

LEBARON, FRANCIS NEWTON, b Framingham, Mass, July 26, 22; m 53; c 1. NEUROCHEMISTRY, LIPID CHEMISTRY. *Educ:* Mass Inst Technol, BS, 44; Boston Univ, MA, 48; Harvard Univ, PhD(biochem), 51. *Prof Exp:* Asst biochemist, McLean Hosp, Mass, 52-53 & 54-57, assoc biochemist, 57-64; from assoc prof to prof biochem, Sch Med, Univ NMex, 64-83, chmn dept, 71-78; RETIRED. *Concurrent Pos:* USPHS fel, McLean Hosp, Waverley, Mass, 51-52 & Maudsley Hosp, 53-54; res assoc, Harvard Med Sch, Harvard Univ, 56-59, assoc, 59-64, tutor, 57-64; vis scholar, Mass Inst Technol, 74-75. *Mem:* AAAS; Am Soc Neurochem; Am Soc Biol Chem; Am Inst Nutrit. *Res:* Biochemistry of the nervous system, especially the chemistry of proteins and lipids and their nervous complexes as they occur in mammalian nervous tissues; role of polyunsaturated fatty acids in nervous tissues. *Mailing Add:* PO Box 779 Mashpee MA 02649

LEBARON, HOMER MCKAY, b Barnwell, Alta, May 13, 26; US citizen; wid; c 7. PESTICIDE CHEMISTRY, PESTICIDE EFFECTS. *Educ:* Utah State Univ, BS, 56, MS, 58; Cornell Univ, PhD(chem), 60. *Prof Exp:* Plant physiologist, Va Truck Exp Sta, 60-63; mem field res & develop, Geigy Chem Corp, 64; group leader herbicide res, 64-75, sr staff scientist, 75-84, SR RES FEL, CIBA-GEIGY CORP, 84- *Concurrent Pos:* Exec bd, Weed Sci Soc Am, 65-68, 81-85, vpres, 87-88, pres elect, 88-89, pres, 89-90; vpres, Northeastern Weed Sci Soc, 68-69, pres, 69-70; res comt chmn, Southern Weed Sci Soc, 84-85, prog chmn & pres elect, 85-86, pres, 86-87. *Honors & Awards:* Distinguished Serv Award, Southern Weed Sci Soc, 84; Cert Appreciation Groundwater Res Progs, USDA, 90; Charles A Black Award, Coun Agr Sci & Tech, 91. *Mem:* Fel Weed Sci Soc Am; Am Soc Agron; Entom Soc Am; Am Soc Plant Physiologists; Am Chem Soc; Am Phytopath Soc; Europ Weed Res Soc; Coun Agr Sci & Technol; Sigma Xi; AAAS; Am Inst Biol Sci; Aquatic Plant Mgt Soc. *Res:* Direct and coordinate all basic greenhouse and laboratory studies on herbicides, insecticides, fungicides, and other agricultural chemicals in areas of mode of action, environmental studies, soil interactions, metabolism and residues outside Ciba-Geigy; pesticide fate; weed and pest resistance to control measures. *Mailing Add:* Agr Div Biochem Basic Res & New Technol Ciba-Geigy Corp PO Box 18300 Greensboro NC 27419

LE BEAU, STEPHEN EDWARD, b Escanaba, Mich, Sept 6, 54. POWDER METALS, CONTINUOUS CASTING. *Educ:* Mich Technol Univ, BS, 76; Rensselaer Polytechnic Inst, MS, 78; Univ Wis, PhD(metall eng), 82. *Prof Exp:* Develop engr powder metals, Caterpillar Tractor Co, 78-79; res engr coated steels, USX Corp, 83; Supvr mat processing, Babcock & Wilcox Co, 83-90; DIR RES & DEVELOP MAT PROCESSING, PHOENIX METALS CORP, 90- *Mem:* Am Soc Metals; Am Powder Metall Inst; Am Soc Qual Control; Metall Soc Am Inst Mech Engrs; Soc Automotive Engrs; Soc Mfg Engrs. *Res:* Materials processing and manufacturing; steel powder manufacturing; continuous casting of steel; seamless and welded tubemaking technology. *Mailing Add:* Phoenix Metals Corp 300 Dunn St Plymouth MI 48170

LEBEL, JACK LUCIEN, b Montreal, Que, Sept 16, 33; US citizen; m 60; c 2. RADIOLOGY, RADIATION BIOLOGY. *Educ:* Univ Montreal, DVM, 58; Colo State Univ, MS, 66, PhD(radiation biol), 67; Am Col Vet Radiol, dipl, 68. *Prof Exp:* Asst prof vet med, Univ Montreal, 61-64, assoc prof radiol, 67-68; assoc prof radiol & radiation biol, 68-73, asst dean curric col, 70-72, PROF RADIOL & RADIATION BIOL, COL VET MED & BIOMED SCI, COLO STATE UNIV, 73- *Concurrent Pos:* Consult, Orthop Found Animals, 69-, Rockewell Int, 73- & Wildlife Pharmaceut, 90- *Mem:* Am Vet Med Asn; Am Vet Radiol Soc. *Res:* Bone pathology; biological effects of plutonium contamination; densitometry; pulmonary physiology. *Mailing Add:* Dept Radiol Health Colo State Univ Ft Collins CO 80523

LEBEL, JEAN EUGENE, b Can, Mar 21, 22. MATHEMATICS. *Educ:* McGill Univ, BSc, 44; Univ Toronto, MA, 50, PhD(appl math), 58. *Prof Exp:* Theoret physicist, Newmont Explor, Ltd, Ariz, 52-54; lectr, McGill Univ, 55-57; from asst prof to assoc prof math, Georgetown Univ, 58-65; ASSOC PROF MATH, UNIV TORONTO, 65- *Mem:* Am Math Soc; Can Math Cong. *Res:* Analysis; applied mathematics. *Mailing Add:* 27 Cheston Rd Toronto ON M4S 2X4 Can

LEBEL, NORMAN ALBERT, b Augusta, Maine, Mar 22, 31; m 52; c 3. ORGANIC CHEMISTRY. *Educ:* Bowdoin Col, AB, 52; Mass Inst Technol, PhD, 57. *Prof Exp:* Chemist, Merck & Co, Inc, 52-54; from asst prof to assoc prof, 57-64, chmn dept, 71-78, interim dean, Col Lib Arts, 83-84, PROF CHEM, WAYNE STATE UNIV, 64- *Concurrent Pos:* Sloan Found fel, 61-65; Welch Found lectr, 74. *Mem:* AAAS; Am Chem Soc; Sigma Xi; Royal Soc Chem. *Res:* Stereochemistry and mechanism of elimination reactions; chemistry of nitrones and nitrogen heterocycles; additions to olefins; bridged polycyclic molecules; new synthetic reactions. *Mailing Add:* 277 Chem Bldg Wayne State Univ Detroit MI 48202

LEBEL, ROLAND GUY, b Edmundston, NB, Feb 10, 32; m 64; c 1. PHYSICAL CHEMISTRY, CHEMICAL ENGINEERING. *Educ:* NS Tech Col, BEng, 55; Mass Inst Technol, SM, 56; McGill Univ, PhD(phys chem), 62. *Prof Exp:* Res engr, Fraser Pulp & Paper Co Ltd, 58-59; demonstr, McGill Univ, 59-62; sr develop engr, Anglo Paper Prod Ltd, 63-65; res & develop proj coordr pulp & paper, Reed Inc, 65-69, tech dir, Papeterie Reed Ltd, 69-80, dir res & develop, 80-85; DEAN, SCH FOREST SCI, UNIV MONCTON, 85- *Mem:* Sr mem Can Pulp & Paper Asn. *Res:* Planning, organization and coordination of research and development of projects related to the manufacture of newsprint and other paper products. *Mailing Add:* Sch Forest Sci Univ Moncton 165 Blvd Hebert Edmundston NB E3V 2S8 Can

LEBEL, SUSAN, b July 5, 59; Can citizen; m. MOLECULAR BIOLOGY. *Educ:* Montreal Univ, Que, BSc, 82, PhD(molecular biol), 87. *Prof Exp:* Postdoctoral fel, 87-88, SR SCIENTIST, BIOTECHNOL RES INST, BIOMIRA INC, 89- *Concurrent Pos:* Natural Sci & Eng Res Coun Can vis res fel, 87-88. *Mem:* Am Soc Cell Biol. *Mailing Add:* Biomira Inc 6100 Royalmount Ave Montreal PQ H4P 2R2 Can

LEBEN, CURT (CHARLES), b Chicago, Ill, July 7, 17; m 44; c 2. PLANT PATHOLOGY. *Educ:* Ohio Univ, BS, 40; Univ Wis, PhD(plant path), 46. *Prof Exp:* Asst bot, Ohio Univ, 39-40; asst, Univ Wis, 42, asst plant path, 42-46, res assoc, 46-49, asst prof, 49-55; plant pathologist, Eli Lilly & Co, Ind, 55-57, head agr res labs, 57-59; prof bot & plant path & assoc chmn dept, Ohio State Univ, 59-67, actg chmn, 67-68, prof plant path, Agr Res & Develop Ctr, 67-88; RETIRED. *Mem:* AAAS; fel Am Phytopath Soc; Sigma Xi. *Res:* Antibiotics and antibiosis in relation to plant diseases; microbiology; epiphytic microorganisms; bacterial and decay diseases; forest tree diseases; biological control. *Mailing Add:* 923 Thorne Ave Wooster OH 44691

LEBENBAUM, MATTHEW T(OBRINER), b Portland, Ore, Nov 29, 17; m 42; c 2. ELECTRONICS ENGINEERING. *Educ:* Stanford Univ, BA, 38; Mass Inst Technol, MS, 45. *Prof Exp:* Asst elec eng, Stanford Univ, 38-39 & Mass Inst Technol, 39-41; asst syst planning engr, Am Gas & Elec Corp, 41-42; res assoc, Radio Res Lab, Harvard Univ, 42-45; dir, Appl Electronics Div, Airborne Instruments Lab, Cutler-Hammer, Inc, 45-; RETIRED. *Mem:* Fel Inst Elec & Electronics Engrs; Int Union Radio Sci. *Res:* Radio receivers, especially microwave; methods and techniques of noise measurement; intermediate frequency amplifiers; application of electronic methods to power systems; radio astronomy instrumentation; low noise devices. *Mailing Add:* 80 Whitehall Blvd Garden City NY 11530

LEBENSOHN, ZIGMOND MEYER, b Kenosha, Wis, Sept 8, 10; m 40, 79; c 4. PSYCHIATRY. *Educ:* Northwestern Univ, BS, 30, MB, 33, MD, 34; Am Bd Psychiat & Neurol, dipl, 41. *Prof Exp:* Instr neurol, Med Sch, George Washington Univ, 36-41; PROF CLIN PSYCHIAT, MED SCH, UNIV HOSP, GEORGETOWN UNIV, 41-; CHIEF EMER, DEPT PSYCHIAT, SIBLEY MEM HOSP, 76- *Concurrent Pos:* Med officer, St Elizabeth's Hosp, 35-39; mem staff, Doctors Hosp, 41-74; consult, Vet Admin, 46-49, US Naval Hosp, Bethesda, Md, 52-, US Info Agency, 58-61, Walter Reed Army Hosp, Washington, DC, 58-66, 77- & NIMH, 68-; consult med adv panel, Fed Aviation Agency, 61-66; chief, Dept Psychiat, Sibley Mem Hosp, 57-75. *Honors & Awards:* Jacobi Soc Award, 83. *Mem:* Fel Am Psychiat Asn; fel AMA; Am Psychopath Asn; Asn Res Nerv & Ment Dis. *Res:* Legal aspects of psychiatry; psychiatric hospital design; trans-cultural psychiatry. *Mailing Add:* 2015 R St NW Washington DC 20009

LEBENTHAL, EMANUEL, b Jerusalem, Israel, Apr 12, 36; US citizen; c 5. GASTROENTEROLOGY, PEDIATRICS. *Educ:* Hebrew Univ, Israel, MD, 64. *Prof Exp:* Asst med, Children's Hosp, Boston, 72-74, assoc, 74-76; assoc prof pediat, 76-80, PROF PEDIAT GASTROENTEROL, STATE UNIV NY, BUFFALO, 80-; CHIEF PEDIAT GASTROENTEROL, CHILDREN'S HOSP BUFFALO, 80-, DIR, INT INST INFANT NUTRIT & GASTROINTESTINAL DIS, 84- *Concurrent Pos:* Consult to clin staff, Roswell Park Mem Inst, 77- *Honors & Awards:* Int Prize Mod Nutrit, Int Dairy Fedn, Switz, 84. *Mem:* Soc Pediat Res; NAm Soc Pediat Gastroenterol; Am Pancreatic Asn; Am Asn Study Liver Dis; Am Soc Clin Nutrit; Am Inst Nutrit. *Res:* Impact of the ontogeny of the gut on feeding the compromised and premature infant, specifically the ontogeny of the pancreas; determinants that affect acture diarrhea to develop into chronic diarrhea in children. *Mailing Add:* Dept Gastroenterol Children's Hosp Buffalo 219 Bryant St Buffalo NY 14222

LEBER, PHYLLIS ANN, b Scranton, Pa, 49; m 76; c 1. REACTION MECHANISMS. *Educ:* Albright Col, Pa, BS, 76; Univ NMex, PhD(chem), 81. *Prof Exp:* Jr chemist, Am Color & Chem Co, Reading, Pa, 71-76; asst prof org chem, Pomona Col, Claremont, Calif, 81-82; ASST PROF ORG CHEM, FRANKLIN & MARSHALL COL, PA, 82- *Concurrent Pos:* Vis asst prof org chem, Univ NMex, 83. *Mem:* Am Chem Soc; Sigma Xi. *Res:* Physical organic chemistry as it relates to the elucidation of reaction mechanisms of thermal unimolecular isomerizations, both uncatalyzed and catalyzed. *Mailing Add:* 1849 Wheatland Ave Lancaster PA 17603

LEBER, SAM, b Rockford, Ill, Nov 15, 25; m 52; c 2. METALLURGY. *Educ:* Univ Ill, BS, 48, MS, 49. *Prof Exp:* X-ray crystallographer, Horizons Inc, 50-55; metallurgist, Lamp Metals & Components Dept, Gen Elec Co, 55-66, supvr struct eval unit, 66-71, mgr metals eval subsect, Refractory Metals Dept, 71-86; RETIRED. *Mem:* Am Soc Metals; Am Inst Mining, Metall & Petrol Engrs. *Res:* Physical metallurgy; x-ray, optical and electron metallurgy. *Mailing Add:* 2495 Deborah Dr Beachwood OH 04122

LEBERMAN, PAUL R, b New York, Mar 1, 04; m 32. UROLOGY. *Educ:* NY Univ, BS, 25; Univ Pa, MS, 27, MD, 31; Am Bd Urol, dipl, 42. *Prof Exp:* Assoc prof clin urol & asst prof urol, Grad Sch Med, 51-66, PROF CLIN UROL, UNIV PA, 66-, CHIEF UROL SURG, OUTPATIENT DEPT, HOSP, 51-, EMER PROF UROL, 72- *Concurrent Pos:* Chief urol, Philadelphia Gen Hosp, 59-; consult, US Naval Hosp, Philadelphia, 60- *Mem:* Fel Am Col Surg; fel Royal Soc Med; fel Int Soc Urol; fel Am Acad Pediat; fel Pan-Pac Surg Asn; Sigma Xi. *Res:* Urological surgery. *Mailing Add:* Warwick 17th St & Locust St Philadelphia PA 19103

LEBERMANN, KENNETH WAYNE, b Davenport, Iowa; m 62; c 2. FOOD SCIENCES. *Educ:* Univ Ill, Urbana, BS, 59, PhD(food sci), 64; Univ Calif, Davis, MS, 61. *Prof Exp:* Scientist, CPC Int, 64-66; group leader cereals, 66-67, sr group leader bakery prod, 67-69, sect mgr bakery prod, 69-70, sect mgr pet foods, 70-75, mgr pet food res, 75-78, asst dir, 78, dir, 78-79, VPRES RES, QUAKER OATS CO, 79- *Mem:* Inst Food Technologists; Sigma Xi. *Res:* Industrial research related to new product development, pet food research. *Mailing Add:* Quaker Oats Co 617 W Main St Barrington IL 60010

LEBHERZ, HERBERT G, b San Francisco, Calif, July 27, 41; div. BIOCHEMISTRY. *Educ:* San Francisco State Univ, BA, 64, MA, 66; Univ Wash, PhD(biochem), 70. *Prof Exp:* From res assoc to sr res assoc cell biol, Swiss Fed Inst Technol, 71-75; cancer res sci biochem, Roswell Park Mem Inst, 75-76; assoc prof, 76-80, PROF BIOCHEM, SAN DIEGO STATE UNIV, 80- *Mem:* Am Heart Asn; Am Soc Biol Chemists; AAAS. *Res:* Elucidation of the mechanisms involved in the regulation of protein synthesis and protein degradation in developing adult and diseased organisms. *Mailing Add:* Dept Chem San Diego State Univ 5300 Campanile Dr San Diego CA 92182

LEBIEDZIK, JOZEF, b Feb 13, 40; US citizen; m 69; c 1. QUANTITATIVE MICROSCOPY, MATERIALS SCIENCE. *Educ:* Pa State Univ, BS, 70, MS, 72, PhD(solid state sci), 75. *Prof Exp:* Res asst quant micros, Mat Res Lab, Pa State Univ, 70-75, res assoc, 75-76; vpres res & develop quant micros, Lemont Sci Inc, 76-86; PRES, MODERN INSTRUMENTATION TECHNOL, INC, 80- *Mem:* Microbeam Anal Soc; Am Vacuum Soc. *Res:* Microbeam analysis on automated scanning electron microscopes or electron probe systems; laboratory automation; surface analysis systems - instrumentation. *Mailing Add:* 2434 30th St Boulder CO 80301-1232

LEBIEN, TUCKER W, b Minneapolis, Minn, Dec, 8, 48; m. LABORATORY MEDICINE, PATHOLOGY. *Educ:* NDak State Univ, BS, 70, MS, 73; Univ Nebr, PhD(med microbiol-immunol), 77. *Prof Exp:* Teaching asst, Dept Med Microbiol, Nebr Med Ctr, 74-76; postdoctoral res fel, Dept Lab Med & Path, Univ Minn, 77-79, from instr to assoc prof, 79-89, assoc dir, Pathobiol Grad Prog, 85-87, DIR GRAD STUDIES, PATHOBIOL GRAD PROG, UNIV MINN, 88-, PROF, DEPT LAB MED & PATH, 89- *Concurrent Pos:* New investr award, Nat Cancer Inst, 80-82; organizing fac, Fermentation Biotechnol Ctr, Univ Minn, 83-, dir, NIH Immunol Training Prog, 89-; assoc ed, J Immunol, 84-88, Blood, 91-93; vis assoc prof path, Lab Molecular Immunol, Harvard Med Sch, 86. *Honors & Awards:* McFadden Lectr, Nebr Med Ctr, 87; Stohlman Mem Award, Leukemia Soc Am, 89. *Mem:* Am Asn Immunologists; Am Asn Cancer Res; Am Soc Hemat; Am Soc Microbiol; Sigma Xi; Fedn Am Scientists; AAAS. *Res:* Neutral endopeptidase; differentiative programs of lymphoid progenitor cells. *Mailing Add:* Dept Lab Med & Path-UMHC Univ Minn 420 Delaware St SE Box 609 Minneapolis MN 55455

LEBLANC, ADRIAN DAVID, b Salem, Mass, May 21, 40; m 66; c 2. RADIOLOGICAL PHYSICS, NUCLEAR MEDICINE. *Educ:* Univ Mass, BA, 62; Iowa State Univ, MS, 66; Univ Kans, PhD(radiation biophys), 72; Am Bd Health Physics, cert; Am Bd Radiol, cert in nuclear med physics. *Prof Exp:* HEALTH PHYSICIST, HOSP & RADIOL PHYSICIST, DEPT NUCLEAR MED, METHODIST HOSP, 66-; ASST PROF MED, BAYLOR COL MED, 72- *Concurrent Pos:* Radiation physicist, Vet Admin Hosp, 67- *Mem:* Health Physics Soc; Soc Nuclear Med; Am Asn Physicists in Med; Sigma Xi. *Res:* Coronary blood flow; neutron activation analysis; x-ray fluorescence; health physics. *Mailing Add:* 1701 Herman Dr Apt 707 Houston TX 77004

LEBLANC, ARTHUR EDGAR, b Moncton, NB, Sept 29, 23; US citizen; m 52; c 2. GEOLOGY, PALYNOLOGY. *Educ:* Univ Mass, Amherst, BS, 52, MS, 54. *Prof Exp:* Paleontologist, Shell Oil Co, Tex, 54-62 & Calif, 62-65; sr res scientist, Res Ctr, Pan Am Petrol Corp, Okla, 65-69; proj palynologist, 69-77, SR PROJ GEOLOGIST, GULF OIL RES & DEVELOP CO, 77- *Mem:* AAAS; Am Asn Petrol Geologists; Soc Econ Paleontologists & Mineralogists; Am Asn Stratig Palynologists. *Res:* Stratigraphic Paleozoic palynology; Mesozoic and Cenozoic stratigraphic palynology. *Mailing Add:* 11622 Spriggs Way Houston TX 77024

LEBLANC, DONALD JOSEPH, b Shirley, Mass, Jul 3, 42; m 66; c 4. PLASMID BIOLOGY, MICROBIAL PHYSIOLOGY. *Educ:* St Michael's Col, BA, 64; Fordham Univ, MS, 66; Univ Mass, Amherst, PhD(microbiol), 70. *Prof Exp:* Postdoctoral, Georgetown Univ Sch Med & Dent, 70-72, staff fel, Nat Inst Health, 72-75, sr staff fel, 75-77, res microbiologist, plasmid biol, 77-81, sect head, 81-88; PROF MICROBIOL, UNIV TEX HEALTH SCI CTR, SAN ANTONIO, 88- *Concurrent Pos:* Consult, Food & Drug Admin, 78, vis scientist, Am Soc Microbiol Minority Student Sci Careers Support Prog & mem, Am Soc Microbiol Culture Collection Subcomt, Pub Affairs Con, 87-, Ed bd, Appl & Environ Microbiol, 88-90. *Honors & Awards:* Dr John C Hartnett lectr, St Michael's Col, 78. *Mem:* NY Acad Sci; Am Soc Microbiol; Am Asn Advan Sci. *Res:* Molecular, genetic, and biochemical techniques to identification and characterization of chromosomal and extrachromosomal genetic traits. *Mailing Add:* Dept Microbiol Univ Tex Health Sci Ctr 7703 Floyd Curl Dr San Antonio TX 78284

LEBLANC, FRANCIS ERNEST, b North Sydney, NS, June 10, 35; m 61; c 3. NEUROPHYSIOLOGY, NEUROSURGERY. *Educ:* St Francis Xavier Univ, BSc, 55; Univ Ottawa, MD, 59; Univ Montreal, MSc, 62, PhD(neurophysiol), 64; FRCS(C), 68. *Prof Exp:* Intern, Montreal Gen Hosp, Que, 59-60, jr asst resident surg, Queen Mary Vet Hosp, Montreal, 60-61; res asst, Neurol Sci Lab, Univ Montreal, 61-62, lectr physiol, Univ, 61-64; demonstr neurol, McGill Univ, 64-67, lectr neurosurg, 67-68, asst prof, 68-70; from asst porf to assoc prof, 71-79, PROF SURG, UNIV CALGARY, 80-, ASSOC DEAN CLIN SERV, 83-; CHIEF, DIV NEUROSURG, FOOTHILLS HOSP, CALGARY, 74- *Concurrent Pos:* Med Res Coun fel, Univ Montreal, 61-64, res fel, 62-64; clin fel & chief resident, Montreal Neurol Inst, 66-67, res scholar, 67-70, asst neurosurgeon, 68; consult, Queen Mary Vet Hosp, Montreal, 67; vis neurosurgeon, Royal Victoria Hosp, Montreal, 68; consult neurosurgeon, Foothills Hosp, 71-74. *Mem:* Can Neurosurg Soc; Cong Neurol Surg; Asn Acad Surg; fel Am Col Surg. *Res:* Cerebrovascular physiology; epilepsy; movement disorders. *Mailing Add:* Div Neurosci Univ Calgary Fac Med 3330 Hospital Dr NW Calgary AB T2N 4N1 Can

LEBLANC, GABRIEL, b Montreal, Que, June 24, 27; m 68; c 1. GEOPHYSICS, SEISMOLOGY. *Educ:* Univ Montreal, BA, 52; L'Immaculee-Conception, Montreal, LPh, 53; Boston Col, MSc, 58, SThL, 60; Pa State Univ, PhD(geophys), 66. *Prof Exp:* Res asst seismol, Pa State Univ, 63-66; assoc prof geophys, Laval Univ, 66-71; sr res scientist, Seismol Div, Dept Energy, Mines & Resources, Can, 71-76; SR STAFF CONSULT & SR SEISMOLOGIST, WESTON GEOPHYS CORP, 77- *Mem:* Am Geophys Union; Can Geophys Union; Can Asn Physicists; Seismol Soc Am. *Res:* Regional tectonics and local seismicity; seismic risk analysis; strong motion; induced seismicity; local arrays. *Mailing Add:* Weston Geophys Corp PO Box 550 Westboro MA 01581

LEBLANC, JACQUES ARTHUR, b Quebec, Que, Aug 23, 21; nat US; m 51; c 3. PHYSIOLOGY. *Educ:* Laval Univ, BA, 43, BSc, 47, PhD(physiol), 51. *Prof Exp:* Physiologist human physiol, Defence Res Bd, Can Dept Nat Defence, 49-56; PROF HUMAN PHYSIOL, FAC MED, LAVAL UNIV, 58- *Mem:* AAAS; Am Physiol Soc; Fedn Am Socs Exp Biol; Soc Exp Biol & Med; Fr-Can Asn Advan Sci. *Res:* Amines in stress conditions; tranquilizers; mast cells; basic and applied work in environmental physiology. *Mailing Add:* Dept Physiol Fac Med Laval Univ Quebec PQ G1K 7P4 Can

LEBLANC, JERALD THOMAS, b Baton Rouge, La, Mar 10, 43; m 68. PHOTOGRAPHIC SCIENCE, ORGANIC CHEMISTRY. *Educ:* Birmingham-Southern Col, BS, 65; Fla State Univ, PhD(org chem), 70. *Prof Exp:* SR CHEMIST RES LABS, EASTMAN KODAK CO, 70-, PROF MP TECH STAFF. *Mem:* Soc Photog Scientists & Engr. *Res:* Silver halide chemistry and physics; organic dye synthesis. *Mailing Add:* Eastman Kodak Co Bldg 30 Rochester NY 14652-3403

LEBLANC, LARRY JOSEPH, b New Orleans, La, July 21, 47; m 69; c 2. OPERATIONS RESEARCH, TRANSPORTATION SYSTEMS. *Educ:* Loyola Univ La, New Orleans, BS, 69; Northwestern Univ, Evanston, IL, MS, 71, PhD(opers res), 73. *Prof Exp:* From asst prof to assoc prof opers res, Southern Methodist Univ, 73-80; assoc prof, 80-88, PROF MGT, VANDERBILT UNIV, 88- *Concurrent Pos:* Consult, Mississippi Chem Corp, Dr Pepper Co, Trailways, McDermott, WFAA Radio, Pan Technol, Urban Systa, US Army Inventory Res Off, US Dept Justice, Port Everglades Steel Co, Eaton Corp, Carbon/Graphite Group & John Hamburg & Assoc, Inc, 77-; NSF grants, 80-81 & 82-83; chmn transp sci sect, Opers Res Soc Am, 84-86; grant, US Dept Transp; coun mem, telecommun spec interest group, Opers Res Soc Am, 88-90; vis prof, Linkoping Inst Tech, Sweden, 80, Technion, Israel, 84, Univ Ulm, Ger, 81, 82, 84 & 85, Ecole Centrale Paris, 87, Univ Chile, 78 & 88, Neth Orgn Appl Sci Res, 91. *Mem:* Opers Res Soc Am; Inst Mgt Sci. *Res:* Computer implementation techniques for telecommunication networks, traffic management, production and inventory control; applied optimization techniques. *Mailing Add:* Owen Grad Sch Mgt Vanderbilt Univ Nashville TN 37203

LEBLANC, LEONARD JOSEPH, b Moncton, NB, Nov 6, 37; m 59; c 3. SOLID STATE PHYSICS. *Educ:* St Joseph's Univ, NB, BSc, 59; Univ Notre Dame, PhD(physics), 64. *Prof Exp:* From asst prof to assoc prof physics, Univ Moncton, 64-71, vdean fac sci, 69-74, dean fac sci & eng, 75-80, PROF PHYSICS, UNIV MONCTON, 71-, VPRES ACAD, 80- *Mem:* Am Asn Physics Teachers; Can Asn Physicists. *Res:* Optical and photoelectric properties of metals in the vacuum ultraviolet. *Mailing Add:* Dept Physics & Math Univ de Moncton Moncton NB E1A 3E9 Can

LE BLANC, MARCEL A R, b Gravelbourg, Sask, Mar 25, 29; m 62; c 3. PHYSICS, SUPERCONDUCTIVITY. *Educ:* Univ Ottawa, BA, 49; Univ Sask, BA, 52, MA, 54; Univ BC, PhD(physics), 58. *Prof Exp:* Res assoc physics, Stanford Univ, 59-63; asst prof elec eng, Univ Southern Calif, 63-68; chmn, dept physics, 80-86, PROF PHYSICS, UNIV OTTAWA, 68- *Concurrent Pos:* Asst prof, San Jose State Col, 61-62; consult, Varian Assocs, 61-62 & Spectromagnetic Industs, 62-63; mem tech staff & consult, Aerospace Corp, 63-68. *Mem:* Am Phys Soc; Can Asn Physicists. *Res:* Photonuclear cross-sections; nuclear polarization; low temperature physics; superconductivity. *Mailing Add:* Dept Physics Univ Ottawa Ottawa ON K1N 6N5 Can

LEBLANC, NORMAN FRANCIS, b Boston, Mass, June 28, 26; m 52; c 4. SCIENCE ADMINISTRATION, TECHNICAL MANAGEMENT. *Educ:* Tufts Univ, BS, 47; Mass Inst Technol, PhD(anal chem), 50. *Prof Exp:* Res chemist, Hercules Aerospace Co, 50-56, res chemist solid rocket res & develop, 56-58, dir progs, 58-68, dir develop, Polymers Dept, 68-73, dir, Fibers Div, 73-81, group dir plastics, Hercules, Inc, 81-83, vpres technol, Hercules Aerospace Co, 84-89; RETIRED. *Mem:* Am Chem Soc. *Mailing Add:* Box 426 Chadds Ford PA 19317

LEBLANC, OLIVER HARRIS, JR, b Beaumont, Tex, Nov 14, 31; m 56; c 1. PHYSICAL CHEMISTRY. *Educ:* Rice Univ, BA, 53; Univ Calif, PhD, 57. *Prof Exp:* PHYS CHEMIST, RES & DEVELOP CTR, GEN ELEC CO, 57- *Mem:* AAAS; Am Chem Soc. *Res:* Electrochemistry; inorganic chemistry; membrane biophysics. *Mailing Add:* 1173 Phoenix Ave Schenectady NY 12308

LEBLANC, ROBERT BRUCE, b Alexandria, La, Jan 28, 25; m 68; c 7. INORGANIC CHEMISTRY. *Educ:* Loyola Univ, BS, 47; Tulane Univ, MS, 49, PhD(chem), 50. *Prof Exp:* Asst prof chem, Tex A&M Univ, 50-52; res specialist, Org Res Dept, Dow Chem Co, 52-63, sr textile specialist, 63-67; textile chem develop mgr, ADM Chem Div, Ashland Oil Co, 67-68; res mgr, Nat Cotton Coun, 68-70; PRES, LEBLANC RES CORP, 70- *Concurrent Pos:* Mem, Info Coun Fabric Flammability. *Mem:* Am Chem Soc; Am Asn Textile Chemists & Colorists; Am Soc Testing & Mat; Nat Fire Protection Asn. *Res:* Textile chemistry and phosphorus chemistry; flammability and fire retardance of textiles and plastics. *Mailing Add:* LeBlanc Res Corp PO Box 391 Tallulah LA 71284-0391

LEBLANC, ROGER M, b Trois-Rivieres, Quebec, Jan 5, 42; c 4. SURFACE CHEMISTRY, BIOLOGICAL MEMBRANES. *Educ:* Laval Univ, Can, BSc, 64, DSc(phys chem), 68. *Prof Exp:* Fel phys chem, Davy Faraday Res Lab, Royal Inst, London, Eng, 68-70; chmn, Dept Chem-Biol, Univ Que, 71-75, dir, Group Biophys Res, 78-81, chmn, Ctr Photobiophys, 81-91, PROF PHYS CHEM, DEPT CHEM-BIOL, UNIV QUE, TROIS-RIVIERES, 70- *Concurrent Pos:* Mem, cell biol & genetics selection comt, Natural Sci & Eng Res Coun Can, 80-82, pres, 82-83, Anal & Phys Chem Selection Comt, 88-90, pres, 90-91. *Honors & Awards:* Vincent Award, Fr Can Asn Advan Sci, 78; Noranda Award, Chem Inst Can, 82; Barringer Award, Spectros Soc Can, 83. *Mem:* Chem Inst Can; Am Chem Soc; Biophys Soc; Europ Photochem Asn; Brit Biophys Soc; Am Soc Photobiol. *Res:* The interaction of chlorophyll (a) with itself and with various chloroplast components; the specific interaction in a two dimensional array is being studied with monolayer and photophysics techniques. *Mailing Add:* Ctr Photobiophys Res Univ Que CP 500 Trois-Rivieres PQ G9A 5H7 Can

LEBLANC, RUFUS JOSEPH, b Erath, La, Oct 12, 17; m 40; c 4. GEOLOGY. *Educ:* La State Univ, BS, 39, MS, 41. *Prof Exp:* Asst geologist, La State Univ, 41-43; geologist, Miss River Comn, US War Dept, 44-46, chief geol sect, 47-48; sr res geologist, Shell Oil Co, 48-52, mgr dept geol res, 53-56, sr geologist, Tech Servs, 57-60, sr staff geologist, Explor Dept, Offshore Div, 61-65, mem staff explor training dept, clastic sediments, 65-86; OWNER, LEBLANC SCH CLASTIC SEDIMENTS, HOUSTON, TX, 86- *Concurrent Pos:* Mem comt fundamental res occurrence and recovery petrol, Am Petrol Inst, 52-57; assoc ed, J Am Asn Petrol Geologists. *Honors & Awards:* Sidney Powers Medal, Am Asn Petrol Geologists, 88. *Mem:* Fel Geol Soc Am; hon mem Soc Econ Paleont & Mineral; hon mem Am Asn Petrol Geologists. *Res:* Fundamental research in stratigraphy and sedimentology; exploration for oil and gas; quaternary geology; exploration training; environmental geology gulf coast. *Mailing Add:* 3751 Underwood St Houston TX 77025

LEBLOND, CHARLES PHILIPPE, b Lille, France, Feb 5, 10; m 36; c 4. ANATOMY. *Educ:* Univ Nancy, Lic es S, 32; Univ Paris, MD, 34; Univ Montreal, PhD(iodine metab), 42; Univ Sorbonne, DSc, 45. *Hon Degrees:* DSc, Acadia Univ, 72, Mc Gill Univ, 82, Univ Montreal, 85, York Univ, 86. *Prof Exp:* Asst histol, Med Sch Paris, 34-35; Rockefeller fel, Sch Med, Yale Univ, 36-37; asst, Lab de Synthese Atomique, Paris, 38-40; lectr histol & embryol, 41-43, from asst prof to assoc prof, 43-48, chmn dept, 57-75, PROF ANAT, MCGILL UNIV, 48- *Honors & Awards:* Wilson Medal, Am Soc Cell Biol, 82. *Mem:* Am Asn Anat; Am Asn Cancer Res; fel Royal Soc; fel Royal Soc Can. *Res:* Histological localization of vitamin C; uptake of iodine by thyroid; tracing of radio elements and labelled precursors of nucleic acids, proteins and glycoproteins by means of radioautography. *Mailing Add:* Dept Anat 3640 University St Montreal PQ H3A 2B2 Can

LEBLOND, PAUL HENRI, b Quebec, Can, Dec 30, 38; m 63; c 3. PHYSICAL OCEANOGRAPHY. *Educ:* Laval Univ, BA, 57; McGill Univ, BSc, 61; Univ BC, PhD(physics), 64. *Prof Exp:* Nat Res Coun Can fel, Inst Meereskunde, Kiel, Ger, 64-65; from asst prof to assoc prof physics, 65-75, HEAD & PROF PHYSICS & OCEANOG, UNIV BC, 75- *Concurrent Pos:* Vis assoc prof, Simon Fraser Univ, 70; vis scientist, Inst Oceanology, USSR Acad Sci, Moscow, 73-74; vis prof, Laval Univ, 79-80; dir, Seaconsult Marine Res Ltd; assoc dean, Fac Sci, Univ BC, 83-85; vis prof, Univ Marseille. *Honors & Awards:* President's Prize, Can Meteorol Oceanog Soc, 81. *Mem:* Can Meteorol Oceanog Soc; Am Geophys Union; Am Meteor Soc; Int Soc Cryptozool; fel Royal Soc Can, 82; Nat Marine Coun, 88-90. *Res:* Surface, internal, planetary waves; ocean currents; tides; estuarine circulation; cryptozoology; theoretical biology. *Mailing Add:* Dept Oceanog Univ BC Vancouver BC V6T 1W5 Can

LEBO, GEORGE ROBERT, b Chadron, Nebr, Sept 27, 37; m 58; c 2. PHYSICS, RADIO ASTRONOMY. *Educ:* Wheaton Col, BS, 59; Univ Ill, MS, 60; Univ Fla, PhD(physics), 64. *Prof Exp:* Res assoc radio astron, 64-65, asst prof, 65-77, ASSOC PROF PHYSICS & ASTRON, UNIV FLA, 77- *Mem:* Am Astron Soc; Am Geophys Union; Am Phys Soc. *Res:* Study of decametric radiation from the planets, particularly Jupiter. *Mailing Add:* Dept Physics & Astron Univ Fla Gainesville FL 32611

LEBOFSKY, LARRY ALLEN, b Brooklyn, NY, Aug 31, 47; m 80; c 1. PLANETARY SCIENCES. *Educ:* Calif Inst Technol, BS, 69; Mass Inst Technol, PhD(earth & planetary sci), 74. *Prof Exp:* Res assoc planetary sci, Jet Propulsion Lab, 75-77; from res assoc to assoc planetary Sci, 77-89, SR RES SCIENTIST, UNIV ARIZ, 89- *Mem:* Am Astron Soc; Int Astron Union; Am Geophys Union; Meteoritical Soc; Nat Sci Teachers Asn; Asn Educ Teachers Sci. *Res:* Remote sensing of the visual, near infrared and thermal infrared spectra of asteroids and satellites for the study of composition; related studies of laboratory reflection spectra. *Mailing Add:* Lunar & Planetary Lab Univ Ariz Tucson AZ 85721

LEBOUTON, ALBERT V, b La Salle, Ill, July 10, 37; m 59; c 3. MICROSCOPIC ANATOMY, CELL BIOLOGY. *Educ:* San Diego State Col, BS, 60; Univ Calif, Los Angeles, PhD(anat), 66. *Prof Exp:* Asst prof anat, Univ Calif, Los Angeles, 66-72, actg head dept, 72-74, ASSOC PROF ANAT, UNIV ARIZ, 72- *Mem:* AAAS; Am Inst Biol Sci; Fedn Am Sci; Am Soc Cell Biol; Am Asn Anat. *Res:* Radioautography, radiobiochemistry, and immunocytochemistry of protein metabolism and growth in the liver. *Mailing Add:* Dept Anat Univ Ariz Col Med 1501 N Campbell Ave Tucson AZ 85724

LEBOVITZ, NORMAN RONALD, b New York, NY, Sept 27, 35; m 71; c 2. APPLIED MATHEMATICS, ASTROPHYSICS. *Educ:* Univ Calif, Los Angeles, AB, 56; Univ Chicago, MS, 57, PhD(physics), 61. *Prof Exp:* Moore instr math, Mass Inst Technol, 61-63; from asst prof to assoc prof, 63-69, PROF MATH, UNIV CHICAGO, 69- *Concurrent Pos:* Sloan Found fel, 67; Guggenheim Found fel, 77; managing ed, Soc Indust & Appl Math. *Mem:* Am Astron Soc; Am Math Soc; Soc Indust & Appl Math. *Res:* Stability theory; bifurcation theory; rotating fluid masses; singular perturbation theory. *Mailing Add:* Dept Math Univ Chicago 5734 S University Ave Chicago IL 60637

LEBOVITZ, ROBERT MARK, b Scranton, Pa, May 6, 37; m; c 3. NEUROPHYSIOLOGY, BIOMATHEMATICS & BIOENGINEERING. *Educ:* Calif Inst Technol, BS, 59, MS, 60; Univ Calif, Los Angeles, PhD(neurophysiol), 67. *Prof Exp:* Mem tech staff, Hughes Aircraft Co, 59-62; resident consult, Dept Math, Rand Corp, 67; NSF fel, Sch Med, NY Univ, 67-69; res assoc neural models, Ctr Theoret Biol, State Univ NY Buffalo, 69; staff scientist auditory & visual info processing, Recognition Equip, Inc, 69-70; assoc prof, 70-80, PROF NEUROPHYSIOL, UNIV TEX HEALTH SCI CTR, DALLAS, 80- *Concurrent Pos:* Consult, Dept Path & Eng Sci, Rand Corp, 69, Recognition Equip, Inc, 69, Neurosyst, Inc, 71, Dallas Epilepsy Asn, 71, Energy Conversion Devices, 71 & Equitable Environ Health, 71-; chief technical officer, Centra-Guard Inc & Neighborhood Coop Patrol, 69-71; NSF fel, Univ Tex Regents, 70-72; adj prof, Inst Technol, Southern Methodist Univ, 71-74; prof bd, Home Health Serv, Dallas, 74-; NIH grants, 71-, Food & Drug Admin grant, 74-, Nat Inst Environ Health

Sci grant, 77- & Off Naval Res Contracts, 78-; Sloan Consortium scholar, 74-75. *Mem:* Am Physiol Soc; NY Acad Sci; Biophys Soc; Inst Elec & Electronics Engrs; Soc Neurosci. *Res:* Neural networks and neural modelling; neurophysiology of epilepsy and behavior; modification and control of behavior via drugs and implanted or extraneous brain stimulating arrays; microwave interactions with the nervous system and behavior; electronic medicine. *Mailing Add:* Dept Physiol Univ Tex Health Sci Ctr 5323 Harry Hines Blvd Dallas TX 75235

LEBOW, IRWIN L(EON), b Boston, Mass, Apr 27, 26; m 51; c 3. ELECTRONICS. *Educ:* Mass Inst Technol, SB, 48, PhD(physics), 51. *Prof Exp:* Mem staff, Lincoln Lab, Mass Inst Technol, 51-60, assoc group leader, 60-65, group leader, 65-70, assoc div head, 70-75; chief scientist & assoc dir technol, Defense Commun Agency, 75-81; vpres eng, Am Satellite Co, 81-84; vpres, Syst Res & Appln Corp, 84-87, CONSULT, 87- *Concurrent Pos:* Mem, Defense Commun Agency Sci Adv Corp, Radio Eng Adv Comt, Voice Am. *Mem:* Fel Am Phys Soc; fel Inst Elec & Electronics Engrs; AAAS. *Res:* Communication systems, including, satellite communications, command and control systems and information processing systems; author on information technology. *Mailing Add:* 2800 Bellevue Terrace NW Washington DC 20007-1366

LEBOW, MICHAEL DAVID, b Detroit, Mich, June 24, 41; m 69; c 2. BEHAVIORAL MEDICINE, HEALTH PSYCHOLOGY. *Educ:* Univ Calif, Los Angeles, BS, 64; Univ Utah, MA, 67, PhD(psychol), 69. *Prof Exp:* Asst prof psychol, Univ Man, 69-72; asst prof med psychol, Med Sch Dartmouth, Univ, 72-74; assoc prof, 74-79, PROF PSYCHOL, UNIV MAN, 79- *Concurrent Pos:* Chmn, Psychol Intervention Subcomt, Task Force Treatment Obesity, Can Health & Welfare, 86- *Mem:* Sigma Xi; Am Psychol Asn; Soc Behav Med; Can Psychol Asn; Asn Advan Behav Therap. *Res:* Treatment programs for the obese; attitudes, perceptions and practices of children, adolescents and adults towards the obese. *Mailing Add:* Dept Psychol Univ Manitoba Winnipeg MB R3T 2N2 Can

LEBOWITZ, ELLIOT, b Monticello, NY, June 19, 41; m 63; c 4. NUCLEAR MEDICINE, MEDICAL DEVICES. *Educ:* Columbia Univ, BA, 61, PhD(chem), 67. *Prof Exp:* Res assoc chem kinetics & asst chemist, Dept Chem, Brookhaven Nat Lab, NY, 67-69, assoc chemist, Dept Appl Sci, 69-75; asst to vpres & actg dir drug regulatory affairs, New England Nuclear Corp, 75-76, div mgr radiopharmaceut div, 76-78, div mgr, Med Diag Div, 78-82; dir, new technol res, DuPont Corp, 82-85; dir, 85-87, VPRES, RES & DEVELOP, BARD CARDIOSURG DIV, 87- *Res:* Medical technology. *Mailing Add:* Bard Cardiosurgery Div Billerica MA 01821

LEBOWITZ, JACOB, b Brooklyn, NY, Oct 20, 35; m 78. MOLECULAR BIOLOGY, BIOCHEMISTRY. *Educ:* Brooklyn Col, BS, 57; Purdue Univ, PhD(phys chem), 62. *Prof Exp:* Res fel biophys chem, Calif Inst Technol, 62-66; from asst prof to assoc prof biochem, Syracuse Univ, 66-74; assoc prof, 74-77, PROF MICROBIOL, MED CTR, UNIV ALA, BIRMINGHAM, 77- *Concurrent Pos:* Career develop award, NIH, 72-77; scholar award, Am Cancer Soc, 82-83. *Mem:* AAAS; Am Chem Soc; Am Soc Microbiol; Am Soc Biol Chemists. *Res:* Analysis of RNA polymerase-promoter interactions; structural transitions in supercoiled DNA in relation to biological activity. *Mailing Add:* Dept Microbiol 520 CHSB Univ Ala Med Ctr Birmingham AL 35294

LEBOWITZ, JACOB MORDECAI, b New York, NY, Mar 21, 36; m 65; c 2. NUCLEAR PHYSICS. *Educ:* Yeshiva Univ, BA, 57; Columbia Univ, MA, 60, PhD(physics), 65. *Prof Exp:* From instr to asst prof, 59-70, ASSOC PROF PHYSICS, BROOKLYN COL, 70- *Mem:* Am Phys Soc; Sigma Xi. *Res:* Nuclear forces; fission. *Mailing Add:* 138 W Beech Sts Long Beach NY 11561

LEBOWITZ, JOEL LOUIS, b Taceva, Czech, May 10, 30; nat US; m 53. STATISTICAL MECHANICS, MATHEMATICAL PHYSICS. *Educ:* Brooklyn Col, BS, 52; Syracuse Univ, MS, 55, PhD(physics), 56. *Hon Degrees:* Dr Hon Causa, Ecole Polytech Fed Lausanne, 77. *Prof Exp:* NSF res fel, Yale Univ, 56-57; asst prof physics, Stevens Inst Technol, 57-59; from asst prof to prof physics, Grad Sch Sci, Yeshiva Univ, 57-77; GEORGE WILLIAM HILL PROF MATH & PHYSICS & DIR CTR MATH SCI RES, RUTGERS UNIV, 77- *Concurrent Pos:* Vis prof, Sch Med, Cornell Univ; Guggenheim fel, 76-77; mem, Sci Comt Sci Matters, Inst des Hautes Etudes Scientifiques, Bures-sur-Yvette, France, 79-82, Inst Theoret Physics; ed-in-chief J Statist Physics, 75, ed/co-ed Annals NY Acad Sci, Collective Phenomena, 80- *Honors & Awards:* A Cressy Morrison Award in Natural Sci, NY Acad Sci, 86. *Mem:* Fel Nat Acad Sci; NY Acad Sci (pres-elect, 77-79, pres, 79); fel Am Phys Soc; Am Math Soc; fel AAAS; Int Union Pure & Appl Physics (secy, 82-84, pres, 85-87). *Res:* Statistical mechanics of equilibrium and nonequilibrium processes; theory of liquids; biomathematics and mathematical economics. *Mailing Add:* Dept Math Hill Ctr Rutgers Univ New Brunswick NJ 08903

LEBOWITZ, MICHAEL DAVID, b New York, NY, Dec 21, 39; m 60; c 3. EPIDEMIOLOGY, PULMONARY DISEASES. *Educ:* Univ Calif, Berkeley, AB, 61, MA, 65; Univ Wash, PhC, 69, PhD(epidemiol), 71. *Prof Exp:* Pub health statistician, Alameda County Health Dept, 62-63; biostatistician, Calif Dept Pub Health, 67; res assoc environ health, Univ Wash, 67-71; asst prof internal med, 71-75, assoc prof, 75-80, asst dir respiratory sci, 74-85, PROF INTERNAL MED, COL MED, UNIV ARIZ, 80-, ASST DIR SPECIALIZED CTR RES, 71-, ASSOC DIR, DIV RESPIRATORY SCI, 85- *Concurrent Pos:* Partic, NSF-Japan Soc Promotion Sci Coop Sci Group in Air Pollution, 69-70; prin investr grants, NIH, 71-, Environ Protection Agency, 77-, Food & Drug Admin, 76-81, Elec Power Res Inst, 83-; consult, NIH & Nat Heart & Lung Inst, 72-, Environ Protection Agency, 76-, Ital Nat Res Coun, 77-, Polish & Hungarian NIH, 81, Sci Adv Bd, 84-; mem & chmn, Pima Co Air Qual Adv Coun, 75-; mem, epidemiol study sect, NIH, 75-78, behav study sect, 77; assoc prog dir, NIH Inst Training Prog, 77-; Fogarty sr int fel, 78-79; sr fel, Univ London, Postgrad Cardiothoracic Inst, Brompton, 78-79; vis prof, Harvard, 78 & 83, Univ Wash,

78, Univ Kans, 78, 82 & 85, Univ Pisa, 79, 80, 82 & 85-, Univ Groningen, 79 & 85, Univ Ill, 79, Univ Crakow, 81 & 86, Polish Acad Sci, 81 & 86, Univ London, 82, Univ Utah, 82, Univ Padua, 82, 86 & 88, Univ Rome, 82 & 85, Univ Catania, 82, NY Univ, 84-, Univ Goteborg, 84 & 88, Johns Hopkins, 84, Yale, 85, Univ Otago, 89, Fed Univ Rio de Janiero, 89; preceptor, NIH fel, Am Thoracic Soc, Fogarty Ctr fel & CNR fel, 78-; co-chmn, comt indoor pollutants, Nat Res Coun, Nat Acad Sci, 79-81; WHO adv, 79-; WHO/EURO work group, 82-; Ariz Gov Spec Environ Investr, 87-; mem, Epidemiol Coun, Am Health Asn. *Mem:* Int Epidemiol Asn; Am Epidemiol Soc; Am Thoracic Soc; Europ Respiratory Soc; Soc Epidemiol Res; fel Am Col Chest Physicians; hon mem Hungarian Soc Med Hyg; Am Pub Health Asn; AAAS; Int Soc Environ Epidemiol. *Res:* Pulmonary and chronic disease epidemiology; etiology and natural history of pulmonary diseases and other chronic diseases; air pollution health effects research. *Mailing Add:* Ariz Med Ctr Sect Pulmonary Dis Univ Ariz Col Med 1501 N Campbell Ave Tucson AZ 85724

LEBOY, PHOEBE STARFIELD, b Brooklyn, NY, July 29, 36; m 84. GENE EXPRESSION, BONE BIOLOGY. *Educ:* Swarthmore Col, AB, 57; Bryn Mawr Col, PhD(biochem), 62. *Hon Degrees:* MA, Univ Pa, 71. *Prof Exp:* Res assoc biochem, Bryn Mawr Col, 61-63; res assoc, Sch Med, 63-66, from instr to assoc prof, 65-76, PROF BIOCHEM, SCH DENT MED, UNIV PA, 76- *Concurrent Pos:* NATO fel, Weizmann Inst Sci, 66-67; NIH res grant, Univ Pa, 68-, & NIH res career develop award, 71-76; vis prof, Univ Calif, San Francisco, 79-80; chairperson, Grad Group Molecular Biol, Univ Pa, 84-87; Fogarty fel, Univ Oxford, 89-90. *Mem:* Am Soc Biol Chem; Am Soc Bone Mineral Res; Am Soc Cell Biol. *Res:* Role of gene expression in tissue mineralization; molecular biology of bone formation. *Mailing Add:* Dept Biochem Sch Dent Med Univ Pa Philadelphia PA 19104-6003

LE BRETON, GUY C, b Miami, Fla, Oct 1, 46. PHARMACOLOGY. *Educ:* Univ Chicago, BS, 68, PhD(pharmacol), 73. *Prof Exp:* Postdoctoral trainee, Dept Pharmacol, 73-75, from instr to assoc prof, 75-85, PROF, DEPT PHARMACOL, UNIV ILL, 85- *Concurrent Pos:* Fac develop award, Pharmaceut Mfrs Asn, 80-82; mem bd trustees, Am Asn Accreditation Lab Animal Care, 82-86; estab investr, Am Heart Asn, 82-87; field ed, J Pharmacol & Exp Therapeut, 82-; mem, Comt Pub Info, Am Soc Pharmacol & Exp Therapeut, 86-88 & Coun Thrombosis, Am Heart Asn. *Mem:* Sigma Xi; Am Soc Pharmacol & Exp Therapeut; AAAS. *Res:* Drug development; thrombosis and anti-thrombotic drugs; blood platelet: aggregating agents, aggregation, prostaglandins, prostaglandin endoperoxides, thromboxanes, thromboxane antagonists, thromboxane receptor, calcium, ion fluxes, cAMP; fluorescent cation probes; cellular secretion; cellular contraction; fatty acid metabolism. *Mailing Add:* Dept Pharmacol Rm 768 CME Univ Ill Col Med 835 S Wolcott Ave Chicago IL 60612

LEBRETON, PIERRE ROBERT, b Chicago, Ill, Sept 17, 42; m 69; c 2. BIOPHYSICAL CHEMISTRY. *Educ:* Univ Chicago, BS, 64; Harvard Univ, MA, 66, PhD(chem physics), 70. *Prof Exp:* Fel physics, Phys Inst, Univ Freiburg, Ger, 70-71; fel chem, Jet Propulsion Lab, Calif Inst Technol, 71-73; from asst prof to assoc prof, 73-87, PROF CHEM, UNIV ILL, CHICAGO, 87- *Concurrent Pos:* NIH grants, 80-83, 87-90; Am Cancer Soc grants, 76-80, 84 & 91-93; res assoc, Nat Ctr Sci Res, Univ Louis Pasteur, Strasbourg, France, 80; Petrol Res Fund Am Chem Soc, grant, 87-91. *Mem:* Am Chem Soc; Int Soc Quantum Biol; Soc Appl Spectros; Biophysical Soc; Am Asn Cancer Res. *Res:* Photoelectron, theoretical quantum mechanical, and time-resolved fluorescence probes of nucleic acid electronic structure, and of nucleic acid interactions with mutagens and carcinogens. *Mailing Add:* Dept Chem Univ Ill Chicago IL 60680

LEBRUN, ROGER ARTHUR, b Providence, RI, May 26, 46. ENTOMOLOGY, INVERTEBRATE PATHOLOGY. *Educ:* Providence Col, AB, 68; Cornell Univ, MSc, 73, PhD(invert path), 77. *Prof Exp:* ASSOC PROF INVERT PATH, UNIV RI, 77- *Concurrent Pos:* Fel, Inst Pasteur; Eli Lilly fel. *Mem:* Soc Invert Path; Entom Soc Am; Sigma Xi. *Res:* Protozoan and fungal pathogens of insects; ciliate parasites of medically important insects; fungal pathogens of coleoptera; host-parasite relationships; microbial ecology of pathogens. *Mailing Add:* Dept Plant Path & Entom Univ RI Kingston RI 02881

LEBSOCK, KENNETH L, b Brush, Colo, Oct 19, 21; m 43; c 2. AGRONOMY. *Educ:* Mont State Col, BS, 49; NDak State Univ, MS, 51; Iowa State Col, PhD(plant breeding), 53. *Prof Exp:* Wheat breeding & genetics, 53-72, AGR ADMINR, AGR RES SERV, USDA, 72- *Mem:* Fel Am Soc Agron. *Mailing Add:* ARS-USDA 220 Second S St Ste 100 Minneapolis MN 55401

LE CAM, LUCIEN MARIE, b Croze, France, Nov 18, 24; m 52; c 3. MATHEMATICAL STATISTICS. *Educ:* Univ Paris, Lic, 45; Univ Calif, PhD, 52. *Prof Exp:* Statistician, Elec of France, 45-50; instr math & asst, Statist Lab, 50-52, instr & jr res statistician, 52-53, from asst prof to prof statist, 53-73, chmn dept, 61-65, PROF STATIST & MATH, UNIV CALIF, BERKELEY, 73- *Concurrent Pos:* Sloan Found fel, 57-58; dir, Ctr Math Res, Univ Montreal, 72, mem adv comt, 74-80; mem, C Appl Theor Stat, NRC, 87-90. *Mem:* Am Math Soc; Inst Math Statist (pres, 72-73); Int Statist Inst; fel Am Acad Arts & Sci; AAAS; Am Statist Asn. *Res:* General statistics; asymptotic methods in statistical decision theory; limit theorems in probability theory. *Mailing Add:* 101 Kensington Rd Kensington CA 94707-1011

LECAR, HAROLD, b Brooklyn, NY, Oct 18, 35; m 58; c 2. BIOPHYSICS. *Educ:* Columbia Univ, AB, 57, PhD(physics), 63. *Prof Exp:* biophysicist, Lab Biophys, Nat Inst Neurol & Commun Dis & Stroke, 63-; PROF, DEPT MOLECULAR & CELL BIOL, UNIV CALIF, BERKELEY. *Concurrent Pos:* Fel commoner, Churchill Col, Cambridge Univ, 75-76; regents lectr, Univ Calif, San Diego, 82. *Mem:* AAAS; Am Phys Soc; Biophys Soc; Sigma Xi. *Res:* Masers; biophysics of excitable membranes. *Mailing Add:* Dept Molecular Cell Biol Div Neurobiol Univ Calif Donner Lab Berkeley CA 94720

LECAR, MYRON, b Brooklyn, NY, Apr 10, 30. ASTROPHYSICS. *Educ:* Mass Inst Technol, BS, 51; Case Inst Technol, MS, 53; Yale Univ, PhD(astron), 63. *Prof Exp:* Lectr astrophys, Yale Univ Observ, 62-65; LECTR ASTROPHYS, COL OBSERV, HARVARD UNIV, 65-; ASTRONR, SMITHSONIAN ASTROPHYS OBSERV, 65- *Concurrent Pos:* Astronr, Inst Space Studies, NASA, 62-65. *Mem:* Am Astron Soc; Fedn Am Scientists; Royal Astron Soc; Sigma Xi. *Res:* Dynamics of the solar system; stellar dynamics and galactic structure; cosmology. *Mailing Add:* Ctr Astrophys Harvard Col Observ & Smithsonian Astrophys Observ 60 Garden St Cambridge MA 02138

LECCE, JAMES GIACOMO, b Williamsport, Pa, Jan 11, 26; m 50. MICROBIOLOGY. *Educ:* Dartmouth Col, BA, 49; Pa State Univ, MS, 51; Univ Pa, PhD(microbiol), 53. *Prof Exp:* Instr prev med, Sch Vet Med, Univ Pa, 53-55; from asst prof to assoc prof microbiol, 55-63, PROF MICROBIOL, NC STATE UNIV, 63-, PROF ANIMAL SCI, 76- *Mem:* Am Soc Microbiol. *Res:* Rotavirus; passive immunity; enteric diseases. *Mailing Add:* Box 7615 Microbiol NC State Univ Raleigh NC 27695-7615

LECH, JOHN JAMES, b Passaic, NJ, June 21, 40. PHARMACOLOGY. *Educ:* Rutgers Univ, Newark, BS, 62; Marquette Univ, PhD(pharmacol), 67. *Prof Exp:* From instr to asst prof, 67-74, assoc prof pharmacol, 74-80, PROF PHARMACOL & TOXICOL, MED COL WIS, 80- *Concurrent Pos:* Am Heart Asn grant, Med Col Wis, 72-75, Sea grant, 71-75. *Mem:* AAAS; Soc Toxicol; Am Fisheries Soc; Am Soc Pharmacol & Exp Therapeut. *Res:* Cardiac triglyceride metabolism; metabolism of foreign compounds by fish. *Mailing Add:* Med Col Wis 8701 Watertown Plank Rd Milwaukee WI 53226

LECHEVALIER, HUBERT ARTHUR, b Tours, France, May 12, 26; nat US; m 50; c 2. MICROBIOLOGY. *Educ:* Laval Univ, MS, 48; Rutgers Univ, PhD(microbiol), 51. *Hon Degrees:* DSc, Laval Univ, 83. *Prof Exp:* From asst prof to assoc prof, 51-66, PROF MICROBIOL, WAKSMAN INST MICROBIOL, RUTGERS UNIV, NEW BRUNSWICK, 66- *Concurrent Pos:* Exchange scientist, Acad Sci USSR, 58-59; USPHS spec fel, Pasteur Inst, Paris, 61-62. *Honors & Awards:* Lindback Award, 76; Charles Trom Award, 82; Bergey Award, 89. *Mem:* Soc for Indust Microbiol; hon mem Fr Soc Microbiol. *Res:* Morphology, classification and products of actinomycetes, including antibiotics; history of microbiology. *Mailing Add:* 28 Juniper Lane Piscataway NJ 08854

LECHEVALIER, MARY P, b Cleveland, Ohio, Jan 27, 28; m 50; c 2. MICROBIOLOGY, TAXONOMY. *Educ:* Mt Holyoke Col, BA, 49; Rutgers Univ, MS, 51. *Prof Exp:* Res microbiologist, E R Squibb & Sons, 60-61; res assoc, 62-75, from asst res prof to assoc res prof, 75-85, RES PROF MICROBIOL, RUTGERS UNIV, PISCATAWAY, 85- *Honors & Awards:* Charles Thom Award, Soc Indust Microbiol, 82. *Mem:* AAAS; Brit Soc Gen Microbiol; Am Soc Microbiol; Soc Indust Microbiol; N Am Mycological Asn. *Res:* Classification, ecology, physiology and natural products of actinomycetes; microbial transformations; nitrogen fixation. *Mailing Add:* Waksman Inst Microbiol Rutgers Univ PO Box 759 Piscataway NJ 08855-0759

LECHLEIDER, J W, b Brooklyn, NY, Feb 22, 33; m 55; c 2. MATHEMATICS, COMPUTER SOFTWARE. *Educ:* Cooper Union, BME, 54; Polytech Inst Brooklyn, MEE, 57, PhD(elec eng), 65. *Prof Exp:* Engr, Gen Elec Co, 54-55; mem tech staff, Bell Tel Labs, 55-65, supvr transmission studies, Outside Plant & Underwater Systs Div, 65-67, head outside plant eng dept, 67-70, head loop transmission maintenance eng dept, 70-76, HEAD, MDF SOFTWARE DESIGN DEPT, BELL LABS, 76- *Mem:* Sr mem Inst Elec & Electronics Engrs; Am Math Soc; Sigma Xi. *Res:* Electromagnetic theory; communication theory; transmission theory. *Mailing Add:* 4 Harding Terr Morristown NJ 07960

LECHNER, BERNARD J, b New York, NY, Jan 25, 32; m 53. RESEARCH MANAGEMENT, TELEVISION SYSTEMS. *Educ:* Columbia Univ, BSEE, 57. *Prof Exp:* Mem tech staff, RCA Labs, 57-62, group head, 62-77, lab dir, 77-83, staff vpres, 83-87; CONSULT, 87- *Honors & Awards:* David Sarnoff Award, RCA Corp, 62; Frances Rice Darne Award, Soc Info Display, 71 & Beatrice Winner Award, 83. *Mem:* Fel Inst Elec & Electronics Engrs; fel Soc Info Display (vpres, 76-78, pres, 78-80); Soc Motion Picture & TV Engr. *Res:* Video-tape recording; high speed digital computer circuits; tunnel diodes; display devices and systems; ferroelectrics; electroluminescence; magnetic thin films; instrumentation; liquid crystals; digital television systems; TV receivers; TV tuning systems; TV broadcast equipment; cable TV systems; high definition TV. *Mailing Add:* 98 Carson Rd Princeton NJ 08540

LECHNER, JAMES ALBERT, b Danville, Pa, Aug 6, 33; m 56; c 3. APPLIED STATISTICS, RELIABILITY. *Educ:* Carnegie Inst Technol, BS, 54; Princeton Univ, PhD(math statist), 59. *Prof Exp:* Instr math, Princeton Univ, 57-58; sr mathematician, Res Labs, Westinghouse Elec Corp, Pa, 60-63, adv mathematician, Aerospace Div, Md, 63-67; mem tech staff, Res Anal Corp, Va, 67-71; MATH STATISTICIAN, STATIST ENG DIV, NAT INST STANDARDS & TECHNOL, 71- *Concurrent Pos:* Instr, Carnegie Tech, Univ Md & George Washington Univ. *Mem:* Am Statist Asn. *Res:* Probability; theory of reliability; systems analysis; stochastic processes. *Mailing Add:* Statist Eng Div Nat Standards & Technol Gaithersburg MD 20899

LECHNER, JOHN FRED, b Holyoke, Colo, Oct 27, 42; m. CELL & MOLECULAR BIOLOGY, TOXICOLOGY. *Educ:* Cornell Univ, BS, 64; Hahnemann Med Col, PhD(microbiol), 70. *Prof Exp:* Grad teaching asst, Ind Univ, Bloomington, Iowa, 64-65, microbiol training fel, 65-66; res asst, Div Genetics, Hahnemann Med Col, Philadelphia, 66-69, from instr to sr instr, 69-71, asst prof, Div Genetics & mem, Grad Sch Fac, 71-73; res assoc, Mass Inst Technol, Cambridge, 73-75; res fel, W Alton Jones Cell Sci Ctr, Lake Placid, NY, 75-76; re investr, Pasadena Found Med Res, 76-78; asst dir, Prostate Cancer Lab, Pasadena, 78-79; chief, Cell Cult & Media Br, Ctr Infectious Dis, Centers Dis Control, Atlanta, 84-85; expert scientist, Nat Cancer Inst, NIH,

79-83, sr staff fel, 83-84, chief, In Vitro Carcinogenesis Sect, Lab Human Carcinogenesis, 85-91, dep lab chief, 90-91; ASST DIR CELL & MOLECULAR BIOL, INHALATION TOXICOL RES INST, ALBUQUERQUE, NMEX, 91- *Concurrent Pos:* Nat res serv award, Pub Health Serv, 75; mem, Handicapped Employee Adv Comt, NIH, 82 & Cellular Physiol Grant Rev Study Sect, 82-84, Prostate Cancer Task Force, Nat Cancer Inst, 86-89; chmn, Comt Carcinogenesis Initiative, Prostate Cancer Working Group, 86-87 & Tumor Tissue Request Rev Comt, Nat Dis Interchange, 87-90; consult, Organogenesis, Inc, Mass, 88-90 & Clonetics Inc, Calif, 89-91. *Mem:* Am Asn Cancer Res; Am Soc Cell Biol; Tissue Cult Asn; AAAS; Am Asn Pathologists; Fedn Am Socs Exp Biol. *Res:* Cell biology-growth factor, nutritional and hormonal control of growth and differentiation; in vitro carcinogenesis of human epithelial cells. *Mailing Add:* Inhalation Toxicol Res Inst PO Box 5890 Albuquerque NM 87185

LECHNER, JOSEPH H, b Boston, Mass, Nov 13, 51; m 81; c 3. DNA SEQUENCE, MAMMALIAN STRUCTURAL PROTEINS. *Educ:* Roberts Wesleyan Col, BS, 72; Univ Iowa, PhD(biochem), 77. *Prof Exp:* Teaching fel, Dent Sch, Northwestern Univ, 78-79; from asst prof to assoc prof, 79-87, PROF CHEM, MT VERNON NAZARENE COL, 87- *Mem:* Am Chem Soc; Am Sci Affil. *Res:* Determination of nucleic acid sequences at or near sites of radiation-induced mutations. *Mailing Add:* 800 Martinsburg Rd Mt Vernon OH 43050

LECHNER, ROBERT JOSEPH, b Danville, Pa, Oct 16, 31; m 55; c 7. APPLIED MATHEMATICS, INFORMATION SCIENCE. *Educ:* Carnegie Inst Technol, BS, 52, MS, 53; Harvard Univ, PhD(appl math), 63. *Prof Exp:* Sr engr, Sylvania Electronics Corp, 55-57, res engr, 57-59, adv res engr, 59-61, eng specialist, 61-63, sr eng specialist, 63-70; sr engr, Honeywell Info Systs, 70-75; sr engr, C S Draper Lab, 75-76; assoc prof elec eng, Northeastern Univ, 76-; PROF, UNIV LOWELL. *Mem:* Inst Elec & Electronics Engrs; Asn Comput Mach. *Res:* Applications of modern algebra to the information sciences; information processing and communications systems; programming languages; software engineering; computer architecture. *Mailing Add:* Comput Sci Dept Univ Lowell WL 205 One University Ave Lowell MA 01854

LECHOWICH, RICHARD V, b Chicago, Ill, June 23, 33; m 57, 83; c 5. FOOD SCIENCE, FOOD MICROBIOLOGY. *Educ:* Univ Chicago, AB, 52, MS, 55; Univ Ill, PhD(food sci), 58. *Prof Exp:* Microbiologist, Am Meat Inst Found, Ill, 52-55; res asst food sci, Univ Ill, Urbana, 55-58; res microbiologist, Continental Can Co, Inc, Ill, 58-63; asst prof food sci, Mich State Univ, 63-66, from assoc prof to prof, 66-71; prof food sci & technol & head dept, Va Polytech Inst & State Univ, 71-81; mgr microbiol, Gen Foods Corp, 81-87; exec vpres, ABC Res Lab, Fla, 87-89; DIR, NAT CTR FOOD SAFETY & TECHNOL, 89- *Mem:* AAAS; Int Asn Milk, Food & Environ Sanitarians; fel Inst Food Technologists; Am Soc Microbiologists; Brit Soc Appl Bact. *Res:* Food safety mechanisms of bacterial spore formation and germination; chemical composition and thermal resistance phenomena of bacterial spores; food poisoning microorganisms, especially Clostridium botulinum; thermal and radiation resistances of microorganisms. *Mailing Add:* Nat Ctr Food Safety & Technol 6502 S Archer Ave Summit IL 60501

LECHOWICZ, MARTIN JOHN, b Chicago, Ill, Feb 23, 47; m 77. PLANT ECOLOGY, EVOLUTIONARY BIOLOGY. *Educ:* Mich State Univ, BA, 69; Univ Wis-Madison, MSc, 73, PhD(plant ecol), 76. *Prof Exp:* Lectr ecol, Dept Bot, Univ Wis-Madison, 75; asst prof, 76-80, assoc prof plant ecol, 81-88, PROF, MCGILL UNIV, 88- *Concurrent Pos:* Nat Res Coun Can operating grant, 77-80, strategic grant, 80-86; Can Jour, Bot, 83, Evolutionary Biol, 87; Environ Can contracts, 79-82; Atmospheric Environ Serv grant, 81- *Mem:* Ecol Soc Am; Can Bot Asn; Soc Study Evolution; Am Soc Naturalists; Europ Soc Evolutionary Biol. *Res:* Physiological ecology and evolutionary ecology of plants, and particularly resource uptake and allocation; plant herbivore interactions. *Mailing Add:* Dept Biol 1205 Ave Dr Penfield Montreal PQ H3A 1B1 Can

LECHTENBERG, VICTOR L, b Butte, Nebr, Apr 14, 45; m 67; c 4. AGRONOMY. *Educ:* Univ Nebr, BS, 67; Purdue Univ, PhD(agron), 71. *Prof Exp:* From instr to prof agron, 69-82, assoc dir, Agr Exp Sta, 82-89, EXEC ASSOC DEAN AGR, PURDUE UNIV, WEST LAFAYETTE, 89- *Honors & Awards:* Ciba-Geigy Award, Am Soc Agron. *Mem:* Fel Am Soc Agron; fel Crop Sci Soc Am (pres); Am Forage & Grassland Coun; AAAS; Coun Agr Sci & Technol. *Res:* Factors that affect forage crop quality and utilization; environmental physiology of forage crops; genetic improvement of crop quality. *Mailing Add:* Office of Dean AGAD Bldg Purdue Univ West Lafayette IN 47907

LECHTMAN, MAX D, b Providence, RI, Apr 24, 35; m 62; c 3. MICROBIOLOGY. *Educ:* Univ RI, AB, 57; Univ Mass, MS, 59; Univ Southern Calif, PhD(microbiol), 68. *Prof Exp:* Microbiologist, Douglas Aircraft Co, Calif, 61-62 & Res & Develop Div, Magna Chem Co, 62-64; instr microbiol, Univ Southern Calif, 64; microbiol consult, Garrett Corp, 64-65, microbiologist, AiRes Mfg Co Div, 65-67; mem tech staff aerospace microbiol, Autonetics Div, NAm, Rockwell Corp, Anaheim, 67-71; INDUST CONSULT, 71- *Concurrent Pos:* Instr, Calif Community Col Syst, 71- *Mem:* Am Soc Microbiol; Soc Indust Microbiol. *Res:* Microbial cytology, cytochemistry and physiology; bioluminescent bacteria for detection of toxic chemicals; laboratory and medical instrumentation. *Mailing Add:* 8641 Delray Circle Westminster CA 92683

LECK, CHARLES FREDERICK, b Princeton, NJ, June 20, 44. ORNITHOLOGY, ECOLOGY. *Educ:* Muhlenberg Col, BS, 66; Cornell Univ, PhD(vert zool), 70. *Prof Exp:* Vis res assoc, Smithsonian Trop Res Inst, 68-69; dir ecol prog, 74-78, ASSOC PROF ECOL, RUTGERS UNIV, NEW BRUNSWICK, 70- *Concurrent Pos:* Vis fac mem, West Indies Lab, Smithsonian Trop Res Inst, 70-73. *Mem:* Wilson Ornith Soc; Cooper Ornith Soc; Am Ornith Union; Sigma Xi; Asn Trop Biologists. *Res:* Avian ecology and landscape ecology and conservation; tropical biology. *Mailing Add:* Dept Zool Rutgers Univ New Brunswick NJ 08903

LECKIE, FREDERICK ALEXANDER, b Dundee, Scotland, Mar 26, 29; m 57; c 3. ENGINEERING MECHANICS, MATERIALS ENGINEERING. *Educ:* Univ St Andrew, BSc, 49; Stanford Univ, MS, 55, PhD(eng mech), 58. *Prof Exp:* Civil engr, Mott, Hay & Anderson, Westminister, London, 49-51; res asst mech, Tech Hochsch, Hannover, Ger, 57-58; lectr, Univ Cambridge, 58-68; prof eng, Univ Leicester, 68-78; prof theoret & appl mech & mech eng, Univ Ill, Urbana-Champaign, 78-88; PROF MECH ENG, UNIV CALIF, SANTA BARBARA, 88- *Mem:* Am Soc Mech Engrs; Am Acad Mech. *Res:* Properties of load bearing mechanical components operating at elevated temperatures; creep rupture and fractures of materials at elevated temperatures. *Mailing Add:* Dept Mech Eng Univ Calif Santa Barbara CA 93106

LECKLITNER, MYRON LYNN, b Canton, Ohio, June 16, 42; c 1. NUCLEAR MEDICINE, MEDICAL ULTRASOUND. *Educ:* Pa State Univ, BS, 64; Univ Ala, Tuscaloosa, BS, 70, Birmingham, MD, 74; Am Bd Nuclear Med, cert, nuclear med, 82. *Prof Exp:* Resident med, Lloyd Noland Hosp, Birmingham, 74-77, nuclear med, 77-79; asst prof radiol, Univ Tex, San Antonio, 79-83; assoc prof radiol & dir diag imaging, 83-86, PROF, UNIV SALA, 86-, SR SCIENTIST, CANCER CTR, 84- *Concurrent Pos:* Mem, Acad Coun, Soc Nuclear Med, 81-; vis prof, Univ Nuevo Leon, Monterrey, Mex, 83, Univ Oxford, UK, 85, 88, & Royal Postgrad Med Sch, Univ London, UK, 85; Am Col Nuclear Physicians, 86-88 & mem bd regents & treas, chmn finance comt, 86-88. *Mem:* Soc Nuclear Med; Am Col Nuclear Med; Am Inst Ultrasound Med; Am Col Radiol; AMA; Radiol Soc Am; Am Col Nuclear Physicians; fel Am Col Nuclear Med. *Res:* Basic science and clinical investigations involving human biokinetics and quality control of radiotracers; relation of physiology and pathophysiology to nuclear medicine. *Mailing Add:* Dept Radiol Univ SAla 2451 Fillingim St Mastin 301 Mobile AL 36617

LECKONBY, ROY ALAN, b Bethlehem, Pa, Aug 1, 49; m 71; c 2. PHYSICAL ORGANIC CHEMISTRY, SYNTHETIC ORGANIC CHEMISTRY. *Educ:* Hamilton Col, AB, 71; Univ Rochester, MS, 74, PhD(chem), 76. *Prof Exp:* Fel res, Robert A Welch Found, Rice Univ, 76-77; sr res chemist, Akron, Ohio, 77-82, plant chemist, La Porte, Tex, 82-86, QUAL ASSURANCE MGR, GOODYEAR TIRE & RUBBER CO, APPLE GROVE, WV, 86- *Mem:* Am Soc Qual Control. *Res:* Organic synthesis; age resistors for rubbers and plastics; monomers; oxidation of organic chemicals. *Mailing Add:* Goodyear Tire & Rubber Co State Rte 2 Apple Grove WV 25502

LECKRONE, DAVID STANLEY, b Salem, Ill, Nov 30, 42; m 64; c 2. ASTROPHYSICS. *Educ:* Purdue Univ, BS, 64; Univ Calif, Los Angeles, MA, 66, PhD(astron), 69. *Prof Exp:* Astrophysicist, 69, SR ASTROPHYSICIST, LAB ASTRON & SOLAR PHYSICS, GODDARD SPACE FLIGHT CTR, NASA, 69- *Mem:* Am Astron Soc; Int Astron Union. *Res:* Ultraviolet stellar spectroscopy and photometry from space vehicles; magnetic and chemically peculiar stars; stellar atomspheres; abundances of the elements in astronomical objects; instrumentation for space astronomy. *Mailing Add:* Code 681 NASA-Goddard Space Flight Ctr Greenbelt MD 20771

LECLAIRE, CLAIRE DEAN, b Huron, SDak, Aug 26, 10; m 41. ORGANIC CHEMISTRY. *Educ:* SDak State Sch Mines, BS, 33; Univ Minn, PhD(org chem), 39. *Prof Exp:* Chemist, Coal Res Lab, Carnegie Inst Technol, 38-41 & Rohm and Haas Co, Pa, 41-42; rubber fel, Mellon Inst, 42-47; res chemist, Firestone Tire & Rubber Co, 47-49; asst dir rubber res proj, Univ Minn, 49-51; sect leader appl res, Adhesives & Coatings Div, Minn Mining & Mfg Co, 51-58; sr res chemist, Int Latex Corp, 58-69 & Standards Brands Chem Industs, Inc, 69-76; res chemist, Reichhold Polymers Inc, 76-81; consult, Int Playtex Inc, 81-83; RETIRED. *Mem:* AAAS; Am Chem Soc. *Res:* Organic synthesis; structure; polymerization; properties and utilization of high polymers; bituminous coals; elastomers; resins; adhesives. *Mailing Add:* 134 Lakeview Ave Dover DE 19901

LECOURS, MICHEL, b Montreal, Que, Aug 1, 40; m 66; c 3. COMMUNICATIONS SYSTEMS, SIGNAL PROCESSING. *Educ:* Univ Montreal, BScA, 63; Univ London, DIC & PhD(electronics), 67. *Prof Exp:* Engr commun, Bell Can, 63; head, Elec Eng Dept, 75-77, PROF ELEC ENG, UNIV LAVAL, 67-, vdean, Fac Sci & Eng, 77-85. *Concurrent Pos:* Mem sci staff, Bell Northern Res, 71-72; consult, Lab-Volt, Inc, 82-; vis res, NTT-ECL, Japan, 86. *Honors & Awards:* Can Award, Excellent Technol Transfer, 86; Annual Mert Award, Ecole Polytechnique Montreal Alumni Asn, 87. *Mem:* Can Soc Elec Eng; sr mem Inst Elec & Electronics Engrs. *Res:* Performance of digital communications systems (including mobile) in non-gaussian noise, interference and fading; signal processing in the radio and audio frequency bands. *Mailing Add:* Fac Sci & Eng Laval Univ Quebec PQ G1K 7P4 Can

L'ECUYER, JACQUES, b St-Jean, Que, Mar 6, 37; m 59; c 3. APPLIED & NUCLEAR PHYSICS. *Educ:* Col St Jean, BA, 56; Univ Montreal, BSc, 59, MSc, 61, PhD(physics), 66. *Prof Exp:* Lectr, Univ Montreal, 61-63 & Univ Sherbrooke, 63-64; asst prof physics, Laval Univ, 64-67; Nat Res Coun Can fel, Oxford Univ, 67-69; from asst prof to assoc prof physics, Univ Montreal, 69-73, prof, 73-86; pres, Quebec Univ Coun, 81-88; VPRES TEACHING & RES, UNIV QUE, 88- *Concurrent Pos:* Mem bd gov, Univ Montreal, 76-79; mem, Quebec Univs Coun, 77-81 & Hong Kong Coun Acad Accreditation, 88- *Mem:* Can Asn Physicists; Fr-Can Asn Advan Sci. *Res:* Experimental nuclear physics. *Mailing Add:* 10890 Berri Montreal PQ H3L 2H5 Can

L'ECUYER, MEL R, b Concordia, Kans, June 4, 36; m 62. MECHANICAL ENGINEERING, HEAT TRANSFER. *Educ:* Purdue Univ, BS, 59, MS, 60, PhD(mech eng), 64. *Prof Exp:* Res engr, Jet Propulsion Ctr, 60-64, from asst prof to assoc prof, 64-76, PROF MECH ENG, PURDUE UNIV, 76- *Concurrent Pos:* Sr eng specialist, LTV Aerospace Div, Ling-Temco-Vought, Inc, Tex, 68- *Mem:* Am Inst Aeronaut & Astronaut; Am Soc Mech Engrs; Am Soc Eng Educ. *Res:* Mass transfer cooling; two-phase flow; propulsion gas dynamics. *Mailing Add:* Dept Mech Eng Purdue Univ West Lafayette IN 47907

LECZYNSKI, BARBARA ANN, b Lowell, Mass, Aug 27, 54. HUMAN & ENVIRONMENTAL MONITORING, QUALITY ASSURANCE. *Educ:* Univ Lowell Mass, BS, 76; Univ Conn, MS, 79. *Prof Exp:* Math statician, US Dept Labor, Bur Labor Statist, 78-80; appl math, Eastman Kodak Co, 80-84; proj mgr, Battelle Mem Inst, 84-89; proj mgr, Wash Consult Group, 89-90; SR CONSULT, DAVID C COX & ASSOCS, 90- *Mem:* Am Statist Asn; Am Soc Qual Control. *Res:* Participated on several multi-disciplinary projects with applications to environmental monitoring in particular human monitoring exposure based assessments and asbestos abatement studies; consumer preference and marketing strategies and economic indicators; developed sampling designs and quality assurance plans for field studies; provided experimental designs for industrial manufactures research projects. *Mailing Add:* 8497 Lazy Creek Ct Springfield VA 22153

LEDBETTER, HARVEY DON, b Pierson, Ill, June 26, 26; m 47; c 3. MATERIALS SCIENCIE ENGINEERING. *Educ:* Univ Ariz, BS, 49, MS, 50; Univ Tenn, PhD(chem), 54. *Prof Exp:* Asst gen chem, Univ Tenn, 50-52; res group leader, Dow Chem USA, 53-67, res supvr, 67-71, tech mgr, Designed Prod Dept, 71-76, lab dir, Cent Res plastics Lab, 76-81; dir technol res & develop, 81-83, sr proj mgr, 83-89, RES SCIENTIST, CENTRAL RES, 90- *Mem:* AAAS; Am Chem Soc; Soc Plastics Engrs; Sigma Xi; Soc Advan Math & Process Eng; Fiber Soc. *Res:* Polymer nucleation and stabilization; vacuum processes; preparation and physics of composite structures; fiber processing. *Mailing Add:* Lake Pinehurst Villas No 14 Pinehurst NC 28374

LEDBETTER, JEFFREY A, IMMUNOTHERAPY. *Prof Exp:* RES FEL, DEPT IMMUNOTHER, ONCOGEN, 81- *Mailing Add:* Dept Immunother Oncogen 3005 First Ave Seattle WA 98121

LEDBETTER, JOE O(VERTON), b New Hope, Ala, Feb 1, 27; m 47. ENVIRONMENTAL HEALTH, CIVIL ENGINEERING. *Educ:* Univ Ala, BSCE, 50; Univ Tex, MS, 58, PhD(civil eng), 63. *Prof Exp:* Proj engr, Ala Hwy Dept, 50-51; resident engr, Tex Hwy Dept, 51-56, from instr to assoc prof, 56-71, PROF CIVIL ENG, UNIV TEX, AUSTIN, 71- *Concurrent Pos:* Pres, US Pub Health Serv, 50- *Mem:* Am Indust Hyg Asn; Health Physics Soc. *Res:* Air pollution control; sampling, evaluation, and abatement; disposal of solid radioactive and hazardous wastes; radiological and industrial health engineering; air pollution from wastewater treatment. *Mailing Add:* Univ Tex 8-6 Cockrell Hall Austin TX 78712-1104

LEDBETTER, MARY LEE STEWART, b Monterrey, Mex, Aug 30, 44; US citizen; m 66; c 2. CELL BIOLOGY, BIOCHEMICAL GENETICS. *Educ:* Pomona Col, BA, 66; Rockefeller Univ, PhD(genetics), 72. *Prof Exp:* Res assoc microbiol, Dartmouth Med Sch, 72-75, instr, 75-78, res assoc psychiat, 77-79, res asst prof biochem, 79-80; asst prof, 80-85, ASSOC PROF BIOL, COL HOLY CROSS, 85- *Concurrent Pos:* USPHS res trainee, Sch Med, NY Univ, 72; Damon Runyon-Walter Winchell fel, 72-73; Leukemia Soc Am Fel, Dartmouth Med Sch, 73-75 & 77-78; USPHS, NCI fel, Dartmouth Med Sch, 75-77; mem cell biol adv panel, NSF, 86-90. *Mem:* Am Soc Cell Biol; AAAS; Sigma Xi; Asn Women Sci. *Res:* Somatic cell genetics; study of metabolic regulation in cultured mammalian cells; cell communication through gap junctions. *Mailing Add:* Dept Biol PO B Col Holy Cross Worcester MA 01610

LEDBETTER, MYRON C, b Ardmore, Okla, June 25, 23. CELL BIOLOGY, BOTANY. *Educ:* Okla State Univ, BS, 48; Univ Calif, MA, 51; Columbia Univ, PhD, 58. *Prof Exp:* Fel plant physiol, Boyce Thompson Inst, 53-57, asst plant anatomist, 57-60; guest investr & fel, Rockefeller Inst, 60, res assoc, 61; res assoc, Harvard Univ, 61-65; cell biologist, Brookhaven Nat Lab, 65-74, sr cell biologist, 74-89; RETIRED. *Concurrent Pos:* Guest assoc, Brookhaven Nat Lab, 58-60; fel training grant prog, USPHS, 60-61; consult, Celanese Res Corp, 81-82; bd mgr, NY Bot Garden, 82-84; adj prof, City Univ NY, 82- *Mem:* AAAS; Am Soc Cell Biol; Bot Soc Am; Electron Micros Soc Am (pres, 78); Am Inst Biol Sci; Torrey Bot Club (pres, 84-85). *Res:* Morphology of physiologically dwarfed tree seedlings; feeding damage to plant tissues by lygus bugs; histopathology of ozone on plants; distribution of fluorine in plants; plant fine structure, microtubules in plants; plant cell walls; electron microscopy; structure of macromolecular complexes by cluster ion bombardment. *Mailing Add:* PO Box 145 Port Jefferson NY 11777-0145

LEDBETTER, STEVEN R, EXPERIMENTAL BIOLOGY. *Prof Exp:* SR RES SCIENTIST, CANCER & INFECTIOUS DIS, UPJOHN CO, 84- *Mailing Add:* Upjohn Co 301 Henrietta St Kalamazoo MI 49001

LEDBETTER, W(ILLIAM) B(URL), b El Paso, Tex, Sept 15, 34; c 4. CIVIL ENGINEERING. *Educ:* Tex A&M Univ, BS, 56; Univ Tex, Austin, PhD(civil eng), 64. *Prof Exp:* From asst prof to assoc prof, 64-71, from asst res engr to assoc res engr, 64-71, PROF CIVIL ENG & RES ENGR, TEX A&M UNIV, 71- *Concurrent Pos:* Chmn Transp Res Bd Comt, Nat Res Coun-Nat Acad Sci, 70-79. *Honors & Awards:* K B Woods Award, Transp Res Bd, Nat Acad Sci, 76. *Mem:* Nat Soc Prof Engrs; Am Soc Civil Engrs; Am Soc Eng Educ; Am Soc Testing & Mat; Am Concrete Inst; Am Soc Qual Control. *Res:* Quality management; project management; construction materials; concrete materials engineering. *Mailing Add:* Dept Civil Eng Clemson Univ Main Campus Clemson SC 29634

LEDDY, JAMES JEROME, b Detroit, Mich, July 28, 29; m 51; c 9. INDUSTRIAL CHEMISTRY. *Educ:* Univ Detroit, BS, 51; Univ Wis, PhD(chem), 55. *Prof Exp:* Chemist inorg res, Dow Chem USA, 55-59, proj leader, 59-60, proj leader, Electrochem & Inorg Res Lab, 60-63, sr res chemist, 63-67, assoc res scientist, 67-71, res scientist, Electrochem & Inorg Res Lab, 71-73, res scientist, Mich Div Inorg Res, 73-80, tech mgr, Chlor-Alkali Technol Ctr, 80-85, SR RES SCIENTIST, CENTRAL RES, DOW CHEM USA, 85- *Concurrent Pos:* Asst prof, Assumption Univ, 59-60. *Honors & Awards:* Herbert H Dow Medal, 87. *Mem:* AAAS; Am Chem Soc; Electrochem Soc; Am Ceram Soc. *Res:* Chemistry of less familiar elements, especially titanium, zirconium and hafnium; coordination compounds; unfamiliar oxidation states; amalgam chemistry; electrochemistry; inorganic polymers; industrial inorganic and electrochemistry; chlor-alkali; engineering ceramics and material science. *Mailing Add:* 311 Cherryview Dr Midland MI 48640-5559

LEDDY, JOHN PLUNKETT, b New York, NY, Sept 10, 31; m 56; c 3. IMMUNOLOGY. *Educ:* Fordham Univ, BA, 52; Columbia Univ, MD, 56. *Prof Exp:* From intern to resident internal med, Boston City Hosp, Harvard Med Serv, 56-59; USPHS trainee hemat, Med Ctr, Univ Rochester, 59-60; sr instr, 62-64, from asst prof to assoc prof med, 64-73, PROF MED & MICROBIOL, MED CTR, UNIV ROCHESTER, 73-, DIR CLIN IMMUNOL UNIT, 70- *Concurrent Pos:* Res med officer immunochem, Walter Reed Army Inst Res, 60-62; Nat Found fel, 62-64; NIH res grant, 65-; sr investr, Arthritis Found, 65-70; dir, USPHS Training Grant, 70-; prog dir, NIH Specialized Res Ctr grant immunol, 78- *Mem:* Am Soc Clin Invest; Am Asn Immunologists. *Res:* Biology of complement system in man; erythrocyte autoantibodies in human diseases. *Mailing Add:* Dept Med Univ Rochester Med Ctr 601 Elmwood Ave Box 695 Rochester NY 14642

LEDDY, SUSAN, b Jersey City, NJ, Feb 23, 39; m 72; c 2. NURSING. *Educ:* Skidmore Col, BS, 60; Boston Univ, MS, 65; NY Univ, PhD(nursing), 73. *Prof Exp:* Instr nursing, Mt Auburn Hosp, 62-65 & New Rochelle Hosp, 65-66; asst prof, Columbia Univ, 66-70 & Pace Univ, 73-75; consult nursing educ, Nat League Nursing, 75; chairperson & prof nursing, Mercy Col, 76-81; dean & prof nursing, 81-84, DEAN, COL HEALTH SCI, UNIV WYO, 84- *Concurrent Pos:* Independent consult, 75- *Mem:* Nat League Nursing (vpres, 85-87). *Res:* Biological rhythms; nursing education and curriculum; health care delivery. *Mailing Add:* Dir Nursing Widener Univ Chester PA 19013

LEDEEN, ROBERT, b Denver, Colo, Aug 19, 28; m 82. BIOCHEMISTRY, NEUROBIOLOGY. *Educ:* Univ Calif, Berkeley, BS, 49; Ore State Univ, PhD(org chem), 53. *Prof Exp:* Fel, Univ Chicago, 53-54; res chemist, Mt Sinai Hosp, New York, 56-59; res chemist, Albert Einstein Col Med, 59-61, from asst prof to assoc prof biochem in neurol, 62-75, prof biochem in neurol, 75-91; PROF NEUROSCI, UNIV MED & DENT NJ, 91- *Concurrent Pos:* Dep chief ed, J Neurochem, 82-88; mem, Neurol Sci Study Sect, NIH, 76-80, Nat Mult Sclerosis Soc Study Sect, 89- *Honors & Awards:* Jacob Javits Award, 87; Alexander von Humboldt Prize, 88. *Mem:* AAAS; Am Chem Soc; Am Soc Biol Chem; Am Soc Neurochem; Int Soc Neurochem; NY Acad Sci. *Res:* Biochemistry of the nervous system; gangliosides and other lipids of the nervous system; myelin lipids; myelin enzymology and pharmacology. *Mailing Add:* Dept Neurosci Univ Med & Dent NJ 185 S Orange Ave Newark NJ 07103-2757

LEDER, FREDERIC, b New York, NY, Nov 1, 39; m 71; c 2. CHEMICAL ENGINEERING. *Educ:* Queens Col, BS, 61; Columbia Univ, BS, 61; Yale Univ, MS, 63, PhD(chem eng), 65. *Prof Exp:* Res engr, Esso Res & Eng Co, Exxon Corp, Linden, 65-68, sr res engr, 68-73, gas separations res assoc, 73; dir explor res, Occidental Res Corp, Irvine, 74-76; dir res, 76-85, MANAGING PARTNER, SPECIALTY CAPITAL GROUP, CITIES APPL RES & TECHNOL CTR, 85- *Mem:* Am Inst Chem Eng; Sigma Xi. *Res:* Phase equilibria at high pressure; gas separations. *Mailing Add:* 4219 E 87th St Tulsa OK 74137

LEDER, IRWIN GORDON, b New York, NY, June 16, 20; m 45; c 3. BIOCHEMISTRY. *Educ:* Brooklyn Col, AB, 42; NY Univ, MS, 47; Duke Univ, PhD(biochem), 51. *Prof Exp:* Res assoc, Duke Univ, 51; USPHS fel, NY Univ, 51-52 & Yale Univ, 52-53; asst, Pub Health Res Inst, Inc, NY, 53-54; BIOCHEMIST, NAT INST ARTHRITIS, METAB & DIGESTIVE DIS, BETHESDA, 54- *Mem:* Fedn Am Socs Exp Biol. *Res:* Niacin metabolism; pyridine nucleotide synthesis; intermediary carbohydrate metabolism; enzymology. *Mailing Add:* Nat Inst Arthritis Metab & Digestive Dis Bldg 8A Rm 2A24 NIH Bethesda MD 20892

LEDER, LEWIS BEEBE, b Brooklyn, NY, Oct 28, 20; m 48; c 1. EXPERIMENTAL PHYSICS. *Educ:* Univ Idaho, BS, 43. *Prof Exp:* Instr physics, Williams Col, 43-44; physicist gas flow, Inst Gas Technol, 44-45; physicist, Dielectric Studies, Indust Condenser Corp, 45-46; physicist electron heat controls, Wheelco Instruments Co, 46-48; physicist nuclear physics, Inst Nuclear Studies, 48-51; physicist electron physics, Nat Bur Standards, 51-62; physicist, Appl Res Lab, Philco Corp, 62-65; physicist electron scattering & superconductivity, Ford Sci Lab, 65-69; res scientist, Xerox Corp, 69-83; RETIRED. *Mem:* Am Vacuum Soc; Am Phys Soc. *Res:* Electron scattering; superconductivity; thin films preparation and properties; photoreceptors. *Mailing Add:* 1505 Oaks Chapel Hill NC 27514

LEDER, PHILIP, b Washington, DC, Nov 19, 34; m 59; c 3. MOLECULAR GENETICS. *Educ:* Harvard Col, AB, 56, Med Sch, MD, 60. *Hon Degrees:* DSc, Yale Univ, 84, Mt Sinai Col Med, 85. *Prof Exp:* Lab chief molecular genetics, Nat Inst Child Health & Human Develop, NIH, 72-80; PROF & CHMN GENETICS, HARVARD MED SCH, 80- *Honors & Awards:* Richard Lounsberry Award, Nat Acad Sci; Nat Medal Sci,89. *Mem:* Nat Acad Sci; Inst Med-Nat Acad Sci; Genetics Soc Am; Am Acad Arts & Sci; Am Soc Biol Chemists. *Res:* Molecular biology and genetics. *Mailing Add:* Dept Genetics Harvard Med Sch 25 Shattuck St Boston MA 02115

LEDERBERG, ESTHER MIRIAM, b New York, NY, Dec 18, 22; m 46, 68. GENETICS, MICROBIOLOGY. *Educ:* Hunter Col, AB, 42; Stanford Univ, MA, 46; Univ Wis, PhD(genetics), 50. *Prof Exp:* Proj assoc genetics, Univ Wis, 50-59; res geneticist, Med Sch, 59-68, res assoc, 68-71, sr scientist, 71-74, res prof, 74-85, EMER PROF MED MICROBIOL, STANFORD UNIV, 85- *Concurrent Pos:* Fulbright fel, Australia, 57; Am Cancer Soc Sr Dernham fel, 68-70; dir, Plasmid Ref Ctr, 76-86; consult, Molecular Biol Comput Res Resource, 86- *Honors & Awards:* Co-recipient, Pasteur Award, Soc Ill Bact, 56. *Mem:* Fel AAAS; Genetics Soc Am; Brit Soc Gen Microbiol; Sigma Xi. *Res:* Genetics of microorganisms; lysogenicity; bacterial recombination and transformation; DNA repair; phase variation of Flagellar antigens in Salmonella; R plasmids. *Mailing Add:* Dept Med Microbiol No 5402 Stanford Univ Stanford CA 94305-2499

LEDERBERG, JOSHUA, b NJ, May 23, 25; m 68; c 2. GENETICS. *Educ:* Columbia Univ, BA, 44; Yale Univ, PhD(microbiol), 47. *Hon Degrees:* ScD, Yale Univ, 60, Columbia Univ, 67, Univ Wis, 67, Albert Einstein Col Med, 70, Mt Sinai, 79, Rutgers Univ, 81 & NY Univ, 84; MD, Univ Turin, 69 & Tufts Univ, 85; LLD, Univ Pima, 79; DLitt, Jewish Theol Seminary, 79. *Prof Exp:* From asst prof to prof genetics, Univ Wis, 47-58, prof med & genetics & chmn dept, 58-59; prof genetics & chmn dept, Med Sch, Stanford Univ, 59-78; PRES, ROCKEFELLER UNIV, NEW YORK, 78- *Concurrent Pos:* Dir, Ctr Advan Study Behav Sci, 75-81; mem, Annual Reviews, Inc & bd of trustees, 72, Natural Resources Def Coun, 72-84; sci consult, Cetus Corp, Berkeley, Calif, 71-; coun mem, Inst Med-Nat Acad Sci, 78-81; bd dir, Charles Babbage Indust, Palo Alto, Calif, 78-, Chem Indust Inst Toxicol, NC, 80-, Inst Sci Info, Philadelphia, 62-, US Defense Sci Bd, 81- & Proctor & Gamble, 84-; mem bd, Revson Found, 86-, Dreyfus Found, 83-, Carnegie Corp, 85-, NYC Partnership, 85- *Honors & Awards:* Nobel Prize, 58; Nat Medal Sci, 89. *Mem:* Nat Acad Sci; Inst Med-Nat Acad Sci; foreign mem The Royal Soc. *Res:* Molecular genetics and evolution; science policy; computer science. *Mailing Add:* Rockefeller Univ New York NY 10021-6399

LEDERBERG, SEYMOUR, b New York, NY, Oct 30, 28; m 59; c 2. MICROBIOLOGY, GENETICS. *Educ:* Cornell Univ, BA, 51; Univ Ill, PhD(bact), 55. *Prof Exp:* Am Cancer Soc fel, Univ Calif, 55-57, vis asst prof bact, 57-58; from asst prof to assoc prof, 58-66, chmn microbiol sect, 70-78, coordr, grad progs biol med, 78-85, PROF BIOL, BROWN UNIV, 66-, ASSOC DEAN, GRAD STUDIES BIOL MED, 85- *Concurrent Pos:* USPHS fel, Inst Biol Phys Chem, Paris, 65-66; consult, Nat Inst Gen Med Sci, 70-74, Gen Med & Sci Adv Coun, Cystic Fibrosis Found, 72-76, Comt Genetic Screening Cystic Fibrosis, Nat Acad Sci-Nat Res Coun, 74-75 & Nat Inst Arthritis, Metab & Digestive Diseases, 75-78; vis scholar, Harvard Law Sch; lectr, Law Sch Ctr for Law & Health Sci, Boston Univ, 73-77, adj prof, pub health law, 77- *Mem:* Am Soc Microbiol; Genetics Soc Am; Am Soc Human Genetics; Environ Mutagen Soc. *Res:* Human, microbial and viral genetics; cystic fibrosis; yeast and bacterial biochemistry; chromosomal, microtubule and cell surface functions in cell cycles; environmental mutagens; intracellular pathogens. *Mailing Add:* Biol & Med Box G-J364 Brown Univ Providence RI 02912

LEDERER, C MICHAEL, b Chicago, Ill, June 6, 38; m 70; c 2. NUCLEAR CHEMISTRY. *Educ:* Harvard Univ, AB, 60; Univ Calif, Berkeley, PhD(nuclear chem), 64. *Prof Exp:* Dir isotopes proj, Lawrence Berkeley Lab, 64-78, head, Info & Data Anal Dept, 78-80, DEP DIR, RADIATION MEASUREMENT, NUCLEAR ENG DEPT, ENERGY INST, UNIV CALIF, BERKELEY, 80- *Concurrent Pos:* Mem, US Nuclear Data Comt, 70-78. *Mem:* AAAS; Am Phys Soc; Am Nuclear Soc; Sigma Xi. *Res:* Energy production, use and environmental effects; experimental nuclear structure physics. *Mailing Add:* UERG Bldg T-9 Univ Calif Berkeley CA 94720

LEDERER, JEROME F, b New York, NY, Sept 26, 02; c 2. AEROSPACE SAFETY. *Educ:* NY Univ, BS, 24, ME, 25. *Prof Exp:* Aeronaut engr, US Air Mail Serv, 26-27, dir aeronaut tech, 27-29; chief engr, Aero Ins Underwriters, 29-40 & 44-48; dir safety bur, Civil Aeronaut Bd, 40-42, airlines war training inst, 42-44; tech dir, Flight Safety Found, 48-67; dir manned space flight safety, NASA, 67-70, dir safety, 70-72; RETIRED. *Concurrent Pos:* James Jackson Cabot prof lectr, Norwich Univ, 39; dir, Cornell-Guggenheim Aviation Safety Ctr, 50-68; adj prof, Inst Safety & Systs Mgt, Univ Southern Calif, 74-; emer pres, Flight Safety Found. *Honors & Awards:* Arthur Williams Award, 53; Von Baumhauer Medal, 54; Monsanto Award, 57; Guggenheim Medal, 61; Laura Tabor Barbour Medal, 64; Wright Bros Award, 65; Amelia Earhart Medal; Gargarin Medal, Soviet Fedn Cosmonauts. *Mem:* Nat Acad Eng; Soc Automotive Engr; Am Soc Mech Engrs; fel Royal Aeronaut Soc; hon mem Airline Pilots Asn; Air Traffic Controllers Asn; fel Aviation Med Asn; hon fel Am Inst Aeronaut & Astronaut. *Res:* Aviation and space safety. *Mailing Add:* 468-D Calle Cadiz Laguna Hills CA 92653

LEDERER, WILLIAM JONATHAN, US citizen; m 75; c 2. CARDIAC MUSCLE, CELLULAR & MOLECULAR PHYSIOLOGY. *Educ:* Harvard Univ, BA, 70; Yale Univ, PhD(physiol), 75, MD, 76. *Prof Exp:* Intern internal med, Univ Wash, 76-77; fel physiol, Oxford Univ, 77-79; from asst prof to assoc prof physiol, 79-88, PROF PHYSIOL, UNIV MD, 88- *Concurrent Pos:* Vis researcher, Univ Col London, 81-90; estab investr, Am Heart Asn, 81-86, mem, Basic Sci Coun. *Mem:* Soc Gen Physiologists; Biophys Soc; NY Acad Sci; AAAS; Am Physiol Soc; Physiol Soc London. *Res:* Mammalian heart muscle to determine how it functions at the cellular and molecular level; links between the electrical activity and the ion transport function of cardiac cells and how these properties relate to the force generated by the heart; cellular control of calcium, sodium and pH. *Mailing Add:* Dept Physiol Sch Med Univ Md 660 W Redwood St Baltimore MD 21201

LEDERIS, KAROLIS (KARL), b Lithuania, Aug 1, 20; m 52; c 2. PHARMACOLOGY, ENDOCRINOLOGY. *Educ:* Bristol Univ, BSc, 58, PhD(pharmacol), 61, DSc(endocrinol), 68. *Prof Exp:* Jr fel pharmacol, Bristol Univ, 61-63, lectr, 63-66, sr lectr, 66-68, reader, 68-69; prof, 69-89, EMER PROF PHARMACOL & THERAPEUT, MED SCH, UNIV CALGARY, 89- *Concurrent Pos:* Wellcome Trust & Ger Res Asn fel, Univ Kiel, 61-62; NSF fel, Univ Calif, Berkeley, 67-68; ed, Pharmacol, Int J Exp & Clin Pharmacol; vis prof, Vilnius Univ, Lithuania, USSR, 76, Univ Bristol, Eng, 79, Kyoto Univ, Japan, 80 & Univ Santiago & Valdivia, Chile, 82; chmn, grants comt prog grants, Med Res Coun Can, 81-, mem, 83-84, exec, Med Res Coun Can, 83-90; int peer review comt, Networks Ctr of Excellence, 89-90. *Honors & Awards:* Fel, Royal Soc Can, 87; UpJohn Award in Pharmacol, 90. *Mem:* Int Brain Res Orgn; Endocrine Soc; Brit Soc Endocrinol; Brit Pharmacol Soc; UK Physiol Soc; Can Physiol Soc; Can Pharmacol Soc; Can Biochem Soc; NY Acad Sci; Am Soc Pharmacol & Exp Therapeut; Fedn Am Socs Exp Biol. *Res:* Hypothalamo-neurohypophyseal system, mechanisms of hormone storage and secretion; central nervous system of teleosts and amphibians; chemistry, pharmacology and gene expression of urotensin peptides, corticotropin releasing hormones and neurohypophyseal hormones. *Mailing Add:* Dept Pharmacol & Therapeut Univ Calgary Med Sch 3330 Hosp Dr NW Calgary AB T2N 4N1 Can

LEDERMAN, DAVID MORDECHAI, b Bogota, Colombia, May 26, 44; m 67; c 2. BIOMEDICAL ENGINEERING. *Educ:* Univ Los Andes, BEng, 66; Cornell Univ, BSc, 66, MEng, 67, PhD(aerospace eng), 73. *Prof Exp:* From prof appl math, Fac Arts & Sci, to dir div biomed eng, Fac Eng, Univ Los Andes, 72-73; sr staff mem, Avco-Everett Res Lab Inc, Avco Corp, 73-76, prin res scientist, 76-79, chmn, Med Res Comt, 79-81, consult, 81-82; CHIEF EXEC OFF & CHMD BD, ABIOMED, INC, 81- *Concurrent Pos:* Dir, Salem State Col Found Bd, Salem, Ma, 84- *Mem:* AAAS; NY Acad Sci; Sigma Xi; Am Phys Soc; Am Soc Artificial Internal Organs; Europ Soc Artificial Organs; Int Soc Artificial Organs; Am Heart Asn. *Res:* Development of clinical cardiovascular devices; prosthetic heart valves; blood-compatible biomaterials; hemodynamics and thrombosis; laser physics and applications to medicine; plastics technology; artificial hearts. *Mailing Add:* PO Box 426 Marblehead MA 01945-0426

LEDERMAN, FRANK L, b Buffalo, NY, Aug 19, 49; m 84. PHYSICS, MATHEMATICS. *Educ:* Carnegie-Mellon Univ, BS & MS, 71; Univ Ill, Urbana, PhD(physics), 75. *Prof Exp:* Physicist, solid state physics & ultrasonics, Gen Elec Corp Res & Develop Ctr, 75-78, mgr ultrasound imager prog, 78-80, mgr energy systems mgt branch, 81-82, mgr power electronics systsbr, 84-87; VPRES, RES & DEVELOP, NORANDA INC, 88- *Mem:* Am Phys Soc; Sigma Xi. *Res:* Solid state physics; signal processing; medical imaging; power circuits, integrated power electronics, lighting systems, motor and drives; metallurgy; metals processing; forestry; computing. *Mailing Add:* Noranda Tech Ctr 240 Hymus Blvd Pointe Claire PQ H9R 1G5 Can

LEDERMAN, LEON MAX, b New York, NY, July 15, 22; m 45, 81; c 3. NUCLEAR PHYSICS, INSTRUMENTS. *Educ:* City Col NY, BS, 43; Columbia Univ, AM, 48, PhD(physics), 51. *Hon Degrees:* DSc, City Col NY, 81, 85, Northern Ill Univ, 83, Univ Chicago, 83, Ill Inst Technol, 87, Lake Forest Col, Carnegie-Mellon Univ, 88, Aurora Univ, 89, Univ Ill, 89, Univ Pittsburgh, 90, Bradley Univ, 90; DHL, Columbia & Rush Univ, 89. *Prof Exp:* Res assoc, Columbia Univ, 51-52, from asst prof to prof physics, 52-89; dir, 79-89, EMER DIR, FERMILAB, BATAVIA, 89-; PROF PHYSICS, UNIV CHICAGO, 89-, FRANK E SULZBERGER PROF, 89- *Concurrent Pos:* Assoc dir, Nevis Labs, 53, dir, 62-79; Guggenheim & Ford Found fel, 58-59, Ernest Kempton Adams fel, 61 & NSF fel, 67; mem, High Energy Physics Adv Panel, Atomic Energy comn, 67-70; collabr, Res Nevis Lab Prog, Brookhaven Nat Lab, Europ Ctr Nuclear Res, Fermilab & State Univ NY, Stony Brook; US rep, Int Comt Future Accelerators; Higgins chair physics, Columbia Univ, 72; mem bd dirs, Weizmann Inst Sci, Israel, 88-, physics & astron, Nat Acad Sci, 88-, Mus Sci Indust, Chicago, 89-; chmn, Aspen Ctr Physics, Colo, 89-; mem Secy Energy Adv Bd, Washington, DC, 90- *Honors & Awards:* Nobel Prize in Physics, 88; Nat Medal Sci, President Lyndon Johnson, 65; Wolf Prize, 82. *Mem:* Nat Acad Sci; fel Am Phys Soc; fel Am Acad Arts & Sci; fel AAAS (pres-elect, 90). *Res:* Properties and interactions of elementary particles; author of over 200 publications. *Mailing Add:* Fermi Nat Accelerator Lab PO Box 500 Batavia IL 60510

LEDERMAN, PETER B, b Weimar, Ger, Nov 16, 31; US citizen; m 57; c 2. CHEMICAL & ENVIRONMENTAL ENGINEERING. *Educ:* Univ Mich, BSE, 53, MS, 57, PhD(chem eng), 61. *Prof Exp:* Jr technologist chem eng, Shell Oil Co, 53; technologist, Cent Res Div, Gen Foods Corp, 56; instr, Univ Mich, 59-61; engr, Esso Res Labs, 61-63, engr chem develop, Esso Res & Eng Co, 63-65, sr chem engr, 66; assoc prof chem eng, Polytech Inst Brooklyn, 66-72; dir indust waste treatment, Res & Develop, US Environ Protection Agency, 72-76; mgr tech develop & res, Cottrell Environ Sci, 76-78, vpres & gen mgr, 78-80; vpres hazardous/toxic mats mgt, 80-83, vpres & gen mgr, Spill Prev & Emergency Response Div, 83-88, VPRES & SR ADV, ROY F WESTON, 88- *Concurrent Pos:* Teaching fel, Univ Mich, 55-59; lectr, Columbia Univ, 65-67; panel arbitrators, Am Arbit Asn. *Honors & Awards:* Silver Medal, US Environ Protection Agency, 76; Larry K Cecil Award Environ Eng, Am Inst Chem Engrs, 87. *Mem:* Fel Am Inst Chem Engrs; Am Chem Soc; Am Soc Eng Educ; Nat Soc Prof Engrs; Air Pollution Control Asn; dipl Am Acad Environ Engrs; Am Soc Mech Engrs. *Res:* Environmental studies; solid and hazardous material management; computer application; process optimization; polymers; mass transfer; pollution prevention; waste minimization; emergency response. *Mailing Add:* 17 Pittsford Way New Providence NJ 07974

LEDFORD, BARRY EDWARD, b Denver, Colo, Feb 27, 42; m 65; c 2. BIOCHEMISTRY, MOLECULAR BIOLOGY. *Educ:* Univ Colo, Boulder, BA, 63, MS, 65; Fla State Univ, Tallahassee, PhD(chem), 71. *Prof Exp:* Teaching asst, Univ Colo, Boulder, 63-64, res asst, 64-65; jr scientist, Lawrence Livermore Nat Lab, 65-66; teaching asst, Fla State Univ, Tallahassee, 66-67, USPHS biochem trainee, Dept Chem, 67-71; USPHS postdoctoral fel, Biol Div, Oak Ridge Nat Lab, 71-73; from asst prof to assoc prof biochem, 73-84, PROF, DEPT BIOCHEM & MOLECULAR BIOL, MED UNIT SC, CHARLESTON, 84-, ASSOC DEAN, COL GRAD STUDIES & DIR, PROG MOLECULAR & CELL BIOL & PATHOBIOL, 89- *Mem:* Am Soc Biochem & Molecular Biol; AAAS; Am Soc Cell Biol; Soc Develop Biol. *Mailing Add:* Dept Biochem & Molecular Biol Med Univ SC 171 Ashley Ave Charleston SC 29425

LEDFORD, RICHARD ALLISON, b Charlotte, NC, June 30, 31; m 57; c 5. FOOD MICROBIOLOGY. *Educ:* NC State Univ, BS, 54, MS, 56; Cornell Univ, PhD(food sci), 62. *Prof Exp:* Dir, NY State Food Lab, NY State Dept Agr & Mkts, 61-64; from asst prof to assoc prof, 64-80, chmn dept, 72-77, assoc dir, Inst Food Sci, 75-77, PROF FOOD SCI, CORNELL UNIV, 80-, CHMN DEPT, 85-, DIR, INST FOOD SCI, 88- *Honors & Awards:* Nordica Award, 87. *Mem:* Am Soc Microbiol; Inst Food Technologists; Am Dairy Sci Asn. *Res:* Microbiological aspects of food science, especially food fermentations and analytical methods. *Mailing Add:* 1396 Ellis Hollow Rd Ithaca NY 14850-9601

LEDFORD, THOMAS HOWARD, b Macon, Ga, Aug 24, 42; m 65; c 2. ORGANIC CHEMISTRY. *Educ:* Univ Ga, BS, 64; Univ Fla, PhD(chem), 73. *Prof Exp:* Res chemist polymer intermediates, Tenn Eastman Res Labs, Kingsport, Tenn, 64-68; res chemist fuels processing, Exxon Res & Develop, 73-80, sr res chemist, 80-88; SR TECH DEVELOP CHEMIST, ROBICON CORP, 88- *Concurrent Pos:* Adj assoc prof chem, La State Univ, 81- *Mem:* Am Chem Soc. *Res:* Polymers; reactions in strong acids; fuel processing chemistry; sulfur chemistry. *Mailing Add:* 2322 W High Meadow Ct PO Box 2226 Baton Rouge LA 70816

LEDIAEV, JOHN P, b Goorgan, Iran, Sept 8, 40; US citizen; div; c 1. MATHEMATICS. *Educ:* Occidental Col, BA, 63; Univ Calif, Riverside, MA, 65, PhD(noether lattices), 67. *Prof Exp:* Asst prof, 67-71, ASSOC PROF MATH, UNIV IOWA, 71- *Mem:* Am Math Soc. *Res:* Structure; representation and embedding of Noether lattices; primary decomposition in multiplicative lattices; semi-prime operations in Noether lattices. *Mailing Add:* Dept Math Univ Iowa Iowa City IA 52240

LEDIG, F THOMAS, b Dover, NJ, Aug 13, 38; div; c 3. EVOLUTION. *Educ:* Rutgers Univ, BS, 62; NC State Univ, MS, 65, PhD(genetics), 67. *Prof Exp:* Lectr, Yale Univ, 66-67, from asst prof to prof forest genetics, 67-82, dir, 79-88; SR SCIENTIST, INST FOREST GENETICS, USDA FOREST SERV, 88- *Concurrent Pos:* Res geneticist, Inst Trop Forestry, Rio Piedres, PR, 78, Keystone Nat Policy Dialogue Biodiversity, 89-90; consult, Nat Res Coun Strategies to Manage Forest Genetic Resources, 89-, UN/FAO Biotechnol for Develop Countries, 89- *Mem:* AAAS; Soc Am Foresters; Soc Conserv Biol; Int Soc Trop Foresters. *Res:* Genecology, population genetics, and evolution; physiological mechanisms of adaptation in wild populations; quantitative genetics and breeding. *Mailing Add:* Inst Forest Genetics USDA-Forest Serv PO Box 245 Berkeley CA 94701

LEDIN, GEORGE, JR, b Seekirchen, Austria, Jan 28, 46; US citizen; m 68; c 2. COMPUTER SCIENCE, STATISTICS. *Educ:* Univ Calif, Berkeley, BS, 67. *Hon Degrees:* JD, Univ San Francisco, 82. *Prof Exp:* Statistician & mathematician, 65-70, lectr math & comput sci, 68-74, sr res assoc statist & math biosci, 70-75, asst prof comput sci, 75-79, ASSOC PROF COMPUT SCI, UNIV SAN FRANCISCO, 80-, CHMN DEPT, 76- *Concurrent Pos:* Consult comput sci & statist, 66-; US rep, Int Fedn Info Processing, 72-74. *Mem:* AAAS; Am Statist Asn; Asn Comput Mach; Math Asn Am; NY Acad Sci. *Res:* Algorithmic languages; programming methodology; heuristic programming; pattern recognition; robotics; mathematical models for biosciences; number theory; combinatorics; graph theory; game theory; information theory; algebra. *Mailing Add:* Sonoma State Univ 1801 E Cotati Ave Darwin 124 Rohnert Park CA 94928

LEDINKO, NADA, b Girard, Ohio, Dec 16, 25. VIROLOGY. *Educ:* Ohio State Univ, BS, 46; Pa State Col, MS, 49; Yale Univ, PhD(microbiol), 52. *Prof Exp:* Res asst virol, Yale Univ, 52-53; Nat Found Infantile Paralysis fel, Walter & Eliza Hall Inst, Australia, 53-55; virologist, Pub Health Res Inst, 56-62; USPHS fel, Carnegie Inst Genetics Res Unit & Salk Inst Biol Studies, 63-65; assoc investr & NIH res career develop awardee, Putnam Mem Hosp Inst Med Res, Bennington, Vt, 65-71; PROF BIOL, UNIV AKRON, 71- *Mem:* AAAS; Am Asn Path & Bact; Tissue Cult Asn; Am Soc Microbiol; Am Asn Cancer Res. *Res:* Genetical and biochemical aspects of viral growth. *Mailing Add:* Dept Biol Univ Akron Akron OH 44325

LEDLEY, BRIAN G, b Brisbane, Australia, Sept 29, 28; m 58; c 1. MAGNETOSPHERIC PHYSICS, REMOTE SENSING. *Educ:* Univ Queensland, BSc, 50; Univ Birmingham, PhD(physics), 55. *Prof Exp:* Asst lectr physics, Univ Birmingham, 55-57; res assoc nuclear physics, Fermi Inst, Univ Chicago, 57-60; res fel cosmic rays, Univ Sydney, 60-62; Nat Acad Sci-NASA res fel space sci, 62-63; sect head, 63-70, PHYSICIST, NASA, 70- *Concurrent Pos:* Gowrie Overseas scholar, 52-55. *Mem:* Am Geophys Union. *Res:* Geomagnetic field measurements using satellites; magnetometer instrumentation; remote optical sensing instrumentation. *Mailing Add:* Code 696 Goddard Space Flight Ctr Greenbelt MD 20771

LEDLEY, FRED DAVID, b Washington, DC, Nov 27, 54; m 76; c 2. PEDIATRICS, METABOLISM. *Educ:* Univ Md, College Park, BS, 74; Georgetown Univ, MD, 78. *Prof Exp:* Resident pediat, Harvard Med Sch, Children's Hosp, 78-81, fel genetics, 81-83; fel genetics, Mass Inst Technol, 81-83; from assoc to asst prof cell biol, 83-89, ASSOC PROF CELL BIOL & PEDIATRICS, BAYLOR COL MED, 89- *Concurrent Pos:* Fel, Am Can Soc, 81-83; Janeway scholar, Harvard Med Sch, Children's Hosp, 81; asst investr, Howard Hughes Med Inst, 85- *Honors & Awards:* Upjohn Achievement Award, Georgetown Univ, 78. *Mem:* Am Soc Human Genetics; Soc Pediat Res; Soc Inherited Metab Dis. *Res:* The molecular cloning and characterization of genes for pathways of intermediary metabolism, particularly those involved in metabolic disease in humans. *Mailing Add:* Dept Cell Biol Baylor Col Med Houston TX 77030

LEDLEY, ROBERT STEVEN, b New York, NY, June 28, 26; m 49; c 2. BIOPHYSICS, COMPUTER SCIENCE. *Educ:* NY Univ, DDS, 48; Columbia Univ, MA, 49. *Prof Exp:* Res physicist, Radiation Lab, Columbia Univ, 48-50, instr physics, 49-50; vis scientist, Nat Bur Standards, 51-52, physicist, External Control Group, Electronic Comput Lab, 53-54; opers res analyst, Opers Res Off, Strategic Div, Johns Hopkins Univ, 54-56; assoc prof elec eng, Sch Eng, George Washington Univ, 57-60; instr pediat, Sch Med, Johns Hopkins Univ, 60-63; prof elec eng, Sch Eng & Appl Sci, George Washington Univ, 68-70; PROF PHYSIOL, BIOPHYS & RADIOL, MED CTR, GEORGETOWN UNIV, 70-; PRES & RES DIR, NAT BIOMED RES FOUND, 60- *Concurrent Pos:* Consult mathematician, Data Processing Systs Div, Nat Bur Standards, 57-60; mem staff, Nat Acad Sci-Nat Res Coun, 57-61; pres, Digital Info Sci Corp, 70-75; affil rep, Pattern Recognition Soc. *Honors & Awards:* Nat Inventors Hall of Fame, 90. *Mem:* Soc Math Biophys; Inst Elec & Electronics Engrs; Biophys Soc; NY Acad Sci; Pattern Recognition Soc. *Res:* Applications of computers to medical instrumentation; computer software systems and applications in medicine and biology; computer aids to medical diagnosis; computer information science; medical imaging; pattern recognition; medical informatics. *Mailing Add:* Dept Physiol Lr-3 Preclin Sci Georgetown Univ 37th & O St NW Washington DC 20057

LEDLEY, TAMARA SHAPIRO, b Washington, DC, May 18, 54; m; c 2. EARTH SYSTEM SCIENCE, CLIMATOLOGY. *Educ:* Univ Md, BS, 76; Mass Inst Technol, PhD(meteorol), 83. *Prof Exp:* Res assoc climat, Rice Univ, 83-85, asst res scientist climat/earth syst sci, 85-90, assoc res scientist, 90, SR FAC FEL CLIMAT/EARTH SYST SCI, RICE UNIV, 90- *Concurrent Pos:* Prin investr, NSF grants, 85- & Tex Advan Technol Prog grants, 87-; mem, Alaska SAR Facil Archive Working Team, 88 & McMurdo SAR Facil Sci Working Team, 90; coordr, Nat Week Educ, Union Concerned Scientists, 89; consult, Sci Connection Elem Sci Curric Proj, Houston Mus Natural Sci, 89-90, Broader Perspectives Inc, 90 & Eastern Res Group Inc, 90- *Mem:* AAAS; Am Meteorol Soc; Am Geophys Union; Oceanog Soc; Sigma Xi. *Res:* Understanding the role of the polar regions in shaping climate on a wide range of time scales by examining how atmosphere-sea ice-ocean interactions influence climate change. *Mailing Add:* Dept Space Physics & Astron Rice Univ PO Box 1892 Houston TX 77251-1892

LEDNEY, GEORGE DAVID, b Sharon, Pa, June 25, 37. BIOLOGY, RESEARCH ADMININISTRATION. *Educ:* Youngstown Univ, BS, 60; Univ Notre Dame, PhD(biol), 65. *Prof Exp:* Asst prof radiation biol, Med Units, Univ Tenn, Memphis, 65-67, assoc prof, 70-73; head, Div Immunol, 73-88, PROG MGR, MICROBIOL-IMMUNOL, ARMED FORCES RADIOBIOL RES INST, 88- *Concurrent Pos:* Nat Cancer Inst fel, 65-67; Am Cancer Soc grants, 68 & 70. *Mem:* Radiation Res Soc; Transplantation Soc; Asn Gnotobiotics; Int Soc Exp Hemat; Soc Exp Biol Med; Am Soc Microbiol; Reticuld Endothelial Soc. *Res:* Radiation biology; radiation and immune functioning; bone marrow transplantation; wound trauma. *Mailing Add:* Dept Exp Hematol Armed Forces Radiobiol Res Inst Bethesda MD 20814-5145

LEDNICER, DANIEL, b Antwerp, Belg, Oct 15, 29; nat US; m 56; c 2. ORGANIC CHEMISTRY. *Educ:* Antioch Col, BS, 52; Ohio State Univ, PhD(chem), 55. *Prof Exp:* Sr chemist, G D Searle & Co, 55-56; res assoc, Duke Univ, 56-58; Esso Res & Develop Co fel, Univ Ill, 58-59; chemist, Upjohn Co, 59-73, sr scientist, 73-76; dir chem res, Mead Johnson & Co, 76-80; dir Med Chem & Pharm, Adria Labs, 80-84; PHARM MGR, ABC LABS, 84- *Mem:* Am Chem Soc. *Res:* Stereochemistry; medicinal chemistry; hypotensives; analgesics; pharmaceutical analysis. *Mailing Add:* 826 Bowie Rd Rockville MD 20852-1023

LEDOUX, ROBERT LOUIS, b Marieville, Que, Apr 19, 33; m 61; c 2. MINERALOGY. *Educ:* Univ Montreal, BSc, 56; Laval Univ, MScA, 60; Purdue Univ, PhD(mineral), 64. *Prof Exp:* From assoc prof to prof mineral, 64-74, PROF PETROL & MINERAL, LAVAL UNIV, 74- *Mem:* Mineral Asn Am; Can Asn Mineral; Geol Asn Can. *Res:* Infrared studies of layered silicates. *Mailing Add:* 711 rue Moreau Ste Foy Quebec PQ G1V 3A5 Can

LEDSOME, JOHN R, b Bebington, Eng, June 18, 32; m 57; c 3. PHYSIOLOGY, MEDICINE. *Educ:* Univ Edinburgh, MB, ChB, 55, MD, 62. *Prof Exp:* Lectr physiol, Univ Leeds, 59-68; PROF PHYSIOL, 68-, HEAD, PHYSIOL, UNIV BC, 80- *Concurrent Pos:* USPHS int fel, 64-65, res grant, 66-68; Med Res Coun Eng res grant, 65-68; Med Red Coun Can res grant, 68-91; BC Heart Found res grant, 72-91. *Mem:* Can Physiol Soc; Brit Physiol Soc; Am Physiol Soc. *Res:* Control of the cardiovascular system; function of left atrial receptors; atrial natriuretic peptide. *Mailing Add:* Dept Physiol Univ BC Vancouver BC V6T 1W5 Can

LEDUC, ELIZABETH, b Rockland, Maine, Nov 19, 21. CELL BIOLOGY. *Educ:* Univ Vt, BS, 43; Wellesley Col, MA, 45; Brown Univ, PhD(biol), 48. *Prof Exp:* Res assoc biol, Brown Univ, 48-49; instr & assoc anat, Harvard Med Sch, 49-53; from asst prof to assoc prof, 53-64, PROF BIOL, BROWN UNIV, 64-, DEAN DIV BIOL & MED, 73- *Concurrent Pos:* Mem adv coun, Nat Inst Gen Med Sci, 72-76. *Mem:* AAAS; Am Soc Cell Biol; Soc Francaise de Microscopie Electronique; Histochem Soc; Am Soc Exp Path. *Res:* Histophysiology and pathology of the liver; cellular mechanism in antibody production; ultrastructural and cytochemical effects of cancer chemotherapeutic compounds on normal and neoplastic cells. *Mailing Add:* Div Biol & Med Sci Brown Univ Box G Providence RI 02912

LEDUC, GERARD, b Verdun, Que, Sept 7, 34; m 59; c 3. FISHERIES, BIOCHEMISTRY. *Educ:* Univ Montreal, BSc, 58, MSc, 60; Ore State Univ, PhD(fisheries), 66. *Prof Exp:* Biologist, Que Wildlife Serv, 63-66; asst prof biol sci, 66-72, chmn dept, 69-72, ASSOC PROF BIOL SCI, SIR GEORGE WILLIAMS UNIV, 72- *Res:* Fisheries problems in water pollution, mainly the long-term effects of sublethal concentrations of toxicant; artificial streams. *Mailing Add:* Dept Biol Sci Concordia Univ Sir G Williams 1455 Demaisonneuve Blvd Montreal PQ H3G 1M8 Can

LE DUC, J-ADRIEN MAHER, b Can, July 2, 24; nat US; m 49; c 2. ELECTROCHEMISTRY, INORGANIC CHEMISTRY. *Educ:* Sir George William Univ, BSc, 47; Polytech Inst New York, MS, 51; Univ Dijon, DSc(phys inorg chem), 53. *Prof Exp:* Anal paper chemist, Howard Smith Paper Co, 43-44; anal paper chemist dyes & colors, L B Holliday & Co, 45; res chemist zirconium alloys, St Lawrence Alloys & Metals Co, 46; consult chemist, Milton Hersey Co, 47; res consult chemist & lab dir, Wyssmont Co, 47-49; sr inorg chemist, J T Baker Chem Co, 49-50; sr electroplating chemist, MacDermid, Inc, 53-54; head electrolytic sect, Diamond Shamrock Chem Corp, 54-56, group leader res, Electrochem Div, 56-59; supvr explor & process res, M W Kellogg Co, 59-62, dir electrochem res, 62-70, mgr electrochem res & develop, 63-70; CONSULT CHEM & ELECTROCHEM TECHNOL, LICENSING & PROCESS TECHNOL EXCHANGE RES ADMIN, 70- *Concurrent Pos:* Ed, Int Electrochem Progress & Chemicals Today; dir, Int Electrochem Inst, Millburn, NJ & Industrialists Int Inc, Summit, NJ. *Honors & Awards:* Order of Merit of Res & Invention Award, Paris, 68, Gold Medal Award Res & Invention, 75. *Mem:* Sr mem Am Chem Soc; Electrochem Soc; NY Acad Sci; Soc Indust Chem; Sigma Xi. *Res:* Inorganic and organic chemicals; electrochemistry; non-ferrous electrometallurgy; metal finishing molten salts; licensing negotiations and technical exchange; research and invention management; aquisitions and ventures. *Mailing Add:* 189 Parsonage Hill Rd Short Hills NJ 07078

LE DUC, JAMES WAYNE, b Long Beach, Calif, Nov 23, 45; m 70; c 3. EPIDEMIOLOGY, MEDICAL ENTOMOLOGY. *Educ:* Calif State Univ, Long Beach, BS, 67; Univ Calif, Los Angeles, MS, 72, PhD(pub health), 77. *Prof Exp:* Med entomologist, Smithsonian Inst, Washington, DC, 67-68; team leader entom & mammal, 68-69; entomologist, Walter Reed Army Inst Res, 69-71, chief virol & entom, Arbovirus Prog, 73-75; COMDR ARBOVIROL, US ARMY MED RES UNIT, BELEM, BRAZIL, 77- *Honors & Awards:* Paul A Siple Award, US Army Sci Conf, 74. *Mem:* Am Soc Trop Med & Hyg; Am Mosquito Control Asn; Soc Vector Ecologists. *Res:* Arbovirology; medical entomology; infectious disease epidemiology; ecological aspects of disease maintenance. *Mailing Add:* US Army Med Res Unit-Belem Brazil APO Miami FL 34030

LEDUC, SHARON KAY, b Hattiesburg, Miss, Apr 28, 43. STATISTICS. *Educ:* Eastern Ill Univ, BS, 65; Univ Mo-Columbia, MA, 67, PhD(statist), 71. *Prof Exp:* Prof atmospheric sci, Univ Mo, 67-89; prof geog, Univ NC, 90-93; Nat Environ Satellite, Data & Info Serv, statistician, Ctr Environ Assessment Serv, Nat Ocean & Atmospheric Admin, Mo, 74- 88, ENVIRON PROTECTION AGENCY, NC, 88- *Mem:* Am Statist Asn; Am Meteorol Soc; Sigma Xi. *Res:* Statistical analysis of atmospheric data for climate change and monitoring. *Mailing Add:* US Environ Protection Agency MD-80 Research Triangle Park NC 27111

LEDUY, ANH, b Vietnam, Feb 6, 46; Can citizen; m 77; c 2. BIOCHEMICAL ENGINEERING, APPLIED MICROBIOLOGY. *Educ:* Univ Sherbrooke, BScA, 69, MScA, 72; Univ Western Ont, PhD(biochem eng), 75. *Prof Exp:* Res assoc chem eng, Univ Sherbrooke, 75-77; from asst to assoc prof, 77-85, PROF CHEM ENG, LAVAL UNIV, 85- *Concurrent Pos:* Consult, 81- *Mem:* Can Soc Chem Eng; Am Inst Chem Engrs; Am Soc Microbiol; Can Soc Mech Eng; NY Acad Sci. *Res:* Utilization of microorganisms; enzymes systems in the production of biomass and precious metabolites from abundant raw materials, waste materials and industrial agricultural by-products. *Mailing Add:* Dept Chem Eng Laval Univ Ste Foy PQ G1K 7P4 Can

LEDWELL, THOMAS AUSTIN, b PEI, June 13, 38; m 64; c 4. MECHANICAL ENGINEERING. *Educ:* NS Tech Col, BE, 60, ME, 65; Univ Waterloo, PhD(mech eng), 68. *Prof Exp:* Res coordr, Nat Res Coun, 75-77; tech adv, Renewable Energy Policy, 77-78, head res coord, 78-80, sr tech adv, Conserv & Renewable, Energy, Mines & Resources, 80-82, chief, Demonstration Prog, 82-85; CONSULT, 85- *Concurrent Pos:* Coordr renewable energy res & develop, Fed Govt Can, 76-77, coordr conserv res & develop, 78-80. *Mem:* Combustion Inst. *Res:* Thermodynamics; engine design and development; applied mathematics; energy research and development. *Mailing Add:* 1922 Marquis Ave Gloucester Ottawa ON K1J 8J2 Can

LEDWITZ-RIGBY, FLORENCE INA, b New York, NY, Feb 14, 46; m 68; c 2. REPRODUCTIVE ENDOCRINOLOGY. *Educ:* City Col NY, BS, 66; Case-Western Reserve Univ, MS, 68; Univ Wis-Madison, PhD(endocrinol & reprod physiol), 72. *Prof Exp:* Res fel reprod physiol, Sch Med, Dept Physiol, Univ Pittsburgh, 72-74, vis asst prof physiol, Dept Biol, 74-75; from asst prof to assoc prof, 75-87, PROF PHYSIOL, DEPT BIOL SCI, NORTHERN ILL UNIV, 87- *Concurrent Pos:* Vis res prof, dept genetics, Univ Ill, Chicago, 87-88. *Mem:* Am Physiol Soc; Soc Study Reprod; Endocrine Soc. *Res:* Endocrine and physiological control mechanisms of ovarian cell function and differentiation. *Mailing Add:* Dept Biol Sci Northern Ill Univ DeKalb IL 60115

LEE, ADDISON EARL, b Maydelle, Tex, June 18, 14; m 37; c 2. BOTANY, SCIENCE EDUCATION. *Educ:* Stephen F Austin State Teachers Col, BS, 34; Agr & Mech Col Tex, MS, 37; Univ Tex, PhD(bot), 49. *Prof Exp:* Lab instr, Stephen F Austin State Teachers Col, 33-36 & Agr & Mech Col Tex, 36-37; from instr to assoc prof bot, Univ Tex, Austin, 46-59, morphologist & asst dir, Plant Res Inst, 49-53, prof, 59-80, EMER PROF SCI EDUC & BIOL & DIR, SCI EDUC CTR, UNIV TEX, AUSTIN, 80- *Concurrent Pos:* High sch teacher, Tex, 34-36; vis prof, Univ Va, 57-58. *Honors & Awards:* Robert H Carleton Award, Nat Sci Teachers Asn, 75. *Mem:* Nat Sci Teachers Asn (pres, 67); hon mem Nat Asn Biol Teachers (pres, 73); Nat Asn Res Sci Teaching. *Res:* Experimental plant morphology. *Mailing Add:* 4505 Balcones Dr Austin TX 78731

LEE, ALFRED M, b Bloomington, Ind, Aug, 23, 51. TELECOMMUNICATIONS POLICY, TECHNOLOGY ASSESSMENT. *Educ:* Univ Ill, BSEE, 73; Cornell Univ, MS, 75, PhD(civil eng & pub policy), 81. *Prof Exp:* Res specialist mobile commun, Cornell Univ, 75-78, res asst electronic message transfer, 78-80, fel, Prog Sci, Technol & Sci, 80-82; telecommun policy analyst, 84-90, SR POLICY ADV, NAT TELECOMMUN & INFO ADMIN, US DEPT COM, 90- *Concurrent Pos:* Mem, Transp Res Bd, Subcomt Telecommun & Transp, Trade-offs, 80-84; assoc ed, Inst Elec & Electronics Engrs Technol & Soc Mag, 81-84. *Mem:* Inst Elec & Electronics Engrs; Sigma Xi. *Res:* Telecommunications policy; transportation-communications trade-offs and the social impact of developments in telecommunications; mobile communications; electronic message transfer. *Mailing Add:* Nat Telecommun & Info Admin H4725 US Dept Commerce Washington DC 20230

LEE, ALFRED TZE-HAU, b Hong Kong, July 22, 39; US citizen; m 70; c 2. ANALYTICAL CHEMISTRY, ORGANIC CHEMISTRY. *Educ:* Univ Calif, Berkeley, BS, 63; Univ Calif, Los Angeles, PhD(chem), 68. *Prof Exp:* Instr chem, East Los Angeles Col, 67-68; PROF CHEM, CITY COL OF SAN FRANCISCO, 68- *Mem:* Am Chem Soc. *Res:* Oxidation-state diagrams; pulse polarography; analytical methods in general, vacuum-ultraviolet spectra of olefins. *Mailing Add:* 154 Hernandez Ave San Francisco CA 94127

LEE, AMY SHIU, b Canton, China, Aug 5, 47; US citizen; m 72; c 2. MOLECULAR BIOLOGY, BIOCHEMISTRY. *Educ:* Univ Calif, Berkeley, BA, 70; Calif Inst Technol, MSc, 72, PhD(biophysics, molecular biol), 75. *Prof Exp:* Res asst bact, Univ Calif, Los Angeles, 70-71; teaching asst biol, Calif Inst Technol, 71-74, res fel, 74-77, sr res fel, 78-79; from asst to assoc prof, 79-87, PROF BIOCHEM, SCH MED, UNIV SOUTHERN CALIF, 88- *Concurrent Pos:* Am Cancer Soc fel, Calif Inst Technol, 75-76, sr res fel, 78-69, NIH Pub Health Serv fel, 77-78; NIH study sect, 80-88; fac res award, Am Cancer Soc, 83-88. *Honors & Awards:* Merit Award, NIH, 88. *Mem:* AAAS. *Res:* DNA sequence organization and gene expression in cell cycle regeneration eukaryotes; recombinant DNA technology. *Mailing Add:* Dept Biochem Univ Southern Calif Los Angeles CA 90033

LEE, ANTHONY, b Canton, China, Dec 8, 41; US citizen; m 69; c 2. PLASMA PHYSICS. *Educ:* Drexel Univ, BS, 66; Stevens Inst Technol, MS, 69, PhD(physics), 71. *Prof Exp:* Fel plasma physics, Univ Sask, 71-73; adj asst prof, Univ S Fla, 73-75, vis asst prof physics, 75-76; SR ENGR, WESTINGHOUSE ELEC CORP, RES & DEVELOP CTR, 76- *Mem:* Am Phys Soc; Inst Elec & Electronics Engrs. *Res:* Nonlinear plasma wave theory; linear and nonlinear low frequency waves in plasmas with density, potential and temperature gradients; catalytic turbulent heating of plasmas as supplementary tokamac heating. *Mailing Add:* Elec Eng Res Labs Gen Motors Warren MI 48090

LEE, ANTHONY L, b Qingdao, China, Nov 16, 34; US citizen; m 62; c 2. CATALYSIS, THERMODYNAMICS. *Educ:* Univ Calif, Berkeley, BS, 58; Mo Sch Mines, MS, 61. *Prof Exp:* Res asst, Calif Inst Technol, 58-59; chemist, Stepan Chem Co, 60; chem engr, Inst Gas Technol, 61-66, supvr fundamental properties res, 66-69, supvr catalytic processing, 69-78, sr chem engr, 78-85, asst dir, Catalyst Develop, 85-88, ASSOC DIR, GAS PROCESSING & CATALYSIS, INST GAS TECHNOL, 88- *Mem:* Am Inst Chem Engrs. *Res:* Transport and thermodynamic peoperties of hydrocarbons; coal gasification research; methanation, water-gas shift, hydrotreating, hydrocracking, and steam reforming catalysis; C1 chemistry; hydroforming. *Mailing Add:* Inst Gas Technol 3424 S State St Chicago IL 60616

LEE, ARNOLD (YING-HO), electrical engineering, for more information see previous edition

LEE, ARTHUR CLAIR, b Abilene, Kans, Aug 3, 23; m 51; c 5. OPHTHALMOLOGY, RADIATION RESEARCH. *Educ:* Colo State Univ, DVM, 52, MS, 63, PhD(radiation biol), 70. *Prof Exp:* Practitioner vet med, 52-60; Morris Found fel, 60-62; vet radiologist, AEC Proj, Colo State Univ, Foothills Campus, 63-66, vet sect leader, Collab Radiol Health Lab, USPHS, 64-90; RETIRED. *Mem:* AAAS; Am Vet Med Asn; Radiation Res Soc Am; Am Soc Vet Ophthal. *Res:* Radiation effects on canine growth and development; ocular lesions as a result of age at exposure; cataractogenesis from heavy charged particles. *Mailing Add:* 1908 Mohawk St Ft Collins CO 80525-1526

LEE, BENEDICT HUK KUN, b Hong Kong, Oct 17, 40; Can citizen; m 66; c 2. MECHANICAL ENGINEERING, AERONAUTICS. *Educ:* McGill Univ, BEng, 63, MEng, 64, PhD(mech eng), 66. *Prof Exp:* Res assoc, McGill Univ, 66-67; SR RES SCIENTIST, NAT RES COUN, 67- *Mem:* Am Inst Aeronaut & Astronaut; Acoust Soc Am. *Res:* Fluid mechanics; gas dynamics; acoustics and aerodynamics; aeroelasticity; structural dynamics. *Mailing Add:* Nat Res Coun Bldg U66 Montreal Rd Ottawa ON K1A 0R6 Can

LEE, BERNARD S, b China, Dec 14, 34; US citizen; m 63; c 3. CHEMICAL ENGINEERING, PHYSICAL CHEMISTRY. *Educ:* Polytech Inst Brooklyn, BChE, 56, PhD(chem eng), 60. *Prof Exp:* Mem staff, Arthur D Little, Inc, 60-65; supvr, mgr & dir coal gasification, 65-75, asst vpres process res, 75-76; vpres, 76-77, exec vpres, 77-78, PRES, INST GAS TECHNOL, 78- *Concurrent Pos:* Lectr, Am Inst Chem Engrs, 81. *Mem:* Fel Am Inst Chem Engrs; Am Chem Soc; Am Inst Mining, Metall & Petrol Engrs; Sigma Xi; Am Gas Asn. *Res:* Energy conversion processes for its production of synthetic fuels from coal, lignite, peat, oil shale, biomass, urban and industrial wastes and efficient energy utilization systems involving solar energy and fuel cells. *Mailing Add:* Inst Gas Technol 3424 St State St Chicago IL 60616

LEE, BURNELL, b Hearne, Tex, Dec 8, 55; m 78; c 2. COATINGS FORMULATIONS, GENERAL FORMULATION. *Educ:* Southwest Mo State Univ, BS, 80; Univ Ill, Champaign-Urbana, PhD(org chem), 87. *Prof Exp:* Sr res chemist, 87-90, RES SPECIALIST, ETHYL CORP, 90- *Mem:* Am Chem Soc. *Res:* Synthesis and application polyimide polymers for coatings, molding, and composite manufacture; applications of proprietary additives in a variety areas including ceramic coatings, cosmetics, agricultural products, and foams. *Mailing Add:* PO Box 14799 Baton Rouge LA 70898

LEE, BURTRAND INSUNG, b Seoul, Korea, Jan 20, 52; US citizen; m 79; c 1. ANALYTICAL CHEMISTRY, THERMODYNAMICS & MATERIAL PROPERTIES. *Educ:* Southern Col, Collegedale, BA, 76; Western Mich Univ, MS, 79; Univ Fla, Gainesville, PhD(mat eng), 86. *Prof Exp:* Chemist, Biospherics Inc, 76-78; asst prof mat sci, 86-91, ASSOC PROF CERAMIC ENG, CLEMSON UNIV, 91- *Concurrent Pos:* Lectr, State Univ NY, 80; prin investr, Petrol Res Fund, Am Chem Soc, 87-88 & NSF, 88-; panel comt mem, Nat Sci & Technol Ctr, 87-88; vis prof, Norweg Inst Technol, 89; Fulbright Found Scholar, 89. *Mem:* Am Chem Soc; Am Ceramic Soc; Mat Res Soc. *Res:* Surface interactions of ceramic and polymeric materials; new methods or new materials for engineering and electronic applications; chemical processing or ultrastructure processing of ceramic materials; sol-gel processing. *Mailing Add:* Dept Ceramic Eng Olin Hall Clemson Univ Clemson SC 29634-0907

LEE, BYUNGKOOK, b Korea, Feb 7, 41; m 64; c 1. X-RAY CRYSTALLOGRAPHY, BIOLOGICAL STRUCTURE. *Educ:* Seoul Nat Univ, BS, 61; Cornell Univ, PhD(phys chem), 67. *Prof Exp:* USPHS res fel, Yale Univ, 69-70; from asst prof to assoc prof chem, Univ Kans, 70-83; expert, 80-87, RES CHEMIST, NIH, BETHESDA, MD, 87- *Mem:* Am Chem Soc; Am Crystallog Asn. *Res:* Biothermodynamics; computer modelling of biological macromolecules. *Mailing Add:* NIH Dort-PSL Bldg 12A Rm 2007 Bethesda MD 20892

LEE, C(HIA) H(UAN), b China, Oct 1, 19; m 53; c 4. ELECTRICAL ENGINEERING. *Educ:* Chiao Tung Univ, China, BS, 42; Cornell Univ, MS, 49, PhD(elec eng), 51. *Prof Exp:* Design engr, Cent Elec Mfg Works, China, 42-47; distribution engr, Canton Power Co, 47-48; develop engr, Reliance Elec & Eng Co, Ohio, 51-55; asst prof elec eng, Polytech Inst Brooklyn, 55-58; fel engr, Westinghouse Elec Corp, 58-64; sr engr specialist, Airesearch Mfg Co, 64-82 ENG CONSULT, AIRESEARCH MFG CO, 83- *Honors & Awards:* Cert Recognition, NASA, 73. *Mem:* Inst Elec & Electronics Engrs; Am Soc Naval Engrs; China Elec Eng Soc. *Res:* Electric machinery; electric power systems; electromagnetic devices; electronic components; circuit theory; land, sea and air transportation; electronic power converter; control systems. *Mailing Add:* 30584 Ganado Dr Rancho Palos Verdes CA 90274-6222

LEE, CHARLES ALEXANDER, b New York, NY, Aug 28, 22; m 53; c 1. PHYSICS. *Educ:* Rensselaer Polytech Inst, BEE, 44; Columbia Univ, PhD(physics), 54. *Prof Exp:* Res assoc molecular beam spectros, Columbia Univ, 52-53; mem tech staff semiconductors, Bell Tel Labs, 53-67; PROF ELEC ENG, CORNELL UNIV, 67- *Mem:* Am Phys Soc; Inst Elec & Electronics Engr; Sigma Xi. *Res:* Solid state device physics; molecular beam spectroscopy. *Mailing Add:* Six 212 Giles St Ithaca NY 14850

LEE, CHARLES NORTHAM, b Syracuse, NY, Jan 13, 25; m 52; c 4. ENGINEERING, FORESTRY. *Educ:* Syracuse Univ, BS, 49, BCE, 57, MCE, 59. *Prof Exp:* Instr civil eng, Syracuse Univ, 57-59; asst prof forest mgt, 59-63, assoc prof forest eng, 64-68, PROF FOREST ENG, STATE UNIV NY COL ENVIRON SCI & FORESTRY, 68-, DIR COMPUT CTR, 64- *Concurrent Pos:* Consult geotech engr, 57-; NSF fel Sci Fac MIT, 63-65. *Mem:* Am Soc Civil Engrs; Asn Comput Mach. *Res:* Land locomotion under off-highway conditions; modulation techniques applied to roadway design; information coding content and transformations for analysis and design of engineering systems. *Mailing Add:* Dept Forest Eng Col Environ Sci & Forestry State Univ NY Syracuse NY 13210

LEE, CHARLES RICHARD, b Tarrytown, NY, Dec 3, 42; m 65; c 3. SOIL CHEMISTRY, PLANT NUTRITION. *Educ:* Univ Tampa, BS, 64; Clemson Univ, MS, 65, PhD(agron), 68. *Prof Exp:* Res scientist, Can Dept Agr, 68-73; RES SOIL SCIENTIST, US ARMY CORPS OF ENGRS, ENVIRON LAB, WATERWAYS EXP STA, 73- *Mem:* Am Soc Agron; Can Soc Soil Sci. *Res:* Land treatment of wastewater; heavy metal uptake by marsh plants; soil fertility; minor elements; influence of soil acidity on potato production; plant growth in high zinc, aluminum or manganese media; restoration of problem soil. *Mailing Add:* Environ Lab US Army Engr Waterways Exp Sta PO Box 631 Vicksburg MS 39180-0631

LEE, CHE-HUNG R, METABOLISM. *Prof Exp:* PRIN INVESTR, METAB RES DIV, CCRD NAVY MED RES INST, 84- *Mailing Add:* Metab Res Div CCRD Navy Med Res Inst Bethesda MD 20814

LEE, CHEN HUI, b Taipei, Taiwan, Dec 2, 29; m 62; c 2. FORESTRY. *Educ:* Nat Taiwan Univ, BS, 53; Mich State Univ, MS, 60, PhD(forestry), 66. *Prof Exp:* Asst, Nat Taiwan Univ, 54-59, instr, 59-62; res asst, Mich State Univ, 62-66; from asst prof to assoc prof, 66-77, PROF FORESTRY, UNIV WIS-STEVENS POINT, 77- *Concurrent Pos:* US Forest Serv res grants, 77 & 79. *Mem:* Soc Am Foresters; Chinese Soc Forestry; Japanese Forestry Soc. *Res:* Forest genetics and tree improvement, especially tree physiology, pine leaf anatomy and wood quality. *Mailing Add:* Col Natural Res Univ Wis Stevens Point WI 54481

LEE, CHENG-CHUN, b Youngchow, China, May 24, 22; nat US; m 59; c 2. TOXICOLOGY. *Educ:* Nat Cent Univ, China, BS, 45, MS, 48; Mich State Univ, MS, 50, PhD(physiol), 52. *Prof Exp:* Asst vet med, Nat Cent Univ, China, 45-48; asst physiol, Mich State Univ, 49-51; from pharmacologist to sr pharmacologist, Eli Lilly & Co, 52-62; from sr pharmacologist to prin pharmacologist, Midwest Res Inst, 62-67, head pharmacol & toxicol, 67-76, asst dir, 76-77, assoc dir, 77-78, dep dir, biol sci div, 78-79; SR SCI ADV, HEALTH & ENVIRON REV DIV, OFF TOXIC SUBSTANCES, ENVIRON PROTECTION AGENCY, 79- *Concurrent Pos:* Lectr, Univ Mo-Kansas City, 65-66 & Med Ctr, Univ Kans, 66-79; spec consult antimalarials & drug develop, WHO, 80, 81, 89; prof lectr, Med Ctr, George Washington Univ, 81- *Honors & Awards:* Bronze Medal & Special Achievement & Contribs, Environ Protection Agency, 79, 81, 88 & 90. *Mem:* Am Physiol Soc; Am Soc Pharmacol & Exp Therapeut; Soc Toxicol; Am Col Toxicol; NY Acad Sci. *Res:* Safety evaluation; drug metabolism and disposition; antimalarials; antineoplastics; pharmaceuticals; chemicals; drug development; mechanism of drug action and toxicity; liver and renal functions; author or coauthor of over 100 publications. *Mailing Add:* Health & Environ Rev Div (TS-796) Environ Protection Agency 401 M St SW Washington DC 20460

LEE, CHENG-SHENG, US citizen. AQUACULTURE & MARICULTURE, REPRODUCTIVE PHYSIOLOGY. *Educ:* Nat Taiwan Univ, BS, 70, MS, 72; Univ Tokyo, PhD(aquacult), 79. *Prof Exp:* Res assoc, 79-81, shrimp prog mgr, 81-84, FINFISH PROG MGR, OCEANIC INST, 84- *Concurrent Pos:* Aquatic biologist, Tungkang Marine Lab, 73-76. *Mem:* World Aquacult Soc; Asian Fisheries Soc; Am Fisheries Soc. *Res:* Induction of maturation and spawning of marine finfish; evaluating optimal environment conditions for food organisms and early life stages of finfish. *Mailing Add:* Oceanic Inst PO Box 25280 Honolulu HI 96825

LEE, CHEUK MAN, b China, Feb 22, 29; m; c 2. ORGANIC CHEMISTRY, PHARMACEUTICAL CHEMISTRY. *Educ:* Univ Hong Kong, BSc, 54, MSc, 57; Univ Mich, PhD(pharmaceut chem), 60. *Prof Exp:* ASSOC RES FEL, ABBOTT LABS, 60 - *Mem:* Am Chem Soc. *Res:* Synthesis of organic compounds of biological activities; heterocyclic chemistry; anti-biotics. *Mailing Add:* Abbott Labs One Abbott Park Rd Abbott Park IL 60064-3500

LEE, CHI-HANG, b Vinh Long, SVietnam, Jan 1, 39; nat US; m 64; c 2. NATURAL PRODUCTS CHEMISTRY, FOOD SCIENCE. *Educ:* Southern Ill Univ, Carbondale, BA, 60; Rutgers Univ, New Brunswick, PhD(natural prod chem), 66. *Prof Exp:* Res asst, Rutgers Univ, 61-65, res assoc, 66; sr chemist, Gen Foods Corp, 67-71, from res specialist to sr res specialist, 72-78; sr res scientist, RJR Foods, 78-80; biochem mgr, 80-85, chem dir, 85-87, DIR ANALYTICAL SERV, DEL MONTE CORP, 88- *Concurrent Pos:* Vis prof, King's Col, 73-77; mem, adv comt, Econ Affairs, Repub China, 77-89. *Honors & Awards:* Chairman's Award, Gen Foods Corp, 77; Agr Comn Appreciation Award, Taiwan, Rep China, 89. *Mem:* Am Chem Soc; Am Sci Affil (pres, 82). *Res:* Carbohydrates; flavors; sweeteners; food chemistry, biochemistry and chemical analysis. *Mailing Add:* Del Monte Corp Res Ctr 205 N Wiget Lane Walnut Creek CA 94598

LEE, CHI-HO, b Taitung, Taiwan, July 2, 41; US citizen; m 72; c 1. PHARMACOLOGY. *Educ:* Kaohsiung Med Col, Taiwan, BS, 67; Univ Tokyo, MS, 72; Sch Med Sci, Cornell Univ, PhD(pharmacol), 76. *Prof Exp:* Fel pharmacol, Roche Inst Molecular Biol, 75-76 & Med Col, Cornell Univ, 76-78; fel, 78-79; staff researcher I, 79-81, STAFF RESEARCHER II, CARDIOVASC PHARMACOL, SYNTEX RES, 81- *Concurrent Pos:* Mem, Am Heart Asn, High Blood Pressure Coun. *Mem:* AAAS; Western Pharmacol Soc; Am Soc Hypertension. *Res:* Cardiovascular pharmacology; antihypertensive drugs; evaluation of drug mechanisms; cerebral and peripheral blood vessels; heart; hemodynamics. *Mailing Add:* Inst Pharmacol Syntex Res 3401 Hillview Ave Palo Alto CA 94304

LEE, CHI-JEN, b Yi-Lan, Taiwan, Feb 8, 36; US citizen; m 60; c 3. BIOCHEMICAL BASIS ON IMMUNOGENICITY OF BACTERIAL POLYSACCHARIDES, BACTERIAL POLYSACCHARIDE VACCINES. *Educ:* Nat Taiwan Univ, BS, 57; Johns Hopkins Univ, ScD, 66. *Prof Exp:* Pharmacist, China Chem & Pharmaceut Co, Taiwan, 59-62; res assoc, dept biochem, Johns Hopkins Univ, 66-67; res assoc, Rockefeller Univ, New York City, 67-68, asst prof, 68-73; sr staff fel, Nat Inst Child Health & Human Develop, NIH, Bethesda, Md, 73-74; SUPVRY RES CHEMIST, CTR BIOLOGICS, FOOD & DRUG ADMIN, 74- *Concurrent Pos:* Mem, Nat Reconstruct Comt Med & Health, 78; chmn, 3rd Pneumococcal Workshop, Food & Drug Admin, Bethesda, Md, 79, chmn, Polysaccharide Vaccine Comt; mem bd dirs, Chinese Med & Health Asn, Washington, DC, 80; mem bd advisors, Dept Biochem, Col Med, Nat Cheng Kung Univ, 83-84, vis prof, 84; thesis dir, dept microbiol, Nat Cheng Kung Univ, 83-84 & 87-; referee, CRC Press, Inc, Boca Raton, Fla, 88. *Mem:* Am Asn Immunologists; Am Soc Biol Chemists. *Res:* Characterization of group 19 pneumolysins and cloning of their ply genes have been studied to examine the relationship of ply to virulence; inactivated pneumolysin is conjugated to pneumococcal polysaccharide to form a polysaccharide-protein conjugate to develop a more effective pneumococcal vaccine. *Mailing Add:* Ctr Biologics Food & Drug Admin Bldg 29 Rm 405 8800 Rockville Pike Bethesda MD 20892

LEE, CHIN OK, b Choong-buk, Korea, June 8, 39; US citizen; m 69; c 2. CARDIAC ELECTROPHYSIOLOGY. *Educ:* Seoul Nat Univ, BS, 65, MS, 67; Ind Univ, PhD(physiol), 73. *Prof Exp:* Res fel biochem, Atomic Energy Res Inst, 67-68; fel cardiac electrophysiol, Univ Chicago, 72-76; from asst prof to assoc prof, 76-86, PROF CARDIAC ELECTROPHYSIOL, CORNELL UNIV MED COL, 86- *Concurrent Pos:* Estab investr, Am Heart Asn; Adv consult, Site Visit, NIH, 80; vis prof, Nat Defense Med Ctr, Taipei, Repub China, 88; mem, res peer rev comt, NY Heart Asn, 88-; overseas vis fel, Brit Heart Found, 90. *Honors & Awards:* Louis N Katz Basic Sci Res Prize, 74; Pfizer Award Outstanding Invest, 88. *Mem:* Am Physiol Soc; Biophys Soc; NY Acad Sci; AAAS. *Res:* Cardiac cellular electophysiology; regulation of intracellular Na and Ca ions in heart muscle; intracellular application of ion-selective microelectrodes, intracellular application of ion sensitive dyes. *Mailing Add:* Dept Physiol Med Col Cornell Univ 1300 York Ave New York NY 10021

LEE, CHIN-CHIU, b Hunan, China, Aug 10, 34; m 64; c 1. MICROBIOLOGY, ELECTRON MICROSCOPY. *Educ:* Taiwan Norm Univ, BSc, 55; Loyola Univ, MS, 64; La State Univ, PhD(parasitol), 68. *Prof Exp:* Biol teacher, Taiwan Prov Agr Sch, 55-56 & High Sch, Taiwan, 56-58; asst instr biol, Taiwan Norm Univ, 58-59 & Nanyang Univ, 59-61; res technologist biochem, La State Univ, 63, res assoc parasitol, 64; ASSOC PROF BIOL, KING'S COL, PA, 68-, CHMN DEPT, 77- *Mem:* Am Soc Parasitol; Electron Micros Soc Am; AAAS. *Res:* Medical parasitology; studies on the physiological and ultrastructural aspects of parasites, particularly of parasitic nematodes. *Mailing Add:* Dept Biol King's Col 133 N River St Wilkes-Barre PA 18711

LEE, CHING TSUNG, b Taiwan, July 1, 37; m 67; c 2. QUANTUM OPTICS, MATHEMATICAL PHYSICS. *Educ:* Nat Taiwan Univ, BS, 62; Rice Univ, MA, 65; PhD(physics), 67. *Prof Exp:* Welch Found fel, Tex A&M Univ, 67-68; NASA fel, Rice Univ, 68-69; assoc prof, 69-73, PROF PHYSICS & MATH, ALA A&M UNIV, 73- *Mem:* Am Phys Soc; Optical Soc Am. *Res:* Superradiance; free-electron laser; squeezed states. *Mailing Add:* Dept Physics Ala A&M Univ Normal AL 35762

LEE, CHING-LI, b Taiwan, Repub China, Mar 30, 42; US citizen; m 71; c 3. IMMUNOLOGY. *Educ:* Chung-Hsing Univ, Taiwan, BS, 69; Wayne State Univ, MS, 72, PhD(biochem), 75. *Prof Exp:* Res assoc, Dept Immunol, Mayo Clinic, 75-77; res scientist, 77-79, sr cancer res scientist, 79-83, RES ASST PROF, ROSWELL PARK MEM INST, 81-, ASSOC CANCER RES SCIENTIST, 83- *Mem:* Am Chem Soc; Am Asn Immunologists; Am Asn Cancer Res. *Res:* Antigenic structure of globular proteins with various biochemical, chemical and physical approaches; isolation and characterization of human tumor associated antigens (enzymes) or tumor specific antigens and producing specific antibodies for clinical application. *Mailing Add:* Diag Immunol Res & Biochem Roswell Park Mem Inst 666 Elm St Buffalo NY 14263

LEE, CHING-TSE, b Sinchu, Taiwan, China, May 21, 40; m 69; c 2. ANIMAL BEHAVIOR, BEHAVIOR MEDICINE. *Educ:* Nat Taiwan Univ, BS, 63; Bowling Green State Univ, MA, 67, PhD(psychol), 69. *Prof Exp:* Fac assoc psychol, Univ Tex-Austin, 69-71; from asst prof to assoc prof, 71-82, PROF PSYCHOL, BROOKLYN COL, CITY UNIV NEW YORK, 82- *Concurrent Pos:* Fac res award, City Univ New York, 71, 73, 74; fel, Dept Health Educ & Welfare, 74. *Honors & Awards:* Nat Sci Coun Award, 79 & 80. *Mem:* Animal Behav Soc; Am Psychol Asn; AAAS; Behav Genetics Asn. *Res:* Investigation of animal communication processes through olfaction and hormonal determinants of the production of olfactory signals; effects of neonatal hormones on behavioral differentation; mathematical models applied to animal behavior; biofeedback and behavior medicine; Chinese medicine theories; states of consciousness. *Mailing Add:* Dept Psychol Brooklyn Col City Univ New York Brooklyn NY 11210

LEE, CHING-WEN, b Yunnan, China, Nov 19, 21; US citizen; m 51; c 1. ENGINEERING MECHANICS, SOLID MECHANICS. *Educ:* Nat Inst Technol, Chungking, China, BS, 44; Ill Inst Technol, MS, 56, PhD(mech), 58. *Prof Exp:* Staff mem, Res & Develop Lab, Int Bus Mach Corp, Endicott, NY, 59-60; asst prof eng mech, Case Inst Technol, 60-62; assoc prof, 62-66, PROF ENG SCI & MECH, UNIV TENN, KNOXVILLE, 66- *Mem:* Am Soc Mech Engrs; Am Soc Eng Educ; Am Acad Mech. *Res:* Mechanics of deformable solids; elasticity; plates and shells; thermal stresses. *Mailing Add:* 8300 Bennington Dr Knoxville TN 37919

LEE, CHI-YU GREGORY, b Taiwan, China, Apr 19, 45; US citizen; m; c 2. EXPERIMENTAL BIOLOGY, OBSTETRICS & GYNECOLOGY. *Educ:* Nat Taiwan Univ, China, BSc, 67; Calif Inst Technol, MSc, 71, PhD(chem), 72. *Prof Exp:* Postgrad res chemist, Dept Chem, Univ Calif, San Diego, 72-75; sr staff fel, Lab Animal Genetics, Nat Inst Environ Health Sci, NIH, Research Triangle Park, NC, 76-81; DIR ANDROLOGY, ACUTE CARE UNIT, UNIV BC, 81-, PROF OBSTET-GYNEC, 89- *Concurrent Pos:* Vis prof, Chem Ctr, Univ Lund, Sweden, 75; res asst prof, Dept Biochem, Univ NC, Chapel Hill, 77-81. *Res:* Applications of biotechnology; sperm antigen-based immunocontraceptive vaccines; new tumor markers for early diagnosis and monitoring of cancer patients; monoclonal antibodies against human proteins-hormones and clinical applications. *Mailing Add:* Acute Care Unit Univ BC Vancouver BC V6T 2B5 Can

LEE, CHOI CHUCK, organic chemistry, for more information see previous edition

LEE, CHONG SUNG, b Seoul, Korea, Sept 4, 39; nat US; m 72; c 2. MOLECULAR BIOLOGY. *Educ:* Seoul Nat Univ, BS, 64; Calif Inst Technol, PhD(chem), 70. *Prof Exp:* Grad res asst biophys, Calif Inst Technol, 65-69; fel biochem, Harvard Med Sch, 69-72; asst prof, 72-78, ASSOC PROF MOLECULAR GENETICS, UNIV TEX, AUSTIN, 78- *Concurrent Pos:* Jane Coffin Childs Mem Fund for Med Res fel, 70-72; vis assoc prof, Harvard Med Sch, 82-83. *Mem:* Genetics Soc Am; Am Soc Cell Biol; Am Soc Biochem & Molecular Biol; Korean Chem Soc. *Res:* Molecular genetics of the rosy locus in Drosophila melanogaster. *Mailing Add:* Dept Zool Univ Tex Austin TX 78712

LEE, CHOONG WOONG, b Pyunganpuk-Do, Korea, May 3, 35; m 64; c 4. TELECOMMUNICATION SYSTEMS, EDTV & HDTV SIGNAL PROCESSING. *Educ:* Seoul Nat Univ, BS, 58, MS, 60; Univ Tokyo, Dr Eng, 72. *Prof Exp:* Lectr, Dept Electronics Eng, Seoul Nat Univ, 64-71, from asst prof to assoc prof, 71-81, PROF, DEPT ELECTRONICS ENG, SEOUL NAT UNIV, 81- *Concurrent Pos:* Res assoc, Commun Systs Lab, Nat Res Inst Defense, Korea, 58-64; vis res fel, Dept Elec Eng, Univ Sydney, 63 & Univ Tokyo, 69-71; chmn, Bd Utilization Radio Waves, Ministry Commun, Korea, 86-89; pres, Korean Soc Med & Biol Eng, 88 & Korean Inst Telematics & Electronics Engrs, 89; trustee, Korea Telecommun, 91. *Honors & Awards:* Dongbaik Order of Merit. *Mem:* Fel Inst Elec & Electronics Engrs; Inst Electronics, Info & Commun Engrs Japan; Korean Inst Telematics & Electronics. *Res:* Communication systems especially wide band am, fm demodulators; edtv and hdtv signal processing; medical electronics. *Mailing Add:* 141-97 Sadang 3 Dong Dong-jack Gu Seoul 151-093 Republic of Korea

LEE, CHOUNG MOOK, b Pyungtek, Korea, Oct 3, 35; m 65; c 2. FLUID MECHANICS, OCEAN ENGINEERING. *Educ:* Seoul Nat Univ, BS, 58; Univ NDak, BS, 61; Univ Calif, Berkeley, MEng, 63, PhD(naval archit), 66. *Prof Exp:* Res scientist hydrodyn, David Taylor Res Ctr, 66-82; sci officer fluid mech, Off Naval Res, 82-86; VPRES, POHANG INST SCI & TECHNOL, 86- *Concurrent Pos:* Adj prof, George Washington Univ, 72-73; vpres, Korea Res Inst Ships, 78-79. *Honors & Awards:* Linnard Prize, Soc Naval Architechts & Marine Engrs, 75. *Mem:* Soc Naval Architects & Marine Engrs; Am Soc Mech Engrs; Soc Naval Architects Japan; Sigma Xi. *Res:* Theoretical, mumerical and experimental investigation of ship hydrodynamics; water waves; stability and dynamics of floating and submerged bodies; body-wave interactions; resistance of ships. *Mailing Add:* Dept Mech Eng Pohang Inst Sci & Technol PO Box 125 Pohang 790-600 Republic of Korea

LEE, CHUAN-PU, b Tsing-Tao, China, Sept 24, 31. BIOCHEMISTRY, PHYSICAL CHEMISTRY. *Educ:* Nat Taiwan Univ, BS, 54; Ore State Univ, PhD(biochem), 61. *Hon Degrees:* DPhil, Univ Stockholm, 78. *Prof Exp:* Instr chem, Nat Taiwan Univ, 54-56; res assoc biochem, Ore State Univ, 60-61; Johnson Found res fel, Univ Pa, 61-63; Jane Coffin Childs Mem Fund Med Res fel physiol chem, Wenner-Gren Inst, Stockholm, 63-65, docent, 65-66; mem staff, Johnson Found, Univ Pa, 66-75, from assoc prof to prof biochem, 70-75; PROF BIOCHEM, SCH MED, WAYNE STATE UNIV, 75-, DISTINGUISHED PROF BIOCHEM, 90- *Concurrent Pos:* USPHS career develop award, 68-73; ed, Biochimica et Biophysica Acta & Biochimica et Biophysica Acta Review on Bioenergetics, 73-; ed, Current Topics in Bioenergetics, 81- *Honors & Awards:* Silver Medal, Chinese Chem Soc, 55; Merck Index Award, 60. *Mem:* AAAS; Chinese Chem Soc; Am Soc Biol Chem; Biophys Soc; NY Acad Sci. *Res:* Reaction mechanisms of electron and energy transfer in oxidative phosphorylation; neuromuscular diseases and mitochondrial metabolism. *Mailing Add:* Dept Biochem Sch Med Wayne State Univ Detroit MI 48201

LEE, CHUNG, b Shanghai, China, Sept 18, 36; US citizen; m 65; c 2. REPRODUCTIVE ENDOCRINOLOGY, NUTRITION. *Educ:* Nat Taiwan Univ, BS, 59; WVa Univ, MS, 66, PhD(nutrit & endocrinol), 69. *Prof Exp:* USPHS fel, Albany Med Col, 69-71; assoc obstet & gynec, 71-74, from asst prof to assoc prof, 74-85, PROF UROL, MED SCH, NORTHWESTERN UNIV, CHICAGO, 85-, PROF CELL, MOLECULAR & STRUCT BIOL, 87-, DIR, UROL RES LAB, 74- *Concurrent Pos:* Prin investr, Am Cancer Soc grant, 73-74, Nat Inst Child Health & Human Develop, 77-91, Abbott Labs grant, 78-84, Elsa Univ Pardee Found grant, 82-84, Nat Inst Diabetes & Digestive & Kidney Dis grant, 87-92; res consult, Matpath, 78, Abbott Labs, 78-, Travenol-Baxter, 83-86, Upjohn Co, 85, Lilly Res Labs, 88-, NuClin Diag, 90; lectr, Cook Col Grad Med Sch, 77-80; treas & chair, Finance Comt, Soc Basic Urol Res, 88-90; pres, Chicago Chap, Soc Chinese Bioscientists Am, 88-; mem, Cancer Prevention Comt, Ill Div, Am Cancer Soc, 88-; vis scholar, Vets Gen Hosp, Taipei, Tawiwan, 89. *Mem:* Endocrine Soc; Soc Study Reproduction; Am Asn Cancer Res; Am Physiol Soc; Am Soc Cell Biol; Am Urol Asn; Nat Kidney Found. *Res:* Hormonal regulation of breast and prostate cancer; mechanism of sex steroid action; cancer and hormones; protein analysis and indexing. *Mailing Add:* Dept Urol Northwestern Univ Med Sch Chicago IL 60611

LEE, CHUNG N, b Sinuiju, Korea, Nov 7, 31. MATHEMATICS. *Educ:* Seoul Nat Univ, BA, 54; Univ Va, MA, 57, PhD, 59. *Prof Exp:* From instr to asst prof, 60-68, ASSOC PROF MATH, UNIV MICH, ANN ARBOR, 68- *Mem:* Am Math Soc. *Res:* Algebraic topology; transformation groups; topology of manifolds. *Mailing Add:* Dept Math Pohang Inst Sci PO Box 125 Pohang Republic of Korea

LEE, CLARENCE EDGAR, b San Jose, Calif, Aug 18, 31; m 76; c 3. MATHEMATICAL PHYSICS, COMPUTATIONAL PHYSICS. *Educ:* Univ Calif, Berkeley, BA, 53; Cornell Univ, MA, 62; Univ Colo, Boulder, PhD(physics), 73. *Prof Exp:* Staff mem, Los Alamos Sci Lab, 53-77; prof nuclear sci & res assoc, dept nuclear eng, Tex A&M Univ, 77-85; sr proj scientist, Technadyne, 85-86; SR SCIENTIST, JTA INC, 86- *Concurrent Pos:* Pres, CTM Systs & Software, Inc, 85-; adj prof nuclear & chem eng, Univ NMex-Albuquerque, 86- *Mem:* Am Physics Soc; Am Nuclear Soc; Sigma Xi; Inst Elec & Electronics Engrs. *Res:* Theoretical, nuclear, plasma, solid state, chemical reactor, radiation, computational and applied physics; hydrodynamics, chemical metallurgy and kinetics, heat transfer; computer sciences and numerical analysis; fuel technology. *Mailing Add:* 12805 Arroyo de Vista Dr NE Albuquerque NM 87111

LEE, CRAIG CHUN-KUO, b Taichung, Taiwan, Oct 31, 54; m 88. WASTE TREATMENT, ENVIRONMENTAL CONSULTING. *Educ:* Nat Cent Univ, Taiwan, BS, 76, Ill Inst Technol, MS, 81; Syracuse Univ, PhD(chem eng), 86. *Prof Exp:* Postdoctoral assoc, Dept Chem Eng, Syracuse Univ, 86-87; sr chem engr environ res & develop, Recra Environ Inc, 87-89; SR CHEM ENGR, PIGMENT PILOT PLANT, MOBAY CORP, 90- *Mem:* Am Inst Chem Engrs; Am Chem Soc. *Res:* Hydrometallurgical solvent extraction, heavy metal removal and recovery; electro-coagulation to remove suspended particles or oils from aqueous solutions; sludge drying and stabilization processes. *Mailing Add:* 107 Churchill Ct Summerville SC 24484

LEE, D(ON) WILLIAM, b Everett, Pa, Oct 30, 27; m 55; c 6. CERAMICS, METALS. *Educ:* Bethany Col, BS, 51; Mass Inst Technol, ScD(ceramics), 58. *Prof Exp:* Asst ceramics, Mass Inst Technol, 54-58; mem res staff, Cent Res Dept, E I du Pont de Nemours & Co, 58-63; sr staff mem, Res & Develop Div, Arthur D Little, Inc, 63-67, mgr mat res, 67-, vpres, 75-; CONSULT. *Mem:* Am Ceramic Soc. *Res:* Physical properties of solids; thermal, mechanical and electrical properties of materials; research and development organizational studies; product and process development; high performance material technology; strategic assessment and utilization of material technology. *Mailing Add:* Arthur D Little Inc 15 Acorn Park Cambridge MA 02140

LEE, DAEYONG, b Ham Nam, Korea, June 16, 33; US citizen; m 62; c 3. MECHANICAL & COMPUTER-AIDED ENGINEERING. *Educ:* Ripon Col, BA, 58; Mass Inst Technol, BS, 58, MS, 62, ScD(metall), 65. *Prof Exp:* Res asst metall, Mass Inst Technol, 61-65, staff mem, 65-66; res staff, Gen Elec Res & Develop Ctr, Schenectady, 66-83; PROF, RENSSELAER POLYTECH INST, TROY, NY, 83- *Concurrent Pos:* Mech Metall, Ladish Co, 58-60; adj prof, Rensselaer Polytech Inst, Troy, 81-83. *Mem:* Am Inst Mining, Metall & Petrol Engrs; Am Soc Metals; Am Soc Mech Eng; Am Deep Drawing Res; Soc Plastics Engrs. *Res:* Mechanical metallurgy; plasticity theory, mechanics, constitutive equations, nuclear materials, fracture, friction and lubrication, materials processing; computer-aided engineering. *Mailing Add:* 33 Cobble Hill Rd Loudonville NY 12211

LEE, DAH-YINN, b Tsing-tao, China, June 4, 34; m 62; c 2. CIVIL ENGINEERING, HIGHWAY MATERIALS. *Educ:* Cheng Kung Univ, Taiwan, BSc, 58; Iowa State Univ, PhD(civil eng), 64. *Prof Exp:* Res assoc, Eng Res Inst, 64-65; from asst prof to assoc prof, 65-78, PROF CIVIL ENG, IOWA STATE UNIV, 78- *Concurrent Pos:* Comt mem, Hwy Res Bd, Nat Acad Sci-Nat Res Coun. *Mem:* Am Soc Testing & Mat; Am Soc Civil Engrs; Am Concrete Inst. *Res:* Asphalt durability; aggregates used for asphalt mixtures; waste materials in construction; pavement recycling; foamed asphalt; sulfur in construction. *Mailing Add:* Dept Civil Eng 476 Town Engr Iowa State Univ Ames IA 50011

LEE, DAISY SI, b Peiping, China, July 21, 34; US citizen; m 66; c 2. PEDIATRICS, ALLERGY. *Educ:* Okla Baptist Univ, BA, 56; Bowman Gray Sch Med, MD, 61; Am Bd Pediat, dipl; Am Bd Allergy & Immunol, dipl, 75. *Prof Exp:* Intern med, Georgetown Div, Washington Gen Hosp, 61-62; resident pediat, St Luke's Hosp Ctr, New York, 62-64, NY Heart Asn fel med, 64-65; fel, Inst Nutrit Sci, Columbia Univ & Dept Med, St Lukes Hosp Ctr, 65-66; fel pediat, Sch Med, Stanford Univ, 66-69; pediatrician, Ctr Develop & Learning Disorders, 69-72, asst prof, 69-76, dir pediat allergy prog, 72-76, ASST CLIN PROF PEDIAT, MED CTR, UNIV ALA, BIRMINGHAM, 76- *Mem:* NY Acad Sci; Sigma Xi; Am Acad Allergy; Am Col Allergists. *Mailing Add:* 1025 S 18th St Suite 303 Birmingham AL 35205

LEE, DANIEL DIXON, JR, b Dillon, SC, Sept 27, 35; m 58; c 4. ANIMAL NUTRITION. *Educ:* Clemson Univ, BS, 57, MS, 64; NC State Univ, Raleigh, PhD(biochem & nutrit), 70. *Prof Exp:* Res supvr biochem, NC State Univ, 67-70; from asst prof to assoc prof mineral metab, Dept Animal Indust, Southern Ill Univ, 70-86, asst dean res, Sch Agr, 76-86; head, 87-91, PROF DAIRY SCI, CLEMSON UNIV, 91- *Mem:* Am Soc Animal Sci. *Res:* Trace mineral metabolism; nonprotein nitrogen utilization; wintering of cattle on crop residues and feeding of recycled animal wastes to ruminants. *Mailing Add:* Dept Dairy Sci Clemson Univ 119 Poole Agr Ctr Clemson SC 29634-0363

LEE, DAVID ALLAN, b Ft Smith, Ark, Nov 7, 37; m 60; c 2. APPLIED MATHEMATICS. *Educ:* Univ Mo-Columbia, BSEE, 59; Brown Univ, ScM, 61; Ill Inst Technol, PhD(mech), 63. *Prof Exp:* Res mathematician, Air Force Aerospace Res Labs, 63-71, dir, Appl Math Res Lab, 71-75, head, Dept Math & Computer Sci, USAF Inst Technol, 75-85; DIR, RES & DEVELOP/PROCUREMENT COST ANALYSIS DIV, OFF SECY DEFENSE, 85- *Concurrent Pos:* Vis prof, von Karman Inst Fluid Dynamics, Rhode-St-Genese, Belg, 69-70; sr exec fel, Sr Exec Fel Prog, J F Kennedy Sch, Harvard Univ, 83. *Res:* Econometrics of forecasting development and procurement costs of major defense acquisitions. *Mailing Add:* OASD(PA&E) Rm 2D278 Pentagon Washington DC 20301-1800

LEE, DAVID CHARLES, b Manchester, Eng, June 10, 50. MICROBIOLOGY, IMMUNOLOGY. *Educ:* Stanford Univ, BS, 73; Univ Wash, PhD(biochem), 79. *Prof Exp:* Postdoctoral fel, Dept Biol Chem, Wash Univ, 79-81, Dept Pharmacol, 81-83, res assoc, 83; sr scientist, Oncogen, Seattle, Wash, 83-85; asst prof, 85-91, DIR, NUCLEIC ACID CORE FAC, LINEBERGER COMPREHENSIVE CANCER CTR, UNIV NC, CHAPEL HILL, 89-, ASSOC PROF, DEPT MICROBIOL & IMMUNOL, 91- *Concurrent Pos:* Helen Hay Whitney Found fel, 80-83; core mem, Cancer Cell Biol Prog, Lineberger Comprehensive Cancer Ctr, Univ NC, 85-, mem, Protein Eng & Molecular Genetics Prog, 91- *Res:* Regulation of transforming growth factors; immunoregulatory effects of the interferons; transgenic expression of TGFa and related growth factors. *Mailing Add:* Lineberger Comprehensive Cancer Ctr 237H Sch Med Univ NC Campus Box 7295 Chapel Hill NC 27599-7295

LEE, DAVID K H, BIOCHEMISTRY. *Educ:* McGill Univ, BSc, 69; Queen's Univ, PhD(biochem), 73. *Prof Exp:* Postdoctoral fel, McGill Univ, 73-76, prof asst, 76-77, assoc scientist, Royal Victoria Hosp, 77; sr scientist, Wyeth-Ayerst Res, 77-80, res assoc, 80-86, group leader, 86-88, sect head, 88-89; RES MGR, DEPT BIOCHEM & PHARMACOL, R W JOHNSON PHARMACEUT RES INST, 89- *Mem:* Am Soc Pharmacol & Exp Therapeut; NY Acad Sci; AAAS; Soc Neurosci; Soc Exp Biol & Med. *Mailing Add:* R W Johnson Pharmaceut Res Inst Welsh & McKean Rd Springhouse PA 19477

LEE, DAVID LOUIS, b Oakland, Calif, Oct 19, 48; m 75; c 2. CHEMISTRY. *Educ:* Univ Calif, Berkeley, BS, 70, PhD(chem), 76; Univ Ill, Urbana, MS, 72. *Prof Exp:* Res assoc chem, Univ Calif, San Francisco, 76-77; sr chemist, Cordova Chem Co, 77-80; res chemist, Stauffer Chem Co, 80-82, sr res chemist, 83-84, group supvr, 85-87; GROUP SUPVR, ICI AMERICAS, 87- *Mem:* Am Chem Soc. *Res:* Defining the structure-activity space of new classes of herbicides. *Mailing Add:* ICI Americas 1200 S 47th St Richmond CA 94804-4610

LEE, DAVID MALLIN, b Brooklyn, NY, Jan 18, 44; m 66; c 5. EXPERIMENTAL NUCLEAR PHYSICS, NUCLEAR SAFEGUARDS. *Educ:* Manhattan Col, BS, 66; Univ Va, PhD(physics), 71. *Prof Exp:* Res assoc physics, Univ Va, 71-74; MEM STAFF, LOS ALAMOS NAT LAB, 74- *Concurrent Pos:* US expert, Int Atomic Energy Agency, 80-81. *Mem:* Am Phys Soc; Sigma Xi. *Res:* Medium energy nuclear physics; position sensitive detectors; beam line instrumentation. *Mailing Add:* MS H846 MP4 Los Alamos Nat Lab Los Alamos NM 87545

LEE, DAVID MORRIS, b Rye, NY, Jan 20, 31; m 60; c 2. PHYSICS. *Educ:* Harvard Univ, AB, 52; Univ Conn, MS, 55; Yale Univ, PhD(physics), 59. *Prof Exp:* From instr to assoc prof, 59-68, PROF PHYSICS, CORNELL UNIV, 68- *Concurrent Pos:* Guggenheim fel, 66-67 & 74-75; guest assoc physicist, Brookhaven Nat Lab, 66-67; Japan Soc Prom Sci fel, 77; lectr, Peking Univ, 81; vis prof, Univ Fla, 74-75, Univ Calif, San Diego, 88, AAAS Sect B Electorate Nominating Comt, 90-; chmn, Nat Res Coun Comt Fundamental Constants & Standards, 90- *Honors & Awards:* Sir Francis Simon Mem Prize, British Inst Physics, 76; Oliver Buckley Prize, Am Phys Soc, 81. *Mem:* Nat Acad Sci; fel Am Acad Arts & Sci; fel AAAS; fel Am Phys Soc. *Res:* Low temperature physics with emphasis on quantum fluids and solids, and superconductivity; solid helium three and solid helium four; normal and superfluid phases of liquid helium three; spin polarized hydrogen gas; liquid helium three helium four mixtures; magnetic resonance; ultrasonics; magnetism. *Mailing Add:* Lab Atomic & Solid State Physics Cornell Univ Dept Physics Ithaca NY 14853

LEE, DAVID OI, Hong Kong, China, Feb 5, 40; US citizen; div; c 1. HEAT TRANSFER, ELECTRO MAGNETIC-SEISMIC GEOPHYSICS. *Educ:* Tex A&M Univ, BS, 62, MS, 64. *Prof Exp:* Mem tech staff, 67-90, SR MEM TECH STAFF, SANDIA NAT LABS, 90- *Mem:* Soc Petrol Engrs; Sigma Xi. *Res:* Heat transfer and fluid mechanics experimental and analytical research; system analysis including economic analysis of solar systems; instrumentation development for enhanced oil recovery; development of seismic tehniques for small event detection; development of electromagnetic techniques for sensing of tunnels and contamination plumes. *Mailing Add:* 6409 Quemado NE Albuquerque NM 87109

LEE, DAVID ROBERT, b Grand Forks, NDak, May 9, 45; m 75; c 5. HYDROLOGY, RADIOECOLOGY. *Educ:* Univ NDak, BS, 68, MS, 72; Va Polytech Inst & State Univ, PhD(zool), 76. *Prof Exp:* Res asst prof earth sci & biol, Univ Waterloo, 76-79; RES OFFICER HYDROL & RADIOECOL, ATOMIC ENERGY CAN, CHALK RIVER, ONT, 79- *Concurrent Pos:* Adj prof, Dept Earth Sci, Univ Waterloo, Ont, 79- *Mem:* Am Geophys Union; Am Soc Limnol & Oceanog. *Res:* Groundwater contaminant flux to surface waters. *Mailing Add:* Environ Res Br Chalk River Nuclear Labs Chalk River ON K0J 1J0 Can

LEE, DAVID WEBSTER, b Wenatchee, Wash, Dec 10, 42; m 72; c 2. PLANT EVOLUTION, PLANT STRUCTURE & FUNCTION. *Educ:* Pac Lutheran Univ, BS, 66; Rutgers Univ, MS, 68, PhD(bot), 70. *Prof Exp:* Res assoc bot & microbiol, Ohio State Univ, 70-72; lectr, Univ Malaya, Kuala Lumpur, 73-76; maite de conf assoc, Univ Montpellier II, 77-78; asst prof, 80-82, ASSOC PROF BIOL, FLA INT UNIV, 82- *Concurrent Pos:* Indo-Am fel, 84-85; field res, Cent Am & Southeast Asia. *Mem:* Am Bot Soc; Soc Trop Biol. *Res:* Evolution and adaptation of plants in humid tropical forests. *Mailing Add:* Dept Biol Sci Fla Int Univ Miami FL 33199

LEE, DER-TSAI, b Taipei, China, Apr 5, 49; m 74; c 2. COMPUTER SCIENCE, COMPUTATIONAL GEOMETRY. *Educ:* Nat Taiwan Univ, BS, 71; Univ Ill, MS, 76, PhD(comput sci), 78. *Prof Exp:* Res asst computer sci, Univ Ill, 74-78; from asst prof to assoc prof, 78-86, PROF ELEC ENG & COMPUTER SCI, NORTHWESTERN UNIV, 86- *Concurrent Pos:* Consult, Gen Elec Co, 77 & 79, IBM Corp, 82 & USDA, 85; prin investr, NSF, 79-; vis prof, Academia Sinica, 84-85,; ed, Algorithmica, 85-; managing ed, Int J Computational Geom & Applns, 90- *Mem:* Inst Elec & Electronics Engrs; Asn Comput Mach. *Res:* Design and analysis of algorithms; computational geometry and data structures; very large scale integration systems; computer graphics. *Mailing Add:* Dept of Elec Eng & Comput Sci Northwestern Univ Evanston IL 60208

LEE, DIANA MANG, b Mukden, China; US citizen; m 60; c 1. LIPOPROTEINS. *Educ:* Nat Taiwan Univ, BS, 55; Utah State Univ, MS, 60; Univ Okla, PhD(biochem), 67. *Prof Exp:* Chemist anal chem, Yung-Kang Cement Corp, 55-57; univ asst, chem eng, Nat Taiwan Univ, 57; supvr in chem, Presby-St Luke's Hosp, 61-64, trainee lipoproteins, 64-67; sr investr, 67-71, asst mem, 71-75, ASSOC MEM LIPOPROTEINS, OKLA MED RES FOUND, 75-; ASSOC PROF BIOCHEM, UNIV OKLA SCH MED, 76- *Concurrent Pos:* Res assoc, dept biochem, Univ Okla Sch Med, 68-71, asst prof, 72-76, assoc prof, 76-; mem, credential comt, Coun Arteriosclerosis, Am Heart Asn, 73-75; assoc ed, Artery, 75-; reviewer, Biochem & Biophys, 77-83, 87-; consult, NIH, 79-81. *Mem:* Am Chem Soc; Sigma Xi; AAAS; Am Oil Chemists Soc; NY Acad Sci; Am Soc Biol Chemists. *Res:* Structural aspects of human plasma lipoproteins and apolipoproteins, particularly in low density lipoproteins and apolipoprotein B, their properties, the oxidative and proteolytic effects, and their relationship with atherosclerosis. *Mailing Add:* Lipoprotein & Atherosclerosis Res Prog Okla Med Res Found 825 NE 13th St Oklahoma City OK 73104

LEE, DO IK, b Chinnampo, Korea, Mar 6, 37; nat US; m 70; c 1. LATEX TECHNOLOGY, POLYMER SCIENCE. *Educ:* Seoul Nat Univ, BS, 59; Columbia Univ, MS, 64, EngScD(chem eng), 67. *Prof Exp:* Res chem engr, Dow Chem USA, 67-72, res specialist, 72-75, sr res specialist, 75-79, assoc scientist, 79-82, sr assoc scientist, 82-88, RES SCIENTIST, DOW CHEM USA, 88- *Honors & Awards:* Coating & Graphic Arts Div Award, Tech Asn Pulp & Paper Indust; Charles W Engelhard Medallion. *Mem:* Am Inst Chem Engrs; Am Chem Soc; fel Tech Asn Pulp & Paper Indust; Sigma Xi; AAAS; Polymer Soc Korea. *Res:* Rheology of disperse systems; coating rheology; colloid science; paper coating; emulsion polymerization; inverse emulsion polymerization; suspension polymerization; latex technology; polymer morphology; polymerization kinetics; structured latex technology; particle packing. *Mailing Add:* Dow USA 1604 Bldg Midland MI 48674

LEE, DOH-YEEL, HUMAN GENETICS. *Prof Exp:* RES ASSOC, DEPT NUTRIT, UNIV FLA, GAINESVILLE, 90- *Mailing Add:* Dept Food Sci & Human Sci Univ Fla Gainesville FL 32611

LEE, DO-JAE, b Namwon, Korea, Jan 24, 28; US citizen; m 57; c 2. PHYSICAL ORGANIC CHEMISTRY. *Educ:* Long Beach State Col, BS, 60; San Diego State Col, MS, 64; Univ Calif, San Diego, PhD(chem), 67. *Prof Exp:* RES CHEMIST, TOMS RIVER CHEM CORP, 68- *Mem:* Am Chem Soc. *Res:* Development of new dyestuff and economic process for plant production. *Mailing Add:* 34 Oakside Dr Toms River NJ 08755

LEE, DONALD EDWARD, mineralogy, for more information see previous edition

LEE, DONALD GARRY, b Midale, Sask, June 21, 35; m 59; c 3. PHYSICAL CHEMISTRY, ORGANIC CHEMISTRY. *Educ:* Univ Sask, BA, 58, MA, 60; Univ BC, PhD(chem), 63. *Prof Exp:* Instr chem, Camrose Lutheran Col, 62-65; res assoc, Harvard Univ, 65-66; prof chem, 67-91, PRES LUTHER COL, UNIV REGINA; ASSOC PROF, PAC LUTHERAN UNIV, 66- *Concurrent Pos:* Vis scholar, Univ Oslo, 72-73; vis prof, Stanford Univ, 80-81. *Mem:* Chem Inst Can; Am Chem Soc. *Res:* Oxidation mechanisms; protonation studies; heavy oil and coal research. *Mailing Add:* Dept Chem Univ Regina Regina SK S4S 0A2 Can

LEE, DONALD JACK, b Goldendale, Wash, Jan 28, 32; m 58; c 3. NUTRITIONAL BIOCHEMISTRY. *Educ:* Wash State Univ, BS, 58, MS, 60; Univ Ill, PhD(nutrit, biochem), 65. *Prof Exp:* Assoc prof food sci & technol, Food Protection Sect, Ore State Univ, 65-75; asst dir, Agr Res Ctr, 75-84, team leader, Lesotho Farming Systs Proj, 84-86, DEPT CHAIR, FOOD SCI & HUMAN NUTRIT, WASH STATE UNIV, 86- *Concurrent Pos:* USPHS res grant, 66-75. *Mem:* AAAS; Am Inst Nutrit; Sigma Xi. *Res:* Nutritional biochemistry, especially lipid metabolism; toxicity and carcinogenicity of natural compounds. *Mailing Add:* Food Sci & Human Nutrit Wash State Univ Pullman WA 99164-6376

LEE, DONALD WILLIAM, b Buffalo, NY, Nov 4, 47; m 78; c 2. FLUID MECHANICS, APPLIED MECHANICS. *Educ:* Clarkson Col, BS, 69, MS, 73; Univ Mich, PhD(appl mech), 77. *Prof Exp:* Engr, Ford Motor Co, 69-70; teaching fel mech eng, Clarkson Col Technol, 70-71; res asst appl mech, Univ Mich, 71-76; res assoc fluid mech, 77-81, res staff mem, 81-89, GROUP LEADER APPL PHYS SCI, OAK RIDGE NAT LAB, 89- *Concurrent Pos:* Instr gen sci, Wayne State Univ, 75-76; adj assoc prof, NC State Univ, 87-; mem, Low-level Radioactive Waste Tech Resource Group for 40CFR 193 & Low-level Radioactive Waste Peer Rev Panel for Dept Energy Order 5820-2A, 88-; secy, Air & Radiation Mgt Comt Environ Eng Div, Am Soc Civil Eng, 89-90, vchmn, 90- *Mem:* Am Soc Civil Engrs; Am Soc Mech Engrs; Sigma Xi. *Res:* Environmental fluid dynamics of surface water and groundwater; environmental impact assessment of energy technologies; low-level radioactive waste management. *Mailing Add:* Oak Ridge Nat Lab PO Box 2008 Oak Ridge TN 37831-6045

LEE, DONG HOON, b Seoul, Korea, Nov 17, 38; m 68; c 2. LIE GROUPS & LIE ALGEBRAS, TOPOLOGICAL GROUPS. *Educ:* Seoul Nat Univ, BS, 61; Tulane Univ, PhD(math), 67. *Prof Exp:* From asst prof to assoc prof, 67-80, PROF MATH, CASE WESTERN RESERVE UNIV, 81- *Concurrent Pos:* Vis prof math, Seoul Nat Univ, 76-77. *Mem:* Am Math Soc. *Res:* Representation theory of lie groups and lie algebras. *Mailing Add:* Math Dept Case Western Reserve Univ Cleveland OH 44106

LEE, DOUGLAS HARRY KEDGWIN, b Bristol, Eng, Feb 22, 05; nat US; m 52; c 1. ENVIRONMENTAL SCIENCES. *Educ:* Univ Queensland, MSc, 27; Univ Sydney, MB & BS, 29, dipl trop med, 33, MD, 40; FRACP, 40; Am Bd Indust Hyg, dipl. *Hon Degrees:* MD, Univ Queensland, 86. *Prof Exp:* Med officer, Commonwealth Dept Health, Australia, 30-33; prof physiol, King Edward VII Col Med, Singapore, 35-36; prof physiol, Univ Queensland, 36-48, dean fac med, 38-42; prof physiol climat & lectr environ med, Johns Hopkins Univ, 48-55; chief res br, Off Qm Gen, 55-58; assoc sci dir res, Qm Res & Eng Command, 58-60; chief occup health res & training facility, USPHS, 60-66, assoc dir, Nat Inst Environ Health Sci, 66-73; RETIRED. *Concurrent Pos:* Consult, US Qm Corps, 47-55 & Food & Agr Orgn, UN, 47-60; Cutter lectr, Sch Pub Health, Harvard Univ, 50; adj prof, NC State Univ, 68-74; consult, Mt Sinai Med Sch, 74-76. *Mem:* AAAS; Am Physiol Soc; fel NY Acad Sci. *Res:* Climatic physiology, effects of climate on man and animals and application to clothing, housing and tropical development; occupational and environmental health. *Mailing Add:* Two Hobson Ct Chapel Hill QLD 4069 Australia

LEE, E(RNEST) BRUCE, b Brainerd, Minn, Feb 1, 32; m 54; c 6. CONTROL ENGINEERING. *Educ:* Univ NDak, BS, 55, MS, 56; Univ Minn, PhD(mech eng), 60. *Prof Exp:* Sr res scientist, Honeywell Inc, 56-60, vis scientist, Res Inst Advan Studies, 60-61; sr res scientist, Honeywell Inc, 61-63; assoc prof, 63-66, head dept, 76-81, PROF ELEC ENG, UNIV MINN, MINNEAPOLIS, 66- *Mem:* Soc Indust & Appl Math; Inst Elec & Electronics Engrs. *Res:* Learning systems; differential equations; optimal control theory. *Mailing Add:* Dept Elec Eng Univ Minn 123 Church St 139 EE Minneapolis MN 55455

LEE, E(UGENE) STANLEY, b Hopei, China, Sept 7, 30; US citizen; m 57, 83; c 5. OPERATIONS RESEARCH, COMPUTER SCIENCE. *Educ:* Chieng-Cheng Inst Tech Taiwan, BChE, 53; Univ NC, MS, 57, Princeton Univ, PhD(chem eng), 62. *Prof Exp:* Res engr, Phillips Petrol Co, Okla, 60-66; from asst prof chem eng to assoc prof, 66-71, PROF INDUST ENG, KANS STATE UNIV, 71- *Concurrent Pos:* NSF grant, 71-; vis prof, Univ Southern Calif, Los Angeles, 72-76; ed, Energy Sci & Technol, 77-; assoc ed, J Math Anal & Appln & Math with Appln, assoc ed, Comput & Math with Appln; off Air Res Grantee, USDA. *Mem:* Soc Indust & Appl Math; Am Inst Chem Engrs; Opers Res Soc Am; Inst Indust Engrs. *Res:* Optimization theory; applied mathematics; quasilinearization and invariant imbedding; systems engineering; set theory; expert systems. *Mailing Add:* Dept Indust Eng Kans State Univ Manhattan KS 66502

LEE, EDWARD HSIEN-CHI, b Taiwan, Aug 31, 35; m 67; c 2. PLANT PHYSIOLOGY, CELL BIOLOGY. *Educ:* Nat Taiwan Univ, BS, 59; Univ Kans, MA, 66; Univ Okla, PhD(bot), 69. *Prof Exp:* Lab instr gen bot & taxon, Nat Taiwan Univ, 61-64; teaching asst gen bot & physiol, Univ Okla, 66-69; assoc prof cellular physiol, genetics & microbiol, Cent Methodist Col, Mo, 69-78; PLANT PHYSIOLOGIST, AGR RES SERV, USDA, 78- *Mem:* AAAS; Am Soc Plant Physiol; Plant Growth Regulator Soc Am. *Res:* Environmental stress, air pollution, photosynthesis and tissue culture. *Mailing Add:* Climate Stress Lab Bldg 001 Rm 206 USDA-Agr Res Serv 10300 Baltimore Ave Beltsville MD 20705-2350

LEE, EDWARD KYUNG CHAI, physical chemistry, for more information see previous edition

LEE, EDWARD PRENTISS, b Tulsa, Okla, Oct 3, 42. PHYSICS. *Educ:* Calif Inst Technol, BS, 64; Univ Chicago, MS, 66, PhD(physics), 68. *Prof Exp:* Mem staff plasma physics, Inst Advan Study, Princeton, NJ, 68-70; staff physicist plasma physics, Lawrence Livermore Lab, 70-82; STAFF PHYSICIST ACCELERATOR PHYSICS, LAWRENCE BERKELEY LAB, 82- *Mem:* Fel Am Phys Soc. *Res:* High current charged particle beams; controlled thermonuclear fusion; astrophysics; particle accelerators. *Mailing Add:* Lawrence Berkeley Lab One Cyclotron Rd Berkeley CA 94618

LEE, ELLEN SZETO, b Hong Kong; Brit citizen; c 1. MODELING & SIMULATION, DESIGN. *Educ:* Mt Holyoke Col, BA, 77; Univ Calif, Berkeley, MS, 79, PhD(elec eng & computer sci), 82. *Prof Exp:* Mem tech staff, Bell Commun Res, Inc, 84-86, MEM TECH STAFF, AT&T BELL LABS, 82-84, 86- *Concurrent Pos:* Vis lectr, Univ Calif, Berkeley, 81-82 & Rutgers Univ, 87. *Mem:* Inst Elec & Electronics Engrs; NY Acad Sci. *Res:* Computer-aided design for all aspects of electrical designs: digital and analog circuits, very large scale integration circuits, switch-mode power supplies; circuit theory; object-oriented computer programming. *Mailing Add:* 4501 Torino Pl Plano TX 75093

LEE, EMERSON HOWARD, b Okmulgee, Okla, Feb 23, 21; m 48; c 4. PHYSICAL CHEMISTRY. *Educ:* Univ Tex, BS, 52, PhD(chem), 55. *Prof Exp:* Chemist, Darco Div, Atlas Powder Co, 46-50; res engr, Develop & Res Dept, Continental Oil Co, 54-56; res chemist, Monsanto Co, 56-59, res specialist, 59-60, group leader, 60-65, scientist, 65-82; CONSULT, 82- *Mem:* Am Chem Soc; Sigma Xi. *Res:* Surface chemistry and catalysis. *Mailing Add:* 48 Beaver Dr St Louis MO 63141

LEE, ERASTUS HENRY, b Southport, Eng, Feb 2, 16; m; c 4. PLASTICITY, CONTINUUM MECHANICS. *Educ:* Cambridge Univ, UK, BA, 37, MA, 43; Stanford Univ, PhD(mech eng), 40. *Prof Exp:* Exp sci officer, Ordnance Bd, War Off, UK, 41-43 & armaments res dept, Ministry Supply, UK, 43-46; asst dir tech eng, Dept Atomic Energy, UK, 46-48; from assoc prof to prof appl math, Brown Univ, 48-62, chmn div appl math, 53-58; prof appl mech & aero eng, Stanford Univ, 62-81; REDFERN PROF ENG, RENSSELAER POLYTECH INST, 81- *Honors & Awards:* Timoshenko Medal, Am Soc Mech Engrs, 76. *Mem:* Nat Acad Eng; fel Am Soc Mech Engrs; fel Inst Mech Eng UK; fel Am Acad Mech; fel Soc Eng Sci. *Res:* plasticity analysis and constitutive equations. *Mailing Add:* Dept Mech Eng Aero Eng & Mech Rensselaer Polytech Inst Troy NY 12180-3590

LEE, ERIC KIN-LAM, b Hong Kong, June 25, 48; m 72; c 1. CHEMICAL ENGINEERING, POLYMER CHEMISTRY. *Educ:* NC State Univ, BS, 70, MS, 72, PhD(chem eng), 76. *Prof Exp:* Proj mgr, Bend Res, Inc, 77-84; sr res eng, DuPont Co, 84-86; DIR, MEMBRANE RES & TECHNOL DEVELOP, SEPRACOR INC, 86- *Concurrent Pos:* Res fel, Max Planck Inst Biophys, 76-77. *Mem:* Am Inst Chem Engrs; Am Chem Soc; NAm Membrane Soc. *Res:* Research and development of synthetic membranes, membrane materials, separation processes and process systems for high value industrial and bioprocessing applications; design and engineering of membrane fabrication equipment; composite membranes bioprocessing applications. *Mailing Add:* Sepracor Inc 33 Locke Dr Marlboro MA 01752

LEE, ERNEST Y, BIOCHEMISTRY, MOLECULAR BIOLOGY. *Prof Exp:* PROF BIOCHEM, DEPT BIOCHEM & MOLECULAR BIOL, SCH MED, UNIV MIAMI, 67- *Mailing Add:* Biochem & Molecular Biol Dept Sch Med Univ Miami PO Box 016129 Miami FL 33101

LEE, EUN SUL, b Gongju, Korea, Sept 19, 34; US citizen; m 64; c 2. SURVEY SAMPLING, DEMOGRAPHIC METHODS. *Educ:* Seoul Nat Univ, BA, 57; Univ Ky, MA, 64; NC State Univ, PhD(exp statist & sociol), 70. *Prof Exp:* Res assoc statist anal, NC Bd Higher Educ, 66-69; res biometrician, 69-72, from asst prof to assoc prof, 72-87, PROF BIOMET & DEMOG, SCH PUB HEALTH, UNIV TEX HEALTH SCI CTR, HOUSTON, 87- *Concurrent Pos:* Vis prof, dept sociol, Utah State Univ, 75; UN Fund Pop Activities, Pop & Develop Inst, Seoul Nat Univ, 76; fel hist med, Univ Cincinnati, Nat Endowment Humanities, 80. *Mem:* Am Sociol Asn; Am Statist Asn; Pop Asn Am; Am Pub Health Asn; Int Union Sci Study Pop. *Res:* Ethnic differentials in mortality, fertility and health behavior; analysis of community nutritional status; changing cardiovascular mortality and morbidity trends; sample survey design. *Mailing Add:* Sch Pub Health Univ Tex Health Sci Ctr PO Box 20186 Houston TX 77025

LEE, FANG-JEN SCOTT, b Taipei, Taiwan, Apr 20, 57; US citizen; m 84; c 1. BIOCHEMISTRY, MICROBIOLOGY. *Educ:* Nat Taiwan Univ, BS, 80; NC State Univ, MS, 84, PhD(biotechnol & microbiol), 86. *Prof Exp:* Res fel, Dept Genetics, Harvard Med Sch & Dept Molecular Biol, Mass Gen Hosp, 87-90; SR STAFF, LAB CELLULAR METAB, NAT HEART, LUNG & BLOOD INST, NIH, 90- *Concurrent Pos:* Consult, Yung-Shin Pharmaceut Industs Co Ltd, 86-90. *Mem:* Am Soc Biochem & Molecular Biol; Am Soc Microbiol; AAAS; Protein Soc; Sigma Xi. *Res:* Investigation of protein processing and signal transduction. *Mailing Add:* Dept Cellular Metab Nat Heart Lung & Blood Inst NIH Rm 5N307 Bldg 10 Bethesda MD 20892

LEE, FLOYD DENMAN, b Hays, Kans, Apr 27, 38. NUCLEAR PHYSICS. *Educ:* Univ Kans, BS, 60, PhD(physics), 66. *Prof Exp:* Instr physics, Univ Kans, 65-66; Nat Acad Sci-Nat Res Coun assoc, 66-68; ASSOC PROF PHYSICS, MONT STATE UNIV, 68- *Mem:* Am Asn Physics Teachers; Am Phys Soc. *Res:* Low-energy nuclear research with Van-de-Graaf accelerators; nuclear structure. *Mailing Add:* Dept Physics AJM Johnson Hall Mont State Univ Bozeman MT 59717

LEE, FRED C, US citizen; c 2. POWER CONVERSION. *Educ:* Nat Cheng Kung Univ, Taiwan, BS, 68; Duke Univ, Durham, MS, 72, PhD(elec eng), 74. *Prof Exp:* Teaching asst, Duke Univ, 70-72, res asst, Spacecraft Systs Res Lab, 72-77; from asst prof to prof, 77-86, DIR, VA POLYTECH ENG CTR, VA POLYTECH INST & STATE UNIV, 85-, JAMES S TUCKER PROF, 86-; DIR, TECHNOL DEVELOP CTR POWER ELECTRONICS, 87- *Concurrent Pos:* Assoc ed, Inst Elec & Electronics Engrs Trans Power Electronics, 85-; mem, Power Electronics Coun, Inst Elec & Electronics Engrs, 85-87; bd dirs, Zytec Corp, 86- & adv bd, Power Integrations Inc, 88- *Honors & Awards:* Ralph R Teeter Award, Soc Automotive Eng, 85; William E Newell Power Electronics Award, Inst Elec & Electronics Engrs, Power Electronics Soc, 89. *Mem:* Fel Inst Elec & Electronics Engrs; Inst Elec & Electronics Engrs Indust Applications Soc; Inst Elec & Electronics Engrs Power Electronics Soc (vpres, 88-89); Brit Inst Elec Engrs. *Res:* Power conversion; power devices; high frequency resonant converters; distributed power systems; power hybrids; space power systems; nonlinear control; design optimization; system modeling; analysis and simulation. *Mailing Add:* Elec Eng Dept Va Polytech Inst & State Univ 340 Whittemore Blacksburg VA 24061

LEE, FREDERICK STRUBE, b Baltimore, Md, Dec 26, 27; m 52; c 3. PHYSICAL CHEMISTRY. *Educ:* Johns Hopkins Univ, AB, 50; Brown Univ, PhD(chem), 58. *Prof Exp:* Res chemist, Agr Div, W R Grace Co, 58-60 & Anal & Phys Div, 60-61; prof chem, Baltimore Jr Col, 61-71, dir gen studies, 71-76, dir sci, bus & technol studies, 76-79, PROF MATH, COMMUNITY COL BALTIMORE, 61- *Mem:* Sigma Xi. *Res:* X-ray crystallography; inorganic synthesis; physical inorganic chemistry. *Mailing Add:* 9126 Winands Rd Owings Mills MD 21117

LEE, GARRETT, b San Francisco, Calif, June 23, 46. CARDIOVASCULAR RESEARCH, LASER MEDICINE. *Educ:* Univ Calif, Berkeley, BA, 68; Univ Calif, Davis, MD, 72. *Prof Exp:* Internship med, Duke Univ Med Ctr, 72-73; residency med, Univ Calif, Davis, 73-75, fel cardiol, 74-76, asst prof med, 76-83; dir Laser Res Lab, Cedars Med Ctr, Fla, 83-84; dir res, Western Heart Inst, San Francisco, 84-85; DIR RES, NORTHERN CALIF HEART & LUNG INST, 86- *Concurrent Pos:* Med dir, Asprin Myocardial Infarction Study, UADV Calif Davis, 75-78; med consult, Calif Comn Peace Officer Standards, 75-80; chmn, cardiovascular Curriculum, Univ Calif, Davis, 79-83, dir Cardiac Cath Lab, 78-83; bd dir, Am Heart Assoc, 81-83; Counr, Am Fedn Clin Res, 77-81. *Mem:* Am Col Cardiol; Am Soc Laser Med & Surg; Am Col Clin Pharmacol; Am Col Angiol; Am Fedn Clin Res; Am Heart Assoc. *Res:* Cardiovascular pharmacology and interventional cardiology including bisers and heart disease, angioscopy, balloon angioplasty and thrombolytic therapy in accute myocardial infarction. *Mailing Add:* 900 Alice St Oakland CA 94607

LEE, GARTH LORAINE, physical chemistry; deceased, see previous edition for last biography

LEE, GARY ALBERT, b Scottsbluff, Nebr, May 18, 41; m 62; c 3. WEED SCIENCE. *Educ:* Univ Wyo, BS, 64, MS, 65, PhD(agron), 71. *Prof Exp:* Instr weed sci, Univ Wyo, 65-71, from asst prof to assoc prof, 71-75; asst dir agr res, 79-80, head, Dept Plant, Soil & Entomol Sci, 80-86, PROF WEED SCI, UNIV IDAHO, 75-, ASSOC DEAN RES & DIR, IDAHO AGR EXP STA, 86- *Concurrent Pos:* Consult, US Borax Res Corp, 75-; pres, Western Soc Weed Sci; bd dirs, WSSA, 77-79, ICIA, 80-86, WRAC, 86-; secy, WAAESD, 87-88, pres 90; bd dirs, Ctr Appl Sci Technol, 90. *Honors & Awards:* Fel, Western Soc Weed Sci, 79. *Mem:* Weed Sci Soc Am; Am Soc Sugarbeet Technologists; Soc Range Mgt; Int Crop Improv Asn. *Res:* Mechanisms of herbicide selectivity in agronomic crops and perennial weed control; population dynamics of weeds in agronomic crops and rangeland; influence of herbicides on the metabolism of weed species; biological control and integrated weed management systems. *Mailing Add:* Idaho Agr Exp Sta Univ Idaho Col Agr Moscow ID 83843

LEE, GEORGE C, b Peiping, China, July 17, 32; m 61; c 2. STRUCTURAL ENGINEERING, BIOMECHANICS. *Educ:* Nat Taiwan Univ, BSE, 55; Lehigh Univ, MS, 58, PhD(civil eng), 60. *Prof Exp:* Res fel civil eng, Lehigh Univ, 56-57, from res asst to res assoc, Frit* Eng Lab, 57-61; asst prof civil eng, 61-63, assoc prof, 63-67, actg chmn, Dept Civil Eng, 70-71, chmn, 72-77, dir grad studies, 71-78, dir socio-eng prog, 71-78, dir, Health Instrument & Device Inst, 84-85, PROF ENG & APPL SCI, STATE UNIV NY, BUFFALO, 67-, DEAN ENG & APPL SCI, 78-, ASSOC DIR, CALSPAN-STATE UNIV NY, BUFFALO RES CTR, 83- *Concurrent Pos:* Spec eng consult, Struct Dynamics Dept, Bell Aerosysts Co, 65-; NIH grant & sr res fel, Dept Physiol, Harvard Univ, Sch Pub Health, 69-70; head eng mech sect, NSF, 77-78. *Honors & Awards:* Adams Mem Award, Am Welding Soc, 74; Superior Accomplishment Award, NSF, 77. *Mem:* Am Soc Civil Engrs; Am Soc Eng Educ; Am Welding Soc; Sigma Xi; AAAS. *Res:* Buckling and stability analysis of structural members, frames, plates and shells; ultimate strength design; respiratory mechanics and lung elasticity; earthquake engineering; cold regions engineering. *Mailing Add:* 288 Countryside Lane Williamsville NY 14221

LEE, GEORGE FRED, environmental engineering, aquatic chemistry, for more information see previous edition

LEE, GEORGE H, II, b Ithaca, NY, Feb 26, 39; m 64; c 2. FORENSIC TOXICOLOGY. *Educ:* Rensselaer Polytech Inst, 61, PhD(phys chem), 65. *Prof Exp:* Res assoc, Cornell Univ, 65-67; res chemist, Res Ctr, Hercules Inc, Del, 67-71; Sr res chemist, Dept Phys & Biol Sci, Southwest Res Inst, 71-73; assoc found scientist, Southwest Found Res & Educ, 73-77; sr res scientist fire technol, Southwest Res Inst, 77-81, sr res scientist, US Army Fuels & Lubricants Res Lab, 81-86; assoc chief, Forensic & Documentation Div, Air Force Drug Testing Lab, 86-88, CHIEF, VOLATILE ORGANICS FUNCTION, OCCUP & ENVIRON HEALTH DIRECTORATE, BROOKS AFB, TEX, 88- *Mem:* NY Acad Sci; Soc Forensic Toxicologists. *Res:* Analysis of potable and non-potable waters, soils and tissues for toxic contaminants. *Mailing Add:* 11107 Whispering Wind San Antonio TX 78230

LEE, GLENN RICHARD, b Ogden, Utah, May 18, 32; m 69; c 2. INTERNAL MEDICINE, HEMATOLOGY. *Educ:* Univ Utah, BS, 53, MD, 56. *Prof Exp:* Intern med, Boston City Hosp, 56-57, asst resident, 57-58; clin fel hemat, 60-61, res fel, 61-63, from instr to assoc prof, 63-73, assoc dean acad affairs, 73-76, PROF MED, UNIV UTAH, 73-, DEAN, COL MED, 78- *Mem:* Am Fedn Clin Res; Am Soc Hemat; Am Col Physicians; Am Soc Clin Invest. *Res:* Clinical and experimentally induced abnormalities in heme biosynthesis; physiologic consequences of copper deficiency; iron metabolism. *Mailing Add:* Dept Int Med Univ Utah Sch Med Vet Admin Med Ctr 500 Foothill Dr Salt Lake City UT 84148

LEE, GLORIA, NEUROLOGY. *Prof Exp:* ASST PROF NEUROSCI, DEPT NEUROL, BRIGHAM & WOMEN'S HOSP, 86- *Mailing Add:* Dept Neurol Brigham & Women's Hosp 75 Francis St Boston MA 02115

LEE, GORDON M(ELVIN), b Minneapolis, Minn, Jan 3, 17; m 41; c 4. ELECTRICAL ENGINEERING. *Educ:* Univ Minn, BEE, 38; Univ Mo, MS, 39; Mass Inst Technol, DSc(elec eng), 44. *Prof Exp:* Asst elec eng, Univ Mo, 38-39; asst elec engr, Mass Inst Technol, 39-44, mem staff, Div Indust Coop, 44-45; tech dir elec eng & secy-treas, Cent Res Labs, Inc, Sargent Industs, 45-73, pres, 73-81; CONSULT, 81- *Concurrent Pos:* Mem, Nat Defense Res Comt, 44; lectr, Univ Minn, 48. *Honors & Awards:* Thompson Mem Prize, Inst Elec & Electronics Engrs, 46. *Mem:* AAAS; Am Nuclear Soc; Inst Elec & Electronics Engrs. *Res:* Remote handling equipment; properties of dielectrics; high-speed oscillography; development of high-speed micro-oscillograph and remote handling equipment. *Mailing Add:* Wacouta Beach Red Wing MN 55066

LEE, GRETA MARLENE, b El Paso, Tex. CELL BIOLOGY, VIDEO MICROSCOPY. *Educ:* Univ Mo, Columbia, BS, 71; E Tenn State Univ, MS, 83; Duke Univ, PhD(zool), 89. *Prof Exp:* FEL, DEPT CELL BIOL & ANAT, UNIV NC, CHAPEL HILL, 89- *Mem:* Am Soc Cell Biol. *Res:* Cell motility and plasma membrane structure; movements of individual molecules in membranes using specific colloidal gold probes and video enhanced contrast microscopy. *Mailing Add:* Dept Cell Biol & Anat Univ NC 108 Taylor Hall Chapel Hill NC 27599-7090

LEE, GRIFF C, b Jackson, Miss, Aug 17, 26; m 50; c 3. OFFSHORE DESIGN, OFFSHORE CONSTRUCTION. *Educ:* Tulane Univ, BE, 48; Rice Univ, MS, 51. *Prof Exp:* Civil engr, Humble Oil & Refining Co, 48-54; prin engr & design engr, 54-66, chief engr, 66-75, group vpres, 75-78, vpres res & develop, McDermott Inc, 78-83; PRES, GRIFF C LEE INC, 83- *Concurrent Pos:* Mem, Marine Bd Nat Res Coun, Offshore Comt Am Petrol Inst, Adv Comt-Offshore Technol Detnorske Veritas, Welding Res Coun, Tech Panel Offshore Installations for Lloyd's Register Shipping, Comt Offshore Platforms & Bd Adv, Tulane Univ, 80- *Mem:* Nat Acad Eng; Am Soc Civil Engrs; Am Concrete Inst; Am Welding Soc; Soc Petrol Eng. *Res:* Advanced engineering technology; offshore construction for the petroleum industry. *Mailing Add:* Griff Lee Inc PO Box 70787 New Orleans LA 70172

LEE, H(O) C(HONG), b Seoul, Korea, Aug 2, 33; m 65; c 2. MECHANICAL ENGINEERING. *Educ:* Univ Bridgeport, BS, 57; Rensselaer Polytech Inst, MME, 59, PhD(mech eng), 62. *Prof Exp:* Asst prof mech eng, Rensselaer Polytech Inst, 62-68; staff engr, IBM Corp, 68-70, adv engr, 70-77, sr engr, 77-89, SR TECH STAFF MEM, IBM CORP, 89- *Concurrent Pos:* Consult, Mech Tech, Inc, 62-65 & Gen Elec Co, 65-68; adj assoc prof, Rensselaer Polytech Inst, 68-70. *Mem:* Am Soc Mech Engrs. *Res:* Dynamics of structural elements; rotor dynamics. *Mailing Add:* Eight Tudor Dr Endicott NY 13760

LEE, HARLEY CLYDE, b Bellville, Ohio, Nov 7, 01; m 32. MINERALOGY, CHEMISTRY. *Educ:* Ohio State Univ, BEM, 27. *Prof Exp:* Chief chemist, Dolomite Inc, 26-30; res asst mineral, US Steel Corp, 30-31; dir res ceramics, Basic Dolomite Inc, 31-41; supt metall, Tech Dept, Basic Magnesium Inc, 41-45; vpres mineral technol, Basic Inc, 45-74, dir, 53-79, consult mineral technol, 74-79; RETIRED. *Concurrent Pos:* Dir, Elgin Electronics Inc, 60-79. *Mem:* Fel Am Ceramic Soc; Am Chem Soc; fel Mineral Soc Am; Geol Soc Am; Am Inst Mining, Metall & Petrol Engrs. *Res:* Res: Refractories, slags, calcium and magnesium compounds; steel and magnesium production; mineral processing; silicate chemistry and high temperature chemistry of refractory oxides. *Mailing Add:* One Watergate Apt 15-b 111 Gulfstream Ave Sarasota FL 34236

LEE, HAROLD HON-KWONG, b China, Jan 31, 34; m 66; c 2. DEVELOPMENTAL BIOLOGY, BIOTECHNOLOGY. *Educ:* Okla Baptist, AB, 56; Univ Tenn, MS, 58, PhD(embryol), 65. *Prof Exp:* USPHS fel, Carnegie Inst, 65-67; PROF BIOL, UNIV TOLEDO, 75-, DIR, MASTER LIB STUDIES, 88- *Concurrent Pos:* Am Cancer Soc, NIH grants, Rockefeller, United Nations grants, Lolor Found, res grants. *Mem:* Soc Develop Biologists; AAAS. *Res:* Cell interactions and fertilization development of reproduction; biotechnology; tissue culture. *Mailing Add:* Dept Biol Univ Toledo 2801 W Bancroft Toledo OH 43606

LEE, HARVEY S, b China, Feb 7, 49; US citizen. DYNAMICS OF VEHICLES ON GUIDEWAY. *Educ:* Newark Col Eng, BSc, 72; Ohio State Univ, MSc, 74. *Prof Exp:* MECH ENGR, US DEPT TRANSP, 75- *Mem:* Am Soc Mech Engrs. *Res:* Vehicle dynamics as it relates to safety; response of railroad vehicles to track irregularities that could lead to derailments; safety considerations of magnetically levitated trains. *Mailing Add:* 51 Appleton St Apt 10 Arlington MA 02174

LEE, HARVIE HO, b Chiang-si, China, Aug 17, 37; US citizen; m 65; c 3. METALLURGY, CORROSION. *Educ:* Nat Taiwan Univ, BS, 59; Va Polytech Inst, MS, 63; Mass Inst Technol, PhD(metall), 71. *Prof Exp:* Res metallurgist coating & corrosion, Inland Steel Res Labs, 63-67; res asst corrosion & metall, Corrosion Lab, Mass Inst Technol, 67-71; SR RES ENGR CORROSION & METALLIC COATINGS, INLAND STEEL RES LABS, 67- *Mem:* Nat Asn Corrosion Engrs; Am Soc Metals; Am Soc Testing & Mat. *Res:* Stress corrosion cracking of high strength low alloy steels; development of new hot dip metallic coatings with improved corrosion resistance. *Mailing Add:* Inland Steel Res Labs 3001 E Columbus Dr East Chicago IN 46312

LEE, HAYNES A, b Johnson City, Tenn, Oct 14, 32; m 59; c 3. LASERS, GLASS TECHNOLOGY. *Educ:* Emory & Henry Col, BS, 54; State Univ NY Col Ceramics, Alfred Univ, MS, 61. *Prof Exp:* Glass technologist, Thatcher Glass Mfg Co, NY, 61-63; glass technologist, Owens-Ill Inc, 63-66, glass scientist, 66-68, chief laser scientist, 68-72; gen mgr optical prod, Laser Inst Am, 73-80, gen mgr, 81-; RETIRED. *Concurrent Pos:* Bd dir, Laser Inst Am, 75-80. *Mem:* Am Ceramic Soc; Sigma Xi; Laser Inst Am; Optical Soc Am. *Res:* Electronic pheonmena in glasses, particularly laser phenomena. *Mailing Add:* 5845 Viramar Rd Toledo OH 43611

LEE, HENRY C, b China, Nov 22, 38; US citizen; m 63; c 2. FORENSIC SCIENCE, BIOCHEMISTRY. *Educ:* John Jay Col NY, BS, 72; NY Univ, MS, 74, PhD(biochem), 75. *Hon Degrees:* DSc, Univ New Haven, 90. *Prof Exp:* Asst prof, 75-76, assoc prof & dir forensic sci, 76-78, PROF FORENSIC SCI, UNIV NEW HAVEN, 78-, DIR, CTR FOR APPLIED RES, 76-; DIR, FORENSIC SCI LAB, 77-; CHIEF, CONN STATE FORENSIC SCI LAB, 79- *Concurrent Pos:* Consult, Conn State Police Forensic Lab, 75; vis prof, Seton Hall Univ, 76, Northeastern Univ, 77-79; vis fac, Yale Univ, 78; res grant, Univ New Haven, 78 & 79; chief, Conn State Forensic Sci Lab, 79-; ed, Forensic Sci, 81- *Honors & Awards:* Distinguished Criminalist Award, Am Acad Forensic Sci, 88; John Dondero Award, Int Asn Identification, 89. *Mem:* NY Acad Sci; fel Am Acad Forensic Sci; AAAS; Am Soc Testing & Mat; Am Soc Crime Lab Dirs; Int Found Sci; distinguished mem Int Asn Identification. *Res:* Protein biosynthesis; blood individualization, forensic science and crime scene investigation; forensic chemistry; DNA typing. *Mailing Add:* Forensic Sci Lab 294 Colony St Meriden CT 06450-2098

LEE, HENRY JOUNG, b Seoul, SKorea, Nov 17, 41; US citizen; m 69; c 3. ANTI-INFLAMMATORY STEROIDS. *Educ:* Seoul Nat Univ, BS, 64, MS, 66; Okla State Univ, PhD(biochem), 71. *Prof Exp:* Instr food technol, Seoul Women's Col, 66-67; res asst biochem, Okla State Univ, 67-71 & res assoc, Mt Sinai Sch Med, 71-73; from asst prof to assoc prof, 73-82, PROF MED CHEM, COL PHARM, FLA A&M UNIV, 82- *Concurrent Pos:* Vis scientist, Rockefeller Univ, 79; prin investr, NIH, 79-, grant reviewer, 85; consult, Taisho Pharmaceut Co, 85, Sandoz Pharmaceut Co, 86- *Mem:* Sigma Xi; Am Chem Soc; Am Soc Biol Chemists. *Res:* Chemical synthesis and evaluation of new anti-inflammatory steroids without adverse effects. *Mailing Add:* Col Pharm Fla A&M Univ Tallahassee FL 32307

LEE, HOONG-CHIEN, b Hong Kong, Aug 12, 41; m 65; c 3. PHYSICS. *Educ:* Nat Taiwan Univ, BSc, 63; McGill Univ, MSc, 67, PhD(physics), 69. *Prof Exp:* Tech collabr physics, Brookhaven Nat Lab, 67-68; assoc res officer, 68-85, SR RES OFFICER THEORET PHYSICS, CHALK RIVER NUCLEAR LABS, ATOMIC ENERGY CAN, LTD, 85- *Mem:* Am Phys Soc; Can Asn Physicists. *Res:* Theoretical nuclear physics; structure of nuclei; electromagnetic and weak interaction; structure and decay of elementary particles and nuclei. *Mailing Add:* Physics Div Chalk River Labs Chalk River ON K0J 1J0 Can

LEE, HSI-NAN, b Taiwan, July 15, 46; m 74; c 2. ATMOSPHERIC SCIENCE, MATHEMATIC NUMERICAL TECHNIQUES. *Educ:* Col Chinese Cult, BS, 70; Univ Utah, MS, 73, PhD(meteorol), 77. *Prof Exp:* Res asst meteorol, Univ Utah, 71-77; asst meteorologist atmospheric sci, Brookhaven Nat Lab, 77-79, assoc meteorologist, 79-80; res asst prof, Dept Meteorol, Univ Utah, 80-82. *Mem:* Sigma Xi; Am Meteorol Soc. *Res:* Advanced numerical modeling study in environmental air pollution; large scale atmospheric wave structure and nonlinear interaction in wave number frequency space; mathematic numerical techniques; numerical techniques for solving partial and ordinary differential equation. *Mailing Add:* 20943 Devonshire St Chatsworth CA 91311

LEE, HSIN-YI, b Hsin-chu, Taiwan; m; c 1. DEVELOPMENTAL BIOLOGY. *Educ:* Nat Taiwan Univ, BS, 59; Oberlin Col, MA, 64; Univ Minn, Minneapolis, PhD(zool), 67. *Prof Exp:* Res assoc tissue cult, Cardiovasc Inst, Michael Reese Hosp & Med Ctr, 68; from asst prof to assoc prof biol, 68-78, PROF ZOOL, RUTGERS UNIV, CAMDEN, 78-, PROF BIOL & CHMN DEPT, 84- *Mem:* AAAS; Am Soc Zoologists. *Res:* Neural tube formation. *Mailing Add:* Dept Biol Rutgers Univ Camden NJ 08102

LEE, HUA, b Taipei, Taiwan, Sept 30, 52; m 76; c 2. ACOUSTIC MICROSCOPY, IMAGING ALGORITHM DESIGN. *Educ:* Nat Taiwan Univ, BS, 74; Univ Calif, Santa Barbara, MS, 78, PhD(elec eng), 80. *Prof Exp:* Asst prof elec eng, Univ Calif, Santa Barbara, 80-83; from asst prof to assoc prof elec eng, Univ Ill, 83-90; PROF ELEC ENG, UNIV CALIF, SANTA BARBARA, 90- *Concurrent Pos:* Presidential young investr award, 85. *Mem:* Inst Elec & Electronics Engrs; Acoust Soc Am. *Res:* All aspects of the imaging technology; high-resolution high-speed imaging techniques; imaging system optimization; radar and sonar imaging; signal analysis and processing; biomedical imaging, high resolution gerome sequencing; computer vision and non-destructive evaluation. *Mailing Add:* Dept Elec & Computer Eng Univ Calif Santa Barbara Santa Barbara CA 93106

LEE, HUA-TSUN, b Nanking, China, May 11, 37; m 63; c 4. MATHEMATICS. *Educ:* Tunghai Univ, Taiwan, BS, 59; Univ Pittsburgh, PhD(math), 71. *Prof Exp:* Asst physics, Tunghai Univ, Taiwan, 61; asst physics, Univ Pittsburgh, 61-64, asst math, 65-67 & 68-69, instr biostatist, Grad Sch Pub Health, 67-68; asst prof, 69-74, ASSOC PROF MATH, POINT PARK COL, 74- *Mem:* Math Asn Am. *Res:* Summability methods of infinite series. *Mailing Add:* Eight Coral Dr Pittsburgh PA 15238

LEE, HULBERT AUSTIN, b Chelsea, Que, June 17, 23; m 47; c 5. GEOLOGY. *Educ:* Queen's Univ, Ont, BSc, 49; Univ Chicago, PhD(geol), 53. *Prof Exp:* Geologist, Geol Surv Can, 50-69; CONSULT GEOLOGIST & PRES, LEE GEO-INDICATORS LTD, 69- *Concurrent Pos:* Vis lectr, Univ NB, 64-65. *Mem:* Fel Geol Soc Am; Can Inst Mining & Metall. *Res:* Correlation of quaternary events around Hudson Bay, the Tyrrell Sea and Keewatin ice divide; quaternary studies in New Brunswick; esker and till methods of mineral exploration now firmly established and extensively used in the exploration industry; kimberlite petrology, engineering terrain analysis of Ontario. *Mailing Add:* Lee Geo-Indicators Ltd Ten Alexander St Box 68 Stittsville ON K2S 1A2 Can

LEE, HYUNG MO, b Tanchon, Korea, Sept 27, 26; US citizen; m 59; c 2. MEDICINE, SURGERY. *Educ:* Keijo Imp Univ, BS, 45; Seoul Nat Univ, MD, 49. *Prof Exp:* Res fel surg, Med Col Va, 59-61; from instr to assoc prof, 63-70, PROF SURG, MED COL VA, 70-, CHMN DIV VASCULAR GEN SURG, 76- *Concurrent Pos:* Dir, Clin Transplant Prog. *Mem:* Am Col Surg; Am Soc Nephrology; Transplantation Soc; NY Acad Sci. *Res:* Renal homotransplantation. *Mailing Add:* Dept Surg MCV Scis Va Commonwealth Univ Box 163 Richmond VA 23298

LEE, I P, b Rupl Korea, Dec 25, 35; US citizen; m 62; c 4. MOLECULAR BIOLOGY, ENDOCRINOLOGY. *Educ:* Pac Lutheran Univ, BA, 59; Univ Wash, MS, 69, PhD(pharmacol), 71. *Prof Exp:* Res pharmacologist, Nat Cancer Inst, NIH, 69-72, Nat Inst Environ Health Serv, 72-85; RES PHARMACOLOGIST MOLECULAR TOXICOL, DEPT HEALTH & HUMAN SERV, FOOD & DRUG ADMIN, 85- *Concurrent Pos:* Vis prof, Fed Tech Univ & Univ Zurich, Inst Toxicol, 75-76, Med Sch, Cath Univ, 83, Med Sch, Yonsei Univ, 84, Beijing Polytech Univ, 87. *Mem:* Am Soc Pharmacol & Exp Therapeut; Soc Toxicol; Sigma Xi. *Res:* Metabolism and toxicology of chemical carcinogens in male reproductive tissues. *Mailing Add:* Dept Health & Human Serv & Molecular Toxicol Food & Drug Admin HFF-162 Washington DC 20204

LEE, INSUP, b Seoul, Korea, Mar 15, 55; c 2. COMPUTER & INFORMATION SCIENCES. *Educ:* Univ NC, Chapel Hill, BS, 77; Univ Wis, Madison, MS, 78 & PhD(comput sci), 83. *Prof Exp:* ASSOC PROF COMPUT SCI, UNIV PA, 83- *Mem:* Inst Elec & Electronics Engrs; Am Comput Mach. *Res:* Distributed real-time computing. *Mailing Add:* Dept Comput & Info Sci Univ Penn Moore Sch Elec Eng Philadelphia PA 19104

LEE, I-YANG, b Nanking, China, Dec 21, 46; m 72; c 2. NUCLEAR PHYSICS. *Educ:* Nat Taiwan Univ, BSc, 68; Univ Pittsburgh, PhD(physics), 74. *Prof Exp:* Physicist nuclear physics, Lawrence Berkeley Lab, 75-77; PHYSICIST NUCLEAR PHYSICS, OAK RIDGE NAT LAB, 77- *Mem:* Am Phys Soc. *Mailing Add:* Oak Ridge Nat Lab Bldg 6000 MS 371 PO Box 2008 Oak Ridge TN 37831-6371

LEE, JA H, b Hamyang, S Korea, Apr 25, 25; US citizen; m; c 4. PLASMA PHYSICS. *Educ:* Kyungpook Nat Univ, Korea, BS, 48; George Peabody Col, Nashville, MS, 61; Vanderbilt Univ, Nashville, MS, 62, PhD(physics), 64. *Prof Exp:* Prof physics, Kyungpook Nat Univ, Daegu, Korea, 65-67; res assoc physics, NASA/Nat Acad Sci, 67-69; sr res assoc, Vanderbilt Univ, Nashville, 69-73, res assoc prof, 73-78, res prof physics, 78-83; SR RES SCIENTIST, NASA LANGLEY RES CTR, 83- *Concurrent Pos:* Prin investr grants, Langley Res Ctr, NASA, 69- *Honors & Awards:* Group Achievement Award, NASA, 81. *Mem:* Am Phys Soc; Inst Elec & Electronics Engrs. *Res:* Pulsed high beta plasma; solar pumped laser and laser pumping source development. *Mailing Add:* 37 East Governor Dr Newport News VA 23602-7405

LEE, JAMES A, b Troy, NY, July 11, 25; m 46; c 1. HUMAN ECOLOGY, PUBLIC HEALTH. *Educ:* Union Col, BS, 49; Cornell Univ, MS, 51; George Washington Univ, MPh, 69, PhD, 70; Univ Sarajevu, 78. *Honor Degrees:* LLD, Penn Col, 79. *Prof Exp:* Sr res biologist, NH, 51-56; tech asst dir, State Conserv Dept, Minn, 56-61, from dep comnr to comnr, State Dept Resources Develop, 61-63; scientist adminr, USPHS, Washington, DC, 63-66, asst environ health to asst secy health & sci affairs, Dept HEW, 67-69, dir human ecol, 69-70; DIR ENVIRON, HEALTH & SCI, THE WORLD BANK, 70- *Concurrent Pos:* Vis prof, Sch Med, Cornell Univ, 76-; co-chmn, Int Trop Dis Res Prog & Int Diarrheal Dis Control Prog; Woodrow Wilson vis fel, 81-; clin prof, Sch Med, Univ Miami, 84- *Mem:* Am Col Prev Med; Ecol Soc Am; Am Soc Trop Med & Hyg; Royal Soc Trop Med & Hyg; Am Acad Health Admin; Sigma Xi. *Res:* Environmental, public health, and socio-cultural aspects of international economic development; human ecology with emphasis on multi-environmental causation of diseases; social anthropology and medical sociology; natural resources planning and management. *Mailing Add:* 112 Cove Rd Williamsburg VA 23185

LEE, JAMES B, b Ware, Mass, June 30, 30; m 64; c 2. ENDOCRINOLOGY, METABOLISM. *Educ:* Col Holy Cross, AB, 51; Jefferson Med Col, MD, 56. *Prof Exp:* Intern med, St Vincent Hosp, Worcester, Mass, 56-57; resident, Pa Hosp, Philadelphia, 57-58; resident, Georgetown Univ Hosp, 58-59; USPH res fel renal metab, Peter Bent Brigham Hosp, Boston, 59-62; dir res metab & endocrinol, St Vincent Hosp, Worcester, 62-68; assoc prof med & chief sect exp med, Sch Med, St Louis Univ, 68-71; med dir, Skilled Nursery Fac, Erie County Med Ctr, 81-87; PROF MED, SCH MED, STATE UNIV NY, BUFFALO, 71-, CHMN, HUMAN INSTNL REV BD, 82- *Concurrent Pos:* Mass Heart Asn res grant, 62-; USPHS res grant, 62-, develop training grant, 64-; asst prof, Georgetown Univ Hosp, 63-68. *Mem:* Am Physiol Soc; Am Soc Clin Invest; Am Fedn Clin Res; Endocrine Soc; Am Heart Asn. *Res:* Hypertension; in vitro metabolism of kidney cortex and medulla related to sodium excretion; isolation and indentification of the renal prostaglandes, from renal medulla and their deficny as a cause of essential hypertension. *Mailing Add:* Sch Med State Univ NY 3435 Main St Buffalo NY 14214

LEE, JAMES C, b Shanghai, China, Dec, 16, 41; US citizen; m 69; c 2. HUMAN BIOLOGY CHEMICAL & GENETICS. *Educ:* Hope Col, BA, 66; Case Western Reserve Univ, PhD(biochem), 71. *Prof Exp:* Postdoctoral fel, Grad Dept Biochem, Brandeis Univ, Boston, Mass, 71-76; from asst prof to prof biochem, St Louis Univ, Mo, 76-90; PROF & ROBERT A WELCH CHAIR CHEM, DEPT HUMAN BIOL CHEM & GENETICS, UNIV TEX MED BR, GALVESTON, 90- *Concurrent Pos:* Mem, Molecular & Cellular Biophys Study Sect, NIH, 81-85 & spec study sect, 89, 90 & 91; vis prof, State Univ NY, Stony Brook, 86. *Mem:* Am Chem Soc-Div Biol Chem; AAAS; Biophys Soc; Am Soc Biol Chemists; Am Soc Cell Biol. *Res:* Elucidate the regulatory mechanisms of biological functions at the molecular level; gene expressions; supramacromolecular assembly; enzyme activity. *Mailing Add:* Human Biol Chem & Genetics Dept Univ Tex Med Br Galveston TX 77550

LEE, JAMES KELLY, mechanical engineering, control systems, for more information see previous edition

LEE, JAMES NORMAN, b Santa Monica, Calif, Dec 20, 56; m 79; c 3. MEDICAL & MAGNET IMAGING. *Educ:* Univ Utah, BA, 80, MS, 82; Duke Univ, PhD, 86. *Prof Exp:* Res assoc, Duke Univ, 86-88; ASST PROF RADIOL, UNIV UTAH, 88- *Mem:* Soc Magnetic Resonance Med; Soc Magnetic Resonance Imaging. *Res:* New techniques in magnetic resonance angiography. *Mailing Add:* Dept Radiol Med Imaging Res Lab AC 213 Sch Med Univ Utah Salt Lake City UT 84132

LEE, JAMES WILLIAM, geology, for more information see previous edition

LEE, JANG Y, CARDIOVASCULAR. *Prof Exp:* RES INVESTR, DEPT CARDIOVASC PHARMACOL, ABBOTT LABS, 84- *Mailing Add:* Dept Cardiovasc Pharmacol Abbott Labs Dept 47-C AP-10 Abbott Park IL 60064

LEE, JEAN CHOR-YIN WONG, b Canton, China, Aug 26, 41; m 67; c 2. BIOCHEMISTRY. *Educ:* Chung Chi Col, Hong Kong, dipl, 62; Univ Nebr, Lincoln, PhD(chem), 67. *Prof Exp:* Res instr biochem, Col Med, Univ Nebr, Omaha, 67-70; fel chem & res assoc, Univ Nebr, Lincoln, 70-86; GROUP

LEADER, HARRIS LABS, INC, 86-; ADJ ASST PROF, UNIV NEBR, LINCOLN, 87- *Honors & Awards:* Lewis E Harris Award for Excellence, 89. *Mem:* Am Chem Soc; Am Asn Pharmaceut Sci; Clin Ligand Assay Soc. *Res:* Biomembranes, structure and transport; analysis of pharmaceuticals. *Mailing Add:* PO Box 80837 Harris Lab Inc Lincoln NE 68501

LEE, JEFFREY STEPHEN, b Salt Lake City, Utah, Aug 25, 44; m 69; c 2. INDUSTRIAL HYGIENE, OCCUPATIONAL HEALTH. *Educ:* Univ Utah, BS, 67; Univ Calif, Berkeley, MPH, 71, PhD(environ health), 80. *Prof Exp:* Indust hygienist, Nat Inst Occup Safety & Health, 69-76; dep dir, Health Response Team, Occup Safety & Health Admin, US Dept Labor, 76-79; asst prof, 80-87, DIR INDUST HYG, ROCKY MT CTR ENVIRON HEALTH, UNIV UTAH, 79-, ASSOC PROF, DEPT FAMILY & PREV MED, MED SCH, 87- *Concurrent Pos:* Chmn, Am Conf Govt Indust Hygienists. *Mem:* Am Acad Indust Hyg; Am Indust Hyg Asn; Am Conf Gov Indust Hyg; Am Pub Health Asn; Int Occup Hyg Asn (pres). *Res:* Industrial hygiene and occupational health; asbestos; quantification and toxicology of particulates specifically cadmium; industrial hygiene problems related to mining and new energy development. *Mailing Add:* Dept Family & Prev Med Univ Utah Salt Lake City UT 84112

LEE, JEN-SHIH, b Kwangtung, China, Aug 22, 40; nat US; m 66; c 3. ENGINEERING MECHANICS. *Educ:* Nat Taiwan Univ, BS, 61; Calif Inst Technol, MS, 63, PhD(aeronaut, math), 66. *Prof Exp:* Asst res engr, Univ Calif, San Diego, 66-69; from asst prof to assoc prof, 69-83, PROF BIOMED ENG, UNIV VA, 84-, CHAIR DEPT, 88- *Concurrent Pos:* San Diego County Heart Asn advan res fel, 66-69; USPHS res grant, Univ Va, 71; Nat Heart & Lung Inst res career develop award, 75-80. *Mem:* Am Physiol Soc; Biomed Eng Soc; Am Soc Mech Eng; Microcirc Soc. *Res:* Hemodynamics; pulmonary mechanics and edema; indicator dilution technique as applied to microcirculation and transcapillary exchange in microvessels. *Mailing Add:* Dept Biomed Eng Univ Va Health Sci Stacey Hall Charlottesville VA 22908

LEE, JOE, b Shanghai, China, Mar 2, 39; US citizen; m 70; c 2. TRANSPORTATION ENGINEERING, TRAFFIC ENGINEERING. *Educ:* Nat Taiwan Univ, BSc, 61; Asian Inst Technol, MEng, 66; Ohio State Univ, PhD(transp & traffic eng), 71. *Prof Exp:* Installation officer maintenance, Chinese Air Force, 61-62; field engr construct, BES Eng Corp, 62-64; res asst, Asian Inst Technol, 64-66 & Ohio State Univ, 66-71; from asst prof to assoc prof civil eng, 71-81, PROF CIVIL ENG, UNIV KANS, 81-, DIR, TRANSP CTR, 77- *Concurrent Pos:* Prin investr, Ctr Res, Inc, Univ Kans, 74-; consult, Land Plan Eng, 78-; vis prof, Nat Chiao Tung Univ, Taiwan, 83- *Mem:* Inst Transp Engrs; Am Soc Civil Engrs; Am Soc Eng Educ; Am Road & Transp Builders Asn. *Res:* Traffic flow dynamics; highway design; traffic safety; traffic signal operation and general systems theory. *Mailing Add:* Dept Civil Eng Univ Kans Lawrence KS 66045

LEE, JOHN A N, b Coventry, UK, Dec 23, 34; US citizen; m 88; c 5. PROGRAMMING LANGUAGES, SOFTWARE ENGINEERING. *Educ:* Univ Nottingham, BSc, 55, PhD(appl sci), 58. *Prof Exp:* Asst engr, Freeman, Fox & Partners, 57-59; asst prof civil eng, Queens Univ, Kingston, Ont, 59-64; prof computer sci, Univ Mass, Amherst, 64-74; PROF COMPUTER SCI, VA TECH, 74- *Concurrent Pos:* Dir, Comput Ctr, Queens Univ, Kingston, Ont, 60-64; head, Dept Computer Sci, Univ Mass, Amherst, 64-69, assoc dir, Comput Ctr, 64-74; chair, Standards Comt, Asn Comput Mach, 64-73, coun mem, 82-84; vis prof, Univ Denver, 70-71; head, Dept Computer Sci, Va Tech, 79-80; fac assoc, Santa Teresa Lab, IBM Corp, San Jose, 80-81; ed & ed-in-chief, Am Fedn Info Processing Socs, 80-91; dir, Ctr Innovative Technol, Inst Info Technol, Herndon, Va, 87-91; ed-in-chief, Inst Elec & Electronics Engrs Computer Soc, 91- *Honors & Awards:* Outstanding Contrib Award, Asn Comput Mach, 81; Cert Distinguished Serv, US Dept Defense, 83. *Mem:* Asn Comput Mach (vpres, 84-86); Inst Elec & Electronics Engrs Computer Soc; Am Fedn Info Processing Socs. *Res:* Programming languages: history, compilers and design; software engineering: formal specifications, testing, integration history of computing. *Mailing Add:* Dept Computer Sci Va Tech 133 McBryde Hall Blacksburg VA 24061-0106

LEE, JOHN ALEXANDER HUGH, b Isle of Wight, Eng, Oct 10, 25; m 49; c 3. EPIDEMIOLOGY. *Educ:* Univ Edinburgh, BSc, 47, MB, ChB, 49, MD, 55; Univ London, DPH, 52. *Prof Exp:* Fel epidemiol, London Sch Hyg & Trop Med, 52-55; mem sci staff, Social Med Res Unit, Med Res Coun, London Hosp, 55-66; PROF EPIDEMIOL, UNIV WASH, 66- *Mem:* Am Epidemiol Soc; Brit Soc Social Med; Brit Med Asn; Int Epidemiol Asn. *Res:* Epidemiology of neoplastic diesease. *Mailing Add:* Dept Epidemiol & Int Health Univ Wash Seattle WA 98195

LEE, JOHN C, IMMUNOLOGY. *Prof Exp:* ASST DIR, DEPT CELL SCI, SMITH KLINE & FRENCH LABS, 89- *Mailing Add:* Dept Immunol Smith Kline & French Labs L-101 PO Box 1539 King of Prussia PA 19406-0939

LEE, JOHN CHAESEUNG, b Seoul, Korea, July 29, 41; US citizen; m 71; c 1. NUCLEAR ENGINEERING. *Educ:* Seoul Nat Univ, BS, 63; Univ Calif, Berkeley, PhD(nuclear eng), 69. *Prof Exp:* Sr engr nuclear eng, Westinghouse Elec Corp, 69-73; sr engr, Gen Elec Co, 73-74; from asst prof to assoc prof, 74-81, actg chmn, Dept Nuclear eng, 86, PROF NUCLEAR ENG, UNIV MICH ANN ARBOR, 81- *Concurrent Pos:* Consult, Adv Comt Reactor Safeguards, US Nuclear Regulatory Comn, 75- & Los Alamos Nat Lab, 77-88; vis scientist, Germany, 81-82. *Mem:* Fel Am Nuclear Soc; AAAS. *Res:* Nuclear reactor physics; reactor kinetics; fuel management; reactor safety analysis; power plant simulation and control. *Mailing Add:* Dept Nuclear Eng Univ Mich Ann Arbor MI 48109-2104

LEE, JOHN CHEUNG HAN, b China, Dec 1, 45; m 77; c 1. INDUSTRIAL MICROBIOLOGY, INFECTIOUS DISEASES-ANTIBOTICS. *Educ:* Rutgers Univ, BA, 67; Long Island Univ, MS, 72; St John's Univ, PhD(microbiol), 79. *Prof Exp:* Sect head, Julius Schmid, Inc, 67-73; supvr, Miles Pharmaceut, 79-; STUART PHARMACEUT, WILMINGTON, DEL; DEPT IMMUNOL, SMITH KLINE & FRENCH. *Mem:* Am Soc Microbiol; NY Acad Sci; Soc Indust Microbiol; Sigma Xi. *Res:* In vitro activity of new antibiotics; effect of antibiotics on functions of macrophages. *Mailing Add:* L-101 At 709 Swedeland Rd Swedeland PA 19479

LEE, JOHN CHUNG, b Shanghai, China, Mar 2, 37; US citizen; m 63; c 2. BIOCHEMISTRY. *Educ:* Taylor Univ, AB, 61; Purdue Univ, West Lafayette, MSc, 64, PhD(molecular biol), 67. *Prof Exp:* Res assoc biochem, Mass Inst Technol, 67-69; from asst prof to assoc prof, Med Sch, 69-85, PROF BIOCHEM, UNIV TEX HEALTH SCI CTR, SAN ANTONIO, 85- *Concurrent Pos:* USPHS res grant, Univ Tex Health Sci Ctr San Antonio, 71- *Mem:* Am Soc Cell Biol; Am Chem Soc; Am Soc Biol Chem; AAAS. *Res:* Structure and function of nucleic acids and of ribosomes differentiation. *Mailing Add:* Dept Biochem Univ Tex Health Sci Ctr San Antonio TX 78284-7760

LEE, JOHN D(AVID), b Barrie, Ont, Aug 24, 24; nat US; m 49; c 3. AERODYNAMICS. *Educ:* Univ Toronto, BSc & MSc, 50, PhD(aerophys), 52. *Prof Exp:* Asst prof & res assoc, 53-55, assoc prof aeronaut eng, 55-59, dir aerodyn lab, 55-68, PROF AERONAUT & ASTRONAUT ENG, OHIO STATE UNIV, 59-, DIR AERONAUT & ASTRONAUT RES LAB, 68- *Concurrent Pos:* Consult, Naval Ord Lab, Wright Air Develop Ctr, Fluidyne Eng Corp, Sandia Corp & US Army. *Mem:* Am Inst Aeronaut & Astronaut; Sigma Xi. *Res:* Hypersonic fluid mechanics; boundary layer phenomena; low density flows; transonic high Reynolds number flows. *Mailing Add:* Dept Aeronaut & Astronaut Eng Ohio State Univ Columbus OH 43210

LEE, JOHN DENIS, b Trinidad, WI, Apr 22, 29; m 58; c 3. METEOROLOGY. *Educ:* Fla State Univ, BS, 70, MS, 71, PhD(meteorol), 73. *Prof Exp:* Fel, Nat Ctr Atmospheric Res, 73-74; asst prof meteorol, Pa State Univ, 74-78; UN expert meteorol educ & training, Saudi Arabia, 82-86; RES ASSOC, PA STATE UNIV, 86- *Concurrent Pos:* Lectr meteorol, UN Develop Prog for Advan Training in Meteorol in Eng-speaking Caribbean Territories, 78-81. *Mem:* Am Meteorol Soc; Sigma Xi; NY Acad Sci. *Res:* Numerical modeling of cooling tower plumes and urban pollution; time series analysis; numerical modeling of flow over obstacles. *Mailing Add:* 447 Kemmerer Rd State College PA 16801

LEE, JOHN HAK SHAN, b Hong Kong, Sept 7, 38; Can citizen; m 62; c 2. COMBUSTION, EXPLOSIONS. *Educ:* McGill Univ, BSc, 60, PhD(eng), 65; Mass Inst Technol, MSc, 62. *Prof Exp:* Res asst mech eng, Mass Inst Technol, 60-62; jr res officer, Nat Res Coun, 62-63; lectr, 62-64, from asst prof to assoc prof, 64-66, PROF MECH ENG, MCGILL UNIV, 73- *Concurrent Pos:* Mem, comt studies hazardous substances, Nat Res Coun, 84- *Honors & Awards:* Silver Medal, Combustion Inst, 80 & Dionizy Smolenski Medal, 88. *Mem:* Order Engrs Quebec; Am Phys Soc; Combustion Inst. *Res:* Cause, effects, prevention and mitigation of accidental explosions in the production, transport, storage of flammable gases, liquids, organic, metallic and coal dusts; hydrogen combustion and vapor explosion problems pertaining to nuclear reactor accident. *Mailing Add:* Dept Mech Eng Rm 459 McGill Univ 817 Sherbrooke St W Montreal PQ H2A 2K6 Can

LEE, JOHN JOSEPH, b Philadelphia, Pa, Feb 23, 33; m 56; c 2. MARINE MICROBIOLOGY, PROTOZOOLOGY. *Educ:* Queens Col, NY, BS, 55; Univ Mass, MA, 57; NY Univ, PhD(biol), 60. *Prof Exp:* Asst prof, NY Univ, 61-66; from asst prof to assoc prof, 66-72, PROF BIOL, CITY COL NEW YORK, 72- *Concurrent Pos:* Res fel & dir, Living Foraminifera Lab, Am Mus Natural Hist, 60-68, res assoc, 70-; dir, Marine Microbiol Ecol Lab, Inst Oceanog, City Univ New York, 68-; res assoc, Lamont-Doherty Geol Observ, 70- & Philadelphia Acad Natural Sci, 85- *Mem:* Fel AAAS; Soc Protozool; Phycol Soc Am; Am Soc Microbiol; Am Micros Soc. *Res:* Cytology, fine structure, life history, ecology, cultivation and nutrition of foraminifera; algal endosymbiosis in foraminifera, meiofauna, benthic marine food webs and diatom assemblages. *Mailing Add:* Biol Convent Ave 138 City Col New York New York NY 10031

LEE, JOHN K, CELL & DEVELOPMENTAL BIOLOGY. *Prof Exp:* POSTDOCTORAL FEL, CELL & DEVELOP BIOL DEPT, HARVARD UNIV, 89- *Mailing Add:* Cell & Develop Biol Dept Harvard Univ 16 Divinity Ave Cambridge MA 02138

LEE, JOHN NORMAN, b Schenectady, NY, Dec 2, 44; m 68; c 2. PHYSICS. *Educ:* Union Col, Schenectady, BS, 66; Johns Hopkins Univ, MA, 68, PhD(physics), 71. *Prof Exp:* Res asst physics, Johns Hopkins Univ, 69-71; physicist, Electronics Res & Develop Command, Harry Diamond Labs, 71-80; SUPVR RES PHYSICIST, NAVAL RES LAB, 80- *Honors & Awards:* Kingslake Medal & Prize, Soc Photo-Optical Instrumentation Engrs. *Mem:* Am Phys Soc; sr mem Inst Elec & Electronics Engrs. *Res:* Signal processing using acousto-optics and surface acoustic wave devices; radiation damage in optical materials and components; spectroscopy of optical and magnetic materials; optical devices for signal processing; optical computing. *Mailing Add:* Naval Res Lab Code 6530 Washington DC 20375

LEE, JOHN WILLIAM, b Sydney, Australia, Apr 7, 35; m 60; c 2. CHEMICAL PHYSICS, BIOPHYSICS. *Educ:* Univ New South Wales, BSc, 56, PhD(phys chem), 60. *Prof Exp:* Res assoc biochem, McCollum-Pratt Inst, Johns Hopkins Univ, 61-63; staff scientist, New Eng Inst Med Res, 63-69; assoc prof, 69-75; PROF BIOCHEM, UNIV GA, 75- *Mem:* Am Chem Soc; Am Phys Soc. *Res:* Positron annihilation in matter; radiation chemistry; energy exchange processes in chemical and biological systems; bioluminescence; chemiluminescence; radiation physics. *Mailing Add:* Dept Biochem 622 Grad Studies Univ Ga Athens GA 30602

LEE, JOHN YUCHU, b Tai-Ho, China, Jan 25, 48; nat US; m 76; c 2. ORGANIC CHEMISTRY, ORGANOMETALLICS. *Educ:* Nat Cheng-Kung Univ, Taiwan, BS, 70; SDak State Univ, MS, 74; Vanderbilt Univ, PhD, 78. *Prof Exp:* Asst, SDak State Univ, 73-74 & Vanderbilt Univ, 74-78; res assoc bioorg chem, chem dept, Tex A&M Univ, 78-79, Robert Welch fel organometallics, 79-80; res chemist indust & specialty chem, res & develop, 80-88, sr res chem, 88-89, RES SPECIALIST, NEW PROJ DEVELOP, ETHYL CORP, 89- *Concurrent Pos:* Chem reagent officer, Chinese Air Force Acad, 70-71; prin & prog dir Baton Rouge Chinese Sch, 85-86; pres, Chinese Asn, Baton Rouge, 89-90. *Mem:* Sigma Xi; fel Am Inst Chemists; Am

Chem Soc; Royal Soc Chem; Japanese Chem Soc; Chinese Am Chem Asn; AAAS; Soc Francaise Chimie. *Res:* Synthesis, isolation and characterization of pharmaceutical, agricultural, surfactant and detergent intermediates, as well as process improvement; bromine chemicals; chemicals from biomass; alkylation products; flame retardants; catalysis; polymer chemistry; advanced materials. *Mailing Add:* Tech Ctr Ethyl Corp 8000 GSRI Ave Baton Rouge LA 70820

LEE, JON H(YUNKOO), b Seoul, Korea, Mar 5, 34; US citizen; m 70; c 4. APPLIED MATHEMATICS, COMPUTER SIMULATION. *Educ:* Seoul Nat Univ, BS, 56; Ohio State Univ, MS, 58, PhD(chem eng), 62. *Prof Exp:* Chem engr, USAF Mat Lab, 62-64, res engr, Aerospace Res Labs, 64-75, RES SCIENTIST, FLIGHT DYNAMICS LAB, WRIGHT-PATTERSON AFB, OHIO, 75- *Concurrent Pos:* Adj prof, Dept Chem Eng, Univ Dayton, 62. *Honors & Awards:* Gen Foulois Res Award, 87. *Mem:* Am Phys Soc; Sigma Xi; Am Acad Mech; Soc Indust & Appl Math. *Res:* Fluid dynamics; structural vibration; turbulence; computational mechanics; dynamical systems. *Mailing Add:* 10661 Putnam Rd Englewood OH 45322

LEE, JONATHAN K P, b Kiangsu, China, July 13, 37; m 67; c 3. NUCLEAR PHYSICS. *Educ:* McGill Univ, BEng, 60, MSc, 62, PhD(nuclear physics), 65. *Prof Exp:* Nat Res Coun Can overseas fel, 65-66; res asst, Univ Toronto, 66-68; from asst prof to assoc prof, 68-81, PROF NUCLEAR PHYSICS, MCGILL UNIV, 81-, DIR, FOSTER RADIATION LAB, 79- *Mem:* Can Asn Physicists; Am Phys Soc; Oper Sci Apliquees. *Res:* Nuclear structure studies; laser spectroscopy. *Mailing Add:* Foster Radiol Lab McGill Univ 3610 University St Montreal PQ H3A 2B2 Can

LEE, JONG SUN, b Suwon, Korea, July 10, 32; m 58; c 3. MICROBIOLOGY. *Educ:* Univ Calif, Berkeley, BA, 58; Ore State Univ, MS, 62, PhD(microbiol), 63. *Prof Exp:* From asst prof to prof food microbiol, Ore State Univ, 63-80, prof food sci & tech, 80-; AT FISHERY INDUST, TECHNOL CTR, KODIAK, ALASKA. *Mem:* AAAS; Am Soc Microbiol; Inst Food Technol; Brit Soc Appl Bact; Sigma Xi. *Res:* Microbiology of seafoods. *Mailing Add:* FITC SFOS Univ Alaska 202 Center St Kodiak AK 99615

LEE, JORDAN GREY, b Lafayette, La, Feb 28, 14. EXPERIMENTAL BIOLOGY. *Educ:* La State Univ, BS, 34, MA, 35; Univ Mo, PhD(chem), 40. *Prof Exp:* Asst prof chem, Okla State Univ, 40-43; from asst prof to prof, 43-79, chmn dept, 62-67, EMER PROF BIOCHEM, LA STATE UNIV, 79- *Mem:* Fel AAAS; Sigma Xi; Am Chem Soc; Am Inst Nutrit. *Mailing Add:* La State Univ 451 Nelson Baton Rouge LA 70808

LEE, JOSEPH CHUEN KWUN, b Chungking, China, Oct 6, 38; US citizen; m. PATHOLOGY, CELL BIOLOGY. *Educ:* Univ Hong Kong, MB, BS, 64; Univ Rochester, PhD(path), 70; FRCP(C), 71, Am Bd Pathol, 72, FRCP(A), 85, MRC(path), 86. *Prof Exp:* Intern, Hong Kong, 64; rotating intern, St Francis Hosp, New York, 66; resident path, New York Hosp-Cornell Med Ctr, 66-67, Toronto Gen Hosp, Univ Toronto, 70-71 & Princess Margaret Hosp, Ont Cancer Inst Can, 71-72; asst prof to assoc prof path & oncol, Univ Rochester Med Ctr, 72-80; vis prof, NIH, 80; res fel, Armed Forces Inst Pathol, 81; dean fac med, 86-89, PROF PATHOL, CHINESE UNIV, HONG KONG, 82- *Mem:* AAAS; Am Soc Cell Biol; Soc Anal Cytol; Am Asn Path; Electron Micros Soc Am. *Res:* Experimental pathology of iron metabolism; structure of chromosomes; nasopharyngeal carcinoma. *Mailing Add:* Fac Med Chinese Univ Hong Kong Shatin New Territories Hong Kong

LEE, JOSHUA ALEXANDER, b Rocky Ford, Ga, Oct 30, 24; m 56; c 2. GENETICS. *Educ:* San Diego State Col, AB, 50; Univ Calif, PhD(genetics), 58. *Prof Exp:* Technician, Univ Calif, 51-53, asst, 54-56; GENETICIST, AGR EXP STA, NC STATE UNIV, 58-, PROF CROP SCI, 71- *Mem:* Crop Sci Soc Am; Soc Study Evolution. *Res:* Genetical problems pertaining to the improvement of domesticated cotton species. *Mailing Add:* 118 West Lake Dr Sylvania GA 30467

LEE, JUI SHUAN, b Hopei, China, Aug 14, 13; US citizen. PHYSIOLOGY. *Educ:* Tsinghua Univ, Peking, BS, 36; Univ Minn, Minneapolis, PhD(physiol), 53. *Prof Exp:* Res fel, Biol lab, Sci Soc China, 36-38; res fel physiol, 39-41, from lectr physiol to assoc prof, Col Med, Nat Cent Univ, Chengtu, 42-47; res fel physiol, Univ Minn, Minneapolis, 53-56; res assoc biochem nutrit, Grad Sch Pub Health, Univ Pittsburgh, 56-57; from res assoc to assoc prof physiol, Univ Minn, Minneapolis, 57-89; RETIRED. *Mem:* Am Physiol Soc; Microcirculatory Soc. *Res:* Mechanism of water absorption and secretion by the mammalian small intestine; lymph flow and capillary pressure; tissue fluid pressure of the small intestine during water transport. *Mailing Add:* 2721 Gaset Pl Santa Ana CA 92724

LEE, JUNE KEY, b Seoul, Korea, Aug 9, 43; US citizen; m 70; c 3. ENGINEERING MECHANICS. *Educ:* Han-Yang Univ, BS, 65; Tenn Technol Univ, MS, 70; Univ Tex, Austin, PhD(eng mech), 76. *Prof Exp:* Res asst comput mech, Res Inst, Univ Ala, Huntsville, 70-73; asst instr, Univ Tex, 75-76; asst prof eng, Drexel Univ, 76-77; from asst prof to assoc prof, 77-86, PROF MECH, OHIO STATE UNIV, 86- *Concurrent Pos:* Prin investr, NSF & Ohio State Univ, 78-; co-prin investr various grants, Dept Energy, 80-82, Ohio State Univ, 80-83 & NASA, 81-83; consult, 80- *Mem:* Am Acad Mech; Soc Eng Sci; Sigma Xi; Am Soc Mech Eng; Am Soc Metals. *Res:* Theory and application of the finite element method in applied mechanics; numerical methods; geotechnical engineering; sheet metal forming. *Mailing Add:* Dept Eng Mech Ohio State Univ 155 W Woodruff Columbus OH 43210

LEE, KAH-HOCK, b Jan 28, 41; US citizen; m 67; c 3. INORGANIC CHEMISTRY. *Educ:* Nanyang Univ, BS, 64; Georgetown Univ, PhD(inorg chem), 70. *Prof Exp:* Environ chemist, DC Dept Environ Serv, 69-80; PRES, ADVAN INVESTMENT MGT CO, 77- *Concurrent Pos:* Res fel, Dept Chem, Grad Sch, Georgetown Univ, 71-73. *Mem:* Am Chem Soc. *Res:* Environmental toxic trace metals; effect of x-ray film developer on radiation protection; heteropoly inorganic anions exchange mechanisms. *Mailing Add:* 947 Cedar Ridge Ct Orange Park FL 32065-6798

LEE, KAI NIEN, b New York, NY, Oct 19, 45; m 71; c 2. ENVIRONMENTAL MANAGEMENT, SCIENCE POLICY. *Educ:* Columbia Univ, AB, 66; Princeton Univ, PhD(physics), 71. *Prof Exp:* Soc Sci Res Coun res training fel, 71-72; asst res social scientist, Inst Govt Studies, Univ Calif, Berkeley, 72-73; res asst prof, Prog Social Mgt of Technol & Dept Polit Sci, Univ Wash, 73-75, from asst prof to assoc prof Inst Environ Studies & Dept Polit Sci, 75-91; PROF & DIR ENVIRON STUDIES, WILLIAMS COL, 91- *Concurrent Pos:* Mem int adv bd, Policy Sci, 74-; White House fel, US Dept Defense, 76-77; mem Naval Res adv comt, 78-; mem adv panel radioactive waste disposal, Off Tech Assessment, 78-81; mem, Environ Studies Bd, 80-82, Bd Radioactive Waste Mgt, Nat Acad Sci, 83-; Kellogg Nat fel, 80-83; mem, Northwest Power Planning Coun, 83-87; vis prof, Kyoto Univ, Japan, 90-91. *Honors & Awards:* White House Fel, 76; Kellogg Fel, 80. *Mem:* Fel Soc Religion in Higher Educ; AAAS; Am Polit Sci Asn. *Res:* Energy and environmental policy and politics: energy, fish and wildlife, global climate change, environmental conflict and dispute settlement; influence of technological change on American life. *Mailing Add:* Ctr Environ Studies Williams Col Williamstown MA 01267

LEE, KAI-FONG, US citizen; m 71; c 3. ANTENNA THEORY & DESIGN, APPLIED ELECTROMAGNETICS. *Educ:* Queen's Univ, Can, BSc, 61, MSc, 63; Cornell Univ, PhD(elec eng), 66. *Prof Exp:* Res asst, Nat Radio Astron Observ, 66; post doctorate fel, Univ Calif, San Diego, 66-67; vis scientist, Nat Ctr Atmospheric Res, 68-69; from asst prof to assoc prof appl phys, Catholic Univ Am, 67-72; sr resident res assoc, Nat Res Coun-Nat Oceanic & Atmospheric Admin, 72-73; lectr electronics, Chinese Univ Hong Kong, 73-77, sr lectr, 77-83, reader, 84-; head dept electronic eng, City Polytech Hong Kong, 84-85; prof elec eng, Univ Akron, 85-88; PROF & CHMN, ELEC ENG, UNIV TOLEDO, 88- *Concurrent Pos:* Vis reseacher, Univ Calif, Los Angeles, 75. *Mem:* Sr mem Inst Elec & Electronics Engrs; fel Inst Elec & Electronics Engrs; Chartered Eng UK; Am Phys Soc; Sigma Xi. *Res:* Theory and design of microstrip, helical and reflector antennas; theory of plasma waves and instabilities. *Mailing Add:* Dept Elec Eng Univ Toledo 2801 W Bancroft St Toledo OH 43606-3390

LEE, KAI-LIN, b Nanking, China, Sept 16, 35; m 64. BIOCHEMISTRY, ENDOCRINOLOGY. *Educ:* Nat Taiwan Univ, BS, 60; Tulane Univ, PhD(biochem), 66. *Prof Exp:* Teaching asst bot, Nat Taiwan Univ, 61-62; res asst biochem, Tulane Univ, 62-66, fel biochem endocrinol, 66-67, instr, 67-68; Hoffmann-La Roche fel, Biochem, Biol Div, 68-69, res assoc, 69-70, BIOCHEMIST, BIOL DIV, OAK RIDGE NAT LAB, 70- *Mem:* AAAS; Endocrine Soc; Am Chem Soc; Am Soc Biol Chemists. *Res:* Hormonal regulation of metabolic processes. *Mailing Add:* Biol Div Oak Ridge Nat Lab PO Box Y Oak Ridge TN 37830

LEE, KANG IN, b Korea, Nov 2, 46; US citizen; m 75; c 2. SYNTHETIC POLYMER CHEMISTRY, PHOTOCHEMISTRY OF POLYMERS. *Educ:* Murray State Univ, BA, 70; State Univ NY Buffalo, MA, 72; Polytech Inst New York, PhD(polymer sci), 76. *Prof Exp:* Res assoc, Inst Polymer Sci, Univ Akron, 76-77; res scientist polymer chem, Cent Res Labs, Firestone Tire & Rubber Co, 77-80; sr mem tech staff polymer sci, Fundamental Res Lab, Gen Tel & Electronics Labs, 80-84; sr res specialist polymer sci, 84-90, RES FEL, MONSANTO CHEM CO, 91- *Concurrent Pos:* Res assoc, Inst Polymer Sci, Univ Akron, 76-77. *Mem:* Am Chem Soc; Sigma Xi; Korean Scientist & Engrs Am. *Res:* Synthesis and characterization of new polymers for microelectronics application; cationic and ziegler natta polymerizations; new photoresist developments; conducting polymers, electron beam sensitive polymers and polymers for optical fiber sensor applications. *Mailing Add:* Monsanto Co Bldg 123 730 Worcester St Springfield MA 01151

LEE, KATHRYN ADELE BUNDING, b Chicago, Ill; m 86. MOLECULE-SURFACE INTERACTIONS, SPECTROSCOPY. *Educ:* Univ Chicago, BA, 71; City Univ NY, PhD(phys org chem), 80. *Prof Exp:* Nat Res Coun fel, Nat Bur Standards, 80-82; vis scientist, IBM, 82-84; res asst, Georgetown Univ, 84-85; res chemist, Naval Res Lab, 85-87; SR SCIENTIST, SC JOHNSON, 87- *Mem:* Am Chem Soc; Sigma Xi; Electrochem Soc. *Res:* Molecule-surface interactions and reactions; use of in situ spectroscopic and electrochemical techniques; Raman and Fourier transform infrared spectroscopies; optical second harmonic generation. *Mailing Add:* S C Johnson 525 Howe St Racine WI 53403

LEE, KEENAN, b Huntington, NY, Nov 20, 36; m 66; c 2. GEOLOGY. *Educ:* La State Univ, Baton Rouge, BS, 60, MS, 63; Stanford Univ, PhD(geol), 69. *Prof Exp:* Geophys trainee, Cuban Stanalind Oil Co, Pan Am Petrol Corp, 57-58; geologist, Mobil Oil Libya, Ltd, 63-66; asst prof, 69-74, ASSOC PROF GEOL, COLO SCH MINES, 74- *Mem:* AAAS; Geol Soc Am; Am Soc Photogram. *Res:* Hydrogeology; remote sensing. *Mailing Add:* Dept Geol Colo Sch Mines Golden CO 80401

LEE, KEN, b West Orange, NJ, Sept 5, 53; m; c 1. FOOD SCIENCE & TECHNOLOGY. *Educ:* Rutgers Univ, BS, 75; Univ Mass, Amherst, PhD(food sci & nutrit), 80. *Prof Exp:* From asst prof to assoc prof food sci, 80-90, PROF & CHMN, DEPT FOOD SCI & TECHNOL, COL AGR, OHIO STATE UNIV, 90- *Concurrent Pos:* Conf chmn, Midwest Food Processing Conf, 84; counr, Inst Food Technologists, 84-; chmn, Tech Adv Comt, Ctr Dairy Res. *Mem:* Sigma Xi; Inst Food Technologists; Am Asn Cereal Chemists; Am Inst Nutrit; Am Dairy Sci Asn. *Res:* Over 40 publications in peer-review research journals and over 30 published abstracts; mineral bioavailability from cured meats; analysis of nutrient inhibitors; mineral binding by dietary fiber; oxidized cholesterol compounds in foods; nitrate metabolism and analysis in foods; anti-nutrients in tea; hydrocolloids in dairy foods. *Mailing Add:* Dept Food Sci & Technol Ohio State Univ 2121 Fyffe Rd Columbus OH 43210-1097

LEE, KENNETH, b San Francisco, Calif, July 3, 37; m 59; c 2. SOLID STATE PHYSICS. *Educ:* Univ Calif, Berkeley, AB, 59, PhD(physics), 63. *Prof Exp:* Res physicist, Varian Assocs, 63-68; mgr & res staff mem, IBM Res Lab, San Jose, Calif, 68-83; dir memory technol, Southwall Technol, Palo Aalto, Calif,

83-84; vpres, 84-86, SR VPRES PROD DEVELOP, DOMAIN TECHNOL, MILPITAS, CALIF, 84- *Concurrent Pos:* Consult, Lawrence Radiation Lab, Univ Calif, 63-68; comt mem, Magnetism & Magnetic Mat Conf, 64-65, prog comt 71 & adv comt, 71-74, 74-77 & 80-83, prog co-chmn, 75, steering comt, 77, 78 & 81, prog comt, 79. *Honors & Awards:* Centennial Medal, Inst Elec & Electronics Engrs, 84. *Mem:* Fel Am Phys Soc; sr mem Inst Elec & Electronics Engrs; Sigma Xi. *Res:* Antiferromagnetism; crystal defects and motion of nuclei in solids; electron and nuclear magnetic resonance; magnetism in thin films; amorphous magnetism; thin film technology; magnetic recording; author of over 50 technical articles and over 20 reports on magnetic recording technology, storage technologies and magnetic bubble materials. *Mailing Add:* 20587 Debbie Lane Saratoga CA 95070

LEE, KEUN MYUNG, b Korea, Oct 17, 45; m 74; c 1. ACOUSTIC EMISSION, SOFTWARE ENGINEERING. *Educ:* Seoul Nat Univ, BS, 67; State Univ NY Stony Brook, PhD(astrophys), 79. *Prof Exp:* Systs analyst x-ray spectromet, Princeton Gamma Tech, Inc, 79-81; SR SCIENTIST ACOUST EMISSION, PHYS ACOUST CORP, 81- *Mem:* Inst Elec & Electronics Engrs. *Res:* Industrial applications and theory of x-ray spectrometry; signal processing; acoustic emission; computer applications of acoustic signal processing. *Mailing Add:* Hewlett Packard Bldg II PO Box 10490 Palo Alto CA 94303

LEE, KING C, m; c 3. PHARMACOLOGY. *Educ:* Eastern Ky Univ, BS, 78; Univ Ky, Lexington, PhD(pharmacol), 83. *Prof Exp:* Res asst, Dept Pharmacol, Col Med, Univ Ky, 80-83, postdoctoral fel, Dept Physiol & Biophys, 83-85; supvr & sr pharmacologist cardiovasc pharmacol, Wyeth-Ayrst Res, 86-87; RES INVESTR, DEPT CARDIOVASC PHARMACOL, STERLING RES GROUP, 87- *Concurrent Pos:* Co-chmn, Cardiovasc Res Subcomt, Wyeth-Ayrst Res, 86-87; mem, Antioxidant Proj Team, Diag Image Discovering Team & Magnetic Resonance Develop Team, Sterling Res Group, 87- *Mem:* AAAS; Am Physiol Soc; Am Soc Pharmacol & Exp Therapeut; NY Acad Sci. *Res:* Therapeutic agents for ischemic-reperfusion, congestive heart failure and hypertension, glaucoma and ocular ischemia; diagnostic agents for magnetic resonance and x-ray imagings. *Mailing Add:* Dept Cardiovasc Pharmacol Sterling Res Group 81 Columbia Turnpike Rensselaer NY 12144

LEE, KIUCK, b Hamhung, Korea, Jan 15, 22; c 5. NUCLEAR PHYSICS. *Educ:* Seoul Nat Univ, BS, 47, MS, 49; Fla State Univ, PhD(physics), 55. *Prof Exp:* Res assoc physics, Fla State Univ, 55-56 & Argonne Nat Lab, 56-57; from asst prof to assoc prof, 57-68, PROF PHYSICS, MARQUETTE UNIV, 68- *Concurrent Pos:* Res assoc, Argonne Nat Lab, 60. *Mem:* Am Phys Soc. *Res:* Nuclear structure studies on deformed nucleus, fission and the superheavy nucleus. *Mailing Add:* Marquette Univ Dept Physics 1515 W Wisconsin Ave Milwaukee WI 53233

LEE, KOK-MENG, b Singapore, Aug 20, 52; m 82; c 1. DYNAMIC SYSTEMS & CONTROL, DESIGN & MANUFACTURING AUTOMATION. *Educ:* State Univ NY, Buffalo, BS, 80; Mass Inst Technol, SM, 82, PhD(mech eng), 85. *Prof Exp:* Res asst, Fluid Power Control Lab, Mass Inst Technol, 80-85; asst prof, 85-90, ASSOC PROF MECH ENG, GA INST TECHNOL, 90- *Concurrent Pos:* Consult, Milliken Textile Co, 87 & 88; prin investr, NSF, 88-, Ga Tech, Gen Motors & Ford, 89-; NSF presidential young investr, 89; Sigma Xi jr fac award, Ga Inst Technol, 89. *Mem:* Am Soc Mech Engrs; Inst Elec & Electronics Engrs; Instrument Soc Am; Am Soc Eng Educ. *Res:* Dynamic system modeling, control and automation; mechatronics and their application to intelligent control systems and manufacturing automation. *Mailing Add:* Sch Mech Eng Ga Inst Technol Atlanta GA 30332-0405

LEE, KOTIK KAI, b Chungking, China, May 30, 41; US citizen; m 67; c 2. MATHEMATICAL PHYSICS. *Educ:* Chung-Yuan Col, BSc, 64; Univ Ottawa, MSc, 67; Syracuse Univ, PhD(physics), 72. *Prof Exp:* Res asst physics, Syracuse Univ, 68-72, instr, 72-73; asst prof, Rio Grande Col, 73-74; asst physics, 65-67, vis prof math, Univ Ottawa, 74-75; scientist, Lab Laser Energetics, Univ Rochester, 77-78; physicist, Santa Barbara Res Ctr, 78-79; scientist, lab laser energetics, Univ Rochester, 79-; AT GEN ELEC CO. *Mem:* Am Math Soc; Am Phys Soc; Can Asn Physicists; Soc Indust & Appl Math; Int Soc Gen Relativity & Gravitation. *Res:* Global structures of spacetimes, singularities in general relativity, gravitational collapse, quantization of the gravitational field, cosmology, mathematical foundations of quantum field theory and statistical mechanics, plasma physics, laser physics and astrophysics. *Mailing Add:* Dept Elec & Comput Eng Univ Colo Colorado Springs CO 80933-7150

LEE, KUO-HSIUNG, b Taiwan, Jan 4, 40; m 68; c 2. MEDICINAL CHEMISTRY, NATURAL PRODUCTS CHEMISTRY. *Educ:* Kaohsiung Med Col, Taiwan, BS, 61; Kyoto Univ, MS, 65; Univ Minn, Minneapolis, PhD(med chem), 68. *Prof Exp:* Postdoc scholar chem, Univ Calif, Los Angeles, 68-70; from asst prof to assoc prof, 70-77, PROF MED CHEM, SCH PHARM, UNIV NC, CHAPEL HILL, 77-, DIR, NATURAL PROD LAB, 83- *Concurrent Pos:* USPHS res grants, Univ NC, 71-; Am Cancer Soc grant, 75-81, 86-; US Army Med Res Acquisition Agency contract, 83-87; ad hoc mem, Br Study Sect, Conf & Sem Prog, NIH-Fogarty Int Ctr, 83; mem, Study Sect Psychopharmacol, NIMH, 83-84; Chem Study Sect, NIH, 85, 86, 88 & 89, Bio-org & Natural Prod Study Sect, 86, 90 & Phys Biochem Study Sect, 88; mem, Develop Therapeut Contracts Rev Comt, Nat Cancer Inst, 84-88; reviewer, natural prod related grant applications, NSF, 85; Hollingsworth fac scholar award, 88; consult, Genelabs, Inc, 88- & Sphinx Biotechnol Corp, 90- *Honors & Awards:* Taite O Soine Mem Award, 90. *Mem:* Am Chem Soc; Am Soc Pharmacog; fel Am Asn Pharm Scientists; fel Acad Pharmaceut Sci; AAAS; fel Chem Soc; Soc Synthetic Org Chem Japan. *Res:* Isolation, structure determination, synthesis, structure-activity relationships and mechanisms of action of plant antitumor, antiviral, and anit-AIDS agents; antibiotics-isolation and structural determination; anti-inflammatory and anti-fungal agents; antimalarial agents and insect antifeedants; Oriental herbal medicines; over 220 research articles in various journals. *Mailing Add:* Sch Pharm Univ NC Chapel Hill NC 27599-7360

LEE, KWANG, b Seoul, Korea, Jan 18, 42; m 70; c 2. POULTRY NUTRITION & MANAGEMENT. *Educ:* Seoul Nat Univ, BS, 64; Southern Ill Univ, MS, 69; Mich State Univ, PhD(poultry sci & nutrit), 73. *Prof Exp:* PROF POULTRY SCI & NUTRIT, UNIV ARK, PINE BLUFF, 73- *Mem:* Poultry Sci Asn; Korean Soc Animal Sci; World Poultry Sci Asn; AAAS. *Res:* Production efficiency and economics of egg production as influenced by nutrition and management. *Mailing Add:* Dept Agr Univ Ark Pine Bluff AR 71601

LEE, KWANG SOO, b Seoul, Korea, Feb 1, 18; m 40; c 3. PHARMACOLOGY. *Educ:* Keijo Imp Univ, Korea, MD, 42, PhD, 45; Johns Hopkins Univ, 56. *Prof Exp:* Asst prof pharmacol, Seoul Nat Univ, 48-49; instr, Jefferson Med Col, 49-50, assoc, 50-51, from asst prof to assoc prof, 51-56; prof pharmacol, State Univ NY Downstate Med Ctr & Col Physicians & Surgeons, Columbia Univ, 62-80; PROF PHARMACOL, CHUNG-ANG UNIV, SEOUL, REPUB S KOREA, 80- *Honors & Awards:* Pres Award, Repub Korea. *Mem:* Am Soc Pharmacol & Exp Therapeut. *Res:* Cardiac metabolism; mechanisms of drug actions. *Mailing Add:* Dept Pharmacol Chung-Ang Univ 221 Heuksuk Dong, Dongjak KU Seoul Republic of Korea

LEE, KWANG-YUAN, petrology, sedimentology; deceased, see previous edition for last biography

LEE, KWOK-CHOY, immunology, for more information see previous edition

LEE, KYU TAIK, b Taegu, Korea, Sept 17, 21; m 44; c 2. MEDICINE, PATHOLOGY. *Educ:* Severence Union Med Col, MD, 43; Wash Univ, PhD(path), 56. *Prof Exp:* Physician in chief, Presby Gen Hosp, Taegu, Korea, 50-53; prof med & chmn dept, Kyung-Pook Nat Univ, 56-60; assoc prof, 60-66, PROF PATH, ALBANY MED COL, 66-, ASSOC DEAN FOR GRAD STUDIES & RES, 76-, DIR, GEOG PATH DIV, 60-, DIR SECT MOLECULAR BIOL & PATH, 68- *Concurrent Pos:* NIH res grants, 56-; dir, Saratoga Conf Molecular Biol & Path, 68; managing ed, Exp & Molecular Path, 69; mem comt comp path, Nat Res Coun, 69-71, mem comt path, 70-71; mem nutrit study sect, NIH, 69-73. *Mem:* Am Heart Asn; Am Soc Exp Path; Int Acad Path; Am Soc Cell Biol; Int Soc Cardiol. *Res:* Ultrastructural and biochemical aspects of atherosclerosis and thrombosis; geographic aspects of atherosclerosis. *Mailing Add:* Dept Path & Cytol 6605 Heidi Ct Albany NY 22101

LEE, KYUNG NO, BIOMEDICINE. *Prof Exp:* ASST SCIENTIST, DEPT BIOMED SCI, SAMUEL ROBERTS NOBLE FOUND INC, 85- *Mailing Add:* Biomed Dept Samuel Roberts Noble Found Inc PO Box 2180 Ardmore OK 73402

LEE, L(AWRENCE) H(WA) N(I), b Shanghai, China, Jan 5, 23; nat US; m 48; c 1. ENGINEERING MECHANICS, STRUCTURAL ENGINEERING. *Educ:* La Univ Utopia, China, BS, 45; Univ Minn, MS, 47, PhD(struct eng & mech), 50. *Prof Exp:* Engr, Lin-Hu Reconstruct Asn, 45-46; instr math & mech, Univ Minn, 49; from asst prof to assoc prof eng mech, 50-60, prof eng sci, 60-69, PROF AEROSPACE & MECH ENG, UNIV NOTRE DAME, 69- *Concurrent Pos:* Consult, Bendix Corp, 53-75, Gen Motors Corp, 60-63 & 73, Rock Island Arsenal, US Army, 76 & Dodge Div, Reliance Elec Corp, 80. *Honors & Awards:* Struct Mech Res Award, Off Naval Res & Am Inst Aeronaut & Astronaut, 71. *Mem:* Am Soc Mech Engrs; Am Soc Civil Engrs; Soc Eng Sci; Am Soc Eng Educ. *Res:* Inelastic stability; dynamic plasticity; elasticity; experimental stress analysis and dynamic stability. *Mailing Add:* Dept Aerospace & Mech Eng Univ Notre Dame Notre Dame IN 46556

LEE, LELA A, b Gorman, Tex, Sept 7, 50; m 84; c 1. DERMATOLOGY. *Educ:* Rice Univ, BA, 72; Univ Tex Southwestern Med Sch, MD, 76; Am Bd Internal Med, cert, 79; Am Bd Dermat, cert, 83. *Prof Exp:* Fel immunodermat, 83-85, ASST PROF DERMAT & MED, UNIV COLO SCH MED, 85- *Concurrent Pos:* Prin investr, biol cutaneous lupus, NIH grant, 85-90, photosens cutaneous lupus, Vet Admin grant, 91-; William Reed travelling fel, Am Acad Dermat & Col Physicians Philadelphia, 83-84; mem bd dirs, Soc Invest Dermat, 85-87, officer, 89-95; mem, Am Acad Dermat Task Force Immunopath, 88-91; chief assoc ed, J Invest Dermat, 90-92. *Honors & Awards:* Stelwagon Award, Am Acad Dermat, 83-84. *Mem:* Soc Invest Dermat; Am Fedn Clin Res; Am Acad Dermat. *Res:* Clinical and basic science studies concerning the definition and pathogenesis of neonatal lupus and of cutaneous lupus in adults. *Mailing Add:* Dept Dermat Univ Colo Sch Med 4200 E Ninth Ave Box B-153 Denver CO 80262

LEE, LESTER TSUNG-CHENG, polymer chemistry, organic chemistry, for more information see previous edition

LEE, LIENG-HUANG, b Fukien, China, Nov 6, 24; m 49; c 4. ORGANIC CHEMISTRY. *Educ:* Amoy (Xiamen) Univ, BSc, 47; Case Inst Technol, MSc, 54, PhD(chem), 55. *Prof Exp:* Jr chemist, Nantou Sugar Factory, China, 47-48; asst res chemist, Chia-Yee Solvent Works, 48-51 & Rain Stimulation Res Inst, 51-52; res assoc, Case Inst Technol, 55-56; lectr, Tunghai Univ, 56-57; vis prof, Taiwan Prov Norm Univ, 57-58; res org chemist, Dow Chem Co, 58-63, sr res chemist, 63-68; SR SCIENTIST, XEROX CORP, 68- *Concurrent Pos:* Consult, Union Res Inst, Taiwan, China, 57-58; adv ed, J Adhesion, 71-; invited lectr, Chinese Acad Sci, 79, 86 & 87; course organizer, Fundamentals Adhesion, 81-; hon adv, Fujian Res Inst light indust, 86-, Res Ctr Solid Lubrication, Chinese Acad Sci, 87-; mem comt, Reliability of Adhesive Bonds in Severe Environs, Nat Res Coun, 84; vis prof chem, Xiamen Univ, China, 86-; hon prof, Lanzhou Inst, Chem Phys Chinese Acad Sci, 88- *Honors & Awards:* Mabery Prize, 54. *Mem:* Am Chem Soc; Sigma Xi; Am Phys Soc. *Res:* Polymer friction and wear; adhesion and surface chemistry; electrophotography. *Mailing Add:* 796 John Glen Blvd Webster NY 14580

LEE, LIH-SYNG, b China, Oct 28, 45; m 74; c 1. BIOCHEMISTRY, BIOPHYSICS. *Educ:* Nat Taiwan Univ, BS, 68; Yale Univ, MPh, 71, MS, 72, PhD(chem), 74. *Prof Exp:* Res scientist biochem & fel, Roswell Park Mem Inst, 74-76; STAFF ASSOC CARCINOGENESIS, COLUMBIA UNIV, 76- *Mem:* NY Acad Sci; AAAS; Am Chem Soc. *Res:* Chemical carcinogenesis; tumor promotion; growth factors; hormone receptors; cell culture; transport; enzymology; DNA metabolism; cell cycle; gene transfer; membrane biophysics; drug design. *Mailing Add:* 22 Van Wyck Dr Princeton NJ 08550

LEE, LINWOOD LAWRENCE, JR, b Trenton, NJ, Aug 5, 28; m 57; c 1. NUCLEAR PHYSICS. *Educ:* Princeton Univ, AB, 50; Yale Univ, MS, 51, PhD(physics), 55. *Prof Exp:* Asst physicist, Argonne Nat Lab, 54-59, assoc physicist, 60-65; vis asst prof physics, Univ Minn, 59-60; prof physics, State Univ NY, Stony Brook, 65-68, dir, Nuclear Struct Lab, 65-84; RETIRED. *Mem:* Fel Am Phys Soc; fel AAAS. *Res:* Experimental studies of nuclear structure and spectroscopy; nucleon transfer reactions; near barrier heavy ion reactions. *Mailing Add:* Eight Johns Rd Setauket NY 11733

LEE, LLOYD LIEH-SHEN, b China, Jan 25, 42; US citizen. STATISTICAL MECHANICS, POLYMER TECHNOLOGY. *Educ:* Nat Taiwan Univ, BS, 63; Northwestern Univ, PhD(chem eng), 71. *Prof Exp:* Res engr polymer, Du Pont Chem Co, 71-72; researcher liquids, Univ Paris, 72; mgr textiles, Tashing Chem Co, 73-75; asst prof, 76-80, ASSOC PROF CHEM ENG & MAT SCI, UNIV OKLA, 80- *Mem:* Am Chem Soc. *Res:* Natural gas properties; electrolyte solutions; perturbation theory for liquid structure and liquid thermodynamics; Monte Carlo and molecular dynamics simulations; turbulence; high-speed spinning of polymeric filaments. *Mailing Add:* Dept Chem Eng 202 W Boyd Rm 23 Univ Okla Univ Okla 660 Parrington Oval Norman OK 73019

LEE, LONG CHI, b Kaohsiung, Taiwan, Oct 19, 40; m 67; c 2. EXPERIMENTAL PHYSICS. *Educ:* Taiwan Normal Univ, BS, 64; Univ Southern Calif, MA, 67, PhD(physics), 71. *Prof Exp:* From res asst to res assoc, 67-72, res staff physicist, 72-77, adj asst prof physics, Univ Southern Calif, 77; physicist, 77-78, sr physicist, Stanford Res Inst Int, 79-81; PROF ELEC & COMPUT ENG, SAN DIEGO STATE UNIV, 82- *Concurrent Pos:* Vis scientist, Univ Kaiserslautern, Ger, 75. *Honors & Awards:* SDSU Meritorious Performance & Prof Award, 86, 88, 90. *Mem:* Am Phys Soc; Inter-Am Photochemical Soc; Inst Elec & Electronics Engrs; Am Geophys Union; Int Soc Optical Eng. *Res:* Photoionization and photodissociation processes of small molecules and radicals using vacuum ultraviolet radiation; photodestruction processes of atmospheric positive and negative ions; molecular processes in electrical discharges; optical characteristics of small aerosol particles; molecular processes in chemical etching; reaction kinetics. *Mailing Add:* Dept Elec & Comput Eng San Diego State Univ San Diego CA 92182-0190

LEE, LOU-CHUANG, b Taiwan, China, Apr 20, 47; c 2. PLASMA PHYSICS, SPACE PHYSICS. *Educ:* Nat Taiwan Univ, BS, 69; Calif Inst Technol, MS, 72, PhD(physics), 75. *Prof Exp:* Res assoc physics, Goddard Space Flight Ctr, NASA, 75-77; vis asst prof, Inst Phys Sci & Technol, Univ Md, 77-78; assoc prof, 78-86, PROF, GEOPHYS INST, UNIV ALASKA, 86- *Concurrent Pos:* Toray vis scholar, 86; Distinguished Fulbright Scholar, 88. *Honors & Awards:* Moore Prize, 87. *Mem:* Am Phys Soc; Am Geophys Union; AAAS. *Res:* Magnetospheric physics. *Mailing Add:* Geophys Inst Univ Alaska Fairbanks AK 99775-0800

LEE, LU-YUAN, AIR WAYS DISEASE, SMOKING & HEALTH. *Educ:* Univ Wis, PhD(physiol), 75. *Prof Exp:* ASSOC PROF PHYSIOL, MED CTR, UNIV KY, 78- *Mailing Add:* Dept Physiol & Biophys Univ Ky Med Ctr Lexington KY 40536

LEE, LYNDON EDMUND, JR, b Islip, NY, Aug 11, 12; m 43; c 3. SURGERY, PHARMACOLOGY. *Educ:* Duke Univ, BS, 37, MD, 38. *Prof Exp:* Asst cardiol, Univ Va, 38; resident obstet & gynec, Duke Univ Hosp, 40; intern surg, Med Col Va, 41; instr, Med Sch, Univ Mich, 42-43, instr pharmacol & surg, 42-47; instr post-grad educ comt, State Med Asn, Tenn, 47-49; dir cancer control, PR Dept Health, 49-54; instr pharmacol & surg, Med Sch, Univ Mich, 54-57; coordr res, 57-69, dir surg servs, 65-69, asst chief med dir res & educ, 69-71, asst chief med dir prof serv, 71-76, special asst chief med dir, US Vet Admin, 76-77, chief staff, Vet Admin Hosp, Richmond, Va & assoc dean & prof surg, Med Col Va, 76-78; SURG, US VET ADMIN, WASHINGTON, DC, 78- *Concurrent Pos:* Nat Res Coun fel, 38-40 & 41-42; assoc physician, Blue Ridge Sanatorium, Charlottesville, Va, 38; physician, Am Hosp in Brit, Oxford, Eng, 41 & Pondville State Hosp Cancer, Walpole, Mass; assoc surg, Mass Gen Hosp, Boston; prof, Sch Med, Univ PR, 52-54; dir surg, Wayne County Gen Hosp, 54-57; mem, Med Sci Div, Nat Acad Sci, 69-76 & White House Fed Coun Sci & Technol, 69-76 Consult, Smithsonian Inst, 58-76, Nat Adv Cancer Coun, NIH, 58-76 & Training Grants Rev Bd & Cancer Chemother Nat Serv Ctr, Nat Cancer Inst, 58-76; mem Nat Adv Coun Child Health & Human Develop, NIH. *Mem:* AAAS; Am Pub Health Asn; Pub Health Cancer Asn Am; fel Am Col Surgeons; fel Int Soc Surg. *Res:* Analgesics and sedatives; neoplasms. *Mailing Add:* 4111 Saul Rd Chevy Chase View Kensington MD 20895

LEE, M(ONHE) HOWARD, b Pusan, Korea, May 21, 37; US citizen; m 67; c 1. THEORETICAL PHYSICS. *Educ:* Univ Pa, BS, 59, PhD(physics), 67. *Prof Exp:* Fel physics, Theoret Physics Inst, Univ Alta, 67-69; res assoc, Dept Physics & Mat Sci Ctr, Mass Inst Technol, 69-71; NIH res grant & investr biomat, Health Sci & Technol, Mass Inst Technol & Harvard Univ, 71-73; from asst prof to assoc prof, 73-85, PROF PHYSICS, UNIV GA, 85- *Concurrent Pos:* Guest lectr, Inst Theoret Physics, Univ Leuven, Belg, 76; grants, NATO, 76-78, Air Force Off Sci Res, 77-78, Dept Energy, 77-87, Off Naval Res, 85-89, NSF, 87-, Army Res Off, 88-; Fulbright-Hays Sr Res Scholar Award, Univ Louvain, Belg, 78-79; co-chmn, 17th Eastern Theoret Physics Conf, Athens, Ga, 79; vis prof, Dept Physics, Seoul Nat Univ, Korea, 80. *Mem:* Am Phys Soc. *Res:* Many-body theory; statistical mechanics; biophysics of membrane transport. *Mailing Add:* Dept Physics Univ Ga Athens GA 30602

LEE, MARIETTA Y W T, b Canton, China, Mar 3, 43; m 69; c 1. DNA REPLICATION. *Educ:* Nazareth Col, BSc, 65; New York Univ, MS, 67; Univ Miami, PhD(biol), 73. *Prof Exp:* Asst prof, 81-87, ASSOC PROF MED, UNIV MIAMI, 87- *Concurrent Pos:* Estab investr, Am Heart Asn, 84-89; mem, ASBMB comt Equal Opportunity Women, 88-91; mem, NIH Biochem Study Sect, 90-94. *Mem:* Am Soc Univ Prof; Am Soc Biol Chemists. *Res:* Molecular cloning of DNA polymerases delta and epsilon; inhibition of human DNA polymerases by antiviral drugs; mechanism of DNA replication. *Mailing Add:* Dept Med PO Box 016960-R57 Univ Miami Miami FL 33101

LEE, MARTIN ALAN, b Bromley, Eng, Oct 9, 45; US citizen. ASTROPHYSICS. *Educ:* Stanford Univ, BSc, 66; Univ Chicago, PhD(physics), 71. *Prof Exp:* NATO fel astrophys, Max Planck Inst Extraterrestrial Physics, WGer, 71-72, res assoc, 72-73; res assoc, Lab Astrophysics & Space Res, Univ Chicago, 73-74; asst prof physics, Washington Univ, 74-79; res scientist astrophys, 79-84, RES ASSOC PROF, UNIV NH, 84- *Concurrent Pos:* Ed, Advances in Space Res, 84; vis prof extraterrestrial physics, Max Planck Inst, WGer, 85; prin investr, NSF grant. *Mem:* Am Astron Soc; Am Geophys Union. *Res:* Energetic particle transport and plasma processes in the solar-terrestrial environment including ion shock acceleration, solar modulation of galactic cosmic rays, plasma instabilities and wave propagation. *Mailing Add:* Space Sci Ctr De Meritt Hall Univ NH Durham NH 03824-3568

LEE, MARTIN J G, b Kings Lynn, Eng, Mar 16, 42. CONDENSED MATTER PHYSICS. *Educ:* Cambridge Univ, BA, 63, MA & PhD(physics), 67. *Prof Exp:* Instr, 67-69, asst prof physics, James Franck Inst & Dept Physics, Univ Chicago, 69-74; assoc prof, 74-79, PROF PHYSICS, UNIV TORONTO, 79- *Mem:* Can Asn Physicists. *Res:* Experimental and theoretical study of Fermi surfaces and electronic structure of metals; photo field emission phenomena; high temperature superconductivity. *Mailing Add:* McLennan Phys Lab 60 St George St Toronto ON M5S 1A7 Can

LEE, MARTIN JEROME, b Bayonne, NJ, May 24, 43; m 67; c 1. BIOCHEMISTRY. *Educ:* Rutgers Univ, New Brunswick, BA, 65, MS, 68, PhD(biochem), 69. *Prof Exp:* Nat Inst Gen Med Sci fel, Univ Wis-Madison, 69-70, res assoc biochem, Enzyme Inst, 70-71; sr res assoc, Pharmacia Fine Chem, Inc, 71-73; staff scientist, Technicon Instruments Corp, 73-80; dir appl res, Coulter Diag, 80-; AT ROCKLAND MEDI-LABS. *Mem:* Biophys Soc; Am Chem Soc; Sigma Xi; NY Acad Sci. *Res:* Bioenergetics; chromatography; separational techniques and instrumentation; automated cytochemistry, immunology, clinical chemistry and enzymology. *Mailing Add:* Great Smokie Diag Lab 18 A Regent Park Blvd Asheville NC 28806

LEE, MARTIN JOE, US citizen. ACCELERATOR PHYSICS, MICROWAVES. *Educ:* Univ Calif, BS, 60; NY Univ, MS, 62; Stanford Univ, PhD(elec eng), 67. *Prof Exp:* Microwave engr commun, Bell Tel Lab, 60-62; microwave engr accelerator eng, Stanford Linear Accelerator Ctr, 62-67; accelerator physicist, Brookhaven Nat Lab, 67-69; accelerator theorist, 69-78, dep chief, Pep Theory Group, Accelerator Physics, Stanford Linear Accelerator Ctr, 78-87; CONSULT, 87- *Concurrent Pos:* Spec consult, Electron Storage Ring Corp, 78- *Mem:* Inst Elec & Electronics Engrs. *Mailing Add:* Stanford Linear Accelerator Ctr Stanford Univ Stanford CA 94305

LEE, MATHEW HUNG MUN, b Hawaii, July 28, 31; m 58; c 3. PHYSICAL MEDICINE & REHABILITATION. *Educ:* Johns Hopkins Univ, AB, 53; Univ Md, MD, 56; Univ Calif, MPH, 62; Am Bd Phys Med & Rehab, dipl, 66. *Prof Exp:* Resident, Inst Phys Med & Rehab, Med Ctr, NY Univ, 62-64, NY State Health Dept assignee, Rehab Serv, 64-65, from asst prof to assoc prof rehab med, 65-73, dir educ & training, Dept Rehab Med, 66-68, assoc dir, 68, DIR DEPT REHAB MED, GOLDWATER MEM HOSP, 68-, PROF REHAB MED, SCH MED, NY UNIV, 73- *Concurrent Pos:* Assoc vis physician, Goldwater Mem Hosp, 65-68, vis physician, 68-, chief electrodiag unit, 66-, vpres med bd, 69-70, pres, 71-; asst clin prof, Col Dent, NY Univ, 66-69, clin asst prof, 69-70, clin assoc prof, 70-; consult, Daughters of Israel Hosp, New York, 65-72, Bur Adult Hyg, 65- & Human Resources Ctr, 66-; asst attending physician, Hosp, NY Univ, 68-; World Rehab Fund consult, Gordon Seagrave & Maryknoll Hosps, Korea, 69; attend physician, Bellevue Hosp Ctr, 71-; consult, US Dept Interior. *Mem:* AAAS; fel Am Acad Phys Med & Rehab; fel Am Col Physicians; Pan-Am Med Asn; fel Am Pub Health Asn. *Mailing Add:* Rusk Inst Rehab Med New York NY 10016

LEE, MAY D-MING (LU), b China, May 12, 49; US citizen; m 74; c 2. NATURAL PRODUCTS ISOLATION, STRUCTURE ELUCIDATION. *Educ:* Univ BC, BS, 72; Univ Ill, Urbana, MS, 74, PhD(chem), 76. *Prof Exp:* Res fel chem, Harvard Univ, 76-77; RES CHEMIST, LEDERLE LABS, 77- *Mem:* Am Chem Soc; AAAS. *Res:* Isolation, screening and structure elucidation of novel antitumor and antibacterial agents from fermentation; screening methodology. *Mailing Add:* 19 Stuart Rd Monsey NY 10952

LEE, MELVIN, b New York, NY, Jan 5, 26; m 49; c 4. NUTRITION, BIOCHEMISTRY. *Educ:* Univ Calif, Los Angeles, BA, 47; Univ Calif, Berkeley, MA, 52, PhD(nutrit), 58. *Prof Exp:* From instr to asst prof prev med, Sch Med, Univ Calif, San Francisco, 58-67, asst prof biochem, 63-67, lectr dent, 61-67; prof nutrit & dir, Sch Home Econ, 67-74, PROF NUTRIT, SCH FAMILY & NUTRIT SCI, UNIV BC, 74- *Concurrent Pos:* USPHS res fel, 66; sci exchange fel, Japan Soc Promotion, 81; vis scientist, Gunma Univ, Japan, 86. *Mem:* AAAS; Am Inst Nutrit; Soc Environ Geochem & Health; Soc Nutrit Latin Am; Can Soc Nutrit Sci. *Res:* Relation of diet to metabolic patterns; factors influencing growth; maternal alcohol and fetal development. *Mailing Add:* Sch Family Nutrit Sci Univ BC Vancouver BC V6T 1W5 Can

LEE, MEN HUI, b Taiwan, Apr 2, 33; US citizen. CANCER RESEARCH. *Educ:* Nat Taiwan Univ, BS, 56; Utah State Univ, MS, 63; Columbia Univ, PhD(biochem), 68. *Prof Exp:* Postdoctoral biochem, Chem Dept, Columbia Univ, 68-69; NIH postdoctoral, Dept Biol Chem, Harvard Univ, 69-71; res assoc pharmacol, Dept Pharmacol, Yale Univ, 71-80; PROF BIOCHEM, DEPT AGR CHEM, NAT TAIWAN UNIV, 80- *Concurrent Pos:* Spec fel,

Leukemia Soc Am, Inc, 74-76; vis prof, Div Oncol-Hemat, Ohio State Univ, 88-89. *Mem:* Am Soc Pharmacol & Exp Therapeut; Am Asn Cancer Res. *Res:* Cancer: fields of chemotherapy, tumor immunology involving lymphokine-activated killer cells and enzymology related to cancer. *Mailing Add:* Agr Chem Dept Biochem Lab Nat Taiwan Univ Sect 4 Roosevelt Rd Taipei 10764 Taiwan

LEE, MERLIN RAYMOND, evolutionary biology; deceased, see previous edition for last biography

LEE, MING T, b Taipei, Taiwan, Aug 15, 40; US citizen; m 70; c 2. HYDRAULICS, WATER RESOURCES. *Educ:* Nat Taiwan Univ, BS, 63, MS, 66; Univ Cincinnati, MS, 68; Purdue Univ, PhD(civil eng), 72. *Prof Exp:* Res engr, Hydraul Lab, Water Resources Planning Comn, 66-67; hydraul engr, Vogt, Iver & Assoc, Ohio, 68; res engr hydraul, Agr Econ Dept, Univ Ill, 72-75; assoc prof scientist, 75-91, PROF SCIENTIST HYDRAUL, ILL STATE WATER SURV, 91- *Mem:* Am Geophys Union; Soil Conserv Soc Am; Am Soc Civil Engrs; Sigma Xi. *Res:* Soil erosion; sediment transport; lake hydrology; watershed erosion control; hydrologic computer modeling. *Mailing Add:* 2204 Griffith Dr Champaign IL 61820

LEE, MING-LIANG, b Tainan, Taiwan, June 26, 36; m 65; c 3. MEDICAL GENETICS, BIOCHEMISTRY. *Educ:* Nat Taiwan Univ, MD, 62; Univ Miami, PhD(biochem), 69. *Prof Exp:* Asst prof, Sch Med, Univ Miami, 72-76; chief fel med genetics, Sch Med, Johns Hopkins Univ, 76-77; ASSOC PROF MED GENETICS & CHIEF DIV, RUTGERS MED SCH, 77- *Mem:* Am Soc Biol Scientist; Am Soc Med Genetics. *Mailing Add:* CN-19 Dept Pediat Rutgers Med Sch New Brunswick NJ 08903

LEE, MIN-SHIU, b Taipei, Taiwan, June 30, 40; US citizen; m 66; c 2. PHYSICAL CHEMISTRY, POLYMER SCIENCE & TECHNOLOGY. *Educ:* Nat Taiwan Univ, BS, 62; NMex Highlands Univ, MS, 66; Case Western Reserve Univ, PhD(macromolecular sci), 69. *Prof Exp:* Res fel chem, Chinese Ord Res Inst, 62-64; teaching asst, NMex Highlands Univ, 64-65, res fel hot atom chem, Inst Sci Res, 65-66; res asst polymer res, Case Western Reserve Univ, 66-69; res chemist, FMC Corp, 69-73, sr res chemist, 73-76; sr res scientist, Jelco labs, 76-80, mgr mat & process, Critikon Inc, 80-83, staff consult, Johnson & Johnson, 83-86; SR RES SCIENTIST, BECTON DICKINSON & CO, 86- *Mem:* Sigma Xi; Am Chem Soc; NY Acad Sci; AAAS; Am Asn Univ Prof; Soc Plastics Engrs; Am Soc Testing & Mat; Asn Advan Med Instrumentation. *Res:* Medical products improvement and development; collagen research and development; chemical process improvement; new polymer products development; textile and non-woven applications; polymer characterization; polymer modification and application; process/quality control; biomaterial development; colloidal macromolecular phenomena. *Mailing Add:* 9642 Meadow Woods Lane Spring Valley OH 45370

LEE, MINYOUNG, b Seoul, Korea, Aug 11, 38; US citizen; m 66; c 2. MATERIALS SCIENCE. *Educ:* Seoul Nat Univ, BS, 61; Providence Col, MS, 67; Brown Univ, PhD(mat sci), 71. *Prof Exp:* MEM TECH STAFF MAT, CORP RES & DEVELOP CTR, GEN ELEC CO, 71- *Concurrent Pos:* Vis scientist, Cavendish Lab, Cambridge Univ, 76. *Mem:* Am Inst Mining, Metall & Petrol Engrs; Soc Mech Engrs. *Res:* Development of very hard materials primarily for cutting tools and wear parts; advanced materials processing technology; tribology; sensors for factory automation. *Mailing Add:* Corp Res & Develop Ctr Gen Elec Co PO Box 8 Schenectady NY 12309

LEE, NANCY L, BIOLOGICAL SCIENCE. *Prof Exp:* PROF MICROBIOL, UNIV CALIF, 65- *Mailing Add:* Biol Sci Dept Univ Calif Santa Barbara CA 93106

LEE, NANCY M, PHARMACOLOGY. *Educ:* Southwestern Univ, Georgetown, Tex, BS, 63; Univ Tex, Austin, PhD(biochem), 67. *Prof Exp:* Res asst, Univ Tex, Austin, 63-67; postdoctoral fel, Dept Biochem, Northwestern Univ, Evanston, Ill, 67; res biochemist III-VI, Dept Pharmacol, Univ Calif, San Francisco, 68-72, asst res biochemist, 72-78, assoc res pharmacologist, Dept Pharmacol & Langley Porter Psychiat Inst, 78-84, adj assoc prof, 84-87, prof, 87-89; PROF, DEPT PHARMACOL, UNIV MINN MED SCH, MINNEAPOLIS, 89- *Mem:* Sigma Xi; Soc Neurosci; Am Soc Pharmacol & Exp Therapeut; AAAS; Int Asn Women Bioscientist; Soc Chinese Bioscientists Am. *Res:* Pharmacology. *Mailing Add:* Dept Pharmacol Univ Minn 3-249 Millard Hall 435 Delaware St SE Minneapolis MN 55455

LEE, NANCY ZEE-NEE MA, b Shanghai, China, Oct 28, 40; US citizen; m 65; c 2. BIOCHEMISTRY. *Educ:* Southwestern Univ, BS, 63; Univ Tex, PhD(chem), 67. *Prof Exp:* Fel biochem, Northwestern Univ, 68; RES BIOCHEM PHARMACOL, UNIV CALIF, SAN FRANCISCO, 69- *Concurrent Pos:* Res biochem, Bay Area Heart Res, 70-72. *Mem:* Sigma Xi; Soc Neurosci. *Res:* Biochemical mechanism for narcotic addiction. *Mailing Add:* Dept Pharmacol 3-249 Millard Hall Univ Minn 435 Delaware St SE Minneapolis MN 55455

LEE, NORMAN K, b Frankfort, Ind, Feb 3, 34; m 56; c 2. APPLIED MATHEMATICS. *Educ:* Hanover Col, BA, 56; Vanderbilt Univ, MA, 58; Purdue Univ, PhD(bionucleonics), 69. *Prof Exp:* Asst prof, 58-66, admin asst dept, 79-82, ASSOC PROF MATH, BALL STATE UNIV, 69- *Concurrent Pos:* NSF fel, 63-64, partic, Acad Year Inst, 62-63; sr instr, Somerset Community Col, 68-69. *Mem:* AAAS. *Res:* Mathematical models. *Mailing Add:* 401 W Main Ridgeville IN 47380

LEE, PANG-KAI, physical chemistry, for more information see previous edition

LEE, PATRICK A, b Hong Kong, Sept 8, 46; m 69; c 2. CONDENSED MATTER THEORY. *Educ:* Mass Inst Technol, BS, 66, PhD(physics), 70. *Prof Exp:* J W Gibbs instr physics, Yale Univ, 70-72; mem tech staff, Bell Labs, 72-73; asst prof physics, Univ Wash, Seattle, 73-74; mem tech staff, Bell Labs, 74-82; PROF PHYSICS, MASS INST TECHNOL, 82- *Honors & Awards:* Oliver Buckley Condensed Matter Physics Prize, 91. *Mem:* Nat Acad Sci; fel Am Phys Soc. *Res:* Disordered electronic systems; quantum transport in small structures; theory of high temperature superconductivity. *Mailing Add:* Dept Physics Mass Inst Technol Cambridge MA 02139

LEE, PAUL D, b Ina, Ill, Feb 15, 40; m 61; c 2. ASTRONOMY, ASTROPHYSICS. *Educ:* Univ Ill, BS, 63, MS, 65, PhD, 68. *Prof Exp:* Asst prof, 68-73, ASSOC PROF PHYSICS & ASTRON, LA STATE UNIV, BATON ROUGE, 73- *Mem:* Royal Astron Soc; Am Astron Soc. *Res:* Spectrophotometry of stellar and nonstellar objects; stellar atmospheres and chemical abundances in stars. *Mailing Add:* Dept Physics & Astron La State Univ Baton Rouge LA 70803-4001

LEE, PAUL L, b China, June 12, 44; US citizen; m 72; c 2. MEDICINE. *Educ:* Calif Inst Technol, BS, 67, PhD(physics), 71. *Prof Exp:* Res fel physics, Calif Inst Technol, 71-73, lectr physics, Calif State Univ, Long Beach, 73-75, from asst prof to assoc prof, 75-83, PROF PHYSICS, CALIF STATE UNIV, NORTHRIDGE, 83- *Concurrent Pos:* Vis assoc, Calif Inst Technol, 76-83; vis fac, Jet Propulsion Lab, Caltech, 83-84, MTS, 85- *Mem:* Sigma Xi; Am Phys Soc. *Res:* Image processing and medical physics; nuclear magnetic resonance imaging protocols and special hardware; algorithms for analyzing X-ray angiograms; data analysis; numerical methods. *Mailing Add:* Dept Physics & Astron Calif State Univ Northridge CA 91330

LEE, PETER CHUNG-YI, b Hankow, Hupei, China, Sept 29, 34; m 61; c 1. ENGINEERING MECHANICS. *Educ:* Cheng Kung Univ, Taiwan, BS, 57; Rutgers Univ, MS, 61; Columbia Univ, MS, 65, DEngSc(eng mech), 65. *Prof Exp:* Sloan vis fel, 65-66, from asst prof to assoc prof, 66-76, PROF CIVIL ENG, PRINCETON UNIV, 76- *Honors & Awards:* C B Sawyer Mem Award, 80. *Mem:* Am Soc Mech Engrs; assoc mem Am Soc Civil Engrs; Acoust Soc Am; Inst Elec & Electronics Engrs. *Res:* Theory of elasticity; vibrations and wave propagation in elastic solids and piezoelectric crystals; effects of initial stresses and accelerations; temperature changes on the vibrations of elastic and crystal plates. *Mailing Add:* Dept Civil Eng & Opers Res Princeton Univ Princeton NJ 08544

LEE, PETER E, b Trinidad, WI, Oct 18, 30; m 60; c 2. PLANT VIROLOGY. *Educ:* Univ Man, BSc, 54; Univ Wis, MSc, 59, PhD(entom), 61. *Prof Exp:* Jr res entomologist, Univ Calif, Berkeley, 61; res officer virus-vector studies, Can Dept Agr, 61-65; from asst prof to assoc prof biol, Carleton Univ, 65-77, prof, 77-; RETIRED. *Mem:* Electron Micros Soc Am. *Res:* Characterization of leafhopper-transmitted viruses and insect viruses; virus purification and electron microscopy. *Mailing Add:* 38 Inurik Crescent Kanata ON K2L 1A2 Can

LEE, PETER H Y, b Chungking, China, Apr 20, 39. FLUID MECHANICS, PLASMA. *Educ:* Nat Taiwan Univ, BS, 61; Tech Univ Aachen, Ger, dipl, 67; Calif Inst Technol, PhD(aeronaut), 73. *Prof Exp:* Mem tech staff, TRW, 73-76; physicist, Lawrence Livermore Lab, 76-84; staff mem, Los Alamos Nat Lab, 84-91; SR SCIENTIST, OCEAN TECHNOL DEPT, TRW SPACE & DEFENSE, 91- *Mem:* Am Phys Soc. *Mailing Add:* TRW Space & Design One Park Bldg R 1 Rm 1008 Redondo Beach CA 90278

LEE, PETER VAN ARSDALE, b San Francisco, Calif, Mar 31, 23; m 51; c 4. PHARMACOLOGY, MEDICINE. *Educ:* Stanford Univ, AB, 44, MD, 47. *Prof Exp:* Intern, San Francisco Hosp, 46-47; asst resident path, Stanford Univ Hosps, 49-50, resident med, 50-51; clin asst, Col Med, State Univ NY, 51-52; instr pharmacol, Sch Med, Stanford Univ, 52-54, asst prof, 54-55; asst prof & asst dean, 55-58, assoc prof pharmacol, 58-67, assoc prof med, 60-67, assoc dean, 58-60, admis officer, 60-65, prof pharmacol, 67-80, PROF MED, SCH MED, UNIV SOUTHERN CALIF, 67-, PROF FAMILY & PREV MED, 80- *Concurrent Pos:* Resident, King's County Hosp, NY, 51-52; consult, Commonwealth Fund, 58-59; vis fel, Brit Asn Study Med Educ, 71-72. *Mem:* AAAS; Am Fedn Clin Res; Asn Am Med Cols; Brit Asn Study Med Educ; fel Royal Soc Med. *Res:* Medical education; clinical pharmacology. *Mailing Add:* Dept Family Med Univ Southern Calif 1420 San Pablo St Los Angeles CA 90033

LEE, PETER WANKYOON, b Seoul, Korea, May 30, 39; US citizen; m 67; c 2. POWDER METALLURGY FORMING, THIN METALLIC COATING PROCESSES. *Educ:* Seoul Nat Univ, Korea, BS, 61; Marquette Univ, BS, 66, MS, 68; Drexel Univ, PhD(mat eng), 72. *Prof Exp:* Designer, Dowha Consult, Seoul, Korea, 61-62; res metallurgist, 68-69, from res specialist to sr res specialist, 72-86, RES SCIENTIST, TIMKEN CO, CANTON, OHIO, 86- *Concurrent Pos:* Chmn, Powder Metall Comt, Am Soc Metals Int, 82-86, Mat Shaping Tech Div, 86-90, Conf Comt, Near Net Shape Mfg, 88 & 90, mem, Tech Div Bd, 86-; ed, Proc Am Soc Metals Int Conf on Rapidly Solidified Mat, 86 & 88. *Honors & Awards:* Cert Recognition, NASA, 89. *Mem:* Fel Am Soc Metals Int; Am Powder Metall Inst; Asn Iron & Steel Engrs. *Res:* Research activities in the areas of metal forming processes, powder metallurgy and surface modifications in order to improve performance and/or to reduce the manufacturing cost. *Mailing Add:* 1922 Dunkeith Dr NW Canton OH 44708-1977

LEE, PHILIP RANDOLPH, b San Francisco, Calif, Apr 17, 24; m 53; c 4. INTERNAL MEDICINE. *Educ:* Stanford Univ, AB, 45, MD, 48; Univ Minn, MS, 56; Am Bd Internal Med, cert, 56. *Hon Degrees:* ScD, MacMurray Col, 67. *Prof Exp:* Asst prof clin phys med & rehab, Sch Med, NY Univ, 55-56; clin instr med, Sch Med, Stanford Univ, 56-59, asst clin prof, 59-67; asst secy health & sci affairs, 67-69, chancellor, 69-72, PROF SOCIAL MED, MED CTR, UNIV CALIF, SAN FRANCISCO, 69-, DIR, INST HEALTH POLICY STUDIES, 72- *Concurrent Pos:* Mem dept internal med, Palo Alto Med Clin, Calif, 56-65; consult, Pub Health Serv, USPHS, 58-63; dir health serv, Off Tech Coop & Res, AID, 63-65; dep asst secy health & sci affairs, Dept HEW, 65, asst secy, 65-69; co-dir, Inst Health & Aging, Sch Nursing, Univ Calif, San Francisco, 80-; pres bd dirs, World Inst Disability, 84-; mem comt pop, Nat Res Coun-Nat Acad Sci, 83-86; mem, Nat Coun

Health Planning & Develop, Dept HEW, 78-80, US Pub Health Serv adv comt, 78, Nat Comn Smoking & Pub Policy, 77-78; mem adv bd, Scripps Clin & Res Found, 80- *Honors & Awards:* Hugo Schaefer Medal, Am Pharmaceut Asn, 76. *Mem:* Inst Med-Nat Acad Sci; AAAS; AMA; Am Pub Health Asn; Am Fedn Clin Res; Am Col Physicians; Am Geriat Soc; Asn Am Med Col. *Res:* Arthritis and rheumatism, especially Rubella arthritis; cardiovascular rehabilitation; academic medical administration; health policy; author of over 10 books and many technical journal articles. *Mailing Add:* Inst Health Policy Studies Sch Med Univ Calif 1326 3rd Ave San Francisco CA 94143

LEE, PING, b Taiwan, Apr 6, 50; US citizen; m 78; c 1. PLASMA PHYSICS. *Educ:* Univ Calif, Berkeley, BS, 73; Mass Inst Technol, PhD(physics), 76. *Prof Exp:* STAFF MEM X-RAY PHYSICS, LOS ALAMOS NAT LAB, 77-; PRIN SCIENTIST, GA TECHNOLOGIES INC, 88- *Mem:* Am Phys Soc. *Res:* Plasmas generlate by laser-matter interaction; measurement and analysis of x-rays produced in high temperature and high density plasmas. *Mailing Add:* L-626 Lawrence Livermore Nat Lab PO Box 808 Livermore CA 94550

LEE, PING-CHEUNG, b Hong Kong, Sept 10, 39. PEDIATRICS. *Educ:* Univ Hong Kong, BSc, 63; Fla State Univ, MS, 65, PhD, 68. *Prof Exp:* Res assoc, Dept Chem, Fla State Univ, Tallahassee, 68-72; res assoc, Dept Biochem, Med Sch, Northwestern Univ, 72-74, instr, Dept Biochem, 74-75; asst prof, Dept Surg & Biochem, Med Col Ohio, Toledo, 75-79; assoc prof, Dept Pediat, State Univ NY, Buffalo, 79-89; RES PROF & DIR GASTROINTESTINAL LABS, DEPT PEDIAT, MED COL WIS, MILWAUKEE, 89- *Concurrent Pos:* Res investr, Steroid Hormone Lab, Wesley Mem Hosp, Chicago, Ill, 72-75; vis investr, Digestive Dis Br, Inst Arthritis, Metab & Digestive Dis, NIH, 78; dir, Gastrointestinal & Nutrit Lab, Children's Hosp, Buffalo, NY, 79-89; asst ed, J Pediat Gastroenterol & Nutrit, 82-90. *Mem:* Am Soc Biochem & Molecular Biol; Am Gastroenterol Asn; Soc Exp Biol & Med; Am Pancreatic Asn. *Mailing Add:* Dept Pediat Med Col Wis MACC Fund Res Ctr 8701 Watertown Plank Rd Milwaukee WI 53226

LEE, PUI KUM, b Peking, China, June 22, 16; US citizen; m 41; c 4. OPTICS. *Educ:* Lingnam Univ, BS, 40; Columbia Univ, MS, 49. *Prof Exp:* Asst chem, Nat Kwangsi Univ, China, 40-41; res chemist, Inst Indust Res, 41-42; mgr, China Chem Corp, 42-46; res assoc chem, Columbia Univ, 49-56; sr chemist reprography, Cent Res Labs, 3M Co, 56-63, res specialist imaging, 63-67, sr res specialist, 67-82; RETIRED. *Honors & Awards:* IR-100 Award, Indust Res Mag, 78. *Mem:* Am Chem Soc; Soc Photog Sci & Eng; Optical Soc Am. *Res:* Imaging optics. *Mailing Add:* 2240 Midland Grove Rd 305 St Paul MN 55113

LEE, RALPH EDWARD, b Gilliam, Mo, July 1, 21; m 42; c 5. COMPUTER SCIENCE. *Educ:* Mo Valley Col, BS, 42; Univ Mo, MS, 49; Ind Univ, MA, 53. *Prof Exp:* From instr to assoc prof, Sch Mines, 46-59, PROF MATH, UNIV MO, ROLLA, 59-, DIR COMPUT CTR, 60- *Concurrent Pos:* NSF fel, Nat Bur Standards, 59. *Mem:* Data Processing Mgt Asn; Asn Comput Mach; Soc Indust & Appl Math; Asn Educ Data Systs. *Res:* Numerical analysis; matrix computations. *Mailing Add:* 211 Math Comput Sci Univ Mo-Rolla Rolla MO 65401

LEE, RAY H(UI-CHOUNG), b Canton, China, Mar 28, 18; nat US; c 4. ELECTRICAL ENGINEERING, MATHEMATICS. *Educ:* Nat Cent Univ, China, BS, 41; Stanford Univ, MA, 45, EE, 46, MS, 47. *Prof Exp:* Design engr, Cent Radio Works China, 41-44; founder & prin, Honolulu Trade Sch, 47-51; engr-specialist, Boeing Airplane Co, 51-55; res specialist & chief mathematician, Chromatic TV Labs & Auktometric Corp, 55-63; staff mem, David Sarnoff Res Ctr, Radio Corp Am, NJ, 64-66; staff mem & mgr, Tex Instruments, 66-69; chief engr, Liquid Crystal Display, Backman Instruments, 69-73; CONSULT PROD DEVELOP & STRATEGIC PLANNING, 73- *Mem:* Inst Elec & Electronics Engrs; Soc Info Display. *Res:* Processings; energy; communication. *Mailing Add:* 23203 Park Esperanza Calabasas Park CA 91302

LEE, RAYMOND CURTIS, b Dallas, Tex, Oct 22, 29; m 53; c 2. PHYSICAL CHEMISTRY. *Educ:* Rice Univ, BA, 50, MA, 53; Univ Tenn, PhD(chem), 61. *Prof Exp:* Tech engr, Gen Elec Co, 56-61; asst prof chem, San Jose State Col, 61-63; res specialist, Lockheed Missiles & Space Co, 63-66; mem tech staff, Aerospace Corp, 66-70; SCIENTIST, SCI APPLICATIONS, INC, 70- *Mem:* Am Chem Soc; Am Nuclear Soc; Am Inst Aeronaut & Astronaut; Am Sci Affil. *Res:* Reactor engineering, in-pile experiment design; testing and evaluation for aircraft nuclear propulsion reactor; nuclear safety evaluation for nuclear rocket project; systems analysis; computer systems modeling and simulation; advanced development engineering. *Mailing Add:* 1014 Venhorst Rd Colorado Springs CO 80918

LEE, RICHARD FAYAO, b Shanghai, China, July 13, 41; US citizen; m 70; c 2. ENVIRONMENTAL CHEMISTRY, BIOLOGICAL OCEANOGRAPHY. *Educ:* San Diego State Col, BA, 64, MA, 66; Univ Calif, San Diego, PhD(marine biol), 70. *Prof Exp:* Res assoc biochem, Pa State Univ, 71-72 & Scripps Inst Oceanog, 72-73; RES PROF OCEANOG, SKIDAWAY INST OCEANOG, 74- *Concurrent Pos:* Lectr oceanog, San Diego State Univ, 71-73; mem adv comt, Marine Resources Res Group Biol Accumulators, UN Food & Agr Orgn, 74-; consult, Exxon Corp, 75. *Mem:* Am Chem Soc; Am Soc Limnol & Oceanog; Sigma Xi. *Res:* Fate of petroleum hydrocarbons in the marine food web; role of lipids in the ecology of marine zooplankton; aquatic toxicology. *Mailing Add:* Skidaway Inst Oceanog PO Box 13687 Savannah GA 31416

LEE, RICHARD J, b Minot, NDak, July 23, 44; m 60. SOLID STATE PHYSICS. *Educ:* Univ NDak, BSEd, 66; Colo State Univ, PhD(physics), 70. *Prof Exp:* Instr physics, Lake Regional Jr Col, 66; res asst, Colo State Univ, 68-70; asst prof, Purdue Univ, Ft Wayne, 70-74; MEM STAFF, US STEEL CORP, 74- *Mem:* Am Phys Soc; Am Asn Physics Teachers; Am Chem Soc. *Res:* Theory of quantum solids; phase transition; light scattering. *Mailing Add:* 3498 Meadowgate Dr Murrysville PA 15668

LEE, RICHARD K C, b Honolulu, Hawaii, Oct 2, 09; m 52; c 3. PUBLIC HEALTH. *Educ:* Tulane Univ, MD, 33; Yale Univ, DrPH, 38. *Hon Degrees:* DSc, Tulane Univ, 73. *Prof Exp:* Instr anat, Sch Med, Tulane Univ, 33-35; intern, Hotel Dieu Hosp, New Orleans, 35-36; dep comnr health, Territory Hawaii, 36-43, dir pub health, 43-53; pres, Hawaii Bd Health, 53-60; dir health, Hawaii Dept Health, 60-62; dir pub health & med activities, Univ Hawaii, 62-65, dean, Sch Pub Health, 65-69, prof, 62-69, exec dir, Res Corp, 70-80, EMER PROF PUB HEALTH & EMER DEAN, SCH PUB HEALTH, UNIV HAWAII, 69-; RES FEL, E W POP INST, E W CTR, 80- *Concurrent Pos:* Lectr, Univ Hawaii, 37-55; WHO fel, 52; mem US deleg, Western Pac Regional Comt Meetings, WHO, 52-65, chief US rep, Manila, 66 & Taiwan, 67, mem US deleg, Assembly Meetings, 57-61; Dept State specialist, Int Educ Exchange Prog, Far East, 55; mem, Western Interstate Comn Higher Educ, Western Ment Health Coun, 60-; coordr-consult health & med prog, East-West Ctr Inst Tech Interchange, 62-; mem task force pub health & med educ, World Affairs of New York & Food Found, 65-; consult, WHO, Manila, 67, Alexandria, UAR, 69, Am Pub Health Asn, Korea, 69, People's Repub China, 78 & Bangkok, 79; chief div environ health & occup med & med coord dept commun med, Straub Clin, 69-71; consult, Water Qual Mgt Prog, City & County of Honolulu, 69-71 & Southern Calif Coastal Water Res Proj, 69-; actg dir, Cancer Ctr Hawaii, 71-73; mem, Nat Adv Coun, Nat Inst Aging, NIH, 75-77; chmn adv bd, Health Manpower Planning Proj, Hawaii, 75-77; prof dir gerontol develop prog, Hawaii, 78-; coordr, US-Japan Conf Agr, Hawaii, 83, Tokyo, Japan, 84, Honolulu, 85 & Japan, 86; consult, Peoples Repub China, 86, Indonesia, 87, Washington DC, 88, Taiwan, 89; mem ctr aging, Univ Hawaii, 87. *Honors & Awards:* Samuel J Crumbine Award, Kans State Univ Interfraternity Coun, 63; Diamond Jubilee Award, Univ Hawaii, 82. *Mem:* AAAS; AMA; Am Pub Health Asn (vpres, 62-63). *Res:* Public health administration; international health activities in the Pacific and Asian areas of the world. *Mailing Add:* E W Pop Inst 1777 E W Cent Rd Honolulu HI 96848

LEE, RICHARD NORMAN, b Waukegan, Ill, Nov 3, 39; m 64; c 2. ATMOSPHERIC CHEMISTRY. *Educ:* Park Col, BA, 61; Univ Kans, PhD(chem), 68. *Prof Exp:* Asst prof chem, St Norbert Col, 66-72; RES SCIENTIST, ATMOSPHERIC SCI DEPT, BATTELLE PAC NORTHWEST LABS, 72- *Mem:* Am Chem Soc. *Res:* Reaction kinetics; environmental chemistry; chemical analysis; atmospheric pollutants and tracers. *Mailing Add:* Lake Powell Villas Page AZ 86040

LEE, RICHARD SHAO-LIN, b Looshan, China, July 19, 29; US citizen; m; c 3. FLUID MECHANICS. *Educ:* Nat Taiwan Univ, BS, 52; NC State Univ, MS, 55; Harvard Univ, PhD(eng & appl sci), 60. *Prof Exp:* From asst prof to assoc prof mech eng, NC State Univ, 60-64; assoc prof, 64-68, chmn dept, 69-75, PROF ENG, STATE UNIV NY, STONY BROOK, 68- *Concurrent Pos:* Vis prof, Univ Queensland, Australia, 70-71; guest prof, Univ Karlsruhe, WGer, 77-78; vis res scientist, Nuclear Res Ctr Karlsruhe, WGer, 81-82; guest prof, Univ Erlangen-Nuremberg, WGer, 85; prof & dir, Inst Appl Mech, Nat Taiwan Univ, Taiwan, 86- *Honors & Awards:* Alexander von Humboldt Award, 80. *Mem:* Am Soc Mech Engrs; Am Inst Aeronaut & Astronaut; Combustion Inst. *Res:* Fire research; combustion gasdynamics; flow instability; vartex flows; biomedical fluid mechanics; synovial lubrication; artificial heart valves; two-phase suspension flows; laser-Doppler anemometry for dispersed flows; particle migration in laminar and turbulent flows. *Mailing Add:* Dept Mech Eng State Univ NY Stony Brook NY 11794-2300

LEE, ROBERT BUMJUNG, b Seoul, Korea, Jan 24, 37; US citizen; m 65; c 2. TRANSPORTATION & CIVIL ENGINEERING. *Educ:* Seoul Nat Univ, BSCE, 61; Polytech Inst Brooklyn, MSTP, 69, PhD(transp eng), 73. *Prof Exp:* Hwy engr civil, Madigan-Hyland Eng Co, 64-67; traffic engr, Port of NY Authority, 67-69; res assoc traffic & transp, Polytech Inst Brooklyn, 69-73, asst prof, 73-75; sr traffic engr, Louis Berger Int Inc, 75-80; vpres consult engr, 80-81, PRES, URBITRAN ASSOCS INC, 81- *Mem:* Am Soc Civil Engrs; Sigma Xi. *Res:* Transportation; transit; environmental impacts. *Mailing Add:* Urbitran Assocs Inc 71 W 23rd St New York NY 10038

LEE, ROBERT E, b Rochester, NY, May 22, 32; m 57; c 4. ELECTRICAL & BIOMEDICAL ENGINEERING. *Educ:* Univ Rochester, BS, 54, MS, 62, PhD(elec eng), 66. *Prof Exp:* Instr thermodyn, Univ Buffalo, 56-57; proj engr, Gen Rwy Signal Co, 57-58; asst prof thermodyn, fluids, calculus & eng, Rochester Inst Technol, 58-62; tech specialist, Bausch & Lomb, Inc, 65-66; asst prof elec eng, Univ WVa, 66-68; ASSOC PROF ELEC ENG, ROCHESTER INST TECHNOL, 68- *Concurrent Pos:* Consult, Eastman Kodak, 81. *Mem:* Am Soc Eng Educ; Inst Elec & Electronics Engrs; Optical Soc Am; Sigma Xi. *Res:* Visual systems, specifically latency to response of human pupil for contraction and dilation; effects of intensity, adaptation and size of change on latency. *Mailing Add:* One Lomb Mem Dr Rochester NY 14623

LEE, ROBERT E, JR, b Albany, NY, Sept 21, 36; m 60; c 2. PHYSICAL CHEMISTRY, AIR POLLUTION. *Educ:* Siena Col, BS, 58; George Washington Univ, MEA, 64; Univ Cincinnati, MS, 67, PhD(phys chem), 69. *Prof Exp:* Chemist, US Army Biol Labs, Md, 58-62; scientist, Melpar, Inc, Va, 62-64; RES CHEMIST, US ENVIRON PROTECTION AGENCY, 64- *Mem:* AAAS; Air Pollution Control Asn; Am Chem Soc. *Res:* Air pollution chemistry. *Mailing Add:* US EPA Nat Environ Res Ctr Research Triangle Park NC 27711

LEE, ROBERT JEROME, b New York, NY, Nov 2, 14; m 41; c 2. VETERINARY MEDICINE. *Educ:* NY Univ, BS, 35; Kans State Univ, DVM, 39. *Prof Exp:* Vet med off, Fed Meat Grading Br, USDA, 49-58, head, Regulatory Sect, Fed Poultry Inspection, 58-61, Poultry Prod Sect, 61-64, training off, 64-68; chief, Meat & Poultry Inspection Sect, Md Dept Agr, 68-80; Staff veterinarian, Flow Labs, 80-; VPRES, AM CRITICAL CARE, MCGRAW PARK, IL. *Concurrent Pos:* Mem, Expert Comt Food Hyg Comt, Codex Alimentarius, WHO-UN Food & Agr Org, 65-68, alt deleg, Poultry Comt, 66-68; mem, Food Hyg Comt, US Animal Health Asn, 70-80, chmn, 78-80 & mem animal welfare comt, 81- *Mem:* Nat Asn State Meat & Food Inspection Dirs (pres, 70-73); Am Asn Food Hyg Vets (secy, 73-78, exec vpres, 78-); Nat Asn Fed Vets (secy-treas, 60-61); Am Vet Med Asn. *Mailing Add:* 1125 Laurel Wood Dr McLean VA 22102

LEE, ROBERT JOHN, b Worcester, Mass, July 2, 29; m 60; c 3. MEDICAL PHYSIOLOGY. *Educ:* Adelphi Univ, AB, 58; Yeshiva Univ, MS, 59; State Univ NY, PhD(physiol), 67. *Prof Exp:* Teacher, NY Schs, 58-59; teaching asst surg, State Univ NY, 59-63; sr scientist, Geigy Res Labs, 66-70; res group leader, Squibb Inst Med Res, 70-77; dir dept pharmacol, 77-78, dir pharmaceut res, 78-81, VPRES, RES & DEVELOP, DUPONT CRITICAL CARE, 81- *Mem:* AAAS; Am Soc Pharmacol & Exp Therapeut; Sigma Xi; Am Col Cardiol; Am Heart Asn. *Res:* Mechanical aspects of cardiovascular physiology; ventricular function. *Mailing Add:* DuPont Merck Pharmaceut P26-1354 4301 Lancaster Pike Barley Mill Plaza Wilmington DE 19880-0026

LEE, ROBERT W, b Cedar Rapids, Iowa, Feb 28, 31; m 51; c 3. EXPERIMENTAL PHYSICS. *Educ:* Mich State Univ, BS, 53, MS, 55. *Prof Exp:* Sr res physicist, Res Labs, Gen Motors Corp, 55-88; RETIRED. *Honors & Awards:* AIP Award, Indust Applns Physics, 85. *Mem:* Sigma Xi. *Res:* Permanent magnets; electro-optics; gas diffusion in solids. *Mailing Add:* 3090 Myddleton Dr Troy MI 43084

LEE, ROBERTO, b Shanghai, China, Jan 10, 37; m 63; c 2. CHEMICAL ENGINEERING. *Educ:* Univ Ill, BS, 58; Purdue Univ, MS, 60, PhD(chem eng), 64. *Prof Exp:* Engr, Corning Glass Works, 60; res engr, E I du Pont de Nemours & Co, 61; sr res chem engr & prin engr specialist, 63-76, engr supt, 77-79, engr mgr, 80-82, ENGR GROUP CONSULT, MONSANTO CO, 82- *Mem:* Fel Am Inst Chem Engrs; Am Chem Soc. *Res:* Reaction engineering; biochemical and chemical process design; chemical process research and development. *Mailing Add:* Monsanto Co 800 N Lindbergh Blvd St Louis MO 63167

LEE, ROLAND ROBERT, b Cleveland, Ohio, July 18, 54. DIAGNOSTIC RADIOLOGY, MAGNETIC RESONANCE IMAGING & SPECTROSCOPY. *Educ:* Calif Inst Technol, BS, 75; Univ Calif, Berkeley, MA, 77; Univ Calif, Los Angeles, MD, 85. *Prof Exp:* Physicist laser fusion, Lawrence Livermore Nat Lab, 75-77; teaching asst physics, Univ Calif Berkeley, 75-77, res asst superconductivity, 77-81; intern, Harbor-Univ Calif Los Angeles Med Ctr, 85-86; resident diag radiol, Harvard Med Sch, Brigham & Women's Hosp, 86-90; clin & res fel magnetic resonance imaging spectros, Long Beach Mem Magnetic Resonance Ctr, Huntington Mem Med Res Inst, 90-91; CLIN INSTR & RES FEL NEURORADIOL, DEPT RADIOL, UNIV CALIF, SAN FRANCISCO, 91- *Mem:* Am Col Radiol. *Res:* Clinical diagnostic radiology; neuroradiology and magnetic resonance imaging; magnetic resonance spectroscopy; medical image processing. *Mailing Add:* Dept Radiol Neuroradiol Sect Univ Calif San Francisco CA 94143

LEE, RONALD NORMAN, b Springfield, Mo, Oct 21, 35; m 59; c 1. SURFACE PHYSICS, MATERIALS SCIENCE. *Educ:* Univ Ill, BS, 58, MS, 60; Brown Univ, PhD(physics), 65. *Prof Exp:* Res assoc physics, Coord Sci Lab, Univ Ill, 60 & Brown Univ, 64-65; fel phys chem, Battelle Mem Inst, Ohio, 65-68; physicist, US Naval Ord Lab, 68-74, PHYSICIST, MATS DIV, NAVAL SURFACE WARFARE CTR, 74-, TECH LEADER, SURFACE SCI GROUP, 79- *Mem:* Am Phys Soc; Am Vacuum Soc; AAAS; Sigma Xi. *Res:* Surface science; composite materials interface properties, carbon fiber surface properties, corrosion, bio-corrosion; physics of electron spectroscopies; physics of scanning tunneling; microscopy, battery electrode chemistry; energetic materials. *Mailing Add:* 54 Shaw Ave Silver Spring MD 20904

LEE, RONALD S, b Ames, Iowa, Dec 29, 38; m 60; c 2. SOLID STATE PHYSICS. *Educ:* Luther Col, Iowa, BA, 61; Iowa State Univ, PhD(physics), 67. *Prof Exp:* Asst prof, 67-74, ASSOC PROF PHYSICS, KANS STATE UNIV, 74- *Mem:* Am Phys Soc; Am Asn Univ Professors. *Res:* Shock waves in chemically reacting media; radiation effects in solids. *Mailing Add:* Lawrence Livermore Lab L-281 PO Box 808 Livermore CA 94550

LEE, RONNIE, b China, Nov 6, 42. TOPOLOGY. *Educ:* Chinese Univ Hong Kong, BS, 65; Univ Mich, PhD(math), 68. *Prof Exp:* Mem, Inst Advan Studies, 68-70; asst prof, 70-73, ASSOC PROF MATH, YALE UNIV, 73-, DIR UNDERGRAD STUDIES, 77- *Concurrent Pos:* Sloan Found fel, 73. *Res:* Differential topology. *Mailing Add:* Dept Math Yale Univ New Haven CT 06520

LEE, SAMUEL C, b Hong-Chow, China, May 4, 37; US citizen. ELECTRICAL ENGINEERING, COMPUTER SCIENCE. *Educ:* Nat Taiwan Univ, BS, 60; Univ Calif, Berkeley, MS, 63; Univ Ill, Urbana, PhD(elec eng), 65. *Prof Exp:* Mem tech staff elec eng, Bell Labs, Murray Hill, 65-67; assoc prof, NY Univ, 67-70 & Univ Houston, 70-75; PROF ELEC ENG & COMPUTER SCI, UNIV OKLA, 75- *Concurrent Pos:* Consult, Bell Labs, 67-70 & NAm Aircraft Co, Conn, 68-69; vis assoc prof, Baylor Col Med & asst neurophysiologist, Methodist Hosp, Houston, 72-75. *Mem:* Inst Elec & Electronics Engrs; Asn Comput Mach; Am Soc Eng Educ. *Res:* Digital systems; logical design; pattern recognition; artificial intelligence. *Mailing Add:* Sch Elec Eng & Comput Sci Rm 219 Univ Okla 202 W Boyd Norman OK 73019

LEE, SHAW-GUANG LIN, b Miao-Li, Taiwan, Oct 9, 44; US citizen; m 68; c 3. BIOCHEMISTRY. *Educ:* Nat Taiwan Univ, BS, 67; Northwestern Univ, PhD(biochem), 72. *Prof Exp:* Fel biochem, Northwestern Univ, 73-75; res biochemist molecular virol, Abbott Labs, 75-81; prin scientist, Meloy Lab, 82-84; ASSOC DIR, BIOTECHNOL & MICROBIOL, WYETH-AYERST RES, 88- *Mem:* Am Chem Soc. *Res:* AIDS vaccine developmental research; lyme disease vaccine; strombolytic agents; fibrinolysis. *Mailing Add:* Dept Biotechnol & Microbiol Wyeth-Ayerst Res PO Box 8299 Philadelphia PA 19101-1245

LEE, SHIH-SHUN, b Taiwan, May 25, 36; US citizen; m 69; c 2. MOLECULAR GENETICS, CANCER CHEMOTHERAPY. *Educ:* Nat Taiwan Univ, BS, 59, MS, 64; Mont State Univ, PhD(genetics), 69. *Prof Exp:* Instr agron, Prov Taiwan Agr Col, 64-66; res asst genetics, Mont State Univ, 66-69; res assoc cancer res, Stehlin Found, 73-74, assoc dir cancer res, 75-81;

CHMN TISSUE CULT DEPT, BURZYNSKI RES INST, 81- *Concurrent Pos:* Fels, Indiana Univ, 69-70 & M D Anderson Hosp & Tumor Inst, 70-73. *Mem:* Sigma Xi; NY Acad Sci; AAAS. *Res:* Tumor tissue culture; tumor chemotherapy; induction of cancer cell differentiation. *Mailing Add:* 2707 Cane Field Dr Sugar Land TX 77479

LEE, SHIH-YING, b Peking, China, Apr 30, 18. MECHANICAL ENGINEERING, ENGINEERING EDUCATION. *Educ:* Mass Inst Technol, ScD, 45. *Prof Exp:* Res engr, Mass Inst Technol, 47-52, fac mem, 52-74, prof mech eng, 66-74, EMER PROF MECH ENG, MASS INST TECHNOL, 74-; PROF & CHIEF EXEC OFFICER, SENTRA SYSTS, 74- *Mem:* Nat Acad Eng. *Mailing Add:* Setra Systs Inc 45 Nagog Park Acton MA 01720

LEE, SHUI LUNG, b Canton, China, Sept 15, 38; m 69; c 2. ORGANIC CHEMISTRY. *Educ:* Univ Western Australia, BS, 65, PhD(org chem), 69. *Prof Exp:* Fel org chem, McMaster Univ, 69-71, Queens Univ, Can, 71-74; sr chemist, Aldrich Chem Co, 74-75; sr res chemist, pigments div, Chemetron Corp, 75-78; chemist, 78-80, SR CHEMIST, GANES CHEMICALS, 80- *Mem:* Am Chem Soc. *Res:* Medicinal and fine organic chemicals. *Mailing Add:* One Toby Terr Towaco NJ 07082-1413

LEE, SHUISHIH SAGE, b Soo-chow, China, Jan 5, 48; m 73; c 2. EXPERIMENTAL PATHOLOGY, ANATOMIC PATHOLOGY. *Educ:* Nat Taiwan Univ, MD, 72; Univ Rochester, PhD(path), 76. *Prof Exp:* Intern path, Med Ctr, Univ Rochester, 76-78, resident, 78-79; RESIDENT PATH, NORTHWESTERN MEM HOSP, 79-; PATHOLOGIST, PARKVIEW MEM HOSP, 81- *Concurrent Pos:* Clin asst prof, Dept Path, Med Sch, Ind Univ, 89- *Mem:* Int Acad Pathol; Am Asn Pathologists; Am Soc Clin Pathologists; Col Am Pathologists; Int Acad Cytol; Electron Micros Soc Am. *Res:* Synthesis of ferritin in rat liver and hepatoma cells. *Mailing Add:* Dept Path Parkview Mem Hosp Ft Wayne IN 46805

LEE, SHUNG-YAN LUKE, b China, Sept 10, 38; US citizen; m 62; c 2. ORGANIC CHEMISTRY. *Educ:* Univ Wis, BS, 59; Ohio State Univ, MS, 62, PhD(phys org chem), 66. *Prof Exp:* RES CHEMIST PHOTOG SYST, E I DU PONT DE NEMOURS & CO, INC, 66- *Mem:* Am Chem Soc. *Res:* Investigations of photopolymerization systems. *Mailing Add:* 2716 Marklyn Dr Wilmington DE 19810-2345

LEE, SHYH-YUAN, b Yuinin, Taiwan, Nov 17, 43. ACCELERATOR PHYSICS, NUCLEAR PHYSICS. *Educ:* Univ Taiwan, BS, 63; State Univ NY, Stony Brook, MS, 69 & PhD(physics), 72. *Prof Exp:* Res assoc, Univ Paris Orsay, 76-77 & Univ Wash Seattle, 77- 78; asst prof physics, State Univ NY, Stony Brook, 78-84; assoc physicist, Brookhaven Nat Labs, 84-85, physicist, 85-90; PROF PHYSICS, IND UNIV, 90- *Mem:* Am Phys Soc; Nat Geog Soc. *Res:* Nonlinear physics; accelerator and spin physics. *Mailing Add:* Dept Physics Ind Univ Bloomington IN 47405

LEE, SI DUK, b Ham Hung, Korea, Jan 2, 32; US citizen; m 57; c 3. ENVIRONMENTAL SCIENCES, BIOLOGICAL CHEMISTRY, TOXICOLOGY. *Educ:* Seoul Nat Univ, BS, 55; Univ Md, MS, 59, PhD(biochem), 62. *Prof Exp:* Res assoc biochem, Med Ctr, Duke Univ, 61-62, NIH fel, 62-63, Am Heart Asn adv res fel, 63-64; res chemist, USPHS, 64-65, supvry res chemist, 65-67, chief biochem unit, 67-69, chief biochem sect, 69-71, dep chief, Nat Air Pollution Control Admin, 69-71; dep chief biol effects br, Nat Environ Res Ctr, Environ Protection Agency, 71-73; SR SCI HEALTH ADV, US ENVIRON PROTECTION AGENCY, 73- *Concurrent Pos:* Adj asst prof, Dept Biol, Col Med, Univ Cincinnati, 67-80 & Dept Environ Health, 75-80; adj assoc prof, Duke Univ Med Sch, 82-87; vis scholar, Harvard Univ, 87- *Honors & Awards:* Bronze Medal, US Environ Protection Agency, 80 & 84. *Mem:* AAAS; Am Chem Soc; Am Col Toxicol; Air Pollution Control Asn; Sigma Xi; Soc Toxicol. *Res:* Effects of air pollutants on metabolism; lipid metabolism; effects of pollutants on aging; effects of sulfur dioxide on subcellular metabolism; environmental management. *Mailing Add:* US Environ Protection Agency ECAO MD52 Research Triangle Park NC 27711-0001

LEE, SIDNEY, chemical engineering, physical chemistry, for more information see previous edition

LEE, SIN HANG, b Hong Kong, Nov 17, 32; US citizen; m 58; c 2. SURGICAL PATHOLOGY, HISTOTECHNISTRY. *Educ:* Wuhan Med Col, China, MD, 56. *Prof Exp:* Postgrad bact, Sichuan Med Col, Chengou, China, 56-57, asst lectr, 57-61; demonstr path, Univ Hong Kong, 61-63; intern clin, South Baltimore Gen Hosp, 63-64; resident path, NY Hosp, Cornell Med Ctr, 64-67; fel, Mem Hosp Cancer & Allied Dis, 67-68; asst prof, McGill Univ, Montreal, Can, 68-71; assoc prof, 71-73, ASSOC CLIN PROF PATH, YALE UNIV, NEW HAVEN, CONN, 73-; PATHOLOGIST, HOSP ST RAPHAEL, NEW HAVEN, CONN, 73- *Concurrent Pos:* Guest prof, Tongji Med Univ Hankow, Wuhan, China, 84- *Mem:* Col Am Pathologists; Int Acad Path; Am Asn Pathologists; NY Acad Sci; AAAS; Royal Col Physicians & Surgeons Can. *Res:* Histochemical localization of enzyme activities; cytochemical assays of steroid receptors in breast cancer and other target cells; mycoplasma pneumoniae antigen for the detection of specific membrane antibodies in patients for early diagnosis of infection; awarded two patents. *Mailing Add:* Dept Path Hosp St Raphael 1450 Chapel St New Haven CT 06511

LEE, SIU-LAM, b Macao, China, Oct 3, 41; m 82; c 3. INSECT ECOLOGY, EVOLUTION. *Educ:* Chung Chi Col, Chinese Univ, Hong Kong, BSc, 62; Oberlin Col, AM, 63; Cornell Univ, PhD(entom), 67. *Prof Exp:* ASST PROF BIOL, UNIV LOWELL, NORTH CAMPUS, 75- *Concurrent Pos:* NSF res award, 69-72. *Mem:* Am Entom Soc; Animal Behav Soc; Bee Res Asn; Sigma Xi. *Res:* Learning ability of fruit fly; behavioral ecology of the leaf-cutter bee; effects of ginseng, Panax ginseng, on mamalian blood cells and malaria infecton; the flora of temperate deciduous forests. *Mailing Add:* Dept Biol Univ Lowell North Campus Lowell MA 01854

LEE, SOOK, solid state physics; deceased, see previous edition for last biography

LEE, STANLEY L, b Newburgh, NY, Aug 27, 19; m 47; c 3. INTERNAL MEDICINE, HEMATOLOGY. *Educ:* Columbia Univ, AB, 39; Harvard Univ, MD, 43. *Prof Exp:* Intern, Mt Sinai Hosp, NY, 43-44, resident med, 46-48, Georg Escherich fel path, 48-49, asst, 49-53, asst attend hematologist, 53-59; assoc prof, State Univ Downstate Med Ctr, 59-68, dean fac, 78-79, actg pres, 79-81, dean, Col Med & vpres acad affairs, 81-82, PROF MED, STATE UNIV DOWNSTATE MED CTR, 68-; CHIEF HEMAT, BROOKDALE HOSP MED CTR, 82- *Concurrent Pos:* Dir hemat, Maimonides Med Ctr, Brooklyn, 59-71; treas, Int Cong Hemat, NY, 65-68; dir med, Jewish Hosp & Med Ctr, Brooklyn, 71-77. *Mem:* Am Soc Hemat; Am Rheumatism Asn; Soc Human Genetics; Am Fedn Clin Res; fel Am Col Physicians. *Res:* Systematic lupus erythematosus; leukemia. *Mailing Add:* Dept Hemat Brookdale Hosp Brooklyn NY 11212

LEE, STEVE S, b Taiwan, Nov 2, 48; US citizen; m 79; c 2. HERBAL PROCESS, ORIENTAL FOLK MEDICINE. *Educ:* Kaohsiung Med Col, Taiwan, BS, 72; Duquesne Univ, MS, 84. *Prof Exp:* Lab mgr, Grad Sch Pub Health, Univ Pittsburgh, 85-87; PLANT MGR, SUNRIDER INT, 87- *Res:* Extraction of Chinese folk medicinal herbs; evaluation of the extract particularly in antitumor and antibacteria. *Mailing Add:* 463 N 100 E Orem UT 84057

LEE, STUART M(ILTON), b New York, NY, Apr 14, 20; m 48; c 3. MATERIALS SCIENCE. *Educ:* Long Island Univ, BS, 41; Univ Nev, MS, 47; Fla State Univ, PhD(chem), 53. *Prof Exp:* Chemist anal & testing, NY Testing Labs, 41-42; chemist anal develop, Gen Dyestuffs Corp, 42-43; chief chemist org synthesis, Trinity Res Found, 49-50; sr res chemist & proj leader org res, Allied Chem Corp, 52-59; res chemist, Aerojet-Gen Corp, 59-61; mgr chem res & develop, Electro-Optical Systs, Inc, Xerox Corp, 61-64; sr tech specialist, Autonetics Div, NAm Rockwell, Calif, 64-71; sr staf scientist, Ford Aerospace & Commun Corp, 71-85; tech dir, Soc Advan Mat & Process Engrs, 89-90; RETIRED. *Concurrent Pos:* Ed, Soc Advan Mat & Process Engrs J, 79-; ed-in-chief, Technomic Publ Co, 83-, ed, Encycl Composites, 85-; consult, SRI Int, GTE, Siemens, Teledyne & others. *Honors & Awards:* Nat Meritorious Bronze Award, Soc Advan Mat & Process Engrs, 82. *Mem:* Am Chem Soc; fel Soc Advan Mat & Process Engrs. *Res:* Electronic materials; composites, biomaterials; instrumental failure analysis; thin films; space materials; electronic and laser organics; high temperature and organic polymers; approximately 70 technical publications including 14 patents; contributing author or editor of 14 books; acetylenic and petrochemical derivatives. *Mailing Add:* 3718 Cass Way Palo Alto CA 94306

LEE, SUE YING, b Schenectady, NY, Jan 11, 40; m 73; c 1. VERTEBRATE MORPHOLOGY. *Educ:* State Univ NY Albany, BS, 61, MS, 63; Univ Ill, Urbana, PhD(zool), 68. *Prof Exp:* Instr vert morphol & human anat, Univ Ill, Chicago, 67-69; asst prof, 69-72, assoc prof, 73-77, PROF VERT MORPHOL & HUMAN ANAT, HUMBOLDT STATE UNIV, 78- *Concurrent Pos:* Consult, Int Union Conserv Nature & Natural Resources, World Wildlife Fund, 74-75. *Mem:* AAAS; Am Soc Zool; Am Inst Biol Sci; Western Soc Naturalists; Soc Vert Paleont; Am Asn Anatomists. *Res:* Reproductive biology; ultrastructure of fetal membranes. *Mailing Add:* Dept Biol Sci Humboldt State Univ Arcata CA 95521

LEE, SUK YOUNG, b Seoul, Korea, June 18, 40; US citizen; m 66; c 2. SOIL SCIENCE, ENVIRONMENTAL CHEMISTRY. *Educ:* Univ Sask, MS, 68; Univ Wis, PhD(soil sci), 73. *Prof Exp:* Fel, Univ Wis, 74-75, Univ S Fla, 75-76 & Tex A&M Univ, 76-77; RES SCIENTIST ENVIRON SCI, OAK RIDGE NAT LAB, 77- *Mem:* Am Soc Agron; Soil Sci Soc Am; Clay Minerals Soc. *Res:* Transport of trace elements and radio nuclides, such as plutonium and uranium, in environment. *Mailing Add:* 12 Monaco Lane Oak Ridge TN 37830

LEE, SUN, b Seoul, Korea, June 2, 20; US citizen; m 45; c 6. SURGERY. *Educ:* Seoul Nat Univ, MD, 45. *Prof Exp:* From instr to asst prof, Univ Pittsburgh, 57-64; assoc prof surg, 68-74, PROF EXP SURG, UNIV CALIF, SAN DIEGO, 74-; ASSOC, SCRIPPS CLIN & RES FOUND, 64- *Concurrent Pos:* Surg fel, Univ Pittsburgh, 55-57. *Honors & Awards:* Gold Medal, Pioneer Exp Microsurg, Ger; Gold Medal, Lombardo Surg, Italy. *Mem:* Int Microsurg Soc; Int Proctol Soc. *Res:* Development of organ transplant in the rat to study transplantation immunology and associated physiology; techniques of heart-lung, liver, spleen, pancreas, testicle, kidney and stomach transplantation and allied microsurgical techniques in rats. *Mailing Add:* 6462 Cardeno Dr La Jolla CA 92038

LEE, SUNG J, applied mathematics; deceased, see previous edition for last biography

LEE, SUNG MOOK, b Seoul, Korea, Mar 2, 33; m 58; c 3. THEORETICAL MECHANICS, SNOW & COLD ENVIRONMENTAL RESEARCH. *Educ:* Yonsei Univ, BSc, 55; Ohio State Univ, MSc, 59, PhD(crystal dynamics), 65. *Prof Exp:* Teacher, Hansung Boy's High Sch, Korea, 54-55; asst prof physics, Denison Univ, 61-65; from asst prof to assoc prof, 65-72, PROF PHYSICS, MICH TECHNOL UNIV, 72-, DIR, KEWEENAW RES CTR, 76-, DEAN RES & GRAD SCH, 88- *Concurrent Pos:* NATO sr fel sci, 74; vis sr res fel, Inst Sound & Vibration Res, Univ Southampton, Eng, 80 & 81. *Mem:* Am Phys Soc; Int Glaciological Soc. *Res:* Vibrational analysis of periodic systems, crystal lattices, and molecules; wave propagation in solids; mechanical properties of solids; mechanics; acoustics; optics; snow and ice. *Mailing Add:* Grad Sch Mich Technol Univ Houghton MI 49931

LEE, SUNGGYU, b Kangjin-Kun, Korea, Mar 11, 52; m 80; c 3. PROCESS DEVELOPMENT, FUEL SCIENCE. *Educ:* Seoul Nat Univ, BS, 74, MS, 76; Case Western Reserve Univ, PhD(chem eng), 80. *Prof Exp:* From asst prof to assoc prof, 80-88, ROBERT IREDELL PROF CHEM ENG & DEPT HEAD, UNIV AKRON, 88- *Concurrent Pos:* Prin investr, Univ Akron, 80-; prof consult, 80- *Honors & Awards:* Louis A Hill Award, 87. *Mem:* Am Inst Chem Engrs; Am Chem Soc; Sigma Xi. *Res:* Process development for the manufacture of clean liquid and solid fuels; coal desulfurization and characterization; manufacture of enzymes; manufacture and kinetics of polymers and polymerization; engineered plastics; author of 6 books, 44 journal articles and 23 papers. *Mailing Add:* Dept Chem Eng Univ Akron Akron OH 44325

LEE, T(IEN) P(EI), b Nanking, China, Sept 8, 33; m 63; c 2. ELECTRICAL ENGINEERING. *Educ:* Taiwan Norm Univ, BS, 57; Ohio State Univ, MS, 59; Stanford Univ, PhD(elec eng), 63. *Prof Exp:* MEM TECH STAFF, BELL TEL LABS, 63- *Mem:* Inst Elec & Electronics Engrs; Sigma Xi. *Res:* Microwave electronics; microwave solid state devices; varactor diodes; parametric amplifiers; semiconductor lasers and related optical communication. *Mailing Add:* Dist Res Mgr Bell Commun Res Rm NVC3Z-369 Navesink Res & Eng Ctr Red Bank NJ 07701

LEE, T(HOMAS) S(HAO-CHUNG), b Soochow, China, Nov 18, 31; m 60; c 2. ELECTRICAL ENGINEERING. *Educ:* Nat Taiwan Univ, BS, 54; Univ Minn, Minneapolis, MS, 56, PhD(elec eng), 61. *Prof Exp:* From instr to asst prof, 57-66, ASSOC PROF ELEC ENG, UNIV MINN, MINNEAPOLIS, 66- *Concurrent Pos:* Consult, mil prod group, Honeywell Regulator Co, 61-62, aero div, 63-64 & US Naval Res Lab, 65- *Mem:* Am Phys Soc; Am Geophys Union. *Res:* Acoustics; explosive phenomena; gas-dynamics; systems; electromagnetism; interplanetary phenomena. *Mailing Add:* Dept Elec Eng 139 Elec Eng Bldg Univ Minn 123 Church St SE Minneapolis MN 55455

LEE, TED C K, b Seoul, Korea, Dec 3, 40; US citizen; m 25; c 3. KINETICS, CHEMICAL MODIFICATION OF PROTEINS. *Educ:* Korea Univ, BS, 65; Okla State Univ, PhD(biochem), 71. *Prof Exp:* Res assoc protein chem, Rockefeller Univ, 71-73; asst prof protein chem, Med Col, Howard Univ, 73-78; vis asst prof protein chem, Cornell Univ, 78-81; sr res scientist, Revlon Health Care Group, 81-85; prin scientist, Rorer Group, 85-90; RES FEL PROCESS DEVELOP, RHONE-POULENC RORER INC, 90- *Mem:* Am Soc Biochem & Molecular Biol; Am Chem Soc. *Res:* Properties of protein molecules; development of novel methods for isolation of proteins; formulation of proteins. *Mailing Add:* Div Biotechnol Develop Rorer Biotechnol Inc 680 Allendale Rd King of Prussia PA 19406

LEE, TEH HSUN, b Shaoshin, China, Mar 25, 17; m 45; c 1. BIOCHEMISTRY. *Educ:* Chekiang Univ, BS, 38; Univ Mich, PhD, 54. *Prof Exp:* Res assoc, Sch Med, Univ Ore, 54-55; res assoc, Sch Med, Yale Univ, 55-60, asst prof exp med, 60-62; sr biochemist, Merck, Sharp & Dohme, 62-64; asst dir, Vet Admin Human Protein Hormone Bank, Vet Admin Hosp, 64-66, chief Protein Hormone Res Lab, 66-84; vis assoc prof biochem, Albert Einstein Col Med, 67-84; RETIRED. *Concurrent Pos:* Assoc prof, Sch Med, Univ Colo, Denver, 64-66. *Mem:* Am Chem Soc; Am Soc Biol Chem; Endocrine Soc. *Res:* Pituitary hormones. *Mailing Add:* 196 Bradley Rd Scarsdale NY 10583

LEE, TEH-HSUANG, b Shanghai, China, Aug 15, 36; m 61; c 2. SOLID STATE PHYSICS. *Educ:* Nat Taiwan Univ, BS, 58; Purdue Univ, West Lafayette, PhD(physics), 67. *Prof Exp:* SR PHYSICIST, RES LABS, EASTMAN KODAK CO, 67- *Mem:* Am Phys Soc. *Res:* Optical properties of solids; semiconductors; magnetic semiconductors. *Mailing Add:* 760 High Tower Way Webster NY 14580

LEE, TEN CHING, MEMBRANES, PHOSPHOLIPIDS. *Educ:* Tulane Univ, New Orleans, PhD(biochem), 67. *Prof Exp:* BIOCHEMIST, UNIV TENN. *Mailing Add:* Oak Ridge Assoc Univ PO Box 117 Oak Ridge TN 37830

LEE, THERESA, b Beijing, China; US citizen; m 70; c 1. RECOMBINANT DNA, BIOTECHNOLOGY. *Educ:* Nat Taiwan Univ, BS, 62; Univ Pittsburgh, MS, 64; Wash Univ, PhD(biochem), 68. *Prof Exp:* Fel biochem, Med Sch, Univ Wis, 68; Harvard Univ, 68-70; fel, Johns Hopkins Univ, 70-71; prof staff microbiol, Med Sch, 71-77; res chemist molecular biol, Nat Cancer Inst, NIH, 77-84; chemist, Chem Div, Food & Drug Admin, 84-88; PROG OFF, DIV PRECLIN RES, NAT INST DRUG ABUSE, 88- *Mem:* Am Soc Biol Chemists; AAAS; Am Soc Microbiologists; NY Acad Sci; Am Chem Soc. *Res:* Recombinant DNA techniques; molecular biology; proteins; biochemistry; virology. *Mailing Add:* Div Preclin Res Parklawn Bldg Rm 10A31 Nat Inst Drug Abuse 5600 Fishers Lane Rockville MD 20857

LEE, THOMAS HENRY, b Shanghai, China, May 11, 23; nat US; m 48; c 3. PHYSICS. *Educ:* Nat Chiao-Tung Univ, China, BS, 46; Union Col, MS, 50; Rensselaer Polytech Inst, PhD(elec eng, physics), 54. *Prof Exp:* Eng analyst, Gen Elec Co, Pa, 54-55, sr res engr, 55-59, mgr eng res, 59-67, mgr lab opers, 67-71, mgr tech resources, 71-74, mgr strategic planning opers, 74-77, staff exec power systs technol oper, 77-80; vis res prof, 79-80, prof elec eng & assoc dir, Energy Lab, 80-82, dir, Lab Electromagnetic & Electronic Systs, 82-84, co dir, 84, dir, Int Inst Appl Systs Anal, 84-87, Philip Sporn prof, 82-88, EMER PROF ENERGY PROCESSING, MASS INST TECHNOL, 88-; PRES, CTR QUAL MGT, 89- *Concurrent Pos:* Adj prof, Rensselaer Polytech Inst, 54-55; lectr, Univ Pa, 59-61 & Lehigh Univ, 61-62. *Mem:* Nat Acad Eng; Am Phys Soc; fel Inst Elec & Electronics Engrs; Power Eng Soc (pres); Am Vacuum Soc; Sigma Xi; Swiss Acad Eng Sci; fel AAAS. *Res:* Electron physics; gaseous discharges; magnetohydrodynamics; ultra high vacuum technology; electrical systems; plasma physics. *Mailing Add:* Beacon Hill 44 Chestnut St Boston MA 02108

LEE, THOMAS W, b New Britain, Conn, Sept 12, 37; m 82; c 2. COMPARATIVE PHYSIOLOGY. *Educ:* Bates Col, BS, 59; Duke Univ, MA, 61; Rice Univ, PhD(biol), 64. *Prof Exp:* Lectr biol, Rice Univ, 64-65; from asst prof to assoc prof, 65-72, asst to chmn, 74-79, actg chmn dept, 69-74, PROF BIOL, CENT CONN STATE UNIV, 72- *Concurrent Pos:* Vis prof biol, Wesleyan Univ, 74-85. *Mem:* AAAS; Am Soc Zool; Sigma Xi; Ecol Soc Am. *Res:* Embryonic development of amphibians; nitrogen metabolism in invertebrate animals. *Mailing Add:* Dept Biol Cent Conn State Univ New Britain CT 06050

LEE, TIEN-CHANG, b Nantou, Taiwan, July 1, 43; m 69; c 2. GEOPHYSICS. *Educ:* Nat Taiwan Univ, BS, 65; Univ Southern Calif, PhD(geophys), 74. *Prof Exp:* Fel marine geophys, Woods Hole Oceanog Inst, 73-74; asst prof, 74-79, ASSOC PROF GEOPHYS, UNIV CALIF, RIVERSIDE, 79- *Mem:* Am Geophys Union; Soc Explor Geophysicists; Geol Soc Am. *Res:* Terrestrial heat flow, electrical exploration and micro-earthquake. *Mailing Add:* Dept Earth Sci Univ Calif Riverside CA 92521

LEE, TONG-NYONG, b July 22, 27; US citizen; m 59; c 3. PLASMA PHYSICS, ATOMIC PHYSICS. *Educ:* Seoul Nat Univ, BS, 50; Univ London, PhD(physics), 59. *Prof Exp:* Asst prof physics, Seoul Nat Univ, 60-63; assoc prof appl physics, Cath Univ Am, 64-70; res physicist plasma physics & optical sci, Naval Res Lab, 70-88; PROF PHYSICS, POHANG INST SCI & TECHNOL, 88- *Mem:* Am Phys Soc. *Res:* Short wavelength laser generation; plasma physics and spectroscopy of high temperature; high density plasma and solar flare study. *Mailing Add:* Pohang Inst Sci & Technol PO Box 125 Pohang City Kyungbuk 790-600 Republic of Korea

LEE, TONY JER-FU, b Hualien, Taiwan, Nov 10, 42; US citizen; m 78; c 2. PHARMACOLOGY. *Educ:* Taipei Med Col, Taiwan, BS, 67; WVa Univ, PhD(pharmacol), 73. *Prof Exp:* From asst prof to assoc prof, 75-87, PROF PHARMACOL, SCH MED, SOUTHERN ILL UNIV, SPRINGFIELD, 87- *Concurrent Pos:* Fel, Univ Calif, Los Angeles, 73-75; mem, High Blood Pressure Coun, Am Heart Asn & Stroke Coun, Soc Neurol Sci; Am Heart Asn grant, NIH. *Mem:* Am Soc Pharmacol & Exp Therapeut; Soc Neurosci. *Res:* Cerebral vessel innervation in health and disease. *Mailing Add:* Dept Pharmacol PO Box 19230 Springfield IL 62794-9230

LEE, TSUNG DAO, b China, Nov 25, 26; m 50; c 2. THEORETICAL PHYSICS. *Educ:* Univ Chicago, PhD(physics), 50. *Hon Degrees:* DSc, Princeton Univ, 58; LLD, Chinese Univ Hong Kong, 69; ScD, City Col New York, 78, Bard Col, 84, Peking Univ, 85, Drexel Univ, 86, Univ Bologna, 88, Columbia Univ, 90, Adelphi Univ, 91. *Prof Exp:* Res assoc astrophys, Univ Chicago, 50; res assoc physics, Univ Calif, 50-51; mem, Inst Advan Study, 51-53; from asst prof to prof, Columbia Univ, 53-60; prof, Inst Advan Study, 60-63; ENRICO FERMI PROF PHYSICS, COLUMBIA UNIV, 63-, UNIV PROF, 84- *Concurrent Pos:* Hon prof, Univ Sci & Technol China, 81, Jinan Univ, 82, Fudan Univ, 82, Quinghua Univ, 84, Peking Univ, 85 & Nanjing Univ, 85, Nankai Univ, 86, Shanghai Jiao Tong Univ, 87, Suzhou Univ, 87, Zhejiang Univ, 88. *Honors & Awards:* Nobel Prize in Physics, 57; Einstein Award Sci, 57; Loeb Lectr, Harvard Univ, 64; Order of Merit, Grande Ufficiale, Repub Italy, 86. *Mem:* Nat Acad Sci; Acad Sci China; Am Acad Arts & Sci; Am Philos Soc. *Res:* Field theory; statistical mechanics; gravity; particle physics. *Mailing Add:* Dept Physics Columbia Univ New York NY 10027

LEE, TSUNG TING, b Anhwei, China, Mar 21, 23; m 50; c 3. PLANT HORMONE, GROWTH REGULATION. *Educ:* Nat Cent Univ, China, BS, 47; Univ Wis, MS, 59, PhD(plant physiol), 62. *Prof Exp:* Plant physiologist, London Res Ctr, Can Dept Agr, 68-88; RETIRED. *Concurrent Pos:* Hon lectr, Univ Western Ont, 76- *Mem:* Am Soc Plant Physiol; Plant Growth Regulator Soc Am; Sigma Xi; Int Asn Plant Tissue Cult; Can Soc Plant Physiol. *Res:* Plant growth regulators; auxin metabolism, concerning regulation of conjugation and oxidation of IAA. *Mailing Add:* 30 Runny Mede London ON N6G 1Z8 Can

LEE, TSUNG-SHUNG HARRY, b Taipei, Taiwan, June 7, 43; m 68; c 1. NUCLEAR PHYSICS. *Educ:* Taiwan Norm Univ, BS, 65; Nat Tsing-Hua Univ, MS, 67; Univ Pittsburgh, PhD(physics), 73. *Prof Exp:* Res assoc physics, Bartol Res Found, 73-75; res assoc physics, 75-77, asst physicist, 77-81, PHYSICIST, ARGONNE NAT LAB, 81- *Mem:* Am Phys Soc. *Res:* Intermediate-energy nuclear physics. *Mailing Add:* Physics Div Argonne Nat Lab Argonne IL 60439

LEE, TUNG-CHING, b Szechwan, China, Oct 28, 41; US citizen; m 70; c 2. FOOD SCIENCE & TECHNOLOGY, FOOD SAFETY. *Educ:* Tung-Hai Univ, Taiwan, BS, 63; Univ Calif, Davis, MS, 66, PhD(agr chem), 70. *Prof Exp:* Res asst, Univ Calif, 65-70; sr food technologist, Hunt-Wesson Foods, Inc, 70-72; from asst prof to assoc prof, 72-79, PROF FOOD SCI, UNIV RI, 79- *Concurrent Pos:* Vis prof, Inst Biochem, Cluj, Romania, 75 & Grad Inst Food Sci, Nat Taiwan Univ, 78-79; adv, Food Indust Res & Develop Adv Comt, Repub China, 76-; consult, many US & int food co, 76- *Mem:* Fel Inst Food Technologists; Am Chem Soc; Am Inst Nutrit; Am Soc Microbiol. *Res:* Nutritional and safety aspects of food processing; Maillard browning reaction; carotenoids and vitamins; food extrusion; marine food technology; biotechnological applications in food technology; fish nutrition; fishfeed technology. *Mailing Add:* Dept Food Sci & Nutrit Univ RI Kingston RI 02881

LEE, TYRONE YIU-HUEN, b China, Sept 6, 44; Can citizen; m 76. NEUROPHARMACOLOGY, PSYCHOPHARMACOLOGY. *Educ:* Univ Calif, Los Angeles, BSc, 68; Univ Toronto, MSc, 72, PhD(pharmacol), 75. *Prof Exp:* Res fel pharmacol, Ont Ment Health Found, 75-78; res assoc, 77-78, lectr pharmacol, 78-79, ASST PROF PHARMACOL & PSYCHIAT, UNIV TORONTO, 79-, GRAD FAC, SCH GRAD STUDIES, 80-; RES SCIENTIST & ASST DIR, PSYCHOPHARMACOL SECT, CLARK INST PSYCHIAT, 80- *Honors & Awards:* Paul Christie Mem Award, Ont Ment Health Found, 78. *Mem:* Soc Neurosci. *Res:* Study of the mechanism of action of anti-psychotic drugs in the central nervous system; radio-receptor binding analysis of post-mortem human brains in the study of schizophrenia, Parkinson's disease and Huntington's chorea; effect of chronic neuroleptic treatment in animals and tardive dyskinesia; interaction of neuronal systems in affective disorders; animal models of schizophrenia. *Mailing Add:* 36 Henry St Toronto ON M5T 1X2 Can

LEE, TZOONG-CHYH, b Taiwan, Jan 2, 36; m 62; c 3. ORGANIC CHEMISTRY, BIO-ORGANIC CHEMISTRY. *Educ:* Yamagata Univ, Japan, BSc, 63; Tohoku Univ, Japan, MSc, 65; Australian Nat Univ, PhD(med chem), 68. *Prof Exp:* USPHS fel, Sloan-Kettering Inst Cancer Res, 68-71; Damon Runyon Res fel, 69-70; res assoc org chem, 71-75, assoc, 75-80; RES MGR, LEA RONAL, INC, 80- *Mem:* Am Chem Soc. *Res:* Nitrogen heterocyclic chemistry; organic synthesis; structure-activity relationship; imaging chemicals. *Mailing Add:* 272 Buffalo Ave Lea Ronal Inc Freeport NY 11520

LEE, VING JICK, b Columbus, Ohio, July 28, 51; m 74; c 2. CHEMISTRY, NATURAL PRODUCTS CHEMISTRY. *Educ:* Ohio State Univ, BA, 71; Univ Ill, Urbana, MS, 73, PhD(chem), 75. *Prof Exp:* Res assoc chem, Univ Ill, Urbana, 71-75; teaching assoc, 71-73; NIH res assoc, Harvard Univ, 75-77; sr res group leader, 87-89, RES CHEMIST, LEDERLE LABS, AM CYANAMID CO, 77-, HEAD DEPT CHEM, INFECTIOUS DIS & MOLECULAR BIOL RES, 90- *Concurrent Pos:* Mem, Med Chem Study Sect, NIH. *Mem:* Am Chem Soc; Int Soc Heterocyclic Chem. *Res:* Antibiotics, antivirals and natural products. *Mailing Add:* Infectious Dis Res Sect Lederle Labs Bldg 65A-302A Pearl River NY 10965

LEE, VIN-JANG, b Honan, China, Feb 14, 37; div; c 1. HETEROGENEOUS CATALYSIS, QUANTUM THEORY. *Educ:* Ord Eng Col, Taiwan, Dipl eng, 53; Notre Dame Univ, MS, 58; Univ Mich, PhD(chem eng), 63. *Prof Exp:* Chem engr, 26th Arsenal, Repub China, 52-57; res specialist, Monsanto Chem Co, 64-65; res specialist, Univ Mo, Columbia, 65-68, assoc prof chem eng, 68-74; vis prof, Dept Chem, Univ Calif, Los Angeles, 72-73; consult catalysis, Libby Corp, 74-77; pres, Lee Securities & Investment Co, 75-80 & Econo Trading Corp, 80-81; PRES, CYBERDYNE, INC, 81- *Mem:* Am Phys Soc; Am Chem Soc; Am Inst Chem Eng. *Res:* Surface physics; catalysis and kinetics; tunneling in catalysis; physical foundations of quantum theory. *Mailing Add:* Cyberdyne Inc 1045 Ocean Ave Suite 2 Santa Monica CA 90403

LEE, VIRGINIA ANN, b Grand Rapids, Mich, Oct 30, 22. BIOCHEMISTRY. *Educ:* Univ Ill, BS, 44; Univ Colo, MS, 46, PhD(biochem), 52. *Prof Exp:* Instr biochem, Sch Med, Univ Colo, Denver, 55-59, asst prof, 59-67; asst prof, 67-78, ASSOC PROF FOOD SCI & NUTRIT, COLO STATE UNIV, 78- *Mem:* Am Dietetics Asn; Soc Nutrit Educ; Sigma Xi. *Res:* Nutrition. *Mailing Add:* 1405 Luke Ft Collins CO 80524

LEE, WAI-HON, b Haiphong, Vietnam, Apr 29, 42; US citizen; m 68; c 2. OPTICAL PHYSICS, COMMUNICATIONS SCIENCE. *Educ:* Mass Inst Technol, BSc, 65, MSc, 67, DSc, 69. *Prof Exp:* Assoc prin eng coherent optics, Electronic Syst Div, Harris Corp, 69-73; staff mem res optical sci, Palo Alto Res Ctr, Xerox Corp, 73-81; mgr laser imaging systs, Xidex Corp, 81-; PRES, HOETRON INC, SUNNYVALE, CALIF, 88- *Mem:* Inst Elec & Electronics Engrs; Optical Soc Am; Soc Photo-Optical Instrumentation Engrs; Soc Photog Scientists & Engrs. *Res:* Applications of grating structures for testing optical surfaces and scanning laser beam; optical methods for storing digital information at high density. *Mailing Add:* Hoetron Inc 776 Palomar Ave Sunnyvale CA 94086

LEE, WARREN FORD, b Harriston, Ont, Aug 25, 41; m 66; c 3. AGRICULTURAL ECONOMICS. *Educ:* Univ Toronto, BSA, 63; Univ Ill, MS, 67; Mich State Univ, PhD(agr econ), 70. *Prof Exp:* Credit adv, Farm Credit Corp, 63-65; assoc prof agr finance, 70-80, PROF AGR ECON & RURAL SOCIOL, OHIO STATE UNIV, 80- *Concurrent Pos:* Economist, Econ Br, Agr Can, 75-76. *Mem:* Am Agr Econ Asn. *Res:* Agricultural credit and finance; farm firm growth; rural capital markets; bank structure and performance; financial institutions. *Mailing Add:* Dept Agr Econ 103 Agr Admin Bldg Main Campus Ohio State Univ Columbus OH 43210

LEE, WEI-KUO, b Hopei, China, Apr 29, 43; US citizen; m. PETROCHEMICAL SEPARATIONS, SOLID FLUIDIZATION TECHNOLOGY. *Educ:* Nat Taiwan Univ, BS, 65; Univ Houston, PhD(chem eng), 71. *Prof Exp:* Unidel postdoctoral fel, Univ Del, 72-73; res engr, Celanese Res Co, 73-77; staff engr, Exxon Res & Eng Co, 77-86; sr res engr, Shell Develop Co, 86-88; RES STAFF MEM, UNION CARBIDE CHEM & PLASTICS CO, 88- *Concurrent Pos:* Adj prof, Chem Eng Dept, NJ Inst Technol, 75-81; vchmn, Chinese Inst Engrs-USA, 91- *Mem:* Am Inst Chem Engrs; Soc Plastics Engrs; Soc Rheology; Am Inst Physics; Sigma Xi. *Res:* Applications of fluid mechanics and rheology in various plastics technology and petrochemical engineering problems, including fibers, films, wire and cable insulations, oil field fluids, solid suspensions, composites, multiphase dispersions, fluidized solids and powders. *Mailing Add:* 328 Eileen Way Bridgewater NJ 08807

LEE, WEI-LI S, b Kiangsi, China, Feb 14, 45; m 70; c 2. BIOCHEMISTRY. *Educ:* Tunghai Univ, BS, 66; State Univ NY, Buffalo, MA, 69, PhD(biol), 72. *Prof Exp:* Asst biol sci, State Univ NY, Buffalo, 66-72; res assoc immunochem, Col Physicians & Surgeons, Columbia Univ, 72-75; instr dermat med, 75-82, ASST PROF DERMAT, STATE UNIV NY DOWNSTATE MED CTR, 83- *Concurrent Pos:* Dermat Found grant, Soc Investigative Dermat, 76, Benjamin Zohn Res & Cult Fund grant, 83. *Mem:* Sigma Xi; NY Acad Sci; Soc Investigative Dermat; Am Fedn Clin Res. *Res:* Identification of specific extracellular and cell surface factors (enzymes) derived from major skin microflora, propionibacterium acnes and studies of their role in leukocytes chemotaxis and chemiluminescence in order to elucidate the mechanisms of inflammation in acne. *Mailing Add:* Dept Dermat PO Box 46 State Univ NY Brooklyn NY 11203

LEE, WEI-MING, b Kiangsu, China, June 11, 36; m 62. PHYSICAL CHEMISTRY, POLYMER CHEMISTRY. *Educ:* Nat Taiwan Univ, BS, 57; Southern Ill Univ, MA, 61; Univ Ill, PhD(phys chem), 64. *Prof Exp:* Teacher chem & math, Univ High Sch, Taiwan Norm Univ, 59; res assoc theoret chem kinetics, Univ Calif, Santa Barbara, 64-65; sr res engr styrene molding polymers res & develop, Dow Chem USA, 65-74, staff mem olefin plastics res & develop, 74-76, staff mem plastics res & develop, 78-80, assoc scientist, 81-83, SR ASSOC SCIENTIST, ADVAN COMPOSITES LAB, CENTRAL RES, DOW CHEM CO, 84- *Concurrent Pos:* Vis prof chem eng, Mich Tech Univ, 80-81. *Mem:* Am Chem Soc; Soc Advan Mat & Process Eng. *Res:* Mechanical properties of polymeric materials and plastic foams; computer simulation of chemical and physical processes; fundamental studies of polyurethanes; composite materials science. *Mailing Add:* Central Res Dow Chem Co Bldg 1702 Midland MI 48640

LEE, WILLIAM CHIEN-YEH, b London, Eng, July 20, 32; m 64; c 2. ELECTRICAL ENGINEERING. *Educ:* Chinese Naval Acad, Taiwan, BSc, 54; Ohio State Univ, MS, 60, PhD(elec eng), 63. *Prof Exp:* Mem tech staff commun, Bell Labs, 64-79; sr scientist & mgr, Defense Commun Div, ITT, 79-84; VPRES, PACTEL CELLULAR, 85- *Concurrent Pos:* Publ chmn, Inst Elec & Electronics Engrs Transactions on Vehicular Technol, 79-; mem, Nat Commun Forum Overseas Coun, 85-; affil mem, Univ Calif, Irvine, 85-; affil mem, Univ Calif, Davis, 85- *Mem:* Brit Inst Elec Engrs; fel Inst Elec & Electronics Engrs; Sigma Xi. *Res:* Wave propagation in anisotropic medium; antennas; signal fading; communication systems, particularly those relating to the ultrahigh frequency and x-band regions; author of three books on mobile communications and more than one hundred technical articles. *Mailing Add:* Pactel Three Park Plaza Box 19707 Irvine CA 92714

LEE, WILLIAM HUNG KAN, b Kwangsi, China, Oct 6, 40; m 66; c 2. GEOPHYSICS. *Educ:* Univ Alta, BSc, 62; Univ Calif, Los Angeles, PhD(planetary & space physics), 67. *Prof Exp:* Asst res geophysicist, Univ Calif, Los Angeles, 67; RES GEOPHYSICIST, US GEOL SURV, 67- *Concurrent Pos:* Secy, Heat-Flow Comt, Int Union Geod & Geophys, 63-65; ed, Geophys Monogr 8, Am Geophys Union, 65; guest lectr, Stanford Univ, 69; translation bd mem, Am Geophys Union, 75-78; co-ed, Chinese Geophys J, 78. *Mem:* AAAS; Am Geophys Union; Seismol Soc Am; Soc Explor Geophys; Sigma Xi. *Res:* Terrestrial heat-flow; thermal evolution of the planets; computer modeling of geologic processes; earthquake seismology. *Mailing Add:* US Geol Surv MS 977 345 Middlefield Rd Menlo Park CA 94025

LEE, WILLIAM JOHN, b Lubbock, Tex, Jan 16, 36; m 61; c 2. PETROLEUM ENGINEERING. *Educ:* Ga Inst Technol, PhD(chem eng), 63. *Prof Exp:* Sr res specialist, Esso Prod Res Co, 62-68; assoc prof petrol eng, Miss State Univ, 68-71; tech adv, Exxon Co, USA, 71-77; NOBLE CHAIR & PROF PETROL ENG, TEX A&M UNIV, 77- *Concurrent Pos:* Sr vpres, S A Holditch & Assocs, 80-; lectr, Soc Mining, Metall & Petrol Engrs, 70- & Am Asn Petrol Geologists, 77- *Honors & Awards:* Distinguished Lectr, Soc Petrol Engrs, 78, Reservoir Eng Award, 86; Distinguished Mem Award, 87. *Mem:* Soc Petrol Engrs. *Res:* Pressure transient testing; low permeability gas well analysis. *Mailing Add:* Dept Petrol Eng Tex A&M Univ College Station TX 77843

LEE, WILLIAM ORVID, b Brigham City, Utah, July 2, 27; m 51; c 4. FIELD CROPS. *Educ:* Utah State Univ, BS, 50, MS, 54; Ore State Univ, PhD(farm crops), 65. *Prof Exp:* Soil scientist, Bur Reclamation, US Dept Interior, 50-51; agronomist, Sci & Educ Admin, Agr Res, USDA, Utah, 51-54 & Wyo, 54-56, res agronomist, 56-83; RETIRED. *Mem:* Weed Sci Soc Am. *Res:* Crop science; control of weeds in forage and turf seed crops; legumes and grasses. *Mailing Add:* 1538 NW 12th Corvallis OR 97701

LEE, WILLIAM ROSCOE, b Little Rock, Ark, Feb 14, 30; m 53; c 3. GENETICS. *Educ:* Univ Ark, BSA, 52; Univ Wis, MS, 53, PhD(genetics, entom), 56. *Prof Exp:* Asst, Univ Wis, 52-56; from asst prof to assoc prof entom, Univ NH, 56-63; asst prof zool, Univ Tex, Austin, 63-67; assoc prof, 67-73, PROF ZOOL & PHYSIOL, LA STATE UNIV, BATON ROUGE, 73- *Concurrent Pos:* Res exec for H J Muller, Ind Univ, 62-63; dir, Inst Mutagenesis, La State Univ, Baton Rouge. *Mem:* Genetics Soc Am; Radiation Res Soc; Environ Mutagen Soc. *Res:* Radiation genetics; mechanisms of mutagenesis; recombinant DNA; southern blot experiments with both cloned and synthetic probes; relation between DNA adducts in the germ cells of Drosophila melanogaster and changes in DNA of induced mutants. *Mailing Add:* Dept Zool & Physiol La State Univ Baton Rouge LA 70803

LEE, WILLIAM STATES, b Charlotte, NC, June 23, 29; m 51; c 3. ENGINEERING. *Educ:* Princeton Univ, BS, 51. *Hon Degrees:* DEng, The Citadel, 85; Dr, Univ SC, 86, Univ NC, Johnson C Smith Univ & Davidson Col, 88, Clemson Univ, 89. *Prof Exp:* Mem staff, Duke Power Co, 55-62, eng mgr, 62-65, vpres eng, 65-71, sr vpres, 71-75, exec vpres, 76-78, pres & chief operating officer, 78-82, chmn & chief exec officer, 82-89, CHMN & PRES, DUKE POWER CO, 89- *Concurrent Pos:* Mem, US Comt Large Dams, 63-; trustee, Queens Col; mem bd dir, Edison Elec Inst; chmn bd, Inst Nuclear Power Opers, 79-82; pres, World Asn Nuclear Operators, 89. *Honors & Awards:* George Westinghouse Gold Medal, Am Soc Mech Engrs, 72; Gantt Award, AMA; Walter Zinn Award, Am Nuclear Soc, 80. *Mem:* Nat Acad Eng; fel Am Soc Mech Engrs; Nat Soc Prof Engrs; Am Nuclear Soc; fel Am Soc Civil Engrs. *Res:* Safety and operating performance of nuclear plants. *Mailing Add:* Duke Power Co 422 S Church St Charlotte NC 28242

LEE, WILLIAM THOMAS, b Hartford, Conn, July 18, 58; m. MICROBIOLOGY. *Educ:* George Washington Univ, BA, 80, MForensic Sci, 82; Johns Hopkins Univ, PhD(immunol), 86. *Prof Exp:* ASST INSTR, DEPT MICROBIOL, UNIV TEX HEALTH SCI CTR, DALLAS, 87- *Mem:* Am Asn Immunologists. *Mailing Add:* Dept Microbiol Univ Tex Southwestern Med Ctr 5323 Harry Hines Blvd Dallas TX 75235

LEE, WILLIAM WAI-LIM, b Shanghai, China, Aug 6, 48; US citizen. RESOURCE MANAGEMENT, ENVIRONMENTAL ENGINEERING. *Educ:* Tulane Univ, BSE, 69; Univ Mich, MSE, 70; Mass Inst Technol, SMCE, 72; Univ Mich, ScD(resources systs mgt), 77. *Prof Exp:* Proj engr water reuse, Los Angeles County Sanit Dists, 70-72; proj engr decision anal, Woodward-Clyde Consults, 77-79; asst prof, civil & urban eng, Univ Pa, 79-82; proj dir, R F

Weston, Inc, 82-85; AT DEPT NUCLEAR ENG, UNIV CALIF, BERKELEY, 86- *Mem:* Am Soc Civil Engrs; AAAS; Am Geophys Union; Am Nuclear Soc; Sigma Xi. *Res:* Applicability of quantitative analytical techniques in resources and environmental problems; analysis of nuclear waste management. *Mailing Add:* Dept Nuclear Eng Univ Calif Berkeley CA 94720

LEE, WILLIAM WEI, b San Francisco, Calif, May 17, 23; m 47; c 3. MEDICINAL CHEMISTRY. *Educ:* Univ Calif, BS, 47; Univ Minn, PhD(org chem), 52. *Prof Exp:* Jr chemist org anal chem, Shell Develop Co, 47-48; asst org chem, Univ Minn, 48-51; org res chemist, Cent Res Dept, Monsanto Chem Co, 52-54; assoc chemist org chem, SRI INT, 54-56, sr org chemist, 56-77, prog dir, Synthetic Cancer Drugs, 78-87; RETIRED. *Mem:* Am Chem Soc; Sigma Xi; Radiation Res Soc. *Res:* Allylic and acetylenic compounds; nucleosides; amino acids and alkylating agents; enzyme chemistry; active halogen compounds; folic acid antagonists; heterocyclic chemistry; chemotherapy, particularly cancer chemotherapy; radiosensitizing agents. *Mailing Add:* 991 N California Ave Palo Alto CA 94303-3407

LEE, WONYONG, b Korea, Dec 29, 30; m 61; c 1. HIGH ENERGY PHYSICS. *Educ:* Calif Inst Technol, BS, 57; Univ Calif, Berkeley, PhD(physics), 61. *Prof Exp:* Res assoc physics, Lawrence Radiation Lab, Univ Calif, 61-62; res assoc, 62-64, from asst prof to assoc prof, 64-72, PROF PHYSICS, COLUMBIA UNIV, 72- *Concurrent Pos:* Sloan Found fel, 65-67. *Mem:* Am Phys Soc. *Res:* High energy experimental physics. *Mailing Add:* Dept Physics Columbia Univ New York NY 10027

LEE, WOONG MAN, b Seoul, Korea, Dec 3, 38; US citizen. PATHOLOGY. *Educ:* Seoul Nat Univ, BS, 60, MD, 64; Am Bd Path, cert, 76. *Prof Exp:* From instr to asst prof path, Albany Med Col, NY, 70-79; PATHOLOGIST, GLEN FALLS HOSP, NY, 79- *Concurrent Pos:* Asst residency, Albany Med Ctr Hosp, 67-70, residency, 70-72, asst attend pathologist, 74-79; attend pathologist, Vet Admin Hosp, Albany, NY, 74-79. *Res:* Anatomical aspects and drug treatment of atherosclerosis. *Mailing Add:* Dept Path Glens Falls Hosp Glens Falls NY 12801

LEE, WOOYOUNG, b Pusan, Korea, Jan 2, 38; m 66; c 2. CHEMICAL ENGINEERING. *Educ:* Seoul Nat Univ, BS, 61; Univ Wis-Madison, MS, 64, PhD(chem eng), 66. *Prof Exp:* Fel chem eng, Univ Wis, 66; res engr, Mobil Res & Develop Corp, 66-69, sr res engr, 69-74, res assoc, 74-77, mgr, reforming & spec process develop, 77-80, synthetic fuels develop, 80-84, mgr, Chem Prods Dept, 84-86, MGR POLYOLEFINS PROCESS, MOBIL RES & DEVELOP CORP, 86-; MGR, EDISON RES LAB, MOBIL CHEM CO, 86- *Mem:* Am Inst Chem Engrs; Am Chem Soc; Soc Polymer Engrs; Soc Chem Indust. *Res:* Conversion of oxygenates to hydrocarbons; aromatics and olefin upgrading process development; fluid bed catalytic cracking; kinetics and reaction engineering; lube additives and synthetic hydrocarbon fluids; polyethylene process product. *Mailing Add:* Mobil Res & Develop Corp PO Box 240 Edison NJ 08818

LEE, Y C, b Hong Kong, Mar 30, 48; Can citizen; m 81; c 2. DATABASES, COMPUTATIONAL GEOMETRY. *Educ:* Simon Fraser Univ, BSc, 77; Univ NB, MSc, 80, PhD(surv eng), 86. *Prof Exp:* Asst prof, 77-91, ASSOC PROF GIS, DEPT SURV ENG, UNIV NB, 91- *Concurrent Pos:* Consult, Universal Systs Ltd, Fredericton, NB, 87-, Can Int Develop Agency, 88-, NB Geog Info Corp, 90-; mem, working group GIS, Int Soc Photogram & Remote Sensing, 89-, Comn Urban Cartog, Int Cartog Asn, 89-, Tech Comt Geomatics, Can Gen Standards Bd, 89- *Honors & Awards:* Bauch & Lomb Photogram Award, Can Inst Surv & Mapping, 82, ACDS Graphics Award, 91. *Mem:* Can Inst Surv & Mapping; Can Cartog Asn; Inst Elec & Electronics Engrs Computer Soc. *Res:* Geographic information systems and automated cartography, particularly in areas of data models, data structures, geometric processing, visualization, user interfaces and spatial search algorithms. *Mailing Add:* Dept Surv Eng Univ NB PO Box 4400 Fredericton NB E3B 5A3

LEE, YAT-SHIR, b Kwangtung, China. INORGANIC CHEMISTRY. *Educ:* Nat Taiwan Univ, BS, 58; Kent State Univ, MS, 67, PhD(chem), 71. *Prof Exp:* MEM TECH STAFF LIQUID CRYSTAL RES, HUGHES AIRCRAFT CO, 72- *Concurrent Pos:* Fel, Harvard Univ, 72. *Mem:* Am Chem Soc. *Res:* Structure, hydrodynamics and electro-optical effects of liquid crystals. *Mailing Add:* 30318 Benecia Laguna Niguel CA 92677

LEE, YIEN-HWEI, b Taiwan, China, Oct 20, 37; m 68; c 3. PHARMACOLOGY. *Educ:* Nat Taiwan Univ, MD, 63; Univ Calif, Los Angeles, PhD(pharmacol), 68. *Prof Exp:* Res asst pharmacol, Univ Calif, Los Angeles, 64-66, res pharmacologist, 66-68, res assoc, Neuropsychiat Inst, 66-68; sr investr pharmacol, Searle Res Lab, G D Searle & Co, 68-72; sect head, Abbott Labs, 72-76; staff, Chicago Med Sch Hosp, 76-77; staff physician, St Joseph Hosp, Chicago, 77-78; fel sect hemat, Rush-Presby & St Luke's Hosp, Chicago, 78-80; Am Cancer Soc fel clin oncol, Sect Hemat & Oncol, Cook County Hosp, Chicago, 80-81; consult specialist, Bethesda Res Lab, Gaithersburg, 81-82; attend hematologist, Georgetown Univ, Washington, DC, 82-83; attend hematologist oncologist, Shady Grove Adventist Hosp, Rockville, Md, 82-85; pres, L&L Clin, Inc, Rockville, Md, 83-85; CONSULT PHYSICIAN, SPRUCE MED CTR, PHILADELPHIA, PA, 85-; MED DIR, LEE CLIN, VILLANOVA, PA, 85- *Concurrent Pos:* Attend hematologist oncologist, Greater Beltsville Laurel Med Ctr, Md, 83-85 & Leland Hosp, Riverdale, Md, 84-85; attend physician, Jefferson Park Hosp, Philadelphia, Pa, 85- *Mem:* AAAS; Am Soc Pharmacol & Exp Therapeut; Sigma Xi; Am Chem Soc; Am Ctr Chinese Med Sci. *Res:* Gastrointestinal pharmacology and physiology; biochemical pharmacology; cancer chemotherapy; experimental leukemia; hematology; oncology. *Mailing Add:* 155 S Spring Mill Rd Villanova PA 19085-1408

LEE, YIM TIN, b Canton, China, June 14, 50. ATOMIC PROCESS IN PLASMA, LAB X-RAY LASER RESEARCH. *Educ:* Univ Calif, Berkeley, BA, 72; Univ Calif, San Diego, PhD(physics), 77. *Prof Exp:* Fel physics, Univ Calif, San Diego, 77-79; RES PHYSICIST, LAWRENCE LIVERMORE NAT LAB, 79- *Concurrent Pos:* Consult, Physical Dynamics Inc, 76-77. *Mem:* Am Phys Soc; AAAS; Fusion Energy Asn. *Res:* Atomic processes in laser produced plasma and soft x-ray lasing. *Mailing Add:* Lawrence Livermore Lab PO Box 808 L-298 Livermore CA 94550

LEE, YING KAO, b Shanghai, China, Dec 14, 32; US citizen; m 61; c 3. POLYMER CHEMISTRY. *Educ:* Tai Tung Univ, BSc, 52; Univ Cincinnati, PhD(chem), 61. *Prof Exp:* Res chemist, Tex-US Chem Co, 60-63, proj leader, 65; res chemist, E I Du Pont de Nemours & Co, Inc, 65-68, staff chemist, 68-70, res assoc, 70-76, res fel, 76-87, sr res fel, 86- 89, dept fel, 89-90, DUPONT FEL, MARSHALL RES & DEVELOP LAB, E I DU PONT DE NEMOURS & CO, INC, 90- *Mem:* Am Chem Soc; Sigma Xi. *Res:* Polymers or polymeric systems used in coating field; high temperature polymer for electronics applications; crosslinking chemistries. *Mailing Add:* Marshall Res & Develop Lab E I Du Pont de Nemours & Co Inc Philadelphia PA 19146

LEE, YONG YUNG, b Kyungpook, Korea, Feb 12, 36; US citizen; m 63; c 2. ACCELERATOR PHYSICS. *Educ:* Kyung-Pook Nat Univ, BS, 58, MS, 60; Univ Mich, PhD(physics), 64. *Prof Exp:* Res assoc physics, Univ Wis, 64-67; asst prof, State Univ NY, Stony Brook, 67-71; PHYSICIST ACCELERATOR PHYSICS, BROOKHAVEN NAT LAB, 71- *Mem:* Fel Am Phys Soc. *Res:* Accelerator physics. *Mailing Add:* Bldg 911B Brookhaven Nat Lab Upton NY 11973

LEE, YOUNG HIE, b Seoul, Korea, Jan 12, 46; m 72; c 2. CHEMICAL ENGINEERING, BIOCHEMICAL ENGINEERING. *Educ:* Seoul Nat Univ, BS, 71; Purdue Univ, MS, 74, PhD(chem eng), 77. *Prof Exp:* PROF CHEM ENG, DREXEL UNIV, 78- *Concurrent Pos:* Prin investr, NSF proj, 78-90. *Mem:* Am Inst Chem Engrs; Am Chem Soc. *Res:* Transport phenomena; gas liquid reaction; waste water treatment; biomass utilization. *Mailing Add:* Dept Chem Eng Drexel Univ 32nd & Chestnut St Philadelphia PA 19104

LEE, YOUNG JACK, b Seoul, Korea, Feb 25, 42; m 67; c 3. STATISTICS. *Educ:* Seoul Nat Univ, BSE, 64; Ohio State Univ, MS, 72, PhD(statist), 74. *Prof Exp:* Instr electronics eng, Korean Air Force Acad, 67-69; asst prof statist, Univ Md, College Park, 74-79; MATH STATISTICIAN, NIH, 79- *Mem:* Am Statist Asn; Biomet Soc. *Res:* Nonparametric/robust design of experiment and statistical analysis in hypothesis testing, ranking and selection and estimation; applications of statistics to social science and life science; method in clinical trial, statistical design and analysis for carcinogenesis and mutagenesis bioassays. *Mailing Add:* NIH Bldg EPN Rm 630 Bethesda MD 20892

LEE, YOUNG-HOON, b Korea, Sept 18, 35; m 65; c 1. SOLID STATE PHYSICS. *Educ:* Dong-Guk Univ, BS, 61, MS, 63; State Univ NY Albany, PhD(physics), 72. *Prof Exp:* Sr asst physics, Dong-Guk Univ, 63-66; fel physics, State Univ NY, Albany, 73-78; MEM RES STAFF, IBM THOMAS J WATSON RES CTR, 78- *Mem:* Am Phys Soc; Am Vacuum Soc. *Res:* Defects in solids and electron spin resonance; electronics engineering; microelectronics fabrication techniques; plasma processings. *Mailing Add:* IBM T J Watson Res Ctr PO Box 218 Yorktown Heights NY 10598

LEE, YOUNG-JIN, b Seoul, Korea, Nov 22, 46; m 72; c 2. CHEMISTRY, ORGANIC CHEMISTRY. *Educ:* Millikin Univ, BA, 68; Univ Rochester, MS, 70; State Univ NY, Albany, PhD(chem), 74. *Prof Exp:* Fel chem, Cornell Univ, 74-75; chemist, Union Carbide Corp, 75-87; CHEMIST, CHEVRON CHEM CO, 87. *Mem:* Am Chem Soc. *Res:* Syntheses and process development of agricultural chemicals; new pesticides. *Mailing Add:* Chevron Chem Co 940 Hensley St Danville CA 94506

LEE, YUAN CHUAN, b Taiwan, China, Mar 30, 32; m 58; c 1. BIOCHEMISTRY. *Educ:* Nat Taiwan Univ, BS, 55, MS, 57; Univ Iowa, PhD(biochem), 62. *Prof Exp:* Res assoc biochem, Univ Iowa, 62 & Univ Calif, Berkeley, 62-65; from asst prof to assoc prof, 65-74, PROF BIOL, JOHNS HOPKINS UNIV, 74- *Concurrent Pos:* Vis prof, Kyoto Univ, Academia Sinica, Taipei & Beijing Med Univ. *Honors & Awards:* Merit Award, NIH. *Mem:* Am Chem Soc; Am Soc Biol Chem; Japanese Biochem Soc; Chinese Biochem Soc; Am Soc Cell Biol. *Res:* Complex carbohydrates; carbohydrate receptors. *Mailing Add:* Dept Biol Johns Hopkins Univ 3400 N Charles St Baltimore MD 21218

LEE, YUAN TSEH, b Hsinchu, Taiwan, Nov 29, 36; m 63; c 3. CHEMISTRY. *Educ:* Nat Taiwan Univ, BS, 59; Nat Tsing Hua Univ, Taiwan, MS, 61; Univ Calif, Berkeley, PhD(chem), 65. *Prof Exp:* From asst prof to prof chem, Univ Chicago, 68-74; PROF CHEM, UNIV CALIF, BERKELEY, 74- *Honors & Awards:* Nobel Prize in Chem, 86; Ernest O Lawrence Award, US Dept Energy, 81; Harrison Howe Award, Rochester Sect, 83; Peter Debye Award, Am Chem Soc, 86; Nat Medal Sci, 86. *Mem:* Nat Acad Sci; fel Am Phys Soc; fel Am Acad Arts Sci; Am Chem Soc. *Res:* Chemical kinetics; reaction dynamics; laser chemisty and molecular interaction. *Mailing Add:* Dept Chem Univ Calif Berkeley CA 94720

LEE, YUEN SAN, b Taipei, Taiwan, Oct 13, 39; m 67; c 2. FOODS, BIOCHEMISTRY. *Educ:* Nat Taiwan Univ, BS, 62; Utah State Univ, MS, 65; Univ Md, College Park, PhD(food sci, biochem), 68. *Prof Exp:* CHEMIST, COMN PUB HEALTH, DC GOVT, 68- *Concurrent Pos:* Prof, Univ DC, 75- *Mem:* Inst Food Technol; Am Chem Soc; Am Dietetic Asn. *Res:* Method development in the determination of pesticides in meat, milk and water; quality control of detecting adulteration in meat and meat products for consumer protection; heavy metals in foods. *Mailing Add:* Five Maplewood Ct Greenbelt MD 20770

LEE, YUE-WEI, b San-Tung, China, Mar 9, 46; m 72; c 2. ORGANIC CHEMISTRY, MEDICINAL CHEMISTRY. *Educ:* Calif State Univ, Sacramento, MS, 73; Columbia Univ, MS, 75, PhD(chem), 78. *Prof Exp:* Res grad asst, Columbia Univ, 74-78; RES STAFF SYNTHESIS & RES CHEMIST, RES TRIANGLE INST, 78- *Mem:* Am Chem Soc. *Res:* Synthesis of steroidal hormone for contraceptive purpose; isolation and structure determination of medicinal component from natural resources; countercurrent chromatography and its applications in natural products research. *Mailing Add:* 105 Highland Dr Chapel Hill NC 27514

LEE, YUNG, b Inchon, Korea, May 11, 32; Can citizen; m 60; c 1. MECHANICAL ENGINEERING. *Educ:* Seoul Nat Univ, BEng, 59, MEng, 61; Univ Liverpool, PhD(mech eng), 64. *Prof Exp:* Lectr mech eng, Liverpool Polytech Inst, 64-65; res officer, Chalk River Nuclear Lab, Atomic Energy Can, Ltd, 65-67; from asst prof to assoc prof mech eng, 67-73, PROF MECH ENG, UNIV OTTAWA, 73-; DIR, OTTAWA-CARLETON INST MECH & AERONAUT ENG, 84- *Mem:* Can Soc Mech Engrs; Eng Inst Can; Asn Prof Engrs Ont. *Res:* Fluid flow; heat transfer. *Mailing Add:* Dept Mech Eng Univ Ottawa Ottawa ON K1N 6N5 Can

LEE, YUNG-CHANG, b Canton, China, Nov 7, 35; US citizen. PHYSICS. *Educ:* Nat Taiwan Univ, BSc, 55; Univ Md, PhD(physics), 63. *Prof Exp:* Mem tech staff, Bell Tel Lab, 61-67; ASSOC PROF PHYSICS, STATE UNIV NY, BUFFALO, 67 - *Concurrent Pos:* Vis prof, Nat Tsing Hua Univ & Nat Taiwan Univ, 73-74; physicist, Lawrence Livermore Lab, 78-79. *Mem:* Am Phys Soc. *Res:* Solid state physics and quantum optics, including superradiance in thin crystal films; excitons in thin films; Anderson localization and Thouless' maximum resistance; interaction of electromagnetic radiation with plasmas; parametric coupling in plasmas; two dimensional crystalline order; high temperature superconductivity; meissner attractive forces on dirty superconductors; density functional theory; clustered hubbard model. *Mailing Add:* Physics Dept State Univ NY Buffalo Amherst NY 14260

LEE, YUNG-CHENG, b Taiwan, Repub China, Feb 5, 56; US citizen; m 83; c 1. ELECTRONIC PACKAGING, HEAT TRANSFER. *Educ:* Nat Taiwan Univ, BS, 78; Univ Minn, MS, 82, PhD(mech eng), 84. *Prof Exp:* Res asst, Univ Minn, 80-84; mem tech staff, AT&T Bell Labs, Murray Hill, 84-89; ASST PROF, DEPT MECH ENG, UNIV COLO, BOULDER, 89- *Concurrent Pos:* NSF presidential young investr award, 90. *Mem:* Am Soc Mech Engrs; Soc Mfg Engrs. *Res:* Low-cost prototyping and manufacturing of electronic multichip modules; three dimensional packaging for portable supercomputers; optoelectronic packaging; mechatronics; plasma-aided manufacturing. *Mailing Add:* Dept Mech Eng Univ Colo Boulder CO 80309-0427

LEE, YUNG-KEUN, b Seoul, Korea, Sept 26, 29; m 58; c 4. NUCLEAR PHYSICS. *Educ:* Johns Hopkins Univ, BA, 56; Univ Chicago, MS, 57; Columbia Univ, PhD(physics), 61. *Prof Exp:* Res scientist, Columbia Univ, 61-64; from asst prof to assoc prof, 64-71, PROF PHYSICS, JOHNS HOPKINS UNIV, 71- *Mem:* Am Phys Soc. *Res:* Nuclear beta decay; nuclear reactions; Mossbauer effects; intermediate energy physics. *Mailing Add:* Dept Physics Johns Hopkins Univ Baltimore MD 21218

LEECH, GEOFFREY BOSDIN, b Montreal, Que, Aug 28, 18; m 46; c 1. RESEARCH ADMINISTRATION. *Educ:* Univ BC, BASc, 42; Queen's Univ, Ont, MSc, 43; Princeton Univ, PhD(petrol, econ geol), 49. *Prof Exp:* Field asst, Geol Surv Can, 40-41 & 42; geologist, Int Nickel Co Can, Ltd, Ont, 43-46; chief party, BC Dept Mines, 47-48; geologist, Geol Surv Can, Dept Energy, Mines & Resources, 49-72, head, Econ Geol Subdiv, 73-78, dir, Econ Geol Div, 79-82; RETIRED. *Concurrent Pos:* Assoc secy gen, Int Asn on Genesis of Ore Deposits, 78-84. *Mem:* Fel Geol Soc Am; fel Soc Econ Geol; fel Royal Soc Can; Can Inst Mining & Metall; fel Geol Asn Can. *Res:* Regional metallogeny; mineral resource evaluation; problems of resource adequacy. *Mailing Add:* 1113 Greenlawn Crescent Ottawa ON K2C 1Z4 Can

LEECH, H(ARRY) WILLIAM, b Triadelphia, WVa, Apr 28, 40; m 66, 88; c 5. PHYSICS, COMPUTER SCIENCE. *Educ:* WVa Univ, BS, 62, MS, 64; Univ Md, PhD(physics), 77. *Prof Exp:* From instr to asst prof physics & math, W Liberty State Col, 64-68; mem tech staff computer sci, Comput Sci Corp, 73-77; asst prof physics, Southern Conn State Col, 77-78; asst prof physics, Bethany Col, 78-80; assoc prof, 80-87, chmn, 84-86, DEAN ENG TECHNOLS & COMMUN, JEFFERSON TECH COL, 86-, PROF ENG TECHNOLS, 87- *Mem:* Am Asn Physics Teachers. *Res:* Cosmic ray astrophysics; celestial mechanics; orbit determination; scientific applications of computer science. *Mailing Add:* West Liberty State Col West Liberty WV 26074

LEECH, ROBERT LELAND, b Chicago, Ill, Oct 21, 38; m 61; c 4. ELECTRICAL ENGINEERING, COMPUTER SCIENCE. *Educ:* US Mil Acad, BS, 60; Purdue Univ, West Lafayette, MSE, 66; Duke Univ, AM, 76, PhD(elec eng), 77. *Prof Exp:* Proj officer & comdr, Automatic Commun Syst Off, US Army Electronics Command, Ft Monmouth, NJ, 63-64; instr & opers officers, Acad Comput Ctr, US Mil Acad, 66-67, from instr to asst prof, Off Dean Educ Syst, Dept Elec, 67-69; chief automatic data processing sect, opers group, US Army War Col, Carlisle Barracks, Pa, 70-72; chief comput opers info systs div, Supreme Hqs Allied Powers Europe, 72-75; assoc prof comput sci, dept elec eng, 77-78, asst dean, Acad Automation, US Mil Acad, 78-81; SR ENG MGR, IBM CORP, 81- *Mem:* Inst Elec & Electronics Engrs; Asn Comput Mach. *Res:* Computer systems architecture; computer performance evaluation; military computer command and control system. *Mailing Add:* IBM Corp PO Box 12195 J45/062 Research Triangle Park NC 27709

LEECH, STEPHEN H, b Mar 27, 42; m; c 4. MICROBIOLOGY, CLINICAL IMMUNOLOGY. *Educ:* Univ Edinburgh, MB, 65; Univ London, PhD(immunol), 76; FRCP(C), 71. *Prof Exp:* House surgeon, Edinburgh Royal Infirm, Scotland, 65-66; royal internship, Mem Res Ctr & Hosp Univ Tenn, Knoxville, 66-67; resident internal med, Univ Minn, Mayo Clin, 67-69; Minneapolis, 69-70; fel allergy & immunol, Royal Victoria Hosp, Montreal,

70-72; clin res fel, Tumor Immunol Unit, Univ Col Hosp, London, 72-76; chief, Sect Allergy & Clin Immunol, Med Ctr, La State Univ, New Orleans, 76-87, from asst prof to assoc prof, Dept Med, 76-87; DIR, IMMUNOL LAB, SENTARA NORFOLK GEN HOSP, VA, 87-, DIR, SENTARA SEROL LAB, 88-; PROF CLIN IMMUNOL, DEPT IMMUNOL & MICROBIOL, EASTERN VA MED SCH, 88- Concurrent Pos: House physician, St Mary's Hosp, Knoxville, 67; dir, Allergy Clin, Charity Hosp, La State Univ, New Orleans, 76-87, dir, B Cell Clin, 81-87; dir, Immunocytogenetics Lab, Med Ctr, La State Univ, 79-87, assoc prof, Dept Biomet & Genetics, 85-87; mem, Nat Histocompatability Comt, United Network Organ Sharing. Mem: Am Acad Allergy & Immunol; Am Soc Histocompatability & Immunogenetics; Am Fedn Clin Res; Am Asn Immunologists; Am Soc Transplant Physicans; Clin Immunol Soc. Mailing Add: Microbiol & Clin Immunol Dept Eastern Va Med Sch 1113 Llewellyn Mews Norfolk VA 23507

LEED, RUSSELL ERNEST, b Denver, Pa, Dec 24, 15. NUCLEAR CHEMISTRY, CHEMICAL PLANT ENGINEERING. Educ: Franklin & Marshall Col, BS, 37; Univ Md, MS, 40, PhD(phys chem), 41. Prof Exp: From instr to asst prof chem, Va Mil Inst, 41-48; assoc prof, Va Polytech Inst, 48-51; chemist, Prod Div, Atomic Energy Comn, Oak Ridge Opers, 51-73, dep dir, 58-73, asst mgr uranium enrichment prog, Dept Energy, 73-80; RETIRED. Mem: Am Chem Soc. Res: Solutions-electrochemistry; war gases; physical properties of alloys; thermal conductivity; phase diagrams; process development; nuclear weapons and separation of uranium isotopes; gas centrifuge and gaseous diffusion processes. Mailing Add: 125 Nesper Rd Oak Ridge TN 37830

LEEDER, JOSEPH GORDEN, b Oil City, Pa, July 4, 16; m 39; c 2. DAIRY CHEMISTRY. Educ: Ohio State Univ, BS, 38; Univ Vt, MS, 40; Pa State Univ, PhD, 44. Prof Exp: Dir labs & Supvr zone qual control, Nat Dairy Prod Co, Ohio, 44-46; chief res chemist, Ramsey Labs, 46-48; from assoc prof to prof dairy indust, dept animal sci, Rutgers Univ, 48-65, res prof food sci, dept food sci, 65-81; RETIRED. Mem: Am Dairy Sci Asn; Inst Food Technologists. Res: Immobilized enzymes for whey utilization; food emulsions; chemistry of butter oil used in deep-fat frying; fat demulsification in ice cream; use of corn syrup in ice cream. Mailing Add: 687 Yarborough Way No 6 Jamesburg NJ 08831

LEEDHAM, CLIVE D(OUGLAS), b London, Eng, Nov 1, 28; m 55; c 2. AUTOMOTIVE ELECTRONICS, COMPUTERS. Educ: Univ London, BSc, 49; Mass Inst Technol, SM, 55; Purdue Univ, PhD(elec eng), 63. Prof Exp: Asst lectr elec eng, Univ London, 48-49; apprentice, Metrop Vickers Elec Co Ltd, 49-51; jr engr, 51-52; teaching asst, Mass Inst Technol, 53-55; instr, Purdue Univ, 58-62; asst prof, Univ Calif, 62-65; consult, Delco Electronics Corp, 63-65; staff engr, 65-76, bus develop mgr digital systs, 76-83, regional mgr int sales, 83- 90, BUS DEVELOP MGR, DIGITAL SYSTS, DELCO ELECTRONICS CORP, 90- Concurrent Pos: Consult, USAF, 61-62. Res: Automatic control; signal processing; acoustics; digital computers. Mailing Add: Delco Electronics Corp 6767 Hollister Ave Goleta CA 93117

LEEDOM, JOHN MILTON, b Peoria, Ill, Oct 18, 33; m 56; c 2. INTERNAL MEDICINE, INFECTIOUS DISEASES. Educ: Univ Ill, BA, 55, BS, 56, MD, 58; Am Bd Internal Med, dipl, 67 & 74. Prof Exp: Resident med, Univ Ill Res & Educ Hosps, 59-60 & 61-62; from asst prof to assoc prof, 62-76, HASTINGS PROF MED, SCH MED, UNIV SOUTHERN CALIF, 76-, CHIEF, DIV INFECTIOUS DIS, DEPT MED, 75- Concurrent Pos: Res fel, Univ Ill Res & Educ Hosps, 60-61; officer res proj, Epidemic Intel Serv, USPHS, Infectious Dis Lab, Univ Southern Calif, 62-64, consult health facil construction div, Health Serv & Ment Health Admin. Mem: Am Fedn Clin Res; Am Soc Microbiol; Infectious Dis Soc Am; Western Soc Clin Invest; Western Asn Physicians. Res: Infectious disease, particularly viral and bacterial diseases of the central nervous system; AIDS and AIDS treatment. Mailing Add: Dept Med Univ Southern Calif 2025 Zonal Ave Los Angeles CA 90033

LEEDS, J VENN, JR, b Wharton, Tex, Oct 26, 32; m 56; c 2. ELECTRICAL ENGINEERING, ENVIRONMENTAL ENGINEERING. Educ: Rice Univ, BA, 55, BSEE, 56; Univ Pittsburgh, MSEE, 60, PhD(elec eng), 63; JD, Univ Houston, 72. Prof Exp: Sr engr, Bettis Atomic Lab, Westinghouse Elec Corp, 56-63; asst prof elec eng, Rice Univ, 63-65, from asst prof to prof elec & environ eng, 65-90, master, Sid W Richardson Col, 70-76, EMER PROF ELEC & ENVIRON ENG, RICE UNIV, 90- Concurrent Pos: Consult, Esso Prod Res Co Div, Exxon, 63-68 & Geospace Corp, 68-72; mem safety & licensing panel, US Nuclear Regulation Comn, 71-78; consult, var ins co and law firms, 72- Mem: Inst Elec & Electronics Engrs; Sigma Xi. Res: Design and analysis of large, complex systems; applied mathematics; interactions of law and engineering; applied mathematics; electrical events; fire and explosions; artificial intelligence. Mailing Add: 10807 Atwell Houston TX 77096

LEEDS, MORTON W, b Brooklyn, NY, Dec 18, 16; m 45; c 1. ORGANIC CHEMISTRY. Educ: Polytech Univ NY, BS, 38, MS, 39, PhD(org chem), 44. Prof Exp: Chemist, Bio-Med Res Lab, NY, 35-39; sr chemist, Res Lab, Interchem Corp, 39-45, head develop dept amino acids, Biochem Div, NJ, 45-48; sr chemist, E I du Pont de Nemours & Co, Inc, 48-50; asst chief chemist, Schwarz Labs, NY, 50-52; head appln & develop, Res Labs, Air Reduction Co, 52-56, supvr org chem develop & res, 56-60, from asst dir to assoc dir chem res, 60-71; mgr clin develop, Ciba-Geigy Pharmaceut Co, 71-77, asst dir med prods mgt, Med Res Div, 77-82; EXEC DIR, MED-CHEM ASSOC, 82- Concurrent Pos: Permanent adj prof, Kean Col, 71-80. Mem: Am Chem Soc; fel Am Inst Chem; NY Acad Sci; Sigma Xi. Res: Organic synthesis and development; acetylene and pharmaceutical chemistry; petrochemicals. Mailing Add: 100 Chestnut Hill Dr Murray Hill NJ 07974

LEEDY, CLARK D, b Chicago, Ill, June 3, 33; m 56; c 5. SOILS SCIENCE. Educ: Purdue Univ, BS, 55; NMex State Univ, MS, 64, MA, 66; Tex A&M Univ, PhD(educ), 74. Prof Exp: Exten Soils Specialist, NMex State Univ, 57-71; div leader, WVa Univ, 73-76; ASSOC SOIL SCIENTIST, UNIV NEV, RENO, 76- Mem: Sigma Xi. Res: Forms, rates, methods and timing of phosphorous fertilizer application on alfalfa yield and quality. Mailing Add: Dept Plant Sci Univ Nev Reno NV 89557

LEEDY, DANIEL LONEY, b North Liberty, Ohio, Feb 17, 12; m 45, 88; c 2. URBAN WILDLIFE, OUTDOOR RECREATION. Educ: Miami Univ, AB, 34, BSc, 35; Ohio State Univ, MSc, 38, PhD(wildlife mgt), 40. Prof Exp: Lab asst geol & zool, Miami Univ, 34-35; asst leader, Ohio Wildlife Res Unit, Ohio State Univ, 40-42; leader, Ohio Unit, US Fish & Wildlife Serv, 45-48, coordr, Coop Wildlife Res Unit Prog, 49-57, chief br wildlife res, 57-63; chief div res & educ, Bur Outdoor Recreation, 63-65, water resources res scientist, Off Water Resources Res, US Dept Interior, 65-74; SR SCIENTIST, NAT INST URBAN WILDLIFE, 75- Concurrent Pos: Mem, res coun comts & panels, Nat Acad Sci, 57-74 & surface mining panel, 78-79. Honors & Awards: Conserv Award, Am Motors Corp, 58; Aldo Leopold Award, Wildlife Soc, 83. Mem: Wildlife Soc (pres, 53, exec secy, 54-57); Am Fisheries Soc; Am Ornithologists Union; Wilson Ornith Soc. Res: Wildlife ecology; socioeconomics of fish and wildlife and recreation; natural resources training and employment; wildlife-land use relationships; water resources; urban wildlife and ecology; ecologic impacts of water development, surface mining, highways, and electric utilities in relation to wildlife. Mailing Add: 12401 Ellen Ct Silver Spring MD 20904

LEEF, AUDREY V, b Hoboken, NJ, July 15, 22; m 47; c 4. MATHEMATICS. Educ: Montclair State Col, BA, 43; Stevens Inst Technol, MS, 47; Rutgers Univ, EdD, 76; Drew Univ, MDiv, 85. Prof Exp: Teacher & dept chairperson math, Millburn High Sch, NJ, 43-48; ASSOC PROF MATH, MONTCLAIR STATE COL, UPPER MONTCLAIR, NJ, 66- Mem: Nat Coun Teachers Math; Asn Women Math; Asn Women Educ; Am Asn Univ Women. Res: Mathematics education; math anxiety. Mailing Add: 24 Overlook Rd Mountain Lakes NJ 07046

LEEF, JAMES LEWIS, b San Francisco, Calif, Mar 6, 37; m 64; c 4. CRYOBIOLOGY, IMMUNOLOGY. Educ: Univ Calif, San Francisco, BA, 67; Univ Tenn, PhD(biol), 74. Prof Exp: sr investr cryobiol & head Malaria Res Dept, 76-82, DIR, BIOMED RES INST, 82-; EXEC DIR, BIOMED RES INC, 82- Concurrent Pos: Consult, Sci & Indust Res & Develop Co, 67-69; fel, Univ Ill, 73-76; guest scientist, Navy Med Res Inst, 76- Mem: Soc Cryobiol; Tissue Cult Asn; Am Asn Tissue Banks; NY Acad Sci; AAAS. Res: Malariology; mechanisms of freezing injury; study of various developmental stages of malaria and schistosomiasis parasites as antigens in developing a malaria and schistosomiasis vaccine and preservation of these forms at low temperatures. Mailing Add: Biomed Res Inst 12111 Parklawn Dr Rockville MD 20852

LEE-FRANZINI, JULIET, b Paris, France, May 18, 33; US citizen; m 64; c 1. EXPERIMENTAL PHYSICS. Educ: Hunter Col, BA, 53; Columbia Univ, MA, 57, PhD(physics), 60. Prof Exp: Res assoc physics, Columbia Univ, 60-62; res fel astrophys, Nat Acad Sci-Nat Res Coun, 62-63; from asst prof to assoc prof elem particle physics, 63-74, PROF PHYSICS, STATE UNIV NY, STONY BROOK, 74- Concurrent Pos: State Univ NY Res Found grant, 63-66; vis assoc physicist, Brookhaven Nat Lab, 64-; AEC grant, 66-72; NSF grant, 72-; vis prof, Cornell Univ, 80-81. Mem: Fel Am Phys Soc. Res: Elementary particle physics; weak interactions. Mailing Add: 26 Meadowlark Rd Ithaca NY 14850

LEEG, KENTON J(AMES), b Spokane, Wash, Aug 31, 11. PLASTICS CHEMISTRY, ENGINEERING. Educ: Univ Calif, AB, 35; Univ Southern Calif, PhD(chem), 40. Prof Exp: Dir res, Baker Oil Tools, Inc, 40-51; pres, Lebec Chem Corp, 51-55; pres, Rez-Coat of Calif, 55-64; CONSULT PLASTICS, 64- Concurrent Pos: Lectr, Univ Southern Calif, 40-50. Mem: Am Chem Soc. Res: Plastics; plastics finishes; high and ultra high temperature plastics; solar heating. Mailing Add: PO Box 1423 Rancho Santa Fe CA 92067

LEE-HAM, DOO YOUNG, b Seoul, Korea, Mar 31, 32; US citizen; m 66; c 2. PHARMACOLOGY, TOXICOLOGY. Educ: Mercer Univ, BA, 57; Cath Univ Am, MS, 61, PhD(physiol), 66. Prof Exp: Res scientist, Microbiol Assoc, Inc, 62-66; spec lectr physiol & biochem, Sungshin Womans Univ & Ewha Womans Univ, 66-67; sr scientist, Melpar, Inc, 67-69; PHYSIOLOGIST, DIV ONCOL & RADIOPHARM DRUG PROD, FOOD & DRUG ADMIN, 69- Mem: NY Acad Sci. Res: Basic and applied cell physiology; in vitro and in vivo testing of drugs and chemicals for their carcinogenic potential, mutagenicity and cell transformation. Mailing Add: Food & Drug Admin HFD-150 Parklawn Bldg 5600 Fishers Lane Rockville MD 20857

LEEHEY, PATRICK, b Waterloo, Iowa, Oct 27, 21; m 44; c 6. APPLIED MECHANICS. Educ: US Naval Acad, BSc, 42; Brown Univ, PhD(appl math), 50. Prof Exp: Proj officer, US Off Naval Res, DC, 51-53, prog officer, David Taylor Model Basin, 53-56, design supt, Puget Sound Naval Shipyard, Wash, 56-58, head ship silencing br, Bur Ships, DC, 58-63, head acoustics & vibration lab, David Taylor Model Basin, 63-64; assoc prof mech eng, Mass Inst Technol, 64-67, from assoc prof to prof naval archit, 64-71, prof appl mech, 71-79; Liaison scientist, 84-85; PROF MECH & OCEAN ENGR, OFF NAVAL RES, LONDON, 79- Concurrent Pos: Fulbright lectr, Austria, 77-78. Honors & Awards: Gold Medal, Am Soc Naval Eng, 62. Mem: Am Math Soc; fel Acoust Soc Am; Am Soc Naval Eng. Res: Hydrodynamics; hydrofoil craft development; unsteady airfoil theory; supercavitating flow theory; acoustics; ship silencing; underwater acoustics, boundary layer noise; mathematics; hyperbolic partial differential equations; singular integral equations; boundary layer stability. Mailing Add: Dept Mech Eng Rm 3-262 Mass Inst Technol Cambridge MA 02139

LEE-HUANG, SYLVIA, b Shanghai, China, July 14, 30; US citizen; m 57; c 3. BIOCHEMISTRY, MOLECULAR BIOLOGY. *Educ:* Nat Taiwan Univ, BS, 52; Univ Idaho, MS, 57; Univ Pittsburgh, PhD(biophys), 61. *Prof Exp:* NIH fel microbiol, Sch Med, Univ Pittsburgh, 61-62; res assoc chem physics, Sloan-Kettering Inst, 62-64; instr biochem, Med Col, Cornell Univ, 64-66; res scientist, 66-67, from instr to asst prof, 67-70, res assoc prof, 70-71, ASSOC PROF BIOCHEM, SCH MED, NY UNIV, 71- *Concurrent Pos:* Prin investr, 72- *Mem:* AAAS; Am Soc Biol Chemists; Biophys Soc; Harvey Soc; NY Acad Sci. *Res:* Molecular mechanism of transmission and expression of genetic information; control and mechanism of differentiation and development; erythropoietin and the regulation of red cell production; mechanism of anti-HIV action of plant proteins and polycyclic compounds. *Mailing Add:* Dept Biochem NY Univ Sch Med New York NY 10016

LEEKLEY, ROBERT MITCHELL, organic chemistry; deceased, see previous edition for last biography

LEELA, SRINIVASA (G), b Mysore, India. MATHEMATICS. *Educ:* Osmania Univ, India, BSc, 55, MSc, 57; Marathwada Univ, India, PhD(math), 65. *Prof Exp:* Lectr math, Women's Col, Kurnool, India, 59-65; instr, Calgary Univ, 65-66; asst prof, Univ RI, 66-68; assoc prof, 68-73, PROF MATH, STATE UNIV NY COL GENESEO, 73- *Mem:* Am Math Soc; Math Asn Am. *Res:* Qualitative analysis in differential equations; stability theory. *Mailing Add:* Dept Math State Univ NY Col Geneseo Geneseo NY 14454

LEELING, JERRY L, b Ottumwa, Iowa, July 11, 36; m 57; c 3. PHARMACOLOGY. *Educ:* Parsons Col, BS, 59; Univ Iowa, MS, 62, PhD(biochem), 64. *Prof Exp:* Res biochemist, 64-70, SR RES SCIENTIST, TOXICOL DEPT, MILES LABS, INC, 70-, SECT HEAD, 73- *Mem:* Am Chem Soc; AAAS; Soc Toxicol; Am Soc Pharmacol & Exp Therapeut. *Res:* Analytical toxicology. *Mailing Add:* Toxicol Dept Miles Inc Elkhart IN 46515

LEEMAN, SUSAN EPSTEIN, b Chicago, Ill, May 9, 30; m 57; c 3. PHYSIOLOGY. *Educ:* Goucher Col, BA, 51; Radcliffe Col, MA, 54, PhD, 58. *Prof Exp:* Instr physiol, Harvard Med Sch, 58-59; fel neurochem, Brandeis Univ, 59-62, sr res assoc biochem, 62-68, adj asst prof, 66-68, asst res prof, 68-71; asst prof physiol, Lab Human Reprod & Reprod Biol, Harvard Med Sch, 72-73, assoc prof, 73-80; PROF PHYSIOL, MED SCH, UNIV MASS, 80-, DIR INTERDEPARTMENTAL NEUROSCI PROG, 84- *Concurrent Pos:* USPHS career develop award, 62-; mem, Endocrinol Study Sect, Div Res Grants, NIH, 81; Albert Heritage med res vis prof, 81. *Honors & Awards:* Astwood Award, 81; Van Dyke Award, 82; Louis & Bert Freedman Found Award, 82. *Mem:* Nat Acad Sci; Endocrine Soc; Soc Neurosci; AAAS; Am Physiol Soc. *Res:* Neuroendocrinology. *Mailing Add:* 139 Park St Newton MA 02158

LEEMANN, CHRISTOPH WILLY, b Basel, Switz, Jan 12, 39; m 69; c 2. EXPERIMENTAL & ACCELERATOR PHYSICS. *Educ:* Univ Basel, PhD(nuclear physics), 69. *Prof Exp:* Res asst nuclear physics, Univ Basel, 63-69, res assoc, 69-70; fel, 70-72, sr fel accelerator physics, 72-73, STAFF SCIENTIST ACCELERATOR PHYSICS, LAWRENCE BERKELEY LAB, UNIV CALIF, 73- *Concurrent Pos:* Sabbatical leave, Europ Orgn Nuclear Res, Geneva, Switz, 80-81. *Mem:* AAAS. *Res:* Design and development of particle accelerators and related devices; beam cooling techniques; colliding beam devices; relativistic heavy ion accelerators. *Mailing Add:* CEBAF 12000 Jefferson Ave Newport News VA 23606

LEEMING, DAVID JOHN, b Victoria, BC, June 8, 39; m 66; c 3. MATHEMATICS. *Educ:* Univ BC, BSc, 61; Univ Ore, MA, 63; Univ Alta, PhD(math), 69. *Prof Exp:* from Instr to assoc prof, 63-86, 75-86, PROF MATH, UNIV VICTORIA, BC, 86- *Concurrent Pos:* Course writer, Open Learning Inst, 79. *Mem:* Math Asn Am; Can Math Soc. *Res:* Approximation theory; error bounds for interpolation schemes; rational approximation. *Mailing Add:* Dept Math Univ Victoria Victoria BC V8W 2Y2 Can

LEENHEER, MARY JANETH, b Zeeland, Mich, Mar 4, 54; m 81. ORGANIC GEOCHEMISTRY. *Educ:* Calvin Col, BS, 76; Univ Mich, MS, 78, PhD(oceanic sci), 81. *Prof Exp:* RES GEOCHEMIST, CITIES SERV CO, 81- *Mem:* Geochem Soc; Sigma Xi; Int Asn Great Lakes Res. *Res:* Chromatographic techniques to characterize organic constituents present in petroleum and recent and ancient sediments in order to better understand depositional environments and post-depositional alteration processes. *Mailing Add:* Dept Atmos/Ocean Sci 2455 Hayward Ann Arbor MI 48109

LEENHOUTS, LILLIAN S, engineering, for more information see previous edition

LEEP, HERMAN ROSS, b Louisville, Ky, Oct 21, 40; m 75. MANUFACTURING, ENGINEERING ECONOMY. *Educ:* Univ Louisville, BSME, 63; Univ Del, MMAE, 67; Purdue Univ, PhD(indust eng), 79. *Prof Exp:* Asst prof graphics, 73-79 & indust eng, 79-83, ASSOC PROF INDUST ENG, UNIV LOUISVILLE, 83-; MECH ENGR, E I DU PONT DE NEMOURS, 63- *Concurrent Pos:* Lectr, GE, 86-88. *Mem:* Inst Indust Engrs; Soc Mfg Engrs; Robotics Int. *Res:* Machining of composite materials; machining of titanium alloys; measurement of tool wear; measurement of cutting forces; measurement of surface finish. *Mailing Add:* Dept Indust Eng Univ Louisville Louisville KY 40292-9966

LEEPER, DENNIS BURTON, b Glendale, Calif, May 3, 41; m; c 5. RADIATION BIOLOGY, HYPERTHERMIC ONCOLOGY CANCER. *Educ:* Univ Iowa, BS, 64, PhD(radiation biol), 69. *Prof Exp:* From asst prof to assoc prof, 70-80, PROF RADIATION ONCOL, THOMAS JEFFERSON UNIV HOSP, 80- *Concurrent Pos:* Atomic Energy Comn fel, Colo State Univ, 69-70; affil grad fac, 70-72; prin investr, NIH Res Grants, 72-; consult radiation biol, Franklin Inst, Pa, 76-82; adj assoc prof biomed eng, Univ Pa, 72-80; chmn, NAm Hyperthermia Group, 81-85; mem, Training

Grant Study Sect, NCI, 82-83, Diag Radiol & Nuclear Med Study Sect, NIH, 83-87; counr biol, Radiation Res Soc, 83-86; assoc ed, Int J Radiation Biol, 83-86, Int J Hyperthermia, 85- *Honors & Awards:* Sci Award, Am Cancer Soc, 86. *Mem:* Radiation Res Soc; Am Asn Cancer Res; Cell Kinetics Soc; AAAS; Am Soc Thermalapeut Radiol Oncol; NAm Hyperthermia Group; Sigma Xi; NY Acad Sci. *Res:* Interaction of radiation, hyperthermia, and anti-cancer drugs in mammalian cells in culture and in normal and tumor tissues in vivo; cell cycle kinetics; experimental radiation oncology. *Mailing Add:* Dept Radiation Oncol & Nuclear Med Thomas Jefferson Univ Hosp Philadelphia PA 19107-5004

LEEPER, HAROLD MURRAY, b Akron, Ohio, July 14, 20; m 42; c 3. POLYMER CHEMISTRY. *Educ:* Univ Akron, BS, 42, MS, 47. *Prof Exp:* Rubber technologist, US Eng Bd, 42-44; polymer chemist rubber, Govt Rubber Labs, 44-47; res chemist polymers, Wm Wrigley Jr Co, 47-54; group leader, Monsanto Co, 54-70; res scientist, 70-91, PRIN SCIENTIST POLYMERS, ALZA CORP, 80- *Mem:* Am Chem Soc; Soc Plastics Engrs; AAAS. *Res:* Chemistry, physics, and technology of rubbers and plastics. *Mailing Add:* Alza Corp 1040 Gest Dr Mountain View CA 94040

LEEPER, HERBERT ANDREW, JR, b Lewistown, Pa, Nov 5, 42; m 77; c 1. SPEECH PATHOLOGY, COMMUNICATIVE DISORDERS. *Educ:* Bloomsburg State Univ, BS, 63; Purdue Univ, MS, 66, PhD(speech path), 69. *Prof Exp:* Speech pathologist, Jewish Hosp, St Louis, 69-71; asst prof, Univ Okla Med Ctr, 71-74; assoc prof, Okla State Univ, 74-77; ASSOC PROF SPEECH PATH, UNIV WESTERN ONT, 77- *Concurrent Pos:* Lectr, Fac Dent, Univ Western Ont, 79-; consult, Cleft Palate Team, Thames Valley Children's Ctr, 79-; clin asst prof otolaryngol, Fac Med, Univ Western Ont, 88- *Mem:* Fel Am Speech Lang Hearing Asn; Am Cleft Palate Asn (secy, 74-77); Acoust Soc Am; Am Asn Phonetic Sci; Can Asn Speech Lang Pathologists & Audiologists; Sigma Xi. *Res:* Description of the physiological, acoustical, and perceptual characteristics of the speech produced by children and adults with cranio-facial anomalies, laryngeal dysfunction and deafness. *Mailing Add:* Dept Commun Dis Elborn Col Univ Western Ont London ON N6G 1H1 Can

LEEPER, JOHN ROBERT, b Hackensack, NJ, July 12, 47; m 73; c 2. ENTOMOLOGY. *Educ:* Carthage Col, BA, 69; Univ Hawaii, MS, 71, PhD(entom), 75. *Prof Exp:* Res assoc entom, Tree Fruit Res Ctr, Wash State Univ, 75-77; asst prof entom, NY State Agr Exp Sta, Cornell Univ, 77-80; sr res biologist, Du Pont Exp Sta, Du Pont Inc, 80-85, licensing prod mgr, New Bus Ventures, 85-88, MGR US PROD DEVELOP, RICE HERBICIDES, AG PROD DEPT, DU PONT INC, 88- *Mem:* Entom Soc Am. *Res:* Tree fruit entomology and insecticide resistance. *Mailing Add:* DuPont Japan Ltd Tsukuba Lab 25-2 Kannondai 1-chome Tsukuba-shi Ibaragi Profecture 305 Japan

LEEPER, RAMON JOE, b Princeton, Mo, Apr 1, 48; m 76; c 1. PLASMA PHYSICS, HIGH ENERGY NUCLEAR PHYSICS. *Educ:* Mass Inst Technol, SB, 70; Iowa State Univ, PhD(high energy nuclear physics), 75. *Prof Exp:* Res assoc high energy nuclear physics, Ames Lab, US Dept of Energy, 75-76; mem tech staff Plasma Physics, 76-86, SUPVR DIAGNOSTICS DIV, SANDIA NAT LABS, 86- *Concurrent Pos:* Guest scientist, Argonne Nat Lab, 71-76. *Mem:* Am Phys Soc; Sigma Xi. *Res:* Particle beam induced controlled thermonuclear fusion; neutron physics; fusion plasma diagnostic techniques; neutron production of inertially confined high temperature fusion plasmas; high intensity pulsed neutron sources; high current ion beams; meson spectroscopy. *Mailing Add:* Diagnostics Div 1267 Sandia Nat Labs Albuquerque NM 87185

LEEPER, ROBERT DWIGHT, nuclear medicine, endocrinology; deceased, see previous edition for last biography

LEEPER, ROBERT WALZ, b Waterloo, Iowa, Apr 28, 15; m 42; c 3. ORGANIC CHEMISTRY. *Educ:* Univ Iowa, BA, 36, MS, 38; Iowa State Univ, PhD(org chem), 42. *Prof Exp:* Nat Defense Res Coun fel, Iowa State Univ, 42-43; org chemist, Gelatin Prod Corp, Mich, 43-45; dir res, Boyle-Midway Div, Am Home Prod Co, NJ, 45-48; org chemist, Pineapple Res Inst, Hawaii, 48-66, head chem dept, 61-66; chem consult & group leader, 69-77, sr scientist & leader, Agr Div, Amchem Prod, Inc, 77-78; sr scientist, Union Carbide Agr Prod, 78-81; RETIRED. *Concurrent Pos:* Vis prof, Mich State Univ, 55; vis scientist, NZ, 62. *Mem:* Am Chem Soc; Sigma Xi. *Res:* Organometallics; synthesis of potential herbicides and plant growth regulators. *Mailing Add:* 47 Woodside Ave Chalfont PA 18914

LEERBURGER, BENEDICT ALAN, b New York, NY, Jan 2, 32; m 58; c 2. SCIENCE WRITING, JOURNALISM. *Educ:* Colby Col, BA, 54. *Prof Exp:* Asst ed, Prod Eng Mag, 54-59; sci ed, Grolier, Inc, 59-61; ed, Cowles Ed Corp, 61-68; proj dir, CCM Info Sci, Inc, 68-70; vpres & ed dir, Nat Micro-Publ Corp, 70-72; dir publ, NY Times, 72-74; publ, Kraus-Thomson Org, Ltd, 74-76; CONSULT COMPUT SCI, PHYS SCI & GEN SCI COMMUN, 79- *Concurrent Pos:* Consult, Sci Digest, 59-61, Cross, Hinshaw & Lindberg, Inc, 65-67, Storrington Printing & Pub Co, Inc, 67-70 & NSF Deep Freeze Prog, Antarctica, 67; ed-in-chief, McGraw-Hill Book Co, 76-79; free lance writer, 80- *Mem:* Nat Asn Sci Writers; Am Hist Asn; Nat Sci Teachers Asn; Am Soc Journalists & Authors. *Res:* Physical science; Antarctica; American scientific history. *Mailing Add:* 338 Heathcote Rd Scarsdale NY 10583

LEE-RUFF, EDWARD, b Shanghai, China, Jan 4, 44; Can citizen; m 69; c 3. CHEMISTRY. *Educ:* McGill Univ, BSc, 64; PhD(org chem), 67. *Prof Exp:* Nat Res Coun fel, Columbia Univ, 67-69; from asst prof to assoc prof, 74-84, PROF CHEM, YORK UNIV, 85- *Mem:* Am Chem Soc; Chem Inst Can. *Res:* Organic photochemistry; reactions of strained molecules; organic chemistry. *Mailing Add:* Dept Chem York Univ Downsview ON M3J 1P3 Can

LEES, ALISTAIR JOHN, b Preston, Eng, July 12, 55; m 79; c 2. PHOTOCHEMISTRY, SPECTROSCOPY. *Educ:* Univ Newcastle, BSc, 76, PhD(chem), 79. *Prof Exp:* Fel phys chem, Univ Southern Calif, 79-81; asst prof, 81-86, ASSOC PROF INORG CHEM, STATE UNIV NY, BINGHAMTON, 86- *Concurrent Pos:* Consult, IBM Corp, 85-86, Int Paper Corp, 85- *Mem:* Am Chem Soc; Royal Soc Chem. *Res:* Photochemistry and spectroscopy of transition metal compounds; organometallic chemistry; homogeneous catalysis; kinetics and mechanism; low-temperature spectroscopy and photochemistry. *Mailing Add:* Dept Chem State Univ NY Binghamton NY 13901

LEES, DAVID ERIC BERMAN, b Boston, Mass, July 22, 50; m 77; c 3. IMAGE PROCESSING. *Educ:* Oakland Univ, BS, 72; Univ Rochester, MS, 74, PhD(optics), 79. *Prof Exp:* Prin develop engr, Honeywell, 79-82; systs engr, Automatix, Inc, 82-86, SYSTS ENGR SPARTA INC, 86- *Concurrent Pos:* Prin investr, NSF Phase I Small Bus Innovation Res Award, 84. *Mem:* Optical Soc Am; Soc Photo-Optical Instrumentation Engrs. *Res:* Machine vision for industrial applications; vision guided robots; optical gauging; speckle imaging, phase-only imaging. *Mailing Add:* 57 Gleason Rd Lexington MA 02173

LEES, GEORGE EDWARD, b Pittsburgh, Pa, Feb 7, 48; m 80. VETERINARY INTERNAL MEDICINE, VETERINARY NEPHROLOGY & UROLOGY. *Educ:* Colo State Univ, BS, 70, DVM, 72; Univ Minn, MS, 79. *Prof Exp:* Intern small animal med & surg, Sch Vet Med, Univ Calif, Davis, 75-76; resident small animal med, Col Vet Med, Univ Minn, St Paul, 76-79, clin asst prof small animal med, 79-80; assoc prof, 80-86, PROF SMALL ANIMAL MED, COL VET MED, TEX A&M UNIV, 86- *Concurrent Pos:* Chmn bd, Am Col Vet Internal Med, 91-92. *Mem:* Am Col Vet Internal Med (vpres, 88-90, pres-elect, 89-90, pres, 90-91). *Res:* Diagnosis, treatment and prevention of spontaneous diseases of the urinary system in dogs and cats. *Mailing Add:* 2600 Rustling Oaks Bryan TX 77802

LEES, HELEN, biochemistry, for more information see previous edition

LEES, MARJORIE BERMAN, b New York, NY, Mar 17, 23; m 46; c 3. NEUROCHEMISTRY, NEUROIMMUNOLOGY. *Educ:* Hunter Col, BA, 43; Univ Chicago, MS, 45; Harvard Univ, PhD(med sci), 51. *Prof Exp:* Asst, Univ Chicago, 43-45; asst, Col Physicians & Surgeons, Columbia Univ, 45-46; Am Cancer Soc res fel, 51-53; res asst, McLean Hosp, 53-55, asst biochemist, 55-58, assoc biochemist, 58-62; sr res assoc pharmacol, Dartmouth Med Sch, 62-66; assoc biochemist, McLean Hosp, 66-76; BIOCHEMIST, EUNICE KENNEDY SHRIVER CTR, 76-, DIR, BIOCHEM DIV, 90-; PROF BIOCHEM, NEUROL DEPT, MED SCH, HARVARD UNIV, 85- *Concurrent Pos:* Instr neuropath, Harvard Med Sch, 55-59, res assoc, 59-62 & 66-71, prin res assoc, 71-75, sr res assoc, 75-85; chief ed, J Neurochem, 86-90; mem, Nat Adv Coun, Nat Inst Neurol & Communicative Disorders & Stroke, 79-82; Javits invest award, NIH, 84-91, 91-97. *Mem:* Am Soc Biol Chem; Am Soc Neurochem (treas, 75-81, pres, 83-85); Soc Neurosci; Am Soc Neuropath; Int Soc Neurochem. *Res:* Chemistry of the nervous system; myelin and demyelinating diseases; neuroimmunology; brain proteins. *Mailing Add:* E K Shriver Ctr 200 Trapelo Rd Waltham MA 02254

LEES, MARTIN H, b London, Eng, May 11, 29; m 59; c 3. PEDIATRICS, CARDIOLOGY. *Educ:* Univ London, MB, BS, 55. *Hon Degrees:* FRCP, 80. *Prof Exp:* Assoc prof, 62-71, PROF PEDIAT, MED SCH, UNIV ORE, 71- *Mem:* Am Pediat Soc. *Res:* Pediatric cardiology; newborn and infant cardiopulmonary physiology and pathophysiology. *Mailing Add:* Dept Pediat Univ Ore Med Sch Portland OR 97201

LEES, NORMAN DOUGLAS, b Providence, RI, Sept 16, 45; m 81; c 3. MICROBIOLOGY, MOLECULAR BIOLOGY. *Educ:* Providence Col, AB, 67; Northwestern Univ, PhD(microbiol), 73. *Prof Exp:* Teaching asst biol sci, Northwestern Univ, 67-73; asst prof, 73-80, ASSOC PROF BIOL, IND UNIV-PURDUE UNIV, INDIANAPOLIS, 80- *Concurrent Pos:* Grants, Ind Univ-Purdue Univ, Indianapolis, 74 & Biomed Sci Res, 75 & 77. *Mem:* Am Soc Microbiol; Sigma Xi. *Res:* Role of sterols in biological membranes. *Mailing Add:* Dept Biol KB 337 1125 E 38th St Indianapolis IN 46205

LEES, ROBERT S, b New York, NY, July 16, 34; m 60; c 4. MEDICINE, BIOCHEMISTRY. *Educ:* Harvard Univ, AB, 55, MD, 59; Am Bd Internal Med, dipl. *Prof Exp:* Intern surg, Mass Gen Hosp, Boston, 59-60, asst resident med, 61-62; hon asst registr cardiol, Nat Heart Hosp, Eng. 62-63; staff assoc & attend physician med, Nat Heart Inst, 63-66; asst prof med & attend physician, Rockefeller Univ, 66-69; dir clin res ctr, 69-74, assoc prof, 69-71, PROF HEALTH SCI TECHNOL, HARVARD UNIV & MASS INST TECHNOL, 71-; CHIEF, DIV PERIPHERAL VASCULAR DIS & DIR MED RES, NEW ENG DEACONESS HOSP, 82- *Concurrent Pos:* USPHS res fel med, 60-61; Dalton scholar, Harvard Med Sch, 60; USPHS fel, 62-63; fel coun arteriosclerosis, Am Heart Asn, 65-; assoc in med, Peter Bent Brigham Hosp; asst in med, Mass Gen Hosp. *Mem:* Am Heart Asn; Am Fedn Clin Res; Am Soc Clin Invest; Am Soc Pharmacol & Exp Therapeut. *Res:* Cardiology, especially ischemic heart disease; lipid and lipoprotein metabolism. *Mailing Add:* New Eng Deaconess Hosp 185 Pilgram Rd Boston MA 02215

LEES, RONALD EDWARD, b Carlisle, UK, Feb 14, 35; Can citizen; m 62; c 2. OCCUPATIONAL MEDICINE. *Educ:* Glasgow Univ, MBChB, 58, DPM, 62, MD, 67. *Prof Exp:* Med officer health, Govt St Lucia, Wis, 62-65; field med officer, Rockefeller Found, NY, 65-68; FROM ASST PROF TO PROF EPIDEMIOL, QUEEN'S UNIV, 69-, PROF FAMILY MED, 75-, DIR, OCCUP MED, 78- *Res:* Tropical health-infant malnutrition, schistosomiasis, parasitic and infectious disease control; environmental and occupational health-epidemiologic studies in effect of toxins, noise induced hearing loss, effects of shift work and stress; occupational health problems in underdeveloped countries. *Mailing Add:* Occup Health & Safety Resource Ctr Abramsky Hall 2nd Floor Queen's Univ Kingston Kingston ON K7L 3N6 Can

LEES, RONALD MILNE, b Sutton, Eng, Oct 28, 39; Can citizen; m 62; c 2. MOLECULAR SPECTROSCOPY. *Educ:* Univ BC, BSc, 61, MSc, 65; Bristol Univ, PhD(physics), 67. *Prof Exp:* Nat Res Coun Can fel, Nat Res Coun, Ottawa, 66-68; assoc prof, 68-77, dept chmn, 81-88, PROF PHYSICS, UNIV NB, FREDERICTON, 77- *Concurrent Pos:* Nat Res Coun Can grant, Univ NB, Fredericton, 68-, vis assoc prof, Physics Dept, Univ BC, Vancouver, 74-75; prin investr, Centres of Excellence in Molecular & Interfacial Dynamics, Fed Networks of Centres Excellence Prog. *Mem:* Can Asn Physicists; Am Asn Physics Teachers; Optical Soc Am; Soc Photo-Optical Instrumentation Engrs. *Res:* Atomic and molecular physics; molecular spectroscopy. *Mailing Add:* Dept Physics Univ NB Fredericton NB E3B 5A3 Can

LEES, SIDNEY, b Philadelphia, Pa, Apr 17, 17; m 46; c 3. ENGINEERING. *Educ:* City Col New York, BS, 38; Mass Inst Technol, SM, 48. *Hon Degrees:* ScD, Mass Inst Technol, 50. *Prof Exp:* Observer, US Weather Bur, 38-40; engr, US Signal Corps, 40-43; res assoc aeronaut, Mass Inst Technol, 47-50, asst prof, 50-57; consult instrumentation, 57-59; vpres, United Res, Inc, 59; pres, Lees Instrument Res, Inc, 59-63; prof eng, Dartmouth Col, 62-66; SR STAFF MEM & HEAD BIOENG DEPT, FORSYTH DENT CTR, BOSTON, 66- *Concurrent Pos:* Chmn, Joint Automatic Control Conf, 65 & Res Conf Instrumentation Sci, 71; vis scientist, Univ Amsterdam Dent Sch, 75; chmn, Conf Ultrasonics, 78; joint chmn, NE Doppler Conf, 81; adj prof, Northeastern Univ. *Mem:* Am Phys Soc; Sigma Xi; Inst Elec & Electronics Engrs; Am Acoust Soc. *Res:* Ultrasonics; bioinstrumentation; measurement systems and components; control systems; geophysical instrumentation. *Mailing Add:* 50 Eliot Mem Rd Newton MA 02158

LEES, THOMAS MASSON, b New York, NY, June 16, 17; m 43; c 2. BIOCHEMISTRY. *Educ:* Long Island Univ, BS, 39; Iowa State Univ, MS, 42, PhD(biophys chem), 44. *Prof Exp:* Res chemist, Am Distilling Co, 44-46; anal res chemist, Pfizer, Inc, 46-80; RETIRED. *Mem:* Am Chem Soc. *Res:* Fermentative production of glycerol; antibiotics development, production and identification; analysis of medicinal compounds. *Mailing Add:* 35 Woodridge Circle Gales Ferry CT 06335

LEES, WAYNE LOWRY, b Washington, DC, July 18, 14; m 39; c 2. EXPERIMENTAL PHYSICS, ENGINEERING PHYSICS. *Educ:* Swarthmore Col, BA, 37; Harvard Univ, MA, 40, PhD(physics), 49. *Prof Exp:* Asst, Bartol Res Found, Pa, 39-40; physicist, Geophys Lab, Wash, DC, 42-44, Nat Bur Standards, 44-46, Tracerlab, Inc, 49-50, Metall Proj, Mass Inst Technol, 50-54, Nuclear Metals, Inc, 54-58, Instrumentation Lab, Mass Inst Technol, 58-65 & Electronics Res Ctr, NASA, Cambridge, 65-70; assoc prof math, Wash Tech Inst, 71-72; proj engr, Design Automation, Inc, Lexington, Mass, 72-73; staff mem, lab phys sci, P R Mallory & Co, Inc, Burlington, Mass, 74-80; mem staff, Duracell Int, Inc, 80-82; RETIRED. *Mem:* AAAS; Am Phys Soc; Fedn Am Sci; Inst Elec & Electronics Engrs. *Res:* Quasistatic electrical systems and dielectric properties; physics of high pressures and metals; electrode phenomena; electron and ion transport; engineering physics. *Mailing Add:* 29 Tower Rd Lexington MA 02173

LEESE, JOHN ALBERT, b Manchester, Md, Dec 6, 32; m 59; c 4. REMOTE SENSING, NUMERICAL MODELING. *Educ:* Pa State Univ, BS, 57; Fla State Univ, MS, 59; Univ Mich, PhD(meteorol), 64. *Prof Exp:* Meteorologist res & develop, Air Force Cambridge Labs, 59-61; lectr, Univ Mich, 61-64; dept mgr res & develop, Atmospheric Sci Dept, IBM Corp, 64-69; div chief res & develop, Nat Environ Satellite Serv, 69-75, assoc dir data processing, 75-78, dep dir admin, 78-82; sr scientist 0dmin, World Meteorol Orgn, Geneva, Switz, 82-88; DIR RES, INST NAVAL OCEANOG, 88- *Concurrent Pos:* Meteorologist, USNR, 58-78. *Mem:* Am Meteorol Soc; Oceanog Soc. *Res:* Digital data processing and quantitative information extraction techniques of environmental satellite data for input to numerical models of the atmosphere and the oceans. *Mailing Add:* Int Naval Oceanog Stennis Space Center MS 39529-5005

LEESER, DAVID O(SCAR), b El Paso, Tex, Aug 3, 17; m 45; c 2. FORENSIC MATERIALS ENGINEERING. *Educ:* Univ Tex, BS, 43; Ohio State Univ, MS, 50. *Prof Exp:* Metallurgist, Bradley Mining Co, Idaho, 43-44; res engr, Battelle Mem Inst, 44-50; assoc metallurgist, Argonne Nat Lab, 50-54; staff metallurgist & chief mat sect, Atomic Power Develop Assocs, Inc, 54-61; chief scientist, Missile Div, Chrysler Corp, 61-68, chief metallurgist, Amplex Div, 68-75; mgr, Eng Mat Lab, Burroughs Corp, 75-86; CONSULT ENGR, 87- *Concurrent Pos:* Reactor mat engr, Nuclear Power Dept, Detroit Edison Co, 54-61; mem, Atomic Indust Forum; mem welding forum, US AEC, 54-64, high temperature nuclear fuel comt, 57-61; US del, World Metall Cong, 57, Int Conf Peaceful Uses of Atomic Energy, Geneva, 58 & Int Atomic Energy Agency Conf, Vienna, 61. *Honors & Awards:* Award, Off Sci Res & Develop; Award, Nat Adv Comt Aeronaut, 44; Apollo Achievement Award, NASA, 69. *Mem:* Am Soc Testing & Mat; Sigma Xi; Am Soc Mech Engrs; Am Mgt Asn; Am Soc Metals. *Res:* Evaluation of aerospace designs with regard to conventional and nonconventional materials applications and advanced aerospace requirements; ground and launch support equipment; materials for high-speed computer and electromechanical business machine systems under development; failure analysis in each category. *Mailing Add:* 11515 N 91st St Suite 151 Scottsdale AZ 85260

LEESON, CHARLES ROLAND, b Halifax, Eng, Jan 26, 26; m 54; c 5. ANATOMY. *Educ:* Cambridge Univ, BA, 47, MB, BChir, 50, MA, 50, MD, 59, PhD(anat), 71. *Prof Exp:* Lectr anat, Univ Col SWales, 55-58; assoc prof, Dalhousie Univ, 58-61; assoc prof anat & histol, Queen's Univ, Ont, 61-63; prof anat, Univ Iowa, 63-66; prof anat & chmn dept, Univ Mo, Columbia, 66-78; PROF ANAT, SCH BASIC MED SCI, UNIV ILL, 78- *Concurrent Pos:* Vis prof anat, London Hosp Med Col, Eng, 73-74. *Mem:* Anat Soc Gt Brit & Ireland; Am Asn Anat. *Res:* Post natal development, particularly in marsupials and rodents and with reference to certain organ systems. *Mailing Add:* Dept Anat Univ Ill Col Med 506 S Mathews Urbana IL 61801

LEESON, LEWIS JOSEPH, b Paterson, NJ, Apr 26, 27; m 53; c 3. BIOPHARMACEUTICS & PHARMACO-KINETICS. *Educ:* Rutgers Univ, BS, 50, MS, 54; Univ Mich, PhD(pharmaceut chem), 57. *Prof Exp:* Intern pharm, Mack Drug Co, 50-51; pharmacist, Silver Rod Drugs, 51-52; asst, Rutgers Univ, 52-54 & Univ Mich, 55-56; res chemist, Lederle Labs, Am Cyanamid Co, 57-67; proj leader pharmaceut, Union Carbide Res Inst, 67-69; asst dir pharmaceut develop, Geigy Chem Corp, 69-71, asst dir, 71-73, dir, 73-78, sr dir pharm res & develop, 78-80, sr res fel biopharm, 80-84, DISTINGUISHED RES FEL, BIOPHARM, CIBA-GEIGY PHARMACEUT CO, 84- *Concurrent Pos:* Relief pharmacist, Frieds Pharm, 52-54. *Mem:* Am Chem Soc; Am Pharmaceut Asn; fel Acad Pharmaceut Sci; Sigma Xi; fel Am Asn Pharmaceut Sci. *Res:* Pharmaceutical product development; application of physical chemical techniques for developing various pharmaceutical dosage forms; biopharmaceutics; pharmacokinetics. *Mailing Add:* Pharm Res & Develop Ciba-Geigy Corp 556 Morris Ave Summit NJ 07901

LEESON, THOMAS SYDNEY, b Halifax, UK, Jan 26, 26; m 52; c 3. ANATOMY. *Educ:* Cambridge Univ, BA, 46, MA, 49, MD & BCh, 50, MD, 59, PhD, 71. *Prof Exp:* Asst lectr anat, Univ Wales, 55-57; from asst prof to assoc prof, Univ Toronto, 57-63; head dept, 63-82, PROF ANAT, UNIV ALTA, 82- *Mem:* Am Asn Anat; Brit Asn Clin Anatomists; Anat Soc Gt Brit & Ireland; Electron Microscopy Soc Am. *Res:* Electron microscopy, histology and embryology; Pancreatic centroacinar cells as they relate to acinar and ductular cells and to insular cells the latter associated with paracrine secreatin and hormonal control of exocrine secretion. *Mailing Add:* Dept Anat Univ Alta Edmonton AB T6G 2G3 Can

LEESTMA, JAN E, b Flint, Mich, Nov 30, 38; m 61; c 2. PATHOLOGY, NEUROPATHOLOGY. *Educ:* Hope Col, BA, 60; Univ Mich, MD, 64; Northwestern Univ, MBA, 85. *Prof Exp:* Resident & intern path, Univ Colo Med Sch, Denver, 64-67; fel neuropath, Einstein Med Col & instr path, Univ Colo Med Sch, Denver, 67-68; from asst prof to assoc prof path, Sch Med, Northwestern Univ, 71-86; prof path & neurol, Univ Chicago, 86-87; ASSOC MED DIR, CHICAGO NEUROSURG CTR & CHICAGO INST NEUROSURG & NEURORESEARCH, 87- *Concurrent Pos:* Consult, DC Gen Hosp, Washington, DC, 69-71, Nat Naval Med Ctr, Bethesda, Md, 69-70, Vet Admin Lakeside Hosp & Vet Admin North Chicago Hosp, Ill, 71-, Baxter-Travenol Labs, Morton Grove, Ill, 73-76, Great Lakes Naval Hosp, Ill, 74- & 80 W Suburban Hosp, Oak Park, Ill, 76-85; attend physician, Northwestern Mem Hosp, Chicago, 71-, Children's Mem Hosp, 82-; asst med examr, Cook County Off Med Examr, Chicago, 77-; dean student, Div Biol Sci, Pritzker Sch Med, Univ Chicago, 86-87. *Mem:* Am Asn Neuropathologists; Sigma Xi; AAAS; NY Acad Sci. *Res:* Experimental neurology; neurological degenerative disease; central nervous system tissue; brain tumors; computerized data analysis; electron microscopy forensic medicine; electron microscopy, forensic neuropathology. *Mailing Add:* Chicago Neurosurgical Ctr Chicago Inst Neurosurg & Neuroresearch 428 W Deming Pl Chicago IL 60614

LEETCH, JAMES FREDERICK, b Butler, Pa, Sept 27, 29; m 58; c 2. MATHEMATICS. *Educ:* Grove City Col, BS, 51; Ohio State Univ, MA, 57, PhD(math), 61. *Prof Exp:* From asst prof to assoc prof, 61-71, PROF MATH, BOWLING GREEN STATE UNIV, 71- *Mem:* Math Asn Am. *Res:* Mathematical analysis. *Mailing Add:* 19 Darlyn Dr Bowling Green OH 43402

LEETE, EDWARD, b Leeds, Eng, Apr 18, 28; nat US; m 54, 76; c 7. ORGANIC CHEMISTRY. *Educ:* Univ Leeds, BSc, 48, PhD(chem), 50, DSc, 65. *Prof Exp:* Goldsmith fel, Nat Res Coun Can, 51-52, res fel, 52-54; from instr to asst prof org chem, Univ Calif, Los Angeles, 54-58; from asst prof to assoc prof, 58-63, PROF ORG CHEM, UNIV MINN, MINNEAPOLIS, 63- *Concurrent Pos:* Mem med chem study sect, NIH, 62-65; Alfred P Sloan fel, 62-65; consult, Philip Morris Res Ctr, Richmond, Va, 74-; vis prof agr, Univ Kyoto, 90. *Honors & Awards:* Guggenheim Fel, Univ Oxford, Eng, 65; First Phytochemistry Prize & Medal, 90. *Mem:* Am Chem Soc; Am Soc Pharmacog; fel Royal Soc Chem; Soc Pharmaceut Bottling Res; Sigma Xi; fel AAAS; Phytochem Soc NAm. *Res:* Biosynthesis of natural substances, especially alkaloids; synthesis of heterocyclic compounds; isolation of enzymes from plants; use of radioactive and stable isotopes. *Mailing Add:* Dept Chem 207 Pleasant St SE Univ of Minn Minneapolis MN 55455

LEEVY, CARROLL M, b Columbia, SC, Oct 13, 20; m 56; c 2. MEDICINE, NUTRITION. *Educ:* Fisk Univ, AB, 41; Univ Mich, MD, 44. *Hon Degrees:* DSc, NJ Inst of Technol, 73; DHH, Dr Hum, Fisk Univ, 81. *Prof Exp:* Intern med, Jersey City Med Ctr, 44-45, resident, 45-48, dir clin invest & outpatient dept, 48-58; res assoc, Harvard Univ, 58-59; assoc prof, 59-62, actg chmn dept med, 66-68, PROF MED, COL MED NJ, 62-, DIR DIV HEPATIC METAB & NUTRIT, 59-; CHMN & PHYSICIAN IN CHIEF, NJ MED SCH, 75-; SCI DIR, SAMMY DAVIS JR NAT LIVER INST, 84- *Concurrent Pos:* USPHS spec res fel, 58-59; consult, US Naval Hosp, St Albans, 48; consult & mem med adv comt, Vet Admin Hosp, East Orange, NJ, 64; physician-in-chief, Martland Hosp, 66-68; mem dean's comt, East Orange Vet Admin Hosp, 66-68, chief med, 66-71; mem clin cancer training comt, NIH, 69-73; consult, Food & Drug Admin, 70- *Honors & Awards:* Mod Med Award, 72. *Mem:* AAAS; Soc Exp Biol & Med; Nat Med Asn; fel AMA; fel Am Col Physicians; Asn Am Physicians; Am Asn Study Liver Dis (pres, 70); Int Asso for Study of Liver (pres, 72-76). *Res:* Pathogenesis of cirrhosis of alcoholics; factors which control hepatic desoxyribonucleic acid synthesis and regeneration; mechanism of portal hypertension and malutilization of vitamins and proteins; immunologic reactivity and liver disease. *Mailing Add:* Col Med 100 Bergen St Newark NJ 07109

LEE-WHITING, GRAHAM EDWARD, b Iroquois Falls, Ont, Can, Mar 2, 26; m 52. CHARGED-PARTICLE OPTICAL SYSTEMS, PHYSICAL BOUNDARY-VALUE PROBLEMS. *Educ:* Univ Toronto, BASc, 48, MA, 49; Bristol Univ, PhD (theoret physics), 52. *Prof Exp:* Theoret physicist, 52-69, head, Theoret Physics Br, 69-91, EMER RESEARCHER, ATOMIC ENERGY CAN RES CO, 91- *Mem:* Can Asn Physicists; Am Phys Soc. *Res:* The theory of focusing and dispersive systems for charged particles; beta spectrometers using magnetic and or electric fields. *Mailing Add:* Chalk River Labs AECL Res Chalk River ON K0J 1J0 Can

LEFAR, MORTON SAUL, b New York, NY, Apr 11, 37; m 61; c 2. ORGANIC CHEMISTRY. *Educ:* Brooklyn Col, BS, 58, MA, 62; Rutgers Univ, PhD(chem), 65. *Prof Exp:* Res chemist, Inst Environ Med, Med Sch, NY Univ, 58-60; sect head org chem div nutrit, Food & Drug Admin, Washington, DC, 66-68; sr scientist, Warner-Lambert Res Inst, 68-69; mgr anal chem, Rhodia, Inc, 69-74; dir qual control, Hoechst-Roussel Pharmaceut, Inc, 75-78; lectr chem, Rutgers Univ, 79-81; consult pharmacuet, 79-81; Merrill-Lynch, 80-89; VPRES, EPOLIN INC, 89- *Mem:* Am Pharmaceut Asn; Am Chem Soc; Sigma Xi. *Res:* Natural products chemistry; analytical chemistry; photochemistry; environmental health; analytical methods development on new pharmaceuticals; new business ventures. *Mailing Add:* Cherry Tree Lane Chester NJ 07930

LEFAVE, GENE M(ARION), chemistry, engineering, for more information see previous edition

LEFCOE, NEVILLE, b Montreal, Que, July 19, 25; m 54; c 4. PHYSIOLOGY. *Educ:* McGill Univ, BSc, 46; Vanderbilt Univ, MD, 50; FRCP(C), 56. *Prof Exp:* From instr to assoc prof, 57-72, PROF MED, UNIV WESTERN ONT, 72- *Mem:* Am Fedn Clin Res; Can Soc Clin Invest. *Res:* Pulmonary physiology, chiefly exercise physiology, cellular mechanisms in bronchial smooth muscle and the domestic microenvironment. *Mailing Add:* Dept Med Victoria Hosp London ON N6A 4G5 Can

LE FEBVRE, EDWARD ELLSWORTH, environmental chemistry, analytical chemistry, for more information see previous edition

LE FEBVRE, EUGENE ALLEN, b St Paul, Minn, Oct 18, 29; m 66; c 2. CONSERVATION ECOLOGY. *Educ:* Univ Minn, BS, 52, AP, 53, MS, 58, PhD(zool), 62. *Prof Exp:* Teaching asst ornith & zool, Univ Minn, Minneapolis, 53-59, res fel ornith, Mus Natural Hist, 60-61, res assoc, 61-66; asst prof, 66-72, ASSOC PROF ZOOL, SOUTHERN ILL UNIV, 72- *Concurrent Pos:* Res assoc, NIH grant, 60-65, co-prin investr, 63-65; NSF res grant, Midway Island, 69-73 & 80-83, Int Documentation & Commun Ctr grant, 81-82, Nat Register Archives gant, 86, Southern Ill Univ, Carbondale grant, 86-87; co-prin investr, Nat Register Archives. *Mem:* AAAS; Am Ornith Union; Cooper Ornith Soc; Ecol Soc Am; Am Inst Biol Sci; Brit Ornith Union; Sigma Xi. *Res:* Conservation biology; habitat requirements of birds; environmental toxicology; biological diversity. *Mailing Add:* Dept Zool Southern Ill Univ Carbondale IL 62901

LEFEBVRE, MARIO, b Montréal, Que, Sept 20, 57. STOCHASTIC PROCESSES, STOCHASTIC CONTROL THEORY. *Educ:* Univ Montréal, BSc, 79, MSc, 80; Univ Cambridge, Eng, PhD(math), 84. *Prof Exp:* Lectr math, Royal Mil Col St-Jean, 84-85; asst prof, 85-90, ASSOC PROF MATH, ÉCOLE POLYTECH MONTRÉAL, 90- *Mem:* Soc Indust & Appl Math; Can Statist Soc. *Res:* Applied probability and stochastic control theory; electrical engineering applications; pure science, such as biology, chemistry and physics. *Mailing Add:* Dept Appl Math Ecole Polytech Montreal PQ H3C 3A7 Can

LEFEBVRE, PAUL ALVIN, b Washington, DC, Mar 12, 50; m 81. DEVELOPMENTAL BIOLOGY. *Educ:* Univ Va, BA, 72; Yale Univ, MPhil, 78, PhD(biol), 80. *Prof Exp:* Fel, Mass Inst Technol, 80-81; asst prof, 82-88, ASSOC PROF GENETICS & CELL BIOL, UNIV MINN, 88- *Mem:* Genetics Soc Am; Am Soc Microbiol. *Res:* Regulation of expression of genes for flagellar proteins in Chlamydomonas; genetics and molecular biology of nitrate reductase. *Mailing Add:* Dept Genetics & Cell Biol 250 Biosci Ctr 1445 Gortner Ave St Paul MN 55108

LEFEBVRE, RENE, b Verdun, Que, Apr 5, 23; m 51; c 2. MEDICINE. *Educ:* Col Montreal, BA, 44; Univ Montreal, MD, 50; Univ Pa, DSc(med), 54. *Prof Exp:* From asst prof to assoc prof, 60-70, PROF PATH, FAC MED, UNIV MONTREAL, 70-; pathologist, Hotel Dieu Hosp, 54-89; RETIRED. *Mem:* Can Asn Path (pres-elect, 65-66, pres, 66-). *Res:* Pathology of kidney. *Mailing Add:* 850 38th Ave Iachine PQ H8T 2C3 Can

LEFEBVRE, RICHARD HAROLD, b Detroit, Mich, Dec 11, 33; m 59; c 3. SCIENCE EDUCATION. *Educ:* Univ Mich, BS, 57; Univ Kans, MS, 61; Northwestern Univ, PhD(geol), 66. *Prof Exp:* Asst prof geol, Univ Ga, 65-67; geologist, US Geol Surv, 75-85; from asst prof to assoc prof, 67-73, chmn dept, 70-75 & 85-88, PROF GEOL, GRAND VALLEY STATE UNIV, 75- *Mem:* Geol Soc Am; Nat Asn Geol Teachers; Nat Sci Teachers Asn. *Res:* Flood basalts of the northwestern United States; remote sensing of Holocene basaltic lava flows, especially of Craters of the Moon National Monument, Idaho. *Mailing Add:* Dept Geol Grand Valley State Univ Allendale MI 49401

LEFEBVRE, YVON, b Montreal, Que, June 20, 31; m 62; c 3. ORGANIC CHEMISTRY, INFORMATION SCIENCE. *Educ:* Col Stanislas, BA, 50; Univ Mont, BSc, 53, MSc, 55, PhD(chem), 57. *Prof Exp:* Res chemist, Ayerst Labs, 58-75, group leader, 68-75, dir, Info Dept, 75-88, mat coord, 88-89; TECH SUPPORT EXEC, DERWENT INC, MCLEAN, VA, 91- *Res:* Steroids chemistry; progestational agents; estrogens; oxidation of furan derivatives in steroid and non-steroid series. *Mailing Add:* Derwent Inc 1313 Dolly Madison Blvd McLean VA 22102

LEFER, ALLAN MARK, b New York, NY, Feb 1, 36; m 59; c 4. CARDIOVASCULAR PHYSIOLOGY, PHARMACOLCOGY. *Educ:* Adelphi Univ, BA, 57; Western Reserve Univ, MA, 59; Case West Res Univ, PhD(physiol), 62. *Prof Exp:* Instr physiol, Case Western Reserve Univ, 62-64; from asst prof to prof, Sch Med, Univ Va, 64-72; PROF PHYSIOL & CHMN DEPT, JEFFERSON MED COL, THOMAS JEFFERSON UNIV, 74- *Concurrent Pos:* USPHS fel, 62-64; estab investr, Am Heart Asn, 68-73; vis prof & USPHS sr fel, Hadassah Med Sch, Hebrew Univ, Israel, 71-72; mem comt pub affairs, Fedn Am Socs Exp Biol; ed, Circulatory Shock; consult, Task Group on Shock, NIH; mem, Int Study Group Res Cardiac Metab & Pancreatic Study Group; mem, coun basic sci, Am Heart Asn, circulation coun; mem, Study Sect Pharmacol, NIH; vis prof, Wellcome Found, 85-86; managing ed, Eicosanoids, 88- *Mem:* Am Physiol Soc; Cardiac Muscle Soc; Soc Exp Biol & Med; Am Soc Pharmacol & Exp Therapeut; Reticuloendothelial Soc; Shock Soc (pres, 83-84); Int Soc Heart Res. *Res:* Cardiovascular effects of adrenal hormones; humoral regulation of myocardial contractility; corticosteroid pharmacology; experimental myocardial infarction; metabolic alterations in shock; pathogenesis of circulatory shock; prostaglandins and thromboxanes; pharmacology of coronary circulation; leukotrienes as mediators of disease states; PAF and other lipid mediators; endothelial function. *Mailing Add:* Jefferson Med Col Dept Physiol Thomas Jefferson Univ Philadelphia PA 19107

LEFEVER, ROBERT ALLEN, b York, Pa, May 29, 27; m 46; c 3. SOLID STATE CHEMISTRY, MATERIALS SCIENCE. *Educ:* Juniata Col, BS, 50; Mass Inst Technol, PhD(inorg chem), 53. *Prof Exp:* Res chemist, Linde Co, 53-56; sr scientist, Va Inst Sci Res, 56-58; mem tech staff, Hughes Res Labs, 58-59, head chem physics group, 59-61; staff mem, Gen Tel & Electronics Labs, Inc, 61-63; supvr, Mat Res Div, Sandia Labs, 63-74; dir mat preparation, Sch Eng, Univ Southern Calif, 74-77; mgr, Process Eng Dept, 77-80, PLANT MGR, FERRITE MEMORY CORE PLANT, AMPEX CORP, 80- *Concurrent Pos:* Consult, Spectrotherm Corp, 74-77 & Luxtron Corp, 77- *Mem:* Am Phys Soc; Am Chem Soc; Am Ceramic Soc; fel Am Inst Chemists; Sigma Xi; Am Asn Crystal Growth. *Res:* Single crystal growth; growth mechanisms and characterization; sintering processes and mechanisms; ferrites; garnets; metal and rare earth oxides; semiconductors; phosphors; thermoelectrics. *Mailing Add:* 1940 S Broadway Grand Junction CO 81503

LEFEVRE, GEORGE, JR, b Columbia, Mo, Sept 13, 17; m 43, 72; c 3. GENETICS, CYTOGENETICS. *Educ:* Univ Mo, AB, 37, AM, 39, PhD(genetics), 49. *Prof Exp:* Asst zool, Columbia Univ, 41-42; res biologist, Oak Ridge Nat Lab, 46-47; instr zool, Univ Mo, 47-48; from asst prof to assoc prof biol, Univ Utah, 49-56; prog dir genetic biol, NSF, 56-59; dir biol labs, Harvard Univ, 59-65; chmn, Calif State Univ, Northridge, 65-79, prof, 65-84, emer prof biol, 84-90; RETIRED. *Concurrent Pos:* Consult, NSF, 59-62 & NIH, 62-66, 79-83; ed, Genetics, 76-81. *Mem:* Genetics Soc Am (treas, 72-75); Sigma Xi. *Res:* Radiation genetics of Drosophila melanogaster; comparative mutagenesis; cytogenetics. *Mailing Add:* Dept Biol Calif State Univ 18111 Nordhoff St Northridge CA 91330

LEFEVRE, HARLAN W, b Great Falls, Mont, May 19, 29; div; c 8. NUCLEAR PHYSICS. *Educ:* Reed Col, BA, 51; Univ Idaho, MS, 57; Univ Wis, PhD(physics), 61. *Prof Exp:* Physicist, Hanford Atomic Prod Oper, Gen Elec Co, Wash, 51-58; assoc prof, 61-71, PROF PHYSICS, UNIV ORE, 71- *Concurrent Pos:* Consult, Lawrence Livermore Lab, Univ Calif, 62- *Mem:* Am Phys Soc. *Res:* Experimental nuclear physics; nuclear reactions; fast neutron spectrometry; scanning microscopy and analysis. *Mailing Add:* Dept Physics Univ Ore Eugene OR 97403

LEFEVRE, MARIAN E WILLIS, b Washington, DC, Jan 21, 23; m 48; c 3. PHYSIOLOGY. *Educ:* Iowa State Univ, BS, 44; Univ Pa, MS, 47; Univ Louisville, PhD(physiol), 69. *Prof Exp:* Assoc, 68-73, asst prof physiol, Mt Sinai Sch Med,73-78; scientist, Brookhaven Nat Lab, 78-85; RETIRED. *Concurrent Pos:* Res collabr, Brookhaven Nat Lab, 68-75, assoc scientist, 75-78. *Mem:* Am Physiol Soc; Am Gastroenterol Asn; Reticuloendothelial Soc; Am Soc Cell Biol; Soc Exp Biol & Med. *Res:* Structure and function of multicellular membranes; ion transport and metabolism; intestinal barrier function. *Mailing Add:* 15 Agassiz Rd Woods Hole MA 02543

LEFEVRE, PAUL GREEN, b Baltimore, Md, Dec 27, 19; m 48; c 4. CELL PHYSIOLOGY. *Educ:* Johns Hopkins Univ, AB, 40; Univ Pa, PhD(physiol, zool), 45. *Prof Exp:* Asst zool, Univ Pa, 43-45; from instr to asst prof physiol, Col Med, Univ Vt, 45-49, assoc prof physiol & biophys, 49-52; asst to chief med br, AEC, 52-55; scientist, Med Res Ctr, Brookhaven Nat Lab, 55-60; prof pharmacol, Sch Med, Univ Louisville, 60-68; prof, 68-83, EMER PROF PHYSIOL & BIOPHYS, HEALTH SCI CTR, STATE UNIV NY, STONY BROOK, 84- *Concurrent Pos:* Bd gov, bio-sci info exchange, 53-55. *Mem:* AAAS; Soc Gen Physiol; Am Physiol Soc; Biophys Soc; Am Soc Cell Biol. *Res:* Mechanisms, kinetics and model systems for cell membrane mediated transport; phospholipid-carbohydrate complexing. *Mailing Add:* 15 Agassiz Woods Hole MA 02543

LEFF, ALAN R, b Pittsburgh, Pa, May 23, 45. MEDICINE. *Educ:* Oberlin Col, AB, 67; Univ Rochester, MD, 71; Am Bd Internal Med, cert, 76. *Prof Exp:* Med officer, Tuberc Br, Ctr Dis Control, USPHS, 72-74; intern internal med, Univ Mich Hosp, 71-72 & House Off II, III internal med, 74-76; clin fel pulmonary dis, Cardiovasc Res Inst, Univ Calif, San Francisco, 76-77, postdoctoral res fel, 77-79; asst prof med, Pulmonary Sect, Pritzer Sch Med, Univ Chicago, 79-85, asst prof, Comt Clin Pharmacol, Div Biol Sci, 83-85, dir, Pulmonary Med Serv, Dept Med, Sect Pulmonary & Crit Care Med, 84-87, assoc prof med & Comt Clin Pharmacol, 85-89, assoc prof anesthesia & crit care, 88-89, PROF MED, ANAESTHESIA & CRIT CARE & COMT PHARMACOL, UNIV CHICAGO, 89-, HEAD SECT PULMONARY & CRIT CARE MED, 89- *Concurrent Pos:* Asst tuberc control officer, Bd Health, Chicago, 72-74; assoc attend physician, Cook County Hosp, 73-74; consult, Tuberc Prog, Dept Pub Health, San Francisco, 77-79 & adv respiratory dis, Dept Pub Health, Chicago, 80-84; dir, Pulmonary Function Labs, Univ Chicago Hosps & Clins, 79-87, Pulmonary Exercise Lab, 83-87; dir, Respiratory Serv, Hyde Park Community Hosp, 82-89; standing counr, Am Fedn Clin Res, 83-87; mem prog comt, Am Thoracic Soc, 86-, chmn, Respiratory Struct & Function Assembly, 88-89; head, Sect Pulmonary & Crit Care Med, Univ Chicago & Michael Reese Hosps, 87-89; mem joint prog comt, Am Thoracic Soc-Am Physiol Soc, 88-89. *Mem:* Am Soc Clin Invest; Cent Soc Clin Invest; Am Thoracic Soc; Am Col Physicians; Sigma Xi; Am Soc Internal Med; Int Union Tuberc; Am Fedn Clin Res; Am Physiol Soc; Am Soc Pharmacol & Exp Therapeut. *Mailing Add:* Dept Med Univ Chicago 5841 S Maryland Ave Box 98 Chicago IL 60637

LEFF, HARVEY SHERWIN, b Chicago, Ill, July 24, 37; m 58; c 4. THERMAL PHYSICS. *Educ:* Ill Inst Technol, BS, 59; Northwestern Univ, MS, 60; Univ Iowa, PhD(physics), 63. *Prof Exp:* Res assoc physics, Case Inst Technol, 63-64, from asst prof to assoc prof, 64-71; assoc prof & chmn dept phys sci, Chicago State Univ, 71-75, prof physics, 75-79; scientist, Oak Ridge Assoc Univ, 79-83; PROF & CHAIR, PHYSICS DEPT, CALIF STATE UNIV, POMONA, 83- *Concurrent Pos:* Vis prof physics, Col Sci & Eng, Harvey Mudd Col, 77-78; workshop leader, Pre-Col Teachers Energy Ideas & Physics of Toys, 86-; physic coordr, Inst Teaching & Learning, Calif State Univ. *Mem:* Am Asn Physics Teachers; Am Phys Soc; Sigma Xi. *Res:* Thermal efficiency of heat engines; Maxwell's Demon; physics of light bulbs; physics of toys; analysis of energy-related topics relevant to energy policymaking, including industrial energy conservation, energy emergency conservation potential and strategies, US natural gas information system, heat pump development; connections between entropy, forces and disorder; entropy production and efficiency for cyclic processes. *Mailing Add:* 538 E Bishop Pl Claremont CA 91711

LEFF, JUDITH, b Vienna, Austria, July 6, 35; US citizen; m 61; c 3. MICROBIOLOGY, ANALYTICAL CHEMISTRY. *Educ:* Sorbonne, Lic natural sci, 58, PhD(photobiol), 61. *Prof Exp:* Jr researcher photobiol seed germination, Nat Ctr Sci Res, Paris, 60-61, res assoc photobiol, 61; fel biol, Brandeis Univ, 62-63; fel pharmacol, Sch Med, Tufts Univ, 65, res assoc, 66-67; res assoc plant morphogenesis, Manhattan Col, 67-71, NY Univ, 71-72 & Hebrew Univ, Jerusalem, 72-73; NIH spec res fel, Albert Einstein Col Med, 74-79; appln chemist, Farrand Optical, 79-82; FOOD TECHNOL, JLN ASSOCS, 83- *Concurrent Pos:* Nat Res Serv award, Albert Einstein Col Med, 76-77. *Mem:* Sigma Xi; Am Soc Microbiol. *Res:* Photobiology; chloroplast development; molecular biology; microbiology; nucleic acids as tools for solving physiological or developmental questions; microbiology of crown gall; replication of mitochondrial DNA in yeast; fluorescence spectroscopy; food and ingredient technology for the Kosher market. *Mailing Add:* 5829 Liebig Ave Bronx NY 10471

LEFFAK, IRA MICHAEL, b New York, NY, Oct 13, 47; m 69; c 1. BIOCHEMISTRY. *Educ:* City Col New York, BS, 69; City Univ New York, PhD(biochem), 76. *Prof Exp:* Res fel biochem, Princeton Univ, 76-78; ASST PROF BIOCHEM, WRIGHT STATE UNIV, 70- *Concurrent Pos:* NIH fel, 76-78. *Mem:* AAAS. *Res:* Molecular biology of development; cell differentiation. *Mailing Add:* Dept Biol Chem Wright State Univ Colonel Glenn Hwy Dayton OH 45435

LEFFALL, LASALLE DOHENY, JR, b Tallahassee, Fla, May 22, 30; m 56; c 1. SURGICAL ONCOLOGY. *Educ:* Fla A&M Univ, BS, 48; Howard Univ, MD, 52; Am Bd Surg, dipl, 58. *Hon Degrees:* DSc, Georgetown Univ, 84, Fla A&M Univ, 87, Clark Univ, 89; LHD, Meharry Med Col, 88. *Prof Exp:* Intern, Homer G Phillips Hosp, St Louis, 52-53; resident, Freedmen's Gen Hosp, Washington, DC, 53-57 & Mem Sloan Kettering Cancer Ctr, NY, 57-59; from asst prof to assoc prof surg, 62-70, asst dean, 64-70, actg dean, 70, PROF SURG & CHMN DEPT, COL MED, HOWARD UNIV, 70- *Concurrent Pos:* Pvt pract med, Washington, DC, 62-; mem staff, Howard Univ Hosp; asst ed, J Nat Med Asn, 55-57, actg co-ed, 64-65, consult ed, 73-; consult, St Elizabeth's Hosp, 66-76, Walter Reed Army Med Ctr, 71-, Nat Cancer Inst, 72- & Am Cancer Soc, 76-77; prof lectr surg, Georgetown Univ, 70-; mem, Med & Sci Comt, Am Cancer Soc, 70, Surg Training Comt, Nat Inst Gen Med Sci, 71-72, Diag Res Activ Group, Nat Cancer Inst, 72-75, Comt Study Surg Serv US, Am Col Surgeons & Am Surg Asn, 72-74, Tissue & Organ Biol Sect, President's Panel Biomed Res, 75, US Comt & Nat Orgn Comt, 13th Int Cancer Cong, 79 & Nat Cancer Adv Bd, 80; vis prof & lectr, numerous US & foreign univs, 71- *Honors & Awards:* William H Sinkler Mem Award, Nat Med Asn, 72; St George Medal, Am Cancer Soc, 77; Thomas Wyatt Turner Award, 82; Florence Nightingale Award, 82; W Montague Cobb Mem Lectr, Nat Med Asn, 84; James H Jackson Award, 84; Roger L Brooke Distinguished Lectr, Brooke Army Med Ctr, 86; Robert Wilson Kitchen, Jr Award, Am Cancer Soc, 86; James Ewing Lectr & Medal, Soc Surg Oncol, 87. *Mem:* Inst Med-Nat Acad Sci; Soc Surg Oncol (secy, 74-76, pres, 78-79); Am Cancer Soc (pres, 78-79); Am Surg Asn; Am Chem Soc; AMA; fel Am Col Surgeons (secy, 83-). *Res:* Cancer diseases; polyps and cancer of the colorctum; breast cancer; head and neck cancer; soft tissue sarcomas; author of more than 100 technical publications. *Mailing Add:* Dept Surg Suite 4B02 Howard Univ Hosp Washington DC 20060

LEFFEK, KENNETH THOMAS, b Nottingham, Eng, Oct 15, 34; m 58; c 2. PHYSICAL ORGANIC CHEMISTRY. *Educ:* Univ London, BSc, 56, PhD(chem), 59. *Prof Exp:* Nat Res Coun Can fel, 59-61; from asst prof to assoc prof, 61-72, dean grad studies, 72-90, PROF CHEM, DALHOUSIE UNIV, 72- *Concurrent Pos:* Leverhulme vis fel, Univ Kent, Canterbury, 67-68; pres, Atlantic Can Chap, Royal Soc Arts, 88-91. *Mem:* Fel Chem Inst Can (vpres, 85-86, pres, 86-87); Royal Soc Chem; Royal Soc Arts. *Res:* Kinetics and mechanisms of organic reactions; primary and secondary kinetic deuterium isotope effects. *Mailing Add:* Dept Chem Dalhousie Univ Halifax NS B3H 3J5 Can

LEFFEL, CLAUDE SPENCER, JR, b Pearisburg, Va, Dec 21, 21; m 80; c 2. PHYSICS. *Educ:* St John's Col, Md, BA, 43; Johns Hopkins Univ, PhD(physics), 60. *Prof Exp:* Tutor math & physics, St John's Col, Md, 46-50; sr physicist, Appl Physics Lab, Johns Hopkins Univ, 60-83; RETIRED. *Mem:* AAAS; Am Phys Soc; Sigma Xi. *Res:* Plasma physics; nuclear physics; applied physics; atmospheric physics; cryogenics. *Mailing Add:* 838 Glenn Allen Dr Baltimore MD 21229

LEFFEL, ROBERT CECIL, b Woodbine, Md, Apr 26, 25; m 59; c 2. PLANT BREEDING, GENETICS. *Educ:* Univ Md, BS, 48; Iowa State Univ, MS, 50, PhD, 52. *Prof Exp:* Res agronomist, Agr Res Serv, 52-57; assoc prof agron, Univ Md, 57-62; investigations leader, 62-72, chief plant nutrit lab, Agr Res Serv, 72-75, chief, Cell Cult & Nitrogen Fixtion Lab, 75-76; staff scientist oilseed crop prod, Nat Prog Staff, 76-83, RES AGRONOMIST, AGR RES SERV, USDA, 83- *Concurrent Pos:* FAO consult on soybeans, Yugoslavia,

83, India, 87; exec secy, Soybean Res Adv Inst, 82-84. *Mem:* Corp Sci Soc Am; Am Soybean Asn. *Res:* Soybean, forage crop and clover genetics; breeding and production; enhancing nitrogen metabolism of soybean; high protein soybeans. *Mailing Add:* Bldg 011-BARC-W Agr Res Serv USDA Beltsville MD 20705

LEFFELL, W(ILL) O(TIS), b Tazewell Co, Va, Dec 11, 12; m 41; c 2. ELECTRICAL ENGINEERING. *Educ:* Washington & Lee Univ, BS, 34; Univ Tenn MS, 39. *Prof Exp:* Asst, 36-39, from instr to prof, 40-77, EMER PROF ELEC ENG, UNIV TENN, KNOXVILLE, 77- *Mem:* Inst Elec & Electronics Engrs. *Res:* Physics; mathematics; magnetism; power generation; transmission and distribution; electrical apparatus and machinery. *Mailing Add:* 3539 Miser Station Rd Louisville TN 37777

LEFFERT, CHARLES BENJAMIN, b Logansport, Ind, May 22, 22; m 45. ENERGY CONVERSION, CHEMICAL PHYSICS. *Educ:* Purdue Univ, BS, 43; Univ Pittsburgh, MS, 57; Wayne State Univ, PhD(chem eng), 74. *Prof Exp:* Chem engr, Res Dept, Union Oil Co Calif, 43-49; chem engr, Pittsburgh Consol Coal Co, 49-52; asst physics, Univ Pittsburgh, 52-56; sr res physicist, Res Labs, Gen Motors Corp, 57-70; res asst, Res Inst Eng Sci, 70-74, ASSOC PROF CHEM ENG, WAYNE STATE UNIV, 74-, DIR COL ENG ENERGY CTR, 74- *Mem:* Am Phys Soc; Am Inst Chem Engrs; Sigma Xi. *Res:* Chemical engineering. *Mailing Add:* 1302 Wren Wood Troy MI 48084

LEFFERT, HYAM LERNER, b New York, NY, May 11, 44. CELL BIOLOGY. *Educ:* Univ Rochester, BA, 65; Brandeis Univ, MA, 67; Albert Einstein Col Med, MD, 71. *Prof Exp:* Fel, Salk Inst Biol Studies, 71-72; res assoc, 72-73, asst res prof cell biol, 73-80; ASSOC PROF MED, SCH MED, UNIV CALIF, SAN DIEGO, 80- *Concurrent Pos:* Res grant, Nat Cancer Inst, NSF & Diabetes Asn Southern Calif, 74; consult cell biol, Dept Nutrit Path, Mass Inst Technol, 76; Nat Heart & Lung Inst, 73- & Dept Med, Vet Admin Hosp, Dallas, 74-; res grant, Nat Cancer Inst, 76-80, Nat Inst Arthritis, Metab & Digestive Dis, 80- & Nat Inst Alcohol Abuse & Alcoholism, 80- *Mem:* Int Study Group Carcinoembryonic Proteins. *Res:* Mechanism of liver regeneration and differentiation in mammals. *Mailing Add:* Dept Med M-013-H Univ Calif San Diego Med Sch La Jolla CA 92093

LEFFEW, KENNETH W, b Louisville, Ky, Nov 5, 50; m 69; c 3. CHEMICAL ENGINEERING, POLYMER ENGINEERING. *Educ:* Univ Louisville, BS, 73, MEng, 73, PhD(chem eng), 81. *Prof Exp:* RES ASSOC, E I DU PONT DE NEMOURS & CO, 73- *Concurrent Pos:* Adj prof, Univ Louisville, 78-80. *Mem:* Am Inst Chem Engrs. *Res:* Application of advanced process control to new process developments in large industrial projects. *Mailing Add:* Du Pont Exp Sta Bldg 304 Wilmington DE 19898

LEFFINGWELL, JOHN C, b Evanston, Ill, Feb 16, 38; m 60; c 3. ORGANIC CHEMISTRY. *Educ:* Rollins Col, BS, 60; Emory Univ, MS, 62, PhD(org chem), 63. *Prof Exp:* Res assoc org chem, Columbia Univ, 63-64; res chemist, Org Chem Div, Glidden Co, Fla, 64-65 & R J Reynolds Tobacco Co, 65-70; head flavor res, R J Reynolds Industs Inc, 70-73, head flavor develop, R J Reynolds Tobacco Co, 73-75; vpres, Aromatics Int, 75-77; vpres res & develop, Sunkist Soft Drinks, sr vpres, 78-85; VPRES, FRANCHISE BEVERAGE PROD, DEL MONTE CORP, 85-; EXEC VPRES, FOXFIRE FARMS, INC, 86-; PRES, LEFFINGWELL & ASSOCS, 90- *Concurrent Pos:* NIH fel, 63-64; dir, Foxfire Farms, Inc, 85- *Honors & Awards:* Philip Morris Award for Distinguished Achievement in Tobacco Sci, Philip Morris Inc & Tobacco Sci, 74. *Mem:* Am Chem Soc; Royal Soc Chem; NY Acad Sci; Inst Food Technologists; AAAS; Soc Soft Drink Technologists. *Res:* Natural products; flavor chemistry; olfaction; consumer products. *Mailing Add:* Foxfire Farms Inc Rte 1 Box 22 Arborhill Rd Canton GA 30114

LEFFINGWELL, THOMAS PEGG, JR, b San Marcos, Tex, June 27, 26; m 60; c 1. CELL BIOLOGY. *Educ:* Southwest Tex State Col, BS, 48, MA, 49; Univ Tex, Austin, PhD(biol sci), 70. *Prof Exp:* Asst dept head & instr aerospace physiol & radiobiol, USAF Sch Aerospace Med, Gunter AFB, 51-53; res assoc physiol, Univ Tex, 53-55; radiobiologist, Nat Hq, Fed Civil Defense Admin, 55-56; res physiologist & group leader, Radiobiol Lab, Univ Tex, Austin & USAF Sch Aerospace Med, 56-65; RES SCIENTIST, CELL RES INST, UNIV TEX, AUSTIN, 65- *Mem:* AAAS; Electron Micros Soc Am; Am Soc Cell Biologists. *Res:* Biosynthesis and physicochemical characteristics of extracellular protein-polysaccharides; matrix materials, antigenic and other recognition factors on the cell surface; specific receptor sites for biological molecules. *Mailing Add:* 8102 Hillrise Dr Austin TX 78759

LEFFLER, AMOS J, b New York, NY, Sept 9, 24; m 49; c 3. INORGANIC CHEMISTRY, PHYSICAL CHEMISTRY. *Educ:* Brooklyn Col, BS, 49; Univ Chicago, PhD(inorg chem), 53. *Prof Exp:* Chemist, Callery Chem Co, 52-55; res chemist, Stauffer Chem Co, 55-60; res assoc chem, Arthur D Little, Inc, 60-65; assoc prof, 65-75, PROF CHEM, VILLANOVA UNIV, 75- *Concurrent Pos:* Am Inst Chemists Award, Brooklyn Col, 49; sr res assoc, Nat Acad Sci-Nat Res Coun, 73. *Mem:* Am Chem Soc; Royal Soc Chem. *Res:* Inorganic and physical chemistry, especially boron and metallorganic chemistry, catalysis and metal oxides and fluorine chemistry. *Mailing Add:* Dept Chem Villanova Univ Villanova PA 19085

LEFFLER, CHARLES WILLIAM, b Cleveland, Ohio, May 21, 47; m 68; c 1. CARDIOVASCULAR PHYSIOLOGY, PERINATAL PHYSIOLOGY. *Educ:* Univ Miami, BS, 69; Univ Fla, MS, 71, PhD(zool), 74. *Prof Exp:* Teaching asst zool, Univ Fla, 69-73, coun fel, 73-74, fel, 74-76; asst prof physiol & biophys, Univ Louisville, 76-77; from asst prof to assoc prof, 77-86, PROF PHYSIOL & BIOPHYS, HEALTH SCI CTR, UNIV TENN, MEMPHIS, 86-, PROF PEDIAT, 89, DIR, LAB RES NEONATAL PHYSIOL, 90- *Concurrent Pos:* Estab investr, Am Heart Asn, 82-87. *Mem:* Am Physiol Soc; Soc Exp Biol & Med; AAAS; Sigma Xi; Am Heart Asn; Soc Pediat Res. *Res:* Pulmonary hemodynamics; pharmacology; autacoids in control of perinatal circulation; cerebral hemodynamics in the newborn. *Mailing Add:* Dept Physiol & Biophys 894 Union Ave NA427 Memphis TN 38163

LEFFLER, ESTHER BARBARA, b Clearfield, Pa, Feb 1, 25. PHYSICAL CHEMISTRY. *Educ:* Pa State Univ, BS, 45; Univ Va, PhD(chem), 50. *Prof Exp:* Asst chemother, Stanford Res Labs, Am Cyanamid Co, 45-46; instr chem, Randolph-Macon Woman's Col, 49-53; from asst prof to prof chem, Sweet Briar Col, 53-66, chmn dept, 56-59 & 60-66; from asst prof to prof chem, Calif State Polytech Univ, 67-88, actg chmn dept, 73-74, assoc dean, Sch Sci, 78-84; CONSULT, 88- *Concurrent Pos:* Res assoc & vis lectr, Stanford Univ, 66-67; res assoc, Oxford Univ, 74-75; resident dir, Calif State Univ Int Prog in UK, 74-75. *Mem:* Sigma Xi. *Res:* Software development for science education. *Mailing Add:* 19950 Esquiline Ave Walnut CA 91789

LEFFLER, HARRY REX, b Rensselaer, Ind, Sept 20, 42; m 71; c 3. PLANT PHYSIOLOGY, PLANT GENETICS. *Educ:* Iowa State Univ, BS, 64; Purdue Univ, MS, 67, PhD(plant physiol), 70. *Prof Exp:* Res assoc agron, Univ Ill, 70-71; res assoc hort, Purdue Univ, 71-72; PLANT PHYSIOLOGIST, COTTON PHYSIOL & GENETICS UNIT, AGR RES SERV, USDA, 72- *Concurrent Pos:* Assoc ed, Agron J, 82-85, Dekalb-Pfizer, 85-88. *Mem:* Am Soc Agron; Am Soc Plant Physiologists; Crop Sci Soc Am. *Res:* Physiological genetics of seed development. *Mailing Add:* 3100 Sycamore Rd Dekalb IL 60115

LEFFLER, JOHN EDWARD, b Brookline, Mass, Dec 27, 20; m 52. CHEMISTRY. *Educ:* Harvard Univ, BS, 42, PhD(org chem), 48. *Prof Exp:* Res assoc chem, Harvard Univ, 42-44 & Univ Chicago, 44-45; res assoc rocket fuels, USN Proj, Mass Inst Technol, 45-46; du Pont fel, Cornell Univ, 48-49; mem fac, Brown Univ, 49-50; from asst prof to assoc prof, 50-59, PROF CHEM, FLA STATE UNIV, 59- *Mem:* Am Chem Soc; fel AAAS. *Res:* Reaction rate theory; polar and radical reactions; reactive intermediates of organic chemistry; peroxides; reactions of adsorbed organic compounds. *Mailing Add:* Dept Chem Fla State Univ Tallahassee FL 32306

LEFFLER, MARLIN TEMPLETON, b College Corner, Ind, Feb 28, 11; m 33; c 3. ORGANIC CHEMISTRY. *Educ:* Miami Univ, AB, 32; Univ Ill, MA, 33, PhD(org chem), 36. *Prof Exp:* Asst chemist, Univ Ill, 33-35; res chemist, Abbott Labs, Ill, 36-46, head org dept, 46-49, asst dir res, 49-50, assoc dir, 50-57, dir chem & agr res, 57-59, dir res liaison, 59-71; consult, 71-80; RETIRED. *Mem:* Am Chem Soc. *Res:* Local anesthetics; antiseptics; chemotherapy; optical activity of organic deuterium compounds; organic medicinal chemistry; biochemistry; overseas technological advances. *Mailing Add:* 102 Whispering Sands Dr Sarasota-Siesta Key FL 34242

LEFFORD, MAURICE J, b London, Eng, Nov 27, 30; US citizen. IMMUNOLOGY, MICROBIOLOGY. *Educ:* Univ London, Eng, BS, 55. *Prof Exp:* Resident internal med, Royal Victoria Hosp, Bournemouth, 57-59 & path, St Bartholomew's Hosp, London, 59-63; mem, Unit Lab Studies Tuberc, Brit Med Res Coun, Royal Postgrad Med Sch, 63-71; asst mem, Trudeau Inst, NY, 71-75, assoc mem, 75-79; assoc prof, 79-85, PROF, DEPT IMMUNOL & MICROBIOL, SCH MED, WAYNE STATE UNIV, 85- *Concurrent Pos:* Mem, Sci Group Leprosy, Nat Inst Allergy & Infectious Dis, 76, Sci Working Group Immunol Leprosy, WHO, 76-80 & US Tuberculosis Panel, US-Japan Coop Med Sci Prog, 84-88; chmn, Mycobact Div, Am Soc Microbiol, 84. *Mem:* Am Asn Immunologists; Am Soc Microbiol; Int Leprosy Asn. *Mailing Add:* Immunol & Microbiol Dept Wayne State Univ Sch Med 540 E Canfield Ave Detroit MI 48201

LEFKOWITZ, IRVING, b New York, NY, July 8, 21; m 55; c 2. SYSTEMS & CONTROL ENGINEERING. *Educ:* Cooper Union Sch Eng, BChE, 43; Case Inst Technol, MS, 55, PhD(control eng), 58. *Prof Exp:* Instrument engr, Calvert Distilling Co, 44-47; instrument engr, J E Seagram & Sons, Inc, 47-51, head instrumentation res, 51-53; res assoc instrumentation eng, Case Western Reserve Univ, 53-58, from asst prof to assoc prof eng, 58-65, dir, Control Indus Systs Prog, 58-85, chmn, Dept Systs Eng, 70-74 & 80-83, prof systs eng & chmn eng, 65-87, EMER PROF SYSTS ENG, CASE WESTERN RESERVE UNIV, 87- *Concurrent Pos:* NATO fel, 62-63; res fel, Int Inst Appl Systs Anal, 74-75; chmn, systs eng comt, Int Fedn Automatic Control, 78-81; chmn, Control Systs Soc Comt on Indust Systs Control, 84-85. *Honors & Awards:* Control Heritage Award, Am Automatic Control Coun; Inst Elec & Electronics Engrs, Control Systs Soc & Man & Cybernets Soc. *Mem:* fel AAAS; fel Inst Elec & Electronics Engrs; Int Fedn Automatic Control; Sigma Xi. *Res:* Hierarchical computer control; control of industrial processes; energy conservation through integrated systems control. *Mailing Add:* Dept Systs Eng Case Western Reserve Univ University Circle Cleveland OH 44106

LEFKOWITZ, ISSAI, solid state physics; deceased, see previous edition for last biography

LEFKOWITZ, LEWIS BENJAMIN, JR, b Dallas, Tex, Dec 18, 30; m 61; c 3. MEDICINE. *Educ:* Denison Univ, BA, 51; Univ Tex Southwest Med Sch Dallas, MD, 56. *Prof Exp:* USPHS res fel med, Univ Tex Southwest Med Sch Dallas, 59-60, instr, 60-61; USPHS res fel, Univ Ill, 61-62, USPHS trainee infectious dis, 62-65; from asst prof to assoc prof, 65-78, PROF PREV MED, SCH MED, VANDERBILT UNIV, 78-, ASST PROF MED, 71- *Concurrent Pos:* Asst clin prof internal med, Meharry Med Col, 66-78, clin prof family & community med, 78-81, assoc clin prof internal med, 78-85; consult, US Army Hosp, Ft Campbell, Ky, 69-85. *Mem:* Am Col Prev Med; Am Pub Health Asn. *Res:* Epidemiology and pathogenesis of infectious diseases; health care delivery. *Mailing Add:* Dept Prev Med Vanderbilt Univ Sch Med Nashville TN 37232

LEFKOWITZ, ROBERT JOSEPH, b New York, NY, Apr 15, 43; m 63; c 5. MOLECULAR PHARMACOLOGY, MEDICAL SCIENCE. *Educ:* Columbia Univ, BA, 62, MD, 66; Am Bd Internal Med, dipl. *Prof Exp:* From intern to jr asst resident, Columbia Presbyterian Med Ctr, NY, 66-68; clin & res assoc, Nat Inst Arthritis & Metab Dis, 68-70; sr asst resident, Mass Gen Hosp, Harvard Univ, 70-71; fel cardiol, 71-73; assoc prof med & asst prof biochem, 73-77, prof med, Med Ctr, 77-82, JAMES B DUKE PROF MED, DUKE UNIV MED CTR, 82-, PROF BIOCHEM, 85- *Concurrent Pos:* Estab investr, Am Heart Asn, 73-76; investr, Howard Hughes Med Inst, 76- *Honors*

& Awards: Janeway Prize, 66; John J Abel Award, Am Soc Pharmacol & Exp Therapeut, 78; Ernst Oppenheimer Mem Award, Endocrine Soc, 82; Gordon Wilson Medal, Am Clin & Climatol Asn,82; Lita Annenberg Hazen Award, 83; Outstanding Res award, Int Soc Heart Res, 85; Goodman & Gilman Award, Am Soc Pharmacol & Expert Therapeut, 86; Gairdner Found Int Award, 88; Novo Nordisk Biotechnol Award, 90; Res Award, Asn Am Med Col, 90; Basic Res Prize, Am Heart Asn, 90. Mem: Nat Acad Sci; Am Soc Biol Chem; Asn Am Physicians; Am Heart Asn; Am Soc Pharmacol & Exp Therapeut; Am Soc Clin Invest; Am Acad Arts & Sci; Endocrine Soc. Res: Molecular pharmacology of drug and hormone receptors; author of numerous articles. Mailing Add: Med Ctr Duke Univ PO Box 3821 Durham NC 27710

LEFKOWITZ, RUTH SAMSON, b Cincinnati, Ohio, Oct 7, 10; m 40; c 2. MATHEMATICS. Educ: Hunter Col, BA, 30; Columbia Univ, MA, 60, EdD(math educ), 66. Prof Exp: Sec Sch teacher math, New York City Bd Educ, 38-59; from asst prof to assoc prof, Bronx Community Col, 60-67; from assoc prof to prof, 67-76, chmn dept, 73-75, EMER PROF MATH, JOHN JAY COL CRIMINAL JUSTICE, CITY UNIV NEW YORK, 76- Mem: AAAS; Math Asn Am; NY Acad Sci. Res: Impact of the computer on the study of mathematics. Mailing Add: 900 W 190th St New York NY 10040

LEFKOWITZ, STANLEY A, b Philadelphia, Pa, Aug 5, 43; m 82. PHYSICAL INORGANIC CHEMISTRY. Educ: Temple Univ, AB, 65; Princeton Univ, PhD(chem), 70. Prof Exp: Asst to vchancellor Urban Affairs, City Univ New York, 70-73; asst dir instruct develop, Queens Col, 73-75; VPRES, FALAMWOOD CORP & MOCATTA METALS CORP, 75- Concurrent Pos: Environ consult, NY State Temp Comn Powers Local Govt, 72-73; consult, Prof Exam Serv, 74-75 & Guana Island Hotel Corp, 76-; dir, Iron Mountain Depository Corp, 79-87. Mem: Am Phys Soc; Fedn Am Scientists. Res: Alternative techniques for the extraction, refining and analysis of precious metals; the development and design of a solar-wind energy installation on Guana Island in the British Virgin Islands. Mailing Add: 60 E Eighth St Apt 10D New York NY 10003

LEFKOWITZ, STANLEY S, b New York, NY, Nov 26, 33; m 78; c 3. MICROBIOLOGY, VIROLOGY. Educ: Univ Miami, BS, 55, MS, 57; Univ Md, PhD(plant path), 61; Am Bd Med Microbiol, dipl, 74. Prof Exp: Fel viral oncol, Variety Childrens Res Found, 61-64, res assoc, 64-65; from asst prof to assoc prof virol, Med Col Ga, 65-69; assoc prof, 72-78, assoc dean, 75-81, actg chairperson, Microbiol Dept & res coordr, 78-81, PROF VIROL, SCH MED, TEX TECH UNIV, 78- Concurrent Pos: Assoc scientist, Sloan Kettering Inst Cancer Res, 77-81; mem, Clin Cancer Educ Comt, NIH, 78-81. Mem: AAAS; fel Am Acad Microbiol; Am Soc Microbiol; Tissue Cult Asn; Soc Exp Biol & Med; Reticuloendothelial Soc (treas); NY Acad Sci; Am Asn Immunol. Res: Viral oncogenesis including its physical and biological implications; properties of interferon; cancer immunology; effects of highly abused drugs on immunity. Mailing Add: 3801 67th St Lubbock TX 79413

LEFORT, HENRY G(ERARD), b Mineola, NY, Apr 4, 28; m 55; c 3. CEARMICS ENGINEERING. Educ: Clemson Col, BCerE, 52; Univ Ill, MS, 57, PhD(ceramic eng), 60. Prof Exp: Ceramic engr, Nat Bur Standards, 52-55; res assoc ceramic eng, Univ Ill, 55-60; chemist, Lawrence Radiation Lab, Univ Calif, 60-62; ASSOC PROF CERAMIC ENG, CLEMSON UNIV, 62- Mem: Am Ceramic Soc; Nat Inst Ceramic Engrs. Res: Ceramic structural adhesives for high temperature use; ceramic coatings; procelain enamels; nuclear ceramics. Mailing Add: Dept Ceramic Eng Clemson Univ Main Campus Clemson SC 29634

LEFRAK, EDWARD ARTHUR, b Newark, NJ, April 21, 43; m 73; c 5. CARDIAC, VASCULAR & THORACIC SURGERY. Educ: State Univ NY, Buffalo, BA, 65; Ind Univ Sch Med, MD, 69; Am Bd Surg, cert, 76; Am Bd Thoracic Surg, cert, 78. Prof Exp: Intern surg, Baylor Col Med Affil Hosp, Houston, Tex, 69-70; resident gen surg, 70-75; resident cardiopulmonary surg, Univ Ore Med Sch, Portland, 75-77; DIR CARDIAC SURG, FAIRFAX HOSP, VA, 77-, MED DIR CARDIAC TRANSPLANTATION PROG, 86- Concurrent Pos: Asst clin prof surg, Georgetown Univ Sch Med, 78-; mem Coun Cardiovasc Surg, Am Heart Asn, 82. Mem: Fel Am Col Cardiol; fel Am Col Surgeons; AMA; fel Am Col Chest Physicians; fel Int Col Surgeons; Int Soc Heart Transplantation; Am Asn Thoracic Surg. Res: Cardiac & vascular surgery; heart transplantation. Mailing Add: 3301 Woodburn Rd No 301 Annandale VA 22003

LEFRANCOIS, LEO, b Bristol, Conn, Jan 6, 56. EXPERIMENTAL BIOLOGY. Educ: Colo State Univ, BS, 78; Wake Forest Univ, PhD(immunol), 82. Prof Exp: Med technologist, Bristol Hosp, Conn, 78-79; teaching asst, Dept Microbiol & Immunol, Bowman Gray Sch Med, Winston-Salem, NC, 81-82; res assoc, Dept Immunol, Scripps Clin & Res Found, La Jolla, Calif, 82-86; SR RES SCIENTIST III, DEPT CELL BIOL, UPJOHN CO, 90- Concurrent Pos: Assoc ed, J Immunol; adv coun mem, Midwest Autumn Immunol Conf. Mailing Add: Dept Cell Biol Upjohn Co Kalamazoo MI 49001

LEFTIN, HARRY PAUL, b Beverly, Mass, Oct 23, 26; m 54; c 3. PHYSICAL ORGANIC CHEMISTRY, INDUSTRIAL CHEMISTRY. Educ: Boston Univ, AB, 50, PhD(chem), 55. Prof Exp: Res fel, Mellon Inst, 54-59; res chemist, 59-60, supvr chem res, 60-67, sr res assoc, Res & Develop Lab, 67-73, MGR RES, PULLMAN KELLOGG, 73- Concurrent Pos: Instr, Fairleigh Dickinson Univ, 60-75; ed, Catalysis Rev, 67-85; assoc dir res, M W Kellogg Co, 85-86; consult, 86- Mem: Am Chem Soc; Catalysis Soc NAm; Sigma Xi. Res: Heterogeneous catalysis; chemisorption; electronic infrared and nuclear magnetic resonance spectra of molecules in the adsorbed state; petroleum and petrochemical process development; gas phase kinetics. Mailing Add: 2314 Lexford Lane Houston TX 77080

LEFTON, LEW EDWARD, b Albuquerque, NMex, Sept 24, 60; m 88. NONLINEAR DIFFERENTIAL EQUATIONS, NONLINEAR ANALYSIS. Educ: NMex Inst Mining & Technol, BS, 82; Univ Ill, Urbana, MS, 86, PhD(math), 87. Prof Exp: Vis asst prof math, Univ Calif, Riverside,

87-89; ASST PROF MATH, UNIV NEW ORLEANS, 89- Mem: Am Math Soc; Math Asn Am. Res: Existence, multiplicity, and qualitative behavior of solutions to nonlinear ordinary and partial differential equations; applying both numerical techniques and analytic methods from functional analysis; elliptic partial differential equations. Mailing Add: Dept Math Univ New Orleans New Orleans LA 70148

LEFTON, PHYLLIS, b Neptune, NJ, Feb 10, 49; m 79; c 1. MATHEMATICS, NUMBER THEORY. Educ: Barnard Col, BA, 71; Columbia Univ, MA, 72, MPhil & PhD(math), 75; Jewish Theol Sem, BHL, 75. Prof Exp: Teaching asst calculus, Columbia Univ, 70-73; instr math, Belfer Grad Sch & Stern Col, Yeshiva Univ, 75-77; asst prof, 77-82, assoc prof math, 82-87, PROF MATH & COMPUT SCI, MANHATTANVILLE COL, 87- Mem: Am Math Soc; Math Asn Am; Asn Women in Math; Nat Coun Teachers Math. Res: Algebraic number theory; analytic number theory; group representation theory; theory of polynomials and field theory. Mailing Add: Dept Math Manhattanville Col Purchase NY 10577

LEGAL, CASIMER CLAUDIUS, JR, b Farrell, Pa, Feb 3, 15; m 38; c 4. INORGANIC CHEMISTRY. Educ: Thiel Col, BS, 37. Prof Exp: Anal chemist soap & raw mat, Lever Bros Co, 37-42; res chemist & supvr res lab & chem eng dept, Davison Chem Corp, 42-47, supvr res eng, 47-56; supvr agr chem res, W R Grace & Co, 56-65, mgr fertilizer res, 65-74, mgr indust res, 74-78, mgr, Environ Dept, 78-85; RETIRED. Mem: Am Chem Soc; Am Inst Chem Engrs. Res: Fertilizer; superphosphate; wet process phosphoric acid; silica gel; land reclamation. Mailing Add: 2460 S Rifle St Aurora CO 80013

LEGAN, SANDRA JEAN, b Cleveland, Ohio, Sept 10, 46. REPRODUCTIVE PHYSIOLOGY, NEUROENDOCRINOLOGY. Educ: Univ Mich, BS, 67, MS, 70, PhD(physiol), 74. Prof Exp: Lab asst, Geigy Co, Basel, Switz, 67-68; instr physiol, Univ Mich, 71-72, teaching fel, 72-73; NIH fel physiol, Emory Univ, 74-75; NIH fel, Reproductive Endocrinol Prog, Univ Mich, Ann Arbor, 75-77, res assoc, 77-79; asst prof, 79-84, ASSOC PROF PHYSIOL, UNIV KY, LEXINGTON, 84- Mem: Am Physiol Soc; Endocrine Soc; Soc Study Reproduction; AAAS; Soc Study Fertil. Res: Neuroendocrine control of gonadotrophin secretion, specifically how modulations in steroid concentrations, environmental stimuli and neural input are transduced into endocrine events in the hypothalamo-hypophyseal axis. Mailing Add: Dept Physiol Univ Ky Lexington KY 40536-0084

LEGARE, RICHARD J, b Central Falls, RI, Dec 27, 34; m 57; c 2. POLYMER CHEMISTRY, CHEMICAL KINETICS. Educ: Providence Col, BS, 56; Univ Minn, MS, 60, PhD(phys chem), 62. Prof Exp: Sr res chemist, Allegany Ballistics Lab, Md, 62-71; staff scientist, Bacchus Works, 71-76 & Fibers Technol Ctr, Research Triangle Park, NC, 76-78, STAFF SCIENTIST, FIBERS TECHNOL CTR, HERCULES INC, 78- Mem: Am Chem Soc. Res: High speed kinetics; biophysical and polymer chemistry; rocket propellants; high temperature resins; composite materials. Mailing Add: 2619 Country Club Dr Conyers GA 30208

LEGASPI, ADRIAN, b Mexico City, Mex, Jul 13, 52; m 80; c 1. ONCOLOGY. Educ: Army Med Sch, Mex, MD, 76. Prof Exp: Res fel, surgery, Cornell Univ NY Hosp, 83-85; clin fel, surgery oncol, mem Sloan Kettering Cancer Ctr Med Col, 85-87; ASST PROF SURG, UNIV MIAMI SCH MED, 87- Mem: Europ Soc Parenteral & Enteral Nutrit; Asn Acad Sug. Res: Treatment of cancer by means of surgery; nutritional aspects of patient support before and after surgery; injury and hospitalization on protein metabolism. Mailing Add: Dept Surg Univ Miami Sch Med PO Box 016310 Miami FL 33138

LEGATES, JAMES EDWARD, b Milford, Del, Aug 1, 22; m 44; c 4. ANIMAL GENETICS. Educ: Univ Del, BS, 43; Iowa State Col, MS, 47, PhD, 49. Prof Exp: Asst, Iowa State Col, 48-49; from asst prof to prof animal indust, NC State Univ, 49-56, actg head dairy husb sect, 55-58, head animal breeding sect, 58-70, William Neal Reynolds prof animal sci & genetics, 56-86, dean sch agr & life sci, 71-86, EMER WILLIAM NEAL REYNOLDS PROF & DEAN, NC STATE UNIV, 86- Concurrent Pos: Consult agr prog, Rockefeller Found, Colombia, 59; consult, Exp Sta Div, USDA, 59-65; vis prof, Nat Inst Animal Sci, Copenhagen, 63 & State Agr Univ, Wageningen, 71. Honors & Awards: Borden Award, Am Dairy Sci Asn, 67, Award of Honor, 89; J Rockefeller Prentice Animal Breeding & Genetics Award, Am Soc Animal Sci, 77. Mem: AAAS; Biomet Soc; Am Soc Animal Sci; Am Dairy Sci Asn. Res: Selection in dairy cattle; genetics of mastitis resistance; quantitative inheritance in mice. Mailing Add: Rte 3 Box 717 Franklinton NC 27525

LEGAULT, ALBERT, b Hull, Que, June 7, 19; m 57. SYSTEMATIC BOTANY, PHYTOGEOGRAPHY. Educ: Univ Montreal, BA, 48, BPed, 53, BSc, 55, MSc, 55; Yale Univ, MSc, 59. Prof Exp: Teacher biol, Montreal-St Louis Col, Montreal, 50-57; researcher palynol, Serv Biogeog, Prov of Que, 61-62; from asst prof to assoc prof, 62-76, PROF BOT, UNIV SHERBROOKE, 76- Mem: Can Bot Asn; Int Asn Plant Taxon; French-Can Asn Advan Sci. Res: Floristics of southeastern, arctic and subarctic Quebec. Mailing Add: Dept Biol Univ Sherbrooke 2500 Blvd de Universite Sherbrooke PQ J1K 2R1 Can

LEGECKIS, RICHARD VYTAUTAS, b Panevezys, Lithuania, Jan 28, 41; US citizen; m 67; c 2. PHYSICAL OCEANOGRAPHY, REMOTE SENSING. Educ: City Univ New York, BS, 65; Fla Inst Technol, MS, 68; Fla State Univ, PhD(phys oceanog), 74. Prof Exp: Space engr, Grumman Corp, 65-70; res assoc sci, Fla State Univ, 74; assoc oceanog, Nat Res Coun, 74-75; SCIENTIST OCEANOG, NAT ENVIRON SATELLITE SERV, NAT OCEANIC & ATMOSPHERIC ADMIN, 75- Concurrent Pos: Nat Res Coun grant, 74. Mem: Am Geophys Union. Res: Ocean currents and temperature fronts; application of satellite remote sensing to ocean studies. Mailing Add: 15004 Whitegate Rd Silver Spring MD 28904

LEGENDRE, LOUIS, b Montreal, Que, Feb 16, 45; m 67. BIOLOGICAL OCEANOGRAPHY, NUMERICAL ECOLOGY. *Educ:* Univ Montreal, BSc, 67; Dalhousie Univ, PhD(oceanog), 71. *Prof Exp:* NATO fel oceanog, Marine Sta Villefranche-sur-Mer, Univ Paris, 71-73; res assoc, 73-74, asst prof, 74-77, assoc prof, 77-81, PROF OCEANOG, LAVAL UNIV, 81- *Concurrent Pos:* Secy gen interuniv group res oceanog, Que, Laval, McGill & Montreal Univs, 77-79; res coordr, 80-86, vpres, 89-; mem Can nat comt for sci, comt oceanog res, Nat Res Coun Can, 78-79; mem Pop Biol Comt, Nat Sci Eng Res Coun Can, 80-83; mem Comt Perfect, Inst Ocean Paris Monaco, 81-; mem, working group 73, Sci Comt Oceanic Res, 83-87, working group 86, Sci Comt Oceanic Res, 88-; mem, Strategic Panel Oceans, Natural Sci & Eng Res Coun Can, 85-87, chmn, 88-89, group chair Life Sci, 89-; mem, Comt Res Centres, FCAR Found One, 88-89, chair Comt Actions struct, 90-; mem, Int Exec Comt, Group on Aquatic Productivity, 88-, Sci Cultural Coun, French Univ Pac, 90-, Int Sci Comt Mar Sta Ctr, Nat Res Sci France, 90-; mem, bd Environ Policy, Royal Soc, 90- *Honors & Awards:* Leo-Pariseau Award, French-Can Asn Advan Sci, 85. *Mem:* Am Soc Limnol & Oceanog. *Res:* Marine primary production; physiological ecology of photosynthesis in marine phytoplankton; numerical analysis of ecological data sets. *Mailing Add:* Dept Biol Laval Univ Quebec PQ G1K 7P4 Can

LEGENDRE, PIERRE, b Montreal, Que, Oct 5, 46; m 69; c 2. NUMERICAL ECOLOGY, COMMUNITY ECOLOGY. *Educ:* Univ Montreal, BA, 65; McGill Univ, MSc, 69; Univ Colo, PhD(biol), 71. *Prof Exp:* Postdoctoral genetics, Genetiska Institutionen, Lunds Universität, Lund, Sweden, 71-72; res assoc environ sci, Univ Que, Montreal, 72-73, res dir environ sci, 73-80, prof physics, 80; assoc prof, 80-84, PROF BIOL, UNIV MONTREAL, 84- *Concurrent Pos:* Res assoc, Nat Sci & Eng Res Coun, 77-80; expert, Environ Training & Mgt Africa, USA, Workshop in Togo, 83; invited prof, Montpellier, France, 85, Louvain-la-Neuve, Belg, 87, 88 & 89; chmn, Theme 5, ECOTHAU Res Prog, Montpellier, France, 85-90 & Grant Selection Comt Pop Biol-18, Nat Sci & Eng Res Coun, 89-90; dir, Advan Res Workshop Numerical Ecol, France, 86; mem bd dirs, Classification Soc NAm, 86-89; res fel, Killam Prog, Can Coun, 89-91, selection comt mem, 90-93. *Honors & Awards:* Michel-Jurdant Prize Environ Sci, Asn Canadienne-Francaise pour l'Advan des Sci, 86. *Mem:* Classification Soc NAm. *Res:* Mathematically analyzing the organization of ecological communities through space and integrating spatial structures into population and community models in order to increase their predictive power; application to several types of ecosystems (aquatic and terrestrial); author of several textbooks on numerical ecology, that have established the foundations of this new sub-discipline. *Mailing Add:* Dept Biol Sci Univ Montreal CP 6128 Succursale A Montreal PQ H3C 3J7 Can

LEGER, ROBERT M(ARSH), b Foochow, China, June 3, 21; US citizen; m 44, 78; c 3. ELECTRICAL ENGINEERING. *Educ:* Antioch Col, BS, 44; Ill Inst Technol, MS, 50, PhD(elec eng), 55. *Prof Exp:* Final test foreman, Collins Radio Co, 43-44, qual control engr, 44-46; asst elec eng, Ill Inst Technol, 47-48, instr, 48-52, asst prof, 52-53; from electronics engr to mgr info systs, Convair Astronaut Div, Gen Dynamics/Convair, 53-70, eng specialist, Gen Dynamics Electronics, 71-86; RETIRED. *Mem:* Sr mem Inst Elec & Electronics Engrs; Sigma Xi. *Res:* Information handling systems; guidance and tracking systems. *Mailing Add:* 6517 Altair Ct San Diego CA 92120

LEGERTON, CLARENCE W, JR, b Charleston, SC, July 8, 22; m 58; c 3. GASTROENTEROLOGY. *Educ:* Davidson Col, BS, 43; Med Col SC, MD, 46. *Prof Exp:* Instr med, Duke Univ Hosp, 51-53; instr, 56-58, asst, 58-60, from asst prof to assoc prof, 61-70, chief div gastroenterol, 66-70, dir Gastroenterol Div, 70-87, PROF MED, MED UNIV SC, 70- *Concurrent Pos:* Consult, Vet Admin Hosp, Charleston, 66-; hon consult, Addenbrooke's Hosp, Sch Med, Cambridge Univ, Eng, 78; nat liaison comt, Am Gastroenterol Asn, 79-; med adv bd, US Ther Chang Index, 80-; Nat Digestive Dis Adv Bd (NIH), 84-87, chmn, Comt Digestive Dis Res Ctr, NIH, 85-87. *Mem:* Am Gastroenterol Asn; Am Fedn Clin Res; Am Col Physicians (gov, 78-82); Am Col Gastroenterol. *Res:* Mechanism of pain in peptic ulcer; effects of anticholinergic drugs on gastric motility and secretion; medical therapy of acid-peptic disease. *Mailing Add:* Dept Med Med Univ SC Charleston SC 29425

LEGG, DAVID ALAN, b Elwood, Ind, Sept 7, 47; m 79; c 1. MATHEMATICAL ANALYSIS. *Educ:* Purdue Univ, BS, 69, MS, 70, PhD(math), 73. *Prof Exp:* from asst prof to assoc prof, 74-80, PROF MATH, IND UNIV-PURDUE UNIV, FT WAYNE, 84- *Mem:* Am Math Soc; Math Asn Am; Sigma Xi. *Res:* Approximation theory. *Mailing Add:* 17424 Lochner Rd Spencerville IN 46788

LEGG, IVAN, b New York, NY, Oct 15, 37; m 62; c 2. BIOINORGANIC CHEMISTRY. *Educ:* Oberlin Col, BA, 60; Univ Mich, MS, 63, PhD(inorg chem), 65. *Prof Exp:* Res assoc inorg chem, Univ Pittsburgh, 65-66; from asst prof to prof chem, Wash State Univ, 66-78, assoc biochem, 75-78, chmn dept chem, 78-87; DEAN, COL SCI & MATH, AUBURN UNIV, 87- *Concurrent Pos:* NIH spec fel, Harvard Med Sch, 72-73. *Mem:* Am Chem Soc. *Res:* Use of metal ions to probe structure-function relationships in metalloenzymes; development of models for metal ions binding sites in proteins. *Mailing Add:* Ext Col Auburn Univ Auburn AL 36849

LEGG, JAMES C, b Kokomo, Ind, Sept 17, 36; m 73; c 3. ATOMIC PHYSICS, NUCLEAR PHYSICS. *Educ:* Ind Univ, BS, 58; Princeton Univ, MA, 60, PhD(physics), 62. *Prof Exp:* Instr physics, Princeton Univ, 61-62; res assoc, Rice Univ, 62-63, asst prof, 63-67; assoc prof, 67-73, dir, Nuclear Sci Lab, 72-83, PROF PHYSICS, KANS STATE UNIV, 73-, HEAD DEPT, 87- *Mem:* Am Phys Soc; AAAS; Am Asn Physics Teachers. *Res:* Atomic and molecular collisions. *Mailing Add:* Dept Physics Kans State Univ Manhattan KS 66506

LEGG, JOHN WALLIS, b Minter City, Miss, Sept 20, 36; m 56; c 3. PHYSICAL CHEMISTRY. *Educ:* Miss Col, BS, 58; Univ Fla, MS, 60, PhD, 64. *Prof Exp:* Chemist, Shell Oil Co, Tex, 58; asst, Univ Fla, 58-60; asst prof, Miss Col, 60-62; asst, Univ Fla, 62-64, instr, 63; assoc prof, 64-71, PROF CHEM, MISS COL, 71-, HEAD DEPT, 82- *Concurrent Pos:* Vis prof, George Peabody Col, 66. *Mem:* Am Chem Soc; Am Sci Affil. *Res:* Adsorption at solid surfaces and heterogeneous catalysis, specifically reactions over thorium oxide catalysts, primarily of the alcohols; dielectric properties of freon hydrates; coal powders and slurries. *Mailing Add:* Dept Chem Miss Col Clinton MS 39058

LEGG, JOSEPH OGDEN, b Tex, Oct 16, 20; m 44. SOIL SCIENCE. *Educ:* Univ Ark, BS, 50, MS, 51; Univ Md, PhD(soil fertil), 57. *Prof Exp:* Soil scientist, USDA, 51-79; RETIRED. *Concurrent Pos:* USDA exchange scientist to USSR, 63-64; adj prof, Agron Dept, Univ Ark, 80- *Mem:* Fel AAAS; Soil Sci Soc Am; Am Soc Agron; Int Soc Soil Sci; Coun Agr Sci & Technol. *Res:* Nitrogen transformations in soils; biological nitrogen fixation; soil organic matter. *Mailing Add:* 2400 W New Hope Rd Rogers AR 72756

LEGG, KENNETH DEARDORFF, b Ogdensburg, NY, Feb 19, 43; m 67. ANALYTICAL CHEMISTRY. *Educ:* Union Col, BS, 64; Mass Inst Technol, PhD(chem), 69. *Prof Exp:* Vis prof chem, Univ Southern Calif, 74-75; asst prof chem, Calif State Univ, Long Beach, 69-74, assoc prof, 75-78; prin scientist, Instrumentation Lab Inc, 78-81, mgr, anal res, 81-82, dir res, 82-83; DIR RES, ALLIED HEALTH & SCI PRODS, 83- *Mem:* Am Asn Clin Chem; Am Chem Soc; Electrochem Soc. *Res:* Study of fast photophysical processes using laser excitation; biomedical instrumentation; electrogenerated chemiluminescence; ion selective electrodes, amperometric sensors and biomedical instrumentation. *Mailing Add:* 29 Jefferson Rd Wellesley MA 02181

LEGG, MERLE ALAN, b San Francisco, Calif, Feb 26, 26. PATHOLOGY. *Educ:* Univ Puget Sound, BS, 48; McGill Univ, MD, 52; Am Bd Path, dipl, 57. *Prof Exp:* Intern, Michael Reese Hosp, Chicago, 52-53; resident & chief resident path, Mallory Inst Path, Boston City Hosp, 55-57; instr path, Boston Univ Med Sch, 55-57, Harvard Med Sch, 56-57 & 58-66; asst resident & resident path, New Eng Deaconess Hosp, 53-55, clin assoc path, Harvard Med Sch, 66-70, asst clin prof, 70-74, chmn, Dept Path, 75-78 & 79-88, ASSOC PROF PATH, HARVARD MED SCH, NEW ENG DEACONESS HOSP, 74-, EMER CHMN, DEPT PATH, 88- *Concurrent Pos:* Assoc staff pathologist, New Eng Deaconess Hosp, 57-60, staff pathologist, 61-; assoc staff pathologist, New Eng Baptist Hosp, 60-63, staff pathologist, 63-, chmn, Dept Path, 74-76 & 79-88, pres med staff, 80-81 & 82-83; consult-lectr path, US Naval Hosp, Chelsea, 62-74; assoc path, Peter Bent Brigham Hosp, Boston, 73-77; mem, Coun Anat Path, Am Soc Clin Pathologists, 74-80, chmn, 78-80; consult path, Cambridge Hosp, 75-87, Children's Hosp Med Ctr, Boston, 76-, Boston Vet Admin Hosp, 77-, West Roxbury Vet Admin Med Ctr, 81-, Waltham Hosp, 84-87; mem, Comt Anat Path, Col Am Pathologists, 78-80, Cancer Comt, 80-81 & Path Comt, Radiation Ther Oncol Group, 82-90; chief path, Brooks Hosp, Boston, 79-80; assoc staff path, Hahnemann Hosp, Brighton, 81-82; hon prof, Xi'an Med Univ, People's Repub China, 88, hon spec consult, 89; referring consult staff, Champlain Valley Physicians Hosp Med Ctr, Plattsburgh, NY, 90- *Honors & Awards:* Distinguished Serv Award, Am Soc Clin Pathologists, 81, Commissioner's Medal, 88. *Mem:* AMA; AAAS; Am Soc Clin Pathologists; fel Col Am Pathologists; Int Acad Path; Am Asn Pathologists. *Res:* Pancreatic endocrine and exocrine tumors; pathology of diabetes mellitus; pulmonary interstitial disease; definitions of pulmonary tumors and their behavior; biliary tract tumors. *Mailing Add:* Dept Path New Eng Deaconess Hosp 185 Pilgrim Rd Boston MA 02215

LEGG, THOMAS HARRY, b Kamloops, BC, May 4, 29; m 57; c 2. PHYSICS, RADIO ASTRONOMY. *Educ:* Univ BC, BASc, 53; McGill Univ, MSc, 56, PhD(physics), 60. *Prof Exp:* Radar engr, Can Aviation Electronics Ltd, 53-54; sci officer radio physics, Defense Res Bd Can, 56-57; SR RES OFFICER, HERZBERG INST ASTROPHYS, NAT RES COUN CAN, 60- *Mem:* AAAS; Am Astron Soc; Can Asn Physicists; Royal Astron Soc Can; Inst Elec & Electronics Engrs; Sigma Xi. *Res:* Microwave diffraction; electronic circuitry; radio interferometry. *Mailing Add:* Herzberg Inst Astrophys Nat Res Coun Can 100 Sussex Dr Ottawa ON K1A 0R6 Can

LEGGE, NORMAN REGINALD, b Edmonton, Alta, Apr 20, 19; US citizen; m 42; c 5. THERMOPLASTIC ELASTOMERS, SYNTHETIC ELASTOMERS. *Educ:* Univ Alta, BSc, 42, MSc, 43; McGill Univ, PhD(phys chem), 45. *Prof Exp:* Proj leader, Polysar, Sarnia, Can, 45-51; dir res & develop, Ky Synthetic Rubber Corp, 51-55; mgr, Elastomers Res, Shell Develop, 55-61; dir, Elastomers Res Lab, Rubber Div, Shell Chem Co, 61-64, mgr res & develop, 65-69, mgr res & develop, Polumer Div, 69-71, mgr, Shell Chem Co, 72-78; CONSULT POLYMER RES & DEVELOP, 79- *Concurrent Pos:* Tech ed, 85- *Honors & Awards:* Charles Goodyear Medalist, Rubber Div, Am Chem Soc, 87. *Mem:* Fel AAAS; Am Chem Soc; Soc Plastic Engrs. *Res:* Elastomers, thermoplastic elastomers, polymerization systems for polydienes and polystyrenes, elastomer latices; initiating and high energy explosives, technological forecasting. *Mailing Add:* 19 Barkentine Rd Rancho Palos Verdes CA 90274

LEGGE, THOMAS NELSON, b Erie, Pa, Sept 23, 36; m 60; c 2. ZOOLOGY, LIMNOLOGY. *Educ:* Edinboro State Col, BS, 59; Miami Univ, MAT, 62; Univ Vt, PhD(zool), 69. *Prof Exp:* Instr biol, Northwestern Mich Col, 62-64; from asst prof to assoc prof, 67-70, dept chmn, 85-87, PROF BIOL, EDINBORO UNIV PA, 70- *Mem:* Am Soc Limnol & Oceanog; Int Asn Gt Lakes Res; Int Asn Theoret & Appl Limnol. *Res:* Physical and biological limnology, especially the distribution and ecology of calanoid copepods; ecology of small reservoirs. *Mailing Add:* Dept Biol Edinboro Univ Pa Edinboro PA 16444

LEGGETT, JAMES EVERETT, b West Union, WVa, Oct 20, 26; m 48; c 3. PLANT PHYSIOLOGY. *Educ:* Glenville State Col, AB, 49; Univ Md, College Park, MS, 54, PhD(bot), 65. *Prof Exp:* Plant physiologist, USDA, 53-70; From assoc prof to prof agron, Univ KY, 70-80; plant physiologist, USDA, 80-86; CONSULT, 86- *Mem:* AAAS; Am Soc Plant Physiologists; Soil Sci Soc Am; Am Soc Agron; Crop Sci Soc Am. *Res:* Ion transport; application of ion transport mechanism to growth of plants and relationships to other environmental factors. *Mailing Add:* Dept Agron Univ Ky N106D Agr Ctr Lexington KY 40506

LEGGETT, ROBERT DEAN, b Midvale, Ohio, Aug 2, 29; m 51; c 6. METALLURGICAL ENGINEERING. *Educ:* Ohio State Univ, BMetE & MSc, 52; Carnegie Inst Technol, PhD(metall eng), 59. *Prof Exp:* Engr, Bettis Lab, Westinghouse Elec Corp, 52-55; res asst, Metals Res Lab, Carnegie Inst Technol, 58-59; sr engr, Hanford Labs, Gen Elec Co, 59-64, tech specialist, 64-65; res assoc irradiation effects in metals, Pac Northwest Labs, Battelle Mem Inst, 65-70 & WADCO Corp, 70-71; res assoc, Westinghouse-Hanford Co, 72-76, mgr, LMR fuel develop, 76-88, LMR Progs, 88-89; RETIRED. *Concurrent Pos:* Pvt consult, 89- *Mem:* Am Soc Metals; fel Am Nuclear Soc. *Res:* Basic mechanisms of irradiation behavior of materials, especially fissionable metals; corrosion of single crystals and bi crystals of stainless steel; hot water corrosion of uranium base alloys and stainless steel; irradiation behavior of mixed oxide; stainless steel clad fuel elements. *Mailing Add:* 2113 Harris Ave Richland WA 99352

LEGGETT, WILLIAM C, b Orangeville, Ont, June 25, 39; m 64; c 2. FISH ECOLOGY, POPULATION DYNAMICS. *Educ:* Waterloo Univ Col, BA, 62; Univ Waterloo, MSc, 65; McGill Univ, PhD(zool), 69. *Prof Exp:* Res scientist fisheries, Essex Marine Lab, 65-70, res assoc, 70-78; from asst prof to assoc prof, 70-79, chmn dept, 81-85, PROF BIOL, MCGILL UNIV, 79-, DEAN FAC SCI, 86- *Concurrent Pos:* Pres, Group Interuniv Oceanog Res Que, 86-; assoc ed, J Am Fisheries Soc, 76-78 & Am J Fisheries Aquatic Sci, 80-85; mem comt prof cert, Am Fisheries Soc, 76-78; mem bd dirs, Memphremagog Conserv Inc, 76-80; mem grants adv comt, Can Nat Sportsmans Fund, 77-81; mem, grant selection comt pop biol, Natural Sci & Eng Res Coun Can, 78-81, chmn 81-82, mem, grant selection comt oceans, 84-86, chmn, 85-86; pres & chmn bd, Huntsman Marine Lab, 80-83, chmn bd, 86- *Honors & Awards:* Stevenson Lectr, 87; Fry Medal, Can Soc Zoologists, 90; Award of Excellence for Fisheries Educ, Am Fisheries Soc, 90. *Mem:* Can Soc Zoologists; Am Fisheries Soc; Am Soc Limnol & Oceanog; Am Soc Naturalists. *Res:* Life history strategies in fishes; reproductive ecology of fish; environmental regulation of migration in fish; larval fish ecology; regulation of mortality in fish; lake ecosystem ecology. *Mailing Add:* Dept Biol McGill Univ 1205 Ave Dr Penfield Montreal PQ H3A 1B1 Can

LEGLER, DONALD WAYNE, b Minneapolis, Minn, Oct 2, 31; m 57; c 4. IMMUNOLOGY, PHYSIOLOGY. *Educ:* Univ Minn, BS, 54, DDS, 56; Univ Ala, PhD(physiol), 66. *Prof Exp:* From instr pedodont to assoc prof oral biol, Sch Dent, Univ Ala, Birmingham, 63-71,asst dean, 71-74, prof oral biol & chmn dept, 71-80, asst dean admin affairs, 74-80; assoc dean, advan educ & res, Sch Dent, Univ Minn, 80-83; DEAN, COL DENT, UNIV FLA, 83- *Concurrent Pos:* NIH trainee, 62-66; Swed Med Res Coun fel, 67-68. *Mem:* Am Dent Asn; fel Am Col Dent; Am Soc Microbiol. *Res:* Comparative immunology; germ free research; preventive dentistry. *Mailing Add:* Dean Dent Univ Fla Health Sci Ctr Gainesville FL 32610

LEGLER, JOHN MARSHALL, b Minneapolis, Minn, Sept 9, 30; m 52; c 3. ZOOLOGY. *Educ:* Gustavus Adolphus Col, BA, 53; Univ Kans, PhD(zool), 59. *Prof Exp:* Asst human anat, Gustavus Adolphus Col, 52-53; asst zool, Univ Kans, 53-57, asst cur herpet, Mus Natural Hist, 55-59, asst instr zool, 58; res asst physiol execise lab, Univ Kans, 58-59, asst instr herpet, 59; vis prof zool, Univ New Eng, Armidale, NSW, Australia, 72-74, 76-77 & 80; from asst prof to assoc prof, 59-69, PROF ZOOL, UNIV UTAH, 69-, CUR HERPET, 59-; CUR REPTILES & AMPHIBIANS, UTAH MUS NATURAL HIST, 69- *Concurrent Pos:* Var individual res grants, 59-; res assoc, Gorgas Mem Lab, Panama, 64- & Los Angeles County Mus Natural Hist, 75- *Mem:* Soc Study Evolution; Am Soc Ichthyologists & Herpetologists; fel Herpetologists League (pres, 68-70); Sigma Xi; Brit Herpet Soc. *Res:* Herpetology; the biology of chelonians, the turtles of Middle America and Australia; biosystematics; evolution; ecology and morphology. *Mailing Add:* Dept Biol Univ Utah 201 S Biol Bldg Salt Lake City UT 84112

LEGLER, WARREN KARL, b Hiawatha, Kans, Apr 28, 30; m 52; c 3. COMPUTER SCIENCE, ELECTRICAL ENGINEERING. *Educ:* Univ Kans, BS, 52, PhD(elec eng), 69; Mass Inst Technol, MS, 60. *Prof Exp:* Physicist, US Naval Ord Test Sta, 52-63; instr elec eng, Med Ctr, Univ Kans, 63-68, instr comput sci, 68-70, asst prof physiol, 70-80; SYSTS ANALYST, DIT-MCO INT, 80- *Mem:* Inst Elec & Electronics Engrs; Asn Comput Mach; Sigma Xi. *Res:* Application of computers to medical research, practice and teaching. *Mailing Add:* 1630 Illinois St Lawrence KS 66044

LEGNER, E FRED, b Chicago, Ill, Oct 17, 32; m 60; c 1. ENTOMOLOGY, ECOLOGY. *Educ:* Univ Ill, Urbana, BS, 54; Utah State Univ, MS, 58; Univ Wis, PhD(entom), 61. *Prof Exp:* Asst entom, Univ Wis, 61-62; from asst entomologist to assoc entomologist, 62-75, ENTOMOLOGIST, UNIV CALIF, RIVERSIDE, 75-, ASSOC PROF ENTOM, 70-, PROF BIOL CONTROL, 73- *Concurrent Pos:* Consult, Africa, Australasia, SAm, Mid-E, Micronesia, WI & Europe, 62, 63 & 65-75; USPHS grants, 64-70 & NSF, 72-74. *Mem:* Entom Soc Am; Int Orgn Biol Control; Entom Soc Can; Am Mosquito Control Asn; Sigma Xi. *Res:* Population dynamics of arthropods and their biological control; behavior of parasitic hymenoptera. *Mailing Add:* Div Biol Control Univ Calif Riverside CA 92521

LEGOFF, EUGENE, b Passaic, NJ, Aug 18, 34; m 60; c 2. ORGANIC CHEMISTRY. *Educ:* Rutgers Univ, BS, 56; Cornell Univ, PhD(org chem), 59. *Prof Exp:* Fel, Harvard Univ, 59-60; fel org chem, Mellon Inst, 60-65; assoc prof, 65-78, PROF ORG CHEM, MICH STATE UNIV, 78- *Mem:* Am Chem Soc. *Res:* Synthesis of pseudoaromatics, non-benzenoid aromatics, heteroannulenes organic conductors, porphyrins new synthetic methods. *Mailing Add:* Dept Chem Mich State Univ East Lansing MI 48824

LEGRAND, DONALD GEORGE, b Springfield, Mass, Apr 3, 30; m 51; c 4. PHYSICAL CHEMISTRY. *Educ:* Boston Univ, BA, 52; Univ Mass, PhD(chem), 59. *Prof Exp:* Res chemist, Mallinckrodt Chem Works, 52; asst prof chem, Univ Mass, 58-59; RES CHEMIST, RES LABS, GEN ELEC CO, 59- *Mem:* Am Chem Soc; Am Phys Soc; Soc Rheol. *Res:* Polymer physics; surface physics; rheo-optics. *Mailing Add:* Gen Elec Res Lab PO Box 1088 Schenectady NY 12301

LEGRAND, FRANK EDWARD, b Mayfield, Okla, Dec 18, 26; m 49; c 5. GENETICS, ECOLOGY. *Educ:* Okla State Univ, BS, 59; NDak State Univ, PhD(plant breeding), 63. *Prof Exp:* Exten agronomist, 63-79, PROF AGRON, OKLA STATE UNIV, 74-, DIR, OKLA PEDIGREED SEED SERV, 79- *Mem:* Am Soc Agron. *Res:* Genetic studies of wheat in relation to the inheritance of several quantitative and qualitative characters. *Mailing Add:* Dept Agron Rm 368 North Agron Hall Okla State Univ Stillwater OK 74078-0507

LEGRAND, HARRY E, b Concord, NC, May 19, 17; m 45; c 2. HYDROGEOLOGY. *Educ:* Univ NC, BS, 38. *Prof Exp:* Geol aide, US Geol Surv, 38-40, geologist, Ground Water Br, 46-49, dist geologist, 49-56, consult geologist, 56-59, res geologist, 59-60, chief radiohydrol sect, 60-62, res geologist, 62-74; CONSULT HYDROLOGIST, 74- *Mem:* AAAS; Geol Soc Am; Am Inst Prof Geologists; Am Geophys Union; Am Water Works Asn; Nat Water Well Asn. *Res:* Contamination and geochemistry of ground water; ground water geology; ground water in igneous and metamorphic rocks; pollution and ground waste disposal. *Mailing Add:* 331 Yadkin Dr Raleigh NC 27609

LEGROW, GARY EDWARD, b Toronto, Ont, Mar 9, 38; m 63; c 3. CHEMISTRY. *Educ:* Univ Toronto, BA, 60, MA, 62, PhD(organosilicon chem), 64. *Prof Exp:* Res assoc metall organosiloxanes, dept chem, Univ Sussex, 64-65; res chemist, 65-68, group leader organo-functional silicon chem, 68-70, group leader resins & chem res, 70-73, sr group leader resins res, 73-77, assoc res scientist resins res, 77-80, assoc res scientist basic mat res, 80-83, assoc res scientist, 83-85, RES SCIENTIST ADVAN CERAMICS, DOW CORNING CORP, 85- *Concurrent Pos:* Lectr, Mich State Univ, 66-68. *Mem:* Am Chem Soc; Sigma Xi. *Res:* Synthesis, kinetics and molecular rearrangements of organo-functional silanes; influence of the proximity of silicon on the reactivity of organic functions; silicon resin process; computer modelling of organosilicon reactions. *Mailing Add:* 1213 Wildwood Midland MI 48640-3105

LEGTERS, LLEWELLYN J, b Clymer, NY, May 23, 32; m 56; c 2. PREVENTIVE MEDICINE, TROPICAL PUBLIC HEALTH. *Educ:* Univ Buffalo, BA & MD, 56; Harvard Univ, MPH, 61. *Prof Exp:* Rotating intern, Akron Gen Hosp, Ohio, 56-57; surgeon, 82nd Airborne Div, US Army, 57-58, 504th Infantry, Ger, 58-59, prev med officer, 8th Infantry Div, Ger, 59-60 & John F Kennedy Ctr Mil Assistance, 63-66, chief, Walter Reed Amy Inst Res Field Epidemiol Surv Team, Vietnam, 66-68, prev med officer, US Army Training Ctr, Ft Ord, Calif, 68-70, command & gen staff col, Ft Leavenworth, Kans, 70-71, chief volar eval group, Training Ctr, Infantry & Ft Ord, 71-72, chief ambulatory health serv, Silas B Hays Army Hosp, Ft Ord, 72-74, US Army War Col, Carlisle Barracks, Pa, 74-75, chief health & environ div, Off Surg Gen, 75-77, comdr surg & US Army Med Dept Activity, XVIII Airborne Corps, Ft Bragg, NC, 77-78; sr med consult, Enviro Control Inc, 79-80; PROF & CHMN DEPT PREV MED & BIOMED, UNIFORMED SERV UNIV HEALTH SCI, BETHESDA, MD, 80- *Concurrent Pos:* La State Univ fel trop med & parasitol, Cent Am, 63. *Mem:* AAAS; fel Am Col Prev Med; Am Soc Trop Med & Hyg; Am Pub Health Asn; NY Acad Sci. *Res:* Epidemiology of infectious diseases, especially malaria, other tropical infectious diseases. *Mailing Add:* 10301 Grosvenor Pl No 1410 Rockville MD 20852

LEHAN, FRANK W(ELBORN), b Los Angeles, Calif, Jan 26, 23; m 44; c 1. ELECTRICAL ENGINEERING. *Educ:* Calif Inst Technol, BSEE, 44. *Prof Exp:* Chief telemetry sect, Jet Propulsion Lab, Calif Inst Technol, 44-49, telecommun sect, 49-52 & electronics res, 53-54; sr staff mem, Ramo-Wooldridge Corp, 54-56, assoc dir electronics res & develop staff, Guided Missile Res Div, 56-58; exec vpres & pres, Space Gen Corp, Calif, 58-66; consult, 66-67; asst secy res & technol, Dept Transp, Washington, DC, 67-69; dir, Syst Develop Corp, 71-83; RETIRED. *Concurrent Pos:* Mem res & develop bd, Dept Defense, 48-52, Gov Coun Ocean Resources, 65-66, President's Health Manpower Comt, 65-67 & naval warfare panel, President's Sci Adv Group, 65-; consult, 69-; chmn elec intel panel, Defense Sci Bd, 70-71. *Mem:* Nat Acad Eng; fel Inst Elec & Electronics Engrs. *Res:* Technical management. *Mailing Add:* 1696 E Valley Rd Santa Barbara CA 93108

LEHENY, ROBERT FRANCIS, b New York, Ny, Dec 8, 38; m 62; c 2. APPLIED PHYSICS. *Educ:* Univ Conn, BS, 60; Columbia Univ, MS, 63, DrEngrSc, 66. *Prof Exp:* Engr electronic systs, Sperry Gyroscope Co, 60-61; res asst, Columbia Univ, 62-66, asst prof, 66-67; mem tech staff semiconductor res & optical properties semiconductors, Bell Labs, Inc, 67-84; dist res mgr, High Speed Device Res Group, 84-87, DIV MGR, ELECTRONIC SCI & TECHNOL RES, BELL COMMUN RES, INC, 87- *Honors & Awards:* LEOS Traveling Lectr Inst Elec & Electronics Engrs , 87-88. *Mem:* AAAS; Am Phys Soc; fel Inst Elec & Electronics Engrs, 91; Sigma Xi; NY Acad Sci; Optical Soc Am. *Res:* Optical properties of semiconductors; plasma physics; electromagnetic radiation; carrier transport in semiconductor; optoelectronic device research. *Mailing Add:* 176 Fox Hill Dr Little Silver NJ 07739

LEHISTE, ILSE, b Tallinn, Estonia, Jan 31, 22. SPEECH PERCEPTION, GENERAL PHONETICS. *Educ:* Univ Hamburg, Ger, PhD, 48; Univ Mich, Ann Arbor, PhD(ling), 59; Univ Lund, Sweden, Phd(philos),82. *Hon Degrees:* Dr, Univ Essex, Eng, 77. *Prof Exp:* Lectr ling, Univ Hamburg, Ger, 48-49; assoc prof, Kans Wesleyan Univ, Salina, 50-51 & Detroit Inst Technol, 51-56; res assoc phonetics, Commun Sci Lab, Univ Mich, 57-63; assoc prof slavic ling, 63-65, chmn, 65-71, prof ling, 65-87, chmn ling, 65-71, EMER PROF LING, OHIO STATE UNIV, 87- *Concurrent Pos:* Vis prof ling, Univ

Cologne, 65, Univ Calif, Los Angeles, 66, Univ Vienna, Austria,74 & tokyo Univ, 80; Guggenheim fel, 69-70 & 75-76. *Mem:* Ling Soc Am (pres, 80); fel Acoust Soc Am; Int Soc Phonetic Sci; Asn Advan Baltic Studies (pres, 74-76); Mod Lang Asn; fel, Japan Soc Prom Sci. *Res:* Acoustic analysis of spoken language; acoustic manifestation of syntactic structure; suprasegmental structure of various languages. *Mailing Add:* Dept Ling Ohio State Univ 1841 Millikin Rd Columbus OH 43210

LEHMAN, ALFRED BAKER, b Cleveland, Ohio, Mar 21, 31. MATHEMATICS. *Educ:* Ohio Univ, BS, 50; Univ Fla, PhD(math), 54. *Prof Exp:* Instr math, Tulane Univ, 54; mem staff, Acoust Lab & Res Lab Electronics, Mass Inst Technol, 55-57; asst prof math, Case Inst Technol, 57-61; vis mem, Math Res Ctr, Univ Wis, 61-63; res assoc, Rensselaer Polytech Inst, 63; res mathematician, Walter Reed Army Inst Res, 64-67; PROF MATH & COMP SCI, UNIV TORONTO, 67- *Concurrent Pos:* Vis prof, Univ Toronto, 65-67. *Mem:* Math Asn Am; Soc Indust & Appl Math. *Res:* Combinatorics. *Mailing Add:* Dept Math Univ Toronto Toronto ON M5S 1A1 Can

LEHMAN, AUGUST F(ERDINAND), b Waukesha, Wis, July 10, 24; c 3. HYDRODYNAMICS, FLUID MECHANICS. *Educ:* Agr & Mech Col, Tex, BS, 50; Pa State Univ, MS, 54. *Prof Exp:* Engr, Ord Res Lab, Pa State Univ, 51-53, group leader, 53-54, proj leader, 54-62, asst prof, 55-59, assoc prof, 59-62; head water tunnel div, Oceanics, Inc, 62-77; PRES, A F LEHMAN ASSOCS INC, 77- *Mem:* Sigma Xi. *Res:* Hydrodynamics and low speed aerodynamics; marine propulsion; flow visualization; cavitation; tip vortices; drag reduction: polymers, suction, heating, body shape; design and construction of water and wind tunnels and special instrumentation and equipment. *Mailing Add:* A F Lehman Assocs Inc PO Box 27 Centerport NY 11721-0027

LEHMAN, DAVID HERSHEY, b Lancaster, Pa; m; c 1. GEOLOGY, PETROLEUM GEOCHEMISTRY. *Educ:* Franklin & Marshall Col, AB, 68; Univ Tex, Austin, PhD(geol), 74. *Prof Exp:* Res geologist, Exxon Co, 74-77, exploration geologist, 77-78, prod geologist, 79; proj supvry geologist, North Alaska, 80-82; dist prod geologist, Kingsville Dist, 83-85; opers geol mgr, 85-86, opers tech mgr, STex, 87-88. *Mem:* Am Asn Petrol Geol; Geol Soc Am. *Res:* Structural geology; tectonics; radiometric age determinations; petroleum geochemistry; petroleum geology; regional geology. *Mailing Add:* 5110 Walnut Hills Kingwood TX 77345

LEHMAN, DENNIS DALE, b Youngstown, Ohio, July 14, 45; div; c 2. CHEMICAL INSTRUMENTATION, PHYSIOLOGICAL CHEMISTRY. *Educ:* Ohio State Univ, BSc, 67; Northwestern Univ, MS, 68, PhD(chem), 73. *Prof Exp:* From asst prof to assoc prof, 68-81, PROF CHEM, LOOP COL, 81- *Concurrent Pos:* Vis scholar, Dept Chem, Northwestern Univ, 73-, vis assoc prof, 77-81, vis prof, 81- & lectr, Med Sch, 81-89. *Mem:* Am Chem Soc; Sigma Xi; AAAS. *Res:* The use of organometallic complexes as catalyst for a variety of inorganic reactions; the reaction of transition metal complexes with small molecules and structural studies on the resulting products. *Mailing Add:* Dept Chem Northwestern Univ Evanston IL 60208

LEHMAN, DONALD RICHARD, b York, Pa, Dec 13, 40; m 62. THEORETICAL NUCLEAR PHYSICS. *Educ:* Rutgers Univ, BA, 62; Air Force Inst Technol, MS, 64; George Washington Univ, PhD(physics), 70. *Prof Exp:* Proj scientist nuclear physics, Air Force Off Sci Res, 64-68; instr physics, George Washington Univ, 69-70; Nat Acad Sci-Nat Res Coun res assoc nuclear physics, Nat Bur Standards, 70-72; from asst prof to PROF PHYSICS, GEORGE WASHINGTON UNIV, 82-, DEPT CHMN, 87-, DIR, CTR NUCLEAR STUDIES, 90- *Concurrent Pos:* Guest worker, Nat Bur Standards, 72-89; vis staff mem & collab, Los Alamos Nat Lab, 74-; co-prin investr, Energy Dept Contract/Grant, George Washington Univ, 79-86, prin investr, 86- *Mem:* Fel Am Phys Soc. *Res:* Nuclear few-body problem; photonuclear physics; intermediate energy physics; hypernuclei; scattering theory. *Mailing Add:* Dept Physics George Washington Univ Washington DC 20052

LEHMAN, DUANE STANLEY, b Berne, Ind, Jan 18, 32; m 55; c 3. INORGANIC CHEMISTRY. *Educ:* Wheaton Col, Ill, BS, 54; Ind Univ, PhD(chem), 59. *Prof Exp:* Res chemist, 58-65, proj leader, Chem Dept Res Lab, 65-67, group leader, Chem Eng Lab, lab dir, chem eng lab, Dow Chem Co, 71-76; res mgr high impact polystyrene, 76-78, DIR, RES & DEVELOP RECRUITING & RES DEVELOP, 78- *Mem:* Am Chem Soc; Sigma Xi. *Res:* Coordination chemistry; brine chemistry; inorganic process research; basic refractories; new product development. *Mailing Add:* 704 Linwood Dr Midland MI 48640

LEHMAN, ERNEST DALE, b Woodward, Okla, Mar 2, 42; m 65; c 3. BIOCHEMISTRY. *Educ:* Northwestern State Col, Okla, BS, 65; Okla State Univ, PhD(biochem), 71. *Prof Exp:* Res assoc biochem, Okla State Univ, 71-72; NIH fel, Case Western Reserve Univ, 72-74; sr res biochem, 74-78, res fel, 78-87, SR RES FEL, DEPT VIRUS & CELL BIOL RES, MERCK, SHARP & DOHME RES LABS, 87- *Mem:* AAAS; Soc Complex Carbohydrates. *Res:* Biochemistry and function of glycoproteins; isolation and identification of bacterial and viral antigens. *Mailing Add:* Dept Virus & Cel Biol Res Merck Sharp & Dohme Res Labs West Point PA 19486

LEHMAN, EUGENE H, b New York, NY, Jan 26, 13; m 61; c 4. MATHEMATICAL STATISTICS. *Educ:* Yale Univ, BA, 33; Columbia Univ, MA, 37; NC State Univ, PhD(math statist), 61. *Prof Exp:* Res assoc math, Univ Alaska, 49-51; asst prof, Univ Fla, 55-57 & Univ San Diego, 57-58; consult statistician, Los Angeles, 61-64; consult biostatistician, Cedars of Lebanon Hosp, 64-66; assoc prof math, Northern Mich Univ, 66-69; prof math, Mo Southern Col, 69-70; prof statist, Univ Que, Trois-Rivieres, 70-76; prof math, Univ Nat du Rwanda, 76-78; prof statist, Concordia Univ, Montreal, 78-79; PROF MATH, CÉGEPS DU QUÉBEC, 81- *Concurrent Pos:* Corresp abstractor, Math Rev, 60-; referee, La Rev Can de Statist; teaching Dumai, Indonesia, 83-84. *Mem:* Am Math Soc; Am Statist Asn; Asn Can-French Advan Sci; Soc Statist Can; World Coun Gifted & Talented Children. *Res:* Children in mathematics. *Mailing Add:* Four Viburnum Ave Pointe Claire PQ H9R 5A7 Can

LEHMAN, GRACE CHURCH, b Mt Holly, NJ, June 10, 41. ZOOLOGY, ENDOCRINOLOGY. *Educ:* Drew Univ, AB, 63; Ind Univ, Bloomington, PhD(zool), 67. *Prof Exp:* USPHS fel, Univ Mich, Ann Arbor, 67 & 68-70, univ fel, 70-71, res assoc, 71-74, RES INVESTR ZOOL, UNIV MICH, ANN ARBOR, 74- *Mem:* Am Soc Zool. *Res:* Endocrine interactions; influence of thyroid activity on reproduction; comparative and developmental endocrinology; reproductive biology of the amphibia. *Mailing Add:* 400 Maynard Ann Arbor MI 48104

LEHMAN, GUY WALTER, b Walkerton, Ind, Sept 21, 23. THEORETICAL PHYSICS. *Educ:* Purdue Univ, BSEE, 48, MS, 50, PhD(physics), 54. *Prof Exp:* Jr engr electronics, Eastman Kodak Co, 48; asst physicist, Cornell Aeronaut Lab, 51; asst physics, Purdue Univ, 51-54; res specialist, Res Dept, Atomics Int Div, NAm Aviation, Inc, 54-62, group leader theoret physics, Sci Ctr, 63-67, mem tech staff, 67-70; PROF PHYSICS, UNIV KY, 70- *Mem:* Fel Am Phys Soc. *Res:* Solid state; mathematical physics; electronic structure; statistical mechanics; electromagnetic theory; lattice dynamics. *Mailing Add:* Dept Physics Univ Ky Lexington KY 40506

LEHMAN, HARVEY EUGENE, b Yuhsien, China; US citizen; m 58. DEVELOPMENTAL BIOLOGY, EMBRYOLOGY. *Educ:* Maryville Col, Tenn, BA, 41; Univ NC, MA, 44; Stanford Univ, PhD(embryol), 48. *Prof Exp:* From asst prof to assoc prof, 48-59, chmn dept, 62-67, PROF ZOOL, UNIV NC, 59-, CHMN DEPT, 76- *Concurrent Pos:* Fel, Univ Berne, 52-53; chg exp embryol course, Bermuda Biol Sta, 60-75; vis prof zool, Univ Vienna, 76. *Mem:* AAAS; Soc Develop Biol; Am Soc Zoologists; Am Micros Soc; Am Soc Cell Biol. *Res:* Rhabdocoele parasitology; amphibian pigmentation; nuclear transplantation in Triton; hybridization in Echinoderms; tissue culture; cell migration and differentiation of the neural crest; invertebrate larvae and metamorphosis; cytochemistry of embryonic differentiation. *Mailing Add:* Dept Zool Univ NC Chapel Hill NC 27599-3280

LEHMAN, HUGH ROBERTS, b Ft Leavenworth, Kans, Jan 25, 21; m 45; c 3. CHEMICAL ENGINEERING, PHYSICS. *Educ:* The Citadel, BS, 41; Ohio State Univ, MSc, 47; Univ Calif, Berkeley, PhD(chem eng), 51. *Prof Exp:* Proj officer, Power Plant Lab, Wright Field, US Air Force, 45-46, Air Proving Ground, Fla, 47-48 & Armed Forces Spec Weapons Proj, 51-52; staff mem nuclear weapons, Los Alamos Sci Lab, 52-55; dep chief anal div, Air Force Spec Weapons Ctr, 55-58; br chief, Air Force Intel Ctr, 58-62; staff asst nuclear disarmament, US AEC, 62-64; STAFF MEM NUCLEAR WEAPONS, LOS ALAMOS NAT LAB, 64- *Concurrent Pos:* Mem consult panel, Ballistics Missile Re-entry Systs, 66-68 & defense technol steering group, AEC. *Mem:* Am Nuclear Soc; Am Inst Aeronaut & Astronaut. *Res:* Employment and effects of nuclear weapons; vulnerability of targets, nuclear and conventional. *Mailing Add:* Los Alamos Nat Lab PO Box 1663 Los Alamos NM 87545

LEHMAN, ISRAEL ROBERT, b Tauroggen, Lithuania, Oct 5, 24; US citizen; m 59; c 3. BIOCHEMISTRY. *Educ:* Johns Hopkins Univ, AB, 50, PhD(biochem), 54. *Hon Degrees:* MD, Univ Gothenberg, 87. *Prof Exp:* Am Cancer Soc fel, 55-57; instr microbiol, Wash Univ, 57-59; from asst prof to assoc prof, 59-66, chmn dept, 74-79 & 84-86, PROF BIOCHEM, SCH MED, STANFORD UNIV, 66- *Concurrent Pos:* Assoc ed, J Biol Chem, 71-74, 81-83, 89-; William Hume prof, Sch Med, Stanford Univ, 79. *Mem:* Nat Acad Sci; fel Am Acad Arts & Sci; Am Soc Biol Chem. *Res:* Nucleic acid metabolism; biochemistry of virus infection. *Mailing Add:* Dept Biochem Sch Med Stanford Univ Stanford CA 94305

LEHMAN, JOE JUNIOR, b Versailles, Mo, July 1, 21; m 43; c 4. ORGANIC CHEMISTRY. *Educ:* Bethel Col, Kans, AB, 43; Wash State Univ, MS, 47, PhD, 49. *Prof Exp:* From instr to assoc prof, 49-64, PROF CHEM, COLO STATE UNIV, 63- *Concurrent Pos:* Res fel, Midwest Res Inst, 58-59; vis prof, US Naval Acad, 61-62. *Mem:* AAAS; Am Chem Soc; Am Soc Microbiol. *Res:* Organic synthesis; modification of compounds by microorganisms; steric acceleration of hydrolytic reactions. *Mailing Add:* Dept Chem Colo State Univ Ft Collins CO 80523

LEHMAN, JOHN MICHAEL, b Abington, Pa, June 19, 42; m; c 2. EXPERIMENTAL PATHOLOGY, VIROLOGY. *Educ:* Philadelphia Col Pharm & Sci, BS, 64; Univ Pa, PhD(path), 70. *Prof Exp:* NIH fel, Wistar Inst Anat & Biol, 70; NIH fel, 70-71, from instr to assoc prof, 71-80, prof path, Med Sch, Univ Colo Med Ctr, Denver; PROF, DEPT MICROBIOL & IMMUNOL, ALBANY MED COL. *Concurrent Pos:* Vis staff mem, Los Alamos Nat Lab, 72- *Mem:* Am Soc Microbiol; Tissue Cult Asn; Am Asn Cancer Res; Am Asn Exp Path; Am Soc Cell Biol. *Res:* Tumor biology and virus transformation with oncogenic DNA viruses. *Mailing Add:* Dept Microbiol & Immunol Albany Med Col Albany NY 12208

LEHMAN, JOHN THEODORE, b Taylor, Pa, Oct 13, 52; m 74; c 2. LIMNOLOGY, ECOLOGY. *Educ:* Yale Univ, BS & MS, 74; Univ Wash, PhD(zool), 78. *Prof Exp:* Asst prof limnol, 78-80, asst prof biol, 80-88, PROF BIOL, UNIV MICH, 88-, MEM, GREAT LAKES RES DIV, 80- *Mem:* Am Soc Limnol & Oceanog; Int Asn Theoret & Appl Limnol; Phycol Soc Am; Sigma Xi. *Res:* Aquatic ecology; population dynamics of phytoplankton and zooplankton; mathematical models and numerical simulations of biological and chemical processes. *Mailing Add:* Div Biol Sci Natural Sci Bldg Univ Mich Ann Arbor MI 48109

LEHMAN, MEIR M, b Karlsruhe, Ger, Jan 24, 25; UK citizen; m 53; c 5. SOFTWARE ENGINEERING, SOFTWARE DEVELOPMENT PROCESS. *Educ:* Imp Col, BSc Hons, 53, PhD(math), 57; London Univ, DSc(computer sci). *Prof Exp:* Logic designer, Ferantti Ltd, 56-57; head, Sci Dept, Israel Ministry Defense, 57-66; res staff mem & mgr, Res Div, IBM, 66-72; prof computer sci, 72-84, head dept, 79-84, EMER PROF, IMP COL, 84-, SR RES FEL, 89- *Concurrent Pos:* Chmn, dir & consult, Imp Software Technol Ltd, 82-87; dir, ICST&N, DOC, ESF Proj, 91- *Mem:* Fel Inst Elec & Electronics Engrs; fel Brit Computer Soc; Asn Comput Mach; fel Royal Acad Eng; fel Inst Elec Engrs UK. *Res:* Software engineering; software development process & its support. *Mailing Add:* Dept Comput Imp Col 180 Queens Gate London SW7 2BZ England

LEHMAN, R SHERMAN, b Ames, Iowa, Jan 25, 30; div; c 6. MATHEMATICS. *Educ:* Stanford Univ, BS, 51, MS, 52, PhD, 54. *Prof Exp:* Prob analyst, Comput Lab, Ballistic Res Labs, Aberdeen Proving Ground, 55-56; Fulbright res grant, Univ Göttingen, 56-57; from asst prof to assoc prof, 58-66, PROF MATH, UNIV CALIF, BERKELEY, 66- *Concurrent Pos:* Consult, Rand Corp, 54-65. *Mem:* Am Math Soc; Math Asn Am; Asn Symbolic Logic. *Res:* Numerical analysis and computing; number theory. *Mailing Add:* Dept Math Univ Calif Berkeley CA 94720

LEHMAN, RICHARD LAWRENCE, b Portland, Ore, Nov 7, 29; m 63; c 4. BIOPHYSICS, ENVIRONMENTAL HEALTH. *Educ:* Univ Ore, BS, 51, MA, 53; Univ Calif, Berkeley, PhD(biophys), 63. *Prof Exp:* High sch instr math & sci, Calif, 53-56; physicist, Lawrence Radiation Lab, Univ Calif, 57-64, asst prof biophys & nuclear med, med ctr, Univ Calif, Los Angeles, 64-68; res physicist, lab nuclear sci, Mass Inst Technol, 68-71; dept dir Off Ecol, 71-81, PHYS SCIENTIST, CLIMATE ANALYSIS CTR, NAT WEATHER SERV, NAT OCEANIC & ATMOSPHERIC ADMIN, 81- *Concurrent Pos:* Vis scientist, Am Inst Biol Sci, 61-; vis scientist, Swiss Fed Inst Technol, 63 & Cambridge Univ, 66. *Honors & Awards:* Wellcome Trust Award, 66. *Mem:* AAAS; Am Geophys Union; Health Physics Soc. *Res:* Climate data applications; definition of forecast; contigent probability distributions; climate impact assessment. *Mailing Add:* Climate Analysis Ctr Nat Oceanic & Atmospheric Admin Washington DC 20233

LEHMAN, ROBERT HAROLD, b Duncannon, Pa, Nov 15, 29; m 52; c 1. PHYSIOLOGY, ECOLOGY. *Educ:* Bloomsburg State Col, BS, 60; Univ Okla, MNS, 65, PhD(physiol, ecol), 70. *Prof Exp:* Asst prof biol, Longwood Col, 66-70, assoc prof bot, 70-74, asst dean, 74-79, dean, continuing studies, 79-86, ASSOC PROF BIOL, LONGWOOD COL, 74-, DEAN, GRAD SCH, 86- *Concurrent Pos:* Consult, environ waste mgt & planning. *Mem:* Sigma Xi. *Res:* Allelopathic effects of caffeoylquinic acids and scopolin on vegetational patterning; land reclamination; use of sledge as fertilizers; biochemistry. *Mailing Add:* Ala A&M Univ PO Box 998 Normal AL 35762

LEHMAN, ROGER H, b Neosho, Wis, Apr 24, 21; m 57; c 3. MEDICINE, OTOLARYNGOLOGY. *Educ:* Univ Wis, BA, 42, MD, 44. *Prof Exp:* Resident otolaryngol, Vet Admin Ctr, Wood, Wis, 48-51; resident ophthal, Milwaukee County Gen Hosp, 51-52; consult, Vet Admin Ctr, Wood, 54-78; PROF OTOLARYNGOL, MED COL WIS, 66- *Concurrent Pos:* Chief otolaryngol, Vet Admin Med Ctr, Milwaukee County Hosp, Wis, 60-88, Milwaukee Children's Hosp, 71-88 & Froedtept Mem Lutheran Hosp, 80-88. *Mem:* AMA; Am Col Surg; Am Laryngol, Rhinol & Otolaryngol Soc; Soc Univ Otolaryngol; Am Acad Otolaryngol; Sigma Xi. *Mailing Add:* Vet Admin Med Ctr Milwaukee WI 53193

LEHMAN, THOMAS ALAN, b Berne, Ind, Jan 12, 39; m 61; c 2. CHEMICAL EDUCATION. *Educ:* Bluffton Col, BS, 61; Purdue Univ, PhD(chem), 67. *Prof Exp:* Asst prof chem, Bluffton Col, 66-69; assoc prof chem & physics, Nat Univ Zaire, 71-73; assoc prof, 73-81, PROF CHEM, BETHEL COL, 81- *Concurrent Pos:* Res assoc, Univ NC Chapel Hill, 69-70; vis scientist, Nat Inst Environ Health Sci, 79-80. *Mem:* Am Chem Soc; Am Soc Mass Spectrometry. *Res:* Gaseous ion/molecule reactions; ion cyclotron resonance spectrometry; author and co-author scientific publications and book. *Mailing Add:* Box 435 North Newton KS 67117

LEHMAN, WILLIAM FRANCIS, plant genetics, agronomy; deceased, see previous edition for last biography

LEHMAN, WILLIAM JEFFREY, b June 20, 45; m 82; c 2. MUSCLE BIOCHEMISTRY & BIOPHYSICS, ELECTRON MICROSCOPY. *Educ:* State Univ NY, Stony Brook, BS, 66; Princeton Univ, PhD(biol), 69. *Prof Exp:* Postdoctoral res fel, Biol Dept, Brandeis Univ, 69-72; higher sci officer, Zool Dept, Oxford Univ, 73; from asst prof to assoc prof, 73-90, PROF PHYSIOL, DEPT PHYSIOL, BOSTON UNIV SCH MED, 90- *Concurrent Pos:* Estab investr, Am Heart Asn, 82; Whitaker award, 83-84. *Mem:* Am Heart Asn; Soc Gen Physiologists; Biophys Soc; Biochem Soc. *Res:* Mechanism of regulation of muscle contraction. *Mailing Add:* Dept Physiol Boston Univ Sch Med 80 E Concord St Boston MA 02118

LEHMANN, A(LDO) SPENCER, b Los Angeles, Calif, Sept 23, 16; m 43; c 2. CHEMISTRY, CHEMICAL ENGINEERING. *Educ:* Stanford Univ, AB, 38; Brown Univ, PhD(chem), 41. *Prof Exp:* Res chemist, Brown Univ, 40-42, 43-45 & Naval Res Lab, 42-43; sr engr, Tenn Eastman Corp, Oak Ridge, Tenn, 45-46; chemist, Shell Develop Co, 46-50, supvr develop, 50-52, tech rep, 52-53, asst to pres, 53-54, mgr, Tech Dept, Wood River Refinery, Shell Oil Co, 54-58, asst mgr, NY Tech Dept, 58-62, process supt, 62-63, chief technologist, 63-64, refinery supt, Houston Refinery, 64-66, refinery mgr, Wilmington Refinery, Calif, 66-68, gen mg, Tech Depts, 68-73, gen mgr, Res Orgn & Facil, 73-76; RETIRED. *Concurrent Pos:* Dir, Fallbrook Pub Utilites Dist, 83-, ACWA/Joint Powers Ins Authority, 86-; bd pres, San Diego Blood Bank, 87. *Honors & Awards:* Cert of Appreciation, Am Petrol Inst. *Mem:* Am Petrol Inst; Am Chem Soc; Sigma Xi; Am Inst Chem Engrs. *Res:* Infrared spectroscopy; development of reaction for producing metallic potassium; chemistry of uranium; design of special equipment for uranium recovery and processing; development of processes for production of petrochemicals; petroleum process design; research laboratory design and construction. *Mailing Add:* 1917 Santa Margarita Dr Fallbrook CA 92028

LEHMANN, ELROY PAUL, b Tigerton, Wis, June 22, 28; m 51; c 2. RESOURCE MANAGEMENT, PETROLEUM GEOLOGY. *Educ:* Univ Wis, BS, 50, MS, 51, PhD(geol), 55. *Prof Exp:* Asst prof geol, Wesleyan Univ, 52-59, actg chmn, 55-57; paleontologist, Mobil Oil Can, Libya, 59-60, sr paleontologist, 60-61, geol lab supvr, 61-63, staff geologist, Mobil Oil Libya Ltd, 63-65, sr staff geologist, Mobil Latin Am Inc, 65-67, sr res geologist, Mobil Res & Develop Corp, 67-69, chief geoscientist, Int Div, Mobil Oil Corp, 69-72, sr staff explorationist, 72-74, explor mgr, Mobil Oil Libya Ltd, 74-78; explor mgr new areas, Mobil Explor & Producing Serv, Inc, 78-79; pres & gen mgr, Mobil Explor Egypt, Inc, 79-83, gen mgr, Mobil Alaska Explor, 83-86; PETROL CONSULT, 86- *Concurrent Pos:* Mem educ comt, Am Geol Inst, 55-57; Fulbright lectr, Karachi, 58-59. *Mem:* AAAS; fel Geol Soc Am; Am Asn Petrol Geologists; Soc Econ Paleontologists & Mineralogists; Libya Petrol Explor Soc (treas, 64, pres, 65); Sigma Xi. *Res:* Geoscience aspects of energy resource identification and evaluation. *Mailing Add:* PO Box 616 Addison TX 75001

LEHMANN, ERICH LEO, b Strasbourg, France, Nov 20, 17; nat US; m 39; c 3. MATHEMATICAL STATISTICS. *Educ:* Univ Calif, MA, 42, PhD(math statist), 46. *Hon Degrees:* DSc, Univ Leiden, 85. *Prof Exp:* From asst prof to assoc prof math, 42-54, prof, 54-89, chmn dept, 73-76, EMER PROF STATIST, UNIV CALIF, BERKELEY, 89- *Concurrent Pos:* Vis assoc prof, Columbia Univ, 50 & Stanford Univ, 51; vis lectr, Princeton Univ, 51; ed, Ann Math Statist, 53-55; Guggenheim fel, 55, 66 & 79. *Honors & Awards:* R A Fisher Award, 88. *Mem:* Nat Acad Sci; Am Statist Asn; Inst Math Statist; Int Statist Inst; Am Acad Arts & Sci; hon fel Royal Statist Soc. *Res:* Theories of testing hypotheses and of estimation; nonparametric statistics. *Mailing Add:* Dept Statist Univ Calif Berkeley CA 94720

LEHMANN, GILBERT MARK, b Libertyville, Ill, Aug 4, 33; m 58; c 1. GAS DYNAMICS, HEAT TRANSFER. *Educ:* Valparaiso Univ, BS, 55; Ill Inst Technol, MS, 57; Purdue Univ, PhD(jet propulsion), 66. *Prof Exp:* Dean, Col Eng, 72-79, PROF MECH ENG, VALPARAISO UNIV, 56- *Mem:* Am Soc Mech Engrs; Am Soc Eng Educ. *Res:* Internal ballistics of solid propellant rockets. *Mailing Add:* Dept Mech Eng Valparaiso Univ Valparaiso IN 46383

LEHMANN, HEINZ EDGAR, b Berlin, Ger, July 17, 11; nat Can; m 40; c 1. PSYCHIATRY. *Educ:* Univ Berlin, MD. *Hon Degrees:* LLD, Univ Calgary. *Prof Exp:* From asst prof to assoc prof, 52-65, prof, 65-81, chmn dept, 70-74, EMER PROF PSYCHIAT, McGILL UNIV, 81- *Concurrent Pos:* Dir med educ & res, Douglas Hosp, 48-; vis prof, Univ Cincinnati, 58-; dep commr res, Off Mental Health, State NY, 81- *Honors & Awards:* Lasker Award, Am Pub Health Asn, 57; McNeill Award, Can Psychiat Asn, 69, 70 & 74. *Mem:* Life fel Am Psychiat Asn; Can Psychiat Asn; Can Ment Health Asn; fel Am Col Neuropharmacol; fel Int Col Neuropsychopharmacol. *Res:* Diagnosis and therapy of psychotic conditions; effects of drugs on mental processes. *Mailing Add:* Dept Psychiat McGill Univ 1033 Pine Ave W Montreal PQ H3A 1A1 Can

LEHMANN, HERMANN PETER, b London, Eng, June 24, 37. CLINICAL BIOCHEMISTRY. *Educ:* Univ Durham, BSc, 59, PhD(phys chem), 64. *Prof Exp:* Weizmann fel, Weizmann Inst Sci, 65-66; Volkswagen Found fel, Max Planck Inst, Mulheim Ruhr, WGer, 66-67; res assoc, Radiation Lab, Univ Notre Dame, 67-69; NIH sr fel biochem, Univ Wash, 69-71; from asst prof to assoc prof, 71-82, PROF PATH, LA STATE UNIV MED CTR, NEW ORLEANS, 82- *Concurrent Pos:* Vis scientist, Charity Hosp La, New Orleans, 71-; consult, Vet Admin Hosp, New Orleans, 73-; assoc mem, Am Soc Clin Path, 87. *Honors & Awards:* Distinguished Serv Award, Am Soc Clin Path, 87-; Outstanding Contrib Educ Award, Am Asn Clin Chem, 90. *Mem:* Am Chem Soc; The Chem Soc; Am Asn Clin Chem; Brit Asn Clin Biochem; Am Soc Clin Path; Sigma Xi. *Res:* Clinical chemistry; molecular pathology. *Mailing Add:* Dept Path La State Univ Med Ctr New Orleans LA 70112

LEHMANN, JOHN R(ICHARD), b Oak Park, Ill, Mar 24, 34; m 56; c 2. COMPUTER SYSTEMS DESIGN, RESEARCH ADMINISTRATION. *Educ:* Univ Ill, BS, 56, MS, 58, PhD(elec eng), 64. *Prof Exp:* Asst-assoc prog dir, Eng Systs Prog, Div Eng, NSF, 63-67, prog dir, Develop Computer Uses Prog, Off Comput Activ, 67-70, prog dir, Computer Systs Design Prog, Div Computer Res, 70-86, prog dir, Computer Res Equip Prog, Div Computer Res, 82-86, prog dir, Microelectronic Systs Archit Prog, 86-87, DEP DIV DIR, MICROELECTRONIC INFO PROCESSING SYSTS DIV, NSF, 87- *Concurrent Pos:* Mem, Simulation Coun, 65-75. *Mem:* Inst Elec & Electronics Engrs; Am Soc Eng Educ; Inst Elec & Electronics Engrs Computer Soc; Sigma Xi. *Res:* Computer systems design; microcomputer applications; electronics. *Mailing Add:* Microelectronic Info Processing Div NSF Washington DC 20550

LEHMANN, JUSTUS FRANZ, b Koenigsberg, Ger, Feb 27, 21; nat US; m 43; c 3. PHYSICAL MEDICINE. *Educ:* Univ Frankfurt, MD, 45. *Prof Exp:* Asst physician internal med, Univ Frankfurt, 45-46; res asst, Max Planck Inst Biophys, 46-48; asst physician internal med, Univ Frankfurt, 48-51; asst prof med, Mayo Clinic, 51-55; asst prof & assoc dir dept, Ohio State Univ, 55-57; PROF PHYS MED & CHMN DEPT PHYS MED & REHAB, UNIV WASH, 57- *Concurrent Pos:* Fel phys med, Mayo Clin, 51-55. *Mem:* Biophys Soc; AMA; Am Asn Electromyog & Electrodiag; Am Acad Phys Med & Rehab; Am Cong Rehab Med. *Res:* Biophysics of physical agents used in medicine; rehabilitation. *Mailing Add:* Univ Wash RJ-30 Seattle WA 98195

LEHMANN, WILMA HELEN, b Chicago, Ill, Nov 14, 29. VERTEBRATE MORPHOLOGY. *Educ:* Mundelein Col, BA, 51; Northwestern Univ, MS, 54; Univ Ill, PhD(zool), 61. *Prof Exp:* Res asst allergy, Med Sch, Northwestern Univ, 54-56; asst prof zool, Pa State Univ, 61-64; asst prof natural sci, Mich State Univ, 64-67; from asst prof to assoc prof, 67-73, PROF BIOL, NORTHEASTERN ILL UNIV, 73-, CHAIRPERSON, DEPT BIOL, 83- *Concurrent Pos:* Indexer, Evolution, 60-66; NSF instnl res grant, 63-64; vis res assoc, Argonne Nat Lab, 69-70. *Mem:* Fel AAAS; Soc Study Evolution; Am Soc Zool; Sigma Xi. *Res:* Comparative vertebrate anatomy; adaptive radiation of primates and rodents; functional mammalian anatomy; functional morphology, gross and microscopic, of bone; glaucoma. *Mailing Add:* Dept Biol Northeastern Ill Univ Bryn Mawr & St Louis Ave Chicago IL 60625

LEHMER, DERRICK HENRY, b Berkeley, Calif, Feb 23, 05; m 28; c 2. MATHEMATICS. *Educ:* Univ Calif, AB, 27; Brown Univ, ScM, 29, PhD(math), 30. *Prof Exp:* Asst, Brown Univ, 28; Nat Res Coun fel, Calif Inst Technol, 30-32; res worker, Inst Advan Study, 33-34; from instr to asst prof math, Lehigh Univ, 34-38 & 39-40; Guggenheim fel, Cambridge Univ, 38-39; from asst prof to prof, 40-72, EMER PROF MATH, UNIV CALIF,

BERKELEY, 72- *Concurrent Pos:* Nat Res Coun fel, Stanford Univ, 30-32; mathematician, Aberdeen Proving Ground, Md, 45-46; dir, Inst Numerical Analysis, 51-53; adv panel, NSF, 52- *Mem:* AAAS; Soc Indust & Appl Math; Am Math Soc (vpres, 53); Math Asn Am; Asn Comput Mach (vpres, 54-57). *Res:* Theory of numbers; computing devices; mathematical tables and other aids to computation. *Mailing Add:* 1180 Miller Ave Berkeley CA 94708

LEHMKUHL, DENNIS MERLE, b Pierre, SDak, Aug 22, 42; m 65; c 3. ENTOMOLOGY, ECOLOGY. *Educ:* Univ Mont, BA, 64, MS, 66; Ore State Univ, PhD(entom), 69. *Prof Exp:* From asst prof to assoc prof, 74-80, PROF BIOL, UNIV SASK, 80- *Concurrent Pos:* Ecol & taxon consult. *Mem:* NAm Benthological Soc; Entom Soc Can. *Res:* Taxonomy and biology of Ephemeroptera; ecology of rivers; arctic and northern aquatic insects, especially ecological adaptations and limiting factors and the resulting zoogeographical implications. *Mailing Add:* Dept Biol Univ Sask Saskatoon SK S7N 0W0 Can

LEHMKUHL, L DON, b Lodgepole, Nebr, Jan 2, 30; m 53; c 3. CLINICAL NEUROPHYSIOLOGY, HEAD INJURY REHABILITATION. *Educ:* Univ Nebr, BS, 53; Univ Iowa, MS, 58, PhD, 59. *Prof Exp:* Instr physiol, Univ Iowa, 59-60; sr instr, Case Western Reserve Univ, 60-66, asst prof phys ther, 60-68, assoc prof, 68-70, asst prof physiol, 66-70; asst dir, Am Med Assoc, Dept Allied Med Prof & Serv, Chicago, 70-76; ASSOC DIR RES, INST REHAB & RES, BAYLOR COL MED, 76- *Concurrent Pos:* Mem voc rehab admin adv panel phys ther, Dept Health, Educ & Welfare, 64-68; assoc ed, J Am Phys Ther Asn, 82-85. *Honors & Awards:* Lucy Blair Award, Am Phys Ther Asn, 83; Worthingham Fel, Am Phys Ther Asn, 84. *Mem:* Sigma Xi; Am Phys Ther Asn; AAAS; Am Physiol Soc; Int Soc Electromyographic Kinesiol; Soc Behav Kinesiol (pres, 77-81); Am Cong Rehab Med; Nat Head Injury Found. *Res:* Functional capacity of peripheral circulation by venous occlusion plethysmography; spread of excitation in cardiac muscle and pacemaker electrophysiology using cultured heart cells; physiological effects of heat and cold; neurophysiologic profile of patients with brain injury or spinal cord injury; disorders of motor control; costs and outcomes of head injury rehabilitation. *Mailing Add:* 3015 Winslow Houston TX 77025

LEHMKUHLE, STEPHEN W, visual electrophysiology, visual psychophysics, for more information see previous edition

LEHN, WILLIAM LEE, b Spring Valley, Ill, Mar 17, 32; m 54; c 4. MATERIALS ENGINEERING, CHEMISTRY. *Educ:* Univ Ill, BS, 54; Univ Rochester, PhD(chem), 58. *Prof Exp:* Chemist polymers, E I du Pont de Nemours, 58-60; res chemist, Air Force Mat Lab, 60-61, group leader, 61-67, tech area mgr, 67-81, mat engr coatings, 81-90; CONSULT, 91- *Honors & Awards:* Sky Lab Achievement Award, NASA, 74. *Mem:* Am Chem Soc; Res Soc Am; Am Inst Aeronaut & Astronaut; Sigma Xi. *Res:* Materials; coatings; protective and functional coatings and materials for aircraft and spacecraft. *Mailing Add:* 450 Deauville Dr Dayton OH 45429

LEHNE, RICHARD KARL, b Newark, NJ, Nov 18, 20; m 45; c 4. ORGANIC CHEMISTRY. *Educ:* Muhlenberg Col, BS, 41; Yale Univ, PhD(org chem), 49. *Prof Exp:* Process develop chemist, Gen Aniline & Film Co, 49-52; from assoc dir to dir res & develop, Wildroot Co, 52-59; mgr hair prod, Colgate-Palmolive Co, 59-63; dir res & develop, Mennen Co, 63-66; dir consumer prod res & develop, Cyanamid Int, 66-71; dir regulatory affairs, Church & Dwight Co, Inc, 71-82; RETIRED. *Mem:* AAAS; Am Chem Soc; Soc Cosmetic Chemists (secy, 60-63). *Res:* Emulsion technology and viscosity versus stability; effects of phenolic additives. *Mailing Add:* 24 Hillcrest Rd Martinsville NJ 08836

LEHNER, ANDREAS FRIEDRICH, b Mt Vernon, NY, Mar 29, 53; c 1. CHROMATOGRAPHY, MASS SPECTROSCOPY. *Educ:* Manhattan Col, Riverdale, NY, BS, 75; Pa State Univ, University Park, MS, 78; Hershey Med Ctr, Pa, PhD(biochem), 83. *Prof Exp:* Res fel, Dept Biochem, Hershey Med Ctr, 82-83; res fel, Dept Med, Med Col Ga, 83-84, asst prof, Dept Endocrinol, 84-87, res asst, Dept Nutrit, 87-88, res fel, Dept Cell & Molecular Biol, Med Col Ga, 88-90; LAB DIR, AUGUSTA TESTING LAB, 87- *Concurrent Pos:* Prin investr, MCG Res Inst grant, 83-84 & NIH grant-Endocrinol Sect, 84-87. *Res:* Patterns of mouse mammary tumor virus integration in mammary carcinogenesis; DNA-small molecule interactions as they relate to patterns of activity in bioactive chemicals; mechanisms of action of an antineoplaston; recombinations during intergeneric mating in microorganisms. *Mailing Add:* Augusta Testing Lab PO Box 3293 Augusta GA 30904

LEHNER, GUYDO R, b Chicago, Ill, Apr 14, 28. TOPOLOGY. *Educ:* Loyola Univ, Ill, BS, 51; Univ Wis, MS, 53, PhD, 58. *Prof Exp:* Instr math, Univ Wis-Milwaukee, 57-58; from instr to assoc prof, 58-68, PROF MATH, UNIV MD, COLLEGE PARK, 68- *Mem:* Am Math Soc; Math Asn Am. *Res:* Abstract spaces; continua; point set topology. *Mailing Add:* Dept Math Univ Md College Park MD 20742

LEHNER, PHILIP NELSON, b NH, July 5, 40; m 67; c 2. ANIMAL BEHAVIOR, ECOLOGY. *Educ:* Syracuse Univ, BS, 62; Cornell Univ, MS, 64; Utah State Univ, PhD(animal behav), 69. *Prof Exp:* Biologist, Bur Sport Fisheries & Wildlife, 62; res asst, Cornell Univ, 62-64 & Smithsonian Inst, 64-65; biologist, USPHS, 65; from asst prof to assoc prof, 69-83, PROF ANIMAL BEHAV, COLO STATE UNIV, 83- *Concurrent Pos:* NIH grant, Colo State Univ, 69-; consult, Stearns-Roger Corp, 70-80; pres, Animal Behav Assocs, Inc, 83- *Mem:* AAAS; Animal Behav Soc; Soc Exp Anal Behav; Wildlife Soc; Am Vet Soc Animal Behav. *Res:* Animal behavior, its description, analysis and the effects of environmental variables; wild and domestic species; fish and wildlife; zoology. *Mailing Add:* Dept Biol Colo State Univ Ft Collins CO 80523

LEHNERT, JAMES PATRICK, b Mecosta, Mich, Mar 29, 36. ZOOLOGY. *Educ:* Univ Mich, BS, 58, MA, 61; Univ Ill, PhD(parasitol), 67. *Prof Exp:* From asst prof to prof biol, Ferris State Col, 67-87; RETIRED. *Mailing Add:* 626 Woodward Big Rapids MI 49307

LEHNERT, SHIRLEY MARGARET, b London, Eng, June 2, 34; m 61; c 2. RADIOBIOLOGY. *Educ:* Univ Nottingham, BSc, 55; Univ London, MSc, 58, PhD(biophys), 61. *Prof Exp:* Fel, Univ Rochester, 61-63; res biophysicist, Montreal Gen Hosp, 63-65; sci serv officer, Defence Bd Can, 65-67; res assoc phys biol, Sloan-Kettering Inst Cancer Res, 68-71; asst prof radiol, Radiol Res Lab, Col Physicians & Surgeons, Columbia Univ, 71-74; asst prof therapeut radiol, 74-82, ASSOC PROF RADIATION ONCOL, MCGILL UNIV, 82- *Concurrent Pos:* Vis scientist, Inst Gustav Roussy, Paris, France, 85. *Mem:* AAAS; Radiation Res Soc. *Res:* Biological and biochemical effects of ionizing radiation. *Mailing Add:* Dept Radiation Oncol Montreal Gen Hosp 1650 Cedar Ave Montreal PQ H3Q 1A4 Can

LEHNHOFF, HENRY JOHN, JR, internal medicine, for more information see previous edition

LEHNHOFF, TERRY FRANKLIN, b St Louis, Mo, July 7, 39; m 60; c 4. MECHANICAL ENGINEERING, ENGINEERING MECHANICS. *Educ:* Univ Mo-Rolla, BS, 61, MS, 62; Univ Ill, Urbana, PhD(theoret & appl mech), 68. *Prof Exp:* Res engr, Caterpillar Tractor Co, 62-65; from asst prof to assoc prof mech eng, 68-77, res assoc, Rock Mech Res Ctr, 70-80, PROF MECH & AEROSPACE ENG, UNIV MO- ROLLA, 77- *Concurrent Pos:* Consult, Detroit Tool, 76-85, Eaton Corp, 80-82 & Rockwell Int, 81-82; pres, Enmeco, 83- *Mem:* Sigma Xi; Am Soc Mech Engrs; Am Soc Metals. *Res:* Thermal stresses; mechanical structures; solid mechanics; rock mechanics; finite elements; pressure vessel design. *Mailing Add:* Dept Mech Aerospace Eng & Eng Mech Univ Mo Rolla MO 65401

LEHOCZKY, JOHN PAUL, b Columbus, Ohio, June 29, 43; m 66; c 2. STATISTICS. *Educ:* Oberlin Col, BA, 65; Stanford Univ, MS, 67, PhD(statist), 69. *Prof Exp:* from asst prof to assoc prof, 69-81, PROF STATIST, CARNEGIE-MELLON UNIV, 81-, DEPT HEAD, 84- *Concurrent Pos:* Area ed, Mgt Sci; assoc ed, J Real-Time Systs. *Mem:* Opers Res Soc Am; fel Inst Math Statist; fel Am Statist Asn; Inst Mgt Sci; AAAS; Inst Elec & Electronics Engrs; Sigma Xi; Int Statistical Inst. *Res:* Applied probability theory; stochastic processes and their application to computer, communication, and repair systems; real-time computer systems. *Mailing Add:* Dept Statist Carnegie Mellon Univ Pittsburgh PA 15213

LEHOUX, JEAN-GUY, b St Severin, Que, Jan 9, 39; m 63; c 3. BIOCHEMISTRY, ENDOCRINOLOGY. *Educ:* Univ Montreal, BSc, 63, MSc, 67, PhD(biochem), 69. *Prof Exp:* Chief chemist, Cyanamid Can Ltd, 63-65; res asst biochem, Univ Montreal, 65-69, lectr med, 69-71; from asst prof to assoc prof obstet & gynec, 71-81 PROF, DEPT BIOCHEM, OBSTET & GYNEC, SHERBROOKE UNIV, 81-, HEAD, CLIN ENDOCRINOL LAB, 74-, CHMN, 80-, HEAD DEPT BIOCHEM, 80- *Concurrent Pos:* Biochemist, Hosp Maisonneuve, 69-70; Med Res Coun Can fels, Fac Med, Univ Montreal, 69-70 & Dept Zool, Univ Sheffield, 70-71; Med Res Coun Que & Med Res Coun Can grant, Univ Sherbrooke, 71-74; Med Res Coun Can scholar, 74. *Mem:* Brit Soc Endocrinol. *Res:* Studies on steroid hydroxylation with a special interest to aldosterone regulation. *Mailing Add:* Fac Med Univ Sherbrooke Sherbrooke PQ J1K 2R1 Can

LEHOVEC, KURT, b Ledvice, Czech, June 12, 18; US citizen; m 52; c 4. SEMI-CONDUCTING DEVICES, OPTO-ELECTRONICS. *Educ:* Prague Charles Univ, BS, 38, MS, 40, PhD(physics), 41. *Prof Exp:* Head res lab, Physics Inst, Prague Univ, 42-45, res fel, 45- 46; res fel, US Signal Corps, Ft Monmouth, NJ, 47-52; dir semiconductor res & develop, Sprague Elec Co, 52-66; PRES, INVENTORS & INVESTORS, 67-; EMER PROF ELECTRONICS, UNIV SOUTHERN CALIF, 88- *Concurrent Pos:* Prof electronics, Univ Southern Calif, 71-88; adj prof, Williams, Col, 67, Univ Calif, Irvine, 80; consult, var co, 67- *Mem:* Fel Am Phys Soc; fel Inst Elec & Electronics Engrs. *Res:* Crystal structure; phase diagrams; defects; energy levels; transistors; solar cells. *Mailing Add:* Dept Elec Eng & Mat Sci Univ Southern Calif University Park MC0483 Seaver Sci Ctr 522 Los Angeles CA 90089

LEHR, CARLTON G(ORNEY), b Boston, Mass, Sept 11, 21; m 47; c 3. ELECTRICAL ENGINEERING. *Educ:* Mass Inst Technol, BS, 43, MS, 48. *Prof Exp:* Mem staff, Div Indust Coop, Mass Inst Technol, 46-47, asst elec eng, 47-48; sr engr, Microwave & Power Tube Div, Raytheon Co, 48-54, mem staff, Microwave Group, Res Div, 54-58, mgr, 58-64; staff engr, Smithsonian Astrophys Observ, 64-76; consult, 76-79; sr radar & optical engr, Ford Aerospace & Commun Corp, 79-86; RETIRED. *Concurrent Pos:* Lectr, Northeastern Univ, 63-70. *Res:* Satellite tracking; lasers; mathematics. *Mailing Add:* Five Childs Rd Lexington MA 02173-4501

LEHR, DAVID, b Sadagura, Austria, Mar 22, 10; US citizen; div; c 2. CARDIOVASCULAR PHARMACOLOGY, MEDICINE. *Educ:* Univ Vienna, Austria, BA, 29, MD, 35. *Prof Exp:* Asst pharmacol, Univ Vienna, 34-48; instr, Univ Lund, 38-39; pharmacologist & res assoc, Path Dept, Newark Beth Israel Hosp, NJ, 39-42; from instr to assoc prof pharmacol, 41-54, prof & chmn, Dept Physiol & Pharmacol, 56-64, prof & chmn, dept pharmacol, 64-79, ASSOC PROF MED, NY MED COL, 49-, EMER PROF PHARMACOL, 80- *Concurrent Pos:* Asst vis physician, Metrop Hosp, Welfare Island, NY, 42-54, vis physician, 54-75; asst attend physician, Flower & Fifth Ave Hosps, 44-49, assoc attend physician, 49-75; vis physician, Bird S Coler Hosp, 54-75; Claud Bernard prof, Inst Exp Med & Surg, Univ Montreal, 61; mem rev comt, Health Res Coun New York, 61-65, vchmn panel neurol & psychiat dis, 61-65; chmn ad hoc comt use of new therapeut agents & procedures in human beings, Assoc Med Schs, NY, 67-; mem, Coun Arteriosclerosis, Am Heart Asn; co-chmn, Coun on Drugs, Am Col Nutrit, 80-; consult ed, J Am Col Nutrit, 82. *Mem:* Fel AAAS; fel Am Col Physicians; fel Am Col Cardiol; Soc Exp Biol & Med; Am Soc Pharmacol & Exp Therapeut; Am Soc Arteriosclerosis; Am Soc Exp Path; Sigma Xi; NY Acad Sci; Harvey Soc; Int Soc Heart Res. *Res:* Cardiology; hypertension; arteriosclerosis; toxicity of sulfonamides; sulfonamide mixtures; chemotherapy; experimental cardiovascular necrosis; parathyroid hormone interrelations; tissue electrolytes. *Mailing Add:* 79 Lloyd Rd Montclair NJ 07042

LEHR, GARY FULTON, b Rockville Centre, NY, July 16, 52; m 81; c 3. NEW TECHNOLOGY EXPLORATION & EVALUATION. *Educ:* Manhattanville Col, AB, 75; Brown Univ, PhD(chem), 81. *Prof Exp:* Res assoc, Dept Chem, Columbia Univ, 79-81; staff scientist res, Cent Res & Develop Dept, 81-84, staff specialist, Comput Consult, 84-89, TECH LEADER, PROTOTYPING GROUP, DU PONT FIBERS, E I DU PONT DE NEMOURS & CO, INC, 89- *Mem:* Am Chem Soc; AAAS. *Res:* Study of reaction mechanisms involving free radical and carbene intermediates, including radical-radical, radical-molecule photochemical, organometallic and autoxidation reactions. *Mailing Add:* 122 Chatham Pl Wilmington DE 19810

LEHR, HANNS H, b Sadagora, Austria, Jan 1, 08; nat US; m 34; c 3. PHARMACEUTICAL CHEMISTRY. *Educ:* Univ Vienna, PhD(chem), 31, MPharm, 32. *Prof Exp:* Asst org chem, Univ Vienna, 30-32; managing dir pharmaceut lab, Salvatorapotheke, Vienna, 33-38; asst pharmacol, Paris, 38-40 & French Pub Health Serv, 40; asst biochem, Univ Aix Marseille, 40-42; asst org chem, Univ Basle, Switz, 43-46; sr res chemist, Hoffmann-La Roche, Inc, 46-66, asst to vpres chem res, 66-68, asst dir, 68-73, consult, 73-76; RETIRED. *Mem:* AAAS; Am Chem Soc; fel Am Inst Chemists. *Res:* Organic chemistry; biochemistry; chemotherapy; antibiotics; synthetic drugs; research administration. *Mailing Add:* Ten Tuers Pl Upper Montclair NJ 07043

LEHR, JAY H, b Teaneck, NJ, Sept 11, 36; m 57, 78; c 2. HYDROLOGY, GROUNDWATER GEOLOGY. *Educ:* Princeton Univ, BSE, 57; Univ Ariz, PhD(hydrol), 62. *Prof Exp:* Hydrol field asst, Groundwater Br, US Geol Surv, NY, 55-56; res assoc hydrol, Univ Ariz, 59-62, from instr to asst prof, 62-64; asst prof, Ohio State Univ, 64-67; EXEC DIR, NAT WATER WELL ASN, 67- *Concurrent Pos:* Ed, Ground Water, 66-; ed-in-chief, Water Well J, 72-; adj prof, The Ohio State Univ. *Res:* Groundwater model studies utilizing consolidated porous medias; groundwater pollution; groundwater and surfacewater law; water well construction techniques. *Mailing Add:* 6375 Riverside Dr Dublin OH 43017

LEHR, MARVIN HAROLD, b Brooklyn, NY, Mar 17, 33; m 56; c 4. POLYMER PHYSICS, POLYMER CHEMISTRY. *Educ:* Reed Col, BA, 54; Yale Univ, MS, 55, PhD(kinetics), 59. *Prof Exp:* Res chemist, B F Goodrich Res Ctr, 59-61, sr res chemist, 61-66, res assoc, 66-73, sr res assoc, 73-78, RES FEL CORP RES, B F GOODRICH RES & DEVELOP CTR, 78- *Res:* Viscoelastic-fracture behavior; polymer morphology; relation of polymer structure to properties; structure-property studies on polymer blends and composites; relationships of physical, mechanical and rheological properties to composition, microstructure and morphology of miscible and immiscible mixtures. *Mailing Add:* B F Goodrich Res & Develop Ctr Brecksville OH 44141-3289

LEHR, ROLAND E, b Quincy, Ill, Nov 7, 42; m 70; c 2. ORGANIC CHEMISTRY. *Educ:* Princeton Univ, AB, 64; Harvard Univ, AM, 66, PhD(chem), 69. *Prof Exp:* From asst prof to assoc prof chem, Univ Okla, 68-80, actg chair, 87-88, interim dean, Col Arts & Sci, 89-90, PROF CHEM, UNIV OKLA, 80- *Concurrent Pos:* Res grants, NASA, 68-69, Am Chem Soc, 68-70 & NIH, 77- *Mem:* Am Chem Soc; Am Asn Cancer Res; Sigma Xi. *Res:* Chemical carcinogenesis of polycyclic aromatic hydrocarbons and their DNA adducts. *Mailing Add:* Dept Chem Univ Okla Norman OK 73019

LEHRER, GERARD MICHAEL, b Vienna, Austria, May 29, 27; US citizen; c 2. NEUROLOGY. *Educ:* City Col New York, BS, 50; NY Univ, MD, 54. *Prof Exp:* Res asst neurol, Col Physicians & Surgeons, Columbia Univ, 53-54; intern, Mt Sinai Hosp, NY, 54-55, asst resident neurologist, 55-57, resident, 58; asst neurol, Sch Med, Wash Univ, 58-60; from asst attend neurologist to assoc attend neurologist, Mt Sinai Hosp, NY, 60-68; assoc prof, 66-67, PROF NEUROL, MT SINAI SCH MED, 67-, DIR DIV NEUROCHEM, 66-; ATTEND NEUROLOGIST, MT SINAI HOSP, NY, 68-; ATTEND NEUROLOGIST, ST VINCENT HOSP, NV, 76- *Concurrent Pos:* NIH trainee, Mt Sinai Hosp, NY, 56-58; NIH spec trainee neurochem & res fel pharmacol, Sch Med, Wash Univ, 58-61; consult, Preclin Psychopharmacol Res Rev Comt, NIMH, 65-69; res collabr, Brookhaven Nat Lab, NY, 67-69; consult neurologist, Vet Admin Hosp, Bronx, NY, 67-75, staff neurologist, 76- *Mem:* Fel Am Acad Neurol; Am Asn Neuropath; Int Soc Neurochem; Soc Neurosci; Am Neurol Asn; Am Soc Neurochem. *Res:* Brain maturation and metabolism; molecular mechanisms of central nervous system differentiation and disease, especially demyelination. *Mailing Add:* Dept Neurol Mt Sinai Sch Med 1200 Fifth Ave New York NY 10029-5208

LEHRER, HAROLD Z, b New York, NY, Aug 22, 27. RADIOLOGY. *Educ:* Columbia Univ, AB, 47; NY Univ, MD, 53; Am Bd Radiol, dipl, 60. *Prof Exp:* Am Cancer Soc fel, 58; asst adj radiologist, Beth Israel Hosp & Med Ctr, New York, 59-63; instr neuroradiol, NY Univ-Bellevue Med Ctr, 63-65; from asst to assoc prof, Sch Med, Tulane Univ, 65-67; ASSOC PROF NEURORADIOL, NY MED COL, 68-; DIR DEPT RADIOL, BIRD S COLER HOSP, NY, 73- *Concurrent Pos:* NIH spec fel neuroradiol, NY Univ-Bellevue Med Ctr, 63-65. *Mem:* AAAS; Radiol Soc NAm; Am Roentgen Ray Soc; Am Soc Neuroradiol. *Res:* Neuroradiology, especially analysis of clinical data mathematically. *Mailing Add:* 89 River St Box M705 Hoboken NJ 07030

LEHRER, HARRIS IRVING, b Boston, Mass, May 28, 39. BIOCHEMISTRY, IMMUNOCHEMISTRY. *Educ:* Brandeis Univ, BA, 60, PhD(biochem), 65. *Prof Exp:* Sr biochemist, Monsanto Corp, 68-69; sr scientist immunochem, Ortho Diag, 69-76, group leader, 76-77; supvr, Sherman-Abrams Lab, 77-78; SR IMMUNOCHEMIST, ICL SCI, 78- *Concurrent Pos:* NIH fel, Marine Biol Lab, Woods Hole, 65, Univ Palermo, 65-67 & Brandeis Univ, 67-68. *Mem:* AAAS; Am Asn Clin Chem; NY Acad Sci; Am Chem Soc. *Res:* Development of new immunochemical techniques and their application to diagnostic testing. *Mailing Add:* 127 Corona Ave Long Beach CA 90803

LEHRER, PAUL LINDNER, b Chicago, Ill, Feb 9, 28; m 53; c 4. PHYSICAL GEOGRAPHY. *Educ:* Univ Cincinnati, BS, 49; Ohio State Univ, MA, 51; Univ Nebr, PhD(geog), 62. *Prof Exp:* Instr geog, Ohio Univ, 56-59; from instr to asst prof, Univ Wis-Milwaukee, 60-66; assoc prof, 66-69, PROF GEOG, UNIV NORTHERN COLO, 69- *Concurrent Pos:* NSF sci fac fel, Univ Witwatersrand, 64-65. *Mem:* Asn Am Geogr; Sigma Xi. *Res:* Soils and regional geography of Subsaharan Africa. *Mailing Add:* Dept Geog Univ Northern Colo Greeley CO 80639

LEHRER, PAUL MICHAEL, b New York, NY, Aug 30, 41; m 65; c 2. RELAXATION THERAPY. *Educ:* Columbia Col, AB, 63; Harvard Univ, PhD(clin psychol), 69. *Prof Exp:* Clin instr, psychol, Tufts Univ Sch Med, 68-70; asst prof, psychol, Rutgers Univ, 70-72; asst prof, 72-79, ASSOC PROF, PSYCHIAT, ROBERT WOOD JOHNSON MED SCH, 79- *Concurrent Pos:* Vis prof, Univ London, 80-81. *Mem:* Am Psychol Asn; Biofeedback Soc Am; Soc Psychophys Res; Soc Behav Med; Asn Adv Behav Ther. *Res:* Psychophysiological research on the effects of relaxation therapy on psychosomatic disease; psychophysiology of asthma, headache, back pain and anxiety; nature and treatment of stage fright. *Mailing Add:* Dept Psychiat R W Johnson Med Sch 675 Hoes Lane Piscataway NJ 08854

LEHRER, ROBERT N(ATHANIEL), b Sandusky, Ohio, Jan 17, 22; m 45; c 2. INDUSTRIAL & SYSTEMS ENGINEERING. *Educ:* Purdue Univ, BS, 45, MS, 47, PhD(indust eng), 49. *Prof Exp:* Asst & instr indust eng, Purdue Univ, 46-49; asst prof, Ore State Col, 49-50; assoc prof, Ga Inst Technol, 50-54, prof, 54-58, res assoc, 50-58; prof & chmn dept, Technol Inst, Northwestern Univ, Ill, 58-62; UNESCO expert, Guadalajara & Guanajuato, 62-63; assoc dir, Sch Indust Eng, 63-66, PROF INDUST ENG, GA INST TECHNOL, 63-, DIR, SCH INDUST ENG, 66- *Concurrent Pos:* Consult opers res, 50-; ed-in-chief, J Indust Eng, 53-62; adv indust & systs eng sem, Japan, 59 & 62; adv indust eng, Eindhoven & Dutch Ministry Educ, 62-; Nat Acad Sci workshop panel indust & technol res, Indonesia, 71; consult ed indust eng & mgt sci ser, Reinhold Publ Corp; mem adv bd mil personnel supplies, Nat Acad Sci-Nat Res Coun. *Honors & Awards:* Outstanding Indust Eng Award, Am Inst Indust Engrs, 57. *Mem:* Opers Res Soc Am; Am Soc Eng Educ; fel Am Inst Indust Engrs (vpres, 60); Inst Mgt Sci. *Res:* Work simplification; operations research and management science; management of improvement. *Mailing Add:* Sch Indust & Systs Eng 765 First Dr Atlanta GA 30332-0205

LEHRER, SAMUEL BRUCE, b New Britain, Conn, Apr 1, 43; m 71; c 4. ALLERGY, IMMUNOLOGY. *Educ:* Upsala Col, BS, 66; Temple Univ, PhD, 71. *Prof Exp:* Lab technician, Microbiol Dept, State Lab Hartford, 65; researcher, Univ Lausanne, Switz, 69; fel, Scripps Clin & Res Found, La Jolla, 71-75; from asst prof to assoc prof, 75-83, PROF MED, SCH MED, TULANE UNIV, 83- *Concurrent Pos:* NIH fel, 74; Nat Inst Allergy & Infectious Dis young investr award, 78-81; Am Lung Asn grant, 78-80; consult, Food & Drug Admin, 78-80; adj assoc prof microbiol & immunol, Tulane Univ, 80- *Mem:* Am Soc Microbiol; Am Acad Allergy; Am Asn Immunologists; Am Thoracic Soc; Col Int Allergologicum. *Res:* Allergic disease in man; environmental and host factors regulating IgE response; molecular aspects of allergenicity. *Mailing Add:* Tulane Univ Sch Med 1700 Perdido St New Orleans LA 70112

LEHRER, SHERWIN SAMUEL, b New York, NY, Apr 2, 34; m 60; c 2. BIOCHEMISTRY. *Educ:* Univ Pittsburgh, BS, 56; Univ Calif, Berkeley, PhD(chem), 61. *Prof Exp:* Staff scientist thin magnetic films, Lincoln Lab, Mass Inst Technol, 61-62; fel biochem, Brandeis Univ, 63-66; res assoc, Retina Found, Mass, 66-70; SR STAFF SCIENTIST BIOCHEM, BOSTON BIOMED RES INST, 70- *Concurrent Pos:* USPHS res grant, Retina Found, Mass & Boston Biomed Res Inst, 67-; assoc, Harvard Med Sch, 68- *Mem:* AAAS; NY Acad Sci; Am Soc Biol Chem; Am Chem Soc; Biophys Soc. *Res:* Application of fluorescence techniques to protein conformation and interactions; muscle protein interactions. *Mailing Add:* Muscle Res Dept Boston Biomed Res Inst 20 Staniford St Boston MA 02114

LEHRER, WILLIAM PETER, JR, b Brooklyn, NY, Feb 6, 16; m 45; c 1. ANIMAL SCIENCE, ANIMAL NUTRITION. *Educ:* Pa State Univ, BS, 41; Univ Idaho, MS, 46 & 54; Wash State Univ, PhD(animal nutrit, chem), 51; Blackstone Sch of Law, LLB, 72, JD, 74; Pepperdine Univ, MBA, 75. *Prof Exp:* Mgt trainee, Swift & Co, WVa, 41-42; US Army Air Corps, 42-43; mgr livestock farm, NY, 44-45; from asst prof & asst animal husbandman to assoc prof animal husb & assoc animal husbandman, Univ Idaho, 46-60, prof, 60; dir nutrit, Albers Milling Co, 60-62, dir nutrit & res, 62-74, dir nutrit & res, Albers Milling Co & John W Eshelman & Sons, 74-76; dir nutrit & res, Milling Div, Carnation Co, 76-81; RETIRED. *Concurrent Pos:* Mem, Comt Animal Nutrit & Comt Dog Nutrit, Nat Acad Sci-Nat Res Coun & Tech Comt, Western Livestock Range Livestock Nutrit, USDA; mem nutrit coun, Am Feed Mfrs, 62-81, chmn, 69-70; mem, Res Adv Coun, US Brewers Asn, 69-81 & Adv Coun Calif State Polytech Univ, Pomona, 69-81. *Honors & Awards:* WAP Award, Agr-Bus Award, 64. *Mem:* Fel AAAS; Am Inst Nutrit; Sigma Xi; fel Am Soc Animal Sci; Inst Food Technologists; Coun Agr Sci & Technol; Am Soc Agr Engrs; Am Registry Prof Animal Scientists. *Res:* Animal production and nutrition; reproduction and growth of beef cattle and dairy cattle, dogs, horses, sheep and swine. *Mailing Add:* Rocking L Ranch 12180 Rimrock Rd Hayden ID 83835

LEHRMAN, GEORGE PHILIP, b New York, NY, Nov 28, 26; m 48; c 3. PHARMACY ADMINISTRATION. *Educ:* Univ Conn, BS, 50, PhD, 55; Purdue Univ, MS, 52. *Prof Exp:* Asst pharm, Purdue Univ, 50-52, instr chem, 52-53; mkt analyst, Mead Johnson & Co, 55-57; res chemist, Am Cyanamid Co, 57-59; head develop, Cent Pharmacal Co, Ind, 59-61; pharmaceut develop mgr, Baxter Labs, 61-62; dir labs, Conal Pharmaceut, 62-64; vpres, Owen Labs, 64-67; assoc prof pharm, Univ Okla, 67-75; ASST DEAN COL PHARM, UNIV NMEX, 75- *Concurrent Pos:* Res fel, Univ Conn, 53-55. *Mem:* Am Chem Soc; Am Pharmaceut Asn; Soc Cosmetic Chemists. *Res:* Product development; drug law. *Mailing Add:* 8431 Palo Duro NE Albuquerque NM 87111

LEHRSCH, GARY ALLEN, b Altoona, Pa, June 16, 54; m 81; c 3. SOIL PHYSICS, SOIL MANAGEMENT. *Educ:* Pa State Univ, BS, 76, MEPC, 79, MS, 81, PhD(soil physics), 85. *Prof Exp:* Res assoc, Miss State Univ, 80-86; soil scientist, Nat Sedimentation Lab, 86-87, SOIL SCIENTIST, SOIL & WATER MGT RES UNIT, AGR RES SERV, USDA, 87- *Concurrent Pos:* Grad res asst, Pa State Univ, 80; affil asst prof, Univ Idaho, 88-; res asst prof, Utah State Univ, 88- *Mem:* Sigma Xi; Am Soc Agron; Am Soc Surface Mining & Reclamation; Soil Sci Soc Am; Soil & Water Conserv Soc. *Res:* Quantification of effects of tillage, compaction, freezing, thawing, water content, climate and time on soil aggregate stability; improvement of crop and irrigation management systems; estimation of soil interrill erodibility from properties of original soil matrix. *Mailing Add:* USDA Agr Res Serv Soil & Water Mgt Res Unit 3793 N 3600 E Kimberly ID 83341-5076

LEHTO, MARK R, b Longview, Wash, Sept 29, 56. HUMAN FACTORS ENGINEERING, SAFETY ENGINEERING. *Educ:* Ore State Univ, BS, 78; Purdue Univ, MSIE, 80; Univ Mich, PhD(eng), 85. *Prof Exp:* Sr res engr, J M Miller, Inc, 83-86; ASST PROF INDUST ENG, PURDUE UNIV, 86- *Concurrent Pos:* Vis asst prof indust eng, Univ Mich, 89-90; lectr, Nordic Inst Advan Training Occup Health, Turkey & Finland, 89; NSF presidential young investr award, 89; NEC fac fel, 91. *Mem:* Sr mem Human Factors Soc; Indust Eng Soc. *Res:* Safety engineering; human factors; computer aided design methods that address product safety problems early in the design process; author of several books. *Mailing Add:* Sch Indust Eng Purdue Univ West Lafayette IN 47907

LEI, DAVID KAI YUI, b Macau, July 30, 44; m 66; c 1. NUTRITION. *Educ:* Univ London, BS, 68; Univ Guelph, MS, 70; Mich State Univ, PhD(human nutrit), 73. *Prof Exp:* Res asst nutrit, Mich State Univ, 70-73; res assoc hemat, Wayne State Univ, 74-75; from asst prof to assoc prof nutrit, Miss State Univ, 75-80; assoc prof, 80-88, PROF NUTRIT, UNIV ARIZ, 88- *Mem:* Am Inst Nutrit; Am Dietetic Asn; Sigma Xi; Am Heart Asn. *Res:* Trace mineral metabolism; lipoprotein metabolism. *Mailing Add:* Dept Nutrit & Food Sci Univ Ariz Tucson AZ 85721

LEI, SHAU-PING LAURA, b Taipei, Taiwan, Oct 7, 53; US citizen; m 80; c 1. AGRICULTURAL & FOOD CHEMISTRY. *Educ:* Nat Taiwan Univ, BS, 76, MS, 80; Univ Calif, Los Angeles, PhD(molecular biol), 85. *Prof Exp:* Teaching assoc analytical chem, Nat Taiwan Univ, 76-78; proj dir, Int Genetic Eng Inc, 86-89 & Trigen Inc, 89-90; DIR PROTEIN PURIFICATION, XOMA CORP, 90- *Res:* Host-vector development for recombinant DNA; gene cloning and expression in bacteria system and chimeric antibody, fab fragment; fabs and lgg purification from mammalian; yeast and bacteria systems in pilot scale. *Mailing Add:* 11452 Clarkson Rd Los Angeles CA 90064

LEIBACH, FREDRICK HARTMUT, b Kitzingen, Ger, Sept 21, 30; US citizen; m 61; c 3. BIOCHEMISTRY, ENDOCRINOLOGY. *Educ:* Southwest Mo State Col, BS, 59; Emory Univ, PhD(biochem), 64. *Prof Exp:* Nat Acad Sci-Nat Res Coun res assoc, Ames Res Ctr, NASA, 64-67; assoc prof endocrinol, 76-79, ASSOC PROF BIOCHEM, MED COL GA, 67-, PROF CELL & MOLECULAR BIOL, 79- *Mem:* AAAS; Am Chem Soc; Am Soc Biol Chemists; Am Physiol Soc. *Res:* Enzymes in protein and amino acid metabolism; peptidases, transpeptidases and esterases; protein turnover; membrane transport of organic solutes. *Mailing Add:* Dept Cell & Molecular Biol Med Col Ga Augusta GA 30912-3331

LEIBACHER, JOHN W, b Chicago, Ill, May 28, 41; m 76. ASTROPHYSICS, RESEARCH ADMINISTRATION. *Educ:* Harvard Univ, PhD(astron), 71. *Prof Exp:* Postdoctoral, Univ Colo, 70-72; res scientist astrophys, Laboratorie de Physique Stellaire at Planetaire, 72-75; res scientist astrophys, Lockheed Res Labs, 75-82; ASTROR ASTROPHYS, NAT SOLAR OBSERV, 82-, DIR ASTROPHYS, 88- *Mem:* Am Astron Soc; Int Astron Union. *Res:* Solar physics; helioseismology. *Mailing Add:* Nat Solar Observ 950 N Cherry Ave Tucson AZ 85719

LEIBBRANDT, VERNON DEAN, b McCook, Nebr, Oct 31, 44; m 67; c 2. ANIMAL NUTRITION. *Educ:* Univ Nebr, BS, 66; Iowa State Univ, PhD(animal nutrit), 72. *Prof Exp:* Asst prof animal nutrit, Univ Fla, 75-78; asst prof, 78-80, ASSOC PROF ANIMAL NUTRIT, UNIV WIS-MADISON, 80- *Concurrent Pos:* Fel, Res Div, Cleveland Clin Found, 72-75. *Mem:* Am Soc Animal Sci. *Res:* Husbandry and nutritional aspects of swine production. *Mailing Add:* 7317 Branford Lane Madison WI 53706

LEIBEL, WAYNE STEPHAN, b Aug 5, 51; US citizen. EVOLUTIONARY GENETICS, MOLECULAR SYSTEMATICS. *Educ:* Dartmouth Col, AB, 73; Yale Univ, MPhil, 75, PhD(biol), 79. *Prof Exp:* ASSOC PROF BIOL, LAFAYETTE COL, 83- *Mem:* Soc Study Evolution; Am Soc Ichthyologists & Herpetologists; Am Soc Zoologists; Sigma Xi. *Res:* Evolution at the molecular level; speciation and diversification of neotropical freshwater fishes. *Mailing Add:* Biol Dept Lafayette Col Easton PA 18042

LEIBHARDT, EDWARD, b New Rome, Wis, Oct 13, 19; m 61; c 2. SPECTROSCOPY. *Educ:* Northwestern Univ, BA, 54, PhD(astron), 59. *Prof Exp:* PRES, DIFFRACTION PROD, INC, 51- *Mem:* Optical Soc Am. *Res:* Developed ruling engine for producing interferometrically ruled diffraction gratings; holographic gratings. *Mailing Add:* 9416 W Bull Valley Rd Woodstock IL 60098

LEIBHOLZ, STEPHEN W, b Berlin, Ger, Jan 28, 32; US citizen; m 58; c 3. INFORMATION & COMMUNICATION SCIENCES. *Educ:* NY Univ, AB, 52. *Prof Exp:* Res asst & teaching fel physics, NY Univ, 52-53; tutor, Queens Col, 53-54; res assoc electronics, Adv Group Electron Devices, US Dept Defense, 54-56; prin engr, Repub Aviation Corp, 57-60; sr mem tech staff, Auerbach Corp, 60-64, prog mgr opers res & anal, 64-66, mgr syst design & anal, 66-67; PRES, CHIEF EXEC OFFICER, ANALYTICS, 67- *Concurrent Pos:* Mem var adv panels, Dept Defense, 70-; mem, Simulation Coun; ed, Mil Oper Res Monogra. *Mem:* Inst Elec & Electronics Engrs; Soc Indust & Appl Math; Opers Res Soc Am; Mil Oper Res Soc. *Res:* Applied mathematics and operations research in areas of information systems and military systems, especially statistical problems; systems architecture and engineering in areas of information, communications, automation control and surveillance systems for industry and government. *Mailing Add:* Analytics 2500 Maryland Rd Willow Grove PA 19090

LEIBMAN, KENNETH CHARLES, b New York, NY, Aug 7, 23; m 46; c 2. BIOCHEMICAL PHARMACOLOGY. *Educ:* Polytech Inst Brooklyn, BS, 43; Ohio State Univ, MSc, 48; NY Univ, PhD(biochem), 53. *Prof Exp:* Org res chemist, Nat Lead Co, 43-44; asst chem, Ohio State Univ, 46-48; instr, Univ Louisville, 48-49; fel oncol, Univ Wis, 53-54, proj assoc, 54-55; specialist tracer techniques, US Tech Coop Mission, India, 55-56; from instr to assoc prof, 56-68, PROF PHARMACOL, COL MED, UNIV FLA, 68- *Concurrent Pos:* Ed, Drug Metab & Disposition, 72- *Mem:* AAAS; Am Soc Pharmacol & Exp Therapeut. *Res:* Drug metabolism; enzymology; toxicology. *Mailing Add:* 4545 Hwy 346 Archer FL 32618

LEIBMAN, LAWRENCE FRED, b Bronx, NY, Sept 10, 47. ORGANIC CHEMISTRY. *Educ:* City Univ New York, PhD(chem), 76. *Prof Exp:* Fel org biochem, Columbia Univ, 75-76; chemist, Am Cyanamid Co, 76-80; MEM STAFF, BASF WYANDOTTE CORP, 80- *Mem:* Am Chem Soc; Sigma Xi. *Res:* Organic reaction mechanisms. *Mailing Add:* BASF Wyandotte Corp 33 Riverside Ave Rensselaer NY 12144

LEIBO, STANLEY PAUL, b Pawtucket, RI, Apr 8, 37; m 61; c 2. CRYOBIOLOGY, EMBRYOLOGY. *Educ:* Brown Univ, AB, 59; Univ Vt, MS, 61; Princeton Univ, MA, 62, PhD(biol), 63. *Prof Exp:* Res assoc, Oak Ridge Nat Lab, 63-64, USPHS res fel, 64-65, staff biologist, Biol Div, 65-80; vpres, Res & Develop Div, Rio Vista Int, 81-; AT DEPT BIOMED SCI, UNIV GUELPH. *Concurrent Pos:* Lectr, Oak Ridge Grad Sch Biomed Sci, Univ Tenn, 69-80; mem adv bd, Am Type Cult Collection, 71-74; vis scientist health sci & technol, Mass Inst Technol, 74; mem sci staff, Inst Immunol, Basel, Switz, 77-78, vis scientist, 80, 82 & 84; mem fac, UNESCO-ICLA-ICRO Training Course & Roving Seminars on Deep Freeze Preservation of Mouse Strains, Neth, 75, Denmark, Hungary & Poland, 76 & Czech, Italy & Yugoslavia, 77; adj prof, Health Sci Ctr, Univ Tex, 83-; mem fac, Univ Wis, 88; adv mem, Colombian Ctr Fertil & Steril, 85-; adj assoc sci, Southwest Found Biomed Res, Tex, 86-88; adj prof biomed eng, Univ Tex, Austin, 87-; adj res prof, Ctr Cryobiol Res, State Univ NY, Binghamton, 88-90; bd dirs & exec comt, Hubbs-Sea World Res Inst, Calif, 86-88; consult, UN Food & Agr Orgn, Rome, 87-; mem comt Basic Sci Found Med Assisted Conception, Inst Med-Nat Acad Sci, 87-88. *Mem:* AAAS; Soc Study Reproduction; Soc Cryobiol (vpres, 78-80 & 83-85, pres, 85-); Int Cell Res Orgn. *Res:* Cryobiology and physiology of mammalian embryos, erythrocytes, lymphocytes and tissue-culture cells; bovine reproduction and embryology; micromanipulation of embryos; immunology and cytogenetics of bovine embryos; biology of bacteriophage; cryobiology of bacteriophage, proteins and algae. *Mailing Add:* Dept Biomed Sci Univ Guelph Bldg 165 Guelph ON N1G 2W1 Can

LEIBOVIC, K NICHOLAS, b Plunge, Lithuania, June 14, 21; m 43; c 3. NEUROSCIENCES, BIOPHYSICS. *Educ:* Cambridge Univ, 43; London Univ, 52. *Prof Exp:* Mathematician, Dulwich Col, Eng, 46-53 & Courtaulds, 53-56; proj leader indust math, Brit Oxygen Res & Develop Co, 56-60; sr mathematician, Westinghouse Res Labs, 60-63; prin mathematician, Cornell Aeronaut Lab, 63-64; assoc prof biophys, 64-74, asst dir, Ctr Theoret Biol, 67-68, PROF BIOPHYS, STATE UNIV NY, BUFFALO, 74- *Concurrent Pos:* Lectr, Norwood Col, Eng, 52-53; mem math adv coun, Battersea Col Technol, 59-60; vis prof, Univ Calif, Berkeley, 69, Hadassah Med Sch, Hebrew Univ, 71 & Inst Ophthal, London, 78; prog dir, Neurosci Res Prog, Mass Inst Technol, 78-79; vis scholar, Harvard Univ, 79-80. *Mem:* AAAS; Soc Neurosci; Biophys Soc; Asn Res Vision & Ophthal. *Res:* Information processing in the nervous system; electrophysiology and psychopyhsics of vision; nervous system theory; mathematical models in biology; industrial mathematics; applications to chemical, mechanical and electrical engineering; computers and operations research. *Mailing Add:* 105 High Park Blvd State Univ NY Buffalo NY 14226

LEIBOVICH, SIDNEY, b Memphis, Tenn, Apr 2, 39; m 62; c 2. FLUID DYNAMICS, APPLIED MATHEMATICS. *Educ:* Calif Inst Technol, BS, 61; Cornell Univ, PhD(theoret mech), 65. *Prof Exp:* NATO fel, London, 65-66; from asst prof to assoc prof thermal eng, 66-78, prof, 78-88, SAMUEL B ECKERT PROF MECH & AEROSPACE ENG, CORNELL UNIV, 88- *Concurrent Pos:* Sr vis fel, Math Inst, Univ St Andrews, 77; assoc ed, Soc Indust & Appl Math J Appl Math, 72-75, J Appl Mech, 76-83, J Fluid Mech, 82-; adv, Mech Eng & Appl Mech, Nat Sci Found, 84-85; chmn, Appl Mech Div, Am Soc Mech Engrs, 87-88; chmn, Div Fluid Dynamics, Am Phys Soc, 88-89, US Nat Comt Theoret & Appl Mech, 90- *Mem:* Fel Am Phys Soc; Soc Indust & Appl Math; fel Am Soc Mech Engrs; Am Geophys Union; AAAS. *Res:* Fluid mechanics, particularly dynamics of vortex flows, geophysical fluid dynamics, hydrodynamic stability, and wave propagation phenomena in fluids. *Mailing Add:* 999 Cayuga Heights Rd Ithaca NY 14857

LEIBOWITZ, GERALD MARTIN, b New York, NY, Feb 17, 36; m 63; c 4. MATHEMATICAL ANALYSIS. *Educ:* City Col New York, BS, 57; Mass Inst Technol, SM, 59, PhD(math), 63. *Prof Exp:* Instr math, Mass Inst Technol, 63; from instr to asst prof, Northwestern Univ, 63-68; assoc dir, Comt on Undergrad Prog in Math, Math Asn Am, Calif, 68-69; ASSOC PROF MATH, UNIV CONN, 69- *Concurrent Pos:* Asst engr, Ford Instrument Co, 57. *Mem:* Am Math Soc; Math Asn Am. *Res:* Functional analysis; Banach algebras. *Mailing Add:* Dept Math Univ Conn Storrs CT 06269-3009

LEIBOWITZ, JACK RICHARD, b Bridgeport, Conn, July 21, 29; m; c 2. LOW TEMPERATURE PHYSICS, SUPERCONDUCTIVITY. *Educ:* NY Univ, BA, 51, MS, 55; Brown Univ, PhD(physics), 62. *Prof Exp:* Physicist, Signal Corps Eng Labs, 51-54 & Electronics Corp Am, 55-56; res physicist,

Lincoln Labs, Mass Inst Technol, 56-61 & Westinghouse Res Lab, 61-64; asst prof physics, Univ Md, College Park, 64-69; assoc prof physics, 69-73, chmn, dept art, 82-86, PROF PHYSICS, CATH UNIV AM, 74-, ASSOC DEAN GRAD STUDIES, 88- *Concurrent Pos:* Consult, Nat Broadcasting Co, 79- *Mem:* Fel Am Phys Soc; Sigma Xi. *Res:* Superconductivity; ultrasonic interactions in solids; intermediate and mixed states; Fermi surfaces; electron-phonon interaction; excitation spectra of inhomogeneous superconductors; physical acoustics of solids. *Mailing Add:* Dept Physics Cath Univ Am Washington DC 20064

LEIBOWITZ, JULIAN LAZAR, b New York, NY, Dec 14, 47. VIROLOGY, PATHOLOGY. *Educ:* Alfred Univ, BA, 68; Albert Einstein Col Med, PhD(cell biol), biol), 74, MD, 75. *Prof Exp:* Med scientist trainee virol, Albert Einstein Col Med, 70-74, med scientist trainee med, 74-75; path resident, Univ Calif, San Diego, 75-77, USPHS fel neuropath & virol, 77-79, asst prof, 79-83; asst prof, 83-85, ASSOC PROF PATH, UNIV TEX, HEALTH SCI CTR, HOUSTON, 85- *Mem:* AAAS; Am Soc Microbiol; Am Soc Virol. *Res:* Animal virology; virus induced demyelinating disease. *Mailing Add:* Dept Path Lab Med Univ Tex Health Sci GR PO Box 20036 Houston TX 77225

LEIBOWITZ, LEONARD, b New York, NY, Feb 5, 31; m 76; c 3. THERMOPHYSICAL PROPERTIES. *Educ:* NY Univ, AB, 51, MS, 54, PhD(chem), 56. *Prof Exp:* Chemist, Pigments Dept, E I du Pont de Nemours & Co, 56-58; asst chemist, Argonne Nat Lab, 58-61, assoc chemist, 61-72, chemist, 72-88, SR CHEMIST, CHEM TECHNOL DIV, ARGONNE NAT LAB, 88- *Mem:* Am Chem Soc; Sigma Xi; AAAS. *Res:* Experimental and theoretical determination of thermodynamic and transport properties of materials, particularly at high temperature. *Mailing Add:* Chem Technol Div Argonne Nat Lab Argonne IL 60439

LEIBOWITZ, LEWIS PHILLIP, b Chicago, Ill, June 22, 42; m 67; c 3. ADVANCED SPACE SYSTEMS, LARGE OPTICAL SYSTEMS. *Educ:* Northwestern Univ, BS, 64; Univ Calif, San Diego, MS, PhD(eng sci), 69. *Prof Exp:* Mem tech staff atmospheric entry technol, Calif Inst Technol, Jet Propulsion Lab, 69-75, team leader geothermal energy, 75-77, mgr advan solar technol, 77-80; prob mgr, NASA, Dept Defense Exploratory Technol, 80-83; mgr, Technol Appl, Int Power Technol, Inc, 83-85; ADVAN SYSTS LEADER, LOCKHEED MISSLES & SPACE CO, 85- *Concurrent Pos:* Team leader, oil explor assessment, Jet Propulsion Lab, Calif Inst Technol, 77. *Honors & Awards:* New Technol Award, NASA, 76. *Mem:* Am Phys Soc; Sigma Xi. *Res:* Dynamic performance of distributed communication and data processing networks, system definition and performance analysis of advanced space systems for communications, space surveillance, astrophysics investigation and robotic applications; developed experiments for the testing of large antennas and deployable optical systems in space; energy systems; solar and thermal technology; developed comprehensive system/cost effectiveness model that permits trade-off of advanced technology, reliability and maintenance. *Mailing Add:* 141 Durazno Way Portola Valley CA 94028

LEIBOWITZ, MARTIN ALBERT, b New York, NY, Oct 15, 35. APPLIED MATHEMATICS, OPERATIONS RESEARCH. *Educ:* Columbia Univ, BA, 56; Harvard Univ, MA, 57, PhD(appl math), 61. *Prof Exp:* Staff scientist, Int Bus Mach Res Ctr, 60-63; mem tech staff, Bellcomm, Inc, Washington, DC, 63-66; assoc prof eng, 66-73, ASSOC PROF APPL MATH & STATIST, STATE UNIV NY, STONY BROOK, 73- *Mem:* Opers Res Soc Am; Asn Comput Mach. *Res:* Random processes with application to control theory; guidance and communications; scientific programming. *Mailing Add:* 15 Charles St New York NY 10014

LEIBOWITZ, MICHAEL JONATHAN, b Brooklyn, NY, May 14, 45; m 66; c 2. VIROLOGY, PLASMIDS. *Educ:* Columbia Univ, AB, 66; Albert Einstein Col Med, PhD(molecular biol), 71, MD, 73. *Prof Exp:* Intern, Barnes Hosp, Wash Univ, St Louis, 74; res assoc, Nat Inst Gen Med Sci & guest worker, Lab Biochem Pharmacol, Nat Inst Arthritis, Metab & Digestive Dis, NIH, 74-76, sr staff fel, Lab Biochem Pharmacol, 76-77; from asst prof to assoc prof, 82-86, PROF MOLECULAR GENETICS & MICROBIOL, UNIV MED & DENT NJ, ROBERT WOOD JOHNSON MED SCH, 86-, DIR, DNA SYNTHESIS NETWORK LAB, NJ CTR ADVAN BIOTECHNOL & MED, 86- *Concurrent Pos:* Dir, Lore Curric Molecular & Cell Biol, Rutgers Univ & Univ Med & Dent NJ; Alexandrine & Alexander L Sinsheimer Scholar, 80-83; ad hoc mem, Genetics Study Sect, NIH, 84, mem, Biomed Sci Rev Group, Subcomt 7, 86. *Mem:* Am Soc Cell Biol; Am Chem Soc; Genetics Soc Am; Am Soc Microbiol; Am Soc Virol; AAAS. *Res:* Molecular basis of the interaction of viruses and plasmids with eukaryotic host cells, mainly in Saccharomyces cevevisiae; molecular genetics of pneumocystis carinii. *Mailing Add:* Dept Molecular Genetics & Microbiol UMDNJ Robert Wood Johnson Med Sch 675 Hoes Lane Piscataway NJ 08854-5635

LEIBOWITZ, SARAH FRYER, b White Plains, NY, May 23, 41; m 66; c 3. PSYCHOPHARMACOLOGY, NUTRITION. *Educ:* NY Univ, BA, 64, PhD(physiol psychol), 68. *Prof Exp:* USPHS fel & guest investr, 68-70, asst prof, 70-78, ASSOC PROF NEUROPHARMACOL, ROCKEFELLER UNIV, 78- *Concurrent Pos:* Alfred P Sloan found award, 77-79. *Honors & Awards:* First Prize, Div Psychopharmacol, Am Psychol Asn, 69. *Mem:* AAAS; fel Am Psychol Asn; NY Acad Sci; Am Soc Pharmacol & Exp Therapeut; Soc Neurosci; Asn Res Neurol & Ment Dis; Sigma Xi; fem Am Psychol Soc; fel Acad Behav Med Res. *Res:* Study of neurochemical mechanisms in the brain which regulate behavioral and physiological responses. *Mailing Add:* Rockefeller Univ 1230 York Ave New York NY 10021

LEIBSON, IRVING, b Wilkes Barre, Pa, Sept 28, 26; m 50. CHEMICAL ENGINEERING. *Educ:* Univ Fla, BChE, 45, MS, 47; Carnegie Inst Technol, 49, DSc(chem eng), 52. *Prof Exp:* Chem engr, Humble Oil & Refining Co, 52, staff engr, 57-59, supv engr, 59-61; mgr process engr, Rexall Chem Co, 61-63, develop mgr, 63-65, dir res & develop, 65-67, dir com develop, 67, gen mgr acrylonitrile-butadiene-styrene plastic div, 67-69; vpres, Dart Indust Chem Group, 69-74, mgr com ventures & investment dept, 74-75; mgr process & environ sci develop, 75-78, VPRES & MGR RES & ENG, BECHTEL GROUP INC, 78- *Concurrent Pos:* Lectr, Univ Md; prof, Rice Univ, 58; mem, Eng Manpower Comn; assoc, Coal Indust Adv Bd, Int Energy Agency, 80- & World Coal Study, 79-80. *Mem:* Am Chem Soc; Am Inst Chem Eng; fel Am Inst Chem. *Res:* Unit operations. *Mailing Add:* 2920 SE Dunes Dr Apt 410 Stuart FL 34996

LEIBSON, PAUL JOSEPH, b Chicago, Ill, June 15, 52; m; c 1. IMMUNOLOGY. *Educ:* Univ Ill, Urbana, BS, 74; Univ Chicago, PhD(immunol), 79, MD, 81; Am Bd Pediat, cert, 86. *Prof Exp:* Intern & resident, Health Sci Ctr, Univ Colo, Denver, 81-84; fel, Nat Jewish Hosp & Res Ctr, Denver, 84-86; ASST PROF, DEPT IMMUNOL, MAYO CLIN, 86- *Concurrent Pos:* Course chmn immunol, Mayo Med Sch. *Honors & Awards:* Henry Kaplan Award, 88. *Mem:* Am Asn Immunologists; Soc Leukocyte Biol. *Res:* Author of numerous publications. *Mailing Add:* Dept Immunol Mayo Clin 301 Guggenheim Bldg Rochester MN 55905

LEIBU, HENRY J, b Schlesiengrube, Ger, Apr 22, 17; c 2. CHEMISTRY. *Educ:* Swiss Fed Inst Technol, ChemE, 42, ScD(chem), 45. *Prof Exp:* Instr & res assoc indust chem, Swiss Fed Inst Technol, 45-49; chemist res plant develop, Polychem Dept, E I du Pont de Nemours & Co, Inc, 49-59, sales develop, Europe, SAm & Australia, 59-66, sr res chemist res & develop elastomers, 66-73, tech sales & develop, 73-85; RETIRED. *Concurrent Pos:* Consult, elastomers, med prod, high performance liquid chromatography. *Mem:* Am Chem Soc. *Res:* Elastomers; urethanes. *Mailing Add:* 4905 Threadneedle Rd Wilmington DE 19807

LEIBY, CLARE C, JR, b Ashland, Ohio, May 4, 24; m 52; c 5. GRAVITATION & ELECTROMAGNETISM, BLACK-HOLES. *Educ:* Mass Inst Technol, SB, 54; Univ Ill, MS, 58. *Prof Exp:* Physicist, Air Force Cambridge Res Labs, 54-56; res assoc, Univ Ill, 60-61 & Sperry Rand Res Ctr, 61-64; res physicist, Air Force Cambridge Res Labs, 64-76; RES PHYSICIST, ROME AIR DEVELOP CTR, 76- *Concurrent Pos:* Consult, Leghorn Labs. *Mem:* Sigma Xi. *Res:* Gravitation theory, gravitation and its relationship to electromagnetism; extremely dense astrophysical objects, such as black holes, neutron and white dwarf stars; experimental and theoretical research in laser interactions with molecules; molecular beams; atomic clocks; neutron/white dwarf stars. *Mailing Add:* 229 Old Billerica Rd Bedford MA 01730

LEIBY, ROBERT WILLIAM, b Allentown, Pa, Apr 2, 49; m 74. ORGANIC CHEMISTRY. *Educ:* Albright Col, BS, 71; Lehigh Univ, MS, 73, PhD(org chem), 75. *Prof Exp:* Res assoc org chem, Dartmouth Col, 75-76; med chemist, Purdue Frederick Co, 76-77; vis asst prof chem, Hampden-Sydney Col, 77-78; vis asst prof chem, Duke Univ, 78-80; mem fac, dept chem, Univ Wis-Whitewater, 80-85; Univ RI, 85-87; ASST PROF CHEM, SOUTHERN ILL UNIV, 87- *Mem:* Am Chem Soc; Sigma Xi. *Res:* Organic mass spectroscopy; heterocyclic synthesis; synthesis of pharmaceutical agents particularly central nervous system agents and antineoplastic agents: organic rearrangements and mechanisms; development of new synthetic techniques and reagents. *Mailing Add:* 700 N Buchana St Edwardsville IL 62025

LEICH, DOUGLAS ALBERT, b Paterson, NJ, Jan 26, 47; m 72; c 1. NUCLEAR COSMOCHEMISTRY, MASS SPECTROMETRY. *Educ:* Colgate Univ, BA, 68; Calif Inst Technol, PhD(physics), 74. *Prof Exp:* Asst res physicist, Univ Calif, Berkeley, 73-76; dep div leader, 86-89, CHEMIST NUCLEAR CHEM, LAWRENCE LIVERMORE LAB, UNIV CALIF, 76- *Concurrent Pos:* Mem, Lunar & Planetary Sample Team, Lunar & Planetary Inst, 81-83. *Mem:* Meteoritical Soc; Am Phys Soc; Am Chem Soc. *Res:* Isotopic abundance variations in materials and their uses in interpreting natural and anthropagenic processes. *Mailing Add:* Nuclear Chem Div Mail Code L-232 PO Box 808 Livermore CA 94551

LEICHNER, GENE H(OWARD), b Richmond, Ind, Nov 14, 29; m 59; c 1. ELECTRICAL ENGINEERING. *Educ:* Univ Ill, BS, 51, MS, 55, PhD(elec eng), 58. *Prof Exp:* Elec engr electronics, Ballistic Res Lab, Aberdeen Proving Ground, Md, 51-52; elec engr, Control Systs Lab, Univ Ill, 55-56, asst, Digital Comput Lab, 56-58, assoc prof elec eng, 58-67, dir eng, Intersci Res Inst, 65-67; leader systs eval, RCA Instnl Systs, Systs Control, Inc, Palo Alto, 67-69, mgr eng, 69-72, prog mgr, Comput Systs Develop, 72-85; VPRES, EPIC ENG INC, 85- *Mem:* Inst Elec & Electronics Engrs; Asn Comput Mach. *Res:* Applications of computers to business and scientific problems; computer system performance analysis and prediction; on line computer applications. *Mailing Add:* Epic Eng Inc 5150 El Camino Real Suite A30 Los Altos CA 94022

LEICHNER, PETER K, b W Berlin, Ger, Jan 24, 39; m 86; c 1. DOSIMETRY, TOMOGRAPHY. *Educ:* Univ Calif, Riverside, BA, 64; Univ Kans, PhD (physics), 69. *Prof Exp:* Asst prof physics, McMurry Col, Abilene, Tex, 69-71; asst prof, Univ Ky, Lexington, 71-77; from instr to asst prof, 77-82, ASSOC PROF ONCOL, JOHNS HOPKINS UNIV MED SCH, 83- *Concurrent Pos:* Invited lectr, Radiol Soc NAm, 80-88, Soc Nuclear Med, 85-, Am Col Nuclear Physicians, 86-, Univ London, Royal Postgrad Sch, 86, Nat Cancer Inst, 87, Am Soc Therapeut Radiol & Oncol, 87; assoc ed, Antibody Immunoconjugates & Radiopharmaceuts, 87- *Honors & Awards:* R S Landauer Mem Award & Lectr, 85. *Mem:* Am Phys Soc; Am Asn Physicists Med; Soc Nuclear Med; Am Soc Therapeut Radiol & Oncol. *Res:* Quantitative medical imaging and image analysis; tomographic reconstruction algorithms; radiation dosimetry for radiolabeled cancer therapy agents. *Mailing Add:* Dept Radiol Univ Nebr Med Ctr 600 S 42nd St Omaha NE 68198-1045

LEICHNETZ, GEORGE ROBERT, b Buffalo, NY, Oct 15, 42; m 67; c 3. NEUROANATOMY. *Educ:* Wheaton Col, Ill, BS, 64; Ohio State Univ, MS, 66, PhD(anat), 70. *Prof Exp:* Instr anat, Ohio State Univ, 69-70; from asst prof to assoc prof, 70-85, PROF ANAT, MED COL VA, VA COMMONWEALTH UNIV, 85- *Concurrent Pos:* A D Williams res grant,

Med Col Va, Va Commonwealth Univ, 71-72; NSF grant, 78-86 & 90- *Mem:* Am Asn Anatomists; Soc Neurosci; Am Sci Affil. *Res:* Comparative neuroanatomy of primates; connections from cerebral cortex to brainstem pre-oculomotor nuclei; central nervous system connections related to pain mechanisms. *Mailing Add:* Dept Anat Med Col Va Box 709 Va Commonwealth Univ Richmond VA 23298-0709

LEICHTER, JOSEPH, b Feb 4, 32; US citizen. NUTRITION, FOOD SCIENCE. *Educ:* Cracow Col, Poland, BS, 56; Univ Calif, Berkeley, MS, 66, PhD(nutrit), 69. *Prof Exp:* Chemist, Pharmaceut Plant, Cracow, Poland, 56-57; chemist, Ministry Com & Indust, Haifa, Israel, 57-60; chemist, Anresco Lab, Calif, 60-65; from asst prof to assoc prof, 69-82, PROF NUTRIT, UNIV BC, 82- *Mem:* Am Inst Nutrit; Can Soc Nutrit Sci. *Res:* Folic acid metabolism; effect of protein-calorie malnutrition on carbohydrate digestion and absorption; effect of dietary lactose on intestinal lactase activity; effect of maternal alcohol consumption on growth and development of offspring. *Mailing Add:* Human Nutrit Div Sch Family & Nutrit Sci Univ BC Vancouver BC V6T 1W5 Can

LEID, R WES, b Walla Walla, Wash, May 25, 45; m 66; c 2. INFLAMMATION, BIOCHEMICAL MECHANISMS OF VIRAL SUPPRESSION OF ALVEOCAR MACROPHAGE FUNCTION. *Educ:* Cent Wash Univ, BA, 68 & MS, 70; Mich State Univ, PhD(microbiol), 73. *Prof Exp:* Fel immunol, Harvard Med Sch, 73-76, instr med, 76-77; asst prof path, Mich State Univ, 77-80; assoc prof, 80-85, PROF MICROBIOL & PATH, WASH STATE UNIV, 85- *Concurrent Pos:* Consult, WHO, 78; reviewer, Am J Vet Res, 80-83; vis prof, Walter & Eliza Hall Inst, Melbourne, 85. *Mem:* Am Asn Immunol; Am Asn Pathologists; AAAS; Reticulendothelial Soc; Am Soc Biochem & Molecular Biol. *Res:* Pulmonary inflammation; immediate hypersensitivity; eosinophil, neutrophil, alveolar macrophage and mast cell and basophilic function; platelet biochemistry, molecular interactions of the complement cascade. *Mailing Add:* Dept Vet Microbiol & Path Lab Molecular & Cellular Inflammation Wash State Univ Pullman WA 99164-7040

LEIDER, HERMAN R, b Detroit, Mich, Jan 14, 29; m 60; c 2. SOLID STATE CHEMISTRY, PHYSICAL CHEMISTRY. *Educ:* Wayne State Univ, BS, 51, PhD(chem), 54. *Prof Exp:* Asst, Wayne State Univ, 51-52, res assoc, 52-54; aeronaut res scientist, Solid State Physics Br, Chem Mat Sec, Nat Adv Comt Aeronaut, 54-56; chemist, 56-70, group leader, Hydrides Group, 70-76, & Chem Compatibility Group, 76-81, SECT LEADER PHYS CHEM, LAWRENCE LIVERMORE LAB, UNIV CALIF, 81- *Mem:* Am Phys Soc; Am Chem Soc; AAAS. *Res:* Alkali halides; compatibility of materials; color center; luminescence; radiation effects; hydrides. *Mailing Add:* Dept Chem Lawrence Livermore Lab Livermore CA 94550

LEIDERMAN, P HERBERT, b Chicago, Ill, Jan 30, 24; m 47; c 4. PSYCHIATRY. *Educ:* Calif Inst Technol, MS, 49; Univ Chicago, MA, 49; Harvard Med Sch, MD, 53. *Prof Exp:* Asst psychol, Univ Chicago, 48-49; intern med, Beth Israel Hosp, 53-54; resident neurol, Boston City Hosp, 54-56; resident psychiat, Mass Gen Hosp, 56-57; res fel, Mass Ment Health Ctr, 57-58; assoc, Harvard Med Sch, 58-63; assoc prof, 63-68, PROF PSYCHIAT, MED SCH, STANFORD UNIV, 68- *Concurrent Pos:* Consult, USPHS & Nat Res Coun. *Mem:* AAAS; Am Psychosom Soc; Soc Res Child Develop; Am Acad Child Psychiat; Sigma Xi. *Res:* Child development; psychology; transcultural psychiatry; psychophysiology. *Mailing Add:* 828 Lathrop Dr Stanford CA 94305

LEIDERSDORF, CRAIG B, b Mineola, NY, Mar 12, 50; m 84. COASTAL ENGINEERING, ARCTIC ENGINEERING. *Educ:* Stanford Univ, BS, 72; Univ Calif, Berkeley, MS, 75. *Prof Exp:* Res assoc hydraul res, Univ Calif, Berkeley, 75-76; coastal engr, Swan Wooster Eng Co Ltd, 76-77 & Tetra Tech, Inc, 77-80; EXEC VPRES COASTAL ENG, TEKMARINE, INC, 80- *Concurrent Pos:* Guest lectr, Univ Calif, Berkeley, 82, 84, 85. *Mem:* Am Soc Civil Engrs; Marine Technol Soc; Am Shore & Beach Preserv Asn. *Res:* Coastal engineering and coastal oceanography, including field data acquisition, assessment of coastal stability and design of slope protection; arctic engineering relating to offshore structures and sea ice. *Mailing Add:* Coastal Frontiers Inc 9424 Eton Ave SW Chatsworth CA 91311

LEIDHEISER, HENRY, JR, b Union City, NJ, Apr 18, 20; m 44; c 2. PHYSICAL CHEMISTRY. *Educ:* Univ Va, BS, 41, MS, 43, PhD(phys chem), 46. *Prof Exp:* Res worker, Nat Adv Comt Aeronaut Proj, Univ Va, 43-45, res assoc, Cobb Chem Lab, 46-49; proj dir, Va Inst Sci Res, 49-52, mgr lab, 52-58, dir res, Lab, 58-60, dir res, Inst, 60-68; prof chem & dir, Zettlemoyer Ctr Surface Studies, Lehigh Univ, 68-83; chair prof, Alcoa, 83-90, dept chmn, 88-89; RETIRED. *Concurrent Pos:* Mem adv comt, Oak Ridge Nat Lab; chmn, Gordon Conf Corrosion, 64; NATO sr scientist fel, Cambridge Univ, 69; consult, Marshall Space Flight Ctr, NASA, 71-; Von Humboldt Award, 85-86. *Honors & Awards:* Award, Oak Ridge Inst Nuclear Studies, 48; J Shelton Horsley Res Prize, Va Acad Sci, 49; Young Auth Prize, Electrochem Soc; Silver Medal, Am Electroplaters' Soc, 78; Arch T Colwell Award, Soc Automotive Engrs, 78; Whitney Award, Nat Asn Corrosion Engrs, 83; Humboldt Award, 85; Silver Medal, SAfrican Corrosion Inst, 86; Electrode Position Div Res Award, Electrochem Soc, 87; Mattiello Award, Fedn Socs Coatings Technol, 90; Uhlig Award, Electrochem Soc, 91. *Mem:* Am Chem Soc; Electrochem Soc; Nat Asn Corrosion Engrs; fel AAAS. *Res:* Corrosion; surface science; electrodeposition; Mossbauer spectroscopy; polymer coatings; long-term food storage; paint adherence. *Mailing Add:* 822 Carnoustie Dr Venice FL 34295

LEIDY, BLAINE I(RVIN), b Conemaugh, Pa, Aug 15, 23; m 57; c 1. MECHANICAL ENGINEERING. *Educ:* Univ Pittsburgh, BS, 51, MS, 57, PhD(mech eng), 62. *Prof Exp:* From instr to asst prof, Univ Pittsburgh, 51-62, assoc prof mech eng, 62-90; RETIRED. *Concurrent Pos:* Consult, Westinghouse Elec Corp, 54-57 & 69-73. *Mem:* Am Soc Eng Educ; Am Soc Mech Engrs. *Res:* Heat transfer; fluid mechanics; thermodynamics; energy conservation in small industries under federally supported grant. *Mailing Add:* Dept Mech Eng Sch Eng Univ Pittsburgh Pittsburgh PA 15261

LEIDY, ROSS BENNETT, b Newark, Ohio, June 1, 39; m 71; c 2. AGRICULTURAL & FOOD CHEMISTRY. *Educ:* Tex A&M Univ, BS, 63, MS, 66; Auburn Univ, PhD(biochem), 72. *Prof Exp:* Res asst radiation biol, Radiation Biol Lab, Tex A&M Univ, 65-66; instr & lab supvr, Biol Br, Microbiol Lab, US Army Chem Ctr & Sch, Ft McClellan, Ala, 66-68; supvr pesticide chem, Lab Div, NC Dept Human Resources, 73-74; SR RES SCIENTIST, PESTICIDE RESIDUE RES LAB, NC STATE UNIV, 74- *Concurrent Pos:* NIH res assoc, Dept Animal Sci, NC State Univ, 72-73; consult, pesticide residues in air, soils & surfaces in home & working environ. *Mem:* Sigma Xi; NY Acad Sci; Am Chem Soc. *Res:* Methodology and analyses of pesticide residues on plant and animal products and in air relating to the laboratory's research projects. *Mailing Add:* Pesticide Residue Res Lab 3709 Hillsborough St NC State Univ Raleigh NC 27607

LEIER, CARL V, b Bismarck, NDak, Oct 20, 44; m 70; c 3. BIOLOGY, MEDICINE. *Educ:* Creighton Univ, BS, 65; Creighton Univ Col Med, MD, 69; Am Bd Internal Med, cert & dipl, 73; cardiovascular subspecialty, Am Bd Internal Med, cert, 78. *Prof Exp:* Intern, 69-70, instr med, 71-73, chief instr, 73-74, clin instr, Div Cardiol, 74-76, from asst prof to assoc prof, 76-84, PROF MED & PHARMACOL, OHIO STATE UNIV COL MED, 84-, DIR, DIV CARDIOL, 86- *Concurrent Pos:* Mem, Internship Selection Comt, dept med, Ohio State Univ Col Med, 73-74, hosp procedures; comt, Ohio State Univ Hosps, 73-74, pharmacol & therapeut comt, 76-80; res comt, Cent Ohio Heart Chapter, Am Heart Asn, 77-84, mem, bd trustees, 79-; fac mem, grad sch, Ohio State Univ Col Med, 80-, dir res, div cardio, 80-83, James W Overstreet prof med, 83- *Mem:* Am Col Physicians; fel Am Col Clin Pharmacol; Am Fed Clin Res; Am Col Cardiol; AAAS; Am Soc Clin Invest; Int Soc Heart Res. *Mailing Add:* Div Cardiol Col Med Ohio State Univ Hosp Rm 669 1654 Upham Dr Columbus OH 43210

LEIES, GERARD M, b Chicago, Ill, Aug 19, 18. NUCLEAR PHYSICS. *Educ:* Loyola Univ, Ill, BS, 40; Univ Calif, MA, 53; Georgetown Univ, PhD(physics), 62. *Prof Exp:* Asst tech dir, Air Force Tech Appln Ctr, 62-73 command tech dir, 73-88; RETIRED. *Mem:* Am Phys Soc; Am Geophys Union. *Res:* Nuclear weapon test detection. *Mailing Add:* 472 Carmine Dr Cocoa Beach FL 32931

LEIF, ROBERT CARY, b New York, NY, Feb 27, 38; m 63; c 2. IMMUNOHEMATOLOGY, BIOMEDICAL ENGINEERING. *Educ:* Univ Chicago, BS, 59; Calif Inst Technol, PhD(chem), 64. *Prof Exp:* Fel, Univ Calif, Los Angeles, 64-66; res assoc microbiol, Sch Med, Univ Southern Calif, 66-67; asst prof chem & biochem, Fla State Univ, 67-71; assoc scientist, Papanicolaou Cancer Res Inst, 71-72, sr scientist, 72-81; PRIN SCIENTIST, COULTER ELECTRONICS, 81- *Concurrent Pos:* Consult, Int Equip Corp, 65-73, Xerox Corp, 66-67, Damon Eng, 69-73, Solid State Radiation, Calif, 67-73, Coulter Electronics, 75- & Photometrics, 75-; res scientist, Dept Microbiol, Univ Miami, 71-72, adj asst prof microbiol, 73-76 & adj asst prof biomed eng, 74-76, assoc prof microbiol & biomed eng, 76-, assoc prof oncol, 80- *Mem:* AAAS; Sigma Xi; Biomed Eng Soc; Am Chem Soc; Am Soc Cytol. *Res:* Cellular differentiation; cytology automation; clinical chemistry instrumentation; cytology specimen preparation; computer based instrumentation; cytophysical and histochemical techniques to separate, purify and analyze heterogeneous cell populations; identification of biological activities with cell morphology. *Mailing Add:* 1030 Mariposa Ave Coral Gables FL 33146

LEIFER, CALVIN, b New York, NY, Mar 4, 29; m 63; c 2. EXPERIMENTAL PATHOLOGY, ELECTRON MICROSCOPY. *Educ:* NY Univ, BA, 50, DDS, 54; State Univ NY Buffalo, PhD(exp path), 71. *Prof Exp:* Pvt pract, 58-65; USPHS fels, State Univ NY Buffalo, 65-70; assoc prof, 70-78, PROF PATH, SCH DENT, TEMPLE UNIV, 78- *Concurrent Pos:* Pres, Philadelphia Sect, Am Asn Dent Res, 85-90. *Mem:* Am Dent Asn; hon mem Sigma Xi; AAAS; Int Asn Dent Res; Am Acad Oral Path; NY Acad Sci. *Res:* Ultrastructural and biochemical alterations of the rat parotid gland following single and multiple doses of x-irradiation; ultrastructural and histochemical features of tumors of the salivary glands in humans; ultrastructure of human dental pulp. *Mailing Add:* Dept Path Temple Univ Sch Med Philadelphia PA 19140

LEIFER, HERBERT NORMAN, b New York, NY, Jan 30, 25; m 48; c 3. SOLID STATE PHYSICS. *Educ:* Univ Calif, Los Angeles, BA, 48, PhD(physics), 52. *Prof Exp:* Asst physics, Univ Calif, Los Angeles, 48-51, res engr, 51-52; res assoc, Res Lab, Gen Elec Corp, 52-55; staff scientist, Lockheed Missiles & Space Co, 55-59, mgr solid state electronics dept, 61-62; mgr basic physics, Fairchild Semiconductor Corp, 62-65; staff scientist, Electro optical Lab, Autonetics Div, NAm Aviation, Inc, Calif, 65-67 & High Energy Laser Lab, TRW Systs Group, 68-72; sr staff scientist, Rand Corp, 72-79; sr staff scientist, Rocketdyne Div, Rockwell Int, 79-90; RETIRED. *Concurrent Pos:* Mem staff, Physics Lab, Ecole Normale Superieure, Paris, 60. *Mem:* Am Phys Soc. *Res:* Semiconductors; thermoelectric effects; electron-acoustic interactions; electrooptic effects; lasers. *Mailing Add:* 16557 Park Lane Circle Los Angeles CA 90049

LEIFER, LARRY J, BIOMEDICAL ENGINEERING. *Educ:* Stanford Univ, BS, 62, MS, 63, PhD(biomed eng), 69. *Prof Exp:* Staff mem, Ames Res Ctr, NASA, 69-72; NASA res exchange fel, Man-Vehicle Lab, Mass Inst Technol, 73; asst prof biomed systs anal, Swiss Fed Inst Technol, Zurich, 73-76; assoc prof, 76-82, PROF MECH ENG & DIR, CTR DESIGN RES, STANFORD UNIV, 82- *Concurrent Pos:* Co-founder, Ohlone Int Corp, 89- *Res:* Developed laboratory and curriculum for programmable electromechanical systems design; develop basic design theory and methodology through application of knowledge-based engineering technology to a wide range of industrial machine design problems. *Mailing Add:* Mech Eng Design Div Stanford Univ Stanford CA 94305-4021

LEIFER, LESLIE, b New York, NY, Apr 13, 29; m 57; c 1. PHYSICAL CHEMISTRY. *Educ:* City Col New York, BS, 50; Univ Kans, PhD(phys chem), 59. *Prof Exp:* Res assoc nuclear & inorg chem, Mass Inst Technol, 56-59; asst prof chem, Clark Univ, 59-60; res assoc & staff mem, Lab Nuclear Sci, Mass Inst Technol, 61-63; assoc prof, Boston Col, 63-66; PROF CHEM, MICH TECHNOL UNIV, 66- *Concurrent Pos:* Travel awards, US AEC, Stockholm, 62, Australia, 63, Stockholm, 71; res award, Mich Technol Univ, 70. *Mem:* AAAS; Am Chem Soc; Sigma Xi. *Res:* Solution physical chemistry; Mossbauer spectroscopy; quantum chemistry; energy storage materials. *Mailing Add:* Dept Chem Mich Technol Univ Houghton MI 49931

LEIFER, ZER, b Brooklyn, NY, May 24, 41; m 72; c 5. GENETIC TOXICOLOGY, POLYAMINE BIOSYNTHESIS. *Educ:* Yeshiva Univ, BA, 63; Harvard Univ, MA, 65; NY Univ, PhD(microbiol), 72. *Prof Exp:* Fel microbiol, NY Univ, 72-74, Queens Col, City Univ New York, 74-76; res asst prof, 76-81, RES ASSOC PROF MICROBIOL, NY MED COL, 81- *Concurrent Pos:* Mem, DNA Repair Deficient Bacterial Assay Work Group, Genetic Toxicol Prog, Environ Protection Agency, 78-81. *Mem:* Am Soc Microbiol; Am Chem Soc; AAAS; Environ Mutagen Soc; Sigma Xi. *Res:* Development and utilization of microbiol assay systems for the detection of environment mutagens and carcinogens; biosynthesis and biological role of polyamines. *Mailing Add:* Dept Basic Sci NY Col Podiatric Med 53 E 124th St New York NY 10035

LEIFIELD, ROBERT FRANCIS, b St Louis, Mo, Jan 29, 28; m 52; c 7. INORGANIC CHEMISTRY. *Educ:* St Louis Univ, BS, 52, MS, 59. *Prof Exp:* Chemist, Great Lakes Carbon Co, 52-56; chemist, Mallinckrodt Chem Works, 56-58, supvr metall & ceramics, 58-61, group leader process develop, 61-62, res chemist, 62-66, assoc mgr res, 66-69, res & develop mgr, Calsicat Div, 69-70, tech mgr, 70-77, dir res & develop, chem div, 77-85, dir new prod develop, Catalysts & Performance Chem Div, Mallinckrodt Inc, 85-86; PRES, LEIFIELD INC, 86- *Mem:* Catalysis Soc; Am Chem Soc; Licensing Execs Soc. *Res:* Column and thin layer chromatography; analytical reagents; product and process research and development; uranium metallurgical chemistry; heterogeneous catalysis; licensing consultation. *Mailing Add:* 7433 Hibbard Lane St Louis MO 63123-2015

LEIFMAN, LEV JACOB, analysis, probability; deceased, see previous edition for last biography

LEIGA, ALGIRD GEORGE, b New York, NY, Mar 25, 33; m 55; c 4. PHYSICAL CHEMISTRY. *Educ:* NY Univ, BA, 55, MS, 60, PhD(phys chem), 63. *Prof Exp:* Res scientist, Dept Chem, NY Univ, 62-64; sr scientist, Mat Sci Lab, NY, 64-73, mgr mat develop, Xeroradiography, Pasadena, 73-77; corp res & develop staff, Xerox Corp, Palo Alto, 77-81; mgr mat technol opers, Pasadena, 81-87, mgr supplies bus ctr, Monrovia, 88-89, MGR MAT TECHNOL & MFG, XEROX MED SYSTS, MONROVIA, CALIF, 89- *Concurrent Pos:* Councilman, City Claremont, Calif, 90-94. *Mem:* Am Chem Soc; Optical Soc Am; Soc Photog Scientists & Engrs; Am Phys Soc. *Res:* Vacuum ultraviolet photochemistry and spectroscopy; decomposition reactions of solids; materials development for electrophotography. *Mailing Add:* 3790 Elmira Ave Claremont CA 91711

LEIGH, CHARLES HENRY, b Southport, Lancashire, Jan 10, 26, US citizen; m 49; c 3. RE-ENTRY PHYSICS & CHEMISTRY, TECHNICAL MANAGEMENT & CONSULTING. *Educ:* Victoria Univ Manchester, UK, BS, 47, PhD(chem), 52. *Prof Exp:* Chemist transuranics, Harwell, UK Atomic Energy, 48-50; vis scientist chem, Brookhaven Nat Lab, 52-54; chemist thermodyn, M W Kellogg Co, 54-57; chief phys sci, avco-RAD, 57-62; chief tech proj, NASA-ERC, 62-70; SR SCIENTIST KINETICS, HYDRO-QUE RES INST, 70- *Concurrent Pos:* Consult, Air Force Spec Weapons Ctr, 58-59, Fed Aviation Agency, 65-66, Univ Que, 71-72, Can Elec Asn, 72-73. *Res:* Thermodynamics and chemical kinetics, high temperature reactions; chemical reaction mechanisms; radiochemistry; hydrocarbon conversion reactions; transuranic chemistry; atomic energy, free radical reactions, plasma physics and heat transfer; electro-optics, infrared, missile technology. *Mailing Add:* Hydro-Que Res Inst Varennes PQ J3X 1S1 Can

LEIGH, DONALD C, b Toronto, Ont, Feb 25, 29; nat US; m 52; c 3. CONTINUUM MECHANICS. *Educ:* Univ Toronto, BASc, 51; Cambridge Univ, PhD(math), 54. *Prof Exp:* Sr aerophys engr, Gen Dynamics/Ft Worth, 54-56; supvr tech comput, Curtiss-Wright Corp, 56-57; lectr mech eng, Princeton Univ, 57-58, asst prof aerospace & mech sci, 58-65; assoc prof eng mech & math, 65-68, chmn dept eng mech, 72-80, PROF ENG MECH, UNIV KY, 68-, ASSOC DEAN ENG, 83- *Mem:* Soc Natural Philos; Soc Rheol; Am Acad Mech; Am Soc Mech Engrs; Nat Soc Prof Engrs. *Res:* Continuum mechanics; systems engineering; pressure vessels. *Mailing Add:* Dept Eng Mech Univ Ky Lexington KY 40506-0046

LEIGH, EGBERT G, JR, b Richmond, Va, July 27, 40; m 68; c 2. ECOLOGY, POPULATION GENETICS. *Educ:* Princeton Univ, AB, 62; Yale Univ, PhD(biol), 66. *Prof Exp:* Actg instr biol, Stanford Univ, 66; asst prof, Princeton Univ, 66-72; BIOLOGIST, SMITHSONIAN TROP RES INST, 69- *Mem:* Ecol Soc Am; Am Soc Naturalists; Paleont Res Inst. *Res:* Ecological aspects of population genetics; patterns of evolution in communities and in individual species; evolutionary biology; physiognomy and trophic organization of tropical rain forests. *Mailing Add:* Smithsonian Trop Res Inst Box 2072 Balboa Panama

LEIGH, RICHARD WOODWARD, b New York, NY, Apr 26, 42. ENERGY TECHNOLOGY, ELECTRIC UTILITY ANALYSIS. *Educ:* Oberlin Col, AB, 65; Columbia Univ, PhD(physics), 73. *Prof Exp:* Res assoc physics, City Col New York, 72-73, adj asst prof, 73-75; res fel, Lab Spectrosc Hertzienne, Ecole Normale Superieure, Paris, 75-77; asst scientist, Brookhaven Nat Lab, 77-80, Assoc scientist energy syst, 80-87; ASSOC PROF PHYSICS, PRATT INST, 87- *Concurrent Pos:* Consult energy conserv & utility anal, US and developing countries. *Mem:* AAAS; Am Phys Soc. *Res:* Energy technologies, especially heat exchanger materials and design; conservation, storage and solar; technical, economic and infrastructural requirements and benefits; electric utility planning methods and systems; coherent optics. *Mailing Add:* Dept Math & Sci Pratt Inst Brooklyn NY 11205

LEIGH, THOMAS FRANCIS, b Loma Linda, Calif, Mar 6, 23; m 54; c 2. ENTOMOLOGY, PEST MANAGEMENT. *Educ:* Univ Calif, BS, 49, PhD(entom), 56. *Prof Exp:* Res asst entom, Univ Calif, 52-54; asst prof, Univ Ark, 54-58; from asst entomologist to assoc entomologist, Univ Calif-Davis, 58-68; ENTOMOLOGIST, UNIV CALIF, 68- *Concurrent Pos:* NIH, USDA & NSF-Int Biol Prog & Rockefeller Found grants; consult, Commonwealth Sci & Indust Res Orgn, Australia, Int Atomic Energy Agency, Food & Agr Orgn UN & pvt indust. *Mem:* AAAS; Entom Soc Am; Ecol Soc Am; Mex Soc Entom; Am Registry Prof Entomologists. *Res:* Biology, ecology and control of cotton insects; insect resistance in crop plants; plant nutrition and insect abundance. *Mailing Add:* 242 Pine St Shafter CA 93263

LEIGHLY, HOLLIS PHILIP, JR, b St Joseph, Ill, May 28, 23; m 51; c 2. PHYSICAL METALLURGY. *Educ:* Univ Ill, BS, 48, MS, 50, PhD(metall eng), 52. *Prof Exp:* Res metallurgist, Bendix Aviation Corp, 52-54; res metallurgist, Denver Res Inst, 54-60, asst prof metall & chmn dept, Univ Denver, 58-60; assoc prof, 60-70, PROF METALL ENG, UNIV MO-ROLLA, 70- *Concurrent Pos:* Sabbatical, Dept Phys Metall, Birmingham Univ, 67-68 & Oak Ridge Nat Lab, 74-75; NATO fel, Univ Guelph, 74; res fel, Univ East Anglia, Norwich, Eng, 79-80 & 87. *Mem:* AAAS; Am Soc Metals; Am Inst Mining, Metall & Petrol Engrs; fel Inst Metallurgists; Sigma Xi. *Res:* Recrystallization; nuclear reactor materials; radiation damage; electron microscopy; positron annihilation. *Mailing Add:* Dept Metall Eng Univ Mo Rolla MO 65401-0249

LEIGHT, WALTER GILBERT, b New York, NY, Nov 19, 22; m 48; c 2. OPERATIONS RESEARCH, STANDARDS ENGINEERING. *Educ:* City Col New York, BS, 42. *Prof Exp:* High sch instr, NY, 42; res meteorologist, US Weather Bur, 46-53; from sci analyst to dir opers anal div, Opers Eval Group & sr sci analyst, Systs Eval Group, Ctr Naval Anal, 53-70; prog mgr decision systs, Tech Anal Div, Nat Bur Standards, 71-74, chief, Off Consumer Prod Safety, Nat Inst Standards & Technol, 74-78, chief, Prod Safety Technol Div, 78-81, chief, Off Standards Info Anal & Develop, 81- 82, prog mgr, Standards Code & Info, 82-86, asst assoc dir, indust & standards, 87-88; DEP DIR, OFF STANDARDS SERV, NAT INST STANDARDS & TECHNOL, 89- *Concurrent Pos:* Chmn, Covoss Comn Int Standardization, Am Soc Testing & Mat. *Honors & Awards:* Bronze Medal, Dept Com, 79. *Mem:* Fel AAAS; Opers Res Soc Am; Am Meteorol Soc; Standards Eng Soc; Am Soc Testing & Mats. *Res:* Decision systems; criminal justice; search and rescue; nuclear safeguards; military systems; extended forecasting; consumer product safety; standards and trade. *Mailing Add:* 9416 Bulls Run Pkwy Bethesda MD 20817

LEIGHTON, ALEXANDER HAMILTON, b Philadelphia, Pa, July 17, 08; m; c 2. PSYCHIATRIC EPIDEMIOLOGY, CULTURAL ANTHROPOLOGY. *Educ:* Princeton Univ, BA, 32; Cambridge Univ, MA, 34; Johns Hopkins Univ, MD, 36. *Hon Degrees:* AM, Harvard Univ, 66; SD, Acadia Univ, 74. *Prof Exp:* Social Sci Res Coun fel field work among Navajos & Eskimos, Columbia Univ, 39-40; Guggenheim fel, 46-47; dir, Southwest Proj, Cornell Univ, 48-53, dir, Prog Social Psychiat, 55-66, prof sociol & anthrop, Col Arts & Sci, 47-66; prof social psychiat, Med Col, 56-66; prof social psychiat & head dept behav sci, 66-75, EMER PROF SOCIAL PSYCHIAT, HARVARD SCH PUB HEALTH, 75-; PROF PSYCHIAT & COMMUNITY HEALTH EPIDEMIOL, DALHOUSIE UNIV, 75- *Concurrent Pos:* Prof, Sch Indust & Labor Rels, 47-52; consult Bur Indian Affairs, US Dept Interior, 48-50; mem bd dirs, Social Sci Res Coun, 48-58, chmn comt psychiat & social sci, 50-58; dir, Stirling Co Proj, 48-75; consult, Surgeon Gen Adv Comt Indian Affairs, 56-59; tech adv, Milbank Mem Fund, 56-63; fel, Ctr Advan Study Behav Sci, 57-58; mem expert adv panel ment health, WHO, 57-75; Thomas W Salmon Mem Lectr, NY Acad Med, 58; mem sub-panel behav sci, President's Sci Adv Comt, 61-62; consult, Peace Corps, 61-63; reflective fel, Carnegie Corp NY, 62-63; vis lectr, Cath Univ Louvain, 71; mem comt effects of herbicides in Vietnam, Nat Acad Sci, 71-73; mem, Consult Comt Ment Health Res, Dept Nat Health & Welfare, Can, 82- *Honors & Awards:* Human Rels Award, Am Soc Advan Mgt, 46; La Pouse Award, Am Pub Health Asn, 75; Ment Health Asn Res Achievement Award, 75; Nat Health Scientist Award, Can, 75-84; Malinowski Award, Soc Appl Anthrop, 84. *Mem:* Fel AAAS; fel Am Psychiat Asn; Asn Am Indian Affairs; fel Am Anthrop Asn; Am Psychopath Asn; hon fel Royal Col Psychiatrists, UK. *Res:* Social and cultural change; social psychiatry; psychiatric epidemiology. *Mailing Add:* Dept Community Health Dalhousie Univ Halifax NS B3H 4H7 Can

LEIGHTON, ALVAH THEODORE, JR, b Portland, Maine, Apr 17, 29; m 53; c 4. GENETICS, PHYSIOLOGY. *Educ:* Univ Maine, BS, 51; Univ Mass, MS, 53; Univ Minn, PhD(poultry genetics & physiol), 60. *Prof Exp:* Asst poultry genetics, Univ Mass, 51-52 & Univ Minn, 55-59; assoc prof, 59-71, PROF POULTRY SCI, VA POLYTECH INST & STATE UNIV, 71- *Mem:* Sigma Xi; World Poultry Sci Asn; Am Genetic Asn; Poultry Sci Asn. *Res:* Reproductive physiology and management of turkey populations. *Mailing Add:* Dept Poultry Sci Va Polytech Inst & State Univ Blacksburg VA 24061

LEIGHTON, CHARLES CUTLER, b Boston, Mass, June 27, 38; m 63; c 3. PHARMACEUTICAL RESEARCH & DEVELOPMENT. *Educ:* Colby Col, BA, 60; Harvard Med Sch, MD, 64. *Prof Exp:* Prof educ writer, Merck & Co, 65-66, gen mgt trainee, 66-67, mgr data coord & dir data & info servsm 67-71, sr dir regulatory affairs, 71-74, exec dir regulatory affairs & med admin, 74-78, vpres regulatory affairs & med admin, 78-87, SP VPRES REGULATORY AFFAIRS WORLDWIDE, MERCK & CO, 87-; exec dir regulatory affairs & med admin, 74-78; vpres regulatory affairs & med admin, 78-87; sr vpres regulatory affairs worldwide, 87-89, SR VPRES ADMIN, PLANNING & SCI POLICY, MERCK & CO, 90- *Concurrent Pos:* Pres, Drug Info Asn, 73-74; ed, Drug Info J, 78-87; cong comn on Fed Drug Approval Process, 81-82; trustee, Int Life Sci Inst, 84-; dir, Indust Biotechnol Asn, 88-; mem, Nat Comt Rev Current Procedure Approval New Drugs for Cancer & AIDS, 89-90. *Mem:* Am Med Asn; Drug Info Asn; Am Soc Clin Pharmacol & Therapeut; Am Heart Asn; Int Life Sci Inst; Am Med Writers Asn. *Mailing Add:* Merck & Co Res Labs West Point PA 19486

LEIGHTON, DOROTHEA CROSS, medicine, psychiatry; deceased, see previous edition for last biography

LEIGHTON, FREDERICK ARCHIBALD, b Nov 4, 48. VETERINARY MEDICINE. *Educ:* Cornell Univ, AB, 70; Univ Sask, DVM, 79; NY State Col Vet Med, PhD(exp path), 84. *Prof Exp:* Teacher, Crescent Collegiate, Robert's Arm, Nfld, 72-73; res asst wildlife ecol, Bald Eagle Proj, Besnard Lake, Sask, 74; student asst wildlife dis studies, Dept Vet Path, Western Col Vet Med, Univ Sask, 75-78, instr, 79-80, assoc prof, 84-88, PROF & HEAD, DEPT VET PATH, WESTERN COL VET MED, UNIV SASK, 88-, SUPVR, ELECTRON MICROS LAB, 85- *Concurrent Pos:* Hon vis res fel, Dept Biochem, Mem Univ Nfld, St John's, 83, hon vis assoc prof, 85; mem mgt comt, Wildlife Health Fund, 84-, chmn, 87-; mem, Toxicol Group, Univ Sask, 84-, Computer Coord Comt, Western Col Vet Med, 85-87, 88-89 & Electron Micros User's Comt, 86-90. *Honors & Awards:* F W Schofied Prize in Path, 78; Pfizer Award, 78. *Mem:* Wildlife Dis Asn; Am Col Vet Pathologists; AAAS; Am Asn Pathologists; US & Can Acad Path; Can Vet Med Asn; Can Asn Vet Pathologists (vpres, 87-88, pres, 88-89). *Mailing Add:* Vet Path Dept Western Col Vet Med Univ Sask Saskatoon SK S7N 0W0

LEIGHTON, FREEMAN BEACH, b Champaign, Ill, Dec 19, 24; c 4. GEOLOGY. *Educ:* Univ Va, BS, 46; Calif Inst Technol, MS, 49, PhD(geol), 51. *Prof Exp:* From asst prof to prof geol, Whittier Col, 50-78; PRES, LEIGHTON & ASSOCS, INC, 60- *Concurrent Pos:* Dir undergrad res prog, NSF, 59-65; mem eng geol qual bd, City Los Angeles, 62-70; state-of-the-art reviewer, US Geol Surv/HUD/Asn Bay Area Govts, 72-74; adj res prof, Whittier Col, 78- *Honors & Awards:* Claire P Holdredge Award, Nat Asn Eng Geologists, 67. *Mem:* AAAS; Geol Soc Am; Am Asn Petrol Geologists; Am Asn Prof Geologists; Nat Asn Geol Teachers. *Res:* Active faulting; environmental planning; landslides; hillside development; geomorphology; engineering geology. *Mailing Add:* 1151 Duryea Ave Irvine CA 92714

LEIGHTON, HENRY GEORGE, b London, Eng, May 2, 40; Can citizen; m 62; c 3. METEOROLOGY. *Educ:* McGill Univ, BS, 61, MS, 64; Univ Alta, PhD(nuclear physics), 68. *Prof Exp:* Res assoc nuclear physics, R J van de Graaff Lab, Holland, 68-70; vis asst prof, Univ Ky, 70-71; res assoc, 71-72, asst prof, 72-80, ASSOC PROF METEOROL, MCGILL UNIV, 80- *Mem:* Can Meteorol Soc; Am Meteorol Soc. *Res:* Atmospheric radiation; cloud physics and cloud chemistry. *Mailing Add:* Dept Meteorol McGill Univ 805 Sherbrooke St Montreal PQ H3A 2K6 Can

LEIGHTON, JOSEPH, b New York, NY, Dec 13, 21; m 46; c 2. PATHOLOGY, ONCOLOGY. *Educ:* Columbia Univ, AB, 42; Long Island Col Med, MD, 46. *Prof Exp:* From assoc prof to prof path, Sch Med, Univ Pittsburgh, 46-71; chmn dept, Med Col Pa, 71-87, prof path, 71-89; MEM STAFF, PERALTA CANCER RES INST, 89- *Concurrent Pos:* Intern, Mt Sinai Hosp, NY, 46-47; resident path anat, Mass Gen Hosp, Boston, 48-49; resident clin path, USPHS Hosp, Baltimore, 50, exp pathologist, Path Lab, Nat Cancer Inst, 51-56, consult, Nat Serv Ctr, USPHS, 59-; mem coun, Gordon Res Confs, 63-66, chmn, Gordon Res Conf Cancer, 63. *Mem:* Am Soc Exp Path; Soc Develop Biol; Am Asn Path & Bact; Tissue Cult Asn; Am Asn Cancer Res. *Res:* Experimental surgical pathology; development of matrix methods and histophysiologic gradient methods for tissue culture; pathogenesis of tumor invasion and metastasis; cancer research; tissue culture; cancer matastasis. *Mailing Add:* Peralta Cancer Res Inst 333 E Eighth St Oakland CA 94606

LEIGHTON, MORRIS WELLMAN, b Champaign, Ill, June 17, 26; m 47; c 3. EXPLORATION GEOLOGY. *Educ:* Univ Ill, BS, 47; Univ Chicago, MS, 48, PhD(geol), 51. *Prof Exp:* Res geologist, Jersey Prod Res Co, 51-58, geol sect head, 58-61, geologist-in-chg Europ study group, Esso Mediter, 61-63, sr res geologist, Jersey Prod Res Co, 63-64, geol adv, Esso Explor Inc, 64-68, asst explor mgr, Esso Standard Oil Ltd, Australia, 69-70, explor mgr, Esso Australia Ltd, 70-72, div mgr, Esso Prod Res Co, 72-74, chief geologist, Esso InterAmerica, 74-83; CHIEF, ILL STATE GEOL SURV, 83- *Concurrent Pos:* Statistician & vpres, Asn Am State Geologists; mem adv comt, US Geol Surv, mem, Hydrocarbon Res Drilling Subcomt, Nat Acad Sci; mem, Interagency Dept Energy/US Geol Surv/NSF Coun Continental Sci Drilling; distinguished lectr, Am Asn Petrol Geologists, 90. *Mem:* Am Asn Petrol Geologists; fel Geol Soc Am. *Res:* Petroleum geology; basin studies; sedimentary and igneous petrology; basin and play assessment; estimating hydrocarbon potential. *Mailing Add:* Ill State Geol Surv 615 E Peabody Dr Natural Resource Bldg Champaign IL 61820

LEIGHTON, ROBERT BENJAMIN, b Detroit, Mich, Sept 10, 19; m 43, 77; c 4. TELESCOPE DESIGN, MILLIMETER-WAVE ASTRONOMY. *Educ:* Calif Inst Technol, BS, 41, MS, 44, PhD(physics), 47. *Prof Exp:* Mem res staff, 43-45, res fel, 47-49, from asst prof to prof, 49-85, William L Valentine prof, 85-86, EMER PROF PHYSICS, CALIF INST TECHNOL, 86- *Honors & Awards:* Rumford Award, Nat Acad Arts & Sci, 86; James Craig Watson Award, Nat Acad Sci, 88. *Mem:* Nat Acad Sci; AAAS; Am Phys Soc; Am Astron Soc; Am Acad Arts & Sci. *Res:* Millimeter-wave; submillimeter; infrared-astronomy. *Mailing Add:* Calif Inst Technol Mail Code 405-47 Pasadena CA 91124

LEIGHTON, TERRANCE J, b Twin Falls, Idaho, Oct 14, 44. BIOCHEMISTRY. *Educ:* Ore State Univ, BSc, 66; Univ BC, PhD, 70. *Prof Exp:* Postdoctoral fel, Dept Biochem & Biophys, Univ Calif, Davis, 70-72; asst prof, Dept Microbiol, Med Sch, Univ Mass, Worcester, 72-74; from asst prof to prof microbiol, 74-88, PROF BIOCHEM & MOLECULAR BIOL, DEPT MOLECULAR & CELL BIOL, UNIV CALIF, BERKELEY, 89- *Concurrent Pos:* Guest Ger Govt, Max Planck Inst Molecular Genetics, Berlin, WGer, 76; Alexander von Humboldt-Stiftung fel, 80-83; sr fel, Max Planck Soc, 84- *Mem:* AAAS; Sigma Xi; Biophys Soc; Am Soc Biochem & Molecular Biol; Am Chem Soc; Am Soc Microbiol; fel Am Inst Chem; Soc Indust Microbiol. *Res:* Genetic engineering of high performance bioprocess systems; molecular genetic regulation of Bacillus subtilis development; molecular genetic regulation of Bacillus subtilis translational initiation and termination; environmental mutagenesis, carcinogenesis and the role of flavonols and flavonol glycosides as dietary anticarcinogens. *Mailing Add:* Biochem Dept Univ Calif 401 Barker Hall Berkeley CA 94720

LEIGHTON, WALTER (WOODS), mathematics; deceased, see previous edition for last biography

LEIK, JEAN, MALE GAMETOGENESIS, TERATOLOGY. *Educ:* Univ Wis, PhD(cytol), 65. *Prof Exp:* RES ASST, DEPT PATH, UNIV WASH, 83- *Mailing Add:* Dept Biol Univ Wash SM-20 Seattle WA 98195

LEIKIN, JERROLD BLAIR, b Chicago Ill, Aug 28, 54; m 82; c 2. TOXICOLOGY, EMERGENCY MEDICINE. *Educ:* Univ Iowa, BS, 76; Chicago Med Sch, MD, 80. *Prof Exp:* ATTEND MED TOXICOL, COOK COUNTY HOSP, 85-; ASST PROF MED, RUSH MED COL, 88- *Concurrent Pos:* Chmn, Res & Educ, Am Col Emergency Physicians, 87- *Mem:* Am Med Assoc; Am Acad Clin Toxicol; Am Col Physicians; Am Col Emergency Physicians. *Res:* Toxicology and emergency medicine. *Mailing Add:* 1037 Edgebrook Lane Glencoe IL 60022

LEIMANIS, EUGENE, b Koceni, Latvia, Apr 10, 05; m 42; c 6. APPLIED MATHEMATICS. *Educ:* Univ Latvia, Mag Math, 29; Univ Hamburg, Dr rer nat(math), 47. *Prof Exp:* Asst math, Univ Latvia, 29-35, privat-docent, 35-37, docent, 37-44; docent, Univ Greifswald, 44-45; assoc prof, Baltic Univ, Ger, 46-48; from asst prof to prof, 49-74, EMER PROF MATH, UNIV BC, 74- *Concurrent Pos:* Fel, Summer Res Inst, Can Math Cong, 53 & 55; sci adv, Lockheed Missles & Space Co, Palo Alto, Calif, 62. *Mem:* Am Math Soc; Math Asn Am; London Math Soc; Asn Advan Baltic Studies; Am Astron Soc; NY Acad Sci; fel AAAS. *Res:* Differential equations; non-linear and celestial mechanics. *Mailing Add:* 3839 Selkirk St Vancouver BC V6H 2Z2 Can

LEIMGRUBER, RICHARD M, PHYSICAL SCIENCE. *Prof Exp:* SR RES SPECIALIST, PHYS SCI CTR, MONSANTO CO, 83- *Mailing Add:* Phys Sci Ctr Monsanto Co 700 Chesterfield Village Pkwy Chesterfield MO 63198

LEIMKUHLER, FERDINAND F, b Baltimore, Md, Dec 31, 28; m 56; c 6. INDUSTRIAL ENGINEERING, OPERATIONS RESEARCH. *Educ:* Loyola Col, Md, BS, 50, Johns Hopkins Univ, BEng, 52, Dr Eng, 62. *Prof Exp:* Engr, E I du Pont de Nemours & Co, 52-57; res assoc & instr indust eng, Johns Hopkins Univ, 57-61; assoc prof, 61-66, head, Sch Indust Eng, 69-74, PROF INDUST ENG, PURDUE UNIV, 66-, HEAD, SCH INDUST ENG, 81- *Concurrent Pos:* Vis prof, Univ Calif, Berkeley, 68-69, 90-91; Fulbright prof, Univ Ljubjana, Yugoslavia, 74-75. *Honors & Awards:* Distinguished lectr, Am Soc Info Sci, 71-72. *Mem:* Opers Res Soc Am; Inst Mgt Sci; Am Inst Indust Engrs; Am Soc Eng Educ. *Res:* Library operations research; engineering economic analysis; transportation of highly radioactive materials; stochastic system theory; manufacturing systems. *Mailing Add:* Sch Indust Eng Purdue Univ Lafayette IN 47907

LEIN, ALLEN, b New York, NY, Apr 15, 13; m 41; c 2. ANIMAL PHYSIOLOGY. *Educ:* Univ Calif, Los Angeles, BA, 35, MA, 38, PhD(endocrinol), 40. *Prof Exp:* Instr surg res, Sch Med, Ohio State Univ, 41-42, res assoc aviation physiol, Res Found, 42-43; from asst prof to prof physiol, Med Sch, Northwestern Univ, 47-68, dir student affairs, 60-63, dir honors prog med educ, 62-68, asst dean, Med Sch & dir med scientist training prog, 64-68, asst dean, Grad Sch, 66-68; assoc dean & prof med, Univ Calif, San Diego, 68-73, prof reproductive med, 73-84, assoc dean grad studies health sci, 74-77 & 81-82, prog dir med scientist training, 75-77, prog dir health professions hon prog, 79-80, actg assoc dean acad affairs, 80-82, EMER PROF REPRODUCTIVE MED, UNIV CALIF, SAN DIEGO, 84- *Concurrent Pos:* Asst prof physiol, Med Sch, Vanderbilt Univ, 46-47; vis prof chem, Calif Inst Technol, 54-55; consult, Vet Admin Res Hosp, Chicago, Ill, 64-68; Scholar in residence, Bellagio Study Ctr, Rockefeller Found, 77. *Honors & Awards:* Guggenheim Fel, Col France, Paris, 58-59. *Mem:* Am Physiol Soc; AAAS; Am Inst Biol Sci; Soc Exp Biol & Med; Endocrine Soc; Fedn Am Soc Exp Biol; Sigma Xi. *Res:* Endocrine regulation of reproductive function, thyroid function and carbohydrate and fat metabolism. *Mailing Add:* Dept Reproductive Med T-002 Univ Calif La Jolla CA 92093

LEINBACH, F HAROLD, b Ft Collins, Colo, Jan 7, 29; wid; c 2. PHYSICS. *Educ:* SDak State Univ, BS, 49; Calif Inst Technol, MS, 50; Univ Alaska, PhD(geophys), 62. *Prof Exp:* Geophysicist, Geophys Inst, Alaska, 50-53 & 56-62; asst prof physics, Univ Iowa, 62-66; instr, 78-79, PHYSICIST, SPACE ENVIRON LAB, NAT OCEANOG & ATMOSPHERIC ADMIN, UNIV COLO, 66- *Concurrent Pos:* Actg dir, Space Environ Lab, 82-86. *Mem:* Am Astron Soc; Am Geophys Union; Am Asn Physics Teachers; Sigma Xi. *Res:* High latitude ionospheric absorption of cosmic radio noise; solar cosmic rays and their interaction with the ionosphere; solar physics; space physics; laboratory plasma physics. *Mailing Add:* 2015 Kohler Dr Boulder CO 80303

LEINBACH, RALPH C, JR, b Esterly, Pa, Nov 24, 28; m 51; c 4. ALLOYS. *Educ:* Lehigh Univ, BS, 54. *Prof Exp:* Metallurgist, Atomic Power Div Res Ctr, Babcock & Wilcox Co, Alliance, Ohio, 54; metallurgist, Res & Develop Lab, Electronics & Magnetics, Carpenter Technol Corp, Reading, Pa, 55, melting metallurgist, 56, melting metallurgist, Bridgeport, Conn plant, 57-61, plant metallurgist, 61-65, mgr, Mill Metallurgy, 65-68, chief metallurgist, 68-70, asst vpres metall, Carpenter Technol Corp, 70-71, vpres metall, 71-75, vpres tech, 75-76, div vpres tech, 76-79, group vpres, Carpenter Steel Div, 79-82, sr vpres technol, eng & purchasing, 82-87; RETIRED. *Concurrent Pos:* Mem bd dirs, Metal Prop Coun Inc, 80; Mem, Interim Core Group Mkt Develop Comt, Am Iron & Steel Inst, 85. *Honors & Awards:* Regional Tech Award, Am Iron & Steel Inst, 63, Spec Achievement Cert, 67; Bradley Stoughton Award, Am Soc Metals, 71. *Mem:* Fel Am Soc Metals; Am Inst Metall Engrs; Am Iron & Steel Inst; Soc Automotive Engrs; Soc Metall Engrs; Am Welding Soc; Metals Soc; Am Vacuum Soc. *Res:* Stainless steel making utilizing vacuum treatment; vacuum induction melting of specialty steels and alloys; specialty steel melting; US and Canadian patents; technical publications. *Mailing Add:* 2404 Bell Dr Reading PA 19609

LEINEN, MARGARET SANDRA, b Chicago, Ill, Sept 20, 46; m; c 1. PALEOCEANOGRAPHY, PALEOCLIMATE. *Educ:* Univ Ill, BS, 69; Ore State Univ, MS, 75; Univ RI, PhD(oceanog), 80. *Prof Exp:* Marine scientist, Univ RI, 80-82, from asst res prof to assoc res prof, 82-88, PROF RES & ASSOC DEAN, UNIV RI, 88- *Mem:* Geochem Soc; fel Geol Soc Am; Am Geophys Union. *Res:* Deep sea sedimentary processes; history of atmospheric circulation. *Mailing Add:* Grad Sch Oceanog Univ RI Narragansett RI 02882-1197

LEINHARDT, THEODORE EDWARD, solid state physics; deceased, see previous edition for last biography

LEININGER, HAROLD VERNON, b Baton Rouge, La, June 18, 25; m 50; c 1. MICROBIOLOGY. *Educ:* La State Univ, BS, 48, MS, 51. *Prof Exp:* Lab technician, Dairy Improv Ctr, La State Univ, 51; food & drug inspector, Food & Drug Admin, 51, from bacteriologist to dir biol warfare proj, 52-63, res microbiologist, 63-71, dir, Minneapolis Ctr Microbiol Invest, 71-80; CONSULT MICROBIOL, FOOD, DRUGS & COSMETICS, 80- *Concurrent Pos:* Proj officer, Test Site, AEC, Nev, 57. *Mem:* Asn Off Analytical Chemists; Am Soc Microbiologists; Inst Food Technologists; Int Asn Milk, Food & Environ Sanit. *Res:* Microbiological research in food toxicity, decomposition, natural flora, and sanitation. *Mailing Add:* 3200 Voss Dr El Paso TX 79936

LEININGER, PAUL MILLER, b Pa, Oct 29, 11; m 37; c 3. PHYSICAL CHEMISTRY. *Educ:* Univ Pa, BS, 32 & 34, MS, 36, PhD(phys chem), 39. *Prof Exp:* Asst instr chem, Univ Pa, 34-38; chemist, E I du Pont de Nemours & Co, 39-49; asst prof chem, Lafayette Col, 49-54; from assoc prof to prof, 54-77, chmn dept, 69-77, EMER PROF CHEM, ALBRIGHT COL, 78- *Concurrent Pos:* Lectr, Sch Nurses, Reading Hosp, Pa, 54-61; mem teaching staff exp prog teacher educ, Temple Univ, 55-60; consult, George W Bollman Co. *Mem:* AAAS; Am Chem Soc; Am Soc Metals; Am Leather Chemists Asn; NY Acad Sci. *Res:* Case hardening and heat treatment of metals in molten salts; new uses and processes for sodium cyanide and related cyanogen compounds; reaction kinetics in solutions; chemical education; electrolytic dissociation and technology of felting. *Mailing Add:* 1726 Hampden Blvd Reading PA 19604

LEININGER, ROBERT IRVIN, b Cleveland, Ohio, May 11, 19; m 42; c 4. BIOMEDICAL ENGINEERING, POLYMER CHEMISTRY. *Educ:* Cleveland State Univ, BChE, 40; Case Western Reserve, MS, 41, PhD, 43. *Prof Exp:* Instr chem eng, Fenn Col, 40-43; res chemist, Monsanto Chem Co, 43-48; prin chemist, Battelle Mem Inst, 48-51, asst chief rubber & plastics div, 51-60, chief polymer res sect, 60-65, mgr, 65-69; tech adv, Korean Inst Sci & Technol, Seoul, 69-70; prof dir biomat, Dept Biol, Environ & Chem, 70-73, mem res coun, 74-80, dir res coun, Columbus Labs, Battelle Mem Inst, 80-84; dir, Nat Ctr Biomed Infrared Spectros, 83-85; CONSULT, BIOMAT, 86- *Concurrent Pos:* Mem adv comt, Div Technol Devices, Nat Heart & Lung Inst, 72-73; mem adv comt, Nuclear Powered Artificial Heart Prog, Energy Res & Develop Admin, 75-76; mem, Surg & Bioeng Study Sect, 76-79, chmn, Biomat Adv Panel, NIH, 80- *Honors & Awards:* IR-100 Indust Res Award, 72; Clemson Award, Soc Biomat, 81. *Mem:* Int Soc Artificial Organs; Am Soc Artificial Internal Organs; Soc Biomat; Am Chem Soc; NY Acad Sci; Sigma Xi. *Res:* Biomaterials; biomedical engineering; polymer chemistry. *Mailing Add:* 1973 Milden Rd Columbus OH 43221

LEINROTH, JEAN PAUL, JR, b Utica, NY, July 4, 20; m 46; c 3. CHEMICAL ENGINEERING. *Educ:* Cornell Univ, BMechEng, 41; Mass Inst Technol, SM, 48, ScD(chem eng), 63. *Prof Exp:* Trainee, Standard Oil Co, Ohio, 41-42; asst job engr, M W Kellogg Co, NY, 42-43; proj engr, Union Carbide Chem Co, 48-56, proj leader, 56-59; instr thermodyn, Mass Inst Technol, 60-61, vis assoc prof, 63-64; assoc prof chem eng, Cornell Univ, 64-71; vis prof, Mass Inst Technol, 71-72; process dir, 72-80, mgr spec projs, John Brown E&C, 80-85; vis prof, Univ Conn, 86-88; RETIRED. *Concurrent Pos:* Consult, Union Carbide Chem Co, 64-68, Develop Sci, 71-72, Gen Elec, Sterling Org, John Brown E&C, Vanderbilt Chem Corp, Syntex, Velsicol Westex, 87-; vis prof, Mass Inst Technol, 88-89. *Mem:* Am Chem Soc; Am Inst Chem Engrs. *Res:* Chemical kinetics; thermodynamics; staged operations; computer applications; desalination. *Mailing Add:* 33 Millstone Rd Wilton CT 06897

LEINWEBER, FRANZ JOSEF, b Berlin, Ger, Jan 18, 31; US citizen; m 60; c 2. DRUG METABOLISM. *Educ:* Univ Tuebingen, Dr rer nat(biol), 56. *Prof Exp:* Fel biochem, Tex A&M Univ, 57-60 & Johns Hopkins Univ, 60-63; res assoc, Univ Tenn, 63-65; sr scientist, McNeil Labs, Inc, 65-69; sr scientist, Warner-Lambert Res Inst, 69-77; SR SCIENTIST, HOFFMANN-LA ROCHE INC, 77- *Mem:* Am Soc Pharmacol & Exp Therapeut. *Res:* Photoperiodism and biological clocks; enzymology, intermediary metabolism and metabolic regulation of sulfur amino acid biosynthesis in bacteria and molds; drug metabolism and separation methods; enzymatic mechanisms of drug biotransformation. *Mailing Add:* Dept Drug Metab Hoffmann-La Roche Inc Nutley NJ 07110

LEIPNIK, ROY BERGH, b Los Angeles, Calif, May 6, 24; m 44; c 3. MATHEMATICAL ANALYSIS, MATHEMATICAL PHYSICS. *Educ:* Univ Chicago, SB, 45, SM, 48; Univ Calif, PhD(math), 50. *Prof Exp:* Asst math statist & econ, Univ Chicago, 45-46; from asst to assoc math, Univ Calif, 46-48; fel, Sch Math, Inst Advan Study, 48-50; asst prof, Univ Wash, 50-57; sr res scientist, Naval Weapons Ctr, Calif, 57-75; PROF APPL MATH & MEM ALGEBRA INST, UNIV CALIF, SANTA BARBARA, 75- *Concurrent Pos:* Fulbright res prof, Univ Adelaide, 55, 63 & 68; lectr, Univ Calif, Los Angeles, 59-; prof, Univ Fla, 61-62, 64-65 & 70; consult, Decisional Controls Assocs & Commun Res Labs. *Mem:* Am Math Soc; Math Asn Am; Inst Math Statist; Inst Elec & Electronics Engrs; Soc Indust & Appl Math. *Res:* Operator analysis; mathematical physics; control systems; stochastic processes; information theory; plasma physics; solid state physics; microprocessor design; transportation theory; allocation and assignment algorithms; recursive algorithms; differential equations; engineering mechanics. *Mailing Add:* Santa Barbara Math Dept Goleta CA 93106

LEIPOLD, MARTIN H(ENRY), b Englewood, NJ, July 4, 32; m 58; c 3. CERAMICS. *Educ:* Rutgers Univ, BS, 54; Ohio State Univ, MS, 55, PhD(ceramics), 58. *Prof Exp:* Res assoc ceramics, Ohio State Univ Res Found, 55-58; sr scientist, Jet Propulsion Lab, Calif Inst Technol, 58-62, res specialist, 62-67; assoc prof mat sci, Col Eng, Univ Ky, 67-74; MEM TECH STAFF, JET PROPULSION LAB, 74- *Concurrent Pos:* Adj prof, Univ Calif, Los Angeles, 76-84; mgr, Historically Black Cols & Univ Initiative, 87-90.

Honors & Awards: Except Serv Award, NASA, 89. *Mem:* Fel Am Ceramic Soc; Am Inst Ceramic Engrs; Am Soc Metals. *Res:* Properties, fabrication and structure of ceramics and semi-conductor materials; photovoltaic power systems; IR sensors; focal plane arraxs. *Mailing Add:* 1118 Sheraton Dr La Canada CA 91011

LEIPPER, DALE F, b Salem, Ohio, Sept 8, 14; m 42; c 4. SEA-AIR INTERACTION. *Educ:* Wittenberg Univ, BS 37; Ohio State Univ, MA, 39; Univ Calif, PhD(oceanog), 50. *Hon Degrees:* DSc, Wittenberg Univ, 68. *Prof Exp:* Weight & balance engr, Consol Aircraft, Calif, 40; sch teacher, Calif, 40-41; oceanogr, Scripps Inst, Univ Calif, 46-49; from assoc prof to prof oceanog & head dept, Tex A&M Univ, 49-68; chmn, 68-80, CHMN, DEPT OCEANOGR, NAVAL POSTGRAD SCH, MONTEREY, 80- *Concurrent Pos:* Head dept, Tex A&M Univ, 50-64, assoc exec dir, Tex A&M Found, 53-54, trustee, Univ Corp Atmospheric Res, 59-65; dir, World Data Ctr Oceanog, 57-60; consult, Comt Sci & Astronaut, US House Rep, 60; mem joint panel sea-air interaction, Nat Acad Sci, 60-62; adj prof oceanog, Naval Postgrad Sch, 80-88. *Mem:* Am Meteorol Soc; Am Soc Limnol & Oceanog (pres, 58); Am Soc Oceanog (pres, 67); Am Geophys Union; Oceanog Soc. *Res:* Coastal fog forecasting; analysis of sea temperature variations; use of the bathythermograph; interaction between ocean and atmosphere; physical oceanography; marine meteorology. *Mailing Add:* 716 Terra Ct Reno NV 89506-9606

LEIPUNER, LAWRENCE BERNARD, b Long Beach, NY, May 27, 28; m 48; c 3. HIGH ENERGY & ELEMENTARY PARTICLE PHYSICS. *Educ:* Univ Pittsburgh, BS, 50; Carnegie Inst Technol, MS, 54, PhD(physics), 62. *Prof Exp:* SR RES PHYSICIST, BROOKHAVEN NAT LAB, 55- *Concurrent Pos:* Vis prof, Yale Univ, 67-68. *Mem:* Fel Am Phys Soc. *Res:* Lepton and quark experiments. *Mailing Add:* Dept Physics 510A Brookhaven Nat Lab Upton NY 11973

LEIPZIGER, FREDRIC DOUGLAS, b New York, NY, Aug 26, 29; m 51; c 2. ANALYTICAL CHEMISTRY. *Educ:* Univ Conn, BA, 51; Univ Mass, MS, 53, PhD(chem), 56. *Prof Exp:* Res assoc chem, Gen Elec Co, 55-62; head anal chem dept, Sperry Rand Res Ctr, 62-66; mgr anal serv, Ledgemont Lab, Kennecott Copper Corp, 66-81; PRES, NORTHERN ANALYTICAL LAB, INC, 82- *Mem:* Am Chem Soc; Soc Appl Spectros; Am Soc Mass Spectrometry; Am Asn Crystal Growth. *Res:* Electron microscopy; mass spectrometry; atomic absorption; automated analyses; process control. *Mailing Add:* Northern Analytical Lab Three Northern Blvd Amherst NH 03031-2313

LEIS, BRIAN NORMAN, b Kitchener, Ont, Oct 25, 47; m 69; c 3. FRACTURE MECHANICS, DAMAGE MECHANICS. *Educ:* Univ Waterloo, BASc, 71, MASc, 72, PhD(civil eng), 79. *Prof Exp:* Scientist appl mech, Battelle Columbus Lab, 74-79, staff scientist, 79-81, sr scientist, 81-85, RES LEADER APPL MECH, BATTELLE COLUMBUS LAB, 85- *Concurrent Pos:* Adj prof mech eng, Ohio State Univ, 80-83; subcomt chmn, Am Soc Testing & Mat, 81-85. *Mem:* Am Soc Testing & Mat; Am Soc Mech Engrs; Am Inst Metall Engrs. *Res:* Damage mechanics in materials and structures, including model simulation, with emphasis on fatigue, fracture and environmental degradation; granted one patent. *Mailing Add:* Battelle Columbus Lab Columbus OH 43201

LEIS, DONALD GEORGE, b Jeannette, Pa, Aug 26, 19; m 45; c 3. ORGANIC CHEMISTRY. *Educ:* St Vincent Col, BS, 41; Univ Notre Dame, MS, 42, PhD(org chem), 45. *Prof Exp:* Chemist, Carbide & Carbon Chem Co, 46-55; group leader, 55-63, mgr mkt develop-cellular prod, 63-71, com mkt mgr, 71-75, SR MKT CONSULT, SILICONES & URETHANES DIV, UNION CARBIDE CORP, 75- *Mem:* AAAS; Nat Fire Protection Asn; Soc Plastics Engrs; Am Chem Soc; Soc Plastics Indust. *Res:* Urethane products; polyethers; polyglycols; alkylene oxides; alkylene oxide derivatives; ethylene oxide; propylene oxide; urethanes; surfactants; lubricants and coatings. *Mailing Add:* 11 Coulter St No 28 Old Saybrook CT 06475

LEIS, JONATHAN PETER, b Brooklyn, NY, Aug 17, 44; m 70; c 2. NUCLEIC ACID ENZYMOLOGY, VIROLOGY. *Educ:* Hofstra Univ, BA, 65; Cornell Univ, PhD(biochem), 70. *Prof Exp:* Fel develop biol & cancer, Albert Einstein Col Med, 70-73; asst prof surg, microbiol & immunol, Med Ctr, Duke Univ, 74-79; assoc prof, 79-86, PROF BIOCHEM, MED SCH, CASE WESTERN RESERVE UNIV, 86- *Concurrent Pos:* Damon Runyon res fel, 71; res career develop awards, NIH, 74-79. *Mem:* Am Soc Biol Chemists; Am Soc Microbiol; Am Soc Virol. *Res:* Control of expression of eukaryotic genes; biochemical mechanisms of replication of Retro viruses; author of over 61 publications. *Mailing Add:* Dept Biochem Med Sch Case Western Reserve Univ 2119 Abington Rd Cleveland OH 44106

LEISE, ESTHER M, b Washington, DC, May 13, 53. INVERTEBRATE NEUROANATOMY, LARVAL NEUROBIOLOGY. *Educ:* Univ Md, BS, 75; Univ Wash, PhD(zool), 83. *Prof Exp:* Teaching asst histol, Dept Zool, Univ Wash, 75-83; postdoctoral res assoc, Dept Zool, Univ Calif, Davis, 83-88; postdoctoral res assoc, Dept Biol, Ga State Univ, 88-90; asst researcher, Pac Biomed Res Ctr, Univ Hawaii, Honolulu, 90-91; ASST PROF NEUROBIOL, DEPT BIOL, UNIV NC, GREENSBORO, 91- *Mem:* Soc Neurosci; AAAS; Am Soc Zoologists. *Res:* Neuroanatomy of crayfish central nervous systems; neuronal control of settlement and metamorphosis in marine invertebrate larvae, particularly molluscan veligers. *Mailing Add:* Dept Biol Univ NC Greensboro NC 27412-5001

LEISERSON, LEE, b Toledo, Ohio, Mar 30, 16; m 43, 71; c 6. ORGANIC CHEMISTRY, PHYSICAL CHEMISTRY. *Educ:* Antioch Col, BS, 37; Univ NC, MA, 40, PhD(org chem), 41. *Prof Exp:* Res chemist, Eastman Kodak Co, NY, 41-45; fel, Va Smelting Co, NC, 45-47; chemist, Am Cyanamid Co, NJ, 47-51; chief org chemist, Liggett & Myers Tobacco Co, 51-55; chemist & res adminr, Air Force Off Sci Res, Washington, DC, 55-62; chief, Chem Div, Off Saline, 62-74, consult, 74-76; field serv coordr, Environ Protection Agency, 76-80; RETIRED. *Concurrent Pos:* Consult. *Mem:* AAAS; Am Chem Soc.

Res: Surface active agents; organic synthesis; development of natural products; turpentine and tall oil separation processes; reactions in liquid sulfur dioxide; tobacco; water and aqueous solutions; environmental chemistry. *Mailing Add:* 200 Tabernacle Rd F-46 Black Mountain NC 28711-2592

LEISMAN, GILBERT ARTHUR, b Washington, DC, May 12, 24; m 52. BOTANY. *Educ:* Univ Wis, BS, 49; Univ Minn, MS, 52, PhD, 55. *Prof Exp:* Asst plant physiol, Univ Wis, 49-50; asst bot, Univ Minn, 50-55; from asst prof to assoc prof, 55-64, PROF BIOL, EMPORIA STATE UNIV, 64- *Concurrent Pos:* Mem, World Orgn Paleobot, Int Union Biol Sci. *Mem:* AAAS; Bot Soc Am; Nat Asn Biol Teachers; Int Asn Plant Taxon. *Res:* Coal ball plants; morphology of pteridosperm leaves and fructifications; plant succession and soil development of mine dumps. *Mailing Add:* Dept Biol Emporia State Univ 1200 Coml St Emporia KS 66801

LEISS, ERNST L, b Ger, July 7, 52. SOFTWARE SYSTEMS. *Educ:* Univ Waterloo, Can, MMath, 74; Tech Univ Vienna, Austria, Dipl Ing, 75, Dr Techn, 76. *Prof Exp:* Asst prof, 79-85, PROF COMPUT SCI, UNIV HOUSTON, 85-, DIR, RES COMPUTATION LAB, 85- *Concurrent Pos:* Vis prof, var univs, Brazil & Italy; Nat lectr, Asn Comput Mach, 91-92. *Mem:* Sr mem Inst Elec & Electronics Engrs; Asn Comput Mach; Soc Explor Geophysicists. *Res:* Vector and parallel computing; seismic data processing; data security; databases; automata theory. *Mailing Add:* Dept Comput Sci Univ Houston Houston TX 77204-3475

LEISS, JAMES ELROY, b Youngstown, Ohio, June 2, 24; m 45; c 4. PHYSICS. *Educ:* Case Inst Technol, BS, 49; Univ Ill, MS, 51, PhD(physics), 54. *Prof Exp:* Lab asst, Gen Elec Co, 48-49; asst physics, Univ Ill, 49-54; dir, Ctr Radiation Res, Nat Bur Standards, 54-88; dir, Off High Energy & Nuclear Physics, Dept Energy; RETIRED. *Mem:* Am Phys Soc. *Res:* Nuclear physics, especially photonuclear reactions and photomeson reactions; design of particle accelerators. *Mailing Add:* Rte 2 Box 142C Broadway VA 22815

LEISSA, A(RTHUR) W(ILLIAM), b Wilmington, Del, Nov 16, 31; m 53; c 2. VIBRATIONS, BUCKLING. *Educ:* Ohio State Univ, BME & MSc, 54, PhD(eng mech), 58. *Prof Exp:* Assoc engr, Sperry Gyroscope Co, 54-55; res assoc, Res Found, 55-56, from instr to assoc prof, 56-64, PROF ENG MECH, OHIO STATE UNIV, 64-, RES FOUND SUPVR, 62- *Concurrent Pos:* Mech engr, Ralph & Curl Engrs, 54-58; fac assoc, Boeing Airplane Co, 57; consult, NAm Aviation, Inc, 58-64, Battelle Mem Inst, 64-, Kaman Nuclear, 68-70, Medtronic, Inc, 84-85; vis prof, Swiss Fed Inst Technol, 72-73; assoc ed, Appl Mech Reviews, 85-, J Vibration Acoust, 89-; vis prof, USAF Acad, 85-86; chmn, Orgn Comt, Pan Am Congress Appl Mech, 86-89; res fel, Japan Soc Prom Sci, 90. *Mem:* Assoc fel Am Inst Aeronaut & Astronaut; Am Soc Eng Educ; Int Asn Shell Struct; fel Am Soc Mech Engrs; fel Am Acad Mech(pres 87-88). *Res:* Elasticity; plates and shells; vibration of continuous systems; buckling; numerical methods for solving boundary value and eigenvalue problems; composite structures. *Mailing Add:* Dept Eng Mech Ohio State Univ 155 W Woodruff Columbus OH 43210

LEISTER, HARRY M, b Quakertown, Pa, Mar 3, 41; m 62; c 4. PHYSICAL CHEMISTRY. *Educ:* Pa State Univ, BS, 63; Drexel Univ, MS, 65; Temple Univ, PhD(phys chem), 70. *Prof Exp:* Res chemist, E I du Pont de Nemours & Co, 69-70; chemist, Amchem Div, Union Carbide Corp, 71-73, group leader, 73-81, SCIENTIST, AMCHEM PROD, INC, 81- *Mem:* Am Chem Soc. *Res:* Organic coatings; inorganic coatings. *Mailing Add:* 51 Hendricks St Ambler PA 19002

LEISURE, ROBERT GLENN, b Cromwell, Ky, Jan 29, 38; m 62. SOLID STATE PHYSICS. *Educ:* Western Ky Univ, BS, 60; Wash Univ, PhD(physics), 67. *Prof Exp:* Res scientist, Boeing Sci Res Lab, 67-70; from asst prof to assoc prof, 70-78, chmn, Dept Physics, 84-90, PROF PHYSICS, COLO STATE UNIV, 78- *Concurrent Pos:* Vis scientist, Univ Paris VI, 78-79; sr vis fel, St Andrews Univ, Scotland, 83. *Honors & Awards:* US-France Exchange Scientist Award. *Mem:* Am Phys Soc; Sigma Xi. *Res:* Ultrasonics; propagation of ultrasound in solids; metal hydrides. *Mailing Add:* Dept Physics Colo State Univ Ft Collins CO 80523

LEITCH, GORDON JAMES, GASTROINTESTINAL PHYSIOLOGY. *Educ:* Univ Chicago, PhD(physiol), 64. *Prof Exp:* PROF MED PHYSIOL, MOREHOUSE SCH MED, 77-, CHMN DEPT, 78- *Res:* Diarrhea pathophysiology; alcohol pathophysiology. *Mailing Add:* Dept Physiol Morehouse Sch Med 720 Westview Dr SW Atlanta GA 30310-1495

LEITCH, LEONARD CHRISTIE, b Ottawa, Ont, Aug 22, 14; m 42; c 2. ORGANIC CHEMISTRY. *Educ:* Univ Ottawa, BSc, 35; Laval Univ, DSc, 49. *Prof Exp:* Res chemist, Mallinckrodt Chem Works, 37-46; res chemist, Div Chem, Nat Res Coun Can, 46-70, prin res chemist, 70-80; HON SR SCIENTIST, DEPT CHEM, UNIV OTTAWA, 88- *Mem:* Chem Inst Can; Royal Soc Chem; Sigma Xi. *Res:* Synthetic drugs and plant hormones; synthesis of organic compounds with stable isotopes; reaction mechanisms; insecticide. *Mailing Add:* 64 Blackburn Ave Ottawa ON K1N 8A5 Can

LEITE, RICHARD JOSEPH, b Fremont, Ohio, Mar 8, 23; m 55; c 3. PRELIMINARY DESIGN OF SPACECRAFT & SPACEFLIGHT PAYLOADS, DEVELOPMENT OF MINIATURIZED SENSORS & INSTRUMENTS. *Educ:* Univ Notre Dame, BNS, 45, BSE, 47; Univ Mich, MSE, 48, PhD(aero eng), 56. *Prof Exp:* Res assoc, Univ Mich, Ann Arbor, 48-56, res engr, 58-71; sr engr, Booz-Allen Appl Res, Inc, 56-58; sr staff engr, Bendix Aerospace Corp, 71-72; sr scientist, KMS Fusion, Inc, 72-77; STAFF MGR, ENG & TEST DIV, TRW INC, REDONDO BEACH, CALIF, 77- *Concurrent Pos:* Lectr, Univ Mich, Ann Arbor, 58-60, prin investr, 63-71, res consult, 68-71. *Mem:* Sigma Xi. *Res:* Electrical systems design and development of spacecraft; qualification testing and electrical integration of systems components and systems; laser fusion fuel pellet development and insertion technology; spaceflight mass spectrometer development; upper atmosphere composition measurement pioneer; experimental demonstration of stability criteria for tube flow; one United States patent. *Mailing Add:* 6742 Abbottswood Dr Rancho Palos Verdes CA 90274-3018

LEITER, EDWARD HENRY, b Columbus, Ga, Apr 17, 42; m 64. CELL BIOLOGY. *Educ:* Princeton Univ, BS, 64; Emory Univ, MS, 66, PhD(biol), 68. *Prof Exp:* NIH trainee, Univ Tex, Austin, 68-71; asst prof biol, Brooklyn Col, 71-74; from assoc staff scientist to staff scientist, 74-89, SR STAFF SCIENTIST, JACKSON LAB, 90- *Concurrent Pos:* Nat Inst Arthritis & Metab Dis res grant, 74-; Juvenile Diabetes Found grant, 76- *Mem:* Tissue Cult Asn; Am Soc Cell Biol; Endocrine Soc; Am Diabetes Asn; Am Asn Immunologists; Histochem Soc. *Res:* Function of normal and diabetic pancreatic endocrine cells in vitro; genetic, viral, and environmental parameters producing pancreatic pathologies in the mouse; immunology of type 1 diabetes. *Mailing Add:* Jackson Lab Bar Harbor ME 04609

LEITER, ELLIOT, b Brooklyn, NY, May 24, 33; m 63; c 3. MEDICINE, UROLOGY. *Educ:* Columbia Univ, AB, 54; NY Univ, MD, 57. *Prof Exp:* Asst urol, Johns Hopkins Hosp, 58-59; asst urol, NY Univ, 60-63; instr, Columbia Univ, 66-67; dir urol, Beth Israel Hosp & Med Ctr, 78-84; from asst prof to assoc prof, 66-78, PROF UROL, MT SINAI SCH MED, 78-; ATTEND UROL, MT SINAI HOSP, 78- *Concurrent Pos:* Fel, NY Univ, 60-61, USPHS trainee, 61-62, fel hypertensive renal group, 63; asst vis surgeon, Bellevue Hosp, NY, 63 & Greenpoint Hosp, Brooklyn, 64; asst attend urologist, Mt Sinai Hosp, NY, 63-69, assoc attend urologist, 69- *Mem:* Asn Acad Surg; Am Urol Asn; Am Col Surg; Soc Univ Urol; Soc Pediat Urol. *Res:* Renal disease; pediatric urology; kidney transplantation; hypertension. *Mailing Add:* 109 E 38th St New York NY 10016

LEITER, HOWARD ALLEN, b Mt Gilead, Ohio, Feb 16, 18; m 52; c 3. PHYSICS. *Educ:* Miami Univ, AB, 40; Univ Ill, AM, 42, PhD(physics), 49. *Prof Exp:* Asst physics, Univ Ill, 40-42; mem staff, Radiation Lab, Mass Inst Technol, 42-45; asst physics, Univ Ill, 45-48, res assoc, 48-49; physicist, Res Labs, Westinghouse Elec Corp, 49-58; sr engr, Labs, Int Tel & Tel Corp, 58-69; assoc prof physics, Tri-State Col, 70-75; instr, Inventive Indust, 75-77; instr, Int Tel & Tel Corp, 78-83; RETIRED. *Concurrent Pos:* Guest lectr, Off-campus Grad Prog, Purdue Univ, 59-60. *Mem:* Am Phys Soc. *Res:* Charged particle scattering; microwave components; interaction of electromagnetic radiations with matter; infrared detectors and systems; image tubes; cryogenic equipment; field emission microscopy; satellite instrumentation. *Mailing Add:* 2703 Capitol Ave Ft Wayne IN 46806

LEITER, JOSEPH, b New York, NY, May 14, 15; m 39; c 2. BIOCHEMISTRY. *Educ:* Brooklyn Col, BS, 34; Georgetown Univ, PhD(biochem), 49. *Prof Exp:* Jr chemist org & fibrous mat, Nat Bur Standards, 35-38; carcinogenesis, Nat Cancer Inst, 38-40, asst chemist, 40-42, assoc chemist chemother, 46-47, chemist, 47-49, from sr chemist to sr scientist & chief biochem sect, Lab Chem Pharmacol, 49-55, scientist dir & asst chief lab activ, Cancer Chemother Nat Serv Ctr, 55-63, chief, Ctr, 63-65, assoc dir libr opers, Nat Libr Med, 65-83; CONSULT, 83- *Mem:* AAAS; Soc Pharmacol & Exp Therapeut; Am Chem Soc; Am Asn Cancer Res. *Res:* Carcinogenesis, production of tumors with chemical agents, air dust; chemotherapy of cancer; drug metabolism, effect of chemical agents on enzymes in normal and malignant tissues; biomedical library and information systems. *Mailing Add:* 4814 Essex Ave Chevy Chase MD 20815

LEITH, ARDEAN, b Warsow, NY, Mar 21, 47; m. BIOLOGICAL COMPUTING. *Educ:* Rensselaer Polytech Inst, BS, 68, MS, 86; Univ Rochester, PhD(biol), 72. *Prof Exp:* Fac mem biophys, Nat Univ Malaysia, 72-76 & cell biol, Univ Pertanian Malaysia, 78-82; sr res assoc cell biol, Worcester Polytech Inst, Mass, 76-78; asst prof cell biol, Univ Guam, USA, 82-85; RES SCIENTIST, HEALTH RES INC, ALBANY, NY, 85- *Mem:* Asn Comput Mach; Inst Elec & Electronics Engrs; Am Chem Soc; AAAS. *Res:* Visualization of cellular structure; display techniques for tomographic and confocal microscopy data; mathematical modeling of biological processes; computer graphics applications in biology. *Mailing Add:* Wadsworth Labs Empire State Plaza PO Box 509 Albany NY 12201

LEITH, CARLTON JAMES, b Madison, Wis, Sept 24, 19; m 41; c 2. GEOLOGY. *Educ:* Univ Wis, BA, 40, MA, 41; Univ Calif, PhD(geol), 47. *Prof Exp:* Asst geol, Univ Calif, 41-42; from jr mineral economist to asst mineral economist, Mineral Prod & Econ Div, US Bur Mines, 42-43; geologist, Standard Oil Co, Tex, 46; asst geol, Univ Calif, 46-47; from instr to asst prof geol, Univ Ind, 47-49; chief petrog unit, US Engrs Testing Lab, 49-51; geologist, Standard Oil Co, Calif, 51-60 & Holmes & Narver, Inc, 60-61; assoc prof, 61-65, prof geol, 65-80, emer prof geosci, NC State Univ, 80, Head dept, 67-80; RETIRED. *Mem:* Am Asn Petrol Geol. *Res:* Engineering geology; sedimentary petrology; areal geology; gravity and magnetics. *Mailing Add:* 17960 Tanleaf Lane Salinas CA 93907

LEITH, CECIL ELDON, JR, b Boston, Mass, Jan 31, 23; m 42; c 3. TURBULENCE, ATMOSPHERIC SCIENCES. *Educ:* Univ Calif, AB, 43, PhD(math), 57. *Prof Exp:* Physicist, Lawrence Radiation Lab, Univ Calif, 46-68; sr scientist, Lawrence Livermore Nat Lab, 68-78, dir, Atmospheric Anal & Prediction Div, 78-81, sr scientist, Nat Ctr Atmospheric Res, 81-83, physicist, 83-90, EMER PHYSICIST, LAWRENCE LIVERMORE NAT LAB, DEPT ENERGY, 90- *Concurrent Pos:* Mem, Int Comn Dynamic Meteorol, Int Asn Meteorol & Atmospheric Physics, 72-80 & Int Comn Climate, 78-80; mem joint organizing comt, Global Atmospheric Res Prog, World Meteorol Orgn & Int Counc Sci Unions, 76-80, officer, Joint Sci Comt, World Climate Res Prog, 81-83; chmn, Comt Atmospheric Sci, Nat Res Coun, 78-80. *Honors & Awards:* Meisinger Award, Am Meteorol Soc, 67, Carl-Gustaf Rossby Res Medal, 81. *Mem:* Fel AAAS; fel Am Phys Soc; fel Am Meteorol Soc; Am Math Soc. *Res:* Computational fluid dynamics; statistical hydrodynamics; turbulence. *Mailing Add:* Lawrence Livermore Nat Lab PO Box 808 Livermore CA 94550

LEITH, DAVID W G S, b Glasgow, Scotland, Sept 5, 37; m 62; c 3. HIGH ENERGY PHYSICS. *Educ:* Univ Glasgow, BSc, 59, PhD(natural philos), 62. *Prof Exp:* Glasgow Univ res fel physics, Europ Orgn Nuclear Res, Geneva, Switz, 62-63, staff physicist, 63-66; assoc prof, 66-70, PROF PHYSICS, LINEAR ACCELERATOR CTR, STANFORD UNIV, 70- *Mem:* Fel Am

Phys Soc; Brit Inst Physics & Phys Soc. *Res:* Strong interaction physics with emphasis on scattering experiments and investigations of resonance properties, their classification and the associated phenomenological analysis; study of electroweak interaction via the production and decay of z boson. *Mailing Add:* 754 Mayfield Ave Stanford Univ Stanford CA 94305

LEITH, EMMETT NORMAN, b Detroit, Mich, Mar 12, 27; m 56; c 1. ELECTRO-OPTICS. *Educ:* Wayne State Univ, BS, 49, MS, 52, PhD, 79. *Prof Exp:* Lab instr physics, Wayne State Univ, 51-52; asst eng res inst, 52-56, res assoc, 56-59, assoc res engr, 59-65, assoc prof, 65-68, PROF ELEC ENG, UNIV MICH, ANN ARBOR, 68- *Honors & Awards:* Gordon Mem Award, Soc Photo-Optical Instrumentation Eng, 65, Liebmann Award, Inst Elec & Electronics Engrs, 67; Daedalion Award, 68; Stuart Ballantine Medal, Franklin Inst, 69; R W Wood Prize, Optical Soc Am, 75; Holly Medal, Am Soc Chem Engrs, 76; Inventor Year Award, Asn Adv Invention & Innovation, 76; Nat Medal Sci, 79; Ivestr Medal, Optical Soc Am; Dennis Gabor Medal, Soc Photo-Optical Instrumentation Eng, 84, Gold Medal, 89. *Mem:* Nat Acad Eng; fel Optical Soc Am; fel Inst Elec & Electronics Engrs; fel Soc Photo-Optical Instrument Engrs. *Res:* Wavefront reconstruction; electronic physics; electromagnetics; radar; resonant cavity design; data processing; optical system design; coherent optics; interferometry; holography. *Mailing Add:* Dept Elec & Comput Eng Univ Mich Ann Arbor MI 48109-2122

LEITH, JOHN DOUGLAS, b Grand Forks, NDak, Apr 20, 31; m 57; c 2. PATHOLOGY. *Educ:* Lehigh Univ, BA, 52; Univ Pa, MD, 56; Univ Wis, PhD(cytol), 64; Am Bd Path, cer anat & clin path, 74, cert radioisotopic path, 75. *Prof Exp:* Intern, Med Ctr, Univ Calif, San Francisco, 56-57; asst zool, Univ Wis, 59-60, NSF fel, 60-63; Nat Cancer Inst spec fel, 63-64; asst prof anat & cell biol, Med Sch, Univ Pittsburgh, 64-67; from asst prof to assoc prof biol, Univ Wis-Oshkosh, 67-71; resident path, Peter Bent Brigham Hosp, Boston, 71-74; assoc pathologist, 75-91, ACTG CHIEF PATHOLOGIST, BROCKTON HOSP, 91- *Concurrent Pos:* Am Cancer Soc Inst res grant, 65-66, Health Res Serv Found res grant, 66-67, NSF grant, 68-70, Univ Wis res grants 68-70. *Mem:* Col Am Path; Am Soc Clin Path; Sigma Xi. *Mailing Add:* Dept Path Brockton Hosp Brockton MA 02402-3395

LEITH, THOMAS HENRY, philosophy of science, geophysics; deceased, see previous edition for last biography

LEITH, WILLIAM CUMMING, b Kimberley, BC, Apr 15, 25; m 50; c 3. MECHANICAL ENGINEERING, POLLUTION CONTROL. *Educ:* Univ BC, BAppSc, 48, MAppSc, 49; McGill Univ, PhD(mech eng), 60. *Prof Exp:* Jr engr, Dominion Eng Works, Que, 49-50 & Cominco-Trail, BC, 51-52; mech res engr, Dominion Eng Works, Que, 53-61; sr res scientist, Hydronautic Inc, Md, 61-62; design engr,Cominco-Trail, BC, 62-64 & H G Acres Co, Ont, 64-67; res assoc prof nuclear eng, Univ Wash, 67-73; MECH ENGR, COMINCO LTD, 73- *Concurrent Pos:* Consult, wood chip refining. *Honors & Awards:* Duggan Prize & Medal, Eng Inst Can, 59. *Mem:* Am Soc Mech Engrs. *Res:* Design of devices for access to blood circulatory systems such as cannulas, fistulas and catheters; pollution control; scrubbing of gases; uranium enrichment by gas centrifuge; cavitation correlated to vibration white finger in loggers' hands and minimum energy model of wood chip refining pulp/ paper making. *Mailing Add:* Cominco Ltd Eng Dept Trail BC V1R 4L4 Can

LEITMAN, MARSHALL J, b Yonkers, NY, Jan 16, 41. APPLIED MATHEMATICS, CONTINUUM PHYSICS. *Educ:* Rensselaer Polytech Inst, BS, 62; Brown Univ, PhD(appl math), 65. *Prof Exp:* Res assoc appl math, Brown Univ, 65-66; asst prof, 66-71, assoc prof, 71-81, PROF MATH, CASE WESTERN RESERVE UNIV, 81- *Concurrent Pos:* Vis asst prof, Cath Univ Louvain, 70-71. *Mem:* Soc Natural Philos; Soc Indust & Appl Math. *Res:* Mechanics; viscoelasticity. *Mailing Add:* Dept Math Case Western Reserve Univ University Circle Cleveland OH 44106

LEITMANN, G(EORGE), b Vienna, Austria, May 24, 25; nat US; m 55; c 2. MECHANICS. *Educ:* Columbia Univ, BS, 49, MA, 50; Univ Calif, PhD(eng sci), 56. *Hon Degrees:* DSc, Technische Univ, Vienna, Univ Paris; DIng, Technische Univ, Darmstadt. *Prof Exp:* Physicist, Naval Ord Test Sta, 50-55, head aeroballistics anal sect, 55-57; from asst prof to assoc prof eng sci, 57-63, chmn, Div Appl Mech, 71-72, Univ Ombudsman, 68-70, PROF ENG SCI, UNIV CALIF, BERKELEY, 63-, ASSOC DEAN, COL ENG, 81-, HUGHES CHAIR MECH ENG, 90- *Concurrent Pos:* Consult, Martin Co, 57-58 & Lockheed Missiles & Space Co, 58-66; ed, J Math Analysis Appl. *Honors & Awards:* Pendray Aerospace Lit Award, Am Inst Aeronaut & Astronat, 77; Levy Medal, Franklin Inst, 81; Mech & Control of Flight Award, Am Inst Aeronaut & Astronaut; Alexander von Humboldt Sr Scientist Award, 81. *Mem:* Nat Acad Eng; corresp mem Int Acad Astronaut; Acad Sci Bologna; fel Am Inst Aeronaut & Astronaut; Arg Acad Eng. *Res:* Exterior ballistics of rockets and astrodynamics; variational problems in mechanics and astronautics; optimal control of dynamic systems; game theory; control of uncertain systems; applications to economics, engineering. *Mailing Add:* Dept Mech Eng Univ Calif Berkeley CA 94720

LEITNER, ALFRED, b Vienna, Austria, Nov 3, 21; m 48; c 3. MATHEMATICAL THEORY OF WAVE PROPAGATION. *Educ:* Univ Buffalo, BA, 44; Yale Univ, MS, 45, PhD(physics), 48. *Prof Exp:* Res scientist, Courant Inst, NY Univ, 47-51; from asst prof to prof physics, Mich State Univ, 51-67; prof, 67-87, EMER PROF PHYSICS, RENSSELAER POLYTECH INST, 87- *Concurrent Pos:* Vis prof physics & Guggenheim fel, Aachen Technische Hochschule, Ger, 58-59; vis prof physics, Rensselaer Polytech Inst, 64 & US Mil Acad, West Point, 83-85; res assoc, Proj Physics, Harvard Univ, 65-66, consult, 66-67; Ger exchange fel, Deutsches Mus, Munich, 77-78. *Mem:* Fel Am Phys Soc; Am Asn Physics Teachers. *Res:* Mathematical theory of wave propagation, boundary value problems and special functions; production of educational films demonstrating physical phenomena for students of physics; history of physics. *Mailing Add:* 1201 Eighth Terr N Naples FL 33940

LEITNER, PHILIP, b Peking, China, June 16, 36; US citizen; m 60; c 2. VERTEBRATE ZOOLOGY. *Educ:* St Mary's Col, Calif, BS, 58; Univ Calif, Los Angeles, MA, 60, PhD(zool), 61. *Prof Exp:* Jr res zoologist, Univ Calif, Los Angeles, 61-62; from instr to assoc prof, 62-76, chmn dept, 70-76, PROF BIOL, ST MARY'S COL, CALIF, 76- *Concurrent Pos:* NIH res grant, 63-65, NSF res grants, 65-70. *Mem:* AAAS; Am Soc Zoologists; Soc Study Evolution; Am Soc Mammalogists. *Res:* Environmental physiology of mammals, especially physiological responses to temperature and photoperiod. *Mailing Add:* Dean Sci St Mary's Col Moraga CA 94575

LEITZ, FRED JOHN, JR, b Portland, Ore, Feb 2, 21; m 45; c 3. PHYSICAL CHEMISTRY. *Educ:* Reed Col, BA, 40; Univ Calif, PhD(phys chem), 43. *Prof Exp:* Instr chem, Univ Calif, 43-44; sr res chemist, Monsanto Chem Co, Ohio, 44-46; sr chemist, Oak Ridge Nat Lab, Tenn, 46-48; chemist, Radiochem & Reactor Metall Res, Hanford Works, Gen Elec Co, Wash, 48-56; nuclear engr, Atomic Power Develop Assocs, Mich, 56-58; develop proj engr, Atomic Power Equip Dept, Gen Elec Co, 58-64, mgr fast reactor core eng & test, Adv Prod Oper, 64-66, mgr steam reactor technol, 66-68; consult to dir, Battelle Northwest Lab, 69-70; sr staff scientist, Westinghouse Hanford Co, 70-76, mgr planning & anal, 76-79, staff mgr technol, 79-84; CONSULT, 84- *Mem:* Am Chem Soc; Am Nuclear Soc. *Res:* Heavy element and fission product chemistry; nuclear fuel cycle development; fast and steam cooled reactor design and technology. *Mailing Add:* 10411-140th Ave E Puyallup WA 98374

LEITZ, FREDERICK HENRY, b Hastings, Mich, Nov 20, 28; m 70. PHARMACOLOGY. *Educ:* Kalamazoo Col, BA, 52; Univ Calif, Los Angeles, PhD(chem), 62. *Prof Exp:* Res grant, Inst Org Chem, Royal Inst Technol, Sweden, 62-63; res scientist, Lamont Geol Observ, 63-65; staff fel, Lab Chem Pharmacol, Nat Heart Inst, 65-70; SR PRIN SCIENTIST, SCHERING CORP, 70- *Mem:* Am Chem Soc; Am Soc Pharmacol & Exp Therapeut; NY Acad Sci. *Res:* Drug metabolism, bioavailability, bioequivalency and pharmacokinetics. *Mailing Add:* Schering Corp 86 Orange St Bloomfield NJ 07003

LEITZ, VICTORIA MARY, b Yorkshire, Eng. BIOCHEMISTRY, LABORATORY MEDICINE. *Educ:* Oxford Univ, BA, 64, DPhil(clin chem), 68. *Prof Exp:* Res scientist human genetics, Med Res Coun, Oxford, Eng, 67-68; chemist neurochem, Sect Child Neurol, Nat Inst Neurol Dis & Stroke, NIH, 68-70; mgr develop chem & clin chem, Becton-Dickinson, NJ, 71-72; mgr diag chem & clin chem, Electro-Nucleonics, Inc, 74-79, dir tech serv, 79-81, dir mkt, 81-84, vpres int sales & mkt, 84-89, DIR CLIN CHEM, PHARMACIA DIAG, ELECTRO-NUCLEONICS, INC, 90- *Mem:* Am Asn Clin Chem; Nat Comt Clin Lab Standards; Biomed Mkt Asn. *Mailing Add:* 196 N Mountain Ave Montclair NJ 07042

LEITZEL, JAMES ROBERT C, b Shenandoah, Pa, May 27, 36; m 65; c 2. MATHEMATICS. *Educ:* Pa State Univ, BA, 58, MA, 60; Ind Univ, PhD(math), 65. *Prof Exp:* Asst prof math, Bloomsburg State Col, 59-63; asst prof, 65-69, ASSOC PROF MATH, OHIO STATE UNIV, 69- *Concurrent Pos:* Chair, Comt Math Educ Teachers, Math Asn Am. *Mem:* Am Math Soc; Math Asn Am; AAAS; Nat Coun Teachers Math; Asn Women Math. *Res:* Algebra, especially class field theory and algebraic function fields. *Mailing Add:* Dept Math Ohio State Univ Columbus OH 43210-1174

LEITZEL, JOAN PHILLIPS, b Valparaiso, Ind, July 2, 36; m 65; c 2. MATHEMATICS. *Educ:* Hanover Col, AB, 58; Brown Univ, AM, 61; Univ Ind, PhD(algebra), 65. *Prof Exp:* Instr math, Oberlin Col, 61-62; from asst prof to assoc prof, 65-83, PROF MATH, OHIO STATE UNIV, 83- *Concurrent Pos:* Vchmn math dept, Ohio State Univ, 73-, assoc provost, 85-90; div dir, NSF, 90- *Mem:* Am Math Soc; Math Asn Am. *Res:* Field theoretical proofs for cohomological results in class field theory. *Mailing Add:* Dept Math Ohio State Univ Columbus OH 43210

LEITZMANN, CLAUS, b Dahlenburg, Ger, Feb 6, 33; m 57; c 4. BIOCHEMISTRY, NUTRITION. *Educ:* Capital Univ, BS, 62; Univ Minn, MS, 64, PhD(biochem), 67. *Prof Exp:* Nat Inst Gen Med Sci res asst molecular biol inst, Univ Calif, Los Angeles, 67-69; vis prof biochem, Mahidol Univ, Thailand, 69-71; chief labs, Anemia & Malnutrit Res Ctr, Thailand, 71-74; assoc, 74-78, PROF INST NUTRIT, UNIV GIESSEN, 78-, DIR INST NUTRIT, 90- *Concurrent Pos:* Mem, Trop Inst, Univ Giessen, 74- *Honors & Awards:* Zabel prize, 87. *Mem:* AAAS; Inst Soc Nutrit; Am Soc Clin Nutrit; Am Inst Nutrit. *Res:* Nutrition in developing countries; interaction of nutrition and infection; adaptations to changes in food intake; hunger and satiety; obesity; dietary fibers; vegetarianism. *Mailing Add:* Inst Nutrit Univ Giessen Giessen 6300 Germany

LEIVO, WILLIAM JOHN, b New Castle, Pa, Sept 11, 15; m 39; c 2. PHYSICS. *Educ:* Carnegie Inst Technol, BS, 39, MS, 45, DSc(physics), 48. *Prof Exp:* Supt bldg construct, Matthew Leivo & Sons, Inc, Pa, 33-35 & 39-42; from instr to asst prof physics, Carnegie Inst Technol, 42-55; PROF PHYSICS, OKLA STATE UNIV, 55- *Mem:* Fel Am Phys Soc; Am Asn Physics Teachers. *Res:* Color centers in crystals; radiation effects in solids; optics; solid state physics; semiconducting diamond; ESR studies of blood cell membranes. *Mailing Add:* Dept Physics Okla State Univ Stillwater OK 74078

LEJA, J(AN), b Grodzisko, Poland, May 27, 18; m 47; c 6. SURFACE CHEMISTRY, METALLURGY. *Educ:* Univ London, BSc, 45; Univ Krakow, dipl Ing, 47; Cambridge Univ, PhD(surface chem), 54. *Hon Degrees:* Dr, Marie Curie-Sklodowska Univ, Poland, 76. *Prof Exp:* Res metallurgist, Southwest Africa Co, Eng, 47-49, reduction officer, SAfrica, 49-52; res fel colloid sci, Cambridge Univ, 54-57; from asst prof to prof metall, Univ Alta, 57-65; prof, 65-83, EMER PROF METALL, UNIV BC, 83- *Mem:* Brit Inst Mining & Metall; fel Can Inst Chem; Can Inst Mining & Metall. *Res:* Surface chemistry; infrared spectroscopy of adsorption; effluent control; dissolution of metals; corrosion. *Mailing Add:* Dept Mining & Mineral Process Eng Univ BC 6350 Stores Rd Vancouver BC V6T 1W5 Can

LEJA, STANISLAW, b Grodzisko, Poland, Jan 3, 12; nat US; m 39, 87; c 4. MATHEMATICS. *Prof Exp:* Teacher high schs, Palestine & Eng, 45-51; from asst to instr, Cornell Univ, 53-57; asst prof, 57-67, EMER PROF MATH, WESTERN MICH UNIV, 82- *Mem:* Am Math Soc; Math Asn Am. *Res:* Real variable; Fourier analysis. *Mailing Add:* 2205 Thurber Ct Orlando FL 32821-6785

LEKEUX, PIERRE MARIE, b Liege, Belg, Apr 7, 54; m 78; c 2. VETERINARY PHYSIOLOGY, BOVINE DISEASES. *Educ:* Univ Liege, Belg, DVM, 78; Univ Utrecht, Neth, 84. *Prof Exp:* Pres, Comp Respiratory Soc, 88-89; PROF PHYSIOL, UNIV LIEGE, 86-; DIR, LAB FUNCTIONAL INVEST, 88- *Concurrent Pos:* Assoc ed-in-chief, Annals of Med Vet, 86 & Pratipue Vet Equine, 91; pres equine, Equine Res Funds, 88-; secy gen bovine, World Asn Buiatrics, 90- *Mem:* Comp Respiratory Soc; Am Physiol Soc; World Asn Buiatrics. *Res:* Physiological, pathophysiological and pharmacological studies of the cardio-pulmonary function in large animals. *Mailing Add:* Univ Liege Bat Sart Tilman Liege B-4000 Belgium

LEKLEM, JAMES ERLING, b Rhinelander, Wis, Aug 1, 41; m 67; c 2. NUTRITION. *Educ:* Univ Wis, BS, 64, MS, 66, PhD(nutrit), 73. *Prof Exp:* Proj assoc clin oncol, Univ Wis, 66-71, res assoc, 73-75; from asst prof to assoc prof, 75-85, PROF NUTRIT, ORE STATE UNIV, 85- *Honors & Awards:* Borden Award, Am Home Econ Found, 85. *Mem:* Sigma Xi; Am Inst Nutrit. *Res:* Vitamin B6; metabolism of tryptophan; nutrient relationship to cancer etiology; obesity; diabetes. *Mailing Add:* 1625 NW Woodland Dr Corvallis OR 97330

LELACHEUR, ROBERT MURRAY, b Ottawa, Ont, Oct 12, 20; US citizen; m 46; c 4. PHYSICS. *Educ:* Mt Allison Univ, BSc, 42; Dalhousie Univ, MSc, 47; Univ Va, PhD(physics), 49. *Prof Exp:* Physicist, Nat Res Coun Can, 49-53; mem tech staff, Bell Labs, NJ, 53-58; asst supt eng, 58-62, dir mat & chem processes res & develop, NY, 62-66, MGR DEVELOP & MFG ENG, WESTERN ELEC CO, INC, READING, 66- *Mem:* Am Phys Soc; Inst Elec & Electronics Engrs; Sigma Xi. *Res:* Materials properties and processing; semiconductor device engineering. *Mailing Add:* 2200 Monroe Ave West Wyomissing PA 19609

LELAND, FRANCES E(LBRIDGE), b Chicago, Ill, Apr 22, 32. PHYSICAL CHEMISTRY. *Educ:* Swarthmore Col, BA, 54; Northwestern Univ, PhD(phys chem), 59. *Prof Exp:* Instr chem, Brooklyn Col, 59-61; from asst prof to assoc prof, 62-73, PROF CHEM, MACMURRAY COL, 73- *Mem:* Am Chem Soc. *Res:* Molecular quantum mechanics. *Mailing Add:* Dept Chem MacMurray Col Jacksonville IL 62650

LELAND, HAROLD R(OBERT), b Eau Claire, Wis, Apr 18, 31; m 58; c 2. ELECTRICAL ENGINEERING, SYSTEMS ANALYSIS. *Educ:* Univ Wis, BS & MS, 54, PhD(elec eng), 58. *Prof Exp:* Var res & supvry positions, Cornell Aeronaut Lab,Inc, 58-63, staff scientist, 63-66, asst head, Syst Res Dept, 66-70, head, 70-71, vpres & dir elec syst group, 71-73, vpres com develop group, 73-76, pres, 74-76, vpres electronics & systs group, 76-78, vpres & gen mgr, Advan Technol Ctr, 78-83, PRES, CALSPAN CORP, 83- *Mem:* Am Inst Aeronaut & Astronaut; Inst Elec & Electronics Engrs. *Res:* Automatic controls and pattern recognition; electronic warfare; military systems analysis; mathematical modeling of large systems. *Mailing Add:* 198 Bridle Path Williamsville NY 14221

LELAND, STANLEY EDWARD, JR, b Chicago, Ill, Aug 1, 26; m 50; c 3. PARASITOLOGY. *Educ:* Univ Ill, BS, 49, MS, 50; Mich State Univ, PhD(parasitol), 53. *Prof Exp:* Asst parasitol, Mich State Univ, 50-53; from assoc parasitologist to parasitologist, Univ Ky, 53-60, prof animal path, 60-63; assoc parasitologist, Univ Fla, 63-67; actg head dept infectious dis, 70-72, PROF PARASITOL, KANS STATE UNIV, 67-, ASSOC DIR AGR EXP STA, 75- *Concurrent Pos:* Coop agent animal dis & parasite res div, USDA, 53-59; consult, Eli Lilly Co, 62-66; vis prof parasitol, Ahmadu Bello, Univ, Zaria, Nigeria, 72-73; USDA Comt of Nine, 85-88. *Honors & Awards:* Col Vet Med Res Award, Kans State Univ, 71 &. *Mem:* Am Soc Parasitol. *Res:* Electrophoresis; drug testing; pathology; physiology; biochemistry; in vitro cultivation; immunology as related to parasitology published in over 120 scientific papers and nine book chapters. *Mailing Add:* Agr Exp Sta-Waters Hall Kans State Univ Manhattan KS 66506

LELAND, THOMAS W, JR, chemical engineering; deceased, see previous edition for last biography

LELAND, WALLACE THOMPSON, b Minn, Jan 21, 22; m 43; c 4. LASERS. *Educ:* Univ Minn, BEE, 43, PhD(physics), 50. *Prof Exp:* Head, Instrument Develop Dept, Carbide & Carbon Chem Corp, 46-47; MEM STAFF NUCLEAR RES, LOS ALAMOS NAT LAB, 50- *Mem:* Am Phys Soc; Sigma Xi. *Res:* Mass spectroscopy and nuclear reactions; high energy lasers. *Mailing Add:* MS E532 Los Alamos NM 87545

LELE, PADMAKAR PRATAP, b Chanda, India, Nov 9, 27; m 59; c 2. BIOMEDICAL ULTRASONICS. *Educ:* Bombay Univ, BS, 49, MD, 50; Oxford Univ, DPhil(biophys), 55. *Prof Exp:* Vis scientist, NIH, 58-59; tech dir med acoust & neurosurg assoc, Mass Gen Hosp, Boston, 59-69; assoc prof, 69-71, PROF EXP MED, MASS INST TECHNOL, 72-; PROF, HARVARD-MASS INST TECHNOL HEALTH SCI & TECHNOL, 78- *Concurrent Pos:* Physiol assoc, Harvard Med Sch, 59-69; consult, NIH, 60- & Harvard Univ Hosps, 70-; mem, diag radiol study sect, NIH; assoc ed, Ultrasound in Med & Biol, 78-, In Vivo, 86-; Rockefeller Found fel. *Honors & Awards:* Hist Med Ultrasound Pioneer Award, World Fedn Ultrasound Med & Biol, 88; Joseph H Holmes Pioneer Award, Am Inst Ultrasound Med, 88. *Mem:* Fel Am Inst Ultrasound Med; fel Acoust Soc Am; Am Soc Clin Hyperthermic Oncol; Inst Elec & Electronics Engrs; NAm Hyperthermia Group-Radiation Res Soc; Bioelectromagnetics Soc; Int Clin Hyperthermic Soc; Europ soc Hyperthermic Oncol; Am Soc Physicists Med. *Res:* Non-ionizing radiations; ultrasound and microwaves; applications in diagnostic and therapeutic medicine, industry and home, and safety; development and evaluation of local hyperthermia in cancer therapy, ultrasonic surgery and diagnosis, safety; biomedical engineering; heat transfer. *Mailing Add:* Mass Inst Technol Rm E17-434 Cambridge MA 02139

LELE, SHREEDHAR G, b Varanasi, India, Apr 19, 31; m 66; c 2. ELECTRICAL ENGINEERING, PHYSICS. *Educ:* Banaras Hindu Univ, India, MSc, 52; Univ Mich, MSE, 62, PhD(elec eng), 66. *Prof Exp:* Lectr physics, Banaras Hindu Univ, India, 52-60; res asst elec eng, Univ Mich, 61-66; DIR, ENG PROG, UNIV MASS, BOSTON, 85- *Concurrent Pos:* Consult, 80- *Mem:* Inst Elec & Electronics Engrs. *Res:* Ionospheric physics; microwave tubes; design of electron guns and solid-state devices. *Mailing Add:* Boston Harbor Campus Univ Mass Sci Bldg 3rd Floor Rm 110 Boston MA 02125

LELEIKO, NEAL SIMON, b Brooklyn, NY, Oct 26, 46; m 67; c 2. PEDIATRIC GASTROENTEROLOGY, HEPATOLOGY-NUTRITION. *Educ:* Brooklyn Col, BS, 67; NY Med Col, MD, 71; MIT, PhD(biochem & metab), 79. *Prof Exp:* Dir, Gen Clin Res Ctr, 87-90, ASSOC PROF PEDIAT, MT SINAI, 87- *Mem:* AAAS; Am Acad Pediat; NAm Soc Pediat Gastroenterol; Soc Pediat Res; Am Fedn Clin Res. *Res:* Molecular biology of gene nutrient interactions; cost of illness in children and adults; treatment and cause of inflammatory bowel disease. *Mailing Add:* Div Pediat Gastroenterol Mt Sinai Hosp Fifth Ave & 100th St New York NY 10029

LE LEVIER, ROBERT ERNEST, b Los Angeles, Calif, Nov 7, 23; m 45; c 3. THEORETICAL PHYSICS. *Educ:* Univ Calif, Los Angeles, PhD(physics), 51. *Prof Exp:* Mem staff, Lawrence Radiation Lab, Univ Calif, 51-57 & Rand Corp, 57-71; mem staff, 71-80, CHIEF SCIENTIST, R&D ASSOCS, 80- *Mem:* Am Phys Soc. *Res:* Ionospheric physics; nuclear physics; geophysics. *Mailing Add:* 961 Jacon Way Pacific Palisades CA 90272

LELLINGER, DAVID BRUCE, b Chicago, Ill, Jan 24, 37; m 63; c 2. TAXONOMIC BOTANY. *Educ:* Univ Ill, AB, 58; Univ Mich, MS, 60, PhD(bot), 65. *Prof Exp:* CUR FERNS, US NAT HERBARIUM, SMITHSONIAN INST, 63- *Concurrent Pos:* Ed-in-chief, Am Fern Soc, 66-84; Nat Geog Soc & Smithsonian Res Found explor & res grants, 71, 74; hon assoc Curator Pteridophytes, Mus Nat Costa Rica; ed, Pteridologia, 85- *Mem:* Int Asn Plant Taxon; Brit Pterid Soc. *Res:* Taxonomy of ferns and fern allies, especially those of the New World tropics. *Mailing Add:* US Nat Herbarium NHB 166 Smithsonian Inst Washington DC 20560

LELLOUCHE, GERALD S, b New York, NY, June 21, 30; m 58; c 1. THEORETICAL PHYSICS, NUCLEAR ENGINEERING. *Educ:* Purdue Univ, BS, 52; NC State Col, PhD(nuclear eng), 60. *Prof Exp:* Jr engr, Brookhaven Nat Lab, 52-55, asst nuclear eng, 60-64, assoc physicist, 64-68, physicist, 68-74; prog mgr, Probabilistics & Statist, Elec Power Res Inst, 74-80, sr prog mgr code develop & validation, 80-86; mgt consult, S Levy Inc, 86-90; PRES, TECH DATA SERV, 90- *Mem:* Am Chem Soc; Am Inst Chem Eng; Am Nuclear Soc. *Res:* Reactor kinetics, nonlinear dynamics; thermal hydraulics, two-phase flow; probabilistics, risk analysis. *Mailing Add:* 6252 N Lakewood Chicago IL 60660

LELONG, MICHEL GEORGES, b Casablanca, Morocco, Mar 20, 32; US citizen; m 59; c 3. PLANT TAXONOMY. *Educ:* Univ Algiers, baccalaureat, 50; Northwestern State Col, La, BS, 59, MS, 60; Iowa State Univ, PhD(syst bot), 65. *Prof Exp:* Assoc prof, 65-77, PROF BIOL, UNIV SALA, 77- *Mem:* Am Soc Plant Taxon; Int Asn Plant Taxon. *Res:* Systematics of Panicum subgenus Dichanthelium of North America; flora of the Mobile Bay region. *Mailing Add:* Dept Biol Sci Univ SAla 307 Univ Blvd Mobile AL 36688

LEM, KWOK WAI, b Canton, China, July 14, 52; US citizen; c 2. POLYMER SYNTHESIS, MOLECULAR COMPOSITE. *Educ:* Univ Toronto, Ont, Can, 76; Polytech Inst NY, MSc, 80; PhD(polymer sci eng), 83. *Prof Exp:* Chem specialist, Can Hanson Ltd, Toronto, Ont, 76-77; polymer chemist, Schenectady Chemicals Can, Ont, 77-78; res engr, Corp Technol, Allied Corp, 83-85; res engr, gas separation membrane, Corp Technol, Allied-Signal Inc, 85-86, res engr, polymer blends, 86-87, sr res engr, polymer alloys & composites, Corp Technol, 87-89, SR RES ENGR, ARMORS & COMPOSITES, CORP RES & TECHNOL, ALLIED-SIGNAL INC, 89- *Concurrent Pos:* Adj lectr, Polytech Univ, NY, 91. *Mem:* Soc Plastics Engrs; Soc Physics; Soc Rheology; Sigma Xi. *Res:* Polymer processing and rheology; reactive extrusion and injection molding; advanced polymer blends and synthesis; electrochemical and membrane technologies; processing-structure-property relations in polymer materials; advanced armor materials; dynamic behavior of advanced materials; impact dynamics; degradation and stability of materials; flammability materials. *Mailing Add:* 11 Old Coach Rd Randolph NJ 07869

LEMAIRE, IRMA, PHARMACOLOGY. *Prof Exp:* PROF, DEPT PHARMACOL, UNIV OTTAWA, 90- *Mailing Add:* Pharmacol Dept Health Sci Ctr Sch Med Univ Ottawa 451 Smyth Rd Ottawa ON K1H 8M5 Can

LEMAIRE, PAUL J, b Colchester, Vt, Aug 11, 53; m 86; c 2. OPTICAL FIBER RESEARCH & DEVELOPMENT. *Educ:* Mass Inst Technol, BS, 75, PhD(ceramics), 80. *Prof Exp:* RES SCIENTIST, AT&T BELL LABS, 80- *Honors & Awards:* Purdy Award, Am Ceramic Soc, 84. *Mem:* Am Ceramic Soc; Mat Res Soc. *Res:* Optical fiber research and development; optical loss; defects in glasses; waveguide design; optical fiber processing and manufacturing; hermetic coatings; fiber reliability and hydrogen-glass reactions. *Mailing Add:* AT&T Bell Labs Rm 6C-326 600 Mountain Ave Murray Hill NJ 07974-0636

LEMAISTRE, CHARLES AUBREY, b Lockhart, Ala, Feb 10, 24; m 52; c 4. INTERNAL MEDICINE, EPIDEMIOLOGY. *Educ:* Univ Ala, BA, 43; Cornell Univ, MD, 47. *Hon Degrees:* LLD, Austin Col, 70 & Univ Ala, 71; DSc, Univ Dallas, 78 & Southwestern Univ, 81. *Prof Exp:* From instr to asst prof internal med, Med Col, Cornell Univ, 51-54; assoc prof, Sch Med, Emory Univ, 54-59, prof prev med & chmn dept, 57-59; prof, Univ Tex Southwestern Med Sch, Dallas, 59-66, assoc dean, 65-66; vchancellor health affairs, Univ Tex, Austin, 66-68; from exec vchancellor to chancellor-elect, 68-70, chancellor, Univ Tex Syst, 71-78; PRES, UNIV TEX M D ANDERSON

CANCER CTR, 78- *Concurrent Pos:* Mem human ecol study sect, NIH, 62-65; mem, Surgeon Gen Adv Comt Smoking & Health, 63-64; mem, Gov Comt Eradication Tuberc, 63-64; mem comt res tobacco & health, AMA Educ & Res Found, 64-66; mem, Nat Citizens Comn Int Coop, 65-; mem Surgeon Gen emergency health preparedness adv comt, Dept Health, Educ & Welfare, 67, consult, Div Physician Manpower, 67-70; mem, President's Comn White House Fel, 71; mem, Comn Non-Traditional Study, 71-73; mem joint task force continuing competence pharm, Am Pharmaceut Asn-Am Asn Col Pharm, 73-74; mem bd comnr, Nat Comn Accrediting, 73-76; trustee, Biol Humanics Found, Dallas, 73-; mem, Nat Coun Educ Res, 73-75, consult, 75; chmn subcomt diversity & pluralism, Nat Coun Educ Res, 73-75; mem, United Negro Col Fund Develop Coun, 74-78; mem, Nat Adv Coun, Inst Serv Educ, 74-77. *Mem:* Am Cancer Soc (pres, 86-87). *Res:* Chest diseases. *Mailing Add:* Dept Tex M D Anderson Cancer Ctr 1515 Holcomb 1515 Holcomb Houston TX 77030

LE MAISTRE, CHRISTOPHER WILLIAM, b Moradabad, India, Aug 20, 38; Australian citizen; m 63; c 2. AUTOMATION & ROBOTICS. *Educ:* Univ Adelaide, BS, 63 & 64; Rensselaer Polytech Inst, PhD(mat eng), 72. *Prof Exp:* Exp officer, Australian Defence Sci, 64-72, sr res scientist, 72-79; assoc dir, Mfg Ctr, 79-84, DIR, CTR INDUST INNOVATION, RENSSELAER POLYTECH INST, 84-, ASST DEAN ENG, 84- *Concurrent Pos:* Res & develop rep, Australian High Comn, London, 74-77; head int progs, Australian Defence Sci, Canberra, 77-78, head lab progs, 78-79. *Mem:* Am Ceramic Soc; Am Soc Metals; Am Soc Eng Educ. *Res:* Near net shape (sintering); structural studies of carbon fibers; composite design and fabrication; automation and robotics; intellectual property. *Mailing Add:* Joseph St Troy NY 12180

LEMAL, DAVID M, b Plainfield, NJ, Feb 20, 34; m 63; c 4. ORGANIC CHEMISTRY. *Educ:* Amherst Col, AB, 55; Harvard Univ, PhD, 59. *Prof Exp:* From instr to asst prof chem, Univ Wis, 58-65; from assoc prof to prof, 65-81, ALBERT W SMITH PROF CHEM, DARTMOUTH COL, 81- *Concurrent Pos:* A P Sloan Found res fel, 68-70; trustee, Gordon Res Conf, 73-79 & chmn bd, 77-78; mem exec comt, Fluorine Div, Am Chem Soc, 80-83. *Mem:* Am Chem Soc. *Res:* Unusual species, stable and short-lived, in organic chemistry; organic reaction mechanisms; organic photochemistry; organofluorine chemistry. *Mailing Add:* Dept Chem Dartmouth Col Hanover NH 03755

LEMAN, ALLEN DUANE, b Peoria, Ill, Jan 15, 44. VETERINARY MEDICINE. *Educ:* Univ Ill, BS, 66, DVM, 68, PhD(physiol), 74. *Prof Exp:* Instr, Univ Ill, 69-75; ASSOC PROF VET MED, UNIV MINN, 75- *Concurrent Pos:* Dir, Nat Pork Producers Coun Res Comt, 72-; prog chmn, Int Pig Vet Cong, 73-; ed, Dis Swine, Iowa State Press, 74-; dir, Soc Study of Breeding Soundness & Nat Swine Improv Fedn, 75- *Mem:* Am Asn Swine Practitioners (pres, 75); Am Vet Med Asn; Soc Study Reproduction; Am Soc Animal Sci; Soc Study Breeding Soundness. *Res:* Causes of swine infertility; causes of lameness in boars; maximum economic returns from pork production. *Mailing Add:* 1620 Superior Ave Webster City IA 50595

LEMANN, JACOB, JR, b New Orleans, La, Aug 31, 29; m; c 3. EXPERIMENTAL BIOLOGY. *Educ:* Univ Calif, Berkeley, AB, 50; Univ Buffalo, MD, 54; Am Bd Internal Med, cert, 63 & 74. *Prof Exp:* Intern, Mass Mem Hosp, Boston, 54-55, res fel med, renal & metabol dis, 57-59 & clin fel, 60-61; asst resident, New Eng Med Ctr, Boston, 59-60; instr med, Boston Univ Sch Med, 61-63; from asst prof to assoc prof med, Marquette Sch Med & assoc dir, Clin Res Ctr, 63-68; assoc prof, Boston Univ Sch Med, 68-70; assoc prof, 70-71, CHIEF NEPHROLOGY DIV, MED COL WIS, MILWAUKEE, 70-, PROF MED, 71- *Concurrent Pos:* Chief, Renal Sect, Boston Univ Sch Med, 68-70; mem, Gen Med B Study Sect, NIH, 70-74 & 82-84; mem, Coun Med Adv Comt, Kidney Found Wis, 70- *Mem:* AAAS; fel Am Col Physicians; Am Fedn Clin Res; AMA; Am Physiol Soc; Am Soc Clin Invest; Am Soc Bone & Mineral Res; Am Soc Nephrology; Asn Am Physicians; Int Soc Nephrology. *Res:* Medicine. *Mailing Add:* Froedtert Mem Lutheran Hosp Med Col Wis 9200 W Wisconsin Ave Milwaukee WI 53226

LEMANSKI, LARRY FREDERICK, b Madison, Wis, June 5, 43; m 66; c 2. DEVELOPMENT BIOLOGY, IMMUNOELECTRON MICROSCOPY. *Educ:* Univ Wis, BS, 66; Ariz State Univ, MS, 68, PhD(zool), 71. *Prof Exp:* Fel biol & biochem, Univ Pa, 71-75; asst prof anat, Univ Calif, San Francisco, 75-77; from assoc prof to prof anat, Univ Wis, 77-83; PROF & CHMN ANAT & CELL BIOL, STATE UNIV NY, 83- *Concurrent Pos:* Prin investr, NIH, 76-; estab investr award, Am Heart Asn, 76-81, Louis Katz res prize, 78; distinguished sci examr, Bhopal Univ, India, 84-; dir, Cell & Molecular Training Prog, 87-90. *Mem:* Am Soc Cell Biol; AAAS; Asn Anat Chairmen; NY Acad Sci; Soc Develop Biol; Sigma Xi; Am Heart Asn. *Res:* Embryonic heart development using cellular and molecular biology approaches; early heart development and the initiation and maintenance of normal heart function in vertebrate embryos. *Mailing Add:* Dept Anat & Cell Biol State Univ NY Health Sci Ctr Syracuse NY 13210

LEMANSKI, MICHAEL FRANCIS, b Cleveland, Ohio, Nov 16, 46. HETEROGENEOUS CATALYSIS. *Educ:* Univ Dayton, BS, 69; Ohio State Univ, MS, 72, PhD(inorg chem), 75. *Prof Exp:* Sr res chemist, Diamond Shamrock Corp, 75-78; RES SCIENTIST, BP AM CORP, 78- *Concurrent Pos:* Vis researcher, Ctr Catalytic Sci & Technol, Univ Del, 80. *Mem:* Am Chem Soc. *Res:* Heterogeneous catalysis, including selective oxidation and selective reduction of small molecules. *Mailing Add:* Bf Chem R&DD Saltend Hull HU12 8DS England

LEMASTER, EDWIN WILLIAM, b Perryton, Tex, Apr 27, 40; m 64; c 1. SOLID STATE PHYSICS. *Educ:* West Tex State Univ, BS, 62; Tech Tech Univ, MS, 66; Univ Tex, PhD(physics), 70. *Prof Exp:* Asst prof physics, Gen Motors Inst, 64-66; asst prof physics, Pan Am Univ, 70-, chmn, phys sci dept, 73-; dean sci & technol, NMex Highlands Univ, 86-88; PROF PHYS SCI, EDINBURG UNIV. *Mem:* Am Phys Soc; Am Asn Physics Teachers. *Res:* Metalammonia solution properties; amorphous semiconductors; remote sensing of vegetative canopies; mathematical modeling. *Mailing Add:* Edinburg Univ 1201 W University Dr Edinburg TX 78539

LEMASTERS, JOHN J, b Newark, Ohio, May 29, 47; m 80; c 3. CELL BIOLOGY, ANATOMY. *Educ:* Yale Univ, BA, 69; Johns Hopkins Univ, MD, 75, PhD(cell biol & anat), 75. *Prof Exp:* Teaching asst, Dept Anat, Sch Med, Johns Hopkins Univ, 71-72; teaching asst neuroanat, Dept Cell Biol, Southwestern Med Sch, Univ Tex Health Sci Ctr, 73, asst prof, 75-77; from asst prof to assoc prof, 77-85, PROF CELL BIOL & ANAT, LAB CELL BIOL, SCH MED, UNIV NC, CHAPEL HILL, 85- *Concurrent Pos:* Dir, Grad Studies in Anat, Univ NC, 77-84, Electron Microscope Lab, 83-90, Confocal Imaging Facil, 89-; estab investr, Am Heart Asn, 82-87; vis prof, Div Gastroenterol & Sect Transplantation Surg, Mayo Grad Sch Med, 87; prin investr, Off Naval Res, 88-91, USPHS, 89-94 & 90-94, NSF, 90-92; mem, Coun Circulation, Am Heart Asn. *Mem:* AAAS; Am Asn Study Liver Dis; Am Asn Anatomists; Am Heart Asn; Am Soc Biochem & Molecular Biol; Biophys Soc; Electron Micros Soc Am. *Res:* Rescue of injured myocytes; liver preservation for transplantation; laser scanning confocal microscope; mechanisms of cell death in hepatocytes. *Mailing Add:* Lab Cell Biol Dept Cell Biol & Anat Univ NC Sch Med Campus Box 7090 236 Taylor Hall Chapel Hill NC 27599-7090

LEMASURIER, WESLEY ERNEST, b Wash, DC, May 3, 34; m 63; c 3. GEOLOGY. *Educ:* Union Col, BS, 56; Univ Colo, MS, 62; Stanford Univ, PhD(geol), 65. *Prof Exp:* Geologist, US Geol Surv, 61-64; asst prof geol, Cornell Univ, 64-68; assoc prof, 68-76, PROF GEOL, DIV NAT & PHYS SCI, UNIV COLO, DENVER, 76- *Mem:* AAAS; Geol Soc Am; Am Geophys Union; Int Asn Volcanol & Chem Earth's Interior; Sigma Xi. *Res:* Subglacial volcanism; petrology and tectonic relationships of volcanism in Antarctica. *Mailing Add:* Geol Dept Univ Colo Box 172 1200 Larimer St Denver CO 80204

LEMAY, CHARLOTTE ZIHLMAN, b Ft Worth, Tex, June 30, 19; m 44; c 3. PHYSICS, SOLID STATE PHYSICS. *Educ:* Tex Christian Univ, AB, 40; Mt Holyoke Col, MA, 41; La State Univ, PhD(physics), 50. *Prof Exp:* Res physicist, Monsanto Chem Co, 43-44; instr physics, Mt Holyoke Col, 45-46; instr physics, La State Univ, 46-48, res asst, 48-50; engr, Tex Instruments, Inc, 52-53, 55-57; res physicist, Stanford Res Inst, 53-54; engr, Westinghouse Elec Corp, 58-60 & Int Bus Mach Corp, 60-63; from asst prof to assoc prof sci, 63-69, prof physics & chmn dept, 69-77 & 85-89, EMER PROF PHYSICS, WESTERN CONN STATE UNIV, 90-; OWNER, SESAME STUDIOS, 89- *Concurrent Pos:* Otis Skinner fel physics, 40-41. *Mem:* Am Phys Soc; Am Asn Physics Teachers; sr mem Inst Elec & Electronics Engrs; Sigma Xi; Am Soc Eng Educ; sr mem Soc Women Engrs; Optical Soc Am. *Res:* Dielectric liquids; transistors; fiber optics; invention world's first silicon and silicon carbide transistors. *Mailing Add:* 60 Chestnut Ridge Rd Mt Kisco NY 10549

LEMAY, HAROLD E, JR, b Tacoma, Wash, May 28, 40; m 64; c 2. CHEMICAL EDUCATION. *Educ:* Pac Lutheran, BS, 62; Univ Ill, MS, 64, PhD(inorg chem), 66. *Prof Exp:* From asst prof to assoc prof, Univ Nev, Reno, 66-78, vchmn dept, 74-76, chmn dept, 84-85, assoc chmn, 85-90, PROF CHEM, UNIV NEV, RENO, 78- *Concurrent Pos:* Vis prof, Univ NC, Chapel Hill, 77-78, Univ Col Wales, 78 & Univ Calif Los Angeles, 89-90. *Mem:* Am Chem Soc; Sigma Xi. *Res:* Preparation and characterization of coordination compounds; reactions of coordination compounds in the solid phase; writer of chemistry textbooks. *Mailing Add:* Dept Chem Univ Nev Reno NV 89557

LE MAY, I(AIN), metallurgy, failure analysis, for more information see previous edition

LEMAY, JEAN-PAUL, b St Hyacinthe, Que, July 4, 23; m 52; c 2. ANIMAL PHYSIOLOGY, ANIMAL BREEDING. *Educ:* Classical Col St Hyacinth, BA, 45; Univ Montreal, BSA, 49; Univ Mass, MSc, 51; Laval Univ, PhD, 67. *Prof Exp:* Mem artificial insemination unit, Classical Col St Hyacinthe, 48-49; from instr to prof animal sci, Res Sta La Pocatiere, Que, 51-62; PROF ANIMAL SCI, LAVAL UNIV, 62- *Mem:* Am Soc Animal Sci; Can Soc Animal Prod. *Res:* Early weaning of sheep; histophysiology of sperm atogenesis and ovogenesis in sheep; sterility in dairy cattle; physiology of reproduction in dairy cattle, sheep and goat. *Mailing Add:* Dept Animal Sci Laval Univ Fac Agr Ste Foy PQ C1K 7P4 Can

LEMBACH, KENNETH JAMES, b Rochester, NY, June 16, 39; m 65; c 2. MONOCLONAL ANTIBODIES, IMMUNE REGULATION. *Educ:* Mass Inst Technol, BS, 61; Univ Pa, PhD(biochem), 66. *Prof Exp:* USPHS fel, Mass Inst Technol, 66-68, res assoc biochem, 68-69; from asst prof to assoc prof, Sch Med, Vanderbilt Univ, 69-79; plasma prod res sect head, 79-82, prin staff scientist, 82- 86, mgr, cell physiol res, 86-88, MGR, CELL & MOLECULAR BIOL RES, CUTTER BIOL, MILES INC, 89- *Concurrent Pos:* US Nat Cancer Inst res grants, 71-74 & 75-79. *Mem:* AAAS; Tissue Cult Asn; Am Soc Cell Biol; Am Soc Biochem & Molecular Biol. *Res:* Immune regulation and therapies; mammalian cell expression. *Mailing Add:* Cellular & Molecular Biol Res Cutter Biol Miles Inc Berkeley CA 94710

LEMBECK, WILLIAM JACOBS, b Kansas City, Mo, Aug 29, 28; m 60; c 1. MICROBIOLOGY, ACADEMIC ADMINISTRATION. *Educ:* La State Univ, BS, 50, MS, 56, PhD(bact), 62. *Prof Exp:* Supvry bacteriologist, US Army Chem Corps, Pine Bluff Arsenal, Ark, 57-59; from asst prof to assoc prof biol, Ark Agr & Mech Col, 61-62; asst prof bact, McNeese State Col, 62-65, assoc prof microbiol, 65-66; asst prof bot, Baton Rouge, 66-68, assoc prof biol & head div sci, 68-75, PROF BIOL, LA STATE UNIV, EUNICE, 75- *Mem:* Am Inst Biol Sci; Sigma Xi. *Res:* Effects of herbicides on normal soil microflora; biological catalysis of herbicides in soil; effects of herbicides on cellulose decomposition by Sporocytophaga myxococcoides. *Mailing Add:* Div Sci La State Univ PO Box 1129 Eunice LA 70535

LEMBERG, HOWARD LEE, b Queens, NY, July 29, 49; m 70; c 2. CHEMICAL PHYSICS, COMPUTER SCIENCES. *Educ:* Columbia Univ, BA, 69; Univ Chicago, PhD(chem physics), 73. *Prof Exp:* Res chem physics, Bell Labs, 73-75; asst prof chem, Univ NC, 75-78; mem tech staff, Bell Labs, 78-81, supvr, 81-84, DIST MGR, BELLCORE, 84- *Mem:* Inst Elec &

Electronics Engrs; Am Phys Soc; Soc Photo Instrument Engr. *Res:* Statistical mechanics of liquids; electronic structure of surfaces; fluctuations and instabilities in chemical systems; computer protocols; communications networks; optical networks; optical fiber-subscriber loop networks. *Mailing Add:* Bellcore 2M289 445 South St Morristown NJ 07962

LEMBERG, LOUIS, b Chicago, Ill, Dec 27, 16; m 39. PHYSIOLOGY. *Educ:* Univ Ill, BS, 38, MD, 40; Am Bd Internal Med, dipl, 50, recert, 74; Am Bd Cardiovasc Dis, dipl, 55. *Prof Exp:* Intern, Mt Sinai Hosp, Chicago, Ill, 40-41, res, 45-48; PROF CLIN CARDIOL, SCH MED, UNIV MIAMI, 69- *Concurrent Pos:* Dir cardiol, Dade County Hosp, 55-57; attend specialist, Vet Admin Hosp, 55-64; chief staff, Nat Children's Cardiac Hosp, Miami, 55-66; attend cardiologist, Mercy & Cedars of Lebanon Hosp; chief div electrophysiol, 56-74, dir coronary care unit, Jackson Mem Hosp, 68-74; chief div cardiol, Mercy Hosp, 76- mem coun clin cardiol, Am Heart Asn; ed-in-chief, Accel for Nurses, 83-85, Current Concepts Cardiovasc Dis, 85-87; Endowed chair, Cardiol, "Louis Lemberg" Chair Cardiol, Univ Miami. *Honors & Awards:* Savage Award, 60; Luis Guerrero Mem Award, Philippines. *Mem:* Hon mem Philippine Med Asn; fel Am Col Physicians; Am Col Chest Physicians; Am Col Cardiol; NY Acad Sci. *Res:* Cardiology. *Mailing Add:* Div Cardiol (D39) Univ Miami PO Box 016960 Miami FL 33101

LEMBERGER, AUGUST PAUL, b Milwaukee, Wis, Jan 25, 26; m 47; c 7. PHARMACEUTICS. *Educ:* Univ Wis, BS, 48, PhD(pharm), 52. *Prof Exp:* Sr chemist pharmaceut res, Merck & Co, Inc, 52-53; from instr to prof pharm, Univ Wis-Madison, 53-69, coordr exten serv, 65-69; prof pharm & dean, Col, Univ Ill Med Ctr, 69-80; PROF PHARM & DEAN, SCH PHARM, UNIV WIS-MADISON, 80- *Concurrent Pos:* Mem & secy, Wis Pharm Internship Comn, 65-69; consult, Dept Health, Educ & Welfare, 72-74; mem, Am Coun Pharmaceut Educ, 78-84, bd trustees, Am Pharmaceut Soc, 85-88; mem, Tech Adv Coun Ill Dept Pub Health Drug Substitution law, 78-80. *Honors & Awards:* Kiekhofer Award, Univ Wis, 57. *Mem:* Am Pharmaceut Asn; fel Acad Pharm Res & Sci; fel AAAS; fel Am Asn Pharmaceut Scientists; Am Asn Cols Pharmacy; Am Soc Hosp Pharmacists. *Mailing Add:* 7439 Cedar Creek Trail Madison WI 53717

LEMBERGER, LOUIS, b Monticello, NY, May 8, 37; m 59; c 2. CLINICAL PHARMACOLOGY. *Educ:* Long Island Univ, BS, 60; Albert Einstein Col Med, PhD(pharmacol), 64, MD, 68. *Prof Exp:* Fel pharmacol, Albert Einstein Col Med, 64-68; med intern, Metropolitan Hosp Ctr-NY Med Col, 68-69; pharmacol & toxicol res assoc clin pharmacol, Lab Clin Sci, NIMH, 69-71; clin pharmacologist, 71-75, chief, 75-78; dir clin pharmacol, 78-89, CLIN RES FEL, LILLY LAB CLIN RES, LILLY RES LABS, 89- *Concurrent Pos:* Dir, Clin Pharmacol Training Prog, Sch Med, Ind Univ, 72-75, asst prof pharmacol & med, 72-73, assoc prof, 73-77, prof pharmacol, med & psychiat, 77-; assoc prof, Grad Fac, 75-77, prof, 77-; adj prof clin pharmacol, Ohio State Univ, 75-; chmn, Second World Conf Clin Pharmacol, Int Union Clin Pharmacol. *Mem:* Am Soc Pharmacol & Exp Therapeut (pres, 87-88); Am Soc Clin Pharmacol & Therapeut (pres, 83-84); Am Col Neuro Psychopharm; Am Col Physicians; Sigma Xi. *Res:* Drug metabolism and drug-drug interactions; synthesis and metabolism of biogenic amines; biochemical mechanisms of drug action; psychopharmacology; marihuana & cannabinoids. *Mailing Add:* Lilly Lab for Clin Res Lilly Res Labs Indianapolis IN 46202

LEMBKE, ROGER ROY, b Clayton Co, Iowa, Apr 24, 40; m 69; c 2. PHYSICAL CHEMISTRY, RADIATION CHEMISTRY. *Educ:* Luther Col, AB, 62; Univ Nebr, Lincoln, MS, 66; Wayne State Univ, PhD(phys chem), 73; Univ Evansville, Ind, MCSE, 84. *Prof Exp:* Instr chem, Hastings Col, 65-69; guest scientist, Hahn-Meitner Inst, Berlin, 73-74; res assoc, Univ Fla, 75; asst prof, Cornell Col, 75-76; assoc prof, 76-78, PROF CHEM & CHMN DEPT, CENT METHODIST COL, 78- *Mem:* Am Chem Soc; Asn Comput Mach. *Res:* Radiolysis and photolysis; rate constants and mechanisms. *Mailing Add:* Dept Chem Cent Methodist Col Fayette MO 65248

LEMBO, NICHOLAS J, b Boston, Mass, Aug 18, 29; m 56; c 4. ANALYTICAL CHEMISTRY. *Educ:* Boston Col, BS, 51; Teachers Col City Boston, EdM, 52; Northeastern Univ, MS, 62. *Prof Exp:* Instr pub schs, Mass, 52-56; instr phys sci, 56-59, asst prof chem, 59-65, assoc prof chem, Boston State Col, 65-; AT DEPT SCI, BUNKER HILL COMMUNITY COL. *Concurrent Pos:* Lectr, Lincoln Col, Northeastern Univ, 53-65. *Res:* Improvement of science teaching in the elementary schools. *Mailing Add:* Dept Sci Bunker Hill Community Col New Rutherford Ave Boston MA 02129

LEMCOE, M M(ARSHALL), b St Louis, Mo; m 51; c 2. CIVIL ENGINEERING. *Educ:* Wash Univ, BS, 43, MS, 49; Univ Ill, PhD(civil eng), 57. *Prof Exp:* Struct engr, Curtiss-Wright Corp, 43-46; mem staff, Res Found & Dept Civil Eng, Wash Univ, 47-51; supvr aeroelasticity & spec consult, Southwest Res Inst, 51-52, mgr, Strength Anal Sect, Dept Struct Res, 52-61; supvr exp mech & sr tech specialist, Atomics Int Div, NAm Rockwell Corp, Calif, 61-71; TECH ADV, COLUMBUS DIV, BATTELLE MEM INST, 71- *Honors & Awards:* Award, Curtiss-Wright Corp, 45; IR-100 Award, 76. *Mem:* Soc Exp Stress Anal; Sigma Xi. *Res:* Experimental stress analysis; structures; pressure vessels; high temperature materials technology and strain gage technology; high temperature behavior of structures. *Mailing Add:* 12990 Camino Rainllette San Diego CA 92128

LE MEE, JEAN M, b June 4, 31; US citizen; m 64. ENGINEERING. *Educ:* Carnegie-Mellon Univ, MS, 59, PhD(mech eng), 63. *Prof Exp:* Design engr, James Gordon & Co Ltd, Eng, 55-58; teaching asst eng, Carnegie-Mellon Univ, 59-61; res engr, Lawrence Radiation Lab, Univ Calif, Berkeley, 60 & Westinghouse Res Labs, 62-64; assoc prof, 64-80, PROF MECH ENG, COOPER UNION SCH ENG & SCI, 80- *Mem:* Inst Elec & Electronics Engrs; Sigma Xi. *Res:* Control systems; semiconductor devices. *Mailing Add:* 16 Mevan Ave Tributary Woods Englewood NJ 07631

LE MEHAUTE, BERNARD J, b St Brieuc, France, Mar 29, 27; m 53; c 2. HYDRODYNAMICS, COASTAL ENGINEERING. *Educ:* Univ Rennes, Baccalaureat, 47; Univ Toulouse, lic es sc, 51; Univ Grenoble, Dr es Sc(hydrodyn), 57. *Prof Exp:* Res engr, Neyrpic-Sogreah, France, 53-57; assoc prof hydrodyn, Polytech Sch, Montreal, 57-59; res prof, Queen's Univ, Ont, 59-61; mem tech staff, Nat Eng Sci Co, 61-62; mem sr staff, 62-64, assoc dir hydrodyn, 64-66; vpres, Tetra Tech, Inc, 66-70, sr vpres, 66-78, mem bd, 66-84; prof & chmn ocean eng, 78-83, PROF APPL MARINE PHYSICS, UNIV MIAMI, ROSENSTIEL SCH MARINE & ATMOSPHERIC SCI, 84- *Concurrent Pos:* Mem, nat sea grant rev panel, 70-78; mem, coastal eng res bd, Dept Defense, 82-88; mem, marine bd, Nat Res Coun, 89- *Honors & Awards:* Int Coastal Eng Award, Am Soc Civil Engrs, 79; Creative Award, NSF, 81. *Mem:* Nat Acad Eng; Am Soc Civil Eng; Marine Technol Soc; Int Asn Hydraul Res. *Res:* Hydrodynamics and hydraulic and coastal engineering, ranging from theoretical fluid mechanics to physical oceanography applied to the design of engineering structures for water power, coastal harbors and offshore drilling. *Mailing Add:* Univ Miami Rosenstiel Sch Marine & Atmospheric Sci 4600 Rickenbacker Causeway Miami FL 33149

LEMENT, BERNARD S, b Boston, Mass, Feb 11, 17; m 42; c 3. FAILURE ANALYSIS, ACCIDENT RECONSTRUCTION. *Educ:* Mass Inst Technol, BS, 38, ScD(metall), 49. *Prof Exp:* Metallurgist testing, NY Testing Labs, 38-39; res asst tool steels, Mass Inst Technol, 40; assoc metallurgist army ord, Watertown Arsenal, 40-46; instr physics, Univ Mass, 46-47; res staff mem dimensional stability, Mass Inst Technol, 47-49; asst prof metall, Univ Notre Dame, 49-51; mem res staff electron microscopy, Mass Inst Technol, 51-57; proj dir res & develop, ManLabs, Inc, 57-67; MAT ENG CONSULT, LEMENT & ASSOCS, 67- *Mem:* Fel Am Soc Metals; Am Inst Metall Engrs; Am Welding Soc; Am Soc Testing & Mat; Am Soc Safety Engrs. *Mailing Add:* 24 Graymore Rd Waltham MA 02154

LEMESHOW, STANLEY ALAN, b Brooklyn, NY, Jan 29, 48; m 72; c 2. SAMPLING, EXPERIMENTAL DESIGN. *Educ:* City Col New York, BBA, 69; Univ NC, Chapel Hill, MSPH, 70; Univ Calif, Los Angeles, PhD(biostatist), 76. *Prof Exp:* Res asst, Dept Prev Med, NY Med Col, 68-69; statist supvr, Health Res Training Prog, New York City Dept Health, 69; anal statistician, Comn Off, Nat Ctr Health Statist, USPHS, 70-72; sr statisticin, Sch Pub Health, Univ Calif, Los Angeles, 74-75; asst prof, 76-80, ASSOC PROF BIOSTATISTS, UNIV MASS, 80- *Concurrent Pos:* Dir, Coord Ctr for Multicenter Clin Trial of Hyperbaric Oxygen in Treatment of Burn Injuries, Univ Mass, 77-78; prog dir, Biopharmceut Res Unit, Div Pub Health, Univ Mass, Amherst, 78- *Mem:* Am Statist Asn; Soc Epidemiol Res; Am Pub Health Asn. *Res:* Sampling; variance estimation in complex sampling designs; sample size determination and logistic regression analysis; medical and other applied health sciences. *Mailing Add:* Div Pub Health Sch Health Sci Univ Mass Amherst MA 01003

LEMESSURIER, WILLIAM JAMES, b Pontiac, Mich, June 12, 26; m 53; c 3. STRUCTURAL ENGINEERING. *Educ:* Harvard Univ, AB, 47; Mass Inst Technol, SM, 53. *Prof Exp:* Founder, Goldberg-LeMessurier, 52-61, founder & partner, LeMessurier Assoc, Inc, 61-73, chmn & chief exec officer, Sippican Consult Int, Inc, 73-85, CHMN & CHIEF EXEC OFFICER, LEMESSURIER CONSULT, INC, 85- *Concurrent Pos:* Asst prof, Mass Inst Technol, 52-56, assoc prof, 64-67, sr lectr, 76-77; assoc prof, Harvard Grad Sch Design, 56-61, lectr, 73-, adj prof, 82-; Masonry Res Adv Coun, Sci Adv Comt, Nat Ctr Earthquake Eng Res 500 Allied Prof Medal, Am Inst Architects, 68. *Honors & Awards:* Special Award, Am Inst Steel Construction, 72. *Mem:* Nat Acad Eng; fel Am Soc Civil Engrs; fel Am Concrete Inst; hon mem Am Inst Architects. *Res:* Structural engineering design; precast concrete high rise housing system; staggered truss system for high rise steel structures; tuned mass damper system used to reduce tall building motion; structural stability. *Mailing Add:* LeMessurier Consult Inc 1033 Massachusetts Ave Cambridge MA 02238-5388

LEMIEUX, RAYMOND URGEL, b Lac la Biche, Alta, June 16, 20; m 48; c 6. VIROLOGY, PHYSIOLOGY. *Educ:* Univ Alta, BSc, 43; McGill Univ, PhD(org chem), 46. *Hon Degrees:* DSc, Univ NB, 67, Laval Univ, 70, Univ Ottawa, 75, Waterloo Univ, 80, Mem Univ, 81, Univ de Que, 82, Queens Univ, 83, McGill Univ, 84, Univ de Sherbrooke, 86, McMaster Univ, 86; Doctorate, Univ de Provence, France, 72; LLD, Univ Calgary, 79; Dr Philos, Univ Stockholm, 88. *Prof Exp:* Res assoc carbohydrate chem, Ohio State Univ, 46-47; asst prof org chem, Univ Sask, 47-49; res officer chem natural prod, Prairie Regional Lab, Nat Res Coun, 49-54; prof chem, chmn dept & vdean fac pure & appl sci, Ottawa Univ, Can, 54-61; prof, 61-85, EMER & UNIV PROF ORG CHEM, UNIV ALTA, 85- *Concurrent Pos:* Merck lectr, 56; Folkers lectr, 58; Karl Pfister lectr, 68; Purves lectr, 70; pres & dir res, Raylo Chem Ltd, Alta, 66-76; pres, Chembiomed Ltd, 77-78, mem bd, 77-78 & 83-84, hon bd mem, 86-, chmn, Sci Adv Comt, 90- *Honors & Awards:* Medal, Chem Inst Can, 64; C S Hudson Award, Am Chem Soc, 66; Medal of Serv, Order of Can, 68; Haworth Medal Chem Soc, Eng, 78; Killam Prize, Can Coun, 81; Medal Hon, Can Med Asn, 85; Gairdner Found Int Award, 85; Rhone-Poulenc Award, Royal Soc Chem, 89; LeSueur Award, Soc Chem Indust, 89; King Faisal Int Prize Sci, 90. *Mem:* Royal Soc Chem; fel Chem Inst Can; fel Royal Soc Can. *Res:* Stereochemistry; conformational analysis; carbohydrate chemistry; synthesis and conformation; especially antibiotics and blood group determinants. *Mailing Add:* Dept Chem Univ Alta Edmonton AB T6G 2G2 Can

LEMING, CHARLES WILLIAM, b Cutler, Ill, Nov 5, 43; m 65; c 1. PHYSICS. *Educ:* Eastern Ill Univ, BS, 65; Mich State Univ, MS, 67, PhD(physics), 70. *Prof Exp:* PROF PHYSICS, HENDERSON STATE UNIV, 70- *Concurrent Pos:* Res assoc, Fac Develop Prog, NSF, 77 & Student Sci Training Prog, 78, proj dir, 79; res assoc, Carbondale Mining Technol Ctr, 81; mem proj staff, NSF Consortium upper level physics software. *Mem:* AAAS; Am Asn Physics Teachers. *Res:* Optical devices; radon detection; science museum programs for teacher enhancement; textbooks which integrate computer methods into undergraduate physics courses. *Mailing Add:* Box 7605 Henderson State Univ Arkadelphia AR 71923

LEMIRE, ROBERT JAMES, b Toronto, Ont, Mar 12, 45; m 68. SOLUTION CHEMISTRY. *Educ:* Univ Toronto, BSc, 68, MSc, 71, PhD(chem), 75. *Prof Exp:* Researcher fel chem, Univ Ky, 75-77; ASST RES OFFICER CHEM, WHITESHELL NUCLEAR RES ESTAB, ATOMIC ENERGY CAN LTD, 77- *Mem:* Chem Inst Can; Am Chem Soc. *Res:* Properties of aqueous and non-aqueous solutions; complexation; solvent extraction. *Mailing Add:* Atomic Energy Can Ltd Pinawa MB R0E 1L0 Can

LEMIRE, RONALD JOHN, b Portland, Ore, Apr 20, 33; c 5. TERATOLOGY, PEDIATRICS. *Educ:* Univ Wash, MD, 62. *Prof Exp:* Intern, King County Hosp, Seattle, Wash, 62-63; NIH fel teratology & embryol, Univ Wash, 63-65, asst resident pediat, 65-67; chief resident, Children's Orthop Hosp & Med Ctr, 67-68; from asst prof to assoc prof, 68-77, PROF PEDIAT, UNIV WASH, 77- & DIR INPATIENT SERV, CHILDREN'S HOSP & MED CTR, SEATTLE. *Mem:* Soc Pediat Res; Teratology Soc. *Res:* Neuroembryology; neuroteratology. *Mailing Add:* Dir Inpatient Serv Childrens Hosp & Med Ctr 4800 Sand Pt Way NE PO Box C-5371 Seattle WA 98105

LEMISH, JOHN, b Rome, NY, July 4, 21; m 46; c 5. ECONOMIC GEOLOGY, GEOCHEMISTRY. *Educ:* Univ Mich, BS, 47, MS(geol), 48, PhD, 55. *Prof Exp:* Geologist, US Geol Surv, 48, 49-51; instr geol, Univ Mich, 53-55; from asst prof to prof geol, Iowa State Univ, 55-91; CONSULT, 91- *Concurrent Pos:* Mem comts, Hwy Res Bd, Nat Acad Sci-Nat Res Coun, 58-70; chmn, State Mining Bd, 64-73, chmn publ comt, Am Geol Inst; mem adv comt, Iowa Coal Proj, 74-79; mem tech adv comt, Nat Gas Surv, Fed Power Comn, 75-81; NSF fel, 52-53. *Mem:* AAAS; fel Geol Soc Am; Geochem Soc; Am Asn Petrol Geol; Am Inst Mining, Metall & Petrol Eng. *Res:* Weathering studies of concrete; behavior of carbonate aggregates in Portland cement concrete; aggregate-cement reactions; physical and chemical phenomena related to ore deposition; structural geology; trace elements in Pennsylvania shales; occurrence of coal deposits in Iowa; coal exploration; occurrence of deep coal in Iowa; geology of Forest City Basin; pore properties of carbonate rocks in Iowa suitable for desulfurization by burning coal in fluidized bed combustion. *Mailing Add:* Dept Geol Sci Iowa State Univ Ames IA 50011

LEMKAU, PAUL V, b Springfield, Ill, July 1, 09; m 34; c 5. EPIDEMIOLOGY, PSYCHIATRY. *Educ:* Baldwin-Wallace Col, AB, 31; Johns Hopkins Univ, MD, 35. *Hon Degrees:* DSc, Baldwin-Wallace Col, 51; DPH, Dickinson Col, 58. *Prof Exp:* Res asst to prof & dept chmn ment hyg, Sch Hyg & Pub Health, Johns Hopkins Univ, 39-75; RETIRED. *Concurrent Pos:* Dir Ment Health Div, Md State Dept Health, 43-47; consult, Vet Admin Hosp, Perry Point, Md, 46-52, WHO, 59-75; pres, Md Psychiat Soc, 50; dir, New York City, Community Ment Health Bd, 55-57. *Mem:* Am Psychiat Asn (vpres, 61); Am Pub Health Asn. *Res:* Epidemiology of mental illnesses in Baltimore and in Croatian Republic of Yugoslavia; administration of public mental health programs. *Mailing Add:* PO Box 178 Lusby MD 20657

LEMKE, CALVIN A(UBREY), b Waco, Tex, Aug 25, 21; m 48; c 1. CIVIL ENGINEERING. *Educ:* Agr & Mech Col, Tex, BS, 43, MS, 61, PhD, 88. *Prof Exp:* Instr math, Baylor Univ, 52-56; asst prof, 56-65, ASSOC PROF CIVIL ENG, LA TECH UNIV, 65- *Res:* Structures and highways; highway culverts; soils. *Mailing Add:* PO Box 235 Ruston LA 71273

LEMKE, CARLTON EDWARD, b Buffalo, NY, Oct 11, 20; m 55; c 2. MATHEMATICS. *Educ:* Univ Buffalo, BA, 49; Carnegie Inst Technol, MA, 51, PhD(math), 53. *Prof Exp:* Instr math, Carnegie Inst Technol, 52-54; res assoc anal, Knolls Atomic Power Lab, Gen Elec Co, NY, 54-55; engr Radio Corp Am, NJ, 55-56; from asst prof to prof math, 56-67, FORD FOUND PROF MATH, RENSSELAER POLYTECH INST, 67- *Mem:* Am Math Soc; Soc Indust & Appl Math; Opers Res Soc Am; Economet Soc; Math Asn Am. *Res:* Algebra; mathematical programming; probability and statistics; operations research. *Mailing Add:* Dept Math Amos Eaton Bldg Rm 301 Rensselaer Polytech Inst Troy NY 12180-3590

LEMKE, DONALD G(EORGE), b Chicago, Ill, Mar 25, 32; m 55; c 4. MECHANICAL ENGINEERING. *Educ:* Ill Inst Technol, BS, 55; Univ Pa, MS, 60, PhD(mech eng), 70. *Prof Exp:* Jr engr, Int Harvester Co, 54-55; engr, Teletype Corp, 55-58 & Westinghouse Elec Corp, 58-59; appl mech specialist, Dyna/Struct, Inc, 60-61; res engr, Advan Space Proj Dept, Gen Elec Co, 62-63; assoc scientist, Missile Div, Chrysler Corp, 63-64, res mgr aero ballistics & mech, 64-65; asst prof eng mech, Mich Technol Univ, 65-68, asst prof mech eng, 68-77, ASSOC PROF MECH ENG, UNIV ILL, CHICAGO CIRCLE, 77- *Mem:* Am Soc Mech Engrs; Am Inst Aeronaut & Astronaut. *Res:* Machine mechanics; dynamics; structural mechanics. *Mailing Add:* Dept Civil Eng Univ Ill Chicago IL 60680

LEMKE, JAMES UNDERWOOD, b Grand Rapids, Mich, Dec 26, 29; m 53; c 3. MAGNETIC RECORDING, MAGNETIC MATERIALS. *Educ:* Ill Inst Technol, BS, 59; Northwestern Univ, MS, 60; Univ Calif, Santa Barbara, PhD, 66. *Prof Exp:* Electronics engr prod develop, Temco, 51-53; vpres eng, AV Mfg Co, 53-56; assoc to tech vpres, Armour Res Found, 57-60; dir magnetic res, Bell & Howell Res Labs, 60-68; pres & founder, Spin Physics Inc, 68-82; res fel, Eastman Kodak Res Labs, 82-86; PRES & FOUNDER, REC PHYSICS INC, 86- *Concurrent Pos:* Adj prof, Univ Calif, San Diego, 84- *Mem:* Nat Acad Eng; Inst Elec & Electronics Engrs; Am Phys Soc; AAAS; Am Asn Physics Teachers. *Res:* Physics of magnetic recording process; development of related components, materials, processes, devices and circuits. *Mailing Add:* Rec Physics Inc 2326 India St San Diego CA 92101

LEMKE, PAUL ARENZ, b New Orleans, La, July 14, 37; div; c 2. GENETICS, MICROBIOLOGY. *Educ:* Tulane Univ, BS, 60; Univ Toronto, MA, 62; Harvard Univ, PhD(biol), 66. *Prof Exp:* Instr biol, Tulane Univ, 62-63; sr microbiologist, Eli Lilly & Co, Ind, 66-72; assoc prof biol sci, Carnegie-Mellon Univ, 72-79; prof bot, plant path, microbiol & head dept, 79-85, PROF MOLECULAR GENETICS, AUBURN UNIV, 85- *Concurrent Pos:* Instr, Franklin Col, 67; sr fel, Carnegie-Mellon Inst Res, 72-79; Alexander von Humboldt Award, 77-78; vis prof, Ruhr Univ, WGer, 77-78 & Emory Univ, 85-86. *Honors & Awards:* Porter Award, Soc Indust Microbiol. *Mem:* Am Soc Microbiol; Genetics Soc Am; Mycol Soc Am; Bot Soc Am; Soc Indust Microbiol (treas, 76-78, pres-elect, 78-79, pres, 79-80); fel Am Acad Microbiol. *Res:* Genetics and viruses of fungi; plasmid DNA in fungi; immunochemistry of fungal viruses and double-stranded RNA; biosynthesis of antibiotics; cytoplasmic inheritance in fungi; flourescent staining of fungal nuclei and chromosomes; gene cloning in fungi. *Mailing Add:* Dept Bot & Microbiol Auburn Univ Auburn AL 36849

LEMKE, RONALD DENNIS, b St Paul, Minn, Apr 27, 41; m 67; c 2. THERMAL MODELS USED TO QUANTIFY COOKING METHODS, PSYCHOMETRIC MODELS. *Educ:* Mankato State Univ, BS, 66; Col St Thomas, MBA, 80. *Prof Exp:* Engr, Honeywell, 66-69; eng mgr, Thermoking Div, Westinghouse, 69-80; res & develop mgr, Despatch Inc, 80-85; vpres eng & opers, Appl Vision Systs, 85-87; DIR ENG, STEIN INC, 87- *Mem:* Am Soc Heating Refrig & Air Conditioning Engrs. *Res:* Development of products that utilize different heat transfer methods for the preparation of food products; forced convection, latent heat and liquid immersion heat transfer. *Mailing Add:* 2608 Hull Rd Huron OH 44839

LEMKE, THOMAS FRANKLIN, b Tremont, Pa, July 28, 42; m 65. ORGANIC CHEMISTRY, CORROSION. *Educ:* Wake Forest Univ, BS, 64; Marshall Univ, MS, 66; Lehigh Univ, PhD(chem), 68. *Prof Exp:* Biochemist, Med Res Labs, Edgewood Arsenal, 68-70; asst prof chem, Marshall Univ, 70-72; tech serv specialist, 72-80, MEM STAFF MKT DEVELOP, INCO ALLOYS INT INC, 80- *Mem:* Am Chem Soc; Nat Asn Corrosion Engrs; Sigma Xi; Am Soc Metals. *Res:* Synthesis of heterocyclic compounds of medicinal interest; corrosion of nickel base alloys. *Mailing Add:* Inco Alloys Int Inc PO Box 1958 Huntington WV 25720

LEMKE, THOMAS LEE, b Waukesha, Wis, June 1, 40; m 63; c 3. MEDICINAL CHEMISTRY. *Educ:* Univ Wis, BS, 62; Univ Kans, PhD(med chem), 66. *Prof Exp:* Res assoc org chem & patent liaison, Upjohn Co, Mich, 66-70; from asst prof to assoc prof, 70-84, PROF PHARM, UNIV HOUSTON, 84- *Mem:* Am Chem Soc; Am Pharmaceut Asn; Am Asn Col Pharm; Am Asn Pharmaceut Scientists. *Res:* Heterocyclic chemistry; anticancer agents; Favorskii rearrangement; drugs for mental disease; cardiovascular agents. *Mailing Add:* Col Pharm Univ Houston 4800 Calhoun Rd Houston TX 77204-5515

LEMKEY, FRANKIN DAVID, b Oak Park, Ill, Jan 6, 37; m 78; c 3. MATERIALS SCIENCE ENGINEERING, METALLURGY & PHYSICAL METALLURGICAL ENGINEERING. *Educ:* Univ Mich Ann Arbor, BSE, 60; Univ Oxford, Eng, DPhil, 73. *Prof Exp:* Adj prof eng, Dartmouth Col, 82-88; VIS PROF METALL ENG MAT, UNIV CONN, 88- *Concurrent Pos:* Sr consult scientist, United Technologies Res Ctr, 60-; counr, Mat Res Soc, 78-82; chmn, Exec Bd Rev, Metall Trans, Am Inst Mech Engrs, 80-81; expert, DWG mem, Univs Space Res Asn, 80-88; vis, Accreditation Bd Eng & Technol, 85-; dir, Microgravity Sci & Appln Div, NASA Hq, 88-89; pres exec exchange fel, White House, 88. *Honors & Awards:* Grossman Author's Award, Am Soc Metals Int, 70. *Mem:* Fel Am Soc Metals Int; Mat Res Soc; Am Inst Mech Engrs. *Res:* Melt grown metallic and ceramic composites for high temperature applications; discovered Raney type nickel catalysts from RSR atomization of powders; shock compaction of ferrous alloy powders; high temperature austenitic stainless steel alloys for stirling engine components; high temperature oxidation and corrosion of alloys together with thermochemical properties evaluations. *Mailing Add:* MS 75 United Technologies Res Ctr East Hartford CT 06108

LEMLICH, ROBERT, b Brooklyn, NY, Aug 22, 26; m. CHEMICAL ENGINEERING, BUBBLES & FOAM. *Educ:* NY Univ, BChE, 48; Polytech Inst Brooklyn, MChE, 51; Univ Cincinnati, PhD(chem eng), 54. *Prof Exp:* Chem res engr, Gen Chem Div, Allied Chem & Dye Corp, 48-49; from asst prof to prof, 52-85, EMER PROF CHEM ENG, UNIV CINCINNATI, 85- *Concurrent Pos:* Res grants, Res Corp, Procter & Gamble, USPHS, HEW & NSF, 54-81, 85-88; Fulbright lectr, Israel Inst Technol, 58-59 & Univ Arg, 66; fel, Grad Sch, Univ Cincinnati, 71-, chmn fels, 76-78. *Honors & Awards:* Sigma Xi award, 69. *Mem:* Fel AAAS; Am Chem Soc; Am Soc Eng Educ; fel Am Inst Chem Engrs. *Res:* Foam fractionation and properties. *Mailing Add:* Dept Chem Eng Univ Cincinnati Cincinnati OH 45221-0171

LEMM, ARTHUR WARREN, b Biloxi, Miss, Mar 9, 52; m 72; c 5. NEW MATERIALS DEVELOPMENT. *Educ:* Marquette Univ, BS, 74; Univ Wis-Milwaukee, MS, 90. *Prof Exp:* Supv chemist, Cerac Inc, 74-77; lab technician, RTE Corp, 77-81, res & develop mat engr, 81-86; SUPVR, ANALYTICAL LAB, COOPER POWER SYSTS, 86- *Mem:* Am Chem Soc; Inst Elec & Electronics Engrs; Am Soc Testing & Mat; Soc Plastics Engrs; Am Soc Metals; Mat Res Soc. *Res:* Investigation of new materials development and how their utilization can be used to improve and optimize the performance and reduce costs for products of the electrical and electronic industries. *Mailing Add:* Cooper Power Systs 11131 Adams Rd Franksville WI 53126

LEMMERMAN, KARL EDWARD, b Willoughby, Ohio, May 30, 23; m 46; c 3. PHYSICAL CHEMISTRY. *Educ:* Oberlin Col, AB, 47; Cornell Univ, PhD(chem), 51. *Prof Exp:* Asst gen chem, Cornell Univ, 47-50; res chemist, Procter & Gamble Co, 51-88; RETIRED. *Mem:* Am Chem Soc. *Res:* Kinetics of gas-phase photochemical reactions; complex inorganic electrolytes; surfactant solutions; colloids. *Mailing Add:* 1952 Compton Rd Cincinnati OH 45231

LEMMING, JOHN FREDERICK, b Dayton, Ohio, Oct 31, 43. NUCLEAR PHYSICS. *Educ:* Univ Dayton, BS, 66; Ohio Univ, MS, 68, PhD(physics), 72. *Prof Exp:* Fel physics, Ohio Univ, 72-74; SR PHYSICIST NUCLEAR SPECTROS, MONSANTO RES CORP, MOUND LAB, 74- *Concurrent Pos:* Nuclear infor res assoc, Nat Acad Sci, Nat Res Coun Comt Nuclear Sci, 72-74. *Mem:* Am Inst Physics; Am Phys Soc. *Res:* Nuclear safeguards. *Mailing Add:* 4137 Woodedge Dr Bellbrook OH 45305

LEMMON, DONALD H, b Sugar Grove, Pa, Oct 19, 35; m 56; c 1. SPECTROSCOPY. *Educ:* Univ Pittsburgh, PhD(chem), 66. *Prof Exp:* Fel, State Univ NY, Stony Brook, 66-67; SR ENGR, WESTINGHOUSE RES CTR, 67- *Mem:* Am Chem Soc; Coblentz Soc. *Res:* Infrared, Raman and nuclear magnetic resonance spectroscopy; mass spectrometry. *Mailing Add:* 5525 Floral Ave Verona PA 15147

LEMMON, RICHARD MILLINGTON, b Sacramento, Calif, Nov 24, 19; m 49; c 3. RADIATION CHEMISTRY. *Educ:* Stanford Univ, AB, 41; Calif Inst Technol, MS, 43; Univ Calif, PhD(chem), 49. *Prof Exp:* Res chemist, Calif Inst Technol, 43-45; fel, Med Sch, Univ Calif, 49-50; USPHS fel, Fed Inst Tech, Switz, 50-51; RES CHEMIST, LAWRENCE BERKELEY LAB, UNIV CALIF, 51- *Concurrent Pos:* Guggenheim fel, Helsinki, 65; assoc dir, Lab Chem Biodynamics, Univ Calif, 57-84. *Honors & Awards:* Mosher Award, Am Chem Soc, 89. *Mem:* AAAS; Am Chem Soc; Radiation Res Soc. *Res:* Radiochemistry; hot-atom chemistry; radiation decomposition of organic compounds; chemical evolution. *Mailing Add:* 298 Los Altos Dr Berkeley CA 94708

LEMNIOS, A(NDREW) Z, b Newburyport, Mass, Nov 23, 31; m 54; c 2. AERONAUTICAL ENGINEERING, APPLIED MECHANICS. *Educ:* Mass Inst Technol, BS, 53, MS, 54; Univ Conn, PhD(appl mech), 67, Harvard Univ, Cert, 83. *Prof Exp:* Asst, Aeroelastic Struct & Res Labs, Mass Inst Technol, 53-54; res engr, Res Labs, United Aircraft Corp, 54-61; sr anal engr, Kaman Corp, 61-63, res proj mgr, 63-65, chief fluid mech res, 65-69, chief res engr, 69-76, dir res & technol, 76-89, ASST VPRES RES & TECHNOL, KAMAN AEROSPACE CORP, BLOOMFIELD, CONN, 90- *Concurrent Pos:* Instr, Western New Eng Col, 57-; adj fac, Univ Mass, 77-; mem aeronaut adv comt, NASA, 78-84; mem adv comt, Rotary Wing Technol, Rennselaer Polytech Inst, Univ Md & Ga Inst Tech. *Mem:* Am Inst Aeronaut & Astronaut; Am Helicopter Soc. *Res:* Aeroelastic behavior of rotating structures; structural dynamics and vibrations; structures and structural mechanics; computer modeling. *Mailing Add:* 144 Primrose Dr Longmeadow MA 01106

LEMNIOS, WILLIAM ZACHARY, b Athens, Greece, Sept 13, 25; US citizen; m 54; c 4. PHYSICS, ELECTRICAL ENGINEERING. *Educ:* Mass Inst Technol, BS, 49; Univ Ill, MS, 51. *Prof Exp:* Staff mem systs anal, 52-65, asst group leader, 64-65, group leader, 65-69, assoc div head, 69-83, DIV HEAD RADAR MEASUREMENTS DIV, LINCOLN LAB, MASS INST TECHNOL, 83- *Mem:* AAAS; Am Phys Soc; sr mem Inst Elec & Electronics Engrs; Sigma Xi; Am Inst Aeronaut & Astronaut. *Res:* Computer systems and simulation; radar systems. *Mailing Add:* Lincoln Lab Mass Inst Technol 36 Independence Ave Lexington MA 02173

LEMOINE, ALBERT N, JR, b Nelson, Nebr, Apr 13, 18; m 40; c 3. OPHTHALMOLOGY. *Educ:* Univ Kans, AB, 39; Wash Univ, MD, 43; Am Bd Ophthal, dipl. *Prof Exp:* Teaching & res fel ophthal, Harvard Med Sch, 45-46; chmn dept, 50-80, PROF OPHTHAL, SCH MED, UNIV KANS MED CTR, KANSAS CITY, 50- *Mem:* AAAS; Asn Res Vision & Ophthal; fel Am Col Surg; Sigma Xi; Am Acad Ophthal & Otolaryngol. *Res:* Surgical and applied anatomy of the eye and orbit. *Mailing Add:* Dept Ophthal Sud 2 Col Health Univ Kans 39th St & Rainbow Blvd Kansas City KS 66103

LEMON, EDGAR ROTHWELL, b Buffalo, NY, Aug 22, 21; m 44; c 3. SOIL SCIENCE. *Educ:* Cornell Univ, BS, 43, MS, 49; Mich State Univ, PhD(soil physics), 54. *Prof Exp:* Prof agron, Tex A&M Univ, 51-56; prof agron, Cornell Univ, 56-80; soil scientist, Agr Res Serv, USDA, 51-80; RETIRED. *Concurrent Pos:* Guggenheim & Fulbright fel, Australia, 62-63; USSR-US exchange scientist, 69; Dept Sci & Indust Res fel, NZ, 70-71; consult, Univ Guelph, 83-85. *Honors & Awards:* Soil Sci Award, Am Soc Agron, 72; Biometeor Award, Am Meterol Soc, 90. *Mem:* Fel AAAS; fel Am Soc Agron; Soil Sci Soc Am. *Res:* Applied physics, particularly physical processes in the environment of agricultural crops. *Mailing Add:* 67 Ricardo Niagara-on-the-Lake ON L0S 1J0 Can

LEMON, LESLIE ROY, b Greenville, SC, Jan 19, 47; m 68; c 3. METEOROLOGICAL RADAR SYSTEMS DESIGN, SEVERE CONVECTIVE STORM METEOROLOGY. *Educ:* Univ Okla, BS, 70. *Prof Exp:* Meteorologist, Severe Storms Res, Nat Severe Storms Lab, Nat Oceanic & Atmospheric Admin Comn Corps, 69-70, mem staff oceanog data collection, 70-73, res meteorologist, 73-76, meteorologist, Nat Severe Storms Forecast Ctr, 76, res meteorologist, Tech Develop Unit, 76-81; MGR, NEXRAD OPERS COMPATIBILITY ASSURANCE SOC, NEXRAD RADAR DEVELOP, UNISYS SURVEILLANCE & FIRE CONTROL SYSTS, UNISYS CORP, 81- *Concurrent Pos:* Consult & lectr severe storms. *Mem:* Am Meteorol Soc; Nat Weather Asn. *Res:* Understanding and documenting severe storm structure and evolution and tornado genesis, as well as operational application of conventional and meteorological Doppler radar to the warning services. *Mailing Add:* 16416 Cogan Dr Independence MO 64055

LEMON, PETER WILLIAN REGINALD, b London, Ont, Mar 15, 51; m 79; c 2. EXERCISE PHYSIOLOGY, NUTRITION. *Educ:* McMaster Univ, BA & BPE, 73; Univ Windsor, MS, 75; Univ Wis- Madison, PhD(exercise physiol), 79. *Prof Exp:* From asst prof to assoc prof, 79-87, PROF EXERCISE PHYSIOL, KENT STATE UNIV, 87- *Concurrent Pos:* Vis scientist, McMaster Univ, 89. *Mem:* Fel Am Col Sports Med; Am Physiol Soc; Can Asn Sport Sci. *Res:* Protein/amino acid metabolism during both heavy and resistance and endurance exercise; quantify the protein/amino acid requirements of active individuals and to identify the mechanisms responsible for any increased needs. *Mailing Add:* Appl Physiol Res Lab Kent State Univ Kent OH 44242

LEMON, ROY RICHARD HENRY, b Birmingham, Eng, July 13, 27; Can citizen; m 59. GEOLOGY. *Educ:* Univ Wales, BSc, 51; Univ Toronto, MA, 53, PhD, 55. *Prof Exp:* Geologist, Ghana Geol Surv, 56-57; asst cur invert paleont, Royal Ont Mus, 57-58, 59-61, assoc cur, 61-67; staff geologist, Texaco Oil Co, Trinidad, WI, 67-68; PROF GEOL, FLA ATLANTIC UNIV, 68-, CHMN DEPT, 77- *Concurrent Pos:* Nat Res Coun res grant, 63-66; asst prof, Queen's Univ, Ont, 58-59; assoc prof, Univ Toronto, 62-67. *Mem:* Geol Soc Am; Am Asn Petrol Geol. *Res:* Pliocene and Pleistocene geology and faunas of the west coast of South America and the Caribbean; world wide Pleistocene sea level changes; origin of sedimentary phosphates. *Mailing Add:* Dept Geol Fla Atlantic Univ Boca Raton FL 33431

LEMONDE, ANDRE, b Saint-Liboire, Que, May 30, 21; m 53; c 2. BIOCHEMISTRY, PHYSIOLOGY. *Educ:* Univ Montreal, BA, 42; Laval Univ, BS, 47, ScD(biol), 51. *Prof Exp:* Demonstr physiol, Laval Univ, 47-51; hon fel biochem & entom, Cornell Univ, 51-52; from asst prof to assoc prof, 52-66, head dept, 76-81, PROF BIOCHEM, SCH MED, LAVAL UNIV, 66- *Mem:* AAAS; Am Physiol Soc; Can Biochem Soc; Nutrit Soc Can. *Res:* Comparative biochemistry and physiology. *Mailing Add:* Dept Biochem Laval Univ Sch Med Quebec PQ J1K 7P4 Can

LEMONE, DAVID V, b Columbia, Mo, Apr 16, 32; m 55; c 2. INVERTEBRATE PALEONTOLOGY, PALEOBOTANY. *Educ:* NMex Inst Mining & Technol, BS, 55; Univ Ariz, MS, 59; Mich State Univ, PhD(geol), 64. *Prof Exp:* Geologist, Stanolind Oil & Gas Co, 55-56 & Tex Co, 58-59; assoc prof geol, SMiss Univ, 61-64; assoc prof, 64-77, PROF GEOL, UNIV TEX, EL PASO, 77- *Mem:* Fel AAAS; Am Paleont Soc; Soc Econ Paleontologists & Mineralogists; Paleont Soc Japan; Geol Soc Am. *Res:* Paleophycology; stratigraphic paleontology; systematic invertebrate paleontology and paleobotany; palynology; paleoecology; numerical taxonomy. *Mailing Add:* Univ Tex PO Box 3 El Paso TX 79968

LEMONE, MARGARET ANNE, b Columbia, Mo, Feb 21, 46; m 76; c 2. CONVECTIVE STORMS. *Educ:* Univ Mo, AB, 67; Univ Wash, PhD(atmospheric sci), 72. *Prof Exp:* Fel, Advan Study Prog, 72-73, SCIENTIST, NAT CTR ATMOSPHERIC RES, 73- *Mem:* Am Meteorol Soc; AAAS; Am Geophys Union. *Res:* Structure and dynamics of atmospheric boundry layer and its interaction with cumulus clouds; structure and dynamics of cumulus and cumulonimbus clouds and mesoscale convective systems and their interaction with the environment and larger-scale flow. *Mailing Add:* Nat Ctr Atmospheric Res PO Box 3000 Boulder CO 80307

LEMONICK, AARON, b Philadelphia, Pa, Feb 2, 23; m 50; c 2. HIGH ENERGY PHYSICS. *Educ:* Univ Pa, BA, 50; Princeton Univ, MA, 52, PhD, 54. *Prof Exp:* Instr physics, Princeton Univ, 53-54; asst prof, Haverford Col, 54-57, assoc prof & chmn dept, 57-61; assoc prof, Princeton Univ, 61-64, assoc dir, Princeton-Pa Accelerator, 61-67, assoc chmn dept, 67-69, dean grad sch, 69-73, dean fac, 73-89, PROF PHYSICS, PRINCETON UNIV, 64- *Concurrent Pos:* NSF sci fac fel, Univ Calif, Berkeley, 60-61. *Mem:* Fel Am Phys Soc; Am Asn Physics Teachers. *Mailing Add:* Physics Dept Princeton Univ PO Box 708 Princeton NJ 08544

LEMONS, JACK EUGENE, b St Petersburg, Fla, Jan 20, 37; m 62; c 2. BIOMATERIALS, MATERIALS ENGINEERING. *Educ:* Univ Fla, BS, 63, MS, 64, PhD(metall, chem, physics), 68. *Prof Exp:* Owner & operator, J E Lemons Gen Repair & Mach Shop, 55-60; res assoc metall & mat, Univ Fla, 63-64, asst, 64-68; res metall & head, Phys Metall, Eng Div, Southern Res Inst, Ala, 68-70; asst prof interdisciplinary studies, Clemson Univ, 70-71; from instr to assoc prof, 71-73, prof eng & chmn dept, 77-90, PROF, BIOMAT & SURG, DIV RES, UNIV ALA, BIRMINGHAM, 90- *Concurrent Pos:* NIH spec fel, Med Sch, Univ Ala, 71-73. *Honors & Awards:* I Lew Mem Award, Am Acad Implant Dent, 85. *Mem:* Am Soc Metals; Am Inst Mining, Metall & Petrol Engrs; Soc Biomat; Int Asn Dent Res; Orthod Res Soc. *Res:* Properties of materials for applications in physiological environments; interfacial interactions between synthetic biomaterials and tissues. *Mailing Add:* Dept Biomat SDB49 Univ Ala Birmingham AL 35294

LEMONS, THOMAS M, b Indianapolis, Ind, Sept 15, 34; m 59; c 2. REFLECTOR DESIGN, LIGHTING PRODUCT DESIGN. *Educ:* Purdue Univ, BS, 56. *Prof Exp:* Mgr illumination eng, GTE-Sylvania Lighting Prods, 56-70; PRES, TLA-LIGHTING CONSULTS, INC, 70- *Concurrent Pos:* Co-founder & vpres, ARC Sales, Inc, 79-, lectr, TLA Lighting Consults, Inc, 80- & Mass Inst Technol, 84-; group mgr, design & applications, Illuminating Eng Soc, 82-84; expert witness, Tech Adv Serv for Attorneys, 84-; consult, Qualite Sports Lighting, Inc, 85- & Wilmette Park Dist, Ill, 86- *Honors & Awards:* Distinguished Serv Award, Illum Eng Soc, 83. *Mem:* Illum Eng Soc; Nat Soc Prof Engrs; Int Comn Illum; Soc Motion Picture & TV Engrs; fel US Inst Theatre Technol. *Res:* Energy efficient products and lighting systems; combine the latest light sources with improved optical systems to achieve unique results; author of over sixty technical papers; nine patent awards. *Mailing Add:* TLA Lighting Consults Inc 72 Loring Ave Salem MA 01970

LEMONTT, JEFFREY FIELDING, b New York, NY, July 1, 44; m 70; c 2. MOLECULAR GENETICS, TUMOR CELL DRUG RESISTANCE. *Educ:* Rensselaer Polytech Inst, BS, 65; Univ Calif, Berkeley, MBiorad, 67, PhD(biophys), 70. *Prof Exp:* Res fel genetics, Nat Res Coun Can, 70-72 & Nat Inst Med Res, 73-74; res staff mem yeast genetics, Oak Ridge Nat Lab, 74-81; sr scientist, Integrated Genetics, Inc, 82-89; SR SCIENTIST, GENZYME CORP, 89- *Concurrent Pos:* Lectr, Oak Ridge Grad Sch Biomed Sci, Univ Tenn, 74-81. *Mem:* Genetics Soc Am. *Res:* Yeast genetics and molecular biology; mechanisms of mutagenesis and DNA repair in yeast; expression of foreign genes in yeasts; yeast transformation; recombinant DNA technology; detection of pathogenic fungi; mechanisms of multidrug resistance in human tumor cell lines; bacterial and mammalian protein expression systems. *Mailing Add:* 165 Fairway Dr West Newton MA 02165

LEMOS, ANTHONY M, b Arlington, Mass, Aug 31, 30; m 53; c 3. THEORETICAL PHYSICS, SOLID STATE PHYSICS. *Educ:* Boston Col, AB, 52; Univ Chicago, MS, 56; Ill Inst Technol, PhD(physics), 64. *Prof Exp:* Instr physics, Lake Forest Col, 58-63; assoc prof, 64-71, PROF PHYSICS, ADELPHI UNIV, 71- & CHMN DEPT, 77- *Concurrent Pos:* Res collabr,

Brookhaven Nat Lab, 65- *Mem:* Am Phys Soc. *Res:* Theoretical understanding of the phenomena surrounding the F-center; color centers in the alkali halides. *Mailing Add:* Dept Physics Adelphi Univ Grad Sch Arts & Sci Garden City NY 11530

LEMP, JOHN FREDERICK, JR, b Alton, Ill, May 25, 28; m 53; c 3. VIROLOGY, CELL BIOLOGY. *Educ:* Univ Ill, BS, 51; Nat Registry Microbiologists, Regist. *Prof Exp:* Bacteriologist, Com Solvents Corp, Ind, 51, fermentation supt & microbiologist, Ill, 53-57; microbiologist, Pilot Plants Div, US Army Biol Labs, Ft Detrick, Md, 57-61, prin investr, Process Develop Div, 61-63; sr microbiologist & asst br chief, Biol Ctr, 63-71; proj mgr retrovirus, Electro-Nucleonics Inc, Md, 71-72; DIR, CELL SCI LAB, BIOTECHNOLOGIES, INC, COLUMBIA MD, 72-, SR SCIENTIST, 89- *Mem:* Am Soc Microbiol; NY Acad Sci; Sigma Xi. *Res:* Fermentation; purification microbial products, B-12, riboflavin and penicillin; bacitracin, alcohols, fungal amylase, continuous sterilization and culture, pH control, polarographic dissolved oxygen; mammalian tissue culture; virus propagation and purification; electrophoresis; human interferons; human lymphokines; acquired immundeficiency syndrome virus diagnostic tests; HIV-1, HIV-2 and HBLV research and development. *Mailing Add:* 14 W Broad Way Lovettsville VA 22080

LEMPER, ANTHONY LOUIS, b Buffalo, NY, June 3, 39; m 62; c 4. ORGANIC CHEMISTRY. *Educ:* Univ Buffalo, BA, 60; State Univ NY Buffalo, PhD(org chem), 66. *Prof Exp:* Res chemist, Res Div, Goodyear Tire & Rubber Co, 65-66, sr res chemist, 66; res chemist, Corp Res Ctr, Hooker Chem Corp, 66-68, sr res chemist, 68-71, res group leader, 71-75, res mgr, 75-78; DIR PROD RES APPL & SALES SERV, FMC CORP, 78- *Mem:* Soc Plastics Engrs; Am Chem Soc. *Res:* Inorganic industrial chemicals and flame retardant chemicals; engineering thermoplastics research and development; polyvinyl chloride chemistry and technology; organophosphorus chemistry; organofluorine chemistry; inorganic industrial and flame retardant chemicals. *Mailing Add:* 8413 Golfview Dr Orland Park IL 60462-2848

LEMPERT, JOSEPH, b North Adams, Mass, July 3, 13; m 41; c 3. MAGNETOHYDRODYNAMICS. *Educ:* Mass Inst Technol, BS, 35; Stevens Inst Technol, MS, 42. *Prof Exp:* Engr, Lamp Div, 36-44, sect engr, 44-53, mgr eng sect, Electronic Tube Div, 53-56, adv develop, 56-58, sect mgr camera tubes, 58, res engr, 58-66, adv engr, Westinghouse Res & Develop Ctr, 66-78; CONSULT ENG, 78- *Mem:* Am Phys Soc; fel Inst Elec & Electronics Engrs. *Res:* Photoemission; photoconductivity; secondary emission; thin films; vacuum tube electronics; electronic imaging; x-ray image intensification; x-rays; storage techniques; electron beams; electron beam welding; thermionic emission; magnetohydrodynamics. *Mailing Add:* 140 Spring Grove Rd Pittsburgh PA 15235

LEMPERT, NEIL, b New York, NY, Nov 25, 33; m 62; c 5. BIOLOGY, CHEMISTRY. *Educ:* Hamilton Col, BS, 54. *Prof Exp:* Asst instr, 60-65, instr, 67-68, asst prof, 68-72, assoc prof, 72-78, PROF SURG, ALBANY MED COL, 78- *Concurrent Pos:* Resident exp surg, Albany Med Ctr, 60-61; res fel surg, Mary Imogene Bassett Hosp & Clin, Cooperstown, NY, 63-64; consult surg, Vet Admin Med Ctr, 67-; dir, Histocompatibility Lab, Albany Med Col, 72- *Mem:* Cryobiol Soc; Asn Acad Surg; Transplantation Soc; Am Soc Transplant Surgeons; Cent Surg Asn. *Res:* Transplantation and preservation of tissues and organs; basic immunology of organ transplantation. *Mailing Add:* Dept Surg Albany Med Col 47 New Scotland Ave Albany NY 12208

LEMPICKI, ALEXANDER, b Warsaw, Poland, Jan 26, 22; nat US; m 52; c 2. PHYSICS. *Educ:* Imp Col, Univ London, MSc, 52, PhD, 60. *Prof Exp:* Res physicist, Electronic Tube Co, Ltd, Eng, 49-54; head quantum physics group, 65-72, MGR ELECTROOPTICS LAB, GOVT TECHNOL CTR, GEN TEL & ELECTRONICS LABS, INC, 73-; RES PROF, CHEM & PHYSICS DEPT, BOSTON UNIV. *Concurrent Pos:* Mem adv subcomt electrophys, NASA, 69-71. *Mem:* Fel Am Phys Soc; fel Optical Soc Am. *Res:* Electroluminescence; optical properties of solids; spectroscopy and molecular structure of organo metallic complexes; optical maser materials, particularly liquid luminescence; luminescence and structure of glasses; luminescent solar collectors; semiconductors; spectroscopy of transition metal ions; scintillator materials. *Mailing Add:* 303 A Commonweatlh Ave Boston MA 02115

LEMPKE, ROBERT EVERETT, b Dover, NH, Nov 27, 24; m 49; c 4. SURGERY. *Educ:* Yale Univ, MD, 48. *Prof Exp:* Intern surg, Johns Hopkins Hosp, Baltimore, 48-49; resident, Med Ctr, Ind Univ, Indianapolis, 49-51; med officer, Army Med Res Lab, Ft Knox, Ky, 51-53; resident, Med Ctr, Ind Univ, Indianapolis, 53-55; assoc chief of staff med res, Vet Admin Hosp, Indianapolis, 56-71; from instr to assoc prof, 56-65, PROF SURG, SCH MED, IND UNIV, INDIANAPOLIS, 65-; CHIEF SURG, VET ADMIN HOSP, INDIANAPOLIS, 59- *Concurrent Pos:* Vis prof surg, Jinnah Postgrad Med Ctr, Karachi, Pakistan, 64-65. *Mem:* Soc Surg Alimentary Tract; Cent Surg Asn; Am Col Surg; Asn Vet Admin Surgeons. *Res:* Diseases of the alimentary tract. *Mailing Add:* 1481 W Tenth St Indianapolis IN 46202

LEMYRE, C(LEMENT), b Shawinigan, Que, Apr 2, 34; m 62; c 4. ELECTRICAL ENGINEERING. *Educ:* Laval Univ, BScEng, 57; Univ London, PhD(transistors), 62. *Prof Exp:* From asst prof to assoc prof, Laval Univ, 62-69; secy fac sci & eng, Univ Ottawa, 70-73, chmn dept, 71-78, dir, Coop Educ Progs, 80-83, asst dean acad affairs, 86-88, ASSOC PROF ELEC ENG, UNIV OTTAWA, 69- *Concurrent Pos:* Asn Orgn Stages France fel, 65-66. *Mem:* Sr mem Inst Elec & Electronics Engrs; fel Eng Inst Can; Can Soc Elec & Computer Eng; Am Soc Eng Educ. *Res:* Characterization of transistors. *Mailing Add:* Dept Elec Eng Univ Ottawa Ottawa ON K1N 6N5 Can

LENA, ADOLPH J, b Latrobe, Pa, Oct 10, 25. METALLURGY. *Educ:* Pa State Univ, BS, 48; Carnegie Mellon Inst, MS, 51, ScD, 52. *Prof Exp:* Pres & chief operating officer, 86-90, SPEC ASST TO CHIEF EXEC OFFICER, CARPENTER TECHNOL CORP, 90- *Mem:* Am Soc Metals Int. *Mailing Add:* Carpenter Technol Corp PO Box 14662 Reading PA 19612-4662

LENARD, ANDREW, b Balmazujvaros, Hungary, July 18, 27; US citizen; m 53; c 2. MATHEMATICAL PHYSICS. *Educ:* State Univ Iowa, BA, 49, PhD(physics), 53. *Prof Exp:* Res assoc physics, Columbia Univ, 55-57; res staff mem, Plasma Physics Lab, Princeton Univ, 57-65; PROF MATH PHYSICS, IND UNIV, BLOOMINGTON, 66- *Concurrent Pos:* Mem, Inst Haute Etudes Sci, 79-80. *Mem:* Am Phys Soc; Am Math Soc. *Res:* Kinetic theory; statistical mechanics; fundamental problems of quantum physics; mathematical problems related to physics. *Mailing Add:* Dept Physics & Math Ind Univ Bloomington IN 47405

LENARD, JOHN, b Vienna, Austria, May 17, 37; US citizen; m 59; c 4. VIROLOGY, PHYSIOLOGY. *Educ:* Cornell Univ, BA, 58, PhD(biochem), 64. *Prof Exp:* Res assoc biochem, Cornell Univ, 63-64; fel biol, Univ Calif, San Diego, 64-65, Am Heart Asn advan res fel, 65-67; asst prof biochem, Albert Einstein Col Med, 67-68; assoc, Sloan-Kettering Inst Cancer Res, 68-72; assoc prof, 73-76, PROF PHYSIOL & BIOPHYS, RUTGERS MED SCH, UNIV MED & DENT NJ, 76- *Concurrent Pos:* Am Heart Asn estab investr, 70-72; adj asst prof biol, Hunter Col, 70-73. *Mem:* Am Soc Biol Chemists; Am Soc Cell Biol; Am Soc Microbiol; Biophys Soc; Am Soc Virol. *Res:* Structures of biological membranes and enveloped viruses; entry and assembly of enveloped viruses; function of viral glycoproteins; endocrinology of lower eukaryotes; insulin action. *Mailing Add:* Dept Physiol & Biophys Univ Med & Dent NJ-Robert Wood Johnson Med Sch 675 Hoes Lane Piscataway NJ 08854

LENARZ, WILLIAM HENRY, b Sacramento, Calif, Sept 18, 40; m. FISH BIOLOGY, BIOSTATISTICS. *Educ:* Humboldt State Univ, BS, 63; Univ Wash, MS, 66, PhD(fisheries), 69. *Prof Exp:* Fishery biologist, La Jolla, 68-76, FISHERY BIOLOGIST, SOUTHWEST FISHERIES CTR, MARINE FISHERIES SERV, TIBURON, 76- *Concurrent Pos:* Sci adv, US Deleg Int Comn Conserv of Atlantic Tunas, 70-74 & Pac Fisheries Mgt Coun, 77-88. *Mem:* Am Statist Asn; Biometric Soc; Sigma Xi; AAAS; Ecol Soc Am; Am Inst Fishery Res Biologists. *Res:* Dynamics of exploited populations of fish. *Mailing Add:* Tiburon Fisheries Lab 3150 Paradise Dr Tiburon CA 94920

LENCHNER, NATHANIEL HERBERT, b New York, NY, Aug 28, 23; m 59; c 3. PROSTHODONTICS. *Educ:* NY Univ, BA, 43, DDS, 50. *Prof Exp:* Instr dent, Col Dent, NY Univ, 50-55; asst clin prof prev dent, Sch Dent & Oral Surg, Columbia Univ, 74-78; DENT CONSULT, WHALEDENT INT, IPCO, 77- *Concurrent Pos:* Asst attend dentist, Long Island Col Hosp, 64-65; assoc ed, J Prosthetic Dent, 77-; adj assoc prof biomed eng, Sch Eng & Archit, Univ Miami, 80- *Mem:* Study Group Advan Dent Diag (pres, 68-69); Am Prosthodontic Soc; Am & Int Asn Dent Res; Am Dent Asn; fel Acad Gen Dent. *Res:* Biomedical engineering relative to dental devices; electrosurgery; the true effect of wave forms on cutting and coagulation. *Mailing Add:* 104-20 Queens Blvd Forest Hills NY 11375

LENDARIS, GEORGE G(REGORY), b Helper, Utah, Apr 2, 35; m 58; c 2. SYSTEMS SCIENCE, SYSTEMS DESIGN FACILITATOR. *Educ:* Univ Calif, Berkeley, BS, 57, MS, 58, PhD(elec eng), 61. *Prof Exp:* Sr staff scientist adaptive flight control systs, Gen Motors Defense Res Labs, 61-63, sr res engr, 63-69; assoc prof systs sci & chmn fac, Ore Grad Ctr Study & Res, 69-71; PROF SYSTS SCI, SYSTS SCI PHD PROG, PORTLAND STATE UNIV, 71- *Concurrent Pos:* NSF fel, 60-61; Mem, Gov Tech Adv Comt, Ore, 70-72; consult to pres, Ore State Senate, 71; mem, Ore State Senate Task Force Econ Develop, 72-73; vis scientist, Johnson Space Ctr, NASA, 73-74; Nat Acad Sci res fel, 73-74; vis scholar, Eng & Econ Systs Dept, Stanford Univ, 78. *Mem:* Fel Inst Elec & Electronics Engrs; Soc Gen Systs Res; Pattern Recognition Soc; Asn Transpersonal Psychol; Sigma Xi; Int Neural Network Soc. *Res:* Developing methodologies for assisting teams of people to carry out systems design, engineering; analysis of social and human systems; models for complex systems; structural modeling; artificial intelligence; neural networks with application to conceptual graph knowledge systems. *Mailing Add:* 1717 SW Park Ave Apt 1305 Portland OR 97201

LENDER, ADAM, US citizen; c 2. DIGITAL COMMUNICATIONS, SPEECH COMPRESSION. *Educ:* Columbia Univ, BS, 54, MS, 56; Stanford Univ, PhD(elec eng), 72. *Prof Exp:* Mem tech staff, Bell Tel Labs, Murray Hill, NJ, 54-60; proj engr, Int Tel & Tel Labs, 60-61; head, Advan Develop, Gen Tel Lenkurt Labs, San Carlos, 61-84; SR CONSULT SCIENTIST, LOCKHEED PALO ALTO RES LABS, 84- *Concurrent Pos:* Chmn, Data Commun Systs Comt, Inst Elec & Electronics Engrs, Commun Soc, 72-76, mem bd gov, 77-79 & 82-84; adj prof elec eng, Santa Clara Univ, 76-; ed-in-chief, Trans Commun, Inst Elec & Electronics Engrs, 78-84; J Selected Areas Commun, 83-84, sr tech ed, Commun Mag, 87- *Honors & Awards:* Centennial Medal, Inst Elec & Electronics Engrs, 84. *Mem:* Fel Inst Elec & Electronics Engrs; assoc fel Am Inst Aeronaut & Astronaut. *Res:* Digital communications; invented correlative, duobinary or partial response used worldwide for efficient, fast digital communications applied to high density magnetic disk recording for computers; 30 US patents. *Mailing Add:* Lockheed Palo Alto Res Labs 3251 Hanover St 0-9150 B 251 Palo Alto CA 94304

LENEL, FRITZ (VICTOR), b Kiel, Germany, July 7, 07; nat US; m 43; c 5. METALLURGY. *Educ:* Univ Heidelberg, PhD, 31. *Prof Exp:* Fel, Univ Goettingen, 31-33; metallurgist, Charles Hardy, Inc, NY, 33-37 & Delco-Moraine Div, Gen Motors Corp, Ohio, 37-47; from asst prof to prof, 47-73, chmn dept, 65-69, EMER PROF METALL ENG, RENSSELAER POLYTECH INST, 73- *Honors & Awards:* Powder Metall Pioneer Award, Am Powder Metall Inst, 84. *Mem:* Fel Am Soc Metals; Am Inst Mining, Metall & Petrol Engrs; Brit Inst Metals; fel Am Soc Testing & Mat. *Res:* Powder metallurgy. *Mailing Add:* Dept Mat Eng Rensselaer Polytech Inst Troy NY 12181

LENER, WALTER, b New York, NY, Mar 20, 25. ENTOMOLOGY. *Educ:* NY Univ, BA, 48, MA, 50, PhD(biol), 57; Rutgers Univ, MS, 60. *Prof Exp:* Instr biol, State Univ NY Col Oneonta, 50-51, sci consult, New Paltz, 51-52, from instr to prof biol, Geneseo, 52-64, coordr biol sci, 62-64; PROF BIOL,

NASSAU COMMUNITY COL, 64- *Concurrent Pos:* Res grant, Res Found, State Univ NY, 63-64, 65-67; fel trop med, Sch Med, La State Univ, 64; NSF res grant, 66-68; consult. *Mem:* AAAS; Ecol Soc Am; Animal Behav Soc; NY Acad Sci; Sigma Xi; Entom Soc Am. *Res:* Investigating the physiology, genetics and ethology of large milkweed bug, Oncopeltus fasciatus. *Mailing Add:* Dept Biol Nassau Community Col Garden City NY 11530

LENES, BRUCE ALLAN, b White Plains, NY, April 4, 49; m 74; c 3. HEMATOLOGY. *Educ:* Union Col, Union Univ, BS, 71; Albany Med Col, Union Univ, MD, 75. *Prof Exp:* Resident int med, Shands Teaching Hosp & Clin, Univ Fla, 75-78; fel hematol, Georgetown Univ Hosp, Washington, DC, 78-80; fel blood banking, NIH, Bethesda Md, 80-81; MED DIR, AM RED CROSS BLOOD SERV, MIAMI, 81-; ASST PROF MED & PATH, SCH MED, UNIV MIAMI, 81- *Concurrent Pos:* Clin asst prof med, Sch Med, Georgetown Univ, 80- *Mem:* Sigma Xi; Am Med Asn; Am Asn Blood Banks. *Res:* Clinical research in hematology and blood banking; special emphasis on pheresis and hemolytic anemia. *Mailing Add:* Am Red Cross Blood Serv SFla Region PO Box 013201 Miami FL 33101

LENEY, LAWRENCE, b New York, NY, Dec 14, 17; m 45; c 5. WOOD SCIENCE, MICROSCOPY. *Educ:* State Univ NY Col Forestry, Syracuse, BS, 42, MS, 48, PhD, 60. *Prof Exp:* Instr wood tech, State Univ NY Col Forestry, Syracuse, 46-52; asst prof, Univ Mo, 52-60; from assoc prof to prof wood & fiber sci, Col Forest Resources, Univ Wash, 60-83; RETIRED. *Mem:* AAAS; Forest Prod Res Soc; Tech Asn Pulp & Paper Indust; Int Asn Wood Anat; Soc Wood Sci & Technol. *Res:* Wood anatomy; microtechnique; machining wood; photomicrography of woody tissue; seasoning and preservation of wood; pulp and paper fiber analysis. *Mailing Add:* 2101 E First Ave Ellensburg WA 98926

LENFANT, CLAUDE J M, b Paris, France, Oct 12, 28; US citizen; m 49; c 5. PHYSIOLOGY. *Educ:* Univ Rennes, BS, 48; Univ Paris, MD, 56. *Hon Degrees:* DSc, State Univ NY, Buffalo, 88. *Prof Exp:* From res asst to dir res, Ctr Marie Lannelongue, France, 54-57; res fel, Univ Buffalo, 57-58; res fel, Columbia Univ, 58-59; asst prof physiol, Univ Lille, 59-60; from instr to prof med, physiol & biophys, Univ Wash, 61-72; assoc dir lung progs & actg assoc dir collab res & develop prog, Nat Heart & Lung Inst, 70-72, actg chief, Pulmonary Res Br, 72-74, dir, Div Lung Dis, Nat Heart, Lung & Blood Inst, 72-80, dir, Fogarty Int Ctr & assoc dir, Int Res, 81-82, DIR, NAT HEART, LUNG & BLOOD INST, NIH, 82- *Concurrent Pos:* Fulbright fel, 56-58; assoc dir, Inst Respiratory Physiol & staff physician, Firland Sanitorium, Seattle, 61-68; mem physiol study sect, NIH, 69-70; hon prof, Nat Yang-Ming Med Col, Taipei, Taiwan, 80, Peruvian Univ, Lima, Peru, 81; bd gov, US Israel Binat Sci found, 90-93. *Honors & Awards:* Thesis Prize, Univ Paris, France, 56; Distinguished Serv Award, Am Heart Asn, 83; Forrest M Bird Contributory Award, Am Respriatory Ther Found, 85; Breath of Life Award, Cystic Fibrosis found, 88; Brotherhood Award, Asn Black Cardiologists, 90. *Mem:* Inst Med Nat Acad Sci; Am Physiol Soc; Am Fed Clin Res; Fr Physiol Soc; Asn Am Physicians; hon fel Am Col Chest Physicians; hon fel Am Heart Asn; hon mem Royal Soc Med; USSR Acad Med Sci; hon mem French Cardiol Soc; Am Soc Clin Invest; Int Fed Med Electronics; Am Soc Zool; Soc Exp Med & Biol; Undersea Med Soc; NY Acad Sci. *Res:* Respiratory physiology, especially in gas exchange; comparative physiology related to the development and environmental adaptation of the respiratory system; author or co-author of 192 scientific publications. *Mailing Add:* Nat Heart Lung & Blood Inst NIH Bethesda MD 20892

LENG, DOUGLAS E, b Kitchener, Ont, May 28, 28; m 55; c 3. CHEMICAL ENGINEERING. *Educ:* Queen's Univ, Ont, BSc, 51, MSc, 53; Purdue Univ, PhD(chem eng), 56. *Prof Exp:* Chem engr, Benzene Prod Lab, 56-61, res engr, Process Fundamentals Lab, 62-63, sr res engr, 63-70, assoc scientist, Dow Interdisciplinary Groups Eng, 70-74, res scientist, 75-82, PRES & DIR, CENT RES ENG LAB, 82- *Concurrent Pos:* Mem adv bd, Queen's Univ, Kingston, Ont, 85-88; mem panel m, bd assessment, Nat Bur Standards/Nat Res Coun, 85-88. *Mem:* Am Chem Soc; Am Inst Chem Engrs. *Res:* Multiphase behavior; coalescence and dispersion; mixing, micromixing. *Mailing Add:* 1714 Sylvan Lane Midland MI 48640

LENG, EARL REECE, b Williamsfield, Ill, June 12, 21; m 44; c 3. GENETICS, RESEARCH ADMINISTRATION. *Educ:* Univ Ill, BS, 41, MS, 46, PhD(agron), 48. *Prof Exp:* Spec asst agron, 41-42, asst plant genetics, 46-48, from asst prof to assoc prof, 48-58, from asst dir to assoc dir int prog, 69-73, crop specialist, USAID, 75-77; prof, 58-77, EMER PROF AGRON, UNIV ILL, URBANA, 77- *Concurrent Pos:* Fulbright sr res fel, Max Planck Inst, Ger, 61; consult, Fed Govt Yugoslavia & USAID, 60-61; res adv, USAID & Uttar Pradesh Agr Univ, India, 64-66; adv, USAID & Midwest Univs Consortium for Int Activities, Indonesia, 71; consult, Food & Agr Orgn, Thailand, 71; consult, Int Coffee Orgn, 72; consult, UNDP/FAO, Yugoslavia, 73, 75; consult, World Bank, Malaysia, 74, USAID & Pac Consults, Sudan, Jamaica & Mauretania, 77-78; prof dir, INTSORMIL, Univ Nebr, Lincoln, 79-84. *Mem:* Crop Sci Soc Am; Am Soc Agron. *Res:* Comparative international agriculture; genetics and breeding of maize; breeding systems; evolution of maize and relatives; international soybean improvement; international agricultural development, emphasis on major cereal crops. *Mailing Add:* SE 181 Arcadia Shores Rd Shelton WA 98584

LENG, MARGUERITE LAMBERT, b Edmonton, Alta, Can, Sept 25, 26; m 55; c 3. AGRICULTURAL BIOCHEMISTRY, ANALYTICAL BIOCHEMISTRY. *Educ:* Univ Alta, BSc, 47; Univ Sask, MSc, 50; Purdue Univ, PhD(biochem), 56. *Prof Exp:* Ed asst chem & physics, Nat Res Coun Can, 47-48, anal chemist, 48-49; sr chemist, Allergy Res Lab, Univ Mich Hosp, Ann Arbor, 50-53; anal chemist, Agr Dept, 56-59, registr specialist, Agr-Org Dept, 66-73, sr regist specialist, 73-80, RES ASSOC INT REGULATORY AFFAIRS, HEALTH & ENVIRON SCI, DOW CHEM CO, MIDLAND, MICH, 80- *Mem:* Fel Am Chem Soc; Sigma Xi. *Res:* Pesticides, their toxicology, metabolism, residues, analytical methods and realistic evaluation of hazard to the environment; meaningful communication of scientific information; international regulation of hazardous chemicals. *Mailing Add:* 1714 Sylvan Lane Midland MI 48640

LENGEL, ROBERT CHARLES, b Key West, Fla, Oct 1, 57; m 89. VIBROACOUSTIC MEASUREMENT & ANALYSIS, MEASUREMENT SYSTEM ENGINEERING & DEVELOPMENT. *Educ:* Va Polytech Inst & State Univ, BS, 80; Univ Tex, Austin, MS, 91. *Prof Exp:* Mech design engr, Geophys Serv, Inc, Tex Instruments, Inc, 80-81; acoust engr, David Taylor Res Ctr, Dept Navy, 81-82, sr systs engr, 82-87; STAFF ENGR & SCIENTIST, TRACOR APPL SCI, 87- *Mem:* Acoust Soc Am; Inst Noise Control Engrs. *Res:* Application of advanced signal processing to problems in acoustics and vibration; vibroacoustic measurement systems engineering; noise control engineering. *Mailing Add:* Tracor Appl Sci 6500 Tracor Lane Austin TX 78725-2050

LENGEMANN, FREDERICK WILLIAM, b New York, NY, Apr 8, 25; m 50; c 2. PHYSICAL BIOLOGY. *Educ:* Cornell Univ, BS, 50, MNS, 51; Univ Wis, PhD(dairy husb), 54. *Prof Exp:* Res assoc radiation biol, Univ Tenn, 54-55, asst prof chem, 55-59; assoc prof, 59-67, prof phys biol, 67-88, EMER PROF PHYSIOL, NY STATE VET COL, CORNELL UNIV, 88- *Concurrent Pos:* biochemist, Div Biol & Med, US Atomic Energy Comn, 62; Consult, FAD-IAEA, Vienna, 66 & 76. *Mem:* Fel AAAS; Am Dairy Sci Asn; Am Inst Nutrit; Coun Agr Sci & Technol. *Res:* Environmental contamination; fission product and mineral metabolism; milk secretion; mineral absorption; bone calcification. *Mailing Add:* Dept Physiol NY State Col Vet Med Cornell Univ Ithaca NY 14853

LENGYEL, BELA ADALBERT, b Budapest, Hungary, Oct 5, 10; US citizen; m 42; c 2. LASERS, MATHEMATICAL PHYSICS. *Educ:* Pamany Univ Budapest, PhD (math), 35. *Prof Exp:* Res fel, Harvard Univ, 35-36; asst actuary, Astra Insurance Co, Budapest, 37-38; asst statistician, Worcester State Hosp, Mass, 38-39; instr math, Rensselaer Polytech Inst, NY, 39-42; instr phys, City Col New York, 42-43; asst prof phys, Univ Rochester, NY, 43-46; physicist, Navy Dept, Washington, DC, 46-52; sr staff physicist, Hughes Res Labs, Calif, 52-63; prof, 63-77, emer prof physics, State Univ, Northridge, 77; RETIRED. *Concurrent Pos:* Vis prof, Universidad Nat LaPlata, Arg, 70, Lund Tech Univ, Sweden, 71, Eidgenössiche Tech Hochschule, Zürich, Switz, 72; independent indust consult, 78-85. *Mem:* Am Math Soc; Am Phys Soc; Inst Elec & Electronics Engrs. *Res:* Functional analysis; math statistics; engineering applications of electromagnetic theory; author three books on lasers. *Mailing Add:* 28 Sequoia Tree Lane Irvine CA 92715

LENGYEL, G(ABRIEL), b Budapest, Hungary, Apr 30, 27; US citizen. ELECTRICAL ENGINEERING, OPTICAL COMMUNICATIONS. *Educ:* Budapest Tech Univ, BASc, 49; Univ Toronto, PhD(elec eng), 63. *Prof Exp:* Demonstr math, Budapest Tech Univ, 49-50; res engr, Elec Power Res Inst, Budapest, 50-56; proj engr, E B Eddy Co, Que, 56-58; develop engr, Sangamo Co, Ont, 58-59; sr res fel appl & solid state physics, Ont Res Found, 59-66; assoc prof, 66-76, PROF ELEC ENG, UNIV RI, 76- *Concurrent Pos:* Res asst, Univ Toronto, 60-63; mem assoc comt elec insulation, Nat Res Coun Can, 60-64. *Mem:* Inst Elec & Electronics Engrs; Am Phys Soc; Optical Soc Am. *Res:* Semiconductor lasers and optical modulations. *Mailing Add:* Dept Elec Eng Kelley Hall Univ RI Kingston RI 02881

LENGYEL, ISTVAN, b Kaposvar, Hungary, July 12, 31; US citizen. ORGANIC CHEMISTRY, HISTORY OF SCIENCE. *Educ:* Eotvos Lorand, Univ Budapest, dipl, 55; Mass Inst Technol, PhD(org chem), 64. *Prof Exp:* Res chemist, G Richter Pharmaceut Co, 54-55; sci co-worker geochem, Geophys Res Inst Hungary, 55-56; lab chemist, Kundl Tirol Austria Pharmaceut Co, 57-58; res asst biochem, Sch Med, Johns Hopkins Univ, 58-59; res assoc org synthesis, Mass Inst Technol, 59-64; fel, Munich Tech Univ, 64-65; res assoc mass spectrometry, Mass Inst Technol, 65-67; from asst prof to assoc prof, 67-73, PROF CHEM, ST JOHN'S UNIV, NY, 73-, CHMN DEPT, 85- *Concurrent Pos:* NAS vis scholar, Univ Budapest, 73. *Honors & Awards:* Sr Awardee, Alexander Von Humboldt Found, Germany, 73-74. *Res:* History of organic chemistry. *Mailing Add:* Dept Chem St John's Univ Grand Central & Utopia Jamaica NY 11439

LENGYEL, JUDITH ANN, b Rochester, NY, May 15, 45; m 83; c 2. DEVELOPMENTAL BIOLOGY, MOLECULAR BIOLOGY. *Educ:* Univ Calif, Los Angeles, BA, 67, MA, 68; Univ Calif, Berkeley, PhD(molecular biol), 72. *Prof Exp:* Fel molecular biol, Univ Calif, Berkeley, 72-73; fel cell biol, Mass Inst Technol, 73-75; from asst prof to assoc prof, 76-87, PROF BIOL, UNIV CALIF, LOS ANGELES, 87- *Concurrent Pos:* Prin investr, NIH, 76- & NSF, 81- *Mem:* Sigma Xi; Soc Develop Biol; Am Soc Cell Biol; Genetics Soc Am. *Res:* Molecular and genetic analysis of genes required to establish the body plan during early embryogenesis in Drosophila. *Mailing Add:* Dept Biol Univ Calif Los Angeles CA 90024

LENGYEL, PETER, b Budapest, Hungary, May 24, 29; US citizen; m 56; c 2. BIOCHEMISTRY. *Educ:* Budapest Tech Univ, Dipl, 51; NY Univ, PhD(biochem), 62. *Prof Exp:* Instr biochem, Sch Med, NY Univ, 62-63, asst prof, 63-65; assoc prof molecular biophys, 65-69, PROF MOLECULAR BIOPHYS & BIOCHEM, YALE UNIV, 69- *Concurrent Pos:* NIH spec fel, Pasteur Inst, Paris, 63-64. *Mem:* Am Soc Virol; Am Soc Biolchem & Molecular Biol. *Res:* Protein biosynthesis; nucleic acid and protein metabolism of animal cells and viruses; interferon defense mechanism; oncogenes. *Mailing Add:* Dept Biophys & Biochem Yale Univ Box 6666 New Haven CT 06511

LENHARD, JAMES M, CELL BIOLOGY, PHYSIOLOGY. *Prof Exp:* RES ASST, MED SCH, WASH UNIV, 89- *Mailing Add:* Cell Biol & Physiol Dept Med Sch Wash Univ 4566 Scott Ave St Louis MO 63110

LENHARD, JOSEPH ANDREW, b Detroit, Mich, June 18, 29; m 83; c 2. HEALTH PHYSICS, NUCLEAR PHYSICS. *Educ:* Vanderbilt Univ, BA, 53, MS, 57; Am Bd Health Physics, dipl, 60. *Prof Exp:* Health physicist radiation protection, Oak Ridge Opers Off, 57-61, sr health physicist broad nuclear safety, US Atomic Energy Coun, 61-67, dir safety & environ control div, 67-72, dir res, energy res & develop admin, 72-77, dir res, Dept Energy, 77-89;

MGT & TECH CONSULT, 89- *Concurrent Pos:* Charter mem, Sr Exec Serv US Govt, 79. *Mem:* Health Physics Soc. *Res:* Research administration; physical, life and engineering sciences. *Mailing Add:* 125 Newell Lane Oak Ridge TN 37830

LENHARDT, MARTIN LOUIS, b Elizabeth, NJ, Dec 14, 44; m 66; c 11. AUDIOLOGY, SPEECH & HEARING SCIENCES. *Educ:* Seton Hall Univ, BS, 66, MS, 68; Fla State Univ, PhD(audiol, speech sci), 70. *Prof Exp:* Nat Inst Neurol Dis & Stroke fel, Johns Hopkins Univ, 70-71; asst prof, 71-75, ASSOC PROF OTORHINOLARYNGOL, MED COL VA, VA COMMONWEALTH UNIV, 71-, ASSOC PROF PEDIAT DENT, 76- *Concurrent Pos:* Mem staff adj fac, Va Inst Marine Sci, Col William & Mary, 80- *Mem:* Acoust Soc Am; Am Audiol Soc; Animal Behav Soc; Asn Res Otolaryngol. *Res:* Psychological and physiological acoustics; speech communication; bioacoustics and linguistics. *Mailing Add:* Med Col Va Box 168 MCV Sta Richmond VA 23298

LENHART, JACK G, b Bremen, Ohio, May 6, 29; m 51; c 3. FLUID CONTROLS, THERMODYNAMIC SYSTEMS. *Educ:* Case Inst Technol, BSME, 51. *Prof Exp:* Mat res engr, NAm Aviation, 51-54; proj eng, TRW Equip Lab, 56-67; sr proj engr, chief engr, mgr ground support dept, mgr contracts admin, Accessories Div, Parker-Hannifin, 67-75; chief engr, Scott & Fetzer-Meriam Instrument, 75-78; DIR ENG, TELEDYNE REPUB MFG, 78- *Mem:* Nat Fluid Power Asn; Am Met Soc; Am Soc Mech Eng. *Mailing Add:* 2000 Winchester Rd Lyndhurst OH 44124

LENHERT, ANNE GERHARDT, b Lynchburg, Va, Apr 1, 36; m 67; c 2. ORGANIC CHEMISTRY. *Educ:* Hollins Col, BA, 58; Univ NMex, MS, 63, PhD(chem), 65. *Prof Exp:* Res fel, Univ NMex, 64-65; asst prof chem, Cent Mo State Col, 65-67; ASST PROF CHEM, KANS STATE UNIV, 67- *Mem:* Am Chem Soc; Sigma Xi; Int Heterocyclic Chem. *Res:* Synthesis of heterocyclic ring systems as potential purine and pteridine antagonists; anti-cancer agents and anti-radiation drugs. *Mailing Add:* Dept Chem Kans State Univ Manhattan KS 66506

LENHERT, DONALD H, b Winfield, Kans, Nov 25, 34; m 67; c 2. COMPUTER ENGINEERING. *Educ:* Kans State Univ, BS, 56; Syracuse Univ, MS, 58; Univ NMex, PhD(elec eng), 66. *Prof Exp:* Syst engr, Gen Elec Co, NY, 56-58; res engr, Dikewood Corp, NMex, 60-62; res & teaching assoc elec eng, Univ NMex, 62-66; from asst prof to assoc prof elec eng, 66-81, PROF ELEC & COMPUT ENG, KANSAS STATE UNIV, 81- *Concurrent Pos:* Consult Air Force Spec Weapons Ctr, NMex, 60-62 & Am Inst Prof Educ, 77-83; vis prof, Intel Corp, 83-84, Teletronix, 83, Motorola, 84 & Delco Electronics, 84. *Mem:* Inst Elec & Electronics Engrs; Nat Soc Prof Engrs. *Res:* Microprocessor systems; microprocessor applications; testing of digital systems; testing of analog to digital converters. *Mailing Add:* Dept Elec & Comput Eng Kans State Univ Durland Hall Manhattan KS 66506

LENHERT, P GALEN, b Dayton, Ohio, July 31, 33; m 56; c 2. CRYSTALLOGRAPHY, COMPUTER SCIENCE. *Educ:* Wittenburg Univ, AB, 55; Johns Hopkins Univ, PhD(biophys), 60. *Prof Exp:* USPHS res fel chem crystallog, Oxford Univ, 60-61; asst prof physics, Wittenburg Univ, 61-64; from asst prof to assoc prof, 64-82, PROF PHYSICS, VANDERBILT UNIV, 82- *Concurrent Pos:* USPHS grants, 62-63, 64-75, NSF grants, 72-76, 78-81; vis scientist, WPAFB Mats Lab, 82-90. *Mem:* AAAS; Am Crystallog Asn. *Res:* Determination of molecular structures by x-ray crystallographic methods; phase changes and modulated structures; polymer fiber diffraction. *Mailing Add:* Dept Physics Vanderbilt Univ Box 1807 Sta B Nashville TN 37235

LENHOFF, HOWARD MAER, b North Adams, Mass, Jan 27, 29; m 54; c 2. INVERTEBRATE ZOOLOGY, HISTORY & PHILOSOPHY OF SCIENCE. *Educ:* Coe Col, BA, 50; Johns Hopkins Univ, PhD(biol), 55. *Hon Degrees:* DSc, Coe Col, 76. *Prof Exp:* USPHS fel, Loomis Lab, Nat Cancer Inst, 54-56; actg chief, Biochem Sect, Armed Forces Insts Path, 56-57; assoc consult res, George Washington Univ, 57-58; fel, Dept Terrestrial Magnetism, Carnegie Inst Technol, 58; assoc prof biol, Univ Miami, Fla, 59-65, prof, 66-69, dir lab quant biol, 63-69; assoc dean sch biol sci, 69-71, dean grad div, 71-73, PROF DEVELOP & CELL BIOL, FAC RES FACIL, UNIV CALIF, IRVINE, 69- *Concurrent Pos:* Vis lectr, Howard Univ, 57-58; investr, Biochem Labs, Howard Hughes Med Inst, 58-63; USPHS career develop award, 65-69; vis scientist, Polymer Lab, Weizmann Inst Sci, Israel, 68-69; vis prof, Hebrew Univ Jerusalem, 70, 71, 77-78; vis prof chem eng, Israel Inst Technol, 73-74, social ecol, Ben Gurion Univ, Beersheba, Israel, 81; vis sr res fel, Jesus Col, Oxford Univ, 88. *Honors & Awards:* Hon mem, Soc Phys & Nat Hist, Geneva Swiss Acad Sci, 90. *Mem:* Am Soc Biol Chem; Am Soc Cell Biol; Am Chem Soc; Biophys Soc; Soc Develop Biol; Hist Sci Soc. *Res:* Invertebrate biology; chemoreception; symbiosis; cellular differentiation; immobilized enzymes; enzyme immunoassays; history of biology. *Mailing Add:* Fac Res Facil Univ Calif Irvine CA 92717

LENIART, DANIEL STANLEY, b Norwich, Conn, Jan 5, 43; m 71. PHYSICAL CHEMISTRY. *Educ:* The Citadel, BS, 64; Cornell Univ, PhD(phys chem), 69. *Prof Exp:* Fel, Varian Assocs, 69-70, appln engr, 70-75, mgr EPR res & develop, 75-80; Gas Chromatography-Mass Spectros, Hewlett Packard, 81-85; IMMUNOL DIAG INSTRUMENTATION, TECHNICON INSTRUMENT CORP, 86- *Mem:* Am Phys Soc. *Res:* Study of relaxation phenomena using the techniques of electron spin resonance, electron nuclear double resonance and electron-electron double resonance. *Mailing Add:* Six Guernsey Rd Brookfield CT 06805-1951

LENKE, ROGER RAND, b Brooklyn, NY, April 6, 46. MATERNAL FETAL MEDICINE, MEDICAL GENETICS. *Educ:* Columbia Univ, MD, 71. *Prof Exp:* Intern med, Roosevelt Hosp, 71-72; resident obstet & gynec, Columbia Presby, NY, 72-76; fel neurol, Mass Gen Hosp, 79; fel maternal & fetal med, Univ Southern Calif, Los Angeles, 79-81; dir prenatal diag, Univ Wash, Seattle, 82-85; DIR MATERNAL & FETAL MED, MED COL OHIO, TOLEDO & PROF OBSTETS & GYNEC, 85- *Mem:* Am Col Obstetricians & Gynecologists; Am Inst Ultrasound Med; Am Soc Human Genetics; Soc Perinatal Obstets; AMA; Int Soc Fetal Med & Surg. *Res:* High risk obstetrics and prenatal diagnosis. *Mailing Add:* Mass Gen Hosp Burnham Eight Fruit St Boston MA 02114

LENKER, SUSAN STAMM, b Bridgeport, Conn, Nov 13, 45; m 68; c 2. MATHEMATICAL LOGIC, STATISTICS. *Educ:* Western Conn State Col, BS, 69; Univ Colo, MA, 70; Univ Mont, PhD(math), 75. *Prof Exp:* Asst prof math, Univ Louisville, 75-76; asst prof oper res, 76-88, assoc prof software systs, Dept Info Systs & Anal 89-91, ASSOC PROF MATH, DEPT MATH, CENT MICH UNIV, 91- *Mem:* Am Statist Asn; Opers Res Soc Am; Am Inst Decision Sci; Am Math Soc; Sigma Xi. *Res:* Data base theory; decision theory including fuzzy sets; cluster analysis, statistics; category theory. *Mailing Add:* Dept Math Cent Mich Univ Mt Pleasant MI 48859

LENKOSKI, L DOUGLAS, b Northampton, Mass, May 13, 25; m 52; c 4. MEDICINE, PSYCHIATRY. *Educ:* Harvard Univ, AB, 48; Western Reserve Univ, MD, 53. *Prof Exp:* Fel psychiat, Yale Univ, 55-56; teaching fel, 57-60, from instr to assoc prof, 60-69, dir dept, Cleveland Metrop Gen Hosp, 69-76; chmn dept, 70-86, PROF PSYCHIAT, SCH MED, CASE WESTERN RESERVE UNIV, 69-, ASSOC DEAN, 82- *Concurrent Pos:* Consult, DePaul Maternity & Infant Home, 58-67, & Cleveland Ctr on Alcoholism, 58-61; actg dir dept psychiat, Univ Hosps Cleveland, 62-66, assoc dir dept, 66-69, dir dept, 69-86; consult, Cleveland Vet Admin Hosp, 65-; chief of staff, Univ Hosps Cleveland, 82-90; dir, Substance Abuse Ctr, Case Western Reserve Univ, 90- *Mem:* AAAS; Am Col Psychiat; fel Am Psychiat Asn; Am Psychoanal Asn. *Res:* Psychiatric education; community mental health planning. *Mailing Add:* Dept Psychiat Sch Med Case Western Reserve Univ Cleveland OH 44106

LENLING, WILLIAM JAMES, b Madison, Wis, Jan 16, 61; m 87. THERMAL SPRAY COATINGS. *Educ:* Univ Wis-Madison, BS, 85, MS, 87. *Prof Exp:* Mat engr plasma coating res & develop, Sandia Nat Labs, 88-90; mat engr plasma coating res & develop, Fisher-Barton, Inc, 87-90; RES MGR, THERMAL SPRAY TECHNOLOGIES, 91- *Concurrent Pos:* Mem, Mat Sci Comt, Thermal Spray Div, Am Soc Metall, 90- *Mem:* Am Soc Metall. *Res:* Material science coating development of thermal spray coatings; develop coatings for a wide variety of industrial applications; author of four publications and two patents. *Mailing Add:* Thermal Spray Technologies 201 Frederick St Watertown WI 53094

LENN, NICHOLAS JOSEPH, b Chicago, Ill, Nov 26, 38; m 64; c 3. NEUROLOGY, ANATOMY. *Educ:* Univ Chicago, SB, 59, MS & MD, 64, PhD(anat), 67. *Prof Exp:* Res assoc neuroanat, NIH, 64-66; asst prof pediat & med, Univ Chicago, 70-74; asst prof neurol, Univ Calif, Davis, 74-80, asst prof pediat, 76-80; MEM FAC NEUROL, SCH MED, UNIV VA, CHARLOTTESVILLE, 80- *Mem:* AAAS; Am Asn Anat; Am Acad Neurol. *Res:* Synaptic organization of mammalian brain. *Mailing Add:* Dept Neurol State Univ NY HSC T12-20 Stony Brook NY 11794-8121

LENNARTZ, MICHELLE R, MEDICINE. *Prof Exp:* DIR MED, WASH UNIV, 91- *Mailing Add:* Med Dept Sch Med Wash Univ 660 S Euclid Ave Box 8051 St Louis MO 63110

LENNARZ, WILLIAM J, b New York, NY, Sept 28, 34; m; c 3. BIOCHEMISTRY, CELL BIOLOGY. *Educ:* Pa State Univ, BS, 56; Univ Ill, PhD(chem), 59. *Prof Exp:* Postdoctoral fel, Harvard Univ, 59-62; from asst prof to prof, Dept Biol Chem, Sch Med, Johns Hopkins Univ, 62-83; Robert A Welch prof chem & chmn, Dept Biochem & Molecular Biol, M D Anderson Cancer Ctr, Univ Tex, 83-89, prof, 84-89; LEADING PROF & CHMN, DEPT BIOCHEM & CELL BIOL, STATE UNIV NY, STONY BROOK, 89-, DIR, INST CELLULAR & DEVELOP BIOL, 90- *Concurrent Pos:* NSF fel, 69-60, NIH fel, 60-62; mem, Physiol Chem Study Sect, NIH, 74-78, ad hoc mem, Molecular Biol Study Sect, 80-81, mem, Pathobiol Chem Study Sect, 82-86; vis prof biochem, Sch Med, WVa Univ, 82, Burroughs Wellcome vis prof, State Univ, NY, 83, Univ PR, 88; mem, Grad Prog Cell & Develop Biol, Genetics Prog, 89- & Acad Standards Coun, 90- *Mem:* Am Chem Soc; Am Soc Biochem & Molecular Biol (pres-elect, 88-89); Am Soc Microbiol; Sigma Xi; Am Soc Cell Biol; Soc Complex Carbohydrates; Am Soc Zoologists. *Res:* Cancer research; author of numerous scientific publications. *Mailing Add:* Dept Biochem & Cell Biol State Univ NY 450 Life Sci Bldg Stony Brook NY 11794-5215

LENNETTE, EDWIN HERMAN, b Pittsburgh, Pa, Sept 11, 08; m 30; c 2. EPIDEMIOLOGY, EXPERIMENTAL PATHOLOGY. *Educ:* Univ Chicago, BS, 31, PhD(hyg & bact), 35; Rush Med Col, MD, 36. *Prof Exp:* Instr bact, Univ Chicago, 36-37, res assoc, 37-38; instr path, Wash Univ, 38-39; mem staff, Int Health Div, Rockefeller Found, 39-46; chief med-vet div, Camp Detrick, Md, 46-47; chief biomed lab, 73-78, chief, 47-78, interim dir, Walton Jones Cell Sci Ctr, 81, EMER CHIEF VIRAL & RICKETTSIAL DIS LAB, CALIF STATE DEPT HEALTH SERV, 78-,. *Concurrent Pos:* Lectr, Sch Pub Health, Univ Calif, 47-78, lectr, Univ, 48-58; consult physician, Highland-Gen Hosp, 48-80; consult, Sixth Army Surgeon, 48-; assoc mem comn influenza, Armed Forces Epidemiol Bd, 48-51, mem, 51-73, mem comn rickettsial dis, 51-73; dep dir & chief lab serv prog, Calif State Dept Health, 72-73; mem, Armed Forces Epidemiol Bd, Off Surgeon Gen, 70-76, pres, 73-76; mem adv panel, Naval Biol Lab, 48-56; dir regional lab, Influenza Study Prog, WHO, 49-75, mem expert adv panel virus dis, 51-, mem expert adv panel zoonoses, 52-62, mem WHO sci group on virus dis, 66-75, mem sci adv comt to WHO team, EAfrican Virus Res Inst, 71-; mem viral rickettsial registry, Am Type Cul Collection, 49-65; coordr sect XII, sect res prog, NIH, 51-56; mem virus & rickettsial study sect, NIH, 51-53, chmn, 52-53, mem med lab serv adv comt, 68-72; consult, NIH, 51-, co-chmn microbiol & immunol study sect, Div Res Grants & Fels, 53-54, chmn, 55-56,; mem adv comt poliomyelitis vaccine eval ctr, Nat Found Infantile Paralysis, 53-54; chmn bd sci counsr, Nat Inst Allergy & Infectious Dis, 57-61, mem training grant comt, 60-61, chmn, 62, mem panel respiratory & related viruses, 60-63, mem comt vaccine develop, 63-68, mem & chmn subcomt rubella virus, 65-

66, mem nat adv allergy & infectious dis coun, 63-66; consult physician, Peralta Hosp, 57-; mem microbiol panel study manpower needs in basic health sci, Fedn Am Soc Exp Biol, 60-61; chmn ad hoc comt rubella vaccine, Nat Inst Neurol Dis & Blindness, 63-64; mem microbiol panel, Wooldridge Comt, White House, 64; mem & chmn panel virus dis, US-Japan Coop Med Sci Prog, Off Sci & Technol, White House & Off Int Res, 65-69, mem US del, US Dept State, 70-76; mem sci adv comt, Hastings Found, 66-76; consult, Univ Tex, MD Anderson Hosp & Tumor Inst, 66-77; mem solid tumor-virus segment, Spec Virus-Cancer Prog, Nat Cancer Inst, 66-72, consult Nat Cancer Inst, 67-73; founding mem, Am Biol Coun, 68-70; mem bd dir, Rush Med Col, 70-74; consult, Bur of Biologics, Fed Drug Admin & mem, Bur Panel Viral & Rickettsial Vaccines, 73- *Honors & Awards:* Bronfman Award & Prize Achievement Pub Health, Am Pub Health Asn, 69; Wyeth Award Clin Microbiol & Prize, Am Soc Microbiol, 76. *Mem:* Fel Am Pub Health Asn, 61-77; hon fel Am Soc Clin Path, 72; Soc Gen Microbiol; Tissue Cult Asn (pres, 76-78); fel Royal Soc Trop Med & Hyg. *Res:* Virology; clinical, epidemiologic and immunologic research on viral and rickettsial diseases, including poliomyelitis, enteroviruses, respiratory disease, Q fever and virus-cancer relationships. *Mailing Add:* Calif Pub Health Found 2001 Addison St Suite 210 Berkeley CA 94704-1103

LENNEY, JAMES FRANCIS, b St Louis, Mo, Oct 11, 18; m 42, 73; c 2. BIOCHEMISTRY. *Educ:* Wash Univ, BA, 39; Mass Inst Technol, PhD(gen physiol), 46. *Prof Exp:* Asst zool, Wash Univ, 41; asst, Mass Inst Technol, 42-43; res chemist, 44-45; res biochemist, Fleischmann Labs, 46-49, head, Enzyme Dept, 49-56; sect head, Union Starch & Ref Co, 56-63; assoc prof, 64-74, PROF PHARMACOL, UNIV HAWAII, 74- *Honors & Awards:* Fel, AAAS. *Mem:* Am Soc Pharmacol & Exp Therapeut; AAAS; Am Chem Soc. *Res:* Biochemical pharmacology; enzyme and protein chemistry. *Mailing Add:* Dept Pharmacol Univ Hawaii Sch Med Honolulu HI 96822

LENNON, EDWARD JOSEPH, b Chicago, Ill, Aug 2, 27; m 73; c 3. INTERNAL MEDICINE. *Educ:* Univ Ill, BA, 47, MA, 48; Northwestern Univ, MD, 52. *Prof Exp:* Intern med, Milwaukee County Hosp, Wis, 52-53, resident internal med, 55-58; from instr to assoc prof, 58-68, from assoc dean to dean, 68-84, PROF MED, MED COL WIS, 68- *Concurrent Pos:* Fel, Mass Mem Hosp, Boston, 60-61; dir clin res ctr, Milwaukee County Hosp, 61-68, chief renal serv, 63-70. *Mem:* AAAS; Am Soc Clin Invest; Am Physiol Soc; Am Fedn Clin Res; Am Diabetes Asn. *Res:* Endocrine disorders; renal disease; acid-base metabolism. *Mailing Add:* PO Box 26509 Milwaukee WI 53226

LENNON, JOHN W(ILLIAM), b Columbus, Ohio, Oct 21, 17; m 44; c 7. CERAMIC ENGINEERING. *Educ:* Ohio State Univ, BCerE, 40, MSc, 43. *Prof Exp:* Ceramic engr, Stupakoff Ceramic & Mfg Co, Pa, 40-42; asst ceramic eng, Ohio State Univ, 42-43; ceramic engr, Isolantite Inc, NJ, 43-44; fel, Mellon Inst, 45-46; ceramic engr, Orefraction, Inc, 46-49; eng exp sta, Ohio State Univ, 49-51; res engr, Battelle Mem Inst, 51-82; RETIRED. *Mem:* Am Ceramic Soc; Mineral Soc Am. *Res:* High frequency electrical insulation; special oxide compositions; high dielectrics; heat shock compositions; enamels; refractory coatings; glass; ceramic microstructure; electrodeposition of ceramic coatings; ferroelectric and ferromagnetic ceramics. *Mailing Add:* 942 Lambeth Rd Columbus OH 43220

LENNON, PATRICK JAMES, b Amsterdam, NY, July 24, 50. HOMOGENEOUS CATALYSIS. *Educ:* State Univ NY at Binghamton, BA, 72; Brandeis Univ, PhD(org chem), 77. *Prof Exp:* Res fellowship, Oxford Univ, 77-79; SR RES SPEC, MONSANTO CO, 80- *Mem:* Am Chem Soc; Royal Soc Chem. *Res:* Homogeneous catalysis of organic reactions by transition metal complexes; organosilicon chemistry; design a synthesis of enzyme inhibitors. *Mailing Add:* 7540 Wydown Blvd No 3W Clayton MO 63105

LENNON, VANDA ALICE, b Sydney, Australia, Aug 1, 43; m 75. NEUROIMMUNOLOGY. *Educ:* Univ Sydney, MB, BS, 66; Univ Melbourne, PhD(immunol), 73. *Prof Exp:* Res asst nuclear med, Univ Sydney, 66; from jr intern to asst med resident, Montreal Gen Hosp, 66-68; fel immunol, Walter & Eliza Hall Inst Med Res, 68-72; res assoc, Salk Inst Biol Studies, 72-73, asst res prof, 73-77; CONSULT NEUROL & IMMUNOL, MAYO CLINIC, 78-, assoc prof neurol & immunol, Mayo Grad Sch Med, 78-83, PROF NEUROL & IMMUNOL, MAYO GRAD SCH MED, 83- *Concurrent Pos:* Assoc adj prof, Dept Neurosci, Univ Calif, San Diego, 77-78. *Mem:* Am Acad Neurol; Sigma Xi; Am Asn Immunol; Soc Neurosci; Am Soc Clin Invest; Am Asn Neuropath. *Res:* Autoimmunity to antigens of central and peripheral nervous systems and muscle; identification of neural antigens on small cell lung cancer; immunologic studies of patients with neurological and paraneoplastic diseases of presumed autoimmune basis. *Mailing Add:* Neuroimmunol Lab Depts Neurol & Immunol Mayo Clinic Rochester MN 55905

LENNOX, ARLENE JUDITH, b Cleveland, Ohio, Dec 3, 42. MEDICAL PHYSICS, ELEMENTARY PARTICLE PHYSICS. *Educ:* Notre Dame Col, Ohio, BS, 63; Univ Notre Dame, MS, 73, PhD(physics), 74. *Prof Exp:* Teacher, Marymount High Sch, 63-64, Regina High Sch, 64-65 & Shrine High Sch, 65-69; res assoc physics, Fermilab, 74-77; prof physics, NCent Col, 77-80; staff physicist, 80-86, DEPT HEAD, FERMILAB NEUTRON THERAPY FACIL, 86- *Concurrent Pos:* Vis physicist, Fermilab, 78-80. *Mem:* Am Phys Soc; AAAS; Am Asn Physics Teachers; Am Asn Physicists Med; Am Soc Therapeut Radiol & Oncol. *Res:* Experiments to study backward peak in pi-p elastic scattering; experiments to measure pion form factor; p-p colliding beams; neutron therapy physics; medical uses for proton linacs. *Mailing Add:* Fermilab MS 301 PO Box 500 Batavia IL 60510-0500

LENNOX, DONALD HAUGHTON, b Toronto, Ont, June 7, 24; m 47; c 2. HYDROLOGY. *Educ:* Univ Toronto, BA, 49; Univ Alta, MSc, 60. *Prof Exp:* Tech Off, Occup Health Lab, Dept Nat Health & Welfare Can, 50-57; asst res officer, Res Coun Alta, 57-61; head groundwater div, 61-68; maritime res sect, 68-70, head groundwater subdiv, 70-72; chief, Hydrol Res Div, 72-79,

dir, Nat Hydrol Res Inst, 79-85, SPEC ADV, INLAND WATERS DIRECTORATE, ENVIRON CAN, 85- *Mem:* Geol Soc Am; Geol Asn Can; Nat Water Well Asn. *Res:* Application of geophysical techniques to shallow groundwater exploration; investigation of analytical methods for the determination of aquifer and well characteristics. *Mailing Add:* Inland Waters Directorate Dept Environ Environ Can Ottawa ON K1A 0E7 Can

LENNOX, ROBERT BRUCE, b New Orleans, La, June 5, 57; Can citizen; m 85. ELECTROCHEMISTRY, INTERFACIAL CHEMISTRY. *Educ:* Univ Toronto, BSc, 79, MSc, 81, PhD(chem), 85. *Prof Exp:* Res assoc chem, Imp Col, Univ London, 85-87; ASST PROF CHEM, MCGILL UNIV, 87- *Mem:* Electrochem Soc; Chem Inst Can; Amer Chem Soc. *Res:* Interfacial reactivity, organized assembly chemistry, bioelectrochemistry, organic thin films, bioelectrocatalysis. *Mailing Add:* Dept Chem McGill Univ 801 Sherbrooke St W Montreal PQ H3A 2K6 Can

LENNOX, ROBERT BRUCE, b New Orleans, La, June 5, 57; Can citizen; m 85. INTERFACIAL CHEMISTRY, BIOELECTROCHEMISTRY. *Educ:* Univ Toronto, BSc, 79; MSc, 81, PhD(chem), 85. *Prof Exp:* Postdoctoral fel chem, Imp Col, Univ London, 85-87; ASST PROF CHEM, MCGILL UNIV, 87- *Mem:* Chem Inst Can; Am Chem Soc; Electrochem Soc. *Res:* Interfacial chemistry; organic chemistry; thin film; monolayers; bioorganic mechanisms; biosensors; electrochemistry. *Mailing Add:* 801 Sherbrooke St W Montreal PQ H3A 2K6 Can

LENNOX, WILLIAM C(RAIG), b Mount Forest, Ont, May 22, 37; m 61. MECHANICS. *Educ:* Univ Waterloo, BASc, 62, MSc, 63; Lehigh Univ, PhD(mech), 66. *Prof Exp:* From asst prof to assoc prof, 66-71, chmn dept, 76-77 & 79-82, PROF CIVIL ENG, UNIV WATERLOO, 71-, DEAN ENG, 82- *Concurrent Pos:* Vis prof, Col Petrol & Minerals, Saudi Arabia, 70-71 & Harvey Mudd Col, 77-79. *Mem:* Am Inst Aeronaut & Astronaut; Am Soc Eng Educ; Am Acad Mech. *Res:* Stochastic processes; nonlinear mechanisms; stochastic processes; nonlinear mechanics; ice research. *Mailing Add:* Dept Civil Eng Univ Waterloo Waterloo ON N2L 3G1 Can

LE NOBLE, WILLIAM JACOBUS, b Rotterdam, Netherlands, July 19, 28; nat US; m 71; c 3. ORGANIC CHEMISTRY. *Educ:* Advan Tech Sch, Neth, BS, 49; Univ Chicago, PhD, 57. *Prof Exp:* Res chemist, Indust Lab, Rohm & Haas Co, 57; instr chem, Rosary Col, 58; NSF fel & res asst, Purdue Univ, 58-59; from asst prof to assoc prof, 59-69, PROF ORG CHEM, STATE UNIV NY STONY BROOK, 69- *Concurrent Pos:* Sr ed, J Org Chem. *Honors & Awards:* Humboldt Sr Scientist Award. *Mem:* Am Chem Soc. *Res:* Chemical kinetics, mechanisms and equilibria in liquid systems under high pressure; face selectivity. *Mailing Add:* Dept Chem State Univ NY Stony Brook NY 11794

LENOIR, J(OHN) M, chemical engineering, for more information see previous edition

LENOIR, WILLIAM BENJAMIN, b Miami, Fla, Mar 14, 39; m 64; c 1. GEOPHYSICS, ELECTRICAL ENGINEERING. *Educ:* Mass Inst Technol, SB, 61, SM, 62, PhD(elec eng), 65. *Prof Exp:* Asst elec eng, Mass Inst Technol, 62-64, instr, 64-65, asst prof, 65-67; Ford fel eng, 65-66; SCIENTIST-ASTRONAUT, JOHNSON SPACE CTR, NASA, 67- *Mem:* AAAS; Am Geophys Union. *Res:* Microwave studies of planetary atmospheres; propagation of partially polarized waves. *Mailing Add:* Booz-Allen & Hamilton 4330 East-West Hwy Bethesda MD 20814

LENOIR, WILLIAM CANNON, JR, b Loudon, Tenn, Sept 22, 29; m 56; c 3. BOTANY. *Educ:* Maryville Col, BS, 51; Univ Ga, MS, 62, PhD(bot), 65. *Prof Exp:* Instr high sch, Tenn, 57-59; asst prof biol, 60-62, from asst prof to assoc prof bot, 62-73, PROF BOT, COLUMBUS COL, 73-, CHMN DIV SCI & MATH, 73- *Mem:* Am Inst Biol Sci; Bot Soc Am. *Res:* Role of light in morphogenesis; organogenesis in pine; morphogenesis of the leaf of Lygodium japonicum. *Mailing Add:* Dept Biol Columbus Col Columbus GA 31907

LENON, HERBERT LEE, b Battle Creek, Mich, June 8, 39; m 62; c 3. FISH BIOLOGY. *Educ:* Albion Col, AB, 61; Wayne State Univ, MS, 64; Mich State Univ, PhD(fisheries), 68. *Prof Exp:* Asst prof, 67-74, ASSOC PROF FISHERIES BIOL & ICHTHYOL, CENT MICH UNIV, 74- *Mem:* Am Fisheries Soc; Nat Audubon Soc; Nat Wildlife Fedn. *Res:* Freshwater fish population dynamics; management evaluation. *Mailing Add:* Dept Biol Cent Mich Univ Mt Pleasant MI 48859

LENOX, RONALD SHEAFFER, b Lancaster, Pa, Jan 25, 48; m 71; c 1. POLYMER CHEMISTRY. *Educ:* Juniata Col, BS, 69; Univ Ill, PhD(org chem), 73. *Prof Exp:* Asst prof chem, Wabash Col, Crawfordsville, Ind, 73-79; res unit mgr, 79-90, SR PRIN SCIENTIST, ARMSTRONG WORLD INDUSTS, 90- *Concurrent Pos:* Ad cont, dept chem, Univ Southern Miss. *Mem:* Am Chem Soc. *Res:* Synthesis of natural products; particularly insect pheromones; new synthetic reactions; sulfonyl azide chemistry; polymer blends and alloys; electrostatic discharge products. *Mailing Add:* Innovation Armstrong World Indust Lancaster PA 17604

LENSCHOW, DONALD HENRY, b LaCrosse, Wis, July 17, 38; m 64; c 2. METEOROLOGY. *Educ:* Univ Wis, BS, 60, MS, 62, PhD(meteorol), 66. *Prof Exp:* SCIENTIST, NAT CTR ATMOSPHERIC RES, 66-; AFFIL PROF, COLO STATE UNIV, 74- & UNIV COLO, 89- *Mem:* Fel Am Meteorol Soc. *Res:* Atmospheric boundary layer; airborne turbulence measurements and airplane research instrumentation. *Mailing Add:* 95 Pawnee Dr Boulder CO 80303

LENSTRA, HENDRIK W, b Zaadax, Neth, Apr 16, 49. NUMBER THEORY. *Educ:* Univ Amsterdam, PhD(math), 77. *Prof Exp:* PROF MATH, UNIV AMSTERDAM, 78- *Honors & Awards:* Fulkerson Prize, AMS & Parisenne Soc, 85; Royal Dutch Acad Ser Prize. *Mem:* Am Math Soc; Dutch Math Soc. *Res:* Algorithmic number theories which interface with computer sciences and algebraic number theories. *Mailing Add:* Dept Math Univ Calif Berkeley CA 94720

LENTINI, EUGENE ANTHONY, b Boston, Mass, July 6, 29; m 51; c 4. PATHOLOGY. *Educ:* Boston Univ, AB, 51, MA, 55, PhD(myocardial metab), 58. *Prof Exp:* Instr physiol, Med Sch, Univ Ore, 58-64; asst prof, Med Col Va, 64-68; assoc prof physiol, Albany Col Pharm, 68-75; assoc prof, Dept Physiol & Pharmacol, Philadelphia Col Obsteopath Med, 77-81; scientist, Vet Admin Hosp, Philadelphia, 81-82; res assoc prof surg, Pa Med Col, 82-84; CONSULT BIOHAZARD MED, 85- *Concurrent Pos:* Nat Heart & Lung Inst fel, 56-58; Heart & Lung res awards, 60-65 & 69-72; vis prof, Mass State Col & Univ Lowell, 75-77; adj prof physiol, Sch Vet Med, Univ Penn, 81-84; grant awards, Nat Heart, Lung & Blood Inst, Am Heart Asn, Ore Heart, Va Heart, Am Osteop Asn & AMA. *Mem:* Sigma Xi; NY Acad Sci; AAAS; Am Physiol Soc; Am Heart Asn; Am Asn Univ Prof. *Res:* Bioelectronics, electronic micrometer, chart viewer; biophysics determination of oxygen diffusion coefficient through heart muscle; biochemical interrelation between ventricular dynamics and oxidative metabolism; effects of metabolic inhibitors on endogenous substrate; analysis of endogenous lipids and glycogen; physiological myocardial contract; substrate utilization; myocardial infarct model; vascular effects of catheteryction; oncology and smooth muscle dynamics; 40 publications. *Mailing Add:* 221 Canterbury Dr Broomall PA 19008

LENTON, PHILIP A(LFRED), b Detroit, Mich, Aug 13, 19; m 42; c 2. CHEMICAL ENGINEERING. *Educ:* Wayne State Univ, BSChE, 41; Mich State Col, MSChE, 43. *Prof Exp:* Chem engr res, Girdler Corp, 43-44; prod supvr gas mfr, Houdaille Hershey Corp, 44-45; mem staff chem eng res, Wyandotte Chem Corp, 45-49, sect head, BASF Wyandotte Corp, 49-56, develop engr, 56-63, acquisition specialist, 63-69, corp planning coordr, 69-77, sr indust engr, 77-80; RETIRED. *Mem:* Am Chem Soc; Am Inst Chem Engrs. *Res:* Process for manufacturing of lubricating grease; development of various organic and inorganic processes, including sodium carboxymethylcellulose and sodium alkyl aryl sulfonate; evaluation of business opportunities, including possible corporate acquisitions; long-range corporate planning. *Mailing Add:* 1920 Dacosta St Dearborn MI 48128

LENTZ, BARRY R, b Philadelphia, Pa, Sept 2, 44; m 66; c 3. BLOOD COAGULATION, CELLULAR FUSION. *Educ:* Univ Pa, BA, 66; Cornell Univ, PhD, 73. *Prof Exp:* Vis scientist biophysics, Weitmann Inst Sci, 72; NIH fel biophys, Dept Biochem, Univ Va Sch Med, 73-75; from asst prof to assoc prof, 75-88, PROF, DEPT BIOCHEM, UNIV NC, CHAPEL HILL, 88- *Concurrent Pos:* Estab investr, Am Heart Asn, 79-84. *Mem:* Am Chem Soc; Am Heart Asn; Am Soc Biochem & Molecular Biol; AAAS; N Am Thermal Anal Soc; Biophys Soc. *Res:* Physical chemistry used for the solution of biologically relevant problems; platelet-derived membranes in blood coagulation; poly ethylene glycol induced cellular fusion. *Mailing Add:* 179 Tradescant Dr Chapel Hill NC 27514

LENTZ, CHARLES WESLEY, b Mt Pleasant, Mich, May 6, 24; m 47; c 7. CHEMISTRY, SILICON CHEMISTRY. *Educ:* Mich State Univ, BS, 46. *Prof Exp:* Chemist, Mich Chem Corp, 46-52 & Columbia-Southern Div, Pittsburgh Plate Glass Co, 52-55; chemist, Dow Corning Corp, 55-61, supvr develop, 61-68, mgr develop, 68-70, mgr res, 70-75, mgr life sci res & develop, 75- 77, dir health & environ sci, 77-86; RETIRED. *Concurrent Pos:* Mem comt MC-B5, Hwy Res Bd, Nat Acad Sci-Nat Res Coun, 67- *Honors & Awards:* Sigma Xi Award, 65. *Res:* Study of silica as a reinforcing agent for silicone rubber, silicate minerals and the silicate structure changes that occur in portland cement during hydration. *Mailing Add:* 5105 Foxcroft Midland MI 48640

LENTZ, CLAUDE PETER, b Westerham, Sask, Nov 27, 19; m 45; c 4. FOOD REFRIGERATION. *Educ:* Univ Sask, BSc, 49; Univ Toronto, MASc, 51. *Prof Exp:* Prin res officer, Nat Res Coun Can, 49-53, head, Food Technol Sect, Div Biol Sci, 53-79; RETIRED. *Concurrent Pos:* Consult, 80-81. *Honors & Awards:* W J Eva Award, Can Inst Food Technol, 67; John Labatt Award, Chem Inst Can, 78. *Mem:* Can Inst Food Technol; Am Soc Heating, Refrig & Air-Conditioning Engrs. *Res:* Food preservation by refrigeration and related heat and mass transfer problems; waste treatment. *Mailing Add:* 19 David Dr Nepean ON K2G 2M8 Can

LENTZ, GARY LYNN, b Hollywood, Calif, July 15, 43; m 65; c 6. ECONOMIC ENTOMOLOGY. *Educ:* Univ Mo-Columbia, AB, 65; Iowa State Univ, PhD(entom), 73. *Prof Exp:* Res assoc entom, Iowa State Univ, 68-72; asst prof, Univ Ariz, 72-74; asst prof, 74-80, ASSOC PROF ENTOM, AGR EXP STA, UNIV TENN, 80- *Mem:* Entom Soc Am. *Res:* Pest management of cotton and soybean insects. *Mailing Add:* WTenn Exp Sta 605 Airways Blvd Jackson TN 38301

LENTZ, MARK STEVEN, b Madison, Wis, July 3, 49; m 80. ENERGY CONSERVATION, SPECIALTY ENVIRONMENT DESIGN. *Educ:* Univ Wis-Madison, BSME, 78. *Prof Exp:* Proj engr, Affil Engrs, Inc, 76-81 & Stanley Consults, 81-83; sr mech engr, Donohue & Assocs, Inc, 83-91; SR PROJ ENGR, PSJ ENG, INC, 91- *Concurrent Pos:* Mem, TC 9.8-Large Bldg Air Conditioning Appln, Am Soc Heating, Refrigerating & Air Conditioning Engrs, Inc, 84-88, corresp mem, 88-; mem, TC 9.2-Indust Air Conditioning, Am Soc Heating, Refrigerating & Air Conditioning Engrs, Inc, 85-89; mem, Prog Comt, Am Soc Heating, Refrigerating & Air Conditioning Engrs, Inc, 89-92, corresp mem, TG 9.LS-Lab Systs, 89- & TC 5.7-Evaporative Cooling, 90-, consult, Standing Comt-Energy Conserv Bldgs, 91-95. *Honors & Awards:* Energy Award, Am Soc Heating, Refrigerating & Air Conditioning Engrs, 88. *Mem:* Am Soc Heating, Refrigerating & Air Conditioning Engrs. *Res:* Advanced systems design for high-tech laboratory and specialty environment employing variable-volume ventilation and evaporative cooling; author of three publications. *Mailing Add:* 437 Center Walk Kohler WI 53044

LENTZ, PAUL JACKSON, JR, b Niagara Falls, NY, Oct 10, 44; m 68; c 2. BIOCHEMISTRY, X-RAY CRYSTALLOGRAPHY. *Educ:* Univ Alaska, BS, 66; Purdue Univ, PhD(molecular biol), 71; Univ Miami, MD, 84. *Prof Exp:* Fel biol, Wallenberg Lab, Uppsala Univ, Sweden, 71-75; scholar chem, Univ Mich, 75-78; asst prof biol, Kings's Col, 78-82; med resident, Univ Mich Hosps, 84-87; MED PRACT, 87- *Concurrent Pos:* NIH grant, 71-73; lectr chem, Univ Mich, 76-78. *Mem:* Am Crystallog Asn; AAAS; Sigma Xi; AMA; Am Acad Family Phys. *Res:* X-ray crystallographic structure determination of proteins, nucleic acids and viruses. *Mailing Add:* 304 State St Adrian MI 49221

LENTZ, PAUL LEWIS, b Indianapolis, Ind, May 26, 18; m 43; c 2. MYCOLOGY. *Educ:* Butler Univ, AB, 40; Univ Iowa, MS, 42, PhD(mycol), 53. *Prof Exp:* Asst bot lab, Butler Univ, 38-40, bact lab, 40; asst mycol, Univ Iowa, 40-42, 46-47; assoc mycologist, Plant Indust Sta, USDA, 47-56, mycologist, Plant Sci Res Div, 56-72, instr advan educ sci, Grad Sch-Found, 58-71, chief, Mycol Lab, Sci & Educ Admin-Age Res, 72-83; RETIRED. *Mem:* Bot Soc Am; Mycol Soc Am; Int Soc Plant Taxon; Sigma Xi. *Res:* Basidiomycete taxonomy, anatomy, morphology and biology; Aphyllophorales; National Fungus Collections. *Mailing Add:* Five Orange Ct Greenbelt MD 20770

LENTZ, THOMAS LAWRENCE, b Toledo, Ohio, Mar 25, 39; m 61; c 3. ACETYLCHOLINE RECEPTOR, RABIES VIRUS. *Educ:* Yale Univ, MD, 64. *Prof Exp:* From instr to asst prof anat, Sch Med, Yale Univ, 64-69, assoc prof cytol, 69-74, assoc prof cell biol, 74-85, PROF CELL BIOL, SCH MED, YALE UNIV, 85-, ASST DEAN ADMIS, 76- *Mem:* AAAS; Am Soc Cell Biol; Soc Neurosci; NY Acad Sci. *Res:* Characterization of functional domains on the acetylcholine receptor; identification of cellular receptors for rabies virus. *Mailing Add:* Dept Cell Biol Yale Univ Sch Med 333 Cedar St New Haven CT 06510

LENZ, ALFRED C, b Olds, Alta, Jan 6, 29; m 54; c 2. GEOLOGY. *Educ:* Univ Alta, BSc, 54, MSc, 56; Princeton Univ, PhD(paleont), 59. *Prof Exp:* Paleontologist, Calif Standard Co, 59-64; from asst prof to assoc prof, 64-75, prof paleont & stratig, 75-80, PROF GEOL, UNIV WESTERN ONT, 80- *Concurrent Pos:* Lectr, Univ Alta, 60-61. *Mem:* Int Palaeont Asn; Paleont Soc; Can Palaeont Asn. *Res:* Lower Paleozoic biostratigraphy; Devonian stratigraphy and paleontology; graptolite biostratigraphy; Upper Silurian and Lower Devonian brachiopods. *Mailing Add:* Dept Geol Univ Western Ont Middlesex Coll London ON N6A 5B7 Can

LENZ, ARNO T(HOMAS), b Fond du Lac, Wis, Sept 22, 06; m 32; c 3. ENGINEERING. *Educ:* Univ Wis, BS, 28, MS, 30, CE, 37, PhD(hydraul eng), 40. *Prof Exp:* From instr to assoc prof hydraul & sanit eng, 28-48, prof hydraul eng, 48-77, chmn dept civil eng, 58-72, dir hydraul model tests, 45-58, EMER PROF HYDRAUL ENG, UNIV WIS-MADISON, 77- *Concurrent Pos:* Consult, 51- *Mem:* Am Soc Civil Engrs; Am Meteorol Soc; Nat Soc Prof Engrs; Am Water Works Asn; Am Geophys Union. *Res:* Hydrology of rain fall-runoff relations; model tests of hydraulic structures; oil tests on v-notch weirs. *Mailing Add:* 930 Cornell Ct Madison WI 53705

LENZ, CHARLES ELDON, b Omaha, Nebr, Apr 13, 26. ENGINEERING, MATHEMATICS. *Educ:* Mass Inst Technol, SB, 51, SM, 53; Cornell Univ, PhD, 57; Univ Calif, MS, 71. *Prof Exp:* Engr, Gen Elec Co, 49-56; consult, Assoc Univs, 56; sr staff engr, Avco Corp, 58-60; mem tech staff, Armour Res Found, 60-62; sr scientist, Autonetics Div, NAm Aviation, Inc, 62-69; consult lectr, Univ Nebr, 73-77; sr engr, Control Data Corp, 77-80; staff develop engr, USAF, Offutt AFB, 80-84; res engr, Union Pac Syst, 84-87; MEM TECH STAFF, MITRE CORP, 88- *Concurrent Pos:* Prof, Univ Hawaii, 66-68; guest lectr, Univs Hawaii & Minn, Cornell Univ & Col Aeronaut, Cranfield, Eng; lectr, Univ Calif; IEEE Nat Feedback-Control Comt. *Mem:* Inst Elec & Electronics Engrs; Instrument Soc Am; Am Soc Civil Engrs. *Res:* solving significant electronic, economic and other engineering problems with simulation, novel hardware and new mathematical techniques later incorporated into computer programs of broad utility. *Mailing Add:* 5016 Western Ave Omaha NE 68132

LENZ, GEORGE H, b Irvington, NJ, Oct 9, 39; m 61; c 2. NUCLEAR PHYSICS. *Educ:* Rutgers Univ, AB, 61, MS, 63, PhD(physics), 67. *Prof Exp:* Asst prof physics, Univ Va, 67-71; assoc prof, 71-76, WHITNEY-GUION PROF PHYSICS, SWEET BRIAR COL, 76-, CHMN DEPT, 71- *Mem:* Am Phys Soc; Am Asn Physics Teachers. *Res:* Analogue states; compound nucleus and direct reactions; Coulomb energy systematics. *Mailing Add:* Box 668 Amherst VA 24521

LENZ, GEORGE RICHARD, b Chicago, Ill, Nov 22, 41; m 70; c 3. RESEARCH ADMINISTRATION, BIOENGINEERING & BIOMEDICAL. *Educ:* Ill Inst Technol, BS, 63; Univ Chicago, MS, 65, PhD(chem), 67; Northwestern Univ, MBA, 83. *Prof Exp:* Nat Cancer Inst fel, Yale Univ, 67-69; res investr, 69-71, sect head, G D Searle & Co, 71-85; DIR HEALTH CARE RES & DEVELOP, BOC GROUP TECH CTR, MURRAY HILL, NJ, 85- *Mem:* AAAS; Am Chem Soc; Royal Soc Chem. *Res:* Photochemistry; medicinal chemistry. *Mailing Add:* BOC Group Tech Ctr 100 Mountain Ave Murray Hill NJ 07974

LENZ, PAUL HEINS, b Newark, NJ, Mar 29, 38; m 60; c 4. PHYSIOLOGY, ENDOCRINOLOGY. *Educ:* Franklin & Marshall Col, BS, 60; Rutgers Univ, MS, 64, PhD(endocrinol), 66. *Prof Exp:* Asst prof, 66-70, assoc prof physiol, 70-80, PROF BIOL SCI, FAIRLEIGH DICKINSON UNIV, 80- *Concurrent Pos:* Univ res grants, 68-71, Eli Lilly grant, 70; Ciba grants, 70-71. *Mem:* Endocrine Soc; Am Oil Chemists' Soc; Am Asn Clin Chemists; Am Heart Asn. *Res:* Development of micro-chemical techniques; hormonal and biochemical control of lipid metabolism; platelet aggregation and its control. *Mailing Add:* Dept Biol Sci Fairleigh Dickinson Univ Madison NJ 07940

LENZ, ROBERT WILLIAM, b New York, NY, Apr 28, 26; m 53; c 4. ORGANIC CHEMISTRY. *Educ:* Lehigh Univ, BS, 49; Inst Textile Technol, MS, 51; State Univ NY, PhD(polymer chem), 56. *Prof Exp:* Res chemist, Chicopee Mfg Corp, 51-53; res chemist, Polymer Res Lab, Dow Chem Co, 55-61, Eastern Res Lab, 61-63; asst dir, Fabric Res Labs, Inc, 63-66; assoc prof, 66-69, PROF POLYMER SCI & ENG, UNIV MASS, AMHERST, 69- *Concurrent Pos:* Vis prof, Univ Mainz, Ger, 72-73, Royal Inst Technol,

Stockholm, Sweden, 75, Univ Freiburg, Ger, 79-80 & Japan Soc Prom Sci, 79, Univ Pisa, Italy, 87. *Honors & Awards:* Humboldt Prize, 79. *Mem:* Am Chem Soc; Am Inst Chem Eng; Mat Res Soc. *Res:* Monomer and polymer synthesis; kinetics and mechanism of polymerization; structure-property relations of polymers; reactions of polymers; new polymeric materials and applications; biopolymers. *Mailing Add:* Polymer Sci & Eng Dept Univ Mass Amherst MA 01003

LENZEN, K(ENNETH) H(ARVEY), civil engineering, engineering mechanics; deceased, see previous edition for last biography

LEO, ALBERT JOSEPH, b Winfield, Ill, Sept 29, 25; m 47; c 3. MEDICAL CHEMISTRY. *Educ:* Pomona Col, BA, 48; Univ Chicago, MS, 49, PhD(chem), 52. *Prof Exp:* Res assoc med chem, 68-71, DIR, MED CHEM PROJ, POMONA COL, 71-, ADJ ASST PROF, 81- *Mem:* Sigma Xi; Am Chem Soc. *Res:* Database of parameters useful in drug design, toxicological and environmental fate studies. *Mailing Add:* 311 Armsley Sq Ontario CA 91762-1606

LEO, GERHARD WILLIAM, b Frankfurt, WGer, Jan 31, 30; US citizen; m 68; c 2. GEOCHEMISTRY, MINERALOGY-PETROLOGY. *Educ:* Stanford Univ, BS, 51, PhD(geol), 61. *Prof Exp:* Actg mineral, Stanford Univ, 56; geologist, US geol surv, Menlo Park Calif, 57-59; vis lectr, Univ Bahia, Salvador, 59-61; RES GEOLOGIST, US GEOL SURV, WASHINGTON, DC, 61- *Mem:* Fel Geol Soc Am; Am Geophys Union; Am Soc Testing & Mat. *Res:* Investigations of early paleozoic; metamorphosed plutonic and volcanic rocks in the northern Appalachians. *Mailing Add:* US Geol Surv Nat Ctr 928 Reston VA 22092

LEON, ARTHUR SOL, b Brooklyn, NY, Apr 26, 31; m 56; c 3. MEDICAL RESEARCH, NUTRITION. *Educ:* Univ Fla, BS, 52; Univ Wis-Madison, MS, 54, MD, 57. *Prof Exp:* Intern, Henry Ford Hosp, Detroit, 57-58; fel internal med, Lahey Clin, Boston, 58-60; fel cardiol, Sch Med, Univ Miami & Jackson Mem Hosp, 60-61; chief gen med & cardiol, 34th Gen Hosp, US Army, France, 61-64, cardiol consult, US Armed Forces, France, 61-64, res cardiologist, Dept Cardiorespiratory Dis, Walter Reed Army Inst Res, 64-67, mem med eval team, Gemini & Apollo Projs, 66-67; dir clin pharmacol, Roche Spec Treatment Unit, Newark Beth Israel Med Ctr, 67-73; assoc prof, 73-80, PROF, LAB PHYSIOL HYG, DIV EPIDEMIOL, SCH PUB HEALTH, UNIV MINN, MINNEAPOLIS, 80-, DIR APPL PHYSIOL, NUTRIT SECT, 73- *Concurrent Pos:* Res assoc, Dept Clin Pharmacol, Hoffmann-La Roche Inc, 67-73; from instr to assoc prof, Col Med & Dent NJ, 67-73; chief med serv, 322nd Gen Hosp, US Army Reserve, Newark, 67-73; sr investr multiple coronary risk factor intervention and lipid research clinic trials, Univ Minn, Minneapolis, 73-; chief med cardiol & prof serv, 551st Army Hosp, Ft Snelling, Minn, 73-86; pres Hennepin Div, Am Heart Asn, 83-84. *Honors & Awards:* William G Anderson Award, Am Alliance & Health Phys Educ, 81. *Mem:* Am Col Cardiol; Am Col Chest Physicians; Am Physiol Asn; Am Soc Pharmacol & Exp Therapeut; Am Col Sports Med (vpres, 77-79); Am Geriat Soc; Am Asn Cardiac & Pulmonary Rehab; Am Pub Health Asn; Am Heart Asn. *Res:* Prevention of coronary heart disease by risk factor modification; metabolic and cardiovascular effects of exercise; exercise testing; effects of exercise conditioning; evaluation of new cardiovascular and lipid-lowering drugs. *Mailing Add:* Div Epidemiol 1-210 MOOS Tower 515 Delaware St SE Minneapolis MN 55455

LEON, B(ENJAMIN) J(OSEPH), b Austin, Tex, Mar 20, 32; m 54; c 4. ELECTRICAL ENGINEERING. *Educ:* Univ Tex, BS, 54; Mass Inst Technol, SM, 57, ScD, 59. *Prof Exp:* Mem staff, Lincoln Lab, Mass Inst Technol, 54-59; tech staff, Hughes Aircraft Co, 59-62; from assoc prof to prof elec eng, Purdue Univ, W Lafayette, 62-80; chmn dept, Univ Ky, Lexington, 80-84, prof elec eng, 80-88; sr staff officer, Nat Res Coun, 88-90; PROF ELEC & COMPUT ENG, UNIV SOUTHWESTERN LA, 90- *Concurrent Pos:* Ed, Trans Circuit Theory, Inst Elec & Electronics Engrs, 67-69; consult ed, Holt, Rinehart & Winston Series Elec Eng, Electronics & Systs, 67-73; Rome Air Develop Ctr fel & vis prof, Cornell Univ, 68-69; elec engr, Defense Commun Agency, 75-76; Consult, Westinghouse Telecommun, 80; vis prof, Southern Methodist Univ, 86-87. *Honors & Awards:* Centennial Medal, Inst Elec & Electronics Engrs. *Mem:* fel AAAS; fel Inst Elec & Electronic Engrs (vpres, Educ Div, 79-80). *Res:* Communications systems, circuit and system theory. *Mailing Add:* Univ Southwestern La Box 43890 Lafayette LA 70504-3890

LEON, H(ERMAN) I, b Chicago, Ill, Mar 27, 24; m 46; c 2. ENGINEERING. *Educ:* Univ Calif, BS, 47, MS, 48, PhD(eng), 55. *Prof Exp:* Systs analyst, Commun Div, Ramo-Wooldridge Corp, 55-56, assoc mgr reentry systs dept, Space Tech Labs, Inc, Div, Thompson Ramo Wooldridge, Inc, 56-62; dir systs res & technol, ITT Fed Labs, 62-63; asst dir systs anal, Apollo Spacecraft Systs Anal Prog, TRW Systs Group, 63-68, asst mgr plans, Houston Opers, 68-70, mgr electronic systs lab, Washington Opers, McLean, 70-75, asst mgr eng, 72-75, sr staff planning div, 75-78, PROJ MGR, OIL SHALE DEVELOP, TRW ENERGY SYSTS GROUP, 78- *Concurrent Pos:* Lectr, Univ Calif, Los Angeles, 50-60. *Mem:* Am Inst Aeronaut & Astronaut. *Res:* Reentry and space dynamics; nose cone vulnerability; fuzing; instrumentation techniques; military space systems; manned spacecraft systems; energy systems and policy analysis; shale oil development. *Mailing Add:* 1528 N Ivanhoe Arlington VA 22205

LEON, HENRY A, b San Francisco, Calif, Sept 25, 28; m 58; c 3. ENVIRONMENTAL PHYSIOLOGY, AEROSPACE BIOLOGY. *Educ:* Univ Calif, Berkeley, BS, 52, PhD(physiol), 60. *Prof Exp:* Nat Cancer Inst fel, Wenner-Gren Inst, Stockholm, Sweden, 60-61; Milton res fel path, Harvard Med Sch, 61-62; res scientist aerospace biol, 62-81, PAYLOAD PROJ SCIENTIST, AMES RES CTR, NASA, 81- *Concurrent Pos:* Mem staff, Mass Gen Hosp, Boston, 61-62. *Mem:* Am Physiol Soc; Aerospace Med Asn. *Res:* Effect of space cabin environments on blood elements; stress and the control of liver protein synthesis; nutrition and stress. *Mailing Add:* 2371 Richland Ave San Jose CA 95125

LEON, KENNETH ALLEN, b New York, NY, Nov 19, 37; m 63; c 2. FISH BIOLOGY. *Educ:* Ohio State Univ, BS, 60; Col William & Mary, MS, 63; Univ Wash, PhD(fisheries mgt), 70. *Prof Exp:* Biol consult, Ichthyol Assocs, 70-71; res biologist, Tunison Lab Fish Nutrit, US Bur Sport Fisheries & Wildlife, 71-74; prin biologist, Fish Rehab, Enhancement & Develop Div, Juneau, 75-88, REGIONAL BIOLOGIST, ALASKA DEPT FISH & GAME, FISHERIES REHAB ENHANCEMENT DIV, DOUGLAS, 88- *Mem:* Am Fisheries Soc. *Res:* Enhancement and rehabilitation of salmonid species, specialty salmon incubation and hatchery design. *Mailing Add:* Alaska Dept Fish & Game PO Box 20 Douglas AK 99824-0020

LEON, MELVIN, b Brooklyn, NY, Sept 2, 36; m 63; c 2. THEORETICAL PHYSICS. *Educ:* Univ Md, BS, 57; Cornell Univ, PhD(physics), 61. *Prof Exp:* Imp Chem Industs res fel & NSF fel theoret physics, Univ Birmingham, 61-63; res physicist, Carnegie Inst Technol, 63-66; asst prof physics, Rensselaer Polytech Inst, 66-72; MEM STAFF, LOS ALAMOS NAT LAB, 72- *Mem:* Fel Am Phys Soc. *Res:* Exotic atoms; muon spin rotation; muon-catalyzed fusion. *Mailing Add:* Los Alamos Nat Lab MP-DO MS-H844 Los Alamos NM 87545

LEON, MICHAEL ALLAN, b New York, NY, Nov 23, 47; m 70; c 2. EARLY LEARNING, NEUROBIOLOGY. *Educ:* Brooklyn Col, BS, 68; Univ Chicago, PhD(biopsychol), 72. *Prof Exp:* From asst prof to assoc prof psychol, McMaster Univ, 72-80; assoc prof psychobiol, 80-84, PROF PSYCHOBIOL, UNIV CALIF, IRVINE, 84- *Concurrent Pos:* Assoc ed, Develop Psychobiol, 85- *Mem:* Soc Neurosci; Int Soc Develop Psychobiol; Asn Chemoreception Sci. *Res:* Neurobiology of early learning. *Mailing Add:* Dept Psychobiol Univ Calif Irvine CA 92717

LEON, MYRON A, b Troy, NY, July 13, 26. IMMUNOLOGY. *Educ:* Columbia Univ, BS, 50, PhD(biochem), 54. *Prof Exp:* Assoc surg res, 53-64, ASSOC HEAD PATH RES, ST LUKE'S HOSP, 64-74; PROF IMMUNOL, SCH MED, WAYNE STATE UNIV, 74- *Concurrent Pos:* Fel, Univ Lund, Sweden, 58. *Mem:* Am Asn Immunol; AAAS; Am Soc Microbiol. *Res:* Immunochemistry; mechanisms of natural resistance to infection; complement; myeloma proteins; lymphocyte stimulation. *Mailing Add:* Dept Immunol Wayne State Univ Sch Med Detroit MI 48201

LEON, RAMON V, b Holguin, Oriente, Cuba, Sept 29, 48; m 79. RELIABILITY THEORY, STOCHASTIC INEQUALITIES. *Educ:* Fla State Univ, BS, 72, MS, 76, PhD(statist), 79; Tulane Univ, MS, 75. *Prof Exp:* Vis instr statist, Fla State Univ, 78-79; asst prof statist, Rutgers Univ, 79-81; MEM TECH STAFF STATIST, BELL LABS, 81- *Honors & Awards:* Ralph A Bradley Award, Fla State Univ, 79. *Mem:* Inst Math Statist; Am Statist Asn; Am Soc Quality Control; Inst Environ Sci. *Res:* Reliability theory and the mathematics and statistics associated with it; characterizations of distributions; stochastic inequalities; probability modeling of systems. *Mailing Add:* PO Box 494 Holmdel NJ 07733

LEON, ROBERT LEONARD, b Denver, Colo, Jan 18, 25; m 47; c 4. MEDICINE, PSYCHIATRY. *Educ:* Univ Colo, MD, 48. *Prof Exp:* Intern, Univ Hosp, Ann Arbor, Mich, 48-49; resident psychiat, Med Ctr, Univ Colo, 49-52; resident child psychiat, State Dept Health, Conn, 52-53; asst dir child psychiat, Greater Kansas City Ment Health Found, 53-54; prof psychiat, Southwest Med Sch, Univ Tex, 57-67; PROF PSYCHIAT & CHMN DEPT, UNIV TEX HEALTH SCI CTR SAN ANTONIO, 67- *Concurrent Pos:* Chief ment health serv, USPHS, Mo, 54-57; consult, Bur Indian Affairs, 62-67; consult regional off VI, NIMH, 57-73, mem psychiat training rev comt, 70-74; consult, Audie Murphy Mem Vet Hosp, 73- *Mem:* Fel Am Psychiat Asn; fel Am Col Psychiatrists (pres, 87-88); Am Orthopsychiat Asn; fel Am Acad Child Psychiat; AMA; fel Am Asn Soc Psychiatrists (pres 90-92). *Res:* Social psychiatry. *Mailing Add:* Dept Psychiat Univ Tex Health Sci Ctr 7703 Floyd Curl Dr San Antonio TX 78284

LEON, SHALOM A, b Sofia, Bulgaria, Apr 7, 35; m 62; c 3. BIOCHEMISTRY, RADIOBIOLOGY. *Educ:* Hebrew Univ, Jerusalem, MSc, 60, PhD(pharmacol), 64. *Prof Exp:* Jr res asst pharmacol, Med Sch, Hebrew Univ, Jerusalem, 60-64; res assoc biochem, Ind Univ, 65-67; MEM BIOSCI STAFF, ALBERT EINSTEIN MED CTR, 68-, DIR, RADIATION RES LAB, 79-; ASSOC PROF RADIOBIOL, SCH MED, TEMPLE UNIV, 79- *Mem:* Am Asn Cancer Res; AAAS; Radiation Res Soc; Am Chem Soc; NY Acad Sci; Am Asn Immunol. *Res:* Mechanism of antibiotic action; biosynthesis of nucleic acids and proteins; use of radioactive isotopes in clinical research and diagnosis; relationship between structure and biological activity of toxins from microorganisms; effect of radioprotective agents against ionizing radiation. *Mailing Add:* Radiation Res Lab Albert Einstein Med Ctr Philadelphia PA 19141

LEONARD, A(NTHONY), b June 2, 38; US citizen; m 60; c 2. ENGINEERING. *Educ:* Calif Inst Technol, BS, 59; Stanford Univ, MS, 60, PhD(nuclear eng), 63. *Prof Exp:* Mem tech staff, Rand Corp, 63-66; from asst prof to assoc prof mech eng, Stanford Univ, 66-73; res scientist, NASA Ames Res Ctr, 75-85; PROF AERONAUT ENG, CALIF INST TECHNOL, 85- *Concurrent Pos:* Lectr, Calif Inst Technol, 65-66; consult, Gen Elec Co; NASA Ames Res Ctr sr fel, 73-75. *Honors & Awards:* Edward Teller Award, 63. *Mem:* Am Phys Soc; Soc Indust & Appl Math. *Res:* Nuclear reactor theory; particle transport theory; turbulence theory; numerical fluid mechanics. *Mailing Add:* 301-46 Grad Aeronaut Labs Calif Inst Technol Pasadena CA 91125

LEONARD, ARNOLD S, b Minneapolis, Minn, Oct 26, 30; m 50; c 4. SURGERY. *Educ:* Univ Minn, Minneapolis, BA, 52, BS, 53, MD, 55, PhD(surg path), 63. *Prof Exp:* Univ fel, 56-63, from asst prof to assoc prof, 63-73, PROF SURG, UNIV MINN, MINNEAPOLIS, 73- *Mem:* Am Soc Artificial Internal Organs; Am Soc Exp Path; Int Soc Hist Med; Soc Univ Surg; Am Pediat Surg Asn; Sigma Xi. *Res:* Gastrointestinal physiology; hypothalamic stimulation and study of gastric secretion; transplantation; extracorporeal organ perfusion; pediatric surgery; computer technology. *Mailing Add:* Univ Minn Hosp Minneapolis MN 55455

LEONARD, B(ENJAMIN) F(RANKLIN), (III), b Dobbs Ferry, NY, May 12, 21; m 50; c 2. MINERALOGY & PETROLOGY. *Educ:* Hamilton Col, BS, 42; Princeton Univ, MA, 47, PhD(geol), 51. *Prof Exp:* Geol field asst, Geol Surv Nfld, 42; from jr geologist to geologist, 43-62, GEOLOGIST-IN-CHARGE, ORE MICROS LAB, US GEOL SURV, 62- *Concurrent Pos:* Vis prof, 67-68, adj prof, Colo Sch Mines, 90-; mem, Int Comn Ore Micros, 68-70, vchmn, 82-86; regional counr NAm, Int Asn Genesis of Ore Deposits, 84-89. *Honors & Awards:* Meritorious Serv Award, US Dept Interior, 88. *Mem:* Fel Mineral Soc Am; fel Geol Soc Am; fel Soc Econ Geol; Soc Geol Appl Mineral Deposits; Mineral Asn Can; Asn Explor Geochemists. *Res:* Ore deposits, especially gold, iron and tungsten; geology of central Idaho and northwest Adirondacks; ore minerals; rock-forming minerals; geochemical and biogeochemical exploration. *Mailing Add:* Cent Mineral Resources Br US Geol Survey Box 25046 Stop 905 Fed Ctr Denver CO 80225

LEONARD, BILLIE CHARLES, b Purdy, Mo, Mar 11, 34; m 54. INSTRUCTIONAL BEHAVIOR, INSTRUCTIONAL TECHNOLOGY. *Educ:* Southwest Mo State Univ, BSE, 57; Univ Mo, Columbia, MEd, 63, PhD(educ admin), 68. *Prof Exp:* Supt, Purdy R-11 Sch, 63-66; asst sch admin, Univ Mo, Columbia, 67-68, dir, Ctr Educ Improvement, 68-75; dir, Diversified Educ Serv Corp, 76-79; prin, La Middle Sch, 79-83; dir, Comput Educ Mgt Systs, 83-85; dir statewide med educ res develop, 85-87, DIR, INST RES & DEVELOP, KIRKSVILLE COL OSTEOPATHIC MED, 87-; PROF & CHAIR, DEPT EDUC LEADERSHIP, WRIGHT STATE UNIV. *Concurrent Pos:* From asst prof to prof educ, Univ Mo, Columbia, 68-76; consult, Educ Ctr, 75-76; vis prof, Sch Educ, Seattle Univ, 80, Col Teacher Educ, Northwest La State Univ, 84 & Northwest Mo State Univ, 86; prof med educ, Kirksville Col Osteopathic Med, 86; Asn Supervision & Curriculum Develop; Nat Coun Univ Res Adminrs. *Mem:* Nat Soc Study Educ; Am Asn Sch Adminrs; Am Educ Res Asn. *Res:* Comparative studies of varying instructional methodologies in the teaching of science and other biomedical cognates. *Mailing Add:* Wright State Univ 372 Millett Hall Dayton OH 45435

LEONARD, BOWEN RAYDO, JR, b Houston, Tex, Mar 7, 26; div; c 2. PHYSICS. *Educ:* Tex Western Col, BS, 47; Univ Wis, MS, 49, PhD(physics), 52. *Prof Exp:* Asst, Univ Wis, 47-51; physicist, Hanford Labs, Gen Elec Co, 52-53, sr scientist, 53-57, mgr exp physics res, 57-64; mgr exp physics res, Pac Northwest Lab, Battelle Mem Inst, 65-67, sr staff scientist, 67-82; RETIRED. *Concurrent Pos:* Mem, nuclear cross sect adv group, Atomic Energy Comn, 57-63, ad-hoc mem, 69-, mem cross sect eval working group, 60- *Mem:* Fel Am Phys Soc; Am Nuclear Soc; Sigma Xi. *Res:* Neutron cross section measurements; nuclear physics; x-ray scattering; slow neutron in-elastic scattering studies of solids and liquids. *Mailing Add:* 212 S Morain St Kennewick WA 99336

LEONARD, BRIAN PHILLIP, b Melbourne, Australia, June 4, 36; m 64; c 2. PLASMA PHYSICS, FLUID MECHANICS. *Educ:* Univ Melbourne, BMechE, 58; Cornell Univ, MAeroE, 61, PhD(aerospace eng), 65. *Prof Exp:* Asst aerospace eng, Cornell Univ, 61-64, asst elec eng, 64-65, vis asst prof, 65-66; sr lectr aeronaut eng, Royal Melbourne Inst Technol, 67; lectr appl math, Monash Univ, Australia, 67-68; Air Force Off Sci Res assoc plasma physics, Columbia Univ, 69-70; asst prof eng sci, Richmond Col NY, 70-76; assoc prof, City Univ New York, 76-82; PROF MECH ENG, UNIV AKRON, 82- *Mem:* Am Phys Soc; Am Inst Aeronaut & Astronaut; Am Nuclear Soc. *Res:* High temperature gas dynamics; shock wave structure; magnetically driven shock waves; applied mathematics; control systems; hydrodynamics and ship stability and control; thermonuclear fusion. *Mailing Add:* Dept Mech Eng Univ Akron Leigh Hall Rm 201F Akron OH 44325-3903

LEONARD, BYRON PETER, b Morgan City, La, Feb 26, 25; m 46. PHYSICS. *Educ:* Southwestern La Univ, BS, 43; Univ Tex, MA, 52, PhD, 53. *Prof Exp:* Proj engr, US Naval Ord Test Sta, 46-47; instr physics, Southwestern La Univ, 48-50 & Univ Tex, 50-53; chief nuclear res & develop, Gen Dynamics Corp, 53-59; sr staff engr, Space Technol Labs, 59-60; dir satellite-missile observation syst prog, 60-65, vpres & gen mgr, Man Orbiting Lab, Systs Eng Off, 65-68, VPRES & GEN MGR EL SEGUNDO TECH OPERS & GROUP VPRES, PROGS GROUP, AEROSPACE CORP, 68- *Mem:* Am Nuclear Soc; Am Inst Aeronaut & Astronaut; Am Chem Soc. *Res:* Nuclear shielding; radiation effects to materials and operating components; radiation hazards of fission products released to the atmosphere; design of research reactors; design and use of satellite systems, particularly for surveillance applications. *Mailing Add:* 2600 W Farewell Ave Chicago IL 60645

LEONARD, CHARLES BROWN, JR, b Woodbury, NJ, May 28, 34; m 55; c 2. BIOCHEMISTRY. *Educ:* Rutgers Univ, AB, 55; Univ Md, MS, 57, PhD(biochem), 63. *Prof Exp:* Asst, 55-58, from instr to assoc prof, 58-76, dir off admis, 75-77, asst dean recruitment & admis, 77-85, PROF BIOCHEM, DENT SCH, UNIV MD, BALTIMORE, 76-, CHMN DEPT, 85- *Concurrent Pos:* Consult, Dr H L Wollenweber, clin pathologist, 59-61. *Mem:* AAAS; Am Chem Soc; NY Acad Sci; Am Inst Chem; Am Asn Dent Sch; Sigma Xi. *Res:* Amino acid incorporation into rat liver ribosomes; effect of divalent ions on structure of rat liver RNA; effect of o,p'-DDD on cellular metabolism; metabolic products of o,p'-DDD. *Mailing Add:* Univ Md Sch Dent Baltimore MD 21201

LEONARD, CHARLES GRANT, b Detroit, Mich, Apr 7, 39; m 64; c 2. ASTRONOMY. *Educ:* Eastern Mich Univ, BS, 63; Wayne State Univ, MA, 66. *Prof Exp:* Instr physics, Wis State Univ-Whitewater, 66-68; ASST PROF ASTRON & PHYSICS, JACKSON COMMUNITY COL, 68- *Mem:* Am Asn Physics Teachers. *Mailing Add:* Dept Math & Eng Jackson Community Col 211 Emmons Rd Jackson MI 49201

LEONARD, CHESTER D, soils; deceased, see previous edition for last biography

LEONARD, CHRISTIANA MORISON, b Boston, Mass, Jan 22, 38; m 59, 82; c 2. NEUROANATOMY, PSYCHOLOGY. *Educ:* Radcliffe Col, BA, 59; Mass Inst Technol, PhD(psychol), 67. *Prof Exp:* USPHS trainee, Rockefeller Univ, 67-70, res assoc, 70-71, asst prof neuropsychol, 71-74; asst prof anat, Mt Sinai Sch Med, 74-76; assoc prof, 76-86, PROF NEUROSCI, COL MED, UNIV FLA, 86- *Mem:* AAAS; Soc Neurosci; Sigma Xi; Am Anat Asn; Animal Behav Soc. *Res:* Neurological basis of behavior. *Mailing Add:* Dept Neurosci Box J244 JHM Health Ctr Col Med Univ Fla Gainesville FL 32610

LEONARD, DAVID E, b Greenwich, Conn, Dec 28, 34; m 57; c 2. ENTOMOLOGY. *Educ:* Univ Conn, BS, 56, MS, 60, PhD(entom), 64. *Prof Exp:* Asst entomologist, Conn Agr Exp Sta, 64-69, assoc entomologist, 69-70; assoc prof, 70-76, prof entom, Univ Maine, Orono, 76-; AT DEPT ENTOM, UNIV MASS. *Concurrent Pos:* Co-ed, Annals Entom Soc Am. *Mem:* Entom Soc Am; Entom Soc Can; Ecol Soc Am; AAAS. *Res:* Biosystematics, biology and ecology of insects; host-parasite relationships. *Mailing Add:* Dept Entom Univ Mass Amherst MA 01003

LEONARD, EDWARD (FRANCIS), b Paterson, NJ, July 6, 32; m 55; c 5. CHEMICAL ENGINEERING. *Educ:* Mass Inst Technol, BS, 53; Univ Pa, MS, 55, PhD(chem eng), 60. *Prof Exp:* Res engr, Barrett Div, Allied Chem Corp, 53-55; instr chem eng, Univ Pa, 55 & 57-58; from asst prof to assoc prof, 58-67, chmn bioeng comn, 65-68, PROF CHEM ENG, COLUMBIA UNIV, 67-, DIR, ARTIFICIAL ORGANS RES LABS, 68- *Concurrent Pos:* Resident eng pract, Ford Found, 64-65; consult, Mt Sinai St Luke's Hosp, NY, Procter & Gamble Co & Baxter Labs; mem bd, Assoc Univs Inc, 71-78. *Honors & Awards:* Allan P Colburn Award, Am Inst Chem Engrs, 69. *Mem:* Fel Am Inst Chem Engrs; Am Soc Artificial Internal Organs (pres, 72); Biomed Eng Soc. *Res:* Heat, mass, momentum transport in fluid systems; distributed parameter chemical systems; transient behavior of chemical process systems; design of transport devices in medicine, particularly for immunotherapy and cell separation. *Mailing Add:* Dept Chem Eng Columbia Univ New York NY 10027

LEONARD, EDWARD CHARLES, JR, b Burlington, NC, Aug 21, 27; m 52; c 1. POLYMER CHEMISTRY. *Educ:* Univ NC, BS, 47, PhD(chem), 51; Univ Chicago, MBA, 74. *Prof Exp:* Asst, Univ NC, 47-50; sr res chemist, Res Dept, Bakelite Co, 51-56, group leader, Union Carbide Plastic Co, 56-64; res mgr, Borden Chem Co, 64-67; mgr indust chem prod lab, Res & Develop Div, Kraft, Inc, 67-73, tech dir, Humko Sheffield Chem Co Div, Kraft, Inc, 73-77, vpres res & develop, Humko Sheffield Chem Co Div, 77-80, VPRES RES & DEVELOP, HUMKO CHEM DIV, WITCO CHEM CORP, 80-, GEN MGR, VPRES & OFFICER, 83- *Concurrent Pos:* Vpres & mem Bd Dirs, Enenco, Inc, 74- *Mem:* Am Chem Soc. *Res:* Synthetic surface active agents; ionic polymerizations; graft polymers; fatty acids; homogeneous catalysis; chemical economics. *Mailing Add:* Humko Chem Div Witco Chem Corp PO Box 125 Memphis TN 38101

LEONARD, EDWARD H, b Berwick, Maine, Feb 21, 19; m 51; c 1. ANALYTICAL CHEMISTRY. *Educ:* Dartmouth Col, AB, 42; Tufts Univ, MA, 54; Univ NH, MS, 61. *Prof Exp:* Res & develop engr, Elec Res Lab, Simplex Wire & Cable Co, 42-46, Eng Dept, 46-51; head sci dept high sch, NJ, 51-60, sci coord, 60-64; ASSOC PROF PHYSICS & NATURAL SCI, WORCESTER STATE COL, 64- *Mem:* AAAS; Am Chem Soc; Am Asn Physics Teachers; Nat Sci Teachers Asn. *Res:* Design and development of apparatus and aids for the teaching of physical science. *Mailing Add:* 184 Holden St Holden MA 01520

LEONARD, EDWARD JOSEPH, b Boston, Mass, Mar 20, 26; m 56; c 3. MEDICINE. *Educ:* Harvard Med Sch, MD, 49. *Prof Exp:* Investr, Nat Heart Inst, 53-69; investr, 69-73, head tumor antigen sect, Biol Br, 73-76, HEAD IMMUNOPATH SECT, LAB IMMUNOBIOL, NAT CANCER INST, 76- *Concurrent Pos:* From instr to assoc clin prof, George Washington Univ, 57-74. *Mem:* Am Fedn Clin Res; Soc Gen Physiologists; Am Asn Immunol. *Res:* Tumor immunology. *Mailing Add:* Immunopath Sect Nat Cancer Inst Frederick MD 21702

LEONARD, ELLEN MARIE, b New York, NY, Nov 28, 44. NUCLEAR PHYSICS, PLASMA PHYSICS. *Educ:* Univ Mich, BS, 66, MS, 68, PhD(plasma physics), 73. *Prof Exp:* STAFF MEM, LOS ALAMOS NAT LAB, 73- *Mem:* Am Phys Soc; Am Nuclear Soc; AAAS; Inst Elec & ElectronicS eNGRS. *Res:* Weapon physics advanced concepts. *Mailing Add:* 102 Monte Rey Dr N Los Alamos NM 87544

LEONARD, FREDERIC ADAMS, physiology, for more information see previous edition

LEONARD, HENRY SIGGINS, JR, b Needham, Mass, Oct 12, 30; m 54; c 1. MATHEMATICS. *Educ:* Mich State Univ, BS, 52; Harvard Univ, AM, 53, PhD(math), 58. *Prof Exp:* From asst prof to assoc prof math, Carnegie Inst Technol, 58-68; asst chmn dept math sci, 75-78, PROF MATH, NORTHERN ILL UNIV, 68- *Concurrent Pos:* Prin investr, NSF grants, 59-70; vis assoc prof, Univ Ill, Urbana, 67-68; vis fel, Yale Univ, 73-74; vis scholar, Univ Chicago, 80-81; vis, Univ Manchester, Eng, 87-88. *Mem:* Am Math Soc; Math Asn Am. *Res:* Theory of groups of finite order. *Mailing Add:* Dept Math Northern Ill Univ DeKalb IL 60115-2888

LEONARD, JACK E, b Chickasha, Okla, Feb 6, 43; m 65; c 3. CHEMISTRY. *Educ:* Harvard Univ, AB, 65; Southern Methodist Univ, BD, 67; Calif Inst Technol, PhD(chem & biol), 71. *Prof Exp:* Asst prof chem, State Univ NY, 61-75 & Tex A&M Univ, 75-81, assoc res scientist, 82-83; sr environ scientist, Indianapolis Ctr Advan Res, Inc, 85-90; PRES, ENVIRON MGT INST, 90- *Concurrent Pos:* Sr res chemist, Allied Corp, 80; vis assoc prof chem, Univ Tex, El Paso, 81-82; fac mem, Blinn Col, 83-85. *Mem:* Am Chem Soc; AAAS. *Res:* Physical organic chemistry from mechanisms of photochemical and electrochemical reactions to laser synthesis of catalysts to mathematical group and graph theory; environmental chemical policy. *Mailing Add:* Environ Mgt Inst 5610 Crawfordsville Rd Suite 15 Indianapolis IN 46224

LEONARD, JACQUES WALTER, b Montreal, Que, Aug 7, 36; m 63; c 2. POLYMER CHEMISTRY, PHYSICAL CHEMISTRY. *Educ:* Univ Montreal, BSc, 60, MSc, 61, PhD(chem), 64. *Prof Exp:* Can Nat Res Coun fel, Univ Leeds, 64-66; from asst prof to assoc prof chem, Laval Univ, 66-75, dept dir, 78-81, vdean res, Fac Sci & Eng, 87-89, PROF CHEM, LAVAL UNIV, 75- *Concurrent Pos:* Vis prof, Univ Sussex, Eng, 77-78; vis fel, Inst Charles Sadron, Nat Ctr Sci Res, Strasbourg, France, 90, Univ De Bordeaux, France, 91. *Mem:* Fel Chem Inst Can; Am Chem Soc. *Res:* Kinetics and thermodynamics of polymerizations in solution; effect of the medium on the equilibrium of reversible cyclizations, homo- and copolymerizations; thermodynamics of polymer solutions and binary liquid mixtures. *Mailing Add:* Dept Chem Laval Univ Quebec PQ G1K 7P4 Can

LEONARD, JAMES JOSEPH, b Schenectady, NY, June 17, 24; m 54; c 4. INTERNAL MEDICINE. *Educ:* Georgetown Univ, MD, 50. *Prof Exp:* From intern to jr asst resident med, Georgetown Univ Hosp, 50-52; asst resident med serv, Boston City Hosp, Mass, 52-53; resident, Pulmonary Dis Div, DC Gen Hosp, 54-55; instr, Sch Med, Georgetown Univ, 55-56; instr, Med Sch, Duke Univ, 56-57; asst prof & dir, Div Cardiol, Georgetown Univ Serv, DC Gen Hosp, 57-59; from asst prof to assoc prof med, Univ Tex Med Br, 59-62; dir cardiopulmonary lab, 61-62; assoc prof med & dir cardiac diag lab, Ohio State Univ, 62-63; assoc prof med & dir div cardiol, Sch Med, Univ Pittsburgh, 63-67, actg chmn dept med, 70-71, prof med, 67-77, chmn dept, 71-77; PROF MED & CHMN DEPT, UNIV HEALTH SCI, 77- *Concurrent Pos:* Washington Heart Asn fel cardiol, Georgetown Univ Hosp, 53-54; Am Trudeau Soc fel, Pulmonary Dis Div, DC Gen Hosp, 54-55; NIH cardiac trainee, Duke Univ Hosp, 56-57; med officer, DC Gen Hosp, 55-56, chief cent heart sta, 57-59; attend cardiol, Mt Alto's Vet Hosp, DC, 57-59. *Mem:* Asn Am Physicians; Asn Prof Med; Asn Univ Cardiologists; Sigma Xi; Am Col Physicians. *Res:* Cardiopulmonary physiology. *Mailing Add:* Dept Med Uniformed Serv Univ 4301 Jones Bridge Rd Bethesda MD 20814

LEONARD, JANET LOUISE, b Ames, Iowa, Feb 24, 53. NEUROETHOLOGY, INVERTEBRATE ZOOLOGY. *Educ:* Univ Wis, Madison, BS, 73, PhD(zool), 80. *Prof Exp:* Asst prof, Univ Maine, Orono, 80-81; fel med physiol, Univ Calgary, Can, 81-85; fel neurosci, Univ Calif, San Diego, 85; ASST PROF ZOOL, UNIV OKLA, 86- *Concurrent Pos:* Vis asst prof, Hatfield Marine Sci Ctr, Ore State Univ, Newport, 89- *Mem:* Animal Behav Soc; Am Soc Zoologists; Int Soc Neuroethologists; Soc Neurosci. *Res:* Ethology and neuroethology of invertebrates; behavioral organization; mating systems; neuroethology of opisthcbranchs; coelenterate behavior. *Mailing Add:* Dept Zool Univ Okla Norman OK 73019

LEONARD, JOHN ALEX, b Swindon, Eng, Dec 13, 37; m 61; c 2. INDUSTRIAL CHEMISTRY. *Educ:* Univ London, BSc, 59, PhD(chem), 62. *Prof Exp:* Res chemist polymers, Shell Develop Co, Calif, 63-66; from sr scientist catalysis to bus planning, Imperial Chem Indust, UK & USA, 66-74; res adv, C-I-L, Inc, 74-77, technol & agreement mgr, 77-84, Ventures Mgr, 84-88, CHEM BUS MGR, ICI CAN, 88- *Concurrent Pos:* Fel, Harvard Univ, 62-63. *Mem:* Am Chem Soc; Chem Soc Can; Royal Soc Chem. *Res:* Catalytic, electrochemical and biological processes and research management. *Mailing Add:* ICI Can 90 Sheppard Ave E PO Box 200 Sta A North York ON M2N 6H2 Can

LEONARD, JOHN EDWARD, b Great Falls, Mont, Apr 18, 18; div; c 2. ORGANIC CHEMISTRY, PHYSICAL CHEMISTRY. *Educ:* Antioch Col, BS, 42; Ohio State Univ, PhD(chem), 49. *Prof Exp:* Res engr, Battelle Mem Inst, 42-46; res fel, Calif Inst Technol, 49-52; res scientist, Beckman Instruments, Inc, 52-56, chief proj engr, 56-62, sr scientist, 62-66, mgr appl res, Med Develop Activ, 66-69; chief scientist, Int Biophys Corp, 69-71; consult electrochem sensors & instrumentation, 71-78; RES DIR, BROADLEY-JAMES CORP, SANTA ANA, 78- *Mem:* Am Chem Soc; AAAS; NY Acad Sci. *Res:* Analytical instruments, particularly electrochemical, for chemical research and industrial use; biomedical engineering; medical instrumentation research. *Mailing Add:* PO Box 278 Freeland WA 98249

LEONARD, JOHN JOSEPH, b Philadelphia, Pa, Feb 12, 49; m 72; c 2. PHYSICAL ORGANIC CHEMISTRY. *Educ:* Drexel Univ, BS, 72, PhD(phys org chem), 72. *Prof Exp:* Res assoc chem, Univ Pa, 72-73; sr res chemist, Arco Chem Co, 73-78, supvr catalyst res, 78-79, mgr catalyst res, 79-85, mgr res & develop, 85-89, MGR CORP PLANNING, ARCO CHEM CO, DIV ATLANTIC RICHFIELD CO, 89- *Concurrent Pos:* Adj prof math, Drexel Univ Evening Div, 73-80. *Mem:* Am Chem Soc; Int Catalysis Soc; AAAS. *Res:* Kinetics and mechanisms of organic reactions especially catalysis of organic oxidation reactions(heterogeneous and homogeneous catalysis); spectroscopy of organic molecules. *Mailing Add:* 37 S Hillcrest Springfield PA 19064-2413

LEONARD, JOHN LANDER, b Jamaica, NY, Oct 20, 35; m 65; c 1. MATHEMATICS. *Educ:* Carnegie Inst Technol, BS, 57; Univ Calif, Santa Barbara, MA, 63, PhD(math), 66. *Prof Exp:* Opers analyst, Comput Dept, Gen Elec Co, 59-60, mem tech staff, Tech Mil Planning Oper, 60-61; asst math, Univ Calif, Santa Barbara, 61-63, 64-66; asst prof math, 66-76, LECTR MATH, UNIV ARIZ, 76- *Honors & Awards:* Fulbright Lectr, Peru, 73, Intercountry Fulbright Lectr, Columbia, 73. *Mem:* Math Asn Am; Sigma Xi. *Res:* Graph theory, extremal problems, connectivity; real function theory; mathematical analysis. *Mailing Add:* Dept Math Univ Ariz Tucson AZ 85721

LEONARD, JOHN W, SCIENCE ADMINISTRATION. *Prof Exp:* SR EXEC CONSULT, MORRISON-KNUDSEN CO, INC. *Mem:* Nat Acad Eng. *Mailing Add:* Morrison-Knudsen Co Inc One Morrison-Knudsen Plaza, PO Box 7808 Boise ID 83729

LEONARD, JOSEPH THOMAS, b Scranton, Pa, Aug 8, 32; m 58; c 4. FUEL SCIENCE. *Educ:* Univ Scranton, BS, 54; Pa State Univ, University Park, PhD(fuel technol), 59. *Prof Exp:* Res asst chem, Pa State Univ, University Park, 54-59; RES CHEMIST FUELS, NAVAL RES LAB, WASHINGTON, DC, 59- *Mem:* Am Chem Soc. *Res:* Electrostatic charging of hydrocarbon liquids and fuels; suppression of evaporation of hydrocarbons and smoke abatement techniques. *Mailing Add:* Naval Res Lab Code 6180 Washington DC 20390

LEONARD, JOSEPH WILLIAM, b Pottsville, Pa, Dec 24, 30; m 52; c 4. MINING ENGINEERING. *Educ:* Pa State Univ, BS, 52, MS, 58. *Prof Exp:* Asst to div supt coal mining, Philadelphia & Reading Coal & Iron, Pottsville, Pa, 52-54; asst pre engr, United Elec Coal Co, Chicago, Ill, 54-56; res asst coal prep, Pa State Univ, 56-58; res engr coal mining, US Steel Corp, Monroeville, Pa, 58-61; dir bur, 61-81, dean, 78-81, PROF MINING, WVA UNIV, 74-, WILLIAM N POUNDSTONE RES PROF, 81- *Concurrent Pos:* Consult, Pa Elec Co, Johnstown, 71-, Cortix, Bochum, WGer, 78- *Honors & Awards:* Howard N Eavenson Award, Am Inst Mining Engrs, 69. *Mem:* Fel Am Inst Chemists; Am Inst Mining Engrs; Am Mining Cong; Sigma Xi. *Res:* Mining; coal reserve analysis; coal preparation including design; coal utilization. *Mailing Add:* 3195 Burnham Ct Lexington KY 40503

LEONARD, KATHLEEN MARY, b Grand Rapids, Mich, Aug 14, 54. GROUND WATER STUDIES, FATE OF CONTAMINANTS. *Educ:* Univ Wis-Milwaukee, BS, 83, MS, 85; Univ Ala, Huntsville, PhD(environ eng), 90. *Prof Exp:* VPRES, OPTECHNOL, INC, 89-; ASST PROF CIVIL ENG, UNIV ALA, HUNTSVILLE, 91- *Mem:* Am Soc Civil Engrs; Soc Women Engrs; Water Pollution Control Fedn. *Res:* Using optical fibers for remote chemical sensing of environmental systems, specializing in ground water applications. *Mailing Add:* Optechnol Inc 287 Blooming Acres Lane Gurley AL 35748

LEONARD, KURT JOHN, b Holstein, Iowa, Dec 6, 39; m 61; c 3. PLANT DISEASE EPIDEMIOLOGY, PLANT HOST-PARASITE GENETICS. *Educ:* Iowa State Univ, BS, 62; Cornell Univ, PhD(plant path), 68. *Prof Exp:* Res plant scientist, Agr Res Serv-USDA, NC State Univ, 68-88; DIR, CEREAL RUST LAB, AGR RES SERV-USDA, UNIV MINN, 88- *Concurrent Pos:* Mem coun, Am Pythopathology Soc, 82-85; ccounr, Int Soc Plant Path, 83- *Mem:* Fel Am Phytopath Soc; Int Soc Plant Path; Mycol Soc Am; Brit Soc Plant Path. *Res:* Epidemiology and genetics of cereal rust diseases; population genetics of host-parasite interactions in plant diseases. *Mailing Add:* USDA-Agr Res Serv Cereal Rust Lab Univ Minn St Paul MN 55108

LEONARD, LAURENCE, b New York, NY, Jan 9, 32; m 58; c 4. PHYSICAL METALLURGY, MATERIALS SCIENCE. *Educ:* Mass Inst Technol, SB, 54, SM, 56, ScD(metall), 62. *Prof Exp:* Asst prof metall, Case Western Reserve Univ, 62-69; group supvr, SKF Industs, Inc, 69-71; PRIN SCIENTIST, FRANKLIN RES CTR, 71- *Concurrent Pos:* Adj assoc prof, Drexel Univ, 78-; adj assoc prof, Great Valley Grad Ctr, Pa State Univ, 90- *Mem:* Am Soc Metals. *Res:* Materials failure analysis; physical metallurgy of rolling contact bearings; scanning electron microscopy; metal embrittlement; x-ray diffraction; residual stresses; phase transformations; heat treatment; nondestructive testing; wear monitoring by oil analysis. *Mailing Add:* Franklin Res Ctr 2600 Monroe Blvd Norristown PA 19403

LEONARD, MARTHA FRANCES, b New Brunswick, NJ, May 10, 16. PEDIATRICS. *Educ:* NJ Col Women, BSc, 36; Johns Hopkins Univ, MD, 40. *Hon Degrees:* MS, Yale Univ, 79. *Prof Exp:* Intern, Baltimore City Hosp, 40-41; asst resident med, Vanderbilt Univ Hosp, 42-43; asst resident pediat, New York Hosp, 43-46; pvt pract, 46-60; fel, Child Study Ctr, Yale Univ, 60-62, from instr to prof pediat, 79-86; RETIRED. *Honors & Awards:* Winslow Award, 88. *Mem:* Am Acad Pediat; Asn Ambulatory Pediat; World Assoc Infant Psychiatry & Allied Disciplines. *Res:* Normal and deviant child development; effects of deprivation; failure to thrive; child abuse; developmental impact of conditions such as genetic, metabolic and endocrine disorders. *Mailing Add:* Child Study Ctr Yale Univ New Haven CT 06510

LEONARD, NELSON JORDAN, b Newark, NJ, Sept 1, 16; wid; c 4. ORGANIC BIOCHEMISTRY. *Educ:* Lehigh Univ, BS, 37; Univ Oxford, BSc, 40, DSc, 83; Columbia Univ, PhD(org chem), 42. *Hon Degrees:* ScD, Lehigh Univ, 63; Dr, Adam Michiewicz Univ, 80; DSc, Univ Ill, 88. *Prof Exp:* Sci consult & spec investr, Field Intel Agency Tech, US Army Dept & US Dept Com, Europ Theatre, 45-46; fel & res asst chem, Univ Ill, Urbana, 42-43, instr, 43-44, assoc, 45-47, from asst prof to prof chem, 47-86, head, Div Org Chem, 54-63, prof chem & biochem, 73-86, Reynold C Fuson prof chem & mem, Ctr Advan Study, 81-86, EMER R C FUSON PROF, UNIV ILL, 86- *Concurrent Pos:* Mem, Comt Med Res, 44-46; ed, Org Syntheses, 51-58, ed-in-chief, 56, pres, bd dirs, 80-88; Am-Swiss Found lectr, 53, 70; Guggenheim Mem Found fel, 59, 67; mem prog comt basic phys sci, Alfred P Sloan Found, 61-66; Stieglitz lectr, 62; mem educ adv bd & bd of selection, John Simon Guggenheim Mem Found, 69-88; Edgar Fahs Smith Mem lectr, Univ Pa, 75; Arapahoe lectr, Univ Colo, 79; Calbiochem-Beohring lectr, Univ Calif, San Diego, 81. *Honors & Awards:* Synthesis Award, Am Chem Soc, 63, Edgar Fahs Smith Award, 75, Roger Adam Award, 81; Synethetic Org Chem Mfrs Award, 70. *Mem:* Nat Acad Sci; Am Acad Arts & Sci; Am Chem Soc; Royal Soc Chem; Swiss Chem Soc; Ger Chem Soc; Am Soc Biol Chemists; AAAS; Am-Can Soc Plant Physiol; foreign mem Polish Acad Sci; hon mem Pharmaceut Soc Japan. *Res:* Structure, synthesis and biological activity of cytokinins; modification of nucleic acid bases; fluorescent probes of coenzyme, enzyme binding and nucleic acid structures; intramolecular interactions. *Mailing Add:* Dept Chem & Biochem Univ Ill 1209 W Calif St Urbana IL 61801-3731

LEONARD, RALPH AVERY, b Louisburg, NC, Mar 2, 37; m 58; c 3. SOIL CHEMISTRY. *Educ:* NC State Univ, BS, 59, PhD(soil chem), 66; Purdue Univ, MS, 62. *Prof Exp:* Instr soil sci, NC State Univ, 62-66; RES SOIL SCIENTIST, USDA, 66- *Mem:* Am Chem Soc; Soil Sci Soc Am; Am Soc Agron; Sigma Xi. *Res:* Physical chemistry of soils; fate of pesticides in soil and water; soil chemical aspects of waste disposal and utilization on the land. *Mailing Add:* 224 Franklin Ave River Forest IL 60305

LEONARD, REID HAYWARD, b Littleton, NH, Aug 28, 18; m 46; c 3. CHEMISTRY. *Educ:* Univ Vt, BS, 40; Univ WVa, MS, 42; Univ Wis, PhD(biochem), 47. *Prof Exp:* Asst, Exp Sta, Univ WVa, 40-42; asst, Univ Wis & Forest Prod Lab, US Forest Serv, 43-45; res chemist, Salvo Chem Corp, Wis, 46-47; res chemist, Newport Industs, 47-56; consult biochemist, 56-; RETIRED. *Mem:* Am Chem Soc. *Res:* Chemistry of wood; sugars from wood; lignin; levulinic acid; kidney stones; blood lipids; gas chromatography. *Mailing Add:* 537 Brent Lane Pensacola FL 32503

LEONARD, ROBERT F, b Oceanside, NY, Aug 21, 34; m 64; c 3. PHOTOLITHOGRAPHY, MICROLITHOGRAPHY. *Educ:* Hofstra Univ, BA, 66; Worchester Polytech Inst, MS, 81. *Prof Exp:* Dir res, Litho Chem & Supply Co, 68-69; mgr printing prod, Rogers Corp, 69-76; res mgr printing prod, 76-79 & photopolymer applns, 79-84, DIR RES, PROD DEVELOP, PHILIP A HUNT CHEM CORP, 84-; DIR RES, PROD DEVELOP, OCG MICROELECTRONICS MAT, INC, 90- *Mem:* Am Chem Soc; Am Inst Chemists; Am Soc Testing Mat; Electrochem Soc; Soc Photographer Scientists & Engrs; Soc Photo-Optical Instrumentation Engrs. *Res:* Photolithographic chemicals and processes; microelectronics. *Mailing Add:* 24 Lens Ave Dayville CT 06241

LEONARD, ROBERT GRESHAM, b Roanoke, Va, Jan 27, 37; m 60; c 2. MECHANICAL ENGINEERING, CONTROL ENGINEERING. *Educ:* Va Polytech Inst, BS, 60, MS, 65; Pa State Univ, PhD(mech eng), 70. *Prof Exp:* Eng trainee, Gen Elec Co, Va, 56-59; instr mech eng, Va Polytech Inst, 60-65 & Pa State Univ, 65-70; from asst prof to assoc prof mech eng, Purdue Univ, 70-78; PROF MECH ENG & ASST DEPT HEAD, VA POLYTECH INST & STATE UNIV, 78- *Concurrent Pos:* Lectr, Purdue Univ, 67, asst dir, Ray W Herrick Labs, 76-78. *Honors & Awards:* Homer Addams Award, Am Soc Heating Refrig & Air-Conditioning Engrs, 75. *Mem:* Am Soc Mech Engrs. *Res:* Automatic controls; dynamic systems modeling; simulation; fluid power systems; parameter identification. *Mailing Add:* Tech-Develop Voice Control Systs 2620 Teakwood Plano TX 75075

LEONARD, ROBERT STUART, b Berkeley, Calif, Jan 20, 30; m 56; c 2. GEOPHYSICS, AERONOMY. *Educ:* Univ Nev, BS, 52, MS, 53; Univ Alaska, PhD(geophys), 61. *Prof Exp:* Res asst auroral studies, Geophys Inst, Univ Alaska, 53-58, instr, 58-60; radio physicist, SRI Int, 61-62, sr ionospheric physicist, 62-69, prog mgr, 69-72, asst dir, 72-77, dir, radio physics lab, 77-87; CONSULT, 87- *Mem:* Int Union Radio Sci; Am Geophys Union; Am Phys Soc. *Res:* Chemical seeding in the ionosphere; transionospheric propagation; ionospheric disturbances. *Mailing Add:* PO Box 450 Lakehead CA 96051

LEONARD, ROBERT THOMAS, b Providence, RI, Dec 18, 43; div; c 1. PLANT PHYSIOLOGY. *Educ:* Univ RI, BS, 65, MS, 67; Univ Ill, Urbana, PhD(biol), 71. *Prof Exp:* Fel plant physiol, Univ Ill & Purdue Univ, 71-73; from asst prof to prof plant physiol, Univ Calif, Riverside, 73-82, vchmn dept, 78-82, ASSOC DEAN, GRAD DIV & RES DEVELOP, 85- *Mem:* Am Soc Plant Physiologists; Am Inst Biol Sci; AAAS. *Res:* Physiology and biochemistry of ion transport in plants. *Mailing Add:* Dept Bot & Plant Sci Univ Calif Riverside CA 92521

LEONARD, ROY J, b Central Square, NY, Aug 17, 29; c 2. GEOTECHNICAL ENGINEERING. *Educ:* Clarkson Col Technol, BSCE, 52; Univ Conn, MS, 54; Iowa State Univ, PhD(civil eng), 58. *Prof Exp:* Asst prof civil eng, Univ Del, 57-59; from asst prof to assoc prof, Lehigh Univ, 59-66; PROF CIVIL ENG, UNIV KANS, 66- *Concurrent Pos:* NSF res grants, 59 & 62, sci fac fel, 63-65; pres, Alpha-Omega Geotech, Inc, Kansas City, Kans. *Mem:* Asn Eng Geologists; Soc Mining Engrs; Am Soc Testing Mats; Int Soc Found Eng & Soil Mechanics; fel Am Soc Civil Engrs. *Res:* Applied soil and rock mechanics; foundation engineering; tunnels; earth dams and conduits. *Mailing Add:* Dept Civil Eng Univ Kans 2006 Engineering Lawrence KS 66045

LEONARD, SAMUEL LEESON, zoology, for more information see previous edition

LEONARD, STANLEY LEE, b Oakland, Calif, Aug 27, 26; wid; c 4. PLASMA PHYSICS. *Educ:* Principia Col, BS, 47; Univ Calif, PhD(physics), 53. *Prof Exp:* Physicist, Radiation Lab, Univ Calif, 52-53; instr physics, Principia Col, 53-55, asst prof, 55-56; mem tech staff, Ramo-Wooldridge Corp, 56-59 & Space Technol Labs, Inc, 59-60; mem tech staff, 60-64, head, plasma radiation dept, Plasma Res Lab, 64-73, head, chem physics dept, Chem & Physics Lab, 73-74, DIR PHOTOVOLTAIC SYSTS, ENERGY SYSTS DIRECTORATE, AEROSPACE CORP, 74- *Mem:* Am Phys Soc; Int Solar Energy Soc. *Res:* Analysis of terrestrial photovoltaic applications. *Mailing Add:* 2617 Via Carrillo Palos Verdes Estates CA 90274

LEONARD, THOMAS JOSEPH, b Watertown, Mass, July 27, 37; m 65; c 1. DEVELOPMENTAL GENETICS. *Educ:* Clark Univ, AB, 62; Ind Univ, PhD(microbiol), 67. *Prof Exp:* NIH fel, Harvard Univ, 67-68; assoc prof mycol, Univ Ky, 68-74; PROF BOT & GENETICS, UNIV WIS-MADISON, 74- *Mem:* AAAS; Genetics Soc Am; Mycol Soc Am; Brit Mycol Soc. *Res:* Physiology and genetics of fungi as applied to development; genetics and physiological aspects of cell differentiation. *Mailing Add:* Dept Bot Univ Wis 132 Birge Hall 430 Lincoln Dr Madison WI 53706

LEONARD, WALTER RAYMOND, b Scott Co, Va, July 5, 23; m 51; c 2. ZOOLOGY, PHYSIOLOGY. *Educ:* Tusculum Col, BA, 46; Vanderbilt Univ, MS, 47, PhD(zool), 49. *Prof Exp:* Asst prof biol, 49-50, assoc prof & acting chmn dept, 50-53, REEVES PROF BIOL & CHMN DEPT, WOFFORD COL, 54- *Res:* Respiratory metabolism of Allomyces arbuscula; effects of activity on growth in hydra. *Mailing Add:* Dept Biol Wofford Col Spartanburg SC 29301

LEONARD, WARREN J, b Washington, DC, Feb 28, 52; m; c 2. CELL BIOLOGY. *Educ:* Princeton Univ, AB, 73; Stanford Univ, MD, 77; Am Bd Internal Med, dipl, 80; Am Bd Allergy & Immunol, dipl, 83. *Prof Exp:* Residency internal med, Barnes Hosp, St Louis, 78-80; res assoc, Sch Med, Wash Univ, 80-81; sr staff fel, Metab Br, Nat Cancer Inst, NIH, 81-85, sr staff fel, Cell Biol & Metab Br, Nat Inst Child Health & Human Develop, 85-87, med officer res, Cell Biol & Metab Br, 87-91, CHIEF, SECT PULMONARY & MOLECULAR IMMUNOL, OFF DIR, INTRAMURAL RES PROG, NAT HEART, LUNG & BLOOD INST, NIH, BETHESDA, 91- *Mem:* Sigma Xi; Am Asn Immunologists; Am Soc Clin Invest. *Res:* Cell biology; two patents; numerous publications. *Mailing Add:* Cell Biol & Metab Br NICHD NIH 9000 Rockville Pike Bldg 18T Rm 101 Bethesda MD 20892

LEONARD, WILLIAM F, b Hampton, Va, Jan 18, 38; m 58, 75; c 5. ELECTRICAL ENGINEERING. *Educ:* Univ Va, BSEE, 60, MSEE, 63, ScD(elec eng), 66. *Prof Exp:* Aerospace technologist, NASA-Langley Res Ctr, Va, 60-66; from asst prof to prof elec eng, Southern Methodist Univ, 66-84, assoc dean & dir, Grad Div, 79-81 & 82-84, dean ad interim, 81-82; DIR, CORP RES & ENG, ROCKWELL INT, 84- *Concurrent Pos:* Consult, WTex Eng, 57-59, Nuclear Systs Inc, 70-71, Marlow Indust, 77- & Varo Semiconductor, 77-78. *Mem:* Sigma Xi; Inst Elec & Electronics Engrs; Am Soc Eng Educ. *Res:* Physical electronics; bulk and surface electronic transport studies in solids; fabrication of III-V compound and alloy semiconductors; infrared photodetectors. *Mailing Add:* 9411 Shady Valley Dr Dallas TX 75238

LEONARD, WILLIAM J, JR, b Ravenna, Ohio, Apr 28, 36. PHYSICAL CHEMISTRY. *Educ:* Kent State Univ, BS, 58; Purdue Univ, PhD(chem), 63. *Prof Exp:* Fel phys polymer chem, Stanford Univ, 63-65; chemist, Shell Develop Co, 65-72; dir polymer sci, 72-78, DIR TECH EVAL, DYNAPOL, INC, 78- *Mem:* AAAS; Am Chem Soc; NY Acad Sci; Inst Food Technologists. *Res:* Protein conformation; polymer chain statistics; solution thermodynamics; optical rotatory dispersion; liquid crystals. *Mailing Add:* 18 Carson St San Francisco CA 94114

LEONARD, WILLIAM WILSON, b Portland, Maine, May 1, 34; m 61; c 2. MATHEMATICS. *Educ:* Univ Tampa, BS, 60; Univ SC, MS, 63, PhD(math), 65. *Prof Exp:* Asst prof math, Susquehanna Univ, 64-65; from asst prof to assoc prof, 65-74, PROF MATH, GA STATE UNIV, 74-, MEM, URBAN LIFE FAC, 77- *Mem:* Am Math Soc; Math Asn Am; Math Soc France. *Res:* Module theory; homological algebra. *Mailing Add:* Box 3770 RFD 1 Poland Spring ME 04274

LEONARDS, G(ERALD) A(LLEN), b Montreal, Que, Apr 29, 21; m 45; c 2. CIVIL ENGINEERING. *Educ:* McGill Univ, BSCE, 43; Purdue Univ, MSCE, 48, PhD(soil mech), 52. *Hon Degrees:* DSc, McGill Univ, Montreal, Can. *Prof Exp:* Lectr mech, McGill Univ, 43-46; from instr to prof soil mech, 46-64, head sch civil eng, 64-68, PROF CIVIL ENG, PURDUE UNIV, 68- *Concurrent Pos:* Mem adv bd, joint hwy res proj, Ind State Hwy Dept, 58-72, dir, 64-68; res award, Hwy Res Bd, Nat Acad Sci-Nat Res Coun, 65. *Honors & Awards:* Norman Medal, Am Soc Civil Engrs, 65, Karl Terzaghi Award, 89. *Mem:* Nat Acad Eng, 88; Fel Am Soc Civil Engrs; Int Soc Soil Mech & Found Engrs; hon mem Geotech Soc Columbia SAm. *Res:* Engineering properties of soils; foundation engineering; geotechnical aspects of earthquake engineering; soil-structure interactions; earth dams. *Mailing Add:* Sch Civil Eng Purdue Univ West Lafayette IN 47907

LEONBERGER, FREDERICK JOHN, b Washington, DC, Sept 25, 47; m 70; c 2. INTEGRATED OPTICS, FIBER OPTICS. *Educ:* Univ Mich, BSE, 69; Mass Inst Technol, MS, 71, PhD(elec eng), 75. *Prof Exp:* Staff mem, Lincoln Lab, Mass Inst Technol, 75-81, group leader, 81-84; mgr photonics, United Technol Res Ctr, 84-90, GEN MGR, UNITED TECHNOL PHOTONICS, 91- *Mem:* Fel Optical Soc Am; fel Inst Elec & Electronics Engrs; Inst Elec & Electronics Engrs Laser & Electrooptics Soc (pres, 88). *Res:* Photonic device research, with emphasis on integrated optics, fiber optics and optoelectronic devices and their applications to communication, sensing and signal processing. *Mailing Add:* United Technol Photonics Silver Lane East Hartford CT 06108

LEONE, CHARLES ABNER, b Camden, NJ, July 13, 18; m 41; c 3. IMMUNOLOGY, RADIATION BIOLOGY. *Educ:* Rutgers Univ, BS, 40, MS, 42, PhD, 49. *Prof Exp:* Asst zool, Rutgers Univ, 40-42, instr, 46-49; from asst prof to prof, Univ Kans, 49-68; prof biol & dean grad sch, Bowling Green State Univ, 68-71, vprovost res & grad studies, 71-75; vpres & vprovost, 75-79, vpres res & grad studies, 79-81, PROF ZOOL, UNIV ARK, FAYETTEVILLE, 81- *Concurrent Pos:* Resident res assoc, Argonne Nat Lab, 55, consult, 59-62; adj prof, Med Col Ohio, Toledo, 69-75. *Mem:* Fel AAAS; Am Asn Immunologists; Sigma Xi. *Res:* Immunochemistry; radiation biophysics; comparative serology among arthropods, mollusks and mammals. *Mailing Add:* Dept Zool Univ Ark Fayetteville AR 72701

LEONE, FRED CHARLES, b New York, NY, Aug 3, 22; m 45; c 7. STATISTICS. *Educ:* Manhattan Col, BA, 41; Georgetown Univ, MS, 43; Purdue Univ, PhD(math statist, educ), 49. *Prof Exp:* Instr, Georgetown Univ, 42-43; instr, Purdue Univ, 43-44 & 46-49; from instr to prof math, Case Western Reserve Univ, 49-66, dir statist lab, 51-65, actg chmn dept, 63-65; prof statist & indust eng, Univ Iowa, 66-73; exec dir & secy-treas, 73-83, EXEC DIR & SECY, AM STATIST ASN, 84- *Concurrent Pos:* Fulbright prof, Univ Sao Paulo, Brazil, 68-69; ed, Technometrics, 63-68; NAm ed, Statist Theory & Methods Abstracts, 69-73. *Mem:* Fel AAAS; fel Am Soc Qual Control; fel Am Statist Asn; Sigma Xi; Math Asn Am. *Res:* Experimental design and statistics applied to engineering; order statistics, especially in analysis of variance. *Mailing Add:* 201 E Wayne Ave Silver Spring MD 20901

LEONE, IDA ALBA, b Elizabeth, NJ, Apr 28, 22. POLLUTION BIOLOGY. *Educ:* Rutgers Univ, BS, 44, MS, 46. *Prof Exp:* Asst plant path, Col Agr, 46-50, res assoc, 50-58, asst res specialist, 58-70, assoc res prof, 70-76, prof plant biol, 76-87, PROF II DEPT PLANT PATH, COOK COL, RUTGERS UNIV, 87- *Concurrent Pos:* Consult, NY State Environ Protection Bur, 75-76, US Dept Interior, NY State Dept Transp, Pa Power & Light Co, Niagara Mohawk Power Co, Rohm & Haas, Cambridge Mass Landfill Revegetation Comn, Cabot Corp; lectr, univ & inst, India, 77, China, 85; dir, NJ Jr Acad Sci. *Mem:* Sigma Xi; Am Phytopath Soc; Am Soc Plant Physiologists; Air Pollution Control Asn; NY Acad Sci; Indian Soc Air Pollution Control. *Res:* Effect of air pollution; nutritional, physiological and environmental factors on plant growth; plants as sources of air pollution; undergraduate and graduate courses in air pollution effects; effect of cooling-tower or de-icing salt spray on crops; phytotoxicity of anaerobic landfill gases; role of mycorrhizae in adapting woody species to landfill conditions. *Mailing Add:* 876 Rayhon Terr Rahway NJ 07065

LEONE, JAMES A, b Braddock, Pa, Dec 11, 37; m 61; c 1. PHYSICAL CHEMISTRY, INSTRUMENTATION. *Educ:* Univ Cincinnati, BS, 61; Johns Hopkins Univ, MA, 63, PhD(phys chem), 65. *Prof Exp:* Res assoc, Univ Notre Dame, 65-67; from asst prof to assoc prof phys chem, 67-74, dir med technol, 74-77, ASSOC PROF CHEM & COMPUT SCI, CANISIUS COL, 77- *Concurrent Pos:* Vis assoc prof, Va Polytech Inst & State Univ, 75-76. *Mem:* Am Chem Soc; Sigma Xi; Soc Appl Spectros. *Res:* Radiation chemistry; ESR; on-line minicomputers; minicomputer and microprocessor interfacing; minicomputers and microprocessors in instrumentation automation. *Mailing Add:* Comput Sci Canisius Col Buffalo NY 14208

LEONE, LUCILE P, b Ohio, 02. HEALTH ADMINISTRATION. *Educ:* Univ Del, BA, 24; Johns Hopkins, BS, 27; Teachers Col, Columbia, MS, 29. *Prof Exp:* Staff nursing, Johns Hopkins, 27-29; staff nursing, Univ Minn, 29-41; comt, Student Nursing Serv, USPHS, 41-42, dir student nursing, Cadet Corp Prog, 42-48, asst nursing, 48-66; assoc dean nursing, Tex Womans Col, 77-82; RETIRED. *Mem:* Inst Med-Nat Acad Sci. *Mailing Add:* 1400 Geary Blvd San Francisco CA 94109

LEONE, RONALD EDMUND, b New York, NY, Aug 11, 42. ORGANIC CHEMISTRY. *Educ:* Northwestern Univ, BA, 64; Princeton Univ, MA, 67, PhD(org chem), 70. *Prof Exp:* Fel org chem, Yale Univ, 69-71; SR RES CHEMIST, EASTMAN KODAK CO, 71- *Mem:* Am Chem Soc; Sigma Xi. *Res:* Aspects of physical organic chemistry including organic reaction mechanisms and nuclear magnetic resonance spectroscopy; synthesis of compounds for photographic applications including sensitizing dyes, silver halide fogging agents, couplers, interlayer scavengers, and latent image stabilizers; photographic chemistry. *Mailing Add:* 755 Corwin Rd Rochester NY 14610

LEONE, STEPHEN ROBERT, b New York, NY, May 19, 48. CHEMICAL PHYSICS. *Educ:* Northwestern Univ, BA, 70; Univ Calif, Berkeley, PhD(phys chem), 74. *Prof Exp:* Asst prof chem, Univ Southern Calif, 74-76; physicist, Nat Bur Standards, 76-78; FEL, NAT BUR STANDARDS, 86- *Concurrent Pos:* Adj asst prof chem, Univ Colo, 76-80, adj prof, 80-; Alfred P Sloan Found fel, 77. *Honors & Awards:* Nobel Laureate Signature Award, Am Chem Soc, Pure Chem Award, Coblentz Award; Gold Medal, Dept Com; Arthur S Flemming Award, 85; Herbert P Broida Prize, Am Phys Soc, 89. *Mem:* Fel Am Phys Soc; Am Chem Soc; Am Inst Physics; Sigma Xi; fel Optical Soc Am; fel AAAS. *Res:* Laser-excited chemical reactions; kinetics and spectroscopic investigations of excited states using specific laser excitation; energy transfer and dynamical processes of small gas phase molecules; photodissociation; new laser development; ion molecule reaction dynamics; surface dynamics. *Mailing Add:* Joint Inst for Lab Astrophysics Univ Colo Boulder CO 80309-0440

LEONG, JO-ANN CHING, b Honolulu, Hawaii, Mar 15, 42; c 2. VIROLOGY. *Educ:* Univ Calif, Berkeley, BA, 64; Univ Calif, PhD(microbiol), 71. *Prof Exp:* Sr res asst virol, Dept Surg, Stanford Univ Sch Med, 65-67; from teaching assoc microbiol to res biochemist, Univ Calif, San Francisco, 71-73, res fel biochem, 73-75; from asst prof to assoc prof, 75-85, PROF MICROBIOL, ORE STATE UNIV, 85- *Concurrent Pos:* Dernham fel, Am Cancer Soc, Calif Div, 73-75; Giannini Found fel,73. *Honors & Awards:* Res Award, Sigma Xi, 90. *Mem:* Am Soc Microbiologists; AAAS; Soc Gen Microbiol; NY Acad Sci; Am Asn Cancer Res; Am Soc Virol. *Res:* Virus-cell interactions; tumor virology. *Mailing Add:* Dept Microbiol Ore State Univ Nash Hall Corvallis OR 97331

LEONG, KAM CHOY, b Honolulu, Hawaii, Dec 17, 20; m 50; c 3. BIOCHEMISTRY, POULTRY NUTRITION. *Educ:* Wash State Univ, BS, 49, MS, 50; Univ Wis, PhD(biochem, poultry), 58. *Prof Exp:* Asst, Wash State Univ, 48-50; jr animal husbandman, Univ Hawaii, 51-54; asst, Univ Wis, 54-57; fel, Wash State Univ, 57-58; jr poultry scientist, 58-61; res chemist, Bur Com Fisheries, 61-65; nutritionist, Milling Co Div, Carnation Co, 65-81, asst dir nutrit, 81-86; RETIRED. *Mem:* Am Poultry Sci Asn; Am Inst Nutrit. *Res:* Amino acids; enzymes; vitamins; protein; metabolizable energy. *Mailing Add:* 410 S Las Flores Dr Nipomo CA 93444

LEONHARD, FREDERICK WILHELM, physics, for more information see previous edition

LEONHARD, WILLIAM E, b Middletown, Pa, Dec 9, 14. CIVIL & ELECTRICAL ENGINEERING. *Educ:* Pa State Univ, BS, 36; Mass Inst Technol, MS, 40. *Hon Degrees:* LLD, Pepperdine Univ, 87. *Prof Exp:* From lt to lt colonel, US Army Corp Engrs, 36-51; dir construct, USAF, 52-56, dep commdr, missile & space div, 56-61, chief staff-brigadier gen, hq systs command, 61-64; dir, Tittan III prog, United Technol Corp, 64-66; sr vpres & gen mgr, Parsons Corp, 66-74, pres, 74-75, pres & chief exec officer, 75-78, chmn, pres & chief exec officer, 78-90; RETIRED. *Honors & Awards:* George Washington Award, Inst Advan Eng, 84. *Mem:* Nat Acad Eng; Inst Advan Eng. *Mailing Add:* 1265 S Orange Grove Blvd No 1 Pasedena CA 91105

LEONHARDT, EARL A, b Council Bluffs, Iowa, Apr 18, 19; m 41; c 3. MATHEMATICS. *Educ:* Union Col, Nebr, BA, 50; Univ Nebr, ME, 52, PhD(sec educ, math), 62. *Prof Exp:* High sch instr, Nebr, 51-52; from instr to prof math, Union Col, Nebr, 52-90; RETIRED. *Concurrent Pos:* Mem, Nat Coun Teachers Math. *Mem:* Math Asn Am. *Mailing Add:* Dept Math Union Col 3800 S 48th St Lincoln NE 68506

LEONORA, JOHN, b Milwaukee, Wis, Jan 30, 28; m 52; c 2. ENDOCRINOLOGY. *Educ:* Univ Wis, BS, 49, MS, 54, PhD(zool), 57. *Prof Exp:* Asst endocrinol, Univ Wis, 52-57; from instr to assoc prof, 59-69, PROF MED, SCH MED, LOMA LINDA UNIV, 69-, CO-CHMN DEPT PHYSIOL & PHARMACOL, 74- *Concurrent Pos:* NIH fel, Univ Wis, 57-59. *Honors & Awards:* Res Award Sigma Xi. *Mem:* AAAS; NY Acad Sci; Endocrine Soc; Sigma Xi. *Res:* Hypothalamic-parotid endocrine axis; relationship of dentinal fluid movement to dental caries. *Mailing Add:* Dept Physiol Loma Linda Univ Sch Med Loma Linda CA 92354

LEONTIS, T(HOMAS) E(RNEST), b Plainfield, NJ, Mar 13, 17; m 54; c 3. PHYSICAL METALLURGY. *Educ:* Stevens Inst Technol, ME, 38; Carnegie Inst Technol, MS, 42, DSc(phys metall), 46. *Prof Exp:* Instr chem, Stevens Inst Technol, 38-39; res metallurgist, Vanadium Corp Am, 39-41; grad fel, Carnegie Inst Technol, 41-44; metallurgist, Dow Chem Co, 44-51, sect chief metall lab, 51-57, asst to dir metall labs, 57-62, proj planning mgr, Dow Metal Prod Co Div, 62-65, mgr govt bus, 65-68, mgr tech & govt laison, 68-71; sr tech adv & assoc mgr, Battelle-Columbus, 71-75, mgr, Magnesium Res Ctr, 75-82; RETIRED. *Mem:* AAAS; Am Soc Metals (secy, 66-68, vpres, 70, pres, 71); Am Inst Mining, Metall & Petrol Engrs; Am Soc Metals Found Educ & Res (pres, 72). *Res:* Oxidation of metals; age hardening; powder metallurgy extrusion; extrusion of metals; high temperature magnesium alloys; magnesium alloy development; melting, casting and solidification; die casting; metal protection; surface finishing. *Mailing Add:* 3590 Hythe Ct Columbus OH 43220

LEOPOLD, ALDO CARL, b Albuquerque, NMex, Dec 18, 19; c 3. PLANT PHYSIOLOGY, AGRONOMY. *Educ:* Univ Wis, BA, 41; Harvard Univ, MA, 47, PhD(biol), 48. *Prof Exp:* Plant physiologist, Hawaiian Pineapple Co, Hawaii, 48-49; from asst prof to prof hort, Purdue Univ, 49-75; grad dean & asst vpres res, Univ Nebr, 75-77; distinguished scientist, 77-78, WILLIAM C CROCKER SCIENTIST, BOYCE THOMPSON INST, ITHACA, NY, 78- *Concurrent Pos:* Carnegie vis prof, Univ Hawaii, 62; mem panel regulatory biol, NSF, 65, sr policy analyst, 74-75; bd govs, Am Inst Biol Sci & Am Soc Gravitational Space Biol, 84; mem bd agr & renewable resources, Nat Res Coun, 75-78; adj prof, Cornell Univ, 78-, Univ Fla, 88- *Mem:* Fel AAAS; Bot Soc Am; Am Soc Plant Physiol (vpres, 59, pres, 65); Crop Sci Soc Am; Am Soc Gravitational Space Biol (vpres, 88, pres, 89). *Res:* Plant growth and development; seed viability; senescence and aging. *Mailing Add:* Boyce Thompson Inst Ithaca NY 14853

LEOPOLD, BENGT, b Valbo, Sweden, Dec 23, 22; m 45; c 3. PAPER CHEMISTRY, PULP & PAPER TECHNOLOGY. *Educ:* Royal Inst Tech, Sweden, BChem Eng, 47, MS, 49, PhD(org chem), 52. *Prof Exp:* Sr res chemist, Columbia-Southern Chem Corp, Ohio, 52-53; mgr pioneering res div, Indust Cellulose Res Ltd, Can Inst Paper Co, 53-58; mgr basic res div, Mead Corp, Ohio, 58-60, assoc dir res, 60-61; PROF PULP & PAPER RES & DIR, EMPIRE STATE PAPER RES INST, STATE UNIV NY COL ENVIRON SCI & FORESTRY, SYRACUSE UNIV, 61-, CHMN DEPT PAPER SCI & ENG, 74- *Concurrent Pos:* Ed, J Tech Asn Pulp & Paper Indust, 66- *Mem:* AAAS; Am Chem Soc; fel Tech Asn Pulp & Paper Indust; Paper Indust Mgt Asn; Can Pulp & Paper Asn. *Res:* Fiber physics; structure of lignin; mechanical properties of wood fibers; cellulose-water interactions. *Mailing Add:* Dept Forestry State Univ NY Col Environ Sci Forestry Syracuse NY 13210

LEOPOLD, DANIEL J, b NY. ELECTRONIC & OPTICAL MATERIALS. *Educ:* Rochester Inst Technol, BS, 77; Wash Univ, MA, 79, PhD(physics), 83. *Prof Exp:* Postdoctoral fel physics, Harvard Univ, 83-84; SCIENTIST PHYSICS, MCDONNELL DOUGLAS RES LABS, 84- *Concurrent Pos:* Adj assoc prof, Physics Dept, Wash Univ, 90- *Mem:* Am Phys Soc; Mat Res Soc. *Res:* Optical and electronic properties of semiconductor quantum wells; molecular beam epitaxy; thin film amorphous semiconductors and conductive polymers; photonic devices. *Mailing Add:* McDonnell Douglas Res Labs MC 111 1041 PO Box 516 St Louis MO 63166-0516

LEOPOLD, DONALD JOSEPH, b Ft Thomas, Ky, July 13, 56; m 80; c 2. FOREST ECOLOGY, DENDROLOGY. *Educ:* Univ Ky, BS, 78, MSF, 81; Purdue Univ, PhD(forest ecology), 84. *Prof Exp:* Res assoc, Univ Ga, Athens, 85; ASSOC PROF DENDROL, COL ENVIRON SCI & FORESTRY, STATE UNIV NY, SYRACUSE, 85- *Mem:* Ecol Soc Am; Soc Am Foresters; Torrey Bot Club; Soc Conserv Biologists; Int Asn Veg Sci. *Res:* Vegetation responses to disturbance (natural or man-induced); restoration techniques for disturbed plant communities; rare plant management. *Mailing Add:* State Univ NY Col Envrion Sci & Forestry Syracuse NY 13210

LEOPOLD, ESTELLA (BERGERE), b Madison, Wis, Jan 8, 27. BOTANY. *Educ:* Univ Wis, PhB, 48; Univ Calif, Berkeley, MS, 50; Yale Univ, PhD(bot), 55. *Prof Exp:* Asst res hydrologist, Tree Ring Res Lab, Univ Ariz, 51; teaching asst, Dept Plant Sci, Yale Univ, 52-53, Dept Zool, 54; res botanist, Paleont & Stratig Br, US Geol Survey, Denver, Colo, 55-76; prof bot & forest resources & dir, Quaternary Res Ctr, 76-82, prof, Dept Bot & Col Forest Resources, 82-89, PROF BOT & ENVIRON STUDIES, UNIV WASH, SEATTLE, 89- *Concurrent Pos:* Res asst, Genetics Exp Sta, Smith Col, 52; mycologist, Forest Prod Labs, Madison, Wis, 52; Jr Sterling scholar, Yale Univ, 53-54, Sheffield Sci Sch scholar, 54-55; NSF travel grant to Spain, 57, Poland, 61, Eng, 76, USSR, 82, Can, 87; adj prof, Dept Biol, Univ Colo, 67-76; NSF grants, 68-69, 79-81, 82-83 & 91; vis prof, Dept Bot & Inst Environ Studies, Univ Wis-Madison, 71-72; mem, McIntyre Stennis Coop Forestry Res Adv Comt, 74-82; mem, US Nat Comt, Int Union Quaternary Res, 76-78, vchmn, 78-82, chmn, 82-87; mem, Environ Studies Bd, Nat Acad Sci, 77-80,

Climate Res Bd, 83; assoc ed, Quaternary Res, 80-83; mem, Comt on Climate, AAAS, 83-84; mem, Mt St Helens Sci Adv Bd, US Forest Serv, 86-89; mem bd, Friends of the Earth Found, 87-89. *Honors & Awards:* Co-recipient, Conservationist of Year Award, Colo Wildlife Fedn, 69; Keep Colo Beautiful Ann Award, 79. *Mem:* Nat Acad Sci; fel AAAS; Am Quaternary Asn (pres-elect, 80-82, pres, 82-84); Bot Soc Am; Ecol Soc Am; fel Geol Soc Am; Sigma Xi. *Res:* Late Cenozoic paleobotany, palynology, paleoecology and paleoclimate; pollen and spore floras of late Cenozoic age in Wyoming, Idaho, Washington, Colorado and Alaska; palynology research in late quaternary deposits of Connecticut, Washington & California; Upper Cretaceous pollen and spore floras of Alabama and Wyoming; history of western grasslands; forest history of Washington; history of Pacific Northwest forest associations; climate and vegetation patterns since glaciation. *Mailing Add:* Bot Dept Univ Wash Seattle WA 98195

LEOPOLD, LUNA BERGERE, b Albuquerque, NMex, Oct 8, 15; m 40; c 2. GEOMORPHOLOGY. *Educ:* Univ Wis, BS, 36; Harvard Univ, PhD(geol), 50; Univ Calif, Los Angeles, MA, 45. *Hon Degrees:* DrGeog, Univ Ottawa, 70; DSc, Iowa Wesleyan Univ, 72, Univ Wis, 80, St Andrews Univ, Scotland, 81, Univ Murcia, Spain, 88. *Prof Exp:* From jr engr to assoc engr, Soil Conserv Serv, USDA, NMex, 36-40; assoc engr, US Eng Off, Los Angeles, 41-42; assoc engr, bur reclamation, US Dept Interior, Washington, DC, 46-47; head meteorologist, Pineapple Res Inst, 47-50; hydraul engr, US Geol Surv, 50-56, chief hydrologist, 56-66, sr res hydrologist, 66-72; PROF GEOL, UNIV CALIF BERKELEY, 72- *Honors & Awards:* Nat Medal of Sci, 91; Dept Interior Distinguished Serv Award, 58; Bryan Award, Geol Soc Am, 58; Veth Medal, Royal Neth Geog Soc, 63; Liege Univ Medal, 66; Cullum Medal, Am Geog Soc, 68; Rockefeller Pub Serv Award, 71; Busk Medal, Royal Geog Soc, London, 84. *Mem:* Nat Acad Sci; Am Soc Civil Eng; Geol Soc Am (pres, 71); Am Philos Soc; Am Acad Arts & Sci. *Res:* Hydrology of arid regions; rainfall characteristics; river morphology, erosion and sedimentation. *Mailing Add:* Dept Geol Univ Calif Berkeley CA 94720

LEOPOLD, REUVEN, b Arad, Rumania, May 5, 38; US citizen; m 62; c 3. OCEAN ENGINEERING, HYDRODYNAMICS. *Educ:* Mass Inst Technol, BS, 61, Marine Mech Engr, 65, PhD(eng), 77; George Washington UNiv, MBA, 77. *Prof Exp:* Res scientist hydrodyn, Hydronautics Inc, 61-62; res engr, Mass Inst Technol, 62-66; dir ship eng & design, Shipbldg Div, Litton Industs, 66-71; tech dir surface ship & submarine design, Naval Ship Eng Ctr, 72-78; vpres advan systs aerospace, Govt Prod Div, Pratt & Whitney Aircraft Group, 78-81; mgr bus develop & strategic plan, Mil Engines Oper, Gen Elec Co , 81-84; chmn bd & chief exec officer, N K F Engr Inc, 84-87; chmn bd & pres, Syntek Eng & Comput Systs Inc, 87-89; CHMN BD J J H, INC, 87- *Concurrent Pos:* Mem vis comt, Mass Inst Technol, 73-76 & 87-; assoc mem, Defense Sci Bd, 75-84; task force mem, Atlantic Coun US, 76-78; mem, Chief Naval Opers Sr Adv Bd, 79- *Honors & Awards:* Harold E Saunders Award, Am Asn Nuclear Engrs, 86. *Mem:* Fel Soc Naval Architects & Marine Engrs; Am Soc Naval Engrs; Sigma Xi; Am Defense Preparedness Asn. *Res:* Computers; materials; ship design; marine engineering. *Mailing Add:* 4001 N Fairfax Dr No 400 Arlington VA 22203

LEOPOLD, ROBERT L, b Philadelphia, Pa, Oct 5, 22; m 44; c 3. PSYCHIATRY. *Educ:* Harvard Univ, AB, 43; Univ Pa, MD, 46. *Prof Exp:* Intern neurol, Grad Hosp, 47-50, from instr to assoc prof psychiat, 50-68, clin psychiat, 68-69, PROF COMMUNITY PSYCHIAT & COMMUNITY MED, DIV GRAD MED, SCH MED, UNIV PA, 68-, CHMN DEPT COMMUNITY MED, 71-, DIR DIV COMMUNITY PSYCHIAT, 65- *Concurrent Pos:* Resident, Philadelphia Psychoanal Inst, 49-55; fel, Psychiat Inst Pa, 50-51; resident, Univ Hosp, Univ Pa, 51-52; psychiat consult, Am Friends Serv Comt, 56-; sr psychiat consult, Peace Corps, 61-67; dir, WPhiladelphia Community Ment Health Consortium, 67-72; psychiatrist-in-chief, Philadelphia Psychiat Ctr, 80- *Mem:* Am Sci Affil; Am Psychoanal Asn; fel Am Psychiat Asn; AMA. *Res:* Community psychiatry; psychoanalysis. *Mailing Add:* 21 Piersol Bldg 3400 Spruce St Philadelphia PA 19104

LEOPOLD, ROBERT SUMMERS, b Dayton, Ohio, June 21, 15; m 43; c 3. ORGANIC CHEMISTRY. *Educ:* Miss State Univ, BS, 37; Univ NC, MA, 40; Univ Fla, PhD(chem), 42. *Prof Exp:* Instr, Va Mil Inst, 42-43; asst prof chem, Ga Inst Tech, 46; assoc prof chem, Fla State Univ, 46-49; mem staff, US Naval Dent Sch, 49-52; head chemist, Naval Med Field Res Lab, 52-57, head personnel protection div, 59-62; assoc prof chem, The Citadel, 63-80; CONSULT, 80- *Mem:* Am Chem Soc. *Res:* Biochemistry. *Mailing Add:* 225 Sea Myrtle Ct, Kiawah Johns Island SC 29455

LEOPOLD, ROGER ALLEN, b Redwood Falls, Minn, Mar 23, 37; m 88; c 2. CYTOLOGY, EMBRYOLOGY. *Educ:* Concordia Col, Minn, BA, 62; Mont State Univ, PhD(entom), 67. *Prof Exp:* Res asst stress physiol, Mont State Univ, 62-67; res leader, 76-78, lead scientist, 86-88, RES ENTOMOLOGIST, BIOSCI RES LAB, AGR RES SERV, USDA, 67- *Concurrent Pos:* Adj prof zool, NDak State Univ, 71- *Mem:* AAAS; Entom Soc Am; Am Soc Zoologists; Soc Cryobiol. *Res:* Insect reproductive physiology and development; cryobiology. *Mailing Add:* Bio Sci Res Lab USDA Agr Res Serv PO Box 5674 Fargo ND 58105

LEOPOLD, WILBUR RICHARD, III, b Paterson, NJ, July 26, 49; m 73; c 1. EXPERIMENTAL CHEMOTHERAPY, TUMOR BIOLOGY. *Educ:* Univ Ill, BS, 71, MS, 73; Univ Wis, PhD(oncol), 81. *Prof Exp:* Chem engr, Exxon Co, 71-72 & 73-75; res oncologist, Southern Res Inst, 81-; AT WARNER-LAMBERT PARK-DAVIS PHARMACEUT RES DIV. *Mem:* Am Chem Soc; AAAS; Am Assoc Cancer Res. *Res:* Model development for cancer therapy; evaluation of anticancer drugs; chemical carcinogenesis and toxicological evaluations. *Mailing Add:* 2229 Loch Highland Dexter MI 48130-9597

LEOSCHKE, WILLIAM LEROY, b Lockport, NY, May 2, 27; m 56; c 2. NUTRITION. *Educ:* Valparaiso Univ, BA, 50; Univ Wis, MS, 52, PhD(biochem), 54. *Prof Exp:* Proj assoc biochem, Univ Wis, 54-59; from asst prof to assoc prof, 59-69, PROF CHEM, VALPARAISO UNIV, 69- *Concurrent Pos:* Consult, Mink Specialties Co, Ill, 55-; mem, Nat Res Coun Sub-comt Fur Animal Nutrit. *Mem:* Am Chem Soc; Am Asn Univ Prof; Sigma Xi. *Res:* Biochemistry and nutrition of mink; fundamental nutritional requirements of mink; mink diseases of nutritional origin; composition of blood and urine of mink. *Mailing Add:* Dept Chem Valparaiso Univ Valparaiso IN 46383

LEOVY, CONWAY B, b Hermosa Beach, Calif, July 16, 33; m 58; c 4. METEOROLOGY. *Educ:* Univ Southern Calif, BA, 54; Mass Inst Technol, PhD(meteorol), 63. *Prof Exp:* Meteorologist, Rand Corp, Calif, 63-69; assoc prof atmospheric sci, 69-74, dir, 86-89, PROF ATMOSPHERIC SCI & GEOPHYS & ADJ PROF ASTRON, UNIV WASH, 74- *Concurrent Pos:* Mem, Comt on Atmospheric Sci, Nat Acad Sci, 72-75 & comt on Lunar & Planetary Exploration, 74-76; ed, J Atmos Sci; Solar Syst Explor Comt, NASA, 84-87. *Honors & Awards:* NASA Outstanding Sci Achievement Award, 72. *Mem:* Am Meteorol Soc; Am Geophys Union; AAAS. *Res:* Dynamics, radiation and photochemistry of earth and planetary atmospheres. *Mailing Add:* Dept Atmospheric Sci Univ Wash Seattle WA 98195

LEPAGE, RAOUL, b Detroit, Mich, Mar 5, 38; m 61; c 2. RANDOM PROCESSES, SEQUENTIAL ANALYSIS. *Educ:* Mich State Univ, BS, 61, MS, 62; Univ Minn, PhD(math statist), 67. *Prof Exp:* Instr statist & probability, Columbia Univ, 65-66; asst prof statist & probability, Columbia Univ, NY, 67-70; vis assoc prof statist & probability, Univ Colo, 71; assoc prof, 72-76, PROF STATIST & PROBABILITY, MICH STATE UNIV, 77- *Concurrent Pos:* Consult statist. *Mem:* Inst Math Statist. *Res:* Isolating and proving significant properties of random processes; nonstandard statistical questions of an applied character and the interface between probability, statistics and computing. *Mailing Add:* A413 Wells Hall-Statist Mich State Univ East Lansing MI 48824

LE PAGE, WILBUR R(EED), b Kearney, NJ, Nov 16, 11; c 1. ELECTRICAL ENGINEERING. *Educ:* Cornell Univ, EE, 33, PhD(elec eng), 41; Univ Rochester, MS, 39. *Prof Exp:* Instr elec eng, Univ Rochester, 33-38; res engr advan develop sect, Photophone Div, Radio Corp Am Mfg Co, 41-42; res physicist radiation lab, Johns Hopkins Univ, 42-46; sr res engr, Stromberg-Carlson Co, 46-47; assoc prof, 47-50, chmn dept, 56-74, PROF ELEC & COMPUTER ENG, SYRACUSE UNIV, 50- *Mem:* Fel Inst Elec & Electronics Engrs; Asn Comput Mach. *Res:* Network theory; applied mathematics; education; computer applications. *Mailing Add:* Dept Elec Eng Link Hall Syracuse Univ Syracuse NY 13244

LEPARD, DAVID WILLIAM, b Newmarket, Ont, Nov 1, 37; m 61; c 2. MOLECULAR PHYSICS. *Educ:* Univ Toronto, BA, 59, MA, 60, PhD(physics), 64. *Prof Exp:* Asst prof physics, Mem Univ, 64-65; fel spectros sect, Div Pure Physics, Nat Res Coun Can, 65-67; from asst prof to assoc prof physics, Brock Univ, 72-82; SYSTS ANALYST, ENERGY, MINES & RESOURCES, 82- *Concurrent Pos:* Nat Res Coun Can res grants, 64-65, 67-; Ont Dept Univ Affairs res grants, 68-69. *Mem:* Can Asn Physicists; Am Phys Soc. *Res:* Theory; infrared and Raman spectra of polyatomics; electronic spectra of diatomics. *Mailing Add:* 3303 33rd St NW Calgary AB T2L 2E7 Can

LEPESCHKIN, EUGENE, b Kazan, Russia, Apr 15, 14; nat US; m 49; c 3. CARDIOLOGY. *Educ:* Univ Vienna, MD, 39. *Prof Exp:* Asst physiol, Univ Vienna, 39-40; asst, Balneolog Inst, Bad Nauheim, Ger, 40-42; asst, I Med Clin, Vienna, 42-44; cardiologist, Hosp Team 1064, UNRRA, Munich, Ger, 45-47; from asst prof to prof exp med, 47-65, prof, 65-79, EMER PROF MED, COL MED, UNIV VT, 79- *Concurrent Pos:* Nat Heart Inst res career award, 62-; res cardiologist, Life Ins Hosp, Bad Nauheim, Ger, 40-42; chief cardiographer, Goesbbriand Hosp, Burlington, 52-62, consult, 62-79; consult, Middlebury Hosp, 52-68; estab investr, Am Heart Asn, 53-58, mem basic sci clin cardiol coun, Comt Standard Electrocardiograph Vectorcardiograph Leads, 53-67. *Honors & Awards:* Einthoiven Medal, 19th Cong Electrocardiol, Budapest, 78. *Mem:* Am Physiol Soc; Am Col Cardiol. *Res:* Physiology and pathology of the heart and circulation, especially electrophysiology of the heart, arrhythmias, electrocardiography, magnetocardiography and phonocardiography. *Mailing Add:* PO Box 80755 San Diego CA 92138

LEPIE, ALBERT HELMUT, b Malapane, Ger, Aug 6, 23; US citizen; m 56; c 1. PHYSICAL CHEMISTRY. *Educ:* Aachen Tech Univ, MS, 58; Munich Tech Univ, PhD(chem), 61. *Prof Exp:* Res chemist, Ger Inst Res Aeronaut, 61-63 & US Naval Propellant Plant, Md, 63-64; RES CHEMIST, NAVAL WEAPONS CTR, 64- *Concurrent Pos:* Mem, Interagency Chem Rocket Propulsion Group; fel, Naval Weapons Ctr, 90. *Honors & Awards:* Jannaf Cert of Recognition, 85; William B McLean Award, 88. *Mem:* AAAS; Sigma Xi; Am Chem Soc. *Res:* Performance calculations of propellants; hypergolic ignitions; mechanical behavior of polymers; advanced testing methods. *Mailing Add:* 121 Desert Candles Ridgecrest CA 93555

LEPLAE, LUC A, b Hammemille, Belgium, Nov 27, 30; m 59; c 4. THEORETICAL SOLID STATE PHYSICS. *Educ:* Cath Univ Louvain, Lic en Theoret Physics, 55; Univ Md, PhD(physics), 62. *Prof Exp:* Res assoc physics, Inst Theoret Physics, Naples, Italy, 62-66; res assoc, 66-67, vis asst prof, 67-68, asst prof, 68-77, ASSOC PROF PHYSICS, UNIV WIS-MILWAUKEE, 77- *Res:* Application of the Boson method to superconductivity, superfluidity, magnetism and phase transitions; many body problem; solid state physics. *Mailing Add:* Dept Physics Univ Wis Milwaukee WI 53201

LEPLEY, ARTHUR RAY, b Peoria, Ill, Nov 1, 33; m 85; c 4. PHYSICAL ORGANIC CHEMISTRY. *Educ:* Bradley Univ, AB, 54; Univ Chicago, SM, 56, PhD(chem), 58. *Prof Exp:* Res assoc org chem, Univ Munich, 58-59 & Univ Chicago, 59-60; asst prof chem, State Univ NY Stony Brook, 60-65; from assoc prof to prof chem, Marshall Univ, 68-88; QUAL ASSURANCE OFFICER, LABS ADMIN, DEPT HEALTH & MENT HYG, DIV LICENSURE CERT & TRAINING, MD, 88- *Concurrent Pos:* NSF fel,

58-59; USPHS gen med fel, 60; vis prof, Univ Utah, 69-71; guest worker, Lab Chem Phys, Nat Inst Arthritis, Metabolism & Digestive Dis, Md, 75-76; consult, Interox Res & Develop Labs, Widnes, Eng, 85; Resources Conserv, Bellevue, Wash, 86. *Mem:* AAAS; Am Chem Soc; Am Inst Chem; Sigma Xi. *Res:* Flow nuclear magnetic resonance; microprocessor application in chemistry; hydroxy radical oxidations; direct alpha alkylation of tertiary amines; free radical intermediates; nuclear magnetic resonance emission spectroscopy; chemically induced dynamic nuclear polarization; word processing; quality assurance; laboratory database. *Mailing Add:* Labs Admin PO Box 2355 Baltimore MD 21203

LEPOCK, JAMES RONALD, b Fairmont, WVa, Oct 20, 48; m 70; c 3. BIOPHYSICS, MEMBRANES. *Educ:* WVa Univ, BS, 70, MS, 72; Pa State Univ, PhD(biophys), 76. *Prof Exp:* Fel radiobiol, New Eng Med Ctr, Tufts Univ, 76-77; from asst prof to assoc prof, 77-87, PROF PHYSICS, UNIV WATERLOO, 87- *Concurrent Pos:* Med Res Coun Can grants, 82-85; Nat Sci & Eng Res Coun Can grant, 78-, NIH grant, 85- *Mem:* AAAS; Biophys Soc; Radiation Res Soc; NAm Hyperthermia Group; Biophys Soc Can. *Res:* Membrane biology; spin labeling and electron spin resonance; mammalian cell tissue culture; hypothermia and hyperthermia, radiation biology; fluorescence spectroscopy. *Mailing Add:* Dept Physics Univ Waterloo Waterloo ON N2L 3G1 Can

LEPOFF, JACK H, b Portland, Maine, July 22, 23; m 47; c 2. PHYSICS. *Educ:* Univ NH, BS, 43; Columbia Univ, MA, 48. *Prof Exp:* Electronic scientist, Nat Bur Standards, 49-50, Naval Res Lab, 50-51, Nat Bur Standards, 51-53 & Naval Ord Lab, 53-54; sr staff mem, Motorola, Inc, 54-59; eng specialist, Sylvania Electronic Defense Lab, Gen Tel & Electronics Corp, 59-65; diode appln mgr, HPA Div, 65-73, APPLNS ENGR, HEWLETT PACKARD CO, 73- *Mem:* Sigma Xi; Inst Elec & Electronics Engrs. *Res:* Microwaves; semiconductors. *Mailing Add:* Hewlett-Packard Co Microwave Semiconductor Div Mail Stop 90TJ 350 W Trimble Rd San Jose CA 95131

LEPORE, JOHN A(NTHONY), b Philadelphia, Pa, Feb 19, 35; m 59; c 4. CIVIL ENGINEERING, APPLIED MECHANICS. *Educ:* Drexel Inst, BSCE, 57; Univ Pa, MS, 61, PhD(appl mech), 67. *Prof Exp:* Nuclear engr, NY Ship Bldg Corp, 57-61; supv engr missile & space div, Gen Elec Co, 61-68; asst prof civil eng, Dept Civil & Urban Eng, Univ Pa, 68-71, Winterstein asst prof, 71-78, from assoc prof to prof, 78-86, PROF & UNDERGRAD CHAIR, DEPT SYST, UNIV PA, 87- *Concurrent Pos:* Danforth assoc. *Mem:* Am Soc Civil Engrs; Am Soc Eng Educ; Am Acad Mech; Earthquake Eng Res Inst. *Res:* Stability of dynamic systems; applied mathematics; random processes; earthquake, wind and ocean engineering; disaster mitigation; solar energy applications. *Mailing Add:* 113 Towne Bldg Univ Pa Philadelphia PA 19104

LEPORE, JOSEPH VERNON, b Detroit, Mich, Oct 9, 22; div. PHYSICS. *Educ:* Allegheny Col, BS, 43; Harvard Univ, PhD(physics), 48. *Prof Exp:* Instr physics, Princeton Univ, 43-44; physicist, Tenn Eastman Corp, 44-46; AEC fel, Inst Advan Study, 48-50; asst prof physics, Ind Univ, 50-51; lectr, Univ Calif, Berkeley, 54-65, staff sr scientist physics, Lawrence Berkeley Lab, Univ Calif, 51-81; RETIRED. *Concurrent Pos:* Adv to test dir, Nev Test Site, 57. *Mem:* Am Phys Soc. *Res:* Nuclear physics; quantum field theory; scattering theory. *Mailing Add:* 712 Moraga Rd Moraga CA 94556-2301

LEPOUTRE, PIERRE, b Roubaix, France, July 28, 33; Can citizen; m 62; c 3. POLYMER CHEMISTRY, PAPER SCIENCE. *Educ:* Sch Advan Indust Studies, Lille, BSc, 57; NC State Univ, MSc, 60, PhD(chem eng), 68. *Prof Exp:* Chem engr, Olegum, France, 57-58, Rohm & Haas France, 60-63; res engr, Int Cellulose Res, 63-66; res engr, Consol Bathurst, 68-71; res engr, 71-78, head polymer sect, 78-82, DIR, APPL SURFACE SCI DIV, PULP & PAPER RES INST, 82- *Honors & Awards:* Coating & Graphic Arts Div Award, Tech Asn Pulp & Paper Indust. *Mem:* Fel Tech Asn Pulp & Paper Indust; Can Tech Asn Pulp & Paper Indust. *Res:* Chemical modification of cellulose; adhesion; polymer latexes; paper coating. *Mailing Add:* Pulp & Paper Res Inst Can 570 St John Blvd Pointe Claire PQ H9R 3J9 Can

LEPOVETSKY, BARNEY CHARLES, b Ridgway, Pa, Jan 17, 26; m 50; c 2. SCIENCE ADMINISTRATION. *Educ:* Ohio State Univ, BSc, 49, MSc, 51, PhD(microbiol), 54; Ohio Northern Univ, JD, 63. *Prof Exp:* Prof microbiol, Ohio Northern Univ, 54-64; scientist adminstr, NIH, 64-69, dep assoc dir extramural progs, Nat Inst Dent Res, 69-71, chief, Off Collab Res, 71-75, CHIEF RES MANPOWER BR, NAT CANCER INST, 75- *Mem:* Int Acad Law & Sci Res. *Res:* Law. *Mailing Add:* NIH Bldg BL Rm 424 Bethesda MD 20205

LEPOW, MARTHA LIPSON, b Cleveland, Ohio, Mar 28, 27; m 58; c 3. PEDIATRICS, INFECTIOUS DISEASES. *Educ:* Oberlin Col, BA, 48; Case Western Reserve Univ, MD, 52. *Prof Exp:* Intern & resident pediat, Case Western Reserve Univ, 52-56, sr instr & asst prof, 58-67; fel infectious dis, Cleveland Metro Gen Hosp, 56-58; from assoc prof to prof pediat, Sch Med, Univ Conn, 67-78; PROF PEDIAT, ALBANY MED COL, 78-, dir clin studies ctr, 78-87, VCHMN DEPT, 82- *Concurrent Pos:* Mem study sects, NIH, 72-76; mem comt infectious dis, Am Acad Pediat, 85- *Mem:* Am Pediat Soc; Am Asn Immunol; Infectious Dis Soc Am; emer mem Soc Pediat Res; Am Acad Pediat; Am Soc Microbiol. *Res:* Clinical vaccine evaluation. *Mailing Add:* 73 Bentwood Ct Albany NY 12203

LEPOWSKY, JAMES IVAN, b New York, NY, July 5, 44. LIE THEORY & FINITE GROUP THEORY. *Educ:* Harvard Univ, AB, 65; Mass Inst Technol, PhD(math), 70. *Prof Exp:* Lectr & res assoc math, Brandeis Univ, 70-72; asst prof math, Yale Univ, 72-77; assoc prof, 77-80, PROF MATH, RUTGERS UNIV, 80- *Concurrent Pos:* Mem, Sch Math, Inst Advan Study, 75-76, 80, 85 & 87-88; Alfred P Sloan fel, 76-78; vis assoc prof math, Univ Paris, 78; assoc ed, Am Math Soc, 80-85; mem, Math Sci Res Inst, 83-84; Guggenheim fel, 87-88. *Mem:* Am Math Soc; Math Asn Am; Am Phys Soc. *Res:* Conformal field theory; representations of infinite-dimensional Lie algebras and vertex operator algebras; interactions with other branches of mathematics and physics; finite group theory; string theory; combinations. *Mailing Add:* Dept Math Rutgers Univ New Brunswick NJ 08903

LEPP, CYRUS ANDREW, b Brooklyn, NY, Aug 11, 46; m 72; c 2. CLINICAL BIOCHEMISTRY, CLINICAL CHEMISTRY. *Educ:* Syracuse Univ, BS, 68, PhD(biochem), 74. *Prof Exp:* Lab technician clin chem, Nassau Hosp, 68-69; sr biochemist clin chem & biochem, 74-80, mgr, develop, Corning Med & Sci, Corning Glass Works, 80-85, MGR, REAGENT SYSTS RES & DEVELOP, CIBA-CORNING DIAG, 85- *Mem:* AAAS; Am Asn Clin Chem; Am Chem Soc; NY Acad Sci. *Res:* Electrophoretic separations of isoenzymes and hemoglobins; development of specific isoenzyme assay procedures; development of immunologic assays, clinical chemistry reagents, clinical chemistry controls, bloodgas reagents & controls. *Mailing Add:* CIBA-Corning Diag Corp 132 Artino St Oberlin OH 44074

LEPP, HENRY, b Russia, Mar 4, 22; m 52; c 4. GEOLOGY. *Educ:* Univ Sask, BSc, 44; Univ Minn, PhD(geol), 54. *Prof Exp:* Geologist & mining engr, Consol Mining & Smelting Co, Ltd, 44-46; mining engr, N W Byrne, Consult Mining Engr, 46-48; geologist, Aluminum Labs, Ltd, 48-50 & Freeport Sulphur Co, 54; from asst prof to prof, Univ Minn, Duluth, 54-64; PROF GEOL & CHMN DEPT, MACALESTER COL, 64- *Concurrent Pos:* Consult, Que Cartier Mining, 56-58, Roberts Mining Co, 61-64, Potlach Corp, 68. *Mem:* Geol Soc Am; Soc Econ Geol. *Res:* Mining geology; sedimentary iron formations, particularly the geochemistry of these formations and its possible relation to the evolution of the atmosphere. *Mailing Add:* Dept Geol Macalester Col St Paul MN 55105-1899

LEPPARD, GARY GRANT, b Medicine Hat, Can, Aug 6, 40; m 70; c 2. CELL BIOLOGY, LIMNOLOGY. *Educ:* Univ Sask, BA, 62, BA hons, 63, MA, 64; Yale Univ, MS, 66, MPhil, 67, PhD(biol), 68. *Prof Exp:* RES SCIENTIST BIOL, CAN DEPT ENVIRON, 71- *Concurrent Pos:* NATO Sci fel, Fac Med, Univ Paris, 68, Inst Pharmacol, Univ Milan, 69; Nat Res Coun Sci fel, Fac Sci, Univ Laval, 69-70, Biochem Lab, Nat Res Coun Can, 70-71; sci res exec mem, Prof Inst Pub Serv Can, 71-73; adj prof, Dept Biol, Univ Ottawa, 74-75; NATO sci award, 81; Comn Europ Communities award, 82; prof biol, McMaster Univ, Hamilton, Ont, 88- *Mem:* Sigma Xi; Can Fedn Biol Soc; Can Soc Cell Biol; Int Asn Great Lakes Res; Int Soc Limnol; Soc Environ Toxicol & Chem. *Res:* Physico-chemical and ecotoxicological relationships between living cells, nutrients and contaminants in aquatic ecosystems. *Mailing Add:* 226 Simon Dr Burlington ON L7N 1X9 Can

LEPPELMEIER, GILBERT WILLISTON, b Cleveland, Ohio, Aug 19, 36; m 59, 84; c 3. PHYSICS. *Educ:* Yale Univ, BS, 57; Univ Calif, Berkeley, PhD(physics), 65. *Prof Exp:* Jr physicist neutron physics, Lawrence Livermore Lab, 57-59; NATO fel physics, Free Univ Brussels, 65-66; lectr physics, Univ Sussex, 66-67; physicist laser physics, Lawrence Livermore Lab, 67-75; sr scientist laser-fusion, Lab for Laser Energetics, Univ Rochester, 75-77; physicist magnetic-fusion, Lawrence Livermore Lab, 77-85; scientist, Inst Tech Physics, Nuclear Res Ctr, Karlsruhe, Ger, 85-86; SR SCIENTIST, TECH RES CENT, FINLAND, 86-, RES PROF, 90- *Mem:* Am Phys Soc; Inst Elec & Electronics Engrs; Europ Phys Soc. *Res:* Application of small computers to instrument development and control; space instruments. *Mailing Add:* VTT/INS PO Box 107 Espoo 02151 Finland

LEPPER, MARK H, b Washington, DC, 1917. MEDICINE. *Educ:* George Washington Univ, MD, 44. *Prof Exp:* Intern, Sibley Hosp Washington, DC, 41-42; fel med, George Washington, 42-43; res med, Gallinger Hosp, 43-44, med staff, 46-50; med staff, George Washington Hosp, 46-50; consult staff, Walter Reed Hosp, 48-50; med supt, Munic Contagious Dis Hosp, Chicago, 50-52; clin assoc prof prev med, 50-52, assoc prof med charge prev med, 52-55, head dept, Ill, 55-66; exec vpres, prof & acad affairs, Presby St Lukes Hosp, Chicago, 66-73, attend staff, 56-82, vpres prog eval, 76-82; RETIRED. *Concurrent Pos:* Instr med, George Washington, 46-48, asst med, 49-50, prev med, Ill, 55-; prof med & prev med, Rush, 66-, dean, 70-73. *Mem:* Nat Acad Sci. *Mailing Add:* Goodwin House 4800 Fillmore Ave No 319 Alexandria VA 22311

LEPPI, THEODORE JOHN, b Mountain Iron, Minn, May 30, 33; m 59; c 3. ANATOMY, MEDICAL SCHOOL ADMINISTRATION. *Educ:* Albion Col, BA, 59; Yale Univ, PhD(anat), 63. *Prof Exp:* From asst prof to assoc prof anat, Sch Med, Univ NMex, 66-71; assoc dean, Sch Med, Univ Minn, Duluth, 71-77, prof Biomet Anat & Chmn Dept, 71-88, dir admis, 71-77, 83-89; ASSOC DEAN ADMIS & PROF ANAT & CELL BIOL, TEX COL OSTEOP MED, 90- *Concurrent Pos:* Staff fel, Lab Exp Path, Nat Inst Arthritis & Metab Dis, 63-66; guest lectr, Sch Med, Georgetown Univ, 63-64,; asst prof lectr, Sch Med, George Washington Univ, 65-66; Lederle Med Fac Award, 68-71; vis scientist, Pac Biomed Res Ctr, Univ Hawaii, 78-79; vis prof, Sch Med Univ Hawaii, 88. *Mem:* Am Asn Anat; Am Asn Clin Anat; Sigma Xi. *Res:* Effects of hormones on connective tissues; histochemistry and cytochemistry of glycosaminoglycans; immunohistochemistry of fibronection. *Mailing Add:* 108 Sundance Ct Weatherford TX 76087

LEPPLA, STEPHEN HOWARD, b Oak Park, Ill, Feb 1, 41; m 80. BACTERIAL TOXINS, PROTEIN PURIFICATION. *Educ:* Calif Inst Technol, BS, 63; Univ Wis-Madison, PhD(biochem), 69. *Prof Exp:* NIH fel dept molecular biol, Univ Calif, Berkeley, 69-71; res assoc, div biol & med sci, Brown Univ, 71-73; res chemist, US Army Med Res Inst Infectious Dis, 74-89; RES CHEMIST, NAT INST DENT RES, 89- *Mem:* Am Soc Microbiol; Am Chem Soc; Am Soc Cell Biol. *Res:* Study of bacterial protein toxins to discover mechanisms of action, structure-function relationships; vaccine design; employment of methods of protein purification; immunochemical characterizations using monoclonal antibodies; gene cloning, and eukaryotic cell culture. *Mailing Add:* Bldg 30 Rm 309 NIH Bethesda MD 20892-0030

LEPPLE, FREDERICK KARL, b Newark, NJ, July 1, 44; m 75; c 2. CHEMICAL OCEANOGRAPHY, GEOCHEMISTRY. *Educ:* Univ Miami, BS, 67, MS, 71; Univ Del, PhD(marine studies), 75. *Prof Exp:* Res chemist org polymers, Air Reduction Corp, 67-68; res chemist marine aerosols, 74-89, PROG MGR, CHIEF NAVAL OPERS, NAVAL RES LAB, 89- *Concurrent Pos:* Res assoc, Nat Acad Sci-Nat Res Coun, Naval Res Lab, 74-76. *Mem:* AAAS; Am Chem Soc; Am Geophys Union. *Res:* Chemistry, transport and effects of aerosols in the marine environment; marine geochemistry and oceanography. *Mailing Add:* 2613 Woodlawn Lane Alexandria VA 22306

LEPS, THOMAS MACMASTER, b Keyser, WVa, Dec 3, 14; m 40; c 1. CIVIL ENGINEERING, GEOTECHNICAL ENGINEERING. *Educ:* Stanford Univ, AB, 36; Mass Inst Technol, MS, 39. *Prof Exp:* Jr engr dams, US Corps Engrs, 39-41; asst engr dams, US Bur Reclamation, 41-42; chief civil engr dams & power plants, Southern Calif Edison Co, 46-61; chief engr dams & foundations, Shannon & Wilson, 61-63; CONSULT CIVIL ENGR DAMS & POWER PLANTS, THOMAS M LEPS, INC, 63- *Concurrent Pos:* Mem & chmn exec comt, Soil Mech & Found Div, Am Soc Civil Engrs, 55-61; mem peer group mem comt, Nat Acad Eng, 75-77; mem exec comt, US Comt Large Dams, 76-, vchmn, 78-80. *Honors & Awards:* Cert of Appreciation, Am Soc Civil Engrs, 61. *Mem:* Nat Acad Eng; Am Soc Civil Engrs. *Res:* Soil mechanics and seismologic engineering. *Mailing Add:* PO Box 2228 Menlo Park CA 94026-2228

LEPSE, PAUL ARNOLD, b Seattle, Wash, Mar 18, 37; m 61; c 2. ORGANIC CHEMISTRY. *Educ:* Seattle Pac Col, BS, 58; Univ Wash, PhD(org chem), 62. *Prof Exp:* NSF fel, Univ Munich, 62; from asst prof to assoc prof, 63-72, PROF CHEM, SEATTLE PAC UNIV, 72- *Mem:* AAAS; Am Chem Soc; Sigma Xi. *Res:* Organic reaction mechanisms; carbene chemistry. *Mailing Add:* Dept Chem Seattle Pac Univ Seattle WA 98119

LEPSELTER, MARTIN P, b New York, NY, Nov 24, 29. ENGINEERING. *Prof Exp:* Dir, Advan Very Large Scale Integration Develop Lab, AT&T Bell Labs, Murray Hill, NJ, 57-87; PRES, LEPTON INC, 87- *Honors & Awards:* Daniel C Hughes Jr Mem Award, Int Soc Hybrid Microelectronics; Jack A Morton Award, Inst Elec & Electronics Engrs, 79. *Mem:* Nat Acad Eng; fel Inst Elec & Electronics Engrs. *Mailing Add:* 25 Sweetbriar Rd Summit NJ 07901

LEPSON, BENJAMIN, b New York, NY, Mar 4, 24; m 48; c 1. MATHEMATICAL ANALYSIS, COMPUTATIONAL MATHEMATICS. *Educ:* Yale Univ, BS, 43, MS, 44; Columbia Univ, PhD(math), 50. *Prof Exp:* Lab asst physics, Yale Univ, 43-44; math physicist, Naval Ord Lab, 44-46; lectr math, Columbia Univ, 46-48; mem, Inst Advan Study, 50-52; mathematician, Off Naval Res, 52-53; asst prof math, Cath Univ Am, 53-54; head, Res Comput Ctr, 54-61, Numerical Anal Br, 61-65, math consult, Nucleonics Div, 65-66, Nuclear Physics Div, 66-67, Math & Info Sci Div, 67-69, Space Sci Div, 69-74, res mathematician, Math Res Ctr, 74-75, res mathematician & actg head, appl math staff, 75-76, consult, Res Comput Ctr, 76-79, mathematician & comput sci consult, Mgt Info Div, 79-83, Chief Staff Office, 83-87, COMMAND SUPPORT DIV, US NAVAL RES LAB, 87- *Concurrent Pos:* Asst instr math, Stanford Univ, 47, Univ Ill, 48; lectr, Univ Md, 52-53 & Am Univ, 52-53, 56-57; lectr & res assoc, Cath Univ Am, 54-64, adj prof, 64-74; vis prof math & statist, Univ Md, 74-75. *Mem:* Am Math Soc; Math Asn Am; Inst Math Statist; Inst Elec & Electronics Engrs; Comput Soc; Sigma Xi. *Res:* Complex function theory, especially entire and meromorphic functions; Dirichlet type series; real function theory; potential theory; computer science; numerical analysis; applied mathematics; probability; mathematical statistics; application of mathematics, computers and statistics to physical sciences. *Mailing Add:* 9429 Curran Rd Silver Spring MD 20901

LE QUESNE, PHILIP WILLIAM, b Auckland, NZ, Jan 6, 39; m 65, 90; c 2. ORGANIC CHEMISTRY. *Educ:* Univ Auckland, MSc, 61, PhD(chem), 64. *Hon Degrees:* DSc, Univ Auckland, 79. *Prof Exp:* Res assoc org chem, Oxford Univ, 64-65; res assoc, Univ BC, 65-66, teaching fel, 66-67; asst prof, Univ Mich, Ann Arbor, 67-73; assoc prof org chem, 73-78, chmn dept, 79-87, PROF CHEM & MED CHEM, NORTHEASTERN UNIV, 78- *Mem:* Am Chem Soc; Phytochem Soc NAm; The Chem Soc; assoc NZ Inst Chem; Am Soc Pharmacog. *Res:* Natural product chemistry, especially steroids, alkaloids, terpenoids, fungal metabolites; comparative phytochemistry; physiologically active compounds. *Mailing Add:* Dept Chem Northeastern Univ Boston MA 02115-5096

LEQUIRE, VIRGIL SHIELDS, b Maryville, Tenn, June 15, 21; m 46; c 4. PATHOLOGY. *Educ:* Maryville Col, BA, 42; Vanderbilt Univ, MD, 46. *Prof Exp:* NIH res asst, 49-50, from asst prof to assoc prof anat, 50-66, actg chmn dept path, 71-73; PROF PATH & CELL BIOL, SCH MED, VANDERBILT UNIV, 66- *Concurrent Pos:* USPHS sr fel. *Honors & Awards:* Thomas Jefferson Award. *Mem:* AAAS; Am Asn Anat; Am Fedn Clin Res; Am Soc Exp Path; Sigma Xi; NY Acad Sci. *Res:* Lipoprotein transport; fat embolization; membrane structure. *Mailing Add:* Dept Path Vanderbilt Univ Sch Med Nashville TN 37232

LERBEKMO, JOHN FRANKLIN, b Alta, Can, Dec 8, 24; m 49; c 4. GEOLOGY. *Educ:* Univ BC, BASc, 49; Univ Calif, Berkeley, PhD(geol), 56. *Prof Exp:* Asst prof geol, 56-59, assoc prof sedimentary geol, 59-68, PROF SEDIMENTARY GEOL, UNIV ALTA, 68- *Mem:* Geol Soc Am; Int Asn Sedimentol; Geol Asn Can; Can Soc Petrol Geologists. *Res:* Sedimentary petrology; detrital sediments; magnetostratigraphy. *Mailing Add:* Dept Geol Univ Alta Edmonton AB T6G 2E3 Can

LERCH, IRVING A, b Chicago, Ill, June 29, 38; m 63. MEDICAL PHYSICS. *Educ:* US Mil Acad, BS, 60; Univ Chicago, SM, 66, PhD(med physics), 69. *Prof Exp:* Res assoc med physics, Univ Chicago, 69-72; first officer, Int Atomic Energy Agency, Vienna, 73-75; PROF & SR PHYSICIST, NY UNIV MED CTR, 76- *Concurrent Pos:* Consult med radiation physics, Int Atomic Energy Agency & WHO, 76-86, various corps, 76- *Mem:* Am Phys Soc; Am Asn Physicists in Med; Radiation Res Soc. *Res:* Diagnostic radiological physics as applied to problems in image quality; cell kinetic and modelling studies as applied to problems in radiation therapy; dosimetry in radiation oncology physics; radiofrequency-induced tissue hyperthermia; computers in medicine and computer-mediated telecommunications. *Mailing Add:* Dept Radiation Oncol NY Univ Med Ctr New York NY 10016

LERCHER, BRUCE L, b Milwaukee, Wis, June 7, 30; m 60; c 2. MATHEMATICAL LOGIC. *Educ:* Univ Wis, BS, 51, MS, 52; Pa State Univ, PhD(math), 63. *Prof Exp:* Instr math, Univ Rochester, 59-62; asst prof, 62-67, ASSOC PROF MATH, STATE UNIV NY, BINGHAMTON, 67- *Mem:* Math Asn Am; Asn Symbolic Logic. *Res:* Combinatory logic. *Mailing Add:* Dept Math State Univ NY Binghamton NY 13902-6000

LE RICHE, WILLIAM HARDING, b Dewetsdorp, SAfrica, Mar 21, 16; Can citizen; m 43; c 5. MEDICINE. *Educ:* Univ Witwatersrand, BSc, 37, MB, BCh, 43, MD, 49; Harvard Univ, MPH, 50; FRCP(C), 72. *Prof Exp:* Med officer, Union Health Dept, 44-49, epidemiologist, 50-52; consult epidemiol, Dept Nat Health & Welfare, Ottawa, Can, 52-54; res med officer, Physicians Serv, Inc, 54-57; res assoc pub health, 57-59, prof, 59-62, prof & head dept epidemiol & biomet, Sch Hyg, 62-75, prof epidemiol, 75-81, EMER PROF EPIDEMIOL, DEPT PREV MED, FAC MED, UNIV TORONTO, 82- *Concurrent Pos:* Carnegie fel, Bur Educ & Social Res, SAfrica, 37-39; consult, Physicians Serv Inc, 57-66, Toronto & Can Forces Med Coun, 69- *Honors & Awards:* Defries Medal, Can Pub Health Asn, 81. *Mem:* Can Med Asn; fel Am Col Physicians. *Res:* Child growth, nutrition and infectious diseases; medical care studies; epidemiology; cardiovascular disease; education in public health and preventive medicine; hospital infections. *Mailing Add:* Fac Med McMurrich Bldg 12 Queen's Park Crescent W Toronto ON M5S 1A8 Can

LERMAN, ABRAHAM, b Harbin, China, Nov 14, 35; US citizen; c 1. TECHNICAL MANAGEMENT. *Educ:* Hebrew Univ, Israel, MSc, 60; Harvard Univ, PhD(geol), 64. *Prof Exp:* Lectr geol, Johns Hopkins Univ, 64, asst prof, 64-65; asst prof, Univ Ill, Chicago, 65-68; sr scientist, Weizmann Inst, Israel, 66-69; res scientist chem limnol, Can Centre Inland Waters, Can Dept Environ, 69-71; assoc prof, 71-75, PROF GEOL SCI, NORTHWESTERN UNIV, 75- *Concurrent Pos:* Guggenheim fel, 76; vis prof, Inst Aquatic Sci, Swiss Fed Inst Technol, Duebendorf, 76-77 & Univ Karlsruhe, 79, 81 & 87; consult, Basalt Waste Isolation Proj, US Dept Energy, 81-87, mem, Mat Rev Bd, 85-88; mem, peer rev panels nuclear repository projs, Argonne Nat Lab, 83-; underground injection controls, US Environ Protection Agency, 89- *Mem:* AAAS; Geochem Soc; Am Chem Soc; fel Geol Soc Am; Am Geophys Union; Sigma Xi. *Res:* Global biogeochemical cycles; water and sediment geochemistry; transport processes; nuclear wastes; surface and ground water quality. *Mailing Add:* Dept Geol Sci Northwestern Univ Evanston IL 60208

LERMAN, CHARLES LEW, b Elizabeth, NJ, Apr 23, 48; m 71; c 1. BIO-ORGANIC CHEMISTRY. *Educ:* Yale Univ, BS, 69; Harvard Univ, AM, 70, PhD(org chem), 74. *Prof Exp:* Asst prof chem, Juniata Col, 74-76; asst prof chem, Haverford Col, 76-81; SR RES CHEMIST, ICI AMERICAS, 81- *Mem:* Am Chem Soc; AAAS; Sigma Xi. *Res:* Enzyme mechanisms and model systems; specificity of molecular interactions; mechanism of 5-aminolevulinic acid dehydratase; model for the active site of ribonuclease; NMR studies of biochemical systems. *Mailing Add:* Stuart Pharmaceut Div ICI Americas Wilmington DE 19897

LERMAN, LEONARD SOLOMON, b Pittsburgh, Pa, June 27, 25; m 74; c 3. MOLECULAR BIOLOGY. *Educ:* Carnegie Inst Technol, BS, 45; Calif Inst Technol, PhD(chem), 50. *Prof Exp:* Asst org chem & explosives, Explosives Res Lab, Carnegie Inst Technol, 45; asst chem, Calif Inst Technol, 45-49; Schenley fel, Univ Chicago, 49-51; instr pediat, Univ Colo, 51-52, asst prof, 52-53, from asst to prof biophys, 53-65; prof molecular biol, Vanderbilt Univ, 65-77; prof & chmn dept biol sci, State Univ NY, Albany, 77-84; dir diagnostics, Genetics Inst, 84-87; SR LECTR, DEPT BIOL, MASS INST TECHNOL, 87- *Concurrent Pos:* USPHS Res Career Award, 63-65, mem, Nat Sci Found Adv Panel, 65-68; NIH study sect, 69-73; Guggenheim fel, 71-72; mem, Health & Environ Adv Bd, Dept Energy, 87-; bd, Radiation Effects Res, Nat Res Coun, 88-; subcomt Human Genome, DOE-NIH, 89-; ed, Genomics, 90- *Mem:* Nat Acad Sci; AAAS; Soc Human Genetics. *Res:* Aspects of the physical nature of DNA as related to human genetics; mutagenesis; structure of the nucleus; recognition of genetic variations. *Mailing Add:* Dept Biol Rm 56-731 Mass Inst Technol Cambridge MA 02139

LERMAN, MANUEL, b New York, NY, Feb 5, 43; m 75; c 2. RECURSION THEORY. *Educ:* City Col New York, BS, 64; Cornell Univ, PhD(math logic), 68. *Prof Exp:* Instr math, Mass Inst Technol, 68-70; asst prof, Yale Univ, 70-73; assoc prof, 73-76, PROF MATH, UNIV CONN, 76- *Concurrent Pos:* Vis prof, Univ Ill, Chicago Circle, 75-76 & Univ Chicago, 80; mem, Math Sci Res Inst, Berkeley, 90. *Mem:* Am Math Soc; Asn Symbolic Logic. *Res:* Recursive function theory; recursive model theory. *Mailing Add:* Dept Math Univ Conn Storrs CT 06269

LERMAN, MICHAEL ISAAC, b Korosten, USSR, Sept 21, 32; US citizen; m 75; c 2. MOLECULAR CLONING OF HUMAN TUMOR SUPPRESSOR GENES & GENES CAUSING ALZHEIMERS DISEASE. *Educ:* First Moscow Med Sch, MD, 57; Acad Med Sci, Moscow, PhD(biochem), 61, DSc(molecular biol), 68. *Prof Exp:* Asst prof biochem, Dept Biochem, First Moscow Med Sch, 60-62; sr scientist, Inst Molecular Biol, Acad Sci, Moscow, 62-64 & Bach Inst Biochem, 64-66; sr scientist, Inst Biol Med Chem, Acad Med Sci, Moscow, 66-78, dir, Lab Molecular Pathobiol, 68-79; vis scientist, 80-87, expert, 87-90, RES CHEMIST, LAB IMMUNOL, NAT CANCER INST, FCRDC, NIH, 90- *Concurrent Pos:* Consult, Diag Div, Abbott Labs, 89- *Mem:* Am Soc Biol Chemists; AAAS; Am Soc Human Genetics; Genetic Soc Am; Int Mammalian Genome Soc. *Res:* Molecular biology of protein biosynthesis; molecular biology of aging; molecular genetics of human cancers; molecular cloning of disease genes. *Mailing Add:* Nat Cancer Inst-FCRDC Bldg 560 Rm 12-26 Frederick MD 21702

LERMAN, SIDNEY, b Montreal, Can, Oct 6, 27; m 57; c 2. OPHTHALMOLOGY, CHEMISTRY. *Educ:* McGill Univ, BSc, 48, MD, CM, 52; Univ Rochester, MS, 61. *Prof Exp:* Dir ophthalmic res, Univ Rochester, 57-68, asst prof biochem, 61-68, assoc prof ophthal, 62-68; prof ophthal & biochem & dir exp ophthal, McGill Univ, 68-73; prof ophthal, Sch Med, Emory Univ, 75-88; PROF OPHTHAL, NY MED COL, 88- *Concurrent Pos:* Chmn, Int Conf Ophthalmic Biochem, Woods Hole, Mass, 64-72; consult, Bausch & Lomb, 66-69, Nat Patent Develop Corp, 68-73 & Alza Corp, 69-72; adj prof chem, Ga Inst Technol, 75- *Honors & Awards:* Award in Ophthal & J B Cramer Mem Award, Rochester Acad Med, 58; Parker Heath Mem Award, 85. *Mem:* Am Asn Res Vision & Ophthal; Am Chem Soc; Am Soc Biol Chem; Am Soc Photobiol; Int Soc Ocular Toxicol (pres, 91). *Res:* Ophthalmic biochemistry and photobiology. *Mailing Add:* Eye Res Lab Rm 41 NY Med Col 100 Grasslands Rd Valhalla NY 10595

LERMAN, STEPHEN PAUL, b Philadelphia, Pa, Oct 3, 44; m; c 1. IMMUNOLOGY. *Educ:* Philadelphia Col Pharm & Sci, BS, 66; Hahnemann Med Col, MS, 70, PhD(microbiol), 73. *Prof Exp:* Fel, 72-74, asst res scientist, 74-75, res asst prof path, Sch Med, NY Univ, 75-78; ASSOC PROF IMMUNOL & MICROBIOL, WAYNE STATE UNIV SCH MED, 78-, DIR FLOW CYTOMETRY LAB, 86- *Concurrent Pos:* Spec fel, Leukemia Soc Am, 76-78. *Mem:* Am Asn Immunologists; Am Soc Microbiol; fel Am Acad Microbiol. *Res:* Tumor immunology. *Mailing Add:* Wayne State Univ Sch Med Dept Immunol & Microbiol 540 E Canfield Ave Detroit MI 48201

LERMAN, STEVEN I, b Bronx, NY, Nov 14, 44; m 65; c 3. OCCUPATIONAL SAFETY & HEALTH, ASBESTOS. *Educ:* Queen's Col, City Univ New York, BS, 65; Adelphi Univ, MS, 72. *Prof Exp:* Chemist, Con Edison NY, 65-86; sr chemist, NY Power Authority, 86-88; CONSULT CHEM, SIL CONSULT, 88- *Concurrent Pos:* Cong sci counr, Am Chem Soc, 81-83; subcomt chmn, Am Soc Testing & Mat, 83-; instr, Inst Asbestos Awareness, 88-90, Tall Oaks Publ, 89- & FED Training Ctr, 91-; consult, Nat Inst Standards & Technol, US Dept Com, 88- *Mem:* Fel Am Inst Chemists; Nat Asbestos Coun; Am Chem Soc; Am Soc Testing & Mat. *Res:* Environmental laboratories; quality assurance programs for compliance; industrial health and safety; computer systems. *Mailing Add:* Three Allan Gate Plainview NY 11803

LERNER, AARON BUNSEN, b Minneapolis, Minn, Sept 21, 20; m 45; c 4. DERMATOLOGY, BIOCHEMISTRY. *Educ:* Univ Minn, BA, 41, MS, 42, MB & PhD(physiol chem), 45, MD, 45; Am Bd Dermat, dipl, 53. *Prof Exp:* Asst physiol chem, Univ Minn, 41-45; Am Cancer Soc fel, Sch Med, Western Reserve Univ, 48-49; asst prof dermat, Med Sch, Univ Mich, 49-52; assoc prof, Univ Ore, 52-55; assoc prof, 55-57, chmn dept, 58-85, PROF, DEPT DERMAT, YALE UNIV SCH MED, 58- *Honors & Awards:* Myron-Gordon Award, 69; Stephen Rothman Award, 71; Dome Lectr, 80; Lita Annenberg Hazen Award, 81. *Mem:* Sr mem Inst Med-Nat Acad Sci; Soc Invest Dermat; Am Acad Dermat; Am Soc Biol Chemists; Sigma Xi. *Res:* Plasma proteins associated with disease; metabolism of phenylalanine and tyrosine; biochemistry of melanin pigmentation; mechanism of endocrine control of pigmentation; malignant melanomas; cryoglobulins; biochemistry of skin; author or co-author of numerous publications. *Mailing Add:* Dept Dermat Yale Univ Sch Med 333 Cedar St PO Box 3333 New Haven CT 06510

LERNER, ALBERT MARTIN, b St Louis, Mo, Sept 3, 29; div; c 4. INTERNAL MEDICINE. *Educ:* Wash Univ, BA, 50, MD, 54; Am Bd Internal Med, dipl, 61. *Prof Exp:* Intern, Barnes Hosp, St Louis, Mo, 54-55; lab investr, Nat Inst Allergy & Infectious Dis, 55-57; asst resident, Harvard Med Serv, Boston City Hosp, Mass, 57-58; sr asst resident, Barnes Hosp, Mo, 58-59; res assoc biol, Mass Inst Technol, 62-63; assoc prof med & assoc microbiol & path, Col Med, Wayne State Univ, 63-67; assoc med & path & dir bact lab, Detroit Gen Hosp, 64-69, clin consult bact lab, 69-; prof med, 67-82, chief, Hutzel Hosp Med Unit, 70-82, CLIN PROF MED, COL MED, WAYNE STATE UNIV, 82- *Concurrent Pos:* Res fel med, Thorndike Mem Lab, Boston City Hosp & Harvard Med Sch, 59-62; fel, Med Found Greater Boston, Inc, 60-63; consult, Vet Admin Hosp, Allen Park, Mich, 63-82. *Mem:* Fel Am Col Physicians; Am Soc Clin Invest; Inf Dis Soc Am; dipl mem Pan-Am Med Asn; Asn Am Physicians; Am Fedn Clin Res. *Res:* Infectious diseases. *Mailing Add:* 31000 Lahser Birmingham MI 48010

LERNER, B(ERNARD) J, b Brooklyn, NY, Apr 28, 21; m 51; c 3. CHEMICAL ENGINEERING. *Educ:* Cooper Union, BChE, 43; Univ Iowa, MS, 47; Syracuse Univ, PhD(chem eng), 49. *Prof Exp:* Instr chem eng, Univ Iowa, 46-47; res engr, Inst Indust Res, Syracuse Univ, 47-48; asst prof, Univ Tex, 49-54; group leader, Gulf Res & Develop Co, 54-59; consult, Dominion Gulf Co, 59-63; pres, Patent Develop Assocs, Inc, Pa, 63-68; vpres res & dir chem eng res, MK Res & Develop Co, Pa, 68-70; PRES, BECO ENG CO, 70- *Concurrent Pos:* Pvt consult chem engr 59-80; Consult, Monsanto Chem Co, 52-53 & Maurice A Knight Co, 63-66. *Mem:* Am Chem Soc; Am Inst Chem Engrs; Air Pollution Control Asn. *Res:* Mass transfer; two-phase fluid flow; air pollution control. *Mailing Add:* Beco Eng Co PO Box 443 Oakmont PA 15139

LERNER, DAVID EVAN, b Kansas City, Mo, Mar 21, 44. MATHEMATICAL PHYSICS. *Educ:* Haverford Col, BA, 64; Univ Pittsburgh, PhD(math), 72. *Prof Exp:* Instr math, Univ Pittsburgh, 72-73; res assoc physics, Syracuse Univ, Relativity Group, 73-75; asst prof, 75-79, ASSOC PROF MATH, UNIV KANS, 80- *Concurrent Pos:* Math Inst, Univ Oxford, 76-77. *Mem:* Am Phys Soc; Am Math Soc. *Res:* Application of Lie groups and differential geometry to general relativity; associative memory systems. *Mailing Add:* Dept Math Univ Kans 12th St Lawrence KS 66045

LERNER, EDWARD CLARENCE, b Brooklyn, NY, Sept 10, 24. THEORETICAL PHYSICS. *Educ:* Mass Inst Technol, BS, 45, PhD(physics), 52. *Prof Exp:* Mem staff, Lincoln Lab, Mass Inst Technol, 52-58; from assoc prof to prof physics, Univ SC, 57-; RETIRED. *Mem:* AAAS; Am Phys Soc. *Res:* Electrodynamics; field theory; classical and quantum dynamics. *Mailing Add:* 12 Shadow Creek Ct Columbia SC 29209

LERNER, HARRY, electrochemistry, for more information see previous edition

LERNER, JOSEPH, b Wilkes-Barre, Pa, Jan 16, 42; m 63; c 2. BIOCHEMISTRY. *Educ:* Rutgers Univ, BS, 63, PhD(biochem), 67. *Prof Exp:* Sr res investr biochem, Eastern Utilization Res & Develop Div, USDA, 67-68; from asst prof to assoc prof biochem, Univ Maine, Orono, 68-77, prof, 77-, chmn dept, 78-83; PROF CHEM, COL ARTS & SCI, TENN TECH UNIV, COOKEVILLE, 84-, DEAN, 84- *Concurrent Pos:* Coe Res Fund grant, 68-69; Hatch Fund grant, 69-; NIH res grant, 73-82; res assoc, Dept Avian Sci, Univ Calif, Davis, 74; fac fel acad higher admin, Univ NH, 82-82; pres, Tenn Coun Arts & Sci Deans, 88. *Mem:* Am Chem Soc; Am Inst Nutrition; NY Acad Sci; Sigma Xi; AAAS. *Res:* Intestinal absorption of amino acids in chicken; metabolism of small intestine; genetic aspects of transport processes; separation of nucleotide derivatives by column chromatography. *Mailing Add:* Col Arts & Sci Tenn Tech Univ Box 5065 Cookeville TN 38505

LERNER, JULES, b Englewood, NJ, Oct 24, 41; m 69. GENETICS, CYTOLOGY. *Educ:* Bowdoin Col, BA, 63; Johns Hopkins Univ, PhD(biol), 67. *Prof Exp:* Asst prof, 67-71, assoc prof, 71-77, PROF BIOL, NORTHEASTERN ILL UNIV, 77- *Mem:* AAAS. *Res:* Developmental biology and genetics. *Mailing Add:* Dept Biol Northeastern Ill Univ 5500 N Saint Louis Ave Chicago IL 60625

LERNER, LAWRENCE ROBERT, b New York, NY, Mar 17, 43; m 64; c 2. ORGANIC CHEMISTRY. *Educ:* City Col New York, BS, 64; Mich State Univ, PhD(org chem), 68. *Prof Exp:* Res chemist org pigments, E I du Pont de Nemours & Co, 68-72; group leader, Harmon Colors Corp, 72-73, supvr org pigments, 73-81, mgr process control & develop, 81-85, TECH MGR, MOBAY CHEM CORP, 85- *Concurrent Pos:* Adj asst prof, County Col Morris, NJ, 73-74. *Mem:* Sigma Xi; Am Chem Soc; AAAS; Inter-Soc Color Coun. *Res:* Synthesis of colored organic pigments; study of the effects of structure on the photostability, color and physical properties of organic pigments. *Mailing Add:* 176 Grove Terr Livingston NJ 07039

LERNER, LAWRENCE S, b New York, NY, Mar 10, 34; m 59. SOLID STATE PHYSICS, HISTORY OF SCIENCE. *Educ:* Univ Chicago, AB, 53, MS, 55, PhD(physics), 62. *Prof Exp:* Staff mem, Labs Appl Sci, Univ Chicago, 58-60; physicist, Hughes Res Labs, 62-65 & Hewlett-Packard Labs, Calif, 65-67; res scientist, Lockheed Palo Alto Res Lab, 67-69; assoc prof, 69-73, dir, Gen Honors Prog, 76-80, PROF PHYSICS & ASTRON, CALIF STATE UNIV, LONG BEACH, 73- *Concurrent Pos:* Danforth assoc, 75-; mem, Nat Fac Humanities Arts & Sci 78-; foreign mem, Ctr Hist of Ideas in the Anglo-Am World Sorbonne; mem, Calif Curric Framework & Criteria Comt Sci, 88-89. *Mem:* AAAS; Am Phys Soc; Am Asn Physics Teachers; Hist Sci Soc; Sigma Xi. *Res:* Fermi surfaces of metals and semimetals; preparation and properties of ternary compound semiconductors; semiconductor physics; influence of non-scientific philosophical movements on early scientific revolution. *Mailing Add:* Dept Physics & Astron Calif State Univ Long Beach CA 90840

LERNER, LEON MAURICE, b Chicago, Ill, Feb 2, 38; m 59; c 3. BIOCHEMISTRY. *Educ:* Ill Inst Technol, BS, 59, MS, 61; Univ Ill, PhD(biochem), 64. *Prof Exp:* Res assoc biochem, Col Med, Univ Ill, 64-65; from instr to assoc prof, 65-80, PROF BIOCHEM, STATE UNIV NY HEALTH SCI CTR, BROOKLYN, 80- *Mem:* AAAS; Am Chem Soc; Sigma Xi. *Res:* Potential nucleic acid antimetabolites; nucleoside analogs; chemistry and biochemistry of carbohydrates. *Mailing Add:* 450 Clarkson Ave Brooklyn NY 11203

LERNER, LEONARD JOSEPH, b Roselle, NJ, Sept 26, 22. ENDOCRINOLOGY, REPRODUCTION & FERTILITY. *Educ:* Rutgers Univ, BS, 43, AB, 51, MS, 53, PhD(zool), 54. *Prof Exp:* Pharmacist, 46-51; asst, Bur Biol Res, Rutgers Univ, 53-54; endocrinologist, Wm S Merrell Co, 54-58; head endocrine res, Squibb Inst Med Res, 58-71; dir endocrinol, Gruppo Lepetit Spa, 71-77; RES PROF DEPTS OBSTET & GYNEC & PHARMACOL, THOMAS JEFFERSON MED COL, 77- *Concurrent Pos:* Assoc mem, Bur Biol Res, Rutgers Univ; vis prof obstet & gynec, Hahnemann Med Sch; mem, Steering Comt Int Study Group, Steroid Hormones; mem, Breast Cancer Task Force Comt & Concept Rev Comt, Nat Cancer Inst; consult, pharmaceut co; mem, Animal Res & Experimentation Comt, NY Acad Sci; Comt Animals in Res, Soc Study Reprod, co-ed res steroids; chmn & vchmn sect Biol & Med, NY Acad Sci; mem bd dirs, Am Diabetes Soc, NJ affil; lectr pharmacol, Sch Nursing, Univ Pa, 90- *Honors & Awards:* Cain Mem Award, Am Asn Cancer Res, 89. *Mem:* Endocrine Soc; Am Physiol Soc; fel NY Acad Sci; Am Fertil Soc; Soc Study Reproduction; Am Asn Cancer Res; Soc Exp Biol Med; fel AAAS. *Res:* Hormone antagonists; fertility control; pregnancy, ova transport and reproduction; placenta; prostaglandins; ovulation; steroids; endocrine-tumor relationships and anti-cancer research; hormone treatment pre- and post- natally effects on hormonal and behavior responses; endocrine biochemistry; central nervous system-endocrine system relationship; adrenal physiology; growth and development; diabetes, atherosclerosis and endocrine in relation to stress, hormone, blood lipids; anti-inflammation; diabetes and reproduction, diabetes and cardiovascular system; awarded 35 patents; author of numerous publications. *Mailing Add:* Thomas Jefferson Univ Med Col Dept Pharmacol 1020 Locust St Philadelphia PA 19107

LERNER, LOUIS L(EONARD), b Chicago, Ill, Feb 25, 15; m 49; c 1. COSMETIC CHEMISTRY, MEDICAL & HEALTH SCIENCES. *Educ:* Cent YMCA Col, BS, 42. *Prof Exp:* Chem asst, Universal Merchandise Co, 34, assoc chemist, 35-37; chief chemist, Russian Duchess Labs, 37; pres & dir res, LaLerne Labs, 37-40; dir res & prod, Consol Royal Chem Corp, 40-46; exec vpres & dir res & prod, Allied Home Prods Corp, 46-49; vpres & dir res, Kalech Res Labs, 49-50; vpres & dir res & new prod develop, Bymart, Inc, 50-52; sr scientist, Personal Care Div, Gillette Co, 52-74; consumer protection specialist, 74, PHYS SCIENTIST, FED TRADE COMN, US GOVT, 75- *Concurrent Pos:* Dir, Brokers, Inc, 37-40; instr, Cent YMCA Col, 42-44; dir, Allied Home Prods Corp, Ill, 47-49; vpres, Phil Kalech Co, 49-50; dir, AD Prods Corp, 64-66; consult, Seaquist Valve Co, 74-; sect ed, Chem Bull, Am Chem Soc, 71-77, consult ed, 78-80, ed, 91- *Mem:* Fel AAAS; Am Chem Soc; Soc Cosmetic Chem; NY Acad Sci; Nat Asn Sci Writers; Soc Dyers & Colorists, UK. *Res:* Product development, exploratory research; pharmaceuticals, proprietaries, cosmetics, detergents, emulsions, waving compositions, dyes and pigments; chemistry of polymers, proteins and enzymes; mechanical devices; surface chemistry; consumer products; environmental sciences; biology. *Mailing Add:* 900 N Lake Shore Dr Chicago IL 60611

LERNER, MELVIN, b Milwaukee, Wis, June 8, 17; m 68; c 2. ANALYTICAL CHEMISTRY. *Educ:* City Col NY, BS, 37. *Prof Exp:* Res chemist, Univ Patents, Inc, 38-39; chemist, Hoover & Strong, Inc, 39-41; chemist, US Customs Lab, 41-50, from asst chief chemist to chief chemist, 50-67; dir tech serv, US Customs, 67-87; CONSULT, 87- *Mem:* AAAS; Am Chem Soc; Am Soc Testing & Mat. *Res:* Analytical chemistry of marijuana; narcotics and dangerous drugs; sampling of bulk materials; instrumental analytical chemistry. *Mailing Add:* 4017 Flamingo Dr El Paso TX 79902-1313

LERNER, MICHAEL PAUL, b Los Angeles, Calif, May 2, 41; m 65; c 2. VIROLOGY, CELL BIOLOGY. *Educ:* Univ Calif, Los Angeles, BA, 63; Kans State Univ, MS, 67; Northwestern Univ, PhD(microbiol), 70. *Prof Exp:* Nat Inst Neurol Dis & Stroke fel, Univ Calif, Los Angeles, 70-71, asst res biologist neurochem, Ctr Health Sci, 72-73; ASSOC PROF MICROBIOL, HEALTH SCI CTR, UNIV OKLA, 73- *Concurrent Pos:* Fel, NATO Advan Study Inst, Italy, 72. *Mem:* Am Soc Microbiol. *Res:* Mammalian cell biochemistry and development. *Mailing Add:* Dean McGee Eye Inst 608 Stanton L Young Blvd Oklahoma City OK 73104

LERNER, MOISEY, b Kiev, USSR, May 15, 31; US citizen; wid; c 2. SOFTWARE SYSTEMS, ELECTRICAL ENGINEERING. *Educ:* Polytech Inst, USSR, MS, 53; Pedagogical Inst, BA, 54; Power Eng Inst, USSR, MS, 57, PhD(elec eng), 61. *Hon Degrees:* Docent, Supreme Cert Comn, Moscow, USSR, 64. *Prof Exp:* Asst prof physics, Inst Elec Commun, USSR, 62-68; asst prof elec eng, Air Force Mil Eng Acad, USSR, 68-73; DIR RES & DEVELOP, SANFORD PROCESS CORP, NATICK, MA, 75- *Concurrent Pos:* Pres, Tomin Corp, Mass, 82; patent agent. *Mem:* Am Electroplaters Soc; Inst Elec & Electronics Engrs. *Res:* Low voltage hardcoating of aluminum; designing large electrical and software systems; bottom line double entry accounting system; heat dissipation at non-sinusoidal voltages; topology. *Mailing Add:* Tomin Corp PO Box 82-206 Wellesley MA 02181

LERNER, NARCINDA REYNOLDS, b Brooklyn, NY, Oct 10, 33; m 59. CHEMICAL KINETICS. *Educ:* Hofstra Univ, BA, 56; Univ Chicago, MS, 59, PhD(chem), 62. *Prof Exp:* Mem tech staff, Hughes Res Labs Labs, Calif, 62-63; res scientist, Lockheed Palo Alto Res Lab, 66-70; RES SCIENTIST, AMES RES CTR, NASA, 70- *Mem:* AAAS; Am Phys Soc. *Res:* Paramagnetic resonance; electron nuclear double resonance; crystalline field theory; crystal preparation; electrical properties of polymers; polymer degradation; abiotic synthesis of organic compounds. *Mailing Add:* Ames Res Ctr NASA MS 230-3 Moffett Field CA 94035

LERNER, NORMAN CONRAD, b New York, NY; c 2. ENGINEERING ECONOMICS, INVESTMENT FEASIBILITY ANALYSIS. *Educ:* Mass Inst Technol, BS, 57; Columbia Univ, MBA, 61; Am Univ, PhD(math econ), 68. *Prof Exp:* Proj mgr, ITT/RCA Corp, 57-65 & Mitre Corp, 65-68; dir, Command Systs Div, Computer Sci Div, 68-71; PRES, TRANSCOM, INC, 69- *Concurrent Pos:* Spec asst to dir, Exec Off Pres, US Off Telecommun Policy, 71-72; assoc prof mgt sci, George Washington Univ; rep, US-Peoples Repub China Telecommun Protocol; partic, World Admin Radio Conf, Presidential Task Force Commun Policy, Cong Off Technol Assessment-Int Telecommun, NASA Joint Study Group-Satellite Commun Underdevelop Countries & FCC Cable TV Adv Panel; mem, US State Dept Comt Ctr Telecommun Develop, Int Telecommun Union. *Mem:* Nat Soc Prof Engrs; Am Econ Asn; Inst Elec & Electronics Engrs; Nat Asn Bus Economists. *Res:* Economic, financial and market development aspects of high technology industries in the US and overseas, particularly telecommunications and energy; extensive writing addresses the privatization of telecommunications in developing countries. *Mailing Add:* 4527 Pickett Rd Fairfax VA 22042

LERNER, PAULINE, b Baltimore, Md, July 4, 48; div. NEUROCHEMISTRY. *Educ:* Goucher Col, BA, 69; Univ Md, PhD(chem), 73. *Prof Exp:* Chemist, NIH, 74-80; CHEMIST, FOOD & DRUG ADMIN, 80- *Mem:* AAAS; Asn Women Sci; Am Chem Soc; Am Soc Neurochem; Soc Neurosci. *Res:* Regulatory work in the area of nutrition and health. *Mailing Add:* Food & Drug Admin HFF-204 200 C St SW Washington DC 20204

LERNER, RICHARD ALAN, b Chicago, Ill, Aug 26, 38; m 66; c 3. IMMUNOLOGY. *Educ:* Stanford Univ, MD, 64. *Prof Exp:* Intern med, Stanford Univ, 64-65; assoc cell biol, Wistar Inst, 68-70; assoc immunol, 70-71, assoc mem, 71-73, PROF DEPT CHEM & DIR, RES INST, SCRIPPS CLIN, LA JOLLA, CALIF, 74- *Concurrent Pos:* USPHS grant, Scripps Clin & Res Found, 65-68; consult, Nat Cancer Inst, 72- *Mem:* Nat Acad Sci; Am Soc Path; Biophys Soc; Am Soc Microbiol; Am Soc Immunol. *Res:* Molecular medicine; differentiation. *Mailing Add:* Dept Molecular Biol Res Inst Scripps Clin 10666 N Torrey Pines Rd La Jolla CA 92037

LERNER, RITA GUGGENHEIM, b New York, NY, May 7, 29; m 54; c 2. INFORMATION SCIENCE. *Educ:* Radcliffe Col, AB, 49; Columbia Univ, MA, 51, PhD(chem phys), 56. *Prof Exp:* Res assoc chem phys, Columbia Univ, 56-64; staff physicist info sci, Am Inst Physics, 65, dep dir, Info Anal & Retrieval Div, 66-67; dir labs, Dept Biol Sci, Columbia Univ, 68; mgr planning & develop, Am Inst Physics, 69-73, mgr spec projs, 73-79, mgr mkt, 79-84, mgr books, 85-90; CONTRIB ED, VCH PUBL, INC, 91- *Concurrent Pos:* Info sci ser ed, Hutchinson Ross Publ Co, 73-85. *Mem:* Am Phys Soc; Am Chem Soc; AAAS; Am Soc Info Sci; NY Acad Sci; Sigma Xi. *Res:* Information systems and information retrieval; applications of new technology to publishing and distribution of information. *Mailing Add:* Winding Rd S Chauncey Ardsley NY 10502

LERNER, ROBERT GIBBS, b Brooklyn, NY, Mar 3, 36; m 58; c 3. HEMATOLOGY, HEMOSTASIS. *Educ:* NY Univ, AB, 56, MD, 60. *Prof Exp:* Teaching asst, NY Univ Sch Med, 61-62; instr, Univ Southern Calif Sch Med, 65-67; from asst prof to assoc prof, 67-80, PROF MED & CHIEF HEMAT, NY MED COL, 80- *Concurrent Pos:* Prin investr var grants & projs, NY Med Col, 70-; consult, Food & Drug Admin, 75-78; bd dirs, Island Peer Rev Orgn, 90-; mem, Coun Thrombosis, Am Heart Asn. *Mem:* Fel Am Col Physicians; Am Soc Hemat; Nat Hemophilia Found. *Res:* Hematology, specifically hemostasis and thrombosis; clinical aspects of the diagnosis and treatment of hemorrhagic and thrombotic. *Mailing Add:* NY Med Col Valhalla NY 10595

LERNER, SAMUEL, b Providence, RI, Jan 21, 08; m 39; c 1. ENGINEERING. *Educ:* Brown Univ, BSc, 30; Syracuse Univ, MS, 33; Turin Polytech, DrEng, 57. *Prof Exp:* Instr hydraul, Syracuse Univ, 31-33; instr civil eng, from asst prof to prof, 41-76, dir construct planning, 63-76, EMER PROF CIVIL ENG, BROWN UNIV, 73- *Concurrent Pos:* Owner, Lerner Assocs, 73- *Mem:* Am Soc Civil Engrs; Am Concrete Inst; Int Asn Bridge & Struct Eng. *Res:* Building construction. *Mailing Add:* 14 Cooke St Pawtucket RI 02860

LERNER, SIDNEY ISAAC, occupational medicine; deceased, see previous edition for last biography

LEROI, GEORGE EDGAR, b London, Eng, June 23, 36; US citizen; m 73; c 4. CHEMICAL PHYSICS. *Educ:* Univ Wis, BA, 56; Harvard Univ, AM, 58, PhD(chem), 60. *Prof Exp:* Res assoc chem, Univ Calif, Berkeley, 60-62; lectr, Princeton Univ, 62-64, asst prof, 64-67; assoc prof, 67-72, PROF CHEM, MICH STATE UNIV, 72- *Concurrent Pos:* Guest Prof, Lab Phys Chem, Swiss Fed Inst Technol, 74-75; Japan Soc Prom Sci vis prof, 77; Syncrotron Ultraviolet Radiation Facil fel, US Nat Bur Standards, 81-82; guest res, Brookhaven Nat Lab, 85; Ford Found fel, 52-56; mem, Phys Chem Div, Am Chem Soc. *Honors & Awards:* Coblentz Prize, 72. *Mem:* Am Phys Soc; Am Chem Soc. *Res:* Molecular spectroscopy and structure; vacuum ultraviolet, visible, infrared, far infrared, Raman; ion/molecule chemistry; laser spectroscopy and photochemistry. *Mailing Add:* Dept Chem Mich State Univ East Lansing MI 48824-1322

LEROITH, DEREK, b Cape Town, SAfrica, Jan 3, 45; m 79; c 3. MEDICINE, DIABETES. *Educ:* Univ Cape Town, MB, ChB, 67, PhD(med), 73. *Prof Exp:* Registr med, Univ Cape Town, SAfrica, 72-75; sr registr med, Middlesex Med Sch, London, England, 75; sr lectr med, Univ Ben Gurion, Israel, 76-79; assoc prof med, Univ Cincinnati, 83-84; vis scientist diabetes, 79-83, SR INVESTR DIABETES, NIH, 84- *Mem:* Fel Am Col Physicians; Am Endocrine Soc. *Res:* Evolutionary origins of the vertebrate endocrine systems; hormonal substances in invertebrates; brain insulin receptors. *Mailing Add:* Diabetes Br NIH Bldg 10 Rm 85243 Bethesda MD 20892

LEROUX, EDGAR JOSEPH, b Ottawa, Ont, Jan 23, 22; m 44; c 3. INSECT ECOLOGY, ENTOMOLOGY. *Educ:* Carleton Univ, Can, BA, 50; McGill Univ, MSc, 52, PhD(entom), 54. *Hon Degrees:* DSc, McGill Univ, 73; DU, Univ Ottawa, 86. *Prof Exp:* Asst entomologist, Fruit Insect Invests, Sci Serv, 49-50; res scientist, Res Br, Can Dept Agr, 50-62, res coordr, 65-68, asst dir gen, 68-75, dir gen, 75-78, asst dep minister res, 78-87; RETIRED. *Concurrent Pos:* Demonstr, MacDonald Col, McGill Univ, 50-51, asst, 53-54, lectr, 58-62, assoc prof, 62-65, hon prof, 70-71; mem orchard protection comt, Info & Res Serv, Que Dept Agr, 59-63, adv comt entom probs, Defense Res Bd, Dept Nat Defense, 64-67, panel experts integrated pest control, Food & Agr Orgn, 66-71; sci ed, Can J Plant Sci, 65-68; dir, Biol Coun Can, 66-70, pres, 70-71; off cor entom, Commonwealth Inst Biol Control, 67-73; Can rep, Int Soc Hort Sci, 67-73; mem, World Hort Coun, 68-70; pres Biol Coun Can, 70-71, Entom Soc Can, 69-70 & Entom Soc Que, 65-66; dir, Can Soc Zool, 68-70; negotiated grants comt & adv comt biol, Nat Res Coun, 73-78; vchmn, Agr Stabilization Bd & Agr Prod Bd, 77-80, Can Agr Res Coun, 77-87; Co-chmn, Can/USSR Agr Working Group & Can/Romania Comt Cooperation Agr, 81-87; chmn, comt agr, Orgn Environ Coop & Develop, 78-80, Fed Interdept Comt Pesticides, 78-87, assoc comt biotechnol, Nat Res Coun, 83-87; mem, coord comt, Can Agr Servs, 75-87, interdept comt, Can Ministry State Sci & Technol, 78-84, indept panel, Energy Mines & Resources Can, 78-87; mem, Nat Res Coun Can, 80-86, Nat Biotechnol Adv Comt, 83-87. *Honors & Awards:* Jubilee Medal, Governor Gen Can, 77; Armand Frappier Medal, 84; Gold Medal, Entom Soc Can, 86; Golden Award, Can Feed Indust Asn, 86. *Mem:* Fel Entom Soc Can (pres, 69-70); French-Can Asn Advan Sci; fel Agr Inst Can; Can Soc Zoologists; Asn Advan Sci Can (hon treas, 70-72). *Res:* Insect ecology; integrated pest control; morphology; toxicology; author of over 75 publications. *Mailing Add:* 27 Keppler Cres Nepean ON K2H 5Y1 Can

LEROUX, EDMUND FRANK, b Muskegon Heights, Mich, Mar 8, 25; m 49; c 3. HYDROLOGY. *Educ:* Mich State Univ, BS, 48. *Prof Exp:* Geologist, US Geol Surv, 49-60, chief, Manpower Sect, 61-64, asst dist chief hydrol, 64-; RETIRED. *Mem:* Am Geophys Union; Int Asn Hydrogeologists; AAAS. *Res:* Ground water temperature; hydrology of glacial terrain in a semiarid climate. *Mailing Add:* 1675 Illinois SW Huron SD 57350-2469

LEROUX, PIERRE, b Quebec City, Can, Aug 18, 42; m 64; c 2. MATHEMATICS. *Educ:* Univ Montreal, BSc, 64, MSc, 66, PhD(math), 70. *Prof Exp:* PROF MATH, UNIV QUE, MONTREAL, 71-, DIR, COMBINATORICS & MATH INFO LAB, 90- *Mem:* Can Math Soc; Am Math Soc; Soc Indust & Appl Math. *Res:* Enumerative combinatorics and special functions. *Mailing Add:* Math Info Dept Univ Que Montreal PQ H3C 3P8 Can

LEROY, ANDRË FRANÇOIS, b Philadelphia, Pa, Sept 30, 33. ANALYTICAL CHEMISTRY, PHYSICAL CHEMISTRY. *Educ:* Yale Univ, BE, 56; Calif Inst Technol, MS, 57; Harvard Univ, AM, 65, PhD(eng), 67. *Prof Exp:* Engr, Radiol Health Res Activ, USPHS, 58-60, res chemist, Northeastern Radiol Health Lab, Mass, 63-68; engr, NIH, Md, 69-78; mem staff, Sci Off, Am Embassy, Paris, France, 78-80; CHIEF, ANALYTICAL METHODS, BIOMED ENG & INSTRUMENTATION BR, DIV RES SERV, NIH, 80- *Concurrent Pos:* Dir, Tech Equip Seminars, US Dept Com, France, 80, Spain, Italy, Greece, 81 & France & Belg, 83; coordr, instrumentation & biomed eng sci res, NIH-Nat Inst Health & Med Res, France, 83-; div res serv rep, Fogarty Int Ctr, 84-; coordr, NIH-CGR Collab Res Prog in Magnetic Resonance Imaging, 86-; mem, Bd Sci Counrs, Life Sci Div, French AEC, 85-86. *Honors & Awards:* Clemens Herschel Prize; Commendation Medal, Pub Health Serv, 87. *Mem:* Am Chem Soc; Am Inst Chem Engrs; Sigma Xi. *Res:* Physical chemistry of transition metal complexes; analytical chemical and ultra-trace level isolation, characterization and quantitation of metal complex species and kinetics of protein binding and their transformations in biological systems and natural waters; high spatial resolution analysis of elements by instrumental micro analysis (electron- probe); WDX and EDX, chromatography, neutron activation analysis. *Mailing Add:* 11705 College View Dr Wheaton MD 20902

LEROY, DONALD JAMES, chemical kinetics, science policy; deceased, see previous edition for last biography

LEROY, EDWARD CARWILE, b Elizabeth City, NC, Jan 19, 33; m 60; c 2. RHEUMATOLOGY. *Educ:* Wake Forest Univ, Winston-Salem, BS, 55; Univ NC, Chapel Hill, MS, 58, MD, 60; Am Bd Internal Med, dipl, 67. *Prof Exp:* Med internship & asst residency, Presby Hosp, Columbia-Presby Med Ctr, NY, 60-62; clin assoc, Nat Heart Inst, Bethesda, Md, 62-65; fel arthritis, Columbia Univ Col Physicians & Surgeons, 65-66, assoc, 66-67, from asst prof to assoc prof, Dept Med, 67-75, dir, Div Rheumatic Dic, 71-75; PROF MED & DIR, DIV RHEUMATOLOGY & IMMUNOL, DEPT MED, MED UNIV SC, CHARLESTON, 75-, CHMN, DEPT PHYS MED & REHAB, 82- *Concurrent Pos:* NIH spec fel, 65-67; asst attend physician, Presby Hosp, NY, 67-70, assoc attend physician & dir, Edward Daniels Faulkner Arthritis Clin, 70-75, mem & chmn, var comts, Am Rheumatism Asn, 67-, Am Col Rheumatology, 89- & Am Col Physicians, 75-80; vis prof & lectr, numerous US & foreign univs, 69-91; Fogarty sr int fel endothelial cell biol, US Dept Health & Human Serv, 82-83; sabbatical, Corpus Christi Col, Cambridge Univ, Eng, 82-83. *Mem:* Sigma Xi; fel Am Rheumatism Asn; Am Col Rheumatology; fel Am Col Physicians; AAAS; Am Asn Immunologists. *Res:* Rheumatology; immunology; author of numerous medical publications. *Mailing Add:* Arthritis Clin & Res Ctr Med Univ SC 171 Ashley Ave Charleston SC 29425

LEROY, ROBERT FREDERICK, b Passaic, NJ, July 24, 50; m 74; c 2. NEUROLOGY, ELECTROENCEPHALOGRAPHY. *Educ:* Brown Univ, AB, 72; Pa State Univ, MD, 77. *Prof Exp:* Intern, Dept Med, Baltimore City Hosp, 77-78; resident neurol, Dept Neurol, Sch Med, Yale Univ, 78-81, fel, Merritt Pulnam Epilepsy Found Am, 81-82; ASST PROF & DIR CLIN NEUROPHYSIOL, DEPT NEUROL, HEALTH SCI CTR, SOUTHWESTERN MED SCH, UNIV TEX, 82- *Concurrent Pos:* Consult, Dallas Tex Vet Admin, Med Ctr, 82. *Mem:* Am Acad Neurol. *Res:* Electroencephalography and general clinical neurophysiology as it pertains to epilepsy and its treatments. *Mailing Add:* 2606 Augusta Lane Arlington TX 76012-4247

LEROY, ROBERT JAMES, b Ottawa, Ont, Sept 30, 43; m 67; c 4. INTERMOLECULAR FORCES, MOLECULAR CLUSTERS. *Educ:* Univ Toronto, BSc, 65, MSc, 67; Univ Wis-Madison, PhD(chem), 71. *Prof Exp:* From asst prof to assoc prof, 72-82, PROF CHEM, UNIV WATERLOO, 82- *Concurrent Pos:* A P Sloan Found fel, 74, sr vis fel, Sci Res Coun UK, 76 & J S Guggenheim Mem Found fel, 79; invited prof, Univ Paris-S Orsay, France, 82; dir, Guelph-Waterloo Ctr Grad Work Chem, 82-85. *Honors & Awards:* Rutherford Mem Medal, Royal Soc Can, 84. *Mem:* Can Asn Physicists; fel Chem Inst Can; Am Phys Soc; Am Chem Soc. *Res:* Empirical methods for determining intermolecular forces in simple systems; understanding and predicting the properties of simple molecules and molecular clusters. *Mailing Add:* Guelph-Waterloo Ctr Grad Work Chem Univ Waterloo Waterloo ON N2L 3G1 Can

LEROY, RODNEY LASH, b Ottawa, Ont, Nov 15, 41; m 83; c 6. PHYSICAL CHEMISTRY, ELECTROCHEMISTRY. *Educ:* Univ Toronto, BSc, 64, MA, 65, PhD(phys chem), 68; McGill Univ, dipl 78, MBA, 83. *Prof Exp:* Fel phys chem, Univ Colo & Yale Univ, 68-70; assoc scientist chem, Noranda Res Ctr, 70-72, group leader electrochem, 72-78, prin scientist electrochem, 78-79, prog mgr res, 78-79; tech dir, Electrolyser Inc, 80-85, exec vpres, 85-88; MGR ENERGY & PROD APPLICATIONS LAB, NORANDA TECHNOL CTR, 88- *Mem:* Fel Chem Inst Can; Electrochem Soc; Nat Asn Corrosion Engrs; Int Asn Hydrogen Energy; Soc Petrol Engrs; Soc Explor Geophysicists; Petrol Soc. *Res:* Corrosion research; electrometallurgy of non-ferrous metals, especially copper and zinc; hydrogen production by electrolysis of water; petroleum engineering. *Mailing Add:* 240 Hymus Blvd Pointe Claire PQ H9R 1G5 Can

LERSTEN, NELS R, b Chicago, Ill, Aug 6, 32; m 58; c 3. BOTANY. *Educ:* Univ Chicago, BS, 58, MS, 60; Univ Calif, Berkeley, PhD(bot), 63. *Prof Exp:* From asst prof to assoc prof, 63-70, PROF BOT, IOWA STATE UNIV, 70- *Mem:* Bot Soc Am; Sigma Xi. *Res:* Systematic and developmental anatomy of angiosperms; embryology of flowering plants. *Mailing Add:* Dept Bot Iowa State Univ Ames IA 50011-1020

LES, EDWIN PAUL, b Adams, Mass, Dec 28, 23; m 67; c 2. LABORATORY ANIMAL SCIENCE. *Educ:* Northeastern Univ, BS, 52; Ohio State Univ, MS, 53; PhD(genetics), 59. *Prof Exp:* Assoc staff scientist, Jackson Lab, 59-60; biologist, Biol Div, Oak Ridge Nat Lab, 60-62; staff scientist, Jackson Lab, 62-75, sr staff scientist, 75-; RETIRED. *Mem:* Am Asn Lab Animal Sci. *Res:* Effect of environment on reproduction, growth and survival of laboratory mice; mouse husbandry techniques and practices; laboratory animal ecology. *Mailing Add:* Jackson Lab 600 Main St Bar Harbor ME 04609

LESAGE, LEO G, b Concordia, Kans, Apr 15, 35; m 58; c 2. NUCLEAR ENGINEERING, PHYSICS. *Educ:* Univ Kans, BS, 57; Stanford Univ, MS, 62, PhD(nuclear eng), 66; Univ Chicago, MBA, 81. *Prof Exp:* Nuclear engr, Fast Breeder Reactor Develop, 66-81, DIR, APPL PHYSICS DIV, ARGONNE NAT LAB, 81- *Concurrent Pos:* US mem, Comt on Reactor Physics, Nuclear Energy Agency, 81- *Mem:* Fel Am Nuclear Soc. *Res:* Fast reactor physics; fast reactor critical experiments. *Mailing Add:* Eng Physics Div Argonne Nat Lab 9700 S Cass Ave Bldg 207 Argonne IL 60439

LESCARBOURA, JAIME AQUILES, b Barcelona, Spain, Aug 29, 37; US citizen; m 57; c 2. PETROLEUM ENGINEERING. *Educ:* Univ Kans, BS, 59, PhD(chem eng), 67; Univ Wis-Madison, MS, 61. *Prof Exp:* Engr, Cardon Refinery, Shell Oil Co Venezuela, 61-63; res scientist, Continental Oil Co, 67-76, sr res scientist, 76-82, staff eng, 82-84; res assoc, 84-90, SR STAFF ENGR, CONOCO, INC, 91- *Concurrent Pos:* Am Oil Found fel, 66-67. *Mem:* Soc Petrol Engrs; Soc Rheology. *Res:* Flow of Newtonian and non-Newtonian fluids; falling cylinder viscometer for non-Newtonian fluids; turbulent flow drag reduction by addition of polymers; well testing; formation evaluation; rheology of crosslinked gels, drilling muds, waxy crudes and viscous crudes; well stimulation by blasting. *Mailing Add:* 2702 Shadowdale Houston TX 77043

LESER, ERNST GEORGE, b Mineola, NY, May 3, 43; m 69; c 3. ORGANIC CHEMISTRY. *Educ:* Bucknell Univ, BS, 65; Fordham Univ, PhD(org chem), 70. *Prof Exp:* Res chemist, Jackson Lab, 69-74, prod supvr, Chamber Works, 74-84, sr res chemist, Jackson Lab, 84-90, RES ASSOC, JACKSON LAB, E I DU PONT DE NEMOURS & CO, INC, 90- *Mem:* Am Chem Soc. *Res:* Supervision of dyes and intermediates production; fluorochemical research; tetraethyl lead process assistance. *Mailing Add:* Chambers Works E I du Pont de Nemours & Co Inc Wilmington DE 19898

LESER, RALPH ULRICH, b Bloomington, Ind, Dec 31, 05; m 66; c 2. INTERNAL MEDICINE. *Educ:* Ind Univ, AB, 27, MD, 30; Am Bd Internal Med, dipl, 44. *Prof Exp:* Intern, Philadelphia Gen Hosp, Pa, 30-32; fel internal med, Mayo Clin, 34-37; from instr to assoc med, Ind Univ, Indianapolis, 38-50, asst prof, 50-67, assoc prof med, Sch Med, 67-88; RETIRED. *Concurrent Pos:* Vis physician, Marion County Gen Hosp, 38-, chief, Diag Clin, 46-58; vis physician, Methodist, St Vincent's & Community Hosps. *Mem:* AMA; fel Am Col Physicians. *Res:* Cardiology; diseases of metabolism; gastroenterology. *Mailing Add:* 5434 Ashurst St Indianapolis IN 46220

LESH, THOMAS ALLAN, b Chicago, Ill, Aug 6, 29; m 79. PHYSIOLOGY. *Educ:* Mich State Univ, BS, 51; Ind Univ, PhD(physiol), 68. *Prof Exp:* Assoc ed, Howard W Sams & Co, Inc, Ind, 55-63; USPHS cardiovasc trainee, Bowman Gray Sch Med, 68-70; asst prof physiol, Med Ctr, Univ Ark, Little Rock, 70-72; asst prof, 72-77, ASSOC PROF PHYSIOL & HEALTH SCI, BALL STATE UNIV & MUNCIE CTR MED EDUC, 77- *Mem:* Asn Am Physiol Soc; Sigma Xi. *Res:* Control of blood flow. *Mailing Add:* Dept Physiol & Health Sci Ball State Univ Muncie IN 47306

LESHER, DEAN ALLEN, b Endicott, NY, Feb 18, 27; m 49; c 3. CLINICAL PHARMACOLOGY, NEPHROLOGY. *Educ:* Colgate Univ, AB, 48; Univ Wis, PhD(pharmacol), 56; Univ Buffalo, MD, 62. *Prof Exp:* Asst pharmacol, Univ Mich, 51-52; asst, Univ Wis, 52-55; instr, Univ Buffalo, 55-58, assoc, 58-62; from intern to resident, Henry Ford Hosp, Detroit, 62-63, assoc med, 64-71; assoc dir med res, Lederle Labs, 71-73; assoc dir, Ciba Geigy Corp, 74-77, dir clin pharmacol, 77-78, sr dir clin res, 78-80, sr res fel, 80-84, dir gen drugs, 84-86, ASSOC DIR CLIN RES, CIBA GEIGY CORP, 86- *Mem:* Am Soc Clin Invest; Int Soc Nephrol; Am Soc Nephrol; Am Soc Clin Pharmacol & Therapeut; fel Am Col Clin Pharm; Soc Clin Trials; Drug Info Asn. *Res:* Renal transport of acids and bases; renal potassium transport; diuretics; renal disease; hemodialysis; peritoneal dialysis; renal transplantation; drug evaluation: anti inflammatory, central nervous system, antibiotic, ophthalmic; drug intoxication; anti-tumor agents. *Mailing Add:* CIBA-Geigy Corp 556 Morris Ave Summit NJ 07901

LESHER, GARY ALLEN, b Chicago, Ill, June 29, 50; m 73; c 2. PHARMACOLOGY, DRUG METABOLISM. *Educ:* Carroll Col, Wis, BS, 72; Purdue Univ, West Lafayette, MS, 75, PhD(pharmacol), 77. *Prof Exp:* ASST PROF PHARMACOL, SCH PHARM, UNIV MD, 77- *Mem:* Am Asn Col Pharm. *Res:* Narcotic drug dependence and drug interactions in narcotic dependent animals, effects of narcotics on drug metabolism; effects of environmental contaminents on drug metabolism; toxicology. *Mailing Add:* Dept Basic & Health Sci Ill Col Optom 3241 S Michigan Ave Chicago IL 60616

LESHER, GEORGE YOHE, pharmaceutical & synthetic organic chemistry; deceased, see previous edition for last biography

LESH-LAURIE, GEORGIA ELIZABETH, b Cleveland, Ohio, July 28, 38; m 69. DEVELOPMENTAL BIOLOGY. *Educ:* Marietta Col, BS, 60; Univ Wis, MS, 61; Case Western Reserve Univ, PhD(biol), 66. *Prof Exp:* Instr biol, Case Western Reserve Univ, 65-66; asst prof biol sci, State Univ NY Albany, 66-69; from asst prof to assoc prof biol, Western Reserve Col, Case Western Reserve Univ, 69-77, asst dean, 73-76; chmn, Dept Biol, Cleveland State Univ, 77-81, PROF BIOL, 77-, dean col grad studies, 81-86, interim provost, 89-90, DEAN COL ARTS & SCI, CLEVELAND STATE UNIV, 86- *Concurrent Pos:* NY State Res Found res fel, 66-67; USPHS instnl grant, 70-71; Am Cancer Soc grants, 68-71 & 77-80; Res Corp grant, 71; Am Cancer Soc instnl grant, 73; Am Heart Asn Grant, 82-83; Wright fel, Bermuda Biol Sta, 84. *Mem:* AAAS; Am Soc Zool; Soc Develop Biol; NY Acad Sci; Am Soc Cell Biol. *Res:* Study of the neural control of developmental events in cnidarian systems and the role of nematocyst products on the mammalian cardiovascular system. *Mailing Add:* Dept Biol Cleveland State Univ Cleveland OH 44115

LESHNER, ALAN IRVIN, b Lewisburg, Pa, Feb 11, 44; m 69; c 2. PSYCHOPHYSIOLOGY. *Educ:* Franklin & Marshall Col, AB, 65; Rutgers Univ, MS, 67, PhD(psychol), 69. *Prof Exp:* From asst prof to prof psychol, Bucknell Univ, 69-81; proj mgr, NSF, 80-82, dep exec dir, NSB comn on precol educ, 82-83, dep dir, Div Behav & Neurol Sci, 83-85, exec dir, Behav & Social Sci, 85-87; dir, Off S & T Ctrs Develop, 88; dep dir, 88-90, ACTG DIR, NIMH, 90- *Concurrent Pos:* NIMH res grant, 70-71 & 78-80; NSF res grant, 70-72 & 75-77; Fulbright lectr, Weizmann Inst, Israel, 77-78; vis scientist, Wis Regional Primate Res Ctr, 76-77. *Mem:* AAAS; fel Am Psychol Asn; Soc Neurosci; fel NY Acad Sci. *Res:* Biological basis of behavior; current and emerging science and technology policy issues; precollege and college level mathematics and science education. *Mailing Add:* Off Actg Dir NIMH 5600 Fishers Lane Rm 17-99 Rockville MD 20857

LESIKAR, ARNOLD VINCENT, b Galveston, Tex, Nov 3, 37; m 76. CHEMICAL PHYSICS, PHYSICS. *Educ:* Rice Univ, BS, 58; Calif Inst Technol, PhD(physics), 65. *Prof Exp:* Res asst physics, Calif Inst Technol, 60-65; asst physics, Tech Univ, Munich, Ger, 65-66; PROF PHYSICS, ST CLOUD STATE UNIV, 66- *Concurrent Pos:* Vis prof chem physics, Cath Univ Am, 78- *Mem:* Sigma Xi; Am Chem Soc; Am Asn Physics Teachers; Am Phys Soc. *Mailing Add:* Dept Physics St Cloud State Univ St Cloud MN 56301

LESINSKI, JOHN SILVESTER, b Philadelphia, Pa, Mar 29, 13; m 40; c 4. OBSTETRICS & GYNECOLOGY. *Educ:* Jagellonian Univ, Poland, MD, 39; Johns Hopkins Univ, MPH, 67. *Prof Exp:* Dep dir, Nat Res Inst Mother & Child Health, Warsaw, 52-65; chmn obstet & gynec, Med Acad, Warsaw & Inst Postgrad Training Physicians, Warsaw, 58-65; assoc prof mother & child health, obstet & gynec, Johns Hopkins Univ, 67-88; RETIRED. *Concurrent Pos:* Expert, Mother & Child Health, WHO, 60- *Honors & Awards:* Medal Exemplary Serv Health, Govt Poland, 62, Order Polonia Restituta, 50 & 63. *Mem:* Fel Am Acad Reproductive Med; fel Am Pub Health Asn; fel Am Col Obstetricians & Gynecologists; fel Royal Soc Health; fel Am Col Prev Med. *Res:* High risk factors in reproductive failure; sequelae of induced abortion; study of reproductive performance of individuals born prematurely or with low birth weight. *Mailing Add:* 5204 W Melrose St Chicago IL 60641-5397

LESKO, KEVIN THOMAS, b Apr 30, 56. FISSION PROPERTIES. *Educ:* Leland Stanford, Jr Univ, BS, 78; Univ Wash, PhD(physics), 83. *Prof Exp:* Postdoctoral fel, Argonne Nat Lab, 83-85; fel, 85-87, STAFF SCIENTIST II & PHYSICIST, NUCLEAR SCI DIV, LAWRENCE BERKELEY LAB, 87- *Concurrent Pos:* Tandem Accelerator Operator, Stanford Nuclear Physics Lab, 76-78; teaching asst, Stanford Physics Dept, Stanford Univ, 77-78; res asst, Nuclear Physics Lab, Univ Wash, 78-83, Accelerator Operator Instr, 81-83. *Mem:* Am Phys Soc. *Res:* Author of several scientific journals. *Mailing Add:* Lawrence Berkeley Lab Cyclotron Rd Bldg 88 Berkeley CA 94720

LESKO, PATRICIA MARIE, b Oakland, Calif, Jan 19, 47. POLYMER CHEMISTRY. *Educ:* Rice Univ, BA, 68, MS, 72, PhD(org chem), 73. *Prof Exp:* Fel org chem, Syntex Corp, 73-75; CHEMIST POLYMER CHEM, RES LABS, ROHM & HAAS CO, 75- *Mem:* AAAS; Am Chem Soc. *Res:* Emulsion polymers; coatings; polymeric controlled release formulations; corrosion. *Mailing Add:* PO Box 219 Spring House PA 19477

LESKO, STEPHEN ALBERT, b Cassandra, Pa, Dec 30, 31; m 81. BIOCHEMISTRY. *Educ:* Ind Univ, Pa, BS, 59; Univ Md, PhD(biochem), 65. *Prof Exp:* Instr, Johns Hopkins Univ, 65-68, res assoc, 68-73, asst prof, 73-82, assoc prof biophys, 82-90, ASSOC PROF ENVIRON HEALTH SCI, JOHNS HOPKINS UNIV, 82-, ASSOC PROF BIOCHEM, 90- *Mem:* AAAS; Am Chem Soc; Am Asn Cancer Res. *Res:* Chemical carcinogenesis; nucleic acid chemistry and biology; oxygen toxicity; chromosome topography in interphase nuclei; relationship between DNA damage and cellular responses in the carcinogenic process; quantification of gene expression at the single cell level; gene mapping. *Mailing Add:* Dept Biochem Johns Hopkins Univ 615 N Wolfe St Baltimore MD 21205

LESKOWITZ, SIDNEY, b New York, NY, Nov 15, 22; m 48; c 3. IMMUNOLOGY. *Educ:* City Col New York, BS, 43; Columbia Univ, MA, 48, PhD(chem), 50. *Prof Exp:* Res asst microbiol, Col Physicians & Surgeons, Columbia Univ, 50-54; from res assoc to asst prof bact & immunol, Harvard Med Sch, 54-70; PROF PATH, MED SCH, TUFTS UNIV, 70- *Concurrent Pos:* Asst immunol, Mass Gen Hosp, 54-57, assoc, 57-70. *Mem:* AAAS; Am Asn Immunol; Soc Exp Biol & Med. *Res:* Delayed hypersensitivity; induction of immunologic tolerance; structure of antigens; antigen processing and presentation. *Mailing Add:* Dept Path Tufts Med Sch 136 Harrison Ave Boston MA 02111

LESLEY, FRANK DAVID, b El Paso, Tex, Dec 20, 44; m 67; c 2. ANALYSIS. *Educ:* Stanford Univ, BS, 66; Univ Calif, San Diego, MA, 68, PhD(math), 70. *Prof Exp:* From asst prof to assoc prof, 70-77, PROF MATH, SAN DIEGO STATE UNIV, 77- *Mem:* Am Math Soc. *Res:* Boundary behavior of conformal mappings, including minimal surfaces and approximation theory. *Mailing Add:* Dept Math San Diego State Univ 5500 Campanik Dr San Diego CA 92182

LESLIE, CHARLES MILLER, b Lake Village, Ark, Nov 8, 23; m 46; c 3. MEDICAL ANTHROPOLOGY. *Educ:* Univ Chicago, PhB, 49, MA, 50, PhD(anthrop), 59. *Prof Exp:* Instr anthrop, Southern Methodist Univ, 50-51; instr, Univ Minn, 54-56; from instr to assoc prof, Pomona Col, 56-65; vis prof, Univ Wash, 65; assoc prof, Case Western Reserve Univ, 66-67; chmn dept, Univ Col, NY Univ, 67-71, prof anthrop, 67-76; PROF, CTR SCI & CULT, UNIV DEL, 76- *Concurrent Pos:* NSF fel, Sch Oriental & African Studies, Univ London, 62-63; res assoc, Dept Anthrop, Univ Chicago, 74-75; NSF res grant, 74, grant, 76; vis prof, Univ Calif, Berkeley, 87; med Anthrop ed, Social Sci & Med, 77- *Mem:* Fel AAAS; fel Am Anthrop Asn; fel Royal Anthrop Inst Gt Brit & Ireland; Asn Asian Studies; Soc Med Anthrop; Latin Am Studies Asn. *Res:* World view and social change in India and Latin America; comparative study of medical systems. *Mailing Add:* Ctr Sci & Cult Univ Del Newark DE 19716

LESLIE, GERRIE ALLEN, b Red Deer, Alta, Nov 19, 41; m 65; c 2. IMMUNOLOGY, IMMUNOCHEMISTRY. *Educ:* Univ Alta, BSc, 62, MSc, 65; Univ Hawaii, PhD(microbiol), 68. *Prof Exp:* From asst prof to assoc prof microbiol, Sch Med, Tulane Univ, 70-74; assoc prof, 74-81, PROF MICROBIOL & IMMUNOL, ORE HEALTH SCI UNIV, 81- *Concurrent Pos:* USPHS fel, Col Med, Univ Fla, 68-70; adj assoc prof microbiol, Sch Med, Tulane Univ, 74-80; res affil, Delta Regional Primate Res Ctr, Covington, La; NL Tartat res fel, 77 & 82. *Mem:* Am Asn Immunol; Am Soc Microbiol; Am Soc Zoologists; NY Acad Sci. *Res:* Phylogeny of immunoglobulin structure and function; regulation of the immune response; secretory immunologic system; immunoglobulin D. *Mailing Add:* Immunol Consults Lab Inc 13301 Knaus Rd Lake Oswego OR 97034

LESLIE, JAMES, b Belfast, Ireland, Apr 25, 34; m 64; c 3. PHARMACEUTICAL CHEMISTRY. *Educ:* Queen's Univ, Belfast, BSc, 56, PhD(chem), 59. *Prof Exp:* Fel, Okla State Univ, 59-61, asst prof & res assoc, 61-62; asst prof chem, Wash Col, 62-63; asst prof, 63-66, assoc prof med chem, 66-79, ASSOC PROF PHARM, SCH PHARM, UNIV MD, BALTIMORE, 79- *Concurrent Pos:* NIH res grant, 64, vis, Dept Clin Physics & Bioeng, Western Regional Hosp Bd, Glasgow, Scotland, 71-72. *Mem:* Am Chem Soc; Am Asn Cols Pharm. *Res:* Kinetics of processes of biological interest; drug analysis. *Mailing Add:* Univ Md Sch Pharm 20 N Pine St Baltimore MD 21201

LESLIE, JAMES C, b Berlin, Pa, July 14, 33; div; c 4. MATERIALS SCIENCE, CHEMICAL ENGINEERING. *Educ:* Pa State Univ, BS, 56; Ohio State Univ, MS, 58, PhD(chem eng), 64. *Prof Exp:* Asst chem anal, Pa State Univ, 55-56; eng exp sta, Ohio State Univ, 56-57, Dept Chem Eng, 57-58, fel, 58-59, res asst chem eng, Res Found, 59-62; sr res engr, Allegany Ballistics Lab, Hercules Inc, 62-66, group supvr mat res, 66-67, composite mat group suvpr, 67, group supvr, Adv Fiber Struct Group, 67-69, dept head mat res, 69-71, dept head advan composite mat struct, Prod & Process Develop, 71-72, western regional mgr, Advan Composites, 72-75, mgr advan compos, Reliable Mfg, 75-76; OWNER, JIMLESLIE CONSULTS & SALES REPS, 76-; OWNER & GEN MGR, ADVAN COMPOSITE PROD & TECHNOL, 81- *Concurrent Pos:* Instr, Frostburg State Col, 62-63 & WVa Univ Exten Serv, Allegany Ballistics Lab, 64-65; mgr mkt & eng, Advan Composite Pipe & Tube, 77-81. *Mem:* Assoc fel Am Inst Aeronaut & Astronaut; Am Inst Chem Eng; Soc Aerospace Mat & Process Eng; Soc Advan Mat & Process Eng (first vpres, 73, pres, 74); Solar Mesosphere Explor. *Res:* Research and development; consultant; prototype and process development in many areas of advanced composite materials fabrication; tracer techniques; development of graphite fibers; advanced composite materials & manufacturing techniques. *Mailing Add:* Advan Composite Prods & Technol 15602 Chemical Lane Huntington Beach CA 92649-1507

LESLIE, JAMES D, b Toronto, Ont, July 6, 35; m 64; c 3. SOLID STATE PHYSICS. *Educ:* Univ Toronto, BASc, 57; Univ Ill, MS, 60, PhD(physics), 63. *Prof Exp:* Asst prof, 63-68, ASSOC PROF PHYSICS, UNIV WATERLOO, 68- *Mem:* Am Phys Soc; Can Asn Physicists. *Res:* Low temperature physics; far infrared spectroscopy; superconductivity; electron tunneling. *Mailing Add:* Dept Physics Univ Waterloo Waterloo ON N2L 3G1 Can

LESLIE, JEROME RUSSELL, b Luverne, Minn, Oct 11, 39; m 64; c 3. WORD PROCESSING & ELECTRONIC DELIVERY OF NEWS RELEASES, PHOTOGRAPHY. *Educ:* SDak State Univ, BS, 62, MS, 90. *Prof Exp:* Reporter-photogr, Watertown Pub Opinion, SDak, 62-63; reporter-photogr & deskman, Sioux City J, Iowa, 63-73; state ed, Brookings Daily Reporter, SDak, 73-78; AGR NEWS ED, SDAK STATE UNIV, 78- *Mem:* Int Agr Commun Educ. *Res:* Evaluating electronic transfer and other news delivery methods; publications on plant science, animal science, veterinary science, dairy science and economics. *Mailing Add:* SDak State Univ Box 2231 Brookings SD 57007

LESLIE, JOHN FRANKLIN, b Dallas, Tex, July 2, 53; m 76; c 2. FUNGAL GENETICS. *Educ:* Univ Dallas, BA, 75; Univ Wis, Madison, MS, 77, PhD(genetics), 79. *Prof Exp:* Fel trainee, NIH, Lab Genetics, Univ Wis, Madison, 76-79; fel res affil, Dept Biol Sci, Stanford Univ, Calif, 79-81; res microbiologist genetics, Int Mineral & Chem Corp, 81-84; asst prof, 84-90, ASSOC PROF PLANT PATH, KANS STATE UNIV, MANHATTAN, 90- *Concurrent Pos:* Tech adv, Inst Christian Resources, San Jose, Calif, 81-; Assoc ed, Phytopath, 88-90. *Mem:* Genetics Soc Am; Mycol Soc Am; Am Soc Microbiol; Brit Mycol Soc; Soc Gen Microbiol; Am Phytopath Soc. *Res:* Molecular, classical and population genetics of filamentous fungi, especially Neurospora and Fusarium; stability of transformed DNA in fungi, genetics of mycotoxins in F-moniliforme; genetics of nit mutants and vegetative compatibility. *Mailing Add:* Dept Plant Path Throckmorton Hall Kans State Univ Manhattan KS 66506-5502

LESLIE, PAUL WILLARD, b Peekskill, NY, Apr 23, 48. POPULATION GENETICS, DEMOGRAPHY. *Educ:* Bucknell Univ, BA, 70; Pa State Univ, MA, 72, PhD(anthrop), 77. *Prof Exp:* Asst prof anthrop, Univ Tex, Austin, 76-78; ASST PROF ANTHROP, STATE UNIV NY, BINGHAMTON, 78- *Mem:* AAAS; Am Asn Phys Anthropologists; Human Biol Coun; Pop Asn Am; Soc Study Social Biol. *Res:* Population genetics, demography of small populations, mathematical modeling and computer simulation; interactions among the social, demographic, and genetic structures of human populations. *Mailing Add:* Dept Anthrop State Univ NY Binghamton NY 13901

LESLIE, RONALD ALLAN, b Welland, Ont, Oct 11, 48; m 74; c 1. CEREBROVASCULAR ANATOMY, NEUROANATOMY. *Educ:* Brock Univ, Ont, BSc, 70; Cambridge Univ, PhD(neurobiol), 75. *Prof Exp:* Fel neurosci, McMaster Univ, 75-76; asst prof, 76-82, ASSOC PROF ANAT, DALHOUSIE UNIV, 82- *Concurrent Pos:* Res fel, Oxford Univ, 83-84. *Mem:* Am Asn Anatomists; Soc Neurosci; Can Asn Anatomists; Can Asn Neurosci. *Res:* Neuroanatomy and neuropharmacology of brainstem autonomic centers and somato sensory cortex. *Mailing Add:* Dept Anat Dalhousie Univ Halifax NS B3H 4H6 Can

LESLIE, STEPHEN HOWARD, b New York, NY, Nov 6, 18; m 43; c 2. MEDICINE. *Educ:* NY Univ, BS, 38, MD, 42. *Prof Exp:* Clin asst, 46-49, clin instr med, 49-54, asst prof, 54-69, ASSOC PROF CLIN MED, SCH MED, NY UNIV, 69- *Concurrent Pos:* Fel, Sch Med, NY Univ, 44-46. *Mem:* Endocrine Soc; Am Diabetes Asn; Am Fedn Clin Res. *Res:* Metabolic diseases; endocrinology. *Mailing Add:* 160 E 38th St 33B New York NY 10016-2651

LESLIE, STEVEN WAYNE, b Franklin, Ind, Jan 23, 46; m 70; c 1. PHARMACOLOGY. *Educ:* Purdue Univ, BS, 69, MS, 72, PhD(pharmacol), 74. *Prof Exp:* Asst prof, 74-81, ASSOC PROF PHARMACOL, UNIV TEX, AUSTIN, 81- *Mem:* Sigma Xi; AAAS. *Res:* Investigations concerning the role of cellular organelles in calcium-mediated termination mechanisms in secretory tissues and the effects of various drugs on these termination mechanisms. *Mailing Add:* Col Pharmacol Univ Tex Austin TX 78712

LESLIE, THOMAS M, b Philadelphia, Pa, Nov 11, 54; m 82. PHOTOCHEMISTRY, SYNTHETIC ORGANIC CHEMISTRY. *Educ:* Rider Col, BS, 76; Univ Notre Dame, PhD(chem), 80. *Prof Exp:* Mem tech staff, Bell Tel Labs, 80-85; STAFF SCIENTIST, CELANESE RES CO, 85- *Mem:* Am Chem Soc; NY Acad Sci. *Res:* Mechanism of photochemical reactions via reaction intermediates produced in laser flash photolysis; study of the effect of subtle changes in chemical constitution on materials exhibiting liquid crystal phases; liquid crystal side chain polymers; nonlinear optical materials. *Mailing Add:* Dept Chem Univ Ala Huntsville AL 35899

LESLIE, WALLACE DEAN, b Dacoma, Okla, Nov 9, 22; m 48; c 3. ANALYTICAL CHEMISTRY. *Educ:* Northwestern State Col, Okla, BS, 47; Okla State Univ, MS, 50. *Prof Exp:* Asst chem, Okla State Univ, 47-49; instr, Northwestern State Col, Okla, 49-51; anal chemist, Mfg Dept, Continental Oil Co, 51-52, from assoc res chemist to sr res chemist, 52-62, res group leader, res & develop dept, 62-77, dir, anal res sect, 77-85; RETIRED. *Mem:* Am Chem Soc. *Res:* Analytical research and development; petroleum and petroleum products; petrochemicals. *Mailing Add:* 11 Forest Rd Ponca City OK 74604

LESLIE, WILLIAM C(AIRNS), b Dundee, Scotland, Jan 6, 20; nat US; m 48; c 1. METALLURGY. *Educ:* Ohio State Univ, BMetE, 47, MSc, 48, PhD(metall), 49. *Prof Exp:* Res assoc, Res Found, Ohio State Univ, 47-49; metallurgist, US Steel Corp, 49-53; assoc dir res, Thompson Prod Inc, 53-54; metallurgist, US Steel Corp, 54-57, sr scientist, Fundamental Res Lab, 57-63, asst dir phys metall, 63-69, mgr phys metall, E C Bain Lab Fundamental Res, 69-73; prof mat eng, 73-85, EMER PROF MAT ENG, UNIV MICH, 85- *Concurrent Pos:* Battelle vis prof, Ohio State Univ, 64-65; chmn, Int Conf Strength Metals & Alloys, Calif, 70; vis prof Melbourne Univ, Australia, 79; NSF-Coun Sci & Indust Res exchange scientist, India, 81; distinguished alumnus lectr, Ohio State Univ, 84. *Honors & Awards:* Krumb Lectr, Am Inst Mining, Metal & Petrol Engrs, 67, & Howe lectr, 82; Andrew Carnegie Lectr, Am Soc Metals, 70, Campbell Lectr, 71, Sauveur Lectr & Jeffries Lectr, 75; Garofalo Lectr, Northwestern Univ, 77. *Mem:* Fel & hon mem Am Soc Metals; fel Am Inst Mining Metall & Petrol Engrs (vpres, 75); fel Metals Soc Gt Brit; Iron Steel Inst Japan. *Res:* Physical metallurgy, especially of steels. *Mailing Add:* RR 2 Box 416 Palmyra VA 22963

LESNAW, JUDITH ALICE, b Chicago, Ill, July 30, 40. VIROLOGY, MOLECULAR BIOLOGY. *Educ:* Univ Ill, BS, 62, MS, 64, PhD(cell biol), 69. *Prof Exp:* Res assoc virol, Univ Ill, 69-74; from asst prof to assoc prof, 74-88, PROF BIOL SCI, UNIV KY, 88- *Mem:* Am Soc Microbiol; Am Soc Virol; Am Soc Biochem & Molecular Biol. *Res:* Structure and function of viral proteins and RNA; replication of RNA viruses; defective interfering particles. *Mailing Add:* Sch Biol Sci L P Markey Cancer Ctr Univ Ky 800 Rose St Lexington KY 40536-0093

LESNER, SHARON A, b Lorain, Ohio, April 1, 51. REHABILITATIVE AUDIOLOGY. *Educ:* Hiram Col, BA, 73; Kent State Univ, MA, 75; Wayne State Univ, MA, 76; Ohio State Univ, PhD(audiol), 79. *Prof Exp:* PROF AUDIOL, UNIV AKRON, 79- *Mem:* Am Speech, Lang & Hearing Asn; Acad Rehab Audiol; Acoust Soc Am; Alexander Graham Bell Asn Deaf; Am Auditory Soc. *Res:* Visual, auditory and audio-visual reception of speech; evoked potentials including auditory and visual; hearing aid use; central auditory processing. *Mailing Add:* Speech & Hearing Ctr Univ Akron Akron OH 44325-3001

LESNIAK, LINDA, b Gary, Ind, Aug 14, 48. MATHEMATICS. *Educ:* Western Mich Univ, BA, 70, MA, 71, PhD(math), 74. *Prof Exp:* Asst prof, La State Univ, Baton Rouge, 74-78; asst prof to assoc prof math, Western Mich Univ, 78-85; ASSOC PROF MATH, DREW UNIV, 85- *Honors & Awards:* Fulbright Award, 90. *Mem:* Am Math Soc; Sigma Xi; Math Asn Am; NY Acad Sci; Asn Women in Math. *Res:* Extremal problems in graph theory; generalized degree conditions. *Mailing Add:* Dept Math & Comput Sci Drew Univ Madison NJ 07940

LESPERANCE, PIERRE J, b Montreal, Que, Aug 16, 34; m 60; c 3. INVERTEBRATE PALEONTOLOGY. *Educ:* Univ Montreal, BSc, 56; Univ Mich, MS, 57; McGill Univ, PhD(geol), 61. *Prof Exp:* Geologist, Dept Natural Resources, Que, 60-61; from asst prof to assoc prof, 61-71, chmn dept, 75-79, PROF GEOL, UNIV MONTREAL, 71- *Mem:* Geol Soc Can; Am Asn Petrol Geol; Soc Econ Paleont & Mineral; Paleont Soc; Brit Palaeont Asn. *Res:* Low and middle Paleozoic field mapping in Quebec; paleontology and biostratigraphy of Upper Ordovician to Lower Devonian trilobites and brachiopods. *Mailing Add:* Dept Geol Univ Montreal Montreal PQ H3C 3J7 Can

L'ESPERANCE, ROBERT LOUIS, mining engineering, geology, for more information see previous edition

LESSARD, JAMES LOUIS, b Eau Claire, Wis, Mar 9, 43; m 65; c 2. BIOCHEMISTRY. *Educ:* Marquette Univ, BS, 65, PhD(biochem), 70. *Prof Exp:* Fel, Roche Inst Molecular Biol, Nutley, NJ, 69-71; res scholar biochem, Children's Hosp Res Found, 71-72; assoc prof, 72-79, ASSOC PROF RES PEDIAT, MED SCH, UNIV CINCINNATI, 79-, ASST PROF BIOL CHEM, 74- *Concurrent Pos:* Fel pharmacol-morphol, Pharmaceut Mfrs Asn Found, Cincinnati, Ohio, 72-74. *Mem:* Am Chem Soc; AAAS; Sigma Xi. *Res:* Regulatory processes in development; cell motility; immunochemistry. *Mailing Add:* 8579 Hallridge Ct Cincinnati OH 45231

LESSARD, JEAN, b East-Broughton, Que, Apr 29, 36; div; c 2. ORGANIC ELECTROCHEMISTRY. *Educ:* Laval Univ, BA, 56, BSc, 60, PhD(org chem), 65. *Prof Exp:* Nat Res Coun Can fel, Imp Col, Univ London, 65-67; asst res officer org chem, Nat Res Coun Can, 67-69; from asst prof to assoc prof, 69-76, PROF ORG CHEM, UNIV SHERBROOKE, 76- *Mem:* Chem Inst Can; Royal Soc Chem; Am Chem Soc; Electrochem Soc. *Res:* Transition metals; electrochemistry and photochemistry used to study new methods of effecting organic reactions or new organic reactions, investigation of the mechanism, scope and synthetic utility of these reactions. *Mailing Add:* Dept Chem Univ Sherbrooke Sherbrooke PQ J1K 2R1 Can

LESSARD, RICHARD R, b Lowell, Mass, Mar 15, 43; m 66; c 2. PETROLEUM PRODUCTS RESEARCH. *Educ:* Lowell Technol Inst, BS, 66; Univ Maine, MS, 68, PhD(chem eng), 70. *Prof Exp:* Proj staff engr, Exxon Res & Eng Co, 70-76, sect head, 76-79, lab dir res & develop, 80-82, res coordr, 82-84, mgr, 84-89, OIL SPILL TECHNOL CONSULT, EXXON RES & ENG CO, 89- *Concurrent Pos:* Mem, Fossil Energy Res Working Group III, Dept Energy, 80. *Mem:* Am Inst Chem Engrs; Sigma Xi. *Res:* Coordinator of technical studies in support of Alaska oil spill cleanup. *Mailing Add:* 12 Buckley Hill Rd Morristown NJ 07960

LESSARD, ROGER ALAIN, b East Broughton, Que, Sept 11, 44; m 67; c 2. OPTICS. *Educ:* Univ Laval, BS, 69, DSc(optics), 73. *Prof Exp:* Res officer lasers, Gentec Co, 71-72; from lectr to assoc prof 72-82, PROF OPTICS, LAVAL UNIV, 78- *Concurrent Pos:* Invited prof, Tianjin Univ, People's Repub China, 82 & 84. *Mem:* Can Asn Advan Sci; Can Asn Physicists; Optical Soc Am; Am Phys Soc; Soc Photo Optical Eng. *Res:* Holography and optical information processing; optical memories recording media. *Mailing Add:* Dept Physics Univ Laval Quebec PQ G1K 7P4 Can

LESSE, HENRY, psychiatry, neurophysiology; deceased, see previous edition for last biography

LESSELL, SIMMONS, b Brooklyn, NY, May 25, 33; m 55; c 4. NEUROLOGY, OPHTHALMOLOGY. *Educ:* Amherst Col, BA, 54; Cornell Univ, MD, 58. *Prof Exp:* Intern med, Cornell Univ, 58-59; resident neurol, Univ Vt, 59-60; resident ophthal, Mass Eye & Ear Hosp, 63-66; assoc prof neurol, 67-70, PROF ANAT, NEUROL & OPHTHAL, SCH MED, BOSTON UNIV, 70- *Concurrent Pos:* Physician, NIH, 59-60; lectr, Sch Med, Tufts Univ, 66-; vis surgeon & dir dept ophthal, Boston City Hosp; vis surgeon, Univ Hosp; consult ophthal, Vet Admin Hosp & Tufts-New England Med Ctr. *Mem:* Asn Res Vision & Ophthal. *Res:* Optic neuropathies, clinical and experimental; histochemistry and experimental pathology of the optic nerve. *Mailing Add:* Mass Eye & Ear Inst 243 Charles St Boston MA 02114

LESSEN, MARTIN, b New York, NY, Sept 6, 20; m 48; c 3. FLUID MECHANICS, SOLID MECHANICS. *Educ:* City Col New York, BME, 40; NY Univ, MME, 42; Mass Inst Technol, ScD(mech eng), 48. *Prof Exp:* Mech engr, Navy Dept, 41-46; asst fluid mech, Mass Inst Technol, 48; aeronaut res scientist, Nat Adv Comt Aeronaut, 48-49; prof aero eng, Pa State Col, 49-53; prof appl mech, Univ Pa, 53-60; prof mech & aerospace sci & chmn dept, 60-70, Yates mem prof eng, 68-83, EMER YATES MEM PROF ENG, UNIV ROCHESTER, 83- *Concurrent Pos:* NSF sr fel shock waves & instabilities, Cambridge Univ, 66-67; Nat Acad Sci exchange visitor, USSR, 67; IBM Corp pure sci div grant; consult, RCA Corp, NJ, Rochester Appl Sci Assocs, GE Corp, Philadelphia & Adv Comt Energy Div, Oak Ridge Nat Lab; liaison scientist, US Off Naval Res, London, 76-79; Vollmer Fries fel, Rensselaer Polytech Inst, 78; founding chmn, Energetics Prof Div, Am Soc Mech Engrs, 63, 67. *Mem:* Fel Am Phys Soc; fel Am Soc Mech Engrs; Am Soc Eng Educ; Ger Soc Appl Math & Mech; Am Inst Aeronaut & Astronaut; fel AAAS. *Res:* Fluid mechanics; thermodynamics and heat transfer; vibrations; hydrodynamic stability and transition to turbulence; continuum mechanics; thermoelasticity; biomechanics; biophysics; plasma dynamics; field theory. *Mailing Add:* 12 Country Club Dr Rochester NY 14618

LESSEPS, ROLAND JOSEPH, b New Orleans, La, Aug 13, 33. DEVELOPMENTAL BIOLOGY. *Educ:* Spring Hill Col, BS, 58; Johns Hopkins Univ, PhD(biol), 62. *Prof Exp:* From asst prof to assoc prof, 67-81, chmn dept, 78-83, PROF BIOL, LOYOLA UNIV, LA, 81- *Concurrent Pos:* Vis prof, Roman Cath Univ, Nijmegen; mem bd trustees, Am Killfish Asn, 86- *Mem:* AAAS; Am Soc Zool; Electron Micros Soc Am; Soc Develop Biol; Nat Asn Biol Teachers; Am Killfish Asn. *Res:* Morphogenetic movements of embryonic cells; electron microscopy of the cell surface; time-lapse filming of cell movements in living embryos. *Mailing Add:* Dept Biol Loyola Univ 6363 St Charles Ave New Orleans LA 70118

LESSHAFFT, CHARLES THOMAS, JR, pharmacy; deceased, see previous edition for last biography

LESSIE, THOMAS GUY, b New York, NY, Dec 14, 36; m 62; c 3. MICROBIAL PHYSIOLOGY. *Educ:* Queens Col, NY, BS, 58; Harvard Univ, AM, 61, PhD(biol sci), 63. *Prof Exp:* Res asst microbiol, Haskins Labs, NY, 58-59; NIH fels biochem, Oxford Univ, 63-65 & biol sci, Purdue Univ, 65-67; res assoc microbiol, Univ Wash, 67-68; asst prof, 68-74, assoc prof, 74-81, PROF MICROBIOL, UNIV MASS, AMHERST, 81- *Concurrent Pos:* NSF grants, 68-70 & 85-88; NIH grants, Inst Arthritis & Metab Dis, 68-74, Inst Gen Med Sci, 74-81 & 81-85, Inst Allergy & Infectious Dis, 87-; vis prof, dept biol sci, Purdue Univ, 74, dept microbiol, Med Col Va, 81, dept biol, Yale Univ, 88; Environ Protection Agency grant, 88-91. *Mem:* Am Soc Microbiol; Am Chem Soc. *Res:* Biochemical genetics of Pseudomonas cepacia with emphasis on regulatory mechanisms governing carbohydrate and amino acid metabolism; roles of transposable gene-activating elements in evolution of new metabolic functions. *Mailing Add:* Dept Microbiol Univ Mass Amherst MA 01003

LESSIN, LAWRENCE STEPHEN, b Washington, DC, Oct 14, 37; m 62; c 3. HEMATOLOGY, ONCOLOGY. *Educ:* Univ Chicago, MD, 62. *Prof Exp:* Instr med, Univ Pa, 65-67; Nat Heart Inst spec fel hematol med, Inst Cell Path, Paris, France, 67-68; asst prof med, Sch Med, Duke Univ, 68-70; assoc prof, 70-74, PROF MED PATH & DIR HEMAT & ONCOL, SCH MED, GEORGE WASHINGTON UNIV, 74- *Concurrent Pos:* Consult, Nat Heart, Lung & Blood Inst & US Naval Med Ctr, 74- & Walter Reed Army Med Ctr, 75- *Mem:* Am Col Physicians; Am Soc Hemat; Am Fedn Clin Res; Int Soc Hemat; Am Soc Clin Oncol; Sigma Xi. *Res:* Red cell membrane structure in hemolytic anemias; red cell rhelogy; pathophysiology of sickle cell disease; preleukemia (myelodysprasia). *Mailing Add:* 2150 Pennsylvania Ave NW Washington DC 20037

LESSING, PETER, b Englewood, NJ, June 15, 38; m 65; c 2. ENVIRONMENTAL GEOLOGY. *Educ:* St Lawrence Univ, BS, 61; Dartmouth Col, MA, 63; Syracuse Univ, PhD(geol), 67. *Prof Exp:* Asst prof geol, St Lawrence Univ, 66-71; environ geologist, WVa Geol Surv, 71-73, chief, Geol Div, 80-85, head, Environ Geol Sect, 73-89, SR RES GEOLOGIST, WVA GEOL SURV, 89- *Concurrent Pos:* Adj prof, WVa Univ, 73- *Mem:* Geol Soc Am; Int Asn Eng Geol; Hist Earth Sci Soc. *Res:* Geologic field mapping; environmental geology investigations; history of geology; landslide evaluation; geologic hazard studies; hydrology and water use. *Mailing Add:* WVa Geol Surv PO Box 879 Morgantown WV 26507-0879

LESSIOS, HARILAOS ANGELOU, b Thessaloniki, Greece, Mar 4, 51; m 83; c 2. EVOLUTION. *Educ:* Harvard Univ, BA, 73; Yale Univ, MPhil, 76, PhD(biol), 79. *Prof Exp:* STAFF BIOLOGIST, SMITHSONIAN TROP RES INST, 79- *Mem:* AAAS; Soc Study Evolution. *Res:* Evolution and ecology of marine organisms. *Mailing Add:* Smithsonian Trop Res Inst Unit 0948 APO Miami FL 34002-0948

LESSLER, JUDITH THOMASSON, b Charlotte, NC, Oct, 10, 43; m 70; c 2. SURVEY RESEARCH METHODS. *Educ:* Univ NC, Chapel Hill, AB, 66, PhD(biostatist), 74; Emory Univ, Ga, MAT, 67. *Prof Exp:* Statistician, 74-78, dept mgr, 80-84, SR STATISTICIAN, RES TRIANGLE INST, 78- *Concurrent Pos:* Prin investr, NSF grant, 79-81; adj asst prof, Biostatist Dept, Univ NC, 81-; serv fel, Nat Ctr Health Statist, 84-85; bd mem, Am Statist Asn, 85-87. *Mem:* Am Statist Asn; Am Pub Health Asn; AAAS. *Res:* Survey research methods; statistical treatment of nonsampling errors and multiframe-multiplicity estimators; application of cognitive psychology to survey design. *Mailing Add:* Res Triangle Inst PO Box 12194 Rm 407-Hill Bldg Research Triangle Park NC 27709

LESSLER, MILTON A, b New York, NY, May 18, 15; m 43; c 3. PHYSIOLOGY. *Educ:* Cornell Univ, BS, 37, MS, 38; NY Univ, PhD(biochem cell physiol), 50. *Prof Exp:* Technician cardiac res, NY State Health Dept, 40-42; from asst prof to prof, 51-85, EMER PROF PHYSIOL, COL MED, OHIO STATE UNIV, 85- *Concurrent Pos:* Nat Cancer Inst fel, NY Univ-Washington Square Col, 49-50; NSF fac fel, Univ Mich, Ann Arbor, 58-59; vis lectr, Am Physiol Soc, 62-66; consult, Yellow Springs Instrument Co, 65-; ed-in-chief, Ohio J Sci, 74-81. *Mem:* Fel AAAS; fel NY Acad Sci; Am Physiol Soc; Am Asn Cancer Res; Am Soc Cell Biol. *Res:* Cell physiology; effects of environmental pollutants on the hemopoietic system; cellular radiobiology; erythropoiesis and lead poisoning. *Mailing Add:* 1814 Riverside Dr B Columbus OH 43212

LESSMAN, GARY M, b Hillsboro, Ill, July 15, 38. SOIL FERTILITY. *Educ:* Southern Ill Univ, BS, 60, MS, 62; Mich State Univ, PhD(soil sci), 67. *Prof Exp:* Exten agronomist, Purdue Univ, 67-68; asst prof, 69-77, ASSOC PROF AGRON, UNIV TENN, KNOXVILLE, 77- *Mem:* Sigma Xi; Am Soc Agron. *Res:* Micronutrient nutrition. *Mailing Add:* Dept Plant & Soil Sci Univ Tenn Knoxville TN 37996

LESSMANN, RICHARD CARL, b New York, NY, Oct 14, 42; m 65; c 3. MECHANICAL ENGINEERING, FLUID MECHANICS. *Educ:* Syracuse Univ, BSME, 64; Brown Univ, ScM, 66, PhD(eng), 69. *Prof Exp:* Asst prof, 69-75, ASSOC PROF MECH ENG & APPL MECH, UNIV RI, 75- *Mem:* Am Inst Aeronaut & Astronaut; Am Phys Soc; Sigma Xi. *Res:* Turbulent flows; boundary layer theory; heat transfer. *Mailing Add:* 35 Courtland Dr Narragansett RI 02882

LESSNER, HOWARD E, b Philadelphia, Pa, Feb 28, 27; m 57; c 3. INTERNAL MEDICINE, ONCOLOGY. *Educ:* Univ Pa, MD, 53. *Prof Exp:* Jr asst resident, Jackson Mem Hosp, 54-55, sr asst resident, 55-56, clin fel, 56-57; res fel, Nat Heart Inst, Barnes Hosp, 57-58, clin fel, Nat Cancer Inst, 58-59; from instr to prof med, 59-75, PROF ONCOL, UNIV MIAMI, 74- *Concurrent Pos:* Dir, Comprehensive Cancer Ctr, 72-74, clin dir, 74- *Mem:* Am Col Physicians; AMA; Am Soc Clin Oncol; Am Asn Cancer Res; AAAS. *Mailing Add:* 8950 N Kendall Dr No 410 Miami FL 33176

LESSO, WILLIAM GEORGE, b Cleveland, Ohio, Mar 23, 31; m 52; c 5. OPERATIONS RESEARCH. *Educ:* Univ Notre Dame, BSME, 53; Xavier Univ, Ohio, MBA, 63; Case Inst Technol, MS, 66, PhD(opers res), 67. *Prof Exp:* Design engr, Clevite Corp, 53-58; proj engr, Flight Propulsion Div, Gen Elec Co, 58-64; assoc prof mech eng, 67-72, PROF MECH ENG, UNIV TEX, AUSTIN, 72- *Mem:* Opers Res Soc Am; Inst Mgt Sci. *Res:* Application of operations research to industrial and economic problems. *Mailing Add:* Dept Mech Eng Univ Tex Austin TX 78712

LESSOFF, HOWARD, b Boston, Mass, Sept 23, 30; m 59; c 2. SOLID STATE SCIENCE. *Educ:* Northeastern Univ, BS, 53, MS, 57. *Prof Exp:* Staff engr, Radio Corp Am, Mass, 57-60, sr staff mem, 61-64; staff mem, Bell Tel Labs, 60-61; aerospace technologist, Electronic Res Ctr, NASA, 64-70; supvry physicist, 70-75, BR HEAD ELECTRONIC MAT, NAVAL RES LAB, 75- *Concurrent Pos:* Lectr, Lincoln Col, Northeastern Univ, 57-70; consult, Datacove Corp, NJ, 69- *Mem:* AAAS; Inst Elec & Electronics Engrs; Sigma Xi. *Res:* Crystal growth; semiconductor materials; microwave and optic properties; solid state physics; magnetic materials. *Mailing Add:* Electronic Mat Br Code 6820 Naval Res Lab Washington DC 20375-5000

LESSOR, ARTHUR EUGENE, JR, analytical chemistry, crystallography; deceased, see previous edition for last biography

LESSOR, DELBERT LEROY, b 1941; US citizen; m 62; c 2. PHYSICS, APPLIED MATHEMATICS. *Educ:* Ft Hays State Univ, BS, 62; Kans State Univ, PhD(physics), 67. *Prof Exp:* Temp asst prof physics, Kans State Univ, 66-67; sr res scientist phys sci, 67-80, sr res scientist eng physics, 80-89, STAFF SCIENTIST, APPL PHYSICS CTR, PAC NORTHWEST LABS, BATTELLE MEM INST, 89- *Mem:* Am Phys Soc; Sci Res Soc NAm; Sigma Xi; Bioelectromagnetics Soc; Am Vacuum Soc. *Res:* Electromagnetic field computation in industrial and instrument configurations; air filtration theory; nuclear particle transport; nuclear reactor theory; nuclear reactor instrumentation; geothermal chemistry; fluid flow calculation; optics theory; bioelectromagnetics effects; low energy electron diffraction analysis. *Mailing Add:* 2229 Davison Richland WA 99352

LESSOR, EDITH SCHROEDER, b Chicago, Ill, Aug 5, 30; wid; c 2. ANALYTICAL CHEMISTRY, GENERAL CHEMISTRY. *Educ:* Valparaiso Univ, BS, 52; Indiana Univ, Bloomington, PhD(anal chem), 55. *Prof Exp:* Instr chem, Ulster Community Col, 64-65; lectr, Harpur Col, State Univ New York, Binghamton, 65-67; from asst prof to assoc prof, Mt St Mary Col, NY, 67-76, chmn, Div Natural Sci & Math, 73-84, chmn dept, 68-88,

PROF CHEM, MT ST MARY COL, NY, 76-, CHMN DIV NATURAL SCI, 88- *Mem:* AAAS; Am Chem Soc; Sigma Xi. *Res:* Spectrophotometry of organic analytical reagents and analytical chemistry of water pollution control. *Mailing Add:* Div Natural Sci Mt St Mary Col Newburgh NY 12550-3494

LESTER, CHARLES TURNER, b Covington, Ga, Nov 10, 11; m 36; c 2. ORGANIC CHEMISTRY. *Educ:* Emory Univ, AB, 32, MA, 34; Pa State Univ, PhD(org chem), 41. *Prof Exp:* Teacher high sch, Ga, 34-35; instr chem, Emory Jr Col, 35-39; res chemist, Calco Div, Am Cyanamid Corp, NJ, 41-42; from asst prof to assoc prof, chmn, Dept Chem, 54-57, vpres grad studies, 70-74, vpres arts & sci, 74-78, PROF CHEM, EMORY UNIV, 50-, DEAN GRAD SCH, 57-, EXEC VPRES & DEAN FAC, 78- *Concurrent Pos:* Mem bd dirs, Oak Ridge Assoc Univs, 62-65; mem, Ga Sci & Technol Comn, 63-72; mem bd dirs, Atlanta Speech Sch, 65-71; mem exec comt, Coun Grad Schs US, 65-68; mem bd dirs, Southeastern Educ Lab, 66-68; bd trustees, Reinhardt Col, 66-72; vchmn, Ocean Sci Ctr, Atlantic Comn, 67-72; mem bd trustees, Huntingdon Col, 68; chief acad progs br, Bur Higher Educ, 69-70; chmn elect, Coun Grad Sch, 73, chmn, 74; coun mem, Oak Ridge Assoc Univs, 58-62, 70-79. *Honors & Awards:* Herty Medal, 65. *Mem:* Am Chem Soc; Sigma Xi; Am Inst Chem. *Res:* Sterically hindered ketones; indigosol dyes; biphenyl mercaptan; oxetanones; alkyl aryl ketones; antimicrobial compounds. *Mailing Add:* 281 Chelsea Circle Decatur GA 30030

LESTER, DAVID, biochemistry, pharmacology; deceased, see previous edition for last biography

LESTER, DONALD THOMAS, b New London, Conn, Aug 26, 34; m 62; c 2. FORESTRY. *Educ:* Univ Maine, BS, 55; Yale Univ, MF, 57, PhD(forest genetics), 62. *Prof Exp:* From asst prof to prof forestry, Univ Wis-Madison, 62-77; SUPVR, FOREST BIOL, FORESTRY RES DIV, CROWN ZELLERBACH CORP, 77- *Mem:* Sigma Xi. *Res:* Tree breeding; genecology. *Mailing Add:* 3883 W King Edward Ave Vancouver BC V6S 1M9

LESTER, GEORGE RONALD, b War Eagle, WVa, Sept 6, 34; m 56; c 4. AIR PURIFICATION, CATALYSIS. *Educ:* Berea Col, BA, 54; Univ Ky, MS, 56, PhD(chem), 58. *Prof Exp:* Chemist, Universal Oil Prod Co, 58-63, assoc res coordr, UOP, Inc, 63-74, mgr appl catalysis, 74-76, dir mat res, 76-83; sr res scientist, Allied-Signal Engineered Mat Res Ctr, 83-90; RES FEL, ALLIED-SIGNAL RES & TECHNOL, 90- *Concurrent Pos:* Chair, Gordon Res Conf Catalysis, 91. *Mem:* AAAS; Am Chem Soc; Faraday Soc; Sigma Xi; Catalysis Soc; Soc Automotive Eng. *Res:* Conductivity of nonaqueous solutions; adsorption of gases on solids; heterogeneous catalysis; petrochemical processes; material science; automotive exhaust catalysis; solar systems; energy conservation; catalytic combustion; automotive gas turbine engines; electric power plant catalytic combustion; electrocatalysts for fuel cells for power plants; catalytic destruction of chemical warfare agents; catalytic incineration of hazardous pollutants. *Mailing Add:* 209 Vine Park Ridge IL 60068

LESTER, HENRY ALLEN, b New York, NY, July 4, 45; c 2. NEUROBIOLOGY, BIOPHYSICS. *Educ:* Harvard Col, AB, 66; Rockefeller Univ, PhD(biophys), 71. *Prof Exp:* Res fel molecular neurobiol, Inst Pasteur, 71-73; PROF BIOL, CALIF INST TECHNOL, 73- *Concurrent Pos:* Res grants, NIH Res Career Develop Award, 77-82, Alfred P Sloan Res fel, 74-76, & Sen Jacob Javits investr, NIH, 85-; vis prof, Dept Biol Chem, Hebrew Univ, Israel, 80-81; mem, Physiol Study Sect, 85-89. *Mem:* Soc Neurosci; Biophys Soc; Soc Gen Physiologists; fel AAAS. *Res:* Excitable membranes; molecular neuroscience. *Mailing Add:* Div Biol Calif Inst Technol Pasadena CA 91125

LESTER, JOHN BERNARD, b San Diego, Calif, Mar 11, 45; m 72; c 2. ASTRONOMY, ASTROPHYSICS. *Educ:* Northwestern Univ, BA, 67; Univ Chicago, MS, 69, PhD(astron), 72. *Prof Exp:* Lectr physics, Univ Wis-Milwaukee, 69-71; presidential intern astron, Smithsonian Astrophys Observ, 72-73, physicist, 73-76; asst prof, 76-81, ASSOC PROF ASTRON, UNIV TORONTO, 81- *Mem:* Am Astron Soc; Astron Soc Pac; Int Astron Union. *Res:* High dispersion stellar spectroscopy; stellar abundances; ultraviolet astronomy; infrared spectroscopy. *Mailing Add:* Erindale Col Univ Toronto Mississauga ON L5L 1C6 Can

LESTER, JOSEPH EUGENE, b Bay City, Tex, July 2, 42; m 59; c 2. PHYSICAL CHEMISTRY. *Educ:* Rice Univ, BA, 64; Univ Calif, Berkeley, PhD(chem), 68. *Prof Exp:* Asst prof chem, Northwestern Univ, Evanston, 67-74; mem staff, GTE Labs, 73-77, sr res chemist, 78-81, res assoc, Gulf Sci & Technol, 81-85, MGR MAT CHARACTERIZATION, GTE LABS, GULF RES & DEVELOP, 85- *Mem:* Am Chem Soc; Am Phys Soc; Catalysis Soc. *Res:* Kinetics and mechanisms of surface reactions; catalysis; extended x-ray absorption spectroscopy; extended x-ray absorption fine structure. *Mailing Add:* GTE Labs 40 Sylvan Rd Waltham MA 02154

LESTER, LARRY JAMES, b Bay City, Tex, July 15, 47; m 69, 78, 83; c 1. POPULATION GENETICS. *Educ:* Univ Tex, Austin, BA, 69, PhD(pop genetics), 75. *Prof Exp:* From asst prof to assoc prof, 75-91, PROF BIOL, UNIV HOUSTON-CLEAR LAKE, 91- *Mem:* Sigma Xi; World Aquacult Soc; Crustacean Soc. *Res:* Genetics of aquaculture species. *Mailing Add:* 2700 Bay Area Blvd Univ Houston-Clear Lake Houston TX 77058

LESTER, RICHARD GARRISON, b New York, NY, Oct 24, 25; m 53; c 2. RADIOLOGY. *Educ:* Princeton Univ, AB, 46; Columbia Univ, MD, 48. *Prof Exp:* From instr to assoc prof radiol, Univ Minn, 54-61; prof & chmn dept, Med Col Va, 61-65; prof radiol & chmn dept, Duke Univ, 65-76; PROF RADIOL, MED SCH, UNIV TEX, HOUSTON, 76-, CHMN RADIOL, 77- *Concurrent Pos:* Mem comt acad radiol, Nat Acad Sci, 66; mem steering comt, Soc Chmn Acad Radiol Dept, 67; mem bd trustees, Am Bd Radiol; mem bd trustees, Meharry Med Col, 75- *Mem:* AMA; Am Roentgen Ray Soc; Am Col Radiol; Soc Pediat Radiol (secy-treas, 58-62); regent Am Col Chest Physicians. *Res:* Cardiovascular and pediatric radiology. *Mailing Add:* Box 1980 Norfolk VA 23501

LESTER, ROBERT LEONARD, b New Haven, Conn, Aug 21, 29; m 54; c 2. BIOCHEMISTRY. *Educ:* Yale Univ, BS, 51; Calif Inst Technol, PhD(biochem), 56. *Prof Exp:* Asst prof biochem, Univ Wis, 58-60; from asst prof to assoc prof, 60-68, chmn dept, 74-83, PROF BIOCHEM, MED SCH, UNIV KY, 68- *Concurrent Pos:* Res fel, Inst Enzyme Res, Univ Wis, 55-58; NIH res grants, 60-; vis res biologist, Univ Calif, San Diego, 69-70. *Mem:* AAAS; Am Soc Biol Chemists; Am Chem Soc; Am Soc Microbiol; Fedn Am Sci. *Res:* Lipid metabolism. *Mailing Add:* Univ Ky Med Sch Lexington KY 40536

LESTER, ROGER, b Brooklyn, NY, Dec 26, 29; m 54; c 2. MEDICINE. *Educ:* Princeton Univ, AB, 50; Yale Univ, MD, 55. *Prof Exp:* From intern to resident med, Col Med, Univ Utah, 55-57, resident, 59-60; NIH fel, 56-59; fel Thorndike Mem Lab, Harvard Univ, 60-62; asst prof, Sch Med, Univ Chicago, 62-65; from asst prof to prof med, Sch Med, Boston Univ, 65-73; PROF GASTROENTEROL & CHIEF DIV, SCH MED, UNIV PITTSBURGH, 73- *Concurrent Pos:* NIH career develop award, 63-73, res grant, 65- *Mem:* Am Fedn Clin Res; Am Asn Study Liver Dis; Am Soc Clin Invest; Am Gastroenterol Asn; Int Asn Study Liver. *Res:* Fetal hepatic and intestinal function; effect of alcohol and liver disease on sexual function. *Mailing Add:* Div Gastroenterol UAMS Slot No 567 4301 W Markham St Little Rock AR 72205

LESTER, WILLIAM ALEXANDER, JR, b Chicago, Ill, Apr 24, 37; m 59; c 2. COLLISION DYNAMICS. *Educ:* Univ Chicago, BS, 58, MS, 59; Cath Univ, PhD(chem), 64. *Prof Exp:* Proj asst physics, lab molecular struct & spectra, Univ Chicago, 57-59; asst chem, Wash Univ, 59-60 & Cath Univ, 60-62; phys chemist, Phys Chem Div, Nat Bur Standards, 61-64; proj assoc, Theoret Chem Inst, Univ Wis-Madison, 64-65, asst dir, 65-68; mem permanent prof staff, IBM Res Lab, 68-75, mem tech planning staff, T J Watson Res Ctr, IBM Corp, Yorktown Heights, NY, 75-76, mgr molecular interactions group, IBM Res Lab, San Jose, Calif, 76-78; CONSULT, 78- *Concurrent Pos:* Lectr, Univ Wis-Madison, 66-68; Secy & treas, Wisc sect, Am Chem Soc, 67-68, treas, div comput in chem, 74-77, vchmn, div phys chem, 77, chmn elec, 78, chmn, 79; ed, Proc of Conf on Potential Energy Surface in Chem, 71; mem, res eval panel, Off Sci Res, USAForce, 74-78, US Nat Comt, Int Union Pure & Appl Chem, 76-79, Nat Res Coun Panel for Chem physics, Nat Bur Standards, 80-83, Comt to survey chem sci, Nat Acad Sci, 82-84, exec bd, Nat Orgn Black Chemists & Chem Engrs, 84-87, comt on recommendations, US Army Basic Sci Res, 84-87; consult, NSF, 76-77; exec bd, Nat Orgn Bl Chemists & Chem Engrs, 84-87; vchmn, div chem physics, Am Phys Soc, 85, chmn, 86; mem, Sci Yr adv bd, World Bk, Inc, 89-, external adv comt, Sci & Technol Ctr Res in Parallel Computation, NSF, 89- & Fed Networking Adv Comt, 91-; consult, Teltech, Inc, 91- *Honors & Awards:* Percy L Julian Award, Nat Orgn Black Chemists & Chem Engrs, 79. *Mem:* Am Chem Soc; fel Am Phys Soc; Sigma Xi; Nat Orgn Bl Chemists & Chem Engrs; fel AAAS. *Res:* Molecular quantum mechanics and molecular collision theory. *Mailing Add:* Dept Chem Univ Calif Berkeley CA 94720

LESTER, WILLIAM LEWIS, b Webster City, Iowa, July 21, 32; m 64; c 5. MICROBIOLOGY. *Educ:* San Jose State Col, BA, 58; Univ Calif, Davis, PhD(microbiol), 68. *Prof Exp:* Lab technician pharmacol, Univ Calif, Davis, 62-66; supvr res & develop, Cutter Labs, 68-70; from asst to assoc prof, 70-79, PROF MICROBIOL HUMBOLDT STATE UNIV, 79- *Concurrent Pos:* Nat Oceanic & Atmospheric Admin sea grant, Samoa & Calif, 70-73; bd dirs, Redwood Health Consortium, 73-75; univ rep, Conf Assist Undergrad Sci Educ, 75-; health manpower coordr, Humboldt State Univ, 76-83, mem Acad Senate, 89- *Mem:* AAAS; Am Soc Microbiol; Wildlife Soc; Am Soc Allied Health Prof. *Res:* Biodegradation of kraft pulp mill effluent; microbial ecology; marine bioassays utilizing echino embryo. *Mailing Add:* Dept Biol Humboldt State Col Arcata CA 95521

LESTINGI, JOSEPH FRANCIS, b Long Island, NY, Apr 24, 35; m 57; c 4. ENGINEERING MECHANICS, STRUCTURAL ENGINEERING. *Educ:* Manhattan Col, BCE, 57; Va Polytech Inst, MS, 59; Yale Univ, DEng(solid mech), 66. *Prof Exp:* Instr eng mech, Va Polytech Inst, 57-59 & Pa State Univ, 59-60; struct res engr, Elec Boat Div, Gen Dynamics Corp, Conn, 60-65; sr mech engr, Battelle Mem Inst, Ohio, 65-67; from asst prof to prof civil eng, Univ Akron, 67-78; prof eng mech & chmn, Dept Math & Eng Mech, 78-79, Gen Motors Inst, 78-79, prof mech eng & head dept, 79-83; DEAN ENG & PROF MECH ENG, MANHATTAN COL, RIVERDALE, NY, 83- *Concurrent Pos:* Res fel, Am Soc Civil Engrs; consult eng staff, Gen Motors Corp, Warren, Mich, 79-; Danforth assoc. *Honors & Awards:* Western Elec Fund Award, Am Soc Eng Educ, 75. *Mem:* Fel Am Soc Civil Engrs; fel Am Soc Mech Engrs; Am Soc Eng Educ; Am Acad Mech; Sigma Xi; Soc Automotive Engrs. *Res:* Computer assisted design; computer assisted manufacturing; finite element methods; shock and vibration analysis; computer methods. *Mailing Add:* 37 Indianfield Ct Mahwah NJ 07430

LESTON, GERD, b Germany, Sept 19, 24; nat US; m 50; c 2. ORGANIC CHEMISTRY. *Educ:* City Col New York, BS, 48; Purdue Univ, MS, 49, PhD(chem), 52; Univ Pittsburgh, BS, 81. *Prof Exp:* Chemist, Koppers Co Inc, 52-54, sr chemist, 54-58, group mgr, 58-66, sr group mgr, 67-72, sr proj scientist, Koppers Co Inc, 72-85; CONSULT, 85- *Mem:* Am Chem Soc; Sigma Xi; Catalysis Soc. *Res:* Synthetic organic chemistry, particularly phenol chemistry, aromatic substitution; aromatic alkylation and dealkylation; hydrogenation; aromatic acylation; ultraviolet stabilizers; antioxidant synthesis and testing; homogenous and heterogenous catalysis; pesticides; drugs; separation techniques; organics-salt complexes. *Mailing Add:* 1219 Raven Dr Pittsburgh PA 15243-1241

LESTOURGEON, WALLACE MEADE, b Alexandria, La, Jan 16, 43; m 86; c 2. MOLECULAR BIOLOGY. *Educ:* Univ Tex, Austin, BS, 66, PhD(cell biol), 70. *Prof Exp:* NIH fel oncol, McArdle Lab Cancer Res, 70-74, asst scientist, 74-78; from asst prof to assoc prof, 78-86, PROF MOLECULAR BIOL, VANDERBILT UNIV, 86- *Concurrent Pos:* Prin investr, NSF grants, 75, 78, 81, 85 & 88; dir, Cell Biol Prog, NSF, 83-84. *Mem:* AAAS; Am Soc Cell Biol; Sigma Xi. *Res:* Molecular biological, biochemical and physical chemical studies on the structure of 40's nuclear ribonucleoprotein particles and their role in RNA splicing and in the modulation of information flow in eucaryotes; gene regulation. *Mailing Add:* Dept Molecular Biol Vanderbilt Univ Nashville TN 37235

LESTRADE, JOHN PATRICK, b New Orleans, La, Mar 25, 49; m 74; c 2. ATMOSPHERIC PHYSICS, SPACE PHYSICS. *Educ:* La State Univ, New Orleans, BS, 71; Purdue Univ, MS, 72; Rice Univ, MS, 76, PhD(space physics), 78. *Hon Degrees:* French Lang Master, Univ Aix-Marseille, France, 68. *Prof Exp:* Design scientist nuclear reactors, Westinghouse-Bettis Labs, 73-74; res assoc planetary atmospheres, Rice Univ, 78-80; asst prof physics, Tex A&M Univ, 80-84; ASST PROF PHYSICS, MISS STATE UNIV, 84- *Mem:* Am Astron Soc. *Res:* High-energy astrophysics; gamma-ray burster modelling; radiative transfer in various media-clouds, soil, oceans. *Mailing Add:* Dept Physics Miss State Univ PO Box 5328 Miss State MS 39762

LESTZ, SIDNEY J, US citizen. FUEL ENGINEERING. *Educ:* Pa State Univ, BS, 57, MS, 59. *Prof Exp:* Asst, Dept Petrol & Natural Gas Eng, Pa State Univ, 58-59; asst proj engr, Wright Aeronaut Div, Curtiss Wright Corp, 61-64; sr res engr, Exxon Res & Eng Co, 64-70; sr res engr, 70-71, mgr fuels & lubricants eng, 71-78, DIR, US ARMY FUELS & LUBRICANTS RES LAB, SOUTHWEST RES INST, 78-, DIR MERADCOM FUELS & LUBRICANTS PROG, 76- *Mem:* Combustion Inst; Soc Automotive Engrs; Coord Res Coun; Sigma Xi. *Res:* Wider boiling range fuels; synthetic fuels; alternate fuels; synthetic lubricants; universal hydraulic power transmission fluid development; safety fuels and fluids technology. *Mailing Add:* 622 Briar Oak San Antonio TX 78216

LESURE, FRANK GARDNER, b Camden, SC, Jan 28, 27; m 63; c 2. GEOLOGY. *Educ:* Va Polytech Inst, BS, 51; Yale Univ, MS, 52, PhD(geol), 55. *Prof Exp:* GEOLOGIST, US GEOL SURV, 55- *Honors & Awards:* Meritorious Serv Award, US Dept Interior, 85. *Mem:* AAAS; Soc Econ Geol; Mineral Soc Am; Geol Soc Am; Geochem Soc. *Res:* Geology of Oriskany iron deposits in Virginia, uranium deposits in southeastern Utah, gold deposits in North Carolina and Georgia, mica pegmatites of southeastern United States; mineral resources of eastern wilderness areas; geochemistry. *Mailing Add:* US Geol Surv 954 Nat Ctr Reston VA 22092

LE SURF, JOSEPH ERIC, b London, Eng, July 21, 29; Can citizen; m 52; c 3. PHYSICAL CHEMISTRY. *Educ:* Univ London, BSc, 50 & 51. *Prof Exp:* Sci officer corrosion, Royal Naval Sci Serv, 51-57; sr sci officer, UK Atomic Energy Authority, 57-64; head, Syst Mat Br, Atomic Energy Can LTD, 64-78; tech dir, London Nuclear Ltd, 78-84, PRES & CHIEF EXEC OFFICER, LONDON NUCLEAR LTD & LONDON NUCLEAR SERV INC, 84- *Mem:* Nat Asn Corrosion Engrs; Am Nuclear Soc; Can Nuclear Asn; Can Nuclear Soc. *Res:* Marine corrosion; corrosion, material selection, for nuclear decontamination processing plants and nuclear power plants. *Mailing Add:* 3529 Yale Cres Niagara Falls ON L2J 3C4 Can

LETARTE, JACQUES, b Montreal, Que, Aug 19, 34; m 60; c 2. PEDIATRICS, ENDOCRINOLOGY. *Educ:* Univ Montreal, BA, 57, MD, 62. *Prof Exp:* Resident med, Notre Dame Hosp, Montreal, 62; resident pediat, St Justine Hosp, 63-64; resident, Royal Postgrad Med Sch, London, 68; assoc prof, 69-80, PROF PEDIAT, UNIV MONTREAL, 80- *Concurrent Pos:* Mead-Johnson fel pediat, Univ Montreal, 63-64; res fel biochem, Children's Hosp, Zurich, 64-65; Queen Elizabeth II res fel, Can, 64-68; res fel, Clin Biochem Inst, Geneva, 65-67; res fel metab, Royal Postgrad Med Sch, London, 68-69; Med Res Coun Can fel, 68-69, scholar, 69-74; dir, Pediat Res Ctr, Hosp Ste-Justine, 82-86, med dir, 87-89; assoc med dir, Schering Can Inc, 89- *Mem:* AAAS; Can Soc Clin Invest; Soc Pediat Res; Endocrine Soc. *Res:* Hormonal regulation of carbohydrate metabolism; hyperammonemia in children; lipid and carbohydrate metabolism in children. *Mailing Add:* Schering Can Inc 3535 Trans-Can Pointe Claire PQ H9R 1B4 Can

LETARTE, MICHELLE VINLAINE, b Quebec, Oct 12, 47. MOLECULAR IMMUNOLOGY, LEUKEMIA. *Educ:* Laval Univ, BSc, 68; Univ Ottawa, PhD(biochem), 72. *Prof Exp:* From asst prof to assoc prof med biophys, immunol & pediat, Univ Toronto, 75-86; SR INVESTR, RES INST, HOSP SICK CHILDREN, 80-; TERRY FOX RES SCIENTIST, NAT CANCER INST, 85-; PROF, DEPT IMMUNOL, MED BIOPHYS & PEDIAT, UNIV TORONTO, 87- *Concurrent Pos:* Centennial Studentship, Nat Res Coun Can, 68-72; Nat Cancer Inst scholarship, Med Res Coun Can, 75-81, associateship, 81- *Mem:* Am Asn Histocompatibility Testing; Am Asn Immunologists. *Res:* Structure/function of glycoproteins of childhood leukemia; CDIO/CALLA and endoglin and integrins are current major focus/adhesion to endothelial cells (primary distribution of endoolin, a putative adhesion molecule). *Mailing Add:* Hosp Sick Children 555 University Ave Toronto ON M5G 1X8 Can

LETBETTER, WILLIAM DEAN, neurophysiology, neuroanatomy, for more information see previous edition

LETCHER, DAVID WAYNE, b Dover, NJ, May 5, 41; m 63; c 4. METEOROLOGY, CLIMATOLOGY. *Educ:* Rutgers Univ, BS, 63; Univ Nebr, MS, 65; Cornell Univ, PhD(meteorol), 71. *Prof Exp:* Grad asst agr climat, Univ Nebr, 63-65; grad asst meteorol, Cornell Univ, 65-68; from asst prof to assoc prof meteorol, 68-81, coordr acad comput, 81-85, ASSOC PROF METEOROL, PHYSICS DEPT, TRENTON STATE COL, 85- *Concurrent Pos:* Vis prof, Pub Serv Elec & Gas Co, Newark, NJ, 77-78; coadj prof atmospheric sci, Rutgers Univ & Mercer County Community Col. *Mem:* Air Pollution Control Asn; Am Meteorol Soc; Sigma Xi; Data Processing Mgt Asn; Soc Info Mgt. *Res:* Atmospheric dispersion modeling of power plant stacks used in district heating; severe weather in New Jersey. *Mailing Add:* Dept Bus Admin Trenton State Col Hillwood Lakes Cn4700 Trenton NJ 08650

LETCHER, JOHN HENRY, III, b Wilkes-Barre, Pa, July 18, 36; m 60; c 2. PHYSICS, COMPUTER SCIENCE. *Educ:* Univ Tulsa, BS, 57 & 58; Univ Mo, MS, 59, PhD(physics), 63. *Hon Degrees:* DEng Tech, Toulane Col, 62. *Prof Exp:* Mem staff, Advan Electronics Techniques Div, McDonnell Corp, 63-64; mem staff, Cent Res Dept, Monsanto Co, 64-68; vpres systs & res, Data Res Corp, 68-70; PRES, SYNERGISTIC CONSULTS, INC, 70- *Concurrent Pos:* Mem staff, Dept Comput Sci, Southern Methodist Univ, Dallas, 75-79; prof comput sci, Univ Tulsa. *Mem:* Sigma Xi. *Res:* Computer software; hardware systems development; quantum physics and chemistry; medical physics-magnetic resonance imaging. *Mailing Add:* 7421 S Marion Ave Tulsa OK 74136

LETCHER, STEPHEN VAUGHAN, b Chicago, Ill, Dec 13, 35; m 59; c 2. PHYSICS. *Educ:* Trinity Col, BS, 57; Brown Univ, PhD(physics), 64. *Prof Exp:* From asst prof to assoc prof, 63-75, PROF PHYSICS, UNIV RI, 75- *Mem:* Am Phys Soc; Am Asn Physics Teachers; Acoust Soc Am. *Res:* Physical acoustics; physics of fluids. *Mailing Add:* Dept Physics Univ RI Kingston RI 02881

LETEY, JOHN, JR, b Carbondale, Colo, June 13, 33; m 55; c 3. BIOPHYSICS. *Educ:* Colo State Univ, BS, 55; Univ Ill, PhD, 59. *Prof Exp:* Asst agron, Univ Ill, 55-59; asst prof, Univ Calif, Los Angeles, 59-64; assoc prof, soil physics, Univ Calif, Riverside, 64-68, chmn div environ sci, 68-75, chmn dept soil sci & environ sci, 75-80, dir Univ Calif Kearney Found Soil Sci, 80-85, PROF SOIL PHYSICS, UNIV CALIF, RIVERSIDE, 68- *Honors & Awards:* Soil Sci Award, Soil Sci Soc Am, 70. *Mem:* Fel Am Soc Agron; Soil Sci Soc Am; Sigma Xi. *Res:* Soil aeration; soil-water-plant relationships; soil wettability, infiltration; environmental pollutants. *Mailing Add:* Dept Soil & Environ Sci Univ Calif Riverside CA 92521

LETIZIA, GABRIEL JOSEPH, b New Rochelle, NY, Mar 15, 50; m 76; c 2. RESEARCH ADMINSTRATION, PHARMACOLOGY. *Educ:* Herbert H Lehman Col, City Univ, NY, BA, 76, MA, 79; Southwestern Univ, PhD(health sci), 84. *Prof Exp:* Qual control chemist, Purdue Frederick Co, Inc, 76-77; pigment chemist, Paul Uhlich & Inc, 77-79; tech dir, Alvin Last Pharmaceut & Cosmetics, Inc, 79-82; dir, AMA Labs, Inc, 82-84; EXEC DIR, WELLS LABS, INC, 84- *Mem:* Asn Consult Chem & Chem Engrs; AAAS; Soc Cosmetic Chemists; Am Soc Microbiol; Am Chem Soc; NY Acad Sci. *Res:* Testing criteria with respect to HILV-III/LAV Acquired Immune Deficiency Syndrome (AIDS), in tissue culture and inanimate surfaces; patent application for medical device treating common cold; clinical studies; toxicology; chemistry; microbiology; tissue culture. *Mailing Add:* 25-27 Lewis Ave Jersey City NJ 07306

LETKEMAN, PETER, b Winkler, Man, Feb 12, 38; m 64; c 3. CHEMISTRY. *Educ:* Univ Man, BSc, 60, MSc, 61, PhD(chem), 69. *Prof Exp:* Teacher high sch, Man, Can, 61-63; lectr, 63-66, from asst prof to assoc prof, 66-76, dept head, 72-80, PROF CHEM, BRANDON UNIV, 76-, DEAN SCI, 82- *Concurrent Pos:* Mem sci curric coun, Dept Educ, Man, 68-; grant, Univ Calif, Riverside, 70; consult, Christie Sch Supplies, Man, 70-; mem bd gov & senate, Brandon Univ, 73-77; pres, Western Man Sci Fair, 76; judge-in-chief, Can Wide Sci Fair, 75; mem staff, Tex A&M Univ, 77-78. *Honors & Awards:* Fel, Chem Inst Can. *Mem:* Chem Inst Can; AAAS. *Res:* The polarography and nuclear magnetic resonance of metal complexes in aqueous media; environmental research with regard to water and soil analysis; determination of stability constants of metal complexes. *Mailing Add:* Dean Sci Brandon Univ Brandon MB R7A 6A9 Can

LETO, SALVATORE, b Borgetto, Sicily, Nov 28, 37; US citzen; m 64; c 2. ANDROLOGY, CLINICAL CHEMISTRY. *Educ:* City Col New York, BS, 61; Georgetown Univ, PhD(biol), 67. *Prof Exp:* Staff res fel, Nat Inst Child Health & Human Develop, NIH, 67-71; supvr clin lab, Idant Corp, NY, 71-72, dir clin lab, Baltimore, 72-73; DIR, CLIN LAB WASHINGTON FERTIL STUDY CTR, 73- *Mem:* Am Fertil Soc; Am Physiol Soc; Am Soc Andrology; Sigma Xi; Am Asn Tissue Banks; AAAS. *Res:* Human male fertility; sperm cryo-preservation; immuno-infertility; endocrinology of reproduction. *Mailing Add:* 2600 Virginia Ave Suite 500 Washington DC 20037

LETOURNEAU, BUDD W(EBSTER), mechanical & nuclear engineering; deceased, see previous edition for last biography

LETOURNEAU, DUANE JOHN, b Stillwater, Minn, July 12, 26; m 47; c 3. PLANT BIOCHEMISTRY. *Educ:* Univ Minn, BS, 48, MS, 51, PhD(agr bot), 54. *Prof Exp:* Asst, Univ Minn, 48-53; asst prof agr chem & asst agr chemist, Univ Idaho, 53-58, assoc prof & assoc agr chemist, 58-63, prof agr biochem & agr biochemist, 63-73, actg head dept agr biochem & soils, 61-62, asst dept head, 89-90, PROF BIOCHEM & BIOCHEMIST, UNIV IDAHO, 73-, SECY FAC, 90- *Concurrent Pos:* Resident res assoc, USDA, 64-65; vis prof, Bot Dept, Univ Sheffield, Eng, 73; vis scientist, Nat Res Coun, Saskatoon, Can, 81. *Mem:* Fel AAAS; Am Soc Plant Physiol; Am Chem Soc; Am Phytopath Soc; Mycol Soc Am. *Res:* Plant biochemistry; plant cell culture techniques. *Mailing Add:* Off Fac Secy Univ Idaho Moscow ID 83843

LETOURNEUX, JEAN, b Que, Mar 23, 35; m 70. THEORETICAL NUCLEAR PHYSICS. *Educ:* Laval Univ, BSc, 59; Oxford Univ, DPhil(physics), 62. *Prof Exp:* Ciba fel, Inst Theoret Physics, Copenhagen, 62-64; res assoc physics, Univ Va, 64-65, asst prof, 65-66; from asst prof to assoc prof, 66-74, PROF PHYSICS, UNIV MONTREAL, 74- *Mem:* Am Phys Soc; Can Asn Physicists. *Res:* Nuclear theory. *Mailing Add:* Dept Physics Univ Montreal PO Box 6128 Montreal PQ H3C 3J7 Can

LETSINGER, ROBERT LEWIS, b Bloomfield, Ind, July 31, 21; m 43; c 3. ORGANIC CHEMISTRY. *Educ:* Mass Inst Technol, BS, 43, PhD(org chem), 45. *Prof Exp:* Asst, Mass Inst Technol, 43-45, res assoc, 45-46; res chemist, Tenn Eastman Corp, 46; from instr to prof, 46-88, chmn dept, 72-75, C H HALL PROF CHEM, NORTHWESTERN UNIV, 88- *Concurrent Pos:* Guggenheim fel, 56; mem, NIH Fel Rev Panel, 65-69; med chem study sect, NIH, 71-75. *Honors & Awards:* Rosenstiel Award, 85; Humboldt Sr Scientist Award, 88. *Mem:* Nat Acad Sci; Am Soc Biol Chemists; AAAS; fel Japan Soc Prom Sci; Am Chem Soc; Am Acad Arts & Sci. *Res:* Bioorganic chemistry; synthesis of polynucleotides and nucleotide analogs; photochemistry; organoboron and organoalkali metal compounds. *Mailing Add:* Dept Chem Northwestern Univ Evanston IL 60201

LETT, GREGORY SCOTT, b Denver, Colo, April 20, 58; m 83; c 3. SCIENTIFIC COMPUTING, OPTIMIZATION. *Educ:* Univ Colo, BA, 82. *Prof Exp:* Staff engr, Martin Marietta Astronaut Group, 83-85; consult, Software Develop Lab, 85-86; STAFF ENGR, MARTIN MARIETTA ASTRONAUT GROUP, 86- *Mem:* Am Math Soc; Math Asn Am. *Res:* Numerical linear algebra; simulation; stochastic estimation and optimization; numerical analysis of groundwater flow; parallel algorithms for fast transforms; applied and computational math; aerospace industry. *Mailing Add:* 11536 Green Circle Conifer CO 80433

LETT, JOHN TERENCE, b London, Eng, Dec 23, 33; m 56; c 1. BIOPHYSICS, RADIATION BIOLOGY. *Educ:* Univ London, BSc, 56, PhD(phys org chem), 60. *Prof Exp:* Sr lectr, Inst Cancer Res, Univ London, 56-67; PROF RADIOL & RADIATION BIOL, GRAD SCH, COLO STATE UNIV, 68- *Concurrent Pos:* Res assoc, Univ Calif, 61; vis scientist, Oak Ridge Nat Lab, 64. *Mem:* Radiation Res Soc; Brit Biophys Soc; Biophys Soc; Brit Asn Radiation Res. *Res:* DNA structure of the chromosome; repair of radiation damage to cellular DNA; radiation and aging. *Mailing Add:* Dept Radiol & Radiation Biol Colo State Univ Ft Collins CO 80523

LETT, PHILIP W(OOD), JR, b Newton, Ala, May 4, 22; m 48; c 3. MECHANICAL ENGINEERING. *Educ:* Auburn Univ, BME, 43; Univ Ala, MS, 47; Univ Mich, PhD(mech eng), 50; Mass Inst Technol, ScM, 61. *Prof Exp:* Instr, Univ Mich, 48-50; proj engr, Eng Div, Chrysler Corp, 50-54, asst chief engr, 54-58, chief engr, 58-61, operating mgr, Chrysler Defense Eng, 61-73, prog mgr, 73-76, gen mgr, Sterling Defense Div, 76-79, vpres eng, Chrysler Defense Div, 79-82; vpres res & eng, Gen Dynamics Land Systs, Warren, Mich, 82-87; PRES, PWL INC, 87- *Concurrent Pos:* Mem, Eng Coun, Auburn Univ, contrib author to the Int Defense Rev, Gen Dynamics, US Army; chmn, Vehicle Technol Sect, Tank-Automotive Div, Am Defense Preparedness Asn; consult, Gen Dynamics, US Army. *Honors & Awards:* Cheonsu Medal, Repub Korea; Silver Medal, Am Defense Preparedness Asn. *Mem:* Nat Acad Eng; Soc Automotive Engrs; Asn US Army; Am Defense Preparedness Asn. *Res:* Analytical and experimental studies of dynamic stability of surface vehicles employing models in wind tunnels and full scale instrumented vehicles on roads; computer simulation of business systems using industrial dynamics techniques; vehicular systems and surface mobility. *Mailing Add:* PWL Inc 1330 Oxford Birmingham MI 48009

LETTERMAN, GORDON SPARKS, b St Louis, Mo, Aug 17, 14; m 47; c 1. SURGERY, PLASTIC SURGERY. *Educ:* Wash Univ, AB, 37, BS, 40, MD, 41; Am Bd Surg, dipl; Am Bd Plastic Surg, dipl. *Prof Exp:* Asst surg, Wash Univ, 43-48; instr, George Washington Univ, 49-53, from assoc to asst prof, 53-60, from assoc clin prof to clin prof, 60-64, prof surg, 64-89, EMER PROF SURG, GEORGE WASHINGTON UNIV, 89- *Concurrent Pos:* Mem, Int Cong Plastic Surgeons, 55. *Mem:* Am Soc Plastic & Reconstruct Surg; Am Asn Plastic Surg; Int Soc Aesthetic Plastic Surg; Am Soc Aesthetic Plastic Surg; fel Am Col Surgeons. *Res:* Plastic surgery. *Mailing Add:* 5101 River Rd Bethesda MD 20816

LETTERMAN, HERBERT, b Brooklyn, NY, Oct 8, 36; m 57; c 4. ANALYTICAL CHEMISTRY, LABORATORY MANAGEMENT. *Educ:* City Col New York, BS, 58; Brooklyn Col, MA, 62; Seton Hall Univ, MS, 67, PhD(analytical chem) 73. *Prof Exp:* Analytical chemist, Brooklyn Jewish Hosp, NY, 58-59; analytical chemist, Ciba Pharmaceut Co, NJ, 59-63; group leader phys chem res & develop, 63-66, head qual control, 66-78, mgr qual serv, 78-86, MGR PROD STABILITY DEVELOP, BRISTOL-MYERS PROD DIV, 86- *Mem:* Sigma Xi; Am Chem Soc; Acad Pharmaceut Sci; sr mem Am Soc Qual Control. *Res:* Quality control; analytical method development. *Mailing Add:* 44 Delaware Ave New Providence NJ 07974

LETTIERI, THOMAS ROBERT, b Scranton, Pa, Sept 9, 14, 52. OPTICS, MICROMETROLOGY. *Educ:* Univ Miami, BS, 73; Univ Rochester, MS, 76, PhD(optics), 78; Univ Md, MGA, 87. *Prof Exp:* Physicist liquids, Nat Bur Standards, 78-79; PHYSICIST OPTICS, NAT INST STANDARDS & TECHNOL, 79- *Concurrent Pos:* Consult, Develop Proj, Chinese Univ; tech monitor, Agency Int Develop Projs, Egypt & India. *Honors & Awards:* IR-100 Award, 86. *Mem:* Sigma Xi; Optical Soc Am. *Res:* Microparticle measurements; light scattering; dimensional metrology. *Mailing Add:* Met-A117 Nat Inst Standards & Technol Gaithersburg MD 20899

LETTON, JAMES CAREY, b Paris, Ky, June 9, 33; m 56; c 3. PHARMACEUTICAL CHEMISTRY, ORGANIC SYNTHESIS. *Educ:* Ky State Col, BS, 55; Univ Ill, Chicago Med Ctr, PhD(chem), 71. *Hon Degrees:* LHD, Ky State Univ. *Prof Exp:* Prod foreman, Julian Labs, 57-62, supt prod, Smith Kline & French Labs, 62-64, res & develop chemist, Julian Res Inst, 64-69; instr org chem, Triton Col, 68-70; assoc prof org chem, Ky State Univ, 70-73, chmn, Dept Chem, 71-75, prof org chem, 73-75; ORG CHEMIST, PROCTER & GAMBLE CO, 76- *Concurrent Pos:* Nat Pres, Ky State Univ Alumni Asn, 78-84. *Honors & Awards:* Percy L Julian Award, Nat Orgn Prof Advan Black Chemists & Chem Engrs, 89. *Mem:* Fel Am Inst Chemists; Am Chem Soc; hon Soc Pharmaceut Sci. *Res:* Medicinal chemistry, especially beta amino ketones and analgesic properties; morphine-like compounds; steroid synthesis-carbohydrate chemistry; surface active and nonionics-synthesis. *Mailing Add:* Proctor & Gamble Co 11530 Reed Hartman Hwy Cincinnati OH 45241

LETTS, LINDSAY GORDON, b Warragul, Australia, Jan 9, 48; m 69; c 3. RESEARCH ADMINISTRATION. *Educ:* Monash Univ, BSc, 71; Sydney Univ, PhD(pharmacol), 80. *Prof Exp:* Tutor pharmacol, Sydney Univ, 76-80; res scientist, Royal Col Surgeons Eng, 80-82; sr res fel pharmacol, Merck

Frosst Can Inc, 82-87; DIR, BOEHRINGER INGELHEIM PHARMACEUT, 87- *Concurrent Pos:* Mem, bd, Nat Inst Community Health Educ, Quinnipiac Col, Conn, 90-; sect ed, Prostaglandins, 86-; adj assoc prof, Yale Univ Sch Med, 91-; ed, Mediators of Inflammation, 91- *Mem:* Am Thoracic Soc; Soc Leukocyte Biol; NY Acad Sci; Am Heart Asn. *Res:* Inflammation research, including arachidonic acid metabolites, especially prostaglandins and leukotrienes; pulmonary inflammation including cellular influx, mediator release, airway responsiveness, asthma; elucidation of role of integrins during inflammatory responses; author of numerous publications. *Mailing Add:* Boehringer Ingelheim Pharmaceut Inc 90 E Ridge Rd PO Box 368 Ridgefield CT 06877

LETTVIN, JEROME Y, b Chicago, Ill, Feb 23, 20; m 47; c 3. NEUROPHYSIOLOGY. *Educ:* Univ Ill, BS, 42, MD, 43. *Prof Exp:* Intern neurol, Boston City Hosp, 43-44; physiologist, Dept Psychol, Univ Rochester, 47-48; neuropsychiatrist & physiologist, Manteno State Hosp, 48-51; PROF COMMUN PHYSIOL, DEPTS BIOL, ELEC ENG & COMPUT SCI, MASS INST TECHNOL, 66-, NEUROPHYSIOLOGIST, LAB ELECTRONICS, 51- *Concurrent Pos:* Lectr neurol, Harvard Med Sch, 75- *Mem:* Am Physiol Soc. *Res:* Experimental epistemology. *Mailing Add:* Dept Biol Mass Inst Technol 77 Mass Ave Rm 20C-014 Cambridge MA 02139

LEU, MING C, b Taoyuan, Taiwan, Apr 27, 51; US citizen; m 78; c 3. ROBOTICS, MANUFACTURING AUTOMATION. *Educ:* Nat Univ Taiwan, BS, 72; Pa State Univ, MS, 77; Univ Calif, Berkeley, PhD(mech eng), 81. *Prof Exp:* Res asst surface friction, Pa State Univ, 75-77; res asst vibration & control, Univ Calif, Berkeley, 77-81; asst prof robotics & automation, Cornell Univ, 81-87; PROF & CHAIR ROBOTICS & AUTOMATION, NJ INST TECHNOL, 87- *Concurrent Pos:* Consult, Moog Inc, 81-87, AT&T, 88-; presidential young investr award, NSF, 85; exec comt mem, Prod Eng Div, Am Soc Mech Engrs, 86-89, chmn, 89-90; prog chmn, Japan-US Symp on Flexible Automation, 90- *Honors & Awards:* Wood Award, Forest Prod Res Soc, 81. *Mem:* Am Soc Mech Engrs; Inst Elec & Electronics Engrs; Soc Automotive Engrs; Soc Mfg Engrs; Int Soc Prod Engrs. *Res:* Motion planning and control; sensors and actuators; geometric modeling; robotics; automated assembly; author of over 100 technical publications. *Mailing Add:* Three Hamilton Pl Pine Brook NJ 07058

LEU, RICHARD WILLIAM, b Argonia, Kans, Jan 5, 35; m 60; c 2. MICROBIOLOGY, IMMUNOLOGY. *Educ:* Northwestern State Col, Okla, BS, 60; Univ Okla, MS, 63, PhD(microbiol, immunol), 70. *Prof Exp:* USPHS res training fel pediat & path, Med Sch, Univ Minn, Minneapolis, 70-74; MEM STAFF, NOBLE FOUND, 74-, HEAD IMMUNOL SECT, 76- *Res:* Cellular immunity; effector molecules associated with macrophage inhibition, proliferation and activation; role of cytophilic antibody in cellular immunity; localized immunity in the lung. *Mailing Add:* PO Box 2180 Ardmore OK 73402

LEUBNER, GERHARD WALTER, b Walton, NY, Aug 31, 21; m 44; c 3. ORGANIC CHEMISTRY. *Educ:* Union Col, BS, 43; Univ Ill, PhD(chem), 49. *Prof Exp:* Chemist, Winthrop Chem Co, 43-45; asst, Univ Ill, 45-46; res assoc, Eastman Kodak Co, 48-; RETIRED. *Mem:* Am Chem Soc. *Res:* Patent information storage and retrieval systems. *Mailing Add:* 151 Upland Dr Rochester NY 14617

LEUBNER, INGO HERWIG, b Prittlbach, Ger, Apr 9, 38; m 69; c 2. PHYSICAL CHEMISTRY. *Educ:* Munich Tech Univ, Dipl, 63, PhD(phys chem), 66. *Prof Exp:* Ger Res Asn res fel phys chem, Munich Tech Univ, 66-68; Welch Found fel & lectr photochem, Tex Christian Univ, 68-69; SR RES CHEMIST, RES LABS, EASTMAN KODAK CO, 69- *Mem:* Fel Soc Imaging Sci & Technol; Am Chem Soc; Sigma Xi. *Res:* Photochemistry of organic and inorganic compounds; crystal formation (nucleation). *Mailing Add:* 23 Willowview Dr Penfield NY 14526

LEUCHTAG, H RICHARD, b Breslau, Ger, June 2, 27; US citizen; m 55; c 1. MEMBRANE BIOPHYSICS, ION CHANNEL THEORY. *Educ:* Univ Calif, Los Angeles, BA, 50, MA, 55; Ind Univ, PhD(physics), 74. *Prof Exp:* Instr, Don Bosco Tech High Sch, 61-62, Univ San Diego Col Men, 62-63, San Diego State Col, 63-65 & Ind Univ-Purdue Univ, Indianapolis, 65-70; res assoc, Biophys Lab, Phys Dept, NY Univ, 72-74; assoc ed, Physics Today, Am Inst Physics, 74-78; res scientist, Dept Physiol & Biophys, Univ Tex Med Br, 78-82; asst prof biol, 82-87, ASSOC PROF BIOL, TEX SOUTHERN UNIV, 87- *Concurrent Pos:* Physicist, Western Elec Co, 66; consult, dept physiol & biophys, NY Univ Med Ctr, 78; secy, Int Conf Structure & Function Excitable Cells, 81; vis scientist, Mat Res Lab, Pa State Univ, 90. *Mem:* Am Phys Soc; Biophys Soc; Soc Neurosci; Soc Indust & Appl Math; Soc Gen Physiologists; Sigma Xi. *Res:* Physical basis of excitability in channels and membranes; ferroelectric-superionic transition hypothesis; measurement of noise, admittance and impedance in axons; monitored retrievable disposal of high-level radioactive waste. *Mailing Add:* Dept Biol Tex Southern Univ 3100 Cleburne Houston TX 77004

LEUCHTENBERGER, CECILE, cytology, for more information see previous edition

LEUNG, ALBERT YUK-SING, b Hong Kong, May 24, 38; nat US; m 68; c 2. PHARMACOGNOSY, BIOMEDICAL INFORMATION SERVICES. *Educ:* Nat Taiwan Univ, BS, 61; Univ Mich, Ann Arbor, MS, 65, PhD(pharmacog), 67. *Prof Exp:* NIH res chemist, Med Ctr, Univ Calif, 67-69; res supvr microbial protein prod, Bohna Eng & Res, Inc, 69-71; tech dir chem & microbiol consult, Sci Res Info Serv, Inc, 71-74; dir res & develop, Dr Madis Labs, Inc, 74-77; CONSULT NATURAL PRODS, 77- *Concurrent Pos:* Lily Found fel, 63-67. *Mem:* Am Chem Soc; Am Soc Pharmacog; NY Acad Sci; Sigma Xi. *Res:* isolation of active principles from plants and microorganisms; retrieval and dissemination of biomedical information, especially from Chinese sources. *Mailing Add:* 35 Cumberland Rd Glen Rock NJ 07452

LEUNG, ALEXANDER KWOK-CHU, b Hong Kong, Oct 1, 48; Can citizen; m 75; c 4. GENERAL PEDIATRICS. *Educ:* Univ Hong Kong, MBBS, 73; Royal Col Physicians London & Royal Col Surgeons Eng, DCH, 77; Royal Col Physicians & Surgeons Ireland, DCH, 79, MRCPI, 78; MRCP(UK), 80; FRCP(C), 79, FAAP, 80, Am Bd Pediat Endocrinol & Pediat, dipl, 86; World Univ, PhD, 88, FRCP(G), 89, FRC, 90, FRCPI, 90. *Prof Exp:* Intern, Univ Hong Kong, 73-74; lectr surg, 74; lectr pediat, Univ Queensland, 77; endocrine fel, Univ Calgary, 78-80; consult pediat, Calgary Gen Hosp, 80-81 & Grace Hosp, 80-82; clin asst prof, 82-90, CLIN ASSOC PROF PEDIAT, UNIV CALGARY, 90-; CONSULT PEDIAT, FOOTHILLS PROV HOSP & ALTA CHILDRENS HOSP, 80- *Honors & Awards:* Gold Medal Award, ABI, 87; Physician Recognition Award, AMA, 87 & 90; Prep Fel Award, Am Acad Pediat, 87 & 90. *Mem:* Fel Royal Soc Med; fel Royal Soc Health; fel Can Pediat Soc; fel Am Acad Pediat; fel Royal Col Physicians Can; fel Royal Col Physicians Ireland; fel Royal Col Physicians Eng; fel Royal Col Physicians Glasgow. *Res:* General pediatrics; author of over 250 publications in various fields of pediatrics. *Mailing Add:* 330 Mkt Hall Prof Bldg 4935 40 Ave NW Calgary AB T3A 2N1 Can

LEUNG, BENJAMIN SHUET-KIN, b Hong Kong, June 30, 38; US citizen; m 64; c 3. ENDOCRINOLOGY, ONCOLOGY. *Educ:* Seattle Pac Col, BS, 63; Colo State Univ, PhD(biochem), 69. *Prof Exp:* Res asst steroid hormones, Pac Northwest Res Found, 63-66; from asst prof to assoc prof surg, Med Sch, Univ Ore, dir Hormone Receptor Lab, Clin Res Ctr Lab, 71-76; sr res scientist, Dept Surg, Cedars-Sinai Med Ctr, Los Angeles, 76-78; assoc prof, 78-85, PROF & DIR, HORMONE RES LAB, DEPT OBSTET & GYNEC, UNIV MINN, 78-, CHIEF, DIV REPRODUCTIVE CELL BIOL, 84-, DIR GRAD STUDIES, 90- *Concurrent Pos:* NIH & Ford Found res fel reprod endocrinol, Med Sch, Vanderbilt Univ, 69-71; Med Res Found Ore grant, Med Sch, Univ Ore, 71-72; Am Cancer Soc Ore Div res grants, 72-74; Cammack Trust Fund grant, 74-75; NIH grants, 75-79; assoc oncologist, Div Surg Oncol, Univ Calif, Los Angeles, 76-78; Med Res Found Ore res grant, 76-77; Nat Cancer Inst res grant, 77-91; Minn Med Found res grant, 78-79; ad hoc consult & grant rev, NSF, NIH, Nat Cancer Inst, Nat Inst Alcohol Abuse & Alcoholism; ad hoc ed & referee sci manuscripts, J Nat Cancer Inst, J Biol Chem & J Steroid Biol; grant-in-aid, Univ Minn, 81-82; from assoc prof to prof, Dept Animal Physiol, Univ Minn, 82-; Nat Inst Alcohol Abuse & Alcoholism grant, 89-92. *Mem:* Am Soc Biol Chemists; AAAS; Endocrine Soc; Soc Gynec Invest; Am Asn Cancer Res. *Res:* Fetal growth; cancer growth; growth factors; oncogens. *Mailing Add:* Dept Obstet & Gynec Box 395 UMHC Minneapolis MN 55455

LEUNG, CHARLES CHEUNG-WAN, b Hong Kong, June 27, 46; US citizen; m 73; c 1. SEMICONDUCTOR PROCESSING, DEVICE PHYSICS. *Educ:* Univ Hong Kong, BSc, 69; Univ Chicago, MS, 71, PhD(physics), 76. *Prof Exp:* Sr scientist, Corning Glass Works, 76-80; sr staff engr, Motorola, 80-81; SR MEM TECH STAFF, AVANTEK, 81- *Mem:* Am Phys Soc; Am Chem Soc; Am Vacuum Soc; Soc Info Display. *Res:* Electro-optic materials; glassification; carbon; vacuum deposition; thin film; surface science. *Mailing Add:* Bipolarics 108 Albright Way Los Gatos CA 95030

LEUNG, CHRISTOPHER CHUNG-KIT, b Hong Kong, Jan 3, 39; m 70; c 2. EMBRYOLOGY, IMMUNOLOGY. *Educ:* Howard Univ, BSc, 64; Jefferson Med Col, PhD(anat, embryol), 69. *Prof Exp:* Res asst, Sch Med, Univ Rochester, 64-65; from instr to assoc prof pediat, Jefferson Med Col, Thomas Jefferson Univ, 69-74, instr anat, 69-75, instr, Col Allied Health Sci, 70-75, res assoc prof pediat, 74-75; asst prof anat, Univ Kans Med Ctr, 75-79; assoc prof anat, Sch Med, La State Univ, 79-85; ASSOC PROF ANAT, NJ MED SCH, 85- *Concurrent Pos:* NIH fel, Stein Res Ctr, Thomas Jefferson Univ, 65-69; NIH res grant, 79-89; mem Ad Hoc Study Sect, NIH, 85. *Mem:* Teratol Soc; Am Asn Anatomists; Am Asn Immunologists; Am Soc Cell Biol. *Res:* Teratology; immunopathology; cell biology. *Mailing Add:* Dept Anat NJ Med Sch Univ Med & Dent Newark NJ 07103

LEUNG, CHUNG NGOC, b Macao, Hong Kong, May 12, 56; m. THEORETICAL PHYSICS, ELEMENTARY PARTICLE PHYSICS. *Educ:* Univ Minn, BS, 77, PhD(physics), 83. *Prof Exp:* Res assoc, Fermi Nat Accelerator Lab, 83-85; res assoc, Purdue Univ, 85-86 & 87-89; res assoc, Max Planck Inst Physics & Astrophys, 86-87; ASST PROF, UNIV DEL, 89- *Res:* Elementary particle theory; dynamical symmetry breaking in gauge theories; fermion mass problem; physics of massive neutrinos; phenomenology of elementary particles; chaos and nonlinear dynamical systems; self-organized critical phenomena. *Mailing Add:* Dept Physics & Astron Univ Del Newark DE 19716

LEUNG, DAVID WAI-HUNG, b Hong Kong, Aug 2, 51; m 82; c 2. BIOCHEMISTRY. *Educ:* Whittier Col, BA, 73; Univ Ill, Urbana-Champaign, PhD(biochem), 78. *Prof Exp:* Res fel biochem, Univ BC, 78-80; res fel, 80-81, scientist, 81-88, SR SCIENTIST, GENENTECH INC, 88- *Res:* Isolation and cloning of genes for growth factors and their receptors; optimization of the cloned gene products in bacteria, yeast and tissue culture using recombinant DNA technology. *Mailing Add:* Genentech Inc 460 Pt San Bruno Blvd S San Francisco CA 94080

LEUNG, DONALD YAP MAN, b New York, NY, Oct 1, 49; m 78; c 2. KAWASAKI DISEASE, ATOPIC DERMATITIS. *Educ:* Johns Hopkins Univ, BA, 70; Univ Chicago, PhD(biochem), 75, MD, 77. *Prof Exp:* Intern pediat, Children's Hosp, Boston, 77-78; resident pediat, 78-79; fel allergy-immunol, 79-81; clin fel pediat, Harvard Med Sch, 77-79, instr pediat, 81-85; from asst prof to assoc prof pediat, 83-89; ASSOC PROF PEDIAT, UNIV COLO, 89- *Concurrent Pos:* Dir allergy, Children's Hosp, 87-89; head, Div Allergy-Immunol, Nat Jewish Ctr Immunol & Respiratory Med, 89- *Mem:* Am Fedn Clin Res; Am Acad Allergy & Immunol; Am Asn Immunologists; Soc Pediat Res; AAAS. *Res:* Mechanisms of allergic diseases including atopic dermatitis, asthma and food allergy, IgE responses; immunopathogenesis of Kawasaki disease; use of immunomodulatory agents in the treatment of allergic disorders. *Mailing Add:* Dept Pediat K926 Nat Jewish Ctr Immunol & Resp Med 1400 Jackson St Denver CO 80206

LEUNG, FREDERICK C, b Hong Kong, Dec 1, 52; US citizen; m 78; c 2. MOLECULAR BIOLOGY, ZOOLOGY. *Educ:* Univ Calif, Berkeley, BA, 74, PhD(endocrinol), 78. *Prof Exp:* Fel neuroendocrinol, dept physiol, Mich State Univ, 78-80; sr res biochemist animal physiol, Merck, Sharp & Dohme Res Labs, 80-84, res fel, 84-85; SR RES SCIENTIST CELL BIOL, DEPT BIOL & CHEM, BATTELLE, PAC NORTHWEST LABS, 85- *Concurrent Pos:* Adj prof, dept animal sci, Rutgers Univ, 82-85. *Mem:* Endocrinol Soc; Am Physiol Soc; NY Acad Sci; Am Soc Zool. *Res:* Hormonal regulation of growth; hypothalamic regulation of anterior pituitary function; structure & function of growth hormone and its receptor; eukaryotic cell expression and gene insertion. *Mailing Add:* Battelle Pac Northwest Labs PO Box 999 Richland WA 99352

LEUNG, IRENE SHEUNG-YING, b Hong Kong, July 10, 34. MINERALOGY. *Educ:* Univ Hong Kong, BA, 57; Ohio State Univ, MA, 63; Univ Calif, Berkeley, PhD(geol), 69. *Prof Exp:* Res staff geologist, Yale Univ, 69-71; asst prof, 71-77, ASSOC PROF GEOL, LEHMAN COL, 77- *Mem:* Sigma Xi; Mineral Soc Am; Am Geophys Union; Geochem Soc; Asian Environ Soc. *Res:* X-ray investigation of mineral inclusions in natural diamonds; magmatic crystallization and sector-zoning in crystals; deformation structures and glide mechanisms in deformed minerals. *Mailing Add:* Dept Geol & Geog City Univ NY Lehman Col Bedford Park Blvd W Bronx NY 10468

LEUNG, JOSEPH YUK-TONG, b Hong Kong, June 25, 50; m 73. COMPUTER SCIENCE. *Educ:* Southern Ill Univ, Carbondale, BA, 72; Pa State Univ, PhD(comput sci), 77. *Prof Exp:* Asst prof computer sci, Va Polytech Inst & State Univ, 76-77; asst prof computer sci, Northwestern Univ, Evanston, 77-81, assoc prof eng & computer sci, 81-85; prof computer sci prog, Univ Tex, Dallas, 85-90; CHMN DEPT COMPUTER SCI & ENG, UNIV NEBR, LINCOLN, 90- *Mem:* Asn Comput Mach; Inst Elec & Electronics Engrs. *Res:* Operating systems; scheduling theory; analysis of algorithms; data structure; computational complexity. *Mailing Add:* Dept Computer Sci & Eng Univ Nebr-Lincoln Lincoln NE 68588-0115

LEUNG, JULIA PAULINE, monoclonal antibody, cancer biology, for more information see previous edition

LEUNG, KAM-CHING, b Hong Kong, June 16, 35; m 63; c 2. ASTRONOMY, ASTROPHYSICS. *Educ:* Queen's Univ, Ont, BSc, 61; Univ Western Ont, MA, 63; Univ Pa, PhD(astron), 67. *Prof Exp:* Nat Acad Sci-Nat Res Coun res fel astron, Inst Space Studies, NASA, 68-70; asst prof physics, 70-72, assoc prof, 72-78, PROF PHYSICS & ASTRON, UNIV NEBR, LINCOLN, 78- *Concurrent Pos:* NSF res grant, Univ Nebr, Lincoln, 70-71, 75, 81-82, 85-87 & 87-89, NASA grant, 86-87, dir observ, 72-75; staff assoc astron sect, NSF, 75; mgt specialist, Off Nuclear Energy, ERDA, 77; sr assoc, Off Energy Res, Dept Energy. *Mem:* Fel AAAS; Int Astron Union; Am Astron Soc. *Res:* Stellar photometry and spectroscopy; intrinsic variable stars; binary stars. *Mailing Add:* Dept Physics & Astron Univ Nebr Lincoln NE 68588-0111

LEUNG, KA-NGO, b Canton, China. ELECTRICAL ENGINEERING, SYSTEMS DESIGN. *Educ:* Chinese Univ Hong Kong, BS, 68; Univ Akron, MS, 71; Univ Calif, Los Angeles, PhD(physics), 75. *Prof Exp:* Asst prof physics, James Madison Univ, Va, 75-78; staff physicist, 78-88, SR STAFF PHYSICIST, LAWRENCE BERKELEY LAB, 88- *Mem:* Fel Am Phys Soc; Sigma Xi. *Res:* Development of positive and negative ion sources for neutral beam heating in fusion reactors and particle accelerators. *Mailing Add:* Lawrence Berkeley Lab Bldg 4 Berkeley CA 94720

LEUNG, LAI-WO STAN, b Macau, Can, July 21, 52; m 79; c 1. NEUROPHYSIOLOGY, BEHAVIOR & NEURAL ACTIVITY. *Educ:* Calif State Univ, Northridge, BSc, 73; Univ Calif, Berkeley, PhD(biophysics), 78. *Prof Exp:* Teaching assoc physiol & biophysics, Univ Calif, Berkeley, 77-78; fel physiol psychol, Univ Western Ont, London, Can, 78-79, fel brain res, Int Brain Res Orgn & UNESCO fel, Inst Med Physics, Utrecht, Holland, 79-80; asst prof, 80-86, ASSOC PROF CLIN NEUROL SCI, PHYSIOL & PSYCHOL, UNIV WESTERN ONT, LONDON, CAN, 86- *Concurrent Pos:* Univ res fel, Natural Sci & Eng Res Coun, 80-; mem biopsychol study sect, NIH, 87- *Mem:* Soc Neurosci. *Res:* Analyses of normal and abnormal neuronal activities in the cerebral cortex in vivo and in vitro. *Mailing Add:* Dept Clin Neurol Sci Univ Western Ont London ON N6A 5C5 Can

LEUNG, PAK SANG, b Shanghai, China, June 8, 35; US citizen; m 65; c 2. COLLOID CHEMISTRY. *Educ:* Nat Taiwan Univ, BSc, 57; Columbia Univ, MA, 62, PhD(phys chem), 67. *Prof Exp:* Dyes lab asst, Imp Chem Industs, 57-59; demonstr chem, Hong Kong Baptist Col, 59-61; res scientist, Brookhaven Nat Lab-Columbia Univ, 66-67; RES SCIENTIST, UNION CARBIDE CORP, TARRYTOWN, 67- *Concurrent Pos:* Mem chem adv bd, Harriman Col, 74- *Mem:* Sigma Xi; Am Chem Soc. *Res:* Surface and collidal chemistry; polymer composite; polymer processing; clinical chemistry. *Mailing Add:* 15 Woodland Rd Highland Mills NY 10930

LEUNG, PETER, b New York, NY, Apr 12, 55. DRUG METABOLISM, PHARMACOKINETICS. *Educ:* Johns Hopkins Univ, BA, 77; State Univ NY, PhD(pharmacol), 83; Am Bd Toxicol, dipl, 87. *Prof Exp:* Postdoctoral fel, Col Med, Tex A&M Univ, 83-86; sr scientist, Schering-Plough Corp, 86-89; staff toxicologist, Med Toxicol Br, Calif Dept Food & Agr, 89-91; STAFF TOXICOLOGIST, CALIF DEPT PESTICIDE REGULATION, 91- *Concurrent Pos:* Am Soc Pharmacol & Exp Therapeut travel award, 84; res assoc, Dept Med Pharmacol & Toxicol, Col Med, Tex A&M Univ, College Station, 86. *Mem:* Soc Toxicol; Am Col Toxicol; Am Soc Pharmacol & Exp Therapeut. *Res:* Metabolism and disposition of cyanide in the presence and absence of cyanide antagonists; new prophylactic and therapeutic treatments for cyanide intoxication; erythrocytes as drug carriers for purified enzymes. *Mailing Add:* Med Toxicol Br Calif Dept Pesticide Regulation 1220 N St Sacramento CA 95814

LEUNG, PHILIP MAN KIT, b Hong Kong, June 22, 33; Can citizen; m 59; c 2. MEDICAL PHYSICS, BIOPHYSICS. *Educ:* Univ Toronto, BASc, 60; McMaster Univ, MSc, 61; Univ Toronto, PhD(biophys), 67. *Prof Exp:* Lectr, Ryerson Inst Technol, Can, 61-62; physicist, BC Cancer Inst, Can, 67-68; SR PHYSICIST, ONT CANCER INST, 68- *Concurrent Pos:* Fel, Univ BC, 67-68; consult physicist, Orillia Soldier Mem Hosp, Can, 69-70 & Can Soc Radiol Technicians, 69-78; ed, Physics in Med & Biol, 72-75; investr, Children's Cancer Study Group, Nat Cancer Inst, USPHS, 76-77 & 77-78; lectr, Dept Med Biophys, Univ Toronto, 79- *Res:* Radiation dosimetry; radiotherapy treatment techniques and new equipment associated with radiation oncology. *Mailing Add:* Ont Cancer Inst 500 Sherbourne St Toronto ON M4X 1K9 Can

LEUNG, PHILIP MIN BUN, b Canton, China, July 31, 32; c 3. NUTRITION, BIOCHEMISTRY. *Educ:* Nat Chung Hsing Univ, Taiwan, BSc, 56; McGill Univ, MSc, 59; Mass Inst Technol, PhD(nutrit biochem), 65. *Prof Exp:* Res asst, McGill Univ, 56-59; res asst, Mass Inst Technol, 59-65; group leader biochem res, Med Sci Res Lab, Miles Labs Inc, Ind, 65-67; asst res nutritionist to assoc res nutritionist, 67-86, NUTRITIONIST, SCH VET MED, UNIV CALIF , DAVIS, 86 - *Mem:* AAAS; Am Inst Nutrit; Inst Food Technol; NY Acad Sci; Soc Neurosci. *Res:* Nutrition and biochemistry of amino acid imbalance; nutritional regulation of protein intake and metabolism; influence of nutrition, especially amino acid balance on food intake regulation. *Mailing Add:* 1100 Radcliffe Dr Davis CA 95616

LEUNG, SO WAH, b China, Nov 2, 18; nat US; m 57; c 1. DENTISTRY, PHYSIOLOGY. *Educ:* McGill Univ, DDS, 43, BSc, 45; Univ Rochester, PhD(physiol), 50; FRCDent(C). *Prof Exp:* Intern dent, Royal Victoria Hosp, Montreal, 43-44; from assoc prof to prof physiol, Sch Dent, Univ Pittsburgh, 50-61, head dept, 52-61, prof dent res & dir grad educ, 57-61; prof oral biol, Sch Dent & lectr physiol, Sch Med, Univ Calif, Los Angeles, 61-62; dean sch dent, 62-77, PROF ORAL BIOL, UNIV BC, 62- *Concurrent Pos:* Mem comt dent, Nat Acad Sci-Nat Res Coun, 57-61; consult, Colgate-Palmolive Co, 58-62; mem dent study sect, NIH, 59-63; consult, Nat Bd Dent Exam, 60-66 & Lever Bros, 63-65; mem assoc comt dent res, Nat Res Coun Can, 63-68, exec comt, 65-68; mem, Nat Dent Exam Bd Can, 65-67, chmn exam comt, 67-71; chmn res comt, Asn Can Fac Dent, 68-70, pres, 70-72. *Mem:* Am Dent Asn; fel Am Col Dent; fel Int Col Dent; NY Acad Sci; Sigma Xi. *Res:* Salivary chemistry; oral calculus formation; physiology of salivary glands. *Mailing Add:* 4510 NW Marine Drive Vancouver BC V6R 1B8 Can

LEUNG, WAI YAN, PHYSICS. *Prof Exp:* Asst prof, Dept Chem, Fla Int Univ, 89-91; ASSOC PROF, DEPT CHEM, AMES LAB, IOWA STATE UNIV, 91- *Mem:* Am Phys Soc. *Res:* Service structures. *Mailing Add:* Dept Physics Spedding Hall Ames Lab Rm 224 Iowa State Univ Ames IA 50011

LEUNG, WING HAI, b Hong Kong, July 29, 37; m 65; c 3. SURFACE CHEMISTRY. *Educ:* Univ Hong Kong, BSc, 63; Univ Miami, MS, 70, PhD(phys chem), 74. *Prof Exp:* Res assoc, State Univ NY, Buffalo, 74-76; sr chemist res, GAF Corp, Binghamton, NY, 76-77; res scientist, Clinton Corn Corp, Iowa, 77-78; asst prof, Hampton Inst, 78-82, ASSOC PROF CHEM, HAMPTON UNIV, 82- *Mem:* Am Chem Soc; Sigma Xi; Am Geophys Union. *Res:* Surface phenomena and kinetic studies of crystal growth; the structure and interaction at solid-solution interfaces. *Mailing Add:* Hampton Univ Box 6422 Hampton VA 23668

LEUNG, WOON F (WALLACE), b Hong Kong, Jan 25, 54; m 78; c 2. FLUID DYNAMICS, SOLID-LIQUID SEPARATION. *Educ:* Cornell Univ, BS, 77; Mass Inst Technol, MS, 78, ScD(mech eng), 81. *Prof Exp:* Res engr, flow in porous media, Gulf Res & Develop Co, 81-84; proj leader, flow near well-bore & transient testing, Schlumberger, 84-86; SR RES SCI, CENTRIFUGE DYNAMICS, BIRD MACH CO, 86- *Concurrent Pos:* Chmn, Centrifuge Network; tech session chmn, Am Filtration Soc Ann meetings, 90, 91, conf chmn, 93. *Honors & Awards:* Cedric Ferguson Medal, Soc Petrol Engrs, 87. *Mem:* Soc Mech Engrs; Soc Petrol Engrs; Am Filtration Soc; Am Inst Aeronaut & Astronautics; Soc Rheology. *Res:* Physical-chemical hydrodynamics for industrial applications, solid-liquid separation in industrial centrifuge, suspension rheology, membrane filtration and inclined lamella settlers; fluid flow and compaction-consolidation in porous media; oil and gas production through wellbore perforations; petroleum reservoir engineering; develop and use computer models and experimental test rigs in research and developmental study; develop innovative dynamic balancing methods for rotating machinery under process condition; ten published technical papers and over 200 in-house technical reports. *Mailing Add:* Bird Mach Co 100 Neponset St S South Walpole MA 02071

LEUPOLD, HERBERT AUGUST, b Brooklyn, NY, Jan 6, 31. PHYSICS, MATERIALS SCIENCE. *Educ:* Queens Col, NY, BS, 53; Columbia Univ, AM, 58, PhD(physics), 64. *Prof Exp:* Fel physics, Lawrence Radiation Lab, Livermore, Calif, 64-67; RES PHYSICIST, ELECTRONICS RES DEVELOP COMMAND, US ARMY, 67- *Concurrent Pos:* Lectr physics, Queens Col, NY, 57 & Monmouth Col, 67-70, 84-85, & Univ Dayton 84, 86, 88 & lectr chem, Trenton State Col, 83. *Honors & Awards:* Commendation, Inst for Explor Res, US Army, 69 & Electronics Technol & Devices Lab, US Army, 72; Harold Jacobs Award, 87. *Mem:* Am Phys Soc; Sigma Xi; Inst Elec & Electronics Engrs; Inst Elec & Electronics Engrs Magnetics Soc. *Res:* Magnetism; semiconductors; superconductivity; thermodynamics; cryogenics; magnetic circuit design; magnetic materials. *Mailing Add:* 26 B Stony Hill Gradens Eatontown NJ 07724

LEUSCHEN, M PATRICIA, b Iowa, June 3, 43; m 63; c 4. PERINATOLOGY, NEUROENDOCRINOLOGY. *Educ:* Creighton Univ, BS, 65, MS, 67; Med Ctr, Univ Nebr, Omaha, MS, 74, PhD(anat), 76. *Prof Exp:* Res asst pediat, 80-82, ASST PROF PEDIAT & RES DIR, DIV NEWBORN MED, MED CTR, UNIV NEBR, OMAHA, 82-; CLIN ASST PROF, CREIGHTON UNIV, OMAHA, 89- *Concurrent Pos:* Chair, Med Sci Interdepartmental Area Grad Prog, Univ Nebr Med Ctr, 89-, Exec Grad Comt, Univ Nebr Syst, 90-91. *Mem:* Am Asn Anatomists; Am Soc Cell Biol;

Soc Neurosci. *Res:* Cerebral microvasculature and intracranial hemorrhage in premature infants; B-endorphin and related neurohormones and stress in neonates, ultrastructural studies including morphometry; association with perinatal asphyxia and/or apnea; interaction between prostaglandin synthesis and neuroendocrine axis. *Mailing Add:* Dept Pediat Div Newborn Med Univ Nebr Med Ctr 3001 Douglas 7th Floor N Omaha NE 68131

LEUSSING, DANIEL, JR, b Cincinnati, Ohio, Oct 8, 24; m 57; c 3. ANALYTICAL CHEMISTRY. *Educ:* Univ Cincinnati, BA, 45; Univ Ill, MS, 47; Univ Minn, PhD(chem), 53. *Prof Exp:* Instr anal chem, Univ Minn, 51-52; chemist, Am Cyanamid Co, 53; instr anal chem, Mass Inst Technol, 53-55; from instr to asst prof, Univ Wis, 55-60; chemist, Nat Bur Standards, 60-62; from asst prof to assoc prof, 62-70, PROF CHEM, OHIO STATE UNIV, 70- *Mem:* Am Chem Soc. *Res:* Physical chemistry of aqueous solutions; coordination chemistry; metal mercaptide complexes; Schiff base complexes; kinetics. *Mailing Add:* Dept Chem Ohio State Univ Columbus OH 43210

LEUTENEGGER, WALTER, b Winterthur, Switz, Oct 18, 41; US citizen. BIOLOGICAL ANTHROPOLOGY, PRIMATOLOGY. *Educ:* Univ Zurich, PhD(biol anthrop), 69. *Prof Exp:* Anthropologist, Bern Natural Mus Hist, Switz, 67-69; sci res asst biol anthrop, Univ Zurich, 69-71; from asst prof to assoc prof, 71-83, PROF BIOL ANTHROP, UNIV WIS-MADISON, 83- *Concurrent Pos:* Affil scientist, Wis Regional Primate Res Ctr, 71- *Mem:* Am Asn Phys Anthropologists; Am Soc Naturalists; Am Soc Primatologists; Soc Vert Paleont. *Res:* Functional anatomy of the primate locomotor apparatus; determinants of behavioral and morphological sexual dimorphism; reconstruction of early hominid social organization and behavior. *Mailing Add:* Dept Anthrop Univ Wis 5240 Soc Sci Bldg 1180 Observatory Dr Madison WI 53706

LEUTERT, WERNER WALTER, applied mathematics; deceased, see previous edition for last biography

LEUTGOEB, ROSALIA ALOISIA, chemistry; deceased, see previous edition for last biography

LEUTHEUSSER, H(ANS) J(OACHIM), b Eisenach, Ger, Feb 1, 27; Can citizen; m 55; c 3. FLUID MECHANICS, TRIBOLOGY. *Educ:* Karlsruhe Univ, Dipl Ing, 52; Univ Toronto, MASc, 57, PhD(mech eng), 61. *Prof Exp:* Asst hydraul eng, Theodor-Rehbock Lab, Karlsruhe Univ, 51-52; field engr, Oulujoki Oy, Helsinki, Finland, 52-53; sr engr, Friedrich Buchner, Wuerzburg, Ger, 53-54; instr fluid mech, dept mech eng, 55-57, lectr, 57-62, from asst prof to assoc prof, 63-70, assoc chmn dept, 77-79, assoc chmn dept & coordr grad studies, 84-88, PROF FLUID MECH, DEPT MECH ENG, UNIV TORONTO, 70- *Concurrent Pos:* Consult engr, 57-; sabbatical leaves, Inst Mech Statist of Turbulence, Univ Aix Marseille, 66-67, Inst Hydromech, Karlsruhe Univ, 75-76, Univ Santiago & Aristotle Univ, Thessaloniki, 83, Univ Victoria, BC, 89, Univ Shanghai & Tokyo, 90. *Mem:* Am Soc Civil Engrs; Int Asn Hydraul Res. *Res:* Fundamental fluid mechanics and applications; turbulence; fluid elasticity and transients; biomechanics; building aerodynamics; tribology. *Mailing Add:* Dept Mech Eng Univ Toronto Toronto ON M5S 1A4 Can

LEUTZE, WILLARD PARKER, b Burlington, Vt, Mar 2, 27; m 51; c 2. GEOLOGY. *Educ:* Syracuse Univ, BS, 51, MS, 55; Ohio State Univ, PhD(geol), 59. *Prof Exp:* Geologist, US Geol Surv, 54-55; asst geol, Ohio State Univ, 55-58; instr geol & soil sci, Earlham Col, 58-60; geologist, Texaco, Inc, 60-66; biostratigrapher, Atlantic Richfield Co, 66-71, sr geologist, 71-75; sr geologist, Stone Oil, 75-77; INDEPENDENT CONSULT, 77- *Mem:* Asn Soc Econ Paleont & Mineral; Soc Prof Well Log Analysts; Soc Independent Earth Scientists. *Res:* Paleontology, particularly foraminifera, arthropods and echinoderms; stratigraphy of Upper Silurian and of Gulf Coast; subsurface stratigraphy of south Louisiana. *Mailing Add:* Box 52641 OCS Lafayette LA 70501

LEUTZINGER, RUDOLPH L(ESLIE), b Dallas Center, Iowa, June 17, 22; m 50; c 6. MECHANICAL ENGINEERING, AEROSPACE ENGINEERING. *Educ:* Iowa State Univ, BS, 43; Univ Mich, MS, 52; Univ Iowa, PhD(mech eng), 76. *Prof Exp:* Stress analyst, Douglas Airplane Co, 44-46 & McDonnell Airplane Co, 46 & 47; instr aeronaut eng, Iowa State Univ, 47-50; res assoc dynamics, Aeronaut Res Ctr, Univ Mich, 51; res engr, Midwest Res Inst, Mo, 51-53; asst prof aeronaut eng & appl mech, Univ Kans, 54-56; assoc prof aerodyn, Agr & Mech Col Tex, 56-58; assoc prof thermodyn, Univ Mo-Rolla, 58-60; assoc prof eng & chmn dept, Univ Mo-Kansas City, 62-76; assoc prof mech eng, 76-79, EMER PROF MECH ENG, UNIV MO-COLUMBIA, 79-; DIR, AERO TURB MFG, 79-; CONSULT ENGR, LEUTZINGER CONSULT ASSOCS, 79- *Concurrent Pos:* Consult, Boeing Airplane Co, 57, Gas Turbine Div, Westinghouse Elec Corp, 56 & McDonnell Aircraft Corp, 58, 59 & 62; assoc mem grad fac & eng admin coun, Agr & Mech Col Tex, 58; mem grad fac, Univ Mo, 59; NSF inst res grants, 61, 62 & 64; Kans City Regional Coun Higher Educ grant, 64-65; vis lectr, Col Eng, Univ Iowa, 69-; NASA fel propulsion, 73 & 77. *Mem:* Soc Eng Sci; Am Inst Aeronaut & Astronaut (secy-treas, 42 & 75-78); Am Soc Eng Educ; Nat Soc Prof Engrs. *Res:* Structural mechanics and dynamics; gas dynamics; flow fields in turbomachinery, especially three-dimensional and boundary layers; vehicle design and analysis; internal and external aerodynamics of ducts and bodies. *Mailing Add:* Leutzinger Consult Assocs 1521 N Holder Independence MO 64050

LEUZE, REX ERNEST, b Sabetha, Kans, Mar 7, 22; m 48; c 3. CHEMICAL ENGINEERING, PROCESS CHEMISTRY. *Educ:* Kans State Univ, BS, 44; Univ Tenn, Knoxville, MS, 56. *Prof Exp:* Anal chemist, Monsanto Chem Co, Ill, 44-45; anal chemist, Clinton Labs, 45-47; chem engr. Tech Div, Oak Ridge Nat Lab, 47-49, develop group leader inorg fluorides, Chem Tech Div, 49-54, develop group leader transuranium element chem, 54-63, asst chief, Chem Develop Sect, 63-72, asst chief, Pilot Plant Sect, 72-76, sect head, Exp Eng Sect, 76-81, head, Pilot Plant, 81-87; RETIRED. *Concurrent Pos:* Consult, radiochem opers safety, 89- *Mem:* Am Nuclear Soc; Am Chem Soc; fel Am Inst Chem; Sigma Xi; AAAS. *Res:* Ion exchange and solvent extraction, especially of the transuranium elements, neptunium through fermium; preparation and properties of concentrated colloids of metal oxides and hydroxides; nuclear fuel reprocessing and waste treatment. *Mailing Add:* 517 W Fifth Ave Lenoir City TN 37771

LEV, MAURICE, b St Joseph, Mo, Nov 13, 08; m 47; c 2. PATHOLOGY. *Educ:* NY Univ, BS, 30; Creighton Univ, MD, 34; Northwestern Univ, MA, 66; Am Bd Path, dipl path anat, 41, dipt clin path, 43. *Hon Degrees:* LHD, De Paul Univ, 81. *Prof Exp:* From instr to assoc prof path, Col Med, Univ Ill, 39-41; asst prof, Sch Med, Creighton Univ, 46-47; from assoc prof to prof, Sch Med, Univ Miami, 51-57; prof, 57-77, EMER PROF PATH, MED SCH, NORTHWESTERN UNIV, CHICAGO, 77-; DIR CONGENITAL HEART DIS RES & TRAINING CTR, HEKTOEN INST MED RES, 57- *Concurrent Pos:* Pathologist, Chicago State Hosp, 40-42; pathologist & dir res labs, Mt Sinai Hosp, Miami Beach, 51-57; career investr & educr, Chicago Heart Asn, 66-; consult, Children's Mem Hosp, Chicago, 57-; prof lectr, Univ Chicago, 59-; lectr, Col Med, Univ Ill, 63-, Chicago Med Sch, Univ Health Sci, 70- & Stritch Sch Med, Loyola Univ, 71-; distinguished prof pediat, Rush Med Col, 74-, distinguished prof int med, 75-, distinguished prof path, 77-; lectr, Cook County Grad Sch Med, 77. *Mem:* Am Soc Clin Path; Am Asn Path & Bact; AMA; fel NY Acad Sci. *Res:* Cardiac pathology; pathology of congenital heart disease and of conduction system. *Mailing Add:* Deborah Heart & Lung Ctr Brown Mills NJ 08015

LEVAN, MARIJO O'CONNOR, b Detroit, Mich, Oct 27, 36; m 59; c 3. MATHEMATICS. *Educ:* Spring Hill Col, BS, 59; Univ Ala, MA, 61; Univ Fla, PhD(math), 64. *Prof Exp:* From instr to asst prof math, Univ Fla, 62-67; asst prof, Southeast Mo State Col, 67-69; assoc prof, 69-74, actg chmn, 78-79, chmn, 79-84, PROF MATH, EASTERN KY UNIV, 74- *Mem:* Am Math Soc. *Res:* Number theory; partition functions; translated geometric progressions; pseudo perfect numbers; additive distributive functions. *Mailing Add:* Dept Math Eastern Ky Univ Richmond KY 40475

LEVAN, MARTIN DOUGLAS, JR, b Chattanooga, Tenn, Aug 30, 49; m 77; c 2. ADSORPTION, FLUID MECHANICS. *Educ:* Univ Va, BS, 71; Univ Calif, Berkeley, PhD(chem eng), 76. *Prof Exp:* Sr res engr, Amoco Prod Co, Standard Oil Co, 76-78; asst prof, 78-83, ASSOC PROF CHEM ENG, UNIV VA, 83- *Concurrent Pos:* Mem, Nat Prog Comt Adsorption & Ion Exchange, Am Inst Chem Engrs, 80-, vchmn, 83-85 & chmn, 85-87, mem, Nat Prog Comt Interfacial Phenomena, 82-87; consult, Amoco Prod Co, 84-, Amoco Corp, 86-87; Fulbright sr scholar, Univ Porto, Portugal, 85-86. *Mem:* Am Inst Chem Engrs; Am Chem Soc. *Res:* Adsorption and fluid mechanics; fixed-bed adsorption; adsorption equilibria; low Reynolds number hydrodynamics; free surface flows; computer applications in chemical engineering. *Mailing Add:* Dept Chem Eng Univ Va Thronton Hall Charlottesville VA 22901

LEVAN, NHAN, b Quang Yen, Vietnam, Nov 6, 36; m 60; c 2. SYSTEM SCIENCE. *Educ:* Univ New Eng, Australia, BSc, 60; Univ New South Wales, MSc, 62; Monash Univ, Australia, PhD(elec eng), 66. *Prof Exp:* Lectr elec eng, Monash Univ, Australia, 65-66; asst prof syst sci, 67-73, assoc prof, 73-79, PROF ELEC ENG, SCH ENG & APPL SCI, UNIV CALIF, LOS ANGELES, 79- *Mem:* AAAS; Inst Elec & Electronics Engrs; Sigma Xi. *Res:* System theory; circuit theory; distributed parameter systems; applied functional analysis and scattering systems; control theory and applications. *Mailing Add:* 7732 Boelter Hall Univ Calif Los Angeles CA 90024

LEVAND, OSCAR, b Parnu, Estonia, Nov 3, 27; US citizen; div; c 1. ORGANIC CHEMISTRY, BIOCHEMISTRY. *Educ:* Miss State Col, BS, 54; Purdue Univ, MS, 58; Univ Hawaii, PhD(org chem), 63; Univ Minn, Minneapolis, MPH, 70. *Prof Exp:* Jr res chemist, Mead Johnson Co, Ind, 54-56; res chemist, Knoll Pharmaceut Co, NJ, 58-59; fel NIH, 62-63; res chemist, Dole Co, Hawaii, 63-68; consult, Air Pollution Control Prog, Govt of Guam, 70-74; asst prof, 74-80, ASSOC PROF CHEM, UNIV GUAM, 80- *Mem:* Am Chem Soc; Sigma Xi. *Res:* Air and water chemistry. *Mailing Add:* Dept Chem Univ Guam Sta Mangilao GU 96923

LEVANDER, ORVILLE ARVID, b Waukegan, Ill, Apr 6, 40; m 81; c 2. NUTRITION. *Educ:* Cornell Univ, BA, 61; Univ Wis-Madison, MS, 63, PhD(biochem), 65. *Prof Exp:* Res fel biochem, Col Physicians & Surgeons, Columbia Univ, 65-66; res assoc, Sch Public Health, Harvard Univ, 66-67; res chemist, Food & Drug Admin, 67-69, RES CHEMIST, USDA HUMAN NUTRIT RES CTR, 69- *Concurrent Pos:* Mem, Nat Res Coun Comt Biol Effects Environ Pollutants, 74-77, mem subcomt nutrit, Safe Drinking Water Comt, 77-79; temp adv, Environ Health Criteria Doc on Selenium, WHO, 77-87; mem, Comt Animal Nutrit, Agr Bd, Nat Res Coun, 79-83, Comt Dietary Allowances, Food & Nutrit Bd, 80-85; William Evans vis fel, Univ Otago, Dunedin, NZ, 82; mem, US Nat Comt Int Union of Nutrit Sci, 85-; travel fel, Danish Med Res Coun, 87; Burroughs Wellcome vis prof nutrit, Ore State Univ, Corvallis, 87. *Honors & Awards:* Osborne & Mendel Award, Am Inst Nutrit, 86. *Mem:* AAAS; Am Inst Nutrit; Am Chem Soc; Am Soc Clin Nutrit. *Res:* Toxicology and nutrition of selenium; pharmacology of heavy metals; trace mineral nutrition; vitamin E; drug metabolism; lead poisoning; tropical parasitic diseases; malaria. *Mailing Add:* USDA Human Nutrit Res Ctr Beltsville MD 20705

LEVANDOWSKI, DONALD WILLIAM, b Stockett, Mont, Dec 20, 27; m 55; c 2. GEOLOGY. *Educ:* Mont Col Mineral Sci & Technol, BS, 50; Univ Mich, MS, 52, PhD(mineral), 56. *Hon Degrees:* Geol Engr, Mont Col Mineral Sci & Technol, 68. *Prof Exp:* Res geologist, Calif Res Corp, Standard Oil Co, Calif, 55-64, staff asst to mgr explor res, Chevron Res Co, 64-65, geophysicist, Western Opers, Inc, 65-67; assoc prof, 67-75, assoc head dept, 70-76, actg head dept, 76-78, head dept, 78-88, PROF GEOSCI, PURDUE UNIV, 75- *Mem:* Fel Geol Soc Am; AAAS; Am Asn Petrol Geologists; fel Geol Asn Can; Am Soc Photogrammetry. *Res:* Mineral deposits; remote sensing; igneous and metamorphic petrology; geophysics. *Mailing Add:* 3711 Capilano Dr West Lafayette IN 47906

LEVANDOWSKY, MICHAEL, b Knoxville, Tenn, Aug 15, 35. MARINE ECOLOGY, BEHAVIOR ETHOLOGY. *Educ:* Antioch Col, AB, 61; Columbia Univ, Ma, 65, PhD(biol), 70; NY Univ, MS, 73. *Prof Exp:* Instr biol, Bard Col, 67-69; instr, Bronx Community Col, 69-70; asst prof biol, 70-71, RES SCIENTIST, HASKINS LABS, PACE UNIV, 70- *Concurrent Pos:* Nat Sci Found sci fac fel, Courant Inst Math Sci, NY Univ, 71-72; asst prof biol, York Col, NY, 73-74; mem citizens' adv comt on resource recovery for borough of Brooklyn, 81-; distinguished lectr, NE Algal Soc, 86; bd dir, The River Proj, 89. *Mem:* Soc Protozool; Phycol Soc Am; Water Pollution Control Fedn; Am Soc Limnol & Oceanog; Am Soc Microbiol; Am Soc Photobiol. *Res:* Mathematical models in ecology; microbial ecology; marine biology; sensory physiology and behavior of Protista; resource recovery. *Mailing Add:* Haskins Labs Pace Univ 41 Park Row New York NY 10038

LEVASSEUR, MAURICE EDGAR, b Rivière-du-Loop, Que, June 30, 53; m 82; c 2. OCEANOGRAPHY. *Educ:* Univ Laval, Can, BS, 79, MS, 84; Univ BC, PhD(oceanog), 90. *Prof Exp:* Biologist, 85-90, RESEARCHER, FISHERIES & OCEANS, CAN, 90- *Mem:* Am Soc Limnol & Oceanog. *Res:* Ecological and physiological aspects of marine algae; physiological effects of different N sources upon microalgae. *Mailing Add:* Fisheries & Oceans Maurice-Lamontagne Inst PO Box 1000 Mont-Joli PQ G5H 3Z4 Can

LEVCHUK, JOHN W, b Hudson, NY, Jan 13, 42; m 66; c 2. PARENTERAL TECHNOLOGY, HOSPITAL PHARMACY. *Educ:* Philadelphia Col Pharm & Sci, BS, 63, MS, 68; Univ Ariz, MEd, 73, PhD(pharm), 77. *Prof Exp:* Pharmacist, USPHS Indian Hosp, Gallup, NMex, 64-66, comn officer, USPHS, 64; asst prof drug info & poison control, State Univ NY, Buffalo, 67-69; dir, Poison Control & Drug Info Ctr, Buffalo Children's Hosp, 67-69; from asst prof to assoc prof hosp pharm & sterile prod, Univ NMex, 69-79; assoc prof sterile prod & hosp pharm, Univ Alta, 79-83; assoc prof sterile prod, Univ Tenn, 83-87; CONSUMER SAFETY OFFICER, OFF COMPLIANCE, DIV MFG PROD QUAL, STERILE DRUGS BR, US FOOD & DRUG ADMIN, 87- *Concurrent Pos:* Regional coordr residency progs, Can Soc Hosp Pharmacists, 79-83; mem, Training Comt, Parenteral Drug Asn, 83- *Mem:* Parenteral Drug Asn; Am Soc Hosp Pharmacists. *Res:* Microbiological quality of pharmaceutical processing facilities, activities and dosage forms; experimental aerobiology; packaging of sterile dosage forms. *Mailing Add:* FDA Sterile Drugs Br HFD-322 5600 Fishers Lane Rockville MD 20857

LEVEAU, BARNEY FRANCIS, b Denver, Colo, Oct 2, 39; m 61; c 3. PHYSICAL MEDICINE. *Educ:* Univ Colo, BS, 61, MS, 66; Mayo Clin, RPT, 65; Pa State Univ, PhD(phys educ), 73. *Prof Exp:* Teacher math & sci, Colo Springs Sch Dist, 61-63; from asst prof to assoc prof phys educ, WChester State Col, 66-70; from asst prof to assoc prof phys ther, Sch Med, Univ NC, Chapel Hill, 72-85; PROF & CHMN, DEPT PHYS THER, SOUTHWEST MED CTR, UNIV TEX, 85- *Mem:* Am Phys Ther Asn; Am Col Sports Med; Int Soc Biomech; Am Asn Health, Phys Educ & Recreation. *Res:* Biomechanics as it applies to physical therapy and physical education; sports medicine. *Mailing Add:* Southwest Med Ctr 5233 Harry Hines Blvd Dallas TX 75235-8876

LEVEEN, HARRY HENRY, b Woodhaven, NY, Aug 10, 16; c 2. SURGERY. *Educ:* Princeton Univ, BA, 36; NY Univ, MD, 40; Univ Chicago, MS, 47; Am Bd Surg, dipl. *Prof Exp:* Instr & res assoc, Univ Chicago, 45-47; instr surg, Col Med, NY Univ, 47-50; assoc prof physiol, Sch Med, Loyola Univ, Ill, 50-55; assoc prof surg, Chicago Med Sch, 55-56; assoc prof surg, Col Med, State Univ NY, Downstate Med Ctr, 57-59, prof, 60-79; prof, 79-89, EMER PROF SURG, MED UNIV SC, 89- *Concurrent Pos:* Assoc prof, NMex Mil Inst, 52-55; chief surgeon, Vet Admin Hosp, 57-78. *Mem:* Soc Exp Biol & Med; Am Physiol Soc; Int Soc Surg; fel Am Col Surg; NY Acad Med. *Res:* Surgical physiology; radiofrequency thermotherapy; cirrhosis. *Mailing Add:* 321 Convederate Cir Charleston SC 29407

LEVEILLE, GILBERT ANTONIO, b Fall River, Mass, June 3, 34; m 81; c 3. NUTRITION, BIOCHEMISTRY. *Educ:* Univ Mass, BVA, 56; Rutgers Univ, MS, 58, PhD(nutrit), 60. *Prof Exp:* Biochemist, US Army Med Res & Nutrit Lab Colo, 60-66; assoc prof nutrit biochem, Univ Ill, Urbana, 66-69, prof, 69-71; prof food sci & human nutrit & chmn dept, Mich State Univ, 71-80; dir nutrit & health sci, Gen Foods Corp, 80-86; VPRES RES & TECH SERV, NABISCO BISCUIT CO, 86- *Honors & Awards:* Res Award, Poultry Sci Asn, 65; Mead Johnson Res Award, Am Inst Nutrit, 71. *Mem:* AAAS; Am Inst Nutrit; Am Soc Clin Nutrit; Am Chem Soc; Poultry Sci Asn; Inst Food Technologists. *Res:* Lipid metabolism; protein and amino acid nutrition and metabolism; atherosclerosis; obesity. *Mailing Add:* VPres Res & Tech Serv Nabisco Biscuit Co PO Box 1944 East Hanover NJ 07936-1944

LEVELTON, B(RUCE) HARDING, b Bella Coola, BC, June 18, 25; m 50; c 3. CHEMICAL ENGINEERING. *Educ:* Univ BC, BASc, 47, MASc, 48; Tex A&M Univ, PhD(chem eng), 51. *Prof Exp:* Asst res engr, BC Res Coun, 51-54, assoc res engr, 54-58, res engr, 58-64, assoc head div appl chem, 64-66; PRIN, B H LEVELTON & ASSOCS LTD, 66-, PRES, 66-; VPRES, CHATTERTON PETROCHEM CORP, 84- *Concurrent Pos:* Spec lectr, Univ BC, 57-65. *Mem:* Am Inst Chem Engrs; Nat Asn Corrosion Engrs; Air Pollution Control Asn; Chem Inst Can; Forest Prod Res Soc; Am Water Works Asn. *Res:* Treatment and beneficiation of industrial minerals; utilization of wood wastes by carbonization; corrosion of metals in chemical industry and in marine service; corrosion of copper in potable waters; environmental technology; solid waste disposal; toxic and hazardous waste disposal. *Mailing Add:* 6130 St Clair Pl Vancouver BC V6N 2A5 Can

LEVEN, ROBERT MAYNARD, b Chicago, Ill, Nov 7, 55; m 82; c 2. CELL BIOLOGY, HEMATOLOGY. *Educ:* Wash Univ, BA, 77; Univ Pa, PhD(anat), 82. *Prof Exp:* Postdoctoral fel thrombosis, Med Sch, Temple Univ, 82-84; staff scientist, res med, Lawrence Berkeley Lab, 84-90; ASST PROF ANAT & MED, RUSH MED COL, CHICAGO, ILL, 90- *Concurrent Pos:* Instr anat, Univ Calif, San Francisco, 86; consult, Amgen, Inc, 89-90. *Mem:* Am Soc Cell Biol; Int Soc Exp Hemat. *Res:* Hormonal and microenvironmental factors that control the formation of blood platelets from megakaryocytes using in vitro cell culture models. *Mailing Add:* Dept Anat Rush Med Col 600 S Paulina Chicago IL 60612

LEVENBERG, MILTON IRWIN, b Chicago, Ill, Nov 5, 37; div; c 2. MASS SPECTROMETRY, COMPUTER SCIENCE. *Educ:* Ill Inst Technol, BS, 58; Calif Inst Technol, PhD(chem), 65. *Prof Exp:* Sr chem physicist, Abbott Labs, 65-73, assoc res fel, 73-84, sect head, 84-90, MGR, ABBOT LABS, 90- *Mem:* Am Chem Soc; Sigma Xi; Am Soc Mass Spectrometry. *Res:* Computer applications to instrumentation; instrumentation; electronics; mass spectrometry; nuclear magnetic resonance spectroscopy. *Mailing Add:* Abbott Labs D-418 AP-9 Abbott Park IL 60064-3500

LEVENBOOK, LEO, b Kobe, Japan, Dec 29, 19; nat US; m 50; c 1. BIOCHEMISTRY. *Educ:* Univ London, BSc, 41; Cambridge Univ, PhD(biochem), 49. *Prof Exp:* Asst insect biochem, Cambridge Univ, 46-50; fel, Harvard Univ, 50-51; res assoc biochem genetics, Inst Cancer Res, Philadelphia, 51-54; asst prof biochem, Jefferson Med Col, 54-58; biochemist, Nat Inst Diabetes, Metab & Digestive Dis, NIH, 58-86; RETIRED. *Concurrent Pos:* Lectr, Haverford Col, 53-54. *Mem:* Am Soc Biol Chemists. *Res:* Insect physiology and biochemistry. *Mailing Add:* 5100 Dorset Ave No 308 Chevy Chase MD 20815

LEVENE, CYRIL, b Gateshead, Eng, May 27, 26; m 52; c 3. ANATOMY. *Educ:* Queen's Univ Belfast, MB, BCh & BAO, 48, MD, 60. *Prof Exp:* Demonstr anat, Queen's Univ Belfast, 51-52, asst lectr, 52-54; lectr human anat, Univ Col WI, 54-65, sr lectr anat, 65-67; assoc prof, Univ Western Ont, 67-69; assoc prof, 69-74, prof anat, Div Morphol Sci, 74-88, EMER PROF ANAT, DEPT ANAT, FAC MED, UNIV CALGARY, 88- *Concurrent Pos:* WHO fel human genetics, 66. *Mem:* Hon mem Can Asn Anat. *Res:* Medical education. *Mailing Add:* Dept Anat Univ Calgary 3300 Hospital Dr NW Calgary AB T2N 4N1 Can

LEVENE, HOWARD, b New York, NY, Jan 17, 14. MATHEMATICAL STATISTICS. *Educ:* NY Univ, BA, 41; Columbia Univ, PhD(math statist), 47. *Prof Exp:* Exten lectr zool & math statist, 47-48, from instr to prof math statist & biomet, 48-70, prof math statist & genetics, 70-82, chmn, Dept Statist, 76-82, EMER PROF MATH STATIST & SPECIAL LECTR, COLUMBIA UNIV, 82- *Mem:* AAAS; Am Math Soc; Soc Study Evolution; Biomet Soc; Am Soc Human Genetics (vpres, 63); Am Soc Naturalists (pres, 76); Inst Math Statist; Am Statist Soc. *Res:* Mathematical genetics; nonparametric tests; biometrics; population genetics and evolution. *Mailing Add:* Dept Statist Columbia Univ New York NY 10027

LEVENE, JOHN REUBEN, b Hull, Eng, Dec 7, 29; m 59; c 2. OPTOMETRY, PHYSIOLOGICAL OPTICS. *Educ:* City Univ, London, dipl ophthalmic optics, 54; Ind Univ, MS, 62; Oxford Univ, PhD(biol sci), 66; Pa Col Optom, OD, 81. *Prof Exp:* Asst prof optom & physiol optics, Univ Houston, 62-63; lectr optom, City Univ, London, 65-67; from assoc prof to prof optom, Ind Univ, Bloomington, 67-75, chmn physiol optics prog & dir low vision clin, 70-75; DEAN FAC & PROF OPTOM, SOUTHERN COL OPTOM, MEMPHIS, TENN, 75- *Honors & Awards:* Obrig Labs Mem Award, 69. *Mem:* Am Acad Optom (vpres, Brit chap, 66); Brit Soc Hist Sci; Royal Micros Soc; Optom Hist Soc (vpres, 69); Am Asn Hist Med. *Res:* Pathological processes concerning vision; history of visual science. *Mailing Add:* 3936 Galloway Dr Memphis TN 38111

LEVENE, RALPH ZALMAN, b Winnipeg, Man, May 17, 27; nat US; m 54; c 2. OPHTHALMOLOGY. *Educ:* Univ Man, MD, 49; NY Univ, DSc(ophthal), 57; Am Bd Ophthal, dipl, 55. *Prof Exp:* Intern, Winnipeg Gen Hosp, Can, 49-50, resident ophthal, 51-55; from instr to assoc prof, Med Sch, NY Univ, 55-73; PROF OPHTHAL, UNIV ALA, BIRMINGHAM, 73- *Mem:* AMA; Asn Res Vision & Ophthal; Am Acad Ophthal & Otolaryngol; NY Acad Med; Sigma Xi. *Res:* Clinical and basic science aspects of glaucoma. *Mailing Add:* 2008 Brookwood Med Ctr Dr Suite 209 Birmingham AL 35209

LEVENSON, ALAN IRA, b Boston, Mass, July 25, 35; m 60; c 2. PSYCHIATRY. *Educ:* Harvard Univ, AB, 57, MD, 61, MPH, 65; Am Bd Psychiat & Neurol, Dipl, 67. *Prof Exp:* Intern, Univ Hosp, Ann Arbor, 61-62; resident in psychiat, Mass Ment Health Ctr, Boston, 62-65; staff psychiatrist, NIMH, 65-66, dir servs div, 67-69; head dept, 69-89, PROF PSYCHIAT, COL MED, UNIV ARIZ, 69- *Concurrent Pos:* Chief exec officer, Palo Verde Mental Health Serv, 71- *Mem:* Fel Am Pub Health Asn; fel Am Psychiat Asn; fel Am Col Psychiat; Group Advan Psychiat; fel Am Col Mental Health Admin. *Res:* Organization and delivery of mental health services. *Mailing Add:* Dept Psychiat Ariz Health Sci Ctr Univ Ariz Tucson AZ 85724

LEVENSON, HAROLD SAMUEL, b Allentown, Pa, July 12, 16; m 38; c 3. FOOD SCIENCE. *Educ:* Lehigh Univ, BSChE, 37, MS, 39, PhD(chem physics), 41. *Prof Exp:* Asst chem, Lehigh Univ, 37-41; res chemist, Gen Foods Corp, NJ, 41-46, chief chemist, Maxwell House Div, Calif, 46-51, res mgr, NJ, 51-64, dir coffee res, Tech Ctr, NY, 65-78; RETIRED. *Mem:* Am Chem Soc; Inst Food Technol; Sigma Xi. *Res:* Antioxidants; food spoilage; kinetics of saponification; hydrocaffeic acid and esters as antioxidant for edible materials; coffee technology. *Mailing Add:* 5577 Inverness Ave Santa Rosa CA 95404

LEVENSON, JAMES B, b San Francisco, Calif, Aug 22, 44; c 2. PLANT ECOLOGY, GEOGRAPHIC INFORMATION SYSTEMS. *Educ:* Ind State Univ, Terre Haute, BS, 71, MA, 73; Univ Wis-Milwaukee, PhD(bot), 76. *Prof Exp:* Res assoc, Univ Wis-Milwaukee, 76-77; asst prof ecol, Saginaw Valley State Col, 77-79; asst environ sci, 79-84, ECOLOGIST, ARGONNE NAT LAB, 84- *Concurrent Pos:* Co-investr, NSF grant, 78-80. *Mem:* AAAS; Am Inst Biol Sci; Ecol Soc Am; Sigma Xi. *Res:* Interactions and resultant impacts of man-dominated systems on remnant ecosystem patches; identification, description and quantification of natural areas; application of ecological concepts to regional assessments of projected energy scenarios for federal agencies; development of natural resource data bases and spatial display systems for assessment of energy-related impacts to natural and man-dominated ecosystems. *Mailing Add:* Energy & Environ Systs Argonne Nat Lab Bldg 362/3D 9700 S Cass Ave Argonne IL 60439

LEVENSON, LEONARD L, b San Francisco, Calif, Sept 18, 28; m 57; c 3. PHYSICS. *Educ:* Univ Calif, Berkeley, AB, 52, MS, 55; Univ Paris, PhD(physics), 68. *Prof Exp:* Physicist, US Naval Ord Test Sta, 52; res engr, Univ Calif, Berkeley, 52-58, physicist, Lawrence Radiation Lab, 58-62; physicist, Nuclear Res Ctr, Saclay, France, 62-68; from asst prof to assoc prof physics, Univ Mo-Rolla, 68-76, prof, 76-81, dir grad ctr mat res, 75-81; chmn dept, 81-84, PROF, DEPT PHYSICS & ENERGY SCI, UNIV COLO, COLORADO SPRINGS, 84- *Concurrent Pos:* Vis prof, Kyoto Univ, Japan, 86-87 & 88-89, Univ Houston, 89. *Mem:* Inst Elec & Electronics Engrs; Am Phys Soc; Am Vacuum Soc; Sigma Xi. *Res:* Gas-surface interactions; thin films; surface physics; electronics materials. *Mailing Add:* Dept Physics & Energy Sci Univ Colo PO Box 7150 Colorado Springs CO 80933

LEVENSON, MARC DAVID, b Philadelphia, Pa, May 28, 45; m 71. LASERS, QUANTUM ELECTRONICS. *Educ:* Mass Inst Technol, BS, 67; Stanford Univ, MS, 68, PhD(physics), 72. *Prof Exp:* Res fel non-linear optics, Gordon McKay Lab, Harvard Univ, 71-74; asst prof physics, Univ Southern Calif, 74-77, assoc prof physics & elec eng, 77-79; head mgr, Optical Storage, 87-88, MEM RES STAFF, IBM RES LAB, 79- *Concurrent Pos:* Alfred P Sloan fel, 75-77; Joint Inst Lab Astrophys vis fel, Univ Colo, 78-79. *Honors & Awards:* Adolph Lomb Award, Optical Soc Am, 76. *Mem:* Am Phys Soc; Inst Elec & Electronics Engrs; Optical Soc Am. *Res:* Development and application of new techniques of laser spectroscopy to problems in atomic, molecular and condensed matter physics; application of optical and laser techniques to electronics manufacturing; quantum optics; optical memories; photolithography. *Mailing Add:* IBM Res Lab K31/802(D) 650 Harry Rd San Jose CA 95120-6099

LEVENSON, MILTON, b St Paul, Minn, Jan 4, 23; m 50; c 5. CHEMICAL ENGINEERING. *Educ:* Univ Minn, BChE, 43. *Prof Exp:* Jr engr, Houdaille-Hershey Corp, 44; asst engr, Oak Ridge Nat Labs, 44-48; from assoc engr to assoc lab dir energy & environ, Argonne Nat Lab, 48-73; dir nuclear power, Elec Power Res Inst, 73-81; exec eng, Bechtel Power Corp, San Francisco, 81-90, vpres, Bechtel Int, 83-90; CONSULT, 90- *Honors & Awards:* Robert E Wilson Award, Am Inst Chem Engrs, 75. *Mem:* Nat Acad Eng; fel Am Inst Chem Engrs; fel Am Nuclear Soc (pres); AAAS. *Res:* Water reactor technology; fuel cycle technology; breeder reactor development. *Mailing Add:* 21 Politzer Dr Menlo Park CA 94025

LEVENSON, MORRIS E, b New York, NY, Nov 13, 14; m 43. MATHEMATICS. *Educ:* NY Univ, PhD(math), 48. *Prof Exp:* Asst, Duke Univ, 37-38; instr math, NY Univ, 43-44; mathematician, David Taylor Model Basin, 44-46; instr math, Cooper Union, 46-49; from instr to assoc prof, 49-71, PROF MATH, BROOKLYN COL, 71- *Mem:* Am Math Soc; Math Asn Am; Sigma Xi. *Res:* Nonlinear vibrations. *Mailing Add:* 160 W 73rd St New York NY 10023

LEVENSON, ROBERT, b New York, NY. CELL BIOLOGY. *Prof Exp:* ASSOC PROF CELL BIOL, SCH MED, YALE UNIV, 89- *Mailing Add:* Cell Biol Dept Sch Med Yale Univ 333 Cedar St New Haven MD 06510

LEVENSON, STANLEY MELVIN, b Dorchester, Mass, May 25, 16; m 42; c 2. SURGERY, SURGICAL RESEARCH. *Educ:* Harvard Univ, AB, 37, MD, 41; Am Bd Nutrit, dipl, 52; Am Bd Surg, dipl, 57. *Prof Exp:* Surg house officer, Beth Israel Hosp, Boston, Mass, 41-42; resident burn serv & res assoc surg, Boston City Hosp, 42-43; surg scientist, Med Nutrit Lab, Univ Chicago, 47-49; from asst resident to sr asst resident surg, Med Col Va, 50-52; chief dept surg metab & physiol, Walter Reed Army Inst Res, 56-61, from assoc dir to dir dept germfree res, 56-61, dir div basic surg res, 61; PROF SURG, ALBERT EINSTEIN COL MED, 61-, DEP DIR RES SURG, COL MED, 67- *Concurrent Pos:* Res fel med, Thorndike Mem Lab, Harvard Univ, 44-47; NIH res career award, 62-; chmn subcomt burns & radiation injury, Food & Nutrit Bd, Nat Res Coun, 49-50, comt on trauma, 56-; dir surg metab lab & clin assoc prof, Georgetown Univ, 59-61; res assoc physiol, Sch Pub Health, Harvard Univ, 41-; consult, Walter Reed Army Inst Res, 61-63; Am Surg Asn rep, Nat Res Coun-Nat Acad Sci, 71-75. *Honors & Awards:* Jonathan E Rhoads lectr, Am Soc Parenteral & Enteral Nutrit, 78; Arnold M Seligman Mem, Sinai Hosp Baltimore, 79; McCollum Award, Am Soc Clin Nutrit, 83. *Mem:* Am Surg Asn; Am Burn Asn; Am Inst Nutrit; Am Soc Clin Nutrit; Am Col Surgeons; Am Soc Parenteral & Enteral Nutrit. *Res:* Metabolic and clinical response to trauma; wound healing; germfree life; infection; burns; neoplasia; nutrition; radiation. *Mailing Add:* Dept Surg Albert Einstein Col Med 1300 Morris Park Ave Bronx NY 10461-1975

LEVENSPIEL, OCTAVE, b Shanghai, China, July 6, 26; nat US; m 52; c 3. CHEMICAL ENGINEERING. *Educ:* Univ Calif, BS, 47; Ore State Col, MS, 49, PhD(chem eng), 52. *Hon Degrees:* Doctorate, ENSIC, Nancy, France, 87. *Prof Exp:* Jr res engr, Inst Eng Res, Univ Calif, 51-52; asst prof chem eng, Ore State Col, 52-54; asst & assoc prof, Bucknell Univ, 54-58; assoc & prof, Ill Inst Technol, 58-68; PROF CHEM ENG, ORE STATE UNIV, 68- *Concurrent Pos:* NSF sr fel, Cambridge Univ, 63-64, Fulbright fel, 68-69; vis prof, Univ NSW, Australia, 76, Denmarks Tech Univ, 77, Univ Groningen, Neth, 84, Univ Sydney, Australia, 85. *Honors & Awards:* 3M Lectureship Award, Am Soc Eng Educ, 66; Wilhelm Award, Am Inst Chem Engrs, 79; Danckwerts Award, London, 88. *Mem:* Am Chem Soc; Am Inst Chem Engrs. *Res:* Chemical reactor design; fluidization. *Mailing Add:* Dept Chem Eng Ore State Univ Corvallis OR 97331-2702

LEVENSTEIN, HAROLD, b Philadelphia, Pa, June 28, 23; m 47; c 3. LASER RADAR SYSTEMS, FIRE CONTROL SYSTEMS. *Educ:* Cooper Union Sch Eng, BEE, 43; Polytech Inst Brooklyn, MS; Columbia Univ, MS, 79. *Prof Exp:* Engr mgr radar & control systs, various co, 43-64; dept head systs & opers res, Div Cutler Hammer, Ail, 64-71; prog mgr radar systs, Missile & Surface Radar Div, RCA, 71-73; tech dir electrooptics systs, Military Systs Div, Perkin Elmer Corp, 73-88; CONSULT, ELECTROOPTICS, 88- *Concurrent Pos:* Mem ASW comt, Navy/NSIA, 56-60; adj asst prof statist models, dept marine sci, C W Post Ctr, Long Island Univ, 68-71; consult radar, Electronics Div, W L Maxson, 71; lectr statist, Univ Conn, Storrs, 83; speaker, NAS/AF Study Hi Energy Laser Optics, 84; mem adv bd, Defense Intel Agency, 88- *Mem:* Inst Elec & Electronics Engrs; Opers Res Soc Am; Am Statist Asn; Soc Indust & Appl Math; Am Inst Aeronaut & Astronaut; Int Soc Optical Eng. *Res:* Design development of radar and control systems for guidance and navigation; network theory for multivariable systems; modular optical (laser) radar for space applications. *Mailing Add:* Seven Golden Heights Rd Danbury CT 06811

LEVENSTEIN, IRVING, b Fair Lawn, NJ, Aug 14, 12; m 37; c 2. ENDOCRINOLOGY, TOXICOLOGY. *Educ:* NY Univ, BA, 34, MSc, 36, PhD, 38. *Prof Exp:* Instr biol sci, NY Univ, 36-40; PRES & DIR, LEBERCO LABS, 42- *Concurrent Pos:* Res fel, Nat Comt Maternal Health, 38-42. *Mem:* Am Soc Zoologists; Am Asn Anatomists; Am Pharmaceut Asn; Soc Toxicol; Am Chem Soc. *Res:* Hormones; histology; anatomy. *Mailing Add:* 890 Westminster Ave Hillside NJ 08002

LEVENTHAL, BRIGID GRAY, b London, Eng, Aug 31, 35; US citizen; m 62; c 4. PEDIATRICS, ONCOLOGY. *Educ:* Univ Calif, Los Angeles, BA, 55; Harvard Univ, MD, 60. *Prof Exp:* Sr investr leukemia serv, Med Br, Nat Cancer Inst, 65-73, head chemoimmunother sect, Pediat Oncol Br, 73-76; CHIEF PEDIAT ONCOL DIV, JOHNS HOPKINS HOSP, 76- *Concurrent Pos:* Fel pediat, Harvard Univ, 60-62; fel, Boston Univ, 62-63; fel med, Tufts Univ, 63-64; fel hemat, Nat Cancer Inst, 64; assoc prof oncol & pediat, Johns Hopkins Univ. *Mem:* Soc Pediat Res. Am Soc Hemat; Am Asn Cancer Res; Am Soc Clin Oncol; Am Soc Clin Invest. *Res:* Pediatric hematology and oncology. *Mailing Add:* Div Pediat Oncol Johns Hopkins Hosp CMSC 720 Rutland Ave Baltimore MD 21205

LEVENTHAL, CARL M, b New York, NY, July 28, 33; m 62; c 4. NEUROLOGY, NEUROPATHOLOGY. *Educ:* Harvard Univ, AB, 54; Univ Rochester, MD, 59. *Prof Exp:* Intern med, Johns Hopkins Hosp, 59-60, asst res physician, 60-61; asst resident neurol, Mass Gen Hosp, 61-62, fel neuropath, 62-63, resident, 63-64; assoc neuropathologist, Nat Inst Neurol Dis & Blindness, 64-66, neurologist, Nat Cancer Inst, 66-68, asst to dep dir sci, NIH, 68-74, actg dep dir sci, 73-74; dep dir, Bur Drugs, Food & Drug Admin, 74-77; dep dir, Nat Inst Arthritis, Metab & Digest Dis, 77-81; DIV DIR, NAT INST NEUROL DIS & STROKE, 81- *Concurrent Pos:* Fel, Johns Hopkins Univ, 59-61; fel, Harvard Univ, 61-64; instr, Georgetown Univ, 64-66, asst prof, 67-74; med officer, USPHS, 64- *Mem:* Am Acad Neurol; Am Asn Neuropath; Am Neurol Asn; Asn Res Nervous & Ment Dis. *Res:* Government research administration; clinical neuropathology. *Mailing Add:* Nat Inst Neurol Dis & Stroke Bethesda MD 20892

LEVENTHAL, EDWIN ALFRED, b Brooklyn, NY, Jan 26, 34; m 56; c 2. SOLID STATE PHYSICS. *Educ:* Cornell Univ, BEng Phys, 56; Polytech Inst Brooklyn, MS, 59; NY Univ, PhD(physics), 63. *Prof Exp:* Sr physicist, Philips Labs Div, NAm Philips Co, 61-70; ed & publ, Med Instrument Reports, 70-74; dir, systs planning, Friesen Int, 74-84; DIR, MKT INFO SYSTS, AM MED INT, 84- *Mem:* Am Phys Soc; Asn Advan Med Instrumentation; NY Acad Sci. *Res:* Materials handling and information processing in hospital management and design; medical equipment planning. *Mailing Add:* 23723 Kivik St Woodland Hills CA 91367

LEVENTHAL, HOWARD, b Brooklyn, NY, Dec 7, 31; m 54; c 2. HUMAN EMOTION, COGNITIVE PSYCHOLOGY. *Educ:* CUNY Queens Col, BS, 52; Univ NC Chapel Hill, MS, 54, PhD(psychol), 56. *Prof Exp:* Sr asst sci psychol, US Public Health Serv, 56-58; asst prof psychol, Yale Univ, 58-64, assoc prof psychol, 64-67; prof psychol, Univ Wisc, Madison, 67-88; PROF PSYCHOL, RUTGERS UNIV, 88- *Concurrent Pos:* Prof sociol, Univ Wis-Madison, 74-, assoc, Clin Cancer Ctr, 80-; mem, Behav Med Study Sect, NIH, 86-; chair, Dept Psychol, Univ Wis-Madison, 87-88; mem adv bd, Acad Behav Med Res, 88- *Mem:* Fel AAAS; fel Am Psychol Asn; Acad Behav Med Res. *Res:* Focus on common sense views of illness and how these representations affect health and illness behaviors and emotional reactions to health crises; decision process and the social and biological factors affecting it at different times in life. *Mailing Add:* Inst Health Policy & Aging 30 College Ave New Brunswick NJ 08903

LEVENTHAL, JACOB J, b Brooklyn, NY, Dec 18, 37; m 62. ATOMIC PHYSICS, MOLECULAR PHYSICS. *Educ:* Wash Univ, BS, 60; Univ Fla, PhD(physics), 65. *Prof Exp:* Res assoc physics & chem, Brookhaven Nat Lab, 65-67, assoc chemist, 67-68; from asst prof to assoc prof, 68-77, PROF PHYSICS, UNIV MO-ST LOUIS, 77- *Mem:* Am Phys Soc. *Res:* Interactions of positive ions with neutral molecules; spectroscopic observations of excited state production in low energy atomic and molecular collision processes. *Mailing Add:* Dept Physics Univ Mo 8001 Nat Bridge Rd St Louis MO 63121

LEVENTHAL, LEON, b New York, NY, Jan 25, 22; wid; c 4. RADIOCHEMISTRY, CHEMICAL ENGINEERING. *Educ:* Univ Calif, BS, 42; Va Polytech Inst, BS, 44; Univ Calif, Los Angeles, MS, 48. *Prof Exp:* Control chemist, Richfield Oil Corp, Calif, 42-43; res chemist, Metall Lab, Chicago, 44-45; jr chem engr, Oak Ridge Nat Lab, 45; chem engr, Atomic Bomb Lab, Los Alamos Sci Lab, 45-46; res chemist, Radiol Defense Lab, San Francisco Naval Shipyard, 47-49; res radiochemist, Tracerlab, Inc, 49-50, sr chemist, 50-57, dept head, 57-59, div mgr tech serv, 59-67, vpres, 75-81, gen mgr, LFE Environ Anal Labs, Div LFE Corp, 67-90, tech dir, EAL Corp, 81-90; tech dir, TMA/Norcal, 88-90; RETIRED. *Concurrent Pos:* Prof engr, State Calif Nuclear Eng; consult, 90- *Mem:* Am Chem Soc; fel Am Health Phys Soc; fel Am Nuclear Soc; fel Am Inst Chem. *Res:* Nuclear, plutonium and semimicro chemistry; plutonium metallurgy; complex compounds of zinc with zinc 65; general radiochemistry of radiological defense and radioactive waste problems; fission products; environmental and fallout studies; particle analysis; mass spectrometry of plutonium and uranium; applications of radioisotopes to science and industry; transuranium nuclides in the environment; decontamination and decommissioning; radioactive and hazardous waste disposal. *Mailing Add:* 1511 Arch St Berkeley CA 94708

LEVENTHAL, MARVIN, b New York, NY, Dec 4, 37; m 61; c 2. ASTROPHYSICS, ATOMIC PHYSICS. *Educ:* City Col New York, BS, 58; Brown Univ, PhD(physics), 64. *Prof Exp:* Res assoc physics, Yale Univ, 63-67, asst prof, 67-68; MEM TECH STAFF, BELL LABS, 68- *Mem:* Fel Am Phys Soc; Am Astronom Soc. *Res:* Precision measurements of atomic physics quantities which have bearing on quantum electrodynamics; experimental and theoretical gamma ray astronomy; laboratory astrophysics. *Mailing Add:* Bell Labs Rm 1E-349 Murray Hill NJ 07974

LEVENTHAL, STEPHEN HENRY, b New York, NY, Apr 2, 49; m 71; c 2. NUMERICAL ANALYSIS, RESERVOIR SIMULATION. *Educ:* Rutgers Univ, BA, 69; Univ Md, MA, 71; PhD(math), 73. *Prof Exp:* Res math, Naval Surface Weapons Ctr, 73-77; res math, Gulf Res & Develop Co, Gulf Oil, 77-79, sr res math, 79-81, supvr math, 81-83, dir reservoir simulation, 83-85; staff res, 85-89, SR STAFF RES MATH, SHELL DEVELOP CO, SHELL OIL, 89- *Mem:* Soc Indust & Appl Math; Soc Petrol Engrs. *Res:* Development of high order numerical methods and state of the art reservoir simulation; founder of OCI method; principal developer of Gulf Oil's Black Oil Simulator; one of the principal developers of Shell Oil's multi-purpose simulator. *Mailing Add:* Shell Develop Co PO Box 481 Houston TX 77001

LEVEQUE, WILLIAM JUDSON, b Boulder, Colo, Aug 9, 23; m 49, 70; c 1. MATHEMATICS. *Educ:* Univ Colo, BA, 44; Cornell Univ, MA, 45, PhD(math), 47. *Prof Exp:* Benjamin Peirce instr math, Harvard Univ, 47-49; from instr to prof, Univ Mich, Ann Arbor, 49-70, chmn dept, 67-70; prof math, Claremont Grad Sch, 70-77; exec dir, Am Math Soc, 77-88; RETIRED. *Concurrent Pos:* Fulbright res scholar, 51-52; Sloan res fel, 57-60; exec ed, Math Rev, 65-66; chmn, Conf Bd Math Sci, 73-74. *Mem:* Am Math Soc; Math Asn Am; fel AAAS. *Res:* Theory of numbers. *Mailing Add:* 77 Ives St No 51 Providence RI 02906

LEVER, ALFRED B P, b London, Eng, Feb 21, 36; m 63, 87; c 3. INORGANIC CHEMISTRY. *Educ:* Univ London, BSc & ARCS, 57, dipl, Imp Col & PhD(chem), 60. *Prof Exp:* Hon res asst, Univ Col, London, 60-61, hon res assoc, 61-62; lectr chem, Inst Sci & Tech, Univ Manchester, 62-66; vis lectr, Ohio State Univ, 67; assoc prof, 67-72, PROF CHEM, YORK UNIV, 72- *Concurrent Pos:* Ed, Coord Chem Rev, 66-; prog chmn, XIVth Int Conf Coord Chem, Toronto, 72; vis prof, Calif Inst Technol, 76-77, Sydney Univ, 78, Univ Calabria, 83 & Univ Pavia, 89. *Honors & Awards:* Alcan Lecture Award, 81. *Mem:* Am Chem Soc; Chem Inst Can; Royal Soc Chem; fel Jan Prom Sci. *Res:* Inorganic electronic spectroscopy; solar energy conversion; phthalocyanine chemistry; electrochemistry. *Mailing Add:* Dept Chem York Univ North York ON M3J 1P3 Can

LEVER, CYRIL, JR, b Abington, Pa, June 5, 29; m 61; c 2. ORGANIC CHEMISTRY. *Educ:* Pa Mil Col, BS, 53. *Prof Exp:* Asst treas & asst dir res, 53-57, pres & dir res, 57-90, PRES & CHIEF EXEC OFFICER, C LEVER CO, INC, 90- *Mem:* Soc Am Mil Eng. *Res:* Dyes and colors for paper. *Mailing Add:* C Lever Co Inc Lever Bldg 736 Dunks Ferry Rd Bensalem PA 19020-6575

LEVER, JULIA ELIZABETH, b Montreal, Que, Nov 23, 45. BIOCHEMISTRY, MOLECULAR BIOLOGY. *Educ:* McGill Univ, BS, 66; Univ Calif, PhD(biochem), 71. *Prof Exp:* Vis res scientist, Roche Inst Molecular Biol, 74; vis scientist staff, Dept Cell Regulation, Imp Cancer Res Fund Lab, 74-77; asst res prof, Salk Inst Biol Studies, 77-79; from asst prof to assoc prof, 79-86, PROF, DEPT BIOCHEM & MOLECULAR BIOL, MED SCH, UNIV TEX, 86- *Concurrent Pos:* Prin investr, Nat Cancer Inst, 76-77; mem fac, Grad Sch Biomed Sci, Univ Tex, 79- *Mem:* Soc Biol Chemists; Am Physiol Soc. *Res:* Biochemistry; molecular biology; numerous publications. *Mailing Add:* Biochem & Molecular Biol Dept Univ Tex Med Sch PO Box 20708 Houston TX 77225

LEVER, REGINALD FRANK, b Birmingham, Eng, July 5, 30; div; c 1. MATERIALS SCIENCE. *Educ:* Oxford Univ, BA, 51, MA, 54. *Prof Exp:* Sci officer, Serv Electronics Res Lab, Baldock, Eng, 51-57; sr sci officer, UK Atomic Energy Agency Indust Group, Lancashire, 57-58; staff mem, Res Div, Philco Corp, 58-60; staff mem, Thomas J Watson Res Ctr, 60-70, STAFF MEM COMPONENTS DIV, IBM CORP, 70- *Mem:* Am Phys Soc. *Res:* Growth of crystals from vapor by chemical deposition; gaseous diffusion; semiconductors; surfaces; silicon device processing; material analysis by MeV ion backscattering; plasma etching; process modeling. *Mailing Add:* IBM GTD D62G Bldg 300-47A Rte 52 Hopewell Junction NY 12533

LEVER, WALTER FREDERICK, b Erfurt, Ger, Dec 13, 09; US citizen; m 40, 71; c 4. ELECTRON MICROSCOPY. *Educ:* Univ Leipzig, Ger, MD, 34. *Hon Degrees:* MD, Frei Univ, West Berlin, 84. *Prof Exp:* Res fel dermat, Mass Gen Hosp, 38-44; from instr to assoc clin prof, Harvard Med Sch, 44-59; prof, 59-75, chmn dept, 61-75, EMER PROF DERMAT, TUFTS UNIV MED SCH, 75- *Concurrent Pos:* Lectr, Harvard Med Sch, 59-75; dermatologist-in-chief, New Eng Med Ctr Hosps, 59-78; hon dermatologist, Mass Gen Hosp, 59-78; dir dermat, Boston City Hosp, 61-74. *Honors & Awards:* Dohi Medal, Japanese Dermat Soc, 63. *Mem:* Am Acad Dermat; Am Dermat Asn; Am Soc Dermatopathology (past pres); Soc Investigative Dermat (past pres). *Res:* Pathology of the skin, especially appendage tumors; blistering skin diseases. *Mailing Add:* Im Kleeacker 29 D-7400 Tubingen 07071-75355 Germany

LEVERANT, GERALD ROBERT, b Hartford, Conn, June 18, 40; m 62; c 2. MATERIALS SCIENCE ENGINEERING. *Educ:* Rensselaer Polytech Inst, BMetE, 62, PhD(metall), 66. *Prof Exp:* Sr res scientist, Res Lab, United Aircraft Corp, 66, res assoc, Mat Eng & Res Lab, Pratt & Whitney Aircraft Div, 66-68, sr res assoc, 68-74, group leader, 74-77; mgr metall, Southwest Res Inst, 77-81, asst dir, 81-85, dir mat sci, 85-90, DIR MAT & MECH, SOUTHWEST RES INST, 90- *Honors & Awards:* Henry Marion Howe Gold Medal, Am Soc Metals, 70. *Mem:* Am Soc Metals; Am Inst Mining, Metall & Petrol Engrs. *Res:* Relation of metallurgical structure to mechanical properties. *Mailing Add:* Southwest Res Inst PO Drawer 28510 San Antonio TX 78228-0510

LEVERE, RICHARD DAVID, b Brooklyn, NY, Dec 13, 31; m 78; c 3. INTERNAL MEDICINE, HEMATOLOGY. *Educ:* State Univ NY, MD, 56. *Prof Exp:* From intern to asst resident med, Bellevue Hosp, 56-58; resident, Kings County Hosp, 60-61; instr med, State Univ NY, 62-63; res assoc biochem, Rockefeller Inst, 62-63, asst prof, 64-65; from asst prof to prof med, State Univ NY Downstate Med Ctr, 65-77, chief hemat sect, 70-77; PROF MED & CHMN DEPT, NEW YORK MED COL, 77- *Concurrent Pos:* Fel hemat, State Univ NY, 61-62; NIH grant, 65-82; adj prof, Rockefeller Univ, 73-; dir med serv, Westchester County Med Ctr, 78- *Mem:* AAAS; Am Soc Clin Invest; Am Fedn Clin Res; Am Soc Hemat; Am Col Physicians; Sigma Xi. *Res:* Control mechanisms in heme and porphyrin synthesis; metabolism of normal and abnormal hemoglobins; diseases of porphyrin metabolism. *Mailing Add:* Dept Med NY Med Col Munger Pavilion Valhalla NY 10595

LEVERE, TREVOR HARVEY, b London, Eng, March 21, 44; Can & Brit citizen; m 66; c 2. HISTORY OF CHEMISTRY, SCIENCE & ARCTIC EXPLORATION. *Educ:* Oxford Univ, BA, 66, MA, 69, DPhil(mod hist & hist sci), 69. *Prof Exp:* Lectr, 68-69, from asst prof to assoc prof, 69-81, dir inst hist & philos sci, 81-86, PROF HIST SCI, UNIV TORONTO, 81- *Concurrent Pos:* Killam fel, Can Coun, 75-77; John Simon Guggenheim Found fel, 83; vis fel, Clare Hall & vis scholar, Scott Polar Res Inst, Cambridge Univ, 83-84. *Mem:* Foreign mem Dutch Soc Sci; Can Soc Hist & Philos Sci; corresp mem Int Acad Hist Sci; fel Royal Soc Can; fel Royal Geog Soc; Hist Sci Soc. *Res:* History of sciences from eighteenth to twentieth centuries; social and cultural contexts; science and Romanticism; science and Arctic exploration. *Mailing Add:* Inst Hist & Philos Sci Victoria Col Univ Toronto 73 Queen's Park Crescent E Toronto ON M5S 1K7 Can

LEVERENZ, HUMBOLDT WALTER, b Chicago, Ill, July 11, 09; m 40; c 4. SOLID STATE SCIENCE. *Educ:* Stanford Univ, AB, 30. *Prof Exp:* Res chemico-physicist, Radio Corp Am, 31-54, dir phys & chem lab, 54-57, asst dir res, 57-59, dir, 59-61, assoc dir, RCA Labs, 61-66, staff vpres, Res & Bus Eval, 66-68, staff vpres & chmn educ aid comt, RCA Corp, 68-74; RETIRED. *Concurrent Pos:* Advan mgt prog, Bus Sch, Harvard Univ, 58; mem, Mat Adv Bd, Nat Acad Sci, 64-68; mem conf comt, Nat Conf Admin Res, 64-68. *Honors & Awards:* Brown Medal, Franklin Inst, 54. *Mem:* Nat Acad Eng; fel AAAS; Am Chem Soc; fel Am Phys Soc; fel Inst Elec & Electronic Engrs; fel Optical Soc Am. *Res:* Syntheses and applications of solids used in electronics; phosphors; secondary-emitters; photoconductors; semiconductors; nonmetallic magnetic materials; scotophors; crystals used in electronics. *Mailing Add:* 2240 Gulf Shore Blvd N Apt K4 Naples FL 33940

LEVERETT, DENNIS HUGH, b Cleveland, Ohio, June 22, 31; c 7. DENTISTRY, CLINICAL RESEARCH. *Educ:* Ohio State Univ, DDS, 56; Harvard Univ, MPH, 68; Am Bd Dent Pub Health, dipl. *Prof Exp:* Intern dent, USPHS, 56-57, dent officer, 57-60; pvt pract gen dent, 60-66; pub health dentist, NMex Dept Pub Health, 66-67; res fel ecol dent, Sch Dent Med, Harvard Univ, 67-69; resident dent pub health, Mass Dept Pub Health & Sch Dent Med, Harvard Univ, 68-69; exec dir, Ctr Community Dent Health, Portland, Maine, 69-73; CHMN DEPT COMMUNITY DENT, EASTMAN DENT CTR, 73-; PROF DENT RES, SCH MED & DENT, UNIV ROCHESTER, 84- *Concurrent Pos:* Consult, Bio-Dynamics, Inc, Mass, 68, Maine Dept Health & Welfare, 69-70, Mass Dept Pub Health, 69, Southern Maine Comp Health Asn, Inc, 70-73, Genesee Valley Group Health Asn, 73, Rochester Regional Med Prog, 76-77 & Health Econ Group, Inc, 78-; lectr dent ecol, Sch Dent Med, Harvard Univ, 69-73; clin instr social dent, Sch Dent Med, Tufts Univ, 70-73; clin assoc prof prev med & community health & clin dent, Sch Med & Dent, Univ Rochester, 73-84; adj prof dent hyg, Monroe Community Col, 73-82; dir dent care progs, Portland, Maine Model Cities, 69-73, chmn health task force, 70-72; mem prof adv comt, Portland City Health Dir, 69-73; mem med adv comt, Southern Maine Comprehensive Health Asn, 69-73; mem attend staff, Maine Med Ctr, Portland; mem courtesy staff, Mercy Hosp, Portland, Maine, 70-73; mem bd dirs, Smilemobile, Monroe County, NY, 73-77; mem dent hyg adv comt, Monroe Community Col, 73-85; dent dir, Monroe County Health Dept, 73-; mem bd dir, Westside Health Serv, Rochester, NY, 75-, chmn prog comt, 76-; co-prin investr, USPHS res grants, 74-77 & 78-81; prin investr, Nat Inst Dent Res grants, 75-79, 77-81 & 82-; lectr dent, var hosps & orgn, 73-81; ed, J Pub Health Dent, 87- *Mem:* Am Asn Pub Health Dent; Int Asn Dental Res; Am Asn Dental Res; Am Pub Health Asn; Europ Orgn Caries Res. *Res:* Clinical trials of therapeutic and preventive agents, including adhesive sealants and fluorides; evaluation of third party payment mechanisms, dental care delivery systems and post-doctoral dental education. *Mailing Add:* Dept Community Dent 625 Elmwood Ave Rochester NY 14620

LEVERETT, M(ILES) C(ORRINGTON), b Danville, Ill, Dec 18, 10; m 38. NUCLEAR ENGINEERING. *Educ:* Kans State Col, BS, 31; Univ Okla, MSE, 32; Mass Inst Technol, ScD(chem eng), 38. *Prof Exp:* Asst chemist, Marathon Paper Mills Co, Wis, 32-33 & Phillips Petrol Co, Okla, 33-35; sr res engr, Humble Oil & Ref Co, Tex, 38-42; assoc div dir, Metall Lab, Univ Chicago, 42-43, div dir, 43-48; res assoc, Humble Oil & Ref Co, 48-49; tech dir nuclear engine propulsion aircraft proj, Fairchild Eng & Airplane Corp, 49-51; mgr eng, Aircraft Nuclear Propulsion Dept, Gen Elec Co, 51-56 & Develop Labs, 56-61, mgr res & eng, Nuclear Reactor Dept, 61-67, safety & qual mgr, Nuclear Energy Div, 67-71 & Nuclear Safety & Boiling Water Reactor Qual Assurance, 71-76; NUCLEAR CONSULT ENGR, 76- *Concurrent Pos:* Div dir, Monsanto Chem Co, 43-48 & Carbide & Chem Co, Tenn, 43-48; leader US deleg, Int Standardization Orgn Meeting Reactor Safety Stand, 58, 60; mem res adv comt nuclear energy processes, NASA, 59-60. *Honors & Awards:* Robert E Wilson Award, Am Inst Chem Engrs, 76. *Mem:* Nat Acad Engrs; fel Am Nuclear Soc (vpres, 59-60, pres, 60-61); Am Inst Mining, Metall & Petrol Engrs; Am Inst Chem Engrs; assoc fel Am Inst Aeronaut & Astronaut; Am Phys Soc. *Res:* Petroleum emulsions; flow of fluids in porous solids; nuclear reactors; reactor materials; reactor safety; nuclear physics. *Mailing Add:* 15233 Via Pinto Monte Sereno CA 95030

LEVERT, FRANCIS E, b Tuscaloosa, Ala, Mar 28, 39; m 65; c 3. INSTRUMENTATION DEVELOPMENT, HEAT TRANSFER. *Educ:* Tuskegee Inst, BS, 64; Univ Mich, MS, 66; Pa State Univ, PhD(nuclear eng), 71. *Prof Exp:* From asst prof to assoc prof nuclear engr & mech engr, Tuskegee Inst, 66-72, head mech eng dept, 72-73; indust fel reactor anal, Commonwealth Edison Elec Co, 73-74; nuclear engr, Appl Physics Div, Argonne Nat Lab, 74-79; chief scientist, Technol Energy Corp, 79-85; VPRES, KEMP CORP, 85- *Concurrent Pos:* Consult, Technol Energy Corp, 85-; mem, Am Nuclear Soc-Nuclear Educ & Comt Disadvantaged Youth, 72-74, Currie Comt Eng Educ Disadvantaged, Pa State Univ, 69-71. *Mem:* Am Soc Mech Engrs; Nat Soc Prof Engrs; Am Nuclear Soc; Am Soc Mech Engrs. *Res:* Nuclear reactor noise analysis, neutron and gamma ray detector development; thermionic converters and instrumentation development for the fossil power fuel power industry; author of 63 technical papers. *Mailing Add:* 1909 Matthew Lane Knoxville TN 37923

LEVERTON, WALTER FREDERICK, b Imperial, Sask, Dec 24, 22; m 48; c 2. SOLID STATE PHYSICS, MATERIALS SCIENCE. *Educ:* Univ Sask, BS, 46, MS, 48; Univ BC, PhD(physics), 50. *Prof Exp:* Asst prof elec eng, Univ Minn, 50-51; asst div mgr semiconductors, Res Div, Raytheon Co, 51-60; group vpres develop, Aerospace Corp, 60-79; CONSULT, 79- *Concurrent Pos:* Mem, Defence Commun Agency Sci Adv Group, 74-79. *Mem:* Am Phys Soc; fel Inst Elec & Electronics Engrs. *Res:* Cathode materials; semiconductors; 1061 Glenhaven Dr. *Mailing Add:* Pacific Palisades CA 90272

LEVESQUE, ALLEN HENRY, b Jewett City, Conn, Nov 1, 36; m 60; c 3. COMMUNICATION THEORY, COMMUNICATION SYSTEMS DEVELOPMENT. *Educ:* Worcester Polytech Inst, BSEE, 59; Yale Univ, MEng, 60, DEng, 65. *Prof Exp:* Sr engr commun, Sylvania Appl Res Lab, 60-62; res asst elec eng, Sch Eng, Yale Univ, 63-65; eng specialist commun res & develop, Sylvania Appl Res Lab, 65-66, eng specialist, Sylvania Commun Systs Lab, 66-69, sr mem tech staff, GTE Labs, Inc, 69-74, sr eng specialist, Eastern Div, 74-81, sr eng specialist commun res & develop, 81-84, sr scientist, Commun Systs Div, GTE Sylvania Inc, 84-88, SR SCIENTIST, GTE GOVT SYSTS CORP, 88- *Concurrent Pos:* Adj prof elec eng, Northeastern Univ, 78-81. *Mem:* Sr mem Inst Elec & Electronics Engrs; assoc mem Sigma Xi. *Res:* Information theory; algebraic coding theory; communication systems development; computer communications; digital signal processing; communication networks; authored book on error-control coding. *Mailing Add:* Elec Def Commun Div GTE Govt Systs Corp 100 First Ave Waltham MA 02254-1191

LEVESQUE, CHARLES LOUIS, b Manchester, NH, Feb 16, 13; m 38; c 3. ORGANIC CHEMISTRY. *Educ:* Dartmouth Col, AB, 34, AM, 36; Univ Ill, PhD(org chem), 39. *Prof Exp:* Instr anal chem, Dartmouth Col, 34-36; sr chemist, Resinous Prod & Chem Co, 39-41, group leader, 41-45, lab head, 45-48; res supvr, Rohm & Haas Co, 48-69, asst dir res, 69-71; prof appl sci & dir, Eve Sch, Ursinus Col, 71-79, dean continuing educ, 79-81; RETIRED. *Mem:* Am Chem Soc; Sigma Xi. *Res:* Structures of vinyl polymers; polyester resins and raw materials; new organic synthesis; surface active agents; pharmaceuticals. *Mailing Add:* Normandy Farms Estates Box 1108 Apt F-304 Blue Bell PA 19422

LEVESQUE, RENE J A, b St-Alexis, Que, Oct 30, 26; div; c 3. NUCLEAR PHYSICS. *Educ:* Sir George Williams Col, BSc, 52; Northwestern Univ, PhD(physics), 57. *Prof Exp:* Res assoc physics, Univ Md, 57-59; from asst prof to assoc prof, Univ Montreal, 59-67, dir, Lab Nuclear Physics, 65-69, dir, dept physics, 68-73, vdean res fac arts & sci, 73-75, dean fac arts & sci, 75-78, prof physics, 67-87, vpres res, 78-85, vpres res & planning, 85-87; PRES, ATOMIC ENERGY CONTROL BD, 87- *Concurrent Pos:* Asst ed, Can J Physics, 73-75; vpres, Can-France-Hawaii Telescope Corp, 79, pres, 80; pres, Asn Sci, Eng & Technol Community Can, 80. *Honors & Awards:* Queen Elizabeth Jubilee Medal. *Mem:* Can Asn Physicists (pres, 76-77); Natural Sci & Eng Res Coun Can (vpres, 81-86). *Res:* Nuclear spectroscopy; nuclear reactions at low energy. *Mailing Add:* Atomic Energy Control Bd PO Box 1046 Ottawa ON K1P 5S9 Can

LEVETIN AVERY, ESTELLE, b Boston, Mass, Mar 24, 45; m 74; c 2. MYCOLOGY, BOTANY. *Educ:* State Col Boston, BS, 66; Univ RI, PhD(bot & mycol), 71. *Prof Exp:* Lab instr & teaching asst bot, Univ RI, 69-71, asst prof, Exten Div, 71-72, fel res assoc, Dept Plant Path, 71-72; asst prof physiol, Mt St Joseph Col, 72; asst prof, 72-78, ASSOC PROF BOT, UNIV TULSA, 78- *Concurrent Pos:* Consult, Joint Res Prog, Allergy Clin Tulsa, Inc, 75-76. *Mem:* Mycol Soc Am; Bot Soc Am; Int Asn Aerobiol; Brit Mycol Soc; Pan Am Aerobiol Asn. *Res:* Physiology and development of fungi; distribution of fleshy fungi in Oklahoma; distribution of air-borne fungi and pollen in Tulsa County; allergenic spores and pollen; indoor air bioaerosols. *Mailing Add:* Fac Biol Sci Univ Tulsa 600 S College Tulsa OK 74104

LEVEY, DOUGLAS J, b Boston, Mass, Sept 20, 57; m 88. ECOLOGY, BEHAVIOR-ETHOLOGY. *Educ:* Earlham Col, BA, 79; Univ Wis, MS, 82, PhD(zool), 86. *Prof Exp:* Archie Carr fel zool, 87-88, ASST PROF ORNITH, UNIV FLA, 88- *Mem:* Ecol Soc Am; Am Ornithologists Union; Asn Trop Biol; Animal Behav Soc. *Res:* Community structure and co-evolution of fruit eating birds and fruiting plants in the tropics; digestive physiology of frugivores; seed dispersal systems. *Mailing Add:* Dept Zool Univ Fla Gainesville FL 32611

LEVEY, GERALD SAUL, b Jersey City, NJ, Jan 9, 37; m 61; c 2. INTERNAL MEDICINE, ENDOCRINOLOGY. *Educ:* Cornell Univ, AB, 57; NJ Col Med, MD, 61. *Prof Exp:* Intern med, Jersey City Med Ctr, 61-62, resident, 62-63; resident, Mass Gen Hosp, Boston, 65-66; clin assoc endocrinol, Nat Inst Arthritis & Metab Dis, 66-68; sr investr endocrinol, Nat Heart & Lung Inst, 69-70; assoc prof med, 70-73, Sch Med, Univ Miami, Fla, prof, 73-; CHMN, DEPT MED, SCH MED, UNIV PA, PHILADELPHIA. *Concurrent Pos:* NIH fel biochem, Med Sch, Harvard Univ, 63-65; consult med, Vet Admin Hosp, Miami, Fla, 70-; investr, Howard Hughes Med Inst, 71-; lecture, Channel 10, WTLG. *Mem:* Am Soc Clin Invest; Am Col Physicians; Am Thyroid Asn; Am Fedn Clin Res; Soc Exp Biol & Med. *Res:* Mechanism of hormone action; cyclic adenosine monophosphate. *Mailing Add:* Univ Pittsburgh Sch Med 1218 Scaife Hall Pittsburgh PA 15261

LEVEY, GERRIT, physical chemistry; deceased, see previous edition for last biography

LEVEY, HAROLD ABRAM, b Boston, Mass, Aug 14, 24; m 59; c 2. ENDOCRINE PHYSIOLOGY. *Educ:* Harvard Univ, AB, 47; Univ Calif, Los Angeles, PhD(zool), 53. *Prof Exp:* Jr & asst res physiol chemist, Univ Calif, Los Angeles, 53-56; from instr to asst prof, 56-64, ASSOC PROF PHYSIOL, COL MED, STATE UNIV NY DOWNSTATE MED CTR, 64- *Concurrent Pos:* USPHS fel, 53-; China Med Bd vis prof physiol, Fac Med, Univ Singapore, 66-67. *Mem:* AAAS; Am Physiol Soc; Endocrine Soc; NY Acad Sci; Harvey Soc. *Res:* Pituitary chemistry and physiology; pituitary-thyroid interrelationships; factors influencing metabolism of endocrine organs; electrophysiology of thyroid. *Mailing Add:* Dept Physiol State Univ NY Downstate Med Ctr Brooklyn NY 11203

LEVI, ANTHONY FREDERIC JOHN, b London, Eng, Feb 3, 59; m 85. ELECTRONIC & OPTO-ELECTRONIC DEVELOPMENT. *Educ:* Univ Sussex, England, BS, 80; Univ Cambridge, England, PhD(physics), 84. *Prof Exp:* Mem technol staff physic, 84-88, DISTINGUISHED MEM TECHNOL STAFF PHYSICS, AT&T BELL LABS, 88- *Mem:* Am Physical Soc. *Res:* Experimental and theoretical study of nonequilibrium electron transport in unipolar and bipolar semiconductor transistor structures; electron dynamics in quantized systems; exploration of new materials for electronic and opto-electronic device applications. *Mailing Add:* Rm 1E450 AT&T Bell Labs 600 Mountain Ave Murray Hill NJ 07974

LEVI, BARBARA GOSS, b Washington, DC, May 5, 43; m 66; c 2. ARMS CONTROL. *Educ:* Carleton Col, BA, 65; Stanford Univ, MS, 67, PhD(physics), 71. *Prof Exp:* Lectr physics, Fairleigh Dickinson Univ, 70-76; lectr physics, Ga Inst Technol, 77-80; mem res staff, Bell Labs, 82-83; mem res staff, Ctr Energy & Environ Studies, Princeton Univ, 81-82 & 83-87; assoc ed, 87-88, SR ASSOC ED, PHYSICS TODAY MAG, AM INST PHYSICS, 89- *Concurrent Pos:* Mem, gov coun, Fedn Am Scientists, 86-88; mem task force energy, Am Asn Univ Women, 75-77; consult, Off Technol Assessment, US Cong, 76-; vis prof, Rutgers Univ, 88-89; chair, Forum Physics & Soc, Am Phys Soc, 88-89, mem educ comt, 89- *Mem:* Am Phys Soc; Am Asn Physics Teachers. *Res:* Writing news of current physics research; nuclear arms control problems. *Mailing Add:* 20 N Point Dr Colts Neck NJ 07722

LEVI, DAVID WINTERTON, b Berryville, Va, Sept 2, 21; m 47; c 2. POLYMER CHEMISTRY. *Educ:* Randolph-Macon Col, BS, 43; Va Polytech Inst, MS, 51, PhD(chem), 54. *Prof Exp:* From instr to assoc prof chem, Va Polytech Inst, 46-59; SUPVRY CHEMIST, PICATINNY ARSENAL, DOVER, 59- *Mem:* Am Chem Soc. *Res:* Solution properties of high polymers; polymer-energetic compatibility; adhesives; thermal degradation of polymers. *Mailing Add:* Two Oak Hill Dr Succasunna NJ 07876-2006

LEVI, ELLIOTT J, b Brooklyn, NY, June 12, 40; m 64; c 2. ORGANIC CHEMISTRY, PHYSICAL CHEMISTRY. *Educ:* City Col New York, BS, 61; Univ Cincinnati, PhD(chem), 66. *Prof Exp:* Chief chemist, Apollo Chem Corp, 66-68; group leader, Chem Systs Inc, 68-70; res mgr chem, Drew Chem Corp, 70-80, dir res & develop, 81-87; dir res planning, Ashland Chem Corp, 85-87; RETIRED. *Concurrent Pos:* Adj asst prof, Upsala Col, 67-71. *Mailing Add:* 1307 Mercedes St Teaneck NJ 07666-2130

LEVI, ENRICO, b Milano, Italy, May 20, 18; US citizen; m 41. ENERGY CONVERSION, PLASMA PHYSICS. *Educ:* Israel Inst Technol, BSc, 41, Ing, 42; Polytech Inst Brooklyn, MEE, 56, DEE, 58. *Prof Exp:* Foreman elec shop, Shipwrights & Engrs Ltd, Israel, 42-44; mech engr, Palestine Elec Co, 44-45; sect head elec eng, Mouchly Eng Co, 45-48; lectr, Israel Inst Technol, 48-55; fel, Microwave Res Inst, Polytech Inst Brooklyn, 56-57; sr scientist, Elec & Electronic Res Found, Westbury, 57-58; from assoc prof to prof, 58-88, EMER PROF ELECTROPH, POLYTECH UNIV NY, 88-; PRES & TREAS, ENRICO LEVI, INC, 76- *Concurrent Pos:* Consult, Lever Bros, Israel, 48-55, Hudson Paper Mill Co, 54-55, Am Mach & Foundry Co, 60-62, Westinghouse Elec Astronuclear Labs, Pa, 62-64, Gen Appl Sci Labs, 65, Van Karman Inst Fluid Dynamics, 66, Consol Edison Co, New York, 72, Long Island Lighting Co, 73-74, US Dept Energy, 76-78, Lawrence Livermore Lab, 78, Argonne Nat Lab, 78, Nasa Lewis Res Ctr, 80 & North-Hills Electronics, 84-; mem, Israel Govt Comt, 50-51 & Elec Wire Standard Comt, Israel, 50-55; Technion vis prof, 80-81 & 84; Lady Davis fel award, 80. *Honors & Awards:* Charles J Hirsch Award, Inst Elec & Electronics Engrs, 80. *Mem:* Inst Elec & Electronics Engrs; Sigma Xi. *Res:* Electromechanical power conversions; magnetic amplifiers; automatic control; linear electric propulsion; variable speed drives, electric power. *Mailing Add:* 110 20 71st Rd Forest Hills NY 11375

LEVI, HERBERT WALTER, b Frankfurt am Main, Ger, Jan 3, 21; nat US; m 49; c 1. ARACHNOLOGY, SYSTEMATICS. *Educ:* Univ Conn, BS, 46; Univ Wis, MS, 47, PhD(zool), 49. *Hon Degrees:* AM, Harvard Univ, 70. *Prof Exp:* From instr to assoc prof bot & zool, Exten Div, Univ Wis, 49-56; from asst cur to assoc cur, Mus, 55-66, mem fac educ, Univ, 64-66, lectr biol, 64-70, PROF BIOL, HARVARD UNIV, 70-, AGASSIZ PROF ZOOL, 72-, CUR ARACHNOL, MUS COMP ZOOL, 66- *Concurrent Pos:* Secy, Rocky Mountain Biol Lab, 59-65; vpres, Ctr Int Document Arachnol, 65-68, pres, 80-83; vis prof, Hebrew Univ Jerusalem, 75; hon curator, Univ Panama Mus Invertebrates. *Mem:* Fel AAAS; Am Arachnol Soc (pres, 79-81); Soc Syst Zool; Am Ecol Soc; Am Inst Biol Sci; Am Micros Soc; Soc Study Evolution; Soc Syst Zool; Am Soc Zool; Centre Int de Doc Arachnologique. *Res:* Evolution; systematic zoology; spiders and other arachnids; animal transplantation; systematic studies of orb-weaving spiders in the family Araneidae and Tetragnathidae. *Mailing Add:* Mus Comp Zool Harvard Univ Cambridge MA 02138-9706

LEVI, IRVING, b Winnipeg, Man, Dec 15, 14; m 44; c 4. MEDICINAL CHEMISTRY. *Educ:* Univ Man, BSc, 38, MSc, 39; McGill Univ, PhD(chem), 42. *Prof Exp:* Carnegie Corp res fel, McGill Univ, 42-43, res assoc, 44-46, lectr, 46-47; sr res chemist, Charles E Frosst & Co, 48-68; PRES, ALMEDIC DIV, RHOING LTD, 68- *Concurrent Pos:* Civilian with Can Govt, 40-44. *Mem:* Am Chem Soc; fel Chem Inst Can. *Res:* Organic synthesis; carbohydrates; synthetic analgesics and sedatives; antibiotic and cancer chemotherapy; amino acids and derivatives; steroids and hormones; medicinal applications of natural products and derivatives. *Mailing Add:* Almedic Div Rhoing Ltd 4900 Cote Vertu Rd St Laurent PQ H4S 1J9 Can

LEVI, MICHAEL PHILLIP, b Leeds, Eng, Feb 5, 41; m 66; c 2. FOREST PRODUCTS. *Educ:* Univ Leeds, BS, 61, PhD(biophys), 64. *Prof Exp:* Fulbright travel scholar & res fel wood prod path, Sch Forestry, Yale Univ, 65; res fel, NC State Univ, 65-66; Sci Res Coun-NATO res fel, Univ Leeds, 66-67; sr biologist, Timber Res & Develop Lab, Hickson & Welch, Eng, 67-68, head res wood preservation, 68-71; assoc prof forestry, 71-77, PROF WOOD PAPER SCI, NC STATE UNIV, 77- *Mem:* Forest Prod Res Soc; Royal Soc Chem; Am Phytopath Soc. *Res:* Wood preservation; mode of action of fungicides; wood deterioration by fungi; wood as fuel. *Mailing Add:* Agr Exten Serv 216 Ricks Box 7602 NC State Univ Raleigh NC 27695

LEVI, ROBERTO, b Milano, Italy, Mar 2, 34; m 62; c 2. PHARMACOLOGY. *Educ:* Univ Florence, MD, 60. *Prof Exp:* Asst pharmacol, Univ Florence, 60-61; from asst prof to assoc prof, 66-77, PROF PHARMACOL, MED COL, CORNELL UNIV, 77- *Concurrent Pos:* Fulbright travel fel pharmacol & exp therapeut, Sch Med, Johns Hopkins Univ, 61-63; sr res fel electrophysiol, Univ Florence, 63-66; prin investr, USPHS grant, 67-68; co-investr, NIH grant, 67-69; prin investr, NY Heart Asn grant, 68-71, 71-73 & 74-76; Nat Inst Gen Med Sci grant, 74-81; fel, Polachek Found Med Res, 73-76; vis prof pharmacol, Col Physicians & Surgeons, Columbia Univ, 77-78. *Honors & Awards:* Alberico Benedicenti Prize, 64; J Murray Steele Prize, 70. *Mem:* Am Soc Pharmacol & Exp Therapeut; Harvey Soc. *Res:* Cardiovascular pharmacology; heart electrophysiology; neuropharmacology; immunopharmacology. *Mailing Add:* Dept Pharmacol Cornell Univ Med Col 1300 York Ave New York NY 10021

LEVIALDI, STEFANO, b Rome, Italy, Nov 6, 36; m 85; c 3. PARALLEL PROCESSING, MULTICOMPUTER ARCHITECTURES. *Educ:* Marconi Col, cert advan electronics, 61. *Prof Exp:* Lectr electronics, Univ Genoa, 61-65 & Univ Naples, 66-68; sr researcher image processing, Ital Nat Coun Res, 68-81; prof computer sci, Univ Bari, Italy, 81-83; PROF COMPUTER SCI, UNIV ROME, 83- *Concurrent Pos:* Assoc ed, Computer Vision, Graphics & Image Processing & Signal Processing, 79-, Pattern Recognition, 80-, Pattern Recognition Lett, 82-, Image & Vision Comput, 83- & J Parallel & Distrib Comput, 84-; co-ed, J Visual Lang & Comput, 89- *Mem:* Fel Inst Elec & Electronics Engrs; Int Asn Pattern Recognition (vpres, 90-). *Res:* Image analysis and understanding, algorithms, languages, architectures; visual languages; iconic interfaces; scientific visualization; computational metaphors. *Mailing Add:* Info Sci Univ Rome Via Salaria 113 Rome 00198 Italy

LEVICH, CALMAN, b Iowa City, Iowa, May 26, 21; m 46; c 4. RADIATION BIOPHYSICS, ACCIDENT ANALYSIS. *Educ:* Morningside Col, BS, 49; Cath Univ Am, PhD(physics), 66. *Prof Exp:* Biophysicist, Naval Med Res Inst, 50-61; proj dir, Armed Forces Radiobiol Res Inst, 61-67; assoc prof physics, Cent Mich Univ, 67-68; chmn dept, Seton Hall Univ, 68-70; prof, 70-83, chmn dept, 75-83, EMER PROF PHYSICS, CENT MICH UNIV, 83-; PRES, CALEB ASSOCS, 83- *Concurrent Pos:* Mem, legislative off of sci adv, State of Mich, 80-81, radiation adv bd, 82-90, governor's task force on high level radiation waste, 83-86; mem, Mich Indoor Radon Task Force, 87-90. *Mem:* AAAS; Biophys Soc; Radiation Res Soc; Am Asn Physics Teachers; Sigma Xi; Health Physics Soc. *Res:* Mechanical properties of muscle; radiation biophysics; reactor operator education; radiation safety and transport of radioactive materials. *Mailing Add:* PO Box 546 Pentwater MI 49449-0546

LEVIE, HAROLD WALTER, b Augusta, Ga, Jan 17, 49. SURFACE PHYSICS. *Educ:* William Marsh Rice Univ, BA, 71, MS, 73, PhD(mat sci), 76. *Prof Exp:* Physicist, Phys Sci Lab, US Army Missile Command, 71; MAT SCIENTIST SURFACE TECHNOL, INORG MAT DIV, LAWRENCE LIVERMORE LAB, 75- *Concurrent Pos:* Instr corrosion eng, Nat Asn Corrosion Engrs, 75. *Mem:* Nat Asn Corrosion Engrs; Sigma Xi. *Res:* Analysis and characterization of solid surfaces; kinetics of surface reactions and interface formation. *Mailing Add:* 4279 Amherst Way Livermore CA 94550

LEVIEN, LOUISE, b New York, NY, Mar 23, 52; m 84; c 2. PETROPHYSICS, CRYSTALLOGRAPHY. *Educ:* Brown Univ, ScB, 74; State Univ NY, Stony Brook, MS, 75, PhD(geochem), 79. *Prof Exp:* Weizmann fel, Calif Inst Technol, 79-81; res geologist, 81-84, res specialist, 84-91, SR RES SPECIALIST, EXXON PROD RES CO, 91- *Concurrent Pos:* Mem, Am Geol Inst Women Geoscientists Comt, 78-80, chmn, 80; mem educ & human resources comt, Am Geophys Union, 80-84. *Mem:* Am Asn Petrol Geologists; Am Geophys Union; Mineral Soc Am; Soc Prof Well Log Analysts. *Res:* Geochemistry and geophysics of hydrocarbon reservoirs; relationship of elastic properties and crystal chemistry of minerals. *Mailing Add:* Exxon Prod Res Co PO Box 2189 Houston TX 77252-2189

LEVIEN, ROGER ELI, b Brooklyn, NY, Apr 16, 35; m 60; c 2. SYSTEMS ANALYSIS, INFORMATION SCIENCE. *Educ:* Swarthmore Col, BS, 56; Harvard Univ, MS, 58, PhD(appl math), 62. *Prof Exp:* Engr, Rand Corp, 60-67, head syst sci dept, 67-71, mgr, Wash Domestic Progs, 71-74; proj leader, Int Inst Appl Systs Anal, Austria, 74-75, dir, 75-81; dir strategic systs anal, 82-85, VPRES, STRATEGY OFF, XEROX CORP, 85- *Concurrent Pos:* Adj prof, Univ Calif, Los Angeles, 70-74. *Honors & Awards:* Austrian Ehrenkreuz First Class, Sci & Art. *Mem:* Asn Comput Mach; Inst Elec & Electronics Engrs; Opers Res Soc Am; Asn Pub Policy Anal & Mgt. *Res:* Systems analysis; operations research; research and development management; information sciences; strategic planning. *Mailing Add:* 28 Fresh Meadow Rd Weston CT 06883

LEVI-MONTALCINI, RITA, b Torino, Italy, Apr 22, 09; Italian & US citizen. GROWTH FACTORS. *Educ:* Univ Turin, MD, 40. *Hon Degrees:* Dr, Univ Uppsala, Sweden, 77, St Mary's & Notre Dame's Col, 80; PhD, Wash Univ Med Sch, St Louis, Mo, 82, Univ London, Eng, 87, Univ Buenos Aires, 87, Loyola Univ, Chicago, 87 & Biophys Inst, Univ Brazil, 87, Harvard Univ, 89 & Univ Urbino, Italy, 90. *Prof Exp:* Asst prof anat, Univ Turin, 45-47; res assoc, Inst Zool, Wash Univ, 47-56, assoc prof, 56-58, prof neurobiol, Inst Biol, 58-77, EMER PROF NEUROBIOL, INST BIOL, WASH UNIV, 77-; AT INST NEUROBIOL, COMN NATURAL RESOURCES, NAT RES COUN, ROME, 89- *Concurrent Pos:* Dir, Neurobiol Res Ctr, Comn Natural Resources, Nat Res Coun, 61-69, dir, Cellular Biol Lab, 69-79, researcher/guest prof, 79-89, guest prof, Inst Neurobiol, 89; Fogarty Scholar, Wash, DC, 78; mem, Int Sci Adv Bd, Int Acad Biomed & Drug Res, Belgium, 90, Nat Comt Bioethics, Italy, & Nat Comn Unesco, Italy, 90. *Honors & Awards:* Nobel Prize Med/Physiol, 86; Max Weinstein Award, Cerebral Palsy Found, 62; Harvey Lectr, 65; Feltrinelli Int Prize Med, 69; US Nat Medal Sci, 87; Golden Plate Award, Am Acad Achievement, 70; Ibico-Reggino Award Biol Sci, 70; Int St Vincent Award, 79; Knights of Humanity Award, Int Philanthrop Soc, 79; Gold Medal Sci, Rome, 86; Thudicum Award & Lectr, Eng, 87; Gold Medal, Ministry Pub Health, Rome, 88. *Mem:* Nat Acad Sci; AAAS; Soc Develop Biol; Am Asn Anatomists; hon mem Tissue Cult Asn; Sigma Xi; Am Acad Arts & Sci; Am Philos Soc; hon mem Am Soc Zoologists; hon mem Am Med Women's Asn. *Res:* Experimental neurology; effect of a nerve growth factor isolated from the mouse salivary gland on the sympathetic nervous system and of an antiserum to the nerve growth factor; study of other specific growth factors. *Mailing Add:* Inst Neurobiol Nat Res Coun Viale Marx 15 Rome 00156 Italy

LEVIN, AARON R, b Johannesburg, SAfrica, Mar 19, 29; m 55; c 3. PEDIATRICS, CARDIOLOGY. *Educ:* Univ Witwatersrand, BSc, 48, MBBCh, 53, MD, 68; Royal Col Physicians & Surgeons, dipl child health, 60. *Prof Exp:* Intern, Edenvale Hosp, SAfrica, 54-55; sr intern, Johannesburg Fever Hosp, 55; pediat intern, Coronation Hosp, 55-56; pediat registr, 56-60; pediat registr, Charing Cross Hosp, Eng, 61; gen pract, 62-63; instr pediat, Med Ctr, Duke Univ, 64-66; from asst prof to assoc prof, 66-74, PROF PEDIAT, MED CTR, CORNELL UNIV, 74- *Concurrent Pos:* NIH fel cardiol, Med Ctr, Duke Univ, 64-66; attend physician, Pediat Intensive Care Unit, NY Hosp-Cornell Med Ctr; fel Royal Col Physicians, 81. *Mem:* Fel Am Acad Pediat; Soc Pediat Res; Am Pediat Soc; Am Heart Asn. *Res:* Pediatric cardiology, specifically related to studies of pressure-flow dynamics in various forms of congenital heart disease; extra cardiac factors in congenital heart disease; right ventricular hypertrophy at cellular level. *Mailing Add:* Pediat Cardiopulmonary Lab NY Hosp-Cornell Med Ctr New York NY 10021

LEVIN, ALFRED A, b Chicago, Ill, May 18, 28; m 56; c 4. RESEARCH ADMINISTRATION, AGRICULTURAL & FOOD CHEMISTRY. *Educ:* Univ Ill, Urbana, BS, 51; Loyola Univ, Chicago, MS, 62. *Prof Exp:* Chemist, Leaf Brands Inc, 51-53 & Wallace A Erickson & Co, 53-55; res chemist, 55-70, mgr labels & petitions, 70-75, dir govt compliance, 76-78, dir staff & support progs, 78-80, DIR TOXIC SUBSTANCES CONTROL, VELSICOL CHEM CORP, 80- *Mem:* Am Chem Soc; Am Inst Chemists. *Res:* Synthesis of chemicals with intended pesticidal properties. *Mailing Add:* 8242 Ridgeway Skokie IL 60076

LEVIN, ANDREW ELIOT, b Newton, Mass, Mar 9, 54; m. EXPERIMENTAL BIOLOGY. *Educ:* Princeton Univ, AB, 76; Univ Wis-Madison, 84. *Prof Exp:* Res technician, Genetics Unit, Mass Gen Hosp, 76-77; fel, Dept Cellular & Develop Biol, Harvard Univ, 84-87; FOUNDER & PRES, IMMUNETICS, INC, 87- *Concurrent Pos:* Consult ed, Encycl Sci Instruments; adv bd mem, Am Chem Soc; consult, Soviet biotechnol & biomed res technol. *Mem:* Am Soc Cell Biol. *Res:* Immunochemistry and immunoassays; protein purification and characterization; cell culture; monoclonal antibodies and hybridoma production; immunofluorescence microscopy; four patents. *Mailing Add:* Immunetics Inc 380 Green St Cambridge MA 02139

LEVIN, BARBARA CHERNOV, b Providence, RI, May 5, 39; m 61; c 2. TOXICOLOGY, FIRE SCIENCES. *Educ:* Brown Univ, BA, 61; Georgetown Univ, PhD(microbial genetics), 73. *Prof Exp:* Res asst endocrinol, Sch Med, Johns Hopkins Univ, 62-63; teaching asst biol, Georgetown Univ, 68-73; fel molecular biol, NIH, 73-75, staff fel, NIH, 75-78; res biologist, NBS, 78-82, group leader, NBS, 82-85, PROJ LEADER FIRE TOXICOL, NAT INST STANDARDS & TECHNOL, 85- *Concurrent Pos:* Mem, comt develop toxicity test method to assess combustion prod, Nat Bur Standards, 78-82, sci officer numerous grants, Extramural Res, 78-; lectr environ toxicol, grad sch, NIH, 83-91; mem, comt toxicity complex mixtures, Nat Acad Sci, 84-88; chmn, Tech Adv Group to Int Standards Orgn on Toxic Hazards in Fire, 84-; counr, Am Col Toxicol, 89-91. *Mem:* Soc Toxicol; Am Chem Soc; Sigma Xi; Am Col Toxicol; Am Soc Microbiol; Am Soc Testing & Mat. *Res:* Toxicology of combustion products; assessment of acute inhalation toxicity of combustion products; development of model to predict toxicity through examination of actions of individual and combined fire gases. *Mailing Add:* Bldg 224 Rm A363 Nat Inst Standards & Technol Gaithersburg MD 20899

LEVIN, BARRY EDWARD, b Brooklyn, NY, May 1, 42. NEUROBIOLOGY, NEUROLOGY. *Educ:* Emory Univ, MD, 67; Am Bd Psychiat & Neurol, dipl. *Prof Exp:* Instr & chief resident neurol, Cornell Med Sch, 71-72; clin assoc, Nat Inst Neurol Dis & Blindness, 72-74; asst prof neurol & psychiat, Dartmouth Med Sch, 74-77; ASSOC PROF NEUROSCI, COL MED NJ, 77- *Concurrent Pos:* Grantee, Vet Admin Res & Educ grant, 74-; dir lab of neuropharmacol & dept neurosci, Col Med NJ, 77-; staff neurologist, Vet Admin Hosp, East Orange, NJ, 77-; attend neurologist, Martland Hosp, Col Med NJ, 78- *Mem:* Soc Neurosci; Am Acad Neurol. *Res:* Metabolism, axonal transport and rhythms of catecholamines in health and disease. *Mailing Add:* Neurol Serv NJ Med Sch Vet Admin Med Ctr East Orange NJ 07019

LEVIN, BRUCE, b New York, NY, Mar 14, 48; m 70; c 2. MATHEMATICAL STATISTICS. *Educ:* Columbia Univ, AB, 68; Harvard Univ, MA, 72, PhD(appl math). 74. *Prof Exp:* Data analyst & comput programmer, Albert Einstein Col Med, 66-72; prof math statist & biostatist, 74-83, ASSOC PROF CLIN PUB HEALTH BIOSTATIST, COLUMBIA UNIV, 83- *Concurrent Pos:* Consult, Statistica, Inc, 78- *Mem:* Am Statist Asn; Inst Math Statist; Sigma Xi. *Res:* Statistical inference and data analysis. *Mailing Add:* 39 Claremont Ave No 42 New York NY 10027

LEVIN, EDWIN ROY, b Philadelphia, Pa, Nov 4, 27; m 51; c 3. SOLID STATE SCIENCE. *Educ:* Temple Univ, AB, 49, MA, 51, PhD(physics), 59. *Prof Exp:* Asst physics, Temple Univ, 49-51; physicist, Frankford Arsenal, US Army, 51-63; mem tech staff, RCA Labs, 63-87; CONSULT, ELECTRON MICROS APPLICATIONS, 87- *Concurrent Pos:* Secy Army res & study fel, Cavendish Lab, Cambridge Univ, 61-62; guide prof, World Univ, 73- *Mem:* AAAS; Am Phys Soc; Electron Micros Soc Am. *Res:* Solid state physics; theory of dielectrics; photoconductivity; quantum electronics; analysis of solid materials for electronics, including electron microscopy and related methodologies. *Mailing Add:* Micco Assocs PO Box 3263 Princeton NJ 08543

LEVIN, EUGENE (MANUEL), b New York, NY, Aug 14, 34; m 60; c 3. PHYSICS. *Educ:* Univ Vt, BA, 56; Columbia Univ, MA, 59; NY Univ, PhD(physics), 67. *Prof Exp:* PROF PHYSICS, YORK COL, NY, 67- *Mem:* Am Asn Physics Teachers. *Res:* Excited states and fluorescence properties of organic molecules; applications of fluorescence techniques to charged particle dosimetry. *Mailing Add:* Dept Physics York Col City Univ NY Jamaica NY 11451

LEVIN, FRANK S, b Bronx, NY, Apr 14, 33; m 55, 73; c 4. NUCLEAR PHYSICS. *Educ:* Johns Hopkins Univ, AB, 55; Univ Md, PhD(physics), 61. *Prof Exp:* Res assoc physics, Rice Univ, 61-63 & Brookhaven Nat Lab, 63-65; temp res assoc, Atomic Energy Res Estab, Eng, 65-67; assoc prof, 67-77, PROF PHYSICS, BROWN UNIV, 77- *Concurrent Pos:* Exchange scientist, US-India Exchange Scientists Prog, 73; sr vis fel, UK Sci Res Coun, 74; founder, Am Phys Soc Topical Group Few Body Systs & Multiparticle Dynamics. *Honors & Awards:* Alexander von Humboldt Sr US Scientist Award, 79-80. *Mem:* Fel Am Phys Soc. *Res:* Nuclear reaction theory; scattering theory; few-body problems; molecular structure; muon catalyzed fusion. *Mailing Add:* Dept Physics Brown Univ Providence RI 02912

LEVIN, FRANKLYN KUSSEL, b Terre Haute, Ind, June 28, 22; m 46; c 3. EXPLORATION GEOPHYSICS. *Educ:* Purdue Univ, BS, 43; Univ Wis, PhD(physics), 49. *Prof Exp:* Physicist, Sam Labs, Columbia Univ, 43-44, Carbide & Carbon Chem Corp, 44-46; asst physics, Univ Wis, 46-47; physicist, Carter Oil Co, 49-53; asst dir, Hudson Labs, Columbia Univ, 53-54; physicist, Carter Oil Co, 54-58; physicist, Jersey Prod Res Co, Standard Oil Co, 58-59, res assoc, 59-63; sr res assoc, 63-64, sr res assoc, Esso Prod Res Co, 65-67, res scientist, 67-73, sr res scientist, Exxon Prod Res Co, 73-86; CONSULT, 87- *Concurrent Pos:* Lectr, Univ Tulsa, 58-63; ed, Geophys, 69-71. *Honors & Awards:* Robert Earll McConnell Award, 81; Reginald Fessenden Award, Soc Explor Geophys, 84; Maurice Ewing Medal, Soc Explor Geophys, 88. *Mem:* AAAS; Seismol Soc Am; Am Geophys Union; Acoust Soc Am; Soc Explor Geophys; Europ Asn Explor Geophys; Inst Elec & Electronics Engrs. *Mailing Add:* 802 W Forest Dr Houston TX 77079

LEVIN, GEOFFREY ARTHUR, b Los Alamos, NMex, Dec 7, 55; m 81; c 1. PLANT SYSTEMATICS. *Educ:* Pomona Col, BA, 77, Univ Calif, Davis, MS, 80, PhD(bot), 84. *Prof Exp:* Asst prof bot, Ripon Col, 82-84; CUR BOT, SAN DIEGO NATURAL HIST MUS, 84- *Concurrent Pos:* Adj prof bot, San Diego State Univ, 85-; vis prof biol, Univ San Diego, 87, 89- *Honors & Awards:* Jesse M Greenman Award, 87. *Mem:* Bot Soc Am; Am Soc Plant Taxonomists; Sigma Xi; Am Inst Biol Sci. *Res:* Systematics of Euphorbiaceae; evolution of dicotyledonous foliar morphology; Mexican flora. *Mailing Add:* Natural Hist Mus PO Box 1390 San Diego CA 92112-1390

LEVIN, GERSON, b Philadelphia, Pa, Oct 27, 39; m 69. MATHEMATICS. *Educ:* Univ Pa, AB, 61; Univ Chicago, MS, 62, PhD(math), 65. *Prof Exp:* NSF fel, Univ Ore, 66, vis asst prof math, 66-67; asst prof, NY Univ, 67-74; asst prof, 74-76, ASSOC PROF MATH, BROOKLYN COL, 76- *Res:* Commutative rings and homological algebra. *Mailing Add:* 420 12th St Brooklyn NY 11215

LEVIN, GIDEON, b Mazkeret Ratia, Israel, Apr 6, 36; US citizen; m 63; c 2. PHYSICAL ORGANIC CHEMISTRY, PHOTOCHEMISTRY. *Educ:* Israel Inst Technol, BSc, 60; Purdue Univ, West Lafayette, MSc, 65; State Univ NY Col Environ Sci & Forestry, PhD(chem), 71. *Prof Exp:* Chemist polymers, Dow Corning Corp, 65-67; res assoc photochem, Upsala Univ, 72; res assoc photochem, Col Environ Sci & Forestry, State Univ NY, 72-75, sr res assoc, 75-; AT DEPT MAT RES, WEIZMANN INST SCI. *Concurrent Pos:* Vis scientist, Weizmann Inst Sci, 78- *Mem:* Am Chem Soc. *Res:* Mechanism of photochemical reaction initiated by flash of light which includes conversion of light energy to chemical energy and photo-oxidation and photoreduction of organic and organo metallic molecules which have biological significance. *Mailing Add:* Dept Mat Res Weizmann Inst Sci Rehovot 76100 Israel

LEVIN, GILBERT VICTOR, b Baltimore, Md, Apr 23, 24; m 53; c 3. ENVIRONMENTAL HEALTH, ENGINEERING. *Educ:* Johns Hopkins Univ, BE, 47, MS, 48, PhD(sanit eng), 63. *Prof Exp:* Jr asst sanit engr, State Dept Health, Md, 48-50; asst sanit engr, Dept Pub Health, Calif, 50-51; pub health engr, DC, 51-56; vpres, Resources Res, Inc, 56-63; dir spec res, Hazleton Labs, Inc, 63-65; dir life systs div, 65-67; PRES, BIOSPHERICS INC, 67- *Concurrent Pos:* Res asst biochem, Schs Med & Dent, Georgetown Univ, 52-61; clin asst prof, 53-60; biochemist, Dept Sanit Eng, DC, 62-63; consult, Dept Interior, 63-71; NASA planetary quarantine adv, 65-74; NASA experimenter, Mariner 9, 71 & Viking Mission to Mars, 76; trustee, Johns Hopkins Univ, 82-85. *Honors & Awards:* IR-100 Indust Res Mag, 75;

Necomb Cleveland Prize, AAAS, 77. *Mem:* Am Soc Civil Eng; Am Water Works Asn; fel Am Pub Health Asn; Water Pollution Control Fedn; NY Acad Sci; Am Inst Biol Sci. *Res:* Inventor PhoStrip process for wastewater phosphorus removal; Lev-O-Cal L-sugar noncaloric sweetener; life sciences; applied biology; water supply; waste disposal; sanitary biology; environmental sanitation; life detection techniques; public health and medical microbiology; low caloric sweeteners instrumentation; space biology. *Mailing Add:* Biospherics Inc 12051 Indian Creek Ct Beltsville MD 20705

LEVIN, HAROLD LEONARD, b St Louis, Mo, Mar 11, 29; m 54; c 3. GEOLOGY, PALEONTOLOGY. *Educ:* Univ Mo, AB, 51, MA, 52; Wash Univ, PhD(paleont), 56. *Prof Exp:* Geologist, Standard Oil Co Calif, 56-61; from asst prof to assoc prof, 61-71, chmn, dept earth & planetary sci, 73-76, PROF PALEONT, WASH UNIV, 71-, ASSOC DEAN, COL ARTS & SCI, 76- *Concurrent Pos:* Res grants, Wash Univ, 61-72; consult ecol serv, Mo Bot Garden, 73-75; auth. *Mem:* AAAS; Soc Econ Paleont & Mineral; Paleont Soc. *Res:* Foraminifera, Coccolithophoridae and related microfossils; biostratigraphy of microorganisms; geological education. *Mailing Add:* Dept Earth Wash Univ St Louis MO 63130

LEVIN, HARVEY STEVEN, b New York, NY, Dec 12, 46; m 68; c 1. NEUROPSYCHOLOGY. *Educ:* City Col, Univ NY, BA, 67; Univ Iowa, MA, 71, PhD(clin psychol), 72. *Prof Exp:* Fel, Dept Neurol, Univ Iowa, 72-73; intern clin psychol, Ill Masonic Med Ctr, 73-74; from asst prof to assoc prof, 74-84, PROF NEUROPSYCHOL, UNIV TEX MED BR, 84- *Concurrent Pos:* Consult, Dept Neurol, Univ Hosps, Iowa, 73-74; vis lectr, dept psychol, Univ Mo, Columbia, 81; vis prof, dept neurosurg, Univ Pa, 81-; prin investr, Neuropsychol Sect, Nat Inst Neurol & Commun Dis & Stroke Prog Proj, 75-; investr, Int Study Group Pharmacol Memory, 78-; ed, Cortex, 81-, J Clin & Exp Neuropsychol, 83-; Develop Neuropsychol, 84- & Brain Injury, 85-; mem, Vet Admin Merit Rev Bd, 84-; mem, med prof adv bd, Nat Head Injury Found, 85-; Jacob K Javits Neurosci Investr Award, 84. *Honors & Awards:* Caveness Award, Nat Head Injury Found, 85. *Mem:* AAAS; fel Am Psychol Asn; Soc Neurosci; Acad Aphasia. *Res:* Recovery from brain injury in children and adults; cholinergic augmentation in dementia of the Alzheimer type; visual perception in patients with focal brain lesions. *Mailing Add:* Div Neurosurg E17 Univ Tex Med Br Galveston TX 77550

LEVIN, IRA WILLIAM, b Washington, DC, Sept 20, 35; m 61; c 2. CHEMICAL PHYSICS. *Educ:* Univ Va, BS, 57; Brown Univ, PhD(chem), 61. *Prof Exp:* Res instr chem, Univ Wash, 61-62; guest worker, NIH, 63-65, staff fel, 65-66, res chem, Phys Biol Lab, 66-72, actg chief, Lab Chem Physics, 84-85, RES CHEMIST, LAB CHEM PHYSICS, NIH, 72-, DEP CHIEF, 87-, CHIEF, SECT MOLECULAR BIOPHYS, 79- *Concurrent Pos:* Lectr, Georgetown Univ, 64-65; assoc mem grad fac chem, 74-75. *Honors & Awards:* Lippincott Award, 85. *Mem:* Coblentz Soc (pres, 77-78); fel Am Phys Soc; Biophys Soc; Am Soc Biol Chemists. *Res:* Vibrational spectroscopy; absolute intensities; molecular dynamics and structure; spectra; spectroscopy of biomembranes. *Mailing Add:* Lab Chem Physics NIH Bethesda MD 20892

LEVIN, IRVIN, b Baltimore, Md, Dec 18, 12; m 49; c 3. PHYSICAL CHEMISTRY. *Educ:* Johns Hopkins Univ, BS, 35; Univ Md, MS, 40, PhD(chem), 48. *Prof Exp:* Res chemist, Nat Dairy Prod Corp, Md, 35-42; phys chemist, Signal Corps, US Army, Camp Evans, NJ, 42-45; instr chem, Univ Md, 45-46, res assoc, 48-50; chief dept biophys instrumentation, Army Med Serv Grad Sch, Walter Reed Army Med Ctr, 50-55, dir instrumentation div, Walter Reed Army Inst Res, 55-76; res assoc, Am Univ, 78-79; RETIRED. *Concurrent Pos:* Consult, Power Condenser & Electronics Corp, Washington, DC, 49-50; consult, US Army Inst Dent Res, Washington, 76-77, Lab Tech Develop, Nat Heart & Lung Inst, Bethesda, 77-78 & Nuclear Support Serv, Inc, 79-80. *Mem:* Am Chem Soc; Am Phys Soc; Sigma Xi. *Res:* Electrochemistry; photochemistry; design of laboratory apparatus; photovoltaic behavior of chemical substances. *Mailing Add:* 1404 Billman Lane Wheaton MD 20902

LEVIN, JACK, b Newark, NJ, Oct 11, 32; m 75. INTERNAL MEDICINE, HEMATOLOGY. *Educ:* Yale Univ, BA, 53, MD, 57; Am Bd Internal Med, dipl, 65, recert, 74. *Prof Exp:* Chief resident & instr, Yale Univ, 64-65; from instr to assoc prof, Johns Hopkins Univ, 65-78, prof med, div hemat, 78-82; PROF LAB MED & MED, UNIV CALIF SCH MED, SAN FRANCISCO, 82-; DIR HEMAT LAB & BLOOD BANK, VET ADMIN MED CTR, SAN FRANCISCO, 82- *Concurrent Pos:* Fel med, Sch Med, Johns Hopkins Univ, 62-64; Markle scholar acad med, 68-73; mem corp, Marine Biol Lab, 65-; physician chg hemat out-patient clin, Johns Hopkins Hosp, 67-71, 76-; consult, Vet Admin Hosp, Baltimore, Md, 68-; atten physician Lab Med & Internal Med, Univ Calif Med Ctr, San Francisco, 86-; dir, Flow Cytometry Facil, VA Med Ctr, San Francisco, 87-90; res career develop awardee, USPHS, 70-75. *Honors & Awards:* Frederik B Bang Award, 86. *Mem:* Int Soc Hemat; Int Soc Exp Hemat; fel Am Col Physicians; Am Soc Hemat; Am Soc Clin Invest; Sigma Xi. *Res:* Blood coagulation, platelets; thrombopoiesis; endotoxin and endotoxemia; Shwartzman phenomenon; thrombocytosis; invertebrate blood coagulation; megakaryocytopoiesis. *Mailing Add:* Clin Path Serv 113A Vet Admin Hosp 4150 Clement St San Francisco CA 94121

LEVIN, JACOB JOSEPH, b New York, NY, Dec 21, 26; m 52; c 3. MATHEMATICAL ANALYSIS. *Educ:* City Col New York, BEE, 49; Mass Inst Technol, PhD, 53. *Prof Exp:* Instr math, Mass Inst Technol, 52-53; instr, Purdue Univ, 53-55; vis lectr, Mass Inst Technol, 55-56; staff mem, Lincoln Lab, 56-63; assoc prof, 63-66, PROF MATH, UNIV WIS-MADISON, 66- *Concurrent Pos:* NSF sr fel, Univ Calif, Los Angeles, 70-71; vis prof, Univ BC, 77-78. *Mem:* Am Math Soc; Soc Indust & Appl Math. *Res:* Differential equations; integral equations. *Mailing Add:* 1110 Frisch Rd Madison WI 53711

LEVIN, JEROME ALLEN, b Washington, DC, Aug 25, 39; m; c 3. MEDICAL INFORMATICS. *Educ:* Philadelphia Col Pharm & Sci, BSc, 61; Univ Mich, PhD(pharmacol), 66. *Prof Exp:* Res assoc pharmacol, State Univ NY Downstate Med Ctr, 66-68; asst prof, Med Col Ohio, 68-74, interim chmn, 73-75, assoc prof pharmacol, 74-79, PROF PHARMACOL, MED COL OHIO, 79-, ASSOC DEAN ACAD RESOURCES, 87-, DIR COMPUTER LEARNING RESOURCE CTR, 88- Concurrent Pos: USPHS fel, State Univ NY Downstate Med Ctr, 66-68; Am Heart Asn res grant, Med Col Ohio, 69-75, USPHS res grant, 70-76. Mem: AAAS; Am Heart Asn; Am Soc Pharmacol & Exp Therapeut. Res: Computer applications in medicine. Mailing Add: Deans Off Med Col Ohio CS 10008 Toledo OH 43699

LEVIN, JOSEPH DAVID, b New York, NY, Feb 7, 18; m 47; c 2. INDUSTRIAL MICROBIOLOGY. Educ: Queens Col, NY, BS, 41. Prof Exp: Tech aide, E R Squibb & Sons, 47-50, res asst, 50-53, res asst supvr, Squibb Div, Olin Mathieson Chem Corp, 53-59, res scientist, 59-64, sr res scientist, 64-69, lab supvr, Inst Med Res, Squibb Corp, New Brunswick, 68-82; RETIRED. Mem: NY Acad Sci; Am Soc Microbiol. Res: Analytical microbiology; test and develop microbiological assays of antibiotics including traces in mammalian tissues; test and development methods for pharmaceutical preservative efficacy; co-patentee, diagnostic aid for fungal infection. Mailing Add: 244 Benner St Highland Park NJ 08904

LEVIN, JUDITH GOLDSTEIN, b Brooklyn, NY, Nov 8, 34; m 57; c 2. BIOCHEMISTRY, VIROLOGY. Educ: Barnard Col, Columbia Univ, BA, 55; Harvard Univ, MA, 57; Columbia Univ, PhD(biochem), 62. Prof Exp: Sr scientist molecular biol viruses, Nat Cancer Inst, 69-73, sr scientist, Lab Molecular Genetics, 73-86, HEAD, UNIT VIRAL GENE REGULATION, NAT INST CHILD HEALTH & HUMAN DEVELOP, 86- Concurrent Pos: Nat Heart Inst res fel biochem genetics, 62-69; USPHS fel, 63-65; Am Heart Asn advan res fel, 66-68; consult lab path, Nat Cancer Inst, 69; estab investr, Am Heart Asn, 69-74. Mem: AAAS; Am Soc Biochem & Molecular Biol; Am Chem Soc; Am Soc Microbiol; Am Soc Virol. Res: Molecular genetics of retrovirus replication: correlation of gene structure with functional activity; regulated expression of viral genetic information in mammalian cells, in particular translational control mechanisms; translational control mechanisi. Mailing Add: Lab Molecular Genetics Nat Inst Child Health & Human Develop NIH Bethesda MD 20892

LEVIN, KATHRYN J, b Lawrence, Kans, Feb 25, 44; m 69. SOLID STATE PHYSICS THEORY. Educ: Univ Calif, Berkeley, BA, 66; Harvard Univ, PhD(physics), 70. Prof Exp: Res assoc physics, Univ Rochester, 70-72; asst res physicist, Univ Calif, Irvine, 72-75; asst prof, 75-85, PROF PHYSICS, UNIV CHICAGO, 85- Mem: Am Phys Soc. Res: Exotic superconductivity disordered systems. Mailing Add: James Franck Inst Univ Chicago 5640 Ellis Ave Chicago IL 60637

LEVIN, LEONID A, b USSR, Nov 2, 48. ALGORITHMIC COMPLEXITY. Educ: Moscow Univ, 72; Mass Inst Technol, PhD(math), 79. Prof Exp: PROF MATH & COMPUT SCI, BOSTON UNIV, 80- Concurrent Pos: Vis scientist comput sci, Mass Inst Technol, 78-; vis MacKey prof, UC Berkeley, 86; vis prof, Calif Inst Technol, 87. Res: Foundations of mathematics, statistics and computer science; algorithmic complexity with applications to randomness and information theories, inductive inference, functional analysis, combinatorics and graph theory, mathematical logic, theory of computations; randomness and information. Mailing Add: 150-3 Kenrick St Brighton MA 02135

LEVIN, MARTIN ALLEN, b Philadelphia, Pa, Aug 14, 49; m 77; c 3. MICROANATOMY. Educ: Rutgers Univ, BA, 71, MS, 73; Ohio Univ, PhD(zool), 77. Prof Exp: Teaching asst biol, Rutgers Univ, 71-73; teaching assoc biol, Ohio Univ, 74-77; from asst prof to assoc prof, 78-88, PROF BIOL, EASTERN CONN STATE COL, 88- Mem: Electron Micros Soc Am. Res: Ultrastructural studies of the abdominal muscles of terrestrial and semiterrestrial amphipods: correlating structure to locomotory function. Mailing Add: Eastern Conn State Univ Eastern Conn State Col 83 Windham St Willimantic CT 06226-2295

LEVIN, MICHAEL H(OWARD), b New York, NY, Sept 25, 36; m; c 1. ENVIRONMENTAL SCIENCES, ENGINEERING ECOLOGY. Educ: Univ Vt, BS, 58; Rutgers Univ, MS, 60, PhD(bot), 64. Prof Exp: Res assoc taxon, NY Bot Garden, 63-64; cur, Greene-Nieuwland Herbarium & asst prof biol, Univ Notre Dame, 64-66; asst prof bot & cur herbarium, Univ Man, 66-68; asst prof landscape archit & regional planning, Univ Pa, 68-73; PRES, ENVIRON RES ASSOCS, INC, 70-, DIR RES, 73- Concurrent Pos: Adj prof agr & natural resources, Del State Col, 79- Mem: Fel AAAS; Ecol Soc Am; Sigma Xi. Res: Environmental sciences and geotechnical investigations; ecology of altered communities and ecosystems; ecological management; application of gradient analysis to terrestrial communities; wetlands ecology; hydrobiology, hydrology and water resources; wood science; research and testing of natural and manmade materials; health and safety evaluations; environmental studies and surveys; testing and laboratory services; engineering and planning. Mailing Add: Environ Res Assocs Inc 414 Mill Rd Havertown PA 19083-3740

LEVIN, MORRIS A, b New York, NY, May 15, 34; m 57; c 2. MICROBIOLOGY. Educ: Univ Chicago, BS, 59; Univ RI, PhD(microbiol), 70. Prof Exp: Microbiologist aerobiol, Dept Defense, 57-66; microbiologist marine microbiol, Dept Health Educ & Welfare, 66-70; MICROBIOLOGIST HEALTH EFFECTS, ENVIRON PROTECTION AGENCY, 70- Concurrent Pos: Adj prof civil eng & microbiol, Univ RI, 75. Mem: Sigma Xi. Res: Quantitating of microorganisms in the environment, dose-response relationships and epidemiological considerations correlating the public health effects of exposure to microbial populations under natural conditions; forecasting, trend analysis of environmental problems, genetic engineering. Mailing Add: 14405 Woodcrest Dr Rockville MD 20853

LEVIN, MORTON LOEB, b Russia, Aug 25, 03; div; c 2. PREVENTIVE MEDICINE, PUBLIC HEALTH. Educ: Univ Md, MD, 30; Johns Hopkins Univ, DrPH, 34. Prof Exp: Intern & asst resident, Mt Sinai Hosp, Baltimore, Md, 30-32; asst dispensary physician, Johns Hopkins Hosp, 32-33; comnr health, Ottawa County, Mich, 34-35; instr epidemiol, Sch Hyg & Pub Health,

Johns Hopkins Univ, 35-36; assoc physician, Roswell Park Mem Inst, NY, 36-39; from asst dir to dir div cancer control, State Dept Health, NY, 39-47, asst comnr med serv, 47-60; chief, Dept Epidemiol, Roswell Park Mem Inst, 60-67; vis prof epidemiol, Sch Hyg & Pub Health, Johns Hopkins Univ, 67-; VIS PROF. Concurrent Pos: Dir, Comn Chronic Illness, 50-51. Honors & Awards: Biggs Medal, 60; Haven Emerson Award, 62; John Snow Award, 78. Mem: Am Epidemiol Soc; Am Pub Health Asn; AAAS. Res: Epidemiology of malignant tumors. Mailing Add: 4089 Nescoset Hwy Centereach NY 11720

LEVIN, MURRAY LAURENCE, b Boston, Mass, Nov 14, 35; m 61; c 2. INTERNAL MEDICINE, NEPHROLOGY. Educ: Harvard Col, AB, 57; Tufts Univ, MD, 61. Prof Exp: Intern med, Beth Israel Hosp, Boston, Mass, 61-62, resident, 62-64; assoc, 66-69, from asst prof to assoc prof, 69-80, PROF MED, MED SCH, NORTHWESTERN UNIV, CHICAGO, 80-, CHIEF, SECT NEPHROL, MED CTR, 86- Concurrent Pos: Res fel renal dis, Univ Tex Southwestern Med Sch Dallas, 64-66; NIH res fel, 65-66; Chicago Heart Asn res grants, 66-70 & 73-75; Nat Inst Arthritis & Metab Dis res grant, 67-70; attend physician, Vet Admin Lakeside Hosp, Chicago, Ill, 66-, chief renal sect, 72-76, chief med serv, 76-85; adj staff, Passavant Mem Hosp, 68-, assoc attend physician, 75-80; attending physician, Northwestern Mem Hosp, 80-; sec-tres, Cent Soc Clin Res, 82-87; chief, Sect Nephrology/ Hypertension, Northwestern Univ Med Ctr, 86- Mem: AAAS; Am Fedn Clin Res; Int Soc Nephrol; Am Soc Nephrol. Res: Salt and water metabolism; uremia; membrane transport; calcium and phosphorus metabolism. Mailing Add: Sect Nephrol Dept Med Northwestern Univ Med Sch 303 E Chicago Ave Chicago IL 60611

LEVIN, NATHAN, pharmaceutical chemistry; deceased, see previous edition for last biography

LEVIN, NORMAN LEWIS, b Hartford, Conn, Mar 31, 24; m 50; c 2. ZOOLOGY, PARASITOLOGY. Educ: Univ Conn, BS, 48, MS, 49; Univ Ill, PhD(zool, parasitol), 56. Prof Exp: Asst zool, Univ Ill, 53-56, instr, 56-57; asst prof biol, Westminster Col, Mo, 57-60; from instr to assoc prof, 60-76, PROF BIOL, BROOKLYN COL, 76-, DEP CHMN DEPT, 84- Mem: Fel AAAS; Am Soc Zool; Am Soc Parasitol; Am Soc Trop Med & Hyg; Am Micros Soc; Am Inst Biol Sci. Res: General taxonomy; morphology; life cycles; interrelationship of larval trematodes and marine snails. Mailing Add: Dept Biol Brooklyn Col Brooklyn NY 11210

LEVIN, ROBERT AARON, b New York, NY, July 25, 29; m 55; c 4. CLINICAL CHEMISTRY. Educ: St John's Univ, NY, BS, 51, MS, 55. Prof Exp: Res toxicologist, Norwich Pharmacal Co, 55-58, sr researcher clin path & toxicol, 58-62, unit leader clin path, 62-80 & 80-89, MGR PATH, ANATOMIC & CLIN PATH SECT, NORWICH-EATON PHARMACEUT, 90- Concurrent Pos: Sci adv, Med Technol Dept, State Univ NY Agr & Tech Col Morrisville, Broome Tech Col, State Univ NY, Canton; adj prof, State Univ NY, Utica. Mem: Am Chem Soc; Am Asn Clin Chem; Am Soc Vet Clin Pathologists. Res: Automation and computerization of chemical technics; drug safety assessment; establishing effects on clinical pathology parameters; veterinary hematology. Mailing Add: Clin Path Unit Norwich-Eaton Pharmaceut Norwich NY 13815

LEVIN, ROBERT E, b Boston, Mass, Dec 1, 30; c 2. MICROBIOLOGY, FOOD SCIENCE. Educ: Los Angeles State Col, BS, 52; Univ Southern Calif, MS, 54; Univ Calif, Davis, PhD(microbiol), 63. Prof Exp: Asst prof microbiol, Ore State Univ, 63-64; from asst prof to assoc prof, 64-77, PROF FOOD SCI, UNIV MASS, AMHERST, 77- Concurrent Pos: NIH res grant, 65-68. Mem: Am Soc Microbiol; Inst Food Technologists; Soc Cryobiol. Res: Microbiological sulfate reduction; yeast cytology; psychophilic bacteria; enzymology; molecular taxonomy. Mailing Add: Dept Food Sci Univ Mass Amherst MA 01002

LEVIN, ROBERT EDMOND, b Orange, Calif, Oct 11, 31; m 58; c 2. LIGHT & RADIOMETRIC OPTICS. Educ: Stanford Univ, BS, 53, MS, 54, Engr, 56, PhD(elec eng), 60. Prof Exp: Assoc prof elec eng, Calif State Univ, San Jose, 58-63; SR SCIENTIST LIGHT & RADIATION, GTE SYLVANIA INC, 63- Concurrent Pos: Consult engr, 58-63; instr continuing educ, Northeastern Univ, 68-82; contrib ed, McGraw-Hill, 71-75; adj prof elec eng, Univ NH, 83-; adj assoc prof archit, Rensselaer Polytech Inst, 90- Mem: Optical Soc Am; sr mem Inst Elec & Electronics Engrs; fel Illum Eng Soc; Am Soc Photobiol; Am Soc Eng Educ; Soc Motion Picture & TV Engrs; Sigma Xi. Res: Control and application of non-ionizing radiation in photobiological, photochemical and visual systems; radiometric optics. Mailing Add: GTE Sylvania Inc 60 Boston St Salem MA 01970

LEVIN, ROBERT HAROLD, b Chicago, Ill, Nov 1, 15; m 41; c 4. ORGANIC CHEMISTRY. Educ: Univ Ill, AB, 37; Univ Wis, PhD(org chem), 41. Prof Exp: Chem libr asst, Univ Ill, 34-36; asst chem, Univ Wis, 37-41; res chemist, Upjohn Co, Mich, 41-46, group leader chem res, 46-52, head dept chem, 52-58, asst dir res, 58-68; vpres res, Richardson-Merrell, Inc, 68-78; RES/MGT CONSULT, 78- Concurrent Pos: Mem subcomt steroid nomenclature, Nat Res Coun, 50-55; mem coun, Gordon Res Conf. Mem: AAAS; Am Chem Soc; Sigma Xi. Res: Chemistry of steroids, especially the cortical hormones; biomedical research and new drug development long range planning for pharmaceutical research; international pharmaceutical product licensing. Mailing Add: 11127 Jardin Pl Cincinnati OH 45241

LEVIN, ROBERT MARTIN, b New York, NY, Apr 6, 45; m 67; c 2. PHARMACOLOGY. Educ: Albright Col, BS, 67; Univ Pa, MS, 69, PhD(pharmacol), 74. Prof Exp: Fel, Dept Pharmacol, Med Col Pa, 74-76, instr pharrmacol, 76-78; res assoc, Div Urol, Univ Pa, 78-79, from res asst prof to res assoc prof, 79-87, assoc prof pharmacol, 83-87, RES PROF PHARMACOL & UROL, DIV UROL, UNIV PA, 87- Concurrent Pos: Dir, Urol res, Univ Pa, 78-; pharmacologist, term appt, Vet Med Ctr 84- Mem: Am Urol Asn; Am Soc Pharmacol & Exp Therapeuts; Am Fertil Soc; Urodynamics Soc; Int Soc Artificial Organs. Res: Smooth muscle plasticity in response to pathological situations. Mailing Add: Div Urol Univ Pa Hosp 3006 Ravdin Courtyard Bldg 3400 Spruce St Philadelphia PA 19104

LEVIN, ROGER L(EE), b Clearfield, Pa, Mar 21, 36; m 60; c 3. MATERIALS SCIENCE. *Educ:* Pa State Univ, BS, 58; Yale Univ, MEng, 61; Northwestern Univ, PhD(mat sci), 63. *Prof Exp:* USN, 58-70, anal officer, sonal opers anal, Naval Test & Eval Detachment, Fla, 63-65 sonar proj officer, Accoust Warfare Proj Off, Naval Ship Systs Command, DC, 56-70; mgr undersea warfare progs, Hydrospace Res Corp, 70-72; PRES, MAR INC, 72-, TECH DIR, 77- *Mem:* Am Soc Naval Engrs; Am Oceanic Orgn; Am Defense Preparedness Asn; US Naval Inst. *Res:* Underwater acoustics. *Mailing Add:* Four Cleveland Ct Rockville MD 20850

LEVIN, RONALD HAROLD, b San Francisco, Calif, Sept 26, 45; m 69; c 2. ORGANIC CHEMISTRY. *Educ:* Case Western Reserve Univ, BS, 67; Princeton Univ, PhD(chem), 70. *Prof Exp:* Fel chem, Univ Freiburg, 70-71 & Calif Inst Technol, 71-72; asst prof chem, Harvard Univ, 72-77; MEM STAFF, IBM CORP, 78- *Mem:* Am Chem Soc; Chem Soc London; Sigma Xi. *Res:* Reactive intermediates; thermal and photochemical transformations; applications of magnetic resonance; electrophotography. *Mailing Add:* 460 Oakwood Pl Boulder CO 80302

LEVIN, ROY, b New York, NY, 1948. PROGRAMMING ENVIRONMENT, DISTRIBUTED SYSTEMS & OPERATING SYSTEMS. *Educ:* Yale Univ, BS, 70; Carnegie Mellon Univ, PhD(computer Sci), 77. *Prof Exp:* Mem tech staff, 84-88, SR CONSULT ENGR, SYSTS RES CTR, DIGITAL EQUIP CORP, 88- *Concurrent Pos:* Chmn, Spec Interest Group Oper Systs, Asn Comput Mach, 87-91. *Mem:* Asn Comput Mach; Inst Elec & Electronics Engrs. *Mailing Add:* Systs Res Ctr Digital Equip Corp 130 Lytton Ave Palo Alto CA 94301

LEVIN, S BENEDICT, b New Orleans, La, July 9, 10; m 36; c 2. REMOTE SENSING, TECHNOLOGY TRANSFER. *Educ:* Columbia Univ, AB, 31, BS, 32, EM, 33, PhD(geol), 48. *Prof Exp:* Mining geologist, Central Am Mines, 34-37; instr geol, Hunter Col, 37-42; geol engr mineral explor, US Bur Mines, 42-45; res dir, US Army Electronics Command, 45-60; dir, Inst Explor Res, 60-68; asst dir res, Off Secy Defense, 68-70; exec vpres, Earth Satellite Corp, 70-76; prof eng & appl sci, George Washington Univ, 76-79; RETIRED. *Concurrent Pos:* Mem, Solid State Sci Panel, Nat Acad Sci, 49-68, chmn Panel, Space Info, Nat Acad Eng, 74-75; chmn, Defense Ctr Res, 69-70; mem, Fed Coun Sci & Technol, Acad Sci & Eng, 69-70 & adv coun, Technol Transfer, NASA, 78-80; consult, earth resources, 79- *Honors & Awards:* Medal, Antarctic Serv, NSF, 65. *Mem:* Fel Geol Soc Am; fel Am Geophys Union; Sigma Xi; fel Am Soc Photogram & Remote Sensing; fel Mineral Soc Am. *Res:* Application of remote sensing from earth satellites and aircraft to resource exploration and development; solid state physics. *Mailing Add:* Lake Waramaug New Preston CT 06777

LEVIN, SAMUEL JOSEPH, b Detroit, Mich, Sept 19, 35; m 63; c 2. BIOCHEMISTRY. *Educ:* Wayne State Univ, BA, 58, PhD(chem), 61; Am Bd Clin Chem, dipl. *Prof Exp:* Res assoc chem, Col Med, Wayne State Univ, 55-61; scientist, Warner Lambert Pharmaceut Co, 61-62; from instr to asst prof biochem, Div Grad Studies, Med Col, Cornell Univ, 63-66; from asst prof to assoc prof biochem, Sch Dent, Univ Mo-Kansas City, 67-74, from asst prof to assoc prof, Sch Med, 68-74; asst dir clin path, Michael Reese Hosp, 74-76, mem, Michael Reese Inst, 74-76, DIR, DIV BIOCHEM, MICHAEL REESE HOSP, 77-, ASSOC DIR, CLIN PATH, 83-; ASST PROF, PATH, UNIV ILL, 89- *Concurrent Pos:* Asst attend biochemist, Mem Hosp Cancer & Allied Dis, 62-66; assoc, Sloan-Kettering Inst, 63-66; chief biochemist, Dept Path, Kansas City Gen Hosp, 66-74. *Mem:* AAAS; Am Asn Clin Chem; Clin Lab Mgt Asn; Sigma Xi. *Res:* Clinical biochemistry. *Mailing Add:* Div Biochem two Blum Michael Reese Hosp Chicago IL 60616

LEVIN, SEYMOUR A(RTHUR), b Newark, NJ, Sept 16, 22; m 48; c 4. CHEMICAL ENGINEERING. *Educ:* Johns Hopkins Univ, BE, 43. *Prof Exp:* Res asst, Columbia Univ, 43-45; dept head, Union Carbide Nuclear Co Div, 45-68, HEAD LONG RANGE PLANNING, NUCLEAR DIV, UNION CARBIDE CORP, 68- *Mem:* Am Chem Soc; AAAS; Nat Soc Prof Engrs. *Res:* Design and analysis of isotope separation process. *Mailing Add:* 956 W Outer Dr Oak Ridge TN 37830

LEVIN, SEYMOUR R, b Chicago, Ill, Apr 27, 34; m 57; c 3. INTERNAL MEDICINE. *Educ:* Univ Ill, BS, 56, MD, 61; Am Bd Internal Med, dipl internal med, 70 & endocrinol, 73. *Prof Exp:* Intern, Cook Co Hosp, Chicago, 61-62; resident, Wadsworth Vet Admin Hosp, Los Angeles, 62-65; physician, US Army Hosp, Ft Carson, 65-67; res fel endocrinol, Univ Calif, San Francisco, 67-69, asst res physician, 69-73; DIR DIABETES CLIN & CHIEF METAB UNIT, WADSWORTH VET ADMIN HOSP, 73-; PROF MED, UNIV CALIF, LOS ANGELES, 81- *Concurrent Pos:* Endocrine Div, Univ Calif, Los Angeles, 73-; assoc prof med, Univ Calif, Los Angeles, 75-81; grants, Vet Admin, 73-91 & NIH Tug grant assoc inv, 83-92. *Mem:* Fel Am Col Physicians; Am Fedn Clin Res; Am Diabetes Asn; Endocrine Soc. *Res:* Studies of insulin secretion and mechanisms of secretion by the endocrine pancreas. *Mailing Add:* Wadsworth Vet Admin Hosp 691/111K Los Angeles CA 90073

LEVIN, SIDNEY SEAMORE, b Philadelphia, Pa, Mar 29, 29; m 62; c 2. PHYSIOLOGY, PHARMACOLOGY. *Educ:* Univ Pittsburgh, BS, 51, MS, 53, PhD(biol sci), 55. *Hon Degrees:* MA, Univ Pa, 71. *Prof Exp:* RES ASSOC, HARRISON DEPT SURG RES, SCH MED, UNIV PA, 58- *Mem:* AAAS; NY Acad Sci. *Res:* Adrenal output in shock; effect of hypertension on adrenal cortical steroids; cytochrome P-450. *Mailing Add:* Dept Surg Stemmler Hall Univ Pa Philadelphia PA 19104-6070

LEVIN, SIMON ASHER, b Baltimore, Md, Apr 22, 41; m 64; c 2. MATHEMATICS, BIOLOGY. *Educ:* Johns Hopkins Univ, BA, 61; Univ Md, PhD(math), 64. *Hon Degrees:* DS, Eastern Mich Univ, 90. *Prof Exp:* Asst math, Univ Md, 61-62; NSF fel biomath, Univ Calif, Berkeley, 64-65; asst prof math, Cornell Univ, 65-70, assoc prof appl math, 71-77, assoc prof ecol & systs & theoret & appl math, 72-77, chmn sect ecol & systs, 74-79, dir, Ecosysts Res Ctr, 80-87, dir, Ctr Environ Res, 87-90, PROF APPL MATH & ECOL, CORNELL UNIV, 77-, CHARLES A ALEXANDER PROF BIOL SCI, 85-, DIR, PROG THEORET & COMPUT BIOL, 90- *Concurrent Pos:* Res assoc, Univ Md, College Park, 64; co-chmn biomath, Gordon Res Conf, 70, chmn theoret biol & biomath, 71; vis prof, Univ Md, College Park, 68, Univ Wash, Seattle, 73- 74, Weizmann Inst, Rehovot, Israel, 77 & 80, Univ BC, Vancouver, 79-80, Stanford Univ, 88; assoc ed, Ecol & Ecol Monographs, Ecol Soc Am, 73-75, ed, 75-77; assoc ed, Theoret Pop Biol, 76-84; managing ed, Lecture Notes Biomath, 73-, J Appl Math, Soc Indust & Appl Math, 75-79 & Biomathematics, 76-; adv ed, J Math Biol, 73-76, ed, 76-79, managing ed, 79-; ed, Lect on Math in Life Sci, 74-79; mem US comt, Israel Environ, 75-; coun mem, Ecol Soc Am, 75-77, Soc Indust & Appl Math, 77-79; adv ed, J Theoret Biol, 76-; consult ed, Evolutionary Theory, 76-, Math Intelligencer, 77-84 & Int J Math Modeling, 79-; mem adv comt, Environ Sci Div, Oak Ridge Nat Lab, 78-81; Guggenheim fel, 79-80; vchmn math, Comt Concerned Scientists, 79-; co-dir, autumn course on ecol, Trieste, Italy, Inter Atomic Energy Agency, UNESCO, 82, 86 & 90; sci panel, Hudson River Found, 82-, chmn, 85-86, bd dir, 86; mem, ed adv coun, Nat Res Modeling, 84-; dir, Ctr Environ Res, 87-; ed-in-chief, Ecol Appln, 88-, assoc ed, Bull Math Biol, 88-; mem, Comn Life Sci, Nat Res Coun, Nat Acad Sci, 83-89, Comt Release Genetically Eng Organisms into Environ, 86-87, bd biol, 83-89, chmn, Subcomt Ecol & Ecosyst, 86-87; mem, Health & Environ Res Adv Comt, Dept Energy, 86-90; mem bd dir, Hudson River Found, 86-; mem, Comn Ecol, Int Union Conserv Nature & Natural Resources, 87- *Honors & Awards:* Lansdowne Lectr, Univ Victoria, BC, 81; Grace Kimball Mem Lectr, Wilkes Col, Pa, 86; H J Oosting Mem Lectr, Duke Univ, 87; MacArthur Award, Ecol Soc Am, 88. *Mem:* Soc Math Biol (pres, 87-89, vpres, 89-91); Am Math Soc; Am Soc Naturalists; Ecol Soc Am (pres-elect, 89-90, pres, 90-91); Soc Indust & Appl Math; Brit Ecol Soc. *Res:* Theoretical ecology; mathematical and computational models of ecological and evolutionary processes; biological growth and spread; landscape models in relation to disturbance and global change; terrestrial, intertidal, and marine ecosystems. *Mailing Add:* Sect Ecol & Syst 347 Corson Hall Cornell Univ Ithaca NY 14853

LEVIN, SIMON EUGENE, b Philadelphia, Pa, Nov 29, 20; m 48; c 2. TOXICOLOGY, INDUSTRIAL HYGIENE. *Educ:* Philadelphia Col Pharm, BS, 41; Pa State Col, MS, 42, PhD, 49. *Prof Exp:* Bacteriologist, La Wall & Harrison Res Labs, 38-41; lab asst bact, Pa State Col, 41-42; bioassayist, La Wall & Harrison Res Labs, 42-43 & 46; asst bact, Pa State Col, 46-49; res assoc chemother, E R Squibb & Sons, 49-50; head div biol, La Wall & Harrison Res Labs, 50-56; pres, Huntingdon Farms, Inc, West Conshohocken, 57-74; dir life sci div, Am Standards Testing Bur, 74-77; OCCUP HEALTH CONSULT, NJ STATE DEPT LABOR & INDUST, 77- *Concurrent Pos:* Dir, Syndot Labs, 57-74; consult, Decker Corp, 59-70; comn radiation protection, State NJ, 78- *Mem:* AAAS; Am Indust Hyg Asn; Toxicol Soc; Am Chem Soc; NY Acad Sci. *Res:* Medical and industrial pharmacology and toxicology. *Mailing Add:* Beaver Hill Apts 309 Florence Ave N205 Jenkintown PA 19046-2602

LEVIN, VICTOR ALAN, b Milwaukee, Wis, Nov 22, 41; m 63; c 2. CANCER, NEUROLOGY. *Educ:* Univ Wis-Madison, BS, 63, MD, 66. *Prof Exp:* Intern med, St Louis City Hosp, Washington Univ, 66-67; staff assoc chem pharm, Nat Cancer Inst, 67-69; resident neurol, Mass Gen Hosp, 69-71, Nat Inst Neurol Dis & Stroke fel, 71-72; from instr to assoc prof neurol, 72-81, prof neuro-oncol & pharmacol chem, Univ Calif, San Francisco, 81-88; PROF & CHMN DEPT NEURO-ONCOL, UNIV TEXAS, 88- *Honors & Awards:* Fac Res Award, Am Cancer Soc, 77-81; Ann & Jason Faber Award, 88. *Mem:* Am Acad Neurol; Am Asn Cancer Res; Am Soc Clin Oncol. *Res:* Pharmacology and pharmacokinetics of brain tumor chemotherapeutic agents and experimental brain tumor chemotherapy. *Mailing Add:* Dept Neuro-oncol MD Anderson Cancer Ctr 1515 Holcombe Blvd Houston TX 77030

LEVIN, WAYNE, b New York, NY, Feb 29, 40; m 62; c 2. PROTEIN CHEMISTRY, CHEMICAL CARCINOGENESIS. *Educ:* Ithaca Col, BA, 62; Univ Ill, MS, 64. *Prof Exp:* Biochemist, Burroughs Wellcome & Co, 65-70; biochemist, Dept Biochem, Hoffmann-La Roche Inc, 70-71, sr scientist, 71-74, group chief, 74-78, sect head, 78-85, mem, Roche Inst Molecular Biol, 85-87, distinguished res leader, Dept Protein Biochem, 87-90, DIR, DEPT PROTEIN BIOCHEM, HOFFMAN-LA ROCHE INC, 90- *Concurrent Pos:* Mem, Study Sect Chem Path, NIH, 77-79; distinguished lectr, Col Vet Med, Tex A&M Univ, 85-86; vis prof, Univ BC, 89-90; adj prof, Rutgers Univ, 89- *Honors & Awards:* Achievement Award, Acad Pharmaceut Sci, 79; Bernard B Brodie Award, Am Soc Pharmacol & Exp Therapeut, 88. *Mem:* Am Soc Biochem & Molecular Biol; Am Soc Pharmacol & Exp Therapeut; Am Asn Cancer Res; Soc Toxicol; NY Acad Sci; AAAS. *Res:* Purification and characterization of membrane-bound proteins involved in the biotransformation of drugs and chemical carcinogens, structure-activity relationships and immunochemical characterization of proteins. *Mailing Add:* Dept Protein Biochem Hoffmann-La Roche Inc Nutley NJ 07110

LEVIN, WILLIAM COHN, b Waco, Tex, Mar 2, 17; m 41; c 2. INTERNAL MEDICINE. *Educ:* Univ Tex, BA, 38, MD, 41. *Hon Degrees:* Dr, Univ Montpellier, 80. *Prof Exp:* From instr to assoc prof internal med, 44-65, dir hemat res lab & blood bank, 46-74, PROF INTERNAL MED, UNIV TEX MED BR GALVESTON, 65-, PRES, 74- *Concurrent Pos:* Consult, USPHS Hosp, Nassau Bay, 52; dir clin res ctr, John Sealy Hosp, 62-75. *Honors & Awards:* Ordre des Palmes Academiques, 81. *Mem:* Am Fedn Clin Res; Am Soc Hemat; AMA; fel Am Col Physicians; fel Int Soc Hemat. *Res:* Hematology; immunology; oncology. *Mailing Add:* Off Pres Univ Tex Med Br Galveston TX 77550

LEVIN, ZEV, b Haifa, Israel, Dec 17, 40; US citizen; m 65; c 2. ATMOSPHERIC SCIENCES, CLOUD PHYSICS. *Educ:* Calif State Univ, Los Angeles, BS, 66; Univ Wash, PhD(atmospheric sci), 70. *Prof Exp:* Res meteorologist, Univ Calif, Los Angeles, 70-71; head, Dept Geophys & Planetary Sci, 85-87, PROF ATMOSPHERIC SCI, TEL AVIV UNIV, ISRAEL, 71-, VPRES RES DEVELOP & DEAN RES, 87- *Concurrent Pos:*

Vis sr scientist atmospheric sci, Nat Ctr Atmospheric Res, 76-77; sr res assoc, Nat Res Coun, Ames Res Ctr, NASA, 81, fac fel, 85. *Mem:* Am Meteorol Soc; Am Geophys Union; Sigma Xi. *Res:* Formation of clouds and precipitation; cloud electrifications; atmospheric aerosols; ice nucleation. *Mailing Add:* Dept Geophys & Planetary Sci Tel Aviv Univ Ramat Aviv 69978 Israel

LEVINE, AARON WILLIAM, b New York, NY, July 14, 43; m 64; c 3. PHYSICAL CHEMISTRY, POLYMER CHEMISTRY. *Educ:* Yeshiva Univ, BA, 63; City Col New York, MA, 66; Seton Hall Univ, PhD(org chem), 70. *Prof Exp:* Teacher, High Schs, NY, 63-66; res chemist, M&T Chem, Inc, 66-69; teaching asst chem, Seton Hall Univ, 69; mem tech staff, David Sarnoff Res Ctr, Subsid SRI Int, 69-84, head org mat & lithography res, RCA Labs, 84-87, head, Thin Film & Org Mat Res, 87-89, HEAD, ADVAN MAT RES, DAVID SARNOFF RES CTR, SUBSID SRI INT, 90- *Mem:* Am Chem Soc; fel Am Inst Chemists. *Res:* Liquid crystals; kinetics of organic reactions in solution; polymer syntheses and reaction; organic electrochemistry; microlithography. *Mailing Add:* David Sarnoff Res Ctr CN 5300 Princeton NJ 08543-5300

LEVINE, ALAN STEWART, b New York, NY, Aug 11, 44; m 67; c 2. HEMATOLOGY, BIOCHEMISTRY. *Educ:* Monmouth Col, NJ, BS, 66; Univ Del, PhD(chem), 71. *Prof Exp:* Teaching asst chem, Univ Del, 66-71; res assoc, Sch Pharm, Univ Kans, 71-72; from staff fel to sr staff fel, 72-77, health scientist adminr & dep br chief, 77-86, CHIEF, BLOOD DIS BR, NAT HEART LUNG BLOOD INST, NIH, 86- *Mem:* Am Soc Hemat; Am Chem Soc; Biophys Soc; AAAS; NY Acad Sci. *Res:* Molecular mechanism of human red blood cell sickling; diseases of the red blood cell. *Mailing Add:* Fed Bldg Rm 5A12 Nat Heart Lung & Blood Inst NIH Bethesda MD 20892

LEVINE, ALFRED MARTIN, b Brooklyn, NY, Apr 5, 41; m 65; c 2. QUANTUM OPTICS & DISSAPATIVE SYSTEMS. *Educ:* Cooper Union, BEE, 61; Princeton Univ, MA, 64, PhD(elec eng), 66. *Prof Exp:* Mem res staff, Plasma Physics Lab, Princeton Univ, 66; scientist, Gas Lab, Ionizatti, Frascati, Italy, 66-68; mem tech staff, Bell Labs, 68-70; PROF ENG SCI, COL STATEN ISLAND, CITY UNIV NEW YORK, 70-, MEM DOCTORAL FAC PHYSICS, CITY UNIV, 71- *Concurrent Pos:* Vis scientist, Weizmann Inst, Israel, 86. *Mem:* Am Phys Soc; Inst Elec & Electronics Engrs; Optical Soc Am; Fulbright Fel, 86. *Res:* noise in laser systems; quantum optics; four wave mixing; computer modelling of environmental systems; dissipation in quantum mechanical systems. *Mailing Add:* Dept Appl Sci Col Staten Island 130 Stuyvesant Pl Staten Island NY 10301

LEVINE, ALLEN STUART, b Newark, NJ, Aug 1, 49; m 73; c 1. FOOD INTAKE, NUTRIENT ABSORPTION. *Educ:* Rutgers Univ, BA, 70; Univ Minn, MS, 73 & PhD(nutrit), 77. *Prof Exp:* From asst prof to assoc prof, 81-86, PROF FOOD SCI & NUTRIT, UNIV MINN, 86-, PROF SURG, 87-; ASSOC DIR RES, VET ADMIN MED CTR, 87- *Concurrent Pos:* Res chemist, Vet Admin Med Ctr, 78-87. *Honors & Awards:* Mead Johnson Award, Am Inst Nutrit, 85. *Mem:* Am Inst Nutrit; Soc Neurosci. *Res:* Regulation of food intake; investigation of the interactions of neuropeptides and monoamines in the control of feeding; emphasis on the role of the endogenous opioids in the initiation of feeding; importance of specific opiate receptors; role of regulatory peptides in energy expenditure. *Mailing Add:* Neuroendocrine Res Lab Vet Admin Med Ctr 151 One Veterans Dr Minneapolis MN 55417

LEVINE, ALVIN SAUL, b Hamlet, NC, Aug 29, 25; m 51; c 4. VIROLOGY. *Educ:* Wake Forest Col, BS, 48; Univ NC, MSPH, 50; Rutgers Univ, PhD(microbiol), 54. *Prof Exp:* Res asst biochem, Duke Univ, 50-51; instr bact & immunol, Harvard Med Sch, 56-58; from asst prof to assoc prof microbiol, 58-64, PROF MICROBIOL & IMMUNOL, SCH MED, IND UNIV, INDIANAPOLIS, 64-; PROF LIFE SCI & DIR TERRE HAUTE CTR MED EDUC, IND STATE UNIV, TERRE HAUTE, 71-; ASST DEAN, IND UNIV SCH MED, 80- *Concurrent Pos:* Res fel microbiol, Rutgers Univ, 51-54; teaching fel bact & immunol, Harvard Med Sch, 54-56; Fulbright vis prof, Univ West Indies, 67-68; asst dean, Sch Med, Ind Univ, Indianapolis, 81- *Mem:* Fel Am Soc Microbiol; Am Asn Path; Am Asn Immunol; Am Asn Cancer Res. *Res:* Infectious diseases; viral oncology; RNA viruses; biochemical, biophysical and immunological studies. *Mailing Add:* Terre Haute Ctr Med Educ Sch Med Ind State Univ Terre Haute IN 47809

LEVINE, ARNOLD DAVID, b Brooklyn, NY, Oct 24, 25; m 62. THEORETICAL PHYSICS. *Educ:* Columbia Univ, PhD(physics), 58. *Prof Exp:* Asst prof physics, WVa Univ, 57-60; asst prof, Wayne State Univ, 60-62; from asst prof to assoc prof, 62-71, PROF PHYSICS, WVA UNIV, 71- *Concurrent Pos:* Consult, Columbia Liquified Natural Gas Corp, 71-73; consult, Am Gas Asn, currently. *Mem:* Combustion Inst; Am Phys Soc; Am Asn Physics Teachers; Sigma Xi. *Res:* Meson physics; quantum field theory; non-equilibrium thermodynamics; fluid dynamics; combustion. *Mailing Add:* Dept Physics WVa Univ Morgantown WV 26506

LEVINE, ARNOLD J, b Brooklyn, NY, July 30, 39. BIOLOGY. *Educ:* State Univ NY, BA, 61; Univ Pa, PhD(microbiol), 66. *Prof Exp:* From asst prof to prof biochem, Princeton Univ, 68-79; chmn & prof microbiol, Sch Med, State Univ NY, Stony Brook, 79-83; HARRY C WIESS PROF MOLECULAR BIOL & CHMN, DEPT BIOL, PRINCETON UNIV, 84- *Concurrent Pos:* Mem, Human Cell Biol Panel, NSF, 71-72, Genetics & Biol Panel, 72-73; Camille & Henry Dreyfus Found teacher-scholar, 72-77; assoc ed, Virol, 73-74, ed, 74-84; mem, Virus Cancer Prog Sci Rev Comt, Nat Cancer Inst, 76-77, bd sci counr, Div Cancer Biol & Diag, 86-90; panel mem, Basic Virol-Cell Biol Task Force, Nat Inst Allergy & Infectious Dis, 77; mem, Papovirus Study Group, Int Comt Taxonomy of Viruses, 77-; mem, Biochem-Cell Biol Panel, Am Heart Asn, 78-80; mem, Microbiol Sect, Nat Bd Med Examnrs, 83-84; assoc ed, J Cellular & Molecular Biol, 84-88; ed-in-chief, J Virol, 84-; alt counr, Div S, DNA Viruses, Am Soc Microbiol, 86-87; Gen Motors vis prof, Univ Southern Calif Cancer Ctr, 89; Am Cancer Soc scholar, 90-91; Rosie & Max Varon vis prof, Dept Immunol, Weizmann Inst, Rehovot, Israel, 90-91; John Simon Guggenheim mem fel, 91. *Honors & Awards:* Merit Award, Nat Cancer Inst, 89; Susan Swerling Lectr, Harvard Med Sch, 90. *Mem:* Nat Acad Sci; NY Acad Sci; Fedn Am Socs Exp Biol; Am Soc Biol Chemists; Sigma Xi; Am Soc Microbiol; AAAS. *Res:* DNA replication; animal virology; tissue culture systems for the study of gene expression and the regulation of the cell cycle; genetics of higher organisms. *Mailing Add:* Dept Molecular Biol Princeton Univ Princeton NJ 08544

LEVINE, ARNOLD MILTON, b Preston, Conn, Aug 15, 16; m 41; c 3. COMMUNICATION ENGINEERING. *Educ:* Tri-State Col, BS, 39; Univ Iowa, MS, 40. *Hon Degrees:* DSc, Tri-State Col, 60. *Prof Exp:* Head, Sound Dept, Columbia Broadcasting Co, 40-42; from asst engr to vpres & dir missile & space systs, Int Tel & Tel Corp, 42-71, vpres & gen mgr, ITT Aerospace, 71, vpres & tech dir, ITT Gilvillan Inc, 71-74, sr scientist, Int Tel & Tel Corp, 74-86; RETIRED. *Mem:* Fel Inst Elec & Electronics Engrs; Am Inst Navig; fel Inst Advan Eng. *Res:* Research and development in communication and missile guidance navigation; time division systems; pulse code modulation; psuedo noise modulation; pulse and continuous wave missile guidance; radar research and development; fiber optics. *Mailing Add:* 10828 Fullbright Ave Chatsworth CA 91311

LEVINE, ARTHUR SAMUEL, b Cleveland, Ohio, Nov 1, 36; m 59; c 3. MOLECULAR VIROLOGY, ONCOLOGY. *Educ:* Columbia Univ, AB, 58; Chicago Med Sch, MD, 64, Am Bd Pediat, dipl, 70; Am Bd Hemat-Oncol, dipl, 76. *Prof Exp:* Intern & resident pediat Univ Minn Hosps, 64-66, USPHS fel hemat & genetics, Univ Minn, Minneapolis, 66-67; staff fel oncol, Div Cancer Treatment, Nat Cancer Inst, NIH, 67-70, sr investr molecular virol & oncol, 70-75, chief, Pediat Br, 75-82, SCI DIR, NAT INST CHILD HEALTH & HUMAN DEVELOP, NIH, 82- *Concurrent Pos:* Vis lectr, Cold Spring Harbor Lab, NY, 73; vis prof, Benares Hindu Univ, India, 74, Univ Minn, Minneapolis, 74, Hebrew Univ, Israel, 81, Univ Bologna, Italy, 89 & Univ Calabria, Italy, 90; prof pediat, Uniformed Servs Univ of Health Sci, 83-; prof med & pediat, Georgetown Univ, 75-; ed-in-chief, New Biologist, 88- *Honors & Awards:* Karon Mem Lectr, Univ Southern Calif, Los Angeles, 83; Seham Lectr, Univ Minn, Minneapolis, 83; Meritorious Serv Medal, USPHS, 87. *Mem:* Am Soc Clin Invest; Soc Pediat Res; Am Asn Cancer Res; Am Soc Hemat; AAAS; Am Soc Microbiol. *Res:* Molecular genetics of SV40 and adenovirus-SV40 hybrids; mechanism of viral oncogenesis; DNA repair and mutagenesis; oncology. *Mailing Add:* NIH Bldg 31 Rm 2A50 Bethesda MD 20892

LEVINE, BARRY FRANKLIN, b Brooklyn, NY, Sept 5, 42; m 68. LASERS. *Educ:* Polytech Inst Brooklyn, BS, 63; Harvard Univ, PhD(physics), 69. *Prof Exp:* PHYSICIST, BELL LABS, 68-, DEPT HEAD, 77- *Mem:* Am Phys Soc. *Res:* Experimental and theoretical nonlinear optics of crystals and liquids; coherent Raman scattering; optical picosecond spectroscopy of surfaces; novel high speed semiconductor devices (phototransistors, functional element tests, lasers, photodetectors). *Mailing Add:* 22 Bear Brook Lane Livingston NJ 07039

LEVINE, BERNARD BENJAMIN, b New York, NY, Nov 8, 28. IMMUNOLOGY, MEDICINE. *Educ:* City Col New York, BS, 50; NY Univ, MD, 54. *Prof Exp:* From asst prof to assoc prof, 62-70, PROF MED, MED CTR, NY UNIV, 70-, DIR ALLERGY, 62- *Concurrent Pos:* Res fel path, Med Ctr, NY Univ, 60-62. *Mem:* Am Asn Immunol; Soc Exp Biol & Med; Am Soc Clin Invest; Am Acad Allergy. *Res:* Immunopathology; hypersensitivity; antigenicity; immune response; allergy. *Mailing Add:* NY Univ Sch Med 566 First Ave New York NY 10016

LEVINE, CHARLES (ARTHUR), b Des Moines, Iowa, Dec 25, 22; m 48; c 2. PHYSICAL CHEMISTRY, ELECTROCHEMISTRY. *Educ:* Iowa State Col, BS, 47; Univ Calif, PhD(chem), 51. *Prof Exp:* Asst, Univ Calif, 48-49, asst, Radiation Lab, 49-51; res chemist, 51-65, assoc scientist, Dow Chem Co, 65-86; SR SCIENTIST, OMNI-TECH, INT, 86- *Mem:* AAAS; Electrochem Soc; Am Chem Soc; Am Phys Soc; Am Inst Chem Eng. *Res:* Nuclear chemistry; radiation chemistry; electrochemistry. *Mailing Add:* 124 Buena Vista Ave Santa Cruz CA 95062

LEVINE, D(ONALD) J(AY), b Brooklyn, NY, Oct 10, 21; m 46; c 2. ELECTRICAL ENGINEERING. *Educ:* City Col New York, BEE, 43; Polytech Inst Brooklyn, MEE, 52. *Prof Exp:* Sr asst, Microwave Res Inst, Polytech Inst Brooklyn, 46-48; dir, Radio Transmission & Anti-Submarine Warfare Lab, Int Tel & Tel Corp, 48-65; vpres & mgr, Transmission Systs Div, Commun Systs Inc, Comput Sci Corp, 65-67; dept head, Network Eng & Anal Dept, Commun Div, Mitre Corp, 67-73; dir systs eng, Page Commun Eng, Vienna, Va, 73-74; dir commun systs, Litton-Amecom, College Park, 74-75; dir commun eng, Aerospace Corp, Washington, DC, 75-76; vpres eng, Kings Electronics Co, Inc, Tuckahoe, NY, 76-82 & Am Nucleonics Corp, Westlake Village, Calif, 82-83; chief, Broadcast Syst Eng Div, USIA/Voice Am (SES-4), Washington, DC, 84-86; consult, USIA/Voice Am (SES-4), Washingto, DC, 86-88 & Int Broadcast Syst, Inc, 88-89; pres, Int Broadcast Systs, Inc, 88; electronics engr, USN TELCOM, 89-90; CONSULT, USN TELCOM, 90- *Honors & Awards:* Scott Helt Mem Award, Outstanding Contrib Inst Elec & Electronics Engrs Broadcast Tech Trans, 89. *Mem:* Nat Soc Prof Engrs; sr mem Inst Elec & Electronics Engrs; Armed Forces Commun & Electronics Asn. *Res:* Microwave components, systems, antennas and antenna systems; radio communications; line of sight and troposcatter systems; switched telecommunications systems; network management planning and analysis; economic engineering for fixed, mobile, surface, air, space and submarine environments; operations analysis; shortwave/ mediumwave broadcast systems engineering. *Mailing Add:* 7420 Westlake Terr Apt 609 Bethesda MD 20817

LEVINE, DANIEL, b New York, NY, July 21, 20; m 57; c 5. ELECTRICAL ENGINEERING. *Educ:* Univ Mich, BS, 41, MS, 42; Ohio State Univ, MSc, 48, PhD(elec eng), 55. *Prof Exp:* Electronics engr, Aircraft Radiation Lab, Wright-Patterson AFB, Ohio, 46-51, br tech consult, Aerial Reconnaissance Lab, 53-54; sr engr, Goodyear Aircraft Corp, Ariz, 54-56; consult engr, 56-61;

CONSULT SCIENTIST, LOCKHEED MISSILES & SPACE CO, 61- *Mem:* Optical Soc Am; Am Soc Photogram; Inst Elec & Electronics Engrs. *Res:* System design of aerospace reconnaissance and mapping equipments; instrumentation for photographic and radar stereoanalysis; analogue simulators for radar trainers and guidance equipments; digital simulators for radar detection and tracking systems. *Mailing Add:* 1043 Enderby Way Sunnyvale CA 94087

LEVINE, DAVID MORRIS, b Boston, Mass, Dec 15, 39; m 65; c 2. BEHAVIORAL SCIENCES, HEALTH EDUCATION. *Educ:* Brandeis Univ, AB, 59; Univ Vt, MD, 64; Johns Hopkins Univ, MPH, 69, SCD, 72; Nat Bd Med Examr, dipl, 65; Am Col Prev Med, dipl, 71; Pan Am Med Assoc, dipl. *Prof Exp:* Intern, Montefiore Hosp, Pittsburgh, 64-65; resident, Waltham Hosp, Mass, 65-66; US Army Med Corp, 66-68; resident prev med, 68-70, assoc prof pub health, med educ & internal med, 72-81, PROF BEHAV SCI, HEALTH EDUC, JOHNS HOPKINS UNIV, 81-, DIR MANPOWER STUDIES, CTR HEALTH SERV RES & DEVELOP, 72- *Concurrent Pos:* Fel pub health serv, Sch Hyg & Pub Health, Johns Hopkins Univ, 68-71; consult, Nat Ctr Health Serv Res, 72- & Am Asn Med Col, 73-; mem study sect, Nat Heart-Lung Inst, NIH, 75-, Vet Admin Mert Rev, 80- *Mem:* AAAS; Am Pub Health Asn; Am Fedn Clin Res; Am Col Prev Med; Pan Am Med Asn. *Res:* Health behavior, health education and health promotion; health care manpower-services, health education strategies in managing chronic disease process and outcome of medical education. *Mailing Add:* Dept Med Johns Hopkins Sch Med 1830 E Monument St Baltimore MD 21205

LEVINE, DONALD MARTIN, b Boston, Mass, Oct 17, 29. ZOOLOGY, PARASITOLOGY. *Educ:* Univ Vt, BA, 51; Univ RI, MS, 53; Univ Pa, PhD(zool), 58. *Prof Exp:* USPHS fel, 58-60; helminthologist, Liberian Inst, Am Found Trop Med, 60-62; assoc prof, 62-74, PROF BIOL SCI, WILLIAM PATERSON COL NJ, 74- *Mem:* Am Soc Trop Med & Hyg; Am Inst Biol Sci. *Res:* Immunology and ecology of parasitic infections. *Mailing Add:* Dept Biol Sci William Paterson Col NJ Wayne NJ 07470

LEVINE, DUANE GILBERT, b Baltimore, Md, July 5, 33; m 57; c 6. CHEMISTRY, PHYSICS. *Educ:* Johns Hopkins Univ, BES, 56, MS, 58. *Prof Exp:* Combustion, electrochem & petrol researcher, Exxon Res & Eng Co, 59-68, head air pollution control res & develop, Automotive Emission Res Sect, 68-70, adv logistics, Exxon Corp, 70-71, mgr petrol fuels res & develop, Fuels Prod Qual Res Lab, Exxon Res & Eng Co, 71-74, mgr petrol process eng, Gasoline & Lubes Process Eng Div, 74-76, gen mgr synthetic fuels res & develop, Baytown Res & Develop Div, 76-78, exec dir, Corp Res-Sci Labs, 78-89, MGR SCI & STRATEGY DEVELOP, EXXON CORP, 89- *Concurrent Pos:* Mem, Eng & Tech Res Comt, Am Petrol Inst, 71-74; mem, Air Pollution Res Adv Comt, joint comt US Govt, Petrol Indust & Automotive Indust, 71-74; participant UN/Indust-Sponsored Conf Environ Mgt, Versailles, France, 84; chmn, Rene Dubos Int Forum Managing Hazardous Mat, New York, 87, Forum Global Urbanization, 88; mem, Adv Comt, Calif Inst Technol, Johns Hopkins Univ, Rene Dubos Ctr. *Mem:* AAAS; fel Am Inst Chemists; Am Inst Chem Engrs; Am Chem Soc; Int Combustion Inst; Sigma Xi; NY Acad Sci. *Res:* Solid state sciences; surface sciences; optics; catalysis; materials; theoretical and mathematical sciences; biosci; engineering sciences; laser chemistry; polymer sciences; emulsion chemistry; chemical physics. *Mailing Add:* 225 E John W Carpenter Freeway Irving TX 75062-2298

LEVINE, ELLIOT MYRON, b Brooklyn, NY, June 16, 37; m 59; c 3. CELL BIOLOGY, CELL CULTURE. *Educ:* Queens Col, NY, BS, 57; Yale Univ, PhD(biochem), 61. *Prof Exp:* Sr asst scientist biochem, Nat Inst Arthritis, Metab & Digestive Dis, 61-63; from assoc to asst prof cell biol, Albert Einstein Col Med, 63-72; coordr res training, 74-80, from asst prof to assoc prof,72-84, PROF, WISTAR INST, 84- *Concurrent Pos:* NIH fel, Albert Einstein Col Med, 63-64, NIH spec res fel, 64-65, NIH career develop award, 68-72; NSF res grants, Albert Einstein Col Med & Wistar Inst, 70-; NIH res grants, Wistar Inst, 72-; mem grad groups cell biol, genetics & pathol, Univ Pa, 75-, Lung Cell Comt, Am Type Cult Col, 76-; staff mycoplasma detection course, W Alton Jones Cell Sci Ctr, 77-; mem, Cell Biol Study Sect, NIH, 78-82, chmn, 80-82. *Mem:* Fel AAAS; Tissue Cult Asn (pres, 90-92); Sigma Xi. *Res:* Mycoplasma detection, cellular senesence and differentiation in cultured cells, especially vascular endothelial and smooth muscle cells. *Mailing Add:* Wistar Inst 3601 Spruce St Philadelphia PA 19104-4268

LEVINE, EUGENE, b Brooklyn, NY, Jan 11, 25; m 48; c 3. ANALYTICAL STATISTICS, OPERATIONS RESEARCH. *Educ:* City Col New York, BBA, 48; NY Univ, MPA, 50; Am Univ, PhD(pub admin), 60. *Prof Exp:* Statistician, New York City Dept Health, 47-50; chief, Manpower Anal & Resources Br, Div Nursing, USPHS, 50-78, dep dir, Div Health Prof Anal, 78-80; ASSOC, LEVINE ASSOC, 80- *Concurrent Pos:* Res consult, Sch Nursing, Georgetown Univ, 84- *Mem:* Am Pub Health Asn; Nat League Nursing; Am Statist Asn. *Res:* Health manpower analysis; problems of health services organization and delivery; psychometric analysis into problems of job satisfaction; career choice and motivation; evaluation of health care programs. *Mailing Add:* 8135 Inverness Ridge Rd Potomac MD 20854

LEVINE, GEOFFREY, b Washington, DC, Sept 2, 42; m 70; c 3. NUCLEAR PHARMACY, HEALTH PHYSICS. *Educ:* Temple Univ, BS, 65, MS, 67; Northwestern Univ, PhD(civil eng & environ health), 78. *Prof Exp:* Clin asst prof pharmaceut, 72-80, coord, prog radiopharm, 72-83, asst prof radiol, 72-83, assoc prof, sch pharm, 85, ASSOC PROF RADIOL, SCH MED, UNIV PITTSBURGH, 83-; CLIN PROF NUCLEAR MED, ALLEGHENY COUNTY COMMUNITY COL, 84- *Concurrent Pos:* Grants, Am Cancer Soc, Union Carbide Corp, Soc Nuclear Med & others, 74-; dir nuclear pharm, Univ Health Ctr, Presbyterian-Univ Hosp, 72-87, radiopharm adv nuclear pharm, 72-, radiation safety comt, 75-, radiopharmacist, 72-, clin asst med staff, 73-83, assoc med staff, 84-, nuclear med res comt, 85-; consult, Shadyside Hosp, 74-75, Ames Labs, 75, Charleston Area Med Ctr, 79-80, NEN-Dupont Radiopharm Div, 84, Am Pharm Asn, 86-, Mallinckrodt-NeoRx Monoclonal Antibody Develop Prog, 88-90; pharm staff, Montefiore Hosp, 75-, radiation res comt, 78-; mem, Human Use Subcomt Radiation Safety Comt, Univ Pittsburgh, 85-, Radioactive Drug Res Comt, 85-; dir nuclear pharm, Central Imaging Serv, Inc, 85-; assoc mem, Pittsburgh Cancer Inst, 87- *Mem:* Health Physics Soc; Soc Nuclear Med (secy-treas, 74); Sigma Xi; Am Pharmaceut Asn; AAAS; Health Physics Soc; Am Soc Hosp Pharmacists. *Res:* Drug interactions; radioactive pharmaceuticals and radioactive monoclonal antibodies for tumor detection; author of numerous publications; cost-benefit risk analysis; inventory control modeling of radiopharmaceuticals; radiopharmacology. *Mailing Add:* Presbyterian-Univ Hosp Dept Radiol Nuclear Pharm De Soto at O'Hara Sts Pittsburgh PA 15213

LEVINE, HAROLD, b New York, NY, Mar 24, 22; m 47. APPLIED MATHEMATICS. *Educ:* City Col New York, BS, 41; Cornell Univ, PhD(physics), 44. *Prof Exp:* Res fel physics, Harvard Univ, 45-54; assoc prof, 55-70, PROF MATH, STANFORD UNIV, 70- *Concurrent Pos:* Lectr, Harvard Univ, 52-54; consult, Lawrence Radiation Lab, Univ Calif. *Mem:* Am Phys Soc. *Res:* Boundary value problems of classical field theories, particularly acoustics, electrodynamics and hydrodynamics. *Mailing Add:* Dept Math Stanford Univ Stanford CA 94305

LEVINE, HAROLD, b Lynn, Mass, Dec 14, 28; m 61. MATHEMATICS. *Educ:* Univ Chicago, PhD(math), 57. *Prof Exp:* Fulbright fel & Ger Res Asn grant, Univ Bonn, 57-59; instr math, Yale Univ, 59-60; from asst prof to assoc prof, 60-70, PROF MATH, BRANDEIS UNIV, 70- *Mem:* Am Math Soc; Math Asn Am. *Res:* Differential topology. *Mailing Add:* Dept Math Brandeis Univ Waltham MA 02254-9110

LEVINE, HARVEY ROBERT, b New York, NY, Sept 15, 31; m 56; c 2. PARASITOLOGY, MEDICAL ENTOMOLOGY. *Educ:* City Col New York, BS, 53; Univ Mass, MS, 55, PhD(entom), 58. *Prof Exp:* Instr entom, Univ Mass, 55; from asst prof to prof biol, Bemidji State Col, 58-68; chmn dept biol, 68-76, asst dean acad affairs, Sch Sci, 71-72, PROF BIOL, QUINNIPIAC COL, 68- *Concurrent Pos:* Consult, Trout Unlimited; mem, Mus Natural Hist; bd trustees, Quinnipiac Col, 84-; dir Title II Math, Sci Inst, 89- *Mem:* Am Inst Biol Sci; Am Asn Lab Animal Sci; Soc Vector Ecol; Entom Soc Am; AAAS; Sigma Xi; Am Asn Univ Profs. *Res:* Freshwater insects; medical entomology; lyme disease. *Mailing Add:* Dept Biol Sci Quinnipiac Col Hamden CT 06518

LEVINE, HERBERT JEROME, b Boston, Mass, July 22, 28; m 58; c 2. CARDIOLOGY. *Educ:* Harvard Univ, AB, 50; Johns Hopkins Univ, MD, 54; Am Bd Internal Med, dipl, 63. *Prof Exp:* Intern med, Peter Bent Brigham Hosp, 54-55, sr resident, 58-59; resident, Mass Gen Hosp, 57-58; res fel, Harvard Med Sch, 59-61; chief cardiol serv, New Eng Med Ctr Hosps, 66-88; sr instr, 61-63, from asst prof to assoc prof, 63-70, PROF MED SCH MED, TUFTS UNIV, 70- *Concurrent Pos:* Res fel cardiol, Peter Bent Brigham Hosp, 56-61; consult, Vet Admin Hosp, Mass, 66-; lectr, US Naval Hosp, Mass, 67- *Mem:* Fel Am Fedn Clin Res; fel Asn Univ Cardiol; fel Am Col Cardiol; fel Am Soc Clin Invest; Asn Am Physicians. *Res:* Clinical cardiology; physiology of congestive heart failure; muscle mechanics and energetics in the intact heart. *Mailing Add:* New Eng Med Ctr Hosp 750 Washington St Boston MA 02111

LEVINE, HERMAN SAUL, b Jeannette, Pa, Feb 11, 22; m 47; c 3. PHYSICAL CHEMISTRY, HIGH TEMPERATURE CHEMISTRY. *Educ:* Univ Pittsburgh, BS, 43; Univ Ill, PhD(phys chem), 48. *Prof Exp:* Res asst, Ill State Geol Surv, 44-46; staff mem, NY State Col Ceramics, Alfred Univ, 48-51 & USPHS, R A Taft Sanit Eng Ctr, 51-57; MEM TECH STAFF, SANDIA LABS, 57- *Mem:* Am Chem Soc. *Res:* X-ray spectroscopy. *Mailing Add:* 5501 Vista Sandia NE Albuquerque NM 87111-5782

LEVINE, HOWARD ALLEN, b St Paul, Minn, Jan 15, 42; m 74; c 2. MATHEMATICS. *Educ:* Univ Minn, Duluth, BA, 64; Cornell Univ, MA, 67, PhD(math), 69. *Prof Exp:* Asst prof math, Univ Minn, Minneapolis, 69-73; from asst prof to assoc prof, Univ RI, 73-78; assoc prof, 78-79, prof math, 80-89, DEPT CHAIR, IOWA STATE UNIV, 89- *Concurrent Pos:* Vis scientist, Battelle Advan Studies Ctr, Switz, 71 & 72; Sci Res Coun Gt Brit grant, Univ Dundee, 72; NSF res grant, 74-77 & 78-79; part-time res consult, Naval Underwater Systs Ctr, 77-78; assoc prof, Iowa State Univ, 78-79, Consiglio Nazionale delle Recerche, Italy & Math Sci Res Inst, 83. *Mem:* Am Math Soc; Sigma Xi. *Res:* Partial differential equations; numerical analysis. *Mailing Add:* Dept Math Iowa State Univ Ames IA 50011

LEVINE, HOWARD BERNARD, b Brooklyn, NY, Apr 15, 28; m 67; c 2. COMPUTER LANGUAGES, SCIENTIFIC SOFTWARE. *Educ:* Univ Ill, BS, 50; Univ Chicago, MS, 52, PhD(chem), 55. *Prof Exp:* Res fel chem, Inst Atomic Res, Iowa State Univ, 55-56; chemist, Lawrence Radiation Lab, Univ Calif, 56-62; mem tech staff, NAm Aviation Sci Ctr, 62-70; proj assoc, Theoret Chem Inst & Space Sci & Eng Ctr, Univ Wis, 70-71; prof chem eng, Va Polytech Inst & State Univ, 71-73; prog mgr chem systs, Systs, Sci & Software, 73-76; prin scientist, Jaycor, 76-82; pres, 21st Century Data, Inc, 82-89; SR SOFTWARE ENGR, TELEDYNE RYAN AERONAUT, 89- *Concurrent Pos:* Consult, Tech Adv Bd Supersonic Transport, Dept Com; mem ad hoc comt ozone & environ studies bd, Nat Acad Sci. *Mem:* Am Chem Soc; fel Am Phys Soc; Am Inst Chem Eng. *Res:* Thermodynamics; statistical mechanics; quantum mechanics; spectroscopy; atmospheric chemistry; molecular physics; applied mathematics; chemical kinetics; computer languages. *Mailing Add:* 2817 Luciernaga St Carlsbad CA 92009

LEVINE, IRA NOEL, b Brooklyn, NY, Sept 8, 37. PHYSICAL CHEMISTRY. *Educ:* Carnegie Inst Technol, BS, 58; Harvard Univ, AM, 59, PhD(chem), 63. *Prof Exp:* Res assoc chem, Univ Pa, 63-64; from instr to assoc prof, 64-77, PROF CHEM, BROOKLYN COL, 78- *Concurrent Pos:* Am Chem Soc Petrol Res Fund starter grant, 65-66. *Mem:* Am Chem Soc; Am Phys Soc. *Res:* Quantum chemistry. *Mailing Add:* Dept Chem Brooklyn Col Brooklyn NY 11210

LEVINE, J(OSEPH) S(AMUEL), b San Antonio, Tex, Sept 14, 15; m 55; c 2. PETROLEUM ENGINEERING. *Educ:* Univ Tex, BS, 36; Pa State Col, MS, 38, PhD(petrol eng), 41. *Prof Exp:* Asst & instr petrol eng, Pa State Col, 36-42; sr chemist fluid flow res, Shell Develop Co Div, Shell Oil Co, 46-60, sr exploitation engr, 60-64, staff engr, 64-65, staff res engr, 65-83; RETIRED. *Mem:* Am Inst Mining, Metall & Petrol Engrs; Soc Petrol Engrs. *Res:* Fluid flow; hydrodynamics; fluid flow through porous media; mechanism of displacement of oil by water; secondary recovery of oil. *Mailing Add:* 5614 Jackwood St Houston TX 77096-1106

LEVINE, JACK, b Philadelphia, Pa, Dec 15, 07; m 38. MATHEMATICS. *Educ:* Univ Calif, Los Angeles, AB, 29; Princeton Univ, PhD(math), 34. *Prof Exp:* Asst math, Univ Calif, Los Angeles, 29-30; instr, Princeton Univ, 30-35; from instr to assoc prof, 35-47, PROF MATH, NC STATE UNIV, 47- *Concurrent Pos:* Res analyst, US Dept War, 42-43. *Mem:* Am Math Soc; Math Asn Am. *Res:* Differential geometry; tensor analysis; combinatorial analysis; particle dynamics. *Mailing Add:* Dept Math NC State Univ Raleigh NC 27695-8205

LEVINE, JEFFREY, b Brooklyn, NY, Feb 7, 45; m 66. MATHEMATICS. *Educ:* State Univ NY Stony Brook, BS, 66; Rutgers Univ, New Brunswick, PhD(math), 70. *Prof Exp:* Asst prof math, Monmouth Col, NJ, 69-71; asst prof math, State Univ NY Col Geneseo, 71-80; MEM STAFF, MC DONNELL DOUGLAS CORP, 80- *Mem:* Am Math Soc; Math Asn Am. *Res:* Ring theory. *Mailing Add:* 761 LaFeil Dr Manchester MO 63021

LEVINE, JEROME PAUL, b New York, NY, May 4, 37; m 58; c 3. TOPOLOGY. *Educ:* Mass Inst Technol, BS, 58; Princeton Univ, PhD(math), 62. *Prof Exp:* Instr math, Mass Inst Technol, 61-63; NSF fels, 63-64; from asst prof to assoc prof math, Univ Calif, Berkeley, 64-66; assoc prof, 66-69, chmn dept, 74-76 & 88-90, PROF MATH, BRANDEIS UNIV, 69- *Concurrent Pos:* NSF postdoctoral fel, 63-64; Sloan Found fel, 66-68. *Honors & Awards:* Humboldt Prize, Ger. *Mem:* Am Math Soc. *Res:* Differential topology; knot theory. *Mailing Add:* Dept Math Brandeis Univ Waltham MA 02154

LEVINE, JERRY DAVID, b Mount Vernon, NY, June 27, 52; m 77; c 1. ENVIRONMENTAL PROTECTION. *Educ:* State Univ NY, Stony Brook, BS, 74; Polytech Inst NY, MS, 76. *Prof Exp:* Asst engr, Ebasco Ser, Inc, 76-77, assoc engr, 77-78, engr, 78-80, sr engr, 80-84, prin engr, Envirosphere Co, 84-87; NUCLEAR-ENVIRON ENGR, PLASMA PHYSICS LAB, PRINCETON UNIV, 87- *Mem:* Am Nuclear Soc. *Res:* Nuclear fusion safety studies; review of nuclear safety and environmental aspects of design and operation of Tokamak devices, including the Tokamak Fusion Test Reactor and the planned Compact Ignition Tokamak. *Mailing Add:* One Ivy Way Dayton NJ 08810-1420

LEVINE, JOEL S, b Brooklyn, NY, May 14, 42; m 68; c 1. GEOCHEMISTRY, GEOPHYSICS. *Educ:* Brooklyn Col, BS, 64; NY Univ, MS, 67; Univ Mich, MS, 73, PhD, 77. *Prof Exp:* Res scientist atmospheric sci, Goddard Inst Space Studies, 64-70, SR RES SCIENTIST, ATMOSPHERIC SCI DIV, LANGLEY RES CTR, NASA, 70- *Concurrent Pos:* Instr physics & dir astron observ, Brooklyn Col, 64-70; res scientist atmospheric sci, Geophys Sci Lab, NY Univ, 64-70; consult, Mars Aeronomy, Viking Proj NASA, 72-76, Comt Planetary Biol & Chem Evolution, Space Sci Bd, Nat Res Coun-Nat Acad Sci, 78-81 & Va Dept Ed Sci Dir, 85-; prin guest investr, Orbiting Astron Observ-Copernicus, 74-76, Int Ultraviolet Explorer, 81-83; res adv, Sch Eng, Old Dominion Univ, 77-, adj assoc prof, dept geol sci, 85-; lectr, Col William & Mary, 76- & Tidewater Ctr, Univ Va, 82-; prin investr, Global Troposheric Chem Photochem Processes, 77-, Atmospheric Chem Exp, NASA Storm Hazards Proj, 79-82 & Photochem & Geochem Early Earth, 83-, Global Biomass Burning, 87-; mem NASA Life Sci Adv Comt, 85-88, Sci Steering Comt, Origins Solar Systs Prog, 87-, Space Sci & Appln Adv Comt, 88-90, Ctr Explor Prog Scientists, 88-; ed, Photochem of Atmosphers: Earth, Other Planets & Comets, 85, Spare Opportunities for Tropospheric Chem Res, 87, Global Biomass Burning: Atmospheric, Climate & Biospheric Implications, 91. *Honors & Awards:* Halpern Award Photochem, NY Acad Sci, 82; Medal Exceptional Sci Achievement, NASA, 83. *Mem:* Am Geophys Union; AAAS; Int Soc Study Origin Life. *Res:* Origin, evolution, physics and chemistry of planetary atmospheres; atmospheric photochemistry; biogeochemical cycling; global climate change; origin & evolution of life. *Mailing Add:* Theoret Studies Br Atmospheric Sci Div NASA Langley Res Ctr Hampton VA 23665-5225

LEVINE, JON DAVID, b New York, NY, Mar 20, 45. MEDICAL SCIENCES, INTERNAL MEDICINE. *Educ:* Univ Mich, BS, 66; Yale Univ, PhD(neurobiol), 72; Univ Calif, San Francisco, MD, 78. *Prof Exp:* FEL RHEUMATOL & CLIN IMMUNOL, UNIV CALIF, SAN FRANCISCO, 81-, FEL CLIN PHARMACOL & THERAPEUT, 82- *Concurrent Pos:* Asst prof med, Univ Calif, San Francisco, 84-; Hartford Found fel. *Mem:* Am Soc Clin Invest. *Res:* Mechanisms of pain and anlagesia and application of research in this area to the diagnosis and mangement of clinical pain; pathophyisology of inflammatory joint disease; rheumatology; neurobiology; clinical pharmacology. *Mailing Add:* Dept Med & Neurosci Univ Calif Rm U-426 Box 0724 San Francisco CA 94143-0724

LEVINE, JON HOWARD, b Toronto, Ont, July 13, 41; m 64; c 3. ENDOCRINOLOGY. *Educ:* Univ Toronto, MD, 65, MSc, 69; Royal Col Physicians & Surgeons Can, FRCP(C), 71; Am Bd Internal Med, cert endocrinol, 77. *Prof Exp:* Instr, Vanderbilt Univ, 71-73; from asst prof to assoc prof endocrinol, 73-82, PROF MED, MED UNIV SC, 82- *Concurrent Pos:* Fel, Med Res Coun Can, 71-73. *Mem:* Am Fedn Clin Res; Endocrine Soc; Can Soc Endocrin & Metab. *Res:* Medical education; pituitary regulation of adrenal steroidogenesis; clinical problem solving techniques by physicians. *Mailing Add:* 1900 Patterson St Nashville TN 37203

LEVINE, JOSEPH H, b Mineral Wells, Tex, Apr 16, 26; m 50; c 1. RISK MANAGEMENT, PRODUCT ASSURANCE. *Educ:* Southern Methodist Univ, BS, 50, MS, 58. *Prof Exp:* Proj engr, Gen Dynamics Corp, 56-62; chief, Reliability Div, Johnson Space Ctr, NASA, 62-86; OWNER, ENG CONSULT SERV, 86- *Concurrent Pos:* Mem, Nuclear Regulation Comn Invest Group, 86. *Mem:* Assoc fel Am Inst Aeronaut & Astronaut; Nat Asn Consults. *Res:* Manufacturing processes relative to risk; process failure modes and effects analysis technique applied to solid propulsion improvement program and advanced solid propulsion program. *Mailing Add:* 3722 Montvale Houston TX 77059

LEVINE, JULES DAVID, b New York, NY, June 24, 37; m 66; c 2. MATERIALS SCIENCE ENGINEERING. *Educ:* Columbia Univ, BS, 59; Mass Inst Technol, PhD(physics, nuclear eng), 63. *Prof Exp:* Mem tech staff surface & mat res, David Sarnoff Res Ctr, RCA Labs, 63-73, proj mgr flat panel TV, 73-76, proj mgr cathode res & develop, 76-79; BR MGR SOLAR CELL DEVELOP, TEXAS INSTRUMENTS, 79- *Concurrent Pos:* Vis lectr elec eng, Princeton Univ, 71-72 & 74-75. *Mem:* Fel Inst Elec & Electronics Engrs; sr mem Am Vacuum Soc. *Res:* Physical processes and engineering of surfaces; thin films; semiconductors; electron emitters; display and power tubes; thermionic energy conversion; high voltage phenomena; electron beams; varistors; vacuum science and technology; fabricates novel solar cells made from miniature single crystal silicon spheres mounted in a planar matrix. *Mailing Add:* Texas Instruments MS 147 PO Box 225936 Dallas TX 75265

LEVINE, JULES IVAN, b Brooklyn, NY, Apr 17, 38; m 62; c 2. HEALTH SCIENCES, MEDICAL ADMINISTRATION. *Educ:* Univ Va, BEE, 60, PhD(biomed eng), 72; Johns Hopkins Univ, MS, 68. *Prof Exp:* Sr engr aerospace electronics, Westinghouse Elec Corp, 63-68; from asst prof to assoc prof pediat, 72-86, asst vpres health affairs, 78-86, PROF HEALTH AFFAIRS & ASSOC VPRES HEALTH SCI, UNIV VA, 86- *Concurrent Pos:* Consult, Health Resources Admin, Hyattsville, Md, Dept Health, Educ & Welfare, 74-; assoc dean, Sch Med, Univ Va, 74-78. *Mem:* Soc Col & Univ Planning; Am Asn Med Clin. *Res:* Planning and evaluation of health resources and the health care delivery system. *Mailing Add:* Univ Va Med Ctr Box 492 Charlottesville VA 22908

LEVINE, LAURENCE, b New York, NY, July 10, 26; m 51; c 3. CELL BIOLOGY. *Educ:* NY Univ, BA, 49; Univ Wis, MA, 52, PhD(zool, biochem), 55. *Prof Exp:* Asst parasitol, Univ Wis, 50-55; from instr to prof biol, Wayne State Univ, 55-91, coordr freshman biol, 68-86, undergrad officer, 68-91, EMER PROF BIOL, WAYNE STATE UNIV, 91- *Mem:* AAAS; Sigma Xi. *Res:* Cell contractility; chromosome motion; chromosome structure and function; mechanism of meiosis. *Mailing Add:* Dept Biol Wayne State Univ Detroit MI 48202

LEVINE, LAWRENCE, b Hartford, Conn, July 18, 24. IMMUNOCHEMISTRY. *Educ:* Univ Conn, BA, 48; Univ Mich, MS, 50; Johns Hopkins Univ, DSc(microbiol), 53. *Prof Exp:* Instr microbiol, Johns Hopkins Univ, 53-54; res scientist, Div Labs & Res, State Dept Health, NY, 54-57; from asst prof to assoc prof, 57-70, PROF BIOCHEM, BRANDEIS UNIV, 70- *Res:* Blood proteins and their immunol properties. *Mailing Add:* Dept Biochem Brandeis Univ Waltham MA 02254

LEVINE, LAWRENCE ELLIOTT, b Chelsea, Mass, June 23, 41; m 65; c 7. APPLIED MATHEMATICS. *Educ:* Rensselaer Polytech Inst, BS, 63; Univ Md, PhD(appl math), 68, Stevens Inst Technol, MEng, 77. *Prof Exp:* Asst prof, 68-72, assoc prof, 68-77, PROF MATH, STEVENS INST TECHNOL, 77- *Mem:* Am Math Asn; Asn Develop Comput Based Instrnl Systs. *Res:* Fluid dynamics; partial differential equations; perturbation methods; CAI in mathematics. *Mailing Add:* Dept Pure & Appl Math Stevens Inst Technol Hoboken NJ 07030

LEVINE, LEO MEYER, b Brooklyn, NY, May 26, 22; m 49; c 3. MATHEMATICS. *Educ:* City Col New York, BS, 42; NY Univ, PhD(math), 60. *Prof Exp:* Asst physicist, Signal Corps Labs, Eatontown, NJ, 42-43; sr physicist, Mat Lab, NY Naval Shipyard, 47-59; from asst res scientist to assoc res scientist, Courant Inst Math Sci, NY Univ, 59-63, from asst prof to assoc prof, 63-70; assoc prof, 70-81, PROF MATH, QUEENSBOROUGH COMMUNITY COL, 81- *Concurrent Pos:* Consult, Radio Corp Am, 61-62. *Mem:* Am Math Soc; Math Asn Am. *Res:* Applied mathematics; ordinary and partial differential equations; acoustics; electromagnetic theory. *Mailing Add:* 138-21 77th Ave Flushing NY 11367

LEVINE, LEON, b Brooklyn, NY, Jan 6, 34; m 66; c 2. POLYMER CHEMISTRY. *Educ:* Brooklyn Col, BS, 56; Polytech Inst New York, PhD(org chem), 63. *Prof Exp:* Res & develop chemist polymers, Foster Grant Co, 63-66, Gaylord assoc, 66-67; res & develop chemist polymers, Sun Chem Co, 67-68; res & develop chemist dent mat, Warner Lambert Co, 68-72; res & develop chemist polymers, Nat Patent Develop Corp, 72-76; res & develop chemist polymers, Loctite Corp, 76-80; sr chemist, Coats & Levine, Inc, 81-82; sr chemist, Richardson Polymer Corp, 82-89; CONSULT, 89- *Concurrent Pos:* Consult, L & E Assocs, 82. *Mem:* Am Chem Soc. *Res:* Adhesives; photopolymerizations; do it yourself products; dental materials; hydrophilic polymers; suspension polymerization. *Mailing Add:* 109 S Main St No B1 West Hartford CT 06107-2526

LEVINE, LEONARD, b Atlantic City, NJ, Jan 28, 29; m 52; c 2. NEUROPHYSIOLOGY. *Educ:* Rutgers Univ, BS, 50; Columbia Univ, PhD(physiol), 59. *Prof Exp:* Instr physiol, Columbia Univ, 57-60; from asst prof to assoc prof, Univ Va, 61-66; PROF PHYSIOL, PAC UNIV, 66-, PROF PHARMACOL, 76- *Concurrent Pos:* USPHS fel physiol, Columbia Univ, 59-60; fel biophys, Univ Col, Univ London, 60-61; USPHS res grants, 62-65, 67-68 & 70-72; res grant proposal evaluator, Regulatory Biol Prog, NSF, 78- *Mem:* AAAS; Am Physiol Soc; Biophys Soc; Am Soc Zool; Am Soc Pharmacol & Exp Therapeut; Sigma Xi. *Res:* Electrophysiology and pharmacology of ocular tissues; trophic interrelations between nerve and muscle tissues. *Mailing Add:* Col Optom Pac Univ 2043 Col Way Forest Grove OR 97116

LEVINE, LEONARD P, b Newark, NJ, July 24, 32; m 54; c 1. HUMAN-COMPUTER INTERFACING. *Educ:* Queens Col, NY, BS, 54; Syracuse Univ, MS, 56, PhD(physics), 60. *Prof Exp:* Engr, Sperry-Gyroscope Co, 59-60; sr scientist, Honeywell Res Ctr, 60-64, prin res scientist, 64-66; PROF ELEC ENG & COMPUT SCI, UNIV WIS-MILWAUKEE, 66- *Mem:* Asn Comput Mach. *Res:* Human and machine interfacing; system to system interfacing; small machine system design; computer teaching techniques. *Mailing Add:* Dept Elec Eng & Comput Sci Univ Wis Milwaukee WI 53201

LEVINE, LOUIS, b New York, NY, May 14, 21. GENETICS, ANIMAL BEHAVIOR. *Educ:* City Col New York, BS, 42, MS, 47; Columbia Univ, MA, 49, PhD(zool), 55. *Prof Exp:* From instr to assoc prof, 55-67, PROF BIOL, CITY COL NEW YORK, 68- *Concurrent Pos:* NSF grants, 60-; AEC grant, 63- *Mem:* Fel AAAS; Animal Behav Soc; Am Genetic Asn; Genetics Soc Am; Am Soc Naturalists. *Res:* Genetics of animal behavior and population genetics. *Mailing Add:* Dept Biol City Col NY New York NY 10031

LEVINE, MAITA FAYE, b Cincinnati, Ohio, Oct 17, 30. MATHEMATICS. *Educ:* Univ Cincinnati, BA, 52, BE, 53, MAT, 66; Ohio State Univ, PhD(math educ), 70. *Prof Exp:* Teacher, High Sch, Ohio, 53-63; instr, 63-70, from asst prof to assoc prof, 70-85, PROF MATH, UNIV CINCINNATI, 85- *Concurrent Pos:* NSF res grant, 74, 85. *Mem:* Math Asn Am; Am Educ Res Asn; Nat Coun Teachers Math; Asn Women Math; Am Asn Univ Prof (vpres, 86-88). *Res:* Relationship between mathematical competence and mathematical confidence; mathematical modeling; reasons why qualified women do not pursue mathematical careers; applications of computers and graphics calculators in the undergraduate curriculum; applications of mathematics to politics. *Mailing Add:* 1106 Lois Dr Cincinnati OH 45237

LEVINE, MARK DAVID, b Cleveland, Ohio, May 26, 44; m. ENERGY ANALYSIS, CHEMISTRY. *Educ:* Princeton Univ, BA, 66; Univ Calif, Berkeley, PhD(chem), 75. *Prof Exp:* Staff scientist, Ford Found Energy Policy Proj, 72-73; sr policy analyst, Stanford Res Inst, 74-78; PROG LEADER, LAWRENCE BERKELEY LAB, 78- *Concurrent Pos:* Fulbright scholar; Woodrow Wilson Found scholar; mem, bd dirs, Ctr Clean Air Policy, adv bd, Int Inst Energy Conserv. *Mem:* Int Asn Energy Economists; Am Soc Heating, Refrig & Air Conditioning Engrs. *Res:* Comprehensive analysis and policy studies of the People's Republic of China; analysis of energy efficiency standards and guidelines for commericial buildings with application in developing countries; development of techniques to improve energy demand forecasting in the US, particularly for the building sector; analysis of energy issues related to global climate change. *Mailing Add:* Lawrence Berkeley Lab Bldg 90 Rm 3124 Berkeley CA 94720

LEVINE, MARTIN, b Brooklyn, NY, Oct 27, 25; m 60; c 2. ENGINEERING, EDUCATION. *Educ:* City Col New York, BSEE, 50; Univ Pittsburgh, MLitt, 50, MEd, 60; Univ Mich, PhD(higher educ), 69. *Prof Exp:* Proj engr, Air Res & Develop, 50-53; mem fac, Pa State Univ, 53-63 & Harrisburg Area Community Col, 65-68; PROF ELEC TECHNOL, VA WEST COMMUNITY COL, 68- *Mem:* Inst Elec & Electronics Engrs; Am Soc Eng Educ. *Res:* Student-work interface. *Mailing Add:* Dept Eng Technol Va West Community Col PO Box 14065 Roanoke VA 24038

LEVINE, MARTIN DAVID, b Montreal, Que, Mar 30, 38; m 61; c 2. COMPUTER VISION. *Educ:* McGill Univ, BEng, 60, MEng, 63; Univ London, DIC & PhD(control theory), 65. *Prof Exp:* From asst prof to assoc prof, 65-77, PROF ELEC ENG, McGILL UNIV, 77-, DIR, CTR INTELLIGENT MACH. *Concurrent Pos:* Vis prof comput sci, Hebrew Univ, Jerusalem, Israel, 79-80; Am Soc Eng Educ-Ford Found fel, 72; assoc ed, Comput Vision, Graphics & Image Processing; tech staff mem, Image Processing & Jet Propulsion Labs, Pasadena, Calif, 72-73; assoc ed, Trans Pattern Anal & Mach Intel, Inst Elec & Electronics Engrs. *Mem:* Fel Inst Elec & Electronics Engrs; Pattern Recognition Soc; Int Asn Pattern Recognition. *Res:* Computer vision; biomedical image processing; artificial intelligence; intelligent robotics. *Mailing Add:* Dept Elec Eng McGill Univ 3480 University St Montreal PQ H3A 2A7 Can

LEVINE, MELVIN MORDECAI, b Richmond, Va, Nov 20, 25; m 50; c 3. NUCLEAR ENGINEERING & REACTOR PHYSICS. *Educ:* Mass Inst Technol, BS, 46; Univ Va, PhD(physics), 55. *Prof Exp:* Instr physics, Pa State Univ, 46-48; physicist, Babcock & Wilcox Co, 55-59; physicist, Brookhaven Nat Lab, 59-88; RETIRED. *Mem:* Fel Am Nuclear Soc. *Res:* Nuclear reactor safety research and applications, including neutronics and thermal-hydraulic phenomena; computational methods for reactor physics and engineering problems. *Mailing Add:* Two Meadow Lane Saranac Lake NY 12983

LEVINE, MICHAEL S, b Brooklyn, NY, Sept 22, 44; c 1. ANIMAL PHYSIOLOGY. *Educ:* Queens Col, BA, 66; Univ Rochester, PhD(physiol psychol), 70. *Prof Exp:* Fel neurophysiol, Brain Res Inst, 70-72, asst res neurophysiologist, 72-76, lectr psychol, 75-76, from asst prof to assoc prof psychiat, 76-85, PROF PSYCHIAT, UNIV CALIF, LOS ANGELES, 85- *Concurrent Pos:* Consult neurophysiologist, Hereditary dis Found, 75. *Mem:* Soc Neurosci; Am Psychol Asn; Am Asn Anatomists; Sigma Xi. *Res:* Neurophysiology and neuroanatomy of basal ganglia in mature, developing and aging animals; role of basal ganglia in regulation of behavior; development and prediction of learning ability in developing animals. *Mailing Add:* Ment Retardation Res Ctr Dept Psychiat Univ Calif 760 Westwood Plaza Los Angeles CA 90024

LEVINE, MICHAEL STEVEN, b Los Angeles, Calif, Mar 5, 55; m 85; c 1. DEVELOPMENTAL BIOLOGY. *Educ:* Univ Calif, Berkeley, BA, 76; Yale Univ, PhD(molecular biol), 81. *Prof Exp:* Fel, Univ Basel, Switz, 82-83 & Univ Calif, Berkeley, 83-84; from asst prof to assoc prof, 84-88, PROF BIOL, COLUMBIA UNIV, 88- *Concurrent Pos:* Sloan fel, 85. *Res:* The control of gene expression during early embryonic development; DNA binding proteins and regulatory switch genes; the developmental regulation of eukaryotic promoters by crudely localized positional cues and morphogen gradients. *Mailing Add:* Bonner Hall Rm 2425 Univ Calif San Diego 9500 Gilman Dr La Jolla CA 92093

LEVINE, MICHAEL W, b New York, NY, Mar 10, 43; m 69; c 2. VISUAL SCIENCE, SENSORY PROCESSES. *Educ:* Mass Inst Technol, BS, 65, MS, 67; Rockefeller Univ, PhD(biophysics), 72. *Prof Exp:* Res asst mech eng, Mass Inst Technol, 65-67; proj engr, Lion Res Corp, Newton, Mass, 67; grad fel biophysics, Rockefeller Univ, 67-72; from asst prof to assoc prof, 72-85, PROF PSYCHOL, UNIV ILL, CHICAGO, 85- *Concurrent Pos:* Assoc prof bioeng, Univ Ill Chicago, 81-84; vis scholar, Northwestern Univ, 81; vis prof, Univ Sydney, Australia, 87-88. *Mem:* AAAS; Asn Res Vision & Ophthal; Sigma Xi; Soc Neurosci. *Res:* Visual system; firing patterns of retinal ganglion cells; statistics of neural discharges; sensation and perception; author of various publications. *Mailing Add:* Dept Psychol M/C 285 Univ Ill Chicago Box 4348 Chicago IL 60680

LEVINE, MICHEAL JOSEPH, b Oak Park, Ill, Dec 1, 40; m 68; c 1. PHYSICS, DATA ACQUISITION. *Educ:* Yale Univ, BS, 62, MS, 64, PhD(physics), 68. *Prof Exp:* SR PHYSICIST NUCLEAR PHYSICS, BROOKHAVEN NAT LAB, 68- *Concurrent Pos:* Consult, High Voltage Eng Corp, 72-75; guest physicist, Max Planck Inst Nuclear Physics, Heidelberg, Ger, 75, Ctr d'Etudes Nucleaires, Saclay, France, 80-81. *Mem:* Am Phys Soc; Inst Elec & Electronics Engrs. *Res:* Study of nuclear reactions induced by relativistic heavy ions; development of magnetic spectrometers and associated focal plane detectors; development of data acquisition architectures. *Mailing Add:* Brookhaven Nat Lab Bldg 510A Upton NY 11973

LEVINE, MYRON, b Brooklyn, NY, July 28, 26; m 50; c 2. GENETICS, VIROLOGY. *Educ:* Brooklyn Col, BA, 47; Ind Univ, PhD(zool), 52. *Prof Exp:* Res assoc microbiol, Univ Ill, 54-56; asst to assoc biologist, Brookhaven Nat Lab, 56-61; assoc prof, 61-66, PROF HUMAN GENETICS, SCH MED, UNIV MICH, ANN ARBOR, 66- *Concurrent Pos:* Am Cancer Soc fel, Johns Hopkins Univ, 53-54; Commonwealth Fund fel, Univ Geneva, 66-67; ed, J Virol, 72-76; vis scientist, Imp Cancer Res Fund, London, 73-74 & Cambridge Univ, 82; chmn grad prog cell & molecular biol in health sci, Univ Mich, 74-90; mem & chmn genetic basis of dis rev comt, Nat Inst Gen Med Sci, NIH, 75-79; sr fel, Soc Fels, Univ Mich, 82-85; Clare Hall life fel, Cambridge Univ, 82-; chmn, educ comt, Genetics Soc Am; vis scientist, Weizmann Inst Sci, Israel, 83 & Inst Sci Res Cancer, France; vis prof, biol dept, Harbin Normal Univ, Harbin, People's Repub China. *Mem:* Genetics Soc Am; Am Soc Microbiol; Am Soc Virol. *Res:* Genetics and regulation of gene expression of animal viruses; herpesvirus genetics, latency and biology. *Mailing Add:* Dept Human Genetics Univ Mich Sch Med Ann Arbor MI 48109-0618

LEVINE, NATHAN, b Brooklyn, NY, Aug 7, 30; m 53; c 2. COMMUNICATIONS ENGINEERING. *Educ:* Mass Inst Technol, BS, 52; Univ Ill, MS, 54, PhD(physics), 57. *Prof Exp:* Mem tech staff, Bell Tel Labs, 57-61, supvr re-entry physics, 61-64, dept head, 64-83, dir anti-missile systs res, 68-71, dir toll transmission eng ctr, 71-83, dir educ ctr, 83-86, dir transmission facil planning, 86-89, DIR, NETWORK SERV PERFORMANCE, BELL TEL LABS, 89- *Mem:* Inst Elec & Electronics Engrs. *Res:* System studies and design of integrated network planning tools for the evolution of communication networks; network services performance evaluation. *Mailing Add:* Network Serv Performance Bell Telephone Labs Holmdel NJ 07733

LEVINE, NORMAN DION, b Boston, Mass, Nov 30, 12; wid. PARASITOLOGY, PROTOZOOLOGY & HUMAN ECOLOGY. *Educ:* Iowa State Col, BS, 33; Univ Calif, PhD(zool), 37; Am Bd Med Microbiol, cert pub health & med lab parasitol. *Hon Degrees:* DSc, Univ Ill, Urbana, 89. *Prof Exp:* Asst zool, Univ Calif, 33-37; asst animal parasitologist, Univ Ill, 37-41, assoc animal path, 41-42, from asst prof to assoc prof vet parasitol, 46-53, ass to dean col vet med, 47-57, prof, 53-83, EMER PROF VET PARASITOL & VET RES, COL VET MED, UNIV ILL, URBANA, 83- *Concurrent Pos:* Mem, Nat Res Coun, 56-62; mem bd gov, Am Bd Microbiol, 59-64; sr mem, Ctr Zoonoses Res, 60-74, prof zool, 65-77, dir, Ctr Human Ecol, Univ Ill, Urbana, 68-74; vis prof, Univ Hawaii, 62, Santa Catalina Marine Biol Lab, 72 & J Hopkins Marine Sta, 80; mem comt health sci achievement award prog, NIH, 65-66, mem trop med & parasitol study sect, 65-69, chmn, 66-69, mem animal resources adv comt, 71-75; ed, J Protozool, 65-71. *Mem:* AAAS; Am Soc Parasitol; hon mem Soc Protozool (secy, 52-58, vpres, 58-59, pres, 59-60, actg secy, 60-62); hon mem Micros Soc Am (pres, 69-70); fel Am Acad Microbiol; Sigma Xi; hon mem, World Asn Advan Vet Parasitol. *Res:* Protozoan and roundworm parasites of domestic and wild animals; malaria and other insect-borne diseases. *Mailing Add:* Col Vet Med Univ Ill Urbana IL 61801

LEVINE, O ROBERT, EXPERIMENTAL BIOLOGY. *Prof Exp:* PROF PEDIAT, UNIV MED & DENT NJ, 72- *Mailing Add:* NJ Med Sch Univ Med & Dent NJ 185 S Orange Ave Rm F-576 Newark NJ 07103

LEVINE, OSCAR, b Brooklyn, NY, Feb 6, 23; m 48; c 2. PHYSICAL CHEMISTRY. *Educ:* City Col New York, BS, 43; Columbia Univ, AM, 48; Georgetown Univ, PhD(chem), 57. *Prof Exp:* Nat Adv Comt Aeronaut, Ohio, 48-52; chemist, USN Res Lab, 52-58; chemist, chem & mat res, Gillette Safety Razor Co, Boston, 58-85; RETIRED. *Concurrent Pos:* Vpres res & develop lubricant coatings, Ro-59, Inc, 85- *Mem:* Am Chem Soc. *Res:* Chemistry and physics of solid and liquid surfaces and interfaces; lubrication; adhesion. *Mailing Add:* 43 Connolly St Randolph MA 02368

LEVINE, PAUL HERSH, b New York, NY, Sept 27, 35; m 63. THEORETICAL PHYSICS, APPLIED PHYSICS. *Educ:* Mass Inst Technol, BS, 56; Calif Inst Technol, MS, 57, PhD(theoret physics), 63. *Prof Exp:* Sr scientist, Jet Propulsion Lab, Calif Inst Technol, 63-64; chief scientist, Astrophys Res Corp, 64-72; chief scientist, Megatek Corp, 72-82; CONSULT PHYSICIST, 82- *Mem:* Am Phys Soc. *Res:* Ionospheric physics; over-the-horizon radar; quantum many-body problem; exploding wire phenomena; electron field emission; radiative transport; electromagnetic propagation; navigation and communication systems analysis; minicomputer applications; electroencephalography; psychobiology of consciousness. *Mailing Add:* PO Box 8827 Incline Village NV 89450

LEVINE, PAUL HOWARD, b New York, NY, Sept 11, 37; m 60; c 3. VIRAL ONCOLOGY, INTERNAL MEDICINE. *Educ:* Cornell Univ, BA, 59; Univ Rochester, MD, 63. *Prof Exp:* Intern internal med, Strong Mem Hosp, 63-64; resident fel oncol, Roswell Park Mem Inst, 64-66; resident internal med, Univ Colo, 66-68; RES INVESTR VIRAL ONCOL, NAT CANCER INST, 68- *Concurrent Pos:* Co-chmn immunol group, Nat Cancer Inst, 71-72, chmn immunol-epidemiol segment, Virus Cancer Prog, 72-75, head clin studies sect, Viral Leukemia & Lymphoma Br, 74-75, chmn clin adv group, Div Cancer Cause & Prev, 76-81, head clin studies sect, Lab Viral Carcinogenesis, 78-82, sr investr, Epidemiol & Biostatists, 82-; clin asst prof med, George Washington Univ, Med Ctr, 78- *Mem:* Am Asn Cancer Res; Am Col Physicians; Am Col Epidemiol; AAAS. *Res:* Epidemiology of oncogenic viruses, particularly Epstein-Barr virus and HTLV-I; viral immunol, application of assays to cancer etiology, diagnosis, treatment. *Mailing Add:* Epidemiol & Biostatists EPN 434 Nat Cancer Inst Bethesda MD 20892

LEVINE, PHILLIP J, b Providence, RI, Jan 7, 34; m 55; c 2. PHARMACY. *Educ:* Univ RI, BS, 55; Univ Md, MS, 57, PhD(pharm), 63. *Prof Exp:* Instr pharm, Sch Pharm, Univ Md, 57-63; from asst prof to assoc prof, 63-70, PROF PHARM, COL PHARM, DRAKE UNIV, 70-, COORDR CONTINUING EDUC PROG PHARM, 77- *Concurrent Pos:* Consult, Dr Salsbury's Labs, Charles City, Iowa, 65-70; dir, Coop IV Additive Proj, 67-69; chmn, Mayor's Task Force on Drugs, Des Moines, Iowa, 69-70; consult, Gov, State of Iowa, 70-72. *Mem:* Am Pharmaceut Asn. *Res:* Development of topical anesthetic suspensions to test their applicability to long duration of anesthesia in dental patients; product development in area of suspension and formulations. *Mailing Add:* Dept Pharm Drake Univ 25th St University Des Moines IA 50311

LEVINE, RACHMIEL, b Poland, Aug 26, 10; nat US; m 43; c 2. ENDOCRINOLOGY. *Educ:* McGill Univ, BA, 32, MD, 36. *Hon Degrees:* MD, Univ Ulm, 69; ScD, Northwestern, 85, McGill, 87. *Prof Exp:* Asst dir dept metab & endocrine res, Michael Reese Hosp, 39-42, dir, 42-58, chmn dept med & dir med educ, 52-60; prof & chmn dept, NY Med Col, 60-70; med dir, 70-78, EMER MED DIR, CITY OF HOPE MED CTR, 82- *Concurrent Pos:* Williams fel, Michael Reese Hosp, 36-37, res fel, 37-39; Endocrine Soc Upjohn scholar, 57; Guggenheim Found fel, 71-72; consult, NSF, 56-59 & 70-; pres, Int Fedn Diabetes, 67-70; mem bd dirs, Found Fund Psychiat Res. *Honors & Awards:* Thompson Award, Am Geriat Soc, 71; Gairdner Found Award, 71. *Mem:* Nat Acad Sci; Am Physiol Soc; Soc Exp Biol & Med; Endocrine Soc; Am Diabetes Asn (pres, 64-65); fel Am Acad Arts & Sci. *Res:* Hormonal control of metabolism; mode of action of insulin; diabetes. *Mailing Add:* City Hope Med Ctr Duarte CA 91010

LEVINE, RANDOLPH HERBERT, b Denver, Colo, Nov 20, 46; m 70; c 2. ASTROPHYSICS, SOLAR PHYSICS. *Educ:* Univ Calif, Berkeley, AB, 68; Harvard Univ, AM, 69, PhD(physics), 72. *Prof Exp:* Vis scientist solar physics, High Altitude Observ, Nat Ctr Atmospheric Res, Boulder, Colo, 72-74; Res fel solar physics, Ctr Astrophys, Harvard Col Observ, 74-75, res assoc solar physics, 75-81, lectr astron, 77-81; sr scientist & dir comput, atmospheric & environ res, 81-82, mgr software eng, 82-85, MKT EXEC, DIGITAL EQUIP CORP, 85- *Mem:* Am Astron Soc; Am Geophys Union; Int Astron Union; Am Phys Soc; Inst Elec & Electronics Engrs. *Res:* Scientific computing; design and development of products. *Mailing Add:* 50 Carver Rd Newton Highlands MA 02161

LEVINE, RHEA JOY COTTLER, b Brooklyn, NY, Nov 26, 39; m 60; c 3. CELL BIOLOGY, CYTOCHEMISTRY. *Educ:* Smith Col, AB, 60; NY Univ, MS, 63, PhD(biol), 66. *Prof Exp:* Lab instr biol, Sch Com, Acct & Finance, Wash Sq Col, NY Univ, 63-64; res assoc neuropath, Sch Med, Univ Pa, 68-69; from asst prof to assoc prof, 69-80, PROF ANAT, MED COL PA, 80- *Concurrent Pos:* A H Robins Co fel biochem res, Manhattan State Hosp, Ward's Island, New York, 66; USPHS fel, Sch Med, Yale Univ, 66-68; Nat Heart & Lung Inst grant, Pa Muscle Inst, 73-; Nat Inst Neurol Commun Dis & Stroke career develop award, 74-79; Nat Inst Gen Med Sci res grant, 75-81; NSF grants, 79-80, 80-81 & 85-86; mem, Cardiovasc & Pulmonary Study Sect, Div Res Grants, NIH, 80-84; co-ed, Basic Biol Muscle; trustee, Stockton State Col, Pomona, NJ, 83-; Nat Inst Arthritis & Metab Dis res grant, 84-87; Biol Instrumentation grant, 87; reviewer, J Cell Biol, Am J Physiol Sci & Biophys J, 75-; reviewer grants, NIH, NSF & Vet Admin, 75- *Mem:* AAAS; Histochem Soc; Am Asn Anat; NY Acad Sci; Soc Gen Physiol; Biophys Soc; Am Soc Cell Biol; Sigma Xi. *Res:* Ultrastructure; muscle structure and function; comparative aspects of immunohistochemistry and cytochemistry. *Mailing Add:* Dept Anat/EPPI Div Med Col Pa 3200 Henry Ave Philadelphia PA 19129

LEVINE, RICHARD JOSEPH, b New York, NY, Nov 12, 39; m 69; c 2. OCCUPATIONAL MEDICINE. *Educ:* Princeton Univ, AB, 60; Calif Inst Technol, MS, 64, St Louis Univ, MD, 76; Harvard Univ MPH, 76. *Prof Exp:* Intern med, Grady Mem Hosp, 71-72; epidemiologist, Ctr Dis Control, Epidemic Intell serv, 72-75; sr med scientist, Ctr Occup & Environ Health, Stanford Res Inst, 76-77; CHIEF EPIDEMIOL, CHEM INDUST INST TOXICOL, 77- *Concurrent Pos:* Asst state epidemiologist, Ala State Health Dept, 72-73; epidemiologist, Cholera Res Lab, Dacca, Bangladesh, 73-75; partic, Working Group Asbestos, Int Agency Cancer Res, 77; adj asst prof, Dept Family Commun Med, Div Occup Med, Duke Univ, 78-, assoc prof, 83-; adj assoc prof, Dept Epidemiol, Univ NC Sch Pub Health, 84- *Mem:* Soc Epidemiol Res; Am Occup Med Asn; fel Am Col Occup Med. *Res:* Epidemiology of cholera and mass hysteria; effects of occupation on male reproduction. *Mailing Add:* 1509 Pinecrest Rd Six Davis Dr Durham NC 27705

LEVINE, RICHARD S, b Pittsburgh, Pa, Jan 14, 47; m 69; c 1. CORROSION CONTROL & WATER TREATMENT, ENVIRONMENTAL TESTING. *Educ:* Carnegie-Mellon Univ, BS, 68; Univ Ill, MS, 71. *Prof Exp:* Chemist, Univ Ill, 71-73; PRES, INDUST CORROSION MGT, INC, 73- *Mem:* Am Chem Soc; Am Soc Testing & Mat; Am Water Works Asn; Asn Off Anal Chemists. *Res:* Corrosion control and water treatment in central air conditioning and heating systems; environmental testing; author of numerous publications. *Mailing Add:* Indust Corrosion Mgt Inc 1152 Rte 10 Randolph NJ 07869

LEVINE, ROBERT, b Boston, Mass, July 30, 19; m 50; c 3. ORGANIC CHEMISTRY. *Educ:* Dartmouth Col, BA, 40, MA, 42; Duke Univ, PhD(org chem), 45. *Prof Exp:* Asst, Dartmouth Col, 40-42; asst, Duke Univ, 42-45; chemist, Mathieson Chem Corp, NY, 45-46; from instr to assoc prof, 46-59, PROF CHEM, UNIV PITTSBURGH, 59- *Concurrent Pos:* Consult, Monsanto Co, 52-62, Schering Corp, 59-63, Reilly Tar & Chem Corp, 64-66, FMC Corp, 65-67, Columbia Org Chem Co, 70-73, Pressure Chem Co, 70-75, Fike Chem Inc, 71-74 & Mallinckrodt Chem Works, 74-75. *Mem:* Am Chem Soc; Int Asn Heterocyclic Chem; NY Acad Sci; Israel Chem Soc; Sigma Xi. *Res:* Heterocyclic nitrogen chemistry, including pyridine, pyrazine, pyrimidine and triazine; synthesis of organic fluorine compounds; chemistry of organometallic compounds; synthesis of potential medicinals. *Mailing Add:* 121 Virginia Rd Pittsburgh PA 15237

LEVINE, ROBERT, b New York, NY, Nov 10, 26; m 54; c 2. PEDIATRIC CARDIOLOGY. *Educ:* City Col New York, BS, 48; Western Reserve Univ, MD, 54. *Prof Exp:* From intern to resident pediat, State Univ NY Upstate Med Ctr, 54-57; from instr to assoc prof, Col Physicians & Surgeons, Columbia Univ, 62-72; PROF PEDIAT & DIR PEDIAT CARDIOL, NJ MED SCH, COL MED & DENT NJ, 72- *Concurrent Pos:* NIH trainee pediat cardiol, Col physicians & Surgeons, Columbia Univ, 59-61 & NIH fel cardiorespiratory physiol, 61-62; NY City Health Res Coun career scientist award, 62-72; John Polachek Found fel, 68-69; prin investr, NIH Grad Training Prog Pediat Cardiol, Columbia Univ, 70-72, responsible investr, Nat Heart & Lung Inst-SCOR, Col Physicians & Surgeons, 71-72. *Mem:* Am Acad Pediat; Am Pediat Soc; Am Physiol Soc. *Res:* Cardiorespiratory physiology. *Mailing Add:* Univ Med & Dent NJ Med Sch 185 S Orange Ave Newark NJ 07130-2714

LEVINE, ROBERT ALAN, b New York, NY, June 12, 32; m 56; c 3. MEDICINE, PHARMACOLOGY. *Educ:* Cornell Univ, AB, 54, MD, 58; Am Bd Gastroenterol, cert. *Prof Exp:* Intern med, NY Hosp-Cornell Med Ctr, 58-59, asst resident, 59-60; clin fel med, Liver Study Unit, Sch Med, Yale Univ, 61-62, res fel, 62-63; from asst chief to chief metab div, Army Med Res & Nutrit Lab, Fitzsimons Gen Hosp, 63-65; chief div gastroenterol, Brooklyn-Cumberland Med Ctr, 65-71, assoc prof med, 69-71; PROF MED, STATE UNIV NY UPSTATE MED CTR, 71-, CHIEF DIV GASTROENTEROL, STATE UNIV HOSP, 71- *Concurrent Pos:* Clin fel gastroenterol, NY Hosp-Cornell Med Ctr, 60-61. *Mem:* Am Soc Pharmacol & Exp Therapeut; Am Fedn Clin Res; Am Gastroenterol Asn; Am Asn Study Liver Dis. *Res:* Basic and clinical research in gastroenterology, metabolism and pharmacology; cyclic adenosine 3', 5'-monophosphate in vivo and in vitro; isolated perfused rat liver; chronic hepatitis; hormone regulation of gastrointestinal function. *Mailing Add:* Dept Med State Univ NY Health Sci Ctr Rm 6416 750 E Adams St Syracuse NY 13210

LEVINE, ROBERT JOHN, b New York, NY, Dec 29, 34; m 87; c 2. INTERNAL MEDICINE, MEDICAL ETHICS. *Educ:* George Washington Univ, MD, 58; Am Bd Internal Med, dipl, 65. *Prof Exp:* Intern internal med, Peter Bent Brigham Hosp, Boston, Mass, 58-59, asst resident, 59-60; clin assoc clin pharmacol, Nat Heart Inst, 60-62; resident internal med, Vet Admin Hosp, West Haven, Conn, 62-63; investr clin pharmacol, Nat Heart Inst, 63-64; from instr to assoc prof internal med & pharmacol, Yale Univ, 64-73, chief sect clin pharmacol, 66-74, dir physician's assoc prog, 73-75, PROF INTERNAL MED & LECTR PHARMACOL, SCH MED, YALE UNIV, 73- *Concurrent Pos:* Clin asst, Yale-New Haven Hosp, 64-65, asst attend physician, 65-68, attend physician, 68-; clin investr, Vet Admin Hosp, West Haven, Conn, 64-66, attend physician, 66-; mem myocardial infarction comt, Nat Heart & Lung Inst, 69-72; ed, Clin Res, Am Fedn Clin Res, 71-76; consult, Nat Comn Protection Human Subj Biomed & Behav Res, 74-78; mem lipid metab adv comt, Nat Heart, Lung & Blood Inst, 77-79; ed, IRB: Review Human Subjects Res, 79-; vchmn, Comn Fed Drug Approval Process, 81-82; mem, Adv Comt Aids Prog, US Dept Health & Human Serv, 89- *Mem:* Am Soc Pharmacol; Am Soc Clin Pharmacol & Therapeut; Am Soc Clin Invest; fel Am Col Physicians; Am Soc Law & Med (pres, 89-90); fel Hastings Ctr. *Res:* Writing, teaching and consulting in the field of medical ethics; concentrating on research involving human subjects; the doctor-patient relationship and care of the dying patient. *Mailing Add:* Dept Internal Med Yale Univ Sch Med New Haven CT 06510

LEVINE, ROBERT PAUL, b Brooklyn, NY, Dec 18, 26; m 69. GENETICS. *Educ:* Univ Calif, Los Angeles, AB, 49, PhD(genetics), 51. *Hon Degrees:* AM, Harvard Univ, 57. *Prof Exp:* Instr biol, Amherst Col, 51-53; from asst prof to prof, Harvard Univ, 53-78, chmn dept, 67-70; PROF GENETICS, MED SCH, WASHINGTON UNIV, 78- *Concurrent Pos:* NSF sr fel, 63-64. *Mem:* AAAS; Genetics Soc Am; Sigma Xi; Soc Gen Physiol; Am Soc Cell Biol. *Res:* Genetic specification of membrane structure. *Mailing Add:* Wash Univ Sch Med Box 8031 St Louis MO 63110

LEVINE, ROBERT S(IDNEY), b Des Moines, Iowa, June 4, 21; m 47, 70; c 5. FIRE RESEARCH, FIRE PROTECTION ENGINEERING. *Educ:* Iowa State Col, BS, 43; Mass Inst Technol, SM, 46, ScD(chem eng), 49. *Prof Exp:* Assoc res dir, Rocketdyne Div, Rockwell Int, 49-66; chief liquid rocket res & technol, Off Advan Res & Technol, NASA, 66-74; CHIEF, FIRE RES RESOURCES DIV, NAT BUR STANDARDS, 74- *Concurrent Pos:* Mem subcomt combustion, Nat Adv Comt Aeronaut, 58; asst prof heat & mass transfer, Univ Calif, Los Angeles, 70-74; prof combustion, George Washington Univ, 77. *Mem:* Am Chem Soc; Am Inst Aeronaut & Astronaut; Nat Fire Protection Asn; Combustion Inst (vpres, 70, pres, 74-78); Soc Fire Protection Engrs; Am Inst Aeronaut & Astronaut. *Res:* Combustion and combustion stability in liquid rocket engines; combustion phenomena and heat transfer in liquid rocket engines; mathematical modeling of growth of unwanted fire in buildings. *Mailing Add:* 19017 Threshing Pl Gaithersburg MD 20879

LEVINE, RUTH R, b New York, NY; m 53. PHARMACOLOGY. *Educ:* Hunter Col, BA, 38; Columbia Univ, MA, 39; Tufts Univ, PhD(pharmacol), 55. *Prof Exp:* From instr to asst prof pharmacol, Sch Med, Tufts Univ, 55-58; from asst prof to prof, 58-65, UNIV PROF PHARMACOL, SCH MED, BOSTON UNIV, 72-, CHMN DIV MED & DENT SCI, GRAD SCH, 64-, ASSOC DEAN, SCH MED, 81- *Mem:* Am Soc Pharmacol & Exp Therapeut (secy-treas, 75); Biophys Soc; Acad Pharmaceut Sci; Am Chem Soc; fel AAAS; Sigma Xi. *Res:* Pharmacokinetics mechanisms of transport of drugs across biological barriers, particularly the intestinal epithelium; biochemical, histological and physiological factors influencing intestinal absorption; environmental toxicology. *Mailing Add:* Div Med & Dent Sci Boston Univ Sch Med Boston MA 02118

LEVINE, SAMUEL, b Brooklyn, NY, Jan 21, 21; m 53; c 3. PHYSICAL CHEMISTRY. *Educ:* Brooklyn Col, BA, 46; Columbia Univ, MA, 52, PhD(chem), 55. *Prof Exp:* Electrochemist, Arc Anodying & Plating Co, 46-47; phys chemist thermodyn, Nat Bur Standards, 47-51, proj leader, Macromolecular Properties Unit, Northern Regional Res & Develop Div, 55-58; assoc prof chem, Western Ill Univ, 58-59; chemist, Dow Chem Co, 59-61; prof chem & dir sci, Delta Col, 61-64; PROF CHEM & DEAN, SAGINAW VALLEY STATE COL, 64- *Mem:* AAAS; Am Chem Soc; Sigma Xi. *Res:* Physical chemistry of polymers; thermodynamics; kinetics. *Mailing Add:* 1604 E Canterbury TR AP Mt Pleasant MI 48858-2597

LEVINE, SAMUEL GALE, b Malden, Mass, Nov 1, 28; m 53; c 4. ORGANIC CHEMISTRY. *Educ:* Tufts Univ, BS, 50; Harvard Univ, MA, 52, PhD(org chem), 54. *Prof Exp:* Res assoc, Forrestal Res Ctr, Princeton Univ, 53-54; res chemist, Walter Reed Army Inst Res, 54-56; res chemist, Eastern Regional Res Br, USDA, 56-60; sr chemist, Natural Prod Lab, Res Triangle Inst, 60-64; assoc prof, 64-68, PROF CHEM, NC STATE UNIV, 68- *Concurrent Pos:* Consult, Res Triangle Inst, 64-; Weizmann fel, Weizmann Inst Sci, 71-72. *Mem:* AAAS; Am Chem Soc. *Res:* New methods in organic synthesis; stereochemistry and conformational analysis; structure determination and synthesis of natural products; chromiumtricarbonyl complexes in organic synthesis. *Mailing Add:* Dept Chem NC State Univ Raleigh NC 27650

LEVINE, SAMUEL HAROLD, b Hazlehurst, Ga, Nov 30, 25; m 55; c 3. NUCLEAR ENGINEERING, REACTOR PHYSICS. *Educ:* Va Polytech Inst, BS, 47; Univ Ill, MS, 48; Univ Pittsburgh, PhD(physics), 54. *Prof Exp:* Instr physics, Va Polytech Inst, 49-50; sr scientist, Bettis Atomic Power Lab, Westinghouse Elec Corp, 54-55, supv scientist, 55-57, mgr, 57-59; physicist in charge, Gen Atomic Div, Gen Dynamics Corp, 59-61; group physicist, Rocketdyne Div, NAm Aviation, Inc, 61-62; lab head nuclear sci, Northrop Space Labs, 62-68; dir nuclear reactor facil, 68-68, PROF NUCLEAR ENG, PA STATE UNIV, UNIV PARK, 68- *Concurrent Pos:* Lectr, Univ Calif, Los Angeles, 64-68; consult, Int Atomic Energy Agency, 77-, PP&L, 88- *Honors & Awards:* Invention Award, NASA, 73. *Mem:* Am Phys Soc; fel Am Nuclear Soc. *Res:* In-core fuel management; neutron detection; experimental reactor physics; neutron radiography; nuclear reactor fuel management; dosimetry. *Mailing Add:* 231 Sackett Bldg PA State Univ University Park PA 16802

LEVINE, SAMUEL W, b Dallas, Tex, May 15, 16; m 44; c 1. PHYSICAL CHEMISTRY. *Educ:* Agr & Mech Col Tex, BS, 38, MS, 41; Mass Inst Technol, PhD(phys chem), 48. *Prof Exp:* Combustion engr, Lone Star Gas Co, Tex, 38-39; instr thermodyn, Agr & Mech Col Tex, 40-41; assoc chemist, Atlantic Refining Co, 48-51; dir develop labs, Fisher Sci Co, 51-53; assoc dir res & develop, Fairchild Camera & Instrument Corp, 53-55, dir res & eng, Graphic Equip Div, 55-59, dir res & eng, Defense Prod Div, 59-61, tech dir, Corp, NY, 61-70; vpres technol, Varadyne, Inc, Calif, 70-72; vpres corp develop, Datel Systs, 72-76, vpres Semi Alloys, 76-85; RETIRED. *Mem:* Am Chem Soc; Optical Soc Am; Inst Elec & Electronics Engrs; NY Acad Sci; fel Am Inst Chemists. *Res:* X-ray spectroscopy; emission spectroscopy; petroleum reservoir characteristics; thermodynamic properties of hydrocarbons; radar systems research and development; instrumentation physics; radioactive tracers; photogrammetry instrumentation; corporate technical management; semiconductors; integrated circuits. *Mailing Add:* 11 Melby Lane East Hills NY 11576

LEVINE, SAUL, b Montreal, Que, May 31, 38; m 62; c 3. PSYCHIATRY. *Educ:* McGill Univ, BSc, 59, MD & CM, 63; Stanford Univ, dipl psychiat, 68; FRCP(C), 69. *Prof Exp:* SR PSYCHIATRIST, HOSP FOR SICK CHILDREN, 71-; PROF PSYCHIAT & ASSOC DIR, CHILD IN THE CITY PROG, UNIV TORONTO, 77- *Mem:* Fel Am Orthopsychiat Asn; fel Am Psychiat Asn; Can Psychiat Asn; Am Soc Adolescent Psychiat; Int Cong Social Psychiat. *Mailing Add:* Sunnybrook Med Ctr 2075 Bayview Ave Toronto ON M4V 3M5 Can

LEVINE, SEYMOUR, b New York, NY, Mar 13, 25; m 45; c 2. PATHOLOGY, NEUROPATHOLOGY. *Educ:* NY Univ, BA, 46; Chicago Med Sch, MB, 47, MD, 48. *Prof Exp:* Pathologist, St Francis Hosp, Jersey City, NJ, 56-64; pathologist & chief labs, Ctr Chronic Dis, Bird S Coler Hosp, 64-77; chief neuropath, Westchester County Med Ctr, 77-87; PROF PATH, NY MED COL, 87- *Concurrent Pos:* Consult, Vet Admin Hosp, Montrose, NY, 77- *Mem:* Am Asn Path; Soc Exp Biol & Med; Am Soc Exp Path; Am Asn Neuropath (pres, 68-69). *Res:* Demyelinating diseases; autoimmune disease. *Mailing Add:* Dept Path NY Med Col Valhalla NY 10595

LEVINE, SEYMOUR, b New York, NY, Jan 25, 25; m 49; c 3. PSYCHOPHYSIOLOGY. *Educ:* NY Univ, PhD(psychol), 52. *Prof Exp:* Res assoc, Queens Col, NY, 51-52; asst prof, Boston Univ, 52-53; lectr, Northwestern Univ, 54-56; asst prof psychiat, Med Sch, Ohio State Univ, 56-60; assoc prof, 62-69, PROF PSYCHOL, SCH MED, STANFORD UNIV, 69- *Concurrent Pos:* USPHS fel, 53-55; res assoc, Inst Psychosom & Psychiat, Michael Reese Hosp, 55-56; consult, Nat Cancer Inst, 57-; Found Fund Res Psychiat fel, Dept Neuroendocrinol, Inst Psychiat, Maudsley Hosp, London, 60; consult, Nat Inst Child Health & Human Develop, 66-67, mem neuropsychol res, 67-70, consult, Nat Comt Causes & Prev Violence, 68. *Honors & Awards:* Hoffheimer Res Award, 61. *Mem:* Am Psychol Asn; Endocrine Soc; Int Soc Develop Psychobiol. *Res:* Infantile experience development physiology and endocrinology. *Mailing Add:* NY Med Col Path Westchester Co Med Ctr Valhalla NY 10595

LEVINE, SEYMOUR, b Chicago, Ill, Apr 30, 22; m 43, 66; c 2. VIROLOGY. *Educ:* Univ Chicago, BS, 43; Univ Ill, MS, 45, PhD(bact), 49. *Prof Exp:* Asst bact, Med Sch, Univ Ill, 45-49; from instr to asst prof biophys, Univ Colo, 51-56; res biologist, Lederle Labs, Am Cyanamid Co, 56-65; sr res scientist, Upjohn Co, Mich, 65-71; from assoc prof to prof microbiol, Sch Med, Wayne State Univ, 71-89; RETIRED. *Concurrent Pos:* Nat Res Coun AEC fel, Univ Colo, 49-50; Case Western Reserve Univ, 50-51. *Mem:* Am Soc Microbiol; Tissue Cult Asn; Am Acad Microbiol; Soc Exp Biol & Med. *Res:* Viral-host cell interactions; tissue culture; viral replication; viral interference and interferon. *Mailing Add:* 34223 Hillside Dr Paw Paw MI 49079

LEVINE, SOLOMON LEON, b Schenectady, NY, Jan 7, 40; m 60; c 3. ANALYTICAL CHEMISTRY. *Educ:* Rensselaer Polytech Inst, BS, 61; Univ RI, PhD(anal chem), 66. *Prof Exp:* Sr assoc engr, Components Div, 65-66, sr assoc chemist, 66-68, staff chemist, 68-69, proj chemist, 69-72, develop chemist, 72-74, adv chemist, 74-79, SR CHEMIST, IBM CORP, 79- *Mem:* Sigma Xi; Electrochem Soc; Soc Electroanal Chem; Am Chem Soc; Am Electroplaters & Surface Finishing Soc. *Res:* Spectroscopy, absorption and emission; electroanalytical chemistry; technical management. *Mailing Add:* 12 Denver Ct E Endicott NY 13760

LEVINE, STEPHEN ALAN, b Brooklyn, NY, Dec 24, 38; m 61; c 4. ORGANIC CHEMISTRY, INFORMATION SYSTEMS. *Educ:* City Col New York, BS, 61; Purdue Univ, PhD(org chem), 66; Marist Col, MS(info systs), 87. *Prof Exp:* Res chemist, Acme Shellac Prod Co, 61; chemist, 66-67, sr chemist, 67-73, res chemist, 73-79, SR RES CHEMIST, TEXACO RES CTR, 79- *Concurrent Pos:* Adj Prof, Marist Col, 88- *Mem:* Fel Am Inst Chem; Sigma Xi; NY Acad Sci. *Res:* Polymer chemistry; computer information system design and development; process research; lubricant additive synthesis. *Mailing Add:* Texaco Res Ctr PO Box 509 Beacon NY 12508

LEVINE, SUMNER NORTON, b Boston, Sept 5, 23; m 52; c 1. PHYSICAL CHEMISTRY. *Educ:* Brown Univ, BS, 46; Univ Wis, PhD(phys chem), 49. *Prof Exp:* Instr phys chem, Univ Chicago, 49-50; sr res fel, Columbia Univ, 50-54; dir res labs, US Vet Admin Hosp, East Orange, NJ, 54-56; mgr chem & physics lab, Gen Eng Labs, Am Mach & Foundry Co, 56-58; sr staff scientist, Surface Commun Div, Radio Corp Am, 58-60, head solid state devices & electronics, 60-61; chmn dept, 61-67, PROF MAT SCI, STATE UNIV NY STONY BROOK, 61- *Concurrent Pos:* Childs fel, Univ Chicago, 49; Runyan fel, Columbia Univ, 52; lectr, Atomic Indust Forum, 56, Albert Einstein Med Col, 57 & Grad Div, Univ Conn, 57-58; instr, Grad Div, Brooklyn Col, 60 & City Col New York, 60; vis prof & dir urban res, Grad Ctr, City Univ New York, 67-68; ed-in-chief, Advan in Biomed Eng & Med Physics, J Socio-Econ Planning Sci & J Biomed Mat Res; NSF guest lectr, Berlin Acad Sci. *Honors & Awards:* Clemson Award, Soc Biol Mats Res, 73; Dan Forth Lectureship, Am Res Col, 64. *Mem:* Am Chem Soc; Electrochem Soc; Sigma Xi; sr mem Inst Elec & Electronics Engrs; Inst Mgt Sci. *Res:* Biophysical investigation of reaction mechanisms and isotopes; semiconductor physics; solid state high frequency devices; thermoelectric materials and devices; energy conversion techniques; superconductors. *Mailing Add:* PO Box D Setauket NY 11785

LEVINE, WALTER (GERALD), b Detroit, Mich, Dec 18, 30; m 55; c 3. PHARMACOLOGY. *Educ:* Wayne State Univ, BS, 52, MS, 54, PhD(physiol, pharmacol), 58. *Prof Exp:* Res assoc physiol & pharmacol, Wayne State Univ, 54-56 & 57-58, asst, 56-57; from asst prof to assoc prof, 61-76, PROF PHARMACOL, ALBERT EINSTEIN COL MED, 76- *Concurrent Pos:* Fel pharmacol, Albert Einstein Col Med, 58-61; USPHS career develop award. *Mem:* Am Soc Pharmacol & Exp Therapeut; NY Acad Sci; AAAS; Int Soc Study Xenobiotics. *Res:* Biochemical pharmacology; drug metabolism and disposition; regulation of the hepatic metabolism of azo dye carcinogens. *Mailing Add:* Dept Molecular Pharmacol Col Med Yeshiva Univ Bronx NY 10461

LEVINE, WILLIAM SILVER, b Brooklyn, NY, Nov 19, 41; m 63; c 2. ELECTRICAL ENGINEERING. *Educ:* Mass Inst Technol, SB, 62, SM, 65, PhD(elec eng), 69. *Prof Exp:* Asst, Mass Inst Technol, 65-69; from asst prof to assoc prof, 69-81, PROF ELEC ENG, UNIV MD, COLLEGE PARK, 81- *Concurrent Pos:* Res engr, Data Technol Inc, Mass, 62-64; consult, BTS Inc, 82-; vis scientist, Nat Inst Res in Informatics & Automation, France, 85-86. *Honors & Awards:* Distinguished Mem, Inst Elec & Electronics Engrs Control Systs Soc, 90. *Mem:* Fel Inst Elec & Electronics Engrs; Soc Ind & Appl Math. *Res:* Optimal controls and systems with special emphasis on the theories of optimal feedback control and system identification and the application of this theory to biological and transportation systems. *Mailing Add:* Dept Elec Eng Univ Md College Park MD 20742

LEVINGER, BERNARD WERNER, b Berlin, Ger, Sept 3, 28; nat US; m 54; c 3. MATHEMATICS. *Educ:* Lehigh Univ, BS, 48; Mass Inst Technol, MS, 50; NY Univ, PhD(math), 60. *Prof Exp:* Asst metallurgist, Armour Res Found, Ill Inst Technol, 51-52; res metallurgist, Tung-Sol Elec, Inc, 52-57; res engr, Labs, Gen Tel & Electronics Corp, 57-62; asst prof math, Case Western Reserve Univ, 62-68; ASSOC PROF MATH, COLO STATE UNIV, 68- *Mem:* Am Math Soc; Math Asn Am; Soc Indust & Appl Math. *Res:* Matrix theory; numerical analysis; group theory. *Mailing Add:* Math & Statist Dept Colo State Univ Ft Collins CO 80523

LEVINGER, JOSEPH S, b New York, NY, Nov 14, 21; m 43; c 4. PHYSICS. *Educ:* Univ Chicago, BS, 41, MS, 44; Cornell Univ, PhD(physics), 48. *Prof Exp:* Jr physicist, Metall Lab, Univ Chicago, 42-44; physicist, Franklin Inst, 45-46; asst, 46-48, instr physics, Conell Univ, 48-51; from asst prof to prof, La State Univ, 51-61; Avco vis prof, Cornell Univ, 61-64; PROF PHYSICS, RENSSELAER POLYTECH INST, 64- *Concurrent Pos:* Guggenheim fel,

57-58; Fulbright travel grant, 72-73; assoc prof, Univ Paris, 72-73. *Mem:* Fel Am Phys Soc. *Res:* Theoretical physics: specialities, the few-nucleon problem and nuclear photoeffect. *Mailing Add:* Dept Physics Rensselaer Polytech Inst Troy NY 12181

LEVINGS, CHARLES SANDFORD, III, b Madison, Wis, Dec 1, 30; c 4. GENETICS. *Educ:* Univ Ill, BS, 53, MS, 56, PhD(agron), 63. *Prof Exp:* Res instr, 62-64, from asst prof to assoc prof, 64-72, PROF GENETICS, NC STATE UNIV, 72- *Concurrent Pos:* Assoc ed, Current Genetics, 80-, Develop Genetics, 80-85, Maydica, 80-, Plant Physiol, 80-87, Plant Molecular Biol, 86, Science, 88- *Mem:* Nat Acad Sci; Am Genetic Asn; Genetics Soc Am; Am Soc Plant Physiologists; Am Soc Agron; AAAS; Int Soc Molecular Biol; Crop Sci Soc Am. *Res:* Autotetraploid genetics; maize biochemical genetics; higher plants; extrachromosomal inheritance; mitochondria and mitochondrial genomes; molecular genetics. *Mailing Add:* Dept Genetics Box 7614 NC State Univ Raleigh NC 27695-7614

LEVINGS, COLIN DAVID, b Victoria, BC, May 23, 42; m 68; c 2. BIOLOGICAL OCEANOGRAPHY, FISHERIES ECOLOGY. *Educ:* Univ BC, BSc Hons, 65, MSc, 67; Dalhousie Univ, PhD(oceanog), 73. *Prof Exp:* Field biologist technician marine fish ecol, Int Pac Halibut Comn, Seattle, Wash, 62-63; scientist, Fisheries Res Bd Can, Pac Biol Sta, Nanaimo, BC, 67-68; res scientist & prog head, 72-83, SECT HEAD, HABITAT RES SECT, BIOL SCI BR, W VANCOUVER LAB, 83- *Concurrent Pos:* Res assoc, Dept Zool, Univ BC; assoc ed, Can J Fish Aquat Sci, 89-; vis sci, Inst Marine Res, Bergen, Norway, 89; vis lecr, Univ Tsukuba, Japan, 90. *Mem:* Can Soc Zoologists; Pac Estuarine Res Soc, (pres, 88-); Estuarine Res Fedn. *Res:* Ecology of marine and estuarine benthos; community structure at disrupted habitats; ocean dumping and dredging; coastal fish habitats and food webs; ecology of fjords; juvenile salmonid ecology; river habitats of fishes; aquaculture siting in coastal areas. *Mailing Add:* Dept Fisheries & Oceans W Vancouver Lab 4160 Marine Dr West Vancouver BC V7V 1N6 Can

LEVINGS, WILLIAM STEPHEN, geology, for more information see previous edition

LEVINS, RICHARD, b New York, NY, June 1, 30; m 50; c 3. POPULATION BIOLOGY, MATHEMATICAL BIOLOGY. *Educ:* Cornell Univ, AB, 51; Columbia Univ, PhD(zool), 65. *Hon Degrees:* MPH, Harvard Univ, 75. *Prof Exp:* Res assoc pop genetics, Univ Rochester, 60-61; assoc prof biol, Univ PR, 61-66; from assoc prof to prof math biol, Univ Chicago, 67-75; JOHN ROCK PROF POP SCI, SCH PUB HEALTH, HARVARD UNIV, 75- *Concurrent Pos:* Farmer, 51-56; NIH res grant, 63-66; consult agr ecol prog, Cuban Acad Sci, 64-65; NSF res grant, 64-66. *Mem:* Am Acad Arts & Sci; Am Soc Naturalists. *Res:* Ecology and genetics; complex systems; agriculture. *Mailing Add:* Dept Pop Sci Harvard Sch Pub Health Boston MA 02115

LEVINSKAS, GEORGE JOSEPH, b Tariffville, Conn, July 8, 24; m 46; c 3. TOXICOLOGY, ENVIRONMENTAL HEALTH. *Educ:* Wesleyan Univ, AB, 49; Univ Rochester, PhD(pharmacol), 53; Am Bd Toxicol, dipl. *Prof Exp:* Res assoc biol sci, USAEC, Univ Rochester, 52-53; dept occup health, Grad Sch Pub Health, Univ Pittsburgh, 53-54, res assoc & lectr, 54-56, asst prof appl toxicol, 56-58; res pharmacologist, Cent Med Dept, Am Cyanamid Co, 58, chief indust toxicologist & dir environ health lab, 59-71; mgr prod eval, Dept Med & Health Sci, Monsanto Co, 71, mgr environ assessment & toxicol, 72-77, dir, 78-85, sr toxicol consult, 86-91; RETIRED. *Concurrent Pos:* Fel, Acad Toxicol Sci. *Mem:* Soc Toxicol; Am Chem Soc; Am Indust Hyg Asn; Environ Mutagen Soc; Am Soc Pharmacol & Exp Therapeut; fel AAAS; Sigma Xi. *Res:* Pharmacology and toxicology of boron compounds; organic phosphates; industrial chemicals; food additives; insecticides; chemistry of bone mineral. *Mailing Add:* 526 Fairways Circle Creve Coeur MO 63141-7554

LEVINSKY, NORMAN GEORGE, b Boston, Mass, Apr 27, 29; m 56; c 3. ANIMAL PHYSIOLOGY. *Educ:* Harvard Univ, AB, 50, MD, 54. *Prof Exp:* Intern & resident med, Beth Israel Hosp, Boston, Mass, 54-56; clin assoc, Nat Heart Inst, 56-58; NIH spec fel med, Boston Univ Hosp, 58-60; from instr to assoc prof, 60-68, Wesselhoeft prof, 68-72, WADE PROF MED & CHMN DIV, SCH MED, BOSTON UNIV, 72-; DIR EVANS MEM DEPT CLIN RES, PREV MED & PHYSICIAN-IN-CHIEF, UNIV HOSP, 72- *Concurrent Pos:* Asst dir, Univ Med Serv, Boston City Hosp, 61-68, dir, 68- *Mem:* Am Fedn Clin Res; Am Soc Clin Invest; Asn Am Physicians; Am Soc Nephrol; Am Heart Asn; Asn Prof Med (secy-treas, 84, pres, 88-89); Am Col Physicians. *Res:* Renal physiology and medical research. *Mailing Add:* Boston Univ Med Ctr 75 E Newton St Boston MA 02118

LEVINSKY, WALTER JOHN, b Meadville, Pa, Sept 16, 20; m 48; c 4. MEDICINE. *Educ:* Allegheny Col, BS, 42; Temple Univ, MD, 45, MS, 52; Am Bd Internal Med, dipl, 54 & 74. *Prof Exp:* Intern med, Hamot Hosp, Erie, Pa, 45-46; resident path, Univ Hosp, 48-49, resident internal med, 49-52, instr, Sch Med, 52-54, assoc, 54-58, from asst prof to assoc prof internal med, 58-74, CLIN PROF MED, SCH MED, TEMPLE UNIV, 74- *Concurrent Pos:* Chief dept med, Northeastern Hosp, Philadelphia, Pa, 54-58. *Mem:* Sr mem Am Fedn Clin Res; fel Am Col Physicians; fel Royal Soc Med. *Res:* Internal medicine; clinical research. *Mailing Add:* Dept Internal Med Temple Univ Sch Med Philadelphia PA 19140

LEVINSON, ALFRED ABRAHAM, b Staten Island, NY, Mar 31, 27. EXPLORATION GEOCHEMISTRY. *Educ:* Univ Mich, BS & MS, 49, PhD(mineral), 52. *Prof Exp:* Asst, 50-52, res assoc, Univ Mich, 52-53; asst prof mineral, Ohio State Univ, 53-56; mineralogist, Dow Chem Co, 56-62; sr res geologist, Gulf Res & Develop Co, 62-67; PROF GEOL, UNIV CALGARY, 67- *Concurrent Pos:* Lectr, Univ Houston, 57-59; exec ed, Geochimica et Cosmochimica Acta, 67-70; ed, Proc Apollo 11 & Second Lunar Sci Conf. *Mem:* Fel Mineral Soc Am; Geochem Soc; fel Geol Soc Am; Mineral Asn Can; Sigma Xi. *Res:* General mineralogy and geochemistry with industrial application; environmental geochemistry; relation of geochemistry to health; exploration geochemistry. *Mailing Add:* Dept Geol Univ Calgary Calgary AB T2N 1N4 Can

LEVINSON, ALFRED STANLEY, b Portland, Ore, Aug 27, 32; m 58; c 3. ORGANIC CHEMISTRY. *Educ:* Reed Col, BA, 54; Wesleyan Univ, MA, 57; Ind Univ, PhD(org chem), 63. *Prof Exp:* Res assoc chem, Ind Univ, 62-63; from asst prof to assoc prof, 63-73, PROF CHEM, PORTLAND STATE UNIV, 73- *Mem:* AAAS; Am Chem Soc. *Res:* Isolation and characterization of natural products; organic synthesis. *Mailing Add:* Dept Chem Portland State Univ Box 751 Portland OR 97207

LEVINSON, ARTHUR DAVID, b Seattle, Wash, Mar 31, 50; m 78; c 2. MOLECULAR BIOLOGY. *Educ:* Univ Wash, BS, 72; Princeton Univ, PhD(biochem), 77. *Prof Exp:* NIH res fel, Univ Calif, San Francisco, 77-78, Am Cancer Soc, sr res fel, 78-80; sr scientist, 80-83, dir, Dept Cell Genetics, 87-89, vpres res technol, 89-90, STAFF SCIENTIST, GENENTECH, INC, 83-, VPRES, RES, 90- *Mem:* AAAS; Am Soc Microbiol; Am Soc Biochem & Molecular Biol. *Res:* Role of cellular genes in tumor development; regulation of mammalian gene expression; molecular genetics. *Mailing Add:* Genentech Inc 460 Pt San Bruno Blvd South San Francisco CA 94080

LEVINSON, CHARLES, b San Antonio, Tex, Dec 31, 36; m 67; c 2. CELL PHYSIOLOGY. *Educ:* Univ Tex, BA, 58; Trinity Univ, MA, 60; Rutgers Univ, PhD(physiol), 64. *Prof Exp:* Nat Cancer Inst fel, Med Col, Cornell Univ, 65-66; sr cancer res scientist, Roswell Park Mem Inst, 66-68; assoc prof, 68-72, PROF PHYSIOL, MED SCH, UNIV TEX, SAN ANTONIO, 72- *Mem:* Biophys Soc; Soc Gen Physiol; Am Physiol Soc; Sigma Xi. *Res:* Membrane phenomena; ion transport in tumor cells. *Mailing Add:* Dept Physiol Univ Tex Health Sci Ctr 7703 Floyd Curl Dr San Antonio TX 78284-7756

LEVINSON, DAVID W, b Chicago, Ill, Feb 24, 25; m 49; c 3. PHYSICAL METALLURGY. *Educ:* Ill Inst Technol, BS, 48, MS, 49, PhD(metall eng), 53. *Prof Exp:* Res metallurgist, Armour Res Found, 53-57, supvr non-ferrous metall res, 57-59, asst dir metals res, 59-62, sci adv, Metals & Ceramics Div, IIT Res Inst, 62-64; assoc head dept mat eng, 64-67, actg dean col eng, 67-69, prof, 64-87, PROF EMER METALL, UNIV ILL, CHICAGO CIRCLE, 87- *Concurrent Pos:* Consult, IIT Res Inst, 64-, Mat Tech Lab, Wright-Patterson AFB, 65-, Fotofabrication Corp & Specialloy, Inc; adj prof, Univ Ariz, Tucson, 90- *Mem:* Am Soc Metals; Am Inst Mining, Metall & Petrol Engrs; Am Soc Eng Educ; Sigma Xi. *Res:* High temperature alloy development; coatings for thermal control of surfaces; metallurgical transformations in alloys in thin film form; binary and ternary phase equilibria in metallic systems; stress relaxation in high carbon steels, fatigue and fracture toughness of resulfurized steels. *Mailing Add:* Dept Mat Sci & Eng Univ Ariz Tucson AZ 85712

LEVINSON, GILBERT E, b New York, NY, Jan 25, 28; m 50; c 2. MEDICINE, CARDIOLOGY. *Educ:* Yale Univ, AB, 48; Harvard Med Sch, MD, 53. *Prof Exp:* Intern med, Harvard Med Serv, Boston City Hosp, 53-54, asst resident, 54-55, chief resident, Thorndike Mem Ward, 58-59; from asst prof to prof med, Univ Med & Dent NJ, NJ Med Sch, 59-76; chief med, St Vincent Hosp, Worcester, Mass, 76-85; chief med, Worchester City Hosp, 85-91; PROF MED, MED SCH, UNIV MASS, 76-; CHIEF MED, ST VINCENT HOSP, WORCHESTER, MASS, 91- *Concurrent Pos:* Teaching fel, Harvard Med Sch, 54-55; Nat Heart Inst res fel, Thorndike Mem Lab, Boston City Hosp, 57-59; Nat Heart & Lung Inst res career develop award, 67-70; assoc dir, T J White Cardiopulmonary Inst, B S Pollak Hosp, Jersey City, NJ, 61-71; consult, USPHS Hosp, Staten Island, NY, 63-; estab investr, Union County Heart Asn, NJ, 61-66 & 70-75. *Mem:* Am Fedn Clin Res; fel Am Col Cardiol; Am Physiol Soc; Am Soc Clin Invest; fel Am Col Physicians. *Res:* Hemodynamics in valvular heart disease; indicator-dilution theory and methodology; cardiopulmonary blood volumes; relations between myocardial performance and metabolism. *Mailing Add:* Dept Med St Vincent Hosp Worcester MA 01604

LEVINSON, JOHN Z, b Odessa, Ukraine, Apr 21, 16; US citizen; m 41; c 3. PSYCHOLOGY, COMMMUNICATIONS ENGINEERING. *Educ:* Univ Toronto, BA, 39, MA, 40, PhD(physics), 48. *Prof Exp:* Demonstr physics, Univ Toronto, 46-48; res assoc, Mass Inst Technol, 48-50; asst prof physics, Alfred Univ, 50-54; sr physicist, Transistor Products, 54-56; mem tech staff, Bell Tel Lab, 56-71; prof, 71-87, EMER PROF PSYCHOL, UNIV MD, 87 - *Concurrent Pos:* Prin investr, Nat Eye Inst grant, Univ Md, 76 -; consult, NIH Study Section, Vision B, 76-80; vis res scholar, Cambridge Univ, England, 67-68 & Syracuse Univ, 77-78. *Mem:* Fel Optical Soc Am; Asn Res Vision & Opthalmol; fel AAAS. *Res:* Temporal and spatial aspects of human vision studied by psychophysical methods; processes that underlie vision, emphasising how they limit speed and acuity. *Mailing Add:* Dept Psychol Univ Md College Park MD 20742

LEVINSON, LIONEL MONTY, b Johannesburg, SAfrica, Mar 12, 43. SOLID STATE PHYSICS, CERAMIC ENGINEERING. *Educ:* Univ Witwatersrand, BSc, 65, MSc, 66; Weizmann Inst Sci, PhD(solid state physics), 70. *Prof Exp:* Physicist, Gen Elec Corp Res & Develop, 70-79; MGR, ELECTRONIC MAT & DEVICES, GE CORP RES, 80- *Concurrent Pos:* Fel, Am Ceramic Soc. *Mem:* Am Phys Soc. *Res:* Electronic ceramics; varistors; electronic packaging. *Mailing Add:* Gen Elec Corp Res & Develop PO Box 8 Schenectady NY 12301

LEVINSON, MARK, b Brooklyn, NY, June 12, 29; m 53; c 2. HISTORY OF TECHNOLOGY, MECHANICS. *Educ:* Polytech Inst Brooklyn, BAeroE, 51, MS, 60; Calif Inst Technol, PhD, 64. *Prof Exp:* Asst appl mech, Polytech Inst Brooklyn, 51-52, instr math, 56, sr asst appl mech, 59; stress analyst, Foster-Wheeler Corp, 57-58; asst prof mech eng, Ore State Col, 60-61; assoc prof, Clarkson Col Technol, 64-66; assoc prof theoret & appl mech, WVa Univ, 66-67; prof eng mech, McMaster Univ, 67-80; A O WILLEY PROF MECH ENG, UNIV MAINE, ORONO, 80- *Mem:* AAAS; Am Soc Mech Engrs; Am Inst Aeronaut & Astronaut; Soc Indust & Appl Math; Soc Hist Technol. *Res:* Theory of elasticity; continuum mechanics; elastic stability; structural dynamics; history of early aviation. *Mailing Add:* 19301 89th Pl W Edmonds WA 98020

LEVINSON, SIDNEY BERNARD, b Russia, July 4, 11; nat US; m 65; c 2. CHEMISTRY, CHEMICAL ENGINEERING. *Educ:* City Col New York, BS(chem) & BS(eng), 32, Chem Engr, 33. *Prof Exp:* Consult, Protective Coatings, Joachim Res Labs, 33-36; pres, Indust Consult Labs, 36-42; vpres & tech dir, Adco Chem Co, 42-48; supt & tech dir, Garland Co, 48-52; vpres & tech dir, D H Litter Co, 52-73, pres, David Litter Labs, Inc, DBA D/L Labs, 74-83; OWNER, SIDLEV ASSOCS, 83- *Concurrent Pos:* dir, Artists Tech Res Inst, 65-74. *Honors & Awards:* PaVaC Award & lectr; Roy H Kienle Award, NY Soc Testing & Mat; Nat Asn Corrosion Engrs; Fedn Socs Coatings Technol. *Res:* Protective coatings; thermosetting and reinforced plastics, sealants and allied products; evaluation of raw materials; formulation; testing of finished products; certification; preparation of specifications and manuals, personnel training, industry surveys, investigation of failures and legal assistance. *Mailing Add:* 20B John Adams St Cranbury NJ 08512

LEVINSON, SIMON ROCK, b Buffalo, NY, Dec 2, 46; m 69; c 2. MEMBRANE TRANSPORT PHENOMENA, BIOELECTRIC PHENOMENA. *Educ:* Calif Inst Technol, BS, 68; Univ Cambridge, PhD(physiol), 75. *Prof Exp:* Fel chem, Calif Inst Technol, 76-79; asst prof, 79-84, ASSOC PROF PHYSIOL, UNIV COLO HEALTH SCI CTR, 84- *Mem:* Biophys Soc; Soc Gen Physiologists; Soc Neuroscience. *Res:* Molecular mechanisms which underlie the electrical excitation phenomena in nerve and muscle cells; mechanisms of the voltage-sensitive sodium channel and the molecular structures which mediate its complex function. *Mailing Add:* Dept Physiol C240 Med Sch Univ Colo Denver CO 80262

LEVINSON, STANLEY S, m; c 3. PATHOLOGY. *Educ:* Boston Univ, BA, 64; Univ Calif, MS, 67, PhD(physiol), 70; Am Bd Clin Chem, cert. *Prof Exp:* Res assoc, Dept Nutrit, Mass Inst Technol, 70-73; res fel med, Clin Chem Lab, Mass Gen Hosp, 73-74; adj asst prof, Dept Path & assoc, Dept Biochem, Sch Med, Wayne State Univ, 81-86, adj assoc prof, Dept Path & assoc, Dept Biochem, 88-89; DIR CLIN CHEM & IMMUNOL, LAB SERV, VET ADMIN MED CTR, UNIV LOUISVILLE, 89-, ASST PROF, DEPT PATH, 89- *Concurrent Pos:* Lectr clin chem, Grad Sch Pharm & Health, Northeastern Univ, 78-80; dir clin lab, Joslin Found Diabetic Unit, Brookline Hosp, 74-80. *Mem:* Am Asn Immunol; Nat Acad Clin Biochem; Clin Ligand Assay Soc; AAAS; Am Asn Clin Chem; Soc Complex Carbohydrates. *Res:* Methods of assay of antigens via immunochemical techniques; methods of assaying circulating immune complexes, rheumatoid factors and idiotypic antibodies; regulation of the immune response through molecules that stimulate and inhibit the synthesis of circulating immune complexes and rheumatoid factors by affecting lymphocytes; clinical correlates with measurement of circulating immune complexes and complement function tests; testing for apolipoproteins and clinical studies related to lipid metabolism as risk factors for cardiovascular disease; numerous publications. *Mailing Add:* 800 Zorn Ave Louisville KY 40206-1499

LEVINSON, STEPHEN, b New York, NY, Sept 27, 44. SPEECH PROCESSING, AUTOMATIC SPEECH. *Educ:* Harvard Univ, BA, 66; Univ RI, MS, 72, PhD(elec eng), 74. *Prof Exp:* Instr comput sci, Yale Univ, 74-76; ELEC ENGR, AT&T BELL LAB, 76- *Mem:* Fel Inst Elec & Electronics Engrs; fel Acoust Soc Am. *Mailing Add:* Ling Res Dept 2D-446 Bell Lab 600 Mountain Ave Murray Hill NJ 07974

LEVINSON, STEVEN R, b Brooklyn, NY, Oct 13, 47. ANALYTICAL CHEMISTRY, PHOTOGRAPHY. *Educ:* Rensselaer Polytech Inst, BS, 68, PhD(anal chem), 73. *Prof Exp:* SR RES CHEMIST, PHOTOG RES DIV, KODAK RES LABS, EASTMAN KODAK CO, 73- *Concurrent Pos:* Instr, Rochester Inst Technol, 76- *Mem:* Am Chem Soc; Sigma Xi; Soc Photog Scientists & Engrs. *Res:* Research and development of photographic materials. *Mailing Add:* 102 Mountain Rd Rochester NY 14625-1819

LEVINSON, STUART ALAN, b Detroit, Mich, Oct 29, 20; m 47; c 3. PALEONTOLOGY, GEOLOGY. *Educ:* Wayne State Univ, BS, 47; Washington Univ, AM, 49, PhD(geol), 51. *Prof Exp:* Asst geol, Washington Univ, 47-51; sr geologist, Humble Oil & Refining Co, 51-64; res supvr, Esso Prod Res Co, 64-66, res assoc, 66-90; RETIRED. *Concurrent Pos:* Instr, Washington Univ, 50-51. *Mem:* AAAS; Paleont Soc; Soc Econ Paleontologists & Mineralogists (vpres, 57); Geol Soc Am; Am Asn Petrol Geologists. *Res:* Invertebrate paleontology, micropaleontology and zoology. *Mailing Add:* 5150 Hidalgo No 705 Houston TX 77056-0806

LEVINSON, WARREN E, b Brooklyn, NY, Sept 28, 33; m 65. MICROBIOLOGY, IMMUNOLOGY. *Educ:* Cornell Univ, BS, 53; Univ Buffalo, MD, 57; Univ Calif, Berkeley, PhD(virol), 65. *Prof Exp:* Assoc prof, 65-70, PROF MICROBIOL, MED CTR, UNIV CALIF, SAN FRANCISCO, 70- *Concurrent Pos:* Am Cancer Soc fel tumor viruses, Univ Col, Univ London, 65-67; R W Johnson Health Policy fel, 80-81. *Res:* Tumor viruses. *Mailing Add:* Dept Microbiol Univ Calif Med Ctr San Francisco CA 94143

LEVINSTEIN, HENRY, physics; deceased, see previous edition for last biography

LEVINTHAL, CHARLES F, b Cincinnati, Ohio, July 6, 45; m 73; c 2. NEUROSCIENCES. *Educ:* Univ Cincinnati, AB, 67; Univ Mich, MA, 68, PhD(exp psychol), 71. *Prof Exp:* From asst prof to assoc prof, 71-87, PROF PSYCHOL, HOFSTRA UNIV, 87- *Mem:* Am Psychol Asn; Soc Neuroscience; Soc Psychophysiol Res; NY Acad Sci. *Res:* Studies of hemisphere specialization among bilingual individuals. *Mailing Add:* Dept Psychol Hofstra Univ 1000 Fulton Ave Hempstead NY 11550

LEVINTHAL, CYRUS, biophysics; deceased, see previous edition for last biography

LEVINTHAL, ELLIOTT CHARLES, b Brooklyn, NY, Apr 13, 22; m 44; c 4. PHYSICS. *Educ:* Columbia Univ, BA, 42; Mass Inst Technol, MS, 43; Stanford Univ, PhD(physics), 49. *Prof Exp:* Proj engr, Sperry Gyroscope Co, NY, 43-46; res assoc nuclear physics, Stanford Univ, 46-48; res physicist, Varian Assocs, 49-50, res dir, 50-52; chief engr, Century Electronics & Instruments, Inc, 52-53; pres, Levinthal Electronic Prod, Inc, 53-61; assoc dean res affairs, Sch Med, Stanford Univ, 71-74; dir Instrumentation Res Lab, 61-81; dir, defense sci off, Defense Advan Res Agency, Dept Defense, 81-83; PROF MECH ENG & ASSOC DEAN RES, SCH ENG, STANFORD UNIV, 83- *Concurrent Pos:* Adj prof genetics, Sch Med, Stanford Univ, 61-81. *Mem:* AAAS; fel Am Phys Soc; sr mem Inst Elec & Electronics Eng; Optical Soc Am; Biomed Eng Soc; Sigma Xi. *Res:* Measurements of nuclear moments; applications of computers to image processing and medical instrumentation; exobiology and planetary sciences. *Mailing Add:* 59 Sutherland Dr Atherton CA 94027-6430

LEVINTHAL, MARK, b Brooklyn, NY, Mar 3, 41; m 62, 90; c 2. MICROBIAL GENETICS, MOLECULAR EVOLUTION. *Educ:* Brooklyn Col, BS, 62; Brandeis Univ, PhD(biol), 66. *Prof Exp:* Fel genetics, Johns Hopkins Univ, 66-68; staff fel genetics lab molecular biol, Nat Inst Arthritis & Metal Dis, 68-72; ASSOC PROF BIOL, PURDUE UNIV, 72- *Concurrent Pos:* NIH fel, 66-68 & 72-77. *Mem:* Am Soc Microbiol; Sigma Xi; Genetics Soc Am; AAAS; Italian Molecular Biol Soc; Soc Study Evolution. *Res:* Regulation of enzyme synthesis of biosynthetic pathways and its relationship to general metabolic controls in bacteria; regulatory mechanisms-their evolution and contribution to general evolutionary theory. *Mailing Add:* Dept Biol Sci Purdue Univ West Lafayette IN 47907

LEVINTON, JEFFREY SHELDON, b New York, NY, Mar 20, 46; m 79; c 2. ECOLOGY, PALEONTOLOGY. *Educ:* City Col New York, BS, 66; Yale Univ, MPhil, 69, PhD(paleoecol), 71. *Prof Exp:* From instr to assoc prof paleoecol, 70-83, PROF ECOL & EVOLUTION, STATE UNIV NY STONY BROOK, 83- *Concurrent Pos:* State Univ NY Stony Brook Res Found fel & grant-in-aid, 71; managing ed, Am Naturalist, 74-75; vis prof, Uppsala Univ, Sweden, 81 & Univ Cambridge, 84; Guggenheim fel, 84-85; chmn, Panel Hudson River Found, 86-90, assoc ed Ecol, 86-89, assoc ed Ecol Appln, 90. *Mem:* Ecol Soc Am; Am Soc Naturalists; Soc Study Evolution; AAAS. *Res:* Marine benthic ecology; paleoecology; fossil population dynamics; benthic deposit feeder-detritus-microbial interactions. *Mailing Add:* Dept Ecol & Evolution State Univ NY Stony Brook NY 11794

LEVINTOW, LEON, b Philadelphia, Pa, Nov 10, 21; m 46; c 4. BIOCHEMISTRY, VIROLOGY. *Educ:* Haverford Col, AB, 43; Jefferson Med Col, MD, 46. *Prof Exp:* Intern, Jefferson Hosp, Philadelphia, Pa, 46-47; chief of lab, US Army Hepatitis Res Ctr, Ger, 47-49; biochemist, Nat Cancer Inst, 49-56, asst chief lab cell biol, Nat Inst Allergy & Infectious Dis, 56-61, asst chief lab biol viruses, 61-65; PROF MICROBIOL, SCH MED, UNIV CALIF, SAN FRANCISCO, 65-, CHMN, DEPT MICROBIOL & IMMUNOL, 80- *Concurrent Pos:* Res fel, Biochem Res Lab, Mass Gen Hosp, Boston, 51-52. *Mem:* Am Soc Microbiol; Am Chem Soc; Am Soc Biol Chemists. *Res:* Biochemistry of viruses. *Mailing Add:* Dept Microbiol & Immunol Univ Calif Sch Med San Francisco CA 94143

LEVIS, ALEXANDER HENRY, b Yannina, Greece, Oct 3, 40; m 70; c 2. MECHANICAL ENGINEERING. *Educ:* Ripon Col, BA, 63; Mass Inst Technol, BS & MS, 65, ME, 67, ScD(mech eng), 68. *Prof Exp:* Res asst control systs, Eng Projs Lab, Mass Inst Technol, 63-65; engr, Christina Lab, E I du Pont de Nemours & Co, Inc, 65; res asst transp systs, Electronic Systs Lab, Mass Inst Technol, 65-68; from asst prof to assoc prof elec eng, Polytech Inst New York, 68-74; sr engr, 73-76, dept mgr, Systs Control Inc, 76-81; res scientist, Mass Inst Technol, 79-91; PROF ELEC, COMPUTER & SYST ENG, GEORGE MASON UNIV, 90- *Concurrent Pos:* Consult, Sweet Assocs Ltd, 86-; pres, Control Syst Soc, Inst Elec & Electronics Engrs, 87, distinguished mem, 87. *Mem:* Fel Inst Elec & Electronics Engrs; AAAS; Sigma Xi; Am Inst Aeronaut & Astronaut; Dir, Am Automatic control Coun, 86-87. *Res:* Mathematical organization theory, distributed intelligence systems command and control; socio-economic systems modeling. *Mailing Add:* 10607 Springvale Ct Great Falls VA 22066

LEVIS, C(URT) A(LBERT), b Ger, Apr 16, 26; nat US; m 58; c 3. ELECTROMAGNETISM. *Educ:* Case Inst Technol, BS, 49; Harvard Univ, AM, 50; Ohio State Univ, PhD(elec eng), 56. *Prof Exp:* Studio engr, Radio Sta WSRS, Inc, 48-49; res assoc, Antenna Lab, 50-56, assoc supvr, 56-61, from asst prof to prof elec eng, 56-85, dir, Antenna Lab, 61-69, EMER PROF, OHIO STATE UNIV, 85- *Concurrent Pos:* Sr fel, Nat Ctr Atmospheric Res, 76-77; guest vis, Inst Telecommun Sci, 85-86. *Mem:* Sr mem Inst Elec & Electronics Engrs. *Res:* Radiowave propagation; antennas; electromagnetic theory. *Mailing Add:* Dept Elec Eng Ohio State Univ 2015 Neil Ave Columbus OH 43210-1272

LEVIS, DONALD J, b Cleveland, Ohio, Sept 19, 36; c 2. CLINICAL PSYCHOLOGY, BEHAVIORAL THERAPY. *Educ:* John Carroll Univ, BSS, 58; Kent State Univ, MA, 60; Emory Univ, PhD(psychol), 64. *Prof Exp:* USPHS fel clin psychol, Lafayette Clin, Detroit, 64-65; res psychologist psychobiol, 65-66; from asst prof to assoc prof clin psychol, 65-72, PROF CLIN PSYCHOL, STATE UNIV NY, BINGHAMTON, 72- *Concurrent Pos:* Lectr, Emory Univ, 63-64; adj asst prof, Wayne State Univ, 65-66; dir res & training clin, Univ Iowa, 70-72; dir & developer clin psychol training prog, State Univ NY, Binghamton, 72-81; dir & developer research & training clin, 73-76; adj prof, Col Med, Upstate Med Ctr, Syracuse, 79- *Mem:* Fel Am Psychol Asn; Psychonomic Soc; Asn Advan Behavior Ther; Soc Psychophysiol Res; Sigma Xi. *Res:* Developing the theoretical model and applied therapeutic behavioral technique of implosive (flooding) therapy; decoding of traumatic memories motivating psychopathology; author of over 80 scientific articles. *Mailing Add:* Dept Psychol State Univ NY Binghamton NY 13901

LEVIS, WILLIAM WALTER, JR, b Chicago, Ill, May 14, 18; m 41; c 2. ORGANIC CHEMISTRY. *Educ:* Univ Fla, BS, 41. *Prof Exp:* Res chemist, Fla Chem Indust, 41-42; sr res chemist, Sharples Chem, Inc, 42-52; sr res chemist, BASF Wyandotte Corp, 52-55, sect head, 55-56, res supvr, 56-84, sr res assoc, 84-86; RETIRED. *Mem:* Am Chem Soc. *Res:* Organic synthesis; catalysis; hydrogenation; amination; oxyalkyation. *Mailing Add:* 2069 SE 37th Ct Circle Ocala FL 32671

LEVI-SETTI, RICCARDO, b Milan, Italy, July 11, 27; m 59; c 2. PALEONTOLOGY. *Educ:* Univ Pavia, Dr, 49. *Prof Exp:* Asst physics, Univ Pavia, 49-51 & Univ Milan, 52-56; res assoc, Inst, 56-57, from asst prof to assoc prof, Univ, 57-64, PROF PHYSICS, ENRICO FERMI INST, UNIV CHICAGO, 65- *Concurrent Pos:* Guggenheim fel, 63-64. *Mem:* Fel Am Phys Soc; Mat Res Soc. *Res:* Imaging microanalysis of materials by secondary ion mass spectrometry at high lateral resolution; studies of metal alloys, ceramics, minerals, biomaterials; development of new microanalytical instrumentation. *Mailing Add:* Enrico Fermi Inst Univ Chicago 5630 Ellis Ave Chicago IL 60637

LEVISON, MATTHEW EDMUND, b New York, NY, May 18, 37; m 66; c 2. MEDICAL SCIENCE, HEALTH SCIENCES. *Educ:* Columbia Univ, BA, 58; State Univ NY, MD, 62. *Prof Exp:* Asst instr med, Downstate Med Ctr, State Univ NY, 65-67; asst physician, NY Hosp, 67-69; instr, Med Col, Cornell Univ, 68-69; clin instr, Downstate Med Ctr, State Univ, NY, 69-70; from asst prof to assoc prof med & chief, 70-77, PROF MED & CHIEF INFECTIOUS DIS DIV, MED COL PA, 77- *Concurrent Pos:* Attend physician & chief infectious dis unit, Queens Hosp Ctr, Long Island Jewish Med Ctr affil, 69-70; attend staff, Philadelphia Vet Admin Hosp, 70- *Mem:* Am Soc Microbiol; Am Fedn Clin Res; Infectious Dis Soc Am; fel Am Col Clin Pharmacol; fel Am Col Physicians. *Res:* Anaerobic bacteria, the pathogenesis of the renal concentrating defect in experimental pylonephritis and the pathogenesis of experimental endocarditis; antimerotical pharmacodynamics. *Mailing Add:* Med Col Pa 3300 Henry Ave Philadelphia PA 19129

LEVISON, SANDRA PELTZ, b New York, NY, Apr 20, 41; m 66; c 2. NEPHROLOGY. *Educ:* Hunter Col, NY, BA, 61; New York Univ, MD, 65. *Prof Exp:* Asst instr med, State Univ NY, Downstate, 66-67 & 68-70; asst instr med, New York Univ Med Ctr, 67-68; from instr to assoc prof med, 70-81, dir, Hypertension Ctr, 74, Nephrology Fel Training Prog, 76 & Hemodialysis Serv, 78, chief hypertension, renal & dialysis, 78, PROF MED, MED COL PA, 81- *Concurrent Pos:* Pres, Women in Nephrology; bd mem, Am Diabetes Asn; med review bd, Nat Kidney Found, Greater Del Valley Chap. *Mem:* Am Soc Nephrology; Int Soc Nephrology; fel Am Col Physicians. *Res:* Elucidation of the renal concentrating defect in experimental infective Pyelonephritis; effects of exercise on blood pressure of adolescents; comparing blood pressures in infants and children of toxemic, hypertensive and normal mothers; geriatric renal disease. *Mailing Add:* Med Col Pa 300 Henry Ave Philadelphia PA 19129

LEVISON, WILLIAM H(ENRY), b Cincinnati, Ohio, Mar 21, 36; m 66; c 2. ENGINEERING PSYCHOLOGY. *Educ:* Mass Inst Technol, BS, 58, MS, 60, ScD, 64. *Prof Exp:* Sr Scientist, 64-88, DIV SCIENTIST, BOLT BERANEK & NEWMAN, INC, 64-; DIV SCIENTIST, BBN LABS INC, 64. *Mem:* Inst Elec & Electronics Engrs; Am Inst Aeronaut & Astronaut. *Res:* Manual control systems and human operator modelling. *Mailing Add:* 19 Phinney Rd Lexington MA 02173

LEVIT, EDITHE J, b Wilkes-Barre, Pa, Nov 29, 26; m 52; c 2. MEDICINE, MEDICAL ADMINISTRATION. *Educ:* Bucknell Univ, BS, 46; Woman's Med Col Pa, MD, 51. *Hon Degrees:* DMS, Med Col Pa, 78; DSc, Wilkes Univ, 90. *Prof Exp:* Intern med, Philadelphia Gen Hosp, 51-52, fel endocrinol, 52-53, clin instr, 53-57, dir med educ, 57-61; asst dir, 61-67, secy & assoc dir, 67-75, vpres & secy, 75-77, pres & chief exec officer, 77-86, EMER PROF & LIFE MEM BD, NAT BD MED EXAMR, 87- *Concurrent Pos:* Bd dirs, Philadelphia Elec Co, 80-, Germantown Savings Bank, Philadelphia, 79-; consult, women in med, Josiah Macy Jr Found, 66-76, Comt Pract Fed Ct, US Judiciary, 77, Off Technol Assessment, US Cong, 78-79; mem sci coun, Nat Lib Med Bd, 81-, adv coun Inst Nuclear Power Opers, Atlanta, 88- *Honors & Awards:* Commonwealth Comt of Woman's Med Col Award, 70; Distinguished Serv Award, Fedn State Med Bds, 87. *Mem:* Inst Med-Nat Acad Sci; AMA; master Am Col Physicians; Asn Am Med Cols. *Res:* Evaluation of professional competence in medicine. *Mailing Add:* 1910 Spruce St Philadelphia PA 19103

LEVIT, LAWRENCE BRUCE, b Cleveland, Ohio, Sept 24, 42; m 67; c 1. PHYSICS. *Educ:* Case Western Reserve Univ, BS, 64, PhD(physics), 71. *Prof Exp:* Res assoc physics, Case Western Reserve Univ, 66-69; asst prof, La State Univ, Baton Rouge, 69-74; MKT MGR, DIV HIGH ENERGY PHYSICS, LECROY RES SYST CORP, 74- *Mem:* Am Phys Soc; Am Inst Physics. *Res:* Ultrahigh energy physics research using cosmic rays as a particle source. *Mailing Add:* Leroy Res Syst Corp 5912 Storebridge Rd Pleasanton CA 94566

LEVIT, ROBERT JULES, b San Francisco, Calif, Aug 17, 16; m 43, 55; c 3. LOGIC, COMPUTER SCIENCE. *Educ:* Calif Inst Technol, BS, 38, MS, 39; Univ Calif, PhD(math), 41. *Prof Exp:* Asst math, Univ Calif, 40-41; from asst prof to assoc prof, Univ Ga, 46-53; vis asst prof, Mass Inst Technol, 54-55; mem staff, Appl Sci Div, Int Bus Mach Corp, 55-57; from asst prof to prof math, 57-72, EMER PROF MATH, SAN FRANCISCO STATE UNIV, 72- *Mem:* Am Math Soc; Math Asn Am; Asn Symbolic Logic. *Res:* Foundations of mathematics; algebra; number theory, analysis. *Mailing Add:* 100 Bay Pl No 1902 Oakland CA 94610

LEVITAN, ALEXANDER ALLEN, b Boston, Mass, Oct 19, 39; c 3. MEDICAL ONCOLOGY, MEDICAL HYPNOSIS. *Educ:* Cornell Univ, BA, 59; Univ Rochester, MD, 63; Univ Minn, MPH, 70; Am Bd Internal Med, dipl, 71 & 77, dipl oncol, 73. *Prof Exp:* Med resident, Harvard Med Serv, Boston City Hosp, 64-65 & med br, Nat Cancer Inst, NIH, Bethesda, MD, 65-67; instr microbiol, Univ Minn, Minneapolis, 68-70, clin assoc prof, Dept Family Pract, 75-89; PVT PRACT, INTERNAL MED & ONCOL, MINNEAPOLIS & FRIDLEY, MINN. *Mem:* Am Fedn Clin Res; fel Am Soc Clin Hypnosis; fel Am Col Physicians; Am Soc Prev Oncol; fel Soc Clin & Exp Hypnosis. *Mailing Add:* 7260 Univ Ave NE Suite 235 Fridley MN 55432

LEVITAN, HERBERT, b Brooklyn, NY, Apr 25, 39; m 64; c 2. NEUROBIOLOGY, MEMBRANE BIOPHYSICS. *Educ:* Cornell Univ, BEE, 62, PhD(phys biol), 65. *Prof Exp:* NIH fel neurophysiol, Brain Res Inst, Univ Calif, Los Angeles, 65-67, anatomist, Anat Dept, 67; NIH fel, Lab Neurophysiol Cellulaire Ctr Etude Physiol Nerveuse, Paris, France, 68-70, Lab Neurophysiol, NIMH, 70, Lab Neurobiol, Nat Inst Child Health & Human Develop, 70-72; assoc prof, 72-83, PROF, DEPT ZOOL, UNIV MD, COL PARK, 83- *Concurrent Pos:* Instr neurobiol, Marine Biol Lab, Woods Hole, 74; neurophysiologist, Lab Neurosci Gerontol Res Ctr, Nat Inst Aging, 79-82; Fulbright Hayes Fel, 87-88; prog dir, NSF, Washington, DC, 90-92. *Mem:* Soc Neurosci; Am Physiol Soc; Soc Gen Physiologists; Am Soc Cell Biol. *Res:* Physico-chemical and biophysical mechanisms underlying the effects of drugs on the physiology of nerves and muscles. *Mailing Add:* Dept Zool Univ Md College Park MD 20742

LEVITAN, MAX, b Tverai, Lithuania, Mar 1, 21; nat US; m 47; c 3. GENETICS, ANATOMY. *Educ:* Univ Chicago, AB, 44; Univ Mich, MA, 46; Columbia Univ, PhD(zool), 51. *Prof Exp:* Statistician, USPHS, 44-45; asst zool, Columbia Univ, 46-49; assoc prof genetics, Va Polytech Inst, 49-55; from asst prof to assoc prof anat, Woman's Med Col Pa, 55-62, prof anat & med genetics, 62-66; prof biol & chmn dept, George Mason Col, Univ Va, 66-68; assoc prof, 68-70, PROF ANAT, MT SINAI SCH MED, CITY UNIV NY, 70- *Concurrent Pos:* Seminar assoc, Columbia Univ, 58-; spec lectr, Univ Pa, 62-63; adj prof anat & genetics, George Wash Univ & Sch Med, Univ Va, 66-68, assoc ed, Evolution, 77-79; auth. *Honors & Awards:* Just Lectr, Howard Univ, 68. *Mem:* Am Asn Anatomists; Am Soc Naturalists; Am Soc Human Genetics; Genetics Soc Am; Soc Study Evolution; Soc Study Social Biol. *Res:* Cytogenetics; population genetics of linked loci; chromosome breakage; cytoplasmic inheritance; medical genetics; author of textbook. *Mailing Add:* 1212 Fifth Ave New York NY 10029

LEVITAN, MICHAEL LEONARD, b Brooklyn, NY, Sept 12, 41; div; c 2. MATHEMATICS. *Educ:* Rensselaer Polytech Inst, BS, 62; Univ Minn, MS, 66, PhD(math), 67. *Prof Exp:* Asst prof math, Drexel Univ, 67-70; asst prof, 70-74, ASSOC PROF MATH, VILLANOVA UNIV, 74- *Mem:* Am Math Soc; Math Asn Am; Am Asn Univ Professors. *Res:* Probability theory; Markov processes; operations research; statistics; math anxiety. *Mailing Add:* Dept Math Sci Villanova Univ Villanova PA 19085

LEVITAN, RUVEN, b Kaunas, Lithuania, Mar 12, 27; US citizen; m 49; c 3. INTERNAL MEDICINE, GASTROENTEROLOGY. *Educ:* Hebrew Univ, Israel, MD, 53; Am Bd Internal Med, dipl & cert gastroenterol. *Prof Exp:* Resident, Mt Sinai Hosp, NY, 56-57; resident, Beth Israel Hosp, Boston, 58-59; dir gastroenterol res, New Eng Med Ctr Hosps, 64-68; assoc prof, 68-70, PROF MED, ABRAHAM LINCOLN SCH MED, UNIV ILL MED CTR, 70- *Concurrent Pos:* Spec fel med neoplasia, Mem Ctr Cancer & Allied Dis, NY, 57-58; fel gastroenterol & res fel med, Mass Mem Hosps & Sch Med, Boston Univ, 59-61, sr res fel, Mass Mem Hosps, 61-62; from asst prof to assoc prof, Sch Med, Tufts Univ, 64-69; lectr, Sch Med, Boston Univ, 65-68; pres, Chicago Soc Gastroenterol; chief gastroenterol sect, Vet Admin West Side Hosp, 68-77. *Mem:* Fel Am Col Physicians; Am Physiol Soc; Am Gastroenterol Asn; Am Asn Study Liver Dis; Am Soc Clin Invest. *Res:* Water electrolyte absorption from the intestine; hormonal influences on absorption; lymphomas, including involvement of liver and gastrointestinal tract. *Mailing Add:* Dept Med Gastroenterol Univ Ill Abraham Lincoln Sch Med 4709 Golf Rd Suite 1000 Skokie IL 60076

LEVITAS, ALFRED DAVE, b New York, NY, Mar 27, 20; m 43; c 1. PHYSICS. *Educ:* Syracuse Univ, BA, 47, MS, 50, PhD(physics), 58. *Prof Exp:* Res engr solid state physics, Sylvania Elec Corp, 53-55 & Sprague Elec Corp, 55-56; physicist, Honeywell Res Ctr, 56-58; prof physics, State Univ NY, Albany, 58-; RETIRED. *Concurrent Pos:* Consult, Naval Res Lab, 60-63. *Mem:* Am Phys Soc. *Res:* Solid state and statistical physics; thermodynamics. *Mailing Add:* 420 Sand Creek Rd No 604 SUNY at Albany 1400 Walsh Ave Albany NY 12205

LEVITICUS, LOUIS I, b Aalten, Neth, July 4, 31; Dutch & US citizen; m 82; c 4. TRACTOR-OFF ROAD VEHICLE DEVELOPMENT, ENGINE DEVELOPMENT. *Educ:* Technion, Israel Inst Technol, BSc, 60, MSc, 63; Purdue Univ, PhD(agr eng), 69. *Prof Exp:* Sr res engr, Stevens Inst Technol, 69-71; consult & instr agr eng, Technion, Israel Inst Technol & Israel Defense Forces, 71-75; chief engr, Tractor Test Lab, 75-87, assoc dir, Ctr Agr Equip, 87-90, PROF AGR ENG, UNIV NEBR, 75-, SUPV TEST & DEVELOP, NEBR POWER LAB, 90- *Mem:* Int Soc Terrain Vehicle Systs; Soc Automotive Engrs; Am Soc Agr Engrs; Asian Soc Agr Engrs. *Res:* Off-road locomotion, oriented on agricultural and military soil-wheel interaction; alternate fuel use in diesel engines, alcohol, ethanol, ETBE, natural gas, vegetable oils; instrumentation for handicapped farmers. *Mailing Add:* Rm 207 L W Chase Hall Univ Nebr Lincoln NE 68583-0726

LEVITON, ALAN, b Brooklyn, NY, June 17, 38; m 63; c 2. PUBLIC HEALTH, EPIDEMIOLOGY. *Educ:* NY Univ, AB, 59; State Univ NY, MD, 63; Harvard Univ, SM, 71. *Prof Exp:* Intern, King's County Hosp, Brooklyn, 63-64, resident, 64-65; officer epidemiol, Epidemic Intel Serv, Ctr Dis Control, USPHS, 65-67; resident neurol, Wash Univ, St Louis, 67-70; teaching fel epidemiol, Sch Pub Health, 70-71, from instr to asst prof, 71-78, ASSOC PROF NEUREL, HARVARD UNIV SCH MED, 78- *Concurrent Pos:* Consult, biomet & field studies, Nat Inst Neurol & Commun Dis & Stroke, NIH, 75-84, mem, proj rev A comt, Neurol Dis Prog, 76-80 & adv comt, Stroke & Trauma Prog, 85. *Mem:* Am Acad Neurol; Soc Epidemiol

Res; Am Pub Health Asn; Am Neurol Asn; Child Neurol Soc; Am Col Epidemiol. *Res:* Using epidemiological techniques to search for antecedents of neurologic handicaps in children. *Mailing Add:* Children's Hosp 300 Longwood Ave Boston MA 02115-5747

LEVITON, ALAN EDWARD, b Brooklyn, NY, Jan 11, 30; m 52; c 2. SYSTEMATIC ZOOLOGY, ZOOGEOGRAPHY. *Educ:* Stanford Univ, AB, 49, AM, 53, PhD, 60. *Prof Exp:* From asst cur to assoc cur, 57-62, chmn dept, 62-83, CUR HERPET & CHMN COMPUT SERV, CALIF ACAD SCI, 83- *Concurrent Pos:* Assoc cur div syst biol, Stanford Univ, 62-63, lectr, 62-70; adj prof biol sci, San Francisco State Univ, 67-; exec dir, Pac Div AAAS, 79- *Mem:* Fel AAAS; Soc Syst Zool; Soc Study Amphibians & Reptiles; Am Soc Ichthyol & Herpet; Geol Soc Am; Hist Sci Soc; Hist Earth Sci Soc. *Res:* Herpetology of Asia; Tertiary paleogeography; phylogeny and taxonomy of reptiles. *Mailing Add:* Dept Herpet Calif Acad Sci San Francisco CA 94118

LEVITSKY, LYNNE LIPTON, b Columbia, SC, May 14, 42; m 67; c 3. ENDOCRINOLOGY, PEDIATRICS. *Educ:* Bryn Mawr Col, BA, 62; Yale Univ, MD, 66. *Prof Exp:* Intern pediat, Bronx Munic Hosp Ctr, 66-67; resident, Childrens Hosp Philadelphia, 67-68; fel endocrinol & metab, Sch Med, Univ Md, 68-70; asst prof pediat, Sch Med, Univ Ill, 70-73; dir div pediat endocrinol, Michael Reese Hosp Med Ctr, 73-86; from asst prof to assoc prof, 73-85, PROF, PEDIAT PRITZKER SCH MED, UNIV CHICAGO, 55- *Mem:* Soc Pediat Res; Endocrine Soc; Lawson Wilkins Pediat Endocrine Soc. *Res:* Carbohydrate metabolism; diabetes; fetal and neonatal metabolism and endocrinology. *Mailing Add:* Wyler Childrens Hosp 5841 Maryland Ave Chicago IL 60637

LEVITSKY, MYRON, b New York, NY, June 22, 30. MECHANICAL ENGINEERING. *Educ:* Cooper Union, BME, 51; NY Univ, MS, 64, PhD(mech eng), 69. *Prof Exp:* Res engr, Heat & Mass Flow Analyzer Lab, Columbia Univ, 51-53; instr mech eng, NY Univ, 53-54; proj engr, Consumer's Union, 62-63; LECTR & ASSOC PROF MECH ENG, CITY COL NEW YORK, 65- *Concurrent Pos:* Assoc res scientist, NY Univ, 65-70, vis mem, Courant Inst, 77-78; NSF res grant, City Col New York, 71-72. *Mem:* AAAS; Am Soc Mech Engrs; Am Phys Soc; Am Soc Eng Educ; Am Acad Mech. *Res:* Elasticity theory; heat transfer; thermal stresses in chemically hardening media; applications to concrete and plastic molding. *Mailing Add:* 392 Central Park W New York NY 10025

LEVITSKY, SIDNEY, b New York, NY, Mar 3, 36; m 67; c 3. CARDIOVASCULAR SURGERY, SURGERY. *Educ:* Albert Einstein Col Med, MD, 60; Bd Surg & Bd Thoracic Surg, dipl, 68. *Prof Exp:* Instr surg, Sch Med, Yale Univ, 64-66; chief surg, Third Surg Hosp, Vietnam, 66-67; thoracic surgeon, Valley Forge Army Hosp, 67-68; sr investr cardiac surg, Nat Heart Inst, NIH, 68-70; assoc prof surg, 70-75, PROF SURG & PHARMACOL, COL MED, UNIV ILL, 75-, CHIEF, DIV CARDIOTHORACIC SURG, MED CTR, 74-, LECTR SURG, COOK COUNTY, GRAD SCH, 70- *Concurrent Pos:* Estab investr, Am Heart Asn, 71; attend surgeon, Cook County Hosp, 73-; sr consult, West Side Vet Hosp, 75- *Mem:* Soc Univ Surgeons; Am Physiol Soc; Soc Thoracic Surgeons; Am Asn Thoracic Surg; Asn Acad Surg; Sigma Xi; Am Surg Asn. *Res:* Thoracic surgery; non-invasive methods of monitoring myocardial contractility; intra-operative protection of myocardium; myocardial ischemia and metabolism. *Mailing Add:* Dept Surg Univ Ill Med Ctr PO Box 6998 Chicago IL 60680

LEVITT, ALBERT P, b Lynn, Mass, Jan 17, 24; m 51; c 3. MATERIALS SCIENCE, MECHANICAL ENGINEERING. *Educ:* Harvard Univ, AB, 44, MS, 47. *Prof Exp:* Mech engr, Pratt & Whitney Aircraft Proj, Harvard Univ, 44-46, engr jet engine compressor res, 47-51; mech engr, US Naval Ord Lab, Md, 51-52; guided missile design engr, Rocket Br, Res & Develop Div, Off Chief Ord, Pentagon, 52-54; consult, High Temperature Missile Mat Probs, 54-55, chief, High Temperature Mat Br, Metals Lab, 54-66, chief, Interdisciplinary Res Lab, 66-68, staff adv composites, 68-70, chief, Metal Matrix Composites Group, Army Mat & Mech Res Ctr, 70-85, CHIEF, PROCESSING RES BR, ARMY MAT TECHNOL LAB, 85- *Concurrent Pos:* Army rep panels, Mat for Guided Missiles, Mat Adv Bd, Nat Acad Sci, 55-56, Ceramics, 57, Alloys for Use Elevated Temperatures, 56-57, Aircraft Appln, 58-59 & Plasma Phenomena, 59-60; mem bd, US Civil Serv Exam for Engrs New Eng Area, 58-; mem, Inorg Non-Metallic Mat Panel, Tripartite Tech Coop Prog, 60-63; consult, Army Mat Prob, Bell Tel Labs, NJ, 62; chmn, Achievement Awards Comt, Army Mat & Mech Res Ctr, 62-; alt mem, Composites for Turbines Res Panel, US Air Force, 65; co-developer, magnesium alloy-graphite fiber composites, 71; guest lectr, Sandia Corp, Ohio State Univ, Univ NH, NY Univ, Brown Univ & Univ Calif, Los Angeles, Continuing Educ Inst; mem, Dod Metal Matrix Composites Steering Comt. *Mem:* Am Soc Mech Engrs; Am Soc Testing & Mat. *Res:* Developing new and improved materials for high temperature service in army weapons; metals, ceramics, whisker and reinforced composites; research and development of advanced metal matrix composites for army aircraft, bridging, vehicles and weapon systems; production of graphite fiber reinforced metals; metal processing research and development; two patent awards. *Mailing Add:* 75 Lovett Rd Newton Centre MA 02159

LEVITT, BARRIE, b Brooklyn, NY, Aug 19, 35; m 68; c 1. PHARMACOLOGY, INTERNAL MEDICINE. *Educ:* State Univ NY Downstate Med Ctr, MD, 59. *Prof Exp:* Rotating intern, Mt Sinai Hosp, New York, 59-60, resident med, 60-63; fel pharmacol, State Univ NY Downstate Med Ctr, 63-64; fel, Med Col, Cornell Univ, 64-65, from instr to asst prof pharmacol, 65-69; asst prof med, New York Med Col, 69-70, assoc prof med & pharmacol & dir, Div Clin Pharmacol, 70-80; PROF MED & CARDIOL, ALBERT EINSTEIN COL MED, 80- *Concurrent Pos:* NY Heart Asn sr investr, Med Col, Cornell Univ, 66-69; consult, Bur Drugs, US Food & Drug Admin, 71-; clin prof med, Albert Einstein Col Med. *Mem:* Am Soc Pharmacol & Exp Therapeut; Am Heart Asn; Sigma Xi. *Res:* Clinical and cardiovascular pharmacology; cardiology. *Mailing Add:* 1100 Park Ave New York NY 10028

LEVITT, DAVID GEORGE, b Minneapolis, Minn, May 9, 42; m 64; c 2. PHYSIOLOGY. *Educ:* Univ Minn, BS, 66, MD & PhD(physiol), 68. *Prof Exp:* Assoc prof, 68-77, PROF PHYSIOL, UNIV MINN, MINNEAPOLIS, 77- *Res:* Theoretical transport processes across membranes and in capillary beds; intestinal absorption; microcirculation in skeletal muscle. *Mailing Add:* Physiol 6-255 Millard Hall Univ Minn Med Sch 435 Delaware St S E Minneapolis MN 55455

LEVITT, GEORGE, b Newburg, NY, Feb 19, 25; m 50; c 4. ORGANIC CHEMISTRY. *Educ:* Duquesne Univ, BS, 50, MS, 52; Mich State Univ, PhD, 57. *Prof Exp:* Res chemist, Exp Sta, E I du Pont de Nemours & Co, Inc, 56-63, res chemist, Stine Lab, 63-66, res chemist, Exp Sta, 66-68, sr res chemist, 68-80, res assoc, 81-86; RETIRED. *Concurrent Pos:* Instr, Del Tech & Community Col, 75-80. *Honors & Awards:* Quadrennial Award Pesticide Res, Swiss Soc Chem Industs, 82; Nat Agr Award Excellence, Nat Agr Mkt Asn, 87 & 88; Am Chem Soc Award Creative Invention, Corp Assocs Am Chem Soc, 89. *Mem:* Am Chem Soc; Int Union Pure & Appl Chem; AAAS; Sigma Xi. *Res:* Organic syntheses; herbicides, fungicides, medicinals; pesticides; heterocyclic compounds; synthesis, characterization and identification of novel organic compounds for biological evaluation; developed programs to define and optimeze chemical structure- biological activity relationships; sulfonylurea herbicides; heterocyclics; exploratory process research. *Mailing Add:* 110 Downs Dr Greenville DE 19807-2556

LEVITT, ISRAEL MONROE, b Philadelphia, Pa, Dec 19, 08; m 37; c 2. ASTRONOMY. *Educ:* Drexel Univ, BS, 32; Univ Pa, MA, 37, PhD(astron), 48. *Hon Degrees:* DSc, Temple Univ, 58, Drexel Univ, 58, Philadelphia Col Pharm, 63. *Prof Exp:* Engr, Abrasive Co, 29-30; astronr, Franklin Inst, 33-39, asst dir, Fels Planetarium, 39-48, dir, 48-70, vpres, Inst, 70-72; EXEC DIR, MAYOR'S SCI & TECHNOL COUN, PHILADELPHIA, PA, 72- *Concurrent Pos:* Engr, Eclipse Exped, Franklin Inst, 32, asst assoc dir astron, photog & seismol, 38-48 & assoc dir astron & seismol, 49-70; astronr, Cook Observ, Univ Pa, 35-46; mem, Air Pollution Control Bd, Philadelphia, 64-, chmn, 66- *Honors & Awards:* Cert Recognition, NASA, 77; Joseph Priestly Award, Spring Garden Inst, 63; Samuel S Fels Medal, 70. *Mem:* AAAS; fel Am Astronaut Soc; Am Inst Aeronaut & Astronaut; Am Astron Soc; Brit Astron Soc. *Res:* Lunar studies; scientific museum and planetarium operation; technology transfer. *Mailing Add:* 3900 Ford Rd Apt 19-D Philadelphia PA 19131

LEVITT, JACOB, plant physiology; deceased, see previous edition for last biography

LEVITT, LEONARD SIDNEY, inorganic chemistry, physical organic chemistry, for more information see previous edition

LEVITT, LEROY P, b Plymouth, Pa, Jan 8, 18; m 71; c 4. PSYCHIATRY. *Educ:* Pa State Univ, BS, 39; Chicago Med Sch, MD, 43; Inst Psychoanal, cert, 59. *Prof Exp:* Pvt pract, 49-66; prof psychiat & dean, Chicago Med Sch, 66-73; dir dept ment health, State of Ill, 73-76; vpres med affairs, 76-81, chmn, dept psychiat, 81-88, DIR, MED EDUC, MT SINAI HOSP MED CTR, 88-; PROF PSYCHIAT, RUSH MED COL, 76- *Concurrent Pos:* Consult, Chicago Am Red Cross, 50-54, Asn Family Living, 50- 54 & Nat Coun Aging, 52-; mem, Mayor's Comn Aging, 60- & Gov Comn Ment Health Planning Bd, 66-; pres, Chicago Bd Health, 78-82. *Honors & Awards:* Chicagoan Year Award in Med, 71; Laughlin Award, Am Col Psychoanalysts. *Mem:* Fel Am Psychiat Asn; fel Am Psychoanal Asn; fel Acad Psychoanal; Am Col Psychiat; Am Col Psychoanal (1st vpres, 81-82, pres, 84-85). *Res:* Process of aging; medical education and administration and study of personality of medical students; psychoanalysis; geriatric psychiatry; mental health. *Mailing Add:* Mt Sinai Hosp California & 15th Sts Chicago IL 60608

LEVITT, MARVIN FREDERICK, b New York, NY, Dec 9, 20; c 2. NEPHROLOGY. *Educ:* Cornell Univ, BA, 41; NY Univ, MD, 44. *Prof Exp:* Res asst med, Mt Sinai Sch Med, 50-53; asst attend physician, Mt Sinai Hosp, 53-60; CHIEF DIV NEPHROLOGY, DEPT MED, MT SINAI SCH MED, 60-, PROF MED, 68- *Concurrent Pos:* Mem cardio-vascular renal panel, Mayor's Res Coun, 69-72; mem sci adv bd, NY State Kidney Dis Inst, 69-72; emer mem, Nat Heart Inst Training Comt; chmn med adv bd, NY Kidney Dis Found. *Mem:* NY Acad Sci; Am Soc Clin Invest; Am Fedn Clin Res; fel Am Col Physicians; Asn Am Physicians. *Mailing Add:* Mt Sinai Sch Med Ctr 19 E 98th St Suite 7A New York NY 10029

LEVITT, MELVIN, b Chicago, Ill, Mar 13, 25; div; c 1. NEUROBIOLOGY. *Educ:* Roosevelt Univ, BS, 49, MA, 53; Mich State Univ, PhD(psychol), 58. *Prof Exp:* Res asst neurol & psychiat, Med Sch, Northwestern Univ, 52-54; res assoc neurophysiol, Rockefeller Inst, 61; assoc anat, Sch Med, Univ Pa, 61-65, asst prof anat & mem, Inst Neurol Sci, 65-70; assoc prof physiol, Bowman Gray Sch Med, 70-90; RETIRED. *Concurrent Pos:* USPHS fel, Inst Neurol Sci, Sch Med, Univ Pa, 57-61. *Mem:* Am Physiol Soc; Am Asn Anat; Soc Neurosci; Int Asn Study Pain. *Res:* Dysesthesias of central neural origin in subhumans. *Mailing Add:* 724 Chester Rd Winston-Salem NC 27104-1706

LEVITT, MICHAEL D, b Chicago, Ill, May 10, 35; m 56; c 3. GASTROENTEROLOGY. *Educ:* Univ Minn, BS, 58, MD, 50. *Prof Exp:* Intern, Univ Minn Hosp, 60-61; resident, Boston Univ Hosp, 61-64; resident, Beth Israel Hosp, Boston, 64-65; fel gastroenterol, Boston City Hosp, 65-68; from asst to assoc prof, 68-74, PROF MED, MED SCH, UNIV MINN, 74- *Concurrent Pos:* Guest lectr, Gastroenterol Res Group, 72; counr, Am Fedn Clin Res, 74-76; consult med, Minneapolis Vet Admin Hosp, 74- *Mem:* Am Fedn Clin Res; Am Soc Clin Invest; Am Gastroenterol Soc. *Res:* Studies employing gas to investigate gastrointestinal physiology and studies of serum and urinary isoamylases. *Mailing Add:* 150 Malcolm Ave SE Minneapolis MN 55414

LEVITT, MORTON, NEUROCHEMISTRY. *Prof Exp:* SR RES ASSOC, NY STATE PSYCHIAT INST, 81- *Mailing Add:* Dept Neurochem NY State Psychiat Inst 722 W 168th St New York NY 10032

LEVITT, SEYMOUR H, b Chicago, Ill, July 18, 28; div; c 3. RADIOTHERAPY. *Educ:* Univ Colo, BA, 50, MD, 54. *Prof Exp:* Instr radiation ther & radiol, Med Sch, Univ Mich, 61-62; asst radiotherapist, Sch Med & Dent, Univ Rochester, 62-63; assoc prof radiation ther & chief div, Sch Med, Univ Okla, 63-66; prof radiol & chmn div radiation ther, Med Col Va, 66-70; PROF THERAPEUT RADIOL & HEAD DEPT, UNIV MINN, MINNEAPOLIS, 70- *Concurrent Pos:* Consult radiother, Vet Admin Hosp, Minneapolis; trustee, Am Bd Radiol; mem, Am Joint Comt; pres, Soc Chmn Acad Radiol Oncol Prog, 74-76, mem bd dirs, 76-78. *Mem:* Fel Am Col Radiol; Am Radium Soc (pres, 83); Soc Nuclear Med; Radiol Soc NAm; Am Soc Therapeut Radiol (pres, 78-79). *Res:* Experimental and clinical radiation therapy; radiation biology. *Mailing Add:* Dept Therapeut Radiol Univ Minn Hosps Minneapolis MN 55455

LEVITZ, HILBERT, b Lebanon, Pa, Nov 13, 31; m 80. MATHEMATICS, COMPUTER SCIENCE. *Educ:* Univ NC, BA, 53; Pa State Univ, PhD(math), 65. *Prof Exp:* Instr math, Williams Col, 65; asst prof, NY Univ, 65-69; assoc prof math, 69-84, PROF COMPUTER SCI, FLA STATE UNIV, 84- *Mem:* Am Math Soc; Asn Symbolic Logic; Asn Computing Mach. *Res:* Mathematical logic; concrete systems of ordinal notations. *Mailing Add:* Dept Computer Sci Fla State Univ Tallahassee FL 32306

LEVITZ, MORTIMER, b New York, NY, May 11, 21; m 47; c 2. BIOCHEMISTRY, ENDOCRINOLOGY. *Educ:* City Col New York, BS, 41; Columbia Univ, MA, 47, PhD(org chem), 51. *Prof Exp:* Res assoc steroid biochem, Col Physicians & Surgeons, Columbia Univ, 51-52; res assoc, 52-56, from asst prof to assoc prof, 56-67, PROF OBSTET & GYNEC, MED CTR, NY UNIV, 67- *Concurrent Pos:* NIH res career award, 62-72; consult, Endocrine Study Sect, NIH, 66-70 & 73-75, Clin Sci Study Sect, 81-85, chmn, 83-85; ed, Endocrinol, 83-87, assoc ed-in-chief, 86-87. *Mem:* Am Chem Soc; Am Soc Biol Chemists; Endocrine Soc; Soc Gynec Invest. *Res:* Estrogen metabolism and mechanisms of action in pregnancy and cancer. *Mailing Add:* NY Univ Med Ctr 550 First Ave New York NY 10016

LEVITZKY, MICHAEL GORDON, b Elizabeth, NJ, Jan 3, 47; m 85; c 1. PULMONARY PHYSIOLOGY, CARDIOVASCULAR PHYSIOLOGY. *Educ:* Univ Pa, BA, 69; Albany Med Col, Union Univ, PhD(physiol), 75. *Prof Exp:* Instr physiol, Albany Med Col, Union Univ, 74-75; from asst prof to assoc prof, 75-85, PROF PHYSIOL, MED CTR, LA STATE UNIV, 85- *Concurrent Pos:* Consult, NIH grants, 74-76 & 79-80, prin investr, 76-81 & 82-86; chmn, Acad Studies Comt, La State Univ, 81-89, Curric Comt, 82-88, 90-; mem, Basic Sci Coun, Am Heart Asn; adj prof pediat, Tulane Univ Med Ctr, 90- *Mem:* Am Physiol Soc; Sigma Xi; Soc Exp Biol & Med; NY Acad Sci; Am Thoracic Soc. *Res:* Cardiopulmonary physiology, particularly in those factors that control pulmonary blood flow. *Mailing Add:* Dept Physiol Med Col La State Univ 1901 Perdido St New Orleans LA 70112-1393

LEVKOV, JEROME STEPHEN, b New York, NY, June 12, 39; m 70. PHYSICAL CHEMISTRY, FORENSIC SCIENCES. *Educ:* City Col New York, BS, 61; Univ Pa, PhD(phys chem), 67. *Prof Exp:* Swiss Copper Inst fel, Swiss Fed Inst Technol, 67-68; asst prof gen & phys chem, Drexel Univ, 68-69; asst prof, 70-80, PROF GEN & PHYS CHEM, IONA COL, 80- *Mem:* N Am Thermal Anal Soc; Am Chem Soc. *Res:* Transport properties in electrolyte solutions; structure of solutions of electrolytes in solvents of low dielectric constatn; polymorphic transitions; forensic chemistry; computers in chemistry education. *Mailing Add:* 3801 Hudson Manor Terr Riverdale NY 10463-1111

LEVOW, ROY BRUCE, b Richmond, Va, June 3, 43; m 62; c 2. COMPUTER SCIENCE. *Educ:* Univ Pa, AB, 64, PhD(appl math), 69. *Prof Exp:* Sci programmer, Atlantic-Richfield Co, 64-65; asst prof math, Univ Hawaii, 69-70; from asst prof to assoc prof math, 70-80, chmn dept, 74-78, assoc prof comput & info syst, 80-87, spec asst comput & commun planning, 85-86, ASSOC PROF COMPUT SCI, FLA ATLANTIC UNIV, 87- *Concurrent Pos:* Consult comput aids, classification & retrieval, 75- & prof develop & comput personnel, 79- *Mem:* Asn Comput Mach; Inst Elec & Electronics Engrs. *Res:* Software engineering; CASE tools; data communications; database systems; operating systems; computer science education; programming languages and programming environments; information retrieval. *Mailing Add:* Dept Comput Sci Fla Atlantic Univ Boca Raton FL 33431

LEV-RAN, ARYE, b Leningrad, USSR, June 7, 30; US citizen; m 68; c 4. DIABETES, INTERNAL MEDICINE. *Educ:* First Leningrad Med Sch, MD, 53; Cand Sci, Inst Physiol, Acad Sci, Leningrad, 59, DSci, 64. *Prof Exp:* Jr res scientist, 56-64, sr res scientist endocrinol, Dept Gen Endocrinol, Inst Ob/Gyn, Acad Med Sci, Leningrad, USSR, 64-67; dir Central Endocrinol Lab of Sick Fiend, Tel Aviv, Israel, 67-75, sr lectr, Tel Aviv Univ Sch Med, 69-75; staff physician, Scripps Clin, La Jolla, CA, 77-81; RES SCIENTIST, DEPT DIABETES & ENDOCRINOL, CITY HOPE MED CTR, DUARTE, CA, 81- *Concurrent Pos:* Clin prof med, Univ Southern Calif, 82- *Mem:* Am Diabetes Asn; Endocrine Soc; Europ Asn Study Diabetes; fel Am Col Physicians; AAAS. *Res:* Pathogenesis of diabetes; epidemial growth factor in pathology; EGF receptors in carcinogenosis. *Mailing Add:* City Hope Nat Med Ctr Duarte CA 91010

LEVY, ALAN, b New York, July 25, 37; m 62; c 2. POLYMER CHEMISTRY. *Educ:* City Col New York, BS, 58; Purdue Univ, PhD(chem), 62. *Prof Exp:* Sr res chemist, Cent Res Lab, Allied Chem Corp, NJ, 62-66; sr res scientist org polymer chem, 66-70, prin scientist & group leader, 70-75, mgr, Polymer Dept, 75-77, assoc dir res, 77-78, dir develop, 78-81, vpres res & develop, 81-84, vpres mech prod & new technol, 84-86, VPRES RES & NEW TECHNOL, ETHICON INC, 86- *Honors & Awards:* Johnson Medal, 81. *Mem:* AAAS; Am Chem Soc; NY Acad Sci. *Res:* Biomedical materials; polymer and synthetic organic chemistry. *Mailing Add:* Ethicon Inc Rte 22 Somerville NJ 08876

LEVY, ALAN B, b San Francisco, Calif, Apr 12, 45; m 69; c 2. ORGANOMETALLIC CHEMISTRY. *Educ:* Univ Calif, Berkeley, BS, 67; Univ Colo, Boulder, PhD(chem), 71. *Prof Exp:* Fel chem, Purdue Univ, 71-74; asst prof chem, State Univ NY Stony Brook, 74-80; sr res chemist, Allied Corp, 80-85, res assoc, 85-90, RES SCIENTIST, ALLIED-SIGNAL CORP, 91- *Mem:* Am Chem Soc; Sigma Xi; AAAS. *Res:* The use of organoboranes and organocopper reagents for the development of new synthetic methods; the total synthesis of natural products; homogenous and heterogenous catalysis in organic synthesis; process control and the modeling of chemical processes. *Mailing Add:* Allied Corp 101 Columbia Rd PO Box 1021 Morristown NJ 07962

LEVY, ALAN C, b Baltimore, Md, Feb 24, 30; m 56; c 2. PHYSIOLOGY, TOXICOLOGY. *Educ:* Univ Md, BS, 52; George Washington Univ, MS, 56; Georgetown Univ, PhD(physiol), 58. *Prof Exp:* Instr physiol, Sch Med, Howard Univ, 58-60; sect head, Dept Endocrinol, William S Merrell Co, 60-67; dir labs, Woodard Res Corp, 67-69; group chief, Hoffmann-La Roche Inc, 69-74, sect head, dept toxicol & path, 74-; AT MICROBIOL ASSOCS INC. *Mem:* Am Physiol Soc; Reticuloendothelial Soc; Endocrine Soc; NY Acad Sci; Soc Toxicol; Sigma Xi. *Res:* Inflammation; anti-inflammation; adrenal cortex; neuroendocrinology; lipid metabolism; acute and chronic toxicology; teratology. *Mailing Add:* 9405 Winterset Dr Potomac MD 20854

LEVY, ALAN JOSEPH, b New York, NY, Oct 30, 55; m. MECHANICAL BEHAVIOR OF MATERIALS, DYNAMICS. *Educ:* State Univ, NY, BS, 77; Columbia Univ, MS, 79, MPhil, 81, PhD(eng mech), 82. *Prof Exp:* Asst prof, 82-88, ASSOC PROF MECH, SYRACUSE UNIV, 88- *Concurrent Pos:* Fac res fel mech, Mat Technol Lab, US Army, 87 & 88. *Honors & Awards:* Outstanding Young Men in Am, 83. *Mem:* Am Soc Mech Eng; Soc Indust & Appl Math; Am Acad Mech; Sigma Xi. *Res:* Mechanical behavior of materials and dynamical systems; development of phenomenological and physically based models and their use in the study of flow and failure of nonlinearly viscous and degrading material; debonding of inclusions and inhomogeneitics; dynamics of the nonlinear plastic impact oscillator. *Mailing Add:* Dept Mech & Aerospace Eng Syracuse Univ Syracuse NY 13244

LEVY, ALLAN HENRY, b New York, NY, Nov 2, 29; m 61; c 2. COMPUTER SCIENCES, VIROLOGY. *Educ:* Columbia Univ, AB, 49; Harvard Med Sch, MD, 53. *Prof Exp:* From intern to asst resident, Harvard Med Serv, Boston City Hosp, 53-55; clin assoc, Nat Cancer Inst, 55-57; from instr to asst prof microbiol, Johns Hopkins Univ, 59-65; assoc prof virol & comput sci, Baylor Col Med, 65-71, prof comput sci, 71-73, prof virol & epidemiol, 73-75; PROF CLIN SCI & PROF COMPUT SCI, 75-, HEAD, DEPT MED INFO SCI, COL MED, UNIV ILL, URBANA, 83- *Concurrent Pos:* Res fel, Sch Med, Johns Hopkins Univ, 57-59; USPHS res career develop award, 60-65; consult div hosp & med facil, Bur State Serv, USPHS, 65-70; mem, adv comt to dir, Div Res Grants, NIH. *Mem:* Am Fedn Clin Res; Am Med Info Asn. *Res:* Artificial intelligence in medicine; hospital information systems; general applications of digital computers to medicine and biology; hypertext and information retrieval techniques in medicine. *Mailing Add:* 190 Med Sci Bldg Univ Ill Urbana IL 61801

LEVY, ARTHUR, b New York, NY, Sept 29, 21; m 49; c 4. COMBUSTION KINETICS, ATOMOSPHERIC CHEMISTRY. *Educ:* Queen's Col, BS, 43; Univ Minn, MS, 48. *Prof Exp:* Chemist, Los Alamos Nat Lab, 44-46; aeronaut res scientist, Nat Adv Comt Aeronaut, 48-50; phys chemist, Brookhaven Nat Lab, 50-51; prin phys chemist, Columbus Labs, Battelle Mem Inst, 51-59, asst chief, 56-69, fel, 69-71, sr fel, 71-73, sr res leader, 73-76, mgr combustion, 76-79, res leader, 79-85; RETIRED. *Concurrent Pos:* Consult, 85- *Mem:* Am Chem Soc; Combustion Inst; Air Pollution Control Asn. *Res:* Kinetics of hydrogen and hydrocarbon oxidation; combustion chemistry; kinetics of radiation and ionic reactions; boron hydride chemistry; induced reactions; flame structure; air pollution kinetics; coal-oil combustion and environmental assessments; synthetic fuel combustion. *Mailing Add:* 614 Farrington Dr Worthington OH 43085

LEVY, ARTHUR LOUIS, b Bridgeport, Conn, Aug 2, 17; m 43; c 1. ANALYTICAL CHEMISTRY, CLINICAL CHEMISTRY. *Educ:* Univ Mo, AB, 38; Yale Univ, PhD(phys chem), 48. *Prof Exp:* From instr to asst prof phys chem, Rensselaer Polytech Inst, 48-54; chemist, Hodgkins Dis Res Lab, 54-58, CHIEF CHEMIST, ST VINCENT'S HOSP, 58- *Concurrent Pos:* Ford Found fel, 53-54; dir labs, New York Dept Health, 64- *Honors & Awards:* Van Slyhe Award, Am Asn Clin Chem, 87. *Mem:* AAAS; Am Asn Clin Chem; Am Chem Soc; NY Acad Sci; Asn Clin Sci. *Res:* Electrolyte solutions; immunochemistry of Hodgkins disease; enzymes; standards and methodologies in clinical chemistry including automation and data processing. *Mailing Add:* 15 La Mesa Ave East Chester NY 10707

LEVY, ARTHUR MAURICE, b New York, NY, Nov 20, 30; c 3. CARDIOLOGY. *Educ:* Harvard Univ, BA, 52; Cornell Univ, MD, 56; Am Bd Internal Med, dipl, 66. *Prof Exp:* Intern, Cornell Med Div, Bellevue Hosp, 56-57, resident med, 57-58; resident, 58-59, from instr to assoc prof med, Col Med, 63-76, asoc prof pediat, 69-77, PROF MED, COL MED, UNIV MT, 76-, PROF PEDIAT, 77- *Concurrent Pos:* NIH fel cardiol, Col Med, Univ Vt, 59-60; Nat Heart Inst res fel, 59-60; trainee cardiol, Harvard Med Sch, Boston Children's Hosp, 62-63; teaching scholar, Am Heart Asn, 66-71; fel coun clin cardiol, Am Heart Asn, 69- *Mem:* Am Fedn Clin Res; fel Am Col Physicians; fel Am Col Cardiol. *Res:* Clinical electrophysiology. *Mailing Add:* Cardiol Dept Med Ctr Hosp Vt Burlington VT 05401

LEVY, BARNET M, b Scranton, Pa, Jan 13, 17; m 40. HISTOPATHOLOGY. *Educ:* Univ Pa, AB, 38, DDS, 42; Med Col Va, MS, 44; Am Bd Oral Path, dipl. *Prof Exp:* Instr bact, path & clin dent, Med Col Va, 42-44; asst prof bact & path, Wash Univ, 44-47, assoc prof path. 47-49; prof dent & dir res & postgrad studies, Sch Dent & Oral Surg, Columbia Univ, 49-57; PROF PATH, UNIV TEX DENT BR HOUSTON, 57-, DIR, DENT SCI INST, 64- *Concurrent Pos:* Assoc attend dent surgeon, Presby Hosp, New York, 49-57; consult-instr, US Naval Hosp, St Albans, NY, 52-57; USPHS Hosp, Staten

Island, 51-57; consult, Vet Admin Hosp, Bronx, 54-57, Houston, 57-, Univ Tex M D Anderson Hosp & Tumor Inst, 57-; mem Nat Res Coun, 52; chmn dent study sect, NIH, 57-62, training grants comt, 62-67; mem adv comt dent, Comt Int Exchange Persons; pres, Am Bd Oral Path, 65-66, ed, J Dent Res, 76-; adj prof anat, Sch Vet Med, Tex A&M Univ; vis prof, Facultad De Ontologia, Univ Nat Autonoma De Mexico, Mexico City, 78- *Honors & Awards:* Isaac Schour Mem Award, Int Asn Dent Res, 75. *Mem:* Am Soc Exp Path; Soc Exp Biol & Med; Am Asn Cancer Res; fel Am Acad Oral Path (pres, 69-70); Int Asn Dent Res (pres, 65-66). *Res:* Experimental pathology; inflammation; immunopathology and oncology. *Mailing Add:* Dept Pub Health & Hyg Sch Dent Univ Calif Box 0754 San Francisco CA 94143

LEVY, BERNARD, JR, b New Orleans, La, Oct 27, 24; m 51; c 3. MECHANICAL ENGINEERING. *Educ:* Univ Nebr, BS, 45, MS, 48. *Prof Exp:* Engr, 48-57, mgr var activ, 57-82, MGR, STEAM GENERATOR PROG, ADVAN REACTORS DIV, WESTINGHOUSE ELEC CORP, 80- *Mem:* Am Soc Mech Engrs; Sigma Xi. *Res:* Design and development of nuclear reactor plants for naval application. *Mailing Add:* 2669 Strathmore Lane Bethel Park PA 15102

LEVY, BORIS, b New York, NY, Nov 24, 27; m 56; c 3. PHOTOGRAPHIC CHEMISTRY. *Educ:* NY Univ, BA, 48, MS, 50, PhD(phys chem), 55. *Prof Exp:* Res chemist, Sylvania Elec Co, 50-51; sr res chemist, Radio Corp Am, 55-56; sr scientist, Westinghouse Elec Corp, 56-60; sr res chemist, Socony Mobil Oil Co, 60-65; mgr, Imaging Mat Res & Develop, Polaroid Corp, 65-89; RES PROF CHEM, BOSTON UNIV, 89- *Concurrent Pos:* Assoc prof, Trenton Jr Col, 62-65; assoc ed, Photog Sci & Eng, 75. *Mem:* Am Chem Soc; fel Soc Photog Scientists & Engrs. *Res:* Radiotracers; surface chemistry; electrokinetics; photoconductivity; photoelectron emission from semiconductors; spectral sensitization; energy and electron transfer reactions across phase boundaries; photographic emulsion preparation and characterization; preparation of novel image rector layers in diffusion transfer photography; kinetics of photo-induced processes; photovoltaic solar energy conversion. *Mailing Add:* 14 Waltham Rd Wayland MA 01778-1112

LEVY, CHARLES KINGSLEY, b Boston, Mass, Dec 25, 24; div; c 3. RADIATION ECOLOGY. *Educ:* George Washington Univ, BSc, 48, MSc, 51; Univ NC, Chapel Hill, PhD(physiol), 56. *Prof Exp:* Instr physiol, Vassar Col, 56-58; staff scientist, Worcester Found Exp Biol, 58-62; assoc prof radiol & biol, 62-70, PROF BIOL, BOSTON UNIV, 70- *Concurrent Pos:* Res collabr, Brookhaven Nat Lab, 57-61; Am Physiol Soc fel, Boston Univ, 58; staff scientist, Worcester Found Exp Biol, 58-62; consult, Mass Gen Hosp, 62-; consult bioinstrumentation, NASA, 67-; Fulbright prof zool, Univ Nairobi, 69-70; proj dir avian radioecol nuclear reactor site, AEC, Dept Energy, 73-78. *Mem:* Am Physiol Soc; Radiation Res Soc; Soc Gen Physiol. *Res:* Effect of high energy particulate radiation mammalian systems; dose-rate phenomena and responses of sensory and neural tissues to ionizing radiation; biological impact of reactor effluents on free ranging populations of wild birds; kinship in voles by radionuclide tagging and whole-body gamma spectroscopy. *Mailing Add:* Dept Biol Boston Univ Boston MA 02215

LEVY, DANIEL, b New York, NY, Nov 27, 40; m 68. BIOCHEMISTRY, MEMBRANES. *Educ:* City Col New York, BS, 61; Brandeis Univ, MS, 63, PhD(chem), 65. *Prof Exp:* Res biochemist, Univ Calif, Berkeley, 67-68; assoc prof, 74-80, PROF BIOCHEM, SCH MED, UNIV SOUTHERN CALIF, 80- *Concurrent Pos:* NIH fel biochem, Univ Calif, Berkeley, 65-67; NIH res grant, 73-; vis prof biochem, Univ Basel, Switzerland, 77-78. *Mem:* AAAS; Am Soc Biol Chem; Am Chem Soc. *Res:* Membrane structure and function; mechanism of hormone action. *Mailing Add:* Dept Biochem Univ Southern Calif 2025 Zonal Ave Los Angeles CA 90033

LEVY, DAVID ALFRED, b Washington, DC, Aug 27, 30; m 51; c 3. IMMUNOLOGY, ALLERGY. *Educ:* Univ Md, BS, 52, MD, 54; Am Bd Internal Med, cert, 62; Am Bd Allergy & Immunol, cert, 74. *Prof Exp:* From intern to chief resident med, Univ Hosp, Baltimore, Md, 54-59; physician, Pulmonary Dis Serv, Fitzsimons Gen Hosp, Denver, 59-61; staff physician, Chest Serv, Vet Admin Hosp, Baltimore, 61-62; USPHS fel, Sch Med, 62-66, asst prof radiol sci, 66-68, from assoc prof to prof radiol sci & epidemiol, 68-73, PROF BIOCHEM & EPIDEMIOL, SCH HYG & PUB HEALTH, JOHNS HOPKINS UNIV, 73-, PROF PATHOBIOL, 80- *Concurrent Pos:* Fogarty Sr Int fel, Col de France, Paris, 76. *Mem:* AAAS; Am Asn Immunol; fel Am Acad Allergy; Am Soc Trop Med & Hyg; Soc Exp Biol & Med. *Res:* Mechanisms of allergic reactions; mechanisms of immunotherapy for allergic diseases; alpha-antitrypsin and pulmonary disease; parasite immunology. *Mailing Add:* 11 Quai Saint Michel 75005 Paris France

LEVY, DAVID EDWARD, b Washington, DC, May 10, 41; m 67. NEUROLOGY. *Educ:* Harvard Univ, AB, 63; Harvard Med Sch, MD, 68; Am Bd Internal Med, dipl, 72; Am Bd Psychiat & Neurol, dipl, 75. *Prof Exp:* From intern to resident, New York Hosp, 68-72; fel & instr, 72-75, asst prof, 75-80, ASSOC PROF NEUROL, MED COL, CORNELL UNIV, 80- *Concurrent Pos:* Asst attend neurologist, New York Hosp, 75-80, assoc attend neurologist, 80-; teacher-scientist award, Andrew W Mellon Found, 75; estab investr, Am Heart Asn, 78. *Mem:* Soc Neurosci; Fel Am Col Physicians; Am Acad Neurol; Am Neurol Asn; Fel Am Heart Asn. *Res:* Brain carbohydrate and energy metabolism in cerebral ischemia; prediction of outcome from stroke and coma. *Mailing Add:* 450 E 63rd St 12F Ctr New York NY 10021

LEVY, DEBORAH LOUISE, b Minneapolis, MN, Nov 3, 50. PSYCHOLOGY. *Educ:* Univ Chicago, BA, 72, PhD(psychol), 76. *Prof Exp:* Res assoc, Univ Chicago, 72-76; intern clin psychol, NY Hosp, Cornell Med Ctr, 76-77; fel, Menninger Found, 77-79; asst unit chief, Ill State Psychiat Inst, 79-81, RES SCIENTIST & SUPVR CLIN TEACHING, 79-87; dir psychophysiol, Hillside Hosp, Glen Oaks, NY, 87-90; CO-DIR, PSYCHOL LAB, MCLEAN HOSP, 90-; ASSOC PROF, DEPT PSYCHIAT, HARVARD MED SCH, 91- *Concurrent Pos:* Res assoc, Dept Psychiat, Univ Chicago, 80-87; prin investr, res scientist develop award, 81-86. *Honors & Awards:* Kay Menninger Sci Day Award, Menninger Found, 79. *Mem:* AAAS; Am Psychol Asn; NY Acad Sci; Am Psychopath Asn; Soc Neurosci; Sigma Xi. *Res:* Genetics of the major psychoses. *Mailing Add:* McLean Hosp 115 Mill St Belmont MA 02178

LEVY, DONALD HARRIS, b Youngstown, Ohio, June 30, 39; m 64; c 3. CHEMICAL PHYSICS, SPECTROSCOPY. *Educ:* Harvard Univ, BA, 61; Univ Calif, Berkeley, PhD(chem), 65. *Prof Exp:* From asst prof to assoc prof, 67-78, chmn dept, 83-85, PROF CHEM, DEPT OF CHEM & PHYS SCIS, UNIV CHICAGO, 78- *Concurrent Pos:* NIH fel, Cambridge Univ, 65-66; NATO fel, 66-67; Alfred P Sloan fel, 67-73; DuPont fac fel, 69-70; Guggenheim fel, 75-76; mem rev panel chem physics, Nat Bur Standards, 81-84 & chem adv comt, NSF, 82-85; Sigma Xi nat lectr, 81-83; chmn molecular spectros tech group, Optical Soc Am, 82-84; assoc ed, J Chem Physics, 83-; Nato fel, Cambridge Univ, 66-67, Alfred P Sloan fel, 67-73, DuPont Fac fel, 69-70, Guggenheim fel, 75-76. *Honors & Awards:* Albert Noyes lectr, Rochester Univ; H H King lectr, Kans State; Plyler Prize, Am Phys Soc, 87; Bourke lectr, Royal Soc Chem. *Mem:* Nat Acad Sci; fel AAAS; fel Am Phys Soc; fel Am Acad Arts & Sci. *Res:* Optical spectroscopy in supersonic molecular beams; spectroscopy and photochemistry of vander waals molecules; laser induced fluorescence spectroscopy; energy transfer; spectroscopy of porphyrin and related molecules. *Mailing Add:* James Franck Inst Univ Chicago 5640 S Ellis Ave Chicago IL 60637

LEVY, DONALD M(ARC), b Lynbrook, NY, Mar 27, 35; m 57; c 2. DIGITAL SIGNAL PROCESSING, COMMUNICATION SYSTEMS. *Educ:* Univ Wis, BS, 56, PhD(elec eng), 65; Mass Inst Technol, MS, 58. *Prof Exp:* Mem tech staff, Hycon Eastern, Inc, 56-57; res asst & staff mem, Instrumentation Lab, Mass Inst Technol, 57-58; instr elec eng, Univ Wis, 59-61; staff engr, Commun Systs Dept, Int Bus Mach Corp, 61-63; instr elec eng, Univ Wis, 63-65; from asst prof to assoc prof info eng, Univ Iowa, 65-79; SUPVR SIGNAL PROCESSING SYSTS, WESTERN DEVELOP LABS, FORD AEROSPACE & COMMUN CORP, 79- *Mem:* Inst Elec & Electronics Engrs. *Res:* Statistical communication theory; topological network theory; bioengineering; digital signal processing. *Mailing Add:* Ford Areospace & Communications 3939 Fabian Way Palo Alto CA 94303

LEVY, EDWARD KENNETH, ENERGY CONVERSION, POWER GENERATION. *Educ:* Univ Md, BS, 63; Mass Inst Technol, SM, 64, ScD, 67. *Prof Exp:* PROF MECH ENG, LEHIGH UNIV, 67- *Concurrent Pos:* Prof assoc, Nat Acad Eng. *Mem:* Am Soc Mech Engrs; Am Inst Chem Engrs; Am Nuclear Soc. *Res:* Fluid mechanics, heat transfer and applied thermodynamic aspects of energy with emphasis on power generation systems. *Mailing Add:* Energy Res Ctr Packard Lab 19 Lehigh Univ Bethlehem PA 18015

LEVY, EDWARD ROBERT, b New York, NY, Oct 3, 27; m 51; c 4. ORGANIC CHEMISTRY. *Educ:* City Col New York, BS, 49; Univ Kans, PhD(org chem), 63. *Prof Exp:* Asst instr chem, Univ Kans, 49-50 & 51-53; res chemist, Glyco Prod, Inc, 53-57; process chemist, Chemagro Corp, 57-65, asst supvr, Process Develop Lab, 65-66, supvr, 66-68, asst mgr, 68-70, mgr, 70-73, PRIN CHEMIST, AGR DIV, MOBAY CHEM CORP, 73- *Mem:* AAAS; Am Chem Soc; Sigma Xi. *Res:* Organophosphorus insecticides; carbamates; chelating agents; synthesis and process development. *Mailing Add:* Mobay Chem Corp PO Box 4913 Kansas City MO 64120

LEVY, ELINOR MILLER, b New York, NY, Mar 18, 42; m 62; c 2. IMMUNE REGULATION, PSYCHONEUROIMMUNOLOGY. *Educ:* Brandeis Univ, BA, 63; Emory Univ, PhD(biophysics), 72. *Prof Exp:* Res asst biophysics, Univ BC, 73-75; res assoc, 75-76, from instr to asst prof, 76-84, ASSOC PROF IMMUNOL, SCH MED, BOSTON UNIV, 84- *Concurrent Pos:* Study sect mem, Nat Inst Alcohol Abuse & Alcoholism, 90- *Mem:* Am Asn Immunologists; AAAS. *Res:* Immune suppression and T-cell differentiation defects in the acquired immunodeficiency syndrome; psychosocial and neuroendocrine modulation of the immune response in humans. *Mailing Add:* Sch Med Boston Univ 80 E Concord St Boston MA 02118

LEVY, EUGENE HOWARD, b New York, NY, May 6, 44; m 67; c 3. ASTROPHYSICS, PLANETARY GEOPHYSICS. *Educ:* Rutgers Univ, AB, 66; Univ Chicago, PhD(physics), 71. *Prof Exp:* Fel physics & astron, Univ Md, 71-73; asst prof, Bartol Res Found, Franklin Inst, 73-75; from asst prof to assoc prof, 75-83, assoc dept head, 81-83, PROF PLANETARY SCI & DIR, LUNAR & PLANETARY LAB, UNIV ARIZ, 83-, MEM FAC APPL MATH, 81-, HEAD DEPT PLANETARY SCI, 83- *Concurrent Pos:* Ctr Theoret Physics fel, Univ Md, 71-73; mem comt planetary & lunar explor, Nat Acad Sci, 76-79, chmn, 78-82; partic, Comet Halley Sci Working Group, NASA, 77-78; mem space sci bd, Nat Acad Sci, 79-82, co-chmn bd study, Explor Primitive Solar Syst Bodies, 78, chmn comt, Planetary & Lunar Explor, 79-82 & mem Steering group, 84-86; mem Ad Hoc Panels, Space Sci Steering Comt, NASA, 79-80, Solar Syst Explor Comt, 80-83, Solar Syst Explor Div Mgt Coun, 83-85 & Space & Earth Sci Adv Comt, 85-88; sci consult, Rockwell Corp, 80; chmn, Fields & Particles Panel, Int Comet Mission Rev Comt, NASA, mem, Theory Panel, & Rev Panel on Origins of Plasmas in Earth's Neighborhood, 80; mem, Comprehensive & Coord Sci Prog Int Tech Panel on Comets, 80-82; mem NASA deleg, Int Coop in Invest of Halley's Comet, Italy, 81, Joint Working Group on Near-Earth Space, Moon & Planets, USSR, 81 & head US deleg, Nat Acad Sci-Europ Sci Found Joint Working Group on Coop in Planetary Explor, 82-84; mem exec comt, Univs Space Sci Working Group, Asn Am Univs, 82-; mem adv bd, Int Conf on Cometry Explor, Budapest, 82; mem, Study Panel on Renewing US-Soviet Coop in Space Sci, 84; mem, NASA-LPI Comt on Future Space-Sta Sci Projs, 85- & NASA Space-Sta Sci User's Working Group, 85-88; distinguished vis scientist, Jet Propulsion Lab, Calif Inst Technol, 85-91; Mars Explor Strategy Advan Group, NASA, 86; mem, Ariz Theoret Astrophysics Prog, 85-; dir, Ariz Space Grant Col Consortium, 89-; chmn, Comet Rendezvous & Asteroid Flyby Rev Panel, NASA, 86, mem, Mars Rover Sample Return Sci Working Group, 87-89, Planetary Systs Sci Working Group, 88-, Lunar & Planetary Geophys Rev Panel, 88-90, Origins of Solar Systs Progs Rev Panel, 90-91; chmn, Adv Comt Int Coop Mars Sample Return, Space Sci Bd, Nat Acad Sci, 86-88, mem, Comt Coop USSR on Planetary Sci, 88-89, Astron & Astrophysics Surv Comt, Sci Opportunities Panel, 89-90; conf convenor & chmn, Protostars & Planets III, Int Conf Formation Stars & Planetary Systs,

90; mem, Study Panel Robotic Explor of Moon & Mars, US Cong Off Technol Assessment, 91. *Honors & Awards:* Distinguished Pub Serv Medal, NASA, 83; Alexander von Humboldt-Stiftung Sr Scientist Award, Fed Repub Ger, 89. *Mem:* Am Phys Soc; Am Astron Soc; Am Geophys Union; Int Astron Union; Sigma Xi; AAAS. *Res:* Theoretical astrophysics and solar system studies; magnetohydrodynamics; space and solar physics; planetary and geophysics; magnetic field generation; physical processes associated with the origin of the solar system; techniques for the observational discovery and study of other planetary systems. *Mailing Add:* Dept Planetary Sci Lunar & Planetary Lab Univ Ariz Tucson AZ 85721

LEVY, GABOR BELA, b Budapest, Hungary, July 16, 13; nat US; m 38; c 2. CHEMISTRY. *Educ:* Karlsruhe Tech Univ, Dipl Ing, 38; Inst Divi Thomae, PhD(chem), 53. *Prof Exp:* Asst physics, NY Univ, 38-41; sr res chemist & sect head, Schenley Labs, Inc, 42-50, head anal & phys chem res, 50-55; head chem div, Consumers Union US, 55-57; asst to pres, Photovolt Corp, 57-64, sr vpres, 64-82; CONSULT, 82- *Concurrent Pos:* Adj prof, Polytech Inst NY; consult ed, Int Sci Commun. *Mem:* AAAS; Am Chem Soc; Am Asn Clin Chem; Sigma Xi. *Res:* Applied colloid chemistry; spectrophotometry; polarography; electron microscopy; swelling of casein; determination of antibiotics; physical methods in organic chemistry; enzymes; optical rotation. *Mailing Add:* 11 Bossy Lane Wilton CT 06897

LEVY, GEORGE CHARLES, b Brooklyn, NY, June 4, 44; m 79; c 2. COMPUTER METHODS. *Educ:* Syracuse Univ, AB, 65; Univ Calif, Los Angeles, PhD(chem), 68. *Prof Exp:* Mem res staff, Gen Elec Corp, 68-73; from assoc prof to prof chem, Fla State Univ, 73-81; PROF CHEM, SYRACUSE UNIV, 81-, PROF SCI & TECHNOL, 85- *Concurrent Pos:* Alfred P Sloan res fel, 75-77; Camille & Henry Dreyfus teacher-scholar, 76-81; dir res resource multi-nuclear, Nuclear Magnetic Resonance & Data Processing, NIH, 81-90; ed, Comput Enhanced Spectros J, 82-87; adj prof radiol, State Univ NY Upstate Med Ctr, 84-; founder & chmn, New Methods Res Inc, 83- *Mem:* Am Chem Soc; Sigma Xi; Asn Comput Mach. *Res:* Nuclear magnetic resonance spectroscopy and computer methods in chemistry; chemical and biophysical applications of carbon-13, nitrogen-15, and other nuclei nuclear magnetic resonance; statistical expert systems. *Mailing Add:* Chem Dept Syracuse Univ Sci Technol Ctr Syracuse NY 13244-4100

LEVY, GERALD FRANK, b Paterson, NJ, June 20, 38; m 60; c 4. ECOLOGY. *Educ:* Bowling Green State Univ, BS, 60, MA, 61; Univ Wis, PhD(bot), 66. *Prof Exp:* Tech asst, Univ Wis, 63-65; asst prof bot & zool, Univ Wis, Marinette Campus, 65-67; from asst prof to assoc prof biol & ecol, 67-77, PROF BIOL SCI, OLD DOMINION UNIV, 78- *Concurrent Pos:* Bot consult, Animal Ecol Proj, 68-69; vpres, Environ Consult, Inc, 73-81, chmn bd, 81-82; mem sci adv bd, Va Mus Nat Hist, 85-87; pres, Gerald F Levy & Assocs Inc. *Honors & Awards:* Pres Award, Va Wildlife Fedn, 79. *Mem:* Ecol Soc Am; Sigma Xi. *Res:* Phytosociology; tick ecology research; small mammals; transpiration; terpine emission; ecology of red heart disease; natural hydrocarbon emissions effects on air quality; longleaf pine regeneration; fire ecology. *Mailing Add:* Dept Biol Sci Old Dominion Univ Norfolk VA 23529

LEVY, GERHARD, b Wollin, Germany, Feb 12, 28; nat US; m 58; c 3. PHARMACOLOGY. *Educ:* Univ Calif, BS, 55, PharmD, 57. *Hon Degrees:* Dr, Univ Uppsala, 75, Phila Col Pharm & Sci, 79, Long Island Univ, 81, DSc, Univ Ill, 86. *Prof Exp:* Res pharmacist, Med Ctr, Univ Calif, 57-58; from asst prof to assoc prof pharm, 58-64, prof biopharmaceut, 64-72, actg chmn dept, 59-60, chmn, 66-70; DISTINGUISHED PROF PHARMACEUT, SCH PHARM, STATE UNIV NY BUFFALO, 72- *Concurrent Pos:* Vis prof, Hebrew Univ Jerusalem, 66-; consult, Bur Drugs, Food & Drug Admin, 71-73; mem comt probs drug safety, Nat Acad Sci-Nat Res Coun, 71-75; vis prof, Univ Rochester, 72-73; grad prof, Victorian Col Pharm, Melbourne, Australia, 73- *Honors & Awards:* Richardson Pharm Award, 57; McKeen Cattell Distinguished Achievement Award Clin Pharmacol, Am Col Clin Pharmacol, 78; Host-Madsen Medal, Int Pharmaceut Fedn, 78; Oscar B Hunter Mem Award, Am Soc Clin Pharmacol & Therapeut, 82; Volwiler Res Achievement Award, Am Asn Col Pharm, 82; Sidney Riegelman lectr, Univ Calif, San Francisco, 83; Am Col Clin Pharm Therapeut Frontiers Lectr Award, 83; Takeru Higuchi Res Prize, Acad Pharmaceut Sci, 83; Kenneth L Waters lectr, Univ Ga Col Pharm, 84; Vis Prof, Basic Med Sci, 85-86. *Mem:* Inst Med-Nat Acad Sci; Am Chem Soc; fel Am Pharmaceut Asn; Am Soc Pharmacol & Exp Therapeut; fel AAAS. *Res:* Biopharmaceutics; clinical pharmacology; pharmacokinetics. *Mailing Add:* Dept Pharmaceut State Univ NY Sch Pharm Amherst NY 14260

LEVY, HANS RICHARD, b Leipzig, Ger, Oct 22, 29; nat US; m 60; c 1. BIOCHEMISTRY. *Educ:* Rutgers Univ, BSc, 50; Univ Chicago, PhD(biochem), 56. *Prof Exp:* USPHS fel, Ben May Lab, Univ Chicago, 56-58 & Hammersmith Hosp, London, Eng, 58-59; from instr to asst prof biochem, Ben May Lab, Univ Chicago, 59-63; from asst prof to assoc prof, 63-71, PROF BIOCHEM, SYRACUSE UNIV, 71- *Mem:* AAAS; Am Soc Biol Chem; Am Chem Soc; Am Asn Univ Prof. *Res:* Mechanisms of action and regulation of enzymes; protein structural features that determine pyridine nucleotide dehydrogenase coenzyme specificity, especially for various glucose 6-phospate dehydrogenases. *Mailing Add:* Biol Res Labs Dept Biol Syracuse Univ Syracuse NY 13244-1220

LEVY, HARRIS BENJAMIN, b Philadelphia, Pa, Nov 29, 28; m 62; c 1. DATA ANALYSIS. *Educ:* Univ Pa, BS, 50; Univ Calif, PhD(chem), 53. *Prof Exp:* Teaching asst chem, Univ Calif, Berkeley, 50-51, res asst, 51-53; SR CHEMIST, LAWRENCE LIVERMORE NAT LAB, 53- *Res:* General radiochemical research; plowshare applications of nuclear explosives; statistical data analysis; migration of radionuclides in groundwater; nuclear waste management; radiochemical analysis of nuclear explosion debris. *Mailing Add:* Nuclear Chem Div Lawrence Livermore Lab PO Box 808 Livermore CA 94551

LEVY, HARVEY LOUIS, b Augusta, Ga, Oct 3, 35; m 61; c 2. MEDICINE, PEDIATRICS. *Educ:* Med Col Ga, MD, 60. *Prof Exp:* Intern pediat, Boston City Hosp, 60-61; asst resident path, Columbia-Presby Med Ctr, 61-62; asst resident pediat, Johns Hopkins Hosp, 64-65; chief resident, Boston City Hosp, 65-66; from instr to asst prof, 68-77, ASSOC PROF NEUROL, HARVARD MED SCH, 77- *Concurrent Pos:* NIH fel neurol, Harvard Med Sch, 66-68; consult, Walter E Fernald Sch Ment Retardation, 67-; lectr, Grad Sch Dent, Boston Univ, 68-; prin investr, Mass Dept Pub Health, 69-75; dir, Mass Metab Dis Prog, 75-; assoc, Ctr Human Genetics, Harvard Med Sch, 71-; assoc neurologist & pediatrician, Mass Gen Hosp, 77-; dir, Inborn Errors of Metab-Phenylketonuria Prog, Children's Hosp Med Ctr, 78- *Mem:* Fel Am Acad Pediat; Soc Pediat Res. *Res:* Inborn errors of metabolism; biochemical and genetic disorders. *Mailing Add:* Mass Gen Hosp Boston MA 02114

LEVY, HARVEY MERRILL, b Pittsburgh, Pa, May 12, 28; m 57; c 1. BIOCHEMISTRY, PHYSIOLOGY. *Educ:* Univ Calif, Los Angeles, BA, 50, PhD(biochem), 55. *Prof Exp:* Asst res biochemist, Army Med Res Lab, Ky, 54-56; assoc res biochemist, Brookhaven Nat Lab, 56-58; asst prof pharmacol, Sch Med, NY Univ, 58-60, from asst prof to prof physiol & biophys, 60-71; PROF PHYSIOL & BIOPHYS, STATE UNIV NY STONY BROOK, 71- *Mem:* Am Chem Soc; Am Soc Biol Chemists; Harvey Soc; Biophys Soc; Soc Gen Physiol. *Res:* Muscle biochemistry; enzymology; kinetics. *Mailing Add:* Dept Physiol & Biophys State Univ NY Stony Brook NY 11794

LEVY, HILTON BERTRAM, b New York, NY, Sept 21, 16; m 42; c 2. VIROLOGY. *Educ:* City Col New York, BS, 35; Columbia Univ, MA, 36; Polytech Inst Brooklyn, PhD(biochem), 46. *Prof Exp:* Chief chemist, Gen Sci Labs, NY, 37-41; res biochemist, Mem Hosp Cancer & Allied Dis, 41-46; res biochemist, Overly Biochem Res Found, 46-52; HEAD SECT MOLECULAR VIROL, NAT INST ALLERGY & INFECTIOUS DIS, 52- *Concurrent Pos:* Prof, Med Sch, Howard Univ. *Mem:* Soc Exp Biol & Med; Am Asn Immunol; Soc Gen Physiol; Am Soc Biol Chem; Infectious Dis Soc; Soc Biol Response Modifiers. *Res:* Cancer; nucleic acid metabolism; infectious diseases; virus reproduction; interferon action and induction; treatment of neoplastic and viral diseases. *Mailing Add:* Sect Molecular Virol Nat Inst Allergy & Infect Dis Bethesda MD 20014

LEVY, JACK BENJAMIN, b Savannah, Ga, Jan 17, 41; m 63; c 2. ORGANIC CHEMISTRY. *Educ:* Duke Univ, AB, 62, NC State Univ, MS, 64, PhD(chem), 67. *Prof Exp:* From asst prof to assoc prof, 68-73, prof chem, 73-86, WILL S DELOACH PROF CHEM, UNIV NC, WILMINGTON, 86-, CHMN DEPT, 75- *Mem:* AAAS; Am Chem Soc; Am Inst Chemists. *Res:* Synthesis of heterocyclic organophosphorus compounds; synthesis and spectral properties of new phenoxaphosphine derivatives; chemical education. *Mailing Add:* Dept Chem Univ NC Wilmington Wilmington NC 28403-3297

LEVY, JERRE MARIE, b Birmingham, Ala, Apr 7, 38; m 69; c 2. PSYCHOBIOLOGY. *Educ:* Univ Miami, BA, 62, MS, 66; Calif Inst Technol, PhD(psychobiol), 69. *Prof Exp:* Res tech neuropsychol, Vet Admin Hosp, Denver, 69-70; fel psychol, Univ Colo, 70-71; fel biochem, Ore State Univ, 71-72; from asst prof to assoc prof psychol, Univ Pa, 72-77; assoc prof, 77-82, PROF BIOPSYCHOL, UNIV CHICAGO, 82- *Concurrent Pos:* Prin investr, NSF grant, 75-77 & NIH grant, 77-79 & Spencer Found grant, 79-88; consult ed, J Exp Psychol: Human Perception & Performance, 75-84; bd assoc eds, Brain & Cognition, 82-, Neuropsychol, 88-, J Neurosci, 90- *Mem:* Soc Exp Psychologists; Int Neuropsychol Symp. *Res:* Cerebral asymmetry and cognitive function; evolution and genetics of human brain, especially hemispheric lateralization and correlated behaviors; variations in human lateralization patterns. *Mailing Add:* Dept Psychol Univ Chicago Chicago IL 60637-1588

LEVY, JOSEPH, b New Haven, Conn, June 30, 13; m 41; c 3. O GANIC CHEMISTRY. *Educ:* Yale Univ, BS, 35, PhD(org chem), 38. *Prof Exp:* Asst chem, Yale Univ, 35-38; res chemist, Polyxor Chem Co, 39-40; dir org res, Ernst Bischoff Co, 40-46; res chemist, Nopco Chem Co, 46-50; sr res assoc, Chem Div, Universal Oil Prod Co, 50-72, sr res assoc, Res Ctr, UOP Inc, Des Plaines, 72-78; RETIRED. *Concurrent Pos:* Adj prof chem, Fla Atlantic Univ, 80- *Mem:* Am Chem Soc; Sigma Xi. *Res:* Directed and conducted exploratory research on synthesis and process development of wide range of both new and known organic products; granted over 50 US patents plus associated foreign patents for resulting inventions; organic synthesis; fine organics. *Mailing Add:* 123 Grantham-B Century Village E Deerfield Beach FL 33442

LEVY, JOSEPH BENJAMIN, b Manchester, Eng, Feb 23, 23; nat US; m 48; c 3. PHYSICAL CHEMISTRY, ORGANIC CHEMISTRY. *Educ:* Univ NH, BS, 43; Harvard Univ, MA, 45, PhD(chem), 48. *Prof Exp:* Mem sci staff, Columbia Univ, 47-49; res chemist, US Naval Ord Lab, 49-56 & Atlantic Res Corp, 56-65; PROF CHEM, GEORGE WASHINGTON UNIV, 65- *Mem:* Am Chem Soc. *Res:* Thermal decomposition of nitrate esters; reactions of free radicals; chemistry of rocket propellants; fluorine chemistry. *Mailing Add:* 7610 Honestway Bethesda MD 20817-5520

LEVY, JOSEPH VICTOR, b Los Angeles, Calif, Apr 7, 28; m 54; c 2. PHYSIOLOGY, PHARMACOLOGY. *Educ:* Stanford Univ, BA, 50; Univ Calif, Los Angeles, MS, 56; Univ Wash, PhD(pharmacol), 59. *Prof Exp:* Asst physiol, Stanford Univ, 51-53, asst pharmacol, 54-56; asst, Univ Wash, 56-57; pharmacologist, Surg Res Labs, Presby Hosp, 60; sr res pharmacologist, Res Labs, Presby Med Ctr, 60-65; dir, Lab Pharmacol & Exp Therapeut, Pac Presby Med Ctr, 61-89; assoc prof, Sch Med Sci, 69-77, assoc prof, Sch Dent, 72-85, PHARMACOL COURSE DIR, SCH DENT, UNIV OF THE PAC, 81-, CLIN PROF PHYSIOL & PHARMACOL, SCH DENT, 85- *Concurrent Pos:* NIH res trainee pharmacol, 57-58 & anesthesiol, 58-59; Am Heart Asn res fel, 59-60; Nat Heart Inst, res career prog scientist, 65-70; mem drug interaction panel, Am Pharmaceut Asn, 73-85; mem, Coun Basic Res, Am Heart Asn; mem res comt, Calif Heart Asn, 75-77; mem, Hypertenison Task Force, Nat Heart Inst, NIH, 76-77; consult, WHO & UN Develop Prog,

81; mem expert adv panel geriat drugs, US Pharmacopea, 85-; mem, Pharmacol & Therapeut Comt, Pac Presby Med Ctr, 72- *Mem:* Western Pharmacol Soc; Soc Exp Biol & Med; Cardiac Muscle Soc; Am Soc Clin Pharmacol & Therapeut; Am Soc Pharmacol & Therapeut; Am Chem Soc. *Res:* Physiology and pharmacology; hypertension; diabetes; prostaglandins; inflammation; vascular; immunopharmacology; thrombosis; platelets. *Mailing Add:* Dept Physiol & Pharmacol Sch Dent Univ of the Pacific 2155 Webster St San Francisco CA 94115

LEVY, JULIA GERWING, b Singapore, May 15, 34; nat Can; m 55, 69; c 3. MICROBIOLOGY. *Educ:* Univ BC, BA, 55; Univ London, PhD(bact), 58. *Hon Degrees:* Doctorate, Univ Ottawa, 89; Doctorate, Mt St Vincents Univ, 90. *Prof Exp:* From instr to assoc prof, 58-74, PROF MICROBIOL, UNIV BC, 74-, MEM CANCER RES UNIT, 77- *Concurrent Pos:* Vpres, Res & Develop, Quadralogic Technol Inc. *Honors & Awards:* Killiam Sci Prize; Biely Sci Award. *Mem:* Am Asn Immunologists; fel Royal Soc Can. *Res:* Characterization of antigenic determinants on natural antigens and the effect of these determinants on the cellular immune response; photoimmunotherapy; a study of immunotoxins to which photosensitizers have been conjugated. *Mailing Add:* Dept Bact & Immunol Univ BC Vancouver BC V6T 1W5 Can

LEVY, LAWRENCE S, b Cleveland, Ohio, Oct 2, 33; m 61; c 2. ALGEBRA. *Educ:* Juilliard Sch Music, BS, 54, MS, 56; Univ Ill, MA, 58, PhD(math), 61. *Prof Exp:* Instr math, Univ Ill, 61; from asst prof to assoc prof, 61-71, PROF MATH, UNIV WIS-MADISON, 71- *Mem:* Am Math Soc. *Res:* Structure of associative rings and their modules. *Mailing Add:* 815 Van Vleck Univ Wis-Madison Madison WI 53706

LEVY, LEO, b New York, NY, July 11, 28; m 57; c 2. PSYCHOLOGY, PREVENTIVE MEDICINE. *Educ:* City Col New York, BS, 50, MA, 51; Univ Wash, PhD(psychol), 58; Harvard Univ, SMHyg, 64. *Prof Exp:* Instr psychol, Univ Mich, 58-60; adminr & chief psychologist, Pueblo Guid Ctr, Pueblo, Colo, 60-63; dir planning & eval, Ill Dept Ment Health, 64-69; asst prof psychiat, 65-69, assoc prof prev med, 69-75, PROF PUB HEALTH & PREV MED RES, UNIV ILL MED CTR, 75- *Concurrent Pos:* NIMH fels, Univ Mich, 58-60 & Harvard Univ, 63-64; Fulbright Hays res grant, State Univ Leiden, 72-73; vis assoc prof psychiat, McMaster Univ, 69-71; Fogerty Sr Int fel, Inst Psychiat, London, 79-80; vis prof, St George's Hosp Med Sch, London, 79-80. *Mem:* Am Asn Univ Prof; AAAS; Am Psychol Asn; Am Pub Health Asn. *Res:* Promotion and maintenance of mental health; social planning; drug abuse; social ecology; psychosocial epidemiology; problems of urban mental health. *Mailing Add:* Sch Pub Health Univ Ill Med Ctr PO Box 6998 Chicago IL 60680

LEVY, LEON BRUCE, b New York, NY, July 20, 37; m 68; c 2. INDUSTRIAL ORGANIC CHEMISTRY. *Educ:* NY Univ, BA, 58; Harvard Univ, AM, 59, PhD(chem), 62. *Prof Exp:* Res chemist, Clarkwood Res Lab, Celanese Chem Co, 62-65; sr res chemist, Tech Ctr, 65-71, res assoc, 72-90; SR RES ASSOC, TECH CTR, HOECHST CELANESE CORP, 90- *Mem:* Catalysis Soc. *Res:* Kinetics and mechanisms of homolytic organic reactions; vapor phase oxidations of hydrocarbons; heterogeneous catalysis. *Mailing Add:* Hoechst Celanese Chem Group Box 9077 Corpus Christi TX 78469-9077

LEVY, LEON SHOLOM, b Perth Amboy, NJ, June 28, 30; m 54; c 3. COMPUTER SCIENCE. *Educ:* Yeshiva Univ, BA, 52; Harvard Univ, SM, 55, ME, 57; Univ Pa, PhD(comput sci), 70; Fairleigh Dickinson Univ, MBA, 86. *Prof Exp:* Engr, RCA Corp, 55-58; sr staff engr, Hughes Aircraft Co, 58-63; mgr comput & displays sect, Aerospace Corp, 63-66; systs architect, IBM Corp, 66-67; asst prof statist & comput sci, Univ Del, 70-74, assoc prof, 74-80; prof computer sci, Ben Gurion Univ, 83-84; DISTINGUISHED MTS, BELL LABS, 79-83 & 84- *Concurrent Pos:* Consult, Ling Proj, Univ Pa, 70-72. *Mem:* Asn Comput Mach; Inst Elec & Electronics Eng. *Res:* Relationship of machines and their languages; relationship of functional and structural aspects of computers; software engineering and software economics; artificial intelligence. *Mailing Add:* 2D Dorado Dr Convent Hill NJ 07961

LEVY, LEONARD ALVIN, b New York, NY, Aug 19, 35; m 60; c 2. PODIATRIC DERMATOLOGY, AGING. *Educ:* NY Univ, BA, 56; NY Col Podiat Med, DPM, 61; Columbia Univ Sch Pub Health, MPH, 67. *Prof Exp:* Dean & vpres podiat med, Calif Col Podiat Med, 67-74, dean podiat med, State Univ NY, Stony Brook, 74-76, consult podiat med, Univ Tex Health Sci Ctr, 76-81, DEAN PODIAT MED, UNIV OSTEOPATH MED & HEALTH SCI, 81- *Concurrent Pos:* Mem review comt, US Pub Health Serv Adv Comt, 68-72, clin assoc prof dermat, Stanford Univ Sch Med, 70-74, chmn & mem test comt, Nat Bd Podiat Med examrs, 77-, governing coun mem, Am Pub Health Asn, 78-80 & 84-86, mem test comt dermat, Nat Bd Podiat Med Examrs, 81-; mem, spec med adv group, Dept Veterans Admin, 90- *Mem:* Am Podiat Med Asn; Am Pub Health Asn; Am Acad Dermat; Am Asn Col Podiat Med; Asn Am Med Col; Nat Rural Health Asn. *Res:* Podiatric medical education; public health dermatological problems of the foot and ankle; health promotion and prevention; geriatrics. *Mailing Add:* Col Podiat Med & Surg Univ Osteopath Med & Health Sci 2150 Grand Ave Des Moines IA 50312

LEVY, LOUIS, b Brooklyn, NY, Feb 1, 23; m 54; c 5. PHARMACOLOGY. *Educ:* Univ Iowa, BS, 49, MS, 51, PhD(pharmacol), 54. *Prof Exp:* Asst chem, Syracuse Univ, 49-50; asst pediat, Univ Iowa, 50-52; instr pharmacol, 53-54; asst prof, Med Sch, Georgetown Univ, 54-55; asst prof, Col Med, Univ Cincinnati, 55-59; sr pharmacologist, Riker Labs, Calif, 59-71; from assoc prof to prof pharmacol, Sch Med, Univ Calif, Los Angeles, 71-85; consult, 85-88; STAFF TOXICOLOGIST, DEPT HEALTH SERVS, STATE CALIF, 88- *Concurrent Pos:* Res collabr, Brookhaven Nat Lab, 53-55. *Mem:* AAAS; Am Soc Pharmacol & Exp Therapeut; NY Acad Sci. *Res:* Biochemical pharmacology. *Mailing Add:* Dept Health Servs State Calif 1405 N San Francisco Blvd No 300 Burbank CA 91504

LEVY, LOUIS A, b New York, NY, Mar 6, 41; m 67; c 1. ORGANIC CHEMISTRY. *Educ:* City Col New York, BS, 61; Univ Colo, PhD(chem), 66. *Prof Exp:* Proj leader synthesis, Int Flavors & Fragrances, Inc, 65-66; scientist, Nat Air Pollution Control Admin, USPHS, 66-67; SCIENTIST, LAB ENVIRON CHEM, NAT INST ENVIRON HEALTH SCI, 67- *Mem:* Am Chem Soc; The Chem Soc. *Res:* Chemistry and synthesis of chemicals of environmental concern; NMR spectroscopy. *Mailing Add:* Nat Inst Environ Health Sci Lab Molecular Biophys PO Box 12233 Research Triangle Park NC 27709

LEVY, M(ORTON) FRANK, b New York, NY, May 31, 25; div; c 3. ORGANIC CHEMISTRY. *Educ:* Queens Col, NY, BS, 50; Columbia Univ, MA, 51; Yale Univ, PhD(chem), 56. *Prof Exp:* Group leader org synthesis & anal develop, Argus Chem Co, 55-60; group leader org synthesis, Harchem Div, Wallace & Tiernan, NJ, 60-64; SR CHEMIST, MAT SCI LAB, IBM CORP, SAN JOSE, 64- *Concurrent Pos:* Res assoc, Univ Calif, Berkeley, 75-76. *Mem:* Am Chem Soc; AAAS. *Res:* Utilization of new raw materials in organic synthesis; photosensitive materials; dibasic acids; dye chemistry; preparation of radiolabeled compounds; new polymers; application of computers to chemistry. *Mailing Add:* 101 Pinta Ct Los Gatos CA 95032-6331

LEVY, MARILYN, b New York, NY, Apr 3, 22. PHOTOGRAPHIC ENGINEERING, CHEMISTRY. *Educ:* Hunter Col, AB, 42. *Prof Exp:* Chemist anal-drugs, NY Quinine & Chem Co, 42-43; chemist lacquer formulation, Roxalin Flexible Finishes, 43-46, chemist lacquer mfg, Valspar Corp, 46-48; inspector chem, NY Quartermaster Proc Agency, 51-52; res chemist photog process, US Army Electronics Command, 53-74; chief photog optics div, Photog Eng, US Army Combat & Surveillance Lab, 75-80; CONSULT, 80- *Concurrent Pos:* Mem & US Army rep, Am Nat Standards Inst, 62-; US Army adv, NATO Photog Standards Comt & Air Standardization Coord Comt, 78- *Mem:* Fel Soc Photog Scientists & Engrs; Am Chem Soc. *Res:* Photographic processing; non-conventional photographic systems; aerial photography; photographic sensitometry; color photography; rapid processing; 26 patents and 20 papers in photo science field. *Mailing Add:* 56 Cheshire Sq Little Silver NJ 07739

LEVY, MARK B, b Hong Kong, China, June 19, 51. MATHEMATICS. *Educ:* Cooper Union, BA, 74. *Prof Exp:* Sr prof, Fed Aviation Admin, 83-85; SYST ENG, FED AVIATION ADMIN, 85- *Mem:* Am Math Soc; Am Comput Mach. *Res:* Multi parallel computing operating systems. *Mailing Add:* PO Box 23286 Washington DC 20026

LEVY, MARTIN J LINDEN, b Philadelphia, Pa, June 19, 25; m 58; c 3. MEDICAL SCIENCES, RESEARCH ADMINISTRATION. *Educ:* Pa State Univ, BS, 47; NJ Inst Technol, MS, 56; Stevens Inst Technol, DrSci, 63. *Prof Exp:* Specialist engr, Jet Engine Div, Gen Elec, 56-58; PROF ENG, NJ INST TECHNOL, 58- *Concurrent Pos:* Adj prof, NJ Med Sch, Univ Med & Dent, 80-82; med staff affil, St Barnabas Med Ctr, NJ & bd dir, Southern Inst Biomed Sci, 80-82. *Mem:* Am Col Cryosurg; NY Acad Sci; Int Soc Cryosurg. *Res:* Biomedical sciences; application of engineering and technical methods to instrumentation for diagnosis and therapy. *Mailing Add:* NJ Inst Technol 323 High St Newark NJ 07102

LEVY, MATTHEW NATHAN, b New York, NY, Dec 2, 22; m 46; c 3. PHYSIOLOGY, BIOMEDICAL ENGINEERING. *Educ:* Western Reserve Univ, BS, 43, MD, 45. *Prof Exp:* From instr to asst prof physiol, Western Reserve Univ, 49-53; from asst prof to assoc prof, Albany Med Col, 53-57; dir res, St Vincent Charity Hosp, Cleveland, Ohio, 57-67; assoc prof, 61-68, PROF, PHYSIOL & BIOMED ENG, CASE WESTERN RESERVE UNIV, 68-; CHIEF DEPT INVESTIGATIVE MED, MT SINAI HOSP, 67- *Concurrent Pos:* Res fel, Western Reserve Univ, 48-49; assoc prof, Case Inst Technol, 63-67; assoc ed, Circulation Res, 70-74; sect ed, Am J Physiol, 75-, ed, Am J Physiol, Heart & Circulatory Physiol, 76-81. *Honors & Awards:* Lederle Med Fac Award, 55-57; Shanes Mem Lectureship, 72; Wiggers Award, Am Physiol Soc, 83; Merit Award, NIH, 86. *Mem:* Am Physiol Soc; Am Heart Asn. *Res:* Cardiovascular physiology. *Mailing Add:* Dept Investigative Med Mt Sinai Med Ctr Cleveland OH 44106

LEVY, MICHAEL R, b Los Angeles, Calif, Aug 8, 35; m 62; c 2. CELLULAR BIOLOGY. *Educ:* Univ Calif, Los Angeles, BS, 57, MA, 59, PhD(zool), 63. *Prof Exp:* USPHS fel, 63-65, trainee, 65-66; teaching fel, Univ Mich, 66-67; from asst prof to assoc prof, 67-74, PROF BIOL, SOUTHERN ILL UNIV, EDWARDSVILLE, 74- *Concurrent Pos:* USPHS res grants, 69-80. *Mem:* AAAS; Am Soc Cell Biol. *Res:* Regulation of protein degradation; proteolytic enzymes; lysosomes. *Mailing Add:* Dept Biol Southern Ill Univ Edwardsville IL 62026

LEVY, MOISES, b Panama, Apr 8, 30; US citizen; m 83. LOW TEMPERATURE SOLID STATE PHYSICS, PHYSICAL ACOUSTICS. *Educ:* Calif Inst Technol, BS, 52, MS, 55; Univ Calif, Los Angeles, PhD(physics), 63. *Prof Exp:* Res chemist, Speciality Resins, Inc, 53-54; mem tech staff, Semiconductor Div, Hughes Aircraft Co, 56-58; asst prof solid state physics, Univ Pa, 64-65; asst prof ultrasonic invest solid state, Univ Calif, Los Angeles, 65-70; assoc prof, 71-73, chmn dept, 75-78, PROF PHYSICS, UNIV WIS-MILWAUKEE, 73- *Concurrent Pos:* NATO postdoc fel, 63-64; chmn, Ultrasonics Symp, 74 & 83, prog comt mem, 72-; vis prof, Int Vis Info Serv, 72, 76 & 82, Univ Sao Paolo, 79 & 83, Techmon Univ, 85-86; Lady Doris fel, 85-86. *Mem:* Fel Am Phys Soc; Acoustical Soc Am; Ultrasonic, Ferroelec & Frequency Control Soc. *Res:* Experimental investigation of electron phonon interaction in superconductors and normal metals; spin phonon in magnetic superconductors and rare earth metals; surface wave investigation of superconducting films and magnetic films. *Mailing Add:* Dept Physics Univ Wis Milwaukee WI 53201

LEVY, MORRIS, b Chicago, Ill, May 22, 44; m 74; c 1. EVOLUTIONARY BIOLOGY, BIOSYSTEMATICS. *Educ:* Univ Ill, Chicago Circle, 67; Yale Univ, MPh, 72, PhD(ecol, evolution), 73. *Prof Exp:* Asst prof, 73-79, ASSOC PROF BIOL SCI, PURDUE UNIV, 79- *Concurrent Pos:* Dir, Kriebel Herbarium, Purdue Univ, 73-; NSF grant, Res Prog Biomed Sci, 75; NSF grant, 78 & 79. *Mem:* Bot Soc Am; Soc Study Evolution; Soc Am Naturalists; AAAS. *Res:* Systematics and biochemical ecology of plants; evolution of hybrid and polyploid species; population biology of weeds; host plant-fungal pathogen co-evolution; pollinetion ecology. *Mailing Add:* Dept Biol Sci Purdue Univ Lilly Hall West Lafayette IN 47907

LEVY, MORTIMER, b Rochester, NY, July 7, 24; m 50; c 2. RESEARCH ADMINISTRATION. *Educ:* Cornell Univ, BSEE, 49; Columbia Univ, MA, 51. *Prof Exp:* Physicist, Xerox Corp, 54-57, sect leader, 58-61; dir appl res, Mat Res Corp, 61-63; res scientist, 63-64, sr scientist, 64-67, mgr explor res, 67-73, mgr process sect, 73-75, mgr, Process Element Sect, 75-78, MGR MAT, PROCESSES & CORP STAFF, XEROX CORP, 78- *Mem:* Soc Photog Sci & Eng. *Res:* Electrostatic photography. *Mailing Add:* 105 Towpath Lane Rochester NY 14618

LEVY, NELSON LOUIS, b Somerville, NJ, June 19, 41; m 74; c 6. IMMUNOLOGY, NEUROSCIENCES. *Educ:* Yale Univ, BA & BS, 63; Columbia Univ, MD, 67; Duke Univ, PhD(immunol), 73. *Prof Exp:* Intern surg, Univ Colo Med Ctr, 67-68; res assoc virol & immunol, NIH, 68-70; resident neurol, Duke Univ Med Ctr, 71-72, from asst prof to assoc prof immunol, 76-80; vpres, Pharmaceut Res, Abbott Labs, 81-84; CEO, CORETECHS CORP, 84- *Concurrent Pos:* Mem gastrointestinal cancer study group, Nat Cancer Inst, 74-78; study sects, NIH, Nat Mult Sclerosis Soc, 75-80; bd dir, Bionica Pty, Ltd, Intek Diagnostics, SASHA, Inc, Heybach Enterprises, MedVac, Inc, Quantum Group, Inc. *Mem:* Am Asn Cancer Res; Am Asn Immunologists; Soc Neuroscience; Pharmaceut Mfrs Asn; Drug Info Asn; Licensing Execs Soc. *Res:* Immunologic and non-immunologic defenses against human cancer; pathogenesis and etiology of multiple sclerosis; neurologic control of the immune system. *Mailing Add:* Coretechs Corp 1391 Concord Dr Lake Forest IL 60045

LEVY, NEWTON, JR, b Tampa, Fla, Oct 10, 35; m 61; c 2. INORGANIC CHEMISTRY, TECHNICAL MANAGEMENT. *Educ:* Univ Fla, BSCh, 61, PhD(kinetics), 64. *Prof Exp:* Res chemist, Wash Res Ctr Div, W R Grace & Co, 64-67, sr chemist, 67-69, res supvr, 69-72; sect head refractories, Martin Marietta Labs, 72-74, mgr fuel additives, Refractories Div, 74-76, mgr prod develop, Refractories Div, 77-81, VPRES SALES-CHEM, MARTIN MARIETTA CHEM, 81- *Res:* Preparation, fabrication, characterization and applications of reactive, fine sized ceramic oxide powders; properties and uses of magnesium oxide; chemical treatment of oils for combustion; chemical market studies; marketing specialty chemicals. *Mailing Add:* Martin Marietta Chem Exec Plaza II Hunt Valley MD 21030

LEVY, NORMAN B, b New York, NY, May 28, 31; m 58, 70; c 4. PSYCHIATRY, PSYCHOSOMATIC MEDICINE. *Educ:* NY Univ, BA, 52; State Univ NY Downstate Med Ctr, MD, 56. *Prof Exp:* Res physician & teaching fel med, Sch Med, Univ Pittsburgh, 57-58; dir med serv, US Air Force Hosp, Ashiya, Japan, 58-60; resident physician psychiat, Kings County Hosp Ctr, Brooklyn, NY, 60-63; from instr to asst prof med & psychiat, State Univ NY Downstate Med Ctr, 63-73, dir continuing educ psychiat, 74-76, presiding officer fac, Col Med, 75-76, assoc prof, 73-79, prof psychiat, 79-80, assoc dir, Med Psychiat Liaison Serv, 72-80; PROF PSYCHIAT, MED & SURG, NY MED COL & DIR, LIAISON PSYCHIAT DIV, WESTCHESTER COUNTY MED CTR, 80- *Concurrent Pos:* NIMH career teacher award, 66; vis prof psychiat & med, Univ Hawaii, 81; consult psychiat educ, NIMH, 74-; examr psychiat, Am Bd Psychiat & Neurol, 74-; assoc ed, Int J Psychiat Med, 77-78; assoc ed, General Hosp Pschiat & ed of sect, Liaison Rounds, 79- *Mem:* Fel Am Col Physicians; fel Am Psychiat Asn; fel Int Col Psychosom Med; Sigma Xi; fel Am Col Psychiatrists. *Res:* Effects of psychological stresses on kidney transplant rejections; psychological adaptation to hemodialysis; psychiatry and the changing role of males in society; attitudes of students and physicians on informing patients of their fatal diagnosis; use of fluoxetine in renal failure. *Mailing Add:* Psychiat Inst NY Med Col Valhalla NY 10595

LEVY, NORMAN STUART, b Detroit, Mich, July 17, 40; m 64; c 4. OPHTHALMOLOGY, GLAUCOMA. *Educ:* Case Western Reserve, MD, 65; Univ Chicago, PhD(ophthal), 75. *Prof Exp:* Asst prof ophthal & pediat, 72-75, ASST PROF FAMILY MED, UNIV FLA, 81- *Concurrent Pos:* Chief ophthal, Vet Admin Hosp, Gainesville, 73-75 & Vet Admin Med Ctr, Lake City, 76-81. *Mem:* Am Acad Ophthal; Asn Res Vision & Ophthal; Am Col Surgeons; Kerato Refractive Soc. *Res:* Mechanism of damage in the disease, glaucoma, and method of diagnosis and early treatment. *Mailing Add:* 7106 NW 11th Pl Gainesville FL 32605-3192

LEVY, PAUL, b New York, NY, May 25, 41; m 65; c 1. APPLIED MATHEMATICS. *Educ:* Rensselaer Polytech Inst, BS, 63, MS, 65, PhD(math), 68. *Prof Exp:* Asst prof math, NY Univ, 67-74; asst prof math, NY Inst Technol, 74-77; ASST PROF MATH, STATE UNIV NY MARITIME COL, 77- *Mem:* Am Math Soc; Soc Indust & Appl Math. *Res:* Investigation of problems in wave propagation and elasticity. *Mailing Add:* 604 Second St Brooklyn NY 11215

LEVY, PAUL F, b New York, NY, Dec 9, 34; m 59, 73; c 4. ANALYTICAL CHEMISTRY, INSTRUMENTATION. *Educ:* City Col New York, BS, 59; Columbia Univ, MA, 61, PhD(anal chem), 65. *Prof Exp:* Lectr chem, City Col New York, 59-65; proj engr, 65-68, sr proj engr, 68-69, supvr appln lab, 69-73, prod mgr thermal anal, 73-75, prod mgr liquid chromatog, 75-78, develop mgr, 78-85, MKT DEVELOP MGR, BIOMED PROD DEPT, E I DU PONT DE NEMOURS & CO, 85- *Mem:* Am Chem Soc; Am Asn Clin Chemists. *Res:* Theory, applications, design and development of thermal analysis and other material characterization instrumentation; coulostatic impulse-chain and other forms of polarography; electrochemical instrumentation; clinical and biomedical instrumentation; development management. *Mailing Add:* 4818 Hogan Dr 20381 E I du Pont de Nemours & Co Wilmington DE 19808-1715

LEVY, PAUL SAMUEL, biostatistics, epidemiology, for more information see previous edition

LEVY, PAUL W(ARREN), b Chicago, Ill, Mar 17, 21; m 44; c 4. SOLID STATE PHYSICS. *Educ:* Univ Chicago, BS, 43; Carnegie Inst Technol, PhD, 54. *Prof Exp:* Jr physicist, Metall Lab, Univ Chicago, 43-44; physicist beta-ray spectros, Oak Ridge Nat Lab, 44-48; assoc physicist, 52-58, PHYSICIST RADIATION DAMAGE INSULATORS REACTIVE MAT & MINERALS, BROOKHAVEN NAT LAB, 58- *Concurrent Pos:* Consult to indust, civilian & mil agencies, 55-; adj prof, Adelphi Univ, 66-; adj prof geol, Univ Pa, 76- *Mem:* Fel Am Phys Soc; Optical Soc Am; Mat Res Soc. *Res:* Nuclear physics; luminescence of solids; optical and defect properties of solids; optical spectrophotometry; radiation effects in glasses, insulators, scintillators, metals, explosives and propellants; minerals, especially for radioactive waste applications; crystal growth; geoscience applications of solid state physics; thermoluminescence of solids and applications to dosimetry, mineralogy, mineral exploration and archaeology. *Mailing Add:* Bldg 480 Brookhaven Nat Lab Upton NY 11973

LEVY, PETER MICHAEL, b Frankfurt, Ger, Jan 10, 36; US citizen; m 65; c 2. SOLID STATE PHYSICS. *Educ:* City Col New York, BME, 58; Harvard Univ, MA, 60, PhD(appl physics), 63. *Prof Exp:* Res assoc physics, Lab Electrostatics & Physics of Metals, Grenoble, France, 63-64; res assoc, Univ Pa, 64-66; asst prof, Yale Univ, 66-70; assoc prof, 70-75, chmn dept, 76-82, PROF PHYSICS, NY UNIV, 75- *Concurrent Pos:* NSF fel, 58-62; fel, Nat Ctr Sci Res, France, 63-64; Air Force Off Sci Res grant, Yale Univ & NY Univ, 67-72; NSF grants, NY Univ, 72-75, 75-79, 79-82, 82-86 & 87; Fulbright-Hays res scholar, France, 75-76; res exchange scientist, NSF-Nat Ctr Sci Res, France, 75-76 & 83-84. *Honors & Awards:* Vermeil Medal, Soc Advan Progress, Paris, 78. *Mem:* Am Phys Soc; fel NY Acad Sci. *Res:* The magneto-transport properties of Kondo lattice systems; mixed-valence and heavy fermions; magnetoresistivity of rare earth metallic compounds; orbital effects in rare earth compounds; anisotropy in disordered magnetic systems; spin glasses; long range interactions between local movements in metals; the magnetic and transport properties of metallic multilayered structures (superlattices). *Mailing Add:* Dept Physics NY Univ New York NY 10003

LEVY, RALPH, b London, Eng, Apr 12, 32; US citizen; m 59; c 2. MICROWAVE THEORY, CIRCUIT THEORY. *Educ:* Cambridge Univ, MA, 53; Univ London, PhD(appl sci), 66. *Prof Exp:* Mem sci staff microwave eng, Gen Elec Co, Stanmore, Eng, 53-59, Mullard Res Labs, Redhill, 59-64; lectr elec eng, Univ Leeds, 64-67; vpres res microwave eng, Microwave Develop Labs, 67-84, VPRES ENG, KW MICROWAVE INC, 84- *Concurrent Pos:* Consult, Decca Radar Ltd & Gen Elec Co, 64-67, Weinschel Eng, 65-66. *Mem:* Fel Inst Elec & Electronics Engrs; mem Int Elec Engrs. *Res:* Microwave passive components; distributed circuit theory; military microwave systems. *Mailing Add:* R Levy Assoc 1897 Caminito Velasco La Jolla CA 92037

LEVY, RAM LEON, b Samokov, Bulgaria, Oct 7, 33; US citizen; m 58; c 3. SEPARATION SCIENCE, CHEMICAL INTRUMENTATION. *Educ:* Israel Inst Technol, BSc, 61, MSc, 63; Univ Man, PhD(anal chem), 67. *Prof Exp:* Res assoc, Dept Plant Sci, Univ Man, 63-67; assoc chemist, Midwest Res Labs, Kansas City, 67-68; SCIENTIST, POLYMER CHEMISTRY, MCDONNELL DOUGLAS RES LABS, MCDONNELL DOUGLAS CORP, 68- *Concurrent Pos:* Affil dir, Sch Continuing Prof Educ, Washington Univ, 70-72; vis lectr, Chem Dept, St Louis Univ, 75. *Mem:* Am Chem Soc; Am Soc Mass Spectrometry. *Res:* Chemical mechanisms of polymer aging; detection of stress and fatigue induced molecular phenomena in polymers by infrared spectroscopy; chemluminescence of polymers; chromatographic characterization of oligimers; incorporation of molecular probes in polymer networks; thermal degradation of polymers. *Mailing Add:* 1622 Parquet Ct St Louis MO 63146-4319

LEVY, RENE HANANIA, b Casa Blanca, Morocco, Sept 30, 42; US citizen; m 64; c 3. PHARMACODYNAMICS. *Educ:* Univ Bordeaux, Baccalaureat, 60; Univ Paris, Pharm, 65; Univ Calif, San Francisco, PhD(pharm, pharmaceut chem), 70. *Prof Exp:* Intern, Hosps of Paris, Hopital Corentin Celton, 64-66; asst prof pharm, Col Pharm, 70-74, assoc prof, 74-77, PROF PHARMACEUT, SCH PHARM & PROF NEUROL SURG, SCH MED, UNIV WASH, 77- *Mem:* Am Epilepsy Soc; Fedn Int Pharmaceutique; Am Pharmaceut Asn; Acad Pharmaceut Sci. *Res:* Pharmacokinetic evaluation of anticonvulsants prior to efficacy testing in primates; clinical pharmacology of new antiepileptic drugs; kinetics of drug metabolites. *Mailing Add:* Dept Pharm Univ Wash Seattle WA 98195

LEVY, RICARDO BENJAMIN, b Quito, Ecuador, Jan 11, 45; US citizen; m 67; c 2. SURFACE CHEMISTRY, CHEMICAL ENGINEERING. *Educ:* Stanford Univ, BSc, 66, PhD(chem eng), 72; Princeton Univ, MA, 67. *Prof Exp:* Gen mgr mfg, Sudam Cia Ltda, 67-69; res eng chem physics, Exxon Res & Eng Co, 72-74; vpres, 74-77, EXEC VPRES CONSULT RES & DEVELOP, CATALYTICA ASSOCS, INC, 77- *Concurrent Pos:* Prof chem eng, Inst Politecnico Nac, Quito, Ecuador, 67-69. *Mem:* Am Inst Chem Engrs; Am Chem Soc; Faraday Soc; Catalysis Soc. *Res:* Reactivity of solid surfaces in catalytic reactions; new materials for catalysis; chemical vapor transport. *Mailing Add:* Catalytica Assocs Inc 430 Ferguson Dr Bldg 3 Mountain View CA 94043-5214

LEVY, RICHARD, b Brooklyn, NY, June 29, 44; m 69; c 1. MEDICAL ENTOMOLOGY, CONTROLLED RELEASE TECHNOLOGY. *Educ:* Univ Fla, BS, 67 & BS, 68, MS, 69, PhD(entom), 71. *Prof Exp:* Res asst med entom, Dept Entom & Nematol, Univ Fla, 71-74; tech & training consult pest control, Orkin Exterminating Co, 74; res entomologist med entom, WFla Arthropod Res Lab, 74-75; RES ENTOMOLOGIST MED ENTOM, LEE COUNTY MOSQUITO CONTROL DIST, 75- *Concurrent Pos:* Consult, US & overseas. *Mem:* Am Registry Prof Entomologists; Entom Soc Am; Am Mosquito Control Asn; Int Orgn Biol Control Noxious Animals & Plants; Soc Invert Path; Controlled Release Soc; AAAS. *Res:* Biological and chemical control of insects of medical and veterinary importance; US patents. *Mailing Add:* Lee County Mosquito Control Dist PO Box 06005 Ft Myers FL 33906

LEVY, RICHARD ALLEN, MEDICATION MANAGEMENT. *Educ:* Univ Del, PhD(pharmacol), 70. *Prof Exp:* VPRES SCI AFFAIRS, NAT PHARMACEUT COUN, 81- *Mailing Add:* Nat Pharmaceut Coun 1894 Preston White Dr Reston VA 22091

LEVY, ROBERT, b Montreal, Que, Apr 12, 38; m 64; c 1. BIOCHEMISTRY, CLINICAL CHEMISTRY. *Educ:* McGill Univ, BS, 59, PhD(biochem), 65, Am Bd Clin Chem, dipl, 80. *Prof Exp:* NIH fel, Vet Admin Hosp/Univ Mo-Kansas City, 64-66; chief chemist, Vet Admin Hosp, Washington, DC, 66-67; asst prof neurobiol, Psychiat Inst, Univ Md, Baltimore City, 67-71; DIR LAB SERV, PATH DEPT, CHURCH HOSP CORP, 71- *Concurrent Pos:* Asst prof, George Washington Univ, 66-67; guest lectr, Towson State Col, 69. *Mem:* AAAS; Am Asn Clin Chem. *Res:* Neurochemistry of membranes; neurotransmitters; neuroenzymology; clinical enzymology. *Mailing Add:* 4010 Carthage Rd Randallstown MD 21133

LEVY, ROBERT AARON, b El Paso, Tex, Nov 15, 26; m 56; c 4. SOLID STATE PHYSICS. *Educ:* Univ Tex, BS, 47, MA, 48; Univ Calif, Berkeley, MA, 50, PhD(physics), 55. *Prof Exp:* Physicist, US Naval Radiol Defense Lab, 50-53, asst physics, Univ Calif, 53-55; physicist, Tex Instruments, Inc, 55-57; proj engr, Motorola, Inc, 57-59; mem tech staff, Hughes Aircraft Co, 59-60; physiscist, Nat Co, 61-62; assoc prof, Univ Cincinnati, 63-69; CONSULT PHYS PROB ENERGY & ENVIRON, 69- *Concurrent Pos:* Assoc fac mem, Ariz State Univ, 58-59 & Univ Southern Calif, 60-61; consult, US Naval Radiol Defense Lab, 59; fel, Israel AEC, 62-63. *Mem:* AAAS; fel Am Phys Soc; Am Asn Univ Prof. *Res:* Magnetic resonance spectroscopy; quantum electronics. *Mailing Add:* 1617-D N Mesa St El Paso TX 79902-3527

LEVY, ROBERT EDWARD, b Cincinnati, Ohio, May 23, 39; m 70; c 2. TECHNOLOGY PLANNING, ADVANCED PROCESS CONTROL. *Educ:* Cornell Univ, BChE, 62; Univ Calif, Berkeley, PhD(chem eng), 67. *Prof Exp:* Actg instr process control, Univ Calif, Berkeley, 65-66; engr & mgr, Exxon Corp, 67-86; independent consult, 86-87; VPRES & DIR TECHNOL DEVELOP, M W KELLOGG CO, 87- *Mem:* Am Inst Chem Engrs; Sigma Xi (pres, 91); Licensing Exec Soc. *Res:* Plan and manage all technology development activities for the M W Kellogg Company; the technologies are primarily chemical and refining processes. *Mailing Add:* 2211 Golden Pond Dr Kingwood TX 77345

LEVY, ROBERT I, b Bronx, NY, May 3, 37; m 58; c 4. MEDICINE, BIOCHEMISTRY. *Educ:* Cornell Univ, BA, 57; Yale Univ, MD, 61. *Prof Exp:* Intern med, Yale-New Haven Med Ctr, 61-62, resident, 62-63; clin asst med res, Nat Heart Lung & Blood Inst, 63-65, chief resident med, 65-66, dep clin dir, 68-69, chief clin serv, Molecular Dis Br, 69-73, chief lipid metab br, 70-74, dir, Div Heart & Vascular Dis, 73-75, inst dir, 66-81, head sect lipoproteins, 75-81; prof med, dean & vpres, Sch Med, Tufts Univ, 81-83; vpres health sci, Columbia Univ, 83-84, prof med, Col Physicians & Surgeons, 83-87, sr adv to Univ, 84-87; PRES, SANDOZ RES INST, 88- *Concurrent Pos:* Mem coun arteriosclerosis, Am Heart Asn; surgeon, USPHS, 63-66; mem & chmn, numerous comts, councils & bd, 70-; spec consult, anti-lipid drugs, Food & Drug Admin, 73-83; coordr, Cardiovasc Portion, US-USSR Agreement Health & Med Sci, 75-81; adj prof med, Col Physicians & Surgeons, Columbia Univ, 89- *Honors & Awards:* Arthur S Flemming Award, 75; Humanitarian Award, Assoc Health Found, 76; Am Asn Clin Chem Award, 79; Donald D Van Slyke Award Clin Chem, 80; Albert Lasker Spec Pub Health Award, 80; Roger J Williams Award Prev Nutrit, 85; Humana Heart Found Award, 88. *Mem:* Inst Med-Nat Acad Sci; fel NY Acad Sci; Am Soc Clin Invest; fel Am Col Cardiol; Am Fedn Clin Res; fel Soc Behav Med; Am Inst Nutrit; Am Soc Clin Pharmacol & Therapeut; Asn Am Physicians; Int Soc Cardiol; Asn Univ Cardiologists; Int Soc Hypertension; Am Heart Asn; Am Soc Clin Nutrit. *Res:* Lipid metabolism; lipid transport; atherosclerosis; hyperlipoproteinemia; lipoproteins; preventive cardiology; clinical nutrition; nutrition education. *Mailing Add:* Sandoz Res Inst East Hanover NJ 07936

LEVY, ROBERT S(AMUEL), b New York, NY, May 28, 20; m 54; c 2. ENGINEERING. *Educ:* City Col New York, BME, 40; Polytech Inst Brooklyn, MME, 43, DrAeroEng, 46. *Prof Exp:* Mech Engr, NY Naval Shipyard, 41-46; struct engr, Repub Aviation Div, Fairchild-Hiller Corp, 46-59, develop engr, Fairchild Industs, 59-65, proj eng manned spaced vehicles, 65-66, eng prog mgr supersonic transp, 66-71, mgr design eng, 71-75, dir eng, 75, mgr tech eng, 76, mgr new systs res, 77-78, tech specialist, Fairchild Repub Co, 79-85; ADJ ASSOC PROF, AIRCRAFT DESIGN, POLYTECHNIC UNIV, 85- *Concurrent Pos:* Civilian with Nat Adv Comt Aeronaut, 44. *Mem:* Assoc fel Am Inst Aeronaut & Astronaut; Sigma Xi. *Res:* Aircraft structures. *Mailing Add:* 11 Brookside Dr Huntington NY 17743

LEVY, ROBERT SIGMUND, b Fresno, Calif, Nov 3, 21; m 52; c 1. BIOCHEMISTRY, BIOLOGICAL PSYCHIATRY. *Educ:* Univ Calif, Berkeley, AB, 48, AM, 52; Univ Southern Calif, PhD(biochem, nutrit), 57. *Prof Exp:* Asst zool, Univ Calif, Berkeley, 49-52; asst biochem & nutrit, Sch Med, Univ Southern Calif, 55-57; from asst prof to assoc prof biochem, 57-72, PROF BIOCHEM, SCH MED, UNIV LOUISVILLE, 72-, DIR, LAB BIOL PSYCHIAT, 78- *Concurrent Pos:* Assoc psychiat & emergency med, Sch Med, Univ Louisville; fel coun arteriosclerosis, Am Heart Asn. *Mem:* Asn Multidiscipline Educ Health Sci; Sigma Xi; Am Chem Soc; Am Soc Biol Chem; Soc Neurosci. *Res:* Isolation of unusual peptides from blood and hemodialysates of schizophrenic patients; detection of enkephalins and endorphins in biological fluids by radioimmunoassay and radioreceptor assay; biochemistry of brain function and role of neuropeptides; reversal of atherosclerosis by lipoproteins in cell culture. *Mailing Add:* 1147 Restrevor Circle Louisville KY 40205

LEVY, RONALD FRED, b St Louis, Mo, Dec 11, 44; m 66; c 1. TOPOLOGY. *Educ:* Wash Univ, AB, 66, AM, 70, PhD(math), 74. *Prof Exp:* Asst prof math, Goucher Col, 74-75; instr math, Wash Univ, 75-76; asst prof, 76-81, ASSOC PROF MATH, GEORGE MASON UNIV, 81- *Mem:* Am Math Soc; Math Asn Am. *Res:* Compact Hausdorff spaces; almost-P-spaces; linearly ordered topological spaces. *Mailing Add:* Dept Math George Mason Univ 4400 Univ Dr Fairfax VA 22030

LEVY, SALOMON, b Jerusalem, Apr 4, 26; US citizen; m 51; c 2. MECHANICAL ENGINEERING. *Educ:* Univ Calif, Berkeley, BS, 49, MS, 51, PhD(mech eng), 53. *Prof Exp:* Mgr systs eng, Gen Elec Co, 66-68, mgr design eng dept, 68-71, gen mgr nuclear fuel dept, 71-73, gen mgr boiling water reactor systs dept, 73-75, gen mgr boiling water reactor opers, 75-77; PRES, S LEVY INC, ENG CONSULT, 77- *Concurrent Pos:* Consult, Assoc Midwestern Univ, 76-81, Brookhaven Nat Lab, 77- & Elec Power Res Inst, 77-84 & Elec Power Res Inst, 77-85; dir, Iowa Elec, 85- *Honors & Awards:* Heat Transfer Mem Award, Am Soc Mech Engrs, 66. *Mem:* Nat Acad Eng; fel Am Soc Mech Engrs. *Res:* Heat transfer and fluid flow, particularly two-phase flow and boiling heat transfer; nuclear reactor power plant design and analysis. *Mailing Add:* S Levy Inc 3425 S Bascom Ave Campbell CA 95008

LEVY, SAMUEL C, b Far Rockaway, NY, Jan 5, 37; m 58; c 2. ELECTROCHEMISTRY, ELECTROANALYTICAL CHEMISTRY. *Educ:* Hofstra Col, BA, 58; Iowa State Univ, PhD(inorg chem), 62. *Prof Exp:* STAFF MEM, SANDIA NAT LABS, 62- *Concurrent Pos:* Treas, Battery Div, Electrochem Soc, 90-92. *Mem:* Am Chem Soc; Electrochem Soc. *Res:* Chemical to electrical energy conversion; mechanism of electrochemical reactions; lithium battery research and development. *Mailing Add:* Sandia Labs Div 2523 PO Box 5800 Albuquerque NM 87185-5800

LEVY, SAMUEL WOLFE, b Montreal, Que, Feb 26, 22; m 67; c 2. CLINICAL CHEMISTRY. *Educ:* McGill Univ, BSc, 49, PhD(physiol), 54; Univ Sask, MSc, 51. *Prof Exp:* Multiple Sclerosis Soc Can res fel biochem, McGill-Montreal Gen Hosp Res Inst, 54-56; res assoc, Hotel-Dieu Hosp, Montreal, 56-61; dir, Dept Biochem, Queen Mary Vet Hosp, Montreal, 61-79; DIR, DEPT BIOCHEM, QUEEN ELIZABETH HOSP, MONTREAL, 79- *Concurrent Pos:* Dom-Prov Health grant, 56-61; Dept Vet Affairs grant, 64-71. *Honors & Awards:* Ames Award, Can Soc Clin Chem, 75. *Mem:* Chem Inst Can; Can Soc Clin Chem (pres, 70-71); Am Asn Clin Chem. *Res:* Lysosomal enzymes in blood in inflammation disease; effects of heparin in vivo on enzymes and lipid in blood; serum ribonuclease; assay, properties and alterations in disease. *Mailing Add:* Dept Biochem Queen Elizabeth Hosp Montreal PQ H4A 3L6 Can

LEVY, SANDER ALVIN, METALLURGY. *Educ:* Lehigh Univ, BS & BA, 62, PhD(metal eng), 65. *Prof Exp:* Mem tech staff, Bell Tel Labs, Murray Hill, NJ, 65-66; casting res, Pittman Dun Res Labs, Frankford Arsenal, US Army & teacher welding metall, Drexel Inst Technol, 66-68; res scientist, Metall Lab, Reynolds Metals Co, 68-70, supvr, Ingot Casting Technol Sect, 70-78, mgr, Dept Ingot Casting Technol & Metall Serv, 78-82, mgr, Dept Mfg Technol, 82-86, tech specialist, 87-90, SR METALLURGIST, METALL LAB, REYNOLDS METALS CO, 90- *Res:* Metallurgy; ingot casting technology; molten metal quality for aluminum alloys; diamond machine of aluminum. *Mailing Add:* Reynolds Metal Co 401 E Canal St Richmond VA 23261

LEVY, STUART B, b Wilmington, Del, Nov 21, 38; m; c 2. MOLECULAR BIOLOGY, MICROBIOLOGY. *Educ:* Williams Col, AB, 60; Univ Pa, MD, 65. *Prof Exp:* Intern & med resident, Mt Sinai Hosp, NY, 65-67, res fel, Dept Cellular Biol, 66-67; from asst prof to assoc prof med, molecular biol & microbiol, 71-80, PROF MED, MOLECULAR BIOL & MICROBIOL, MED SCH, TUFTS UNIV, BOSTON, 80- *Concurrent Pos:* Res fel, Dept Microbiol, Univ Milan, Italy, 62 & Dept Microbiol, Keio Univ, Tokyo, 64; publiker nutrit fel, Kenyatta Nat Hosp, Nairobi, 64; staff assoc, Nat Inst Arthritis & Metab Dis, NIH, Bethesda, Md, 67-70; vis prof, Dept Path, Univ Padua, Italy, 70 & Pasteur Inst, Paris, France, 76; fel hemat, New Eng Med Ctr, Boston, Mass, 70-71; collabr, E African Viral Inst, Entebbe, Uganda, 71; res career develop award, 72-77; staff physician, NE Med Ctr Hosp, Boston, Mass, 76-; staff scientist, Cancer Res Ctr, Med Sch, Tufts Univ, 76-; sci adv, Biomed Res Ctr, Univ Nat Pedro Henriquez Urena, Santo Domingo, DR, 77-83; consult, Food & Drug Admin, Washington, DC, 78-80 & 85-87; adv, Fate of the Earth, Inc, 81-; pres, Alliance for the Prudent Use of Antibiotics, 81- & Boston Blood Club, 1984 overseas vis, Bd Postgrad Med Educ, Royal Melbourne Hosp, Australia, 83-84; gen chmn, Int Task Forces on Use of Antibiotics Worldwide, Fogarty Int Ctr, NIH, 83-86; mem, subcomt, Gram-Negative Facultatively Anaerobic Rods, Am Soc Metals, 85-88, Subcomt on Plasmid Ref Ctr Collection, Comt on Genetic & Molecular Microbiol, Am Soc Microbiol, 86, Subcomt Health & Antibiotic Resistance, Environ Protection Agency, 88- & Comt Environ Microbiol, Am Soc Metals, 89-; lectr, Am Soc Metals Found Microbiol, 89-90. *Mem:* Am Asn Cancer Res; Am Soc Biochem & Molecular Biol; Am Soc Clin Invest; Am Soc Hemat; Am Soc Microbiol; Infectious Dis Soc Am. *Res:* Antibiotic resistance. *Mailing Add:* Molecular Biol & Microbiol Dept Sch Med Tufts Univ 136 Harrison Ave Boston MA 02111

LEW, CHEL WING, b San Antonio, Tex, Dec 9, 35; m 59; c 4. CHEMISTRY. *Educ:* Tex A&M Univ, BS, 60. *Prof Exp:* Technician chem, 60-61, asst res chemist, 61-65, res chemist, 65-73, SR RES CHEMIST, SOUTHWEST RES INST, 73- *Mem:* Sigma Xi. *Res:* Microencapsulation. *Mailing Add:* 9218 Old Homestead San Antonio TX 78230

LEW, GLORIA MARIA, b Kingston, Jamaica, March 7, 34; US citizen. BIOCHEMISTRY, PHARMACOLOGY. *Educ:* Mt St Vincent Col, BA, 56; Boston Col, MS, 58; Univ Calif, Berkeley, PhD(zool), 72. *Prof Exp:* Chmn, Sci Dept, Alpha Jr Col, 64-66; instr biol, Cardinal Cushing Col, 66-68, asst prof, 68-69; asst prof, 72-76, ASSOC PROF ANAT, MICH STATE UNIV, 76- *Mem:* Am Soc Zoologists; Am Soc Neurosci; Am Physiol Soc; Int Soc Chronobiol; Am Asn Anatomists. *Res:* Circadian rhythms in catecholamine metabolism; effects of estrogen on catecholamine metabolism in genetic hypertension; biochemistry and ultrastructure of pineal gland effects of cocaine and PCP on neurochemistry of developing rats; molecular effects of drugs & hormones on neuroblastomas. *Mailing Add:* Dept Anat Mich State Univ East Lansing MI 48824

LEW, HIN, b Vancouver, BC, Apr 18, 21; m 59; c 3. PHYSICS. *Educ:* Univ BC, BA, 40; Univ Toronto, MA, 42; Mass Inst Technol, PhD(physics), 48. *Prof Exp:* Jr res physicist acoust, Nat Res Coun Can, 42-45; res assoc atomic beams, Mass Inst Technol, 48-49; assoc & sr res officer Atomic Beams & Molecular Spectros, 49-85, GUEST SCIENTIST, NAT RES COUN CAN, 85- *Mem:* Am Phys Soc; Can Asn Physicists. *Res:* Hyperfine structure of atoms and molecules by the atomic beam magnetic resonance method; spectra and structure of molecular ions. *Mailing Add:* Herzberg Inst Astrophys Nat Res Coun Can Ottawa ON K1A 0R6 Can

LEW, JOHN S, b New York, NY, Sept 9, 34; m 63, 75; c 2. APPLIED MATHEMATICS. *Educ:* Yale Univ, BS, 55; Princeton Univ, PhD(physics), 60. *Prof Exp:* C L E Moore instr math, Mass Inst Technol, 62-64; asst prof appl math, Brown Univ, 64-70; RES STAFF MEM MATH SCI, T J WATSON RES CTR, IBM CORP, 70- *Mem:* Math Asn Am; Soc Indust & Appl Math; Sigma Xi. *Res:* Applied analysis, especially asymptotic expansions. *Mailing Add:* T J Watson Res Ctr IBM Corp Yorktown Heights NY 10598

LEWANDOS, GLENN S, b Dallas, Tex. ORGANIC CHEMISTRY, ORGANOMETALLIC CHEMISTRY. *Educ:* Southern Methodist Univ, BS, 67; Univ Tex, Austin, PhD(chem), 72. *Prof Exp:* Asst prof chem, Sul Ross State Univ, 72-76; vis asst prof, Univ Tex, Austin, 76-77; from asst prof to assoc prof 77-85, PROF CHEM, CENT MICH UNIV, 85- *Concurrent Pos:* Res Corp grant, 78-80, Am Chem Soc grant, 81-84, 88- *Mem:* Am Chem Soc; Royal Soc Chem. *Res:* Synthesis and reactivity of organometallic pi complexes; catalysis by transition metals; crown ethers. *Mailing Add:* Dept Chem Cent Mich Univ Mt Pleasant MI 48859

LEWANDOWSKI, JOHN JOSEPH, b Pittsburgh, Pa, Dec 17, 56; m 83; c 1. STRUCTURE PROPERTY RELATIONSHIPS, DEFORMATION PROCESSING. *Educ:* Carnegie Mellon Univ, BS, 79, ME, 81, PhD(metall eng & mat sci), 83. *Prof Exp:* Hertz Found fel metall eng & mat sci, Carnegie Mellon Univ, 81-84; NATO postdoctoral fel, Univ Cambridge, UK, 84-86; asst prof metall eng & mat sci, 86-90, ASSOC PROF MAT SCI & ENG, CASE WESTERN RESERVE UNIV, 90- *Concurrent Pos:* Vis scientist, Univ Cambridge, UK, 86 & Wright Patterson AFB, 87; mem, Star-Bast Nat Comt, Nat Acad Sci, 89-91; NSF presidential young investr award, 89- *Honors & Awards:* Bradley Stoughton Award, Am Soc Metals Int, 89. *Mem:* Am Soc Metals Int; Metall Soc; Mat Res Soc; Soc Advan Mat & Process Eng. *Res:* Effects of microstructure on deformation and fracture of materials, including metals, ceramics, and composites; effects of high pressure on deformation and fracture, including deformation processing. *Mailing Add:* 3636 Traynham Rd Shaker Heights OH 44122

LEWANDOWSKI, MELVIN A, b Chicago, Ill, Dec 8, 30; m 54; c 1. CHEMICAL ENGINEERING. *Educ:* Northwestern Univ, BS, 54; Univ Chicago, MBA, 74. *Prof Exp:* Res engr, Int Minerals & Chem Corp, 57-61, purchasing agt, 61-65, mgr eng & distrib, 65-74; exec vpres, Chinhae Chem Co, Seoul, Korea, 75-78; corp staff vpres res & develop, Int Minerals & Chem Corp, 78-80, vpres chem group, 80-83, develop fertilizer group, 83-85; consult, Lindsay Int Sales, 85-86; vpres & gen mgr, Richmond Lox, 86-88; RETIRED. *Mem:* Am Inst Chem Engrs; Am Chem Soc. *Res:* Administration. *Mailing Add:* Six N 455 Neva Terr Itasca IL 60143

LEWARS, ERROL GEORGE, Can citizen. ORGANIC CHEMISTRY. *Educ:* LondonUniv, BSc, 64; Univ Toronto, PhD(chem), 68. *Prof Exp:* Fel chem, Harvard Univ, 68-70, Univ Western Ont, 70-72; fel, 72-73, asst prof, 73-78, ASSOC PROF CHEM, TRENT UNIV, 78- *Mem:* Am Chem Soc. *Res:* Synthetic organic chemistry; compounds of theoretical interest. *Mailing Add:* Dept Chem Trent Univ Petersborough ON K9J 7B8 Can

LEWBART, MARVIN LOUIS, b Philadelphia, Pa, May 28, 29; m 57; c 4. BIOCHEMISTRY. *Educ:* Philadelphia Col Pharm & Sci, BSc, 51, MSc, 53; Jefferson Med Col, MD, 57; Univ Minn, PhD(biochem), 61. *Prof Exp:* Intern pharm, Jefferson Med Col, 51-52, res assoc biochem, 53-57; intern med, Lankenau Hosp, 57-58; fel biochem, Mayo Found, Univ Minn, 58-61; USPHS spec res fel, Univ Basel, 61-62; res assoc, Jefferson Med Col, 62-67, from asst prof to assoc prof med, 67-75; assoc med dir, Franklin Mint Corp, 74-82; dir, Steroid Lab, Crozer-Chester Med Ctr, 75-90; CLIN ASSOC PROF MED, HAHNEMANN MED COL, 75- *Concurrent Pos:* USPHS res fel, 59-61. *Mem:* AMA; Am Chem Soc; AAAS. *Res:* Steroid chemistry and metabolism. *Mailing Add:* Steroid Lab 81 Great Valley Pkwy Suite 700 Malvern PA 19355

LEWELLEN, ROBERT THOMAS, b Nyssa, Ore, Apr 27, 40; m 62. GENETICS, PLANT BREEDING. *Educ:* Ore State Univ, BS, 62; Mont State Univ, PhD(genetics), 66. *Prof Exp:* Asst agronomist, Mont State Univ, 65-66; RES GENETICIST, AGR RES SERV, USDA, 66- *Mem:* Am Soc Agron; Crop Sci Soc Am; Am Phytopath Soc; Am Soc Sugar Beet Technol. *Res:* Genetics of disease resistance in sugar beet, Beta vulgaris, and development of resistant lines. *Mailing Add:* Agr Res Sta USDA 1636 E Alisal St Salinas CA 93905

LEWELLEN, WILLIAM STEPHEN, b Reedy, WVa, Aug 7, 33; m 58; c 2. FLUID DYNAMICS. *Educ:* WVa Univ, BS, 57; Cornell Univ, MAeroE, 59; Univ Calif, Los Angeles, PhD(eng), 64. *Prof Exp:* Mem tech staff, Space Tech Labs, Inc, 59-60; mem tech staff, Aerospace Corp, 60-64, mgr fluid dynamics sect, 64-66; vis assoc prof aeronaut & astronaut, Mass Inst Technol, 66-67, assoc prof, 67-72; sr consult, Aeronaut Res Assoc Princeton, Inc, 72-79, vpres fluid mech, 79-82, sr vpres, 82-86; SR CONSULT, CRT INC, 87- *Concurrent Pos:* Consult, Aerojet Gen Corp, 68-70; assoc ed, Am Inst Aeronaut & Astronaut J, 78-79; adv comt, Mech & Aerospace Eng, WVa Univ, 81- *Mem:* AAAS; Am Inst Aeronaut & Astronaut; Am Meteorol Soc. *Res:* Energy conversion; fluid dynamics of vortex flows; micrometeorology; computer modeling of turbulent transport; pollutant dispersal. *Mailing Add:* Div Calif Res & Technol Inc 50 Washington Rd Princeton NJ 08543

LEWENZ, GEORGE F, b Berlin, Ger, Aug 29, 20; US citizen; m 55; c 4. ORGANIC CHEMISTRY. *Educ:* Western Reserve Univ, BS, 47, MS, 52. *Prof Exp:* Aeronaut res scientist, NASA, 48-53; chemist, Texaco, Inc, 53-58 & Esso Res & Eng Co, NJ, 59-68; res logician, 68-71, sr info chemist, 71-73, res specialist, Dow Chem Co, 73-86; RETIRED. *Mem:* Am Chem Soc; fel Am Inst Chemists. *Res:* Organic synthesis; abstracting, indexing and information science. *Mailing Add:* 2305 Burlington Dr Midland MI 48640

LEWERT, ROBERT MURDOCH, b Scranton, Pa, Sept 30, 19; m 48; c 2. MEDICAL PARASITOLOGY, IMMUNOLOGY. *Educ:* Univ Mich, BS, 41; Lehigh Univ, MS, 43; Johns Hopkins Univ, ScD(parasitol), 49. *Prof Exp:* Asst instr biol, Lehigh Univ, 42-43; instr zool, Cols of Seneca, 43-44; instr parasitol, Dept Bact & Parasitol, 48-52, asst prof, 52-54, microbiol, 54-56, assoc prof, 57-61, prof microbiol, 61-83, prof molecular genetics & cell biol, 83-85, EMER PROF, UNIV CHICAGO, 86- *Concurrent Pos:* Fulbright res fel, Philippines, 61; Guggenheim fel, 61; vis prof, Inst Hyg, Univ Philippines, 61 & 63-65; consult, Surg Gen, US Army, 59-75 & clin parasitol, Hines Vet Admin Hosp, 75-80; mem, Comn parasitic diseases, Armed Forces Epidemiol Bd, 59-66, parasitol study sect, NIH Trop Med, 65-69, Am Bd Microbiol, 65-68. *Mem:* AAAS; fel Am Acad Microbiol; Am Soc Parasitol; Sigma Xi; Royal Soc Trop Med & Hyg; Am Soc Trop Med & Hyg. *Res:* Host parasite relationships with emphasis on immunity, tolerance and immunopathology of schistosomiasis; histochemical and cytochemical studies of parasite effects on host; immunity and invasiveness of helminths; schistosomiasis. *Mailing Add:* 37 Henry Mtn Rd Brevard NC 28712

LEWIN, ALFRED S, b Chicago, Ill, Apr 25, 51; m 72; c 3. MITOCHONDRIA, MEMBRANE PROTEIN SYNTHESIS. *Educ:* Univ Chicago, AB, 73, PhD(biol), 78. *Prof Exp:* Acad asst biochem, Biocenter Univ, Basel, 78-81; ASST PROF, DEPT CHEM, IND UNIV, 81- *Concurrent Pos:* Europ Molecular Biol Orgn fel, 78-; jr fac res award, Am Cancer Soc, 82. *Mem:* Am Soc Biol Chemists; Am Soc Microbiol; Genetics Soc Am; AAAS. *Res:* Biosynthesis of mitochondrial enzymes; transport of proteins from the cytoplasm; splicing of mitochondrial messenger RNA. *Mailing Add:* Dept Med Microbiol Univ Fla Col Med Gainesville FL 32610

LEWIN, ANITA HANA, b Bucarest, Rumania, Oct 27, 35; m 56; c 2. PHYSICAL ORGANIC CHEMISTRY, MOLECULAR MODELLING. *Educ:* Univ Calif, Los Angeles, BS, 59, PhD(phys org chem), 63. *Prof Exp:* Res asst prof chem, Univ Pittsburgh, 64-66; from asst prof to assoc prof chem, Polytech Inst Brooklyn, 66-74; fel, 74-75, SR CHEMIST, RES TRIANGLE INST, 75- *Mem:* Am Chem Soc. *Res:* Reaction mechanisms; conformational analysis; hindered rotation; retinoids-synthesis and properties; organic synthesis; synthesis of radiolabeled compounds; molecular modeling; receptor binding; alkaloid synthesis; structure activity correlations. *Mailing Add:* Chem & Life Sci Group PO Box 12194 Research Triangle Park NC 27709

LEWIN, JOYCE CHISMORE, b Ilion, NY, Nov 13, 26; m 50. MICROBIOLOGY. *Educ:* Cornell Univ, BS, 48; Yale Univ, MS, 50, PhD(bot, microbiol), 53. *Prof Exp:* Guest res worker biol, Lab, Nat Res Coun Can, 52-55; res assoc marine biol, Woods Hole Oceanog Inst, 56-60; asst res biologist, Scripps Inst, Univ Calif, San Diego, 60-65; from asst prof to prof, 65-74, RES PROF OCEANOG, UNIV WASH, 74- *Mem:* Am Soc Limnol & Oceanog; Phycol Soc Am; Am Inst Biol Sci; Marine Biol Asn UK; Int Phycol Soc. *Res:* Culture of marine microalgae, especially diatoms; physiology and nutrition of marine diatoms; physiology and ecology of surf diatom blooms. *Mailing Add:* Dept Oceanog WB-10 Univ Wash Seattle WA 98195

LEWIN, LAWRENCE M, b New York, NY, Mar 3, 32; m 58; c 3. BIOCHEMISTRY, CHEMICAL PATHOLOGY. *Educ:* Mass Inst Technol, BS, 53; Cornell Univ, PhD(biochem), 59. *Prof Exp:* Res asst biochem, Cornell Univ, 53-54; res asst biochem, NY State Agr Exp Sta, 56-59; res assoc, Mass Inst Technol, 59-61; NIH fel, Weizmann Inst, 61-62; from asst prof to assoc prof biochem, Schs Med & Dent, Georgetown Univ, 63-71; assoc prof, 72-79, PROF BIOCHEM, DEPT CHEM PATH, SACKLER MED SCH, TEL AVIV UNIV, 79-, HEAD DEPT, 89- *Mem:* Israel Biochem Soc; Am Chem Soc; Israel Soc Chem Path; Israel Microbiol Soc; Israel Fertil Soc. *Res:* Metabolic functions of vitamins, particularly carnitine; male reproduction tract and infertility lipids. *Mailing Add:* Dept Chem Path Tel Aviv Univ Ramat Aviv Israel

LEWIN, LEONARD, b Southend, Eng, July 22, 19; US citizen; m 43; c 2. ELECTRICAL ENGINEERING, MATHEMATICS & EDUCATION. *Hon Degrees:* DSc, Univ Colo, 67. *Prof Exp:* Sci officer radar, Brit Admiralty, 41-45; sr engr microwaves, Standard Telecommun Labs, 46-50, dept head, 50-60, asst mgr transmissions, 60-66, sr prin res electromagnetic theory, 67-68; prof, 68-86, coordr telecommun prog, 74-86, EMER PROF ELEC ENG, UNIV COLO, BOULDER, 87- *Concurrent Pos:* Consult, Standard Telecommun Labs, 68-, Medion Ltd, 70-, Westinghouse Corp, 71, Nat Bur Standards, 78-, Mass Inst Technol Lincoln Labs, 85-; Sci Res Coun grants, UK, 73 & 75; Fulbright fel, 82. *Honors & Awards:* Premium Awards, Brit Inst Elec Engrs, 52 & 60; Microwave Prize & W G Baker Award, Inst Elec & Electronics Engrs, 62; Prestige Lectr, Nat Inst Elec Eng, NZ, 87. *Mem:* Fel Brit Interplanetary Soc; Brit Inst Elec Engrs; fel Inst Elec & Electronics Engrs. *Res:* Electromagnetic theory; wave propagation; waveguides and antennas; mathematics; mathematical applications to engineering. *Mailing Add:* Dept Elec Eng Camput Box 425 Univ Colo Boulder CO 80309

LEWIN, RALPH ARNOLD, b London, Eng, Apr 30, 21; m 69. PHYCOLOGY. *Educ:* Cambridge Univ, BA, 42, MA, 46, ScD, 72; Yale Univ, MSc, 49, PhD, 50. *Prof Exp:* Spec lectr phycol, 50-51, instr bot, Yale Univ, 51-52; asst res off biol, Maritime Regional Lab, Nat Res Coun Can, 52-55; investr phycol, NIH grant, Marine Biol Lab, Woods Hole, 55-59; assoc prof marine biol, 59-67, PROF EXP PHYCOL, SCRIPPS INST OCEANOG, UNIV CALIF, 67- *Concurrent Pos:* Mem, Corp Marine Biol Lab, Woods Hole. *Honors & Awards:* Darbaker Prize, Bot Soc Am, 58. *Mem:* Soc Exp Biol; Marine Biol Asn UK; Brit Phycol Soc; Phycol Soc Am (pres, 70). *Res:* Experimental phycology; microbiology; microbial genetics; marine biology. *Mailing Add:* Scripps Inst Oceanog A-00202 Univ Calif La Jolla CA 92093

LEWIN, SEYMOUR Z, b New York, NY, Aug 16, 21; m 43; c 2. PHYSICAL & ANALYTICAL CHEMISTRY. *Educ:* City Col New York, BS, 41; Univ Mich, MS, 42, PhD(chem), 50. *Prof Exp:* Lectr chem, Univ Mich, 47; from instr to assoc prof, 51-60, PROF CHEM, NY UNIV, 60- *Concurrent Pos:* Belg-Am Educ Found fel, 62; hon prof, Inst Quimico Sarria, Barcelona, Spain, 62; ed, Art & Archeol Tech Abstr, 66-69; consult, US Army Chem Corp, Smithsonian Inst, Food & Drug Admin & Warner-Lambert Co; instrumentation ed, J Chem Educ, 60-68. *Honors & Awards:* A Cressy Morrison Prize, NY Acad Sci, 56; Kasimir Fajans Prize, 58. *Mem:* AAAS; Am Chem Soc; Soc Appl Spectros; Am Inst Chemists; fel NY Acad Sci; fel Am Inst Chemists. *Res:* Crystal growth; spectroscopy; instrumentation; materials of art and archaeology; polymorphism; solid state chemistry; stone decay and preservation. *Mailing Add:* Dept Chem NY Univ Four Washington Place New York NY 10003

LEWIN, VICTOR, b San Francisco, Calif, Sept 8, 30; m 50; c 3. ECOLOGY. *Educ:* Univ Calif, AB, 53, PhD(zool), 58. *Prof Exp:* Asst zool, Univ Calif, 54-58; from asst prof to assoc prof, 58-70, dir, 58-80, CUR HERPET, MUS ZOOL, UNIV ALBERTA, 80-, PROF ZOOL, 70- *Concurrent Pos:* Asst cur, Mus Vert Zool, Univ Calif, 55-56. *Honors & Awards:* Painton Award, Cooper Ornith Soc, 65. *Mem:* Wildlife Soc; Am Soc Mammalogists; Cooper Ornith Soc; Am Ornithologists Union; Can Soc Wildlife & Fishery Biol. *Res:* Wildlife ecology; ecology of game birds and mammals, particularly reproductive anatomy and physiology of gallinaceous birds; effects of chlorinated hydrocarbon residues on birds; ecology of exotic game bird species. *Mailing Add:* PO Box 97 Heriot Bay BC V0P 1H0 Can

LEWIN, WALTER H G, b The Hague, Netherlands, Jan 29, 36; m 59, 81; c 4. HIGH ENERGY ASTROPHYSICS. *Educ:* Univ Delft, Ir, 60, Dr(physics), 65. *Prof Exp:* Res assoc physics, Univ Delft, 59-66; fel space res & asst prof, 66, assoc prof, 68-74, PROF PHYSICS, MASS INST TECHNOL, 74- *Concurrent Pos:* Recipient Guggenheim Fel, 84; Alexander von Humboldt award, 84, 86 & 91; Distinguished Spring Lectr, Princeton Univ, 86. *Honors & Awards:* Outstanding Sci Achievement Award, NASA, 78. *Mem:* Int Astron Union; Am Astron Soc; Am Phys Soc. *Res:* Radioactive isotope applications; nuclear and atomic physics; x-ray astronomy; high-altitude ballooning; satellite observations, orbital solar observatory-7, small astronomy satellite-3, high energy astronomy observatory-1; astrophysics; radar ocean surveillance satellite; Ginga. *Mailing Add:* MIT 37-627 Cambridge MA 02139

LEWINSON, VICTOR A, b New York, NY, 1918; m 57; c 2. OPERATIONS RESEARCH. *Educ:* Harvard Col, AB, 39; Columbia Univ, MA, 45, PhD(chem), 50. *Prof Exp:* Asst chem, Columbia Univ, 39-42; res scientist & sect leader, Manhattan Proj, 42-45; res fel chem, Calif Inst Technol, 50-51; fel, Mellon Inst, 51-54; analyst opers res, Nat Acad Sci, 54-61; mem prof staff, Arthur D Little, Inc, 61-86; RETIRED, 86- *Res:* Freight transportation, especially maritime and railroad. *Mailing Add:* Arthur D Little Inc Acorn Park Cambridge MA 02140-2390

LEWIS, A(LBERT) D(ALE) M(ILTON), b Paoli, Ind, May 20, 20; m 46; c 1. CIVIL ENGINEERING, STRUCTURAL ENGINEERING. *Educ:* Purdue Univ, BS, 41, MS, 51. *Prof Exp:* Field engr oil refinery construct, M W Kellogg Co, 41-44, 46-47 & Gulf Oil Corp, 47-49; instr struct eng, Purdue Univ, 51-52; design engr, Standard Oil Co, Calif, 52-54; assoc prof, 54-85, EMER PROF STRUCT ENG, PURDUE UNIV, WEST LAFAYETTE, 85- *Concurrent Pos:* Consult, Truss Bridge Res Proj, Northwestern Univ; vis assoc prof struct eng, Univ Calif, Los Angeles, 71. *Mem:* Am Soc Civil Engrs; Am Soc Eng Educ; Am Concrete Inst; Am Rwy Eng Asn; Asn Comput Mach; Soc Exp Mech; Nat Soc Prof Engrs; Am Inst Steel Construct. *Res:* Structural analysis; structural design; design optimization; digital computers; experimental mechanics. *Mailing Add:* 107 Sylvia St West Lafayette IN 47906

LEWIS, AARON, b Calcutta, India, Oct 14, 45; US citizen. BIOPHYSICS. *Educ:* Univ Mo, BS, 66; Case Western Reserve Univ, PhD(phys chem), 70. *Prof Exp:* Instr phys chem, Case Western Reserve Univ, 70; NIH fel, 70-72, instr, 71-72, asst prof, 72-76, ASSOC PROF BIOPHYS, CORNELL UNIV, 76- *Concurrent Pos:* Sloan fel, 74-76; vis prof, Calif Inst Technol, 77 & Hebrew Univ, 79-80; Guggenheim fel, 79-80. *Mem:* AAAS; Am Chem Soc; Biophys Soc; Asn Res Vision & Ophthal; Am Photobiol Soc. *Res:* Molecular mechanism of ion gates and pumps, specifically bacteriorhodopsin and rhodopsin. *Mailing Add:* Dept Appl Physics Gergman Bldg Hebrew Univ Jerusalem Israel

LEWIS, ALAN ERVIN, b Milwaukee, Wis, Feb 1, 36; m 61; c 2. MEDICINE. *Educ:* Univ Wis-Madison, BS, 57; Marquette Univ, MD, 60. *Prof Exp:* Intern, Hosp Univ Pa, 60-61; resident, Med Ctr, Univ Mich, 61 & 63-65; from instr to sr instr internal med, Hahnemann Med Col & Hosp, 67-71, from asst prof to assoc prof med, 71-87. *Concurrent Pos:* Fel endocrinol, Sch Med, Tufts Univ, 65-67. *Mem:* Am Diabetes Asn; Am Fedn Clin Res; Am Col Physicians; Endocrine Soc. *Mailing Add:* 1450 S Dobson Rd No 202 Mesa AZ 85202-4774

LEWIS, ALAN GRAHAM, b Pasadena, Calif, Mar 14, 34; m 57; c 2. BIOLOGICAL OCEANOGRAPHY, ZOOLOGY. *Educ:* Univ Miami, BSc, 56, MSc, 58; Univ Hawaii, PhD(zool), 61. *Prof Exp:* Asst prof zool, Univ NH, 61-64; from asst prof to assoc prof, 64-76, PROF OCEANOG & ZOOL, UNIV BC, 76- *Mem:* Am Geophys Union; Sigma Xi. *Res:* Ecology of marine plankton. *Mailing Add:* Dept Oceanog Univ BC 6270 Univ Blvd Vancouver BC V6T 1W5 Can

LEWIS, ALAN JAMES, b Green Bay, Wis, June 2, 43; m 75; c 1. ISLAND BIOGEOGRAPHY, FLORISTIC BIOGEOGRAPHY. *Educ:* Wis State Univ-Eau Claire, BS, 68; Rutgers Univ, PhD(plant ecol), 71. *Prof Exp:* Teaching asst gen biol, Rutgers Univ, New Brunswick, 68-69, NSF fel, 69-70, teaching asst gen biol & plant ecol, 70-71; asst prof biol, Kean Col NJ, 71-74; vis prof, Swarthmore Col, 75-76; asst prof, Mercyhurst Col, 76-78; PROF ECOL, UNIV MAINE, MACHIAS, 78- *Concurrent Pos:* NSF fel, Rutgers Univ, 69; res assoc, Univ Houston, 72; Fulbright scholar, Univ Papua, New Guinea, 85. *Mem:* AAAS; Am Inst Biol Sci; Ecol Soc Am; Sigma Xi. *Res:* Floristic distributions of Maine; island biogeography of Maine. *Mailing Add:* Div Sci & Math Univ Maine 90 Brien Ave Machias ME 04654

LEWIS, ALAN JAMES, b Newport, Gwent, UK. PHARMACY. *Educ:* Southampton Univ, Hampshire, BSc, 67; Univ Wales, Cardiff, PhD(pharmacol), 70. *Prof Exp:* Postdoctoral fel biomed sci, Univ Guelph, Ont, Can, 70-72; res assoc, Lung Res Ctr, Yale Univ, 72-73; sr pharmacologist, Organon Labs, Ltd, Lanarkshire, Scotland, 73-79; res mgr immunoinflammation, 79-82, assoc dir exp therapeut, 82-85, dir exp therapeut, 85-87, asst vpres exp therapeut, 87-89, VPRES RES, AM HOME PROD, WYETH-AYERST RES, 89- *Concurrent Pos:* Ed, Allergy Sect, Agents & Actions & Int Arch Pharmacodyn Ther; reviewer, J Petrol Technol, Biochem Pharmacol, Can J Physiol Pharmacol, Europ J Pharmacol & J Pharmaceut Sci; vpres, Mid-Atlantic Pharmacol Soc, 91- *Mem:* Pulmonary Res Asn; Inflammation Res Asn (pres, 86-88); Am Soc Pharmacol & Exp Therapeut; Pharmaceut Mfrs Asn; Am Rheumatism Asn. *Res:* Mechanisms and treatment of inflammatory diseases including arthritis and asthma; cardiovascular pharmacology; metabolic disorders; central nervous system pharmacology; osteoporosis. *Mailing Add:* CN 8000 Princeton NJ 08543-8000

LEWIS, ALAN LAIRD, b Holyoke, Mass; m; c 2. OPTOMETRY. *Educ:* Mass Col Optom, BSc, 65, OD, 70; Ohio State Univ, PhD(physiol optics), 71. *Prof Exp:* Optometrist, USN, 65-68; from asst prof to assoc prof, 72-85, PROF PHYSIOL OPTICS, COL OPTOM, STATE UNIV NY, 85- *Concurrent Pos:* Mem, US Nat Comt, Comn Int de L'Eclairage, 73-; vis researcher, Nat Bur Standards, 79-80; mem, Int Res Group Colour Vision Deficiencies. *Mem:* Am Acad Optom; fel Illum Eng Soc; Asn Res Vision & Opthal; Optic Soc Am. *Res:* Clinical aspects of color vision; biological effects of optical radiation; accommodation; visibility of objects as a funciton of illuminance; discomfort glare. *Mailing Add:* Col Optom State Univ NY 100 E 24th St New York NY 10010-3677

LEWIS, ALEXANDER D(ODGE), mechanical engineering; deceased, see previous edition for last biography

LEWIS, ALLEN ROGERS, b Ithaca, NY, Aug 11, 47; m 69; c 2. BEHAVIORAL ECOLOGY. *Educ:* Cornell Univ, BS, 69; Univ Del, MS, 71; Univ Rochester, MS, 77, PhD(biol), 79. *Prof Exp:* Marine exten specialist, Col Marine Studies, Univ Del, 71-74; asst prof, 78-81, assoc prof, 81-86, PROF BIOL, UNIV PR, 86- *Concurrent Pos:* Ed, Carribean J Sci, 81-90. *Mem:* AAAS; Asn Trop Biol; Ecol Soc Am; Soc Study Amphibians & Reptiles; Soc Study Evol; Herpetologists League; Am Soc Naturalists. *Res:* Analysis of problems in social behavior and ecology utilizing terrestrial vertebrates in Puerto Rico. *Mailing Add:* Dept Biol Univ PR Mayaguez PR 00709

LEWIS, ALVIN EDWARD, b New York, NY, Nov 21, 16; m 43; c 2. BIOSTATISTICS, PHYSIOLOGY. *Educ:* Univ Calif, Los Angeles, AB, 38; Stanford Univ, AM, 39, MD, 44. *Prof Exp:* Asst path, Stanford Univ, 47-48; clin instr & chief path sect, AEC Proj, Univ Calif, Los Angeles, 49-53; dir labs, Mt Zion Hosp, 53-66; prof path, Mich State Univ, 66-72; prof path & chmn dept, Med Sch, Univ SAla, 72-74; from prof to emer prof path, Univ Calif, 74-87; RETIRED. *Concurrent Pos:* Am Cancer Soc fel, Stanford Univ, 48-49; vis physician, Los Angeles County Harbor Hosp, 49-53; attend physician, Wadsworth Gen Hosp, 50-53; asst clin prof, Med Ctr, Univ Calif, San Francisco, 59-66. *Mem:* AAAS; Am Physiol Soc; AMA; fel Col Am Path. *Res:* Hepatic function tests; plasma volume and distribution. *Mailing Add:* 21 Woodgreen Ct Santa Rosa CA 95409

LEWIS, ANTHONY WETZEL, b Welch, WVa, Nov 4, 42; m 63; c 2. VERTEBRATE ECOLOGY, PUBLIC HEALTH. *Educ:* Loma Linda Univ, BA, 66, MPH, 74; Ariz State Univ, MS, 69, PhD(zool), 73. *Prof Exp:* Chmn dept biol & health, Mountain View Col, Philippines, 72-77; ASST PROF BIOL, LOMA LINDA UNIV, 77- *Concurrent Pos:* Traineeship, USPHS, 71-72. *Mem:* Am Soc Mammalogists; Soc Study Amphibians & Reptiles. *Res:* Vertebrate population ecology. *Mailing Add:* 5165 Harriet Circle Riverside CA 92505

LEWIS, ARMAND FRANCIS, b Fairhaven, Mass, May 22, 32; m 58; c 2. PHYSICAL CHEMISTRY, MATERIALS SCIENCE. *Educ:* Southeastern Mass Univ, BS, 53; Okla State Univ, MS, 55; Lehigh Univ, PhD(chem), 58. *Prof Exp:* Res asst rheology, Lehigh Univ, 58-59; res chemist, Cent Res Div, Am Cyanamid Co, 59-63, sr res chemist, Plastics & Resins Div, 63-64, group leader polymer physics & adhesion, 64-69, proj leader noise control mat, 70-71; sr res assoc, Lord Corp, 71-73, sr mat scientist, 73-81; res assoc, Kendall Co, 84-89; VIS SCHOLAR, TEXTILE SCI DEPT, SOUTHEASTERN MASS UNIV, 90- *Honors & Awards:* Union Carbide Award, Am Chem Soc, 63. *Mem:* Am Chem Soc; Soc Rheology (treas, 66-). *Res:* Polymer physics; rheology; surface chemistry and adhesion; dynamic mechnical properties of polymers; glass transition phenomena in polymeric systems; polymer to metal adhesion and fracture of adhesive joints; vibration and noise control materials; rubber chemicals; engineering composites; marine materials. *Mailing Add:* Textile Sci Dept Southeastern Mass Univ North Dartmouth MA 02747

LEWIS, ARNOLD D, b Philadelphia, Pa, May 6, 20; m 45; c 2. ANALYTICAL CHEMISTRY. *Educ:* Philadelphia Col Pharm, BS, 40; Polytech Inst Brooklyn, MS, 47. *Prof Exp:* Control chemist, Hance Bros & White, 40-41; pilot plant chemist, United Gas Improv Corp, 41-43; asst scientist to Dr E A H Friedheim, 43-44; jr scientist, G D Res Inst, 44-47; scientist, Dept Org Chem, Warner-Lambert Co, 47-54, sr scientist, 54-63, sr res assoc, Chem Res Div, 63-64, dir anal & phys chem, Prof Prod Group, 64-77; consult, 78-81; dir tech affairs, S S T Corp, 81-86; RETIRED. *Mem:* Am Chem Soc. *Res:* Organic synthesis of heterocycles; infrared and ultraviolet absorption spectrophotometry; microanalysis; gas, paper, thin-layer and column chromatography; chemical safety; proton magnetic resonance spectroscopy. *Mailing Add:* 42 Intervale Rd Livingston NJ 07039-2756

LEWIS, ARNOLD LEROY, II, b Portland, Ore, Sept 24, 52. INSTRUMENTATION, SPECTROSCOPY. *Educ:* Pac Lutheran Univ, BA, 75; Ore State Univ, PhD(anal chem), 81. *Prof Exp:* Teaching asst chem, Ore State Univ, 75-78, res asst, 78-81; sr chemist, Exxon Nuclear Idaho Co, Inc, 81; SR SCIENTIST, WESTINGHOUSE IDAHO NUCLEAR CO INC, 81- *Mem:* Am Chem Soc. *Res:* Laser atomic fluorescence studies of laser microprobe plumes; microcomputer interfacing to analytical instrumentation; methods development of analytical techniques in the nuclear industry. *Mailing Add:* No 195552 c/o Aranco Box 6011 Dhahran 31311 Saudi Arabia

LEWIS, ARTHUR B, b Forest, Miss, Nov 21, 01; m 30; c 4. PHYSICS, SOLID STATE PHYSICS. *Educ:* Univ Miss, BA, 23, MA, 25; Johns Hopkins Univ, PhD(physics), 30. *Prof Exp:* Asst prof math & physics, Univ Miss, 25-26; jr physicist, Nat Bur Standards, 26-30, asst physicist, 30-36; from assoc prof to prof physics, 36-57, from assoc prof to prof math, 36-47, chmn, dept physics & astron, 52-57, prof astron, 52-57, dean, Col Liberal Arts, 69-71, prof math, 69-71, EMER PROF PHYSICS & ASTRON & DEAN, COL LIB ARTS, UNIV MISS, 71- *Concurrent Pos:* Consult, Solid State Div, Oak Ridge Nat Lab, 51-68. *Mem:* Am Phys Soc; Am Asn Physics Teachers; AAAS. *Res:* Electrical instruments and measuring techniques; effects of neutron bombardment on the resistivity of precipitation-hardening alloys. *Mailing Add:* 301 Wash Ave Washington Oxford MS 38655

LEWIS, ARTHUR EDWARD, b Jamestown, NY, Jan 11, 29; m 53; c 3. EARTH SCIENCE. *Educ:* St Lawrence Univ, BS, 50; Calif Inst Technol, MS, 55, PhD(geol), 58. *Prof Exp:* Sr engr, Curtiss Wright Corp, 58-60; scientist, Hoffman Sci Ctr, 60-62; mem tech staff, Fairchild Semiconductor, Calif, 62-67; geologist, Lawrence Livermore Lab, Univ Calif, 67-69, group leader, Plowshare Prog, Peaceful Appln Nuclear Explosives, 69-73, proj leader, Oil Shale, 73-84, FOSSIL ENERGY GROUP LEADER, LAWRENCE LIVERMORE LAB, UNIV CALIF, 84- *Concurrent Pos:* Consult, Fed Energy Agency, 74. *Honors & Awards:* Peele Award, Am Inst Mining & Metall Engrs, 74. *Res:* Processes for recovery of oil from oil shale; energy resource development. *Mailing Add:* Lawrence Livermore Lab PO Box 808 Livermore CA 94550

LEWIS, AUSTIN JAMES, b Poole, Eng, Nov 29, 45; m 73; c 1. ANIMAL NUTRITION. *Educ:* Univ Reading, BSc, 67; Univ Nottingham, PhD(nutrit), 71. *Prof Exp:* Assoc animal nutrit, Iowa State Univ, 71-74; res assoc, Univ Nebr, 74-75; asst prof, Univ Alta, 75-77; from asst prof to assoc prof, 77-85, PROF ANIMAL NUTRIT, UNIV NEBR, 85- *Mem:* Am Soc Animal Sci; Am Inst Nutrit. *Res:* Nutritional requirements of swine, especially proteins and amino acids. *Mailing Add:* Dept Animal Sci Univ Nebr Lincoln NE 68583-0908

LEWIS, BENJAMIN MARZLUFF, b Scranton, Pa, Oct 7, 25; div; c 3. PHYSIOLOGY. *Educ:* Univ Pa, MD, 49. *Prof Exp:* Res fel med, Harvard Univ, 50-52; actg chmn dept, 70-71, from asst prof to assoc prof, 56-62, PROF INTERNAL MED, WAYNE STATE UNIV, 62- *Concurrent Pos:* Teaching fel med, Harvard Univ, 52-53; fel, Grad Sch Med, Univ Pa, 53-55; consult, Vet Admin Hosp, 56. *Mem:* Fel Am Col Physicians; Am Physiol Soc; Am Soc Clin Invest. *Res:* Pulmonary physiology, particularly gas diffusion and pulmonary circulation. *Mailing Add:* Dept Med Wayne State Univ Sch Med Detroit MI 48201

LEWIS, BERNARD, b London, Eng, Nov 1, 1899; nat US; m 34; c 1. CHEMICAL ENGINEERING, COMBUSTION SCIENCE. *Educ:* Mass Inst Technol, BS, 23; Harvard Univ, MA, 24; Cambridge Univ, PhD(phys chem), 26. *Hon Degrees:* ScD, Cambridge Univ, 53. *Prof Exp:* Demonstr phys chem, Cambridge Univ, 25-26; Nat Res Coun fel, Univ Berlin & Univ Minn, 26-29; phys chemist, US Bur Mines, 29-42, chief, Explosives & Phys Sci Div, 46-53; pres, Combustion & Explosives Res, Inc, 53-87; CONSULT, ENERGY SYSTS ASSOCS, 87- *Concurrent Pos:* Dir res powder & explosives, Ord Dept, US Army, 51-52; consult, US Army, USN, USAF & Nat Bur Standards; mem, Sci Adv Comt, Ord Corps, Aberdeen Proving Ground, Md, Combustion Comt, Nat Adv Comt Aeronaut & Fire Res Conf, Nat Acad Sci; pres, Comt high temperature, Int Union Pure & Appl Chem; US ed, J Combustion & Flame; co-ed, Phys Measurements in Gas Dynamics & Combustion, 54 & Combustion Processes, 56; consult, tech adv comt, US Dept Interior, numerous industs & res insts. *Honors & Awards:* Lewis Gold Medal, Combustion Inst, 58; Gold Medal, Ital Thermotech Asn, 61; Pittsburgh Award, Am Chem Soc, 74; Orleans Medal, France, 75; Bordeaux Medal, France, 81; Tecknior Israel Medal, 82. *Mem:* Emer mem, Am Chem Soc; Am Phys Soc; fel Am Inst Aeronaut & Astronaut; Combustion Inst (pres, 54-66, hon pres, 66-); fel NY Acad Sci; fel Inst Chem. *Res:* Chemical kinetics of gas reactions; thermodynamics of explosives; flame propagation; ignition; combustion in jet propulsion; propellants; detonation; internal combustion engines; oxidation of hydrocarbons; fuels, interior ballistics; combustion and flame phenomena; explosion hazard prevention in industry; nuclear power plant safety; author of book on combustion. *Mailing Add:* 5863 Malborough Ave Pittsburgh PA 15217

LEWIS, BERTHA ANN, b Lewisville, Minn, Oct 21, 27. CARBOHYDRATE CHEMISTRY, FOOD SCIENCE. *Educ:* Univ Minn, BChem, 49, MS, 54, PhD(biochem), 57. *Prof Exp:* Res fel biochem, Univ Minn, St Paul, 57-65, res assoc, 65-67; assoc prof design & environ anal, 67-70, assoc dean. Col Human Ecol & asst dir, Cornell Agr Exp Sta, 74-80, ASSOC PROF, DIV NUTRIT SCI, CORNELL UNIV, 70- *Concurrent Pos:* Prin investr, NIH, 71-74 & 80-82, Nat Cancer Inst, 78-89 & NSF, 79-82; mem, US Dept Agr Comt Regional Res & Home Econ Sub-Comt Agr Exp Stas Comt on Policy, 76-79, & Nat Agr Res Comt, 80-85; vis prof, Dept Chem, Univ BC, 84 & 85. *Mem:* Am Chem Soc; Inst Food Technologists; Soc Complex Carbohydrates; Fiber Soc; Am Asn Cereal Chemists. *Res:* Carbohydrate chemistry and biochemistry; chemistry of glycoproteins; protein structure and functionality; dietary fiber; anti-nutrients in food. *Mailing Add:* Div Nutrit Sci Cornell Univ 117 Savage Ithaca NY 14853-6301

LEWIS, BRIAN KREGLOW, b SAfrica, Sept 2, 32; US citizen; m 53; c 6. HUMAN PHYSIOLOGY, COMPUTER BASED INSTRUCTION. *Educ:* Ohio State Univ, BS, 54; Tufts Univ, PhD(physiol), 71. *Prof Exp:* Res assoc physiol, Sch Med, Tufts Univ, 71, May Inst Med Res, Jewish Hosp Cincinnati, 71-74 & Col Med, Univ Cincinnati, 74-75; from asst prof to assoc prof health sci, Grand Valley State Col, 75-81; assoc prof physiol, 81-84, PROF & CHMN DEPT PHYSIOL, DIR, COMPUT SCI, PONCE SCH MED, 87- *Concurrent Pos:* Adj asst prof physiol, Col Med, Univ Cincinnati, 72-75; comput consult & prog bus, 84- *Mem:* Endocrine Soc; Soc; Study Reprod; Study Fertil; Sigma Xi. *Res:* Developing computer simulations for use in physiology instruction for medical and PhD students; develop software and training for small businesses. *Mailing Add:* Dept Physiol Ponce Sch Med PO Box 7004 Ponce PR 00732-7004

LEWIS, BRIAN MURRAY, b Oxford, UK, June 20, 43; UK & NZ citizen; m 67; c 2. GALAXIES, RADIO ASTRONOMY MEASUREMENTS. *Educ:* Adelaide Univ, Australia, BS, 65; Australian Nat Univ, PhD(astron), 70. *Prof Exp:* Res asst astron, Jodrell Bank, Univ Manchester, UK, 69-71; dir teaching-pub rels-res, Carter Observ, Wellington, NZ, 73-81; STAFF SCIENTIST RES, ARECIBO OBSERV, PR, 82- *Concurrent Pos:* Secy, Nat Comt Astron, NZ, 73-80; vis fel, Nat Radio Astron Observ, 79-80; vis prof, Cornell Univ, 84, 87. *Mem:* Australian Astron Soc; Am Astron Soc; Int Astron Union. *Res:* Properties of galaxies, the precision of velocity estimates and their use in Tully-Fisher relation; missing mass in galaxies; identification of OH-IR stars; properties of circumstellar envelopes. *Mailing Add:* Arecibo Observ PO Box 995 Arecibo PR 00613

LEWIS, C S, JR, b Muskogee, Okla, July 19, 20. INTERNAL MEDICINE. *Prof Exp:* DIR INT STUDIES INTERNAL MED, MED COL, UNIV OKLA, TULSA. *Mem:* Inst Med-Nat Acad Sci. *Mailing Add:* Dept Internal Med Univ Okla Med Col Tulsa OK 74129

LEWIS, CAMERON DAVID, b Staunton, Va, June 1, 20; m 45; c 2. ORGANIC CHEMISTRY. *Educ:* Univ Buffalo, AB, 42; Univ Ill, AM, 45, PhD(org chem), 47. *Prof Exp:* Asst chemist, Ill Geol Surv, 42-46; asst chem, War Prod Bd Prog, Univ Ill, 46-47; chemist, 47-74, SR RES CHEMIST, E I DU PONT DE NEMOURS & CO, WVA, 74- *Mem:* Am Chem Soc. *Res:* Analytical test methods; polymer intermediates. *Mailing Add:* 1908 Applewood Dr Hagerstown MD 21740-6764

LEWIS, CARMIE PERROTTA, b New Castle, Pa, June 9, 29; m 65. HISTOLOGY. *Educ:* Thiel Col, BS, 51; Univ NH, MS, 53; Univ Wis, PhD(anat, zool), 56. *Prof Exp:* Res asst, Univ Wis, 53-56; Am Asn Univ Women res fel, Cambridge Univ, 56-57; lectr embryol & histol, Fac Med, Queen's Univ, Ont, 57-58; instr anat, Sch Med, Yale Univ, 58-61; asst radiobiologist, Brookhaven Nat Lab, 61-64, res collabr, 64-67; assoc prof, 67-74, PROF BIOL, SUFFOLK COUNTY COMMUNITY COL, 74- *Concurrent Pos:* USPHS res fel, 61-64; asst prof, Queens Col, NY, 64-67; Mellon fel, Ctr Univ NY-Community Col Proj, 88. *Mem:* Am Asn Anat. *Res:* Radiobiology; endocrines of reproduction. *Mailing Add:* Dept Biol Suffolk Community Col Ammerman Ca Selden NY 11784

LEWIS, CHARLES E, b Kansas City, Mo, Dec 28, 28; m 63; c 4. MEDICINE, PREVENTIVE MEDICINE. *Educ:* Harvard Med Sch, MD, 53; Univ Cincinnati, MS, 57, ScD(prev med), 59. *Prof Exp:* House officer med, Univ Kans Hosps, 53-54; resident occup med, Eastman Kodak Co, 58-59, plant physician, Tex Div, 59-60; asst prof epidemiol, Col Med, Baylor Univ, 60-61; assoc prof med, Med Ctr, Univ Kans, 61-62, prof prev med, 62-69; prof social med, Harvard Med Sch, 69-70; PROF PUB HEALTH, UNIV CALIF, LOS ANGELES, 70-, PROF MED, 72, PROF NURSING, 74- *Concurrent Pos:* Fel prev med, Kettering Lab, Univ Cincinnati, 56-58; USPHS trainee, 57-58; dir, Kans Regional Med Prog, 67-69. *Honors & Awards:* Ginsberg Prize, 54; Glasier Award, 88. *Mem:* Am Pub Health Asn; Asn Am Physicians; Am Col Physicians. *Res:* Medical care and education. *Mailing Add:* Nursing B-558 Factor Bldg Univ Calif 405 Hilgard Ave Los Angeles CA 90024

LEWIS, CHARLES J, b Park River, NDak, May 20, 27; m 50; c 2. ANIMAL SCIENCE. *Educ:* Utah State Univ, BS, 52; Iowa State Univ, MS, 54, PhD(animal sci), 56. *Prof Exp:* Dir nutrit, Kent Feeds, Inc, 56-58, vpres & nutritionist, 58-67, mem, Bd Dirs, 60-67; prof animal sci & head dept, SDak State Univ, 67-68; EXEC VPRES, GRAIN PROCESSING CORP & KENT FEEDS, INC, 68- *Mem:* AAAS; Am Inst Biol Sci; Am Soc Animal Sci; Inst Food Technologists; Poultry Sci Asn Am. *Res:* Animal nutrition and research. *Mailing Add:* Seven Colony Dr Muscatine IA 52761

LEWIS, CHARLES JOSEPH, b New York, NY, Sept 18, 17; m 67; c 2. MATHEMATICS. *Educ:* Georgetown Univ, AB, 41, MS, 45; Brown Univ, PhD(math), 57. *Prof Exp:* Instr math, Georgetown Univ, 43-44 & St Peters Col, 45; instr, Fordham Univ, 54-56, asst prof, 56-65, chmn dept, 58-65; from assoc prof to prof math, Monmouth Col, NJ, 65-89; PROF MATH, GEORGIAN COURT COL, NJ, 89- *Concurrent Pos:* NSF fac fel math, Harvard Univ, 59-60; chmn dept math, Monmouth Col, NJ, 68-74. *Mem:* Am Math Soc; Math Asn Am. *Res:* Complex function theory; extremal problems; growth of entire functions; generalized potential theory; special functions and differential equations of mathematical physics. *Mailing Add:* Eight Timothy Lane Tinton Falls NJ 07724

LEWIS, CHARLES WILLIAM, b New York, NY, Oct 29, 20; wid; c 1. PHYSICAL CHEMISTRY. *Educ:* City Col, New York, BS, 41; Polytech Inst New York, PhD(chem), 50. *Prof Exp:* Res chemist, Res Labs, Westinghouse Elec Corp, 49-58; assoc dir basic res, Int Resistance Co, 58-65; staff scientist, PPG Industs, Inc, 65-84; RETIRED. *Mem:* Am Chem Soc. *Res:* Polymer chemistry; solid and liquid dielectrics; kinetics and mechanism of organic reactions; physics of thin films; mechanical behavior of polymers. *Mailing Add:* 2400 McGinley Rd Monroeville PA 15146

LEWIS, CLARK HOUSTON, b McMinnville, Tenn, Nov 6, 29; m 58; c 2. FLUID MECHANICS, GAS DYNAMICS. *Educ:* Univ Tenn, BSME, 51, MS, 59, PhD(viscous flow), 68. *Prof Exp:* Supvr theoret gas dynamics, Aerophys Div, Aro Inc, Arnold Eng Develop Ctr, Tenn, 51-68; prof aerospace eng, Va Polytech Inst & State Univ, 68-84; PRES, VRA, INC, 84- *Mem:* Assoc fel Am Inst Aeronaut & Astronaut; Am Phys Soc; Am Soc Mech Engrs. *Res:* Physical gas dynamics; high-speed viscous flows; chemically reacting flows; thermophysical gas properties; numerical methods in engineering. *Mailing Add:* 716 Dickerson Lane Blacksburg VA 24060

LEWIS, CLAUDE IRENIUS, b Stanley, NC, Apr 21, 35; m 56; c 2. ANALYTICAL CHEMISTRY. *Educ:* Duke Univ, BS, 57; Va Polytech Inst, MS, 59, PhD(chem), 62. *Prof Exp:* Res chemist, Texaco, Inc, 61-62 & E I du Pont de Nemours & Co, 62-65; res chemist, 65-66, sr res chemist, 66-70, supvr anal develop, 70-76, DIR QUAL ASSURANCE, LORILLARD CORP, 70- *Concurrent Pos:* Instr, Guilford Col, 67-70. *Mem:* AAAS; Am Chem Soc; Am Inst Chem Eng; Am Soc Quality Control; Sigma Xi. *Res:* Cigarette tobacco technology; tobacco smoke chemistry; polyester fiber technology; alkyl benzene synthesis; synthesis of polycyclic aromatic compounds. *Mailing Add:* 3001 Shadyhawn Dr Greensboro NC 27408

LEWIS, CLIFFORD JACKSON, b Altoona, Pa, Aug 18, 12. INORGANIC CHEMISTRY, METALLURGICAL ENGINEERING. *Educ:* Franklin & Marshall Col, BS, 33; Univ Pittsburgh, BS, 44; Pa State Univ, MA, 37. *Prof Exp:* Tech dir, Warner Co, 45-51; sr res fel, Mellon Inst, 51-54; res dir, Res Inst, Colo Sch Mines, 55-70; environ consult indust wastes & sulfur oxides control, 70-73; environ consult, 73-80, DIR ENVIRON SERV, NAT LIME ASN, 80-; TECH DIR, STEEL BROS CAN, LTD, 85- *Mem:* Am Chem Soc; Am Inst Chem Engrs; Air Pollution Control Asn. *Res:* Municipal waste treatment; control of sulfur oxides emission by wet scrubbing and metals recovery by hydrometallurgical processes. *Mailing Add:* 2446 Otis Ct Edgewater CO 80214

LEWIS, CORNELIUS CRAWFORD, b Appomattox, Va, May 24, 21; m 49. AGRONOMY, SOIL SCIENCE. *Educ:* Va State Col, BS, 42; Mich State Univ, MS, 45; Univ Mass, PhD(agron), 48. *Prof Exp:* Prof agron, Ft Valley State Col, 47-48; head dept agr, WVa State Col, 48-49; prof agron, Univ Md, 49-50; anal chemist, New York Testing Lab, 50-51; soil specialist, USDA For Serv, Liberia, WAfrica, 51-53; head dept plant industs, Agr & Tech Col, NC, 54-56; head dept agr, Grambling Col, 56-63; PROF AGRON & NUCLEAR SCI, VA STATE COL, 63- *Mem:* AAAS; Am Chem Soc; Am Soc Agron; Soil Sci Soc Am; Sigma Xi. *Res:* Field crops; soil fertility; plant nutrient relationship, particularly fertility levels and nutrient requirements for economic crops. *Mailing Add:* Box 43 Va State Univ Petersburg VA 23803

LEWIS, CYNTHIA LUCILLE, b Los Angeles, Calif, Nov 26, 48; m 72; c 1. INVERTEBRATE BIOLOGY, DEVELOPMENTAL BIOLOGY. *Educ:* Calif Polytech State Univ, BS, 70; Univ Alta, PhD(zool), 75. *Prof Exp:* Lab instr physiol & invert embryol, Ore State Univ, 70-71; lab instr invert zool, marine biol & gen biol, Univ Alta, 71-75; fel develop biol & anomalies, Dent Inst, NIH, 75-77; asst prof zool & biol, San Diego State Univ, 77-78; asst prof biol & chem, Point Loma Col, 78-79; res assoc, Scripps Inst Oceanog, 79-81; ASST PROF, BIOL DEPT, POINT LOMA COL, 82- *Concurrent Pos:* Fel, Smithsonian Inst, Washington, DC, 73-75. *Mem:* Sigma Xi; Am Soc Zoologists; Grad Women Sci. *Res:* Physiological mechanisms, morphological changes and morphogenetic movements involved in invertebrate development; larval development, physiological ecology and settlement of invertebrate larvae. *Mailing Add:* 4267 Calle DeVida San Diego CA 92124

LEWIS, DANIEL MOORE, b Barnesville, Ohio, Oct 1, 45; m 70; c 3. ALLERGIES & HYPERSENSITIVITY DISEASES, PULMONARY IMMUNOLOGY. *Educ:* Ohio State Univ, BS, 67; WVa Univ, MS, 69, PhD(microbiol), 74. *Prof Exp:* Res fel immunol, Mayo Clin & Found, Rochester, Minn, 74-76; res asst prof immunol, Ohio State Univ, Columbus, 76-80; from adj asst prof to adj assoc prof, 82-89, ADJ PROF MICROBIOL, WVA UNIV, MORGANTOWN, 89-; IMMUNOLOGIST, USPHS, MORGANTOWN, 80- *Concurrent Pos:* Sect chief, immunol sect, DRDS, Nat Inst Occup Safety & Health, Morgantown, WVa, 91- *Mem:* Am Asn Immunologists. *Res:* Immunologic aspects of occupational lung diseases with special interest in allergy and occupational asthma; development of assays to evaluate the inflammatory potential of organic dusts found in agricultural work sites and development of assays for the measurement of airborne allergens in the worksite. *Mailing Add:* Immunol Sect Nat Inst Occup Safety & Health 944 Chestnut Ridge Rd Morgantown WV 26505

LEWIS, DANIEL RALPH, b Camden, Ark, Oct 31, 44. MATHEMATICS. *Educ:* La State Univ, Baton Rouge, 66, MS, 68, PhD(math), 70. *Prof Exp:* Asst prof math, Va Polytech Inst & State Univ, 70-72; asst prof math, Univ Fla, 72-77; assoc prof math, Ohio State Univ, 77-; AT DEPT MATH, TEX A&M UNIV. *Mem:* Am Math Soc. *Res:* Functional analysis. *Mailing Add:* Dept Math Tex A&M Univ College Station TX 77843

LEWIS, DANNY HARVE, b Decatur, Ala, Apr 9, 48; m 68; c 2. POLYMER CHEMISTRY. *Educ:* Univ NAla, BS, 69; Univ Ala, PhD(chem), 73. *Prof Exp:* Res chemist textile fibers, E I du Pont de Nemours & Co Inc, 73-75; sr chemist polymer chem, Southern Res Inst, 75-77, head, Biomat Sect, 77-80, head, Biosysts Div, 80-82; VPRES NEW PROD DEVELOP, STOLLE RES & DEVELOP CORP, 82- *Concurrent Pos:* Consult, NIH. *Mem:* Sigma Xi; Controlled Release Soc (pres-elect). *Res:* Controlled-release delivery systems; biomaterials for dental and orthopedic use; synthesis and characterization of new polymers; physical properties of polymers; polymers for fiber spinning, polymers as adhesives and membranes. *Mailing Add:* Stolle Res & Dev Corp PO Box 5310 Decatur AL 35601

LEWIS, DAVID EDWIN, b Tailem Bend, SAustralia, Nov 21, 51; m 78. NATURAL PRODUCT SYNTHESIS. *Educ:* Univ Adelaide, BSc, 72, PhD(org chem), 80. *Prof Exp:* Res assoc chem, Univ Ark, 77-78, lectr, 79-80; vis asst prof, Univ Ill, Urbana-Champaign, 80-81; ASST PROF CHEM, BAYLOR UNIV, 81- *Mem:* Royal Australian Chem Inst; Am Chem Soc. *Res:* Total synthesis of natural products; development of new synthetic strategy; applications of cycloaddition reactions to the total synthesis of terpenoid compounds. *Mailing Add:* Dept Chem SDak State Univ Brookings SD 57006

LEWIS, DAVID HAROLD, b New York, NY, Dec 22, 25; m 47, 63; c 2. CARDIOLOGY. *Educ:* Columbia Univ, AB, 44, MD, 47. *Prof Exp:* Intern med, Bellevue Hosp, New York, 47-48; intern, Kings County Hosp, 48-49, resident, 49-50; from instr to asst prof physiol, Sch Med, Univ Pa, 50-57, assoc cardiol, Grad Sch Med, 55-63; guest investr, First Surg Dept, Univ Goteborg, 63-78; guest investr, 78-80, CHIEF CLIN RES, CLIN RES CTR, UNIV HOSP, LINKOPING UNIV, 80- *Concurrent Pos:* Chief hemodynamics sect, Div Cardiol, Philadelphia Gen Hosp, 55-63; estab investr, Am Heart Asn, 57-62. *Mem:* Am Physiol Soc; Am Fedn Clin Res; Am Heart Asn. *Res:* Cardiovascular physiology. *Mailing Add:* Clin Res Ctr Univ Hosp Linkoping Univ S-581 85 Linkoping Sweden

LEWIS, DAVID KENNETH, b Poughkeepsie, NY, Feb 11, 43; m 64; c 3. PHYSICAL CHEMISTRY. *Educ:* Amherst Col, AB, 64; Cornell Univ, PhD(phys chem), 70. *Prof Exp:* From asst prof to prof chem, 69-88, assoc dean fac, 85-86, chmn chem, 87-88, CHARLES A DANA PROF CHEM, COLGATE UNIV, 88-, DIR, DIV NAT SCI & MATH, 82-85 & 88- *Concurrent Pos:* Vis sr res fel, Univ Colo/Nat Oceanic & Atmospheric Admin, 77-78; environ mgt coun, Madison County, NY, 81-86; mem Cent NY Task Force Hazardous & Toxic Waste, 82-84; vis prof chem, Syracuse Univ, 86, Univ NC, 87. *Mem:* Am Chem Soc; Sigma Xi. *Res:* Chemical kinetics and energy transfer in gases at high temperatures; ultra-high resolution molecular spectroscopy; atmospheric chemistry and physics; innovative teaching methods. *Mailing Add:* Dept Chem Colgate Univ Hamilton NY 13346

LEWIS, DAVID KENT, b Madison, Wis, June 11, 38; m 62; c 3. FOREST MANAGEMENT, FOREST ECONOMICS. *Educ:* Univ Minn, BS, 60; Yale Univ, MF, 66; Univ Oxford, D Phil, 76. *Prof Exp:* Forester, Ore, 63-65, silviculturist, Forestry Res Ctr, 67-76, FOREST ECONOMIST, RES & DEVELOP, WEYERHAEUSER CO, WASH, 76-, NEW TITLE INDUST HYGIENE MGR. *Mem:* Soc Am Foresters; Am Econ Asn. *Res:* Economics of producing timber crops. *Mailing Add:* Dept Forestry Rm 008C Agr Hall Okla State Univ Stillwater OK 74078

LEWIS, DAVID S(LOAN), JR, b North Augusta, SC, July 6, 17; m 41; c 4. AERONAUTICAL ENGINEERING. *Educ:* Ga Tech Univ, BS, 39. *Hon Degrees:* DSc, Clarkson Col Technol, 71; LLD, St Louis Univ. *Prof Exp:* Aerodynamicist, Martin Co, 39-46; chief aerodyn, McDonnell Aircraft Corp, 46-52, chief preliminary design, 52-55, mgr sales, 55-56, mgr all projs, 56-57, vpres, 57-59, sr vpres, 59-61, sr vpres opers, 60-61, exec vpres, 61-62, pres, 62-67, pres, McDonnell Douglas Corp, 67-70; chmn bd & chief exec officer, 70-86, DIR, GEN DYNAMICS CORP, 86- *Concurrent Pos:* Mem subcomt highspeed aerodyn & subcomt stability & control, Nat Adv Comt Aeronaut, 51-57. *Honors & Awards:* Collier Trophy, Pres Ford, 76; Wright Bros Trophy; Sands of Time Award, 77; Distinguished Achievement Award, Wings Club, 83. *Mem:* Nat Acad Eng; fel Am Inst Aeronaut & Astronaut. *Res:* Aerodynamics; high speed flight characteristics; space mechanics. *Mailing Add:* PO Box 300 Leary GA 31762

LEWIS, DAVID THOMAS, b Downing, Mo, Sept 27, 35; m 68; c 1. AGRONOMY, SOIL MORPHOLOGY. *Educ:* Univ Maine, BS, 60, MS, 62; Univ Nebr, PhD(agron), 71. *Prof Exp:* Instr soil sci, Dept Agron, Univ Maine, 60-62; soil scientist, Soil Conserv Serv, USDA, 62-67; instr agron, 67-71, asst prof soil classification, 71-75, assoc prof, 75-80, PROF SOIL CLASSIFICATION, DEPT AGRON, UNIV NEBR, 80- *Mem:* Soil Sci Soc Am; Soil Conserv Soc Am; Sigma Xi. *Res:* Studies relating to the genesis and classification of soils and to the solution of problems that relate to proper correlation of survey mapping units. *Mailing Add:* Dept Agron Rm 228 Keim Hall Univ of Nebr Lincoln NE 68583

LEWIS, DAVID W(ARREN), b Salem, Ohio, June 16, 30; m 53; c 4. MECHANICS. *Educ:* Rice Univ, BA, 52, BS, 53, MS, 55; Northwestern Univ, PhD(mech), 58. *Prof Exp:* Instr mech eng, Northwestern Univ, 55-57; asst prof, US Naval Postgrad Sch, 58-60; staff engr, IBM Corp, 60-63; assoc prof mech eng, 63-71, PROF MECH ENG & BIOMED ENG, UNIV VA, 71- *Mem:* Am Soc Eng Educ; Am Soc Mech Engrs. *Res:* Mechanics, especially elasticity and kinematics. *Mailing Add:* Dept Mech & Aerospace Eng Univ Va Charlottesville VA 22901

LEWIS, DENNIS ALLEN, b Morristown, NJ, Dec 25, 42; m 70; c 2. ORGANIC CHEMISTRY. *Educ:* St Peters Col, BS, 64; Univ Conn, PhD(org chem), 72. *Prof Exp:* Instr & sr instr, Nuclear Weapons Employ Div, Ft Sill, Okla, 70-71; asst prof, 72-76, ASSOC PROF CHEM, ROSE-HULMAN INST TECHNOL, 76- *Concurrent Pos:* Instr, US Army Reserve Sch, Ft Benjamin Harrison, Ind, 73- *Mem:* Am Chem Soc; Sigma Xi. *Res:* Synthesis of small-ring compounds via photochemical reactions involving carbene and nitrene intermediates; investigation of chemiluminescent systems. *Mailing Add:* Dept Chem Rose-Hulman Inst Technol Terre Haute IN 47803

LEWIS, DENNIS OSBORNE, organic chemistry, for more information see previous edition

LEWIS, DONALD EDWARD, b Ironton, Ohio, May 14, 31; m 57; c 1. ELECTRICAL ENGINEERING. *Educ:* Univ Cincinnati, Ohio, EE, 54; Ohio State Univ, Columbus, MSc, 57, PhD(elec engr), 64. *Prof Exp:* Instr elec engr, Ohio State Univ, 57-59; engr, USAF, 59-65; PROF ELEC ENG, UNIV DAYTON, OHIO, 65- *Concurrent Pos:* Consult, Mead Corp, 79-80, Systs Res Labs, Inc, 77-, Raytheon Serv Co & Systs & Appl Sci Corp, 80- *Mem:* Inst Elec & Electronics Engrs; Am Soc Eng Educ. *Res:* Automatic control; convolution and transform theory; signal processing; image processing; special television-video techniques and apparatus. *Mailing Add:* Dept Elec Eng Univ Dayton Dayton OH 45469

LEWIS, DONALD EVERETT, b Paducah, Tex, July 3, 31; m 68; c 2. BIOCHEMISTRY, SCIENCE EDUCATION. *Educ:* Abilene Christian Col, BS, 52; Fla State Univ, MS, 54, PhD(biochem), 57. *Prof Exp:* From assoc prof to prof chem, Queen's Col, NC, 57-66; assoc prof, 66-68, PROF CHEM, ABILENE CHRISTIAN COL, 68- *Concurrent Pos:* Vis assoc prof, Abilene Christian Col, 63-64; vis prof chem, Univ Tex, Austin, 80-81. *Mem:* Am Chem Soc. *Res:* Synthesis and biological assay of amino acid analogues. *Mailing Add:* 2541 Campus Courts Abilene TX 79601

LEWIS, DONALD HOWARD, b Stamford, Tex, May 31, 36; m 60; c 3. FISH PATHOLOGY, MICROBIOLOGY. *Educ:* Univ Tex, Austin, BA, 59; Southwest Tex State Univ, MA, 64; Tex A&M Univ, PhD(vet microbiol), 67. *Prof Exp:* Res assoc, 66-68, asst prof, 69-75, assoc prof, 75-79, PROF MICROBIOL, TEX A&M UNIV, 79-, ACTG HEAD VET MICROBIOL & PARASITOL, 81- *Concurrent Pos:* Consult, TerEco Corp, 75. *Mem:* AAAS; Am Soc Microbiol; Am Fisheries Soc; Soc Invert Path; World Maricult Soc. *Res:* Microbial diseases and immune mechanisms of aquatic animals; role of microflora upon host welfare; antibiotic resistance; molecular biology. *Mailing Add:* Dept Vet Microbiol Col Vet Med Tex A&M Univ College Station TX 77843

LEWIS, DONALD JOHN, b Adrian, Minn, Jan 25, 26; m 53. MATHEMATICS. *Educ:* Col St Thomas, BS, 46; Univ Mich, MS, 49, PhD(math), 50. *Prof Exp:* Instr math, Ohio State Univ, 50-52; NSF fel, Inst Adv Study, 52-53; from asst prof to assoc prof, Univ Notre Dame, 53-61; assoc prof, 61-63, PROF MATH, UNIV MICH, ANN ARBOR, 63-, CHMN, 84- *Concurrent Pos:* NSF sr fel, Manchester & Cambridge Univs, 59-61; sr vis fel, Cambridge Univ, 65-69; vis fel, Brasenose Col, Oxford. 69; guest prof, Heidelberg Univ, 79-80 & 83. *Honors & Awards:* Humboldt Stiftung Sr Award, 80 & 82. *Mem:* Am Math Soc; Math Asn Am. *Res:* Diophantine equations; finite fields; algebraic number theory. *Mailing Add:* Dept Math Univ Mich Main Campus Ann Arbor MI 48109-1003

LEWIS, DONALD RICHARD, b New Leipzig, NDak, May 18, 20; m 43; c 2. ARCHAEOMETRY, ARCHAEOLOGICAL CHEMISTRY. *Educ:* Univ Wis-Madison, BS, 42, MS, 47, PhD(chem), 48. *Prof Exp:* Ballistics supvr, Hercules Powder Co, 42-46; asst, Univ Wis, 46-48; res chemist, Shell Develop Co, div Shell Oil Co, 48-58, staff res chemist, 58-70, group leader, 70-75, proj leader, 75-80; FAC ASSOC, CTR ARCHAEOL RES, UNIV TEX, SAN ANTONIO, 80-, LECTR, ISOTOPE GEOCHEM, 87-, FAC ASSOC, CTR GROUNDWATER RES, 87- *Concurrent Pos:* Exchange scientist, Shell Develop Co & Royal Dutch Shell, Amsterdam, Neth, 56-57; assoc ed, comt clay minerals, Nat Acad Sci-Nat Res Coun, 57-59; prin scientist, USAF-Advan Res Proj Agency, 65-66; consult & dir, Nuclear Monitoring Syst & Mgt Corp, 80-; lectr archaeometry, chem & geochem, div earth & phys sci & div behav & cult sci, Univ Tex, San Antonio, 80-; res fel, Tex Archaeol Res Lab, Univ Tex, Austin, 88- *Mem:* Fel AAAS; Am Chem Soc; Am Phys Soc; fel Mineral Soc Am; Geochem Soc; Instrument Soc Am; Geol Soc Am. *Res:* Thermoluminescence dating; anthrosol land use; trace element and stable isotope studies for environmental characterization and paleodiet studies; trace element geochemistry; isotope geochemistry and geochronology; groundwater geochemistry. *Mailing Add:* 9219 Lasater San Antonio TX 78250-2418

LEWIS, DOUGLAS SCOTT, b Dayton, Ohio, Aug 10, 51; m 80; c 4. ANIMAL PHYSIOLOGY. *Educ:* Univ Ga, BS, 73; Mich State Univ, PhD(biochem), 78. *Prof Exp:* Postdoctoral physiol, Univ Tex Health Sci Ctr, San Antonio, 78-80; asst res scientist, 80-86, ASSOC RES SCIENTIST, SOUTHWEST FOUND BIOMED RES, 87- *Concurrent Pos:* Asst prof, Dept Physiol, Univ Tex Health Sci Ctr, San Antonio, 84- *Mem:* Am Physiol Soc; Am Heart Asn. *Res:* Mechanisms by which nutritional factors regulate lipid metabolism during growth and development; investigate endocrine regulation of fat cell and hepatic metabolism in preweaning infants using a non-human primate. *Mailing Add:* Dept Physiol & Med Southwest Found Biomed Res San Antonio TX 78228-0147

LEWIS, EDWARD B, b Wilkes-Barre, Pa, May 20, 18; m 46; c 2. GENETICS. *Educ:* Univ Minn, BA, 39; Calif Inst Technol, PhD(genetics), 42. *Hon Degrees:* Dr, Univ Umea, Sweden, 81. *Prof Exp:* From instr genetics to assoc prof genetics, 46-56, prof biol, 56-66, THOMAS HUNT MORGAN PROF BIOL, CALIF INST TECHNOL, 66- *Concurrent Pos:* Rockefeller Found fel, Sch Bot, Cambridge Univ, 48-49; mem, Nat Adv Comt Radiation, 58-61; guest prof, Univ Copenhagen, 75-76. *Honors & Awards:* Morgan Medal, Genetics Soc Am; Gairdner Found Int Award, 87; Wolf Found Prize in Med, 89; Rosenstiel Award, 90; Nat Medal Sci, 90. *Mem:* Nat Acad Sci; AAAS; Am Acad Arts & Sci; Genetics Soc Am; Am Phil Soc; foreign mem Royal Soc London; hon mem Genetical Soc Gt Brit. *Res:* Developmental genetics; somatic effects of ionizing radiation. *Mailing Add:* Biol Div 156-29 Calif Inst Technol Pasadena CA 91125

LEWIS, EDWARD LYN, b Aberystwyth, UK, Oct 9, 30; m 59; c 3. OCEANOGRAPHY. *Educ:* Univ London, BSc, 51, MSc, 58, PhD(physics), 62. *Prof Exp:* Physicist, Mullard Res Labs, 52-56; Harwell res fel, Univ London, 56-59; res assoc microwave electronics, Univ BC, 59-62; RES SCIENTIST, OCEAN SCI & SURV, DEPT FISHERIES & OCEANOG, CAN, 62. *Mem:* Am Geophys Union; Glaciol Soc. *Res:* Arctic oceanography; ice physics; energy exchange ocean-atmosphere; arctic instrument development. *Mailing Add:* 3904 Bedford Rd Victoria BC V8N 4K5 Can

LEWIS, EDWARD R(OBERT), structural & experimental mechanics; deceased, see previous edition for last biography

LEWIS, EDWARD SHELDON, b Berkeley, Calif, May 7, 20; m 55; c 2. CHEMISTRY, CHEMICAL DYNAMICS. *Educ:* Univ Calif, BS, 40; Harvard Univ, MA, 47, PhD(chem), 47. *Prof Exp:* Nat Res Coun fel, Univ Calif, Los Angeles, 47-48; from asst prof to prof, 48-90, chmn dept, 80-85, EMER PROF CHEM, RICE UNIV, 90- *Concurrent Pos:* Vis prof, Univ Southampton, 57; chmn dept chem, Rice Univ, 65-67, 81-86; Guggenheim fel, 67; vis prof, Phys Chem Lab, Oxford Univ, 67-68, Univ Col, Dublin, 78. *Honors & Awards:* Southwest Regional Award, Am Chem Soc. *Mem:* Am Chem Soc; Royal Soc Chem; AAAS. *Res:* Mechanism of reactions of organic compounds, es0ecially diazonium salts, hydrogen isotope effects, methyl transfers and organo-phosphorus chemistry. *Mailing Add:* Dept Chem Rice Univ PO Box 1892 Houston TX 77251-1892

LEWIS, EDWIN REYNOLDS, b Los Angeles, Calif, July 14, 34; m 60; c 2. BIOENGINEERING. *Educ:* Stanford Univ, AB, 56, MS, 57, PhD(elec eng), 62. *Prof Exp:* Mem res staff neural modeling, Lab Automata Res, Gen Precision, 61-67; PROF BIOENG, UNIV CALIF, BERKELEY, 67- *Mem:* Fel Inst Elec & Electronics Engrs; Acoust Soc Am; Soc Neurosci; AAAS; Asn Res Otolaryngol; Sigma Xi. *Res:* Applications of engineering analytical tools to problems in neurobiology; network models of dynamical biological systems; morphology and physiology of vestibular and auditory systems. *Mailing Add:* 1047 Overlook Rd Berkeley CA 94708

LEWIS, FORBES DOWNER, b New Haven, Conn, Apr 15, 42. COMPUTER SCIENCE. *Educ:* Cornell Univ, BS, 67, MS, 69, PhD(comput sci), 70. *Prof Exp:* Asst prof, Harvard Univ, 70-75; assoc prof, State Univ NY, Albany, 75-78; assoc prof comput sci & chmn dept, 78-82, PROF COMPUT SCI, UNIV KY, 83- *Mem:* Asn Comput Mach; Asn Symbolic Logic; Inst Elec & Electronics Engrs. *Res:* CAD algorithms for VLSI; computational complexity. *Mailing Add:* Dept Comput Sci 917 Patterson Off Tower Lexington KY 40506

LEWIS, FRANCIS HOTCHKISS, JR, b Milwaukee, Wis, Aug, 14, 37; div; c 4. PHYSICS, MECHANICAL ENGINEERING. *Educ:* Stevens Inst Technol, ME, 59; Stanford Univ, PhD(physics), 64; Univ San Francisco, JD, 74. *Prof Exp:* Res asst prof physics, Univ Wash, 64-66; physicist, Lawerence Livermore Lab, Univ Calif, 66-81. *Concurrent Pos:* Patent atty, Lewis & Lewis, 76-89, Hewlett Packard, 89-90. *Mem:* Am Phys Soc; Sigma Xi. *Res:* Theoretical nuclear and particle physics; solid state physics. *Mailing Add:* 401 Grand Ave No 100 Oakland CA 94610-5054

LEWIS, FRANK HARLAN, b Redlands, Calif, Jan 8, 19; m 45, 68, 84; c 2. BOTANY. *Educ:* Univ Calif, Los Angeles, BA, 41, MA, 42, PhD(bot), 46. *Prof Exp:* Asst instr bot, Univ Calif, Los Angeles, 42-44, instr, 46-47; Nat Res Coun fel, John Innes Hort Inst, London, 47-48; from asst prof to prof, 48-82, EMER PROF BOT, UNIV CALIF, LOS ANGELES, 82- *Concurrent Pos:* Teaching fel, Calif Inst Technol, 43-44; Guggenheim fel, 54-55; consult, NSF, 58-69; chmn, Dept Bot, Univ Calif, Los Angeles, 59-62, dean, Div Life Sci, 62-82; vpres, Int Orgn Biosyst, 64-69, pres, 69-75; ed, Evolution, 72-74. *Honors & Awards:* Award of Merit, Bot Soc Am, 72. *Mem:* Fel AAAS; Am Inst Biol Sci; Am Soc Naturalists (pres, 71); Am Soc Plant Taxonomists (pres, 69); Soc Study Evolution (secy, 53-58, vpres, 59, pres, 61); Bot Soc Am; Genetics Soc Am. *Res:* Mechanisms of evolution; systematics of flowering plants. *Mailing Add:* Dept Biol Univ Calif Los Angeles CA 90024

LEWIS, FRANK LEROY, b Wurzburg, Ger, May 11, 49; m 86; c 2. ELECTRICAL ENGINEERING. *Educ:* Rice Univ, BA & MEE, 71; Univ WFla, MS, 77; Ga Inst Technol, PhD(elec eng), 81. *Prof Exp:* From asst prof to assoc prof, 81-90, PROF, GA INST TECHNOL, ATLANTA, 90- *Concurrent Pos:* Assoc ed, J Circuits, Systs, Signal Processing, 86. *Honors & Awards:* Frederick E Terman Award, Am Soc Eng Educ, 89. *Mem:* Sr mem Inst Elec & Electronics Engrs Control Systs Soc; Soc Indust & Appl Math; AAAS; Sigma Xi. *Res:* Optimal control systems; nonlinear control; robotics; implicit systems; author of various publications. *Mailing Add:* Automation & Robotics Res Inst Univ Tex Arlington 7300 Jack Newell Blvd S Ft Worth TX 76118

LEWIS, FREDERICK D, b Boston, Mass, Aug 12, 43; m 68; c 2. PHOTOCHEMISTRY. *Educ:* Amherst Col, BA, 65; Rochester Univ, PhD(chem), 68. *Prof Exp:* USPHS res fel chem, Columbia Univ, 68-69; from asst prof to assoc prof, 69-79, PROF, NORTHWESTERN UNIV, 79-, ASSOC DEAN, 89- *Concurrent Pos:* Fel, Dreyfus Found, 73-78 & Sloan Found, 75-77; consult, 3M Corp; Assoc ed, J Phys Org Chem, 87- *Mem:* Am Chem Soc; InterAm Photochem Soc; Europ Photochemical Asn. *Res:* Organic photochemistry; radical ions; free radicals; cycloaddition reactions; exciplexes. *Mailing Add:* Dept Chem Northwestern Univ Evanston IL 60208-3113

LEWIS, GEORGE CAMPBELL, JR, b Williamsburg, Ky, Mar 25, 19; m 45; c 6. OBSTETRICS & GYNECOLOGY, ONCOLOGY. *Educ:* Haverford Col, BS, 42; Univ Pa, MD, 44; Am Bd Obstet & Gynec, dipl, 53; Gyn Oncol, 80. *Prof Exp:* Intern med, Hosp Univ Pa, 44-45, resident obstet & gynec, 47-50, instr, Sch Med, Univ Pa, 50-53, instr radium ther, 51-63, res asst, 53-56, asst prof obstet & gynec, 56-63; prof obstet & gynec & chmn dept, Hahnemann Med Col & Hosp, 62-73, dir div gynec oncol, 71-73; PROF GYNEC ONCOL & DIR DIV, JEFFERSON MED COL, 73- *Concurrent Pos:* Am Cancer Soc fel gynec oncol, Hosp Univ Pa, 50-52; consult lectr, US Naval Hosp Philadelphia, 56-77; consult, Lankenau Hosp & Philadelphia Gen Hosp, 62-, Am Oncol Hosp, 63- & Magee Mem Hosp Rehab Ctr, 68-; mem div gynec oncol, Am Bd Obstet & Gynec; chmn, Gynec Oncol Grp, 75-89. *Mem:* Am Cancer Soc; Soc Gynec Oncol (pres, 69); Am Gynec Soc; Am Asn Obstet & Gynec; Am Col Obstet & Gynec. *Res:* Etiology, early diagnosis and evaluation of modes of therapy of gynecologic oncology. *Mailing Add:* Jefferson Med Col 1025 Walnut St Philadelphia PA 19107-5041

LEWIS, GEORGE EDWARD, b Lorain, Ohio, Oct 27, 08; m 37, 61; c 3. CONTINENTAL STRATIGRAPHIC PALEONTOLOGY, STRATEGIC & TERRAIN INTELLIGENCE. *Educ:* Yale Univ, PhB, 30, PhD, 37. *Prof Exp:* Instr geol, Yale Univ, 38-43, asst prof, 43-45; geologist, US Geol Surv, 44-79; RETIRED. *Concurrent Pos:* Cur, Peabody Mus, Yale Univ, 39-45; vis prof, Univ Guayaquil, Ecuador, 42-43; comdr, USNR, 49-79. *Mem:* Fel Geol Soc Am; Soc Vert Paleont. *Res:* Continental stratigraphy; Permo-Triassic vertebrates; Cenozoic mammals, primates; strategic and terrain intelligence. *Mailing Add:* 155 Brentwood St Lakewood CO 80226

LEWIS, GEORGE EDWIN, b Decatur, Ga, Jan 6, 33; m 56. ORGANIC CHEMISTRY. *Educ:* Emory Univ, AB, 52, MS, 53; Fla State Univ, PhD(chem), 58. *Prof Exp:* Res asst, Ga Inst Technol, 58-59; asst prof chem, La State Univ, 59-66; assoc prof, 66-80, PROF CHEM, JACKSONVILLE UNIV, 80- *Mem:* Am Chem Soc; The Chem Soc. *Res:* Mechanisms of organic reactions. *Mailing Add:* Div Sci & Math Jacksonville Univ Jacksonville FL 32211-3393

LEWIS, GEORGE MCCORMICK, b Los Angeles, Calif, Sept 14, 40; m 64; c 3. THEORETICAL PHYSICS. *Educ:* Stanford Univ, BA, 61; Univ Southern Calif, MA, 64, PhD(math), 70. *Prof Exp:* Asst prof, 67-72, assoc prof, 72-79, PROF MATH, CALIF POLYTECH STATE UNIV, SAN LUIS OBISPO, 79- *Mem:* Am Math Soc; Math Asn Am. *Res:* Synthetic differential geometry. *Mailing Add:* Dept Math Calif State Polytech Univ San Luis Obispo CA 93407

LEWIS, GEORGE R(OBERT), b Kansas City, Mo, June 26, 24; m 54; c 3. CHEMICAL ENGINEERING. *Educ:* Ohio State Univ, BChE, 48, MSc, 49, PhD(chem eng), 51. *Prof Exp:* Res engr cellophane process, E I du Pont de Nemours & Co, 51-53, chem develop supvr, Film Dept, 53-55; res engr, Paperboard Res Labs, Mead Corp, Dayton, 55-60, staff consult, 60-64, prod mgr, 64-69, tech dir, Paperboard Prod Div, 69-88; CONSULT, 88- *Mem:* Tech Asn Pulp & Paper Indust; Am Inst Chem Engrs. *Res:* Industrial wastes; cellophane process; pulp and paper; recycled paperboard; paperboard products. *Mailing Add:* 9450 Sugar Bend Trail Dayton OH 45458-3863

LEWIS, GLENN C, b Oakley, Idaho, July 13, 20; m 56; c 6. SOIL CHEMISTRY. *Educ:* Univ Idaho, BS, 46, MS, 49; Purdue Univ, PhD(soils), 62. *Prof Exp:* Anal agr chem, Univ Idaho, 47-52, from asst prof agr chem to assoc prof soils, 52-67, prof soils, 67-85; RETIRED. *Mem:* Soil Sci Soc Am; Int Soc Soil Sci; AAAS. *Res:* Chemical and mineralogical studies on slick spot soils; water quality, including effects of irrigation water quality on soil characteristics; phosphorus reactions in calcareous soils; mineralogical studies on loess. *Mailing Add:* Dept Plant & Soil Sci Univ Idaho Moscow ID 83844

LEWIS, GORDON, b Cincinnati, Ohio, Apr 7, 33; m 58; c 3. CERAMIC ENGINEERING. *Educ:* Alfred Univ, BS, 56, PhD(ceramics), 63. *Prof Exp:* Ceramic engr, Carborundum Co, NY, 57-58; res assoc chem, Univ Kans, 62-64; assoc prof ceramic eng, 64-73, PROF CERAMIC ENG, UNIV MO, ROLLA, 73- *Mem:* Am Ceramic Soc; Nat Inst Ceramic Engrs; Am Soc Eng Educ. *Res:* High temperature chemistry; phase equilibria and vaporization behavior in oxide systems; thermogravimetric behavior and phase identification of high alumina refractory cements. *Mailing Add:* Dept Ceramic Eng Clemson Univ Clemson SC 29634-0907

LEWIS, GORDON DEPEW, b Charlottesville, Va, July 22, 29; m 54. FOREST ECONOMICS, POLICY. *Educ:* Va Polytech Inst, BS, 51; Duke Univ, MFor, 57; Mich State Univ, PhD(forest econ), 61. *Prof Exp:* Asst prof forest econ, Univ Mont, 59-62; proj leader, Southeastern Forest Exp Sta, US Forest Serv, 62-66, economist, Washington, DC, 66-67, br chief, 68-71, proj leader, Rocky Mountain Forest Exp Sta, 71-77, prog mgr, western environ forestry res, Rocky Mountain Forest Exp Sta, 77-81, ASST DIR, SOUTHEASTERN FOREST EXP STA, US FOREST SERV, 81- *Mem:* Soc Am Foresters; Am Econ Asn. *Res:* Economic evaluations of alternative methods of exploiting natural resources for regional development consistent with the maintenance of the quality of rural and wildlife environments. *Mailing Add:* Southeastern Forest Exp Sta Forest Serv USDA PO Box 2680 Asheville NC 28802

LEWIS, GWYNNE DAVID, b Hackensack, NJ, June 12, 28; m 60; c 1. PLANT PATHOLOGY. *Educ:* Rutgers Univ, BS, 51; Purdue Univ, MS, 53; Cornell Univ, PhD, 58. *Prof Exp:* Asst plant path, Purdue Univ, 51-53 & Cornell Univ, 53-58; from asst prof to assoc prof, 58-70, PROF PLANT PATH, RUTGERS UNIV, NEW BRUNSWICK, 70- *Honors & Awards:* Bronze Medal, Am Rhododendron Soc, 75. *Mem:* Am Phytopath Soc. *Res:* Diseases of vegetable crops; plant nematology; control of plant and vegetable diseases. *Mailing Add:* Dept Plant Path Rutgers Univ New Brunswick NJ 08903

LEWIS, H(ERBERT) CLAY, b Newton, Mass, Aug 7, 13; m 49. CHEMICAL ENGINEERING. *Educ:* Bowdoin Col, AB, 34; Mass Inst Technol, MS, 37; Carnegie Inst Technol, ScD(chem eng), 42. *Prof Exp:* Chem engr, Humble Oil & Ref Co, Tex, 37-40; res assoc, Nat Defense Res Comt, 41-42; from instr to asst prof chem eng, Univ Ill, 42-45; res assoc, Mass Inst Technol, 45-46; from asst prof to assoc prof chem eng, Ga Inst Technol, 46-53, prof, 53-80; RETIRED. *Concurrent Pos:* Vis prof, Imp Col, Univ London, 60-61; fel, Ctr Advan Eng Study, Mass Inst Technol, 71-72. *Mem:* Am Chem Soc; Am Inst Chem Engrs. *Res:* Chemical technology. *Mailing Add:* 212 Winnona Dr Decatur GA 30030

LEWIS, H(AROLD) RALPH, b Chicago, Ill, June 7, 31; m 61; c 2. PHYSICS. *Educ:* Univ Chicago, AB, 51, SB, 53; Univ Ill, MS, 55, PhD(physics), 58. *Prof Exp:* Res assoc physics, Univ Heidelberg, 58-60; instr, Princeton Univ, 60-63; mem staff, 63-75, assoc group leader, 75-81, dep group leader, 81-83, LAB FEL, LOS ALAMOS NAT LAB, 83- *Concurrent Pos:* Ger Acad Exchange Serv fel, Univ Heidelberg, 58-59. *Mem:* Am Phys Soc. *Res:* Plasma physics; nulcear spectroscopy; superconductivity. *Mailing Add:* Los Alamos Nat Lab Los Alamos NM 87545

LEWIS, HAROLD WALTER, b Keene, NH, May 7, 17; m 46; c 2. NUCLEAR PHYSICS. *Educ:* Middlebury Col, BS, 38; Univ Buffalo, AM, 40; Duke Univ, PhD(physics), 50. *Prof Exp:* Vis instr & res assoc, 46-49, from asst prof to prof, 49-86, vprovost, 63-80, dean fac, 69-80, chmn dept, 81-86, univ distinguished serv prof, 80-86, EMER PROF PHYSICS, DUKE UNIV, 86- *Concurrent Pos:* Dean arts & sci, Duke Univ, 63-69. *Mem:* Fel Am Phys Soc; Am Asn Physics Teachers. *Mailing Add:* 1708 Woodburn Rd Durham NC 27705-5723

LEWIS, HAROLD WARREN, b New York, NY, Oct 1, 23; m 47; c 2. PHYSICS. *Educ:* NY Univ, AB, 43; Univ Calif, AM, 44, PhD(physics), 48. *Prof Exp:* Asst prof physics, Univ Calif, 48-53; mem tech staff, Bell Tel Labs, NJ, 51-56; from assoc prof to prof physics, Univ Wis, 56-64; PROF PHYSICS, UNIV CALIF, SANTA BARBARA, 64- *Concurrent Pos:* Mem staff, Inst Advan Study, 47-48 & 50-51; dir, Quantum Inst, Univ Calif, Santa Barbara, 69-73. *Mem:* Am Phys Soc; Sigma Xi. *Res:* Theoretical physics. *Mailing Add:* Dept Physics Univ Calif Santa Barbara Santa Barbara CA 93106

LEWIS, HARVYE FLEMING, b Hodge, La, Dec 24, 17. NUTRITION. *Educ:* La Polytech Inst, BS, 38; Univ Tenn, MS, 42; Iowa State Col, PhD(nutrit), 50. *Prof Exp:* Teacher high sch, La, 38-40; nutritionist, State Dept Pub Health, Tenn, 42; res assoc, Agr Exp Sta, La State Univ, 43-47, asst nutritionist, 50-52; assoc prof food & nutrit, Fla State Univ, 52-65; prof food & nutrit, sch home econ, La State Univ, 65-82; RETIRED. *Mem:* AAAS; Am Dietetic Asn; Am Home Econ Asn; Inst Food Technologists; Am Inst Nutrit; Sigma Xi. *Res:* Vitamin content of foods; nutritional requirements; food patterns and nutritional health of children. *Mailing Add:* 11275 Mollylea Dr Baton Rouge LA 70815-5247

LEWIS, HENRY RAFALSKY, b Yonkers, NY, Nov 19, 25; m 57; c 3. PHYSICS. *Educ:* Harvard Univ, AB, 48, MA, 49, PhD(physics), 56. *Prof Exp:* Mem staff opers res, Opers Eval Group, Mass Inst Technol, 51-53, 56; group head quantum electronics, David Sarnoff Res Ctr, RCA Corp, 57-66, dir, electronic res lab, 66-70; vpres res & develop, Itek Corp, 70-73; pres, Optel Corp, 73-74; group vpres & dir, Dennison Mfg Corp, 74-82, sr vpres & dir, 82-85, vchmn & dir, 86-91; RETIRED. *Concurrent Pos:* Dir, Delphany Syst, Randolph, Mass, 80, AOI Inc, Lowell, Mass, 87-, Genzyme Corp, Cambridge, Mass, 87- *Mem:* Am Phys Soc; Inst Elec & Electronics Engrs; NY Acad Sci; Sigma Xi. *Res:* Paramagnetic resonance; quantum electronics; operations research; molecular beams. *Mailing Add:* 35 Clover St Belmont MA 02178

LEWIS, HERMAN WILLIAM, b Chicago, Ill, July 10, 23; c 2. GENETICS, ZOOLOGY. *Educ:* Univ Ill, BS, 47, MS, 49; Univ Calif, PhD(genetics), 53. *Prof Exp:* USPHS res fel, Univ Calif, 52-54; asst prof biol, Mass Inst Technol, 54-61; prof life sci & chmn dept, Mich State Univ, 61-62; prog dir genetic biol, NSF, 62-66, head, Cellular Biol Sect, 66-77, sr scientist, 77-84, dep dir, Div Molecular Biosci, 84-86, dep exec dir, US-Israel Binat Sci Found, 86-; RETIRED. *Honors & Awards:* Mendel Medal. *Mem:* AAAS; Biophys Soc; Genetics Soc Am; Am Soc Cell Biol. *Res:* Biochemical, physiological and molecular genetics; biophysics and cytology of genetic material; human cell biology. *Mailing Add:* One Gristmill Ct Apt 204 Baltimore MD 21208

LEWIS, HOMER DICK, b Covington, Ky, Oct 4, 26; m 48; c 4. METALLURGY, NUCLEAR ENGINEERING. *Educ:* Univ Cincinnati, MetE, 52; Univ NMex, MS, 64, MSc, 71. *Prof Exp:* Staff mem uranium casting, Los Alamos Sci Lab, Univ Calif, 52-57; res engr, Boeing Airplane Co, 57-58; staff mem, Los Alamos Sci Lab, Univ Calif, 58-86; RETIRED. *Concurrent Pos:* Co-prin investr, Liquid Metal Fast Breeder Reactor Fuels Properties, 75-78, mem, Nat Task Group, Los Alamos Lab Rep, 77-81; sect leader, Solidification Tech, 81-86. *Mem:* Am Soc Metals; Soc Advan Mat & Process Eng. *Res:* Packing behavior of particulate solids; small particle statistics; physics of particulate systems; powder metallurgy; carbon and graphite research and development; electrical and thermal transport properties of plutonium and plutomium alloys and compounds; solidification process; thermochemistry of uranium, plutonium compounds; solidification processes. *Mailing Add:* PO Box 644 320 East Dr Bayfield CO 81122

LEWIS, IRA WAYNE, b Hillsboro, Tex, Sept 22, 50. CONTINUUM THEORY, GEOMETRIC TOPOLOGY. *Educ:* Univ Houston, BSc, 72; Tex A&M Univ, MSc, 74; Univ Tex, Austin, PhD(math), 77. *Prof Exp:* Vis lectr, 77-79, from asst prof to assoc prof, 79-89, PROF MATH, TEX TECH UNIV, 89- *Concurrent Pos:* Vis asst prof, Tulane Univ, 79-80, Univ Tex, Austin, 80, Univ Ala, Birmingham, 81, Univ Ky, 82 & vis assoc prof, Auburn Univ, 84-85; Nat Acad Sci exchange scientist, Inst Math, Polish Acad Sci, 81 & Stephan Banach Ctr, 84-85; invited lectr, Inst Math, Hanoi, S R Vietnam, 86. *Mem:* Am Math Soc; Am Inst Aeronaut & Astronaut; Astron Soc Pac; AAAS; Sigma Xi; Am Math Asn. *Res:* Continuum theory and geometric topology including indecomposable continua, classification of homogeneous continua, embeddings of tree-like continua, continuous decompositions and homogeneous embeddings of continua in manifolds; manifolds. *Mailing Add:* Dept Math Tex Tech Univ Lubbock TX 79409-1042

LEWIS, IRVING JAMES, b Boston, Mass, July 9, 18; m 41; c 3. HEALTH & PUBLIC POLICY. *Educ:* Harvard Univ, AB, 39; Univ Chicago, AM, 40. *Prof Exp:* Res fel, Brookings Inst, 41; with US Govt, 42 & 46-55, dep chief, Int Div, Bur Budget, 55-57, dept head, Inter-govt Comn Europ Migration, Geneva, Switz, 57-59, dep chief, Int Div, Bur Budget, 59-65, chief, Health & Welfare Div, 65-67, dep asst dir, Bur Budget, 67-68, dep adminr health serv & ment health admin, Dept HEW, 68-70; prof community health, 70-90, EMER PROF CMMUNITY HEALTH, ALBERT EINSTEIN COL MED, 90- *Concurrent Pos:* WHO fel. *Honors & Awards:* Except Serv Award, Bur Budget, 64; Career Serv Award, Nat Civil Serv League, 69. *Mem:* Inst Med Nat Acad Sci; Am Soc Pub Adminr; Am Pub Health Asn; Asn Am Med Cols. *Res:* Public health & epidemiology. *Mailing Add:* 3310 N Leisure World Blvd Apt 623 Silver Spring MD 20906

LEWIS, IRWIN C, b New York, NY. PYROLYSIS CHEMISTRY, ELECTRON SPIN RESONANCE. *Educ:* City Col New York, BS, 53; Univ Kans, PhD(org chem), 57. *Prof Exp:* Assoc phys-org, Pa State Univ, 58-60; CORP FEL, UNION CARBIDE CORP, 60- *Mem:* Am Chem Soc; Sigma Xi; Am Carbon Soc. *Res:* Studies of pyrolysis of aromatic hydrocarbons, characterization and reactions of carbonaceos materials, coal and petroleum chemistry, electron spin resonance of aromatic radicals, liquid crystal and polymerization in pitch, carbon fibers. *Mailing Add:* Union Carbide Corp PO Box 6116 Cleveland OH 44101-1116

LEWIS, J(OHN) E(UGENE), b St John, NB, Apr 11, 41; m 64; c 3. ELECTRICAL ENGINEERING. *Educ:* Univ NB, Fredericton, BScE, 64; Univ BC, PhD(elec eng), 68. *Prof Exp:* Nat Res Coun Can fel, Univ Southampton, 68-69; from asst prof to assoc prof, 74-80, chmn dept, 80-89, PROF ELEC ENG, UNIV NB, FREDERICTON, 80- *Concurrent Pos:* Dir, Cadmi Microelectronics, Inc, 86- *Mem:* Inst Elec & Electronics Engrs; Brit Inst Elec Engrs; Int Microwave Power Inst. *Res:* Industrial applications of microwaves to materials processing and process control; microwave measurement of nonelectrical quantities; low-loss waveguides. *Mailing Add:* Elec Eng Dept Univ NB PO Box 4400 Fredericton NB E3B 5A3 Can

LEWIS, J(ACK) R(OCKLEY), b Eureka, Kans, July 30, 20; m 52; c 1. METALLURGY. *Educ:* Stanford Univ, BS, 47, PhD(metall), 51. *Prof Exp:* Res assoc metall, Stanford Univ, 47-50; metallurgist, Fairchild Engine & Airplane Corp, 50-51; engr, Gen Elec Co, 51-57, supvr fuel element develop, 57-60, consult engr, 60-61; group leader mat develop, Atomics Int Div, NAm Aviation, Inc, 61-67; mgr advan mat, Rocketdyne Div, Rockwell Int, 67-85; CONSULT ADVAN MAT TECHNOL, 85- *Mem:* Am Inst Mining, Metall & Petrol Engrs; fel Am Soc Metals; Am Inst Chem Engrs; Am Ceramics Soc. *Res:* Nuclear reactor and radioisotope materials; liquid metals; high-temperature materials; intermetallic compounds and cermets; rocket engine materials and processes; composite materials; structrual ceramics. *Mailing Add:* 11300 Yarmouth Ave Granada Hills CA 91344

LEWIS, JACK A, b Brooklyn, NY, Apr 8, 39; m 68; c 2. PLANT PATHOLOGY, SOIL MICROBIOLOGY. *Educ:* Brooklyn Col, BS, 60; Rutgers Univ, PhD(microbiol), 65. *Prof Exp:* MICROBIOLOGIST, USDA, 65- *Mem:* Am Soc Microbiol; Am Phytopath Soc. *Res:* Biological control of soil-borne plant pathogenic fungi; microbial decomposition of natural materials in soil. *Mailing Add:* USDA ARS BPDL Rm 273 Barc-W 0300 Baltimore Ave Beltsville MD 20705-2350

LEWIS, JAMES BRYAN, molecular biology, for more information see previous edition

LEWIS, JAMES CHESTER, b Kalamazoo, Mich, Jan 31, 36; m 57; c 3. WILDLIFE ECOLOGY. *Educ:* Univ Mich, BS, 57; Mich State Univ, MS, 63; Okla State Univ, PhD(wildlife ecol), 74. *Prof Exp:* Biologist aide, Mich Game Div, 57-59; dist biologist, Tenn Game Div, 59-60, res proj leader game mgt, 60-64, res supvr, 64-67; asst unit leader, Okla Coop Wildlife Res Unit, 67-77, from asst prof to assoc prof life sci, Sch Biophys Sci, Okla State Univ, 67-77; TECH ED, US FISH & WILDLIFE SERV, COLO STATE UNIV, 77- *Concurrent Pos:* Consult, Nat Audubon Soc, 80. *Mem:* Wildlife Soc. *Res:* Endangered species research; deer and turkey management; ecology of wildlife rabies; mourning dove and sandhill crane behavior and ecology. *Mailing Add:* US Fish & Wildlife Serv Endangered Species PO Box 1306 Albuquerque NM 87103

LEWIS, JAMES CLEMENT, b Lewisville, Minn, Aug 10, 15; m 39; c 3. BIOCHEMISTRY. *Educ:* Univ Minn, BCh, 36; Ore State Col, MS, 39, PhD(soils, agr chem), 40. *Prof Exp:* Analyst, Univ Minn, 36-37; asst chemist animal nutrit, Exp Sta, Ore State Col, 40-41; biochemist, Western Regional Res Lab, Bur Agr & Indust Chem, USDA, 41-53, biochemist, Western Regional Res Lab, Agr Res Serv, 53-75; RETIRED. *Mem:* AAAS; Am Chem Soc; Am Soc Microbiol; Am Soc Biol Chem. *Res:* Microbial biochemistry; trace elements in plant and microbiology; bacterial spores. *Mailing Add:* One Harvard Circle Berkeley CA 94708-2206

LEWIS, JAMES EDWARD, intellectual property management, technology transfer, for more information see previous edition

LEWIS, JAMES KELLEY, b Waco, Tex, Oct 24, 24; m 49; c 4. RANGE SCIENCE. *Educ:* Colo State Univ, BS, 48; Mont State Univ, MS, 51. *Prof Exp:* Asst prof, 50-58, ASSOC PROF ANIMAL SCI, SDAK STATE UNIV, 58- *Mem:* Soc Range Mgt; Am Soc Animal Sci; Ecol Soc Am; Wildlife Soc; Brit Grassland Soc. *Res:* Structure, function, measurement, manipulation, uses and systems analysis of range ecosystems; range animal nutrition and management; coupling of range and agronomic ecosystems. *Mailing Add:* RR 5 Box 341A Carthage MO 64836

LEWIS, JAMES LABAN, III, b Nashville, Tenn, Sept 17, 42; m 69; c 3. ELECTRICAL ENGINEERING. *Educ:* Vanderbilt Univ, BE, 63; Princeton Univ, MSE, 65; Purdue Univ, PhD(elec eng), 69. *Prof Exp:* Instr elec eng, Purdue Univ, 66-67; mem tech staff, TRW Systs Group, 69-75, head signal design sect, TRW Defense & Space Systs Group, 75-77, SR STAFF ENGR, TRW DEFENSE & SPACE SYSTS GROUP, 78- *Concurrent Pos:* Japan Soc Prom Sci fel, Kyoto Univ, 71-72; lectr, Loyola Marymount Univ, 77- *Mem:* Inst Elec & Electronics Engrs; Inst Math Statist. *Res:* Systems engineering; communication theory; stochastic processes; digital systems. *Mailing Add:* TRW Electronics & Def Space Comm Div One Space Park Redondo Beach CA 90278

LEWIS, JAMES PETTIS, b Omaha, Nebr, Apr 21, 33; m 85; c 4. LIQUEFIED NATURAL GAS & CRYOGENIC TECHNOLOGY, FACILITY SAFETY FOR HAZARDOUS MATERIALS. *Educ:* Calif Inst Technol, BSME, 55. *Prof Exp:* Dist engr, Richfield Oil Corp, 55-62; asst chief engr, Cosmodyne Corp, 62-69; tech dir, Distrigas Corp, 69-72; proj mgr, Transco Energy Co, 72-78; PRES, PROJ TECH LIAISON ASSOCS, 78- *Concurrent Pos:* Chair, Hydrocarbon Processing Comt, Am Soc Mech Engrs. *Mem:* Am Soc Mech Engrs; Am Inst Chem Engrs. *Res:* Identification of life limiting mechanisms and life extension measures, including monitoring programs for hazardous facilities; non destructive composite material testing; failure mechanisms in brittle insulating materials. *Mailing Add:* 7803 Aleta Dr Spring TX 77379

LEWIS, JAMES VERNON, b Neligh, Nebr, May 2, 15; div; c 3. ENVIRONMENTAL SCIENCES. *Educ:* Univ Calif, AB, 37, MA, 39, PhD(math), 42; Univ NMex, MCRP, 84. *Prof Exp:* Asst, Univ Calif, 39-42; jr physicist, USN, Calif, 42-43; mathematician, Radiation Lab, Univ Calif, 43-45; mathematician, Aberdeen Proving Ground, 45-53; assoc prof, 53-80, EMER PROF MATH, UNIV NMEX, 80- *Concurrent Pos:* Asst prof, Univ Nev, 46-47. *Mem:* Am Planning Asn; Math Asn Am; Sigma Xi; Fedn Am Scientists. *Res:* Urban planning; iterative methods for decision making in urban planning; calculuc of variations. *Mailing Add:* 3401 Mars Rd NE Albuquerque NM 87107

LEWIS, JAMES W L, b Natchez, Miss, May 3, 38; m 61; c 3. MOLECULAR PHYSICS, FLUID PHYSICS. *Educ:* Univ Miss, BS, 60, MS, 64, PhD(physics), 66. *Prof Exp:* Physicist, US Naval Weapons Lab, 61-62 & ARO, Inc, 66-68; UK Sci Res Coun fel physics, Queen's Univ, Belfast, 68-69; assoc prof physics, Space Inst, Univ Tenn, 66-77; physicist, Aro, Inc, 69-80; PROF PHYSICS, SPACE INST, UNIV TENN, 77-; PHYSICIST, CALSPAN, INC, 81- *Mem:* Am Inst Aeronaut & Astronaut; Am Phys Soc. *Res:* Vibrational relaxation processes in gases; molecular processes in hypersonic flow phenomena; molecular and atomic beam collision processes using high temperature shock tube source; raman-rayleigh scattering in gases; condensation processes in gases; nonlinear optics. *Mailing Add:* Dept Physics Univ Tenn Space Inst Tullahoma TN 37388

LEWIS, JANE SANFORD, b Pasadena, Calif, Dec 26, 18; m 42; c 4. NUTRITION. *Educ:* Pomona Col, BA, 40; Cornell Univ, MS, 42; Univ Calif, Los Angeles, MPH, 66, DrPH(nutrit), 69. *Prof Exp:* Home economist, Wilson & Co, Ill, 42-43; anal chemist, Nat Defense Res Coun, Northwestern Univ, 43-45; anal chemist, Nat Defense Res Coun Proj, Calif Inst Technol, 45-46; technician, Nutrit Lab, Sch Pub Health, Univ Calif, Los Angeles, 65; nutritionist, Head Start Prog, Fedn Settlements & Recreation Ctr, Calif, 66; technician, Nutrit Lab, Sch Pub Health, Univ Calif, Los Angeles, 67; assoc prof, 68-73, PROF HOME ECON, CALIF STATE COL, LOS ANGELES, 73- *Concurrent Pos:* Chmn, Task Force, Calif Nutrit Coun, 70- *Mem:* Am Pub Health Asn; Am Dietetic Asn; Soc Nutrit Educ; Am Inst Nutrit; Am Home Econ Asn. *Res:* Nutritional status of children of varying backgrounds; effect of oral contraceptives and anticonvulsant drugs on nutritional status; food habits of various ethnic groups, anthropometric measurements of Oriental children. *Mailing Add:* Dept Family Studies & Consumer Sci Calif State Univ 5151 State University Dr Los Angeles CA 90032

LEWIS, JASPER PHELPS, b Danville, Va, Nov 8, 17; m 50. CHEMISTRY, BIOCHEMISTRY. *Educ:* Univ Va, BSChem, 46; Univ Louisville, MS, 58; Med Col Ga, PhD(biochem), 66. *Prof Exp:* Res chemist biochem, Sch Med, Univ Va, 46-50; clin chemist, Vet Admin Hosp, Bay Pines, Fla, 50-52, clin chemist, Louisville, Ky, 52-60 & res chemist, St Louis, Mo, 60-62; BASIC SCIENTIST ERYTHROPOIESIS RES, VET ADMIN HOSP, AUGUSTA, 62- *Concurrent Pos:* Res prof, Med Col Ga, 67- *Mem:* AAAS; Am Soc Hemat; Am Soc Biol Chemists; Am Chem Soc; Am Asn Clin Chemists; Int Soc Hematology. *Res:* Erythropoiesis regulatory factors. *Mailing Add:* Vet Admin Hosp Downtown Div & Med Col Ga Forest Hills Div Augusta GA 30910

LEWIS, JERRY PARKER, b Terre Haute, Ind, Sept 20, 31; m 56; c 4. MEDICINE. *Educ:* James Millikin Univ, 52; Univ Ill, BS, 53, MD, 56. *Prof Exp:* Asst prof med, Univ Ill, 64-67; assoc prof, 67-69, chief div hemat & oncol, 67-80, lectr clin path & vet med, 68-74, chief staff, Med Ctr, 79-80, actg chmn, Dept Int Med, 80-82, PROF MED, UNIV CALIF, DAVIS, 69-, PROF PATH, 78-, CHIEF DIV HEMAT & ONCOL, 82- *Concurrent Pos:* NIH fel hemat, Presby-St Luke's Hosp, Chicago, 61-63, res fel, 63-65; spec chief clin hemat & chief spec hemat, Presby-St Luke's Hosp, Chicago, 65-67; consult, David Grant Hosp, Travis, AFB, Calif, 68- *Mem:* Am Soc Hemat; Am Soc Clin Oncol; fel Am Col Physicians; Soc Exp Hemat; Am Soc Human Genetics. *Res:* Leukemia; cytogenetics; toxicity of laetrile; molecular biology of oncogenes. *Mailing Add:* Sch Med Univ Calif Davis CA 95616

LEWIS, JESSE C, b Vaughan, Miss, June 26, 29; m 59; c 1. MATHEMATICS, COMPUTER SCIENCES. *Educ:* Univ Ill, MS, 55, MA, 59; Syracuse Univ, PhD(math), 66. *Prof Exp:* Instr math, Southern Univ, 55-57 & Prairie View Agr & Mech Col, 57-58; asst prof, Jackson State Col, 59-61; res asst, Comput Ctr, Syracuse Univ, 63-66; dir, Comput Ctr & chmn, Dept Comput Sci, Jackson State Univ, 66-80, prof math & chmn Div Natural Sci, 67-80, assoc dean comput serv & prof comput sci, 80- 84; VPRES ACAD AFFAIRS, NORFOLK STATE UNIV, 84- *Concurrent Pos:* Consult, Comt Undergrad Prog Math, Jackson State Col; mem eval panel sci comput, Nat Bur Standards, 80-83. *Mem:* Math Asn Am; Am Math Soc; Asn Comput Mach; Asn Educ Data Systs. *Res:* Computer study of permanents of n-square (0,1)-matrices with k l's in each row and column. *Mailing Add:* Norfolk State Univ Acad Affairs Norfolk VA 23504

LEWIS, JESSICA HELEN, b Harpswell, Maine, Oct 26, 17; m 46; c 5. MEDICINE. *Educ:* Goucher Col, AB, 38; Johns Hopkins Univ, MD, 42. *Prof Exp:* Intern, Hosp Women, Baltimore, Md, 42-43; asst resident, Univ Calif Hosp, 43-44; res fel, Thorndike Mem Lab & Harvard Univ, 44-46; res assoc physiol, Univ NC, 48-55, res assoc med, 55-58, res assoc prof, 58-70, res prof med, 70-77, PROF MED, UNIV PITTSBURGH, 77-, VPRES, CENT BLOOD BANK. *Concurrent Pos:* USPHS res fel, Univ NC, 44-46; asst med, Boston City Hosp, 44-46; res assoc, Med Sch, Emory Univ, 46-47; assoc med, Med Sch, Duke Univ, 51-55; staff mem, Presby-Univ Hosp, 55-; dir res, Cent Blood Bank Pittsburgh, 69-75, vpres, 75- *Mem:* Am Soc Hemat; World Fedn Hemophilia; Am Physiol Soc; Am Soc Clin Invest; Am Fedn Clin Res. *Res:* Blood coagulation; enzyme and protein chemistry; comparative hematology. *Mailing Add:* Dept Med Univ Pittsburgh Cent Blood Bank 812 Fifth Ave Pittsburgh PA 15219

LEWIS, JOHN BRADLEY, b Ottawa, Ont, Jan 12, 25; m 80; c 3. MARINE BIOLOGY. *Educ:* McGill Univ, BSc, 40, MSc, 50, PhD(zool), 54. *Prof Exp:* Asst marine biol, Inst Marine Sci, Univ Miami, 51-54; dir, Bellairs Res Inst, 54-71, assoc prof, 61-69, dir, Pedpath Mus, 71-83, PROF MARINE SCI, MCGILL UNIV, 69-, DIR, INST OCEANOG, 83- *Mem:* Can Soc Zool; Int Soc Reef Studies. *Res:* Tropical marine ecology and physiology; tropical marine organisms and coral reef ecology. *Mailing Add:* Dept Biol McGill Univ Stewart Biol Bldg 1205 Docteur Penfield Montreal PQ H3A 1B1 Can

LEWIS, JOHN E, b Riverside, Calif, Mar 30, 39; m 66; c 2. AUTOMATIC TEST. *Educ:* Calif State Univ, BS, 64; Univ Southern Calif, MS, 75. *Prof Exp:* Qual mgr, Interstate Electronics, 65-67; sr engr, USN, 67-90; CONSULT, 90- *Mem:* Sr mem Inst Elec & Electronics Engrs; fel Inst Automotive Engrs. *Res:* Automatic test and test software. *Mailing Add:* Nine Alcoba Irvine CA 92714

LEWIS, JOHN G(ALEN), b Panora, Iowa, Oct 26, 20; m 52; c 3. CHEMICAL ENGINEERING. *Educ:* Iowa State Univ, BS, 43; Univ Mich, MS, 51, PhD(chem eng), 54. *Prof Exp:* Staff engr, Pennwalt Corp, 43-46; chem engr, BASF Wyandotte Corp, 47-50; assoc res engr, Univ Mich, 54-58; asst dir, Res Div, Lear Siegler, Inc/Fabricated Prod Group, 58-62; consult engr, 62-74, KMS Fusion, 74-82; RETIRED. *Mem:* Am Chem Soc; Am Inst Chem Engrs; Nat Soc Prof Engrs; Am Nuclear Soc. *Res:* Promotion of chemical reactions with gamma radiation; nuclear fuel reprocessing; design, construction of vacuum and atmospheric furnaces and their application to refractories and graphite; heat exchange and thermal stress analysis; design and construction of organic and inorganic chemical processes and related facilities. *Mailing Add:* 1340 Burgundy Rd Ann Arbor MI 48105

LEWIS, JOHN HUBBARD, b Jamestown, NY, Apr 13, 29; m 56; c 4. GEOLOGY. *Educ:* Allegheny Col, BS, 56; Univ Colo, PhD(geol), 65. *Prof Exp:* From instr to assoc prof geol, Colo Col, 58-74, chmn dept, 70-78, prof, 74-80; CONSULT, 80- *Concurrent Pos:* Lectr, Exten Div, Univ Colo, 58-66; dir, NSF Sec Sci Training Prog, Colo Col, 65-67; US Antarctic res partic, Tex Tech Col, 67-68. *Res:* Sedimentary petrology; petrology and diagenesis of upper Cambrian rocks of Colorado; structural geology. *Mailing Add:* 918 N Royer Colorado Springs CO 80903

LEWIS, JOHN L, JR, b San Antonio, Tex, June 5, 29; m 55; c 3. OBSTETRICS & GYNECOLOGY. *Educ:* Harvard Univ, BA, 52, MD, 57; Am Bd Obstet & Gynec, dipl, 67, cert, 79. *Prof Exp:* Clin assoc endocrinol br, Nat Cancer Inst, 59-61, sr investr surg br, 65-67; assoc prof, 68-71, PROF OBSTET & GYNEC, MED COL, CORNELL UNIV, 71-; ATTEND SURGEON, GYNEC SERV, MEM HOSP CANCER & ALLIED DIS & JAMES EWING HOSP, 90- *Concurrent Pos:* Sr investr clin ctr, NIH, 65-67; assoc attend gynecologist, Francis Delafield Hosp, 67; assoc attend obstetrician & gynecologist, Presby Hosp, NY, 67; assoc attend obstetrician & gynecologist, New York Lying-in Hosp, 68-71, attend obstetrician & gynecologist, 71-; attend surgeon, Mem Hosp Cancer & Allied Dis, 68-, chief gynec serv, 68-90; assoc, Sloan-Kettering Inst Cancer Res, 68-73, mem, 73-; assoc prof, Col Physicians & Surgeons, Columbia Univ, 67, lectr, 68-; dir, Am Bd Obstet & Gynec, 70-76 & Div Gynec Oncol, 70-76; consult Am joint comt cancer staging & end result reporting. *Mem:* Soc Gynec Invest; AMA; Soc Surg Oncol; Am Radium Soc; Am Asn Cancer Educ. *Res:* Gynecologic cancer; hormonal, immunologic and therapeutic aspects of gestational trophoblastic neoplasms. *Mailing Add:* Mem Hosp Cancer & Allied Dis 1275 York Ave New York NY 10021

LEWIS, JOHN MORGAN, b Joliet, Ill, June 5, 20; m 44; c 3. ANIMAL HUSBANDRY, ANIMAL SCIENCE & NUTRITION. *Educ:* Univ Ill, BS, 43. *Prof Exp:* Asst supt, 43-59, actg supt, 59-62, assoc prof animal sci, Dixon Springs Exp Sta, Univ Ill, Urbana, 62-; RETIRED. *Mem:* Am Soc Animal Sci. *Res:* Sheep breeding, feeding and management. *Mailing Add:* RR 2 Box 305 Metropolis IL 62960

LEWIS, JOHN RAYMOND, b Philadelphia, Pa, July 25, 18; m 42; c 1. POLYMER CHEMISTRY. *Educ:* Franklin & Marshall Col, BS, 42. *Prof Exp:* Chemist, Naval Stores Div, Res Ctr, Hercules Inc, 42-44, shift supvr explosives dept, Sunflower Ord Works, Kans, 44-45, chemist, Naval Stores Div, 45-49, res chemist, 49-55, res supvr, 55-59, res mgr, Plastics & Elastomers Div, 59-64, res assoc, Cent Res Div, 64, mgr develop, Res Dept, 64-69, venture projs, New Enterprise Dept, 69-75, mgr planning & acquisitions, New Enterprise Dept, 75-77, mgr corp acquisitions, 77-83; RETIRED. *Concurrent Pos:* Consult, 83- *Mem:* Am Chem Soc; Financial Analysts Asn; Sigma Xi; Com Develop Asn. *Res:* Commercial development; polymers, energy and raw materials; acquisitions. *Mailing Add:* 118 Dickinson Lane Wilmington DE 19807

LEWIS, JOHN REED, b Ottawa, Kans, Dec 27, 15; m 38; c 3. PHARMACOLOGY. *Educ:* Ottawa Univ, AB, 37; Mich State Col, MS, 40; Univ Mich, PhD, 49. *Prof Exp:* Asst chem exp sta, Mich State Col, 39-41; supvr biol control, Frederick Stearns & Co, 41-45; sr biologist, Sterling-Winthrop Res Inst, 47-53, assoc dir sect coord & integration, 53-60; asst secy coun drugs, AMA, 60-64, assoc dir, 64-72, sr scientist, dept drugs, 72-81, consult, div drugs, 81-82; RETIRED. *Mem:* Am Soc Pharmacol & Exp Therapeut; Soc Toxicol; Am Soc Clin Pharmacol Therapeut; NY Acad Sci; Drug Info Asn. *Res:* Vitamin assays; pharmacology of sympathomimetics and analgesics; diuretics; anticholinesterases; coordination of research projects in development of new drugs; medical writing. *Mailing Add:* 6337 Ravenwood Dr Sarasota FL 34243

LEWIS, JOHN SIMPSON, b Trenton, NJ, June 27, 41; m 64; c 6. GEOCHEMISTRY, METEORITICS. *Educ:* Princeton Univ, AB, 62; Dartmouth Col, MA, 64; Univ Calif, San Diego, PhD, 68. *Prof Exp:* From asst prof to assoc prof chem, earth & planetary sci, Mass Inst Technol, 68-80, prof planetary sci, 80-82; PROF PLANETARY SCI, UNIV ARIZ, 82- *Concurrent Pos:* Mem, Working Group Outer Planet Probe Sci, NASA-Ames Res Ctr, 74-, NASA Phys Sci Comt, 75-78 & Space Sci Bd spec panels outer solar syst & explor Venus, Nat Acad Sci-Nat Res Coun; chmn, Uranus Sci Adv Comt, NASA-Jet Propulsion Lab, 74-75, mem, Sci Adv Group Outer Solar Syst; consult, Aerospace Div, Martin-Marietta Corp, 72 & Avco Systs Div, Avco Corp; Guggenheim lectr, Nat Air & Space Mus, Smithsonian Inst, 73; sci lectr, Div Planetary Sci, Am Astron Soc, 74, Space Sci Bd, Nat Acad Sci, 80-82. *Honors & Awards:* J B Macelwayne Award, Am Geophys Union, 76. *Mem:* AAAS; Am Chem Soc; Int Astron Union; Am Astron Soc. *Res:* Composition, structure and origin of planetary atmospheres; atmosphere-lithosphere interactions; application of thermodynamics to problems of composition and origin of meteorites; exploitation of extraterrestrial resources. *Mailing Add:* LPL Univ Ariz Tucson AZ 85721

LEWIS, JONATHAN JOSEPH, b Johannesburg, SAfrica, May 23, 58; m 90. GROWTH FACTOR SIGNAL TRANSDUCTION. *Educ:* Witwatersrand Univ, MB Bch, 82, PhD(cell biol), 90; FRCS(E), 87. *Prof Exp:* Resident surg, Witwatersrand Univ Sch Med, 83-87; postdoctoral assoc surg & cell biol, 87-90, CHIEF RESIDENT SURG, YALE UNIV SCH MED, 90- *Mem:* Am Soc Cell Biol; Am Asn Cancer Res; AAAS; AMA; NY Acad Sci. *Res:* Growth factor signal transduction; parietal cell signal transduction. *Mailing Add:* Dept Surg Yale Univ Sch Med New Haven CT 06510

LEWIS, KATHERINE, CARCINOGENESIS, INTERMEDIARY METABOLISM. *Educ:* Temple Univ, PhD(biochem), 51. *Prof Exp:* ASST DEAN EDUC, UNIV MED & DENT NJ, 59- *Mailing Add:* Dept Biochem Univ Med & Dent NJ 100 Bergen St Newark NJ 07103

LEWIS, KENNETH D, b Newark, NJ, Aug 11, 49; m 78; c 2. ENGINEERING, MATHEMATICS. *Educ:* Rutgers Univ, AB, 71; Lehigh Univ, MS, 72; Stanford Univ, MSE, 74; Univ Ill, Urbana-Champaign, AM & PhD(nuclear eng), 82; Trinity Theological Seminary, ThD, 86. *Prof Exp:* PRIN ENGR, MARTIN-MARIETTA AVLIS DIV, LAWRENCE LIVERMORE NAT LAB. *Concurrent Pos:* Vis asst prof math, Univ Pac, Stockton, Calif; youth pastor, Stockton, Calif. *Mem:* Sigma Xi; Health Physics Soc. *Res:* Nuclear safety and nuclear criticality safety; photolysis using ceramic p-n junctions; numerical analysis. *Mailing Add:* L-468 Lawrence Livermore Nat Lab Livermore CA 94550

LEWIS, L GAUNCE, JR, b Boston, Mass, Sept 14, 49; m 84. MATHEMATICS, ALGEBRAIC TOPOLOGY. *Educ:* Harvard Col, AB, 71; Univ Chicago, MS, 76, PhD(math), 78. *Prof Exp:* Asst prof, Univ Mich, Ann Arbor, 78-81; asst prof, 81-84, ASSOC PROF MATH, SYRACUSE UNIV, NY, 85- *Concurrent Pos:* Alexander von Humboldt fel, 89-90. *Mem:* Am Math Soc; Math Asn Am; Sigma Xi. *Res:* Equivariant homotopy theory; generalized cohomology theories; stable category; group actions on rings; representation theory. *Mailing Add:* 9A4 Sunrise Terr Oswego NY 13126-1838

LEWIS, LAWRENCE GUY, b Logan, Utah, July 28, 41; m 64; c 4. MATHEMATICS, INFORMATION SCIENCE. *Educ:* Univ Utah, BA, 65; Ind Univ, PhD(math), 69. *Prof Exp:* Fel math, Grad Ctr, City Univ New York, 69-70; asst prof math, Univ Utah, 70-73; mgr licensees, 73-78, dir mgt info serv, Ireco Chem, 78-82; PRES, SOFTWARE FIRST INC, 83-; VPRES, INFO TECHNOL PARTNERS, 91- *Concurrent Pos:* Adj assoc prof math, Univ Utah, 86- *Mem:* AAAS; Data Processing Mgt Asn; Planning Exec Group. *Res:* Ideal boundaries and information systems. *Mailing Add:* 3200 Terrace View Rd Salt Lake City UT 84109

LEWIS, LENA ARMSTRONG, b Lancaster, Pa, July 12, 10. PHYSIOLOGY. *Educ:* Lindenwood Col, AB, 31; Ohio State Univ, MA, 38, PhD(physiol), 40. *Hon Degrees:* LLD, Lindenwood Col, 52. *Prof Exp:* Asst biochem, Sch Med, Johns Hopkins Univ, 31-32; bacteriologist & technologist, Gen Hosp, Lancaster, Pa, 32-36; asst physiol, Ohio State Univ, 36-41; mem res staff & supvr electrophoresis lab, Cleveland Clin Found, 43-45, mem staff & supvr, Electrophoresis Lipoprotein Lab, 45-75, CONSULT & EMER STAFF MEM, LAB MED & RES, CLEVELAND CLIN FOUND, 75- *Concurrent Pos:* Spec fel endocrinol, Cleveland Clin Found, 41-43; adj prof, Cleveland State Univ, 71-74, clin prof chem, 74-87; fel arteriosclerosis coun, Am Heart Asn. *Honors & Awards:* Award Outstanding Contrib Clin Chem in Field Lipids & Lipoproteins, Am Asn Clin Chem, 74. *Mem:* AAAS; Am Physiol Soc; Soc Exp Biol & Med. *Res:* Relation of adrenal to electrolyte metabolism; changes in plasma proteins in endocrine disease hypertension; factors regulating lipid and protein metabolism, especially their relation to atherosclerosis; electrophoresis in physiology; electrophoresis of lipoproteins. *Mailing Add:* Cleveland Clin Found 9500 Euclid Ave Cleveland OH 44195

LEWIS, LEROY CRAWFORD, b Pocatello, Idaho, Mar 18, 40; m 62; c 2. PROCESS ANALYTICAL CHEMISTRY. *Educ:* Col Idaho, BS, 62; Ore State Univ, PhD(phys chem), 68. *Prof Exp:* Sr res chemist, Idaho Nuclear Corp, 68-71; sr res chemist, Allied Chem Corp, 71-72, group supvr, 72-74, sect leader, 74-76, br mgr, 76-79; br mgr, Exxon Nuclear Idaho Co, 79-84; BR MGR, WESTINGHOUSE IDAHO NUCLEAR CO, 84- *Mem:* Am Chem Soc; Sigma Xi. *Res:* Nuclear fuel reprocessing chemistry; chemical waste handling chemistry; actinide chemistry; electrochemistry; analytical chemistry. *Mailing Add:* Westinghouse Idaho Nuclear Co PO Box 4000 Idaho Falls ID 83403-5210

LEWIS, LESLIE ARTHUR, b Castries, St Lucia, WI, May 17, 40; m 68; c 2. GENETICS, MICROBIOLOGY. *Educ:* Univ Toronto, BSA, 63, MSA, 64; Columbia Univ, PhD(genetics), 68. *Prof Exp:* NIH fel, Mich State Univ, 68-69; from lectr to assoc prof, 69-85, PROF BIOL, YORK COL, NY, 85- *Concurrent Pos:* Univ Paris, France, 72-73. *Mem:* Genetics Soc Am; AAAS; Am Soc Microbiol. *Res:* Non-reciprocal recombination in the fungus Sordaria; genetic basis of resistance to aminoglycoside antibiotics. *Mailing Add:* York Col 9420 Guy R Brewer Blvd Jamaica NY 11433-1126

LEWIS, LOWELL N, b Kingston, Pa, July 9, 31; m 53; c 3. PLANT PHYSIOLOGY. *Educ:* Pa State Univ, BS, 53; Mich State Univ, MS, 58, PhD(hort, biochem), 60. *Prof Exp:* Asst horticulturist, Univ Calif, Riverside, 60-65, assoc prof hort, 66-70, PROF PLANT PHYSIOL, UNIV CALIF,

OAKLAND, 70-, ASSOC DEAN RES, COL NATURAL & AGR SCI, 71- & ASSOC VPRES, DIV AGR & NATURAL RESOURCES, 81- Concurrent Pos: Guggenheim res fel, Mich State Univ-AEC Plant Res Lab, 67-68. Mem: Am Soc Hort Sci; Am Soc Plant Physiol; Japanese Soc Plant Physiol; Sigma Xi. Res: Hormonal regulation of plant cell development, especially senescence and abscission. Mailing Add: Univ Calif 300 Lakeside Dr Oakland CA 94612-3560

LEWIS, LYNN LORAINE, b Terra Alta, WVa, Mar 2, 29; m 54; c 4. ANALYTICAL CHEMISTRY. Educ: WVa Wesleyan Col, BS, 50; Marshall Univ, MS, 52; Univ Tenn, PhD(chem), 55. Prof Exp: Engr semiconductors, Westinghouse Elec Corp, 55-56; sr res chemist, Res Lab, US Steel Corp, Pa, 56-66; asst head, 66-75, HEAD ANALYTICAL CHEM DEPT, GEN MOTORS RES LAB, 76-. Mem: Am Chem Soc; Am Soc Testing & Mat; Soc Appl Spectros. Res: Behavior and determination of gases in metals; analysis of metals; instrumentation for chemical analysis. Mailing Add: Res Lab Gen Motors Corp Warren MI 48090

LEWIS, MARC SIMON, b Cleveland, Ohio, Oct 30, 26; wid; c 3. BIOCHEMISTRY, BIOPHYSICS. Educ: Western Reserve Univ, BS, 46, MS, 47; Georgetown Univ, PhD(biochem), 55. Prof Exp: Guest scientist, Nat Inst Arthritis & Metab Dis, 52-55; USPHS fel, 55-57; biochemist, Nat Inst Arthritis & Metab Dis, 57-58, biochemist, Nat Inst Dent Res, 58-62, head sect ophthal chem, Nat Inst Neurol Dis & Blindness, 62-70, sr res investr, lab vision res, Nat Eye Inst, 70-78; BIOPHYSICIST, BIOMED ENG & INSTRUMENTATION, NAT CTR RES RESOURCES, NIH, 78- Concurrent Pos: Lectr, Found Advan Educ Sci Grad Prog, NIH, dept biochem, George Washington Univ, Sch Med. Mem: Am Chem Soc; NY Acad Sci; Biophys Soc; Am Soc Biochem & Molecular Biol; Undersea & Hyperbaric Med Soc; AAAS. Res: Applications of analytical ultracentrifugation to physical biochemistry and biophysics, particularly to the thermodynamics of the interactions of molecules of biological interest; mathematical modeling for the study of such systems. Mailing Add: Biomed Eng & Instrumentation Prog NIH Bethesda MD 20892

LEWIS, MARGARET NAST, b Baltimore, Md, Aug 20, 11. PHYSICS. Educ: Goucher Col, AB, 31; Johns Hopkins Univ, PhD(physics), 37. Prof Exp: Asst physics, Vassar Col, 37-38; Am Asn Univ Women Berliner fel, Univ Calif, 38-39, fel, Crocker Radiation Lab, 39-40; Howell fel, 40-42; instr, Vassar Col, 42-43; instr physics, Univ Pa, 43-48, assoc physics res, 53-54; lectr, Boston Univ, 48-50; physicist, Nat Bur Standards, 50-52; asst prof res, Brown Univ, 54-58; assoc prof, Univ Mass, 58-61; res fel, 61-70, ASSOC, HARVARD COL OBSERV, 70- Concurrent Pos: Radioisotopes res, Mass Mem Hosp, 48-49 & Haverford Col, 52-54; nat consult, Schlesinger Libr, Radcliffe Col. Mem: Am Phys Soc; Sigma Xi. Res: Spectroscopy; atomic structure. Mailing Add: 30 Fernald Dr Cambridge MA 02138

LEWIS, MARIAN L MOORE, b Decatur, Ga, Mar 5, 37. EXPERIMENTAL BIOLOGY. Educ: Ga State Col Women, BA, 59; Univ Ariz, MS, 68; Univ Houston, PhD(biophys sci), 79. Prof Exp: Technician & res asst, Commun Dis Ctr, 59-64; res scientist, Dept Virol & Epidemiol, Col Med, Baylor Univ, Tex Med Ctr, 64-66; res asst, Dept Virol, M D Anderson Hosp & Tumor Inst, 68-71; res analyst, Dept Virol, Northrop Serv, Inc, 71-73; CHIEF, BIOREACTOR LAB, KENNETH E JOHNSON RES CTR, UNIV ALA, 89- Mem: Assoc fel Am Inst Aeronaut & Astronaut; Am Soc Microbiol; NY Acad Sci; AAAS; Sigma Xi; Int Soc Thrombosis & Haemostasis; Am Heart Asn; Am Soc Cell Biol; Am Soc Space & Gravitational Biol; Inst Advan Studies Life Support. Res: Experimental biology; numerous publications. Mailing Add: BioReactor Lab Univ Ala Johnson Res Ctr Sci Bldg Rm 360 Huntsville AL 35899

LEWIS, MARION JEAN, b Windsor, Ont, Sept 21, 25. IMMUNO-HAEMATOLOGY, MAPPING THE HUMAN GENOME. Educ: Univ Man, BA, 60. Hon Degrees: DSc, Univ Winnipeg, 86. Prof Exp: From asst prof to assoc prof, 73-84, SCIENTIST IMMUNOHEMAT, RH LAB, DEPT PEDIAT, UNIV MAN, 44-, PROF IMMUNOHEMAT, DEPT PEDIAT, 84- Concurrent Pos: Mem, Med Adv Comt, Children's Hosp Winnipeg Res Found, 82-; chmn, Int Soc Blood Transfusion, Working Party Terminology for Red Cell Surface Antigens, 83-90; prof immunogenetics, Dept Human Genetics, Univ Man, 86- Honors & Awards: Karl Landsteiner Mem Award, Am Asn Blood Banks, 71; La Médaille de la Ville de Paris, Paris, France, 87. Mem: Human Genome Orgn; hon fel Can Col Med Geneticists; Int Soc Blood Transfusion; Can Soc Immunol. Res: Description, distribution, expression, inheritance of red cell antigens; genetic linkage and mapping of blood group genes; red cell immunization in haemolytic disease of the newborn; international reference laboratory; 160 publications. Mailing Add: Rh Lab 735 Notre Dame Ave Winnipeg MB R3E 0L8

LEWIS, MARK HENRY, b Boston, Mass, Feb 5, 50; m 77; c 1. PSYCHOPHARMACOLOGY. Educ: Bowdoin Col, BA, 72; Western Mich Univ, MA, 75; Vanderbilt Univ, PhD(psychol), 80. Prof Exp: FEL, BIOL SCI RES CTR, MED SCH, UNIV NC, 80- Mem: Sigma Xi; Soc Neurosci. Res: Neuropharmacology of oxidative metabolites of pherothiazine anti-psychotic drugs both in vivo and in vitro; dopamine receptor supersensitivity; function ascorbic acid in brain. Mailing Add: Biol Sci Res Ctr CB 7250 Univ NC Chapel Hill NC 27599

LEWIS, MICHAEL ANTHONY, b South Bend, Ind, Oct 20, 48; m 74; c 1. AQUATIC ECOLOGY, FISH BIOLOGY. Educ: Gannon Col, BS, 70; Pa State Univ, MS, 72; Ariz State Univ, PhD(zool), 77. Prof Exp: Instr environ biol, Scottsdale Community Col, 77-78; AQUATIC BIOLOGIST, PROCTER & GAMBLE CO, 78- Mem: Sigma Xi; Am Fisheries Soc; Am Inst Fishery Res Biologists. Res: Effects of pollutants on stream primary productivity, biotic survival and community structure and diversity; fish life history studies; fish ecology. Mailing Add: Environ Safety Div A11-OB Ivory Dale Tech Ctr 5299 Spring Grove Ave Cincinnati OH 45217

LEWIS, MICHAEL EDWARD, b Chicago, Ill, Nov 9, 51; m 81. NEUROPHARMACOLOGY, HISTOCHEMISTRY. Educ: George Washington Univ, BA, 73; Clark Univ, MA, 75, PhD(psychol), 77. Prof Exp: Guest worker neurochem, Sect Intermediary Metab, Lab Develop Neurobiol, Nat Inst Child Health & Human Develop, NIH, 77; fel behav neurochem, Psychol Lab, Cambridge Univ, 77-79; instr, Europ Div, Univ Col, Univ Md, 79; res psychologist, Sect Biochem & Pharmacol, Biol Psychiat Br, NIMH & Nat Inst Drug Abuse, 80-81; RES INVESTR, MENT HEALTH RES INST, UNIV MICH, 81- Concurrent Pos: Felix & Elizabeth Brunner fel, Ment Health Found, London, 77-79; Twinning grant, Europ Training Prog in Brain & Behav Res, Europ Sci Found, Strasbourg, 79-; Wellcome res fel, The Wellcome Trust, London, 79; John G Searle clin pharmacol fel, 81. Res: Histochemical and biochemical analysis of receptors and endogenous ligands; pharmacology of neuropeptides and psychoactive drugs; recovery of function after brain damage; neuropsychology. Mailing Add: Cephalon Inc 145 Brandywine Pkwy West Chester PA 19380

LEWIS, MILTON, b New York, NY, Dec 30, 21; m 43; c 3. NUCLEAR ENGINEERING. Educ: Univ Wash, BS; Univ Calif, PhD(chem), 50. Prof Exp: Field serv consult, Off Sci Res Develop, 43-46; asst, Univ Calif, 46-48; chemist, Gen Elec Co, 48-51, chg pile coolant studies, 51-54, supvr, Nonmetallic Mat Develop, 54-56, sr engr prog, 56-62, mgr chem & metall, 62-67; mgr chem & metall, Douglas United Nuclear, Inc, 67-68, fuel & target technol, 68, asst chief, Mat Br, Donald W Douglas Labs, 68-70, mgr, Betacel Prog, 70-74; pres, Columbia Engrs Serv, Inc, Wash, 74-80; SR RES SCIENTIST, BATTELLE PAC NORTHWEST LABS, 82- Concurrent Pos: Vis lectr, Univ Calif, Los Angeles, 60-61; consult, 80- Mem: Am Nuclear Soc. Res: Mechanism of irreversible reactions; analytical chemistry of fission products; corrosion in aqueous media; radiation effects on materials; safety of nuclear processes. Mailing Add: 2600 Harris Ave Richland WA 99352

LEWIS, MORTON, b Oak Park, Ill, June 28, 36; m 63; c 3. ORGANIC CHEMISTRY, ADHESIVES & COATINGS. Educ: Purdue Univ, BS, 58; Univ Chicago, PhD(org chem), 62. Prof Exp: Res chemist, Swift & Co, 62-74, head, Specialty Chem Res Div, Swift & Co Div, 74-76, head, Tech Prod Div, Unitech Chem, Inc Div, Esmark Inc, 76-81; PRIN CHEMIST, WILSON SPORTING GOODS CO, 81- Mem: Am Chem Soc; Sigma Xi; Royal Soc Chem; NY Acad Sci. Res: Synthesis and reactions of steroid derivatives; products of fats and oils; surface active agents; quaternary ammonium salts and organophosphorous compounds; flame retardants; plastics adhesives and coatings; specialty chemicals; ultraviolet cured monomers oligomers and coatings; epoxidized oils and epoxy resins. Mailing Add: Wilson Sporting Goods Co 2233 West St River Grove IL 60171

LEWIS, NATHAN SAUL, b Los Angeles, Calif, Oct 20, 55. ELECTROCHEMISTRY, PHOTOELECTROCHEMISTRY. Educ: Calif Inst Technol, BS, 77, MS, 77; Mass Inst Technol, PhD(inorg chem), 81. Prof Exp: Res asst chem, Calif Inst Technol, 74-77; Mass Inst Technol, 77-81; from asst prof to assoc prof chem, Stanford Univ, 81-88; assoc prof, 88-90, PROF CHEM, CALIF INST TECHNOL, 91- Concurrent Pos: Div ed, J Electrochem Soc, 84-90; consult, Inst Defense Anal, 85-90; Alfred P Sloan Found fel, 85-87. Honors & Awards: Am Chem Soc Award in Pure Chem, 90; Fresenos Award, 90. Mem: Am Chem Soc; Electrochem Soc; Int Soc Electrochem. Res: Electrochemistry of semiconductor surfaces; scanning tunneling microscopy in electrochemistry; inorganic complexes use as electrocatalyst. Mailing Add: 127-72 Dept Chem Calif Inst Technol Pasadena CA 91125

LEWIS, NEIL JEFFREY, b New York, NY, Feb 10, 45; m; c 2. MEDICINAL CHEMISTRY, ORGANIC CHEMISTRY. Educ: City Col New York, BS, 66; Univ Kans, PhD(med chem), 72. Prof Exp: NIH res assoc, Ohio State Univ, 72, from asst prof to assoc prof med chem, Div Med Chem, Col Pharm, 72-82, dir, Environ Chem Anal Lab, 77-82; dir drug develop & assoc dir res prog, Muscular Dystrophy Asn, 82-87; dir res develop, 87-88; VPRES RES, JACOBUS PHARMACEUT CO, 88-; ADJ PROF, BUR BIOL RES, RUTGERS UNIV, 89- Concurrent Pos: Lady Davis vis prof, 79-80; mem, US Environ Protection Agency Human Health Effects Study Sect, 81-85, Int Comt Neuromuscular Dis Drug Develop, 84-88; NJ Munic Environ Adv Comt, 91- Mem: Am Chem Soc; Am Pharmaceut Asn; Sigma Xi; NY Acad Sci; Soc Clin Trials. Res: Chemotherapeutics; environmental carcinogenesis and toxicology; immunochemotherapy; antiviral agents; drug development; neuromuscular diseases; clinical trials. Mailing Add: Jacobus Pharmaceut Co PO Box 5290 Princeton NJ 08540

LEWIS, NINA ALISSA, b Princeton, NJ, Oct 5, 54. COMPUTER SECURITY, COMPUTER NETWORKS. Educ: Univ Calif, Santa Barbara, BS, 75, MS, 85. Prof Exp: PRIN SOFTWARE ENGR, UNISYS DEFENSE SYSTS, INC, 85- Mem: Asn Comput Mach; Inst Elec & Electronics Engrs; Inst Elec & Electronics Engrs Computer Soc. Res: Computer security risk analysis; intrusion detection; computer penetration analysis; network security. Mailing Add: 1070 Miramante Dr No 5 Santa Barbara CA 93109

LEWIS, PAUL HERBERT, b New York, NY, Jan 19, 24; m 55; c 3. PHYSICAL CHEMISTRY. Educ: Columbia Univ, AB, 47, MA, 48; Iowa State Col, PhD(chem), 52. Prof Exp: Chemist paints, E I du Pont de Nemours & Co, Inc, 48; petrol chemist, Texaco Inc, 52-85; RETIRED. Mem: Am Chem Soc. Res: X-ray analysis; catalysts. Mailing Add: 1600 Tawakoni Lane Plano TX 75075-6728

LEWIS, PAUL KERMITH, JR, b Monticello, Ark, Jan 24, 31; m 55; c 3. MEAT SCIENCE. Educ: Okla State Univ, BS, 53; Univ Wis, MS, 55, PhD, 58. Prof Exp: Res asst animal husb & biochem, Univ Wis, 53-57; from asst prof to assoc prof animal indust, 57-68, PROF ANIMAL SCI, UNIV ARK, FAYETTEVILLE, 68- Mem: Am Soc Animal Sci; Am Meat Sci Asn; Inst Food Technologists; Coun Agr Sci & Technol. Res: Pre-slaughter stress and storage life of beef and pork; sensory characteristics of beef and pork; cholesteral comparatives of beef. Mailing Add: Dept Animal & Poultry Sci Univ Ark Fayetteville AR 72701

LEWIS, PAUL WELDON, b Dallas, Tex, Jan 31, 43; m 65; c 2. MATHEMATICS. *Educ:* NTex State Univ, BA, 65, MS, 66; Univ Utah, PhD(math), 70. *Prof Exp:* Asst prof, 70-74, ASSOC PROF MATH, NTEX STATE UNIV, 74- *Mem:* Am Math Soc. *Res:* Vector measures; functional analysis; operators on function spaces. *Mailing Add:* Dept Math NTex State Univ Denton TX 76203

LEWIS, PETER ADRIAN WALTER, b Johannesburg, SAfrica, Oct 3, 32; US citizen; m 60; c 2. STATISTICS. *Educ:* Columbia Univ, BA, 54, BS, 55, MS, 57; Univ London, PhD(statist), 64. *Prof Exp:* Res staff mem statist, Int Bus Mach Res Labs, 55-71; PROF STATIST & OPERS RES, NAVAL POSTGRAD SCH, 71- *Concurrent Pos:* NIH spec fel, Imp Col, Univ London, 69-70. *Mem:* Inst Math Statist; Royal Statist Soc; Am Statist Asn. *Res:* Stochastic process; applications of statistics in computer applications. *Mailing Add:* Dept Oper Res Naval Postgrad Sch Code OR/LW Monterey CA 93940

LEWIS, PHILIP M, b New York, NY, May 30, 31; m 53; c 2. COMPUTER SCIENCE. *Educ:* Rensselaer Polytech Inst, BEE, 52; Mass Inst Technol, SM, 54, ScD, 56. *Prof Exp:* From instr to asst prof elec eng, Mass Inst Technol, 54-59; mem tech staff, Gen Elec Res & Develop Ctr, 59-69, consult automata theory & software design, 59, mgr, 69-78, mgr computer sci br, 78-87; LEAD PROF & CHAIR, COMPUTER SCI DEPT, STATE UNIV NY, STONY BROOK, 87. *Concurrent Pos:* Consult, Epsco, Inc, 55, Lincoln Labs, Mass Inst Technol, 55-56, Hycon Eastern Inc, 56-57 & Sanders Assocs, Inc, 58; adj prof, Rensselaer Polytech Inst, 60-; managing ed, J Comput, Soc Indust & Appl Math, 71-; Coolidge fel, Gen Elec Res & Develop Corp, 77- *Mem:* Asn Comput Mach; Soc Indust & Appl Math; fel Inst Elec & Electronics Engrs. *Res:* Theory of information processing, including compiler design, retrieval and self organization; automata theory; abstract languages. *Mailing Add:* Computer Sci Dept State Univ NY Stony Brook NY 11794-4400

LEWIS, PHILLIP ALBERT, education administration, for more information see previous edition

LEWIS, RALPH WILLIAM, b Marion, Mich, May 21, 11; m 37; c 2. STRUCTURE OF KNOWLEDGE. *Educ:* Mich State Col, BS, 34, MS, 37, PhD(plant path), 45. *Prof Exp:* From instr to asst prof bot, 37-44, from asst prof to prof, 44-80, EMER PROF BIOL, MICH STATE UNIV, 80- *Concurrent Pos:* Fel, Calif Inst Technol, 47; NIH spec res fel, Instituto Superiore Sanita, Rome, 58-59. *Mem:* Fel AAAS; Am Soc Naturalists; Nat Sci Teachers Asn; Nat Asn Biol Teachers; Inst Biol. *Res:* Study of the structure of biological knowledge. *Mailing Add:* Col Natural Sci Mich State Univ East Lansing MI 48824-1031

LEWIS, RANDOLPH VANCE, b Powell, Wyo, Apr 8, 50; m 72; c 2. ENDOCRINOLOGY, PROTEIN CHEMISTRY. *Educ:* Calif Inst Technol, BS, 72; Univ Calif, San Diego, MS, 74, PhD(biochem), 78. *Prof Exp:* Asst, Roche Inst Molecular Biol, 78-80; from asst prof to assoc prof biochem, 80-89, HEAD MOLECULAR BIOL, UNIV WYO, 86-, PROF MOLECULAR BIOL, 89- *Mem:* Am Chem Soc; Am Soc Biol Chemists. *Res:* Peptide hormones of adrenal medulla; protein sequencing; chemical structures of spider silks; protein and peptide purification methods. *Mailing Add:* Univ Wyo PO Box 3944 Laramie WY 82071-3944

LEWIS, RICHARD JOHN, b Chicago, Ill, Jan 20, 35; m 61; c 3. HEMATOLOGY, CLINICAL PHARMACOLOGY. *Educ:* Univ Notre Dame, BS, 56; Northwestern Univ, Chicago, MD, 60. *Prof Exp:* Rotating intern, Cook County Hosp, Chicago, 60-61; med intern & resident, Columbia Div, Bellevue Hosp, 61-63; med resident, Presby Hosp, New York, 63-64; chief nuclear med lab, USPHS Hosp, San Francisco, 66-68; clin instr, Med Ctr, Univ Calif, San Francisco, 68-70, asst prof med, 70-75; ASST CLIN PROF MED, MED SCH, UNIV CALIF, LOS ANGELES, 75- *Concurrent Pos:* Fel hemat, Montefiore Hosp, 64-65; NIH res fel, Med Sch, Univ Wash, 65; clin investr hemat, Vet Admin Hosp, San Francisco, 70-74. *Mem:* Int Soc Hemat; Soc Nuclear Med; Am Fedn Clin Res; NY Acad Sci; Am Soc Hemat. *Res:* Metabolism of coumarin anticoagulant drugs. *Mailing Add:* 2200 Santa Monica Blvd Santa Monica CA 90404

LEWIS, RICHARD THOMAS, b East Cleveland, Ohio, Jan 9, 43; div; c 1. CARBON CHEMISTRY, THERMAL ANALYSIS. *Educ:* Case Western Reserve Univ, BS, 64; Univ Chicago, PhD(phys chem), 70. *Prof Exp:* Staff scientist chem, Carbon Prod Div, Union Carbide Corp, 70-74, group leader, 74-78, res scientist, 78-80, sr res scientist, 80-82, res assoc, 82-89; RES ASSOC, UCAR CARBON CO, 89- *Mem:* Am Chem Soc; NAm Thermal Anal Soc. *Res:* Surface chemistry; physical chemistry and chemistry of carbonization. *Mailing Add:* Ucar Carbon Co 12900 Snow Rd Parma OH 44130

LEWIS, ROBERT ALLEN, b Dunkirk, NY, July 27, 43; m 69. ORGANIC CHEMISTRY. *Educ:* Carnegie Inst Technol, BS, 65; Princeton Univ, MS, 67, PhD(org chem), 69. *Prof Exp:* Postdoctoral, Mass Inst Technol, 69-70; res assoc, Chevron Res Co, Richmond, 70-82, prod develop mgr, Chevron Cent Labs, Rotterdam, Neth, 82-86, tech mgr, Orogil, Nevilly-Seine, France, 86-90, UNIT MGR, CHEVRON RES & TECHNOL CTR, CHEVRON CORP, RICHMOND, CALIF, 90- *Mem:* Am Chem Soc; Soc Automotive Engrs. *Res:* Synthesis, evaluation and chemical process definition of fuel and lubricant additives. *Mailing Add:* 1298 Grizzly Peak Blvd Berkeley CA 94708

LEWIS, ROBERT ALLEN, b New York, NY, Feb 2, 45. EXPERIMENTAL BIOLOGY. *Educ:* Yale Univ, BA, 67; Univ Rochester, MD, 71. *Prof Exp:* Instr med, Harvard Med Sch, 77-79, from asst prof to assoc prof med, 79-87; sr vpres & dir, Basic Res, Syntex Res, 86-89; ASSOC CLIN PROF MED, STANFORD UNIV, 86-; EXEC VPRES & DIR, BASIC RES & DRUG EVAL, SYNTEX RES, 89- *Concurrent Pos:* Rheumatologist & allergist, Med & Pediat Serv, USAF Med Ctr, 74-76; asst physician, Robert B Brigham Hosp, 77-79, Robert B Brigham Div & jr assoc med, Peter B Brigham Div, Brigham & Women's Hosp, 80-82; Allergic Dis Acad Award, NIH, 80-85; rheumatologist & immunologist, Brigham & Women's Hosp, 82-86; adj prof med, Univ Calif, San Francisco, 86- *Mem:* Am Fedn Clin Res; Am Asn Immunologists; Am Acad Allergy & Immunol; Collegium Int Allergologicum. *Res:* Leukotriene structure, function and metabolism; regulation of prostagland in D2 Synthesis from mast cells; regulation of the mast cell secretory response. *Mailing Add:* Syntex Corp Res Div 3401 Hillview Ave MS-R1-240 R6E-1 Palo Alto CA 94304

LEWIS, ROBERT DONALD, agronomy; deceased, see previous edition for last biography

LEWIS, ROBERT EARL, b Richmond, Ind, Dec 1, 29; m 52. ENTOMOLOGY, VERTEBRATE ZOOLOGY. *Educ:* Earlham Col, AB, 52; Univ Ill, MS, 56, PhD(entom), 59. *Prof Exp:* From asst prof to assoc prof zool, Am Univ, Beirut, 59-67; assoc prof, 67-71, PROF ENTOM, IOWA STATE UNIV, 71- *Concurrent Pos:* Consult, US Naval Med Res Unit, Egypt, 63-; grants, Off Naval Res, 66-71 & NIH, 64-67; ARS, 67- *Mem:* Entom Soc Am; Am Entom Soc; Am Soc Mammal; Soc Syst Zool; Royal Entom Soc London. *Res:* Siphonaptera of the world, their host relationships and zoogeography. *Mailing Add:* 306 21st St Ames IA 50010

LEWIS, ROBERT EDWIN, JR, b Meridian, Miss, Mar 11, 47. IMMUNOPATHOLOGY, PATERNITY TESTING. *Educ:* Univ Miss, Oxford, BA, 69, MS, 73; Univ Miss, Jackson, PhD(immunol & path), 76. *Prof Exp:* Instr path & anesthesiol, 76-77, asst prof anesthesiol, 77-85, asst prof, 77-84, ASSOC PROF PATH, UNIV MISS MED CTR, 84-, CO-DIR, TISSUE TYPING LAB & CLIN IMMUNOPATH LAB & DIR, PATERNITY TESTING LAB, 81- *Concurrent Pos:* Sr ed, Path & Immunopath Res & Immunol Res, dep ed-in-chief, Pathobiol. *Mem:* Fel Royal Soc Health (UK); Am Asn Pathologists; Am Asn Immunologists; Can Soc Immunol; Reticuloendothelial Soc; Am Soc Microbiol. *Res:* Cellular immunology; natural killer cell morphology, levels and function in human patients with leukemias, lymphomas and in renal allotransplant recipients; mechanisms of Natural Killer and natural cytotoxicity cell actions in allograft rejection and antitumor immunity; Interleukin-2 in immunomodulation of tumors in man and animals. *Mailing Add:* Dept Path Univ Miss 2500 N State St Jackson MS 39216-4505

LEWIS, ROBERT FRANK, b Wis, Dec 13, 20; m 45; c 2. PUBLIC HEALTH. *Educ:* Univ Calif, BS, 50, MPH, 53; Univ Mich, PhD(public health statist), 58. *Prof Exp:* Epidemiol res, Commun Dis Ctr, USPHS, Ga, 49-52; health analyst, San Joaquin Local Health Dist, Calif, 53-54; state health analyst, Calif, 54-56; res assoc cerebral palsy, Univ Mich, 58; from asst prof to prof biomet, Sch Med, Tulane Univ, 58-67, head div, 60-67; prof human med, Col Human Med, Mich State Univ, 67-75; PROF, DEPT EPIDEMIOL & BIOSTATIST, SCH PUB HEALTH, UNIV SC, 78- *Concurrent Pos:* Consult, Div Radiol Health, NIH, 66-70, div manpower intel, 71; chief, Mich Ctr Health Statist, 67-69; consult, Nat Ctr Health Servs Res, 75-80, 85-, US Cong Off Technol & Assessment, 77-78. *Mem:* Fel AAAS. *Res:* Public health and medicine, especially development of methodology in health research in rural settings. *Mailing Add:* Dept Pub Health Univ SC Columbia SC 29208

LEWIS, ROBERT GLENN, b Morehead City, NC, Nov 11, 37; m 60; c 3. ORGANIC CHEMISTRY. *Educ:* Univ NC, BS, 60; Univ Wis, PhD(org chem), 64. *Prof Exp:* Res chemist, Chemstrand Res Ctr, Inc, 64-69, group leader and res specialist, 69-71, sect chief, 71-79, CHIEF, METHODS DEVELOP BR, ENVIRON MONITORING SYSTS LAB, ENVIRON PROTECTION AGENCY, 79- *Concurrent Pos:* NSF fel, 61; NIH fel, 61-64. *Mem:* Am Chem Soc; Sigma Xi; AAAS; Am Soc Testing & Mat; Air & Waste Mgt Asn. *Res:* Environmental chemistry; environmental toxicology; air pollution analysis; reaction mechanisms; organic photochemistry; ultraviolet-visible absorption and luminescence spectroscopy; organic analyses; pesticide chemistry and analysis; mass spectrometry. *Mailing Add:* Atmospheric Res & Exposure Assessment Lab MD-44 US Environ Protection Agency Triangle Park NC 27711

LEWIS, ROBERT MILLER, b Flushing, NY, May 20, 37; m 58; c 2. IMMUNOLOGY. *Educ:* Wash State Univ, DVM, 61. *Prof Exp:* Intern, Angell Mem Animal Hosp, 61-62; res fel, Harvard Med Sch, 62-65; from instr to sr instr surg, Sch Med, Tufts Univ, 65-67, from asst prof to assoc prof, 67-75, dir lab animal sci, 69-75; PROF PATH & CHMN DEPT, NY STATE COL VET MED, CORNELL UNIV, 75- *Concurrent Pos:* Res assoc path, Angell Mem Animal Hosp, 62-63, assoc pathologist, 65-67; affil in med, 68-75; consult surg res, New Eng Med Ctr Hosp, 62-65, mem spec sci staff, 66-75; chief vet serv, 70-75; asst path, Harvard Med Sch, 65-68, clin asst, 68-76. *Honors & Awards:* Mary Mitchell Award Outstanding Res, 61. *Mem:* Am Vet Med Asn; Am Col Vet Pathologists; Int Acad Path; Am Asn Lab Animal Sci; Am Soc Vet Clin Pathologists; Sigma Xi. *Res:* Investigations on the etiology and pathogenesis of spontaneous immunologic diseases of animals which mimic human diseases. *Mailing Add:* Dept Path Col Vet Med Cornell Univ Ithaca NY 14853

LEWIS, ROBERT MINTURN, b Hempstead, NY, Aug 23, 24; m 53; c 3. FISH BIOLOGY. *Educ:* Cornell Univ, BS, 54, MS, 56. *Prof Exp:* Res fishery biologist, Striped Bass Prog, Mid-Atlantic Coastal Fisheries Res Ctr, 56-63, Menhaden Prog, 63-78, RES FISHERY BIOLOGIST, ATLANTIC ESTUARINE FISHERIES CTR, NAT MARINE FISHERIES SERV, 63- *Mem:* Am Fisheries Soc; Am Inst Fisheries Res Biol. *Res:* Effects of environmental conditions on larval and juvenile marine fishes; population dynamics of marine fishes; electrophoretic studies of fish protein; analysis of large scale tagging; analysis and identification of off-shore and estuarine larval fish populations; fecundity of Atlantic and Gulf menhaden. *Mailing Add:* 1611 Front St Beaufort NC 28516-2315

LEWIS, ROBERT RICHARDS, JR, b New Haven, Conn, Mar 7, 27; m 50; c 4. THEORETICAL PHYSICS. *Educ:* Univ Mich, BS, 50, MS, 53, PhD(physics), 54. *Prof Exp:* Asst prof physics, Univ Notre Dame, 54-58; from asst prof to assoc prof, 58-65, PROF PHYSICS, UNIV MICH, ANN

ARBOR, 65- *Concurrent Pos:* Mem, Inst Advan Study, 56-58. *Mem:* Am Phys Soc. *Res:* Quantum theory; angular correlation theory; parity nonconservation in atoms; partial coherence theory. *Mailing Add:* Dept Physics Univ Mich Ann Arbor MI 48109

LEWIS, ROBERT TABER, b New York, NY, May 12, 32; m 58; c 1. PHYSICS. *Educ:* Alfred Univ, BS, 54; Univ Calif, Berkeley, PhD(solid state physics), 64. *Prof Exp:* Res physicist, 63-72, SR RES PHYSICIST, CHEVRON RES CO, 72- *Mem:* Am Phys Soc. *Res:* Magnetism; heterogenous catalysts; x-ray photoelectron spectroscopy. *Mailing Add:* 100 Chevron Way Rm 71-7654 Richmond CA 94802

LEWIS, ROBERT WARREN, b Mansfield, Ohio, Feb 4, 43; m 78; c 2. ELECTRICAL ENGINEERING, OPTICS. *Educ:* Univ Cincinnati, BSEE, 66; Univ Mich, MSE, 68, MA, 69, PhD(elec eng), 73. *Prof Exp:* Res asst coherent optics, Willow Run Labs, Univ Mich, 69-72; res assoc radar, Environ Res Inst Mich, 72-73; assoc sr res scientist optics & computerized tomography, Gen Motors Res Labs, 78-87; ADJ PROF, OAKLAND COMMUNITY COL, 90- *Concurrent Pos:* Educ consult & imaging consult. *Mem:* Optical Soc Am; Inst Elec & Electronics Engrs; Sigma Xi. *Res:* Optics and electrical engineering; computerized tomographic mapping of temperature induced refractive index fields in combusting mixtures. *Mailing Add:* 247 Whims Ct Rochester MI 48306

LEWIS, ROGER ALLEN, b Wellington, Kans, June 1, 41; m 62; c 3. BIOCHEMISTRY. *Educ:* Phillips Univ, BA, 63; Ore State Univ, PhD(biochem), 68. *Prof Exp:* Res assoc pyrimidine nucleotide metab, Stanford Univ, 68-69; from asst prof to assoc prof, 69-82, PROF BIOCHEM, UNIV NEV, RENO, 82- *Mem:* Am Chem Soc; Sigma Xi; AAAS; Am Soc Pharmacol & Exp Therapeut; Am Soc Biochem Molecular Biol. *Res:* Purine deoxynucleotide biosynthesis and its control; toxicology of pesticides with respect to nucleotide metabolism and DNA and/or RNA synthesis. *Mailing Add:* Dept Biochem Univ Nev Reno NV 89557

LEWIS, ROSCOE WARFIELD, animal nutrition, biochemistry; deceased, see previous edition for last biography

LEWIS, ROY STEPHEN, b Oakland, Calif, Aug 10, 44; m 66; c 3. METEORITICS. *Educ:* Univ Calif, Berkeley, AB, 67, PhD(atmospheric & space sci), 73. *Prof Exp:* SR RES ASSOC METEORITICS, DEPT CHEM, UNIV CHICAGO, 73- *Honors & Awards:* Except Sci Achievement Medal, NASA. *Mem:* Meteoritical Soc; fel AAAS. *Res:* Isotopic composition and elemental abundances of noble gases in meteorites and other samples. *Mailing Add:* Enrico Fermi Inst Univ Chicago 5630 S Ellis Ave Chicago IL 60637

LEWIS, RUSSELL J, b Liberty Road, Ky, Jan 23, 29; m 54; c 2. SOIL CHEMISTRY. *Educ:* Univ Ky, BS, 56, MS, 57; NC State Col, PhD(soils), 61. *Prof Exp:* Int Atomic Energy Agency fel chem, Univ NC, 61-62; from asst prof to assoc prof soil chem, Univ Tenn Knoxville, 62-80. *Mem:* Am Soc Agron; Soil Sci Soc Am; Clay Minerals Soc; Sigma Xi. *Res:* Surface chemistry of colloids; ion exchange, fixation and nutrient availability. *Mailing Add:* Univ Tenn Dept Plant & Soil Sci Knoxville TN 37901-1071

LEWIS, RUSSELL M(ACLEAN), b New York, NY, June 20, 30; m 57; c 4. TRAFFIC ENGINEERING, TRANSPORTATION PLANNING. *Educ:* Trinity Col, Conn, BS, 52; Rensselaer Polytech Inst, BCE, 53, MCE, 59; Purdue Univ, PhD, 62. *Prof Exp:* From instr to assoc prof civil eng, Rensselaer Polytech Inst, 57-68; assoc, Byrd, Tallamy, MacDonald & Lewis, Div Wilbur Smith & Assocs, 68-70, partner, 71-72, sr assoc, 72-80; CONSULT ENGR, 80- *Concurrent Pos:* Ford Found fel, Purdue Univ, 60-62; affil, Transp Res Bd, Nat Res Coun-Nat Acad Sci; mem construct & maintenance subcomt, Nat Comt Uniform Traffic Control Devices, 79- *Mem:* Am Soc Civil Engrs; Inst Traffic Engrs; Am Soc Eng Educ; Am Soc Training & Develop. *Res:* Highway accident analysis; highway safety research; traffic and parking studies; expert witness testimony in accident cases; training programs for transportation agencies. *Mailing Add:* 8313 Epinard Ct Annandale VA 22003-4441

LEWIS, SHELDON NOAH, b Chicago, Ill, July 1, 34; m 57; c 3. PHYSICAL CHEMISTRY, ORGANIC CHEMISTRY. *Educ:* Northwestern Univ, BA & MS, 56; Univ Calif, Los Angeles, PhD(phys & org chem), 59. *Prof Exp:* NSF fel, Univ Basel, 59-60; sr chemist, Rohm and Haas Co, 60-61, group leader org chem, 61-63, lab head, 63-68, res supvr, 68-73, dir specialty chem res, 73-74, gen mgr, DCL Lab AG, Switz, 74-75, dir, European Labs, France, 75-76, corp dir res polymers, resins & monomers worldwide, 76-78; vpres res & develop, 78, group vpres & dir, 78-84, EXEC VPRES & DIR, CLOROX CO, 84- *Mem:* Am Chem Soc; Indust Res Inst; Soc Chem Indust. *Res:* Reaction mechanisms; organic synthesis; process development; agricultural chemicals; polymers and surface coatings; leather, paper, textile, cosmetic and petroleum chemicals; plastics and modifiers; ion exchange resins; adhesives; building products. *Mailing Add:* 3711 Rose Ct Lafayette CA 94549

LEWIS, SHERRY M, NUTRITION. *Prof Exp:* ANALYTICAL NUTRITIONIST, NAT CTR TOXICOL RES, 87- *Mailing Add:* Nat Ctr Toxicol Res MC 916 Nat Ctr Toxicol Res Dr Jefferson AR 72079

LEWIS, SILAS DAVIS, b Gastonia, NC, June 26, 30; m 62; c 1. ORGANIC CHEMISTRY. *Educ:* Wake Forest Col, BS, 52; Ga Inst Technol, PhD(org chem), 59. *Prof Exp:* Sr chemist, Atlast Chem Industs, Inc, 59-63; assoc prof chem, Del Valley Col, 63-66; assoc prof chem, Augusta Col, 66-88; RETIRED. *Mem:* Am Chem Soc; Sigma Xi. *Res:* Polyphenyls; Ullmann reaction; nitro and nitrato compounds and explosives. *Mailing Add:* 1760 Kissing Bower Rd Augusta GA 30904

LEWIS, SIMON ANDREW, b Welling Kent, Eng, Apr 18, 48; Can citizen; m 76. EPITHELIAL TRANSPORT, ELECTROPHYIOLOGY. *Educ:* Univ BC, BSc, 70, MSc, 71; Univ Calif, Los Angeles, PhD(physiol), 75. *Prof Exp:* Res assoc physiol, Univ Calif, Los Angeles, 75-76; res assoc physiol, Med Br, Univ Tex, 76-77; from asst prof to assoc prof physiol, Yale Med Sch, 77-87; PROF PHYSIOL & BIOPHYS, UNIV TEX, 87- *Mem:* Biophys Soc. *Res:* Mechanisms of salt and water transport across epithelial cells membranes using electrophysiological methods. *Mailing Add:* Dept Physiol & Biophys Univ Tex Med Br Basic Sci Bldg Rte F41 Galvaston TX 77550

LEWIS, STANDLEY EUGENE, b Twin Falls, Idaho, Nov 15, 40; m 65; c 2. ENTOMOLOGY, PALEONTOLOGY. *Educ:* Univ Nebr, Omaha, BA, 62, MA, 64; Wash State Univ, PhD(entom), 68. *Prof Exp:* Teaching asst biol, Univ Nebr, Omaha, 63-64; teaching asst zool-entom, Wash State Univ, 64-68; from asst prof to assoc prof biol, 68-78, PROF BIOL, ST CLOUD STATE COL, 78- *Concurrent Pos:* Res asst mosquito control, Adams County Abate Dist, 65-; res consult, N States Power Co, 75-77; res assoc, Sci Mus Minn; Sigma Xi, Geol Soc Am & St Cloud instnl grants, 68, 70, 75, 77, 83 & 85. *Mem:* Sigma Xi; Paleont Soc Am; Entom Soc Am. *Res:* Paleobiology, specifically paleoentomology; tertiary insect site in US; fossil insect studies, Miocene sites: Wash And Idaho, Oligocene sites: Mont, Cretaceous sites: Minn; fossil bison kill site in Central Minn. *Mailing Add:* Dept Biol St Cloud State Univ St Cloud MN 56301

LEWIS, STEPHEN ALBERT, b Sodus, NY, Sept 9, 42; m 68; c 1. PLANT NEMATOLOGY. *Educ:* Pa State Univ, BS, 64; Rutgers Univ, MS, 69; Univ Ariz, PhD(plant path), 73. *Prof Exp:* Sales rep, Stand Oil Calif, 65-66; asst prof, 73-77, ASSOC PROF NEMATOL, DEPT PLANT PATH & PHYSIOL, CLEMSON UNIV, 77- *Mem:* Soc Nematologists; Sigma Xi. *Res:* Host-parasite relations of the phytoparasitic nematodes, Hoplolaimus columbus and Criconemoides xenoplax on field crops and peach trees, respectively; nematode-mycorrhizae-rhizobium relationships; gnotobiotic culture of nematodes. *Mailing Add:* Dept Plant Path Long Hall Clemson Univ Clemson SC 29631

LEWIS, STEPHEN B, b Berkeley, Calif, Mar 9, 40; m 72; c 3. ENVIRONMENTAL MEDICINE. *Educ:* Univ Calif, Berkeley, BA, 62; Wash Univ, MD, 66. *Prof Exp:* Intern med serv & chmn, Dept Med, Michael Reese Hosp, Chicago, 66-67, asst resident med, 67-69; postdoctoral fel, Dept Physiol, Sch Med, Vanderbilt Univ, 69-72, instr, Dept Med & res assoc, Dept Physiol, 71-72; endocrinologist & asst dir, Clin Invest Ctr & head, Endocrinol & Metab Br, Med Serv, Naval Regional Med Ctr, Oakland, Calif, 72-78, dir, Clin Invest Ctr & head, Endocrinol Br, 78-83; clin asst prof med, Univ Calif, San Francisco, 77-83 & Tissue Bank Stem Cell Res, Naval Med Res Inst, 83-86; dep dir, Environ Med Dept, 86-90; CAPT, US NAVAL MED CORPS, 79- *Concurrent Pos:* Spec res fel, NIH, 71-72. *Mem:* Am Diabetes Asn (pres, 81-83); Am Fed Clin Res; Soc Exp Biol & Med; Endocrine Soc; AAAS; Am Physiol Soc; Europ Asn Study Diabetes. *Res:* Blood and blood substitutes; wounds, sepsis and shock; hypothermia and non-freezing cold injury; readiness planning and material; post injury enhancement. *Mailing Add:* Environ Med Dept Naval Med Res Inst Stop 11 Bethesda MD 20814-5055

LEWIS, STEPHEN ROBERT, b Mt Horeb, Wis, Aug 26, 20; m 48; c 2. PLASTIC SURGERY. *Educ:* Carroll Col, Wis, BA, 41; Marquette Univ, MD, 44. *Prof Exp:* Instr surg, 50-53, from asst prof to assoc prof plastic & maxillofacial surg, 53-61, asst dean med, 58-62, chief staff, 69-73, dir postgrad educ, 56-80, PROF SURG, UNIV TEX MED BR GALVESTON, 61-, CHIEF PLASTIC SURG, 61- *Concurrent Pos:* Consult, St Mary's Infirmary, Galveston, 54-, Galveston County Mem Hosp, 54-, USAF, 57- & USPHS Hosp, 58-; mem, Am Bd Plastic Surg, 66-72, chmn, 71-72. *Mem:* Am Soc Plastic & Reconstruct Surg (-pres, 56-); fel Am Col Surg; AMA; Am Asn Plastic Surg; Sigma Xi. *Res:* Tissue culture studies on human skin; burns; multiple studies in systemic and local problems; lymphatics in Lymphedema; congenital deformities of face, neck and hands. *Mailing Add:* Dept Surg Univ Tex Med Br Galveston TX 77550

LEWIS, STEVEN CRAIG, b Anderson, Ind, Dec 30, 43; m 69. CHEMICAL CARCINOGENESIS, BIOSTATISTICS. *Educ:* Ind Univ, BA, 70, PhD(toxicol), 75; Am Bd Toxicol, dipl, 80, recert, 85 & 90. *Prof Exp:* Res asst biochem, Ind Univ Med Sch, 65-71, res fel toxicol, 71-75, supvr, Statist Cancer Res Unit, 72-75; toxicologist, 75-79, sr toxicologist & unit head, 79-82, TOXICOL ASSOC, EXXON CORP, 82- *Concurrent Pos:* Feature ed, Neurotoxicol, 79-; vchmn & chmn, toxicol comt, Am Petrol Inst, 83-; mem, sci rev group, NY Dept Environ Conserv, 85- *Mem:* Europ Soc Toxicol; Soc Toxicol; Am Indust Hyg Asn. *Res:* Chemical carcinogenesis; biostatistics; risk analysis; neurotoxicology; toxicology; product safety. *Mailing Add:* Mettlers Rd CN2350 East Millstone NJ 08875-2350

LEWIS, STEVEN M, b Washington, DC, June 9, 48. RESPIRATORY PHYSIOLOGY, SIMULATION. *Educ:* Calif Inst Technol, BS, 69; Univ Wash, PhD(biophys), 74. *Prof Exp:* From asst prof to assoc prof biomed eng, Univ Southern Calif, University Park, 76-89; MEM TECH STAFF, AEROSPACE CORP, 89- *Mem:* Biomed Eng Soc; Am Physiol Soc; Inst Elec & Electronics Engrs; Asn Comput Mach. *Mailing Add:* Aerospace Corp 2350 El Segundo Blvd MS MI-107 El Segundo CA 90245-4691

LEWIS, SUSANNA MAXWELL, b Boston, Mass; m 86; c 1. DNA RECOMBINATION. *Educ:* Tufts Univ, BS, 76; Mass Inst Technol, PhD(biol), 85. *Prof Exp:* Postdoctoral fel biochem, Stanford Univ, 85-86; staff fel, Lab Molecular Biol, NIH, 86-88; SR RES FEL BIOL, CALIF INST TECHNOL, 88- *Res:* Mechanism of immunoglobulin and T cell receptor gene rearrangement. *Mailing Add:* Div Biol 156-29 Calif Inst Technol Pasadena CA 91125

LEWIS, T(HOMAS) SKIPWITH, b Bluefield, WVa, Nov 21, 36; m 62; c 2. ELECTRICAL ENGINEERING. *Educ:* Va Polytech Inst, BS, 59; Univ Va, MS, 64, ScD(elec eng), 67. *Prof Exp:* Engr, Air Arm Div, Westinghouse Elec Corp, 59-62; instr elec eng, Univ Va, 64-67; sr res engr, United Aircraft Res Labs, 67-70; adj asst prof elec eng, Univ Hartford, 67-70, assoc prof, 70-81, dean, Col Eng, 71-81; asst vpres, 81-82, vpres, 82-85, SR VPRES, HARTFORD STEAM BOILER, 85- *Mem:* Inst Elec & Electronics Engrs. *Res:* Microwave engineering and antennas; microwave properties of materials. *Mailing Add:* 41 N Parker Rd Hebron CT 06248

LEWIS, THEODORE, b New York, NY, Apr 9, 24. CHEMICAL ENGINEERING, POLYMER CHEMISTRY. *Educ:* Rensselaer Polytech Inst, BChE, 46, MChE, 48; Princeton Univ, PhD(chem eng), 55; NY Univ, MBA, 71. *Prof Exp:* Res engr chem & chem eng, Esso Res Eng Co, Standard Oil Co, NJ, 54-59, sr mkt develop engr, Enjay Chem Co, 59-65, assoc synthetic elastomers dept, 65-71; assoc prof, Col Bus, Fairleigh Dickinson Univ, Rutherford, pharmaceut/chem prog & admin asst to dean, 71-74; INDUST, GOVT & UNIV CONSULT, 74- *Concurrent Pos:* Adj fac, Fairleigh Dickinson Univ & Adelphi Univ, 74- *Mem:* Am Chem Soc; Am Inst Chem Engrs. *Res:* Textile fibers; synthetic elastomers; petrochemicals; products and process research; market development of thermosets, synthetic elastomers and special industries. *Mailing Add:* PO Box 177 Roselle NJ 07203

LEWIS, THOMAS BRINLEY, b Cleveland, Ohio, Nov 3, 38; m 65; c 3. BIOTECHNOLOGY. *Educ:* John Carroll Univ, BS, 60, MS, 62; Mass Inst Technol, PhD(physics), 65. *Prof Exp:* Res assoc phys chem, Cornell Univ, 65-66; sr res chemist, group leader, Cent Res Dept, Monsanto Co, 66-72, proj mgr, Rubber Chem Div, 72-77, mgr Com Develop, 77-78, dir results Mgt Corp Res & Develop Staff, 78-80, dir Corp Res Labs, 80-83, gen mgr, Rubber Chem Div, 83-86, dir Res & Develop, 87-88; VPRES, CELGENE CORP, 88- *Mem:* Am Phys Soc; Sigma Xi; Am Chem Soc; AAAS; NY Acad Sci. *Res:* Management of programs in biotechnology; speciality chemicals; plastics and resins and separations and the environment; dynamic mechanical properties of composite materials, polymers and rubber; management of programs in biotechnology; specialty chemicals; plastics and resins and separations. *Mailing Add:* Celgene Corp F Powder Horn Dr Warren NJ 07059

LEWIS, TRENT R, b Baltimore, Md, Feb 3, 32; m 65; c 3. PUBLIC HEALTH & EPIDEMIOLOGY, ENVIRONMENTAL TOXICOLOGY. *Educ:* Univ Md, BS, 54, MS, 57; Mich State Univ, PhD(nutrit), 61. *Prof Exp:* Res asst dairy sci, Univ Md, 55-57; res instr, Mich State Univ, 57-61; asst prof animal sci, Univ Maine, 61-63; res chemist, R A Taft Sanit Eng Ctr, USPHS, 63-68; chief chronic & explor toxicol, Nat Air Pollution Control Admin, 68-71; chief, Explor Toxicol Sect, Nat Inst Occup Safety & Health, 71-77 & Exp Toxicol Br, 78-87; dir, Toxicol & Microbiol Div, US Environ Protection Agency, 88-89; PVT CONSULT, 90- *Concurrent Pos:* Mem, Am Conf Ind Hygienists, 75-; lectr & consult, 77- *Mem:* Soc Toxicol; Sigma Xi; Am Conf Govt Indust Hygienists. *Res:* Occupational toxicology; environmental toxicology; mammalian cardiopulmonary physiology; biochemical mechanisms; development or modification of standard toxicity testing regimens; mutagenic chemical agents via reproductive and teratogenic methodology; development of dose-response criteria. *Mailing Add:* 9270 Silva Dr Cincinnati OH 45251

LEWIS, TREVOR JOHN, b Vancouver, BC, Jan 19, 40; m 64; c 3. GEOPHYSICS. *Educ:* Univ BC, BASc, 63, MSc, 64; Univ Western Ont, PhD(geophys), 75. *Prof Exp:* RES SCIENTIST GEOTHERMAL STUDIES, DEPT ENERGY MINES & RESOURCES, GEOL SURV CAN, 64- *Mem:* Geol Asn Can; Can Geophys Union; Am Geophys Soc; Can Geothermal Energy Asn; Int Heat Flow Comt. *Res:* Geothermal studies, thermal structure of the earth. *Mailing Add:* Pac Geosci Ctr PO Box 6000 Sidney BC V8L 4B2 Can

LEWIS, URBAN JAMES, b Flagstaff, Ariz, Apr 28, 23; m 50; c 2. ENDOCRINOLOGY. *Educ:* San Diego State Col, BA, 48; Univ Wis, MS, 50, PhD(biochem), 52. *Prof Exp:* NIH fel, Med Nobel Inst, Stockholm, 52-53; instr biochem & biochemist, Am Meat Found, Univ Chicago, 53-54; sr biochemist, Merck & Co, Inc, 54-61; mem, Scripps Clin & Res Found, 61-82; MEM, WHITTIER INST, 82- *Mem:* Am Soc Biol Chem; Am Chem Soc; Endocrine Soc. *Res:* Proteolytic enzymes; pituitary hormones. *Mailing Add:* Whittier Inst Diabetes & Endocrinol 9894 Genesee Ave La Jolla CA 92037

LEWIS, VANCE DE SPAIN, b Los Angeles, Calif, June 26, 09; m 36; c 2. PHYSICS. *Educ:* Univ Calif, Berkeley, BA, 33, MA, 40; Univ Southern Calif, PhD(educ), 54. *Prof Exp:* From asst prof to prof, 46-64, EMER PROF PHYSICS, CALIF POLYTECH STATE UNIV, SAN LUIS OBISPO, 72- *Concurrent Pos:* Assoc dean, Sch Sci & Math, Calif Polytech State Univ, San Luis Obispo, 68-72. *Mem:* Am Phys Soc. *Res:* Optics; statistical analysis. *Mailing Add:* 17050 Arnold Dr F-113 Riverside CA 92508

LEWIS, W DAVID, b Towanda, Pa, June 24, 31; m 86; c 3. HISTORY OF TECHNOLOGY. *Educ:* Pa State Univ, BA, 52, MA, 54; Cornell Univ, PhD(hist), 61. *Prof Exp:* Instr pub speaking, Hamilton Col, 54-57; fel coordr, Eleutherian Mills-Hagley Found, Inc, Wilmington & lectr hist, Univ Del, 59-65; from assoc prof to prof, State Univ NY, Buffalo, 65-71; HUDSON PROF HIST & ENG, AUBURN UNIV, 71- *Concurrent Pos:* Dir, Nat Endowment Humanities Proj Technol, Human Values & Southern Future, Auburn Univ, 74-; fel Nat Humanities Inst, Univ Chicago, 78-; grants, State Univ NY, Auburn Univ, Eleutherian Mills Hist Libr, Nat Endowment for The Humanities & Delta Airlines Found. *Mem:* Soc Hist Technol. *Res:* History of technology, particularly history of iron and steel industry and aerospace history. *Mailing Add:* 7008 Haley Ctr Auburn Univ Auburn AL 36830

LEWIS, WALLACE JOE, b Smithdale, Miss, Oct 30, 42; m 65; c 2. ENTOMOLOGY. *Educ:* Miss State Univ, BS, 64, MS, 65, PhD(entom), 68. *Prof Exp:* ENTOMOLOGIST, SOUTHERN GRAIN INSECTS RES LAB, ENTOM RES DIV, AGR RES SERV, USDA, 67- *Concurrent Pos:* Asst prof, Univ Fla, 70- *Mem:* Entom Soc Am; Sigma Xi. *Res:* Ecological and physiological relationships between parasitic insects and their hosts; development of methods for the use of parasitic insects for control of insect pests. *Mailing Add:* Grain Insects Coastal Plain Exp Sta Agr Res Serv USDA Tifton GA 31794

LEWIS, WALTER HEPWORTH, b Carleton Place, Ont, June 26, 30; m 57; c 2. BOTANY. *Educ:* Univ BC, BA, 51, MA, 54; Univ Va, PhD(bot), 57. *Prof Exp:* Asst prof biol & dir herbarium, Stephen F Austin State Col, 57-61, assoc prof biol, 61-64; assoc prof, 64-69, PROF BIOL, WASH UNIV, 69- *Concurrent Pos:* Guggenheim fel, 63-64; dir herbarium, Mo Bot Garden, 64-

72, sr botanist, 72- *Honors & Awards:* Horsley Res Award, Va Acad Sci, 57. *Mem:* Bot Soc Am; Am Soc Plant Taxon; Asn Trop Biol; Int Asn Plant Taxon; Int Orgn Biosyst; fel Linnean Soc London. *Res:* Cytotaxonomy of Rosa, the Rubiaceae, palynotaxonomy of angiosperms and southern flora; medical plants; allergy. *Mailing Add:* Dept Biol Wash Univ Lindell-Skinker Blvd St Louis MO 63130

LEWIS, WILFRED BENNETT, physics; deceased, see previous edition for last biography

LEWIS, WILLARD DEMING, physics, academic administration; deceased, see previous edition for last biography

LEWIS, WILLIAM E(RVIN), b Hagerstown, Md, Sept 10, 40; m 58; c 5. COMPUTER PERFORMANCE EVALUATION, ANALYTICAL MODELING. *Educ:* Johns Hopkins Univ, BES, 62; Northwestern Univ, Evanston, MS, 64, PhD(indust eng), 66. *Prof Exp:* Assoc prof, 65-80, chmn dept, 80-85, PROF COMPUT SCI, ARIZ STATE UNIV, 80-, ASST DEAN ENG, 85- *Concurrent Pos:* Consult, Good Samaritan Hosp, 66-68; Gen Elec Info Systs, 69-71, Honeywell Info Systs, 71-79; eval analyst, Phoenix Alcohol Safety Action Proj, 71-73; staff mem, Intel, 79-80; staff mem, Ariz Dept Health Serv, 87- *Mem:* Am Inst Indust Engrs; Asn Comput Mach. *Res:* Application of operations research and computer techniques to industrial problems; computer systems design. *Mailing Add:* 1213 E Loyola Tempe AZ 85287-5206

LEWIS, WILLIAM JAMES, b Talahassee, Fla, Feb 11, 45; m 82; c 3. MATHEMATICS. *Educ:* La State Univ, Baton Rouge, BS, 66, PhD(math), 71. *Prof Exp:* Instr math, La State Univ, 71; asst prof, 71-77, vchmn dept, 80-83, ASSOC PROF MATH, UNIV NEBR-LINCOLN, 77-, CHAIR, 88- *Mem:* Am Math Soc; Math Asn Am; Am Asn Union Profs. *Res:* Commutative algebra; valuation theory; ring theory. *Mailing Add:* Dept Math Univ Nebr Lincoln NE 68583

LEWIS, WILLIAM MADISON, b Faison, NC, Nov 26, 22; m 43; c 4. FISHERIES. *Educ:* NC State Col, BS, 43; Iowa State Col, MS, 48, PhD(zool), 49. *Prof Exp:* Sci bact aide, USDA, 42; asst prof, 49-60, PROF ZOOL, SOUTHERN ILL UNIV, 60-, DIR, COOP FISHERIES LAB, 49-, CHMN DEPT ZOOL, 72- *Mem:* Am Fisheries Soc. *Res:* Aquaculture and fish management. *Mailing Add:* Rte 2 Box 290 Saison NC 28341

LEWIS, WILLIAM MASON, b Ithaca, NY, Aug 13, 29; m 57; c 5. WEED SCIENCE. *Educ:* Tex A&M Univ, BS, 52; Univ Minn, MS, 56, PhD(plant genetics), 57. *Prof Exp:* Asst agron & plant genetics, Univ Minn, 52-56; from instr to assoc prof, 56-69, PROF CROP SCI & WEED SCI EXT SPECIALIST, NC STATE UNIV, 69- *Concurrent Pos:* Vis prof, Univ Ill, 74. *Mem:* Am Soc Agron; Crop Sci Soc Am; Weed Sci Soc Am; Int Turf Grass Soc. *Res:* Agronomy; turf weed control. *Mailing Add:* Dept Crop Sci NC State Univ Raleigh NC 27695-7620

LEWIS, WILLIAM PERRY, b Swatow, China, Aug 12, 29; US citizen; m 51; c 2. MEDICAL MICROBIOLOGY. *Educ:* Univ Redlands, BS, 51; Univ Calif, Los Angeles, PhD(infectious dis), 62. *Prof Exp:* Asst res parasitologist & instr parasitol, Sch Pub Health, Univ Calif, Los Angeles, 62-69; ASSOC PROF PATH, SCH MED, UNIV SOUTHERN CALIF & CHIEF MED MICROBIOLOGIST, LOS ANGELES COUNTY-UNIV SOUTHERN CALIF MED CTR, 69- *Mem:* AAAS; Am Soc Trop Med & Hyg; Am Soc Microbiol; Am Soc Parasitol. *Res:* Immunology of parasitic diseases, especially toxoplasmosis, amebiasis and filariasis; diagnostic bacteriology, parasitology and immunology. *Mailing Add:* Microbiol Lab Los Angeles County- Univ Southern Calif Med Ctr 1200 N State St Los Angeles CA 90033

LEWISTON, NORMAN JAMES, b Perry, Iowa, Oct 8, 38. PEDIATRICS, CELL BIOLOGY. *Educ:* Iowa State Univ, BS, 60; Univ Iowa, MD, 65. *Prof Exp:* Fel allergy, 72-73, fel respiratory med, 73-74, asst prof, 74-82, assoc pediat, 82-88, PROF PEDIAT, SCH MED, STANFORD UNIV, 88- *Concurrent Pos:* Chmn, Consumer Focus, Cystic Fibrosis Found, 83-86. *Mem:* Am Fedn Clin Res; Am Thoracic Soc; Am Acad Allergy; Am Acad Pediat; Am Col Chest Physicians. *Res:* Lung transplantation; humoral response to chronic respiratory infection. *Mailing Add:* Children's Hosp 520 Sand Hill Rd Palo Alto CA 94304

LEWONTIN, RICHARD CHARLES, b New York, NY, Mar 29, 29; m 47; c 4. GENETICS, POPULATION BIOLOGY. *Educ:* Harvard Univ, AB, 51; Columbia Univ, MA, 52, PhD(zool), 54. *Prof Exp:* Reader biomet, Columbia Univ, 53-54; asst prof genetics, NC State Col, 54-58; from asst prof to prof biol, Univ Rochester, 58-64; prof biol, Univ Chicago, 64-73; PROF BIOL, HARVARD UNIV, 73- *Concurrent Pos:* NSF fel, 54-55, sr fel, 61-62 & 71-72; lectr, Columbia Univ, 59, sem assoc, 59-61; Fulbright fel, 61-62; co-ed, Am Naturalist, Am Soc Nat, 65. *Mem:* AAAS; fel Am Acad Arts & Sci; Genetics Soc Am; Soc Study Evolution (pres, 70). *Res:* Population genetics, ecology and evolution. *Mailing Add:* Mus Comp Zool Harvard Univ Cambridge MA 02138

LEWY, ALFRED JAMES, b Chicago, Ill, Oct 12, 45. PSYCHIATRY, NEUROSCIENCE. *Educ:* Univ Chicago, MD, 73, PhD(pharmacol), 77. *Prof Exp:* PROF PSYCHIAT & OPHTHAL, ORE HEALTH SCI UNIV, 86-, PROF PHARMACOL, 88- *Mem:* Sigma Xi; Am Psychiat Asn; Am Col Neuropsychopharmacol; Sleep Res Soc; Soc Biol Rhythm. *Res:* Chronobiology research in psychiatry; bright light treatment of sleep and mood disorders; clinical and basic pineal melatonin research. *Mailing Add:* Ore Health Sci Univ Portland OR 97201

LEX, R(OWLAND) G(ARBER), JR, b Philadelphia, Pa, Dec 29, 24; m 52; c 2. ELECTRICAL ENGINEERING. *Educ:* Univ Pa, BS, 49, MS, 50. *Prof Exp:* Elec engr, Leeds & Northrup Co, 49-55, group chief, 55-58, sect head, 58-62, mgr, Develop Div, 62-68, gen mgr, Digital Equip Div, 68-69, mgr, Eng Coord & Serv Dept, 69-72, gen mgr, Recorder & Test Instrument Div, 72-76, dir, Develop & Eng Dept, Instrument Group, 76-87, vpres res develop & eng, 87-90; RETIRED. *Mem:* Inst Elec & Electronics Engrs; fel Instrument Soc Am. *Res:* Application of digital techniques including computers to measurement and control of processes. *Mailing Add:* 1306 E Butler Pike Ambler PA 19002

LEY, ALLYN BRYSON, b Springfield, Mass, Dec 5, 18; m 43, 67; c 2. MEDICINE. *Educ:* Dartmouth Col, AB, 39; Columbia Univ, MD, 42; Am Bd Internal Med, dipl. *Prof Exp:* Asst med, Med Col, Cornell Univ, 47-49; instr, 51-52, asst dir, Sloan-Kettering Div, 54-55, cancer coord, 54-63, from asst prof to assoc prof, 54-63, PROF MED, MED COL, CORNELL UNIV, 63-, CLIN DIR UNIV HEALTH SERV, 71- *Concurrent Pos:* Asst, Boston City Hosp, 49-51; dir blood bank, Mem Hosp, 51-63, dir hemat labs, 55-63; asst attend physician, NY Hosp, 54-63, attend physician & dir ambulatory serv, 63-69; consult, Manhattan Vet Admin Hosp, 58-60, Hosp Spec Surg, 58-71 & Mem Sloan Kettering Cancer Ctr, 71-; assoc vis physician, Bellevue Hosp, 60-67; chief staff, SS Hope, 69-70; attend physician, Tompkins Community Hosp, 71- *Mem:* AAAS; Harvey Soc; Am Soc Hemat; Am Fedn Clin Res; Am Col Physicians. *Res:* Immunohematology; erythrocyte biochemistry; medical education and care. *Mailing Add:* Cornell Univ Health Serv Ten Central Ave Ithaca NY 14853

LEY, B JAMES, b New York, NY, May 26, 21; m 42; c 2. ELECTRICAL ENGINEERING. *Educ:* NY Univ, BEE, 42, MEE, 48. *Prof Exp:* Elec engr, Gen Elec Co, 42-44; from instr to prof elec eng, NY Univ, 46-73; prof elec eng, Manhattan Col, 73-; RETIRED. *Concurrent Pos:* Grant, Ford Found, 61; indust consult. *Mem:* Inst Elec & Electronics Engrs; Am Soc Eng Educ. *Res:* Active networks and analog and digital computers; computer aided design; alternate energy sources. *Mailing Add:* Manhattan Col Bronx NY 10471

LEY, HERBERT L, JR, EXPERIMENTAL BIOLOGY. *Prof Exp:* PRES CONSULT FIRM, HERBERT LEY FIRM, 70- *Mailing Add:* PO Box 2047 Rockville MD 20847-2047

LEYBOURNE, A(LLEN) E(DWARD), III, b Jacksonville, Fla, Aug 26, 34; m 54; c 2. CHEMICAL ENGINEERING. *Educ:* Univ Fla, BS, 56, PhD(chem eng), 61; Pa State Univ, MS, 58. *Prof Exp:* Res Assoc petrol res, Pa State Univ, 56-58, instr chem, Univ Fla, 58-59; res chemist, Atlantic Refining Co, 60; sr engr, Am Oil Co, 61-62; supvr textile develop, Textile Div, Monsanto Co, 63-71; plant mgr, Texfi Industs, Inc, 71-77, dir prod eng, 77-80; DIR ENG, INTERPINE LUMBER CO, 80- *Mem:* Am Chem Soc; Am Inst Chem Eng; Am Asn Textile Technol. *Res:* Polymer textiles. *Mailing Add:* Eng Tech SS Box 5137 Univ Southern Miss Southern Sta Box 5137 Hattiesburg MS 39406

LEYDA, JAMES PERKINS, b Youngstown, Ohio, Oct 2, 35; m 67; c 3. PHARMACY, PHARMACEUTICAL CHEMISTRY. *Educ:* Ohio Northern Univ, BS, 57; Ohio State Univ, MS, 59, PhD(pharm), 62. *Prof Exp:* Develop chemist, Lederle Labs Div, Am Cyanamid Co, 62-66, mgr prod develop, Int Med Res & Develop, 66-69; mgr new prod develop, Merrell Int Div, Richardson-Merrell, 69-76, dir, 76-81, head, Dept Pharm Res & Develop, Merrell Dow Res Inst, 81-84, com develop, US area, 84-89, ASSOC DIR DRUG REG AFFAIRS, MARION MERRELL DOW PHARM, 90- *Honors & Awards:* Lunsford Richardson Award, 60. *Mem:* AAAS; Am Pharmaceut Asn; NY Acad Sci; Am Asn Pharm Scientist; Sigma Xi; Am Soc Hosp Pharmacists; Am Soc Microbiol. *Res:* Drug delivery systems; antibiotics; cardiovascular agents and commercial liaison. *Mailing Add:* 10597 Tanager Hills Dr Cincinnati OH 45249

LEYDEN, DONALD E, b Gadsden, Ala, June 26, 38; m 61; c 2. ANALYTICAL CHEMISTRY. *Educ:* Kent State Univ, BS, 60; Emory Univ, MS, 61, PhD, 64. *Prof Exp:* Res assoc, Univ NC, 64-65; from asst prof to assoc prof chem, Univ Ga, 65-76; Phillipson prof chem, Univ Denver, 76-81; PROF CHEM, COLO STATE UNIV, 82- *Mem:* Am Chem Soc; Soc Appl Spectros; Mat Res Soc. *Res:* Chemically modified surfaces; ion-exchange; applications of nuclear magnetic resonance to the study of chemical systems of analytical importance. *Mailing Add:* Phillip Morris USA Res & Develop PO Box 26583 Richmond VA 23261-6583

LEYDEN, RICHARD NOEL, b Santa Monica, Calif, Nov 14, 48. ORGANOMETALLIC CHEMISTRY, POLYMER CHEMISTRY. *Educ:* Univ Calif, Los Angeles, BS, 71, PhD(chem), 75. *Prof Exp:* Res asst polymer res, Univ Witwatersrand, 75-76; res asst inorg chem, Calif Inst Technol, 76-77; TECH STAFF POLYMER RES, HUGHES AIRCRAFT CO, 77- *Mem:* Am Chem Soc; Sigma Xi. *Res:* Polymers, especially with semiconducting or electrical properties; organometallic polymers; organic metals; ultra high pressure chemistry. *Mailing Add:* 22024 Alta Dr Topanga CA 90290

LEYDORF, GLENN E(DWIN), b Perrysburg, Ohio, June 7, 14; m 43; c 2. ELECTRONICS. *Educ:* Univ Toledo, BE, 42; Univ Md, MS, 54. *Prof Exp:* Instr elec eng & physics, Univ Toledo, 42-44; from instr to assoc prof elec eng, 46-57, PROF ELEC ENG, US NAVAL ACAD, 57- *Mem:* Inst Elec & Electronics Engrs; Am Soc Eng Educ. *Res:* Electronic circuit applications to metastable atom studies. *Mailing Add:* Dept Elec Eng US Naval Acad Annapolis MD 21402

LEYENDECKER, PHILIP JORDON, phytopathology; deceased, see previous edition for last biography

LEYLAND, HARRY MOURS, organic chemistry, for more information see previous edition

LEYMASTER, GLEN RONALD, b Aurora, Nebr, Aug 7, 15; m 42; c 3. MEDICINE. *Educ:* Univ Nebr, AB, 38; Harvard Univ, MD, 42; Johns Hopkins Univ, MPH, 50; Am Bd Prev Med, dipl. *Prof Exp:* Intern, Boston City Hosp, 42-43, asst resident, 43, resident, 44; clin instr, Sch Med, Johns Hopkins Univ, 44-46; from instr to asst prof bact, Sch Hyg & Pub Health, 46-48; assoc prof pub health & prev med, Sch Med, Univ Utah, 48-50, prof prev med & head dept, 50-60; assoc secy, Coun Med Educ & Hosps, AMA, 60-63; pres, dean, prof prev med & assoc prof med, Med Col, Pa, 64-70; dir dept undergrad med educ, AMA, 70-75; exec dir, 75-81, exec vpres, Am Bd Med Specs, 81-82; RETIRED. *Concurrent Pos:* Clin asst, Harvard Med Sch, 42-44; asst prof & dir univ health serv, Univ Utah, 50-60; med educ adv, US Dept State, Int Coop Admin, Thailand, 57-58. *Mem:* AMA; Col Prev Med. *Res:* Clinical and epidemiological character of influenza and of data regarding encephalitis; experimental immunity and epidemiology of mumps; industrial toxicology; epidemiology of gastroenteritis; medical education. *Mailing Add:* 100 W New England St Worthington OH 43085

LEYON, ROBERT EDWARD, b Newton, Mass, July 28, 36; m 62; c 2. ANALYTICAL CHEMISTRY, ATOMIC SPECTROSCOPY. *Educ:* Williams Col, BA, 58; Princeton Univ, MA, 60, PhD(chem), 62. *Prof Exp:* Instr chem, Princeton Univ, 61-62; from instr to asst prof, Swarthmore Col, 62-69; asst prof, 69-72, chmn dept, 79-83, 88-90, ASSOC PROF CHEM, DICKINSON COL, 72- *Concurrent Pos:* Res assoc, Univ NC, 67-68 & Colo State Univ, 75-76, Univ Tex, Austin, 83-84. *Mem:* Am Chem Soc; Soc Appl Spectros. *Res:* Graphite furnace spectroscopy; trace metal analysis by atomic absorption. *Mailing Add:* Dept Chem Dickinson Col Carlisle PA 17013

LEYSE, CARL F(ERDINAND), b Kewaunee, Wis, Feb 11, 17; m 46; c 2. MECHANICAL ENGINEERING, PHYSICS. *Educ:* Univ Wis, BS, 48. *Prof Exp:* Phys sci aide rocket res, Naval Res Lab, 48; mech eng, Argonne Nat Lab, 48-51; chief eng res sect, Atomic Energy Div, Phillips Petrol Co, 51-56; tech dir, Internuclear Co, 56-59; asst mgr, Nuclear Dept, Res Div, Curtiss-Wright Corp, 59-60; mem staff, Atomic Energy Div, Gen Dynamics Corp, 60-61; pres, Internuclear Co, 61-63; mgr, Nuclear & Radiation Safety Dept, Reon Div, Aerojet Nuclear Systs Co, Sacramento, 63-67, mgr, Nuclear Safety, 67-72, staff scientist, Aerojet Nuclear Co, Idaho Falls, 72-76; mgr Tech Develop, EG&G Idaho Inc, 76-84; RETIRED. *Mem:* Am Nuclear Soc. *Res:* Nuclear reactor development and design. *Mailing Add:* 2860 Holly Pl Idaho Falls ID 83402-4632

LEYSIEFFER, FREDERICK WALTER, b Milwaukee, Wis, Jan 30, 33; m 64; c 3. MATHEMATICS, PROBABILITY. *Educ:* Univ Wis-Madison, BA, 55, MA, 56; Univ Mich, PhD(math), 64. *Prof Exp:* From asst prof to assoc prof, 64-82, assoc head dept, 69-76, chmn dept, 81-87, PROF STATIST, FLA STATE UNIV, 82- *Concurrent Pos:* Vis lectr, Sheffield Univ, Sheffield, Eng, 73-74, Leverhulme Commonwealth-Am fel, 73. *Mem:* Am Math Soc; Am Statist Asn; Math Asn Am; Inst Math Statist; AAAS; Sigma Xi. *Res:* Probability theory; stochastic processes; sampling theory; environmental statistics. *Mailing Add:* Dept Statist B-167 Fla State Univ Tallahassee FL 32306

LEYSON, JOSE FLORANTE JUSTININANE, b Philippines, Aug 17, 46. HUMAN SEXUALITY, URODYNAMICS. *Educ:* Cebu Inst Technol, BS, 65, MD, 70; Am Bd Urol, dipl. 79. *Prof Exp:* Fel entom, US Agency Int Develop, 70-71; fel sexuality, Johns Hopkins Hosp, 76; fel urodynamics, Yale Univ Hosp, 78; fel spinal cord, 76-77, CLIN CHIEF, SPINAL CORD INJURY UNIT, VET ADMIN HOSP, ORANGE, NJ, 77-; ASSOC PROF UROL, NJ MED SCH, 88- *Concurrent Pos:* Vis prof urodynamics, Yale Univ Hosp, 79; fel sex educ & parenthood, Am Univ, Washington, DC, 81; consult, Urodynamics, Bronx Vet Admin Hosp, NY, 79, Forum, Essence & MS Quarterly Mag, 78- & AMA Archives Internal Med; dir, Urodynamics & Sex Clin, Vet Admin Hosp, NJ, 79- & Sexual Dysfunction Ctr, Newark, NJ, 80-; prof Sexuality, Fairleigh Dickinson Univ, NJ, 80-85. *Mem:* Philippines Asn Sexologists Am (pres, 80-82); Philippine Med Asn Am; Am Urol Asn Inc. *Res:* Sexuality for both abled-bodied and disabled persons; drugs or electrical stimulations to produce erection in impotent patients; ways to produce urination in patients with paralysis and spinal defects due to injury or birth defects; laser erectiometer; erection pacemaker; author one book. *Mailing Add:* Spinal Cord Injury Unit Vet Admin Hosp Med Ctr East Orange NJ 07019

LEZNOFF, CLIFFORD CLARK, b Montreal, Que, May 30, 40; m 63; c 3. ORGANIC CHEMISTRY, PHTHALOCYANINES. *Educ:* McGill Univ, BSc, 61, PhD(org chem), 65. *Prof Exp:* Fel org chem, Northwestern Univ, 64-65; Nat Res Coun Can overseas fel, Cambridge Univ, 65-67; from asst prof to assoc prof, 67-79, PROF ORG CHEM, YORK UNIV, 80-, CHAIR CHEM, 90- *Concurrent Pos:* Vis prof, Weizmann Inst Sci, 73-74, Australian Nat Univ, 80-81 & Univ BC, 87-88. *Mem:* Am Chem Soc; Chem Inst Can. *Res:* Polymer supports in organic synthesis; synthesis of chiral compounds, pheromones, phthalocyanines and flourinated heterocyclic compounds; phthalocyanines in photodynamic therapy of cancer, multinuclear phthalocyanines. *Mailing Add:* Dept Chem York Univ Downsview ON M3J 1P3 Can

LHERMITTE, ROGER M, b Pontchartrain, France, May 28, 20; US citizen; m 45; c 2. PHYSICAL METEOROLOGY. *Educ:* Univ Paris, MS, 51, DrSci(meteor), 54. *Prof Exp:* Physicist meteorol res, Meteorol Nat, France, 46-60; Air Force Cambridge Res Lab, 60-63; head res br, Nat Oceanic & Atmospheric Admin, 63-70; PROF PHYS METEOROL, UNIV MIAMI, 70- *Concurrent Pos:* Mem, Active Microwave Workshop, NASA, 74-75; chmn, Nat Ctr Atmospheric Res Adv Panel, 74-; secretariat, Thunderstorm Res Int Prog; mem, Coun Atmospheric Elec, Am Meteorol Soc. *Honors & Awards:* Second Half Century Award, Am Meteorol Soc, 76. *Mem:* Fel Am Meteorol Soc; Int Union Geodesy & Geophys; Int Union Radio Sci. *Res:* Use of Doppler radars for the observation and study of atmospheric motion; the observation of three dimensional motion fields inside these systems by use of three Doppler radars operated simultaneously. *Mailing Add:* 4600 Rickenbacker Causeway Univ Miami Miami FL 33149

L'HEUREUX, JACQUES (JEAN), b Trois-Rivieres, Que, Dec 20, 39; div; c 3. COSMIC RAY PHYSICS, ASTROPHYSICS. *Educ:* Univ Montreal, BSc, 61; Univ Chicago, MSc, 62, PhD(physics), 66. *Prof Exp:* Res assoc physics, Univ Chicago, 66-69; asst prof physics, Univ Ariz, 69-77; SR RES ASSOC, UNIV CHICAGO, 78- *Mem:* Fel Am Phys Soc; Am Geophys Union. *Res:* Primary cosmic ray electrons; solar modulation of cosmic rays; primary heavy nuclei at high energies. *Mailing Add:* Enrico Fermi Inst Lab Astrophys & Space Res Univ Chicago 933 E 56th St Chicago IL 60637

L'HEUREUX, MAURICE VICTOR, b Lewiston, Maine, May 23, 14; m 46; c 5. BIOCHEMISTRY. *Educ:* Col of the Holy Cross, BS, 36, MS, 37; Yale Univ, PhD(biochem), 44. *Prof Exp:* Control chemist, Stokely Bros-Van Camp, Inc, Ind, 40-41; assoc, 46-49, from asst prof to assoc prof, 49-59, PROF BIOCHEM, STRITCH SCH MED, LOYOLA UNIV CHICAGO, 59-, EMER PROF, 79- *Mem:* Am Soc Biol Chem. *Res:* Lipid metabolism; modifying and regulatory effects of parathyroid hormone, calcitonin and vitamin D upon calcium metabolism. *Mailing Add:* 304 W Kenliworth Ave Villa Park IL 60181-2524

LHILA, RAMESH CHAND, b Rangoon, Burma, India. PRESSURE SENSITIVE ADHESIVES, POLYMER BLENDS. *Educ:* Calcutta Univ, BS, 72, BTech, 75; Univ Akron, PhD(polymer sci), 83. *Prof Exp:* Res chemist, Tuck Industs, New Rochelle, NY, 82, group leader res & develop, 82-83, asst dir res & develop, 83-87, tech dir gen prod, 87-88; DIR RES, TESA TUCK INC, ROCHELLE, NY, 88- *Concurrent Pos:* Hon instr, St Xavier's Col, Calcutta, India, 72. *Mem:* Am Chem Soc; Am Chem Soc Polymer Chem Div; Am Chem Soc Rubber Div; fel Plastics & Rubber Inst; Soc Plastics Engrs. *Res:* Pressure sensitive adhesive tapes based on rubber, acrylics and silicones by various methods including solution, hot melt and water-based coatings; coating technology; emulsion and solution polymerisation; polymer blends and rubber-modified high impact plastics. *Mailing Add:* One LeFevre Lane New Rochelle NY 10801

LHOTKA, JOHN FRANCIS, b Butte, Mont, Dec 13, 21; m 51. HISTOCHEMISTRY, MICROANATOMY. *Educ:* Univ Mont, BA, 42; Northwestern Univ, MS, 47, MB, 49, MD, 51, PhD(anat), 53. *Prof Exp:* Asst anat, Northwestern Univ, 46-49; mem clin staff, Minneapolis Gen Hosp, 50-51; asst prof microanat, 51-54, assoc prof, 54-69, prof anat, med sch, 69-87, EMER PROF ANAT SCI, UNIV OKLA, 87- *Concurrent Pos:* VPres, Introgene Found, 75-83, pres, 83-84. *Mem:* Am Asn Anatomists; Histochem Soc; Int Acad Path; Am Chem Soc; Am Soc Zoologists; Am Inst Chemists. *Res:* Polysaccharide histochemistry (theoretical and applied) in aging, in human embryo and fetus; argyrophilia (neurofibrillar and reticular); heavy metal histochemistry in toxicology; applied histochemistry in selected lower animals; aging in the mouse. *Mailing Add:* Anat Sci Univ Okla Health Sci Ctr PO Box 26901 Oklahoma City OK 73190

LI, C(HING) C(HUNG), b Changshu, Kiangsu, China, Mar 30, 32; m 61; c 2. PATTERN RECOGNITION & IMAGE PROCESSING, COMPUTER VISION. *Educ:* Nat Taiwan Univ, BS, 54; Northwestern Univ, MS, 56, PhD(elec eng), 61. *Prof Exp:* Asst prof , 59-60, 61-62, assoc prof, 62-67, PROF ELEC ENG, UNIV PITTSBURGH, 67-, PROF COMPUT SCI, 77- *Concurrent Pos:* Vis assoc prof, Univ Calif, Berkeley, 64, vis prin scientist, Alza Corp, Palo Alto, Calif, 70; prin investr NSF res grants, 75-81, 85-87, Commonwealth Pa, Dept Health, 77-79, Western Pa Advan Technol Ctr, 83-84, 86-88; chmn, Biocybernetics Tech Comt, Inst Elec & Electronics Engrs Systs, Man & Cybernetics Soc, 72-79; ed adv bd, J Cybernetics & Info Sci, 76-79; exec comt, Pattern Anal & Mach Intel Tech Comt, Inst Elec & Electronics Engrs, Comput Soc, 81-84; consult, Westinghouse Res Develop Ctr, 81; mem, Ctr Multivariate Anal, 82-87; fac res partic, Dept Energy, Pittsburgh Energy Technol Ctr, 82, 83, 85 & 88; biomed pattern recognition tech comt, Int Asn Pattern Recognition, 83-, chmn, 87-90; assoc ed, Pattern Recognition, 85-; Health Res & Serv Found, 85-86; mem, Ctr Parallel & Distrib Intelligent Systs, 86-; sabbatical leave, Lab Info & Decision Systs, Mass Inst Technol, 88. *Mem:* Sigma Xi; fel Inst Elec & Electronics Engrs; Pattern Recognition Soc; Biomed Eng Soc; NY Acad Sci; AAAS. *Res:* Integral pulse frequency modulated control systems; stability of nonlinear systems; adaptive and learning systems; modelling of physiological control systems; biomedical pattern recognition; image processing; computer vision. *Mailing Add:* Dept Elec Eng Univ Pittsburgh Pittsburgh PA 15261

LI, CHE-YU, b Honan, China, Nov 15, 34; m 61; c 3. MATERIALS SCIENCE & ENGINEERING. *Educ:* Nat Taiwan Univ, BSE, 54; Cornell Univ, PhD(chem eng), 60. *Prof Exp:* Res assoc, 60-62, from asst prof to assoc prof, 62-72, PROF MAT SCI & ENG, CORNELL UNIV, 72- *Concurrent Pos:* Mem staff, US Steel Res Ctr, 65-66; mem staff, Argonne Nat Lab, 69-71, consult, Nuclear Mat, Electronic Packaging. *Mem:* Am Phys Soc; Am Inst Mining, Metall & Petrol Engrs. *Res:* Mechanical behavior; radiation damage; surface and interface; development of a state variable description of mechanical properties of crystalline solids; micro-mechanical testing; nuclear materials, electronic packaging, high-temperature engineering alloys. *Mailing Add:* 241 Bard Hall-Mat Sci Cornell Univ Main Campus Ithaca NY 14853

LI, CHIA-CHUAN, b Taipei, Taiwan, Dec 29, 46; US citizen; m 84; c 2. ELECTRO-OPTICAL SYSTEMS, INFRARED DETECTORS. *Educ:* Nat Taiwan Univ, BS, 69; Rutgers Univ, MS, 74; Univ Mich, PhD(mat eng), 77; Pepperdine Univ, MBA, 86. *Prof Exp:* Sr engr, Gen Atomic Co, 77-84; syst engr, Electro-Optical & Data Systs, Hughes Aircraft Co, 84-85; PROG MGR, ELECTRO-OPTICAL CTR, ROCKWELL INT CORP, 85- *Concurrent Pos:* Lectr, San Diego City Col, 80-83. *Mem:* Metall Soc; Sigma Xi; Am Soc Metals. *Res:* Metallurgy, friction and wear; high temperature materials and coatings; infrared focal plane array materials; signal processing electronics. *Mailing Add:* 11 Meadow Wood Dr Trabuco Canyon CA 92679

LI, CHIA-YU, b Shanghai, China, May 5, 41; m 69; c 3. ELECTROANALYTICAL CHEMISTRY. *Educ:* Taiwan Normal Univ, BS, 63; Univ Louisville, MS, 67; Wayne State Univ, PhD(anal chem), 72. *Prof Exp:* Res fel electrochem, Univ Ariz, 72-73; from asst prof to assoc prof, 73-84, PROF CHEM, E CAROLINA UNIV, 84- *Mem:* Am Chem Soc; Sigma Xi. *Res:* Electrochemistry of organic and biological model compounds; computer-controlled electrochemical instrumentation; electrochemical detecting techniques for high performance liquid chromatography; electrochemical detection of trace elements. *Mailing Add:* Dept Chem E Carolina Univ Greenville NC 27858

LI, CHING CHUN, b Tientsin, China, Oct 27, 12; nat US; m 41; c 2. POPULATION GENETICS, BIOMETRICS. *Educ:* Nanking Univ, BS, 36; Cornell Univ, PhD(plant breeding), 40. *Prof Exp:* Plant breeder, Agr Exp Sta, Yenching Univ, 36-37; asst prof, Agr Col, Nat Kwangsi Univ, 42-43; prof genetics & biomet, Agr Col, Nanking Univ, 43-46; prof agron & head dept, Peking Univ, 46-50; from res fel to asst prof, Univ Pittsburgh, 51-58, from assoc prof to prof, 58-75, head dept, 69-75, univ prof human genetics, Grad Sch Pub Health, 75-82; RETIRED. *Mem:* Am Soc Human Genetics (pres, 60); AAAS; Int Statist Inst; Biomet Soc; fel Am Statist Asn; Academia Sinica. *Res:* Biometry; design of experiments; population genetics; path analysis. *Mailing Add:* Dept Human Genetics Univ Pittburgh Grad Sch Pub Health Pittsburgh PA 15261

LI, CHIN-HSIU, b Taiwan, China, Jan 3, 38; m 69; c 2. ENGINEERING MECHANICS, THERMAL TECHNOLOGY. *Educ:* Nat Cheng Kung Univ, BS, 61; Nat Cent Univ, MS, 64; Brigham Young Univ, MS, 66; Univ Mich, PhD(eng mech), 68. *Prof Exp:* Res assoc fluid mech, Case Western Reserve Univ, 69-71; vis assoc prof, Nat Cheng Kung Univ, 71-72; res assoc, Univ Mich, 72; assoc sr res engr, Gen Motors Res Labs, 72-76, sr res engr, eng mech, 76-81, staff res engr, Mech Res Dept, 81-85, sr staff res engr, Eng Mech Dept, 85-88, ENGINE RES DEPT, GEN MOTORS RES LABS, 88- *Honors & Awards:* McCuen Award, Gen Motors Labs, 84. *Mem:* Soc Automotive Engrs; Am Soc Mech Engrs. *Res:* Hydrodynamic stability; fluid flow and heat transfer; lubrication theory; rotor dynamics; mechanics of automotive components; computer-aided engine design, ceramic engines. *Mailing Add:* Engine Res Dept Gen Motors Res Labs Warren MI 48090

LI, CHI-TANG, b Ningtu, Kiangsi, China, Oct 16, 34; m 62; c 4. PHYSICAL CHEMISTRY, CRYSTALLOGRAPHY. *Educ:* Nat Taiwan Univ, BS, 55; Univ Louisville, MS, 59; Mont State Univ, PhD(chem), 64. *Prof Exp:* Scientist, 64-67, sr scientist, Adv Mat Res Sect, Owens-Ill Tech Ctr, 67-78; supvr fuel-cell mat res, Inst Gas Technol, 78-80, prin engr, 80-82; RES SPECIALIST, ADV CERAMICS PROG, DOW CORNING, 83- *Mem:* Am Chem Soc; Am Crystallog Asn; Am Ceramic Soc. *Res:* Crystal structure and chemistry; studies of glass ceramic materials, research on high temperature materials; fuel cell research; fiber and composite development. *Mailing Add:* Mat Res 1510 Timber Dr Midland MI 48640

LI, CHOU H(SIUNG), b Haining, China, June 8, 23; US citizen; m 53; c 2. PHYSICAL METALLURGY, STATISTICS. *Educ:* Chiao Tung Univ, BS, 44; Purdue Univ, MS, 49, PhD(phys metall), 51. *Prof Exp:* Metallurgist, Radio Corp Am, 51-59; sr scientist, Shockley Transistor Corp, 59-60; mgr semiconductors, Gen Instrument Corp, 60-62; sr res scientist, Grumman Aerospace Corp, 62-77; staff technologist, Singer Co, 78-79; dir, Res & Develop, Semi-Alloys, 80-81; PRES, LINTEL TECHNOL, INC, 81- *Concurrent Pos:* Ed, Chinese Inst Engrs J, 65-68; NASA Skylab consult specialist, 72-75; adj prof mat sci, State Univ NY, Stony Brook, 77-85. *Honors & Awards:* David Gessner Prize, Am Soc Eng Educ, 55 & 56; NASA New Technol Innovation Award, 77. *Mem:* Sr mem Inst Elec & Electronics Engrs; sr mem Am Soc Qual Control; Am Inst Mining, Metall & Petrol Engrs; Am Phys Soc; Chinese Inst Engrs. *Res:* Automation; ceramic bonding and coating; space manufacturing; semiconductors and thin films; solidification and related phenomena; powder packing, pressing and sintering; particle-surface interactions; friction; physics of failures; reliability; computer programming; artificial intelligence. *Mailing Add:* 379 Elm Dr Roslyn NY 11576

LI, CHUNG-HSIUNG, b Chia-Yi, Taiwan, Repub China, May 10, 39; US citizen; m 76; c 2. HEAT TRANSFER, FLUID MECHANICS. *Educ:* Tunghai Univ, BS, 64; Univ NH, MS, 69; Univ Ill, Chicago, PhD(eng), 76. *Prof Exp:* Specialist eng, Sargent & Lundy Engrs, 74-79; sr engr, C-E Air Preheater Co, 79-89; PROF, TAIWAN UNIV, 89- *Mem:* Nat Soc Prof Engrs; Am Inst Chem Engrs; Am Soc Mech Engrs; Soc Indust & Appl Math; Soc Eng Sci; Combustion Inst. *Res:* Development of new technologies; heat transfer and fluid mechanics. *Mailing Add:* 130 Highland Ave Wellsville NY 14895

LI, FREDERICK P, b China, May 7, 40; US citizen; m 72; c 1. CANCER ETHOLOGY. *Educ:* Univ NY, Rochester, BA, 60, MD, 65; Georgetown Univ, MA, 69. *Prof Exp:* EPIDEMIOLOGIST, NAT CANCER INST, 67-; ASSOC PROF MED, HARVARD MED SCH, 80- *Mem:* Am Soc Clin Oncol; Am Asn Cancer Res. *Res:* Identification of persons at high risk of cancer; genetic and environmental causes of cancer susceptability. *Mailing Add:* 44 Binney St Boston MA 02115

LI, GEORGE SU-HSIANG, b Chunking, China, Oct 24, 43; m 71; c 1. ORGANIC CHEMISTRY. *Educ:* Cheng Kung Univ, BS, 65; Purdue Univ, PhD(org chem), 71. *Prof Exp:* Res assoc, Med Chem Dept, Purdue Univ, 71-73; sr res chem, 74-77, res assoc, 77-86, RES SCIENTIST II, POLYMER RES, BP CO, OHIO, 87- *Mem:* Am Chem Soc. *Res:* Synthesis of latex based polymers; exploration of novel polymers with high heat resistance and barrier characteristics; polymer modification and alloying. *Mailing Add:* 4440 Warrensville Ctr Rd Cleveland OH 44128

LI, HONG, b China, Feb 25, 62. SURFACE PHYSICS, EPITAXIAL GROWTH. *Educ:* Zhejiang Univ, China, BS, 82; Univ Wis, MS, 84, PhD(physics), 88. *Prof Exp:* Res asst surface sci, Lab Surface Studies, Univ Wis, 84-88; RES ASSOC, DEPT MAT SCI & ENG, STATE UNIV NY-STONY BROOK, 88- *Mem:* Am Phys Soc. *Res:* Surface science; structure of surfaces, interfaces and epitaxial thin films; processing and characterization of epitaxially grown thin metal films on metals and semiconductors; superlattices and strained layer materials; development and application of surface analytical spectroscopies. *Mailing Add:* Dept Mat Sci & Eng State Univ NY Stony Brook NY 11794-2275

LI, HSIN LANG, b Shangtung, China, Sept 25, 30; US citizen; m 58. POLYMER SCIENCE. *Educ:* Univ Mich, Ann Arbor, BS, 54; Univ Ill, Urbana, MS, 55, PhD(mech eng), 59. *Prof Exp:* Res asst prof, Univ Ill, 59-60; sr res engr, E I Du Pont de Nemours & Co Inc, 60-64; SR RES ASSOC, ALLIED-SIGNAL, INC, 64- *Honors & Awards:* Delmonte Award, Soc Advan Mat & Process Eng, 89. *Mem:* Sigma Xi; Am Phys Soc; Fibers Soc. *Res:* Heat transfer; fluid dynamics; viscoelastic properties of polymers; polymer processing; fundamentals of polymer deformation; rheology of polymers; amorphous alloys; textile fibers; texturing of fibers; spectra armors. *Mailing Add:* Bldg CRL Allied-Signal Inc Morristown NJ 07960

LI, HSUEH MING, b Taiwan, Rep China, Oct 25, 39; US citizen; m 63; c 2. POLYMER CHEMISTRY. *Educ:* Tunghai Univ, Taiwan, BS, 62; Southern Methodist Univ, MS, 66; Polytech Inst Brooklyn, PhD(polymer chem), 71. *Prof Exp:* Fel x-ray diffraction, Polytech Inst Brooklyn, 70-72; res assoc polymer chem, Midland Macromolecular Inst, 72-73; res chemist polymer res, 73-79, sr res chemist, 79-85, RES ASSOC, ETHYL CORP, 85- *Mem:* Am Chem Soc. *Res:* Opacifying plastic pigment; polymeric flame retardants based on phosphazene-synthesis and evaluation; synthesis, characterization and mechanism of linear and cyclic phosphonitrillic chloride oligomers; advanced composites, specialty glasses, high-tech ceramics. *Mailing Add:* Ethyl Corp 8000 GSRI Ave 14799 Baton Rouge LA 70820-7497

LI, HUNG CHIANG, b Kinhwa, China, Dec 10, 21; US citizen; m 57; c 2. STATISTICS, ANALYTICAL MATHEMATICS. *Educ:* Univ Chekiang, BS, 46; Mich State Univ, MS, 64; Purdue Univ, PhD(statist), 69. *Prof Exp:* Asst math, Taiwan Normal Univ, 47-50; instr, Nat Taiwan Univ, 50-52; asst prof, Taiwan Inst Technol, 52-55; assoc prof, Tunghai Univ, 55-62; PROF STATIST, UNIV SOUTHERN COLO, 69- *Concurrent Pos:* Consult, Univ Southern Colo, 69- *Mem:* Inst Math Statist; Am Math Soc; Math Asn Am; Sigma Xi. *Res:* Multivariate analysis, particularly interested in normal distributions and the asymptotic expansions for distributions of characteristic roots of normal populations. *Mailing Add:* 87 Massarri Rd Pueblo CO 81001

LI, J(AMES) C(HEN) M(IN), b Nanking, China, Apr 12, 25; m 50; c 3. MATERIALS SCIENCE, MECHANICAL ENGINEERING. *Educ:* Nat Cent Univ, China, BS, 47; Univ Wash, MS, 51, PhD, 53. *Prof Exp:* Res chemist, Sch Med, Univ Wash, 51-53; res chemist & fel, Univ Calif, 53-55; supvr res proj, Mfg Chemists Asn, Carnegie Inst Technol, 55-56; phys chemist, Res Labs, Westinghouse Elec Corp, 56-57; scientist, Fundamental Res Lab, US Steel Corp, 57-60, sr scientist, 60-64, staff scientist, 64-69; mgr strength physics dept, Mat Res Ctr, Allied Chem Corp, 69-71; ALBERT ARENDT HOPEMAN PROF ENG, UNIV ROCHESTER, 71- *Concurrent Pos:* Vis prof, Columbia Univ, 64-65, adj prof, 65-71; consult, Mat Res Ctr, Allied Chem Corp, 71-79, NSF, 75-77 & US Steel Corp, 76; vis prof, Ruhr Universität Bochum, Germany, 78-79; vis scientist, Naval Res Lab, 84-85, 88; Alexander von Humboldt sr award, 78-79. *Honors & Awards:* Mathewson Gold Medal, Metall Soc, 72; Robert Mehl Gold Medal, Am Inst Mining, Metall & Petrol Engrs & Inst Metals lectr, 78; Lu Tse-Hon Medal, Chinese Soc Mat Sci, 88; Acta Metallurgica Gold Medal, Am Soc Metals Int, 90. *Mem:* Fel Am Phys Soc; fel Am Inst Mining, Metall & Petrol Engrs; fel Am Soc Metals; Mat Res Soc; Am Soc Mech Engrs. *Res:* Dislocations and defects; plastic deformation; amorphous and polymeric materials; microstructural interactions; equilibrium and non-equilibrium phenomena. *Mailing Add:* Dept Mech Eng Univ Rochester Rochester NY 14627

LI, JAMES C C, MEDICINE. *Educ:* Nat Taiwan Univ, BS, 61, MS, 63; Boston Univ Sch Med, PhD(biochem), 71. *Prof Exp:* Asst res fel, Academia Sinica, Taiwan, China, 63-67; res fel microbiol & molecular genetics, Harvard Med Sch, 70-72; res assoc exp biol, Worcester Found, 72-74; res assoc, Dept Microbiol & Molecular Genetics, Harvard Med Sch, 74-76, prin assoc, 76-79; clin chemists, Dept Path, New Eng Deaconess Hosp, 80-82; ASSOC DIR, MED RES LAB, HEBREW REHAB CTR AGED, 82-; PRIN ASSOC, DEPT MED, HARVARD MED SCH, 84- *Res:* Biochemistry and cell biology in medical research; all types cell culture including cultivation of cells from primary and secondary cultures; experienced in cell culture techniques such as growing cells in monolayer; roller bottle or spinner bottle; semisolid agar colonizing; cell volume determination. *Mailing Add:* 26 Aidon Ct Brookline MA 02146

LI, JANE CHIAO, b Shanghai, China, May 1, 39; US citizen; m 63; c 2. APPLIED STATISTICS. *Educ:* Hunter Col, BS, 63; Rutgers Univ, MS, 65, PhD(statist), 71. *Prof Exp:* Res chemist, Endo Labs, Long Island, NY, 61-63; statist consult comput sci & statist, Rutgers Univ, 74-78; sr statistician, Paramins Div, Exxon Chem Co, 78-79, sect head world-wide product testing, 79-81; mgr, 81-86, SR MGR, DATA ANALYSIS & PROCESSING, UOP, INC, 86- *Concurrent Pos:* Course dir, Ctr Prof Achievement, 86-; co-chair, Am Inst Chem Engrs Symposium 87 & 88. *Mem:* Sigma Xi; Am Statist Asn; Inst Math Statist; Am Soc Qual Control. *Res:* Design and analysis of mixture experiments with process variables; statistical design of experiments and data and uses in research and development manufacturing; statistical process control for continuous and batch chemical processes; database management for petroleum and chemical processes. *Mailing Add:* UOP Inc 25 E Algonquin Rd Des Plaines IL 60017-5017

LI, JEANNE B, b New York, NY, Apr 15, 44. BIOCHEMISTRY, PHYSIOLOGY. *Educ:* Vassar Col, AB, 66; Harvard Univ, PhD(biochem), 71. *Prof Exp:* Res fel physiol, Harvard Med Sch, 71-73; asst prof physiol, 73-79, RES ASSOC PEDIAT, HERSHEY MED CTR, PA STATE UNIV, 79-, SR APPL CHEMIST, WATERS DIV. *Concurrent Pos:* Nat Cancer Inst & Muscular Dystrophy Asn fels, Harvard Med Sch, 71-73; Am Diabetes Asn grant, Hershey Med Ctr, Pa State Univ, 74-76. *Mem:* Am Physiol Soc; AAAS. *Res:* Regulation of protein synthesis and degradation in mammalian tissues. *Mailing Add:* Sr Application Chemist Waters Div Millipore 34 Maple St Milford MA 01757

LI, JOHN KONG-JIANN, b Taiwan, China, Aug 28, 50; US citizen; m 74; c 2. BIOMEDICAL INSTRUMENTATION. *Educ:* Univ Manchester, BSc, 72; Univ Pa, MSEng, 74, PhD(bioeng), 78. *Prof Exp:* Instr physics, Cent Found High Sch, London, 72; res fel bioeng, Univ Pa, 73-77, thesis supvr, 78-79; ASST PROF ELEC ENG, RUTGERS UNIV, 79-, ADJ ASST PROF SURG & BIOENG, UNIV MED & DENT NJ, RUTGERS MED SCH, 81- *Concurrent Pos:* Biomed engr cardiol, Presby Univ Pa Med Ctr, 77-79; prin investr, Rutgers Univ, NSF, 80-; vchmn, Eng Med & Biol, Princeton Sect, Inst Elec & Electronics Engrs, 80-; vis scientist, Fedn Am Soc Exp Biol, 81- *Mem:* Inst Elec & Electronics Engrs; Am Physiol Soc; NY Acad Sci; AAAS; Sigma Xi. *Res:* Cardiovascular dynamics; biomedical instrumentation; diagnostic cardiology; comparative physiology; physiological controls. *Mailing Add:* Dept Elec Eng Rutgers Univ Piscataway NJ 08854

LI, KAM W(U), b China, Feb 16, 34; m 56; c 2. MECHANICAL ENGINEERING. *Educ:* Chu Hai Col, Hong Kong, BSME, 57; Colo State Univ, MSME, 61; Okla State Univ, PhD, 65. *Prof Exp:* Asst prof, Tex A&I Univ, 65-67; from asst prof to assoc prof mech eng, 67-73, PROF MECH ENG, N DAK STATE UNIV, 73-, ASSOC DEAN ENG COL, 89- *Concurrent Pos:* Consult engr, Scott Eng Sci Corp, 67-68; NSF Instnl Fund grants, NDak State Univ, 67-68 & 71; Dept Defense grant, 69; Am Soc Eng Educ-Ford Found resident fel, Northern States Power Co, Minneapolis, Minn, 71-72; eng consult, Chas T Main Inc, Boston, 73-80; USDA res grants, 74-78 & 76-78; consult, Ctr Prof Advancement, NJ, 82-83. *Mem:* Am Soc Mech Engrs; NY Acad Sci. *Res:* Heat transfer; fluid dynamics; thermodynamics; power generation; thermal system design; applied mathematics; energy models; new power generation systems. *Mailing Add:* Dept of Mech Eng NDak State Univ Fargo ND 58102

LI, KE WEN, chemical engineering, for more information see previous edition

LI, KELVIN K, b Kwantung, China, Mar 25, 34; US citizen; m 65; c 4. PHYSICS. *Educ:* McGill Univ, BEng, 58; Mass Inst Technol, PhD(physics), 64. *Prof Exp:* PHYSICIST, BROOKHAVEN NAT LAB, 65- *Mem:* Am Phys Soc. *Res:* Elementary particle interactions. *Mailing Add:* Dept of Physics Brookhaven Nat Lab Upton NY 11973

LI, KUANG-PANG, b Kwang-tung, China, Oct 11, 38; c 3. ANALYTICAL CHEMISTRY. *Educ:* Nat Taiwan Univ, BS, 61; Univ Ill, MS, 68, PhD(anal chem), 70. *Prof Exp:* Lectr chem, Kaohsiung Prov Inst Technol, Taiwan, 64-65; res assoc, Ariz State Univ, 70-72; res assoc, Univ Ill, 72-73; asst prof chem, Univ Fla, 73-80; ASSOC PROF, DEPT CHEM, UNIV LOWELL, 80- *Mem:* Am Chem Soc; NY Acad Sci; AAAS. *Res:* Metallic ion transport in biomembranes; membrane interactions of carcinogenic polynuclear aromatics; theoretical and practical developments of chromatographic methods; excitation mechanism in inductively coupled plasma (ICP). *Mailing Add:* Dept Chem Univ Lowell Lowell MA 01854-1996

LI, KUN, b Kunming, China, Nov 20, 23; m 51; c 2. CHEMICAL ENGINEERING. *Educ:* Nat Southwest Assoc Univ, China, BEng, 45; Carnegie Inst Technol, MS, 49, DSc, 52. *Prof Exp:* Res chemist, Petrol Res Lab, Carnegie Inst Technol, 52-55, supvr & sr res chemist, 55-56; sr res engr, Jones & Laughlin Steel Corp, 56-58, res assoc, 58-62; assoc prof chem eng, 62-64, PROF CHEM ENG, CARNEGIE-MELLON UNIV, 64- *Concurrent Pos:* Consult, Jones & Laughlin Steel Corp. *Mem:* Am Chem Soc; Am Inst Chem Engrs; Am Inst Mining, Metall & Petrol Engrs. *Res:* Fluid flow; kinetics of high-temperature processes. *Mailing Add:* 112 Alpine Circle Pittsburgh PA 15215-1902

LI, LI-HSIENG, b Peking, China, Dec 31, 33; m 60; c 2. BIOCHEMISTRY. *Educ:* Nat Taiwan Univ, BS, 55; Va Polytech Inst, MS, 62, PhD(biochem), 64. *Prof Exp:* Res assoc, Ind Univ, 64-65; sr res scientist, 65-73, SR SCIENTIST, UPJOHN CO, 88- *Mem:* Am Asn Cancer Res; Am Asn Biol Chemists. *Res:* mechanism of action of anticancer agent; immmunity and cancer; cell biology; experimental therapeutic and pharmacology. *Mailing Add:* Upjohn Co Cancer Res Kalamazoo MI 49001

LI, LING-FONG, b Fukien, China, Apr 17, 44. THEORETICAL HIGH ENERGY PHYSICS. *Educ:* Nat Taiwan Univ, BS, 65; Univ Pa, MS, 67, PhD(physics), 70. *Prof Exp:* Res assoc physics, Rockefeller Univ, 70-72; res assoc, Stanford Linear Accelerator Ctr, Stanford Univ, 72-74; from asst prof to assoc prof, 74-83, PROF PHYSICS, CARNEGIE-MELLON UNIV, 83- *Mem:* Fel Am Phys Soc; Sigma Xi. *Res:* Unified theories of weak and electromagnetic interactions in relation to the fundamental structure of the elementary particles. *Mailing Add:* Dept Physics Carnegie-Mellon Univ Pittsburgh PA 15213

LI, LU KU, b Honan, China, Apr 26, 36; m 61; c 1. BIOCHEMISTRY, PROTEIN CHEMISTRY. *Educ:* Nat Taiwan Univ, BS, 58; Princeton Univ, PhD(biol), 64. *Prof Exp:* Res asst biol, Princeton Univ, 63-64; res assoc chem, Cornell Univ, 64-66; instr ophthal, 66-68, assoc, 68-69, asst prof, 69-74, RESEARCHER OPHTHAL, COLUMBIA UNIV, 74- *Mem:* Sigma Xi; Am Chem Soc; Am Soc Biol Chemists. *Res:* Maturation of lens fiber cells and its relation to cataractogenesis; the subunits interactions of lens proteins; vision and opthalmology. *Mailing Add:* Five Hemlock Lane Glen Cove NY 11542-1432

LI, MING CHIANG, b Ningpo, China, June 18, 35; US citizen; m 65; c 2. PHYSICS, MATHEMATICS. *Educ:* Peking Univ, BS, 58; Univ Md, PhD(physics, math), 65. *Prof Exp:* Lectr physics, Norm Col Inner Mongolia, China, 58-61; res asst, Univ Md, 64-65; fel & mem sci, Inst Advan Study, 65-67; asst prof, 68-72, ASSOC PROF PHYSICS, VA POLYTECH INST & STATE UNIV, 72- *Concurrent Pos:* Sr tech staff, Mitre Coop, 82-83; sr physicist, 83-88, group leader, Naval Res Lab, 88- *Mem:* Am Phys Soc; Inst Elec & Electronics Engrs. *Res:* Atomic and molecular physics; interferometry; laser optics; electronics counter measures radar development. *Mailing Add:* 11415 Bayard Dr Mitchellville MD 20716

LI, NORMAN CHUNG, b Foochow, China, Jan 13, 13; nat US; m 37; c 5. PHYSICAL CHEMISTRY. *Educ:* Kenyon Col, BS, 33; Univ Mich, MS, 34; Univ Wis, PhD(chem), 36. *Prof Exp:* Prof chem, Anhwei Univ, China, 36-38; lectr, Yenching Univ, 38-40; from assoc prof to prof, Cath Univ, China, 40-46; from asst prof to assoc prof, St Louis Univ, 46-52; prof, 52-82, distinguished serv prof, 78-82, EMER PROF, DUQUESNE UNIV , 82-; RES PROF CHEM, CATHOLIC UNIV AM, 82- *Concurrent Pos:* Consult, Argonne Nat Lab, 56-58; vis scientist, NIH, 62; tech asst expert, Int Atomic Energy Agency, 64; vis prof, Tsing Hua Univ, China, 64; adv chem res ctr, Nat Taiwan Univ, 66-; consult, Nat Res Coun of Repub China, 74-75; adv, Inst Chem, Nat Tsing Hua Univ, Repub China, 74-; sabbatical leave, res chemist, Naval Res Lab, 81-82; spec chair prof, Nat Tsing Hua Univ, Repub China, 84-87. *Mem:* Fel AAAS; Am Chem Soc; Sigma Xi. *Res:* Nuclear magnetic resonance studies of hydrogen and metal binding; hydrogen bonding in coal and asphaltenes; research on metalloproteins; stability of liquid fuels. *Mailing Add:* Dept of Chem Duquesne Univ Pittsburgh PA 15219

LI, NORMAN N(IAN-TZE), b Shanghai, China, Jan 14, 33; US citizen; m 63; c 2. MATERIALS SCIENCE & POLYMER ENGINEERING. *Educ:* Taiwan Nat Univ, BS, 54; Wayne State Univ, MS, 57; Stevens Inst Technol, ScD(chem eng), 63. *Prof Exp:* Chem engr, Shinlin Paper & Pulp Co, 53 & Parke-Davis & Co, 56; instr chem, Newark Col Eng, 61-63; res engr, Exxon Res & Eng Co, 63-66, sr res engr, 66-70, res assoc, 70-77, sr res assoc, 77-81, head, Separation Sci Group, 76-81; dir separations res, UOP Inc, 81-84; dir, separation sci & technol, 84-88, dir, chem & process technol, 88-90, DIR, ENG MAT & PROCESS TECHNOL, ALLIED SIGNAL RES CTR, 90- *Concurrent Pos:* Vis lectr, Newark Col Eng, 63-67; consult, Bell Aerosysts Co, 67; chmn, Gordon Res Conf Separations & Purification, 73; chmn, Gordon Res Conf Transport Phenomena in Membranes, 75; chmn, Eng Found Int Conf Separations, 84 & 87; chmn, Int Cong Membranes & Membrane Processes, 90. *Honors & Awards:* Am Chem Soc Award in Separation Sci & Technol, 88. *Mem:* Nat Acad Engrs; fel Am Inst Chem Engrs; Am Chem Soc; NY Acad Sci; Am Inst Chem. *Res:* Mass transfer; surface chemistry; interfacial phenomena; transport through membranes; separation techniques; catalysis; material engineering. *Mailing Add:* Allied-Signal Res Ctr PO Box 5016 Des Plaines IL 60017-5016

LI, PEI-CHING, b Kiangsu, China, Nov 2, 19; m 45; c 3. CHEMICAL ENGINEERING. *Educ:* Nat Southwest Assoc Univ, China, BE, 45; Univ Rochester, MS, 55, PhD(chem eng), 59. *Prof Exp:* Asst chem eng, Nat Southwest Assoc Univ, China, 44-46; teacher sci, Chungking Women's Norm Sch, 46-47; chemist & engr, Taiwan Sugar Corp, 47-53; asst chem eng, Univ Rochester, 54-57, res assoc glass, 58-59; mem res staff, Raytheon Co, 59-64; res scientist, Am Standard Corp, 64-65; res scientist ceramics div, IIT Res Inst, 65-68; ADV ENGR, IBM CORP, 68- *Mem:* Am Ceramic Soc; Am Chem Soc; Sigma Xi. *Res:* Physical properties of molten glass, particularly enamel glass and binary system of borates; solid state reactions of ferrites; pyrolytic high temperature materials; chemical vapor deposition and plasma enhanced chemical vapor deposition dielectric films. *Mailing Add:* 12466 Beechgrove Ct Moorpark CA 93021-3108

LI, PEN H (PAUL), b China, May 4, 33; US citizen; m 63; c 2. HORTICULTURE, PLANT PHYSIOLOGY. *Educ:* Ore State Univ, PhD(hort & plant physiol), 63. *Prof Exp:* PROF HORT & PLANT PHYSIOL, UNIV MINN, 63- *Concurrent Pos:* Vis prof, Int Potato Ctr, 73, Inst Low Temperature Sci, Hookkaido Univ, 76, Peking Agr Univ, 80 & Inst Plant Physiol, USSR Acad Sci, 87. *Honors & Awards:* Dow Chem Co Award, Am Soc Hort Sci, 65, Alex Laurie Award, 66. *Mem:* Fel Am Soc Hort Sci; Am Soc Plant Physiologists; Potato Asn Am; Soc Cryobiol; Am Soc Agron. *Res:* Plant hardiness and stress physiology. *Mailing Add:* Dept of Hort Sci Univ of Minn St Paul MN 55108

LI, PETER WAI-KWONG, b Hong Kong, Apr 18, 52; m 82; c 3. GEOMETRIC ANALYSIS, PARTIAL DIFFERENTIAL EQUATION. *Educ:* Calif State Univ, BA, 74; Univ Calif, Berkeley, MA, 77, PhD(math), 79. *Prof Exp:* Res mem, Inst Advan Study, 79-80; from asst prof to assoc prof math, Stanford Univ, 80-83; assoc prof math, Purdue Univ, 83-85; prof math, Univ Utah, 85-89 & Univ Ariz, 89-91; PROF MATH, UNIV CALIF, IRVINE, 91- *Concurrent Pos:* Prin investr, NSF, 80-; vis asst prof, Univ Calif, San Diego, 80 & 81; Sloan fel, Alfred P Sloan Found, 82; res mem, Math Sci Res Inst, 83; Guggenheim fel, John Simon Guggenheim Found, 89; ed, Rocky Mountain J Math, 89-91 & Proc Am Math Soc, 91- *Mem:* Am Math Soc. *Res:* Interplay between the geometry, topology and the analysis of geometrical objects. *Mailing Add:* Dept Math Univ Calif Irvine CA 92717

LI, SEUNG P(ING), b Hong Kong, Jan 15, 32; m 57; c 3. PHYSICS, ENGINEERING. *Educ:* Univ Hong Kong, BSc, 54; Princeton Univ, MSE, 60; Univ Colo, PhD(physics), 66. *Prof Exp:* Demonstr gen eng, Univ Hong Kong, 54-57; engr, Scott & Wilson Consult Engrs, 57-59; asst lectr physics, Chung Chi Col, Hong Kong, 60-63, lectr & chmn dept, 65-68; from asst prof to assoc prof, 68-75, PROF ELEC & ELECTRONICS ENG, CALIF STATE POLYTECH UNIV, POMONA, 75- *Mem:* Inst Elec & Electronics Engrs; Sigma Xi. *Res:* Semiconductor physics and devices; electromagnetic theory. *Mailing Add:* 1791 Via Palomares San Dimas CA 91773

LI, SHENG-SAN, b Hsin-Chu, Taiwan, Dec 10, 38; m 64; c 3. ELECTRICAL ENGINEERING. *Educ:* Cheng Kung Univ, Taiwan, BS, 62; Rice Univ, MS, 66, PhD(elec eng), 68. *Prof Exp:* Engr, China Elec Mfg Co, Taiwan, 63-64; teaching asst elec eng, Rice Univ, 64-67; from asst prof to assoc prof, 68-78, PROF ELEC ENG, UNIV FLA, 78- *Concurrent Pos:* Electronic engr, Nat Bur Standards, 75-76; consult, Battelle Columbus Labs, Ohio, 75-77 & Harris Semiconductors Inc, Fla, 78-, Hughes Res Labs, 85- *Mem:* Am Phys Soc; sr mem Inst Elec & Electronics Engrs; Am Soc Testing & Measurements; Electrochem Soc. *Res:* Semiconductor device physics; transport phenomena in semiconductors; photoelectric effects in semiconductors and devices; defect and recombination properties in semiconductor materials and devices; solar cells and photodetectors; high-speed devices. *Mailing Add:* 231 Benton Hall Univ Fla Gainesville FL 32611

LI, SHIN-HWA, b Taipei, Taiwan, Apr 8, 58; m 83. CHEMICAL VAPOR DEPOSITION, III-V SEMICONDUCTORS. *Educ:* Nat Cent Univ, Taiwan, BS, 80; Univ SWLa, Lafayette, 85; Univ Utah, Salt Lake City, PhD(mat sci eng), 90. *Prof Exp:* Liaison officer, Repub China Marine Corps, 80-82; qual control engr, Ko-sheng Enterprises, Ltd, Taiwan, 82-83; res asst, Univ SWLa, Lafayette, 83-85; res asst, 85-90, RES ASSOC, UNIV UTAH, SALT LAKE CITY, 90- *Mem:* Inst Elec & Electronic Engrs; Minerals, Metals & Mat Soc. *Res:* Organometallic vapor-phase epitaxy reaction mechanisms; decomposition of the precursors for epitaxy of many III-V materials; mass spectrometric methods used to analyze the reaction mechanisms. *Mailing Add:* Univ Utah 304 Emro Salt Lake City UT 84112

LI, SHU-TUNG, connective tissue biochemistry, for more information see previous edition

LI, STEVEN SHOEI-LUNG, b Taiwan, China, Oct 20, 38; m 67; c 2. GENETICS, BIOCHEMISTRY. *Educ:* Nat Taiwan Univ, BS, 61, MS, 63; Univ Mo, PhD(genetics), 68. *Prof Exp:* Res assoc, Univ Tex, Austin, 68-70; res assoc, Stanford Univ, 70-74; assoc prof, Mt Sinai Sch Med, 74-77; RES GENETICIST, NAT INST ENVIRON HEALTH SCI, NIH, 77- *Concurrent Pos:* Adj prof biochem & chem, Univ NC, Chapel Hill, 87- *Mem:* AAAS; Am Soc Biochem & Molecular Biol; Genetics Soc Am. *Res:* Biochemical genetics; structure, regulation and evolution of eukayotic genes and proteins. *Mailing Add:* Nat Inst Environ Health Sci NIH Research Triangle Park NC 27709

LI, SU-CHEN, b Taipei, Taiwan, June 8, 35; US citizen; m 62; c 2. BIOCHEMISTRY. *Educ:* Nat Taiwan Univ, BS, 58; Univ Okla, PhD(biochem), 65. *Prof Exp:* From asst prof to assoc prof, 72-80, PROF BIOCHEM, SCH MED, TULANE UNIV, 80- *Concurrent Pos:* Career Develop Award, NIH, 75-80. *Mem:* Am Soc Biol Chemists; AAAS; Soc Complex Carbohydrates. *Res:* Biochemical studies of glycoconjugates and glycosidases. *Mailing Add:* Dept Biochem Med Sch Tulane Univ 1430 Tulane Univ New Orleans LA 70112

LI, TAO PING, b Szechwan, China, Nov 16, 20; m 48; c 3. ORGANIC CHEMISTRY. *Educ:* Nat Szechwan Univ, China, BS, 41; Univ Tex, Austin, MA, 59, PhD(org chem), 60. *Prof Exp:* Fel, Univ Tex, Austin, 60-61; sr proj chemist, Am Oil Co, Ind, 61-64; res specialist, 64-66, from group leader to sr group leader, 66-82, SCI FELLOW, MONSANTO CO,82- *Mem:* AAAS; Am Chem Soc; Catalysis Soc; NY Acad Sci. *Res:* Chemical kinetics, heterogeneous catalysis and chemistry of metal organic compounds. *Mailing Add:* 295 Heather Crest Dr Chesterfield MO 63017-2855

LI, THOMAS M, EXPERIMENTAL BIOLOGY. *Educ:* Univ Ill, BS, 72, PhD(biophys chem), 76. *Prof Exp:* Res fel, Inst Cancer Res, Philadelphia, 76-78; res & develop mgr, Miles Labs, Inc, 78-83; res & develop dir, 3M Diag, 83-84; develop mgr, Syntex Med Diag, 84-86; develop mgr, 86-88, sr develop mgr, 88-90, VISTA PROG CHEM MGR, SYVA CO, 90- *Mem:* Sigma Xi; NY Acad Sci; Am Chem Soc; Biophys Soc; Protein Soc; Am Asn Clin Chem; Am Soc Biochem & Molecular Biol; fel Am Inst Chemists; fel Nat Acad Clin Biochem. *Res:* Experimental biology; numerous publications. *Mailing Add:* Develop Dept Syva Co 900 Arastradero Rd Palo Alto CA 94304

LI, TIEN-YIEN, b Hunan, China, June 28, 45; m 71; c 1. MATHEMATICS. *Educ:* Nat Tsing-Hua Univ, BS, 68; Univ Md, PhD(math), 74. *Prof Exp:* Instr math, Univ Utah, 74-76; from asst prof to assoc prof, 76-82, PROF MATH, MICH STATE UNIV, 82- *Concurrent Pos:* Vis assoc prof math, Res Ctr, Univ Wis, 78-79; vis prof, Res Inst Math Sci, Kyoto Univ, Japan, 87-88. *Mem:* Am Math Soc; Soc Indust & Appl Math. *Res:* Differential equations, dynamical systems and numerical analysis. *Mailing Add:* Dept Math Mich State Univ East Lansing MI 48824

LI, TING KAI, b Nanking, China, Nov 13, 34; US citizen; m 60; c 2. MEDICINE, BIOCHEMISTRY. *Educ:* Northwestern Univ, AB, 55; Harvard Med Sch, MD, 59; Mass Inst Technol, 60-61. *Prof Exp:* House officer, Peter Bent Brigham Hosp, 59-60, asst med, 60-63, jr assoc, 63-65; instr med, Harvard Med Sch, 65-67, assoc, 67-69; dep dir div biochem, Walter Reed Army Inst Res, 69-71; prof, 71-80, John B Hickam prof, 80-85, DISTINGUISHED PROF MED & BIOCHEM, SCH MED, IND UNIV, INDIANAPOLIS, 85- *Concurrent Pos:* Helen Hay Whitney Found fel, 60-64; Med Found Boston fel, 64-68; Markle scholar acad med, 67-73; asst med, Harvard Med Ach, 60-63, res assoc biochem, 63-65; chief med resident, Peter Bent Brigham Hosp, 65-66; guest scientist, Nobel Med Inst, Sweden, 68. *Honors & Awards:* Res Excellence Award, Res Soc on Alcoholism; Jellinck Award; James B Isaacson Award, Res Substance Abuse. *Mem:* Am Chem Soc; Am Soc Clin Invest; Endocrine Soc; Am Soc Biol Chem; Am Inst Nutrit; Asn Am Physicians. *Res:* Enzymology; metabolism; chemical basis of biological specificity; alcohol metabolism. *Mailing Add:* Dept Med Emerson Hall 421 Ind Univ Sch Med 545 Barnhill Dr Indianapolis IN 46223

LI, TINGYE, b Nanking, China, July 7, 31; m 56; c 2. ELECTRICAL ENGINEERING, PHYSICS. *Educ:* Univ Witwatersrand, BSc, 53; Northwestern Univ, MS, 55, PhD(elec eng), 58. *Prof Exp:* Mem tech staff, AT&T Bell Labs, 57-67, head, Repeater Tech Res Dept, 67-76, head, Lightwave Media Res Dept, 76-84, HEAD, LIGHTWAVE SYSTS RES DEPT, AT&T BELL LABS, 84- *Honors & Awards:* W R G Baker Prize, Inst Elec & Electronics Engrs, 75, David Sarnoff Award, 79. *Mem:* Nat Acad Eng; fel Optical Soc Am; fel AAAS; Chinese Inst Engrs-USA; fel Inst Elec & Electronics Engrs; Sigma Xi. *Res:* Optical communications; lasers and coherent-wave optics; electromagnetic field theory; antennas and propagation; microwave theory and techniques. *Mailing Add:* Crawford Hill Lab AT&T Bell Labs Holmdel NJ 07733

LI, WEN-CH'ING WINNIE, b Taiwan, Dec 25, 48; c 2. COMBINATURICS & FINITE MATHEMATICS, NUMBER THEORY. *Educ:* Nat Taiwan Univ, BS, 70; Univ Calif, Berkeley, PhD(math), 74. *Prof Exp:* Asst prof math, Harvard Univ, 74-78 & Univ Ill, Chicago, 78-79; assoc prof, 79-84, PROF MATH, PA STATE UNIV, 84- *Concurrent Pos:* Mem, Inst Advan Study,

Princeton, 78 & 84; Alfred Sloan fel, 81-83; vis prof, Univ Paris, Orsay, 85-86. *Mem:* Am Math Soc. *Res:* Automorphic forms; representation theory; number theory; combinatorics. *Mailing Add:* Dept Math University Park PA 16802

LI, WEN-HSIUNG, b Ping-Tung, Taiwan, Sept 22, 42; US citizen; m 75; c 3. EVOLUTIONARY GENETICS, MOLECULAR EVOLUTION. *Educ:* Chung-Yuang Col Sci & Eng, Taiwan, BE, 65; Nat Cent Univ, Taiwan, MS, 68; Brown Univ, PhD(appl math), 72. *Prof Exp:* Proj assoc, Univ Wis-Madison, 72-73; from asst prof to assoc prof, 73-84, PROF POP GENETICS, UNIV TEX, HOUSTON, 84- *Concurrent Pos:* Assoc ed, Genetics, 86- *Mem:* Genetics Soc Am; AAAS; Am Soc Human Genetics; Soc Study Evolution. *Res:* Molecular evolution; biomathematics; human population genetics; evolution of DNA sequences, duplicate genes and pseudogenes, cytochrome C genes and pseudogenes and lactate dehydrogenase genes and pseudogenes; structure and evolution of apolipoprotein genes; mathematical theory of population genetics. *Mailing Add:* Dept Pop Genetics Univ Tex PO Box 20334 Houston TX 77225

LI, WU-SHYONG, b Taipei, Taiwan, Aug 20, 43; m 75; c 2. PHYSICAL ORGANIC CHEMISTRY. *Educ:* Nat Taiwan Univ, BS, 66; Kent State Univ, MS, 69; Univ Minn, PhD(org chem), 73. *Prof Exp:* Fel, Ohio State Univ, 73-75; sr chemist plastic, Rohm & Haas Co, 75-78; sr chemist, 78-80, res specialist, 80-88, SR RES SPECIALIST, 3M CO, 89- *Mem:* Am Chem Soc. *Res:* Reaction mechanism, kinetics, polymers and UV curing. *Mailing Add:* 3M Co 3M Ctr Bldg 209-BS-01 Maplewood MN 55144

LI, YAO TZU, b Beijing, China, Feb 1, 14. SCIENCE EDUCATION. *Educ:* Mass Inst Technol, ScD, 39. *Prof Exp:* CHMN & TREAS, SETRA SYSTS INC, 68- *Concurrent Pos:* Prof control & guidance, Mass Inst Technol, 70-79; dir, Innovation Ctr, 72-79. *Mem:* Fel Nat Acad Eng. *Mailing Add:* Setra Systs Inc 45 Nagog Park Acton MA 01720

LI, YING SING, b Kwangtung, China, July 26, 38; US citizen; m 68; c 4. STRUCTURAL CHEMISTRY, MOLECULAR SPECTROSCOPY. *Educ:* Cheng Kung Univ, BS, 60; Univ Kansas, PhD(chem), 68. *Prof Exp:* Res asst, Taiwan Sugar Exp Sta, 61-63; res assoc, Princeton Univ, 68-70 & Univ SC, 70-75 & 76-82; assoc prof, Benedict Col, 78-82; ASSOC PROF RES & TEACHING, MEMPHIS STATE UNIV, 82- *Concurrent Pos:* Oak Ridge Nat Lab, 90-91. *Mem:* Am Chem Soc; Sigma Xi. *Res:* Microwave infrared and Raman spectra conformations; structures of cyclic, fluoro, and organometalic compounds; chemical bonding; intermolecular interactions; matrix isolations infrared; surface-enhanced Raman scattering. *Mailing Add:* Dept Chem Memphis State Univ Memphis TN 38152

LI, YUAN, b Ningpo, China, Sept 15, 36. SOLID STATE PHYSICS, HIGH ENERGY PHYSICS. *Educ:* Nat Taiwan Univ, BS, 58; Ind Univ, PhD(physics), 66. *Prof Exp:* Res assoc physics, Rutgers Univ, 65-68, asst prof, 65-77; assoc prof, Tuskegee Inst, 75-77; ASSOC PROF PHYSICS, RUTGERS UNIV, 77- *Concurrent Pos:* NSF res grant, 76-77. *Mem:* Am Phys Soc; Am Asn Physics Teachers. *Res:* Crystallography and lattice dynamics. *Mailing Add:* Dept of Physics NCAS Rutgers Univ Newark NJ 07102

LI, YU-TEH, b Hsin-Chu City, Formosa, Apr 1, 34; m 62; c 2. BIOCHEMISTRY. *Educ:* Nat Taiwan Univ, BS, 57, MS, 60; Univ Okla, PhD(biochem), 63. *Prof Exp:* From instr to asst prof biochem, Sch Med, Univ Okla, 63-66; chief, Delta Regional Primate Res Ctr, 66-85, PROF BIOCHEM, SCH MED, TULANE UNIV, 74- *Concurrent Pos:* Fel biochem, Sch Med, Univ Okla, 63-64; Nat Cancer Inst grant, 64-66; NSF grant, 68-; NIH grant, 71-; USPHS res career develop award, 71-76; Javits Neurosci Investr Award, 84-91, 91-98. *Mem:* Am Soc Neurochem; Am Soc Biol Chem. *Res:* Biochemical studies on glycoconjugates and various glycosidases. *Mailing Add:* Dept Biochem Sch Med Tulane Univ New Orleans LA 70112

LI, ZI-CAI, b Sichnam, China, Mar 14, 39; Can citizen; m; c 2. NUMERICAL METHODS, IMAGE TRANSFORMATION. *Educ:* Qing-Hua Univ, China, BA, 63; Univ Toronto, PhD(appl math), 86. *Prof Exp:* Assoc prof, Shanghai Inst Comput Technol, China, 72-80; ASSOC RES PROF COMPUTER SCI, CONCORDIA UNIV & CTR INFO, MONTREAL, 87- *Mem:* Soc Indust & Appl Math; Inst Elec & Electronic Engrs. *Res:* Numerical methods for partial differential equations, in particular the combined methods and applied them into pattern recognitions and image processing; over 70 publications and two monographs. *Mailing Add:* 1550 de Maisonneuve Blvd W Montreal PQ H3G 1N2 Can

LIAN, ERIC CHUN-YET, b Tainan Hsien, Taiwan, Nov 11, 38; m 73; c 2. HEMATOLOGY. *Educ:* Nat Taiwan Univ, MD, 64. *Prof Exp:* Res fel hemat, Univ Miami, 69-71; res assoc hemostasis, Harvard Med Sch, 71-73; asst prof med, Sch Med, 73-76, assoc prof med, Sch Med, 78-87, COMPREHENSIVE HEMOPHILIA CTR, UNIV MIAMI, 76-, DIR HEMOSTASIS LAB, 76-, PROF MED, SCH MED, 87- *Concurrent Pos:* Prin investr biochem, immunol & physiol of antihemophilic factor, 76- & pathogenesis of thrombotic thrombocytopenic purpura, 78- *Mem:* Fel Am Col Physicians; AAAS; Am Fedn Clin Res; Am Soc Hemat; Am Heart Asn. *Res:* Thrombosis and hemostasis; hematology. *Mailing Add:* Vet Admin Hosp 1201 NW 16th St Miami FL 33125

LIAN, SHAWN, b Taipei, Taiwan, Mar 2, 58. MATERIALS SCIENCE ENGINEERING. *Educ:* Nat Taiwan Univ, BS, 82. *Prof Exp:* Res assoc, Chung Sun Acad Res Ctr, 84-87; RES ASSOC, UNIV TEX AUSTIN, 87- *Mem:* Am Phys Soc; Electrochem Soc. *Res:* Photo-enhanced chemical vapor deposition to investigate low temperature si processing which is the main trend for future ultralarge-scale integration and heterostructure devices. *Mailing Add:* 2501 Lake Austin Blvd No A104 Austin TX 78703

LIANG, CHANG-SENG, b Fukien, China, Jan 6, 41; US citizen; m 68; c 3. CARDIOLOGY. *Educ:* Nat Taiwan Univ, MD, 65; Boston Univ, PhD(pharmacol), 71. *Prof Exp:* Instr med, Boston Univ Sch Med, 73-74, from asst prof to assoc prof med & pharmacol, 73-82; assoc prof med, 82-86, PROF MED, UNIV ROCHESTER MED CTR, 86- *Concurrent Pos:* Prin investr, NIH res grants, 77-, study sect reviewer, NIH, 81-85. *Mem:* Am Physiol Soc; Am Soc Pharmacol & Exp Therapeut; Am Heart Asn; Am Soc Clin Invest; Am Fedn Clin Res. *Res:* Circulatory control and neurohumoral regulation of the cardiovascular system in heart failure; receptor pharmacology; changes of membrane signal transduction; hemodynamic measurements. *Mailing Add:* Cardiol Unit 601 Elmwood Ave Box 679 Rochester NY 14642

LIANG, CHARLES C, b Nanking, China, June 9, 34; m 61; c 3. PHYSICAL CHEMISTRY, ANALYTICAL CHEMISTRY. *Educ:* Nat Taiwan Univ, BS, 56; Baylor Univ, PhD(phys chem), 62. *Prof Exp:* Res chemist, Houdry Process & Chem Co, 62-63; from asst prof to assoc prof phys chem, WVa Inst Technol, 63-65; sr staff mem electrochem, P R Mallory Co, Inc, 65-73, assoc tech dir batteries, Lab Phys Sci, 73-77; dir, VP Technol, Wilson Greatbatch, LTD, 77-79; pres, Electrochem Industs, Inc, 79-82; PRES, OMNION ENTERPRISE, INC, 82- *Honors & Awards:* IR 100 Award, Indust Res Mag, 71. *Mem:* Am Chem Soc; Electrochem Soc. *Res:* Mechanisms and kinetics of electrode processes; chemical thermodynamics; analytical techniques; solid state chemistry. *Mailing Add:* 9460 Greiner Rd Clarence NY 14031

LIANG, CHARLES SHIH-TUNG, b Peking, China, Dec 10, 40; US citizen; m 65; c 2. AERONAUTICAL & ASTRONAUTICAL ENGINEERING. *Educ:* Univ Ill, BS, 62, PhD(elec eng), 68; Harvard Univ, SM, 63. *Prof Exp:* Asst elec eng, Univ Ill, 63-68; ENG STAFF SPECIALIST, FT WORTH DIV, GEN DYNAMICS CORP, 69- *Mem:* AAAS; Inst Elec & Electronics Engrs; Int Union Radio Sci. *Res:* Electromagnetic scattering research as related to radar signature analysis and target identification; radar antenna design and development; advanced technology aircraft design and development; radar cross-section measurement techniques. *Mailing Add:* General Dynamics Corp Ft Worth Div Grant Lane PO Box 748 Ft Worth TX 76101

LIANG, CHING YU, physics; deceased, see previous edition for last biography

LIANG, EDISON PARK-TAK, b Canton, China, July 22, 47; nat US; m 71; c 3. ASTROPHYSICS, PLASMA RADIATION. *Educ:* Univ Calif, Berkeley, BA, 67, PhD(physics), 71. *Prof Exp:* Res assoc, Univ Tex, Austin, 71-73; res assoc & assoc instr astrophys & relativity, Univ Utah, 73-75; asst prof astrophys, Mich State Univ, 75-76; asst prof physics, Stanford Univ, 76-79; group leader, physics dept, 83-90, assoc div leader, 88-90, PHYSICIST, LAWRENCE LIVERMORE NAT LAB, 80-; PROF SPACE PHYSICS & ASTRON, RICE UNIV, 91- *Concurrent Pos:* Lectr & vis scholar, Ctr for Space Sci & Astrophys, Stanford Univ, 80-90. *Mem:* Sigma Xi; Am Astronom Soc; fel Am Phys Soc. *Res:* Plasma radiation; astrophysics of compact objects (x-ray and gamma ray sources); galaxy formation; laser-plasma interactions; relativity and cosmology; laser cooling. *Mailing Add:* Dept Space Physics & Astron Rice Univ PO Box 1892 Houston TX 77251

LIANG, GEORGE H, b Beiging, China, Oct 1, 34; m 63; c 1. PLANT GENETICS, PLANT BREEDING. *Educ:* Taiwan Prov Col Agr, BS, 56; Univ Wyo, MS, 61; Univ Wis, PhD(agron), 65. *Hon Degrees:* Prof, Jiangsn Agri Col, 84; Henan Agri Univ, 84; Chinese Acad Scis, China, 87. *Prof Exp:* Agronomist, Taiwan Prov Res Inst Agr, 58-59; from asst prof to assoc prof, 64-76, PROF PLANT GENETICS & CYTOGENETICIST, KANS STATE UNIV, 77-, CHMN DEPT GENETICS, 82 - *Concurrent Pos:* Sabbatica, Univ Calif-Davis, 81; consult, World Bank, 83. *Mem:* Am Soc Agron; Crop Sci Soc Am; Am Genetic Asn; Sigma Xi; Genetics Soc Can. *Res:* Quantitative genetics in plant species; cytogenetics and breeding aspects in cultivated crops; somatic cell genetics. *Mailing Add:* Dept Agron Throckmorton Hall Kans State Univ Manhattan KS 66506

LIANG, ISABELLA Y S, b Kwellin, China; US citizen. CARDIAC VASCULAR DISEASES. *Educ:* Hong Kong Chinese Univ, BS, 66; Hong Kong Univ, PhD(physiol), 79. *Prof Exp:* HEALTH SCI ADMINR, NIH, 90- *Mem:* Am Heart Asn; Sigma Xi; Am Physiol Soc; Am Soc Molecular Biol & Med. *Mailing Add:* Cardiac Dis Br NIH Fed Bldg Rm 3C06 Bethesda MD 20892

LIANG, JOSEPH JEN-YIN, b China; US citizen; m 65; c 2. MATHEMATICS. *Educ:* Nat Taiwan Univ, BA, 58; Univ Detroit, MA, 62; Ohio State Univ, PhD(math), 69. *Prof Exp:* Res fel math, Calif Inst Technol, 69-70; from asst prof to assoc prof, 70-77, PROF MATH, UNIV S FLA, 78- *Concurrent Pos:* Vis asst prof, Ohio State Univ, 72; vis assoc, Calif Inst Technol, 75 & 77; vis res prof, Nat Tsing Hua Univ, Taipei, Taiwan, Repub China, 83. *Mem:* Am Math Soc; Math Asn Am. *Res:* Number theory; coding theory; algorithms. *Mailing Add:* Dept of Math Univ of S Fla Tampa FL 33620

LIANG, KAI, b Hunan, China, Mar 23, 34; m 64. PHYSICAL CHEMISTRY. *Educ:* Nat Taiwan Univ, BS, 56; Univ Utah, PhD(phys chem), 64. *Prof Exp:* Res asst phys chem, Univ Utah, 60-64; sr res chemist, 64-70, MEM SR STAFF, COLOR PHOTOG DIV, RES LABS, EASTMAN KODAK CO, 70- *Mem:* Am Chem Soc. *Res:* Diffusion and transport processes; mathematical modelling of diffusion kinetics; color photography. *Mailing Add:* Imaging Mechanism R L 7/820 Eastman Kodak Co Rochester NY 14650-2102

LIANG, KENG-SAN, b Tainan, Taiwan, Dec 17, 43; m 69; c 2. MATERIALS SCIENCE, SOLID STATE PHYSICS. *Educ:* Nat Taiwan Univ, BS, 66; Stanford Univ, MS, 70, PhD(appl physics), 73. *Prof Exp:* From assoc scientist to scientist mat sci, Xerox Corp, 73-78; STAFF PHYSICIST MAT SCI, CORP RES LABS, EXXON RES & ENG CO, 78- *Mem:* Am Phys Soc; Am Vacuum Soc. *Res:* Amorphous solids, x-ray diffraction, x-ray photoelectron spectroscopy, electronic structure, thin films. *Mailing Add:* 226 Windmill Ct Bridgewater NJ 08807

LIANG, SHOU CHU, b Foochow, China, May 14, 20; m 50; c 2. CHEMISTRY. *Educ:* Cent Univ, China, BS, 42; Princeton Univ, MA, 46, PhD(chem), 47. *Prof Exp:* Instr, Teacher High Sch, China, 42; anal chemist, China Match Raw Mat Mfg Co, 42-44; asst, Princeton Univ, 45-47, Int Nickel Co fel, 47-48; res chemist, Merck & Co, NJ, 48-49; fel, Nat Res Coun Can, 49-51, asst res officer II, 51-53; group leader, Res Lab, Dom Tar & Chem Co, 53-56; res engr, Consol Mining & Smelting Co Can, Ltd, 56-64, head gen metall res, 64-70; gen mgr, 70-85, CONSULT, COMINCO ELECTRONIC MAT INC, 85- *Mem:* Am Soc Testing & Mat; Am Chem Soc; Electrochem Soc; NY Acad Sci; Chem Inst Can; AAAS. *Res:* Flotation of ores; preperation of inorganic reagents; chemical method of analysis; surface catalysis; heterogeneity of catalyst surfaces for chemisorption; fast drying paint; metallurgy; semiconductors. *Mailing Add:* S 4206 Helena St Spokane WA 99203

LIANG, SHOUDAN, b Fuzhou, China, Feb 10, 61. CORRELATED CONDENSED MATTER SYSTEMS, COMPUTATIONAL PHYSICS. *Educ:* Peking Univ, BS, 82; Univ Chicago, PhD(physics), 86. *Prof Exp:* Res assoc physics, Princeton Univ, 86-88 & Univ Ill, Urbana-Champaign, 88-90; ASST PROF PHYSICS, PA STATE UNIV, 90- *Mem:* Am Phys Soc. *Res:* Theoretical condensed matter physics include quantum antiferromagnets, new algorithms for overcoming slow dynamics in spin glasses and simulated annealing, random growth and computational physics. *Mailing Add:* 104 Davey Lab Pa State Univ University Park PA 16802

LIANG, SHOUDENG, b Yantai, China, Mar 9, 59. VACUUM TECHNOLOGY & APPLICATIONS, SURFACE CHARACTERIZATIONS & PROPERTY ANALYSIS. *Educ:* Qufu Normal Univ, China, BS, 81; Univ Ill-Chicago, MS, 87; Univ Wis-Madison, MS, 90. *Prof Exp:* Researcher catalysis & phys chem, Dalian Inst Chem Physics, China, 82-85; res asst surface chem, Chem Dept, Univ Ill-Chicago, 86-88; res asst surface sci, Chem Dept, 89, proj asst X-ray lithography, 89-90, RES ASST X-RAY LITHOGRAPHY, CTR X-RAY LITHOGRAPHY, UNIV WIS, 90- *Mem:* Am Vacuum Soc; Mat Res Soc. *Res:* Surface chemistry of thin films and interfaces of metals/semiconductors and gas/metals; application of synchrotron radiation to materials research and scanning x-ray spectromicroscopy; x-ray lithography from synchrotron radiation sources. *Mailing Add:* Chem Dept Univ Wis 1101 University Ave Madison WI 53706

LIANG, SHU-MEI, b Taiwan, Feb 4, 49; US citizen; m; c 1. CYTOKINE RESEARCH, STRUCTURE-FUNCTION STUDIES. *Educ:* Nat Taiwan Univ, BS, 71; Univ Ark, PhD(biochem), 78. *Prof Exp:* Vis fel res, Div Bact Prod, BOB, Food & Drug Admin, 77-80, staff fel, Div Biochem & Biophysics, 80-83, res chemist res & regulatory work, Div Virol, CBER, 86-88; scientist res & develop, Dept Protein Chem, Biogen, SAm, 83- 85; res chemist, Div Virol, 86-88, DIV CYTOKINE BIOL, CBER, FOOD & DRUG ADMIN, 88- *Mem:* Am Soc Biochem & Molecular Biol; Protein Soc; Chinese Biochem Soc. *Res:* Structure-function relationships of cytokines especially interleukin-2 and the regulation of immune response by thiol compounds; purification and characterization of proteins especially membrane proteins and recombinant DNA derived proteins. *Mailing Add:* Div Cytokine Biol Bldg 29A Rm 3C22 Food & Drug Admin 8800 Rockville Pike Bethesda MD 20817

LIANG, TEHMING, b Taiwan, Apr 14, 45; m; c 1. BIOCHEMISTRY. *Educ:* Nat Taiwan Univ, BS, 68; Univ Chicago, PhD(biochem), 73; Univ Miami, MD, 87. *Prof Exp:* Res assoc biochem, Ben May Lab Cancer Res, Univ Chicago, 73-76, res asst prof, 76-77; sr res biochem, Merck Inst Therapeut Res, 77-81, res fel, 81-87; resident internal med, Robert Wood Johnson Med Sch, 87-88, RESIDENT DERMAT, UNIV CHICAGO, 89- *Mem:* Am Soc Biol Chemists; Endocrine Soc; Soc Neurosci; Am Med Asn. *Res:* Molecular mechanism of hormone action. *Mailing Add:* Dept Dermatol Univ Chicago 5841 S Maryland Ave PO Box 409 Chicago IL 60637

LIANG, TUNG, b Peking, China, June 7, 32; m 58; c 2. OPERATIONS RESEARCH, AGRICULTURAL ENGINEERING. *Educ:* Nat Taiwan Univ, BS, 56; Mich State Univ, MS, 63; NC State Univ, PhD(biol eng), 67. *Prof Exp:* Assoc prof, 68-76, PROF AGR ENG, UNIV HAWAII, 76- *Mem:* Am Soc Agr Engrs. *Res:* Agricultural system modeling and optimization; development of natural resource information system. *Mailing Add:* Dept Agr Eng Univ Hawaii 3050 Maile Way Honolulu HI 96822

LIANG, WEI CHUAN, b Shanghai, China, Nov 23, 36; m 70; c 2. ORGANIC CHEMISTRY. *Educ:* Kalamazoo Col, BA, 61; Case Western Reserve Univ, MS, 66; Ohio State Univ, PhD(org chem), 72. *Prof Exp:* Res chemist, Lubrizol Corp, 61-67; fel org chem res, Ga Inst Technol, 72-74; res chemist, Union Carbide Corp, 74-77, group leader process develop, 77-87, res scientist, 82-87; PRIN SCIENTIST, RHÔNE POULENC INC, 87- *Mem:* Am Chem Soc. *Res:* Exploratory syntheses; pesticide process research; process research and development. *Mailing Add:* Rhone Poulenc Inc CN 7500 Cranbury NJ 08512

LIANG, YOLA YUEH-O, b Taiwan, Feb 12, 47; m 75; c 2. ANALYTICAL CHEMISTRY. *Educ:* Nat Taiwan Normal Univ, BS, 70; Univ Kans, PhD(chem), 78. *Prof Exp:* RES LEADER ANAL CHEM, DOW CHEM CO, 81- *Mem:* Am Chem Soc; Asn Women Sci. *Res:* Analytical chemistry; organic electrochemistry; neurochemistry; enzyme kinetics; analytical separations using gas chromatography and liquid chromatography. *Mailing Add:* 386 Mt Sequoia Pl Clayton CA 94517

LIANIDES, SYLVIA PANAGOS, b Lynn, Mass, Sept 2, 31; div; c 3. PHYSIOLOGY, BIOCHEMISTRY. *Educ:* Tufts Univ, BS, 53, PhD(physiol), 59. *Prof Exp:* Res biologist, US Naval Radiol Defense Lab, 59-60; lectr biol, Col Notre Dame, Calif, 62-71; instr biol sci, De Anza Col, 71-73; instr ecol, Chabot Col, 73-75; chmn, Biol Dept, 81-83, PROF ANAT & PHYSIOL, WEST VALLEY COL, SARATOGA, CALIF, 75- *Mem:* AAAS; Sigma Xi; Nat Asn Biol Teachers; Nat Sci Teachers Asn. *Res:* Hormonal and environmental influences upon mitochondrial oxidative phosphorylation; effects of environmental cold on lipid and carbohydrate metabolism; radiation physiology; anatomy. *Mailing Add:* PO Box 2334 Saratoga CA 95070

LIAO, HSUEH-LIANG, b Silo, Taiwan, Jan 24, 41; US citizen; m 52; c 2. ANALYTICAL CHEMISTRY. *Educ:* Cheng Kong Univ, BS, 65; Drexel Univ, MS, 69; Georgetown Univ, PhD(chem), 72. *Prof Exp:* Sr res assoc chem, Northeastern Univ, 72-74; sr res chemist, Norwich Pharmacal Co, Morton-Norwich Prod Inc, 74-76; sr res scientist, Bristol Myers Co, 76-77; GROUP LEADER & SCIENTIST, LEDERLE LABS, AM CYANAMID CO, 77- *Mem:* Am Chem Soc; Am Inst Chem Eng. *Res:* Analytical methods development; chromatographic methods of separation (HPLC, GC & TLC) and quantitation; analytical and physical chemistry of drug compounds; solution thermodynamics. *Mailing Add:* 83 Ridge Rd New City NY 10956-2856

LIAO, MARTHA, b Leeds, Eng, Feb 9, 48; US citizen; m 91. SOMATIC CELL GENETICS, RECOMBINANT DNA. *Educ:* Bryn Mawr Col, BA, 70; Univ Pa, PhD(chem), 74. *Prof Exp:* Fel chem, Univ Denver, 74-75; fel, 75-79, inst fel, 79-86, SR FEL GENETICS, HEALTH SCI CTR, ELEANOR ROOSEVELT INST CANCER RES, UNIV COLO, 86-; ASSOC PROF, DEPT PEDIAT, UNIV COLO HEALTH SCI CTR, 86- *Concurrent Pos:* Am res scholar, Comt Scholarly Commun with People's Republic China, Nat Acad Sci, 81-82; NIH fel, 76-79. *Mem:* Am Cell Biol; Am Soc Human Genetics; AAAS. *Res:* Human gene mapping using chinese hamster/human cell hybrids; using recombinant DNA techniques to obtain DNA markers from specific human chromosome. *Mailing Add:* Eleanor Roosevelt Inst Cancer Res 1899 Gaylord St Denver CO 80206

LIAO, MEI-JUNE, INTERFERON SCIENCE. *Educ:* Nat Tsing-Hua Univ, BS, 73; Yale Univ, MPh, 77, PhD(phys biochem), 80. *Prof Exp:* Assoc biochem & biophys, Mass Inst Technol, 80-83; sr scientist, Bioresponse Modifier Group, Interferon Sci, Inc, 83-84, head, Cellular Immunol Group, 84-85, dir cell biol, 85-86, DIR, RES & DEVELOP, INTERFERON SCI, INC, 87- *Mem:* Am Soc Biochem & Molecular Biol; Int Soc Interferon Res; Soc Chinese Bioscientists Am. *Res:* Purification and characterization of natural and recombinant human interferon proteins; natural interferon induction systems; mechanism of interferon action; interferon receptor purification and characterization; production of monoclonal and polyclonal antibodies; development of immunochemical assay techniques; state of the art protein chemistry techniques and cytokine assays and cDNA cloning and expression of human cytokines; preparation of PLA and response to the Food and Drug Administration for the approval of natural and recombinant interferon product. *Mailing Add:* Interferon Sci Inc 783 Jersey Ave New Brunswick NJ 08901

LIAO, PAUL FOO-HUNG, b Philadelphia, Pa, Nov 10, 44; m 68; c 2. ENGINEERING PHYSICS. *Educ:* Mass Inst Technol, BS, 66; Columbia Univ, PhD(physics), 73. *Prof Exp:* Res assoc, Radiation Lab, Columbia Univ, 72-73; mem tech staff physics, Bell Labs, 73-80, head Quantum Electronics Res Dept, 80-83; div mgr, physics & optical sci res, Bellcore, 84-87, div mgr, Photonic Sci & Tech Res, 87-89, asst vpres, Solid State Sci & Tech, 89-90, ASST VPRES, NETWORK SYSTS RES, BELLCORE, 90- *Concurrent Pos:* Chmn, Joint Coun Quantuim Electronics, 66; ed Acad Press, Quantum Electronics; ed, J Optical Soc Am, 88. *Mem:* Fel Am Phys Soc; fel Optical Soc Am; fel Inst Elec & Electronics Engrs. *Res:* Communications systems; nonlinear optics. *Mailing Add:* Bellcore Red Bank NJ 07701

LIAO, PING-HUANG, b Taipei, Taiwan, Jan 16, 46; Can citizen; m 71; c 3. SANITARY & ENVIRONMENTAL ENGINEERING, OTHER ENVIRONMENTAL EARTH & MARINE SCIENCES. *Educ:* Tunghai Univ, BSc, 69; Univ Neb, MS, 73 & PhD(chem), 76. *Prof Exp:* Fel, 75-78, RES SCIENTIST, UNIV BC, 78- *Concurrent Pos:* Instr, Columbia Col, 81. *Res:* Fermentation biotechnology, biomass utilization and bioenergy production from renewable resources; waste management. *Mailing Add:* Dept Bio-Resource Eng Univ BC 2357 Main Hall Vancouver BC V6T 1W5 Can

LIAO, SHU-CHUNG, b Tainan, Taiwan, Oct 18, 39; m 71. PHYSICAL-ANALYTICAL CHEMISTRY. *Educ:* Nat Taiwan Univ, BSc, 63; Univ Western Ont, PhD(phys chem), 70. *Prof Exp:* Res assoc polymer chem, Univ Mich, Ann Arbor, 70-71, human nutrit prog, Sch Pub Health, 72-73; SR RES ASSOC, CLIMAX MOLYBDENUM CO, MICH, 73- *Mem:* Am Chem Soc; Am Asn Textile Chemists & Colorists (secy, 80-). *Res:* Physical chemistry of molecular complexes and macromolecules; food chemistry and nutrition; instrumental analysis; analytical chemistry. *Mailing Add:* Battelle 505 King Ave Rm 7325B Columbus OH 43201-2693

LIAO, SHUEN-KUEI, b Morioka, Japan, June 27, 40; Can citizen; m 72; c 2. CANCER, IMMUNOLOGY. *Educ:* Tunghai Univ, Taiwan, BSc, 64; McMaster Univ, PhD(immunol), 71. *Prof Exp:* Demonstr, fel histol, Univ Toronto, 70-73; prof asst cancer, Hamilton Clin, Ont Cancer Found, 73-74; lectr, Dept Pediat, 74-76, asst prof, 76-80, ASSOC PROF PATH & PEDIAT, SCH MED, MCMASTER UNIV, 80- *Concurrent Pos:* Mem staff lab med, Henderson Gen Hosp, Hamilton, 74-; res grants, Med Res Coun Can, Ont Cancer Treatment & Res Found, 74 & Nat Cancer Inst Can, 81-; Ont Cancer Fund res associateship, 74-84. *Mem:* AAAS; Can Soc Cell Biol; Am Asn Cancer Res; Can Soc Immunol; NY Acad Sci; Int Soc Pigment Cell. *Res:* Cancer immunology; cell biology. *Mailing Add:* 36 Mountain Brow Blvd Hamilton ON L8T 1A3 Can

LIAO, SHUTSUNG, b Tainan, Taiwan, Jan 1, 31; m 60; c 4. BIOCHEMISTRY, ENDOCRINOLOGY. *Educ:* Nat Taiwan Univ, BSc, 54, MSc, 56; Univ Chicago, PhD(biochem), 61. *Prof Exp:* From asst prof to assoc prof, 64-71, PROF DEPT BIOCHEM & MOLECULAR BIOL, BEN MAY INST, UNIV CHICAGO, 72- *Concurrent Pos:* NIH res grant, 63-; Am Cancer Soc res grant, 74-; mem study sect, NIH; assoc ed, Cancer Res. *Honors & Awards:* Sci Achievement Award, Taiwanese-Am Found. *Mem:* Am Soc Biol Chem; Endocrine Soc. *Res:* Mechanism of hormone action; control of gene expression; enzymology. *Mailing Add:* Ben May Inst Univ Chicago 5841 S Maryland Chicago IL 60637

LIAO, SUNG JUI, b Changsha, China, Nov 15, 17; nat US; m 53; c 4. PHYSICAL MEDICINE. *Educ:* Hsiang Ya Med Col, China, MD, 42; Nat Cent Univ, China, MPH, 44; London Sch Hyg & Trop Med, Univ London, DPH, 46, dipl bact, 47; Am Bd Phys Med & Rehab, dipl, 58. *Prof Exp:* Asst prof prev med, Sch Med, Yale Univ, 50-54; res assoc & assoc res prof bact, Col Med, Univ Utah, 49-50; dir phys med & rehab, Waterbury Hosp, 57-73; clin assoc prof rehab med, 71-82, CLIN PROF ORAL & MAXILLO FACIAL SURG, NY UNIV, 78-; LECTR REHAB MED, BOSTON UNIV SCH MED, 73- *Concurrent Pos:* Milbank Mem fel prev med, Sch Med, Yale Univ, 47-49; clin fel phys med, Mass Gen Hosp, Boston, 55-57; consult physiatrist, Middlesex Mem Hosp, Middletown, Conn, 57-60 & St Raphael Hosp, New Haven, Conn, 72-82; med dir, Waterbury Area Rehab Ctr, 57-62; dir phys med & rehab, St Mary's Hosp, 57-67 & Danbury Hosp, 57-69; attend physiatrist, Waterbury Hosp, 57-90, hon physiatrist, 91-; med consult, Waterbury & Danbury, Conn State Div Voc Rehab, 63-72, chief admin med consult, 69-73; assoc clin prof, Sch Med, Boston Univ, 67-73; hon consult biomech, NY Univ Inst Rehab Med, 69-76; chmn ad hoc comt acupuncture, Conn State Med Soc; secy, Am Acad Acupuncture, Inc; pres, Res Inst Acupuncture & Chinese Med; consult, Rhode Island State Bd Acupuncture, 80-84. *Mem:* sr fel Am Col Physicians; sr fel Royal Soc Med; sr fel Am Acad Phys Med & Rehab; sr mem Sigma Xi. *Res:* Excitability and conduction nerve and muscle; biomedical engineering; acupuncture; themography; pain. *Mailing Add:* 66 Skyline Dr Middlebury CT 06762

LIAO, TA-HSIU, b Taipei, Taiwan, Feb 22, 42; m; c 2. PROTEIN CHEMISTRY, ENZYMOLOGY. *Educ:* Nat Taiwan Univ, BS, 64; Univ Calif, Los Angeles, PhD(biol chem), 69. *Prof Exp:* Postdoctoral biochem, Moore-Stein Lab, Rockefeller Univ, 72-73, asst prof biochem, 73-74; from asst prof to prof biochem, Biochem Dept, Okla State Univ, 74-85; PROF BIOCHEM, INST BIOCHEM, NAT TAIWAN UNIV, 85- *Concurrent Pos:* Vis prof, A Kornberg's Lab, Stanford Univ, 81. *Mem:* Am Soc Biochem & Molecular Biol. *Res:* Protein chemistry; enzymology; deoxyribonuclease; structure and function relationships of proteins; physical methods of characterization of biological macromolecules. *Mailing Add:* Biochem Dept Nat Taiwan Univ Col Med No 1 Jen Ai Rd 1st Sect Taipei 10018 Taiwan

LIAO, TSUNG-KAI, b Chiayi, Taiwan, Aug 1, 23; m 63; c 3. ORGANIC CHEMISTRY, PHARMACEUTICAL CHEMISTRY. *Educ:* Nat Taiwan Univ, BS, 52; Wesleyan Univ, MA, 57; Univ Kans, PhD(chem), 60. *Prof Exp:* Asst chem, Nat Taiwan Univ, 53-55; fel, Wesleyan Univ, 55-57; asst, Univ Kans, 57-60; res assoc, Univ Mich, 60-61; assoc chemist, Midwest Res Inst, 61, sr chemist, 61-77; dir, molecular electronics, Carnegie-Mellon Inst Res, Carnegie-Mellon Univ, 78-87; SR CHEMIST, MIDWEST RES INST, 87-, CHEM CONSULT, 87- *Concurrent Pos:* Fel, Res Inst, Univ Mich, 60-61; vis prof, Midwest Res Inst, vis specialist of Nat Sci Coun, Repub China & Lectr of Sixth Tamkang Chair, Tamkang Col, 76; mem, Contract Develop Therapeut Comt, Nat Cancer Inst, 85-87. *Mem:* Am Chem Soc; Sigma Xi. *Res:* Synthesis of biologically active organic compounds; organic semiconductors; chemistry of nitrogen heterocyclic compounds; polymer chemistry, reverse osmosis composite membranes; high temperature lubricants; zeroshrink thermosetting polymers; polymer chemistry. *Mailing Add:* 1317 E 101 St Terrace Kansas City MO 64131

LIAU, ZONG-LONG, b Taipei, Taiwan, Aug 25, 50; US citizen; m; c 2. SEMICONDUCTOR LASERS, INTEGRATED OPTOELECTRONICS. *Educ:* Nat Taiwan Univ, BS, 72; Calif Inst Technol, PhD(appl physics), 79. *Prof Exp:* Vis scientist, Bell Labs, 77-78; STAFF MEM, LINCOLN LAB, MASS INST TECHNOL, 78- *Mem:* Optical Soc Am; Böhmische Phys Soc. *Res:* Physics and technology of semiconductor devices; materials science and engineering; atomic phenomena in compound semiconductor surfaces; fabrication of diode lasers, miniature mirrors, microlenses, integrated micro-optical and optoelectronic systems; issued 9 patents. *Mailing Add:* Lincoln Lab Mass Inst Technol PO Box 73 Lexington MA 02173-9108

LIAUW, KOEI-LIANG, b Indonesia, May 4, 35; US citizen; m 61; c 1. ORGANIC CHEMISTRY. *Educ:* Nanyang Univ, Singapore, BSc, 60; Univ Calif, Berkeley, MS, 62, PhD(chem), 64. *Prof Exp:* Res chemist, Gen Chem Div, Allied Chem Corp, 64-66; sr res chemist, Mobil Chem Co, Div Mobil Oil Corp, 66-68; PROJ LEADER CENT RES, TECH CTR, WITCO CHEM CORP, 69- *Mem:* Sigma Xi; Am Chem Soc. *Res:* Textile treating agents; paper sizings; process development; synthetic organic chemistry; organotin chemistry; stabilizers for polyvinyl chloride and polyolefins; flame retardants; vapor phase catalysis. *Mailing Add:* 285 W Steven Ave Wyckoff NJ 07481

LIAW, JYE REN, b Hsin-Chu, Taiwan, May 12, 46; m 72; c 2. NUCLEAR ENGINEERING, RADIATION PHYSICS. *Educ:* Nat Tsing Hua Univ, Taiwan, BS, 68; Univ Ore, Eugene, MS, 71; Ore State Univ, PhD(nuclear eng), 75. *Prof Exp:* Res asst physics, Univ Ore, Eugene, 71-73; res asst nuclear eng, Ore State Univ, 73-75; asst prof nuclear eng, Univ Okla, 75-80; consult, Los Alamos Sci Lab, 77-80; mem staff, Appl Physics Div, 80-89, MEM STAFF, REACTOR ANAL DIV, ARGONNE NAT LAB, 90- *Mem:* Am Phys Soc; Am Nuclear Soc; Sigma Xi; Nat Soc Prof Engrs. *Res:* Nuclear reactor design and analysis; radiation transport and dosimetry; nuclear fission product data evaluation; Van de Graff accelerator and nuclear reactor experiments; high vacuum technology; nuclear fuel reprocessing. *Mailing Add:* Reactor Anal Div Argonne Nat Lab Argonne IL 60439

LIBAN, ERIC, b Vienna, Austria, June 20, 21; nat US; m 54; c 3. APPLIED MATHEMATICS. *Educ:* NY Univ, BA, 48, MS, 49, PhD(math), 57. *Prof Exp:* Instr math, Long Island Univ, 49 & NY Univ, 49-50; asst, Ind Univ, 50-51; mathematician, Naval Res Lab, 51-52; assoc mathematician, Proj Cyclone, Reeves Instrument Corp, NY, 52; sr dynamics engr, Repub Aviation Corp, 52-55; staff mem analog comput & consult ctr, Dian Labs, Inc, 55-58; assoc prof eng sci, Pratt Inst, 58-61; res scientist, Grumman Aircraft Eng Corp, 61-67; assoc prof, 67-71, PROF MATH, YORK COL, 71-, CHMN DEPT, 75- *Concurrent Pos:* Lectr, Univ Md, 50; consult, Avco Res & Develop Corp, Mass, 59-60 & Comput Systs, Inc, NJ, 59-61; adj lectr, Polytech Inst Brooklyn, 62-65; adj prof, Adelphi Univ, 66-67; adj assoc prof, Queens Col, 67-68. *Mem:* Am Math Soc; Asn Comput Mach. *Res:* Applications and methods of simulation on analog computers; logical design of computing systems; automata studies; theory of servo and feedback systems; information and communication theory; operations research. *Mailing Add:* Dept Math CUNY York Col 94-20 Guy Brewer Blvd Jamaica NY 11451

LIBBEY, LEONARD MORTON, b Boston, Mass, Apr 17, 30; m 71. FOOD SCIENCE. *Educ:* Univ Mass, BVA, 53; Univ Wis, MS, 54; Wash State Univ, PhD(food technol), 61. *Prof Exp:* From asst prof to assoc prof, 61-81, PROF FOOD SCI & TECHNOL, ORE STATE UNIV, 81- *Mem:* AAAS; Inst Food Technologists; Am Chem Soc; Am Oil Chemists Soc; Sigma Xi; Am Soc Mass Spectrometry. *Res:* Food chemistry; chromatographic and spectrometric analysis, especially gas chromatography and mass spectrometry. *Mailing Add:* Dept Food Sci & Technol Ore State Univ Corvallis OR 97331

LIBBEY, WILLIAM JERRY, b Grand Rapids, Minn, Mar 18, 42; m 64; c 2. POLYOLEFINS, ENGINEERING PLASTICS. *Educ:* Carleton Col, BA, 64; Univ Wis, PhD(org chem), 69. *Prof Exp:* Res chemist, Continental Oil Co, 68-72, sr res chemist, 72-76, res group leader, Conoco, Inc, 77-81, sect dir, 82-85; SR RES ASSOC, E I DU PONT DE NEMOURS & CO, INC, 85- *Mem:* Am Chem Soc. *Res:* Carbonium ion chemistry; thermal rearrangements; alkyl halide chemistry; Fischer-Tropsch chemistry; polyolefins; Ziegler-Natta catalysis; hydrocarbon pyrolysis; high density polyethylene; engineering plastics. *Mailing Add:* Polymer Prod Dept E I Du Pont de Nemours & Co, Inc Bldg 323 Exp Sta Wilmington DE 19880-0323

LIBBY, CAROL BAKER, b South Kingstown, RI, Apr 20, 49; m 69; c 1. ENZYMOLOGY. *Educ:* Pa State Univ, BS, 71, PhD(org chem), 75. *Prof Exp:* Asst prof chem, Oberlin Col, 74-75; vis asst prof chem, Kenyon Col, 75-77; asst prof chem & physics, Skidmore Col, 77-79; mem staff, A E Staley Mfg Co, 79-82; mem staff, Best Foods, Res & Engr Ctr, 83-85; vis asst prof, Univ Maine, Farmington, 86-87; ASST PROF CHEM, COLBY COL, 85- *Concurrent Pos:* Res assoc, State Univ NY, 78. *Mem:* Am Chem Soc; AAAS; Asn Women Sci. *Res:* Enzyme catalyzed reactions, especially mechanism and isolation of enzymes with emphasis on carbohydrate hydrolases and glycoproteins. *Mailing Add:* Dept Chem Colby Col Waterville ME 04901

LIBBY, PAUL A(NDREWS), b Mineola, NY, Sept 4, 21; m 55; c 2. AERONAUTICAL ENGINEERING. *Educ:* Polytech Inst Brooklyn, BAE, 42, MS, 47, PhD, 49. *Prof Exp:* Design engr, Chance Vought Aircraft Co, Conn, 42-43; instr aeronaut eng, Polytech Inst Brooklyn, 43-44, 46-49, from asst prof to prof, 49-64, asst dir aerodyn lab, 59-64; assoc dean grad studies, 67-72, PROF FLUID MECH, UNIV CALIF, SAN DIEGO, 64- *Concurrent Pos:* Consult, Gen Bronze Corp, 48, NAm Aviation, Inc, 55, Gen Elec Co, 56, Gen Appl Sci Labs, Inc, 56-72, Avco, 76 & Systs Sci & Software, 76-; mem, Fluid Dynamics Panel, Adv Group Aerospace Res & Develop, NATO, 60-72, Fluid Mech Adv Comn, NASA, 63-69, Air Force Systs Command Scramjet Panel, 64-65 & Nat Acad Sci Adv Comn on Scramjet, 65-; corresp mem, Eng Sci Sect, Int Acad Astronaut, Int Astronaut Fedn, 66-; Guggenheim fel, 72-73. *Honors & Awards:* Royal Soc Guest Fel, 82-83. *Mem:* Fel Am Inst Aeronaut & Astronaut; Am Phys Soc. *Res:* Combustion theory; turbulent flow; turbulent combustion. *Mailing Add:* Univ Calif San Diego Box 109 La Jolla CA 92093

LIBBY, PAUL ROBERT, b Torrington, Conn, Sept 2, 34; m 59; c 3. BIOCHEMISTRY. *Educ:* Yale Univ, BS, 56; Univ Chicago, PhD(biochem), 62. *Prof Exp:* Fel biochem, Univ Calif, Davis, 62-63; sr cancer res scientist, 63-72, ASSOC CANCER RES SCIENTIST, ROSWELL PARK MEM INST, 72-; ASSOC RES PROF, DEPT PHARMACOL, STATE UNIV NY, 80- *Concurrent Pos:* res prof, Niagara Univ, NY, 78-; From asst to assoc res prof, Dept Physiol, State Univ NY, 71-80. *Mem:* AAAS; Am Asn Cancer Res; Endocrine Soc; Am Chem Soc; Am Soc Cell Biol; Sigma Xi. *Res:* Biochemical mechanisms of chemical carcinogenesis. *Mailing Add:* Roswell Park Mem Inst 666 Elm St Buffalo NY 14203

LIBBY, PETER, b Berkeley, Calif, Feb 13, 47; m 75; c 2. CELL BIOLOGY, CARDIOVASCULAR MEDICINE. *Educ:* Univ Calif, Berkeley, BA, 69; Univ Calif, San Diego, MD, 73. *Prof Exp:* Res physician, Peter Bent Brigham Hosp, 73-76; res fel physiol, Med Sch, Harvard Univ, 76-79; cardiol fel, Brigham & Women's Hosp, 79-80; from asst prof to assoc prof med, 80-90, ASSOC PROF PHYSIOL, SCH MED, TUFTS UNIV, 88- *Concurrent Pos:* S A Levine fel, Am Heart Asn, 76-77; Nat Heart, Lung & Blood Inst nat res serv award, 76-77; fel, Med Found, 80-82; estab investr, Am Heart Asn. *Mem:* Am Heart Asn; Am Soc Cell Biol; AAAS; Am Fedn Clin Res; Am Soc Physiol; Am Soc Clin Invest; Am Col Cardiol; Am Asn Pathologists; Am Asn Immunologists. *Res:* Cellular, molecular, and biochemical aspects of cardiovascular diseases; atherogenesis; arterial wall biology. *Mailing Add:* Vascular Med Unit Brigham & Women's Hosp 75 Francis St Boston MA 02115

LIBBY, WILLARD GURNEA, b Eugene, Ore, July 18, 29; m 65; c 3. GEOLOGY. *Educ:* Ore State Col, BS, 51; Northwestern Univ, MS, 59; Univ Wash, PhD(geol), 64. *Prof Exp:* Explor geologist, Stand Oil Co Calif, 56-59; asst prof geol, Univ SC, 63-65; lectr, Univ BC, 65-68; from asst prof to assoc prof, San Diego State Col, 69-71; PETROLOGIST, GEOL SURV WESTERN AUSTRALIA, 71- *Mem:* Geol Soc Australia. *Res:* Igneous and metamorphic petrology; geochronology. *Mailing Add:* Geol Surv of Western Australia Mineral House 100 Plain St Perth 6004 Australia

LIBBY, WILLIAM JOHN, (JR), b Oak Park, Ill, Sept 10, 32; m 56; c 3. FORESTRY, GENETICS. *Educ:* Univ Mich, BS, 54; Univ Calif, Berkeley, MS, 59, PhD(genetics), 61. *Prof Exp:* NSF fel genetics, NC State Col, 61-62; asst prof forestry, 62-67, assoc prof forestry & genetics, 67-72, PROF FORESTRY & GENETICS, UNIV CALIF, BERKELEY, 72- *Concurrent Pos:* Pack lectr, Yale Univ, 67; L T Murray distinguished vis lectr forest resources, Univ Wash, 68; Fulbright res scholar, NZ Forest Res Inst, Univ Canterbury, Australian Forest Res Inst, 71. *Mem:* Soc Am Foresters; Genetics Soc Am. *Res:* Quantitative genetics of forest trees; gene conservation; vegetation propagation of conifers; maturation of woody plant meristens. *Mailing Add:* Dept Forestry Univ Calif Berkeley CA 94720

LIBELO, LOUIS FRANCIS, b Brooklyn, NY, Oct 12, 30; m 54; c 4. THEORETICAL PHYSICS. *Educ:* Brooklyn Col, BS, 53; Univ Md, MS, 56; Rensselaer Polytech Inst, PhD(physics), 64. *Prof Exp:* Engr physics, Md Electronics Co, 54; proj engr, Ahrendt Instrument Co, 55; physicist, Opers Res Off, Johns Hopkins Univ, 57-58; asst prof physics, Am Univ, 65-68, adj prof, 68-73; adj prof, State Univ NY, Albany, 73-79; RES PHYSICIST, HARRY DIAMOND LAB, 80- *Concurrent Pos:* Res physicist, US Naval Surface Weapons Ctr, 64-80; consult physicist, L & L Assocs, 76-, Entron Inc & Lutech Inc, 78-81. *Honors & Awards:* Hinman Award, 87. *Mem:* Inst Elec & Electronic Engrs; Am Phys Soc; Electromagnetic Soc; Sigma Xi; NY Acad Sci. *Res:* Scattering theory for finite size targets and by apertures; theory of cooperative phenomena in solids; theory of nonlinear phenomena in insulators; microwaves and electromagnetic theory; interaction, coupling and generation. *Mailing Add:* L&L Assocs 9413 Bulls Run Pkwy Bethesda MD 20817

LIBERA, RICHARD JOSEPH, b Thorndike, Mass, Aug 26, 29; m 54; c 2. MATHEMATICS. *Educ:* Am Int Col, BA, 56; Univ Mass, MA, 58; Rutgers Univ, PhD(math), 62. *Prof Exp:* Instr math, Rutgers Univ, 60-62; from asst prof to assoc prof, 62-73, PROF MATH, UNIV DEL, 73- *Mem:* Am Math Soc; Math Asn Am; Polish Math Soc. *Res:* Geometric function theory. *Mailing Add:* Dept of Math Univ of Del Newark DE 19711

LIBERATORE, FREDERICK ANTHONY, b Framingham, Mass, Dec 11, 44; m 68; c 4. BIOLOGICAL RADIOSOTOPES. *Educ:* Mass State Col Framingham, BA, 70; Univ NH, PhD(biochem), 74. *Prof Exp:* Fel, Ohio State Univ, 74-76; mem staff, Sigma Chem Co, 76-78; MEM STAFF, DUPONT/NEW ENG NUCLEAR CORP, 78- *Mem:* AAAS; Sigma Xi; Am Chem Soc. *Res:* Labeling proteins with radioactive isotopes including iodine; protein cross-linking; bifunctional chelators; protein purification and characterization; ELISA and RIA. *Mailing Add:* Treble Cove Rd 49 Liberty Dr North Billerica MA 01862-3276

LIBERLES, ARNO, b Aschaffenburg, Ger, July 7, 34; US citizen; m 66; c 2. ORGANIC CHEMISTRY. *Educ:* Univ Mass, BS, 56; Yale Univ, MS, 59, PhD(chem), 60. *Prof Exp:* Fel chem, Col of France, 60-61; res chemist, W R Grace & Co, 61-62; from asst prof to prof chem, Fairleigh Dickinson Univ, 62-81. *Mem:* Am Chem Soc; Sigma Xi. *Res:* Theoretical organic chemistry. *Mailing Add:* Teaneck Campus Fairleigh Dickinson Univ Teaneck NJ 07666

LIBERMAN, ALLEN HARVEY, b Memphis, Tenn, Sept 15, 43. ELECTRICAL ENGINEERING. *Educ:* Rensselaer Polytech Inst, BEE, 65; Carnegie Inst Technol, MSEE, 66; Univ Detroit, DEng(elec eng), 68. *Prof Exp:* Asst prof comput design, Univ Detroit, 67-69, assoc dir comput eng, 69; vpres comput res & develop, Mgt Sci Inc, 69-70; vpres & dir comput res & develop, Nat Info Serv, Inc, 69-74; vpres finance, DBX Inc, Newton, 74-81; PARTNER, TRADE QUOTES INC, 81- *Concurrent Pos:* Technician, Digital Electronics Inc, 62-63; engr, Fairchild Camera & Instrument Corp, 64-67 & Hell Corp, 65; consult, Burroughs Corp, 68-69; adj prof & grant, Univ Detroit, 69-70. *Mem:* Am Soc Eng Educ; Inst Elec & Electronics Engrs. *Res:* Digital computer applications in personal identification and high density photographic memories; multi-user/multi-task minicomputer operating systems. *Mailing Add:* V P Trade Quotes Inc 675 Massachusetts Ave Cambridge MA 02139

LIBERMAN, ARTHUR DAVID, b Newark, NJ, Oct 13, 40; m 68; c 2. HIGH ENERGY PHYSICS. *Educ:* Dartmouth Col, AB, 62; Harvard Univ, MA, 63, PhD(physics), 69. *Prof Exp:* Res assoc high energy physics, Linear Accelerator Lab, Univ Paris, 69-70; adj asst prof particle physics, Univ Calif, Los Angeles, 70-74; res physicist, High Energy Physics Lab, Stanford Univ, 74-80; AT SCHLUMBERGER-DOLL RES CTR, 80- *Mem:* Am Phys Soc. *Res:* The study of gamma rays and entirely neutral final states in the annihilation interactions at electron-positron storage rings by utilizing the Crystal Ball, a large solid angle, good energy resolution, highly modularized NaI(Tl) detector. *Mailing Add:* Schlumberger-Doll Res Ctr Old Quarry Rd Ridgefield CT 06877

LIBERMAN, IRVING, b New York, NY, June 24, 37. LASERS, OPTICAL PROCESSING & DEVICES. *Educ:* City Col New York, BEE, 58; Northwestern Univ, MSEE, 60, PhD(elec eng), 65. *Prof Exp:* MGR OPTICAL PROCEDURES, WESTINGHOUSE ELEC CORP, SCI & TECHNOL CTR, 63- *Concurrent Pos:* Indust staff mem, Los Alamos Nat Lab, 74-80. *Mem:* Am Phys Soc; Optical Soc Am; Sigma Xi. *Res:* Development of optically pumped solid state lasers and transverse electrically excited gas lasers; design, installation, alignment, and evaluation of optical systems of carbondioxide lasers for laser fusion experiments. *Mailing Add:* Westinghouse Sci Technol Ctr 1310 Beulah Rd Pittsburgh PA 15235

LIBERMAN, MARTIN HENRY, b Los Angeles, Calif, Oct 5, 36; m 67; c 1. ANALYTICAL & POLYMER CHEMISTRY. *Educ:* Univ Calif, Los Angeles, BS, 58; Fla State Univ, PhD(phys chem), 68. *Prof Exp:* Asst, US Army Med Res & Nutrit Lab, Denver, Colo, 60-62; res assoc, Stanford Univ, 68-72; CHEMIST, US CUSTOMS SERV, 74- *Mem:* Am Chem Soc; Royal Soc Chem; AAAS. *Res:* Analytical chemistry; polymer chemistry. *Mailing Add:* US Customs Lab 630 Sansome St Rm 1429 San Francisco CA 94111

LIBERMAN, ROBERT PAUL, b Newark, NJ, Aug 16, 37; m 61, 73; c 5. PSYCHIATRY, CLINICAL PSYCHOLOGY. *Educ:* Dartmouth Col, AB, 59; Dartmouth Med Sch, dipl med, 60; Univ Calif, MS, 61; Johns Hopkins Univ, MD, 63. *Prof Exp:* Intern internal med, Bronx Munic Hosp Ctr, Albert Einstein Col Med, 63-64; res scientist, NIMH, 68-70; from asst prof to assoc prof, 70-76, PROF PSYCHIAT, SCH MED, UNIV CALIF, LOS ANGELES, 77-; DIR, PROG CLIN RES, CAMARILLO-NEUROPSYCHIAT INST RES PROG, 70-; CHIEF, REHAB MED SERV, BRENTWOOD VA MED CTR, 80- *Concurrent Pos:* NIMH res grants, 67-; consult various insts & govt, 70-; consult, Ventura County, Los Angeles County Ment Health Dept & Calif State Dept Ment Health, 72-; Fogarty sr res int fel, NIH, 75-76; assoc ed, J Appl Behav Anal, 76-77 & Schizophrenia Bull, 80-86; mem med staff, var hosps, 70-; prin investr, Mental Health Clin Res Ctr, 77- & proj dir, Rehab Res & Training Ctr, 80-85; mem, Res Rev Comt, NIMH, 79-81. *Honors & Awards:* First Prize, Int Rehab Film Festival, 83. *Mem:* Am Psychiat Asn; Asn Advan Behav Ther. *Res:* Experimental analysis of behavior in clinical psychiatry and psychology; interactions between drug effects and behavior modification; community mental health; behavior therapy; schizophrenia. *Mailing Add:* Dept of Psych B7-349 Univ of Calif Los Angeles 405 Hilgara Ave Los Angeles CA 90024

LIBERTA, ANTHONY E, b La Salle, Ill, May 17, 33; m 60; c 2. MYCOLOGY. *Educ:* Knox Col, Ill, AB, 55; Univ Ill, Urbana, MS, 59, PhD(bot), 61. *Prof Exp:* Res mycologist, Ill State Natural Hist Surv, 61; assoc prof, 61-67, PROF MYCOL, ILL STATE UNIV, 67- *Concurrent Pos:* NSF grants, 63-71 & 81- *Mem:* Mycol Soc Am. *Res:* Effects of soil disturbance and surface-mining on endomycorrhizae; ecological relationships of prairie plants and endomycorrhizae. *Mailing Add:* Dept Biol Ill State Univ Normal IL 61761-6901

LIBERTI, FRANK NUNZIO, b Warsaw, NY, Nov 2, 39; m 66; c 2. POLYMER CHEMISTRY, PHYSICAL CHEMISTRY. *Educ:* Rensselaer Polytech Inst, BChE, 61, PhD(phys chem), 67. *Prof Exp:* Develop chemist, 67-69, specialist prod develop, 69-73, mgr qual assurance, 73-75, specialist advan develop, 75-76, specialist prod develop, 76-79, mgr process technol, 79-85, MGR PRODUCT DEVELOP PROGS, PLASTICS BUS GROUP, GEN ELEC CO, 85- *Mem:* Soc Plastics Engrs. *Res:* Stabilization of polymers; flame retardant polymers; solid state of polymers; polymer crystallinity; thermal analysis of polymers. *Mailing Add:* Gen Elec Plastics Gen Elec Co Lexan Lane Mt Vernon IN 47620

LIBERTI, JOSEPH POLLARA, b Passaic, NJ, Nov 2, 37. CELL BIOLOGY. *Educ:* Fairleigh Dickinson Univ, BS, 59; Loyola Univ, MS, 62, PhD(biochem), 64. *Prof Exp:* Res fel biochem, Univ Minn, 64-65; res fel endocrinol, Mem Sloan-Kettering Cancer Inst, 65-66, res assoc, 66-67; from asst prof to assoc prof, 67-75, PROF BIOCHEM, MED COL VA, RICHMOND, 75- *Concurrent Pos:* Instr biochem, Cornell Univ Med Ctr, 66-67; vis prof, Oxford Univ, 77-78; vis scientist, Sloan-Kettering Cancer Inst, 83-84. *Mem:* Endocrine Soc; Am Chem Soc; Am Soc Biochem & Molecular Biol. *Res:* Regulation of cell growth and proliferation: molecular actions of lactogenic hormones, particularly post-receptor signalling events. *Mailing Add:* Dept Biochem Box 614 Med Col Va Sta Med Col Va Richmond VA 23298

LIBERTI, PAUL A, b Lyndhurst, NJ, Mar 18, 36; m 61; c 4. IMMUNOLOGY, PHYSICAL BIOCHEMISTRY. *Educ:* Columbia Col, AB, 59; Loyola Univ, Ill, MS, 61; Stevens Inst Technol, PhD(phys chem), 66. *Prof Exp:* From instr to prof biochem, Jefferson Med Col, 67-84; PRES & CHIEF SCIENTIST, IMMUNICON CORP, PA, 83- *Concurrent Pos:* Lectr, Fairleigh Dickenson Univ, 64-67 & Temple Univ, 67-; res fel phys chem, Stevens Inst Technol, 66; Nat Inst Allergy & Infectious Dis res career develop award, 73; lectr, Fairleigh Dickinson Univ, 64-67 & Temple Univ, 67-; adv ed, Immunochemistry, 76-81; adj prof biochem, Jefferson Med Col, 84- *Honors & Awards:* Ottens Res Award, 69. *Mem:* Am Asn Immunol; Am Asn Biol Chem; Am Asn Clin Chem. *Res:* complement proteins; development of immune specific MRI contrast agents; development of protein ferrofluids and high magnetic gradient separation systems; viral immunity; development of the immune selective plasma component removal medical device; development of new immune diagnostic technology. *Mailing Add:* Immunicon Corp 1310 Masons Mill Bus Park Huntingdon Valley PA 19006

LIBERTINY, GEORGE ZOLTAN, b Szolnok, Hungary, June 14, 34; m 56; c 2. MECHANICAL ENGINEERING, MATERIALS SCIENCE. *Educ:* Univ Strathclyde, BSc, 59; Bristol Univ, PhD(mech eng), 64. *Prof Exp:* Res & develop engr, English Elec Co Ltd, Eng, 59-60; from asst prof to assoc prof mech eng, Univ Miami, 63-68; assoc prof, Ill Inst Technol, 68-71; sr res engr, 71-73, prin res eng assoc, Advan Testing Methods Dept, 73-78, PRIN RES ENG ASSOC, AUTOMOTIVE SAFETY OFF, FORD MOTOR CO, 78- *Concurrent Pos:* Mem, US Adv Comt, Int Standard Orgn Fluid Power, 68-71, Eng Educ Comt, Soc Automotive Engrs, 73-86, chmn, 84-86, Am Soc Mech Engrs, 74-, chmn, 87-; assoc ed, J Vibration, Stress & Reliability in Design, 82-86; adj prof mech eng, Univ Mich, 83-; consult, safety, mfg & design. *Honors & Awards:* R R Teetor Award, Soc Automotive Engrs, 67, Forest R McFarland Award, 83. *Mem:* Fel Am Soc Mech Engrs; Sigma Xi; Soc Automotive Engrs; Soc Exp Stress Anal; Am Soc Eng Educ. *Res:* Fatigue of metals; static and dynamic fractures due to multiaxial stress-strain systems; nondestructive testing; experimental stress analysis; high pressure engineering; design; safety risk analysis; special transducer. *Mailing Add:* Ford Motor Co 330 Town Center Dr 500 & 300 Dearborn MI 48121

LIBET, BENJAMIN, b Chicago, Ill, Apr 12, 16; m 39; c 4. PHYSIOLOGY. *Educ:* Univ Chicago, BS, 36, PhD(physiol), 39. *Prof Exp:* Asst physiol, Univ Chicago, 37-39; instr, Albany Med Col, 39-40; res assoc physiol & biochem, Inst Pa Hosp, 40-43; instr physiol, Sch Med, Univ Pa, 43-44; mat engr, Personal Equip Lab, US Air Force, Ohio, 44-45; from instr to asst prof physiol, Univ Chicago, 45-48; staff physiologist, Kabat-Kaiser Inst, 48-49; from asst to prof physiol, 62-84, EMER PROF PHYSIOL, MED SCH, UNIV CALIF, SAN FRANCISCO, 84- *Concurrent Pos:* Consult, Mt Zion Neurol Inst, 56-; vis scientist, Japan Soc for Promotion of Sci, 79; fel, Commonwealth Fond, 56-57 & 64; scholar, Bellagio Study, Ctr Rockfeller Found, 77. *Mem:* Fel AAAS; Am Physiol Soc; Soc Neurosci; Int Brain Res Orgn. *Res:* Neurophysiology; electrical and metabolic aspects of neural function; synaptic mechanisms; cerebral mechanisms in conscious experience. *Mailing Add:* Dept of Physiol S-762 Sch Med Univ Calif San Francisco CA 94143-0444

LIBOFF, ABRAHAM R, b Paterson, NJ, Aug 27, 27; m 52; c 1. MEDICAL PHYSICS, BIOPHYSICS. *Educ:* Brooklyn Col, BS, 48; NY Univ, MS, 52, PhD(physics), 64. *Prof Exp:* Jr physicist, Naval Ord Lab, Md, 48-50; sr physicist, Metall Res Lab, Sylvania Elec Prod, Inc, 51-58; res asst cosmic ray lab, NY Univ, 59-64, assoc res scientist, 64-68, assoc dir environ radiation lab, 68-69, sr res scientist & proj coordr, Biophys Res Lab, 69-72; PROF PHYSICS & CHMN DEPT, OAKLAND UNIV, 72-, DIR MED PHYSICS PROG, 73- *Concurrent Pos:* Adj assoc prof physics, Hunter Col, NY, 68-72. *Mem:* Am Phys Soc; Biophys Soc; Am Geophys Union; Am Asn Physicists in Med; Bioelec Repair & Growth Soc (secy, 81). *Res:* Physics of collagenous tissues; biophysics of growth and development; electrically induced osteogenesis; sea-level cosmic ray ionization; environmental radiation; acoustic detection of nucleonic cascades; pyroelectric properties of bone. *Mailing Add:* Dept Physics Oakland Univ Rochester MI 48309

LIBOFF, RICHARD L, b New York, NY, Dec 30, 31; m 54; c 2. THEORETICAL PHYSICS. *Educ:* Brooklyn Col, AB, 53; NY Univ, PhD(physics), 61. *Prof Exp:* Res asst appl math, Courant Inst Math Sci, NY Univ, 56-61, asst prof physics, NY Univ, 62-64; assoc prof, 64-69, PROF ELEC ENG & APPL PHYSICS, CORNELL UNIV, 69- *Concurrent Pos:* Chief consult, NRA, Inc, 63-65; prin investr, Off Naval Res contract, 66-76, Air Force Off Sci Res, 77-81 & Army Res Off, 83-; Solvay fel, Univ Brussels, 71; vis prof physics, Univ Paris, Orsay, 79 & Tel Aviv Univ, 84 & 85; consult, Battelle Columbus Lab, 83 & 85; Fulbright scholar, 84. *Mem:* AAAS; fel Am Phys Soc; sr mem Int Elec & Electronic Engrs. *Res:* Kinetic theory; quantum mechanics; short wavelength lasing; fusion physics; dense recombining plasma; strongly coupled plasmas and fluids; condensed-matter theory; semiconductor transport and superlattice theory; applied mathematics with emphasis on classical and quantum chaos; author of 3 technical books. *Mailing Add:* Appl Physics & Elec Engr Phillips Hall Cornell Univ Ithaca NY 14853

LIBONATI, JOSEPH PETER, b Philadelphia, Pa, Nov 16, 41; m 69; c 3. CLINICAL MICROBIOLOGY. *Educ:* St Joseph's Col, Pa, BS, 63; Duquesne Univ, MS, 65; Univ Md, Baltimore, PhD(microbiol), 68. *Prof Exp:* From instr to asst prof med in clin microbiol, Sch Med, 68-77, SPEC LECTR MICROBIOL, SCH DENT, UNIV MD, BALTIMORE, 69-; CHIEF, DIV MICROBIOL, LABS ADMIN, MD STATE DEPT HEALTH & MENTAL HYG, 77- *Mem:* AAAS; Am Soc Microbiol. *Res:* Enteric bacterial diseases; pathophysiology; immunologic response and vaccine development. *Mailing Add:* 3801 Juniper Rd Baltimore MD 21218

LIBOVE, CHARLES, b New York, NY, Nov 7, 23; m 51; c 2. STRUCTURES, FAILURE ANALYSIS. *Educ:* City Col NY, BCE, 44; Univ Va, MS, 52; Syracuse Univ, PhD(mech eng), 62. *Prof Exp:* Aeronaut res scientist, Nat Adv Comt Aeronaut, 44-53; appl mathematician, Brush Labs, 53-55; assoc prof aeronaut eng, Tri-State Col, 55-58; from instr to assoc prof, 58-67, PROF MECH & AEROSPACE ENG, SYRACUSE UNIV, 67- *Concurrent Pos:* postdoctoral fel, NSF, Nottingham Univ, 67-68; consult, Pratt & Whitney Aircraft Co, 79 & 80. *Mem:* Am Soc Civil Engrs; assoc fel Am Inst Aeronaut & Astronaut; fel Am Soc Mech Engrs; Struct Stability Res Coun. *Res:* Stress analysis of swept wings, sandwich plates, composite thin-walled beams, corrugated plates and microelectronic packaging; elastic stability. *Mailing Add:* Dept Mech & Aerospace Eng Syracuse Univ Syracuse NY 13244-1240

LIBOW, LESLIE S, b New York, NY; m 73; c 2. GERIATRICS, INTERNAL MEDICINE. *Educ:* Brooklyn Col, BA, 54; Chicago Med Sch, MD, 58. *Prof Exp:* Intern, Mt Sinai Hosp New York, 58-59, resident, 63-64; resident, Bronx Vet Admin Hosp, 59-60; clin assoc bio-med psychiat, NIH, 60-62, res assoc, 62-63; chief geriat med, Mt Sinai City Hosp Ctr, Elmhurst, NY, 64-75; from asst prof to assoc prof med, Mt Sinai Sch Med, NY, 67-75; assoc prof, 75-78, prof med, State Univ NY Stony Brook, 78-; med dir, Jewish Inst Geriat Care, New Hyde Park, NY & chief geriat med, Long Island Jewish Hillside Med Ctr, NY, 75-; CLIN DIR LONG TERM CARE DEPT, GERIAT & ADULT DEVELOP, MT SINAI SCH MED & CHIEF MED SERV, JEWISH HOME & HOSP AGED, NY. *Concurrent Pos:* Consult to dir, Nat Inst Aging, Bethesda, 76-; consult, NIH, 75- *Honors & Awards:* Kent Award, Geront Soc Am, 81; Mascher-Manning Award, Am Geriat Soc, 87. *Mem:* Geront Soc; Am Geriat Soc; Am Col Physicians; AAAS. *Res:* Diseases of late life; brain and behavioral changes; human aging; thyroid disease; health care delivery. *Mailing Add:* Dept Geriat & Adult Develop Mt Sinai Sch Med One Gustave L Levy Pl New York NY 10029

LIBOWITZ, GEORGE GOTTHART, b Brooklyn, NY, June 18, 23; m 49, 86; c 2. SOLID STATE CHEMISTRY, MATERIALS SCIENCE. *Educ:* Brooklyn Col, BA, 45, MA, 50; Cornell Univ, PhD(phys chem), 54. *Prof Exp:* Chemist, Chromium Corp Am, 45-46, R Kann Chem Lab, 47-48 & Picatinny Arsenal, US Dept Army, 49; asst physics, Cornell Univ, 49-53; sr engr chem, Sylvania Elec Prod, Inc, 54; res assoc, Tufts Univ, 54-57; res supvr, Atomics Int Div, NAm Aviation, Inc, 57-61; sect head, Mat Sci Lab, Aerospace Corp, 61-63; staff scientist, Ledgemont Lab, Kennecott Copper Corp, 63-73; mgr, Solid State Chem Dept, Mat Res Ctr, Allied-Signal Corp, 73-78, mgr, Inorg & Solid State Chem Dept, Corp Res Ctr, 78-80, sr scientist, 80-86; CONSULT, G G LIBOWITZ, INC, 86- *Concurrent Pos:* Assoc ed, Solid State Ionics, 80-84, Mat Letters, 82-88; co ed, Mat Sci & Technol Series, 80-86; consult, Dept Energy & Environ, Brookhaven Nat Lab, 81-82. *Mem:* AAAS; Am Chem Soc; fel Am Phys Soc; Sigma Xi; NY Acad Sci; Int Asn Hydrogen Energy; Mat Res Soc. *Res:* Solid state chemistry; metal hydrides and metal-hydrogen systems; nonstoichiometric compounds; thermodynamic properties of solids. *Mailing Add:* G G Libowitz Inc PO Box 392 Morristown NJ 07960

LIBSCH, JOSEPH F(RANCIS), b Rockville, Conn, May 7, 18; m 41; c 3. METALLURGY, MATERIALS. *Educ:* Mass Inst Technol, SB & SM, 40, ScD(metall), 41. *Prof Exp:* From asst prof to prof metall, 46-54, dir magnetic mat lab, 46-55, head dept metall, 55-69, dir Mat Res Ctr, 62-69, Alcoa Found prof metall eng, 66-83, vpres res, 69-83, ALCOA FOUND EMER PROF METALL ENG, LEHIGH UNIV, 83- *Concurrent Pos:* Consult, Bullard Co, Conn, 46-47, Lepel High Frequency Labs, Inc, NY, 46-, Quaker State Metals Co, 56-58 & Smith Kline & French Labs, 56-73; consult ed metall ser, Chilton Publ Co, 58-60; bd mem, Pa Sci & Eng Found; mem, Gov Sci Adv Coun, Pa; bd mem, Eng Joint Coun; Alco-Richards prof, Lehigh Univ, 67-83; vpres & pres, Am Soc Metals, 72-74, mem bd educ & res, 74-79; chmn, Alt Energy Res, US Nat World Energy Conf. *Honors & Awards:* RR & EE Hillman Award, Lehigh Univ, 65. *Mem:* Fel hon mem Am Soc Metals; Am Inst Mining, Metall & Petrol Engrs; Am Soc Eng Educ; Sigma Xi. *Res:* Metallurgy of induction heating; materials selection. *Mailing Add:* Whitaker Lab Lehigh Univ Bethlehem PA 18015

LIBURDY, ROBERT P, b Detroit, Mich, Oct 23, 47; m. EXPERIMENTAL BIOLOGY. *Educ:* Brown Univ, PhD(biochem), 75; Univ Northern Colo, MBA, 80. *Prof Exp:* Grad teaching fel introductory biol, molecular biophysics & biochem pharmacol, Dept Biol & Med Sci, Brown Univ, 69-74; chief, Environ Health Serv, USAF Clin, Electronic Syst Div, Hanscom AFB, Mass, 80-81; asst prof environ med, Inst Environ Med, NY Univ Med Ctr, New York, NY, 81-84; RES STAFF SCIENTIST, LAWRENCE BERKELEY LAB, RES MED & RADIATION BIOPHYS DIV, BIOELECTROMAGNETIC RES FACIL, 84- *Concurrent Pos:* Prin investr, Electromagnetic Radiation Bioeffects Res Prog, Radiation Sci Div, USAF Sch Aerospace Med, Brooks AFB, 75-80; prin investr, USAF Proj, 75-80, Off Naval Res Proj, 81-87, Div Res Resources, NIH, 84-85, Liposome Technol, Inc, Menlo Park, Calif, 86-, Dept Energy Proj, 88- & NIH Proj, High-Field NMR Bioeffects, 91-; prog environ health sci, Grad Sch Arts & Sci, NY Univ, Washington Sq, New York, 81-84; course instr, Health Effects of Nonionizing Electromagnetic Radiation, Prog Environ Health Sci & Grad Sch Arts & Sci, NY Univ, Washington Sq, New York, 83-84; adj prof mech eng, Col Eng, Clemson Univ, SC, 87-; co-prin investr, Dept Energy Proj, Off Health & Environ Res, 88- *Mem:* Am Asn Immunologists; Am Chem Soc; Am Heart Asn; Am Soc Biochem & Molecular Biol; Bioelectromagnetics Soc; Biophys Soc; Inst Elec & Electronics Engrs; NY Acad Sci; Radiation Res Soc; Sigma Xi. *Res:* Electromagnetic field interactions with biological systems; liposome drug delivery; response of cellular systems to hyperthermia; atherogenesis. *Mailing Add:* Lawrence Berkeley Lab Res Med & Radiation Biophys Univ Calif Bioelectromagnetics Res Facil Bldg 74 One Cyclotron Rd Berkeley CA 94720

LICARI, JAMES JOHN, b Norwalk, Conn, July 22, 30; m 48. ORGANIC CHEMISTRY. *Educ:* Fordham Univ, BS, 52; Princeton Univ, PhD, 55. *Prof Exp:* Res chemist, Am Cyanamid Co, 55-57; res proj chemist, Am Potash & Chem Corp, 57-59; sr res engr, NAm Aviation, Inc, 59-61, supvr org chem, 61-67, group scientist, Res & Eng Div, NAm Rockwell Corp, 67-70, supvr chem lab, 70-72, MGR MICROCIRCUIT ENG LABS, ROCKWELL INT CORP, ANAHEIM, 72- *Concurrent Pos:* Asst prof, Fordham Univ, 55-56; lectr, Cal State Univ, Fullerton. *Mem:* Am Chem Soc; Int Soc Hybrid Microelec. *Res:* Materials and processes for microelectronics. *Mailing Add:* 15711 Arbela Dr Whittier CA 90603

LICEAGA, CARLOS ARTURO, b San Juan, PR, Nov 20, 58; m 80; c 3. RELIABILITY MODELING, FAULT-TOLERANT. *Educ:* Univ PR, BS, 81; Col William & Mary, MS, 84. *Prof Exp:* RES COMPUTER ENGR, NASA-LANGLEY RES CTR, 79- *Concurrent Pos:* Instr, Physics Lab, Univ PR, 80 & Electronics Col & Computer Prog, 81; asst instr computer eng, Carnegie-Mellon Univ, 85; computer sci instr, Thomas Nelson Community Col, 87-88; consult computer engr, Compass Consults Corp, 90. *Mem:* Inst Elec & Electronics Engrs; Digital Equip Computer Users Soc. *Res:* Automation of reliability and availability modeling of life- critical fault-tolerant computer systems; software reliability engineering and modeling; fault-tolerant hardware and software research and design. *Mailing Add:* NASA Langley Res Ctr MS 130 Hampton VA 23665-5225

LI-CHAN, EUNICE, b Hong Kong, Oct 23, 53; Can citizen; m 76; c 2. FOOD PROTEIN CHEMISTRY, CHEMICAL MODIFICATION. *Educ:* Univ BC, BSc, 75, PhD(food sci), 81; Univ Alta, MSc, 77. *Prof Exp:* Killam fel food chem, Can Coun, 80-82; res asst, 82-83, RES ASSOC FOOD CHEM, DEPT FOOD SCI, UNIV BC, 83- *Mem:* Am Chem Soc; Inst Food Technologists; Can Inst Food Sci & Technol; NY Acad Sci. *Res:* Properties of food proteins, including the relationship of molecular structure to function; improvement in nutritional and functional properties by chemical or enzymatic modification. *Mailing Add:* Dept Food Sci Univ BC 6650 NW Marine Dr Vancouver BC V6T 1W5 Can

LICHSTEIN, HERMAN CARLTON, b New York, NY, Jan 14, 18; m 42; c 2. MICROBIOLOGY, MEDICAL EDUCATION. *Educ:* NY Univ, AB, 39; Univ Mich, MSPH, 40, ScD(bact), 43; Am Bd Microbiol, dipl. *Prof Exp:* Asst bact, Univ Mich, 40-43; instr, Univ Wis, 43-46; Nat Res Coun fel, Cornell Univ, 46-47; from assoc prof to prof, Univ Tenn, 47-50; from assoc prof to prof, Univ Minn, 50-61; dir dept, 61-78, dir grad studies, 62-78 & 81-83, prof microbiol, 61-84, FEL, GRAD SCH, UNIV CINCINNATI 65-, EMER PROF MICROBIOL, 84- *Concurrent Pos:* Lewis lectr; Novy lectr; consult, Carbide & Carbon Chem Corp, Oak Ridge Nat Lab, 48-54 & Vet Admin Hosp, 57-65; consult ed, Life Sci Series, Burgess Pub Co, 60-67; mem sci fac fel panel, NSF, 60-63, Microbiol Training Comt, Nat Inst Gen Med Sci & Adv Bd Methods Biochem Anal; mem microbiol fels rev comt, NIH; trustee, Asn Med Sch Microbiol chmn, 74-77; mem, Linacre Col, Oxford Univ. *Honors & Awards:* Herman C Lichstein Distinguished Lectr endowed, 87. *Mem:* Fel AAAS; Am Soc Biol Chem; hon mem Am Soc Microbiol; Soc Gen Microbiol; fel Am Acad Microbiol. *Res:* Microbial physiology and metabolism. *Mailing Add:* Dept Microbiol & Molecular Genetics Col Med Univ Cincinnati Cincinnati OH 45267-0524

LICHT, ARTHUR LEWIS, b Hartford, Conn, Dec 18, 34; m 58. PHYSICS. *Educ:* Brown Univ, BSc, 57; Univ Md, PhD(physics), 63. *Prof Exp:* Physicist, Nat Bur Standards, 57-59; res physicist, NASA, 59-61 & US Naval Ord Lab, 61-70; ASSOC PROF, DEPT PHYSICS, UNIV ILL CHICAGO CIRCLE, 70- *Concurrent Pos:* Asst prof, Univ Md, 63-65; mem sch math, Inst Adv Study, 65-66. *Mem:* Am Phys Soc. *Res:* Space physics; quantum field theory. *Mailing Add:* Dept Physics Univ Ill Chicago IL 60680

LICHT, PAUL, b St Louis, Mo, Mar 12, 38; m 63. ZOOLOGY, ENDOCRINOLOGY. *Educ:* Washington Univ (Mo), AB, 59; Univ Mich, MS, 61, PhD(zool), 64. *Prof Exp:* From asst prof to assoc prof, 64-73, PROF ZOOL & CHMN DEPT, UNIV CALIF, BERKELEY, 73- *Concurrent Pos:* Lalor Found grant, 67-68; NSF grants, 64-90; consult ed, Col Div, McGraw-Hill Book Co, 68-; chmn, Comp Endocrinol Div, Am Soc Zoologists. *Honors & Awards:* Grace Pickford Award. *Mem:* Fel AAAS; Soc Study Reproduction; Am Soc Zoologists; Am Soc Ichthyologists & Herpetologists. *Res:* Comparative physiology and evolution of pituitary hormones with special reference to reproduction. *Mailing Add:* 940 Contra Costa Dr El Cerrito CA 94530

LICHT, STUART LAWRENCE, b Boston, Mass, July 24, 54. SOLAR ENERGY & ENERGY STORAGE, ELECTROCHEMISTRY & ANALYTICAL CHEMISTRY. *Educ:* Wesleyan Univ, BA(chem) & BA(physics), 76, MA, 80; The Weizmann Inst Sci, PhD(chem), 86. *Prof Exp:* CARLSON CHAIR & ASSOC PROF CHEM, CLARK UNIV, 88- *Concurrent Pos:* Vis asst prof chem, Northeastern Univ, 85-86; vis scientist & fel chem, Mass Inst Technol, 85-86. *Honors & Awards:* Delek Energy Award, Delek Corp, 83; Elad Res Excellence Award, Weizmann Inst, 85; Weizmann-Bantrell Award, Bantrell Found of Israel, 86. *Mem:* Am Chem Soc; Electro Chem Soc; Sigma Xi; Mat Res Soc. *Res:* Highest efficiency photoelectrochemical solar cells, novel materials for electrochemical energy storage, sulfur chemistry, analytical and environmental methods, and fundamental studies in the structure of electrolytes (pH, conductivity, and microelectrochemistry) and in electron correlation. *Mailing Add:* Dept Chem Clark Univ 950 Main St Worcester MA 01610

LICHT, W(ILLIAM), JR, b Cincinnati, Ohio, Sept 29, 15; m 42. CHEMICAL ENGINEERING. *Educ:* Univ Cincinnati, ChE, 37, MS, 39, PhD(chem eng), 50. *Prof Exp:* Asst, 37-39, from instr to prof, 39-85, head dept, 54-68, EMER PROF CHEM ENG, UNIV CINCINNATI, 85- *Concurrent Pos:* Consult govt & var indust concerns, 42-; vis prof, Univ Minn, 68 & 72. *Honors & Awards:* Award, Am Inst Chem Engrs, 72. *Mem:* Am Soc Eng Educ; fel Am Inst Chem Engrs; Air Pollution Control Asn. *Res:* Properties of azeotropic mixtures; drying of gases and refrigerants; adsorption in dessicant beds; dewpoint indicators; mechanics of drops; air pollution control; dust collection; design of systems; mathematical modelling particulate collection; fuel ethanol production. *Mailing Add:* Dept Chem Univ Cincinnati Cincinnati OH 45221-0171

LICHTBLAU, IRWIN MILTON, b Woodmere, NY, May 11, 36; m 65. CHEMICAL ENGINEERING. *Educ:* Princeton Univ, BSE, 58; Yale Univ, MEng, 60, DEng, 63. *Prof Exp:* Sr res engr, Chevron Res Corp, 63-69, asst mgr systs develop & appln, Western Opers Inc, 69-71, mgr comput opers, Comput Serv Dept, 71-74, sr eng assoc, Chevron Res Corp, sr staff econ analyst, Anal Div, 76-80, CONSULT, CORP DEVELOP, STANDARD OIL CALIF, 80- *Mem:* Am Inst Chem Engrs; Sigma Xi. *Res:* Petroleum process design; high pressure technology; compressibility of gas mixtures at high pressures and temperatures. *Mailing Add:* 1096 Upper Happy Valley Rd Lafayette CA 94549

LICHTEN, WILLIAM LEWIS, b Philadelphia, Pa, Mar 5, 28; m 50; c 3. PHYSICS. *Educ:* Swarthmore Col, BA, 49; Univ Chicago, MS, 53, PhD(physics), 56. *Prof Exp:* NSF fel, 56-57; res physicist, Radiation Lab, Columbia Univ, 57-58; from asst prof to assoc prof physics, Univ Chicago, 58-64; dir undergrad studies, 69-71, PROF PHYSICS, YALE UNIV, 64-, PROF PHYSICS & ENG & APPL SCI, 75- *Concurrent Pos:* Mem bd dirs, Nat Asn Metric Educ. *Mem:* Fel Am Phys Soc; Am Asn Physics Teachers; Optical Soc Am. *Res:* Psychology of perception; biophysics; chemical physics; optics; laser spectroscopy; atomic physics; science education. *Mailing Add:* Dept Physics Yale Univ Box 6666 New Haven CT 06511-8167

LICHTENBAUM, STEPHEN, b Brooklyn, NY, Aug 24, 39; m 61; c 5. NUMBER THEORY. *Educ:* Harvard Univ, AB, 60, AM, 61, PhD(math), 64. *Prof Exp:* Lectr math, Princeton Univ, 64-67; from asst prof to assoc prof, Cornell Univ, 67-73, chmn dept math, 79-82, prof math, 73-90; PROF MATH, BROWN UNIV, 90- *Concurrent Pos:* Guggenheim fel, John Simon Guggenheim Mem Found, 73-74. *Mem:* Am Math Soc. *Res:* Algebraic number theory and algebraic geometry, particularly the study of the values of zeta and L-functions. *Mailing Add:* Dept Math Brown Univ Providence RI 02912

LICHTENBERG, ALLAN J, b Passaic, NJ. NON-LINEAR DYNAMICS. *Educ:* Harvard Univ, AB, 52; Mass Inst Technol, MS, 54; Oxford, DPhil, 61. *Prof Exp:* From asst prof to assoc prof, 61-72, PROF ELEC ENG & COMP SCI, UNIV CALIF, BERKELEY, 72- *Concurrent Pos:* Chmn energy & resources group, Univ Calif, Berkeley; Guggenheim fel, 65, NSF fel, 84. *Mem:* Fel Am Phys Soc. *Res:* Plasma physics and engineering; non-linear dynamics; energy conservation and related problems. *Mailing Add:* Elec Eng & Comput Sci Dept Univ Calif Berkeley CA 94720

LICHTENBERG, BYRON KURT, b Stroudsburg, Pa, Feb 19, 48; m 70; c 2. BIOENGINEERING, BIOMEDICAL ENGINEERING. *Educ:* Brown Univ, ScB, 69; Mass Inst Technol, MS, 75, ScD, 79. *Prof Exp:* Res scientist, Mass Inst Technol, 78-84; CHIEF SCIENTIST, PAYLOAD SYSTS INC, 89- *Honors & Awards:* Haley Spaceflight Award, Aerospace Indust Asn Am. *Mem:* Sigma Xi. *Res:* Human adaptation to spaceflight particularly in the field of the inner ear system; human-machine interface, performance and habitation in spaceflight. *Mailing Add:* 728 Wolfsnase Cresent Virginia Beach VA 23454

LICHTENBERG, DON BERNETT, b Passaic, NJ, July 2, 28; m 54; c 2. THEORETICAL PHYSICS. *Educ:* NY Univ, BA, 50; Univ Ill, MS, 51, PhD(physics), 55. *Prof Exp:* Res assoc physics, Ind Univ, 55-57; guest prof, Univ Hamburg, 57-58; from asst prof to assoc prof, Mich State Univ, 58-63; physicist, Linear Accelerator Ctr, Stanford Univ, 62-63; assoc prof, 63-66, PROF PHYSICS, IND UNIV, BLOOMINGTON, 66- *Concurrent Pos:* Vis

prof, Tel-Aviv Univ, 67-68, Imp Col, Univ London, 71 & Oxford Univ, 79-80, Univ Wash, 86-87, Univ Turin, 87-88; sr fel, Sci Res Coun, UK, 79-80; Fulbright travel grant, Italy, 87-88. *Mem:* Fel Am Phys Soc. *Res:* Physics of the elementary particles. *Mailing Add:* Dept of Physics Ind Univ Bloomington IN 47405

LICHTENBERGER, DENNIS LEE, b Elkhart, Ind, Sept 30, 47; m 68; c 2. SURFACE SCIENCE, CATALYSIS. *Educ:* Ind Univ, BS, 69; Univ Wis-Madison, PhD(chem), 74. *Prof Exp:* Fel, Univ Ill, Champaign-Urbana, 74-76; from asst prof to assoc prof, 76-87, PROF CHEM, UNIV ARIZ, TUCSON, 87- *Concurrent Pos:* Counr, Am Chem Soc, 89- *Honors & Awards:* Sci Award, Eastman Kodak, 74. *Mem:* AAAS; Am Chem Soc; NY Acad Sci; Am Vacuum Soc. *Res:* Study of the behavior of organometallic molecules, molecules on surfaces, and catalysts through the development of high resolution gas phase photoelectron spectroscopy, ultra high vacuum surface spectroscopy and scanning tunneling microscopy. *Mailing Add:* Dept Chem Univ Ariz Tucson AZ 85721

LICHTENBERGER, GERALD BURTON, b St Louis, Mo, Jan 14, 45; m 73; c 3. MEDICAL DEVICE TECHNOLOGIES, MEDICAL IMAGING. *Educ:* Mass Inst Technol, BS, 66, MS, 67; Yale Univ, PhD(elec eng), 72. *Prof Exp:* Consult appln statist, IBM Res, 70-72; mem tech staff ocean systs res, Bell Labs, 72-75; prin scientist, dir & co-founder comput systs, Xybion Corp, 75-79; pres & founder, Systs of the Future, Inc, 79-86; vpres, strategic planning, Pentax, 86-90; PRES & CEO, ISIGHT, INC, 90- *Mem:* Inst Elec & Electronics Engrs. *Res:* Application of state of the art computer technology to diverse disciplines such as interactive data management and analysis in medical research; signal and information processing; array processing; random process modeling, electronic imaging. *Mailing Add:* Heather Hill Way Mendham NJ 07945

LICHTENBERGER, HAROLD V, b Decatur, Ill, Apr 22, 20; m 43; c 5. PHYSICS. *Educ:* Millikin Univ, AB, 42. *Prof Exp:* From mem staff to dir, Idaho Div, Metall Lab & Argonne Nat Lab, 42-56; vpres, Gen Nuclear Eng Corp, 56-61; asst div dir & mgr mfg, Nuclear Div, 61-69, dir nuclear prod mfg div, 69-74, VPRES NUCLEAR FUEL NUCLEAR POWER SYSTS, COMBUSTION ENG, INC, 74- *Mem:* Fel Am Nuclear Soc. *Res:* Design and use of nuclear reactors for production of heat and electrical power. *Mailing Add:* 34 Fox Den Rd West Simsbury CT 06092

LICHTENFELS, JAMES RALPH, b Robinson, Pa, Feb 14, 39; m 61; c 2. PARASITOLOGY, TAXONOMY. *Educ:* Ind Univ Pa, BS, 62; Univ Md, MS, 66, PhD(zool), 68. *Prof Exp:* ZOOLOGIST, ANIMAL PARASITOL INST, AGR RES SERV, USDA, 67-, CUR NAT PARASITE COLLECTION, 71- *Concurrent Pos:* Instr, USDA Grad Sch, 71-77; res assoc, Div Worms, Mus Natural Hist, Smithsonian Inst, Wash, DC, 72-; res affiliate, Div Parasitol, State Mus, Univ Nebr, Lincoln, 72-; mem coun resources, Asn Syst Collections, 75-77; ed, Proc Helm Soc Wash, 83-, asst ed, Systematic Parasitol, 87- *Mem:* Am Soc Parasitol (vpres, 87); Wildlife Dis Asn; Am Micros Soc; Sigma Xi; Am Asn Zool Nomenclature (pres, 86). *Res:* Intra and interspecific variation in parasitic nematodes; effects of host on morphology of parasitic nematodes; identification, classification and description of parasitic nematodes of vertebrates. *Mailing Add:* 12311 Whitehall Dr Bowie MD 20715

LICHTENSTEIN, E PAUL, b Selters, WGer, Feb 24, 15; nat US; m 51; c 2. ENTOMOLOGY. *Educ:* Hebrew Univ, Israel, MSc, 41, PhD(entom, biochem), 48. *Prof Exp:* Lectr biol, Sch Educ, Israel, 41-53; asst prof physiol & anat, Ill Wesleyan Univ, 53-54; proj assoc, 54-56, asst prof, 56-65, PROF ENTOM, UNIV WIS-MADISON, 65, ASSOC DIR, CTR ENVIRON TOXICOL, 72- *Mem:* Entom Soc Am; Am Chem Soc; Soc Toxicol. *Res:* Pesticidal residues and their effect on the biological complex on our environment; factors affecting persistence and breakdown of pesticides in soils, crops and water; naturally occurring toxicants. *Mailing Add:* Dept Entom Univ Wis Madison WI 53706

LICHTENSTEIN, HARRIS ARNOLD, b Houston, Tex, May 7, 41; m 69; c 2. ANALYTICAL CHEMISTRY, MOLECULAR BIOLOGY. *Educ:* Tulane Univ, BA, 63; Univ Houston, BS, 66, MS, 67, PhD(biol), 70. *Prof Exp:* Pres, Spectrix Corp, 69-86; VPRES, KEYSTONE ENVIRON INC, SUBSID KOPPERS CO, INC, 86- *Concurrent Pos:* Chief exec officer & chmn, Intrepid Biomed Inc. *Mem:* AAAS; Am Chem Soc; Sigma Xi. *Mailing Add:* 5701 Woodway No 324 Houston TX 77057

LICHTENSTEIN, LAWRENCE M, b Washington, DC, May 31, 34; m 56; c 3. MEDICINE, IMMUNOLOGY. *Educ:* Univ Chicago, BA, 54, MD, 60; Johns Hopkins Univ, PhD(immunol), 65. *Prof Exp:* Intern Med, 60-61, fel microbiol, 61-65, resident, 65-66, from asst prof to assoc prof, 66-75, PROF MED, SCH MED, JOHNS HOPKINS UNIV, 75- *Mem:* Am Acad Allergy; Am Asn Immunol; Am Fedn Clin Res; Am Soc Clin Invest; Am Asn Physicians. *Res:* Mechanisms of reactions of immediate hypersensitivity and relationship to clinical problems. *Mailing Add:* Johns Hopkins Asthma & Allergy Ctr 301 Bayview Blvd Baltimore MD 21224

LICHTENWALNER, HART K, b Easton, Pa, Oct 1, 23; m 45; c 3. CHEMICAL ENGINEERING. *Educ:* Lafayette Col, BS, 43; Lehigh Univ, MS, 49, PhD(chem eng), 50. *Prof Exp:* Org chemist, Res Labs, Gen Motors Corp, 43-48; chem engr, Silicone Prod Dept, 50-61, eng leader, 61-62, mgr process develop, 62-66, mgr room-temp vulcanising rubber develop, 66-68, mgr res & develop, 68-70, mgr vard prod sect, 70-77, managing dir, Gen Elec Silicones-Europe, 77-80, MGR STRATEGIC PLANNING & VENTURE DEVELOP, GEN ELEC CO, 80- *Mem:* Am Chem Soc; fel Am Inst Chem Engrs; NY Acad Sci. *Res:* Chemical process technology of organosilanes and siloxanes. *Mailing Add:* c/o Gen Elec Siligones Waterford NY 12188

LICHTER, BARRY D(AVID), b Boston, Mass, Nov 29, 31; m 58, 71; c 1. MATERIALS SCIENCE, SCIENCE. *Educ:* Mass Inst Technol, SB, 53, SM, 55, ScD(metall), 58. *Prof Exp:* Asst, Mass Inst Technol, 52-58; metallurgist, Air Force Cambridge Res Ctr, 58-61 & Oak Ridge Nat Lab, 61-62; fel, Lawrence Radiation Lab, Univ Calif, Berkeley, 62-64; assoc prof metall eng, Univ Wash, 64-68; assoc prof mat sci, 68-72, PROF MAT SCI & MGT TECHNOL, VANDERBILT UNIV, 72- *Concurrent Pos:* Tech consult, Boeing Co, 66; centennial fel, Vanderbilt Univ, 74-75; NSF fac fel, 75-76; consult, Off Technol Assessment, 76-78 & Oak Ridge Nat Lab, 78-79. *Mem:* Am Inst Mining, Metall & Petrol Engrs; NY Acad Sci; Nat Asn Corrosion Engrs; Electrochem Soc; Am Soc Metals. *Res:* Corrosion; oxidation thermodynamics; technology and human values; materials policy studies; philosophy and engineering ethics. *Mailing Add:* Box 16 Sta B Vanderbilt Univ Nashville TN 37235

LICHTER, EDWARD A, b Chicago, Ill, June 5, 28; m 52; c 2. PREVENTIVE MEDICINE, COMMUNITY HEALTH. *Educ:* Univ Chicago, PhB, 47; Roosevelt Univ, BS, 49; Univ Ill, MS, 51, MD, 55. *Prof Exp:* Asst physiol, Col Med, Univ Ill, 50-51, resident internal med, 58-61, instr med, 60-61; USPHS fel immunochem, Nat Inst Allergy & Infectious Dis, 61-63, mem staff, 63-66; assoc prof, 66-68, prof health care serv & head dept, Sch Pub Health, 72-79, prof prev med & head dept, 68-86, PROF, COMMUNITY HEALTH SCI, SCH PUB HEALTH, 80-, PROF MED, COL MED, UNIV ILL, 86- *Mem:* Am Pub Health Asn; Soc Clin Res; fel Am Col Prev Med; fel Am Col Physicians. *Res:* Radiation effects on peripheral circulation; chronic pulmonary infections; clinical pharmacology and therapeutic evaluation of antibiotics; immunochemistry; immunogenetics of immunoglobulins and other serum proteins; structure and function of health care services. *Mailing Add:* Dept Med Col Med MC 787 Univ Ill Med Ctr Box 6998 Chicago IL 60680

LICHTER, JAMES JOSEPH, b Algona, Iowa, Apr 29, 39. ENERGY DEMAND FORECASTING, INDUSTRIAL ENERGY USE. *Educ:* Loras Col, BS, 61; Fordham Univ, MS, 63; Duke Univ, PhD(physics), 69; Univ Calif Santa Barbara, MBE, 86. *Prof Exp:* From Physicist to res Physicist US Naval Ord Lab, Md,60-65; res asst, Duke Univ, 65-69; assoc scientist, ITT Fed Elec Corp, 69-77, software specialist, 77-83; ENERGY SPECIALIST, CALIF ENERGY COMN, 87- *Mem:* Am Phys Soc; Nat Asn Bus Economists. *Res:* Forecasts of industrial energy demand; economic models; radiation damage to bases of DNA; molecular biophysics; electron paramagnetic resonance; nuclear magnetic resonance; quantum biochemistry; computer simulation models. *Mailing Add:* 2348 American River Dr No 201 Sacramento CA 95825

LICHTER, ROBERT (LOUIS), b Cambridge, Mass, Oct 26, 41; c 2. ORGANIC CHEMISTRY, EDUCATION ADMINISTRATOR. *Educ:* Harvard Univ, AB, 62; Univ Wis-Madison, PhD(chem), 67. *Prof Exp:* USPHS fel, Brunswick Tech Univ, 67-68; res fel chem, Calif Inst Technol, 68-70; from asst prof to prof chem, Hunter Col, 70-83, chmn dept, 77-82; regional dir grants, Res Corp, 83-86; vprovost res & grad studies, staff Univ New York Stony Brook, 86-89; EXEC DIR, CAMILLE & HENRY DREYFUS FOUND INC, 89- *Concurrent Pos:* Vis scientist, Sandoz Res Lab, 81, Exxon Res & Eng Co, 82; adj prof, Hunter Col, City Univ NY, 83-86. *Mem:* Am Chem Soc; NY Acad Sci. *Res:* Organonitrogen chemistry; nuclear magnetic resonance spectroscopy; application of carbon and nitrogen nuclear magnetic resonance to organic chemistry. *Mailing Add:* Camille & Henry Dreyfus Found Inc 445 Park Ave New York NY 10022

LICHTI, ROGER L, b Milford, Nebr, Aug 27, 45; m 70; c 2. EXPERIMENTAL SOLID STATE PHYSICS. *Educ:* Ottawa Univ, BSc, 67; Univ Ill, MS, 69, PhD(physics), 72. *Prof Exp:* Vis asst prof, Univ Kans, 72-73, res assoc, 73-74; res assoc, Univ Mass, 74-77, vis asst prof, 78-79; ASST PROF PHYSICS, TEX TECH UNIV, 79- *Mem:* Am Phys Soc. *Res:* Magnetic resonance of dilute paramagnetic systems; spin-phonon and spin-spin interactions; structural transitions; impurity centers in semiconductors. *Mailing Add:* Physics Dept Tex Tech Univ Lubbock TX 79409

LICHTIG, LEO KENNETH, b Brooklyn, NY, Oct 20, 53; m 77; c 2. HEALTH CARE COSTS & FINANCING, CASE MIX. *Educ:* Rensselaer Polytech Inst, BS & MS, 74, PhD(commun res), 76. *Prof Exp:* Res asst commun res, Rensselaer Polytech Inst, 74-76; asst prof commun, State Univ NY Albany, 76-77; asst proj mgr, NY State Dept Health, 77-82, assoc proj mgr, 82; policy res specialist, Blue Cross Northeastern, NY, 82-83; vpres, Health Care Res Found, 82-90; CASE MIX ECONOMIST, NETWORK, INC, 90- *Concurrent Pos:* Mem, Ad-hoc Comt, US Dept Health, Educ & Welfare, 79-81; mem, tech adv comt, NY Statewide Planning & Res Coop Syst, 82-, tech rev comt & reimbursement adv group, NY State Long Term Care Case Mix Reimbursement Proj, 84-86; mem, Comt Privacy & Confidentiality, Am Statist Asn, 81-84; chairperson, Inst Rev Bd, Health Care Res Found, 83-90; mem, Tech Adv Group, Health Info Reporting Co, 87-90, Tech Adv Comt, NY State Off Mental Health Case Mix Classification Proj Steering, 86-89, Health Care Financing, 85-87; subcomt Qual & Productivity Measures in Health Care, Am Statist Asn, 88-; adj fac, Grad Prog Health Admin, Russell Sage Col, 86-, Union Col, 91. *Mem:* NY Acad Sci; AAAS; Am Statist Asn; Int Commun Asn. *Res:* Integration of clinical and financial information to address issues affecting the cost and management of health services and public health policy; development of case mix classification systems for specialized patient populations; numerous published articles. *Mailing Add:* 57 Fairlawn Latham NY 12110

LICHTIN, J LEON, b Philadelphia, Pa, Mar 5, 24; m 50; c 2. PHARMACEUTICAL CHEMISTRY, COSMETIC CHEMISTRY. *Educ:* Philadelphia Col Pharm, BS, 44, MS, 47; Ohio State Univ, PhD(pharmaceut chem), 50. *Prof Exp:* Asst prof, Cincinnati Col Pharm, 50-51, assoc prof pharm, 51-55; from assoc prof to prof, 55-71, ANDREW JERGENS PROF PHARM, UNIV CINCINNATI, 72- *Mem:* AAAS; fel Soc Cosmetic Chem. *Res:* Dermatologicals; formulation of pharmaceutical products; cosmetics. *Mailing Add:* Dept of Pharm Univ of Cincinnati Cincinnati OH 45267-0004

LICHTIN, NORMAN NAHUM, b Newark, NJ, Aug 10, 22; m 47; c 3. PHOTO CHEMISTRY, CATALYTIC CHEMISTRY. *Educ:* Antioch Col, BS, 44; Purdue Univ, MS, 45; Harvard Univ, PhD(phys org chem), 48. *Prof Exp:* Teaching fel, Harvard Univ, 45-47; lectr, 47, from instr to prof, 48-73, chmn dept, 73-84, dir, Div Eng & Appl Sci, 83-87, UNIV PROF CHEM, BOSTON UNIV, 73- *Concurrent Pos:* Vis chemist, Brookhaven Nat Lab, 57-58, res collab, 58-70; NSF sr fel, 62-63; guest scientist, Weizmann Inst, 62-63; vis prof, Hebrew Univ Jerusalem, 62-63, 70-71, 72, 73, 76 & 80; assoc ed, Solar Energy, 76-; vis prof, Inst Physics & Chem Res, Wako, Saitama, Japan, 80; sabbatical vis, Solar Energy Res Inst, Golden, Colo, 80. *Honors & Awards:* Coochbehar lectr, Soc Cult Sci, India, 80. *Mem:* Fel AAAS; Am Chem Soc; Sigma Xi; Int Solar Energy Soc. *Res:* Radiation chemistry; atomic nitrogen chemistry; photochemical conversion of solar energy; physical photochemistry; photo-assisted solid catalysis. *Mailing Add:* 195 Morton St Newton Centre MA 02159-1522

LICHTMAN, DAVID, b New York, NY, Feb 7, 27; m 48; c 3. SURFACE PHYSICS. *Educ:* City Col New York, BS, 49; Columbia Univ, MS, 50. *Prof Exp:* Physicist, Airborne Instruments Lab, 50-56; res engr, Sperry Gyroscope Co, 56-62; sr prin res scientist, Honeywell Res Ctr, 62-67; assoc prof, 67-70, PROF PHYSICS, UNIV WIS-MILWAUKEE, 70- *Concurrent Pos:* NATO sr sci fel, 71. *Mem:* AAAS; Am Phys Soc; Am Vacuum Soc. *Res:* Mass spectrometry; beam-surface interactions; thin films; metal-ceramic seals; dark trace tubes; gaseous discharge phenomena; high and ultra-high vacuum; surface physics; electron spectroscopy; photodesorption. *Mailing Add:* Dept of Physics Univ of Wis Milwaukee WI 53201

LICHTMAN, HERBERT CHARLES, b New York, NY, Sept 6, 21; m 46; c 3. INTERNAL MEDICINE, CLINICAL PATHOLOGY. *Educ:* Brooklyn Col, BA, 42; Long Island Col Med, MD, 45; Am Bd Internal Med, dipl, 53; Am Bd Clin Path, dipl, 73. *Prof Exp:* Intern, Long Island Col Serv, Kings County Hosp Ctr, 45-46; asst resident path, Montefiore Hosp, Bronx, NY, 48-49; asst resident med, Long Island Col Div, Kings County Hosp Ctr, 49-50; from instr to prof med, Col Med, State Univ NY Downstate Med Ctr, 51-70; PROF MED, BROWN UNIV, 70 - *Concurrent Pos:* Res fel clin med, Long Island Col Med, 50; clin fel hemat, Col Med, Univ Utah, 50-51; clin asst vis physician, Kings County Hosp, 51-53, assoc attend physician, 53-59, attend physician, 59 -; chief hemat & blood bank, State Univ NY Hosp, 66-70; dir div clin path, Dept Lab Med & chief div lab med, Miriam Hosp, 70-74, physician-in-chief, 74-86. *Mem:* AAAS; Am Soc Hemat; Soc Exp Biol & Med; Am Fedn Clin Res; Harvey Soc; Sigma Xi. *Res:* Hematology; leukemia and malignant lymphona; heme synthesis. *Mailing Add:* Miriam Hosp Dept of Med 164 Summit Ave Providence RI 02906

LICHTMAN, IRWIN A, b New York, NY, Nov 3, 20; m 48; c 1. SURFACE CHEMISTRY, DEFOAMERS. *Educ:* City Col New York, BS, 43; NY Univ, MS, 48, PhD(phys chem), 51. *Prof Exp:* Instr chem, Seton Hall Col, 47-48; asst prof, Community Col, NY, 48-52; sr res chemist, Lever Bros Res Ctr, 52-55; group leader phosphates & detergents, Food Mach & Chem Co, 55-60; sr res chemist, Shell Chem Co, 60-64; mgr phys chem lab, 64-77, group mgr res & develop, process chem div, Diamond Shamrock Chem Co, Morristown, 77-82; RETIRED. *Concurrent Pos:* Vis prof, NJ Inst Technol, Chem Eng Dept, 82-84; consult surface chem, 82- *Mem:* Am Chem Soc; Am Inst Chem. *Res:* Surface and colloid chemistry; defoamers; insecticide decomposition mechanisms; reaction kinetics; mechanism of defoamer action, particularly role of hydrophobic particles. *Mailing Add:* 24 Wenzel Lane Stony Point NY 10980-2310

LICHTMAN, MARSHALL A, b New York, NY, June 23, 34; m 57; c 3. HEMATOLOGY, BIOPHYSICS. *Educ:* Cornell Univ, AB, 55; Univ Buffalo, MD, 60; Am Bd Internal Med, dipl, 67. *Prof Exp:* Resident internal med, Med Ctr, Univ Rochester, 60-63; res assoc epidemiol, Sch Pub Health, Univ NC, 63-65; instr med, Sch Med & chief resident, Med Ctr, 65-66, sr instr med, Sch Med, 66-67, from asst prof to assoc prof med, radiation biol & biophys, 71-74, chief hematol unit, 75-77, assoc dean acad affairs & res, 79-80, sr assoc dean, 80-88, PROF MED, RADIATION BIOL & BIOPHYS, SCH MED, UNIV ROCHESTER, 74-, ACAD DEAN, 88-, CO-CHIEF, HEMATOL UNIT, 77- *Concurrent Pos:* USPHS res fel, Univ Rochester, 67-69; Leukemia Soc scholar, 69-74; from asst physician to sr assoc physician, Strong Mem Hosp, 65-71, sr physician, 74- *Mem:* Fel Am Col Physicians; Am Soc Hemat (vpres, 87, pres, 89); Am Soc Clin Invest; Asn Am Physicians; Am Physiol Soc; Am Soc Cell Biol. *Res:* Biochemical and biophysical studies of human erythrocytes and leukocytes. *Mailing Add:* Dept Hematol Univ Rochester Med Ctr PO Box 706 601 Elmwood Ave Rochester NY 14642

LICHTNER, FRANCIS THOMAS, JR, b Philadelphia, Pa, Mar 21, 53; m 76. AGRICULTURAL CHEMICALS, TRANSPORT PHYSIOLOGIST. *Educ:* Lebanon Valley Col, BS, 75; Cornell Univ, PhD(plant physiol), 79. *Prof Exp:* Asst prof bot, Univ Calif, Davis, 79-82; res scientist, Agr Prod Dept, Exp Sta, 82-86, SR RES SCIENTIST, AGR PROD DEPT, STINE-HASKELL RES CTR, E I DU PONT DE NEMOURS & CO, INC, NEWARK, DEL, 86- *Mem:* Am Soc Plant Physiologists; Weed Sci Soc Am. *Res:* Mechanisms of absorption and translocation of agricultural chemicals, ions and organic nutrients (sugars and amino acids), by plants; herbicide discovery and characterization. *Mailing Add:* Agr Prod Dept E I Du Pont De Nemours & Co Inc Newark DE 19714

LICHTON, IRA JAY, b Chicago, Ill, Sept 18, 28; m 49; c 1. NUTRITION. *Educ:* Univ Chicago, PhB, 47; Univ Ill, BS, 50, MS, 51, PhD(physiol), 54. *Prof Exp:* Res assoc obstet & gynec, Univ Chicago, 54-56; Am Heart Asn res fel cardiovasc physiol, Med Res Inst, Michael Reese Hosp, Chicago, Ill, 56-58; instr physiol, Stanford Univ, 58-62; assoc prof, 62-68, PROF NUTRIT, UNIV HAWAII, 68- *Concurrent Pos:* Vis prof, Hebrew Univ, 88-89. *Mem:* AAAS; Am Physiol Soc; Soc Study Reprod. *Res:* Water and electrolyte metabolism in pregnancy; growth; nutritional status. *Mailing Add:* Dept Food Sci & Human Nutrit Univ Hawaii 1800 East-West Rd Honolulu HI 96822

LICHTWARDT, ROBERT WILLIAM, b Rio de Janeiro, Brazil, Nov 27, 24; US citizen; m 51; c 2. MYCOLOGY. *Educ:* Oberlin Col, AB, 49; Univ Ill, MS, 51, PhD(bot), 54. *Prof Exp:* Fel, NSF, 54-55; res assoc bot, Iowa State Univ, 55-57; from asst prof to assoc prof, 57-65, PROF BOT, UNIV KANS, 65- *Concurrent Pos:* NSF sr fel, 63-64; ed-in-chief, Mycologia, 65-70; chmn, dept bot, Univ Kans, 71-74 & 81-84. *Mem:* AAAS; Bot Soc Am; Mycol Soc Am (pres, 71-72); Mycol Soc Japan. *Res:* Fungi association with arthropods, particularly those inhabiting their guts. *Mailing Add:* Dept of Bot Univ of Kans Lawrence KS 66045-2106

LICINI, JEROME CARL, b 1958. QUANTUM TRANSPORT, SEMICONDUCTOR DEVICES. *Educ:* Princeton Univ, AB, 80; Mass Inst Technol, PhD(condensed matter physics), 87. *Prof Exp:* Res asst, Int Bus Mach Corp, Tucson, 81; vis researcher, AT&T Bell Labs, 84-85; postdoctoral assoc, Mass Inst Technol, 87; ASST PROF PHYSICS, LEHIGH UNIV, 87. *Concurrent Pos:* Consult, Valley Enterprises, Inc, 91- *Mem:* Am Phys Soc. *Res:* New quantum phenomena in ultra-small semiconductor devices; fabrication of sub-micron silicon and gallium-arsenide sub- micron devices; measurement of quantum transport phenomena at ultra-low temperatures. *Mailing Add:* Physics Bldg 16 Lehigh Univ Bethlehem PA 18015

LICK, DALE W, b Marlette, Mich, Jan 7, 38; m 56; c 3. PURE MATHEMATICS, APPLIED MATHEMATICS. *Educ:* Mich State Univ, BS, 58, MS, 59; Univ Calif, Riverside, PhD(math, partial differential equations), 65. *Prof Exp:* Instr math & chmn dept, Port Huron Jr Col, 59-60; asst to comptroller, Mich Bell Tel Co, 60-61; from instr to asst prof math, Univ Redlands, 61-63; asst prof, Univ Tenn, 65-67; asst res mathematician, Dept Appl Math, Brookhaven Nat Lab, 67-68; assoc prof math, Univ Tenn, 68-69; assoc prof & head dept, Drexel Univ, 69-72; vpres acad affairs, Russell Sage Col, 72-74; prof math & dean, Sch Sci & Health Professions, Old Dominion Univ, 74-78; prof math & comput sci & pres, Ga Southern Col, 78-86; PROF MATH & PRES, UNIV MAINE, 86- *Concurrent Pos:* Consult, Union Carbide Corp, AEC, Oak Ridge Nat Lab, 66-67; adj assoc prof, Med Sch, Temple Univ. *Mem:* AAAS; Am Math Soc; Asn Comput Mach; Math Asn Am; Soc Indust & Appl Math; Sigma Xi. *Res:* Singular non-linear hyperbolic second order partial differential equations; non-linear Dirichlet problems; systems of non-linear boundary and initial value problems; partial differential equations and their numerical solution. *Mailing Add:* Univ Maine Orono ME 04469

LICK, DON R, b Marlette, Mich, Sept 3, 34; m 61; c 2. MATHEMATICS. *Educ:* Mich State Univ, BS, 56, MS, 57, PhD(math), 61. *Prof Exp:* Asst prof math, Purdue Univ, 61-63 & NMex State Univ, 63-66; vis assoc prof, Western Mich Univ, 65-66, from assoc prof to prof math, 72-85; PROF & HEAD MATH, EASTERN MICH UNIV, 85- *Concurrent Pos:* NSF res grant, 69-70; US Army Res Off Conf grant, 71-72; vis prof, Univ Calif, Irvine, 72-73 & Calif State Univ, Los Angeles, 72-73. *Mem:* Math Asn Am; Am Math Soc; Am Asn Univ Prof. *Res:* Complex analysis; sets of convergence of series; representation of measurable functions by series; graph theory; connectivity; structural problems; coloring problems. *Mailing Add:* Dept Math Eastern Mich Univ Ypsilanti MI 48197

LICK, WILBERT JAMES, b Cleveland, Ohio, June 12, 33; m 65; c 2. ENVIRONMENTAL ENGINEERING, MARINE SCIENCES. *Educ:* Rensselaer Polytech Inst, BS, 55, MA, 57, PhD(aeronaut eng), 58. *Prof Exp:* Res fel & lectr mech eng, Harvard Univ, 59-61, asst prof, 61-66; sr res fel aeronaut, Calif Inst Technol, 66; assoc prof eng, Case Western Reserve Univ, 66-69, chmn dept earth sci, 73-76, prof geophys & eng, 69-79; chmn dept, 82-84, PROF MECH & ENVIRON ENG, UNIV CALIF, SANTA BARBARA, 79- *Concurrent Pos:* Guggenheim fel, 65; Fulbright fel, 78. *Mem:* Am Geophys Union; Am Soc Mech Eng; Int Asn Great Lakes Res; Soc Indust Appl Math. *Res:* Applied mathematics. *Mailing Add:* Dept Mech Eng Univ Calif Santa Barbara CA 93106

LICKLIDER, JOSEPH CARL ROBNETT, computer science; deceased, see previous edition for last biography

LICKO, VOJTECH, b Banska Stiavnica, Czech, Aug 30, 32; US citizen; m 59; c 1. MATHEMATICAL BIOLOGY. *Educ:* Slovak Univ Bratislava, MS, 54; Czech Acad Sci, CSc(biophys), 63; Univ Chicago, PhD(math biol), 66. *Prof Exp:* Chief radioisotope lab, Inst Endocrinol, Slovak Acad Sci, Bratislava, 54-63; fel math biol, Univ Chicago, 63-66; scientist & assoc prof biophys, Inst Physics, Comenius Univ, Bratislava, 66-68; res fel biomath, Dept Biochem & Biophys, 68-71, res asst, 73-74, res asst biomath, 74-78, assoc adj prof, 78-90, ADJ PROF BIOMATH, CARDIOVASC INST, UNIV CALIF, SAN FRANCISCO, 90- *Concurrent Pos:* Assoc adj prof biomath, Dept Med, Univ Calif, San Francisco, 80-90, dir, Biomath Core Fac, Liver Ctr, 80-; assoc ed, Bull Math Biol, 73-86. *Mem:* Biophys Soc; AAAS; Soc Math Biol; NY Acad Sci. *Res:* Pharmacokinetics and pharmacodynamics; mathematical modeling of biochemical and physiological processes; theory of secretory mechanisms; dynamics of glucose-insulin control in man; kinetics of transport of substances through epithelia. *Mailing Add:* Cardiovasc Res Inst Univ Calif San Francisco CA 94143-0130

LIDDELL, CRAIG MASON, b Sydney, NSW, Australia, June 10, 58; m 89. EPIDEMIOLOGY OF PLANT DISEASE, ECOLOGY OF SOIL FUNGI. *Educ:* Univ Sydney, BSc, 79, Dipl plant path, 81, PhD(plant path), 86. *Prof Exp:* Postdoctoral res asst plant path, Univ Calif-Davis, 86-87 & Univ Wis-Madison, 88-89; ASST PROF PLANT PATH, NMEX STATE UNIV, 89- *Concurrent Pos:* Mem Mycol Comt, Am Phytopath Soc, 90-93. *Mem:* Am Phytopath Soc; Mycol Soc Am; Soil Sci Soc Am. *Res:* Physical ecology of soil fungi; computer modelling of fungal growth and development; computer modelling of plant disease epidemiology; biological control of soilborne diseases of crop plants. *Mailing Add:* Dept Entom Plant Path & Weed Sci NMex State Univ Las Cruces NM 88003

LIDDELL, ROBERT WILLIAM, JR, b Pittsburgh, Pa, Sept 11, 13; m 40; c 3. ORGANIC CHEMISTRY, BIOCHEMISTRY. *Educ:* Univ Pittsburgh, BS, 34, PhD(chem), 40. *Prof Exp:* Chem engr, Swindell-Dressler Corp, 34-35; chemist, Hall Labs, 35-36; res chemist, Hagan Chem & Controls, Inc, 40-55, asst res mgr, 55-63; mgr prod eng, Calgon Corp, 63-70, mgr pilot res & develop, 70-78; CONSULT, 78- *Mem:* Am Chem Soc. *Res:* Water treatment; phosphate chemicals. *Mailing Add:* 20 Calle Lecho Green Valley AZ 85614-1999

LIDDELL, WILLIAM DAVID, b Dayton, Ohio, Sept 17, 51; m 77. PALEOECOLOGY. *Educ:* Miami Univ, BA, 73; Univ Mich, MS, 75, PhD(geol), 80. *Prof Exp:* Asst prof geol & paleont, Earth Sci Dept, Univ New Orleans, 79-81; ASST PROF GEOL & PALEONT, GEOL DEPT, UTAH STATE UNIV, 81- *Mem:* AAAS; Ecol Soc Am; Int Palaeont Asn. *Res:* Paleoecology of ancient, primarily Paleozoic, communities; geology and ecology of modern coral reefs. *Mailing Add:* Dept Geol Utah State Univ Logan UT 84322-4505

LIDDICOAT, RICHARD THOMAS, JR, b Kearsage, Mich, Mar 2, 18; m 39. GEMOLOGY, MINERALOGY. *Educ:* Univ Mich, BS, 39, MS, 40; dipl, Gemol Inst Am, 44; Calif Inst Technol, MS, 44. *Prof Exp:* Asst mineral, Univ Mich, 37-40; instr, 40-41, dir ed, 41-42, 46-49, asst dir, 49-52, exec dir, 52-83, pres, 70-83, CHMN BD, GEMOLOGICAL INST AM, 83- *Concurrent Pos:* Ed, Gems & Gemology, 52-; US deleg, Int Gem Conf, 60-81; hon res staff, Los Angeles Mus Nat Hist, 68. *Honors & Awards:* Robert M Shipley Award, Am Gem Soc, 76; Hanneman Award, 78. *Mem:* Sigma Xi; fel AAAS; fel Geol Soc Am; fel Mineral Soc Am; hon fel Gemol Asn Gt Brit; Gemol Asn Australia (hon vpres); Am Gem Soc. *Res:* Gem identification and grading. *Mailing Add:* Gemol Inst Am 1660 Stewart St Santa Monica CA 90404

LIDDICOET, THOMAS HERBERT, pesticide chemistry; deceased, see previous edition for last biography

LIDDLE, CHARLES GEORGE, b Detroit, Mich, Mar 22, 36; m 60; c 4. VETERINARY MEDICINE, RADIATION BIOLOGY. *Educ:* Mich State Univ, BS, 58, DVM, 60; Univ Rochester, MS, 63. *Prof Exp:* Vet, Pvt Pract, Mich, 60-61; chief radioisotopes div, Fourth Army Med Lab, Vet Corps, US Army, Ft Sam Houston, 61-62, res vet, Walter Reed Army Inst Res, 63-65, chief small animal test, Dept Med Chem, 65, chief radioisotope sect, Army Med Res & Nutrit Lab, Denver, 65-69, lab vet, Navy Prev Med Unit, Viet Nam, 69; chief biophys unit, Twinbrook Res Lab, 70-73, RES VET, EXP BIOL DIV, HEALTH EFFECTS RES LAB, ENVIRON RES CTR, ENVIRON PROTECTION AGENCY, 73- *Mem:* Am Vet Med Asn. *Res:* The effects of microwaves on the immunologic competence of laboratory animals. *Mailing Add:* 4004 Oak Park Rd Raleigh NC 27612

LIDDLE, GRANT WINDER, endocrinology; deceased, see previous edition for last biography

LIDE, DAVID REYNOLDS, JR, b Gainesville, Ga, May 25, 28; m 55; c 4. CHEMICAL PHYSICS, SCIENCE INFORMATION. *Educ:* Carnegie Inst Technol, BS, 49; Harvard Univ, AM, 51, PhD(chem physics), 52. *Prof Exp:* Fulbright scholar & Ramsay mem fel, Oxford Univ, 52-53; res fel, Harvard Univ, 53-54; physicist, Nat Bur Standards, 54-63, chief, Infrared & Microwave Spectros sect, 63-68, dir, Off Standards Ref Data, 68-88; ED-IN-CHIEF, CRC HANDBOOK CHEM & PHYSICS, 88- *Concurrent Pos:* Lectr, Univ Md, 56-66; NSF sr fel, Univ London, 59-60 & Univ Bologna, 67-68; ed, J Phys & Chem Ref Data, 72-; US nat deleg, Comt Data Sci & Technol, Int Coun Sci Unions, 73-81, secy gen, 82-86, pres, 86-90; counr, Am Phys Soc, 76-83; chmn, Comn Symbols, Terminology & Units, Int Union Pure & Appl Chem, 77-81, pres, Phys Chem Div, 83-87; chmn, Am Inst Physics Publ Bd, 78-80 & Comt Chem Databases, 85-89; mem, adv bd, Chem Abstracts Serv, 78-83, Petrol Res Fund, 82-84, adv comt, Eng Info, Inc, 84- & bd gov, Mat Properties Data Network, Inc, 89-; US Adv Comt Int Coun Sci Unions, 87-; Comt on Atomic & Molecular Sci, NAS, NRC, 80-84; vchmn, Joint Comt Atomic & Molecular Phys Data, 89- *Honors & Awards:* Silver Medal, US Dept Com, 65, Gold Medal, 68; Stratton Award, Nat Bur Standards, 68; Presidential Rank Award Meritorious Fed Exec, 86; Herman Skolnik Award, Am Chem Soc, 88. *Mem:* Soc Scholarly Publ; Am Chem Soc; fel Am Phys Soc. *Res:* Free radicals, high temperature, microwave and infrared spectroscopy; molecular structure and spectroscopy; critical data evaluation in the physical sciences; molecular lasers; scientific databases; thermodynamics. *Mailing Add:* 13901 Riding Loop Dr Gaithersburg MD 20878

LIDE, ROBERT WILSON, b Hwanghsien, Shantung, China, June 27, 22; US citizen; m 55; c 3. NUCLEAR PHYSICS. *Educ:* Wake Forest Col, BS, 43; Univ Mich, MS, 50, PhD, 59. *Prof Exp:* Asst prof, 57-65, ASSOC PROF PHYSICS, UNIV TENN, KNOXVILLE, 65- *Mem:* Am Phys Soc. *Res:* Low-energy nuclear physics; gamma-gamma angular correlation; gamma-ray spectroscopy. *Mailing Add:* Dept Physics Univ Tenn Knoxville TN 37916

LIDIAK, EDWARD GEORGE, b La Grange, Tex, Mar 14, 34. GEOLOGY. *Educ:* Rice Univ, BA, 56, MA, 60, PhD(geol), 63. *Prof Exp:* Res scientist, Univ Tex, 62-64; from asst prof to assoc prof geol, 64-80, PROF GEOL & PLANETARY SCI, UNIV PITTSBURGH, 76-, CHMN, DEPT GEOL, 71- *Concurrent Pos:* Geologist, US Geol Surv, Pa, 65- *Mem:* AAAS; Geochem Soc; Geol Soc Am. *Res:* Petrology of island arc volcanic rocks; geology of buried Precambrian rocks of United States; phase equilibria in mineral systems. *Mailing Add:* Dept Geol Univ Pittsburgh Main Campus Pittsburgh PA 15260

LIDICKER, WILLIAM ZANDER, JR, b Evanston, Ill, Aug 19, 32; m 56, 89; c 2. POPULATION BIOLOGY, MAMMALOGY. *Educ:* Cornell Univ, BS, 53; Univ Ill, MS, 54, PhD(zool), 57. *Prof Exp:* From instr to assoc prof, Univ Calif, 57-69, vchmn dept zool, 66-67 & 81-83, from asst cur to assoc cur, Mus, 57-69, assoc dir, Mus Vert Zool, 68-81, actg dir, 74-75, prof zool, 69-89, CUR MAMMALS, UNIV CALIF, BERKELEY, 69-, PROF INTEGRATIVE

BIOL, 89- *Concurrent Pos:* Bd of Trustees, Biosciences Info Serv Inc, 87; assoc res prof, Miller Inst Basic Res Sci, 67-68; hon res fel, Dept Animal Genetics, Univ Col London, 71-72; hon lectr, Dept Biol, Royal Free Hosp Sch Med, London, 71-72; NAm rep steering comt, Int Theriological/ Mammalogical Coun, 78-89; co-chmn, rodent specialist group, Species Survival Comn, Int Union Conserv Nature & Natural Resources, 80-84 & chmn, 84-89; fac mem, Inst Ecology; vis scholar, Savannah River Ecol Lab, Univ Ga, 89-90, div zool, Univ Oslo, Norway, 89-90. *Honors & Awards:* C Hart Marriam Award, Am Soc Mammologists, 86. *Mem:* Am Soc Mammal (2nd vpres, 74-76, pres, 76-78); Ecol Soc Am; Soc Study Evolution; Am Soc Naturalists; Am Soc Zoologists; AAAS. *Res:* Ecology and evolution of mammals. *Mailing Add:* Mus of Vertebrate Zool Univ of Calif Berkeley CA 94720

LIDIN, BODIL INGER MARIA, b Malmo, Sweden, May 31, 39; m 73; c 1. MICROBIOLOGY. *Educ:* Univ Stockholm, MS, 74; Univ Ala, Birmingham, PhD(microbiol), 79. *Prof Exp:* Sr technologist, Inst Gustove-Rossy, Paris, 67-68, Univ Fla, 68-69; res asst, Karolinska Inst, Stockholm, 70-73; fel, Dept Pediat & Infectious Dis, Univ Ala, Birmingham, 79-80 & Wallenberg Lab, Uppsala Univ, 80-81; res assoc dept microbiol, 81-83, RES ASSOC SURG, UNIV ALA, BIRMINGHAM, 84- *Mem:* Am Soc Microbiol. *Res:* Epstine-Barr virus. *Mailing Add:* Dept Surg Rm 776 LHR Univ Ala Birmingham AL 35294

LIDMAN, WILLIAM G, b Rochester, NY, Nov 22, 21; m 43; c 3. METALLURGY, CERAMICS. *Educ:* Univ Mich, BS, 43. *Prof Exp:* Res scientist, NASA, Ohio, 43-52; eng sect head, Sylcor Div, Gen Tel & Electronics Corp, 52-57, eng dept head, 57-60, proj mgr nuclear fuel elements, 60-61; tech dir beryllium mfg, Gen Astrometals Corp, 61-71; mgr Hazleton, Pa Plant & Yonkers NY Div, Cabot Corp, 71-74, group mgr metall res & develop, Kawecki Berylco Industs, 74-80, tech sales mgr develop prods, KBI Div, 80-84, prod mgr, Aluminum Master Alloys, 84-86, DIR, PROD MGT, KB ALLOYS, CABOT CORP, READING, PA, 86- *Mem:* Sigma Xi; Am Soc Metals (treas, 55-57); Am Inst Metall Engrs. *Res:* Sintering mechanism of powder metallurgy products; production methods for manufacturing fuel elements for nuclear reactors and beryllium products; chemical specialty metals and beryllium; non-ferrous materials; refractory metals and ceramics. *Mailing Add:* PO Box 14927 Reading PA 19612-4927

LIDOFSKY, LEON JULIAN, b Norwich, Conn, Nov 8, 24; m 48; c 2. NUCLEAR ENGINEERING, COMPUTER SCIENCE. *Educ:* Tufts Univ, BS, 45; Columbia Univ, MA, 47, PhD(physics), 52. *Prof Exp:* Instr physics, NY State Maritime Col, 48-49; res asst, 49-52, res assoc, 52-59, from asst prof to assoc prof nuclear sci & eng, 59-64, PROF APPL PHYSICS & NUCLEAR ENG, COLUMBIA UNIV, 64- *Concurrent Pos:* Res scholar, Inst Nuclear Physics, Amsterdam, 68-69; consult, Mt Sinai Sch Med, 70-77 & Am Phys Soc Study Group, 76-77. *Mem:* Am Nuclear Soc; Am Phys Soc. *Res:* Radiation transport; nuclear physics. *Mailing Add:* Dept Nuclear Eng Columbia Univ Main Div New York NY 10027

LIDOW, ERIC, b Vilnius, Lithuania, Dec 9, 12; US citizen; m 52; c 4. ELECTRICAL ENGINEERING, SOLID STATE PHYSICS. *Educ:* Tech Univ, Berlin, MS, 37. *Prof Exp:* Chief engr, Selenium Corp Am, 41-44, vpres in charge res & eng, 44-46; PRES & CHMN BD, INT RECTIFIER CORP, 47- *Mem:* Sr mem Inst Elec & Electronics Engrs. *Res:* Photoelectric phenomena; selenium photocells; selenium rectifiers; silicone power devices. *Mailing Add:* Int Rectifier Corp 233 Kansas St El Segundo CA 90245

LIDTKE, DORIS KEEFE, b Bottineau County, NDak, Dec 6, 29; m 51. COMPUTER SCIENCE EDUCATION. *Educ:* Univ Ore, BS, 52, PhD(comput sci educ), 79; Johns Hopkins Univ, MEd, 74. *Prof Exp:* Jr mathematician, Shell Develop Co, 55-59; programmer, Univ Calif, Berkeley, 60-62; asst prof comput, Lansing Community Col, 63-67; educ specialist, Johns Hopkins Univ, 68; asst prof comput sci & math, 68-81, assoc prof computer sci, 81-90, PROF COMPUTER SCI, TOWSON STATE UNIV, 90- *Concurrent Pos:* Vis assoc prof, Univ Ore, 81-83, 85; assoc prog dir, Nat Sci Found, 84; Software Productivity Consortium, 87-88. *Mem:* Asn Comput Mach; Inst Elec & Electronics Engrs, Comput Soc; Nat Educ Comput Conf. *Res:* Impact of computer on society; computer literacy and computer awareness; computers and education. *Mailing Add:* Dept Comput & Info Sci Towson State Univ Baltimore MD 21204

LIDZ, THEODORE, b New York, NY, Apr 1, 10; m 39; c 3. PSYCHIATRY. *Educ:* Columbia Univ, AB, 31, MD, 36; Am Bd Psychiat & Neurol, dipl. *Hon Degrees:* MA, Yale Univ, 51. *Prof Exp:* From instr to assoc prof psychiat, Johns Hopkins Univ, 40-51; prof, 51-77, sterling prof, 77-78, EMER STERLING PROF PSYCHIAT, YALE UNIV, 78- *Concurrent Pos:* Examr, Am Bd Psychiat & Neurol, 46-51; psychiatrist-in-chief, Grace-New Haven Hosp, 51-61 & Yale Psychiat Inst, 51-61; chmn comt educ, Am Psychiat Asn, 52-55; mem study sect res grants, NIMH, 52-56, mem training grants comt, 59-63, career investr, 61-78, mem ment health prog-proj comt, 63-67; consult, Off Surgeon Gen, 58-72; fel, Ctr Advan Study Behav Sci, 65-66; chmn dept psychiat, Sch Med, Yale Univ, 67-69. *Honors & Awards:* Frieda Fromm-Reichmann Award, Acad Psychoanal, 61; William C Menninger Award, Am Col Physicians, 72; Stanley R Dean Award, Am Col Psychiat, 73; Van Gieson Award, NY State Psychiat Inst, 73; Laughlin Award, Am Col Psychoanalysts, 82; Spec Award, Am Family Ther Asn, 88. *Mem:* Am Psychosom Soc (secy-treas, 52-56, pres, 57-58); fel Am Psychiat Asn; fel Am Col Psychoanal (pres-elect, 90, pres, 91); fel Am Col Psychiat; Am Psychoanal Asn; Sigma Xi. *Res:* Schizophrenia; family. *Mailing Add:* 60 Orchard Rd Woodbridge CT 06525

LIE, WEN-RONG, b Taiwan, Repub China, Sept 27, 57; m 86; c 1. MOLECULAR IMMUNOLOGY, IMMUNOGENETICS. *Educ:* Tunghai Univ, BS, 79; Iowa State Univ, PhD(immunobiol), 87. *Prof Exp:* Res asst, Vet Gen Hosp, 79-81; grad res asst, Iowa State Univ, 82-87; res assoc, 82-87, NIH POSTDOCTORAL FEL, WASH UNIV SCH MED, 88- *Concurrent Pos:* Am Asn Immunologists travel award, 89. *Res:* Molecular biology of class I major histocompatibility complex molecules. *Mailing Add:* 12758 Castlebar Dr St Louis MO 63146

LIEB, CARL SEARS, b San Antonio, Tex, May 27, 49; m 80; c 2. EVOLUTIONARY BIOLOGY, HERPETOLOGY. *Educ:* Tex A&M Univ, BS, 71, MS, 73; Univ Calif, Los Angeles, PhD(biol), 81. *Prof Exp:* asst cur, Lab nviron Biol, Univ Tex, El Paso, 81-87, assoc cur, 87-89, Interim dir, Centennial Mus, 89-90, MEM, GRAD FAC, LAB ENVIRON BIOL, UNIV TEX, EL PASO, 83-, ASSOC PROF & COORDR, INTROD BIOL LABS, 90- *Concurrent Pos:* Mus assoc, Natural Hist Mus Los Angeles County, 73-82, res assoc, 82-; vis fac, Brigham Young Univ, 84-85; Herpetology Ed, Southwestern Naturalist, 87-89. *Mem:* Am Soc Ichthyologists & Herpetologists; Herpetologists' League; Soc Study Amphibians & Reptiles; Southwestern Asn Naturalists; Sigma Xi. *Res:* Evolution, biosystematics and evolutionary genetics of vertebrate animals, particularly amphibians and reptiles. *Mailing Add:* Dept Biol Univ Tex El Paso TX 79968-0519

LIEB, ELLIOTT HERSHEL, b Boston, Mass, July 31, 32; m 75; c 2. MATHEMATICAL PHYSICS. *Educ:* Mass Inst Technol, BSc, 53; Univ Birmingham, PhD(physics), 56. *Hon Degrees:* DSc, Univ Copenhagen, 79. *Prof Exp:* Fulbright fel physics, Kyoto Univ, 56-57; res assoc, Univ Ill, 57-58; lab nuclear studies, Cornell Univ, 58-60; staff physicist, Res Lab, IBM Corp, 60-63; assoc prof physics, Belfer Grad Sch Sci, Yeshiva Univ, 63-66; prof, Northeastern Univ, 66-68; prof math, Mass Inst Technol, 68-75; PROF MATH & PHYSICS, PRINCETON UNIV, 75- *Concurrent Pos:* Sr lectr, Univ Col Sierra Leone, 61-62; consult, IBM Corp, 63-65; guest prof, Inst Advan Sci Studies, France, 72-73; Guggenheim Found fel, 72 & 78; vis prof, Inst Advan Study, NJ, 82; ed, Commun Math Phys, Studies Appl Math, Lett Math Phys & Rev Mod Phys; mem bd gov, Inst Math & Applns, 83-87; mem bd trustees, Math Sci Res Inst, 85-89. *Honors & Awards:* Boris Pregel Award, NY Acad Sci, 70; Heineman Prize, Am Phys Soc & Am Inst Physics, 78; Sci Prize, UAP, 85; Birkhoff Prize, Am Math Soc & Soc Indust Appl Math, 88. *Mem:* Nat Acad Sci; fel Am Phys Soc; Austrian Acad Sci; Int Asn Math Physics(pres, 81-84); Royal Danish Acad Sci & Letts, 88. *Res:* Statistical mechanics; field theory; solid state physics; atomic physics; analysis; mathematical physics. *Mailing Add:* Dept of Physics Princeton Univ PO Box 708 Jadwin Hall Princeton NJ 08544-0708

LIEB, MARGARET, b Bronxville, NY, Nov 28, 23. GENETICS. *Educ:* Smith Col, BA, 45; Ind Univ, MA, 46; Columbia Univ, PhD, 50. *Prof Exp:* Asst prof biol, Brandeis Univ, 55-60; vis assoc prof, 60-62, assoc prof, 62-67, PROF MICROBIOL, SCH MED, UNIV SOUTHERN CALIF, 67- *Concurrent Pos:* USPHS fel, Calif Inst Technol, 50-52, Nat Found Infantile Paralysis fel, 52-53; fel, Inst Pasteur, 53-54; French Govt fel, Inst Radium, 54-55; NIH res career award, 62-72; prog dir genetic biol, NSF, 72-73. *Mem:* Fel AAAS; Genetics Soc Am; Am Soc Microbiol. *Res:* Bacteriophage genetics; recombination DNA repair; molecular biology. *Mailing Add:* Dept of Microbiol Sch Med Univ of Southern Calif Los Angeles CA 90033

LIEB, WILLIAM ROBERT, b Chicago, Ill, Aug 31, 40; div. BIOPHYSICS, ANESTHESIOLOGY. *Educ:* Univ Ill, BS, 62, MS, 63, PhD(biophys), 67. *Prof Exp:* Air Force Off Sci Res-Nat Res Coun fel biochem, Univ Manchester, 67-68; res fel, Dept Polymer Sci, Weizmann Inst Sci, 69; inst life sci, Hebrew Univ, Jerusalem, 69-70; staff scientist, Med Res Coun Biophys Unit, King's Col, London, 70-76, res fel, dept biophys, 76-84; RES FEL, BIOPHYS SECT, IMP COL LONDON, 84- *Concurrent Pos:* Prin investr, Med Res Coun grants, 78- *Res:* Molecular mechanisms of anesthesia; transport across biological membranes. *Mailing Add:* Biophys Sect Dept Physics Blackett Lab Imp Col Sci & Technol & Med Prince Consort Rd London SW7 2AZ England

LIEBE, DONALD CHARLES, b Cleveland, Ohio, Nov 16, 42; m 64. PHYSICAL CHEMISTRY. *Educ:* Case Western Reserve Univ, BA, 66, MA, 68, PhD(phys chem), 70. *Prof Exp:* Res assoc chem, Yale Univ, 70-71, NIH res fel, 71-73, asst instr, 74; res investr, G D Searle & Co, 74-79, sect head, Res & Develop Div, 79-83; SECT MGR, PACKAGE DEVELOP, S C JOHNSON & SON, INC, 83- *Mem:* Am Chem Soc. *Res:* Physical chemistry of nucleic acids, protein-nucleic acid interactions; binding to nucleic acids; mechanism of animal virus replication; growth factors; physical pharmacy in pharmaceutical development; emulsion science; novel drug delivery systems; polyene macrolide antibiotic physical chemistry; polymer physical chemistry; packaging science. *Mailing Add:* S C Johnson & Son Inc 1525 Howe St MS 042 Racine WI 53403

LIEBE, RICHARD MILTON, b Norwalk, Conn, May 26, 32; m 55; c 3. GEOLOGY. *Educ:* Bates Col, BS, 54; Univ Houston, MS, 59; Univ Iowa, PhD(geol), 62. *Prof Exp:* Assoc prof geol, Col Wooster, 61-67; PROF GEOL, STATE UNIV NY COL BROCKPORT, 67- *Mem:* Paleont Soc; Nat Asn Geol Teachers. *Res:* Stratigraphic paleontology of the Paleozoic era using conodonts; shallow water sedimentology and coral reef ecology. *Mailing Add:* Dept of the Earth Sci State Univ NY Col at Brockport Brockport NY 14420

LIEBELT, ANNABEL GLOCKLER, b Washington, DC, June 27, 26; div; c 4. MICROSCOPIC ANATOMY, CANCER. *Educ:* Western Md Col, BA, 48; Univ Ill, MS, 55; Baylor Col Med, PhD(anat), 60. *Prof Exp:* Biologist, Path Sect, Nat Cancer Inst, 49-52; asst anat, Col Med, Univ Ill, 52-54; asst, Col Med, Baylor Univ, 54-58, from instr to assoc prof, 58-71; assoc prof cell & molecular biol, Med Col Ga, 71-74; prof anat, Northeastern Ohio Univs Col Med, 74-86; FEL EXPERT, NAT CANCER INST, 82- *Concurrent Pos:* Dir, Kirschbaum Mem Lab, Col Med, Baylor Univ, 62-71, coordr, Micros Anat Teaching Prog, 77, dir, 78-79, chmn, 79-81; bd dirs, Augusta Radiation Ctr, 72-74, Portage Country Children's Serv Ctr, 79-82; bd dirs, Am Cancer Soc, Portage County Unit, 74-82, vpres, 76-77, pres, 77-78, chair, Prof Educ Comt, 74-78; consult, Breast Cancer Task Force, Nat Cancer Inst, 76-80; vis prof, Univ Tokushima Med Sch, Japan. *Mem:* NY Acad Sci; life mem Am Cancer Soc; Am Asn Path; Am Asn Lab Animal Sci; Int Acad Path; Soc Toxicol Path; Am Asn Cancer Res; Sigma Xi; Am Asn Women Sci; Am Asn Anatomists; Soc Toxicol Path. *Res:* Carcinogenesis and aging in inbred mice of several organ systems, especially the endocrine and reproductive (emphasis on mammary gland); biology histopathology; etiology; metastasis, environmental influences; animal models. *Mailing Add:* Registry of Exp Cancers NIH, Nat Cancer Inst, Bldg 41 Rm D311 Bethesda MD 20892

LIEBELT, ROBERT ARTHUR, b Chicago, Ill, Feb 3, 27; m 80; c 5. ANATOMY, ALCOHOLISM. *Educ:* Loyola Univ, Ill, BS, 50; Wash State Univ, MS, 52; Baylor Univ, PhD(anat), 57, MD, 58. *Prof Exp:* Asst, Wash State Univ, 50-52; asst, Col Med, Baylor Univ, 54-57, from instr to prof anat & chmn dept, 57-71; prof cell & molecular biol & exp med & assoc dean curriculum, Med Col Ga, 71-72, provost, 72-74; charter dean, 74-79, provost/dean, 79-82, PROF ANAT, NORTHEASTERN OHIO UNIV COL MED, 74- *Concurrent Pos:* Vis prof, Okayama Univ, 61; dir med educ, St Thomas Hosp, Akron, Ohio & dir, Ignatia Hall Alcoholism Ctr, 82. *Mem:* AAAS; Soc Exp Biol & Med; Am Asn Anat; Am Asn Cancer Res; NY Acad Sci. *Res:* Adipose tissue in obesity; relationship between nutrition and neoplasia; hypothalamus and appetite control; hypothalmic-pituitary relationships in experimental neoplasia; effects of pressure on food intake and body composition; biostereometric analysis for breast cancer; medical education; alcoholism. *Mailing Add:* Dept Anat Northeastern Ohio Univs Col Med 4209 State Rte 44 Box 95 Rootstown OH 44272

LIEBENAUER, PAUL (HENRY), b Cleveland, Ohio, Sept 21, 35; m 62; c 2. EXPERIMENTAL NUCLEAR PHYSICS. *Educ:* Case Western Reserve Univ, BS, 57, MS, 60, PhD(physics), 71. *Prof Exp:* Instr physics, Clarkson Col Technol, 60-62; asst prof, 68-70, ASSOC PROF PHYSICS, STATE UNIV NY COL OSWEGO, 70- *Concurrent Pos:* Consult, NASA, 71-; NSF grant, 72. *Mem:* Am Phys Soc; Am Asn Physics Teachers; Sigma Xi. *Res:* Low energy nuclear physics. *Mailing Add:* Dept Physics State Univ Col Oswego NY 13126

LIEBENBERG, DONALD HENRY, b Madison, Wis, July 10, 32; m 57; c 2. LOW TEMPERATURE PHYSICS, HIGH PRESSURES. *Educ:* Univ Wis, BS, 54, MS, 56, PhD, 71. *Prof Exp:* Asst, Univ Wis, 54-61; staff mem physics, Los Alamos Nat Lab, 61-87; staff mem, NSF, 81-88; STAFF MEM, OFF NAVAL RES, 88- *Concurrent Pos:* Solar-terrestrial res prog dir, NSF, 67-68; app liaison to Geophys Res Bd, Nat Acad Sci & US Comt Solar Terrestrial Res; US coordr, Solar Eclipse 70; sabbbatical leave, prog dir low temp physics, on leave from NSF to Dept of Energy, 86-87. *Mem:* AAAS; Am Astron Soc; Am Phys Soc; Am Geophys Union. *Res:* Low temperature physics, especially superfluidity and helium films; solar physics; magneto optics; high pressure physical measurements; high pressure physics, superconductivity. *Mailing Add:* 5100 Pheasant Ridge Rd Fairfax VA 22030

LIEBENBERG, STANLEY PHILLIP, b Los Angeles, Calif, Oct 16, 45; m 70; c 2. INHALATION TOXICOLOGY, VETERINARY PREVENTIVE MEDICINE. *Educ:* Univ Calif, Davis, BS, 68, DVM, 70. *Prof Exp:* Asst chief, anat & microbiol, dept enlisted inst, US Army Med Dept Vet Sch, 70-73; officer-in-chg, S Jutland Br, Denmark Div, US Army Vet Detachment, Europe, 73-76; vet lab animal officer, US Army Biomed Lab, Aberdeen Proving Ground, Md, 76-81; chief, lab animal & surg serv, Dept Clin Invest, Madigan Army Med Ctr,Tacoma, Wash, 81-84; vet lab animal officer, Walter Reed Army Inst Res, Washington, DC, 84-85; CHIEF, VET SERVS BR, TOXICOL DIV, CHEM RES DEVELOP & ENG CTR, ABERDEEN PROVING GROUND, MD, 85- *Concurrent Pos:* Affil asst prof, div animal med, Univ Wash Sch Med, Seattle, WA, 82-84; lectr, Essex Community Col, Baltimore, Md, 84-85. *Mem:* Am Vet Med Asn; Am Col Lab Animal Med; Am Col Vet Prev Med; Am Asn Lab Animal Sci; Am Soc Lab Animal Practrs. *Res:* Effects of environmental variables and process chemicals upon physiologic parameters in various laboratory animal species. *Mailing Add:* Res & Develop MS L-70 709 Swereland Rd King of Prussia PA 19406

LIEBER, CHARLES SAUL, b Antwerp, Belg, Feb 13, 31; US citizen; m 74; c 3. INTERNAL MEDICINE, NUTRITION. *Educ:* Univ Brussels, MD, 55. *Prof Exp:* Asst resident med, Univ Hosp Brugmann Brussels, Belg, 55-56; instr med, Harvard Med Sch, 61-62, assoc, 62-63; assoc prof, Med Col, Cornell Univ, 63-68; assoc prof, 68-69, PROF MED, MT SINAI SCH MED, 69-; CHIEF SECT LIVER DIS & NUTRIT, VET ADMIN HOSP, 68-; DIR, ALCOHOLISM RES & TREAT CTR, 77- *Concurrent Pos:* Belg Coun Sci Res fel internal med, Med Found Queen Elizabeth, 56-58; Belg-Am Found res fel med, Harvard Med Sch, 58-60; mem fat comt, Food & Nutrit Bd, Nat Acad Sci-Nat Res Coun, 61-67; dir liver dis & nutrit unit, Bellevue Hosp, 63-68; NIH res career develop award, 64-68. *Honors & Awards:* Laureate, Belg Govt, 56; McCollum Award, Am Soc Clin Nutrition, 73; E M Jellinek Mem Award, 77; W S Middleton Award, US Vet Admin, 77; Sci Excellence Award, Res Soc Alcoholism, 80. *Mem:* Am Soc Clin Invest; Am Med Soc Alcoholism (pres, 75); Am Soc Clin Nutrit (pres, 75); Asn Am Physicians; Res Soc Alcoholism (pres, 79). *Res:* Diseases of the liver; nutrition and intermediary metabolism, especially alcoholic cirrhosis, fatty liver, hyperlipemia, hyperuricemia, pathogenesis and treatment of hepatic coma and ascites, and pathophysiology of liver regeneration and drug abuse. *Mailing Add:* Alcohol Res Ctr Bronx Vet Admin Med Ctr Mt Sinai Sch Med 130 W Kingsbridge Rd Bronx NY 10468

LIEBER, MICHAEL, b Brooklyn, NY, Dec 28, 36; m 64; c 3. THEORETICAL PHYSICS. *Educ:* Cornell Univ, AB, 57; Harvard Univ, AM, 58, PhD(physics), 67. *Prof Exp:* Sr scientist, Res & Advan Develop Div, Avco Corp, 63-66, chief sci probs, 66-67; assoc res scientist & adj asst prof physics, NY Univ, 67-70; chmn dept, 83-86, from asst prof to assoc prof, 70-83, PROF PHYSICS, UNIV ARK, 83- *Concurrent Pos:* Prin investr, Dept Energy grant, 80-85; vis mem, Inst Theoret Physics, Univ Calif, Santa Barbara, 88; planetarium lectr, Univ Ark, 72-83, dir, Reach Kit Proj, 74-78. *Mem:* Am Phys Soc; Am Asn Physics Teachers; Sigma Xi. *Res:* Quantum scattering theory; few body problems; quantum electrodynamics and field theory; mathematical methods; atomic collisions; cosmic rays; elementary particles; general relativity and cosmology. *Mailing Add:* Dept of Physics Univ of Ark Fayetteville AR 72701

LIEBER, RICHARD L, b Walnut Creek, Calif; m 80; c 2. MUSCLE MECHANICS & REHABILITATION. *Educ:* Univ Calif, Davis, BS, 78, PhD, 82. *Prof Exp:* Fel, NIH, 78-81; res assoc, Univ Calif, Davis, 81-82; res physiologist, 82-85, ASST PROF SURG, DEPT SURG-ORTHOP & REHAB, UNIV CALIF, SAN DIEGO, 85-; BIOMED ENGR, DEPT ORTHOP RES, VET ADMIN MED CTR, 83- *Concurrent Pos:* Invited speaker, Frontiers Eng Health Care, Inst Elec & Electronic Engrs, 84; lectr, Skelatal Muscle Plasticity Series, Univ Calif, San Diego, Statist in Exp Biol, 85 & 86 & Bioeng Lab, 87 & 88. *Honors & Awards:* Talbot Award, Biophys Soc, 81. *Mem:* Biophys Soc; Inst Elec & Electronics Engrs; Orthopaedic Res Soc; Rehab Eng Soc NAm; Soc Neurosci. *Res:* Characterize skeletal muscle adaptation to altered use; elucidate mechanisms of torque generation during normal movement. *Mailing Add:* Div Orthop Univ Calif San Diego Vet Admin Med Ctr V-151 3350 La Jolla Village Dr San Diego CA 92161

LIEBERMAN, ABRAHAM N, b Brooklyn, NY, July 8, 38; c 4. NEUROLOGY. *Educ:* Cornell Univ, AB, 59; NY Univ Med Sch, MD, 63. *Prof Exp:* Instr, 70-71, from asst prof to assoc prof, 71-80, PROF NEUROL, NY UNIV MED CTR, 80- *Concurrent Pos:* Staff neurologist, US Air Force Hosp, Tachikawa, Japan, 67-69; asst chief neurol, Manhattan Vet Admin Hosp, 70-72; attend physician, Univ Hosp, NY Univ Sch Med, 70-; grants & contracts from var univs, cols & insts, 71-85; dir, Neurol Clin, Bellevue Hosp, 73-; consult, AMA Drug Evaluations, Chicago, Ill, 79- *Mem:* Asn Res Nervous & Ment Disease; Am EEG Soc; Am Soc Clin Pharmacol & Therapeut; fel Am Col Clin Pharmacol; Am Neurol Asn. *Mailing Add:* 650 Frist Ave New York NY 10016

LIEBERMAN, ALVIN, b Chicago, Ill, June 14, 21; m 47; c 2. CHEMICAL ENGINEERING. *Educ:* Cent YMCA Col, BS, 42; Ill Inst Technol, MS, 49. *Prof Exp:* Res assoc metal-ceramics, Alfred Univ, 49-51; res chem engr, IIT Res Inst, 51-63, sect mgr fine particles res, 63-68; dir res, Pac Sci Co, 68-74, vpres, Royco Instruments, Inc, 74-80, vpres, adv develop, HIAC/ROYCO Instrument Div, 80-83; RETIRED. *Concurrent Pos:* Regional ed, Powder Technol, 69-; tech spec, Particle Measuring Systs, 83-85, 87-; chief scientist, HIAC/ROYCO instr div, Pac Sci Co, 85-87. *Honors & Awards:* Hausner Award, Fine Particle Soc, 84; Whitfield Award, Inst Environ Sci, 85. *Mem:* AAAS; Am Chem Soc; Am Inst Chem Engrs; Am Asn Contamination Control (vpres, 62-63); Fine Particle Soc (pres, 77-78); Sigma Xi; Am Asn Aerosol Res; Inst Environ Sci. *Res:* Aerosol studies and application of electronic techniques; cloud physics; dust-free assembly area control and procedures; particle technology for gas/liquid suspensions. *Mailing Add:* 1943 Mt Vernon Ct 309 Mountain View CA 94040

LIEBERMAN, ARTHUR STUART, b Brooklyn, NY, Feb 24, 31; m 56; c 3. REGIONAL LANDSCAPE PLANNING & LANDSCAPE ECOLOGY. *Educ:* Cornell Univ, BS, 52, MS/LD, 58. *Prof Exp:* Teacher high sch, NY, 52-53; from instr to prof, 56-86, EMER PROF ENVIRON QUAL, CORNELL UNIV, 86- *Concurrent Pos:* Adv, Nature Reserves Authority & Ministry Agr, Israel, 71-72; res fel, Technion Israel Inst Technol, Haifa, 75-76, Lady Davis vis prof award, 80-81, prof, Technion, 80-81; chmn & coordr Cornell Tree Crops Res Proj, coordr Multidisciplinary Int Land-Use Planning Prog, Cornell Univ; consult, UN Food & Agr Orgn, Bangolore, India, 85; res dir Cornell (Abroad) Prog Israel, 87- *Mem:* Int Asn Landscape Ecol. *Res:* Physical environmental quality; ecology-based regional land-use planning; regional landscape inventories and information systems for physical planning; analysis and use of vegetation in comprehensive land planning; tree crops (agroforestry) for food and forage on rough marginal lands. *Mailing Add:* Landscape Archit Prog Kennedy Hall Cornell Univ Ithaca NY 14853

LIEBERMAN, BURTON BARNET, b Boston, Mass, Sept 28, 38; m 63; c 2. MATHEMATICS. *Educ:* Harvard Univ, BA, 60; NY Univ, MS, 62, PhD(math), 67. *Prof Exp:* Asst prof, 65-69, ASSOC PROF MATH, POLYTECH UNIV, 69- *Concurrent Pos:* Prin investr & consult, US Golf Asn, 79- *Mem:* Am Math Soc; Sigma Xi; Inst Math Stat; Am Stat Asn; Math Asn Am. *Res:* Ordinary differential equations; random differential equations; robust statistical methods. *Mailing Add:* Dept Math Polytechnic Univ 333 Jay St Brooklyn NY 11201

LIEBERMAN, DANIEL, b Gunnison, Utah, Feb 21, 19; c 1. PSYCHIATRY. *Educ:* Univ Calif, AB, 42, MD, 46. *Prof Exp:* Chief hosp serv, Sonoma State Hosp, Calif, 49-54; supt & med dir, Mendocino State Hosp, 54-60; from chief dep dir to dir, State Dept Ment Health, 60-63; pvt pract, 63-64; comnr ment health, Del Dept Ment Health, 64-67; dir, Jefferson Community Ment Health-Ment Retardation Ctr, Thomas Jefferson Univ, 67-74, prof & actg chmn, Dept Psychiat & Human Behav, 74-76, prof & dir, Psychosom Serv, 76-83, prof & chmn, 83-89, EMER PROF, DEPT PSYCHIAT & HUMAN BEHAV, JEFFERSON MED COL, THOMAS JEFFERSON UNIV, 89- *Concurrent Pos:* Consult forensic psychiat, Calif Superior Courts, 54-63; consult ment hosp serv, Am Psychiat Asn, 60-62; consult forensic psychiat, US Fed Court, 61-; consult, Ment Health Res Inst, Palo Alto, Calif, Nat Inst Alcohol Abuse & Acoholism, 63-64; consult state ment progs, NIMH, 65-68, consult alcohol rev comt, 67-70, consult, Nat Coun Community Health Ctrs, 72-73; consult, Vet Admin Hosp, Coatesville, Pa, 67-89. *Mem:* Fel AAAS; charter fel Am Col Psychiat; life fel Am Asn Ment Deficiency; fel Acad Psychosom Med; fel Am Psychiat Asn; fel Am Asn Social Psych; Sigma Xi. *Res:* Chronic pain; psychosymatic medicine; treatment of stress disorders. *Mailing Add:* Pier 5 No 140 Philadelphia PA 19106

LIEBERMAN, DIANA DALE, b Los Angeles, Calif, Jan 19, 49; m 68; c 1. POPULATION BIOLOGY, TROPICAL FOREST ECOLOGY. *Educ:* Univ Ghana, Legon, BSc, 76, PhD(bot), 79. *Prof Exp:* Demonstr plant ecol, Dept Bot, Univ Ghana, 76-79; vis scholar forest ecol & trop biol, Dept Environ Sci, Univ Va, 80-81; asst prof, 81-86, ASSOC PROF BIOL, DEPT BIOL, UNIV NDAK, 86- *Concurrent Pos:* NSF res grants, 81-86; NASA, res grant, 87-90. *Mem:* Ecol Soc Am; Sigma Xi; Asn Trop Biologists. *Res:* Tree growth rates, age-size relationships and tropical forest dynamics; plant population biology, phenology and seed dispersal; community ecology. *Mailing Add:* Dept Biol Univ NDak PO Box 8238 Grand Forks ND 58202-8238

LIEBERMAN, EDWARD MARVIN, b Lowell, Mass, Feb 10, 38; m 60; c 3. PHYSIOLOGY. *Educ:* Tufts Univ, BS, 59; Univ Mass, MA, 61; Univ Fla, PhD(physiol), 65. *Prof Exp:* Res assoc physiol, Col Med, Univ Fla, 66; asst prof, 68-72, assoc prof, Bowman Gray Sch Med, 72-76; assoc prof, 76-78;

PROF PHYSIOL, SCH MED, E CAROLINA UNIV, 78- *Concurrent Pos:* Swed Med Res Coun fel, Col Med, Univ Uppsala, 66-68. *Mem:* Soc Neurosci; Biophys Soc; Am Heart Asn; Am Physiol Soc; NY Acad Sci. *Res:* Cellular nerve physiology; membrane ion and water transport and metabolism; ultraviolet radiation effects on membranes; Schwann cell axon interactions. *Mailing Add:* Dept Physiol E Carolina Univ Sch Med Greenville NC 27858-4354

LIEBERMAN, EDWIN JAMES, b Milwaukee, Wis, Nov 21, 34; m 59, 88; c 2. PSYCHIATRY, SCIENCE COMMUNICATIONS. *Educ:* Univ Calif, Berkeley, AB, 55; Univ Calif, San Francisco, MD, 58; Harvard Univ, MPH, 63; Am Bd Psychiat & Neurol, dipl, 66. *Prof Exp:* Psychiat fel, Mass Ment Health Ctr, Boston, 59-61; child psychiat fel, Putnam's Children Ctr, Boston, 61-62; psychiatrist & chief, Ctr Child & Family Ment Health, NIMH, 63-70; dir family ther, Hillcrest Children's Ctr, DC, 71-74, dir ment health proj, 72-75, dir family planning proj, Am Pub Health Asn, 75-77; clin assoc prof psychiat, Sch Med, George Washington Univ, 77-87; adj prof, Family & Community Develop, Univ Md, 87-90; CLIN PROF PSYCHIAT, SCH MED, GEORGE WASH UNIV, 90- *Concurrent Pos:* Child psychiat fel, Hillcrest Children's Ctr, DC, 65-66; mem bd dirs, Sex Info & Educ Coun US, 66-69 & 73-76; clin asst prof psychiat, Sch Med, Howard Univ, 67-76; mem bd dirs, Nat Coun Family Rels, 69-73; vis lectr maternal & child health, Harvard Sch Pub Health, 69-73. *Mem:* AAAS; fel Am Psychiat Asn; fel Am Pub Health Asn; fel Am Asn Marriage & Family Therapists; Esperanto League NAm (pres, 72-75). *Res:* Mental health; preventive psychiatry; family planning; nonviolence; Esperantic studies; international language planning. *Mailing Add:* 3900 Northampton St NW Washington DC 20015

LIEBERMAN, GERALD J, b New York, NY, Dec 31, 25; m 50; c 4. OPERATIONS RESEARCH, STATISTICS. *Educ:* Cooper Union, BME, 48; Columbia Univ, AM, 49; Stanford Univ, PhD(statist), 53. *Prof Exp:* Math statistician, Nat Bur Standards, 49-50; from asst prof to prof statist & indust eng, Stanford Univ, 53-67, chmn, dept opers res, 67-75, assoc dean, Sch Humanities & Sci, 75-77, dean res, 77-80, vprovost,77-85, actg vpres & provost, 79, dean, grad studies & res, 80-85, PROF STATIST & OPERS RES, STANFORD UNIV, 67- *Concurrent Pos:* Mem, Maritime Transp Res Bd, Nat Res Coun, 66-71, panel on appl math alternatives for the Navy, 77-89, comt appl & theoret statist, 78-81 & panel appl math for Nat Bur Stanards, 83-89, chmn, 85-89; mem, adv panel math sci, Nat Sci Found, 68-73; mem bd dirs, Am Statist Asn, 74-76; mem bd adv, Naval Postgrad Sch, 76-85; mem, Grad Record Exam Bd, 84-88, panel on Qual Control of Family Assistance Progs, Nat Res Coun, 86-88, bd math sci, Nat Res Coun, 88-, bd trustees, Ctr Advan Study Behav Sci, 90-; fel, Ctr Advan Study Behav Sci, 85-86. *Honors & Awards:* Shewhart Medal, Am Soc Qual Control, 72. *Mem:* Nat Acad Eng; Opers Res Soc Am; fel Am Soc Qual Control (teas, 60-64); fel Am Statist Asn (vpres, 63-64); fel Inst Math Statist; Int Statist Inst; Inst Mgt Sci (pres, 80-87). *Res:* Industrial statistics; quality control; reliability; applied probability models. *Mailing Add:* Dept Opers Res Stanford Univ Stanford CA 94305-4022

LIEBERMAN, HERBERT A, b New York, NY, Aug 6, 20; m 49; c 2. PHARMACEUTICAL CHEMISTRY. *Educ:* Univ Ark, BS, 40; Columbia Univ, AM, 48, BS, 51, MS, 52; Purdue Univ, PhD(pharmaceut chem), 55. *Prof Exp:* Res fel biochem, Beth Israel Hosp, New York, 40-41; chemist, Pine Bluff Arsenal, Ark, 41-43; instr & assoc anal chem, Col Pharm, Columbia Univ, 46-52, res pharmacist, Res Inst, Wyeth Labs, 54-57; mgr pharmaceut prod develop, Isodine Pharmacal Co, 57-61; sr res assoc, Warner Lambert Co, Inc, 61-63, dir pharmaceut res & develop, 63-72, vpres, Personal Prods Div, 72-77, dir develop consumer prods, 77-85; PRES, LIEBERMAN ASSOCS INC, LIVINGSTON, NJ, 85- *Concurrent Pos:* Ed, var pharmaceut journals. *Mem:* Am Chem Soc; Am Pharmaceut Asn; fel Acad Pharmaceut Sci; Sigma Xi; fel Acad Pharmaceut Sci, 72; fel Am Asn Pharmaceut Scientists, 86. *Res:* Industrial pharmacy; pharmaceutical technology, particularly process and product development; analytical methods development for pharmaceutical products. *Mailing Add:* Lieberman Assocs Inc Four Browning Dr Livingston NJ 07039

LIEBERMAN, HILLEL, b Philadelphia, Pa, Jan 24, 42; m 66; c 2. ORGANIC CHEMISTRY, MICROBIAL BIOCHEMISTRY. *Educ:* Temple Univ, BS, 63, MS, 65, PhD(med org chem), 70. *Prof Exp:* Var admin & sci asst dir res, 73-76, asst vpres res, 76-79, vpres res, 79-82, vpres res & develop, 82-86, CHMN, BETZ INC, TREVOSE, 86-, SR VPRES, 87- *Mem:* Am Soc Microbiol; Am Chem Soc; Sigma Xi; Tech Asn Pulp & Paper Indust. *Res:* Development of chemical agents of an antimicrobial and/or antipollution nature to be employed in industrial water systems; development of conceptual information to aid in application of the aforementioned. *Mailing Add:* 3782 Midvale Lane Huntingdon Valley PA 19006

LIEBERMAN, IRVING, b Brooklyn, NY, Oct 25, 21; m 47; c 2. CELL BIOLOGY. *Educ:* Brooklyn Col, BA, 44; Univ Ky, MS, 48; Univ Calif, PhD(bact), 52. *Prof Exp:* Asst prof bact, Miami Univ, 48-49; from instr to asst prof microbiol, Sch Med, Washington Univ, 53-56; from asst prof to prof microbiol, 56-66, PROF ANAT & CELL BIOL, SCH MED, UNIV PITTSBURGH, 66- *Concurrent Pos:* Mem cell biol study sect, USPHS, 60-64, chmn, 71-73; fel engr, Westinghouse Res Ctr, 69- *Mem:* Am Soc Biol Chemists. *Res:* Chemistry. *Mailing Add:* 304 Iroqoiu Bldg Univ Pittsburgh Pittsburgh PA 15261

LIEBERMAN, JACK, b Chicago, Ill, Jan 4, 26; m 55; c 4. PULMONARY DISEASES, ENZYMOLOGY. *Educ:* Univ Calif, Los Angeles, AB, 49; Univ Southern Calif, MD, 54; Am Bd Internal Med, dipl, 62. *Prof Exp:* Intern, Harbor Gen Hosp, 54-55, resident internal med, 55-58; clin investr, Vet Admin Hosp, Long Beach, Calif, 60-63, sect chief internal med, 63-68; assoc clin prof med, Sch Med, Univ Calif, Los Angeles, 68-71; assoc clin prof med, Sch Med, Univ Calif, Irvine, 71-76; assoc dir, Dept Respiratory Dis, City of Hope Med Ctr, 71-76; chief, Respiratory Dis Div, Sepulveda Vet Admin Hosp, 76-88; PROF MED, UNIV CALIF, LOS ANGELES, 77- *Concurrent Pos:* Long Beach Heart Asn res fel, Harbor Gen Hosp, 58-60. *Mem:* AAAS; fel Am Col Physicians; fel Am Col Chest Physicians; Am Fedn Clin Res. *Res:* Cystic fibrosis; emphysema; antitrypsin deficiency; blood test for sarcoidosis. *Mailing Add:* Vet Admin Hosp 16111 Plummer St Sepulveda CA 91343

LIEBERMAN, JAMES, b New York, NY, June 2, 21; m 43; c 1. PUBLIC HEALTH. *Educ:* Middlesex Univ, DVM, 44; Univ Minn, MPH, 47. *Prof Exp:* Sr consult vet, UNRRA, 46; regional milk & food consult, USPHS, Kansas City, 48-50, asst to the chief, Milk & Food Br, DC, 50-51, from asst chief to chief spec proj br, Bur State Serv, 51-52, liaison officer to US Navy, 52, consult, spec regulatory prog, 52-54, detailed epidemiologist, Div Epidemiol & Commun Dis Control, NY State Health Dept, 54-55, training consult, Training Br, Commun Dis Ctr, asst chief training br & chief audiovisual sect, 59-62, chief med audiovisual br dir, Pub Health Serv Audiovisual Fac, Commun Dis Ctr, 62-67, dir, Nat Med Audiovisual Ctr & assoc dir audiovisual & telecommun, Nat Libr Med, 67-70, asst surgeon gen, USPHS, 68-70; vpres, med div, Videorecord Corp Am, 70-73; consult health sci educ & commun, 73-76; DIR, DEPT HEALTH, GREENWICH, CONN, 76- *Concurrent Pos:* Secy, Conf Pub Health Vets, 53-57; consult, WHO, Geneva, 55; chmn, Fed Adv Coun Med Training Aids, 60; mem task force sci commun, Surgeon Gen Conf Health Commun, 62; secy AV Conf Med & Allied Sci, 62-70; pres, Metrop Atlanta Commun Coun, 65; chmn conf biomed commun, NY Acad Sci, 67; mem comt bio-technol, Ga Sci & Technol Comn, 69-70; mem prof adv coun, Nat Easter Seal Soc Crippled Children & Adults, 69-75; vis prof, Sch Med, Hahnemann Univ, 73-; mem bd gov & pres-elect, Conn Inst Health Manpower Resources, 74-78; mem, adv coun, Pub Understanding Sci Prog, NSF, 78-80 & bd gov, US Conf Local Health Officers, 79-87; chairperson, Ment Health Consortium, Darien, Greenwich, New Canaan & Stamford, Inc, 86-87. *Honors & Awards:* Cert Commendation, UNRRA, 46; Letter of Commendation, Surgeon Gen, US Navy, 52; Citation, Nat League Nursing, 62; Citation, Nat AV Asn, 67; Brenda Award, Theta Sigma Phi, 68; Myrtle Wreath Award, Hadassah, 69. *Mem:* Fel Am Pub Health Asn; Asn Mil Surg US; Am Vet Med Asn; NY Acad Sci; Sigma Xi. *Res:* Biomedical communication and education; public health practice - policy and epidemiology; training and administration; comparative medicine. *Mailing Add:* 12 Silver Brook Rd Westport CT 06880

LIEBERMAN, LESLIE SUE, b Rockville Ctr, NY, June 23, 44; c 1. BIOLOGICAL ANTHROPOLOGY. *Educ:* Univ Colo, BA, 65; Univ Ariz, MA, 71; Univ Conn, PhD(biobehav sci), 75. *Prof Exp:* Proj assoc body composition, Human Performance Res Lab, Pa State Univ, University Park, 75-76; asst prof anthrop, 76-81, grad coordr, Ctr Geront Studies, 79-82, ASSOC PROF ANTHROP & PEDIAT, UNIV FLA, 81-, RES EPIDEMIOLOGIST, DIABETES RES, EDUC & TREAT CTR, 79 - *Concurrent Pos:* Fel, Nat Inst Gen Med Sci Human Performance Res Lab & Dept Anthrop, Pa State Univ, 74-75; mem, Coun Nutrit Anthrop, vpres, 80-82; prin investr, NSF grant, 81-83; vis lectr, Am Anthrop Asn, 81, Fla State Univ, London, 85, Univ Zagreb, Yugoslavia, 88; exec comt, Am Anthrop Asn, 87- *Mem:* AAAS; Am Anthrop Asn; Am Asn Phys Anthropologists; Human Biol Coun; Sigma Xi. *Res:* Study of body composition and the effects of nutritional behavior and diet on adaptation and microevolution in human populations; epidemiology of diabetes mellitus. *Mailing Add:* Dept of Anthrop 1350 Tur Univ of Fla Gainesville FL 32611

LIEBERMAN, MELVYN, b Brooklyn, NY, Feb 4, 38; m 61; c 2. PHYSIOLOGY. *Educ:* Cornell Univ, BA, 59; State Univ NY, PhD(physiol), 65. *Prof Exp:* Instr biol, Queen's Col, NY, 60; asst physiol, State Univ NY Downstate Med Ctr, 60-64; res assoc, Div Biomed Eng, Sch Eng & Dept Physiol & Pharmacol, 67-68, from asst prof to prof physiol, 68-88, PROF CELL BIOL, DUKE UNIV, 88-, ASSOC PROF, DEPT MED, 89- *Concurrent Pos:* Nat Heart Inst fel, 64-65; Carnegie Inst fel, 65; Nat Heart Inst fel, Biophys Inst, Brazil, 65-67; Nat Heart Inst spec fel, Med Ctr, Duke Univ, 67-68; lectr, Queen's Col, NY, 63-64; vis investr, Jan Swammerdam Inst, Netherlands, 75; Soc Gen Physiol secy, Nat Res Coun, 69-71, rep, 71-75, pres, 81-82; chmn res rev comt, NC Heart Asn, 75-76 & mem res rev comt, NY Heart Asn, 80-85; co-coordr, US Japan Coop Sci Prog, 74, 88 & US BrazilCoop Sci Prog, 80; Porter Develop Prog, Am Physiol Soc, 74-77, educ mat rev bd, 75-77; assoc ed, Am J Physiol, 81-, Expenentia, 82-90, Physiol Rev, 85-91 & Molecular Cell Biochem 91-; consult, Macy Found, Nat Heart Lung & Blood Inst, NSF, Vet Admin & Am Heart Asn; investr, Am Heart Asn, 71-76, mem physiol study sect, 80-84, Cardiovasc Res Study Comt, 87-90, Fel Rev Comt, 89- & Res Training Study Sect, 90-; coun, cell & gen physiol sect, Am Physiol Soc, 84-87. *Honors & Awards:* Cecil Hall Award, Electron Micros Soc Am, 89. *Mem:* Am Heart Asn; Am Physiol Soc; Biophys Soc; Cardiac Muscle Soc; Int Soc Heart Res; NY Acad Sci; Soc Gen Physiol; Physiol Soc; AAAS; Tissue Cult Asn. *Res:* Electrophysiology of cardiac muscle; regulation of ion transport; cultured heart cells as a model to study myocardial ischemia; mechanisms of cardiac cell injury; cardiotoxicity. *Mailing Add:* Dept Cell Biol Div Physiol Duke Univ Med Ctr Box 3709 Durham NC 27710

LIEBERMAN, MICHAEL A, b New York, NY, Aug 22, 50; m; c 2. MOLECULAR GENETICS. *Educ:* Mass Inst Technol, SB, 72; Brandeis Univ, PhD(biochem), 78. *Prof Exp:* Fel, Dept Biol Chem, Sch Med, Wash Univ, 77-79, res assoc, 79-80; asst prof biochem, Dept Nutrit, Harvard Sch Pub Health, 81-83; ASSOC PROF, DEPT MOLECULAR GENETICS, BIOCHEM & MICROBIOL, COL MED, UNIV CINCINNATI, 83- *Concurrent Pos:* Jr fac award, Am Cancer Soc, 81-84; mem, Study Sect Cellular Biol & Physiol II, NIH, 86-89. *Mem:* Am Soc Biochem & Molecular Biol; Am Soc Cell Biol. *Res:* Numerous publications; molecular genetics. *Mailing Add:* Molecular Genetics Biochem & Microbiol Dept Univ Cincinnati Col Med 231 Bethesda Ave Cincinnati OH 45267-0524

LIEBERMAN, MICHAEL MERRIL, b Chicago, Ill, June 10, 44; div; c 1. MICROBIOLOGY. *Educ:* Univ Chicago, BS, 66, PhD(microbiol), 69. *Prof Exp:* Nat Res Coun res assoc, Ames Res Ctr, NASA, Calif, 69-71; sr res microbiologist, Cutter Labs, Inc, 71-75 & Brooke Army Med Ctr, Tex, 76-83; chief microbiol serv, Dept Clin Invest, Tripler Army Med Ctr, Hawaii, 83-87; CHIEF MICROBIOL & IMMUNOL, DEPT PATH, WM BEAUMONT ARMY MED CTR, EL PASO, TEX, 87- *Mem:* Am Soc Microbiol; Am Asn Immunol. *Res:* Bacterial vaccine development; enzymology of halophilic bacteria; biochemical genetics and metabolic regulation; bacterial antigen and toxin purification. *Mailing Add:* Dept Pathol William Beaumont Army Med Ctr El Paso TX 79920-5001

LIEBERMAN, MICHAEL WILLIAMS, b Pittsburgh, Pa, Apr 20, 41; m 68; c 2. EXPERIMENTAL PATHOLOGY, MOLECULAR BIOLOGY. *Educ:* Yale Univ, BA, 63; Univ Pittsburgh, MD, 67, PhD(biochem), 72; Am Bd Path, cert anat path, 72. *Prof Exp:* Sarah Mellon Scoife fel, 69-70; res assoc, Fels Inst, Temple Health Sci Ctr, 70-72 & Exp Path Br, Nat Cancer Inst,74-76; head, Somatic Cell Genetics Sect, Nat Inst Environ Health Sci, 74-76; assoc prof, Sch Med, Wash Univ, 76-80, prof, dept path & dir grad studies, div biol & biomed sci, 80-84; chmn, dept path, Fox Chase Cancer Ctr, Philadelphia, PA, 84-88; RETIRED. *Concurrent Pos:* Adj asst & assoc prof, dept path, Med Sch, Univ NC, 74-76; assoc pathologist, Barnes Hosp, St Louis, 76-84; mem, Chem Path Study Sect, NIH, 78-; mem, Bd Toxicol, & Environ Health Hazards, Nat Res Coun, 80-84; vchmn, US Nat Comt Int Coun Soc Path, 87-90, counr, Am Asn Pathologists, 88-90. *Honors & Awards:* Warner-Lambert & Parke-Davis Award, Am Asn Pathologists, 81. *Mem:* Am Asn Cancer Res; AAAS; Am Asn Pathologists; Environ Mutagen Soc; Am Soc BioChemists & Molecular Biologists. *Res:* Molecular analysis of disease; molecular biology and gene expression, especially the role of oncogenes in the modulation of cellular gene expression in vitro and in carcinogenesis; chemical carcinogenesis. *Mailing Add:* Dept Path Baylor Col Med One Baylor Plaza Houston TX 77030

LIEBERMAN, MILTON EUGENE, b Chicago, Ill, Aug 30, 34; m 68; c 1. EVOLUTIONARY BIOLOGY, TROPICAL & QUANTITATIVE ECOLOGY. *Prof Exp:* Sr lectr ecol, Dept Zool, Univ Ghana, 74-79; vis prof, Dept Environ Sci, Univ Va, 80-81; RES PROF ECOL, DEPT BIOL, UNIV NDAK, 81- *Concurrent Pos:* NSF res grants, 81-86; res assoc, Mo Bot Garden, St Louis, 80-; sr res assoc, Nat Res Coun, 85-86; res grant, NASA, 87-90. *Mem:* Ecol Soc Am; Sigma Xi; Asn Trop Biologists. *Res:* Ecology of new world tropical forests, tropical marine benthic algal assemblages; reproductive phenology of temperate and tropical fleshy-fruited plants; plant-animal interactions. *Mailing Add:* Dept Biol PO Box 8238 Grand Forks ND 58202-8238

LIEBERMAN, MORTON LEONARD, b Chicago, Ill, Nov 22, 37; m 62; c 2. PHYSICAL CHEMISTRY. *Educ:* Ill Inst Technol, BS, 59, MS, 63, PhD(phys chem), 65. *Prof Exp:* Sr chemist, Res & Develop Labs, Corning Glass Works, 65-68; STAFF MEM TECH RES, SANDIA LABS, 68- *Mem:* Am Chem Soc. *Res:* High-temperature chemistry, thermodynamics; phase transitions; thin films; optical properties; fossil fuels; pyrotechnics and explosives; carbon research. *Mailing Add:* 1316 Paisano Northeast Albuquerque NM 87112-4524

LIEBERMAN, ROBERT, b Columbus, Ohio, Apr 9, 24; m 62; c 3. RADIOCHEMISTRY. *Educ:* Ohio State Univ, BA, 48, MSc, 52. *Prof Exp:* Chemist, Plastics Div, Battelle Mem Inst, 55-58, res scientist, Chem Physics Div, 58-64, sr chemist, 64-67, chief bioassay sect, Southeastern Radiol Health Lab, USPHS, 67-69, chief chem & biol, 69-71; chief phys sci br, Eastern Environ Radiation Lab, 71-74, chief qual assurance sect, 74-79, RES & DEVELOP CHEMIST, EASTERN ENVIRON RADIATION FACIL, ENVIRON PROTECTION AGENCY, 74- *Res:* Polyurethane foams; fission gas release; neutron dosimetry; radiation effects on plastics; use of radiotracers on wear studies; measurement of radionuclides in environmental samples. *Mailing Add:* 3707 Laconia Lane Montgomery AL 36111

LIEBERMAN, SAMUEL VICTOR, b Philadelphia, Pa, Nov 3, 14; m 38; c 2. CHEMISTRY. *Educ:* Univ Pa, BS, 36, MS, 37, PhD(chem), 48. *Prof Exp:* Asst, Sharp & Dohme, Inc, 41-42; res chemist, Wyeth Inst Med Res, 45-47, sr res chemist, 48-55; scientist in charge, Phys Analysis Dept, Prod Div, Bristol-Myers Co, 55-57, dir develop & phys sci, 57-60; consult pharmaceut prod, 61-79; RETIRED. *Mem:* AAAS; Sigma Xi. *Res:* Research, development and testing of pharmaceutical products. *Mailing Add:* 3400 N Ocean Dr No 508 Singer Island FL 33404

LIEBERMAN, SEYMOUR, b New York, NY, Dec 1, 16; m 44; c 1. BIOCHEMISTRY. *Educ:* Brooklyn Col, AB, 36; Univ Ill, MS, 37; Stanford Univ, PhD(chem), 41. *Prof Exp:* Chemist, Schering Corp, 38-39; Rockefeller Found asst, Stanford Univ, 39-41; spec res assoc, Harvard Univ, 41-45; assoc, Sloan-Kettering Inst, 45-50; from asst prof to emer prof biochem, Col Physicians & Surgeons, 50-87, assoc dean, 84-90, ASSOC VPROVOST, COLUMBIA UNIV, 87- *Concurrent Pos:* Mem panel steroids, Comt on Growth, Nat Res Coun, 46-50 & panel endocrinol, 55-56; traveling fel from Mem Hosp, New York to Basel, Switz, 46-47; mem endocrinol study sect, NIH, 58-63, mem, Insts, 59-65, chmn, 63-65, mem gen clin res ctrs comn, 67-70; mem med adv comt, Pop Coun, 61-74; assoc ed, J Clin Endocrinol & Metab, 63-67; prog officer, Ford Found, 74-75; pres, St Luke's-Roosevelt Inst Health Sci, 81- *Honors & Awards:* Ciba Award, Endocrine Soc, 52 & Koch Award, 70; Roussel Prize, 84; Dale Medal, 86. *Mem:* Nat Acad Sci; Am Chem Soc; Am Soc Biol Chem; fel NY Acad Sci; Endocrine Soc (vpres, 67, pres, 74). *Res:* Steroid chemistry and biochemistry; biogenesis and metabolism of steroid hormones; steroid hormone-protein conjugates; steroid sulfates and lipoidal derivatives of steroids. *Mailing Add:* St Luke's-Roosevelt Inst Healt Sci 432 W 58 St New York NY 10019

LIEBERMANN, HOWARD HORST, b Ger, Nov 27, 49; m 79; c 2. THERMODYNAMICS & MATERIAL PROPERTIES, ELECTROMAGNETISM. *Educ:* Polytech Inst New York, BS, 72; Univ Pa, MS, 75, PhD(metall & mat sci), 77. *Prof Exp:* Staff metallurgist amorphous alloys, Corp Res & Develop, Gen Elec Co, 77-81; sr metallurgist amorphous alloys, 82-84, MGR, RES & DEVELOP, ALLIED CORP METGLAS PROD, 85- *Mem:* Sigma Xi; Magnetics Soc; Am Soc Metals; Am Inst Mining, Metall & Petrol Engrs. *Res:* Materials processing; amorphous alloys; magnetic materials; electronic soldering alloys. *Mailing Add:* Allied Corp Metglas Products Six Eastmans Rd Parsippany NJ 07054

LIEBERMANN, LEONARD NORMAN, b Ironwood, Mich, May 14, 15; m 41; c 3. PHYSICS. *Educ:* Univ Chicago, BS, 37, MS, 38, PhD(physics), 40. *Prof Exp:* Instr physics, Wash Univ, 40-41; instr, Univ Kans, 41-43, asst prof, 43-44; prin physicist bur ships, Woods Hole Oceanog Inst, Mass, 44-46; res assoc, marine phys lab, 46-48, assoc prof geophys, 48-54, PROF PHYSICS, UNIV CALIF, SAN DIEGO, 54- *Concurrent Pos:* Guggenheim Found fel, 52-53; dir, Proj Sorrento, 59. *Mem:* Fel Am Phys Soc; fel Acoust Soc Am. *Res:* Ultrasonics; underwater sound; hydrodynamics; properties of liquids; electromagnetic propagation; solid state. *Mailing Add:* Physics-0319 Univ Calif San Diego 9500 Gillman Dr La Jolla CA 92093-0319

LIEBERMANN, ROBERT C, b Ellwood City, Pa, Feb 6, 42; m 64; c 3. GEOPHYSICS. *Educ:* Calif Inst Technol, BS, 64; Columbia Univ, PhD(geophys), 69. *Prof Exp:* Res scientist, Lamont-Doherty Geol Observ, 69-70; res fel geophys, Calif Inst Technol, 70; mem fac, Australian Nat Univ, 70-76; ASSOC PROF, STATE UNIV NY, STONY BROOK, 76- *Concurrent Pos:* Assoc ed, J Geophys Res, 73-76. *Mem:* Fel Royal Astron Soc; Am Geophys Union; Seismol Soc Am. *Res:* Relative excitation of seismic waves by earthquakes and underground explosions; elastic properties of minerals and rocks as a function of pressure and temperature; composition and mineralogy of earth's mantle. *Mailing Add:* Dept Earth Space Sci State Univ NY Stony Brook NY 11794

LIEBERT, JAMES WILLIAM, b Coffeyville, Kans, June 19, 46. ASTRONOMY, ASTROPHYSICS. *Educ:* Univ Kans, BA, 68; Univ Calif, Berkeley, MA, 70, PhD(astron), 77. *Prof Exp:* Res assoc, 76-79, asst prof, 79-86, PROF ASTRON, STEWARD OBSERV, UNIV ARIZ, 86- *Concurrent Pos:* Prin investr, NSF grant, Univ Ariz, 78-80. *Honors & Awards:* Trumpler Prize, Astron Soc Pac, 77. *Mem:* Am Astron Soc; Int Astron Union; Astron Soc Pac. *Res:* Observational stellar astronomy and astrophysics; white dwarf stars. *Mailing Add:* Steward Observ Univ of Ariz Tucson AZ 85721

LIEBES, SIDNEY, JR, b San Francisco, Calif, Dec 13, 29; m 58; c 2. PHYSICS. *Educ:* Princeton Univ, BSE, 52; Stanford Univ, PhD(physics), 58. *Prof Exp:* Instr physics, Princeton Univ, 57-61, asst prof, 61-64; res assoc-physicist, dept genetics, Med Ctr, 64-80, MEM STAFF, DEPT COMPUT SCI, STANFORD UNIV, 80-; AT HEWLETT-PACKARD CO. *Mem:* Am Phys Soc; Am Asn Physics Teachers. *Res:* Experimental atomic and electron physics; gravitation experiments; mass spectrometry; physical microanalysis; techniques applied to bio-medical research; computer imagery processing; Martian Lander imagery. *Mailing Add:* Hewlett-Packard Co 1501 Page Mill Rd 3 U Palo Alto CA 94304

LIEBESKIND, HERBERT, b New York, NY, Nov 24, 21; m 43, 83; c 2. PHYSICAL CHEMISTRY. *Educ:* NY Univ, BS, 41. *Prof Exp:* Asst instr chem, NY Univ, 43-45; from instr to prof, 45-88, asst dean, 68-72, dir admissions & registr, 70-87, dean admissions & records, 72-87, EMER PROF CHEM, COOPER UNION, 88- *Concurrent Pos:* Vis lectr, Stevens Inst Technol, 52-54; vis assoc prof, Yeshiva Univ, 61-62. *Mem:* Am Chem Soc; Am Soc Eng Educ; NY Acad Sci. *Mailing Add:* Dept Chem Cooper Union New York NY 10003

LIEBESKIND, LANNY STEVEN, b Buffalo, NY, Sept 5, 50. ORGANIC CHEMISTRY, ORGANOMETALLIC CHEMISTRY. *Educ:* State Univ NY, Buffalo, BS, 72; Univ Rochester, MS, 74, PhD(chem), 76. *Prof Exp:* NSF fel chem, Mass Inst Technol, 76-77; NIH fel, Stanford Univ, 77-78; ASST PROF CHEM, FLA STATE UNIV, 78-; AT DEPT CHEM, EMORY UNIV. *Mem:* Am Chem Soc. *Res:* Application of organotransition metal chemistry to the solution of problems in synthetic organic chemistry. *Mailing Add:* Dept Chem Emory Univ Atlanta GA 30332-0001

LIEBHARDT, WILLIAM C, b Duluth, Minn, Feb 16, 36; m 61; c 4. SOILS, PLANT PHYSIOLOGY. *Educ:* Univ Wis-Madison, BS, 58, MS, 64, PhD(soils), 66. *Prof Exp:* Agronomist, Stand Fruit Co, 66-68; sr agronomist, Allied Chem Corp, 68-69; from asst prof to assoc prof plant sci, Univ Del, 70-81; from assoc dir to dir res, Rodale Res Ctr, 81-87; DIR, SUSTAINABLE AGR RES & EDCU PROG, DEPT AGR & RANGE SCI, UNIV CALIF, DAVIS, 87- *Res:* Soil fertility; plant nutrition; sustainable agriculture; farming systems. *Mailing Add:* Dept Agr & Range Sci Univ Calif Davis CA 95616

LIEBIG, WILLIAM JOHN, b Huntingdon, Pa, Mar 24, 23; m 78; c 1. NEW PRODUCT DEVELOPMENT, WORLDWIDE DISTRIBUTION. *Educ:* Juanita Col, BS, 43; Augustana Col, 43; Univ Pa, 47; Philadelphia Col Textiles & Sci, MS, 49; New York Univ, MBA, 51. *Prof Exp:* Vpres & gen mgr, Meadox Weaving Co, 54-55; pres, Dormeyer Sales Corp, 55-60; PRES & CHIEF EXEC OFFICER, MEADOX MEDICALS, INC, 61-, DIR, 82- *Concurrent Pos:* Fel, Augustana Col, Rock Island, Ill, 43 & Univ Pa, 47; div mgr, Susquehanna Mills, Inc, 49-54; div sales mgr, Webcor, Inc & Camfield, Inc, 55-60; chmn, Liebig Found, 61-; vpres & dir, Huntingdon Throwing Mills, 67-; dir, Meadox Ltd, UK, 78-, Brazil, 81-, Surgimed A-S, 82-, France, 83-, Deutschland GMBH, 86- *Honors & Awards:* Gold Hektoen Award, AMA, 76. *Mem:* Asn Advan Med Instrumentation; Health Indust Mfr's Asn. *Res:* Cardiac and vascular prosthetics, biological and synthetic, for surgical implant purposes. *Mailing Add:* 112 Bauer Dr Oakland NJ 07436

LIEBLEIN, SEYMOUR, b New York, NY, June 17, 23. AEROSPACE ENGINEERING. *Educ:* City Col New York, BS, 44; Case Inst Technol, MS, 52. *Prof Exp:* Researcher, Nat Adv Comt Aeronauts, Lewis Res Ctr, NASA, 44-57, chief, Flow Anal Br, 57-65, chief, Vertical Takeoff & Landing Propulsion Br, 65-70, div tech asst, Short Takeoff, Landing & Noise Div, 70-74; mgr & owner, Tech Report Serv, 77-; RETIRED. *Honors & Awards:* Gas Turbine Award, Am Soc Mech Engrs, 61; Goddard Award, Am Inst Aeronaut & Astronaut, 67. *Mem:* Am Soc Mech Engrs; assoc fel Am Inst Aeronaut & Astronaut. *Res:* Fluid flow and design in axial flow compressors; aerodynamic performance and design of vertical takeoff and landing propulsion systems; waste-heat systems, space power; technical report writing; wind turbine blades and flow. *Mailing Add:* 3400 Wooster Rd Suite 320 Rocky River OH 44116

LIEBLING, RICHARD STEPHEN, b Brooklyn, NY, Aug 31, 38; m 70. MINERALOGY. *Educ:* Columbia Univ, BA, 60, MA, 61, PhD(mineral), 63. *Prof Exp:* Sr ceramist, Carborundum Co, 63-68; asst prof, 68-73, ASSOC PROF GEOL, HUNTER COL, 73- *Mem:* Mineral Soc Am; Sigma Xi. *Res:* Clay mineralogy of sediments. *Mailing Add:* Dept Geol-Geog CUNY Hunter Col 695 Park Ave New York NY 10021

LIEBMAN, ARNOLD ALVIN, b St Paul, Minn, Mar 5, 31; m 55, 77; c 4. ORGANIC CHEMISTRY, RADIOCHEMISTRY. *Educ:* Univ Minn, BS, 56, PhD(pharmaceut chem), 61. *Prof Exp:* Asst prof biochem, Loyola Univ, La, 61-63; Nat Inst Gen Med Sci res fel chem, Univ Calif, Berkeley, 63-66; asst prof chem, Sch Pharm, Univ Md, 66-68; sr chemist, 68-72, res group chief, 72-80, sr res group chief, 80-81, res sect chief, 82-85, RES INVESTR, HOFFMANN-LA ROCHE, INC, 85- *Mem:* AAAS; Am Chem Soc; Int Isotope Soc. *Res:* Heterocyclic chemistry of natural products; isotopic synthesis, heterocyclic chemistry. *Mailing Add:* Chem Res Dept Hoffmann-La Roche Inc Nutley NJ 07110

LIEBMAN, FREDERICK MELVIN, b New York, NY, July 26. 22; m 48; c 3. PHYSIOLOGY. *Educ:* NY Univ, BA, 42, PhD, 56; Univ Pa, DDS, 47. *Prof Exp:* Asst, 53-56, from instr to assoc prof, 56-65, PROF PHYSIOL & BIOPHYS, COL DENT, NY UNIV, 65-, CHMN DEPT, 69- *Concurrent Pos:* Int Asn Dent Res rep, Int Cong Physiol, Buenos Aires, Arg, 59; dir, Basic Med Sci (Dent), Grad Sch Arts & Sci, NY Univ. *Mem:* AAAS; Am Physiol Soc; Harvey Soc; NY Acad Sci; fel Am Col Dent. *Res:* Peripheral circulation; control of circulation in the dental pulp and oral cavity; functional activity of the muscles of mastication; control of posture and movement; pain; analgesics; narcotic-antagonists; opioid agonists. *Mailing Add:* Dept of Physiol NY Univ Col of Dent New York NY 10010

LIEBMAN, JEFFREY MARK, b Milwaukee, Wis, Nov 7, 46; m 74; c 2. PSYCHOPHARMACOLOGY, PHYSIOLOGICAL PSYCHOLOGY. *Educ:* Oberlin Col, BA, 68; Univ Calif, Los Angeles, PhD(psychol), 73. *Prof Exp:* Res assoc fel psychopharmacol, Sch Med, Univ Calif, San Diego, 73-76; SR STAFF SCIENTIST PSYCHOPHARMACOL, CIBA-GEIGY PHARMACEUT, 76- *Concurrent Pos:* Res fels, Sloan Found, 73-74 & Alcoholism, Drug Abuse & Ment Health Admin, 74-76. *Mem:* Soc Neurosci; Sigma Xi. *Res:* Neurotransmitter mechanisms of behavior; psychopharmacological models of mental disorders. *Mailing Add:* 14 Tall Oaks Dr Summit NJ 07901

LIEBMAN, JOEL FREDRIC, b Brooklyn, NY, May 6, 47; m 70. THEORETICAL CHEMISTRY, INORGANIC CHEMISTRY. *Educ:* Brooklyn Col, BS, 67; Princeton Univ, MA, 68, PhD(chem), 70. *Prof Exp:* NATO fel, Depts Phys & Theoret Chem, Cambridge Univ, 70-71; Nat Res Coun & Nat Bur Standards fel, Inorg Chem Sect, Nat Bur Standards, 71-72; from asst prof to assoc prof, 72-82, PROF, DEPT CHEM, UNIV MD, BALTIMORE COUNTY, 82- *Concurrent Pos:* Ramsay hon fel, Ramsay Mem Fel Trust, 70; consult & contractor, Nat Bur Standards, 72-; unofficial consult, Argonne Nat Lab, 72-75, guest scientist, 75-82; co-ed, Molecular Struct & Energetics, 84-; mem ed adv bd, Methods Stereochem Anal, 85-; consult ed, Struct Chem, 90- *Mem:* Am Chem Soc; Am Phys Soc; Sigma Xi. *Res:* Chemical bonding theory, rules and regularities of molecular geometry and energetics; strain and resonance energy of alicyclic and aromatic hydrocarbons; noble gas and fluorine compounds; thermochemistry of molecular ions; mathematical chemistry; structural chemistry. *Mailing Add:* Dept Chem & Biochem Univ of Md Baltimore County Campus Baltimore MD 21228

LIEBMAN, JON C(HARLES), b Cincinnati, Ohio, Sept 10, 34; m 58; c 3. ENGINEERING, OPERATIONS RESEARCH. *Educ:* Univ Colo, BS, 56; Cornell Univ, MS, 63, PhD(sanit eng), 65. *Prof Exp:* From asst prof to assoc prof environ eng, Johns Hopkins Univ, 65-72; assoc head dept civil eng, 76-78, head, dept civil eng, 78-84, PROF ENVIRON ENG, UNIV ILL, URBANA-CHAMPAIGN, 72- *Mem:* Fel AAAS; Inst Mgt Sci; Am Soc Civil Engrs; Sigma Xi. *Res:* Applications of operations research to the field of environmental engineering. *Mailing Add:* Dept of Civil Eng Univ of Ill Urbana-Champaign Urbana IL 61801-2397

LIEBMAN, JUDITH STENZEL, b Denver, Colo, July 2, 36; m 58; c 3. RESEARCH ADMINISTRATION. *Educ:* Univ Colo, BA, 58; Johns Hopkins Univ, PhD(opers res), 71. *Prof Exp:* Engr data anal, Convair Astronaut, Gen Dynamics, 58-59; programmer eng systs, Gen Elec Co, 63-64; programmer chem, Cornell Univ, 64-65; res asst opers res, Johns Hopkins Univ, 65-71, asst prof & health serv res scholar, 71-72; from asst prof opers res to assoc prof opers res, 72-84, PROF OPERS RES, UNIV ILL, URBANA, 84-, VCHANCELLOR RES, & DEAN GRAD COL, 87- *Concurrent Pos:* Pres, bd dir, E Cent Ill Health Systs Agency, 80-82; mem, NSF adv comt eng, 88-; chair, NSF adv comt data & policy analysis, 90- *Mem:* Opers Res Soc Am (pres, 87-88); Am Inst Indust Engrs; Sigma Xi; Inst Mgt Sci; Am Asn Univ Profs; Am Soc Eng Educ; AAAS. *Res:* Mathematical optimization; model building; applications of operations research in health, energy management and transportation. *Mailing Add:* 420 Swanlund 601 E John Univ of Ill Champaign IL 61820

LIEBMAN, PAUL ARNO, b Pittsburgh, Pa, Aug 1, 33; m; c 1. BIOPHYSICS, PHYSIOLOGY. *Educ:* Univ Pittsburgh, BS, 54; Johns Hopkins Univ, MD, 58. *Prof Exp:* Intern internal med, Barnes Hosp, St Louis, Mo, 58-59; res assoc physiol, 63-65, asst prof, 65-69, assoc prof, 69-76, PROF ANAT, UNIV PA, 76-, PROF OPHTHALMOL, 77- *Concurrent Pos:* Fel biophys, Univ Pa, 59-63. *Mem:* Biophys Soc; Asn Res Vision & Ophthal; Am Soc Neurosci. *Res:* Vision; microspectrophotometry of single visual receptors; transducer mechanism of photoreceptors in vision. *Mailing Add:* Dept Anat Univ Pa Sch Med 36th & Hamilton Wk Philadelphia PA 19104-6058

LIEBMAN, SUSAN WEISS, b New York, NY, Dec 2, 47; m 69; c 2. MOLECULAR GENETICS. *Educ:* Mass Inst Technol, BS, 68; Harvard Univ, MA, 69; Univ Rochester, PhD(biophys), 74. *Prof Exp:* Am Cancer Soc fel, Sch Med & Dent, Univ Rochester, 74-76; from asst prof to assoc prof, 77-87, PROF BIOL, UNIV ILL, CHICAGO CIRCLE, 87- *Concurrent Pos:* USPHS career develop award. *Mem:* AAAS; Genetics Soc Am; Am Soc Microbiol. *Res:* Molecular genetics of yeast, including nonsense suppression, mutators, transposable elements. *Mailing Add:* Dept of Biol Sci Box 4348 Chicago IL 60680

LIEBNER, EDWIN J, b Chicago, Ill, July 12, 21; m 63; c 2. RADIOLOGY. *Educ:* Univ Ill, BS, 44, MD, 46. *Prof Exp:* Resident radiol, Ill Res Hosps, 53-56; from asst prof to assoc prof, 56-66, PROF RADIOL, UNIV ILL HOSP, 66-, ACTG HEAD DEPT RADIOL, UNIV ILL COL MED, 71-; DIR RADIOTHER DIV, ILL RES & EDUC HOSPS, 61- *Concurrent Pos:* Consult radiol, Vet Admin Hosp, Hines, 64- *Mem:* Am Radium Soc; Roentgen Ray Soc; Am Soc Therapeut Radiol; Radiol Soc NAm. *Res:* Therapeutic lymphography; refrigeration and irradiation; therapeutic pediatric radiology. *Mailing Add:* Dept of Radiol Univ of Ill Col of Med PO Box 6998 Chicago IL 60680

LIEBNITZ, PAUL W, b Kansas City, Mo, Jan 18, 35; m 61; c 4. MATHEMATICS. *Educ:* Rockhurst Col, BS, 55; Univ Kans, MA, 57, PhD(math), 64. *Prof Exp:* Asst prof, 61-67, ASSOC PROF MATH, UNIV MO-KANSAS CITY, 67- *Mem:* Am Math Soc; Math Asn Am. *Res:* Topology, theory of retracts. *Mailing Add:* Dept Math Univ Mo Kansas City MO 64110

LIEBOWITZ, HAROLD, b Brooklyn, NY, June 25, 24; m 51; c 3. MECHANICS, AERONAUTICAL ENGINEERING. *Educ:* Polytech Inst Brooklyn, BAeroE, 44, MAeroE, 46, DAeroE, 48. *Prof Exp:* Res asst, Polytech Inst Brooklyn, 45-46, res assoc, 46-47, sr res assoc, 47-48; aeronaut engr, Off Naval Res, 48-50, eng mech scientist, 50-51, physicist, 51-54, aeronaut res engr & tech consult, 54-56, chief tech eng consult & coordr struct mech, 56-59, head struct mech br, 59-60, eng consult, 60-61, head struct mech br, 61-69, eng adv & coordr Polaris prog & dir prog solid propellant mech, 62-68; dean & prof, 68-90, L STANLEY CRANE PROF ENG & APPL SCI, GEORGE WASHINGTON UNIV, 90- *Concurrent Pos:* Vis prof aeronaut eng, actg asst dean grad sch & exec dir exp sta, Univ Colo, 60-61, dir eng curricula study, NSF grant, 60-61; res prof, Cath Univ Am, 62-68; tech adv, US House of Rep, founder & ed-in-chief, J Eng Fracture Mech, & prin investr, NASA res grants, 68-; ed-in-chief, J Comput & Struct, 72-; mem, Adv Comt Space Vehicles, NASA, Inter-Agency Eng Sci Group, Adv Comt Mat, Submarine Struct Res, Submarine Acoustics Comt & Deep Submergence Steering Task Group, US Navy, Ad Hoc Working Group Micromech & Design for Brittle Mat Group, Mat Adv Bd, Nat Acad Sci-Nat Res Coun & Tech Adv Group Cent Activity for Shock, Vibration & Assoc Environ, Dept Defense; consult, Acad Press, Advan Eng Res & Develop Co, CASA/GIFTS, Inc, ESDU Int, Ltd, Pergamon Press, Pratt Whitney Aircraft Co, Reynolds Metals Co; lectr, Asn Govt Civil Engrs, Philippines, 65, Asn Struct Engrs, 65, Pac Meeting Joint Am Soc Mech Engrs & Am Soc Testing Mats, Seattle, 65, Am Inst Aeronauts & Astronauts, Chicago, 66, Am Inst Mining, Metall & Petrol Engrs, Los Angeles, 66, Univ Cincinnati, 66, Univ Md, 66, NY Univ, 67, Syracuse Univ NSF prog, 67, Princeton Univ, 67, Aeronaut Soc India, 69, Am Soc Metals, 69, Duke Univ, 69; mem, NSF Eng Progs Comt, Metal Composites Panel Mat Res Bd, Comput Mech Comt & Mat Adv Bd Struct Mat Design Comt, Nat Res Coun. *Honors & Awards:* Nilakantan Mem lectr, 69. *Mem:* Nat Acad Eng; fel AAAS; fel Am Inst Aeronaut & Astronaut; fel Soc Eng Sci (pres, 72-80); Am Soc Mech Engrs; Am Asn Univ Profs; Am Soc Eng Educ; Marine Technol Soc; Soc Exp Stress Analysis; Am Inst Metall Eng; fel Am Acad Mech; fel Am Soc Metals; Am Soc Testing & Mat; Asn Comput Mach; Soc Mfg Engrs; Sigma Xi. *Res:* Applied mechanics; engineering curricula; astronautics and aeronautics; materials engineering; solid mechanics; solid propellant propulsion; rheology; dynamics; aeronautical missile, space and ship structures; weapons and weapons systems; fundamental engineering research; fracture mechanics; computers and structures; author of more than 130 technical publications. *Mailing Add:* Sch Eng & Appl Sci 2021 K St NW Suite 710 Washington DC 20006

LIEBOWITZ, STEPHEN MARC, PHARMACEUTICAL CHEMISTRY. *Educ:* State Univ NY, Buffalo, BS, 74; Va Commonwealth Univ, PhD(med chem), 80. *Prof Exp:* Fel, Ohio State Univ, 80 & Adria Labs, 80-81; fel, Adria Labs, 80-81; ASST PROF ANALYSIS PHARMACEUT CHEM, UNIV TEX, AUSTIN, 81- *Concurrent Pos:* Prin investr, Robert A Welch Found, 82- *Mem:* Am Chem Soc; NY Acad Sci; AAAS. *Res:* Synthesis of organic molecules to aid in a basic understanding of biochemical processes on a molecular level; new synthetic methodology. *Mailing Add:* Schering Plough Bldg K-11-2 Calloping Hill Rd Kenilworth NJ 07033

LIEBSON, SIDNEY HAROLD, b New York, NY, July 9, 20; m 47; c 2. PHYSICS. *Educ:* City Col New York, BS, 39; Univ Mich, MS, 40; Univ Md, PhD(physics), 47. *Prof Exp:* Physicist, Naval Res Lab, 40-49, head electromagnetics br, 49-55; mgr res & develop, Nuclear Develop Corp Am, 55-59; asst dir physics, Armour Res Found, Ill Inst Technol, 59-60; mgr phys res, Nat Cash Register Co, 60-66; mgr xerographic technol, Xerox Corp, 66-69, sr corp planner, 69-74, mgr mfg Res & Develop, 74-83; CONSULT, 83- *Mem:* Fel Am Phys Soc; Inst Elec & Electronics Eng. *Res:* Solid state phenomena; Geiger counters; electronic circuit analysis and design; discharge mechanism of self-quenching Geiger-Miller counters; scintillation and fluorescence of organics; photoconductivity; manufacturing technologies; research administration. *Mailing Add:* 15 Forestwood Dr Stamford CT 06903

LIECHTY, RICHARD DALE, b Lake Geneva, Wis, Oct 20, 25; m 52; c 3. ENDOCRINE SURGERY. *Educ:* Yale Univ, BA, 50; Northwestern Univ, MD, 54. *Prof Exp:* Assoc prof surg, Med Sch, Univ Iowa, 61-71; assoc dean grad med educ, 84-88, PROF SURG, MED SCH, UNIV COLO, 71- *Mem:* Fel Am Col Surgeons; Am Thyroid Asn; Am Asn Endocrine Surgeons; Western Surg Asn (pres, 86). *Res:* Endocrinology; surgery of Grave's disease; anatomy of parathyroid glands. *Mailing Add:* 4200 E Ninth Ave B-192 Denver CO 80262

LIEDL, GERALD L(EROY), b Fergus Falls, Minn, Mar 2, 33; m 57; c 2. MATERIALS SCIENCE, METALLURGICAL ENGINEERING. *Educ:* Purdue Univ, BS, 55, PhD(metall eng), 60. *Prof Exp:* From instr to assoc prof, 58-73, asst head dept, 69-78, PROF METALL ENG, PURDUE UNIV, WEST LAFAYETTE, 73-, HEAD DEPT, 78- *Mem:* Am Soc Eng Educ; fel Am Soc Metals; Metals Soc; Mat Res Soc. *Res:* Diffraction; electron microscopy; correlations among structure, texture, and properties of crystalline solids and thin films. *Mailing Add:* Sch of Mat Eng Purdue Univ MSEE Bldg West Lafayette IN 47907

LIEF, HAROLD ISAIAH, b New York, NY, Dec 29, 17; m 61; c 5. MARITAL & SEX THERAPY & RESEARCH. *Educ:* Univ Mich, AB, 38; NY Univ, MD, 42; Columbia Univ, cert psychoanal, 50. *Hon Degrees:* MA, Univ Pa, 71. *Prof Exp:* Intern, Queens Gen Hosp, Jamaica, NY, 42-43; resident psychiat, Long Island Med Col, 46-48; res asst, Col Physicians & Surgeons, Columbia Univ, 49-51; from asst prof to prof psychiat, Sch Med, Tulane Univ, 51-67; dir, Div Family Study, 67-81, dir, Marriage Coun Philadelphia & Ctr Study Sex Educ in Med, 68-81, prof, 67-82, EMER PROF PSYCHIAT, SCH MED, UNIV PA, 82- *Concurrent Pos:* Vis prof, Sch Med, Univ Va, 58; pres, Sex Info & Educ Coun of US, 68-70; consult, HEW, 69-76, WHO, 71 & 74, AMA, 70-78 & Psychiat Educ Br, NIMH, 74-75; assoc psychiatrist, 81-83, psychiatrist, Pa Hosp, 83-87, emer psychiatrist, 87-; hon co-pres, World Cong Sexology, 81. *Honors & Awards:* Ann Award Soc Sci, Am Asn Sex Educrs, Counrs & Therapists, 80 & 90. *Mem:* Fel Am Acad Psychoanal (pres, 67-68); fel Am Psychiat Asn; fel Am Col Psychiat; fel Am Col Psychoanal; Am Psychosomatic Soc; Soc Sex Therapists & Researchers; World Asn Sexology (secy, 81-85, vpres, 85-89). *Res:* Marital and sexual relations; sex education in medicine; adult development; psycho-endocrinological-pharmacologic aspects human sexuality; adolescent sexuality. *Mailing Add:* 700 Spruce St Suite 503 Philadelphia PA 19106

LIEGEY, FRANCIS WILLIAM, b Frenchville, Pa, Jan 4, 23; m 47; c 6. MICROBIOLOGY. *Educ:* St Bonaventure Univ, BS, 47, MS, 50, PhD(microbiol), 59. *Prof Exp:* From instr to assoc prof biol, St Bonaventure Univ, 48-64; prof biol, Ind Univ Pa, 64-, chmn dept, 72-90; RETIRED. *Mem:* Am Soc Microbiol. *Res:* Microbial ecology of acid mine streams. *Mailing Add:* 23 Elm St Indiana PA 15701

LIEHR, JOACHIM G, b Namslay, Ger, June 20, 42; US citizen. CANCER RESEARCH, HORMONAL CARCINOGENESIS. *Educ:* Univ Munster, Ger, Vordiplom, 65; Univ Del, PhD, 68. *Prof Exp:* Vis asst prof, Inst Lipid Res, Baylor, 72-74; res chemist, Ciba-Geigy Ltd, Switz, 74-76; asst prof, Univ Tex Med Sch, Houston, 76-85; PROF PHARMACOL, UNIV TEX MED BR, 85- *Concurrent Pos:* Mem, Chem Pathol Study Sect, Nat Cancer Inst, 86-90. *Mem:* Am Asn Cancer Res; AAAS; Am Soc Mass Spectroscopy; Endocrine Soc; Am Soc Biol Chemists. *Res:* Mechanism of estrogen-induced cancer; tumor-preventing action of vitamin C; synthesis of non-carcinogenic estrogens; biochemical and clinical applications of mass spectroscopy. *Mailing Add:* Pharmacol & Toxicol Br Univ Tex Med Br 301 University Blvd Galveston TX 77550-2774

LIELMEZS, JANIS, b Riga, Latvia, June 1, 26; US citizen; m 70. CHEMICAL ENGINEERING. *Educ:* Univ Denver, BS, 54; Northwestern Univ, MS, 56. *Prof Exp:* Engr, Snow, Ice & Permafrost Res Estab, US Army Corps Eng, 56; engr, Shell Develop Co, 57-58, 59, consult chem eng, 58-59; res engr, Inst Mineral Res, Mich Col Mining & Technol, 60-63, asst prof chem eng, 62-63; from asst prof to assoc prof, 63-78, PROF CHEM ENG, UNIV BC, 78- *Concurrent Pos:* Ed chem, Tech Rev, Latvian Engrs Asn, 74-; ed sci & technol, Latvian Encycl, 76- *Honors & Awards:* Sci & Technol Award, World Fedn Free Latvians, 81. *Mem:* Fel NY Acad Sci; Am Chem Soc; Am Inst Chem Engrs; fel Chem Inst Can; Am Asn Univ Profs; Am Inst Chem Engrs. *Res:* Applied and theoretical thermodynamics; applied mathematics; fluid flow; magnetism and phase transformations; magnetocatalytic effect in chemical reactions; transport properties of fluids. *Mailing Add:* Dept of Chem Eng Univ of BC Vancouver BC V6T 1W5 Can

LIEM, KAREL F, b Java, Indonesia, Nov 24, 35; m 65. VERTEBRATE MORPHOLOGY. *Educ:* Indonesia Univ, BSc, 57, MSc, 58; Univ Ill, PhD(zool), 61. *Prof Exp:* Asst prof zool, Leiden Univ, 62-64; from asst prof to assoc prof anat, Univ Ill Col Med, 64-72; HENRY BRYANT BIGELOW PROF, CUR ICHTHYOL & PROF BIOL, HARVARD UNIV, 72- *Concurrent Pos:* Head, Div Vert Anat, Chicago Natural Hist Mus, Ill, 65-72; mem comt Latimeria, Nat Acad Sci, 67; Guggenheim fel, 70-71; mem vis comt, New Eng Aquarium, 74-; trustee, Cohosset Marine Biol Sta, 74-; ed, Copeia & assoc ed, J Morphol, 74- *Mem:* Am Soc Zool; Am Soc Ichthyol & Herpet; Soc Syst Zool; fel Zool Soc London; Neth Royal Zool Soc. *Res:* Evolution of chordate structure; functional anatomy of teleosts; morphology and hydrodynamics of air-breathing teleost blood circulations; sex reversal in teleosts; functional anatomy and evolution of African cichlid fishes. *Mailing Add:* Mus of Comp Zool Harvard Univ Cambridge MA 02138

LIEM, RONALD KIAN HONG, b Lombok, Indonesia, Feb 8, 46; US citizen. NEUROBIOLOGY. *Educ:* Amherst Col, AB, 67; Cornell Univ, MSc, 69, PhD(chem). *Prof Exp:* Assoc prof pharmacol, NY Univ Sch Med, 78-87; assoc prof, 87-90, PROF PATH, ANAT & CELL BIOL, COL PHYSICIANS & SURGEONS, COLUMBIA UNIV, 91- *Mem:* Am Soc Cell Biol; Am Soc Neurosci; Am Soc Neurochem. *Res:* Biochemical studies on the neuronal cytoskeleton, especially with regard to the assembly of neurofilaments and their interactions with other cytoskeletal elements, both in vivo and in vitro. *Mailing Add:* Dept Path Columbia Univ 630 W 168th St New York NY 10038

LIEMOHN, HAROLD BENJAMIN, b Minneapolis, Minn, Feb 2, 35; m 57; c 5. SPACECRAFT-ENVIRONMENT INTERACTIONS. *Educ:* Univ Minn, BA, 56, MS, 59; Univ Wash, PhD(physics), 62. *Prof Exp:* Teaching asst physics, Univ Minn, 56-59; staff mem geo-astrophys, Sci Res Labs, Boeing Co, 59-63; asst prof atmospheric & space sci, Southwest Ctr Advan Studies, 63-66; staff mem Environ Sci Res Labs, Boeing Co, 66-72; res staff math-physics, Pac Northwest, Battelle Mem Inst, 72-77; MGR SPACE PHYSICS, ENG TECHNOL, BOEING AEROSPACE CO, 77- *Concurrent Pos:* Adj asst prof, Southern Methodist Univ, 64-65; vis assoc prof, Univ Wash, 68-; chmn & secy local arrangements, Ann Meeting, Comt Space Res, Seattle, 71; reporter particle-wave interactions, Comn V, Int Asn Geomag & Aeronomy, 71-73. *Mem:* AAAS; Am Phys Soc; Am Geophys Union; Am Asn Physics Teachers; Am Inst Aeronaut & Astronaut. *Res:* Theoretical research in radiation belt physics, electromagnetic waves in magnetoplasma, hydromagnetic waves in magnetosphere, and spacecraft charging and contamination. *Mailing Add:* PO Box 3999 MS 2T-50 Boeing Aerospace Co Seattle WA 98124

LIEN, ERIC JUNG-CHI, b Kaohsiung, Taiwan, Nov 30, 37; m 65; c 2. PHARMACEUTICAL CHEMISTRY. *Educ:* Taiwan Univ, BS, 60; Univ Calif, San Francisco, PhD(pharmaceut chem), 66. *Prof Exp:* Res assoc bio-org chem, Pomona Col, 67-68; from asst prof to assoc prof, 68-76, coordr, 78-84, PROF PHARMACEUT & BIOMED CHEM, SCH PHARM, UNIV SOUTHERN CALIF, 76- *Concurrent Pos:* Mem comt, Develop Therapeut Contract Rev, Nat Cancer Inst, 83-87. *Mem:* AAAS; Am Pharmaceut Asn; Am Chem Soc; Am Asn Cols Pharm; Am Asn Cancer Res. *Res:* Structure-activity relationship and bioorganic chemistry; physical organic chemistry; natural products; antiviral and antitumor agents. *Mailing Add:* Sch Pharm 1985 Zonal Ave Los Angeles CA 90033

LIEN, ERIC LOUIS, b Hammond, Ind, Apr 9, 46; m 69, 87; c 4. BIOCHEMISTRY. *Educ:* Col Wooster, BA, 68; Univ Ill, Urbana-Champaign, MS, 71, PhD(biochem), 72. *Prof Exp:* Fel biochem, Sch Med, Univ Pa, 72-75; sr biochemist, Wyeth-Ayerst Labs, 75-83, mgr, Metab Disorders Sect, 83-87, assoc dir, nutrit res & develop, 87-90, DIR NUTRIT RES, WYETH-AYERST LABS, 90- *Mem:* AAAS; Am Inst Nutrit; Am Soc Parenteral & Enternal Nutrit. *Res:* Infant nutrition; triglyceride absorption; amino acid analysis. *Mailing Add:* Wyeth-Ayerst Labs Box 8299 Philadelphia PA 19101

LIEN, HWACHII, b Taipei, Taiwan, Nov 10, 30; US citizen; m 64; c 3. TURBULENT FLOWS, ENGINEERING INSTRUMENTATION. *Educ:* Nat Taiwan Univ, BS, 53; Kans State Univ, MS, 56; Univ Pa, PhD(appl mech), 62. *Prof Exp:* Staff scientist, Avco Missile Syst Div, 62-67; tech supvr, Gen Appl Sci Lab, 67-68; asst mgr, Avco Syst Div, 68-70, prin scientist, 71-74; GEN MGR ADMIN, HOPAX CO, 74- *Concurrent Pos:* Vis prof, Nat Taiwan Univ, 70-71; tech dir, Yuen Foong Yu Paper Mfg Co, 70-71; consult, Anathon Corp, 79-; pres, Financial Scis Inc, 74-; chmn, Liberty Bank & Trust Co, 77- *Mem:* Am Inst Aeronaut & Astronaut. *Res:* Fluid dynamics and its related diagnostic instrumentation in general; turbulence; chemically reacting flows; magnetohydrodynamics; mechanics of multiphase fluids; electrostatic probes. *Mailing Add:* 28 Berkshire Dr Winchester MA 01890

LIENER, IRVIN ERNEST, b Pittsburgh, Pa, June 27, 19; m 46; c 2. BIOCHEMISTRY, NUTRITION. *Educ:* Mass Inst Technol, BS, 41; Univ Southern Calif, PhD(biochem, nutrit), 49. *Prof Exp:* From instr asst prof to prof, 49-59, EMER PROF BIOCHEM, UNIV MINN, ST PAUL, 89- *Concurrent Pos:* Guggenheim fel, Carlsberg Lab, Copenhagen, Denmark, 57; ed, J Agr & Food Univ, 83-; pres, Int Lectin Soc, 87- *Honors & Awards:* Spencer Award Outstanding Achievement Agr & Food Chem, Am Chem Soc, 77 & 82; Fulbright Award, 90. *Mem:* Fel Venezuelan Asn Advan Sci; Am Chem Soc; Am Soc Biol Chem; Am Inst Nutrit; Sigma Xi. *Res:* Isolation and characterization of antinutritional factors in legumes; structure and mechanism of action of proteolytic enzymes and their naturally-occurring inhibitors. *Mailing Add:* Dept of Biochem Col of Biol Sci Univ of Minn St Paul MN 55108

LIENHARD, GUSTAV E, b Plainfield, NJ, June 21, 38; m 60; c 2. BIOCHEMISTRY. *Educ:* Amherst Col, BA, 59; Yale Univ, PhD(biochem), 64. *Prof Exp:* From asst prof to assoc prof biochem & molecular biol, Harvard Univ, 65-72; assoc prof, 72-75, PROF BIOCHEM, DARTMOUTH MED SCH, 75- *Concurrent Pos:* Res fel biochem, Brandeis Univ, 63-65. *Mem:* Am Soc Biochem & Molecular Biol. *Res:* Mechanisms of insulin action; regulation of transport by hormones. *Mailing Add:* Dept Biochem Dartmouth Med Sch Hanover NH 03756

LIENHARD, JOHN H(ENRY), b St Paul, Minn, Aug 17, 30; m 59; c 2. HISTORY, MECHANICAL ENGINEERING. *Educ:* Ore State Col, BS, 51; Univ Wash, Seattle, MS, 53; Univ Calif, Berkeley, PhD(mech eng), 61. *Prof Exp:* Design engr, Boeing Airplane Co, 51-52; instr mech eng, Univ Wash, Seattle, 55-56; assoc, Univ Calif, Berkeley, 56-61; assoc prof, Wash State Univ, 61-67; prof mech eng, Univ KY, 67-80; PROF MECH ENG, UNIV HOUSTON, 80- *Concurrent Pos:* Distinguished Speakers Bur, Am Soc Mech Engrs, 87- *Honors & Awards:* Charles Russ Richards Mem Award, Am Soc Mech Engrs, 80, Heat Transfer Mem Award, 81. *Mem:* Fel Am Soc Mech Engrs; Am Soc Eng Educ; AAAS; Soc Hist Technol. *Res:* Statistical mechanical modeling of macroscopic systems; thermal systems with emphasis on boiling and other two-phase problems; nuclear thermohydraulics; history of technology; equations of state. *Mailing Add:* Dept Mech Eng Univ Houston Houston TX 77204-4792

LIENK, SIEGFRIED ERIC, b Gary, Ind, Oct 16, 16; m 51; c 1. ENTOMOLOGY. *Educ:* Univ Idaho, BS, 42; Univ Ill, MS, 47, PhD, 51. *Prof Exp:* Entomologist pear psylla control, USDA, 42, Alaska Insect Control Proj, 49; assoc prof fruit invests, 50-70, PROF ENTOM, STATE AGR EXP STA, STATE UNIV NY COL AGR, CORNELL UNIV, 70-, ENTOMOLOGIST, 50- *Mem:* Entom Soc Am. *Res:* Biology, ecology and control of phytophagous mites; biology and control of stone fruit insects. *Mailing Add:* Dept Entom NYS Agr Exp Sta Geneva NY 14456

LIENTZ, BENNET PRICE, b Hollywood, Calif, Oct 24, 42. SIMULATION, INFORMATION SYSTEMS. *Educ:* Claremont Men's Col, BA, 64; Univ Wash, MS, 66, PhD(math), 68. *Prof Exp:* Instr math, Univ Wash, 65-68; sr res scientist, Syst Develop Corp, 68-70; assoc prof indust eng & Air Force Off Sci Res grant, Univ Southern Calif, 70-72; PROF GRAD SCH MGT, UNIV CALIF, LOS ANGELES, 77- *Concurrent Pos:* Dir, Off Admin Info Serv, 78-81. *Mem:* Opers Res Soc Am (secy-treas, 71); Inst Math Statist; Am Statist Asn; Am Math Soc. *Res:* Communication and network analysis; computers and systems analysis; computer security networks. *Mailing Add:* Grad Sch of Mgt 405 Hilgard Los Angeles CA 90024

LIEPA, GEORGE ULDIS, b Oldenburg, Germany, Oct 4, 46; US citizen; m 79; c 2. LIPID METABOLISM. *Educ:* Drake Univ, BA, 68, MA, 70; Iowa State Univ, PhD(molecular biol), 76. *Prof Exp:* Asst instr med & grad physiol, Univ Tex Health Sci Ctr, San Antonio, 76-77, NIH fel, 77-78; ASSOC PROF NUTRIT, TEX WOMAN'S UNIV, 79- *Concurrent Pos:* Assoc ed, J Am Oil Chemists Soc, 85- *Mem:* Sigma Xi; Am Oil Chemists Soc; Am Inst Nutrit; Latvian-Am Asn Univ Profs & Scientists. *Res:* Lipid metabolism; dietary care of trauma patient; metabolism during trauma. *Mailing Add:* 1020 Burning Tree Pkwy Denton TX 76201-1453

LIEPINS, ATIS AIVARS, b Aloja, Latvia, Apr 17, 35; m 60; c 2. STRUCTURAL ENGINEERING, APPLIED MECHANICS. *Educ:* Mass Inst Technol, SB, 57, SM, 60, Engr, 60. *Prof Exp:* Res engr appl mech, Res Labs, United Aircraft Corp, 60-61; prin engr, Dynatech Corp, 61-68; sr staff engr, Littleton Res & Eng Corp, 68-77; ASSOC, SIMPSON GUMPERTZ & HEGER INC, 77- *Mem:* Am Inst Aeronaut & Astronaut; Am Soc Mech Engrs; Am Soc Civil Engrs. *Res:* Static and dynamic response of thin shell structures; propeller induced ship vibration; soil-structure interaction. *Mailing Add:* Simpson Gumpertz & Heger, Inc 297 Broadway Arlington MA 02174

LIEPINS, RAIMOND, b Plavinas, Latvia, May 19, 30; US citizen; m 61; c 4. ORGANOMETALLIC CHEMISTRY, ORGANIC CHEMISTRY. *Educ:* Southern Ill Univ, BA, 54; Univ Minn, MS, 56; Kans State Univ, PhD(org chem), 60. *Prof Exp:* Res chemist, B F Goodrich Co, 60-64; res assoc polymer res, Univ Ariz, 64-66; sr chemist, Res Triangle Inst, 66-77; SECT LEADER MAT SCI & TECHNOL, LOS ALAMOS NAT LAB, 77- *Concurrent Pos:* Lectr, NC State Univ, 76, Clemson Univ, 76 & State Univ NY, New Paltz, 77; prin investr, Indust Org Chem Indust, Res Triangle Inst, Environ Protection Agency, 76-77; consult, 85- *Mem:* Am Vacuum Soc; Am Chem Soc; fel Am Inst Chem; Am Soc Mat. *Res:* Coatings for laser fusion targets; gas phase coating techniques; flame retardance; low pressure plasma applications; high temperature polymers; piezoelectric polymers; organometallic polymers; conducting polymers; liquid crystal polymers; tamper proof coating; magnetic processing. *Mailing Add:* Los Alamos Nat Lab MS E549 Los Alamos NM 87545

LIEPMAN, H(ANS) P(ETER), b Kiel, Germany, Oct 24, 13; nat US; m 46; c 4. AEROSPACE ENGINEERING. *Educ:* Swiss Fed Inst Technol, Dipl, 37; Harvard Univ, MS, 39; Univ Mich, PhD(aeronaut), 53. *Prof Exp:* From instr to asst prof aeronaut eng, Univ Cincinnati, 39-44; sr aerodynamicist, Goodyear Aircraft Corp, 44-46, chief aerodynamicist, 46-49; lectr aeronaut eng, Univ Mich, 49-55, assoc prof, 56-59, dir supersonic wind tunnel, 49-59; asst mgr, Systs Develop Dept & asst prog mgr, Aerosci Lab, TRW Systs Group, Calif, 59-71; educ & technol admin & develop consult, 71-73; mem res staff, Inst for Defense Anal, 73-78; aerospace eng, Dept Defense, 81-87; AEROSPACE CONSULT, 87- *Concurrent Pos:* Consult, Space Technol Labs, 57-59, Appl Physics Lab, Johns Hopkins Univ, 78-80 & Inst Defense Anal, 78-80; vis prof aerospace eng, Iowa State Univ, 80-81. *Mem:* Assoc fel Am Inst Aeronaut & Astronaut. *Res:* Reentry system performance and penetration aids system studies; multiple nozzle plume interactions; rocket exhaust interactions; base heating; supersonic nozzle design; system engineering of space and missile systems; comparative analysis and evaluation of tactical and strategic weapons systems and test data. *Mailing Add:* 672 Serrano Dr #1 San Luis Obispo CA 93405

LIEPMANN, H(ANS) WOLFGANG, b Berlin, Ger, July 3, 14; nat US; m 39, 54; c 2. AERONAUTICS. *Educ:* Univ Zurich, PhD(physics), 38. *Hon Degrees:* DEng, Tech Univ, Aachen, 85. *Prof Exp:* Fel physics, Univ Zurich, 38-39; fel aeronaut, Calif Inst Technol, 39-45, from asst prof to prof, 45-76, Charles Lee Powell prof fluid mech & thermodyn, 76-83, Theodore von Karman prof aeronaut, 83-85, dir, Grad Aeronaut Labs, 72-85, EMER THEODORE VON KARMAN PROF AERONAUT, CALIF INST TECHNOL, 85- *Concurrent Pos:* Foreign fel, Max-Planck Inst, 88. *Honors & Awards:* Ludwig-Prandtl-Ring, Ger Soc Aeronaut & Astronaut, 68; Worcester Reed Warner Medal, Am Soc Mech Engrs, 69; Monie A Ferst Award, Sigma Xi, 78; Fluid Dynamics Prize, Am Phys Soc, 80 & Otto Laporte Award, 85; Fluids Eng Award, Am Soc Mech Engrs, 84; Nat Medal Sci, 86; Guggenheim Medal, 86. *Mem:* Nat Acad Sci; Nat Acad Eng; AAAS; hon fel Am Inst Aeronaut & Astronaut; fel Am Acad Arts & Sci; hon fel Indian Acad Sci; hon mem Am Soc Mech Engrs. *Res:* Laminar instability, transition and turbulence; shock wave boundary layer interaction; transonic flow; aerodynamic noise; fluid mechanics of Helium II. *Mailing Add:* Mail Stop 105-50 Calif Inst Technol Pasadena CA 91125

LIER, FRANK GEORGE, b New York, NY, Feb 19, 13; m 37. BOTANY, ECOLOGY. *Educ:* Columbia Univ, PhD, 50. *Prof Exp:* Asst bot, 46-47, lectr, Sch Gen Studies, 47-50, from asst prof to prof bot, 50-77, prof biol sci, 77-80, EMER PROF BIOL SCI, SCH GEN STUDIES, COLUMBIA UNIV, 80- *Mem:* Bot Soc Am; Torrey Bot Club (pres, 64); Am Inst Biol Sci; NY Acad Sci; Am Bryol & Lichenological Soc. *Res:* Plant morphology; developmental anatomy. *Mailing Add:* 2 Cambridge Court East Old Saybrook CT 06475

LIER, JOHN, b Amsterdam, The Netherlands, Feb 23, 24; US citizen; m 50; c 2. REGIONAL CLIMATOLOGY, TROPOSPHERIC METEOROLOGY. *Educ:* Clark Univ, MA, 63; Univ Calif, Berkeley, PhD(geog), 68. *Prof Exp:* Asst prof geog, San Fransisco State Univ, 65-66 & Univ Hawaii, Hilo, 66-67; PROF GEOG & ENVIRON STUDIES, CALIF STATE UNIV, HAYWARD, 68- *Concurrent Pos:* Instr geog, Univ Sask, Saskatoon, 63. *Mem:* AAAS; Am Geog Soc; Am Meteorol Soc; Asn Am Geogr. *Res:* Meteorology and regional climatology; coastal winds; climate classification; precipitation regimes; agroclimatology; agrometeorology; man-atmosphere interaction; 19th century weather observation. *Mailing Add:* Dept Geog & Environ Studies Calif State Univ Hayward CA 94542

LIES, THOMAS ANDREW, b Oak Park, Ill, Jan 16, 29; m 59; c 2. ORGANIC CHEMISTRY. *Educ:* John Carroll Univ, BS, 49; Univ Chicago, SM, 51; Univ Wis-Madison, PhD(org chem), 58. *Prof Exp:* Res chemist, Wyandotte Chem Corp, 58-59; RES CHEMIST, AM CYANAMID CO, 59- *Mem:* Am Chem Soc. *Res:* Synthesis of pesticides. *Mailing Add:* 893 Cherry Hill Rd Princeton NJ 08540

LIESCH, JERROLD MICHAEL, b Chicago, Ill, Dec 12, 49. NATURAL PRODUCTS STRUCTURE DETERMINATION, MASS SPECTROMETRY. *Educ:* Ill Inst Technol, BS, 71; Univ Ill, MS, 73, PhD(org chem), 75. *Prof Exp:* NIH fel org chem, Mass Inst Technol, 76-77; SR INVESTR, MERCK & CO, INC, 77- *Mem:* Am Chem Soc; Am Soc Mass Spectrometry. *Res:* Structure elucidation; mass spectrometry; proton and carbon magnetic resonance spectrometry. *Mailing Add:* Merck & Co Inc PO Box 2000 R80Y345 Rahway NJ 07065-0900

LIESE, HOMER C, b New York, NY, Oct 25, 31; m 55. MINERALOGY, PETROLOGY. *Educ:* Syracuse Univ, BS, 53; Univ Utah, MS, 57, PhD(mineral), 62. *Prof Exp:* X-ray technician, Kennecott Copper Inc, 60-61; ASSOC PROF GEOL, UNIV CONN, 62- *Mem:* Geol Soc Am; Mineral Soc Am; Soc Appl Spectros. *Res:* Spectroscopy of minerals. *Mailing Add:* Dept Geol Univ Conn Main Campus U-45 354 Mansifield R Storrs CT 06268

LIETH, HELMUT HEINRICH FRIEDRICH, b Kuerten-Steeg, Germany, Dec 16, 25; m 52; c 4. GEOECOLOGY, ECOLOGICAL MODELLING. *Educ:* Univ Cologne, DrPhil, 53; Univ Stuttgart-Hohenheim, PD, 60. *Prof Exp:* Asst bot, Univ Cologne, 54-55, Agr Univ, Stuttgart-Hohenheim, 55-66; prof, Univ Hawaii, 66-67; from assoc prof to prof bot & ecol, Univ NC, Chapel Hill, 67-77; prof, 77-91, EMER PROF BIOL & ECOL, UNIV OSNABRUECK, 91- *Concurrent Pos:* Nat Res fel bot & ecol, Univ Montreal, 60-61; guest prof bot, Cent Univ, Caracas, Venezuela, 61, Univ Tolima, Ibague, Colombia, 63; guest scientist ecol & biophys, Nuclear Res Lab, Julich, Germany, 73-74; govt adv ecol, Portugal, 74; adj prof ecol, Univ NC, Chapel Hill, 77-; guest prof, Waseda Univ, Tokyo, 88- & United Arab Emirates, Univ AC Ain, 90; ed-in-chief, Int J Biometeorol, 87-, Vegetatio, 90-, Task for Veg Sci, 81. *Honors & Awards:* Biometerol Res Found Award, 82; Biometeorol Award, Am Meteorol Soc, 87. *Mem:* Int Biometeorol Soc (pres, 79-84); Int Soc Trop Ecol (pres, 85-91); Inst Asn Ecol (treas, 86-90); Ecol Soc Am. *Res:* Systems ecology, geoecology; inter-disciplinary modelling ecology, economy, sociology, climate and atmosphere, plant relations, net primary productivity, phenology and seasonality; high salinity ecosystems. *Mailing Add:* FB 5 Biol Univ Osnabrueck PO Box 4469 Osnabrueck D 4500 Germany

LIETMAN, PAUL STANLEY, b Chicago, Ill, Mar 24, 34; m 56; c 3. CLINICAL PHARMACOLOGY. *Educ:* Western Reserve Univ, AB, 55; Columbia Univ, MD, 59; Johns Hopkins Univ, PhD(physiol chem), 68. *Prof Exp:* Asst prof pediat, 68-72, asst prof pharmacol, 69-72, assoc prof med, pediat & pharmacol, 72-80, WELLCOME PROF CLIN PHARMACOL & PROF MED, PEDIAT & PHARMACOL, JOHNS HOPKINS UNIV, 80- *Concurrent Pos:* Investr, Howard Med Inst, 68-72. *Mem:* Am Soc Pharmacol & Exp Therapeut; Soc Microbiol; Am Pediat Soc; Soc Pediat Res; Am Acad Pediat. *Res:* Developmental pharmacology; antibiotics; antiviral agents. *Mailing Add:* Johns Hopkins Hosp Dept of Pediat 720 Rutland Ave Baltimore MD 21205

LIETZ, GERARD PAUL, b Chicago, Ill, Dec 10, 37; m 64; c 5. NUCLEAR PHYSICS. *Educ:* DePaul Univ, BS, 59; Univ Notre Dame, PhD(nuclear physics), 64. *Prof Exp:* Exchange asst nuclear physics, Univ Basel, 66-67; asst prof, 67-77, ASSOC PROF PHYSICS, DEPAUL UNIV, 77- *Mem:* Am Phys Soc; Am Asn Physics Teachers; Sigma Xi. *Res:* Energy levels in nuclei. *Mailing Add:* 1527 Lincoln St Evanston IL 60201

LIETZKE, DAVID ALBERT, b Pontiac, Mich, Apr 20, 40; m; c 2. PEDOLOGY, CLAY MINERALOGY. *Educ:* Mich State Univ, BS, 62, MS, 68, PhD(clay mineral geomorphol), 72. *Prof Exp:* Soil scientist, Soil Conserv Serv, Mich, 62-68 & 71-73, Mich Agr Exp Sta, 68-71; asst prof & urban soils specialist, Va Polytech Inst & State Univ, 73-79; assoc prof soils, Univ Tenn, Knoxville, 79-86; INDEPENDENT SOIL CONSULT, 86- *Concurrent Pos:* Consult, soil mapping, soil invests, 85; secy-treas, Nat Soc Consult Soil Scientists, 88-90, exec secy, 90. *Mem:* Am Soc Agron; Soil Sci Soc Am; Nat Soc Consult Soil Scientists (secy-treas, 88-90). *Res:* Processes of soil formation, soil-geomorphic relationships, soil-climatic relationships and fundamental weathering processes of earth materials to form soil parent materials. *Mailing Add:* Rte Three Box 607 Rutledge TN 37861

LIETZKE, MILTON HENRY, b Syracuse, NY, Nov 23, 20; m 43, 65; c 5. PHYSICAL CHEMISTRY. *Educ:* Colgate Univ, BA, 42; Univ Wis, MS, 44, PhD(chem), 49. *Prof Exp:* Asst, Univ Wis, 42-43, instr chem, 43-44; lab foreman, Tenn Eastman Corp, 44-47; asst, Univ Wis, 47-49; res chemist, Oak Ridge Nat Lab, 49-83; prof, 63-89, EMER PROF, UNIV TENN, 89- *Concurrent Pos:* Prof, Univ Tenn, 63- *Mem:* Am Chem Soc; fel NY Acad Sci; fel Am Inst Chemists. *Res:* Electrochemistry, electrodeposition; potential measurements; high temperature solution thermodynamics; corrosion research; phase studies; application of high speed computing techniques to chemical problems. *Mailing Add:* 7600 Twining Dr Knoxville TN 37919-7127

LIEUX, MEREDITH HOAG, b Morgan City, La, Nov 9, 39; m 68. PALYNOLOGY, BOTANY. *Educ:* La State Univ, Baton Rouge, BS, 60, PhD(bot), 69; Univ Miss, MS, 64. *Prof Exp:* Teacher pub schs, Lake Charles & Monroe, La, 60-71; fel geol, 71-72, instr bot, 72-74, asst prof, 74-80, ASSOC PROF BOT, LA STATE UNIV, BATON ROUGE, 80- *Mem:* Am Asn Stratig Palynologists; Int Bee Res Asn; Bot Soc Am; Sigma Xi; Asn Women Sci. *Res:* Holocene spore and pollen studies in the Gulf of Mexico Region; pollen morphology involving light, scanning electron and transmission electron microscopy; applied insect-pollen related studies, or melissopalynology. *Mailing Add:* La State Univ 1627 Louray Dr Baton Rouge LA 70838

LIEW, CHOONG-CHIN, b Malaysia, Sept 2, 37; Can citizen; m 64; c 3. MOLECULAR BIOLOGY, PROTEIN CHEMISTRY. *Educ:* Nanyang Univ, Singapore, BSc, 60; Univ Toronto, MA, 64, PhD(path chem), 67. *Prof Exp:* Guest investr, Rockefeller Univ, 69-70; from asst prof to assoc prof biochem, 70-78, assoc prof med, 78-79, PROF CLIN BIOCHEM & MED, UNIV TORONTO, 79- *Concurrent Pos:* Hon prof biochem, Peking Union Med Col, Zhongshan Med Col, WChina Med Univ, Zheijiang Med Univ, Shanghai Second Med Col, Xian Med Univ, Harabin Med Sci Univ, Shihezi Med Col & Xinjiang Med Col. *Honors & Awards:* Hoffman Mem Prize, 67. *Mem:* Can Biochem Soc; Biochem Soc; Am Soc Cell Biol. *Res:* Gene regulation in eukaryotes; chromosomal proteins; structure and function of cardiac myosin heavy chain genes; correlation of chromatin process and genetically determined heart diseases. *Mailing Add:* Dept Clin Biochem 100 College St Toronto ON M5G 1L5 Can

LIFKA, BERNARD WILLIAM, b Chicago, Ill, Apr 8, 31; m 56; c 4. METALLURGICAL ENGINEERING. *Educ:* Purdue Univ, BS, 58. *Prof Exp:* Engr corrosion, Alcoa Labs, 58-62, sr engr, 63-77, sr engr alloy develop, 77-79, staff engr, 80-81, tech supvr, 81-85, TECH CONSULT, ALCOA LABS, 85- *Honors & Awards:* IR 100 Award, 87; Arthur Vining Davis Award, 88. *Mem:* Am Soc Metals; Nat Asn Corrosion Engrs; Sigma Xi; Am Soc Testing & Mat. *Res:* Alloy development, increased plant production and resistance to corrosion of high-strength, heat-treatable aluminum alloys for aerospace and automotive applications. *Mailing Add:* Alloy Technol Div Bldg C Alcoa Tech Ctr PA 15069

LIFSCHITZ, MEYER DAVID, b Patchogue, NY, Jan 9, 42; m 70; c 3. NEPHROLOGY, EICOSINOID PHYSIOLOGY. *Educ:* Mass Inst Technol, BS, 63; Boston Univ, MS, 66, MD, 67. *Prof Exp:* From asst prof to assoc prof, 73-84, PROF MED, UNIV TEX HEALTH SCI CTR, SAN ANTONIO, 84. *Concurrent Pos:* NIH fel, 71, NIH, Vet Admin & NASA res grants, 73-; chief, Renal Sect, Audie Murphy Vet Admin Hosp, 76-85, actg chief, SDTU, 78-81, assoc chief staff res & develop, 84-; consult, Brooke Army Med Ctr, 79- & St Lukes Lutheran Hosp, 84-; actg chief, Div Nephrology, Univ Tex Health Sci Ctr, San Antonio, 89-91; mem, Kidney & Hypertension Coun, Am Heart Asn. *Mem:* Am Soc Nephrology; Am Physiol Soc; Am Soc Clin Invest; Am Heart Asn. *Res:* Inter-relationships of hormones and kidney function. *Mailing Add:* Dept Med Audie Murphy Vet Admin Hosp 7703 Floyd Curl Dr San Antonio TX 78284

LIFSON, NATHAN, physiology; deceased, see previous edition for last biography

LIFSON, WILLIAM E(UGENE), b Newark, NJ, Apr 17, 21; m 46; c 2. CHEMICAL ENGINEERING. *Educ:* Mass Inst Technol, BS, 41, MS, 42. *Prof Exp:* Res engr petrol prod res, Exxon Res & Eng Co, 46-51, group head, 51-54, sect head, 54-56, asst dir, 56-62, dir, Prod Res Div, 62-64, dir, Chem Res Div, 64-65, dir, Enjay Chem Labs, 65-66, mgr chem planning & coord, 66-69, asst gen mgr, Exxon Eng Technol Dept, 69-80, relocation mgr, 80-81, new facil proj exec, 81-84; RETIRED. *Mem:* Am Chem Soc; Soc Automotive Engrs. *Res:* Petroleum products and petrochemicals; engineering research and development process industries. *Mailing Add:* 365 Long Hill Dr Short Hills NJ 07078

LIGGERO, SAMUEL HENRY, b Amsterdam, NY, Apr 17, 42; m 66; c 2. PHYSICAL ORGANIC CHEMISTRY. *Educ:* Fordham Univ, BS, 64; Georgetown Univ, PhD(chem), 69. *Prof Exp:* NIH fel, Princeton Univ, 69-70; sr lab supvr film develop, Polaroid Corp, 70-71, res group leader film develop, 72-76, mgr process eng, 77-78, sr tech mgr, 79-82, plant mgr, 82-85, mktg dir, 85-87, DIR, IMAGING APPLN, POLAROID CORP, 87- *Honors & Awards:* Chuck Hall Award, 87. *Mem:* The Chem Soc; AAAS; Soc Photog Scientists & Engrs; Am Chem Soc; Sigma Xi. *Res:* Application of physical organic chemistry principles to the development of instant color photographic transparencies employing diffusion transfer processes; application of chemistry, physics and engineering to the manufacture and development of the polaroid 35mm instant slide system. *Mailing Add:* 69 Sheridan Rd Wellesley Hills MA 02181

LIGGETT, JAMES ALEXANDER, b Los Angeles, Calif, June 29, 34; m 60. CIVIL ENGINEERING. *Educ:* Tex Tech Col, BS, 56; Stanford Univ, MS, 57, PhD(civil eng), 59. *Prof Exp:* Engr, Chance Vought Aircraft Corp, 59-60; asst prof hydraul, Univ Wis, 60-61; PROF HYDRAUL, CORNELL UNIV, 61- *Mem:* Am Soc Civil Engrs; Int Asn Hydraul Res. *Res:* Hydraulics; fluid mechanics; free surface flow; circulation and temperature distribution in lakes; groundwater flow; numerical methods. *Mailing Add:* Holister Hall Cornell Univ Ithaca NY 14853

LIGGETT, LAWRENCE MELVIN, b Denver, Colo, June 22, 17; m 43; c 2. ANALYTICAL CHEMISTRY, ORGANIC CHEMISTRY. *Educ:* Cent Col, Iowa, AB, 38; Iowa State Col, PhD(chem), 43. *Prof Exp:* Res chemist, Nat Defense Res Comt, Iowa State Col, 41-43; plant supt, Alkali Chlorates & Perchlorates Cardox Corp, 43-48; supvr inorg res, Wyandotte Chems Corp, 48-55; dir res, Speer Carbon Co, 55-64, vpres & tech dir, 65-67, vpres & gen mgr, Airco Speer Electronics, 67-70, pres, Airco Speer Electronics Div, 70-75, pres, Vacuum Equip & Systs, Airco Temescal Div, Airco Inc, 75-82; RETIRED. *Concurrent Pos:* Bus & tech consult, 82- *Mem:* Am Chem Soc; Electrochem Soc. *Res:* Alkali perchlorate production; nonblack pigments for rubber and paper; carbon and graphite technology; resistors; capacitors and electronic components; technical management. *Mailing Add:* 1856 Piedras Circle Danville CA 94526

LIGGETT, THOMAS MILTON, b Danville, Ky, Mar 29, 44; m 72; c 2. MATHEMATICS. *Educ:* Oberlin Col, AB, 65; Stanford Univ, MS, 66, PhD(math), 69. *Prof Exp:* From asst prof to assoc prof, 69-76, PROF MATH, UNIV CALIF, LOS ANGELES, 76- *Concurrent Pos:* Sloan fel, 73; ed, Annals of Probability, 85-87. *Mem:* Am Math Soc; Math Asn Am; fel Inst Math Statist; Bernoulli Soc. *Res:* Probability theory; interacting particle systems. *Mailing Add:* Dept of Math Univ of Calif 405 Hilgard Ave Los Angeles CA 90024-1555

LIGGETT, WALTER STEWART, JR, b Abington, Pa, Aug 27, 40; m 62; c 3. EXPERIMENTAL DESIGN, APPLIED STATISTICS. *Educ:* Rensselaer Polytech Inst, BS, 61, MS, 64, PhD(math), 67. *Prof Exp:* Prin engr, Submarine Signal Div, Raytheon Co, Portsmouth, 65-73; mathematician, Rand Inst, New York, 73-75; statistician, Div Environ Planning, Tenn Valley Authority, 75-79; MATH STATISTICIAN, STATIST ENG DIV, NAT INST STANDARDS & TECHNOL, 79- *Concurrent Pos:* Mem, Am Statist Asn, Comt Statist & Environ, 80- *Mem:* Inst Math Statist; Soc Indust & Appl Math; Am Statist Asn. *Res:* Statistical planning of studies that involve physical measurements having non-standard error properties such as multiple components, a non constant variance, a non-normal distribution, or serial correlation. *Mailing Add:* Nat Inst Standards & Technol 20418 Shadow Oak Ct Gaithersburg MD 20879

LIGH, STEVE, b Canton, China, Nov 12, 37; US citizen. MATHEMATICS. *Educ:* Univ Houston, BS, 61; Univ Mo-Columbia, MA, 62; Tex A&M Univ, PhD(math), 69. *Prof Exp:* Instr math, Ohio Univ, 62-64, Houston Baptist Col, 65-66 & Tex A&M Univ, 68-69; asst prof, Univ Fla, 69-70; assoc prof, 70-72, PROF MATH, UNIV SOUTHWESTERN LA, 72- *Mem:* Math Asn Am; Am Math Soc. *Res:* Algebra; generalizations of rings; near rings. *Mailing Add:* Dept Math Univ Southwestern La Lafayette LA 70504

LIGHT, ALBERT, b Brooklyn, NY, June 19, 27; m 52; c 2. BIOCHEMISTRY. *Educ:* City Col New York, BS, 48; Yale Univ, PhD(biochem), 55. *Prof Exp:* Fel biochem, Cornell Univ, 55-57; asst res prof, Univ Utah, 57-63; assoc prof, Univ Calif, Los Angeles, 63-65; assoc prof, 65-77, head, Div Biochem, 78-82, PROF, PURDUE UNIV, 77- *Mem:* AAAS; Am Chem Soc; Am Soc Biochem & Molecular Biol. *Res:* Protein chemistry and enzymology; protein folding; relationship of structure to function of biologically active proteins. *Mailing Add:* Dept Chem Purdue Univ Lafayette IN 47923

LIGHT, AMOS ELLIS, pharmacology, nutrition; deceased, see previous edition for last biography

LIGHT, DOUGLAS B, b New York, NY, Apr 9, 56; m 77; c 3. ION CHANNEL REGULATION, OVERWINTERING ADAPTATIONS. *Educ:* Colby Col, BA, 78; Univ Minn, MS & PhD(physiol), 86. *Prof Exp:* Biol teacher, Winslow High Sch, 78-81; teaching asst physiol, Univ Minn, 81-84, res asst physiol, 84-86; postdoctoral fel physiol, Dartmouth Med Sch, 86-88, res assoc, 88-89; ASST PROF BIOL, RIPON COL, 89- *Concurrent Pos:* Instr biol, Sch Life Long Learning, 88. *Honors & Awards:* Award for Res Excellence, Am Physiol Soc, 87, Caroline tum Suden Award, 88. *Mem:* Sigma Xi; Biophys Soc; Am Soc Zoologists; Am Physiol Soc; Soc Gen Physiol & Cryobiol. *Res:* Ion channel regulation, cell volume regulation and overwintering adaptations of invertebrates. *Mailing Add:* Dept Biol Ripon Col 300 Seward St Ripon WI 54971-0248

LIGHT, IRWIN JOSEPH, b Montreal, Que, July 21, 34; m 72; c 2. PEDIATRICS. *Educ:* McGill Univ, BS, 55, MD, 59. *Prof Exp:* Intern med, Royal Victoria Hosp, 59-60, resident, 60-61; resident pediat, Montreal Children's Hosp, 61-63; from asst prof to assoc prof pediat, 65-73, from asst prof to assoc prof obstet & gynec, 68-73, dir newborn clin serv, 73-83, PROF PEDIAT, OBSTET & GYNEC, UNIV CINCINNATI, 73-; RES ASSOC PEDIAT, CHILDREN'S HOSP RES FOUND, 63- *Concurrent Pos:* Clin fel pediat, Cincinnati Gen Hosp & res fel, Univ Cincinnati, 63-65; chmn, Instnl Rev Bd, Children's Hosp Med Ctr, 84- *Mem:* AAAS; Am Acad Pediat; Soc Pediat Res; Am Pediat Soc; Am Fedn Clin Res. *Res:* Neonatal infectious diseases; newborn metabolism. *Mailing Add:* Univ Hosp Cincinnati OH 45267-0541

LIGHT, JOHN CALDWELL, b Mt Vernon, NY, Nov 24, 34; m 78; c 3. CHEMICAL PHYSICS. *Educ:* Oberlin Col, BA, 56; Harvard Univ, PhD(chem), 60. *Prof Exp:* NSF fel, Brussels, 59-61; from instr to assoc prof chem, 61-70, dir, Mat Res Lab, 70-73, chmn, Dept Chem, 80-82, PROF CHEM, UNIV CHICAGO, 70- *Concurrent Pos:* Sloan fel, 66; vis fel, Joint Inst Lab Astrophys, Univ Colo, 76-77; ed, J Chem Phys, 82- *Mem:* AAAS; Am Inst Phys; Am Chem Soc. *Res:* Theoretical studies of elementary gas phase reactions; quantum mechanics and chemical kinetics; scattering theory; computational methods for theoretical chemistry. *Mailing Add:* Dept Chem Univ Chicago Chicago IL 60637

LIGHT, JOHN HENRY, b Annville, Pa, Dec 15, 24; m 50; c 3. MATHEMATICS. *Educ:* Lebanon Valley Col, BS, 48; Pa State Univ, MS, 50 & 57. *Prof Exp:* Res assoc, Ord Res Lab, Pa State Univ, 51-58, eng mech, 58-59; from assoc prof to prof math, Dickinson Col, 59-89; RETIRED. *Concurrent Pos:* Consult, Naval Supply Depot, Mechanicsburg, Pa, 59-62. *Mem:* Am Math Soc. *Res:* Spectroscopy; environmental testing. *Mailing Add:* 619 Belvedre St Carlisle PA 17013

LIGHT, KENNETH FREEMAN, b Detroit, Mich, Jan 22, 22; m 43; c 3. MECHANICAL ENGINEERING. *Educ:* Univ Ill, Urbana, BS, 49; Mich State Univ, MA, 52, PhD(higher educ), 67. *Prof Exp:* Instr mech eng, Mich Technol Univ, 56-60, assoc prof technol, 60-65; prof mech eng & acad vpres, Lake Superior State Col, 65-77; pres, Ore State Technol, 77-86; RETIRED. *Concurrent Pos:* Pres, Lake Superior State Col. *Mem:* Am Soc Eng Educ. *Res:* Mechanical engineering technology; programmed learning; computer assisted instruction. *Mailing Add:* 1635 NE Country Club Ave Gresham OR 97030

LIGHT, KIM EDWARD, Indianapolis, Ind, Sept 21, 51; m 75; c 1. BIOGENIC AMINES, RECEPTORS. *Educ:* Ind State Univ, Terre Haute, BS, 73; Ind Univ, Bloomington, MS, 75, PhD(pharmacol), 77. *Prof Exp:* Fel physiol, Sch Med, Tex Tech Univ, 79; ASST PROF PHARMACOL, UNIV ARK MED SCI, 79-, ASST PROF INTERDISCIPLINARY TOXICOL, 80- *Concurrent Pos:* Collabr, Col Med, Tex Tech Univ Health Sci, 79- *Mem:* Soc Neurosci; Am Physiol Soc; Sigma Xi. *Res:* Investigations into the functions and interactions of membrane receptors for biogenic amines; functional aspects and regulation of H1 and H2 histamine receptors in various tissues. *Mailing Add:* Dept Biopharmaceut Sci Univ Ark Med Sch Little Rock AR 72205

LIGHT, MITCHELL A, geology, economics; deceased, see previous edition for last biography

LIGHT, ROBLEY JASPER, b Roanoke, Va, Nov 8, 35; m 60; c 1. BIOCHEMISTRY, ORGANIC CHEMISTRY. *Educ:* Va Polytech Inst, BS, 57; Duke Univ, PhD(org chem), 61. *Prof Exp:* NSF fel biochem, Harvard Univ, 60-62; instr, 62-63, from asst prof to assoc prof, 62-72, Chmn, 83-90, PROF BIOCHEM, FLA STATE UNIV, 72- *Concurrent Pos:* USPHS res career develop award, 67-72; Alexander von Humboldt US sr scientist award, 77. *Mem:* Am Chem Soc; Am Soc Biochem & Molecular Biol. *Res:* Lipid metabolism, structure, and function; polyketides and other secondary metabolites of microorganisms. *Mailing Add:* Dept of Chem Fla State Univ Tallahassee FL 32306

LIGHT, THOMAS BURWELL, b Dayton, Ohio, July 9, 28; m 51; c 4. PRINTER TECHNOLOGY, MATERIALS SCIENCE. *Educ:* Antioch Col, BS, 51; Ill Inst Technol, MS, 54; Yale Univ, PhD(mat sci), 66. *Prof Exp:* Mem tech staff, Bell Tel Labs, 53-62; mem res staff, Thomas J Watson Res Ctr, 65-81, DEVELOP CONSULT, IBM US DEVELOP STAFF, IBM CORP, 81- *Mem:* AAAS; Am Phys Soc; Inst Elec & Electronics Engrs. *Res:* Deposition, structure and properties of thin films; structure of and crystallization in amorphous materials; structure of oxide layers and interface reactions. *Mailing Add:* 2 Briarcliff Rd Chappaqua NY 10514

LIGHT, TRUMAN S, b Hartford, Conn, Dec 16, 22; m 46, 80; c 3. ANALYTICAL CHEMISTRY. *Educ:* Harvard Univ, SB, 43; Univ Minn, MS, 49; Univ Rome, DrChem, 61. *Prof Exp:* Asst prof chem, Boston Col, 49-59; staff scientist, Res & Adv Develop Div, Avco Corp, 59-64; sr res chemist, Foxboro Co, 64-72; mgr, Chem Analysis & Mat Lab, 72-80, prin res scientist, 80-88; adj prof chem, Boston Col, 87-88; CONSULT, 88-; ADJ PROF CHEM, SUFFOLK UNIV, 91- *Concurrent Pos:* NSF fel, Chem Inst, Univ Rome, Italy, 60-61; consult, Children's Med Ctr, Boston, Mass, 56-60 & Watertown Arsenal, 51-55. *Mem:* Am Chem Soc; Electrochem Soc; Soc Appl Spectros; Instrument Soc Am; Sigma Xi. *Res:* Instrumental methods of analysis; electrochemistry; physical chemistry; materials sciences, water quality and pollution controls. *Mailing Add:* Four Webster Rd Lexington MA 02173-8222

LIGHTBODY, JAMES JAMES, b Detroit, Mich, Mar 1, 39; m 64; c 2. IMMUNOLOGY, BIOCHEMISTRY. *Educ:* Wayne State Univ, BA, 61, BS, 64, PhD(biochem), 66. *Prof Exp:* Res assoc biochem & NIH trainee, Brandeis Univ, 67-69; instr pediat & Swiss Nat Sci Found grant, Univ Bern, 69-70; sr res assoc immunol, Basel Inst Immunol, 70-71; res assoc, Univ Wis, 71-72; asst prof, 72-76, ASSOC PROF BIOCHEM, SCH MED, WAYNE STATE UNIV, 76-, ASSOC IMMUNOL, 73-, CLIN ASSOC PROF INTERNAL MED, 76- *Concurrent Pos:* Vis prof, Mem Sloan-Kettering Cancer Ctr, 73; lectr, Cancer Inst, Cairo Univ, 74. *Mem:* AAAS; Am Asn Immunol; Transplantation Soc. *Mailing Add:* Dept Biochem Wayne State Univ Sch Med Detroit MI 48201

LIGHTERMAN, MARK S, b New York, NY, Jan 17, 60. COMPUTER PROGRAMMING. *Educ:* Syracuse Univ, BS & BA, 82; Univ Miami, MS, 85. *Prof Exp:* VPRES, HMMM CORP, 82- *Concurrent Pos:* Mem bd, Fla Alliance Technol Educ, 89-91, pres, 91- *Mem:* NY Acad Sci; Asn Comput Mach; Am Math Soc; Inst Elec & Electronics Engrs; Am Med Info Asn; Soc Motion Picture & TV Engrs. *Res:* Curriculum enhancement by bringing technology back into the classroom. *Mailing Add:* 9230 SW 59th St Miami FL 33173

LIGHTFOOT, DONALD RICHARD, b Los Angeles, Calif, Aug 8, 40; m 72. BIOCHEMICAL GENETICS. *Educ:* Univ Redlands, BA, 62; Univ Ariz, MS, 67, PhD(biochem), 72. *Prof Exp:* Teacher, Philippine High Sch, Peace Corps, 62-64; res trainee biochem, Med Sch, Univ Ore, 69-71; fel, Univ Calif, Riverside, 71-74; asst prof biochem & nutrit, Va Polytech Inst & State Univ, 74-79; ASSOC PROF BIOL & DIR BIOCHEM/BIOTECHNOL, EASTERN WASH UNIV, 80- *Concurrent Pos:* DNA diagnostics consult, Sacred Heart Med Ctr. *Mem:* AAAS; Am Chem Soc; Soc Exp Biol & Med; Sigma Xi. *Res:* Plant virology; tobacco mosaic virus infection process; gene titration in tobacco species; minor nucleosides; transfer RNA, messenger RNA 5.85 ribosomal RNA and viral RNA NMR structure and function; biotechnology education, turnip yellow mosaic virus structure/function; polymerase chain reaction applications. *Mailing Add:* Dept Biol Eastern Wash Univ Cheney WA 99004

LIGHTFOOT, E(DWIN) N(IBLOCK), JR, b Milwaukee, Wis, Sept 25, 25; m 49; c 5. SEPARATIONS, BIOTECHNOLOGY. *Educ:* Cornell Univ, BS, 47, PhD(chem eng), 51. *Hon Degrees:* Dr, Tech Univ Norway. *Prof Exp:* Chem engr, Chas Pfizer & Co, 50-53; from asst prof to prof biochem eng, 53-80 HILLDALE PROF CHEM ENG, UNIV WIS-MADISON, 80- *Concurrent Pos:* Vis prof, Tech Univ Norway, 62; Stanford Univ & Tech Univ Denmark, 71, Univ Canterbury, NZ, 72; tech consult, 53-88; Erskine fel, Univ Canterbury, NZ, 72. *Honors & Awards:* William H Walker Award, Am Inst Chem Engrs, 75, Food, Pharm & Bioeng Award, 79; Lacey lectr, Calif Inst Technol, 84; Van Winkle lectr, Univ Tex, 84; Goff Smith lectr, Univ Mich, 87. *Mem:* Nat Acad Eng; Royal Norweg Soc Sci & Letters; Am Inst Chem Engr; Am Chem Soc. *Res:* Physical separation techniques; mass transfer; biomedical engineering; author of 14 books and technical articles. *Mailing Add:* Dept Chem Eng Univ Wis 1415 Johnson Dr Madison WI 53706

LIGHTFOOT, RALPH B(UTTERWORTH), b Fall River, Mass, June 19, 13; m 37; c 2. AERONAUTICAL ENGINEERING. *Educ:* Univ RI, BS & ME, 35. *Prof Exp:* Wind tunnel engr, Sikorsky Aircraft Div, United Technologies Corp, 35-37, chief wind tunnel engr, 38-40, chief flight test engr, 40-43, chief flight res, 44-57, chief engr, 57-66, engr mgr, 66-67, sr staff engr, 68-74; CONSULT ENGR, 74- *Concurrent Pos:* Instr, Univ Bridgeport, 39, 77, lectr, 63-65; instr, Bullard Havens Tech Inst, 41-43; lectr, NY Univ, 43; designated eng rep, Fed Aviation Agency, 55-60, mem airworthiness stand eval comt, 66; mem, NASA, 56-68; mem adv coun, Bridgeport Eng Inst, Univ Bridgeport, Housatonic Community Col & Adv Group Aerospace Res & Develop, NATO, 59, 62; navigator, US Power Squadron; lectr, Univ Conn, 62; mem adv coun, Univ RI, 81- *Honors & Awards:* Merit Award, Am Helicopter Soc, 48; Bell Award, 61; Sperry Award, 64; Gold Medal, NY Acad Sci & Cierva Prize, Royal Aeronaut Soc, 65. *Mem:* Nat Soc Prof Engrs; hon fel Am Helicopter Soc (vpres, 50, pres, 51); fel Am Inst Aeronaut & Astronaut; fel NY Acad Sci; fel Royal Aeronaut Soc. *Res:* Aerodynamics; wind tunnel; analytical flight test; flying boats; helicopters; management; airplanes. *Mailing Add:* 55 Eliphamets Lane Chatham MA 02633

LIGHTMAN, ALAN PAIGE, b Memphis, Tenn, Nov 28, 48; m 76; c 2. THEORETICAL ASTROPHYSICS, THEORETICAL PHYSICS. *Educ:* Princeton Univ, AB, 70; Calif Inst Technol, MA, 73, PhD(physics), 74. *Prof Exp:* Res assoc physics, Calif Inst Technol, 74; res assoc astrophysics, Cornell Univ, 74-76; asst prof astron, Harvard Univ, 76-79, lectr astron & phys, 79-88; PROF SCI & WRITING, MASS INST TECHNOL, 88- *Concurrent Pos:* Mem staff, Smithsonian Astrophys Observ, 79-88. *Mem:* Fel Am Phys Soc; Am Astron Soc; fel AAAS; Soc Lit & Sci. *Res:* Theoretical frameworks for analyzing modern gravitation theories; relativistic astrophysics, x-ray astronomy; stellar dynamics; radiation processes; philosophy of science. *Mailing Add:* Dept Physics Mass Inst Technol Cambridge MA 02139

LIGHTNER, DAVID A, b Los Angeles, Calif, Mar 25, 39; m 74; c 2. BIOORGANIC CHEMISTRY, STEREOCHEMISTRY. *Educ:* Univ Calif, Berkeley, AB, 60; Stanford Univ, PhD, 63. *Prof Exp:* NSF fels, Stanford Univ, 63-64 & Univ Minn, 64-65; asst prof chem, Univ Calif, Los Angeles, 65-72; assoc prof, Tex Tech Univ, 72-74; assoc prof, 74-76, dept chmn, 85-88, PROF CHEM, UNIV NEV, RENO, 76-, RC FUSION PROF, 84- *Concurrent Pos:* Assoc ed, Photochem Photobiol, 83-86; fel, Ctr Advan Study, Univ Nev; counr, Am Soc Photobiol; found prof, Univ Nev Reno, 87-90. *Mem:* AAAS; Am Chem Soc; Royal Soc Chem; Am Soc Photobiol; Inter-Am Photochem Soc. *Res:* Photochemistry of biological materials; molecular recognition; synthesis and stereochemistry; circular dichroism and optical rotatory dispersion; phototherapy and jaundice. *Mailing Add:* Dept Chem Univ Nev Reno NV 89557-0020

LIGHTNER, JAMES EDWARD, b Frederick, Md, Aug 29, 37. MATHEMATICS, EDUCATION. *Educ:* Western Md Col, AB, 58; Northwestern Univ, AM, 62; Ohio State Univ, PhD(math, educ), 68. *Prof Exp:* Teacher, Frederick County Bd Educ, Md, 58-62; from instr to assoc prof math, Western Md Col, 62-77, chmn dept, 68-73, dir Jan term, 69-83, PROF MATH & EDUC, WESTERN MD COL, 77-, DIR MATH PROFICIENCY, 83- *Concurrent Pos:* Fed Liaison Rep, Western Md Col, 73-78, coordr int studies, 80-83; consult sch systs & Md State Dept Educ; exec secy, Md Coun Teachers Math, 88- *Mem:* Nat Coun Teachers Math; Math Asn Am. *Res:* Undergraduate mathematics curricula; secondary mathematics curricula and methodology; secondary school geometry; history of mathematics. *Mailing Add:* Dept Math Western Md Col Westminster MD 21157

LIGHTON, JOHN R B, b Johannesburg, SAfrica, Aug 25, 52. ECOLOGICAL PHYSIOLOGY, WATER RELATIONS. *Educ:* Univ Cape Town, BA, 75, BSc, 81, MSc, 84; Univ Calif, Los Angeles, PhD(physiol), 87. *Prof Exp:* Hollaender distinguished postdoctoral fel ecophysiol, Univ Calif, Los Angeles, 87-89; adj asst prof, 89-90; ASST PROF ECOPHYSIOL, UNIV UTAH, 91- *Concurrent Pos:* Guest prof ecophysiol, Univ Zurich, 90. *Res:* Ecological physiology of animals, concentrating on insects, their respiratory and ventilatory physiology. *Mailing Add:* Biol Dept Univ Utah Salt Lake City UT 84112

LIGHTSEY, PAUL ALDEN, b Wray, Colo, Aug 25, 44; m 65; c 1. AERONAUTICAL & ASTRONAUTICAL ENGINEERING, ATMOSPHERIC CHEMISTRY & PHYSICS. *Educ:* Colo State Univ, BS, 66; Cornell Univ, MS, 69, PhD(physics), 72. *Prof Exp:* Res aide atmospheric physics, Colo State Univ, 62-66; physicist, surface physics, Dow Chem-Rocky Flats Div, 66; res asst solid state physics, Cornell Univ, 66-72; lectr physics, Beloit Col, 72-73; asst prof physics & math, Univ Dallas, 73-75; electrician, Great Western Sugar Co, 75-76; assoc prof physics & math, Colo Mountain Col, 76-77; assoc prof & chmn dept physics, Univ Northern Colo, 77-86; PRIN SYSTS ENGR, BALL AEROSPACE SYSTS GROUP, 86- *Concurrent Pos:* Instr, Frontiers Sci Inst, Univ Northern Colo, 80-82; physicist, Nat Oceanic & Atmospheric Admin, 82-84; scientist, Nat Ctr Atmospheric Res, 86- *Mem:* Am Asn Physics Teachers; Sigma Xi; Int Soc Biomechanics in Sports; Soc Photo Instrumentation Engrs; Optical Soc Am; Nat Sci Teachers Asn. *Res:* Use of laser radar for remote sensing characteristics of atmosphere; biomechanical analysis of sports; optical propagation through atmospheric turbulence; electro-optical systems engineering; atmospheric optics. *Mailing Add:* Ball Aerospace PO Box 1062 Boulder CO 80306-1062

LIGHTY, JOANN SLAMA, b Weehawken, NJ, Jan 5, 60. WASTE REMEDIATION, INCINERATION. *Educ:* Univ Utah, BS, 82, PhD(chem eng), 88. *Prof Exp:* Proj engr, Northwest Pipeline Corp, 82-84; ASST PROF CHEM ENG, UNIV UTAH, 88- *Concurrent Pos:* Sr scientist, Reaction Eng Int, 90- *Mem:* Am Inst Chem Eng; Soc Women Engrs; Combustion Inst; Air & Waste Mgt Asn. *Res:* Remediation of contaminated solids by thermal treatment; MSW incineration; fate of metals during incineration; circulating fluidized bed combustion. *Mailing Add:* Dept Chem Eng Univ Utah Salt Lake City UT 84112

LIGHTY, RICHARD WILLIAM, b Freeport, Ill, Nov 8, 33; m 55; c 2. PLANT GENETICS, HORTICULTURE. *Educ:* Pa State Univ, BS, 55; Cornell Univ, MS, 58, PhD(genetics), 60. *Prof Exp:* Geneticist, Longwood Gardens, Pa, 60-67; assoc prof plant sci & coordr Longwood prog ornamental hort, 67-82, DIR MT CUBA CTR STUDY PIEDMONT FLORA, UNIV DEL, 83- *Honors & Awards:* A H Scott Medal & Award; Silver Medal, Mass Hort Soc; Eloise Payne Lequer Medal, Garden Club Am. *Mem:* AAAS; Am Asn Bot Gardens & Arboretums. *Res:* Plant breeding; cytotaxonomy; horticultural taxonomy; floriculture. *Mailing Add:* Mt Cuba Ctr PO Box 3570 Greenville DE 19807

LIGLER, FRANCES SMITH, b Louisville, Ky, June 11, 51; m 72; c 2. BIOSENSORS, ARTIFICIAL BLOOD. *Educ:* Furman Univ, BSc, 72; Oxford Univ, Eng, DPhil. *Prof Exp:* Fel biochem, Univ Tex Health Sci Ctr, San Antonio, 75-76; asst instr immunol, Southwestern Med Sch, 76-78, instr, 78-80; primary scientist immunol, E I Dupont De Nemours & Co, Inc, 80-84, Group Leader & Cellular Immunol, 84-85; SR SCIENTIST, NAVAL RES LAB, 85-, DEP HEAD, CTR BIO-MOLECULAR SCI & ENG, 90- *Concurrent Pos:* From adj asst prof to adj assoc prof, Hahnemann Univ, 81-85. *Honors & Awards:* Am Asn Med Instrumentation Ann Meeting Manuscript Award, 3M. *Mem:* Am Asn Immunologists; AAAS; Am Asn Pathologists; Am Chem Soc. *Res:* Biosensors; immunoassay development; liposome technology; fluorescence; B cell activation. *Mailing Add:* Naval Res Lab Code 6090 Washington DC 20375-5000

LIGOMENIDES, PANOS ARISTIDES, b Pireus, Greece, Apr 3, 28; US citizen; m 73; c 2. SYNERGETIC & NEURAL COMPUTERS, EXPERIMENTAL KNOWLEDGE ENGINEERING. *Educ:* Univ Athens, Greece, dipl physics, 51, MSc, 52; Stanford Univ, MSc, 56, PhD(elec eng & physics), 58. *Prof Exp:* Radio engr radiotel, Greek Tel & Tel Co, 54-55; res & staff engr elec eng & computers, IBM, 58-64; asst prof elec eng, Univ Calif, Los Angeles, 64-69; adj prof elec eng, Stanford Univ, 69-70; PROF COMPUTER ENG, ELEC ENG DEPT, UNIV MD, 71-; VPRES RES, CAELUM RES CORP, 87- *Concurrent Pos:* Tech consult indust & govt, 64-; Orgn Econ Coop & Develop fel, Greece, 65 & 74; Ford Found fel, SAm, 66 & 68; vis prof, Univ Ceara, Brazil, 66, Stanford Univ, 67, Univ Athens, Greece, 68-69 & Polit Univ Madrid, Spain, 82-84; prin investr grants & contracts from indust & govt, 68-; Fulbright prof univs in Greece, 70-71; Salzburg Sem Fel, Salzburg Sem Am Studies, 71; distinguished prof, Elec Eng Dept, Univ Md, 71-72; pres & owner, Computer Eng Consults, 75-81. *Mem:* Sr mem Inst Elec & Electronics Engrs; Int Soc Optical Eng. *Res:* Applied artificial intelligence; neural networks; decisions support technologies; pattern recognition; computer architectures; microcomputer-based systems; synergetic computer applications. *Mailing Add:* 8802 Magnolia Dr Lanham MD 20706

LIGON, JAMES DAVID, b Wewoka, Okla, Feb 2, 39; m 67; c 1. ZOOLOGY. *Educ:* Univ Okla, BS, 61; Univ Fla, MS, 63; Univ Mich, PhD(zool), 67. *Prof Exp:* Asst prof biol, Idaho State Univ, 67-68; from asst prof to assoc prof, 68-77, PROF BIOL, UNIV NMEX, 77- *Mem:* Am Ornith Union; Cooper Ornith Soc. *Res:* Avian ecology and behavior. *Mailing Add:* Dept Biol Univ NMex Albuquerque NM 87131

LIGON, JAMES T(EDDIE), b Easley, SC, Feb 20, 36; m 58; c 3. SOILS & SOIL SCIENCE. *Educ:* Clemson Univ, BS, 57; Iowa State Univ, MS, 59, PhD(agr eng, soil physics), 61. *Prof Exp:* Asst prof agr eng, Univ Ky, 61-66; assoc prof, 66-71, chmn directorate, Water Resources Res Inst, 75-78, actg head, Agr Eng Dept, 84-85, PROF AGR ENG, CLEMSON UNIV, 71- *Mem:* Am Soc Agr Engrs; Am Geophys Union; Sigma Xi; Am Soc Eng Educ. *Res:* Soil drainage and physics; soil, water and plant relationships; hydrologic modeling. *Mailing Add:* Dept of Agr Eng Clemson Univ Clemson SC 29634-0357

LIGON, WOODFIN VAUGHAN, JR, b Farmville, Va, Apr 24, 44. ORGANIC MASS SPECTROMETRY & CHEMISTRY. *Educ:* Longwood Col, BS, 66; Univ Va, PhD(org chem), 70. *Prof Exp:* Vis asst prof org chem, Univ Ill, 72-73; STAFF SCIENTIST ORG MASS SPECTROMETRY, GEN ELEC CORP RES & DEVELOP, 73- *Concurrent Pos:* NIH fel, Univ Ill, Urbana, 71-72. *Mem:* Sigma Xi; Am Soc Mass Spectrometry. *Res:* New modes of ionization for mass spectrometry; secondary ion mass spectrometry; multidimensional gas chromatography. *Mailing Add:* PO Box 8 K1-2A40 Schenectady NY 12301

LIGUORI, VINCENT ROBERT, b Brooklyn, NY, Dec 15, 28; m 49, 52; c 5. MARINE MICROBIOLOGY. *Educ:* St Francis Col, NY, BS, 51; Long Island Univ, MS, 58; NY Univ, PhD(microbiol), 67. *Prof Exp:* Res asst cancer chemother, Sloan-Kettering Inst Cancer Res, 55-56; supvr oncol lab, Vet Admin Hosp, NY, 56-62; staff scientist microbiol, New York Aquarium, 62-65; asst prof biol & marine sci, Long Island Univ, 65-66; res assoc microbiol, Osborn Labs Marine Sci, New York Aquarium, 66-71; lectr, 66-68, assoc prof biol & dep chmn dept biol sci, 71- 73, PROF BIOL, KINGSBOROUGH COMMUNITY COL, CUNY, 73- *Concurrent Pos:* Lectr, Nassau County Mus Natural Hist, 65-68, Richmond Col, 67-70 & Queens Col, 68-69; mem bd dir, Mid Atlantic Natural Sci Coun, Inc, 75-78; mem, Bermuda Biol Sta Res, 75-78. *Mem:* Am Soc Microbiol; Sigma Xi. *Res:* Biological effects of natural products and the mechanism of adhesion in marine invertebrates; role of marine microorganisms in the disease processes of marine animals; aquaculture; invertebrates. *Mailing Add:* 173 Edgewood Ave PO Box 359 Oakdale NY 11769

LIH, MARSHALL MIN-SHING, b Nanking, China, Sept 15, 36; m 62; c 3. CHEMICAL ENGINEERING. *Educ:* Nat Taiwan Univ, BS, 58; Univ Wis-Madison, MS, 60, PhD(chem eng), 62. *Prof Exp:* Res engr, E I du Pont de Nemours & Co, 62-64; mem fac chem eng, Cath Univ Am, 64-74; sr res scientist, Nat Biomed Res Found, 66-76; prog dir, Thermodynamics & Mass Transfer, NSF, 73-76, sect head, Eng Chem & Energetics, 76-79, div dir chem & process eng, 79-87, US SR EXEC SERV, ENG CENTERS, NSF, 79-, DIR CROSS DISCIPLINARY RES, 87- *Concurrent Pos:* Dir, Inst Creative Eng Methodology, 68-70; NSF vis prof & chmn dept chem eng, Nat Taiwan Univ, 70-71; adj lectr, Georgetown Univ Med Sch, 83- *Mem:* Fel Am Inst Chem Engrs; Sigma Xi. *Res:* Kinetics and catalysis; transport processes; color technology; application of mathematics in chemical engineering; biomedical engineering. *Mailing Add:* NSF Eng Centers Div Rm 1121 1800 G St NW Washington DC 20550

LIIMATAINEN, T(OIVO) M(ATTHEW), b Gloucester, Mass, Nov 14, 10; m 50. ENGINEERING PHYSICS, MATHEMATICS. *Prof Exp:* Engr, Gen Elec Co, 41-46 & Sylvania Elec Prod Co, 46-48; proj engr, Nat Bur Standards, 48-53; asst br chief electron devices, Diamond Ord Fuze Labs, US Dept Army, 53-59, br chief microelectronics, 59-63; aerospace engr, Goddard Space Flight Ctr, NASA, Md, 63-66 & Electronics Res Ctr, Cambridge, 66-70; gen engr, Transp Systs Ctr, Dept Transp, Mass, 70-71; consult, 71-80; RETIRED. *Mem:* AAAS; sr mem Inst Elec & Electronics Engrs; NY Acad Sci. *Res:* Semiconductor devices; integrated circuits; microelectronics; high vacuum and gas discharge devices. *Mailing Add:* 1004 Union St Schenectady NY 12308

LIITTSCHWAGER, JOHN M(ILTON), b Alden, Iowa, Oct 24, 34; m 55; c 4. OPERATIONS RESEARCH, INDUSTRIAL ENGINEERING. *Educ:* Iowa State Univ, BS, 55; Northwestern Univ, MS, 61. *Prof Exp:* Engr, foods div, Anderson Clayton & Co, 55-56; consult pub utility, Mid West Serv Co, 56-60; chmn dept, 74-81, PROF INDUST & MGT ENG, UNIV IOWA, 61- *Mem:* Opers Res Soc Am; Inst Mgt Sci; Inst Indust Eng; Asn Comput Mach. *Res:* Legislative districting by computer; mathematical programming; reliability theory. *Mailing Add:* 4104 Eng Bldg Univ Iowa Iowa City IA 52242

LIJEWSKI, LAWRENCE EDWARD, b Milwaukee, Wis, Mar 12, 48; m 75; c 2. MISSILE AERODYNAMICS, COMPUTATIONAL FLUID DYNAMICS. *Educ:* Univ Notre Dame, BSAE, 70, MSAE, 72, PhD(aerospace), 74. *Prof Exp:* Mech engr, Army Aviation Systs Command, 74-77; AEROSPACE ENGR, AIR FORCE ARMAMENT LAB, 77- *Mem:* Am Inst Aeronaut & Astronaut. *Res:* Aerodynamics of aircraft and missiles; computational fluid dynamics and experimental methods to obtain aerodynamic characteristics and fundamental understanding of basic aerodynamic phenomena. *Mailing Add:* US Air Force Armament Lab AFATL/FXA Eglin AFB FL 32542

LIJINSKY, WILLIAM, b Dublin, Ireland, Oct 19, 28; Brit citizen; m 73; c 2. ENVIRONMENT CARCINOGENESIS. *Educ:* Univ Liverpool, BSc, 49, PhD(biochem), 51. *Prof Exp:* From asst prof to assoc prof cancer res, Chicago Med Sch, 55-68; prof biochem, Univ Nebr Med Sch, 68-71; group leader cancer res, Oak Ridge Nat Lab, 71-76; PROG DIR CHEM CARCINOGENESIS, FREDERICK CANCER RES FACIL, NAT CANCER INST, 76- *Concurrent Pos:* AA Noyes fel, Calif Inst Technol, 52-54, Damon Runyon fel, McGill-MGH Res Inst, Montreal, 54-55. *Mem:* Sigma Xi; Am Chem Soc; Am Asn Cancer Res; Am Soc Biol Chemists; Environ Mutagen Soc. *Res:* Detection and identification of environmental carcinogens, especially nitrosamines and polycyclic aromatic compounds; biological testing in animals and other systems; investigation of mechanisms of action of capainogens. *Mailing Add:* 5521 Woodlyn Rd Frederick MD 21702

LIKE, ARTHUR A, b New York, NY. PATHOLOGY. *Prof Exp:* PROF PATH, MED SCH, UNIV MASS, 75- *Mailing Add:* Path Dept Med Sch Univ Mass 55 Lake Ave N Worcester MA 01655

LIKENS, GENE ELDEN, b Pierceton, Ind, Jan 6, 35; m; c 4. AQUATIC ECOLOGY, LIMNOLOGY. *Educ:* Manchester Col, BS, 57; Univ Wis, MS, 59, PhD(zool), 62. *Hon Degrees:* DSc, Manchester Univ, 79, Rutgers Univ, 85, Plymouth State Col, 89, Miami Univ, 90. *Prof Exp:* Asst zool, Univ Wis, 57-61; instr, Dartmouth Col, 61; from proj asst to res assoc, Univ Wis, 62, res assoc meteorol, 62-63; from instr to assoc prof biol sci, Dartmouth Col, 63-69; from assoc prof to prof ecol & syst, Cornell Univ, 69-83, actg chmn sect, 73-74; VPRES & DIR, NY BOT GARDEN INST ECOSYST STUDIES, 83- *Concurrent Pos:* Vis lectr, Univ Wis, 63; vis assoc ecologist, Brookhaven Nat Lab, 68; NATO sr fel, Eng & Sweden, 69; Guggenheim fel, 72-73; mem, US Nat Comt Int Hydrol Decade, 66-70; mem adv panel, US Senate Comt Pub Works, 70-73; US Nat Rep, Int Asn Theoret & Appl Limnol, 70-; mem comt water qual policy, Nat Acad Sci, 73-76, mem asembly life sci, 77-; mem ecol adv comt & sci adv bd, Environ Protection Agency, 74-78; mem biol res comt, Edmund Niles Huyck Preserves & resource adv, NY State Dept Environ Conserv, 74-; vis prof, Ctr Advan Sci, Dept Environ Sci, Univ Va, 78-79; adv, White House, Special Envoy for Acid Rain to President, 85; adj prof, Sect Ecol & Systematics, Cornell Univ, 83-; assoc ed, Am Water Resources Asn, 83-; prof, Dept Biol, Yale, 84-; sr adv, Ctr Energy & Environ Res, Univ PR, 84-; mem, US Environ Protection Agency Steering Comt, State Univ NY, Albany, 84-; prof, Grad Field Ecol, Rutgers Univ, 85- *Honors & Awards:* Am Motors Conserv Award, 69; First G E Hutchinson Award, Am Soc Limnol & Oceanog, 82; NY Acad Sci Award, 86; Int ECI Prize, Limnetic Ecol, 88; Distinguished Serv Award, Am Inst Biol Sci, 90. *Mem:* Nat Acad Sci; Am Polar Soc; Am Soc Limnol & Oceanog (vpres, 75-76, pres, 76-77); Ecol Soc Am (vpres, 78-79, pres, 81-82); Int Asn Theoret & Appl Limnol; hon mem Am Water Resource Asn; Int Asn Great Lakes Res; Int Water Resource Asn; fel AAAS; Am Inst Biol Sci. *Res:* Circulation in lakes using radioactive tracers; meromictic lakes; biogeochemistry and analysis of ecosystems; antarctic and arctic limnology; precipitation chemistry. *Mailing Add:* Inst Ecosyst Studies NY Bot Garden Box AB Millbrook NY 12545

LIKES, CARL JAMES, b Charleston, SC, Sept 11, 16; m 43. PHYSICAL CHEMISTRY. *Educ:* Col Charleston, BS, 37; Univ Va, PhD(phys chem), 41. *Prof Exp:* Instr chem, Univ Va, 41-43; asst prof, Tulane Univ, 43, asst prof, 44-46; prof & head dept, Hampden-Sydney Col, 47-52; proj supvr, Va Inst for Sci Res, 52-58; prof, 58-82, EMER PROF CHEM, COL CHARLESTON, 82- *Mem:* Am Chem Soc. *Res:* Electrophoretic and ultracentrifugal analysis of proteins. *Mailing Add:* 2280 Shore Line Dr Johns Island SC 29455

LIKINS, PETER WILLIAM, b Tracy, Calif, July 4, 36; m 55; c 6. DYNAMICS, CONTROL SYSTEMS. *Educ:* Stanford Univ, BS, 57, PhD(eng mech), 65; Mass Inst Technol, SM, 58. *Prof Exp:* Develop engr, Jet Propulsion Lab, Calif Inst Technol, 58-60; from asst prof to prof eng, Univ Calif, Los Angeles, 64-76, from asst dean to assoc dean, 74-76; from prof & dean to provost, Columbia Univ, 76-82; PRES, LEHIGH UNIV, 82- *Concurrent Pos:* Consult to various industs & govt res agencies, 66-; mem, US President's Coun Adv Sci & Technol. *Mem:* Nat Acad Eng; fel Am Inst Aeronaut & Astronaut. *Res:* Problems of space vehicle dynamics, stability and control. *Mailing Add:* Alumni Bldg 27 Lehigh Univ Bethlehem PA 18015

LIKOFF, WILLIAM, medicine; deceased, see previous edition for last biography

LIKUSKI, ROBERT KEITH, b Hillcrest, Alta, Oct 16, 37; m 71; c 2. BIOMEDICAL ENGINEERING, LIQUID CHROMATOGRAPHY. *Educ:* Univ Alta, BS, 59; Univ Ill, MS, 61, PhD(elec eng), 64. *Prof Exp:* Asst prof elec eng, Univ Tex, Austin, 65-70; staff engr comput memories, Micro-Bit Corp, 70-76; chief engr biomed eng, 76-80, dir res & develop, Berkeley Bio-Eng, Inc, 80-81; ADV DEVELOP MGR, BECHMAN INSTRUMENTS, INC, 81- *Mem:* Inst Elec & Electronics Engrs; Sigma Xi. *Res:* Development of instrumentation for analytical & medical use. *Mailing Add:* 4430 School Way Castro Valley CA 94546

LILENFELD, HARVEY VICTOR, b Brooklyn, NY, Aug 25, 45. PHYSICAL CHEMISTRY. *Educ:* Polytech Inst Brooklyn, BS, 66; Mass Inst Technol, PhD(phys chem), 71. *Prof Exp:* Res assoc, Brookhaven Nat Lab, 71-72; scientist chem, McDonnell Douglas Corp, 72-80. *Mem:* Am Chem Soc; Sigma Xi. *Res:* Laser chemistry; kinetics of gas phase reactions. *Mailing Add:* 2340 Driftwood Pl St Louis MO 63146

LILES, JAMES NEIL, b Akron, Ohio, Apr 25, 30; m 55; c 3. COMPARATIVE PHYSIOLOGY, HUMAN PHYSIOLOGY. *Educ:* Miami Univ, BA, 51; Ohio State Univ, MSc, 53, PhD(insect physiol), 56. *Prof Exp:* Asst prof biol, Univ SC, 56-58; res assoc entom, Ohio State Univ, 58-60; from asst prof to assoc prof, 60-71, PROF ENTOM, UNIV TENN, 71- *Mem:* Am Inst Biol Sci; Am Soc Zool; Geront Soc; Entom Soc Am; Am Col Sports Med; Sigma Xi. *Res:* Aging in insects; insect nutrient utilization. *Mailing Add:* Dept Zool Univ Tenn F-209 Walter Life Sci Bldg Knoxville TN 37916

LILES, SAMUEL LEE, b Texas City, Tex, June 24, 42; m 86; c 1. PHYSIOLOGY. *Educ:* McNeese State Col, BS, 64; La State Univ Med Ctr, New Orleans, PhD(physiol), 68. *Prof Exp:* Instr, 68-70, asst prof, 70-75, ASSOC PROF PHYSIOL, LA STATE UNIV MED CTR, NEW ORLEANS, 75- *Concurrent Pos:* Nat Inst Neurol Dis & Stroke grant, La State Univ Med Ctr, New Orleans, 70- *Mem:* AAAS; Am Physiol Soc; NY Acad Sci; Soc Neurosci; Sigma Xi. *Res:* Regional neurophysiology; electrophysiological correlates between brain neuronal activity and voluntary motor and sensory function. *Mailing Add:* Dept Physiol La State Univ Med Ctr 1100 Florida Ave New Orleans LA 70119

LILEY, NICHOLAS ROBIN, b Halifax, Eng, Dec 17, 36; m 61. ZOOLOGY. *Educ:* Oxford Univ, BA, 59, DPhil(zool), 64. *Prof Exp:* Nat Res Coun Can fel zool, 63-65, from asst prof to assoc prof, 65-78, PROF ZOOL, UNIV BC, 78- *Mem:* Animal Behav Soc; Can Soc Zool. *Res:* Comparative ethology and the evolution of behavior; endocrine mechanisms in control of behavior. *Mailing Add:* Dept Zool Univ BC 2075 Wesbrook Mall 6270 University Blvd Vancouver BC V6T 1Z4 Can

LILEY, PETER EDWARD, b Barnstaple, Eng, Apr 22, 27; m 63; c 2. PHYSICS, CHEMICAL ENGINEERING. *Educ:* Univ London, BSc, 51, PhD(physics), 57; Imp Col, Univ London, dipl, 57. *Prof Exp:* Chem engr, Brit Oxygen Eng, Ltd, 55-57; from asst prof to assoc prof, 57-72, PROF MECH ENG, PURDUE UNIV, WEST LAFAYETTE, 72- *Concurrent Pos:* Mem, Thermophysical Properties comt, Am Soc Mech Engrs, 60-87 & chmn, 71-73. *Res:* Thermodynamic and transport properties of matter, principally fluids; cryogenic engineering; high pressure. *Mailing Add:* Sch of Mech Eng Purdue Univ West Lafayette IN 47907-1101

LILIEN, OTTO MICHAEL, b New York, NY, Apr 26, 24; c 6. GENITOURINARY SURGERY. *Educ:* Jefferson Med Col, MD, 49; Columbia Univ, MA, 60. *Prof Exp:* Lectr zool, Columbia Univ, 56-58; from asst prof urol surg to assoc prof urol, 61-67, PROF UROL, STATE UNIV NY UPSTATE MED CTR, 67-, CHMN DEPT, 63- *Concurrent Pos:* Nat Cancer Inst trainee, 56-58. *Mem:* AMA; Am Urol Asn; fel Am Col Surg. *Res:* Renal and cell physiology. *Mailing Add:* 750 E Adams St Syracuse NY 13210

LILIENFIELD, LAWRENCE SPENCER, b New York, NY, May 5, 27; m 50; c 3. MEDICINE, PHYSIOLOGY. *Educ:* Villanova Col, BS, 45; Georgetown Univ, MD, 49, MS, 54, PhD, 56; Am Bd Internal Med, dipl, 57, 74. *Prof Exp:* Intern med, Georgetown Univ Hosp, 49-50, from jr asst resident to sr asst resident, 50-53, asst chief cardiovasc res lab & attend physician, 56; instr med, Med Sch, 55-57, instr physiol, 56-57, from asst prof to assoc prof med, physiol & biophys, 57-64, PROF PHYSIOL & BIOPHYS, SCHS MED &DENT, GEORGETOWN UNIV, 64-, CHMN DEPT PHYSIOL & BIOPHYS, SCH MED, 63- *Concurrent Pos:* Am Heart Asn res fel, 57; USPHS sr res fel, 59 & res career award, 63; attend physician, DC Gen Hosp, 56 & Vet Admin Hosp, 57; estab investr, Am Heart Asn, 58; consult, USPHS, 65-72; vis prof, Univ Saigon; vis prof, Univ Tel-Aviv, 67-68; assoc, Comt Int Exchange Persons, 71-76. *Mem:* AAAS; Biophys Soc; Am Physiol Soc; Soc Exp Biol & Med; Am Soc Clin Invest. *Res:* Transcapillary exchange; hemodynamics; blood distribution in organs; renal concentrating mechanisms; medical education. *Mailing Add:* Dept Physiol Sch Med Georgetown Univ Washington DC 20007

LILL, GORDON GRIGSBY, b Mt Hope, Kans, Feb 23, 18; m 43; c 2. MARINE GEOLOGY. *Educ:* Kans State Col, BS, 40, MS, 46. *Hon Degrees:* DSc, Univ Miami, 66. *Prof Exp:* Asst chief party, State Hwy Comn, Kans, 41; asst geol, Univ Calif, 46-47; head geophys br, US Off Naval Res, 47-59, earth sci adv, 59-60; corp res adv, Lockheed Aircraft Corp, 60-64; dir proj Mohole, NSF, 64-66; sr sci adv, Lockheed Aircraft Corp, 66-70; dep dir, Nat Ocean Surv, Nat Oceanic & Atmospheric Admin, US Dept Com, 70-79; RETIRED. *Concurrent Pos:* Pvt res, US Nat Mus; geologist, Bikini Sci Resurv, 47; mem mineral surv, Cent & Western Prov, Liberia, 49-50; consult comt geophys & geol res & develop bd, Nat Mil Estab, 47-53; chmn panel oceanog, Int Geophys Year; mem comt, Am Miscellaneous Soc, Proj Mohole, Nat Acad Sci-Nat Res Coun; vchmn, Calif Adv Comn Marine & Coastal Resources, 68-70; mem adv coun, Inst Marine Resources, Univ Calif, 69-71. *Mem:* Fel AAAS; Geol Soc Am; Am Geophys Union; Marine Technol Soc. *Res:* Sedimentary petrology; submarine geology. *Mailing Add:* 3916 Rusthill Pl Fairfax VA 22030

LILL, PATSY HENRY, b Mesa, Ariz, July 6, 43. IMMUNOLOGY, PATHOLOGY. *Educ:* Northwestern Univ, BS, 66; Univ Wis, MST, 72; Chicago Med Sch-Univ Health Sci, PhD(path), 75. *Prof Exp:* Teacher biol, Highland Park High Sch, Ill, 66-70; lab supvr cancer res, Dept Exp Path, Mt Sinai Hosp Med Ctr, 70-75; fel tumor immunology & cancer biol, Frederick Cancer Res Ctr, 75-77; ASST PROF PATH, SCH MED, UNIV SC, 77- *Concurrent Pos:* Prin investr, Univ SC Res & Prod Scholarship grant, 79 & Nat Cancer Inst grant, 79-82; mem, Charles Louis Davis Doctor Vet Med Found. *Mem:* AAAS; Am Asn Pathologists; Am Asn Cancer Res. *Res:* Tumor immunology; effect of physical and chemical carcinogens on syngeneic tumor growth. *Mailing Add:* Dept of Path Univ SC Sch of Med Columbia SC 29208

LILLARD, DORRIS ALTON, b Thompson Station, Tenn, July 17, 36. FOOD CHEMISTRY, BIOCHEMISTRY. *Educ:* Middle Tenn State Univ, BS, 58; Ore State Univ, MS, 61, PhD(food sci), 64. *Prof Exp:* Res fel lipid autoxidation, Ore State Univ, 58-64; asst prof food flavor chem, Iowa State Univ, 64-68; assoc prof, 68-80, PROF FOOD SCI, UNIV GA, 80- *Mem:* AAAS; Am Chem Soc; Am Oil Chem Soc; Inst Food Technol; Am Meat Sci Asn. *Res:* Flavor chemistry of foods; autoxidation of lipids; mycotoxins in foods; food microbiology. *Mailing Add:* Dept of Food Sci Univ of Ga Athens GA 30602

LILLEGRAVEN, JASON ARTHUR, b Mankato, Minn, Oct 11, 38; m 64, 83; c 2. PALEONTOLOGY. *Educ:* Calif State Col Long Beach, BA, 62; SDak Sch Mines & Technol, MS, 64; Univ Kans, PhD(zool), 68. *Prof Exp:* Instr zool, Calif State Col Long Beach, summer, 64; NSF fel paleont, Univ Calif, Berkeley, 68-69; from asst prof to prof zool, San Diego State Univ, 69-75; assoc prof, 76-78, cur, Geol Mus, 76-88, PROF GEOL, UNIV WYO, 78- *Concurrent Pos:* Prog dir, syst biol prog, NSF, 77-78; co-ed, Contrib to Geol, 76-88; assoc dean, Col Arts & Sci, Univ Wyo, 84-85; ed bd, Nat Geog Res, 86-; Humboldt sr scientist, Freie Univ, WBerlin, 88-89. *Honors & Awards:* Alexander von Humboldt Sr US Scientist Award, 88 & 89. *Mem:* Paleont Soc; Soc Vert Paleont (vpres & pres, 84-86); Am Soc Mammal; fel Linnean Soc London. *Res:* Paleogeography; Mesozoic and early Cenozoic mammalian paleontology, comparative anatomy and evolution of mammalian reproduction. *Mailing Add:* Dept Geol & Geophys Univ Wyo Laramie WY 82071-3386

LILLEHOJ, EIVIND B, b Kimballton, Iowa, Aug 11, 28; m 48; c 4. PLANT PHYSIOLOGY, BIOCHEMISTRY. *Educ:* Iowa State Univ, BS, 60, MS, 62, PhD(plant physiol), 64. *Prof Exp:* NIH fel, Carlsberg Lab, Copenhagen, Denmark, 64-65; microbiologist, Northern Regional Res Lab, USDA, 65-89; MICROBIOL CONSULT, 89- *Mem:* Am Soc Plant Physiol; Am Soc Microbiol; Am Inst Biol Sci. *Res:* Fungal physiology; mycotoxins; fermentation; microbial products. *Mailing Add:* PO Box 22 Kimballton IA 51543-0022

LILLEHOJ, HYUN SOON, b Seoul, Korea, Mar 1, 49; m 79; c 2. IMMUNOGENETICS, IMMUNOPARASITOLOGY. *Educ:* Univ Hartford, BS, 74; Univ Conn, MS, 76; Wayne State Univ, PhD(immunol), 79. *Prof Exp:* Staff fel, Nat Inst Allergy & Infectious Dis, NIH, 81-84; RES IMMUNOLOGIST, USDA, 84- *Concurrent Pos:* Assoc ed, Poultry Sci, 88; adj prof, Univ Del, 89. *Mem:* Am Asn Immunologists; Poultry Sci Asn; Am Asn Avian Vet Pathologists. *Res:* Vaccine against Eimeria; monoclonal antibodies detecting avian lymphocytes and lymphokines; immunopathology and immunogenetics of avian coccidiosis; molecular cloning of avian lymphokines. *Mailing Add:* BARC-E Bldg 1040 PDL LPSI USDA Beltsville MD 20705

LILLELAND, OMUND, b Stavanger, Norway, Mar 12, 99; US citizen; m 34; c 2. POMOLOGY. *Educ:* Univ Calif, BS, 21, PhD, 34. *Prof Exp:* Jr pomologist, 26-30, from asst to assoc pomologist, 31-46, pomologist, 46-66, EMER POMOLOGIST, UNIV CALIF, DAVIS, 67-; CONSULT, 67- *Res:* Growth of fruits; thinning of deciduous tree fruits; phosphate and potash nutrition of fruit trees. *Mailing Add:* 40 College Park Davis CA 95616

LILLELEHT, L(EMBIT) U(NO), b Parnu, Estonia, Mar 9, 30; US citizen; m 60; c 2. THERMAL SCIENCES, ENERGY CONVERSION. *Educ:* Univ Del, BChE, 53; Princeton Univ, MSE, 55; Univ Ill, PhD(chem eng), 62. *Prof Exp:* Engr process develop & res, E I du Pont de Nemours & Co, Inc, 54-57; from asst prof to assoc prof chem eng, Univ Alta, 60-66; ASSOC PROF CHEM ENG, UNIV VA, 66- *Concurrent Pos:* Partner, Assoc Environ Consults, 72-; vis assoc prof, Solar Energy Res Inst, 78-79; lectr solar energy, US Int Comn Agency, 78-79. *Mem:* Am Inst Chem Engrs; Am Chem Soc; Int Solar Energy Soc; Sigma Xi; AAAS; Am Solar Energy Soc. *Res:* Multiphase flows; air pollution control; nucleation and condensation of refractory vapors; heat transfer; utilization of solar and other alternative energy resources. *Mailing Add:* Dept Chem Eng Thornton Hall Univ Va Charlottesville VA 22903-2442

LILLER, WILLIAM, b Philadelphia, Pa, Apr 1, 27; m 85; c 3. ARCHAEOASTRONOMY. *Educ:* Harvard Univ, AB, 49; Univ Mich, AM, 50, PhD(astron), 53. *Prof Exp:* Mem meteor exped, Harvard Univ, 47-48, supt, 52-53; asst, McMath-Hulbert Observ, Univ Mich, 52, from instr to assoc prof astron, 53-60; chmn, dept astron, 60-66, prof, 60-70, ROBERT WHEELER WILLSON PROF APPL ASTRON, HARVARD UNIV, 70- *Concurrent Pos:* Guggenheim fel, 64-65; master, Adams House, Harvard Univ, 68-73, head tutor, astron dept, 77-80; vis comnr, Bartol Found, 68-, chmn, 76-79; sr res fel, Isaac Newton Inst, Santiago, Chile, 81- *Mem:* Am Astron Soc; Royal Astron Soc Can; fel AAAS; Am Acad Arts & Sci; Int Astron Union; Brit Astron Asn. *Res:* Photoelectric photometry of planetary nebulae and hot stars; investigation of x-ray sources; spectrophotometry; globular clusters; archaeoastronomy (easter I). *Mailing Add:* Vina del Mar Casilla 437 Chile

LILLESAND, THOMAS MARTIN, b Laurium, Mich, Oct 1, 46; m 68; c 3. REMOTE SENSING. *Educ:* Univ Wis-Madison, BS, 69, MS, 70, PhD(civil eng), 73. *Prof Exp:* Prof remote sensing, State Univ NY, Syracuse, 73-78 & Univ Minn, 78-82; PROF REMOTE SENSING, UNIV WIS-MADISON, 82- *Concurrent Pos:* Consult, 73- *Honors & Awards:* Alan Gordon Award, Am Soc Photogram & Remote Sensing, 79, Talbert Abrams Award, 84, Fennell Award, 88. *Mem:* Am Soc Photogram & Remote Sensing; Am Soc Civil Engrs; Soc Am Foresters. *Res:* Remote sensing and image processing of satellite data for application in agriculture, forestry, water resources and environmental monitoring; space policy. *Mailing Add:* Environ Remote Sensing Ctr Rm 1231 1225 W Dayton St Madison WI 53706

LILLEVIK, HANS ANDREAS, b Sherman, SDak, Feb 4, 16; m 46; c 4. BIOCHEMISTRY. *Educ:* St Olaf Col, BA, 38; Univ Minn, MS, 40, PhD(biochem), 46. *Prof Exp:* Instr biochem, Univ Minn, 42-44; res chemist, Minn Mining & Mfg Co, 44-45; from instr to assoc prof chem & biochem, 46-70, PROF BIOCHEM, MICH STATE UNIV, 70- *Concurrent Pos:* Am Scand Found fel, Carlsberg Lab, Denmark, 47-48. *Mem:* AAAS; Am Chem Soc; Am Soc Biol Chemists; Am Dairy Sci Asn. *Res:* Chemical properties and biological function of proteins and enzymes. *Mailing Add:* 708 Knoll Rd East Lansing MI 48823-2826

LILLEY, ARTHUR EDWARD, b Mobile, Ala, May 29, 28. ASTRONOMY. *Educ:* Univ Ala, BS, 50, MS, 51; Harvard Univ, PhD(radio astron), 54. *Prof Exp:* Physicist, Naval Res Lab, 54-57; asst prof radio astron, Yale Univ, 57-59; assoc prof, 59-63, PROF RADIO ASTRON, HARVARD UNIV, 63-; assoc dir, Harvard-Smithsonian Ctr Astrophys, 72-89, ASTRONOMER-IN-CHARGE, SMITHSONIAN OBSERV, 65- *Concurrent Pos:* Res Corp grant, 57-59; Sloan res fel, 58-60. *Honors & Awards:* Bok Prize, Harvard Univ, 58. *Mem:* AAAS; Int Union Radio Sci; Int Astron Union; Am Astron Soc; Am Phys Soc. *Res:* Spectral line and satellite radio astronomy; radio astronomical navigation techniques. *Mailing Add:* Harvard Univ Ctr Astrophys 60 Garden St Cambridge MA 02138

LILLEY, DAVID GRANTHAM, b Shipley, Eng; US citizen. COMBUSTION AERODYNAMICS, FIRE MODELING. *Educ:* Sheffield Univ, Eng, BSc, 66, MSc, 67, PhD(chem eng), 70. *Prof Exp:* Lectr math, Sheffield Polytech, Eng, 70-73; sr res assoc, Cranfield Inst Technol, Eng, 73-75; vis assoc prof combustion, Univ Ariz, 75-76; assoc prof mech eng, Concordia Univ, Montreal, 76-78; assoc prof, 78-82, PROF MECH ENG, OKLA STATE UNIV, 82- *Mem:* Assoc fel Am Inst Aeronaut & Astronaut; Am Soc Mech Engrs; Inst Fuel; assoc fel Inst Math & Applns. *Res:* Theoretical combustion aerodynamics; computational fluid dynamics; swirling flows; combustor design; numerical methods; finite difference methods; turbulent reacting flows; heat transfer, fires, flames, and computer simulation. *Mailing Add:* Dept Eng Sci Okla State Univ Main Campus Stillwater OK 74078-0545

LILLEY, JOHN RICHARD, b Fall River, Mass, Apr 2, 34; div; c 4. RADIATION FLOW, SHOCK PHYSICS. *Educ:* Univ Calif, Berkeley, AB, 56; Univ Idaho, MSc, 62. *Prof Exp:* Physicist, Gen Elec Co, Wash, 56-62; sr scientist, Western Div, McDonnell Douglas Astronaut Co, 62-72; STAFF MEM, LOS ALAMOS NAT LAB, 72- *Concurrent Pos:* Lectr, Radiol Physics Fel Prog, AEC, 61; collabr, Comm Atomic Energy, Bruyeres-le-Chatel, France, 79-80; adj prof eng, Los Alamos Univ NMex, 85- *Res:* Radiation shielding; reactor physics; statistics; nonlinear optimization; experimental data analysis; thermal analysis; nuclear weapons effects; nuclear weapons design theory. *Mailing Add:* Los Alamos Nat Lab Univ of Calif PO Box 1663 Los Alamos NM 87545

LILLICH, THOMAS TYLER, b Cincinnati, Ohio, Sept 8, 43; m 65; c 2. BACTERIAL PHYSIOLOGY, HOST PARASITE RELATIONS. *Educ:* Miami Univ, AB, 65; NC State Univ, MS, 68, PhD(microbiol), 70. *Prof Exp:* Asst prof oral & cell biol, Univ Ky, 72-75, act chair oral biol, 75, assoc prof oral biol & microbiol & immunol, 75-81, chair, dept oral biol, 80-88, MEM MICROBIOL GRAD FAC, UNIV KY, 74-, PROF ORAL BIOL & MICROBIOL & IMMUNOL, 81-, CHAIR DEPT ORAL HEALTH SCI, 88- *Concurrent Pos:* NIH res fel, Univ Ky, 70-72, & Agr Res Serv contractee, 73-76; vis prof, Roy Dent Col, Arhus Denmark, 78; consult, Div Educ Res Prog, Am Asn Med Sch, 77-85, Bd Educ Training, Am Soc Microbiol, 77-79, Univ Tex Dent Br San Antonio, Ohio State Univ Col Dent, 87; comt, Dent Accreditation, Am Dent Asn, 90-91, SIll Univ Grad Sch, 91. *Mem:* Am Soc Microbiol; Am Asn Dent Sch (chair elect, 75-76, 88-89, chair, 76-77, 89-90); Int Asn Dent Res; Sigma Xi; Am Asn Dent Res. *Res:* Effects of antimicrobials on the oral microflora; emphasis on oral microflora of medically comprised patients; identification and ratios of organisms; control with tropical antimicrobials to reduce systematic disease; host response to microbial challenge. *Mailing Add:* Dept Oral Health Sci Col Dent Univ of Ky Lexington KY 40536-0084

LILLIE, CHARLES FREDERICK, b Indianola, Iowa, Feb 20, 36; wid; c 3. AERONAUTICAL ENGINEERING, ASTRONAUTICAL ENGINEERING. *Educ:* Iowa State Univ, BS, 57; Univ Wis, Madison, PhD(astrophys), 68. *Prof Exp:* Instr eng, NASA Flight Res Ctr, Edwards, Calif, 60-62; teaching asst, Dept Astron, Univ Wis, 62-64, res asst, Washburn Observ, 64-68, proj assoc, Space Astrophys Lab, 68-70; from asst prof to assoc prof physics & astrophys, Univ Colo, 70-77, assoc prof astrogeophys, Attendant Rank & fel Lab Atmospheric & Space Physics, 77-79; SR SYSTS ENGR, FED SYSTS DIV, TRW SPACE & TECH GROUP, REDONDO BEACH, CALIF, 79- *Concurrent Pos:* Prin investr, Voyager Photopolarimeter Exp, 72-79; co-investr, Apollo 17 Ultraviolet Spectrometer Exp, 72-74; team mem, Large Space Telescope Inst Definition Team High Resol Spectrograph, 73-75. *Mem:* Am Astron Soc; AAAS; Soc Photo-Optical Instrumentation Engrs; Int Astron Union; Am Inst Aeronaut & Astronaut; Air Force Asn. *Res:* Surface brightness of the night sky, zodiacal light, diffuse galactic light, interstellar radiation density, extragalactic light; cometary physics, ultraviolet spectroscopy of stars and nebulae; spacecraft design, RF environment, contamination; spacecraft operations and on orbit servicing; optics. *Mailing Add:* 6202 Vista del Mar #364 Playa del Rey CA 90293

LILLIE, JOHN HOWARD, b Oak Park, Ill, Dec 16, 40; m 63; c 2. ANATOMY, DENTISTRY. *Educ:* Univ Mich, DDS, 66, PhD(anat), 72. *Prof Exp:* PROF ANAT, SCH MED & ASSOC PROF, SCH DENT, UNIV MICH, ANN ARBOR, 72-, STAFF MEM, DENT RES INST, 72- *Concurrent Pos:* Admin, Sch Dent. *Mem:* Am Soc Cell Biol; Am Asn Anat. *Res:* Cellular control mechanisms in exocrine secretion and epithelia-connective tissue interactions; features of synthesis and control in the production of basal lamina constituents. *Mailing Add:* Dept Anat & Cell Biol Univ Mich Rm 4818 Med Sch 11 Ann Arbor MI 48109

LILLIE, ROBERT JONES, b Rochester, Minn, Apr 15, 21; m 46; c 2. POULTRY NUTRITION. *Educ:* Pa State Col, BS, 44; Univ Md, MS, 46, PhD(poultry nutrit), 49. *Prof Exp:* Asst poultry dept, Univ Md, 45-47; poultry husbandman, Animal & Poultry Husb Res Br, USDA, 47-72, res animal scientist, Nonruminant Animal Nutrit Lab, Nutrit Inst, Sci & Educ Admin Agr Res, 72-78; RETIRED. *Concurrent Pos:* Mem standard diet subcomt, Nat Res Coun, 54. *Honors & Awards:* Am Poultry Sci Asn Award, 50; Commission Award, US Civil Serv, 76. *Mem:* Am Poultry Sci Asn; Am Inst Nutrit; Worlds Poultry Cong. *Res:* Vitamins, antibiotics, surfactants, arsenicals, unidentified factors, proteins and amino acids; pesticides; reproductive efficiency; air pollutants affecting poultry; trace minerals in swine. *Mailing Add:* Cornwall Manor PO Box 125 Cornwall PA 17016-0125

LILLIEFORS, HUBERT W, b Reading, Pa, June 14, 28; m 80; c 2. STATISTICS. *Educ:* George Washington Univ, BA, 52, PhD(statist), 64; Mich State Univ, MA, 53. *Prof Exp:* Mathematician, Diamond Ord Fuze Labs, 53-55; sr scientist opers res, Lockheed Missile Systs Div, 55-56, opers analyst, Opers Eval Group, 56-57; mathematician opers res, Appl Physics Lab, Johns Hopkins Univ, 57-64; from instr to assoc prof, 62-67, PROF STATIST, GEORGE WASHINGTON UNIV, 67- *Mem:* Am Statist Asn; Inst Math Statist. *Res:* Nonparametric statistics; statistical inference. *Mailing Add:* Dept of Statist George Washington Univ Washington DC 20052

LILLIEN, IRVING, b New York, NY, Feb 2, 29. ORGANIC CHEMISTRY, COMPUTER SCIENCES. *Educ:* Univ Denver, BS, 50; Purdue Univ, MS, 52; Polytech Inst NY, PhD(org chem), 59. *Prof Exp:* Fel org chem, Wayne State Univ, 59-61; asst prof, Georgetown Univ, 61-62; asst prof, Univ Miami, 62-65, Sch Med, 65-67; assoc prof, Marshall Univ, 67-69; assoc prof org chem, 69-80, PROF CHEM, MIAMI-DADE COMMUNITY COL, 80- *Concurrent Pos:* Air Force Off Sci & Res grant, 63-65. *Mem:* AAAS; Am Chem Soc; Royal Soc Chem; Inst Elec & Electronics Engrs. *Res:* Physical-organic chemistry; mechanisms of organic reactions; chemistry and conformation of small and medium size rings; science education and administration. *Mailing Add:* Dept Chem Miami-Dade Community Col 11011 SW 104 St Miami FL 33176

LILLINGTON, GLEN ALAN, b Winnipeg, Can, Oct 20, 26; m 57; c 3. PULMONARY MEDICINE, CONTINUING MEDICAL EDUCATION. *Educ:* Univ Manitoba, BSc, 46, MD, 51; Univ Minn, MS, 57; FRCP, 59; FACP, 67. *Prof Exp:* Asst staff med, Mayo Clin & Found, 57-58; lectr, Univ Manitoba Fac Med, 58-60; asst clin prof, Sch Med, Stanford Univ, 60-73; PROF MED, SCH MED, UNIV CALIF, DAVIS, 73- *Concurrent Pos:* Res assoc, Palo Alto Med Res Found, 65-73; consult, Rand Corp, Santa Monica, 69-70; med dir respiratory therapy, Sch Respiratory Therapy, Foothill Col, 72-73; travelling fel, Webb-Waring Inst, Denver, 73-74; actg chmn, Dept Med, Med Sch, Univ Calif, Davis, 79-80, chief staff, 79-80, dir residency med, Med Ctr, 79-80, prof med, 75-81, chief pulmonary, critical care div, 75-87. *Res:* Experimental emphysema; pulmonary mechanics; differential diagnosis of pulmonary diseases based on roentgenographic patterns; decision analysis in medicine. *Mailing Add:* Med Ctr Univ Calif 4301 X St Sacramento CA 95817

LILLWITZ, LAWRENCE DALE, b Hinsdale, Ill, June 1, 44; m 68; c 5. INDUSTRIAL ORGANIC CHEMISTRY, INDUSTRIAL PROCESS CHEMISTRY. *Educ:* Ill Benedictine Col, BS, 66; Univ Notre Dame, PhD(org chem), 70. *Prof Exp:* Group leader, Chem Div, Quaker Oats Co, 70-77; res chemist, 77-79, staff res chemist, 79-81, sr res chemist, 81-86, ASSOC RES CHEMIST, AMOCO CHEM CO, 86- *Mem:* Am Chem Soc. *Res:* Monomer synthesis; organic reaction mechanisms; homogeneous and heterogeneous catalysis. *Mailing Add:* 773 Crescent Blvd Glen Ellyn IL 60137

LILLY, ARNYS CLIFTON, JR, b Beckley, WVa, June 3, 34; m 56; c 3. PHYSICS. *Educ:* Va Polytech Inst, BS, 57,PhD, 89; Carnegie Inst Technol, MS, 63; Va Polytech Inst, PhD, 89. *Prof Exp:* Res physicist, Gulf Res & Develop Co, Pa, 57-65; res physicist, 65-67, sr scientist, Physics Div, 67-74, assoc prin scientist, 74-81, prin scientist, 81-84, RES FEL, PHILLIP MORRIS RES CTR, 84- *Mem:* Am Phys Soc; Sigma Xi. *Res:* Ion & electron optics; dielectric theory and experiment; electrostatics and organic conduction; space charge in insulators; thermal physics; combustion; laser processing; fluid mechanics; quantum chemistry; theoretical physics. *Mailing Add:* Res Fel Philip Morris Res Ctr PO Box 26583 Richmond VA 23261

LILLY, DAVID J, b Washington, DC, Sept 21, 31; m 56; c 4. AUDIOLOGY. *Educ:* Univ Redlands, BA, 54, MA, 57; Univ Pittsburgh, PhD(audiol), 61. *Prof Exp:* Res assoc, Cent Inst Deaf, St Louis, Mo, 61-64; prof audiol, Univ Iowa, 64-80; PROF OTOLARYNGOL & MAXILLOFACIAL SURG, SPEECH PATH & AUDIOL, UNIV MICH, 80-; AT GOOD SAMARITAN HOSP & MED CTR, PORTLAND, ORE. *Concurrent Pos:* NIH res fel, 61, Nat Inst Neurol Dis & Blindness trainee, 62-63; consult hearing aid res & procurement prog, Vet Admin, 66- *Mem:* Acoust Soc Am; Am Speech & Hearing Asn; Audio Eng Soc. *Res:* Experimental audiology, especially on measurements of acoustic impedance at the tympanic membrane of normal and pathologic ears; speech audiometry; masking; auditory adaptation; bone conduction; audiometric standards and calibration. *Mailing Add:* Dept Audiol Good Samaritan Hosp & Med Ctr 1040 NW 22nd Ave Portland OR 97210

LILLY, DOUGLAS KEITH, b San Francisco, Calif, June 16, 29; m 54; c 3. MESOSCALE DYNAMICS. *Educ:* Stanford Univ, BS, 50; Fla State Univ, MS, 54, PhD(meteorol), 59. *Prof Exp:* Res meteorologist, US Weather Bur, 58-65; prog scientist, Nat Ctr Atmospheric Res, 65-73, sr scientist, 73-82; PROF METEOROL, 82-, DIR COOP INST, MESOSCALE METEOROL STUDIES (CIMMS), UNIV OKLA, 87- *Honors & Awards:* Rossby Medal, Am Meteorol Soc, 86. *Mem:* Fel Am Meteorol Soc. *Res:* Atmospheric convection, turbulence and gravity waves; numerical simulation of meteorological flows. *Mailing Add:* Dept Meteorol Univ Okla Main Campus Norman OK 73019

LILLY, FRANK, b Charleston, WVa, Aug 28, 30. GENETICS, ONCOLOGY. *Educ:* WVa Univ, BS, 51; Univ Paris, PhD(org chem), 59; Cornell UNIV, PhD(biol, genetics), 65. *Prof Exp:* Res fel, 65-74, chmn, Dept Genetics, 76-88, PROF IMMUNOGENETICS & ONCOGENETICS, ALBERT EINSTEIN COL MED, 65- *Concurrent Pos:* New York City Health Res Coun career scientist award, Albert Einstein Col Med, 67-72; mem, Breast Cancer-Virus Working Group, Nat Cancer Inst, 72-79; mem bd dirs, Leukemia Soc Am, 73-78, 83-; mem, Sci Adv Coun, Cancer Res Inst, Inc, 75-; mem sci adv comn, Wistar Inst, 80-; mem, Presidential comn HIV Epidemic, 88-89; mem bd overseers, Jackson Lab, 89- *Mem:* Nat Acad Sci; NY Acad Sci; Genetics Soc Am; Am Asn Immunologists; Am Asn Cancer Res; AAAS. *Res:* Oncogenetics, study of genes which influence susceptibility or resistance to oncogenic agents in mice; immunogenetics. *Mailing Add:* Dept Molecular Genetics Albert Einstein Col Med Bronx NY 10461

LILLY, JOHN RUSSELL, b Milwaukee, Wis, May 23, 29. PEDIATRIC SURGERY. *Educ:* Univ Wis, BS, 51, MD, 54. *Prof Exp:* Fel pediat surg, Hosp Sick Children, Eng, 63-64; chief resident, Children's Hosp, Washington, DC, 64-65, chief organ transplant & surg dir clin res ctr, 70-73; PROF & HEAD DIV PEDIAT SURG, UNIV COLO HEALTH SCI CTR, DENVER, 73-; INTERIM SURGEON-IN-CHIEF, CHILDREN'S HOSP, DENVER, 90- *Concurrent Pos:* Fel transplantation, Univ Colo Health Sci Ctr, 69-70, actg chmn dept surg, 80-84; dir surg res, Res Found, Children's Hosp, DC, 70-73; consult, pediat surg, Denver Gen Hosp, 73-, NIH, DC, 81-, Nat Jewish Hosp, Denver, 86-; dir pediat surg, Rose Med Ctr, Denver, 88- *Mem:* Am Col Surgeons; Am Pediat Surg Asn; Am Acad Pediat; Am Asn Study Liver Disease; Am Surg Asn. *Res:* Pediatric hepatobiliary disease and pediatric thoracic surgery; thoracic surgery of infants and children. *Mailing Add:* Pediat Surg No B323 Children's Hosp 1950 Ogden St Denver CO 80218

LILLY, PERCY LANE, b Spanishburg, WVa, July 14, 27; m 51; c 4. PLANT TAXONOMY. *Educ:* Concord Col, BS, 50; Univ WVa, MS, 51; Pa State Univ, PhD, 57. *Prof Exp:* Instr biol, Salem Col, WVa, 51-53; from asst to assoc prof, 56-64, PROF BIOL, HEIDELBERG COL, 64-, CHMN DEPT, 65- *Concurrent Pos:* Spec field staff mem, Rockefeller Found, Colombia, 68-69. *Mem:* AAAS; Bot Soc Am. *Res:* Plant genetics and microbiology; nitrogen fixation in Azotobacter; tropical botany. *Mailing Add:* Heidelberg Col Tiffin OH 44883

LILLYA, CLIFFORD PETER, b Chicago, Ill, May 23, 37; m 62; c 2. ORGANIC CHEMISTRY. *Educ:* Kalamazoo Col, AB, 59; Harvard Univ, PhD(chem), 64. *Prof Exp:* Staff assoc, 63-64, from asst prof to assoc prof, 64-73, PROF CHEM, UNIV MASS, AMHERST, 73- *Concurrent Pos:* Fel, Woodrow Wilson, 59, NSF, 59-63, Alfred P Sloan Found, 69-71; vis scholar, Univ Calif, Los Angles, 70-71, Stanford Univ, 78. *Mem:* Am Chem Soc; AAAS. *Res:* Organic polymers; liquid crystals; nitramines. *Mailing Add:* Dept Chem Univ Mass Amherst MA 01003

LILLYWHITE, HARVEY B, b Nogales, Ariz, Dec 1, 43; m 67; c 2. COMPARATIVE PYHSIOLOGY, PHYSIOLOGICAL ECOLOGY. *Educ:* Univ Calif, Riverside, BA, 66; Univ Calif, Los Angeles, MA, 67, PhD(zool), 70. *Prof Exp:* Postdoctoral fel zool, Univ Calif, Berkeley, 70-71; from asst prof to prof physiol, Univ Kans, Lawrence, 71-84; PROF ZOOL, UNIV FLA, GAINESVILLE, 84- *Concurrent Pos:* Vis lectr, Monash Univ, Clayton, Victoria, Australia, 75-76; sect ed, Copeia, 78-82, Ecol & Ecol Monographs, 82-86; secy nominating comt, Ecol Soc Am, 78-79; vis scientist, Scripps Inst Oceanog, La Jolla, Calif, 79-80; res fel, San Diego Zoo & Wild Animal Park, 79-80; mem publ policy comt, Am Soc Ichthyologists & Herpetologists, 80-82; res fel, Univ New Eng, Armidale, NSW, Australia, 91. *Mem:* Am Soc Zoologists; Am Physiol Soc; Ecol Soc Am; Am Soc Ichthylogists & Herpetologists; Soc Study Amphibians & Reptiles; AAAS. *Res:* Comparative and ecological physiology of vertebrates, especially amphibians and reptiles; cardiovascular adaptations of reptiles; functional morphology of vertebrate integument; water and thermal relations; ecology of fire disturbance and animal coloration. *Mailing Add:* Dept Zool Univ Fla Bartram Hall Gainesville FL 32611

LIM, ALEXANDER TE, b Manila, Philippines, June 17, 42; US citizen; m 71; c 2. HEAT TRANSFER & THERMAL SCIENCES, AIR CONDITIONING & REFRIGERATION. *Educ:* Univ St Thomas, BS, 64; Duke Univ, MS, 69. *Prof Exp:* Instr, Col Eng, Ateneo de Manila Univ, 64-66; teaching asst undergrad eng, Mech Eng Dept, Duke Univ, 66-69; sr engr, Res & Develop Div, Carrier Corp, 69-76, mgr develop eng, Room Air Opers, 76-80, dir eng, Light Residential Div, 80-85; dir res & develop, ICG Keeprite Corp, 85-86, vpres, 86-87; DIR TECHNOL, INTER-CITY PROD CORP, 87- *Concurrent Pos:* comt mem, Am Soc Heating, Refrig & Air-conditioning Engrs, Comt, 74-79; Comt mem, Am Home Appliance Mfrs Eng Comt, 78-83 & 88-91; comt mem, Air Conditioning & Refrig Inst Packaged Terminal Air Conditioning Eng Comt, 85-91, mem, Res & Technol Comt, 89- 91. *Mem:* Am Soc Mech Engrs; Am Soc Heating, Refrig & Air Conditioning Engrs; Air Conditioning & Refrig Inst. *Res:* Air-side and refrigerant-side heat and mass transfer as applied to air-conditioning and refrigeration; new concepts and cycles for air-conditioning application; granted four patents. *Mailing Add:* Inter-city Prod Corp 1136 Heil Quaker Blvd PO Box 3005 Lavergne TN 37086

LIM, DANIEL V, b Houston, Tex, Apr 15, 48; m 73. PATHOGENIC MICROBIOLOGY, ENVIRONMENTAL MICROBIOLOGY. *Educ:* Rice Univ, BA, 70; Tex A&M Univ, PhD(microbiol), 73. *Prof Exp:* Fel, Baylor Col Med, 73-76; from asst prof to assoc prof microbiol, 76-87, actg chmnbiol, 83-85, PROF MICROBIOL, UNIV SFLA, 87-, DIR, INST BIOMOLECULAR SCI, 88- *Concurrent Pos:* Consult, Pharmacia Diag-Pharmacia AB, 79-86, Life Technols, Inc-Gibco Labs, 82-86, The Conservancy, 80-84; assoc fac mem, Tampa Gen Hosp, 82-88; pres, Micro Concepts Res Corp, 86-; grad fel panel, Nat Res Coun/NSF, 89-; pres, southeastern br, Am Soc Microbiol, 90-91. *Mem:* Am Soc Microbiol; Inter-Am Soc Chemother (vpres, 83-88); AAAS; Sigma Xi; fel Am Acad Microbiol; Am Pub Health Assn. *Res:* The virulence of bacteria and rapid diagnosis of bacterial diseases; group B streptococci, Neisseria gonorrhoeae, and environmental pathogens; invented Lim group B strep broth; author of one book. *Mailing Add:* Dept Biol/Inst Biomolecular Sci LIF136 Univ S Fla Tampa FL 33620-5150

LIM, DAVID J, b Seoul, Korea, Nov 27, 35; US citizen; m 66; c 2. OTOLARYNGOLOGY, ELECTRON MICROSCOPY. *Educ:* Yonsei Univ, Korea, AB, 55, MD, 60. *Prof Exp:* Intern, Nat Med Ctr, Seoul, Korea, 60-61, resident otolaryngol, 61-64; res assoc, 66-67, from asst prof to assoc prof, 67-76, PROF OTOLARYNGOL, COL MED, OHIO STATE UNIV, 76- & PROF ANAT, 77- *Concurrent Pos:* Spec fel otol res, Mass Eye & Ear Infirmary & Harvard Med Sch, 65-66; dir, Otol Res Labs, Ohio State Univ, 67-; mem task force, Am Acad Ophthal & Otolaryngol & Am Bd Otolaryngol, 69-72; consult comt res otolaryngol, Am Acad Ophthal & Otolaryngol, 77-; mem ad hoc adv comt, Commun Disorders Prog, Nat Inst Neurol & Commun Disorders & Stroke, 76-79; mem sci rev comt, Deafness Res Found, 77-80; mem commun sci study sect, NIH, 79; mem, Nat Adv Neurol & Commun Disorders & Stroke Coun, NIH, 79-83; adv-at-large, Comt Hearing, Bioacoust & Biomech, Nat Res Coun, 80-; mem bd dir, Deafness Res Found, 80-; prin investr, various grants & contracts, 69-; Fogarty sr Int & vis scientist, Swed Med Res Coun, Karolinska Inst, 82. *Honors & Awards:* First Award Scientific Exhibit, Am Acad Ophthal & Otolaryngal, 72. *Mem:* Am Acad Otolaryngol; Soc Neurosci; Am Otol Soc; Asn Res Otolaryngol (secy-treas, 73-75, pres, 76-77, past pres & prog chair, 77-78, ed hist, 80-); Barany Soc; Histochem Soc; Am Soc Cell Biol; Soc Mucosal Immunol. *Res:* Investigation of the ear as to the normal function and disorders of hearing and balance with the use of light and electron microscopy; immunocytochemistry; immunochemistry and microbiology. *Mailing Add:* Dept Otolaryngol Otol Res Labs Ohio State Univ 4331 Univ Hosp Clin 456 W Tenth Ave Columbus OH 43210

LIM, EDWARD C, b Seoul, Korea, Nov 17, 32; nat US; m 58; c 2. PHYSICAL CHEMISTRY. *Educ:* St Procopius Col, BS, 54; Okla State Univ, MS, 57, PhD(chem), 59. *Prof Exp:* Instr phys chem, Loyola Univ, Ill, 58-60, from asst prof to prof, 60-68; prof chem, Wayne State Univ, 68-89; GOODYEAR PROF CHEM, UNIV AKRON, 89- *Mem:* Fel Am Phys Soc; Am Chem Soc. *Res:* Molecular electronic spectrosocpy; molecular photophysics. *Mailing Add:* Dept Chem Univ Akron Akron OH 44325-3601

LIM, H(ENRY) C(HOL), b Seoul, Korea, Oct 24, 35; US citizen; m 59; c 3. CHEMICAL & BIOCHEMICAL ENGINEERING. *Educ:* Okla State Univ, BS, 57; Univ Mich, MSE, 59; Northwestern Univ, PhD(chem eng), 67. *Prof Exp:* Process develop engr, Pfizer, Inc, 59-63; from asst prof to prof chem eng, Purdue Univ, 66-87; PROF & CHMN, BIOCHEM ENG & PROF MICROBIOL & MOLECULAR GENETICS, UNIV CALIF, IRVINE, 87- *Concurrent Pos:* Vis scholar, Calif Inst Technol, 77. *Honors & Awards:* Food Pharmaceut & Bio Eng Award, Am Inst Chem Eng, 87. *Mem:* Am Inst Chem Engrs; Am Soc Microbiol; Am Chem Soc; Inst Food Technologists. *Res:* Modelling, optimization and control of chemical and biochemical processes; biological reactor engineering; cellular growth kenetics; optimal operating strategies for fed-batch bioreactors of recombinant cells with regulated promoters; on-line optimization of continuous flow bioreactors with little priori knowledge; engineering of recombinant cell product expression and secretion. *Mailing Add:* Biochem Eng Univ Calif Irvine CA 92717

LIM, HONG SEH, b Hong Kong, Aug 5, 58; m 84; c 2. IMAGE PROCESSING, COMPUTER VISION. *Educ:* Univ Hong Kong, BSc, 81; Stanford Univ, MS(elec eng) & MS(opers res), 83, MS & PhD(elec eng), 87. *Prof Exp:* Res asst, Stanford Univ, 82-87; staff mem, Los Angeles Sci Ctr, IBM, 87-90 & Palo Alto Sci Ctr, 90-91; LANG MGR, CALERA RECOGNITION SYSTS, 91- *Concurrent Pos:* Prin investr, AKM Assocs, 85-87; lectr, Univ Calif, Los Angeles, 89; consult, Univ Calif, San Francisco, 90- *Mem:* Sr mem Inst Elec & Electronics Engrs; Asn Comput Mach; Optical Soc Am; Opers Res Soc Am; Inst Mgt Sci. *Res:* Application of image processing and computer vision to target identification, object recognition, industrial parts identification and manufacturing inspection; character recognition. *Mailing Add:* 26180 Rancho Manuella Lane Los Altos Hills CA 94022

LIM, JAMES KHAI-JIN, b Batavia, Java, March 11, 33; US citizen; m 62; c 3. PHARMACEUTICS. *Educ:* Univ Malaya, Singapore, BPharm, 58; Univ NC, MS, 62, PhD(pharmaceut), 65. *Prof Exp:* From asst prof to assoc prof, 66-76, PROF PHARM, SCH PHARM, WVA UNIV, 76- *Concurrent Pos:* Res fel, Biochem Dept, Univ NC, 65-66; vis scientist, Lipid Dept, Med Div, Oak Ridge, 65-66; dent res, Inst Advan Educ Dent Res, 71. *Mem:* Am Pharmaceut Asn; Am Asn Pharmaceut Sci; Am Asn Col Pharm; Am Asn Dent Res; Malayan Pharmaceut Asn; Soc Cosmetic Chemists. *Res:* Pharmaceutical formulations for solubilization and stability of drugs; caries research involving in vitro pellicle and streptococci plaque; blood cholesterol and triglyceride levels with fiber diets; viscosity measurements of semisolids; tableting formulations. *Mailing Add:* Sch Pharm Health Sci Ctr WVa Univ Morgantown WV 26506

LIM, JOHNG KI, b Seoul, Korea, Feb 12, 30. GENETICS. *Educ:* Univ Minn, BS, 58, MS, 60, PhD(genetics), 64. *Prof Exp:* From asst prof to assoc prof, 63-69, PROF BIOL, UNIV WIS-EAU CLAIRE, 69- *Concurrent Pos:* Vis prof, Dept of Med Genetics, Univ Wis-Madison, 77-78. *Mem:* AAAS; Genetics Soc Am; Environ Mutagen Soc; Sigma Xi. *Res:* Chemical mutagenesis; cytogenetics. *Mailing Add:* Dept of Biol Univ of Wis Eau Claire WI 54701

LIM, KIOK-PUAN, b Rengam, Malaysia, Oct 6, 47; Can citizen; m 74; c 1. BIOLOGICAL CONTROL, INSECT PATHOGENS. *Educ:* Nat Taiwan Univ, BSc, 71; McGill Univ, MSc, 74, PhD(entom), 79. *Prof Exp:* Vector control officer, Vector Control & Res Dept, Ministry Environ, Singapore, 74-75; lab demonstr, Dept Entom, McGill Univ, MacDonald Campus, 75-79; vis fel, Res Inst, Agr Can, London, Ont, 80; RES SCIENTIST, NFLD FOREST RES CTR, CAN FORESTRY SERV, 80- *Concurrent Pos:* Res asst, Dept Entom, McGill Univ, MacDonald Campus, 75-79. *Mem:* Entom Soc Can; Int Orgn Biol Control; Soc Invert Path. *Res:* Biological control of insect pests; biology of parasitic hymenoptera; insect diseases; entomogenous nematodes. *Mailing Add:* Nfld Forestry Ctr Forestry Canada PO Box 6028 St John's NF A1C 5X8 Can

LIM, RAMON (KHE SIONG), b Cebu, Philippines, Feb 5, 33; m 61; c 3. NEUROCHEMISTRY. *Educ:* Univ Santo Tomas, Manila, MD, 58; Univ Pa, PhD(biochem), 66. *Prof Exp:* Intern, Long Island Col Hosp, NY, 59-60; USPHS trainee & fel, Univ Pa, 60-66; asst res biochemist, Ment Health Res Inst, Univ Mich, 66-69; asst prof neurosurg & biochem, Brain Res Inst & Sect Of Neurosurg, Univ Chicago, 69-76, assoc prof neurochem, 76-81; PROF NEUROL & NEUROCHEM, UNIV IOWA, 81-, DIR, DIV NEUROCHEM & NEUROBIOL, DEPT NEUROL, 81- *Concurrent Pos:* NIMH spec res fel, 68-69; Int Soc Neurochem lectureship, China, 86; vis prof, Univ Santo Tomas, Manila, 74, Cent Univ Venezuela, Caracas, 78, Nat Yang-Ming Med Col, Taipei, 87. *Mem:* AAAS; Am Soc Neurochem; Int Soc

Neurochem; Am Soc Biochem Mol Biol; Soc Neurosci; Am Soc Cell Biol; Int Soc Develop Neursci. *Res:* Brain proteins and peptides; tissue culture; growth and maturation of brain cells. *Mailing Add:* Dept Neurology Univ Iowa Iowa City IA 52242

LIM, SUNG MAN, US citizen; m 68; c 2. PLANT PATHOLOGY. *Educ:* Seoul Univ, Korea, MS, 59; Miss State Univ, MS, 63; Mich State Univ, PhD(crop sci & plant path), 66. *Prof Exp:* Agronomist, Crop Exp Sta, Suwon, Korea, 60-61; res asst, Miss State Univ, 61-63 & Mich State Univ, 63-66; res assoc, 67-71, asst prof, 71-77, assoc prof, 77-83, PROF PLANT PATH, UNIV ILL, 83-; PLANT PATHOLOGIST, USDA, 77- *Concurrent Pos:* Assoc ed, Plant Dis; mem, Soybean Germplasm Adv Comt, USDA; assoc ed & sr ed, Phytopath. *Honors & Awards:* Soybean Researcher's Award, Am Soybean Asn; Distinguished Sci Award, Minist Sci & Technol, Korea. *Mem:* Fel Am Phytopath Soc; Am Genetic Asn; Am Soc Agron; Crop Sci Soc Am; fel AAAS. *Res:* Epidemics of plant diseases; genetics of host-pathogen interactions. *Mailing Add:* Dept of Plant Path Univ of Ill Urbana IL 61801

LIM, TECK-KAH, b Malacca, Malaysia, Dec 1, 42; m 66; c 2. THEORETICAL NUCLEAR PHYSICS, THEORETICAL ATOMIC & MOLECULAR PHYSICS. *Educ:* Univ Adelaide, BS, 64, PhD(nuclear physics), 68. *Prof Exp:* Lectr math, Univ Malaya, 68; res assoc nuclear physics, Fla State Univ, 68-70; asst prof physics, 70-75, assoc prof, 75-82, PROF PHYSICS & ATMOSPHERIC SCI, DREXEL UNIV, 82- *Concurrent Pos:* Secy-treas, Topical Group, few-body systs, Am Phys Soc; consult, UN Develop Prog, 85. *Mem:* Fel Am Phys Soc. *Res:* Few-nucleon problem; spin-polarized quantum systems; chemical physics; molecular physics; computers in education. *Mailing Add:* Dept Physics Drexel Univ Philadelphia PA 19104

LIM, TEONG CHENG, b Penang, Malaysia, Oct 4, 39; m 66. APPLIED PHYSICS, ELECTRICAL ENGINEERING. *Educ:* Nat Taiwan Univ, BSc, 63; Ottawa Univ, MSc, 64; McGill Univ, PhD(elec eng), 68. *Prof Exp:* Elec engr, Malayan Racing Asn, 62-63; res asst elec eng, Ottawa Univ, 63-64; electronic engr, Can Marconi Co, Montreal, 65; res asst elec eng, McGill Univ, 65-68; Nat Res Coun Can fel, Imp Col, Univ London, 68-70; mem tech staff, Sci Ctr, NAm Rockwell Corp, 70-74, group leader, 75-80, mgr, Sci Ctr, Rockwell Int, 80-; PRES, AMERASIA TECHNOL, INC. *Mem:* Sr mem Inst Elec & Electronics Engrs; Brit Inst Elec Engrs. *Res:* Physics of ferroelectric and display materials and devices. *Mailing Add:* Amerasia Technol Inc 368 Venus St Thousand Oaks CA 91360

LIM, YOUNG WOON (PETER), b Seoul, Korea, Oct 25, 35; m 68; c 3. SURFACE CHEMISTRY, COLLOID CHEMISTRY. *Educ:* Ohio Wesleyan Univ, AB, 57; Univ Dayton, MS, 63; State Univ NY Col Forestry, Syracuse, PhD(polymer chem), 69. *Prof Exp:* Res chemist, Paper Res Dept, NCR Corp, 69-71, group leader analytical chem, Appleton Papers Div, 71-74; res assoc chem, Tissue & Towel Res & Develop, Am Consumer Prod, Am Can Co, Neenah, Wis, 74-77; PROJ LEADER, CROWN ZELLERBACH CENT RES, CAMAS, WASH, 77- *Mem:* Am Chem Soc; Tech Asn Pulp & Paper Indust. *Res:* Application of surface and colloid chemistry to pulp and paper research and development; morphology of cellulose and synthetic fibers; functional coatings; microencapsulation. *Mailing Add:* 5220 NE 51st St Vancouver WA 98661

LIMA, GAIL M, b Cambridge, Mass, July 2, 57. INVERTEBRATE ZOOLOGY. *Educ:* Wash Univ, St Louis, AB, 79; Tufts Univ, Medford, Mass, MS, 83; Rutgers Univ, New Brunswick, PhD(zool), 87. *Prof Exp:* ASST PROF BIOL, ILL WESLEYAN UNIV, 87- *Mem:* Am Soc Zoologists; Am Malacological Union; Sigma Xi; Western Soc Naturalists. *Res:* Biology of the invertebrates, reproduction and development; metamorphosis of marine invertebrates; relationship of shell morphology to mode of development in marine gastropods; encapsulation in marine and freshwater gastropods. *Mailing Add:* Dept Biol Ill Wesleyan Univ Bloomington IL 61702

LIMA, JOHN J, b New Bedford, Mass, June 19, 40; c 2. CLINICAL PHARMACOKINETICS. *Educ:* Bridgewater State Col, BSEd, 62; Mass Col Pharm, BS, 67; Univ Mich, PharmD, 77. *Prof Exp:* Pharm dir, Al Jordan Health Ctr, 72-75; fel, State Univ NY, Buffalo, 76-77, res assoc, 77-78; ASSOC PROF PHARM, OHIO STATE UNIV, 79- *Mem:* Am Asn Pharmaceut Scientist; Am Col Clin Pharm; Am Asn Cols Pharm. *Res:* Modelling time course; extent and mechanisms associated with drug/hormone induced adaptations, specifically hypersensitivity to adrenergic stimulation folling chronic treatment with certain drugs. *Mailing Add:* Pharm 217 Lloyd Parks Hall Ohio State Univ Main Campus Columbus OH 43210

LIMARZI, LOUIS ROBERT, b Chicago, Ill, Nov 27, 03; c 2. MEDICINE. *Educ:* Univ Ill, BS, 28, MD, 30, MS, 35. *Prof Exp:* Intern, Ill Res & Educ Hosp, Chicago, 30-31; clin asst med, 32-35, clin assoc, 35-40, from asst prof to assoc prof, 40-55, PROF MED & DIR HEMAT SECT, UNIV ILL COL MED, 55- *Concurrent Pos:* Resident, Ill Res & Educ Hosp, Chicago, 32-35, attend physician, 35-; investr, Midwest Coop Chemother Group, USPHS; civilian med consult, Hines Vet Admin Hosp, Ill & Surgeon Gen Off, 40-45, USPHS, 40- & Fed Civil Defense Admin, 50-; mem adv bd, Hemat Res Found, 40-; mem comt civil defense blood & blood derivatives, Ill Civil Defense Orgn, 50-; attend physician, West Side Vet Admin Hosp, 54-; hemat ed, Abstr Bioanal Tech, 61; consult hematologist, Augusta Hosp, Am Orthop Asn, Chicago. *Mem:* AAAS; fel Am Soc Clin Path; fel Col Am Path; fel Am Col Physicians. *Res:* Leukemia; intermediate metabolism of leukemic leukocytes and effects of anti-leukemic agents; idiopathic thrombocytopenic purpura; polycythemia vera. *Mailing Add:* 910 N East Ave Oak Park IL 60302

LIMB, JOHN ORMOND, b Pinjarra, Western Australia. COMMUNICATIONS, ELECTRICAL ENGINEERING. *Educ:* Univ Western Australia, BEE, 63, PhD(elec eng), 67. *Prof Exp:* Engr, res labs, Australian Post Off, 66-67; mem tech staff, 67-71, DEPT HEAD, COMMUN, BELL TEL LABS, 71- *Concurrent Pos:* Ed on commun, Inst Elec & Electronics Engrs. *Honors & Awards:* Leonard G Abraham Award,

Inst Elec & Electronics Engrs, 73. *Mem:* Optical Soc Am; fel Inst Elec & Electronics Engrs; Asn Res Vision & Ophthalmol. *Res:* Visual communications; efficient coding of picture signals; human visual perception; local area networks. *Mailing Add:* Lab Mgr Hewlett Packard Bldg 46g 19046 Pruneridge Ave Cupertino CA 95014

LIMBERT, DAVID EDWIN, b Omaha, Nebr, Oct 21, 42; m 72; c 2. DYNAMIC SYSTEMS, CONTROL ENGINEERING. *Educ:* Iowa State Univ, BS, 64; Case Inst Technol, MS, 65; Case Western Reserve Univ, PhD(control eng), 69. *Prof Exp:* Asst prof to assoc prof, 69-83, PROF MECH ENG, UNIV NH, 83- *Concurrent Pos:* Dir, DEL Eng, 71- *Mem:* Am Soc Mech Engrs; Inst Elec & Electronic Engrs; Air Brake Asn; Sigma Xi. *Res:* Modeling and computer simulation of freight train air brake systems including piping and valves; boundary element simulation of electric fields for electroplating; communication aids for handicapped; electric wheel chair controller design. *Mailing Add:* Dept Mech Eng Kingsbury Hall Univ NH Durham NH 03824

LIMBERT, DOUGLAS A(LAN), b Council Bluffs, Iowa, Feb 6, 48. INSTRUMENTED SYSTEM & COMPONENT TESTING, SYSTEM DYNAMICS. *Educ:* Mass Inst Technol, SB & SM, 70, ScD, 77. *Prof Exp:* Instr mech eng, Mass Inst Technol, 74-76; asst prof mech eng, Ariz State Univ, 77-84; managing engr, 84-90, SR MANAGING ENGR, FAILURE ANALYSIS ASSOCS, INC, 90- *Concurrent Pos:* Assoc ed, Am Soc Mech Engrs Trans: J Dynamic Systs & Control, 84-87. *Mem:* Sigma Xi; Am Soc Mech Engrs; Inst Elec & Electronic Engrs; Soc Automotive Engrs. *Res:* Modeling, dynamics and control of physical systems; advanced ground transportation suspensions; test engineering; instrumentation; automotive engineering; failure analysis. *Mailing Add:* Failure Anal Assocs Inc 1850 W Pinnacle Peak Rd Pheonix AZ 85027

LIMBIRD, LEE EBERHARDT, b Philadelphia, Pa, Nov 27, 48. MOLECULAR BASIS HORMONE ACTION. *Educ:* Univ NC, Chapel Hill, PhD(biochem), 73. *Prof Exp:* PROF PHARMACOL, VANDERBILT UNIV, 83- *Mailing Add:* Dept Pharmacol Vanderbilt Univ Nashville TN 37232-6600

LIMBURG, WILLIAM W, b Buffalo, NY, Nov 9, 35; m 66. ORGANIC POLYMER CHEMISTRY. *Educ:* Univ Buffalo, BA, 59, MA, 62; Univ Toronto, PhD(organosilicon chem), 65. *Prof Exp:* From sr chemist to assoc scientist, 65-66, scientist, 67-73, sr scientist, 73-80, TECH SPECIALIST & PROJ MGR, XEROX CORP, 80- *Mem:* Am Chem Soc; Royal Soc Chem; Soc Photog Scientists & Engrs. *Res:* Synthesis of organometallic compounds; mechanistic and stereochemical studies of molecular rearrangements of carbon-functional silicon-containing compounds; non-silver halide imaging methods; synthesis of organic photoconductive materials; synthesis of novel polysiloxanes. *Mailing Add:* 66 Clearview Dr Penfield NY 14526-2433

LIMERICK, JACK MCKENZIE, SR, b Fredericton, NB, July 16, 10; m 37. CHEMISTRY. *Educ:* Univ NB, BSc, 31, MSc, 34. *Prof Exp:* Res chemist, Fraser Co, 34-37; chief chemist, Bathurst Paper Co, 37-41, supt control dept, 41-44, tech & res dir, Bathurst Paper Co Ltd, 44-67, assoc dir res & develop, Consol-Bathurst Ltd, 67-71; CONSULT, 72- *Concurrent Pos:* Lectr, Royal Tech Inst, Sweden, 52; fel, Chem Inst Can, 53; fel, Tech Asn, Pulp & Paper Indust, 68; consult, Iran, 72-78, pulp, paper & container indust, Brazil, 73- & US & Can, 80- *Honors & Awards:* Award, Tech Asn Pulp & Paper Indust, 59, 82 & 90. *Mem:* Tech Asn Pulp & Paper Indust; Can Pulp & Paper Asn; fel Chem Inst Can; Pulp & Paper Res Inst Can. *Res:* Pulp; paper; containers; author of over 100 publications. *Mailing Add:* 36 E St PH4 Oakville ON L6L 5K2 Can

LIMPERT, FREDERICK ARTHUR, b Frankfort, NY, Feb 4, 21; m 44. HYDROLOGY. *Educ:* Wash State Univ, BSCE, 43. *Prof Exp:* Civil engr, Columbia Basin Proj, Wash Bur Reclamation, 46-61; head hydrol sect & chief hydrologist, Bonneville Power Admin, 61-77; RETIRED. *Concurrent Pos:* Mem, Interagency Adv Comt Water Data, 72- & Coord Coun Water Data Acquisition Methods, 74- *Honors & Awards:* Meritorious Serv Award, US Dept Interior. *Mem:* Fel Am Soc Civil Engrs; Nat Soc Prof Engrs; Western Snow Conf. *Res:* Use of satellite data for determining areal snow cover and cloud classification for areal precipitation. *Mailing Add:* 10701 Pinion Lane Sun City AZ 85373-1831

LIMPERT, RUDOLF, b Neuhaldensleben, Ger, Mar 19, 36; US citizen; m 62; c 6. MECHANICAL ENGINEERING. *Educ:* Wolfenbuettel Univ, Ger, Ing, 58; Brigham Young Univ, BES & MS, 68; Univ Mich, Ann Arbor, PhD(mech eng), 72. *Prof Exp:* Proj engr, Alfred Teves Corp, Ger, 63-65; res asst, Hwy Safety Res Inst, Univ Mich, Ann Arbor, 69-72; safety standards engr, Nat Hwy Traffic Safety Admin, Dept Transp, 72-73; res prof, Univ Utah, 73-81; CONSULT, MOTOR VEHICLE SAFETY, 81- *Mem:* Soc Automotive Engrs. *Res:* Motor vehicle accident reconstruction and cause analysis; product liability research; automotive systems design. *Mailing Add:* 2145 N Oak Lane Provo UT 84604

LIN, ALICE LEE LAN, b Shanghai, China, Oct 28, 37; US citizen; m 62; c 1. SPACE PHYSICS, ELECTRO-OPTICS & ACOUSTO-OPTICS. *Educ:* Univ Calif, Berkeley, AB, 63; George Washington Univ, MA, 74. *Prof Exp:* Res asst physics, Cavendish Lab, Cambridge, Eng, 65-66; info anal specialist, Nat Acad Sci, Washington, DC, 70-71; teaching fel physics, George Washington Univ, Washington, DC, 72-74; physicist, Goddard Space Flight Ctr, NASA, Greenbelt, Md, 75-80; physicist, Nondestructive Eval Br, 80-82, PHYSICIST, MECH MAT DIV, US ARMY MAT TECHNOL LAB, WATERTOWN, MASS, 82- *Honors & Awards:* Mencius Educ Award. *Mem:* Am Phys Soc; Am Soc Nondestructive Testing; Soc Exp Stress Anal; Am Ceramic Soc; NY Acad Sci; AAAS. *Res:* Finite element analysis using numerical methods and computer simulation for studying the behavior of failure analysis in the mechanics of materials for Department of Defense use. *Mailing Add:* 28 Hallett Hill Rd Weston MA 02193

LIN, BENJAMIN MING-REN, US citizen; m 70; c 2. COMPUTER NETWORKING, REAL-TIME SYSTEMS. *Educ:* Taipei Inst Technol, dipl, 61; Univ Wyo, MS, 67; Univ Iowa, PhD(elec eng), 73. *Prof Exp:* Engr, Radio Wave Res Labs, 62-65; design & develop engr, Collins Radio Co, 67-68; engr, Addressograph Multigraph Corp, 68-69; PROF COMPUT SCI, MOORHEAD STATE UNIV, 73- *Concurrent Pos:* Dir, Grad Sch Info Eng, Tamkang Univ, Taiwan, 80-81; vis prof, Shandong Inst Mining & Technol, China, 88. *Mem:* Sigma Xi; Asn Comput Mach; Inst Elec & Electronics Engrs; Comput Soc. *Res:* Application of microprocessors in consumer products and data communications; fault-tolerant computing systems design; computer architecture in artificial intelligence; real-time signature verification. *Mailing Add:* Dept of Comput Sci Moorhead State Univ 11th St S Moorhead MN 56563

LIN, BOR-LUH, b Fukien, China, Mar 4, 35; m 63; c 3. MATHEMATICS. *Educ:* Nat Taiwan Univ, BS, 56; Univ Notre Dame, MS, 60; Northwestern Univ, PhD(math), 63. *Prof Exp:* From asst prof to assoc prof, 63-72, PROF MATH, UNIV IOWA, 72- *Concurrent Pos:* Vis assoc prof, Ohio State Univ, 70-71; vis prof, Univ Calif, Santa Barbara. *Mem:* Am Math Soc. *Res:* Functional analysis, Banach spare theory; minimax theorems. *Mailing Add:* Dept of Math Univ of Iowa Iowa City IA 52242

LIN, CHANG KWEI, b Taiwan, Sept 11, 41; m 67; c 2. PHYCOLOGY. *Educ:* Chung Shim Univ, BSc, 64; Univ Alta, MSc, 68; Univ Wis-Milwaukee, PhD(bot), 73. *Prof Exp:* Asst cur, Acad Natural Sci, Philadelphia, 72-74; assoc res limnol, Univ Mich, Ann Arbor, 80- *Concurrent Pos:* Scientist, WHO, 77-78. *Mem:* AAAS; Phycol Soc Am; Am Soc Limnol & Oceanog; Soc Int Limnol. *Res:* Physio-ecology of algae in general, with particular interest in nutrient limitation to phytoplankton growth in the Great Lakes; pond dynamics and fish culture. *Mailing Add:* Great Lakes Res Div Univ Mich Ann Arbor MI 48109

LIN, CHARLIE YEONGCHING, b Long-Yen, Taiwan, Feb 12, 56; US citizen; m 90. MATERIAL SYNTHESIS & CHARACTERIZATION. *Educ:* Nat Taiwan Univ, BS, 78; Univ Minn, MS, 83, PhD(mat sci & metall), 85. *Prof Exp:* Grad res asst, Dept Chem Eng & Mat Sci, Univ Minn, 81-85, teaching asst, 84-85; res scientist, Eastman Kodak Co, 85-86, sr res scientist, 86-88; VPRES & COFOUNDER, ADVAN DIVERSIFIED TECHNOL, INC, 88- *Concurrent Pos:* Lectr, Univ Calif, Davis, 85; lectr & consult, Advan Diversified Technol, 87; vis prof, var univs, Taiwan & China, 87; consult, Yuen Foong Yu Paper Mfg Co, Tancan Co & Am Electronic Mat, 90-; prin investr, NASA, Dept Defense & NIH, 90- *Mem:* Am Chem Soc; Electrochem Soc; Mat Res Soc. *Res:* Synthesis and characterization of ceramics and inorganic polymer materials and their applications for electronic, microelectronic and optoelectronic devices; thin film technology; ultrastructure processing and sol gel technology; instrument analysis; surface and interface science; corrosion prevention; coating technology; diamond and superconductor technology development. *Mailing Add:* Advan Diversified Technol Inc 13289 Benchley Rd San Diego CA 92130

LIN, CHE-SHUNG, quantum chemistry; deceased, see previous edition for last biography

LIN, CHIA CHIAO, b Foochow, China, July 7, 16; US citizen. APPLIED MATHEMATICS. *Educ:* Nat Tsing Hua Univ, China, BSc, 37; Univ Toronto, MA, 41; Calif Inst Technol, PhD(aeronaut), 44. *Hon Degrees:* LLD, Chinese Univ Hong Kong, 73. *Prof Exp:* Asst, Tsing Hua Univ, China, 37-39; from asst to res engr, Calif Inst Technol, 43-45; from asst prof to assoc prof appl math, Brown Univ, 45-47; from assoc prof to Inst prof, 47-87, EMER INST PROF MATH, MASS INST TECHNOL, 87- *Concurrent Pos:* Guggenheim fels, 53 & 60; mem, Inst Advan Study, Princeton, NJ, 59-60 & 65-66; mem, Comt on Support Res in Math Sci, Nat Acad Sci, 66-68. *Honors & Awards:* John Von Neumann lectr, Soc Indust & Appl Math-Am Math Soc, 67; Otto Laporte Mem lectr, Am Phys Soc, 73; Timoshenko Medal, Am Soc Mech Engrs, 75; Award Appl Math & Numerical Anal, Nat Acad Sci, 77; Fluid Dynamics Prize, Am Phys Soc & US Off Naval Res, 79. *Mem:* Nat Acad Sci; Am Astron Soc; Soc Indust & Appl Math (pres, 72-74); Am Math Soc; fel Am Acad Arts & Sci; Am Philos Soc; Am Phys Soc; fel Inst Aerospace Sci. *Res:* Hydrodynamics; stellar dynamics; astrophysical problems; spiral structure of galaxies; density wave theory developed in great mathematical detail with predictions checked against various astronomical observations. *Mailing Add:* Dept Math Mass Inst Technol Cambridge MA 02139

LIN, CHII-DONG, b Taiwan. ATOMIC PHYSICS. *Educ:* Nat Taiwan Univ, BS, 69; Univ Chicago, MS, 73, PhD(physics), 74. *Prof Exp:* Fel astrophys, Ctr Astrophys, Harvard Col Observ, 74-76; from asst prof to assoc prof, 76-80, PROF, KANS STATE UNIV, 84- *Concurrent Pos:* Sloan fel, 79-83. *Mem:* fel, Am Phys Soc. *Mailing Add:* Dept Physics Kans State Univ Cardwell Hall Manhattan KS 66506

LIN, CHIN-CHUNG, b Taipei, Taiwan, Oct 8, 37; m; c 2. BIOCHEMICAL PHARMACOLOGY. *Educ:* Chung Hsing Univ, Taiwan, BS, 60; Tuskegee Inst, MS, 65; Northwestern Univ, PhD(biochem), 69. *Prof Exp:* Sr scientist, 69-75, prin scientist, 75-79, res fel, 79-82, ASSOC DIR, SCHERING CORP, 83- *Concurrent Pos:* Res fel biochem, Med Sch, Northwestern Univ, 69. *Mem:* AAAS; Am Chem Soc; Am Soc Pharmacol & Exp Therapeut; NY Acad Sci. Res; Drug metabolism and the mechanism of enzymatic hydroxylation. *Mailing Add:* Schering Corp Dept Drug Metab 60 Orange St Bloomfield NJ 07003

LIN, CHING Y, b Taiwan, May 22, 40; Can citizen; m 65; c 2. ANIMAL BREEDING, QUANTITATIVE GENETICS. *Educ:* Nat Chung-Hsing Univ, Taiwan, BS, 63; Iowa State Univ, MS, 71; Ohio State Univ, PhD(dairy sci), 76. *Prof Exp:* Jr specialist agr extension, Taiwan Prov Dept Agr & Forestry, 64-68; res assoc poultry breeding, Dept Animal Sci, Univ Guelph, Can, 76-80; RES SCIENTIST DAIRY CATTLE BREEDING, ANIMAL RES CTR, AGR CAN, 80- *Concurrent Pos:* Vis res fel, Japan, 89-90. *Mem:* Am Soc Animal Sci; Am Dairy Sci Asn; Can Soc Animal Sci. *Res:* Dairy cattle breeding; quantitative genetics; statistical analysis as applied to animal breeding. *Mailing Add:* 2226 Bingham Ottawa ON K1G 2V7 Can

LIN, CHINLON, b Taiwan, Rep of China, Jan 19, 45; m 69; c 2. QUANTUM ELECTRONICS, OPTICAL COMMUNICATION. *Educ:* Nat Taiwan Univ, BS, 67; Univ Ill, MS, 70; Univ Calif, Berkeley, PhD(elec eng), 73. *Prof Exp:* MEM TECH STAFF, LASER SCI RES, BELL LABS, HOLMDEL, 74- *Concurrent Pos:* Assoc ed, J Appl Optics. *Mem:* Inst Elec & Electronics Engrs; fel Optical Soc Am. *Res:* Lasers and quantum electronics; optical fibers and lightwave communications. *Mailing Add:* Bell Commun Res 331 Newman Springs Rd Rm 3X249 Red Bank NJ 07701

LIN, CHIN-TARNG, b Chu-Nan, Taiwan, Dec 11, 38; US citizen; m 83; c 1. CELL BIOLOGY, NEUROBIOLOGY. *Educ:* Nat Taiwan Univ, DDS, 63; Med Br, Univ Tex, Galveston, PhD(cell biol), 75. *Prof Exp:* Doctor & teaching asst path, Dept Path, Col Med, Nat Taiwan Univ, 64-69, instr, 69-75; teaching asst, Div Cell Biol, Med Br, Univ Tex, Galveston, 71-75; assoc prof path, Dept Path, Col Med, Nat Taiwan Univ, 75-78; res assoc, Dept Cell Biol, Baylor Col Med, 78-80, res instr, 80-83; asst prof, Dept Physiol, Col Med, Pa State Univ, 83-87; RES FEL, INST BIOMED SCI, ACAD SINICA, TAIPEI, TAIWAN, ROC, 87- *Concurrent Pos:* Vis prof, Dept Pathol, Col Med, Nat Taiwan Univ, 87-89, PROF PATHOL, 89- *Mem:* Histochem Soc; AAAS; Am Soc Cell Biol; Soc Neurosci. *Res:* Immunochemical approaches to protein synthesis and transport in normal and cancer cells; cell biology; neurobiology; neurotransmitter synthesizing enzymes; immunohistochemistry; immunoelectron microscopy; in situ nucleic acid hybridization; monoclonal hybridoma technique; nasopharyngeal carcinoma tumor biology; stroke animal model. *Mailing Add:* Inst Biomed Sci Academia Sinica Taipei 11529 17033 Taiwan

LIN, CHI-WEI, b Hong Kong, May 16, 37; m 65; c 1. CANCER, BIOCHEMISTRY. *Educ:* Nat Taiwan Univ, BS, 61; Univ Wis-Madison, MS, 65, PhD(biochem), 69. *Prof Exp:* Fel cancer res, Sch Med, Tufts Univ, 69-71, res assoc, 71-72, asst prof path, 72-80; DIR, UROL RES LAB & STAFF MEM, DEPT UROL, MASS GEN HOSP, 79- *Mem:* Biochem Soc; AAAS; Histochem Soc; Am Asn Cancer Res; Sigma Xi. *Res:* Biochemical characteristics of cancer, specifically, the studies of tumor-associated enzymes and isozymes, including histaminase, acid phosphatase and alkaline phosphatase; processes of synthesis and distribution of acid hydrolases and the biogenesis of lysosomes. *Mailing Add:* Dept Urol Mass Gen Hosp Fruit St Boston MA 02114

LIN, CHUN CHIA, b Canton, China, March 7, 30; US citizen. ATOMIC & MOLECULAR COLLISIONS, ELECTRONIC ENERGIES OF SOLIDS. *Educ:* Univ Calif, Berkeley, BS, 51, BA, 52; Harvard Univ, PhD(chem), 55. *Prof Exp:* From asst prof to prof physics, Univ Okla, 55-68; PROF PHYSICS, UNIV WIS, 68- *Concurrent Pos:* Consult & univ retainee, Tex Instruments, Inc, 60-68; consult, Sandia Labs, 76; Alfred P Sloan fel, 62-66; secy-treas, Div Electron & Atomic Physics, Am Phys Soc, 74-77; secy, Gaseous Electronics Conf, 73, chmn, 90-92. *Mem:* Am Phys Soc. *Res:* Atomic and molecular collision processes; radiation of atoms and molecules excited by electron impact and laser irradiation; electronic energy band theory of crystalline solids, impurity atoms in solids, amorphous solids. *Mailing Add:* Dept Physics Univ Wis Madison WI 53706

LIN, DENIS CHUNG KAM, b Hong Kong, July 7, 44; Can citizen; m 69; c 1. ANALYTICAL CHEMISTRY, MASS SPECTROMETRY. *Educ:* Univ Man, BSc, 68, MSc, 70, PhD(chem), 72. *Prof Exp:* Fel, Univ Montreal, 72-74; staff chemist, Battelle Mem Inst, 74-80; AT ETC CORP, 80- *Mem:* Am Chem Soc; Am Soc Mass Spectrometry; Int Asn Forensic Toxicologists. *Res:* Identification and quantification of low levels of drugs and their metabolites in biological samples by mass spectrometry and other techniques; nucleic acid and protein sequencing; pyrolytic reactions. *Mailing Add:* PO Box 26518 New Orleans LA 70186-6518

LIN, DIANE CHANG, b China, Aug 6, 44; US citizen; m 69; c 2. CELL BIOLOGY, CELL MOTILITY. *Educ:* Nat Taiwan Univ, BS, 66; Univ Calif, Los Angeles, PhD(biol), 71. *Prof Exp:* Asst res scientist pharmacol, Univ Calif, San Francisco, 71-74; RES SCIENTIST BIOPHYS, JOHNS HOPKINS UNIV, 74- *Concurrent Pos:* Prin investr, Johns Hopkins Univ, 78- *Mem:* Am Soc Biochem & Molecular Biol; Am Soc Cell Biol. *Res:* Actin-binding proteins from chicken muscles. *Mailing Add:* Dept Biophys Johns Hopkins Univ 3400 N Charles St Baltimore MD 21218

LIN, DONG LIANG, b Taiwan, China, Mar 5, 47; m 71; c 3. PHYSICS, ELECTRICAL ENGINEERING. *Educ:* Nat Taiwan Univ, BS, 69; Columbia Univ, MS, 72, PhD(physics), 75. *Prof Exp:* Res fel physics, Johns Hopkins Univ, 75-77; staff scientist physics, Sci Appln Inc, 77-80; MEM STAFF, BELL LABS, HOLMDEL, NJ, 80- *Mem:* Am Phys Soc; Nat Soc Prof Engrs. *Res:* Atomic physics; plasma physics; solid state physics. *Mailing Add:* Bell Lab 600 Mountain Ave Murray Hill NJ 07974

LIN, DUO-LIANG, b Juian, China, May 16, 30; m 63; c 2. THEORETICAL PHYSICS, CONDENSED MATTER PHYSICS. *Educ:* Taiwan Norm Univ, BSc, 56; Tsing Hua Univ, China, MSc, 58; Ohio State Univ, PhD(physics), 61. *Prof Exp:* Res assoc physics, Yale Univ, 61-64; asst prof, 64-67, ASSOC PROF PHYSICS, STATE UNIV NY BUFFALO, 67- *Concurrent Pos:* Sr vis, Oxford Univ, 70-71; vis prof, Nat Taiwan Univ, 71, Tsing Hua Univ, Peking, 78, Liao Ning Univ, Shen Yang, 81 & Jiaotong Univ, Shanghai, 85; hon prof, Neimonggu Univ, China, 85-; Tokten consult, UN Develop Prog, 86-; adj prof, Univ Sci & Technol China, Hefei, 87-, Shandong Univ, Jinan, 88, Sichuan Univ, Chengdu, 90; adv prof, Chengdu Univ Sci & Technol, 86, Southwestern Jiao, Tong Univ, Emei, 87, Chongqing Univ, 85. *Mem:* Am Phys Soc. *Res:* Electronic, optical and transport properties of semiconductor heterostructures; effects of surfaces and interfaces on magnetic properties; nonlinear optical response and ultrafast processes in polymers; mechanism of high Tc superconductivity. *Mailing Add:* Dept Physics State Univ NY Buffalo NY 14260

LIN, EDMUND CHI CHIEN, b Peking, China, Oct 28, 28; nat US. BIOCHEMISTRY. *Educ:* Univ Rochester, AB, 52; Harvard Univ, PhD, 57. *Prof Exp:* Instr biochem, 57-60, assoc, 60-63, from asst prof to assoc prof, 63-69, PROF MICROBIOL & MOLECULAR GENETICS, HARVARD MED SCH, 69- *Concurrent Pos:* Vis prof, Univ Calif, Berkeley, 72; Guggenheim fel, Pasteur Inst Paris, 69; prof chmn dept, Harvard Med Sch, 73-75; Fogarty Sr Int fel, Univ Paris, VI, 77-78; vis prof biol, Univ Konstanz, Ger, 81; hon res prof, Inst Plant Physiol, Academia Sinica, Shanghai, 80- *Mem:* Am Soc Microbiol; Am Soc Biol Chem. *Res:* Bacterial physiology and genetics and biochemical evolution. *Mailing Add:* Microbiol & Molecular Genetics Harvard Med Sch Boston MA 02115

LIN, FU HAI, b Fukien, China, Feb 15, 28; US citizen; m 56; c 5. MOLECULAR BIOLOGY, NEUROSCIENCES. *Educ:* Nat Taiwan Univ, BS, 53; Univ WVa, MS, 59; Rutgers Univ, PhD(bact), 65. *Prof Exp:* Asst, Univ WVa, 58-59; tech asst biochem, Boyce Thompson Inst, 59-61; res asst, Rutgers Univ, 61-65; asst mem biochem, Albert Einstein Med Ctr, 65-69; sr res scientist, 69-70, assoc res scientist, 70-76, RES SCIENTIST V & HEAD LAB VIRAL BIOCHEM, INST BASIC RES DEVELOP DISABILITY, 76- *Mem:* AAAS; Am Soc Microbiol; Am Chem Soc; Sigma Xi; NY Acad Sci. *Res:* Mitochondrial mutation in Alzheimer disease and related neurodegenerative syndromes; gene expression of animal RNA viruses of slow infection; biochemistry and function of proteins of slow viruses. *Mailing Add:* Inst Basic Res Developmental Disability 1050 Forest Hill Rd Staten Island NY 10314

LIN, GEORGE HUNG-YIN, b Shantung, China, Mar 9, 38; m 69; c 1. PHYSICAL CHEMISTRY. *Educ:* Tunghai Univ, Taiwan, BS, 60; Univ Nev, MS, 65; Univ Calif, Davis, PhD(chem), 69; Am Bd Toxicol, dipl, 82. *Prof Exp:* NSF fel biochem, Univ Wis-Madison, 69-71; Rockefeller fel chem, Univ Calif, Riverside, 71-74; prog specialist, 74-76, staff toxicologist, 74-84, SR SCIENTIST, XEROX CORP, 84- *Concurrent Pos:* Adj fac toxicol, Univ Rochester, 80-82. *Mem:* Soc Toxicol; Environ Mutagen Soc. *Res:* Industrial toxicology; gaseous emissions; trace metal sampling and analysis; trace organic compounds sampling and analysis; inhalation toxicology; biometrics; chemical carcinogenesis; x-ray crystallography; general toxicology; carcinogen risk assessment; structure-activity relationships. *Mailing Add:* Xerox Corp Webster NY 14580

LIN, GRACE WOAN-JUNG, b Taipei, Taiwan; US citizen; m. NUTRITIONAL BIOCHEMISTRY. *Educ:* Nat Taiwan Univ, BS, 59; Tex Woman's Univ, MS, 64; Univ Calif, Berkeley, PhD(nutrit), 71. *Prof Exp:* Res asst, US Naval Med Res Unit #2, 59-62 & Thorndike Mem Lab, Med Sch, Harvard Univ, 64; res fel, Columbia Univ, 69-70; asst res specialist, 74-83, ASSOC RES SPECIALIST, RUTGERS UNIV, 83- *Mem:* Res Soc Alcoholism; Am Inst Nutrit; Int Soc Biomed Res Alcoholism. *Res:* Effects of ethanol on absorption and metabolism of nutrients (amino acids and water soluble vitamins) and on fetal development. *Mailing Add:* Ctr Alcohol Studies Rutgers Univ Piscataway NJ 08854

LIN, H(UA), b Peiping, China, Nov 25, 19; nat US; m 47; c 2. AERONAUTICAL ENGINEERING. *Educ:* Nat Tsing Hua Univ, China, BS, 40; Univ Mich, MS, 44; Mass Inst Technol, ScD(aeronaut eng), 55. *Prof Exp:* Engr, aeronaut dept, Cent Aircraft Mfg Co, China, 40-42; instr, Nat Tsing Hua Univ, 42-43; stress analyst, Stinson Div, Consol Vultee Aircraft Corp, Mich, 44-45; engr, Cincinnati Milling Mach Co, Ohio, 45-47; asst mgr, Far East Develop Corp, NY, 47-49; res aeronaut engr, aeroelastic & struct res lab, Mass Inst Technol, 49-56; res specialist, struct dynamics staff, 56-58, chief dynamics, systs mgt off, 58-59, chief struct tech unit, aerospace div, 59-64, mgr, struct & mat tech dept, Aerospace Group Div, Boeing Co, 65-66, chief missile tech, missile & info systs div, 66-68, chief engr, minuteman prog, missile div, 68-70, dep prog mgr, minuteman prog, Aerospace Group, 70-71, prog mgr hardsite defense, 71-73, prod develop mgr, Boeing Airplane Co, 73-75; dir offensive systs, US Directorate Res & Eng, Dept Defense, 75-78; CHIEF SCIENTIST, BOEING AEROSPACE CO, 78- *Concurrent Pos:* Partic, sr exec prog, Sloan Sch Indust Mgt, Mass Inst Technol, 69. *Mem:* Sr mem Am Astron Soc; fel Am Inst Aeronaut & Astronaut; fel Brit Interplanetary Soc; Am Soc Mech Engrs. *Res:* Aeroelasticity; structural dynamics; structural analysis; steady-state and unsteady aerodynamics; aerodynamic heating and thermal analysis; stability and control; structural flexility and servo-control interaction. *Mailing Add:* 3212 90th Pl Mercer Island WA 98040

LIN, HSIU-SAN, b Nagoya, Japan, March 15, 35; US citizen; m 62; c 3. RADIATION ONCOLOGY, MICROBIOLOGY. *Educ:* Nat Taiwan Univ, MD, 60; Univ Chicago, PhD(microbiol), 68. *Prof Exp:* From asst prof to assoc prof, 71-84, PROF RADIOL, WASH UNIV, ST LOUIS, 84-, ASSOC PROF MICROBIOL, 85- *Concurrent Pos:* Vis scientist, Univ Oxford, Eng, 77-78; consult ed, J Leukocyte Biol. *Honors & Awards:* Res Career Develop Award, Nat Cancer Inst, 74. *Mem:* Am Soc Microbiologists; Am Soc Therapeut Radiol & Oncol; Am Asn Cancer Res; Reticuloendothelial Soc; Sigma Xi. *Res:* Differentiation of monocytes and macrophages; radiobiology of mononuclear phagocytes. *Mailing Add:* Mallinckrodt Inst Radiol Wash Univ Sch Med St Louis MO 63110

LIN, HUNG CHANG, b Shanghai, China, Aug 8, 19; US citizen; m 59; c 2. ELECTRONICS. *Educ:* Chiao Tung Univ, BSEE, 41; Univ Mich, MS, 48; Polytech Inst Brooklyn, DEE(elec eng), 56. *Prof Exp:* Engr, Cent Radio Works of China, 41-44 & Cent Broadcasting Admin of China, 44-47; res engr, RCA, 48-56; mgr appln, CBS Semiconductor Opers, 56-59; lectr, 66-69, vis prof elec eng, 69-71, PROF ELEC ENG, UNIV MD, COLLEGE PARK, 71-; MGR ADVAN DEVELOP, MOLECULAR ELECTRONICS DIV, WESTINGHOUSE CORP, LINTHICUM HEIGHTS, 63-, SR ADV ENGR, AEROSPACE DIV, BALTIMORE, 69- *Concurrent Pos:* Adv engr, Res Lab, Westinghouse Corp, Baltimore, 59-63; adj prof, Univ Pittsburgh, 59-63; vis lectr, Univ Calif, Berkeley, 65-66. *Honors & Awards:* Ebers Award, Inst Elec & Electronic Engrs, Electron Device Soc, 78. *Mem:* Fel Inst Elec & Electronic Engrs; Sigma Xi. *Res:* Semiconductor and integrated circuits. *Mailing Add:* 8 Shindler Ct Silver Spring MD 20903

LIN, JAMES C H, b Macao, Aug 12, 32; wid; c 3. GENETICS, CELL PHYSIOLOGY. *Educ:* Taiwan Prov Norm Univ, BS, 54; Rice Univ, MA, 60, NC State Univ, PhD(genetics), 65. *Prof Exp:* Lab instr zool, Nat Taiwan Univ, 55-57; res asst nuclear med, Methodist Hosp, Houston, Tex, 59-60 & Hermann Hosp, 60; from asst prof to assoc prof, 65-75, PROF BIOL, NORTHWESTERN STATE UNIV, 75- *Concurrent Pos:* Vis prof, Univ Tex, M D Anderson Hosp & Tumor Inst, 80-81. *Mem:* Genetics Soc Am; Sigma Xi; Tissue Culture Asn. *Res:* Chemical mutagenesis; cholinesterase in fire ants; crossing over in Drosophila; nucleolar organizing regions of Chinese hamster ovary cells. *Mailing Add:* Dept of Biol Sci Northwestern State Univ Natchitoches LA 71497

LIN, JAMES CHIH-I, b Dec 29, 42; m 70; c 3. BIOENGINEERING, ELECTRICAL ENGINEERING. *Educ:* Univ Wash, Seattle, BS, 66, MS, 68, PhD(elec eng), 71. *Prof Exp:* Elec engr, Crown Zellerbach Corp, 66-67; teaching & res asst elec eng, Univ Wash, Seattle, 67-71, asst prof rehab med, 71-74, asst dir, Bioelectromagnetic Res Lab, 74; prof elec eng, Wayne State Univ, Detroit, 74-80; dir robotics & automation lab, 82-89, PROF ELEC ENG & HEAD BIOENG, UNIV ILL, CHICAGO, 80- *Concurrent Pos:* Consult, Walter Reed Army Inst Res, 73-75, Battelle Mem Inst, 76-80, SRI Int, 78-79, Arthur D Little, Inc, 80-82 & Ga Tech Res Inst, 84-86, URS Corp, 86-87, CBS, Inc, 88, ACS, Inc, 88-89, Luxtron, Inc, 90-; appointee, Diag Radiol Study Sect, NIH, 81-85, Nat Acad Sci & Int Union Radio Sci, 80-82, Presidential Young Investr Award Panel, NSF, 84-89; vis prof, Nat Yang Ming Med Col, Taipei, 81, Chung Yuan Univ, Taiwan, 81 & 88, Univ Rome, 85 & 88, Shangdong Univ, China, 88; chair, US Am Comn Man & Radiation, Inst Elec & Electronic Engrs, 90-91. *Mem:* Fel Inst Elec & Electronics Engrs; Bioelectromagnetic Soc; Biomed Eng Soc; AAAS; NAm Hyperthermia Group; Robotics Int; Am Soc Engr Educ; Nat Soc Prof Engr. *Res:* Electromagnetic tissue imaging; biological effects and medical applications of electromagnetic fields; hyperthermia for cancer therapy; visual and nonvisual robotic sensing. *Mailing Add:* Dept Bioeng MC063 Univ Ill Box 4348 Chicago IL 60680

LIN, JAMES PEICHENG, b New York, NY, Sept 30, 49. ALGEBRAIC TOPOLOGY. *Educ:* Univ Calif, Berkeley, BS, 70; Princeton Univ, PhD(math), 74. *Prof Exp:* From asst prof to prof math, Univ Calif, San Diego, 74-86. *Concurrent Pos:* Vis prof math, Princeton Univ, 76-77; Sloan Found fel, 77-78; mem, Inst Advan Studies, Hebrew Univ, Jerusalem, Israel, 81-82; vis scholar, Univ Calif, Berkeley, 82, Mass Inst Technol, 83-84 & Univ Neuchatel, Switz, 84; Prin invest, NSF grant. *Mem:* Am Math Soc. *Res:* Algebraic topology, concentrating on finite H-spaces. *Mailing Add:* Dept Math C-012 Univ Calif San Diego Box 109 La Jolla CA 92093

LIN, JEONG-LONG, physical chemistry; deceased, see previous edition for last biography

LIN, JIA DING, b Fuzhou, China, Dec 24, 31; US citizen; m 58; c 3. FLUID MECHANICS, HYDROLOGY AND WATER RESOURCES. *Educ:* Nat Univ Taiwan, BS, 53, Univ Ill, MS, 56; Mass Inst Technol, ScD(hydromech), 61. *Prof Exp:* Res scientist, Hydronautics, Inc, 60-62; asst prof, 62-64, assoc prof, 64-81, PROF CIVIL ENG, UNIV CONN, 81- *Concurrent Pos:* Vis prof, Taiwan Univ, 79-80. *Mem:* Asn Hydrol Res; Am Geophys Union. *Res:* hydrodynamics; hydrology. *Mailing Add:* Dept Civil Eng Univ Conn Main Campus U-37 261 Glenbrook Rd Storrs CT 06268

LIN, JIAN, b Fuzhou, Fujian. MARINE GEOPHYSICS, TECTONOPHYSICS. *Educ:* Univ Sci & Technol, China, BS, 82; Brown Univ, MS, 84, PhD(geophys), 88. *Prof Exp:* Vis scientist res geophys, US Geol Surv, Menlo Park, Calif, 88; ASST SCIENTIST GEOPHYS, WOODS HOLE OCEANOG INST, 88- *Concurrent Pos:* Mem, Crustal Accretion Variables Working Group, US Ridge Prog, 89; prog chmn, Tectonophysics Sect, Am Geophys Union, 92-93. *Mem:* Am Geophys Union; Sigma Xi. *Res:* Mid-ocean ridge dynamics; thermal evolution of the lithosphere; mantle convections; crustal deformation. *Mailing Add:* Dept Geol & Geophys Woods Hole Oceanog Inst Woods Hole MA 02543

LIN, JIANN-TSYH, b Taoyuan, Taiwan, Jan 15, 40; US citizen; m 69; c 2. PLANT HORMONES, GIBBERELLINS. *Educ:* Chung-Hsing Univ, Taiwan, BS, 63; Univ Miss, MS, 67; Drexel Univ, Philadelphia, PhD(biochem), 71. *Prof Exp:* Res assoc, Univ Tenn Med Ctr, 71-72, Hormel Inst, Univ Minn, 72-76 & Harborview Med Ctr, 76-77; RES CHEMIST, WESTERN REGIONAL RES CTR, USDA, 77- *Mem:* Am Chem Soc; Am Soc Plant Physiologists; Plant Growth Regulator Soc Am. *Res:* Biochemistry and analytical chemistry of steroids, lipids and enzymes; biochemistry and analysis of plant hormone, gibberellins, in wheat and apple. *Mailing Add:* USDA Western Regional Res Ctr 800 Buchanan St Albany CA 94710

LIN, JIUNN H, b Taiwan, 1943. DRUG METABOLISM. *Prof Exp:* ASSOC DIR, MERCK SHARP & DOHME, 88- *Mailing Add:* Dept Drug Metab Merck Sharp & Dohme Res Labs West Point PA 19486

LIN, KUANG-FARN, b Taiwan, China, Feb 25, 36; m 58; c 2. POLYMER SCIENCE. *Educ:* Cheng Kung Univ, Taiwan, BSc, 57; NDak State Univ, MS, 63, PhD(polymers, coatings), 69. *Prof Exp:* Asst instr chem, Chinese Naval Acad, 57-59; supt synthetic resins, Yung Koo Paint & Varnish Mfg Co, 59-61; chemist, 63-67, res chemist, 69-73, proj leader, 73-75, sr res chemist, 74-78, RES SCIENTIST, HERCULES RES CTR, 78-, TECH MGR, 79-, RES ASSOC, 84-, VENTURE MGR, 89- *Concurrent Pos:* Asst, NDak State Univ, 63 & 67. *Mem:* Am Chem Soc; Sigma Xi; Tech Asn Pulp & Paper Indust; fel Am Inst Chemists; Am Inst Mining, Metall & Petrol Engrs. *Res:* Structure-property relationship; mineral processing; adhesion, coatings and polymer synthesis; elastomers. *Mailing Add:* Hercules Mkt Ctr Wilmington DE 19894

LIN, KUANG-MING, b Tapei, Taiwan, Mar 10, 32; m 62; c 4. FLUID MECHANICS, DYNAMICS. *Educ:* Nat Taiwan Univ, BS, 56; Auburn Univ, MS, 58; Mich State Univ, PhD(appl mech), 64; Bowling Green State Univ, MBA, 79. *Prof Exp:* Asst res instr fluid mech, Mich State Univ, 58-60; asst prof eng mech, Tri-State Col, 61-63; from asst prof to assoc prof eng sci, Tenn Technol Univ, 64-66; sr res engr, Brown Eng Co, 66-68; specialist reliability eng, ITT Fed Elec Corp, 68; sr staff engr, Corp Res & Develop, 68-73, dep dir planning & develop, Asia-Pac, Dana Int, Div, 73-75, mgr int planning, 75-85, DIR, ASIA-PAC LIAISON, DANA CORP, 85- *Concurrent Pos:* NSF res initiation grant, 65-66. *Mem:* Am Soc Mech Engrs; Soc Automotive Engrs. *Res:* Two-phase flow through porous medias; boundary layer theory; hydrology; aerodynamics; lubrication of porous bearings; control systems for power transmission devices. *Mailing Add:* Dana Corp PO Box 1000 Toledo OH 43697

LIN, KUANG-TZU DAVIS, b Nantou, Taiwan, Aug 12, 40; m 68; c 2. BIOCHEMISTRY, BIOCHEMICAL GENETICS. *Educ:* Nat Taiwan Univ, BM, 66; Univ Wis-Madison, PhD(physiol chem), 71. *Prof Exp:* Rotating intern, Nat Taiwan Univ Hosp, 65-66; surg officer, Tainan Air Force Hosp, 66-67; res asst physiol chem, Univ Wis-Madison, 67-71, proj assoc, 71-74; asst prof, 74-82, assoc prof med biol, mem res ctr, Univ Tenn, 82-86; ASSOC PROF PEDIATRICS, MEHARRY, 88- *Concurrent Pos:* Med staff, dept pediat, William Beaumont Hosp. *Mem:* Soc Exp Biol Med; Am Soc Biol Chemists; Am Bd Med Genetics. *Res:* Structure and function, especially carbonic anhydrase, hemoglobin, protease inhibitors and erythropoietin; amino acid metabolic disorder; pediatrics. *Mailing Add:* Dept Pediat Meharry Med Col 1005 D B Todd Jr Blvd Nashville TN 37208

LIN, KWAN-CHOW, b Hong Kong; Can citizen. BIOLOGICAL WASTEWATER TREATMENT, WATER QUALITY STUDIES. *Educ:* Chu Hai Col Hong Kong, BSc, 66; Univ NB, MScE, 69; Univ Toronto, PhD(environ eng), 74. *Prof Exp:* Civil engr, Domtar Newsprint Ltd, 66-67; res assoc, Univ NB, 69-70, lectr, 74, from asst prof to assoc prof, 74-83, PROF, UNIV NB, 83- *Concurrent Pos:* Design engr, Environ Resources Consult Ltd, 69-70; consult, WHO, Copenhagen, 76, var eng firms & munic in Can, 78-; lectr, Univ Hong Kong, 80. *Mem:* Can Soc Civil Eng; Eng Inst Can; Water Pollution Control Fedn; Can Asn Water Pollution Res & Control; Overseas Chinese Environ Engrs & Scientists Asn. *Res:* Biological wastewater treatment employing aerobic and anaerobic processes for organics, solids and nutrient removal; water quality studies employing laboratory analyses and mathematical modelling of surface and ground water quality and transport. *Mailing Add:* Dept Civil Eng Univ NB PO Box 4400 Fredericton NB E3B 5A3 Can

LIN, LARRY Y H, b China. CIVIL ENGINEERING. *Educ:* Nat Taiwan Univ, BS, 57; WVa Univ, MS, 63, PhD(civil eng), 66; Am Acad Environ Engrs, dipl. *Prof Exp:* Teaching asst, Nat Taiwan Univ, 59-61; res asst assoc, WVa Univ, 61-64; MEM STAFF, ROY WESTON INC, 66- *Mem:* Water Pollution Control Fedn. *Res:* Process design of industrial and municipal wastewater treatment facilities; physical, chemical and biological aspects of water pollution; air and solid waste problems; data analysis and computer programming; energy conservation. *Mailing Add:* 19 Montbard Dr Chaddsford PA 19317

LIN, LAWRENCE I-KUEI, b Fuchou, China, May 21, 48; m 71; c 3. STATISTICS, DATA MANAGEMENT. *Educ:* Nat Chengchi Univ, Taiwan, BC, 70; Univ Iowa, MS, 73, PhD(statist), 79. *Prof Exp:* Res asst, Dept Preventive Med, Univ Iowa, 73, statistician, Iowa Epidemiol Study Pesticides, 73-79; res statistician, 79-87, SR RES STATISTICIAN, BAXTER HEALTHCARE CO, 87- *Mem:* Am Statist Asn; Biomet Soc; Drug Info Asn. *Res:* Discriminant analysis; risk assessment; M-estimator; pharmacokinetics; generalized Linear model. *Mailing Add:* Baxter Healthcare Co Rte 120 & Wilson Round Lake IL 60073

LIN, LEEWEN, b Taipei, Taiwan. BIOCHEMISTRY. *Prof Exp:* STAFF FEL, CBER FOOD & DRUG ADMIN, 90- *Mailing Add:* Div Biochem & Biophys CBER Food & Drug Admin 8800 Rockville Pike Bldg 29 Rm 516 Bethesda MD 20892

LIN, LEU-FEN HOU, b Kwangtung, China; US citizen; m 72; c 2. BIOCHEMISTRY, ORGANIC CHEMISTRY. *Educ:* Nat Taiwan Univ, BS, 67; Univ Minn, PhD(biochem), 72. *Prof Exp:* Instr biochem, Mt Sinai Sch Med, 72-76; res fel, Harvard Univ, 76-79; scientist biochem, E K Shriver Ctr Ment Retardation, Harvard Med Sch, 79-88; NEUROSCIENTIST, SYNERGEN, 88- *Concurrent Pos:* Asst neurol, Mass Gen Hosp, 81- *Res:* Membrane structure, function and biosynthesis. *Mailing Add:* Synergen Corp 1885 33rd St Boulder CO 80301

LIN, MAO-SHIU, b Tainan, Taiwan, June 20, 31. ELECTRICAL ENGINEERING. *Educ:* Nat Taiwan Univ, BSE, 55; Univ Mich, MSE, 58, PhD(elec eng), 64. *Prof Exp:* Asst engr elec mach, Ta-Tung Elec Mfg Co, Taiwan, 55-56; assoc res engr, Electron Physics Lab, Univ Mich, 64-66; from asst prof to assoc prof elec eng, 66-72, chmn, elec & comput eng dept, 76-87, PROF ELEC ENG, SAN DIEGO STATE UNIV, 72- *Mem:* Inst Elec & Electronics Engrs; Am Soc Eng Educ. *Res:* Material science; solid state electronics; quantum electronics; power engineering. *Mailing Add:* Dept of Elec Eng 5402 College Ave San Diego CA 92182

LIN, MICHAEL C, b 1938. HORMONE REGULATION, CELLULAR DIFFERENTIATION. *Educ:* Med Col Ga, PhD(biochem), 66. *Prof Exp:* RES CHEMIST, NIH, 75- *Mem:* Am Soc Biochem & Molecular Biol; Am Chem Soc. *Mailing Add:* 11213 Blackhorse Ct Potomac MD 20854-2019

LIN, MING CHANG, b Hsinpu, Hsinchu, Taiwan, Oct 24, 36; US citizen; m 65; c 3. CHEMICAL KINETICS, LASERS. *Educ:* Taiwan Normal Univ, BSc, 59; Univ Ottawa, Can, PhD(phys chem), 65. *Prof Exp:* Res fel, Univ Ottawa, 65-67; res assoc, Cornell Univ, 67-69; res chemist, Naval Res Lab, 70-74, supvry res chemist, 74-82, sr scientist, 82-88; ROBERT W

WOODRUFF PROF PHYS CHEM, EMROY UNIV, 88- *Concurrent Pos:* Adj prof, Dept Chem, Catholic Univ, Washington, DC, 81-88; Guggenheim fel, 82-83. *Honors & Awards:* Hillebrand Prize, Chem Soc Washington, 75; Phys Sci Award, Washington Acad Sci, 76; Alexander von Humboldt Award, 82. *Mem:* Am Chem Soc; Combustion Inst; Sigma Xi. *Res:* Kinetics of chemical reactions are studies with modern diagnostic tools such as lasers with special emphasis on the elucidation of mechanisms of combustion and planetary reactions, heterogeneous catalytic processes and microelectronic processing chemistry. *Mailing Add:* Dept Chem Emory Univ Atlanta GA 30322

LIN, MOW SHIAH, b Kwangtung, China, June 18, 41; US citizen; m 68; c 3. BIOCHEMISTRY, ORGANIC CHEMISTRY. *Educ:* Tamkang Col, BS, 65; Univ Wyo, PhD(chem), 73. *Prof Exp:* Fel vision, NIH, 73-75; res assoc enzyme, 75-77, ASSOC CHEMIST, BROOKHAVEN NAT LAB, 77- *Mem:* AAAS; Am Chem Soc; Sigma Xi. *Res:* Chemistry of vision; isomerase and bacteriorhodopsin; nuclear engineering; nuclear wastes. *Mailing Add:* 81 Westchester Dr Rocky Point NY 11778

LIN, OTTO CHUI CHAU, b Kwongtang, China, Aug 8, 38; m 63; c 3. POLYMER CHEMISTRY, RHEOLOGY. *Educ:* Nat Taiwan Univ, BS, 60; Columbia Univ, MA, 63, PhD(phys chem), 67. *Prof Exp:* Res chemist, E I Dupont de Nemours & Co, Inc, 67-69, staff chemist, 69-71, res assoc, Marshall Res Lab, Fabrics & Finishes Dept, 71-; DIR GENERAL, MAT RES LAB, ITRI, 83-, VPRES, 85- *Concurrent Pos:* Vis prof, Inst Polymer Sci & actg dean, Col of Eng, Nat Tsing Hua Univ, 78-80; bd dir, Feng Chia Univ, 86-; bd dir, China Tech Consult, Inc, 84- *Mem:* AAAS; Am Chem Soc; Am Inst Chem Engrs; NY Acad Sci; Soc Rheol; Chinese Soc Mat Sci (pres, 86-). *Res:* Physical chemical characterization of polymers; rheological properties of polymers; sedimentation; viscometry; organic coatings; ecological impacts of polymer applications; polymers for electronics applications. *Mailing Add:* Mat Res Lab Itri PO Box 2-32 Hsinchu 30098 Taiwan

LIN, P(EN) M(IN), b China, Oct 17, 28; nat US; m 62; c 3. ELECTRICAL ENGINEERING. *Educ:* Taiwan Univ, BS, 50; NC State Col, MS, 56; Purdue Univ, PhD(elec eng), 60. *Prof Exp:* From instr to asst prof elec eng, Purdue Univ, 56-60; mem tech staff, Bell Tel Labs, NJ, 60-61; from asst prof to assoc prof, 61-74, PROF ELEC ENG, PURDUE UNIV, 74- *Concurrent Pos:* Assoc ed, Trans Circuit Theory, Inst Elec & Electronics Engrs, 71-73. *Mem:* Fel Inst Elec & Electronics Engrs. *Res:* Circuit theory; applications of graph theory; computer-aided circuit analysis. *Mailing Add:* Sch Elec Eng Purdue Univ Lafayette IN 47907

LIN, PAUL C S, b Taiwan; Can citizen; m 76; c 1. HILBERT & BANACH SPACES, ALGEBRAS. *Educ:* Nat Taiwan Norm Univ, BSc, 63; McMaster Univ, MSc, 65; Univ NB, PhD(math), 73. *Prof Exp:* Asst prof math, St Thomas Univ, 78-81; from asst prof to assoc prof, 81-89, PROF MATH BISHOP'S UNIV, 89- *Concurrent Pos:* Res fel, Res Inst Can Math Cong, 77; hon res assoc, Univ NB, 77-85; chmn, Dept Math, St Thomas Univ, 79-81 & Bishop's Univ, 89- *Mem:* Am Math Asn; Math Asn Am; Can Math Soc. *Res:* Normed linear spaces; Hilbert and Banach spaces; operators on Hilbert spaces. *Mailing Add:* Dept Math Bishop's Univ Lennoxville PQ J1M 1Z7 Can

LIN, PAUL KUANG-HSIEN, b Tung-Shih, Taiwan, Nov 12, 46; m 78; c 2. EXPERIMENTAL DESIGN, QUALITY ENGINEERING. *Educ:* Fu-Jen Univ, Taiwan, BS, 70; Brigham Young Univ, 75; Wayne State Univ, PhD(statist), 80. *Prof Exp:* Asst prof statist, Oakland Univ, 80-82; asst prof statist, Western Mich Univ, 82-84; asst prof, 84-87, ASSOC PROF STATIST, UNIV MICH, DEARBORN, 87- *Concurrent Pos:* Consult, Qual Improv. *Mem:* Am Statist Asn; Inst Math Statist; Am Soc Quality Control; Sigma Xi. *Res:* Experimental design; interval estimation and hypothesis testing; Taguchi's methods for quality improvement. *Mailing Add:* Dept Math Statist Univ Mich Dearborn MI 48128-1491

LIN, PI-ERH, b Taiwan, China, Jan 8, 38; m 63; c 3. MATHEMATICAL STATISTICS. *Educ:* Taiwan Norm Univ, BSc, 61; Columbia Univ, PhD(math statist), 68. *Prof Exp:* Consult ment ctr, Columbia Univ, 67-68; asst prof, 68-74, assoc prof, 74-80, PROF STATIST, FLA STATE UNIV, 80- *Concurrent Pos:* Fla State Univ fac res grant, 71-72. *Mem:* Inst Math Statist; Am Statist Asn; Bernoulli Soc. *Res:* Multivariate analysis; statistical inference. *Mailing Add:* Dept of Statist Fla State Univ Tallahassee FL 32306

LIN, PING-WHA, b Canton, China, July 11, 25; m 60; c 2. ENVIRONMENTAL ENGINEERING. *Educ:* Jiao Tong Univ, BS, 47; Purdue Univ, MS, 49, PhD(sanit eng), 51. *Prof Exp:* Engr, Amman & Whitney, NY, 51-54; Ebesco, 54-55, Parsons, Brinkerhoff, Hall & MacDonald, 55-57 & Lockwood Greene Engrs, 57-59; engr, World Health Orgn, 59-60, consult, 62-66; engr, John Graham & Co, NY, 60-61; PROF ENVIRON ENG, TRI-STATE UNIV, 66-79 & 81-; PRES, LIN TECHNOLOGIES, INC, 88- *Concurrent Pos:* Fel, NSF workshop, Mass Inst Technol, 69; Dept Energy indiv grant, 82; proj mgr, WHO, 79-81, consult, 86. *Honors & Awards:* Achievement Award, United Inventors & Scientists of Am, 74. *Mem:* Fel Am Soc Civil Engrs; Am Water Works Asn; Sigma Xi; Am Chem Soc. *Res:* Acid neutralization; metal waste treatment; fly ash utilization; soil stabilization; quenching process; flue gas desulfurization; energy development. *Mailing Add:* 506 S Darling St Angola IN 46703

LIN, RENEE C, LIPID METABOLISM, CULTURE HEPATOCYTES. *Educ:* Univ Wis, Madison, PhD(biochem), 69. *Prof Exp:* ASSOC SCIENTIST MED RES, VET ADMIN MED CTR, 74-; ASST PROF MED, SCH MED, IND UNIV, 85- *Mailing Add:* Vet Admin Med Ctr Ind Univ Sch Med 1481 W Tenth St Indianapolis IN 46202

LIN, RENG-LANG, b Hsin-Chu, Taiwan, Feb 28, 37; m 65; c 2. TOXICOLOGY, PSYCHOPHARMACOLOGY. *Educ:* Nat Taiwan Univ, BS, 59, MS, 63; Okla State Univ, PhD(biochem), 69. *Prof Exp:* Fel, Univ Wis-Madison, 69-71; res scientist, Galesburg State Hosp, Ill, 71-75; res scientist, Ill State Psychiat Inst, 75-78; ASST CHIEF TOXICOLOGIST, OFF MED

EXAMINER, COOK COUNTY, 78- *Mem:* Am Chem Soc; AAAS. *Res:* Forensic toxicology; biochemistry of mental illness; biosynthesis and metabolism of biogenic amines; biochemistry and pharmacology of psychoactive drugs. *Mailing Add:* Toxicol Dept Med Examr Off 1014 W 32nd Pl Chicago IL 60608

LIN, ROBERT I-SAN, b Fukien, China, 42; m; c 4. SCIENCE ADMINISTRATION, RESOURCE MANAGEMENT. *Educ:* Nat Taiwan Univ, BS, 61; Univ Calif, Los Angeles, MS, 65, PhD(biophysics & nuclear med), 68. *Prof Exp:* Res fel chem, Calif Inst Technol, 68-70; clin trainee metab dis, Univ Calif Med Ctr, 70-71; life sci mgr, Gen Tel & Electronic Corp, 71-73; dir enzyme prod, Worthington Biochem Corp, 73-74; vpres technol diag prod, RIA Inc, 74-75; chief scientist, Frito-Lay Inc, 75-82; vpres, Natural Prod Div, Richardson Vicks, 82-85; sr vpres, Makers of Kal, 86-87; sr vpres, Weider Health & Fitness, 88-90; EXEC VPRES, NUTRIT INT, 91- *Concurrent Pos:* Vis prof biochem & molecular biol, Pepperdine Univ, 69-70; trainee biotechnol & nutrit, Mass Inst Technol, 71; vis distinguished prof nutrit & food sci, Tex Woman's Univ, 81-; chmn, First World Cong on Health Significance of Garlic, 90. *Mem:* NY Acad Sci; Sigma Xi; Am Photobiol Soc; Soc Appl Nutrit; Am Agr Econ Asn; Am Mgt Asn; Am Col Nutrit; Am Col Sport Med; Inst Food Technol; Am Chem Soc. *Res:* Biotechnology and genetic engineering: recombinant DNA--hybridoma and subcellular organelle transfer; chemistry and rheology of natural and synthetic polymers; physical chemistry of surfactants and viscosity modifiers; nutrition, aging and degenerative diseases; sport nutrition, herbal sciences and pharmacognosy; management of technological development and industrialization; operation research and econometrics. *Mailing Add:* Six Silverfern Irvine CA 92715

LIN, ROBERT PEICHUNG, b China, Jan 24, 42; US citizen; m 83. SOLAR PHYSICS, HIGH ENERGY ASTROPHYSICS. *Educ:* Calif Inst Technol, BS, 62; Univ Calif, Berkeley, PhD(physics), 67. *Prof Exp:* From asst res physicist to assoc res physicist, Univ Calif, 67-79, res physicist, 79-88, sr fel, 80-88, adj prof astron, 88-90, PROF PHYSICS, UNIV CALIF, 91- *Concurrent Pos:* Prin investr, NASA & ESA Giotto Mission, 79-, ISTP/GGS Wind, 81-, Mars Observer, 87-, Max 91 Balloon, 89- & NSF Antarctic Balloon, 88- *Mem:* Am Geophys Union; Am Astron Soc. *Res:* Solar flares, radio bursts and cosmic rays; interplanetary particles; magnetospheric processes; lunar magnetism; astrophysical x-ray and gamma ray spectroscopy; comets. *Mailing Add:* Space Sci Lab Univ Calif Berkeley CA 94720

LIN, SHAO-CHI, b Canton, China, Jan 5, 25; nat US; m 55. AEROSPACE ENGINEERING. *Educ:* Nat Cent Univ, China, BSc, 46; Cornell Univ, PhD(aeronaut eng), 52. *Prof Exp:* Engr, Bur Aircraft Indust, China, 47-48; asst, Cornell Univ, 48-51, actg instr, 52, res assoc, 52-54, actg asst prof, 54; prin res scientist, Avco-Everett Res Lab, Mass, 55-64; PROF ENG PHYSICS, UNIV CALIF, SAN DIEGO, 64- *Concurrent Pos:* Consult, Aerospace Corp, Avco-Everett Res Lab, Inst Defense Anal & Rand Corp; panel mem re-entry physics, Nat Acad Sci-Nat Res Coun. *Honors & Awards:* Res Award, Am Inst Aeronaut & Astronaut, 66. *Mem:* Am Inst Aeronaut & Astronaut; Am Phys Soc; Am Astronaut Soc; Am Geophys Union. *Res:* Physical gas dynamics; hypersonic flight; reentry physics; laser physics and interaction. *Mailing Add:* 2614 Costebelle Dr La Jolla CA 92037

LIN, SHENG HSIEN, b Sept 17, 37; US citizen; m 70. CHEMICAL KINETICS, CHEMICAL PHYSICS. *Educ:* Nat Taiwan Univ, BS, 59, MS, 61; Univ Utah, PhD(chem), 64. *Prof Exp:* Fel chem, Columbia Univ, 64-65; from asst to assoc prof, 65-72, PROF CHEM, 72, REGENT PROF, ARIZ STATE UNIV, 88- *Concurrent Pos:* A P Sloan fel, 67-69; Guggenheim fel, 71-73; Humboldt Sr US scientist awardee, 79-80 & 88-89; Hon Prof, Nanjing Univ, 88- *Mem:* Am Chem Soc; Acad Sinica. *Res:* Energy transfer; femtosecond processes; optical rotations and the Faraday effect; reaction kinetics; electron transfer; magnetic properties of molecules; multi-photon processes; molecular relaxation processes; theory of time-resolved x-ray diffraction. *Mailing Add:* Dept of Chem Ariz State Univ Tempe AZ 85281

LIN, SHERMAN S, food science, flavor chemistry, for more information see previous edition

LIN, SHIN, b Hong Kong, Feb 14, 45; US citizen; m 69; c 2. BIOCHEMISTRY, BIOPHYSICS. *Educ:* Univ Calif, Davis, BS, 65; San Diego State Univ, MS, 67; Univ Calif, Los Angeles, PhD(biol chem), 71. *Prof Exp:* Fel, Univ Calif, San Francisco, 71-74; from asst prof to assoc prof, 74-82, PROF BIOPHYS, JOHNS HOPKINS UNIV, 82-, PROF BIOL, 89- *Concurrent Pos:* NIH res career develop award, 76-81; chmn biophys, Johns Hopkins Univ, 83- *Mem:* AAAS; Am Soc Biol Chemists; Am Soc Cell Biol; Biophys Soc. *Res:* Biochemical and biophysical studies on cytoskeletal and motile functions of eukaryotic cells, with emphasis on drugs and cellular proteins affecting the assembly and interactions of actin filaments in vivo and in vitro. *Mailing Add:* Dept of Biophys Johns Hopkins Univ Baltimore MD 21218

LIN, SHU, b Nanking, China, May 20, 36; m 63; c 3. ELECTRICAL ENGINEERING, INFORMATION SCIENCES. *Educ:* Nat Taiwan Univ, BS, 59; Rice Univ, MS, 64, PhD(elec eng), 65. *Prof Exp:* Res assoc, Univ Hawaii, 65-66, from asst prof to assoc prof elec eng, 66-73, prof, 73-81, prof elec eng, 86-88; prof elec eng, 82-88, IRMA RUNYON CHAIR PROF, TEX A&M UNIV, 88-; PROF & CHMN, UNIV HAWAII, 88- *Concurrent Pos:* NSF grants, 67-92; Air Force Cambridge Res Lab grant, 70-71, NASA grants, 83-91; vis scholar, Univ Utah, 71-72; vis scientist, IBM Watson Res Ctr, 78-79. *Mem:* Fel Inst Elec & Electronics Engrs; Sigma Xi. *Res:* Coding theory and error control in data transmission systems; coding theory and multi-access communications. *Mailing Add:* Univ Hawaii-Manoa 2500 Campus Rd Honolulu HI 96822

LIN, SHWU-YENG TZEN, b Tainan, Formosa, May 11, 34; m 60; c 3. TOPOLOGY. *Educ:* Nat Taiwan Normal Univ, BSc, 58; Tulane Univ, MS, 62; Univ Fla, PhD(math), 65. *Prof Exp:* Asst math, Inst Math, Academia Sinica, 58-60; instr, Tulane Univ, 61-63; lectr, 64-65, asst prof, 65-71, ASSOC

PROF MATH, UNIV SOUTH FLA, 71- *Concurrent Pos:* Reviewer, Math Rev, Am Math Soc, 68-; Zentralblatt fur Mathmatik, 70- *Mem:* Math Asn Am. *Res:* Topology and relation-theory. *Mailing Add:* Dept of Math Univ of SFla 4202 Fowler Ave Tampa FL 33620

LIN, SIN-SHONG, b Taiwan, Oct 24, 33; m 64; c 3. HIGH TEMPERATURE CHEMISTRY. *Educ:* Nat Taiwan Univ, BS, 56; Nat Tsing-Hua Univ, Taiwan, MS, 58; Univ Kans, PhD(chem), 66. *Prof Exp:* Fel, Northwestern Univ, Evanston, 66-67; RES CHEMIST, ARMY MAT TECH LAB, 67- *Mem:* Am Chem Soc; Am Vacuum Soc; Electrochem Soc; Am Carbon Soc; Soc Advancement Mat & Process Eng. *Res:* Thermodynamics of vaporization processes; material research and development; atmospheric sampling of gases; carbon fiber processing and characterization; electron spectroscopy for chemical analysis & auger electron spectroscopy. *Mailing Add:* 40 Lyons Rd Westwood MA 02090

LIN, SPING, b Canton, China, Sept 8, 18; nat US; m 46; c 2. NEUROCHEMISTRY, PHYSIOLOGY. *Educ:* Sun Yat-Sen Univ, BA, 40; Univ Minn, MS, 50, PhD(entom), 52. *Prof Exp:* Asst entom, Sun Yat-Sen Univ, 40-44, instr, 44-47; res fellow, 54-61, res assoc, 61-63, asst prof, 63-69, ASSOC PROF NEUROL, MED SCH, UNIV MINN, MINNEAPOLIS, 69- *Mem:* Int Soc Neurochem; AAAS; Am Soc Neurochem; Sigma Xi. *Res:* Neurobiology. *Mailing Add:* 1785 Fairview Ave N St Paul MN 55113

LIN, STEPHEN FANG-MAW, b Nantou, Taiwan, Aug 21, 37; m 66; c 1. PHYSICAL CHEMISTRY. *Educ:* Nat Taiwan Univ, BS, 60; Univ Ill, Urbana, MS, 68, PhD(phys chem), 70. *Prof Exp:* ASSOC PROF CHEM, NC CENT UNIV, 70- *Mem:* AAAS; Sigma Xi; Am Chem Soc. *Res:* Conformation and stability of sulfur ring compounds; molecular spectroscopy. *Mailing Add:* 3116 Annadale Rd Durham NC 27705-5466

LIN, STEPHEN Y, b Pingtung, Taiwan, Apr 23, 39; US citizen; m 72; c 2. LIGNIN CHEMISTRY, ORGANIC CHEMISTRY. *Educ:* Nat Taiwan Univ, Taipei, BS, 62; Univ Wash, Seattle, MS, 67; NC State Univ, Raleigh, PhD(chem), 70. *Prof Exp:* Res chemist, Westvaco Corp, 72-76, sr res chemist, 76-79; res assoc, Lignin Chem Res, Am Can Co, 79-80, supvr res, 80-81, sr supv res, 81-82; res mgr, Reed Lignin Inc, 83-88 & Daishowa Chemicals Inc, 89-91; RES MGR, LIGNOTECH US INC, 91- *Concurrent Pos:* Adj prof, NC State Univ. *Honors & Awards:* George Olmsted Award, Am Paper Inst, 71. *Mem:* Tech Asn Pulp & Paper Indust; Am Chem Soc. *Res:* Lignin organic-physical chemistry; modification and application of industrial lignins, including kraft and sulfite lignins; dispersants. *Mailing Add:* Lignotech US Inc 100 Highway 51 S Rothschild WI 54474-1198

LIN, SUE CHIN, b Taipei, China, Nov 8, 36; m 62; c 2. MATHEMATICS. *Educ:* Univ Calif, Berkeley, MA, 64, PhD(math), 67. *Prof Exp:* Asst prof math, Univ Miami, 67-69; mem, Inst Advan Study, 69-71; assoc prof math, 71-86, DIR GRAD STUDIES, UNIV ILL, CHICAGO CIRCLE, 86- *Mem:* Am Math Soc. *Res:* Functional analysis. *Mailing Add:* Dept Math Univ Ill Chicago IL 60680

LIN, SUI, b Wenlin, Zhejiang, China, 1929; Can citizen; m; c 2. HEAT & MASS TRANSFER, FLUID MECHANICS. *Educ:* Ord Eng Col, Taiwan, BS, 53; Univ Karlsruhe, WGer, Dipl-Ing 62, Dr-Ing(mech eng), 64. *Prof Exp:* Res assoc refrig, Inst Refrig Eng, Univ Karlsruhe, WGer, 62-65 & gas dynamics, Inst Fluid Mech & Fluid Mach, 65-69; fel sonic boom, Inst Aerospace studies, Univ Toronto, Ont, 69-70; asst prof thermodynamics & fluid mech, 70-75, assoc prof, 75-81, PROF HEAT TRANSFER & FLUID MECHANICS, DEPT MECH ENG, CONCORDIA UNIV, MONTREAL, 81- *Concurrent Pos:* Sr vis scientist, QIT-Fer et Titane Inc, Sorel, Quebec, 84-85. *Mem:* Eng Inst Can; Can Soc Mech Eng; Am Soc Heating, Refrig & Air Conditioning Engrs; Deutsche Gesellschaft fuer Luft-und Raum-fahrt. *Res:* Heat and mass transfer with phase changes; confined vortex flows; heat pump systems for cold climates; freezing preservation of biological cells. *Mailing Add:* Dept Mech Eng Concordia Univ 1455 de Maisonneuve Blvd W Montreal PQ H3G 1M8 Can

LIN, SUNG P, b Taipei, Taiwan, Apr 18, 37; US citizen; m 66; c 2. FLUID MECHANICS, APPLIED MATHEMATICS. *Educ:* Taiwan Univ, BS, 58; Univ Utah, MS, 61; Univ Mich, PhD(eng mech), 65. *Prof Exp:* Engr, Ministry Econ, China, 58-60; lectr eng mech, Univ Mich, 65-66; from asst prof to assoc prof mech eng, 66-74, chmn fluid & thermal sci group, 78-80, PROF MECH ENG, CLARKSON UNIV, 74-, CHMN APPL MECH PROG, 81- *Concurrent Pos:* NSF initiation grant, 67-69, res grants, 70-72, 74-76, 78-80, 80-85 & 88-91; sr vis, Cambridge Univ, 73-74; Kodak grant, 74-76, consult, Eastman Kodak Co, 77-81; vis prof, Rochester Univ & Stanford Univ, 80-81; Bausch & Lomb grant, 80-81; NASA fel, 83,84, grant, 84-86, ARO, 85- *Mem:* Am Phys Soc; Am Soc Mech Engrs. *Res:* Theory and application of mechanics; fluid and biofluid mechanics; hydrodynamic stability; flow separation; heat transfer; transieut phenomena; film coating technology; aerosol dynamics; atomization. *Mailing Add:* Dept Mech & Aeronaut Eng Clarkson Univ Potsdam NY 13676

LIN, TAI-SHUN, b Fukien, China, Oct 10, 38; m 65; c 1. MEDICINAL CHEMISTRY, ORGANIC CHEMISTRY. *Educ:* Nat Taiwan Norm Univ, BSc, 59; Univ Wash, MS, 65; Western Mich Univ, PhD(org & med chem), 70. *Prof Exp:* Fel org chem, NC State Univ, 70-71; res assoc med chem, Western Mich Univ, 71-74; res assoc, 74-76, asst prof, 76-82, res scientist, 82-85, SR RES PHARMACOL, SCH MED, YALE UNIV, 85- *Concurrent Pos:* Assoc ed, J Carbohydrates, Nucleosides, Nucleotides, 74-81. *Mem:* Am Chem Soc; assoc fel Inst Chem; NY Acad Sci. *Res:* Design, synthesis, mechanism of action and development of anticancer and antiviral agents; pharmacology. *Mailing Add:* Dept of Pharmacol Yale Univ Sch of Med New Haven CT 06510

LIN, TIEN-SUNG TOM, b Taiwan, China, Jan 9, 38; m 66; c 3. PHYSICAL CHEMISTRY. *Educ:* Tunghai Univ, BS, 60; Syracuse Univ, MS, 66; Univ Pa, PhD(phys chem), 69. *Prof Exp:* Res fel, Harvard Univ, 70; from asst prof to assoc prof, 76-86, PROF CHEM, WASH UNIV, 86- *Concurrent Pos:* Scientist-in-Res, Argonne Nat Lab, 80. *Mem:* Am Chem Soc. *Res:* Molecular spectroscopy; photophysical and photochemical processes; structural aspects of organic free radicals. *Mailing Add:* Dept Chem Wash Univ St Louis MO 63130

LIN, TSAU-YEN, b Taiwan, China, July 18, 32; m 67. BIOCHEMISTRY. *Educ:* Nat Taiwan Univ, BS, 55, MS, 57; Univ Calif, Berkeley, PhD(biochem), 65. *Prof Exp:* Instr clin chem, Kaohsiung Med Col, Taiwan, 57-58; res chemist, China Chem & Pharmaceut Co, 58-59; res asst biochem, US Naval Med Res Unit Number 2, Taiwan, 59-61; res assoc, Univ Calif, Berkeley, 65-67, asst res biochemist, 67-68; biochemist, Merck Sharp & Dohme Res Labs, 69-91; SCIENTIST, CTD LABS, 91- *Mem:* AAAS; Am Chem Soc; Am Soc Biol Chemists; Sigma Xi. *Res:* Biosynthesis and function of complex carbohydrates; mechanism and active site structure of enzymes; biochemical characterization of complement. *Mailing Add:* Two Concord Ave Piscataway NJ 08854-5233

LIN, TSUE-MING, b Ping-tung, Taiwan, June 10, 35; US citizen; m 64; c 3. IMMUNOLOGY, MICROBIOLOGY. *Educ:* Nat Taiwan Univ, DVM, 58, dipl pub health, 60; Tulane Univ, MS, 64; Univ Tex Med Br, PhD(microbiol), 68. *Prof Exp:* Teaching asst & lab instr med parasitol, Col Med, Nat Taiwan Univ, 60-62; res assoc, 68-69, from instr to assoc prof pediat, 69-80, res assoc prof pediat, Med Sch, Univ Miami, 80-86; dir res, 86-90, VPRES, RES & DEVELOP, DIAMEDIX CORP, 90- *Concurrent Pos:* Teaching asst & lab instr, Taipei Med Col, 61-62; consult, Cordis Labs, 72-74 & sr staff immunologist to asst dir res & develop, 77-86. *Mem:* AAAS; Am Soc Microbiol; Am Asn Immunol. *Res:* Host-parasite relationship; enzyme-linked immunoassays for infectious and immunological diseases; immunology; human heart autoimmune system; trichinosis; amebiasis; toxoplasmasis; human pregnancy-associated plasma proteins. *Mailing Add:* 8245 Southwest 56 St Miami FL 33155

LIN, TSUNG-MIN, b Chefoo, China, Oct 8, 16; nat US; m 43; c 2. PHYSIOLOGY, PHARMACOLOGY. *Educ:* Nat Tsing Hua Univ, China, BS, 38; Univ Ill, MS, 52, PhD, 54. *Prof Exp:* Asst physiol, Nat Tsing Hua Univ, China, 39-40; asst, Nat Chung Cheng Med Col, 40-41, lectr physiol, 41-43; lectr, Nat Kweiyang Med Col, 43-46, asst prof, 46-48; sr instr, Peking Union Med Col, 48-51; asst prof clin sci, Col Med, Univ Ill, 54-56; pharmacologist, Res Labs, Eli Lilly & Co, 56-69, res scientist, 59-63, sr res scientist, 64-85; CONSULT, MED RES METHODIST HOSP, INDIANAPOLIS, IN, 86- *Concurrent Pos:* Adj prof physiol, Ind Univ Sch Med, 86- *Mem:* Am Physiol Soc; Am Soc Pharmacol & Exp Therapeut; Am Gastroenterol Asn; Am Pancreatic Asn. *Res:* Gastrointestinal physiology and pharmacology. *Mailing Add:* 9146 Compton Ave Indianapolis IN 46240

LIN, TU, b Fukien, China, Jan 18, 41; US citizen; m 67; c 3. INTERNAL MEDICINE, ENDOCRINOLOGY & METABOLISM. *Educ:* Nat Taiwan Univ, Taipei, MD, 66. *Prof Exp:* Intern, Episcopal Hosp, Temple Univ, 67-68; resident med, Berkshire Med Ctr, Union Univ, 68-70; fel endocrinol, Lahey Clin, Boston, 70-71, Roger Williams Gen Hosp, Brown Univ, 71-73; staff physician, Vet Admin Hosp, Salisbury, 73-75; from asst prof to assoc prof, 76-84, PROF ENDOCRINOL, SCH MED, UNIV SC, 84-; CHIEF, ENDOCRINE SECT, WILLIAM JENNINGS BRYAN DORN VET HOSP,74- *Concurrent Pos:* Physician, Richland Mem Hosp, Columbia, SC, 77- *Mem:* Endocrine Soc; Am Fedn Clin Res; Am Soc Andrology; fel Am Col Physicians. *Res:* Male reproductive endocrinology; Leydig cell function including cell membrane receptors, cyclic AMP metabolism, steroid receptors, steroidogenesis, phospholipid turnover and long term cell culture; insulin receptors, insulin-like growth factors I and II receptors of Leydig cells. *Mailing Add:* William Jennings Bryan Dorn Vet Hosp Columbia SC 29201

LIN, TUNG HUA, b Chungking, China, May 26, 11; nat US; m 39; c 3. MECHANICS. *Educ:* Tangshan Col Eng, BS, 33; Mass Inst Technol, SM, 36; Univ Mich, DSc(eng mech), 53. *Prof Exp:* From assoc prof to prof aeronaut eng, Tsing Hua Univ, China, 37-39; from designer to chief engr & prod mgr, Chinese Aircraft Mfg Plant, 39-45; mem, Chinese Tech Mission in Eng, 45-49; from assoc prof to prof aeronaut eng, Univ Detroit, 49-56; prof eng, 56-78, EMER PROF, UNIV CALIF, LOS ANGELES, 78- *Concurrent Pos:* Consult, Continental Motor Corp, Mich, 54-55, Off Ord Res, 58, NAm Aviation Inc, 62-68 & ARA Inc, 66-; prin investr, res projs, Off Sci Res, US Air Force, 55-59, 88-, & NSF, 61-78, US Off Naval Res, 85- *Honors & Awards:* Theodore Van Karman Award, Am Soc Civil Engrs, 88. *Mem:* Nat Acad Eng; Soc Eng Sci; fel Am Acad Mech; Am Soc Civil Engrs; fel Am Soc Mech Engrs. *Res:* Micromechanics; multiaxial stress-strain relations based on microstress fields in polycrystals; fatigue crack initiation mechanism based microstresses; inelastic structures; el asto-plastic analysis of beams, columns and plates; creep analysis of columns and plates. *Mailing Add:* 906 Las Palgas Rd Pacific Palisades CA 90272

LIN, TUNG YEN, b Foochow, China, Nov 14, 11; c 2. STRUCTURAL ENGINEERING. *Educ:* LLD, Chinese Univ Hong Kong, 72, Golden Gate Univ, San Francisco, 82, Jiaotung Univ, Taiwan & Tongji Univ. *Prof Exp:* From engr to chief designer, Ministry of Rwy, China, 33-41; prof bridge eng, Tungchi Univ, 39-41; chief engr, Kung Sing Eng Corp, 41-45; comnr, Taiwan Sugar Rwy, 45-46; from assoc prof to prof civil eng, Univ Calif, Berkeley, 46-76, chmn div struct eng & mech & dir lab, 60-63; chmn bd, 54-88, HON CHMN BD, TY LIN INT, 88- *Concurrent Pos:* Hon prof, Tsinghua Univ & Tongji Univ, Beijing, China, Jiaotung Univ, Shanghai & Omei. *Honors & Awards:* Howard Gold Medal, Am Soc Civil Engrs; Nat Medal Sci, 86; Award of Merit, Am Consult Engrs Coun, 87; Roebling Medal, 90. *Mem:* Nat Acad Eng; fel Am Soc Civil Engrs; Int Asn Bridge & Struct Engrs; Am Concrete Inst; Am Soc Eng Educ. *Res:* Bridge and structural engineering; design of prestressed concrete and steel structures; structural concepts and systems. *Mailing Add:* T Y Lin Int 315 Bay St San Francisco CA 94133

LIN, TUNG-PO, b Fukien, China, Dec 31, 26; nat US; m 56; c 4. MATHEMATICS. *Educ:* Nat Cent Univ, China, BSc, 49; Mass Inst Technol, PhD(phys chem), 58. *Prof Exp:* Res chemist, E I du Pont de Nemours & Co, Del, 58-61; from asst prof to assoc prof math, 61-69, PROF MATH, CALIF STATE UNIV, NORTHRIDGE, 69- *Concurrent Pos:* Consult, IBM Corp, 61-68. *Mem:* Am Math Soc; Math Asn Am. *Res:* Functional analysis; applied mathematics. *Mailing Add:* Dept Math Calif State Univ Northridge CA 91330

LIN, TZ-HONG, b Taiwan, Jan 30, 34; m 69; c 2. ORGANIC CHEMISTRY, RADIOPHARMACEUTICAL RESEARCH. *Educ:* Nat Taiwan Univ, BS, 56; NMex Highlands Univ, MS, 64; Univ Calif, Berkeley, PhD, 69. *Prof Exp:* Teaching & res asst, Dept Chem, NMex Highlands Univ, 61-63; teaching asst, Dept Chem, Univ Calif, Berkeley, 63-64, res asst, Biodynamics Lab, Lawrence Radiation Lab, 64-69; vis asst prof, Dept Chem, La State Univ, 69-71; dir res chem, Medi-Physics, Inc, 71-74, res group leader, 74-78, proj mgr, 78-83, assoc dir, Res & Develop, 83-85; consult, Photon Diag Inc, 85-87; vpres, Imagents Inc, 87-90; VPRES, IMP INC, 90- *Mem:* Soc Nuclear Med; Am Chem Soc; AAAS. *Res:* Research and development of new radiopharmaceuticals; hot atom chemistry; free radical chemistry; antibody research. *Mailing Add:* IMP Inc 8050 El Rio Houston TX 77054

LIN, WEI-CHING, b Taipei, China, Dec 31, 30; m 59. SPACE PHYSICS. *Educ:* Nat Taiwan Univ, BSc, 54; Univ Iowa, MSc, 61, PhD(physics), 65. *Prof Exp:* Res assoc space physics, Univ Iowa, 63-64; asst prof, Dalhousie Univ, 64-68; ASSOC PROF SPACE PHYSICS, UNIV PRINCE EDWARD ISLAND, 68- *Mem:* Am Geophys Union; Am Asn Physics Teachers. *Res:* Galactic and solar cosmic rays. *Mailing Add:* Dept of Physics Univ of Prince Edward Island Charlottetown PE C1A 4P3 Can

LIN, WEN-C(HUN), b Kutien, China, Feb 22, 26; US citizen; m 56; c 3. ELECTRICAL & COMPUTER ENGINEERING. *Educ:* Taiwan Univ, BS, 50; Purdue Univ, MS, 56, PhD(elec eng), 65. *Prof Exp:* Engr, elec lab, Taiwan Power Co, 50-54; engr high voltage lab, Gen Elec Co, 56-59; sr engr, electronic data processing div, Honeywell, Inc, 59-61; instr, Purdue Univ, 61-65; from asst prof to prof syst eng, Case Western Reserve Univ, 65-78; PROF ELEC & COMPUT, UNIV CALIF, DAVIS, 78- *Concurrent Pos:* Autonomous Mobile Robot Syst. *Mem:* Inst Elec & Electronics Engrs. *Res:* Special electronic instrumentation; signal processing; pattern recognition; microcomputers; artificial neuron network. *Mailing Add:* Dept Elec & Comput Eng Univ Calif Davis CA 95616

LIN, WILLY, b Taiwan, China, July 2, 44; m 71; c 2. PLANT PHYSIOLOGY, PLANT MOLECULAR BIOLOGY. *Educ:* Nat Taiwan Norm Univ, BS, 67; Ill State Univ, MS, 72; Univ Ill, Urbana, PhD(biol), 76. *Prof Exp:* Res assoc, Dept Biol, Brookhaven Nat Lab, 76-77; staff scientist, Cent Res & Develop Exp Sta, 77-86, STAFF SCIENTIST, AGR PROD DEPT, E I DU PONT DE NEMOURS & CO, INC, 87- *Mem:* Am Soc Plant Physiologists; Am Inst Biol Sci; Sigma Xi; NY Acad Sci; Int Asn Plant Tissue Culture; AAAS. *Res:* Ion transport mechanism in plant tissues; plant tissue culture and genetic transformation. *Mailing Add:* DuPont Agr Prod E42/4130 E I du Pont de Nemours & Co Wilmington DE 19880-0402

LIN, WUNAN, b Tainan, Taiwan, Aug 1, 42; US citizen; m 71. GEOPHYSICS, ROCK MECHANICS. *Educ:* Cheng-Kung Univ, BSE, 64; Univ Calif, Berkeley, MS, 69, PhD(geophys), 77. *Prof Exp:* Prof asst mining eng, Cheng-Kung Univ, 65-67; GEOPHYSICIST, LAWRENCE LIVERMORE LAB, UNIV CALIF, 77- *Concurrent Pos:* Res asst, Univ Calif, Berkeley, 67-68 & 71-77. *Mem:* Am Geophys Union; Int Soc Rock Mech; Inst Elec & Electronic Engrs. *Res:* Solid earth geophysics; physical properties of rocks at high pressure and high temperature. *Mailing Add:* Lawrence Livermore Lab L-201 PO Box 808 Livermore CA 94550

LIN, Y(U) K(WENG), b Foochow, China, Oct 30, 23; US citizen; m 52; c 4. STRUCTURAL ENGINEERING, APPLIED PROBABILITY. *Educ:* Amoy Univ, BS, 46; Stanford Univ, MS, 55, PhD(struct eng), 57. *Prof Exp:* Stress engr, Vertol Aircraft Corp, Pa, 56-57; prof eng, Imp Col Eng, Ethiopia, 57-58; res engr, Boeing Co, 58-60; asst prof aeronaut eng, Univ Ill, Urbana, 60-62, from assoc prof aeronaut & asstronaut eng to prof, 62-83; CHARLES E SCHMIDT EMINENT SCHOLAR ENG, FLA ATLANTIC UNIV, 83- *Concurrent Pos:* Consult, transport div, Boeing Co, 61, Wichita Div, 62, Gen Dynamics/Convair, 67, US Army Weapons Command, Ill, 72-77, Res Labs, Gen Motors Corp, 75-, TRW Defense & Space Systs, 78- & Brookhaven Nat Lab, 90-; vis prof, Mass Inst Technol, 67-68; sr vis fel, Inst Sound & Vibration Res, Univ Southampton, Eng, 76; NSF sr fel, 67-68. *Honors & Awards:* Alfred M Freudenthal Medal, Am Soc Civil Engrs, 84. *Mem:* Am Inst Aeronaut & Astronaut; Acoust Soc Am; Am Acad Mech; Am Soc Civil Engrs. *Res:* Structural dynamics; random vibrations; systems reliability. *Mailing Add:* Col Eng Fla Atlantic Univ Boca Raton FL 33431-0991

LIN, YEONG-JER, b Taiwan, China, Nov 11, 36; m 66; c 2. METEOROLOGY, ATMOSPHERIC SCIENCES. *Educ:* Nat Taiwan Univ, BS, 59; Univ Wis-Madison, MS, 64; NY Univ, PhD(meteorol), 69. *Prof Exp:* Res asst meteorol, Univ Wis-Madison, 62-64; asst res scientist, NY Univ, 65-69, assoc res scientist, 69; from asst prof to assoc prof, 69-76, PROF METEOROL, ST LOUIS UNIV, 76- *Concurrent Pos:* NSF res grants, St Louis Univ, 70-89. *Mem:* Am Meteorol Soc; Am Geophys Union. *Res:* Dynamical and observational studies of severe local storms; numerical modelling of meso-scale circulation. *Mailing Add:* Dept Earth & Atmospheric Sci St Louis Univ St Louis MO 63103

LIN, YI-JONG, b Feng Yuan, Taiwan, June 19, 44; US citizen; m; c 2. DRUG METABOLISM, TOXICOLOGY. *Educ:* Univ Tokyo, BS, 67, MS, 71, PhD(pharmaceut), 75. *Prof Exp:* Postdoctoral fel pharmacol, Univ Mich, 75-76; res assoc pharmaceut, State Univ NY, Buffalo, 76; res fel pharmacol, Univ Mo, Kansas City, 77-79; sr res pharmacologist, 79-84, group leader, 84-90, SECT LEADER PHARMACOL, COLGATE-PALMOLIVE CO, 90- *Concurrent Pos:* Prin, CACA Mid-Jersey Chinese Sch, 87-88. *Mem:* Am

Pharmaceut Soc; Am Soc Pharmacol & Exp Therapeut. *Res:* Plan and execute projects on pharmacokinetics, drug metabolism, drug delivery, bucal absorption and drug safety evaluation; plan and monitor acute, chronic toxicity, carcinogenicity, clinical pharmacokinetics, mutagenicity and reproductive toxicology studies. *Mailing Add:* Seven Argyle Way Colgate-Palmolive Co 909 River Rd Robbinsville NJ 08691

LIN, YONG YENG, b Feb 2, 33; Taiwan citizen; m 61; c 2. BIO-ORGANIC CHEMISTRY. *Educ:* Nat Taiwan Univ, BSc, 56; Tokyo Kyoiku Univ, MSc, 63; Tohoku Univ, Japan, PhD(org chem), 66. *Prof Exp:* Res assoc chem, Fla State Univ, 66-68; res assoc biochem, Univ Tex Med Br, Galveston, 68-70; res assoc chem, Univ Toronto, 70-72; res scientist chem, 72-81, RES SCIENTIST CHEM, INST BASIC RES, UNIV TEX MED BR, 82- *Mem:* AAAS; Am Chem Soc; Sigma Xi. *Res:* Mechanisms of biological oxidation; enzyme models; application of enzymes and enzyme models to preparative organic chemistry; organic synthesis. *Mailing Add:* 1050 Forest Hill Rd Staten Island NY 10314

LIN, YOU-FENG, b Feng-Shan, Taiwan, July 31, 32; m 60; c 3. TOPOLOGY. *Educ:* Nat Taiwan Normal Univ, BS, 57; Univ Fla, PhD(math), 64. *Prof Exp:* Asst math, Inst Math, Chinese Acad Sci, 56-59; from asst to assoc prof, 64-69, res asst prof, 65-66, PROF MATH, UNIV S FLA, 69- *Concurrent Pos:* Reviewer, Math Rev, Am Math Soc, 65- *Mem:* Am Math Soc; Math Asn Am. *Res:* Topological algebra; structure of topological semigroups; semigroup of measures; topology and relation-theory. *Mailing Add:* Dept of Math Univ of S Fla 4202 Fowler Ave Tampa FL 33620

LIN, YU-CHONG, b Taiwan, Repub China, Apr 24, 35; m 60; c 2. PHYSIOLOGY. *Educ:* Taiwan Norm Univ, BS, 59; Univ NMex, MS, 64; Rutgers Univ, PhD(physiol), 68. *Prof Exp:* Teaching asst biol, Taiwan Norm Univ, 60-62; from asst prof to assoc prof, 69-76, PROF PHYSIOL, SCH MED, UNIV HAWAII, MANOA, 76- *Concurrent Pos:* Fel, Inst Environ Stress, Univ Calif, Santa Barbara, 68-69; physiologist consult, Tripler Army Med Ctr, 79-; vis prof, Nat Yang Ming Med Col, Taipei, Taiwan, 83-, Kosin Med Col, Basan, Korea, 83-, Nat Defense Med Col, Taipei, Taiwan, 89- *Mem:* AAAS; Am Physiol Soc; Fedn Am Socs Exp Biol; Undersea & Hyperbaric Med Soc. *Res:* Cardiovascular research in the area of diving, exercise, and effect of environmental factors. *Mailing Add:* Dept of Physiol Sch Med Univ Hawaii Honolulu HI 96822

LIN, YUE JEE, b Canton, China, Oct 8, 45; US citizen; m 72; c 2. CYTOGENETICS. *Educ:* Nat Taiwan Univ, BS, 67; Ohio State Univ, MS, 72, PhD(genetics), 76. *Prof Exp:* Res asst, Nat Taiwan Univ, 68-69, Taiwan Agr Res Inst, 69-70; teaching assoc genetics & biol, Ohio State Univ, 70-76; asst prof, 76-82, ASSOC PROF GENETICS & CYTOGENETICS, ST JOHN'S UNIV, 82- *Concurrent Pos:* Mem bd dirs, Coun Advan Psychol Professions & Sci, 87-92. *Mem:* Am Soc Cell Biol; Genetics Soc Am; Am Genetic Asn; Sigma Xi. *Res:* Cytogenetics of complex heterozygotes, Rhoeo spathacea; cytogenetics of polyploids; cytogenetic effects of mutagens and environmental chemicals. *Mailing Add:* Dept Biol St John's Univ Jamaica NY 11439

LINAM, JAY H, b Carey, Idaho, Mar 9, 31; m 65; c 2. ENTOMOLOGY. *Educ:* Univ Idaho, BS, 53; Univ Utah, MS, 57, PhD(entom, zool), 65. *Prof Exp:* Asst entomologist, Ecol Res Lab, Univ Utah, 58-59; mgr, Magna Mosquito Abatement Dist, Utah, 60-62; from instr to assoc prof, 65-75, chmn dept life sci, 87-88, PROF BIOL, UNIV SOUTHERN COLO, 75- *Mem:* Entom Soc Am; Am Mosquito Control Asn. *Res:* Taxonomy and biology of mosquitoes of Western United States. *Mailing Add:* Biol Dept Univ Southern Colo Pueblo CO 81001

LINASK, KERSTI KATRIN, b Zittau, Ger, Feb 10, 45; US citizen; m 67; c 2. CELL ADHESION, SIGNAL TRANSDUCTION. *Educ:* Russell Sage Col, BA, 67; Univ Calif, Los Angeles, MA, 68; Univ Pa, PhD(develop biol), 86. *Prof Exp:* Instr biol, Holy Family Col, 70-80; postdoctoral fel develop biol, Thomas Jefferson Univ, 86-89; ASST PROF PEDIAT, UNIV PA, 90- *Concurrent Pos:* Asst prof, Children's Hosp Philadelphia, 90-; mem, Basic Sci Coun, Am Heart Asn. *Mem:* Soc Develop Biol; Am Soc Cell Biol; Int Soc Develop Biol; Am Heart Asn; AAAS. *Res:* Mechanisms underlying cell adhesion systems, growth factor signalling and signal transduction during early avian and mammalian heart development. *Mailing Add:* Div Cardiol Children's Hosp Philadelphia PA 19104

LINAWEAVER, FRANK PIERCE, b Woodstock, Va, Aug 22, 34; m 68; c 2. ENVIRONMENTAL ENGINEERING, CIVIL ENGINEERING. *Educ:* Johns Hopkins Univ, BES, 55, PhD(water resources, sanit eng), 65; Am Acad Environ Engrs, dipl. *Prof Exp:* From jr civil engr to sr civil engr, Bur Water Supply, Baltimore, 55 & 58-61; res staff asst, dept sanit eng & water resources, Johns Hopkins Univ, 61-65, res assoc, dept environ eng sci, 65-66; res staff mem, water resources group, Resources for Future, Inc, DC, 66; White House fel, US Govt, 66-67; assoc prof environ sci, dept environ eng & geog, Johns Hopkins Univ, 67-68; dep dir, Dept Pub Works, City of Baltimore, 68-69, dir, 69-74; consult environ & civil eng, 74-78; partner, Rummel, Klepper & Kahl, Consult Engrs, 78-86; PRES, E A ENG INC, 86- *Concurrent Pos:* Mem, President's Air Qual Adv Bd, 68-71; consult rev panel, URS, Inc, 70-72; vis comt, Sch Archit, Univ Md, 75-78; dir, T Rowe Price Mutual Funds, 79-; trustee, Johns Hopkins Univ, 80-86, 87- *Mem:* Fel Am Soc Civil Engrs; Am Acad Environ Eng; fel AAAS; Nat Soc Prof Engrs; Am Water Works Asn; Am Consult Engs Coun. *Res:* Residential and commercial water use and their impact on urban water systems; urban water management and water resources; street cleaning relation to water pollution control; public works; sanitary environmental engineering; hydrology. *Mailing Add:* E A Corp Ctr 11019 McCormick Rd Hunt Valley MD 21031

LIN-CHUNG, PAY-JUNE, b Tienjin, China; Nat US; m 67; c 2. SEMICOUNDUCTOR PHYSICS, THEORETICAL SOLID STATE PHYSICS. *Educ:* Nat Taiwan Univ, China, BS, 58; Univ Penn, MS, 61, PhD(physics), 65. *Prof Exp:* Res physicist, Inst Metals, Univ Chicago, 64-65; res physicist, Univ Calif, Berkeley, 65-66; asst prof physics, State Univ Northridge Calif, 66-67; RES PHYSICIST, NAVAL RES LAB, 67- *Concurrent Pos:* Vis scientist, Argonne Nat Lab, 71, Cambridge Univ, England, 78-79; vis lectr, Univ Durham, England, 73-74. *Mem:* Am Physical Soc; Mat Res Soc. *Res:* Theoretical investigations of the electronic and lattice vibrational structure and of electro-optical effects in solids; defects, surfaces and superlattices on semiconductors. *Mailing Add:* Naval Res Lab Code 6877 Washington DC 20375-5000

LINCICOME, DAVID RICHARD, b Champaign, Ill, Jan 17, 14; m 41, 53; c 2. PARASITOLOGY, PHYSIOLOGY. *Educ:* Univ Ill, BS & MS, 37; Tulane Univ, PhD(parasitol), 41; Am Bd Med Microbiol, dipl, 65. *Prof Exp:* Asst zool, Univ Ill, 37; asst trop med, Sch Med, Tulane Univ, 34-41; from instr to asst prof zool, Univ Ky, 41-47; asst prof parasitol, Univ Wis, 47-49; sr res parasitologist, E I du Pont de Nemours & Co, 49-54; from asst prof to prof zool, Howard Univ, 55-70; RETIRED. *Concurrent Pos:* USPHS res grants, 58-68; ed, Exp Parasitol, 49- & chmn ed bd, 50-76; guest scientist, Naval Med Res Inst, 55-61; ed, Int Rev Trop Med, 60-; vis scientist, Lab Phys Biol, Nat Inst Arthritis & Metab Dis, 64-65; chmn comt exam & cert, Am Bd Med Microbiol, 72-; dir res, Am Dairy Goat Asn, mem bd dirs, 73-79; ed, Trans, Am Micros Soc, 70-71; dir, Am Dairy Goat Asn, 72-88 & Nat Pygmy Goat Asn; founder & ed, Int Goat & Sheep Res; guest scientist, USDA Exp Sta, Beltsville, Md, 78-; vis scholar, Nat Agr Libr, 90-; dist dir, Natural Colored Wool Growers Asn, 90-; dir & registrar, Jacob Sheep Conservancy. *Mem:* Fel AAAS; Am Soc Parasitol; Helminth Soc (secy, 60, vpres, 61, pres, 68); fel NY Acad Sci; Am Dairy Goat Asn; Nat Pygmy Goat Asn (pres, 79). *Res:* Diagnosis of protozoan and helminthic diseases; amebiasis; taxonomy and systematics of Acanthocephala, Nematoda and Cestoda; epidemiology of tropical diseases; molecular biology of parasitism; nutritional exchange between parasite and host. *Mailing Add:* Frogmoor Farm Rte 1 Box 352 Midland VA 22728

LINCK, ALBERT JOHN, b Portsmouth, Ohio, Aug 18, 26; m 57; c 2. PLANT PHYSIOLOGY. *Educ:* Ohio State Univ, BSc, 50, MSc, 51, PhD(plant physiol), 55. *Prof Exp:* From instr to prof plant physiol, Univ Minn, St Paul, 55-84, asst dir, Minn Agr Exp Sta, 66-71, dean, Col Agr, 71-73, assoc vpres acad admin, 73-84; PROVOST & ACAD VPRES, COLO STATE UNIV, 84- *Mem:* AAAS; Am Soc Plant Physiol; Bot Soc Am; Scand Soc Plant Physiol; Am Inst Biol Sci. *Res:* Translocation of inorganic and organic compounds; mechanism of action of growth regulators. *Mailing Add:* 1406 Springwood Dr Ft Collins CO 80525

LINCK, RICHARD WAYNE, b Los Angeles, Calif, Apr 9, 45; m 72; c 1. CELL BIOLOGY. *Educ:* Stanford Univ, BA, 67; Brandeis Univ, PhD(biol), 72. *Prof Exp:* Fel res microtubules, Med Res Coun, Eng, 71-73; instr, 74-75, ASST PROF, DEPT ANAT, HARVARD MED SCH, 75- *Concurrent Pos:* Fel, Europ Molecular Biol Org, 73; res grant, NIH, 74. *Mem:* Am Soc Cell Biol; Biophys Soc. *Res:* Relation of structure to motility; biochemistry and ultrastructure of microtubule proteins in cilia and flagella; reassembly of such proteins and enzymatic interaction with accessory components; optical diffraction of electron micrographs. *Mailing Add:* Dept Cell Biol & Neuroanat Univ Minn 321 Church St SE Minneapolis MN 55356

LINCK, ROBERT GEORGE, b St Louis, Mo, Nov 18, 38; m 62. INORGANIC CHEMISTRY. *Educ:* Case Western Reserve Univ, BS, 60; Univ Chicago, PhD(chem), 63. *Prof Exp:* Asst prof chem, Univ Calif, 66-72, assoc prof, 72-81; assoc prof, US Naval Acad, 86-88; from asst prof to assoc prof, 81-86, PROF CHEM, SMITH COL, 88- *Honors & Awards:* Catalyst Award, Chem Mfrs Asn. *Mem:* AAAS; Am Chem Soc. *Res:* Rates of inorganic reactions, especially electron-transfer reactions; electronic structure and photochemistry of complex ions. *Mailing Add:* Dept of Chem Smith Col Northampton MA 01063

LINCOLN, CHARLES ALBERT, b Rudyard, Mont, May 13, 39; m 63; c 3. THEORETICAL PHYSICS. *Educ:* Mont State Univ, BS, 62, MS, 64; Univ Va, DSc(eng physics), 69. *Prof Exp:* From instr to assoc prof physics, State Univ NY Col Fredonia, 64-79; RETIRED. *Concurrent Pos:* NDEA fel, Univ Va, 66-69; NSF assistantship, Theoret Inst Physics, Boulder, 68 & Inst Statist Mech & Theoret Thermodyn, Univ Tex, 70; Fulbright exchange prof, Newcastle upon Tyne Polytech, Newcastle/Tyne, Eng, 73-74; adj instr, Walla Walla Community Col, 81-; adj res assoc physics, Whitman Col, 84-; consult electro-acoust, noise control & archit acoust. *Mem:* Am Phys Soc; Am Asn Physics Teachers; Inst Elec & Electronic Eng; Am Sci Affil; Audio Eng Soc; Am Astron Soc; Astron Soc Pac; Int Soc Optical Eng. *Res:* Field theoretic methods in statistical mechanics and fluids; a generalized dynamical formalism of statistical mechanics; information theory and electroacoustics; astrophysics. *Mailing Add:* 2163 Granite Dr Walla Walla WA 99362

LINCOLN, DAVID ERWIN, b Detroit, Mich, Oct 8, 44; m 70. CHEMICAL ECOLOGY. *Educ:* Kalamazoo Col, BA, 71; Univ Calif, Santa Cruz, PhD(biol), 78. *Prof Exp:* Fel, Stanford Univ, 78-80; asst prof, 80-86, ASSOC PROF BIOL, UNIV SC, 87- *Mem:* Ecol Soc Am; Bot Soc Am; AAAS; Entom Soc Am; Int Soc Chem Ecol; Phytochem Soc NAm. *Res:* Environmental and genetic control of secondary chemical production by plants and the roles of these chemicals in plant-herbivore coevolution; effects of rising atmospheric carbon dioxide on plant-herbivore interactions. *Mailing Add:* Dept Biol Univ SC Columbia SC 29208

LINCOLN, JEANNETTE VIRGINIA, b Ames, Iowa, Sept 7, 15. GEOPHYSICS, SOLAR PHYSICS. *Educ:* Wellesley Col, BA, 36; Iowa State Univ, MS, 38. *Prof Exp:* Asst household equip, Iowa State Univ, 37-38, instr, 38-42; physicist, Nat Bur Standards, DC, 42-54, sect chief, Radio Warning Serv, Colo, 59-65, dep chief data serv, Inst Telecommun Sci & Aeronomy, Environ Sci Serv Admin, 65-66, dep chief data serv & chief, Upper Atmosphere Geophys, 66-70, chief data serv & dir, World Data Ctr A, Solar-Terrestrial Physics, Nat Geophys & Solar Terrestrial Data Ctr, Environ Data & Info Serv, Nat Oceanic & Atmospheric Admin, 70-81; RETIRED. *Concurrent Pos:* Mem US preparatory comt study group ionospheric propagation, Int Radio Consult Comt, 59-80; secy, Int Ursigram & World Days Serv, 61-81, mem, US Comn G, Int Sci Radio Union, 63-, secy, Ionospheric Network Adv Group, 69-72, vchmn, 72-81; forecasting reporter, Int Asn Geomagnetism & Aeronomy, 63-67, mem comns IV & V, 67-73; mem working groups 3 & 5, Inter-Union Comn on Solar-Terrestrial Physics, 69-72; Am Geophys Union mem, Am Geophys Union-Int Sci Radio Union Bd of Radio Sci, 69-74, US Nat Comt, Int Union Geodesy & Geophys, 76-79; chmn working group V6, Geophys Indices, 73-81; mem comn 40, Int Astron Union, 76- *Honors & Awards:* Gold Medal, Dept of Com, 73. *Mem:* Fel AAAS; Sigma Xi; fel Am Geophys Union; Am Astron Soc; fel Soc Women Engrs. *Res:* Radio propagation disturbances and forecasts; solar-terrestrial relationships; publication of solar and geophysical data; prediction of solar indices; data center management. *Mailing Add:* 2005 Alpine Dr Boulder CO 80304-3607

LINCOLN, KENNETH ARNOLD, b Oakland, Calif, Oct 1, 22; m 56; c 4. HIGH TEMPERATURE CHEMISTRY, MASS SPECTROMETRY. *Educ:* Stanford Univ, AB, 44, MS, 48, PhD(phys chem), 57. *Prof Exp:* Phys chemist, US Naval Radio Defense Lab, 58-69; RES SCIENTIST, NASA-AMES RES CTR, 70- *Mem:* Am Chem Soc; Am Soc Mass Spectrometry; Am Sci Affil; Combustion Inst; Am Vacuum Soc. *Res:* Thermochemistry of the vaporization of refractory materials; thermokinetics of pulsed energy deposition; development of instrumentation combining lasers and high-speed mass spectrometry for in-situ analyses of short-lived chemical species; space flight spectrometric instrumentation. *Mailing Add:* 2016 Stockbridge Ave Redwood City CA 94061

LINCOLN, LEWIS LAUREN, b Canandaigua, NY, Oct 9, 26; m 49; c 6. PHOTOGRAPHIC CHEMISTRY. *Prof Exp:* Lab technician, 46-60, res chemist, 60-65, sr res chemist, 65-70, RES ASSOC CHEM, RES LABS, EASTMAN KODAK, 70- *Mem:* Am Chem Soc. *Res:* The study and synthesis of photographic sensitizing dyes and addenda. *Mailing Add:* 426 Mount Airy Dr Rochester NY 14617-2164

LINCOLN, RICHARD CRIDDLE, b Boston, Mass, Nov 25, 42; c 2. APPLIED PHYSICS. *Educ:* Cornell Univ, BEP, 66, MS, 68, PhD(mat sci), 71. *Prof Exp:* Instr & res assoc mat sci, Cornell Univ, 70-71; TECH STAFF MEM APPL PHYSICS, SANDIA LABS, 71-, SUPVR, APPL TECHNOL DIV. *Res:* High pressure and high temperature experimental techniques; analysis of nuclear waste management systems. *Mailing Add:* Sandia Labs Div 6341 Albuquerque NM 87185

LINCOLN, RICHARD G, b Portland, Ore, Nov 1, 23; m 46; c 2. PLANT PHYSIOLOGY. *Educ:* Ore State Univ, BS, 49; Univ Calif, Los Angeles, PhD, 55. *Prof Exp:* Plant physiologist, Sugarcane Field Sta, USDA, La, 55-56; PROF BOT, CALIF STATE UNIV, LONG BEACH, 56- *Mem:* AAAS; Am Soc Plant Physiol. *Res:* Plant photoperiodism; inhibition of flowering and extraction of the flowering stimulus; rhythmic phenomena in fungi. *Mailing Add:* Div of Natural Sci Long Beach St Coll Long Beach CA 90840

LIND, ALEXANDER R, cardiovascular physiology; deceased, see previous edition for last biography

LIND, ARTHUR CHARLES, b Chicago, Ill, May 28, 32; m 57; c 3. NUCLEAR MAGNETIC RESONANCE, MATERIALS CHARACTERIZATION. *Educ:* Univ Ill, Urbana, BS, 55; Rensselaer Polytech Inst, PhD(physics), 66. *Prof Exp:* Physicist, Knolls Atomic Power Lab, 58-61 & Watervliet Arsenal, US Army, 63-66; assoc scientist, McDonnell Douglas Res Labs, 66-76, scientist, 76-79, sr scientist, 79-83, prin scientist, 83-88, chief scientist, 88-89, DIR RES, McDONNELL DOUGLAS RES LABS, 89- *Mem:* Am Phys Soc; Am Chem Soc; Int Soc Magnetic Resonance; Soc Advancement Mat & Process; Am Inst Aeronaut & Astronaut; Mat Res Soc. *Res:* Electromagnetic processing of composite materials; nuclear magnetic resonance studies of polymers; theoretical and experimental studies of electromagnetic scattering; propagation of electromagnetic waves in turbulent media; measurement of dielectric properties at high temperatures. *Mailing Add:* 15450 Country Mill Ct Chesterfield MO 63017

LIND, CAROL JOHNSON, b Minneapolis, Minn, Dec 8, 26; div; c 2. MINERAL IDENTIFICATION, CHEMICAL ANALYSIS. *Educ:* Univ Minn, BS, 49. *Prof Exp:* Anal chemist coal anal, Twin City Testing Co, 49-50; jr chemist org res, Julius Hyman Co, 50-52; phys sci tech pollen identification, 63-66, phys sci tech chem res, 66-67, RES CHEMIST, US GEOL SURV, 67 - *Mem:* Am Chem Soc; Am Geophys Union; Am Sci Affil. *Res:* Natural-water chemistry of aluminum, aluminum-silicon, trace metals and manganese and their corresponding natural-organic influences. *Mailing Add:* Water Resources Div Mail Stop 427 US Geol Surv 345 Middlefield Rd Menlo Park CA 94025

LIND, CHARLES DOUGLAS, physical chemistry; deceased, see previous edition for last biography

LIND, DAVID ARTHUR, b Seattle, Wash, Sept 12, 18; m 45; c 4. NUCLEAR PHYSICS. *Educ:* Univ Wash, Seattle, BS, 40; Calif Inst Technol, MS, 43, PhD(physics), 48. *Prof Exp:* Jr aerodynamicist, Boeing Airplane Co, Wash, 42-43; physicist, Appl Physics Lab, Univ Wash, Seattle, 43-45; res fel physics, Calif Inst Technol, 48-50; Guggenheim fel, Nobel Inst Physics, Stockholm, 50-51; asst prof, Univ Wis, 51-56; from assoc prof to prof, 56-83, chmn dept physics & astrophys, 74-78, EMER PROF PHYSICS, UNIV COLO, BOULDER, 83- *Concurrent Pos:* Consult off instnl prog, NSF, 63-66; physicist div res, US AEC, 69-70; mem prog adv comn, Los Alamos Meson Facil, 71-74, chmn users group, 75-76; consult, Los Alamos Nat Lab, 83- *Mem:* Fel Am Phys Soc; Sigma Xi. *Res:* X-rays; crystal diffraction; nuclear spectroscopy; sector focused cyclotron design; charged particle scattering; reaction studies; fast neutron spectroscopy; particle beam optics; physics of snow and avalanche phenomena. *Mailing Add:* 920 Jasmine Circle Boulder CO 80304

LIND, DOUGLAS A, b Arlington, Va, Aug 11, 46. MATHEMATICS. *Educ:* Univ Va, BS, 68; Stanford Univ, MA, 71, PhD(math), 73. *Prof Exp:* PROF MATH, UNIV WASH, 76- *Mem:* Am Math Soc. *Res:* Researches the interplay between symbol dynamics; smooth dynamcal systems and data storage and transmission. *Mailing Add:* Univ Wash GN-50 Seattle WA 98195

LIND, MAURICE DAVID, b Jamestown, NY, July 25, 34; m 62; c 1. PHYSICAL CHEMISTRY, X-RAY CRYSTALLOGRAPHY. *Educ:* Otterbein Col, BS, 57; Cornell Univ, PhD(phys chem), 62. *Prof Exp:* NSF fel, 62-63; res chemist phys chem, Union Oil Co, Calif, 63-66; MEM TECH STAFF, SCI CTR, ROCKWELL INT, 66- *Concurrent Pos:* Vis prof appl physics, Tech Univ Denmark, 85. *Mem:* Am Phys Soc; Am Crystallog Asn; Sigma Xi; Am Asn Crystal Growth. *Res:* X-ray crystal mography; crystal growth. *Mailing Add:* 1690 Stoddard Ave Thousand Oaks CA 91360

LIND, NIELS CHRISTIAN, b Copenhagen, Denmark, Mar 10, 30; Can citizen; m 84; c 4. APPLIED MECHANICS. *Educ:* Royal Tech Univ Denmark, MSc, 53; Univ Ill, PhD(theoret & appl mech), 59. *Prof Exp:* Design engr, Dominia Ltd, Denmark, 53-54; engr, Bell Tel Co, Can, 54-55; field engr, Drake & Merritt Co, Labrador, 55-56; design engr, Fenco, Que, 56; asst stress anal, Univ Ill, 56-57, instr, 57-58, res assoc, 58-59, asst prof theoret & appl mech, 59-60; assoc prof, 60-62, dir, Inst Risk Res, 82-87, PROF CIVIL ENG, UNIV WATERLOO, 62- *Concurrent Pos:* Mem, Can Nat Study Group Math Higher Educ, Orgn for Econ Coop & Develop, 63-65; adv comt nuclear safety, Atomic Energy Control Bd, 81-; vis prof, Univ Laval, 69, Inst Eng, Nat Univ Mex, 75 & 81 & Tech Univ Denmark, 77-78. *Mem:* Fel Am Acad Mech (pres, 71-72); fel Royal Soc Can; fel Can Eng Acad. *Res:* Structural mechanics; theory of design; structural reliability and optimization. *Mailing Add:* Dept Civil Eng Univ Waterloo Waterloo ON N2L 3G1 Can

LIND, OWEN THOMAS, b Emporia, Kans, June 2, 34; m 54, 90; c 2. LIMNOLOGY, WATER RESOURCES. *Educ:* William Jewell Col, AB, 56; Univ Mich, MS, 60; Univ Mo, PhD(zool), 66. *Prof Exp:* Biologist, Parke, Davis & Co, Mich, 56-60; asst prof biol, William Jewell Col, 60-62; res assoc limnol, Univ Mo, 60; from asst prof to assoc prof, 66-79, PROF BIOL, BAYLOR UNIV, 79- *Concurrent Pos:* Prin investr, Off Water Resource & Technol, 71, 73, 76, NSF, 83, 88; consult, US Nat Park Serv, 69, 74-76, Corps Engrs, 79, USAID, 81-83, State of Tex, 82; dir, Inst Environ Studies, 71-76; mem exec comt, Tyler Ecol Award, 74- *Mem:* Am Soc Limnol & Oceanog; Sigma Xi; Int Asn Theoret & Appl Limnol; Brit Freshwater Biol Asn; NAm Lake Mgt Soc. *Res:* Factors governing production of lakes and reservoirs; tropical limnology and water resources of third world countries. *Mailing Add:* Dept Biol Baylor Univ Waco TX 76798-7388

LIND, ROBERT WAYNE, b Ishpeming, Mich, Aug 25, 39; m 64; c 2. THEORETICAL PHYSICS. *Educ:* Mich Technol Univ, BS, 61; Univ Pittsburgh, PhD(physics), 70. *Prof Exp:* Engr, Ford Motor Co, 63-66; res assoc physics, Syracuse Univ, 70-72; sr res assoc, Temple Univ, 72-73; res assoc, Fla State Univ, 73-74; asst prof physics, WVa Inst Technol, 74-78, chmn dept, 76-82; assoc prof, 78-83, PROF PHYSICS & ELECT ENG, UNIV WIS-PLATTEVILLE, 83- *Concurrent Pos:* Res physicist, Naval Res Lab, 87-88. *Mem:* Am Asn Physics Teachers; Int Soc Gen Relativity & Gravitation; AAAS. *Res:* General relativity and electromagnetism; analysis of high frequency radio wave probing of the ionosphere. *Mailing Add:* Dept of Physics Univ Wis-Platteville One University Plaza Platteville WI 53818-3099

LIND, VANCE GORDON, b Brigham City, Utah, Feb 12, 35; m 64; c 11. PHYSICS, ASTROPHYSICS. *Educ:* Utah State Univ, BS, 59; Univ Wis, MS, 61, PhD(elem particles), 64. *Prof Exp:* Eng asst, Edgerton, Germeshausen & Grier, Inc, summer, 59, res asst, 60; res assoc, Univ Mich, 64; from asst prof to assoc prof, 64-75, dept head, 81-88, PROF PHYSICS, UTAH STATE UNIV, 75- *Concurrent Pos:* Woodrow Wilson fel, 59-64; Utah State Univ Res Found grant, 64-66, 72-76 & 78-79; investr, NSF res grant, Utah State Univ, 66-71, 78-80, 80-85; Dept Energy res grant, Utah State Univ, 86- *Mem:* Am Phys Soc; Sigma Xi; Am Solar Energy Soc; Int Solar Energy Soc. *Res:* Basic interactions; elementary particle interactions, meson and nucleon interactions with nuclei, nuclear mass measurements, astronomy and astrophysics; solar energy technology. *Mailing Add:* Dept Physics Utah State Univ Logan UT 84322-4415

LIND, WILTON H(OWARD), b Oakland, Calif, Feb 14, 27; m 51; c 2. CHEMICAL ENGINEERING. *Educ:* Univ Calif, Berkeley, BS, 50, MS, 52; JD, Empire, Col, 77. *Prof Exp:* From asst res engr to sr res engr, Chevron Res Co, 51-78, analyst, 78-80, asst secy, Finance Dept, 80-85, mgr, Legal Process & Secy Dept, Chevron Corp, 85-86; RETIRED. *Mem:* Am Chem Soc; Am Inst Chem Engrs. *Res:* Petrochemical research and development; aromatics chemistry; pilot plant design and operation. *Mailing Add:* PO Box 30004 Walnut Creek CA 94598

LINDAHL, CHARLES BLIGHE, b N Platte, Nebr, Feb 4, 39; m 59, 91; c 2. SYNTHETIC INORGANIC CHEMISTRY, INORGANIC CHEMISTRY. *Educ:* Iowa State Univ, BS, 60; Univ Calif, Berkeley, PhD(chem), 64. *Prof Exp:* Chemist, Ames Lab, 60; res asst, Lawrence Berkeley Lab, 61-64; sr chemist, Eastman Kodak, 64-65; mem tech staff, Rocketdyne Div, Rockwell INt, 65-70; sr chemist, Reheis Chem, 70-71; head new prod develop, 71-72, tech dir, 73-90, GEN MGR & TECH DIR, OZARK-MAHONING CO, 90- *Mem:* Am Chem Soc; Am Asn Dent Res; Int Asn Dent Res; Sigma Xi. *Res:* Inorganic synthesis; fluorides for dental applications; oxidizer chemistry; unusual oxidation states. *Mailing Add:* Ozark-Mahoning Atochem NAm 5101 W 21st St Tulsa OK 74107-2230

LINDAHL, LASSE ALLAN, b Copenhagen, Denmark, Sept 9, 44; m 78; c 2. MOLECULAR GENETICS, RNA. *Educ:* Univ Copenhagen, MSc, 69, PhD(microbiol), 73. *Prof Exp:* Fel molecular biol, Univ Wis-Madison, 73-76; asst prof, Univ Aarhus, Denmark, 76-78; from asst prof to assoc prof, 78-89, PROF BIOL, UNIV ROCHESTER, 89- *Mem:* Am Soc Microbiol; AAAS. *Res:* Molecular basis for the regulation of ribosome synthesis; in vitro DNA recombination site directed mutagenesis, regulatory mutations and biochemical measurements of the synthesis and decay of specific RNA and protein molecules; in vitro RNA synthesis; RNA folding. *Mailing Add:* Dept Biol Univ Rochester Rochester NY 14627

LINDAHL, RONALD, b Detroit, Mich, Aug 11, 48; m 70; c 2. MOLECULAR BIOLOGY, GENERAL BIOLOGY. *Educ:* Wayne State Univ, BA, 70, PhD(biol), 73. *Prof Exp:* Postdoctoral, Argonne Nat Lab, 74-75; prof biol, Univ Ala, 75-89; PROF & CHAIR, BIOCHEM & MOLECULAR BIOL, SCH MED, UNIV SDAK, 89- *Concurrent Pos:* Prin investr, Nat Cancer Inst grant, 79-; univ res prof, Univ Ala, 84-89; reviewer, NSF, 87-90 & NIH, 89-90; co-prin investr, Nat Inst Alcohol Abuse & Alcoholism, 87- *Mem:* Am Soc Biochem & Molecular Biol; Am Asn Cancer Res; AAAS; Int Soc Biomed Res Alcoholism; Sigma Xi. *Res:* Molecular biology of gene expression; aldehyde dehydrogenases; genetic changes in carcinogenesis. *Mailing Add:* Biochem & Molecular Biol Dept Univ SDak Sch Med 414 E Clark St 145 Lee Med Bldg Vermillion SD 57069

LINDAHL, RONALD GUNNAR, b Detroit, Mich, Aug 11, 48; m 70; c 2. BIOCHEMISTRY. *Educ:* Wayne State Univ, BA, 70, PhD(biol), 73. *Prof Exp:* Fel biochem, Argonne Nat Lab, 74-75; from asst prof to assoc prof biol, Univ Ala, 75-84, prof biol & biochem, 84-89; PROF & CHMN, DEPT BIOCHEM & MOLECULAR BIOL, SCH MED, UNIV SDAK, 89- *Concurrent Pos:* Univ Res fel, 84. *Mem:* Am Asn Cancer Res; AAAS; Genetics Soc Am; Asn Southeastern Biologists; Int Soc Biomed Res Alcoholism. *Res:* Biochemical changes during neoplasia; developmental biochemistry; transplacental and perinatal carcinogenesis; genetic regulation of enzyme activity. *Mailing Add:* Dept Biochem & Molecular Biol Univ SDak Sch Med Vermillion SD 57069

LINDAHL, ROY LAWRENCE, b Los Angeles, Calif, Aug 22, 25; m 48, 76; c 7. DENTISTRY. *Educ:* Univ Southern Calif, BS & DDS, 50; Univ Mich, MS, 52; Am Bd Pedodontics, dipl, 56. *Prof Exp:* Asst prof, 52-56, prof, 56-85, dir, continuing educ & dent demonstr pract, 70-83, EMER PROF PEDIAT DENT, SCH DENT, UNIV NC, CHAPEL HILL, 85-; PRES, DELTA DENT PLAN OF NC, 85- *Concurrent Pos:* Mem bd trustees, NC Cerebral Palsy Hosp, 57-63; consult, Womack Army Hosp, Ft Bragg, NC, 60-78; examr, Am Bd Pedodontics, 60-67, chmn, 67. *Mem:* AAAS; Am Soc Dent for Children (from secy to pres, 69-73); Am Dent Asn; Am Acad Pedodontics (vpres, 62-63, pres-elect, 63-64, pres, 64-65); Int Asn Dent Res. *Res:* Pedodontics; effective utilization of dental auxiliary personnel; problems of the handicapped patient; pre-payment dental care programs; health services research-quality assurance. *Mailing Add:* 305 Clayton Rd Chapel Hill NC 27514

LINDAL, GUNNAR F, b Oslo, Norway, Mar 24, 36; US citizen; m 62; c 1. ELECTRICAL ENGINEERING. *Educ:* Stanford Univ, PhD(elec eng), 64. *Prof Exp:* Res assoc, Stanford Electronics Lab, 64-69; SR RES SCIENTIST, JET PROPULSION LAB, 69- *Mem:* Am Astron Soc; Am Geophys Union; Union Radio Sci Int. *Res:* Applied mathematics and computer science; radio propagation and communication; planetary atmospheres and surfaces; participated in the radio occultation measurements of the atmospheres of Mercury, Venus, Mars, Jupiter, Saturn, Uranus, Neptune and Tritan during the Mariner, Pioneer, Viking and Voyager spaceflight missions to these planets and satellites. *Mailing Add:* PO Box 4092 Point Dume CA 90265

LINDAMOOD, JOHN BENFORD, b Galax, Va, Aug 6, 29; m 53; c 1. DAIRY TECHNOLOGY. *Educ:* Va Polytech Inst & State Univ, BS, 53, MS, 55; Ohio State Univ, PhD(educ), 74. *Prof Exp:* Prod mgr, Evaporated Milk Div, Carnation Co, 56-61; asst prof, 74-78, ASSOC PROF FOOD SCI, DEPT FOOD SCI & NUTRIT, OHIO STATE UNIV, 78- *Mem:* Inst Food Technologists; Am Dairy Sci Asn; Int Asn Milk, Food & Environ Sanitarians. *Res:* Milk and milk products. *Mailing Add:* Dept Food Sci & Nutrit Ohio State Univ Main Campus Columbus OH 43210

LINDAU, EVERT INGOLF, b Vaxjo, Sweden, Oct 4, 42. SOLID STATE PHYSICS. *Educ:* Chalmers Univ Technol, Sweden, Civilingenjor, 68, Technol Licentiat, 70, PhD(physics), 71, DrTechnol, 72. *Prof Exp:* Res asst physics, Chalmers Univ Technol, Sweden, 68-71; res scientist, Varian Assocs, 71-72; res assoc physics, 72-74, PROF PHYSICS, STANFORD UNIV, 74- *Mem:* Am Phys Soc; Am Vacuum Soc; Swed Soc Technol. *Res:* Optical and photoemission studies of the electronic structure of materials using synchrotron radiation with emphasis on surface properties; surface states, surface photoemission, physisorbtion, chemisorbtion, surface composition and catalytic activities. *Mailing Add:* Stanford Electronics Lab Stanford Univ Stanford CA 94305

LINDAUER, GEORGE CONRAD, b Queens, NY, Nov 5, 35; m 59; c 3. NUCLEAR ENGINEERING, INFORMATION SCIENCE. *Educ:* Cooper Union, BS, 56; Mass Inst Technol, ScM, 57; Univ Pittsburgh, PhD(mech eng), 62; Long Island Univ, MS, 71. *Prof Exp:* From jr engr to sr engr, Bettis Atomic Power Lab, Westinghouse Elec Corp, 57-64; from asst chem engr to chem engr, Brookhaven Nat Lab, 64-71; PROF NUCLEAR SCI & LIBRN, SPEED SCI SCH, UNIV LOUISVILLE, 71-, PROF MECH ENG, 79- *Mem:* Am Soc Mech Engrs. *Res:* heat transfer; fluid dynamics. *Mailing Add:* Speed Sci Sch Univ Louisville Louisville KY 40292

LINDAUER, IVO EUGENE, b Grand Valley, Colo, Apr 7, 31; m 57; c 2. PLANT ECOLOGY. *Educ:* Colo State Univ, BS, 53, PhD(bot), 70; Univ Northern Colo, MA, 60. *Prof Exp:* Instr biol, Univ Northern Colo, 60-64, asst prof sci, 64-65; res assoc & teaching asst bot, Colo State Univ, 65-67; from asst prof to assoc prof, 67-75, asst dean, Col Arts & Sci, 76-81, PROF BOT, UNIV NORTHERN COLO, 75- *Concurrent Pos:* Tri-Univ Proj grant, NY Univ, 70; US Bur Reclamation grant, proposed Narrows Dam site, 70-72 & 72-75; mem, vpres & pres bd trustees, Colo Nature Conserv; Northwest Colo Wildlife Consortium grant, 81-83; Colo Div Wildlife grant, 84-85. *Mem:* Ecol Soc Am; Nat Asn Biol Teacher (secy-treas, 69-70); Am Inst Biol Sci. *Res:* Analysis of vegetational communities found along flood plains; ecological studies of river bottom ecosystems; ecosystem modeling and assessment of remote sensing vegetation data bases. *Mailing Add:* 1832 23rd Ave Lane Greely CO 80631

LINDAUER, MAURICE WILLIAM, b Millstadt, Ill, Sept 25, 24; m 46; c 3. ANALYTICAL CHEMISTRY, PHYSICAL CHEMISTRY. *Educ:* Wash Univ, AB, 49, AM, 53; Harvard Univ, MEd, 62; Fla State Univ, PhD, 70. *Prof Exp:* Res chemist, Mallinckrodt Chem Works, 52-55 & Am Zinc, Ill, 55-56; res chemist, Nitrogen Div, Allied Chem & Dye Corp, 56-57; assoc prof anal & phys chem, Valdosta State Col, 57-71, prof chem, 71-84, head dept, 81-84; RETIRED. *Concurrent Pos:* NSF sci fac fel, 64-65. *Mem:* Am Chem Soc; Sigma Xi. *Res:* Mossbauer spectroscopy; history of chemistry; thermodynamics and chemical equilibrium. *Mailing Add:* 1401 Miramar St Valdosta GA 31601-3616

LINDBECK, WENDELL ARTHUR, b Rockford, Ill, Sept 28, 12; m 38; c 3. ORGANIC CHEMISTRY. *Educ:* Beloit Col, BS, 36; Univ Wis, PhM, 37, PhD(chem), 40. *Prof Exp:* Teacher, Tenn Jr Col, 40-44; tech coord, Goodyear Synthetic Rubber Corp, Ohio, 44-47; assoc prof & chmn, Natural Sci Div, Univ Ill, 47-49; from assoc prof to prof phys sci & chem, 49-78, EMER PROF CHEM, NORTHERN ILL UNIV, 78- *Concurrent Pos:* NSF sci fac fel, Univ Calif, Berkeley, 60-61; US AEC grant, Argonne Nat Lab, 69-70. *Mem:* Am Chem Soc; Sigma Xi. *Res:* Synthesis of organic compounds. *Mailing Add:* 204 Windsor Dr Dekalb IL 60115

LINDBERG, CRAIG ROBERT, b Edmonton, Alta. STATISTICS & SIGNAL PROCESSING, WAVE PROPAGATION. *Educ:* Univ Alta, BSc, 79; Univ Calif, San Diego, PhD(earth sci), 86. *Prof Exp:* Researcher geophys, Esso Resources Ltd, 79; res asst physics, Mass Inst Technol, 79-80; res asst geophys, Univ Calif, San Diego, 80-86, postgrad geophysicist, 86-88; postdoctoral mem tech staff, 88-90, RESIDENT VISITOR, AT&T BELL LABS, 91- *Concurrent Pos:* Assoc prin investr, Univ Calif, San Diego, 84-88; consult, AT&T Bell Labs, 87; assoc ed, Am Geophys Union, 90- *Mem:* Soc Indust & Appl Math; Am Statist Asn; Inst Elec & Electronic Engrs; Am Geophys Union. *Res:* Statistical signal processing and application to physical problems; robust regression methods; climate change; seismic and speech data analysis; group theory solution of pdes. *Mailing Add:* AT&T Bell Labs Rm 2C-541 600 Mountain Ave Murray Hill NJ 07974-2070

LINDBERG, DAVID ROBERT, b Elgin, Ill, Feb 7, 48; m 68; c 1. PALEOBIOLOGY, EVOLUTIONARY ECOLOGY. *Educ:* San Francisco State Univ, BA, 77; Univ Calif, Santa Cruz, PhD(biol), 83. *Prof Exp:* Dir res, Farallon Res Group, Oceanic Soc, 73-77; res biologist, dept invert zool, Calif Acad Sci, 75-77; sr mus sci, 82-84, asst res paleontologist, 84-86, ASSOC RES PALEONTOLOGIST, CURATION MUS PALEONT, UNIV CALIF, BERKELEY, 84-, ADJ ASSOC PROF, DEPT INTEGRATIVE BIOL, 89- *Concurrent Pos:* Res assoc, dept invert zool, Calif Acad Sci, San Francisco, 77-; co-investr, res prog, US Fish & Wildlife Serv, Univ Calif, Santa Cruz, 79-91; res assoc, dept invert paleont, Nat Hist Mus Los Angeles, Calif, 86-; fel, Calif Acad Sci, 85, Willi Hennig Soc, Stockholm, 88. *Honors & Awards:* Award for Excellence, Pac div, AAAS, 80. *Mem:* Am Soc Zoologists; Am Malacological Union; Paleont Res Inst; Paleont Soc; Soc of Systematic Zoologists. *Res:* The evolution and biology of patellacean limpets; the evolution of brooding and hermaphroditism in marine molluscs; rocky intertidal community structure; molluscan evolution; unites malacologica. *Mailing Add:* Mus Paleont Univ Calif Berkeley CA 94720

LINDBERG, DAVID SEAMAN, SR, academic administration, health sciences; deceased, see previous edition for last biography

LINDBERG, DONALD ALLAN BROR, b Brooklyn, NY, Sept 21, 33; m 57; c 3. PATHOLOGY, COMPUTER SCIENCE. *Educ:* Amherst Col, AB, 54; Columbia Univ, MD, 58; Am Bd Path, dipl, 63. *Hon Degrees:* ScD, Amherst Col, 79 & State Univ NY, Syracuse, 87; LLD, Univ Mo, 90. *Prof Exp:* From instr to prof path, Sch Med, 62-84; DIR NAT LIBR MED, NIH, 84- *Concurrent Pos:* Markle scholar, 64-69; mem, Comput Res & Biomath Study Sect, NIH, 67-71 & Comput Sci & Eng Bd, Nat Acad Sci, 71-73; chmn, CBX Adv Comt, Nat Bd Med Examrs, 71-74, mem, Joint CBX Comt, Nat Bd Med Examrs & Am Bd Internal Med, 74-81; US rep, Comt Comput Med, Int Fedn Info Processing; mem, Biomed Review Libr Comt, Nat Libr Med, 79-80; consult & mem, Peer Review Group, TRIMIS, Dept Defense, 77-84; prof & chmn, dept info sci, Sch Libr & Info Sci, 69-71, dir info sci group, Sch Med, 71-84, dir, Health Serv Res Ctr & Health Care Technol Ctr, Univ Mo, Columbia, 76-80; adj prof path, Sch Med, Univ Md, 84-; mem coun, Inst Med-Nat Acad Sci, 90- *Honors & Awards:* Silver Core Award, Int Fedn Info Processing, 86; Surgeon General's Medallion, USPHS, 89; Nathan Davis Award, AMA, 89. *Mem:* Inst Med-Nat Acad Sci; Am Med Informatics Asn; Col Am Pathologists; Am Asn Artificial Intel; fel Am Col Med Informatics; Sigma Xi; AAAS. *Res:* Information processing; computers in medicine; infectious diseases; author of four books and over 100 technical articles. *Mailing Add:* Nat Libr Med MLN Bldg 38 Rm 2E 17B 8600 Rockville Pike Bethesda MD 20894

LINDBERG, EDWARD E, b Boston, Mass, Aug 16, 38; m 58; c 2. MECHANICAL ENGINEERING. *Educ:* Worcester Polytech Inst, BSME, 60, MSME, 63. *Prof Exp:* Test engr, Scintilla Div, Bendix Corp, 60-61; engr, Alden Res Labs, Worcester Polytech Inst, 61-63; asst prof mech eng, 63-67, ASSOC PROF MECH ENG, WESTERN NEW ENG COL, 67-, DIR COMPUT CTR, 68- *Mem:* Am Soc Mech Engrs; Am Soc Eng Educ; Instrument Soc Am; Sigma Xi. *Res:* Automatic controls; fluid mechanics; thermodynamics; computer sciences. *Mailing Add:* 76 Craiwell Ave West Springfield MA 01089

LINDBERG, GEORGE DONALD, b Salt Lake City, Utah, Feb 9, 25; m 55; c 3. PLANT PATHOLOGY. *Educ:* Ariz State Univ, BS, 50; Okla State Univ, MS, 52; Univ Wis, PhD(plant path), 55. *Prof Exp:* Asst prof, 55-59, assoc prof, 59-70, PROF PLANT PATH, LA STATE UNIV, BATON ROUGE, 70- *Mem:* Am Phytopath Soc. *Res:* Plant virology; diseases of forage crops; abnormalities in the fungi. *Mailing Add:* Dept Plant Path La State Univ Baton Rouge LA 70803-1720

LINDBERG, JAMES GEORGE, b Grand Rapids, Mich, Sept 19, 40; c 3. ORGANIC CHEMISTRY. *Educ:* Kalamazoo Col, BA, 62; Baylor Univ, PhD, 69. *Prof Exp:* From asst prof to assoc prof, 67-78, PROF CHEM, DRAKE UNIV, 78- *Concurrent Pos:* Vis scholar, Stanford Univ, 83-84. *Mem:* Am Chem Soc; Royal Soc Chem. *Res:* Nuclear magnetic resonance spectroscopic studies of steric effects; conformational analysis of cyclohexanones. *Mailing Add:* Dept of Chem Drake Univ Des Moines IA 50311

LINDBERG, JOHN ALBERT, JR, b New York, NY, Apr 19, 34; m 64; c 2. MATHEMATICAL ANALYSIS. *Educ:* Wagner Col, BA, 54; Univ Minn, MA, 57, PhD(math), 60. *Prof Exp:* Instr math, Univ Minn, 58-59 & Yale Univ, 60-62; from asst to assoc prof, 62-72, PROF MATH, SYRACUSE UNIV, 72- *Concurrent Pos:* Res fel math, Yale Univ, 68-69. *Mem:* AAAS; Am Math Soc; Math Asn Am; Sigma Xi. *Res:* Theory of algebraic extensions of Banach algebras and factorization of polynomials over such algebras; inverse producing normed extensions. *Mailing Add:* Dept of Math Room 200 Carnegie Syracuse Univ Syracuse NY 13244

LINDBERG, LOIS HELEN, b Scott Air Force Base, Ill, Sept 1, 32. MEDICAL MICROBIOLOGY. *Educ:* San Jose State Col, AB, 52; Univ Calif, MPH, 58; Stanford Univ, PhD, 67. *Prof Exp:* Jr microbiologist, State Dept Pub Health, Calif, 53-54; instr bact, San Jose State Col, 54-55; assoc pub health, Pub Health Lab, Univ Calif, 55-58; asst prof bact, 58-65, assoc prof biol, 65-70, prof biol, 70-80, assoc dean fac, 78-79, PROF MICROBIOLOGY, SAN JOSE STATE UNIV, 80- *Concurrent Pos:* NSF sci teachers fel, 62, fel, 66 & 67; res assoc, Stanford Univ Med Sch. *Mem:* AAAS; Am Pub Health Asn; Am Soc Microbiol. *Res:* Medical microbiology as related with the pathology and immunology of streptococcal infections. *Mailing Add:* 211 Santa Margarita Apt Menlo Park CA 94025

LINDBERG, R(OBERT) G(ENE), b Los Angeles, Calif, June 9, 24; m 47; c 4. ZOOLOGY. *Educ:* Univ Calif, Los Angeles, BA, 48, PhD(zool), 52. *Prof Exp:* Asst zool, Univ Calif, Los Angeles, 50, jr res biologist radiation ecol, Atomic Energy Proj, 52-55, asst res biologist, 55-59; head bioastronaut lab, Northrop Corp, 59-67, mem sr tech staff, Labs, 67-74; LECTR ENVIRON SCI ENG, UNIV CALIF, LOS ANGELES, 74-, RES BIOLOGIST, LAB NUCLEAR MED & RADIATION BIOL, 75- *Concurrent Pos:* Instr, Art Ctr Col Design, 51-; Pauley fel, Univ Hawaii, 52; consult, Sch Med, Univ Calif, Los Angeles, 59-65 & Am Inst Biol Sci, 67-70; secy, Bd Trustees, BIOSIS, 74-79; mem, Space Biol Adv Panel, Am Inst Biol Sci/NASA, 75-80. *Mem:* AAAS; Ecol Soc Am; Am Soc Zoologists. *Res:* Radiobiology; circadian rhythms; chronobiology; environmental effects of alternative energy systems. *Mailing Add:* Environ Sci & Eng Univ Calif Los Angeles CA 90024

LINDBERG, STEVEN EDWARD, b St Paul, Minn, Oct 17, 42; c 2. ORGANIC CHEMISTRY, POLYMER CHEMISTRY. *Educ:* Gustavus Adolphus Col, BS, 64; Univ Minn, Minneapolis, PhD(org chem), 69. *Prof Exp:* Res chemist, 69-75, RES SUPVR, RES & DEVELOP DEPT, AMOCO CHEM CO, 75- *Mem:* Am Chem Soc; Am Soc Testing & Mat. *Res:* Petroleum additives, lubricant formulation; polymers; tertiary oil recovery; plastics. *Mailing Add:* Amoco Chem Co PO Box 3011 Naperville IL 60566

LINDBERG, STEVEN ERIC, b Waukegan, Ill, May 9, 47; m 69; c 1. GEOCHEMISTRY, ENVIRONMENTAL SCIENCES. *Educ:* Duke Univ, BS, 69; Fla State Univ, MS, 73, PhD(oceanog), 79. *Prof Exp:* Teacher & adv, Antioch Upper Grade Ctr, 69-71; fel chem oceanog, Fla State Univ, 71-74; RES STAFF GEOCHEM, ENVIRON SCI DIV, OAK RIDGE NAT LAB, 74- *Concurrent Pos:* Chmn, Nat Atmosphere Depostion Prog; chmn, Int Conf Heavy Metals Environ; Humboldt-Stiftung fel, Fed Repub Ger, 85-86. *Honors & Awards:* Environ Sci Achievement Award, Oak Ridge Nat Lab, 84; Martin Marietta Tech Achievement Award, 87. *Mem:* AAAS; Am Geologists Union. *Res:* Influence of fossil fuel utilization on geochemical cycles. *Mailing Add:* Environ Sci Div Oak Ridge Nat Lab Bldg 1505 Box 2008 Oak Ridge TN 37831-6038

LINDBERG, VERN WILTON, b Rimbey, Alta, May 5, 49; m 73; c 2. SOLID STATE PHYSICS, PHYSICAL VAPOR DEPOSITION. *Educ:* Univ Alta, BSc, 69; Case Western Reserve Univ, MS, 72, PhD(physics), 76. *Prof Exp:* Asst prof physics, Hartwick Col, 76-79; ASSOC PROF PHYSICS, ROCHESTER INST TECHNOL, 79- *Concurrent Pos:* Vis researcher, Case Western Reserve Univ, 76 & 78; vis res scientist, Kodak Res Labs, 84-85. *Mem:* Am Asn Physics Teachers; Am Vacuum Soc. *Res:* Vacuum deposition of thin films by evaporation and sputtering; adhesion of thin films to polymers, surface morphology, resistivity of thin films; use of glow discharge and ion beam to modify substrate; optical thin films. *Mailing Add:* Dept Physics Rochester Inst Technol Rochester NY 14623-0887

LINDBLAD, WILLIAM JOHN, b Glen Head, NY, Oct 14, 54; m 85; c 2. WOUND HEALING, COLLAGEN BIOCHEMISTRY. *Educ:* Univ Maine, BS, 76; Cleveland State Univ, MS, 77; Univ RI, PhD(pharmacol), 80. *Prof Exp:* Postdoctoral fel, Dept Surg, Med Col Va, 80-81, res assoc, 81-83, from asst prof to assoc prof, 89-90; ASSOC PROF, DEPT PHARMACEUT SCI, WAYNE STATE UNIV, 90- *Mem:* Am Asn Study Liver Dis; NY Acad Sci; Am Soc Pharmacol & Exp Therapeut; Wound Healing Soc. *Res:* Fibrogenic response to tissue injury; developing pharmacologic approaches to control fibrogenesis in pathologic conditions; determine the ability of extracellular matrices to control cellular phenotype. *Mailing Add:* Dept Pharmaceut Sci 721 Shapero Hall Detroit MI 48202

LINDBURG, DONALD GILSON, b Wagner, SDak, Nov 6, 32; m 54; c 3. ANIMAL BEHAVIOR. *Educ:* Houghton Col, BA, 56; Univ Chicago, MA, 62; Univ Calif, Berkeley, PhD(anthrop), 67. *Prof Exp:* Res asst primatol, Nat Ctr Primate Biol, 64-66; res anthropologist, Sch Med, Univ Calif, Davis, 69-72, asst prof anthrop, Univ Calif, Davis, 67-73; chmn & assoc prof, Ga State Univ, 73-75; assoc prof anthrop, Univ Calif, Los Angeles, 75-79; RES BEHAVIORIST, SAN DIEGO ZOO, 79- *Concurrent Pos:* Res anthropologist, Nat Ctr Primate Biol, 66-69; NSF fel, Univ Calif, Davis, 72-75; vis lectr, Univ Calif, Berkeley, 72, Univ Calif, San Diego, 85; exec bd, Int Primatol Soc, 84-86; PI, Inst Mus Serv grant "Conserv lion-tailed macaque," 84-85; mem, Res Adv Bd, Int Soc Endangered Cats, 88-; ed, Zoo Biol, 88- *Honors & Awards:* Nat Zoo Centennial Award for Excellence in Zoo Res, 90. *Mem:* Am Soc Primatology (pres, 84-86); Animal Behav Soc; Int Primatol Soc; Am Asn Phys Anthrop Int Primatol Soc. *Res:* Captive reproduction of exotic mammals; behavioral correlates of steroid hormone excretions during different phases of the reproductive cycle in primates and carnivores. *Mailing Add:* Zool Soc San Diego PO Box 551 San Diego CA 92112

LINDE, ALAN TREVOR, b Lowood, Australia, Feb 13, 38; m 60; c 3. GEOPHYSICS. *Educ:* Univ Queensland, BSc, 59, PhD(physics), 72. *Prof Exp:* Lectr physics, Univ Queensland, 62-72; STAFF SCIENTIST, DEPT MAGNETISM, CARNEGIE INST, 72- *Mem:* Am Geophys Union; Seismol Soc Am; fel Japan Soc Prom Sci. *Res:* Theoretical and observational studies of earthquake source mechanisms to determine properties of the earth's interior and hence to understand the earth's tectonic engine. *Mailing Add:* Carnegie Inst 5241 Broad Branch Rd NW Washington DC 20015

LINDE, HARRY WIGHT, b Woodbridge, NJ, Jan 1, 26; m 56; c 2. ANESTHESIOLOGY. *Educ:* Tufts Col, BS, 50; Mass Inst Technol, PhD(chem), 53. *Prof Exp:* Sr chemist, Res Labs, Air Reduction Co, Inc, 53-56; res assoc anesthesia, Med Sch, Univ Pa, 56-63; group leader, Air Prod & Chem, Inc, 63-65; asst prof anesthesia, 65-70, Northwestern Univ, asst dir, Anesthesia Res Ctr, 67-71, coordr res & sponsored progs, 71-76, from assoc prof to prof anesthesia, 70-91, assoc dean hons prog med educ, 76-91, vchmn res, 77-91, EMER PROF ANESTHESIA, MED SCH, NORTHWESTERN UNIV, CHICAGO, 91- *Concurrent Pos:* Res assoc, Col Med, Univ Ill, 55-56; mem, Comt Admis, Northwestern Univ, 67-72, human subjects rev, 70-76 & res comt, 71-76, gen fac comt, 85-88, chair, 87-88; consult res anesthesia, Vet Admin Lakeside Med Ctr, Chicago, 68-91; assoc staff mem, Chicago Wesley Mem Hosp, 69-72 & Northwestern Mem Hosp, 72-; consult, US Naval Hosp, Great Lakes, 69-78; assoc ed, Yearbk Anesthesia, 70-81. *Mem:* Fel AAAS; Am Chem Soc; Int Anesthesia Res Soc; Am Soc Anesthesiol; Sigma Xi. *Res:* Pharmacology of anesthesia; gas analysis; bioanalytical chemistry. *Mailing Add:* Dept of Anesthesia Northwestern Univ Med Sch Chicago IL 60611

LINDE, LEONARD M, b New York, NY, June 1, 28; m 51; c 4. CARDIOLOGY, PHYSIOLOGY. *Educ:* Univ Calif, BS, 47, MD, 51; Am Bd Pediat, dipl & cert cardiol, 61, dipl & cert pediat, 75. *Prof Exp:* Intern, Morrisania City Hosp, New York, 51-52; sr resident pediat, Children's Hosp, Los Angeles, 52-53 & 55-56; prof pediat & cardiol, Sch Med, Univ Calif, Los Angeles, 57-76, physiol, 59-76; PROF PEDIAT CARDIOL, SCH MED, UNIV SOUTHERN CALIF, 76- *Concurrent Pos:* Fel pediat cardiol, Med Ctr, Univ Calif, Los Angeles, 56-57; consult, Child Cardiac Clin, Los Angeles City Health Dept, 57- & Surg Gen, US Air Force; vis prof, Univ Tokyo, 65-; chief pediat cardiol, St Vincent's Hosp, Los Angeles, 73-86. *Honors & Awards:* Ross Award for Pediat Res, 62. *Mem:* Soc Pediat Res; Am Pediat Soc; Fel Am Acad Pediat. *Res:* Pediatric cardiology; cardiopulmonary physiology; clinical cardiology; psychological aspects of congenital heart disease; cholesterol problems in children. *Mailing Add:* Div Cardiol Children's Hosp-Los Angeles Los Angeles CA 90027

LINDE, PETER FRANZ, b Berlin, Ger, June 9, 26; nat US; m 53; c 4. PHYSICAL CHEMISTRY, ELECTROCHEMISTRY. *Educ:* Reed Col, BA, 46; Univ Ore, MA, 49; Wash State Univ, PhD(chem), 54. *Prof Exp:* Phys chemist, Sandia Corp, 53-57; from asst prof to assoc prof, 57-66, PROF CHEM, SAN FRANCISCO STATE UNIV, 66- *Concurrent Pos:* Ed, Gmelin, Frankfurt, Ger, 81-82. *Mem:* Am Chem Soc; Sigma Xi. *Res:* Electrochemistry of quaternary ammonium compounds; supporting electrolytes in polarography; shock tube measurements; chemometrics. *Mailing Add:* San Francisco State Univ 1600 Holloway San Francisco CA 94132

LINDE, RONALD K(EITH), b Los Angeles, Calif, Jan 31, 40; m 60. SOLIDS, ENVIRONMENTAL SCIENCE. *Educ:* Univ Calif, Los Angeles, BS, 61; Calif Inst Technol, MS, 62, PhD(mat sci), 64. *Prof Exp:* Engr, Litton Systs, Inc, 61; res asst, Calif Inst Technol, 61-64; mat scientist, SRI Int, 64-67, head solid state res, Poulter Labs, 65-67, chmn shock & high pressure physics dept & mgr tech serv, 67-68, chief exec, 68-69, dir phys sci, 68-69; chmn bd & chief exec officer, Envirodyne Industs, Inc, 69-89. *Res:* Environmental engineering; pollution control; solid state physics; properties of materials; physical chemistry; crystallographic phase transformations; shock wave propagation in solids; physics of soilds. *Mailing Add:* 180 E Pearson St Chicago IL 60611

LINDEBERG, GEORGE KLINE, b Spencer, Iowa, June 6, 30; m 54; c 2. SOLID STATE PHYSICS. *Educ:* St Olaf Col, BA, 52; Princeton Univ, PhD(exp physics), 57. *Prof Exp:* Asst physics, Princeton Univ, 56-57; PHYSICIST, MINN MINING & MFG CO, 57- *Mem:* Am Phys Soc. *Res:* Non-equilibrium electronic processes in solids; thermodynamics; physics operations research. *Mailing Add:* 276 W Grove Rd Hudson WI 54016

LINDELL, ISMO VEIKKO, b Viipuri, Finland, Nov 23, 39; m 64; c 2. ELECTROMAGNETIC THEORY. *Educ:* Helsinki Univ Technol, dipl eng, 63, LicTech, 67, DrTech(radio eng), 71. *Prof Exp:* From asst prof to assoc prof radio eng, 62-89, PROF ELECTROMAGNETICS, HELSINKI UNIV TECHNOL, 89- *Concurrent Pos:* Vis prof, Univ Ill, 72-73; vis scientist, Mass Inst Technol, 86-87. *Honors & Awards:* SA Schelkunoff Prize, Inst Elec & Electronic Engrs, Antennas & Propagation Soc, 87. *Mem:* Fel Inst Elec & Electronic Engrs Antennas & Propagation Soc; Int Union Radio Sci. *Res:* Electromagnetic theory. *Mailing Add:* Elec Eng Dept Helsinki Univ Technol Espoo 02150 Finland

LINDELL, THOMAS JAY, b Red Wing, Minn, July 22, 41; div; c 2. MOLECULAR & CELLULAR BIOLOGY. *Educ:* Gustavus Adolphus Col, BS, 63; Univ Iowa, PhD(biochem), 69. *Prof Exp:* Assoc prof pharmacol, Health Sci Ctr, 70-83, ACTG HEAD MOLECULAR & CELLULAR BIOL, UNIV ARIZ, 83- *Concurrent Pos:* USPHS fel biochem, Univ Wash, 68-69 & biochem, biophys & develop biol, Univ Calif, San Francisco, 69-70; assoc ed, J Life Sci. *Mem:* AAAS; Sigma Xi; Am Soc Biol Chemists; Am Soc Microbiol. *Res:* Control of eukaryotic transcription. *Mailing Add:* Dept Molecular Cellular Biol Univ Ariz Life Sci S Tucson AZ 85721

LINDEMAN, ROBERT D, b Ft Dodge, Iowa, July 19, 30; m 54; c 5. INTERNAL MEDICINE, NEPHROLOGY. *Educ:* State Univ NY Col Forestry, Syracuse Univ, BS, 52; State Univ NY, MD, 56. *Prof Exp:* From asst resident to asst instr internal med, State Univ NY Upstate Med Ctr, 57-60; med officer, Okla State Dept Health, 60-62; med officer geront, Baltimore City Hosps, Md, 62-66; asst prof med & prev med, Med Ctr, Univ Okla, 66-68, assoc prof med & physiol, 68-71, assoc prof biostatist & epidemiol, 69-77, prof med & physiol, 71-77, chief renal sect, 67-77; assoc dean, Vet Admin Affairs & prof med, Sch Med Univ Louisville, 77-88; chief staff, Louisville Vet Admin Med Ctr, 77-88; CHIEF GERIAT & EXTENDED CARE, ALBUQUERQUE VET ADMIN MED CTR, 88- *Concurrent Pos:* Clin asst med, Univ Okla, 60-62; instr, Sch Med, Johns Hopkins Univ, 62-66; asst chief res staff, Oklahoma City Vet Admin Hosp, 67-77; assoc ed, The Kidney, 74-77; mem, US Pharmacopeia Comt Rev & chmn, Subcomt Electrolytes, Large Volume Parenterals & Renal Drugs, 75-; prof, Univ NMex, 88- *Mem:* Am Fedn Clin Res; Cent Soc Clin Res; Int Soc Nephrology; Southern Soc Clin Invest; fel Am Col Physicians. *Res:* Renal and electrolyte problems; hypertension; renal and cardiovascular physiology; aging; trace metal metabolism and nutrition. *Mailing Add:* 2513 Myra Pl NE Albuquerque NM 87112

LINDEMANN, CHARLES BENARD, b Staten Island, NY, Dec 17, 46; m 75; c 3. CELL PHYSIOLOGY, BIOPHYSICS. *Educ:* State Univ NY Albany, BS, 68, PhD(biol), 72. *Prof Exp:* Res assoc cell physiol, Pac Biomed Res Ctr, Univ Hawaii, 72-73; res assoc biophys, State Univ NY Albany, 73-74; PROF PHYSIOL, OAKLAND UNIV, 74- *Mem:* Biophys Soc; Am Soc Cell Biol; Soc Study reprod. *Res:* Flagellar motility: the mechanisms of force production and the factors which control motility onset are under investigation in mammalian sperm. *Mailing Add:* Dept Biol Sci Oakland Univ Rochester MI 48309-4401

LINDEMANN, WALLACE W(ALDO), b Bigelow, Minn, Aug 7, 25; m 50; c 6. ELECTRICAL ENGINEERING. *Educ:* Univ Minn, BEE, 50, MS & PhD(elec eng), 55. *Prof Exp:* Asst, Univ Minn, 50-51, res fel, 51-55; prin scientist, Gen Mills, Inc, 55-60; dir solid state res, Control Data Corp, 60-69, gen mgr, 69-79, vpres, Comput Components Div, 79-85; prof elec eng & dir ctr microelectronics & comput sci, Univ Minn, 85-90; RETIRED. *Mem:* Inst Elec & Electronic Engrs. *Res:* Solid state device development; microelectronics including thin film and semiconductor technologies. *Mailing Add:* 227 Lind Hall Univ Minn 207 Church St SE Minneapolis MN 55455

LINDEMANN, WILLIAM CONRAD, b East St Louis, Ill, Aug 31, 48; m 78. SOIL MICROBIOLOGY. *Educ:* Southern Ill Univ, BS, 70; Univ Minn, MS, 74, PhD(soil sci), 78. *Prof Exp:* Res asst, Univ Minn, 72-74, 75-77; tech asst, Res Seeds Inc, 74-75; ASST PROF SOIL MICROBIOL, NMEX STATE UNIV, 78- *Mem:* AAAS; Am Soc Microbiol; Am Soc Agron; Soil Sci Soc Am; Sigma Xi. *Res:* Soil nitrogen fixation; rhizobiology; legume innoculation; legume nutrition. *Mailing Add:* Crop & Soil Sci Box 3Q NMex State Univ Las Cruces NM 88003

LINDEMER, TERRENCE BRADFORD, b Gary, Ind, Feb 17, 36; m 62; c 2. HIGH TEMPERATURE CHEMISTRY, NUCLEAR CHEMISTRY. *Educ:* Purdue Univ, BS, 58; Univ Fla, PhD(metall eng), 66. *Prof Exp:* Mem res staff, Inland Steel Co, 58-61 & Solar Aircraft Co, 61-63; MEM RES STAFF, CHEM TECHNOL DIV, OAK RIDGE NAT LAB, 66- *Mem:* Fel Am Ceramic Soc; Mat Res Soc. *Res:* Thermodynamic and kinetic factors affecting reactor performance of nuclear fuels and fission products; ceramic superconductors; structural ceramics. *Mailing Add:* Oak Ridge Nat Lab PO Box 2008 Oak Ridge TN 37831-6221

LINDEMEYER, ROCHELLE G, b Philadelphia, Pa, June 14, 52. PEDIATRIC DENTISTRY. *Educ:* West Chester State Col, BA, 72; Univ Pittsburgh, DMD, 77. *Prof Exp:* Residency pedodontics, Children's Hosp Philadelphia, 77-79; asst prof oper dent, 79-81, ASSOC PROF ORAL PEDIAT, SCH DENT, TEMPLE UNIV, 81- *Concurrent Pos:* Pvt pract pedodontics, 77-; clin affil, Children's Hosp Philadelphia, 79-, St Christopher's Hosp Children, 82- *Honors & Awards:* Am Acad Gen Dent award, 77. *Mem:* Int Asn Dent Res; Sigma Xi; Am Acad Pediat Dent; Am Dent Asn; Am Soc Dent Children. *Res:* Hormone receptors; periodontitis in the pediatric dental patient. *Mailing Add:* Dept Oral Pediat Sch Dent Temple Univ 3223 N Broad St Philadelphia PA 19140

LINDEN, CAROL D, b Philadelphia, Pa, Oct 1, 49; c 2. CELL MEMBRANE, ENDOCYTOSIS. *Educ:* Univ Calif, Los Angeles, PhD(molecular biol), 74. *Prof Exp:* BIOLOGIST, US ARMY MED RES INST, 79- *Mem:* Am Soc Cell Biol; AAAS; Am Women in Sci. *Mailing Add:* Prog Anal Off US Army Med Res Inst Infectious Dis Fort Detrick Frederick MD 21701

LINDEN, DENNIS ROBERT, b Greeley, Colo, June 22, 42; m 66; c 2. SOIL SCIENCE, HYDROLOGY. *Educ:* Colo State Univ, MS, 68, MS, 70; Univ Minn, PhD(soil), 79. *Prof Exp:* SOIL SCIENTIST, SCI & EDUC ADMIN-AGR RES, USDA, 70- *Mem:* Am Soc Agron; Soil Sci Soc Am. *Res:* Soil physics and hydrology; water and energy transport within soil and exchange with the atmosphere at the soil-atmosphere interface. *Mailing Add:* Soil Sci 439 Borlaug Hall Univ Minn St Paul 1991 Upper Burford Cr St Paul MN 55108

LINDEN, DUANE B, b Toledo, Ohio, June 1, 30; m 67; c 3. PLANT GENETICS, CELL BIOLOGY. *Educ:* Hiram Col, AB, 52; Univ Minn, PhD(plant genetics), 56. *Prof Exp:* Res assoc plant genetics, Univ Minn, 56-57; asst prof genetics, Univ Fla, 57-61; assoc scientist, PR Nuclear Ctr, 61-65; assoc prof biol, 65-69, PROF BIOL, KEAN COL NJ, 69-, CHMN DEPT, 73- *Mem:* AAAS; Am Inst Biol Sci; Genetics Soc Am; Nat Asn Biol Teachers; Inst Soc Ethics & Life Sci. *Res:* Effects of radiation on biological systems; study of paramutagenic systems in maize. *Mailing Add:* 1238 Medinah Dr Ft Myers FL 33919

LINDEN, HENRY R(OBERT), b Vienna, Austria, Feb 21, 22; nat US; m 67; c 2. RESEARCH ADMINISTRATION, ENERGY POLICY & ECONOMICS. *Educ:* Ga Inst Technol, BS, 44; Polytech Inst Brooklyn, MChE, 47; Ill Inst Technol, PhD(chem eng), 52. *Prof Exp:* Chem engr, petrol fuel res, Socony-Vacuum Labs, 44-47; supvr oil gasification, Inst Gas Technol, 47-52 from asst res dir to dir, 52-69, exec vpres, 69-74, pres & trustee, 74-78, pre & mem bd dirs, 76-87; Gunsanlus distinguished prof chem eng, Ill Inst Technol, 87-90, interim pres & chief exec officer, 88-90, interim chmn & chief exec officer, 89-90, MAX MCGRAW PROF ENERGY & POWER ENG & MGT, ILL INST TECHNOL, 90-; EXEC ADV, GAS RES INST, 87- *Concurrent Pos:* From adj assoc prof to adj prof, Ill Inst Tech, 54-78, res prof chem eng, 78-87, prof gas eng, 78-85; chmn, Gordon Res Conf Coal Sci, 65; chief operating officer, Gas Develop Corp, subsid Inst Gas Technol, 65-73, chief exec officer, 73-78, dir, 65-78; pres & mem bd dirs, Gas Res Inst, 76-87. *Honors & Awards:* Coal Res Awards, Am Chem Soc, 59 & 65, H M Storch Award, 67; Oper Sect Award, Am Gas Asn, 56; Walton Clark Medal, Franklin Inst, 72; Bunsen Pettenkofer Ehrentafel Award, Deut Ver des Gas und Wasserfaches, 78; Gas Indust Res Award, Am Gas Asn, 82; Nat Energy Resources Orgn Res Award, 86. *Mem:* Nat Acad Eng; Am Chem Soc; Am Gas Asn; fel Am Inst Chem Engrs; fel Brit Inst Fuel. *Res:* Petroleum properties; petrochemicals; fossil fuel combustion and gasification; synthetic fuels; coal and petroleum pyrolysis and hydrogenolysis; energy economics; energy policy. *Mailing Add:* Ill Inst Technol PH-135 Ten W 33rd St Chicago IL 60616

LINDEN, JAMES CARL, b Greeley, Colo, Sept 12, 42; m 68; c 2. PLANT BIOCHEMISTRY, INDUSTRIAL MICROBIOLOGY. *Educ:* Colo State Univ, BS, 64; Iowa State Univ, PhD(biochem), 69. *Prof Exp:* Alexander von Humboldt stipend, Bot Inst, Univ Munich, 69, fel plant biochem, 71; fel mammalian cell cult, Dept Microbiol, St Louis Univ, 71-72; biochemist, Great Western Sugar Co, 72-76; sr chemist, Adolph Coors Co, 77-78; ASSOC PROF, DEPTS AGR, CHEM ENG, & MICROBIOL, COLO STATE UNIV, 78- *Concurrent Pos:* Prin investr, Colo Res Develop Corp, 84-85; Consult, Dept Biotechnol, Swiss Fed Inst Technol, 80. *Mem:* Am Chem Soc; Sigma Xi; Soc Indust Microbiol; Am Soc Microbiol. *Res:* Fuels from biomass, lignocellulose pretreatment, cellulase enzymology; lignin biochemistry; membrane biochemistry; microbial fermentations acetone and butanol, ethanol and lactic acid; polysaccharide biosynthesis plant cell culture. *Mailing Add:* Depts Agr, Chem Eng & Microbiol Colo State Univ Ft Collins CO 80523

LINDEN, JOEL MORRIS, b Boston, Mass, May 30, 52. PHARMACOLOGY, CARDIOLOGY. *Educ:* Brown Univ, BS, 74; Univ Va, PhD(pharmacol), 78. *Prof Exp:* Res assoc, Dept Pharmacol, 78-80, RES ASST PROF PHYSIOL & PHARMACOL, UNIV VA, 80-; AT OKLA MED RES FOUND, OKLAHOMA CITY. *Res:* Mechanism of action of cardioactive drugs and the involvement of cyclic nucleotides in the control of cardiac contractility. *Mailing Add:* Sch Med Jordan Hall Univ Va Box 449 Charlottesville VA 22908

LINDEN, KURT JOSEPH, b Berlin Ger, Dec 27, 36; US citizen; m 62; c 3. SOLID STATE PHYSICS, ELECTRICAL ENGINEERING. *Educ:* Univ Utah, BS, 59; Mass Inst Technol, MS, 61; Purdue Univ, PhD(elec eng), 66. *Prof Exp:* Engr physics, Air Force Cambridge Res Lab, 63; sr engr infrared, Raytheon Co, 66-76; mgr solid state device activ, laser anal, Div Spectra Physics, 76-84; mgr, 84-88, DIR ELECTRONIC MAT DIV, SPIRE CORP, 89- *Concurrent Pos:* Res asst elec eng, Mass Inst Technol, 59-61, teaching asst, 61-63; instr, Purdue Univ, 63-66, NSF fel, 65; sr lectr, Northeastern Univ, Ctr Continuing Educ, 77-; guest lectr, Mass Inst Technol, 79- *Mem:* Sr mem Inst Elec & Electronic Engrs; Am Phys Soc; Int Soc Optical Engrs. *Res:* Optoelectronic semiconductor materials and devices; infrared detectors and emitters; low energy detectors and diode lasers of gallium arsenide, gallium aluminum arsenide and lead salts; management of research and development and manufacturing activities. *Mailing Add:* 17 Keith Rd Wayland MA 01778-4560

LINDENAUER, S MARTIN, b New York, NY, Dec 10, 32; m 56; c 4. SURGERY. *Educ:* Tufts Univ, MD, 57. *Prof Exp:* From instr to assoc prof, 64-72, asst dean, Med Sch, 74-81, PROF SURG, UNIV MICH, ANN ARBOR, 72- *Concurrent Pos:* Chief surg serv, Vet Admin Hosp, 68-74, chief staff, 74-81. *Mem:* Am Col Surg; Asn Acad Surg; Soc Vascular Surg; Int Cardiovasc Soc; Soc Surg Alimentary Tract. *Res:* Vascular surgery; biliary tract surgery. *Mailing Add:* Dept Surg Rm 2922 G Univ Mich Med Ctr/ Taubman Ctr Ann Arbor MI 48109-0331

LINDENBAUM, S(EYMOUR) J(OSEPH), b New York, NY, Feb 3, 25; m 58. PHYSICS. *Educ:* Princeton Univ, AB, 45; Columbia Univ, MA, 48, PhD(physics), 51. *Prof Exp:* Res assoc, Nevis Cyclotron Lab, Columbia Univ, 47-51; assoc physicist, 51-54, physicist, 54-63, SR PHYSICIST, BROOKHAVEN NAT LAB, 63-; MARK W ZEMANSKY CHAIR PHYSICS, CITY COL NEW YORK, 70- *Concurrent Pos:* Group leader high energy counter res group, Brookhaven Nat Lab, 54-88; vis prof, Univ Rochester, 58-59; dep sci affairs, High Energy Prog, Div Phys Res, Energy Res & Develop Agency, 76-77; vis, Europ Orgn Nuclear Res; consult, Saclay Nuclear Res Ctr. *Mem:* AAAS; fel Am Phys Soc; NY Acad Sci. *Res:* High energy elementary particle interactions; high energy experimental techniques; heavy ion physics search for a quark gluon plasma; search for and probable discovery of glueballs of quantum chromodynamics (with collaborators). *Mailing Add:* Physics Dept Brookhaven Nat Lab Upton NY 11973

LINDENBAUM, SIEGFRIED, b Unna, Ger, July 24, 30; nat US; m 56; c 3. PHYSICAL CHEMISTRY. *Educ:* Rutgers Univ, BS, 52, PhD(chem), 55. *Prof Exp:* Chemist, Oak Ridge Nat Lab, 55-71; assoc prof, 71-76, PROF PHARMACEUT CHEM, UNIV KANS, 76- *Concurrent Pos:* Merz professorship, Frankfort, 90. *Mem:* Fel Acad Pharmaceut Sci. *Res:* Solvent extraction; separations; thermodynamics of electrolyte solutions; physical chemistry of bile; solution calorimetry; protein binding; enzyme inhibitors; analysis of peptides; drug stability. *Mailing Add:* Sch Pharm Univ Kans Lawrence KS 66045

LINDENBERG, KATJA, b Quito, Ecuador, Nov 2, 41; US citizen; m 70, 90; c 2. PHYSICAL CHEMISTRY, CHEMICAL PHYSICS. *Educ:* Alfred Univ, BA, 62; Cornell Univ, PhD(theoret physics), 67. *Prof Exp:* Res assoc & asst prof physics, Univ Rochester, 67-69; lectr chem & res chemist, 69-72, asst prof chem residence, 72-73, from asst prof to assoc prof, 73-81, PROF CHEM, UNIV CALIF, SAN DIEGO, 81- *Concurrent Pos:* Res physicist, Univ Calif, San Diego, 69-71; researcher, Oak Ridge Summer Inst Theoret Biophys, 69-75; consult, Chem Div, Oak Ridge Nat Lab, 75; assoc, La Jolla Inst, 79- *Mem:* Am Phys Soc; Am Chem Soc; Mat Res Soc. *Res:* Theory of stochastic processes with applications to physical and chemical systems; non-equilibrium statistical mechanics. *Mailing Add:* Chem 0340 Univ Calif, San Diego La Jolla CA 92093-0340

LINDENBERG, RICHARD, b Bocholt, Ger, Feb 18, 11; US citizen; m 37; c 1. NEUROPATHOLOGY. *Educ:* Univ Berlin, MD, 44. *Prof Exp:* Chief resident neuropath, Kaiser-Wilhelm Inst Brain Res, 36-39; resident & dir anat lab, Neuropsychiat Hosp, Univ Frankfurt, 45-47; res neuropathologist, Sch Aviation Med, Randolph Field, Tex & Army Chem Ctr, Md, 47-51; dir neuropath & legal med, Md State Dept Health & Ment Hyg, 51-78; RETIRED. *Concurrent Pos:* Clin prof path, Sch Med, Univ Md, 51-; lectr neuro-ophthal, Sch Med, Johns Hopkins Univ, 59-; consult, Greater Baltimore Med Ctr, 59- *Honors & Awards:* Meritorious Award, Am Asn Neuropath, 81. *Mem:* Am Asn Neuropath; Am Soc Clin Path; fel Col Am Path; AMA; World Fedn Neurol. *Res:* Neuropathology of head injury and circulatory disorders; neuro-ophthalmologic pathology; forensic neuropathology. *Mailing Add:* 13801 York Rd, Apt P Five Cockeysville MD 21030

LINDENBLAD, IRVING WERNER, b Port Jefferson, NY, July 31, 29; m 58; c 2. ASTROMETRY, GEODETIC ASTRONOMY. *Educ:* Wesleyan Univ, BA, 50; Colgate Rochester Divinity Sch, MDiv, 56; George Washington Univ, MA, 63. *Prof Exp:* Astronr, US Naval Observ, 53, 58-60, 63-89; CHAPLAIN INTERN, WASH HOSP CTR, 90- *Mem:* Fel The Royal Astron Soc; Am Astron Soc. *Res:* Photographic visual binary stars; motion and magnitude difference of the components of sirius; variation of latitude; rotation and polar motion of the earth; sunspots. *Mailing Add:* 4735 Arlington Blvd Arlington VA 22203-2612

LINDENFELD, PETER, b Vienna, Austria, Mar 10, 25; nat US; m 53; c 2. LOW TEMPERATURE PHYSICS, SUPERCONDUCTIVITY. *Educ:* Univ BC, BASc, 46, MASc, 48; Columbia Univ, PhD(physics), 54. *Prof Exp:* Asst physics, Univ BC, 46-47; asst, Columbia Univ, 48-52, res scientist, 53; vis lectr, Drew Univ, 52-53; from instr to assoc prof, 53-66, PROF PHYSICS, RUTGERS UNIV, 66- *Concurrent Pos:* Dir NSF in-serv insts for high sch teachers, 64-66; regional counr NJ, Am Inst Physics, 63-71; Rutgers Res Coun fel & guest scientist fac sci, Univ Paris-South, 70-71; guest scholar, Kyoto Univ, 82. *Honors & Awards:* Robert A Millikan Medal, Am Asn Physics Teachers, 89. *Mem:* Fel Am Phys Soc; Am Asn Physics Teachers; AAAS; Am Asn Univ Prof. *Res:* Metal-insulator transition audits relation to superconductivity; electric, magnetic, and thermal properties of materials. *Mailing Add:* Dept Physics Rutgers Univ New Brunswick NJ 08903

LINDENLAUB, JOHN CHARLES, b Milwaukee, Wis, Sept 10, 33; m 57; c 4. ELECTRICAL ENGINEERING. *Educ:* Mass Inst Technol, BS, 55, MS, 57; Purdue Univ, PhD(elec eng), 61. *Prof Exp:* From asst prof to assoc prof, 61-72, dir, Ctr Instrnl Develop Eng, 77-81, PROF ELEC ENG, PURDUE UNIV, 72- *Concurrent Pos:* Danforth Assoc, Danforth Found, 66; mem tech staff, Bell Tel Labs, Inc, 68-69. *Honors & Awards:* Helen Plants Award, Frontiers in Educ Conf, 80; Educ Soc Achievement Award, Inst Elec & Electronics Engrs, 84; Chester F Carlson Award, Am Soc Eng Educ, 88. *Mem:* Inst Elec & Electronics Engrs; Am Soc Eng Educ. *Res:* Statistical communication theory; computer engineering; engineering education. *Mailing Add:* 238 Connolly St West Lafayette IN 47906

LINDENMAYER, ARISTID, theoretical biology; deceased, see previous edition for last biography

LINDENMAYER, GEORGE EARL, b Port Arthur, Tex, Aug 22, 40; m 63; c 2. BIOCHEMICAL PHARMACOLOGY. *Educ:* Baylor Univ, BS, 62; Baylor Col Med, MD & MS, 67, PhD(pharmacol), 70. *Prof Exp:* Instr pharmacol, Baylor Col Med, 69-70; staff assoc cardiol, Nat Heart & Lung Inst, 70-72; asst prof pharmacol & med, Baylor Col Med, 72-74, assoc prof cell biophys & med, 74-75; assoc prof pharmacol & med, 75-77, PROF PHARMACOL & ASSOC PROF MED, MED UNIV SC, 77- *Concurrent Pos:* Estab investr, Am Heart Asn, 73-78. *Mem:* Am Soc Pharmacol & Exp Therapeut; Int Study Group Res Cardiac Metab; Am Chem Soc; Biophys Soc; Am Heart Asn; Sigma Xi. *Res:* Information transfer between extracellular and intracellular environments of myocardial cells. *Mailing Add:* Dept of Pharmacol Med Univ SC 171 Ashley Ave Charleston SC 29425

LINDENMEIER, CHARLES WILLIAM, b Ft Collins, Colo, Dec 2, 30; m 58; c 2. THEORETICAL PHYSICS, NUCLEAR PHYSICS. *Educ:* Colo State Univ, BS, 52; Cornell Univ, PhD(theoret physics), 60. *Prof Exp:* Sr physicist, Hanford Labs, Gen Elec Co, 60-63, mgr theoret physics, 63-64; mgr, Pac Northwest Labs, Battelle Mem Inst, 65-70, mgr math & physics res, 70-73; mgr design anal, Laser Enrichment Dept, 74-81, mgr neutron develop, neutron & fuel mgt, 81-90, SR STAFF ENGR, ADVAN NUCLEAR FUELS CORP, 90- *Mem:* Am Phys Soc; Am Nuclear Soc. *Res:* Reactor physics; neutron thermalization; nuclear reactions; computer applications; laser isotope separation. *Mailing Add:* Advan Nuclear Fuels Corp 2101 Horn Rapids Rd Richland WA 99352

LINDENMEYER, PAUL HENRY, b Bucyrus, Ohio, May 4, 21; m 44; c 3. PHYSICAL CHEMISTRY, MATERIALS SCIENCE. *Educ:* Bowling Green State Univ, BS, 44; Ohio State Univ, PhD(chem), 51. *Prof Exp:* Asst, Ohio State Univ, 46-49, res assoc, Res Found, 49-51; res chemist, Visking Co Div, Union Carbide Corp, 51-53, res supvr, 53-57, mgr, Pioneering Res Dept, 57-59; mgr fiber sci, Chemstrand Res Ctr, Inc, 59-69; head mat sci lab, Boeing Sci Res Labs, 69-72, sci adv, Boeing Aerospace Co, 72-73; prog dir, Div Mat Res, NSF, 73-75; sr prin scientist, Boeing Co, 77-88; PRES, DYNAMIC MAT INC, 82- *Concurrent Pos:* Mat res consult, 75-; Royal Soc sr vis fel, Inst Sci & Technol, Univ Manchester, 83-84. *Honors & Awards:* US Sr Scientist Award, Humboldt Found, 75. *Mem:* AAAS; Fiber Soc; Am Chem Soc; Am Phys Soc; Am Crystallog Asn. *Res:* X-ray crystallography; spectroscopy; microscopy; crystal growth and structure of high polymers; materials processing; irreversible thermodynamics and materials properties. *Mailing Add:* 165 Lee St Seattle WA 98109-3104

LINDER, ALLAN DAVID, b Grand Island, Nebr, Sept 27, 25; m 49; c 1. VERTEBRATE ZOOLOGY. *Educ:* Univ Nebr, BSc, 51; Okla State Univ, MSc, 52, PhD(zool), 56. *Prof Exp:* Asst prof zool, Univ Wichita, 56-59 & Southern Ill Univ, 59-60; chmn dept, Idaho State Univ, 60-75, assoc dean, Col Lib Arts, 66-69 & 76-78, prof zool, 60-; RETIRED. *Mem:* Am Soc Ichthyologists & Herpetologists; Soc Study Amphibions & Reptiles; Sigma Xi; Soc Vert Paleont; Wilderness Soc. *Res:* Ichthyology, paleo-ichthyology and herpetology. *Mailing Add:* 33 Brair Bate Terr Pueblo CO 81001

LINDER, BRUNO, b Sniatyn, Poland, Sept 3, 24; nat US; m 53; c 5. THEORETICAL CHEMISTRY, CHEMICAL PHYSICS. *Educ:* Upsala Col, BS, 48; Univ Ohio, MS, 50; Univ Calif, Los Angeles, PhD(chem), 55. *Prof Exp:* Asst chem, Univ Ohio, 48-49; asst chem, Univ Calif, Los Angeles, 50-55, asst res chemist, 55; proj assoc theoret chem, Naval Res Lab, Wis, 55-57; from asst prof to assoc prof, 57-65, assoc chmn dept, 80-83, PROF PHYS CHEM, FLA STATE UNIV, 65- *Concurrent Pos:* Guggenheim fel, Inst Theoret Physics, Univ Amsterdam, 64-65; chmn chem physics prog, Fla State Univ, 71-73 & 75-; vis prof, Hebrew Univ, Jerusalem, 73. *Mem:* Am Phys Soc; Sigma Xi. *Res:* Intermolecular forces; van der Waals dipoles; theory of adsorption; liquid crystal theory; solvent effects on infrared and Raman intensities. *Mailing Add:* Dept of Chem Fla State Univ Tallahassee FL 32306

LINDER, CLARENCE H, b Jan 18, 03; US citizen. ELECTRICAL ENGINEERING. *Educ:* Univ Tex, BS, 23, MS, 25. *Hon Degrees:* DEng, Worchester Polytech Inst, 55 & Clarkson Col, 56; LLD, Union Col, 65 & Lehigh Univ, 72. *Prof Exp:* Mem staff, Gen Elec Co, 24-51, gen mgr large appliance div, 51-53, vpres eng serv, 53-60, vpres exec elec utilities group, 60-61, mem staff, 61-63; RETIRED. *Concurrent Pos:* Dir, Western Union Tel & Western Union Corp; res prof elec eng, Union Col. *Mem:* Nat Acad Eng (vpres, 66-70, pres, 70-73); fel Am Inst Elec Engrs (pres, 60-61); fel Elec & Electronic Engrs (pres, 64); fel Am Soc Mech Engrs; Am Soc Eng Educ; fel AAAS. *Mailing Add:* 1334 Ruffner Rd Schenectady NY 12309

LINDER, DONALD ERNST, b Yoakum, Tex, Oct 4, 38; m 61; c 3. LIQUID CHROMATOGRAPHY, GEL PERMEATION CHROMATOGRAPHY. *Educ:* Sul Ross State Univ, BS, 61; Tex A&M Univ, MS, 64, PhD(chem), 67. *Prof Exp:* ENVIRON PROJ COORDR, CONOCO, INC, 66- *Mem:* Am Chem Soc. *Res:* Liquid chromatography, adsorption, liquid-liquid, ion exchange and gel permeation; large scale preparative gas-liquid chromatography; analytical distillations; environmental sampling and testing; EPA protocol groundwater monitoring. *Mailing Add:* CONOCO Inc PO Drawer 1267 RDE 346 Ponca City OK 74602

LINDER, ERNEST G, b Waltham, Mass, May 16, 02; m 44; c 2. MICROWAVE DEVELOPMENT, SOLAR CELLS. *Educ:* State Univ Iowa, BA, 25, MS, 27; Cornell Univ, PhD(physics), 31. *Prof Exp:* Res assoc, Cornell Univ, 28-32; res physicist, Res Dept, RCA, 32-42 & RCA Labs, 42-55, spec proj mgr, 55-68; RETIRED. *Concurrent Pos:* Comt insulation, Nat Res Coon, 29; deleg, Int Conf Peaceful Usage Atomic Energy, Geneva, Switz, 55; consult, Pub Broadcast Syst, 74-79. *Mem:* fel Am Phys Soc; fel Am Inst Elec Engrs; Union Concerned Scientists; Sigma Xi. *Res:* Electrical discharges in gases, microwaves, magnetrons, electron physics, nuclear and solar batteries; radar, microwave propagation. *Mailing Add:* 16 Colonial Club Dr Apt 205 Boynton Beach FL 33435

LINDER, FORREST EDWARD, b Waltham, Mass, Nov 21, 06; m 84; c 2. STATISTICS. *Educ:* State Univ Iowa, BA, 30, MA, 31, PhD(psychol, math), 32. *Prof Exp:* Tech expert, Div Vital Statist, Bur of Census, 35-42, asst chief, 42-45; asst chief, Med Statist Div, US Navy, 44-46; dep chief, Nat Off Vital Statist, USPHS, 46-47; chief, Demog & Social Statist Br, UN, 47-56; dir nat health surv, USPHS, 56-60, dir, Nat Ctr Health Statist, 66-67; prof biostatist & dir int prog labs pop statist, Univ NC, Chapel Hill, 67-77; pres, Int Inst Vital Regist & Statist, 72-87; RETIRED. *Concurrent Pos:* Consult, Ford Found, India, 62 & 64 & WHO, 66 & 68-71; mem expert adv panel health statist, 67; mem policy res adv comt, Nat Inst Child Health & Human Develop, 67-71; mem res adv comt, AID, 68-83; consult, Pan Am Health Orgn, 68 & 69; chmn, Nat Comt Health & Vital Statist, Nat Ctr Health Statist, 69-72; mem adv comt statist policy, Off Budget & Mgt, US Exec Off, 72; mem world fertility surv steering comt, Int Statist Inst, 72-84; mem, Inter-Am Statist Inst, 45- *Honors & Awards:* Bronfman Prize, Am Pub Health Asn, 67. *Mem:* Fel AAAS; Int Union Sci Study Pop; fel Am Statist Asn; fel Am Pub Health Asn; Pop Asn Am. *Res:* Development of statistical methods for measurement of population change; public health statistics; census and vital statistics methods. *Mailing Add:* 15115 Vantage Hill Rd Silver Spring MD 20906

LINDER, HARRIS JOSEPH, b Brooklyn, NY, Jan 3, 28; m 52; c 4. ZOOLOGY. *Educ:* Long Island Univ, BS, 51; Cornell Univ, MS, 55, PhD(zool), 58. *Prof Exp:* Asst zool, Cornell Univ, 52-57; resident res assoc, Div Biol & Med, Argonne Nat Lab, 57-58; asst prof, 58-63, ASSOC PROF ZOOL, UNIV MD, COLLEGE PARK, 63- *Concurrent Pos:* Contrib ed, Instrnl Media, J Col Sci Teaching, 76-79. *Mem:* AAAS; Am Soc Zool; Am Micros Soc; Am Inst Biol Sci; Soc Study Reproduction; Sigma Xi. *Res:* Comparative invertebrate endocrinology; neurosecretion; experimental studies on earthworm reproduction. *Mailing Add:* Dept of Zool Univ of Md College Park MD 20742

LINDER, JAMES, b Omaha, Nebr, Oct 21, 54. PATHOLOGY, MICROBIOLOGY. *Educ:* Iowa State Univ, BS, 76; Univ Nebr, MD, 80; Am Bd Path, cert anat & clin path, 83, cert cytopath, 89. *Prof Exp:* Teaching asst, Dept Biochem, Iowa State Univ, 75-76, Univ Minn, 76-77; clin path fel, Duke Univ/Caberras Hosp, 81-82; from asst prof to assoc prof path, Univ Nebr, 83-89, dir path residency prog, 83-89, dir regional lab, 89-90, DIR CYTOPATH, DEPT PATH & LAB MED, MED CTR, UNIV NEBR, 83-, DIR SURG PATH, 85-, PROF PATH & MICROBIOL, 89- *Concurrent Pos:* Med examr, Durham, NC, 81-82; path resident, Med Ctr, Duke Univ, 80-82, Univ Nebr, 82-83; mem grad fac, Med Ctr, Univ Nebr, 84-87; consult. *Mem:* Am Fedn Clin Res; Am Soc Cytol; Am Asn Pathologists; fel Col Am Pathologists; fel Am Soc Clin Pathologists; AMA; Sigma Xi. *Res:* Immune disorders; hematologic disorders. *Mailing Add:* Dept Path & Microbiol Med Ctr Univ Nebr 600 S 42nd St Omaha NE 68198

LINDER, JOHN SCOTT, b Baton Rouge, La, May 3, 35. MICROELECTRONICS, SOLID STATE PHYSICS. *Educ:* La State Univ, BS, 56, MS, 60; Univ Ariz, PhD(elec eng), 67. *Prof Exp:* Tech investr chem processing, E I du Pont de Nemours & Co, 56-58; assoc elec eng, La State Univ, 58-60, instr, 62-63; sr engr, comput div, Bendix Corp & Control Data Corp, 60-62; instr elec eng, Univ Ariz, 63-67; asst prof elec eng, Tex A&M Univ, 67-68, assoc prof, 68-79; PROF ELEC ENG, TEX A&I UNIV, 79- *Concurrent Pos:* Mem tech staff, Sandia Corp, 66; consult, Burr Brown Corp, 67-68, missiles & space div, LTV Aerospace Corp & Consoltec Inc, 69-, Teledyne 79- *Mem:* Inst Elec & Electronics Engrs; Am Soc Eng Educ; Am Phys Soc. *Res:* Solid state devices; semiconductor technology; solid state materials; active and distributed synthesis. *Mailing Add:* Col Eng Tex A&I Univ Kingsville TX 78363

LINDER, LOUIS JACOB, b East St Louis, Ill, May 10, 16; m 48; c 3. ANALYTICAL CHEMISTRY & OPTICAL EMISSION SPECTROSCOPY. *Educ:* Wash Univ, AB, 41. *Prof Exp:* Chemist, Eagle-Picher Lead Co, 41-44 & US Army Chem Warfare Serv, 44-46; anal chemist, Alumina & Chem Div, Res Labs, Aluminum Co Am, 46-50, res chemist, 50-72; lab mgr Sch Sci, Southern Ill Univ, Edwardsville, 72-86; RETIRED. *Mem:* Soc Appl Spectros. *Res:* Analytical procedures on aluminous materials; application of optical emission spectroscopy to analysis of alumina, aluminous ores and sodium aluminate slags; spectrographic analysis of gallium oxide and metal. *Mailing Add:* 7907 W Washington St Belleville IL 62223-2317

LINDER, MARIA C, b New York, NY. CHEMISTRY, GENERAL BIOCHEMISTRY. *Prof Exp:* PROF BIOCHEM, CALIF STATE UNIV, 78- *Mailing Add:* Chem & Biochem Dept Calif State Univ Fullerton CA 92634

LINDER, REGINA, b New York, NY, June 21, 45. MICROBIOLOGY, BIOCHEMISTRY. *Educ:* City Col New York, BS, 67; Univ Mass, MS, 69; NY Univ, PhD(microbiol), 75. *Prof Exp:* Asst res scientist, Sch Med, NY Univ, 75-78, asst prof microbiol, 78-82; asst prof, 82-87, ASSOC PROF HEALTH SCI, HUNTER COL, CITY UNIV NEW YORK, 87-, DIR MED LAB SCI PROG, 89- *Mem:* Am Soc Microbiol. *Res:* Investigation of the mechanism of action of bacterial and animal toxins which specifically interact with membrane lipids; studies on the enzyme target of penicillin in bacterial cells. *Mailing Add:* Hunter Col Sch of Health Sci 425 E 25th St New York NY 10010

LINDER, SEYMOUR MARTIN, b New York, NY, Dec 17, 25; m 55; c 2. ORGANIC ANALYTICAL CHEMISTRY, INDUSTRIAL ORGANIC CHEMISTRY. *Educ:* City Col New York, BS, 46; Polytech Inst Brooklyn, MS, 49, PhD(chem), 53. *Prof Exp:* Jr chemist, Hoffmann-La Roche, Inc, 46-51; proj leader, Becco Chem Div, FMC Corp, 53-58 & Org Chem Div, 58-72; dir synthesis res, Alcolac, Inc, 72-80; prin chemist & shift leader, Patapsco Wastewater Treatment Plant, Balto City, 81-90; RETIRED. *Mem:* Am Chem Soc; Am Inst Chemists. *Res:* Chemistry of hydrogen peroxide and peroxy acids; epoxidations; epoxyresins; process development; terpene and medicinal chemistry; insecticides; gas chromatography; specialty organic chemicals; functional monomers; (meth)acrylate esters; organometallic compounds; quaternary salts; copolymerizable surfactants; analysis of wastewater and sludge; determination of primary pollutants by GC and GC/MS; toxicity studies on wastewater treatment plant biomass. *Mailing Add:* 1902 Tadcaster Rd Baltimore MD 21228

LINDER, SOLOMON LEON, b Brooklyn, NY, Mar 13, 29; m 53; c 3. INFRARED SYSTEMS, LASER SYSTEMS. *Educ:* Rutgers Univ, BS, 50; Wash Univ, PhD(physics), 55. *Prof Exp:* Mem tech staff, Bell Tel Labs, Inc, 55-62; sr group engr, McDonnell Aircraft Corp, 62-67, SR GROUP ENGR & TECH SPECIALIST, McDONNELL DOUGLAS ASTRONAUT CO, 67- *Concurrent Pos:* Eve instr, Fairleigh Dickinson Univ, 59-62, Univ Col, Wash Univ, 63-67, Fla Technol Univ, 70-71, Southern Ill Univ, Edwardsville, 74-75 & Univ Col, Wash Univ, 75- *Mem:* Optical Soc Am; sr mem Inst Elec & Electronics Eng. *Res:* Nuclear magnetic resonance; military systems; electrooptics. *Mailing Add:* McDonnell Douglas Aircraft Co Mail Code 1061061 318/101 St Louis MO 63166

LINDERMAN, ROBERT G, b Crescent City, Calif, Feb 2, 39; m 61; c 3. PLANT PATHOLOGY. *Educ:* Fresno State Col, BA, 60; Univ Calif, Berkeley, PhD(plant path), 67. *Prof Exp:* Lab technician plant path, Univ Calif, Berkeley, 64-67, asst res plant pathologist, 67; res plant pathologist, Agr Res Serv, USDA, 67-73; SUPVRY RES PLANT PATHOLOGIST, RES LEADER & COURTESY PROF BOT & PLANT PATH, ORE STATE UNIV, 73- *Mem:* Am Phytopath Soc. *Res:* Ecology of soil-borne fungus plant pathogens; biological control; biological effects of plant residue decomposition in soil; ornamental plant diseases; mycorrhizal fungi. *Mailing Add:* USDA-ARS Hort Crops Res Lab 3420 NW Orchard Ave Corvallis OR 97330

LINDERS, JAMES GUS, b St Catharines, Ont, June 27, 36; m 65; c 3. COMPUTER SCIENCE, MATHEMATICS. *Educ:* Univ Toronto, BASc, 60, MASc, 61; Univ London, DIC & PhD(comput sci), 69. *Prof Exp:* Teaching fel math, St Michael's Col, Univ Toronto, 61-63; lectr comput sci, Ryerson Polytech Inst, 62-65; lectr, Imp Col, Univ London, 65-69; asst prof, Univ Waterloo, 69-77; PROF, UNIV GUELPH, 82-, CHMN DEPT COMPUT & INFO SCI, 80- *Concurrent Pos:* Consult, Dept Energy, Mines & Resources, Can, 67- & Ministry Natural Resources, Ont, 77- *Mem:* Fel Brit Comput Soc; fel Royal Geog Soc; Asn Comput Mach; Inst Elec & Electronics Engrs. *Res:* Computer-aided design; data base design; knowledge base engineering; development of automated cartography; geo referenced data system. *Mailing Add:* Dept of Comput & Info Sci Univ of Guelph Guelph ON N1G 2W1 Can

LINDFORS, KARL RUSSELL, b Saginaw, Mich, July 10, 37; m 58; c 2. PHYSICAL CHEMISTRY. *Educ:* Univ Mich, BS, 59; Univ Wis, PhD(phys chem), 64. *Prof Exp:* Spectroscopist, Tracerlab, 63-64; PROF CHEM, CENT MICH UNIV, 64-, CHMN DEPT, 78 - *Mem:* Am Chem Soc. *Res:* Molecular spectroscopy; species in solution. *Mailing Add:* Dept Chem Cent Mich Univ Mt Pleasant MI 48859

LINDGREN, ALICE MARILYN LINDELL, b Minneapolis, Minn, Jan 31, 37; m 59; c 3. RADIATION BIOLOGY, IMMUNOLOGY. *Educ:* Augsburg Col, BA, 58; Univ Minn, Minneapolis, MS, 61; Univ Iowa, PhD(radiation biol), 70. *Prof Exp:* From instr to assoc prof, 63-81, PROF BIOL, BEMIDJI STATE UNIV, 81- *Concurrent Pos:* Consult, Agassiz Nursing Educ Consortium, 72- & Itasca Nursing Educ Consortium, 81-; vis asst prof radiation biol, Univ Iowa, 75; vis prof radiol, Univ Iowa, 84-85. *Mem:* Sigma Xi; Radiation Res Soc; Cell Kinetics Soc. *Res:* Cell cycle kinetics; effect of radiation on the cell cycle; control of the cell cycle by cyclic nucleotides; response of rat lens epithelial cells to a wound stimulus; lymphocyte blast cell formation (perturbation by drugs and radiation). *Mailing Add:* Sci Div Bemidji State Univ Bemidji MN 56601

LINDGREN, BERNARD WILLIAM, b Minneapolis, Minn, May 13, 24; m 45; c 3. MATHEMATICS. *Educ:* Univ Minn, PhD(math), 49. *Prof Exp:* Instr math, Univ Minn, 43-44, 46-49 & Mass Inst Technol, 49-51; res mathematician, Minn-Honeywell Regulator Co, 51-53; from instr to assoc prof, 53-69, chmn dept, 63-73, PROF STATIST, UNIV MINN, MINNEAPOLIS, 69- *Mem:* Fel Am Statist Asn; Int Statist Inst; Sigma Xi. *Res:* Analysis; probability; statistics. *Mailing Add:* Sch Statist Vincent Hall Univ Minn Minneapolis MN 55455

LINDGREN, CLARK ALLEN, b Green Bay, Wis, Aug 31, 58; c 2. SYNAPTIC PHYSIOLOGY, NEUROMUSCULAR PHYSIOLOGY. *Educ:* Wheaton Col, BS, 80; Univ Wis-Madison, MS, 82, PhD(physiol), 85. *Prof Exp:* Res assoc neurobiol, Dept Neurobiol, Duke Univ Med Ctr, 85-89; ASST PROF BIOL, ALLEGHENY COL, 89- *Concurrent Pos:* Vis lectr, Univ NC, Chapel Hill, 88-89. *Mem:* Soc Neurosci; AAAS. *Res:* Presynaptic mechanisms at the chemical synapse; role of calcium in activating neurotransmitter release; modulation of release by hormones and co-transmitters; disturbance of neuromuscular transmission in Lambert-Eaton myasthenic syndrome. *Mailing Add:* Dept Biol Allegheny Col Meadville PA 16335

LINDGREN, DAVID LEONARD, b St Paul, Minn, Sept 17, 06; m 33; c 3. ENTOMOLOGY. *Educ:* Univ Minn, BS, 30, MS, 31, PhD(entom), 35. *Prof Exp:* Jr entomologist, 35-41, from asst entomologist to entomologist, 41-74, EMER ENTOMOLOGIST & LECTR, CITRUS EXP STA, UNIV CALIF, RIVERSIDE, 74- *Mem:* Fel AAAS; Entom Soc Am; Am Asn Cereal Chemists. *Res:* Insecticides; citrus insects; stored product insects. *Mailing Add:* 4738 Elmwood Ct Riverside CA 92506

LINDGREN, E(RIK) RUNE, b Sodertlje, Sweden, Aug 15, 19; m 63; c 4. THEORETICAL & EXPERIMENTAL FLUID MECHANICS. *Educ:* Tech Col Stockholm, BS, 43; Royal Inst Technol, Sweden, MS, 47, Tekn lic, 56,. *Hon Degrees:* DSc, Royal Inst Technol. *Prof Exp:* Resident assoc, Lumalampan Inc, Sweden, 45-47 & Aeronaut Lab, Royal Inst Technol, Sweden, 47-49; lectr physics & mech, Tech Col Stockholm, 49-51; res fel fluid mech, Royal Inst Technol, Sweden, 51-59, lectr, 59-61; vis asst prof mech, Johns Hopkins Univ, 61-63; assoc prof mech & fluid mech, Okla State Univ, 63-65; PROF ENG SCI, UNIV FLA, 65- *Concurrent Pos:* Consult, Kockums Shipyard & Royal Swedish Naval Bd, 51-59 & Res Lab, Presby Hosp, 62-63; res grants, Swedish State Coun Tech Res, 53-56, Air Res & Develop Command, US Air Force, 56-59, Docent Fluid Dynamics, Royal Inst Technol, Sweden, 59-, David Taylor Model Basin, US Bur Ships, 62-65, NSF, 66-80 & Off Naval Res, 75-80; vis prof mech, Roy Inst Technol, Sweden, 72- *Mem:* Am Phys Soc; Swedish Math Soc. *Res:* Experimental mechanics; cavitation; turbulent transition; structure of shear in flows of Newtonian and non-Newtonian systems, specifically liquid crystals; dynamics of immersed bodies; Theory of inviscid, incompressible fluid dynamics; non-linear mechanics. *Mailing Add:* Dept Eng Sci Univ Fla Aero Bldg Gainesville FL 32611

LINDGREN, FRANK TYCKO, b San Francisco, Calif, Apr 14, 24; m 53. BIOPHYSICS. *Educ:* Univ Calif, Berkeley, BA, 47, PhD(biophys), 55. *Prof Exp:* Res asst biophysicist, 55-56, res assoc biophysicist, 56-67, RES BIOPHYSICIST, DONNER LAB, UNIV CALIF, BERKELEY, 67- *Concurrent Pos:* Assoc ed, Lipids, Am Oil Chemists Soc, 66-76; fel, coun arteriosclerosis, Am Heart Asn; reviewer, NIH, NSF grants. *Mem:* Am Oil Chemists Soc; Sigma Xi; AAAS; Am Heart Asn. *Res:* Physical chemistry and biochemistry of blood lipids and lipo-proteins as they occur in states of health and diseases; instrumentation and engineering necessary to facilitate such investigations. *Mailing Add:* 315 Donner Lab Univ of Calif Berkeley CA 94720

LINDGREN, GORDON EDWARD, b Minneapolis, Minn, Apr 29, 36; m 59; c 3. PHYSICS, MATH. *Educ:* Augsburg Col, BA, 59; Univ SDak, MA, 63; Univ Iowa, PhD(sci educ), 70. *Prof Exp:* Teacher physics/math, Minnetonk High Sch, Excelsior, Minn, 59-61; from instr to assoc prof, 63-77, dean sci & math, 72-80, PROF PHYSICS, BEMIDJI STATE UNIV, 77- *Mem:* Am Asn Physics Teachers; Am Asn Univ Prof; Radiation Res Soc. *Res:* Electron spin resonance; radiation physics. *Mailing Add:* Div Sci/Math Bemidji State Col Bemidji MN 56601

LINDGREN, RICHARD ARTHUR, b Providence, RI, June 2, 40; m 63; c 4. NUCLEAR PHYSICS. *Educ:* Univ RI, BA, 62; Wesleyan Univ, MA, 64; Yale Univ, PhD(nuclear physics), 69. *Prof Exp:* Res assoc nuclear physics, Univ Md, College Park, 69-70; res assoc, Nat Res Coun, Nat Acad Sci, 70-71 & Univ Rochester, 71-73; res physicist nuclear physics, Naval Res Lab, Washington, DC, 73-77; ASSOC PROF, NUCLEAR PHYSICS GROUP, UNIV MASS, AMHERST, 77- *Concurrent Pos:* Instr, George Mason Univ, 73-75; assoc prof, Cath Univ Am, 75-76; consult, Naval Res Lab, 77- & Lawrence Livermore Nat Lab, 81- *Mem:* Sigma Xi; Am Phys Soc. *Res:* Nuclear structure studies using inelastic electron scattering, particularly those nuclear states excited strongly via nuclear magnetization currents; comparison of inelastic proton and electron scattering for high spin stretched states. *Mailing Add:* Inst Nuclear & Particle Physics Univ Va Charlottesville VA 22901

LINDGREN, WILLIAM FREDERICK, b San Mateo, Calif, Dec 23, 42; c 1. MATHEMATICS. *Educ:* SDak Sch Mines & Technol, BS, 64, MS, 66; Southern Ill Univ, PhD(math), 71. *Prof Exp:* Mathematician & analyst, Atomic Energy Div, Phillips Petrol Co, 66-67; assoc prof, 71-80, PROF MATH, SLIPPERY ROCK STATE COL, 80- *Concurrent Pos:* Vis prof math, Va Polytech Inst, 78-79. *Mem:* Am Math Soc; Sigma Xi. *Res:* General topology. *Mailing Add:* Slippery Rock Univ Pa Slippery Rock PA 16057

LINDH, ALLAN GODDARD, b Mason City, Wash, Mar 18, 43; m 71; c 2. SEISMOLOGY, EARTHQUAKE PREDICTION. *Educ:* Univ Calif, BA; Stanford Univ, MS, PhD(geophys). *Prof Exp:* GEOPHYSICIST, US GEOL SURV, 73- *Mem:* AAAS; Soc Explor Geophys; Am Geophys Union. *Res:* Earthquake prediction and estimation of probabilities of earthquake occurrence; quantification of earthquake characteristics. *Mailing Add:* US Geol Surv MS 977 345 Middlefield Rd Menlo Park CA 94025

LINDHEIMER, MARSHALL D, b Brooklyn, NY, June 28, 32; m 58; c 5. INTERNAL MEDICINE, NEPHROLOGY. *Educ:* Cornell Univ, AB, 52; Univ Geneva, BSM, 57, MD, 61. *Prof Exp:* Intern, Rochester Gen Hosp, 61-62; resident, Brooklyn Vet Admin Hosp, 62-63; resident & chief resident, Brookdale Hosp, Brooklyn, 63-64; US Pub Health Serv fel, Boston Univ, 64-66; sr instr med, Case Western Reserve Univ, 66-69; asst prof, Northwestern Univ, 69-70; from asst to assoc prof, 70-76, PROF MED, OBSTET & GYNEC, UNIV CHICAGO, 76- *Concurrent Pos:* Prin investr grants, NIH, 72-; fel, High Blood Pressure Res Coun, Am Heart Asn, 80, Asn Am Physicians, 85. *Honors & Awards:* Chesler Award, 88. *Mem:* Am Physiol Soc; Soc Gynec Invest; fel Am Col Physicians; Int Soc Study Hypertension Pregnancy (secy-treas, 81-87, pres elect, 88); Asn Am Physicians. *Res:* Salt and water physiology and renal disease; renal physiology and hypertension in pregnancy; volume homeostasis and vasopressin in gravid animal models. *Mailing Add:* 950 E 59th St Univ Chicago Hosp & Clins Chicago IL 60637

LINDHOLM, FREDRIK ARTHUR, b Tacoma, Wash, Feb 26, 36; m 59, 69. ELECTRICAL ENGINEERING. *Educ:* Stanford Univ, BS, 58, MS, 60; Univ Ariz, PhD(elec eng), 63. *Prof Exp:* From instr to assoc prof elec eng, Univ Ariz, 60-66; PROF ELEC ENG, UNIV FLA, 66- *Concurrent Pos:* Assoc scientist, Lockheed Corp, 60; sr engr, Motorola Semiconductor Prods, Phoenix, Ariz, 63-66; mem res adv comt electronics, NASA, 68-70; vis prof elec eng, Univ Leuven, Belg, 73-74; consult, Jet Propulsion Lab, Pasadena, Calif, 78-, Los Alamos Nat Lab, 81- & Motorola Bipolar Technol Ctr, Mesa, Ariz, 84- *Honors & Awards:* Awards, Inst Elec & Electronics Engrs, 63 & 65. *Mem:* Fel Inst Elec & Electronics Engrs; Am Phys Soc; Am Asn Physics Teachers. *Res:* Semiconductor device physics, including transistors, diodes, integrated circuits, photovoltaics, photoconductivity and equivalent circuit representations; solar cells. *Mailing Add:* Dept Elec Eng Univ Fla Gainesville FL 32601

LINDHOLM, JOHN C, b Wichita, Kans, Nov 3, 23; m 47; c 4. MACHINE DESIGN, DYNAMICS. *Educ:* Kans State Univ, BS(mech eng) & BS(bus admin), 49; Univ Kans, MS, 56; Purdue Univ, PhD(mach design), 61. *Prof Exp:* Design engr, Gen Elec Co, 49-52; sr engr, Midwest Res Inst, 52-54; instr mech eng, Univ Kans, 54-57 & Purdue Univ, 57-59; assoc prof, Kans State Univ, 60-74, prof mech eng, 74-80, prof eng tech & head dept, 80-87, prof mech eng, 87-; RETIRED. *Concurrent Pos:* Vis prof, Univ Assiut, 64-66; Fulbright prof, Univ S Pac, 87-88. *Mem:* Am Soc Mech Engrs; Soc Exp Stress Anal; Am Soc Eng Educ; Soc Exp Mech. *Res:* Three-dimensional photoelastic stress analysis; mechanical properties materials at intermediate strain rates; kinematic synthesis of mechanisms. *Mailing Add:* 744 Elling Dr Manhattan KS 66502

LINDHOLM, ROBERT D, b Rockford, Ill, June 17, 40; m 62; c 2. PHYSICAL CHEMISTRY. *Educ:* Northern Ill Univ, BS, 63, MS, 64; Univ Southern Calif, PhD(phys chem), 69. *Prof Exp:* Sr res chemist, 68-73, RES ASSOC, EASTMAN KODAK CO, 73- *Mem:* Am Chem Soc. *Res:* Photochemistry of transition-metal complexes; silver halide photochemistry and microelectronics fabrication. *Mailing Add:* Res Lab Eastman Kodak Co Rochester NY 14650

LINDHOLM, ROY CHARLES, b Washington, DC, Mar 8, 37; m 65; c 2. STRATIGRAPHY, SEDIMENTATION. *Educ:* Univ Mich, BS, 59; Univ Tex, MA, 63; Johns Hopkins Univ, PhD(geol), 67. *Prof Exp:* Instr, Johns Hopkins Univ, 65-66; from asst prof to assoc prof, 67-77, chmn geol, 86-89 PROF GEOL, GEORGE WASHINGTON UNIV, 77- *Mem:* Soc Econ Paleont & Mineral. *Res:* Paleozoic carbonate rocks of eastern United States; sequences of carbonate cements; geology of Triassic-Jurassic rocks in Virginia; lacustrine deposits; sedimentology of Cretaceous sandstones in northern Virginia. *Mailing Add:* Dept Geol George Washington Univ Washington DC 20052

LINDHOLM, ULRIC S, b Washington, DC, Sept 11, 31; m 62; c 4. MATERIALS SCIENCE, APPLIED MECHANICS. *Educ:* Mich State Univ, BS, 53, MS, 55, PhD(appl mech), 60. *Prof Exp:* Sr res engr, 60-64, mgr eng mech, 64-71, asst dir, Dept Mec Sci, 71-73, dir, Dept Mat Sci, 73-85, VPRES DIV ENG & MAT SCI, SOUTHWEST RES INST, 85- *Mem:* Fel Am Soc Mech Engrs; Sigma Xi; Am Soc Metals; fel AAAS. *Res:* Applied mechanics, structural dynamics and vibrations; wave propagation; material properties. *Mailing Add:* 6220 Culebra Rd San Antonio TX 78284

LINDHORST, TAYLOR ERWIN, b St Louis, Mo, Aug 11, 28; m 51; c 2. MYCOLOGY. *Educ:* St Louis Col Pharm, BS, 51; Wash Univ, MA, 54, PhD(mycol), 67. *Prof Exp:* Instr pharm, 51-52, resident biol, 52-55, assoc instr, 56-59, asst prof, 59-67, assoc prof, 67-74, PROF BIOL & DEAN STUDENTS, ST LOUIS COL PHARM, 74- *Mem:* Am Soc Pharmacog; Bot Soc Am. *Res:* Mycological studies concerning response and growth variation to antibiotic substances. *Mailing Add:* Dean of Students St Louis Col Pharm 4588 Parkview Pl St Louis MO 63110

LINDLEY, BARRY DREW, b Orleans, Ind, Jan 25, 39; div; c 3. PHYSIOLOGY, BIOPHYSICS. *Educ:* DePauw Univ, BA, 60; Western Reserve Univ, PhD(physiol), 64. *Prof Exp:* From asst prof to assoc prof, 65-84, PROF PHYSIOL, SCH MED, CASE WESTERN RESERVE UNIV, 84-, ASSOC DEAN MED EDUC, 85-, PROF & ACTG CHMN ANAT. *Concurrent Pos:* NSF fel neurophysiol, Nobel Inst Neurophysiol, Karolinska Inst, Sweden, 64-65; Lederle med fac award, 6770; USPHS res career develop award, 71-76; mem, Physiol Study Sect, NIH, 75-79. *Mem:* Am Physiol Soc; Soc Gen Physiologists; Biophys Soc. *Res:* Muscle biophysics; ion and water transport; membrane permeability; irreversible thermodynamics; electrophysiology of nerve, muscle and glandular tissue. *Mailing Add:* Off Med Educ Case Western Reserve Univ Sch Med 2109 Adelbert Rd Cleveland OH 44106-4901

LINDLEY, CHARLES A(LEXANDER), b Union City, Ind, May 12, 24; m 46; c 2. AEROSPACE SYSTEMS ENGINEERING, ALTERNATIVE ENERGY SYSTEMS. *Educ:* Ohio State Univ, BAeroEng & MS, 49; Calif Inst Technol, PhD(aeronaut), 56. *Prof Exp:* Instr, Ohio State Univ, 47-48; eng aid, Nat Adv Comt Aeronaut, 48; compressor design engr, Thompson Aircraft Prod, Inc, 49-52, consult, 52-55; eng specialist, Marquardt Corp, 55-57, mgr engine res, 57-59, res consult, 59-61, chief res consult, 61-63; sr staff engr, Appl Mech Div, Aerospace Corp, 63-65, dir vehicle design, Satellite Systs Div, 65-73; res assoc, Environ Qual Lab, Calif Inst Technol, 73-74; assoc dir advan systs off, Energy Systs Group, 74-78, SR STAFF ENGR, THREAT ANAL OFF, AEROSPACE CORP, 78- *Concurrent Pos:* Lectr & consult wind power & wind resources, Univ Calif, Santa Barbara. *Mem:* AAAS; assoc fel Am Inst Aeronaut & Astronaut; Sigma Xi. *Res:* Aeronautical and space propulsion; air breathing and recoverable boosters; physics and chemistry of the upper atmosphere; satellite systems engineering; wind and solar energy; manned and unmanned space vehicle systems engineering; energy conversion devices; meteorology. *Mailing Add:* 18900 Pasadero Dr Tarzana CA 91356

LINDLEY, CHARLES EDWARD, b Macon, Miss, Dec 21, 21; m 45. ANIMAL HUSBANDRY. *Educ:* Miss State Univ, BS, 46; Wash State Univ, MS, 48; Okla State Univ, PhD, 57. *Prof Exp:* Asst, Wash State Univ, 46-48, asst prof animal husb, 48-51; asst, Okla State Univ, 51-52; chmn dept animal sci, 52-69, PROF ANIMAL SCI, MISS STATE UNIV, 52-, DEAN COL AGR, 69- *Mem:* Am Soc Animal Sci. *Res:* Livestock production and animal breeding. *Mailing Add:* Off of Dean Col of Agr Miss State Univ Drawer AG Mississippi State MS 39762

LINDLEY, KENNETH EUGENE, b Stratton, Colo, Mar 16, 24; m 48; c 4. ELECTRICAL ENGINEERING, MATHEMATICS. *Educ:* Univ Wis, BS, 48, MS, 49; State Univ Iowa, PhD(elec eng), 53. *Prof Exp:* From instr to prof elec eng, SDak State Col, 49-63; prof Physics & Math & Chmn Sci & Math Div, Houghton Col, 63-89; RETIRED. *Concurrent Pos:* Develop consult, Acme Elec Corp, 66- *Mem:* Am Sci Affil; Am Inst Elec & Electronics Engrs; Am Soc Eng Educ. *Res:* Electrical power supplies. *Mailing Add:* Rte 1 Box 43A Houghton NY 14744

LINDMAN, ERICK LEROY, JR, b Seattle, Wash, Mar 20, 38; m 63; c 5. COMPUTATIONAL PLASMA PHYSICS, COMPUTATIONAL ELECTROMAGNETISM. *Educ:* Calif Inst Technol, BS, 60; Univ Calif, Los Angeles, MS, 63, PhD(physics), 64. *Prof Exp:* Res scientist, Univ Tex, Austin, 64-65, asst prof physics, 65-68; physicist, Austin Res Assocs, 68-71; staff mem, Los Alamos Nat Lab, 71-78, assoc group leader laser fusion target design, 78-80, assoc group leader inertial fusion supporting physics, 80-82, group leader, advan concept & Plasma applns, 83-86, staff mem Inertia Fusion, 86-87; SR SCIENTIST, MISSION RES CORP, 87- *Concurrent Pos:* Vis prof, sr vis fel, UK Sci & Eng Res Coun, Blackett Lab, Imp Col, London UK, 82-83. *Mem:* Am Phys Soc. *Res:* Computational theoretical and experimental plasma physics including computer simulation code development with applications to plasma opening switches; particle beam sources; space plasmas and intense laser and particle beam interaction with matter. *Mailing Add:* Mission Res Corp 127 E Gate Dr Los Alamos NM 87544

LINDMAYER, JOSEPH, b Budapest, Hungary, May 8, 29; US citizen; m 55; c 2. SOLID STATE PHYSICS. *Educ:* Williams Col, MS, 63; Aachen Tech Univ, PhD, 68. *Prof Exp:* Scientist, Inst Measurements Tech, Hungarian Acad Sci, 55-56; scientist res ctr, Sprague Elec Co, 57-63, dept head semiconductor physics, 63-68; br mgr, Comsat Labs, Commun Satellite Corp, Clarksburg, 68-74; dir physics lab, Defense Language Inst, 74-76; PRES, SOLAREX CORP, 76-; PRES, QUANTEX CORP. *Concurrent Pos:* Vis lectr, Yale Univ, 68-69. *Mem:* Inst Elec & Electronics Engrs. *Res:* Semiconductor physics, electronics and devices. *Mailing Add:* One Riverwood Ct Potomac MD 20854

LINDNER, DUANE LEE, b Ft Dodge, Iowa, May 7, 50; m 77; c 3. PHYSICAL CHEMISTRY, THERMODYNAMICS. *Educ:* Mass Inst Tehnol, SB, 72; Univ Calif, Berkeley, PhD(chem), 77. *Prof Exp:* Mem tech staff, 77-86, supvr, Chem Div, 86-89, MGR MAT SCI & TECHNOL, SANDIA NAT LABS, 89- *Mem:* Am Chem Soc; Mat Res Soc; Sigma Xi. *Res:* Chemical kinetics and reaction mechanisms of solid-solid and gas-solid systems, particularly metal-hydrogen reactions. *Mailing Add:* Sandia Nat Labs Livermore CA 94551-0969

LINDNER, ELEK, b Budapest, Hungary, June 3, 24; US citizen; m 60; c 1. ANALYTICAL BIOCHEMISTRY, MARINE BIOLOGY. *Educ:* Budapest Tech Univ, Dipl Chem Eng, 46, PhD(biochem), 74. *Prof Exp:* Prof asst agr chem, Budapest Tech Univ, 47-48 & food chem, 48-50; chief chemist, Anal & Res Lab, Elida Cosmetic Factory, 50-51; res chemist, Res Inst, Fatty Oil Chem Indust, 51-56, Res & Develop Div, Lever Bros Co, NJ, 57-61 & Chevron Res Corp, Calif, 61-64; prod mgr, Sawyer Tanning Co, Calif, 64-65; res chemist, Paint Lab, Mare Island Naval Shipyard, 65-73; SUPVRY RES CHEMIST, NAVAL OCEAN SYSTS CTR, 73- *Concurrent Pos:* Res chemist, Naval Ship Res & Develop Ctr, Annapolis, Md, 73. *Mem:* Am Chem Soc. *Res:* Chemistry and biochemistry of food and agricultural products; analytical methods; chemistry of fatty oils; detergents and surfactants; biochemistry of marine organisms. *Mailing Add:* PO Box 82148 San Diego CA 92138-2148

LINDNER, LUTHER EDWARD, b Toledo, Ohio, Aug 6, 42; m 69; c 3. PATHOLOGY. *Educ:* Univ Toledo, BS, 64; Western Reserve Univ, MD, 67, Case Western Reserve Univ, PhD(exp path), 74. *Prof Exp:* From intern to resident, Univ Hosp, Cleveland, 67-72; fel path, Case Western Reserve Univ, 69-72; staff pathologist, William Beaumont Army Med Ctr, 72-74, chief, Anatomic Path, 74-75; asst prof lab med, Univ Nev, 75-82; ASSOC PROF PATH, LAB MED, TEX A&M UNIV, 82- *Concurrent Pos:* Consult path, Reno Vet Admin Hosp, Marlin Vet Admin Hosp, Burleson County Hosp; dir lab, Path Consult Serv, Tex A&M Univ. *Mem:* Am Soc Cytol; Am Soc Clin Path; Col Am Pathologists; US-Can Acad Path. *Res:* Studies of anatomic changes in disease with histochemical correlations and application to diagnosis; sexually transmitted diseases. *Mailing Add:* Dept Path & Lab Med Tex A&M Univ College Station TX 77843-1114

LINDNER, MANFRED, b Chicago, Ill, Oct 21, 19; m 46; c 2. NUCLEAR CHEMISTRY. *Educ:* Northwestern Univ, Ill, BS, 40; Univ Calif, Berkeley, PhD(nuclear chem), 48. *Prof Exp:* Chemist, Hanford Eng Works, Wash, 44-46; res asst, Univ Calif, Berkeley, 46-48; asst prof chem, Wash State Col, 48-51; chemist, Calif Res & Develop Co, 51-53; SR CHEMIST, RADIOCHEM DIV, LAWRENCE LIVERMORE LAB, 53- *Concurrent Pos:* Rothschild fel, Weizmann Inst Sci, 62-63. *Mem:* AAAS; Am Phys Soc. *Res:* Neutron capture cross-sections; nuclear structure. *Mailing Add:* Lawrence Livermore Nat Lab L-234 PO Box 808 Livermore CA 94551

LINDORFF, DAVID EVERETT, b Moline, Ill, Aug 25, 45; m 72; c 2. HYDROGEOLOGY. *Educ:* Augustana Col, BA, 67; Univ Wis-Madison, MA, 69, MS, 71. *Prof Exp:* Geologist, Pa Dept Environ Resources, 71-75; asst geologist, Ill State Geol Surv, 75-80; HYDROGEOLOGIST, WIS DEPT NAT RESOURCES, 80- *Concurrent Pos:* Coordr, Midwest Groundwater Conf, 87. *Mem:* Asn Ground Water Scientists & Engrs. *Res:* Ground-water contamination; hydrogeology of strip mines; siting of sanitary landfills; ground-water standards; ground-water sampling procedures. *Mailing Add:* Dept Natural Resources PO Box 7921 Madison WI 53707

LINDOWER, JOHN OLIVER, b Ashland, Ohio, March 15, 29; m 51; c 3. SCIENCE EDUCATION. *Educ:* Ashland Col, AB, 50; Ohio State Univ, MD, 55, PhD(pharmacol), 68. *Prof Exp:* Gen rotating internship, Miami Valley Hosp, Dayton, Ohio, 55-56, asst surg resident, 58-59; gen med officer, US Army, 56-58; gen med pract, Dayton, Ohio, 59-65; fel pharmacol, Col Med, Ohio State Univ, 65-68, from instr to assoc prof, 68-75; prof & chmn pharmacol, 75-82, from asst dean to assoc dean curric affairs, 76-81, assoc dean acad affairs, 81-87, PROF PHARMACOL & TOXICOL, WRIGHT STATE UNIV, 82-, INTERIM DEAN, 87- *Concurrent Pos:* Coordr curric affairs, Sch Med, Wright State Univ, 75-76; mem, Comn Accrediting, Asn Theol Schs US & Can, 78-80; mem, NCent Res Rev & Adv Comn, Am Heart Asn, 79-81; proj dir, Area Health Educ Ctr, Region IV, Ohio, 82-87; mem, Drug Info Adv Panel Geriat, US Pharmacopoieia, Inc, 85- *Mem:* AMA. *Res:* Clinical pharmacology; cardiovascular research; subcellular and ultrastructural pharmacology. *Mailing Add:* Dept Pharmacol Sch Med Wright State Univ PO Box 927 Dayton OH 45401

LINDQUIST, ANDERS GUNNAR, b Lund, Sweden, Nov 21, 42; m 66, 86; c 3. SYSTEMS AND CONTROL. *Educ:* Royal Inst Technol, Sweden, MS, 67, TeknL, 68, TeknD(optimization & systs theory), 72. *Prof Exp:* Res assoc optimization, Royal Inst Technol, Sweden, 69-72, docent, 72; vis asst prof math, Univ Fla, 72-73; assoc prof, Brown Univ, 73; assoc prof math, Univ Ky, 74-80, prof, 80-83; PROF OPTIMIZATION & SYSTS THEORY, ROYAL INST TECHNOL, 82- *Concurrent Pos:* Affil prof, Washington Univ, St Louis, 89- *Mem:* Soc Indust & Appl Math; fel Inst Elec & Electronic Engrs. *Res:* Stochastic systems theory, control theory and estimation. *Mailing Add:* Dept Math Royal Inst Technol Stockholm 10044 Sweden

LINDQUIST, DAVID GREGORY, b Chicago, Ill, Feb 14, 46; m 73; c 2. ICHTHYOLOGY. *Educ:* Univ Calif, Los Angeles, BA, 68; Calif State Univ, Hayward, MA, 72; Univ Ariz, PhD(zool), 75. *Prof Exp:* Asst prof, 75-81, assoc prof, 81-86, PROF BIOL, UNIV NC, WILMINGTON, 86- *Concurrent Pos:* Fulbright Fel, 88. *Mem:* Am Soc Ichthyologists & Herpetologists. *Res:* Ethology and behavioral ecology of reef fishes; natural history of Southeastern freshwater fishes. *Mailing Add:* Dept Biol Sci Univ NC Wilmington NC 28403-3297

LINDQUIST, EVERT E, b Susanville, Calif, June 26, 35; m 57; c 4. ACAROLOGY, SYSTEMATIC ENTOMOLOGY. *Educ:* Univ Calif, Berkeley, BS, 57, MS, 59, PhD(entom), 63. *Prof Exp:* Res scientist, 62-75, SR RES SCIENTIST, BIOSYST RES CTR, AGR CAN, 75- *Concurrent Pos:* Adj prof, Carleton Univ, 71-83; vis lectr, Ohio State Univ Summer Acarology Prog, 72-88; vis lectr, Univ Nat Auton Mex, 83, 87, col postgrad Mex, 87, Inst Polytech Nat Mex, 87; mem exec comt, Int Cong Acarology, 90- *Honors & Awards:* Acarology Award, Ohio State Univ, 74, Agr Acarology Award, 88. *Mem:* Fel Entom Soc Can; Acarological Soc Am; Sigma Xi; Europ Asn Acarologists. *Res:* Systematics, cladistics and classification of Acari diversi; symbiotic relationships between mites and insects; homology of acarine external structures; faunistics and biogeography of mites. *Mailing Add:* Biosyst Res Centre Agr Can Ottawa ON K1A 0C6 Can

LINDQUIST, RICHARD KENNETH, b Minneapolis, Minn, Oct 2, 42; m 69. ENTOMOLOGY. *Educ:* Gustavus Adolphus Col, BA, 64; Kans State Univ, MS, 67, PhD(entom), 69. *Prof Exp:* Instr entom, Kans State Univ, 68-69; asst prof, 69-74, ASSOC PROF ENTOM, OHIO AGR RES & DEVELOP CTR, 74- *Mem:* Entom Soc Am. *Res:* Biology, ecology and control of insect and mite pests of floral and greenhouse vegetable crops. *Mailing Add:* Entomol 103 Bot/Zool Bldg Ohio State Univ Main Campus Columbus OH 43210

LINDQUIST, RICHARD WALLACE, b Worcester, Mass, May 6, 33; m 57; c 2. PHYSICS. *Educ:* Worcester Polytech Inst, BS, 54; Princeton Univ, AM, 57, PhD(physics), 62. *Prof Exp:* Instr physics, Princeton Univ, 58-60; asst prof, Adelphi Univ, 60-64; res assoc, Univ Tex, 64-65; assoc prof, 65-77, PROF PHYSICS, 77-, CHAIR, WESLEYAN UNIV, 78-84, 86- *Mem:* Am Phys Soc; Am Asn Physics Teachers. *Res:* General relativity; geometrodynamics; gravitational collapse. *Mailing Add:* Dept Physics Wesleyan Univ Middletown CT 06457-6036

LINDQUIST, ROBERT HENRY, b Minneapolis, Minn, Feb 27, 28; m 50; c 2. PHYSICAL CHEMISTRY. *Educ:* Univ Minn, BChem, 49, MS, 50; Univ Calif, PhD(chem), 55. *Prof Exp:* Res chemist, Chevron Res Co, 55-60, sr res chemist, 60-64, sr res assoc, 64-75, asst to pres, 75-78, mgr solar, 78-80; consult corp develop, Chevron Corp, 80-86, mgr corp res & planning dept, Chevron Res Co. 86-88; PRES, LINQUIST CONSULTS, 88- *Concurrent Pos:* Consult, high tech bus profitability. *Mem:* Am Chem Soc; Am Phys Soc. *Res:* Solid state physics; magnetic resonance; physics of ultra-fine particles; heterogeneous catalysis; reaction kinetics; synthetic fuels; alternate energy sources. *Mailing Add:* 225 Arlington Berkeley CA 94707

LINDQUIST, ROBERT MARION, b Cumberland, Wis, Dec 4, 23; c 3. ORGANIC CHEMISTRY, PHOTOGRAPHIC CHEMISTRY. *Educ:* Univ Wis, BS, 44; Univ Minn, PhD, 50. *Prof Exp:* Res chemist photog res, Gen Aniline & Film Corp, 50-56; assoc chemist photog processes, 56-57, staff chemist, 57-62, adv chemist, 62-63, develop chemist, 63-65, SR CHEMIST, IBM CORP, 65- *Honors & Awards:* First Level Invention Award, IBM Corp, 62, Outstanding Contrib Award, 74. *Mem:* Sr mem Am Chem Soc; Sigma Xi; fel Am Inst Chemists; sr mem Soc Photog Scientists & Engrs. *Res:* Electrophotographic processes. *Mailing Add:* 4788 Briar Ridge Trail Boulder CO 80301

LINDQUIST, ROBERT NELS, b Bakersfield, Calif, Sept 29, 42; m 68. BIOCHEMISTRY, ORGANIC CHEMISTRY. *Educ:* Occidental Col, BA, 65; Ind Univ, PhD(chem), 68. *Prof Exp:* Chemist, Shankman Labs, 65-66; res chemist, Shell Develop Co, 68-71; asst prof, 71-75, assoc prof, 75-80, PROF CHEM, SAN FRANCISCO STATE UNIV, 80- *Mem:* AAAS; Am Chem Soc. *Res:* Enzyme and enzyme model reaction kinetics and mechanisms. *Mailing Add:* Dept Chem San Francisco State Univ 1600 Holloway Ave San Francisco CA 94132

LINDQUIST, WILLIAM BRENT, b Ft Frances, Ont, June 23, 53; m 89. NUMERICAL SOLUTION OF PDES, RIEMANN PROBLEMS. *Educ:* Univ Man, BS, 75; Cornell Univ, PhD(physics), 81. *Prof Exp:* Assoc res scientist, Courant Inst Math Sci, NY Univ, 81-85, from res asst prof to res assoc prof, 85-89; ASSOC PROF APPL MATH, STATE UNIV NY, STONY BROOK, 89- *Concurrent Pos:* Consult, Inst Energy Technol, Kjeller, Norway, 85-88. *Mem:* Am Phys Soc; Soc Indust & Appl Math; Am Math Soc; Soc Petrol Engrs. *Res:* Hyperbolic equations; numerical methods for PDE's; flow in porous media; anomalous magnetic moment of the electron. *Mailing Add:* Dept Appl Math & Statist State Univ NY Stony Brook NY 11794-3600

LINDROOS, ARTHUR E(DWARD), b Worcester, Mass, Aug 14, 22; m 44; c 4. CHEMICAL ENGINEERING. *Educ:* Worcester Polytech Inst, BS, 43, MS, 44; Yale Univ, DEng, 49. *Prof Exp:* Chem engr res & develop, Kellex Corp, 49-53; chem engr process develop, Air Reduction Chem Co, NJ, 53-59, mgr develop, 59-61; tech mgr, Cumberland Chem Corp, 61-62; mgr process eng, Air Reduction Chem & Carbide Co, 62-67; mgr process eng, Airco Chem & Plastics Co, 67-69, eng mgr, 69-71; vpres, Techni-Chem Co, 71-76; div supvr, Unit CPC Int, 76-79, assoc mgr eng, 79-85, mgr eng, 85-86, DIR ENG, PENICK CORP, 86- *Mem:* Am Chem Soc; Am Inst Chem Engrs; Sigma Xi. *Res:* Mass transfer; phase equilibria at elevated pressures; nuclear reactor fuel reprocessing; acetylenic chemistry; vinyl monomers; resins and emulsions; polyvinyl alcohol; calcium carbide; lime recovery; chlorinated hydrocarbons; narcotics; pharmaceuticals. *Mailing Add:* 20 N Briarcliff Rd Mountain Lakes NJ 07046

LINDSAY, BRUCE GEORGE, b The Dalles, Ore, Mar 7, 47; m 69; c 2. MATHEMATICAL STATISTICS. *Educ:* Univ Ore, BA, 69; Univ Wash, PhD(biomath), 78. *Prof Exp:* From asst prof to assoc prof, 79-87, PROF STATIST, PA STATE UNIV, 87- *Concurrent Pos:* Prin investr, NSF, 79-; assoc ed, Annals Statist, 85-; vis prof, Johns Hopkins Univ, 87, Cornell Univ, 88 & Yale Univ, 90. *Honors & Awards:* Humboldt Sr Scientist Award, Humboldt Found, Ger, 90. *Mem:* Fel Inst Math Statist; Am Statist Asn; fel Int Statist Inst. *Res:* Statistical methods in semiparametric models, with emphasis on maximum likelihood and minimum distance methods in mixture models and computation. *Mailing Add:* 219 Pond Lab University Park PA 16802

LINDSAY, CHARLES MCCOWN, b Fayetteville, Tenn, July 5, 32; m 55; c 4. MATHEMATICS. *Educ:* Univ of the South, BS, 54; Univ Iowa, MS, 57; George Peabody Col, PhD(math), 65. *Prof Exp:* From instr to assoc prof, 57-71, interim dean, 75-76, PROF MATH, COE COL, 71-, CHMN DEPT, 63- *Mem:* Am Math Soc; Math Asn Am. *Res:* Mathematics education. *Mailing Add:* Coe Col Coe Col 1220 First Ave NE Cedar Rapids IA 52402

LINDSAY, DAVID TAYLOR, b Philadelphia, Pa, Mar 22, 35; m 59; c 1. DEVELOPMENTAL BIOLOGY. *Educ:* Amherst Col, BA, 57; Johns Hopkins Univ, PhD(biol), 62. *Prof Exp:* From asst prof to assoc prof zool, 62-77, MEM FAC, UNIV GA, 77- *Concurrent Pos:* Nat Sci Found res grant develop biol, 63-66. *Mem:* AAAS; Soc Develop Biol; Am Soc Zoologists; NY Acad Sci. *Res:* Mechanisms of cellular differentiation; regulation of protein systhesis in differentiation; biological role of histone proteins; bilateral symmetry. *Mailing Add:* Dept of Zool Univ of Ga Athens GA 30602

LINDSAY, DELBERT W, systematic botany; deceased, see previous edition for last biography

LINDSAY, DEREK MICHAEL, b Belfast, Northern Ireland, Oct 3, 44; m 70. PHYSICAL CHEMISTRY, CHEMICAL PHYSICS. *Educ:* Trinity Col, Dublin, BA, 67; Harvard Univ, PhD(chem), 75. *Prof Exp:* Res asst, Mass Inst Technol, 75-76; res asst chem, Coop Inst Res Environ Sci & Univ Colo, Boulder, 76-78; PROF CHEM, CITY COL NEW YORK, 78- *Mem:* Am Chem Soc; Am Phys Soc. *Res:* Laser fluorescence and electron spin resonance studies of small metal clusters; application of this research to catalysis, surface science and solid state physics; intramolecular perturbations and gas-phase energy transfer processes. *Mailing Add:* Dept Chem City Col Convent Ave & 138th St New York NY 10031

LINDSAY, DWIGHT MARSEE, b Versailles, Ind, June 19, 21; m 43; c 2. MAMMALOGY. *Educ:* Hanover Col, AB, 47; Univ Ky, MS, 49; Univ Cincinnati, PhD, 58. *Prof Exp:* Asst zool, Univ Ky, 47-48, instr, 51-52; from instr to prof, 49-86, chmn dept biol sci, 74-84, EMER PROF BIOL, GEORGETOWN COL, 86- *Res:* Endocrinology; physiology; histology; embryology. *Mailing Add:* Dept Biol Georgetown Col Georgetown KY 40324

LINDSAY, EVERETT HAROLD, JR, b La Junta, Colo, July 2, 31; div; c 3. VERTEBRATE PALEONTOLOGY. *Educ:* Chico State Col, AB, 53, MA, 57; Cornell Univ, MST, 62; Univ Calif, Berkeley, PhD(paleont), 67. *Prof Exp:* From asst prof to assoc prof, 67-80, PROF GEOL, UNIV ARIZ, 80- *Mem:* AAAS; Soc Vert Paleont; Geol Soc Am; Paleont Soc. *Res:* Biostratigraphy; magnetostratigraphy; taxonomy and evolution of small mammal fossils. *Mailing Add:* 2771 N Treat Ave Tucson AZ 85716-2159

LINDSAY, GEORGE EDMUND, b Pomona, Calif, Aug 17, 16. PLANT TAXONOMY. *Educ:* Stanford Univ, BA, 51, PhD, 56. *Prof Exp:* Dir, Desert Bot Garden, Ariz, 39-40; admin asst, Arctic Res Lab, Off Naval Res, 52-53; exec dir, San Diego Mus Natural Hist, 56-63; exec dir, Calif Acad Sci, 63-82; dir, Pepperwood Ranch Natural Preserve, 82-86; RETIRED. *Mem:* AAAS; Cactus & Succulent Soc Am; Int Orgn Succulent Plant Studies; Am Asn Mus; Asn Dirs Sci Mus. *Res:* Taxonomic botany; taxonomy and ecology of Cactaceae and xerophytic plants of Baja California and other parts of Mexico. *Mailing Add:* 574 Arballo Dr San Francisco CA 94132

LINDSAY, GLENN FRANK, b Portland, Ore, June 13, 35; m 65. INDUSTRIAL ENGINEERING, OPERATIONS RESEARCH. *Educ:* Ore State Univ, BSc, 60; Ohio State Univ, MSc, 62, PhD(indust eng), 68. *Prof Exp:* Res assoc, Systs Res Group, Ohio State Univ, 61-65; asst prof opers res, 65-69, ASSOC PROF OPERS RES, NAVAL POSTGRAD SCH, 69- *Mem:* Opers Res Soc Am; Inst Mgt Sci; Am Soc Eng Educ. *Res:* Counter-insurgency small-unit military operations; industrial inspection systems. *Mailing Add:* Dept Opers Naval Postgrad Sch Res Monterey CA 93940

LINDSAY, HAGUE LELAND, JR, b Ft Worth, Tex, Jan 24, 29; m 56; c 4. VERTEBRATE ZOOLOGY. *Educ:* Tex Christian Univ, BA, 49; Univ Tex, MA, 51, PhD(zool), 58. *Prof Exp:* Res scientist, Univ Tex, 54; PROF ZOOL, UNIV TULSA, 56- *Mem:* AAAS; Am Fisheries Soc; Am Soc Ichthyol & Herpet. *Res:* Vertebrate speciation, especially with amphibians; fish distribution and ecology, especially with darters. *Mailing Add:* 8518 E 35th St Tulsa OK 74145

LINDSAY, HARRY LEE, b Cotesfield, Nebr, Sept 3, 25; m 49; c 4. VIROLOGY. *Educ:* Univ Nebr, BS, 50, MS, 52; Univ Wis, PhD(bact), 63. *Prof Exp:* Microbiologist, Gateway Chemurgic Co, 51-52 & Hiram Walker & Sons, Inc, 53-59; microbiologist, Lederle Labs, Am Cyanmid Co, 63-88; RETIRED. *Mem:* Am Soc Microbiol; Am Chem Soc. *Res:* Respiratory viruses, including rhinoviruses and influenza; herpesviruses, antiviral agents; organ cultures; tissue culture and mycoplasma; antibiotic fermentations and culture improvement. *Mailing Add:* 129 Standish Dr Pearl River NY 10965

LINDSAY, HUGH ALEXANDER, b Moose Jaw, Sask, Mar 5, 26; m 56; c 3. PHYSIOLOGY. *Educ:* Univ Western Ont, BSc, 49, MSc, 52; Univ Toronto, PhD(pharmacol), 55; WVa Univ, MD, 73. *Prof Exp:* Asst pharmacol, Univ Toronto, 52-55; from asst prof to assoc prof, 55-70, PROF PHYSIOL, SCH MED, WVA UNIV, 70- *Mem:* AAAS; NY Acad Sci; Am Physiol Soc; Pharmacol Soc Can. *Res:* Growth in congenital cardiovascular disease; osteoporosis. *Mailing Add:* 1133 Van Voorhis Rd Morgantown WV 26505

LINDSAY, JAMES EDWARD, JR, b Denver, Colo, Feb 26, 28; m 49; c 5. ELECTRICAL ENGINEERING. *Educ:* Univ Denver, BS, 53; Univ Colo, Boulder, MS, 58, PhD(elec eng), 60. *Prof Exp:* Res engr, RCA Labs, 53-55; instr elec eng, Univ Denver, 55-56, res engr, Denver Res Inst, 57-58; from instr to assoc prof appl math & elec eng, Univ Colo, 58-62; from asst prof to assoc prof elec eng, Univ Denver, 62-65; res engr, Martin Marietta Co, 66-67; assoc prof geophys & basic eng, Colo Sch Mines, 67-70; res engr, Denver Res Inst, Univ Denver, 70-76; assoc prof, 76-80, PROF ELEC ENG, UNIV WYO, 80- *Mem:* Soc Indust & Appl Math; Inst Elec & Electronics Engrs; Sigma Xi. *Res:* Electromagnetic field theory; antennas and propagation; teaching of graduate and undergraduate courses in electrical engineering, geophysics and applied mathematics. *Mailing Add:* 1938 Riverwood Trails Dr Florissant MO 63031

LINDSAY, JAMES GORDON, JR, b Norfolk, Va, Jan 23, 41; m 62; c 1. NUCLEAR PHYSICS. *Educ:* Va Polytech Inst & State Univ, BS, 64, MS, 66, PhD(physics), 71. *Prof Exp:* From asst prof to assoc prof, 69-77, PROF PHYSICS, APPALACHIAN STATE UNIV, 77- *Mem:* Am Asn Physics Teachers; Am Nuclear Soc; Am Soc Mech Engrs. *Res:* Neutron activation analysis; thermal neutron cross sections. *Mailing Add:* Dept Physics Appalachian State Univ Boone NC 28608

LINDSAY, KENNETH LAWSON, b Springfield, Ill, Aug 26, 25; m 49; c 1. ORGANIC CHEMISTRY. *Educ:* Univ Ill, BS, 48; Univ Minn, PhD(chem), 52. *Prof Exp:* Res chemist, Ethyl Corp, 52-55, develop chemist, 55-61, develop assoc, 61-63, process res supvr, 63-78, sr res assoc, 78-82; AUTOMATED PUBLISHING CONSULT, 85- *Mem:* Am Chem Soc; Am Inst Chem. *Res:* Applied kinetics; organometallic chemistry; chlorine chemistry. *Mailing Add:* PO Box 148 Laie HI 96762-0148

LINDSAY, RAYMOND H, b Perry, Ga, Dec 9, 28; m 85; c 5. BIOCHEMISTRY, PHARMACOLOGY. *Educ:* Jacksonville State Col, BS, 48; Univ Ala, MS, 57, PhD(pharmacol), 61. *Prof Exp:* From asst prof to assoc prof pharmacol, 63-71, from asst prof to assoc prof med, 63-72, PROF MED, MED CTR, UNIV ALA, BIRMINGHAM, 72-, PROF PHARMACOL, 71-; DIR PHARMACOL RES UNIT, VET ADMIN HOSP, 71- *Concurrent Pos:* NIH fel physiol chem, Univ Wis, 60-62, univ fel, 62-63; dir metab res, Vet Admin Hosp, Birmingham, 65-67, asst chief radioisotope serv, 64-72. *Mem:* AAAS; Endocrine Soc; Am Thyroid Asn; Am Chem Soc; Am Physiol Soc; Am Soc Pharmacol & Exp Therapeut. *Res:* Biochemistry, pharmacology and physiology of thyroid function; antithyroid drugs; environmental goitrogens. *Mailing Add:* Vet Admin Hosp 700 S 19th St Birmingham AL 35233

LINDSAY, RICHARD H, b Portland, Ore, Sept 24, 34; m 58; c 6. NUCLEAR PHYSICS. *Educ:* Univ Portland, BS, 56; Stanford Univ, MS, 58; Wash State Univ, PhD(nuclear physics), 61. *Prof Exp:* Teaching assoc physics, Wash State Univ, 60-61; assoc prof, 61-66, PROF PHYSICS, WESTERN WASH UNIV 66- *Mem:* Am Phys Soc; Am Asn Physics Teachers. *Res:* Nuclear reactions with 30 to 65 million electron volts alpha particles; reactions with 14 million electron volts neutrons; theoretical nuclear physics; instrument design. *Mailing Add:* Dept of Physics Western Wash Univ Bellingham WA 98225

LINDSAY, ROBERT, magnetism, for more information see previous edition

LINDSAY, ROBERT CLARENCE, b Montrose, Colo, Nov 30, 36; m 57; c 2. FOOD SCIENCE, FOOD CHEMISTRY. *Educ:* Colo State Univ, BS, 58, MS, 60; Ore State Univ, PhD(food sci), 65. *Prof Exp:* Asst prof food sci, Ore State Univ, 64-69; assoc prof, 69-74, PROF FOOD SCI, UNIV WIS-MADISON, 74- *Honors & Awards:* Dairy Res Found Award, 86; Elected Fel, Inst Food Technologists, 88. *Mem:* Am Chem Soc; Fel inst Food Technologists; Am Dairy Sci Asn; Am Soc Microbiol. *Res:* Flavor chemistry; sensory evaluation of food; enzymic generation of flavor chemicals and biotechnological applications. *Mailing Add:* Dept Food Sci Univ Wis-Madison Madison WI 53706

LINDSAY, ROBERT KENDALL, b Cleveland, Ohio, Aug 13, 34; c 2. ARTIFICIAL INTELLIGENCE, COGNITIVE THEORY. *Educ:* Carnegie-Mellon Univ, BS, 56, PhD(admin), 61; Columbia Univ, MA, 57. *Prof Exp:* From asst prof to assoc prof psychol, Univ Tex, Austin, 60-65; RES SCIENTIST PSYCHOL, UNIV MICH, 65- *Mem:* Am Asn Artificial Intel; AAAS. *Res:* Artificial intelligence and cognitive science, especially verbal and spatial reasoning and their interactions; expert systems; methodological and philosophical aspects of psychological theory and computation. *Mailing Add:* 205 Zina Pitcher Pl Ann Arbor MI 48109

LINDSAY, W(ESLEY) N(EWTON), b Ontario, Calif, Dec 29, 13; m 38. CHEMICAL ENGINEERING. *Educ:* Univ Calif, AB, 35. *Prof Exp:* Chemist, Arabol Mfg Co, NY, 35-43; chief, Spec Prob Sect, Food Mach & Chem Corp, 43-52, chief, Chem Sect, 52-55; assoc prof eng, San Jose State Col, 55-59; eng specialist, Jennings Radio & Mfg Corp, 59-60, mgr, Eng Design Dept, 60-64 & Res Dept, 64-72; indust consult, 72-84; RETIRED. *Concurrent Pos:* Consult, 55-59. *Mem:* Am Chem Soc; Inst Elec & Electronics Engrs. *Res:* Starch, protein and resin adhesives; drying of gases and foodstuffs; high pressure water sprays; recovery of nitrous gases; fluidization and high vacuum techniques; ultrahigh temperature furnaces. *Mailing Add:* 10710 Ridgeview Ave San Jose CA 95127-2643

LINDSAY, WILLARD LYMAN, b Dingle, Idaho, Apr 7, 26; m 51; c 4. SOIL SCIENCE. *Educ:* Utah State Univ, BS, 52, MS, 53; Cornell Univ, PhD(soil sci), 56. *Prof Exp:* Res asst Utah State Univ, 52-53 & Cornell Univ, 53-56; soil chemist, Soils & Fertilizer Res Br, Tenn Valley Authority, 56-60; asst prof agron, 60-62, from assoc prof to prof, 62-70, centennial prof, 70-78, PROF AGRON, COLO STATE UNIV, 74- *Concurrent Pos:* Vis prof, State Agr Univ, Wageningen, Netherlands, 72. *Mem:* Soil Sci Soc Am; fel Am Soc Agron; Int Soc Soil Sci; Sigma Xi. *Res:* Chemical reactions of phosphate in soils; physicochemical equilibria of plant nutrients in soils; chemistry and availability of micronutrients to plants; equilibrium of metal chelates in soils; solubility of heavy metals in soils. *Mailing Add:* Dept of Agron Colo State Univ Ft Collins CO 80523

LINDSAY, WILLIAM GERMER, JR, b Cleveland, Ohio, Nov 22, 28; m 56; c 3. PHYSIOLOGY. *Educ:* Oberlin Col, AB, 51; Univ Pa, MS, 57, PhD(zool), 62. *Prof Exp:* Instr physiol, Albany Med Col, 62-66; asst prof biol, 66-69, assoc prof, 69-75, PROF BIOL, ELMIRA COL, 75- *Mem:* AAAS; Soc Study Reproduction. *Res:* Spermatozoa metabolism; biological limnology. *Mailing Add:* Dept Biol Elmira Col Elmira NY 14901

LINDSAY, WILLIAM TENNEY, JR, b Scranton, Pa, Apr 4, 24; m 51; c 2. CHEMICAL ENGINEERING. *Educ:* Rensselaer Polytech Inst, BChE, 48; Mass Inst Technol, PhD(phys chem), 52. *Prof Exp:* Engr, Procter & Gamble Co, 48; asst, Mass Inst Technol, 49-51, res assoc, 52-53; sr scientist, Atomic Power Div, Westinghouse Elec Corp, 53-54, supv engr, 55-59, fel engr, Res Labs, 59-64, mgr phys chem dept, 64-73, mgr phys & inorg chem dept, Res Labs, 73-77, consult, Res & Develop Ctr, Westinghouse Elec Corp, 77-83,; CONSULT, LINDSAY & ASSOC, 83- *Mem:* Fel AAAS; Am Chem Soc; Am Phys Soc; Electrochem Soc; NY Acad Sci; Am Soc Testing & Mat; Am Inst Chem Engrs. *Res:* Electrolytic solutions; nuclear reactor coolant technology. *Mailing Add:* 47 E Main St Hopkinton MA 01748

LINDSEY, ALTON ANTHONY, b Monaca, Pa, May 7, 07; m 39; c 2. PLANT ECOLOGY. *Educ:* Allegheny Col, BS, 29; Cornell Univ, PhD(bot), 37. *Hon Degrees:* ScD, Allegheny Col, 88. *Prof Exp:* Asst bot, Cornell Univ, 29-33; biologist, Byrd Antarctic Exped, 33-35; asst bot, Cornell Univ, 35-37; instr bot, Am Univ, 37-40; asst prof, Univ Redlands, 40-42 & Univ NMex, 42-47; from asst prof to prof, 47-74, EMER PROF BIOL, PURDUE UNIV, 74- *Concurrent Pos:* Botanist, Purdue Can-Arctic Permafrost Exped, 51, ecologist, Purdue Res Team, Sonoran Desert, 53-54; bot ed, Ecol, Ecol Soc Am, 57-61, managing ed, Ecol & Ecol Monogr, 72-74; dir, Ind Natural Areas Surv, 67-68. *Honors & Awards:* Spec Cong Medal, Ecol Soc Am, 35 & Eminent Ecologist Award. *Mem:* Fel AAAS; Ecol Soc Am; Sigma Xi. *Res:* Indiana vegetation; flood plain ecology; 8 technical and 2 popular books. *Mailing Add:* Dept Biol Sci Purdue Univ W Lafayette IN 47907

LINDSEY, BRUCE GILBERT, b Rockland, Maine, June 22, 49. NEUROPHYSIOLOGY, NEUROANATOMY. *Educ:* Williams Col, BA, 71; Univ Pa, PhD(neuroanat), 74. *Prof Exp:* Fel neurophysiol, Univ Pa, 74-77; FROM ASST PROF TO PROF PHYSIOL & BIOPHYS, UNIV S FLA MED CTR, 77- *Mem:* Soc Neurosci; AAAS; Am Physiol Soc; Int Neural Network Soc. *Res:* Parallel information processing in the nervous system; sensory-motor intergration; brain stem control of breathing. *Mailing Add:* Dept Physiol Univ SFla Tampa FL 33612

LINDSEY, CASIMIR CHARLES, b Toronto, Ont, Mar 22, 23; m 48. ICHTHYOLOGY. *Educ:* Univ Toronto, BA, 48; Univ BC, MA, 50; Cambridge Univ, PhD(zool), 52. *Prof Exp:* Res biologist, BC Dept Game, 52-57; from asst prof to assoc prof zool, Univ BC, 57-66, cur fishes, Inst Fisheries, 52-66; PROF ZOOL, UNIV MAN, 66-; PROF ZOOL, UNIV BC, VANCOUVER. *Concurrent Pos:* Vis prof, Univ Singapore, 62-63, Wallace mem lectr, 63; fisheries consult, Pakistan, 64 & Fiji, 71-72. *Mem:* Am Soc Ichthyologists & Herpetologists; Am Fisheries Soc; fel Royal Soc Can; Can Soc Zoologists (pres, 77-78); Can Soc Environ Biologists (vpres, 74-). *Res:* Meristic variation; taxonomy; zoogeography of northern freshwater fishes; comparison of tropical and temperate fisheries. *Mailing Add:* Zool Dept Univ BC 6270 University Blvd Vancouver BC V6T 1Z4 Can

LINDSEY, DAVID ALLEN, b Nebraska City, Nebr, May 26, 42; m 66; c 2. GEOLOGY. *Educ:* Univ Nebr, BS, 63; Johns Hopkins Univ, PhD(geol), 67. *Prof Exp:* Geologist, US Geol Surv, Colo, 67-74, staff geologist mineral resources, Va, 74-75, geologist, 76-86, CHIEF, BR CENT MINERAL RESOURCES, US GEOL SURV, COLO, 87- *Concurrent Pos:* Geol Soc Am res grant, 65-66. *Mem:* Geol Soc Am; Am Asn Petrol Geol; Soc Econ Geol. *Res:* Glacial deposits, alluvial conglomerates and sandstones; beryllium deposits and volcanic rocks in Utah; intrusive complexes in central Montana; copper in sedimentary rocks; laramide and tertiary tectonics. *Mailing Add:* US Geol Surv MS 905 Fed Ctr Lakewood CO 80225

LINDSEY, DONALD L, b Stockton, Kans, May 25, 37; m 61; c 3. PLANT PATHOLOGY. *Educ:* Ft Hays Kans State Col, BS, 59; Colo State Univ, MS, 62, PhD(plant path), 65. *Prof Exp:* Jr plant pathologist, Colo State Univ, 61-65, asst plant pathologist, 66-69; instr bot, Colo State Col, 66; asst prof, 69-74, assoc prof, 74-81, PROF PLANT PATH, NMEX STATE UNIV, 81- *Mem:* Am Phytopath Soc; Sigma Xi; Soc Nematologist. *Res:* Biological control of plant pathogens; ecology of soil fungi; mine spoil revegetation. *Mailing Add:* PO Box 30001 Las Cruces NM 88003

LINDSEY, DORTHA RUTH, b Kingfisher, Okla, Oct 26, 26. GEROKINESIATRICS, KINESIOTHERAPY. *Educ:* Okla State Univ, BS, 48; Univ Wis, MS, 56; Ind Univ, PED, 63. *Prof Exp:* Instr health, phys educ & recreation, Okla Stae Univ, 48-50; instr, Monticello Col, 51-54; instr, DePauw Univ, 54-56; prof health, phys educ & recreation, Okla State Univ, 56-75; prof, Calif State Univ, Long Beach, 76-88; VIS PROF, UTAH STATE UNIV, 76- *Concurrent Pos:* Consult, Payne County Guid Ctr, 66-71; ed, Fencing Guide, Am Asn Health, Phys Educ & Recreation, 61-62; consult fitness & exercise, Ask the Professors, 82-; ed, Perspectives: J Western Soc, Phys Ed Col Women, 87- *Honors & Awards:* Julian Vogel Mem Award, Am Kinesiotherapy Asn, 88. *Mem:* Am Alliance Health, Phys Ed, Recreation & Dance; Am Kinesiotherapy Asn; Am Col Sports Med; Nat Coun Against Health Fraud. *Res:* Physical education; kinesiotherapy; electromyographical and kinesiological analyses of muscle action; quackery in physical fitness and reducing; therapeutic exercise; gerokinesiatrics. *Mailing Add:* Dept Phys Educ Calif State Univ Long Beach CA 92683-7418

LINDSEY, EDWARD STORMONT, b West Palm Beach, Fla, June 3, 30; m 53; c 2. MEDICINE. *Educ:* Tulane Univ, BS, 51, MD, 58, MMedSci, 68. *Prof Exp:* Intern, Charity Hosp, La, 58-59, resident surg, 59-61 & thoracic surg, 63-64; from instr to assoc prof, 63-76, CLIN ASSOC PROF SURG, TULANE UNIV, 76-, DIR TRANSPLANTATION RES UNIT, 66- *Concurrent Pos:* Resident surg, Southern Baptist Hosp, 61-62; Nat Heart Inst spec fel, Univ Edinburgh, 64-65; consult surg, Charity Hosp La & Keesler Air Force Hosp, 65-; mem adv comt, Nat Transplant Registry, 66-67. *Mem:* Transplantation Soc; Am Col Surg; NY Acad Sci; Asn Advan Med Instrumentation; Am Soc Artificial Internal Organs. *Res:* Thoracic and vascular surgery; transplantation biology. *Mailing Add:* 4440 Magnolia St Suite 410 New Orleans LA 70115

LINDSEY, GEORGE ROY, b Toronto, Ont, June 2, 20; m 51; c 2. MILITARY STRATEGIC ANALYSIS, SYSTEMS ANALYSIS. *Educ:* Univ Toronto, BA, 42; Queen's Univ Ont, MA, 46; Cambridge Univ, PhD(physics), 50. *Prof Exp:* Defence sci officer oper res, Can Defence Res Bd, 50-53; sr oper res officer, Air Defence Command, Royal Can Air Force, 54-59; dir defence systs anal group, Can Dept Nat Defence, 59-61; oper res group leader, Antisubmarine Warfare Res Ctr, Supreme Allied Comdr, Atlantic, Italy, 61-64; sr oper res scientist, 64-67, chief oper res anal estab, Dept Nat Defense, 67-87; SR RES FEL, CAN INST STRATEGIC STUDIES, 87- *Concurrent Pos:* Mem, Can Govt Bicult Develop Prog, 70-71 & Can Comt Int Inst Appl Systs Anal, 73-79; consult, Inst Res Pub Policy, 75-78; head, Can deleg NATO High Level Group Nuclear Planning, 77-87; chmn TTCP Panel, Undersea Warfare, 81-86; vis fel, Can Inst Int Peace & Security, 90- *Honors & Awards:* Award of Merit, Can Opers Res Soc, 84. *Mem:* Opers Res Soc Am; Can Inst Strategic Studies; Int Inst Strategic Studies; Can Oper Res Soc (pres, 61); Can Inst Int Affairs; Royal Can Astron Soc. *Res:* Arms control; advancing military technology and security analysis of strategic problems related to the attainment and preservation of a stable military balance. *Mailing Add:* 55 Westward Way Ottawa ON K1L 5A8 Can

LINDSEY, GERALD HERBERT, b Marshall, Mo, Aug 3, 34; m 58; c 6. MECHANICAL & AERONAUTICAL ENGINEERING. *Educ:* Brigham Young Univ, BES, 60, MS, 62; Calif Inst Technol, PhD(aeronaut eng), 66. *Prof Exp:* PROF AERONAUT ENG, NAVAL POSTGRAD SCH, 65- *Concurrent Pos:* Consult, Chem Systs Div, United Technol, 66- *Mem:* Am Inst Aeronaut & Astronaut. *Res:* Fracture, aircraft fatigue and design; viscoelastic stress and fracture. *Mailing Add:* Dept Aeronaut Naval Postgrad Sch Monterey CA 93940

LINDSEY, JAMES RUSSELL, b Tifton, Ga, Dec 6, 33; m 58; c 4. PATHOLOGY. *Educ:* Univ Ga, BS, 56, DVM, 57; Auburn Univ, MS, 61; Am Col Lab Animal Med, dipl, 67; Am Col Vet Pathologists, dipl, 67. *Prof Exp:* From instr to asst prof, Sch Vet Med, Auburn Univ, 57-61; from instr to asst prof, lab animal med & path, Johns Hopkins Univ Sch Med, 61-67; chmn dept com med, 67-75, prof, 67-86, assoc prof path, 69-85, PROF PATH, UNIV ALA, BIRMINGHAM, 85-, PROF & CHMN COMP MED, SCH MED & DENT, 86- *Concurrent Pos:* Fel path, Johns Hopkins Univ, 61-63; chief, RILAMSAT, Birmingham Vet Admin Hosp, 68-; adj prof, Sch Vet Med, Auburn Univ, 80-; vis scientist & consult lab animal dis, Orgn Health Res, Inst Exp Gerontol & Radiobiol, Rijswijk, Neth & Cent Lab Animal Breeding Fac, Zeist, Neth, 78-79. *Honors & Awards:* Charles River Prize Lab Animal Med, Am Vet Med Asn, 79; T S Williams lectr, Tuskegee Inst, 82. *Mem:* Am Vet Med Asn; Sigma Xi; Int Acad Path; Am Asn Pathologists; Am Thoracic Soc; Am Soc Microbiol; Int Orgn Mycoplasmologists. *Res:* Comparative pathology; mycoplasmal respiratory disease; laboratory animal diseases complicating research. *Mailing Add:* Dept Comp Med Univ Ala Sch Med Birmingham AL 35294

LINDSEY, JULIA PAGE, b Pine Bluff, Ark, Dec 9, 48. MYCOLOGY, PLANT PATHOLOGY. *Educ:* Hendrix Col, BA, 70; Univ Ariz, MS, 72, PhD(plant path), 75. *Prof Exp:* Teaching asst biol, Univ Ariz, 70-72, teaching asst plant path, 72-75; seed analyst seed cert, Ark State Plant Bd, 75-78; ASST PROF BIOL, FT LEWIS COL, 78- *Mem:* Mycol Soc Am. *Res:* A compilation of descriptive, cultural and taxonomic data concerning woodrotting basidiomycetes that decay aspen in North America; identification and taxonomy of plant pathogenic fungi. *Mailing Add:* Dept Biol & Agr Ft Lewis Col Durango CO 81301

LINDSEY, MARVIN FREDERICK, b Stockville, Nebr. PLANT BREEDING. *Educ:* Univ Nebr, BSc, 53, MSc, 55; NC State Univ, PhD(genetics), 60. *Prof Exp:* Asst prof agron, Univ Nebr, 60-63; geneticist, Rockefeller Found, 63-66; asst prof agron, Univ Wis, 66-69; area dir, Dekalb-Pfizer Genetics, 70-; RETIRED. *Mem:* Am Soc Agron; Crop Sci Soc Am. *Res:* Maize breeding and genetics. *Mailing Add:* Box 319 Beaver City NE 68926

LINDSEY, NORMA JACK, b Canton, Tex, June 16, 29. CLINICAL MICROBIOLOGY. *Educ:* Tex Woman's Univ, BA & BS, 51; Univ Calif, MPH, 64; Colo State Univ, PhD(microbiol), 69. *Prof Exp:* Bacteriologist, Dallas Health Dept Lab, Tex, 51-54; microbiologist, Ariz Health Dept Labs, Tucson, 56-65; teaching asst microbiol, Colo State Univ, 66-67; chief of microbiol, NMex Health Labs, 69-70; res microbiologist, Dept Health, Educ & Welfare, 70-73; ASST PROF PATH & ACTG HEAD, MICROBIOL SECT, CLIN LABS, UNIV KANS MED CTR, 73- *Mem:* NY Acad Sci; Am Soc Microbiol; Am Pub Health Asn; AAAS; Sigma Xi. *Res:* Clinical and applied microbiology. *Mailing Add:* 9931 Cedar Dr Overland Park KS 66207

LINDSEY, ROLAND GRAY, b Sylvatus, Va, June 26, 27; m 48; c 3. CHEMICAL ENGINEERING. *Educ:* Univ Del, BChE, 51; Ohio State Univ, MSc, 54, PhD(chem eng), 59. *Prof Exp:* Chem engr, Polychem Div, Dow Chem Corp, 56-59, proj leader polymer res, 59-60; res engr, Fabrics & Finishes Dept, E I Du Pont de Nemours & Co, 60-66, staff engr, 66-85, res assoc, 85-87, sr res assoc, 87; RETIRED. *Concurrent Pos:* Consult, 87-91. *Mem:* Am Chem Soc; Am Inst Chem Engrs. *Res:* Computerized kinetic model description of free radical polymer processes using numerical and Monte Carlo methods. *Mailing Add:* 613 Sherman Rd Springfield PA 19064

LINDSEY, WILLIAM B, b Iowa Park, Tex, July 26, 22; m 48; c 2. ORGANIC CHEMISTRY. *Educ:* Univ Tex, BS, 48; Ind Univ, MA, 49, PhD(chem), 54. *Prof Exp:* Res chemist, E I du Pont de Nemours & Co, Inc, Buffalo, 52-70, staff scientist, Film Dept, Richmond, 70-75, Clinton, Iowa, 75-77, sr res chemist, Polymers & Plastics Dept, 77-82, res assoc, 82-85; RETIRED. *Res:* Organic coatings; polymerization; surface phenomena; polymer stabilization; coating techniques; surface treatment for adhesion; adhesion; film extrusion and orientation; film evaluation; monomer and general organic synthesis; cellulose chemistry. *Mailing Add:* 416 Spring House Rd Camp Hill PA 17011

LINDSKOG, GUSTAF ELMER, b Boston, Mass, Feb 7, 03; m 34; c 2. SURGERY. *Educ:* Mass Agr Col, BS, 23; Harvard Univ, MD, 28; Am Bd Surg, dipl & Am Bd Thoracic Surg, cert, 52. *Prof Exp:* Intern surg, Lakeside Hosp, 28-29; asst surg & path, Sch Med, Yale Univ, 29-30; asst res surgeon, obstetrician & gynecologist, New Haven Hosp, 30-32; Nat Res Coun fel, Mass Gen Hosp, 32-33; from instr to prof, 33-71, EMER PROF SURG, SCH MED, YALE UNIV, 71- *Concurrent Pos:* Res surgeon, New Haven Hosp, 33-34; chmn, Am Bd Surg, 57-58. *Mem:* Soc Univ Surg; Soc Clin Surg; Am Asn Thoracic Surg; Am Surg Asn; fel Am Col Surg; Sigma Xi. *Res:* Thoracic surgery and physiology. *Mailing Add:* 15 Cow Path Lane Woodbridge CT 06525

LINDSLEY, DAN LESLIE, JR, b Evanston, Ill, Oct 13, 25; m 47; c 4. GENETICS. *Educ:* Univ Mo, AB, 47, MA, 49; Calif Inst Technol, PhD(genetics), 52. *Prof Exp:* Nat Res Coun fel biol, Princeton Univ, 52-53; NSF fel, Univ Mo, 53-54; from assoc biologist to biologist, Oak Ridge Nat Lab, 54-67; chmn dept, 77-79, prof biol, 67-91, EMER PROF BIOL, UNIV CALIF, SAN DIEGO, 91- *Concurrent Pos:* NSF sr fels, Univ Sao Paulo, 60-61 & Inst Genetics, Univ Rome, 65-66; USPHS spec fel, Dept Genetics, Div Plant Indust, Commonwealth Sci & Indust Res Orgn, Canberra, Australia, 72-73, Fogarty Int Fel, Dept Develop Genetics, Ctr Molecular Biol, Autonomous Univ Madrid, Spain, 80-81. *Honors & Awards:* T H Morgan Medal, Genetics Soc Am, 89. *Mem:* Nat Acad Sci; Genetics Soc Am (treas, 75-78, vpres, 85, pres, 86); Am Acad Arts & Sci; Lepidopterists Soc. *Res:* Cytogenetics of Drosophila. *Mailing Add:* Dept of Biol Univ Calif San Diego 9500 Gilman Dr La Jolla CA 92093-0322

LINDSLEY, DAVID FORD, b Cleveland, Ohio, May 18, 36; m 60; c 3. NEUROPHYSIOLOGY. *Educ:* Stanford Univ, BA, 57; Univ Calif, Los Angeles, PhD(anat, neurophysiol), 61. *Prof Exp:* Asst prof physiol, Med Sch, Stanford Univ, 63-67; ASSOC PROF PHYSIOL, MED SCH, UNIV SOUTHERN CALIF, 67- *Concurrent Pos:* USPHS fels, Moscow State Univ, 61-62 & Cambridge Univ, 62-63; Lederle med fac award, 64-67; visitor, Max Planck Inst Psychiat, Munich, 71 & 74-75; Guggenheim fel, 74-75. *Mem:* AAAS; Am Physiol Soc; Am Asn Anatomists; Soc Neurosci; Int Brain Res Orgn. *Res:* Central nervous system neurophysiology; behavioral neurophysiology; brain mechanics of attention and perception, using single neurons of central visual system of primates with particular interest in where in the brain and how incoming sensory stimuli become tagged as significant. *Mailing Add:* Dept Physiol USC Med Sch 1333 San Pablo St Los Angeles CA 90033

LINDSLEY, DONALD B, b Brownhelm, Ohio, Dec 23, 07; m 33; c 4. PSYCHOPHYSIOLOGY, NEUROPSYCHOLOGY. *Educ:* Wittenberg Univ, AB, 29; Univ Iowa, MA, 30, PhD, 32. *Hon Degrees:* DSc, Brown Univ, 58, Wittenberg Univ, 59, Trinity Col, Conn, 65 & Loyola Univ Chicago, 68; PhD, Johannes Gutenberg Univ, Mainz, WGer, 77. *Prof Exp:* Instr psychol, Univ Ill, 32-33; Nat Res Coun fel physiol & neuropsychiat, Harvard Med Sch & Mass Gen Hosp, Boston, 33-35; res assoc anat, Sch Med, Western Reserve Univ, 35-38; asst prof psychol, Brown Univ, 38-46; prof, Northwestern Univ, 46-51; chmn dept psychol, 59-62, prof psychol & physiol, 51-77, EMER PROF PSYCHOL & PHYSIOL, UNIV CALIF, LOS ANGELES, 77- *Concurrent Pos:* Dir, Psychol & Neurophysiol Labs, Bradley Hosp, East Providence, RI, 38-46 & Nat Defense Res Comt Proj, Off Sci Res & Develop, Yale Univ Contract, Camp Murphy & Boca Raton AFB, Fla, 43-46; mem sci adv bd, US Air Force, 47-49, chmn human resources comt, 48-49; mem aviation psychol comt, Nat Res Coun, 47-49 & undersea warfare comt, 51-64; consult, Study Sect USPHS, NIMH, 51-54, Nat Inst Neurol Dis & Blindness, 58-61, Nat Inst Gen Med Sci, 65-69, exp psychol comt, NSF, 52-54 & Guggenheim Found, 63-70; mem, Am Inst Biol Sci-NASA Behav Biol Panel, 65-70; mem space sci bd, Nat Acad Sci, 67-70, chmn long-duration missions in space comt, 67-71, mem space med comt, 67-; treas, Int Brain Res Orgn, 67-71; mem sci & technol adv coun, Calif State Assembly, 69-71; mem adv comt psychiat & neurol, Vet Admin Ctr, Wash, DC, 56-58; mem, Brain Res Inst, 61-77; Guggenheim fel, 59. *Honors & Awards:* Distinguished Sci Contrib Award, Am Psychol Asn, 59; William James Lectr, Harvard Univ, 58; Donald B Lindsley Prize, Soc Neurosci, 78; Ralph Gerard Prize, Soc Neurosci, 88; Gold Medal Award for Life Achievement in Psychol Sci, Am Psychol Found, 89. *Mem:* Nat Acad Sci; Am Psychol Asn; AAAS; Soc Neurosci; Am Physiol Soc; hon fel Am Electroencephalog Soc (pres, 65); Soc Psychophysiol Res; Soc Exp Psychol; Am Acad Arts & Sci; Finnish Acad Sci. *Res:* Brain function; emotion; behavior disorders; electroencephalography; neurophysiology; vision and visual perception; brain organization and behavior. *Mailing Add:* Dept Psychol Univ Calif Los Angeles CA 90024-1563

LINDSLEY, DONALD HALE, b Princeton, NJ, May 22, 34. PHASE EQUILIBRIA, GEOTHERMOMETRY. *Educ:* Princeton Univ, AB, 56; Johns Hopkins Univ, PhD(geol), 61. *Prof Exp:* Fel, Geophys Lab, Carnegie Inst, Washington, 60-62, petrologist, 62-70; PROF PETROL, DEPT EARTH & SPACE SCI, STATE UNIV NY STONY BROOK, 70- *Concurrent Pos:* Vis assoc prof, Calif Inst Technol, 69; vis scientist, Univ BC, 76-77; adj prof, Univ Wyo, 90- *Mem:* Mineral Soc Am (vpres, 81, pres, 82); Geol Soc Am; Geochem Soc; Am Geophys Union (vpres, 89-91, pres, 91-); AAAS; Mineralogical Asn Can. *Res:* High-pressure and high-temperature phase relations and thermodynamic solution models of mineral systems; redox reactions in earth; origin of anorthosites and related rocks; origin of lunar magmas. *Mailing Add:* Dept Earth & Space Sci State Univ NY Stony Brook NY 11794-2100

LINDSTEDT, P(AUL) M, b Stromsburg, Nebr, Feb 28, 17; m 42; c 3. CHEMICAL ENGINEERING. *Educ:* Univ Nebr, BS, 39. *Prof Exp:* Trainee eng, Goodyear Tire & Rubber Co, 39-40, jr chem engr, 40-44, sect head, 44-49, asst mgr chem eng & pilot plants, 49-51, mgr, 51-80; RETIRED. *Mem:* Am Inst Chem Engrs. *Res:* Synthetic rubber drying process; process development of elastomers, resins, antioxidants and rubber accelerators. *Mailing Add:* 2830 Hastings Rd Cuyahoga Falls OH 44224

LINDSTEDT-SIVA, K JUNE, b Minneapolis, Minn, Sept 24, 41; m 69. BIOLOGY. *Educ:* Univ Southern Calif, AB, 63, MS, 67, PhD(biol), 71. *Prof Exp:* Asst coordr sea grant progs, Univ Southern Calif, 71; environ specialist, Southern Calif Edison Co, 71-72; asst prof biol, Calif Lutheran Col, 72-73; sci adv, Atlantic Richfield Co, 73-77, sr sci adv, 77-81, mgr, environ sci, 81-86, MGR, ENVIRON PROTECTION, ATLANTIC RICHFIELD CO, 86- *Concurrent Pos:* Consult, Jacques Cousteau, Metromedia Producers Co, 70 & Southern Calif Edison Co, 72; mem task force, Fate & Effects of Oil, Am Petrol Inst, 73-80, chmn biol res subcomt; mem Clearmnon, oil spills comt & opers subcomt, Marine Water Qual Comn, Water Pollution Control Fedn, 75-76; chmn environ subcomt, Marine Indust Group, 81-82; chmn, spills technol comt, 83-85; mem bd trustees, Bermuda Biol Sta Res, 79-, res subcomt, 86-; chmn, Dispersant Use Guidelines Task Force, Am Soc Testing & Mat, 82-; mem bd dir, Southern Calif Acad Sci, 84-, pres, 90-; mem, Nat Sci Bd, 84-90; mem adv bd, Cabrillo Marine Museum, 85-; biol adv coun, Calif State Univ, Long Beach, 81-; mem, Nat Acad Sci Ecol Panel, Outer Continental Shelf Environ Studies Prog Rev, 82-84, Nat Sci Bd Comt, Educ & Human Resources, 85-90, Int Sci, 88-, Polar Regions, 87-90, Sci Judicators, 89 & Biodiversity, 89-90; mem, Univ Southern Calif, Marine Tech Adv Panel, 87-90, bd dir, Univ Southern Calif Oceanog Assoc, 88-90; mem panel ecol risk reduction, Environ Protection Agency, 90; mem panel oil spill res & develop, Nat Acad Sci, 90- *Honors & Awards:* Trident Award Marine Sci, Int Rev Underwater Activities, Ustica, Italy, 70; Award of Merit, Am Soc Testing & Mat, 90. *Mem:* Soc Petrol Indust Biologists (pres); Marine Technol Soc; AAAS; Sigma Xi; Am Inst Biol Sci; fel Am Soc Testing & Mat. *Res:* Chemoreception in aquatic animals, especially chemical control of feeding behavior in sea anemones; effects of oil on marine organisms; oil spill response planning; oil spill cleanup and control; environmental planning and management in industry; use of dispersants in oil spill response. *Mailing Add:* Atlantic Richfield Co 515 S Flower St Los Angeles CA 90071

LINDSTROM, DAVID JOHN, b Ashland, Wis, Mar 1, 45; m 71. GEOCHEMISTRY. *Educ:* Univ Wis-Madison, BS, 66; Univ Chicago, SM, 68; Univ Ore, PhD(chem), 76. *Prof Exp:* Res assoc lunar sci, Goddard Space Flight Ctr, NASA, 75-77; res scientist geochem, Dept Earth & Planetary Sci, Wash Univ, 71-; JOHNSON SPACE CTR. *Res:* Experimental trace element geochemistry; experimental petrology; properties of silicate liquids; extraterrestrial materials processing. *Mailing Add:* SN2 NASA Johnson Space Ctr Houston TX 77058

LINDSTROM, DUAINE GERALD, b Raymond, Wash, Jan 18, 37; m 67. MECHANICAL ENGINEERING. *Educ:* Univ Wash, BS, 59, PhD(nuclear eng), 68; Univ Mich, MS, 60. *Prof Exp:* Physics specialist, Aerojet Nuclear Systs Co, Sacramento, 68-71; lectr nuclear technol, Imperial Col, London, 71-75; assoc prof nuclear eng, Univ Okla, 75-82; ASSOC PROF & PROG COORDR NUCLEAR & CHEM ENG, WASH STATE UNIV, RICHLAND, WA, 82- *Concurrent Pos:* Lectr, Calif State Univ, Sacramento, 69-70. *Mem:* Am Nuclear Soc; Brit Nuclear Energy Soc; Health Physics Soc; Sigma Xi; Am Inst Chem Engrs. *Res:* Radiation transport, radiation shielding and protection; nuclear fuel cycle; reactor operations. *Mailing Add:* Wash State Univ TC 100 Sprout Rd Richland WA 99352

LINDSTROM, EUGENE SHIPMAN, b Ames, Iowa, Jan 12, 23; m 49; c 4. BACTERIOLOGY. *Educ:* Univ Wis, BA, 47, MS, 48, PhD(bact), 51. *Prof Exp:* Asst bact, Univ Wis, 46-51, AEC fel enzyme chem, 51-52; from asst prof to assoc prof, Pa State Univ, 52-64, from asst dean to assoc dean col sci, 66-77, prof bact, 64-68, head Dept Biol, 77-88; RETIRED. *Concurrent Pos:* NSF fel, Univ Minn, 61. *Mem:* Am Soc Microbiol; Am Acad Microbiol; Am Soc Biol Chem & Molecular Biol. *Res:* Bacterial physiology; physiology of Athiorhodaceae; physiology and ecology of photosynthetic bacteria. *Mailing Add:* 236 Ellen Ave State College PA 16801

LINDSTROM, FREDRICK THOMAS, b Astoria, Ore, July 30, 40; m 64; c 2. APPLIED MATHEMATICS. *Educ:* Ore State Univ, BS, 63, MS, 65, PhD(appl math), 69. *Prof Exp:* Res asst, 64-69, asst prof, 69-74, ASSOC PROF STATISTICS & MATH, ORE STATE UNIV, 74- *Mem:* Soc Indust & Appl Math; Am Math Soc; Am Statist Asn; Sigma Xi. *Res:* Mass transport phenomenon, especially in porous and permeable mediums; compartmental analysis and the mathematical modeling of drug distributions in mammalian tissue systems. *Mailing Add:* 6743 Dominion Ct Las Vegas NV 89103

LINDSTROM, GARY J, b Beacon, NY, Aug 4, 39; m 63. SEMICONDUCTOR PROCESS ENGINEERING, PROCUREMENT ENGINEERING. *Educ:* Marist Col, BA, 69. *Prof Exp:* Adv engr, IBM Corp, 64-90; SUBSTITUTE TEACHER MATH & SCI, WAPPINGERS CENT SCH DIST, 90-; SUBSTITUTE TEACHER MATH & SCI, BEACON CITY SCH DIST, 90- *Res:* Science education. *Mailing Add:* 52 Kent Rd Wappingers Falls NY 12590

LINDSTROM, IVAR E, JR, b Milligan, Nebr, Oct 15, 29; m 52; c 2. PHYSICS. *Educ:* Nebr Wesleyan Univ, AB, 50; Univ Ore, MA, 52, PhD(physics), 59. *Prof Exp:* MEM STAFF PHYSICS, LOS ALAMOS NAT LAB, UNIV CALIF, 58- *Mem:* Am Phys Soc; Sigma Xi. *Res:* Explosives, particularly initiation by shock waves; nuclear spectroscopy; solid state physics. *Mailing Add:* 327 Venado St Los Alamos NM 87544

LINDSTROM, JON MARTIN, b Moline, Ill, Oct 9, 45; m 77; c 3. AUTOIMMUNITY. *Educ:* Univ Ill, BA, 67; Univ Calif, San Diego, PhD(biol), 71. *Prof Exp:* Muscular Dystrophy Asn fel, Salk Inst for Biol Studies, 71-73, from asst res prof to assoc res prof, 73-83, assoc prof neurosci & mem, 83-90; TRUSTEE PROF NEUROL SCI & PHARMACOL, MED SCH, UNIV PA, 90- *Concurrent Pos:* Sloan fel; adj prof neurosci, Univ Calif, San Diego, 87-90; mem sci adv comt, MDA, Los Angeles & Calif chap MG Found. *Honors & Awards:* McKnight Neurosci Develop Award; Jacob Javits Award. *Mem:* Soc Neurosci. *Res:* Acetylcholine receptor structure and function; pathological mechanisms in myasthenia gravis. *Mailing Add:* Inst Neurol Sci Univ Pa Med Sch 503 Clin Res Bldg Curie Blvd Philadelphia PA 19104-6142

LINDSTROM, MARILYN MARTIN, b Jacksonville, Fla, Nov 28, 46; m 71; c 2. GEOCHEMISTRY & COSMOCHEMISTRY, IGNEOUS PETROLOGY. *Educ:* Univ Calif, San Diego, BA, 69; Univ Ore, PhD(geochem), 76. *Prof Exp:* Technician geochem, Geol Dept, Univ Ore, 68-69; res assoc, Univ Md, 75-77; res scientist, 77-79, sr res scientist geochem, Dept Earth & Planetary Sci, Wash Univ, 79-86; METEORITE CUR NASA JOHNSON SPACE CTR, 86- *Concurrent Pos:* Prin investr, NSF & NASA grants; assoc ed, Proc Lunar & Planetary Sci Conf; GSA Penrose grants. *Mem:* Geochem Soc; Am Geophys Union; Meteoritical Soc. *Res:* Geochemistry and petrology of igneous rocks and extraterrestrial materials; oceanic volcanic rocks, lunar samples and meteorites; trace element geochemistry; instrumental neutron activation analysis. *Mailing Add:* NASA Johnson Space Ctr Code SN2 Houston TX 77058

LINDSTROM, MERLIN RAY, b New Rockford, NDak, Oct 28, 51; m 72; c 3. COATINGS CHEMISTRY, POLYMER CHEMISTRY. *Educ:* NDak State Univ, BS, 73, PhD(chem), 78. *Prof Exp:* CHEMIST, PHILLIPS PETROL CO, 78- *Mem:* Fedn Coatings Technol; Soc Mfg Engrs. *Res:* Coatings; sulfur chemicals; metal cleaners; adhesives; water soluble resins; electroplating; sealants. *Mailing Add:* 919 King Circle Bartlesville OK 74006

LINDSTROM, RICHARD EDWARD, b Bristol, Conn, June 15, 32; m 52; c 3. PHYSICAL PHARMACY. *Educ:* Univ Conn, BS, 55; Syracuse Univ, MS, 62, PhD(phys chem), 67. *Prof Exp:* Asst prof chem, US Air Force Acad, 62-66 & Salem State Col, 66-68; from assoc prof to prof pharmaceut, Univ Conn, 68-; RETIRED. *Concurrent Pos:* Consult, Vick Chem Co, 74-, US Air Force, 80- & US Army, 81- *Mem:* Am Chem Soc; Am Pharmaceut Asn; Acad Pharmaceut Sci. *Res:* Thermodynamics of solution phenomena via molar volume and solubility data. *Mailing Add:* One Club Circle Tequesta FL 33469

LINDSTROM, RICHARD S, b Cleveland, Ohio, Mar 5, 27; m 53; c 3. HORTICULTURE. *Educ:* Ohio State Univ, BS, 50, MS, 51, PhD(hort), 56. *Prof Exp:* Instr hort, Mich State Univ, 53-56, from asst prof to assoc prof, 56-68; PROF HORT, VA POLYTECH INST & STATE UNIV, 68- *Mem:* Am Soc Hort Sci; Sigma Xi. *Res:* Physiology of floricultural plants including work with growth regulators, nutrition and photoperiodic control. *Mailing Add:* 18 Henzie St Reading MA 01867

LINDSTROM, TERRY DONALD, b Minneapolis, Minn, Sept 23, 51; m 76; c 2. OXIDATIVE DRUG METABOLISM, FREE RADICAL BIOCHEMISTRY. *Educ:* Augsburg Col, BA, 73; Univ Minn, PhD(pharmacol), 77. *Prof Exp:* Fel biochem, Mich State Univ, 77-79; dept head, 87-88, SR RES SCIENTIST BIOCHEM, LILLY RES LABS, ELI LILLY & CO, 79-; ASST PROF PHARMACOL, DEPT PHARMACOL, MED SCH, IND UNIV, 82- *Concurrent Pos:* Reviewer, J Med Chem, 90- *Mem:* AAAS; Am Soc Pharmacol & Exp Therapeut; Int Soc Study Xenobiotics; Int Union Pharmacol. *Res:* Oxidative and reductive metabolism of various drugs in subcellular, cellular and in vivo metabolism models; free radical damage and axidative damage to tissue. *Mailing Add:* Drug Metab & Disposition Lilly Res Labs Indianapolis IN 46285

LINDSTROM, WENDELL DON, b Kiron, Iowa, Feb 7, 27; m 50; c 2. MATHEMATICS. *Educ:* Univ Iowa, AB, 49, MS, 51, PhD(math), 53. *Hon Degrees:* DSc, Kenyon Col, 88. *Prof Exp:* From instr to asst prof math, Iowa State Univ, 53-58; from assoc prof to prof, 58-88, EMER PROF MATH, KENYON COL, 88- *Concurrent Pos:* NSF sci fac fel, Univ Calif, Berkeley, 62-63; vis prof, Robert Col, Istanbul, 68-69; Dana Early Retirement fel, 88- *Mem:* Am Math Soc; Math Asn Am. *Res:* Fields, rings, algebras, differential algebra; algebraic geometry. *Mailing Add:* Box 212 Gambier OH 43022

LINDT, JAN THOMAS, b Amsterdam, Holland, July 8, 42; m 77; c 3. POLYMER ENGINEERING, CHEMICAL ENGINEERING. *Educ:* Delft Univ Technol, MSc, 64, PhD(chem eng), 71. *Prof Exp:* Asst prof chem eng, Delft Univ Technol, 69-71; scientist polymer eng, Shell Res Ltd, 72-77; assoc prof, 78-85, PROF POLYMER ENG, UNIV PITTSBURGH, 85- *Mem:* Soc Rheology; Soc Plastics Engrs; Polymer Proc Soc. *Res:* Plasticating extrusion; reaction injection molding; reactive extrusion of polymers; devolatilization of polymer solutions; polyurethane foaming processes; polymer composites. *Mailing Add:* Dept Metall Eng 848 Benedam Univ Pittsburgh Main Campus Pittsburgh PA 15260

LINDVALL, F(REDERICK) C(HARLES), engineering administration; deceased, see previous edition for last biography

LINDZEN, RICHARD SIEGMUND, b Webster, Mass, Feb 8, 40; m 65; c 2. DYNAMIC METEOROLOGY, APPLIED MATHEMATICS. *Educ:* Harvard Univ, AB, 60, SM, 61, PhD(appl math), 64. *Prof Exp:* Res fel meteorol, Univ Wash, 64 & Univ Oslo, 64-65; res scientist, Nat Ctr Atmospheric Res, 65-68; prof, Univ Chicago, 68-72; prof meteorol, Harvard Univ, 72-83, dir, Ctr Earth & Planetary Physics, 80-83; SLOANS PROF METEOROL, MASS INST TECHNOL, 83-; DISTINGUISHED VIS SCIENTIST, JET PROPULSION LAB, 88- *Concurrent Pos:* Exec mem, Nat Acad Comt Global Atmospheric Res Prog, 68-79; consult, Naval Res Lab, 72-83, Control Data Corp, 77 & NASA, 77-; mem, Nat Acad Assembly Math & Phys Sci, 78-81, NRC Math Sci Educ Bd, 87-, NRC Bd Atmospheric Sci & Climate, 90- *Honors & Awards:* Meisinger Award, Am Meteorol Asn, 68; Macelwane Award, Am Geophys Union, 69; Charney Award, Am Meteorol Soc, 85. *Mem:* Nat Acad Sci; fel Am Acad Arts & Sci; fel Am Meteorol Soc; fel Am Geophys Union. *Res:* Hydrodynamic stability; climatology; upper atmosphere dynamics; general atmospheric circulation; tides. *Mailing Add:* Bldg 54 Rm 1720 Mass Inst Technol Cambridge MA 02139

LINE, JOHN PAUL, b Pontiac, Mich, Mar 2, 29; m 57; c 4. MATHEMATICS, APPLIED MATHEMATICS. *Educ:* Univ Mich, BS, 50, MS, 51. *Prof Exp:* Instr math, Oberlin Col, 55 & Univ Rochester, 55-56; asst prof, 56-62, ASSOC PROF MATH, GA INST TECHNOL, 62- *Mem:* Am Math Soc; Math Asn Am. *Res:* Integral transformations as applied to solution of boundary value problems in partial differential equations. *Mailing Add:* Sch of Math Ga Inst of Technol Atlanta GA 30332

LINEBACK, DAVID R, b Russellville, Ind, June 7, 34; m 56; c 3. CARBOHYDRATE CHEMISTRY. *Educ:* Purdue Univ, BS, 56; Ohio State Univ, PhD(org chem), 62. *Prof Exp:* Res chemist, Monsanto Chem Co, 56-57; fel, Univ Alta, 62-64; from instr to asst prof biochem, Univ Nebr, Lincoln, 64-69; from assoc prof to prof grain sci & indust, Kans State Univ, 69-76; prof food sci & head dept, Pa State Univ, 76-80; PROF FOOD SCI & HEAD DEPT, NC STATE UNIV, 80- *Concurrent Pos:* Regional adv, Wheat Indust Coun, 82-86; mem ed bd, Lebensmittel-Wissenschaft und Technologie, 82-; fel Inst Food Technologists, 82; mem Comt on Food Protection Bd, Nat Res Coun, Nat Acad Sci, 83-85; mem bd dir, League for Int Food Educ, 83-87; mem bd gov, Food Processing Inst, 86-, bd dir, 87-, prog chmn, 88- *Honors & Awards:* Spec Award Merit, Japanese Soc Starch Sci, 85; William F Geddes Mem Lectureship, 88. *Mem:* Am Asn Cereal Chem (pres-elect, 82-83, pres, 83-86); Inst Food Technol (exec comt, 88-); Am Chem Soc; Soc Nutrit Educ; Am Inst Nutrit; Japanese Soc Starch Sci. *Res:* Reaction and structure of carbohydrates; characterization of enzymes of starch hydrolysis and synthesis; cereal chemistry; structure of starch and functionality in food products. *Mailing Add:* Dept Food Sci NC State Univ Box 7624 Raleigh NC 27695-7624

LINEBACK, JERRY ALVIN, b Ottawa, Kans, Oct 25, 38; m 69; c 3. REGIONAL GEOLOGY. *Educ:* Univ Kans, BS, 60, MS, 61; Ind Univ, PhD(geol), 64. *Prof Exp:* From asst geologist to geologist, stratig & areal geol sect, Ill State Geol Surv, 64-81; sr geologist, Robertson Res, 81-86; ASST STATE GEOLOGIST & HYDROL PROG MGR, GA GEOL SURV, 87- *Mem:* Fel Geol Soc Am. *Res:* Ground water quality and protection; ground water resource development and management; Mississippian stratigraphy; regional geology. *Mailing Add:* Georgia Geol Surv Rm 400 Martin Luther King Jr Dr Atlanta GA 30334

LINEBERGER, ROBERT DANIEL, b Dallas, NC, Nov 9, 48; m 71; c 3. HORTICULTURE, PLANT PHYSIOLOGY. *Educ:* NC State Univ, BS, 71; Cornell Univ, MS, 74, PhD(hort), 78. *Prof Exp:* L H Bailey res asst, Dept Floricult, Cornell Univ, 71-77; ASSOC PROF HORT, OHIO STATE UNIV, 77- *Mem:* Am Soc Hort Sci; Am Soc Plant Physiologists; Bot Soc Am. *Res:* Plant cell and tissue culture; freeze preservation of germ plasm; freezing injury to plant cells; plant cell ultrastructure. *Mailing Add:* Dept Hort Clemson Univ Main Campus Clemson SC 29634

LINEBERGER, WILLIAM CARL, b Hamlet, NC, Dec 5, 39; m 79. CHEMICAL PHYSICS. *Educ:* Ga Inst Technol, BEE, 61, MSEE, 63, PhD, 65. *Prof Exp:* Asst prof elec eng, Ga Inst Technol, 65; res physicist atmospheric physics, Aberdeen Res & Develop Ctr, Md, 67-68; res assoc physics, Inst, 68-70, from asst prof to prof, 70-85, E U CONDON DISTINGUISHED PROF CHEM, UNIV COLO, BOULDER, 85-, FEL PHYSICS, JOINT INST LAB ASTROPHYS, 71- *Concurrent Pos:* J S Guggenheim fel. *Honors & Awards:* Broida Prize, Am Phys Soc, 81; Bomenn Michaelson Prize, 87; Meggers Prize, 88. *Mem:* Nat Acad Sci; fel AAAS; Am Chem Soc; fel Am Phys Soc; Sigma Xi. *Res:* Negative ion structure; molecular fluorescence; ion molecule reactions; tunable lasers. *Mailing Add:* Dept of Chem Univ of Colo Box 215 Boulder CO 80309

LINEHAN, JOHN HENRY, b Chicago, Ill, July 8, 38; m 60; c 5. MECHANICAL & BIOMEDICAL ENGINEERING. *Educ:* Marquette Univ, BSME, 60; Rensselaer Polytech Inst, MSME, 62; Univ Wis-Madison, PhD(mech eng), 68. *Prof Exp:* Engr, Knolls Atomic Power Lab, NY, 60-62; instr mech eng, 62-64, asst prof mech & biomed eng, 68-76, assoc prof, 76-78, PROF MECH & BIOMED ENG, MARQUETTE UNIV, 78-; PROF PHYSIOL, MED COL WIS, 79- *Mem:* Am Soc Mech Engrs; Am Physiol Soc; Biomed Eng Soc. *Res:* Multiphase flow; heat and mass transfer; hemodynamics; thermodynamics; lung physiology. *Mailing Add:* Dept Biomed Eng Marquette Univ Vet Admin Med Ctr Res Serv 5000 W National Ave Milwaukee WI 53295-1000

LINEHAN, URBAN JOSEPH, b Brockton, Mass, Oct 13, 11; m 50; c 3. PHYSICAL GEOGRAPHY. *Educ:* Bridgewater State Col, BS, 33; Clark Univ, MA, 46, PhD(geog), 55. *Prof Exp:* Instr geog, Univ Cincinnati, 40-45; from instr to asst prof, Univ Pittsburgh, 45-48; asst prof, Cath Univ Am, 48-56; analyst, US Govt, 56-73; RETIRED. *Res:* Synoptic climatology of Pittsburgh, Pennsylvania; areal and temporal distribution of tornado deaths in the United States; landscapes and off-road recreation of southwestern United States. *Mailing Add:* 13921 Pinetree Dr Sun City West AZ 85375

LINEHAN, WILLIAM MARSTON, b Tulsa, Okla, June 25, 47; m 79; c 2. CANCER RESEARCH, MOLECULAR BIOLOGY. *Educ:* Brown Univ, BA, 69; Univ Okla, MD, 73. *Prof Exp:* HEAD, UROL & ONCOL SECT, SURG BR, NAT CANCER INST, 82- *Concurrent Pos:* Sr investr, Nat Cancer Inst, Nat Inst Health, 82- *Mem:* Am Urol Asn; Soc Univ Surgeons; Am Col Surgeons. *Res:* Study of recessive oncogenes in human renal cell carcinoma; paracrine, endocrine and andocrine effects of genital ovinary tumor produced growth factors; adoptive immunotherapy of advanced malignancies. *Mailing Add:* Nat Cancer Inst Bldg Ten 2B47B Bethesda MD 20892

LINEMEYER, DAVID L, BIOCHEMISTRY REGULATION. *Educ:* Colo State Univ, BS, 71; Univ Wash, MS, 73, PhD(microbiol), 77. *Prof Exp:* staff fel, Lab Tumor Virus Genetics, Nat Cancer Inst, NIH, Bethesda, Md, 77-80, sr staff fel, 80-81; res fel, Dept Biochem Genetics, Merck Sharp & Dohme Res Labs, Rahway, NJ, 81-85, Dept Biochem Fundamental & Exp Res, 85-86, Dept Molecular Pharmacol & Biochem, 86-89, ASSOC DIR, DEPT BIOCHEM REGULATION, MERCK SHARP & DOHME RES LABS, RAHWAY, NJ, 89- *Res:* Author of numerous scientific publications. *Mailing Add:* Dept Biochem Regulation Merck Sharp & Dohme Res Labs R80W27 126 E Lincoln Ave Rahway NJ 07065-0900

LINES, ELLWOOD LEROY, consumer products, general chemistry, for more information see previous edition

LINES, MALCOLM ELLIS, b Banbury, Eng, Apr 26, 36; m 62; c 2. THEORETICAL SOLID STATE PHYSICS. *Educ:* Oxford Univ, BA, 59, MA & DPhil(physics), 62. *Prof Exp:* Fel physics, Magdalen Col, Oxford Univ, 61-63, 65-66; MEM TECH STAFF, BELL LABS, 63-65, 66- *Concurrent Pos:* Consult, Atomic Energy Res Estab, Harwell, Eng, 73. *Mem:* Fel Brit Inst Physics; Am Ceramic Soc; fel Phys Soc Gt Brit; Soc Photo-Optical Instrumentation Engrs. *Res:* Statistical mechanics; magnetism; ferroelectricity; structure of glasses; light scattering. *Mailing Add:* Bell Labs Murray Hill NJ 07974

LINFIELD, WARNER MAX, b Hannover, Ger, Jan 8, 18; nat US; m 45; c 1. SURFACE & SURFACTANT CHEMISTRY, LIPID CHEMISTRY. *Educ:* George Washington Univ, BS, 40; Univ Mich, MS, 41, PhD(pharmaceut chem), 43. *Prof Exp:* Anna Fuller Fund res fel, Northwestern Univ, 43-44; res chemist, Emulsol Corp, Ill, 44-46; group leader, E F Houghton & Co, Pa, 46-52 & Quaker Chem Prod Co, 52-55; dir res, Soap Div, Armour & Co Ill, 55-58, tech dir grocery prod div, 58-63; vpres, Culver Chem Co, 63-65; mgr org chem res, IIT Res Inst, 65-71; res leader, Eastern Regional Res Ctr Div, 71-84; CONSULT, 84- *Concurrent Pos:* Assoc ed, J Am Oil Chemists Soc, 74- *Honors & Awards:* Alton E Bailey Medal, Am Oil Chem Soc. *Mem:* Am Chem Soc; Am Oil Chemists Soc; Am Inst Chem; Sigma Xi. *Res:* Surface-active agents; soaps and detergents; synthesis of germicides; anti-malarial drugs; textile finishing agents; food technology; enzymatic fat splitting. *Mailing Add:* 2010 Bridle Lane Oreland PA 19075-1503

LINFOOT, JOHN ARDIS, b Grand Forks, NDak, May 16, 31; m 55; c 3. MEDICINE, ENDOCRINOLOGY. *Educ:* Univ NDak, BA, 53, BS & MS, 55; Harvard Univ, MD, 57; Am Bd Internal Med, dipl; Am Bd Endocrinol, dipl; Am Bd Nuclear Med, dipl. *Prof Exp:* Fel metab & endocrinol, Univ Utah Hosps, 59-60; sr staff scientist, Lawrence Berkeley Lab, 61-76; sr staff scientist, Donner Lab, Univ Calif, Berkeley, 61-81; CLIN PROF MED, UNIV CALIF, DAVIS, SACRAMENTO MED CTR, 81- *Concurrent Pos:* Consult, Martinez Vet Admin Hosp & Children's Hosp, East Bay; dir endocrine & metab servs, Alta Bates Hosp, 70; dir, Diabetes & Endocrine Inst, Providence Hosp. *Mem:* AAAS; Am Fedn Clin Res; fel Am Col Physicians; Endocrine Soc; Am Diabetes Asn. *Res:* Growth hormone; acromegaly; Cushing's syndrome; diabetic retinopathy; heavy particle pituitary irradiation. *Mailing Add:* Providence Hosp 350 30th St Suite 208 Oakland CA 94609

LINFORD, GARY JOE, b Laramie, Wyo, June 13, 40; m 63, 86; c 3. LASER PHYSICS, NONLINEAR OPTICS. *Educ:* Mass Inst Technol, BS, 62; Univ Utah, PhD(physics), 71. *Prof Exp:* Res asst physics, Electronics Res Lab, Mass Inst Technol, 61-62; mem tech staff laser physics, Laser Technol Dept, Aerospace Group, Hughes Aircraft Co, 63-68, group head, Laser Div, 68-69; teaching asst physics & astron, Univ Utah, 69-71; sect head, Laser Div, Hughes Aircraft Co, 71-74; group head laser res, Laser Fusion Prog, Lawrence Livermore Nat Lab, Univ Calif, 74-82; physicist laser res, Max Planck Inst Quantum Optics, WGermany, 82-83; sect head, appl tech div, 83-84, mgr Advan Technol Dept, TRW, 84-86; chief scientist, 86-90, AREA LEADER ADVAN LASER TECHNOL, OPTICS & DIRECTED ENERGY LAB, 90- *Concurrent Pos:* Guest lectr, Dept Physics, Calif State Univ, 81-82; guest scientist, Max Planck Inst for Plasma Physics, 78-79. *Res:* Inertial confinement laser fusion experiments; harmonic conversion of infrared light to ultraviolet at intensities of up to 8 gigawatt/cubic centimeter; design of laser amplifiers and high power propagation optics; target irradiation experiments and diagnostics; non-linear optics; excimer lasers; chemical lasers; free electron lasers; raman laser physics; phase conjugation; spacecraft telescopes; xenon lasers, xenon flashlamps. *Mailing Add:* TRW Space & Tech Group One Space Park Dr 01 1230 Redondo Beach CA 90278

LINFORD, RULON KESLER, b Cambridge, Mass, Jan 31, 43. MAGNETIC CONFINEMENT SYSTEMS FOR FUSION ENERGY. *Educ:* Univ Utah, BS, 66; Mass Inst Technol, SM & EE, 69, PhD(elec eng), 73. *Prof Exp:* Staff mem, Los Alamos Nat Lab, 73-75, asst group leader, 75-77, group leader, 77-80, asst div leader, 80-81, assoc div leader, 81-86, prog dir, 86-89, div leader, 89-91, PROG DIR, LOS ALAMOS NAT LAB, 91- *Concurrent Pos:* Mem, Magnetic Fusion Adv Comt, 82-86; Ignition Tech Oversight: CIT Steering Comt, 85-90, Int Thermonuclear Exp Reactor Steering Comt, 90; mem var comts, Am Phys Soc Div Plasma Physics, 87- *Mem:* Fel Am Phys Soc; Sigma Xi. *Res:* Fusion energy and accelerator transmution of radioactive waste; plasma theory; fusion product diagnostics; tritium technology; radiation resistant materials; accelerators; nuclear chemistry; neutronics; nuclear engineering; system studies; high voltage pulsed power systems. *Mailing Add:* Los Alamos Nat Lab M/S E529 Los Alamos NM 87545

LING, ALAN CAMPBELL, b London, Eng, July 28, 40. RADIOCHEMISTRY. *Educ:* London Univ, BSc, 63, PhD(chem), 66. *Prof Exp:* Fel chem, Univ Wis-Madison, 66-68; prof, WVa Univ, 68-75; PROF CHEM & DEAN, SCH SCI, SAN JOSE STATE UNIV, 75- *Concurrent Pos:* Proj dir, NASA-Ames Res Ctr, Mountain View, Calif, 75-, US Dept Energy & NSF. *Mem:* Royal Soc Chem, London; Am Chem Soc; Sigma Xi. *Res:* Radiation chemistry; combustion and fire protection; general physical chemistry; chemical education. *Mailing Add:* Sch Sci San Jose State Univ San Jose CA 95192-0099

LING, ALFRED SOY CHOU, b New York, NY, Mar 16, 28; m 54; c 3. CLINICAL RESEARCH, DEVELOPMENT. *Educ:* Princeton Univ, AB, 48; Univ Ill, MSc, 50; Univ Md, PhD(pharmacol), 58, MD, 62. *Prof Exp:* Asst endocrinol & mammal physiol, Univ Ill, 50-53, asst endocrinol, 53-54; asst pharmacol, Sch Med, Univ Md, 55-59; intern med, Pa Hosp, Philadelphia, 62-63, resident internal med, 63-65; res assoc, Rockefeller Univ, 65-67, asst prof, 67-70, adj assoc prof biochem pharmacol, 70-85, adj fac, 85; dep dir, 78-80, vpres, 80-84, sr vpres & dir sci affairs, Ives Labs, 84-86; VPRES CLIN RES, WYETH-AYERST LABS, 86- *Concurrent Pos:* From asst physician to assoc physician, Rockefeller Univ, 65-70, physician, 70-80; assoc dir clin pharmacol, Ciba-Geigy Corp, 70-74; exec dir clin res, Wallace Labs, 74-77; sr assoc dir clin res, Squibb Inst Med Res, 77-78. *Mem:* Am Soc Pharmacol & Exp Therapeut; Sigma Xi; Am Soc Clin Pharmacol; Endocrine Soc; Am

Fedn Clin Res; fel Am Col Pharmacol. *Res:* Role of thyroid gland in brain metabolism; auto-immune aspects of thyroiditis; total body x-irradiation effect on blood volume of intact and adrenalectomized animals; effect of chemo-convulsant agents on brain metabolism; use of new convulsant agent, hexafluorodiethyl ether, in therapy of mentally ill patients; clinical development of new and investigational drugs. *Mailing Add:* Wyeth-Ayerst Pres Clin Res PO Box 8299 Philadelphia PA 19101

LING, CHUNG-MEI, b Chekiang, China, May 5, 31; m 57; c 2. BIOCHEMISTRY. *Educ:* Nat Taiwan Univ, BS, 58; Ill Inst Technol, MS, 62, PhD(biochem), 65. *Prof Exp:* Teaching asst biochem & physiol, Ill Inst Technol, 60-64; res assoc biochem res, Michael Reese Res Found, Chicago, 64-65; asst prof biochem, Ill Inst Technol, 65-68; molecular biologist, Abbott Labs, 68-71, assoc res fel virol, 71-74, res fel, Dept Biochem, 74-76, head, Molecular Biol Lab, 74-77, head, virol lab, 77-; CHMN BD, GEN BIOL CORP, TAIWAN, ROC. *Concurrent Pos:* Adj asst prof, Ill Inst Technol, 68-69. *Mem:* AAAS; Sigma Xi; Am Soc Biol Chemists. *Res:* Molecular biology and immunodiagnostics of hepatitis viruses. *Mailing Add:* Gen Biol Corp # Six Innovation First Rd Sci-Based Indust Pk HsinChu Taiwan

LING, DANIEL, b Wetherden, Eng, Mar 16, 26; Can citizen; m 58; c 2. AUDIOLOGY, COMMUNICATIONS. *Educ:* St John's Col, Univ York, dipl, 50; Victoria Univ, Manchester, dipl, 51; McGill Univ, MS, 66, PhD(human commun dis), 68. *Prof Exp:* Organizer educ deaf, Reading Educ Comt, 55-63; prin, Montreal Oral Sch Deaf, 63-66; DIR RES DEAF CHILDREN, McGILL UNIV, 66-,ASSOC PROF & DIR ORAL REHAB, SCH HUMAN COMMUN DIS, 70-, PROF AURAL HABILITATION & EDUC, 74-; dir, speech & hearing div, Royal Victoria Hosp, 70-; AT DEPT COMMUN DIS, UNIV WESTERN ONT. *Concurrent Pos:* Can Fed Prov Health grants, McGill Univ, 66-; res assoc educ, Cambridge Univ, 55-58; asst prof audiol, Sch Human Commun Dis, McGill Univ, 68-70; hon dir, Coun Children's Audiol Rehab, Ctr Deaf Children, Mexico City, 66-; dir, Speech & Hearing Div, 70- *Mem:* Am Speech & Hearing Asn; Can Speech & Hearing Asn; Am Audiol Soc; Acoust Soc Am. *Res:* Communication development in deaf children; speech production among hearing impaired children; speech recognition using linear and coding amplifiers; diagnostic procedure relative to deafness. *Mailing Add:* Dept Commun Disorders Univ Western Ont London ON N6A 5B9 Can

LING, DANIEL SETH, JR, b Chicago, Ill, Oct 22, 24; m 46; c 2. THEORETICAL PHYSICS. *Educ:* Univ Mich, BSE(physics) & BSE(math), 44, MS, 45, PhD(physics), 48. *Prof Exp:* From asst prof to assoc prof physics, Univ Kans, 48-87, assoc prof astron, 73-87. *Mem:* AAAS; Am Phys Soc. *Res:* Nuclear physics. *Mailing Add:* 3601 W Sixth St Lawrence KS 66049

LING, F(REDERICK) F(ONGSUN), b Tsingtao, China, Jan 2, 27; m 54; c 3. TRIBOLOGY. *Educ:* St John's Univ, China, BS, 47; Bucknell Univ, BS, 49; Carnegie Inst Technol, MS, 51, DSc, 54. *Prof Exp:* Proj mech engr, Carnegie Inst Technol, 52-54, asst prof math, 54-56; from asst prof to prof mech, Rensselaer Polytech Inst, 55-70, William Howard Hart prof rational & tech mech, 70-88, chmn, Dept Mech Eng, Aeronaut Eng & Mech, 74-86; prof mech eng, Columbia Univ, 87-90, dir Columbia Eng Prod Ctr, 87-90; EMER WILLIAM HOWARD HART PROF, RATIONAL & TECH MECH, RENSSELAER POLYTECH INST, 88-, PRES, INST PRODUCTIVITY RES, 90- *Concurrent Pos:* Consult, Southwest Res Inst, 59-65, Gen Elec Co, 60-62, Mitre Corp, 61-63, Alco Prod, Inc, 61-62, Mech Tech, Inc, 62-70 & Wear Sci Corp, 71-; NSF sr fel, 70; vis prof, Univ Leeds, 70-71; Jacob Wallenburg Found, Sweden, 87. *Honors & Awards:* Nat Award, Soc Tribologists & Lubrication Engrs, 70; Mayo D Hersey Award, Am Soc Mech Engrs, 84. *Mem:* Nat Acad Eng; fel Am Soc Mech Engrs; fel Am Soc Lubrication Engrs; NY Acad Sci; fel Am Acad Mech; Am Phys Soc; fel AAAS; Soc Eng Sci; Am Soc Tribologists & Lubrication Engrs; Am Soc Mfg Engrs. *Res:* Surface mechanics and related phenomena; elastic stability. *Mailing Add:* Inst Productivity Res 488 Seventh Ave No 8E New York NY 10018

LING, GEORGE M, b Trinidad, WI, Apr 11, 23; Can citizen; m 55; c 2. PHARMACOLOGY. *Educ:* McGill Univ, BA, 43; Univ BC, MA, 57, PhD, 60. *Prof Exp:* Res asst pharmacol, Univ BC, 57-60, res assoc, 60-61, from asst prof to assoc prof, 61-65; prof pharmacol & chmn dept, Univ Ottawa, 65-75; external examr, Fac Med, 75-76, PROF PHARMACOL, UNIV WI, KINGSTON, JAMAICA, 76-; DIR, DIV NARCOTIC DRUGS, UN VIENNA INT CTR, AUSTRIA, 80- *Concurrent Pos:* Asst res anatomist, Brain Res Inst, Med Ctr, Univ Calif, Los Angeles, 62-63, assoc res pharmacologist, 63-64; head, Ment Health, Drug Dependence Unit, WHO, Geneva, 74-76; dir, UN Div Narcotic Drugs & spec adv UN Fund Drug Abuse Control, Geneva, 76-80. *Mem:* Am Soc Pharmacol & Exp Therapeut; Pharmacol Soc Can; Can Physiol Soc; Can Soc Chemother; Int Soc Chemother. *Res:* Pharmacology of central nervous system; neuro-humoral mediators and mechanisms of action of dependence producing substances and their antagonists behavior. *Mailing Add:* Div Narcotic Drugs UN Vienna Int Ctr PO Box 500 Vienna A-1400 Austria

LING, GILBERT NING, b Nanking, China, Dec 26, 19; m 51; c 3. PHYSIOLOGY. *Educ:* Nat Cent Univ, China, BSc, 43; Univ Chicago, PhD(physiol), 48. *Prof Exp:* Comen fel, Univ Chicago, 48-50; instr physiol optics, Sch Med, Johns Hopkins Univ, 50-53; from asst prof to assoc prof neurophysiol, Univ Ill, 53-57; sr staff scientist, Eastern Pa Psychiat Inst, 57-62; DIR DEPT MOLECULAR BIOL, PA HOSP, 62- *Concurrent Pos:* Mem, Woods Hole Marine Biol Corp. *Mem:* Am Physiol Soc. *Res:* Molecular mechanisms in cell function. *Mailing Add:* 140 Church St Kings Park NY 11754

LING, HAO, b Taichung, Taiwan, Sept 26, 59; US citizen; m 84. ELECTROMAGNETICS, ANTENNAS. *Educ:* Mass Inst Technol, BS, 82; Univ Ill, Urbana-Champaign, MS, 83, PhD(elec eng), 86. *Prof Exp:* Asst prof, 86-90, ASSOC PROF ELECTROMAGNETICS, DEPT ELEC & COMPUTER ENG, UNIV TEX, AUSTIN, 90- *Concurrent Pos:* Engr, Int

Bus Mach Res Lab, 82; NSF presidential young investr award, 87; vis fac mem, Lawrence Livermore Nat Lab, 87; Air Force fel, Rome Air Develop Ctr, Hanscom AFB, 90. *Mem:* Inst Elec & Electronic Engrs; Int Union Radio Sci. *Res:* Electromagnetic scattering; characterization and reduction of the radar cross sections of inlet structures and plumes in low-observable targets. *Mailing Add:* Dept Elec & Comput Eng Univ Tex Austin TX 78712-1084

LING, HARRY WILSON, b Painesville, Ohio, Feb 14, 27; div. INORGANIC CHEMISTRY. *Educ:* Bowling Green State Univ, AB, 50; Ohio State Univ, PhD(inorg chem), 54. *Prof Exp:* Res chemist, Pigments Dept, Res Div, 54-64, tech serv chemist, Sales Div, 64-67, sr res chemist, 67-69, supvr, 69-72, prod mgr, 72-74, MGR TECH SERV, E I DU PONT DE NEMOURS AND CO, INC, 74- *Mem:* Am Chem Soc; Fedn Soc Paint Technol; Sigma Xi. *Res:* Inorganic nitrogen chemistry; elemental silicon; anodic oxidation of metal substrates; electrolytic capacitors; titanium dioxide pigments; pigment colors. *Mailing Add:* 2410 Alister Dr Wilmington DE 19808

LING, HSIN YI, b Taiwan, Dec 5, 30; m 58; c 2. MICROPALEONTOLOGY. *Educ:* Nat Taiwan Univ, BS, 53; Tohuku Univ, MS, 58; Wash Univ, PhD(geol), 63. *Prof Exp:* Instr geol, Nat Taiwan Univ, 54-55; res engr, Res Ctr, Pan Am Petrol Corp, Okla, 60-63; res instr geol oceanog, Dept oceanog, Univ Wash, 63-64, from res asst prof to res assoc prof, 64-74, res prof, 74-78; PROF, GEOL DEPT, NORTHERN ILL UNIV, 78- *Mem:* Fel AAAS; Soc Econ Paleontologists & Mineralogists; Paleont Soc; Paleont Res Inst; Sigma Xi. *Res:* Marine micropaleontology and palynology. *Mailing Add:* Dept Geol Northern Ill Univ De Kalb IL 60115

LING, HUBERT, b Chungking, China, Apr 28, 42; US citizen; m 68; c 2. MICROBIAL GENETICS, PLANT TISSUE CULTURE. *Educ:* Queens Col, BS, 63; Brown Univ, MS, 66; Wayne State Univ, PhD(biol), 69. *Prof Exp:* Assoc prof biol, Univ Del, 69-77; res microbiologist, E I du Pont, 77-80; sterilization scientist, Johnson & Johnson, Ethicon, 80-81; plant scientist tissue cult, Samsen Lab, 81; ASSOC PROF BIOL, COUNTY COL MORRIS, 83- *Concurrent Pos:* Consult, Morton Salt Co, 72-74 & Int Chem Co, 74- *Mem:* Torrey Bot Club; Am Orchid Soc. *Res:* Genetic control of asexual cell fusion in Myxomycetes; ultrastructure of Myxomycetes; control of sporulation in Myxomycetes; tissue culture propagation of North American terrestrial orchids: Platanthera, Cypripedium and Calopogon. *Mailing Add:* Biol Dept County Col Morris 214 Center Grove Rd Randolph NJ 07869

LING, HUEI, b Fukien, China, Feb 24, 34; m 64. COMPUTER ARCHICTECTURE & LOGIC DESIGN. *Educ:* Nat Taiwan Univ, BSc, 57; Univ NB, MSc, 61; Univ Okla, PhD(elec eng), 65. *Prof Exp:* Method supvr, Bell Tel Can, 61; electronic engr, Sundstrand Aviation, 62-63; asst prof, Fla State Univ, 65; res staff mem, San Jose Res Lab, 65-77, mgr logic design & exp eng, 77-80, DEPT MGR COMPUTER ENG, IBM RES CTR, IBM CORP, 80- *Concurrent Pos:* Adj assoc prof, Fairleigh-Dickinson Univ, Teaneck Campus, 67-68, adj prof, 68. *Mem:* Asn Comput Mach; Inst Elec & Electronics Engrs. *Res:* Computer design. *Mailing Add:* Watson Res Ctr Div IBM Corp Box 218 Yorktown Heights NY 10598

LING, HUNG CHI, b Wenchow, China, 1950; US citizen; m 79; c 3. MATERIALS SCIENCE, ELECTRONIC CERAMICS. *Educ:* Mass Inst Technol, BS, 72, ScD, 78; State Univ NY, Stony Brook, MA, 74. *Prof Exp:* Res assoc mat sci, Dept Mat Sci & Eng, Mass Inst Technol, 78-81; DISTINGUISHED MEM TECH STAFF, AT&T BELL LABS ENG RES CTR, PRINCETON, NJ, 81- *Concurrent Pos:* Vis lectr, Shanghai Jiao Tong Univ, Peoples Repub China, 82. *Mem:* Am Soc Metals; Metall Soc Am Inst Mech Engrs; Sigma Xi; Mat Res Soc; Am Ceramic Soc; Inst Elect & Electronics Engrs. *Res:* Solid state phase transformations; thin film materials; electronics ceramics; dielectrics and varistors; superconducting oxides; metal-ceramic interaction; shape memory alloys. *Mailing Add:* AT&T Eng Res Ctr PO Box 900 Princeton NJ 08540

LING, JAMES GI-MING, b Wuhan, China, Oct 4, 30; m 61; c 3. SCIENCE POLICY, CHEMICAL ENGINEERING. *Educ:* Cornell Univ, BChemE, 53; Iowa State Univ, MS, 59; Stanford Univ, MS, 66, PhD(mgt sci), 67. *Prof Exp:* Chem engr, Chambers Works, DuPont Co, 53-54; transp navigator, Mil Airlift Command, 55-58; proj engr, Army Reactors, Atomic Energy Comn, 59-63; syst analyst, Joint Chiefs Staff, 67-70; chief opers, Anal Div, Hq 7th Air Force, 70-71; spec asst res & develop, Off Secy Defense, 71-72, asst plans & progs, Defense Directorate Energy, 73-75; group leader, Energy & Environ Eng, MITRE Corp, 75-78; asst dir, Off Sci & Technol Policies, Exec Off Pres, 81-86; PRES, LING TECHNOL, 86- *Concurrent Pos:* Chmn, Working Group Energy Resources, NATO, 75; adv to chmn, President's Comn Defense Mgt, 86. *Mem:* Inst Mgt Sciences; Am Chem Soc; Am Inst Chem Engrs. *Res:* Planning models for resource allocation; evaluation of research and development management and institutions. *Mailing Add:* 1111 Arlington Blvd W-615 Arlington VA 22209-3205

LING, JOSEPH TSO-TI, b Peking, China, June 10, 19; US citizen; m 44; c 4. AIR & WATER POLLUTION TECHNOLOGY. *Educ:* Hangchow Christian Col, Shanghai, BS, 44; Univ Minn, Minneapolis, MS, 50, PhD(sanit eng, pub health), 52. *Prof Exp:* Dist engr, Nanking-Shanghai RR Syst, 44-47; res asst sanit eng, Univ Minn, 48-52; sr staff engr, Gen Mills, Inc, 53-55; dir, Nat Res Inst Munic Eng, Peking, China, 56-57; prof civil eng, Baptist Univ, Hong Kong, 58-59; mgr, water & sanit eng, 3M Co, 60-65, mgr, sanit & civil eng, 66-69, dir, environ eng & pollution control, 70-74, vpres, environ eng & pollution control, 74-84; EXEC CONSULT, COMMUNITY SERV EXEC PROG, 84- *Concurrent Pos:* Adv, Ohio River Water Sanit Comn, 62-70, Environ Pollution Panel, US Chamber of Com, 67-72 & Tech Contact, Nat Indust Pollution Control Coun, 71-84; President's adv bd air qual, 74-78; mem environ health comt, President's Domestic Policy Rev, 78-80; sci adv bd, US Environ Protection Agency, 84-88; mem, World Environ Ctr, 84-; mem, Am Inst Pollution Prevention, 88- *Honors & Awards:* Edward Cleary Award, Am Acad Environ Engr, 81; First Gold Medal Int Corp Environ Award, World Environ Ctr, 85; Queneau Palladium Medal, Nat Audubon Soc, 90; Leadership Award, Nat Asn Photographic Mfrs, 90. *Mem:* Nat Acad Eng; Am Soc Civil Engrs; Air Pollution Control Asn; Water Pollution Control Fedn; Am Water Works Asn; Am Acad Environ Engr. *Res:* water filtration and related purification processes; biological oxidation and advanced treatment technology for water pollution control; thermo-oxidation, control techniques in air pollution and solid and hazardous waste disposal; waste minimization technologies. *Mailing Add:* 2090 Arcade St St Paul MN 55109

LING, NICHOLAS CHI-KWAN, b Hong Kong, Aug 15, 40; m 71; c 1. PEPTIDE CHEMISTRY, PEPTIDE HORMONES. *Educ:* San Jose State Univ, BS, 64; Stanford Univ, PhD(org chem), 69. *Prof Exp:* Res assoc crystallog, Stanford Univ, 69-70; res assoc biochem, Stalk Inst, 70-73, asst res prof, 74-78, assoc res prof, 79-88; SR MEM, WHITTIER INST, 89- *Mem:* Am Chem Soc; AAAS; Am Soc Mass Spectrometry; Endocrine Soc; Am Soc Biol Chem; Soc Neuroscience. *Res:* Isolation and characterization of peptide hormones; synthesis of peptides by solid phase methodology; peptide sequence determination by mass spectrometry and Edman technique. *Mailing Add:* Whittier Inst 9894 Genesee Ave La Jolla CA 92037

LING, ROBERT FRANCIS, b Hong Kong, Apr 21, 39; US citizen; m 63; c 1. STATISTICS. *Educ:* Berea Col, BA, 61; Univ Tenn, MA, 63; Yale Univ, MPhil, 68, PhD(statist), 70. *Prof Exp:* Asst prof math, E Tenn State Univ, 64-66; from instr to asst prof statist, Univ Chicago, 70-75; assoc prof, 75-76, PROF STATIST, CLEMSON UNIV, 77- *Concurrent Pos:* Vis prof, Owen Grad Sch Bus, 82, Grad Sch Bus, Univ Chicago, 83, Mass Inst Technol, 89, Harvard, 90; assoc ed, J Am Statist Asn, 77-85; vis lectr, Comt Pres Statist Socs, 83-86; mem coun, Classification Soc NAm, 74-77, prog chmn, 82 & 89, bd dirs, 88-90. *Honors & Awards:* Frank Wilcoxon Prize, 84. *Mem:* Fel Am Statist Asn; Classification Soc NAm. *Res:* Applied statistics; statistical computing; interactive data analysis; regression diagnostics. *Mailing Add:* Dept Math Sci Clemson Univ Clemson SC 29634-1907

LING, RUNG TAI, b Taipei, Taiwan, July 28, 43; US citizen. COMPUTATIONAL ELECTROMAGNETICS, COMPUTATIONAL FLUID DYNAMICS. *Educ:* Nat Taiwan Univ, BS, 65; Univ Mo-Columbia, MS, 68; Univ Calif, San Diego, PhD(physics), 72. *Prof Exp:* Lectr physics, San Diego State Univ, 72-73; res fel, Calif Inst Technol, 73-76; sr res scientist, STD Res Corp, 76-78 & R & D Assoc, 78-81; sr res specialist, 81-84, SR TECH SPECIALIST, NORTHROP AIRCRAFT DIV, LOCKHEED CALIF CO, 84- *Mem:* Am Phys Soc; Am Inst Aeronauts & Astronauts. *Res:* Computational physics including numerical solutions and modelling of atomic and molecular physics problems; electromagnetic and acoustic scattering phenomena; fluid dynamics including magnetohydrodynamics and radiation hydrodynamics. *Mailing Add:* 4715 Lasheart Dr La Canada CA 91011

LING, SAMUEL CHEN-YING, b Canton, China, May 7, 29; US citizen; m 57; c 3. PHYSICS. *Educ:* Nat Taiwan Univ, BS, 51; Baylor Univ, MS, 53; Ohio State Univ, PhD(physics), 69. *Prof Exp:* Asst prof physics, Augustana Col, Ill, 59-64; asst prof, 69-74, ASSOC PROF PHYSICS WRIGHT STATE UNIV, 74-; AT C COWE, KOWLOON, HONG KONG. *Mem:* Am Phys Soc. *Res:* Rutherford backing scattering studies of semiconducting thin films on the damage and recovery of crystallinity following ion implantation. *Mailing Add:* c/o Ccowe PO Box 98435 TST Hong Kong Hong Kong

LING, TA-YUNG, b Shianghai, China, Feb 2, 43; US citizen; m 69; c 3. ELEMENTARY PARTICLE PHYSICS. *Educ:* Tunghai Univ, Taiwan, BS, 64; Univ Waterloo, Ont, MS, 66; Univ Wis-Madison, PhD(physics), 71. *Prof Exp:* Teaching asst physics, Univ Waterloo, Can, 65-66; teaching asst physics, Univ Wis-Madison, 66-67, res asst, 67-71; res assoc physics, Univ Pa, 72-75, asst prof, 75-77; from asst prof to assoc prof, 77-83, PROF PHYSICS, OHIO STATE UNIV, COLUMBUS, 83- *Honors & Awards:* Outstanding Jr Investr, Dept Energy, 77. *Mem:* Am Phys Soc. *Res:* Experimental high energy physics: deep inelastic neutrino- nucleon scattering, neutrino masses and mixing, neutrino oscillations, deep inelastic electron-proton scattering, high energy proton-proton collisions. *Mailing Add:* Dept Physics Smith Lab Ohio State Univ Columbus OH 43210

LING, TING H(UNG), b Hwaiyuan, China, Nov 23, 19; US citizen; m 54; c 2. CHEMICAL ENGINEERING. *Educ:* Chekiang Univ, BS, 39; Ohio State Univ, MS, 49; Case Inst Technol, PhD(chem eng), 53. *Prof Exp:* Jr chem engr, 42nd arsenal, Ministry Nat Defense, China, 42-45; rubber compounder & troubleshooter, Dragon Rubber Co, China, 45-48; rubber chemist, Anaconda Wire & Cable Co, 53-55, sr res chemist, 55-63, supvr polymer res, 63-68, mgr, 68-78, SR TECH CONSULT, WIRE & CABLE DIV, ANACONDA CO, 78- *Mem:* Am Chem Soc; Inst Elec & Electronics Engrs. *Res:* Utilization and modification of high polymers in the insulation field, especially for the wire and cable industry; electric properties of high polymers; radiation effects on high polymers. *Mailing Add:* 643 Candlewood Dr Marion IN 46952-5341

LINGAFELTER, EDWARD CLAY, JR, b Toledo, Ohio, Mar 28, 14; m 38; c 5. CHEMISTRY. *Educ:* Univ Calif, BS, 35, PhD(chem), 39. *Prof Exp:* Assoc phys chem, 39-41, from instr to prof, 41-84, assoc dean, Grad Sch, 60-68, EMER PROF CHEM, UNIV WASH, 84- *Mem:* AAAS; Am Chem Soc; Am Crystallog Asn (pres, 74); Sigma Xi. *Res:* Colloidal electrolytes; crystal structure of paraffin-chain compounds; structure of coordination compounds; hydrogen bond. *Mailing Add:* Dept of Chem BG-10 Univ of Wash Seattle WA 98195

LINGANE, JAMES JOSEPH, b St Paul, Minn, Sept 13, 09; m 38; c 4. ANALYTICAL CHEMISTRY. *Educ:* Univ Minn, ChB, 35, PhD(chem), 38. *Hon Degrees:* MA, Harvard Univ, 46. *Prof Exp:* Asst chem, Univ Minn, 35-37, instr & Baker fel, 38-39; instr, Univ Calif, 39-41; instr, Harvard Univ, 41-44, fac instr, 44-46, from assoc prof to prof chem, 46-76; RETIRED. *Concurrent Pos:* Priestley lectr, Pa State Univ, 53. *Honors & Awards:* Gordon Res Conf Award, Am Asn Advan Sci, 52; Fisher Award, Am Chem Soc, 58; Medaile d'Hommage, Universite Libre de Bruxelles, 65. *Mem:* Am Chem Soc; Am Acad Arts & Sci; hon mem Brit Soc Analytical Chem; hon fel Royal Soc Chem. *Res:* Electrochemistry; electroanalysis; instrumental methods of analysis; polarographic analysis; with the dropping-mercury electrode; physiochemical methods of chemical analysis. *Mailing Add:* 94 Adams St Lexington MA 02173

LINGANE, PETER JAMES, b Oakland, Calif, May 12, 40; m 67; c 2. ELECTROCHEMISTRY, HYDROMETALLURGY. *Educ:* Harvard Univ, AB, 62; Calif Inst Technol, PhD(chem), 66. *Prof Exp:* Asst prof chem, Univ Minn, Minneapolis, 66-70; sr chemist, Ledgemont Lab, Kennecott Copper Corp, 70-77; group leader, prod res div, Conoco, 77-84; SUPVR, STANDARD ALASKA PROD CO, 84- *Mem:* Am Chem Soc; Soc Petrol Engrs. *Res:* Chemistry related to the solution mining of nonferrous ore minerals and to enhanced oil recovery, specifically carbon dioxide flooding; kinetics and mechanisms of solution reactions with particular emphasis upon the reactions which surround electrode processes; phase behavior of hydro carbon systems. *Mailing Add:* One Rosamond Villas Churchpath East Sheen SW 14 8 NP England

LINGAPPA, BANADAKOPPA THIMMAPPA, b Mysore, India, Mar 19, 27; nat US; m 53; c 3. MICROBIOLOGY. *Educ:* Benaras Hindu Univ, BSc, 50, MSc, 52; Purdue Univ, PhD, 57. *Prof Exp:* Lectr mycol, Benaras Hindu Univ, 52-53; res asst, Purdue Univ, 53-57; res assoc, Univ Mich, 57-59; res assoc, Mich State Univ, 59-60, asst prof med mycol, 60; Nat Inst Sci India sr res fel, Bot Lab, Univ Madras, 61; asst prof, Mich State Univ, 61-62; from asst prof to assoc prof, 62-68, PROF BIOL, COL HOLY CROSS, 68- *Concurrent Pos:* Vis scientist, Mass Inst Technol, 68-69; vis prof, Inst Gen Bot, Univ Geneva, 69-70; vis scientist, Worc Found Exp Biol, 76-77, Harvard Univ, 88; fac fel, Col Holy Cross, 70 & 88. *Mem:* Mycol Soc Am; Am Soc Microbiol; Sigma Xi; AAAS; Am Phytopath Soc. *Res:* Physiology of fungi; dormancy and germination of spores; methane production by anaerobic fermentation of solid waste; molecular biology. *Mailing Add:* Prof Biol Col Holy Cross Worcester MA 01610

LINGAPPA, JAISRI RAO, b Ann Arbor, Mich, June 11, 59. INFECTIOUS DISEASES. *Educ:* Swarthmore Col, BA, 79; Harvard Univ, PhD(cell biol), 85; Univ Mass, Worcester, MD, 87. *Prof Exp:* Resident internal med, 87-90, INFECTIOUS DIS FEL, DEPT MED, UNIV CALIF, SAN FRANCISCO, 90- *Honors & Awards:* Am Med Women's Asn Award, 87. *Mem:* Sigma Xi. *Mailing Add:* 49 Cragmont Ave San Francisco CA 94116

LINGAPPA, YAMUNA, b Mysore, India, Dec 6, 29; nat US; m 53; c 3. MICROBIOLOGY. *Educ:* Mysore Univ, BSc, 49; Madras Univ, BT, 51; Purdue Univ, MS, 55, PhD, 58. *Prof Exp:* Res assoc, Univ Mich, 57-59 & Mich State Univ, 59-60; sci pool officer, Govt India, 61; RES ASSOC BIOL, COL HOLY CROSS, 63- *Concurrent Pos:* Vis scientist, Inst Bot, Univ Geneva, 69-70; instr human nutrit, Clark Univ & Worcester State Col, 74; res consult, Dept Pub Health, City Worcester; fac adv, Undergrad Res Partic Proj Methane Generation, Col Holy Cross, vis lectr nutrit & world hunger, 78; comnr, Gov's Comn on Status of Women, Mass, 77-79. *Mem:* Mycol Soc Am; Am Soc Microbiol; Sigma Xi; Am Inst Biol Sci. *Res:* Human nutrition; solid waste disposal; physiology of pathogenic fungi; microbial interactions. *Mailing Add:* Four McGill St Worcester MA 01607

LINGELBACH, D(ANIEL) D(EE), b Wilkinsburg, Pa, Oct 4, 25; m 49; c 3. ELECTROMECHANICAL ENERGY CONVERSION. *Educ:* Kans State Col, BSEE, 47, MS, 48; Okla State Univ, PhD, 60. *Prof Exp:* From instr to asst prof elec eng, Univ Ark, 48-55; asst prof, 55-61, assoc prof, 61-79, prof elec eng, 79-86, EMER PROF ELEC & COMPUT ENG, OKLA STATE UNIV, 86- *Honors & Awards:* Charles Schneider Award, Nat Asn Relay Mfrs, 68. *Mem:* Am Soc Eng Educ; Nat Soc Prof Engrs; Inst Elec & Electronics Engrs; Sigma Xi. *Res:* Electric power system modeling and optimization; power systems analysis. *Mailing Add:* Sch of Elec & Comput Eng ES 202 Okla State Univ Stillwater OK 74078

LINGENFELTER, RICHARD EMERY, b Farmington, NMex, Apr 5, 34; m 57; c 2. ASTROPHYSICS, COSMIC RAY PHYSICS. *Educ:* Univ Calif, Los Angeles, AB, 56. *Prof Exp:* Physicist, Lawrence Radiation Lab, Univ Calif, 57-62; assoc res geophysicist, Univ Calif, Los Angeles, 62-66, res geophysicist, Inst Geophys & Planetary Physics, 66-69, prof in residence, Dept Geophys & Space Physics, 69-79 & Dept Astron, 74-79; RES PHYSICIST, CTR ASTROPHYS & SPACE SCI, UNIV CALIF, SAN DIEGO, 79- *Concurrent Pos:* Fulbright res fel geophys & planetary physics, Tata Inst Fundamental Res, Bombay, India, 68-69. *Mem:* Fel Am Phys Soc; Int Astron Union; Am Astron Soc. *Res:* Cosmic ray origins and interactions; gamma ray astronomy; solar flare particle interactions; planetology; radiocarbon variations. *Mailing Add:* Ctr Astrophys & Space Sci Univ Calif San Diego La Jolla CA 92093

LINGG, AL J, b Mt Hope, Kans, Mar 26, 38; m 61; c 3. MICROBIOLOGY. *Educ:* Kans State Univ, BS, 64, MS, 66, PhD(microbiol), 69. *Prof Exp:* Instr, Kans State Univ, 66-68; PROF MICROBIOL, UNIV IDAHO, 69- *Concurrent Pos:* Fulbright lectr, Nepal, 79-80. *Mem:* Sigma Xi; Am Soc Microbiol. *Res:* Environmental microbiology; water quality; fungal insect pathogens. *Mailing Add:* Dept Bact Univ Idaho Moscow ID 83843

LINGLE, SARAH ELIZABETH, b Woodland, Calif, July 22, 55; m 89. PHYSIOLOGY, BIOCHEMISTRY. *Educ:* Univ Calif, Davis, BS, 77; Univ Nebr, MS, 78; Wash State Univ, PhD(agron), 82. *Prof Exp:* Res asst, Univ Nebr, 77-78; teaching asst plant breeding, Wash State Univ, 79-80, res asst, 79-82; res assoc, Fargo, ND, 82-84, PLANT PHYSIOLOGIST, USDA, AGR RES SERV, WESTLACO, TEX, 84-, ACTG RES LEADER, 91- *Mem:* Am Soc Agron; Crop Sci Soc Am; Am Soc Plant Physiologists; AAAS; Sigma Xi. *Res:* Physiology and biochemistry of sucrose accumulation in sugarcane, specifically the transport and metabolism of sucrose and how it relates to the balance between growth and storage in the stalk. *Mailing Add:* USDA Agr Res Serv Weslaco TX 78596

LINGREL, JERRY B, b Byhalia, Ohio, July 13, 35; m 58; c 2. MOLECULAR GENETICS. *Educ:* Otterbein Col, BS, 57; Ohio State Univ, PhD(biochem), 60. *Prof Exp:* From asst prof to prof biol chem, 62-81, PROF & CHMN MOLECULAR GENETICS, BIOCHEM & MICROBIOL, UNIV CINCINNATI, 81- *Concurrent Pos:* Fel biol, Calif Inst Technol, 60-62. *Honors & Awards:* George Rieveschl Award. *Mem:* Am Soc Biol Chemists; Am Soc Cell Biol. *Res:* Regulation of gene expression in animal cells; hemoglobin biosynthesis; messenger RNA; gene structure. *Mailing Add:* Dept Micro & Molecular Genetics Univ Cincinnati Col Med Cincinnati OH 45267-0524

LINGREN, WESLEY EARL, b Pasadena, Calif, Aug 27, 30; m 61; c 2. PHYSICAL CHEMISTRY, OCEANOGRAPHY. *Educ:* Seattle Pac Col, BS, 52; Univ Wash, MS, 54, PhD(electrochem), 62. *Prof Exp:* Instr phys sci, Pasadena Col, 56-58; from asst prof to assoc prof, 62-68, chmn dept, 68-73, PROF CHEM, SEATTLE PAC UNIV, 68-, DIR GEN HONORS, 70-, DEAN SCH SCI, 90- *Concurrent Pos:* Res assoc, US Naval Radiol Defense Lab, 63-69; NSF fel, Yale Univ, 67-68, Solar Energy Res Inst, 84. *Mem:* Am Chem Soc; Sigma Xi. *Res:* Rates of electrode reactions; electroanalytical chemistry; oxidation states of elements in seawater oceanography. *Mailing Add:* 10628 NE 16th St Bellevue WA 98004

LINGWOOD, CLIFFORD ALAN, b Dorset, Eng, Jan 2, 50; m 74; c 3. PROTEIN-GLYCOLIPID INTERACTIONS. *Educ:* Univ Hull, BSc, 71; Univ London, PhD(cell biol), 75. *Prof Exp:* Res fel, dept pathobiol, Univ Wash, Seattle, 75-77; res fel, 77-81, ASST PROF, DEPT BIOCHEM, HOSP SICK CHILDREN, TORONTO, 81-, ASSOC PROF, DEPT MICROBIOL, 87-; ASSOC PROF BIOCHEM, UNIV TORONTO, 83-, ASSOC PROF MICROBIOL, 88- *Concurrent Pos:* Vis scientist, dept biochem oncol, Hutchison Cancer Ctr, Seattle, 75-77. *Mem:* Am Soc Cell Biol; Soc Complex Carbohydrates; NY Acad Sci; Can Biochem Soc. *Res:* Metabolism and function of cell membrane sulfoglycolipids during spermatogenesis fertilization; glycolipid receptors for microorganisms; glycolipid binding by bacterial toxins. *Mailing Add:* Dept Microbiol Hosp Sick Children 555 University Ave Toronto ON M5G 1X8 Can

LINHARDT, ROBERT JOHN, b Passaic, NJ, Oct 18, 53; m 75; c 2. BIOPOLYMER CHEMISTRY, APPLIED ENZYMOLOGY. *Educ:* Marquette Univ, BS, 75; Johns Hopkins Univ, MA, 77, PhD(org chem), 79. *Prof Exp:* Res & teaching asst org chem, dept chem, Johns Hopkins Univ, 75-79; res assoc, dept appl biol, Biochem Eng Labs, Mass Inst Technol 79-81, Johnson & Johnson fel, Whitaker Col Health Sci, Technol & Mgt, 81-82; from asst prof to assoc prof, 82-89, PROF MED CHEM & NATURAL PRODS CHEM, COL PHARM, UNIV IOWA, 89- *Honors & Awards:* Ernst M Marks Award, Johns Hopkins Univ, 79. *Mem:* Am Chem Soc; AAAS; Sigma Xi. *Res:* Bio-organic chemistry and applied enzymology in the study of the structure-activity-relationship of complex polysaccharides; biopolymeric drugs, their preparation, formulation and applications. *Mailing Add:* Dept Med Chem & Natural Prods Univ Iowa Col Pharm Iowa City IA 52242

LINHART, YAN BOHUMIL, b Prague, Czech, Oct 8, 39; US citizen. EVOLUTION, ECOLOGY. *Educ:* Rutgers Univ, New Brunswick, BS, 61; Yale Univ, MF, 63; Univ Calif, Berkeley, PhD(genetics), 72. *Prof Exp:* Jr specialist forest genetics, Sch Forestry, Univ Calif, Berkeley, 63-65; asst specialist, 65-66, from asst prof to assoc prof, 71-83, PROF BIOL, UNIV COLO, BOULDER, 83- *Concurrent Pos:* Res grant, Univ Colo, Boulder, 71-73; Colo Energy Res Inst grant, 75-76; NSF grants, 75-78, 78-80, 81-84 & 85-90; Nat Geog Soc grant, 84-85. *Mem:* AAAS; Am Inst Biol Sci; Brit Ecol Soc; Soc Study Evolution; Soc Am Foresters; Asn Trop Biol; Am Soc Naturalists; Bot Soc Am; Ecol Soc Am. *Res:* Adaptation; population biology; reproductive biology of plants; pollination biology; forest biology; plant biogeography. *Mailing Add:* Dept Biol Univ Colo Boulder CO 80309-0334

LININGER, LLOYD LESLEY, b Iowa City, Iowa, Mar 13, 39; m 59; c 3. MATHEMATICS, MATHEMATICAL STATISTICS. *Educ:* Univ Iowa, PhD(math), 64. *Prof Exp:* Asst prof math, Univ Mo, 64-65; asst to Prof Montgomery, Inst Advan Study, Princeton Univ, 65-67; res instr, Univ Mich, 67-70; ASSOC PROF MATH, STATE UNIV NY ALBANY, 70- *Concurrent Pos:* Statistician, Biometry Sect, NIH, Bethesda, 77-78; statistician, Environ Protection Agency, 85-88. *Mem:* Am Math Soc; Am Statist Asn; Am Pub Health Asn. *Res:* Applications of statistics; biostatistics. *Mailing Add:* Dept Statist Sch Pub Health State Univ of NY Albany NY 12222

LINIS, VIKTORS, mathematics; deceased, see previous edition for last biography

LINK, BERNARD ALVIN, b Columbus, Wis, Mar 23, 41. MEAT SCIENCE, FOOD SCIENCE. *Educ:* Univ Wis-Madison, BS, 62, MS, 64, PhD(meat & animal sci), 68. *Prof Exp:* Res asst meat sci, Univ Wis-Madison, 62-68; Welch Found fel, Tex A&M Univ, 68-72, res scientist meat chem, 70-72, res assoc biochem & biophys, 72-73; res biochemist, 73-77, MGR, PROTEIN RES, CARGILL INC, 77- *Mem:* Am Meat Sci Asn; Sigma Xi; Am Asn Cereal Chemists; Inst Food Technol. *Res:* Soy protein products. *Mailing Add:* 108 Gibson Ave Liftwood Wilmington DE 19803

LINK, CONRAD BARNETT, b Dunkirk, NY, Mar 5, 12; m 40; c 3. HORTICULTURE. *Educ:* Ohio State Univ, BS, 33, MS, 34, PhD(hort), 40. *Prof Exp:* Hybridist, Good & Reese Co, Ohio, 34-35; asst hort, Ohio State Univ, 35-38, exten specialist, 39-40; from instr to asst prof floricult, Pa State Univ, 38-45; horticulturist, Brooklyn Bot Garden, 45-48; prof, 48-82, EMER PROF HORT, UNIV MD, COLLEGE PARK, 82- *Honors & Awards:* Ware Award, Am Soc Hort Sci, L H Vaughan Award. *Mem:* Fel AAAS; fel Am Soc Hort Sci; Bot Soc Am; Am Hort Soc (secy, 48-49); Int Soc Hort Sci. *Res:* Photoperiodism; plant anatomy, nutrition and plant propagation. *Mailing Add:* Dept Hort Holzapfel Hall Univ Md College Park MD 20742-5611

LINK, FRED M, b York, Pa, Oct 11, 04; m 30; c 1. RADIO COMMUNICATIONS ENGINEERING. *Educ:* Penn State Univ, BSEE, 27. *Hon Degrees:* LHD, York Col Penn, 88. *Prof Exp:* Plant Engr, New York Telephone Co, 27-28; mgr, Power Tube Div, Deforest Radio Co, 28-31; owner, pres, Link Radio Corp, 31-51; dir oper, Mobile Radio Div, DuMont Labs Inc, 54-59; consult, Landmobile, radio Div, RCA Corp, Camden, NJ, 59-66; CONSULT, LANDMOBILE RADIO DIV, 66- *Concurrent Pos:*

Consult, EF Johnson Co, 68-, Philip Commun Systs, Cambridge, UK & Australia, 70-, REPCO Inc of Orlando Fla, Decibel Prods Inc, Dallas, 72-, Mobile Radio Technol, Overland Park, Kans, 85- *Mem:* Fel Inst Elec & Electronics Engrs; fel Radio Club Am (pres). *Res:* Producer of FM Land Mobile Radio System. *Mailing Add:* Robin Hill Farms Pittstown NJ 08867

LINK, GARNETT WILLIAM, JR, b Charlottesville, Va, May 7, 45; m 71; c 1. ICHTHYOLOGY. *Educ:* Univ Va, Charlottesville, BA, 67; Univ Richmond, MA, 71; Univ NC, Chapel Hill, PhD(zool), 80. *Prof Exp:* Teacher biol & phys sci, ECarteret High Sch, 75-81; res asst, 81, res asst ichthyol, Inst Marine Sci, 82-83, RES ASST, DEPT SURG, UNIV NC CHAPEL HILL, 83- *Concurrent Pos:* Grad res asst, Inst Marine Sci, Univ NC, 72-80; consult, Marine Occup, Ecarteret High Sch, 80-82, Mentor NC Sch Sci & Math, Durham, 85- *Mem:* Am Soc Ichthylogists & Herpetologists; Asn Southeastern Biologists; Southeastern Fish Coun; Soc Systematic Zool. *Res:* Life-histories and systematics of fishes including Serranids and Percids. *Mailing Add:* Campus Box 7050 Rm 253 Clin Serv Univ NC Chapel Hill Chapel Hill NC 27599-7050

LINK, GORDON LITTLEPAGE, b Charleston, WVa, Feb 9, 32; m 55; c 1. PHYSICAL CHEMISTRY. *Educ:* Col William & Mary, BS, 54; Univ Va, PhD(phys chem), 58. *Prof Exp:* MEM TECH STAFF, BELL LABS, INC, 58- *Mem:* Am Phys Soc; Am Chem Soc. *Res:* Dielectrics. *Mailing Add:* Tingley Rd Brookside NJ 07926

LINK, JOHN CLARENCE, b Iowa, Jan 5, 08; m 36; c 3. PHYSICS. *Educ:* Creighton Univ, AB, 28; Cath Univ Am, MA, 29. *Prof Exp:* Electronic scientist, US Dept Navy, 29-64, consult, Naval Res Lab, 55-64; mgr missile projs, Aerospace Corp, 64-66; mem staff, Avco Missile Systs Div, 66-68; independent consult electromagnetic reflectors, 68-81; RETIRED. *Res:* Countermeasures; electromagnetic reflectors; energy conversion. *Mailing Add:* 6413 Halleck St Forestville MD 20747

LINK, PETER K, b Batavia, Java, Nov 7, 30; US citizen; m 90; c 2. GEOLOGY, METEOROLOGY. *Educ:* Univ Wis, BS, 53, MS, 55, PhD(stratig geol), 65. *Prof Exp:* Geologist, Esso Standard Inc, Libya, 57-58, party chief, 58-59, subsurface geologist, 59-60, regional geologist, 60-61; regional geologist, Humble Oil & Refining Co, Okla, 62-63; res geologist, Atlantic Richfield Co, Dallas, 65-68, sr res geophysicist, 68-70; sr res scientist, Amoco Prod Co, 70-73; CONSULT, 73- *Concurrent Pos:* Adj prof geol, Univ Tulsa, 74-77; found mem, Associated Resource Consult Inc, Tulsa, 79; staff instr, Oil & Gas Consult Int, Inc, Tulsa, 79- *Mem:* Am Asn Petrol Geologists; fel Geol Soc Am. *Res:* Stratigraphy; structure; tectonics; field, regional, well site, subsurface and petroleum geology; research operations; stratigraphic-seismic research exploration; exploration programs; sedimentation; photogeology; minerals and petroleum exploration; author one book. *Mailing Add:* 7106 E Briarwood Dr Englewood CO 80112

LINK, RICHARD FOREST, mathematical statistics, for more information see previous edition

LINK, ROGER PAUL, veterinary pharmacology; deceased, see previous edition for last biography

LINK, WILLIAM B, b Darke, WVa, Mar 25, 28; m 56; c 3. ANALYTICAL CHEMISTRY, ORGANIC CHEMISTRY. *Educ:* Shepherd Col, BS, 53. *Prof Exp:* Med technician, Baker Vet Ctr, Martinsburg, WVa, 55; chemist, US Food & Drug Admin, 55-57, anal chemist, 57-62, supvy chemist, 62-63, supvy anal res chemist, 63-85; RETIRED. *Res:* Chemistry of all color additives used in foods, drugs and cosmetics. *Mailing Add:* 4113 LaMarre Dr Fairfax VA 22030

LINK, WILLIAM EDWARD, b Ironwood, Mich, Jan 24, 21; m 47; c 2. ANALYTICAL CHEMISTRY. *Educ:* Northland Col, BA, 42; Univ Wis, MS, 51, PhD, 54. *Prof Exp:* Asst prof chem, Northland Col, 47-52; group leader, Res Lab, ADM Chem, 54-69, group leader, Res Ctr, Ashland Chem Co, 69-71, mgr, Anal Chem Res & Develop Div, Ashland Oil & Refining Co Ohio, 71-76, res mgr, Chem Prod Div, 76-78; DIR RES & DEVELOP, SHEREX CHEM CO, 79- *Concurrent Pos:* Ed, Off & Tentative Methods, Am Oil Chemists Soc, 71- *Mem:* Am Chem Soc; Am Oil Chemists Soc (pres, 75-76). *Res:* Organic analytical research; fats and oils chemistry; industrial fatty derivatives; specialty chemicals; fatty nitrogen chemicals; industrial fatty derivatives analysis; resin analysis. *Mailing Add:* 6039 Sedgwick Rd Worthington OH 43085

LINKE, HARALD ARTHUR BRUNO, b Bautzen, Ger, Aug 18, 36; m 71; c 1. MICROBIOLOGY. *Educ:* Univ Berlin, BSc, 61; Univ Gottingen, MSc, 63, PhD(biochem, microbiol), 67. *Prof Exp:* Res assoc enzym, Univ Gottingen, 66-67; fel biochem, Rutgers Univ New Brunswick, 67-69; res microbiologist, Allied Chem Corp, 69-72; res assoc, Inst Microbiol, Rutgers Univ, 72-73; from asst prof to assoc prof, 73-85, PROF, DEPT MICROBIOL, NY UNIV DENT CTR, 85- *Concurrent Pos:* Referee, Zentralblatt Bakteriologie II. Abteilung, 66- *Mem:* New York Acad Sci; Am Soc Microbiol; Ger Chem Soc; Am Asn Dental Res; Europ Orgn Caries Res. *Res:* Isolation and characterization of enzymes; utilizing isotope techniques in the study of microorganisms; biosynthesis and biodegradation of chemical and natural compounds; taxonomy of streptococci; etiology of dental caries and periodontal disease; artificial sweeteners. *Mailing Add:* Dept Microbiol NY Univ Dent Ctr 421 First Ave New York NY 10010

LINKE, RICHARD ALAN, b Plainfield, NJ, Feb 15, 46; m 67; c 2. HARDWARE SYSTEMS. *Educ:* Columbia Col, BA, 68, MS, 70, PhD(physics), 72. *Prof Exp:* Mem tech staff radio physics res, Bell Tel Labs, 72-86, head, Lightware Commun Res Dept, 86-89; SR RES SCIENTIST, NEC RES INST, 89- *Honors & Awards:* Traveling Lectr Award, Inst Elec & Electronic Engrs/Lazers & Electro-optics Soc. *Mem:* Fel Optical Soc Am; sr mem Inst Elec & Electronic Engrs. *Res:* Application of optical communications techniques to computing; optical fiber communications systems; development of low noise millimeter wave receivers. *Mailing Add:* Eight Anderson Lane Princeton NJ 08540

LINKE, SIMPSON, b Jellico, Tenn, Aug 10, 17; m 46; c 2. ELECTRICAL ENGINEERING EDUCATION. *Educ:* Univ Tenn, BS, 41; Cornell Univ, MEE, 49. *Prof Exp:* From instr to prof, 46-86, asst dir lab plasma, 67-75, acting dir, 75-76, coordr Elec Eng Grad Studies, 81-84, EMER PROF ELEC ENG, CORNELL UNIV, 86- *Concurrent Pos:* Consult, Philadelphia Elec Co, 56-57, Brookhaven Nat Labs, 76-80 & NMex Pub Serv Comn, 80-82; chief investr, NSF res grant, 61-64, prog mgr, NSF, 71-72; mem, US Nat Comt, Int Conf Large Elec Systs, 63-88; Attwood assoc, US Nat Comt, Int Conference Large Elec Systs, 88- *Mem:* Inst Elec & Electronic Engrs; Sigma Xi. *Res:* Transient stability of synchronous machines; energy conversion, electric energy systems; high voltage direct current transmission; electric power transmission. *Mailing Add:* Sch Elec Eng Phillips Hall Cornell Univ Ithaca NY 14853

LINKE, WILLIAM FINAN, b Ravena, NY, Aug 5, 24; m 49; c 3. PHYSICAL CHEMISTRY. *Educ:* City Col New York, BS, 45; NY Univ, MS, 46, PhD(chem), 48. *Prof Exp:* Asst chem, NY Univ, 45-48, from instr to asst prof, 48-57; group leader phys chem, 57-59, group leader paper chem, 59-64, mgr res & develop paper & film chem, 65-67, tech dir paper chem dept, 67-70, dir res, Indust Chem & Plastics Div, 71, dir, Stamford Res Ctr, 72-79, dir, technol assessment & licensing, 80- 85, DIR, CHEM RES DIV, AM CYANAMID CO, 86- *Concurrent Pos:* Adv comt, Univ Conn. *Mem:* AAAS; Am Chem Soc; Tech Asn Pulp & Paper Indust; Soc Chem Indust; Indust Res Inst. *Res:* Solubilities; phase equilibria; polyelectrolytes; stability of colloids; flocculation; adsorption; mining and paper chemicals; sizing; polymers; monomers; petrochemical processes; refinery catalysts; auto exhaust catalysts. *Mailing Add:* 75 Ridge Crest Rd Stamford CT 06903-3120

LINKENHEIMER, WAYNE HENRY, physiology; deceased, see previous edition for last biography

LINKER, ALFRED, b Vienna, Austria, Nov 23, 19; US citizen; m 54; c 2. BIOCHEMISTRY, CARBOHYDRATE CHEMISTRY. *Educ:* City Col New York, BS, 49; Columbia Univ, PhD(biochem), 54. *Prof Exp:* Assoc biochem, Columbia Univ, 56-59; asst res prof biochem & path, 60-64, assoc res prof biochem, 64-72, ASSOC PROF PATH, COL MED, UNIV UTAH, 64-, RES PROF BIOCHEM, 72- *Concurrent Pos:* Res biochemist, Vet Admin Hosp, Salt Lake City, 60- *Mem:* Am Soc Biol Chemists; AAAS. *Res:* Structure, function and metabolism of the glycosaminoglycans of connective tissue, including studies of heparin, heparitin sulfate, the chondroitin sulfates, hyaluronic acid, and a variety of degradative enzymes isolated from mammalian and bacterial sources. *Mailing Add:* Vet Admin Hosp Res Serv 151 E Foothill Blvd Salt Lake City UT 84148

LINKINS, ARTHUR EDWARD, b Middletown, Ohio, Jan 13, 45; m 74; c 2. BIOLOGY. *Educ:* Dartmouth Col, AB, 67; Univ Mass, Amherst, PhD(bot), 73. *Prof Exp:* Fel plant physiol, Dept Plant Sci, Univ Calif, Riverside, 72-74; asst prof, 74-80, ASSOC PROF, DEPT BIOL, VA POLYTECH INST & STATE UNIV, 80- *Concurrent Pos:* Adj assoc prof, Inst Arctic Biol, Univ Alaska, 80- *Mem:* Am Soc Microbiol; Soil Sci Soc Am; Mycological Soc Am; AAAS; Am Inst Biol Sci. *Res:* Fungal physiological ecology: role of temperature in regulation of physiology and role of temperature, oxygen, and substrat quality in regualtion of fungal associated decomposition of organic matter. *Mailing Add:* Dept Biol Clarkson Univ Potsdam NY 13699-5800

LINKOW, LEONARD I, b Brooklyn, NY, Feb 25, 26; m 52; c 2. DENTAL IMPLANTOLOGY. *Educ:* Long Island Univ, BS, 48; NY Univ, DDS, 52. *Prof Exp:* Assoc attend chief implantology, Jewish Mem Hosp, 70-84; CLIN PROF REMOVABLE PROSTHODONTICS & IMPLANTOLOGY, SCH DENT, TEMPLE UNIV, PHILADELPHIA, PA, 80- *Concurrent Pos:* Instr at over twenty Am Dent Cols, 71-80; vis lectr, Fixed Bridge Dept, Sch Dent, Loyola Univ, Mayville, Ill, 69- *Honors & Awards:* Thomas P Hinman Award, Am Acad Implant Dent, 72, Aaron Gershoff Mem Award, 74. *Mem:* Fel Am Acad Implant Dent (pres, 74-75); fel & hon mem Int Cong Oral Implantology; fel Am Acad Implants & Transplants; fel Royal Soc Med Eng; German Soc Dent Implantology (hon pres); Am Soc Dent Aesthetics. *Res:* Frontiers of dental implantology. *Mailing Add:* 18 E 50th St New York NY 10022-6865

LINMAN, JAMES WILLIAM, b Monmouth, Ill, July 20, 24; m 46; c 4. MEDICINE. *Educ:* Univ Ill, BS, 45, MD, 47; Am Bd Internal Med, dipl, 55, cert hemat, 74. *Prof Exp:* From intern to jr clin instr internal med, Univ Mich, 47-51, instr, 51-52 & 54-55, asst prof, 55-56; from asst prof to assoc prof med, Northwestern Univ, 56-65; from assoc prof to prof internal med, Mayo Grad Sch Med, Univ Minn, 65-72, consult, Div Hemat, Mayo Clin, 65-72; prof med & dir, Osgood Leukemia Ctr, Univ Ore Health Sci Ctr, 72-79, head Div Hemat, 74-78; PROF MED, JOHN A BURNS SCH MED, UNIV HAWAII, 79- *Mem:* Fel Am Col Physicians; Int Soc Hemat; Am Soc Clin Invest; Am Soc Hemat. *Res:* Hematology. *Mailing Add:* Univ Hawaii Hohn A Burns Med Sch 1356 Lusitana St Honolulu HI 96813

LINN, BRUCE OSCAR, b East Orange, NJ, Dec 12, 29; m 51; c 3. MEDICINAL CHEMISTRY, BIOCHEMISTRY. *Educ:* Duke Univ, BS, 52, PhD(org chem), 56. *Prof Exp:* Asst, Duke Univ, 52-54 & Off Naval Res, 53-54; sr chemist, 56-75, RES FEL, MERCK SHARP & DOHME RES LABS, 76 - *Mem:* Am Chem Soc. *Res:* Medicinal and synthetic organic chemistry in human and animal health. *Mailing Add:* 743 Wingate Dr Bridgewater NJ 08807

LINN, CARL BARNES, b Auburn, Nebr, Feb 18, 07; m 32; c 3. ORGANIC CHEMISTRY. *Educ:* Univ Nebr, BS, 29, MSc, 30; Stanford Univ, PhD(chem), 34. *Prof Exp:* Asst, Univ Nebr, 34-35; from res chemist to group leader, Universal Oil Prod Co, 35-64; prin chemist, Midwest Res Inst, 64-68; sr res chemist, C J Patterson Co, 68-74; RETIRED. *Mem:* Fel AAAS; Am Chem Soc. *Res:* Grignard reagent; pyrolysis of hydrocarbons; homogeneous and heterogeneous catalytic reactions of hydrocarbons and derivatives; hydrogen fluoride technology; high pressure and high temperature technology; Friedel-Crafts reactions; carbohydrate chemistry, including catalytic condensation of sugars and derivatives with hydrocarbons and derivatives; chemistry of lactic acid. *Mailing Add:* 8125 Beverly Dr Prairie Village KS 66208

LINN, DEVON WAYNE, b Estherville, Iowa, Oct 9, 29; m 53; c 3. LIMNOLOGY. *Educ:* Mankato State Col, BA, 52; Ore State Univ, MS, 55; Utah State Univ, PhD(fishery biol, statist), 62. *Prof Exp:* Chemist, Mayo Clin, Minn, 52-53; res biologist, Fisheries Res Inst, Univ Wash, 55-58; asst prof biol, Dakota Wesleyan Univ, 62-64; from asst prof to assoc prof, 64-73, chmn dept, 69-73, PROF BIOL, SOUTHERN ORE STATE COL, 73- *Concurrent Pos:* Consult, Northwest Biol Consults, 62-; Peace Corps vol serving as Dep to Chief Fisheries Officer, Fisheries Dept Ministry Agr & Natural Resources, Lilongwe, Malawi, EAfrica, 73-75; vis prof biol, Univ Swaziland, Kwaluseni, Africa, 83-85. *Mem:* Am Sci Affil. *Res:* Physiological effects of radiation; water pollution and abatement; environmental quality and resource management. *Mailing Add:* Dept of Biol Southern Ore State Col Ashland OR 97520-5071

LINN, JOHN CHARLES, b Bellingham, Wash. COMPUTER SCIENCE, SYSTEMS THEORY. *Educ:* Univ Wash, BS, 68; Stanford Univ, MS, 69, PhD(elec eng), 73. *Prof Exp:* Res engr laser commun, Honeywell Inc, 68; instr comput sci, Stanford Univ, 72; MEM TECH STAFF COMPUT SCI, TEX INSTRUMENTS INC, 73- *Mem:* Inst Elec & Electronics Engrs; Asn Comput Mach. *Res:* Computer architecture; algorithms, memory organization; human speech and language. *Mailing Add:* Tex Instruments Inc 305 Mail Sta 8481 PO Box 869 Plano TX 75086

LINN, MANSON BRUCE, plant pathology; deceased, see previous edition for last biography

LINN, STUART MICHAEL, b Chicago, Ill, Dec 16, 40; m 67; c 3. BIOCHEMISTRY. *Educ:* Calif Inst Technol, BS, 62; Stanford Univ, PhD(biochem), 66. *Prof Exp:* Helen Hay Whitney fel, Univ Geneva, 66-68; from asst prof to prof biochem, 68-87, Univ Calif, Berkeley, head, div biochem & molecular biol, 87-90; CONSULT, 90- *Concurrent Pos:* Res grants, USPHS, Univ Calif, Berkeley, 68- & Dept Energy, 70-; Guggenheim fel, 74-75. *Mem:* AAAS; Am Soc Microbiol; Am Soc Biol Chemists. *Res:* Biochemistry of nucleic acids; nucleic acid enzymes. *Mailing Add:* Div Biochem & Molecular Biol Barker Hall Univ of Calif Berkeley CA 94720

LINN, WILLIAM JOSEPH, b Crawfordsville, Ind, July 14, 27; m 56; c 2. ORGANIC CHEMISTRY. *Educ:* Wabash Col, AB, 50; Univ Rochester, PhD(chem), 53. *Prof Exp:* SR RES ASSOC, AGR PROD DEPT, E I DU PONT DE NEMOURS & CO, INC, 53- *Concurrent Pos:* Res assoc, Northwestern Univ, 69-70. *Mem:* Am Chem Soc; AAAS; Catalysis Soc. *Res:* Organometallic compounds; heterogeneous and homogeneous catalysis; catalytic oxidation; process chemistry. *Mailing Add:* Agr Prod Dept Exp Sta E I du Pont de Nemours & Co Inc Wilmington DE 19880-0402

LINNA, TIMO JUHANI, b Tavastkyro, Finland, Mar 16, 37; m 61; c 3. CANCER, IMMUNOLOGY. *Educ:* Univ Uppsala, BMed, 59, MD, 65, PhD(histol), 67. *Prof Exp:* Asst prof histol, Med Sch, Univ Uppsala, 67-71; asst prof, 70-71, assoc prof, 71-78, adv clin immunol, 72-80, prof, 78-80, RES PROF MICROBIOL & IMMUNOL, SCH MED, TEMPLE UNIV, 80-; GROUP LEADER, IMMUNOL CONTROL RES & DEVELOP DEPT, E I DU PONT DE NEMOURS & CO, INC, WILMINGTON, 80- *Concurrent Pos:* USPHS int res fel, Univ Minn, Minneapolis, 68-70, Univ Minn spec res fel, 70; consult immunol, UN Develop Prog/World Bank/WHO spec prog for res & training in tropical dis, WHO, Geneva, Switzerland, 78-79. *Mem:* Am Soc Exp Path; NY Acad Sci; Reticuloendothelial Soc; Swed Royal Lymphatic Soc; Am Asn Immunologists; Am Asn Cancer Res. *Res:* Immunobiology; experimental pathology; tumor immunology; cell kinetics. *Mailing Add:* Med Prod Dept E I du Pont de Nemours & Co Barley Mill Plaza P26/1236 Wilmington DE 19898

LINNARTZ, NORWIN EUGENE, b Fischer, Tex, Apr 9, 26; m 57; c 2. FOREST SOILS, SILVICULTURE. *Educ:* Tex A&M Univ, BS, 53; La State Univ, MF, 59, PhD(soils), 61. *Prof Exp:* Range mgt asst soil conserv serv, USDA, 53-54, range conservationist, 54-57; res asst, Sch Forestry & Agr Exp Sta, 57-60, from asst prof to assoc prof, 61-70, asst dean, Grad Sch, 77-80, PROF FORESTRY, LA STATE UNIV, BATON ROUGE, 70- *Concurrent Pos:* Asst dir, Sch Forestry, Wildlife & Fisheries, 86- *Mem:* Ecol Soc Am; fel Soc Am Foresters; Soil Sci Soc Am. *Res:* Hardwoods silviculture; forest soil-moisture-plant relationships; forest fertilization; forest range. *Mailing Add:* Sch Forestry Wildlife & Fisheries La State Univ Baton Rouge LA 70803-6202

LINNELL, ALBERT PAUL, b Canby, Minn, June 30, 22; m 44; c 5. ASTROPHYSICS. *Educ:* Col Wooster, AB, 44; Harvard Univ, PhD(astron), 50; Amherst Col, MA, 62. *Hon Degrees:* MA, Amherst Col, 62. *Prof Exp:* From instr to prof astron, Amherst Col, 49-66; chmn, Astron Dept, 66-76, PROF PHYSICS & ASTRON, MICH STATE UNIV, 66- *Concurrent Pos:* Mem adv comt, Comput Ctr, Mass Inst Technol, 60-63; mem bd dirs, Asn Univs for Res Astron, 62-65. *Mem:* Int Astron Union; AAAS; Am Astron Soc; Sigma Xi. *Res:* Instrumentation for photoelectric photometry; photometry and theory of eclipsing binaries. *Mailing Add:* 1918 Yuma Trail Okemos MI 48864

LINNELL, RICHARD D(EAN), b Rapid River Twp, Mich, Sept 18, 20; m 58; c 1. AERODYNAMICS. *Educ:* Univ NH, BS, 46; Mass Inst Technol, SM, 48, ScD(aerodyn eng), 50. *Prof Exp:* Aerodyn engr, United Aircraft Corp, 48; aerodyn engr, Mass Inst Technol, 49, sr engr, 50-52; aerodyn engr, Convair Div, Gen Dynamics Corp, 52-55, staff scientist aerodyn, 56-60; actg mgr aerothermodyn, Gen Elec Co, 55-56; Chance Vought prof aeronaut eng, Sch Eng, Southern Methodist Univ, 60-62; analyst, Ctr Naval Anal, 62-79; ENGR, TRACOR INC, 81- *Mem:* Am Phys Soc; Am Inst Aeronaut & Astronaut. *Res:* Systems analysis; fluid mechanics; vehicle design. *Mailing Add:* Box 342 Rte 1 Northwood NH 03261

LINNELL, ROBERT HARTLEY, b Kalkaska, Mich, Aug 15, 22; m 50; c 4. ACADEMIC ADMINISTRATION, INSTITUTIONAL RESEARCH. *Educ:* Univ NH, BS, 44, MS, 47; Univ Rochester, PhD(chem), 50. *Prof Exp:* Instr chem, Univ NH, 47; asst prof, Am Univ Beirut, 50-52, assoc prof & chmn dept, 52-55; vpres, Tizon Chem Co, 55-58, dir, 55-62; assoc prof chem, Univ Vt, 58-61; lab dir, Scott Res Labs, 61-62; prog dir phys chem, NSF, 62-65, staff assoc planning, 65-67, dep dir dept develop prog, 67-69; dean col letters, arts & sci, 69-70, dir off Instl Studies, 70-82, prof chem, 69-85, chmn dept Safety Sci, 82-85, EMER PROF CHEM, UNIV SOUTHERN CALIF, 85-; PRES, HARMONY INST, 85- *Concurrent Pos:* Grants, Res Corp, 50-54 & 58-60, NSF, 59-61, USPHS, 61-62 & Am Petrol Inst, 61-62; consult, Reheis Corp, 58-61, Tizon Chem Co, 58-62, Col Chem Consult Serv, Lake Erie Environ Studies Prog & Environ Protection Agency; grant, Exxon Educ Found, 74-76 & 78-80 & Carnegie Corp, 76-77 & 78-80; consult environ health & safety, 85-; mem bd dirs, Central Calif Chap, Am Lung Asn, 86-, pres, 91- *Mem:* AAAS; Am Chem Soc; Asn Instnl Res; Am Soc Safety Engrs. *Res:* Hydrogen bonds; air pollution energy planning; science and public policy; science manpower; faculty and staff personnel research (salaries, fringe benefits, policies for consulting, intellectual properties and adult education); student and faculty surveys; higher education evaluation and planning; indoor air pollution; asbestos; radon; environmental health and safety; drunk driving. *Mailing Add:* Harmony Inst PO Box 70 Tollhouse CA 93667-0070

LINNEMANN, ROGER E, b St Cloud, Minn, Jan 12, 31; m 51; c 5. RADIOLOGY, NUCLEAR MEDICINE. *Educ:* Univ Minn, Minneapolis, BA, 52, BS & MD, 56; Am Bd Radiol, cert, 64; Am Bd Nuclear Med, cert, 72. *Prof Exp:* Intern, Walter Reed Army Hosp, 56-57; physician, US Army, Europe, 57-61; res assoc radiobiol, Walter Reed Army Hosp, 61-62, resident radiol, 62-65; cmndg officer, Nuclear Med Res Detachment, US Army, Europe, 65-68; asst prof radiol, Univ Minn, Minneapolis, 68; radiologist, Hosp, 68-69, asst prof, 69-74, ASSOC PROF CLIN RADIOL, UNIV PA, 74-; PRES, RADIATION MGT CORP, 69- *Concurrent Pos:* US deleg radiation protection comt & panel experts med aspects nuclear biol & chem warfare, NATO, 65-68; Nat Res Coun James Picker Found res grant radiol, 68-69; nuclear med consult, Philadelphia Elec Co, 68-; mem ad hoc comt med aspects radiation accidents, AEC, 69-; vis assoc prof clin radiol, Northwestern Univ Sch Med, 77-; res scholar, Univ Minn. *Mem:* AMA; Am Col Radiol; Am Nuclear Soc; Am Pub Health Asn; Indust Med Asn. *Res:* Medical aspects of nuclear industry accidents; kidney function studies using isotopes; radiological health. *Mailing Add:* 5301 Tacony St Box 208 Bldg 11 Arsenal Bus Ctr Philadelphia PA 19137-2307

LINNER, JOHN GUNNAR, b St Paul, Minn, Dec 30, 43. ELECTRON MICROSCOPY, IMMUNOCYTOCHEMISTRY. *Educ:* Mankato State Univ, BS, 66, MS, 70; Iowa State Univ, PhD(cell biol), 78. *Prof Exp:* Electron microscopist, 76-77, res assoc, Iowa State Univ, 77-78; postdoctoral fel, neuro & anat, 78-79, instr, 79-80, asst prof, 80-87, CTR DIR, CRYOBIOL, UNIV TEX HEALTH SCI CTR HOUSTON, 86-, ASSOC PROF, 87- *Concurrent Pos:* Consult, LifeCell Corp, 86- *Mem:* Am Soc Cell Biol; Electron Microscopy Soc Am; Histochem Soc. *Res:* Development of equipment and methodologies to achieve ultrarapid cryofixation of biological samples and molecular distillation drying, the application of electron microscopic molecular and elemental probing techniques to cryoprepared samples. *Mailing Add:* Inst Technol Develop & Assessment Grad Sch Biomed Sci GSBS 1 S128 Houston TX 77030

LINNERT, GEORGE EDWIN, b Chicago, Ill, Dec 8, 16; m 37; c 3. METALLURGY OF WELDING. *Prof Exp:* Metallurgist ferrous, Repub Steel Corp, 35-41; res metallurgist, Rustless Iron & Steel Corp, 41-46; res mgr, Armco, Inc, 46-73; head, N Am Off, Welding Inst, 73-85; CHIEF EXEC OFFICER, GML PUBL, 85- *Concurrent Pos:* Emer head, Edison Welding Inst, 85. *Honors & Awards:* Samuel Wylie Miller Mem Medal, Am Welding Soc, 72; Adams lectr, Am Welding Soc. *Mem:* Am Welding Soc (pres, 70-71); fel Am Soc Metals. *Res:* Metallurgy of joining metals by welding, brazing and soldering; author of textbooks. *Mailing Add:* GML Publications 31 Oyster Shell Lane Hilton Head SC 29926

LINNERUD, ARDELL CHESTER, b Whitehall, Wis, Apr 9, 31; m 56. EXPERIMENTAL STATISTICS. *Educ:* Wis State Univ River Falls, BS, 53; Univ Minn, MS, 62, PhD(dairy husb), 64. *Prof Exp:* Res asst dairy husb, Univ Minn, 57-63, consult biomet, 64; fel biomath, 64-67, asst prof statist, 67-75, ASSOC PROF STATIST, NC STATE UNIV, 75- *Concurrent Pos:* Statist consult, Inst for Aerobics Res, 74- *Mem:* Am Dairy Sci Asn; Am Soc Animal Sci. *Res:* Design of experiments and mathematical model building; animal science and exercise physiology. *Mailing Add:* Box 8203 Statsist NC State Univ Main Campus Raleigh NC 27695-8203

LINNOILA, MARKKU, ALCOHOLSIM. *Prof Exp:* ACTG SCI DIR, NAT INST ALCOHOL ABUSE & ALCOHOLISM, NIH, 91- *Mailing Add:* NIH Nat Inst Alcohol Abuse & Alcoholism Sci Dir Actg Bldg 10 Rm 3C103 Bethesda MD 20892

LINNSTAEDTER, JERRY LEROY, b Lindale, Tex, July 25, 37; m 62; c 3. MATHEMATICS. *Educ:* Tex A&M Univ, BA, 59, MS, 61; Vanderbilt Univ, PhD(math), 70. *Prof Exp:* Instr math, Northeastern La State Univ, 61-63 & Vanderbilt Univ, 67-68; assoc prof, 68-71, PROF MATH, ARK STATE UNIV, 71-, CHMN DEPT COMPUT SCI, MATH & PHYSICS, 68- *Concurrent Pos:* Prin investr NASA res grant, Ark State Univ, 69-71; mem, Ark Comn Improving Pub Sch Basic Skill Opportunities. *Mem:* Am Math Soc; Math Asn Am; Sigma Xi. *Res:* Multistage calculus of variations; classical analysis; multi stage bolza problems and related control problems; applications analysis; Zermelo flow problems. *Mailing Add:* Col Sci Ark State Univ Box 877 State University AR 72467

LINOWSKI, JOHN WALTER, b Boston, Mass, July 7, 45; m 72. PHYSICAL CHEMISTRY, ANALYTICAL CHEMISTRY. *Educ:* Boston Col, BS, 67; Canisius Col, MS, 70; Rutgers Univ, PhD(phys chem), 74. *Prof Exp:* Res assoc molecular dynamics, Univ Ill, 74-76; sr res chemist, 76-81, res leader, 81-82, group leader, 82-87, RES MGR, DOW CHEM CO, 87- *Honors & Awards:* Rieman Award 1976. *Mem:* Am Chem Soc; Sci Res Soc NAm; Sigma Xi. *Res:* Dynamic nuclear polarization; molecular dynamics of liquids at high pressure and extreme temperatures; nuclear magnetic resonance; catalysis; process chemistry. *Mailing Add:* 2455 Woodland Estates Dr Midland MI 48640

LINS, THOMAS WESLEY, b Nov 24, 23; US citizen; m 69. MARINE GEOLOGY, STRUCTURAL GEOLOGY. *Educ:* Cornell Univ, BS, 48; Univ Kans, MS, 59, PhD(geol), 69. *Prof Exp:* Dist geologist, Sunray Oil Corp, 50-51; dist geologist, Monsanto Chem Co, 51-57, asst div geologist, 57-60, div geologist, 60-61, res geologist, 61-63; asst prof geol, Lamar Univ, 68-74; from asst prof to assoc prof geol & geog, Miss State Univ, 74-89; RETIRED. *Mem:* AAAS; Geol Soc Am. *Res:* Tectonics, structure of island arcs and trenches, recent sedimentation of the Gulf of Mexico. *Mailing Add:* 33 Stonehinge Lane Malvern PA 19355

LINSAY, ERNEST CHARLES, b Cleveland, Ohio, May 3, 42; m 66; c 3. ORGANIC CHEMISTRY. *Educ:* Yale Univ, BS, 63; Univ Wis-Madison, PhD(org chem), 68. *Prof Exp:* Res chemist, 68-78, sr res chemist, Organics Dept, 78-83, tech supt, 83-86, res & develop supt, 86-87, RES SCIENTIST, COATINGS & ADDITIVES, HERCULES INC, 87- *Mem:* Am Chem Soc; Soc Automotive Engrs. *Res:* Physical organic chemistry; rosin and fatty acids; dispersions and emulsions; rosin-, terpene- and hydrocarbon-based resins; coatings; nitrocellulose; jet engine lucricants; wood preservatives. *Mailing Add:* 43 Slashpine Circle Hockessin DE 19707

LINSCHEID, HAROLD WILBERT, b Goessel, Kans, Sept, 24, 06; m 33; c 3. MATHEMATICAL ANALYSIS. *Educ:* Bethel Col, Kans, BA, 29; Phillips Univ, MEd, 36; Univ Okla, MA, 40, PhD, 55. *Prof Exp:* Prin high sch, Okla, 29-36; instr, Okla Jr Col, 36-38; instr math, Univ Okla, 38-41; instr math & physics, Bluffton Col, 41-43; army specialized training prog, Univ Nebr, 43-44; asst prof math & physics, Eastern NMex Col, 44-46; assoc prof, Col Emporia, 51-58; from assoc prof to prof, 58-77, EMER PROF MATH, WICHITA STATE UNIV, 77- *Mem:* Am Math Soc; Math Asn Am. *Res:* Algebra; geometry; physics; electricity and magnetism. *Mailing Add:* 3701 E Funston Wichita KS 67218

LINSCHITZ, HENRY, b New York, NY, Aug 18, 19; m 64; c 1. PHYSICAL CHEMISTRY. *Educ:* City Col New York, BS, 40; Duke Univ, MA, 41, PhD(chem), 46. *Prof Exp:* Mem staff, Explosives Res Lab, Nat Defense Res Comt, 43; sect leader, Los Alamos Sci Lab, 43-45; fel, Inst Nuclear Studies, Univ Chicago, 46-48; from asst prof to assoc prof chem, Syracuse Univ, 48-57; assoc prof, 57-59, PROF CHEM, BRANDEIS UNIV, 59-, CHMN DEPT, 58- *Concurrent Pos:* Vis scientist, Brookhaven Nat Lab, 56-57; Fulbright vis prof, Hebrew Univ, Israel, 60; mem adv comt space biol, NASA, 60-61, study sect biophys & biophys chem, NIH, 62-66 & comt photobiol, Nat Res Coun, 64-69; Guggenheim fel, Weizmann Inst, 71-72. *Mem:* AAAS; Am Chem Soc; Am Acad Arts & Sci; Fedn Am Scientists. *Res:* Photochemistry; spectroscopy and luminescence of complex molecules; photobiology. *Mailing Add:* Dept Chem Brandeis Univ Waltham MA 02254

LINSCOTT, DEAN L, b Blue Springs, Nebr, Mar 31, 32; m 53; c 7. AGRONOMY. *Educ:* Univ Nebr, BSc, 53, MSc, 57, PhD(agron), 61. *Prof Exp:* Instr agron, Agr Res Serv, USDA, Univ Nebr, 57-61, RES AGRONOMIST, AGR RES SERV, USDA, & PROF FIELD CROPS, CORNELL UNIV, 61- *Mem:* AAAS; fel Weed Sci Soc Am (secy, 84, 85); Soil Sci Soc Am; Am Soc Agron; Crop Sci Soc Am. *Res:* Absorption, translocation and degradation of herbicides; persistence of herbicides; vegetation management; plant protection. *Mailing Add:* Dept Soil Crop & Atmospheric Sci Cornell Univ Ithaca NY 14853-0144

LINSCOTT, WILLIAM DEAN, b Bakersfield, Calif, Apr 23, 30; m 55; c 3. IMMUNOLOGY. *Educ:* Univ Calif, Los Angeles, BA, 51, PhD(infectious dis), 60. *Prof Exp:* From asst prof to prof microbiol, Med Ctr, Univ Calif, San Francisco, 64-82; RETIRED. *Concurrent Pos:* USPHS res fels, Labs Microbiol, Howard Hughes Med Inst, Fla, 60-62 & Div Exp Path, Scripps Clin & Res Found, Calif, 62-64; publ, Linscott's Dir Immunol & Biol Reagents. *Res:* Complement; immunologic unresponsiveness. *Mailing Add:* 4877 Grange Rd Santa Rosa CA 95404

LINSKY, CARY BRUCE, b Chicago, Ill, June 9, 42; m 68; c 2. BIOLOGICAL CHEMISTRY. *Educ:* Univ Wis-Madison, BS, 64; Loyola Univ, PhD(biochem), 71. *Prof Exp:* RES ASSOC, JOHNSON & JOHNSON, 71-, ASST MGR, 80- *Mem:* Am Chem Soc; Sigma Xi; Am Burn Asn. *Res:* Role of inflammatory response and local environment in cutaneous wound healing; cellular components of inflammation; scar formation in surgical wounds; collagen biochemistry; hemostasis; prevention of post surgical adhesions. *Mailing Add:* Dept Clin Res Ethicon Inc Summerville NJ 08876

LINSKY, JEFFREY L, b Buffalo, NY, June 27, 41; m 67; c 2. SPACE PHYSICS, SOLAR PHYSICS. *Educ:* Mass Inst Technol, BS, 63; Harvard Univ, AM, 65, PhD(astron), 68. *Prof Exp:* Res assoc astrophys, 68-69, assoc prof adjoint, 74-79, LECTR, DEPT PHYSICS & ASTROPHYS, & DEPT ASTROGEOPHYS, UNIV COLO, 69-, PROF ADJOINT, DEPT ASTRON, PLANETARY & ATMOSPHERIC SCI, 79- *Concurrent Pos:* Mem, Joint Inst Lab Astrophys, 68-71, fel, 71-; consult, NASA, 72-; astronomer, Lab Astrophys, Nat Bur Standards, 69- *Honors & Awards:* Medal for Except Sci Achievement, NASA, 88. *Mem:* Am Astron Soc; Int Astron Union. *Res:* Radiative transfer; formation of spectral lines in the solar and stellar chromospheres; atmospheres of latetype stars; stellar coronae; ultraviolet and x-ray astronomy from space. *Mailing Add:* Joint Inst Lab Astrophys Campus Box 440 Univ of Colo Boulder CO 80309-0440

LINSLEY, EARLE GORTON, b Oakland, Calif, May 1, 10; m 35; c 2. ENTOMOLOGY. *Educ:* Univ Calif, BS, 32, MS, 33, PhD(entom), 38. *Prof Exp:* Asst entom, Univ Calif, 33-35; agr, Univ Calif, Los Angeles, 35-37; assoc quarantine entomologist, State Dept Agr, Calif, 38-39; instr entom & jr entomologist, 39-43, asst prof & asst entomologist, 43-49, assoc prof & assoc entomologist, 49-53, chmn dept entom & parasitol, 51-59, asst dir, Agr Exp Sta, 60-63, assoc dir, 63-73, dean, Col Agr Sci, 60-73, prof entom & entomologist, 53-73, EMER PROF & ENTOMOLOGIST, AGR EXP STA, UNIV CALIF, BERKELEY, 73- *Concurrent Pos:* Instr, Yosemite Sch Field Nat Hist, 38-41; res assoc, Calif Acad Sci, 39-; secy, Am Comn Entom Nomenclature, 43-48; ed, Pan-Pac Entomologist, 43-50; Guggenheim fel, Am

Mus Natural Hist, 47-48; collabr, US Dept Interior, 53-54; res prof, Miller Inst Basic Res in Sci, 60-62; mem comt insect pests, Nat Res Coun-Nat Acad Sci, 63-69; partic, Galapagos Int Sci Proj, 64; mem Orgn Trop Studies, 68- *Mem:* Fel AAAS; hon mem Entom Soc Am (vpres, 46, 48, pres, 52); Ecol Soc Am; Am Soc Naturalists; Soc Syst Zool. *Res:* Systematic entomology; ecology and taxonomy of Coleoptera and Hymenoptera Apoidea; geographical distribution; mimicry and adaptive coloration; ethology of solitary bees; entomophagous Coleoptera; interrelations of flowers and insects; host specificity. *Mailing Add:* Div Entom & Parasitol Univ Calif Berkeley CA 94720

LINSLEY, JOHN, b Minneapolis, Minn, Mar 12, 25; m 66; c 3. PHYSICS, ASTRONOMY. *Educ:* Univ Minn, BPhys, 47, PhD(physics), 52. *Prof Exp:* Asst prof physics, Univ Va, 51-52; res fel, Univ Minn, 52-54; res assoc, Mass Inst Technol, 54-55, asst prof, 55-58, res assoc, 58-72; adj prof, 72-77, RES PROF PHYSICS, UNIV N MEX, 77- *Mem:* Fel Am Phys Soc; Sigma Xi; Am Astron Soc. *Res:* Origin and behavior of highest-energy cosmic rays by means of experimental and theoretical investigations of extensive air showers. *Mailing Add:* 1712 Old Town Rd NW Albuquerque NM 87104

LINSLEY, RAY K(EYES), JR, civil engineering; deceased, see previous edition for last biography

LINSLEY, ROBERT MARTIN, b Chicago, Ill, Feb 19, 30; div; c 3. INVERTEBRATE PALEONTOLOGY. *Educ:* Univ Mich, BS, 52, MS, 53, PhD(geol), 60. *Prof Exp:* Ford intern geol, 54-55, from instr to asst prof, 55-64, assoc prof & chmn dept, 64-71, dir natural sci course, 62-70, phys sci course, 59-64, prof, 71-78, HAROLD ORVILLE WHITNALL PROF GEOL, COLGATE UNIV, 78- *Concurrent Pos:* Mem Paleont Res Inst. *Mem:* AAAS; Geol Soc Am; Paleont Soc; Soc Study Evolution. *Res:* Evolution, functional morphology; behavior and taxonomy of Gastropoda. *Mailing Add:* Dept of Geol Colgate Univ Hamilton NY 13346

LINSTEDT, KERMIT DANIEL, b Portland, Ore, Nov 6, 40; m 64; c 3. SANITARY ENGINEERING. *Educ:* Ore State Univ, BS, 62; Stanford Univ, MS, 63, PhD(sanit eng), 68. *Prof Exp:* Sanit engr asst, Los Angeles Dept Water & Power, 61-62; from asst prof to prof sanit eng, Dept Civil & Environ Eng, Univ Colo, 67-81; proj mgr, 81-88, REGIONAL OFF MGR, BLACK & VEATCH, DENVER, 88-, PARTNER, 89- *Concurrent Pos:* Consult, Denver Metro Dist, 69-71 & Environ Protection Agency, 78- *Honors & Awards:* Dow Award, Am Soc Eng Educrs, 72; Bedell Award, Water Pollution Control Fedn, 73; Res Div Award, Am Water Works Asn, 78. *Mem:* Water Pollution Control Fedn; Am Water Works Asn; Sigma Xi. *Res:* Treatment methods for water reuse; characterization and treatment of oil shale retort water. *Mailing Add:* 6647 Apache Ct Longmont CO 80503

LINSTONE, HAROLD A, b Hamburg, Germany, June 15, 24; m 46; c 2. CORPORATE PLANNING, FORECASTING. *Educ:* City Col NY, BS, 44; Columbia Univ, MA, 47; Univ Southern Calif, PhD(math), 54. *Prof Exp:* Sr scientist, Hughes Aircraft Co, 49-61 & Rand Corp, 61-63; assoc dir corp develop planning, Lockheed Aircraft Corp, 63-70; dir systs sci PhD prog, 70-77, prof systs sci, 77-86, EMER PROF, PORTLAND STATE UNIV, 86- *Concurrent Pos:* Ed-in-chief, Technol Forecasting & Social Change, 69-; consult, IBM, Electric Power Resource Inst, Atlantic Richfield Co, Weyerhauser, Nero & Assocs, Inc, United Nations Asian-Pacific Ctr Technol Transfer, 70-; pres, Systs Forecasting Inc, 70-; vis prof, Univ Rome, 74, Univ Wash, 77, Univ Calif, Riverside, 85-86, Univ Kiel, 89. *Mem:* Opers Res Soc Am; Inst Mgt Sci. *Res:* Multiple perspectives for decision making; technological forecasting; futures research; corporate planning; risk analysis; systems science; policy analysis. *Mailing Add:* 70 Wheatherstone Ct Lake Oswego OR 97035

LINSTROMBERG, WALTER WILLIAM, b Beaufort, Mo, Oct 30, 12; m 43; c 2. ORGANIC CHEMISTRY. *Educ:* Univ Mo, AB, 37, MA, 50, PhD(chem), 55. *Prof Exp:* Instr chem, Univ Mo, 52-55; from asst prof to prof org chem, Univ Nebr, Omaha, 55-78; RETIRED. *Concurrent Pos:* Vis prof, Utah State Univ, 57; vis prof & res assoc, Univ Nebr, 60. *Mem:* Am Chem Soc; Sigma Xi. *Res:* Pharmaceutical chemistry. *Mailing Add:* 630 S 90th St Omaha NE 68114

LINTNER, CARL JOHN, JR, b Louisville, Ky, July 15, 17; m 48; c 2. PHARMACEUTICAL CHEMISTRY. *Educ:* Univ Ky, BS, 40; Univ Wis, PhD(pharmaceut chem), 50. *Prof Exp:* Res chemist, Upjohn Co, 50-55, sect head, 55-82; CONSULT, 82- *Concurrent Pos:* Adj prof, Fla A&M Univ, Col Pharm, 75- *Honors & Awards:* W E Upjohn Award. *Mem:* AAAS; fel Acad Pharmaceut Sci; Am Pharmaceut Asn. *Res:* Organic synthesis; determination of functional groups; essential oil determination; kinetic studies; chromatography and ion exchange resins; phytochemistry; tablet coatings; ointment bases; chemistry of antibiotics; instrumentation tablet compression; pharmaceutical product stability. *Mailing Add:* 2125 Aberdeen Dr Kalamazoo MI 49008

LINTON, EVERETT PERCIVAL, b St John West, NB, Dec 30, 06; m 36, 80; c 3. PHYSICAL CHEMISTRY. *Educ:* Mt Allison Univ, BSc, 28; McGill Univ, MSc, 30, PhD(chem), 32. *Hon Degrees:* DSc, Acadia Univ, 78. *Prof Exp:* Instr chem, Mt Allison Univ, 28-29; Royal Soc Can fel, Univ Munich, 32-33; chemist, Biol Bd Can, Halifax, NS, 34-36; instr chem, Acadia Univ, 36-41; asst phys chemist, Fisheries Res Bd, Halifax, 41-44; prof chem, Acadia Univ, 44-75, head dept, 66-75; RETIRED. *Mem:* Am Chem Soc. *Res:* Preparation of hydrogen peroxide; measurement of dielectric constants; interaction of neutral molecules; air-drying solids; smokes; colloidal chemistry; dipole moments of amine oxides; drying and smoke curing of fish. *Mailing Add:* PO Box 166 Wolfville NS B0P 1X0 Can

LINTON, FRED E J, b Italy, Apr 8, 38; US citizen; m 90. MATHEMATICS. *Educ:* Yale Univ, BS, 58; Columbia Univ, MA, 59, PhD(math), 63. *Hon Degrees:* MA, Wesleyan Univ, 72. *Prof Exp:* From asst prof to assoc prof, 63-72, chmn dept, 75, PROF MATH, WESLEYAN UNIV, 72- *Concurrent Pos:*

Nat Res Coun res fel, Swiss Fed Inst Technol, 66-67; Izaak Walton Killam sr res fel, Dalhousie Univ, 69-70. *Honors & Awards:* Hon Gold Medal, Polytech Univ Blagoevgrad, Bulgaria, 87. *Mem:* Am Math Soc; Math Asn Am. *Res:* Categorical algebra, a branch of positive speculative philosophy. *Mailing Add:* Wesleyan Univ Middletown CT 06459

LINTON, PATRICK HUGO, psychiatry; deceased, see previous edition for last biography

LINTON, RICHARD WILLIAM, b Scranton, Pa, Apr 17, 51; m 76. SURFACE SPECTROSCOPY, MICROBEAM ANALYSIS. *Educ:* Univ Del, BS, 73; Univ Ill, MS, 75, PhD(chem), 77. *Prof Exp:* From asst prof to assoc prof, 77-88, PROF CHEM, UNIV NC, CHAPEL HILL, 89-, ASST VPRES RES, 86- *Honors & Awards:* Outstanding Young Scientist Award, Microbeam Anal Soc, 90. *Mem:* Am Chem Soc; Microbeam Analysis Soc; Am Soc Testing Mat; Am Soc Mass Spectrometry; Mat Res Soc; Am Vacuum Soc. *Res:* Surface and microprobe techniques for chemical analysis (SIMS, XPS, Auger spectroscopy, SEM, electron and ion microprobe analysis); biological microanalysis; ion beam-surface interactions; environmental analytical chemistry; chemistry and characterization of polymer surfaces; digital imaging. *Mailing Add:* Dept Chem CB#3290 Univ NC Chapel Hill NC 27599-3290

LINTON, THOMAS LARUE, b Carlisle, Tex, July 25, 35; m 61; c 2. FISHERIES. *Educ:* Lamar State Col, BS, 59; Univ Ga, MS, 61; Univ Mich, PhD(fisheries), 66. *Prof Exp:* Res asst zool, Univ Ga, 63-65, res assoc, 65-67, asst prof, 67-70; mem staff, Div Com Sports Fisheries, NC Dept Conserv & Develop, 70-73, mem staff, NC Dept Natural & Econ Resources, 73-80; AT DEPT WILDLIFE & FISHERIES SCI, TEX A & M UNIV. *Concurrent Pos:* Res grants, Ga Game & Fish Comn, 65-68 & US Dept Interior, 66-; Ga rep biol comt, Atlantic State Marine Fisheries Comn, 64-66. *Mem:* AAAS; Am Fisheries Soc. *Res:* Physiology; commercial and sport fisheries; pollution ecology. *Mailing Add:* Dept Wildlife Mgt Tex A&M Univ College Station TX 77843

LINTVEDT, RICHARD LOWELL, b Edgerton, Wis, June 23, 37; m 59; c 3. PHYSICAL INORGANIC CHEMISTRY. *Educ:* Lawrence Univ, BA, 59; Univ Nebr, PhD(inorg chem), 66. *Prof Exp:* Res chemist, Chem Div, Morton Int, 59-62; asst prof inorg chem, 66-71, assoc prof, 71-76, PROF CHEM, WAYNE STATE UNIV, 76-, CHMN DEPT, 83- *Concurrent Pos:* Petrol Res Fund grant, 66-68, 70-73, 76-78 & 84-86; Res Corp grant, 69-71; NSF grant, 76-90; Dept of Energy grant, 78-83. *Mem:* Am Chem Soc; fel AAAS. *Res:* Electronic structure and bonding in inorganic coordination and chelate compounds; electrochemistry; physical inorganic chemistry; magnetochemistry of transition metal complexes. *Mailing Add:* 911 Barrington Grosse Pointe Park MI 48230

LINTZ, JOSEPH, JR, b New York, NY, June 15, 21; m 44; c 3. GEOLOGY. *Educ:* Williams Col, AB, 42; Univ Okla, MS, 47; Johns Hopkins Univ, PhD(geol), 56. *Prof Exp:* Jr geologist, Gen Petrol Corp, 47-48; geologist, Pure Oil Co, 49; from asst prof to assoc prof, 51-65, actg dean, 81-82, prof geol, 65-, EMER PROF GEOL, MACKAY SCH MINES, UNIV NEV, RENO. *Concurrent Pos:* Instr, Mt Lake Biol Sta, Va, 51; asst geologist, Nev Bur Mines, 51-56, assoc geologist, 56-; vis prof, Bandung Tech Inst, 59-61; Nat Acad Sci-Nat Res Coun res assoc, Manned Spacecraft Ctr, Tex, 66-67; consult, Econ Comm Asia & Far East, UN, 71 & 74; consult, Atomic Energy Comn, 71-79. *Mem:* Paleont Soc; Am Asn Petrol Geol; AAAS. *Res:* Remote sensing of environment; petroleum possibilities and Pennsylvanian system of Nevada. *Mailing Add:* 760 Singingwood Dr Reno NV 89509

LIN-VIEN, DAIMAY, Taiwan citizen; m; c 1. INFRARED-SPECTROSCOPTIST, POLYMER SPECTROSCOPIST. *Educ:* Kans State Univ, PhD(anal chem), 88. *Prof Exp:* ASSOC RES CHEMIST ANAL CHEM, SHELL DEVELOPMENT CO, 88- *Mem:* Soc Appl Spectros; Coblenz Soc. *Res:* Analytical applications of fourier transform infrared spectroscopy; fourier transform infrared microspectroscopy; raman spectroscopy; polymer characterization; polymer-polymer interaction; additive-polymer interaction. *Mailing Add:* 2014 Highland Hills Houston TX 77478

LINVILL, JOHN G(RIMES), b Kansas City, Mo, Aug 8, 19; m 43; c 2. ELECTRICAL ENGINEERING, SOLID-STATE ELECTRONICS. *Educ:* William Jewell Col, AB, 41; Mass Inst Technol, SB, 43, SM, 45, ScD(elec eng), 49. *Hon Degrees:* DAppSc, Cath Univ Louvain, 66. *Prof Exp:* Asst prof elec eng, Mass Inst Technol, 49-51; mem tech staff, Bell Tel Labs, Inc, 51-55; asoc prof elec eng, 55-57, chmn dept, 64-80, assoc dean sch eng, 72-80, PROF ELEC ENG, STANFORD UNIV, 57-, PROF INTEGRATED SYSTS & DIR, CTR INTERGRATED SYSTS, 81- *Concurrent Pos:* Co-founder & dir, Telesensory Systs, Inc, 71- *Honors & Awards:* Educ Medal, Inst Elec & Electronic Engrs, 76; John Scott Award, 80; Medal Achievement, Am Electronics Asn, 83; Louis Braille Prize, Deutscher Blindenverband, 84. *Mem:* Nat Acad Eng; fel Inst Elec & Electronics Engrs; fel AAAS; fel Am Acad Arts & Sci. *Res:* Custom integrated circuits and systems as sensory aids for the blind. *Mailing Add:* Dept Elec Eng Stanford Univ Stanford CA 94305-4070

LINZ, ARTHUR, b Barcelona, Spain, Jan 30, 26; US citizen; m 55, 84; c 1. SOLID STATE PHYSICS, MATERIALS SCIENCE. *Educ:* Brown Univ, BS, 46; Univ NC, MS, 50, PhD(physics), 52. *Prof Exp:* Res physicist, titanium div, Nat Lead Co, NJ, 52-58; res physicist, div sponsored res staff, lab insulation res, 58-63, res assoc elec eng, 63-70, sr res assoc elec eng, Crystal Physics Lab, Mat Sci Ctr, Mass Inst Technol, 70-86; TECH CONSULT, 86- *Concurrent Pos:* Vis prof, Univ Nancy, 63; consult, Brookhaven Nat Lab, 69-79; vis sr reader, Univ Bath, 76. *Mem:* Am Phys Soc; Optical Soc Am; Am Asn Crystal Growth; Sigma Xi. *Res:* Materials engineering; single crystal growth; properties of crystals; laser crystals; photodetectors. *Mailing Add:* 9495 Periwinkle Dr Vero Beach FL 32963

LINZ, PETER, b Apatin, Yugoslavia, July 19, 36; US citizen; m 82; c 2. COMPUTER SCIENCE. *Educ:* McGill Univ, BSc, 57; Univ Mich, MS, 60; Univ Wis, PhD(comput sci), 68. *Prof Exp:* Res engr, Dominion Eng Ltd, 57-59; assoc programmer, IBM Corp, 63-65; staff specialist numerical anal, Comput Ctr, Univ Wis, 65-68; asst prof comput sci, NY Univ, 68-70; from asst prof to assoc prof, 70-77, prof math, 77-83, PROF COMPUT SCI UNIV CALIF, DAVIS, 83- *Res:* Numerical analysis; quadrature methods; solution and applications of integral equations; numerical software. *Mailing Add:* Div Computer Sci Univ Calif Davis CA 95616

LINZER, MELVIN, b New York, NY, Aug 5, 37; m 64; c 4. ULTRASOUND, PHYSICAL CHEMISTRY. *Educ:* Brooklyn Col, BS, 57; Princeton Univ, MA, 59, PhD(chem), 62. *Prof Exp:* Res assoc chem, Princeton Univ, 61; Nat Acad Sci-Nat Res Coun fel, 61-63, group leader signal processing & imaging, 78-81, PHYSICAL CHEMIST, NAT BUR STANDARDS, 63-, GROUP LEADER, FRACTURE & DEFORMATION DIV, 81- *Concurrent Pos:* Ed-in-Chief, Ultrasonic Imaging, 79-; cochairperson, Ultrasonic Tissue Signature Working Group, 76-78; chmn, Int Symp on Ultrasonic Imaging and Tissue Characterization, 75-; chmn, Int Symp on Ultrasonic Mat Characterization, 78- *Honors & Awards:* Ross Coffin Purdy Award, Am Ceramic Soc, 75; US Dept Commerce Gold Medal Award, 77; Nat Bur Standards Appl Res Award, 78. *Mem:* Am Phys Soc; Am Inst Ultrasound in Med; Sigma Xi. *Res:* Nondestructive evaluation; ultrasound medical diagnosis; acoustic emission; laser spectroscopy; combustion diagnostics; magnetic resonance spectroscopy; measurement techniques for spectroscopic and materials applications; shock wave structure. *Mailing Add:* Two Fulham Ct Silver Spring MD 20902

LINZER, ROSEMARY, oral microbiology, for more information see previous edition

LINZEY, DONALD WAYNE, wildlife biology, mammalogy, for more information see previous edition

LIOI, ANTHONY PASQUALE, b Pittsburgh, Pa, May 15, 49; m 75; c 5. FLUID MECHANICS. *Educ:* Univ Pittsburgh, BS, 71; Drexel Univ, MS, 76, PhD(biomed eng), 79. *Prof Exp:* Co-prin investr, Univ Utah Artificial Heart Res Lab, 82-85, prin investr, 85-86; CO-PRIN INVESTR, RES SCIENTIST & MGR, MED DEVICES, NU-TECH INDUST, INC, 86- *Concurrent Pos:* Res instr surg, Univ Utah Artificial Heart Res Lab, 79-86. *Res:* Artificial heart research; development of automatic physiolosic control methods and the design and development of a miniature hydraulic pump drive system for the artificial heart; blood pump design. *Mailing Add:* Nu-Tech Industries Inc 5905 Wolf Creek Pike Dayton OH 45426

LIONETTI, FABIAN JOSEPH, b Jersey City, NJ, Mar 3, 18; m 43; c 3. BIOCHEMISTRY. *Educ:* NY Univ, AB, 43, MS, 45; Rensselaer Polytech Inst, PhD(phys chem), 48. *Prof Exp:* From instr to assoc prof biochem, Sch Med, Boston Univ, 49-65; assoc mem, Inst Health Sci, Brown Univ, 65-68; SR INVESTR, CTR BLOOD RES, 68- *Honors & Awards:* Mathewson Medal, Am Inst Metals, 52. *Mem:* Am Soc Biol Chemists; Am Chem Soc; Cryobiol Soc. *Res:* Metabolism and preservation of human blood cells. *Mailing Add:* 800 Huntington Ave Boston MA 02115-6303

LIOR, NOAM, b Mar 11, 40; US citizen; c 3. MECHANICAL ENGINEERING. *Educ:* Technion, Israel, BS, 62, MS, 66; Univ Calif, Berkeley, PhD(mech eng), 73. *Hon Degrees:* MA, Univ PA, 78. *Prof Exp:* Instr, Dept Mech Eng, Technion, 65-66; res asst res eng water desalination, Seawater Conversion Lab, Univ Calif, Berkeley, 66-73; from asst prof to assoc prof, 73-85, chmn Mech Eng Grad Group, 86-90, PROF MECH ENG, DEPT MECH ENG & APPL MECH, UNIV PA, 85- *Concurrent Pos:* Prin investr, NSF grants, Pa Sci & Eng Found grant, US HUD grant, 75- & US Dept Energy grant, 75-; consult, Westinghouse Elec Corp, 77-78, Lawrence Livermore Labs & Solar Energy Res Inst, Argonne Nat Lab ,; mem, Solar Collector Standards Comt, Am Soc Heating, Refrig & Air Conditioning Engrs, 75- *Honors & Awards:* Ralph Teetor Award, 86. *Mem:* Am Soc Mech Engrs; Int Solar Energy Soc; Instrument Soc Am; Am Soc Heating, Refrig & Air Conditioning Engrs; Int Desalination Asn. *Res:* Heat transfer, thermodynamics and fluid mechanics as related to solar energy applications, water desalination, combustion and cooling of electronics; thermo-fluid measurements. *Mailing Add:* Dept Mech Eng & Appl Mech Univ Pa Philadelphia PA 19104-6315

LIOTTA, LANCE, PATHOLOGY. *Prof Exp:* Actg chief, 81-82, CHIEF, LAB PATH, NAT CANCER INST, NIH, 82- *Mailing Add:* NIH Nat Cancer Inst Lab Path Bldg 10 Rm 2A33 Bethesda MD 20892

LIOU, JUHN G, b Taiwan, Dec 28, 39; m 64; c 2. METAMORPHIC PETROLOGY, GEOCHEMISTRY. *Educ:* Nat Twiwan Univ, BS, 62; Univ Calif, Los Angeles, PhD(geol), 70. *Prof Exp:* Teaching & res asst geol, Nat Taiwan Univ, 63-64; teaching asst, Univ Calif, Los Angeles, 65-69, NSF fel, 70; res fel geochem, Nat Res Coun, NASA, Houston, 70-72; asst prof geol, 72-76, ASSOC PROF GEOL, STANFORD UNIV, 76- *Concurrent Pos:* Guggenheim fel, John Simon Guggenheim Found, 78-79. *Honors & Awards:* Mineral Soc Am Award, 79. *Mem:* Geol Soc Am; Am Geophys Union; Mineral Soc Am. *Res:* To understand, through hydrothermal experiments and field observations, the parageneses and compositions of metamorphic minerals zeolites, prehnite, pumpellites and epidote; their imposed physical conditions for low-grade metabasites. *Mailing Add:* Dept of Geol Stanford Univ Stanford CA 94305

LIOU, KUO-NAN, b Tapei, Taiwan, Nov 16, 43; US citizen; m 68; c 2. CLIMATE MODELING, REMOTE SOUNDING. *Educ:* Nat Taiwan Univ, BS, 65; NY Univ, MS, 68, PhD(atmospheric physics), 70. *Prof Exp:* Res asst, NY Univ, 66-70; res assoc, Goddard Inst Space Studies, 70-72; asst prof atmospheric sci, Univ Wash, 72-74; assoc prof, 75-80, PROF ATMOSPHERIC SCI, UNIV UTAH, 80-, DIR, DTR ATMOSPHERIC & REMOTE SOUNDING STUDIES, 87- *Concurrent Pos:* Prin investr, NSF,

Air Force Geophys Lab, 74; vis prof, Univ Calif, Los Angeles, 81; mem adv panel, Int Satellite Cloud Climat Prog, Climate Res Comt, Nat Acad Sci, 84-87; consult, Ames Res Ctr, NASA, 84; vis scholar, Harvard Univ, 85. *Honors & Awards:* David P Gardner Award, Univ Utah, 78. *Mem:* Fel Optical Soc Am; fel Am Meteorol Soc; Am Geophys Union; AAAS. *Res:* Investigation of cloud-radiation processes in weather and climate models for the improvement of prediction; exploration of remote sounding of atmospheric infrared cooling rates from satellites. *Mailing Add:* Dept Meteorol Univ Utah Salt Lake City UT 84112

LIOY, FRANCO, b Gorizia, Italy, May 24, 32; m 66; c 1. PHYSIOLOGY. *Educ:* Univ Rome, MD, 56; Univ Minn, Minneapolis, PhD(physiol), 67. *Prof Exp:* Instr med, Univ Rome, 56-61; instr physiol, Univ Minn, Minneapolis, 66-67; from asst prof to assoc prof, 67-75, PROF PHYSIOL, UNIV BC, 75- *Concurrent Pos:* Mem, Sci Subcomt, Can Heart Found. *Mem:* Can Physiol Soc. *Res:* Cardiovascular physiology; coronary circulation; effect of hyperthermia on circulatory system; sympathetic control of circulation. *Mailing Add:* Dept Physiol Univ BC 2075 Westbrook Mall Vancouver BC V6T 1W5 Can

LIOY, PAUL JAMES, b Passaic, NJ, May 27, 47; m 71; c 1. ENVIRONMENTAL HEALTH, HUMAN EXPOSURE. *Educ:* Montclair State Col, BA, 69; Auburn Univ, MS, 71; Rutgers Univ, MS, 73, PhD(environ sci), 75. *Prof Exp:* Sr air pollution engr, Interstate Sanit Comn, NY, 75-78; from asst prof to assoc prof environ med, Inst Environ Med, NY Univ Med Ctr, 78-85; assoc prof environ & community med, 85-88, DIR, EXPOSURE, MEASUREMENT & ASSESSMENT DIV, UNIV MED & DENT NJ, 85-, PROF ENVIRON & COMMUNITY MED, 88- *Concurrent Pos:* Lectr, Dept Civil & Environ Eng, Polytech Inst NY, 76-78; consult, State NJ, 78-, US Environ Protection Agency Off Res & Develop & Sci Adv Bd, 84-, indust orgn, 80-; mem, Am Conf Govt Indust Hygienists; mem, NJ Clean Air Coun, 81-, chmn, 83-85; dir human exposure, Health Sci Inst, EOHSI-UMDNJ-Rutgers Univ, 86-; mem, Comt on Air Pollution epidemiology, Nat Acad Sci, 84-85, chair, 87-, mem, Comn Air Pollution Exposure Assessment, 90; mem bd environ studies & toxicol, Nat Res Coun, 89-; exed ed, Atmos Environ, 89- *Mem:* Air Pollution Control Asn; fel NY Acad Sci; Int Soc Environ Epidemiol; Am Asn Aerosol Res; Air & Waste Mgt Asn; Int Soc Exposure Anal (treas). *Res:* Atmospheric transport of pollutants; chemical characteristics of inorganic trace elements and organic species; aerosol and gaseous monitoring equipment and techniques; industrial and occupational hygiene; environmental health; human exposure to toxic pollutants and epidemiology; hazardous wastes and multimedia pollution issues. *Mailing Add:* Dept Environ & Community Med Univ Med & Dent NJ 675 Hoes Lane Piscataway NJ 08854

LIPARI, NUNZIO OTTAVIO, b Ali' Terme, Italy, Jan 1, 45. SOLID STATE PHYSICS, MOLECULAR PHYSICS. *Educ:* Univ Messina, Laurea Physics, 67; Lehigh Univ, PhD(physics), 70. *Prof Exp:* Res asst solid state physics, Lehigh Univ, 67-70; res assoc, Univ Ill, 70-72; from asst scientist to scientist sr physics, Webster Res Lab, Xerox Corp, 75-77; MEM RES STAFF, IBM THOMAS J WATSON RES CTR, 77- *Mem:* Fel Am Phys Soc. *Res:* Optical properties of solids; electron-phonon interaction in molecular systems; excitation and impurity states in semiconductors. *Mailing Add:* PO Box 218 Yorktown Heights NY 10598

LIPE, JOHN ARTHUR, b Los Fresnos, Tex, Aug 28, 43; m 64; c 2. HORTICULTURE, PLANT PHYSIOLOGY. *Educ:* Tex A&M Univ, BS, 65, MS, 68, PhD(plant physiol), 71. *Prof Exp:* ASST PROF HORT, TEX A&M UNIV RES & EXTEN CTR, OVERTON, 71- *Honors & Awards:* Award, Am Soc Plant Physiol, 71. *Mem:* Am Soc Hort Sci. *Res:* Plant growth regulation, especially with fruits; role of ethylene in fruit dehiscence. *Mailing Add:* 1906 N Llano Suite 104 Fredricksburg TX 78624

LIPELES, MARTIN, b New York, NY, June 22, 38; m 68; c 2. DIGITAL TYPE, COMPUTER GRAPHICS. *Educ:* Columbia Univ, AB, 60, MA, 62, PhD(physics), 66. *Prof Exp:* Part-time res physicist, Radiation Lab, Columbia Univ, 62-66; mem tech staff, Sci Ctr, Rockwell Int, 66-76; pres, Med Microcomputers Inc, 77-78; dir res & develop, Autologic Inc, 78-83; mgr systs eng, Alpharel Inc, 83-88; COMPUTER CONSULT, 88- *Mem:* AAAS; Am Phys Soc; Am Chem Soc; Sigma Xi; Asn Comput Mach. *Res:* Design of large systems for engineering graphics; software for design of digital type and typesetter design; inelastic, ion-atom collisions at low energies; physics and chemistry of photochemical aerosol formation in the atmosphere. *Mailing Add:* 1476 Warwick Ave Thousand Oaks CA 91360

LIPETZ, LEO ELIJAH, b Lincoln, Nebr, Aug 10, 21; m 47; c 3. BIOPHYSICS, NEUROSCIENCES. *Educ:* Cornell Univ, BEE, 42; Univ Calif, PhD(biophys), 53. *Prof Exp:* Jr elec engr, Radar Lab, Signal Corps, 42-43; mem tech staff, Bell Tel Labs, 43-46; asst physics, Univ Southern Calif, 46-47; instr biophys, Exten Div, Univ Calif, 48; from instr to asst prof ophthal, Ohio State Univ, 54-60, asst prof physiol, 56-61, assoc prof biophys, 60-61, actg chmn div, 65-67, assoc prof physiol, 61-65, chmn div, 67-71, chmn dept, 71-76, prof biophys & res assoc, Inst Res Vision, 65-88, prof, dept zool, 81-88, EMER PROF, DEPT ZOOL, OHIO STATE UNIV, 89- *Concurrent Pos:* Nat Found fel, Johns Hopkins Univ, 53-54; sr fel, NIH, 62-63; mem, Biophys Sci Training Comt, Nat Inst Gen Med Sci, 68-70; res fel, Japan Soc Prom Sci, 81. *Mem:* AAAS; Biophys Soc; Asn Res Vision & Ophthal; Soc Neurosci. *Res:* Biophysics of the visual system; physical basis of behavior; local circuit neuronal networks; optics; information transfer; effects of high energy radiation. *Mailing Add:* Zool Dept 1314 Kinnear Rd Columbus OH 43212

LIPICKY, RAYMOND JOHN, b Cleveland, Ohio, May 3, 33; m 58; c 5. INTERNAL MEDICINE, PHARMACOLOGY. *Educ:* Ohio Univ, AB, 55; Univ Cincinnati, MD, 60. *Prof Exp:* From intern to resident med, Barnes Hosp, St Louis, Mo, 60-62; resident, Strong Mem Hosp, Rochester, NY, 64-65; from asst prof to assoc prof pharmacol, Col Med, Univ Cincinnati, 66-72, from asst prof to assoc prof med, 66-73, prof pharmacol, 72-79, prof med & dir, Div Clin Pharmacol, 73-79; med officer, Div Cardiorenal Drug Prods, 79-81, actg dir, 81-84, DIR, FOOD & DRUG ADMIN, 84- *Concurrent Pos:* Fel pharmacol, Univ Pa, 62-63 & Univ Cincinnati, 63-64; trainee cardiol, Strong Mem Hosp, Rochester, 65-66; mem corp, Marine Biol Lab, Woods Hole, Mass; guest worker, Lab Biophysics, NIH, 79-84. *Mem:* Soc Neurosci; Biophys Soc; Am Physiol Soc; Am Soc Pharmacol & Exp Therapeut; Am Soc Hypertension. *Res:* Ion transport; clinical pharmacology; membrane permeability; bioelectric potentials; hemodynamics. *Mailing Add:* 15201 Apricot Lane Gaithersburg MD 20878

LIPIN, BRUCE REED, b New York, NY, Nov 27, 47; m 71; c 2. PETROLOGY. *Educ:* City Col New York, BS, 70; Pa State Univ, PhD(mineral, petrol), 75. *Prof Exp:* Fel, Geophys Lab, Carnegie Inst Washington, 73-74; Nat Res Coun res assoc fel, 74-75, GEOLOGIST, US GEOL SURV, 75- *Mem:* Mineral Soc Am; Soc Econ Geol. Petrology of chromite deposits. *Res:* Economic geology of ultramafic rocks including chromite, platinum, asbestos and talc; petrology and origin of ultramafic rocks. *Mailing Add:* US Geol Surv MS 954 Nat Ctr Reston VA 22092

LIPINSKI, BOGUSLAW, b Sochaczew, Poland, July 21, 33; US citizen; m 57; c 1. BIOELECTRICITY. *Educ:* Inst Nuclear Res, Warsaw, PhD(biochem), 62; Univ Lodz, Poland, DSc, 71. *Prof Exp:* Vis prof, Vascular Lab, Lemnel Shattuck Hosp, Sch Med, Tufts Univ, 71-76; assoc dir, Vascular Lab, 76-81, DIR, BIOELEC LAB, ST ELIZABETH'S HOSP, TUFTS UNIV SCH MED, 81- *Concurrent Pos:* Asst ed, J Bioelec, 81- *Mem:* AAAS; Int Soc Thrombosis & Hemostosis; Fedn Am Scientists; Int Soc Bioelec (pres, 80-). *Res:* Mechanism of intravascular coagulation and fibrinolysis; effect of nutrition on thrombosis and atherosclerosis; effect of electricity on biological systems in relation to tissues regeneration and healing. *Mailing Add:* H S Res Lab 1101 Beacon St Boston MA 02146

LIPINSKI, CHRISTOPHER ANDREW, b Dundee, Scotland, Feb 1, 44; US citizen; m 69; c 2. ORGANIC CHEMISTRY. *Educ:* San Francisco State Col, BS, 65; Univ Calif, Berkeley, PhD(org chem), 68. *Prof Exp:* Nat Inst Gen Med Sci fel, Calif Inst Technol, 69-70; res sci, 70-74, sr res sci, 74-76, sr res investr, 76-81, prin res investr, 81-86, RES ADV, MED RES LABS, PFIZER CENT RES, 86- *Mem:* Am Chem Soc. *Res:* Medicinal chemistry of gastrointestinal and antidiabetic agents; histamine M2-receptor antagonists; aldose reductase inhibitors; bioisosteusm. *Mailing Add:* Pfizer Cent Res Groton CT 06340-5196

LIPINSKI, WALTER C(HARLES), b Chicago, Ill, Jan 5, 27; m 51; c 1. ELECTRICAL ENGINEERING, NUCLEAR ENGINEERING. *Educ:* Univ Ill, BSc, 50; Ill Inst Technol, MSc, 63, PhD(elec eng), 69. *Prof Exp:* From asst elec engr to assoc elec engr, 50-74, SR ELEC ENGR, ARGONNE NAT LAB, 74- *Concurrent Pos:* US deleg, Int Electrotech Comn, 64-; consult, Adv Comt Reactor Safeguards, US Nuclear Regulatory Comn, 64- *Mem:* Am Nuclear Soc; Inst Elec & Electronics Engrs; Sigma Xi. *Res:* Control engineering; instrumentation; nuclear reactor control and instrumentation; nuclear power plant development; nuclear reactor safety. *Mailing Add:* 9700 S Cass Ave Bldg 208 Argonne Nat Lab Argonne IL 60439

LIPINSKY, EDWARD SOLOMON, b Asheville, NC, Nov 15, 29; m 54; c 2. SYSTEMS ANALYSIS COMMERCIAL DEVELOPMENT. *Educ:* Mass Inst Technol, BS, 52; Harvard Univ, AM, 54. *Prof Exp:* Res assoc, Ohio State Univ, 57-59; prin res scientist, 59-71, assoc div chief bus & tech planning, 71-74, res leader, 75-79, sr res leader res mgt, 79-85, RES LEADER ORG & POLYMER CHEM, BATTELLE COLUMBUS MEM INST, 85- *Mem:* Am Chem Soc; fel AAAS; Sigma Xi; NY Acad Sci; Commercial Develop Asn; Tech Asn Pulp & Paper Indust. *Res:* Chemicals from biomass; renewable resource technology; diffusion-controlled reactions; computer-aided idea generation. *Mailing Add:* Battelle Mem Inst 505 King Ave Columbus OH 43201

LIPKA, BENJAMIN, b New York, NY, Feb 5, 29; m 58; c 3. ORGANIC CHEMISTRY. *Educ:* NY Univ, BA, 49, PhD(org chem), 58. *Prof Exp:* Chemist, Geigy Chem Corp, 59-60; sr chemist, Allied Chem Corp, 60-67; SR DEVELOP CHEMIST, UPJOHN CO, 67- *Mem:* Am Chem Soc. *Res:* Laboratory synthesis and chemical plant production of organic compounds. *Mailing Add:* Upjohn Co 410 Sackett Point Rd North Haven CT 06473

LIPKA, JAMES J, b Highland Park, Mich, Aug 1, 54; m 80. BIOINORGANIC CHEMISTRY. *Educ:* Univ Mich, BS, 76; Columbia Univ, MA, 77, PhD(chem), 82. *Prof Exp:* FEL CHEM, BROOKHAVEN NAT LAB, 81- *Mem:* Am Chem Soc; Sigma Xi; AAAS; NY Acad Sci. *Res:* Methods for labelling biological polymers and assemblies with heavy atoms for the purpose of increasing contrast in high-resolution electron microscopy. *Mailing Add:* 549 36th Ave San Francisco CA 94121

LIPKE, PETER NATHAN, b San Francisco, Calif, June 18, 50; m 71; c 5. CELL ADHESION, CELL SURFACE DEVELOPMENT. *Educ:* Univ Chicago, BS, 73; Univ Calif, Berkeley, PhD(biochem), 76. *Prof Exp:* Fel, Dept Zool, Univ Wis-Madison, 76-78; asst prof, 76-82, assoc prof, 82-89, PROF MOLECULAR CELL BIOL, DEPT BIOL SCI, HUNTER COL, 90- *Mem:* Am Soc Microbiol; AAAS; Am Soc Biochem & Molecular Biol. *Res:* Molecular basis for cell-cell adhesion in eukaryotes, using mating in Saccharomyces cerevisiae as a model; developmental changes in cell surface structure. *Mailing Add:* Dept Biol Hunter Col 695 Park Ave New York NY 10021

LIPKE, WILLIAM G, b Chesterton, Ind, Dec 19, 36; m 57; c 4. PLANT PHYSIOLOGY, PLANT BIOCHEMISTRY. *Educ:* Purdue Univ, BS, 59; Univ Nebr, MS, 62; Tex A&M Univ, PhD(plant physiol), 66. *Prof Exp:* From asst prof to assoc prof plant physiol, 65-74, ASSOC PROF BIOL, NORTHERN ARIZ UNIV, 74-, PLANT PHYSIOLOGIST, 65- *Mem:* AAAS; Am Soc Plant Physiol. *Res:* Plant physiology, especially mineral nutrition; weed science, especially plant enzymes. *Mailing Add:* Box 5640 Northern Ariz Univ Flagstaff AZ 86001

LIPKIN, DAVID, b Philadelphia, Pa, Jan 30, 13; m 42; c 2. NUCLEIC ACID CHEMISTRY, PHOSPHORUS CHEMISTRY. *Educ:* Univ Pa, BS, 34; Univ Calif, PhD(org chem), 39. *Prof Exp:* Petrol res chemist, Res & Develop Dept, Atlantic Ref Co, Pa, 34-36; res fel chem, Univ Calif, 39-42, chemist, 42-43; chemist & group leader, Manhattan Dist, Los Alamos, NMex, 43-46; from assoc prof to prof chem 46-69, chmn dept, 64-70, William Greenleaf Eliot prof chem, 69-81, WILLIAM GREENLEAF ELIOT EMER PROF CHEM, WASHINGTON UNIV, 81- *Concurrent Pos:* Guggenheim fel, 55; trustee, Argonne Univs Asn, 69-71; vis res scientist, John Innes Inst, Norwich, England, 60 & 71, spec consult, 78. *Honors & Awards:* St Louis Award, Am Chem Soc, 70. *Mem:* Am Chem Soc. *Res:* Free radicals; organic phosphorus compounds; nucleic acids; electrochemical synthesis. *Mailing Add:* Dept Chem Washington Univ St Louis MO 63130

LIPKIN, GEORGE, b New York, NY, Dec 31, 30; m 57; c 2. CELL BIOLOGY, DERMATOLOGY. *Educ:* Columbia Univ, AB, 52; State Univ NY Downstate Med Ctr, MD, 55. *Prof Exp:* From instr to assoc prof, 61-74, PROF DERMAT, MED SCH, NY UNIV, 74- *Concurrent Pos:* Prin investr, Nat Cancer Inst res grants, Dermat Found, Med Ctr, NY Univ, 61-85; vis scientist, Univ Zurich, 72-73; dir, Berger Found Cancer Res. *Mem:* Soc Invest Dermat; AAAS; Fed Am Sci; Am Acad Dermat; Harvey Soc; Union Concerned Scientists. *Res:* Biology of malignant melanoma; biologic transformation of malignant cells; endogenous inhibitors of growth. *Mailing Add:* Dept Dermat NY Univ Med Ctr 530 First Ave New York NY 10016

LIPKIN, HARRY JEANNOT, b New York, NY, June 16, 21; m 49; c 2. NUCLEAR PHYSICS, PARTICLE PHYSICS. *Educ:* Cornell Univ, BEE, 42; Princeton Univ, AM, 48, PhD(physics), 50. *Prof Exp:* Mem staff, Radiation Lab, Mass Inst Technol, 42-46; vis res fel reactor physics, AEC, France, 53-54; vis assoc prof physics, Univ Ill, 58-59; from assoc prof to prof, 59-71, HERBERT H LEHMAN CHAIR THEORET PHYSICS, WEIZMANN INST SCI, ISRAEL, 71- *Concurrent Pos:* Res physicist, Weizmann Inst Sci, 52-60, actg head dept physics, 60-61; consult, AEC, Israel, 55-58; vis lectr, Hebrew Univ, Israel, 56-58; vis prof, Univ Ill, 62-63 & Tel Aviv Univ, 65-66; vis prof, Princeton Univ, 67-68; vis scientist, Argonne Nat Lab & Nat Accelerator Lab, Ill, 71-72, 76-77 & 79-81. *Honors & Awards:* Rothschild Prize, Jerusalem, 80. *Mem:* Am Phys Soc; Ital Phys Soc; Phys Soc Israel; Europ Phys Soc; Israel Acad Sci & Humanities. *Res:* Elementary particle theory; theoretical and experimental nuclear structure; collective motion in many particle systems; Beta decay; Mossbauer effect; reactor and particle physics; theoretical physics. *Mailing Add:* Dept Physics Weizmann Inst Sci Rehovot Israel

LIPKIN, LEWIS EDWARD, b New York, NY, Nov 2, 25; m 52; c 2. NEUROPATHOLOGY, COMPUTER SCIENCES. *Educ:* NY Univ, BA, 44; Long Island Col Med, MD, 49; Am Bd Path, dipl & cert anat path & neuropath, 52. *Prof Exp:* From intern med to resident path, Mt Sinai Hosp NY, 49-53; asst prof path & neuropath, State Univ NY Downstate Med Ctr, 56-62; head neuropath, Path Sect, Perinatal Res Br, Nat Inst Neurol Dis & Stroke, 62-72; dir, Div Cancer Biol & Diagnosis, head, Image Processing Unit Off, 72-80, CHIEF IMAGE PROCESSING SECT, LAB MATH BIOL, NAT CANCER INST, NIH, 80- *Concurrent Pos:* Asst pathologist, Kings County Hosp, 56-62; USPHS sr res fel neuropath, Mt Sinai Hosp NY, 55-56; USPHS res grant, 59-62; consult, Nat Inst Neurol Dis & Stroke, 61-62. *Mem:* Asn Res Nerv & Ment Dis; Int Acad Path; Am Asn Neuropath; Asn Comput Mach. *Res:* Computer analysis of microscopic images, especially neuropathologic material; automation of analysis of two dimensional gel electrophoresis; analysis and synthesis of nucleic acid secondary structure. *Mailing Add:* Div Cancer Biol & Diag NIH Nat Cancer Inst Bethesda MD 20205

LIPKIN, MACK, internal medicine, psychiatry; deceased, see previous edition for last biography

LIPKIN, MARTIN, b New York, NY, Apr 30, 26; m 58; c 3. GASTROENTEROLOGY, ONCOLOGY. *Educ:* NY Univ, AB, 46, MD, 50. *Prof Exp:* Instr physiol, Sch Med, Univ Pa, 53-54; from instr to assoc prof, Cornell Univ, 58-78; assoc mem, 72-85, head, Lab Gastrointestinal Cancer Res, 72-90, MEM & ATTEND PHYSICIAN, MEMORIAL SLOAN-KETTERING CANCER CTR, 85-, HEAD, IRVING LAB GASTROINTESTINAL CANCER PREV, 90-; PROF MED, MED COL & GRAD SCH MED SCI, CORNELL UNIV, 78- *Concurrent Pos:* Fel physiol, Med Col, Cornell Univ, 52-53, fel med, 55-58; USPHS res fel, 55-56; NIH res career prog award, 61-71; res collabr, Brookhaven Nat Lab, 58-72; dir gastroenterol res unit, Cornell Med Div, Bellevue Hosp, 58-68; guest investr, Rockefeller Inst, 59-60; assoc attend physician, New York Hosp, 70- & Mem Hosp, 71-; assoc prof, Grad Sch Med Sci, Cornell Univ, 71-78; award lectr, Med Soc State NY, 71; secy, Sect Gastroenterol & Colon & Rectal Surg, 71, vchmn, 72, chmn, Sci Prog Comt, 73; hon pres, Int Acad Pathol Conf colorectal cancer, 81; ann hon lectr, Israel Med Asn & Gastroenterol Soc, 82; vis physician, Rockefeller Univ Hosp. *Mem:* Fel Am Col Physicians; Am Soc Clin Invest; Am Physiol Soc; Am Soc Exp Path; Am Asn Cancer Res. *Res:* Proliferation and differentiation of premalignant and malignant gastrointestinal cells in man. *Mailing Add:* Memorial Sloan Kettering Cancer Ctr 1275 York Ave New York NY 10021

LIPKOWITZ, KENNY BARRY, b Bronx, NY, Apr 1, 50; m 78; c 2. COMPUTATIONAL CHEMISTRY, PHYSICAL ORGANIC CHEMISTRY. *Educ:* State Univ NY, Geneseo, BS, 72; Mont State Univ, PhD(chem), 75. *Prof Exp:* Asst chem, Ohio State Univ, 76-77; from asst prof to assoc prof, 77-90, PROF CHEM, PURDUE UNIV, 90- *Concurrent Pos:* Vis prof, Princeton Univ, 81-82, Univ Calif, San Francisco, 89-90. *Mem:* Am Chem Soc. *Res:* Theoretical studies of organic molecules using quantum mechanical and molecular mechanics methods. *Mailing Add:* Dept Chem Purdue Univ 1125 E 38th St Indianapolis IN 46205

LIPMAN, DAVID J, BIOTECHNOLOGY. *Prof Exp:* DIR, NAT CTR BIOTECHNOL INFO, NAT LIBR MED, NIH, 88- *Mailing Add:* NIH Nat Libr Med Nat Ctr Biotechnol Info Bldg 38A Rm 8N803 8600 Rockville Pike Bethesda MD 20894

LIPMAN, JOSEPH, b Toronto, Ont, Can, June 15, 38; m 62; c 2. MATHEMATICS. *Educ:* Univ Toronto, BA, 60; Harvard Univ, MA, 61, PhD(math), 65. *Prof Exp:* Asst prof math, Queen's Univ, Ont, 65 & Purdue Univ, 66-67; vis asst prof, Columbia Univ, 67-68; from asst prof to assoc prof, 68-72, prof, 72-87, HEAD, MATH DEPT, PURDUE UNIV, WEST LAFAYETTE, 87- *Mem:* Am Math Soc; Can Math Cong. *Res:* Algebraic geometry. *Mailing Add:* Purdue Univ West Lafayette IN 47907

LIPMAN, MARC JOSEPH, b Chicago, Ill, Mar 19, 50; m 81; c 2. GRAPH THEORY. *Educ:* Lake Forest Col, BA, 71; Dartmouth Col, AM, 73, PhD(math), 76. *Prof Exp:* Lectr math, Dartmouth Col, 73; asst prof, 76-83, ASSOC PROF MATH, IND UNIV-PURDUE UNIV, FT WAYNE, 83- *Concurrent Pos:* Assoc, Nat Res Coun, Naval Res Lab, 80. *Mem:* Math Asn Am; Soc Indust & Appl Mathematicians; Asn for Comput Mach; Sigma Xi. *Res:* Graph theory; intelligent systems. *Mailing Add:* Off Naval Res Code 1111 800 N Quincy Arlington VA 22217-5000

LIPMAN, PETER WALDMAN, b New York, NY, Apr 21, 35; m 62. GEOLOGY. *Educ:* Yale Univ, BS, 58; Stanford Univ, MS, 59, PhD(geol), 62. *Prof Exp:* GEOLOGIST, US GEOL SURV, 62- *Concurrent Pos:* NSF fel, Geol Inst, Tokyo, 64-65. *Honors & Awards:* Burwell Award, Geol Soc Am, 83. *Mem:* Geol Soc Am; Mineral Soc Am; Am Geophys Union. *Res:* Petrology and structural geology; volcanology, especially geology of calderas and related ash flows. *Mailing Add:* US Geol Surv MS 910 345 Middlefield Rd Menlo Park CA 94025

LIPNER, HARRY JOEL, b New York, NY, Aug 26, 22; m 49; c 4. REPRODUCTIVE ENDOCRINOLOGY. *Educ:* Long Island Univ, BS, 42; Univ Chicago, MS, 47; Univ Iowa, PhD(physiol), 52. *Prof Exp:* Res assoc thyroid iodine trap, Univ Iowa, 52; Nat Cancer Inst res fel thyroid physiol, 52-54; instr clin path, Chicago Med Sch, 54-55; from asst prof to assoc prof, 55-65, PROF PHYSIOL, FLA STATE UNIV, 65- *Concurrent Pos:* NIH fel & vis prof dept anat, Harvard Med Sch, 69-70; Fulbright vis prof, Ctr Advan Biochem, Indian Inst Sci, Bangalore, India, 74-75; mem Regulatory Biol Rev Panel, NSF, 84-87. *Mem:* AAAS; Endocrine Soc; Sigma Xi; Soc Study Reproduction; Am Physiol Soc. *Res:* Mechanism of ovulation; nonsteroidal gonadal feedback control of Gonadotropin secretion. *Mailing Add:* 2459 Royal Oaks Dr Tallahassee FL 32308

LIPNER, STEVEN BARNETT, b Independence, Kans, Sept 30, 43; m 80. SOFTWARE SYSTEMS, COMPUTER SECURITY. *Educ:* Mass Inst Technol, SB, 65, SM, 66. *Prof Exp:* Assoc dept head, Intel & Info Syst, MITRE Corp, 77-80 & Command & Control Systs, 80-81; GROUP MGR, SECURE SYSTS, DIGITAL EQUIP CORP, 81- *Concurrent Pos:* Chmn comt security & privacy, Inst Elec & Electronics Engrs Comput Soc, 83-84, mem, Nat Res Coun Comt Comput Security, Dept Energy, 87-88, Nat Computer Systs Security & Privacy Adv Bd, 89- *Mem:* Asn Comput Mach; Sigma Xi; Inst Elec & Electronics Engrs Comput Soc. *Res:* Security controls for computer systems; theoretical and practical advances in network security and secure operating systems. *Mailing Add:* Six Midland Rd Wellesley MA 02181

LIPNICK, ROBERT LOUIS, b Baltimore, Md, Sept 9, 41; m 67; c 2. TOXICOLOGICAL MECHANISM, PREDICTIVE TOXICOLOGY. *Educ:* Univ Md, College Park, BS, 63; Brandeis Univ, PhD(org chem), 69. *Prof Exp:* Fel chem, Univ Minn, Minneapolis, 68-72; vis scientist, var African univs, 73-74; res assoc, Sloan-Kettering Inst, 74-79; chemist, 79-80, LEADER STRUCTURE ACTIV GROUP, US ENVIRON PROTECTION AGENCY, 80- *Concurrent Pos:* Vis scientist, Borstel Res Inst, Fed Repub Ger, 86 & Pharmacol Inst, Univ Lund, Sweden, 89; assoc ed, Soc Environ Toxicol, 89-; co-organizer workshop, Environ Protection Agency, 88; invited lectr, Comn Europ Communities, Ispra, Italy, 90; mem, Int Sci Comt, Fourth Int Workshop Quant Struct-Activ Relationship in Environ Toxicol & Chem, Veldhoven, Neth, 90. *Mem:* Am Chem Soc; Soc Environ Toxicol & Chem; Chemometrics Soc. *Res:* Development of quantitative structure-activity relationships for predictive toxicology; estimation of physicochemical, conformational and reactive properties of molecules from chemical structure; quantitative structure activity relationship. *Mailing Add:* 5308 Pender Ct Alexandria VA 22304

LIPO, THOMAS ANTHONY, b Milwaukee, Wis, Feb 1, 38; m 64; c 4. POWER ELECTRONICS, ELECTRICAL MACHINES. *Educ:* Marquette Univ, BEE, 62, MSEE, 64; Univ Wis-Madison, PhD, 68. *Prof Exp:* Grad trainee, Allis-Chalmers Mfg Co, Milwaukee, 62-64; eng analyst, 64; instr, Univ Wis-Milwaukee, 64-66; Nat Res Coun res fel, Univ Manchester Inst Sci & Technol, Eng, 68-69; elec engr, Gen Elec Co, Schenectady, 69-79; prof, Purdue Univ, 79-80; prof, 81-90, W W GRAINGER PROF POWER ELECTRONICS & ELECT MACH, UNIV WIS-MADISON, 90- *Concurrent Pos:* Vis assoc prof, Purdue Univ, 73-74; co-dir, Wis Elec Mach & Power Electronics Consortium, 81-; ed, Inst Elec & Electronic Engrs, Power Electronics Soc Trans, 83-90; dir, Wis Power Electronics Res Ctr, 87-; vis prof, Univ Sydney, 89; chmn, Indust Power Conversion Systs, Inst Elec & Electronic Engrs, Indust Applications Soc, 89- *Honors & Awards:* Outstanding Achievement Award, Inst Elec & Electronic Engrs, Indust Applications Soc, 86; William E Newell Award, Inst Elec & Electronic Engrs, Power Electronics Soc, 90. *Mem:* Fel Inst Elec & Electronic Engrs; Inst Elec & Electronic Engrs Power Eng Soc; Inst Elec & Electronic Engrs Indust Applications Soc; Inst Elec & Electronic Engrs Power Electronics Soc. *Res:* New power electronic circuits, controls and electrical machines for alternating current adjustable speed drives for industrial, commercial and utility applications. *Mailing Add:* Dept Elec & Computer Eng Univ Wis 1415 Johnson Dr Madison WI 53706

LIPOVSKI, GERALD JOHN (JACK), b Coleman, Alta, Jan 28, 44; m 68; c 3. COMPUTER ENGINEERING, ELECTRICAL ENGINEERING. *Educ:* Univ Notre Dame, AB & BSEE, 66; Univ Ill, Urbana, MS, 67, PhD(elec eng), 69. *Prof Exp:* Res fel automata theory, Coordinated Sci Lab, Univ Ill, Urbana, 66-68, asst electronics, 67-68, asst comput archit, 68-69; asst prof elec eng, Univ Fla, 69-76; assoc prof, 76-82, PROF ELEC ENG & COMPUT SCI, UNIV TEX, AUSTIN, 82- *Concurrent Pos:* Consult, Harris Semiconductor, 73-74 & Sycor, 76-77. *Mem:* Asn Comput Mach; Comput Soc of Inst Elec & Electronics Engrs. *Res:* Computer architecture; parallel and distributed computer architectures; data base processor architectures; microcomputer architectures and applications; hardware design languages. *Mailing Add:* Dept of Elec Eng Univ of Tex Austin TX 78712

LIPOWITZ, JONATHAN, b Paterson, NJ, Apr 25, 37; m 60; c 2. ORGANOMETALLIC CHEMISTRY, MATERIALS SCIENCE & CERAMICS ENGINEERING. *Educ:* Rutgers Univ, Newark, BS, 58; Univ Pittsburgh, PhD(chem), 64. *Prof Exp:* Fel, Pa State Univ, University Park, 64-65; res chemist, 65-74, res specialist, 74-75, sr res specialist, 75-79, assoc scientist, 79-87, SCIENTIST, RES DEPT, DOW CORNING CORP, 87- *Mem:* Am Chem Soc; Sigma Xi; Am Ceramic Soc; Mat Res Soc; AAAS; NY Acad Sci. *Res:* Silicone flammability, mechanisms of; effects of silicones on flammability of organic polymers; silicone chemistry and physical properties; characterization of advanced ceramics; preparation of ceramic fibers from polymers. *Mailing Add:* Res Dept Dow Corning Corp Midland MI 48640

LIPOWSKI, STANLEY ARTHUR, organic polymer chemistry; deceased, see previous edition for last biography

LIPOWSKI, ZBIGNIEW J, b Warsaw, Poland, Oct 26, 24; Can citizen; m 46; c 2. PSYCHIATRY. *Educ:* Nat Univ Ireland, MB, BCh & BAO, 53; McGill Univ, dipl, 59. *Hon Degrees:* Dr Med, Helsinki; MA, Dartmouth Col. *Prof Exp:* Demonstr psychiat, McGill Univ, 59-62, lectr, 62-65, from asst prof to assoc prof, 65-71; prof psychiat, Dartmouth Med Sch, 71-83; prof, 83-90, EMER PROF PSYCHIAT, UNIV TORONTO, 90- *Concurrent Pos:* Res fel psychophysiol, Allan Mem Inst Psychiat, 57-58; teaching fel psychiat, Harvard Univ & Mass Gen Hosp, 58-59; clin asst, Allan Mem Inst Psychiat, 59-62, from asst psychiatrist to psychiatrist, 62-71; consult psychiat, Montreal Neurol Inst, 68-71; psychiatrist, Mary Hitchcock Mem Hosp, Hanover, NH, 71-83. *Honors & Awards:* Spec Presidential Commendation, Am Psychiat Asn. *Mem:* Fel Am Psychiat Asn; Am Psychosom Soc; Can Med Asn. *Res:* Psychosomatic medicine and psychopathology related to physical illness. *Mailing Add:* 250 College St Toronto ON M5T 1R8 Can

LIPP, STEVEN ALAN, b Brooklyn, NY, Jan 25, 44; c 2. INORGANIC CHEMISTRY. *Educ:* Brooklyn Col, BS, 65; Univ Calif, Berkeley, PhD(inorg chem), 70. *Prof Exp:* MEM TECH STAFF, DAVID SARNOFF RES CTR, 70- *Honors & Awards:* Achievement Award, RCA Labs, 74, David Sarnoff Award, RCA Corp, 75. *Mem:* Electrochem Soc; Soc Info Display. *Res:* Preparation and evaluation of new cathodoluminescent materials, as well as the design and testing of enhancements for chemical milling; fabrication of active liquid crystal display substrate involving all photolithography and processing. *Mailing Add:* David Sarnoff Res Ctr Princeton NJ 08532

LIPPA, ERIK ALEXANDER, b Minneapolis, Minn, Nov 7, 45; m 80; c 2. OPHTHALMOLOGY, CLINICAL RESEARCH. *Educ:* Calif Inst Technol, BS, 67; Univ Mich, MS, 68, PhD(math), 71; Albert Einstein Col Med, MD, 80. *Prof Exp:* NATO fel math, Oxford Univ, 71-72; asst prof math, Purdue Univ, West Lafayette, 72-78; med intern, NY Univ & Manhattan Vet Admin Hosp, 80-81; ophthal resident, Ill Eye & Ear Infirmary, 81-84; assoc dir clin res, 85-88, DIR CLIN RES, MERCK SHARP & DOHME RES LAB, BLUE BELL, PA, 89- *Concurrent Pos:* Adj clin asst prof, Jefferson Med Col, Philadelphia, Pa, 86-; clin assoc, dept ophthal, Univ Pa, Philadelphia, 87- *Mem:* Am Acad Ophthal; Asn Res Vision & Ophthal; Int Soc Eye Res; Sigma Xi; Europ Glaucoma Soc. *Res:* Ophthalmology; clinical pharmacology and clinical research; analytic number theory, specifically Siegel modular forms of several complex variables and their associated Dirichlet series; applications of mathematics to medicine and pharmaceutical sciences. *Mailing Add:* 1045 Stevens Dr Ft Washington PA 19034-1633

LIPPA, LINDA SUSAN MOTTOW, b Boston, Mass, Apr 9, 51; m 80; c 2. OCULAR PATHOLOGY, OCULAR PHARMACOLOGY. *Educ:* Harvard Univ, AB, 73; Columbia Univ, MD, 77. *Prof Exp:* Intern internal med, St Luke's Hosp & Med Ctr, New York, NY, 77-78; resident ophthal, Montefiore Hosp & Med Ctr & Albert Einstein Col Med, Bronx, NY, 78-80, chief resident, 80-81; fel ophthalmic path, Eye & Ear Infirmary, Univ Ill, Chicago, 81-82; clin instr ophthal, Loyola Univ Med Ctr, Chicago, Ill & Hines Vet Admin Hosp, 82-83, clin asst prof, 83-84; clin asst prof ophthal, Univ Minn & attend opthalmologist & ocular pathologist, St Paul Ramsey Med Ctr, 84-85; ASST SURGEON OPHTHAL, WILLS EYE HOSP, PHILADELPHIA, PA, 85-; CLIN ASST PROF OPHTHAL, JEFFERSON MED COL, PHILADELPHIA, PA, 86- *Concurrent Pos:* Josephine Murray travelling fel, Radcliffe Col Res London, 72; Harvard Bur Study Coun org chem, 73; lab teaching asst, Harkness Eye Inst, 74 & Albert Einstein Col Med, 78; mem, Path Curric Comt, Dept Ophthal, Montefiore Hosp & Med Ctr, Albert Einstein Col Med; rep, Ophthalmologists in Training Comt, Am Acad Ophthal, 78- 81, adv fac, Continuing Educ Comt, 78-81, chmn, Ophthalmologists in Training Comt & Continuing Educ Comt, 80-81; clin attend & ocular pathologist, Hines VA, 81-84; attend ophthalmologist & ocular pathologist, Cook County Hosp, Chicago, Ill, 82-84; vis lectr, Dept Ophthal, Univ Md, 85; attend ophthalmologist, Wills Eye Hosp, 85 & ocular path, Wills Eye Hosp, 85-86. *Mem:* Am Asn Ophthalmic Pathologists; Am Med Asn. *Res:* Retinal embryonal development; inflammatory and neoplastic ocular and adnexal disease; glaucoma; development of pharmacologic models for new compound and vehicle assessment. *Mailing Add:* 1045 Stevens Dr Ft Washington PA 19034-1633

LIPPARD, STEPHEN J, b Pittsburgh, Pa, Oct 12, 40; m 64; c 2. INORGANIC CHEMISTRY, BIOCHEMISTRY. *Educ:* Haverford Col, BA, 62; Mass Inst Technol, PhD(chem), 65. *Prof Exp:* NSF fel, Mass Inst Technol, 65-66; from asst prof to prof chem, Columbia Univ, 66-82; prof 82-88, ARTHUR AMOS NOYES PROF CHEM, MASS INST TECHNOL, 88- *Concurrent Pos:* Consult, Esso Res & Eng Co, 67-73 & John Wiley & Sons, Inc, 81-; Alfred P Sloan Found fel, 68-70; John Simon Guggenheim Mem fel, Sweden, 72; ed, Progress in Inorganic Chem, John E Fogarty Sr Int fel, 78-79; consult, Engelhard Indust, 82, Sun Oil Co, 82, NAXCOR Sci Adv Bd, 87-, Johnson Matthey, 89-; assoc ed, J Am Chem Soc. *Honors & Awards:* Camille & Henry Dreyfus Teacher-Scholar Award, 72; Henry J Albert Award, Int Precious Metals Inst, Eng, 85; Alexander von Humboldt Award, 88; Inorg Chem Award, Am Chem Soc, 87, Remson Award, 88. *Mem:* Nat Acad Sci; Am Chem Soc; Am Crystallog Asn; Biophys Soc; Royal Soc Chem; Am Acad Arts & Sci; fel AAAS. *Res:* Inorganic and organometallic coordination chemistry, especially preparation, structural properties and reactions of transition metal complexes; ligand bridged bimetallic complexes and proteins; metal binding to nucleic acids; platinum antitumor drugs; high coordinate organo-metallic chemistry; bioinorganic chemistry. *Mailing Add:* Rm 18-290 Mass Inst Technol Cambridge MA 02139

LIPPE, ROBERT LLOYD, b New York, NY, May 8, 23; c 2. CHEMISTRY. *Educ:* Yale Univ, BE, 42; Princeton Univ, MS, 47. *Prof Exp:* Eng trainee, Joseph E Seagram & Sons, Inc, Md, 43; jr scientist in charge radiographic res, Manhattan Dist, 45-46; chemist, Standard Varnish Works, 47-51; asst to vpres, Standard Tech Chems, Inc, 51-54; PRES, NOD HILL CHEM CORP, 53-; PRES, ARGUS COATINGS CO, 54- *Mem:* Am Chem Soc; Am Inst Chem Eng; Fedn Am Sci. *Res:* Radiography of explosives; organotitanium compounds; synthetic and natural resins for coatings; industrial organic coatings. *Mailing Add:* 1155 Park Ave New York NY 10028

LIPPEL, KENNETH, b New York, NY, Feb 21, 29; m 61; c 1. ATHEROSCLEROSIS LIPID METABOLISM. *Educ:* City Col New York, BS, 49, MBA, 60; Univ Fla, PhD(biochem), 66. *Prof Exp:* NIH fel biochem, Univ Calif, Los Angeles, 66-68; asst prof dermat & biochem, Sch Med, Univ Miami, 68-70; res biochemist, Lipids Br, Human Nutrit Res Div, Agr Res Serv, USDA, 70-72; HEALTH SCI ADMINR, LIPID METAB BR, DIV HEART & VASCULAR DIS, NAT HEART, LUNG & BLOOD INST, NIH, 72- *Concurrent Pos:* Fel, Coun Arteriosclerosis, Am Heart Asn, 75- *Mem:* AAAS; Am Soc Biol Chem; Am Heart Asn; Am Chem Soc; Fedn Am Socs Exp Biol; Sigma Xi; Am Asn Clin Chem. *Res:* Regulation of fatty acid and lipid metabolism; relationship of lipoprotein metabolism to atherosclerosis; vitamin A metabolism. *Mailing Add:* 7550 Wisconsin Ave Rm 4-A10 Bethesda MD 20892

LIPPERT, BRUCE J, b New York, NY. ENZYME CHEMICALS. *Prof Exp:* SR RES BIOCHEMIST, MARION DOW RES INST, 81- *Mailing Add:* Dept Enzyme Chem Marion Dow Res Inst 2110 E Galbraith Rd Cincinnati OH 45215

LIPPERT, BYRON E, b Los Angeles, Calif, July 1, 29; m 52; c 3. PHYCOLOGY. *Educ:* Univ Ore, BS, 54, MS, 57; Ind Univ, PhD(bot), 66. *Prof Exp:* Instr biol, Eastern Ore Col, 56-59; asst prof, 60-69, ASSOC PROF BIOL, PORTLAND STATE UNIV, 69- *Mem:* Phycol Soc Am; Brit Phycol Soc; Int Phycol Soc; Bot Soc Am; Sigma Xi. *Res:* Morphology, life cycles and sexual reproduction in desmids. *Mailing Add:* Dept of Biol Portland State Univ Portland OR 97207

LIPPERT, LAVERNE FRANCIS, b Deerlodge, Mont, Sept 21, 28; m 50; c 5. PLANT PATHOLOGY. *Educ:* State Col Wash, BS, 50; Univ Calif, Davis, PhD(plant path), 59. *Prof Exp:* Asst plant path, Univ Calif, Davis, 55-58; from asst olericulturist to assoc olericulturist, Univ Calif, Riverside, 58-72, prof veg crops & olericulturist, 72-, vhcmn dept plant sci, 75-; RETIRED. *Mem:* Am Genetic Soc; Am Hort Soc. *Res:* Vegetable crops breeding, especially peppers and melons. *Mailing Add:* 2415 Creekside Lane Anacortes WA 98221

LIPPES, JACK, b Buffalo, NY, Feb 19, 24; m 47; c 3. REPRODUCTION, OBSTETRICS & GYNECOLOGY. *Educ:* Univ Buffalo, MD, 47. *Prof Exp:* Clin instr, 52-60, clin asst, 60-66, assoc prof, 66-75, PROF OBSTET & GYNEC, SCH MED, STATE UNIV NY, BUFFALO & ATTEND ASST, CHILDREN'S HOSP, BUFFALO, 68- *Concurrent Pos:* Consult, World Neighbors Found, Okla, 66-78, Pop Coun, Rockefeller Univ, 59-75, Ortho Pharmaceut Corp, 66-78, Syntex Res, 78-79, WHO Comt Studying Human Reprod, 74-78 & Sterling-Winthrop Pharmaceut Corp grant, 77-78; investr, Upjohn Pharmaceut Co, 75-78 & Prog Appl Res Fertil Regulation, 75-78; chmn, Dept Obstet & Gynec, Deaconess Hosp, Buffalo, 75-81; vis prof obstet & gynec, Charing Cross Hosp, Med Sch, London, 81. *Mem:* Asn Planned Parenthood Physicians; Planned Parenthood Fedn; Am Fertil Soc; Am Col Obstet & Gynec. *Res:* Human oviductal fluid; contraception, especially intrauterine contraception; inventor and researcher of intrauterine contraceptive device known as the Loop; immunology of the genital tract. *Mailing Add:* State Univ NY Health Sci Ctr 3435 Main St Buffalo NY 14214

LIPPINCOTT, BARBARA BARNES, b Raleigh, Ill, Oct 27, 34; m 56; c 3. MICROBIOLOGY, PLANT PHYSIOLOGY. *Educ:* Wash Univ, St Louis, AB, 55, MA, 57, PhD(zool & molecular biol), 59. *Prof Exp:* Jane Coffin Childs Mem Fund Med Res fel physiol genetics, Lab Physiol Genetics, Nat Ctr Sci Res, France, 59-60; res assoc biol sci, 60-80, SR RES ASSOC BIOCHEM, MOLECULAR BIOL & CELL BIOL, NORTHWESTERN UNIV, 80- *Concurrent Pos:* Vis scholar, Univ Calif, Berkeley, 70-71; lectr, Northwestern Univ, 72-73 & 81-; vis scientist, Inst Bot, Univ Heidelberg, 74. *Mem:* Am Soc Microbiol; Sigma Xi. *Res:* Electron spin resonance in biological systems; crown-gall tumor formation; control mechanisms in replication, growth and development. *Mailing Add:* Dept Biochem Molecular & Cell Biol Northwestern Univ Evanston IL 60208

LIPPINCOTT, EZRA PARVIN, b Philadelphia, Pa, Sept 7, 39; m 63; c 3. NUCLEAR PHYSICS, NUCLEAR ENGINEERING. *Educ:* Mass Inst Technol, BS, 61, PhD(nuclear physics), 66. *Prof Exp:* Sr scientist, Battelle-Northwest Labs, 66-72; SR SCIENTIST, NUCLEAR ENG, WESTINGHOUSE ELEC CORP, 72- *Mem:* Am Nuclear Soc; Am Phys Soc; Am Soc Testing & Mat. *Res:* Experimental and theoretical reactor physics; passive and active neutron and gamma-ray dosimetry; data analysis; methods development; nuclear cross section measurement and data file evaluation; standards preparation. *Mailing Add:* Westinghouse Elec Corp PO Box 355 Pittsburgh PA 15230-0355

LIPPINCOTT, JAMES ANDREW, b Cumberland Co, Ill, Sept 13, 30; m 56; c 3. PLANT PHYSIOLOGY. *Educ:* Earlham Col, AB, 54; Wash Univ, AM, 56, PhD, 58. *Prof Exp:* Res assoc plant physiol & lectr bot, Wash Univ, 58-59; Jane Coffin Childs Mem Fund Med Res fel, Lab Phytotron, Nat Ctr Sci Res, France, 59-60; from asst prof to prof biol sci, 60-81, assoc dean biol sci, 80-83, PROF BIOCHEM, MOLECULAR & CELL BIOL, NORTHWESTERN UNIV, 81- *Concurrent Pos:* Vis assoc prof, Univ Calif, Berkeley, 70-71; vis prof, Univ Heidelberg, 74. *Honors & Awards:* Centennial lectr, Mich State Univ Agr Exp Sta, 75; Tanner-Shaughnessy Merit Award, Ill Soc for Microbiol, 81. *Mem:* AAAS; Am Soc Plant Physiol; Bot Soc Am; Am Soc Microbiol; Am Phytopath Soc; Sigma Xi; Am Soc Biol Chemists. *Res:* Crown-gall tumor formation; control mechanisms in replication, growth and development; tumor induction in plants by Agrobacterium tumefaciens. *Mailing Add:* Dept Biochem Molecular & Cell Biol Northwestern Univ Evanston IL 60208

LIPPINCOTT, SARAH LEE, b Philadelphia, Pa, Oct 26, 20; wid. ASTROMETRY. *Educ:* Univ Pa, BA, 42; Swarthmore Col, MA, 50. *Hon Degrees:* DSc, Villanova Univ, 73. *Prof Exp:* Res asst astron, 42-51, res assoc, 52-72, lectr, 61-76, dir, 72-81, prof, 77-81, EMER PROF & EMER DIR, SPROUL OBSERV, SWARTHMORE COL, 81- *Concurrent Pos:* Fulbright fel, France, 53-54; Mem Fr solar eclipse exped to Oland, Sweden, 54; partic vis prof prog, Am Astron Soc, 61-; vpres comn 26, Int Astron Union, 70-73, pres, 73-76; nat lectr Sigma Xi, 72. *Mem:* Am Astron Soc; Int Astron Union; Sigma Xi. *Res:* Parallaxes of nearby stars; double stars; search for planetary companions to nearby stars; stellar masses; chromosphere studies; spicules. *Mailing Add:* 507 Cedar Lane Swarthmore PA 19081

LIPPINCOTT-SCHWARTZ, JENNIFER, b Kans, Oct 19, 52; m 75; c 2. MOLECULAR MEMBRANE BIOLOGY, INTRACELLULAR MEMBRANE TRAFFICKING. *Educ:* Swarthmore Col, BA, 74; Stanford Univ, MS, 79; Johns Hopkins Univ, PhD(biol), 86. *Prof Exp:* Fel, Pharmacol Res Assoc, Nat Inst Gen Med Sci, 86-88; fel, Nat Res Serv Award Prog, 88-90; SR STAFF SCIENTIST, NAT INST CHILD HEALTH & HUMAN DEVELOP, NIH, 90- *Mem:* Am Soc Cell Biologists; AAAS. *Res:* Intracellular membrane transport pathways and the molecular basis for intracellular membrane sorting and the biogenesis of organelles. *Mailing Add:* Cell Biol & Metab Br Nat Inst Child Health & Human Develop NIH 9000 Rockville Pike Bethesda MD 20892

LIPPITT, LOUIS, b New York, NY, Mar 19, 24; m 48; c 4. GEOPHYSICS. *Educ:* City Col New York, BS, 47; Columbia Univ, MA, 53, PhD(geol), 59. *Prof Exp:* Physicist, Columbia Univ, 47-50 & NY Univ, 51-53; geologist, Standard Oil Co, Calif, 54-58; staff engr, Lockheed Missles & Space Co, 58-87; RETIRED. *Concurrent Pos:* Instr, Allan Hancock Col, 69- & Chapman Col, 85- *Mem:* Geol Soc Am; Am Geophys Union. *Res:* Satellite systems; geophysical exploration. *Mailing Add:* Lockheed Missiles & Space Co 696 Raymond Avenue Santa Maria CA 93455

LIPPKE, HAGEN, b Yorktown, Tex, Nov 4, 36; m 58; c 3. ANIMAL NUTRITION. *Educ:* Tex A&M Univ, BS, 59, MS, 61; Iowa State Univ, PhD(animal nutrit), 66. *Prof Exp:* From asst prof to assoc prof ruminant nutrit, 66-74, ASSOC PROF ANIMAL SCI, TEX A&M UNIV, 74- *Mem:* Am Dairy Sci Asn; Am Soc Animal Sci; AAAS. *Res:* Ruminant nutrition; forage utilization by cattle; forage characteristics influencing intake and digestibility. *Mailing Add:* Res & Exten Ctr Tex A&M Univ 1619 Garner Field Rd Uvalde TX 78801

LIPPMAN, ALFRED, JR, b New Orleans, La, Mar 13, 08; m 34; c 3. CHEMICAL ENGINEERING. *Educ:* Tulane Univ, BE, 29. *Prof Exp:* Res chemist, Bay Chem Co, Inc, 29-31; sales rep, Chem Rubber Co, Ohio, 31-34; asst chief assayer, US Mint, La, 32-33; chief chemist, Bay Chem Div, Morton Salt Co, 34-37, from asst supt to mgr, 37-51; gen mgr, Commonwealth Eng Co, Ohio, 51-52; plant mgr, Godchaux Sugars, Inc, 53-56; staff asst to gen mgr, Alumina Div, Reynolds Metals Co, Ark, 56-58; gen dir, Alumina Res Div, 58-71; sr vpres & dir res & develop, Toth Aluminum Corp, New Orleans, 71-80; GEN PARTNER, LM-LIMITED PARTNERSHIP, NEW ORLEANS, 86- *Mem:* Fel Am Ceramic Soc; Am Chem Soc; Am Inst Chem Engrs; Am Inst Mining, Metall & Petrol Engrs; Nat Soc Prof Engrs. *Res:* New uses for hydrogen chloride and common salt; utilization of natural gas and agricultural residues as chemical raw material; manufacturing sugar from cane; refining raw sugar; manufacture of heavy and organic chemicals; products and processes for manufacture and use of aluminas and alumina hydrates. *Mailing Add:* 4613 Purdue Dr Metairie LA 70003

LIPPMAN, GARY EDWIN, b Little Rock, Ark; c 2. MATHEMATICAL ANALYSIS, COMPUTER SCIENCE. *Educ:* San Jose State Col, BA, 63; Univ Calif, Riverside, MA, 65, PhD(math), 70; Univ San Francisco, JD, 78. *Prof Exp:* Asst prof math, Kenyon Col, 70-71; PROF MATH, CALIF STATE UNIV, HAYWARD, 71- *Concurrent Pos:* Vis asst prof math, Univ Tenn, vis assoc prof comput sci & statist, Univ RI, 78-79; vis prof elec eng & computer sci, Univ Calif, Berkeley, 83-88. *Mem:* Am Math Soc; Math Asn Am; Comput Law Soc. *Res:* Fourier analysis; patent law. *Mailing Add:* Dept Math & Comput Sci Calif State Univ Hayward CA 94542-3092

LIPPMAN, MARC ESTES, b Brooklyn, NY, Jan 15, 45. BREAST CANCER. *Educ:* Cornell Univ, BA, 64; Yale Med Sch, MD, 68. *Prof Exp:* Intern, Osler Med Serv, Johns Hopkins Hosp, Baltimore, Md, 68-69, asst resident, 69-70; clin assoc, Leukemia Serv, Nat Cancer Inst, NIH, 70-71, clin assoc, Lab Biochem, 71-73, sr investr, Med Br, 74-88, head, Med Breast Cancer Sect, 76-88; CLIN PROF MED & PHARMACOL, UNIFORMED SERV, UNIV HEALTH SCI, 78-; DIR, VINCENT T LOMBARDI CANCER CTR, WASHINGTON, DC & PROF MED & PHARMACOL, GEORGETOWN UNIV SCH MED, 88- *Concurrent Pos:* Fel endocrinol, Yale Med Sch, New Haven, Conn, 73-74; mem, Merit Rev Bd Oncol, Vet Admin Med Res Serv, 77-81, Endocrine Treatment Comt, Nat Surg Adjuvant Breast Proj, 77-86 & pub affairs comt, Endocrine Soc, 80-81; consult, Dept Pharmacol, George Wash Sch Med, 78-89; co-chmn, Gordon Res Conf on Hormone Action, 84, chmn, 85; treas, Int Cong Hormones & Cancer, 84-; mem, med adv bd, Nat Alliance Breast Cancer Orgn, 86-; mem prog comt, Am Asn Cancer Res, 86, Am Soc Clin Oncol, 87-89 & Am Soc Clin Invest, 88; mem, Stage III Monitoring Comt, Nat Surg Adjuvant Proj Breast & Bowel Cancers, 87-89; chmn, local organizing comt, Am Soc Clin Oncol, 89-90; bd trustees, Am Cancer Soc, Dist of Columbia, 89-92; sci adv bd, Coord Coun Cancer Res, 89-; hon dir, Y-ME, Nat Orgn Breast Cancer Info & Support, 90-; Woodward vis prof, mem Sloan-Kettering, 90. *Honors & Awards:* Mallinckrodt Award, Clin Radioassay Soc, 78; Sidney Sachs Mem Lectr, Case Western Reserve, 85; D R Edwards Lect & Medal, Tenovus Inst, Wales, 85; Gosse Lectr, Dalhousie Univ, Halifax, NS, 87; Transatlantic Medal & Lect, Brit Endocrine Socs, 89; Tiffany Award of Distinction, Komen Found, 89; Barofsky Lectr, Howard Univ, 90; Rose Kushner Mem Lectr, Long Beach Mem Med Ctr, 90; Henrietta Banting Mem Lectr, Long Beach Mem Med Ctr, 90; Edward B Astwood Lect Award, Endocrine Soc, 91; Constance Wood Mem Lectr, Hammersmith Hosp, Eng, 91. *Mem:* Asn Am Physicians; Am Soc Clin Invest; Am Soc Biol Chemists; fel Am Col Physicians; Am Fedn Clin Res; Endocrine Soc; Am Soc Cell Biol; Am Asn Cancer Res; Am Soc Clin Oncol; Metastasis Res Soc. *Res:* Growth regulation of cancer; breast cancer; cancer endocrinology; growth factor receptors. *Mailing Add:* Vincent T Lombardi Cancer Research Ctr Georgetown Univ 3800 Reservoir Rd NW Washington DC 20007

LIPPMANN, DAVID ZANGWILL, b Houston, Tex, July 6, 25; m 69. PHYSICAL CHEMISTRY. *Educ:* Univ Tex, BSc, 47, MA, 49; Univ Calif, Berkeley, PhD(phys chem), 53. *Prof Exp:* Chemist, Reaction Motors, Inc, 54-57, Fulton-Irgon Div, Lithium Corp Am, 57-61 & Proteus, Inc, 61-63; asst prof, 63-69, ASSOC PROF CHEM, SOUTHWEST TEX STATE UNIV, 69- *Mem:* Am Chem Soc. *Res:* Theoretical physical chemistry, especially thermodynamics and statistical mechanics; rocketry and ballistics; properties of gems. *Mailing Add:* Dept Chem Southwest Tex State Univ San Marcos TX 78666

LIPPMANN, HEINZ ISRAEL, b Breslau, Ger, May 21, 08; nat US; m 36; c 3. MEDICINE. *Educ:* Univ Freiburg, BA, 26; Univ Berlin, MD, 31; Univ Genoa, MD, 33. *Prof Exp:* From asst prof to prof, 55-76, EMER PROF REHAB MED, ALBERT EINSTEIN COL MED, 76- *Concurrent Pos:* Assoc attend physician, Montefiore Hosp, Bronx, 44-86; lectr, Columbia Univ, 46-62; chief phys med, Workman's Circle Home for Aged, 51-87; chief peripheral vascular clin, Sydney Hillman Health Ctr, 51-67; chief peripheral vascular clin & vis physician, Bronx Munic Hosp Ctr, 56-; chief attend physician, Prosthetic & Brace Clin, 57-67, chief peripheral vascular clin, 57-, dir amputee ctr, 61-76; consult peripheral vascular dis, Englewood Hosp, NJ & Vet Admin Hosp, East Orange, NJ, 67-; dir rehab med, Jewish Hosp & Rehab Ctr, Jersey City, 74-81; attend physician & chief, Rehab Med Dept, Barnert Mem Hosp, Paterson, NJ, 75-85; consult rehab med, NJ & Vet Admin Hosp, East Orange & St Joseph's Hosp, Patterson NJ, 75- *Honors & Awards:* NAm Roentgen Soc Award, 58; Gold Medal Sci Exhibit, Am Cong Rehab Med, 59; Distinguished Clinician's Award, Am Acad Phys Med & Rehab, 86. *Mem:* Am Heart Asn; fel Am Col Physicians; Am Cong Rehab Med; NY Acad Sci; Am Acad Phys Med & Rehab. *Res:* Vascular physiology; peripheral vascular diseases; prosthetics; rheumatology; geriatrics; musicians disabilities. *Mailing Add:* Dept of Rehab Med Albert Einstein Col of Med New York NY 10461

LIPPMANN, IRWIN, b New York, NY, Dec 30, 30; m 56; c 3. PHARMACEUTICAL CHEMISTRY. *Educ:* Rutgers Univ, BS, 52, MS, 56; Univ Mich, PhD(pharmaceut chem), 60. *Prof Exp:* Instr anal chem, Rutgers Univ, 55-56; Am Found Pharmaceut Educ fel, 56-59; sr res pharmacist, Squibb Inst Med Res, 59-62; asst prof pharm, Med Col Va, 64-65; res assoc, 62-74, assoc dir, 74-79, DIR BIOPHARMACEUT RES, A H ROBINS CO, INC, 79- *Mem:* Am Pharmaceut Asn; Am Chem Soc; Am Asn Pharmaceut Scientists. *Res:* Biopharmaceutics-pharmacokinetics; in vitro correlations; sustained-action drugs; physical pharmacy. *Mailing Add:* PO Box 530 1000 S Grand St Hammonton NJ 02037

LIPPMANN, MARCELO JULIO, b Buenos Aires, Arg, May 27, 39; m 65; c 1. GEOTHERMICS, GROUNDWATER HYDROLOGY. *Educ:* Univ Buenos Aires, MS, 66; Univ Calif, Berkeley, MS, 69, PhD(eng sci), 74. *Prof Exp:* Asst geologist, Arg Geol Serv, 63-66, sedimentologist, 66-67; asst res eng, Dept Civil Eng, Univ Calif, Berkeley, 74-76; STAFF SCIENTIST, EARTH SCI DIV, LAWRENCE BERKELEY LAB, 76-, LEADER GEOTHERMAL GROUP, 83- *Concurrent Pos:* Consult hydrogeologist, Hidrosud SA, Buenos Aires, 67; US tech coordr, Dept Energy, Comn Federal de Electricidad, 77-; Jane Lewis fel, Univ Calif, Berkeley, 69-71. *Mem:* Am Geophys Union; Soc Petrol Engrs. *Res:* Geothermal and groundwater resources; physics and numerical modeling of processes in porous media. *Mailing Add:* Earth Sci Div Lawrence Berkeley Lab Bldg 50E Berkeley CA 94720

LIPPMANN, MORTON, b Brooklyn, NY, Sept 21, 32; m 56; c 3. ENVIRONMENTAL SCIENCES. *Educ:* Cooper Union, BChE, 54; Harvard Univ, SM, 55; NY Univ, PhD(indust hyg), 67. *Prof Exp:* Indust hygienist, USPHS, Ohio, 55-57; indust hygienist, US AEC, NY, 57-62; sr res engr, Del Electronics Corp, 62-64; assoc res scientist aerosol physiol, 64-67, from asst prof to assoc prof, 67-77, PROF ENVIRON MED & DIR, HUMAN EXPOSURE & HEALTH EFFECTS PROG, INST ENVIRON MED, NY UNIV, 77- *Concurrent Pos:* Chmn, Environ Protection Agency Clean Air Sci Adv Comn, 83-87, EPA Indoor Air & Total Human Exposure Adv Comt, 87-; mem, bd sci counr, Nat Inst Occup Safety & Health, 88- *Honors & Awards:* David Sinclair Award, Am Asn Aerosol Res. *Mem:* Am Conf Govt Indust Hygienists; Am Indust Hyg Asn; Int Soc Environ Epidemiol; Am Thoracic Soc; Air Pollution Control Asn; Am Asn Aerosol Res. *Res:* Environmental hygiene; regional deposition and clearance of inhaled particles; sampling and analysis of atmospheric particles; aerodynamic behavior of respirable aerosols; field and laboratory studies of health effects of airborne toxicants. *Mailing Add:* NY Univ Inst Environ Med 550 First Ave New York NY 10016

LIPPMANN, SEYMOUR A, b Brooklyn, NY, Nov 23, 19; m 45; c 3. APPLIED PHYSICS. *Educ:* Cooper Union, BChE, 42. *Prof Exp:* Group leader appl physics, Res Dept, US Rubber Co, 47-60, dept mgr phys res, 60-71, res assoc, 71-75; mgr Tire-Vehicle Systs Labs, Uniroyal Tire Co, Uniroyal, Inc, 75-85; RETIRED. *Concurrent Pos:* Sci consult; sem instr, Tire & Vehicle Dynamics, Soc Automotive Engrs. *Mem:* Am Phys Soc; Fel Soc Automotive Eng; Inst Elec & Electronics Eng; Am Soc Testing & Mat; Sigma Xi. *Res:* Physics of polymeric materials; transmission of noise and vibrations; design of electronic instrumentation for the study of dynamic systems and properties; perception of sound in the presence of background noise; dynamics of the human as a link in control systems; mechanics of laminates and tires. *Mailing Add:* 12767 Lincoln Huntington Woods MI 48070

LIPPMANN, WILBUR, b Galveston, Tex, Sept 6, 30. BIOCHEMICAL PHARMACOLOGY. *Educ:* Tex A&M Col, BS, 51; Univ Tex, MA, 56, PhD(biochem), 61. *Prof Exp:* Res biochemist, Biochem Inst, Univ Tex, 54-56, 58-62; res biochemist, Virus Inst, Univ Calif, Berkeley, 56-58; res biochemist, Univ Tex M D Anderson Hosp & Tumor Inst, 58; res biochemist, Lederle Labs, Am Cyanamid Co, NY, 62-66; head biogenic amine lab, 66-69, DIR DEPT BIOCHEM PHARMACOL, AYERST LABS, CAN, 69- *Concurrent Pos:* Fel, Univ Tex, 61-62. *Mem:* AAAS; Am Chem Soc; Am Soc Pharmacol & Exp Therapeut; Pharmacol Soc Can; NY Acad Sci. *Res:* Biosynthesis and mode of action of the biogenic amines; biochemical mechanisms of action of drugs with respect to cardiovascular, central nervous and gastrointestinal systems; biochemical mechanisms involved with gonadotrophin secretion. *Mailing Add:* 1001 N Harbor Blvd Apt 60 Lahabra CA 90631

LIPPS, EMMA LEWIS, b Alexandria, Va, Feb 8, 19. PLANT ECOLOGY. *Educ:* Wesleyan Col, BA, 40; Emory Univ, MS, 49; Univ Tenn, Knoxville, PhD, 66. *Prof Exp:* Asst biol, James Scott Col, 42-43; from asst to assoc prof, 43-62, dir NSF in-serv insts, 58-61, PROF LIFE & EARTH SCI, SHORTER COL, GA, 62-, CUR LIFE & EARTH SCI, MUS, 70- *Concurrent Pos:* Asst, Univ Tenn, 55-56; Southern Fel Fund fel, 61-62. *Mem:* AAAS; Sigma Xi; Ecol Soc Am; Nat Speleol Soc; Soc Vertebrate Paleont. *Res:* Relationship of present and primeval forests of northwest Georgia to geology and soils, especially the Marshall forest; northwestern Georgia's Pleistocene fossils. *Mailing Add:* Dept Biol/Earth Sci Shorter Col Rome GA 30161

LIPPS, FRANK B, b Baltimore, MD, Mar 13, 33. ATMOSPHERIC SCIENCES. *Educ:* Johns Hopkins Univ, BA, 55, PhD(meteorol), 60. *Prof Exp:* Instr meteorol, Univ Chicago, 62-64; res meteorol, Geophys Fluid Dynamics Lab, Environ Sci Servs Admin, Washington, DC, 64-68; RES METEOROL, GEOPHYS FLUID DYNAMICS LAB, NAT OCEANIC & ATMOSPHERIC ADMIN, PRINCETON, NJ, 68- *Concurrent Pos:* NSF posdoctoral fel, Univ Stockholm, Sweeden, 60- 61. *Mem:* Am Meteorol Soc; AAAS; NY Acad Sci. *Res:* Author of numerous scientific articles. *Mailing Add:* Geophys Fluid Dynamics Lab Princeton Univ PO Box 1919 Princeton NJ 08542

LIPPS, FREDERICK WIESSNER, b Baltimore, Md, Feb 18, 29; m 69; c 3. THEORETICAL PHYSICS, MATHEMATICS. *Educ:* Johns Hopkins Univ, AB, 50, PhD(theoret physics), 56. *Prof Exp:* Meson physics, Lorentz Inst, Holland, 56-57; physicst, cesium-clock, Nat Co, Mass, 57-59, cesium-rocket, Electro- Optical Systs, Calif, 60-63, fusion, Hughes Aircraft Co, DC, 63-64 & Apollo, TRW Syst Inc, 66-69; asst prof math, Tex Southern Univ, 71-73; res scientist solar energy, Energy Lab, 73-84, RES SCIENTIST APPL GEOPHYSICS LAB, ELASTIC LAYER SEISMOGRAMS, UNIV HOUSTON, 85- *Concurrent Pos:* Adj assoc prof, Physics Dept, Univ Houston, 77- *Mem:* Am Phys Soc; Am Math Soc; Bioelectromagnetic Soc; Math Asn Am; Sigma Xi. *Res:* Computer simulation of solar central receiver system optics; computer program development including innovations for shading and blocking, image formation and optimization; elastic wave propagation. *Mailing Add:* 4509 Mimosa St Bellaire TX 77401

LIPPS, JERE HENRY, b Los Angeles, Calif, Aug 28, 39; m 73; c 2. GEOLOGY, INVERTEBRATE PALEONTOLOGY. *Educ:* Univ Calif, Los Angeles, AB, 62, PhD(geol), 66. *Prof Exp:* Asst res geologist invert paleont, Calif Res Corp, 63-65; res geologist, Univ Calif, Los Angeles, 65-67; from asst prof to prof geol, Univ Calif, Davis & Bodega Marine Lab, 67-88; prof paleont, 88-89, PROF INTEGRATIVE BIOL & DIR MUS PALEONT, UNIV CALIF, BERKELEY, 89- *Concurrent Pos:* Res assoc, Los Angeles County Mus, 63-; guest prof, Aarhus Univ, Denmark, 77; prin investr grants, NSF, 68-88 & Nat Park Serv, 70-73; dir, Cushman Found Foraminiferal Res. *Honors & Awards:* Lipps Island, Antarctica (named in honor of). *Mem:* Fel AAAS; fel Geol Soc Am; Paleont Soc; Am Naturalists; Soc Econ Paleont & Mineral; Brit Micropaleont Soc. *Res:* Ecology of Foraminifera; evolutionary biology of protists; marine ecology; geology. *Mailing Add:* Dept Integrative Biol Univ Calif Berkeley CA 94720

LIPPSON, ROBERT LLOYD, b Detroit, Mich, Apr 18, 31; m 72; c 7. MARINE BIOLOGY. *Educ:* Mich State Univ, BS, 63, MS, 64, PhD(zool), 75. *Prof Exp:* Res biologist, Chesapeake Biol Lab, Univ Md, 68-71; fisheries biologist, 71-73, asst coordr water resources, 73-75, res coordr, Environ Assessment Div, Oxford Lab, 75-81, res coordr, Habitat Protection Br, 81, liaison officer, Northeast Region, Oxford Lab, 76-81, ASST REGIONAL DIR, NAT MARINE FISHERIES SERV, 81- *Concurrent Pos:* Adj prof zool, Mich State Univ. *Mem:* Estuarine Res Fedn; Atlantic Estuarine Res Soc; Sigma Xi. *Res:* Population dynamics and physiological-ecology of crustacea. *Mailing Add:* Oxford Lab Nat Marine Fisheries Serv Oxford MA 21654

LIPS, HILAIRE JOHN, b Can, July 20, 18; m 46; c 5. CHEMISTRY. *Educ:* Univ BC, BA, 38, MA, 40; McGill Univ, PhD(agr chem), 44. *Prof Exp:* Biochemist & food info officer, Nat Res Coun Can, 43-84; RETIRED. *Res:* Fats; oils; foods. *Mailing Add:* 15 Fairhaven Way Ottawa ON K1K 0R4 Can

LIPSCHULTZ, FREDERICK PHILLIP, b Los Angeles, Calif, Aug 27, 37. PHYSICS. *Educ:* Stanford Univ, BS, 59; Cornell Univ, PhD(physics), 66. *Prof Exp:* Res assoc physics, Cornell Univ, 62-65; fel, Brookhaven Nat Lab, 65-67; asst prof, 67-72, ASSOC PROF PHYSICS, UNIV CONN, 72- *Concurrent Pos:* Vis fel physics, Univ Nottingham, 76-77. *Mem:* AAAS; Am Phys Soc; Inst Elec & Electronic Engrs. *Res:* Thermal conductivity; low temperature physics; ultrasonics; use of thermal and acoustic properties of materials to investigate microscopic defects in solids; laboratory computer interfacing for research and teaching. *Mailing Add:* Dept Physics U-46 Univ Conn 2152 Hillside Rd Storrs CT 06268

LIPSCHUTZ, MICHAEL ELAZAR, b Philadelphia, Pa, May 24, 37; m 59; c 3. INORGANIC CHEMISTRY, COSMOCHEMISTRY. *Educ:* Pa State Univ, BS, 58; Univ Chicago, SM, 60, PhD(phys chem), 62. *Prof Exp:* NSF and NATO fel, Physics Inst, Berne, 64-65; asst prof chem, 65-68, asst prof geosci, 67-68, assoc prof, 68-73, prof chem & geosci, 73-78, PROF CHEM, PURDUE UNIV, 73-, DIR CHEM OPERS, PRIMELAB, 90- *Concurrent Pos:* Fulbright-Hays scholar, Tel Aviv Univ, 71-72; consult, NASA, 73- & Lunar Planetary Inst, 81-; vis prof chem, Max-Planck Inst Chem, Mainz, Ger, 87; COSPAR/SAFISY Panel Space Sci Experts, 90- *Honors & Awards:* NASA Group Achievement Award, 83; Nininger Meteorite Res Award, 62; Cert of Recognition for Creative Develop Technol, Cert of Spec Recognition, NASA, 79; Cert of Appreciation, Nat Comn Space, 86. *Mem:* AAAS; Am Geophys Union; fel Meteoritical Soc; Geochem Soc; Am Chem Soc; Planetary Soc; Sigma Xi; Int Astron Union. *Res:* Neutron activation, atomic absorption and accelerator mass spectrometric methods for trace and ultratrace analysis; geochemistry; stable isotopes in lunar samples and meteorites; cosmogenic nuclear reactions; high pressure and temperature reactions; author or co-author of over 120 scientific papers; minor planet 2641 Lipschutz named by Int Astron Union. *Mailing Add:* Dept Chem Purdue Univ West Lafayette IN 47907

LIPSCOMB, DAVID M, b Morrill, Nebr, Aug 4, 35; m 78; c 5. AUDIOLOGY. *Educ:* Univ Redlands, BA, 57, MA, 59; Univ Wash, PhD(audiol), 66. *Prof Exp:* Asst prof audiol, WTex State Univ, 60-62; asst prof, Univ Tenn, Knoxville, 62-64 & 66-69, assoc prof, 69-72, prof audiol & speech path, 72-87, dir, Noise Res Lab, 71-87; PRES, CORRECT SERV, INC, 86- *Concurrent Pos:* Consult, various industs & attorneys. *Mem:* Fel Am Speech & Hearing Asn; Acoust Soc Am; Am Auditory Soc; Nat Hearing Conserv Asn. *Res:* Effect of high intensity noise upon the peripheral auditory mechanism. *Mailing Add:* PO Box 1680 Stanwood WA 98292-1680

LIPSCOMB, ELIZABETH LOIS, b Hackensack, NJ, Nov 25, 27; m 49; c 1. BIOCHEMISTRY. *Educ:* Duke Univ, BS, 49; Univ Pittsburgh, MS, 51; NC State Univ, PhD(biochem), 73. *Prof Exp:* RES ASSOC BIOCHEM, NC STATE UNIV, 73- *Mem:* Sigma Xi; NY Acad Sci. *Res:* Catalytic and structural characteristics of the branched-chain amino acid aminotransferase of Salmonella typhimurium; creative application of techniques to elucidate the amino acid sequence is stressed. *Mailing Add:* 3453 Leonard St Raleigh NC 27607

LIPSCOMB, JOHN DEWALD, b Wilmington, Del, Apr 16, 47; m 72; c 1. SPECTROSCOPY ENZYMOLOGY. *Educ:* Amherst Col, BA, 69; Univ Ill, MS, 71, PhD(biochem), 74. *Prof Exp:* Fel, Freshwater Biol Inst, 75-77, from asst to assoc prof, 77-87, PROF BIOCHEM, DEPT BIOCHEM, UNIV MINN, 87- *Concurrent Pos:* Prin investr, NIH res grant, 78- *Mem:* Am Chem Soc; Sigma Xi; Am Soc Biol Chemists. *Res:* Enzyme mechanisms, in particular metalloenzymes such as dioxygenases, monooxygenases iron-sulfur proteins; resonance spectroscopy; chemical modification reactions; transient kinetics. *Mailing Add:* 4-225 Millard Hall Biochem Med Univ Minn 435 Delware Minneapolis MN 55455-0100

LIPSCOMB, NATHAN THORNTON, b Jan 16, 34; US citizen; m 62; c 2. POLYMER CHEMISTRY. *Educ:* Eastern Ky State Col, BS, 56; Univ Louisville, PhD(phys chem), 60. *Prof Exp:* From asst prof to assoc prof, 60-75, PROF CHEM, UNIV LOUISVILLE, 75- *Concurrent Pos:* Consult, ORGI, 88- *Mem:* Am Chem Soc; Sigma Xi. *Res:* Kinetics of polymerization; radiation induced polymerization; polymer characterization; polymer properties. *Mailing Add:* Dept Chem Univ Louisville Louisville KY 40292

LIPSCOMB, PAUL ROGERS, b Clio, SC, Mar 23, 14; m 40; c 2. ORTHOPEDIC SURGERY. *Educ:* Univ SC, BS, 35; Med Col, SC, MD, 38; Univ Minn, MS, 42; Am Bd Orthop Surg, dipl. *Prof Exp:* Intern, Cooper Hosp, NJ, 38-39; resident orthop surg, Mayo Found, Univ Minn, 39-42, from asst prof to prof, 49-69; prof orthop, Surg & Chmn Dept, Sch Med, Univ Calif, Davis, 69-81, chmn, Leadership Fund, 84; CONSULT. *Concurrent Pos:* Consult, Mayo Clin, Minn, St Mary's Hosp Methodist Hosp, 42-69, David Grant US Air Force Med Ctr, Travis AFB, Calif, 70-81, Letterman Gen Hosp, Presidio, San Francisco, 72- & Woodland Clin, Woodland, Calif, 81-86; secy, Am Bd Orthop Surg, 68, pres, 71-73; trustee, Sterling Bunnell Found, 84-; pres, Woodland Clin Res & Educ Found, 84-85. *Mem:* Clin Orthop Soc; AMA; Am Col Surg; Am Acad Orthop Surg; Am Orthop Asn (pres, 74-75). *Res:* Surgery of the hand. *Mailing Add:* 749 Sycamore Lane Davis CA 95616

LIPSCOMB, ROBERT DEWALD, b Tulia, Tex, Dec 29, 17; m 43; c 2. ORGANIC CHEMISTRY, POLYMER CHEMISTRY. *Educ:* Univ Nebr, BS, 40, MS, 41; Univ Ill, PhD(org chem), 44. *Prof Exp:* Lab asst chem, Univ Nebr, 40-41; spec asst, Univ Ill, 41-42, investr, Off Sci Res & Develop & Nat Defense Res Comt, 42-44 & Univ Nebr, 44-45; res chemist, E I du Pont de Nemours & Co, 45-81; RETIRED. *Concurrent Pos:* Consult polymer chem & sci educ, 82- *Mem:* Am Chem Soc; Sigma Xi. *Res:* Chemistry of quinoline and benzoquinoline derivatives; amine bisulfites; organic polysulfides; free radicals; high temperature chemistry; polymerization. *Mailing Add:* 300 Jackson Blvd Wilmington DE 19803

LIPSCOMB, WILLIAM NUNN, JR, b Cleveland, Ohio, Dec 9, 19; m 44, 83; c 3. PHYSICAL CHEMISTRY. *Educ:* Univ Ky, BS, 41; Harvard Univ, MA, 59 Calif Inst Technol, PhD(phys chem), 46. *Hon Degrees:* DSc, Univ Ky, 63, Long Island Univ, 77, Rutgers Univ, 79, Gustavus Adolphus Col, 80, Marietta Col, 81, Miami Univ, 83, Univ Denver, 85 & Ohio State Univ, 91; Dr, Univ Munich, 76. *Prof Exp:* From asst prof to prof phys chem, Univ Minn, 46-59, actg chief div, 52-54, chief, 54-59; prof chem, Harvard Univ, 59-71, chmn dept, 62-65, Abbott & James Lawrence prof, 71-90, EMER ABBOTT & JAMES LAWRENCE PROF CHEM, HARVARD UNIV, 90- *Concurrent Pos:* Guggenheim fel, Oxford, 54-55 & Cambridge, 72-73; NSF sr fel, 65-66; Mem nat comt crystallog, Nat Res Coun, 54-58, 60-63 & 65-67; mem rev comt, Chem Div, Argonne Nat Lab, 56-65; grants, Off Naval Res, 58-77, Off Ord Res, 54-56, NSF, 56-65 & 77-78, Air Force Off Sci Res, 58-64, NIH, 58-, Upjohn Co, 58 & Adv Res Projs Agency, 61-73; distinguished lectr, Howard Univ, 66; Mem nat comt crystallog, Nat Res Coun, 54-58, 60-63, 65-67; mem, adv comt, Ctr Struct Biochem, Brookhaven Nat Labs, 70, bd assocs, Linus Pauling Inst Sci & Med, 77, bd dirs, Dow Chem Co, 82-89 & adv comt, Inst Amorphous Studies, 83-; mem sci adv bd, Robert A Welch Found, 82-, Daltex Med Sci, Inc, 84-, Nova Pharmaceut Corp, 85- & Gensia Pharmaceuticals, Inc, 91-; Harvard lectr, Yale, 72; Centenary lect, Chem Soc, 72; Dreyfus distinguished scholar, Univ Chicago, 80, Probst lectr, Southern Ill Univ at Edwardsville, 80, speaker/session chmn, Conf Quantum Chem in Biomed Sci, NY Acad Sci, 80, invited lectr and speaker at many univs throughout the world. *Honors & Awards:* Nobel Prize in Chem, 76; Howe Award, Am Chem Soc, 58, Peter Debye Award, 73, Remsen Award, 76; Welch Found lectr, Univ Tex, 66; Phillips lectr, Univ Okla & Priestly lectr, Pa State Univ, 67; William Pyle Phillips lectr, Haverford Col, 68; Baker lectr, Cornell Univ & Coover lectr, Iowa State Univ, 69; Weizmann Lectr, Rehovoth, Israel, 74; VantHoff centenary commemoration lectr, Univ Leiden & Gilbert Newton Lewis mem lectr, Berkeley, 74; Renaud lectr, Mich State Univ, 75; Dreyfus distinguished scholar-lectr & John Strauffer Mem lectr, Univ Southern Calif, Los Angeles, 80; Centenary lectr, Chem Soc, 72; Probst lectr, Souther Ill Univ-Edwardsville, 80. *Mem:* Nat Acad Sci; fel Am Phys Soc; Am Chem Soc; Am Crystallog Asn (pres, 55); fel Am Acad Arts & Sci; Sigma Xi; hon mem Chem Soc London; hon fel Royal Soc Chem; Int Acad Quantum Mech Sci; Mineral Soc Am; hon mem Int Asn Bioinorg Scientists; foreign mem Neth Acad Arts & Sci. *Res:* Diffraction studies of crystals and molecules of biochemical interest; relationship between structure and function, including the relationship of three-dimensional structure and mechanisms of enzymes and other proteins; relationship of geometric and electronic structures in theoretical inorganic and organic chemistry. *Mailing Add:* Dept Chem Harvard Univ 12 Oxford St Cambridge MA 02138

LIPSETT, FREDERICK ROY, b Vancouver, BC, Can, Sept 26, 25; m 57; c 2. FLUORESCENCE MEASUREMENTS, CRYSTAL GROWTH & MICROGRAVITY. *Educ:* Univ BC, BApSc, 48, MApSc, 51; Univ London, PhD(physics), 54. *Prof Exp:* Sr res officer elec eng div, 54-87, CONSULT, NAT RES COUN CAN, 87- *Concurrent Pos:* Part time lectr, Carleton Univ, Can, 62-64. *Mem:* Catgut Acoustical Soc; Am Asn Crystal Growth; AAAS; Acoust Soc Am. *Res:* Luminescence; analysis and computer simulation of police patrol operations; floating zone crystal growth. *Mailing Add:* 37 Oriole Dr Gloucester ON K1J 7E8 Can

LIPSEY, SALLY IRENE, b Dec 31, 26; US citizen; m 48; c 3. MATHEMATICS EDUCATION. *Educ:* Hunter Col, AB, 47; Univ Wis, AM, 48; Columbia Univ, DEduc, 65. *Prof Exp:* Asst, Dept Math, Univ Wis, 47-48; high sch teacher, Bd Educ, New York, 48-49; lectr, Hunter Col, 49-53 & Barnard Col, 53-59; asst prof, Bronx Community Col, 59-65; asst prof educ, 65-71, asst prof, 71-78, ASSOC PROF MATH, BROOKLYN COL, 78- *Mem:* Nat Coun Teachers Math; Math Asn Am. *Mailing Add:* 70 E Tenth St New York NY 10003

LIPSHITZ, HOWARD DAVID, b Durban, SAfrica, Oct 30, 55; US citizen; m 86; c 1. DEVELOPMENTAL GENETICS, DEVELOPMENTAL BIOLOGY. *Educ:* Univ Natal, Durban, SAfrica, BSc, 75, BSc Hons, 76; Yale Univ, New Haven, Conn, MPhil, 80, PhD(biol), 83. *Prof Exp:* Postdoctoral res fel, Dept Biochem, Stanford Univ, 83-86; ASST PROF, DIV BIOL, CALIF INST TECHNOL, PASADENA, 86- *Concurrent Pos:* Prin investr, NIH, 87-, March of Dimes Birth Defects Found, 90-; Searle scholar, 88-91. *Mem:* Fel AAAS; Genetics Soc Am. *Res:* Molecular genetics of embryonic pattern formation and morphogenesis in Drosophia Melanogaster; localization of RNA in the egg; specification of germ line; homeotic genes and their regulation. *Mailing Add:* Div Biol 156-29 Calif Inst Technol Pasadena CA 91125

LIPSHITZ, STANLEY PAUL, b Cape Town, SAfrica, Nov 25, 43; Can citizen. ELECTROACOUSTICS, SOUND RECORDING & REPRODUCTION. *Educ:* Univ Natal, BSc, 64; Univ SAfrica, MSc, 65; Univ Witwatersrand, PhD(math), 70. *Prof Exp:* Vis lectr math, Univ Ariz, 67-68; from asst prof to assoc prof, 70-88, PROF APPL MATH & PHYSICS, UNIV WATERLOO, 88- *Mem:* fel Audio Eng Soc (pres, 88-89); Acoust Soc Am; Inst Elec & Electronic Engrs. *Res:* Mathematical, physical, and engineering problems of audio and electroacoustics; transducer design and measurement; digital audio signal processing; stereo and surround sound recording and reproduction. *Mailing Add:* Dept of Appl Math Univ of Waterloo Waterloo ON N2L 3G1 Can

LIPSHULTZ, LARRY I, b Philadelphia, Pa, Apr 24, 42; m 66; c 2. UROLOGY. *Educ:* Franklin & Marshall Col, BS, 60; Univ Pa, MD, 68. *Prof Exp:* Asst instr urol, Univ Pa, 73-74, instr, 74-75; asst prof & clin fel reprod med, 75-77, assoc prof, 77-80, PROF UROL, UNIV TEX MED BR, HOUSTON, 80-; DEPT UROL, BAYLOR COL MED. *Concurrent Pos:* Res scholar, Am Urol Asn, 75-77; adj asst prof, Baylor Col Med, 76- *Mem:* Am Fertil Soc; Am Soc Andrology; Am Urol Asn; Soc Univ Urol. *Res:* The evaluation and diagnosis of reproductive disorders in the male, especially in the field of infertility; androgen binding protein in the human testis and epididymis and evaluation of androgen binding protein as a possible marker of sertoli cell function. *Mailing Add:* Dept Urol Baylor Col Med One Baylor Plaza Houston TX 77030

LIPSHUTZ, NELSON RICHARD, b Philadelphia, Pa, July 14, 42; m 64; c 3. ECONOMICS. *Educ:* Univ Pa, AB, 62, MBA, 72; Univ Chicago, SM, 63, PhD(physics), 67. *Prof Exp:* Res assoc physics, Univ Chicago, 67; from instr to asst prof, Duke Univ, 67-70; mgt res analyst, Mgt & Behav Sci Ctr, Wharton Sch Finance & Commerce, Univ Pa, 70-72; mgr consult, Arthur D Little, Inc, 72-77; PRES, REGULATORY RES CORP, 77- *Concurrent Pos:* Instr, Sch Bus, Northeastern Univ, 86. *Mem:* Am Phys Soc; Inst Mgt Sci; AAAS; NY Acad Sci; Nat Asn Forensic Economists. *Res:* Theory of elementary particles; mathematical analysis of management decision problems; economic analysis of regulated industries; anti-trust economics; business valuation. *Mailing Add:* 24 Radcliff Rd Regulatory Res Corp Waban MA 02168

LIPSICH, H DAVID, b Pittsburgh, Pa, Feb 20, 20; m 46; c 2. MATHEMATICS. *Educ:* Univ Cincinnati, MA, 45, PhD(math), 49; Princeton Univ, MA, 46. *Prof Exp:* From instr to assoc prof, 46-61, vprovost, 67, head dept, 61-77, provost undergrad studies, 67-77, PROF MATH, UNIV CINCINNATI, 61-, DEAN, MCMICKEN COL ARTS & SCI, 77- *Concurrent Pos:* NSF fac fel, 59-60. *Mem:* Am Math Soc; Math Asn Am; Asn Symbolic Logic. *Res:* Mathematical logic; set theory. *Mailing Add:* 427 W Galbraith Rd Cincinnati OH 45215

LIPSICK, JOSEPH STEVEN, b Sharon, Pa, Jan 6, 55; m 78; c 2. MOLECULAR ONCOLOGY. *Educ:* Oberlin Col, BA, 74; Univ Calif, San Diego, PhD(physiol, pharmacol), 81, MD, 82. *Prof Exp:* Asst prof residence molecular path, Univ Calif, San Diego, 86-89; ASSOC PROF, DEPT MICROBIOL, STATE UNIV NY, STONY BROOK, 89- *Mem:* Am Soc Microbiol; AAAS. *Res:* To understand the molecular mechanisms by which nuclear protein products of oncogenes cause leukemia, particulary the myb oncogene, which is highly conserved in evolution. *Mailing Add:* Dept Microbiol Life Sci Bldg SUNY Stoney Brook Campus La Jolla CA 92093

LIPSIG, JOSEPH, b Brooklyn, NY, Dec 13, 30; m 60; c 1. PHYSICAL CHEMISTRY. *Educ:* Brooklyn Col, BA, 50; Polytech Inst Brooklyn, PhD(phys chem), 61. *Prof Exp:* Res assoc, Cornell Univ, 60-62; sr res chemist, Atlantic Ref Co, 62-66; asst prof, 66-68, ASSOC PROF CHEM, STATE UNIV NY COL OSWEGO, 68- *Mem:* Am Chem Soc. *Res:* Catalysis; geochemistry. *Mailing Add:* Dept Chem Oswego State Col Oswego NY 13126

LIPSITT, DON RICHARD, b Boston, Mass, Nov 24, 27; m 53; c 2. PSYCHIATRY, PSYCHOANALYSIS. *Educ:* NY Univ, BA, 49; Boston Univ, MA, 50; Univ Vt, MD, 56; Boston Psychoanal Soc & Inst, cert. 69. *Prof Exp:* Asst, 62-65, from instr to assoc prof psychiat, 74-90, CLIN PROF HARVARD MED SCH, 90- *Concurrent Pos:* Teaching fel psychiat, Harvard Med Sch, 60-62; Dept Health, Educ & Welfare res grant, 66-68; head integration clin, Beth Israel Hosp, Boston, 62-69, asst psychiat, 62-64, assoc, 64-66, dir med psychol liaison serv, 66-69; consult behav sci, Lincoln Lab, Mass Inst Technol, 69-73; mem fac, Boston Psychoanal Soc & Inst, Simmons Col, 71-72 & 81, adj prof, Sch Social Work, 71-80; consult, Dept Psychiat, Cambridge Hosp, 71-; consult psychiatrist, McLean Hosp, 71-; ed, Int J Psychiat Med, 70-79, Gen Hosp Psychiat, 79-; consult, NIMH, 74-80; fac div primary care & family med, Harvard Med Sch, 77- *Mem:* Am Psychiat Asn; Am Psychosom Soc; Asn Acad Psychiat; fel Am Col Psychiat; Am Asn Gen Hosp Psychiat. *Res:* Application of medical psychology to health problems in hospital and community; relationship of varieties of doctor-patient interaction to invalidism and chronicity; psychiatry and primary care; hypochondriasis; factitious illness. *Mailing Add:* Dept Psychiat Harvard Med Sch 25 Shattuck St Brookline MA 02115

LIPSITT, HARRY A(LLAN), b Detroit, Mich, June 7, 31; m 56; c 4. PHYSICAL METALLURGY, MATERIALS SCIENCE. *Educ:* Mich State Univ, BS, 52; Carnegie Inst Technol, MS, 55, PhD(metall), 56. *Prof Exp:* Asst metall, Carnegie Inst Technol, 52-53; proj scientist, Aerospace Res Lab, 56-60, supvr metall res, 60-75, sr scientist, air force mat lab, Wright-Patterson AFB, 75-86; PROF MAT SCI & ENG, DEPT MECH & MAT ENG, WRIGHT STATE UNIV, DAYTON, 86- *Concurrent Pos:* Instr, Univ Dayton, 57-58; sabbatical, Cambridge Univ, 63-64; adj prof, Air Force Inst Technol, 65-71; liaison scientist, Off Naval Res, London, Eng, 67-69; adj prof Eng, Wright State Univ, 76-86. *Honors & Awards:* Award, Aerospace Ed Found, 64. *Mem:* Am Inst Mining, Metall & Petrol Engrs. *Res:* Mechanical metallurgy, especially fatigue, creep, tension and hardness and properties of ceramics; diffusion and strain aging; elasticity, precipitation hardening and electron microscopy; alloy development of intermetallic compounds for structural use; high temperature materials. *Mailing Add:* 1414 Birch St Yellow Springs OH 45387-1308

LIPSITT, LEWIS PAEFF, b New Bedford, Mass, June 28, 29; m 52; c 2. INFANT BEHAVIOR & DEVELOPMENT. *Educ:* Univ Chicago, BA, 50; Univ Mass, MS, 52; Univ Iowa, PhD(child psychol), 57. *Hon Degrees:* MS, Brown Univ, 66. *Prof Exp:* From instr to assoc prof, 57-66, PROF PSYCHOL, BROWN UNIV, 66-, DIR CHILD STUDY CTR, 67- *Concurrent Pos:* Dir child training, Brown Univ, 60-80, prof med sci, 74-; fel, Ctr Advan Study Behav Sci, 79-80, Guggenheim fel behav develop, res unit, St Mary's Hosp, London, 72-73; mem bd sci counr, Nat Inst Child Health & Human Develop, 84-88; mem bd adv, Archives Hist Am Psychol, 86-; consult, Behav Sci Panel, Nat Inst Mental Health, 87-88; mem, Int Conf Infant Studies; USPHS fel, 71; fel, Advan Study in Behav Sci, Stanford Univ, 79-80. *Honors & Awards:* Sauer lectr, Northwestern Univ, 80. *Mem:* Am Psychol Asn; AAAS; Soc Res Child Develop; Am Asn Univ Prof. *Res:* Infant behavior and development, particularly sensory and learning processes of babies; crib death; adolescent suicide; study of behavioral misadventures or hazards. *Mailing Add:* Dept Psychol Brown Univ Providence RI 02912

LIPSITZ, PAUL, b York, Pa, Apr 23, 23; m 48; c 4. ORGANIC CHEMISTRY. *Educ:* Lebanon Valley Col, BS, 44; Univ Cincinnati, MS, 48, PhD(chem), 50. *Prof Exp:* Chemist, E I du Pont de Nemours & Co, 50-59; sr patent agent, Pennsalt Chems Corp, 59-69; SR PATENT AGT, PATENT & LICENCES DEPT, SUN VENTURES, INC, 69- *Mem:* Am Chem Soc; Am Inst Chem. *Mailing Add:* 205 Suffolk Rd Flourtown PA 19031

LIPSITZ, PHILIP JOSEPH, b Piketberg, SAfrica, May 17, 28; m 58; c 3. PEDIATRICS, NEONATAL-PERINATAL. *Educ:* Univ Cape Town, MB, ChB, 52; Royal Col Physicians & Surgeons, dipl child health, 56; Am Bd Pediat, dipl, neonatal-perinatal med, cert. *Prof Exp:* House surgeon, Univ Cape Town, 52; house physician, Somerset Hosp, Cape Town, SAfrica, 53; resident surg house officer, Gen Hosp, Salisbury, SRhodesia, 53; Charles' house physician, St Hosp, London, Eng, 55; resident med officer, Banstead Br, Queen Elizabeth Hosp for Children, 56; house physician, Royal Hosp Sick Children, Edinburgh, Scotland, 56; registr, Prof Unit, Children's Hosp, Sheffield, Eng, 57-58; resident med officer, Red Cross War Mem Children's Hosp, Univ Cape Town, 58; with hosp appointments, Southwest Africa, 62-65; asst prof pediat, Med Col Ga, 65-67; assoc prof, 67-68; assoc prof, Beth Israel Med Ctr & Mt Sinai Sch Med, 68-73; prof pediat, Health Sci Ctr, State Univ NY, Stony Brook, 73-90; dir pediat, South Shore Div, Long Island Jewish-Hillside Med Ctr, Far Rockaway, 73-74; CHIEF, NEONATAL-PERINATAL MED, SCHNEIDER CHILDREN'S HOSP OF LONG ISLAND JEWISH-HILLSIDE MED CTR, NEW HYDE PARK, 74-; PROF PEDIAT, ALBERT EINSTEIN COL MED, 90- *Concurrent Pos:* Fel pediat, Sch Med, Western Reserve Univ, 58-60; clin & res fel pediat & med, Children's Hosp Med Ctr, Harvard Med Sch, 60-61. *Mem:* Royal Col Physicians & Surgeons; Soc Pediat Res; NY Acad Sci; Am Acad Pediat. *Res:* Physiology of the newborn. *Mailing Add:* Schneider Children's Hosp LI Jewish-Hillside Med Ctr 270-05 76th Ave New Hyde Park NY 11040

LIPSIUS, STEPHEN LLOYD, b NeW York, NY, Sept 25, 47. PHYSIOLOGY. *Educ:* State Univ NY, BA, 69, PhD(physiol), 75. *Prof Exp:* Teaching asst, Downstate Med Ctr, 70-75; res fel, 75-78; ASST PROF, DEPT PHYSIOL, LOYOLA UNIV, 78- *Concurrent Pos:* NIH fel, Univ Vt, 76-78. *Mem:* Int Study Group Res Cardiac Metab; Cardiac Electrophysiolic Soc. *Mailing Add:* Dept Physiol Loyola Univ Stritch Sch Med 2160 S First Ave Maywood IL 60153

LIPSKY, JOSEPH ALBIN, b Glen Lyon, Pa, Mar 31, 30; m 57; c 2. PHYSIOLOGY. *Educ:* Pa State Univ, BSc, 51; Ohio State Univ, MSc, 59, PhD(physiol), 61. *Prof Exp:* From asst prof to assoc prof, 61-77, PROF PHYSIOL, COL MED, OHIO STATE UNIV, 77- *Concurrent Pos:* Consult to coun, Nat Bd Dent Exam, 67- *Mem:* AAAS; Fedn Am Socs Exp Biol; Am Physiol Soc; Sigma Xi. *Res:* Carbon dioxide transients and stores; hyperventilation. *Mailing Add:* Dept of Physiol Ohio State Univ Columbus OH 43210

LIPSKY, SEYMOUR RICHARD, biochemistry, physical chemistry; deceased, see previous edition for last biography

LIPSKY, STEPHEN E, b New York, NY, Jan 18, 32; m 79; c 3. SIGNAL PROCESSING, ARTIFICAL INTELLIGENCE. *Educ:* NY Univ, Col Eng, BEE, 53, MEE, 62. *Prof Exp:* Eng Officer, TV eng, US Army Pictorial Ctr, 53-55; design eng microwaves, Prod Res Corp, 55-58; proj leader eng, Fisher Radio Corp, 58-63; corp vpres eng, Polarad Elec Corp, 63-70; dir adv systs, Govt Systs Div, Gen Instrument Corp. 70-79; SR VPRES ENG & CHIEF TECH OFFICER, AEL DEFENSE CORP, 79- *Concurrent Pos:* Eng consult, Electro Acoust Res Labs, 58-60 & Radiometric Div Polarad, 70-77; ed consult, 85-; mem, Comt 1986 Symp, IEEE Ant/Microwave Prop Soc, 85-87; adj univ prof microwaves, Drexel Univ Grad Sch, 87- *Honors & Awards:* Bronze Medal, Armed Forces Commun & Electronics Asn, 53. *Mem:* Fel Inst Elec & Electronic Engrs; Armed Forces Commun & Electronics Asn; British Inst Elec Engrs. *Res:* Monopulse passive direction finding and receiving methods for detection and identification of radar and communications signals; microwave analytic design techniques for antenna systems and associated feed networks; author of publication on Microwave Passive Direction Finding. *Mailing Add:* 1254 Cox Rd Rydal PA 19046

LIPSON, EDWARD DAVID, b Winnipeg, Man, Oct 27, 44; m 66; c 2. PHOTOBIOLOGY, SENSORY TRANSDUCTION. *Educ:* Univ Man, BSc, 66; Calif Inst Technol, PhD(physics), 71. *Prof Exp:* Res fel biol, Calif Inst Technol, 71-74; sr res fel, 74-76; from asst prof to assoc, 76-85, PROF PHYSICS, SYRACUSE UNIV, 85-, DIR GRAD BIOPHYSICS PROG, 83- *Concurrent Pos:* Res fel, Alfred P Sloan Found, 79-83. *Mem:* Biophys Soc; AAAS; Am Soc Photobiol; Am Phys Soc. *Res:* Light-growth responses of the microorganism, Phycomyces, with approaches from genetics, biochemistry and nonlinear systems theory, to elucidate the cellular and molecular mechanisms of sensory transduction and adaptation. *Mailing Add:* Dept Physics Syracuse Univ Syracuse NY 13244-1130

LIPSON, HERBERT GEORGE, b Boston, Mass, July 4, 25; m 51; c 3. SOLID STATE PHYSICS. *Educ:* Mass Inst Technol, BS, 48; Northeastern Univ, MS, 64. *Prof Exp:* Jr physicist metal physics, Sylvania Elec Prods, Inc, 48-50; physicist, Brookhaven Nat Lab, 51; physicist, Naval Res Lab, 51-55; physicist, Lincoln Lab, Mass Inst Technol, 55-58; physicist, Dep Electronic Technol, Rome Air Develop Ctr, Hanscom AFB, 58-90; RETIRED. *Mem:* Am Phys Soc; Mat Res Soc. *Res:* Optical properties of solids; lattice vibrations, impurities and plasma effects in semiconductors; laser and laser window material properties; infrared optical properties of impurities in quartz, radiation effects on quartz for radiation hardened oscillators. *Mailing Add:* 68 Aldrich Rd Wakefield MA 01880

LIPSON, JOSEPH ISSAC, b New York, NY, Apr 19, 27; m 50; c 4. INFORMATION SYSTEMS, COGNITIVE SCIENCE. *Educ:* Yale Univ, BS, 50; Univ Calif, Berkeley, PhD(physics), 56. *Prof Exp:* Asst res physicist, Univ Calif, Berkeley, 56-57 & 60-61 & Univ Alta, 57-60; assoc prof physics & geol, Univ Pittsburgh, 61-64, dir, Curric Design Study Group, Learning Res & Develop Ctr, 64-67; prof sci educ, Nova Univ, 67-69; dir, Sci-Math Div, Learning Res Assoc, 69-70, author & educ consult, 70-71; mem, US Comnr Educ Planning Unit, Nat Inst Educ, 71; assoc dean grad col, Univ Ill, Chicago Circle, 71-72, assoc vchancellor acad affairs, 72-75; vpres acad affairs, Univ Mid-Am, 75-77; dir proj, WICAT, Inc, 77-78; group dir, 81-84; div dir, NSF, 78-81; PROF PHYSICS, CALIF STATE UNIV, CHICO, 84- *Mem:* Am Phys Soc; Sigma Xi; Am Educ Res Asn; Nat Sci Teachers Asn; Nat Coun

Teachers Math; AAAS; Am Asn Physics Teachers; Am Soc Info Sci. *Res:* Development of an applied science of instruction; application to cognitive science to education; role of associative sematic network on problem solving; role of errors and misconceptions in learning. *Mailing Add:* Commun-Design 2319 Mayer Way Chico CA 95926-5300

LIPSON, MELVIN ALAN, b Providence, RI, June 1, 36; m 61; c 4. ORGANIC CHEMISTRY. *Educ:* Univ RI, BS, 57; Syracuse Univ, PhD(org chem), 63. *Prof Exp:* Res chemist, I C I (Organics) Inc, 63 & Eltex Res Corp, 63-64; res supvr org synthesis, Wayland Chem Div, Philip A Hunt Chem Corp, RI, 64-67, res mgr, 67-69; tech dir, Morton Thiokol Inc, 69-72, vpres, 72-82, sr vpres, Tech Opers, 82-85, exec vpres, 85-86, pres, Dynachem Div, 86-89; VPRES, MORTON INT, 89- *Mem:* AAAS; Am Chem Soc; The Chem Soc. *Res:* Amino acids and peptides; chelating agents; photographic chemicals; surface active compounds; polymers; dyestuffs; carbohydrates; photopolymers; photoresists; electroless plating; corrosion inhibitors. *Mailing Add:* Morton Int 2631 Michelle Dr Tustin CA 92680

LIPSON, RICHARD L, b Philadelphia, Pa, July 21, 31; c 2. RHEUMATOLOGY, INTERNAL MEDICINE. *Educ:* Lafayette Col, BA, 52; Jefferson Med Sch, MD, 56; Univ Minn, MSc, 60. *Prof Exp:* Res asst biophys, Mayo Clin & Mayo Found, 62-63; asst prof, 63-85, ASSOC PROF MED, COL MED, UNIV VT, 85- *Concurrent Pos:* Assoc dir rheumatism res unit, Univ Vt, 64-67. *Honors & Awards:* Arnold J Bergen Res Award, Mayo Found, 60. *Mem:* AMA; Am Fedn Clin Res; Am Rheumatism Asn; fel Am Col Physicians. *Res:* Biophysics and bionics of connective tissue, osmotic pressure, viscosity, electrogoniometry, biomechanics; fiber-optics instruments; cancer detection by fluorescence; fluorescent endoscopy; photosensitivity; rheumatology. *Mailing Add:* 233 Pearl St Burlington VT 05401

LIPSON, SAMUEL L(OYD), structural engineering; deceased, see previous edition for last biography

LIPSON, STEVEN MARK, b New York, NY, May 25, 45; m 71; c 2. ENVIRONMENTAL HEALTH, MOLECULAR BIOLOGY. *Educ:* Long Island Univ, BS, 67; C W Post Col, MS, 72; NY Univ, PhD(microbiol), 82. *Prof Exp:* Teacher biol, Frasmus Hall & Prospect Heights High Schs, 67-72; teaching fel biol, NY Univ, 74-75; technologist microbiol, Mem Hosp, 75-76; res assoc biol, NY Univ, 74-80; res assoc biol, Dept Neoplastic Dis, Mt Sinai Med Ctr, 81-83; supvr, Hemat/Oncol Lab, Brooklyn Hosp/Caledonian Hosp, 83-84; chief, Virol Lab, Nassau County Med Ctr, NY, 84-90; DIR, VIROL LAB, NORTH SHORE UNIV HOSP, CORNELL UNIV MED COL, 90- *Concurrent Pos:* Vis lectr microbiol, Adelphi Univ, 76; adj instr biol, Fordham Univ, 81-82; adj asst prof, Fiorello H LaGuardia Community Col & Manhattan Community Col, 83-85; adj prof, biol, C W Post Col, 87- *Mem:* Am Soc Microbiol. *Res:* Epidemiology and rapid identification medically relevant viruses; surface interactions between viruses, cells, polymers, and particulates; polymerase chain reaction in the identification of viruses. *Mailing Add:* Infectious Dis & Immunol North Shore Univ Hosp Cornell Univ Med Col 300 Community Dr Manhasset NY 11030

LIPTAY, ALBERT, b Hampton, Ont, Nov 9, 41; m 67; c 3. HORTICULTURE, PLANT PHYSIOLOGY. *Educ:* Univ Guelph, BSA, 66, MSc, 67; McMaster Univ, PhD(biol), 72. *Prof Exp:* Lectr life sci, Conestoga Col, 72-73; asst prof biol, Camrose Lutheran Col, 73-74; RES SCIENTIST VEG MGT, AGR CAN, 74- *Mem:* Am Soc Hort Sci; Int Soc Hort Sci; Can Soc Hort Sci; Agr Inst Can. *Res:* Vegetable management and physiology; seed germination; seed vigour; plant establishment; growth factors. *Mailing Add:* Res Sta Agr Can Harrow ON N0R 1C0 Can

LIPTON, ALLAN, b New York, NY, Dec 29, 38; m 65; c 3. INTERNAL MEDICINE, ONCOLOGY. *Educ:* Amherst Col, BA, 59; NY Univ, MD, 63; Am Bd Internal Med, dipl, 70. *Prof Exp:* Intern med, Bellevue Hosp, New York, 63-64, resident, 64-65; from asst prof, 71-80, PROF MED CTR, PA STATE UNIV, 80-, CHIEF, DIV ONCOL, DEPT MED, 74- *Concurrent Pos:* Fel hemat, Mem Hosp, New York, 67-68, fel oncol, 68-69; Dernham fel, Salk Inst Biol Studies, 69-71. *Mem:* AAAS; Am Asn Cancer Res; Am Fedn Clin Res. *Res:* Control of growth of normal and malignant cells by serum factors. *Mailing Add:* Dept Med Hershey Med Ctr Pa State Univ Hershey PA 17033

LIPTON, JAMES MATTHEW, b Aug 10, 38; US citizen; m 59; c 2. NEUROSCIENCES. *Educ:* Univ Colo, PhD, 64. *Prof Exp:* PROF PHYSIOL & RES ASSOC PROF ANESTHESIOL, UNIV TEX SOUTHWESTERN MED CTR, DALLAS. *Concurrent Pos:* USPHS fel, Neuropath Lab, Med Sch, Univ Mich, 64-66; USPHS sr fel, Inst Animal Physiol, UK, 70-71; consult neurol, Vet Admin Hosp, Dallas, 74-80; mem staff, Anesthesiol Dept, Southwestern Med Sch, Univ Tex Southwestern Med Ctr, Dallas, 81-, mem, neurol study sect, 86-90. *Mem:* AAAS; Soc Neurosci; Am Physiol Soc; Geront Soc; Soc Exp Biol & Med; Am Soc Anesthesiologists. *Res:* Central control of fever, inflammation and immune regions; pain mechanisms and control. *Mailing Add:* Dept Physiol Univ Tex Southwestern Med Ctr 5323 Harry Hines Blvd KS-140 Dallas TX 75235-9040

LIPTON, MICHAEL FORRESTER, b Huntington, WVa, Oct 28, 50; m 85; c 2. PROCESS DEVELOPMENT. *Educ:* Purdue Univ, BS, 72; Univ Colo, PhD(chem), 76. *Prof Exp:* Res assoc chem, Fordham Univ, 76-78; asst prof, Mich State Univ, 78-80; MEM STAFF, UPJOHN CO, 80- *Concurrent Pos:* Res assoc, Dept Entom, USDA, Mich State Univ, 79-80. *Mem:* Am Chem Soc. *Res:* Process development in the pharmaceutical industry; rapid scale-up and design of chemical routes of synthesis of biologically active molecules. *Mailing Add:* Upjohn Co OU 1510-91-1 Kalamazoo MI 49001

LIPTON, MORRIS ABRAHAM, experimental psychiatry, medicine; deceased, see previous edition for last biography

LIPTON, WERNER JACOB, b Ger, Oct 16, 28; nat US; m 52; c 4. FOOD SCIENCE & TECHNOLOGY, PLANT PHYSIOLOGY. *Educ:* Mich State Univ, BS, 51, MS, 53; Univ Calif, PhD(plant physiol), 57. *Prof Exp:* Asst, Univ Calif, 53-57; sr pant physiologist, Hort Field Sta, USDA, 57-87; CONSULT, 87- *Concurrent Pos:* Assoc ed, Am Soc Hort Sci, 72-76, 78-, chmn postharvest hort sect, 76-77, vpres, W Sect, 81-82, sci ed, 88- *Mem:* AAAS; Am Soc Hort Sci; Am Soc Plant Physiol; Sigma Xi. *Res:* Postharvest physiology of vegetables; emphasis on effects of modified atmospheres and preharvest environmental factors. *Mailing Add:* 4550 E Redlands Fresno CA 93755-5558

LIRA, EMIL PATRICK, b Chicago, Ill, Mar 17, 34; m 58; c 4. ORGANIC CHEMISTRY. *Educ:* Elmhurst Col, BS, 56; Rutgers Univ, PhD(org chem), 63. *Prof Exp:* Chemist, Swift & Co, 56; chemist, Corn Prod Co, 58-59; res chemist, 63-69, Int Mineral & Chem Corp, 63-69, supvr org synthesis, 69-73, mgr org chem, 73-74; dir res, Velsicol Chem Corp, 74-76; DIR AGR-CHEM RES, NORTHWEST IND, INC, CHICAGO LAB, 76- *Mem:* Am Chem Soc; Indust Res Inst. *Res:* Organic research and development with plasticizers, adhesives, polymer additives; plant growth regulators; pesticides; animal health products; synthetic sweeteners; organic processes. *Mailing Add:* 300 Parkway Ave Bloomingdale IL 60108-3008

LIS, ADAM W, b Przemysl, Poland, Jan 5, 25; US citizen; c 4. BIOCHEMISTRY. *Educ:* Univ Ark, BS, 49; Univ Calif, Berkeley, PhD(biochem), 60. *Prof Exp:* Res biochemist, Univ Calif, San Francisco, 60-62; res assoc, Univ Ore Health Sci Ctr, 63-65, asst prof, 65-67, assoc prof nucleic acids, 67-77, dir nucleic acids lab, 66-77; DIR, INTERMEDIARY METAB INST, 77-; PROF BIOCHEM, HUXLEY COL, WESTERN WASH UNIV, BELLINGHAM, WASH, 81- *Concurrent Pos:* Nat Cancer Inst fel, Univ Uppsala, 62-63; ed, Physiol Chem & Physics, 67-80. *Honors & Awards:* Copernicam Medal Med, 73. *Mem:* Brit Biochem Soc; Am Chem Soc; Am Soc Cell Biol; Radiation Res Soc Am; Biophys Soc. *Res:* Minor components in nucleic acids and their function; body fluids analysis in malignant and metabolic diseases; discovery and characterization of Minor Bases in Ribonucleic Acid; pseudouridine, 5-hydroxyuracil; 5-chloro-deuxyuridine and enzymes c-csynthetase; diagnoses of autism and mental deficiencies of metabolic origin; chemistry of stress and its pathology. *Mailing Add:* 1117 SE Umatilla St Portland OR 97202

LIS, ELAINE WALKER, b Denver, Colo, Apr 25, 24; m 58; c 3. NUTRITION, BIOCHEMISTRY. *Educ:* Mills Col, AB, 45; Univ Calif, Berkeley, PhD(nutrit), 60. *Prof Exp:* Asst nutrit, Univ Calif, Berkeley, 56-60, USPHS fel, 60-62; lectr, Portland State Univ, 64-68; assoc prof, 68-81, PROF, CRIPPLED CHILDREN'S DIV, ORE HEALTH SCI UNIV, 81- *Concurrent Pos:* Consult, Crippled Children's Div, Med Sch, Univ Ore, 64-68. *Mem:* Am Asn Univ Prof; Am Asn Ment Deficiency; Soc Nutrit Educ; Am Home Econ Asn. *Res:* Metabolic approach to possible causes of retardation, emotional disturbances or other handicapping conditions with emphasis on inborn errors of metabolism such as phenylketonuria. *Mailing Add:* Crippled Children's Div Ore Health Sci Univ 3181 SW Sam Jackson Park Rd Portland OR 97201

LIS, JOHN THOMAS, b Willimantic, Conn, June 15, 48; m 71; c 1. MOLECULAR GENETICS. *Educ:* Fairfield Univ, BS, 70; Brandeis Univ, PhD(biochem), 75. *Prof Exp:* Fel, Dept Biochem, Stanford Univ, 75-78; asst prof, 78-84, ASSOC PROF BIOCHEM, CORNELL UNIV, 84- *Res:* Relationship between genome structure and gene regulation using the heat shock genes of Drosophila melanogaster and yeast as model systems. *Mailing Add:* Sect Biochem Molecular & Cell Biol Biotech Bldg Cornell Univ Ithaca NY 14853

LIS, STEVEN ANDREW, b Dunkirk, NY, Oct 13, 50; m 76; c 2. OPTICAL COMPUTING, MATERIALS RESEARCH. *Educ:* Fredonia State Univ Col, NY, BS, 72; Princeton Univ, PhD(chem), 77. *Prof Exp:* Sr scientist, Radiation Monitoring Devices, 77-81 & GCA Corp, 81-88; PRIN INVESTR, SPARTA INC, 88- *Concurrent Pos:* Sr prin develop engr, Honewell Electro-Optics Div, 88. *Mem:* Am Phys Soc; Electro-Chem Soc. *Res:* Optical computing systems; holography; materials and device; integrated circuit process development; advanced lithographic equipment. *Mailing Add:* 254 Marked Tree Rd Needham MA 02192

LISAK, ROBERT PHILIP, b Brooklyn, NY, Mar 17, 41; m 64; c 2. NEUROLOGY, IMMUNOLOGY. *Educ:* NY Univ, BA, 61; Columbia Univ, MD, 65. *Prof Exp:* Intern med, Montefiore Hosp & Med Ctr, 65-66; res assoc immunol, Lab Clin Sci, NIMH, 66-68; jr resident med, Bronx Munic Med Ctr, Albert Einstein Col Med, 68-69; resident neurol, Hosp, Sch Med, Univ Pa, 69-72, trainee allergy & immunol, 71-72, from asst prof to prof neurol, 72-87, mem immunol grad group, 75-87; PROF NEUROL & CHMN DEPT, SCH MED, WAYNE STATE UNIV, 87- *Concurrent Pos:* Consult neurol, Vet Admin Hosp, Philadelphia, 72-82; spec consult, Nat Multiple Sclerosis Soc, 75 & Swiss Acad Med, 81; Fulbright-Hays sr res scholar, UK, 78-79. *Mem:* Am Asn Immunologists; Am Fedn Clin Res; Am Acad Neurol; AAAS; NY Acad Sci; Soc Neurosci; Am Neurol Asn. *Res:* Humoral and cell-mediated immunologic mechanisms involved in clinical and experimental diseases of the central and peripheral nervous system and muscle. *Mailing Add:* Dept Neurol Sch Med Wayne State Univ Detroit MI 48201

LISANKE, ROBERT JOHN, SR, material science engineering & properties, for more information see previous edition

LISANO, MICHAEL EDWARD, b Houston, Tex, Oct 6, 42; c 2. REPRODUCTIVE PHYSIOLOGY, ENDOCRINOLOGY. *Educ:* Sam Houston State Univ, BS, 64, MS, 66; Tex A&M Univ, PhD(physiol), 70. *Prof Exp:* Instr biol, Hardin-Simmons Univ, 66-67; asst prof, 70-77, ASSOC PROF PHYSIOL, AUBURN UNIV, 77- *Mem:* Wildlife Soc; Southeastern Asn Fish & Wildlife Agencies. *Res:* Reproductive physiology and endocrinology of economically important game species. *Mailing Add:* Grad Sch Hargis Hall Auburn Univ Auburn AL 36849

LISCHER, LUDWIG F, b Mar 1, 15; US citizen. ENGINEERING. *Educ:* Purdue Univ, BSEE, 37. *Hon Degrees:* DEng, Purdue Univ, 76. *Prof Exp:* Mem staff, 37-64, vpres-in-chg eng & res, Commonwealth Edison Co, 64-80; RETIRED. *Concurrent Pos:* Dir, Chicago Eng & Sci Ctr, Proj Mgt Corp; chmn res adv comt, Elec Power Res Inst; mem var adv comts, NSF, Nat Acad Sci & Nat Acad Eng. *Mem:* Nat Acad Eng; fel Inst Elec & Electronics Engrs; Am Soc Mech Engrs; Am Nuclear Soc. *Res:* Electric utility systems for the development of government and industry energy policy. *Mailing Add:* 441 N Park Blvd Glen Ellyn IL 60137

LI-SCHOLZ, ANGELA, b Hong Kong, Aug 15, 36; US citizen; m 66; c 2. ATOMIC PHYSICS, NUCLEAR PHYSICS. *Educ:* Manhattanville Col, BA, 56; NY Univ, MS, 57, PhD(physics), 63. *Prof Exp:* Jr res assoc nuclear physics, Brookhaven Nat Lab, 60-63; res assoc high energy physics, NY Univ, 63; res assoc nuclear physics, Yale Univ, 63-65; asst prof physics, City Col New York, 65-66; res assoc solid state physics, Univ Pa, 67-70; res assoc nuclear chem, Rensselaer Polytech Inst, 70-72; assoc prof, 72-77, PROF SCI, STATE UNIV NY, EMPIRE STATE COL, 77-; RES PROF, DEPT PHYSICS, STATE UNIV NY, ALBANY, 78- *Concurrent Pos:* Ed, Atomic Data & Nuclear Data Tables, 82- *Mem:* Am Phys Soc. *Res:* Atomic inner shell ionization; microbeam analysis; interaction of nuclei with electromagnetic fields in solids. *Mailing Add:* Dept Physics State Univ NY Albany NY 12222

LISCUM, LAURA, b Boston, Mass, Sept 1, 54. PHYSIOLOGY. *Educ:* Hunter Col, BA, 76; Columbia Univ, MA, 78, PhD, 82. *Prof Exp:* Fac fel, Dept Biol Sci, Columbia Univ, 76-82; postdoctoral fel, Dept Molecular Genetics, Health Sci Ctr, Univ Tex, Dallas, 82-85; ASST PROF, DEPT PHYSIOL, SCH MED, TUFTS UNIV, 85- *Concurrent Pos:* Am Heart Asn estab investr, 87-92. *Honors & Awards:* John S Newberry Prize, 82. *Res:* Author of numerous publications. *Mailing Add:* Dept Physiol Sch Med Tufts Univ 136 Harrison Ave Boston MA 02111

LISELLA, FRANK SCOTT, b Lancaster, Pa, Aug 11, 36; m 58. PUBLIC HEALTH. *Educ:* Millersville State Univ, Millersville, 57; Tulane Univ, MPH, 61; Univ Iowa, PhD(prev med), 70. *Prof Exp:* Sanitarian, Pa Dept Health, 57-64; coordr, Commun Dis Control Proj, USPHS, Fla, 64-66, chief training & consultation, Pesticides Prog, Nat Commun Dis Ctr, 66-68; asst to dir, div community studies, Food & Drug Admin, Ga, 69-70; asst dir, div pesticide community studies, Environ Protection Agency, 70-72; health sci adv, Nat Med Audiovisual Ctr, 72-73; chief, prog develop br, Environ Health Servs Div, Ctr Dis Control, 73-81, asst dir, Chronic Dis Div, 81-84, dir, off biosafety, Ctr Dis Control, Ga, 84-87; HEAD BIOSAFETY OFF & LEADER, HAZARDOUS MAT GROUP, GA TECH RES INST, 87- *Concurrent Pos:* Adj prof, Dekalb Col, 70-75. *Mem:* Nat Environ Health Asn; Am Pub Health Asn; Am Biosafety Asn; Can Asn Biol Safety. *Res:* Epidemiology of acute intoxications involving chemical agents of various types; etiology of self-induced intoxications involving medicants, pesticides and other chemical compounds and measures for prevention of repetitive episodes; control/containment of hazardous chemicals & biologicals. *Mailing Add:* 1777 Mountain Shadow Stone Mountain GA 30087

LISENBEE, ALVIS LEE, b Lamesa, Tex, Dec 3, 40; m 68; c 2. STRUCTURAL GEOLOGY. *Educ:* Univ NMex, BS, 64, MS, 67; Pa State Univ, PhD(geol), 72. *Prof Exp:* Asst geologist, Ark & La Gas Co, 64; chief geologist, Posora Mining Co, Esfahan, Iran, 73-76; dept head, 78-85, PROF GEOL, DEPT GEOL ENG, SDAK SCH MINES, 72- *Concurrent Pos:* Consult, Armco Steel Corp, 79-81, Gulf Oil Corp, 81-82, Exxon Mineral Co, 82, SDak Geol Surv, 82 & Turkish Nat Petrol Co, 85-89; Fullbright prof, Turkey, 83. *Mem:* Geol Soc Am; Sigma Xi. *Res:* The evolution of mountain systems, specifically the timing and types of geological features which evolved in the northern Rocky Mountains of the US and the Taurus Mountain range of southeastern Turkey. *Mailing Add:* Dept Geol Eng SDak Sch Mines Rapid City SD 57701

LISH, PAUL MERRILL, pharmacology; deceased, see previous edition for last biography

LISK, DONALD J, b Buffalo, NY, May 12, 30; m 59; c 4. OCCUPATIONAL EPIDEMIOLOGY, EFFECTS OF SOLID WASTES IN AGRICULTURE. *Educ:* Univ Buffalo, BA, 52; Cornell Univ, MS, 54, PhD (soil chem), 56. *Prof Exp:* PROF TOXICOL & DIR, TOXIC CHEMICALS LAB, NY STATE COL AGR, CORNELL UNIV, 56- *Mem:* Soc Toxicol; Am Chem Soc. *Res:* Fate of toxicants in agriculture and environmental systems; heavy metals; pesticides; industrial toxicants. *Mailing Add:* Toxic Chemicals Lab NY State Col Agr Cornell Univ Tower Rd Ithaca NY 14853-7401

LISK, ROBERT DOUGLAS, b Pembroke, Ont, Nov 10, 34. PHYSIOLOGY, ENDOCRINOLOGY. *Educ:* Queen's Univ, Ont, BA, 57; Harvard Univ, AM, 59, PhD(biol), 60. *Prof Exp:* From instr to prof biol, 60-90, dir prog, neurosci, 82-88, EMER PROF BIOL, PRINCETON UNIV, 90- *Concurrent Pos:* NSF grants, 60-66, 73-85 & 86-; vis asst prof, Sch Med, Univ Calif, Los Angeles, 65; NIH grants, 66-71; mem panel regulatory biol, Nat Sci Found, 72-75. *Mem:* Fel AAAS; Am Soc Zool; Am Asn Anat; Endocrine Soc; Am Physiol Soc. *Res:* Neuroendocrine mechanisms such as sites of action and interactions of hormones on central nervous system for regulation of hormone output by endocrine organs and hormone role in differentiation and triggering of behavioral sequences. *Mailing Add:* Dept Biol Princeton Univ Princeton NJ 08544

LISKA, BERNARD JOSEPH, b Hillsboro, Wis, May 31, 31; m 52; c 2. FOOD TECHNOLOGY. *Educ:* Univ Wis, BS, 53, MS, 56, PhD(dairy & food technol), 57. *Prof Exp:* Asst prof, Univ Fla, 57-59; from asst prof to assoc prof animal sci, 59-65, dir, Food Sci Inst, 68-75, assoc dir, Agr Exp Sta, 72-75, dir & assoc dean, 75-80, PROF FOOD SCI, PURDUE UNIV, 65-, DEAN AGR, 80- *Concurrent Pos:* Sci ed, J Food Sci, 70-80; vchmn, Expert Panel on Food Safety & Nutrit, 71-77. *Honors & Awards:* Babcock Hart Award Jury, Inst Food Technologists, 69-73. *Mem:* Fel Inst Food Technologists; Am Dairy Sci Asn; Am Chem Soc; Sigma Xi. *Res:* Food bacteriology; lactic cultures; bulk handling of milk; milk quality and enzymes; food chemistry; food microbiology; chemical residues in food; pesticide residue analysis. *Mailing Add:* Purdue Univ Smith Hall West Lafayette IN 47906

LISKA, KENNETH J, b Hinsdale, Ill, June 4, 29; m 57; c 3. CHEMISTRY, PHARMACOLOGY. *Educ:* Univ Ill, BS, 51, MS, 53, PhD(med chem), 56. *Prof Exp:* Assoc prof pharmaceut chem, Duquesne Univ, 56-61 & Univ Pittsburgh, 61-69; assoc prof chem, US Int Univ, 69-75; PROF CHEM, MESA COL, 75- *Concurrent Pos:* Chmn, San Diego Sect, Am Chem Soc 75-76. *Mem:* Am Chem Soc; AAAS. *Res:* Synthetic organic medicinal chemistry; author of five books. *Mailing Add:* 2947 Honors Ct San Diego CA 92122

LISKAY, ROBERT MICHAEL, b Apr 16, 48; US citizen. GENETIC RECOMBINATION, SOMATIC CELL GENETICS. *Educ:* Univ Calif, Irvine, BS, 70; Univ Wash, Seattle, PhD(genetics), 74. *Prof Exp:* Res assoc, Univ Colo, 77-80; asst prof, 80-84, ASSOC PROF THERAPEUT RADIOL & HUMAN GENETICS, SCH MED, YALE UNIV, 84- *Concurrent Pos:* Scholar, Leukemia Soc Am, 84- *Mem:* Genetics Soc Am. *Res:* Homologous recombination in mammalian cells, its mode of action and cellular processes that it influences; X-chromosome inactivation in mammals. *Mailing Add:* Dept Therapeut Radiol Yale Sch Med 333 Cedar St New Haven CT 06510

LISKEY, NATHAN EUGENE, b Live Oak, Calif, Apr 26, 37; m 57; c 2. HEALTH SCIENCE. *Educ:* La Verne Col, BA, 59; Ind Univ, Bloomington, MS, 61, HSD(health safety), 69. *Prof Exp:* Teacher pub schs, Calif, 59-65; from asst prof to assoc prof, 65-75, PROF HEALTH SCI, CALIF STATE UNIV, FRESNO, 75- *Concurrent Pos:* USPHS grant, HEW, 68-69; sex therapist, Ctr Coun & Ther, 73- *Mem:* Soc Sci Study Sex; Am Asn Sex Educ & Coun; Nat Coun for Int Health. *Res:* Physical and emotional aspects of behavior relating to accident prevention; human sexuality; sexual behavior of the aged. *Mailing Add:* Dept Health Sci Calif State Univ Fresno Fresno CA 93740

LISKOV, BARBARA H, b Los Angeles, Calif, Nov 7, 39; m 70; c 1. SOFTWARE SYSTEMS. *Educ:* Univ Calif Berkeley, BA, 61; Stanford Univ, MS, 65, PhD(philosophy), 68. *Prof Exp:* Programmer, Mitre Corp, 61-62, Harvard Univ, 62-63; res asst, Stanford Univ, 63-68; tech staff, Mitre Corp, 68-72; PROF COMPUT SCI, MASS INST TECHNOL, 72- *Mem:* Nat Acad Eng; Asn Comput Mach; Inst Elec & Electronics Engrs. *Res:* Programming methodology; distributed computing; programming languages; operating systems; numerous articles, papers and publications. *Mailing Add:* Lab Comput Sci 545 Technol Sq Cambridge MA 02139

LISMAN, FREDERICK LOUIS, b Wilkes-Barre, Pa, Jan 14, 39; m 62; c 3. NUCLEAR CHEMISTRY. *Educ:* Fairfield Univ, BS, 60; Purdue Univ, PhD(nuclear chem), 65. *Prof Exp:* Sr res radiochemist, Idaho Nuclear Corp, 65-70; asst prof, 70-72, chmn dept, 75-79, ASSOC PROF CHEM, FAIRFIELD UNIV, 72- *Mem:* AAAS; Am Chem Soc. *Res:* Fission yield determination; radiochemical separations; mass spectrometric techniques; measurement of fissionable material; mass and charge distribution in low Z fission; chemical separation techniques; energy resources. *Mailing Add:* 353 W Rutland Rd Milford CT 06460

LISMAN, HENRY, b Boston, Mass, July 3, 13; m 38; c 2. MATHEMATICS. *Educ:* Univ Boston, BS, 34, MS, 35, PhD(physics), 39. *Prof Exp:* Asst physics, Univ Boston, 34-35; instr, Northeastern Univ, 40-42; from assoc physicist to physicist, Sig Corps Eng Labs, NJ, 42-47; from instr to assoc prof, Yeshiva Univ, 47-57, prof math, 57-78; RETIRED. *Concurrent Pos:* Consult physicist, US Army Electronics Labs, 49-68. *Res:* Electromagnetic wave propagation. *Mailing Add:* 3777 Independence Ave Bronx NY 10463

LISMAN, PERRY HALL, b Sweetwater, Tex, July 21, 32. RESEARCH ADMINISTRATION, SYSTEMS DEVELOPMENT & ELECTRONICS SYSTEMS. *Educ:* Univ Tex, BA, 61. *Prof Exp:* Res scientist, Defense Res Lab, Univ Tex, Austin, 61-67; sr staff mem, Appl Physics Lab, Johns Hopkins Univ, 67-74; DEP DIR, SYSTS DEVELOP DIV, SRI INT, 74- *Mem:* Sr mem Inst Elec & Electronics Engrs. *Res:* Foreign technology; electronics systems; countermeasures; research and development management; radar. *Mailing Add:* SRI Int 333 Ravenswood Ave Menlo Park CA 94025

LISONBEE, LORENZO KENNETH, b Mesa, Ariz, Nov 25, 14; m 38; c 8. BIOLOGY, SCIENCE EDUCATION. *Educ:* Ariz State Univ, BA, 37, MA, 40, EdD, 63. *Prof Exp:* Teacher sci & dept chmn high schs, Ariz, 40-58; SCI SUPVR, PHOENIX HIGH SCHS, 58-; FAC ASSOC, ARIZ STATE UNIV, 63- *Concurrent Pos:* Consult, Am Geol Inst & Am Inst Biol Sci & Biol Sci Curriculum Study; vis prof, San Jose State Univ, 56; pres, Ariz Sci Teachers Asn, 56, Ariz Acad Sci, 63-64; contribr, Encyclopedia Britannica, 62 & 74; chmn, Ariz Comt Corresp, 84-86. *Mem:* Fel AAAS; Nat Asn Res Sci Teaching; Nat Sci Teachers Asn; Nat Asn Biol Teachers. *Res:* Research in science teaching; desert biology. *Mailing Add:* 2146 E Waston Dr Tempe AZ 85283

LISS, ALAN, b Pittsburgh, Pa, Sept 14, 47; m 71; c 3. MICROBIOLOGY, ENVIRONMENTAL EARTH & MARINE SCIENCES. *Educ:* Univ Calif, Berkeley, BS, 69; Univ Rochester, PhD(microbiol), 73. *Prof Exp:* Fel microbiol, York Univ, 73-74; Nat Cancer Inst fel, Scripps Clin & Res Found, 74-75; asst prof biol, Univ Conn, 75-77; sr staff fel, NIH, 77-79, expert-consult, Nat Inst Allergy & Infectious Dis, Rocky Mountain Labs, 79-82; asst prof biol, State Univ NY, Binghampton, 82-89; res dir, Ecol Eng Assocs, 89-91; PRES, IDEA INC, 91- *Honors & Awards:* Sigrid Juselius Found Award, 75. *Mem:* AAAS; Am Soc Microbiol; Sigma Xi; NY Acad Sci. *Res:* Innovative environmentally appropriate bioremediation methods; mutants; environmental impact analysis. *Mailing Add:* Idea Inc 22 Forest St Providence RI 02906

LISS, IVAN BARRY, b Lebanon, Ky, June 21, 38; m 77; c 2. INORGANIC CHEMISTRY. *Educ:* Georgetown Col, BA, 60; Univ Ky, MA, 63; Univ Louisville, PhD(chem), 73. *Prof Exp:* Teacher chem, Shelby Co High Sch, Ky, 60-65; chemist, Reliance Universal, Inc, 65-67; fel, Univ Mo-Columbia, 73-74; MEM FAC CHEM, BLACKBURN COL, 74- *Mem:* Am Chem Soc; Sigma Xi. *Res:* Synthesis and characterization of chelates of the rare earth and transition metals. *Mailing Add:* 445 Mulberg Dr Christiansburg VA 24073

LISS, LEOPOLD, b Lwow, Poland, Nov 19, 23; nat US; m 48; c 2. NEUROPATHOLOGY. *Educ:* Lwow Gramar Sch, Poland, BA, 41; Univ Heidelberg, MD, 50; Univ Mich, MS, 55. *Prof Exp:* From instr to asst prof neuropath, Univ Mich, 51-60; assoc prof, 60-64, PROF NEUROPATH, OHIO STATE UNIV, 64-, CO-DIR OFF GERIAT MED, COL MED, 77- *Mem:* Am Asn Neuropath; Soc Neuroscience; Int Acad Path; Am Geriat Soc; Asn Res Nerv & Ment Dis; Sigma Xi. *Res:* Clinical and experimental neuropathology; aging brain; Dementia; Alzheimer's disease; alcoholic encephalopathies; aluminum neurotoxicity. *Mailing Add:* 2124 Chardon Rd Columbus OH 43220

LISS, MAURICE, b Boston, Mass, Dec 18, 26. BIOCHEMISTRY, BIOLOGY. *Educ:* Harvard Univ, AB, 49; Tufts Univ, PhD(biochem), 58. *Prof Exp:* Chemist, Peter Bent Brigham Hosp, 49-51; chemist, Mass Dept Pub Safety, 51-53; Am Cancer Soc res fel, enzymol, Brandeis Univ, 58-60; res assoc dermat, Sch Med, Tufts Univ, 61-63, from asst prof to assoc prof, 63-68; assoc prof, 68-73, PROF BIOL, BOSTON COL, 73- *Concurrent Pos:* vis prof, Dept Immunol, Hadassah Med Sch, Jerusalem, 82. *Mem:* Am Chem Soc; Am Soc Biol Chem; AAAS; Am Soc Microbiol. *Res:* Proteins; amino acid metabolism. *Mailing Add:* Dept Biol Boston Col 140 Commonwealth Ave Chestnut Hill MA 02167

LISS, ROBERT H, b Boston, Mass, Nov 2, 36; m 61; c 2. CYTOLOGY, CYTOPATHOLOGY. *Educ:* Tufts Univ, BS, 59; Univ Mass, MA, 61, PhD(cytol, biochem), 64. *Prof Exp:* Head electron micros lab life sci div, 65-76, DIR EXP CELLULAR SCI, ARTHUR D LITTLE, INC, 76-; ASSOC SURG, HARVARD MED SCH, 70-; RES ASSOC CARDIOVASC SURG, CHILDREN'S MED CTR, BOSTON, 75- *Concurrent Pos:* Vis scientist, Tex Heart Inst, 72-; res assoc mat sci & eng, Mass Inst Technol, 77- *Mem:* Am Asn Cancer Res; AAAS; Am Soc Cell Biol; Electron Micros Soc Am; Soc Develop Biol. *Res:* Radioautography; electron microscopy. *Mailing Add:* West Roxbury Vet Admin Ctr 1400 UFW Pkwy West Roxbury MA 02132

LISS, WILLIAM JOHN, b Pittsburgh, Pa, June 18, 47. POPULATION & COMMUNITY ECOLOGY. *Educ:* Pa State Univ, BS, 69; Ore State Univ, MS, 74, PhD(fisheries), 77. *Prof Exp:* Res assoc, 77-78, asst prof, 78-85, ASSOC PROF FISHERIES, ORE STATE UNIV, 85- *Mem:* Sigma Xi. *Res:* Population and community ecology of aquatic and terrestrial organisms; fisheries exploitation theory; effects of toxic substances on aquatic communities; watershed and stream classification. *Mailing Add:* Dept Fisheries & Wildlife Oak Creek Lab Biol Oak Creek Lab Biol Ore State Univ Corvallis OR 97331

LISSAMAN, PETER BARRY STUART, b Durban, SAfrica, Apr 10, 31; US citizen; m 55; c 3. AERODYNAMICS. *Educ:* Univ Natal, BS, 51; Cambridge Univ, MA, 54; Calif Inst Technol, MS, 55, PhD(aeronaut), 66. *Prof Exp:* Designer & struct analyst, Bristol Aircraft Co, Eng, 55-56; res aerodynamicist, Handley-Page Aircraft, Eng, 56-58; asst prof aeronaut, US Naval Postgrad Sch, 58-62, Calif Inst Technol, 62-69 & Jet Propulsion Lab, 68-69; dir continuum mech lab, Northrop Corp, 69-71; VPRES, AEROVIRONMENT INC, 72- *Concurrent Pos:* Consult, McDonnell Douglas Corp, 65-68; distinguished lectr, Am Inst Aeronaut & Astronaut, 72-79; prof, Art Ctr of Design, 76-; Nat Lectr, Sigma Xi Soc, 86-89. *Honors & Awards:* Longstreth Medal, Franklin Inst, 79; Kremer Medal, Royal Aeronaut Soc, 79. *Mem:* Fel Am Inst Aeronaut & Astronaut; Soc Exp Test Pilots. *Res:* Aerodynamics, hydrodynamics, structure and dynamics of aircraft; marine and ground vehicles; wind and marine turbines; automotive aerodynamics; wing, rotor theory; turbulence, diffusion, plume modelling; energy systems; bird flight. *Mailing Add:* AeroVironment Inc 222 E Huntington Dr Monrovia CA 91017-7131

LISSANT, ELLEN KERN, b St Louis, Mo, Nov 4, 22; m 47; c 3. PHYCOLOGY, ENVIRONMENTAL SCIENCE. *Educ:* Wash Univ, St Louis, AB, 44, AM, 46, PhD(bot), 68. *Prof Exp:* Lab instr bot, Wash Univ, St Louis, 43-45; teacher, Webster Groves High Sch, 45-46; asst bot, Wash Univ, St Louis, 46-47; asst herbarium, Stanford Univ, 47; from lectr to prof biol, Fontbonne Col, 60-78; RETIRED. *Concurrent Pos:* Fac assoc, Washington Univ, 79-81; instr, St Louis Community Col, Meramec, 81-85. *Mem:* Bot Soc Am; Phycol Soc Am; Sigma Xi; Int Phycol Soc. *Res:* Palaeobotany; genetics; morphogenetic studies in the geus Erythrocladia Rosenvinge. *Mailing Add:* Rte 1 Box 251 A Clever MO 65631

LISSANT, KENNETH JORDAN, b London, Eng, Aug 6, 20; nat US; m 47; c 3. COLLOID CHEMISTRY. *Educ:* Ottawa Univ, Kans, AB, 41; Wash Univ, St Louis, MS, 43; Stanford Univ, PhD(chem), 47. *Prof Exp:* Asst chem, Ottawa Univ, Kans, 39-41; asst chem, Wash Univ, St Louis, 41-44, instr physics, 43-44; res chemist, Petrolite Corp, 44-65, advan res coordr, 65-68, dir advan res, 68-80; RETIRED. *Concurrent Pos:* Bristol-Myers fel; consult, 85- *Mem:* AAAS; Am Chem Soc. *Res:* Solubilization of liquids; polymerization of unsaturates; foams; surfactants; emulsions; information retrieval; pollution abatement; glyphs. *Mailing Add:* RT 1 Box 251A Clever MO 65631

LISSAUER, DAVID ARIE, b Haifa, Israel, Mar 23, 45; m 68; c 3. ELEMENTARY PARTICLE PHYSICS. *Educ:* Univ Calif, Berkeley, BA, 66, MA, 68, PhD(physics), 71. *Prof Exp:* Lectr physics, Tel Aviv Univ, 72-74, sr lectr, 75-77, assoc prof, 79-81; res assoc, Argonne Nat Lab, 74-75, European Nuclear Res Orgn (CERN), 77-78 & 81-82; PHYSICIST, BROOKHAVEN NAT LAB, 85- *Concurrent Pos:* Prof, Tel-Aviv Univ, 87. *Mem:* Am Phys Soc. *Res:* Lepton production in high energy interactions; relativistic heavy ion collisions. *Mailing Add:* Brookhaven Nat Lab 510 A Upton NY 11973

LISSNER, DAVID, b Rochester, NY, July 25, 31. MATHEMATICS. *Educ:* Mass Inst Technol, BS, 53; Cornell Univ, PhD(math), 59. *Prof Exp:* Design engr, NAm Aviation, Inc, 53-55; Off Naval Res fel math, Northwestern Univ, 59-60; instr, Yale Univ, 60-62; from asst prof to assoc prof, 62-77, PROF MATH, SYRACUSE UNIV, 77- *Mem:* Am Math Soc. *Res:* Ring theory; linear, commutative and homological algebra; algebraic geometry. *Mailing Add:* Dept Math Syracuse Univ Syracuse NY 13244

LIST, ALBERT, JR, b East Orange, NJ, Nov 5, 28; m 53; c 2. PLANT PHYSIOLOGY. *Educ:* Univ Mass, BS, 53; Cornell Univ, MS, 58, PhD(plant physiol), 61. *Prof Exp:* Instr bot, Douglass Col, Rutgers Univ, 61-62, asst prof bot & biol, 62-65; fel, Univ Pa, 65-66, lectr biomet & bot, 66-67; assoc prof biol, 67-74, ASSOC PROF BIOL SCI, DREXEL UNIV, 74- *Concurrent Pos:* NSF res grants, 63-65, 66-67 & 70-72; USPHS grant, 70-73. *Mem:* AAAS; Bot Soc Am; Am Soc Plant Physiol; Am Inst Biol Sci; Soc Develop Biol. *Res:* Developmental botany; control theory for plant root growth; relationships of relative elemental growth rates to bioelectric and membrane properties in roots; air pollution effects on growth. *Mailing Add:* Dept of Biol Sci Drexel Univ 32nd & Chestnut Sts Philadelphia PA 19104

LIST, HARVEY L(AWRENCE), b Brooklyn, NY, Sept 5, 24; m 46; c 2. CHEMICAL ENGINEERING. *Educ:* Polytech Inst Brooklyn, BChE, 50, DChE, 58; Univ Rochester, MS, 50. *Prof Exp:* Process engr chem eng, Esso Res & Eng Co, 50-55; assoc prof, 55-71, prof, 71-80, EMER PROF CHEM ENG, CITY COL NEW YORK, 80-; PRES, LIST ASSOCS, INC, 69- *Concurrent Pos:* Private consult, 55-69; Fulbright scholar, Tunghai Univ, 63-64; tech coordr, Adhesive & Sealant Coun, Ill; ed, Int Petrochem Develop; pres, Rickian Inc, 81- *Mem:* Am Chem Soc; Am Soc Eng Educ; Am Inst Chem Engrs. *Res:* Fluidization of solids; petroleum refining; chemical process economics; international relations; risk analysis. *Mailing Add:* 3501 S Ocean Blvd No 103 Palm Beach FL 33480-5951

LIST, JAMES CARL, b Paducah, Ky, July 6, 26; m 47; c 2. HERPETOLOGY. *Educ:* Notre Dame Univ, BS, 48, MS, 49; Univ Ill, PhD(zool), 56. *Prof Exp:* From instr to asst prof biol, Loyola Univ, Ill, 52-57; from instr to assoc prof, 57-66, prof biol, 66-88, EMER PROF BIOL, BALL STATE UNIV, 88- *Mem:* Am Soc Ichthyol & Herpet; Herpetologists League; Soc Study Amphibians & Reptiles; Sigma Xi. *Res:* Anatomy and ecology of amphibians and reptiles. *Mailing Add:* 7522 W Bethel Muncie IN 47304

LIST, ROLAND, b Frauenfeld, Switz, Feb 21, 29; m 56; c 2. ATMOSPHERIC PHYSICS. *Educ:* Swiss Fed Inst Technol, Dipl phys. *Hon Degrees:* DSc, Nat Swiss Fed Inst Technol, 60. *Prof Exp:* Sect head atmospheric ice formation, Swiss Fed Inst Snow & Avalanche Res, 52-63; prof meteorol, Univ Toronto, 63-82, assoc chmn dept physics, 69-73; dep secy-gen, World Meteorol Orgn, Geneva, Switz, 82-84; PROF METEOROL, UNIV TORONTO, 84- *Concurrent Pos:* Chmn working groups cloud physics & weather modifications, World Meteorol Orgn, 69-82; chmn comt meteorol & atmospheric sci, Nat Res Coun; vis prof, Swiss Fed Inst Technol, 74; bd dir, US Nat Ctr Atmospheric Res, Univ Corp Atmospheric Res, 75-78; mem shuttle sci coun, Univ Space Res Asn, 78-81. *Honors & Awards:* Medal, Univ Leningrad, 70; Patterson Medal, 79. *Mem:* Fel Am Meteorol Soc; Am Geophys Union; Can Asn Physicists; Royal Meteorol Soc; Can Meteorol Soc; Can Geophys Soc; Swiss Phys Soc; Can Acad Sci; fel Royal Soc Can. *Res:* Precipitation physics; cloud dynamics; weather modification. *Mailing Add:* Dept Physics Univ Toronto Toronto ON M5S 1A7 Can

LISTER, CHARLES ALLAN, b Trenton, NJ, Nov 15, 18; m 46; c 3. ELECTRICAL ENGINEERING, ELECTRONICS ENGINEERING. *Educ:* Tufts Univ, BS, 40; Case Western Reserve Univ, MS, 51. *Prof Exp:* Test engr, Gen Elec Co, NY, 40-41, design engr, 41-43, appln engr, 46-47; asst prof elec eng, Swarthmore Col, 47-49; develop engr, Elec Controller & Mfg Div, Square D Co, Cleveland, 49-54, asst supvr, 54-56, supvr new prod develop, 56-62; design specialist, Lockheed Missiles & Space Co, Calif, 62-63, mgr test equip eng, 63-64; mgr spec devices eng, Lockheed Aircraft Serv Co, 64-65; dept head, Res & Develop, Otis Elevator Co, 65-67, mgr prod eng, 67-72, asst to vpres, 72-74, mgr proj admin, 74-75; mgr eng serv, 76-83, sr staff engr, Square D Co, Columbia, SC, 83-84; CONSULT ENGR, 84- *Mem:* Inst Elec & Electronics Engrs; Sigma Xi. *Res:* Electromechanical and electronic systems and devices; high-voltage contactors; electric brakes; lifting magnets; elevator dispatching computer systems; elevator control systems; arc interruption; standards for industrial controls. *Mailing Add:* 3215 Gulf Shore Blvd N Apt 511 Naples FL 33940

LISTER, CLIVE R B, b Uxbridge, Eng, Feb 8, 36. HEAT FLOW, GEODYNAMICS. *Educ:* Cambridge Univ, BA, 59, PhD(geophys), 62, ScD(geophys), 84. *Prof Exp:* Consult geophys, Saclant ASW Res Ctr, 62-63; chief geophysicist, Ocean Sci & Eng Co, 63-65; res asst prof oceanog, 65-68, from asst prof to prof, 68-90, EMER PROF OCEANOG & GEOPHYS, UNIV WASH, 90- *Concurrent Pos:* Mem, Joides Heat Flow Comt, 67-74; Mars Penetrator Sci Comt, 76-77; Williams Evans vis prof, Univ Otago, NZ, 89. *Honors & Awards:* Judd H & Cynthia S Qualline lectr, Univ Tex Austin, 87. *Mem:* Am Geophys Union; Brit Astron Asn. *Res:* Measurement of heat flow through the ocean floor; acoustic sub-bottom profiling; thermal theory applied to geodynamics; aspects of porous convection and water penetration applied to geothermal problems; physics of convection in porous media. *Mailing Add:* Sch Oceanog WB-10 Univ Washington Seattle WA 98195

LISTER, EARL EDWARD, b Harvey, NB, Apr 14, 34; m 56, 83; c 3. RUMINANT NUTRITION. *Educ:* McGill Univ, BS, 55, MS, 57; Cornell Univ, PhD(nutrit), 60. *Prof Exp:* Feed nutritionist, Ogilvie Flour Mills Ltd, 60-65; res scientist, Ottawa, Agr Can, 65-75, dep dir, Animal Res Inst, 75-78, prog specialist, Cent Region, Ottawa, 78-80, dir gen, Atlantic Region Res Br, 80-85, dir gen, Pesticides Directorate, Food Prod & Inspection Br, 85-87, DIR ANIMAL RES CTR, AGR CAN, OTTAWA, 87- *Concurrent Pos:* Assoc ed, Can J Animal Sci, 73-75, ed, 75-78. *Mem:* Can Soc Animal Sci; Am Dairy Sci Asn; Am Soc Animal Sci; Agr Inst Can; Asn Advan Sci Can. *Res:* Nutritional requirements of immature ruminants; feeding and management systems for beef cows and calves; production of beef from dairy breeds of cattle. *Mailing Add:* Dir Animal Res Ctr Ontario ON K1A 0C6 Can

LISTER, FREDERICK MONIE, b Trenton, NJ, May 9, 23; m 54; c 3. MATHEMATICS. *Educ:* Tufts Univ, BS, 47; Univ Mich, MA, 51; Univ Utah, PhD(math), 66. *Prof Exp:* Instr math, Phillips Acad, Mass, 47-49; instr, Western Wash Col Educ, 54-56, from asst prof to assoc prof, 58-67; asst prof, Chico State Col, 57-58; prof, Southern Ore Col, 67-68; assoc prof, 68-69, PROF MATH, CENT WASH STATE COL, 69- *Mem:* Am Math Soc; Math Asn Am. *Res:* Geometric topology; embeddings of 2-spheres in Euclidean 3-space. *Mailing Add:* Cent Wash Univ Ellensburg WA 98926

LISTER, MARK DAVID, b Kansas City, Mo, Aug 12, 53; m; c 1. EXPERIMENTAL BIOLOGY. *Educ:* William Jewell Col, BA, 75; Univ Mo, PhD(chem), 85. *Prof Exp:* Postdoctoral fel lipid enzymol, Univ Calif, San Diego, 85-88; SCIENTIST LIPID METAB, SPHINX PHARMACEUT CO, 89- *Mem:* Am Chem Soc; AAAS; Am Soc Biochem & Molecular Biol. *Res:* Protein purification; enzymatic characterization of phospholipases including development of assays; kinetics; inhibitor studies; chemical and enzymatic synthesis or modification of lipids; lipid second messengers in cell signal transduction. *Mailing Add:* Sphinx Pharmaceut Co Two University Pl PO Box 52330 Durham NC 27717

LISTER, MAURICE WOLFENDEN, b Tunbridge Wells, Eng, Mar 27, 14; nat Can; m 40; c 5. INORGANIC CHEMISTRY. *Educ:* Oxford Univ, PhD, 38, MA, 47. *Prof Exp:* Assoc prof, 53-62, PROF CHEM, UNIV TORONTO, 62- *Res:* Complex inorganic compounds; mechanisms of inorganic reactions; magnetic susceptibilities; thermodynamics of solids. *Mailing Add:* 20 Burnham Rd Toronto ON M4G 1C1 Can

LISTER, RICHARD MALCOLM, b Sheffield, Eng, Nov 14, 28; m 53; c 4. PLANT VIROLOGY. *Educ:* Sheffield Univ, BSc, 49, dipl ed, 50; Cambridge Univ, dipl agr sci, 51; Imp Col Trop Agr, Trinidad, dipl, 52; St Andrews Univ, PhD, 63. *Prof Exp:* Plant pathologist, WAfrican Cocoa Res Inst, 52-56 & Scottish Hort Res Inst, 56-66; assoc prof, 66-72, PROF PLANT VIROL, PURDUE UNIV, WEST LAFAYETTE, 72- *Concurrent Pos:* Fel bot & plant path, Purdue Univ, 63-64; res grants, NSF & USDA. *Honors & Awards:* Ruth Allen Award, Am Phytopathological, 86. *Mem:* Fel Am Phytopathological Soc; Brit Asn Appl Biol. *Res:* Methods in plant virology; properties and interactions of virus-specific products of virus infections; serological techniques; transmission, purification, properties, and relationships of selected plant viruses, especially of cereals, soybeans and fruit plants. *Mailing Add:* Dept Bot & Plant Path Life Sci Bldg Purdue Univ West Lafayette IN 47907

LISTERMAN, THOMAS WALTER, b Cincinnati, Ohio, Dec 21, 38; m 69; c 2. SOLID STATE SCIENCE. *Educ:* Xavier Univ, BS, 59; Ohio Univ, PhD(solid state physics), 65. *Prof Exp:* Sr res physicist, Mound Lab, Monsanto Res Corp, 65-67; asst prof, 67-72, asst provost, 71-73, asst dean sci & eng, 70-71, ASSOC PROF PHYSICS, WRIGHT STATE UNIV, 72- *Concurrent Pos:* Vis res assoc prof elec eng, Univ Cincinnati, 87-88. *Mem:* Inst Elec & Electronics Engrs; Mat Res Soc. *Res:* Electronic properties of materials; positron annihilation; cryogenics; photoacoustic spectroscopy; semiconductor device physics. *Mailing Add:* Dept Physics Wright State Univ Dayton OH 45435

LISTGARTEN, MAX, b Paris, France, May 14, 35; Can citizen; m 63; c 3. DENTISTRY, PERIODONTOLOGY. *Educ:* Univ Toronto, DDS, 59; FRCD(C), 69, Univ Pa, MA, 71. *Prof Exp:* Intern dent, Hosp for Sick Children, Toronto, 59-60; res assoc periodont, Harvard Med Sch, 63-64; from asst prof to assoc prof, Fac Dent, Univ Toronto, 64-68; assoc prof, 68-71, PROF PERIODONT & DIR PERIODONT RES, UNIV PA, 71-, DEPT CHMN, 84- *Concurrent Pos:* Nat Res Coun Can fel periodont, Harvard Med Sch, 60-63; US ed, J Biol Buccale, 72-; oral biol & med study sect, NIH, 80-84; pres, Am Asn Dent Res, 91-92. *Honors & Awards:* Award Basic Res Periodont Dis, Int Asn Dent Res, 73; William J Gies Periodont Award, Am Acad Periodont, 81, Clin Res Award, 87. *Mem:* Fel AAAS; Am Dent Asn; fel Am Acad Periodont; Int Asn Dent Res. *Res:* Ultrastructural investigations of the supporting structures of teeth and associated microbial flora in health and disease. *Mailing Add:* Dept Periodont Sch Dent Med Univ Pa 4001 Spruce St Philadelphia PA 19104

LISTON, RONALD ARGYLE, b Buffalo, NY, Apr 11, 26; m 81; c 9. MECHANICAL ENGINEERING, ENGINEERING MECHANICS. *Educ:* Univ Vt, BS, 49; Univ Mich, MS(mech eng), 58, MS(eng mech), 61; Mich Technol Univ, PhD(eng mech), 73. *Prof Exp:* Proof officer ballistics & automotive, US Army, 50-56, supvr automotive res engr, Tank-Automotive Command, 58-70, res mech engr, Cold Regions Res & Eng Lab, 70-74; res ctr dir, Mich Technol Univ, 74-75; SUPVR RES GEN ENGR, US ARMY COLD REGIONS RES & ENG LAB, 75- *Mem:* Int Soc Terrain-Vehicle Systs (gen secy); Sigma Xi; Am Soc Mech Engrs. *Res:* Off road vehicles; over snow vehicles. *Mailing Add:* US Army Cold Regions PO Box 282 Hanover NH 03775

LISTOWSKY, IRVING, b Vilna, Poland, Dec 21, 35; US citizen; m 63; c 3. BIOCHEMISTRY. *Educ:* Yeshiva Univ, BA, 57; Polytech Inst Brooklyn, PhD(org chem), 63. *Prof Exp:* From instr to assoc prof, 65-78, PROF BIOCHEM, ALBERT EINSTEIN COL MED, 79- *Concurrent Pos:* NIH career develop award, 71-76; sr investr, NY Heart Asn, 67-70. *Mem:* Am Soc Biol Chemists. *Res:* Structure-function relationships of biological substances; iron metabolism, intracellular transport and detoxification mechanisms. *Mailing Add:* Albert Einstein Col Med 1300 Morris Park Ave Bronx NY 10461

LISY, JAMES MICHAEL, b Cleveland, Ohio, Aug 5, 52; m 76; c 2. SPECTROSCOPY, MOLECULAR BEAMS. *Educ:* Iowa State Univ, BS, 74; Harvard Univ, MA, 77, PhD(chem physics), 79. *Prof Exp:* Res assoc, Lawrence Berkeley Lab, 79-81; asst prof, 81-87, ASSOC PROF CHEM, UNIV ILL, 87- *Concurrent Pos:* Alfred P Sloan Res fel, 87-91. *Mem:* Am Chem Soc; Am Phys Soc. *Res:* Structure, bonding and intramolecular energy transfer of small molecular and ion clusters are studied using molecular beam techniques and laser spectroscopy. *Mailing Add:* Sch Chem Sci Univ Ill 175 NL-Box 43 505 S Mathews Urbana IL 61801

LISZT, HARVEY STEVEN, b Newark, NJ, Dec 5, 45; m 73; c 2. RADIO ASTRONOMY, MICROCOMPUTER APPLICATIONS IN DATA PROCESSING. *Educ:* Univ Mass, BS, 67; Princeton Univ, AM, 69, PhD(astron), 74. *Prof Exp:* Res assoc spectros, Princeton Univ Observ, 69-71; asst prof physics, Univ Pittsburgh, 75-76; res assoc, 73-75, assoc scientist, 76-79, SCIENTIST ASTRON, NAT RADIO ASTRON OBSERV, 79-

Concurrent Pos: Res prof, Univ Va, 81- *Mem:* Sigma Xi; Int Astron Union. *Res:* Structure and evolution of interstellar clouds; radiation transport in simple interstellar molecules; structure of the galactic nucleus, interstellar chemistry. *Mailing Add:* Nat Radio Astron Observ Edgemont Rd Charlottesville VA 22903-2475

LIT, ALFRED, b New York, NY, Nov 24, 14; m 47. VISION, EXPERIMENTAL PSYCHOLOGY. *Educ:* Columbia Univ, BS, 38, AM, 43, PhD, 48. *Prof Exp:* Lectr optom, Columbia Univ, 46-48, assoc, 48-49, asst psychol, 49, from asst prof to assoc prof optom, 49-56; res psychologist, Univ Mich, 56-59; head, Human Factors Staff, Systs Div, Bendix Corp, 59-61; res prof, Schnurmacher Inst Vision Resm Col Optom, State Univ NY, 85-86; prof psychol, 61-85, EMER PROF PSYCHOL & MEM ADV BD, EMER COL, SOUTHERN ILL UNIV, 85- *Concurrent Pos:* Res grant, Am Acad Optom, Columbia Univ, 49; mem psychol staff, Off Naval Res Contract, 49-56; lectr, Univ Mich, 57-58; res grants, Eye Inst, USPHS & NSF, 62; mem, Armed Forces-Nat Res Coun Comt Vision; consult, Goodyear Aerospace Corp, Nat Res Coun Comt on Vision, Nat Acad Sci & Spec Study Sect, USPHS; sci referee, Am J Optom, 73. *Honors & Awards:* Sigma Xi-Kaplan Res Award. *Mem:* Fel AAAS; fel Optical Soc Am; fel Am Psychol Asn; hon fel Am Acad Optom; fel NY Acad Sci; fel Soc Eng Psychologists; fel Psychonomic Soc Am; Human Factors Soc; Am Asn Univ Prof; Asn Res Vision & Ophthal. *Res:* Perception; applications of visual psychophysics to ophthalmic clinical practice; spatio-temporal factors influencing visual latency and persistence. *Mailing Add:* Dept Psychol Southern Ill Univ Carbondale IL 62901-6502

LIT, JOHN WAI-YU, b Canton, China, Aug 31, 37; Can citizen; m; c 2. OPTICS. *Educ:* Univ Hong Kong, BSc, 58, dipl Ed, 61. *Hon Degrees:* DSc, Univ Laual, 69. *Prof Exp:* Head physics, Diocesan Boys Sch, 61-64; teacher sci, Quebec High Sch, 64-65; fel optics, Univ Western Ont, 68-69; res assoc, 69-71, from asst prof to assoc prof optics, Univ Laval, 71-77; assoc prof, 77-80, chmn physics dept, 80-86 PROF PHYSICS, WILFRID LAURIER UNIV, 80- *Concurrent Pos:* Consult, various indust & govt labs; assoc ed, Optical Soc Am, 74-79; chmn, Div Optical Physics, Can Asn Physicists, 77-78 & 88-89; adj prof elec eng, Univ Waterloo, 87-, adj prof physics, 80-; mem exec comt, bd dirs, Nat Optics Inst, 88- *Mem:* Can Asn Physicists; fel Optical Soc Am; Inst Elec & Electronic Engrs. *Res:* Fiber and integrated optics; optical sensing; optical instrumentation. *Mailing Add:* Dept Physics & Comput Wilfrid Laurier Univ Waterloo ON N2L 3C5 Can

LITANT, IRVING, organic chemistry, materials science, for more information see previous edition

LITCHFIELD, CARTER, b Pasadena, Calif, Feb 18, 32; m 60. BIOCHEMISTRY. *Educ:* Rensselaer Polytech Inst, BS, 53; Am Inst Foreign Trade, BFT, 57; Tex A&M Univ, PhD(chem), 66. *Prof Exp:* Chemist, Procter & Gamble Co, 53-60; from asst prof to assoc prof lipid biochem, Tex A&M Univ, 60-69; assoc prof lipid biochem, Rutgers Univ, 69-73, from assoc prof to prof biochem, 73-79; WRITER & PUBL, OLEARIUS ED, 79- *Concurrent Pos:* Vis scientist, Fisheries Res Bd Can, 67 & Univ Trondheim, 75 & 79. *Honors & Awards:* Bond Award, Am Oil Chem Soc, 63, 66 & 78. *Mem:* Soc Hist Technol; Am Oil Chem Soc; Soc Indust Archeol. *Res:* Biochemistry of lipids of marine organisms; analysis of natural fat triglyceride mixtures; gas liquid chromatography of lipids; biochemical systematics of lipids; history of lipid biochemistry; history of fats & oils technology. *Mailing Add:* Olearius Editions PO Box H Kemblesville PA 19347

LITCHFIELD, JOHN HYLAND, b Scituate, Mass, Feb 13, 29; m 66; c 1. FOOD SCIENCE, INDUSTRIAL MICROBIOLOGY. *Educ:* Mass Inst Technol, SB, 50; Univ Ill, MS, 54, PhD(food technol), 56. *Prof Exp:* Chief chemist, Searle Food Corp, Fla, 50-51; res food technologist, Swift & Co, Ill, 56-57; asst prof food eng, Ill Inst Technol, 57-60; sr food technologist, 60-61, proj leader bioscience, 61-62, asst chief bioscience res, 62-64, chief biochem & microbiol res, 64-67 & microbiol & environ biol res, 67-68, assoc mgr life sci, Dept Chem Eng, 68-70, mgr biol & med sci sect, Columbus Labs, 70-72, sr tech adv, 73-76, mgr Bioeng & Health Sce Sect, 76-80, prog mgr biol scik 80-81, RES LEADER BIOTECHNOLOGY, COLUMBUS LABS, BATTELLE MEM INST, 81- *Concurrent Pos:* Consult to food indust, 57-60. *Honors & Awards:* Charles Porter Award, Soc Indust Microbiol, 77. *Mem:* Fel AAAS; fel Am Inst Chemists; fel Am Acad Microbiol; fel Soc Indust Microbiol (pres, 70-71); fel Am Pub Health Asn; fel Inst Food Technologists. *Res:* Food processing and preservation; fermentation technology, food, industrial, sanitary and public health microbiology; microbial biochemistry; mass cultivation of microorganisms. *Mailing Add:* Battelle Mem Inst 505 King Ave Columbus OH 43201

LITCHFIELD, WILLIAM JOHN, b Waukegan, Ill, Feb 28, 50; m 72; c 1. BIOCHEMISTRY, CLINICAL CHEMISTRY. *Educ:* Univ Ill, BS, 72, MS, 73; Mich State Univ, PhD(biochem), 76. *Prof Exp:* Fel biophys, Johnson Res Found, Sch Med, Univ Pa, 76-77; RES BIOCHEMIST, INSTRUMENT PRODS DIV, E I DU PONT DE NEMOURS & CO, INC, 77-, Bus Planning Mgr Med Prods Dept. *Concurrent Pos:* NIH fel, 76-77. *Mem:* Am Chem Soc; Am Asn Clin Chem; Reticuloendothelial Soc; Biophys Soc. *Res:* Biophysics; solid state biochemical reactions; free- radical reactions in leukocytes, mitochondria and photosystems; immunology, enzymology, lipid chemistry. *Mailing Add:* 23 Covered Bridge Lane Newark DE 19711-2062

LITCHFORD, GEORGE B, b Long Beach, Calif, Aug 12, 18; m 42; c 2. ELECTRONICS. *Educ:* Reed Col, BA, 41. *Prof Exp:* Head eng sect, Aircraft Radio Dept, Sperry Gyroscope Co, 41-51; asst supvr navig dept, 51-55, head dept aviation systs res, 57-65; HEAD DEPT AVIATION SYSTS CONSULT BUS, Litchord Systs, 65-; GEN PARTNER, LITCHSTREET CO. *Concurrent Pos:* Consult, Dept Transp, Dept Defense, NASA & indust, mem, Radio Tech Comn Aeronaut. *Honors & Awards:* Wright Bros Lect Medal & Citation, Am Inst Aeronaut & Astronaut, 78; Lamme Medal, Inst Elec & Electronics Engrs, 81. *Mem:* Fel Inst Elec & Electronics Engrs; fel Am Inst Aeronaut & Astronaut. *Res:* Inventor of many systems including precision omniranges, allweather landing systems, distance measuring equipment, secondary radar systems; developed navigational and collision avoidance concepts; research into low visibility landing and air traffic control. *Mailing Add:* 32 Cherry Lawn Lane Northport NY 11768

LITHERLAND, ALBERT EDWARD, b Wallasey, Eng, Mar 12, 28; Can citizen; m 56; c 2. NUCLEAR PHYSICS. *Educ:* Univ Liverpool, BSc, 49, PhD(physics), 55. *Prof Exp:* Nat Res Coun Can fel nuclear physics, Atomic Energy Can, Ltd, 53-55, sci officer, 55-60; Atomic Energy Can Ltd vis scientist, Oxford Univ, 60-61, 72-73; sci officer, Atomic Energy Can, Ltd, 61-66; prof, 66-79, UNIV PROF PHYSICS, UNIV TORONTO, 79- *Honors & Awards:* Gold Medal, Can Asn Physicists, 71; Rutherford Medal, Brit Inst Physics, 74; Silver Medal, J Appl Radiation & Isotopes, 80; Guggenheim Fel, 86-87. *Mem:* Fel Am Phys Soc; Royal Soc; Can Asn Physicists; Royal Soc Can. *Res:* Electron induced fission of light elements, accelerator mass spectrometry of 14-C and other isotopes; nuclear spectroscopy of light nuclei using charged particle accelerators; radiative capture of charged particles by nuclei; collective motion in light nuclei; fast neutron spectroscopy. *Mailing Add:* Dept of Physics Univ of Toronto Toronto ON M5S 1A7 Can

LITKE, JOHN DAVID, b Winchester, Mass, May 30, 44; m 66. COMPUTER SCIENCE, PHYSICS. *Educ:* Mass Inst Technol, BS, 65; Johns Hopkins Univ, PhD(physics), 76. *Prof Exp:* Instr physics, Johns Hopkins Univ, 67-75; Bell Lab, 76-80, prin scientist photociruits, 80-84; DEP DIR RES, GRUMMAN DATA SYSTS, 85- *Mem:* Sigma Xi; Am Phys Soc; Inst Elec & Electronic Engrs; Asn Comput Mach. *Res:* Sofware engineering; distributed and fault tolerant systems. *Mailing Add:* 645 Park Ave Huntington NY 11743

LITKE, LARRY LAVOE, b Denver, Colo, June 1, 49; m 69; c 3. ELECTRON MICROSCOPY. *Educ:* Ottawa Univ, BA, 71; Kans State Teachers Col, MS, 73; Univ NDak, PhD(anat), 76. *Prof Exp:* Teaching asst anat, Univ NDak, 73-75, res asst, 75-76, instr, 76, asst prof, 77; instr, Med Col Ohio, 77-78, asst prof anat, 78-82; ASST PROF, ELECTRON MICROSCOPE LAB, HARRINGTON CANCER CTR, AMARILLO, TEX UNIV HEALTH SCIS CTR, LUBBOCK, & TEX TECH REGIONAL ACAD HEALTH CTR, AMARILLO, 82- *Mem:* Sigma Xi; Electron Micros Soc Am; Am Asn Anatomists; AAAS. *Res:* Light and electron microscopy of normal and experimentally induced changes in cell morphology and behavior during early embryonic development; early germ layer formation and cardiovascular and nervous systmes of the chick embryo; development and metastasis of experimentally induced colon carcinoma; ultrastructure of pasteurella haemolytica; ultrastructure of candida albicans. *Mailing Add:* Electron Microscope Lab Harrington Cancer Ctr 1500 Wallace Blvd Amarillo TX 79106

LITMAN, BERNARD, b New York, NY, Oct 26, 20; m 49; c 2. ELECTRICAL ENGINEERING. *Educ:* Columbia Univ, BS, 41, PhD(elec eng), 49; Univ Pittsburgh, MS, 43. *Prof Exp:* Engr mach design, Westinghouse Elec Corp, 41-47; instr elec eng, Univ Pittsburgh, 47; engr fire control & guid, Ambac Industs, Div United Technologies Corp, 48-53, mgr airborne equip, 56-58, res, 59-61, chief engr, 63-83; PRIN SCIENTIST, AVIONICS SYSTS, CULL INC, 83- *Res:* Electromechanical control and computing equipment; weapon control and navigation; inertial guidance; data management systems for aircraft and scientific instruments. *Mailing Add:* Parker Hannifin Corp Gull Electronic Syst Div 70 Corporate Dr Smithtown NY 11787

LITMAN, BURTON JOSEPH, b Boston, Mass, May 8, 35; m 58; c 2. BIOCHEMISTRY, BIOPHYSICS. *Educ:* Boston Univ, BA, 58; Univ Ore, PhD(biophys chem), 66. *Prof Exp:* From asst prof to assoc prof, 68-80, PROF BIOCHEM, SCH MED, UNIV VA, 80-, CHMN DEPT, 84- *Concurrent Pos:* USPHS fel biochem, Sch Med, Univ Va, 66-68. *Mem:* AAAS; Biophys Soc; Am Soc Biol Chemists; Asn Res Vision & Ophthal. *Res:* Structure-function relationships in biological membranes with particular emphasis on the molecular mechanism of vision. *Mailing Add:* Dept Biochem Univ Va Sch Med Charlottesville VA 22908

LITMAN, DAVID JAY, biochemistry, immunochemistry, for more information see previous edition

LITMAN, GARY WILLIAM, b Shoemaker, Calif, June 26, 45; m 70. IMMUNOLOGY, BIOCHEMISTRY. *Educ:* Univ Minn, BA, 67, PhD(microbiol), 72. *Prof Exp:* Res asst microbiol, Univ Minn, 67-68, teaching specialist microbiol & pediat, 68-70, instr pediat & path, 70-72, asst prof path, 72; ASSOC MEM, DEPT MACROMOLECULAR BIOCHEM, SLOAN-KETTERING INST, 72-; ASSOC PROF BIOL, SLOAN-KETTERING DIV, GRAD SCH MED SCI, CORNELL UNIV, 73-; CHMN, DEPT MOLECULAR GENETICS, RES INST, SHOWA UNIV. *Concurrent Pos:* Assoc prof genetics, Sloan-Kettering Div, Grad Sch Med Sci, Cornell Univ, 76-, chmn biol unit, 78-80, assoc prof immunol, 80. *Mem:* AAAS; Am Asn Immunologists; Am Asn Biol Chemists; Am Soc Zoologists; Biophys Soc. *Res:* Evolution of immunoglobulin structure; a typical solubility characteristics of proteins; chemical carcinogenesis; chromosomal proteins. *Mailing Add:* Dept Pediat Univ SFla All Childrens' Hosp 801 Sixth Ave S PO Box 707 St Petersburg FL 33701

LITMAN, IRVING ISAAC, b Chelsea, Mass, Nov 16, 25. FOOD TECHNOLOGY. *Educ:* Univ Mass, BA, 49, MS, 51; Wash State Univ, PhD(food technol), 56. *Prof Exp:* Processed food inspector, Prod & Mkt Admin, USDA, 50-51; food technologist, Gen Prod Div, Qm Food & Container Inst, US Armed Forces, 51-53; asst dairy technologist, Wash State Univ, 53-55; jr res chemist, Univ Calif, 55-56; proj leader, Res Ctr, Gen Foods Corp, 56-62; flavor chemist, Givaudan Corp, 62-64; sect head, Durkee Famous Foods, 64-65; RES DIR, STEPAN FLAVORS & FRAGRANCES, INC, 66- *Mem:* Royal Soc Chem; Inst Food Technologists; AAAS; Sigma Xi; Flavor Chemists Soc. *Res:* Development of synthetic and natural flavorings for food, tobacco and pharmaceuticals. *Mailing Add:* 41 Harriet Dr Princeton NJ 08540-3934

LITMAN, NATHAN, b New York, NY, Nov 22, 46; m 69; c 3. PEDIATRICS, INFECTIOUS DISEASES. *Educ:* Brooklyn Col, BS, 67; Albert Einstein Col Med, MD, 71. *Prof Exp:* Intern, resident & chief resident pediat, Montefiore Hosp & Med Ctr, 71-74; lieutenant comdr pediat, USPHS, 74-76; fel infectious dis, Albert Einstein Col Med, 76-78; ATTEND PEDIAT & INFECTIOUS DIS, MONTEFIORE HOSP & MED CTR, 78- *Mem:* Fel Am Acad Pediat; Infectious Dis Soc Am. *Res:* Infectious etiologies of pediatric diarrhea. *Mailing Add:* Dept Pediat NCent Bronx Hosp Bronx NY 10467

LITOSCH, IRENE, b New York, NY, June 7, 52. PHYSIOLOGICAL CHEMISTRY. *Educ:* New York Univ, Univ Arts & Sci, BA, 74; State Univ NY, Downstate Med Ctr, PhD(pharmacol), 79. *Prof Exp:* FEL RES, SECT PHYSIOL CHEM, BROWN UNIV, 79- *Mem:* NY Acad Sci. *Res:* Mechanism of regulation of intiacularlur calcium. *Mailing Add:* Dept Pharmacol Univ Miami Med Sch PO Box 016189 Miami FL 33101

LITOV, RICHARD EMIL, b New York, NY. PEDIATRIC NUTRITION, TRACE MINERALS. *Educ:* Univ Calif, Davis, BS, 75, PhD(nutrit), 80. *Prof Exp:* Staff res assoc, Med Sch & Primate Res Ctr, Univ Calif, 75-76, pulmonary lab tech clin, Sacramento Med Ctr, 76, res asst res, Davis, 76-80; scientist, 81-85, SR SCIENTIST RES, BRISTOL-MYERS SQUIBB, 85- *Concurrent Pos:* Nutrit consult, Mesker Park Zoo, 86-; reviewer, Am J Clin Nutrit, 86-, Pediat, 90- & USDA Competitive Grants Prog, 90-; sci comt chmn, Evansville Mus Arts & Sci, 87-89, long range planning comt chmn, 90- *Mem:* Inst Food Technologists; Am Soc Clin Nutrit; Am Inst Nutrit. *Res:* Therapeutic biologics; oral rehydration solutions; growth factors; antiinfective formulas; appetite/weight control; mineral bioavailability, fiber and lactoferrin; burn injury; selenium bioavailability and aluminum status. *Mailing Add:* Mead Johnson Res Ctr Bristol-Myers Squibb Evansville IN 47721

LITOVITZ, THEODORE AARON, b New York, NY, Oct 14, 23; m 46; c 2. PHYSICS. *Educ:* Cath Univ, AB, 46, PhD, 50. *Prof Exp:* From asst prof to assoc prof, 50-59, PROF PHYSICS, CATH UNIV AM, 59-, CO-DIR, VITREOUS STATE LAB, 68- *Concurrent Pos:* Consult, Univ Hosp, Georgetown, 50-57. *Mem:* Fel Am Phys Soc; fel Acoust Soc Am; Am Philos Soc. *Res:* Ultrasonic propagation and light scattering in studies of molecular motions in liquids and glasses; development of glasses with unique technical applications. *Mailing Add:* Dept Physics Cath Univ Washington DC 20017

LITSEY, LINUS R, GEOLOGY. *Educ:* Univ Mich, BS, 47; Univ Colo, PhD(geol), 55. *Prof Exp:* Uranium geologist, US Geol Surv, 54-57; geologist, Chevron Oil Co, New Orleans, La, 57-63, dipmeter res, Chevron Res Co, La Habra, Calif, 63-65, dipmeter analyst, Chevron Oil Co, New Orleans, 65-69, digital well log processing, Houston, Tex, 69-72; well log analyst, 72-74, supvr reservoir description unit, Aramco, Arabia, 74-76; formation eval geologist, Chevron Oil Co, Denver, Colo, 76-80; SR CONSULT, SCI SOFTWARE-INTERCOMP. *Mem:* Am Asn Petrol Geologists; Geol Soc Am; Soc Prof Well Log Analysts; Sigma Xi. *Mailing Add:* 6338 S Chase Ct Littleton CO 80123-6818

LITSKY, BERTHA YANIS, b Chester, Pa, Jan 2, 20; m 65; c 2. MICROBIOLOGY, HOSPITAL ADMINISTRATION. *Educ:* Philadelphia Col Pharm, BSc, 42; NY Univ, MPA, 64; Walden Univ, PhD(educ), 74. *Prof Exp:* Head dept bact, Assoc Labs of Philadelphia, 42-44; asst supvr prod, Nat Drug Co, 44-45; res bacteriologist, Univ Pa, 45-50; self-employed, Pa & NY, 50-56; head dept bact, Staten Island Hosp, NY, 56-64; environ microbiol consult, 64-74; NURSE CONSULT, BINGHAM ASSOCS FUND, NEW ENG MED CTR HOSP, 74- *Concurrent Pos:* Mem, Standards Comt, Asn Operating Room Nurses, 75- *Mem:* Am Hosp Asn; Am Soc Microbiol; Am Pub Health Asn; Inst Sanit Mgt; Royal Soc Health. *Res:* Environmental and clinical microbiology; control of cross-infection in hospitals; hospital sanitation, environmental microbiology and administration; antimicrobial agents, antiseptics, disinfectants and germicides; disinfection and sterilization; aseptic practices in the operating room. *Mailing Add:* 9 Kettle Pond Rd Amherst MA 01002

LITSKY, WARREN, b Worcester, Mass, June 10, 24; m 65. BACTERIOLOGY. *Educ:* Clark Univ, AB, 45; Univ Mass, MS, 48; Mich State Univ, PhD(microbiol & biochem), 51. *Hon Degrees:* DSc, Clark Univ, 82. *Prof Exp:* Instr microbiol, Mich State Univ, 51; from asst res prof to res prof bact & pub health, Univ Mass, Amherst, 51-61, Commonwealth prof environ sci & pub health, 61- 90, dir, Inst Agr & Indust Microbiol, 62-84, actg dir, Water Resources Res Ctr, Univ Mass, 68-69, dir, Tech Guid Ctr Indust Environ Control, 68-75, chmn, Dept Environ Sci, 72-85, Environ Health & Sci Progs, 85-86; RETIRED. *Concurrent Pos:* Consult, Ultra High Temperature Pasteurization, USPHS, 61 & Div Water Supply & Pollution Control, 61-67, Div Hosp & Med Facil, HEW, 67-77, biochem aspects watercraft waste disposal devices, Nat Sanitation Found, 67-69, Fed Water Qual Admin, Dept Interior, 67-69, Water Qual Prog, US Environ Protection Agency, 69-84 & Shellfish Prog, US Food & Drug Admin, 73-83; NIH res fel, Oceanog Inst, Fla State Univ, 65; mem task group, Microbiol Criteria Radiation Sterilized Foods, Nat Acad Sci, 65-69, Microbiol Aspects Raw Shucked Oysters, Nat Acad Sci-Nat Res Coun, 67, spec legis comn study recycling solid wastes, State Mass, 70-74, res grant rev panel, US Environ Protection Agency, 71-, lab sect coun, Am Pub Health Asn, 71-74, chmn, lab sect, 75-76, gov coun, Am Soc Microbiol, 72-75, Mass comn nuclear safety, 74-77, US Pharmacopeia expert adv panel microbiol, 76-81, study panel Potomac River Basin, Nat Acad Sci, 76-80 & tech comt, Int Standards Orgn, 76-81; contrib ed, Industrias Lacteus, 67-71; affil prof biol, Clark Univ, 67-82; consult ed, J Environ Health, 70-80; chmn task comt fecal streptococci, 79-81; comt on environ impact statements, Lower Pioneer Valley Planning Comn, 71-; consult, Shellfish Prog, Food & Drug Admin, 74-; mem, Mass Comn Nuclear Safety, 74-; mem, US Comt Israel Environ, 73-; dir, Rush-Hampton Industs, 75- *Honors & Awards:* Difco Award, Lab Sect, Am Pub Health Asn, 77 & 82; Carski Found Award, Am Soc Microbiol, 80; Kimble Methodology Award, Conf Pub Health Lab Dirs, 84. *Mem:* Fel AAAS; fel Am Soc Microbiol; Sigma Xi; fel Am Pub Health Asn; fel Royal Soc Health; Am Acad Sanitarians; fel World Acad NZ; fel Explorers Club. *Res:* Water and sewage pollution; thermal death and vaccine production; fermentations; food and dairy bacteriology; disinfection; marine microbiology. *Mailing Add:* Morrill N 235 Univ Mass Amherst MA 01003-0081

LITSTER, JAMES DAVID, b Toronto, Ont, Can, June, 19, 38; m 65; c 2. SOLID STATE PHYSICS, OPTICS. *Educ:* McMaster Univ, BEng, 61; Mass Inst Tech, PhD(physics), 65. *Prof Exp:* From instr to assoc prof, Mass Inst Technol, 65-75, head, Div Atomic, Condensed Matter & Plasma Physics, Dept Physics, 79-83, PROF PHYSICS, MASS INST TECHNOL, 75-, DIR, FRANCIS BITTER NAT MAGNET LAB, 88- *Concurrent Pos:* Fel, John Simon Guggenheim Mem Found, 71-72; vis prof, Univ Paris, Orsay, 71-72; lectr physics, Harvard Med Sch, 74-85; vis scientist, Risø Nat Lab, Denmark, 78; mem Mat Res Adv Comt, NSF, 78-81, chmn, Condensed Matter Sci Subcomt, 80-81; dir, Ctr Mat Sci & Eng, Mass Inst Technol, 83-88; regional ed, Molecular Crystals & Liquid Crystals, 86-; mem, Solid State Sci Panel, Nat Res Coun, 86-; assoc provost & vpres res, Mass Inst Technol, 91- *Mem:* AAAS; fel Am Phys Soc; fel Am Acad Arts & Sci. *Res:* Magnetism; light scattering; liquid crystals; x-ray scattering using synchrotron radiation. *Mailing Add:* Dept Physics Mass Inst Technol Cambridge MA 02139

LITT, BERTRAM D, biostatistics, for more information see previous edition

LITT, MICHAEL, b New York, NY, Apr 17, 33; m 56. BIOCHEMISTRY. *Educ:* Oberlin Col, BA, 54; Harvard Univ, PhD(chem), 58. *Prof Exp:* Instr chem, Reed Col, 58-62, assoc prof, 64-67; assoc prof biochem, 67-71, PROF BIOCHEM & MED GENETICS, MED SCH, UNIV ORE, 71- *Concurrent Pos:* NIH spec fel, Mass Inst Technol, 62-63; NSF fel, Auckland Univ, 66-67. *Mem:* Am Chem Soc; Am Soc Biol Chemists. *Res:* Structure, function and metabolism of transfer RNA. *Mailing Add:* Dept Biochem & Med Genetics Ore Health Sci Univ 3181 SW Sam Jackson Park Rd Portland OR 97201

LITT, MITCHELL, b Brooklyn, NY, Oct 11, 32; m 55; c 2. BIOENGINEERING, CHEMICAL ENGINEERING. *Educ:* Columbia Univ, AB, 53, BS, 54, MS 56, DEngSc(chem eng), 61. *Prof Exp:* Res engr, Esso Res & Eng Co, NJ, 58-61; from asst prof to assoc prof, 61-71, chmn, Dept Bioeng, 81-90, PROF CHEM ENG, UNIV PA, 71-, PROF BIOENG, 73- *Concurrent Pos:* Vis prof, Weizmann Inst, 79. *Mem:* Am Inst Chem Engrs; Am Soc Eng Educ; Biomed Eng Soc; Am Chem Soc; Int Soc Biorheology; Eng Med Biol Soc; Soc Rheology. *Res:* Application of chemical engineering techniques to biomedical problems; biorheology, with applications to blood and epithelial secretions. *Mailing Add:* Dept Bioengineering Univ Pa Philadelphia PA 19104

LITT, MORTIMER, b Brooklyn, NY, Sept 28, 25; m 54; c 3. IMMUNOLOGY. *Educ:* Columbia Univ, BA, 47; Univ Rochester, MD, 52. *Prof Exp:* Med house officer & asst resident physician, Peter Bent Brigham Hosp, 52-54; instr, 56-59, assoc, 60-65, asst prof, 65-71, asst dean teaching resources, 73-78, ASSOC PROF MICROBIOL & MOLECULAR GENETICS, HARVARD MED SCH, 71-, ASSOC DEAN EDUC PROGS, 79- *Concurrent Pos:* Res fel, Harvard Med Sch, 54-56 & 56-59; Helen Hay Whitney Found fel, 59-63; estab investr, Am Heart Asn, 63-68; asst dir dept bact, Boston City Hosp, 69-77. *Res:* Eosinophil leukocyte. *Mailing Add:* Harvard Med Sch 25 Shattuck St Boston MA 02115

LITT, MORTON HERBERT, b Brooklyn, NY, Apr 10, 26; m 57; c 2. POLYMER CHEMISTRY. *Educ:* City Col New York, BS, 47; Polytech Inst Brooklyn, MS, 53, PhD(polymer chem), 56. *Prof Exp:* Turner & Newall res fel, Manchester Univ, 56-57; res assoc, State Univ NY Col Forestry, Syracuse, 58-60; sr scientist, Cent Res Lab, Allied Chem Corp, 60-64, assoc dir res, 65-67; assoc prof, 67-76, PROF POLYMER SCI, CASE WESTERN RESERVE UNIV, 76- *Mem:* Fel AAAS; Am Chem Soc; NY Acad Sci; fel Am Phys Soc; The Chem Soc. *Res:* Ionic and free radical polymerization mechanisms; organo-fluorine chemistry; polymer mechanical properties; polymer electrical properties; emulsion polymerization. *Mailing Add:* Dept Macromolecular Sci Case Western Reserve Univ Cleveland OH 44106

LITTAUER, ERNEST LUCIUS, b London, Eng, Mar 8, 36; US citizen; m 69. ELECTROCHEMISTRY, METALLURGY. *Educ:* Univ London, BSc, 58, PhD(electrometall), 61. *Prof Exp:* Res scientist, Derby Luminescents Div, Derby Metals, London, 62-63; sr scientist, Lockheed Aircraft Serv Co, 63-67, mgr, electrochem dept, 67-72, mgr, chem dept, 72-84, dir mat sci, res & develop div, 84-90, VPRES & ASST GEN MGR, LOCKHEED MISSILE & SPACE CO, 90- *Concurrent Pos:* Fel corrosion, Battersea Col Technol, 61-62; lectr, Sir John Cass Col & Enfield Col, London, 62-63; chmn, Res Coun Corrosion Comt, Lockheed Aircraft Corp, 66-; lectr electrochem, Univ Santa Clara, 74-76. *Mem:* Electrochem Soc; Am Chem Soc; Am Inst Aeronaut & Astronaut; Mat Res Soc. *Res:* Electrochemistry; energy conversion; process chemistry; analytical, inorganic and plasma chemistry; materials evaluation; chemical and chemical engineering development and design; metallurgy; materials science engineering; nondestructive test technology. *Mailing Add:* 27305 Deer Springs Way Los Altos Hills CA 94022

LITTAUER, RAPHAEL MAX, b Leipzig, Ger, Nov 28, 25; US citizen; m 50; c 2. HIGH ENERGY PHYSICS, ELECTRONICS. *Educ:* Cambridge Univ, MA & PhD(physics), 50. *Prof Exp:* Asst physics, Cambridge Univ, 47-50; res assoc nuclear physics, Cornell Univ, 50-54 & Synchrotron Lab, Gen Elec Co, 54-55; res assoc prof physics, 55-63, res prof, 63-65, chmn dept, 74-77, PROF PHYSICS & NUCLEAR STUDIES, CORNELL UNIV, 65- *Mem:* Am Phys Soc. *Mailing Add:* Newman Lab Cornell Univ Ithaca NY 14853

LITTELL, ARTHUR SIMPSON, biostatistics; deceased, see previous edition for last biography

LITTELL, RAMON CLARENCE, b Rolla, Kans, Nov 18, 42; m 66; c 2. AGRICULTURAL STATISTICS, MATHEMATICAL STATISTICS. *Educ:* Kans State Teachers Col, BS, 64; Okla State Univ, MS, 66, PhD(statist), 70. *Prof Exp:* PROF STATIST. *Concurrent Pos:* Consult. *Mem:* Am Statist Asn; Biomet Soc; fel Am Statist Asn. *Res:* Combining tests of significance; lineas models; experimental design. *Mailing Add:* 3840 NW 35th Pl Gainesville FL 32606

LITTEN, RAYE Z, III, PHYSIOLOGY. *Educ:* Bridgewater Col, BA, 69; Med Col Va, MS, 72, PhD(physiol), 76. *Prof Exp:* Instr, Dept Physiol, Med Col Va, 76, postdoctoral fel, 76-78; res asst prof, Dept Physiol & Biophys, Univ Vt, 78-85; res physiologist, Armed Forces Radiobiol Res Inst, Bethesda, Md, 85-89; PHYSIOLOGIST & PROG OFFICER, TREAT RES BR, DIV CLIN & PREVENTION RES, NAT INST ALCOHOL ABUSE & ALCOHOLISM, ROCKVILLE, MD, 89- *Concurrent Pos:* Numerous grants, univs & asns, 76-90; vis res assoc, Dept Surg Res, Naval Med Res Inst, Bethesda, 85-87; instr, Washington Area Coun Alcoholism & Drug Abuse, 91- *Mem:* Am Physiol Soc; Sigma Xi. *Res:* Biochemistry of contractile proteins from vascular smooth muscle; biochemistry of contractile proteins from hypertrophied hearts; cardiovascular alterations from whole-body irradiation, myosin isoenzymes and pituitary-thyroid function; biochemical markers of alcoholism, nutrition and alcohol-induced pathophysiology; numerous scientific publications. *Mailing Add:* Div Clin & Prev Res NIAAA 5600 Fishers Lane Rm16C-05 Rockville MD 20857

LITTERIA, MARILYN, b Cleveland, Ohio, Aug 9, 31. NEUROENDOCRINOLOGY. *Educ:* Case Western Reserve Univ, BS, 55; Univ Calif, Berkeley, PhD(physiol), 67. *Prof Exp:* Lab technician, Case Western Reserve Univ, 55-60; res fel endocrinol, Scripps Clin & Res Found & Develop Neuroendocrinol Lab, Vet Admin Hosp, San Fernando, 67-69; res assoc reproductive biol, Univ NC, 70-71; instr anat & Sloan fel, Northwestern Univ, 71-72; RES PHYSIOLOGIST & PRIN INVESTR, VET ADMIN MED CTR, NORTH CHICAGO, ILL, 72- *Mem:* Endocrine Soc. *Res:* The role of sex steroids in central nervous system development. *Mailing Add:* Vet Admin Med Ctr North Chicago IL 60064

LITTERST, CHARLES LAWRENCE, b Cleveland, Ohio, 1944. PHARMACOLOGY, TOXICOLOGY. *Educ:* Purdue Univ, BS, 66; Univ Wis, MS, 68, PhD(toxicol), 70. *Prof Exp:* Pharmacologist, Food & Drug Admin, Washington, DC, 70-72; toxicologist, Nat Cancer Inst, 72-87, TOXICOLOGIST, NIAID, BETHESDA, MD, 87- *Concurrent Pos:* Mem, Toxicol Info subcomt HEW comt to coord toxicol & related progs, 74-79; faculty FAES-NIH (Toxicology), 74- *Mem:* Am Soc Pharmacol & Exp Therapeut; Soc Toxicol; Sigma Xi; Am Asn Cancer Res. *Res:* Factors altering hepatic microsomal drug metabolism; toxicology and pharmacology of platinum; toxicology of antineoplastic and antiviral drugs. *Mailing Add:* Develop Therap Br AIDS Div NIAID 6003 Executive Blvd Bethesda MD 20892

LITTLE, A BRIAN, b Montreal, Que, Mar 11, 25; US citizen; m 49, 84; c 6. OBSTETRICS & GYNECOLOGY. *Educ:* McGill Univ, BA, 48, MD, CM, 50; Royal Col Physicians & Surgeons, Can, cert obstet & gynec, 55, FRCPS(C), 57; Am Bd Obstet & Gynec, dipl, 59. *Prof Exp:* Intern, Montreal Gen Hosp, 50-51; asst resident & resident, Boston Lying-in-Hosp & Free Hosp Women, Boston, 51-54; asst obstet, Harvard Med Sch, 55-56, instr obstet & gynec, 56-58, tutor med sci, 57-65, assoc obstet & gynec, 58-63, asst prof, 63-65; prof, 66-72, ARTHUR H BILL PROF OBSTET & GYNEC & DIR DEPT REPRODUCTIVE BIOL, SCH MED, CASE WESTERN RESERVE UNIV, 72- *Concurrent Pos:* Teaching fel obstet & gynec, Harvard Med Sch, 52-54; instr, Sch Nursing, Boston Univ, 51 & 55-57; asst obstetrician outpatients, Boston Lying-in-Hosp, 55-56, asst obstetrician, 56-58, assoc, 58-59, obstetrician & gynecologist, 59-65; sr obstetrician prenatal metab div, USPHS, 55-56; assoc vis surgeon, Boston City Hosp, 55-64, assoc dir dept obstet & gynec, 58-63, dir, 63-65, vis surgeon, 65; asst surgeon, Free Hosp Women, 58-64, mem courtesy staff, 64-65; mem consult staff, Sturdy Mem Hosp, Attleboro, 61-65; chief consult, Hunt Mem Hosp, Danvers, 62-65; mem consult staff, Elliott Community Hosp, Keene, NH, 63-65; dir dept obstet & gynec, Cleveland Metrop Hosp, Ohio, 66-72; assoc obstetrician & gynecologist, Univ Hosps, Cleveland, 66-72, dir dept obstet & gynec, 72- *Mem:* AMA; fel Am Col Obstet & Gynec; Am Gynec Soc; Soc Gynec Invest; fel Am Col Surgeons. *Res:* Steroid mechanism in vivo, in vitro, primarily in reproduction. *Mailing Add:* Dept Obstet & Gynec Royal Victoria Hosp McGill Univ Sch Med 687 Pine Ave W Montreal PQ H3A 1A1 Can

LITTLE, ALEX G, b Atlanta, Ga, Aug 24, 43; m 75; c 2. THORACIC SURGERY. *Educ:* Univ NC, BA, 65; Johns Hopkins Univ, MD, 74. *Prof Exp:* From asst prof to assoc prof surg, Univ Chicago, 81-88; PROF & CHMN SURG, UNIV NEV, 88- *Mem:* Am Col Surgeons; Am Assoc Thoracic Surg; Soc Thoracic Surgeons; Soc Univ Surgeons; Soc Surg Alinentary Tract. *Res:* Pathophysiology of benign esophageol diseases; research in basic mechanisms of esophageal and lung cancer. *Mailing Add:* Dept Surg Univ Nev 2040 W Charleston Bld Suite 601 Las Vegas NV 89102

LITTLE, ANGELA C, b San Francisco, Calif, Jan 12, 20; m 47; c 1. FOOD SCIENCE. *Educ:* Univ Calif, AB, 40, MS, 54, PhD(agr chem), 69. *Prof Exp:* Res asst food sci, 53-56, jr specialist, 56-58, from asst specialist to assoc specialist, 58-69, asst food scientist, 69-71, assoc food scientist, 71-79, lectr food sci, 69-79, from assoc prof to prof, 79-82, EMER PROF, UNIV CALIF, BERKELEY, 85- *Concurrent Pos:* Vis scholar, Univ Wash, Seattle, 76-77. *Mem:* AAAS; Inter-Soc Color Coun; Inst Food Technol. *Res:* Colorimetry; methodology and systems relating to objective measurements; color vision; taste perception related to changes in physiological state; determinants of human food practices-historical, cultural, ecological, religious & etc. *Mailing Add:* Dept of Nutrit Sci Univ of Calif Berkeley CA 94720

LITTLE, BRENDA JOYCE, b Akron, Ohio. MICROBIOLOGICALLY INDUCED CORROSION, COLLOID & INTERFACE CHEMISTRY. *Educ:* Baylor Univ, BS, 67; Tulane Univ, PhD(chem), 83. *Prof Exp:* Microbiologist, Nat Park Serv, 74-76; branch head, 85-86, RES CHEMIST, NAVAL OCEAN RES & DEVELOP ACTIV, 76- *Mem:* Am Chem Soc; Int Humic Substances Soc; Nat Asn Corrosion Engrs; Adhesion Soc; Sigma Xi. *Res:* Factors influencing the absorption of dissolved organic material from natural waters and their impact on adhesion of microorganisms; elucidation of mechanisms for biodeterioration of metals in marine environments. *Mailing Add:* Naval Ocean Res & Develop Activ Code 333 Bay St Louis MS 39529-5004

LITTLE, BRIAN WOODS, b Boston, Mass, Dec 15, 45. NEUROPATHOLOGY & NEUROMUSCULAR PATHOLOGY. *Educ:* Cornell Univ, BA, 67; Univ Vt, MD, 73, PhD(biochem), 77. *Prof Exp:* Resident pathologist, 73- 76, attend pathologist, Med Ctr Hosp Vt, 76-84; asst prof path, Univ Vt, 76-84, asst prof biochem, 80-84; Vet Admin Med Ctr, Northport, NY; asst prof path, State Univ NY, Stony Brook, 84-87; ATTEND PATHOLOGIST, LEHIGH VALLEY HOSP CTR, 87- *Mem:* Am Soc Clin Pathologists; Col Am Pathologists. *Res:* Mammalian nucleic acid metabolism; muscle disease; histochemistry; epidemiology of CNS disorders; pediatric, anatomic and clinical pathology. *Mailing Add:* Neurpath Lab Lehigh Valley Hosp Ctr 1200 S Cedar Crest Blvd Allentown PA 18105

LITTLE, CHARLES DURWOOD, JR, b Denver, Colo, Dec 28, 46. ANATOMY, CELL BIOLOGY. *Educ:* Calif State Polytech Univ, Pomona, BS, 73; Univ Pittsburgh, PhD(anat & cell biol), 77. *Prof Exp:* Res fel, Develop Biol Lab, Mass Gen Hosp, Harvard Med Sch, 77-79; fel, Biol Dept, Univ Calif, San Diego, 79-81; asst prof, 81-87, ASSOC PROF ANAT, UNIV VA, 87- *Concurrent Pos:* Adv panel develop biol, NSF; mem, PBC study sect, NIH; bd trustees, Soc Develop Biol. *Mem:* Developmental Biol Soc; Am Soc Cell Biol. *Res:* Developmental biology of the extracellular matrix; cell surface matrix interactions; double immunolabeling techniques for use in fluorescence and electron microscopy. *Mailing Add:* Dept Anat Univ Va Sch Med Box 439 Charlottesville VA 22908

LITTLE, CHARLES EDWARD, b Kansas City, Kans, Apr 18, 26; m 47; c 3. MATHEMATICS. *Educ:* Univ Kans, AB, 48; Ft Hays Kans State Col, MS, 55; Colo State Col, EdD(math educ), 64. *Prof Exp:* Instr pub sch, Kans, 51-60, supt, 60-61; instr math, Colo State Col, 61-63; from asst prof to assoc prof, 64-68, chmn dept math, 67-70, dean Col Arts & Sci, 74-84, PROF MATH EDUC, NORTHERN ARIZ UNIV, 68- *Concurrent Pos:* Nat Coun Teachers Math. *Mem:* Math Asn Am. *Res:* Training of elementary and secondary mathematics teachers; methods of instruction in mathematics at college level, particularly educational media and techniques for handling large groups; mathematics for the social and behavioral sciences; linear models and statistics. *Mailing Add:* Dept Math Northern Ariz Univ Box 5717 Flagstaff AZ 86011

LITTLE, CHARLES GORDON, b Hunan, China, Nov 4, 24; m 54; c 5. ATMOSPHERIC PHYSICS, REMOTE SENSING. *Educ:* Univ Manchester, BSc, 48, PhD(radio astron), 52. *Prof Exp:* Jr engr, Cosmos Mfg Co, Eng, 44-46; jr physicist, Ferranti, Ltd, 46-47; asst lectr physics, Univ Manchester, 52-53; prof geophys res & dep dir geophys inst, Univ Alaska, 54-58; chief radio astron & arctic propagation sect, Nat Bur Standards, 58-60, upper atmosphere & space physics div, Boulder Labs, 60-62, dir, Cent Radio Propagation Lab, 62-65; dir, Inst Telecommun Sci & Aeronomy, Environ Sci Serv Admin, Nat Oceanic & Atmospheric Admin, 65-67 & Wave Propagation Lab, Environ Reasc Admin, 67-86; sr fel univ corp atmospheric res, Naval Environ Res Prediction Facil, 87-89; George Haltiner res prof, Naval Postgrad Sch, 89-90; RETIRED. *Concurrent Pos:* Consult, US Nat Comt for Int Geophys Year, 57-59. *Honors & Awards:* Cleveland Abbe Award, Am Meteorol Soc, 84. *Mem:* Nat Acad Eng; AAAS; Inst Elec & Electronic Engrs; Am Meterol Soc; Int Union Radio Sci. *Res:* Remote measurement of atmosphere and ocean, using electromagnetic and acoustic waves. *Mailing Add:* 6949 Roaring Fork Trail Boulder CO 80301

LITTLE, CHARLES HARRISON ANTHONY, b Toronto, Ont, May 4, 39; m 64; c 2. TREE PHYSIOLOGY. *Educ:* Univ NB, BScF, 61; Yale Univ, MF, 62, PhD(tree physiol), 66. *Prof Exp:* RES SCIENTIST, CAN FORESTRY SERV, 66- *Mem:* Can Soc Plant Physiol; Sigma Xi. *Res:* Production, movement and transformation of carbohydrate; growth substance regulation of bud and cambial activity. *Mailing Add:* 241 Aberdeen St Fredericton NB E3B 1R6 Can

LITTLE, CHARLES ORAN, b Schulenburg, Tex, July 21, 35; m 55; c 3. ANIMAL NUTRITION, AGRICULTURE. *Educ:* Univ Houston, BS, 57; Iowa State Univ, MS, 59, PhD(animal nutrit), 60. *Prof Exp:* Res asst animal nutrit, Iowa State Univ, 57-60; from asst prof to assoc prof, 60-67, indust res grants, 61-72, Agr Res Serv grant, 64-67, prof animal sci, 67-85, assoc dean res, 69-85, DEAN, COL AGR, UNIV KY, 88-; vchancellor for Res & dir Exp Station, 85-88, DIR AGR EXP STATION & DIR COOP EXTENTION, LA STATE UNIV, 88- *Concurrent Pos:* Assoc dir Ky Agr Exp Sta, 69-; mem & chmn, Southern Regional Res Comt, 72-75; mem, Southern Res Planning Comt, 73-75, chmn, Southern Asn Agr Exp Sta Dirs, 76-77, mem, Exp Sta Comt on Policy, 78- *Honors & Awards:* Distinguished Nutritionist Award, Nat Distillers Feed Res Coun, 64; Outstanding Res Awards, Thomas Poe Cooper & Ky Res Founds, 67. *Mem:* AAAS; Am Soc Animal Sci; Am Inst Nutrit. *Mailing Add:* S-107 Agr Sci Ctr Univ Ky Lexington KY 40546

LITTLE, EDWIN DEMETRIUS, b Orlando, Fla, July 2, 26. ORGANIC CHEMISTRY. *Educ:* Rollins Col, BS, 48; Duke Univ, AM, 53. *Prof Exp:* Sr res chemist, Nitrogen Div, Allied Chem Corp, 53-63, supvry res chemist, 63-66, res assoc, Plastics Div, 66-74, group leader, Specialty Chem Div, 74-79; res assoc, Va Chem Inc, 79-85; RES CHEMIST, MERCK SHARP & DOHME RES LAB, MERCK & CO, INC, 85- *Mem:* Fel Am Inst Chem; Am Chem Soc; NY Acad Sci. *Res:* Heterocyclic nitrogen compounds; epoxide reactions; monomer synthesis. *Mailing Add:* 12 English Village 4B Cranford NJ 07016

LITTLE, ELBERT LUTHER, JR, b Ft Smith, Ark, Oct 15, 07; m 43; c 3. BOTANY, DENDROLOGY. *Educ:* Univ Okla, BA, 27, BS, 32; Univ Chicago, MS & PhD(bot), 29. *Prof Exp:* Asst prof biol, Southwestern Okla State Univ, 30-33; from asst forest ecologist to assoc forest ecologist, Ariz, 34-42; dendrologist, US Forest Serv, Washington, DC, 42-76; collabr, 65-76, RES ASSOC, US NAT MUS NATURAL HIST, SMITHSONIAN INST, WASHINGTON, DC, 76- *Concurrent Pos:* Botanist, Econ Admin, Bogota, 43-45; prod specialist, US Com Co, Mexico City, 45; vis prof, Univ Andes, Venzuela, 53-54 & 60; botanist from Univ Md, Guyana, 55; consult, UN mission, Costa Rica, 64-65 & 67, Ecuador, 65 & 75, Nicaragua, 71 & Okla

Forestry Div 30, 77-78; vis prof, Va Polytech Inst & State Univ, 66-67 & Univ DC, 79. *Honors & Awards:* Barringtom Moore Award, Soc Am Foresters, 86. *Mem:* Fel Explorers Club; fel Soc Am Foresters; Bot Soc Am; Am Soc Plant Taxon; Am Inst Biol Sci; Int Asn Plant Taxon. *Res:* Trees of United States and tropical America, their identification, classification, nomenclature, and distribution; plant ecology. *Mailing Add:* 924 20th St S Arlington VA 22202-2616

LITTLE, GWYNNE H, Birmingham, Ala, June 25, 41. BIOCHEMISTRY OF DEVELOPMENT. *Educ:* Med Col Ga, PhD(biochem), 70. *Prof Exp:* ASSOC PROF BIOCHEM, TEX TECH UNIV, HEALTH SCI CTR, 80- *Mem:* Am Soc Biol Chem; Am Asn Clin Chem; Nat Acad Clin Biochem. *Mailing Add:* Dept Biochem Sch Med Tex Tech Univ Health Sci Ctr Lubbock TX 79430

LITTLE, HAROLD FRANKLIN, b Williamsport, Pa, June 18, 32; m 59; c 1. ENTOMOLOGY, FUNCTIONAL MORPHOLOGY. *Educ:* Lycoming Col, AB, 54; Pa State Univ, MS, 56, PhD(zool), 59. *Prof Exp:* Asst prof biol, WVa Wesleyan Col, 59-63; from asst prof to assoc prof, 63-71, chmn, Div Nat Sci, 68-71 & 73-79, chmn, Biol Dept, 73-78, PROF BIOL, UNIV HAWAII, HILO, 71-, CHMN DEPT, 86- *Mem:* Entom Soc Am. *Res:* Damage to Medfly pupal flight mucles; histology and ultrastructure of tephritid fruit fly pheromone glands. *Mailing Add:* Dept Biol, CAS Univ Hawaii Hilo HI 96720-4091

LITTLE, HENRY NELSON, biochemistry; deceased, see previous edition for last biography

LITTLE, JAMES ALEXANDER, b Detroit, Mich, Dec 8, 22; Can citizen; m 53; c 2. MEDICINE, METABOLISM. *Educ:* Univ Toronto, MD, 46, MA, 50; FRCP(C), 52. *Prof Exp:* Res assoc med, St Michael's Hosp, Univ Toronto, 52-67, clin teacher, 54-63, dir diabetic clin, 54-70, assoc, 63-66, dir clin invest unit, 64-72, from asst prof to assoc prof, 66-74, dir lipid clin, 66-90, PROF MED, ST MICHAEL'S HOSP, UNIV TORONTO, 74-, RES COORDR & SECY, RES SOC, 64- *Concurrent Pos:* Nat Res Coun Can fel biochem, Univ Toronto, 47-49, Can Red Cross fel arthritis, Sunnybrook Dept Vet Affairs Hosp, 51-52; mem, Can Nat Comn, Int Union Nutrit Sci, 71-75; dir div endocrinol, metab & nephrol, Lipid Clin, St Michael's Hosp, 70-73; exec comt, Coun Atherosclerosis, Am Heart Asn, 74-77; comt nutrit & cardiovasc dis, Health Protection Br, Govt Can, 74-76 & Med comt; proj dir, Univ Toronto-Mcmaster Lipid Res Ctr, 72-91; pres, Can Lipoprotein Conf, 89-90. *Mem:* Am Heart Asn; Can Cardiovasc Soc; Am Diabetes Asn; Nutrit Soc Can; Can Soc Clin Invest; Can Atherosclerosis Soc (pres, 87-88). *Res:* Relation between human atherosclerosis, plasma lipoproteins, nutrition and genetic factors; effect of insulin antibodies on diabetic complications. *Mailing Add:* 30 Bond St St Michael's Hosp Toronto ON M5B 1W8 Can

LITTLE, JAMES NOEL, b Kansas City, Mo, July 3, 40; m 64; c 3. ANALYTICAL CHEMISTRY. *Educ:* Univ Kans, BS, 62, Mass Inst Technol, PhD(anal chem), 66. *Prof Exp:* Res chemist, Hercules, Inc, Del, 66-67; sr res chemist, Waters Assocs, Inc, 68-69, mgr chromatography res, 69-71, vpres, 71-81; vpres, 81-82, SR VPRES, ZYMARK CORP, 83- *Concurrent Pos:* Dir, Cyborg Corp, 82-87. *Mem:* Am Chem Soc. *Res:* Separations; chromatography; polymer characterization; analytical methods development; spectroscopy; robotics. *Mailing Add:* Zymark Corp Zymark Ctr Hopkinton MA 01748

LITTLE, JOHN BERTRAM, b Boston, Mass, Oct 5, 29; m 60; c 2. CANCER BIOLOGY, RADIATION BIOLOGY. *Educ:* Harvard Univ, AB, 51; Boston Univ, MD, 55; Am Bd Radiol, dipl, 61, cert nuclear med, 61. *Prof Exp:* Intern med, Johns Hopkins Hosp, 55-56; resident radiol, Mass Gen Hosp, 58-61; USPHS res fel, 61-63, instr physiol, 63-65, from asst prof to assoc prof, 65-75, chmn dept, 80-83, PROF RADIOBIOL, 75-, JAMES STEVENS SIMMONS PROF RADIOBIOL, HARVARD SCH PUB HEALTH, 88- *Concurrent Pos:* Consult Mass Gen Hosp, 65-; Peter Bent Brigham Hosp, 68-; Nat Cancer Inst & Am Cancer Soc fels, Sch Pub Health, Harvard, 68-78, lectr, Med Sch, 68-; chmn, bd sci counr, Nat Inst Environ Health Sci, 82-84; dir, Kresge Ctr Environ Health, 82-; bd sci counselors, Nat Toxicol Prog, Nat Cancer Inst, 87-, outstanding investr grant, 88-; coun deleg med sci, AAAS, 88-, coun affairs comt, 89- *Mem:* AAAS; Am Physiol Soc; Health Physics Soc; Am Asn Cancer Res; Radiation Res Soc (pres-elect, 85, pres, 86-87). *Res:* Cellular and molecular radiation biology with emphasis on mutagenesis; experimental carcinogenesis. *Mailing Add:* Dept Cancer Biol Sch Pub Health Harvard Univ Boston MA 02115

LITTLE, JOHN CLAYTON, b Battle Creek, Mich, Jan 1, 33; m 87; c 4. ORGANIC CHEMISTRY, PROCESS DEVELOPMENT. *Educ:* Univ Calif, BS, 54; Univ Ill, PhD(org chem), 57. *Prof Exp:* Res chemist, 57-62, proj leader, 62-64, group leader, 64-71, res mgr, 71-81, mgr chem technol, agr prod dept, 81-89, FACIL PROJ MGR, WESTERN DIV RES & DEVELOP, DOW CHEM USA, 89- *Concurrent Pos:* Mem, Mich Found Advan Res; sect ed, Chem Abstracts, 64-68. *Mem:* Am Chem Soc; Sigma Xi; The Chem Soc. *Res:* Pilot plant and process development studies; organic syntheses and structure-biological activity relationships; chemical manufacturing, environmental studies; petrochemical processing; chlorination; catalytic oxidation and reduction; Diels-Alder reactions and synthetic methods; fluorination processes; research facilities design and management. *Mailing Add:* Western Div Res & Develop Dow Chem USA PO Box 1398 Pittsburg CA 94565

LITTLE, JOHN DUTTON CONANT, b Boston, Mass, Feb 1, 28; m 53; c 4. OPERATIONS RESEARCH, MANAGEMENT SCIENCE. *Educ:* Mass Inst Technol, SB, 48, PhD(physics), 55. *Prof Exp:* Engr tube develop, Gen Elec Co, 49-50; asst physics, Mass Inst Technol, 51-54; from asst prof to assoc prof opers res, Case Inst Technol, 57-62; assoc prof opers res, Mass Inst Technol, 62-67, dir opers res ctr, 69-75, prof opers res & mgt, 67-68, head Mgt Sci Group, 72-82, head behav & policy scis area, 82-88, George M Bunker prof mgt sci, Sloan Sch, 78-89, INST PROF, MASS INST TECHNOL, 89- *Concurrent Pos:* Pres, Mgt Decision Systs, Inc, 67-80, chmn, 67-85; dir Info Resources Inc, 85-; adv panel of Decision, Risk & Mgt Sci Prog, Nat Sci Found, 86-89; vis prof, Europ Inst Bus Admin, Fontainebleau, France, 89; Philip McCord Morse lectr, Opers Res Soc Am, 89-90. *Honors & Awards:* Charles Coolidge Parlin Award, 78; Kimball Medal, Opers Res Soc Am, 87. *Mem:* Nat Acad Eng; fel AAAS; Opers Res Soc Am (pres, 79-80); Inst Mgt Sci (vpres, 76-79, pres, 84-85); Am Mkt Asn. *Res:* Research on mathematical programming,; queuing theory; marketing; traffic control; decision support systems. *Mailing Add:* 37 Conant Rd RR 3 Lincoln MA 01773-3907

LITTLE, JOHN LLEWELLYN, b Lakewood, Ohio, Feb 17, 19; m 49; c 1. COMPUTER STANDARDS, PHYSICS. *Educ:* Case Inst Technol, BS, 42. *Prof Exp:* Staff mem oper res, US Navy, Anti-Submarine Warfare Oper Res Group, 42-45; proj engr electronics, Reed Res, Inc, 47-48; staff mem oper res, Dept Defense, Res & Develop Bd, 49-51; chief planning dept res, Reed Res, Inc, 51-52; comput specialist, Nat Bur Standards, 53-87; RETIRED. *Concurrent Pos:* Task force coordr, Comput Network Protocol, Nat Comn Libr & Info Sci, 76-77. *Mem:* Asn Comput Mach; Inst Elec & Electronics Engrs. *Res:* Computer hardware standards; coded character sets; collating sequence; transferability of computer programs and data between dissimilar computer-based systems; network protocols. *Mailing Add:* 3534 Woodbine Chevy Chase MD 20815

LITTLE, JOHN RUSSELL, JR, b Cheyenne, Wyo, Oct 23, 30; m 55; c 3. IMMUNOLOGY. *Educ:* Cornell Univ, AB, 52; Univ Rochester, MD, 56. *Prof Exp:* Fel microbiol, 62-64, from asst prof to assoc prof med & microbiol, 69-73, PROF MED & MICROBIOL, SCH MED, WASHINGTON UNIV, 73- *Res:* Medicine and lymphocyte membrane structure and function. *Mailing Add:* Jewish Hosp Washington Univ Sch of Med 216 S Kingsway St Louis MO 63110

LITTLE, JOHN STANLEY, b Fredericton, NB, July 27, 31; m 57; c 1. ORGANIC CHEMISTRY. *Educ:* Univ NB, BS, 52, PhD(org chem), 55. *Prof Exp:* Lord Beaverbrook Overseas Scholar, Univ London, 55-56; res chemist, Can Industs, Ltd, 56-62, group leader fiber res & develop, 62-65; sect leader all process & prod res, Celanese Corp, Can, 65-67, mgr appln res, 67-70, mgr indust prod develop, Celanese Fibers Mkt Co, 70-71, dir spec prod develop, 71-73, dir textile prod develop, 73-78, dir int indust markets, Celanese Corp, NY, 78-81; vpres corp technol, Great Lakes Chem Corp, 81-89; MANAGING DIR, ASSOC OCTEL CO LTD, 89- *Mem:* Fel Chem Inst Can. *Res:* Synthetic fibers; alkaloids and steroids. *Mailing Add:* Assoc Octel Co Ltd PO Box 17 Ellesmere Port Oil Sites Rd South Wirral L65 4HF England

LITTLE, JOHN WESLEY, b Washington, DC, June 24, 41; m 69; c 2. BIOCHEMISTRY. *Educ:* Stanford Univ, BS, 62, PhD(biochem), 67. *Prof Exp:* Sr asst scientist, NIH, 67-69, sr staff fel, 69-72; res fel, Stanford Univ, 73-76, res assoc, Univ Ariz, 77-78; adj asst prof microbiol, 78-80, adj assoc prof molecular & med microbiol, 80-81, asst prof biochem, 82-85, ASSOC PROF BIOCHEM & MOLECULAR & CELLULAR BIOL, 85- *Mem:* AAAS; Am Soc Biol Chemists; Am Soc Microbiol. *Res:* Regulatory system which controls how E coli responds to conditions which damage DNA including the biochemistry of the proteins which control this response. *Mailing Add:* Dept Biochem Univ Ariz Bio Sci W Bldg Tucson AZ 85721

LITTLE, JOSEPH ALEXANDER, b Bessemer, Ala, Mar 16, 18; m 41; c 3. MEDICINE. *Educ:* Vanderbilt Univ, BA, 40, MD, 43. *Prof Exp:* Intern, Vanderbilt Univ Hosp, 43 & 46-47; resident, Childrens Hosp, Univ Cincinnati, 47-48, instr pediat, Col Med, Univ, 48-49; from asst prof to prof, Sch Med, Univ Louisville, 49-62; assoc prof, Sch Med, Vanderbilt Univ, 62-70; PROF PEDIAT & HEAD DEPT, SCH MED, LA STATE UNIV, SHREVEPORT, 70- *Concurrent Pos:* Dir outpatient dept, Childrens Hosp, 48-49, physician in chief, 56-; med dir, State Crippled Children Comn, Ky, 51-54; consult, State Dept Health, Ky, 49-51. *Mem:* AAAS; Am Pediat Soc; Am Acad Pediat; NY Acad Sci. *Res:* Pediatric cardiology. *Mailing Add:* Rte 1 Box 55 Sch Med La State Univ Turning Point Rte 1 Box 328 Sherwood Rd Sewanee TN 37375

LITTLE, JOYCE CURRIE, b Pioneer, La, 1934. COMPUTER ETHICS, SOFTWARE ENGINEERING. *Educ:* Northeast La State Col, BS, 57; San Diego State Col, MS, 63; Univ Md, College Park, PhD(educ admin & computer science), 84. *Prof Exp:* PROF COMPUTER & INFO SCI, TOWSON STATE UNIV, 91-, CHAIR DEPT, 84- *Concurrent Pos:* Consult, Univ Md, College Park, 86- *Mem:* Asn Comput Mach; Inst Elec & Electronics Engrs Computer Soc; Am Asn Univ Professors; Sigma Xi. *Res:* Computer personnel research. *Mailing Add:* Dept Computer & Info Sci Towson State Univ Baltimore MD 21204

LITTLE, MAURICE DALE, b North Grove, Ind, Apr 13, 28; m 55; c 3. MEDICAL PARASITOLOGY. *Educ:* Purdue Univ, BS, 50; Tulane Univ, MS, 58, PhD(parasitol), 61. *Prof Exp:* Microbiologist, Ind State Bd Health, 53-56; from instr to asst prof, Sch Med, 63-68, assoc prof, 68-76, PROF PARASITOL, SCH PUB HEALTH & TROP MED, TULANE UNIV, 76- *Concurrent Pos:* NIH fel, Tulane Univ, 61-63; assoc ed, Am Soc Trop Med & Hyg, 75-84. *Mem:* AAAS; Am Micros Soc; Am Soc Parasitologists (secy-treas, 80-81); Am Soc Trop Med & Hyg; Wildlife Dis Asn. *Res:* Morphology, biology and epidemiology of Strongyloides species; zoonotic helminthiases; soil-transmitted helminths; parasites in sewage sludges. *Mailing Add:* Dept of Trop Med Tulane Univ New Orleans LA 70112

LITTLE, MICHAEL ALAN, b Abington, Pa, Mar 24, 37; m 65; c 2. PHYSICAL ANTHROPOLOGY. *Educ:* Pa State Univ, BA, 62, MA, 65, PhD(anthrop), 68. *Prof Exp:* Asst prof anthrop, Ohio State Univ, 67-70; from asst prof to assoc prof, 71-81, PROF ANTHROP, STATE UNIV NY BINGHAMTON, 81- *Concurrent Pos:* Ohio State Univ res fel, Nuñoa, Peru, 68; State Univ NY Binghamton res fel & grant, Nuñoa, 72; vis assoc prof anthrop & sci coordr, US Int Biol Prog, Human Adaptability Component, Pa State Univ, 72-73; NSF sci equip grants, 74 & 80 & res grant, 78, 80, 82, 85

& 87. *Mem:* Fel AAAS; Am Asn Phys Anthrop (vpres, 88-90); Soc Study Human Biol; Human Biol Coun (secy-treas, 73-75, exec comt, 87-91). *Res:* Biocultural adaptations; human biology; environmental stress; heat and cold adaptation; circadian rhythms; human populations at high altitude; child growth and development; ecology of savanna pastoralists. *Mailing Add:* Dept Anthrop State Univ NY Binghamton NY 13902-6000

LITTLE, PATRICK JOSEPH, b Washington, DC, Oct 6, 58. PSYCHIATRY. *Educ:* Col William & Mary, BS, 81; Va Commonwealth Univ, PhD(pharmacol & toxicol), 89. *Prof Exp:* Lab specialist, Dept Pharmacol & Toxicol, Med Col Va, 81-85; postdoctoral fel, Dept Psychiat, Sch Med, Wash Univ, St Louis, 89-90; POSTDOCTORAL FEL, DEPT PHARMACOL, MED CTR, DUKE UNIV, 90- *Mem:* Am Soc Pharmacol & Exp Therapeut; AAAS; NY Acad Sci; Soc Neurosci. *Res:* Drug dependence. *Mailing Add:* Dept Pharmacol Med Ctr Duke Univ Box 3813 Durham NC 27710

LITTLE, PERRY L, b Ball Ground, Ga, Aug 3, 28; m 51; c 1. NUTRITION, PHYSIOLOGY. *Educ:* Berry Col, BS, 50; Auburn Univ, MS, 57, PhD(path, nutrit, physiol), 66. *Prof Exp:* Teacher high sch Ala, 50-52 & 53-55; res asst poultry sci, Auburn Univ, 55-62; assoc prof, 62-80, PROF POULTRY SCI, SAM HOUSTON STATE UNIV, 80- *Mem:* Poultry Sci Asn; World Poultry Sci Asn. *Res:* Nutrition of parasites which involve poultry, currently doing alligator nutrition. *Mailing Add:* Dept of Agr Sam Houston State Univ Huntsville TX 77341

LITTLE, RANDEL QUINCY, JR, b Richmond, Va, Aug 14, 27; m 49; c 3. ORGANIC CHEMISTRY. *Educ:* Univ Richmond, BS, 48; Univ Mich, MS, 49, PhD(org chem), 54. *Prof Exp:* Res chemist, Am Oil Co, Stand Oil Co, Ind, 53-60, group leader motor oil additives, 60-62, res supvr, 62-68, asst dir lubricants res, 68-74, dir lubricants & agr prod res, 74-83, DIR FUELS RES, AMOCO OIL CO, 83 - *Mem:* Am Chem Soc; Sigma Xi; Soc Automotive Engrs. *Res:* Organic reactions; motor oil additives; lubricants; fuels. *Mailing Add:* 860 W Driveway Glen Ellyn IL 60137

LITTLE, RAYMOND DANIEL, b Superior, Wis, Sept 12, 47; m 72; c 3. ORGANIC CHEMISTRY. *Educ:* Univ Wis-Superior, BS, 69; Univ Wis-Madison, PhD(org chem), 74. *Prof Exp:* Postdoctoral fel, Yale Univ, 74-75; from asst prof to assoc prof, 75-86, PROF ORG CHEM, UNIV CALIF, SANTA BARBARA, 86- *Concurrent Pos:* Vis prof, Univ British Columbia, Vancouver, Can, 87. *Honors & Awards:* Plous Award, Univ Calif, Santa Barbara, 81. *Mem:* Am Chem Soc; Sigma Xi. *Res:* Development of new synthetic methods, especially diyl trapping and electroreductive cyclization; total synthesis of pharmacologically active molecules; mechanistic organic chemistry of thermal, electro and photochemical reactions. *Mailing Add:* Dept Chem Univ Calif Santa Barbara CA 93106

LITTLE, RICHARD ALLEN, b Coshocton, Ohio, Jan 12, 39; m 60; c 3. GENERAL MATHEMATICS, GENERAL COMPUTER SCIENCES. *Educ:* Wittenburg Univ, BSc, 60; Johns Hopkins Univ, MA, 61; Harvard Univ, EdM, 65; Kent State Univ, PhD(math ed), 71. *Prof Exp:* Teacher & coach math, Culver Acad, Ind, 61-65; instr & curric consult math, Nigerian proj, Harvard Univ, 65-67; from instr to assoc prof math, Kent State Univ, Stark, 67-75; PROF MATH & COMPUT SCI, BALDWIN-WALLACE CO, OHIO, 75- *Concurrent Pos:* Scholarship, Gen Motors Corp, Wittenburg, 57-60, Ford Found, Johns Hopkins Univ, 60-61 & NSF, Harvard Univ, 63-64; curric consult math, India Inst Proj, NSF, 67; vis prof math, Ohio State Univ, Columbus, 87-88. *Mem:* Asn Comput Mach; Math Asn Am; Nat Coun Teachers Math. *Res:* Mathematics and computer science education; computer applications in small to medium size businesses. *Mailing Add:* Baldwin-Wallace Col Berea OH 44017

LITTLE, ROBERT, b Jedburgh, Scotland, Mar 3, 39; m 62; c 3. PLASMA PHYSICS. *Educ:* Univ Glasgow, BSc, 61, PhD(physics), 64. *Prof Exp:* Res assoc physics, Cambridge Electron Accelerator, Harvard Univ, 64-73 & Mass Inst Technol, 73-75; PROF STAFF PHYSICS, PRINCETON PLASMA PHYSICS LAB, PRINCETON UNIV, 75- *Res:* Development of large tokomaks for controlled thermonuclear research. *Mailing Add:* Physiol & Endocrinol Med Col Ga Augusta GA 30912

LITTLE, ROBERT COLBY, b Norwalk, Ohio, June 2, 20; m 45; c 2. PHYSIOLOGY, MEDICINE. *Educ:* Denison Univ, AB, 42; Western Reserve Univ, MD, 44, MS, 48. *Prof Exp:* Intern, Grace Hosp, Detroit, Mich, 44-45; resident med, Crile Vet Hosp, Cleveland, 49-50; from asst prof to assoc prof physiol, Univ Tenn, 50-54, assoc prof med, 53-54; dir clin res, Mead Johnson & Co, 54-57; dir Cardio-Pulmonary Labs, Scott & White Clin, Tex, 57-58; prof physiol, Seton Hall Col Med, 58-64, asst prof med, 59-64; prof physiol, chmn dept & asst prof med, Col Med, Ohio State Univ, 64-73; chmn dept physiol, 73-86, prof physiol & med, 73-89, EMER PROF PHYSIOL & MED, MED COL GA, 89- *Concurrent Pos:* USPHS res fel, Western Reserve Univ, 48-49. *Mem:* Am Physiol Soc; Soc Exp Biol & Med; Am Heart Asn; Am Fedn Clin Res; AMA; Sigma Xi. *Res:* Cardiovascular dynamics; heart sounds; clinical physiology; muscle dynamics. *Mailing Add:* Dept of Physiol & Endocrinol Med Col of Ga Augusta GA 30912-3376

LITTLE, ROBERT E(UGENE), b Enfield, Ill, May 24, 33; m 61; c 4. MECHANICAL DESIGN, MECHANICAL METALLURGY. *Educ:* Ohio State Univ, MSME, 60; Univ Mich, PhD(mech eng), 63. *Prof Exp:* Asst prof mech eng, Okla State Univ, 63-65; assoc prof, 65-68, PROF MECH ENG, UNIV MICH-DEARBORN, 68- *Mem:* Am Soc Testing & Mat. *Res:* Modes of failure; fatigue; reliability; planned experiments; composites. *Mailing Add:* Dept Mech Eng Univ Mich 4901 Evergreen Rd Dearborn MI 48128-1491

LITTLE, ROBERT LEWIS, b Monticello, Miss, July 1, 29; m 53; c 2. GEOLOGY. *Educ:* Univ Miss, BA, 51, MS, 59; Univ Tenn, Knoxville, PhD(geol), 69. *Prof Exp:* Instr geol, Univ Miss, 58-59 & Univ Tenn, Knoxville, 59-69; from asst prof to assoc prof, Valdosta State Col, 69-81, head dept, 69-81, prof geol & dept head physics, astron & geol, 81-85; RETIRED. *Mem:* Geol Soc Am; Am Asn Univ Prof; Nat Asn Geol Teachers. *Res:* Areal geology; stratigraphy and structural geology. *Mailing Add:* 711 Northside Dr Valdosta GA 31602

LITTLE, ROBERT NARVAEZ, JR, b Houston, Tex, Mar 11, 13; m 42; c 2. NUCLEAR PHYSICS, SCIENCE EDUCATION. *Educ:* Rice Inst, BA, 35, MA, 42, PhD(physics), 43. *Prof Exp:* Asst seismologist, Shell Oil Co, Tex, 36-42, asst physics, Rice Inst, 43; asst prof, Univ Ore, 43-44; testing supvr, Mil Physics Res Lab, Univ Tex, 44-48, from asst prof to assoc prof physics, 46-55 & res scientist, Nuclear Physics Lab, 55-60, PROF PHYSICS & EDUC, UNIV TEX, 55- *Concurrent Pos:* Chief nuclear physics, Gen Dynamics Corp, 54; mem, Comn Col Physics, 62-66; dir physics educ asst proj, Nat Univ Cent Am, 65-71; vis prof, Univ Valle, Guatemala, 71; Catedratico Ad Honorem, Univ Nac de Edu a Distancia, Madrid, Spain, 81. *Mem:* Am Phys Soc; Am Asn Physics Teachers (pres, 70); Int Group Res on Educ Physics; Cent Am Soc Physics. *Res:* Neutron scattering; reactor physics; development and evaluation of physics teaching methods at all levels. *Mailing Add:* Dept Physics Univ Tex Austin TX 78712

LITTLE, SARAH ALDEN, b Cleveland, Ohio, Apr 27, 59. EXPERIMENTAL JET FLOW, CONVECTION. *Educ:* Stanford Univ, BS, 81; Mass Inst Technol, PhD(marine geophys), 88. *Prof Exp:* Postdoctoral fel, Ctr Water Res, Univ Western Australia, 88-89; consult, Mackie Martin & Assoc, 89-90; VIS INVESTR, WOODS HOLE OCEANOG INST, 90- *Concurrent Pos:* Hon postdoctoral, Dept Math, Univ Western Australia, 89-90. *Mem:* Am Geophys Union; Am Women Sci. *Res:* Developing techniques and applications of dynamical systems analysis to natural systems such as fluid flow, ecology and earthquakes. *Mailing Add:* 37 Conant Rd Lincoln MA 01773

LITTLE, STEPHEN JAMES, b Akron, Ohio, July 5, 39; m 73; c 2. ASTRONOMY, ASTROPHYSICS. *Educ:* Univ Kans, BA, 61, MA, 63; Univ Calif, Los Angeles, PhD(astron), 71. *Prof Exp:* Fac assoc astron, Univ Tex, Austin, 68-70; asst prof, Ferris State Col, Big Rapids, Mich, 70-75; asst prof astron, Wellesley Col, 75-83; ASST PROF ASTRON, BENTLEY COL, 83- *Concurrent Pos:* Consult scientist, Solar Physics Div, Am Sci & Eng, Cambridge, 76-77; sci tech staff, Gen Res Co, 84- *Mem:* Am Astron Soc; Sigma Xi. *Res:* Astronomical spectroscopy and photometry of red giant stars, Ap stars, and planets; solar x-ray physics. *Mailing Add:* 17 Service Dr Wellesley MA 02181

LITTLE, THOMAS MORTON, b Picture Rocks, Pa, July 25, 10; m 38; c 2. BIOMETRICS. *Educ:* Bucknell Univ, AB, 31; Univ Fla, MS, 33; Univ Md, PhD(genetics), 43. *Prof Exp:* Asst entom, Cornell Univ, 31; chief hybridist, W Atlee Burpee Co, Calif, 34-41; asst geneticist, USDA, Md, 41-44; dir crop res, Basic Veg Prod, Inc, 44-49; asst prof biol & hort, Univ Nev, 49-53; exten veg crops specialist, 53-64, exten biometrician, 64-72, EMER EXTEN BIOMETRICIAN, UNIV CALIF, RIVERSIDE, 72- *Concurrent Pos:* Fulbright lectr, Univ Zagreb, 67-68; biometrician, Int Inst Trop Agr, Ibadan, Nigeria, 72-74. *Mem:* Am Soc Hort Sci. *Res:* Design and analysis of agricultural experiments. *Mailing Add:* 1488 Argyle Lane Bishop CA 93514

LITTLE, WILLIAM ARTHUR, b Adelaide, SAfrica, Nov 17, 30; nat US; m 55; c 3. PHYSICS. *Educ:* Univ SAfrica, BSc, 50; Rhodes Univ, SAfrica, PhD, 55; Univ Glasgow, PhD, 57. *Prof Exp:* Nat Res Coun Can fel, Univ BC, 56-58; from asst prof to assoc prof, 58-65, PROF PHYSICS, STANFORD UNIV, 65- *Concurrent Pos:* Alfred P Sloan fel, 59-62; Guggenheim fel, 64-65; invited prof, Univ Geneva, 64-65; NSF sr fel, 71-72; chmn, MMR Technol Inc. *Mem:* Fel Am Phys Soc. *Res:* Organic fluorescence; magnetic resonance; low temperature physics; superconductivity; phase transition; chemical physics; neural network theory. *Mailing Add:* Dept of Physics Stanford Univ Stanford CA 94305

LITTLE, WILLIAM ASA, b Ellenville, NY, July 14, 31; m 52; c 2. OBSTETRICS & GYNECOLOGY. *Educ:* Johns Hopkins Univ, AB, 51; Univ Rochester, MD, 55. *Prof Exp:* Intern obstet & gynec, Barnes Hosp, Washington Univ, 55-56; resident, Columbia-Presby Med Ctr, 56-61; from asst prof to assoc prof, Univ Fla, 61-66; prof obstet & gynec & chmn dept, Univ Miami, 66-; AT AM FERTILITY SOC. *Concurrent Pos:* Josia Macy Jr fel, 57-60; Am Cancer Soc fel, 60-61; consult, US Navy, 61-65; asst chief perinatal res, Nat Inst Neurol Dis & Stroke, 62-64, consult, 64-, Nat Inst Child Health & Human Develop, 63-; Markle scholar, 62-; adv, US Food & Drug Admin, 64- *Mem:* AAAS; Am Col Obstet & Gynec; Am Fertil Soc; Am Pub Health Asn; AMA. *Res:* Placental pathology and transfer; enzymology; fetal pharmacology; teratology; mental retardation; cancer; reproductive biology. *Mailing Add:* Dept Ob/Gyn Univ Miami Sch Med R 116 PO Box 016960 Miami FL 33101

LITTLE, WILLIAM C, b Cleveland, Ohio, May 1, 50; m 75; c 2. CARDIOLOGY, CARDIAC PHYSIOLOGY. *Educ:* Oberlin Col, Ohio, BA, 72; Ohio State Univ, MD, 75. *Prof Exp:* Instr, Univ Ala Sch Med, 80-81; from asst prof to assoc prof med, Univ Tex Health Sci Ctr, 81-86; assoc prof, 86-89, PROF MED, BOWMAN GRAY SCH MED, 89-, CHIEF, CARDIOL SECT, 90- *Concurrent Pos:* Prin investr, NIH, 86-92 & 89-94, mem, Study Sect, 89-90; Am Heart Asn estab investr award, 86; Am Fedn Clin Res young investr award, 91. *Honors & Awards:* Lamport Award, Am Physiol Soc, 87. *Mem:* Am Soc Clin Invest; Am Physiol Soc; Cardiac Syst Dynamics Soc; Asn Univ Cardiologists; Asn Professors Cardiol. *Res:* Cardiac dynamics and pathophysiology of myocardial infarction. *Mailing Add:* Sect Cardiol Bowman Gray Sch Med Medical Center Blvd Winston-Salem NC 27157-1045

LITTLE, WILLIAM FREDERICK, b Morganton, NC, Nov 11, 29; m 58; c 1. ORGANIC CHEMISTRY, INORGANIC CHEMISTRY. *Educ:* Lenoir Rhyne Col, BS, 50; Univ NC, MA, 52, PhD(org chem), 55. *Hon Degrees:* DSc, Lenoir Rhyne Col, 85. *Prof Exp:* Instr chem, Reed Col, 55-56; from instr to assoc prof, 56-65, prof chem & chmn dept, 65-70, UNIV DISTINGUISHED PROF, UNIV NC, CHAPEL HILL, 77- *Concurrent Pos:* Consult, Res Triangle Inst, 56-69; asst to dean, Grad Sch Bus Admin, Univ NC, Chapel Hill, 59-62, vice chancellor develop & pub serv, 73-78; pres, Triangle Univ Ctr Adv Studies, Inc, 82-86; chmn & corp secy, Exec comt Res Triangle Found, 86- *Honors & Awards:* Thomas Jefferson Award, 80. *Mem:* Am Chem Soc; Sigma Xi. *Res:* Organometallic compounds, especially metallocenes, and group VIII metals. *Mailing Add:* Dept of Chem Univ of NC Chapel Hill NC 27599-3290

LITTLE, WINSTON WOODARD, JR, b Gainesville, Fla, Sept 4, 38. NUCLEAR ENGINEERING. *Educ:* Mass Inst Technol, BS, 60, MS, 62, ScD(nuclear eng), 64. *Prof Exp:* Mgr Fast Flux Test Facil, Nuclear Design & Anal Unit, Pac Northwest Labs, Battelle Mem Inst, 66-77; CONSULT SCIENTIST, WESTINGHOUSE HANFORD CO, 77- *Mem:* Am Nuclear Soc. *Res:* Nuclear design of the fast flux test facility. *Mailing Add:* Westinghouse Hanford Co PO Box 999 MS K6-42 Richland WA 99352

LITTLEDIKE, ERNEST TRAVIS, b Logan, Utah, Feb 25, 35; m 60; c 4. ENDOCRINOLOGY, MINERAL METABOLISM. *Educ:* Utah State Univ, BS, 58; Wash State Univ, DVM, 60; Univ Ill, PhD(physiol), 65. *Prof Exp:* Instr anat, Univ Ill, 60-62, NIH fel physiol, 62-64; fel endocrinol, Univ Wis, 64-65; vet med officer physiol, Nat Animal Dis Ctr, 65-85; NAT PROG LEADER, AGR RES SERV, USDA, 85- *Concurrent Pos:* Vis scientist endocrinol, Mayo Clinic, 75-76; adj prof, Col Vet Med, Iowa State Univ, 70- *Mem:* Endocrine Soc; Am Soc Bone & Marine Res; World Vet Anatomists Asn; World Vet Physiologists & Pharmacologists Asn. *Res:* Mineral metabolism in domestic amimals and the diseases that result from mineral imbalances; factors that effect mineral metabolism and the pathogenesis of diseases of mineral metabolism in domestic animals. *Mailing Add:* US Meat Animal Res Ctr PO Box 166 Clay Center NE 68933

LITTLEFIELD, JOHN WALLEY, b Providence, RI, Dec 3, 25; m 50; c 3. GENETICS, ANIMAL PHYSIOLOGY. *Educ:* Harvard Med Sch, MD, 47. *Prof Exp:* From intern to resident med, Mass Gen Hosp, 47-50; from clin & res fel to asst prof med, Harvard Med Sch, 54-66, tutor, 59-65, asst prof pediat, 66-69, prof, 70-73; prof pediat & chmn dept, Johns Hopkins Univ, 74-85, PROF PHYSIOL & CHMN DEPT, 85- *Concurrent Pos:* USPHS fel, Inst Enzyme Res, Univ Wis, 51; Am Cancer Soc scholar, 56-59; tutor, Harvard Med Sch, 59-65; Guggenheim fel, 65-66; Josiah Macy scholar, 79. *Mem:* Nat Acad Sci; Am Soc Human Genetics; Am Soc Clin Invest; Am Pediat Soc; Am Soc Biol Chemists; Soc Pediat Res; Asn Am Physicians. *Res:* Somatic cell and molecular human genetics; developmental biology. *Mailing Add:* Dept Physiol Johns Hopkins Univ Sch Med Baltimore MD 21205

LITTLEFIELD, LARRY JAMES, b Ft Smith, Ark, Feb 7, 38; m 63. PLANT PATHOLOGY. *Educ:* Cornell Univ, BS, 60; Univ Minn, MS, 62, PhD(plant path), 64. *Prof Exp:* Res asst plant path, Univ Minn, 60-64; NSF res fel, Uppsala, 64-65; from asst prof to prof plant path, NDak State Univ, 65-85; PROF & HEAD DEPT PLANT PATH, OKLA STATE UNIV, 85- *Concurrent Pos:* NIH fel, Purdue Univ, 69-70; res fel, Oxford Univ, 73-74. *Mem:* Mycol Soc Am; Brit Mycol Soc; Am Phytopathological Soc. *Res:* Histology of host-parasite relations; fungus physiology; electron microscopy of fungi and diseased plants. *Mailing Add:* Plant Path Dept Okla State Univ Stillwater OK 74078

LITTLEFIELD, NEIL ADAIR, b Santa Fe, NMex, Apr 25, 35; m 60; c 5. TOXICOLOGY. *Educ:* Brigham Young Univ, BS, 61; Utah State Univ, MS, 64, PhD(toxicol), 68. *Prof Exp:* Res assoc air pollution, Univ Utah, 66-67; staff scientist inhalation toxicol, Hazleton Lab, Inc, 67-70; pharmacologist pesticide regulation, Environ Protection Agency, 71-72; toxicologist, 72-79, DIR, DIV CHEM TOXICOL, NAT CTR TOXICOL RES, FOOD & DRUG ADMIN, 79- *Concurrent Pos:* Chmn, Interagency Task Force Inhalation Chronic Toxicity & Carcinogenesis, 74-75; mem, Food & Drug Admin Task Force Aerosol Prod, 75- *Mem:* Sigma Xi. *Res:* Investigations in concepts of long-term, low-dose exposures; extrapolation of animal toxicology data to risk-benefit in man; carcinogenesis. *Mailing Add:* 3404 Millbrook Rd Little Rock AR 72207

LITTLEJOHN, OLIVER MARSILIUS, b Cowpens, SC, Sept 29, 24; m 48; c 2. PHARMACY. *Educ:* Univ SC, BS, 48 & 49; Univ Fla, MS, 51, PhD(pharm), 53. *Prof Exp:* Asst prof pharm & head dept, Southern Col Pharm, 53-56; prof & head dept, Univ Ky, 56-57; DEAN, SOUTHERN SCH PHARM, MERCER UNIV, 57- *Mem:* Fel Am Found Pharmaceut Educ; Sigma Xi. *Res:* Pharmaceutical preservatives. *Mailing Add:* 6485 Bridgewood Valley Rd NW Atlanta GA 30328

LITTLE-MARENIN, IRENE RENATE, b Pilsen, Czech, May 4, 41; US citizen; m 73; c 2. ASTROPHYSICS. *Educ:* Vassar Col, AB, 64; Ind Univ, MA, 66, PhD(astrophys), 70. *Prof Exp:* Fel astron, Ohio State Univ, 70-72; asst prof, Univ Western Ont, 72-73; teaching fel, Ferris State Col, 74; res asst solar x-rays, Am Sci & Eng, 76-77; ASST PROF ASTRON, WELLESLEY COL, 77- *Concurrent Pos:* Sabbatical leave, vis asst prof, Dennison Univ & Ohio State, 80-81; vis scientist, Univ Colo, 80-81; res fel, JILA, 83-84, Univ resident res fel, AFGL, 86-88. *Mem:* Am Astron Soc; Sigma Xi. *Res:* A search for the radioactive element technetium in long-period variable stars; analysis of IRAS; low resolution spectra. *Mailing Add:* Whitin Observ Wellesley Col Wellesley MA 02181

LITTLEPAGE, JACK LEROY, b San Diego, Calif, Apr 14, 35; m 60. BIOLOGICAL OCEANOGRAPHY. *Educ:* San Diego State Col, BA, 57; Stanford Univ, PhD(biol), 66. *Prof Exp:* Asst prof, 65-71, ASSOC PROF BIOL, UNIV VICTORIA, 71- *Concurrent Pos:* Oceanographic consult to mining industs, 71- *Mem:* AAAS; Am Soc Limnol & Oceanog. *Res:* Physiology and ecology of marine zooplankton, especially copepods and euphausids; pollution monitoring; salmonid IHN virus; geothermal aguaculture. *Mailing Add:* Dept of Biol Univ of Victoria Victoria BC V8W 2Y2 Can

LITTLER, MARK MASTERTON, b Athens, Ohio, Sept 24, 39; m 66. TAXONOMY, FUNCTIONAL MORPHOLOGY. *Educ:* Ohio Univ, BS, 61, MS, 66; Univ Hawaii, PhD(marine bot), 71. *Prof Exp:* Chemist, Testing Lab, Ohio State Hwy Dept, 61-64; from asst prof to assoc prof, Univ Calif, Irvine, 70-80, prof biol res, 80-82; chmn cur bot res, 82-87, SR BOTANIST RES, SMITHSONIAN INST, 85- *Concurrent Pos:* Vis prof, Stanford Univ, 73 & 74, Univ SCalif, 73 & 75; mem bd, SCalif Acad Sci, 80-82; distinguished vis scientist, Univ Nebr, Lincoln, 81, Northwestern Univ, 81; assoc ed, Aquatic Bd, 82-85, J Psychol 82-86, ed, Smithsonian Contrib to Bot, 82-87; assoc ed, Phycolog Soc Am, 82-86; adj prof, George Mason Univ, 84-; mem bd vis, Ohio Univ Col Arts & Sci, 85-88; rep, Int Univ Biol Sci, Biol Monitoring Proj, Int Phycol Soc, 85- *Honors & Awards:* Earle C Anthony Innovative Res Award, 73; Durbaker Prize, Bot Soc Am, 84; Medal Merit, Ohio Univ, 85. *Mem:* Ecol Soc Am; Int Phycol Soc; Phycol Soc Am; Int Soc Reef Studies; Am Soc Limnol & Oceanog. *Res:* Man's effect on marine ecosystems; taxonomy, developmental morphology and seasonal cycles of marine benthos and phytoplankton; standing stock, productivity and the physiological ecology of temperate and reef-building benthic organisms. *Mailing Add:* Dept Bot NHB 166 Nat Mus Nat Hist Smithsonian Inst Washington DC 20560

LITTLETON, H(AROLD) T(HOMAS) J(ACKSON), b Parksley, Va, June 28, 21; m 48; c 1. CHEMICAL ENGINEERING. *Educ:* Univ Va, BChE, 43. *Prof Exp:* Chem engr, Naval Res Lab, 43-46; chem engr, Res Div, Polychem Dept, Exp Sta, E I du Pont de Nemour Co, 46-53, res supvr, 53-62, sr res engr, Plastic Dept, 62-69, res assoc, 69-73, lab adminr, Platics Prod & Resins Dept, 74-82; RETIRED. *Mem:* Am Chem Soc. *Res:* Process development on nylon intermediates; high pressure processes; new plastics and plastics processing methods. *Mailing Add:* 320 Walden Rd Sharpley Wilmington DE 19803

LITTLETON, JOHN EDWARD, b Ballston Spa, NY, July 28, 43; m 88. STELLAR EVOLUTION, MOLECULAR SPECTROSCOPY. *Educ:* Cornell Univ, BS, 65; Univ Rochester, PhD(astrophys), 72. *Prof Exp:* Res assoc astrophys, Belfer Grad Sch Sci, Yeshiva Univ, 72-73; res fel, Harvard Col Observ, 73-74, res assoc, 74-75; from asst prof to assoc prof, 75-88, PROF PHYSICS, WVA UNIV, 88- *Concurrent Pos:* Vis assoc prof, Univ Ill, 82; vis assoc res astron, Univ Calif Berkeley, 84, 85, 86; vis scientist, Indiana Univ, 84. *Mem:* Am Astron Soc; Am Asn Physics Teachers; Int Astron Union. *Res:* Plasma fluctuations in astrophysical media; structure of detonation waves in degenerate stellar cores; hydrodynamic properties of stellar atmospheres; molecular bands in cool, stellar atmospheres. *Mailing Add:* Dept Physics WVa Univ Morgantown WV 26506

LITTLETON, PRESTON A, JR, DENTAL RESEARCH. *Prof Exp:* EXEC DIR, AM ASN DENT SCHS, 90- *Mailing Add:* Am Asn Dent Schs 1625 Massachusetts Ave NW Washington DC 20036

LITTLETON, ROBERT T, b Sheridan, Wyo, Jan 14, 16. ENGINEERING GEOLOGY. *Educ:* Ore State Univ, BS, 41. *Prof Exp:* Regional geologist, US Bur Reclanation, 61-80; RETIRED. *Mem:* Geological Soc Am. *Mailing Add:* 624 Don Vincente Dr Boulder City NV 89005

LITTLEWOOD, BARBARA SHAFFER, b Buffalo, NY, Oct 8, 41; m 70. BIOCHEMISTRY, GENETICS. *Educ:* Univ Rochester, BA, 63; Univ Pa, PhD(biochem), 68. *Prof Exp:* NIH trainee, Cornell Univ, 68-70; res assoc biochem, 70-73, res assoc physiol chem, 73-76, RES ASSOC BIOCHEM, UNIV WIS-MADISON, 76- *Concurrent Pos:* Lectr, dept genetics, Cornell Univ, 70 & dept biochem, Univ Wis-Madison, 72. *Mem:* Genetics Soc Am; Am Soc Microbiol. *Res:* Yeast genetics and biochemistry. *Mailing Add:* 5109 Coney Weston Pl Madison WI 53711

LITTLEWOOD, ROLAND KAY, b Mendota, Ill, Nov 26, 42; m 70; c 2. COMPUTER SCIENCE, MOLECULAR BIOLOGY. *Educ:* Univ Ill, Urbana, BS, 64; Cornell Univ, PhD(genetics), 70. *Prof Exp:* NIH fel, Lab Molecular Biol, 70-72, res assoc, 72-78, asst scientist, 78-86, SR INFO PROCESSING CONSULT, BIOPHYS LAB, UNIV WIS-MADISON, 86- *Concurrent Pos:* Proprietor, Digital Comput Appl, 74- *Mem:* Asn Comput Mach. *Res:* Application of computers in the biological sciences. *Mailing Add:* Inst for Molecular Virol 1525 Linden Dr Univ Wis Madison WI 53706

LITTMAN, ARMAND, b Chicago, Ill, Apr 4, 21; m 52; c 3. MEDICINE. *Educ:* Univ Ill, Chicago, BS, 42, MD, 43, MS, 48, PhD(physiol), 51. *Prof Exp:* Intern, Cook County Hosp, Chicago, 44; from clin asst to assoc prof, 46-64, PROF MED, UNIV ILL COL MED, 64-; CHIEF MED SERV, HINES VET ADMIN HOSP, 59- *Concurrent Pos:* Raymond B Allen instructorship award, Univ Ill, 57; US Atomic Energy Comn travel award, 58; resident, Cook County Hosp, Chicago, 48-50; pvt pract, 52-59; attend physician, Res & Educ Hosps, 55-; prof, Cook County Grad Sch Med, 58- *Mem:* AMA; Am Col Physicians; Am Fedn Clin Res; Am Gastroenterol Asn; Sigma Xi. *Res:* Gastroenterology; physiology. *Mailing Add:* Vet Admin Hosp Hines IL 60141

LITTMAN, BRUCE H, b New York, NY, Nov 18, 44; m; c 1. EXPERIMENTAL MEDICINE. *Educ:* Univ Wis, BS, 66; State Univ NY, MD, 70; Am Bd Internal Med, dipl, 75, dipl rheumatology, 78. *Prof Exp:* Intern med, Tufts New Eng Med Ctr, Boston, 70-71; staff assoc tumor & immunol res, Nat Cancer Inst, NIH, Bethesda, Md, 71-73; resident, Tufts New Eng Med Ctr, 73-74; res fel, Robert B Brigham Hosp, Harvard Med Sch, 74-76, asst med, Peter Bent Brigham Hosp, 74-76; postdoctoral fel, Am Cancer Soc, Mass Div, Boston, 74-76; from asst prof to assoc prof med & microbiol, Med Col Va, 76-89; SR ASSOC DIR EXP MED, PFIZER CENTRES, PFIZER INC, GROTON, CONN, 89- *Concurrent Pos:* Vis scientist, Metab Br, Nat Cancer Inst, Bethesda, Md, 81-82; chief, Rheumatology Sect, Med Serv, McGuire Vet Admin Med Ctr, Richmond, 82-89; mem, Immunol Sci Study Sect, Div Res Grants, NIH, 83-87. *Mem:* Fel Am Col Rheumatology; Am Fedn Clin Res; Am Asn Immunologists; fel Am Col Physicians; Sigma Xi. *Res:* Transplantation immunity in dogs; cellular immunology; immunogenetics; nucleotide biochemistry; author of numerous articles, chapters and books. *Mailing Add:* Pfizer Cent Res Pfizer Inc Groton CT 06340

LITTMAN, HOWARD, b Brooklyn, NY, Apr 22, 27; m 55; c 3. CHEMICAL ENGINEERING. *Educ:* Cornell Univ, BChE, 51; Yale Univ, PhD(chem eng), 56. *Prof Exp:* From asst prof to assoc prof chem eng, Syracuse Univ, 56-65; assoc prof, 65-67, PROF CHEM ENG, RENSSELAER POLYTECH INST, 67- *Concurrent Pos:* Resident res assoc, Argonne Nat Lab, 57-59; res asst chem eng, Brookhaven Nat Lab, NY, 57; vis prof, Imperial Col, London,

71-72; Fulbright-Hays lectr, Univ Belgrade, 72; vis prof, Chonnam Nat Univ, Kwangju, Korea. *Mem:* Am Chem Soc; Am Inst Chem Engrs. *Res:* Fluidization and fluid-particle systems. *Mailing Add:* Dept Chem Eng Rensselaer Polytech Inst Troy NY 12181

LITTMAN, WALTER, b Vienna, Austria, Sept 17, 29; US citizen; m 60; c 3. MATHEMATICAL ANALYSIS. *Educ:* Univ NY, BA, 52, PhD(math), 56. *Prof Exp:* Instr math, Univ Calif, Berkeley, 56-58, lectr, 58-59; asst prof, Univ Wis, 59-60; from asst prof to assoc prof, 60-66, PROF MATH, UNIV MINN, MINNEAPOLIS, 66- *Concurrent Pos:* Vis mem, Courant Inst Math Sci, NY Univ, 67-68; vis prof, Mittag-Leffler Inst, Djursholm, Sweden, 74, Chalmers Technol Univ, Gothenburg, Sweden, 75, Hebrew Univ, Jerusalem, Israel, 81-82 & 89. *Mem:* Am Math Soc. *Res:* Partial differential equations; functional analysis; mathematical physics. *Mailing Add:* Dept of Math Univ of Minn Minneapolis MN 55455

LITTMANN, MARTIN F(REDERICK), b Brazil, Ind, Feb 9, 19; m 44; c 2. METALLURGY & PHYSICAL METALLURGICAL ENGINEERING. *Educ:* Univ Cincinnati, ChemE, 41, MS, 43. *Prof Exp:* From jr res engr to sr res engr, Armco Inc, 43-68, prin res assoc, 68-75, prin res engr, 68-75, prin res engr, 75-83; RETIRED. *Concurrent Pos:* Consult, 83- *Mem:* Inst Elec & Electronics Engrs; Am Inst Mining, Metall & Petrol Engrs. *Res:* Deformation and recrystallization orientations in soft magnetic materials; studies of magnetic properties in relation to metallurgy of soft magnetic materials. *Mailing Add:* 1050 Knoll Lane Middletown OH 45042-1669

LITTON, GEORGE WASHINGTON, animal husbandry; deceased, see previous edition for last biography

LITVAK, AUSTIN S, b Staten Island, NY, Dec 1, 33. UROLOGY. *Educ:* Wagner Col, BS, 54; Univ Va, MD, 58; Am Bd Urol, dipl. *Prof Exp:* Intern surg, Univ Va Hosp, 58-59; NIH fel, Surg Br, Nat Cancer Inst, 59-60; USPHS surgeon, Claremore Indian Hosp, Okla, 60-61; resident urol, Hosp Univ Pa, Jefferson Med Col Serv & Philadelphia Gen Hosp, 61-63; pvt pract, 63-73; asst in surg, Div Urol, Hahnemann Med Col, 70-73; asst prof, 73-75, ASSOC PROF SURG, DIV UROL, UNIV KY MED CTR, 76- *Concurrent Pos:* Asst attend urol, Monmouth Med Ctr & Riverview Hosp; assoc attend & div rep, Bayshore Hosp; courtesy staff, Jersey Shore Med Ctr; asst urol staff, Hahnemann Med Col; mem urol staff, Univ Ky Med Ctr 73-78; consult urol, Vet Admin Hosp, 73-75; active staff, St Josephs Hosp, Central Baptist Hosp & Claril County Hosp. *Honors & Awards:* First Prize, Nat Clin Soc, 68. *Mem:* Am Urol Asn; Am Fertil Soc; AMA; fel Am Col Surgeons; fel Int Col Surgeons. *Res:* Embryology of the kidney; acute and chronic prostatitis; lower urinary tract infections in children and adults; diseases of the kidney, urethra and bladder in children. *Mailing Add:* PO Box 4066 Lexington KY 40504

LITVAK, MARVIN MARK, b Newark, NJ, Oct 20, 33; m 63; c 2. RADIATION-PLASMA INTERACTION, QUANTUM ELECTRONIC & OPTICS. *Educ:* Cornell Univ, BEngPhys, 56, PhD(theoret physics), 60. *Prof Exp:* Consult, Avco Corp, 55-60, sr staff mem, Avco-Everett Res Lab, 60-63; group leader, Lincoln Lab, Mass Inst Technol, 63-70; sr radio astronomer, Smithsonian Astrophys Observ, 70-78; MEM TECH STAFF, JET PROPULSION LAB, CALIF INST TECHNOL, 78- *Concurrent Pos:* Lectr, Harvard Col Observ, 70-78; consult, Lincoln Lab, Mass Inst Technol, 77-81. *Mem:* Int Astron Union; Am Phys Soc; Am Astron Soc. *Res:* Magnetohydrodynamic shock waves and power generation; current-carrying superconductors; laser-produced gas breakdown; gas laser characteristics; non-linear propagation effects of lasers; nonlinear properties of semiconductors; interstellar molecules and masers; millimeter-wave and submillimeter-wave radio astronomy and aeronomy. *Mailing Add:* 1525 Espinaso Circle Palas Verdes Estates CA 90274

LITVAN, GERARD GABRIEL, b Vienna, Austria, May 17, 27; Can citizen; m 64; c 1. SURFACE CHEMISTRY. *Educ:* Eotovos Univ, Budapest, Dipl, 52; Univ Toronto, PhD(surface chem), 62. *Prof Exp:* Asst prof phys chem, Inst Phys Chem, Eotovos Univ, 52-55; assoc phys chem, Cent Chem Res Inst, Hungarian Acad Sci, 55-56; res chemist phys polymer chem, Can Industs Ltd, 57-59; PRIN RES OFFICER, INST RES CONSTRUCT, NAT RES COUN CAN, 62- *Concurrent Pos:* Lectr, Chem Dept, Carleton Univ, 66-67; First Int Conf Durability of Bldg Mat & Components; Can Stand Asn Subcomt chmn, 78. *Mem:* Fel Am Ceramic Soc; Am Concrete Inst; Am Chem Soc; fel Chem Inst Can; Am Soc Testing & Mat; Can Standards Asn. *Res:* Phase transitions of substances adsorbed in porous solids; mechanism of cyroinjury and cyroprotection in plant and animal tissue; mechanism of frost action in porous building materials; materials science engineering; corrosion of reinforcing steel in concrete,; rehabilitation of concrete structures. *Mailing Add:* Inst Res Construct Nat Res Coun Can Ottawa ON K1A 0R6 Can

LITWACK, GERALD, b Boston, Mass, Jan 11, 29; m 56, 73; c 3. BIOCHEMISTRY. *Educ:* Hobart Col, BA, 49; Univ Wis, MS, 51, PhD(biochem), 53. *Prof Exp:* from asst prof to prof biochem, Rutgers Univ, 54-64; res assoc prof, Grad Sch Med, Univ Pa & dir biochem, Div Cardiol, Philadelphia Gen Hosp, 60-64; prof biochem, Sch Med, Temple Univ, 64-91, fels res inst, 64-91, dep dir, 78-91, Laura H Carnell prof biochem, 88-91; CHMN, DEPT PHARMACOL & DEP DIR, JEFFERSON CANCER INST, THOMAS JEFFERSON UNIV, 91- *Concurrent Pos:* Nat Found Infantile Paralysis fel, Biochem Lab, Univ Sorbonne, 53-54; trainee, Oak Ridge Inst Nuclear Studies, 55; Nat Inst Arthritis & Metab Dis res career develop award, 63-69; vis prof, Univ Calif, 56; hon prof, Rutgers Univ, 60-64; vis scientist, Univ London, 71 & Univ Calif, 72; mem adv bd, Biochem & Chem Carcinogenesis, Am Cancer Soc, 77-80, chmn, 79, Endocrinol, 80-83, Anticancer Res, 82-; mem, Cell Physiol Panel, NSF, 80-83; counc, Soc Exp Biol & Med, 84-; spec study, Reproductive Endocrine Sect, NIH, 85. *Honors & Awards:* Lalor Found Award, 56; Fac Res Award, Temple Univ, 87. *Mem:* Am Soc Biol Chemists; Am Asn Cancer Res; Endocrine Soc. *Res:* Ligandin; hormonal control of enzyme formation and activity; glucocorticoid receptor. *Mailing Add:* Dept Pharmacol & Jefferson Cancer Inst Jefferson Med Col Thomas Jefferson Univ Tenth & Locust Sts Philadelphia PA 19107

LITWAK, ROBERT SEYMOUR, b New York, NY, Nov 25, 24; c 3. SURGERY. *Educ:* Ursinus Col, BS, 45; Hahnemann Med Col, MD, 49; Am Bd Surg, dipl, 56; Am Bd Thoracic Surg, dipl, 58. *Prof Exp:* Asst surg, Sch Med, Boston Univ, 52; from instr to assoc prof surg, Med Sch, Univ Miami, 56-62; ATTEND SURGEON & CHIEF DIV CARDIOTHORACIC SURG, MT SINAI HOSP, 62-; PROF SURG, MT SINAI SCH MED, 71- *Concurrent Pos:* Consult, Vet Admin Hosp, Coral Gables, Fla, 57- & Variety Children's Hosp, 59-; chief div thoracic & cardiovascular surg, Jackson Mem Hosp, 59-62. *Mem:* Fel Am Col Surg; fel Am Col Chest Physicians; fel Am Col Cardiol; fel NY Acad Sci. *Res:* Cardiovascular physiology; cardiac surgery. *Mailing Add:* One E 100th St New York NY 10029

LITWHILER, DANIEL W, b Ringtown, Pa, Feb, 28, 42; m 66; c 4. OPERATIONS RESEARCH. *Educ:* Fla State Univ, BS, 63, MS, 65; Univ Okla, PhD(indust eng), 77. *Prof Exp:* Opers officer, mgt analyst, comptroller & progs officer, US Air Force, 65-72; plans & issue analyst, staff group, Off Secy Air Force, 82; assoc prof, 72-74 & 77-81, tenure prof, 83-85, PROF MATH & HEAD DEPT MATH SCI, US AIR FORCE ACAD, 86- *Mem:* Opers Res Soc Am. *Res:* Location theory, particularly large regions; military applications of operations research. *Mailing Add:* Dept Math Sci US Air Force Acad Colorado Springs CO 80840

LITWIN, MARTIN STANLEY, b Florence, Ala, Jan 8, 30; m 85; c 4. SURGERY. *Educ:* Univ Ala, BS, 51, MD, 55; Am Bd Surg, dipl, 63. *Prof Exp:* Instr med physiol, Sch Med, Univ Ala, 53; intern surg, Michael Reese Hosp, Chicago, 55-56; asst, Peter Bent Brigham Hosp, Boston, 56-58, jr asst resident, 57 & 58-59, sr asst resident, 60-61, sr resident, 61-62; instr, Harvard Med Sch, 66; from asst prof to prof, 66-77, ASSOC DEAN MED DIR FAC PRACT, ROBERT & VIOLA LOBRANO PROF SURG, SCH MED, TULANE UNIV, 77. *Concurrent Pos:* Surg res fel, Harvard Med Sch, 56-58; surg registr, St Mary's Hosp & Med Sch, London, 59-60; George Gorham Peters fel, Peter Bent Brigham Hosp, Boston, 59-60, chief surg res, 65, jr assoc surg, 65-66; Am Cancer Soc clin fel, 61-62; regr to prof teaching unit, St Mary's Hosp, London, 59-60; clin invetr, Vet Admin Hosp, W Roxbury, 64-66, consult, 66-; adj prof biomed eng, Northeastern Univ, 64-67; mem, Surgeon-Gen Adv Comt Optical Lasers, Working Group Safety Stand Use Lasers, Armed Forces Nat Res Coun Comt Vision, Ad Hoc Initial Rev Group, Nat Ctr Radiol Health & Spec Study Sect Laser & NIH session chmn, Gordon Conf Lasers Biol Med, 65-67; sr vis surgeon, Charity Hosp of La, mem active staff, Tulane Hosp; Nat Heart Inst investr career develop award, 68-72. *Mem:* Fel Am Col Surgeons; Am Surg Asn; Am Asn Surg Trauma; Soc Univ Surgeons; Soc Surg Alimentary Tract; Int Surg Soc; Europ Soc Surg Res. *Res:* Blood rheology; vascular and gastrointestinal surgery; surgical metabolism; blood transfusion and treatment of skin cancer. *Mailing Add:* Tulane Univ Med Sch 1415 Tulane Ave New Orleans LA 70112

LITWIN, STEPHEN DAVID, b New York, NY, Apr 30, 34; m 61; c 3. GENETICS, IMMUNOLOGY. *Educ:* Brooklyn Col, BS, 55; New York Univ, MD, 59. *Prof Exp:* Intern & resident med, Montefiore Med Ctr, 59-64; investr immunol, Rockefeller Univ, 64-67; res hematologist, US Army, 67-69; asst prof, 65-71, assoc prof, 71-79, PROF MED, CORNELL MED COL, NEW YORK HOSP, 79- *Concurrent Pos:* USPHS career develop award, 70-72; mem various study sects, NIH, Ad Hoc Study Sect, Am Cancer Soc, 82. *Mem:* Am Soc Clin Investigation; Am Asn Human Genetics; Am Asn Immunologists; Sigma Xi. *Res:* Genetic control of human immunoglobulin synthesis including the genetic allotypes and subclasses of human immunoglobulin and inherited immunodeficiency diseases. *Mailing Add:* 3 Brevity Ct Binghamton NY 13905

LITYNSKI, DANIEL MITCHELL, b Amsterdam, NY, Mar 13, 43; m 63; c 3. OPTICAL SIGNAL PROCESSING. *Educ:* Rensselaer Polytech Inst, BS, 65, PhD(physics), 78; Univ Rochester, MS, 71. *Prof Exp:* Exec officer B, 2/34 Armor Battalion, Vietnam, US Army, 66-67, commanding officer, 551st Lt Maint Co, Vietnam and H0 Co, US Ord Ctr, 67-68, res physicist, Ballistics Res Labs, 72-73, asst prof physics, US Mil Acad, 74-78, exec officer, 19th Maintenance Battalion, WGer, 78-80, assoc prof, 80-86, prof, dep & actg head, Elec Engr, 86-89, US ARMY, 65-,PROF & HEAD ELEC ENGR & COMPUTER SCI, US MIL ACAD, WEST POINT, NY, 90- *Concurrent Pos:* Vis researcher, Harry Diamond Labs, 74 & 75; lectr, Europ Div, Univ Md, 80; prin investr, US Army Res Off, 81-; adj res prof, Elec Comput & Syst Engr Dept, Rensselaer Polytech Inst, 87-; USMA fel, Indust Col Armed Forces, 88-89. *Mem:* Sigma Xi; Inst Elec & Electronics Engrs; Optical Soc Am; Soc Photo-Optical Instrumentation Engrs; Am Soc Eng Educ; NY Acad Sci; Armed Forces Commun & Electronics Asn. *Res:* Optical signal processing using surface acoustic wave devices; optical matrix processing; optical computing. *Mailing Add:* Dept Elec Eng & Computer Sci US Mil Acad West Point NY 10996-1787

LITZ, LAWRENCE MARVIN, b Chicago, Ill, Oct 22, 21; m 42; c 4. GAS LIQUID SOLID MIXING, MEMBRANE APPLICATIONS & ENGINEERING. *Educ:* Univ Chicago, BS, 42; Ohio State Univ, PhD(phys chem), 48. *Prof Exp:* Res chemist, Tenn Valley Auth, 42-43; res metallurgist, Manhattan Proj, Los Alamos, 43-45; res chemist catalysis, Allied Chem Corp, 47-51; group leader chem metal, Standard Oil Calif, 51-53; group leader high temperature mat, Carbon Div, Union Carbide, 53-62, develop mgr fuel cells, Advan Develop Dept, 62-66, sr group leader membrane technol, Corp Develop, 66-72, gen mgr, Membrane Systs, 72-77, sr develop assoc, 77-86, mgr process chem, Linde Div, 86-90, CORP FEL, UNION CARBIDE INDUST GASES, INC, 86- *Honors & Awards:* Kirkpatrick Award, Chem Eng J, 75. *Mem:* Am Inst Chem Engrs; Am Chem Soc; Electrochem Soc; Soc Mining & Metall Engrs. *Res:* Membrane technology; high temperature materials and processes; electrochemical and process engineering; fuel cells, nuclear chemistry; chemical and physical metallurgy; gas-liquid chemical processing; bioengineering. *Mailing Add:* Union Carbide Indust Gases Inc Linde Div Tarrytown NY 10591

LITZ, RICHARD EARLE, b Presque Isle, Maine, July 3, 44; m 69; c 2. PLANT PATHOLOGY, HORTICULTURE. *Educ:* Dalhousie Univ, BA, 66, MSc, 68; Univ Nottingham, PhD(plant virol), 71. *Prof Exp:* Fel mycol, Univ Durham, 71-73; res officer plant path, Twyford Labs Ltd, 73-76; fel, 76-77, res assoc plant path, 77-78, asst prof, 79-84, ASSOC PROF, FRUIT CROPS DEPT, UNIV FLA, 84- *Concurrent Pos:* Rare Fruits Coun Int grant, 78-; USDA grant, 79-84; Rockefeller Found grants, 79-81 & 81-85. *Mem:* Int Asn Plant Tissue Cult; Sigma Xi; Am Soc Hort Sci; Int Soc Plant Molecular Biol. *Res:* Tissue culture of tropical fruits; disease resistance in tropical fruits. *Mailing Add:* TREC 18905 SW 280th St Homestead FL 33031

LITZENBERGER, LEONARD NELSON, b East Macungie, Pa, Oct 15, 45; m 74; c 1. LASERS. *Educ:* Lehigh Univ, BS, 67; Mass Inst Technol, SM, 69, PhD(physics), 71. *Prof Exp:* Prin res scientist, Avco Everett Res Lab, 71-90; PRIN SCIENTIST, TEXTRON DEFENSE SYSTS, 91- *Mem:* Am Phys Soc. *Res:* Laser research, technology development and applications; laser isotope separation; optical resonators; laser bandwidth control; electron beam generation and transport; high voltage, pulsed power systems; wave phenomena in plasmas. *Mailing Add:* Textron Defense Systs 2385 Revere Beach Pkwy Everett MA 02149

LITZENBERGER, SAMUEL CAMERON, b Calgary, AB, July 21, 14; nat US; m 41. AGRONOMY, FIELD CROPS. *Educ:* Colo State Univ, BS, 37; Mont State Univ, MS, 39; Iowa State Univ, PhD(plant path, crop breeding), 48. *Prof Exp:* From instr to assoc prof agron, Mont State Univ, 39-46, asst Exp Sta, 39-43, agronomist, 43-46; assoc agronomist, Exp Sta, Fla, 48-49; agronomist, Ala Exp Sta, Agr Res Admin, USDA, 49-51; agriculturist, res adv agron & head dept, Agr Technol Serv, US Opers Mission, Int Coop Admin, Nicaragua, 51-57, agron adv, Cambodia, 58-63, agron adv, Food & Agr Off & Chief Div, USAID, Guinea, 64-67, agr res adv & chief prod and breeding, Tunisia, 67-68, chief regional cereals off, NAfrica, 68-69, Food & Agr Off, 69-70, agronomy res specialist & chief, Crops Prod Div, 70-75, SPEC CONSULT, OFF AGR, TECH ASSISTANCE BUR, AID, WASHINGTON, DC, 75- *Concurrent Pos:* Consult, Colo State Univ, 75- *Mem:* Fel AAAS; fel Am Soc Agron; Am Phytopathological Soc; Am Genetic Asn; Am Forestry Asn; Am Hort Soc; fel Crop Sci Soc Am. *Res:* Inheritance and disease resistance of small grains; weed control; seed and soil improvement; crops production and improvement under tropical, subtropcial, temperate and sub-arctic environments; integrated programs for agricultural production and improvement of food crops for developing nations. *Mailing Add:* 1230 Tulip St Longmont CO 80501

LIU, ALICE YEE-CHANG, b Hunan, China, July 12, 48; US citizen; m 78. HEAT SHOCK GENES, CELL AGING. *Educ:* Chinese Univ Hong Kong, BSc Hons, 69; City Univ New York, PhD(pharmacol), 74. *Prof Exp:* Instr pharmacol, Mt Sinai Sch Med, 73; fel, Sch Med, Yale Univ, 73-77; asst prof pharmacol, Med Sch, Harvard Univ, 77-84; assoc prof, 84-89, PROF, DEPT BIOL SCI, RUTGERS STATE UNIV, 89- *Concurrent Pos:* Mem, Pharm Sci Rev Comt, NIQMS, NIH, 84-88; mem, Cell Biol Panel, NSF, 89-93. *Mem:* Am Soc Biochem Molecular Biol; Am Soc Pharmacol Exp Therapeut. *Res:* Signal transduction and gene expression; transcriptional regulation of heat shock genes in cell aging and differentiation; effects of cAMP in the regulation of gene expression. *Mailing Add:* Dept Biol Sci Nelson Biol Bldg Rutgers State Univ New Brunswick NJ 08854-1059

LIU, ANDREW T C, b Hong Kong, July 20, 29; Brit citizen; m 55; c 2. INTERNATIONAL BUSINESS DEVELOPMENT, PETROLEUM ACREAGE ACQUISITION. *Educ:* Ind Univ, Bloomington, BS, 51; Lowell Univ, Mass, MS, 53; London Univ, eng, PhD(org chem), 61; Imp Col London, Eng, DIC, 62. *Prof Exp:* Chemist, A C Lawrence Leather Co, Div Swift Co, 53-55; engr, Textile Res Inst, Beijing, China, 56-58; tech officer, ICI, Ltd, Manchester, Eng, 61-67; res chemist, E I du Pont de Nemours & Co, 67-71; chief exec officer & founder, Marble Industs, Liu Indust Corp, 71-77; prof polymer chem, Ling Nam Univ, 73-74; proj coordr, Conoco, Inc, 77-80, vpres, Continental Overseas Oil, 80-85; chief exec officer, China Ctr Technol Develop, 85-88; DIR POLYMER RES & DEVELOP & SR RES SCIENTIST, DENTSPLY INT, 88- *Mem:* Am Chem Soc; Am Soc Mat; Soc Plastic Engrs; Int Asn Dent Res. *Res:* Biocompatible polymer process synthesis of monomers; photo initiator synthesis; strategic planning of research and development; management of scientific research and technology development; entrepreneurship in chemical products; granted ten patents. *Mailing Add:* Dentsply Int 570 W College Ave PO Box 872 York PA 17405-0872

LIU, BEDE, b Shanghai, China, Sept 25, 34; US citizen; m 59; c 1. ELECTRICAL ENGINEERING. *Educ:* Nat Taiwan Univ, BSEE, 54; Polytech Inst Brooklyn, MEE, 56, DEE, 60. *Prof Exp:* Equipment engr, Western Elec Co, 54-56; mem tech staff commun systs, Bell Tel Labs, 59-62; from asst prof to assoc prof, 62-69, PROF ELEC ENG, PRINCETON UNIV, 69- *Concurrent Pos:* Mem bd dirs, Inst Elec & Electronic Engrs, 84-85. *Honors & Awards:* Centennial Medal, Inst Elec & Electronic Engrs, 85, Tech Achievement Award, Signal Processing Soc, 85, Educ Award, Circuits Systs Soc, 88. *Mem:* Fel Inst Elec & Electronic Engrs; Inst Elec & Electronic Engrs Circuit Systs Soc (pres, 82). *Res:* Signal processing and image processing; communication, circuit and system theory. *Mailing Add:* Dept Elec Eng Princeton Univ Princeton NJ 08540

LIU, BENJAMIN Y H, b Shanghai, China, Aug 15, 34; m 58; c 1. MECHANICAL ENGINEERING. *Educ:* Univ Nebr, BSME, 56; Univ Minn, Minneapolis, PhD(mech eng), 60. *Prof Exp:* From asst prof to assoc prof, 60-69, PROF MECH ENG, UNIV MINN, MINNEAPOLIS, 69- *Concurrent Pos:* Guggenheim fel, 68-69, dir, Particle Technol Lab, 73-; sr US scientist award, Alexander von Humboldt Found, WGermany. *Mem:* Nat Acad Eng; Solar Energy Soc; Am Soc Heat, Refrig & Air-Conditioning Engrs; Air Pollution Control Asn; fel AAAS. *Res:* Terrestrial and space application of solar energy; aerosol science and technology; instrumentation and measurement. *Mailing Add:* Dept Mech Eng Univ Minn Minneapolis MN 55455

LIU, C(HANG) K(ENG), b Soochow, China, Mar 28, 21; nat US; m 51; c 2. MECHANICAL ENGINEERING. *Educ:* Nat Chiao-Tung Univ, China, BS, 43; Univ Ill, MS, 46, PhD(theoret & appl mech), 50. *Prof Exp:* Res assoc theoret & appl mech, Univ Ill, 50-52; ammunition design engr, Picatinny Arsenal, 52; res assoc appl math, Brown Univ, 52-54; from asst prof to prof mech eng, Univ Ala, Tuscaloosa, 63-86; RETIRED. *Concurrent Pos:* Consult, Marshall Space Flight Ctr, NASA, 60-68; fallout shelter analyst, 66- *Mem:* Am Soc Mech Engrs; Am Soc Eng Educ; Soc Eng Sci; Am Inst Aeronaut & Astronaut. *Res:* Radiation safety; fluid mechanics; viscous fluid flow; stability of social and human behavior; heat conduction in solids. *Mailing Add:* 2236 Woodland Rd Tuscaloosa AL 35404

LIU, CHAIN T, CERAMICS ENGINEERING. *Educ:* Nat Taiwan Univ, BS, 60; Brown Univ, MS, 64, PhD(Mat Sci & Eng), 67. *Prof Exp:* Teaching asst, Brown Univ, 62-66; sr res staff mem, 67-82, GROUP LEADER, ALLOYING BEHAV & DESIGN GROUP, METALS & CERAMICS DIV, OAK RIDGE NAT LAB, 83-; CORP FEL, MARTIN MARIETTA ENERGY SYSTS, INC, 85- *Concurrent Pos:* Prin ed, J Mat Res, 90-; mem, Adv Tech Awareness Coun, Am Soc Metals, 84. *Honors & Awards:* Pioneer/Jupiter Award, NASA, 74, Pioneer II Saturn Mission Team Award, 77, Spacecrafts Voyage I & II Team Award, 84; IR 100 Award, 79 & 83; Henry J Albert Award, Int Precious Metals Inst, 80; E O Lawrence Award, US Dept Energy, 88; RD 100 Award, 90; Outstanding Achievement Award to Galileo RTG Team, Dept Energy, 90. *Mem:* Fel Am Soc Metals; hon platinum mem Int Precious Metals Inst. *Res:* Mechanical behavior of metals, alloys and intermetallic compounds; phase transformation; gas-metal interactions; alloy design of high-temperature materials; metal-matrix composites; environmental effects on ductability and fracture in metals and alloys; 14 patents; numerous technical publications. *Mailing Add:* Metals & Ceramics Div Oak Ridge Nat Lab PO Box 2008 Oak Ridge TN 37831-6115

LIU, CHAMOND, b Waltham, Mass, Sept 28, 48. PERFORMANCE OF OPERATING SYSTEMS. *Educ:* Univ Calif, Berkeley, AB, 68; Cornell Univ, MS, 71, PhD(math), 73. *Prof Exp:* Asst prof math, Fordham Univ, 73-79; sr assoc programmer, 79-81, STAFF PROGRAMMER, IBM, CORP, 81- *Concurrent Pos:* NSF grant, 75-76. *Mem:* Am Math Soc; Math Asn Am; Asn Comput Mach. *Res:* Performance and performance methodology of large operating systems including automatic work load characterization, automatic work load generation and architectural design. *Mailing Add:* IBM Bldg 052 Dept 798D PO Box 950 Poughkeepsie NY 12602

LIU, CHAO-HAN, b Kwangsi, China, Jan 3, 39; m 63; c 2. PHYSICS, ELECTRICAL ENGINEERING. *Educ:* Nat Taiwan Univ, BS, 60; Brown Univ, PhD(elec sci), 65. *Prof Exp:* Res assoc, 65-66, from asst prof to assoc prof, 66-74, PROF ELEC ENG, UNIV ILL, URBANA-CHAMPAIGN, 74- *Concurrent Pos:* Chair Prof, Nat Taiwan Univ, 81; Distinguished lect, Nat Sci Coun, Taiwan, 88. *Mem:* Am Phys Soc; Am Geophys Union; fel Inst Elec & Electronics Engrs. *Res:* Ionosphere, plasma and atmospheric physics; wave propagation in plasma and random media; radar remote sensing. *Mailing Add:* Dept Elec Eng Univ Ill at Urbana-Champaign Urbana IL 61801

LIU, CHEN YA, b Shanghsien, China, Sept 21, 24; US citizen; m 56; c 3. MECHANICAL ENGINEERING, APPLIED MATHEMATICS. *Educ:* Cent Univ, China, BS, 48; NY Univ, MME, 55, EngScD, 59. *Prof Exp:* Res engr, Ord Serv, China, 48-52, proj engr, 52-54; instr mech eng, NY Univ, 55-59; asst prof, Carnegie Inst Technol, 59-61; sr res engr, Res Ctr, B F Goodrich Co, 61-64; sr res engr, Columbus Labs, Battelle Mem Inst, 65-69, assoc fel, 69-72, fel, 72-87; CONSULT, 87- *Concurrent Pos:* Consult, Budd Electronics, Inc, 61; lectr, Univ Akron, 62-64. *Mem:* Am Inst Aeronaut & Astronaut; Am Soc Mech Engrs. *Res:* Fluid mechanics and heat transfer involving Newtonian or non-Newtonian fluids; elasticity of orthotropic materials. *Mailing Add:* 3139 Alameda Menlo Park CA 94025

LIU, CHEN-CHING, b Taiwan, Dec 30, 54. ELECTRICAL ENGINEERING. *Educ:* Nat Taiwan Univ, BS, 76, MS, 78; Univ Calif, Berkeley, PhD(comput sci & elec eng), 83. *Prof Exp:* Instr elec, Army Signals & Electronic Sch, Taiwan, 78-80; asst prof, 83-87, ASSOC PROF ELEC ENG, UNIV WASHINGTON, 87- *Concurrent Pos:* Chmn, Expert Syst Task Force, Inst Elec & Electronic Engrs, 87, Seattle Sec, 87-88, Tech Comt on Power Syst, 89. *Honors & Awards:* Presidental Young Investr Award. *Res:* Application of expert systems to electric power systems; develop analytical and computer methods for power systems planning and operation; power electronic circuits analysis. *Mailing Add:* Dept Elec Eng MS/FT-10 Univ Wash Seattle WA 98195

LIU, CHIEN, b Canton, China, Mar 6, 21; m 47; c 4. INFECTIOUS DISEASES, VIROLOGY. *Educ:* Yenching Univ, BS, 42; WChina Union Univ, MD, 47; Am Bd Pediat, dipl, 64. *Prof Exp:* Intern, Ill Masonic Hosp, 46-47; med intern, Garfield Mem Hosp, Washington, DC, 47-48; asst med, Johns Hopkins Univ, 51-52, asst physician, Johns Hopkins Hosp, 49-52; res assoc bact & immunol, Harvard Med Sch, 52-55, assoc, 55-58, asst prof, 58; assoc prof pediat, 58-63, PROF MED & PEDIAT, SCH MED, UNIV KANS, 63- *Concurrent Pos:* Res fel med, Sch Med, Johns Hopkins Univ, 49-51; USPHS res career award, 63-; vis prof, Nat Defense Med Ctr, Taiwan, 66-67; med consult, US Naval Med Res Unit 2, 66-67; Res Career Award, Nat Inst Allergy & Infectious Dis, 63- *Mem:* Soc Pediat Res; Am Soc Microbiol; Am Asn Immunologists; Am Acad Microbiol; Infectious Dis Soc Am. *Mailing Add:* Dept of Med Univ of Kans Sch of Med 39th & Rainbow Blvd Kansas City KS 66103

LIU, CHI-LI, b Taiwan, Aug 20, 52; US citizen. BACTERIOLOGY, BIOCHEMISTRY. *Educ:* Nat Taiwan Univ, BS, 74; Univ Ga, MS, 79, PhD(biochem), 81. *Prof Exp:* Lab instr cell biol, Univ Ga, 79; assoc, Exxon Res & Eng Co, Annandale, NJ, 82-84; staff researcher, Novo Labs Inc, 84-90; GROUP LEADER, ENTOTECH INC, 90- *Mem:* Am Soc Microbiol. *Res:* Industrial microbiology. *Mailing Add:* Entotech Inc/Novo Nordisk A/S 1497 Drew Ave Davis CA 95616

LIU, CHING SHI, b Shanghai, China, July 23, 35. AERONAUTICS, FLUID DYNAMICS. *Educ:* SDak Sch Mines & Technol, BS, 57; Kans State Univ, MS, 58; Northwestern Univ, PhD(mech eng), 61. *Prof Exp:* Design engr, Int Harvester Co, 56-57; res engr, Bendix Corp, 59-60; asst prof gas dynamics, Northwestern Technol Inst, 61-68; ASSOC PROF ENG SCI, STATE UNIV NY BUFFALO, 68- *Concurrent Pos:* Consult, Cook Res Lab, Ill, 61-62; res assoc, Argonne Nat Lab, 63-64; sr res fel, Calif Inst Technol, 65- *Mem:* Am Inst Aeronaut & Astronaut; Am Soc Mech Engrs; Am Soc Eng Educ; Inst Elec & Electronics Engrs. *Res:* Gas dynamics; magneto gas dynamics; plasma physics. *Mailing Add:* Dept Mech Eng SUNY Buffalo Jarvis Hall Buffalo NY 11260

LIU, CHING-TONG, b Tai-Shin, Kiangsu, China, Oct 19, 31; US citizen; m 70; c 4. PHYSIOLOGY, PHARMACOLOGY. *Educ:* Nat Taiwan Univ, BS, 56; Univ Tenn, MS, 59, PhD(physiol), 63. *Prof Exp:* Assoc res biologist pharmacol, Sterling-Winthrop Res Inst, 65-66; asst prof physiol, Baylor Col Med, 66-73; RES PHYSIOLOGIST, US ARMY MED RES INST INFECTIOUS DIS, 73-, CHIEF, DEPT CLIN & EXP PHYSIOL, 84- *Concurrent Pos:* USPHS trainee, 63-65; adj prof physiol, Baylor Col Med, 80- *Mem:* Am Soc Pharmacol & Exp Therapeut; Am Physiol Soc; Soc Exp Biol & Med. *Res:* Cardiovascular and renal physiology; water, electrolyte and lipid metabolism; mechanisms of infectious diseases and toxemias; effect of muscle trauma; dynamic functional changes and systematically integrated responses to certain viral infections in animals. *Mailing Add:* US Army Med Res Inst Infect Dis SGRD-UID-C Fort Detrick Frederick MD 21702-5011

LIU, CHI-SHENG, b Chinan, China, Nov 1, 34; US citizen; m 60; c 3. PHYSICAL ELECTRONICS, MATHEMATICS. *Educ:* Nat Taiwan Univ, BSEE, 57; WVa Univ, MSEE, 62; Univ Ill, Urbana, PhD(elec eng), 68. *Prof Exp:* Elec engr radio, Philco Corp, 60-61; mem eng staff TV, RCA Consumer Electronics Div, 62-69; SR RES SCIENTIST GASEOUS DISCHARGE, WESTINGHOUSE RES LABS, 69- *Concurrent Pos:* David Sarnoff fel, RCA, 65-68. *Mem:* Am Phys Soc. *Res:* Study of high pressure gas discharges; high efficiency arc lamps and gas lasers. *Mailing Add:* 11625 Caminito Magnifica San Diego CA 92131

LIU, CHONG TAN, b Shanghai, China, May 11, 36; US citizen; m 63; c 3. INORGANIC CHEMISTRY. *Educ:* Nat Taiwan Univ, BSc, 56; Univ Pittsburgh, PhD(inorg chem), 64. *Prof Exp:* Sr res chemist, Hooker Chem Corp, 64-71; sr res chemist, 72-78, GROUP SUPVR, STAUFFER CHEM CO, 79- *Mem:* Am Chem Soc. *Res:* Water treatment; industrial chemical processes; metal finishing; plating on plastics; corrosion controls; high temperature chemistry; coordination chemistry. *Mailing Add:* 3 Demarest Mill Ct West Nyack NY 10994

LIU, CHUAN SHENG, b Kwanhsi, China, Jan 9, 39; m 65; c 4. THEORETICAL ASTROPHYSICS, PLASMA PHYSICS. *Educ:* Tunghai Univ, BS, 60; Univ Calif, Berkeley, MA, 64, PhD(physics), 68. *Prof Exp:* Asst prof in residence physics, Univ Calif, Los Angeles, 68-70; vis scientist, Gulf Gen Atomic, Inc, 70-71; mem, Inst Advan Study, 71-74; chmn dept, Physics & Astron, 85-90, PROF PHYSICS, UNIV MD, 90- *Concurrent Pos:* Dir, Theoret Sci Div, GA Technolologies, 81-84. *Mem:* Fel Am Phys Soc. *Res:* Fusion and plasma physics. *Mailing Add:* Dept Physics & Astron Univ Md College Park MD 20712-4111

LIU, CHUI FAN, b Chunking, China, Apr 5, 30; US citizen; m 59; c 3. INORGANIC CHEMISTRY. *Educ:* Univ Ill, AB, 52, PhD(chem), 56. *Prof Exp:* Res chemist, Dow Chem Co, 56-58; instr chem, Univ Conn, 58-60; asst prof, Univ Mich, 60-65; assoc prof, Univ Ill, Chicago Circle, 65-70, prof chem, 70-87; RETIRED. *Concurrent Pos:* NSF & NIH res grants, 62- *Res:* Structure and chemistry of coordination compounds; asymmetric synthesis. *Mailing Add:* 1603 S Highland Ave Arlington Heights IL 60005

LIU, CHUI HSUN, b China, Nov 5, 31; US citizen; m 62. ANALYTICAL CHEMISTRY, INORGANIC CHEMISTRY. *Educ:* Univ Ill, BA, 52, PhD(chem), 57. *Prof Exp:* From asst prof to assoc prof chem, Polytech Inst Brooklyn, 57-65; PROF CHEM, ARIZ STATE UNIV, 65- *Mem:* Am Chem Soc. *Res:* Chemistry, electrochemistry and spectroscopy in molten salts and other nonaqueous solvents; chemistry of coordination compounds; chelating agents in chemical separations and analyses. *Mailing Add:* Ariz State Univ Tempe AZ 85287

LIU, CHUNG LAUNG, b Canton, China, Oct 25, 34; US citizen; m 60; c 1. COMPUTER SCIENCE. *Educ:* Cheng Kung Univ, Taiwan, BSc, 56; Mass Inst Technol, SM, 60, ScD(elec eng), 62. *Prof Exp:* From asst prof to assoc prof elec eng, Mass Inst Technol, 62-72; PROF COMPUT SCI, UNIV ILL, URBANA, 73- *Honors & Awards:* Karl V Karlstron Outstanding Educr Award, Asn Comput Mach, 90. *Mem:* Inst Elec & Electronics Engrs; Asn Comput Mach. *Res:* Theory of computation; combinatorial mathematics. *Mailing Add:* Dept of Comput Sci Univ of Ill 2129 DCL Urbana IL 61801

LIU, CHUNG-CHIUN, b Canton, China, Oct 8, 36; m 67; c 1. CHEMICAL ENGINEERING. *Educ:* Cheng Kung Univ, Taiwan, BS, 59; Calif Inst Technol, MS, 62; Case Western Reserve Univ, PhD(chem eng), 68. *Prof Exp:* Res assoc, Eng Design Ctr, Case Western Reserve Univ, 66; prof chem eng, Univ Pittsburgh, 68-78; PROF CHEM ENG, CASE WESTERN RESERVE UNIV, 78- *Concurrent Pos:* Wallace R Persons Prof sensor technol & control. *Mem:* Am Inst Chem Engrs. *Res:* Electrochemistry; bio-medical engineering; material science. *Mailing Add:* Dept Chem Eng Case Western Reserve Univ Cleveland OH 44106

LIU, CHUNG-YEN, b Canton, China. AERONAUTICAL ENGINEERING. *Educ:* Nat Cheng-Kung Univ, BS, 56; Brown Univ, MS, 58; Calif Inst Technol, PhD(aeronaut), 62. *Prof Exp:* From asst prof to prof eng, 62-, EMER PROF UNIV CALIF, LOS ANGELES,. *Mem:* Am Phys Soc; Am Inst Aeronaut & Astronaut. *Res:* Fluid mechanics. *Mailing Add:* 831 Gregna Greenway No 302 Los Angeles CA 90049

LIU, DARRELL T, b Taiwan, Repub China, May 24, 32; US citizen; m; c 3. BIOCHEMISTRY. *Educ:* Nat Taiwan Univ, BS, 55; Univ Pittsburgh, PhD, 61. *Prof Exp:* Res assoc, Rockefeller Univ, 61-65, asst prof, 65-67; biochemist, Brookhaven Nat Lab, 67-69, sr biochemist, 69-73; res chemist, Develop Immunol Br, Nat Inst Child Health Develop, NIH, 73, sr chief biochem microbial struct, 74; dep dir, Div Bact Prod Chief, Biochem Br, Bur Biol, 75-80, DIR, DIV BIOCHEM & BIOPHYS, OFF BIOL RES, CTR BIOL, EVAL RES, FOOD & DRUG ADMIN, 80- *Concurrent Pos:* Mem, Regulatory Sci Prom Comt, Food & Drug Admin, 79-80, Recombinant DNA Rev Comt, 81-89. *Mem:* Sigma Xi; Am Soc Biol Chemists. *Res:* Human C-reactive protein; prototypic acute phase reactant; employing molecular cloning techniques to explore the mechanism of induction and control of the biosynthesis of this protein at the chromosomal level; technique of homologous gene transfection is being pursued to investigate the possible physiological function of CRP in xenopus; isolation and cloning of the genes coding for the enzymes and substrate proteins associated with a unique coagualtion cascade system form Limulus. *Mailing Add:* Div Biochem & Biophys Ctr Biol Eval & Res Bldg 29 Rm 516 Bethesda MD 20892

LIU, DAVID H(O-FENG), b Chekiang, China, Feb 24, 28; m 56; c 2. CHEMICAL ENGINEERING. *Educ:* Jadavpur Univ, BSc, 51; Univ Pa, MSc, 54, PhD(chem eng), 56. *Prof Exp:* Sr chem engr, Monsanto Chem Co, 55-62; supvr chem eng res, Mobil Chem Co, Tex, 64-66; sr res assoc, Uniroyal Inc, 66-74; mgr process develop, Rhone-Poulenc Inc, 74-82; MGR NEW PROD DEVELOP, J T BAKER, INC, 82- *Mem:* Am Chem Soc; Am Inst Chem Engrs. *Res:* Chemical thermodynamics; reaction kinetics; unit operations, engineering and economic analysis; pollution abatement; petrochemical processes; aroma chemicals; polymers; rubber and specialty chemicals; research specialty chemical, and electronic chemicals. *Mailing Add:* 222 Red School Lane Phillipsburg NJ 08865

LIU, DAVID H W, toxicology, for more information see previous edition

LIU, DAVID SHIAO-KUNG, b Chung-King, China, Aug 27, 40; US citizen; m 66; c 3. FLUID MECHANICS, NUMERICAL METHODS. *Educ:* Cheng-Kung Univ, Taiwan, BS, 62; Univ Calif, Berkeley, MS, 65; New York Univ, PhD(appl math & hydraul), 73. *Prof Exp:* Phys scientist fluid mech, 72-76, SR PHYS SCIENTIST, RAND CORP, 76-; PROF HYDRAUL, NAT CHENG-KUNG UNIV, 77- *Concurrent Pos:* Sr consult, Va Inst Marine Sci, 77-; adv, Sci & Technol adv Group, Taiwan. *Mem:* Am Soc Civil Eng; Int Asn Water Resources. *Res:* Numerical modeling of three-dimensional non-homogeneous geophysical fluid systems; stochastic analysis and control theory of physical systems. *Mailing Add:* 3706 Oceanhill Way Malibu CA 90265

LIU, DICKSON LEE SHEN, b Shantung, China, Apr 6, 35; Can citizen; m 67; c 1. MICROBIOLOGY, WATER POLLUTION. *Educ:* Nat Taiwan Chung Hsin Univ, BSc, 62; Univ BC, MSc, 66, PhD(microbiol), 71. *Prof Exp:* Res scientist marine biochem, BC Res Coun, 66-68; res scientist eutrophication, Can Ctr Inland Waters, 71-72; res scientist wastewaters, Can Wastewater Technol Ctr, 72-75; RES SCIENTIST TOXIC SUBSTANCES, NAT WATER RES INST, 75-; assoc prof, 81-89, PROF, DEPT ENVIRON HEALTH SCI, TULANE MED CTR, NEW ORLEANS, LA, 90- *Concurrent Pos:* Expert, Food Agr Orgn, UN, 68-; adv, Wastewater Technol Ctr, 75-; tech ed, Can Res, 77-83; expert biodegradation, Can Nat Comt on Int Orgn Standards, 77-; mem, Assoc Comt on Sci Criteria Environ Qual, Nat Res Coun Can, 77-80; co-ed, Toxicity Assessment: An Int J, 86-90, ed, Environ Toxicol Water Qual, Int J, 91-; co-chmn, Int Symposia on Toxicity Assessment using Microbial Systs, 83- *Res:* Biodegradation of toxic substances; development of standard procedure for assessing the persistence of new substances in the natural environments; lake and river eutrophication; biological treatment of toxic industrial wastewaters; environmental toxicology; biotechnology. *Mailing Add:* Nat Water Res Inst PO Box 5050 Burlington ON L7R 4A6 Can

LIU, EDWIN H, b Honolulu, Hawaii, Apr 11, 42; m 65. BIOCHEMISTRY. *Educ:* Johns Hopkins Univ, AB, 64; Mich State Univ, PhD(biochem), 71. *Prof Exp:* Res assoc biochem, AEC, Plant Res Lab, Mich State Univ, 71-72; asst prof biol, Univ SC, 72-81; asst ecologist, Savannah River Ecology Lab, 81-86; Newport coordr, Calif Regional Water Qual Control Bd, 86-88; REGIONAL MONITORING COORDR, US ENVIRON PROTECTION AGENCY, REGION IX, 88- *Mem:* Am Chem Soc; Genetics Soc Am; Am Soc Plant Physiologists. *Res:* Biochemical ecology. *Mailing Add:* US Environ Protection Agency Region IX 75 Hawthorne St San Francisco CA 94105

LIU, FOOK FAH, b Calcutta, India, Sept 30, 34; m 66; c 3. HIGH ENERGY PHYSICS. *Educ:* Presidency Col, Calcutta, India, BSc, 56; Purdue Univ, PhD(physics), 62. *Prof Exp:* Res assoc, High-Energy Physics Lab, Stanford Univ, 62-66; asst prof physics, Case Inst Technol, 66-68; staff physicist, Stanford Linear Accelerator Ctr, Stanford Univ, 68-70; asst prof, 70-74, assoc prof, 74-78, PROF PHYSICS, CALIF STATE COL, SAN BERNARDINO, 78- *Mem:* Am Phys Soc. *Res:* Photoinduced reactions at high energies; phenomenology of multibody final states; decay of unstable particles; microcomputer systems. *Mailing Add:* Dept Computer Sci Calif State Univ San Bernadino 5500 University Pkwy San Bernardino CA 92407

LIU, FRANK C, b Shantung, China, Apr 15, 26; US citizen; m 54; c 3. MECHANICAL ENGINEERING. *Educ:* Chekiang Univ, BS, 49; Univ Wash, MS, 53; Univ Tex, PhD(mech eng), 58. *Prof Exp:* Teacher high sch, Taiwan, 49-52; instr mech eng, Univ Tex, 55-57; res engr, Boeing Co, 57-62; tech consult, NASA, Marshall Space Flight Ctr, 62-63, sci asst appl math, 63-67; prof eng mech, Univ Ala, Huntsville, 67-85. *Concurrent Pos:* Distinguished vis professorship, AF Inst Technol, 83-84. *Res:* Dynamics; structural dynamics; orbital mechanics and applied mathematics. *Mailing Add:* Dept Mech Eng Univ Ala 4701 University Dr Huntsville AL 35899

LIU, FRED WEI JUI, b Canton, China, Jan 29, 26; nat US; m 61. PHYSICAL CHEMISTRY. *Educ:* St John's Univ, China, BS, 48; Temple Univ, MA, 50; Lehigh Univ, PhD(chem), 52. *Prof Exp:* Res assoc, Lehigh Leather Inst, Pa, 52-53; chief chemist, Lester Labs, Inc, 53-64; DIR, CONTINENTAL CONSULTS, INC, 64-; PRES, CONTINENTAL TRADING CO, 65- *Mem:* Am Chem Soc; Nat Asn Corrosion Engrs. *Res:* Colloid or surface chemistry; detergents; cleaning and maintenance chemicals formulation; corrosion; water treatment; foreign trade; industrial chemicals. *Mailing Add:* 157 Lake Forest Lane NE Atlanta GA 30342-3209

LIU, FREDERICK F, b Chefoo, China, Apr 19, 19; US citizen; m 46; c 2. SYSTEMS DESIGN & SYSTEMS SCIENCE. *Educ:* Technische Hochschule, Berlin, dipl, 39; Carnegie Inst Technol, BS, 46; Princeton Univ, PhD(sci admin), 51. *Hon Degrees:* DSc, Polytech Univ Inst, China, various univ & insts, 54. *Prof Exp:* Res asst, Princeton Univ, 50-52, res assoc, 52-55; res eng specialist, Rockethyne & Atomics Int, NAm Aviation, Inc, 55-57; dir res, Dresser Dynamics, Inc, 57, exec vpres, 57-59; PRES & SCI DIR, QUANTUM DYNAMICS, INC, 59- *Concurrent Pos:* Vis lectr, Mass Inst Technol, 56 & Cambridge Univ, Eng, 64; vis scientist, Kyoto Univ, Japan, 67-; guest lectr, Technische Univ, Berlin, 70-; vis scientist & prof, Inst Mech, Chinese Acad Sci, 79- *Mem:* Sigma Xi; Am Phys Soc; Am Inst Aeronaut & Astronaut; assoc fel Int Inst Refrigeration; fel Am Inst Elec Eng. *Res:* Extremely fast, dynamic and transient phenomena relating to propulsion, weapon, nuclear and space; low temperature physical phenomena, trans-regine viscosity effects on fluid flow theory, together with the development of a range of modern instrumentation and computing technologies. *Mailing Add:* 17812 Community St Northridge CA 91325

LIU, FU-WEN (FRANK), b Taiwan. POMOLOGY, POSTHARVEST HORTICULTURE. *Educ:* Taiwan Univ, BS, 57; Cornell Univ, MS, 69, PhD(pomol), 74. *Prof Exp:* Horticulturist, Sino-Am Joint Comn Rural Reconstruct, 69-71; from asst prof to assoc prof, 74-87, PROF POMOL, CORNELL UNIV, 87- *Concurrent Pos:* FAO consult, 87-88. *Mem:* Am Soc Hort Sci; Sigma Xi. *Res:* Postharvest physiology with emphasis on the control mechanism of maturation and ripening of fruits and storage methods of fruits. *Mailing Add:* Dept Pomol Cornell Univ 125 Plant Sci Bldg Ithaca NY 14853

LIU, HAN-SHOU, b Hunan, China, March 9, 30; US citizen; m 57; c 2. EARTH PHYSICS, SPACE PHYSICS. *Educ:* Cornell Univ, MS, 62, PhD, 63. *Prof Exp:* Res assoc, Nat Acad Sci, 63-64; SCIENTIST, GODDARD SPACE FLIGHT CTR, NASA, 65- *Honors & Awards:* Apollo Achievement Award, NASA, 69. *Mem:* Fel AAAS; Am Astron Soc; Am Geophys Union; Am Inst Aeronaut & Astronaut; Planetary Soc. *Res:* Physics of the earth; planetary interiors. *Mailing Add:* Goddard Space Flight Ctr Code 921 NASA Greenbelt MD 20771

LIU, HAO-WEN, b China, Aug 20, 26; nat US; m 55; c 5. MECHANICS, MATERIALS SCIENCE. *Educ:* Univ Ill, BS, 54, MS, 56, PhD(appl mech), 59. *Prof Exp:* Asst prof appl mech, Univ Ill, 59-61; sr res fel, Calif Inst Technol, 61-63; assoc prof metall, 63-68, prof mat sci, 68-84, PROF MECH & AERONAUT ENG, SYRACUSE UNIV, 84- *Mem:* Am Soc Mech Engrs; Am Inst Mech Engrs; Am Soc Testing & Mat; Sigma Xi. *Res:* Mechanical behavior and properties of materials and applied mechanics. *Mailing Add:* Dept Mech & Aeronaut Eng Syracuse Univ Syracuse NY 13244-1240

LIU, HENRY, b Peking, China, June 3, 36; m 64. FLUID MECHANICS. *Educ:* Nat Taiwan Univ, BS, 59; Colo State Univ, MS, 63, PhD(fluid mech), 66. *Prof Exp:* From asst prof to assoc prof, 65-77, PROF CIVIL ENG, UNIV MO-COLUMBIA, 77- *Concurrent Pos:* Prin investr water resources res grants, Dept Interior, 66-68, Capsule Pipeline res grants, US Dept Energy, 78-81 & NSF grants, 80-86; vis prof, Univ Melbourne, Australia, 80; prof, Natural Gas Pipeline Co, 83-87, endowed chair, James C Dowell, 88-; bd dirs, US Wind Eng Res Coun, 85-89; chmn, Exec Comt, Aerospace Div, Am Soc Civil Engrs, 89-90. *Honors & Awards:* Aerospace Div Award, Am Soc Civil Engrs; Distinguished Lectr Award, Int Freight Pipeline Soc. *Mem:* Am Soc Civil Engrs; Am Wind Energy Asn; Sigma Xi; US Wind Eng Res Coun; Int Freight Pipeline Soc (pres, 89-93). *Res:* Electrokinetics; exploration of the physics of streaming potential fluctuations and the utilization of this phenomenon to study turbulence characteristics in liquid flows; dispersion of pollutants in river; hydraulic capsule pipeline; wind pressure inside buildings; flow measurement; cherepnov water lifter; hydropower; wind damage mitigation; wind energy utilization. *Mailing Add:* Dept of Civil Eng Univ of Mo Columbia MO 65201

LIU, HOUNG-ZUNG, b China, Jan 23, 31; m 69; c 2. BIOCHEMICAL & PLANT PROTOPLAST GENETICS. *Educ:* Taiwan Prov Col, BS, 53; NDak State Col, MS, 59; Cornell Univ, PhD(genetics, biochem, plant physiol), 64. *Prof Exp:* Asst cytol, Taiwan Agr Res Inst, Taipei, 54-56; asst gen genetics, Cornell Univ, 59-64; assoc prof genetics, 64-69, actg dean, Fac Arts & Sci, 81-82, prof genetics & chmn, Dept Biol Sci, 69-81, DEAN, FAC ARTS & SCI, STATE UNIV NY COL, PLATTSBURGH, 83- *Concurrent Pos:* NIH spec res fel, Marquette Univ, 67-68; res collabr, Brookhaven Nat Lab, 73-74. *Mem:* AAAS; Am Chem Soc; Genetics Soc Am; Int Plant Tissue Asn. *Res:* Tryptophan operon mutants of Escherichia coli and indoleglycerol-phosphate synthetase; plant protoplast fusion and culture; plantlet regeneration. *Mailing Add:* Dept Biol Sci State Univ NY Col Plattsburgh NY 12901

LIU, HSING-JANG, b Kiang-Su, China, Dec 2, 42; m 66; c 3. CHEMISTRY. *Educ:* Nat Taiwan Norm Univ, BSc, 64; Univ NB, Fredericton, PhD(chem), 68. *Prof Exp:* Fel chem, Univ NB, Fredericton, 68-69; res assoc, Columbia Univ, 69-70; teaching & res assoc, Univ NB, Fredericton, 70-71; from asst prof to assoc prof, 71-83, PROF CHEM, UNIV ALTA, 83- *Mem:* Am Chem Soc; fel Chem Inst Can. *Res:* Natural products, isolation, identification and synthesis; development of novel synthetic methods. *Mailing Add:* Dept Chem Univ Alta Edmonton AB T6G 2G2 Can

LIU, HUA-KUANG, b Kueilin, China, Sept 2, 39; m 65; c 2. LASER OPTICS, ELECTROOPTICS. *Educ:* Nat Taiwan Univ, BS, 62; Univ Iowa, MS, 65; Johns Hopkins Univ, PhD(elec eng), 69. *Prof Exp:* Res asst, Univ Iowa, 63-64; instr elec eng, Va Mil Inst, 64-65; res asst, Johns Hopkins Univ & jr instr, Eve Col, 65-69; from asst prof to prof elec eng, Univ Ala, 69-84; SR RES ENG, JET PROPULSION LAB, CALIF INST TECHNOL, 84- *Concurrent Pos:* Consult, Optimal Data Corp, Ala, 69-, NASA Marshall Space Flight Ctr, US Army Missile Command, 73 & Newport Res Corp, Rockwell Int; res grants, Univ Ala, Tuscaloosa, 70-71 & NSF, 71-73 & 75-77 & NASA, 73-78; pres, Lumin, Inc, 75; vis assoc prof elec eng, Stanford Univ, 75-76; vis prof, Nat Taiwan Univ & Univ Wis-Madison, 82-83. *Mem:* Sr mem Inst Elec & Electronics Engrs; fel Optical Soc Am; fel Soc Photo-Instrumentation Engrs. *Res:* Solid-state electronics; nonlinear optical image processing; optical pattern recognition, holography and holographic nondestructive testing; halftone contact screens for printing. *Mailing Add:* Jet Propulsion Lab 4800 Oak Grove Dr Pasadena CA 91109

LIU, J(OSEPH) T(SU) C(HIEH), b Shanghai, China, Nov 9, 34; US citizen; m 64; c 3. FLUID MECHANICS. *Educ:* Univ Mich, BSE, 57, MSE, 58; Calif Inst Technol, PhD(aeronaut), 64. *Prof Exp:* Propulsion engr aerothermodyn group, Gen Dynamics & Convair, 58-59; res assoc aerospace & mech sci, Gas Dynamics Lab, Princeton Univ, 64-66; from asst prof to assoc prof, 66-73, PROF ENG, BROWN UNIV, 73- *Concurrent Pos:* Consult, Space Systs Div, Avco Corp, Mass, 66-67, Systs Div, 69-70; vis, Dept Math, Imperial Col, Univ London, 72-73 & 79-80. *Honors & Awards:* Nat Award, Inst Aeronaut Sci, 58. *Mem:* Am Soc Mech Engrs; Am Phys Soc; Am Meteorol Soc; Am Inst Aeronaut & Astronaut. *Res:* Coherent structures in turbulent shear flows, nonlinear hydrodynamic stability and transition; aeroacoustics; fluidized bed instabilities. *Mailing Add:* Div Eng Brown Univ Box D Providence RI 02912

LIU, JIA-MING, b Taichung, Taiwan, July, 13, 53; US citizen; m 90. OPTICS, ELECTRICAL ENGINEERING. *Educ:* Nat Chiao Tung Univ, BS, 75; Harvard Univ, SM, 79, PhD(appl physics), 82. *Prof Exp:* Asst prof elec eng, State Univ NY, Buffalo, 82-84; sr mem tech staff, GTE Labs, Inc, 83-86; ASSOC PROF ELEC ENG, UNIV CALIF, LOS ANGELES, 86- *Concurrent Pos:* Consult, Jaycor, 87-, Battelle Columbus Div, US Army, 89-90. *Mem:* Fel Optical Soc Am; sr mem Inst Elec & Electronics Engrs Laser & Electro-Optics Soc; Am Phys Soc; Sigma Xi. *Res:* Ultrashort laser pulses and applications; nonlinear optics; optical wave propagation; semiconductor lasers; fiber optics. *Mailing Add:* Elec Eng Dept Univ Calif 56-125B Eng IV Los Angeles CA 90024

LIU, JOHN, b Taiwan. OPHTHALMIC PHARMACOLOGY. *Prof Exp:* ASSOC SCIENTIST, EYE RES INST, 88- *Mailing Add:* Dept Ophthal Pharmacol Eye Res Inst Retina Found 20 Staniford St Boston MA 02114

LIU, JOHN K(UNGFU), b Hankow, China, Aug 22, 30; nat US; m 57. MECHANICS. *Educ:* Univ Pa, BSME, 52; Ill Inst Technol, MS, 57. *Prof Exp:* Struct engr, Shih & Assoc, 48-52; from design engr to proj engr, Int Harvester Co, 52-57; from sr proj engr to actg dir res & develop, Clearing Div, US Industs, 57-60, dir marine tech dept, Tech Ctr, 59-60; mgr marine tech lab, Stromberg Carlson Co Div, Gen Dynamics Co, 60-62; vpres, Force Control, Inc, 62-68; VPRES, CLUTCH DIV, PHILADELPHIA GEAR CORP, 68- *Mem:* Am Soc Mech Engrs; Sigma Xi; Am Soc Inventors. *Res:* Oceanographic instruments and devices; solid and fluid mechanics; marine propulsion equipment; pressure vessel design and development; fluid shear power transmission devices; industrial electronics and transducers. *Mailing Add:* 2749 Paige St Lower Burrell PA 15068

LIU, JOSEPH JENG-FU, b Chiangsi, China, Oct 24, 40; m 71; c 3. CELESTIAL & THEORETICAL MECHANICS, APPLIED MATHEMATICS. *Educ:* Cheng Kung Univ, Taiwan, BS, 62; Auburn Univ, MS, 66, PhD(celestial mech), 71. *Prof Exp:* Teaching asst appl mech, Cheng Kung Univ, 63-64 & Auburn Univ, 66-71; mem res staff astrodyn, Northrop Serv, Inc, Huntsville, Ala, 71-77; mem tech staff Astrodyn Aerospace Defense Command, 77-86; CHIEF, SPACE DIVISION, HQ A F SPACE COMMAND. DOA, 86- *Concurrent Pos:* Honorarian, appl math, Univ Colo, Colorado Springs, 82- *Mem:* Am Inst Aeronaut & Astronaut; Am Astronaut Soc; Am Astron Soc. *Res:* General, special perturbation and semi-analytic theories, their applications for the orbital and attitude motions of an artificial satellite perturbed by conservative and nonconservative forces. *Mailing Add:* Hq Air Force Space Command DOJY Stop 7 Peterson AFB CO 80914-5001

LIU, LEROY FONG, b Tao-yuan, Taiwan, July 28, 49; US citizen; m; c 2. DNA TOPOISOMERASES, CANCER PHARMACOLOGY. *Educ:* Nat Taiwan Univ, Taiwan, BS, 71; Univ Calif, Berkeley, PhD(biophys chem), 77. *Prof Exp:* Postdoctoral fel molecular biol, Harvard Univ, 77-78 & Univ Calif, San Francisco, 78-80; from asst prof to assoc prof, 80-88, PROF BIOCHEM, JOHNS HOPKINS UNIV, 88- *Concurrent Pos:* vis prof, Inst Molecular Biol, Academia Sinica, Taiwan, 86-87. *Mem:* Am Soc Biochem & Molecular Biol; Am Asn Cancer Res; Am Math Soc. *Res:* Biological functions of multiple DNA topoisomerases; DNA topoisomerases as therapeutic targets. *Mailing Add:* Biol Chem Dept Johns Hopkins Univ Sch Med, 725 N Wolfe St Baltimore MD 21205

LIU, LIU, b Shanghai, China, Aug 12, 30; m 56; c 3. SOLID STATE PHYSICS. *Educ:* Univ Taiwan, BS, 54; Univ Chicago, MS, 57, PhD(physics), 61. *Prof Exp:* From asst prof to assoc prof, 61-74, PROF PHYSICS, NORTHWESTERN UNIV, ILL, 74- *Concurrent Pos:* Consult, Argonne Nat Lab, 61-64; Fulbright Sr Res Scholar to France, 75-76. *Mem:* Fel Am Phys Soc. *Res:* Theory of narrow-gap and zero-gap semiconductors and magnetic semiconductors. *Mailing Add:* Dept of Physics Northwestern Univ 633 Clark St Evanston IL 60201

LIU, LON-CHANG, b China; US citizen. PHYSICS, NUCLEAR STRUCTURE. *Educ:* Univ Neuchatel, PhD(physics), 73. *Prof Exp:* Instr, City Univ New York, 73-75, asst prof physics, 75-79; STAFF MEM, LOS ALAMOS NAT LAB, 79- *Mem:* Am Phys Soc; Sigma Xi. *Res:* Theoretical intermediate energy nuclear physics; meson nucleus interaction. *Mailing Add:* Group Inc-11 MS H824 Los Alamos Nat Lab Los Alamos NM 87545

LIU, MATTHEW J P, b Peking, China, July 19, 35; US citizen; m 61; c 2. MATHEMATICS. *Educ:* Lafayette Col, BS & BA, 58; Ill Inst Technol, MS, 61; Ind Univ, PhD(math), 75. *Prof Exp:* From instr to assoc prof, 61-76, PROF MATH, UNIV WIS, STEVENS POINT, 76- *Concurrent Pos:* NSF fel, Ind Univ, 67-68. *Mem:* Math Asn Am; Am Math Soc; Nat Coun Teachers Math. *Res:* Mathematics, summability. *Mailing Add:* 2148 W River Dr Stevens Point WI 54481

LIU, MAW-SHUNG, b Taiwan, Feb 2, 40; m 66; c 1. MEDICAL PHYSIOLOGY. *Educ:* Kaohsiung Med Col, Taiwan, DDS, 64; Univ Ky, MSc, 70; Univ Ottawa, PhD(physiol), 76. *Prof Exp:* Staff dentist & lectr oral surg, Chinese Army Hosp, Kaohsiung Med Col Hosp, Taiwan, 64-68; intern path, Med Ctr, Univ Ky, 68-69; Med Res Coun Can fel, 70-73; Alcoholism & Drug Addiction Res Found Ont res scholar, 73-74; instr physiol, 74-76, asst prof physiol, Sch Med, La State Univ Med Ctr, New Orleans, 76-78; assoc prof physiol, Bowman Gray Sch Med, Wake Forest Univ, Winston-Salem, NC, 78-82; PROF PHYSIOL, SCH MED, ST LOUIS UNIV, 82- *Concurrent Pos:* Consult, NIH, Heart Lung & Blood Inst, 83-; hon prof, Nanjing Med Col, China, 84; vis prof, Beijing Med Univ, China & Zhejiang Med Univ China, 86, Kaohaiung Med Col & Chang Gung Med Col, Taiwan, 89 & 90; hon prof, Hunan Med Univ, China, 86; mem, Surg, Anesthesiol & Trauma Study Sect, NIH, 88-92. *Mem:* Int Soc Heart Res; Shock Soc; Am Physiol Soc. *Res:* Myocardial and hepatic intermediary metabolism in endotoxic and septic shock. *Mailing Add:* Dept Physiol St Louis Univ Sch Med St Louis MO 63104

LIU, MIAN, b Sichuan, People's Republic China, Oct 15, 60. TECTONO PHYSICS, NUMERICAL MODELLING. *Educ:* Nanjing Univ, People's Repub China, BSc, 82; McGill Univ, MSc, 85; Univ Ariz, PhD(geophysics), 89. *Prof Exp:* Postdoctoral fel geophys, Minn Supercomputer Inst, 89-90; POSTDOCTORAL FEL GEOPHYS, PA STATE UNIV, 90- *Mem:* Am Geophys Union. *Res:* Geodynamics; mantle convection; deformation and evolution of lithosphere; numerical modelling. *Mailing Add:* Dept Geosci Pa State Univ State College PA 16802

LIU, MICHAEL T H, b Hong Kong, China, Mar 1, 39; Can citizen; m; c 3. PHYSICAL CHEMISTRY. *Educ:* St Dunstan's Univ, BSc, 61; St Francis Xavier, MA, 64; Univ Ottawa, PhD(phys chem), 67. *Prof Exp:* Technician, Can Celanese Ltd, 61-62; group leader qual control, Chemcell Ltd, 64; Nat Res Coun fel, Univ Reading, 67-68; from asst prof to assoc prof, 68-80, PROF CHEM, UNIV PRINCE EDWARD ISLAND, 80- *Concurrent Pos:* Nat Sci & Eng Res Coun Can grant-in-aid, 68-; Def Res Bd of Can grant-in-aid, 74-76; sabbatical leave, Univ BC, 75 & Univ Geneva, 82; adj prof, Dalhousie Univ, 79-85, Univ Bordeaux, 88-89; ed, CRC Press, Boca Raton, Fla, 87. *Mem:* Fel Chem Inst Can; Inter-Am Photochemical Soc. *Res:* Kinetics and mechanism of chain reaction; unimolecular reaction; carbene chemistry; synthesis of new diazirines and 1,2-hydrogen migration; cyclopropanation of electrophilic and ambiphilic carbenes; thermolysis, photolysis and laser photolysis of dizirines. *Mailing Add:* Dept of Chem Univ of Prince Edward Island Charlottetown PE C1A 4P3 Can

LIU, MING-BIANN, b Chang-Hua, Taiwan, June 22, 42; m 75; c 2. PHYSICAL CHEMISTRY, CHEMICAL ENGINEERING. *Educ:* Cheng-Kung Univ, Taiwan, BS, 68; Ill Inst Technol, PhD(chem), 74, MS, 80. *Prof Exp:* Res & teaching, Univ Kans, 74-75; res, Chem Div, Argonne Nat Lab, 76-78, asst chemist, Chem Eng Div, 78-81; ASST CHEMIST, DOW CHEM CO, 81- *Mem:* Am Chem Soc; Electrochem Soc. *Res:* High temperature materials and technology; plasma surface interaction; gas surface interaction; electrochemical processes and technology; flame retardants. *Mailing Add:* 386 Mount Sequoia Pl Clayton CA 94517-1614

LIU, MING-TSAN, b Taiwan, China, Aug 30, 34; US citizen; m 66; c 3. COMPUTER SCIENCE, COMPUTER ENGINEERING. *Educ:* Cheng Kung Univ, BS, 57; Univ Pa, MS, 61, PhD(elec eng), 64. *Prof Exp:* Asst elec eng, Cheng Kung Univ, 59-60; instr, Moore Sch Elec Eng, Univ Pa, 62-65, asst prof, 65-69; assoc prof, 69-78, PROF COMPUT & INFO SCI, OHIO STATE UNIV, 78- *Concurrent Pos:* Consult, Comput Command & Control Co, Philadelphia, Pa, 64-65, Burroughs Corp, Paoli, Pa, 77 & AT&T Bell Labs, Columbus, Ohio, 82-84; distinguished vis, Comput Soc, Inst Elec & Electronics Engrs, 81-84; ed, Inst Elec & Electronics Engrs Trans Comput, 82-86 & ed-in-chief, 86- *Mem:* Fel Inst Elec & Electronics Engrs; Asn Comput Mach; Sigma Xi. *Res:* Computer architecture; computer networks; distributed processing; microcomputer systems; computer communication. *Mailing Add:* Cis Dept 2036 Neil Ave Ohio State Univ Columbus OH 43210-1277

LIU, PAN-TAI, b Taipei, Taiwan, Sept 22, 41; m 66; c 1. APPLIED MATHEMATICS. *Educ:* Nat Taiwan Univ, BS, 63; State Univ NY, Stony Brook, PhD(appl math), 68. *Prof Exp:* Asst prof, 68-74, assoc prof, 74-80, PROF MATH, UNIV RI, KINGSTON, 80- *Concurrent Pos:* Vis prof, Dept Elec Eng, Nat Taiwan Univ, Taipei, Taiwan, 74-75. *Mem:* Am Math Soc. *Res:* Optimal controls; differential games; stochastic processes. *Mailing Add:* Dept of Math Univ of RI Kingston RI 02881

LIU, PAUL CHI, b Chefoo, China, June 18, 35; m 65; c 1. PHYSICAL OCEANOGRAPHY, COASTAL ENGINEERING. *Educ:* Nat Taiwan Univ, BS, 56; Virginia Polytech Inst, MS, 61; Univ Mich, PhD(oceanic sci), 77. *Prof Exp:* Res phys scientist, US Lake Survey, Army Corps Engrs, 65-71; res phys scientist, Lake Surv Ctr, Nat Ocean Surv, 71-74, OCEANOGR, GREAT LAKES ENVIRON RES LAB, NAT OCEANIC & ATMOSPHERIC ADMIN, 74- *Concurrent Pos:* Vis scholar, Univ Mich, 78- *Mem:* Am Geophys Union; Am Meterol Soc; Am Soc Civil Engrs; Int Asn Great Lakes Res; Soc Indust & Applied Math; Sigma Xi. *Res:* Coastal engineering; air-sea interactions. *Mailing Add:* 2205 Commonwealth Blvd Ann Arbor MI 48105

LIU, PAUL ISHEN, b Taiwan; US citizen. CLINICAL PATHOLOGY. *Educ:* Nat Taiwan Univ, MD, 60; St Louis Univ, PhD(path), 69. *Prof Exp:* Assoc prof path, Med Ctr, Univ Kans, 73-74; assoc dir lab med, Med Col Ga, 74-76; prof & vchmn, Med Univ SC, 76-80; PROF PATH & VCHMN DEPT, UNIV SOUTH ALA, 81- *Mem:* AMA; Col Am Path; Am Soc Clin Path; Asn Clin Sci; Am Soc Microbiol. *Res:* Leukemia; immunology. *Mailing Add:* Univ SAla Med Ctr Mobile AL 36608

LIU, PHILIP L-F, b Fu-Chu, China, Dec 11, 46; m; c 2. HYDRODYNAMICS, COASTAL ENGINEERING. *Educ:* Nat Taiwan Univ, BS, 68; Mass Inst Technol, SM, 71, ScD(hydrodyn), 74. *Prof Exp:* Res asst civil eng, Mass Inst Technol, 69-74; from asst prof to assoc prof environ eng, 74-83, assoc dir, Sch Civil & Environ Eng, 85-86, assoc dean, Col Eng, 86-87, PROF ENVIRON ENG, CORNELL UNIV, 83- *Concurrent Pos:* Justice asst prof, Justice Found, 78-79; Eng Found fel, 79; J S Guggenheim fel, 80; vis assoc, Calif Tech, 80-81; vis scientist, Delft Hydraulics, 87; vis prof, Tech Univ Denmark, 88. *Honors & Awards:* Walter L Huber Prize, Am Soc Civil Engrs, 78. *Mem:* Am Soc Civil Engrs; Am Geophys Union. *Res:* Wave hydrodynamics in coastal engineering; coastal currents and shoreline processes; numerical methods for nonlinear free surface problems; groundwater flow modeling. *Mailing Add:* Sch Civil & Environ Eng Cornell Univ Ithaca NY 14853

LIU, PINGHUI VICTOR, b Formosa, China, Feb 9, 24; nat US; m 59; c 2. MEDICAL MICROBIOLOGY. *Educ:* Tokyo Jikel-kai Sch Med, MD, 47; Tokyo Med Sch, PhD(microbiol), 57; Am Bd Med Microbiol, dipl, 62. *Prof Exp:* Intern, Mercy Hosp, Cedar Rapids, Iowa, 54-55; intern internal med, Louisville Gen Hosp, 55-56; from instr to assoc prof, 57-69, prof microbiol, 69-81, PROF MICROBIOL & IMMUNOL, SCH MED, UNIV LOUISVILLE, 81- *Concurrent Pos:* Res fel microbiol, Sch Med, Univ Louisville, 56-57; USPHS sr res fel, 59-, res career develop award, 62-; mem, Subcomt Pseudomonas & Related Organisms, Int Comt Bact Nomenclature, 63- *Mem:* AAAS; Am Soc Microbiol; Infectious Dis Soc Am; NY Acad Sci. *Res:* Pathogenesis and taxonomy of pseudomonads and related organisms, such as aeromonads and vibrios; extracellular toxins, such as hemolysin, lecithinase and protease; immunities to infections. *Mailing Add:* Microbiol Health Sci Ctr Univ Louisville Louisville KY 40292

LIU, QING-HUO, b Fujian, People's Repub China, Feb 4, 63; m 87. WAVE PROPAGATION, NUMERICAL ANALYSIS. *Educ:* Xiamen Univ, China, BSc, 83, MSc, 86; Univ Ill, Urbana-Champaign, PhD(elec eng), 89. *Prof Exp:* Res asst physics, Xiamen Univ, China, 84-86, teaching asst, 85-86; res asst elec eng, Univ Ill, Urbana-Champaign, 86-88, postdoctoral res assoc, 88-90; RES SCIENTIST ELECTROMAGNETICS, SCHLUMBERGER-DOLL RES, 90- *Mem:* Inst Elec & Electronic Engrs, Antennas & Propagation Soc; Inst Elec & Electronic Engrs, Microwave Theory & Tech Soc; Inst Elec & Electronic Engrs, Geosci & Remote Sensing Soc. *Res:* Large scale computation for the foward and inverse modeling of electromagnetic wave propagation in inhomogeneous media. *Mailing Add:* Schlumberger-Doll Res Old Quarry Rd Ridgefield CT 06877

LIU, RAY HO, b Taiwan, Apr 3, 42; US citizen; m 65; c 3. FORENSIC DRUG URINALYSIS, CRIMINALISTICS. *Educ:* Cent Police Col, Taiwan, LLB, 65; Southern Ill Univ, PhD(chem), 76. *Prof Exp:* Asst prof forensic sci, Univ Ill Chicago, 77-80; mass spectrometrist, US Environ Protection Agency, 80-82; ctr mass spectrometrist, USDA, 82-83; assoc prof, 84-89, PROF FORENSIC SCI, UNIV ALA, BIRMINGHAM, 89- *Concurrent Pos:* Tech dir, Environ Health Res & Testing, Inc, 87-91; drug urinalysis expert witness & resource person, USCG, 88-; lab inspector, Nat Inst Drug Abuses Nat Lab Cert Prog, 88-; ed-in-chief, Forensic Sci Rev J, 89-; dir, Forensic Sci Grad Prog, Univ Ala, Birmingham, 91- *Mem:* Fel Am Acad Forensic Sci; Am Chem Soc; Am Soc Mass Spectrometry. *Res:* Application and development of new approaches for solving existing and emerging problems in forensic sciences, with special emphasis on analytical approaches that may be used for sample differentiation purposes. *Mailing Add:* Dept Criminal Justice Univ Ala Birmingham AL 35294

LIU, ROBERT SHING-HEI, b Shanghai, China, Aug 1, 38; m 67; c 2. ORGANIC CHEMISTRY. *Educ:* Howard Payne Col, BS, 61; Calif Inst Technol, PhD(chem), 65. *Prof Exp:* Res chemist, E I du Pont de Nemours & Co, Inc, 64-68; assoc prof, 68-72, PROF CHEM, UNIV HAWAII, 72- *Concurrent Pos:* UH Fujio Matsuda Scholar, 84-85. *Honors & Awards:* Alfred P Sloan fel, 70-72; John Simon Guggenheim Found fel, 74-75; Merit Award, NIH, 89. *Mem:* Inter-Am Photochem Soc; Am Chem Soc; Am Soc Photobiology. *Res:* Photochemistry of polyenes; energy transfer processes in solutions; bioorganic reaction mechanisms; new geometric isomers of vitamin A and carotenoids; visual pigments: primary processes, analogs and binding sites; bacteriorhodospin analogs. *Mailing Add:* Dept of Chem Univ of Hawaii Honolulu HI 96822

LIU, RUEY-WEN, b Kiangsu, China, Mar 18, 30; US citizen; m 57; c 2. ELECTRICAL ENGINEERING. *Educ:* Univ Ill, BS, 54, MS, 55, PhD(elec eng), 60. *Prof Exp:* From asst prof to assoc prof, 60-66; PROF ELEC ENG, UNIV NOTRE DAME, 66- *Concurrent Pos:* NSF grants 62-63, 64-66 & 71-73; vis assoc prof, Univ Calif, Berkeley, 65-66; vis prof, Nat Taiwan Univ, 69 & Univ Calif, Berkeley, 77-78. *Mem:* Am Math Soc; fel Inst Elec & Electronics Engrs. *Res:* System and network theory; large-scale dynamical systems. *Mailing Add:* Dept of Elec Eng Univ Notre Dame Notre Dame IN 46556

LIU, S(HING) G(ONG), b Soochow, China, Oct 24, 33; m 60; c 3. ELECTRICAL ENGINEERING, APPLIED PHYSICS. *Educ:* Univ Taiwan, BS, 54; NC State Col, MS 58; Stanford Univ, PhD(elec eng), 63. *Prof Exp:* Jr engr, Int Bus Mach Corp, 58-59, assoc engr, 59; asst microwave ferrites, Stanford Univ, 59-63; RES SCIENTIST, RCA LABS, 63- *Mem:* Am Phys Soc; Inst Elec & Electronics Engrs. *Res:* Spin waves in ferrites; microwave and optical frequency devices using semiconductors; ion implantation in gallium arsenide and related III-V compound semiconductors. *Mailing Add:* 48 Braeburn Dr Princeton NJ 08540

LIU, SAMUEL HSI-PEH, b Taiyuan, China, Apr 17, 34; m 61; c 2. THEORETICAL SOLID STATE PHYSICS. *Educ:* Taiwan Univ, BS, 54; Iowa State Univ, 58, PhD(physics), 60. *Prof Exp:* Assoc res mem, Res Lab, IBM Corp, 60-61, res staff mem, 61-64; from assoc prof to prof physics, Iowa State Univ, 64-81; sr res staff mem, 81-89, CORP FEL, OAK RIDGE NAT LAB, 89- *Concurrent Pos:* Vis prof, Ruhr Univ Bochum, WGer, 70, H C Oersted Inst, Copenhagen Univ, 71-72, Univ Calif, Berkeley, 75-76 & 86, Free Univ, Berlin, WGer, 80. *Mem:* Fel Am Phys Soc; AAAS; Sigma Xi. *Res:* Solid state theory; electronic and magnetic properties of metals and metallic compounds; solid surfaces and disordered solids. *Mailing Add:* Oak Ridge Nat Lab Oak Ridge TN 37831-6032

LIU, SI-KWANG, b Kwangsi, China, Dec 1, 25; m 60; c 4. VETERINARY PATHOLOGY. *Educ:* Vet Col Chinese Army, DVM, 49; Univ Calif, Davis, PhD(vet path), 64. *Prof Exp:* Sr vet res & diag, Provincial Taitung Agr Sta, China, 50-55; lectr vet path, Col Agr, Taiwan Univ, 56-59, chief path lab, Univ Vet Hosp, 56-59; res asst path & parasitol, Sch Vet Med, Univ Calif, Davis, 59-64; assoc pathologist, 64-66, cardiopulmonary pathologist, 66-69, asst head, Dept Path, 69-73, SR STAFF MEM, ANIMAL MED CTR, 73- *Concurrent Pos:* Vis prof vet path, Nat Taiwan Univ & vis expert, Chinese Sci Coun, 76-77 & 85-86; sci fel, NY Zool Soc, Bronx Zoo, 79-; clin prof comp path, New York Med Col; consult, Pig Res Inst, Taiwan, Repub China, 85. *Honors & Awards:* Ralston Purina Res Award Cardiovasc Path, 82; Carnation Res Awards in Feline Dis & Nutrit, 84; Beecham Award for Res Excellence, 86; ROC Award in Comp Path, Chinese Histopath Soc, 89. *Mem:* hon mem Am Vet Med Asn; Am Soc Parasitol; Sigma Xi; NY Acad Sci; Int Acad Path; Int Skeletal Soc; Int Cardiovasc Path Soc. *Res:* Cardiovascular pathology in domestic animals as well as zoo animals; comparative pathology in cardiovascular and orthopedic diseases. *Mailing Add:* Animal Med Ctr 510 E 62nd St New York NY 10021

LIU, STEPHEN C Y, b Hunan, China, Feb 24, 27; m 54; c 4. MICROBIOLOGY, IMMUNOLOGY. *Educ:* Taiwan Univ, BSc, 51, MSc, 54; Univ Minn, PhD, 57. *Prof Exp:* Instr plant path, Taiwan Univ, 51-54; from res asst to res assoc, Univ Minn, 54-58; res plant virologist, Nat Res Coun, 58-62, asst mgr Chas Pfizer & Co, Inc, 62-65; from asst prof to assoc prof, 65-74, PROF MICROBIOL, EASTERN MICH UNIV, 74- *Concurrent Pos:* Tech consult, People's Republic China, UN, 80-81. *Mem:* AAAS; Am Phytopathological Soc; Am Soc Microbiol; NY Acad Sci; Sigma Xi. *Res:* Genetics of bacteria; immunology; virology; hydrobiology. *Mailing Add:* 2901 Pebble Park Rd Ann Arbor MI 48108

LIU, SUNG-TSUEN, physical chemistry, inorganic chemistry, for more information see previous edition

LIU, TAI-PING, b Taiwan, Repub of China, Nov 18, 45; m 73. MATHEMATICAL ANALYSIS. *Educ:* Nat Taiwan Univ, BS, 68; Ore State Univ, MS, 70; Univ Mich, PhD(math), 73. *Prof Exp:* From asst prof to assoc prof, 73-81, PROF UNIV MD, COLLEGE PARK, 81- *Concurrent Pos:* Sloan fel; Guggenheim fel. *Mem:* Am Math Soc. *Res:* Nonlinear partial differential equations and mechanics; qualitative behavior of solutions to physical systems such as compressible flow and elastic models; gas dynamics. *Mailing Add:* Dept Math Univ Md College Park MD 20742

LIU, TA-JO, b Taipei, Taiwan, Sept 16, 51. CHEMICAL ENGINEERING. *Educ:* Nat Taiwan Univ, BS, 73; Polytechnic Inst, NY, MS, 77, PhD(chem eng), 80. *Prof Exp:* Res scientist, 79-81, Sr Res Scientist, Eastman Kodak Co, 81-; AT DEPT CHEM, NAT TSING HUA UNIV, TAIWAN. *Res:* Fundamental coating; optimal coating die design for Newtonian, Non-Newtonian fluids; development of special coating techniques. *Mailing Add:* Dept Chem Eng Nat Tsing Hua Univ No 101 Sec 2 Kuang Furd Hsinchu 30083 Taiwan

LIU, TEH-YUNG, b Tainan, Taiwan, May 24, 32; nat US; m 61; c 3. BIOCHEMISTRY. *Educ:* Taiwan Nat Univ, BS, 55; Univ Pittsburgh, PhD(biochem), 61. *Prof Exp:* Res assoc biochem, Rockefeller Univ, 61-65, asst prof, 65-67; biochemist, Biol Dept, Brookhaven Nat Lab, 67-73; sect head biochem microbial struct, Nat Inst Child Health & Human Develop, 73-74; dep dir, Div Bact Prod, 74-79, DIR, DIV BIOCHEM & BIOPHYS, CTR BIOLOGICS EVAL & RES, FOOD & DRUG ADMIN, 80- *Concurrent Pos:* Adj prof, Chem Dept, Catholic Univ Am, Washington, DC; vis prof, Univ Peking, China, Kyoto Univ, Japan, Union Med Col, Peking, China; fel, St Johns Col, Cambridge, Eng. *Mem:* Am Soc Biol Chem. *Res:* Limulus lysate; protein chemistry; human C-reactive protein; Rubella virus. *Mailing Add:* Ctr & Biologics Eval & Res FDA Bldg 29 Rm 516 NIH Bethesda MD 20892

LIU, TING-TING Y, b Taipei, Taiwan, Aug 4, 49; m 75; c 1. STARCHBIOSYNTHESIS. *Educ:* Nat Taiwan Univ, BS, 71; Univ Hawaii, MS, 74; Pa State Univ, PhD(hort), 79. *Prof Exp:* Res asst, Dept Agron, Nat Taiwan Univ, 71-72; grad asst, Dept Hort, Pa State Univ, 74-78, res assoc, 79; tech asst, Dept Dairy Sci, 79-82, res asst, 82-85, RES ASSOC, OHIO AGR RES & DEVELOP CTR, OHIO STATE UNIV, 85- *Mem:* Am Soc Plant Physiologists; Sigma Xi. *Res:* Isolate amyloplasts from corn endosperm to study sugar translocation and starch biosynthesis in plant, and the analysis of research data for animal toxicity studies statistically. *Mailing Add:* 1707 Wessel Dr Worthington OH 43085-4313

LIU, TONY CHEN-YEH, b Fu-Chien, China, July 27, 43; US citizen; m 69; c 2. STRUCTURAL ENGINEERING, CIVIL ENGINEERING. *Educ:* Nat Chung-Hsing Univ, Taiwan, BS, 65; SDak Sch Mines & Technol, MS, 68; Cornell Univ, PhD(civil eng), 71. *Prof Exp:* Struct engr civil eng, Ammann & Whitney Inc, 71; group leader nuclear eng, Gen Atomic Co, 72-76; res engr, Waterways Exp Sta, 76-81, CHIEF MAT ENG, US ARMY CORPS ENGRS, 81- *Concurrent Pos:* Guide prof, World Open Univ, 75. *Honors & Awards:* Wason Res Medal, Am Concrete Inst, 74 & 83. *Mem:* Fel Am Concrete Inst; Am Soc Civil Engrs; Sigma Xi; Am Soc Testing & Mat. *Res:* Design of concrete hydraulic structures, precast concrete structures, repair and rehabilitation of deteriorated concrete structures and thermal stress analysis for mass concrete structures. *Mailing Add:* 1284 Towlston Rd Great Falls VA 22066

LIU, TUNG, b Peking, China, Mar 12, 26; nat US; m 58; c 3. CHEMICAL ENGINEERING. *Educ:* Nankai Univ, Tientsin, BS, 47; Univ Ill, MSc, 51, PhD(chem eng), 53. *Prof Exp:* Chem engr, Pittsburgh Consol Coal Co, 53-55; res chem engr, Monsanto Chem Co, 56-60; res mat engr, Air Force Mat Lab, 60-68; sr res engr, Occidental Petrol Corp, 68-77; RES ENGR, BENDIX RES LABS, 77- *Concurrent Pos:* Chem engr, Garrett Corp, 70-77. *Mem:* Am Chem Soc; Am Inst Chem Engrs; Sigma Xi. *Res:* Friction, lubrication and wear. *Mailing Add:* 16734 Hapton Dr Granger IN 46530

LIU, VI-CHENG, b China, Sept 1, 17; nat US; m 47. AEROSPACE ENGINEERING. *Educ:* Chiao Tung Univ, BS, 40; Univ Mich, MS, 47, PhD(aeronaut eng), 51. *Prof Exp:* Res asst aerodyn, Aeronaut Res Inst, Tsing Hua Univ, China, 40-44, res instr, 44-46; res fel, Ministry Educ, 46-48; res assoc eng, Res Inst, 48-50, res engr, 50-59, PROF AEROSPACE ENG, UNIV MICH, ANN ARBOR, 59- *Mem:* Am Phys Soc; assoc fel Am Inst Aeronaut & Astronaut. *Res:* Upper atmosphere; rocket flight, rarefied gas and ionospheric gas dynamics; thermal diffusion; boundary layer flow; turbulent dispersion; plasma interaction; magnetospheric physics; geophysical Fluid Dynamics. *Mailing Add:* 2104 Vinewood Ann Arbor MI 48104

LIU, WEN CHIH, b Liao-Ning Province, China, Feb 19, 21; US citizen; m 60; c 3. BIOCHEMISTRY. *Educ:* Nat Hu-Nan Univ, BS, 44; Baylor Univ, MS, 53; Univ Wis-Madison, PhD(biochem), 58. *Prof Exp:* Res assoc biochem, Univ Wis, 58-59; res assoc, Univ Calif, Los Angeles, 59-60; sr res chemist, Pfizer, Inc, 60-69; sr res investr microbiol, Squibb Inst Med Res, 69-90; RETIRED. *Mem:* AAAS; Am Chem Soc. *Res:* Natural products; antibiotics; cancer chemotherapy; plant growth substances. *Mailing Add:* Seven Wycombe Way Princeton Junction NJ 08550

LIU, WING KAM, b Hong Kong, May 19, 52; US citizen; m; c 2. FINITE ELEMENTS, COMPUTER AIDED ENGINEERING. *Educ:* Univ Ill, Chicago, BSc, 76; Calif Inst Technol, MSc, 77, PhD(civil eng), 81. *Prof Exp:* Res asst, Univ Ill, Chicago, 74-76 & Calif Inst Technol, 76-80; from asst prof to assoc prof, 80-88, PROF MECH & CIVIL ENG, NORTHWESTERN UNIV, 88- *Concurrent Pos:* Consult, reactor anal & safety div, Argonne Nat Lab, 81-; grants, NSF, Army Res Off & NASA, ONR; fel, Am Soc Mech Engrs, 90. *Honors & Awards:* Melville Medal, Am Soc Mech Engrs, 79, Gold Medal, 85; Ralph Teetor's Award, Soc Automotive Engrs, 83; Thomas J Jaeger Prize, Int Asn Struct Mech Reactor Technol, 89. *Mem:* Am Soc Mech Engrs; Am Soc Civil Engrs; Am Acad Mech; Am Soc Mech Engrs, 85. *Res:* Finite elements; computer simulations; fluid-structure interactions; non-linear and inelastic analysis; computer-aided engineering; liquid storage tanks. *Mailing Add:* Dept Mech Eng Northwestern Univ Evanston IL 60208

LIU, WING-KI, b Hong Kong, Feb 24, 50. SURFACE SCIENCE. *Educ:* Univ Ill, Urbana, BS, 71, MS, 72 & PhD(physics), 75. *Prof Exp:* Fel, 75-76, res assoc, 76-79, vis asst prof, 80-81, res asst prof, 81-85, res assoc prof,85-86, ASSOC PROF , DEPT PHYSICS, UNIV WATERLOO, 86-; vis assoc, Calif Inst Technol, 79-80. *Mem:* Am Phys Soc; Can Assoc Physicists. *Res:* Classical and quantum chemical dynamics; theory of chemical physics of solid surfaces. *Mailing Add:* Dept Physics Univ Waterloo Waterloo ON N2L 3G1 Can

LIU, WINGYUEN TIMOTHY, US citizen. REMOTE SENSING, AIR-SEA INTERACTION. *Educ:* Ohio Univ, BS, 71; Univ Wash, MS, 74, PhD(atmospheric sci), 78. *Prof Exp:* Res assoc atmospheric sci, Univ Wash, 78-79; SR SCIENTIST SATELLITE OCEANOG, JET PROPULSION LAB, CALIF INST TECHNOL, 79- *Honors & Awards:* Medal for Except Sci Achievement, NASA. *Mem:* Am Meteorol Soc; Am Geophys Union. *Res:* Study of the boundary layers, the energy exchanges across the atmosphere-ocean interface and their effects on climate variability. *Mailing Add:* Jet Propulsion Lab MS 300-323 4800 Oak Grove Dr Pasadena CA 91109

LIU, YOUNG KING, b Nanking, China, May 3, 34; US citizen; div; c 2. BIOMECHANICS, BIOMEDICAL ENGINEERING. *Educ:* Bradley Univ, BS, 55; Univ Wis-Madison, MS, 59; Wayne State Univ, PhD(mech), 63. *Prof Exp:* Instr mech, Wayne State Univ, 60-63; lectr, Univ Mich, Ann Arbor, 63-64, asst prof, 64-68; vis asst prof aeronaut & astronaut, Stanford Univ, 68-69; assoc prof, 69-72, prof biomech, Tulane Univ, 72-78; prof & dir, Ctr Mat Res, 78-90, PROF & DIR, BIOMECH LAB, UNIV IOWA, IOWA CITY, 90- *Concurrent Pos:* NIH spec res fel, Stanford Univ, 68-69; biophys consult, US Army Aeromed Res Lab, 72-; NIH res career develop award, 71-76; adv at large, Nat Res Coun, Comt Hearing, Bioacoust & Biomech, NAS, 79- *Mem:* Am Soc Eng Educ; Orthop Res Soc; Am Soc Eng Educ; Sigma Xi; Am Acad Mech; Int Soc Study Lumbar Spine; Int Soc Study Pain. *Res:* Biomechanics, biomaterials and physiologic basis of acupuncture. *Mailing Add:* Ctr for Mat Res Univ Iowa 1212 EB Iowa City IA 52242

LIU, YU, b China; US citizen; m 67; c 1. OPTICS, LASERS. *Educ:* Ga Inst Technol, PhD(physics), 74. *Prof Exp:* Res asst physics, Rice Univ, 62-69; instr physics, Tex Tech Univ, 66-69; teaching asst physics, Ga Inst Technol, 69-74; physicist optics & laser, US Army Missile Res & Develop Lab, 74-76; res physicist laser & optics, US Army Missile Res & Develop Command, 76-79; PHYSICIST LASER, INFRARED & OPTICS, US ARMY AVIATION RES & DEVELOP COMMAND, 79- *Mem:* Asn US Army. *Res:* Research, design and development of laser guidance links; low temperature physics research. *Mailing Add:* 1865 Schoettler Valley Dr Chesterfield MO 63017

LIU, YUNG SHENG, b China, Sept 23, 44; m; c 2. PHYSICS, LASERS. *Educ:* Nat Taiwan Univ, BS, 66; Cornell Univ, PhD(appl physics), 73. *Prof Exp:* Teaching asst physics, Cornell Univ, 68-69, res asst, 70-73; PHYSICIST, GEN ELEC RES CTR, 72-; ADJ PROF PHYSICS, STATE UNIV NY, ALBANY, 86- *Concurrent Pos:* Vis scientist physics, Univ Calif, Los Angeles, 68-; fel, Cornell Univ, 69, Avco fel, 70; consult, United Nations Develop Prog, China, 86. *Mem:* Optical Soc Am; Am Phys Soc; Sigma Xi; AAAS; Inst Elec & Electronics Engrs. *Res:* Laser physics; quantum electronics and optics; laser-matter interactions; semiconductor electronics; author or coauthor of over 50 publications. *Mailing Add:* 2201 Pine Ridge Ct Schenectady NY 12309

LIU, YUNG YUAN, b Taipei, Taiwan, Mar 20, 50; US citizen; m 75; c 2. CORROSION-EROSION. *Educ:* Nat Tsing-Hua Univ, Taiwan, BS, 71; Mass Inst Technol, MS, 76, ScD(nuclear eng), 78. *Prof Exp:* Res asst cogeneration, Dept Nuclear Eng, Mass Inst Technol, 74, teaching asst phys metall, 74-75, struct mech, 75 & radiation effects, 75-76, res asst light water reactor fuel performance, 76-78; assoc staff engr, fast reactor fuel performance, Argonne Nat Lab, 78-82, staff engr & theorist, radiation effects & nuclear mat-casting-solidification, Mat Sci & Technol Div, 82-85, staff engr high temperature corrosion-erosion, tribology, 86-88, STAFF ENGR IRRADIATION PERFORMANCE, MAT & COMPONENTS TECHNOL DIV, ARGONNE NAT LAB, 89- *Concurrent Pos:* Staff engr light water reactor fuel performance, Entropy Ltd, Lincoln, Mass, 77-78; mem, fast reactor fuel performance code comt, US Dept Energy & Fuel Performance Eval Task Force, 78-81; prin investr, Argonne Nat Lab Mat Sci & Technol Div, 78-; consult, Los Alamos Nat Lab, 82-83; staff reviewer, Nuclear Technol, 82-, Am Soc Mech Engrs, 83- *Mem:* Am Soc Metals; AAAS. *Res:* Radiation effects on materials & development of nuclear fuel and breeder materials for fission and fusion reactors. *Mailing Add:* Argonne Nat Lab 9700 S Cass Ave Bldg 212 Argonne IL 60439

LIU, YUNG-PIN, CARCINOGENESIS MECHANISMS. *Educ:* Baylor Univ, PhD(biochem), 69. *Prof Exp:* PROG DIR CARCINOGENESIS MECHANISMS, NAT CANCER INST, 84- *Res:* Biochemical pharmacology. *Mailing Add:* Div Cancer Etiol NIH-Nat Cancer Inst EPN 700 Bethesda MD 20892

LIU, YU-YING, b China, May 16, 44. SYNTHESIS OF LABELLED COMPOUND. *Educ:* Taiwan Normal Univ, BS, 67; Univ Minn, PhD(chem), 72. *Prof Exp:* Fel, Am Health Found, 71-73; sr scientist, 74-85, RES INVESTR, HOFFMANN LAROCHE, INC, 86- *Mem:* Am Chem Soc. *Res:* Natural product isolation; metabolism; quantitative analysis; synthesis of labelled compound of pharmaceutical interest. *Mailing Add:* 570 Wellington Rd Ridgewood NJ 07450-1222

LIU-GER, TSU-HUEI, b Kwei-yang, Kwei-chow, Repub of China, Mar 10, 43; US citizen; m 71; c 2. THEORETICAL PHYSICS. *Educ:* Nat Taiwan Univ, BS, 64; Univ Ore, PhD(physics), 69. *Prof Exp:* Asst prof physics, Portland State Univ, 69-75; PHYSICIST, ELECTROMAGNETIC TRANSIENT PROG, BONNEVILLE POWER ADMIN, US DEPT ENERGY, 75- *Res:* Electromagnetic transient studies of the power systems. *Mailing Add:* Bonneville Power Admin Rte EOHB PO Box 3621 Portland OR 97208

LIUIMA, FRANCIS ALOYSIUS, b Utena, Lithuania, Mar 8, 19; nat US. PHYSICS. *Educ:* Boston Col, MS, 50; St Louis Univ, PhD(physics), 54. *Prof Exp:* ASST PROF PHYSICS, BOSTON COL, 54- *Mem:* Am Phys Soc; Am Asn Physics Teachers. *Res:* Microwave spectroscopy. *Mailing Add:* Dept Physics Boston Col Chestnut Hill MA 02167

LIUKKONEN, JOHN ROBIE, b Oakland, Calif, Oct 23, 42. MATHEMATICAL ANALYSIS. *Educ:* Harvard Univ, BA, 65; Columbia Univ, PhD(math), 70. *Prof Exp:* Asst prof, 70-75, ASSOC PROF MATH, TULANE UNIV, 75- *Res:* Representations of locally compact groups; harmonic analysis on locally compact groups. *Mailing Add:* Dept Math Tulane Univ New Orleans LA 70118

LIUZZO, JOSEPH ANTHONY, b Tampa, Fla, Dec 16, 26; m 51; c 3. FOOD SCIENCE. *Educ:* Univ Fla, BS, 50, MSA, 55; Mich State Univ, PhD(nutrit, biochem), 58. *Prof Exp:* Res microbiologist, Univ Fla, 50-51, asst, 54-55; dir microbiol, Nutrilite Prod, Inc, Calif, 51-53, asst to dir biol res, 53-54; asst, Mich State Univ, 55-58; asst prof biochem, 58-62, assoc prof food sci & technol, 62-69, PROF FOOD SCI, LA STATE UNIV, BATON ROUGE, 69- *Mem:* Am Inst Nutrit; Inst Food Technologists; Am Chem Soc. *Res:* Improved utilization of brown ad milled rice; radiation preservation of foods; utilization of by-products from agricultural commodities. *Mailing Add:* Dept of Food Sci La State Univ Baton Rouge LA 70803

LIVANT, PETER DAVID, b New York, NY, Sept 18, 48. PHYSICAL ORGANIC CHEMISTRY. *Educ:* City Col New York, BS, 69; Brown Univ, PhD(chem), 75. *Prof Exp:* Vis asst prof chem, Univ Ill, Urbana-Champaign, 74-75, res assoc, 75-76; res assoc, Univ Guelph, 76-77; ASST PROF CHEM, AUBURN UNIV, 77- *Mem:* Am Chem Soc. *Res:* Mechanisms of radical reactions; chemistry of hypervalent species; tetracoordinate tetracovalent sulfur compounds; chemically induced dynamic nuclear polarization dependence on magnetic field strength. *Mailing Add:* Dept of Chem Auburn Univ Auburn AL 36849

LIVDAHL, PHILIP V, b Bismarck, NDak, Feb 1, 23; m 44; c 3. PHYSICS. *Educ:* St Olaf Col, BA, 48; Univ Wash, MS, 52. *Prof Exp:* Assoc physicist, Calif Res & Develop Corp, 51-54; physicist, Lawrence Radiation Lab, Calif, 54-57; assoc physicist, Argonne Nat Lab, 57-67; actg dir, Fermi Nat Accelerator Lab, 78-79, physicist, 67-, assoc dir, 79-; RETIRED. *Res:* High energy accelerators; linear accelerators and associated equipment for synchrotron injectors; experimental planning and operation for the zero gradient synchrotron. *Mailing Add:* 137 Harlemen Dr Sequim WA 98382

LIVE, DAVID H, b Philadelphia, Pa, Apr 3, 46. BIOPHYSICS, PHYSICAL CHEMISTRY. *Educ:* Univ Pa, BA, 67; Calif Inst Technol, PhD(chem), 74. *Prof Exp:* Res assoc biophys, Rockefeller Univ, 74-78, asst prof phys biochem, 78-85; assoc prof chem, Emory Univ, 86-91; MEM PROF STAFF, CALIF INST TECHNOL, 91- *Concurrent Pos:* Consult, Jet Propulsion Lab, Calif Inst Technol, 75- *Mem:* Am Chem Soc; AAAS; NY Acad Sci. *Res:* Biophysical applications of magnetic resonance to studying molecular conformation, particularly in peptides and proteins; geochemical investigations by magnetic resonance of terrestrial and lunar samples. *Mailing Add:* Dept Chem Calif Inst Technol Pasadena CA 91125

LIVE, ISRAEL, b Austria, Apr 26, 07; nat US; m 36; c 2. VETERINARY SCIENCE. *Educ:* Univ Penn, VMD, 34, AM, 36, PhD(path), 40; Am Bd Microbiol, dipl. *Prof Exp:* Asst histopath & clin path, 34-37, from instr path to asst prof, 37-46, bact, 46-49, assoc prof, 49-53, PROF MICROBIOL, SCH VET MED, UNIV PA, 53- *Concurrent Pos:* Mem expert comt on brucellosis, WHO; mem comt on brucellosis res, Nat Acad Sci. *Mem:* Fel AAAS; fel Am Acad Microbiol; Am Vet Med Asn; Am Soc Microbiol; Asn; fel Am Col Vet Microbiol. *Res:* Diagnosis of filariasis in dogs; nature of Clostridium chauvoei aggressin; diagnosis, therapy and immunization in brucellosis; staphylococci in animals and man; serological characterization of staphylococci. *Mailing Add:* Dept Microbiol Univ Pa Sch Vet Med Philadelphia PA 19104

LIVELY, DAVID HARRYMAN, b Indianapolis, Ind, Aug 17, 30; m 53; c 2. MICROBIOLOGY. *Educ:* Purdue Univ, BS, 52; Univ Tex, MA, 58, PhD, 62. *Prof Exp:* Mem water pollution serv team, State Bd Health, Ind, 52; bacteriologist, Biol Warfare Labs, US Army, 54-56; asst bacteriologist, Univ Tex, 56-57, res scientist, 57-60; sr microbiologist, 61-63, group leader microbiol, Chem Res Div, 63-66, asst mgr antibiotic develop, Antibiotic Mfg & Develop Div, 66-69, head microbiol res, 69-72, RES ASSOC, ELI LILLY & CO, 72- *Mem:* Fel AAAS; Am Chem Soc; Am Soc Microbiol; Sigma Xi. *Res:* Bacterial endospore; production of biological active compounds by microorganisms; automation in microbiology. *Mailing Add:* Antibiotic Develop Div Eli Lilly & Co Dept Ky 412-Bldg 328/1 Indianapolis IN 46285

LIVENGOOD, DAVID ROBERT, b LaJunta, Colo, Mar 18, 37; m; c 4. BIOPHYSICS, ELECTROPHYSIOLOGY. *Educ:* Butler Univ, BS, 60; Ind Univ, PhD(physiol), 70. *Prof Exp:* Res assoc, Dept Biophysics, Sch Med, Univ Md, 71-73; res physiologist, Dept Neurobiol, 73-79, chief, Radiation Biophysics Div, 79-80, CHMN, DEPT PHYSIOL, ARMED FORCES RADIOBIOL RES INST, 80- *Concurrent Pos:* Grass Found fel, Woodshole Marine Biol Lab, 71; fel, Marine Biol Lab, STI, 76; res consult, Dept Physiol, George Washington Univ, 78-84; adj staff, Dept Physiol, Uniformed Serv Univ Health Sci, 80- & dept physiol & biophys, Georgetown Univ Sch Med. *Mem:* Biophys Soc; Soc Neurosci; Soc Gen Physiologists. *Res:* Biophysical properties of the membranes of nerve and muscle cells. *Mailing Add:* Dept Physiol Nat Naval Med Ctr Armed Forces Radiobiol Res Inst Bethesda MD 20814-5145

LIVENGOOD, SAMUEL MILLER, b Salisbury Pa, Nov 1, 17; m 41; c 3. ORGANIC CHEMISTRY. *Educ:* Juniata Col, BS, 38; Rutgers Univ, MS, 41, PhD(org chem), 43. *Prof Exp:* Instr chem, Rutgers Univ, 40-43; fel, Mellon Inst, 43-55, sr fel, 55-59; asst dir, Res & Develop Dept, Chem Div, Union Carbide Corp, 59-75, assoc dir, 59-79, dir res & develop, Ethylene Oxide Derivatives Div, 79-83; RETIRED. *Mem:* Am Chem Soc; Soc Chem Indust; Sigma Xi. *Res:* Detergents and cosmetics; textile intermediates; humectants; water soluble resins; hydraulic fluids; heat transfer fluids; metalworking fluids. *Mailing Add:* 11769 Woodlea Dr Waynesboro PA 17268

LIVERMAN, JAMES LESLIE, b Brady, Tex, Aug 17, 21; m 43, 59; c 5. PLANT PHYSIOLOGY, BIOCHEMISTRY. *Educ:* Tex A&M Univ, BS, 49; Calif Inst Technol, PhD(plant physiol, bioorganic chem). *Prof Exp:* Fel plant physiol, Calif Inst Technol, 52-53; from asst prof to prof biochem, Agr & Mech Col, Univ Tex, 53-60; biochemist, AEC, 58-59, asst chief biol br, 59-60, chief, 60-64; assoc dir biol div, Oak Ridge Nat Lab, 64-67, asst dir life sci, 67-69, assoc dir biomed & environ sci, 69-72; dir div biomed & environ res, US AEC, 72-75, asst gen mgr biomed & environ res & safety, 73-75, dir, Div Biomed & Environ Res & asst adminr environ & safety, US Energy Res & Develop Admin, 75-77; actg asst secy environ, Dept Energy, 77-78, dep asst secy, 78-79; sr vpres & gen mgr, appl sci div, Litton Bionetics Inc, 79-85, vpres bionetics res, 85-87; vpres prod develop, Organon Teknika Corp, 85-87; Consult, Maxwell Commun Biomed Res & Develop Planning, 87-88; consult & dir tech trf database, Univ Sci Eng & Tech, Inc, 88-90; CONSULT ENVIRON & WASTE MGT, 90- *Concurrent Pos:* Consult agr chemist, 56-58; chmn, Gordon Conf Biochem & Agr, 61; Interim dir, Univ Tenn-Oak Ridge Grad Sch Biomed Sci, 65-66. *Mem:* Fel AAAS; Am Soc Plant Physiol; Am Chem Soc; Radiation Res Soc; Ecol Soc Am; Am Mgt Asn. *Res:* Cell physiology; photoperiodism; radiation in biological systems; immunology; bioengineering; policy science. *Mailing Add:* 5308 Manor Lake Court Rockville MD 20853

LIVERMAN, THOMAS PHILLIP GEORGE, b Salzburg, Austria, June 18, 23; US citizen; m 46; c 2. MATHEMATICS. *Educ:* Univ Pa, MA, 48, PhD(math), 56. *Prof Exp:* Instr math, Univ Del, 46-48; engr, C N R Co, France, 48-49; mathematician, Appl Physics Lab, Johns Hopkins Univ, 51-58; assoc prof, 58-60, chmn dept math, 71-74, 76-85, PROF MATH & ADJ PROF ENG & APPL SCI, GEORGE WASHINGTON UNIV, 71- *Mem:* Soc Indust & Appl Math. *Res:* Functional analysis and applied mathematics; function theory; operational calculus; infinite dimensional generalited functions and fock spaces. *Mailing Add:* Rte 1 Box 1256 Amissville VA 22002

LIVERSAGE, RICHARD ALBERT, b Fitchburg, Mass, July 8, 25; m 54; c 4. DEVELOPMENTAL BIOLOGY, REGENERATION. *Educ:* Marlboro Col, BA, 51; Amherst, Col, AM, 53; Princeton Univ, AM, 57, PhD(biol), 58. *Prof Exp:* Lab instr biol, Amherst Col, 54-55; instr, Princeton Univ, 58-60; from asst prof to assoc prof, 60-69, grad secy dept, 75-77, assoc chmn grad affairs, 78-84, actg chmn, 80-81, PROF ZOOL, UNIV TORONTO, 69- *Concurrent Pos:* Vis investr, Huntsman Marine Lab, NB, Can, 68-71; vis prof, Dept Biophys, Strangeways Res Lab, Cambridge, Eng, 72; grad secy dept, Univ Toronto, 75-77, assoc chmn grad affairs, 78-84, acting chmn, 80-81. *Mem:* Can Soc Zool; Soc Develop Biol; Royal Can Inst; Sigma Xi; Am Soc Zool. *Res:* In vivo and in vitro studies on the role of nerves and endocrine secretions at the cellular-molecular levels in amphibian appendage regeneration. *Mailing Add:* Ramsay Wright Zool Labs Univ of Toronto Toronto ON M5S 1A1 Can

LIVESAY, GEORGE ROGER, b Ashley, Ill, Dec 9, 24. MATHEMATICS. *Educ:* Univ Ill, BS & MS, 48, PhD, 52. *Prof Exp:* Instr math, Univ Mich, 50-56; res assoc, 56-58, from asst prof to assoc prof, 58-69, PROF MATH, CORNELL UNIV, 69- *Res:* Topology. *Mailing Add:* Dept of Math Cornell Univ Ithaca NY 14853

LIVETT, BRUCE G, b Melbourne, Australia, Aug 27, 43; m 76; c 2. NEUROSCIENCES, BIOCHEMICAL PHARMACOLOGY. *Educ:* Monash Univ, BSc, 65, PhD(biochem), 68. *Prof Exp:* Nuffield Dominions demonstr pharmacol, Oxford Univ, 69-71, jr res fel, Wolfson Col, 70-71; Queen Elizabeth II res fel biochem, Monash Univ, 71-73, asst prof, 73-77; from assoc prof to prof med & biochem, Montreal Gen Hosp, McGill Univ, 77-83; READER DEPT BIOCHEM, UNIV MELBOURNE, AUSTRALIA, 83- *Concurrent Pos:* Med Res Coun fel neurosci, McMaster Univ, 75-76, prin investr, 77-; prin investr & mem adv bd, Muscular Dystrophy Asn Can, 78-82; coun mem, Int Soc Neurochem, 87- & Australian Neurosci Soc, 87- *Mem:* Int Brain Res Orgn; Int Soc Neurochem; Soc Neurosci; Australian Neurosci Soc; Australian Biochem Soc; Australian Physiol & Pharmacol Soc; Endocrine Soc Australia; Australian Soc Comput in Learning Tertiary Educ. *Res:* Investigation into the role of neuropeptides as neuromodulators of catecholamine secretion in the endocrine adrenal medulla and nervous system; role of the cellular immunity system in muscular dystrophy. *Mailing Add:* Dept Biochem Univ Melbourne/Parkville Victoria 3052 Australia

LIVIGNI, RUSSELL A, b Akron, Ohio, July 20, 34. POLYMER CHEMISTRY. *Educ:* Univ Akron, BS, 56, PhD(polymer chem), 60. *Prof Exp:* Sr res chemist, Gen Tire & Rubber Co, 61-62, group leader, 62-63, sec head, 63-75, mgr, 75-80, assoc dir, 80-87; VPRES & DIR RES, GENCORP, 88- *Concurrent Pos:* chmn, Gordon Res Conf Elastomers, 78; trustee, Edison Polymer Innovation Corp, 86-; adv mem, Akron Coun Eng & Sci Soc, 88-; mem, NSF Indust Panel Sci & Technol, 90- *Mem:* Am Chem Soc; AAAS. *Res:* Technical programs resulting in advanced and improved products and processes, especially related to polymer based technology; granted 30 patents; author of several publications. *Mailing Add:* Gencorp 2990 Gilchrist Rd Akron OH 44305

LIVIGNI, RUSSELL ANTHONY, b Akron, Ohio, July 20, 34. POLYMER CHEMISTRY. *Educ:* Univ Akron, BSc, 56, PhD(polymer chem), 60. *Prof Exp:* Res scientist polymer chem, Ford Sci Lab, 60-61; sr res chemist, 61-62, group leader polymer characterization & kinetics, 62-63, sect head, 63-69, sect head mat chem & polymer characterization, 69-75, dept mgr polymer & anal chem, 75-80, assoc dir, Res Div, 80-88, VPRES & DIR RES, GENCORP, 88- *Mem:* AAAS; Am Chem Soc. *Res:* Kinetics of free radical polymerization; kinetics and mechanism of anionic polymerization and copolymerization; determination of structure of block copolymers; characterization of polymer molecular weights and distribution; new rubbers by anionic polymerization; plastics and composites. *Mailing Add:* 2291 Manchester Rd Akron OH 44314

LIVINGOOD, CLARENCE SWINEHART, b Elverson, Pa, Aug 7, 11; m 47; c 5. DERMATOLOGY. *Educ:* Ursinus Col, BS, 32; Univ Pa, MD, 36; Am Bd Dermat, dipl, 61. *Hon Degrees:* DSc, Ursinus Col, 82. *Prof Exp:* Asst prof dermat & syphil, Med Sch, Univ Pa, 46-48; prof dermat, Jefferson Med Col, 48-49; prof dermat & syphil, Sch Med, Univ Tex, 49-53; CHMN, DEPT DERMAT, HENRY HOSP, 53- *Concurrent Pos:* Consult, Vet Admin & Surg Gen, US Army; mem comn cutaneous dis, Armed Forces Epidemiol Bd; adv panel med sci, Dept Defense, 55-60; secy gen, Int Cong Dermat, 62; secy, Am Bd Dermat, 63- *Honors & Awards:* Gold Medal, Am Acad Dermatol, 75, Presidental Citation, 87; Cert Meritorious Achievement, Dermatol Found, 75; Stephen Rothman Award, Soc Invest Dermatol, 80; Distinguished Serv Award, AMA, 90. *Mem:* AAAS; Soc Invest Dermat (pres, 55); hon mem Am Dermatol Asn; AMA; NY Acad Sci. *Res:* Epidemiology and treatment of cutaneous bacterial infections; topical corticosteroid therapy of cutaneous disease. *Mailing Add:* Dept Dermatol Henry Ford Hosp Detroit MI 48202

LIVINGOOD, JOHN N B, b Birdsboro, Pa, June 8, 13; m 40; c 3. MATHEMATICS. *Educ:* Gettysburg Col, AB, 34; Univ Pa, AM, 36, PhD(math), 44. *Prof Exp:* Teacher high sch, 36-38; instr math, Gettysburg Col, 38-42; Rutgers Univ, 42-44; mathematician, Nat Adv Comt Aeronaut, 44-47; asst prof math, Rutgers Univ, 47-48; aeronaut res scientist, Nat Adv Comt Aeronaut 48-58 & NASA, 58-73; lectr math, Col Boca Raton, Fla, 73-81; RETIRED. *Mem:* Am Math Soc. *Res:* Theory of numbers; aeronautical research; turbine cooling; nuclear engineering. *Mailing Add:* 20 NW 24th St Delray Beach FL 33444

LIVINGOOD, MARVIN D(UANE), b Corning, Kans, Aug 15, 18; m 47; c 4. CHEMICAL ENGINEERING. *Educ:* Okla State Univ, BS, 38, MS, 40; Mich State Univ, PhD(chem eng), 52. *Prof Exp:* Chem engr, Arzone Prod Co, 41; instr chem eng, Mo Sch Mines, 41-46; asst prof, Mich State Univ, 46-49, res asst prof, Eng Exp Sta, 49-52; res chem engr, E I Dupont de Nemours & Co, Inc, 52-74, environ control supvr, 74-79, sr engr, 79-83; PARTNER, RCI LTD SYST CONSULTS, 83- *Concurrent Pos:* Chmn, Prof Develop Comt, Am Inst Chem Engrs, 80-82, Steering Comt, 80-88, chmn fels, 82-86; secy, Mgt Div, Am Inst Chem Engrs, 83-86. *Mem:* Fel Am Inst Chem Engrs; Nat Comp Graphics Asn. *Res:* Chemical engineering design; economics; small scale pilot plant; safety in handling unstable materials; environmental compliance. *Mailing Add:* 2603 Landor Ave Louisville KY 40205-2333

LIVINGSTON, ALBERT EDWARD, b Hartford, Conn, Feb 28, 36. MATHEMATICAL ANALYSIS. *Educ:* Boston Col, BA, 58, MA, 60; Rutgers Univ, MS, 62, PhD(math), 63. *Prof Exp:* Asst prof math, Lafayette Col, 63-67; from asst prof to assoc prof, 67-75, PROF MATH, UNIV DEL, 75- *Mem:* Math Asn Am; Am Math Soc; Sigma Xi. *Res:* Univalent and multivalent fuctions, particularly the application of methods of extreme point theory and subordination chains to extremal problems in multivalent function theory. *Mailing Add:* Dept of Math Univ of Del Newark DE 19711

LIVINGSTON, CLARK HOLCOMB, b Eau Claire, Wis, Nov 25, 20; m 47; c 2. PLANT PATHOLOGY. *Educ:* Colo Agr & Mech Col, BS, 51, MS, 53; Univ Minn, PhD, 66. *Prof Exp:* Assoc prof bot & plant path, Colo State Univ, 55-85; RETIRED. *Mem:* Am Phytopath Soc; Potato Asn Am. *Res:* Potato diseases, particularly physiology of disease and viruses. *Mailing Add:* 3008 Shore Rd Ft Collins CO 80524

LIVINGSTON, DANIEL ISADORE, b New York City, NY, Oct 15, 19; m 56; c 2. POLYMER PHYSICS, TIRE TECHNOLOGY. *Educ:* City Col New York, BS, 41; Polytech Inst Brooklyn, PhD(phys chem), 50. *Prof Exp:* Dir polymer chem, Gen Latex & Chem Corp, 50-51; scientist, Polaroid Corp, 51-55; sr res engr, Ford Motor Co, 55-57; sr res chemist, Continental Can Co, Ill, 57-59; head, Polymer Physics Sect, Goodyear Tire & Rubber Co, 59-84, res & develop assoc, 85-86; PRES LIVINGSTON ASSOC, 86- *Concurrent Pos:* Assoc ed, Rubber Chem & Technol, 69-72. *Mem:* Am Chem Soc; Am Phys Soc; Tire Soc (pres, 86-88); Am Soc Testing & Mat. *Res:* Physical chemistry of polymers; high polymer synthesis, research and development; radiation effects in polymer systems; polymer physics; tire physics; materials science. *Mailing Add:* 731 Frank Blvd Akron OH 44320-1021

LIVINGSTON, DAVID M, b Cambridge, Mass, Mar 29, 41. MEDICINE. *Educ:* Harvard Univ, AB, 61; Tufts Univ, MD, 65; Am Bd Internal Med, cert, 71. *Prof Exp:* Intern med, Peter Bent Brigham Hosp, Boston, Mass, 65-66, jr resident, 66-67; res assoc, Lab Biochem, Nat Cancer Inst, NIH, 67-69, sr staff fel, 71-72; res fel biol chem, Harvard Med Sch, Boston, Mass, 69-71; assoc med, Brigham & Women's Hosp, Boston, Mass, 73-77, sr assoc med, 76-82; sr clin assoc, Dana-Farber Cancer Inst, 73-77, assoc physician, 77-83, vpres, 89-91, DIR & PHYSICIAN-IN-CHIEF, DANA-FARBER CANCER INST, BOSTON, MASS, 91-; PROF MED, HARVARD MED SCH, BOSTON, MASS, 82- *Concurrent Pos:* Sr investr, Nat Cancer Inst, NIH, 72-73; from asst prof to assoc prof med, Harvard Med Sch, 73-82; vis physician med serv, Brigham & Women's Hosp, 74-85, mem, Med Internship Selection Comt, 75-, physician, 82-; chmn, Med Oncol Fel Selection Comt, Dana-Farber Cancer Inst, 74-, vis physician med oncol, 75-, physician, 83-; mem, Virol Study Sect, Div Res Grants, NIH, 79-83 & 86-88. *Mem:* Inst Med-Nat Acad Sci; Am Fedn Clin Res; Am Soc Microbiol; Am Soc Clin Invest; Am Soc Biol Chemists; Asn Am Physicians. *Res:* Molecular biology of virus-induced neoplastic transformation; control of eukaryotic gene expression and DNA replication; author or co-author of over 90 publications. *Mailing Add:* Harvard Med Sch 12 Stanford Rd Wellesley MA 02181

LIVINGSTON, G E, b Rotterdam, Netherlands, Feb 1, 27; div; c 3. FOOD SCIENCE. *Educ:* NY Univ, BA, 48; Univ Mass, MS, 51, PhD(food technol), 52. *Prof Exp:* Chemist, Bur Chem, NY Produce Exchange, 49; from asst prof to assoc prof food technol, Univ Mass, 51-59; PRES, FOOD SCI ASSOCS, INC, 56-; CHMN, SIERRA SUNSET INC, 88- *Concurrent Pos:* Vis prof, Laval Univ, 54; vis lectr, City Col New York, 59-60; res supvr, Continental Baking Co, 59-62; mem, Adv Bd Mil Personnel Supplies, Nat Acad Sci-Nat Res Coun, 61-64, chmn, Comt Food Serv Systs, 68-71; mgr, Instnl Prod Dept, Morton Frozen Foods Co, 62-65; adj prof, Columbia Univ, 66-72; invitee, White House Conf Food, Nutrit & Health, 69; chmn, Food & Nutrit Coun, Am Health Found, 69-, mem bd sci consults, 69-80; chmn, Panel VII, Nat Conf Food Protection, 71; consult, US Army Natick Labs, 71-83; mem, Bd Govs, Food Update, Food & Drug Law Inst, 71-75; adj prof, Pratt Inst, 73-78, NY Univ, 78-; mem, Food Stability Comn; co-ed, J Food Serv Systs, 80-83. *Honors & Awards:* Sigma Xi Res Award, 57. *Mem:* Am Chem Soc; Soc Food Serv Systs (pres, 81-83); Fel Inst Food Technologists; R&D Assocs; Asn Food & Drug Off. *Res:* Food colorimetry; prepared foods; food service systems; nutritive value; fresh food safety. *Mailing Add:* Food Sci Assocs Inc PO Box 265 Dobbs Ferry NY 10522-0265

LIVINGSTON, HUGH DUNCAN, b Glasgow, Scotland, Nov 12, 40; US citizen; m 65. RADIOCHEMISTRY, OCEANOGRAPHY. *Educ:* Glasgow Univ, BSc, 62, PhD(chem), 66. *Prof Exp:* Res assoc chem, Woods Hole Oceanog Inst, 67-69; res fel, Bowman Gray Sch Med, 69-71; res assoc, 71-73, res specialist, 73-79, SR RES SPECIALIST CHEM, WOODS HOLE OCEANOG INST, 79- *Res:* Studies of artificial radioisotopes in the marine environment. *Mailing Add:* Dept Chem Woods Hole Oceanog Inst Woods Hole MA 02543

LIVINGSTON, JAMES DUANE, b Brooklyn, NY, June 23, 30; m 53; c 3. SUPERCONDUCTIVITY, FERROMAGNETISM. *Educ:* Cornell Univ, BEP, 52; Harvard Univ, MA, 53, PhD(appl physics), 56. *Prof Exp:* Physicist, Gen Elec Corp, Schenectady, NY, 56-89; LECTR MAT SCI, DEPT MAT SCI & ENG, MASS INST TECHNOL, CAMBRIDGE, 89- *Concurrent Pos:* Guest prof, Univ Göttingen, 70; vis prof, Rensselaer Polytech Inst, 87-88. *Mem:* Fel Am Soc Metals Int; fel Am Phys Soc; Inst Elec & Electronics Engrs; AAAS; Metall Soc. *Res:* Superconducting, ferromagnetic and mechanical properties of materials and their relation to microstructure and processing. *Mailing Add:* Mass Inst Technol 13-5014 Cambridge MA 02139

LIVINGSTON, KNOX W, b Atlanta, Ga, Apr 24, 19; m 48; c 1. FORESTRY. *Educ:* Univ SC, BS, 40; Duke Univ, MF, 48. *Prof Exp:* Asst forestry, 48-49, asst forester, 49-63, from asst prof to assoc prof, 63-78, emer assoc prof forestry, Auburn Univ, 78-85; RETIRED. *Mem:* Soc Am Foresters. *Res:* Density, site, growth relations, especially planted southern pine; soil, site relations. *Mailing Add:* 856 Cary Dr Auburn AL 36830

LIVINGSTON, LINDA, b Oklahoma City, Okla, Mar 15, 42; div; c 3. PHYSIOLOGY, BIOPHYSICS. *Educ:* Cent State Univ, BS, 63; Univ Okla, MS, 65; Univ Ala, PhD(biophys), 68. *Prof Exp:* Instr physiol, Med Ctr, Univ Okla, 68-70, asst prof physiol & biophys, 70-71; assoc prof, 71-85, PROF PHYSIOL, UNIV TEX MED SCH HOUSTON, 85- *Concurrent Pos:* Physiologist, Vet Admin Hosp, Oklahoma City, 68-70; res assoc, Dept Med, Med Ctr, Univ Okla, 68-71, Manahan grant, Med Sch, 70-71; USPHS grant, Univ Tex Med Sch Houston, 72-, NIMH res scientist develop award, 73-78; consult, Gen Med Study Sect A, NIH, 72; Res Career Develop award, 72-77; consult, Study Sect, Nat Inst Alcohol Abuse & Alcoholism, 74-78; Fulbright prof, Germany, 83-84. *Mem:* AAAS; Am Fedn Clin Res; Biophys Soc; Soc Exp Biol & Med; Am Physiol Soc; Nat Res Soc Alcoholism. *Res:* Gastrointestinal physiology, biophysics, biochemistry and pathology with emphasis on alcoholism. *Mailing Add:* PO Box 20708-Msmb 4210 Houston TX 77025

LIVINGSTON, MARILYN LAURENE, b High Prarie, Alta, Mar 3, 40; wid. NUMBER THEORY. *Educ:* Univ Alta, BSc, 61, MSc, 63, PhD(math), 66. *Prof Exp:* Asst prof, Western Wash State Col, 66-67; vis asst prof, Ore State Univ, 67-69; from asst prof to assoc prof, 69-78, PROF MATH, SOUTHERN ILL UNIV, 78- *Concurrent Pos:* Mem, Sch Math, Inst Advan Study, Princeton, NJ, 74-75; vis scholar, Univ Mich, Ann Arbor, 86-87. *Mem:* Asn Comput Mach; Soc Indust & Appl Math; Inst Elec & Electronic Engrs. *Res:* Combinatorics; design and analysis of algorithms; parallel algorithms for distributed memory machines. *Mailing Add:* Dept Comput Sci Southern Ill Univ Edwardsville IL 62025-1653

LIVINGSTON, RALPH, b Keene, NH, May 16, 19; div; c 4. CHEMICAL PHYSICS, MAGNETIC RESONANCE. *Educ:* Univ NH, BS, 40, MS, 41; Univ Cinncinnati, DSc, 43. *Prof Exp:* Chemist, metall lab, Univ Chicago, 43-45, assoc dir chem div, 65-75; chemist, Oak Ridge Nat Lab, 45-84, group leader, 75-84; RETIRED. *Concurrent Pos:* Guggenheim & Fulbright fel, France, 60-61; vis prof chem, Cornell Univ, 61-62; prof, Univ Tenn, 64-76; res fel, Union Carbide Corp, 79; consult, 84- *Mem:* Int Soc Magnetic Resonance; AAAS; Am Chem Soc; Am Phys Soc. *Res:* Radiation chemistry; chemical physics; pure quadrupole spectroscopy and electron spin resonance. *Mailing Add:* 144 Westlook Circle Oak Ridge TN 37830

LIVINGSTON, ROBERT BURR, b Boston, Mass, Oct 9, 18; m 54; c 3. NEUROSCIENCES. *Educ:* Stanford Univ, AB, 40, MD, 44. *Prof Exp:* Intern, Stanford Hosp, 43, asst resident, 44; instr physiol, Sch Med, Yale Univ, 46-48; asst prof physiol, Sch Med & dir aeromed res unit, Yale Univ, 50-52; from assoc prof to prof physiol & anat, Univ Calif, Los Angeles, 52-56; dir basic res & sci dir, NIMH & Nat Inst Neurol Dis & Blindness, 56-60, chief lab neurobiol, NIMH, 60-63, chief gen res support br & assoc chief prog planning, Div Res Facil & Resources, NIH, 63-65; chmn dept neurosci, 65-71, PROF NEUROSCI, SCH MED, UNIV CALIF, SAN DIEGO, 65- *Concurrent Pos:* Nat Res Coun sr fel neurol, Inst Physiol, Switz, 48-49; Gruber fel neurophysiol, Switz, France & Eng, 49-50; NIMH sr fel, Gothenburg Univ, 56; res asst, Harvard Med Sch, 47-48; asst to pres, Nat Acad Sci, 51-52; prof lectr, Univ Calif, Los Angeles, 56-59; assoc, Neurosci Res Prog, 63-76, hon assoc, 76-; guest prof, Univ Zurich, 71-72. *Honors & Awards:* Matrix Midland Award, 81; Sachs Mem lectr, Dartmouth Med Sch, 81. *Mem:* Am Physiol Soc; Am Asn Neurol Surg; Am Neurol Asn; Am Asn Anatomists; Soc Neuroscience; Int Physicians Prev Nuclear War. *Res:* Mechanisms relating to higher nervous processes, perception, learning and memory; plasticity of nervous system; three-dimensional analysis and display of neuroanatomical structures; mapping human brains in 3-D at microscopic levels of detail; individual differences in human brain structure-function relations. *Mailing Add:* Dept Neurosci 5-001 Univ Calif at San Diego La Jolla CA 92093

LIVINGSTON, ROBERT SIMPSON, b Summerland, Calif, Sept 20, 14; m 55; c 5. PHYSICS, RESEARCH ADMINISTRATION. *Educ:* Pomona Col, BA, 35; Univ Calif, MA, 41, PhD(physics), 41. *Prof Exp:* Asst physics, Pomona Col, 35-36; asst, Univ Calif, 36-39, res fel, Lawrence Radiation Lab, 39-43; physicist, Tenn Eastman Corp, 43-47; res supt, Carbide & Carbon Corp, 47-50; dir, Electronuclear Div, Oak Ridge Nat Lab, 50-71, dir prog planning & anal, 71-81; CONSULT, 81- *Concurrent Pos:* Consult, Nuclear Physics Panel, Physics Surv Comt, Nat Acad Sci, 69-72; chmn, Ad Hoc Comt Heavy Ion Sources, Nuclear Sci Div, Nat Acad Sci, 72-74; chmn, NSF/Dept Energy study group on the role of electron accelerators in US medium energy nuclear sci, 77-78. *Mem:* Fel Inst Elec & Electronics Engrs; fel Am Phys Soc; fel AAAS; Am Nuclear Soc; Sigma Xi. *Res:* Long range planning of scientific research and development in energy; design of isochronous cyclotrons; heavy particle accelerators; new particle accelerator methods; high intensity ion sources. *Mailing Add:* 7204 Fairlane Dr Powell TN 37849

LIVINGSTON, WILLIAM CHARLES, b Santa Ana, Calif, Sept 13, 27; m 57; c 2. ASTRONOMY. *Educ:* Univ Calif, Los Angeles, AB, 53; Univ Calif, PhD(astron), 59. *Prof Exp:* Observer, Mt Wilson Observ, Carnegie Inst, 51-53; from jr astronr to assoc astronr, Kitt Peak Nat Observ, 59-70, astronr, 70-83; ASTRONR, NAT OPTICAL ASTRON OBSERV, 84- *Mem:* Am Astron Soc; Int Astron Union; Astron Soc India; foreign mem Norweg Acad Sci. *Res:* Solar spectroscopy; solar magnetism; solar cycle studies. *Mailing Add:* PO Box 26732 Tucson AZ 85726

LIVINGSTONE, DANIEL ARCHIBALD, b Detroit, Mich, Aug 3, 27; div; c 5. LIMNOLOGY, PALEOECOLOGY. *Educ:* Dalhousie Univ, BSc, 48, MSc, 50; Yale Univ, PhD(zool), 53. *Prof Exp:* Field collector, NS Mus Sci, summers & demonstr biol, Dalhousie Univ, winters, 47-50; asst zool, Yale Univ, 50-53; Nat Res Coun Can fels, Cambridge Univ, 53-54 & Dalhousie Univ, 54-55; asst prof zool, Univ Md, 55-56; from asst prof to assoc prof, 56-66, PROF ZOOL, DUKE UNIV, 66- *Concurrent Pos:* Spec lectr biogeog, Dalhousie Univ, 54-55; limnologist, US Geol Surv, 56-63; Guggenheim fel, 60-61; mem environ biol panel, Nat Sci Found, 64-, consult Nat Sci Found Polar Prog, 74-76. *Mem:* Ecol Soc Am (ed, Ecol Monogr, 62-66); Am Soc Limnol & Oceanog; Am Soc Nat; Am Quaternary Asn; Sigma Xi; Am Soc Ichthyol & Herpet. *Res:* Pollen analysis; history of lakes; Pleistocene geology of Alaska, Nova Scotia, West, East and Central Africa; geochemistry of hydrosphere; sodium cycle; coring technology; paleoecology; limnology; biogeography of African fishes; distribution of grasses and sedges. *Mailing Add:* Dept Zool Duke Univ Durham NC 22706

LIVINGSTONE, FRANK BROWN, b Winchester, Mass, Dec 8, 28; m 60; c 1. PHYSICAL ANTHROPOLOGY. *Educ:* Harvard Univ, AB, 50; Univ Mich, MA, 55, PhD(anthrop), 57. *Prof Exp:* Nat Sci Found fel, 57-59; from asst prof to assoc prof, 59-68, PROF ANTHROP, UNIV MICH, ANN ARBOR, 68- *Mem:* Am Asn Physical Anthropologists; Am Anthropol Asn. *Res:* Human and population genetics; abnormal hemoglobin; cultural determinants of human evolution. *Mailing Add:* Dept Anthrop 2040E LSA Bldg Univ of Mich Ann Arbor MI 48109-1382

LIZARDI, PAUL MODESTO, molecular biology, parasitology, for more information see previous edition

LJUNG, HARVEY ALBERT, b Greensboro, NC, Oct 26, 05; m 35; c 3. ANALYTICAL CHEMISTRY. *Educ:* Univ NC, BS, 27, MS, 28, PhD(chem), 31. *Prof Exp:* Prof chem, 31-69, acad dean, 46-62, Dana prof, 69-71, EMER DANA PROF, GUILFORD COL, 71- *Concurrent Pos:* Nat Defense Res Comt, 41-42; mem gen chem & qual anal subcomt, Exam Comt, Div Chem Educ, 60-71; chem consult. *Mem:* Emer mem Am Chem Soc. *Res:* Chelation; electrochemistry. *Mailing Add:* 5314 W Friendly Ave Greensboro NC 27410-4349

LJUNGDAHL, LARS GERHARD, b Stockholm, Sweden, Aug 5, 26; m 49; c 2. BIOCHEMISTRY, MICROBIOLOGY. *Educ:* Stockholm Tech Inst, BS, 45; Western Reserve Univ, PhD(biochem), 64. *Prof Exp:* Technician med chem, Karolinska Inst, Univ Sweden, 43-46; res chemist, Stockholm Brewery Co, 47-58; technician biochem, Case Western Reserve Univ, 58-59, sr instr, 64-66, asst prof, 66-67; from mem fac to assoc prof, 67-75, PROF BIOCHEM, UNIV GA, 75- *Concurrent Pos:* Alexander Von Humboldt Sr Scientist Award, 74- *Mem:* Am Soc Microbiol; Am Chem Soc; Brit Biochem Soc; Swed Chem Soc; Am Soc Biochem; Sigma Xi. *Res:* Carbohydrate metabolism, carbon dioxide fixation, and one carbon metabolism inanaerobic microorganism; role of corrinoids, tetrahydrofolate derivatives and properties of enzymes in these processes. *Mailing Add:* Dept Biochem Univ Ga Athens GA 30602

LLAMAS, VICENTE JOSE, b Los Angeles, Calif, Feb 15, 44; m 66, 87; c 4. SOLID STATE PHYSICS. *Educ:* Loyola Univ Los Angeles, BS, 66; Univ Mo-Rolla, MS, 68, PhD(physics), 70. *Prof Exp:* Asst physics, Univ Mo-Rolla, 66-68; from asst prof to assoc prof, 70-84, PROF PHYSICS, NMEX HIGHLANDS UNIV, 84-, CHMN, DEPT PHYSICS & MATH, 75-, CO-DIR, SCI & MATH EDUC CTR, 74- *Concurrent Pos:* Consult, Fermi Nat Accelerator Lab, 70-, Minority Sci Educ Bibliog Proj, AAAS, 75- & NIH; assoc dir, Summer Sci Prog, Stanford Linear Accelerator Ctr, 74-77, dir, 78- *Mem:* Sigma Xi (secy-treas, 71-73, pres, 73-75); Am Phys Soc; Am Asn Physics Teachers; AAAS; Nat Sci Teachers Asn. *Res:* Surface studies of alkali halides in the infrared; atmospheric study of air pollutants. *Mailing Add:* Dept Sci & Sci Technol NMex Highlands Univ Las Vegas NM 87701

LLAURADO, JOSEP G, b Barcelona, Catalonia, Spain, Feb 6, 27; m 58, 66; c 6. NUCLEAR MEDICINE, BIOMEDICAL ENGINEERING. *Educ:* Balmes Inst, Barcelona, BA & BS, 44; Univ Barcelona, MD, 50, PhD, 60; Drexel Univ, MS, 63. *Prof Exp:* Inst med, Sch Med, Univ Barcelona, 50-52; asst med res, Postgrad Med Sch, Univ London, 52-54; asst prof exp surg, Med Sch, Univ Otago, NZ, 54-57; USPHSFound fel, steroid biochem, Col Med Univ Utah, 58-59; assoc prof physiol, Sch Med, Univ Pa, 63-67; prof biomed eng & physiol, Marquette Univ & Med Col, Wis, 67-82; PROF RADIATION SCI, UNIV LOMA LINDA SCH MED, 83-; CHIEF NUCLEAR MED SERV, VET ADMIN HOSP, LOMA LINDA, 83- *Concurrent Pos:* Brit Coun scholar, Postgrad Med Sch, Univ London, 52-54; Hite Found fel exp med, Univ Tex M D Anderson Hosp & Tumor Inst, 57-58; fel, Coun Adv Sci Invests, Spain, 50-52; Rockefeller vis prof, Univ Valle, Colombia, 58; partic, Nat Colloquim Theoret Biol, NASA, Colo, 65; physician, Vet Admin Hosp, Wood, Wis, 67-; US rep, Int Atomic Energy Agency Symp Dynamic Studies Radioisotopes Med, Rotterdam, 70 & Knoxville, Tenn, 74; vis prof, Polytech Univ, Barcelona, Spain, 73 & 75; vis prof, Univ Zulia, Venezuela, 74 & 75 & Univ Padua, Italy, 75; ed, Int J Biomed Comput; consult, Good Samaritan Hosp, Milwaukee, Wis & St Joseph Mem Hosp, West Bend, 79-82; chief ed, Intl Journal Biomed. *Honors & Awards:* Catalan Jocs Florals Prize, Amsterdam, 74 & Caracas, 75. *Mem:* Fel Am Col Nutrit; Am Soc Pharmacol & Exp Therapeut; Catalan Soc Biol; Soc Math Biol; sr mem Inst Elec & Electronics Engrs. *Res:* Radionuclides in cardiology (thallium-201 and analogs); radionuclide (P-32) treatment of pulmonary cancer; computers in nuclear medicine; biomathematics; compartmental analysis of electrolytes. *Mailing Add:* Veterans Hosp No 115 Loma Linda CA 92357

LLEWELLYN, CHARLES ELROY, JR, b Richmond, Va, Jan 16, 22; m 48; c 2. PSYCHIATRY. *Educ:* Hampden-Sydney Col, BS, 43; Med Col Va, MD, 46; Univ Colo, MSc, 53; Am Bd Psychiat & Neurol, dipl, 56. *Prof Exp:* Instr psychiat, Med Col Va, 46-47; assoc, 55-56, asst prof, 56-63, asst dir, Psychiat Outpatient Div, 55-56, head, Psychiat Outpatient Div, 56-76, chief training, Community & Social Psychiat Div, 81-85, ASSOC PROF PSYCHIAT, SCH MED, DUKE UNIV, 63-, HEAD, DIV COMMUNITY & SOCIAL SOCIAL PSYCHIAT DIV, 76-81, 85- *Concurrent Pos:* Partic, NIMH vis fac sem community psychiat, Lab Community Psychiat, Harvard Med Sch, 65-67; prog dir, Duke Study Group, Interuniv Forum Educr Community Psychiat, 67-71; consult, State Dept Social Serv, NC, 55-79; psychiat consult, NC Med Peer Rev Found, 75-79; chief psychiat consult for Peer Rev & Qual Assurance, NC Div Ment Health Serv, 75-78; med consult, Substance Abuse Treatment Ctr, Durham Mental Health Prog, 70- *Mem:* Am Group Psychother Asn; Pan-Am Med Asn; Am Med Asn; Am Asn Marital & Family Therapists; Am Psychiat Asn. *Res:* Community mental health; individual and group psychotherapy; marital and family therapy. *Mailing Add:* Duke Univ Med Ctr Box 3173 Durham NC 27710

LLEWELLYN, GERALD CECIL, b Lonaconing, Md, Feb 8, 40; m 62; c 3. BIONUCLEONICS. *Educ:* Frostburg State Col, BS, 62; Purdue Univ, MS, 66, PhD(bionucleonics), 70. *Prof Exp:* Instr biol chem, Frederick County Bd Educ, Md, 62-66; lectr biol & microbiol, Frederick Community Col, 66-67; from asst prof to assoc prof, 69-77, ASSOC BIOL EDUC, VA COMMONWEALTH UNIV, 77-, DIR, BUR TOXIC SUBSTANCES. *Mem:* Nat Sci Teachers Asn; Sigma Xi. *Res:* Toxicological responses of hamsters to aflatoxin B. *Mailing Add:* 2107 Dresden Rd Richmond VA 23229

LLEWELLYN, J(OHN) ANTHONY, b Cardiff, Wales, Apr 22, 33; m 57; c 3. ENGINEERING SCIENCE, CHEMICAL PHYSICS. *Educ:* Univ Wales, BSc, 55, PhD(chem), 58. *Prof Exp:* Nat Res Coun Can fel, 58-60; res assoc chem, Fla State Univ, 60-61, res assoc chem, Inst Molecular Biophys, 61-62, asst prof chem, 62-64, assoc prof eng sci, Sch Eng Sci, 64-72; PROF DEPT CHEM & MECH ENG, UNIV S FLA, 72-, ASSOC DEAN & DIR ENG COMPUT. *Concurrent Pos:* Scientist astronaut, NASA, 67-68. *Mem:* Am

Chem Soc; Am Inst Aeronaut & Astronaut; Royal Inst Chemists; Am Soc Mass Spectrometry; Am Vacuum Soc. *Res:* Computing applications in medical imaging; theories of reaction rates; computer applications in chemical engineering. *Mailing Add:* Eng Comput Col Eng Univ S Fla Tampa FL 33620

LLEWELLYN, RALPH A, b Detroit, Mich, June 27, 33; m 55; c 4. NUCLEAR PHYSICS. *Educ:* Rose-Hulman Inst Technol, BS, 55; Purdue Univ, PhD(physics), 62. *Prof Exp:* Asst prof physics, Rose-Hulman Inst Technol, 61-64, assoc prof, 64-68, prof, 68-70, chmn dept, 69-70; prof & chmn dept, Ind State Univ, Terre Haute, 70-73; exec secy, Bd on Energy Studies, Nat Acad Sci, Nat Res Coun, 73-74; prof physics & chmn dept, Ind State Univ, Terre Haute, 74-80; dean, Col Arts & Sci, 80-84, PROF PHYSICS, UNIV CENT FLA, 80- *Concurrent Pos:* Mem, NSF Apparatus Develop Workshop, Rensselaer Polytech Inst, 64-65; prof physics & acting chmn dept, St Mary-of-the-Woods Col, 69-70. *Mem:* AAAS; Am Phys Soc; Am Asn Physics Teachers; NY Acad Sci; Forum Physics & Soc (secy-treas). *Res:* Environmental physics, particularly beta and gamma decay; low level radiation in the environment; energy resources, energy and public policy; world energy resources. *Mailing Add:* Dept Physics Univ Cent Fla Orlando FL 32816-0993

LLINAS, MIGUEL, b Cordoba, Argentina, Oct 16, 38; m 63; c 4. NUCLEAR MAGNETIC RESONANCE SPECTROSCOPY, MOLECULAR BIOPHYSICS. *Educ:* Cordoba Nat Univ, Argentina, Licentiate, 63; Univ Calif, Berkeley, PhD(biophysics), 71. *Prof Exp:* Fel assoc, Univ Calif, Berkeley, 71-74; asst, Swiss Fed Inst Technol, Zurich, 74-76; ASSOC PROF CHEM, CARNEGIE-MELLON UNIV, PA, 76- *Mem:* Biophys Soc; Am Chem Soc; AAAS. *Res:* Applications of nuclear magnetic resonance spectroscopy to the study of biological polypeptides; structure and function of siderophores; protein structure and dynamics; conformation of human plasminogen and its interaction with antifibrinolytics. *Mailing Add:* Dept Chem Carnegie-Mellon Univ 4400 Fifth Ave Pittsburgh PA 15213-3876

LLINAS, RODOLFO, b Bogota, Colombia, Dec 16, 34; m 65; c 2. NEUROBIOLOGY, ELECTROPHYSIOLOGY. *Educ:* Pontificial Univ Javeriana, Colombia, MD, 59; Australian Nat Univ, PhD(neurophysiol), 65. *Hon Degrees:* Dr, Univ Salamanca, 85. *Prof Exp:* Instr neurophysiol, Nat Univ Colombia, 59; assoc prof, Univ Minn, 65-66; assoc mem neurobiol, Inst Biomed Res, AMA Educ & Res Found, 66-68, mem, 69-70; prof physiol & biophys & head div neurobiol, Univ Iowa, 70-76; PROF & CHMN, DEPT PHYSIOL & BIOPHYS, SCH MED, NY UNIV, 76-, THOMAS & SUZANNE MURPHY PROF NEUROSCI, 85- *Concurrent Pos:* Res fel psychiat & neurosurg, Mass Gen Hosp, 59-61; fel physiol, Univ Minn, 61-63; res scholar, Australian Nat Univ, 63-65; prof lectr, Col Med, Univ Ill, 67-68, clin prof, 68-71; guest lectr, Wayne State Univ, 67-74; assoc prof, Med Sch, Northwestern Univ, 67-71; mem, Neurol Sci Res Training A Comt, NIH, 71-74 & Neurol A Study Sect, 74-78, panel mem, NIH Task Force on Basic Sci, 78; consult, US Air Force Sch Aerospace Med, 72-75; assoc, Neurosci Res Prog, Mass Inst Technol, 74-84; chief ed, Neurosci, 74-; mem, USA Nat Comt for IBRO, Nat Res Coun, 78-81, actg chmn, 82, chmn, 83-88. *Honors & Awards:* Bowditch Lectr, Am Physiol Soc, 73; Lang Lectr, Marine Biol Lab, Woods Hole, 82; Ralph Gerard Lectr, Univ Calif, Irvine, 87; Albert Einstein Gold Medal Award in Sci, UNESCO, 91. *Mem:* Nat Acad Sci; Am Soc Cell Biol; Soc Neurosci; Int Brain Res Orgn; Biophys Soc; Am Physiol Asn. *Res:* Structural and functional studies of neuronal systems; synaptic transmission in vertebrate and invertebrate forms; evolution and development of the central nervous system. *Mailing Add:* Dept Physiol & Biophys NY Univ Med Ctr New York NY 10016

LLOYD, CHARLES WAIT, endocrinology; deceased, see previous edition for last biography

LLOYD, CHRISTOPHER RAYMOND, b Bridgwater, Eng, Oct 16, 51; m 80. PALEOCLIMATES, TECTONICS. *Educ:* Cambridge Univ, BA, 73; Birmingham Univ, MSc, 74; Lancaster Univ, PhD(paleoclimatol), 77. *Prof Exp:* Ed geosci, Elsevier Sci Publ Co, 79-81; tech ed atmospheric sci, Am Meteorol Soc, 81-85; ELSEVIER INT BULL. *Mem:* Geol Soc Am; Am Meteorol Soc. *Res:* Mid-Cretaceous global tectonics, paleogeography, ocean circulation and temperature, atmospheric circulation and surface climate. *Mailing Add:* Elsevier Int Bull 256 Banbury Rd Oxford OX2 7DE England

LLOYD, DOUGLAS ROY, b Kitchener, Ont, Sept 15, 48; m 74. MEMBRANE SCIENCE, POLYMER SCIENCE. *Educ:* Univ Waterloo, BASc, 73, MASc, 74, PhD(chem eng), 77. *Prof Exp:* Asst prof chem eng, Va Polytech Inst & State Univ, 78-80; asst prof, 80-83, ASSOC PROF CHEM ENG, UNIV TEX, 83- *Concurrent Pos:* Res engr, Union Carbide Can, 69-70; Crane Can Ltd, 70-71; res assoc, Angelstone Ltd, 73; res assoc, Dept Chem Eng, Univ Waterloo, 72, fel, 77-78. *Mem:* Am Chem Soc; Am Inst Chem Engrs; NAm Membrane Soc. *Res:* Synthetic polymeric membranes; membrane separation processes; polymer physics; enzyme engineering. *Mailing Add:* Dept of Chem Engr Univ Tex Austin TX 78712

LLOYD, DOUGLAS SEWARD, b Brooklyn, NY, Oct 16, 39. PUBLIC HEALTH. *Educ:* Duke Univ, AB, 61, MD, 71; Univ NC, MPH, 71. *Prof Exp:* COMNR, CONN STATE DEPT HEALTH SERV, 73- *Concurrent Pos:* Mem courtesy staff, Hartford Hosp, 73-; lectr, Sch Med, Yale Univ, 73- & Univ Conn Health Ctr, 73-; chmn, Asn State & Territorial Health Off Found, McLean, Va; comdr & spec consult to Naval Med Command, US Naval Res; health commentator, WFSB-TV, "House-Call" prog. *Mem:* Asn State & Territorial Health Off (past pres); US Interagency Comt Smoking & Health. *Mailing Add:* 139 Town Farm Rd Farmington CT 06032

LLOYD, EDWARD C(HARLES), b Chicago, Ill, Jan 28, 15; m 38; c 2. ENGINEERING. *Educ:* Univ Calif, BS, 38; Univ Md, MS, 59. *Prof Exp:* Engr, Bur Ships, US Dept Navy, 39-47, head systs & auxiliary mach sect, Design Div, 47-51; asst to chief, Off Basic Instrumentation, Nat Bur Standards, 51-54, chief, Mech Instruments Sect, 54-60, consult, 61-64, chief, Mech Measurements Br, 64-70; CONSULT ENGR, 71- *Concurrent Pos:*

Mem, Am Nat Standards Comt Pressure & Vacuum Gauges & Comt Calibration Instruments. *Mem:* Fel Am Soc Mech Engrs; fel Instrument Soc Am. *Res:* Measurement, data handling and control; electromechanical devices; standards and measurements of pressure, vacuum, vibration and humidity. *Mailing Add:* 7655 Marsh Ave Rosemead CA 91770

LLOYD, EDWIN PHILLIPS, b San Antonio, Tex, Sept 18, 29; m 54; c 1. ENTOMOLOGY. *Educ:* Tex A&M Univ, BS, 51, MS, 52, PhD(entom), 58. *Prof Exp:* RES ENTOMOLOGIST, BOLL WEEVIL RES LAB, AGR RES SERV, USDA, 56-, DIR, 82- *Concurrent Pos:* Sci adv, pilot boll weevil eradication exp, 71-73; adj prof, Miss State Univ, 71- *Honors & Awards:* Superior Serv Award, USDA, 74; Res Award, Miss Entom Asn, 74. *Mem:* Entom Soc Am. *Res:* Cotton insects, specifically the boll weevil. *Mailing Add:* PO Box 1143 Starkville MS 39759

LLOYD, HARRIS HORTON, b Conway, Ark, Nov 14, 37; m 60; c 4. CANCER, CHEMOTHERAPY. *Educ:* Ouachita Baptist Univ, BA & BS, 59; Purdue Univ, PhD(phys chem), 68. *Prof Exp:* Chief, dept chem, 406th Med Lab, US Army Med Command, Japan, 64-67; res chemist, 72-83, head, Math Biol & Data Anal Sect, Southern Res Inst, 72-83; CHIEF, INFO RESOURCES MGT, VA MED CTR, NORTH LITTLE ROCK, ARK, 83-; ADJ PROF COMPUT SCI, UNIV ARK LITTLE ROCK, 87- *Mem:* Cell Kinetics Soc; Am Sci Affil. *Res:* Chemical kinetics; data analysis and mathematical simulation; pharmacokinetics; kinetics of tumor growth and cell killing; design of computer-based information management systems. *Mailing Add:* Vet Admin Med Ctr 2200 Ft Roots Dr North Little Rock AR 72114

LLOYD, JAMES ARMON, endocrinology, ethology; deceased, see previous edition for last biography

LLOYD, JAMES EDWARD, b Oneida, NY, Jan 17, 33; m 58; c 2. EVOLUTIONARY BIOLOGY, INSECT BEHAVIORAL ECOLOGY. *Educ:* State Univ NY Col Fredonia, BS, 60; Univ Mich, MA, 62; Cornell Univ, PhD(entom), 66. *Prof Exp:* Teacher high sch, 60; NSF res assoc syst & evolutionary biol, 66; from asst prof biol sci & entom to assoc prof entom & nematol, 66-74, PROF ENTOM & NEMATOL, UNIV FLA, 74- *Concurrent Pos:* NSF res grant 68, 80; Nat Geog Soc res grant, 80. *Honors & Awards:* Res Award, Sigma Xi, 74. *Mem:* Can Entom Soc; Coleopterist's Soc. *Res:* Function of luminescence in insects; systematics, behavior and ecology of Lampyridae. *Mailing Add:* Dept Entom & Nematol Bldg 970 Univ Fla Gainesville FL 32611

LLOYD, JAMES NEWELL, b Orange, NJ, Oct 20, 32; m 59; c 2. PHYSICS. *Educ:* Colgate Univ, BA, 54; Cornell Univ, PhD(physics), 63. *Prof Exp:* From instr to asst prof physics, 61-70, chmn, dept physics & astron, 73-76, assoc prof, 70-79, PROF PHYS, COLGATE UNIV, 79- *Mem:* Am Phys Soc; Am Asn Physics Teachers. *Res:* Ferromagnetic resonance and transport properties in metals. *Mailing Add:* Dept of Physics & Astron Colgate Univ Hamilton NY 13346

LLOYD, JOHN EDWARD, b Munhall, Pa, Sept 28, 40; m 62; c 2. VETERINARY ENTOMOLOGY. *Educ:* Pa State Univ, BS, 62; Cornell Univ, PhD(entom), 67. *Prof Exp:* Asst prof entom, Pa State Univ, 67-68; from asst prof to assoc prof, 68-76, PROF ENTOM, UNIV WYO, 76-, *Concurrent Pos:* Actg head plant sci dept, Univ Wyo, 85-86. *Mem:* Entom Soc Am; Am Mosquito Control Asn. *Res:* Economic entomology; insects affecting livestock; insects affecting man. *Mailing Add:* Entom Sect Box 3354 Univ Wyo Laramie WY 82071

LLOYD, JOHN WILLIE, III, b Winchester, Va, May 25, 43; m 65; c 2. ENDOCRINOLOGY, PHARMACOLOGY. *Educ:* Shepherd Col, BS, 66; WVa Univ, MS, 69, PhD(endocrinol), 73. *Prof Exp:* Asst prof pharmacol, WVa Univ, 73-74; asst prof endocrinol, Eastern Va Med Sch, 74-78; ASSOC PROF ENDOCRINOL, HOWARD UNIV, 78- *Concurrent Pos:* Dir, Endocrine Serv, Eastern Va Med Sch, 76-78. *Mem:* Sigma Xi; Soc Exp Biol & Med; Endocrine Soc; Am Physiol Soc. *Res:* Endocrinology; Reproductive physiology; hormonal regulation of the adrenal gland and accessory sex organs; prostatic cancer. *Mailing Add:* 518 N Loudoun St Winchester VA 22601

LLOYD, KENNETH OLIVER, b Denbigh, Wales, May 17, 36; US citizen; m 62; c 2. BIOCHEMISTRY, IMMUNOCHEMISTRY. *Educ:* Univ Wales, BSc, 57, PhD(chem), 60. *Prof Exp:* Res assoc microbiol, Columbia Univ, 63-68, asst prof biochem, 68-74; assoc prof, Sch Med, Tex Tech Univ, 74-75; assoc, 75-78, assoc mem, 78-87, MEM, SLOAN-KETTERING CANCER CTR, 87-, CHMN, IMMUNOL PROG, 89- *Concurrent Pos:* Fel, Wash Univ, 60-63; USPHS res career develop award, 68-73; mem, Allergy & Immunol Study Sect, NIH, 84-88. *Honors & Awards:* Philip Levine Award, Am Asn Clin Pathologists, 86. *Mem:* AAAS; Am Chem Soc; Am Asn Immunologists; Soc Complex Carbohydrates; Am Asn Cancer Res. *Res:* Biochemistry, structure and immunochemistry of glycoproteins and glycolipids, particularly tumor antigens. *Mailing Add:* Mem Sloan Kettering Cancer Ctr 1275 York Ave New York NY 10021

LLOYD, L KEITH, b 1941; c 3. UROLOGY. *Educ:* Centenary Col, BS, 62; Tulane Univ Sch Med, MD, 66. *Prof Exp:* Intern, US Pub Health Serv Hosp, Norfolk, Va, 66-67; med officer, 68-70; jr asst resident, Gen Practice Prog, 67-68; asst resident urol, Tulane Univ Sch Med, 70-71, asst resident instr, 70-72, res fel urol, 71-72, resident urol, 72-73, resident instr, 72-74, sr resident urol, 72-74; from asst prof to assoc prof, 74-81, PROF UROL, DEPT SURG, UNIV ALA MED CTR, 81-; DIR UROL REHAB & RES CTR, SPAIN REHAB CTR, BIRMINGHAM, 77- *Concurrent Pos:* Bd dirs & comt educ, Am Spinal Injury Asn, 86. *Mem:* Am Urol Asn; Am Med Asn; Am Col Surgeons; Am Spinal Injury Asn. *Res:* Urodynamics and neurogenic bladder; urologic care in spinal cord injury; male sexual dysfunction; over 60 publications on urology, nuclear and physical medicine and rehabilitation; author of 6 books in urology. *Mailing Add:* Div Urol Univ Ala Med Ctr 606 MEB Birmingham AL 35294

LLOYD, LAURANCE H(ENRY), b Salem, Ore, Apr 24, 15; m 40; c 1. OPERATIONS RESEARCH, ELECTRICAL ENGINEERING. *Educ:* Ore State Univ, BS, 37; Ohio State Univ, MS, 51. *Prof Exp:* Engr, Idaho Power Co, 37-42; proj engr, Control Equip Br, Eng Div, Air Mat Command, Wright-Patterson AFB, 46-47, proj engr, Guided Missiles Br, Intel Dept, 48, chief, Nuclear Energy Br, 48, chief, Sci Br, 50, actg chief & chief engr, Signal Anal Sect, Air Technol Intel Ctr, 53; staff engr, Systs Eng Sect, Missile Test Proj, Radio Corp Am, Patrick AFB, 54; chief eval div, Dept Electronics Intel, Aerospace Technol Intel Ctr, Wright-Patterson AFB, 55-61, actg tech dir, 59-60, aerospace engr, Directorate of Synthesis, Systs Eng Group, 61-69 & Directorate of Opers Res, 69, opers res analyst, Simulation & Anal Div, 69-76; pres, Trident Marine Servs, 72-76; PRES, L-TECH SYSTS, 76- *Mem:* Fel AAAS; sr mem Inst Elec & Electronics Engrs; Opers Res Soc Am. *Res:* System effectiveness evaluation; computer simulation; electrical waveform analysis; electrical measurement theory. *Mailing Add:* 180 Inlets Blvd Nokomis FL 34275

LLOYD, LEWIS EWAN, b Montreal, Que, Feb 3, 24; m 50; c 4. NUTRITION. *Educ:* Macdonald Col, McGill Univ, BSc, 48, MSc, 50, PhD(nutrit), 52. *Hon Degrees:* LLD, Dalhousie Univ, 87. *Prof Exp:* Asst nutrit, Macdonald Col, McGill Univ, 48-52; res assoc, Cornell Univ, 52-53; from asst prof to assoc prof, Macdonald Col, McGill Univ, 53-60, prof animal sci & chmn dept, 60-67; dir sch home econ, Univ Man, 67-70, dean fac home econ, 70-77; dean fac agr, MacDonald Col, 77-85, EMER PROF, MCGILL UNIV, 85-; EMER DEAN UNIV MAN, 85-; SCIENTIST, FOOD RES INST, AGR CAN, OTTAWA, 85- *Concurrent Pos:* Underwood fel, Rowett Res Inst, Scotland, 58-59. *Mem:* Agr Inst Can; Can Inst Food Sci & Technol; Am Inst Nutrit; Nutrit Soc Can. *Res:* Utilization of food nutrients by the body. *Mailing Add:* Box 6, RR 1 Rockland ON K4K 1K7 Can

LLOYD, MILTON HAROLD, b Des Moines, Iowa, Mar 27, 25; m 44; c 3. INORGANIC, CHEMISTRY. *Educ:* Creighton Univ, BS, 50, MS, 54. *Prof Exp:* Chemist prod develop, Tidy House Prod Co, 50-56; sect mgr process res & develop, Oak Ridge Nat Lab, 56-86; RETIRED. *Mem:* Am Chem Soc; AAAS. *Res:* Transuranium element isolation and purification; plutonia sol-gel processes for preparation of advanced reactor fuels; chemical studies of plutonium behavior in reactor fuel reprocessing and waste solutions. *Mailing Add:* 111 Locust Lane Oak Ridge TN 37830

LLOYD, MONTE, b Omaha, Nebr, July 6, 27; m 46, 69; c 4. ANIMAL ECOLOGY, TROPICAL FOREST CONSERVATION. *Educ:* Univ Calif, Los Angeles, AB, 52; Univ Chicago, PhD(zool), 57. *Prof Exp:* NSF fel, Bur Animal Pop, Oxford Univ, 57-59, Brit Nature Conserv res grant, 59-62; asst prof zool, Univ Calif, Los Angeles, 62-67; PROF BIOL, UNIV CHICAGO, 67- *Concurrent Pos:* Ed, Ecol, 68-72. *Mem:* Soc Study Evolution; Am Soc Nat; Ecol Soc Am. *Res:* Dynamics of animal populations and community ecology. *Mailing Add:* Dept Ecol & Evolution Univ Chicago 1101 E 57th St Chicago IL 60637

LLOYD, NELSON ALBERT, b Lorain, Ohio, Oct 12, 26; m 45; c 5. ANALYTICAL CHEMISTRY. *Educ:* Southern Methodist Univ, BS, 50, MS, 51; Okla State Univ, PhD(anal chem), 55. *Prof Exp:* Res chemist, Goodyear Atomic Corp, 54-56; assoc prof anal chem, Northeastern La State Col, 56-61 & Univ Ala, Tuscaloosa, 61-67; chmn Div Natural Sci, Mobile Col, Mobile, 67-70; CHIEF GEOCHEM DIV, GEOL SURV OF ALA, TUSCALOOSA, 70- *Concurrent Pos:* Consult, Tuscaloosa Metall Res Ctr, US Bur Mines, 63-67 & State Oil & Gas Bd, Ala, 66-67. *Mem:* Am Chem Soc. *Res:* Rock analysis, whole rock and trace metals in rock; trace substances in water, heavy metals, pesticides herbicides and various nitrogen species. *Mailing Add:* 209 32nd Ave E Tuscaloosa AL 35404

LLOYD, NORMAN EDWARD, b Oak Park, Ill, Feb 20, 29; m 51; c 8. BIOCHEMISTRY, BIOTECHNOLOGY. *Educ:* Rockjurst Col, BS, 52; Kans State Col, MS, 53; Purdue Univ, PhD(biochem), 56. *Prof Exp:* Assoc chemist, Corn Prods Co, 56-58; cereal chemist, Int Milling Co, 58-59; supvr starch chem res, 60-64, dir sci develop, 64-69, asst res dir, 69-70, supvr chem res, 70-78, dir res & develop, 78-79, vpres tech, 79-82, GROUP DIR, BIOTECHNOL, CLINTON CORN PROCESSING CO, 83- *Mem:* Am Chem Soc; Am Asn Cereal Chem. *Res:* Production and characterization of starches, sweeteners and enzymes; enzyme kinetics and immobilization; food biotechnology. *Mailing Add:* 188 Sawmill Rd Sparta NJ 07871-3004

LLOYD, RAY DIX, b March 10, 30; m 54; c 5. RADIATION PHYSICS, RADIOBIOLOGY. *Educ:* Univ Utah, PhD(biol), 74; Am Bd Health Physics, cert, 68. *Prof Exp:* Res asst prof anat, Radiobiol Div, Dept Anat, Col Med, Univ Utah, 61-79; res assoc prof, 79-84, RES PROF, RADIOBIOL DIV, SCH MED, UNIV UTAH, 84- *Concurrent Pos:* Coun mem, Nat Coun Radiation Protection & Measurements, 80; adj asst prof, Dept Mech & Indust Eng, Univ Utah, 75- *Mem:* Radiation Res Soc; Health Physics Soc; Int Radiation Protection Asn; Am Acad Health Physics. *Res:* Biological effects of ionizing radiation; internal emitters; dose-response models; risk assessment; risk modification by chelation therapy; application of radioactivity to biomedical studies; health physics; reconstruction of radiation dose from Nevada nuclear testing. *Mailing Add:* Radiobiol Div Bldg 586 Univ Utah Salt Lake City UT 84112

LLOYD, RAYMOND CLARE, b Sioux Falls, SDak, July 22, 27; m 49; c 3. NUCLEAR REACTORS, CRITICALITY ANALYSIS. *Educ:* Augustana Col, BA, 49; SDak State Univ, MS, 51. *Prof Exp:* Scientist, Hanford Atomic Plant, Gen Elec Co, 51-65; STAFF SCIENTIST, BATTELLE PAC NORTHWEST LAB, 65- *Mem:* Am Nuclear Soc. *Res:* Reactor physics and criticality analysis. *Mailing Add:* 2068 Hudson Ave Richland WA 99352

LLOYD, ROBERT, b Jackson, Miss, Mar 1, 16; m 49; c 1. CHEMICAL ENGINEERING. *Educ:* Purdue Univ, BS, 38; Temple Univ, MA, 46, PhD(phys chem), 54; Carnegie Mellon Univ, MS, 78. *Prof Exp:* Asst city engr, Gary, Ind, 38-41; chemist, Kingsbury Ord Plant, Ind, 41-42; chem engr, Graham Savage & Assocs, Inc, 45-55; chem engr, Bettis Atomic Power Lab,

Westinghouse Elec Corp, 55-86; RETIRED. *Mem:* Am Chem Soc; Am Inst Chem Engrs. *Res:* Metal complexes; electrochemistry; electroplating procedures; engineering design; nuclear atomic power engineering. *Mailing Add:* 4004 Jane St West Mifflin PA 15122-1640

LLOYD, ROBERT MICHAEL, b Los Angeles, Calif, Dec 26, 38; m 73; c 2. SYSTEMATIC BOTANY. *Educ:* Pomona Col, BA, 60; Claremont Grad Sch, MA, 62; Univ Calif, Berkeley, PhD(bot), 69. *Prof Exp:* Herbarium technician, Univ Calif, Los Angeles, 62-63; asst prof bot, Univ Hawaii, 68-71; from asst prof to assoc prof, 71-81, PROF BOT, OHIO UNIV, 81- *Concurrent Pos:* Prin invstr, NSF grants, 78-80 & 80-83. *Mem:* AAAS; Bot Soc Am; Am Fern Soc (vpres, 78 & 79, pres 80 & 81); Am Soc Plant Taxon; Linnean Soc London. *Res:* Systematics and reproductive biology of ferns; systematics, morphology and evolution of the Pteridophyta; spore germination, breeding systems and genetics. *Mailing Add:* Dept Bot Ohio Univ Athens OH 45701

LLOYD, THOMAS BLAIR, b Reedsville, WVa, Aug 29, 21; m 44; c 3. INDUSTRIAL CHEMISTRY, COLLOID CHEMISTRY. *Educ:* Washington & Jefferson Col, BS, 42; Western Res Univ, MS, 46, PhD(phys chem), 48. *Prof Exp:* Asst prof chem, Muhlenberg Co, 48-54; chem investr, 54-66, res supvr, NJ Zinc Co, 66-83; RES SCIENTIST, LEHIGH UNIV, 83- *Mem:* Am Chem Soc; Sigma Xi. *Res:* Industrial process research, particularly pigments, hydrometallurgy and pollution control; particle dispersion in various media; surface science; PTFE etching; coal. *Mailing Add:* 127 E Bridle Path Rd Bethlehem PA 18017

LLOYD, WALLIS A(LLEN), b Harrisburg, Pa, July 24, 26; m 55; c 4. CHEMICAL ENGINEERING. *Educ:* Pa State Univ, BS, 49; Univ Minn, PhD(chem eng), 54. *Prof Exp:* Res assoc chem eng, Univ Minn, 54-55; design engr, Calif Res Corp Div, Standard Oil Co Calif, 55-56; asst prof chem eng, Pa State Univ, 56-64; RES DIR, CANNON INSTRUMENT CO, 64- *Mem:* Assoc mem Am Inst Chem Engrs; Am Nuclear Soc; Am Soc Testing & Mat; Sigma Xi. *Res:* Heat transfer; chemonuclear research; separation processes; viscometry. *Mailing Add:* 490 Orlando Ave State College PA 16801

LLOYD, WELDON S, b Miami, Fla, July 26, 39; m 60; c 2. CALCIUM & BONE METABOLISM. *Educ:* Boston Univ, BA, 66; Northeastern Univ, MS, 71. *Hon Degrees:* DSc, Boston Univ, 78. *Prof Exp:* Assoc, Harvard Univ, 63-68; instr oral pharmacol, 68-71, assoc pharmacol res, 71-78, ASSOC PROF NUTRIT, BOSTON UNIV, 78- *Concurrent Pos:* Sr lectr, Roxbury Community Col, 80. *Mem:* AAAS. *Res:* Bone disease and hormone related studies; calcium and bone metabolism related studies in health and disease. *Mailing Add:* Sch Grad Dent Boston Univ 100 E Newton St Boston MA 02118

LLOYD, WILLIAM GILBERT, b New York, NY, July 10, 23; m 47; c 3. ORGANIC CHEMISTRY, COAL SCIENCE. *Educ:* Kalamazoo Col, AB, 47; Brown Univ, ScM, 50; Mich State Univ, PhD(org chem), 57. *Prof Exp:* Chemist, Dow Chem Co, 50-60, assoc scientist, 60-62; sr process res specialist, Lummus Co, NJ, 62-67; prof chem, Western Ky Univ, 67-74; sr res scientist, Inst Mining & Minerals Res, Univ Ky, 74-75, chief chemist, 75-77, assoc dir & mgr, Mat Div, 77-80; dean, Ogden Col Sci, Technol & Health, 80-85, prof, 80-88, PROF CHEM, WESTERN KY UNIV, 88- *Concurrent Pos:* Dir, Larox Res Corp, 72- *Mem:* Am Chem Soc. *Res:* Chemistry of coal and coal-derived products; catalysis of organic reactions. *Mailing Add:* Dept Chem Western Ky Univ Bowling Green KY 42101

LLOYD, WINSTON DALE, b Pensacola, Fla, Sept 9, 29; m 58; c 3. ORGANIC CHEMISTRY. *Educ:* Fla State Univ, BS, 51; Univ Washington, PhD(org chem), 56. *Prof Exp:* Org chemist, Dow Chem Co, 56-58 & USDA, 59-62; res prof, 65-66, ASSOC PROF CHEM, UNIV TEX, EL PASO, 62- *Mem:* Am Chem Soc; Sigma Xi. *Res:* Stereochemistry of cyclic dienes; mechanisms of organic chemical reactions; natural products; synthesis. *Mailing Add:* Dept of Chem Univ of Tex El Paso TX 79968

LLUCH, JOSE FRANCISCO, b San Germán, PR, Jan 30, 54; m 79; c 2. CONSTRUCTION MANAGEMENT, ENGINEERING EDUCATION. *Educ:* Univ PR, Mayagüez, BSCE, 75; Ga Inst Technol, MSCE, 76, PhD(construct mgt), 81. *Prof Exp:* Instr civil eng, Univ PR, Mayagüez, 77-79, from asst prof to assoc prof, 81-89, actg dir grad studies, 85-86, asst dean eng, 86-88, DEAN ENG, UNIV PR, MAYAGÜEZ, 88-, PROF CIVIL ENG, 89- *Concurrent Pos:* Construct mgt consult, var construct contractors, 81-88; prin investr & co-prin investr, var externally funded res projs, NSF & others, 81-88. *Mem:* Am Soc Civil Engrs. *Res:* Construction project planning and control; computer applications in construction management. *Mailing Add:* Dean Eng Univ PR-Mayaguez Mayaguez PR 00681-5000

LNENICKA, GREGORY ALLEN, b Cedar Rapids, Iowa, Dec 4, 52; m 83. ELECTROPHYSIOLOGY. *Educ:* Univ Iowa, BA, 76; Univ Va, PhD(biol), 82. *Prof Exp:* Fel Univ Toronto, 82-87; ASST PROF BIOL, DEPT BIOL SCI, STATE UNIV NY, ALBANY, 87- *Mem:* Soc Neurosci; AAAS; NY Acad Sci. *Mailing Add:* Biol Dept State Univ NY 1400 Wash Ave Albany NY 12222

LNENICKA, WILLIAM J(OSEPH), b Hay Springs, Nebr, Oct 16, 22; m 47; c 2. STRUCTURAL ENGINEERING. *Educ:* Univ Nebr, BS, 49; Kansas State Univ, MS; Ga Inst Technol, PhD, 61. *Prof Exp:* Instr civil eng, Univ Nebr, 49-51 & Kans State Univ, 52-54; asst prof eng mech, Univ Okla, 54-57 & La State Univ, 57-58; from asst prof to assoc prof, Ga Inst Technol, 58-68, prof eng mech, 68-, assoc vpres acad affairs, 78-; RETIRED. *Concurrent Pos:* Consult, Aerial Tower Mfg Co, 54-59; mem bd dir, Vinings Chem Co, 64-73, Vinings Leasing Co, 68-73 & Wilroy & Assoc Consult Engrs, 70-78. *Mem:* Am Soc Eng Educ; Soc Am Mil Engrs; Soc Prof Engrs. *Res:* Chief field strength of materials. *Mailing Add:* 3235 Laramie Dr NW Atlanta GA 30339

LO, ANDREW W, US citizen. FINANCIAL ECONOMICS. *Educ:* Yale Univ, BA, 80; Harvard Univ, MA & PhD(econ), 84. *Prof Exp:* Asst prof finance, Wharton Sch, Univ Pa, 84-87, assoc prof, 87-88; assoc prof, 88-91, PROF FINANCE, SLOAN SCH MGT, MASS INST TECHNOL, 91- *Concurrent Pos:* Res assoc, Nat Bur Econ Res, 88-, Olin fel, 88; Batterymarch fel, Batterymarch Financial Mgt, 89. *Mem:* Am Finance Asn; Am Statist Asn; Am Econ Asn; Inst Math Statist; Econometric Soc; Soc Indust & Appl Math. *Res:* Statistical analysis of financial asset pricing models for equities, fixed income securities and derivative products; numerical computation of economic systems for financial forecasting. *Mailing Add:* Sloan Sch Mgt MassInst Technol 50 Memorial Dr Cambridge MA 02139

LO, ARTHUR W(UNIEN), b Shanghai, China, May 21, 16; nat US; m 50; c 2. COMPUTER SCIENCE. *Educ:* Yenching Univ, China, BS, 38; Oberlin Col, MA, 46; Univ Ill, PhD(elec eng), 49. *Prof Exp:* Asst prof elec eng, Mich Col Mining & Technol, 49-50; lectr, City Col New York, 50-51; res & develop engr, Victor Div, Radio Corp Am, 51-52, sr mem tech staff, RCA Labs, 52-60; mgr adv tech develop, Data Systs Div, Int Bus Mach Corp, 60-64; prof, 64-86, EMER PROF ELEC ENG & COMPUT SCI, PRINCETON UNIV, 86- *Mem:* Fel Inst Elec & Electronics Engrs. *Res:* Digital electronics and computer organization. *Mailing Add:* 102 Maclean Circle Princeton NJ 08540

LO, CHENG FAN, b Taichung, Taiwan, Dec 14, 37; m 66; c 3. WOOD CHEMISTRY. *Educ:* Nat Taiwan Univ, BS, 62; Auburn Univ, MS, 66; Ore State Univ, PhD(wood chem), 70. *Prof Exp:* Res asst, Dept Forestry, Auburn Univ, 64-66; wood chemist, Forest Res Lab, Ore State Univ, 66-69; RES CHEMIST, BOISE CASCADE CHEM RES LAB, 69- *Mem:* Am Chem Soc; Tech Asn Pulp & Paper Indust; Forest Prod Res Soc. *Res:* By-products development in wood cellulose and lignin material; technical assistance to paper production. *Mailing Add:* 12309 NE Cassidy Ct Vancouver WA 98685

LO, CHU SHEK, b Hong Kong; US citizen; m 69; c 1. ENDOCRINOLOGY, BIOCHEMISTRY. *Educ:* Nat Taiwan Univ, Taipei, Repub China, BS, 62; Univ Notre Dame, Ind, MS, 65; Ind Univ Med Sch, Indianapolis, PhD(physiol), 72. *Prof Exp:* Sr lab technician Dept Physiol, Sch Med, Univ Miami, 65-66 & Ind Univ, 67-68; fel, Cardiovasc Res Inst, Sch Med, Univ Calif San Francisco, 72-75; asst prof Dept Physiol, Sch Med, Univ Md, Baltimore, 75-77; asst prof, 77-81, ASSOC PROF DEPT PHYSIOL, UNIFORMED SERV UNIV HEALTH SCI, BETHESDA, MD, 81- *Concurrent Pos:* Collaborator, Roche Res Found, Sci Exchange & Biomed, Switzerland, 76; guest worker, Metabolic Dis Br, NIH, 76-78; Mem Grad Affairs Comt, Dept Physiol, Uniformed Servs Univ Health Sci, 85-, Grad Educ Comt, 88-90; mem res comt & peer rev subcomt, Md Affil, Inc, Am Heart Asn, 90- *Honors & Awards:* Sidney C Wener Lectr, Dept Med, Col Physicians & Surgeon, Columbia Univ, 80. *Mem:* Am Physiol Soc; Biophys Soc; Am Soc Cell Biol; Soc Chinese Bioscientists Am. *Res:* Germ-free animal research; cellular aging including whole body electrolytes in young and old rats; effects of mucosal anaerobiosis on galactose transport across the apical membrane of the hamster small intestine; mode of action of thyroid hormone and corticosterone on sodium transport in rat kindney and small intestine; mechanism of action of catecholamine on sodium and potassium transport; molecular biological approches to study the mechanisms of action of hormone(thyroid and gluocorticoid) on membrane transport and membrane biochemistry; mechanisms of action of growth factors on phosphatidylinositol metabolism in smooth muscle cell; signal transduction in diabetics glomerulus. *Mailing Add:* 5304 Elsmere Ave Bethesda MD 20814-4799

LO, CLIFFORD W, b Hempstead, NY, Sept 9, 51; m 84; c 1. PEDIATRICS. *Educ:* Stanford Univ, AB, 72; Univ Hawaii, MD, 77; Univ Calif, Los Angeles, MPH, 81; Mass Inst Technol, ScD(nutrit biochem), 86. *Prof Exp:* CLIN ASSOC & INSTR PEDIAT, HARVARD MED SCH, MASS GEN HOSP, 84-; ASST MED, CHILDREN'S HOSP, BOSTON, 85-, DIR, HOME TPN PROG, 86-, ASSOC DIR, NUTRIT SUPPORT SERV, 90- *Concurrent Pos:* Fulbright scholar, Univ Cambridge, Eng, 89; Royal Soc guest fel, MRC Dunn Nutrit Unit, Cambridge, Eng, 89; lectr nutrit, Inst Health Prof, Mass Gen Hosp, 89-90, adj assoc prof nutrit, 90- *Honors & Awards:* Nat Osteoporosis Found Prize, 90. *Mem:* Am Gastroenterol Asn; NY Acad Sci; Hist Sci Soc; Am Soc Bone & Mineral Res. *Res:* Vitamin D, parathyroid hormone and calcium metabolism; calcium nutrition and bone density in adolescents; total parenteral nutrition in pediatrics; international nutrition. *Mailing Add:* GI Div Children's Hosp 300 Longwood Ave Boston MA 02115

LO, DAVID S(HIH-FANG), b China, Aug 27, 32; US Citizen; m 59; c 2. COMPUTER MEMORY TECHNOLOGY, SWITCH-MODE POWER CONVERSION. *Educ:* Nat Taiwan Univ, BS, 54; Univ Minn, MS, 58, PhD(elec eng), 62. *Prof Exp:* Sr res scientist, Honeywell Res Ctr, 62-64; prin physicist, 64-71, staff physicist, 72-79, SR STAFF SCIENTIST, ELECTRONIC & INFO SYSTS GROUP, UNISYS CORP, 80- *Concurrent Pos:* Lectr, Univ Minn, 62- & adj prof, 82- *Mem:* Inst Elec & Electronics Engrs. *Res:* Electrical properties of ferric oxide semiconductors; ferromagnetic films; magnetic and optical memories; electroluminescent displays; optoelectronics; switch-mode power conversion. *Mailing Add:* Unisys Corp PO Box 64525 St Paul MN 55164-0525

LO, ELIZABETH SHEN, b Shanghai, China, Feb 24, 26; m 50; c 2. ORGANIC CHEMISTRY. *Educ:* St John's Univ, Shanghai, BS, 45; Univ Ill, MS, 47, PhD(chem), 49. *Prof Exp:* Univ Ill fel, 49-50; res chemist, Metalsalsts Corp, 51; J T Baker Chem Co Div, Vick Chem Co, 51-52; M W Kellogg Co, 53-57; Permacel Div, Johnson & Johnson, 57-60; staff chemist, IBM Corp, 60-63; sr res chemist, Thiokol Chem Corp, 65-70; vis fel, Princeton Univ, 71-73; mgr, mat & chem process, Fairchild-PMS Prod, 74-75; mem tech staff, David Sarnoff Res Ctr, RCA, 76-77; chief chemist, Optel Div, Refac Electronics Corp, 79-81; MGR POLYMER MAT, ELECTRO-SCI LABS, INC, PA, 82- *Mem:* Am Chem Soc; Int Soc Hybrid Microelectronics. *Res:* Polymer, rubber and resin chemistry; fluorocarbon polymers; liquid crystals; polymer thick film for electronic industry. *Mailing Add:* 102 Maclean Circle Princeton NJ 08540

LO, GEORGE ALBERT, b Hong Kong, June 26, 34; US citizen; m 57; c 3. PHYSICAL CHEMISTRY, INORGANIC CHEMISTRY. *Educ:* Univ Ore, BA, 57, MA, 60; Wash State Univ, PhD(chem), 63. *Prof Exp:* Mem tech staff, Rocketdyne Div, Rockwell Int Corp, 63-77; sr staff scientist, 77-87, MGR, CHEM DEPT, LOCKHEED PALO ALTO RES LAB, 87- *Mem:* Am Chem Soc; Sigma Xi. *Res:* Chemical kinetics and propulsion; chemistry of inorganic complexes; environment monitor and remediation; polymers and composites. *Mailing Add:* Lockheed Palo Alto Res Lab 3251 Hanover St Palo Alto CA 94304

LO, GRACE S, FIBER RESEARCH. *Prof Exp:* DIR FIBER RES, RALSTON PURINA CO, CHECKERBOARD SQ, 76- *Mailing Add:* Dept Fiber Res Ralston Purina Co Checkerboard Sq St Louis MO 63164

LO, HILDA K, SURGERY. *Educ:* Nat Taiwan Univ, BS, 64; Ill Inst Technol, MS, 68, PhD(biol), 74. *Prof Exp:* Res asst genetics, Dept Biol, Ill Inst Technol, 64-66, microbiol & virol, 66-68; res assoc, Dept Microbiol, Chicago Med Sch, 68-70; fel lipid biochem & microbiol, Ill Inst Technol, 70-74; postdoctoral fel, Div Endocrinol & Metab, Sch Med, Univ Miami, 74-77, instr med, 77-80, res instr, 81-89, RES ASST PROF, DEPT SURG, UNIV MIAMI, 89- & DEPT ORTHOP & REHAB, 90- *Res:* Author of numerous scientific articles, books and chapters. *Mailing Add:* 10901 SW 116th Ave Miami FL 33176

LO, HOWARD H, b Hsinchu, Taiwan, Sept 3, 37; US citizen; m 65; c 2. GEOCHEMISTRY, PETROLOGY. *Educ:* Nat Taiwan Univ, BS, 60; Univ Minn, MSc, 64; Wash Univ, PhD(geochem), 70. *Prof Exp:* Jr geologist, Geol Surv Taiwan, 60-62; res & teaching asst, Univ Minn, 62-64; mine geologist, Opemiska Copper Mines, Ltd, Que, 64-65; instr sci & math, Ottawa Col Inst, Can, 65-67; res asst, Wash Univ, St Louis, 67-70; asst prof, 70-76, ASSOC PROF GEOL SCI, CLEVELAND STATE UNIV, 76- *Concurrent Pos:* Vis res prof, Ctr Volcanology, Univ Ore, 74, Purdue Univ, 78; Ohio Dept Develop grant. *Mem:* Am Geophys Union; Geol Soc Am; Geol Soc China. *Res:* Geochemical and petrological study of the volcanic rocks, igneous rocks and some metamorphic rocks, especially in the modern island arcs and Canadian shield; geochemical study of lake and river waters; environmental quality study including treatment of municipal and industrial wastewater; environmental science. *Mailing Add:* Dept Geol Sci Cleveland State Univ Cleveland OH 44115

LO, KWOK-YUNG, b Nanking, China, Oct 19, 47; US citizen; m 73; c 2. RADIO ASTRONOMY, ASTROPHYSICS. *Educ:* Mass Inst Technol, SB, 69, PhD(physics), 74. *Prof Exp:* Res fel radio astron, Owens Valley Radio Observ, Calif Inst Technol, 74-76; Miller fel basic res sci, Univ Calif, Berkeley, 76-78, asst res astronomer, Radio Astron Lab, 78; sr res fel, Calif Inst Technol, 78-80, asst prof radio astron, 80-86; PROF ASTRON, UNIV ILL, 86- *Concurrent Pos:* Miller fel, 76-79; assoc, Ctr Advan Study, Univ Ill, 91- *Mem:* Am Astron Soc; Int Astron Union. *Res:* Microwave spectroscopy studies of phenomena associated with star formation; studies of the intergalactic medium in nearby groups of galaxies; high angular resolution studies of galactic and extragalactic radio sources by interferometry and very long baseline interferometry techniques; millimeter-wave interferometry; galactic center. *Mailing Add:* Univ Ill 1002 W Green St Urbana IL 61801

LO, MIKE MEI-KUO, b Formosa, China, Sept 21, 36; m 67. PHYSICAL CHEMISTRY. *Educ:* Nat Taiwan Univ, BS, 59; Univ Ill, MS, 65, PhD(chem), 67. *Prof Exp:* SR RES CHEMIST, S C JOHNSON & SON, INC, 67- *Concurrent Pos:* Res fel, Univ Ill. *Mem:* Am Chem Soc; Fine Particle Soc; Am Indust Hyg Asn; Am Soc Testing & Mat. *Res:* Microwave spectroscopy; ultrasonic impedometry; gas chromatography and mass spectroscopy; aerosol science and technology. *Mailing Add:* 3721 Spring Lake Dr Racine WI 53210

LO, THEODORE CHING-YANG, b Shanghai, China, Dec 22, 43; Can citizen; m 74; c 2. MEMBRANE FUNCTIONS, SOMATIC CELL GENETICS. *Educ:* Univ Man, BSc, 69; Univ Toronto, PhD(med biophys), 73. *Prof Exp:* Res fel biochem, Harvard Univ, 73-75; from asst prof to assoc prof, 75-87, PROF BIOCHEM, UNIV WESTERN ONT, 87- *Mem:* Am Soc Biol Chemists; Can Biochem Soc. *Res:* Molecular mechanisms for hexose transport in rat myoblasts; human muscle cells. *Mailing Add:* Dept Biochem Univ Western Ont London ON N6A 5C1 Can

LO, THERESA NONG, b Hai Pong, NViet Nam, Mar 16, 45; US citizen; m 69; c 1. BIOCHEMISTRY. *Educ:* Clarke Col, Dubuque, Iowa, BA, 68; Ind Univ, PhD(biochem), 74. *Prof Exp:* Lab asst dept chem, Clarke Col, 66-68; res asst dept biochem, Ind Univ, 68-73; USPHS trainee, Cardiovas Res Inst, Univ Calif, San Francisco, 73-75; vis fel, Pulmonary Br, 75-77, vis fel, Lab Immunobiol, Nat Cancer Inst, 77-78, res chemist, Lab Cellular Metab, Nat Heart, Lung & Blood Inst, 79-82, res chemist, Lab Chem Pharmacol, 82-88, health sci adminr, Div Blood Dis & Resources, 88-89, HEALTH SCI ADMINR, DIV EXTRAMURAL ACTIV, NAT CANCER INST, NIH, 89- *Mem:* Am Soc Pharmacol & Exp Therapeut; Inflammation Res Asn; Am Soc Cell Biol; Am Soc Biochem & Molecular Biol. *Res:* Enzymology of blood constituents; mechanisms of action peptide cytotoxins; proteases and proteases inhibitors; mechanisms of action of nonsteroidal anti inflammatory drugs; chemotaxis; energy-linked transport processes; author & co-author of numerous articles, in J of biological chemistry and others. *Mailing Add:* 5304 Elsmere Ave Bethesda MD 20814-4799

LO, W(ING) C(HEUK), b Macao, May 20, 24; US citizen; m 57; c 3. CERAMICS, METALLURGY. *Educ:* Lingnan Univ, BS, 47; Mo Sch Mines, BS, 54; Rutgers Univ, PhD(ceramic eng), 60. *Prof Exp:* Assoc ceramic engr, Crane Co, Ill, 54-55; res asst ceramics, Rutgers Univ, 55-59; mem tech staff, Bell Tel Labs, 59-86; RETIRED. *Mem:* Am Ceramic Soc; Nat Inst Ceramic Engrs. *Res:* Evaluation and development of material and process for the fabrication of components for light wave communication. *Mailing Add:* 1466 Locksley Dr Bethlehem PA 18018

LO, WAYNE, semiconductors, electron physics; deceased, see previous edition for last biography

LO, WOO-KUEN, b Hualien, Taiwan, Dec 20, 45; m 72; c 3. LENS RESEARCH. *Educ:* Wayne State Univ, PhD(anat), 78. *Prof Exp:* ASSOC PROF ANAT, MOREHOUSE SCH MED, 87- *Concurrent Pos:* Electron microscopy. *Mem:* Am Soc Cell Biol; Am Asn Anatomists; Asn Res Vision & Ophthal. *Res:* Cell biology of the ocular lens; focusing on structures and functions of intercellular junctions, cell membranes and cytoskeleton, as well as their changes during cataractogenesis. *Mailing Add:* Dept Anat Morehouse Sch Med 720 Westview Dr SW Atlanta GA 30310

LO, Y(UEN) T(ZE), b China, Jan 31, 20; US citizen; m 53; c 2. ELECTRICAL ENGINEERING. *Educ:* Nat Southwest Assoc Univ, BS, 42; Univ Ill, MS, 49, PhD(elec eng), 52. *Prof Exp:* Asst, Radio Res Inst, Tsinghua Univ, Peking, 42-46, instr, Tsinghua & Yenching Univs, 46-48; proj engr, Channel Master Corp, 52-56; from asst prof to prof, 56-90, dir, Electromagnetics Lab, 82-90, EMER PROF ELEC ENG, UNIV ILL, URBANA, 90- *Concurrent Pos:* Consult, Westinghouse Elec Corp, 57-58, Andrew Corp, 63, Am Electronics Lab, 66, Emerson Elec, 68-69, IBM Corp, 69, Raytheon, 69-73, JPL, 81-85, TRW, 85-86, Ford Aerospace, 86 & Lockheed; hon prof, Northwest Telecommunications Eng Inst, PRC; hon prof, Northwest Polytech Univ, PRC. *Honors & Awards:* John T Bolljahn Mem Award, Inst Elec & Electronics Engrs, 64, Centennial Medal, 84; Halli Burt Eng, Educ & Leadership Award, 86. *Mem:* Sigma Xi; Int Union Radio Sci; Nat Acad Engr, 84. *Res:* Antenna; electromagnetic theory; waves in plasma; radio astronomy. *Mailing Add:* Dept Elec Eng & Comp Eng Univ Ill 1406 W Green St Urbana IL 61801

LOACH, KENNETH WILLIAM, b Portsmouth, Eng, Sept 5, 34; m 66; c 2. DATA ANALYSIS, TEACHING. *Educ:* Univ Auckland, BSc, 56, MSc, 58; Univ Wash, PhD(chem), 69. *Prof Exp:* Chemist, Ruakura Animal Res Sta, NZ, 58-60; div plant indust, Commonwealth Sci & Indust Res Orgn, Australia, 60-63; asst prof, 63-73, ASSOC PROF ANAL CHEM, STATE UNIV NY COL PLATTSBURGH, 73- *Concurrent Pos:* NSF grants, Tufts Univ, 71, State Univ NY Col Plattsburgh, 72-73, State Univ NY Res Found fel & grant, 72-73 & 77-78; sr vis fel, Univ Leeds, 77-78; vis assoc prof, Univ Del, 85-86. *Mem:* AAAS; Am Chem Soc; Chemometrics Soc; United Univ Professions. *Res:* Principles of analytical chemistry; chemometrics; chemical computing. *Mailing Add:* Dept Chem State Univ NY Col Plattsburgh NY 12901

LOACH, PAUL A, b Findlay, Ohio, July 18, 34; m 57; c 4. BIOCHEMISTRY, PHYSICAL BIOCHEMISTRY. *Educ:* Univ Akron, BS, 57; Yale Univ, PhD(biochem), 61. *Prof Exp:* Nat Acad Sci-Nat Res Coun fel photosynthesis, Univ Calif, Berkeley, 61-63; from asst prof to assoc prof, 63-73, PROF CHEM, NORTHWESTERN UNIV, 73-, PROF BIOCHEM & MOLECULAR BIOL, 74- *Concurrent Pos:* Res career develop award, NIH, 71-76. *Mem:* AAAS; Am Chem Soc; Am Soc Biol Chem; Biophys Soc; Am Soc for Photobiology. *Res:* Primary photochemistry of photosynthesis; chemistry of porphyrins and metalloporphyrins; biological oxidation and reduction; structure and function in bioenergetic membranes; photochemical models of photosynthesis. *Mailing Add:* Dept of Biochem & Molecular Biol Northwestern Univ 2153 Sheridan Rd Evanston IL 60208-3500

LOADER, CLIVE ROLAND, b Saffron Walden, Eng, Mar 11, 65; NZ citizen. MATHEMATICAL STATISTICS. *Educ:* Univ Canterbury, NZ, BSc, 86; Stanford Univ, PhD(statist), 90. *Prof Exp:* Biometrician, Ministry Agr & Fisheries, 86; MEM TECH STAFF, AT&T BELL LABS, 90- *Mem:* Inst Math Statist; Am Statist Asn. *Res:* Statistical applications of stochastic processes; boundary crossing problems; change points; sequential analysis; scan statistics; goodness of fit tests; confidence bands. *Mailing Add:* AT&T Bell Labs Rm 2C-279 600 Mountain Ave Murray Hill NJ 07974-2070

LOADHOLT, CLAUDE BOYD, b Fairfax, SC, Mar 26, 40; m 63; c 2. BIOSTATISTICS. *Educ:* Clemson Univ, BS, 62, MS, 65; Va Polytech Inst, PhD(statist), 69. *Prof Exp:* Asst exp sta statistician, Clemson Univ, 65-66, asst prof exp statist, 68-70; ASSOC PROF BIOMET, MED UNIV SC, 70- *Mem:* Biomet Soc. *Res:* Statistical consultation in biological and medical research; design of experiments; statistical data processing. *Mailing Add:* Dept of Biomet Med Univ of SC 171 Ashley Ave Charleston SC 29425

LOAN, LEONARD DONALD, b London, Eng, Oct 6, 30; m 55; c 3. POLYMER CHEMISTRY. *Educ:* Univ Birmingham, BSc, 51, PhD(polymer chem), 54. *Prof Exp:* Sci off combustion chem, Royal Aircraft Estab, 54-57; chemist, Arthur D Little Res Inst, 57-59; prin sci off rubber chem, Rubber & Plastics Res Asn, 59-66; mem tech staff polymer chem, 66-74, HEAD PLASTICS CHEM, RES & ENG, AT&T BELL LABS, 74- *Mem:* Am Chem Soc. *Res:* Polymer crosslinking and aging. *Mailing Add:* AT&T Bell Labs 600 Mountain Ave Murray Hill NJ 07974-2070

LOAN, RAYMOND WALLACE, b Ephrata, Wash, Apr 24, 31; m 52; c 4. IMMUNOBIOLOGY. *Educ:* Wash State Univ, BS, 52, DVM, 58; Purdue Univ, MS, 60, PhD(animal path), 61. *Prof Exp:* Instr vet microbiol, Purdue Univ, 58-61; from asst prof to prof vet microbiol, Univ Mo-Columbia, 61-78, chmn dept, 69-78; ASSOC DEAN RES & GRAD INSTR, COL VET MED & PROF VET MICROBIOL & PARASITOL, TEX A&M UNIV, 78- *Mem:* Am Vet Med Asn; Am Asn Immunol; Am Col Vet Microbiol; Am Soc Microbiol; Conf Res Workers Animal Diseases. *Res:* Cell mediated immunity; immunologic aspects of avian leukosis; Bovine Respiratory Disease. *Mailing Add:* Dept Vet Microbiol & Parasitol Tex A&M Univ Col Vet Med College Station TX 77843

LOAR, JAMES M, b Lancaster, Pa, Sept 10, 44. AQUATIC ECOLOGY, FISHERIES BIOLOGY. *Educ:* Gettysburg Col, BA, 66; Temple Univ, MEd, 69; Univ Wyo, PhD(zool), 75. *Prof Exp:* Teacher biol, Cherry Hill High Sch W, NJ, 66-70; RES ASSOC, ENVIRON SCI DIV, OAK RIDGE NAT LAB, 75- *Concurrent Pos:* Prin investr, Oak Ridge Nat Lab, 75- *Mem:* AAAS; Am Fisheries Soc; Ecol Soc Am; Sigma Xi. *Res:* Assessment of impact of nuclear, fossil and hydroelectric energy technologies; responses of aquatic biota to altered flow regimes below hydroelectric projects. *Mailing Add:* 1712 Nighbert Lane Knoxville TN 37922

LOATMAN, ROBERT BRUCE, b Washington, DC, Aug 23, 45; m 69; c 4. COMPUTATIONAL LINGUISTICS, KNOWLEDGE-BASED SYSTEMS. *Educ:* Fordham Col, BA, 67; Fordham Univ, MA, 71, PhD(math), 76. *Prof Exp:* Programmer, Gen Elec Co, 68-69; instr math, Georgetown Univ, 73-76; tech dir, Killalea Assocs Inc, 76-80; computer scientist, 80-84, DIR ARTIFICIAL INTEL DEVELOP, PLANNING RES CORP, 84-, CHIEF SCIENTIST, 90- *Concurrent Pos:* Mem tech staff, Mitre Corp, 78; consult, Phonic Ear Inc, 80. *Honors & Awards:* Parallax Prize, Emhart Corp, 88. *Mem:* Am Asn Artificial Intel; Asn Computational Ling; AAAS. *Res:* Computational linguistics research and development; natural language text understanding systems; knowledge-based systems. *Mailing Add:* 11727 Great Owl Circle Reston VA 22094-1167

LOBAUGH, BRUCE, b Charleroi, Pa, Nov 4, 53; m 83; c 1. BONE & MINERAL METABOLISM, VITAMIN D METABOLISM. *Educ:* Clarion Univ Pa, BS, 75, Pa State Univ, MS, 78, PhD, 81. *Prof Exp:* Res assoc, Duke Univ Med Ctr, 80-83, asst res prof physiol & surg, 83-90, supvr, Surg Endocrinol-Oncol Lab, 83-90, ASST RES PROF MED & CELL BIOL, DIV PHYSIOL, DUKE UNIV MED CTR, 90-, DIR, LIPID LAB & SUPVR, BONE & MINERAL LAB, 90- *Concurrent Pos:* Prin investr, NIH res grants, 82-85 & 87-90. *Mem:* Am Physiol Soc; Am Soc Bone & Mineral Res; Am Fed Clin Res; Adv Mineral Metabolism; AAAS. *Res:* Characterization of the mechanisms underlying regulation of vitamin D metabolism in normal and disease states; changes in vitamin D metabolism which are attendant on the natural aging process. *Mailing Add:* Duke Univ Med Ctr Box 3482 Durham NC 27710

LOBB, DONALD EDWARD, b Saskatoon, Sask, Apr 25, 40. PHYSICS. *Educ:* Univ Sask, BE, 61, MSc, 63, PhD(physics), 66. *Prof Exp:* Nat Res Coun Can overseas fel, 66-67; asst prof, 67-71, assoc prof, 71-87, PROF PHYSICS, UNIV VICTORIA, BC, 87- *Mem:* Inst Elec & Electronics Engrs; Can Asn Physicists. *Res:* Beam optics. *Mailing Add:* Dept Physics Univ Victoria Victoria BC V8W 2Y2 Can

LOBB, R(ODMOND) KENNETH, b Can, Feb 10, 25; m 47; c 2. AERODYNAMICS. *Educ:* Univ Alta, BSc, 47; Univ Toronto, MASc, 48, PhD(aeronaut eng), 50. *Prof Exp:* Aeronaut engr, Naval Ord Lab, 48; res scientist, Can Defense Res Bd, 49; res assoc, Inst Fluid Dynamics, Univ Md, 50-51; aeronaut engr, Naval Ord Lab, 52-53, chief, Hypersonic Wind Tunnel Br, 54-55, Aerophys Div, 56 & Aerodyn Dept, 57-61, assoc head aeroballistics & hydroballistics, 61-74; tech dir, Naval Air Develop Ctr, Dept Defense, 74-81; res prog dir, 81-87, VPRES ADV SYSTS DIV, CTR NAVAL ANALYSES, 87- *Concurrent Pos:* Lectr, Cath Univ Am, 53-54 & Univ Md, 61-63. *Mem:* Am Inst Aeronaut & Astronaut; Int Acad Astronaut. *Res:* Aircraft design; structures; inertial navigation systems; boundary layers; hypersonic wind tunnel research and development; skin friction and heat transfer measurements in hypersonic flows; ballistics range studies; high temperature aerodynamics; underwater acoustics; underwater vehicles; systems engineering. *Mailing Add:* Ctr Naval Analyses Mail Stop 9 4401 Ford Ave Alexandria VA 22302-0268

LOBDELL, DAVID HILL, b Erie, Pa, July 9, 30. PATHOLOGY. *Educ:* Kenyon Col, AB, 52; Univ Mich, MD, 56. *Prof Exp:* From intern to resident path, Bellevue Hosp, New York, 56-59, asst pathologist, 59-60; assoc pathologist, 60-63, DIR LABS, ST VINCENT'S MED CTR, 63-, DIR SCH MED TECHNOL, 63- *Concurrent Pos:* Instr, Sch Med, NY Univ, 59-61, asst clin prof path, 61-69; lectr histol, Fairfield Univ, 64-73. *Mem:* AMA; fel Col Am Path; fel Am Soc Clin Path. *Res:* Osmometry; myeloproliferative disorders. *Mailing Add:* 2800 Main St Bridgeport CT 06606

LOBECK, CHARLES CHAMPLIN, b New Rochelle, NY, May 20, 26; m 54; c 4. PEDIATRICS. *Educ:* Hobart Col, AB, 48; Univ Rochester, MD, 52. *Prof Exp:* From instr to sr instr pediat, Sch Med & Dent, Univ Rochester, 55-58; from asst prof to prof, Sch Med, Univ Wis-Madison, 58-75, chmn dept, 64-74, assoc dean clin affairs, 74-75; dean, Sch Med, Univ Mo-Columiba, 75-; AT SCI CTR, MED SCH, UNIV WIS-MADISON. *Mem:* Sigma Xi. *Res:* Metabolic disease; membrane transport; cystic fibrosis. *Mailing Add:* Med Sch Univ Wis Med Sci Ctr 1300 University Ave Rm 1217 Madison WI 53706

LOBEL, STEVEN A, b Brooklyn, NY, Feb 8, 52; m 78; c 3. CLINICAL IMMUNOLOGY DIRECTOR, IMMUNOPATHOLOGY. *Educ:* Univ Tex, Austin, BA, 73; State Univ NY, Buffalo, MA, 75, PhD(immunol), 77; Am Bd Med Lab Immunol, dipl, 87. *Prof Exp:* Fel, immunol, HHH Ctr Cancer Res, Hadassah, Jerusalem, 77-78, Dept Pathol, Univ Pittsburgh, 78-80; asst prof immunol, Dept Pediat & Microbiol, Univ Ill, Chicago, 80-83, Dept Pathol, Med Col Ga, 83-88; LAB DIR, IMMUNOL, DEPT CLIN PATH, AM MED LABS, 88- *Concurrent Pos:* Fel, Lady Davis Found, 77. *Mem:* Am Asn Immunol; Am Asn Path; Am Soc Clin Path; Am Soc Microbiol; Am Asn Clin Chem; Clin Immunol Soc. *Res:* Cellular immunology of human immunodeficiency virus infection. *Mailing Add:* Dept Clin Path Am Med Labs 11091 Main St Fairfax VA 22030

LOBENE, RALPH RUFINO, b Rochester, NY, Mar 30, 24; m 50. DENTISTRY, PERIODONTOLOGY. *Educ:* Univ Rochester, BS, 44; Univ Buffalo, DDS, 49; Tufts Univ, MS, 62; Am Bd Periodont, dipl, 65. *Prof Exp:* Res chemist, Manhattan Proj, Univ Rochester, 44-45; res chemist, Merck & Co, NJ, 45-46; intern periodont, Eastman Dent Dispensary, Rochester, NY, 49-50; resident oral surg, Strong Mem Hosp, Rochester, 50-51, asst dent surgeon, 53-62, instr dent & clin dent res, Sch Med & Dent, Univ Rochester, 53-62; asst prof periodont, Sch Dent, Univ Pac, 62-63; assoc prof & acad adminr dent asst training prog, 63-73, dean, Forsyth Sch Dent Hygienists, 75-85, DIR ADVAN EDUC & SR MEM STAFF, FORSYTH DENT CTR, 70- *Concurrent Pos:* Res grants, Colgate Palmolive Co, 54-55, 58-59 & 84-88; res grant, Sch Dent, Univ Pac, 62-63; res grant, Gen Elec Co, 62-65; staff dentist, Eastman Dent Dispensary, 53-54, res assoc, 58-62; pvt pract, 53-60; chief dent serv, State Indust & Agr Sch Boys, NY, 54-57; clin asst, Sch Dent Med, Tufts Univ, 61-62, lectr, 63-; asst mem staff, Forsyth Dent Ctr, Northeastern Univ, 63-66, lectr, 63-, head dept clin exp, Inst Res & Advan Study Dent, 66-

80, sr mem staff, 70-91, emer sr mem staff, 91; consult, Dent Res Panel, Gen Elec Co, 63-; vis surgeon, Dept Dent, Boston City Hosp, 64-90; grants, Robert Wood Johnson Found, 73-74 & 76-77, Nat Inst Dent Res, Off Collab Res, 73-75, Merril Richardson Co, 75-77 & Johnson & Johnson Co, 78; mem, Am Dent Asn Clin Cleansing Comt, 74-84; mem, Mass Bd Dent Adv Comt, 78-; fel, Int Col Dent & Am Acad Dent Sci; lectr periodont, Harvard Sch Dent Med, 80; mem ed bd, Clin Preventative Dent, 79-, J Clin Dent, 87-; periodontology lectr, Harvard Sch Dent Med, 80; mem ed rev bd, J Dent Educ, 84-87. *Honors & Awards:* Colgate Palmolive Co Award, Am Dent Asn, 88. *Mem:* Fel AAAS; fel Am Col Dent; Am Dent Asn; Am Acad Oral Med; Am Acad Periodont; fel Int Col Dentists. *Res:* Assessment of periodontal disease and evaluation of the effectiveness of therapeutic methods of treatment of periodontal disease; analytical methods for chemical analysis. *Mailing Add:* 605 Ives Dairy Rd North Miami Beach FL 33179

LOBER, PAUL HALLAM, b Minneapolis, Minn, Sept 25, 19. PATHOLOGY. *Educ:* Univ Minn, Minneapolis, BS, 42, MD, 44, PhD(path), 51; Am Bd Path, dipl, 52. *Prof Exp:* Mem fac, 51-60, surg pathologist, Univ Hosps, 51-74, PROF PATH, MED SCH, UNIV MINN, MINNEAPOLIS, 60-; surg pathologist, Abbott-Northwestern Hosp, 74-89; RETIRED. *Mem:* Am Soc Cytol; fel Am Col Path; Am Asn Pathologists; Int Acad Path; Am Soc Clin Path. *Res:* Coronary heart disease. *Mailing Add:* 1525 W 28th St Minneapolis MN 55408

LOBKOWICZ, FREDERICK, b Prague, Czech, Nov 17, 32; US citizen; m 60; c 2. ELEMENTARY PARTICLE PHYSICS. *Educ:* Swiss Fed Inst Technol, 55, PhD(physics), 60. *Prof Exp:* Res assoc, 60-64, from asst prof to assoc prof, 64-73, PROF PHYSICS, UNIV ROCHESTER, 73- *Concurrent Pos:* Humboldt Found sr fel, 73-74; vis prof, Univ Munich, Ger, 73-74. *Mem:* Fel Am Phys Soc; Swiss Phys Soc. *Res:* Muon and photon interactions. *Mailing Add:* Dept Physics & Astron Univ Rochester Rochester NY 14627

LOBL, THOMAS JAY, b Danville, Va, Oct 20, 44; m 68; c 2. PHARMACEUTICAL CHEMISTRY, REPRODUCTIVE PHYSIOLOGY. *Educ:* Univ NC, Chapel Hill, BS, 66; Johns Hopkins Univ, PhD(org chem), 70. *Prof Exp:* Res fel biochem, Calif Inst Technol, 70-73; Sr res scientist chem & biochem, UpJohn Co, 73-88; Dir, Peptide Res Lab, Immunetech Pharmaceut, 88-90; DIR, CHEM SCI DEPT, TANABE RES LABS USA, 90- *Concurrent Pos:* chair, Kalamazoo Sect, Am Chem Soc, 79, San Diego Sect, 92; Assoc ed, J Andrology & Arch Andrology, 80-84. *Honors & Awards:* President's Award, Am Soc Andrology. *Mem:* Am Chem Soc; AAAS; Sigma Xi; Am Soc Andrology (treas, 81-84); NY Acad Sci; Am Peptide Soc. *Res:* Regulation of male reproduction; spermatogenesis; epididymal function; hormone transport and receptor proteins; male contraception; chemical and biological deaminations; heterocyclic and steroid synthesis; reproductive physiology; peptide synthesis and chemistry; peptide/protein chemistry and biochemistry; signal peptides; peptide secondary structure activity relationships; peptide transport and targeting; peptide membrane interaction, autoimmune diseases, immunology, rheumatoid arthritis, allergy, type I diabetis. *Mailing Add:* Tanabe Res Labs USA 11045 Roselle St San Diego CA 92121

LOBO, ANGELO PETER, b Masindi, Uganda, May 19, 39; m 67; c 2. BIO-ORGANIC CHEMISTRY. *Educ:* Univ Bombay, BSc, 58; Ind Univ, Bloomington, PhD(org chem), 66. *Prof Exp:* Rockefeller Found spec lectr org chem, Makerere Univ Col, Kampala, Uganda, 66-68; res asst chem, Rensselaer Polytech Inst, 68-70; SR RES SCIENTIST, WADSWORTH CTR LABS & RES, NY STATE DEPT HEALTH, 70- *Mem:* Am Chem Soc. *Res:* Synthesis of nucleotide/nucleoside substrates and inhibitors, oligonucleotide synthesis. *Mailing Add:* Wadsworth Ctr Labs & Res NY State Dept of Health Albany NY 12201

LOBO, CECIL T(HOMAS), b Mangalore, India, Sept 22, 34; c 2. ENGINEERING MECHANICS. *Educ:* Gujarat Univ, India, BE, 55; Univ Notre Dame, MS, 60; Purdue Univ, PhD(civil eng), 66. *Prof Exp:* Asst engr, Shah Construct Co, Ltd, India, 56-57; from instr to assoc prof, 63-71, actg chmn dept, 71-72, PROF CIVIL ENG, ROSE-HULMAN INST TECHNOL, 71 - *Concurrent Pos:* Consult, Universal Tank & Iron Works, Inc, Ind. *Mem:* Mem Am Soc Civil Engrs; Am Soc Eng Educ; Sigma Xi; Am Concrete Inst; Prestressed Concrete Inst. *Res:* Structural and soil mechanics. *Mailing Add:* Rose-Hulman Inst Technol Terre Haute IN 47803

LOBO, FRANCIS X, b Aden, UAR, Oct 8, 25; US citizen; m 60; c 3. MICROBIOLOGY. *Educ:* Univ Bombay, BS, 47, MS, 50; Inst Divi Thomae, PhD(exp med, biol), 59; Nat Registry Microbiol, cert. *Prof Exp:* Technician, Path Dept, Worli Gen Hosp, India, 50; control & res microbiologist-chemist, Chemo Pharma Labs Ltd, Worli, Bombay, 50-57; assoc prof sci, 60-70, chmn dept biol, 70-74, PROF BIOL SCI, MARYWOOD COL, 70- *Concurrent Pos:* Consult, Radio Corp Am, 65-; NSF grant, Argonne Nat Lab, 66, resident res assoc, 68-69; fac res partic, Argonne Nat Lab, 67 & St Jude Children Res Hosp, Memphis, Tenn, 70; mem eval team, Pa Dept Educ, 71. *Mem:* Am Soc Microbiol; NY Acad Sci. *Res:* Intestinal microorganisms by enrichment culture techniques; citric acid from a cane-sugar molasses; beef brain extract in controlling staphylococcus infections; etiology of sludge formation in industrial wastes. *Mailing Add:* Dept Sci Marywood Col Scranton PA 18509

LOBO, PAUL A(LLAN), b La Cumbre, Colombia, Oct 10, 28. CHEMICAL ENGINEERING. *Educ:* Mass Inst Technol, SB, 50, SM, 51; Univ Mich, PhD(chem eng), 55. *Prof Exp:* Supvr process develop sect, Petrochem Res Div, Continental Oil Co, Okla, 55-63, European rep, Res Dept, Holland, 63-64, petrolchem coord, Continental Oil Co, Ltd, Eng, 64-65, exec asst to pres, NY, 65-67, mgr develop, Petrochem Dept, 67-68; vpres & gen mgr, Pitt-Consol Chem Co, 68-70; dir bus develop & planning, Tenneco Chem Inc, Piscataway, 71-73, dir, Tech Group, 73-75, dir corp planning, Saddle Brook, NJ, 75-80, vpres planning, 80-85; AT NUODEX, INC. *Concurrent Pos:* Lectr, Univ Okla, 56-63. *Mem:* Am Chem Soc; Nat Soc Prof Engrs; Am Inst Chem Engrs. *Res:* High pressure; petrochemical process development; reaction kinetics. *Mailing Add:* Hul America Inc PO Box 465 Piscataway NY 08855-0465

LOBO, WALTER E(DER), b Brooklyn, NY, Mar 29, 05; m 28; c 5. CHEMICAL ENGINEERING. *Educ:* Mass Inst Technol, BS, 26 & La State Univ, 26. *Prof Exp:* Chief chemist, Cent Agabama Formneto, Cuba, 27; chief chemist & factory supt, Ingenio Manuelita, Palmira, Colombia, 28-29; asst job engr, M W Kellogg Co Div, Pullman Inc NJ, 29-30, chem engr, NY, 30-33, process engr, 33-38, dir, Chem Eng Dept, 38-57; consult chem engr, 57-88; RETIRED. *Concurrent Pos:* Lectr, Columbia Univ, 47-49; pres, United Eng Trustees, 65-67. *Honors & Awards:* Founders Award, Am Inst Chem Engrs, 70. *Mem:* AAAS; Am Soc Mech Engrs; Nat Soc Prof Engrs; Am Inst Chem Engrs; Am Chem Soc. *Res:* Petroleum; petrochemical and chemical processing; furnace design; heat transfer; fluid flow. *Mailing Add:* 497 Lost District Dr New Canaan CT 06840-2017

LOBSTEIN, OTTO ERVIN, b Czech, Apr 12, 22; nat US; m 52; c 3. CLINICAL BIOCHEMISTRY. *Educ:* Univ London, BSc, 45; Smae Inst, Eng, MSF, 45; Northwestern Univ, PhD(biochem), 52; Am Bd Clin Chem, dipl, 55. *Prof Exp:* Asst res chemist, Howards & Sons, Ltd, Eng, 42-46; biochemist, Elgin State Hosp, Ill, 47-48; instr chem, Wesley & Passavant Mem Hosps, Ill, 49-51; res assoc zool, Univ Southern Calif, 52-53; med dir res, Chemtech Labs, 52-62; biochemist-owner, Lobstein Biochem Lab, Calif, 62-64; asst prof chem, Loyola Univ, Calif, 64-65; head, Biochem Dept, St Elizabeth Hosp Med Ctr, 65-76; dir clin chem, 76-82, BIOCHEMIST, COOK COUNTY HOSP, MT SINAI HOSP MED CTR, 82- *Concurrent Pos:* Vis res prof, Univ Redlands, 59-65; secy-treas, Res Found Diseases Eye, 59-65; vis assoc prof, Purdue Univ, 68-73, adj prof, 65-76; prof biochem & path, Rush Presby St Lukes Med Ctr, Chicago, 76- *Honors & Awards:* Sci Award, Cancer Fedn, 79. *Mem:* Fel AAAS; Am Soc Microbiol; sr mem Am Chem Soc; fel Am Asn Clin Chem (secy, 81-83); Am Chem Soc; fel Nat Acad Clin Biochem. *Res:* Biochemical investigation of the crystalline lens of the eye, protein structure and constitution in normal and in cataract lenses of the human and other species; changes in protein with a changed electrolyte environment; clinical investigation of lysozyme in carcinomatosis. *Mailing Add:* 2006 Maple Ave Northbrook IL 60062-5266

LOBUE, JOSEPH, b Union City, NJ, Apr 19, 34; m 59; c 3. PHYSIOLOGY, HEMATOLOGY. *Educ:* St Peter's Col, NJ, BS, 55; Marquette Univ, MS, 57; NY Univ, PhD(physiol), 62. *Prof Exp:* From asst prof to assoc prof, 62-71, PROF BIOL, NY UNIV, 71-, CO-DIR, A S GORDON LAB EXP HEMAT, 67- *Concurrent Pos:* NIH fel, 62; Sigma Xi grant-in-aid, 64-65; Am Cancer Soc grant, 65-66; Nat Cancer Inst grant, 71-73, 75-78 & 79-81; Nat Leukemia Asn grant, 74-76; assoc, Danforth Found, 68-; co-dir, Hemat Training Prog, NIH, 65-75. *Honors & Awards:* Christian R & Mary F Lindback Found Award, NY Univ, 65. *Mem:* AAAS; Soc Exp Biol & Med; Am Asn Anat; Harvey Soc; Am Soc Hemat. *Res:* Mechanisms controlling leukocyte and erythrocyte production and release; pathophysiology and cytokinetics of rodent and avian leukemias. *Mailing Add:* Dept Biol New York Univ Washington Sq Col New York NY 10003

LOBUGLIO, ALBERT FRANCIS, b Buffalo, NY, Feb 1, 38; m 62; c 5. HEMATOLOGY, IMMUNOLOGY. *Educ:* Georgetown Univ, MD, 62. *Prof Exp:* Intern med, Presby Univ Hosp, Pittsburgh, 62-63, resident, 63-65; instr, State Univ NY Buffalo, 67-68, asst prof, 68-69; assoc prof med, Ohio State Univ, 69-73, prof, 73-78; prof med, Univ Mich, 78-; DIR, COMPREHENSIVE CANCER CTR, UNIV ALA, BIRMINGHAM. *Concurrent Pos:* Hemat fel, Thorndike Mem Lab, Boston City Hosp, 65-67; hemat consult, Vet Admin Hosp, Buffalo, 67-69, Dayton, Ohio, 69- *Mem:* Am Fedn Clin Res; Am Soc Hemat; Am Soc Clin Invest. *Res:* Tumor immunology; transplant immunology; human macrophage and lymphocyte functions. *Mailing Add:* Cancer Ctr & Div Hematol-Oncol Univ Ala University Sta Birmingham AL 35294

LOBUNEZ, WALTER, b Ukraine, Nov 22, 20; nat US; m 45; c 1. INDUSTRIAL CHEMISTRY. *Educ:* Univ Pa, MS, 52, PhD(chem), 54. *Prof Exp:* Res assoc immunol, Jefferson Med Col, 54-55; protein chem, Children's Hosp, Univ Penn, 55-59; sr scientist chem, Textile Res Inst, 59-60; res chemist, FMC Corp, 60-67, sr res chemist 67-80, res assoc, 80-83; RETIRED. *Res:* Chemistry of hydrocarbons; protein chemistry; chemistry of cellulose; soda ash processes; Ba and Sr processes. *Mailing Add:* 562 Ewing St Princeton NJ 08540

LOCALIO, S ARTHUR, b New York, NY, Oct 4, 11; m 45; c 4. SURGERY. *Educ:* Cornell Univ, AB, 33; Univ Rochester, MD, 36; Am Bd Surg, dipl, 45. *Hon Degrees:* DSc, Columbia Univ, 42. *Prof Exp:* From instr to assoc prof surg, 45-53, prof clin surg, 53-71, PROF SURG, MED SCH, NY UNIV, 71- *Concurrent Pos:* Asst surgeon, Univ Hosp, 45-47, asst attend surgeon, 47-49, assoc attend surgeon, 49-52, attend surgeon, 52-, clin asst vis surgeon, 49-52, vis surgeon, 52-; Johnson & Johnson distinguished prof surg, NY Univ, 72. *Mem:* Am Asn Surg of Trauma; Am Gastroenterol Asn; Am Col Surg; Am Surg Asn; Sigma Xi. *Res:* Wound healing; surgery of gastro-intestinal disease. *Mailing Add:* Mill Village Rd Deerfield MA 01342

LOCASCIO, SALVADORE J, b Hammond, La, Oct 29, 33; m 54; c 3. HORTICULTURE. *Educ:* Southeastern La Col, BS, 55; La State Univ, MS, 56; Purdue Univ, PhD(plant physiol), 59. *Prof Exp:* Asst prof, 59-65, assoc horticulturist, 65-69, ASSOC PROF HORT, DEPT VEG CROPS, UNIV FLA, 65-, HORTICULTURIST, 69- *Honors & Awards:* Pres Gold Medal Award, 78. *Mem:* Weed Sci Soc Am; Sigma Xi; fel Am Soc Hort Sci. *Res:* Fertilizer and water requirements of vegetables; strawberry culture; chemical weed control for vegetables; teaching of commercial vegetable crops and nutrition of horticultural crops. *Mailing Add:* Dept Veg Crops 1245 Fifield Hall Univ of Fla Gainesville FL 32611

LOCHHEAD, JOHN HUTCHISON, b Montreal, Que, Aug 7, 09; nat US; m 38; c 2. INVERTEBRATE ZOOLOGY. *Educ:* Univ St Andrew's, MA, 30; Cambridge Univ, BA, 32, Bachelor scholar, 33, PhD(zool), 37. *Prof Exp:* With Cambridge Univ Table, Marine Zool Sta, Naples, 34-35; sr cur, Mus Zool, Cambridge Univ, 35-38, instr zool, 36-38; fel by courtesy, Johns Hopkins Univ, 40; asst biologist, Va Fisheries Lab, 41-42; from instr to prof, 42-75,

EMER PROF ZOOL, UNIV VT, 75- Concurrent Pos: Lectr, Col William & Mary, 41-42; instr, Woods Hole Marine Biol Lab, 43-55, mem corp, 44- Mem: Am Soc Zool; Crustacean Soc. Res: Anatomy and physiology of Crustacea including their feeding mechanisms, locomotion, factors controlling swimming positions, responses to light, functions for the blood and related tissues, molting and reproduction. Mailing Add: 49 Woodlawn Rd London SW6 6PS England

LOCHMAN-BALK, CHRISTINA, b Springfield, Ill, Oct 8, 07; wid. INVERTEBRATE PALEONTOLOGY, STRATIGRAPHY. Educ: Smith Col, AB, 29, AM, 31; Johns Hopkins Univ, PhD(geol), 33. Prof Exp: Asst geol, Smith Col, 29-31; Nat Res Coun grant, 34; from instr to assoc prof, Mt Holyoke Col, 35-47; lectr phys sci, Univ Chicago, 47; lectr life sci, NMex Inst Min & Tech, 54; stratig geologist, State Bur Mines & Mineral Resources, 55-57; prof geol, 57-72, EMER PROF GEOL, NMEX INST MINING & TECHNOL, 72- Concurrent Pos: Am Geol Soc Am grants, 36 & 47; NSF grant, 59. Mem: Fel AAAS; Paleont Soc; fel Geol Soc Am; Nat Asn Geol Teachers; Sigma Xi; Am Asn Univ Women. Res: Cambrian paleontology and stratigraphy of the United States. Mailing Add: PO Box 1421 Socorro NM 87801

LOCHMULLER, CHARLES HOWARD, b New York, NY, May 4, 40; m 63; c 3. ANALYTICAL CHEMISTRY. Educ: Manhattan Col, BS, 62; Fordham Univ, MS, 64, PhD(analytical chem), 68. Prof Exp: From asst prof to assoc prof, 69-78, PROF CHEM, DUKE UNIV, 78-, CHMN DEPT, 82- Concurrent Pos: Assoc, Purdue Univ, 67-69; chair, anal div, Am Chem Soc, 83-84. Mem: Am Chem Soc; fel Royal Soc Chem; fel Am Inst Chemists. Res: Factors effecting separation processes; spectroscopy. Mailing Add: P M Gross Chem Lab Duke Univ Durham NC 27706

LOCHNER, JANIS ELIZABETH, b Bethesda, Md, Nov 27, 54. MEMBRANE BIOCHEMISTRY, CELLULAR COMMUNICATION. Educ: Allegheny Col, BS, 76; Univ Ore, PhD(biochem), 81. Prof Exp: RES FEL BIOCHEM, ORE HEALTH SCI UNIV, 81-; ASST PROF CHEM, LEWIS & CLARK COL, 81- Mem: AAAS. Res: Role of the plasma membrane in cellular communication; mechanisms of membrane transduction. Mailing Add: Dept Chem Lewis & Clark Col 0615 SW Palatine Hill Rd Portland OR 97219

LOCHNER, ROBERT HERMAN, b Madison, Wis, Apr 17, 39; m 62; c 4. STATISTICS. Educ: Univ Wis-Madison, BS, 61, MS, 62 & 66, PhD(statist), 69. Prof Exp: Math analyst, A C Electronics Div, Gen Motors Corp, 62-65; from asst prof to assoc prof statist & math, Marquette Univ, 68-85; CONSULT STATIST, SAT POWER ASSOCS, 85- Mem: Am Statist Asn; Inst Math Statist. Res: Statistical methods in reliability and life testing; Bayesian inference. Mailing Add: 2840 S Root River Pkwy Milwaukee WI 53227

LOCHSTET, WILLIAM A, b Port Jefferson, NY, Dec 5, 36; m 82; c 2. PHYSICS. Educ: Univ Rochester, BS, 57, MA, 60; Univ Pa, PhD(physics), 65. Prof Exp: Instr, 65-66; asst prof physics, Pa State Univ, 66-86; ASST PROF PHYSICS, UNIV PITTSBURGH, 86- Mem: Am Phys Soc; AAAS; Sigma Xi; Am Asn Physics Teachers. Mailing Add: Dept of Physics Univ of Pittsburgh Johnstown PA 15904

LOCICERO, JOSEPH CASTELLI, b Ontario Center, NY, 14; m 37; c 2. ORGANIC CHEMISTRY. Educ: Univ Rochester, BA, 36; Pa State Univ, MS, 47, PhD(biochem), 48. Prof Exp: Chemist, Hooker Electrochem Co, NY, 37-43; sr res chemist, Nuodex Prods Co, NJ, 43-45; res chemist, Rohm and Haas Co, 48-52, sr scientist, 52-71; from asst prof to prof chem, Camden County Col, 71-84; RETIRED. Mem: AAAS; Am Chem Soc. Res: Plasticizers, fungicides and insecticides; high pressure reactions; detergents; process development; ion exchange; sugar technology; halogenation; plastics. Mailing Add: 625 Devon Rd Moorestown NJ 08057

LOCK, BRIAN EDWARD, b Yeovil, Eng, Mar 21, 44; m 68; c 3. SEDIMENTOLOGY. Educ: Cambridge Univ, BA, 66, PhD(geol), 69, MA, 70. Hon Degrees: MA, Cambridge Univ, 70. Prof Exp: Consult, Fina Petrol Co, Belg, 69-70; lectr, Rhodes Univ, SAfrica, 70-74, sr lectr, 75-77; assoc prof, 77-80, PROF GEOL, UNIV SOUTHWESTERN LA, 80- Concurrent Pos: Sr res assoc, Exeter Univ, Eng, 75-76; overseas res burser, Coun Sci & Indust Res, Pretoria, 75-76; consult basic appl geol, Superior Oil Co. Mem: Am Asn Petrol Geologists; Soc Econ Paleontologists & Mineralogists; Geol Soc Am; Int Asn Volcanology & Chemistry of the Earth's Interior; Int Asn Sedimentologists. Res: Sedimentology of carbonate rocks; other aspects of sedimentology and stratigraphy; field work in South Africa, Ireland, Spitsbergen, Reunion Island, Canada and United States. Mailing Add: Dept Geol Univ Southwestern La 44530 Lafayette LA 70504

LOCK, COLIN JAMES LYNE, b London, Eng, Oct 4, 33; m 60; c 2. BIOINORGANIC CHEMISTRY, CRYSTALLOGRAPHY. Educ: Univ London, BSc, & ARCS, 54, PhD(inorg chem). Hon Degrees: DSc, Univ London, 87. Prof Exp: Asst exp officer reactor chem, Atomic Energy Res Estab, Eng, 54-57, sci officer, 60; develop chem, Atomic Energy Can Lab, Chalk River, 57-60; asst lectr inorg chem, Imp Col, London, 61-63; from asst prof to assoc prof, 63-73, PROF CHEM, MCMASTER UNIV, 73-, PROF PATH, 84- Honors & Awards: Montreal Medal, Chem Inst Can, 89. Mem: Fel Chem Inst Can; Royal Soc Chem. Res: physical methods of structure determination for heavy transition metal compounds; platinum anti-cancer drugs, technetium radiopharmaceuticals; gold anti-arthritic agents; non-steroidal anti-inflammatory drugs. Mailing Add: Lab Inorg Med ABB-266A McMaster Univ Hamilton ON L8S 4M1 Can

LOCK, G(ERALD) S(EYMOUR) H(UNTER), b London, Eng, June 30, 35; m 59; c 3. MECHANICAL ENGINEERING. Educ: Univ Durham, BSc, 59, PhD(mech eng), 62. Prof Exp: From asst prof to assoc prof, 62-70, PROF MECH ENG, UNIV ALTA, 70- Concurrent Pos: Chmn comt heat transfer, Nat Res Coun Can, 69-71; Sci Coun Can. Honors & Awards: Queen Elizabeth Silver Jubilee Medal, 79. Mem: Am Soc Mech Engrs; fel Eng Inst Can; fel Can Soc Mech Engrs (pres, 77-78). Res: Thermodynamics and heat transfer, especially ice engineering; technology assessment. Mailing Add: Dept Mech Eng Univ Alta Edmonton AB T6G 2G8 Can

LOCK, JAMES ALBERT, b Cleveland, Ohio, Feb 12, 48; m 72. LIGHT SCATTERING. Educ: Case Western Reserve Univ, BS, 70, MS, 73, PhD(physics), 74. Prof Exp: Lectr, Case Western Reserve Univ, 70-74, res assoc physics, 74-78; from asst prof to assoc prof, 78-90, PROF PHYSICS, CLEVELAND STATE UNIV, 90- Mem: Sigma Xi; Optical Soc Am. Res: Light scattering. Mailing Add: Dept of Physics Cleveland State Univ Cleveland OH 44115

LOCK, KENNETH, b Wushi, China, Mar 15, 32; m 54; c 4. ELECTRICAL ENGINEERING, COMPUTER SCIENCE. Educ: Battersea Polytech Inst, BSc, 55; Univ London, MSc, 57; Calif Inst Technol, PhD(elec eng, physics), 62. Prof Exp: From instr to asst prof elec eng, Calif Inst Technol, 59-65; advan programmer, Int Bus Mach Corp, 65-67, vis fel, 67-69; pres, Cyber Data, Inc, Calif, 69-71; mgr design automation, Burroughs Corp, 71-77, dept mgr, MCO div, 77-80; pres, Cybertec, 80-; PRES, ZYBEX. Concurrent Pos: Consult, Jet Propulsion Lab, 62-65. Mem: Inst Elec & Electronics Engrs; Asn Comput Mach. Res: Physical systems on computers; network analysis; switching theory; programming system research; computer design; interactive use of computers in engineering, science and business. Mailing Add: Zybex 10655 Roselle St San Diego CA 92121

LOCKARD, ISABEL, b Brandon, Man, June 27, 15. ANATOMY. Educ: Northwestern Univ, BS, 38; Univ Mich, MA, 42, PhD(anat), 46. Prof Exp: Asst anat, Univ Mich, 42-44 & 47; instr, Univ Pittsburgh, 44-45; instr, Sch Med Georgetown Univ, 47-49, asst prof, 49-52; from asst prof to prof, 52-85, EMER PROF ANAT, MED UNIV SC, 85- Concurrent Pos: Consult, Med Univ SC. Mem: Am Asn Anatomists; Sigma Xi. Res: Neuroanatomy; blood supply of central nervous system. Mailing Add: Dept Anat & Cell Biol Med Univ SC Charleston SC 29425

LOCKARD, J DAVID, b Renovo, Pa, Dec 20, 29; m 51; c 4. BOTANY, SCIENCE EDUCATION. Educ: Pa State Univ, BS, 51, MEd, 55; PhD(bot), 62. Prof Exp: Dept chmn sci dept high sch, Pa, 53-56; consult sci teaching improvement prog, AAAS, 56-58; asst bot, Pa State Univ, 58-61; from asst prof to assoc prof, 61-70, PROF BOT & SCI EDUC, UNIV MD, COLLEGE PARK, 70-, DIR, SCI TEACHING CTR, 62- Concurrent Pos: NSF-AAAS grant, Develop & Maintain Int Clearinghouse Sci & Math Curric Develops, 62-; dir off bio educ, Am Inst Biol Sci, 66-67; dir, NSF-AID Study Improvisation Sci Teaching Mat Worldwide, 68-72; NSF grants, acad year inst sci supvrs, 69-73; dir NSF Impact Study, 74-75; rep, US Nat Comn to UNESCO, 75- Mem: AAAS (vpres, 71); Am Soc Plant Physiol; Int Coun Asn Sci Educ (pres, 73-76); Nat Asn Res Sci Teaching (pres, 72-73); Nat Sci Teachers Asn; Sigma Xi; Nat Asn Biol Teachers. Res: Investigating medicinal and poisonous plants; improving science teaching techniques and equipment; studying science and math curriculum developments internationally; consulting in science education; science writing; studing use of computers in science instructions. Mailing Add: Dept Bot Univ Md College Park MD 20742-5815

LOCKARD, RAYMOND G, b Patricia, Alta, Jan 1, 25; m 51; c 3. HORTICULTURE. Educ: Univ BC, BSA, 49; Univ Idaho, MSc, 51; Univ London, PhD, 56. Prof Exp: Plant physiologist, Can For Aid, Malaysia, 54-59, Ghana, 59 -64; tech expert, Food & Agr Orgn, Philippines, 64-67; assoc prof hort, Univ Ky, 67-72, prof, 72-81; prof crop sci, & crop sci coordr, La State Univ, USAID Prog, Cent Agr Res Inst, Liberia, 81-84; horticulturist, Calif State Polytech Univ, Pomona, USAID Prog, Sanaa, Yemen Arab Repub, 85-88; RETIRED. Mem: Int Soc Hort Sci; Am Soc Hort Sci; Plant Growth Regulator Soc. Res: Chill requirements. Mailing Add: RD 8 Box 8204 Stroudsburg PA 18360-9267

LOCKART, ROYCE ZENO, JR, b Marshfield, Ore, Sept 7, 28; m 51; c 3. MOLECULAR BIOLOGY. Educ: Whitman Col, AB, 50; Univ Wash, MS, 53, PhD(microbiol), 57. Prof Exp: Res fel, Nat Inst Allergy & Infectious Dis, 57-58, bacteriologist, Radiation Br, Nat Cancer Inst, 58-60; from asst prof to assoc prof microbiol, Univ Tex, 60-66; res supvr, E I du Pont de Nemours & Co, Inc, 66-80, biologist, 80-83, patent assoc, 83-; RETIRED. Res: Virus cell interactions, particularly animal viruses and their control by natural means and by chemicals; image analysis of immunological cells; writing biotechnology patents. Mailing Add: 1418 Bucknell Rd Green Acres Wilmington DE 19803

LOCKE, BEN ZION, b New York, NY, Sept 8, 21; m 47; c 4. APPLIED STATISTICS, EPIDEMIOLOGY. Educ: Brooklyn Col, AB, 47; Columbia Univ, MS, 49. Prof Exp: Statistician, NY State Health Dept, 47-56; chief consult sect, Biomet Br, NIMH, 56-66; assoc prof eval & dir res & eval, Community Ment Health Ctr, Temple Univ, 66-67; from asst chief to chief, Ctr Epidemiol Studies, 67-85, CHIEF, EPIDEMIOL PSYCHOPATH BRANCH, NIMH, 85- Honors & Awards: Rema Lapouse Award, Am Pub Health Asn, 90. Mem: AAAS; fel Am Pub Health Asn; Am Statist Asn; Soc Epidemiol Res; fel Am Col Epidemiol. Res: Epidemiology of mental disorders; evaluation of programs designed to prevent and control mental disorders and promote mental health. Mailing Add: EPRB NIMH Rm 10C05 5600 Fishers Lane Rockville MD 20857

LOCKE, CARL EDWIN, JR, b Palo Pinto Co, Tex, Jan 11, 36; m 56; c 2. CORROSION, POLYMER SCIENCE. Educ: Univ Tex, Austin, BS, 58, MS, 60, PhD(chem eng), 72. Prof Exp: Res engr, Continental Oil Co, 59-65; prod engr, R L Stone Co, 65-66; prog res engr, Tracor Inc, 66-68; instr & vis asst prof, Univ Tex, 71-73; from asst to assoc prof, 73-80, prof & dir chem eng & mat sci, 80-86, DEAN ENG, UNIV KANS, 86- Concurrent Pos: Proj dir, & Okla Dept Transp, 75, 76-79. Honors & Awards: D Grant Mickle Award, Transp Res Bd, 88; Eben Junfin Award, Nat Asn Corrosion Engrs, 90. Mem: Am Inst Chem Engrs; Nat Asn Corrosion Engrs; Am Soc Testing & Mat; Am Soc Eng Educ. Res: Corrosion; corrosion in concrete; electrochemistry of corrosion. Mailing Add: 4010 Learned Hall Univ Kans Lawrence KS 66045

LOCKE, CHARLES STEPHEN, statistics, for more information see previous edition

LOCKE, DAVID CREIGHTON, b Garden City, NY, Mar 1, 39; m 62. CHEMICAL SEPARATIONS. *Educ:* Lafayette Col, BS, 61; Kans State Univ, PhD(chem), 65. *Prof Exp:* Res chemist, Esso Res & Eng Co, 65-67; NSF fel, Univ Col Swansea, Univ Wales, 67-68; from asst prof to assoc prof, 68-76, PROF CHEM, QUEENS COL, NY, 76- *Mem:* AAAS; Am Chem Soc; Int Inst Conserv Hist & Artistic Works; NY Acad Sci. *Res:* Analytical chemistry; chemical separations; GC/MS; supercritical fluids. *Mailing Add:* Dept Chem Queens Col Flushing NY 11367-0904

LOCKE, HAROLD OGDEN, b Camden, NJ, Sept 14, 31; m 59; c 2. PHYSICAL CHEMISTRY, ANALYTICAL CHEMISTRY. *Educ:* Wesleyan Univ, BA, 53, MA, 56; Rutgers Univ, PhD(chem), 62. *Prof Exp:* Res chemist, Armstrong Cork Co, 61-65; ANALYTICAL CHEMIST, GAF CORP, 65- *Mem:* Am Chem Soc. *Res:* X-ray crystallography; polymer characterization; surfactants. *Mailing Add:* 816 Prince St Palmer Township PA 18042-2435

LOCKE, JACK LAMBOURNE, b Brantford, Ont, May 1, 21; m 46; c 2. PHYSICS. *Educ:* Univ Toronto, BA, 46, MA, 47, PhD(physics), 49. *Prof Exp:* Demonstr physics, Univ Toronto, 45-47; astrophysicist, Dom Observ, 49-59, chief, Stellar Physics Div, 59-66; radio astronr, Nat Res Coun Can, 66-70, assoc dir, Radio & Elec Eng Div & chief astrophys br, 70-75, dir, Herzberg Inst Astrophysics, 75-85; RETIRED. *Concurrent Pos:* Officer-in-chg, Dom Radio Astrophys Observ, 59-62. *Honors & Awards:* Rumford Medal, AAAS, 71. *Mem:* Am Astron Soc; Can Astron Soc; Int Astron Union. *Res:* Astrophysics; radio astronomy; solar physics; molecular spectra; infrared spectrum of the atmosphere. *Mailing Add:* 2150 Braeside Ave Ottawa ON K1H 7J5 Can

LOCKE, JOHN LAUDERDALE, b Oak Park, Ill, Nov 16, 40; m 77. SPEECH PATHOLOGY, PSYCHOLINGUSITICS. *Educ:* Ripon Col, BA, 63; Ohio Univ, MA, 65, PhD(speech path), 68. *Prof Exp:* Speech pathologist, Vet Admin Ctr, 68-69; prof, Univ Ill, Champaign, 69-80; PROF, DEPT HEARING & SPEECH SCI, UNIV MD, COLLEGE PARK, 80- *Concurrent Pos:* Res fel, Yale Univ, 72-73 & Oxford Univ, 73-74; res consult, Vet Admin Hosp, Danville, Ill, 74-80 & Ft Howard, Md, 80- *Mem:* Am Speech & Hearing Asn; Ling Soc Am; Am Asn Appl Ling. *Res:* Language acquisition and impairment; child phonology; reading and cognition of the deaf; aphasia; speech perception. *Mailing Add:* Mass Gen Hosp River St Boston MA 02114

LOCKE, KRYSTYNA KOPACZYK, b Warsaw, Poland, Dec 2, 26; m 70. BIOCHEMISTRY, ENZYMOLOGY. *Educ:* Wayne State Univ, BS, 53; Western Reserve Univ, MS, 56; Univ Ill, Champaign-Urbana, PhD(lipid chem), 62. *Prof Exp:* Res nutritionist, Atherosclerosis Res Proj, Vet Admin Hosp, Downey, Ill, 56-58; Nat Inst Neurol Dis & Blindness fel biochem, Ment Health Res Inst, Univ Mich, Ann Arbor, 62-64; trainee, Inst Enzyme Res, Univ Wis-Madison, 64-66; proj assoc, 66-69; res biochemist, Biochem Toxicol Br, Div Toxicol, Food & Drug Admin, 69-77; TOXICOLOGIST, TOXICOL BR, HAZARD EVAL DIV, OFF PESTICIDE PROGS, ENVIRON PROTECTION AGENCY, 77- *Mem:* NY Acad Sci; Am Chem Soc. *Res:* Effects of environmental agents on biochemistry and ultrastructure of mitochondria. *Mailing Add:* 6220 Garden Rd Springfield VA 22152

LOCKE, LOUIS NOAH, b Stockton, Calif, Mar 14, 28; m 53; c 1. ANIMAL PATHOLOGY. *Educ:* Univ Calif, AB, 50, DVM, 56. *Prof Exp:* Vet, USPHS, 56-58; wildlife res biologist, Patuxent Wildlife Res Ctr, US Dept Interior, 58-60, histopathologist, 61-75; wildlife pathologist, Nat Wildlife Health Lab, US Fish & Wildlife Serv, 75-89; RETIRED. *Honors & Awards:* Distinguished Serv Award, Wildlife Dis Asn, 84; Merton Rosen Mem Lectr, 85. *Mem:* Wildlife Soc; Am Asn Avian Path; Am Vet Med Asn; Wildlife Disease Asn. *Res:* Wildlife diseases, especially diseases and parasites of the mourning doves, waterfowl; effects of pollutants upon wild birds; lead poisoning in migratory birds. *Mailing Add:* Nat Wildlife Health Res Ctr 6006 Schroeder Dr Madison WI 53711

LOCKE, MICHAEL, b Nottingham, Eng, Feb 14, 29; c 4. INSECT PHYSIOLOGY. *Educ:* Cambridge Univ, BA, 52, MA, 55, PhD, 56, DSc, 76. *Prof Exp:* Lectr zool, Univ WIndies, 56; assoc prof biol, Case Western Reserve Univ, 61-67; prof biol, 67-71, prof zool & chmn dept, 71-85, PROF, UNIV WESTERN ONT, 85- *Concurrent Pos:* Ed, Soc Develop Biol, 62-69; Raman prof, Univ Madras, 69; Killam res fel, 88. *Mem:* AAAS; Am Soc Cell Biol; fel Royal Soc Can. *Res:* Coordination of growth in insects; insect cell development; insect morphogenesis. *Mailing Add:* Dept Zool Univ Western Ont London ON N6A 5B7 Can

LOCKE, PHILIP M, b Rockford, Ill, July 12, 37; m 61; c 2. MATHEMATICS. *Educ:* Bluffton Col, BS, 59; Univ NH, MS, 64, PhD(math), 67. *Prof Exp:* Asst prof math, Mont State Univ, 67-68; asst prof, 68-74, ASSOC PROF MATH, UNIV MAINE, ORONO, 74- *Mem:* Math Asn Am. *Res:* Ordinary differential equations. *Mailing Add:* Dept of Math Univ of Maine Orono ME 04469

LOCKE, RAYMOND KENNETH, b Terre Haute, Ind, July 2, 40; m 70. BIOCHEMISTRY, TOXICOLOGY. *Educ:* Wash Univ, BS, 65. *Prof Exp:* Res asst biochem, Univ Tex, Dallas, 66-67; res chemist, Div Nutrit, Food & Drug Admin, 68-69, Biochem & Metab Sect, Div Pesticides, 69-71 & Metab Br, Div Toxicol, 71-73, res chemist, Biochem Toxicol Br, 73-77, chemist, Div Toxicol, Contaminants & Natural Toxicants Eval Br, 77-79; TOXICOLOGIST, OFF TOXIC SUBSTANCES, US ENVIRON PROTECTION AGENCY, 79- *Mem:* AAAS; Am Chem Soc; NY Acad Sci. *Res:* Biochemical studies of the comparative in vivo and in vitro metabolism of foreign compounds by animals, plants and man. *Mailing Add:* 6220 Garden Rd Springfield VA 22152

LOCKE, STANLEY, b New York City, NY, June 18, 34; m 58; c 3. MATHEMATICS, PHYSICS. *Educ:* NY Univ, BME, 55, MS, 57, PhD(math), 60. *Prof Exp:* Res mathematician, Repub Aviation Corp, 59-60; mem scientific staff, Schlumberger-Doll Res Ctr, 65-86; consult, Stanley Locke & Assoc, 87-88; CONSULT ENGR, TELECO OILFIELD SERV, 88- *Mem:* Sigma Xi; sr mem Inst Elec & Electronic Engrs. *Res:* Solution of mechanical, electro-magnetic, acoustic and nuclear problems arising in the development of new oil field services. *Mailing Add:* 17 Deerwood Ct Norwalk CT 06851

LOCKE, STEVEN ELLIOT, b Englewood, NJ, Dec 2, 45; m 84; c 2. BEHAVIORAL MEDICINE, PSYCHONEUROIMMUNOLOGY. *Educ:* Cornell Univ, AB, 68; Columbia Univ Col Physicians & Surgeons, MD, 72. *Prof Exp:* Fel, psychiat, 74-77, clin instr, 77-79, instr, 80-87, ASST PROF PSYCHIAT, HARVARD MED SCH, 87- *Concurrent Pos:* Intern, Mt Zion Hosp & Med Ctr, San Francisco, Calif, 72-73; resident, McLean Hosp, Belmont, MA, 74-77; fel, Boston Univ Sch Med, 77-79; assoc dir, Psychiat Consult Serv, Beth Israel Hosp, 80-87, dir, Computers in Psychiat, 87-, med student educ in psychiat, 89-; lectr, Beth Israel Hosp, 80-82. *Honors & Awards:* First Prize, Scientific Exhibit, Am Col Emergency Physicians. *Mem:* Biofeedback Soc Am; Am Col Emergency Physicians; fel Am Psychiat Asn; Am Psychosom Soc; Soc Behav Med. *Res:* Interactive computing in psychiatry and medicine; psychophysiology of human emotion; psychoneuroimmunology; behavioral medicine aspects of AIDS and cancer; co-author of five books, 21 papers and book chapters. *Mailing Add:* 27 Camden Rd Auburndale MA 02166

LOCKE, WILLIAM, b Morden, Man, Mar 16, 16; m 45. INTERNAL MEDICINE, ENDOCRINOLOGY. *Educ:* Univ Man, MD, 38; Univ Minn, MS, 47; McGill Univ, DTM, 45. *Prof Exp:* Pres staff, 54-55, partner, Ochsner Clin, 57-81, mem staff, Ochsner Clin & Found Hosp, 50-86, emer head, Sect Endocrinol & Metab, 76-86, TRUSTEE, ALTON OCHSNER MED FOUND, 78- *Concurrent Pos:* Nat Res Coun & Commonwealth Fund fel, Harvard Univ, 48-50; emer clin prof med, Sch Med, Tulane Univ, 87-; sr vis physician, Charity Hosp, New Orleans. *Mem:* AAAS; Am Diabetes Asn; fel Am Col Physicians; Endocrine Soc; Sigma Xi. *Res:* Metabolic diseases. *Mailing Add:* Ochsner Clin 1514 Jefferson Hwy New Orleans LA 70121

LOCKER, JOHN L, b Florence, Ala, Oct 11, 30; m 59; c 8. MATHEMATICS. *Educ:* Auburn Univ, PhD(math), 60. *Prof Exp:* Instr, Auburn Univ, 56-59; mathematician, Redstone Arsenal, 54; assoc prof, 60-70, PROF MATH, UNIV OF NORTHERN ALA, 70- *Concurrent Pos:* Lectr, NSF-Ala Acad Sci Vis Sci Prog, 61-65. *Mem:* Math Asn Am; Nat Coun Teachers Math. *Res:* Statistics; geometry. *Mailing Add:* Univ Northern Ala Box 5051 Florence AL 35630

LOCKETT, CLODOVIA, b Austin, Tex, Jan 23, 13. PHYSIOLOGY, ANIMAL. *Educ:* St Louis Univ, BS, 37, PhD, 52; De Paul Univ, MS, 47. *Prof Exp:* Asst prof biol, LeClerc Col, 47-49; from assoc prof to prof & dir dept, Notre Dame Col, 49-65; prof, 65-86, chmn dept, 68-84, EMER PROF BIOL, UNIV DALLAS, 86- *Concurrent Pos:* Fel, Univ Okla, 61. *Mem:* AAAS. *Res:* Effects of drugs on iodine metabolism in the thyroid; the exchange and growth potential of phosphorus in algae cultures. *Mailing Add:* Dept Biol Univ Dallas Univ Dallas Sta Irving TX 75062-4799

LOCKEY, RICHARD FUNK, b Lancaster, Pa, Jan 15, 40; m; c 2. ALLERGY & IMMUNOLOGY, INTERNAL MEDICINE. *Educ:* Haverford Col, BS, 61; Univ Mich, Ann Arbor, MS, 72; Temple Univ, MD, 65. *Prof Exp:* Asst resident internal med, Univ Hosp, Univ Mich, 66-67, resident, 66-68 & fel allergy & immunol, 69-70; from asst prof to assoc prof med, 73-83, asst dir, Div Allergy & Immunol, 79-82, PROF MED, PEDIAT & PUB HEALTH, COL MED, UNIV SFLA, TAMPA, 83-, DIR, DIV ALLERGY & IMMUNOL, 82- *Concurrent Pos:* Chief, Allergy & Immunol Sect, Carswell AFB Hosp, USAF, Ft Worth, Tex, 71-73 & James A Haley Vet Admin Hosp, Tampa, Fla, 82-; asst chief, Allergy & Immunol, Vet Admin Hosp, Tampa, 73-82. *Mem:* Fel Am Acad Allergy & Immunol (pres-elect, 91-92); fel Am Col Physicians; fel Am Col Chest Physicians; AAAS; Int Asn Aerobiol; Asn Otolaryngol Adminrs. *Res:* Hymenoptera hypersensitivity; imported fire ant; red tide toxin and its mechanism of action on tracheal smooth muscle; important aeroallergens of Florida; immunopharmacology and AIDS; immunity and the aged; asthma. *Mailing Add:* Vet Admin Hosp Div Allergy & Immunol 13000 Bruce B Downs Blvd Tampa FL 33612

LOCKHART, BENHAM EDWARD, b St Vincent, WI, Jan 18, 45; US citizen; m 70; c 2. PLANT PATHOLOGY, AGRICULTURE. *Educ:* Univ WI, Trinidad, BSc, 65; Univ Calif, Riverside, PhD(plant path), 69. *Prof Exp:* Res fels plant path, Univ Nebr, 69-70 & Univ Calif, Berkeley, 70-71; asst prof, Minn Proj, US AID, Rabat, Morocco, 71-76; asst prof, 76-80, ASSOC PROF PLANT PATH, UNIV MINN, ST PAUL, 80- *Mem:* Am Phytopathological Soc; Am Soc Hort Sci; Int Soc Hort Sci. *Res:* Identification properties and control of viruses of vegetable and ornamental crops. *Mailing Add:* 2208 Hendon Ave St Paul MN 55108

LOCKHART, BROOKS JAVINS, b Sandyville, WVa, Feb 8, 20; m 40, 69; c 3. MATHEMATICS. *Educ:* Marshall Col, AB, 37; WVa Univ, MS, 40; Univ Ill, PhD(math), 43. *Prof Exp:* Asst instr math, Univ Ill, 40-43; instr math, Univ Mich, 43-44 & 46-48; from asst prof to prof, 48-79, dean, 62-77, EMER PROF MATH, NAVAL POSTGRAD SCH, 79- *Mem:* Math Asn Am. *Res:* Classical algebraic geometry; numerical analysis; computer programming. *Mailing Add:* Tryst 26105 Carmelo St Carmel CA 93923

LOCKHART, F(RANK) J(ONES), b Austin, Tex, Aug 10, 16; m 45; c 2. CHEMICAL ENGINEERING. *Educ:* Univ Tex, BS, 36, MS, 38; Univ Mich, PhD(chem eng), 43. *Prof Exp:* Jr engr, Humble Oil & Refining Co, Tex, 38-39; from asst to instr chem eng, Univ Mich, 40-43; process engr, Union Oil Co Calif, 43-46; from asst prof to assoc prof chem eng, Southern Calif, 46-52, head dept, 50-52; mgr prod eng dept, Fluor Corp, Ltd, 52-55; head dept, 55-68, PROF CHEM ENG, UNIV SOUTHERN CALIF, 55- *Concurrent Pos:*

Consult, 46-; process engr, Fluor Cor, Ltd, 46. *Mem:* Am Chem Soc; Am Inst Chem Engrs. *Res:* Fluid dynamics; petroleum refining operations; distillation; gas absorption and stripping; design and operation of water cooling towers; liquid-liquid extraction; effect of time and concentration on overall mass transfer coefficients. *Mailing Add:* 648 33rd St Manhattan Beach CA 90266-4698

LOCKHART, HAINES BOOTS, b Crawfordsville, Ind, Oct 29, 20; m 44; c 2. NUTRITION, BIOCHEMISTRY. *Educ:* Wabash Col, AB, 42; Univ Ill, PhD(biochem), 45. *Prof Exp:* Asst, Wabash Col, 41-42; chemist, Univ Ill, 42-45; from res chemist to head baby foods res div, Swift & Co, 45-66, head new foods div, 66-71; sect mgr nutrit res, Quaker Oats Co, 71-80, staff nutritionist, John Stuart Res Labs, 80-85. *Concurrent Pos:* Mem tech adv group, Comt on Nutrit, Am Acad Pediat. *Mem:* Am Chem Soc; Inst Food Technol. *Res:* Amino acids and proteins in nutrition; infant and geriatric nutrition. *Mailing Add:* 333 Sunset Dr Lakewood Crystal Lake IL 60014-5330

LOCKHART, HAINES BOOTS, JR, b Evergreen Park, Ill, Feb 4, 46; m 68; c 2. ENVIRONMENTAL CHEMISTRY, BIOCHEMISTRY. *Educ:* Wabash Col, AB, 67; Univ Nebr, Lincoln, MS, 69, PhD(chem), 73. *Prof Exp:* Sr res biochemist, 72-81, tech assoc, Health & Safety Lab, 81-85, mgr, Health Regulations, 85-86, dir, Environmental Tech Serv, 86-88, DIR, OCCUP HEALTH LAB, EASTMAN KODAK CO, 88- *Mem:* Am Chem Soc; Soc Environ Toxicol & Chem. *Res:* Environmental impact of synthetic chemicals, their biodegradation, photodegradation and bioconcentration in aquatic organisms; risk assessment. *Mailing Add:* Health & Environ Lab Bldg 320 Eastman Kodak Co Kodak Park Rochester NY 14652

LOCKHART, JAMES ARTHUR, plant physiology, theoretical ecology; deceased, see previous edition for last biography

LOCKHART, JAMES MARCUS, b Portsmouth, Ohio, June 11, 48; m 81; c 2. LOW TEMPERATURE PHYSICS, SUPERCONDUCTING ELECTRONICS. *Educ:* Univ Mich, BS, 70; Stanford Univ, MS, 72, PhD(physics), 76. *Prof Exp:* Res affil, Stanford Univ, 76, actg instr physics, 76-77, actg asst prof, 77-80; asst prof physics, Colo Sch Mines, 80-81; sr res assoc, Stanford Univ, 82-83; assoc prof, 83-87, PROF, PHYSICS, SAN FRANCISCO STATE UNIV, 87- *Concurrent Pos:* Fel, Dept Physics, Stanford Univ, 76-78; vis scholar, Stanford Univ, 83- *Mem:* Am Phys Soc; Am Asn Physics Teachers; Sigma Xi. *Res:* Low temperature electrical properties of solids; electron beams; superconductivity, superconducting detectors and electronics, Squids ultra-low; semiconductor device physics; musical acoustics; architectural acoustics; electroacoustics. *Mailing Add:* Physics & Astron Dept San Francisco State Univ 1600 Holloway Ave San Francisco CA 94132

LOCKHART, LILLIAN HOFFMAN, b Columbus, Tex, Oct, 23, 30; m 51; c 3. MEDICINE. *Educ:* Rice Univ, BA, 51; Univ Tex Med Br, Galveston, MA, 55, MD, 57. *Prof Exp:* From asst prof to assoc prof, 63-83, PROF PEDIAT & GENETICS, UNIV TEX MED BR, GALVESTON, 83- *Concurrent Pos:* Fel hemat, Univ Tex Med Br, Galveston, 62-63. *Mem:* Am Acad Pediat. *Res:* Genetics; chromosome disorders. *Mailing Add:* 9 Perthuis Farm Rd La Marque TX 77568

LOCKHART, WILLIAM LAFAYETTE, b Nashville, Tenn, Oct 15, 36; m 60; c 2. INORGANIC CHEMISTRY. *Educ:* Tenn Technol Univ, BS, 58; Univ Miss, MS, 61; Vanderbilt Univ, PhD(inorg chem), 67. *Prof Exp:* Res biochemist, US Food & Drug Admin, 60-63; from asst prof to assoc prof 67-77, chmn dept, 78-82, PROF CHEM, WEST GA COL, 77 - *Res:* Kinetics and mechanisms of inorganic reactions. *Mailing Add:* Dept of Chem W Ga Col Carrollton GA 30118

LOCKHART, WILLIAM RAYMOND, b Carlisle, Ind, Nov 25, 25; m 47; c 4. BACTERIOLOGY. *Educ:* Ind State Teachers Col, AB, 49; Purdue Univ, MS, 51, PhD(bact), 54. *Prof Exp:* Asst bact, Purdue Univ, 50-51; from asst prof to prof bact, Iowa State Univ, 54-90, chmn dept, 60-74; RETIRED. *Mem:* Am Soc Microbiol; fel Am Acad Microbiol; Brit Soc Gen Microbiol. *Res:* Physiology of bacterial growth; numerical taxonomy. *Mailing Add:* 2606 Northwood Dr Ames IA 50010

LOCKRIDGE, OKSANA MASLIVEC, b Czech, Sept 4, 41; US citizen. BIOCHEMICAL PHARMACOLOGY, BIOCHEMICAL GENETICS. *Educ:* Smith Col, BA, 63; Northwestern Univ, Ill, PhD(chem), 71. *Prof Exp:* Fel human genetics, 72-74, res assoc pharmacol, 74-81, RES SCIENTIST, UNIV MICH, ANN ARBOR, 82- *Concurrent Pos:* Prin Investr, Univ Mich, Ann Arbor, 79-88. *Mem:* Am Chem Soc; Asn for Women in Sci; AAAS; Am Soc Biol Chemists. *Res:* Pharmacogenetics; biochemical and structural studies on genetic variants of human pseudocholinesterase. *Mailing Add:* Univ Nebr Med Ctr Eppley Inst 600 S 42nd St Omaha NE 68198-6805

LOCKSHIN, MICHAEL DAN, b Columbus, Ohio, Dec 9, 37; m 65; c 1. RHEUMATOLOGY, IMMUNOLOGY. *Educ:* Harvard Col, AB, 59; Harvard Med Sch, MD, 63. *Prof Exp:* Intern, Second (Cornell) Div Med, Bellevue & Mem Hosp, New York, 63-64; epidemic intel serv officer, Epidemic Intel Serv, Commun Dis Ctr, 64-66; resident, Second (Cornell) Div Med, Bellevue & Mem Hosp, New York, 66-68; fel rheumatol, Columbia Presby Med Ctr, 68-70; from asst prof to prof med, Col Med, Cornell Univ, 82-89; DIR EXTRAMURAL PROG, NAT INST ARTHRITIS & MUSCULOSKELETAL & SKIN DIS, NIH, 89- *Concurrent Pos:* Adj asst prof epidemiol, Sch Pub Health, Univ Pittsburgh, 65-66; assoc scientist & assoc attend physician, Hosp Spec Surg, New York, 70-82, attending physician, 82-; consult rheumatol, Mem Hosp, New York, 70-; mem bd dirs, Arthritis Found, 75-89. *Mem:* Am Col Rheumatism; Am Col Physicians. *Res:* Cellular immunology; clinical rheumatology. *Mailing Add:* Nat Inst Arthritis & Musculoskeletal & Skin Dis NIH Bldg 31-4C32 Bethesda MD 20892

LOCKSHIN, RICHARD ANSEL, b Columbus, Ohio, Dec 9, 37; m 63; c 2. PHYSIOLOGY, DEVELOPMENTAL BIOLOGY. *Educ:* Harvard Univ, AB, 59, AM, 61, PhD(biol), 63. *Prof Exp:* Asst prof physiol, Sch Med, Univ Rochester, 65-75; from assoc prof to prof physiol, 75-83, CHMN DEPT PHYSIOL, ST JOHN'S UNIV, NY, 83- *Concurrent Pos:* NSF fel, Inst Animal Genetics, Univ Edinburgh, 63-64, NIH fel, 64-65. *Mem:* AAAS; Soc Cell Biol; Soc Develop Biol; Gerontol Soc; Am Soc Entom; fel Gerontol Soc Am. *Res:* Destruction of tissues during metamorphosis of insects; early developmental events in insect embryogenesis; cellular differentiation. *Mailing Add:* Dept Biol Sci St John's Univ Grand Central/Utopia Pkwy Jamaica NY 11439

LOCKWOOD, ARTHUR H, b Jan 26, 47; US citizen. MOLECULAR BIOLOGY, CELL BIOLOGY. *Educ:* Carleton Col, BA, 68; Albert Einstein Col Med, PhD(molecular biol), 72. *Prof Exp:* Res fel molecular biol, Dept Biochem Sci, Princeton Univ, 73-74; res scientist cell biol, Sch Med, NY Univ, 75-80; prof anat & cell biol, Med Sch, Univ NC, 80-; AT EINSTEIN MED CTR PEDIAT. *Concurrent Pos:* Arthritis Found fel, 74. *Mem:* AAAS; Am Soc Cell Biol; NY Acad Sci; Am Chem Soc. *Res:* Biology of cytoplasmic microtubules; biochemistry of mitosis, cell senescence. *Mailing Add:* Dept Pediat Albert Einstein Med Ctr Kerman Bldg York & Tabor Rd Philadelphia PA 19141

LOCKWOOD, DAVID JOHN, b Christchurch, NZ, Jan 7, 42; m 79; c 2. SOLID STATE PHYSICS, RAMAN SPECTROSCOPY. *Educ:* Univ Canterbury, BSc, 64, MSc, 66, PhD(physics), 69; Univ Edinburgh, DSc, 78. *Prof Exp:* Teaching fel physics, Univ Canterbury, NZ, 65-69; fel chem, Univ Waterloo, Can, 70-71; res fel physics, Univ Edinburgh, 72-78; sr res officer physics, 78-, SECT HEAD, SURFACE & INTERFACE PHYSICS, NAT RES COUN CAN, 87- *Concurrent Pos:* Univ bursaries, NZ Univ Grants Comt, 66-68; consult vis, Battelle Ctr de Res, Switz, 72-76; tutor, Open Univ, UK, 77-78; consult vis, Univ Paul Sabatier, Toulouse, France, 77-; vis prof, Essex Univ, England, 81-83; prog consult & reviewer, Nat Sci & Eng Res Coun Can, 86-; distinguished vis prof, Ctr Sci Res France, 87-88; mem sci panel, NATO, 87-88, chmn, 89-90; mem Can adv group, NATO Sci Comt, 90- *Mem:* Am Phys Soc; Royal Commonwealth Soc. *Res:* Light scattering studies of structural and magnetic phase transitions, electronic excitations, magnons and phonons in solids; optical properties of semiconductors and superlattices. *Mailing Add:* Microstruct Sci Nat Res Coun Ottawa ON K1A 0R6 Can

LOCKWOOD, DEAN H, b Milford, Conn, June 17, 37; m 58; c 3. BIOCHEMISTRY, MEDICINE. *Educ:* Wesleyan Univ, AB, 59; Johns Hopkins Univ, MD, 63. *Prof Exp:* Intern & resident, Johns Hopkins Univ, 63-65, fel pharmacol, 67-69, asst & assoc prof med, 69-76; staff assoc, NIH, 65-67; PROF MED, UNIV ROCHESTER, 76- *Concurrent Pos:* NIH res develop award, 69-74 & res grant, 69-; mem endocrine merit rev bd, Vet Admin, 76-79; mem, NIH Metab Study Sect, 81- *Mem:* Am Soc Biol Chemists; Am Soc Clin Invest; Endocrine Soc; Am Fedn Clin Res; Am Diabetes Asn. *Res:* Mechanism of action of insulin and glucagon in normal and resistant states; plasma membrane receptors and biological responses emphasized. *Mailing Add:* Dept Med Univ Rochester Endocrine Metab Unit Sch Med Box 693 601 Elmwood Ave Strong Hosp Rochester NY 14642

LOCKWOOD, FRANCES ELLEN, b Passaic, NJ; m 78; c 1. LUBRICANT OXIDATION, TRIBOLOGICAL BEHAVIOR OF LIQUID CRYSTALS. *Educ:* Rensselaer Polytech Inst, BS, 73, Pa State Univ, MS, 76, PhD(chem eng), 78. *Prof Exp:* Assoc sr res engr, Gen Motors Res Labs, 78-80; sr scientist, Martin Marietta Lab, 80-84; dir, 84-85, VPRES PHYS SCI, PENNZOIL PROD CO, 85- *Mem:* Soc Tribologists & Lubrication Engrs; Am Inst Chem Engrs; Soc Automotive Engrs; Am Soc Testing & Mat. *Res:* Lubricant oxidation, tribological behavior of liquid crystals and lubrication of ceramic dry pressing; metal rolling and metal forging lubrication; the fluid state and hydrocarbon oxidation. *Mailing Add:* Pennzoil Co Pennzoil Prod Technol 1520 Lake Front Circle PO Box 7569 The Woodlands TX 77387

LOCKWOOD, GEORGE WESLEY, b Norfolk, Va, June 28, 41; m 65. ASTRONOMY, PLANETARY SCIENCES. *Educ:* Duke Univ, BS, 63; Univ Va, MA, 65, PhD(astron), 68. *Prof Exp:* Astronr, Kitt Peak Nat Observ, 68-73; ASTRONR, LOWELL OBSERV, 73- *Mem:* Am Astron Soc; Int Astron Union; Astron Soc Pac; Am Geophys Union; Sigma Xi. *Res:* Planetary atmospheres; stellar/solar physics; solar-planetary relations; variable stars. *Mailing Add:* Lowell Observ Mars Hill Rd 1400 W Flagstaff AZ 86001

LOCKWOOD, GRANT JOHN, b Byram, Conn, Oct 28, 31; m 56; c 4. EXPERIMENTAL ATOMIC PHYSICS, ELECTRON PHYSICS. *Educ:* Univ Conn, BA, 54, MS, 59, PhD(physics), 63. *Prof Exp:* Res asst physics, Univ Conn, 60-63; STAFF MEM, SANDIA LABS, 63- *Mem:* Am Phys Soc. *Res:* Electronic, atomic and molecular interactions to include ion-atom, ion-molecule, atom-atom and atom-molecule; interaction with surface and solids of ion beams. *Mailing Add:* 7913 Hendrix Ave NE Albuquerque NM 87110

LOCKWOOD, JEFFREY ALAN, b Manchester, Conn, Mar 9, 60; m 82. ECOLOGY, ETHOLOGY. *Educ:* NMex Tech, BS, 82; La State Univ, PhD(entom), 85. *Prof Exp:* Postdoctoral res entom, La State Univ, 85-86; asst prof, 86-90, ASSOC PROF ENTOM, UNIV WYO, 90- *Honors & Awards:* R T Gast Award, Entom Soc Am, 85. *Mem:* Sigma Xi; Entom Soc Am; Orthopterists Soc. *Res:* Insect ecology; behavioral and population ecology of rangeland grasshoppers; studies of chemical ecology and the evolution of insecticide resistance. *Mailing Add:* Dept Plant Soil & Insect Sci Univ Wyo Laramie WY 82071

LOCKWOOD, JOHN ALEXANDER, b Easton, Pa, July 12, 19; m 42; c 3. PHYSICS. *Educ:* Dartmouth Col, AB, 41, Lafayette Col, MS, 43; Yale Univ, PhD(physics), 48. *Prof Exp:* From asst to instr physics, Lafayette Col, 41-44; tech supvr, Tenn Eastman Corp, 44-45; asst physics, Yale Univ, 45-46, asst instr, 46-47, asst, 47-48; from asst prof to assoc prof, 48-58, 74-80, PROF PHYSICS, 58- *Concurrent Pos:* Assoc dir res, Univ NH, 74-80, dir res, 80-82.

Honors & Awards: Sci Award, NASA. *Mem:* AAAS; Am Phys Soc; Am Asn Physics Teachers. *Res:* Development of linear electron accelerators; cosmic ray; nuclear physics; gamma ray astronomy. *Mailing Add:* Space Sci Ctr Inst Earth Oceans & Space Univ NH Durham NH 03824

LOCKWOOD, JOHN LEBARON, b Ann Arbor, Mich, May 28, 24; m 59; c 2. PLANT PATHOLOGY. *Educ:* Mich State Col, BA, 48, MS, 50; Univ Wis, PhD(plant path), 53. *Prof Exp:* Asst prof bot & plant path, Ohio Agr Exp Sta, 53-55; from asst prof to prof, 55-90, EMER PROF BOT & PLANT PATH, MICH STATE UNIV, 90- *Concurrent Pos:* NSF sr fel, Cambridge Univ, 70-71. *Mem:* Fel Am Phytopathological Soc (pres, 84-85); Sigma Xi; Can Phytopathological Soc; Phytopathological Soc Japan. *Res:* Ecology of root-infecting fungi; soybean diseases. *Mailing Add:* Dept Bot & Plant Path Mich State Univ E Lansing MI 48824

LOCKWOOD, JOHN PAUL, b Bridgeport, Conn, Oct 26, 39; m 63; c 2. GEOLOGY. *Educ:* Univ Calif, Riverside, AB, 61; Princeton Univ, PhD(geol), 66. *Prof Exp:* GEOLOGIST, US GEOL SURV, 66- *Concurrent Pos:* Partic, Nat Res Coun-Nat Acad Sci Sci Exchange Prog with USSR, res at Geol Inst Acad Sci, Moscow, 66, res at Dir Volcanol, Bandung, Indonesia, 80-82. *Mem:* Geol Soc Am. *Res:* Petrology, mineralogy and structural features of serpentinites; general geology of the Sierra Nevada Mountains; circum-Pacific distribution of volcanic rocks; Caribbean geology; volcanic hazards; eruptive history and structure of Mauna Loa volcano, Hawaii; volcanic disaster assessments in Indonesia, Italy, Colombia, Cameroon, Northern Marianas Islands, Rwanda, Zaire. *Mailing Add:* US Geol Surv Hawaiian Volcano Observ Hawaii National Park HI 96718

LOCKWOOD, LINDA GAIL, b New York, NY, May 25, 36. ENVIRONMENTAL BIOLOGY, SCIENCE EDUCATION. *Educ:* Columbia Univ, BS, 60, MA, 61 & 65, PhD(bot), 69. *Prof Exp:* Asst prof bot & ecol, Teachers Col, Columbia Univ, 69-73; ASSOC PROF ENVIRON SCI, UNIV MASS, AMHERST, 73- *Concurrent Pos:* Jessie Smith Noyes Found grant environ sci educ, Teachers Col, Columbia Univ, 71-73; prof plant & soil sci & Sch Educ, Univ Mass, Amherst, 73-; co-dir, US Off Educ grant, 74-75; Univ Mass fac res grant, 74-75 & Water Resources Res Ctr grant, 74-75; NSF grants, 75-79. *Mem:* Sigma Xi; AAAS; Scientist's Inst Pub Info; Nat Asn Biol Teachers; Audubon Soc. *Res:* Influence of photoperiod and exogenous nitrogen-containing compounds on the reproductive cycles of the liverwort Cephalozia media Lindb; experimental morphology and physiological ecology; environmental biology, especially physiological ecology, aquatic systems; environmental science education, especially teacher training, history and philosophy of science. *Mailing Add:* Honors Machmert 23 Univ of Mass Amherst MA 01003

LOCKWOOD, ROBERT GREENING, b Faribault, Minn, Jan 12, 28; m 53; c 2. NATURAL & SYNTHETIC ELASTOMERS, SPECIAL ORGANIC COATINGS. *Educ:* Carleton Col, BA, 49; Univ Minn, PhD(org chem), 53. *Prof Exp:* Lab instr inorg & org chem, Univ Minn, 49-53; res chemist, New Prod Develop Lab, Chem Div, Gen Elec Co, 53-54; sr chemist, 54-65, res specialist, 65-81, SR RES SPECIALIST, 3M CO, 81- *Mem:* Am Chem Soc; Sigma Xi. *Res:* Organic synthesis; carboxylic acids and derivatives; condensation polymers; manufacture of alkylated aromatic hydrocarbons and polycarboxylic acids; pressure-sensitive adhesives; release agents. *Mailing Add:* 2 Hingham Circle St Paul MN 55118-1921

LOCKWOOD, WILLIAM RUTLEDGE, b Memphis, Tenn, Apr 10, 29; c 2. PHYSICS, GENERAL OPTICS. *Educ:* Univ Miss, BA, 49, MA, 50; Univ Tenn, Memphis, MD, 57. *Prof Exp:* Intern, Charity Hosp La, New Orleans, 57-58; resident med, Med Ctr, Univ Miss, 59-61, from instr to asst prof, 62-70, asst prof microbiol & path, 66-70, assoc prof med, 70-91; RETIRED. *Concurrent Pos:* USPHS fel, Med Ctr, Univ Miss, 61-64, grant, 64-67; vis instr, Wash Univ, 64; asst dean res & assoc chief staff res, Vet Admin Ctr, 69-73; attend physician, Univ Miss Hosp, 64- *Mem:* Fel Infectious Dis Soc Am; fel Am Col Chest Physicians; Am Soc Trop Med & Hyg; Am Soc Microbiol; fel Am Col Physicians. *Res:* Pathogenesis of acute inflammation; pharmacology of antimicrobial agents. *Mailing Add:* 1200 Meadow Brook Rd Apt 38 Jackson MS 39206

LOCOCK, ROBERT A, b Toronto, Ont, Aug 14, 35; m 61. PHARMACEUTICAL CHEMISTRY, PHARMACOGNOSY. *Educ:* Univ Toronto, BSc, 59, MSc, 61; Ohio State Univ, PhD(pharm), 65. *Prof Exp:* Lectr pharmaceut chem, Univ BC, 61-62; asst pharm, Ohio State Univ, 64-65; asst prof, 65-70, ASSOC PROF PHARM, UNIV ALTA, 70-, ASSOC PROF PHARMACEUT SCI, 74- *Mem:* AAAS; Am Chem Soc; Am Soc Pharmacog; Sigma Xi. *Res:* Chemistry of natural products; phytochemistry; chemotaxonomy; alkaloids and terpenoids. *Mailing Add:* Dept Pharm Univ Alta Edmonton AB T6G 2N8 Can

LOCY, ROBERT DONALD, b Defiance, Ohio, Jan 12, 47; m 69; c 3. PLANT TISSUE CULTURE, PLANT CELL BIOLOGY. *Educ:* Defiance Col, AB, 69; Purdue Univ, PhD(plant biochem), 74. *Prof Exp:* Res assoc biochem, McMaster Univ, 74-76; res fel, Dept Environ Res Lab, Mich State Univ, 76-78; asst prof hort, NC State Univ, 78-82; prof agr, Ind-Purdue, Ft Wayne, 82-83; sr res scientist, NPI, Salt Lake City, Utah, 83-88, mgr, Floral Prod Res, 88-90. *Mem:* Am Soc Plant Physiol; Tissue Culture Asn Am; Sigma Xi; Gamma Sigma Delta. *Res:* Plant biochemistry as applied to crop plant improvement using tissue and cell culture; plant propagation in vitro using automated tissue culture systems. *Mailing Add:* 10908 S Savannah Dr Sandy UT 84094

LODA, RICHARD THOMAS, b Derby, Conn, May 19, 48. SPECTROSCOPY. *Educ:* Waterburg State Tech Col, AAS, 68; Univ Bridgeport, BA, 71; Wesleyan Univ, PhD(phys chem), 80. *Prof Exp:* NIH fel, Chem Dept, Univ Ore, 80-81; CHEMIST RES DEPT, INSTRUMENTAL CHEM ANALYSIS BR, NAVAL WEAPONS CTR, 81-; AT NAVAL RES LAB. *Mem:* Am Chem Soc; Am Phys Soc. *Res:* Application of lasers and spectroscopy to problems of physical and chemical interest; photochemistry; site selection and linewidth phenomena in condensed phase systems; coherent antistokes Raman scattering. *Mailing Add:* Inst for Def Analysesssci & Tech Dir 1801 N Beauregard St Alexandria VA 22311-1772

LODATO, MICHAEL W, b Rochester, NY, June 17, 32; m 59; c 4. OPERATIONS RESEARCH. *Educ:* Colgate Univ, AB, 54; Univ Rochester, MS, 59; Rutgers Univ, PhD(math), 62. *Prof Exp:* Scientist, LFE Monterey Lab, 62-63; mem tech staff, Appl Math Dept, Mitre Corp, Mass, 63-65, head opers anal sub dept, 65-66; sr exec adv, Douglas Aircraft Corp, 66-67, mgr, Info Technol Dept, McDonnell Douglas Corp, 67-68; pres, Macro Systs Assocs, Inc, 68-70; prin bus planner, Xerox Data Systs, 70-71; vpres indust systs, Informatics, Inc, 71-78; exec vpres, Spectrum Int, Inc, 78-80; PRES, MWL INC, 80- *Mem:* Opers Res Soc Am; Asn Data Process Serv Orgn. *Res:* Topology; planning, scheduling and resource allocation; orbital mechanics; production and inventory control; strategic management; author of two books on computer sales and strategic management. *Mailing Add:* 32038 Watergate Ct Westlake Village CA 91361

LODEN, MICHAEL SIMPSON, b Fayette, Ala, Mar 30, 45; m 68; c 1. OLIGOCHAETA, WATER QUALITY. *Educ:* Auburn Univ, BS, 67, MS, 73; La State Univ, PhD(zool), 78. *Prof Exp:* Aquatic biologist, Aquatic Control, Inc, 73-75; asst prof zool, La State Univ, 78-81; environ dir, Jefferson Parish, La, 81-90; DIR ENVIRON RESOURCES, GULF ENG & CONSULTS, INC, 90- *Mem:* Sigma Xi; AAAS; Soc Syst Zool; NAm Benthol Soc; Am Micros Soc. *Res:* Systematics, life histories, ecology and distribution of aquatic Oligochaeta. *Mailing Add:* 917 Trudeau Dr Metairie LA 70003

LODER, EDWIN ROBERT, b Irvington, NJ, Feb 24, 25; m 45; c 4. ANALYTICAL CHEMISTRY. *Educ:* Syracuse Univ, BA, 52; Mass Inst Technol, PhD, 55. *Prof Exp:* Asst chem, Mass Inst Technol, 52-53, asst org microanal, 53-55; chemist, Eastman Kodak Co, 55-59, chief analytical chemist, Maumee Chem, 59-62, dir res serv, 62-65, sect mgr & tech assoc, Gen Aniline & Film Co, NY, 65-66; from dep dir res to dir res, Du Bois Chem Div, W R Grace & Co, 66-70, dir res & vpres, Du Bois Chem Div, Chemed Corp, 70-72, sr vpres corp affairs, 72-73, exec vpres, 73-74, group exec vpres, Du Bois Chem Div, Chemed Corp, 74-85; PRES, DELRAY CHEM, 85- *Concurrent Pos:* Instr, Univ Toledo, 61-62. *Mem:* Fel AAAS; fel Am Inst Chem; Am Chem Soc; Soc Photog Sci & Eng; Am Soc Qual Control; Sigma Xi. *Res:* Electrochemistry; spectroscopy; research management; statistics. *Mailing Add:* 1620 NW 22nd Ave Delray Beach FL 33445

LODEWIJK, ERIC, b Hague, Neth, Nov 15, 40. SYNTHETIC ORGANIC CHEMISTRY. *Educ:* Univ Amsterdam, BSc, 65, PhD(org chem), 68. *Prof Exp:* Res chemist org chem, Syntex Chem, Inc, Bahamas, 69-70, group leader process develop, Chem Div, 70-73, sr res chemist, 73-77, group leader res org chem, 77-79, mgr process res, 79-86, dir res & develop, 86-91, GROUP DIR, PROCESS RES & DEVELOP, SYNTEX CORP, BOULDER, COLO, 91- *Mem:* Am Chem Soc. *Res:* Process development and process research on fine organic chemicals and drugs; synthesis of fluorocorticosteroids and IG steroids; antiflammatory analgesics, beta blockers. *Mailing Add:* Syntex Corp 2075 N 55 St Boulder CO 80301

LODGE, ARTHUR SCOTT, b Liverpool, Eng, Nov 20, 22; m 45; c 3. PHYSICS. *Educ:* Oxford Univ, BA, 45, MA, 48, DPhil(physics), 49. *Prof Exp:* Jr sci officer pile design, Atomic Energy, Montreal Anglo-Can Proj, 45-46 & rheology, Brit Rayon Res Asn, 49-60; lectr math, Inst Sci & Technol, Univ Manchester, 61-63, sr lectr rheology, 63-68; PROF RHEOLOGY, UNIV WIS-MADISON, 68-, CHMN DEPT, 69-, PROF ENG MECH, 68- *Concurrent Pos:* Vis prof chem eng, Univ Wis-Madison, 65-66; vpres, Bannatek Co, Inc. *Honors & Awards:* Bingham Medal, Soc Rheology, 71; Gold Medal, Brit Soc Rheology, 83. *Mem:* Soc Rheology; Brit Soc Rheology; fel Brit Inst Physics & Phys Soc. *Res:* Rheological properties of concentrated polymer solutions; molecular theories of their constitutive equations and stress/optical properties; experimental methods. *Mailing Add:* Dept Eng Univ Wis Madison Mech 2348 Eng Bldg 1415 Johnson Dr Madison WI 53706

LODGE, CHESTER RAY, b McCausland, Iowa, Feb 19, 23; wid. ELECTRICAL ENGINEERING. *Educ:* Iowa State Univ, BS, 43, MS, 49, PhD(eng), 52. *Prof Exp:* Dean eng, Univ Peshawar, Pakistan, 52-53; asst prof, Iowa State Univ, 54; assoc prof, 54-58, PROF ENG, SAN DIEGO STATE UNIV, 58- *Concurrent Pos:* Fulbright lectr, Pakistan, 52-53. *Mem:* Am Soc Eng Educ; Inst Elec & Electronics Engrs. *Res:* Automatic control systems; symmetrical components and applications. *Mailing Add:* Dept Eng San Diego State Univ San Diego CA 92115

LODGE, DAVID MICHAEL, b Athens, Tenn, Apr 1, 57; m 85; c 2. BENTHOS, HERBIVORY. *Educ:* Univ South, BS, 79; Oxford Univ, DPhil(zool), 82. *Prof Exp:* Asst scientist, Ctr Limnol, Univ Wis, 82-85; ASST PROF INVERT ZOOL & ECOL, DEPT BIOL SCI, UNIV NOTRE DAME, 85- *Concurrent Pos:* Exec comn, Am Midland Naturalist, 89-; assoc ed, J NAm Benthological Soc, 90- *Mem:* Am Sci Affil; Am Soc Limnol & Oceanog; Ecol Soc Am; NAm Benthological Soc; Am Inst Biol Sci; Int Soc Limnol. *Res:* Determining the relative importance of biotic and abiotic factors in determining the distribution and abundance of freshwater benthic organisms; predation and herbivory. *Mailing Add:* Dept Biol Sci Univ Notre Dame Notre Dame IN 46556

LODGE, JAMES PIATT, JR, b Decatur, Ill, Feb 4, 26; m 48; c 5. ATMOSPHERIC CHEMISTRY, AIR POLLUTION. *Educ:* Univ Ill, BS, 47; Univ Rochester, PhD(chem), 51. *Prof Exp:* Asst prof chem, Keuka Col, 50-52; chemist, Cloud Physics Lab, Univ Chicago, 52-55; chief chem res & develop sect, Robert A Taft Sanit Eng Ctr, USPHS, Cincinnati, Ohio, 55-61; prog scientist, Nat Ctr Atmospheric Res, 61-74; CONSULT ATMOSPHERIC CHEM, 74- *Concurrent Pos:* Consult, Cook Res Labs, 51-53,; affil prof, La State Univ, 66-69; mem, State Air Pollution Variance Bd, Colo, 66-70; chmn, State Air Pollution Control Comn, Colo, 70-76. *Honors & Awards:* Frank A Chambers Award, Air Pollution Control Asn, 74. *Mem:* Fel AAAS; Am Chem Soc; Am Geophys Union; Am Meteorol Soc; Air & Waste Mgt Asn. *Res:* Air pollution and atmospheric chemistry; microchemical analysis; cloud physics; atmospheric electricity. *Mailing Add:* 801 Circle Dr Boulder CO 80302

LODGE, JAMES ROBERT, b Downey, Iowa, July 1, 25; m 47; c 2. REPRODUCTIVE PHYSIOLOGY. *Educ:* Iowa State Univ, BS, 52, MS, 54; Mich State Univ, PhD(dairy), 57. *Prof Exp:* Asst dairy, Mich State Univ, 54-57; res assoc dairy sci, 57-60, from asst prof to assoc prof, 60-69, PROF PHYSIOL, UNIV ILL, URBANA, 69- *Concurrent Pos:* Res fel, Nat Inst Child Health & Human Develop, 69-70. *Mem:* AAAS; Am Physiol Soc; Soc Study Reproduction; Am Soc Animal Sci; Am Dairy Sci Asn. *Res:* Physiology of reproduction and endocrinology. *Mailing Add:* 328 Mumford Hall Univ Ill 1301 W Gregory Dr Urbana IL 61801

LODGE, MALCOLM A, b Borden, PEI, Can, Mar 16, 39; c 2. COMPUTER SCIENCE, ENGINEERING. *Educ:* NS Tech Col, BEng, 62, MScEng, 69. *Prof Exp:* Sonar engr, Can Forces Dockyard, Halifax, 62-63; prod design engr, Can Westinghouse Co, Ltd, 64-69; prof elec eng, Holland Col, 69-77; res engr, Resource Ventures, Inc, 77-86; PRES, ISLAND TECHNOL, INC, 86- *Concurrent Pos:* Nat Res Coun fel, 64; tech progs officer, Inst Man & Resources, 77-86. *Mem:* Can Wind Energy Asn; Am Wind Energy Asn; Asn Prof Engrs. *Res:* Wind turbine systems; process x-ray applications; biomedical computer applications; energy conservation and supply systems. *Mailing Add:* 201 Water Charlottetown PE C1A 1B1 Can

LODGE, TIMOTHY PATRICK, b Sale, Eng, Apr, 11, 54; m 88. POLYMER SOLUTION DYNAMICS. *Educ:* Harvard Univ, BA, 75; Univ Wis, PhD(anal chem), 80. *Prof Exp:* Nat Res Coun assoc fel, Nat Bur Standards, 81-82; asst prof, 82-88, ASSOC PROF CHEM, UNIV MINN, 88- *Mem:* Am Chem Soc; Am Phys Soc; Soc Rheology; Sigma Xi. *Res:* Conformation and dynamics of macromolecules in solution studies by means of oscillatory flow birefringence, quasi-elastic light scattering, and small angle neutron scattering. *Mailing Add:* Dept Chem Kolthoff & Smith Halls Univ Minn Minneapolis MN 55455

LODHI, MOHAMMAD ARFIN KHAN, b Agra, India, Sept 17, 33; m 65; c 3. NUCLEAR PHYSICS, RENEWABLE SOURCES OF ENERGY. *Educ:* Univ Karachi, BSc, Hons, 52, MSc, 56; Univ London, DIC, 60, PhD(nuclear physics), 63. *Prof Exp:* Lectr math, S M Col, Karachi, 52-59; from asst prof to assoc prof, 63-73, PROF PHYSICS, TEX TECH UNIV, 73- *Concurrent Pos:* Vis res scholar, Bohr Inst Theoret Physics, Copenhagen, Denmark, 62; vis asst prof, Univ Fla, Gainesville, 67, Univ Wyo, Laramie, 68; res assoc, State Univ NY Buffalo, 69; vis assoc prof, Univ Wash, 69-70 & Univ Calif, San Diego, 72; guest scientist, Pinstech, Pakistan, 73 & 76; UN expert & consult, var univs in Pakistan, 81 & 83; consult, Nat Inst Oceanog, Karachi, Pakistan, 84 & 85; vis prof, Bahauddin Zakariya Univ, Pakistan, 84-85; NSF coordr, deleg Int Nathiagali Summer Col, Pakistan; vis prof, Ctr Excellence Anal Chem, Sind Pakistan, 86; guest scientist, Ctr Solar Energy Hyderabad Sin Pakistan, 86-87; vis prof, Middle East Tech Univ, Ankara, Turkey, 87 & Univ Pertanian, Malaysia, 89-90; consult & expert, UN Nat Inst Oceanog, Karachi, Pakistan, 88. *Mem:* Brit Inst Physics; Pakistan Math Soc; fel Phys Soc, UK; Sigma Xi; Am Phys Soc; Am Asn Univ Professors; Am Astronaut Soc; Pakistan Inst Physics; Solar Energy Soc Pakistan. *Res:* High energy electron scattering by nuclei and electromagnetics transitions in nuclei and their role in elucidating nuclear structure; nuclear shell, cluster and resonating group models and their relationship; nuclear nonlocal potential and nuclear systematics; short-range nucleon-nucleon correlations; solar-hydrogen system collection, transduction; extraction of energy from renewable sources, including ocean currents and tides. *Mailing Add:* Dept Physics Tex Tech Univ Lubbock TX 79409

LODISH, HARVEY FRANKLIN, b Cleveland, Ohio, Nov 16, 41; m 63; c 3. BIOCHEMISTRY, MICROBIOLOGY. *Educ:* Kenyon Col, AB, 62; Rockefeller Univ, PhD(biol), 66. *Hon Degrees:* DSc, Kenyon Col, 82. *Prof Exp:* Am Cancer Soc fel biol, Med Res Coun Lab Molecular Biol, Eng, 66-68; from asst prof to assoc prof, 68-76, PROF BIOL, MASS INST TECHNOL, 76-; MEM, WHITEHEAD INST, 82- *Concurrent Pos:* Res career develop award, Nat Inst Gen Med Sci, 71-75; mem, panel develop biol, NSF, 72-; chmn, Gorden Conf on Animal Cells, 76, Red Blood Cells, 85 & Membrane Molecular Biol, 89; Guggenheim fel, 77-78; mem, Whitehead Inst Biomed Res, 82- *Honors & Awards:* Stadie Award, Am Diabetes Asn. *Mem:* Nat Acad Sci; Am Soc Microbiol; Am Chem Soc; Am Soc Biol Chemists; Am Soc Cell Biol; AAAS. *Res:* Structure and function of membrane transport proteins; mechanism and regulation of biosynthesis of membrane proteins; erythropoetin receptor; biosynthesis of cellular and viral membrane proteins; diabetes. *Mailing Add:* Whitehead Inst Biomed Res 9 Cambridge Ctr Cambridge MA 02142

LODMELL, DONALD LOUIS, b Polson, Mont, Aug 27, 39; m 63. VIROLOGY. *Educ:* Northwestern Univ, BA, 61; Univ Mont, MS, 63, PhD(microbiol), 67. *Prof Exp:* Scientist virol, Rocky Mountain Lab, NIH, 67-71; res assoc, Lab Oral Med, NIH, 71-72; sr scientist, 72-81, SCI DIR VIROL, ROCKY MOUNTAIN LAB, NIH, 81- *Concurrent Pos:* fac affil, Dept Microbiol, Univ Mont, 78- *Mem:* Am Soc Microbiol; Am Asn Immunologists; Am Soc Virol. *Res:* Immunological mechanisms of host defense against viral infections of the central nervous system; rabies. *Mailing Add:* Rocky Mountain Lab Hamilton MT 59840

LODOEN, GARY ARTHUR, b Camp Rucker, Ala, May 3, 43. POLYMER CHEMISTRY. *Educ:* Univ NDak, BS, 65; Cornell Univ, PhD(org chem), 69. *Prof Exp:* Fel, Univ Iowa, 69-70; res chemist, 70-73, sr res chemist, 73, res & develop supvr, 73-75, process supvr, 75-77, sr res chemist, 77-82, RES ASSOC, TEXTILE FIBERS DEPT, E I DU PONT DE NEMOURS & CO, INC, 83- *Mem:* Am Chem Soc. *Res:* Spandex chemistry and structure; polyester glycol synthesis and properties; development of new and novel raw materials for spandex yarns. *Mailing Add:* Two Hickory Hill Lane Fisherville VA 22939

LODWICK, GWILYM SAVAGE, b Mystic, Iowa, Aug 30, 17; m 47, 70; c 4. RADIOLOGY, BIOENGINEERING. *Educ:* Univ Iowa, BA, 42, MD, 43. *Prof Exp:* Clin asst prof radiol, Univ Iowa, 52-55, assoc prof, Col Med, 55-56; actg dean, Sch Med, Univ Mo, Columbia, 59, assoc dean, 59-64, prof radiol & chmn dept, 56-78, interim chmn radiol, 80-81, chmn, Dept Radiol, 81-83, res prof, 78-83, EMER PROF, DEPT RADIOL, UNIV MO, COLUMBIA, 83-; RADIOLOGIST, MASS GEN HOSP, 83-; RADIOLOGIST-IN-CHIEF, SPAULDING REHAB HOSP, 86- *Concurrent Pos:* Fel, Armed Forces Inst Path, 51; Nat Inst Gen Med Sci spec fel, 67-68; chief radiol serv, Vet Admin Hosp, Iowa City, 52-55; consult, Ellis Fischel State Cancer Hosp, 59-; mem radiol training comt, Nat Inst Gen Med Sci, 66-70; consult, Jet Propulsion Lab, Calif Inst Technol, 69-73; mem, Comt Radiol, Nat Acad Sci; mem, Comt Radiol, Div Med Sci, Nat Res Coun, 70-75; Sigma Xi res Award, Univ Mo-Columbia, 72; dir, Mid-Am Bone Diag Ctr & Registry; vis prof, Sch Med, Keio Univ, Tokyo, 74 & Univ Turku, Finland, 79; mem, Radiation Study Sect, Div Res Grants, NIH, 76-79, Study Sect Diag Radiol & Nuclear Med, 79-82, chmn, 80-82; vis prof, Harvard Med Sch, 83-; chmn bd sci coun, Nat Libr Med, 87-89; founding fel, Am Col Med Info, 84; distinguished practr, Nat Acad Pract Med, 84. *Honors & Awards:* Founder's Gold Medal, Int Skeletal Soc, 90; Gold Medal, XIII Int Conf Radiol, Madrid, 73; Sakari Mustakallio Medal, Finland, 79. *Mem:* Sr mem Inst Med Nat Acad Sci; AAAS; AMA; Radiol Soc NAm (3rd vpres, 74-75); fel Am Col Radiol. *Res:* Diagnostic radiology; diagnosis and prognosis of bone disease, computer-aided medical diagnosis; automated image analysis and pattern recognition; information systems. *Mailing Add:* One Devonshire Pl Apt 2713 Boston MA 02109

LOE, HARALD, b Steinkjer, Norway, July 19, 26; US citizen; c 2. PERIODONTICS. *Educ:* Thirteen from US & foreign univs, 73-90. *Hon Degrees:* DSc, Univ Gothenburg, Sweden, 73, Cath Univ Leuven, Belg, Univ Athens, Greece & Royal Dental Col, Aarhus, Denmark, 80; Univ Lund, Malmo & Georgetown Univ Sch Dent, 83, Univ Bergen, 85, Univ Md, 86 & Univ Med & Dent NJ, 87. *Prof Exp:* Instr oper dent, Sch Dent, Oslo, 52-55; res assoc, Norweg Inst Dent Res, 56-62; Fulbright res fel & res assoc oral path, Univ Ill, 57-58; univ res fel, Oslo Univ, 59-62; assoc prof periodont, Sch Dent, Oslo Univ, 60-61; vis prof, Hebrew Univ, Jerusalem, 66-67; prof & chmn periodont, Royal Dent Col, Aarhus, Denmark, 62-72, assoc dean & dean elect, 71-72; prof & dir, Dent Res Inst, Univ Mich, 72-74; dean & prof periodont, Sch Dent Med, Univ Conn, 74-82; DIR, NAT INST DENT RES, NIH, BETHESDA, MD, 83- *Concurrent Pos:* Assoc ed, Scand J Dent Res Munksgaard, Copenhagen, Denmark, 62, bd dir, Scand Odontological Act, 65-72, ed, J Periodont Res, 65-85; foreign expert mem, J Indian Dent Asn, 74-82; vis prof periodont, Hebrew Univ, Jerusalem, 66-67; rapporteur, Conf Undergrad Dent Educ Europe, World Health Orgn, 68; mem & chmn, comt postgrad educ, Oslo Dent Soc, 59-61; chmn, Scand Symp Periodont, Aarus, Denmark, 65; secy, World Workshop Periodont, Path Sect, Ann Arbor, 66; consult, comt foreign rel, Am Asn Periodontologists, 67-71, Coun Dent Therapeut, Am Dent Asn, 77-82, Periodont Dis Clin Res Ctr, State Univ NY, Buffalo, 80-81, Coun Int Rel, Am Dent Asn, 83-88, J Am Dent Asn, 79-82, Procter & Gamble Co, 69-76, to the dir, Nat Inst Dent Res, 69-, Naval Dent Sch, Naval Med Command, 85-88, appraisals comt, Ont Coun Grad Studies & Lord Robens Appeal, Brit Soc Dent Res, 85, Med Res Coun, 87; co-chmn, Scan Symp Prosthetics & Periodont, Aarhos, Denmark & Int Conf Periodont Res, Rochester, NY, 69 & chmn, Aarhus Br, 72; chmn & mem, basic res periodont dis award comt, Int Asn Dent Res, 69-74, pres, Periodont Res Group, 70-71; mem, review panel, grants & allocations adv comt, Am Fund Dent Educ, 73-75, comt long range planning, Am Asn Periodontologist, 74-75, comt outreach activ, Univ Conn, dent sch res grant comt & periodont res prog comt, Int Asn Dent Res, 76, organizing comt, Int Conf Periodont Res, Gothenburg, Sweden, & organizing comt, World Workshop Periodont, Chicago, 76-78, nominating comt, Am Asn Dent Res, 76-79, bd dirs, Int Asn Dent Res, 77-82, finance comt, 78-80, gen prog comt, 78-80, int rel comt, 79-82, annual meeting orginizing comt, 79-81, nominating comt, 81-82, hon mem comt, 81-84, comt res periodont, Am Asn Periodontologist, 74-79, comt Orban Prize competition, 76, foreign rel comt, 78- & chmn, 81-82, hom mem comt, 79-82, comt mission & goals, 84-87 & search comt ed J Periodont, 88; electorate nominating comt, Sect Dent, AAAS, 82, 85 & chmn, 85-87; secy, Jury Int Prev Dent Award, Int Dent Fed, 76; mem, prof adv panel prev dent, Johnson & Johnson, 80-81 & other various organizations nationally & internationally. *Honors & Awards:* Peridont Award, William J Gies Found, 78; Int Asn Dent Res Award, 69; Ingv Stokke Prize, 65; Erik Berg Found Prize, 69; Aalborg Dent Soc Prize, 69; Am Soc Prev Dent Int Award, 72; Arthur Merritt Mem lectr, Baylor Univ, 77; William J Gies Award, Am Acad Periodont, 78; Lister Hill Mem lectr, Univ Ala, Birmingham, 83; Goldstein lectr, Emory Univ, 86; Exemplary Serv Award, Surgeon Gen, 88; Swed Dent Soc Int Prize, 88; Award Distinction, Acad Dent Int, 90. *Mem:* Inst Med-Nat Acad Sci; Int Asn Dent Res (pres, 79-81); fel Int Col Dentists; fel Am Col Dentists; fel AAAS; corresp mem Swed Dent Asn; corresp mem Finnish Dent Soc; hon mem Scand Soc Periodont; hon mem Belg Periodont Soc; hon mem Norweg Dent Asn; Int Dent Fed; Int Col Dentists; Sigma Xi. *Res:* Epidemiology, experimental pathology and prevention of peridontal disease; author or coauthor of over 200 scientific articles; author of over 275 publications. *Mailing Add:* Nat Inst Dent Res NIH Bldg 31 9000 Rockville Pike Bethesda MD 20205

LOE, ROBERT WAYNE, plant physiology, for more information see previous edition

LOEB, ALEX LEWIS, b Ithaca, NY, Aug 4, 55. CARDIOVASCULAR PHARMACOLOGY, ENDOTHELIUM DEPENDENT RESPONSES. *Educ:* Beloit Col, BS, 77; George Washington Univ, MS, 81, PhD(pharmacol), 84. *Prof Exp:* Chemist, Nat Bur Standards, 77; consult, Life Sci, Inc, 82-83; teaching fel pharmacol, 84-87, RES ASST PROF, UNIV VA, 87- *Mem:* Am Chem Soc; AAAS; Am Heart Asn. *Res:* Implications of increased DNA synthesis in aorta of hypertensive animals; characterization and properties of endothelium-derived relaxing factor from intact vessels and in cultured cells; interactions between endothelium, smooth muscle and platelets. *Mailing Add:* Ctr Resin Anesthesia Dept Anesthesia Dulles 7 Univ Pa 3400 Spruce St Philadelphia PA 19104-4283

LOEB, ARTHUR LEE, b Amsterdam, Neth, July 13, 23; nat US; m 56. CHEMICAL PHYSICS, DESIGN SCIENCE. *Educ:* Univ Pa, BSCh, 43; Harvard Univ, AM, 45, PhD(chem physics), 49. *Prof Exp:* Mem staff, Bur Study Coun, Harvard Univ, 45-49; mem staff, Div Indust Coop & Lincoln Lab, Mass Inst Technol, 49-58, lectr, 56-58, from asst prof to assoc prof elec eng, 58-63; staff scientist, Ledgemont Lab, Kennecott Copper Corp, 63-73; SR LECTR VISUAL & ENVIRON STUDIES, CUR, CARPENTER CTR & MASTER, DUDLEY HOUSE, HARVARD UNIV, 70- *Concurrent Pos:* Actg head, Dept Chem, Barlaeus Gym, Neth, 46; consult, Mass Inst Technol, 49-, Godfrey Lowell Cabot, Inc, 58 & IBM Corp, NY, 59; mem guest res staff, Univ Utrecht, 54-55. *Honors & Awards:* Golden Door Award, Int Inst Boston; Residency Rockefeller Found Villa Serbelloni Ctr, Scholars & Artists. *Mem:* Acad Mgt; Am Soc Eng Educ; fel Royal Soc Arts; Am Crystallog Asn; fel Am Inst Chem. *Res:* Mathematical crystallography; educational technology; design science; communication of two-dimensional and three-dimensional concepts and patterns. *Mailing Add:* Dept of Visual & Environ Studies Harvard Univ Cambridge MA 02138

LOEB, GERALD ELI, b New Brunswick, NJ, June 26, 48; m 68; c 1. NEUROPHYSIOLOGY, BIOMEDICAL ENGINEERING. *Educ:* Johns Hopkins Univ, BA, 69, MD, 72. *Prof Exp:* Resident surg, Univ Ariz, 72-73; MED OFFICER NEUROPHYSIOL, NAT INST NEUROL & COMMUN DIS & STROKE, 73- *Concurrent Pos:* Fel, Seeing Eye, 69-72; guest res assoc, Artificial Eye Proj, Univ Utah, 71 & adj assoc prof bioengineering, 85-; vis scientist, Univ Calif, San Francisco, 79-83; pres, Biomed Concepts Inc, 81- *Honors & Awards:* Commendation Medal, USPHS. *Mem:* Soc Neuroscience; Biomed Eng Soc; Am Physiol Soc; AAAS. *Res:* Sensorimotor neurophysiology in mammals, chronic unit recording techniques, neural prostheses, information processing by nerve networks. *Mailing Add:* Biomed Eng Queen's Univ Abramsky Hall Kingston ON K7L 3N6 Can

LOEB, JEROD M, b Brooklyn, NY, Oct 21, 49. CARDIOVASCULAR PHYSIOLOGY. *Educ:* City Univ New York, BS, 71; State Univ NY, PhD(physiol), 76. *Prof Exp:* Teaching asst physiol, State Univ NY Downstate Med Ctr, 72-76; res fel med, Harvard Med Sch, 76-77; res assoc physiol, Stritch Sch Med, Loyola Univ, Chicago, 77-79; from asst prof to assoc prof surg & physiol, Med Sch, Northwestern Univ, 79-87; dir, Div Basic Sci, 87-88, Div Biomed Sci, 88-91, ASST PRES SCI & TECHNOL & SECY COUN SCI AFFAIRS, AMA, 91- *Concurrent Pos:* Mem Coun Basic Sci, Am Heart Asn; prin investr, Am Heart Asn Grant-in-Aid, 80-82; career develop award, Schweppe Found, Chicago, Ill, 80-83; prin investr, NIH Grant, 82-; adj prof physiol, Northwestern Univ, 87- *Mem:* Am Physiol Soc; Sigma Xi; NY Acad Sci; Soc Exp Biol & Med; Am Heart Asn. *Res:* Electrophysiologic analysis of normal and abnormal cardiac pacemaker activity; autonomic control of cardiac pacemakers; electrophysiologic mapping of human cardiac arrhythmias. *Mailing Add:* AMA 515 N State St Chicago IL 60610

LOEB, LAWRENCE ARTHUR, b Poughkeepsie, NY, Dec 25, 36; m 58; c 3. CANCER, BIOCHEMISTRY. *Educ:* City Col New York, BS, 57; NY Univ, MD, 61; Univ Calif, Berkeley, PhD(biochem), 67. *Prof Exp:* Intern, Med Ctr, Stanford Univ, 61-62; res assoc biochem, Nat Cancer Inst, 62-64; res assoc zool, Univ Calif, Berkeley, 64-67; asst mem biochem, Inst Cancer Res, 67-69, assoc mem 71-77, mem, 77-78; PROF DEPT PATH, SCH MED, UNIV WASH, 78-, DIR, GOTTSTEIN MEM CANCER RES LABS, 78-, ADJ PROF DEPT BIOCHEM, 78- *Concurrent Pos:* Res grants, Am Cancer Soc, 67-69, Stanley C Dordick Found, 67, NIH & NSF, 69-75; assoc prof, Dept Path, Sch Med & mem biol & molecular biol grad groups, Univ Pa, 67-68. *Mem:* Am Asn Cancer Res (pres, 88-89); Fedn Am Socs Exp Biol; Am Soc Cell Biol; fel Am Col Physicians. *Res:* Fidelity of DNA replication; environmental carcinogenesis; human leukemia; mechanism of catalysis by DNA polymerases; zinc metalloenzymes; lymphocyte transformation. *Mailing Add:* Dept Path SM-30 Univ Wash D-525 H5B Seattle WA 98195

LOEB, LEOPOLD, physical chemistry, for more information see previous edition

LOEB, MARCIA JOAN, b New York, NY, Mar 26, 33; m 53; c 2. INVERTEBRATE PHYSIOLOGY. *Educ:* Brooklyn Col, BA, 53; Cornell Univ, MS, 57; Univ Md, PhD(physiol), 70. *Prof Exp:* Nat Res Coun res assoc physiol & endocrinol of coelenterate develop, Naval Res Lab, 70-72; instr biol & physiol, Northern Va Community Col, 73; res assoc marine biol, Marine Sci Lab, Univ Col N Wales, 74; prof lectr physiol, Am Univ, 75-77; PHYSIOLOGIST, USDA, 77- *Mem:* Am Soc Zoologists; Entom Soc Am; Int Soc Invert Reproduction; Sigma Xi; Tissue Culture Asn. *Res:* Environmental, physiological and endocrine control of strobilation in the Chesapeake Bay sea nettle, Chrysaora quinquecirrha; associated physiological phenomena in Chrysaora quinquecirrha; physiology of settlement in some marine bryzoan larvae; endocrinology and physiology of spermatogenesis in lepidoptera; hormones associated with the testis of lepidoptera. *Mailing Add:* 6920 Fairfax Rd Bethesda MD 20014

LOEB, MARILYN ROSENTHAL, b New York, NY, Feb 26, 30; m 49; c 3. BIOCHEMISTRY OF BACTERIAL CELL SURFACE. *Educ:* Barnard Col, Columbia Univ, BA, 51; Bryn Mawr Col, MA, 55; Univ Pa, PhD(biochem), 58. *Prof Exp:* Res assoc biochem, Univ Pa, 58-59; res assoc, Med Col Pa, 65-68; res assoc biochem, Inst Cancer Res, 68-75; asst res prof microbiol, Med Sch, George Washington Univ, 75-77; prog assoc cell biol prog, NSF, 77-78; from asst prof to assoc prof pediat, 78-89, SCIENTIST, MED SCH, UNIV ROCHESTER, 89- *Mem:* AAAS; Am Soc Microbiol; Sigma Xi. *Res:* Role of outer membrane components in pathogenesis of gram negative bacteria. *Mailing Add:* Dept Pediat Univ Rochester Med Ctr Rochester NY 14642

LOEB, PETER ALBERT, b Berkeley, Calif, July 3, 37; m 58; c 3. MATHEMATICS. *Educ:* Harvey Mudd Col, BS, 59, Princeton Univ, MA, 61; Stanford Univ, PhD(math), 65. *Prof Exp:* Asst prof math, Univ Calif, Los Angeles, 64-68; from asst prof to assoc prof, 68-75, PROF MATH, UNIV ILL, URBANA, 75- *Concurrent Pos:* Grant, Ctr Advan Studies, 71. *Mem:* Am Math Soc. *Res:* Topology; potential theory; non-standard analysis. *Mailing Add:* Dept Math Univ Ill 1409 W Green St Urbana IL 61801

LOEB, VIRGIL, JR, b St Louis, Mo, Sept 21, 21; m 50; c 4. ONCOLOGY, HEMATOLOGY. *Educ:* Wash Univ, MD, 44. *Prof Exp:* From instr to asst prof med, 51-56, asst prof path, 55-78, from asst prof to assoc prof, 56-78, PROF CLIN MED, SCH MED, WASHINGTON UNIV, 78- *Concurrent Pos:* Nat Cancer Inst trainee, Sch Med, Wash Univ, 49-50, Damon Runyan res fel hemat, 50-52; dir, Cent Diag Labs, Barnes Hosp, St Louis, 52-68; NIH grant prin investr, Southeastern Cancer Study Group, 56-77; chmn, cancer clin invest res comt, Nat Cancer Inst, 66-69, consult, 66-, mem, Polycythemia Vera Study Group & Diag Res Adv Group; mem, Oncol Rev Group Vet Admin; mem bd sci counr, Div Cancer Prev & Control, Nat Cancer Inst, 83-; Am Joint Comt Cancer, 82-91. *Mem:* Int Med-Nat Acad Sci; fel Am Col Physicians; Am Asn Cancer Res; Am Soc Clin Oncol; Am Soc Hemat; Am Cancer Soc (pres, 86-87). *Res:* Medical oncology. *Mailing Add:* 6103 Queeny Tower One Barnes Hosp Plaza St Louis MO 63110

LOEBBAKA, DAVID S, b Gary, Ind, Aug 18, 39; m 79; c 3. X-RAY IMAGING IN FLUIDS. *Educ:* Calif Inst Technol, BS, 61; Univ Md, PhD(physics), 67. *Prof Exp:* Assoc res scientist, Univ Notre Dame, 66-68; asst prof high energy physics, Vanderbilt Univ, 68-72; assoc prof, 72-77, PROF PHYSICS, UNIV TENN, 77- *Concurrent Pos:* Fluids Div, Nat Bur Standards, 82-85. *Mem:* Sigma Xi; Am Asn Physics Teachers. *Res:* X-ray imaging in fluid flow. *Mailing Add:* Dept of Geosci & Physics Univ of Tenn Martin TN 38238

LOEBENSTEIN, WILLIAM VAILLE, b Providence, RI, Aug 9, 14; m 49; c 4. PHYSICAL CHEMISTRY, DENTAL RESEARCH. *Educ:* Brown Univ, ScB, 35, ScM, 36, PhD(phys chem), 40. *Prof Exp:* Lab glassblower, Eastern Sci Co, 38-40; res & control chemist, Corning Glass Works, RI, 41; res assoc phys chem, 46-51, chemist, Nat Bur Standards, 51-82; CHEM CONSULT, 82- *Concurrent Pos:* Mem comt com stand, Commodity Stand Div, Off Tech Servs. *Honors & Awards:* Meritorious Serv Medal, US Dept Com, 58. *Mem:* Am Chem Soc; Sigma Xi; Int Asn Dent Res. *Res:* Catalysis; glass technology; physical and chemical adsorption of gases on solid surfaces; kinetics of adsorption from solutions; simplified techniques for the improvement in precision of surface area determinations from adsorption measurements; ortho-para conversion of liquid hydrogen; surface chemistry of teeth and dental materials. *Mailing Add:* 8501 Sundale Dr Silver Spring MD 20910

LOEBER, ADOLPH PAUL, b Detroit, Mich, Feb 1, 20; m 45; c 4. PHYSICS. *Educ:* Wayne State Univ, BS, 41, MA, 49; Univ Chicago, cert, 43; Mich State Univ, PhD(physics), 54. *Prof Exp:* Technician, Metall Dept, Chrysler Corp, 41-42, engr, Truck Dept, 45-47; instr physics, Wayne State Univ, 47-50; asst, Mich State Univ, 50-54; proj engr, Chrysler Corp, 54-55, from asst mgr to mgr physics res, 55-61, mgr phys optics res, Missile Div, 61-64; assoc prof physics, Eastern Mich Univ, 64-66; mgr electrooptics, Missile Div, Chrysler Corp, 66-68; assoc prof, 68-71, PROF PHYSICS, EASTERN MICH UNIV, 71- *Concurrent Pos:* Consult, Missile Div, Chrysler Corp, 64-66. *Mem:* Acoust Soc Am; Indust Math Soc (secy, 57-58); Am Asn Physics Teachers; Optical Soc Am. *Res:* Ultrasonics; physical optics; polarized light. *Mailing Add:* 4035 Antique Lane Bloomfield MI 48013

LOEBER, JOHN FREDERICK, b White Plains, NY, Oct 5, 42; m 62; c 7. MECHANICAL ENGINEERING, ENGINEERING MECHANICS. *Educ:* Lehigh Univ, BS, 64, MS, 65, PhD(appl mech), 68; George Washington Univ, MEA, 86. *Prof Exp:* Engr, 67-71, lead engr, 71-73, mgr methods develop, 75-81, mgr AFC reactor equip design, 81-84, RESIDENT ENGR, DEVELOP APPARATUS REP, KNOLLS ATOMIC POWER LAB, GEN ELEC CO, 84- *Concurrent Pos:* NASA fel, Lehigh Univ, 64-67. *Mem:* Am Soc Mech Engrs. *Res:* Finite element methods of structural analysis including computer program development and graphics; theoretical fracture mechanics. *Mailing Add:* Knolls Atomic Power Lab Bldg D-1 Rm 102 PO Box 1072 Schenectady NY 12301

LOEBL, ERNEST MOSHE, b Vienna, Austria, July 30, 23; nat US; m 50; c 2. CHEMICAL PHYSICS, QUANTUM CHEMISTRY. *Educ:* Hebrew Univ, MSc, 46; Columbia Univ, PhD(chem), 52. *Prof Exp:* Res chemist, Olamith Cement Co, 47; asst chemist, Columbia Univ, 48-50; instr, Rutgers Univ, 50-51; from instr to assoc prof, Polytech Univ, NY, 52-63, head div, 65-73, dep dept head chem, 88-90, prof phys chem, 63-90; RETIRED. *Concurrent Pos:* NSF fel, 63-64; lectr, Esso Res; vis prof, Uppsala Univ, Sweden, 63, Oxford Univ, Eng, 64, Sheffield Univ, Eng, 71 & Hebrew Univ, Jerusalem, 73; dean, Natural Sci & Math, Yeshiva Univ, New York, 80. *Mem:* AAAS; Am Chem Soc; Am Phys Soc; Sigma Xi; Am Asn Univ Professors. *Res:* Theoretical chemistry; quantum theory; polyelectrolytes; solid state; catalysis. *Mailing Add:* 788 Riverside Dr Apt 7E New York NY 10032

LOEBL, RICHARD IRA, b Battle Creek, Mich, Oct 18, 45; m 76. MATHEMATICS. *Educ:* Harvard Univ, AB, 67; Univ Calif, Berkeley, PhD(math), 73. *Prof Exp:* Teaching assoc, Univ Calif, Berkeley, 67-72; actg instr, Univ Calif, Santa Cruz, 72-73; asst prof, 73-79, ASSOC PROF MATH, WAYNE STATE UNIV, 79- *Concurrent Pos:* Res assoc, Univ Calif, Berkeley, 74; fac res award, Wayne State Univ, 75-76. *Mem:* Am Math Soc; Math Asn Am. *Res:* Functional analysis-operator theory. *Mailing Add:* 25319 Scotia Hunting Woods MI 48077

LOEBLICH, ALFRED RICHARD, JR, b Birmingham, Ala, Aug 15, 14; m 39; c 4. MICROPALEONTOLOGY, PALYNOLOGY. *Educ:* Univ Okla, BS, 37, MS, 38; Univ Chicago, PhD(paleont), 41. *Prof Exp:* Instr geol, Tulane Univ, 40-42; assoc cur invert paleont & paleobot, US Nat Mus, 46-57; from sr res paleontologist to res assoc paleontologist, 57-61; sr res assoc, Chevron Oil Field Res Co, Chevron Oil Co Calif, 61-68, sr res assoc, 68-79; RETIRED. *Concurrent Pos:* Adj prof geol, Univ Calif, Los Angeles, 72-82, emer prof, 82-; hon dir, Cushman Found Foraminiferal Res. *Honors & Awards:* Joseph Cushman Award, Cushman Found Foraminiferal Res, 82; Paleont Soc Medal, 82; Raymond C Moore Medal, Soc Econ Paleontologists & Mineralogists, 87. *Mem:* Hon mem Soc Sedimentary Geol; Paleont Soc; Paleont Res Inst; Spanish Paleont Soc; Belgium Geol Soc. *Res:* Stratigraphy; micropaleontology; recent and fossil foraminifera; morphology systematics and classification foraminifera. *Mailing Add:* Dept Earth & Space Sci Univ Calif Los Angeles CA 90024-1567

LOEBLICH, ALFRED RICHARD, III, b New Orleans, La, Mar 2, 41; m 63; c 2. PHYCOLOGY, MARINE BIOLOGY. *Educ:* Univ Calif, Berkeley, AB, 63; Univ Calif, San Diego, PhD(marine biol), 71. *Prof Exp:* Lab helper, Herbarium, Univ Calif, Berkeley, 62-63; teaching asst, Dept Bot, 63-64; lab technician marine biol, Univ Calif, San Diego, 70-71; asst prof & asst cur, Harvard Univ, 71-76, assoc prof & assoc cur, 76-78; ASSOC PROF BIOL, UNIV HOUSTON, 78- *Concurrent Pos:* USPHS fel, Univ Calif, San Diego, 64-70; NIH grant, 72-78; NSF grant, 74-; Mass Sci & Technol Found grant, 75-77; mem, Nomenclature Comt Algae, Int Asn Plant Taxon, 75-; mem, Comt Systs & Evolution, Soc Protozoologists, 77-; mem, Darbaker Prize Comt, Bot Soc Am, 78- *Honors & Awards:* Darbaker Prize, Bot Soc Am, 77. *Mem:* Phycol Soc Am; Soc Protozoologists; Int Phycological Soc; Am Soc Limnol Oceanog; Marine Biol Asn UK. *Res:* Dinoflagellate genetics; characterization of DNA of primitive algae; ultrastructure and physiology of unicellular algae; algal evolution. *Mailing Add:* Dept Biol Univ Houston Univ Park 4800 Calhoun Houston TX 77204

LOEBLICH, HELEN NINA (TAPPAN), b Norman, Okla, Oct 12, 17; m 39; c 4. MICROPALEONTOLOGY, PALEOECOLOGY. *Educ:* Univ Okla, BS, 37, MS, 39; Univ Chicago, PhD(geol), 42. *Prof Exp:* Asst geol, Univ Okla, 37-39; instr, Tulane Univ, 42-43; geologist US Geol Surv, 43-45 & 47-59; hon res assoc paleont, Smithsonian Inst, 54-57; lectr geol, 58-65, assoc res geologist, 61-63, sr lectr, 65-66, vchmn dept, 73-75, prof, 66-84, EMER PROF GEOL, UNIV CALIF, LOS ANGELES, 85- *Concurrent Pos:* Guggenheim fel, 54; hon dir, Cushman Found Foraminiferal Res, 82- *Honors & Awards:* Paleont Soc Medal, 82; Joseph Cushman Award, Cushman Found Foraminiferal Res, 82; Raymond C Moore Medal, Soc Sedimentary Geol, 83. *Mem:* Hon mem Soc Sedimentary Geol; Am Micros Soc; fel Geol Soc Am; Paleont Soc (pres, 85); Paleont Res Inst. *Res:* Micropaleontology; living and fossil foraminiferans, tintinnids, thecamoebians and organic-walled, siliceous and calcareous phytoplankton; morphology, taxonomy, ecology, primary productivity and food chains, evolution and extinctions. *Mailing Add:* Dept Earth & Space Sci Univ Calif Los Angeles CA 90024-1567

LOEBLICH, KAREN ELIZABETH, b Ft Sill, Okla, Oct 10, 44; m 75. ANIMAL BEHAVIOR, ENTOMOLOGY. *Educ:* Univ Calif, Los Angeles, AB, 66, MA, 67; Univ Calif, Davis, PhD(zool), 73. *Prof Exp:* Res assoc entomol, Univ Calif, Davis, 71-72; res assoc, Univ Calif, Riverside, 72-73; res assoc, Univ Hawaii, 74; lectr entomol & zool, San Francisco State Univ, 73-75; res scientist entomol, Agr Div, Upjohn Co, 75-76; lectr ecol & behav, San Diego State Univ, 77-78; MEM STAFF, DEPT ENTOM, UNIV CALIF, DAVIS, 78- UPJOHN CO, 75- *Mem:* Asn Study Animal Behav; Entomol Soc Am; Ecol Soc Am; Sigma Xi; AAAS. *Res:* Behavior and evolution of Diptera: Drosophilidae of Hawaii; insect grooming behavior; integrated pest management, especially of cotton. *Mailing Add:* Dept Entom Univ Calif Davis CA 95616

LOEBNER, EGON EZRIEL, computer science, system intelligence; deceased, see previous edition for last biography

LOECHELT, CECIL P(AUL), b Elfers, Fla, Nov 4, 35; m 56; c 3. CHEMICAL ENGINEERING. *Educ:* Vanderbilt Univ, BE, 56; La State Univ, MS, 62, PhD(adsorption), 64. *Prof Exp:* Instr chem eng, La State Univ, 63-64; sr process design engr, 64-70, SR ECON EVAL ENGR, ETHYL CORP, 70- *Mem:* Am Inst Chem Engrs. *Res:* Mathematical simulation of physical processes; evaluation and design of chemical processes. *Mailing Add:* 5944 Glenwood Dr Baton Rouge LA 70806

LOEFFLER, ALBERT L, JR, b Mineola, NY, Oct 22, 27; m 57; c 3. TURBULENCE. *Educ:* Va Polytech Inst, BS, 49; Iowa State Univ, PhD(chem eng), 53. *Prof Exp:* Res engr, NASA, 54-59; res engr, Grumman Aerospace Corp, 59-60, group leader, 60- 74, staff scientist, res dept, 74-90; ADJ PROF, HOFSTRA UNIV, DOWLING COL, 90- *Mem:* Am Phys Soc; Am Inst Aeronaut & Astronaut; Sigma Xi. *Res:* Turbulence; boundary layers; magnetohydrodynamics; heat transfer; potential flow problems. *Mailing Add:* PO Box 523 Bethpage NY 11714

LOEFFLER, FRANK JOSEPH, b Ballston Spa, NY, Sept 5, 28; m 51; c 4. PHYSICS. *Educ:* Cornell Univ, BS, 51, PhD(physics), 57. *Prof Exp:* Mem staff, Princeton Univ, 57-58; from asst prof to assoc prof, 58-67, PROF PHYSICS, PURDUE UNIV, 67- *Concurrent Pos:* Vis prof, Univ Hamburg, 63-64, Univ Heidelberg, 71 & Univ Hawaii, 85-86. *Mem:* Fel Am Phys Soc; Sigma Xi. *Res:* Elementary particle physics; experimental study of elementary particle interactions at high energy using electronic detection systems; atmospheric physics; investigation of high energy gamma rays and muons from point sources in space; experimental astrophysics using the Haleakala Gamma Ray Observatory and the South Pole GASP facility; development of the GRANDE facility for the study of astrophysical sources and high energy particle interactions. *Mailing Add:* Dept of Physics Purdue Univ Lafayette IN 47907

LOEFFLER, LARRY JAMES, b Beaver Falls, Pa, May 6, 32; m 57; c 2. ORGANIC CHEMISTRY. *Educ:* Princeton Univ, AB, 54, PhD(org chem), 61. *Prof Exp:* USPHS fel, Swiss Fed Inst Technol, 61-62; sr res chemist, Merck Sharp & Dohme Res Labs, Pa, 62-69; spec res fel, NIH, 69-71; asst prof, 71-74, assoc prof, 74-78, PROF MED CHEM, SCH PHARM, UNIV NC, CHAPEL HILL, 78- *Mem:* Am Chem Soc. *Res:* Medicinal chemistry; design and synthesis of compounds of interest as potential drugs; radioimmunoassay development. *Mailing Add:* Sch Pharm Univ NC Chapel Hill NC 27514

LOEFFLER, ROBERT J, b Worcester, Mass, Oct 20, 22; m 56; c 4. BOTANY. *Educ:* Syracuse Univ, BA, 48; Univ Wis, MS, 50, PhD(bot, zool), 54. *Prof Exp:* From asst prof to prof, 54-73, EMER PROF BOT, CONCORDIA COL, MOORHEAD, MINN, 73- *Mem:* Bot Soc Am; Am Inst Biol Sci. *Res:* Pollen analysis of Spiritwood Lake, North Dakota; phytoplankton; plant anatomy and morphology. *Mailing Add:* 704 Eighth St So Moorhead MN 56560

LOEGERING, DANIEL JOHN, b Minn, Mar 11, 43; m 68; c 3. PHYSIOLOGY. *Educ:* St John's Univ, Minn, BS, 65; Univ SDak, Vermillion, MA, 67; Univ Western Ont, PhD(physiol), 70. *Prof Exp:* Instr physiol, Med Col Wis, 69-73; from asst prof to assoc prof, 73-87, PROF PHYSIOL, ALBANY MED COL, 87- *Concurrent Pos:* Wis Heart Asn fel, Med Col Wis, 70-72, NIH spec res fel, 72-73. *Mem:* Reticuloendothelial Soc; Am Physiol Soc. *Res:* Mononuclear phagocyte system function as related to systemic host defense following injury and the cell biology of macrophages. *Mailing Add:* Dept Physiol A-134 Albany Med Col 43 New Scotland Ave Albany NY 12208

LOEHLE, CRAIG S, b Chicago, Ill, Oct 23, 52; m 80; c 2. LIFE HISTORY THEORY, LANDSCAPE ECOLOGY. *Educ:* Univ Ga, BS, 76; Univ Wash, MS, 78; Colo State Univ, PhD(math ecol), 82. *Prof Exp:* Sci programmer, SPSS, Inc, 82-84; postdoctoral ecol, Univ Ga, 84-87; res ecologist, Westinghouse Savannah River Co, 87-91; SR SCIENTIST, ARGONNE NAT LAB, 91- *Mem:* Ecol Soc Am; Int Soc Ecol Modeling; Int Soc Systs Sci. *Res:* Application of mathematics and statistics to ecology, including landscape ecology; simulation modeling; expert systems and artificial intelligence; plant morphology and reproductive strategies; stability theory; author of 80 publications. *Mailing Add:* Argonne Nat Lab 9700 S Cass Ave Argonne IL 60439

LOEHLIN, JAMES HERBERT, b Mussoorie, India, May 23, 34; US citizen; m 75; c 2. PHYSICAL CHEMISTRY, CRYSTALLOGRAPHY. *Educ:* Col Wooster, BA, 56; Mass Inst Technol, PhD(phys chem), 60. *Prof Exp:* Instr chem, Swarthmore Col, 60-61; from instr to asst prof, Col Wooster, 61-64; asst prof, Swarthmore Col, 64-66; from asst prof to assoc prof, 66-77, chmn, 71-74, 81-83 & 86, PROF CHEM, WELLESLEY COL, 77- *Concurrent Pos:* Res assoc, Univ Chicago, 69-70; vis mem fac, Inst Chem, Univ Uppsala, 76-77; vis scholar, Brandeis Univ, 83-84, Cambridge Univ, 84; vis prof, Univ Minn, 90-91. *Mem:* Am Crystallog Asn; Am Phys Soc; AAAS; Sigma Xi; Int Solar Energy Soc; Am Chem Soc. *Res:* Crystallography; molecular structure and solids; energy conversion; intermolecular hydrogen bonds in crystals. *Mailing Add:* Dept of Chem Wellesley Col Wellesley MA 02181

LOEHMAN, RONALD ERNEST, b San Antonio, Tex, Feb 22, 43; m 65, 82; c 2. CERAMICS, SOLID STATE CHEMISTRY. *Educ:* Rice Univ, BA, 64; Purdue Univ, PhD(chem), 69. *Prof Exp:* Res fel mat res, Thermophys Properties Res Ctr, Purdue Univ, 69-70; from asst prof to assoc prof mat eng, Univ Fla, 70-78; staff mem, Sandia Labs, Albuquerque, 77-78; staff scientist, SRI Int, 78-82, staff mem, 82-86, Div Supvr 86-87, DEPT MGR, SANDIA NAT LABS, 87- *Concurrent Pos:* Vis scientist, Univ Rennes, France, 84, Nat Defense Acad, Japan, 88. *Honors & Awards:* Snow Award, Am Ceramic Soc, 84, Fulrath Award, 88. *Mem:* Fel Am Ceramic Soc; AAAS; Nat Inst Ceramic Engrs; Sigma Xi. *Res:* Metal ceramic interfaces and ceramic joining; high temperature materials; electronic properties of materials; glass formation and crystallization; nitrogen ceramics. *Mailing Add:* 1011 Monroe St NE Albuquerque NM 87110

LOEHR, RAYMOND CHARLES, b Cleveland, Ohio, May 17, 31; m 53; c 8. ENVIRONMENTAL ENGINEERING. *Educ:* Case Inst Technol, BS, 53, MS 56; Univ Wis, PhD(sanit eng), 61. *Prof Exp:* From instr civil eng to asst prof, Case Inst Technol, 54-61; from assoc prof civil & sanit eng to prof, Univ Kans, 61-68; prof agr eng & civil eng, Cornell Univ, 68-85, dir environ studies prog, 72-80, Liberty Hyde Bailey prof eng, 81-85; H M ALHARTHY CENTENNIAL CHAIR PROF CIVIL ENG, UNIV TEX, AUSTIN, 85- *Concurrent Pos:* US Pub Health Serv & Environ Protection Agency res grants, 63-; chmn, Technol Assessment & Pollution Control Adv Comt, Sci Adv Bd, Environ Protection Agency, 78-80, & Environ Eng Comt, 82-88; consult to indust. *Honors & Awards:* Water Conserv Award, Nat Wildlife Fedn, 67; Rudolph Hering Award, Am Soc Civil Engrs, 69. *Mem:* Nat Acad Eng; AAAS; Am Soc Civil Engrs; Water Pollution Control Fedn; Asn Environ Eng Professors. *Res:* Environmental health engineering; water and wastewater treatment; hazardous waste treatment; industrial waste management; land treatment of wastes. *Mailing Add:* Univ Tex 8-6 ECJ Hall Austin TX 78712

LOEHR, THOMAS MICHAEL, b Munich, Ger, Oct 2, 39; US citizen; m 65. INORGANIC CHEMISTRY, BIOCHEMISTRY. *Educ:* Univ Mich, Ann Arbor, BS, 63; Cornell Univ, PhD(chem), 67. *Prof Exp:* Asst prof chem, Cornell Univ, 67-68; asst prof chem, 68-74, assoc prof, 74-78, PROF CHEM, ORE GRAD CTR, 78- *Concurrent Pos:* NIH res grant, Ore Grad Ctr, 71-; vis lectr, Portland State Univ, 71-72; vis assoc chem, Calif Inst Technol, 78-79; res grant, NSF, 74-77; mem, NIH Metallobiochem Study Sect, 78-82; actg dept chmn, Ore Grad Ctr, 80-81; chmn, Gordon Res Conf, 87. *Mem:* Am Chem Soc; Soc Appl Spectros; Royal Soc Chem. *Res:* Structural inorganic chemistry; infrared and Raman spectroscopy; metal ion complexes; metallobiochemistry; molecular and electronic structure of metalloproteins; solid state vibrational spectroscopy; analytical applications of Raman spectroscopy. *Mailing Add:* Ore Grad Inst Sci & Technol 19600 NW Von Neumann Dr Beaverton OR 97006-1999

LOEHRKE, RICHARD IRWIN, b Milwaukee, Wis, May 11, 35; m 57; c 2. MECHANICAL ENGINEERING. *Educ:* Univ Wis, BS, 57; Univ Colo, MS, 65; Ill Inst Technol, PhD(mech eng), 70. *Prof Exp:* Tech engr aircraft nuclear propulsion dept, Gen Elec Co, 57-61; res engr, Sundstrand Corp, 61-65; asst prof mech eng, Ill Inst Technol, 70-71; asst prof, 71-76, ASSOC PROF MECH ENG, COLO STATE UNIV, 76- *Mem:* Am Soc Mech Engrs; Am Inst Aeronaut & Astronaut; Sigma Xi. *Res:* Heat transfer; fluid mechanics. *Mailing Add:* 1901 Rangefield Dr Ft Collins CO 80524

LOELIGER, DAVID A, b Scranton, Pa, Mar 1, 39; m 60; c 4. COORDINATION CHEMISTRY, ARCHAEOLOGICAL SOILS. *Educ:* Col Wooster, BA, 61; Univ Chicago, MS, 62, PhD(chem), 65. *Prof Exp:* Asst prof chem, Purdue Univ, 64-67; sr res chemist, Eastman Kodak Co, 67-72; dir, Int Educ Exchange, 81-87, ASSOC PROF CHEM, INT CHRISTIAN UNIV, 72-, RES CONSULT, ARCHEOL RES CTR, 75- *Concurrent Pos:*

Missionary, Am Lutheran Church, 72- *Mem:* Am Chem Soc; Japan Soc Sci Study of Cult Properties. *Res:* Oxidation-reduction and subtitution reactions of transition metal ions and complexes; application of chemical techniques to problems of archaeological interest; chemical analysis of archaeological artifacts and soils. *Mailing Add:* Dept Chem Int Christian Univ Mitaka Tokyo Japan

LOENING, KURT L, b Berlin, Ger, Jan 18, 24; nat US; m 45; c 2. PHYSICAL CHEMISTRY, ORGANIC CHEMISTRY. *Educ:* Ohio State Univ, BS, 44, PhD(chem), 51. *Prof Exp:* From asst ed to sr assoc ed, Chem Abstracts, 51-63, assoc dir, Nomenclature, 63-64, dir, Nomenclature, 64-89; MANAGING DIR, TOPTERM, 90- *Concurrent Pos:* Chmn, Interdiv Comt, Nomenclature & Symbols, Int Union Pure & Appl Chem, 76-87, comt nomenclature, Am Chem Soc, 64-89. *Honors & Awards:* Austin M Patterson-E J Crane Award, 87. *Mem:* AAAS; Am Chem Soc; Am Soc Testing & Mat. *Res:* Acid-catalyzed esterification of organic acids; chemical nomenclature; literature. *Mailing Add:* PO Box 21213 Columbus OH 43221

LOEPPERT, RICHARD HENRY, b Chicago, Ill, Mar 13, 14; m 40; c 1. ORGANIC CHEMISTRY. *Educ:* Northwestern Univ, BS, 35; Univ Minn, PhD(phys chem), 40. *Prof Exp:* Asst, Univ Minn, 35-39; res chemist, Richardson Co, Ill, 39-40; from instr to assoc prof, 40-59, prof, 59-79, EMER PROF CHEM, NC STATE UNIV, 79- *Mem:* Am Chem Soc. *Mailing Add:* 1317 Rand Dr Raleigh NC 27608

LOEPPERT, RICHARD HENRY, JR, b Raleigh, NC, Sept 26, 44; m 89; c 1. SOIL CHEMISTRY. *Educ:* NC State Univ, BS, 66; Univ Fla, MS, 73, PhD(soil sci), 76. *Prof Exp:* Asst county agriculturalist, Agr Extension Serv, Univ Fla, 66-69; asst prof, 79-85, ASSOC PROF SOIL CHEM, TEX A&M UNIV, 85- *Concurrent Pos:* Chmn, Soil Chem Div, Soil Sci Soc Am, 90-91. *Mem:* Soil Sci Soc Am; Am Soc Agron; Am Chem Soc; Clay Minerals Soc; AAAS; Sigma Xi. *Res:* Soil chemical factors influencing availability of plant nutrients under nutrient stress and environmental stress conditions; soil carbonate chemistry; chemistry of soil acidity. *Mailing Add:* Dept Soil & Crop Sci Tex A&M Univ College Station TX 77843

LOEPPKY, RICHARD N, b Lewiston, Idaho, Aug 2, 37; m 65; c 2. PHYSICAL ORGANIC CHEMISTRY. *Educ:* Univ Idaho, BS, 59; Univ Mich, MS, 61, PhD(chem), 63. *Prof Exp:* Instr chem, Univ Mich, 63; NIH fel org chem, Univ Ill, 63-64; from asst prof to assoc prof, 64-80, PROF CHEM, UNIV MO-COLUMBIA, 80- *Concurrent Pos:* Resident vis, Bell Labs, 71-72. *Honors & Awards:* Kasimir Fajans Award, 65. *Mem:* Am Chem Soc; fel AAAS; Am Asn Cancer Res. *Res:* Chemical carcinogenesis and physical organic chemistry; chemical and biochemical transformation of nitrosamines directed at understanding their environmental and biochemical formation, transformation and destruction. *Mailing Add:* 123 Chem Dept Univ Mo Columbia MO 65211

LOESCH, HAROLD CARL, b Tex, Oct 3, 26; m 45; c 4. BIOLOGICAL OCEANOGRAPHY. *Educ:* Tex A&M Univ, BS, 51, MS, 54, PhD(biol oceanog), 62. *Prof Exp:* Prin marine biologist & actg lab dir, Dept Conservation, State Ala, 52-57; assoc res scientist, Tex A&M Res Found, 58-60; shrimp biologist, Food & Agr Orgn, UN, 60-62, fisheries biologist, 62-66, fisheries officer, 67-68, proj mgr & sr resource assessment surveyor, fisheries proj, Bangladesh, 81-85; prof, dept zool & physiol, La State Univ, Baton Rouge, 68-69, prof, dept marine sci & Off Sea Grant Develop, 70-75; expert marine biol, UN develop prog, 76-79; estuarine ecologist, UNESCO, Mexico, 76-79; proj mgr, UN Food & Agr Orgn, Bangladesh, 81-85; RETIRED. *Concurrent Pos:* Consult, Unesco, 79-80, Shrimp Growers Asn Ecuador, 85; vis prof, Org Am States Marine Sci prog, Ecuador, 72. *Mem:* Am Fisheries Soc; Am Soc Limnol & Oceanog; Am Soc Ichthyologists & Herpetologists; fel Int Acad Fishery Sci; Sigma Xi; AAAS; World Maricult Soc. *Res:* Estuarine hydrology and biology; shrimp, spiney lobster and inshore fishes ecology; fisheries statistics. *Mailing Add:* 2140 E Scott St Pensacola FL 32503

LOESCH, JOSEPH G, b Middle Village, NY, May 5, 30; c 3. MARINE BIOLOGY. *Educ:* Univ RI, BS, 65; Univ Conn, MS, 68, PhD, 69. *Prof Exp:* Res asst bluefish migrations, 65-66, asst proj leader, Conn River river herring study, 66-69, shellfish pop studies, 69-76, ANADROMOUS FISH STUDIES, VA, 76-; PROF MARINE SCI, COL WILLIAM & MARY, 69- *Concurrent Pos:* Sr marine scientist, Va Inst Marine Sci; State-Fed fishery mgt comts, 76- *Mem:* Am Fisheries Soc; Atlantic Estuarine Res Soc. *Res:* Marine fisheries; life history studies of anadromous fishes; biometrics and population dynamics of commercially important fishes. *Mailing Add:* Div Biol & Fisheries Sch Marine Sci Va Inst Marine Sci Col William & Mary Gloucester Point VA 23062

LOESCHE, WALTER J, b New Haven, Conn, March 28, 35; m 58; c 3. DENTAL DECAY, PERIODONTAL DISEASE. *Educ:* Yale Univ, BA, 57; Harvard Sch Dent Med, DMD, 61; Mass Inst Technol, PhD(biochem), 67. *Hon Degrees:* Dr, Univ Goteborg, Sweden. *Prof Exp:* Res assoc microbiol, Harvard Sch Dent Med, 61 & 64; mem staff, Forsyth Dent Ctr, 67-69; assoc prof oral biol, 69-74, assoc prof microbiol, 71-74, PROF DENT & ORAL BIOL, UNIV MICH SCH DENT, 74-, DIR RES, 87- *Concurrent Pos:* Assoc nutrit, Mass Inst Technol, 64-66; mem, Nat Affairs Comt, Am Asn Dent Res, 85-88, bd dirs, 85-88; mem, Educ Affairs Comt, Univ Mich Sch Dent, 83-86, mem Grad Studies Comt, Univ Mich Sch Med, 73- *Mem:* Am Soc Microbiol; Int Asn Dent Res; Am Asn Dent Res (pres, 87-88). *Res:* Research intends to demonstrate dental decay is a specific S mutans infection and that advanced periodontal disease is a treatable anaerobic infection; clinical studies are used to document the role of the above bacteria in human decay and periodontal disease. *Mailing Add:* Microbiol M6643 Med Sci 2 Univ Mich Med Sch 1301 Catherine Rd Ann Arbor MI 48109-0620

LOESCHER, WAYNE HAROLD, b Lima, Ohio, Nov 6, 42; m 67; c 1. PLANT PHYSIOLOGY. *Educ:* Miami Univ, BA, 64, MS, 66; Iowa State Univ, PhD(plant physiol), 72. *Prof Exp:* Res assoc plant physiol, Dept Agron, Iowa State Univ, 71-73; plant physiologist, Los Angeles Arboretum, Arcadia, Calif, 73-75; asst prof plant physiol & asst horticulturist, 75-80, ASSOC PROF & ASSOC HORTICULTURIST, WASH STATE UNIV, 80- *Res:* Plant growth and development; plant tissue culture. *Mailing Add:* Dept Hort Wash State Univ Pullman WA 99164-6414

LOESCH-FRIES, LORETTA SUE, b Ventura, Calif, Sept 5, 47; m 76; c 2. PLANT VIROLOGY. *Educ:* Wash State Univ, BS, 69; Univ Wis, PhD(plant path), 74. *Prof Exp:* Res assoc, Dept Plant Path, Univ Fla, 76-77; res assoc, Dept Hort, Univ Wis, 77-80, asst scientist, 80-81; sr res scientist & consult, Agrigenetics, 81-89; asst adj prof, Univ Wis, 87-91; ASST PROF, PURDUE UNIV, 91- *Concurrent Pos:* Grants Rev Panel, Coop State Res Serv, USDA, 85 & 86, Dept of Energy, 88; assoc ed, Molecular Plant-Microbe Interactions, 88- *Mem:* Am Phytopath Soc; Sigma Xi; AAAS; Am Soc Virol; Int Soc Plant Molecular Biol. *Res:* Genome organization and replication of plant viruses; determination of the role of the virus gene products in infection; how plant viruses cause disease and reduce crop yield; control of plant viruses; granted US patent. *Mailing Add:* Dept Bot & Plant Path Purdue Univ 1155 Lilly Hall Life Sci West Lafayette IN 47907-1155

LOESER, EUGENE WILLIAM, b Buffalo, NY, Nov 5, 26; m 55; c 1. MEDICINE. *Educ:* Univ Buffalo, MD, 52. *Prof Exp:* Asst neurol, Columbia Univ, 56-57; asst prof, Univ NC, 57-61; asst prof clin neurol, NY Univ, 64-71; clin assoc prof neurol, Rutgers Med Sch, 71-78; MEM STAFF, KIRKWOOD OUTPATIENT CTR, 78- *Mem:* AMA; Am Acad Neurol. *Res:* Medical neurology. *Mailing Add:* 1000 S Old Dixie Hwy Jupiter FL 33458

LOESER, JOHN DAVID, b Newark, NJ, Dec 14, 35; m 78; c 4. NEUROLOGICAL SURGERY, PAIN MANAGEMENT. *Educ:* Harvard Univ, BA, 57; New York Univ, MD, 61. *Prof Exp:* Asst prof neurosurg,Univ Calif, Irvine, 67-68; med corp, US Army, 68-69; from asst prof to assoc prof, 69-81, PROF NEUROSURG, UNIV WASH, 81- *Concurrent Pos:* Asst dean curric,Univ Wash, 77-82, dir, Multidisciplinary Pain Ctr, 83-; vis prof, Univ N South Wales,80; chief neurosurg,Children's Hosp, Seattle, 85-; Fulbright sr scholar, Australia, 89-90. *Honors & Awards:* Sunderland lect, Australian Pain Soc, 88. *Mem:* Am Pain Soc (treas, 80-85, pres, 86-87); Am Asn Neurol Surgeons; Int Asn Study Pain (secy, 84-90, pres, 90-93); AAAS; Soc Neurosci. *Res:* Clinical and research aspects of chronic pain, pain associated with injuries to nervous system, epidemiology and etiology of low back pain, pediatric neurosurgery, especially congenital malformations. *Mailing Add:* Dept Neurol Surg Univ Wash RC-95 Seattle WA 98195

LOETTERLE, GERALD JOHN, geology; deceased, see previous edition for last biography

LOEV, BERNARD, b Philadelphia, Pa, Feb 26, 28; m 54; c 3. ORGANIC CHEMISTRY, MEDICINAL CHEMISTRY. *Educ:* Univ Pa, BSc, 49; Columbia Univ, MA, 50, PhD(org chem), 52. *Prof Exp:* Instr inorg & org chem, Columbia Univ, 49-51; proj leader, Pennsalt Chem Co, 52-58; group leader, Smith Kline & French Labs, Pa, 58-66, sr investr, 66-67, from asst dir to assoc dir chem, 67-75; dir, Chem Res & Develop Div, USV Pharmaceut Corp, Revlon Health Care Group, 75-80, vpres chem res & develop, 80-83, vpres tech affairs, 83, vpres sci affairs, 83-86; EXEC VPRES, CREATIVE LICENSING INT, 86-; PRES, SCI CONSULT SERV, 87- *Concurrent Pos:* Mem adv bd, Index Chemicus & Intra-Sci Res Found; mem bd dirs, Int Heterocyclic Cong. *Honors & Awards:* Award Outstanding Contrib Med & Org Chem, Am Chem Soc, 74. *Mem:* AAAS; Am Chem Soc; Am Inst Chem; NY Acad Sci. *Res:* Organic synthesis; organic sulfur compounds; medicinal chemistry; nitrogen and sulfur heterocycles; natural products; central nervous system, cardiovascular, asthma, anti-arthritic, anti-ulcer areas; dermatology; patents. *Mailing Add:* 42 Penny Lane Scarsdale NY 10583

LOEW, ELLIS ROGER, b Los Angeles, Calif, Jan 18, 47; m; c 2. VISUAL PHYSIOLOGY, SENSORY BIOPHYSICS. *Educ:* Univ Calif, Los Angeles, BA, 68, MA, 72, PhD(biol), 73. *Prof Exp:* Fel, Vision Unit, Med Res Coun, 73-74, vis fel, 74-75; res fel, 75-77, asst prof, 77-83, ASSOC PROF PHYSIOL, CORNELL UNIV, 83- *Mem:* AAAS; Brit Photobiol Asn; Inst Elec & Electronic Engrs; Asn Res Vision & Ophthal. *Res:* Physiology and biochemistry of visual photoreceptors; biochemistry and biophysics of visual pigments; sensory ecology. *Mailing Add:* Physiol Sect Cornell Univ Ithaca NY 14853

LOEW, FRANKLIN MARTIN, b Syracuse, NY, Sept 8, 39; m 64; c 2. NUTRITIONAL TOXICOLOGY, ANIMAL MEDICINE. *Educ:* Cornell Univ, BS, 61, DVM, 65; Univ Sask, PhD(pharmacol, toxicol), 71. *Prof Exp:* Res scientist, R J Reynolds Tobacco Co, 65-66; res asst, Med Sch, Tulane Univ, 66-67; lectr vet med, Univ Sask, 67-69; MRC fel, Univ Sask, 69-71, dir animal resources, 71-74, prof toxicol, 74-77; chief, Lab Animal Med, Johns Hopkins Univ, 77-82, dir, Div Comp Med, 79-82; DEAN, SCH VET MED, TUFTS UNIV, 82- *Concurrent Pos:* Consult, Am Asn Accreditation Lab Animal Care, 78-81, Human Nutrit Inst, USDA, 78-82, Div Res Resources, NIH & Nat Inst Aging, Howard Hughes Med Inst, 87, Columbia Univ & Univ Pa; chmn, Inst Lab Animal Resources, Nat Acad Sci, 81-87, mem bd trustees, Boston Zool Soc, 84-89; mem, Comt Life Sc, Nat Acad Sci, 81-88; mem, Nat Coun, Div Res Resources, NIH, 88-; Sci Adv Bd, Primate Ctr, Harvard Univ, 88-89; Blue Ribbon Panel, Animal & Plant Health Inspection Serv, USDA, 87-89; bd dir, Ma Biotech Res Inst, 86- & Commonwealth Bioventures Inc, 88- & mem, Pew Fedn, Nat Adv Comn Vet Med Educ, 87-; mem, Mass Govt Adv Comn Sci & Technol, 89-, bd trustees, Mass Biotech Res Inst, 88-, bd dirs, Commonwealth BioVentures, Inc, 88- *Honors & Awards:* Gov-Gen's Medal (Can), 77; NB lectr, Am Soc Microbiol, 84; Armistead Lect, Asn Am Vet Med Cols, 87; Schofield Lectr, Univ Guelph, 87; Smith lectr, Univ Sask, 87; Schalm Lectr, Univ Calif, 89. *Mem:* Soc Toxicol; Am Inst Nutrit; Am Col Lab Animal Med; Am Asn Lab Animal Sci; Royal Soc Med; AAAS; Asn Am Vet Med Cols (pres, 85-86). *Res:* Thiamin deficiency and metabolism;

toxicology of long chain fatty acids; diseases of laboratory animals; animal nutrition, especially in the laboratory and often physiologic states of laboratory animals, including humane care and use; public policy questions regarding animal-related issues of social contention. *Mailing Add:* Sch Vet Med Tufts Univ 200 Westboro Rd North Grafton MA 01536

LOEW, GILDA HARRIS, b New York, NY; c 4. THEORETICAL BIOLOGY, BIOPHYSICS. *Educ:* NY Univ, BA, 51; Columbia Univ, MA, 52; Univ Calif, Berkeley, PhD(chem physics), 57. *Prof Exp:* Res physicist, Lawrence Radiation Lab, Univ Calif, Berkeley, 57-62 & Lockheed Missiles & Space Co, 62-64; assoc quantum biophys, Biophys Lab, Stanford Univ, 64-66; from asst prof to assoc prof physics, Pomona Col, 66-69; res biophysicist & instr biophys, Med Sch, Stanford Univ, 69-79, adj prof genetics, Med Ctr, 74-79; prog dir molecular theory, Life Sci Div, Stanford Res Inst, 79-; PRES, MOLECULAR RES INST, PALO ALTO, CALIF, 79- *Concurrent Pos:* Grants, NSF, 66-, NASA, 69- & NIH, 74-; adj prof, Rockefeller Univ, 79- *Mem:* Biophys Soc; fel Am Phys Soc; Int Soc Magnetic Resonance. *Res:* Molecular orbital and crystal field quantum chemical calculations; models for protein active sites; mechanisms and requirements for specific drug action; theoretical studies related to chemical evolution of life. *Mailing Add:* Molecular Res Inst 701 Welch Rd Suite 213 Palo Alto CA 94304

LOEW, LESLIE MAX, b New York, NY, Sept 2, 47; m 70; c 3. PHYSICAL ORGANIC CHEMISTRY, BIOPHYSICAL CHEMISTRY. *Educ:* City Col New York, BS, 69; Cornell Univ, MS, 72, PhD(chem), 74. *Prof Exp:* Res assoc chem, Harvard Univ, 73-74; from asst prof to assoc prof chem, State Univ NY Binghamton, 79-84; assoc prof physiol, 84-87, PROF PHYSIOL, UNIV CONN, 87- *Concurrent Pos:* Vis scientist, Weizmann Inst Sci, 81-82; Vis assoc prof, Cornell Univ, 84. *Mem:* Am Chem Soc; Biophys Soc; AAAS. *Res:* Organic dye chemistry; biomembranes; theoretical organic chemistry; electrical, adhesive and chemical properties of biomembranes using spectroscopic techniques; microscopy; image processing. *Mailing Add:* Dept Physiol Univ Conn Farmington CT 06030

LOEWE, WILLIAM EDWARD, b Chicago, Ill, Apr 22, 32; m 53; c 3. APPLIED PHYSICS. *Educ:* Univ Chicago, AB, 52; Univ Ill, BS, 53; Ill Inst Technol, MS, 59, PhD(physics), 63. *Prof Exp:* Reactor physicist, Savannah River Lab, E I du Pont de Nemours & Co, 53-54, Savannah River Plant, 54-57; assoc physicist, IIT Res Inst, 57-59, res physicist, 59-62, res physicist group leader, 62-63, mgr nuclear physics, 63-66; adv scientist, Nerva, Astronuclear Lab, Westinghouse Elec Corp, 66-67; sr physicist, Lawrence Livermore Nat Lab, Univ Calif, 67-90; RETIRED. *Concurrent Pos:* Consult, 91- *Mem:* Am Phys Soc. *Res:* Physics of ionized media and radiation transport, applied hydrodynamics, and criticality safety. *Mailing Add:* 1072 Xavier Way Livermore CA 94550

LOEWEN, ERWIN G, b Frankfurt am Maine, Ger, Apr 12, 21; m 52; c 2. MECHANICAL ENGINEERING. *Educ:* NY Univ, BME, 41; Mass Inst Technol, SM, 49, ME, 50. *Prof Exp:* Tech dir, Taft-Peirce Mfg Co, 55-60; head dept metrol, Bausch & Lomb Co, 60-67, dir, Grating & Metrol Labs, 67-84; vpres res, Develop & Eng, Milton Roy, 85-87; EMER PROF, UNIV ROCHESTER, 88- *Honors & Awards:* David Richardson Medal, Optical Soc Am, 84; F W Taylor Medal, Soc Mfg Engrs, 82. *Mem:* AAAS; Am Soc Mech Engrs; Soc Mfg Engrs; Optical Soc Am; Soc Photo Instr Engrs. *Res:* Precision Engineering; metrology; diffraction. *Mailing Add:* 44 Westwood Dr East Rochester NY 14445

LOEWENFELD, IRENE ELIZABETH, b Munich, Ger, June 2, 21; nat US. PHYSIOLOGY. *Educ:* Univ Bonn, PhD(zool), 56. *Prof Exp:* Asst ophthal, Columbia Univ, 58-61, instr, 61-62, res assoc, 62-68; from asst prof to assoc prof, 68-81, PROF OPHTHAL, SCH MED, WAYNE STATE UNIV, 81- *Mem:* Asn Res Vision & Ophthal. *Res:* Neurophysiology; neuroophthalmology; autonomic nervous system; pupil; visual physiology. *Mailing Add:* Three Maker Lane Falmouth MA 02540

LOEWENSON, RUTH BRANDENBURGER, b Zurich, Switz; US citizen; c 2. BIOMETRICS. *Educ:* Univ Minn, Minneapolis, BA, 59, MS, 61, PhD(biomet), 68. *Prof Exp:* From instr to asst prof, Sch Med, Univ Minn, Minneapolis, 65-72, assoc prof neurol & biomet, 72-90; CONSULT, 90- *Concurrent Pos:* Consult statistician, Vet Admin Hosp, Minneapolis, 71-83, FDA Neurol Devices Panel, 83- *Mem:* Am Statist Asn; Biomet Soc; Soc Epidemiol Res; Soc Clin Trials. *Res:* Clinical studies in neurology; clinical trials in neurology. *Mailing Add:* 3320 Louisiana Ave S 408 Minneapolis MN 55426

LOEWENSTEIN, ERNEST VICTOR, b Offenbach am Main, Ger, Sept 3, 31; US citizen; m 61; c 2. OPTICS, SPECTROSCOPY. *Educ:* Cornell Univ, AB, 53; Johns Hopkins Univ, PhD(physics), 60. *Prof Exp:* PHYSICIST, OPTICAL PHYSICS LAB, AIR FORCE CAMBRIDGE RES LABS, 62- *Mem:* Fel Optical Soc Am. *Res:* Optical properties of far infrared materials; optical properties of the atmosphere; Fourier spectroscopy. *Mailing Add:* 57 Hyde St Newton Highlands MA 02161

LOEWENSTEIN, HOWARD, b New York, NY, Jan 1, 24; m 58; c 2. FORESTRY. *Educ:* Colo State Univ, BS, 52; Univ Wis, PhD(soils), 55. *Prof Exp:* Instr soils, Univ Wis, 55-56; asst prof silvicult, Col Forestry, State univ NY, Syracuse, 57-58; from asst prof to prof forest soils, Univ Idaho, 58-89; RETIRED. *Mem:* Soil Sci Soc Am. *Res:* Forest soil-site relationships; forest fertilization; problems of tree seedling establishment; soil microbiology. *Mailing Add:* Wildlife & Range Sci Col Forestry Univ Idaho Moscow ID 83843

LOEWENSTEIN, JOSEPH EDWARD, b Crockett, Tex, Nov 25, 37; m 58; c 2. ENDOCRINOLOGY, INTERNAL MEDICINE. *Educ:* Univ Tex, Austin, BA, 59; Wash Univ, MD, 63. *Prof Exp:* Intern internal med, Barnes Hosp, St Louis, 63-64, resident, 67-69; res assoc, Nat Cancer Inst, 64-66, mem staff, 66-67; instr med, Wash Univ, 70; from asst prof to prof med, 70-84, chief

sect endocrinol, 70-84, CLIN PROF MED, LA STATE UNIV, SHREVEPORT, 84- *Concurrent Pos:* Nat Inst Arthritis & Metab Dis fel metab, Wash Univ, 69-70; consult, US Vet Admin Hosp, Shreveport, 70-; mem, Endocrine & Metab Drugs Adv Comt, US Food & Drug Admin, 80-84, chmn, 82-84. *Mem:* Endocrine Soc; fel Am Col Physicians. *Res:* Physiology of prolactin in humans; kinetics of iodine metabolism in thyroid; metabolic acidosis in diabetes. *Mailing Add:* 2015 Fairfield Suite 2A Shreveport LA 71104

LOEWENSTEIN, MATTHEW SAMUEL, b New York, NY, Dec 3, 41; m 65; c 3. GASTROENTEROLOGY. *Educ:* Union Col, BS, 62; Harvard Med Sch, MD, 67. *Prof Exp:* Intern, Harvard Med Unit, Boston City Hosp, 67-68, jr asst resident, 68-69, dep chief, Salmonella Unit, Ctr Disease Control, 69-70; chief, Enteric Dis Sect, Ctr Dis Control, USPHS, 70-72; sr resident, Harvard Med Unit, Boston City Hosp, 72-73, clin fel, instr & clin res assoc, 73-75, ASST PROF MED, HARVARD MED SCH, 75-; SR RES ASSOC, MALLORY GASTROENTEROL RES LAB, BOSTON CITY HOSP, 76- *Concurrent Pos:* Asst vis physician, Boston City Hosp, 75-80; courtesy staff, Mt Auburn Hosp, 75-77, active staff, 78- *Mem:* Am Gastroenterol Asn; Am Soc Gastrointestinal Endoscopy. *Res:* Clinical use of tumor markers, particularly carcinoembryonic antigen. *Mailing Add:* Mt Auburn Hosp 330 Mt Auburn St Cambridge MA 02138

LOEWENSTEIN, MORRISON, b Kearney, Nebr, Aug 21, 15; m 39; c 3. DAIRY CHEMISTRY, NUTRITION. *Educ:* Univ Nebr, BS, 38; Kans State Col, MS, 40; Ohio State Univ, PhD(dairy tech), 54. *Prof Exp:* Asst dairy, Kans State Col, 38-39; asst supt, Roberts Dairy Co, 39-40; instr dairy, NMex State Col, 40-41; from asst prof to assoc prof dairy, Okla Agr & Mech Col, 47-55; res dir, Crest Foods Co, Inc. 55-66, chmn bd, Sutton Crest Proteins Ltd, Can, 64-66; prof, 66-81, EMER PROF DAIRY SCI, UNIV GA, 81- *Mem:* Am Dairy Sci Asn; Inst Food Technol; AAAS; Sigma Xi. *Res:* Development, modification and compositional control of new and improved dairy products and milk protein concentrates. *Mailing Add:* 173 Beacham Dr Athens GA 30606

LOEWENSTEIN, WALTER B, b Gensungen, Ger, Dec 23, 26; US citizen; m 59; c 2. RESEARCH ADMINISTRATION. *Educ:* Univ Puget Sound, BS, 49; Ohio State Univ, PhD(physics), 54. *Prof Exp:* From asst physicist to sr physicist reactor physics, Argonne Nat Lab, 54-63, head, Fast Reactor Analysis Sect, Reactor Physics Div, 63-66, mgr physics sect, Liquid Metal Fast Breeder Reactor, prog off, 66-68, assoc dir, EBR-II Proj, Argonne Nat Lab, 68-72, actg dir, 72, dir, 72-73; dir, Safety & Analysis Dept, Elec Power Res Inst, 73-81, dep dir, Nuclear Power Div, 81-89; CONSULT, 89- *Concurrent Pos:* Tech adv, US del, Int Conf Peaceful Uses Atomic Energy, Geneva, 58; mem staff, UK Atomic Energy Authority, Dounreay, Scotland, 59; mem, Int Atomic Energy Agency Symp, Vienna, 61, Europ-Am adv comt reactor physics, Atomic Energy Comn, 66-73 & adv comt reactor physics, 66-73; secy-treas, Am Asn Eng Soc, 90. *Mem:* Nat Acad Eng; fel Am Nuclear Soc (vpres, 88-89, pres, 89-90); fel Am Phys Soc. *Res:* Fast reactor physics and related technology, including fast reactor design, analysis and planning of fast critical experiments, fast flux irradiation facilities and conceptual studies; reactor safety; research program development; spare nuclear power; technology transfer. *Mailing Add:* 515 Jefferson Dr Palo Alto CA 94303

LOEWENSTEIN, WERNER RANDOLPH, b Spangenberg, Ger, Feb 14, 26; m 52; 71; c 4. BIOPHYSICS, CELL BIOLOGY. *Educ:* Univ Chile, BSc(physics) & BSc(biol), 45, PhD(physiol), 50. *Prof Exp:* From instr to assoc prof physiol, Univ Chile, 51-57; res zoologist, Univ Calif, Los Angeles, 54-55; from asst prof to prof physiol, Col Physicians & Surgeons, Columbia Univ, 57-71; PROF PHYSIOL & BIOPHYS & CHMN DEPT, SCH MED, UNIV MIAMI, 71- *Concurrent Pos:* Fel neurophysiol, Sch Med & Hosp, Johns Hopkins Univ, 53-54, Kellogg Int fel physiol, 53-55; Block lectr, Univ Chicago, 60; lectr, Royal Swedish Acad Sci, 66; ed, Biochem & Biophys Acta, 67-73; Fulbright distinguished prof, 70; USSR Acad Sci lectr, Leningrad, 75; mem, Biochem, Molecular Genetics & Cell Biol Sect, President's Biomed Res Adv Panel, 77; USAF scientific adv bd, 82-86; ed, Handbook Sensory Physiol, 12 vols, 71-77; ed-in-chief, J Membrane Biol, 69- *Mem:* AAAS; Biophys Soc; Am Physiol Soc; Harvey Soc; fel NY Acad Sci. *Res:* Mechanisms of nerve impulse production and energy conversion at sensory nerve endings; excitation of the nerve cells; biophysics of cellular membranes; intercellular communication. *Mailing Add:* Dept Physiol & Biophys PO Box 016430 Miami FL 33101

LOEWENTHAL, LOIS ANNE, b Middletown, Conn, Oct 31, 26. BIOLOGY. *Educ:* Mt Holyoke Col, AB, 48; Brown Univ, AM, 50, PhD, 54. *Prof Exp:* Asst biol, Brown Univ, 48-53; res assoc zool, Mt Holyoke Col, 50-51; instr animal genetics, Univ Conn, 54-56; from instr to assoc prof zool, 57-74, assoc prof, 74-82, EMER PROF EXP BIOL, UNIV MICH, ANN ARBOR, 82- *Res:* Histology and embryology; skin and hair growth. *Mailing Add:* 55 Miner Brook Dr Middletown CT 06457

LOEWUS, FRANK A, b Duluth, Minn, Oct 22, 19; m 47; c 3. BIOCHEMISTRY. *Educ:* Univ Minn, BSc, 42, MSc, 50, PhD(biochem), 52. *Prof Exp:* Asst agr biochem, Univ Minn, 47-51; res assoc biochem, Univ Chicago, 52-55; chemist, USDA, 55-64; prof cell & molecular biol, Dept Biol, State Univ NY Buffalo, 64-75; agr chemist, Dept Agr Chem, 75-80, prof biochem, 75-89, fel, 80-90, EMER PROF, INST BIOL CHEM, WASH STATE UNIV, 90- *Concurrent Pos:* Ed, Phytochem Soc NAm, 76, 80-84; mem staff, Marine Biol Lab, Woods Hole. *Mem:* Phytochemical Soc NAm (pres, 75-76); AAAS; Am Chem Soc; Am Soc Biol Chem; Am Soc Plant Physiol. *Res:* Intermediary metabolism in plants, mechanisms of enzyme action; biochemistry of natural products. *Mailing Add:* Inst Biol Chem Pullman WA 99164-6340

LOEWUS, MARY W, b Duluth, Minn, Feb 15, 23; m; c 3. BIOCHEMISTRY, ENZYMOLOGY. *Educ:* Univ Minn, PhD(biochem), 53. *Prof Exp:* Assoc scientist, Inst Biol Chem, Wash State Univ, 76-88; RETIRED. *Mem:* Am Fedn Biol Chemists. *Mailing Add:* NE 1700 Upper Dr Pullman WA 99163-4624

LOEWY, ARIEL GIDEON, b Bucharest, Roumania, Mar 12, 25; US citizen; m 51; c 5. PHYSIOLOGY. *Educ:* McGill Univ, BSc, 45, MSc, 47; Univ Pa, PhD, 51. *Prof Exp:* Asst instr, Univ Pa, 47-49; NIH fel & univ res fel phys chem, Harvard Univ, 50-52; Nat Res Coun fel, Cambridge Univ, 52-53; from instr to assoc prof, 53-65, PROF BIOL, HAVERFORD COL, 65-, CHMN DEPT, 57- *Mem:* Am Soc Biol Chem; Am Soc Cell Biol. *Res:* Photosynthesis; protoplasmic streaming and contract; fibrin formation; structural proteins in cellular physiology. *Mailing Add:* Dept of Biol Haverford Col Haverford PA 19041

LOEWY, ARTHUR D(ECOSTA), b Chicago, Ill, Jan 9, 43; m 71; c 1. NEUROANATOMY. *Educ:* Lawrence Univ, BA, 64; Univ Wis-Madison, PhD(anat), 69. *Prof Exp:* Res assoc & instr neuroanat, Univ Chicago, 69-71; res assoc, 71-74, sr res fel neuroanat, Mayo Grad Sch Med, Univ Minn, 74-75; from asst prof to assoc prof, 75-85, PROF ANAT & NEUROBIOL, SCH MED, WASHINGTON UNIV, 85- *Concurrent Pos:* Investr, Am Heart Asn. *Honors & Awards:* Merit Award, Nat Heart, Lung & Blood Inst, 89. *Mem:* Soc Neurosci; Am Physiol Soc. *Res:* Organization of central autonomic pathways; neural control of cardiovascular system. *Mailing Add:* Dept Anat & Neurobiol Sch Med Washington Univ St Louis MO 63110

LOEWY, ROBERT G(USTAV), b Philadelphia, Pa, Feb 12, 26; m 55; c 3. AERONAUTICAL & MECHANICAL ENGINEERING. *Educ:* Rensselaer Polytech Inst, BAE, 47; Mass Inst Technol, MS, 48; Univ Pa, PhD(eng mech), 62. *Prof Exp:* Res asst, Mass Inst Technol, 48; sr vibration engr, Martin Co, Md, 48-49; assoc res engr, Cornell Aero Lab, Buffalo, 49-52, prin engr, 53-55; staff stress engr, Piasecki Helicopter Corp, Pa, 52-53; chief dynamics engr, Vertol Aircraft Corp, Pa, 55-58, chief tech engr, 58-62; from assoc prof to prof mech & aerospace sci, Univ Rochester, 62-74, dean, Col Eng & Appl Sci, 67-74; vpres acad affairs & provost, 74-78, INST PROF MECH & AEROSPACE SCI, RENSSELAER POLYTECH INST, 78- *Concurrent Pos:* Consult various govt agencies & pvt industs, 59-; chief scientist, Dept Air Force, 65-66, dir, Space Sci Ctr, Univ Rochester, 66-71; mem res & eng adv coun, US Post Off Dept, 66-68; mem aviation sci adv group, US Army Aviation Mat Command, 67-71; mem div adv group, Aeronaut Systs Div, US Air Force, 67-69, chmn, 78-84 & Sci Adv Bd, 67-76 & 78-85, vchmn, 71-73, chmn, 73-76; mem mil aircraft panel, President's Sci Adv Coun, 68-72; mem, NASA Aeronaut Comt, Res & Technol Adv Coun, 71-74, chmn aeronaut adv comt, 77-83. *Honors & Awards:* Lawrence Sperry Award, Am Inst Aeronaut & Astronaut, 58; Except Civilian Serv Award, USAF, 66, 75 & 85; Distinguished Pub Serv Medal, NASA, 83; Nikolsky Mem lectr, Am Helicopter Soc, 84. *Mem:* Nat Acad Eng; fel Am Inst Aeronaut & Astronaut; hon fel Am Helicopter Soc; Am Soc Eng Educ; fel AAAS. *Res:* Structural dynamics and aeroelasticity; unsteady aerodynamics; rotorcraft technology. *Mailing Add:* Rensselaer Polytech Inst Troy NY 12181

LOF, JOHN L(ARS) C(OLE), b Denver, Colo, Dec 11, 15; m 48; c 1. ELECTRICAL ENGINEERING. *Educ:* Univ Denver, BS, 38; Mass Inst Technol, SM, 41, EE, 51. *Prof Exp:* Asst elec eng, Mass Inst Technol, 38-45, res assoc, 45-49, instr, 49-52; from asst prof to prof, 52-76, dir, Comput Ctr, 61-76, EMER PROF ELEC ENG, UNIV CONN, 76- *Honors & Awards:* Inst Elec Eng Prize, Inst Elec & Electronics Engrs, 58. *Mem:* Am Soc Eng Educ; Inst Elec & Electronics Engrs; Sigma Xi. *Res:* Electronic computing systems; analog and digital computers; digital differential analyzers. *Mailing Add:* 74 Willington Hill Rd Storrs CT 06268

LOFERSKI, JOSEPH J, b Hudson, Pa, Aug 7, 25; m 49; c 6. SEMICONDUCTORS, PHOTOVOLTAIC ENERGY CONVERSION. *Educ:* Univ Scranton, BS, 48; Univ Pa, MS, 49, PhD(physics), 53. *Prof Exp:* Res assoc physics, Univ Pa, 52-53; res physicist, RCA Labs, 53-60; assoc prof, Brown Univ, 61-66, chmn, Div Eng, 68-74, assoc dean, Grad Sch, 80-83, PROF ELEC ENG, BROWN UNIV, 66- *Concurrent Pos:* Chmn exec comt, Div Eng, Brown Univ, 68-74; vis sr scientist Europ Space Res Orgn, Holland, 67-68; mem res & technol adv comt space power & elec propulsion, NASA, 69-71; mem solar energy panel, comt energy res & develop goals, US Off Sci & Technol, 71-72; consult, Exxon Labs, 73-80, Honeywell Inc, 77-79, & Jet Propulsion Labs, 78-85; exchange fel, US-Poland Acad of Sci, 74-75; pres, Solamat Inc, 77-82; mem, Organizing Comt II through XXII, Inst Elec & Electronic Engrs Photovoltaic Specialists Confs & Sixth Europ Photovoltaic Conf, 84, First thru Fifth Int Photovoltaic Sci & Technol Confs, Japan, China & Australia, 84-90; sci counr, US Embassy, Warsaw, Poland, 85-87, co-dir, Ctr Thin Film Res, 87-; mem, US-Poland Joint Bd for Coop in Sci & Technol, 88-; chmn, Fifth Int Conf Solid Films & Surfaces, Brown Univ, 91. *Honors & Awards:* Freeman Medal, Providence Eng Soc, 74; Wm E Cherry Award, Inst Elec & Electronic Engrs, 81. *Mem:* Fel AAAS; fel Inst Elec & Electronic Engrs; Am Phys Soc; Sigma Xi. *Res:* Semiconductor physics; photovoltaic effect; radiation effects in semiconductors and semiconductor devices; large scale utilization of solar energy; author or coauthor of over 140 publications; thin film deposition and characterization. *Mailing Add:* Div Eng Brown Univ Providence RI 02912

LOFFELMAN, FRANK FRED, b St. Louis, Mo, Nov 29, 25; m 45; c 1. ORGANIC CHEMISTRY. *Educ:* Loyola Univ (Ill), BS, 49; Notre Dame Univ, PhD(chem), 54. *Prof Exp:* Res chemist, 54-65, group leader, 65-76, proj leader, 76-81, MGR, PLASTICS ADDITIVES & FINE CHEM RES & DEVELOP, CHEM RES DIV, AM CYANAMID CO, 81- *Mem:* Am Chem Soc; Sigma Xi; Soc Plastics Engrs; Sigma Xi. *Res:* Organic synthesis; dyestuffs; optical bleaches; light stabilizers for thermoplastics; medicinals; fine chemicals; antioxidants; flame retardants. *Mailing Add:* 1255 Cornell Rd Bridgewater NJ 08807

LOFGREEN, GLEN PEHR, b St David, Ariz, Sept 28, 19; m 45; c 7. ANIMAL NUTRITION. *Educ:* Univ Ariz, BS, 44; Cornell Univ, MS, 46, PhD(animal nutrit), 48. *Prof Exp:* Asst prof animal husb, Mont State Col, 48; from asst prof to assoc prof, Univ Calif, Davis, 48-61, prof, 68-81, emer prof animal husb, 81-; dept animal sci, NMex State Univ; RETIRED. *Concurrent Pos:* Grant, Univ Hawaii, 58-59; consult, USDA; mem subcomt beef cattle nutrit, Nat Res Coun. *Honors & Awards:* Am Feed Mfrs Nutrit Award, 63. *Mem:* Am Soc Animal Sci (vpres, 60); Am Dairy Sci Asn. *Res:* Nutrient requirements and feed evaluation on large domestic animals; calcium and phosphorus metabolism. *Mailing Add:* 1624 Driftwood El Centro CA 92243

LOFGREN, CLIFFORD SWANSON, b St James, Minn, July 29, 25; m 54; c 3. ENTOMOLOGY. *Educ:* Gustavus Adolphus Col, BA, 50; Univ Minn, MS, 54; Univ Fla, PhD(entom), 68. *Prof Exp:* Entomologist, Entom Res Div, Agr Res Serv, 55-57 & Plant Pest Control Div, 57-63, ENTOMOLOGIST, INSECTS AFFECTING MAN RES LAB, USDA, 63-; PROF ENTOM, INST FOOD & AGR SCI, UNIV FLA, 80- *Concurrent Pos:* Asst prof entom & asst entomologist, Univ Fla, 74-80. *Mem:* Entom Soc Am; Am Mosquito Control Asn; Int Union Study Social Insects. *Res:* Methods of controlling mosquitoes, imported fire ants and other insects of medical importance, particularly insecticides and equipment evaluation; studies on resistance, chemosterilants, pheromones and biology. *Mailing Add:* 1321 NW 31st Dr Gainesville FL 32605

LOFGREN, EDWARD JOSEPH, b Chicago, Ill, Jan 18, 14; m 38, 68; c 3. PHYSICS, ACCELERATORS. *Educ:* Univ Calif, AB, 38, PhD(physics), 46. *Prof Exp:* Asst, Univ Calif, 38-40, physicist, Lawrence Radiation Lab, 40-44 & 45-46, group leader, Los Alamos Sci Lab, 44-45; asst prof physics, Univ Minn, 46-48; group leader, 48-73, assoc dir, 73-79, SR STAFF SCIENTIST, LAWRENCE BERKELEY LAB, UNIV CALIF, 79- *Concurrent Pos:* With European Orgn Nuclear Res, 59; mem, High Energy Physics Adv panel, 67-70. *Mem:* AAAS; Am Phys Soc. *Res:* Elementary particle physics; accelerators for particle and heavy-ion physics and for biomedical applications; separation of uranium isotopes; discovery of heavy component of cosmic rays. *Mailing Add:* Lawrence Berkeley Lab Univ Calif Bldg 47 Berkeley CA 94720

LOFGREN, GARY ERNEST, b Los Angeles, Calif, Apr 17, 41; m 65. EXPERIMENTAL PETROLOGY, PLANETARY SCIENCES. *Educ:* Stanford Univ, BS, 63, PhD(geol), 69; Dartmouth Col, MA, 65. *Prof Exp:* Res asst geol, Dartmouth Col, 63-65; res geologist, Cold Regions Res & Eng Lab, 65; teaching asst, Stanford Univ, 65-68; SPACE GEOSCIENTIST, JOHNSON SPACE CTR, NASA, 68- *Concurrent Pos:* Mem, Lunar Sample Preliminary Exam Team, 69-72; team leader, Basaltic Volcanism Study Proj, 76-81; convener, Penrose Conf, Geol Soc Am, 76; adj prof, Univ Houston, 76-87. *Honors & Awards:* Super Achievement Award, NASA, 78. *Mem:* Fel Geol Soc Am; fel Mineral Soc Am; Am Geophys Union; Am Asn Crystal Growth; Sigma Xi; Metall Soc. *Res:* Crystallization properties of silicate melts with emphasis on the kinetics of nucleation and crystal growth; textures of rocks and rock genesis. *Mailing Add:* Johnson Space Ctr NASA SN2 Houston TX 77058

LOFGREN, JAMES R, b West Point, Nebr, May 18, 31; m 62; c 3. PLANT BREEDING, GENETICS. *Educ:* Univ Nebr, BS, 60; NDak State Univ, MS, 62; Kans State Univ, PhD(plant breeding, genetics), 68. *Prof Exp:* Asst agron, NDak State Univ, 60-62; res asst, Kans State Univ, 62-67; asst prof, Northwest Exp Sta, Univ Minn, 67-71; AGRONOMIST-PLANT BREEDER, DAHLGREN & CO, INC, 71- *Mem:* Am Soc Agron; Crop Sci Soc Am. *Res:* Breeding and genetics of sunflowers to improve productivity and quality. *Mailing Add:* Dahlgren & Co Inc 1220 Sunflower St Crookston MN 56716-0609

LOFGREN, KARL ADOLPH, b Killeberg, Sweden, Apr 1, 15; US citizen; m 42; c 2. SURGERY. *Educ:* Harvard Med Sch, MD, 41; Univ Minn, MS, 47; Am Bd Surg, dipl, 53. *Prof Exp:* Intern, Univ Minn Hosp, 41-42; resident surg, Mayo Grad Sch Med, Univ Minn, 42-44 & 46-48; resident, Royal Acad Hosp, Univ Uppsala, 49; asst to staff, Mayo Clin, 49-50, mem surg staff, 50-81; from instr to asst prof, Mayo Grad Sch Med, 51-74, from assoc prof to prof, 74-81, EMER PROF SURG, MAYO MED SCH, 82-; EMER STAFF MEM, MAYO CLIN, 82. *Concurrent Pos:* Head sect peripheral vein surg, Mayo Clin, 66-79, sr consult, 79-81. *Mem:* Fel Am Col Surgeons; Int Cardiovasc Soc; Soc Vascular Surg; Swed Surg Soc; Sigma Xi. *Res:* Peripheral venous disorders. *Mailing Add:* 1001 Seventh Ave NE Rochester MN 55906-7074

LOFGREN, PHILIP ALLEN, b Iowa, July 30, 44; m 77; c 1. NUTRITION, RESEARCH ADMINISTRATION. *Educ:* Iowa State Univ, BS, 66; Cornell Univ, MS, 69, PhD(nutrit), 71. *Prof Exp:* Res asst animal nutrit, Cornell Univ, 66-71; fel nutrit, Univ Calif, Berkeley, 72-73; asst dir, 73-85, DIR GRANT SERVICES, NUTRIT RES, NAT DAIRY COUN, 85- *Mem:* Am Dairy Sci Asn; Inst Food Technologists; Am Oil Chemists Soc; Am Inst Nutrit. *Res:* Human and animal nutrition; unidentified growth factors; nutrient interactions; nutritional physiology of food intake regulation. *Mailing Add:* Nat Dairy Coun 6300 N River Rd Rosemont IL 60018-4233

LOFQUIST, GEORGE W, b Brookhaven, Miss, Oct 6, 30; m 55; c 2. MATHEMATICS, COMPUTER SCIENCES. *Educ:* Univ NC, BS, 52, MEd, 59; La State Univ, Baton Rouge, MS, 63, PhD(math), 67; Univ SFla, MS, 89. *Prof Exp:* Instr math, La State Univ, New Orleans, 59-64 & Baton Rouge, 66-67; from asst prof to assoc prof, 67-73, PROF MATH, ECKERD COL, 73- *Mem:* Am Math Soc; Math Asn Am; Asn Comput Mach. *Res:* Algebra; number theory. *Mailing Add:* Dept Math Eckerd Col PO Box 12560 St Petersburg FL 33733

LOFQUIST, MARVIN JOHN, b Chicago, Ill, Oct 19, 43; m 65; c 2. INORGANIC CHEMISTRY. *Educ:* Augustana Col, BA, 65; Northwestern Univ, PhD(inorg chem), 70. *Prof Exp:* Asst prof chem, Camrose Lutheran Col, 69-73; mem fac, 73-88, PROF & DEPT HEAD, PHYS SCIS DEPT, FERRIS STATE UNIV, 81- *Mem:* Am Chem Soc. *Res:* Kinetics and mechanisms of organometallic transition metal complexes. *Mailing Add:* Dept Phys Sci-Sci 320 Ferris State Col Big Rapids MI 49307

LOFSTROM, JOHN GUSTAVE, b Mason, Wis, June 4, 27; m 52; c 3. ANALYTICAL CHEMISTRY. *Educ:* Northwestern Univ, BS, 50; Univ Wis, PhD(chem), 54. *Prof Exp:* Asst chem, Univ Wis, 50-52; res chemist, Photo Prod Dept, E I du Pont de Nemours & Co, Inc, 53-66, sr res chemist, 66-85. *Mem:* Sigma Xi. *Res:* Instrumental analyses. *Mailing Add:* 58 McGuire St Metuchen NJ 08840

LOFT, JOHN T, b Mankato, Minn, July 21, 32; m 53; c 2. CHEMISTRY. *Educ:* Gustavus Adolphus Col, BA, 54; State Univ Iowa, MS, 56, PhD(chem), 58. *Prof Exp:* Sr res chemist, Sundry Dix Oil Co, 58-65 & Celanese Corp, 65-68; com develop, Celanese Plastics Co, 68-72; projs mgr, Microporus Div, Amerace Corp, 72-77; new bus coordr, Loctite Corp, 77-79, projs mgr, Mkt, 79-83, mgr com develop, 83-86, MGR COM DEVELOP, NEW BUS DEVELOP, LOCTITE CORP, 86- *Concurrent Pos:* Lectr, Chem Dept, Univ Tulsa, 60; chmn, Tulsa Sect, Am Chem Soc, 63-65. *Mailing Add:* 18 Holly Lane Avon CT 06001

LOFTFIELD, ROBERT BERNER, b Detroit, Mich, Dec 15, 19; wid; c 10. ORGANIC CHEMISTRY, BIOCHEMISTRY. *Educ:* Harvard Univ, BS, 41, MA, 42, PhD(org chem), 46. *Prof Exp:* Asst chem, Harvard Univ, 42-44, res assoc, 44-46; res assoc, Mass Inst Technol, 46-48; res assoc, Mass Gen Hosp, 48-56, assoc biochemist, 56-64; assoc, Harvard Med Sch, 56-60, asst prof org chem, 60-64; chmn dept, 64-71 & 78-90, prof, 64-90, EMER PROF BIOCHEM, MED SCH, UNIV NMEX, 90- *Concurrent Pos:* Fel, Brookhaven Nat Lab, 50; Runyon fel, Medinska Nobel Inst, Stockholm, 52-53; Guggenheim fel, Med Res Coun, Cambridge, Eng, 61-62; USPHS sr res fel, Dunn Sch Path, Oxford Univ, 71-72; chief spec assistance div, Off Strategic Serv, 46-47; tutor, Harvard Univ, 48-64; instr, Marine Biol Lab, Woods Hole, 59-62; mem biochem study sect, USPHS, 64-68; mem adv comn pathogenesis of cancer, Am Cancer Soc, 64-67, mem adv comn proteins & nucleic acids, 71-74; Fulbright prof, Abo Akademi, Turku, Finland, 77; mem, Fulbright Adv Comn, 78-81; Fulbright & Heinemann Stiftung fel, Med Univ, Hannover, Ger, 83. *Honors & Awards:* Warren Triennial Prize, 53. *Mem:* Am Chem Soc; Am Soc Biol Chem; Am Asn Cancer Res; Biophys Soc; Am Pub Health Asn. *Res:* Radioactive carbon 14 techniques; organic synthesis; organic reaction mechanisms; protein synthesis. *Mailing Add:* Univ NMex Med Sch Albuquerque NM 87131

LOFTNESS, ROBERT L(ELAND), b Devils Lake, NDak, Feb 14, 22; m 46; c 3. NUCLEAR ENGINEERING. *Educ:* Univ Puget Sound, BS, 43; Univ Wash, MS, 47; Swiss Fed Inst Technol, DSc, 49. *Prof Exp:* Res chemist radiation lab, Univ Calif, 43-45; res engr, US Naval Ord Testing Sta, 45; NAm Aviation, Inc, 49-51; sci attache, Am Embassy, Sweden, 51-53; mgr applns eng, Atomics Int Div, NAm Aviation, Inc, 53-60; com dir, Dynatom, France, 60-61; div dir Atomics Int Div, NAm Aviation, Inc, 61-70; dep dir, Off Atomic Energy Affairs, Bur Int Sci & Tech Affairs, Dept of State, 70-73; dir Washington off, 73-85, EXEC ASST TO PRES, ELEC POWER RES INST, 85- *Mem:* AAAS; Am Nuclear Soc. *Res:* Radiation and nuclear chemistry; dielectric properties of chemical systems; nuclear engineering. *Mailing Add:* 5805 Bent Branch Rd Bethesda MD 20816

LOFTSGAARDEN, DON OWEN, b Big Timber, Mont, July 7, 39; m 62; c 3. MATHEMATICAL STATISTICS. *Educ:* Mont State Univ, BS, 61, MS, 63, PhD(math statist), 64. *Prof Exp:* Res engr, Autonetics Div, NAm Aviation, Inc, 62; statistician, Battelle Mem Inst, 63; instr statist, Mont State Univ, 64-65; asst prof, Western Mich Univ, 65-67; from asst prof to assoc prof, 67-75, chmn dept, 78-79, PROF MATH, UNIV MONT, 75- *Mem:* Am Statist Asn; Inst Math Statist; Math Asn Am; Sigma Xi. *Res:* Statistical inference. *Mailing Add:* Dept of Math Univ of Mont Missoula MT 59801

LOFTUS, JOSEPH P, JR, b Chicago, Ill, Aug 31, 30; m 58; c 6. STATISTICS, PSYCHOLOGY. *Educ:* Cath Univ Am, Wash, DC, BA, 53; Fordham Univ, NY, MA, 56. *Prof Exp:* Res psychologist, Aerospace Med Lab, Wright-Patterson AFB, 59-61; chief, Opers Integ Br, Syst Eng Div, NASA, 64-68, mgr, Prog Eng Off, 68-70, chief, Tech Planning Off, 70-83, ASST DIR PLANS, JOHNSON SPACE CTR, NASA, 83- *Concurrent Pos:* Vchmn, Tech Houston Sect, 73-75, Econ Tech Comt, 72-75, 77-79; Sloan fel, Grad Sch Bus, Stanford Univ, 75-76; chmn, Space Systs Tech Comt, 81, 83, 85, Space Transp Tech Comt, 86-87; gen chmn, Space Technol, Progs Conf, Houston, 86; ed, Orbital Debris From Upper Stage Breakup, Am Inst Aeronaut & Astronaut, Progress in Aeronaut & Astronaut, Vol 121, 89; gen chmn, Am Inst Aeronaut & Astronaut/Dept Defense Orbital Debris Conf, Baltimore, MD, 89; mem, IAA Life Sci Comt, 85-89, secy, Space Policies & Plans Comt, 89-; prog comt, Human Factor Soc, 80. *Honors & Awards:* Commendation Medal, USAF, 62. *Mem:* Int Acad Astronaut; Am Astronaut Soc. *Res:* Orbital debris; launch system technologies; manned spacecraft control and life support systems. *Mailing Add:* NASA Johnson Space Ctr Mail Code AT Houston TX 77058

LOGAN, ALAN, b Newcastle-on-Tyne, Eng, Sept 20, 37; m 62; c 2. PALEOECOLOGY. *Educ:* Univ Durham, BSc, 59, PhD(paleont), 62. *Prof Exp:* Lectr paleont, Univ Leeds, 64-67; asst prof, 67-70, assoc prof, 70-76, PROF GEOL, UNIV NB, ST JOHN, 76- *Concurrent Pos:* Nat Res Coun fel, McMaster Univ, 62-64; vis fel, Univ Calgary. *Mem:* Int Soc Reef Studies. *Res:* Paleontology, paleoecology and ecology of Permian, Triassic and Holocene bivalves and brachiopods; ecology of Holocene coral reefs. *Mailing Add:* Dept of Geol Univ of NB Tucker Park St John NB E2B 4L5 Can

LOGAN, BRIAN ANTHONY, b Newcastle-upon-Tyne, Eng, Dec 22, 38; m 69; c 2. NUCLEAR PHYSICS. *Educ:* Univ Birmingham, BSc, 60, PhD(physics), 64. *Prof Exp:* Res assoc physics, Univ Birmingham, 64-65; lectr, 65-66. from asst prof to assoc prof, 66-81, PROF PHYSICS, UNIV OTTAWA, 81- *Mem:* Can Asn Physicists; Am Phys Soc. *Res:* Nuclear physics. *Mailing Add:* Dept of Physics Univ of Ottawa Ottawa ON K1N 6N5 Can

LOGAN, CHARLES DONALD, b St John, NB, May 15, 24; wid; c 3. WOOD CHEMISTRY. *Educ:* Mt Allison Univ, BSc, 45; McGill Univ, PhD(org chem), 49. *Prof Exp:* From res chemist to sr res chemist, Que & Ont Paper Co, Ltd, 49-65, asst dir, 65-74, dir chem res, 74-84; RETIRED. *Mem:* Am Chem Soc; Am Pulp & Paper Assoc; Brit Paper & Board Makers Asn; Can Res Mgt Asn; Chem Inst Can. *Res:* Vanillin and lignin chemistry; ion exchange chemical recovery; pulp and paper by-product utilization; chemimech pulping. *Mailing Add:* 9 Marlene Dr St Catharines ON L2T 3E7 Can

LOGAN, CHERYL ANN, b Syracuse, NY, Apr 1, 45. ANIMAL BEHAVIOR, NEUROPSYCHOLOGY. *Educ:* Southern Methodist Univ, BA, 67; Univ Calif, San Diego, PhD, 74. *Prof Exp:* Asst prof, 74-79, ASSOC PROF PSYCHOL, UNIV NC, GREENSBORO, 80- *Mem:* Animal Behav Soc; Am Ornith Union; Sigma Xi. *Res:* Animal communication; ecology and evolution of learning; structure and function of birdsong; territorial and reproductive function of mockingbird song. *Mailing Add:* Dept of Psychol Univ of NC Greensboro NC 27412

LOGAN, DAVID ALEXANDER, b Abingdon, Va, Dec 7, 52; m 76. MICROBIOLOGY. *Educ:* Knoxville Col, BS, 75; Univ Tenn, MS, 77, PhD(microbiol), 81. *Prof Exp:* Teaching asst microbiol, Univ Tenn, 75-80; instr biol, Knoxville Col, 80-81; lab asst med microbiol, Univ Calif, Irvine, 82, 83 & 84; ASST PROF BIOL, CELL BIOL & MICROBIOL, DREXEL UNIV, 84- *Mem:* Am Soc Microbiol; Mycol Soc Am. *Res:* Biochemistry of proteinases in fungi. *Mailing Add:* Dept Biol Sci Drexel Univ 32nd & Chestnut St Philadelphia PA 19104

LOGAN, DAVID MACKENZIE, b Toronto, Ont, July 23, 37; m 60; c 3. MOLECULAR BIOLOGY, BIOCHEMISTRY. *Educ:* Univ Toronto, BA, 60, MA, 63, PhD(med biophys), 65. *Prof Exp:* Res assoc biochem, NIH, 65-67; Nat Res Coun Can fel, McMaster Univ, 67-68; asst prof, 68-74, ASSOC PROF MOLECULAR BIOL, YORK UNIV, 68- *Concurrent Pos:* Jane Coffin Childs Mem Fund fel med res, 65-67. *Mem:* AAAS; Biophys Soc; Can Biochem Soc; NY Acad Sci. *Res:* Biochemical and biophysical aspects of nerve-muscle interactions, particularly neurotrophic effects in regeneration and acetylcholine esterases and receptors in normal and dystrophic animals. *Mailing Add:* Dept Biol Sci York Univ 4700 Keele St Downsview ON M3J 1P6 Can

LOGAN, GEORGE BRYAN, b Pittsburgh, Pa, Aug 1, 09; m 39; c 2. PEDIATRICS. *Educ:* Washington & Jefferson Col, BS, 30; Harvard Univ, MD, 34; Univ Minn, MS, 40; Am Bd Pediat, dipl, 41; Am Bd Allergy & Immunol, dipl, 72. *Prof Exp:* From instr to prof pediat, Mayo Grad Sch Med, Univ Minn, 40-73, prof pediat, Mayo Med Sch, 73-75, EMER STAFF, MAYO CLIN, 75- *Concurrent Pos:* Consult, Sect Pediat, Mayo Clin, 40-68, sr consult, 68-75; chmn sub-bd allergy, Am Bd Pediat, 63-66. *Mem:* AAAS; Am Pediat Soc; AMA; Am Acad Allergy & Immunol; Am Acad Pediat (pres, 67-68). *Res:* Allergic and liver diseases in children. *Mailing Add:* 1115 Plummer Circle Rochester MN 55902

LOGAN, H(ENRY) L(EON), b New York, NY, Mar 15, 96; m 21; c 1. ENGINEERING. *Hon Degrees:* LLD, Iona Col, 61. *Prof Exp:* Designer, Cram, Goodhue & Ferguson, NY, 16 & Thomas W Lamb, 16-17; tech dir, L Sonneborn & Sons, 19; asst to chief engr, Holophane Co, Ohio, 19-21, dist mgr, 21-27, eng consult, New York, 27-39, mgr controlens div, 39-46, dir & mgr, Dept Appl Res, 46-50, vpres in charge res, 50-71, chmn bd, 69-71; dir & vpres, Charles Franck Found, 62-72; RES & WRITING, 72- *Concurrent Pos:* Mem exped, Mex, 44; bd consult, Marymount Col, 48-70; dir, Found Sci Relaxation, 52-70; mem, US Nat Comt, Int Comn Illum, 53-; vchmn bd, Holophane Co, 62-69; trustee, Iona Col, 63; mem, Div Eng, Nat Res Coun. *Honors & Awards:* Silver Plate Award, Mex Inst Elec Engrs, 44; Bldg Res Inst Gold Seal Award, 65; Gold Medal Award, Illum Eng Soc, 65. *Mem:* Fel Illum Eng Soc; fel Inst Elec & Electronics Engrs; hon mem Mex Inst Illum; fel Brit Illum Eng Soc. *Res:* Effects of natural solar electromagnetic radiation on people and effects of the artificial radiation which replaced it under modern urban living conditions. *Mailing Add:* Five Hewitt Ave Bronxville NY 10708

LOGAN, JAMES COLUMBUS, b Baltimore, Md. BATTERY RESEARCH & DEVELOPMENT, CORPORATE MANAGEMENT. *Educ:* Johns Hopkins Univ, BES, 68; Harvard Univ, MS, 69, PhD(appl physics), 73. *Prof Exp:* Res fel, Calif Inst Technol, 73-75; br head, Naval Ocean Systs Ctr, 75-79; VPRES, ALTUS CORP, 79- *Mem:* Am Phys Soc; Am Soc Metals; AAAS. *Res:* Research and development of new batteries; lithium batteries and primary lithium batteries. *Mailing Add:* 1421 Tasso St Palo Alto CA 94301

LOGAN, JAMES EDWARD, b Thorndale, Ont, Jan 14, 20; m 82; c 2. CLINICAL CHEMISTRY, HEMATOLOGY. *Educ:* Univ Western Ontario, BSc, 49, PhD(biochem), 52. *Prof Exp:* Sr res asst biochem, Univ Western Ontario, 52-54; chemist, Biol Control Labs, Can Dept Nat Health & Welfare, 54-59, sr biochemist, Clin Labs, 59-73, chief clin chem, 73-77, actg dir, Bur Med Biochem, 77-79, chief clin chem, Lab Ctr Dis Control, 79-84. *Concurrent Pos:* Chmn, Int Fedn Clin Chem Expert Panel, Evaluation Diag Reagent Sets, 79- *Honors & Awards:* Ames Award, Can Soc Clin Chem, 81. *Mem:* Fel Nat Acad Clin Biochem; Can Soc Clin Chem; Am Asn Clin Chem; Can Biochem Soc. *Res:* Chemistry of peripheral nervous system; radioisotope tracer studies; quality control and methodology; hemoglobin; evaluation of diagnostic kits and clinical laboratory instruments; radioimmunoassay; reference methods; trace element analyses. *Mailing Add:* Bur Med Biochem Lab Ctr Dis Control Tunney's Pasture Ottawa ON K1A 0L2 Can

LOGAN, JESSE ALAN, b Pueblo, Colo, June 11, 44; m 70; c 1. POPULATION ECOLOGY, ENTOMOLOGY. *Educ:* Colo State Univ, BS, 67, MS, 69; Wash State Univ, PhD(entom), 77. *Prof Exp:* Res assoc entom, Colo State Univ, 71-72; comput programmer, Wash State Univ, 72-73, res asst, 73-76; res fel, NC State Univ, 76-78; asst prof zool & entom, Colo State Univ, 78-88; ASSOC PROF ENTOM & FORESTRY, VA POLYTECH INST & STATE UNIV, 88- *Mem:* AAAS; Sigma Xi; Entom Soc Am; Am Inst Biol Sci. *Res:* Systems analysis and population dynamics of invertebrate populations; relationship of these organisms to the efficient utilization of renewable natural resources. *Mailing Add:* Dept Entom & Forestry Va Polytech & State Univ Blackburg VA 24061-0324

LOGAN, JOHN MERLE, b Pittsburgh, Pa, July 7, 34; m 82. STRUCTURAL GEOLOGY, TECTONOPHYSICS. *Educ:* Mich State Univ, BS, 56; Univ Okla, MS, 62, PhD(geol), 65. *Prof Exp:* Geologist, Shell Develop Co, 65-67; from asst prof to assoc prof geol & geophys, 67-78, dir, Ctr Tectonophysics,

84-86, PROF EXP ROCK DEFORMATION, DEPT GEOPHYS & GEOL, TEX A&M UNIV, 78- *Concurrent Pos:* Advan Projs Res Agency, US Dept Defense res grant, 71-78, US Geol Surv grant, 71-85 & NSF grant, 71-86; consult, Amoco Prod Co, 67-86 & Los Alamos Nat Lab, 78-86. *Mem:* Assoc Geol Soc Am; assoc Am Geophys Union; AAAS. *Res:* Experimental rock deformation as applied to structural geological problems. *Mailing Add:* Ctr for Tectonophys Tex A&M Univ College Station TX 77843

LOGAN, JOSEPH GRANVILLE, JR, b Washington, DC, June 8, 20; m 44; c 2. PHYSICS. *Educ:* DC Teachers Col, BS, 41; Univ Buffalo, PhD(physics), 55. *Prof Exp:* Physicist aerodyn, Nat Bur Standards, 43-47; physicist aerodyn propulsion, Cornell Aeronaut Lab, Inc, 47-57; head aerophys lab, Space Technol Labs, Inc, 57-59, mgr propulsion res dept, 59-60; dir aerodyn & propulsion lab, Aerospace Corp, 60-67; spec asst to dir res & develop, Western Div, McDonnell Douglas Astronaut Co, 67-69, mgr vulnerability & hardening develop eng, 69-72, chief engr nuclear weapons effects, Western Div, 72-74; pres, Appl Energy Sci, Inc, 74-78; dir physics dept, Calif Polytech Univ, Pomona, 78-79; DIR, URBAN UNIV CTR, UNIV SOUTHERN CALIF, 79- *Mem:* Am Phys Soc; NY Acad Sci. *Res:* New energy systems. *Mailing Add:* 3652 Olympiad Dr Los Angeles CA 90043

LOGAN, JOSEPH SKINNER, b New York, NY, June 4, 32; m 52; c 4. ELECTRICAL ENGINEERING. *Educ:* Cornell Univ, BEE, 55, MS, 56; Stanford Univ, PhD(elec eng), 61. *Prof Exp:* Adv engr, 60-71, sr engr, E Fishkill Facil, 71-83, SR ENGR RES, IBM CORP, 84- *Mem:* Inst Elec & Electronics Engrs; Electrochem Soc; Am Vacuum Soc. *Res:* Semiconductor surface physics and device development; radio frequency sputtering of thin insulator films. *Mailing Add:* IBM Watson Res Ctr 2-148 Box 218 Yorktown Heights NY 10598

LOGAN, KATHRYN VANCE, b Atlanta, Ga, June 12, 46; m 67; c 2. MATERIALS RESEARCH & DEVELOPMENT. *Educ:* Ga Inst Technol, BCerE, 70, MSCerE, 80. *Prof Exp:* From res engr I to res engr II, 70-85. SR RES ENGR, GA INST TECHNOL, 85-, HEAD CERAMIC BR, 89- *Concurrent Pos:* Consult, 81-, prin investr, 82-; res award, Ga Tech, 90. *Honors & Awards:* Monie A Ferst Award, 70; Soc Women Engrs Award, 80. *Mem:* Am Ceramic Soc; Nat Inst Ceramic Engrs; Ceramic Educ Coun; Sigma Xi; Mat Res Soc; Nat Soc Prof Engrs. *Res:* Materials characterization via analytical instrumentation, advanced materials development, microwave ferrites, directionally solidified composites; crystal growth by directional solidification; clay mineralogy; thermite synthesis and forming of titanium diboride. *Mailing Add:* MSTL/Baker Rm 101 Ga Tech Res Inst Atlanta GA 30332

LOGAN, LOWELL ALVIN, b Langley, Ark, Oct 29, 21; m 44. ECOLOGY, PLANT TAXONOMY. *Educ:* Henderson State Col, BS, 43; Univ Ark, MS, 47; Univ Mo, PhD(bot), 59. *Prof Exp:* Instr biol, Ark Polytech Col, 46-49, head dept, 49-60; assoc prof bot, La Polytech Inst, 60-62 & La State Univ, 62-65; prof, Memphis State Univ, 65-67; vpres acad affairs, Southern Ark Univ, Ark, 67-87; RETIRED. *Mem:* AAAS; Ecol Soc Am; Bot Soc Am; Sigma Xi. *Res:* Ecology and distribution of American Beech; local floras; ecological factors affecting vegetation in restricted habitats. *Mailing Add:* 504 Alice St Magnolia AR 71753

LOGAN, R(ICHARD) S(UTTON), b Carthage, Mo, Oct 8, 18; m 48; c 1. CHEMICAL ENGINEERING, TECHNICAL MANAGEMENT. *Educ:* Univ Mo, BS, 47; Okla Agr & Mech Col, MS, 53. *Prof Exp:* Mem staff lubricant & lubricant additives group, Res & Develop Dept, 48-57, Phillips Petrol Co, group leader uranium milling processes, 57-58 & lubricant additives, 58-60, mgr process develop sect, 60-62, mgr catalytic reactions sect, 62-66, mgr refining & separation br, 66-80, dir petrol res, 80, mgr, Petrol & Petrochemical Processes Div, 80-84; RETIRED. *Mem:* Am Inst Chem Engrs; Nat Soc Prof Engrs. *Res:* Petroleum processing; uranium milling; lubricant oils and additives; sulfonation; halogenation; alkylation; catalytic cracking; research and development management. *Mailing Add:* 1808 Skyline Place Bartlesville OK 74006

LOGAN, RALPH ANDRE, b Cornwall, Ont, Sept 22, 26; nat US; m 50; c 9. MATERIALS SCIENCE ENGINEERING. *Educ:* McGill Univ, BSc, 47, MSc, 48; Columbia Univ, PhD(physics), 52. *Prof Exp:* Asst physics, Columbia Univ, 49-52; MEM TECH STAFF, BELL LABS, 52- *Mem:* Fel Am Phys Soc; fel Inst Elec & Electronic Engrs. *Res:* Semiconductor research. *Mailing Add:* AT&T Bell Labs Rm 4-288 Murray Hill NJ 07974

LOGAN, ROBERT KALMAN, b New York, NY, Aug 31, 39. PHYSICS, COMMUNICATIONS. *Educ:* Mass Inst Technol, BS, 61, PhD(physics), 65. *Prof Exp:* Res asst physics, Univ Ill, 65-67; res asst physics, 67-68, asst prof, 68-75, ASSOC PROF PHYSICS, DEPT PHYSICS, UNIV TORONTO, 75-, ASSOC PROF PHYSICS, DEPT MEASUREMENT, EVAL & COMPUT APPLNS, 84- *Res:* Computer applications in education; impact and effect of communication media on science and society. *Mailing Add:* Dept Educ Res & Statist Ont Inst Educ 252 N Bloor St W Toronto ON M5S 1V6 Can

LOGAN, ROWLAND ELIZABETH, b Los Angeles, Calif, Aug 1, 23. PHYSIOLOGY. *Educ:* Univ Calif, AB, 45; Northwestern Univ, MS, 51, PhD(physiol), 54. *Prof Exp:* Instr physiol, Sch Med, WVa Univ, 54-55; instr biol, Bard Col, 56-58; asst prof biol, Gettysburg Col, 58-88; RETIRED. *Mem:* AAAS. *Res:* Cell metabolism; arthropod behavior. *Mailing Add:* HCR 302 Islesboro ME 04848

LOGAN, TED JOE, b Ft. Wayne, Ind, June 22, 31; m 54; c 2. INDUSTRIAL CHEMISTRY. *Educ:* Ind Univ, AB, 53; Purdue Univ, MS, 56, PhD(chem), 58. *Prof Exp:* Res chemist, 58-63, sect head, 63-78, MGR RECRUITING, PROCTER & GAMBLE CO, 78- *Mem:* Am Chem Soc. *Res:* Product development. *Mailing Add:* Procter & Gamble Co PO Box 398707 Cincinnati OH 45239-8707

LOGAN, TERRY JAMES, b Georgetown, Guyana, Feb 6, 43; US citizen; m 73; c 2. SOIL CHEMISTRY. *Educ:* Calif Polytech State Univ, BS, 66; Ohio State Univ, MS, 69, PhD(soil sci), 71. *Prof Exp:* Asst prof soil chem, Ohio Agr Res & Develop Ctr, 71-72; from asst prof to assoc prof, 72-80, PROF SOIL CHEM, OHIO STATE UNIV, 80- *Honors & Awards:* Vis Scientist Award, Am Soc Agron, 84. *Mem:* Fel Soil Sci Soc Am; Fel Am Soc Agron; Int Soil Sci Soc; Soil Conserv Soc Am; Intl Asn Great Lakes Res. *Res:* Non-point sources of pollution; phosphate chemistry of soil and sediments; land disposal of sewage sludge; erosion and sedimentation of agricultural soils. *Mailing Add:* Dept Agron Ohio State Univ 2021 Coffey Rd Columbus OH 43216

LOGCHER, ROBERT DANIEL, b The Hague, Neth, Dec 27, 35; US citizen; m 63; c 3. CIVIL ENGINEERING, COMPUTER SCIENCE. *Educ:* Mass Inst Technol, SB, 58, SM, 60, ScD(civil eng), 62. *Prof Exp:* From asst prof to assoc prof, 62-76, PROF CIVIL ENG, MASS INST TECHNOL, 76- *Concurrent Pos:* Ford Found fel, 62-64; dir & sr consult, Eng Comput Int, Inc, Mass. *Mem:* Am Soc Civil Engrs; Asn Comput Mach; Sigma Xi. *Res:* Application of digital computer to structural design; development of computer-aided design techniques; design process; management of constructed facility projects; information systems. *Mailing Add:* 331 Springs Rd Bedford MA 01730

LOGEMANN, JERILYN ANN, b Berwyn, Ill, May 21, 42. SPEECH PATHOLOGY. *Educ:* Northwestern Univ, Chicago, BA, 63, MS, 64, PhD(speech path), 68. *Prof Exp:* Res assoc, 70-74, from asst prof to assoc prof, 74-83, PROF COMMUN SCI & DIS, NEUROL & OTOLARYNGOL, NORTHWESTERN UNIV, CHICAGO, 83- *Concurrent Pos:* NIH fel, Northwestern Univ, Chicago, 68-70; consult, Downey Vet Admin Hosp, 73-76; assoc attend staff, Northwestern Mem Hosp, 73-; fel, Inst Med Chicago. *Mem:* Am Speech & Hearing Asn; Linguistic Soc Am; Sigma Xi; fel Am Speech Language & Hearing Asn. *Res:* Speech science; laryngeal physiology; voice disorders; language disorders; swallowing physiology. *Mailing Add:* 320 E Superior St Northwestern Univ Chicago IL 60611

LOGERFO, JOHN J, b New York, NY, Feb 12, 18; m 55. ZOOLOGY, MEDICAL TECHNOLOGY. *Educ:* NY Univ, BA, 42; Columbia Univ, MA, 52, EdD, 61. *Prof Exp:* Lab supvr med tech, Lenox Hill Hosp, New York, 41-57; chief biochemist, Clin Lab, S Shore Anal Labs, 57-58; instr biol & gen sci, 58-60, from asst prof to assoc prof, 60-69, PROF BIOL, C W POST COL, LI UNIV, 69-, DIR MED TECHNOL, 63-, CHMN DEPT HEALTH SCI & CHMN PREMED COMT, 71- *Concurrent Pos:* Consult biochemist, St Claires Hosp, New York, 55-58; res assoc, Community Hosp, Glen Cove, NY, 58-64; USPHS res grant, 62-64; prog dir allied health sci traineeship dept, USPHS, 67- *Mem:* AAAS; NY Acad Sci; Am Soc Med Technol; Am Soc Microbiol; Asn Schs Allied Health Professions. *Res:* Hematology; parasitology; comparative anatomy, embryology; chordate vertebrate morphology; experimental embryology. *Mailing Add:* Dept Health Sci CW Post Col Greenvale NY 11548

LOGGINS, DONALD ANTHONY, b Brooklyn, NY, June 13, 51. HEAVY METAL DYNAMICS. *Educ:* City Univ New York, BA, 74; C W Post Col, NY, MPA, 76. *Prof Exp:* Consult urban design, Coun Environ NY, 76-77; vpres, G G Inc, 77-79; staff analyst environ protection & eng, 79-81, HRA-MIS, TASK FORCE NEW YORK, 81- *Concurrent Pos:* Consult, NY Consult Group, 76- *Honors & Awards:* US Environ Protection Agency Award, 79. *Mem:* Nat Audubon Soc; Am Mgt Asn; Am Planning Asn. *Res:* Effects of lead and cadmium in urban soils and flora; urban soil science; environmental design of urban spaces. *Mailing Add:* 723 E 10th St Brooklyn NY 11230

LOGGINS, PHILLIP EDWARDS, b Yorkville, Tenn, Feb 12, 21; m 42; c 2. ANIMAL NUTRITION. *Educ:* Okla State Univ, BS, 52, MS, 53. *Prof Exp:* Instr animal husb, Univ Fla, 53-55, from asst prof to assoc prof, 55-74, PROF ANIMAL HUSB, UNIV FLA, 74-, ANIMAL HUSBANDMAN, AGR EXP STA, 55- *Mem:* Am Soc Animal Sci. *Res:* Animal nutrition; parasitic effect on nutritional requirements; feeding requirements of animals during reproduction. *Mailing Add:* Dept of Animal Sci Univ of Fla Gainesville FL 32611

LOGIC, JOSEPH RICHARD, b Iron Mountain, Mich, Apr 23, 35; m 64. CARDIOVASCULAR PHYSIOLOGY, NUCLEAR MEDICINE. *Educ:* Marquette Univ, MD, 60, MS, 63, PhD(physiol), 64. *Prof Exp:* Intern med, C T Miller Hosp, St Paul, Minn, 60-61; instr physiol, Sch Med, Marquette Univ, 61-64 & 65-66; resident med, Mayo Clin, Rochester, Minn, 64-65; asst prof med, Col Med, Univ Ky, 66-69; assoc prof med & physiol, Univ Tenn, Memphis, 69-73; resident nuclear med, 74, PROF NUCLEAR MED & MED, UNIV ALA MED CTR, BIRMINGHAM, 74- *Res:* Peripheral circulatory failure; adrenergic blockade; electrolyte role in cardiac electrophysiology and contractility; myocardial nuclear medicine. *Mailing Add:* Div Nuclear Med Univ Ala Med Ctr Birmingham AL 35294

LOGIN, ROBERT BERNARD, b Brooklyn, NY, Nov 15, 42; m 71; c 2. ORGANIC CHEMISTRY, POLYMER CHEMISTRY. *Educ:* Brooklyn Col, BA, 66; Purdue Univ, PhD(org chem), 70. *Prof Exp:* Chemist paper specialties, Spring House Lab, Rohm and Haas Co, 70-73; sr chemist, BASF-Wyandotte Corp, 73-74, sect head, 74-75, supvr fiber specialities, 75-80; tech dir, Jordan Chem Co, 80-85; dir surfactants & specialties res & develop, 85-88, DIR POLYMER SCI RES & DEVELOP, GAF CHEM CORP, 88- *Mem:* Am Chem Soc; Am Asn Textile Chemists & Colorists; Soc Cosmetic Chemists. *Res:* Synthesis and applications of polymers and specialties derived from acetylenic-based intermediates. *Mailing Add:* 137 Page Dr Oakland NJ 07436

LOGOTHETIS, ANESTIS LEONIDAS, b Thessaloniki, Greece, June 29, 34; c 2. POLYMER CHEMISTRY. *Educ:* Grinnell Col, BA, 55; Mass Inst Technol, PhD(org chem), 58. *Prof Exp:* Res chemist, Cent Res Dept, 58-66, Elastomers Dept, 66-72, supvr develop, 72-76, div head fluoroelastomer res,

76-83, RES FEL, E I DUPONT DE NEMOURS & CO, INC, 83- *Mem:* Am Chem Soc. *Res:* Fluoropolymers; development of new products. *Mailing Add:* Cent Res & Develop Exp Sta E I du Pont de Nemours & Co Inc Wilmington DE 19880-0293

LOGOTHETIS, ELEFTHERIOS MILTIADIS, b Almyros, Greece; US citizen; m 66; c 2. SOLID STATE DEVICES, ELECTRONIC PROPERTIES OF MATERIALS. *Educ:* Univ Athens, BS, 59; Cornell Univ, MS, 65, PhD(physics), 67. *Prof Exp:* PRIN RES SCIENTIST, DEPT PHYSICS, RES STAFF, FORD MOTOR CO, 67- *Concurrent Pos:* Adj prof physics, Wayne State Univ, 82- *Mem:* Fel Am Phys Soc; Mats Res Soc. *Res:* Electrical and optical properties of semiconductors, metal oxides, layered compounds, ionic materials and high temperature superconductors; defect chemistry and gas/solid interactions; solid state devices such as gas, optical and general automotive sensors. *Mailing Add:* 110 Aspen Birmingham MI 48009

LOGOTHETOPOULOS, J, b Athens, Greece, Mar 12, 18; Can citizen; m 53; c 1. MEDICINE, PHYSIOLOGY. *Educ:* Nat Univ Athens, MD, 41; Univ Toronto, PhD(physiol), 62. *Prof Exp:* From asst prof to assoc prof, 59-64, PROF MED RES, BANTING & BEST DEPT MED RES & DEPT PHYSIOL, UNIV TORONTO, 64- *Concurrent Pos:* Res fel, Postgrad Med Sch, Univ London, 52-56; fel med res, Banting & Best Dept Med Res, Univ Toronto, 56-59. *Mem:* Am Diabetes Asn; Am Soc Exp Path; Can Physiol Soc. *Res:* Structure and function of the thyroid and the pituitary gland; experimental diabetes; structure and function of the islets of Langerhans. *Mailing Add:* Banting & Best Dept Med Res Univ of Toronto Toronto ON M5J 1L6 Can

LOGSDON, CHARLES ELDON, b Mo, May 8, 21; m 48; c 3. PLANT PATHOLOGY. *Educ:* Univ Kansas City, AB, 42; Univ Minn, PhD, 54. *Prof Exp:* Res prof plant path, Univ Alaska, 53-68, plant pathologist, 53-71, assoc dir, Inst Agr Sci, 71-78, prof, 68-78, EMER PROF PLANT PATH, UNIV ALASKA, 78-; CONSULT, 85- *Concurrent Pos:* Pres, Agresources Co, 78-85. *Mem:* AAAS; Am Phytopath Soc. *Res:* Potato and vegetable diseases. *Mailing Add:* PO Box 387 Palmer AK 99645

LOGSDON, DONALD FRANCIS, JR, b Chicago, Ill, Mar 7, 40; m 63; c 4. BIOLOGY, MEDICAL TECHNOLOGY. *Educ:* Northwestern Univ, BA, 61; Trinity Univ, MS, 70; LaSalle Exten Univ, LLB, 72; Colo State Univ, PhD(zool), 75, Thomas A Edison Col, BS, 77, Chapman Col, MAEd, 82, BA, 83, MS, 85. *Prof Exp:* Chief clin lab, 4510th US Air Force Hosp, Luke AFB, Ariz, 66-67; chief radioisotope lab, US Air Force Sch Aerospace Med, 67-70; assoc prof, Dept Life Sci, US Air Force Acad, 70-75; from asst chief to chief & staff biomed scientist, US Air Force Occup Environ Health Lab, 75-78; health sci coordr & asst prof, 78-81, mkt dir, educ prog coordr & prof, Sacramento Area Residence Educ Ctr, Chapman Col, 81-90; INSTR ENGLISH, SACRAMENTO SCH DIST, 90- *Concurrent Pos:* Instr life sci, Am River Col, Sacramento, Calif, 75-79 & Sierra Col, Rocklin, Calif, 77-78, Embry-Riddle Aeronaut Univ, 80, Cosumner River Col, 90. *Mem:* AAAS; Sigma Xi; Am Indust Hyg Asn; Asn Off Anal Chemists; Am Soc Radiol Technologists. *Res:* Effects of radiation on living methods for radiation detection; action of radioprotective drugs; comparison of routine plating versus fluorescent antibody methods for the detection of Beta Streptococcus; medical technology. *Mailing Add:* 7341 Spicer Dr Citrus Heights CA 95621

LOGSDON, JOHN MORTIMER, III, b Cincinnati, Ohio, Oct 17, 37; m 62; c 2. SCIENCE POLICY. *Educ:* Xavier Univ, BS, 60; NY Univ, PhD, 70. *Prof Exp:* Dir, Grad Prog Sci, Technol & Pub Policy, 72-88, DIR SPACE POLICY INST, GEORGE WASHINGTON UNIV, 87-, DIR CTR INT SCI & TECHNOL POLICY, 89- *Mem:* AAAS; Am Inst Aeronaut & Astronaut. *Res:* Evolution, current status and future prospects of the US civilian space program and its relationships with the space programs of other countries; issues of national science and technology policy. *Mailing Add:* Space Policy Inst George Wash Univ Washington DC 20052

LOGUE, J(OSEPH) C(ARL), b Philadelphia, Pa, Dec 20, 20; m 43; c 3. TECHNICAL CONSULTING. *Educ:* Cornell Univ, BEE, 44, MEE, 49. *Prof Exp:* Instr elec eng, Cornell Univ, 44-49, asst prof spec assignment, Brookhaven Nat Lab, 49-51; tech engr, IBM, 51-53, proj engr, 53-55, develop engr, 55-56, mgr mach develop, 56-57, mgr solid state circuit develop, 57-58, mgr explor eng, 58-59, mgr tech develop, 59-63, mgr adv tech systs, 63-64, mgr adv logic tech develop, 64-67, dir corp tech comt, 67-77, mgr technol & design systs, 77-85, dir packaging technol & systs, 86-; CONSULT TO INDUST, 86- *Concurrent Pos:* Secy, Int Solid State Circuits Conf, 57; fel eval comt, Inst Elec & Electronic Engrs, 84-86; mem eval comt, Nat Acad Eng Electronics, 89- *Mem:* Fel Nat Acad Eng; AAAS; fel Inst Elec & Electronic Engrs; Sigma Xi. *Res:* Development and application of new discoveries to advanced digital computers and systems; solid state devices and their applications; electronic aids to aircraft navigation. *Mailing Add:* 52 Boardman Rd Poughkeepsie NY 12603-4228

LOGUE, JAMES NICHOLAS, b Pittston, Pa, June 18, 46; m 72; c 2. EPIDEMIOLOGY. *Educ:* King's Col, BS, 68; Univ Mich, MPH, 71; Columbia Univ, DrPH(epidemiol), 78. *Prof Exp:* Statistician pharmaceut res, Warner-Lambert Res Inst, 70-71, 71-73; sr med biostatistician, Ciba-Geigy Pharmaceut Co, 73-78; sr environ epidemiology mgt consult, Geomet, Inc, 78-80; chief, Epidemiol Studies Br, US Food & Drug Admin Bur Radiol Health, 80-82; DIR ENVIRON HEALTH, PA DEPT HEALTH, 82- *Mem:* Soc Epidemiol Res; Int Epidemiol Asn; Am Pub Health Asn; AAAS. *Res:* Chronic disease; epidemiology; environmental and occupational epidemiology; clinical trials research; mental health research; disaster research. *Mailing Add:* 62 Little Run Rd Camp Hill PA 17011

LOGUE, MARSHALL WOFORD, b Danville, Ky, June 4, 42. CHEMICAL SYNTHESIS. *Educ:* Centre Col Ky, AB, 64; Ohio State Univ, PhD(chem), 69. *Prof Exp:* Res assoc chem, Univ Ill, 69-71; asst prof chem, Univ Md, Baltimore County, 71-77; asst prof chem, NDak State Univ, 77-81; ASSOC

PROF CHEM, MICH TECH UNIV, 81- *Mem:* NY Acad Sci; AAAS; Am Chem Soc; Royal Soc Chem; Sigma Xi. *Res:* Synthetic organic chemistry; bio-organic chemistry; pyrimidines; nucleosides; synthesis of carbohydrates and nucleosides. *Mailing Add:* Dept Chem Mich Tech Univ 1400 Townsend Dr Houghton MI 49931-1295

LOGULLO, FRANCIS MARK, b Wilmington, Del, Dec 19, 39; m 62; c 3. ORGANIC POLYMER CHEMISTRY. *Educ:* Univ Del, BS, 61; Case Inst Technol, PhD(org chem), 65. *Prof Exp:* Res chemist, E I du Pont de Nemours & Co, Inc, 65-70, sr res chemist, 70-77, res assoc, 77-90, REGULATORY AFFAIRS CONSULT, E I DU PONT DE NEMOURS & CO, 90- *Mem:* Am Chem Soc. *Res:* Polymer chemistry; synthetic fibers; chemistry of arynes. *Mailing Add:* Polymer Prod Dept M5624 E I du Pont de Nemours & Co Wilmington DE 19898

LOH, EDWIN DIN, b Suchow, China, Jan 21, 48; US citizen; m; c 1. PHYSICS. *Educ:* Calif Inst Technol, BS, 69; Princeton Univ, PhD(physics), 77. *Prof Exp:* Physicist, US Army Missile Command, 69-71; from instr to asst prof physics, Princeton Univ, 76-87; ASSOC PROF, MICH STATE UNIV, 87- *Res:* Astrophysics. *Mailing Add:* Mich State Univ 106 Physics-Astron Bldg E Lansing MI 48824

LOH, EUGENE C, b Soochow, China, Oct 1, 33; US citizen; c 3. COSMIC RAY, HIGH ENERGY. *Educ:* Va Polytech Inst, BS, 55; Mass Inst Technol, PhD, 61. *Prof Exp:* Res assoc physics, Mass Inst Technol, 61-64, asst prof, 64-65; sr res assoc nuclear studies, Cornell Univ, 65-75; assoc prof, 75-77, PROF PHYSICS & DEPT CHMN, DEPT PHYSICS, UNIV UTAH, 77- *Concurrent Pos:* Vis scientist, Stanford Linear Accelerator Ctr, 80-81. *Mem:* Am Phys Soc; Sigma Xi. *Mailing Add:* Dept Physics 201 Jfb Univ Utah Salt Lake City UT 84112

LOH, HORACE H, b Canton, China, May 28, 36; m 62; c 1. BIOCHEMISTRY, BIOCHEMICAL PHARMACOLOGY. *Educ:* Nat Taiwan Univ, BS, 58; Univ Iowa, PhD(biochem), 65. *Prof Exp:* Lectr biochem pharmacol, Univ Calif, San Francisco, 67, asst res pharmacologist, 67-68; assoc prof biochem pharmacol, Wayne State Univ, 68-70; chief, Drug Dependence Res Ctr, Mendocino State Hosp, Talmage, Calif, 71-72; res specialist, Langley Porte Neuropsychiat Inst, 70-72; from assoc prof to prof, Dept Psychiat & Dept Pharmacol, Sch Med, Univ Calif, San Francisco, 72-88; PROF & HEAD, DEPT PHARMACOL, MED SCH, UNIV MINN, MINNEAPOLIS, 89- *Concurrent Pos:* Fel biochem, Univ Calif, San Francisco, 65-66; USPHS career develop award, 73-78 & 78-83, res scientist award, 83-88, 89-94; mem, Preclin Psychopharmacol Study Sect, NIMH, 77-79 & Basic Psychopharmacol & Neuropsychopharmacol Res Rev Comt, 80-81; consult, US Army Res & Develop, Dept Defense, 80-84; mem, Biomed Res Rev Comt, Nat Inst Drug Abuse, 84-88, chair, 86-88; Wellcome vis prof award pharmacol, Fedn Am Socs Exp Biol & Burroughs Wellcome Labs, 85; chmn, Biochem Subcomt, Biomed Res Rev Comt, Nat Inst Drug Abuse, 86-88, Spec Rev Comt Drug Develop, 89 & Drug Abuse AIDS Res Rev Comt, 89-93. *Honors & Awards:* Humboldt Award for Sr US Scientists, 77; Hamilton Davis Lectr in Neurosci, Sch Med, Univ Calif, Davis, 88; Pfizer Lectr, Med Col Wis, Milwaukee, 88; Merit Award, Nat Inst Drug Abuse, 88. *Mem:* Am Chem Soc; Am Soc Pharm Exp Therapeut; Am Col Neuropsychopharmacol; Soc Chinese Bioscientists Am. *Res:* Opiate receptors; mechanisms of drug tolerance. *Mailing Add:* Dept Pharmacol 3-249 Millard Hall Univ Minn Med Sch 435 Delaware St SE Minneapolis MN 55455

LOH, PHILIP CHOO-SENG, b Singapore, Sept 14, 25; nat US; m 55; c 2. VIROLOGY. *Educ:* Morningside Col, BS, 50; Univ Iowa, MS, 53; Univ Mich, MPH, 54, PhD, 58; Am Bd Microbiol, dipl, 63. *Prof Exp:* Res assoc, Virus Lab, Univ Mich, 58-61; assoc prof, 61-66, PROF VIROL, UNIV HAWAII, 66-, CHMN, 85. *Concurrent Pos:* USPH spec res fel, NIH, 67-68; Eleanor Roosevelt int cancer fel, Int Union Against Cancer, Geneva, 75. *Mem:* Fel AAAS; Am Asn Immunol; Am Soc Microbiol; Tissue Cult Asn; Soc Exp Biol & Med; Am Soc Virol; fel Am Acad Microbiol. *Res:* Biosynthesis and pathobiology of animal viruses at the cellular level and environmental virology. *Mailing Add:* 2552 Peter St Honolulu HI 96816

LOH, ROLAND RU-LOONG, b Shanghai, China, Aug 1, 42; m 73; c 1. HIGH TEMPERATURE SUPERCONDUCTING MATERIALS, STRUCTURAL & ELECTRONIC CERAMICS. *Educ:* Shanghai Iron & Steel Inst, BS, 63; Univ Calif, Berkeley, MS, 85. *Prof Exp:* Lectr math, Iron & Steel Col, Shanghai, 64-65; engr, Shanghai Iron & Steel Co, 66-75, dir res & develop, Dept Ceramics, 76-81; vis scholar, Univ Calif, Berkeley, 82-83, staff scientist, Lawrence Berkeley Nat Lab, 83-85; sr engr, Lambertville Ceramic Mfg Co, 86-88; CHIEF SCIENTIST, HITC SUPERCONCO, DIV ADVAN CERAMETRICS, INC, 88- *Concurrent Pos:* Consult, Shanghai Bao-Shan Iron & Steel Complex, 78-80; prin investr, Dept Energy, 89-90 & NASA, 91; collabr, Los Alamos Nat Lab, 91-92. *Mem:* Am Ceramic Soc; Mat Res Soc. *Res:* Transfering scientific concept and result into production implementation; advanced ceramics and composite materials; author of more than 40 publications; technology transfer and joint venture between United States companies and Far East countries' companies. *Mailing Add:* 20 Cypress Dr Mt Holly NJ 08060

LOHER, WERNER J, b Landshut, Ger, June 27, 29; m 61; c 1. ZOOLOGY. *Educ:* Univ Munich, PhD(zool), 55; Univ London, PhD(entom) & DIC, 59. *Prof Exp:* Asst prof zoophysiol, Univ Tubingen, 60-65, privat docent, 65-67; assoc prof, 67-70, PROF ENTOM, UNIV CALIF, BERKELEY, 70-; DIR, GUMP S PAC BIOL RES STA, MOOREA, FRENCH, POLYNESIA, 85- *Concurrent Pos:* Sr res award, Antilocust Res Ctr, Eng; vis lectr, Glasgow Univ, 67. *Honors & Awards:* A V Humboldt Sr Distinguished Sci Award. *Mem:* AAAS; Animal Behav Soc; Brit Soc Exp Biol; Ger Zool Soc; Acad Scis & Lit, Mainz, 83. *Res:* Hormonal control of reproduction in insects. *Mailing Add:* Dept of Entom Univ of Calif Berkeley CA 94720

LOHMAN, KENNETH ELMO, b Los Angeles, Calif, Sept 11, 97; m 31. GEOLOGY. *Educ:* Calif Inst Technol, BS, 29, MS, 31, PhD(paleont & geol), 57. *Prof Exp:* Chemist, Cert Lab Prod, Inc, Calif, 24-31; geologist, US Geol Surv, 31-67; RES ASSOC, US NAT MUS, SMITHSONIAN INST, 67- *Concurrent Pos:* Chmn, Am Comn Stratig Nomenclature, 59-61. *Mem:* AAAS; fel Geol Soc Am; hon mem Soc Econ Paleont & Mineral (vpres, 61-62); Am Asn Petrol Geol; fel Royal Micros Soc; Am Geophys Union; Sigma Xi. *Res:* Diatoms; paleontology and stratigraphy; stratigraphic nomenclature; paleoecology; petroleum geology; photomicrography. *Mailing Add:* 10608 Huntingshire Lane Fairfax Sta VA 22039

LOHMAN, STANLEY WILLIAM, b Los Angeles, Calif, May 19, 07; m 33; c 3. GEOLOGY. *Educ:* Calif Inst Technol, BS, 29, MS, 38. *Prof Exp:* Asst mineral, Calif Inst Technol, 29-30; from jr geologist to prin geologist, Br Ground Water, Water Resources Div, US Geol Surv, 30-74, dist geologist chg ground water invests, Kans, 37-45 & Colo, 45-51, staff geologist, States in Ark-White-Red Basins, 51-56, br area chief, Rocky Mountain area, 56-59, res geologist, 59-62, staff geologist, 62-81; RETIRED. *Honors & Awards:* Distinguished Serv Award, US Dept Interior, 74; Distinguished Serv in Hydrol Award, Hydrol Div, Geol Soc Am. *Mem:* Fel Geol Soc Am; Am Asn Petrol Geologists; Sigma Xi; fel Am Geophys Union; hon mem Am Water Res Asn. *Res:* Ground-water geology and hydrology. *Mailing Add:* 2060 S Madison Denver CO 80210

LOHMAN, TIMOTHY GEORGE, b Park Ridge, NJ, Dec 10, 40; m 61; c 4. BODY COMPOSITION. *Educ:* Univ Ill, Urbana, BS, 62, MS, 64, PhD(body compos), 67. *Prof Exp:* Res assoc whole-body counting, 67-69, asst prof body compos animals & man, Dept Animal Sci, 69-77, assoc prof phys educ, Univ Ill, Urbana, 77-83; PROF, DEPT EXERCISE & SPORTS SCI, UNIV ARIZ, 84- *Mem:* AAAS; Am Col Sports Med; Soc Study Human Biol Am Alliance Health, Phys Educ & Recreation; Am Acad Phys Educ. *Res:* Exercise physiology; human body composition; physical exercise and body compositional and nutrition. *Mailing Add:* Dept of Exer Sport Sci 114 INA Gittings Bldg, Univ of Ariz Tucson AZ 85715

LOHMAN, TIMOTHY MICHAEL, b Rockville Ctr, NY, June 2, 51; m 78; c 2. BIOPHYSICAL CHEMISTRY, MOLECULAR BIOLOGY. *Educ:* Cornell Univ, Ithaca, AB, 73; Univ Wis, Madison, PhD(phys chem), 77. *Prof Exp:* Res asst biophys chem, Univ Calif, San Diego, 77-79; NIH fel, Univ Ore, 79-81; asst prof biochem & biophys, Tex A&M Univ, 81-85, assoc prof biochem, biophys & chem, 85-90; PROF BIOCHEM & MOLECULAR BIOPHYS, WASHINGTON UNIV, 90- *Concurrent Pos:* Mem molecular & cellular biophys study sect, NIH. *Honors & Awards:* Fac Res Award, Am Cancer Soc. *Mem:* Biophys Soc; Am Soc Biochem & Molecular Biol; AAAS. *Res:* Thermodynamics and kinetics of macromolecular interactions; protein-nucleic acid interactions involved in DNA replication, recombination and control of gene expression. *Mailing Add:* Dept Biochem & Molecular Biophys Sch Med Washington Univ Box 8231 St Louis MO 63110

LOHNER, DONALD J, b Brooklyn, NY, Mar 10, 39. ORGANIC CHEMISTRY. *Educ:* Queens Col, BS, 61; Adelphi Univ, PhD(org chem), 66. *Prof Exp:* Instr chem, Adelphi Univ, 64-66; res chemist, 66-76, res assoc, 76-85, SR RES ASSOC, E I DUPONT DE NEMOURS & CO, INC, 85-, RES FEL, 89- *Mem:* Am Chem Soc. *Mailing Add:* E I du Pont de Nemours & Co IMG Dept Parlin NJ 08859

LOHNES, ROBERT ALAN, b Springfield, Ohio, Feb 5, 37; m 60; c 2. GEOTECHNICAL ENGINEERING, GEOLOGY. *Educ:* Ohio State Univ, BSc, 59; Iowa State Univ, MS, 61, PhD(soil eng, geol), 64. *Prof Exp:* Asst geol, Iowa State Univ, 59-62, instr, 62-64; asst prof, Wis State Univ, River Falls, 64-65; from asst prof to assoc prof, 65-74, PROF CIVIL ENG, IOWA STATE UNIV, 74- *Concurrent Pos:* Vis assoc prof civil eng, Middle East Tech Univ, Ankara, Turkey, 73-74. *Mem:* Am Soc Civil Engrs; Am Geophys Union; Geol Soc Am; Am Rwy Engr Asn; Int Soc Soil Mech & Found Engrs. *Res:* Applied geomorphology, soil fabric as related to mechanical behavior; soil creep and shear strength; engineering properties of tropical soils; quantitative geomorphology; mechanics of bulk solids. *Mailing Add:* Dept Civil Eng Iowa State Univ Ames IA 50011

LOHR, DELMAR FREDERICK, JR, b Madison Co, Va, Sept 9, 34; m. POLYMER CHEMISTRY. *Educ:* Va Polytech Inst, BS, 62; Duke Univ, MA, 63, PhD(chem), 65. *Prof Exp:* Res org chemist, 65-70, sr res scientist, 70-77, assoc scientist, 78-83, RES ASSOC, FIRESTONE TIRE & RUBBER CO, 84- *Mem:* Am Chem Soc; Sigma Xi; Soc Automotive Engrs. *Res:* Synthesis and reactions of aromatic heterocycles, particularly those containing both nitrogen and sulfur; polymer synthesis and characterization. *Mailing Add:* 200 Casterton Ave Akron OH 44303-1517

LOHR, DENNIS EVAN, b Waukegan, Ill, Jan 12, 44. PHYSICAL BIOCHEMISTRY. *Educ:* Beloit Col, BA, 65; Univ NC, Chapel Hill, PhD(biochem), 69. *Prof Exp:* Teacher chem, Peace Corps, Kenya, EAfrica, 70-71; res assoc biochem, Ore State Univ, 72-79; ASST PROF BIOCHEM, ARIZ STATE UNIV, 79- *Res:* Enzymatic investigation of the subunit structure of yeast chromatin and physical characterization of the subunits. *Mailing Add:* Dept Chem Ariz State Univ Tempe AZ 85287

LOHR, JOHN MICHAEL, b Chicago, Ill, June 21, 44. PLASMA PHYSICS. *Educ:* Univ Tex, Austin, BS, 66; Univ Wis-Madison, MS, 67, PhD(nuclear physics), 72. *Prof Exp:* Res assoc plasma physics, Fusion Res Ctr, Univ Tex, Austin, 72-76; STAFF PHYSICIST, GEN ATOMICS CO, SAN DIEGO, 76- *Mem:* Am Phys Soc. *Res:* Tokamak and plasma physics research. *Mailing Add:* Gen Atomics PO Box 85608 San Diego CA 92138

LOHR, LAWRENCE LUTHER, JR, b Charlotte, NC, May 29, 37; m 63; c 1. THEORETICAL CHEMISTRY. *Educ:* Univ NC, BS, 59; Harvard Univ, AM, 62, PhD(chem), 64. *Prof Exp:* Res assoc chem, Univ Chicago, 63-65; res scientist, Sci Lab, Ford Motor Co, 65-68; assoc prof, 68-73, PROF CHEM, UNIV MICH, ANN ARBOR, 73- *Concurrent Pos:* Consult, Ford Motor Co, 68-71 & Bell Tel Labs, 69 & 72; vis prof & scholar, Univ Calif, Berkeley, 74-75; vis scientist, Inst Molecular Sci, Okazaki, Japan, 81, Univ Helsinki, Finland, 84, 88 & Univ Nac Auto Mexico, Mexico City, 89; res fel, Alfred P Sloan, 69-71. *Mem:* Am Phys Soc; Am Chem Soc; AAAS. *Res:* Theories of chemical bonding; interpretation of electronic spectra of molecules and solids with emphasis on properties of transition metal complexes; relativistic quantum chemistry; reaction mechanisms. *Mailing Add:* Dept Chem 1040 Chem Bldg Univ Mich Ann Arbor MI 48109

LOHRDING, RONALD KEITH, b Coldwater, Kans, Jan 1, 41; m 62; c 2. ENERGY POLICY ANALYSIS. *Educ:* Southwestern Col, Kans, BA, 63; Kans State Univ, MA, 66, PhD(statist), 68. *Prof Exp:* Inst math, St John's Col Prep Sch, PR, 63-64; consult statistician, Los Alamos Nat Lab, 68-74, group leader statist & energy assessment, 74-75, energy systs & statist, 75-78, energy systs & econ anal, 78-79, alt div leader, Systs Analysis & Assessment, 78-80, dep assoc dir environ & biosci, 80-81, dep assoc dir int affairs & energy policy, 81-83, asst dir indust & int affairs, 83-88, prog dir energy & tech, 88; PRES & CHIEF EXEC OFFICER, CELL ROBOTICS INC, 88- *Concurrent Pos:* Adj prof, Univ NMex, 69-70; vis prof oper res, Univ Hawaii, 78; res fel, Resource Systs Inst, East-West Ctr, 79. *Mem:* Am Statist Asn; AAAS. *Res:* International technology cooperation, foreign policy and transfer of technology from national labs to US industry; energy policy analysis; energy model evaluation and sensitivity analysis; comparative analysis of energy futures; large data set analysis; computer graphic data displays. *Mailing Add:* 2229 Loma Linda Los Alamos NM 87544

LOHRENGEL, CARL FREDERICK, II, b Kansas City, Mo, Nov 24, 39; m 71. GEOLOGY. *Educ:* Univ Kansas City, BS, 62; Univ Mo-Columbia, MA, 64; Brigham Young Univ, PhD(geol), 68. *Prof Exp:* Res assoc, Marine Inst, Univ Ga, 68-69; from asst prof to assoc prof, 69-81, PROF GEOL & MATH, SNOW COL, 81- *Mem:* Am Asn Petrol Geol; Am Inst Prof Geologists; Nat Asn Geol Teachers. *Res:* Palynology of the Upper Cretaceous of Utah; Upper Cenozoic and modern dinoflagellates of the Georgia coastal plain; Cretaceous stratigraphy of Utah; Upper Cretaceous stratigraphy of Wyoming. *Mailing Add:* Phys Sci Southern Utah State Col Cedar City UT 84720

LOHRMANN, ROLF, b Bissingen-Enz, Ger, Mar 2, 30; m 60. BIO-ORGANIC CHEMISTRY. *Educ:* Stuttgart Tech Univ, dipl(chem), 58, Dr rer nat(chem), 60. *Prof Exp:* Proj assoc, Inst Enzyme Res, Univ Wis, 62-65; sr res assoc, 65-74, ASSOC RES PROF, SALK INST BIOL STUDIES, 74- *Mem:* Ger Chem Soc; Am Chem Soc. *Res:* Prebiotic chemistry; molecular evolution. *Mailing Add:* 5531 Linda Bosa Ave La Jolla CA 92037-7631

LOHSE, CARLETON LESLIE, anatomic pathology; deceased, see previous edition for last biography

LOHSE, DAVID JOHN, b New York, NY, Sept 14, 52; m 78. POLYMER PHYSICS, NEUTRON SCATTERING. *Educ:* Mich State Univ, BS, 74; Univ Ill, PhD(polymer sci), 78. *Prof Exp:* Res asst, Univ Ill, 74-78; res assoc, Nat Bur Standards, 78-80; sr engr, Exxon Chem Co, 80-87, STAFF ENGR, EXXON RES & ENG CO, 88- *Mem:* Am Phys Soc; Am Chem Soc; Sigma Xi. *Res:* Physics of polymer systems, especially the morphology and thermodynamics of polymer blends and solutions and their structure-property relations. *Mailing Add:* 556 Stony Brook Dr Bridgewater NJ 08807

LOHTIA, RAJINDER PAUL, b Bara Pind W Punjab, Oct 9, 38; m 65; c 2. TESTING & QUALITY CONTROL OF CONCRETE, ANALYSIS & DESIGN OF CONCRETE STRUCTURES. *Educ:* Punjab Univ, India, BScEngg, 60; Roorkee Univ, India, MEng, 63; Univ Sask, Saskatoon, Can, Dr, 69. *Prof Exp:* Teaching fel civil eng, Roorkee Univ, India, 60-63, lectr, 63-66, reader, 66-72; head dept, 80-82, PROF CIVIL ENG, REGIONAL ENG COL, KURUKSHETRA, INDIA, 72-, DEAN RES, INDUST LIAISON & PLANNING & DEVELOP, 82- *Concurrent Pos:* Consult struct eng, several govt & semi govt departments, Pub Undertaking, India, 73-; consult civil eng, Govt of Syria, Damascus, 81. *Honors & Awards:* Wason Medal, Am Concrete Inst, 73. *Mem:* Am Concrete Inst. *Res:* Author of more than 70 research papers in the field of structural engineering and concrete technology published in both Indian and foreign journal; structural materials. *Mailing Add:* BA-5 Regional Eng Col Kurukshetra 132119 India

LOHUIS, DELMONT JOHN, b Oostburg, Wis, Jan 24, 14; m 37; c 3. CHEMISTRY. *Educ:* Carroll Col, Wis, BA, 34; Univ Wis, MS, 36. *Prof Exp:* Res dept, Am Can Co, 35-78, asst to corp vpres res & develop, 60-61 & 64-70, dir & vpres res & develop, 61-64. dir corp res & develop staff, 70-75, asst to pres, Tech Air Corp, 75-78; RETIRED. *Mem:* Am Chem Soc; fel Am Inst Chem. *Res:* Container construction materials; pyrolysis and gasification of ligno cellulosic materials; resource recovery from solid wastes; paper based consumer products; specialty chemicals. *Mailing Add:* 9450 River Lake Dr Roswell GA 30075

LOIGMAN, HAROLD, b Philadelphia, Pa, Jan 25, 30; m 51; c 2. CONSTRUCTION MANAGEMENT, GEOTECHNICAL ENGINEERING. *Educ:* Univ Pittsburgh, BS, 51; Univ Pa, MS, 63. *Prof Exp:* Supvry engr, US Army CEngr, 55-68; vpres, Valley Forge Labs, Inc, 68-72; pres, Site Engrs, Inc, 72-90; SR CONSULT, DAY & ZIMMERMANN, INC, 90- *Concurrent Pos:* Active reserve officer, Naval Reserve Construct Forces, 51-82, chief staff, 8th Reserve Naval Construct Regt, 72-74, cmndg officer, 21st Reserve Construct Battalion, 74-76; adj prof eng, Villanova Univ, 68-70, Temple Univ, 75-76, Grad Sch, Villanova Univ, 88-89 & Pa State Univ, 90. *Mem:* Am Soc Civil Engrs; Soc Am Mil Engrs; Nat Soc Prof Engrs; Am Soc Hwy Engrs; Int Soc Soil Mech & Found Engrs. *Mailing Add:* Day & Zimmermann Inc 1818 Market St Philadelphia PA 19103

LOIRE, NORMAN PAUL, b St Louis, Mo, May 7, 27; m 53; c 3. ORGANIC CHEMISTRY. *Educ:* Shurtleff Col, BS, 51; NY Univ, PhD(chem), 60. *Prof Exp:* Chemist, Ciba Pharmaceut Prod Co, 52-54; res chemist, Benger Res Lab, Textile Fibers Dept, E I du Pont de Nemours & Co, 58-62; sr res chemist, Narmco Res & Develop Div, Whittaker Corp, 62-67; sr res chemist,

Chemplex Co, 67-71; lab mgr, Saber Labs, Wheeling, 71-77; SR RES CHEMIST, MORTON THIOKOL INC, WOODSTOCK, ILL, 77- *Mem:* Am Chem Soc. *Res:* Development of new products based on water-borne polymer systems. *Mailing Add:* Morton Thiokol 1275 Lake Ave Woodstock IL 60098-7415

LOIZZI, ROBERT FRANCIS, b Oak Park, Ill, Oct 18, 35; m 60; c 4. PHYSIOLOGY, CELL BIOLOGY. *Educ:* Loyola Univ, Ill, BS, 57; Marquette Univ, MS, 60; Iowa State Univ, PhD(cell biol), 66. *Prof Exp:* Instr physiol, Iowa State Univ, 65-66; from asst prof to assoc prof physiol, 66-78, PROF PHYSIOL, 78-, ASST DIR, RES RESOURCES CTR, 86- *Concurrent Pos:* Assoc dean, Sch Basic Med Sci Col Med. *Mem:* AAAS; Am Soc Cell Biol; Am Physiol Soc. *Res:* Regulation and cell biology of mammary gland; cytoskeleton and milk secretion; hormonal regulation of microtubules; cyclic nucleotides and lactose synthesis; mammary tumor growth; fine structure of secretory processes. *Mailing Add:* Res Resources Ctr (m/c 937) Univ of Ill Box 6998 Chicago IL 60680

LOK, ROGER, b Macao, Oct 19, 43; US citizen; m 70; c 2. ORGANIC CHEMISTRY. *Educ:* Univ Calif, Berkeley, BS, 66; Univ Washington, PhD(org chem), 71. *Prof Exp:* Fel, Dept Pharmacol, Yale Univ, 71-74; RES CHEMIST, EASTMAN KODAK CO, 74- *Mem:* Am Chem Soc; Sigma Xi. *Res:* Organic synthesis; preparation of dyes; enzyme immobilization; affinity chromatography; synthesis of photographically active compounds; study of photographically active compounds. *Mailing Add:* Kodak Park Eastman Kodak Res Labs Rochester NY 14650

LOKAY, JOSEPH DONALD, b Chicago, Ill, Dec 17, 29; m 54; c 4. CHEMICAL ENGINEERING. *Educ:* Ill Inst Technol, BS, 52, MS, 53, PhD(chem eng), 55. *Prof Exp:* Engr, Westinghouse Elec Corp, 55-59 & Argonne Nat Lab, 59-62; mgr res & develop, Continental Can Co, 62-66; STAFF ENGR RES & DEVELOP, GULF OIL CORP, 66- *Mem:* Am Inst Chem Engrs. *Res:* Application of computers to teaching; research and development planning; the commercial evaluation of research and development projects related to petroleum processes and products including synthetic fuels and minerals. *Mailing Add:* Dept Math Univ Pittsburgh 1150 Mt Pleasant Rd Greensburg PA 15601-5898

LOKEN, HALVAR YOUNG, b Oslo, Norway, June 18, 44; US citizen; m 81; c 2. ORGANIC & POLYMER CHEMISTRY. *Educ:* Clark Univ, AB, 66; Brown Univ, PhD(chem), 71. *Prof Exp:* Nat Inst Allergy & Infectious Diseases fel trace org anal, Baker Lab, Cornell Univ, 70-72; res chemist, 72-76, sr res chemist, 76-80, carpet tech supvr, 80-81, mkt rep, 82-84, BUS STRATEGIST & RES ASSOC, E I DU PONT DE NEMOURS & CO, 84- *Concurrent Pos:* chmn, Honeycomb Core Task Force, Suppliers Advan Composites Mfrs Asn. *Mem:* Am Chem Soc; Soc Advan Mat & Process Eng. *Res:* Chemistry and physics of composite materials. *Mailing Add:* 4802 Pennington Ct Linden Heath Wilmington DE 19808

LOKEN, KEITH I, b Sandstone, Minn, Oct 3, 29; m 57; c 3. VETERINARY MICROBIOLOGY. *Educ:* Univ Minn, St Paul, BS, 51 DVM, 53, PhD(vet med), 59. *Prof Exp:* Res fel vet med, 55-58, from instr to assoc prof, 58-71, prof vet microbiol, 71-74, PROF VET BIOL, UNIV MINN, ST PAUL, 74- *Concurrent Pos:* Fulbright res fel, NZ Dept Agr, Ruakura Agr Res Ctr, 65-66. *Mem:* Am Vet Med Asn; Wildlife Dis Asn; Am Soc Microbiol. *Res:* Teaching veterinary microbiology; host-parasite relationships; epidemiology of infectious diseases of animals. *Mailing Add:* Vet Pathobiol Univ Minn 300 B Vet Sci St Paul MN 55108

LOKEN, MERLE KENNETH, b Hudson, SDak, Jan 21, 24; m 47; c 5. NUCLEAR MEDICINE, BIOPHYSICS. *Educ:* Augustana Col, BA, 46; Mass Inst Technol, BS, 48, MS, 49; Univ Minn, PhD(biophys), 56, MD, 62; Am Bd Nuclear Med, cert, 72. *Prof Exp:* Asst physics, Mass Inst Technol, 48-49; asst prof, Augustana Col, 49-51; instr biophys, 53-56, from asst prof to assoc prof, 56-68, dir Div Nuclear Med, 64-87, PROF RADIOL, UNIV MINN, MINNEAPOLIS, 68- *Concurrent Pos:* AMA comt non-mil radiation emergencies. *Mem:* Fel Am Col Radiol; Soc Nuclear Med; Radiol Soc NAm; AMA; Am Col Nuclear Physicians; Sigma Xi. *Res:* Clinical uses of radioisotopes; radiation dosimetry and hazards; effects of radiation on biological systems; applications of radioisotopes as tracer elements in metabolic studies of normal and cancer cells. *Mailing Add:* Box 382 Univ Minn Hosp Minneapolis MN 55455

LOKEN, STEWART CHRISTIAN, b Montreal, Que, Feb 16, 43; m 70; c 2. EXPERIMENTAL HIGH ENERGY PHYSICS. *Educ:* McMaster Univ, BSc, 66; Calif Inst Technol, PhD(physics), 71. *Prof Exp:* Res assoc physics lab nuclear studies, Cornell Univ, 71-74; PHYSICIST, LAWRENCE BERKELEY LAB, UNIV CALIF, 74-, DIR INFO & COMPUT SCIS DIV, 88- *Mem:* Fel Am Phys Soc. *Res:* Measurement of high energy muon-nucleon scattering and experimental studies of rare muon-induced processes at high energies; study of high energy electron-positron interactions. *Mailing Add:* Bldg 50B-2239 Lawrence Berkeley Lab Univ Calif Berkeley CA 94720

LOKENSGARD, JERROLD PAUL, b Saskatoon, Sask, July 30, 40; US citizen; m 65; c 2. ORGANIC CHEMISTRY. *Educ:* Luther Col, Iowa, BA, 62; Univ Wis, Madison, MA, 64, PhD(org chem), 67. *Prof Exp:* NIH fel, 67; res assoc chem, Iowa State Univ, 67; from asst prof to assoc prof, 67-89, chmn dept, 76-79 & 87-90, PROF CHEM, LAWRENCE UNIV, 89- *Concurrent Pos:* Vis assoc prof chem, Cornell Univ, 80-81. *Mem:* Am Chem Soc. *Res:* Natural products synthesis, especially insect defensive compounds; small strained hydrocarbons; novel organic compounds; organic spectroscopy (IR, NMR). *Mailing Add:* Dept Chem Lawrence Univ PO Box 599 Appleton WI 54912

LOKKEN, DONALD ARTHUR, b Tomahawk, Wis, Sept 27, 37; m; c 2. INORGANIC CHEMISTRY, SOLID STATE CHEMISTRY. *Educ:* Univ Wis-Madison, BA, 63; Iowa State Univ, PhD(inorg chem), 70. *Prof Exp:* Chemist, Enzyme Inst, Univ Wis-Madison, 57-59, pesticide chem, Wis Alumni Res Found, 59-63; teaching & res asst, Iowa State Univ, 63-70; ASSOC PROF CHEM, UNIV ALASKA, FAIRBANKS, 70- *Mem:* Am Chem Soc; Am Crystallog Asn; Sigma Xi. *Res:* Inorganic and solid state chemistry; x-ray crystallography; unusual oxidation states. *Mailing Add:* Dept Chem Univ Alaska Fairbanks AK 99775-0520

LOKKEN, STANLEY JEROME, b Fargo, NDak, Sept 22, 31; m 65. PHYSICAL CHEMISTRY. *Educ:* NDak State Univ, BS, 53; Univ Calif, Berkeley, MS, 54; Iowa State Univ, PhD(phys chem), 62. *Prof Exp:* Res chemist, Glidden Co, 62-65 & Continental Oil Co, 65-68; asst prof chem, Univ Wis-Platteville, 68-75; ASST PROF CHEM, BRUNSWICK JR COL, 77- *Concurrent Pos:* Vis asst prof chem, Ill Inst Technol, 75-77. *Mem:* Am Chem Soc. *Res:* Ion exchange theory and techniques; radiochemistry; solution kinetics and mechanisms of reactions; titanium chemistry; phosphate chemistry. *Mailing Add:* 141 Belle Point Pkwy Brunswick GA 31520-2102

LOLLAR, ROBERT MILLER, b Lebanon, Ohio, May 17, 15; m 41; c 2. LEATHER CHEMISTRY, ENVIRONMENTAL CHEMISTRY. *Educ:* Univ Cincinnati, ChE, 37, MS, 38, PhD(leather chem), 40. *Prof Exp:* Develop chemist, Best Foods Corp, Ind, 40-41; assoc prof tanning res & assoc dir, Tanners' Coun Res Lab, 41-58; tech dir, Armour Leather Co Div, Armour & Co, 58-64, dir tech eval, 65-73; tech dir, Leather Indust Am, Univ Cincinnati, 75-86; RETIRED. *Concurrent Pos:* Pres, Lollar & Assocs, Consults, 73- *Honors & Awards:* Alsop Medal, Am Leather Chemists Asn, 54. *Mem:* Am Chem Soc; Am Soc Qual Control; Am Leather Chemists Asn (pres, 66-68); Inst Food Technol; World Mariculture Soc. *Res:* Research administration; statistical quality control; collagen chemistry; industrial biochemistry; marine biology. *Mailing Add:* 5960 Donjoy Dr Cincinnati OH 45242-7508

LOLLEY, RICHARD NEWTON, b Blaine, Kans, May 25, 33; m 59; c 3. PHYSIOLOGY, BIOCHEMISTRY. *Educ:* Univ Kans, BS, 55, PhD(physiol), 61. *Prof Exp:* Pharmacist, Hawk Pharm, Inc, 55-56; asst prof, 66-70, assoc prof, 70-76, assoc mem, Jules Stein Eye Inst, 78-81, PROF ANAT, UNIV CALIF, LOS ANGELES, 76-, MEM, JULES STEIN EYE INST, 81- *Concurrent Pos:* USPHS res fel biochem, Maudsley Hosp, Univ London, 61-62; fel neuropath, McLean Hosp & Harvard Med Sch, 62-65; res pharmacologist, Vet Admin Hosp, 65-71, chief, Lab Develop Neurol, 71-, actg assoc chief staff res, 78-80; Vet Admin res career scientist, 79-; mem, Nat Adv Eye Coun, NIH, 79-84; trustee, Biochem & Molecular Biol Sect, Asn Res Vision & Ophthal, 87-92. *Honors & Awards:* Jules Stein Living Tribute Award, R P Int, 85. *Mem:* Int Soc Neurochemistry; Am Soc Neurochemistry; Soc Neuroscience; Am Asn Anat; Asn Res Vision & Ophthal (pres, 91-92); Int Soc Eye Res. *Res:* Chemical and physiological investigation of retina and regions of the developing brain; quantitative histochemical studies of normal tissues and of regions of the central nervous system afflicted by inherited diseases; animal models of human blindness; role of cyclic nucleotides in photoreceptor cell function and disease. *Mailing Add:* Lab Develop Neurol Vet Admin Med Ctr Sepulveda CA 91343

LOLY, PETER DOUGLAS, b Edmonton, Eng, Mar 7, 41; Can citizen; div; c 2. THEORETICAL PHYSICS, SOLID STATE PHYSICS. *Educ:* Univ London, BSc, 63, PhD(physics) & DIC, 66. *Prof Exp:* Fel, Theoret Physics Inst, Alberta, 66-68; from asst prof to assoc prof, 68-80, assoc head, 89-91, PROF PHYSICS, UNIV MAN, 80- *Concurrent Pos:* Travel fel, Nat Res Coun Can, 75; sabbatical leave, Lab Solid State Physics, Univ Paris-Sud, 75-76 & Physique, Univ Sherbrooke, 81-82. *Mem:* Am Phys Soc; Can Asn Physicists; Brit Inst Physics. *Res:* Spin waves in ferromagnets and anti-ferromagnets; density of states; Fermi surface instabilities; Brillouin zone sums; lattice green functions; spin-peierls and XY systems; real space rescaling; many-body problems; band structure modelling. *Mailing Add:* Dept Physics Univ Man Winnipeg MB R3T 2N2 Can

LOMAN, JAMES MARK, b Waterbury, Conn, Nov 14, 54; m 75; c 3. RADIATION EFFECTS. *Educ:* Villanova Univ, BS, 75; Univ Notre Dame, MS, 77; Univ Del, PhD(physics), 80. *Prof Exp:* Res assoc, Brookhaven Nat Lab, 80-81; eng specialist, Ford Aerospace & Commun Corp, 81-83; STAFF ENGR GEN ELEC CO, 83- *Mem:* Am Phys Soc. *Res:* Experimental radiation effects in electronic devices and insulators; spacecraft charging; radiation effects in geological material for nuclear waste disposal applications. *Mailing Add:* Gen Elec Co PO Box 8555 Philadelphia PA 19101-8555

LOMAN, M LAVERNE, b Stratford, Okla, June 10, 28; m 44; c 1. MATHEMATICS. *Educ:* Univ Okla, BS, 56, MA, 57, PhD(math educ), 61. *Prof Exp:* From asst to instr math, Univ Okla, 56-61; from asst prof to assoc prof, 61-65, PROF MATH, CENT STATE UNIV, OKLA, 66- *Mem:* Math Asn Am; Nat Coun Teachers Math. *Res:* Mathematics education. *Mailing Add:* 2201 Tall Oaks Trail Edmond OK 73034

LOMANITZ, ROSS, b Bryan, Tex, Oct 10, 21; m 47. THEORETICAL PHYSICS. *Educ:* Univ Okla, BS, 40; Cornell Univ, PhD(theoret physics), 51. *Prof Exp:* Teaching asst physics, Univ Calif, Berkeley, 40-42, physicist, Lawrence Radiation Lab, 42-43, teaching asst physics, 46-47; teaching asst, Cornell Univ, 47-49; assoc prof, Fisk Univ, 49; laborer, Okla, 49-54; tutor physics & math, Okla, 54-60; assoc prof physics, Whitman Col, 60-62; asst prof, 62-66, assoc prof, 66-80, PROF PHYSICS, NMEX INST MINING & TECHNOL, 80-, ASSOC PHYSICIST, 62- *Concurrent Pos:* NSF res grant, 63- *Mem:* Am Phys Soc; Am Geophys Union; Math Asn Am. *Res:* Electromagnetic isotope separator; quantum electrodynamics; superconductivity; theoretical plasma physics; theoretical ground water hydrology. *Mailing Add:* Dept Physics NMex Inst Mining & Technol Campus Sta Socorro NM 87801

LOMAS, CHARLES GARDNER, b Ft Peck, Mont; m; c 1. FLUID FLOW INSTRUMENTATION. *Educ:* Univ Md, BS, 57, BS, 64, MS, 75. *Prof Exp:* Engr, Miller Fluid Power, 70-71; asst instr mech engr, Univ Md, 71-77; engr, Dantec Electronics, 77-80; instr eng sci, Lafayette Col, 80-82; asst prof mech eng, Rochester Inst Technol, 85-86; assoc prof fluid power technol, Northampton Community Col, 86-88; ASSOC PROF ENG TECHNOL, CALIF POLYTECH STATE UNIV, 88- *Mem:* Am Soc Mech Engrs; Fluid Power Soc; Instrument Soc Am. *Mailing Add:* Calif Polytech State Univ San Luis Obispo CA 93407

LOMAX, EDDIE, b Atlanta, Ga, Aug 12, 23; m 48. ORGANIC CHEMISTRY. *Educ:* Morehouse Col, BS, 48; Atlanta Univ, MS, 51. *Prof Exp:* Control chemist, 51-53, res chemist, 53-57, lab mgr, 57-71, asst tech dir, 71-73, TECH DIR, PURITAN CHEM CO, 73- *Concurrent Pos:* Sci consult, Atlanta Bd Educ. *Mem:* Am Chem Soc. *Res:* Free radical mechanism in solutions; surfactants, insecticides and disinfectants; floor polishes; cyclobutadiene series. *Mailing Add:* 495 Harlan Rd SW Atlanta GA 30311

LOMAX, HARVARD, b Broken Bow, Nebr, Apr 18, 22; m 43; c 3. ENGINEERING, IED MATHEMATICS. *Educ:* Stanford Univ, BA, 43, MA, 47. *Prof Exp:* RES SCIENTIST, NACA-NASA, AMES RES CTR, 44- *Concurrent Pos:* Teacher, Stanford Univ, 50- *Honors & Awards:* Fluid & Plasma Dynamics Award, Am Inst Aeronaut & Astronaut, 77. *Mem:* Nat Acad Eng; fel Am Inst Aeronaut & Astronaut. *Res:* Subsonic and supersonic dynamics; the upper-astmosphere aerodynamics of hypersonic blunt bodies; the foundations of computational fluid dynamics. *Mailing Add:* Ames Res Ctr Mail Stop 202A-1 Moffett Field CA 94035

LOMAX, MARGARET IRENE, b Roanoke, Va, Nov 13, 38; m 64; c 2. MOLECULAR GENETICS, GENOME ORGANIZATION OF NUCLEAR GENES FOR CYTOCHROME C OXIDASE. *Educ:* Case Western Reserve Univ, BA, 60; Univ Mich, PhD(biol chem), 64. *Prof Exp:* Res assoc biol chem, Univ Mich, Ann Arbor, 64-67, instr, 67-88, asst res sci, Div Biol Sci, 74-85 & Dept Microbiol & Immunol, 85-88, ASST RES SCI, DEPT ANAT & CELL BIOL, UNIV MICH, ANN ARBOR, 88-, DIR, DNA SEQUENCING FACIL, 90- *Concurrent Pos:* Am Cancer Soc fel, 64-66. *Mem:* AAAS; Am Soc Microbiol; Am Soc Biol Chemists; Asn Women Sci. *Res:* Genome organization and molecular evolution of cytochromec oxidase in primates; tissue-specific expression of cytochrome C oxidace nuclear genes. *Mailing Add:* Dept Anat & Cell Biol Univ Mich Med Sch Box 0616 Ann Arbor MI 48109-0616

LOMAX, PETER, b Eng, May 12, 28; m 57; c 3. PHARMACOLOGY. *Educ:* Univ Manchester, MB & ChB, 54, MD, 64, DSc, 71. *Prof Exp:* Resident surgeon neurosurg, Univ Manchester, 54-57, lectr physiol, 57-61, asst prof, 61-69; assoc prof, 69-75, PROF PHARMACOL, SCH MED, UNIV CALIF, LOS ANGELES, 75- *Mem:* AAAS; Am Physiol Soc; Am Soc Pharmacol & Exp Therapeut; Brit Med Asn. *Res:* Pharmacological and immunological studies of drug abuse; role of the central nervous system in cold acclimation; effect of drugs on temperature regulation; experimental epilepsy. *Mailing Add:* Dept Pharmacol Univ Calif at Los Angeles Sch Med Los Angeles CA 90024

LOMAX, RONALD J(AMES), b Stockport, Eng, July 18, 34; m 64; c 2. SOLID STATE DEVICES, COMPUTER SIMULATION. *Educ:* Cambridge Univ, BA, 56, MA & PhD(appl math), 60. *Prof Exp:* From vis asst prof to assoc prof, 61-73, PROF ELEC ENG & COMPUT SCI, UNIV MICH, ANN ARBOR, 73- *Concurrent Pos:* Vis prof, Stanford Univ, 77-78. *Mem:* Inst Elec & Electronics Engrs; Soc Indust & Appl Math; Cambridge Philos Soc. *Res:* Solid-state devices; electron device modeling; finite element method; very large scale integrated circuit design. *Mailing Add:* Dept of Elec Eng & Comput Sci Univ of Mich Ann Arbor MI 48109-2122

LOMBARD, DAVID BISHOP, b Lexington, Mass, June 10, 30; m 52; c 5. EXPERIMENTAL PHYSICS. *Educ:* Northeastern Univ, BS, 53; Pa State Univ, MS, 55, PhD(physics), 59. *Prof Exp:* Sr physicist, Lawrence Livermore Lab, Univ Calif, 59-70; mgr, Atcor, Inc, 70-71; pres, Geo-Resource Assocs, 71-72; vpres, Subcom, Inc, 72-74; prog mgr, NSF, 74; br chief, geothermal energy div, Energy Res Develop Admin, 75-77; br chief, geothermal & biomass progs, 77-83, asst dir, off renewable technol, 83-87, TEAM LEADER, GEOTHERMAL RES, DEPT ENERGY, 87- *Mem:* Sigma Xi; Am Phys Soc; Soc Petrol Engrs. *Res:* Geopressured geothermal energy; neutron physics, fission-to-indium age of neutrons in water; strong shocks in solids; applications of nuclear explosions to natural gas stimulation; oil shale; mining and leaching; disposal of radioactive wastes. *Mailing Add:* Geothermal Div Dept Energy Washington DC 20585

LOMBARD, JULIAN H, b El Paso, Tex, Oct 31, 47. PHYSIOLOGY, ZOOLOGY. *Educ:* Univ Tex, El Paso, BA, 69; Ariz State Univ, MS, 71; Med Col Wis, PhD(physiol), 75. *Prof Exp:* From asst prof to assoc prof, 77-88, PROF PHYSIOL, MED COL WIS, 88- *Concurrent Pos:* Nat Res Serv award, NIH, 75-77; Young Investr res grant, Nat Heart, Lung & Blood Inst, 78-81; estab investr, Am Heart Asn, 85-; mem, Coun High Blood Pressure Res, Am Heart Asn; mem coun, Microcirulatory Soc. *Mem:* Am Physiol Soc; Soc Exp Biol & Med; Sigma Xi; Microcirculatory Soc; Shock Soc. *Res:* Vascular smooth muscle physiology; physiology of the microcirculation; local regulation of blood flow and nervous control of small blood vessels during hemorrhage, low flow states, and hypertension. *Mailing Add:* Dept Physiol Med Col Wis 8701 Watertown Plank Rd Milwaukee WI 53226

LOMBARD, LOUISE SCHERGER, b Wichita, Kans, Nov 20, 21; m 48; c 4. PATHOLOGY. *Educ:* Kans State Univ, DVM, 44; Univ Wis, MS, 47, PhD(path), 50. *Prof Exp:* Assoc vet, Morgan's Animal Hosp, 44-45; diagnostician, Corn States Serum Co, 45-46; instr path, Univ Wis, 46-50; instr, Woman's Med Col Pa, 50-51; res assoc virol, Univ Pa, 51-53, asst prof path, 53-55; biologist, Nat Cancer Inst, 55-57; assoc pathologist, Argonne Nat Labs, Ill, 57-64; assoc prof path, Stritch Sch Med, Loyola Univ Chicago, 64-69; res scientist, Univ Chicago, 69-70, assoc prof path & pharmacol, 70-73; sect head path, Abbott Labs, Abbott Park, Ill, 73-75; assoc path, 75-77, VET PATHOLOGIST, ARGONNE NAT LABS, ILL, 77- *Concurrent Pos:* Consult pathologist, Chicago Zool Park, 60- & Argonne Nat Lab, 64-75. *Mem:* AAAS; Am Soc Exp Path; Wildlife Dis Asn; Am Vet Med Asn; Sigma Xi. *Res:* Neoplasms in animals; chemical carcinogenesis; viral oncology; radiobiology. *Mailing Add:* 373 Borica Dr Danville CA 94526-5457

LOMBARD, PORTER BRONSON, b Yakima, Wash, Feb 6, 30; m 55; c 3. HORTICULTURE. *Educ:* Pomona Col, BA, 52; Wash State Univ, MS, 55; Mich State Univ, PhD(hort), 58. *Prof Exp:* Asst horticulturist, Citrus Exp Sta, Calif, 58-63; assoc prof, 63-70, PROF HORT, ORE STATE UNIV, 70-, SUPT, SOUTHERN ORE EXP STA, 63- *Mem:* AAAS; Am Soc Hort Sci. *Res:* Pear varieties; rootstocks; nutrition, pear fruit bud hardiness and water requirements. *Mailing Add:* Dept Hort Ore State Univ Cordley Hall Rm 2041 Corvallis OR 97331

LOMBARD, RICHARD ERIC, b Brooklyn, NY, May 16, 43; m 67; c 2. MORPHOLOGY. *Educ:* Hanover Col, AB, 65; Univ Chicago, PhD(anat), 71. *Prof Exp:* Res assoc, Mus Vert Zool, Univ Calif, Berkeley, 71; res assoc, Univ Southern Calif, 71-72; asst prof, 72-78, ASSOC PROF ANAT & EVOLUTIONARY BIOL, UNIV CHICAGO, 78-; RES ASSOC, FIELD MUSEUM NATURAL HIST, 81- *Mem:* AAAS; Am Soc Ichthyologists & Herpetologists; Am Soc Zoologists; Soc Study Amphibians & Reptiles; Soc Study Evolution. *Res:* The evolutionary and functional morphology of major adaptive features of lower vertebrates including auditory periphery in frogs, feeding apparatus of frogs and salamanders and the vestibular system in salamanders. *Mailing Add:* Dept Anat Rm 203 Univ Chicago Chicago IL 60637

LOMBARDI, GABRIEL GUSTAVO, b Buenos Aires, Arg, Sept 5, 54. LASERS, NON-LINEAR OPTICS. *Educ:* Univ Chicago, BA, 75; Harvard Univ, PhD(physics), 80. *Prof Exp:* Res assoc, Nat Bur Standards, 80-82; mem tech staff, TRW Inc, 83-84; MEM RES TECH STAFF, NORTHROP RES & TECHNOL CTR, 84- *Concurrent Pos:* Mem energy study group, Forum Physics & Soc, Am Phys Soc, 88- *Mem:* Optical Soc Am; Am Phys Soc. *Res:* Experimental research in optical phase conjugation; stimulated brillouin scattering; stimulated raman scattering; atomic spectroscopy; gas discharge lasers. *Mailing Add:* Northrop Res & Technol Ctr One Res Park Palos Verdes CA 90274

LOMBARDI, JOHN ROCCO, b June 10, 41; US citizen; c 1. PHYSICAL CHEMISTRY. *Educ:* Cornell Univ, AB, 63; Harvard Univ, AM, 66, PhD(chem), 67. *Prof Exp:* Asst prof chem, Univ Ill, 67-72; vis scientist physics, Univ Leiden, Neth, 72-73; vis scientist chem, Mass Inst Technol, 73-75; ASSOC PROF CHEM, CITY COL, CITY UNIV NY, 75- *Mem:* Int Photochem Soc; Sigma Xi. *Res:* Laser spectroscopy; molecular structure; scattering. *Mailing Add:* Dept Chem City Col CUNY New York NY 10021

LOMBARDI, MAX H, b Huanuco City, Peru, Apr 25, 32; m 61; c 3. RADIATION BIOLOGY, NUCLEAR MEDICINE. *Educ:* Univ Lima, BSc & DVM, 58; Cornell Univ, MSc, 61, Am Bd Sci Nuclear Med, cert, 79. *Prof Exp:* From asst prof to assoc prof biochem & nutrit, Vet Col Peru, 60-64; scientist biomed appln & consult, lectr & overall coord progs Latin Am, Oak Ridge Assoc Univs, 64-68, sr scientist & coordr, Radiation Biol & Med Radioisotope Training Progs, Oak Ridge Assoc Univs, 68-77; PROF NUCLEAR MED, HILLSBOROUGH COMMUNITY COL, 77- *Concurrent Pos:* Asst dir in vitro div, Tampa Gen Hosp, 77-79. *Mem:* Soc Nuclear Med; Clin Ligand Assay Soc; Word Fedn Nuclear Med & Biol. *Res:* author of 24 publications in three languages. *Mailing Add:* PO Box 30030 Tampa FL 33630

LOMBARDI, PAUL SCHOENFELD, b Salt Lake City, Utah, Nov 13, 40; m 68; c 3. MICROBIOLOGY, PHYSICAL SCIENCE EDUCATION. *Educ:* Univ Utah, BA, 63, MA, 65; Univ Rochester, PhD(microbiol), 69. *Prof Exp:* Instr, 71-73, asst prof microbiol, Col Med, Univ Utah, 73-78; TEACHER, DAVIS SCH DIST, 85- *Concurrent Pos:* Damon Runyon Mem Fund fel, Swiss Inst Exp Cancer Res, 69-70; Am Cancer Soc fel, Univ Utah, 71-73, NIH grant, 74-76. *Mem:* Am Soc Microbiol; Nat Sci Teachers Asn. *Res:* Cell-virus interactions of polyoma virus in permissive cells; structural proteins of polyoma virions; mycoplasma viruses and their interactions with mammalian cells. *Mailing Add:* 1026 N Oakridge Dr Centerville UT 84014

LOMBARDINI, JOHN BARRY, b San Francisco, Calif, July 2, 41; m 68; c 2. PHARMACOLOGY. *Educ:* St Mary's Col Calif, BS, 63; Univ Calif, San Francisco, PhD(biochem), 68. *Prof Exp:* Fel, Sch Med, Johns Hopkins Univ, 68-72, res assoc pharmacol, 72-73; asst prof, 73-77, ASSOC PROF PHARMACOL, TEX TECH UNIV, HEALTH SCI CTR, 77- *Mem:* Am Soc Pharmacol & Exp Therapeut. *Res:* Function of taurine as a possible neurotransmitter or modulator of nerve impulses; role of taurine in cardiac and retinal tissues; formation, function and regulatory properties of S-adenosylmethionine synthetase. *Mailing Add:* Dept Pharmacol Sch Med Tex Tech Health Sci Ctr Lubbock TX 79430

LOMBARDINO, JOSEPH GEORGE, b Brooklyn, NY, July 1, 33; m 60; c 3. RESEARCH ADMINISTRATION. *Educ:* Brooklyn Col, BS, 54; Polytech Univ, PhD, 58. *Prof Exp:* Sr res investr, 77-79, res adv, 79-86, DIR DEVELOP PLANNING, PFIZER, INC, 86- *Mem:* Am Chem Soc; Int Soc Heterocyclic Chem; Inflammation Res Asn; fel Am Inst Chemists. *Res:* Synthetic organic medicinals; nitrogen heterocycles; anti-inflammatory drugs; immunoregulatory drugs. *Mailing Add:* Pfizer Inc Cent Res Groton CT 06340

LOMBARDO, ANTHONY, b Brooklyn, NY, Jan 4, 39. ORGANIC CHEMISTRY. *Educ:* Queens Col, NY, BS, 61; Syracuse Univ, PhD(org chem), 67. *Prof Exp:* Fel, Univ Calif, Santa Barbara, 67-68; from asst prof to assoc prof, 68-82, PROF CHEM, FLA ATLANTIC UNIV, 82- & CHMN, 83- *Mem:* Sigma Xi. *Res:* Coenzyme models; donor-acceptor complexes; kinetics; spectroscopy. *Mailing Add:* Dept of Chem Fla Atlantic Univ Boca Raton FL 33432

LOMBARDO, PASQUALE, analytical chemistry, for more information see previous edition

LOMBARDO, R(OSARIO) J(OSEPH), b Pawcatuck, Conn, Oct 17, 21; c 3. CHEMICAL ENGINEERING. *Educ:* Univ RI, BS, 43, MS, 47; Pa State Univ, PhD(chem eng), 51. *Prof Exp:* Engr, Hamilton Standard Div, United Aircraft Corp, 43-46; chem engr, 51-56, tech supvr, 56-59, tech supt, 59-60, asst plant mgr, 61, asst dir tech serv lab, 61-64, mgr plants tech sect, Pigments Dept, 65-68, prod mgr, Chem & Pigments Dept, 68-79, MGR MFT SERV, CHEM & PIGMENTS DEPT, E I DU PONT DE NEMOURS & CO, INC, 79- *Mem:* Am Inst Chem Engrs; Sigma Xi. *Res:* Engineering administration; production administration. *Mailing Add:* 1307 Copley Dr Wilmington DE 19803

LOMBOS, BELA ANTHONY, b Apr 22, 31; Can citizen; m 56. MATERIALS SCIENCE, MICROELECTRONICS. *Educ:* Univ Szeged, BSc, 55; Univ Montreal, PhD(spectros), 67. *Prof Exp:* Sci staff, Cent Labs Construct Mat, 55-56, Battelle Mem Inst, 56-59 & Res & Develop, Northern Elec Co, 59-64; fel Nat Ctr Sci Res, France, 67-69; assoc prof, Sir Georges Williams Univ, 69-74; PROF ENG, CONCORDIA UNIV, 74- *Concurrent Pos:* Consult, Northern Elec Co, Ltd, 64-67; Silonex, Inc, 86- *Mem:* Electrochem Soc; Am Soc Crystal Growth. *Res:* Electronic materials sciences; low energy gap semiconductors for photovoltaic infrared detectors; technology of gallium arsenide: semi-insulating gallium arsenide for IC's; semi-magnetic semiconductors. *Mailing Add:* Dept Elec Eng 1455 de Maisonneuve Blvd Montreal PQ H3G 1M8 Can

LOMEDICO, PETER T, MOLECULAR BIOLOGY. *Prof Exp:* SR DIR MOLECULAR BIOL, HOFFMAN-LA ROCHE INC, 89- *Mailing Add:* Dept Molecular Biol Hoffman-La Roche Inc Bldg 102 Nutley NJ 07110

LOMEN, DAVID ORLANDO, b Decorah, Iowa, May 11, 37; m 61; c 1. APPLIED MATHEMATICS. *Educ:* Luther Col, Iowa, BA, 59; Iowa State Univ, MS, 62, PhD, 64. *Prof Exp:* Design specialist, Gen Dynamics & Astronaut, 63-66; from asst prof to assoc prof, 69-74, PROF MATH, UNIV ARIZ, 74 - *Concurrent Pos:* Consult var industs; Marshall Fund Award, Norway-Am Found, 80, 83; vis sr scientist, Norway, 80. *Mem:* Soc Indust & Appl Math; Am Math Soc; Soil Sci Soc Am; Geophys Union; Europ Geophys Soc. *Res:* Modeling water and solute flow in soils; curriculum and software development in mathematics. *Mailing Add:* Dept Math Univ Ariz Tucson AZ 85721

LOMMEL, J(AMES) M(YLES), b Evanston, Ill, Feb 7, 32; m 59; c 2. INFORMATION SCIENCE, METALLURGY. *Educ:* Ill Inst Technol, BS, 53, MS, 54; Harvard Univ, PhD(appl physics), 58. *Prof Exp:* Metallurgist, H M Harper Co, 54-56; metallurgist, Gen Elec Res & Develop Ctr, 57-69, mgr personnel & tech admin, Electronics Sci & Eng, 69-77, mgr info res oper, 77-83, consult, Info Systs, 83-87, mgr, Info Ctr, 88-91, MGR SUPPORT SERV, GEN ELEC RES & DEVELOP CTR, 91- *Concurrent Pos:* Teaching asst, Tufts Univ, 55 & Harvard Univ, 55-56. *Mem:* Inst Elec & Electronics Engrs; Am Inst Mining, Metall & Petrol Engrs. *Res:* Physical metallurgy; magnetic materials and recording; computer systems service; information retrieval. *Mailing Add:* Gen Elec Res & Develop Ctr PO Box 8 Schenectady NY 12301

LOMNITZ, CINNA, b Cologne, Ger, May 4, 25; m 51; c 4. SEISMOLOGY. *Educ:* Univ Chile, CE, 48; Harvard Univ, MS, 50; Calif Inst Technol, PhD(geophys), 55. *Prof Exp:* Res fel seismol, Calif Inst Technol, 55-57; prof geophys, Univ Chile, 57-64, dir inst geophys & seismol, 58-64; assoc res seismologist, Seismog Sta, Univ Calif, Berkeley, 64-68; PROF SEISMOL, INST GEOPHYS, NAT UNIV MEX, 68- *Concurrent Pos:* Consult, Geol Surv, Chile, 58-; vis assoc, Calif Inst Technol & Univ Calif, San Diego, 69- *Mem:* Seismol Soc Am; Am Geophys Union; Sigma Xi. *Res:* Earthquake hazard; creep properties of rocks; viscoelasticity and internal friction in solids; seismicity; structure of the Andes; origin of earthquakes and tsunamis. *Mailing Add:* Geophys Inst Nat Univ Mex Mexico 04510 DF Mexico

LOMON, EARLE LEONARD, b Montreal, Que, Nov 15, 30; nat US; m 51; c 3. PARTICLE & NUCLEAR THEORY. *Educ:* McGill Univ, BSc, 51; Mass Inst Technol, PhD(theoret physics), 54. *Prof Exp:* Res physicist, Can Defence Res Bd, 50-51 & Baird Assocs, Mass, 52-53; Nat Res Coun Can overseas res fel, Inst Theoret Physics, Denmark, 54-55; fel, Weizmann Inst, 55-56; res assoc, Lab Nuclear Studies, Cornell Univ, 56-57; assoc prof theoret physics, McGill Univ, 57-60; assoc prof physics, 60-70, PROF PHYSICS, MASS INST TECHNOL, 70- *Concurrent Pos:* Guggenheim Mem Found fel, 65-66; vis scientist, Los Alamos Nat Lab, 68-; proj dir, Unified Sci & Math for Elem Sch, 71-77; vis prof, Univ Paris, 79-80, & 86-87, Univ Calif, Los Angeles, 83 & Univ Wash, Seattle, 85; adj prof, Louvain-la-Neuve, Belgium, 80; res fel, Univ Col, London, 80; vis scientist, Cern, Geneva, 65-66, KFA, Julich, WGer, 86-90. *Mem:* Am Phys Soc; Can Asn Physicists. *Res:* Nuclear and medium energy physics; field theory. *Mailing Add:* Dept Physics Mass Inst Technol Cambridge MA 02139

LOMONACO, SAMUEL JAMES, JR, b Dallas, Tex, Sept 23, 39; m 68; c 1. MATHEMATICS, COMPUTER SCIENCE. *Educ:* St Louis Univ, BS, 61; Princeton Univ, PhD(math), 64. *Prof Exp:* Asst prof math, St Louis Univ, 64-65 & Fla State Univ, 65-69; res mathematician & comput scientist, Tex Instruments, Inc, 69-71; assoc prof comput sci & math, State Univ NY Albany, 71-80; PROF, DEPT COMPUT SCI, UNIV MD, BALTIMORE COUNTY CAMPUS, 80- *Concurrent Pos:* Indust prof, Southern Methodist Univ, 69-71; actg chmn, Dept Comput Sci, State Univ NY, Albany, 73-74; vis, Inst Defense Anal, Princeton, NJ, 74-76; vis lectr, Dept Math, Princeton Univ, 75-76. *Mem:* Am Math Soc; Asn Comput Mach; Math Asn Am; Soc Indust & Appl Math; Sigma Xi. *Res:* Algebraic topology; higher dimensional knot theory; algebraic coding theory; complexity theory. *Mailing Add:* 10236 Little Brick House Ct Univ Md Baltimore County Campus Ellicott City MD 21043

LOMONT, JOHN S, b Ft Wayne, Ind, Aug 26, 24. MATHEMATICAL PHYSICS. *Educ:* Purdue Univ, MS, 47, PhD(physics), 51. *Prof Exp:* Physicist theoret solid state physics, NAm Aviation, Inc, 51-52; physicist, Res Dept, Michelson Lab, Naval Ord Test Sta, 52-54; physicist, NY Univ, 54-57 & Int

Bus Mach Corp, 57-60; prof math, Polytech Inst Brooklyn, 62-65; PROF MATH, UNIV ARIZ, 65- *Concurrent Pos:* Sabbatical, Courant Inst Math Sci, NY Univ, 71-72. *Mem:* Am Phys Soc; Am Math Soc. *Res:* Applied group theory; quantum field theory; functional analysis. *Mailing Add:* Univ Ariz Bldg 89 Tucson AZ 85721

LONADIER, FRANK DALTON, b Clarence, La, May 6, 32; m 59; c 2. PHYSICAL CHEMISTRY, INORGANIC CHEMISTRY. *Educ:* Northwestern State Col, La, BS, 54; Univ Tex, PhD(phys chem), 59. *Prof Exp:* Res asst, Los Alamos Sci Lab, Univ Calif, 57-58; sr res chemist, 59-61, group leader inorg & nuclear chem, 61-64, sect mgr mat eval, 64-65, sect mgr nuclear develop, 65-67, mgr nuclear prod, 67-69, mgr explosive technol, 69-76, mgr advan devices prod, 76-86, MGR, MFG MOUND LAB, MONSANTO RES CORP, 87- *Concurrent Pos:* Tech Safety Appraisals for Radioactive Mfg. *Mem:* Am Chem Soc; Am Soc Qual Control. *Res:* Actinide elements, particularly uranium and plutonium; inorganic chemistry of polonium; behavior of secondary explosives; environmental pollutant abatement. *Mailing Add:* 221 Estates Dr Dayton OH 45459

LONARD, ROBERT (IRVIN), b Valley Falls, Kans, June 5, 42; m 65; c 1. PLANT TAXONOMY. *Educ:* Kans State Teachers Col, BSE, 64, MS, 66; Tex A&M Univ, PhD(plant taxon), 70. *Prof Exp:* ASST PROF BIOL, PAN AM UNIV, 70- *Mem:* AAAS; Am Soc Plant Taxonomists; Int Asn Plant Taxonomists. *Res:* Flora of south Texas; grass systematics. *Mailing Add:* Dept Biol Pan Am Univ Edinburg TX 78539

LONBERG-HOLM, KNUD KARL, b New York, NY, Sept 22, 31; m 52, 61; c 3. BIOCHEMISTRY. *Educ:* Harvard Univ, BA, 53; Univ Calif, Berkeley, PhD(biochem), 62. *Prof Exp:* Chemist, Hyman Labs, Fundamental Res, Inc, 59-60; biochemist, Cent Res Dept, E I Du Pont de Nemours & Co, Inc, 62-85; CONSULT, 88- *Concurrent Pos:* USPHS fel, Univ Uppsala, 67-69; assoc prof microbiol & immunol, Sch Med, Temple Univ, 76-77; vis prof microbiol & immunol, Hahnemann Univ, 85-88; adj prof, Hahnemann Univ, 88- *Res:* Biochemistry of plasma proteins; virus-cell interaction; biochemical virology. *Mailing Add:* PO Box 95 Lockwood NY 14859

LONCRINI, DONALD FRANCIS, organic chemistry; deceased, see previous edition for last biography

LONDERGAN, JOHN TIMOTHY, b Niagara Falls, NY, Mar 13, 43; m 86; c 3. MEDIUM-ENERGY NUCLEAR THEORY. *Educ:* Univ Rochester, BS, 65; Oxford Univ, DPhil, 69. *Prof Exp:* Res assoc physics, Case Western Reserve Univ, 69-71; Univ Wis, 71-73; from asst prof to assoc prof, 73-82, PROF PHYSICS, IND UNIV, 83-, CHAIR, PHYSICS DEPT, 90- *Concurrent Pos:* Assoc dean, Grad Sch, 84-88; dir, Nuclear Theory Ctr, 85-87; vis prof, Swiss Inst Nuclear Res, 82-83; vis prof, Univ Adelaide, Australia, 89; consult, Los Alamos Nat Lab, 88- *Mem:* Am Phys Soc; Am Asn Univ Professors. *Res:* Intermediate-energy nuclear theory; photonuclear reactions; scattering theory at medium energies; structure of the nucleon. *Mailing Add:* Physic Dept Ind Univ Bloomington IN 47405

LONDON, A(LEXANDER) L(OUIS), b Nairobi, Kenya, Aug 31, 13; US citizen; m 38; c 3. MECHANICAL ENGINEERING. *Educ:* Univ Calif, BS, 35, MS, 38. *Prof Exp:* Engr, Standard Oil Co Calif, 36-37; instr, Univ Santa Clara, 37-38; from instr to prof, 38-78, EMER PROF MECH ENG, STANFORD UNIV, 78- *Concurrent Pos:* Res assoc, Argonne Nat Lab, 55-56. *Mem:* Nat Acad Eng; Am Soc Mech Engrs; Am Soc Eng Educ. *Res:* Heat transfer; thermodynamics; fluid mechanics. *Mailing Add:* Dept Mech Eng Stanford Univ Stanford CA 94305

LONDON, DAVID, b Ardmore, Okla, Feb 27, 53. METAMORPHIC PETROLOGY, FLUID INCLUSION ANALYSIS. *Educ:* Wesleyan Univ, BA, 75; Ariz State Univ, Tempe, MS, 79, PhD, 81. *Prof Exp:* Fel geol, Geophys Lab, Carnegie Inst, 81-83; ASST PROF GEOL, UNIV OKLA, 83- *Concurrent Pos:* Consult exploration, Cabot Mineral Resources, 81-83; prin investr, res grant, US Bur Mines, 83- & NSF, 86- *Mem:* Am Geophys Union; Mineralogical Asn Can; Mineralogical Soc Am. *Res:* Internal evolution of fractionated granite-pegmatite systems; emphasis on crystallization sequences, melt-vapor equilibrium and trace element partitioning and formation of rare-element deposits. *Mailing Add:* Sch Geol & Geophys Univ Okla Norman OK 73019

LONDON, EDYTHE D, b Rome, Italy, Sept 14, 48; US citizen; m 69; c 2. NEUROCHEMISTRY, NEUROPHARMACOLOGY. *Educ:* George Wash Univ, BS, 69; Towson State Univ, MS, 73; Univ Md, PhD(pharmacol), 76. *Prof Exp:* Fel psychopharmacol, Sch Med, Johns Hopkins Univ, 76-78; staff fel, Nat Inst on Aging, 79-81, pharmacologist, 81-82; pharmacologist, 84-85, CHIEF, NEUROPHARMACOLOGY LAB, NAT INST ON DRUG ABUSE, 85- *Honors & Awards:* Mathilde Salowey Award, 87. *Mem:* Soc Neuroscience; AAAS; Am Soc Pharmacol Exp Therapeut; Am Soc Neurochemistry; Int Soc Cerebral Blood Flow & Metab. *Res:* Regional cerebral metabolism and changes in neurotransmitter balance in the aging brain; localization of the actions of psychoactive drugs. *Mailing Add:* NIDA Addiction Res Ctr PO Box 5180 Baltimore MD 21224

LONDON, GILBERT J(ULIUS), b Philadelphia, Pa, May 30, 31; m 52; c 4. METALLURGICAL ENGINEERING. *Educ:* Drexel Inst Technol, BS, 53; Univ Pa, MS, 55, PhD(metall eng), 59. *Prof Exp:* Metallurgist, Aerosci Lab, Gen Elec Co, 56-59; sr res metallurgist & mgr, Mech Metall Lab, Franklin Inst, Pa, 59-70; mgr metall res & develop, Kawecki Berylco Industs, 70-75; BR HEAD STRUCT MAT, NAVAL AIR DEVELOP CTR, 75- *Concurrent Pos:* Adj prof, Drexel Univ. *Mem:* Am Soc Metals; Am Inst Mining, Metall & Petrol Engrs; Am Inst Aeronaut & Astronaut; Sigma Xi. *Res:* Flow and fracture of iron; dispersed hard particle strengthening of metals; beryllium; purification; high purity alloys; micro-strain properties; slip analysis; coextruded composites; beryllium alloys. *Mailing Add:* Naval Air Develop Ctr Code 606D Warminster PA 18974-5000

LONDON, IRVING MYER, b Malden, Mass, July 24, 18; m 55; c 2. MEDICINE. *Educ:* Harvard Univ, AB, 39, MD, 43. *Hon Degrees:* ScD, Univ Chicago, 66. *Prof Exp:* From instr to assoc prof med, Columbia Univ, 47-55; prof & chmn dept, Albert Einstein Col Med, 55-70; PROF MED, HARVARD MED SCH, 72-; PROF BIOL, MASS INST TECHNOL, 69-, GROVER M HERMANN PROF HEALTH SCI & TECH, 77- *Concurrent Pos:* Asst physician, Presby Hosp, NY, 46-52, from asst attend physician to assoc attend physician, 52-55; dir, Harvard-Mass Inst Technol prog in health sci & technol, 69-77, dir, Div Health Sci & Technol, 77-85, dir, Whitaker Col Health Sci Technol & Mgt, 78-82; vis prof, Albert Einstein Col Med, 70-; physician, Peter Bent Brigham Hosp; mem med fel bd & subcomt blood & related probs, Nat Acad Sci-Nat Res Coun, 55-63; res coun mem, Pub Health Res Inst, NY, 58-63; bd sci consult, Sloan-Kettering Inst Cancer Res, 60-72; metab study sect mem, USPHS, 60-63, chmn, 61-63, mem bd sci coun, Nat Heart Inst, 64-68; mem panel biol sci & advan med, Nat Acad Sci, 66-67, bd med, 67-70; mem adv comt to dir, NIH, 66-70; mem, Nat Cancer Adv Bd, 72-76; mem, bd sci coun, Nat Inst Arthritis, Metab & Digestive Dis, 79-83; mem bd dirs, Johnson & Johnson, 82- *Honors & Awards:* Theobald Smith Award in Med Sci, AAAS, 53; Jean Oliver Lectr, State Univ NY, 57; Roger Morris Lectr, Univ Cincinnati, 58; Stuart McGuire Lectr, Med Col Va, 60. *Mem:* Nat Acad Sci; Am Acad Arts & Sci; Asn Am Physicians; Am Soc Clin Invest (pres, 63-64); Am Soc Biol Chem; fel AAAS; Int Soc Hematol; Soc Develop Biol. *Res:* Hemoglobin metabolism; metabolism of erythrocytes; eukaryotic protein synthesis. *Mailing Add:* Mass Inst Technol Harvard-MIT Div Health Sci & Technol Building E25 Rm 551 77 Massachusetts Ave Cambridge MA 02139

LONDON, JULIUS, b Newark, NJ, Mar 26, 17; m 46; c 2. METEOROLOGY. *Educ:* Brooklyn Col, BA, 41; NY Univ, MS, 48, PhD, 51. *Prof Exp:* Meteorologist, US Weather Bur, 42; instr meteorol, US Air Force, 42-47; res assoc meteorol, NY Univ, 48-52, asst prof, 52-56, res assoc prof, 56-59, assoc prof, 59-61; chmn dept astro-geophys, 66-69, prof, 61-87, EMER PROF ASTROPHYS, PLANETARY & ATMOSPHERIC SCI, UNIV COLO, BOULDER, 87- *Concurrent Pos:* Lectr, Columbia Univ, 54-55; vis prof, Pa State Univ, 55; mem, Int Ozone Comn, Int Asn Meteorol & Atmospheric Physics; Max Planck Inst Physics, Gottingen, 58; vis res scientist, Nat Ctr Atmospheric Res, 61-66; mem, Int Ozone Comn, Int Asn Meteorol & Atmospheric Physics, 60-; chmn panel ozone, Nat Res Coun, Nat Acad Sci, 64-65, mem, Comt Human Resources, 78-81; vis prof, Swiss Fed Inst Technol, 67, 74-76; chief US deleg, XVII Gen Assembly, Int Asn Meteorol & Atmospheric Physics, 79; lectr, Chinese Acad Sci, Inst Atmospheric Physics, 80. *Mem:* AAAS; Sigma Xi; Am Geophys Union; Int Radiation Comn (secy, 63-71, pres, 71-79). *Res:* Atmospheric radiation; physics of the atmosphere; ozone. *Mailing Add:* Astrophys Plantetary & Atmospheric Sci Dept Univ Colo Campus Box 391 Boulder CO 80309-0391

LONDON, MARK DAVID, b Brooklyn, NY, May 24, 47; m 70; c 2. IMPACT ASSESSMENT, MANAGEMENT. *Educ:* C W Post Col, Long Island Univ, BS, 70, MS, 74. *Prof Exp:* Biologist, Eng Sci, Inc, 69-72; environ scientist, Woodward-Clyde Consult, 72-76; biologist, Pub Serv Elec & Gas Co, 76-77, lead biologist, 77-79, sr staff biologist, 79-80, sr biologists, 80, environ studies mgr, 82-88; dir, Environ Rev Div, New York Dept City Planning, 88-91; ASST VPRES, ENVIRO-SCI, 91- *Concurrent Pos:* Mem, Twp Denville, NJ Environ Comn, 80-81, chmn, 81-87. *Mem:* Am Soc Testing & Mats; Soc Power Indust Biologists; Am Soc Limnol & Oceanog; Edison Elec Inst Biologist; Am Inst Biol Sci. *Res:* Director and chief reviewer of New York City's City Environmental Quality Review Process reviewing all non-as-of-right-construction in New York City. *Mailing Add:* Enviro-Sci 111 Howard Blvd Mt Arlington NJ 07856

LONDON, MORRIS, BIOCHEMISTRY, ENZYMOLOGY. *Educ:* Ohio State Univ, PhD(physiol), 50. *Prof Exp:* CHIEF CLIN CHEM, BROOKDALE MED CTR, 71- *Res:* Clinical chemistry. *Mailing Add:* Brookdale Hosp Med Ctr Linden Blvd at Brookdale Plaza Brooklyn NY 11212

LONDON, RAY WILLIAM, b Burley, Idaho, May 29, 43. STRESS-CRISIS MANAGEMENT, BEHAVIORAL MEDICINE. *Educ:* Weber State Col, BS, 67; Univ Southern Calif, MSW, 73, PhD(psychol), 76, MBA, 89. *Prof Exp:* Clin trainee & fel, Vet Admin & Children's Hosp,71-74; CONSULT & PVT PRACT, ST JOSEPH HOSP, 73- *Concurrent Pos:* Res affil, Ctr Crisis Mgt, Univ Southern Calif, 87-; adv ed, Int J Clin & Exp Hypn, 81-; clin fac, Univ Calif Irvine Col Med, 78-; fel, Inst Social Scientist Neurobiology & Mental Illness, 78; mem fac, Univ Calif, Los Angeles, Univ Southern Calif, Calif State Univ, 76-86; res assoc, Nat Comt Protection Human Subjects Biomed & Behav Res, 76; res assoc, Bus Adv, Inc, 65-67; dir, Meaning Found, 66-69; mental health liasion, San Bernardino County Social Serv, 68-72; pres, Human Factors Prog, 76-; dir, Human Studies Ctr, 87-; chief exec officer, London Assoc Int, 76-; Erickson Scholar Dipl, Neuropsychology, Med Psychol, Family Psychol, Admin Psychol, Clin Hypn. *Mem:* Int Acad Med & Psychol (pres, 80-); Soc Clin & Exp Hypn (treas, 87-89); Int Psychosomatic Inst; Am Bd Psychol Hypn (pres, 89-); Am Bd Clin Hypn (pres, 90-); Int Consults Found. *Res:* Scientific investigation, integration and application of medical psychology, behavioral medicine, psychophysiology, hypnosis, cognitive, behavioral, organizational and psychosocial data to stress, crisis and human performance issues, problems and policy concerns. *Mailing Add:* 1125 E 17th St Suite E-209 Santa Ana CA 92701-2214

LONDON, ROBERT ELLIOT, b Brooklyn, NY, Oct 25, 46; m 69; c 2. BIOPHYSICAL CHEMISTRY. *Educ:* Brooklyn Col, BS, 67; Univ Ill, MS, 69, PhD(physics), 73. *Prof Exp:* Fel, Los Alamos Nat Lab, 73-75, staff mem biophy chem, 75-83; NMR group leader, Lab Molecular Biophysics, Nat Inst Environ Health Sci, 84- *Mem:* Am Soc Biochem & Molecular Biol; Biophys Soc. *Res:* Nuclear magnetic resonance studies of biologically important molecules. *Mailing Add:* MD 17-05 Nat Inst Environ Health Sci Box 12233 Research Triangle Park NC 27709

LONDON, WILLIAM THOMAS, b New York, NY, Mar 11, 32; m 57; c 4. INTERNAL MEDICINE, ENDOCRINOLOGY. *Educ:* Oberlin Col, BA, 53; Cornell Univ, MD, 57. *Prof Exp:* Intern med, Bellevue Hosp, 57-58, resident, Med Ctr, 58-60; res epidemiologist, Nat Inst Arthritis & Metab Dis, 62-66; res physician, Inst Cancer Res, 66-78; assoc, 66-71, asst prof med, 71-76, assoc prof, 76-78, ADJ PROF MED, SCH MED, UNIV PA, 78-; AT FREDERICK CANCER RES FAC, MD. *Concurrent Pos:* Fel endocrinol, Sloan-Kettering Inst, NY, 60-62; asst, Med Col, Cornell Univ, 60-62; instr, Sch Med, George Washington Univ, 64-; sr res physician, Int Cancer Res, 78- *Mem:* Am Thyroid Asn; Am Asn Cancer Res. *Res:* Susceptibility factors to cancer; variations in host response to hepatitis B infection. *Mailing Add:* Fox Chase Cancer Ctr 7701 Burholme Ave Philadelphia PA 19111

LONE, M(UHAMMAD) ASLAM, b East Punjab, India, Jan 28, 37; m 70; c 3. EXPERIMENTAL NUCLEAR PHYSICS. *Educ:* Punjab Univ, West Pakistan, BSc, 58, MSc, 60; State Univ NY Stony Brook, PhD(physics), 67. *Prof Exp:* Lectr physics, Govt Col, Lahore, Pakistan, 60-62; fel, Ind Univ, Bloomington, 67-68; Nat Res Coun Can fel, 68-70, from res officer physics to assoc res officer, 70-83, SR RES SCIENTIST PHYSICS, CHALK RIVER LABS, ATOMIC ENERGY CAN LTD, 83- *Concurrent Pos:* Mem, Nuclear Energy Agency, Nuclear Data Comt. *Mem:* Can Asn Physicists; Am Phys Soc; Can Radiol Prof Asn. *Res:* Nuclear spectroscopy by gamma ray, neutron, and charged particle induced reactions; investigation of nuclear reaction mechanism; radiation physics, utilization of nuclear radiation for industrial processing; industrial neutron sources. *Mailing Add:* Chalk River Labs Atomic Energy Can Ltd Chalk River ON K0J 1J0 Can

LONERGAN, DENNIS ARTHUR, b West Bend, Ind, May 30, 49; m 80. FOOD SCIENCE, FOOD TECHNOLOGY. *Educ:* Univ Wis-Madison, BS, 71, MS, 75, PhD(food sci), 78. *Prof Exp:* Scientist res, Pillsbury Co, 78-80; asst prof food analysis, Purdue Univ, 80-83; scientist res, Pillsbury Co, 83-90; GOLDEN VALLEY MICRO, WARE FOOD INC, 90- *Mem:* Inst Food Technologists. *Res:* Functionality of casein as a food ingredient; methods of determining water mobility in food; membrane processing of foods. *Mailing Add:* 1825 CR 24 Long Lake MN 55356

LONERGAN, THOMAS A, b Syracuse, NY. BIOLOGICAL SCIENCE. *Prof Exp:* PROF BIOL, UNIV NEW ORLEANS, 88- *Mailing Add:* Biol Sci Dept Univ New Orleans 2000 Lakeshore Dr New Orleans LA 70148

LONEY, ROBERT AHLBERG, b Odebolt, Iowa, June 16, 22; wid; c 3. STRUCTURAL GEOLOGY, PETROLOGY. *Educ:* Univ Wash, BS, 49, MS, 51; Univ Calif, Berkeley, PhD(geol), 61. *Prof Exp:* Geologist, Superior Oil Co, Tex, 51-52 & Wyo, 52-54; GEOLOGIST, US GEOL SURV, 56- *Mem:* Am Geophys Union; Geol Soc Am; Mineral Soc Am; Ger Geol Asn. *Res:* Structural petrology and petrology of mafic-ultramafic complexes and associated terranes; Pacific coastal region. *Mailing Add:* 12112 Foothill Lane Los Altos Hills CA 94022

LONG, ALAN JACK, b Baton Rouge, La, Oct 17, 44; m 66; c 2. FOREST ECOLOGY, FOREST GENETICS. *Educ:* Univ Calif, Berkeley, BS, 67, MS, 71; NC State Univ, PhD(forestry, genetics), 73. *Prof Exp:* Asst prof forest genetics, Pa State Univ, 73-74; res scientist regeneration ecol, Weyerhaeuser Co, 74-79; field sta mgr trop forestry res, Indonesia, 79-80; forestry res field sta mgr, Weyerhaeuser Co, 80-87. *Res:* Technology requisite for plantation establishment and early growth of western conifers; use of clonal material in tree improvement and regeneration programs; root growth of conifer seedlings. *Mailing Add:* Dept Forestry Univ Fla 118 Newins-Ziegler Hall Gainesville FL 32611

LONG, ALAN K, b Burlington, Vt, June 19, 50; m 84; c 1. SYNTHETIC ORGANIC & NATURAL PRODUCTS CHEMISTRY. *Educ:* Yale Univ, BS, 71; Harvard Univ, MA, 76, PhD(chem), 79. *Prof Exp:* RES ASSOC CHEM, DEPT CHEM, HARVARD UNIV, 79- *Concurrent Pos:* Pres, Comput Assisted Synthetic Analysis Group, 83- *Mem:* Am Chem Soc. *Res:* Development of the LHASA computer program for computer-assisted analysis of problems in synthetic organic chemistry; coordination of database expansion for LHASA. *Mailing Add:* Dept Chem Harvard Univ 12 Oxford St Cambridge MA 02138-2902

LONG, ALEXANDER B, b New York, NY, Jan 16, 43; m 66. NUCLEAR ENGINEERING. *Educ:* Williams Col, BA, 64; Univ Ill, Urbana, MS, 66, PhD(nuclear eng), 69. *Prof Exp:* Asst nuclear engr, Argonne Nat Lab, 69-78; mem staff elec power res, Nuclear Safety Anal, 78-; prog mgr, Elec Power Res Inst, 74-83; pres, Expert Ease Systs, 83-90; CONSULT. *Mem:* Am Nuclear Soc; Inst Elec & Electronics Engrs. *Res:* Reactor physics, especially experimental techniques for on line determination of reactor physics parameters; fission physics. *Mailing Add:* 1055 Whitney Dr Menlo Park CA 94025

LONG, ALEXIS BORIS, b New York, NY, Sept 9, 44; m 74; c 3. CLOUD PHYSICS, WEATHER MODIFICATION. *Educ:* Reed Col, BA, 65; Syracuse Univ, MS, 66; Univ Ariz, PhD(atmospheric sci), 72. *Prof Exp:* Res asst cloud physics, Inst Atmospheric Physics, Univ Ariz, 69-72; NSF fel & vis scientist, Div Cloud Physics, Commonwealth Sci & Indust Res Orgn, 72-73; res assoc cloud physics, Coop Inst Res Environ Sci, Univ Colo, 73-75; scientist & head hail suppression group, Nat Hail Res Exp & Convective Storms Div, Nat Ctr Atmospheric Res, 75-79; assoc res scientist, dept meteorol, Texas A & M Univ, 79-81; ASSOC RES PROF, ATMOSPHERIC SCI CTR, DESERT RES INST, RENO, NEV, 81-; PRIN RES SCIENTIST, DIV ATMOSPHERIC RES, COMMONWEALTH SCI & INDUST RES ORGN, AUSTRALIA, 87- *Concurrent Pos:* Mem comt cloud physics, Am Meteorol Soc, 76-82, consult meteorlogist, 84-; assoc ed, J Climate & Appl Meteorol, 85- *Mem:* Am Meteorol Soc; Am Geophys Union; Royal Meteorol Soc; Sigma Xi. *Res:* Cloud physics; precipitation processes in convective and orographic clouds; Doppler radar meteorology and microwave remote sensing; mesometeorological interactions with terrain; precipitation forecasting and modification in mountainous regions; hail measuring systems and methods. *Mailing Add:* Atmospheric Sci Ctr Desert Res Inst Reno NV 89506-0220

LONG, ALTON LOS, JR, b Liberty, Tex, Sept 25, 32; m 55; c 4. ELECTRONICS, MATERIALS TECHNOLOGY. *Educ:* Carnegie Inst Technol, BS, 53, MS, 55; Univ Penn, MS cand, 88. *Prof Exp:* Jr res chemist radiochem, Carnegie Inst Technol, 53-54; unit chief radiation effects, US Army Signal Res & Develop Labs, Ft Monmouth, 57-60, nuclear scientist, 60; develop engr lab, Unisys Defense Systs, 60-61, supvr testing & eval sect, 61-65, staff engr, Adv Develop Dept, 65-70, prog mgr, Comput Microfilm Systs, 70-72, prog mgr, Illiac IV Syst, 72-73, dept mgr, Components Eval, 73-77, prog mgr advan technol, 77-81, dir opers, Spec Devices Div, Systs Develop Corp, Burroughs Corp, 82-87, Mgr, Infusec Prod & Technols, 87-90, PROG MKT MGR, GOV PROG, COMPUTER SYSTS PROD GROUP, UNISYS DEFENSE SYSTS, 90- *Concurrent Pos:* Instr, Monmouth Col, 58-59. *Mem:* Armed Forces Commun & Electronics Asn; Inst Elec & Electronic Engrs; Sigma Xi; Nat Mgt Asn; Inst Cert Prof Mechs. *Res:* Microelectronics; information science; radiation effects on materials; electronic materials; environmental science; physics of failure; radiocarbon dating; applied radiation technology; interconnection and packaging technology. *Mailing Add:* 558 Willis Lane Wayne PA 19087

LONG, ANDREW FLEMING, JR, b Amboy, WVa, Dec 20, 38. MATHEMATICS. *Educ:* WVa Univ, BS, 60, MS, 61; Duke Univ, PhD(math), 65. *Prof Exp:* Asst prof math, St Andrews Presby Col, 65-67; asst prof, 67-75, ASSOC PROF MATH, UNIV NC, GREENSBORO, 75- *Mem:* Math Asn Am; Sigma Xi; Asn Comput Mach. *Res:* Irreducible factorable polynomials over a finite field; number theory; computer software. *Mailing Add:* Dept of Math Univ of NC at Greensboro Greensboro NC 27412

LONG, AUSTIN, b Olney, Tex, Dec 12, 36; m 61; c 2. GEOCHEMISTRY. *Educ:* Midwestern Univ, BS, 57; Columbia Univ, MA, 59; Univ Ariz, PhD(geochem), 66. *Prof Exp:* Res asst geochem, Geochronol Labs, Univ Ariz, 59-63; geochemist, Smithsonian Inst, 63-68; assoc prof geoscience, 68-87, PROF GEOSCIENCE, HYDROL & WATER RESOURCES & CHIEF SCIENTIST, LAB OF ISOTOPE GEOCHEM, UNIV ARIZ, 87- *Mem:* Geochemistry Soc. *Res:* Pleistocene paleoclimatology; radiocarbon dating; stable isotope geochemistry. *Mailing Add:* Dept Geosci Univ Ariz Tucson AZ 85721

LONG, BILLY WAYNE, b Tupelo, Miss, Apr 5, 48; m 72; c 3. PHYSIOLOGY. *Educ:* David Lipscomb Col, Nashville, BA, 69; Univ Miss, MD, 73. *Prof Exp:* Intern & resident med, Univ Miss, 73-75; clin assoc digestive dis, NIH, 75-77; fel gastroenterol, Univ Pa, 77-79, asst prof med, 79-81; ASST PROF MED, UNIV MISS, 81- *Mem:* Fel Am Col Gastroenterol; fel Am Col Physicians; Am Soc Gastrointestinal Endoscopy; AMA. *Res:* Physiology of pancreatic exocrine secretion; physiology of gastrointestinal hormones. *Mailing Add:* 755 Gillespie Pl Jackson MS 39202

LONG, CALVIN H, b Myerstown, Pa, Feb 16, 27; m 54; c 2. ANALYTICAL CHEMISTRY. *Educ:* Univ Miami, BS, 50; Franklin & Marshall Col, MS, 56; Stanford Univ, PhD(chem), 63. *Prof Exp:* Chemist, Armstrong Cork Co, 50-58; res chemist, Chevron Res Co, 63-64; res group leader anal chem, 64-68, sect mgr, 69-78, MGR, KERR-MCGEE CORP, 79- *Mem:* Am Chem Soc. *Res:* Chemical equilibria; mineral benefication. *Mailing Add:* 201 W Penn Rd Lehigh Acres FL 33936

LONG, CALVIN LEE, b NC, Jan 27, 28; m 51; c 3. BIOCHEMISTRY. *Educ:* Wake Forest Col, BS, 48; NC State Col, MS, 51; Univ Ill, PhD, 54. *Prof Exp:* Assoc chemist biochem, Gen Food Corp, 54-57, proj leader, 57-62; res assoc, Harvard Univ, 63 & Col Physicians & Surgeons, Columbia Univ, 64-74; from assoc prof to prof Biochem & Surg, Med Col Ohio, 75-84; ADJ PROF NUTRIT SCI, UNIV ALA, BIRMINGHAM, 84-, DIR RES, BAPTIST MED CTR, 84- *Mem:* AAAS; Am Inst Nutrit; Am Chem Soc; NY Acad Sci; Am Soc Parenteral & Enteral Nutrit. *Res:* Intermediary metabolism and nutritional biochemistry. *Mailing Add:* 701 Princeton Ave SW Birmingham AL 35211

LONG, CALVIN THOMAS, b Rupert, Idaho, Oct 10, 27; m 52; c 2. ELEMENTARY NUMBER THEORY, COMBINATORIAL NUMBER THEORY. *Educ:* Univ Idaho, BS, 50; Univ Ore, MS, 52, PhD(math), 55. *Prof Exp:* Analyst, Nat Security Agency, 55-56; from asst prof to assoc prof, 56-65, chmn dept, 70-78, PROF MATH, WASH STATE UNIV, 65- *Concurrent Pos:* Educ consult, Wash State Dept Educ, 61-67 & NSF, 63-83; vis prof, Univ Jabalpur, India, 65, Univ BC, 72, Clemson Univ, 78-79, Portland State Univ, 79; consult, Educ Comn States, Nat Assessment Educ Progress, 75; educ consult, Rand McNally & Co, 75-77, William Clare, Ltd, 73-77, Wash State Dept Educ, 83, 87-88, 89, NSF, 85-87, Prentice Hall Publ Co, 87-; chair, Comt Rev Guidelines for the Accreditation of Col Math Progs, 74-78, mem, 78-80, mem, Comt Adult Educ, 78-80, Ad Hoc Comt NCATE Guidelines, 79-80, Comt Employment of Math, 80-88, chair, 83 & 85, Comt Math Educ of Teachers, 83-88, mem Ad Hoc Comt Accreditation, 86-90; assoc ed, Math Mag, 86-90; mem, Coun Conf Bd Math Sci, 78-80; mem, Task Force on Post Baccalaureate Educ of Teachers, Nat Coun Teachers of Math, 86-88. *Mem:* Math Asn Am; Nat Coun Teachers Math; Asn Teachers Math; Fibonacci Asn; Am Math Soc. *Res:* Probabilistic and combinatorial number theory and other combinatorial problems. *Mailing Add:* Dept Math Wash State Univ Pullman WA 99164

LONG, CARL F(ERDINAND), b New York, NY, Aug 6, 28; m 55; c 2. CIVIL ENGINEERING, ENGINEERING MECHANICS. *Educ:* Mass Inst Technol, SB, 50, SM, 52; Yale Univ, DEng, 64. *Hon Degrees:* MA, Dartmouth Col, 71. *Prof Exp:* Asst civil eng, Mass Inst Technol, 52-54, res engr, 54; from instr civil eng to assoc prof, 54-70; from assoc dean to dean, 72-84, PROF ENG, THAYER SCH ENG, DARTMOUTH COL, 70-, EMER DEAN & DIRCOOK ENG DESIGN CTR, 84- *Concurrent Pos:* Consult, NH State Water Pollution Comn, 58- & Small Arms Systs Agency, US Army; trustee, Mt Wash Observ, 75-; mem bd overseers, Mary Hitchcock Mem Hosp, 73-; dir, Controlled Environ Corp, Grantham, NH, 75-81, vpres opers, 76-81; pres & dir, OS-Oxygen Processes, Portland, Maine, 79-84; dir, Micro-Tool Co Inc, Fitchburg, MA, 84-; mem, ad hoc vis comt, Eng Coun

Prof Develop, 73-81; mem, vis comt, Mass Bd of Regents of Higher Educ, 84-; pres & dir, Roan Thayer Inc, 87-; dir, Micro-Weigh Systs, Inc, 87-; pres & dir, Hanover Water Works, Inc. *Honors & Awards:* Robert Fletcher Award, Thayer Sch Eng, 85. *Mem:* AAAS; Am Soc Civil Engrs; Am Soc Eng Educ. *Res:* Analytical and experimental investigations of structures and structural elements; planning and decision making for small towns and cities with time-sharing computers. *Mailing Add:* Thayer Sch Eng Dartmouth Col Hanover NH 03755

LONG, CAROLE ANN, b Baltimore, Md, Oct 2, 44; m 80. IMMUNOLOGY. *Educ:* Cornell Univ, AB, 65; Univ Pa, PhD(microbiol & immunol), 70. *Prof Exp:* Fel, Univ Pa Sch Med, 70-73; sr res scientist, Wyeth Labs, Radnor, Pa, 73-75; asst mem, Inst Med Res, 76-77; ASSOC PROF, HAHNEMANN MED COL, 77- *Concurrent Pos:* Res grants, NIH & World Health Orgn. *Mem:* Am Asn Immunologists; Tissue Culture Asn (treas, 76); AAAS; Asn Women Sci; Sigma Xi. *Res:* Murine mammary tumor virus, a model system to study immune responses to virus and virus-induced tumors; possible association of genes in H-2 complex with susceptibility to murine mammary tumor virus. *Mailing Add:* Dept Microbiol & Immunol Hahnemann Univ Sch Med Broad & Vine St Philadelphia PA 19102

LONG, CEDRIC WILLIAM, b Minneapolis, Minn, Mar 4, 37. BIOCHEMISTRY, VIROLOGY. *Educ:* Univ Calif, Los Angeles, BA, 60, MA, 62; Princeton Univ, PhD(biochem), 66. *Prof Exp:* Am Cancer Soc fel biochem, Univ Calif, Berkeley, 66-68; Nat Cancer Inst fel path, Med Sch, NY Univ, 68-69; instr cell biol, 69-70; sr scientist, Flow Labs, Inc, 70-72, head, Cell & Viral Biol Sect, 72-76; head, Biol Type C Viruses Sect, Litton Bionetics, Inc, Frederick Cancer Res Ctr, 76-80; chief, Preclin Trials Sect, 80-86, actg chief, Biol Resources Br, 84-85, actg assoc dir, Biol Response Modifiers Prog, DCT, 85, GEN MGR PROJ OFFICER, FREDERICK CANCER RES & DEVELOP CTR, NAT CANCER INST, 86- *Mem:* AAAS; Am Soc Microbiol; Am Soc Biol Chem. *Res:* Genetic and biochemical aspects of mammalian cell growth; expression of retroviruses; functional aspects of viral proteins; modification of host reponse to tumor cells. *Mailing Add:* Frederick Cancer & Res Develop Ctr Bldg 427 Frederick MD 21702-1201

LONG, CHARLES ALAN, b Pittsburg, Kans, Jan 19, 36; m 60; c 2. ZOOLOGY, GENETICS. *Educ:* Pittsburg State Univ, BS, 57, MS, 58; Univ Kans, PhD(zool), 63. *Prof Exp:* Asst zool, Univ Kans, 59-63; instr, Univ Ill, Urbana, 63-65, asst prof zool & life sci, 65-66; from asst prof to assoc prof, 66-70, PROF BIOL, UNIV WIS-STEVENS POINT, 73- *Concurrent Pos:* Fac fel, Univ Ill, 64; dir, Mus Nat Hist, 68-83, cur mammals, 66-; consult, Lake Mich Proj, Argonne Nat Lab, 74-80 & Ojibway Tribe, Lac de Flambeau, 83-85; Univ Adv Minor Mus Tech, 74-; Fulbright Scholar, 77; Pittsburg State Univ fel, 57-58; vis prof, St Olaf Col, 91. *Mem:* Am Soc Naturalists; Am Soc Mammal; Sigma Xi. *Res:* Vertebrate zoology, particularly systematics and zoogeography of mammals and their evolution; morphology and ecology; variability of mammals; Wyoming and Wisconsin mammals; badgers of the world; fractal geometry and morphology; genetics mammals. *Mailing Add:* Dept Biol Univ Wis Stevens Point WI 54481

LONG, CHARLES ANTHONY, b San Antonio, Tex, Feb 22, 45. CHEMICAL PHYSICS. *Educ:* Carleton Col, BA, 67; Ind Univ, PhD(chem physics), 72; Johns Hopkins Univ, BEE, 82. *Prof Exp:* Fel chem physics, Univ Calif, Riverside, 72-73; asst prof chem, Lake Forest Col, 73-77; res assoc, Brookhaven Nat Lab, 77-79; INSTRUMENTATION SUPVR, JOHNS HOPKINS UNIV, 79- *Concurrent Pos:* NSF res grant, 74. *Honors & Awards:* Roseman Award, 85. *Mem:* Am Phys Soc; Am Chem Soc; Inst Elec & Electronics Engrs. *Res:* Applications of lasers to problems of the chemistry and physics of small molecules; chemical instrumentation of all forms. *Mailing Add:* Dept Chem Johns Hopkins Univ 34th & Charles Baltimore MD 21218

LONG, CHARLES JOSEPH, b Caruthersville, Mo, Dec 25, 35; m 58; c 8. NEUROPSYCHOLOGY. *Educ:* Memphis State Univ, BS, 60, MA, 62; Vanderbilt Univ, PhD(psychol), 67. *Prof Exp:* PROF PSYCHOL, MEMPHIS STATE UNIV, 67- *Concurrent Pos:* Dir, Neuropsychol Training Prog, 72-, Universal Trainer, Memphis State Univ, 72- *Mem:* Sigma Xi; Am Psychol Asn; Int Neuropsychol Asn; Nat Acad Neuropsychologists (treas, 86). *Res:* Neuropsychological assessment and cognitive retraining of head injured, learning and neurologically impaired; study of functional factors influencing chronic pain. *Mailing Add:* 910 Madison Suite 609 Memphis TN 38103

LONG, CLAUDINE FERN, b Nevada, Mo, Sept 10, 38; m 60; c 2. CHEMISTRY. *Educ:* Pittsburg State Univ, BS, 60; Univ Ill-Urbana, MS, 64. *Prof Exp:* Instr biol, 69-70, univ coordr student teachers, Univ Wis, Stevens Pt, 71-75; teacher math & sci, PJ Jacobs Jr High Sch, Stevens Pt, Wis, 76-79; instr biol, Univ Wis, 79-82; leader group nat res, Malaysia, 82-83; SR LECTR CHEM, UNIV WIS, STEVENS PT, 85- *Concurrent Pos:* Teacher sci, W Jr High Sch, Lawrence, Kans, 60-63 & Ben Franklin Jr High Sch, Stevens Pt, Wis, 68; prof, Univ Malaya, Kuala Lumpur, 83; earthwatch researcher, Isle Rhum, Hebrides, Scotland, 85; prin investr serol hyaluronidase, 84- *Mem:* Nat Wildlife Fedn. *Res:* Insecticide resistant houseflies; natural history of birds and mammals; serological properties of hyaluronidase (trematoda); resources of Malaysia (tin, palm oil, pewter, etc); ecology of shore birds; Isle of Rhum, Hebrides, Scotland. *Mailing Add:* Dept Chem Univ Wis Stevens Pt WI 54481

LONG, CLIFFORD A, b Chicago, Ill, Apr 10, 31; m 57; c 4. MATHEMATICS. *Educ:* Univ Ill, BS, 54, MS, 55, PhD(math), 60. *Prof Exp:* From instr to assoc prof, 59-71, PROF MATH, BOWLING GREEN STATE UNIV, 71- *Mem:* Soc Indust & Appl Math; Math Asn Am; Nat Coun Teachers Math. *Res:* Computer graphics; numerical analysis. *Mailing Add:* Dept of Math Bowling Green State Univ Bowling Green OH 43403

LONG, DALE DONALD, b Louisa, Va, Jan 30, 35; m 65; c 2. PHYSICS INSTRUCTION, EXPERIMENTAL PHYSICS. *Educ:* Va Polytech Inst, BS, 58, MS, 62; Fla State Univ, PhD(physics), 66. *Prof Exp:* Instr physics, Va Polytech Inst, 60; instr, Samford Univ, 60-62; asst prof, 67-79, ASSOC PROF PHYSICS, VA POLYTECH INST & STATE UNIV, 79- *Concurrent Pos:* Vis assoc prof, Davidson Col, 88-89. *Mem:* Am Phys Soc; Am Asn Physics Teachers. *Res:* Enhancement of the effectiveness of physics instruction, author of introductory physics textbooks; experimental nuclear physics. *Mailing Add:* Dept Physics Va Polytech Inst & State Univ Blacksburg VA 24061

LONG, DANIEL R, b Redding, Calif, June 9, 38; m 61; c 2. PHYSICS. *Educ:* Univ Wash, PhD(physics), 67. *Prof Exp:* From asst prof to assoc prof, 67-81, PROF PHYSICS, EASTERN WASH UNIV, 81- *Concurrent Pos:* Sloan Found fel, 77. *Mem:* Am Phys Soc; Sigma Xi. *Res:* Electron impact ionization of metastable helium; experimental examination of the mass separation dependence of the gravitational constant. *Mailing Add:* Dept of Physics Eastern Wash Univ Cheney WA 99004

LONG, DARREL GRAHAM FRANCIS, b Yorkshire, Eng, Sept 6, 47; m 73; c 2. CLASTIC & CARBONATE SEDIMENTOLOGY, COAL GEOLOGY. *Educ:* Univ Leicester, Eng, BSc, 69; Univ Western Ont, MSc, 73, PhD(geol), 76. *Prof Exp:* Fel geol, Geol Surv Can, 76-77, res scientist coal geol, 77-81; from asst prof to assoc prof, 81-89, PROF SEDIMENTOL, LAURENTIAN UNIV, SUDBURY, 89- *Mem:* Geol Asn Can; Geol Soc Am; Int Asn Sedimentologists; Soc Econ Paleontologists & Mineralogists; Can Soc Petrol Geologists; Geol Soc Australia. *Res:* Clastic sedimentology of Precambrian sequences in Ontario, Yukon and Northwest Territory Canada; sedimentology and coal bearing sequences in British Columbia, Yukon, Northwest Territory and Ontario; phanerozoic sedimentology and tectonics of the Arctic Islands and Quebec. *Mailing Add:* Dept Geol Laurentian Univ Sudbury ON P3E 2C6 Can

LONG, DARYL CLYDE, b Mason City, Iowa, Aug 19, 39; m 60; c 3. SOIL SCIENCE. *Educ:* Iowa State Univ, BS, 62, MS, 64; Univ Nebr, Lincoln, PhD, 67. *Prof Exp:* Instr soils, Univ Nebr, Lincoln, 64-67; asst prof, 67-77, ASSOC PROF SCI & MATH, PERU STATE COL, 77- *Mem:* Am Soc Agron; Soil Sci Soc Am; Nat Coun Teachers Math; Sigma Xi. *Res:* Mechanics of soil erosion and plant removal of nutrients from soil aggregates. *Mailing Add:* Dept of Sci & Math Peru State Col Peru NE 68421

LONG, DAVID G, SCATTEROMETRY, RADAR. *Educ:* Brigham Young Univ, BS, 82, MS, 83; Univ Southern Calif, PhD(elec eng), 89. *Prof Exp:* Group leader, Jet Propulsion Lab, 83-90; ASST PROF ELEC ENG, BRIGHAM YOUNG UNIV, 90- *Concurrent Pos:* Prin investr, NASA, 89- *Mem:* Inst Elec & Electronic Engrs. *Res:* Spaceborne scatterometry; radar; microwave remote sensing; mesoscale atmospheric dynamics; speech and signal processing; estimation theory. *Mailing Add:* ECEN Dept 459 CB Brigham Young Univ Provo UT 84602

LONG, DAVID MICHAEL, b Shamokin, Pa, Feb 26, 29; c 6. CARDIOVASCULAR SURGERY, THORACIC SURGERY. *Educ:* Muhlenberg Col, BS, 51; Hahnemann Med Col, MS, 54, MD, 56; Univ Minn, PhD(physiol), 65; Am Bd Surg, dipl, 66; Bd Thoracic Surg, dipl, 67. *Prof Exp:* Instr surg, Univ Minn, 65; from asst prof to assoc prof, Chicago Med Sch, 65-67; from assoc prof to prof surg, Abraham Lincoln Sch Med, Univ Ill Med Ctr, 69-73, attend staff & head dir cardiovasc & thoracic surg, Hines, 67-73; CLIN ASSOC PROF RADIOL, UNIV CALIF, SAN DIEGO, 73- *Concurrent Pos:* Assoc prof, Cook County Grad Sch Med, 65-73; assoc attend staff, Cook County Hosp, 65-73; asst dir dept surg res, Hektoen Inst Med Res, 65-68, dir, 68-73; attend staff, W Side Vet Admin Hosp, 66-73; consult, Chicago State Tuberc Sanitarium, 67-72; pvt pract, 73- *Honors & Awards:* First Prize Res, Am Urol Asn, 66. *Mem:* AAAS; Am Asn Thoracic Surg; fel Am Col Cardiol; fel Am Col Chest Physicians; fel Am Col Surg. *Res:* Surgical research; physiology and morphology; cancer chemotherapy; development of the radiopaque compound perfluorocarbon. *Mailing Add:* 10988 Horizon Hill El Cajon CA 92020

LONG, DONLIN MARTIN, b Rolla, Mo, Apr 14, 34; m 59; c 3. NEUROSURGERY, ELECTRON MICROSCOPY. *Educ:* Univ Mo, MD, 59; Univ Minn, PhD(anat), 64. *Prof Exp:* Clin assoc, Surg Neurol Bd, NIH, 65-67; assoc prof neurosurg, Univ Minn Hosps, 67-73; PROF NEUROL SURG & DIR DEPT, SCH MED, JOHNS HOPKINS UNIV, 73- *Concurrent Pos:* Consult neurosurgeon, Vet Admin Hosp, Minneapolis, 67- *Mem:* AAAS; Am Asn Neurol Surg; Cong Neurol Surg; Am Asn Neuropath; Soc Neuroscience. *Res:* Low back pain and brain edema. *Mailing Add:* Dept Neurol Surg Johns Hopkins Univ Sch Med Baltimore MD 21205

LONG, EARL ELLSWORTH, b Akron, Ohio, Mar 27, 19; m 41; c 4. PUBLIC HEALTH LABORATORY ADMINISTRATION. *Educ:* Univ Akron, BSc, 42; Univ Pa, MSc, 47. *Prof Exp:* Asst instr med bact, Sch Med, Univ Pa, 45-48; asst prof bact, Univ Akron, 48-49; dir labs, Akron Health Dept, 49-61; dir labs, Ga Dept Pub Health, 61-82; RETIRED. *Mem:* Am Soc Microbiol; fel Am Pub Health Asn; Asn State & Territorial Pub Health Labs Dirs (pres, 80); Sigma Xi. *Res:* State public health laboratory administration with emphasis on implementation of rapidly changing concepts in service and research. *Mailing Add:* 4669 Canyon Creek Trail NE Atlanta GA 30342

LONG, EDWARD B, b White Plains, NY, Dec 5, 27; m 52, 70; c 3. WETLANDS ECOLOGY. *Educ:* Hamilton Col, BA, 52; Kent State Univ, MS, 71, PhD(biol), 75. *Prof Exp:* Mem staff mkt, Carbon Prod Div, Union Carbide Corp, 6-52-64, proj mgr, New Prod Mkt Develop, 64-69; tech mgr environ prog, Northeast Ohio Areawide Coord Agency, 75-81; ENVIRONM CONSULT, 81- *Mem:* Am Soc Limnol & Oceanog; AAAS. *Res:* Environmental quality of Northeast Ohio. *Mailing Add:* 3140 N Martadale Dr Akron OH 44333

LONG, EDWARD R, b Washougal, Wash, 1942. POLLUTANT-CAUSED BIOLOGICAL EFFECTS. *Educ:* Ore State Univ, BS, 65, MS, 67. *Prof Exp:* Biol Oceanographer, Naval Oceanog Off, 67-73; res biologist, Wapora, Inc, 73-75; MARINE BIOLOGIST, NAT OCEANIC & ATMOSPHERIC ADMIN, 75- *Concurrent Pos:* Lectr, George Washington Univ, 70-75. *Mem:* Marine Technol Soc; Estuarine Res Fedn. *Res:* Administration of technical aspects of marine pollution research, focusing upon measures of biological effects among fish, benthos, birds and mammals; chemical measures with observed biological effects. *Mailing Add:* 415 NE 190th Pl Seattle WA 98155

LONG, EDWARD RICHARDSON, JR, b Annapolis, Md, Sept 1, 41; m 68. MOLECULAR PHYSICS, MATERIALS SCIENCE. *Educ:* Col William & Mary, BS, 63, MS, 67; NC State Univ, PhD(molecular physics, nuclear magnetic resonance), 74. *Prof Exp:* Res scientist human factors, Aeronaut & Space Mech Div, Guid & Control Br, Langley Res Ctr, NASA, 63-67, res scientist solid state physics, Appl Math & Physics Div, Chem & Physics Br, 69-72, res scientist org pollution, Environ & Space Sci Div, Laser & Molecular Physics Br, 72-76, res scientist mat sci, Mat Div, Mat Res Br, 76-80, RES SCIENTIST MAT SCI, MAT DIV, ENVIRON EFFECTS BR, LANGLEY RES CTR, NASA, 80- *Concurrent Pos:* Assoc prof, George Washington Univ, 76- *Mem:* Am Phys Soc. *Res:* Solid state physics and organic chemical physics as applied to pollution spectroscopy and materials science. *Mailing Add:* 233 Chickamanga Pike Hampton VA 23669

LONG, F(RANCIS) M(ARK), b Iowa City, Iowa, Nov 10, 29; m 64; c 4. MICROCIRCUITS. *Educ:* Univ Iowa, BS, 53, MS, 56; Iowa State Univ, PhD(elec eng, biomed electronics), 61. *Prof Exp:* Asst elec eng, Univ Iowa, 55-56; instr, Univ Wyo, 56-58; instr, Iowa State Univ, 58-60; dir bioeng, Univ Wyo, 64-74, head dept elec eng, 77-87, from asst prof to assoc prof, 60-74, PROF ELEC ENG, UNIV WYO, 74- *Concurrent Pos:* Engr, Collins Radio Co, 55, US Naval Air Missile Testing Ctr, Calif, 56 & Good-All Elec Co, 57; NIH spec fel, 72-73; Globe Union Co, 75; cofounder, Wyo Biotelemetry Inc, 78, teaching & consult surface mount technol, 83- *Mem:* Sr mem Inst Elec & Electronic Engrs; Am Soc Eng Educ (vpres, 77-79); Alliance for Eng in Med & Biol (pres, 83-84). *Res:* Instrumentation and system design; system modelling; microcircuit technology; animal biotelemetry; polymer thick film circuits. *Mailing Add:* Dept Elec Eng Box 3295 Univ Sta Laramie WY 82071

LONG, FRANKLIN A, b Great Falls, Mont, July 27, 10; m 37; c 2. SCIENCE POLICY, ARMS CONTROL. *Educ:* Univ Mont, AB, 31, MA, 32; Univ Calif, PhD(phys chem), 35. *Prof Exp:* Instr chem, Univ Calif, 35-36; instr, Univ Chicago, 36-37; from instr to assoc prof chem, Cornell Univ, 37-42, prof chem, 46-79, chmn dept, 50-60, dir prog sci, technol & soc, 69-73, Luce prof sci & soc, 69-79, dir, Peace Studies Prog, 76-79, EMER PROF, CORNELL UNIV, 79- *Concurrent Pos:* Res supvr, Explosives Res Lab, Nat Defense Res Comt, 42-45; consult, Ballistics Res Lab, Dept Army, 53-59; mem sci adv bd, Air Force Off Sci Res, 56-60; mem, Pres Sci Adv Comt, 61 & 64-67; asst dir, US Arms Control & Disarmament Agency, 62-63; mem bd sci & technol for int develop, Nat Acad Sci, 74-77; mem, Indo-US Comn Educ & Cult, 74-82, co-chmn, 77-82; mem adv panel, Policy Res & Analysis Div, NSF, 77-80; mem bd & consult, Carrier Corp, 78-79, Exxon Corp, 69-80; mem bd, Alfred P Sloan Found, 70-83, Arms Control Asn, 71-77, Assoc Univs, Inc, Albert Einstein Peace Prize Found, The Fund for Peace; adj prof, Univ Calif, 88- *Honors & Awards:* Parson's Award, Am Chem Soc, 85; Abelson Prize, AAAS, 89. *Mem:* Nat Acad Sci; AAAS; Am Chem Soc; Am Acad Arts & Sci; Coun Foreign Rel. *Res:* Kinetics of solution reactions; isotopic chemistry; arms control; science and public policy. *Mailing Add:* Sch Social Sci Univ Calif Irvine CA 92717

LONG, GARY JOHN, b Binghamton, NY, Dec 3, 41; m 63; c 1. PHYSICAL INORGANIC CHEMISTRY. *Educ:* Carnegie-Mellon Univ, BS, 64; Syracuse Univ, PhD(chem), 68. *Prof Exp:* From asst prof to assoc prof, 68-82, PROF CHEM, UNIV MO-ROLLA,82- *Concurrent Pos:* Res assoc, Inorg Chem Lab & St John's Col, Oxford Univ, 74-75; res assoc, Atomic Energy Res Estab, Harwell, 75-81; NATO vis prof chem, Univ Padova, Italy, 83; sci & eng res coun fel, Univ Liverpool, Eng, 83-84; vis prof chem, Univ Padova, Italy, 86, 87, 88, Univ Geneva, Switz, 88. *Mem:* Am Chem Soc; fel Royal Soc Chem; Sigma Xi. *Res:* Transition metal inorganic coordination chemistry and solid state chemistry; Mossbauer and electronic spectroscopy; high-pressure optical and infrared spectroscopy; magnetic studies of coupled systems and permanent magnetic materials; x-ray and neutron diffraction studies. *Mailing Add:* Dept Chem em Univ Mo Rolla MO 65401-0249

LONG, GEORGE, b Greenville, Miss, Jan 17, 22; m 51; c 1. CHEMICAL ENGINEERING. *Educ:* Univ Tulane, BE, 44. *Prof Exp:* Res chemist, Div 8, Nat Defense Res Comt, Ohio, 44-45; chief chemist, USAAF, 45-46; res engr, Aluminum Co Am, 46-62; gen coordr res & develop, 62-67, dir, 67-77, mgr dir res & develop, Northern Ill Gas Co, 77-87; CONSULT, NATURAL GAS INDUST, 87- *Mem:* Am Gas Asn; Sigma Xi; fel Am Inst Chem; Chem Mkt Res Asn; Am Chem Soc. *Res:* Process metallurgy of aluminum melting and smelting; aluminum-water explosions; high temperature refractory materials; natural gas utilization, materials and devices for distribution systems, substitute natural gas processes and natural gas combustion; synthetic fuel processes. *Mailing Add:* 24 Sylvia Lane Naperville IL 60540-8014

LONG, GEORGE GILBERT, b Cincinnati, Ohio, July 12, 29; m 52; c 3. INORGANIC CHEMISTRY. *Educ:* Ind Univ, AB, 51; NC State Univ, MS, 53; Univ Fla, PhD(chem), 57. *Prof Exp:* Chemist, Ethyl Corp, 57-58; from asst prof to assoc prof, 58-70, chmn anal chem dept, 69-77, PROF CHEM, NC STATE UNIV, 70- *Mem:* Am Chem Soc. *Res:* Chemistry of group V metalloids-organometalloid compounds; 121-Sb Mossbauer spectroscopy, structure and syntheses; vibrational spectroscopy. *Mailing Add:* Dept Chem NC State Univ Box 8204 Raleigh NC 27650

LONG, GEORGE LOUIS, b Atkin, Minn, Dec 20, 43; m 67; c 5. BIOCHEMISTRY, MOLECULAR BIOLOGY. *Educ:* Pac Lutheran Univ, BA, 66; Brandeis Univ, PhD(biochem), 71. *Prof Exp:* NIH trainee molecular endocrinol sch med, Univ Calif, San Diego, 71-73; asst prof chem, Pomona Col, 73-80; NIH sr fel biochem, Univ Wash, 80-82; scientist, Lilly Res Labs, 82-86; assoc prof, 86-91, PROF BIOCHEM, UNIV VERMONT, 91- *Mem:* Am Soc Molecular & Biol Chem; Sigma Xi; Am Chem Soc; AAAS. *Res:* Comparative enzymology of glycolytic enzymes; molecular biology of hemostasis; bone biochemistry. *Mailing Add:* 92 Hungerford Terr Burlington VT 05401

LONG, H(UGH) M(ONTGOMERY), b Montgomery, Ala, June 28, 24; m 49; c 2. ENHANCED OIL RECOVERY, SUPERCONDUCTING SYSTEMS. *Educ:* Ala Polytech Inst, BS, 47, MS, 49; Oxford Univ, DPhil(physics), 53. *Prof Exp:* Instr math, Auburn Univ, 47-48, res asst, 47-49; res physicist, Linde Div, Oak Ridge Nat Lab, Union Carbide Corp, 54-61, cryogenics consult, 61-71, group leader eng sci, Thermonuclear Div, 71-76, mgr elec energy systs prog, Energy Div, 76-80; assoc prof elec eng, Univ Tenn, Knoxville, 71-80; vpres mkt develop, Vedette Energy Res Inc, 80-81; STAFF EXEC, ENHANCED ENERGY SYSTS, INC, 81- *Concurrent Pos:* Mem, Nat Acad Sci-Nat Res Coun adv panel to Nat Bur Standards Cryogenic Eng Lab, 61-65; US rep, Comt I, Int Inst Refrig, 64-; mem & chmn, US Delegation USSR Scientific & Technol Exchange Superconductivity Power Transmission, 73-79. *Mem:* AAAS; Am Phys Soc; sr mem Inst Elec & Electronics Engrs; NY Acad Sci; Soc Petrol Engrs. *Res:* Low temperature physics; cryogenic engineering; gas liquefaction; low temperature phase equilibria; mechanical properties of materials at low temperatures; superconductivity; power system engineering; energy management. *Mailing Add:* 3551 Lilac Ave Corona Del Mar CA 92625-1661

LONG, HOWARD CHARLES, b Seizholtzville, Pa, Dec 12, 18; m 45; c 3. ACOUSTICS, ATOMIC & MOLECULAR PHYSICS. *Educ:* Northwestern Univ, BS, 41; Ohio State Univ, PhD(physics), 48. *Prof Exp:* Physicist, Naval Ord Lab, 42-45; instr physics, Ohio State Univ, 47-48; asst prof, Washington & Jefferson Col, 48-51; physicist, Naval Ord Lab, 51-52; assoc prof physics & chmn dept, Am Univ, 52-53; prof & chmn dept, Gettysburg Col, 53-59; chmn dept, 63-74, prof, 59-81, EMER PROF PHYSICS, DICKINSON COL, 81- *Concurrent Pos:* Consult, Naval Ord Lab, 54-73. *Mem:* Am Phys Soc; Am Asn Physics Teachers. *Res:* Low period fluctuations in earth's magnetism; environmental noise reduction; air pollution by solid particulates; molecular structure and infrared spectroscopy; electromagnetism. *Mailing Add:* Dept Physics & Astron Dickinson Col Carlisle PA 17013

LONG, JAMES ALVIN, b Porto Alegre, Brazil, July 13, 17; US citizen; wid; c 4. EXPLORATION GEOPHYSICS INTERPRETATIONS & OPERATIONS. *Educ:* Univ Okla, BA, 37. *Prof Exp:* Computer & party chief, Stanolind Oil & Gas Co, 37-46; party chief supvr, United Geophys Corp, area mgr & regional opers mgr, South & Cent Am, 46-62, special tech & res assignments under MB Dobrin & others, 62-67 regional mgr, Latin Am, 67-72; sr geophysicist, Tetra Tech, Inc, 73-74; geophys adv, Yacimientos Petroliferos Fiscales Bolivianos, Santa Cruz, Bolivia, 74-77; int consult geophysicist, 73-84, Peru, US, Australia, Colombia; RETIRED. *Mem:* Soc Explor Geophysicist; fel Explorers Club; Earthwatch. *Res:* Seismic surface sources; special seismic interpretation problems and supervision of operations particularly in South America and Australia. *Mailing Add:* 3951 Gulf Shore Blvd Apt 504 Naples FL 33940-3639

LONG, JAMES DELBERT, b Dover, Okla, Dec 18, 39; c 5. AGRONOMY, HERBICIDE RESEARCH. *Educ:* Okla State Univ, BS, 62; Univ Md, College Park, MS, 67, PhD(hort), 69. *Prof Exp:* Res asst weed control, Univ Md, College Park, 64-67, instr hort, 67-68; res biologist agr chem, 68-79, prod develop mgr, 79-83, RES ASSOC, AGR CHEM DEPT, E I DU PONT DE NEMOURS & CO, INC, WILMINGTON, DEL, 83- *Mem:* Southern Weed Sci Soc; Sigma Xi. *Res:* Control and modification of plant growth through the use of chemicals; new herbicide discovery and development. *Mailing Add:* Stine-Haskell Res Ctr E I Du Pont de Nemours & Co Inc PO Box 30 Newark DE 19711

LONG, JAMES DUNCAN, b Rusk, Tex, Sept 23, 25. ZOOLOGY. *Educ:* Sam Houston State Col, BS, 48, MA, 51; Univ Tex, PhD, 57. *Prof Exp:* Teacher, High Sch, Tex, 48-49 & Pub Schs, 51-52; instr biol, Lamar State Col Technol, 52-53; asst, Univ Tex, 53-56; assoc prof biol & head dept, Ill Col, 56-59; assoc prof, 59-63, dir dept, 63-72, PROF BIOL, SAM HOUSTON STATE UNIV, 63- *Concurrent Pos:* Newsletter ed, Am Mosquito Control Asn. *Honors & Awards:* Pres Citation, Am Mosquito Control Asn. *Mem:* Entom Soc Am. *Res:* Mosquito biology. *Mailing Add:* Dept Biol Sam Houston State Univ Huntsville TX 77341

LONG, JAMES FRANTZ, b Center Valley, Pa, Sept 17, 31; m 56; c 3. PHYSIOLOGY. *Educ:* Mich State Univ, BS, 57, MS, 59; Univ Pa, PhD(physiol), 64. *Prof Exp:* From instr to assoc prof physiol, Albany Med Col, 64-69; prin scientist, dept pharmacol, Schering Corp, 69-75, res fel, 75-; DIR, DEPT GASTROENTEROL, JANSSEN PHARMACEUT, PISCATAWAY, NJ. *Mem:* Am Physiol Soc; Am Gastroenterol Asn. *Res:* Gastrointestinal physiology and pharmacology. *Mailing Add:* Dept Gastroenterol Janssen Res Found 40 Ringsbridge Rd Piscataway NJ 08855-3998

LONG, JAMES WILLIAM, b Boise, Idaho, Aug 26, 43; m 65; c 2. BIOCHEMISTRY. *Educ:* Univ Wash, BS, 65; Univ Calif, Berkeley, PhD(biochem), 69. *Prof Exp:* Res assoc biochem, Purdue Univ, West Lafayette, 70-71, NIH res fel, 71-72, res assoc, 72-73; res assoc, Univ Ore, 73-74; from asst prof to assoc prof chem, Col Great Falls, 74-78; SR INSTR CHEM, UNIV ORE, 78- *Mem:* Am Chem Soc. *Res:* Computers in chemical education; structure-function relationships in enzymes; mechanisms of enzyme action; enzyme model systems; role of metal ions in enzyme catalysis. *Mailing Add:* Dept Chem Univ Ore Eugene OR 97403-1253

LONG, JEROME R, b Lafayette, La, May 17, 35; div; c 2. MAGNETIC AND TRANSPORT PHENOMENA. *Educ:* Univ Southwestern La, BS, 56; La State Univ, MS, 58, PhD(physics), 65. *Prof Exp:* Res engr, Gen Dynamics/Pomona, 58-59; fel metall, Univ Pa, 65-67; asst prof physics, 67-71, ASSOC PROF PHYSICS, VA POLYTECH INST & STATE UNIV, 71- *Concurrent Pos:* Vis prof, Simon Fraser Univ, 78-79, Montana State Univ, 86, Naval Res Lab, 87. *Mem:* Am Phys Soc; Int Elec & Electronics Engrs. *Res:* Transport and magnetic properties of metallic materials; cryophysics; squid susceptometry on layered and or film magnetic and or superconducting materials. *Mailing Add:* Dept Physics Va Polytech Inst & State Univ Blacksburg VA 24061-0435

LONG, JIM T(HOMAS), b Central, SC, Oct 5, 23; m 46; c 1. ELECTRICAL ENGINEERING. *Educ:* Clemson Col, BEE, 43; Ga Inst Technol, MSEE, 49, PhD(elec eng), 64. *Prof Exp:* From instr to assoc prof, 43-67, PROF ELEC ENG & COORDR UNDERGRAD PROG, CLEMSON UNIV, 67- *Concurrent Pos:* Asst, Ga Inst Technol, 48-49, asst prof, 57-64. *Mem:* Am Soc Eng Educ; Inst Elec & Electronics Engrs; Sigma Xi. *Res:* Electronics; network theory; solid state electronics. *Mailing Add:* Dept Elec Eng Clemson Univ Clemson SC 29631

LONG, JOHN A, b Lewistown, Mont, Sept 1, 27; m 49; c 4. AGRONOMY, RESOURCE MANAGEMENT. *Educ:* Univ Idaho, BS, 52; Wash State Univ, MS, 54; Tex A&M Univ, PhD(agron), 61. *Prof Exp:* Asst in agron, NMex State Univ, 54-56; instr, Tex A&M Univ, 56-61; proj leader agron, 61-63, from dir biochem res to dir prod develop, 63-90, CONSULT, O M SCOTT & SONS CO, 90- *Concurrent Pos:* Chmn student interest comt, Southern Weed Control Asn, 59-60, turf sect, Weed Sci Soc Am, 63-64; chmn mem comt, Agr Res Inst, 72-73, mem prog comt, 73-74; pres, Nat Coun Com Plant Breeders, 79-80; chmn, Turf & Garden Com Fertilizer Inst, 87-88. *Res:* Agronomy; horticulture. *Mailing Add:* 1016 Collins Ave Marysville OH 43040

LONG, JOHN ARTHUR, b Kingman, Kans, July 30, 34. CYTOLOGY, ANATOMY. *Educ:* Univ Kans, AB, 56; Univ Wash, PhD(zool), 64. *Prof Exp:* Res assoc anat, Harvard Med Sch, 66-67; asst prof, 67-74, ASSOC PROF ANAT, MED CTR, UNIV CALIF, SAN FRANCISCO, 74- *Mem:* AAAS; Am Soc Zool. *Res:* Fine structure of steroid secreting cells in ovary and adrenal glands. *Mailing Add:* Dept Anat Univ Calif Med Sch 513 Parnassus Ave San Francisco CA 94143

LONG, JOHN FREDERICK, b Napoleon, Ohio, May 30, 24; m 48; c 5. VETERINARY PATHOLOGY. *Educ:* Ohio State Univ, BA, 47, MSc, 48, DVM, 55, PhD(comp neuropath), 66. *Prof Exp:* Res asst animal sci, Ohio Agr Exp Sta, 49-50; diag vet pathologist, Vet Diag Lab, State of Ohio, 55-63; res assoc comp neuropath, 63-64, NIH res fel, 64-66, instr vet path, 66-67, NIH spec res fel comp neuropath, 67-68, asst prof vet path, 68-71, ASSOC PROF VET PATH, OHIO STATE UNIV, 71- *Mem:* Am Vet Med Asn; Am Asn Avian Path. *Res:* Comparative neuropathology; viral encephalomyelitides; use of brain explant culture and germ-free animals in the study of the effects of encephalitogenic agents; demyelinating encephalomyelitides of animals; aluminum encephalopathyy models. *Mailing Add:* Dept Vet Path 325 Goss Lab Ohio State Univ Columbus OH 43210-1358

LONG, JOHN KELLEY, b NY, Dec 12, 21; m 48; c 3. NUCLEAR PHYSICS. *Educ:* Columbia Univ, BS, 42; Ohio State Univ, PhD(physics), 53. *Prof Exp:* Chemist plastics, Hercules Powder Co, 42-45; engr, Wright Field, 47-50; physicist, Battelle Mem Inst, 52-55; physicist, Idaho Div, Argonne Nat Lab, 55-74; Reactor Engr, US Nuclear Regulatory Comm, 74-83; RETIRED. *Concurrent Pos:* Consult, NUS Corp, 91- *Res:* Fast reactor physics; critical experiments; reactor licensing; fast reactor safety test facilities; plutonium toxicity. *Mailing Add:* 227 S 35th West Idaho Falls ID 83402

LONG, JOHN PAUL, b Albia, Iowa, Oct 4, 26; m 50; c 3. PHARMACOLOGY. *Educ:* Univ Iowa, BS, 50, MS, 52, PhD(pharmacol), 54. *Prof Exp:* From asst to instr pharmacol, Univ Iowa, 50-54; res assoc, Sterling-Winthrop Res Inst, 54-56; from asst prof to assoc prof, 56-62, prof, 62-70 & 83-85, head dept, 70-83, CARVOR PROF PHARMACOL, COL MED, UNIV IOWA, 85- *Mem:* Am Soc Pharmacol & Exp Therapeut; Soc Exp Biol & Med. *Res:* Structure-activity relationships of autonomic and anesthetic agents. *Mailing Add:* Dept of Pharmacol Univ of Iowa Col of Med Iowa City IA 52242

LONG, JOHN REED, b Chicago, Ill, Oct 2, 22; m 48, 87; c 2. INDUSTRIAL ENGINEERING, MANUFACTURING ENGINEERING. *Educ:* Northwestern Univ, BS, 47; Iowa State Univ, MS, 48, PhD(chem eng), 51. *Prof Exp:* Atomic Energy Comn asst, Ames Lab, Iowa State Univ, 48-51; sr engr, Hercules, Inc, 51-60, process engr, 61-66, sr process engr, 66-80, supvr process engr, 80-85; RETIRED. *Mem:* Am Chem Soc; Am Inst Chem Engrs; Sigma Xi. *Res:* Process design of chemical plants. *Mailing Add:* 5 Clyth Dr Wilmington DE 19803

LONG, JOHN VINCENT, b San Diego, Calif, Feb 18, 10; m 38; c 4. PHYSICS. *Educ:* Univ Calif, Los Angeles, AB, 37. *Prof Exp:* Lab asst physics, San Diego State Col, 32-35; serv demonstr, Ford Motor Co, Calif, 35-36; res engr, Douglas Aircraft Co, 36-37; geophysicist, Continental Oil Co, Okla, 37-40; res engr, Int Harvester Co, 40, res physicist & asst dir res, 46-51, dir res, Solar Div, 51-80; mem staff, MGL Develop, 80-81; RETIRED. *Mem:* Soc Explor Geophys; Acoust Soc Am; Soc Exp Stress Analysis; assoc Inst Elec & Electronics Engrs. *Res:* Ceramics; metallurgy; vibration and sound; high altitude research; ceramic coatings for high temperature corrosion and oxidation protection of iron, stainless steel; super alloys and refractory metals. *Mailing Add:* 1756 E Lexington Ave El Cajon CA 92019

LONG, JOSEPH POTE, b Baker Summit, Pa, Feb 26, 13; m 42; c 4. OBSTETRICS & GYNECOLOGY. *Educ:* Juniata Col, BS, 34; Jefferson Med Col, MD, 39; Univ Pa, MS, 48. *Prof Exp:* From demonstr to assoc prof obstet & gynec, Jefferson Med Col, Thomas Jefferson Univ, 48-75, clin prof, 75-78, hon clin prof, 78-90; RETIRED. *Mem:* AMA; Am Col Surg; Am Col Obstet & Gynec; Am Fertil Soc; NY Acad Sci. *Mailing Add:* 2209 Douglas Dr Carlisle PA 17013

LONG, KEITH ROYCE, b Lincoln, Kans, Mar 17, 22; m 45; c 5. ENVIRONMENTAL HEALTH. *Educ:* Univ Kans, AB, 51, MA, 53; Univ Iowa, PhD, 60. *Prof Exp:* Asst instr bact, Univ Kans, 52-53, instr bact res, Med Ctr, 53-56; sr bacteriologist & virologist, State Hyg Lab, Col Med, Univ Iowa, 56-57, instr, Inst Agr Med, 57-58, asst bact, 58-60, assoc prof hyg & prev med, 60-69, dir, 74-83, prof prev med, Inst Agr Med & Environ Health, Col Med, 69-86, prof civil eng, 70-86; RETIRED. *Mem:* Am Pub Health Asn; NY Acad Sci. *Res:* Environmental toxicology; epidemiology; pesticides. *Mailing Add:* Dept of Preventive Med Univ Iowa Col Med Iowa City IA 52242

LONG, KENNETH MAYNARD, b Nappanee, Ind, July 10, 32; m 52; c 5. INORGANIC CHEMISTRY, SPELEOLOGY. *Educ:* Goshen Col, BS, 54; Mich State Univ, MA, 60; Ohio State Univ, PhD(chem), 67. *Prof Exp:* Instr, Parochial Sch, Ark, 54-56; instr, High Sch, Mich, 56- 61; from instr to assoc prof, 62-79, asst dean, 71-75, PROF CHEM, WESTMINSTER COL, PA, 79-, CHMN, 83- *Concurrent Pos:* NIH fel, 65-67; fel, Kent State Univ, 79; scholar-in-residence, Northeast Univ Technol, Shenyang, Liaoning, Peoples Repub China. *Mem:* Am Chem Soc; Nat Asn Geol Teachers; Nat Speleol Soc. *Res:* Macrocyclic complexes of transition metals; catalytic properties of transition metal complexes; kinetics; hydrology; geology and mapping of caves. *Mailing Add:* Dept Chem Westminster Col New Wilmington PA 16172

LONG, LARRY L, b St Joseph, Mo, Aug 18, 55; m 84. THIN FILMS, SURFACE PHYSICS. *Educ:* NW Mo State Univ, BS, 77; Univ Mo-Rolla, MS, 82 & PhD(physics), 85. *Prof Exp:* Teaching asst physics, Univ Mo-Rolla, 79-85; sr engr optics & thin films, McDonnell Douglas Corp, 85-88; ASST PROF PHYSICS, PITTSBURG STATE UNIV, 88- *Mem:* Am Phys Soc; Optical Soc Am; Am Asn Physics Teachers; Sigma Xi; Int Soc Optical Eng. *Res:* Thin films; application of surface plasmons to environmental issues; optical constants; application of thin films to thermal batteries; author of various publications. *Mailing Add:* Physics Dept Pittsburg State Univ Pittsburg KS 66762

LONG, LAWRENCE WILLIAM, b Akron, Ohio, Nov 6, 42; m 69. BIOCHEMISTRY. *Educ:* Franklin & Marshall Col, AB, 65; Villanova Univ, PhD(chem), 71. *Prof Exp:* Instr biochem, Thomas Jefferson Univ, 71-73; res scientist, Stevens Inst Technol, 73-74; proj leader chem, 74-77, MGR ALLIED PROD, ANHEUSER-BUSCH INC, 78- *Mem:* Am Chem Soc; Am Soc Brewing Chemists. *Res:* Utilization of brewing residuals. *Mailing Add:* One Bush Pl OSC-1 St Louis MO 63118

LONG, LELAND TIMOTHY, b Auburn, NY, Sept 6, 40; m 70; c 3. GEOPHYSICS, SEISMOLOGY. *Educ:* Univ Rochester, BS, 62; NMex Inst Mining & Technol, MS, 64; Ore State Univ, PhD(geophys), 68. *Prof Exp:* From asst prof to assoc prof, 68-80, PROF GEOPHYS, GA INST TECHNOL, 81- *Concurrent Pos:* Consult Seismol, 78- *Mem:* Am Geophys Union; Seismol Soc Am; Soc Explor Geophys; Sigma Xi. *Res:* Seismic and gravity data acquisition and analysis; earthquake seismology. *Mailing Add:* Sch Earth & Atmospheric Sci Ga Inst Technol Atlanta GA 30332

LONG, LEON EUGENE, b Wanatah, Ind, May 4, 33; m 56; c 2. GEOCHEMISTRY. *Educ:* Wheaton Col, BS, 54; Columbia Univ, MA, 58, PhD(geochem), 59. *Prof Exp:* Geochemist, Lamont Geol Observ, Columbia Univ, 59-60; NSF fel, Oxford Univ, 60-62; from asst prof to assoc prof, 62-75, PROF GEOL, UNIV TEX, AUSTIN, 75- *Mem:* Fel Geol Soc Am; Geochemical Soc; Sigma Xi. *Res:* Isotopic age methods. *Mailing Add:* Dept of Geol Sci Univ of Tex Austin TX 78712

LONG, LYLE NORMAN, b Fergus Fall, Minn, April 7, 54; m 81; c 2. COMPUTATIONAL PHYSICS, PARALLEL PROCESSING. *Educ:* Univ Minn, BME, 76; Stanford Univ, MS, 78; George Washington Univ, DSc, 83. *Prof Exp:* Res asst, Stanford Univ, 77-78; res assoc, George Washington Univ, 78-83; sr aerospace engr, Lockheed-Calif Co, 83-85; sr res scientist, Lockheed Aero Systs Co, 85-89; ASST PROF, PENN STATE UNIV, 89- *Concurrent Pos:* Mem aeroacoustic comt, Am Inst Aeronaut & Astronaut; Mem Aeroacoust Comt & Fluid Dynamics Comt, Am Inst Aeronaut & Astronaut. *Mem:* Am Inst Aeronaut & Astronaut; Am Soc Eng Educ. *Res:* Computational physics; fluid dynamics; unsteady aerodynamics; hypersonic aerodynamics; electromagnetics; parallel processing. *Mailing Add:* 908 Bayberry Dr State College PA 16801

LONG, MAURICE W(AYNE), b Madisonville, Ky, Apr 20, 25; m 50, 63; c 4. ELECTRONICS, PHYSICS. *Educ:* Ga Inst Technol, BEE, 46, MS, 57, PhD(physics), 59; Univ Ky, MSEE, 48. *Prof Exp:* Asst, Eng Exp Sta, Ga Inst Technol, 46-47; instr elec eng, Univ Ky, 47-49; res engr, Eng Exp Sta, Ga Inst Technol, 50-51, asst prof, 51-53, spec res engr, 53-65, head, Radar Br, 55-60, chief, Electronics Div, 59-68, prin res physicist, 65-75, prof elec eng, 68-74, dir res, Eng Exp Sta, Ga Tech Res Inst, 68-75; CONSULT, 75- *Concurrent Pos:* Liaison scientist, Off Naval Res, London, 66-67; mem comt remote sensing prog for earth resources surv, Nat Acad Sci, 77; NASA Space Appln Adv Comt, 83-86. *Mem:* Fel Inst Elec & Electronic Engrs; Acad Electromagnetics. *Res:* Antennas and propagation; radar; electromagnetic scattering from rough surfaces. *Mailing Add:* 1036 Somerset Dr NW Atlanta GA 30327

LONG, MICHAEL EDGAR, b CZ, June 22, 46; m 88; c 1. PHYSICAL CHEMISTRY, TECHNICAL MANAGEMENT. *Educ:* Univ Toledo, BEd, 68; Wayne State Univ, PhD(chem), 73. *Prof Exp:* Fel chem, Cornell Univ, 73-75; RES CHEMIST, EASTMAN KODAK CO, 75- *Concurrent Pos:* NIH fel, Cornell Univ, 74-75. *Res:* Electronic photographic systems development. *Mailing Add:* Eastman Kodak Co Res Labs Bldg 81 Rochester NY 14650

LONG, PAUL EASTWOOD, JR, b Philadelphia, Pa, Oct 9, 42; div; c 2. METEOROLOGY, NUMERICAL ANALYSIS. *Educ:* Drexel Univ, BS, 65, MS, 68, PhD(physics), 70. *Prof Exp:* Mathematician, Philco-Ford Corp, 64-65; fel, Drexel Univ, 70-71; assoc, Nat Weather Serv, 71-73, res

meteorologist, 73-74; res meteorologist, Savannah River Lab, E I du Pont de Nemours & Co, Inc, 74-76; METEOROLOGIST, NAT WEATHER SERV, 76- *Mem:* Am Meteorol Soc; Am Inst Physics. *Res:* Numerical planetary boundary layer modeling. *Mailing Add:* Nat Meteorol Ctr W-N MC 23 Rm 204 WWB Washington DC 20233

LONG, R(OBERT) B(YRON), b Annville, Pa, Feb 18, 23; m 44; c 6. CHEMICAL ENGINEERING. *Educ:* Pa State Col, BS, 44, MS, 47, PhD(chem eng), 51. *Prof Exp:* Res engr, Standard Oil Develop Co, Exxon Res & Eng Co, 50-53, proj leader, 53-57, res assoc, 57-64, sr res assoc, 64-69, sci adv, 69-84; RETIRED. *Concurrent Pos:* Petrol Res Fund Adv Bd, 83-88; pres, Long Consult Inc, 84- *Mem:* Am Chem Soc; Am Inst Chem Engrs. *Res:* Separation processes; solvent extraction; petroleum processing; synthetic fuels; heavy oil processing; hazardous waste treating. *Mailing Add:* 1942 Pucker St Stowe VT 05672-9706

LONG, RAYMOND CARL, b Shattuck, Okla, June 17, 39; m 59; c 4. AGRONOMY. *Educ:* Kans State Univ, BS, 61, MS, 62; Univ Ill, Urbana, PhD(plant physiol), 66. *Prof Exp:* From asst prof to assoc prof, 66-82, PROF, CROP SCI, NC STATE UNIV, 82- *Concurrent Pos:* Vis prof, agron dept, Univ Wis, Madison, 75-76. *Mem:* Am Soc Plant Physiol; Am Soc Agron; Crop Sci Soc Am. *Res:* Biochemistry of growth and senescence of higher plants; nitrogen metabolism; environmental stress and plant growth. *Mailing Add:* Dept of Crop Sci NC State Univ Raleigh NC 27695-7620

LONG, RICHARD PAUL, b Allentown, Pa, Nov 29, 34; m 64; c 2. SUBSURFACE DRAINAGE, FIELD BEHAVIOR OF CLAYS. *Educ:* Univ Cincinnati, CE, 57; Rensselaer Polytech Inst, MSCE, 63, PhD(civil eng), 66. *Prof Exp:* Mgt trainee, Lehigh Struct Steel Co, 57-58; NSF fel res, Rensselaer Polytech Inst, 66-67; from asst prof to assoc prof, 67-78, PROF CIVIL ENG, UNIV CONN, 78-, DEPT HEAD, 77- *Concurrent Pos:* Proj mgr, Storch Engrs, 74; vis assoc prof, Mass Inst Tech, 75; chmn, tech comt, Transp Res Bd, Nat Acad Sci-Nat Res Ctr, 87- *Honors & Awards:* AT&T Award Excellence Eng Educ, Am Soc Eng Educ, 88. *Mem:* Am Soc Civil Engrs; Am Soc Eng Educ. *Res:* Geotechnical engineering; invention of prefabricated underdrain for soils; development of techniques for analyzing field data for settlement of clay; investigation of the process of capping dredged material deposited at shallow ocean sites. *Mailing Add:* Dept Civil Eng Box U-37 Univ Conn Storrs CT 06269-3037

LONG, ROBERT ALLEN, b Kingman, Ariz, Aug 17, 41; m 63; c 3. PHARMACEUTICAL CHEMISTRY, MEDICAL SCIENCES. *Educ:* Portland State Univ, BA, 64; Univ Utah, PhD(org chem), 70. *Prof Exp:* Res chemist, ICN Pharmaceut Inc, Calif, 70-77; clin res scientist, Cardiovasc Sect, 77-83, SR CLIN RES SCIENTIST, CARDIOVASC SECT, MED DIV, BURROUGHS WELLCOME CO, 83- *Mem:* Am Chem Soc; Am Pharmaceut Asn; Acad Pharmaceut Sci; Am Soc Clin Pharmacol & Therapeut; Drug Info Asn. *Res:* Heterocyclic chemistry; nucleic acid chemistry; antiviral and antitumor research; cardiovascular and respiratory research, clinical trials of new drugs; continued medical support for marketed products; project leader for new product development. *Mailing Add:* Burroughs Wellcome Co 3030 Cornwallis Rd Research Triangle Park NC 27709

LONG, ROBERT BYRON, b Annville, Pa, Feb 18, 23; wid; c 6. SEPARATION PROCESSES, HEAVY FUEL CHARACTERIZATION. *Educ:* Pa State Univ, BS, 44, MS, 47, PhD(chem eng), 51. *Prof Exp:* Instr, Pa State Univ, 47-50; engr, Standard Oil Develop, 50-53, group leader, 53-56; res assoc, Exxon Res & Eng Co, 56-64, sr res assoc, 64-69, sci adv, 69-84; PRES, LONG CONSULT, INC, 84- *Mem:* Am Chem Soc; Am Inst Chem Engrs. *Res:* Separation processes; author of 36 publications; holder of 66 US patents. *Mailing Add:* 1942 Pucker St Stowe VT 05672

LONG, ROBERT LEROY, b Renovo, Pa, Sept 9, 36; m 57; c 3. TECHNICAL MANAGEMENT. *Educ:* Bucknell Univ, BS, 58; Purdue Univ, MSE, 59, PhD(nuclear eng), 62. *Prof Exp:* Res assoc exp reactor physics, Argonne Nat Lab, 60-62; reactor specialist nuclear effects br, White Sands Missile Range, NMex, 62-65; from asst prof to prof nuclear eng, Univ NMex, 65-78, asst dean, 72-74, chmn chem & nuclear eng dept, 74-78; mgr, Generation Productivity Dept, GPU Nuclear Corp, 78-79, dir, Reliability Eng Dept, Gen Pub Utilities Serv Corp, 79-80, dir, training & educ, 80-82, vpres, Nuclear Asn Div, 82-87, vpres, Planning & Nuclear Safety Div, 87-89, VPRES, CORP SERV & TMI-Z, GPU NUCLEAR CORP, 89- *Concurrent Pos:* Res partic, Sandia Corp, 65-78; consult, White Sands Missile Range Fast Burst Reactor Facil, 65-78; res assoc nuclear res div, Atomic Weapons Res Estab, Eng, 66-67; assoc reactor engr, Con Edison, NY, 70-71; proj mgr nuclear eng & opers, Elec Power Res Inst, 76-77; mem, US Nuclear Energy Coun. *Mem:* Fel Am Nuclear Soc (pres-elect, 90-91); Soc Risk Analysis. *Res:* Reliability engineering data and applications; experimental reactor physics; fast burst reactors; power reactor technology; engineering teaching methods. *Mailing Add:* GPU Nuclear Corp One Upper Pond Rd Parsippany NJ 07054

LONG, ROBERT RADCLIFFE, b Glen Ridge, NJ, Oct 24, 19; m 63; c 2. METEOROLOGY. *Educ:* Princeton Univ, AB, 41; Univ Chicago, MS, 49, PhD, 50. *Prof Exp:* Meteorologist, US Weather Bur, 46-47; sr investr, hydrodyn lab, Univ Chicago, 49-51; from asst prof to assoc prof meteorol, Johns Hopkins Univ, 51-59, prof fluid mech, 59-; RETIRED. *Concurrent Pos:* Mem adv panel gen sci, US Secy Defense Res & Eng. *Mem:* Am Meteorol Soc. *Res:* Geophysical fluid mechanics; theoretical studies and laboratory models of geophysical phenomena; general circulation of the atmosphere; atmospheric and oceanic flow over barriers. *Mailing Add:* 3031 Willow Green The Meadows Sarasota FL 33580

LONG, RONALD K(ILLWORTH), b Steubenville, Ohio, Dec 5, 32; m 59. ELECTRICAL ENGINEERING. *Educ:* Ohio Wesleyan Univ, BA, 54; Harvard Univ, MS, 56; Ohio State Univ, PhD, 63. *Prof Exp:* Res engr labs, Radio Corp Am, 55; asst, Harvard Univ, 55-56; res engr, NAm Aviation, Inc, 56-57; asst supvr, Antenna Lab, Ohio State Univ, 58-63, from asst prof to assoc prof, 63-69, prof elec eng, 69-80. *Res:* Lasers; atmospheric propagation; infrared techniques; computer data acquisition. *Mailing Add:* Dept Elec Eng Ohio State Univ 2015 Neil Ave Columbus OH 43210

LONG, SALLY YATES, b Moyock, NC, Nov 8, 41; m 73; c 2. EMBRYOLOGY, TERATOLOGY. *Educ:* Col William & Mary, BS, 63; Univ Fla, PhD(anat), 67. *Prof Exp:* Lectr genetics, McGill Univ, 68-70; res assoc teratology, Karolinska Inst, Sweden, 70-71; asst prof, 71-76, asst dean student affairs, 78-81, ASSOC PROF ANAT, MED COL WIS, 76-, ASSOC DEAN STUDENT AFFAIRS, 81- *Concurrent Pos:* NIH fel, McGill Univ, 68-70. *Mem:* Teratology Soc (secy, 77-); Am Asn Anat; Europ Teratology Soc. *Res:* Interactions of genetic and environmental factors in causing malformations, especially cleft palate and limb defects. *Mailing Add:* Dept Anat Med Col Wis 8701 Watertown Plank Rd Milwaukee WI 53226

LONG, SHARON RUGEL, b San Marcos, Tex, Mar 2, 51; m 79; c 2. DEVELOPMENTAL BIOLOGY. *Educ:* Calif Inst Technol, BS, 73; Yale Univ PhD(biol), 79. *Prof Exp:* Res fel, dept biol, Harvard Univ, 78-81; asst prof, 82-87, ASSOC PROF, DEPT BIOL SCI, STANFORD UNIV, 87- *Concurrent Pos:* Shell res found award, 88. *Mem:* Am Soc Plant Physiologists; Genetics Soc Am; Soc Develop Biol; Soc Plant Molecular Biol; Am Soc Microbiol. *Res:* Genetics and developmental biology of symbiotic nitrogen fixation in legumes; role of plasmids in symbiosis; plant cell biology; plant molecular biology. *Mailing Add:* Dept Biol Sci Stanford Univ Stanford CA 94305-5020

LONG, STERLING K(RUEGER), b Petersburg, Va, Mar 11, 27; m 48, 73; c 2. BACTERIOLOGY. *Educ:* Univ Fla, BS, 49, MS, 51; Univ Tex, PhD(bact), 58. *Prof Exp:* Asst instr bot, Univ Miami, 49-50; asst bact & sanit eng, Univ Fla, 50-51; instr bact, Dent Br, Univ Tex, 57-58; assoc indust bacteriologist, Agr Res & Educ Ctr, Univ Fla, 58-74; RES MICROBIOLOGIST, NAT DISTILLERS & CHEM CORP, 74- *Mem:* Am Soc Microbiol. *Res:* Industrial fermentations; clinical and sanitary bacteriology; thermophilic spore forming bacteria; biochemistry. *Mailing Add:* Five E Lake View Dr No 24 Cincinnati OH 45213

LONG, STUART A, b Philadelphia, Pa, Mar 6, 45; m 69; c 3. APPLIED ELECTROMAGNETICS, ANTENNAS. *Educ:* Rice Univ, BA, 67, MEE, 68; Harvard Univ, PhD(appl physics), 74. *Prof Exp:* From asst prof to assoc prof 74-84, PROF ELEC ENG, UNIV HOUSTON, 85-, CHMN DEPT, 81- *Mem:* Fel Inst Elec & Electronic Engrs; Antennas & Propagation Soc; Int Union Radio Sci. *Res:* Applied electromagnetics: antennas; electromagentic methods of nondestructive evaluation; well-logging; subsurface communications; millimeter waveguiding and radiating structures. *Mailing Add:* Dept Elec Eng Univ Houston Houston TX 77204-4793

LONG, TERRILL JEWETT, b Newark, Ohio, Mar 19, 32; m 55; c 4. BOTANY. *Educ:* Ohio Univ, BSAg, 56; Ohio State Univ, MSc, 59, PhD(bot), 61. *Prof Exp:* NIH fel, Oak Ridge Nat Lab, 61-63, res assoc bot, 63-64; asst prof biol, Vanderbilt Univ, 64-65; res assoc biochem, Ohio State Univ, 65-67; from asst prof to assoc prof, 67-83, PROF BIOL, CAPITAL UNIV, 83- *Concurrent Pos:* Consult, C S Fred Mushroom Co, 66-70. *Mem:* AAAS; Bot Soc Am; Mycol Soc Am. *Res:* Physiology and biochemistry of irradiated wheat and mushrooms and related fungi. *Mailing Add:* Dept Biol Capital Univ 2199 E Main St Columbus OH 43209

LONG, THOMAS ROSS, b Lexington, Ky, Nov 6, 29; m 52; c 3. SOLID STATE PHYSICS. *Educ:* Ohio Wesleyan Univ, BA, 51; Case Inst Technol, MS, 53, PhD(physics), 56. *Prof Exp:* Mem tech staff, Bell Tel Labs, Inc, NJ, 56-67, head fundamental studies dept, Bell Labs, Ohio, 67-77; HEAD MAT ENG & CHEM DEPT, BELL LABS, GA, 77- *Mem:* AAAS; Inst Elec & Electronics Engrs; Am Phys Soc. *Res:* Communications device and techniques; memory and logic devices; contact physics; material science. *Mailing Add:* Bell Labs 2E53 2000 Northeast Expressway Norcross GA 30071

LONG, WALTER K, b Austin, Tex, Jan 26, 19; m 50; c 1. HUMAN GENETICS. *Educ:* Univ Tex, BA, 40; Harvard Univ, MD, 43. *Prof Exp:* Res asst cardiol, Thorndike Mem Lab, Boston City Hosp, Mass, 45-48; RES SCIENTIST HUMAN GENETICS, UNIV TEX, AUSTIN, 59-, LECTR ZOOL, 70- *Concurrent Pos:* Life Ins med res fel, 47-48; chief cardiovasc sect, William Beaumont Army Hosp, Ft Bliss, Tex, 51-53. *Mem:* Am Soc Human Genetics; Sigma Xi. *Res:* Relation between sulfahydryl compounds and pharmacology of organic mercurial diuretics; pentose phosphate metabolic pathway in relation to certain human diseases. *Mailing Add:* Dept of Zool Univ of Tex Austin TX 78712

LONG, WALTER KYLE, JR, b Montgomery, Ala, Dec 5, 44. VIROLOGY. *Educ:* Univ Ga, BS, 66; Univ Ill, PhD(microbiol), 72. *Prof Exp:* Fel, Dept Microbiol & Pediat, Univ Ala, Birmingham, 72-75, res assoc, 75-76; from asst prof to assoc prof, Sch Dent, 76-86, ASSOC PROF MICROBIOL, SCH MED, TEMPLE UNIV, 86- *Mem:* AAAS; Am Soc Microbiol; Sigma Xi. *Res:* Effects of antiviral drugs on herpes viruses; oncogenicity of herpes viruses; latency and reactivation of herpes viruses; role of DNA methylation in gene expression. *Mailing Add:* 404 Dresher Rd Townhouse J Horsham PA 19044

LONG, WILLIAM ELLIS, b Minot, NDak, Aug 18, 30; m 55, 71; c 6. HYDROLOGY, GEOMORPHOLOGY. *Educ:* Univ Nev, BS, 57; Ohio State Univ, MSc, 61, PhD(geol), 64. *Prof Exp:* Instr geol, Ohio State Univ, 63-64; explor geologist, Tenneco Oil Co, La, 64-65; from asst prof to assoc prof, 65-72, PROF GEOL, ALASKA METHODIST UNIV, 72-; CHIEF, WATER RESOURCES SECT, ALASKA STATE GEOL SURV, 78- *Concurrent Pos:* Mem, US Antarctic Res Prog, NSF Geol Invest, 63-64; mem discharge prediction glacial melt-water, Off Water Res, 68-70; consult, Shelf Explor Co, 71 & Forest Oil Co, 74-75; investr potential natural landmarks in Alaska, Nat Park Serv, 71; vis lectr, Univ Canterbury, 72. *Honors & Awards:* Long Hills, Antarctica named in honor. *Mem:* Am Groundwater Asn; Am Asn Petrol Geol; Am Inst Prof Geol; Geol Soc Am; Glaciol Soc; Sigma Xi. *Res:* Stratigraphic, geologic and glaciological exploration of Gondwana sequences of Antarctica during International Geophysical Year and following years; stratigraphic and glacial geology; water resources of Alaska. *Mailing Add:* PO Box 1831 Palmer AK 99645

LONG, WILLIAM HENRY, b Decatur, Ala, Sept 20, 28; m 53; c 3. SOILS & SOIL SCIENCE. *Educ:* Univ Tenn, BA, 52; NC State Col, MS, 54; Iowa State Col, PhD, 57. *Prof Exp:* From asst entomologist to prof entom, La State Univ, 57-65; prof, 65-85, DISTINGUISHED SERVICE PROF BIOL SCI, NICHOLLS STATE UNIV, 85- *Concurrent Pos:* Independent agr consult, 65-, UN Food & Agr Orgn, United Arab Repub, 73-74; pres, Long Pest Mgt, Inc, 72-; Consult Entom; entom expert, Int Atomic Energy Agency, 75-76. *Mem:* Entom Soc Am; Am Soc Sugarcane Technologists. *Res:* Development and refinement of sugarcane pest management programs; study of insects, nematodes and other factors which affect sugarcane. *Mailing Add:* PO Box 1193 Thibodaux LA 70301

LONG, WILLIS FRANKLIN, b Lima, Ohio, Jan 30, 34; m 59; c 3. ELECTRICAL ENGINEERING. *Educ:* Univ Toledo, BS, 57, MS, 62; Univ Wis-Madison, PhD(elec eng), 70. *Prof Exp:* Proj engr, Doehler Jarvis, Nat Lead Co, 57, 59-60; asst, Univ Toledo, 60-62, instr elec eng, 62-66; NSF fel, Univ Wis-Madison, 67-68, lectr, 69; mem tech staff, Hughes Res Labs, 69-73; from asst prof to prof elec eng & exten eng, 73-83, chmn exten eng, 80-83, PROF ELEC ENG & EXTEN ENG, UNIV WIS, 85- *Concurrent Pos:* Consult, Hughes Aircraft Co, Los Angeles Dept Power & Water, 73-; Spec Adv Comt, Wis Dept Indust, Labor & Human Rels, 76-77; dir, ASEA Power Systs Ctr, New Berlin, Wis, 83-85. *Mem:* Fel Inst Elec & Electronic Engrs; Int Conf Large High Voltage Elec Systs. *Res:* Analysis, simulation and testing of interconnected AC/DC electric power systems; power electronics switching techniques; continuing education, electric power systems. *Mailing Add:* Dept Eng Prof Develop 432 N Lake St Madison WI 53706-1498

LONG, WILMER NEWTON, JR, b Hagerstown, Md, Apr 24, 18; m 42; c 2. MEDICINE, OBSTETRICS & GYNECOLOGY. *Educ:* Juniata Col, BS, 40; Johns Hopkins Univ, MD, 43. *Prof Exp:* Instr gynec & obstet, Sch Med, Johns Hopkins Univ, 48-65; assoc prof, 65-67; PROF GYNEC & OBSTET, SCH MED, EMORY UNIV, 67- *Concurrent Pos:* Pvt pract obstet, 48-65; med officer in chg obstet & gynec, Navajo Med Ctr, Ft Defiance, Ariz, 53-55. *Mem:* Am Col Obstet & Gynec; AMA. *Res:* Diabetes in pregnancy. *Mailing Add:* 69 Bulter St SE Atlanta GA 30303

LONGACRE, RONALD SHELLEY, b Lindsay, Calif, Aug 15, 41; wid; c 4. PARTICLE PHYSICS. *Educ:* Calif Polytech State Univ, BS, 64; Univ Calif, Berkeley, MA, 68, PhD(physics), 74. *Prof Exp:* Res asst, Dept Physics Elem Particles, Comn L'Etude des Nuages-SACLAY, 74-75; res asst, Northeastern Univ, Boston, 75-78; asst physicist, 78-80, PHYSICIST, BROOKHAVEN NAT LAB, 80- *Res:* Determine Hadronic particle spectrum using three particle decay models; chief tool is the use of partial wave analyses via the Isobar model; model hadronic production in heavy ion collisions. *Mailing Add:* Dept of Physics Brookhaven Nat Lab Upton NY 11973

LONGACRE, SUSAN ANN BURTON, b Los Angeles, Calif, May 26, 41; m 64; c 2. SEDIMENTARY PETROLOGY, PETROLEUM GEOLOGY. *Educ:* Univ Tex, Austin, BS, 64, PhD(geol), 68. *Prof Exp:* Res assoc III, Getty Oil Co, 69-72, res assoc IV, 72-75, res scientist I geol, Explor & Prod Res Lab, 75-76, geol specialist II, Offshore Dist, 76-78, res scientist III explor & prod res, 78-80, prof specialist, 80-84, sr res consult, 84-90, SR SCIENTIST, TEXACO HOUSTON RES CTR, 90- *Concurrent Pos:* Comnr, NAm Comm Stratigraphic Nomenclature, 77-, vchmn, 85-86, chmn, 86-87. *Mem:* Am Asn Petrol Geologists; Geol Soc Am; Soc Econ Paleontologists & Mineralogists. *Res:* Petrology and petrography of carbonate and clastic sediments, particularly those Permian, Jurassic and Cretaceous sediments that accumulated in shallow marine to continental depositional environments. *Mailing Add:* 3901 Briarpark Houston TX 77042

LONGANBACH, JAMES ROBERT, b Akron, Ohio, July 4, 42; m 66; c 2. CHEMISTRY. *Educ:* Univ Akron, BS, 64; Yale Univ, MS, 66, MPh, 67, PhD(chem), 69. *Prof Exp:* Chemist, E I du Pont de Nemours & Co, Inc, 69-71; sr chemist, Res Div, Occidental Petrol Corp, 71-76; prin res chemist, Columbus Labs, Battelle Mem Inst, 76-87; PROJ MGR, MORGANTOWN ENERGY TECHNOL CTR, US ENERGY DEPT, 87- *Mem:* Am Chem Soc. *Res:* Physical-organic, energy and process develop chemistry. *Mailing Add:* PO Box 880 Mail Stop C04 Morgantown WV 26505

LONGCOPE, CHRISTOPHER, b Lee, Mass, Aug 5, 28; m 61; c 3. ENDOCRINOLOGY, REPRODUCTIVE BIOLOGY. *Educ:* Harvard Univ, AB, 49; Johns Hopkins Univ, MD, 53. *Prof Exp:* Intern, Presby Hosp, NY, 53-54; asst resident, 54-55; fel endocrinol, 59-60; asst resident, Johns Hopkins Hosp, 55-56; fel endocrinol, Univ Wash, Seattle, 60-62 & Univ Calif, San Francisco, 62-63; steroid training prog, Worcester Found Exp Biol, 65-66; staff scientist, 66-70, sr scientist, 70-80; PROF OBSTET, GYNEC & MED, MED SCH, UNIV MASS, 80- *Concurrent Pos:* Asst med, Johns Hopkins Univ, 63-64, instr med, 64-65; consult endocrinol, Perry Point Vet Admin Hosp, Md, 63-65; from asst prof med to assoc prof and dir, Endocrine Outpatients Clin, Boston Univ, 68-; mem, Aging Review Comt, Nat Inst Aging, NIH, 73-77 & Breast Cancer Task Force, 80-84. *Mem:* Endocrine Soc; Am Physiol Soc; Soc Exp Biol & Med; Soc Study Reproduction; Am Diabetes Asn. *Res:* Steroid dynamics, their production and mode of action. *Mailing Add:* Dept Obstet & Gynec Med Sch Univ Mass 55 Lake Ave N Worcester MA 01655

LONGENECKER, BRYAN MICHAEL, b Dover, Del, Sept 1, 42; m 63; c 2. IMMUNOLOGY, CELL BIOLOGY. *Educ:* Univ Mo, AB, 64, PhD(zool), 68. *Prof Exp:* Med Res Coun Can fel, 68-71, Nat Cancer Inst Can res grant, 71-73, Nat Cancer Inst Can res scholar immunol, 77-77, ASST PROF IMMUNOL & MEM NAT CANCER INST, UNIV ALTA, 77- *Mem:* AAAS. *Res:* Genetic control of allo-immunocompetence and resistance to virally induced neoplasms. *Mailing Add:* COO & Dir Div Immunotherapeut Biomira Inc 9411 20th Ave Edmonton AB T6N 1E5 Can

LONGENECKER, HERBERT EUGENE, b Lititz, Pa, May 6, 12; m 36; c 4. BIOLOGICAL CHEMISTRY. *Educ:* Pa State Col, BS, 33, MS, 34, PhD(agr biol chem), 36. *Hon Degrees:* ScD, Duquesne Univ, 51; LLD, Loyola Univ, 63; LittD, Univ Miami, 72; DSc, Loyola Univ & Univ Ill, 76. *Prof Exp:* Asst agr & biochem, Pa State Col, 33-35, instr, 35-36; Nat Res Coun fel, Univ Liverpool, 36-37, Univ Cologne, 37-38 & Queen's Univ, Ont, 38; fac mem, Univ Pittsburgh, 38-55, from asst prof to prof, 38-55, dean res natural scis, 44-55, dean grad sch, 46-55; vpres in charge, Univ Ill Med Ctr, 55-60; pres, 60-75, emer pres, Tulane Univ, 75-; mgr dir, Int Trade Mart, 76-79; RETIRED. *Concurrent Pos:* Mem food & nutrit bd, Nat Res Coun, 43-53, chmn comt food protection, 48-53; mem res coun, Chem Corps Adv Bd, 49-65; mem adv panel biol & chem warfare, Off Asst Secy Defense, 53-61; mem nat selection comn Fulbright student awards, 53-55, chmn, Western Europe Sect, 54-55; mem bd gov, Inst Med Chicago, 57-60; mem, Coun Financial Aid to Educ, 64-71; chmn acad bd adv, US Naval Acad, 66-72; dir, A G Bush Found, 69-85; mem panel sci & technol, US House of Rep Comt Sci & Astronaut, 70-73; dir, CPC Int, 66-85, Equitable Life Assurance Soc US, 68-84, United Student Aid Funds, 71-84 & Fed Home Loan Bank Little Rock, 76-79. *Mem:* fel Am Inst Chem; Sigma Xi. *Res:* Nutrition; fat metabolism; research administration. *Mailing Add:* 2717 Highland Ave Birmingham AL 35205

LONGENECKER, JOHN BENDER, b Salunga, Pa, July 8, 30; m 54; c 2. NUTRITION, BIOCHEMISTRY. *Educ:* Franklin & Marshall Col, BS, 52; Univ Tex, MS, 54, PhD(biochem), 56. *Prof Exp:* Res biochemist, E I du Pont de Nemours & Co, Inc, Del, 56-61; group leader, Mead Johnson & Co, Ind, 61-64; PROF NUTRIT & HEAD DIV, UNIV TEX, AUSTIN, 64- *Concurrent Pos:* USPHS grant, 64-71; Allied Health Fel grant, 69-74. *Mem:* Am Chem Soc; Am Inst Nutrit; NY Acad Sci. *Res:* In vivo plasma amino acid studies to evaluate protein and amino acid nutrition; interrelationships among nutrients; nutritional status studies. *Mailing Add:* Dept Home Econ Univ Tex Austin TX 78712

LONGENECKER, WILLIAM HILTON, b Cambridge, Md, Mar 28, 18; m 44. ORGANIC CHEMISTRY. *Educ:* Ohio State Univ, BA, 41; Georgetown Univ, MS, 49. *Prof Exp:* Chemist, Kankakee Ord Works, 42, Universal Oil Prod Co, 43, Armour & Co, 43-44, Toxicity Lab, Univ Chicago, 44, NIH, 46-49, Exp Sta, E I Du Pont de Nemours & Co, Inc, 49-62, Am Petrol Inst, 62-63 & Tech Info Div, Ft Detrick, 63-70; chemist, Nat Agr Libr, USDA, 70-89; RETIRED. *Mem:* AAAS; Am Chem Soc; Sigma Xi. *Res:* Systematic chemical nomenclature; chemical notation systems and machine methods of chemical documentation; bibliography compilations. *Mailing Add:* 11311 Cedar Lane Beltsville MD 20705

LONGERICH, HENRY PERRY, b Du Quoin, Ill, June 20, 40; m 64; c 1. ANALYTICAL CHEMISTRY, COMPUTER SCIENCE. *Educ:* Millikin Univ, BS, 63; Ind Univ, PhD(chem), 67. *Prof Exp:* Asst prof chem, Univ Alaska, 67-72; fel, Dalhousie Univ, 72-74; res assoc, 74-75, res fel, 75-78, asst prof, 78-84, ASSOC PROF EARTH SCI, MEM UNIV NFLD, 84- *Concurrent Pos:* Sessional, comput sci, Mem Univ, 79-82. *Mem:* Am Chem Soc; Spectros Soc Can. *Res:* Real-time on-line computer control and data acquisiton at analytical instrumentation, ICP-MS. *Mailing Add:* Earth Sci Dept Mem Univ St John's NF A1B 3X5 Can

LONGEST, WILLIAM DOUGLAS, b Pontotoc, Miss, Jan 22, 29; m 60. INVERTEBRATE ZOOLOGY. *Educ:* Baylor Univ, BSc, 54, MSc, 56; La State Univ, PhD(invert zool, ecol), 66. *Prof Exp:* Teacher, Parma High Sch, 55-56; instr biol, Northwest Jr Col, 56-59; prof natural sci, Blue Mountain Col, 59-62; instr biol, Memphis State Univ, 62-63; teaching asst zool, La State Univ, 63-65, instr, 65-66; from asst prof to assoc prof, 66-73, PROF BIOL, UNIV MISS, 73- *Mem:* Bot Soc Am; Am Soc Zool. *Res:* Botanical research; foliar embryos of Kalanchoe studied in an explant medium; taxonomy of freshwater Tricladida; study of freshwater triclads in the Florida Parishes of Louisiana. *Mailing Add:* Box 46 Rte 3 Oxford MS 38655

LONGFELLOW, DAVID G(ODWIN), b Akron, Ohio, Nov 16, 42; m 65; c 2. MOLECULAR CARCINOGENESIS, BIOCHEMISTRY. *Educ:* Lynchburg Col, BS, 64; Johns Hopkins Univ, PhD(biol), 72. *Prof Exp:* Damon Runyon res fel breast cancer, Biol Lab, Div Cancer Biol, Nat Cancer Inst, NIH, 72-74, res fel, 74-75, res staff, 75-76, sect head, Molecular Carcinogenesis Sect, 76-79, asst chief, Chem & Phys Br, Div Cancer Etiology, 79-84, CHIEF CHEM & PHYS CARCINOGENESIS BR, DIV CANCER ETIOLOGY, NAT CANCER INST, NIH, 84- *Mem:* Am Asn Cancer Res. *Res:* Chief of an extramural program awarding contracts and grants for research and resource support in the cause and prevention of chemical and physical carcinogenesis; DNA Repair and Chemical Resources programs. *Mailing Add:* Chem & Phys Carcinogenesis Br Nat Cancer Inst Exec Plaza N Suite 700 Bethesda MD 20892

LONGFIELD, JAMES EDGAR, b Mt Brydges, Ont, Mar 12, 25; nat US; m 47; c 3. PHYSICAL CHEMISTRY. *Educ:* Univ Western Ont, BSc, 47, MSc, 48; Univ Rochester, PhD(phys chem), 51. *Prof Exp:* Asst, Univ Rochester, 48-50; res chemist, Res Div, Am Cyanamid Co, 51-57, group leader eng res, 57-62, mgr eng res, 62-72, dir, process eng dept, Chem Res Div, 72-74, dir, Bound Brook Labs, 74-87; RETIRED. *Mem:* Am Chem Soc; Am Inst Chem Eng. *Res:* Vapor phase reactions of organic compounds; reaction kinetics; catalysis; reactor design and mechanism studies. *Mailing Add:* Eight Honey Locust Circle Hilton Head Island SC 29926-2680

LONGHI, JOHN, b White Plains, NY, Oct 12, 46; m 70; c 1. IGNEOUS PETROLOGY, PHYSICAL CHEMISTRY. *Educ:* Univ Notre Dame, BS, 68; Harvard Univ, PhD(geol), 76. *Prof Exp:* Res assoc, Mass Inst Technol, 76-77; res assoc lunar petrol, Univ Ore, 77-80; asst prof, Yale Univ, 80-85, assoc prof, 85-88; RES SCIENTIST, LAMONT-DOHERTY GEOL OBSERV, 88- *Concurrent Pos:* Jr fac fel, Yale Univ, 83-84. *Mem:* Am Geophys Union; Mineral Soc Am; Sigma Xi. *Res:* Origin and evolution of the moon and planets; experimental petrology; physical chemistry of silicates. *Mailing Add:* Lamont-Doherty Geol Observ Palisades NY 10964

LONGHI, RAYMOND, b Plymouth, Mass, Nov 14, 35; m 61; c 3. INORGANIC CHEMISTRY, ORGANIC CHEMISTRY. *Educ:* Univ Mass, BS, 57; Dartmouth Col, MA, 59; Univ Ill, PhD(inorg chem), 62. *Prof Exp:* Res chemist, E I Du Pont de Nemours & Co, Inc, 62-64, sr res chemist, 64-65, res supvr, 65-69, sr supvr tech, 69-71, sr supvr res & develop, 71-74, tech supt, 74-78, res & deveop site mgr, 78-85, mgr, Int Technol Transfer, 85-87, res mgr, 87-89, SR RES FEL, E I DU PONT DE NEMOURS & CO, INC, 89- *Mem:* Am Asn Textile Chemists & Colorists; Am Chem Soc; Sigma Xi. *Res:* Structures of transition metal complexes; reactions of nitrogen oxide; characterization of organic compounds; textile fibers. *Mailing Add:* E I du Pont de Nemours & Co Inc Seaford DE 19973

LONGHOUSE, ALFRED DELBERT, b Dunkirk, NY, Feb 17, 12; m 36; c 2. AGRICULTURAL ENGINEERING. *Educ:* Cornell Univ, BS, 37, MS, 38, PhD(agr eng), 47. *Prof Exp:* Asst agr eng, Cornell Univ, 37-38; instr, WVa Univ, 38-41; spec rep agr educ serv, US Off Educ, DC, 40-41; from asst prof to prof agr eng, West Va Univ, 41-77, head dept, 45-76, emer prof, 76-; RETIRED. *Mem:* Fel Am Soc Agr Engrs. *Res:* Developing undergraduate and graduate education and research programs in agricultural and forest engineering; poultry waste management, utilization, housing and environmental requirements; pollution control and/or elimination for poultry houses; developed courses in forest roads, power and machinery. *Mailing Add:* 11 Lakeview Ave Cassadaga NY 14718

LONGHURST, ALAN R, b Plymouth, Eng, May 3, 25; Can citizen; m 63; c 2. BIOLOGICAL OCEANOGRAPHY, MARINE ECOLOGY. *Educ:* Univ London, BSc, 52, PhD(zool), 62, DSc, 69. *Prof Exp:* Sci officer, West African Fisheries Res Inst, Sierra Leone, 54-57; marine biologist, Fisheries Lab, Wellington, NZ, 57-58; sr sci officer, Fishery Develop & Res Unit, Sierra Leone, 58-60; prin sci officer, Fed Fisheries Serv, Lagos, Nigeria, 60-63; assoc res biologist, Scripps Inst Oceanog, 63-67; dir, Fishery-Oceanog Ctr, Nat Oceanic Atmospheric Admin, 67-71; dep dir, Inst Marine Environ Res, Nat Environ Res Coun, Eng, 71-77; dir, Marine Ecol Lab, NS, 77-79, dir gen, Ocean Sci & Surv, Atlantic, Can Dept Fisheries & Oceans, 79-86, RES SCIENTIST, BEDFORD INST OCEANOG, 87- *Concurrent Pos:* Coordr, Eastern Trop Pac Oceanog Expeditions, 67-70; mem, Group Experts Ocean Variability Intergovt Oceanog Comn/Integrated Global Ocean Sta Syst, 69-71, Food & Agr Orgn Adv Comt Marine Res, 69-74, Dartmoor Nat Park Comt, Devon County Coun, 74-76, UK Delegation UN Conf Law Sea, 74-77; chmn, Continuous Monitoring Biol Oceanog, Sci Comt Oceanic Res/Adv Comt Marine Resources Res, 69-72; secy, Sci Coun Oceanic Res, 79-86. *Res:* Ecology of tropical benthos; population dynamics of tropical demersal fish; descriptive tropical physical oceanography; response to climate changes of marine biota; production and grazing relation in zooplanton in tropical, temperate and arctic oceans; formulation of large scale numerical ecological models; ecology of micoplankton and sub-micron particles. *Mailing Add:* Bedford Inst Oceanog PO Box 1006 Dartmouth NS B2Y 4A2 Can

LONGHURST, JOHN CHARLES, b Napa, Calif, March 18, 47; m 69; c 3. CARDIOVASCULAR PHYSIOLOGY, INTERNAL MEDICINE. *Educ:* Univ Calif, Davis, BS, 69, MD, 73, PhD(physiol), 74; Am Bd Internal Med, dipl, 77. *Prof Exp:* Fac assoc internal med, 78-79, instr internal med & physiol, 79-80, asst prof internal med, Health Sci Ctr, Univ Tex, Dallas, 80-, asst prof ohysiol, 81-; ASSOC PROF, DEPT MED, UNIV CALIF, SAN DIEGO. *Concurrent Pos:* Estab investr, Am Heart Asn, 81-86; mem, Coun Clin Cardiol & Coun Circulation, Am Heart Asn. *Mem:* Am Fedn Clin Res; Am Physiol Soc; fel Am Col Cardiol. *Res:* Neural control of the circulation; exercise physiology; physiology of the coronary circulation. *Mailing Add:* Univ Calif TB No 172 Davis CA 95616

LONGINI, IRA MANN, JR, b Cincinnati, Ohio, Oct 2, 48. EPIDEMIOLOGY. *Educ:* Univ Fla, BS, 71, MS, 73; Univ Minn, PhD(biomet), 77. *Prof Exp:* Assoc fel biomath, Int Ctr Med Res, 77-79; scholar biomet & lectr epidemiol, Dept Epidemiol, Univ Mich, 80-84; ASSOC PROF, DEPT EPIDEMIOL & BIOSTATIST, EMORY UNIV, 84- *Concurrent Pos:* Vis prof biomath, Univ Del Valle, 77-79. *Mem:* Biomet Soc; Soc Math Biol. *Res:* Development of mathmatical and statistical methods in epidemiology; genetics and biology. *Mailing Add:* Sch Pub Health Div Biostatist Emory Univ 1599 Clifton Rd NE Atlanta GA 30329

LONGINI, RICHARD LEON, b US, Mar 11, 13; m 37; c 2. PHYSICS, BIOENGINEERING. *Educ:* Univ Chicago, BS, 40; Univ Pittsburgh, MS, 44, PhD(physics), 48. *Prof Exp:* Physicist, Chicago TV & Res Labs, Inc, 34-35, Akay Electron Co, 35-38 & Wheelco Instruments Co, 38-41; physicist, Westinghouse Elec Corp, 41-51, sect mgr solid state electronics, 51-56, adv physicist, 56-58, sect mgr semiconductors, 58-60, consult physicist, 60-62; prof solid state electronics, 62-75, prof elec eng & urban affairs, 76-78, EMER PROF ELECT ENG & URBAN AFFAIRS, CARNEGIE-MELLON UNIV, 78-, SUPVR, SYSTS ENG LAB, 64- *Mem:* Fel Am Phys Soc; fel Inst Elec & Electronics Engr. *Res:* Solid state and medical electronics; data analysis and automated aids for diagnoses; application of engineering principles to solution of social problems. *Mailing Add:* 100 Norman Dr No 161 Mars PA 16046

LONGLEY, B JACK, b Dousman, Wis, July 19, 13; m 48; c 3. SURGERY. *Educ:* Univ Wis, BS, 34, PhD(pharmacol), 40, MD, 42. *Prof Exp:* Instr, 47-49, ASSOC PROF SURG, SCH MED & ASST DIR TUMOR CLIN UNIV WIS, MADISON, 49-; ASST CHIEF SURG SERV, VET ADMIN HOSP, 50- *Res:* Cardiovascular research. *Mailing Add:* 14 Merlham Dr Madison WI 53705

LONGLEY, GLENN, JR, b Del Rio, Tex, June 2, 42; m 61; c 4. LIMNOLOGY. *Educ:* Southwest Tex State Univ, BS, 64; Univ Utah, MS, 66, PhD(environ biol), 69. *Prof Exp:* Asst prof, 69-77, assoc prof, 77-80, PROF AQUATIC BIOL, SOUTHWEST TEX STATE UNIV, 80-,. *Concurrent Pos:* Res grants, USFWS, US SCS; dir environ consult firm, 80- *Mem:* Nat Water Well Asn; Am Water Res Asn; NAm Benthological Soc; AAAS; Water Pollution Control Fedn; Sigma Xi. *Res:* Use of subterranean fauna as indicators of ground water quality; Edwards Aquifer study; water pollution; heavy metals; organic wastes; pesticides; population dynamics; plankton; groundwater studies. *Mailing Add:* EARDC Southwest Tex State Univ San Marcos TX 78666

LONGLEY, H(ERBERT) JERRY, b Tahoka, Tex, Jan 3, 26; div; c 5. MATHEMATICS. *Educ:* Univ Tex, BS, 46, PhD(physics), 52; Tex Tech Col, BS, 48. *Prof Exp:* Asst prof & res assoc physics, NMex Inst Mining & Technol, 52-54; staff mem, Los Alamos Sci Lab, Univ Calif, 54-71; staff mem, Mission Res Corp, 71-78; CONSULT, 78- *Concurrent Pos:* Mem staff, Los Alamos Nuclear Corp, 70-71. *Mem:* Am Phys Soc. *Res:* Nuclear weapons and weapons testing; hydrodynamics, numerical solutions; radioactive waste storage; nuclear weapons effects; electromagnetic pulse; nuclear physics; linec design; fundamental particles. *Mailing Add:* 5808 Chaparral Circle Farmington NM 87401-4880

LONGLEY, JAMES BAIRD, b Baltimore, Md, June 27, 20; m 44; c 4. HISTOCHEMISTRY. *Educ:* Haverford Col, BSc, 41; Cambridge Univ, PhD(zool), 50. *Prof Exp:* From asst scientist to scientist, Nat Inst Arthritis & Metab Dis, 50-60; assoc prof anat, Sch Med, Georgetown Univ, 60-62; prof anat & chmn dept, Sch Med, Univ Louisville, 62-86; RETIRED. *Concurrent Pos:* USPHS sr res fel, 60-62; instr, Sch Med, Johns Hopkins Univ, 51-52; asst ed, J Histochem & Cytochem, Histochem Soc, 57-64, actg ed, 64-65; ed, Stain Technol, Biol Stain Comn, 73-87. *Mem:* Histochemical Soc; Am Asn Anat; Am Soc Cell Biol; Biol Stain Comn. *Res:* Renal histochemistry, morphology and physiology. *Mailing Add:* Dept of Anat Univ Louisville Health Sci Ctr Louisville KY 40292

LONGLEY, ROBERT W(ILLIAM), b Baltimore, Md, July 7, 25; m 50; c 5. NUTRITION, BIOCHEMISTRY. *Educ:* Loyola Col, Md, BS, 45; George Washington Univ, MS, 55, PhD(biochem), 57. *Prof Exp:* Res asst, Res Lab, Brady Urol Inst, 47-53; technician biochem, George Washington Univ, 53-55, instr biochem, 56; investr, Dorn Lab Med Res, 56-58; asst prof, Med Col Ala, 58-60; biochemist, Cent Res Labs, Gen Mills, Inc, 60-62, res assoc, James F Bell Res Ctr, 62-67; dir food res, Nutrit Div, Mead Johnson Subsidiary, Bristol Myers Co, Ind, 67-68 & Drackett Co, 68-70, dir food prod res, Mead Johnson Res Ctr, 70-71; consult food indust, 71-72; pres, Grist Mill Co, Minn, 72-73; dir res, Delmark Co, Minneapolis, 73-86; RETIRED. *Concurrent Pos:* Mgr res & develop, Camargo Foods Div, Drackett Co, Bristol Myers Co, Ohio, 69-71; mgr spec proj corp res & develop, Joseph Schlitz Brewing Co, Wis, 71. *Mem:* Am Asn Cereal Chemists; Inst Food Technologists; Am Soc Parenteral & Enteral Nutrit. *Res:* Carbohydrate metabolism; diabetes; clinical nutrition. *Mailing Add:* 100 Sweetwater Dr Apple Valley MN 55124

LONGLEY, W(ILLIAM) WARREN, b Paradise, NS, Apr 8, 09; US citizen; m 35, 57; c 3. GEOLOGY. *Educ:* Acadia Univ, BS, 31; Univ Minn, MS & PhD(geol), 37. *Prof Exp:* Instr geol, Dartmouth Col, 35-40; from asst prof to assoc prof geol & geophys, 40-52, prof, 52-77, EMER PROF GEOL, UNIV COLO, BOULDER, 77- *Concurrent Pos:* Consult, Que Dept Mines, 36-50, Kennecott Copper Corp, 45- & Kennco Explor Ltd, 46- *Mem:* Fel Geol Asn Can; fel Geol Soc Am; Soc Econ Geol; Am Asn Petrol Geol; Soc Explor Geophys. *Res:* Photogeology; mineral deposits in pre-Cambrian shield of Canada. *Mailing Add:* 821 Spring Dr Boulder CO 80303

LONGLEY, WILLIAM JOSEPH, b Middleton, NS, May 25, 38; m 63; c 2. REPRODUCTIVE PHYSIOLOGY, ENDOCRINOLOGY. *Educ:* Univ Toronto, BSA, 61, MSA, 63; Univ Mass, PhD(vet animal sci), 67. *Prof Exp:* Lectr physiol, Med Sch, Dalhousie Univ, 67-68, asst prof, 68-73, asst prof path, 73-80; qual control tech prod mgr, Corning Glass Works, 80-81, prod develop mgr, Corning Med & Sci, 81-84; CHEMIST, MARION LABS, 84- *Concurrent Pos:* Endrocrinologist, NS Dept Pub Health, 73-80. *Mem:* Can Soc Clin Chem; Soc Study Reproduction; Sigma Xi. *Res:* Endocrinology of the female, particularly fetal-placental function as related to steroid synthesis; clinical chemistry of various hormones including thyroid and adrenal. *Mailing Add:* Marion Labs PO Box 9627 Park A Kansas City MO 64134

LONGLEY, WILLIAM WARREN, JR, b Hanover, NH, Aug 30, 37; m 60; c 2. COMPUTER SCIENCES, PHYSICS. *Educ:* Univ Colo, BA, 58, PhD(physics), 63. *Prof Exp:* Physicist, Boulder Labs, Nat Bur Standards, 56-59; engr, Denver Div, Martin Co, 59-60, sr engr, 60-63; assoc physicist, Midwest Res Inst, 64-68; asst prof, Upper Iowa Col, Fayette, 68-70, assoc prof physics, 70-81, dir, Comput-Data Processing Ctr, 73-81; assoc prof math & comput sci, St Cloud State Univ, 81-83; ASSOC PROF, PERU STATE COL, 83- *Concurrent Pos:* Fel, Theoret Physics Inst, Univ Alta, 63-64. *Mem:* Fel AAAS; Am Phys Soc; Sigma Xi; Am Econ Asn. *Res:* Computer applications; economic statistics. *Mailing Add:* Peru State Col Admin Bldg 101 Peru NE 68421

LONGLEY-COOK, MARK T, b Tonbridge, Eng, June 29, 43; US citizen; div; c 1. ENVIRONMENTAL ENGINEERING, TRAFFIC ENGINEERING. *Educ:* Cornell Univ, BS, 65, MEng, 66; Univ Ariz, MS, 70, PhD(physics), 72. *Prof Exp:* Postdoctorate fel physics, Inst Atmospheric Physics, 72; physicist, Aircraft Environ Support Off, Naval Air Rework Facil, North Island, 72-76, energy engr, Western Div, Naval Facil Eng Command, 76-79, supvry mech engr, 79-80; principal, Longley-Cook Eng, Inc, 80-88; sr engr, Environ Sci Assoc, San Francisco, 80-89; assoc engr, Phomson Traffic Eng, Inc, Alamed, 80-91; LONGLEY COOK ENG ALAMED, 91- *Concurrent Pos:* Consult, 71- *Mem:* Inst Transp Engrs; Am Planning Asn. *Res:* Community noise; aircraft noise; liquid crystals; instrumentation; analysis; thermal conductivity; programming; atmospheric electricity; diffusion in the atmosphere; air pollution and noise. *Mailing Add:* 875 Portella Ave Coronado CA 94501

LONGMAN, RICHARD WINSTON, b Iowa City, Iowa, Sept 2, 43. CONTROL THEORY, ANALYTICAL DYNAMICS. *Educ:* Univ Calif, Riverside, BA, 65; Univ Calif, San Diego, MS, 67, MA & PhD(aerospace eng), 69. *Prof Exp:* Consult, Rand Corp, 66-69; mem tech staff, Control Systs Res Dept, Bell Tel Labs, NJ, 69-70; from asst prof to assoc prof, 70-79, PROF MECH ENG, COLUMBIA UNIV, 79- *Concurrent Pos:* Managing ed, J Astronaut Sci, 76-84; vis fac, Mass Inst Technol, Univ Bonn & Polytech Darmstadt & Univ Augsburg, WGer; consult, Langley Res Ctr & Goddard Space Flight Ctr, NASA, Europ Space Opers Ctr, Lockheed Missiles & Space Co, Aerospace Systs Div, Naval Res Lab, Martin Marietta Corp, Gen Elec Co, Xerox Res Lab, Marine Environ Corp, Designatronics, Inc & Syst Develop Corp; Alexander von Humboldt res fel, Polytech Darmstadt, WGermany, 77 & 80; sr fel, Nat Res Coun, 90-91. *Honors & Awards:* Dirk Brouwer Award, Am Astronaut Soc. *Mem:* Fel Am Astronaut Soc (vpres, 78-84); fel Am Inst Aeronaut & Astronaut; fel Brit Interplanetary Soc; Am Soc Mech Engrs. *Res:* Dynamics and control: shape control of large flexible spacecraft; control theory; applications of optimal control theory; robot kinematics, dynamics, trajectory optimization, vibration control, robotics in space; satellite dynamics. *Mailing Add:* Dept Mech Eng Columbia Univ New York NY 10027

LONGMIRE, DENNIS B, b Dayton, Ohio, June 4, 44. RUMINANT NUTRITION. *Educ:* Univ Tenn, BS, 66, MS, 69, PhD(animal sci), 73. *Prof Exp:* Dairy res specialist, 73-74, sr ruminant nutritionist, 74-75, ruminant feeds dir, 76-77, DIR INT FEED RES, CENT SOYA CO INC, 78- *Mem:* Am Dairy Sci Asn. *Res:* Management with emphasis on feed processing and production systems. *Mailing Add:* 417 Dearborn St Berne IN 46711

LONGMIRE, MARTIN SHELLING, b Morristown, Tenn, Mar 6, 31. ENGINEERING PHYSICS, GENERAL PHYSICS. *Educ:* Univ Cincinnati, BS, 53; Mass Inst Technol, PhD(phys chem), 61. *Prof Exp:* Res assoc phys chem, Ohio State Univ, 61-62; res fel, Mellon Inst, 62-64; res assoc, Mass Inst Technol, 64-65, physicist, Electronics Res Ctr, NASA, Cambridge, Mass, 65-70; assoc prof physics, Western Ky Univ, 70-88; RETIRED. *Concurrent Pos:* Res physicist, Nat Oceanic & Atmospheric Admin, 72 & Naval Res Lab, 71 & 73-88. *Mem:* AAAS; Am Phys Soc; Am Inst Chem; Sigma Xi. *Res:* Processing of signals from infrared sensors; development of infrared surveillance systems; absorption of solar ultraviolet light by atmospheric contaminants and minor constituents. *Mailing Add:* PO Box 105 Whitesburg TN 37891-0105

LONGMORE, WILLIAM JOSEPH, b La Jolla, Calif, Oct 7, 31; m 53; c 4. BIOCHEMISTRY. *Educ:* Univ Calif, Berkeley, AB, 57; Univ Kans, PhD(biochem), 61. *Prof Exp:* Nat Heart Inst fel metab res, Scripps Clin & Res Found, 61-63, res assoc biochem, 63-66; from asst prof to assoc prof, 66-73, PROF BIOCHEM, SCH MED, ST LOUIS UNIV, 73- *Concurrent Pos:* USPHS res career develop award, 66-76; Fogarty int sr fel, State Univ Utrecht, Neth, 77-78. *Mem:* AAAS; Am Chem Soc; Am Soc Biol Chemists; Sigma Xi. *Res:* Phospholipid metabolism; control mechanisms for lipid metabolism, phospholipid trafficking, especially in lung tissue; pulmonary surtactant. *Mailing Add:* 517 Beaucaire Dr Warson Woods MO 63122

LONGMUIR, ALAN GORDON, b Vancouver, BC, Mar 1, 41; m 62, 89; c 2. CONTROL ENGINEERING. *Educ:* Univ BC, BASc, 64, PhD(elec eng), 68. *Prof Exp:* Control engr, 68-78, mgr metals automation, 78-84, dir mfg systs, 84-88, VPRES RES DEVELOP, KAISER ALUMINUM & CHEM CORP, 88- *Concurrent Pos:* Assoc ed, Automatica, 76-83. *Mem:* Indust Res Inst. *Res:* Control systems engineering; application of computers to industrial process control. *Mailing Add:* Kaiser Aluminum & Chem Corp 6177 Sunol Blvd Pleasanton CA 94566

LONGMUIR, IAN STEWART, b Glasgow, Scotland, Mar 12, 22; m 49; c 4. BIOCHEMISTRY, PHYSIOLOGY. *Educ:* Cambridge Univ, BA, 43, MA & MB, BChir, 48. *Prof Exp:* Res assoc colloid sci, Cambridge Univ, 48-51; prin sci officer, Ministry Supply, Eng, 51-54; sr lectr biochem, Univ London, 54-65; PROF CHEM & BIOCHEM, NC STATE UNIV, 65- *Concurrent Pos:* Ed jour, Brit Polarographic Soc, 57-62; Isaac Ott fel, Univ Pa, 62-63. *Mem:* AAAS; Am Physiol Soc; Int Soc Oxygen Transport to Tissue; Am Soc Biol Chem; Aerospace Med Asn. *Res:* Oxygen transport in blood and tissue; inert gas metabolism. *Mailing Add:* Dept of Biochem NC State Univ Raleigh NC 27695-4622

LONGNECKER, DANIEL SIDNEY, b Omaha, Nebr, June 8, 31; m 52; c 4. PATHOLOGY. *Educ:* Univ Iowa, AB, 54, MD, 56, MS, 62. *Hon Degrees:* MA, Dartmouth Col, 74. *Prof Exp:* From asst to assoc prof path, Univ Iowa, 61-69; assoc prof, Sch Med, St Louis Univ, 69-72; PROF PATH, DARTMOUTH MED SCH, 72- *Concurrent Pos:* NIH spec fel & vis asst prof, Dept Path, Univ Pittsburgh, 65-67; USPHS res grants, Univ Iowa, 67-69, St Louis Univ, 69-71 & Dartmouth Col, 75- *Mem:* Am Soc Clin Path; Int Acad Path; Am Pancreatic Asn; Am Asn Pathologists; Am Asn Cancer Res. *Res:* Biochemical mechanisms of cell injury; pancreatic carcinogenesis; pathology of pancreatic disease and experiemental carcinogenesis in the pancreas; animal models are used to explore the influence of diet, pharmacologic agents, and hormones on pancreatic carcinogenesis. *Mailing Add:* Dept Path Dartmouth Med Sch Hanover NH 03756

LONGNECKER, DAVID EUGENE, b Kendallville, Ind, May 29, 39; m 63; c 3. ANESTHESIOLOGY. *Educ:* Ind Univ, AB, 61, MD, 64. *Prof Exp:* Intern, Blodgett Mem Hosp, Grand Rapids, 64-65; resident anesthesiol, Ind Univ, Indianapolis, 65-68; clin assoc, NIH, 68-70; asst prof, Univ Mo-Columbia, 70-73; from assoc prof to prof anesthesiol, Univ Va, 74-88; R D DRIPPS PROF & CHAIR, DEPT ANESTHESIA, UNIV PA, 88- *Concurrent Pos:* NIH spec res fel, Ind Univ, 67-68; res career develop award, Nat Heart & Lung Inst, 75. *Mem:* Am Soc Anesthesiologists; Inst Anesthesia Res Soc; Am Physiol Soc; Asn Univ Anesthetists. *Res:* Microcirculatory mechanisms during hemorrhagic shock; effect of anesthetics on the microcirculation during normovolemia and hypovolemia. *Mailing Add:* Dept Anesthesia HUP 4N Dulles 3400 Spruce St Philadelphia PA 19104-4283

LONGO, DAN L, b St Louis, Mo, Apr 25, 49; m 71; c 3. IMMUNOLOGY, MEDICAL ONCOLOGY. *Educ:* Washington Univ, AB, 70; Univ Mo, Columbia, MD, 75. *Prof Exp:* Resident internal med, Peter Bent Brigham Hosp, 75-77; fel med, Harvard Med Sch, 75-77; clin assoc med oncol, med br, Nat Cancer Inst, 77-78; clin assoc immunol, Lab Immunol, Nat Inst Allergy & Infectious Dis, 78-80; SR INVESTR ONCOL & IMMUNOL, MED BR, NAT CANCER INST, 80-, DIR, BIOL RESPONSE MODIFIERS PROG, 85- *Concurrent Pos:* asst ed, Am J Clin Nutrit, 81-; ed, Clin Oncol Alert, 85-; assoc ed, Cancer Res & J Nat Cancer Inst, 87-, Yr Bk Oncol & J Immunol, 88-, J Immunotherapy, 89- *Mem:* Am Asn

Immunologists; Am Inst Nutrit; Am Soc Clin Oncol; Am Fedn Clin Res; Am Asn Cancer Res; Am Soc Hemat; Am Soc Clin Invest. *Res:* Thymus function and control of lymphocyte proliferation and gene expression; treatment of lymphorproliferative diseases; biological therapy of human neoplastic infectious and immunological diseases. *Mailing Add:* Biol Response Modifiers Prog Frederick Cancer Res & Develop Ctr Bldg 567 Rm 135 Frederick MD 21702-1201

LONGO, FRANK JOSEPH, b Cleveland, Ohio, Nov 16, 39; m 62; c 6. CELL BIOLOGY. *Educ:* Loyola Univ, BS, 62; Ore State Univ, MS, 65, PhD(cell biol), 67. *Prof Exp:* Asst prof, 70-75, ASSOC PROF ANAT, COL OF MED, UNIV IOWA, 75- *Mem:* AAAS; Am Soc Cell Biol; Am Asn Anat; Soc Study Reproduction. *Res:* Cellular and developmental biology at the fine structural and biochemical levels; comparative pronuclear development and fusion; gametogenesis and fertilization; cell division and differentiation. *Mailing Add:* Dept of Anat Univ of Iowa Iowa City IA 52442

LONGO, FREDERICK R, b Trenton, NJ, May 4, 30; m; c 6. PHYSICAL CHEMISTRY. *Educ:* Villanova Col, BA, 53; Drexel Inst, MS, 58; Univ Pa, PhD(phys chem), 62. *Prof Exp:* Chemist, Am Biltrite Rubber Co, 55-57; assoc prof, 57-68, head dept chem & chem eng, Evening Col, 73-76, PROF CHEM, DREXEL UNIV, 68- *Concurrent Pos:* Sr res assoc, Nat Res Coun, 85-86. *Mem:* Am Chem Soc; Sigma Xi; NY Acad Sci. *Res:* Synthesis and spectral properties of porphyrins; investigation of microemulsions as media for controlled chemical reactions. *Mailing Add:* 6111 N Fairhill St Philadelphia PA 19120-1326

LONGO, JOHN M, b Hartford, Conn, Nov 6, 39; m 64; c 3. SOLID STATE CHEMISTRY, MINERAL REACTIONS. *Educ:* Univ Conn, BA, 61, PhD(inorg chem), 64. *Prof Exp:* Fel, Univ Stockholm, 64-65; chemist, Lincoln Lab, Mass Inst Technol, 65-70; chemist, Corp Res Labs, 70-81, LONG RANGE RES, EXXON PROD RES, 81- *Mem:* Am Chem Soc. *Res:* Preparation and characterization of solid state inorganic materials. *Mailing Add:* 13819 Taylorcrest Houston TX 77079-5814

LONGO, JOSEPH THOMAS, b Ferndale, Mich, Jan 13, 42; m 64; c 2. SOLID STATE PHYSICS. *Educ:* Univ Detroit, BS, 64; Mich State Univ, MS, 66, PhD(solid state physics), 68. *Prof Exp:* Asst, Mich State Univ, 64-68; fel, 68-69, mem tech staff, 69-72, mgr, 72-77, asst dir, 77-78, DIR, NORTH AM ROCKWELL SCI CTR, 78- *Mem:* Am Phys Soc. *Res:* High field magnetoresistance and Hall effect in intermetallic compounds; crystal growth, optical and device properties of narrow gap semiconductors. *Mailing Add:* 712 Kenwood Court Thousand Oaks CA 91360

LONGO, LAWRENCE DANIEL, b Los Angeles, Calif, Oct 11, 26; m 48; c 4. PHYSIOLOGY. *Educ:* Pac Union Col, BA, 49; Loma Linda Univ, MD, 54. *Prof Exp:* Asst prof obstet & gynec, Univ Ibadan, 59-62; asst prof, Univ Calif, Los Angeles, 62-64; lectr physiol, Univ Pa, 64-66; asst prof physiol, 66-68; PROF PHYSIOL, OBSTET & GYNEC, LOMA LINDA UNIV, 68- *Concurrent Pos:* USPHS fel obstet & gynec, Univ Calif, Los Angeles, 59, spec fel physiol, Univ Pa, 64-66 & res career develop award, 66-68, res career develop award, Loma Linda Univ, 68- & grant, 69-; consult, Nat Inst Child Health & Human Develop, 71. *Mem:* AAAS; NY Acad Sci; Am Physiol Soc; Soc Gynec Invest (secy-treas, pres); Perinatal Res Soc. *Res:* Regulation of fetal growth and development; fetal and placental physiology; kinetics of placental transfer of respiratory gases; fetal oxygenation. *Mailing Add:* Div Perinatal Biol Loma Linda Univ Loma Linda CA 92350

LONGO, MICHAEL JOSEPH, b Philadelphia, Pa, Apr 7, 35; m 58; c 3. HIGH ENERGY PHYSICS, SCIENCE EDUCATION. *Educ:* La Salle Col, BA, 56; Univ Calif, Berkeley, PhD(physics), 61. *Prof Exp:* NSF fel physics, Saclay Nuclear Res Ctr, France, 61-62; from asst prof to assoc prof, 62-68, PROF PHYSICS, UNIV MICH, ANN ARBOR, 68- *Mem:* Am Phys Soc; Sigma Xi. *Res:* Nucleon-nucleon interaction at high energies; proportional chambers and scintillation counters; neutrino interactions; magnetic monopoles; science communications; software systems; medical imaging. *Mailing Add:* Dept Physics Univ Mich Ann Arbor MI 48109

LONGOBARDO, ANNA KAZANJIAN, b New York, NY; m 52; c 2. TECHNICAL MANAGEMENT. *Educ:* Columbia Univ, BS, 49, MS, 52. *Prof Exp:* Sr systs engr, Am Bosch Arma Corp, 50-65; res sect head, Sperry Rand Corp, 65-73, mgr eng personnel utilization, 73-77, mgr prog planning, 77-81, mgr planning, 81- 82, dir tech serv, Unisys Corp, 82-89, DIR FIELD ENG, UNISYS CORP, 89- *Concurrent Pos:* Mem bd dirs, Woodward-Clyde Group, Inc, 89- *Mem:* Sr mem Am Soc Mech Engrs; sr mem Am Inst Aeronaut & Astronaut; fel Soc Women Engrs. *Res:* Supervised the Independent Research and Development program and was the strategic planner of a large unit of the Sperry Corporation; supervise the activities of approximately 400 field engineers in sixty locations worldwide who are giving life cycle support to diverse equipments including weather radar systems, radar landing systems, sonar systems and combat systems. *Mailing Add:* 15 Crows Nest Rd Bronxville NY 10708

LONGOBARDO, GUY S, b New York, NY, Oct 23, 28; m 52; c 2. MECHANICAL ENGINEERING, BIOENGINEERING. *Educ:* Columbia Univ, BS, 49, MS, 50, EngScD, 61. *Prof Exp:* Develop engr, E I du Pont de Nemours & Co, 50-52; instr mech eng, Sch Eng, Columbia Univ, 52-61, asst prof mech eng & bioeng, 61-65, dir, Fluid Mech Lab, 63-65; adv engr med info systs, IBM Corp, 65-77, mem corp staff, 77-81; SR FORECASTER, WORLD TRADE CORP, 81- *Concurrent Pos:* Consult, Am Mach & Foundry Co, 61-65 & Case Western Dept Med, 75- *Mem:* Assoc Am Soc Mech Engrs. *Res:* Medical information systems, clinical application of computer technology, operation of the respiratory control system and its unstable modes; medical information systems. *Mailing Add:* 15 Crows Nest Rd Bronxville NY 10708

LONGONE, DANIEL THOMAS, b Worcester, Mass, Sept 16, 32; m 54. ORGANIC CHEMISTRY. *Educ:* Worcester Polytech Inst, BS, 54; Cornell Univ, PhD(org chem), 58. *Prof Exp:* Res assoc org chem, Univ Ill, 58-59; from instr to assoc prof, 59-71, PROF ORG CHEM, UNIV MICH, ANN ARBOR, 71- *Concurrent Pos:* Am Chem Soc-Petrol Res Fund int fel, 67-68; Fulbright scholar, 70-71; vis prof, Univ Cologne, 70-71; vis Calif, Los Angeles, 76. *Mem:* Am Chem Soc. *Res:* Synthetic and mechanistic organic chemistry; bridged aromatic compounds; cyclophane chemistry; monomer synthesis and polymerization. *Mailing Add:* 2307 Chem Bldg Univ Mich Ann Arbor MI 48109-1055

LONGPRE, EDWIN KEITH, b Detroit, Mich, Mar 7, 33; m 65; c 1. SYSTEMATIC BOTANY. *Educ:* Univ Mich, BS, 55, MS, 56; Mich State Univ, PhD(bot), 67. *Prof Exp:* Instr bot, Tex Tech Col, 56-57; assoc prof bot & biol, 65-80, PROF BOT, WESTERN STATE COL COLO, 80- *Concurrent Pos:* Pres-elect, Colo-Wyo Acad Sci, 85-86. *Mem:* Am Soc Plant Taxon; AAAS; Sigma Xi. *Res:* Systematical studies in the tribe Heliantheae of the family Compositae; general cytotaxonomical and floristic studies. *Mailing Add:* Bot Dept Western State Col Gunnison CO 81231

LONGROY, ALLAN LEROY, b Flint, Mich, May 28, 36; m 55; c 3. ORGANIC CHEMISTRY. *Educ:* Univ Mich, AB, 58, MS, 61, PhD(chem), 63. *Prof Exp:* Res fel chem, Brandeis Univ, 62-64; asst prof, Ind Univ, 64-67; asst prof, 67-69, ASSOC PROF CHEM, PURDUE UNIV, 69- *Mem:* Am Chem Soc. *Res:* Organic reaction mechanisms and kinetics; demonstrations in chemistry. *Mailing Add:* Dept Chem Ind Univ-Purdue Univ Ft Wayne IN 46805-1499

LONGSHORE, JOHN DAVID, b Birmingham, Ala, Mar 8, 36; m 64; c 2. PETROLOGY. *Educ:* Emory Univ, BA, 57; Rice Univ, MA, 59, PhD(geol), 65. *Prof Exp:* Teacher, Westminster Schs, Ga, 60-62; PROF GEOL, HUMBOLDT STATE UNIV, 65- *Concurrent Pos:* NASA res grant chem invest Medicine Lake Area, 67-69. *Res:* Chemistry and petrology of igneous rocks. *Mailing Add:* Dept Oceanog Humboldt State Univ Arcata CA 95521

LONGSTRETH, DAVID J, b Phoenix, Ariz, Mar 22, 48. PLANT ECOPHYSIOLOGY. *Educ:* Ariz State Univ, BS 70, MS, 72; Duke Univ, PhD(bot), 76. *Prof Exp:* Fel, Duke Univ & Univ Calif, Los Angeles, 77-79; asst prof, 79-85, ASSOC PROF BOT, LA STATE UNIV, 85- *Mem:* Am Soc Plant Physiologists; Ecol Soc Am; Sigma Xi; AAAS. *Res:* Plant carbon balance in aquatic and semiaquatic environments; salinity effects on plant water relations and photosynthetic response. *Mailing Add:* Dept Bot La State Univ Baton Rouge LA 70803-1705

LONGTIN, BRUCE, b North Fork, Calif, Aug 23, 13; m 53; c 6. THERMODYNAMICS, NUCLEAR & RADIOCHEMISTRY. *Educ:* Univ Calif, BS, 35, MS, 37, PhD(chem), 38. *Prof Exp:* Asst chem, Univ Calif, 35-38, Shell Oil Co fel, 38-39; from instr to assoc prof, Ill Inst Technol, 39-51; from chemist to staff chemist, E I du Pont de Nemours & Co, Inc, 51-78; teaching assoc, Univ SC, Salkehatchie Campus, 78-81, 83-84; RETIRED. *Concurrent Pos:* Assoc chemist, Argonne Nat Labs, 48-49. *Mem:* Am Chem Soc. *Res:* Thermodynamics of industrial processes; thermodynamic properties of solutions; reactor water and water wastes; chemistry, radiolysis and control of impurities in water coolant and moderator of nuclear reactors; rheology and mechanical properties of polymers. *Mailing Add:* 1209 Summerhill Rd North Augusta SC 29841

LONGWELL, JOHN PLOEGER, b Denver, Colo, Apr 27, 18; m 45; c 3. CHEMICAL ENGINEERING, COMBUSTION. *Educ:* Univ Calif, Berkeley, BS, 40; Mass Inst Technol, ScD(chem eng), 43. *Prof Exp:* Asst, Nat Defense Res Comt, Mass Inst Technol, 42-43; chem engr, Exxon Res & Eng Co, 43-55, asst dir, 55-58, head, Spec Proj Unit, 58-73, dir, Cent Basic Res Lab, 59-68, mgr, Corp Res Staff, 68-73, sr sci adv, 73-77; E R Gilliland prof chem eng, 77-88, EMER PROF CHEM ENG MASS INST TECHNOL, 88- *Concurrent Pos:* Mem subcomt combustion, Nat Adv Comt Aeronaut; res adv, Comt Aeronaut Propulsion, NASA; tech adv panel ord, Asst Secy Defense Res & Eng; mem, Aeronaut Adv Comt, Nat Aeronaut & Space Admin, 78-84; chmn, Comt Advan Energy Storage, Nat Res Coun, 78-80, Comt Prod Technologies for Liquid Transp Fuels, Nat Res Coun, 89-90. *Honors & Awards:* Sir Alfred Egerton Medal, Combustion Inst, 74; Chem Eng Pract Award, Am Inst Chem Engrs, 79. *Mem:* Nat Acad Eng; Am Inst Chem Engrs; Am Chem Soc; Combustion Inst (pres). *Res:* Combustion; chemistry; propulsion and propellants; energy technology; coal conversion processes. *Mailing Add:* Mass Inst of Technol 77 Massachusetts Ave Cambridge MA 02139

LONGWELL, P(AUL) A(LAN), b Santa Maria, Calif, Aug 4, 19; m 40; c 2. CHEMICAL ENGINEERING. *Educ:* Calif Inst Technol, BS, 40, MS, 41, PhD(chem eng), 57. *Prof Exp:* Chemist, Shell Oil Co, Calif, 41; instr chem eng, Calif Inst Technol, 41-45; chem engr, US Naval Ord Test Sta, 45-50, head ord processing, 50-51, head, Explosives Dept, 51-54; from instr to assoc prof chem eng, Calif Inst Technol, 55-64; sr staff scientist, Aerojet-Gen Corp, 64-70, CHIEF SCIENTIST, ENVIROGENICS CO, AEROJET-GEN CORP, 70- *Concurrent Pos:* Consult, Aerojet-Gen Corp, 61-64. *Mem:* Am Chem Soc; Am Inst Chem Engrs. *Res:* Applied mathematics in engineering problems; heat, mass and momentum transfer; cryogenic plant processes; desalting plant processes. *Mailing Add:* 1000 Crest Dr Encinitas CA 92024-4042

LONGWORTH, JAMES W, b Stockton Heath, Eng, Sept 16, 38; m 65; c 2. BIOPHYSICS, CHEMICAL PHYSICS. *Educ:* Univ Sheffield, BSc, 59, PhD(biochem), 62. *Prof Exp:* USPHS fel phys chem, Univ Minn, 62-63; mem staff, Bell Tel Labs, 63-65; MEM STAFF, BIOL DIV, OAK RIDGE NAT LAB, 65- *Concurrent Pos:* Mem, US Nat Comt Photobiol, 72-76, chmn, 76-78; prog comt, Int Congr Photobiol, 80; assoc ed, Biophysical J, 79-81, ed, Comments Molecular & Cellular Biophysics, 80- *Mem:* Am Soc Photobiol (pres, 78-79); Biophys Soc; Brit Biochem Soc; Brit Biophys Soc; Am Soc Biol Chem. *Res:* Photophysics and excited state chemistry of proteins, nucleic acids and their synthetic analogues, particularly their luminescent behavior; use of optical methods to study conformation and function of proteins and nucleic acids and their complexes. *Mailing Add:* Dept Physics Ill Inst Tech 3301 S Dearborn St Chicago IL 60616

LONGWORTH, RUSKIN, b Oldham, Eng, Aug 13, 27; m 57; c 4. POLYMER CHEMISTRY, POLYMER PHYSICS. *Educ:* Univ London, BSc, 50, PhD(chem), 56. *Prof Exp:* Asst, Polytech Inst Brooklyn, 52-55; res fel, Univ Leiden, Holland, 55-56; chemist, Vauxhall Motors Ltd, Eng, 56-57; sr res chemist, Polymer Prod Dept, Exp Sta, E I du Pont de Nemours & Co, Inc, 57-85; RETIRED. *Mem:* Am Chem Soc; fel Royal Soc Chem. *Res:* Physical chemistry of polymers, especially rheology, solution properties, ionicpolymers and polyimides. *Mailing Add:* 10 Walnut Ridge Rd Greenville DE 19807

LONGYEAR, JUDITH QUERIDA, b Harrisburg, Pa, Sept 20, 38; c 2. PURE MATHEMATICS. *Educ:* Pa State Univ, BA, 62, MS, 64, PhD(math), 72. *Prof Exp:* Atmospheric physicist, White Sands Missile Range, 66-67; consult, Auerbach Corp, 67-68; asst prof math, Community Col Philadelphia, 68-70; John Wesley Young res assoc, Dartmouth Col, 72-74; from asst prof to assoc prof, 74-83, PROF MATH, WAYNE STATE UNIV, 83- *Mem:* Am Math Soc; Math Asn Am; Am Women in Math; Soc Indust & Appl Math; fel NY Acad Sci. *Res:* Combinatorial mathematics; block designs; Hadamard matrices; transversal theory; tactical configurations. *Mailing Add:* 605 MacKenzie Hall Wayne State Univ 5950 Cass Ave Detroit MI 48202

LONIGRO, ANDREW JOSEPH, b St Louis, Mo, July 22, 36; m 68; c 3. INTERNAL MEDICINE, PHARMACOLOGY. *Educ:* St Louis Univ, BS, 58, MD, 66. *Prof Exp:* Intern-resident, St Louis Univ Hosps, 66-69, fel cardiol, 69-71; from instr to asst prof pharmacol & internal med, Med Col Wis, 71-76; assoc prof, 76-84, PROF INTERNAL MED & PHARMACOL, ST LOUIS UNIV, 84- *Concurrent Pos:* Spec res fel, USPHS, 69-71; res & educ assoc, Vet Admin, 72-74, clin investr, 74-76, prog specialist clin pharmacol, 76-; dir div clin pharmacol, Sch Med, St Louis Univ, 76-; chief clin pharmacol, Vet Admin Hosp, St Louis, 76-87. *Mem:* Am Fedn Clin Res; Am Soc Nephrology; Am Physiol Soc; Am Soc Pharmacol & Exp Therapeut. *Res:* Circulatory control mechanisms; protaglandins, hypertension; renal function. *Mailing Add:* 96 Lake Forest St Louis MO 63117

LONKY, MARTIN LEONARD, b New York, NY, Jan 5, 44; m 66. ELECTRONIC PHYSICS, SOLID STATE PHYSICS. *Educ:* Rensselaer Polytech Inst, BS, 64; Univ Del, MS, 67, PhD(physics), 72. *Prof Exp:* Teaching asst physics, Univ Del, 64-67, res fel, 67-72; Presidential intern chem, US Army Land Warfare Lab, 72, res analyst, 72-73; sr engr electronics, Westinghouse Elec Corp, 73-75, fel engr physics, 76-79; MGR SOLID STATE TECHNOL, QUESTRON CORP, 79- *Mem:* Electrochem Soc; Inst Elec & Electronics Engrs. *Res:* Electron device physics, with emphasis on memory field effect transistors and transparent gate metal-oxide-silicon technology; device fabrication technologies. *Mailing Add:* 27600 Alvesta Pl Rancho Palos Verdes CA 90732

LNNERDAL, BO L, b Linkoping, Sweden, Mar 5, 48; m 74; c 4. INFANT NUTRITION. *Educ:* Univ Uppsala, BSc, 69, MSc, 71, PhD(biochem), 73. *Prof Exp:* Res asst, Dept Biochem, Univ Uppsula, 69-74, res assoc, Inst Nutrit, 74-76; vis asst res nutritionist, 78-80, asst res nutritionist, 80-81, from asst prof to assoc prof, 81-85, PROF NUTRIT, UNIV CALIF, DAVIS, 85- *Concurrent Pos:* Asst prof nutrit, Inst Nutrit, Univ Uppsula, 76- *Honors & Awards:* Henning Throne-Holst's Award, 77. *Mem:* Am Inst Nutrit; Am Soc Clin Nutrit; Soc Exp Biol & Med. *Res:* Composition of breast milk, cow's milk and formulas; trace element metabolism in the perinatal period. *Mailing Add:* Dept Nutrit Univ Calif Davis CA 95616

LONNES, PERRY BERT, b St Paul, Minn, Feb 22, 40; m 65; c 1. ENVIRONMENTAL SCIENCE, ANALYTICAL CHEMISTRY. *Educ:* Univ Minn, St Paul, BS, 63, MS, 65, PhD(environ sci), 72. *Prof Exp:* Instr air anal, Univ Minn, St Paul, 68-70; mgr anal serv & contract res, Environ Res Corp, 70-73; MGR ENVIRON MEASUREMENTS, INTERPOLL INC, 73- *Mem:* Am Chem Soc; Air Pollution Control Asn. *Res:* Characterization of adsorbents to predict gas sampling potentials; gas sampling methodology; gas chromatography; air pollution analytical instrumentation. *Mailing Add:* 5227 Beaver St St Paul MN 55110-6538

LONNGREN, KARL E(RIK), b Milwaukee, Wis, Aug 8, 38; m 63; c 2. PLASMA PHYSICS, ELECTRICAL ENGINEERING. *Educ:* Univ Wis-Madison, BS, 60, MS, 62, PhD(elec eng), 64. *Prof Exp:* Alumni Res Found res asst, Univ Wis, 64; grant, Royal Inst Technol, Sweden, 64-65; from asst prof to assoc prof, 65-72, PROF ELEC ENG, UNIV IOWA, 72- *Concurrent Pos:* Vis scientist, Oak Ridge Nat Lab, 67 & 69, Univ Sask, 71, Inst Plasma Physics, Japan, 72, Math Res Ctr, Univ Wis-Madison, 76-77, Los Alamos Nat Labs, 79 & 80, Inst Space & Astronaut Sci, Japan, 81 & Danish Atomic Energy Comn, 82. *Mem:* Fel Am Phys Soc; fel Inst Elec & Electronics Engrs. *Res:* Nonlinear plasma physics. *Mailing Add:* Dept Elec Eng Univ Iowa Iowa City IA 52242

LONSDALE, EDWARD MIDDLEBROOK, b Kansas City, Mo, July 21, 15; m 41; c 2. ELECTRICAL ENGINEERING. *Educ:* Univ Kans, BS, 36; Univ Iowa, MS, 41, PhD, 52. *Prof Exp:* Dial telephone engr, Southwest Bell Telephone Co, 36-38; TV engr, Midland TV Co, 39-40; radar countermeasures engr, Naval Res Lab, 42-46; prof elec eng, Univ Iowa, 46-56; prof elec eng, Univ Wyo, 56-72; clin engr, 72-80, HEAD BIOMED ENG, ST JOSEPH'S HOSP, 80-, CONSULT CLIN ENG. *Concurrent Pos:* Adj prof elec & comp eng, 82- *Mem:* Am Soc Eng Educ; Am Inst Elec & Electronics Engrs; Asn Advan Med Instrumentation. *Res:* Biomedical instrumentation; radio telemetry from fresh water fish. *Mailing Add:* St Joseph's Hosp 350 N Wilmont Tucson AZ 85711

LONSDALE, HAROLD KENNETH, b Westfield, NJ, Jan 19, 32; m 53; c 2. PHYSICAL CHEMISTRY. *Educ:* Rutgers Univ, BS, 53; Pa State Univ, PhD(chem), 57. *Prof Exp:* Staff mem, Gen Atomic Co, 59-70; prin scientist, Alza Corp, 70-72; vis scientist, Max Planck Inst Biophys, 73; vis prof, Weizmann Inst, 74; pres, Bend Res, Inc, 75-87; CHMN, BEND RES INC, 87- *Concurrent Pos:* Ed, J Membrane Sci, 75-90. *Mem:* Am Chem Soc. *Res:* Transport in synthetic membranes, desalination by reverse osmosis; controlled release of biologically active agents. *Mailing Add:* Bend Res Inc 64550 Research Rd Bend OR 97701

LONSDALE-ECCLES, JOHN DAVID, b Cheshire, Eng, Jan 14, 46; m 73; c 4. PROTEIN BIOCHEMISTRY, CELL BIOLOGY. *Educ:* Queen's Univ Belfast, BSc Hons, 70, PhD(biochem), 74. *Prof Exp:* Res fel biochem, Queen's Univ Belfast, 74-75; sr fel biochem, Univ Wash, Seattle, 75-78, res assoc, Dept Periodont, 78-82; SCIENTIST BIOCHEM, INT LAB RES ANIMAL DIS, NAIROBI, 83-, LAB COORDR, 91- *Concurrent Pos:* Lectr, Ctr Res Oral Biol, Seattle, 78-82; external examr, Univ Nairobi, 87-; consult, WHO, 90- *Mem:* Royal Soc Chem; Biochem Soc; NY Acad Sci; Am Soc Cell Biol; Protein Soc; Biochem Soc Kenya. *Res:* Disecting the biochemical aspects of endocytosis by African tryparosomes and their mechanisms of differentiation from one life cycle stage into another; protenses, protein binases and protein phosphates of the parasites. *Mailing Add:* Dept Biochem Int Lab Res Animal Dis PO Box 30709 Nairobi Kenya

LONSKI, JOSEPH, b Port Jefferson, NY, 43; c 2. DEVELOPMENTAL BIOLOGY. *Educ:* Cornell Univ, BS, 64; Univ Calif, Los Angeles, MA, 66; Princeton Univ, PhD(biol), 73. *Prof Exp:* Instr biol, Southampton Col, 66 & Princeton Univ, 71-72; from asst prof to assoc prof biol, Bucknell Univ, 72-83; CONSULT, 83- *Concurrent Pos:* Stockbroker. *Mem:* Sigma Xi; Am Soc Plant Physiol; Soc Develop Biol. *Res:* Chemotaxis in the myxobacteria and cellular slime molds. *Mailing Add:* 151 Mount Lucas Rd Princeton NJ 08540

LONTZ, ROBERT JAN, b Wilmington, Del, Oct 19, 36; m 62; c 2. PHYSICS. *Educ:* Yale Univ, BSc, 58; Duke Univ, PhD(physics), 62. *Prof Exp:* Asst, Physics Div, US Army Res Off, 62-64, chief, Gen Physics Br, 64-67, assoc dir, 67-73, dir, 73-88; CONSULT, RES FUNDING, RES DEFINITION & MGT, TECH WRITING, 88- *Concurrent Pos:* Dep asst res, Off Undersecy Defense Res & Eng, 78-79. *Res:* Paramagnetic resonance spectroscopy; lasers. *Mailing Add:* 3122 Surrey Rd Durham NC 27707

LONZETTA, CHARLES MICHAEL, b Hazleton, Pa, Jan 28, 50; m 71; c 3. ORGANIC CHEMISTRY, PHYSICAL-ORGANIC CHEMISTRY. *Educ:* Pa State Univ, BS, 71; Harvard Univ, AM, 74, PhD(org chem), 77. *Prof Exp:* Fel phys-org chem, Brandeis Univ, 76-78; RES CHEMIST, ROHM & HAAS CO, 78- *Concurrent Pos:* Head teaching fel, Harvard Exten Sch, 74-78. *Mem:* Am Chem Soc. *Res:* Mechanistic organic chemistry: singlet oxygen formation and reactions; organophosphorus reaction kinetics; free radical reactions; pulsed megawatt infrared laser reaction kinetics; monomer process technology. *Mailing Add:* Rohm & Haas Tex Inc PO Box 672 Tidal Rd Deer Park TX 77536

LOO, BILLY WEI-YU, b Chungking, China, Oct 26, 39; US citizen; m 65; c 2. INSTRUMENT SCIENCE, X-RAY DETECTORS. *Educ:* Univ Mich, Ann Arbor, BSE, 63, MS, 65, PhD(physics & nuclear eng), 72. *Prof Exp:* Asst res physicist high energy physics, Univ Mich, 65-69; SR ENG PHYSICIST, LAWRENCE BERKELEY LAB, UNIV CALIF, 72- *Mem:* Am Phys Soc; AAAS. *Res:* Research and development in medical instrumentation, lung and bone density measurements, and special Si(Li) x-ray detectors for space and nuclear science applications; sampling and analysis of atmospheric aerosols. *Mailing Add:* M/S 70A-3363 Lawrence Berkeley Lab Univ Calif Berkeley CA 94720

LOO, FRANCIS T C, b Tongshan, Hopei, China, July 25, 27; m 58; c 3. SOLID MECHANICS. *Educ:* Nat Taiwan Univ, BS, 52; Syracuse Univ, MS, 59, PhD(mech & aerospace eng), 64. *Prof Exp:* Res assoc mech & aerospace eng, Syracuse Univ, 59-63, asst prof, 63-66; assoc prof mech eng, Clarkson Col Technol, 66-86; PRIN ENGR, ELEC BOAT DIV, GEN DYNAMICS, 86- *Concurrent Pos:* Summer fac res fel, NASA, NSF & US Air Force; assoc ed, J Mech Design, Am Soc Mech Engrs; chmn, RSAFP Comt. *Mem:* Am Soc Mech Engrs; Am Soc Naval Engrs; Soc Naval & Marine Engrs. *Res:* Thermal stresses; stability of shells subjected to concentrated loads and uniform pressure; fracture mechanics; vibration of structures; experimental stress analysis; structural mechanics; finite element method; composite materials. *Mailing Add:* Dept 457 Elec Boat Div Gen Dynamic Groton CT 06340

LOO, MELANIE WAI SUE, b Honolulu, Hawaii, Nov 24, 48. GENETICS. *Educ:* Univ Calif, BA, BS, 70; Univ Wash, PhD(genetics), 74. *Prof Exp:* Proj res assoc genetics, Dept Physiol Chem, Univ Wis, 75-77; ASST PROF BIOL & GENETICS, DEPT BIOL SCI, CALIF STATE UNIV, SACRAMENTO, 77- *Mem:* AAAS. *Res:* Genetic regulation. *Mailing Add:* Dept Biol Sci Calif State Univ Sacramento CA 95819

LOO, TI LI, b Changsha, China, Jan 7, 18; nat US; m 51; c 3. CLINICAL PHARMACOLOGY, CANCER & AIDS CHEMOTHERAPY. *Educ:* Tsing Hua Univ, China, BSc, 40; Oxford Univ, DPhil, 47 & DSc, 85. *Prof Exp:* Asst pharmacol, Oxford Univ, 46-47; fel org chem, Univ Md, 47-51; res assoc, Christ Hosp Inst Med Res, 51-54; supvry chemist, NIH, 55-65; pharmacologist & prof, Dept Develop Therapeut, Univ Tex M D Anderson Hosp & Tumor Inst & prof pharmacol, Univ Tex Med Sch & Grad Biomed Sci, 65-85, Ashbel Smith prof ther, 81-85; RES PROF PHARMACOL, GEORGE WASH UNIV MED CTR, 85- *Concurrent Pos:* Adj prof pharmacol, Univ Houston, 77-85; spec lectr, Japan Soc Clin Pharmacol, 85. *Honors & Awards:* Gottlieb Award, Univ Tex M D Anderson Cancer Ctr, 87. *Mem:* Am Chem Soc; Am Asn Cancer Res; Royal Soc Chem; Am Soc Clin Oncol; Am Soc Clin Pharmacol & Therapeut; Sigma Xi; Am Soc Pharmacol Exp Therapeut. *Res:* Pharmacology of anticancer and anti-AIDS drugs; cancer chemotherapy; metabolism of drugs; chemical structure and biological activities; pharmacokinetics; anti-AIDS chemotherapy. *Mailing Add:* Pharmacol Dept George Wash Univ Med Ctr 2300 Eye St NW Washington DC 20037-2313

LOO, YEN-HOONG, b Honolulu, Hawaii, Dec 19, 14. BIOCHEMISTRY. *Educ:* Barnard Col, BA, 37; Univ Mich, MS, 38, PhD(biochem), 43. *Prof Exp:* Fel biochem, Univ Tex, 43-44; res asst, Univ Ill, 44-51; biochemist, Nat Heart Inst, NIH, 51-52; res biochemist, Eli Lilly & Co, 52-68; res scientist, NY State Inst Basic Res Develop Disabilities, 68-85; RETIRED. *Concurrent Pos:* NIH fel, Cambridge Univ, Eng, 66-67; prin investr, NIH grant, 76-85. *Mem:* Am Soc Neurochem; Am Soc Biol Chemists; fel AAAS. *Res:* Identified neurotoxic metabolite of phenylalanine primarily responsible for brain dysfunction in phenylketonuria (PKU); biochemical, structural and behavioral changes related to retarded brain development in experimental phenylketonuria. *Mailing Add:* 1212 Punahou St Apt 1906 Honolulu HI 96826

LOOFBOURROW, ALAN G, mechanical engineering; deceased, see previous edition for last biography

LOOK, DAVID C, b St Paul, Minn, Dec 19, 38; m 68; c 2. SOLID STATE PHYSICS. *Educ:* Univ Minn, BPhys, 60, MS, 62; Univ Pittsburgh, PhD(physics), 65. *Prof Exp:* Res physicist, Aerospace Res Labs, 66-69; sr res physicist, Univ Dayton, 69-80; SR RES PHYSICIST, WRIGHT STATE UNIV, 80- *Mem:* Am Phys Soc; Am Sci Affil; Inst Elec & Electronics Engrs; Electrochem Soc. *Res:* Transport properties in semiconductor materials and devices; nuclear magnetic resonance; ion implantation; radiation damage. *Mailing Add:* Univ Res Ctr Wright State Univ Dayton OH 45435

LOOK, DWIGHT CHESTER, JR, b Smith Center, Kans, Aug 25, 38; m 60; c 2. THERMAL RADIATIVE HEAT TRANSFER, THERMOPHYSICAL PROPERTIES. *Educ:* Cent Col, Fayette, Mo, BA, 60; Univ Nebr, MS, 62, Univ Okla, PhD(mech & aerosysts), 69. *Prof Exp:* Teaching asst eng physics, Univ Nebr, 60-63; aerosysts engr, Fort Worth Div, Gen Dynamics, 63-67; from asst prof to assoc prof, 69-78, PROF MECH ENG, UNIV MO-ROLLO, 78- *Concurrent Pos:* Adj instr ele maths, Tex Christian Univ, 67; spec instr thermodyn, Univ Okla, 69; co-prin investr, NSF grants, 75- *Honors & Awards:* R R Testor Award, Soc Automotive Engrs, 78. *Mem:* Am Soc Mech Engrs; Am Inst Aeronaut & Astronaut; Am Soc Eng Educ; Int Soc Optical Eng. *Res:* Experimental investigation of thermophysical properties, particularly the reflectance of light from solids and the electromagnetic scattering from small particles; thermodynamics. *Mailing Add:* 203 Mech Eng Bldg Univ Mo Rolla MO 65401

LOOKER, JAMES HOWARD, b Bloomingburg, Ohio, Nov 24, 22; m 46; c 2. ORGANIC CHEMISTRY. *Educ:* Ohio State Univ, BS, 43, PhD(chem), 49. *Prof Exp:* From instr to assoc prof, 50-60, PROF CHEM, UNIV NEBR, LINCOLN, 60- *Concurrent Pos:* NIH spec fel guest prof, Univ Vienna, 63-64. *Mem:* Sigma Xi. *Res:* Flavonoid substances; diazoesters; arylserines; sulfonic esters. *Mailing Add:* Dept Chem Univ Nebr Lincoln NE 68508

LOOKER, JEROME J, b Columbus, Ohio, July 7, 35; m 57; c 3. ORGANIC CHEMISTRY. *Educ:* Kenyon Col, AB, 58; Univ Ill, MS, 60, PhD(org chem), 61. *Prof Exp:* Nat Sci Found fel, Cornell Univ, 61-62; RES CHEMIST, EASTMAN KODAK CO, 62- *Mem:* Am Chem Soc. *Res:* Synthetic organic chemistry. *Mailing Add:* 333 Panorama Terr Rochester NY 14625-2315

LOOKHART, GEORGE LEROY, b North Platte, Nebr, Aug 25, 43; m 63; c 3. ANALYTICAL CHEMISTRY, PHYSICAL BIOCHEMISTRY. *Educ:* Kearney State Col, BS, 68; Univ Wyo, PhD(phys chem), 73. *Prof Exp:* RES ANAL CHEMIST, USDA, US GRAIN MKT RES LAB, 76- *Concurrent Pos:* Teaching internship fel, Chem Dept, Univ Ky, 73-74; fel biochem dept, Univ Mo, Columbia, 74-76. *Mem:* Am Chem Soc; Am Asn Cereal Chemists. *Res:* Develop high pressure liquid chromatographic methods of analysis for protein, estrogens, amino acids and vitamins; develop new electrophoretic methods to fingerprint protein. *Mailing Add:* US Grain Mkt Res Lab 1515 College Ave Manhattan KS 66502

LOOMAN, JAN, b Apeldoorn, Netherlands, Oct 18, 19; Can citizen; m 58; c 2. BOTANY. *Educ:* Univ Wis, MSc, 60, PhD(bot, soils), 62. *Prof Exp:* Technician pasture res, Cent Inst Agr Res, Wageningen, Netherlands, 52-54; technician pasture res, Res Sta, Can Dept Agr, 54-58, res scientist, 62-84; CONSULT, 84- *Mem:* Agr Inst Can; Int Soc Veg Sci. *Res:* Pasture research; classification of plant communities, including lichens and bryophytes, particularly classification in relation to practical application. *Mailing Add:* Res Sta Swift Current SK S9H 3X2 Can

LOOMANS, MAURICE EDWARD, b Wisconsin Rapids, Wis, Aug 10, 33; m 57; c 3. DERMATOLOGY. *Educ:* Hope Col, BA, 57; Univ Wis, MS, 59, PhD(biochem), 62. *Prof Exp:* RES CHEMIST, MIAMI VALLEY LABS, PROCTER & GAMBLE CO, 62- *Mem:* Soc Invest Dermat. *Res:* Keratinization; epidermal cellular control; acne; percutaneous absorption; mediators of inflammation; animal models; rheumatology; arthritis. *Mailing Add:* Procter & Gamble Co Miami Valley Labs Box 398707 Cincinnati OH 45239

LOOMIS, ALBERT GEYER, b Lexington, Mo, Feb 17, 93; m 19, 32; c 3. CHEMISTRY. *Educ:* Univ Mo, AB, 14, AM, 15; Univ Calif, PhD(chem), 19. *Prof Exp:* Instr chem, Univ Ill, 19-20; asst prof, Univ Mo, 20; Nat Res Coun fel, Cryogenic Lab, US Bur Mines, 21-22, phys chemist, 21-28, chemist, Explosives Div, Exp Sta, Pa, 28-29; chief chemist, Gulf Res & Develop Co, 29-35; asst dir, Shell Develop Co, 35-42, assoc dir, 42-45; consult chemist & petrol engr, 45-48; petrol engr, US Bur Mines, 48-63; consult chemist, Loomis Labs, 63-76; GUEST SCIENTIST, CHEM DEPT, UNIV CALIF, BERKELEY, 76- *Concurrent Pos:* Lectr, George Washington Univ, 23-25; coop expert, Int Critical Tables, 25-27; sr indust fel, Mellon Inst, 29-35. *Honors & Awards:* Anthony F Lucas Gold Medal Award, Am Inst Mining, Metall & Petrol Eng, 72. *Mem:* Am Chem Soc; Am Inst Mining, Metall & Petrol Eng. *Res:* Extraction of radium, vanadium and uranium from carnotite ore; liquid ammonia systems; thermodynamic properties of hydrocarbon systems at low temperatures; recovery of helium from natural gas; flame temperatures and explosives; colloid physics of clay dispersions; utilization of redwood products; petroleum production. *Mailing Add:* 85 Parnassus Rd Berkeley CA 94708

LOOMIS, ALDEN ALBERT, b Pittsburgh, Pa, July 22, 34; m 57; c 3. GEOLOGY, OCEANOGRAPHY. *Educ:* Stanford Univ, AB, 56, PhD(petrol, geol), 61. *Prof Exp:* Asst prof geol, San Jose State Col, 60-61; SR SCIENTIST, JET PROPULSION LAB, CALIF INST TECHNOL, 61- *Concurrent Pos:* Assoc prof, Calif State Col Los Angeles, 65-66; consult geoscientist, 69-; eng geologist, State of Calif, 72- *Mem:* Sigma Xi. *Res:* Space applications to oceanography and geology; igneous petrology, volcanology; metamorphic petrology; gravity and crustal structure; geology of moon and Mars; development of experiments for lunar and planetary exploration; engineering and environmental geology; mineral exploration. *Mailing Add:* 1262 E Rubio St Altadena CA 91001

LOOMIS, EARL ALFRED, JR, b Minneapolis, Minn, May 21, 21; m 69; c 4. CHILD DEVELOPMENT, SUBSTANCE DEPENDENCE. *Educ:* Univ Minn, BA, 42, MD, 45; Am Bd Psychiat & Neurol, cert, 51 & 58. *Prof Exp:* Intern internal med & pediat, Evans Mem & Mass Mem Hosp, 45-46; resident psychiat, Western Psychiat Inst, Pittsburgh, 46-48; fel psychiat & child psychiat, Hosp Univ Pa & Inst Pa Hosp, 48-50; instr psychiat, Univ Pa, 49-52; assoc prof, Univ Pittsburgh, 52-56; chief, div child psychiat, St Luke's Hosp, NY, 55-62; prof psychiat & relig, Union Theological Seminary, NY, 56-63; psychiat dir, child & adolescent psychiat, Blueberry Treatment Ctr Seriously Disturbed Children, 63-81; prof psychiat, Dept Psychiat & Health Behav, Med Col Ga, 81-90; RETIRED. *Concurrent Pos:* Consult child psychiat, Gov Bacon Health Ctr, Del, 49-57; res fel in residence child develop, Univ Geneva, Switz, 62-63; lectr, Herbert Hoet Inst, 63-73; attend psychiatrist, Eastern Long Island Hosp, 73-81; psychiat dir, Child Adolescent Prog, Ga Regional Hosp, Augusta, 81-82 & Eugene Talridge Ment Hosp, 82-84. *Mem:* Am Psychiat Asn; Am Psychoanalytic Asn; Am Acad Child Psychiat; Am Soc Alcholism & Other Drug Dependence. *Res:* Play patterns of non-verbal children as indices of ego function and dysfunction; conscience of condoners, abusers and abused in physical and sexual abuse; consequences of parallel and out of phase development of conscience and cognition. *Mailing Add:* 1002 Katherine St No 6 Augusta GA 30904-4481

LOOMIS, FREDERICK B, b Amherst, Mass, Feb 10, 15. PETROLEUM & MINE GEOLOGY. *Educ:* Amherst Univ, BA, 37. *Prof Exp:* Direct mgr, Clark Oil Refining, Milwaukee, Wis, 39-59, geologist & mgr foreign oper, 60-70; mgr Can oper, Petro-Consult, Alta, Can, 70-75; CONSULT GEOLOGIST, 75- *Mem:* Fel Geol Soc Am; Am Asn Petrol Geologists. *Mailing Add:* 2738 S Via Del Bac Green Valley AZ 85614

LOOMIS, GARY LEE, b Baltimore, Md, March 3, 43; m 81. POLYMER SYNTHESIS, FUNCTIONALIZED POLYMERS. *Educ:* Johns Hopkins Univ, MA, 74, PhD(org chem), 75. *Prof Exp:* Res fel org chem, Inst Chem, London, 75-76 & Ecol Normale Superieure, Paris, 76-77; SR RES CHEMIST ORG POLYMER CHEM & BIOMAT, EXP STA, E I DU PONT, CO, 78- *Mem:* Am Chem Soc; Chem Soc London; Sigma Xi. *Res:* Biomaterials, biodegradable polymers, high pressure polymer synthesis, biopolymers, interpenetrating polymer networks, thermoplastic elastomers; organic polymer synthesis with enzymes. *Mailing Add:* E I Du Pont Co PO Box 8023 Wilmington DE 19880-0323

LOOMIS, HAROLD GEORGE, b Erie, Pa, Aug 22, 25; m 47; c 4. NUMERICAL WAVE THEORIES. *Educ:* Stanford Univ, BS, 50; Pa State Univ, MS, 52, PhD(math & physics), 57. *Prof Exp:* Scientist, HRB Singer, 52-55; instr math, Pa State Univ, 55-57; asst prof math, Amherst Col, 57-62; scientist, Nat Oceanic & Atmospheric Admin, 66-82; asst prof math, 63-66, PROF OCEAN ENG, UNIV HAWAII, 82- *Concurrent Pos:* Secy, Tsunami Comn, Int Union Geod & Geophys, 76-82. *Mem:* Soc Indust & Appl Math. *Res:* Numerical hydrodynamics; long and short water wave theories; time series analysis; statistics. *Mailing Add:* 1125 B Ninth Ave Honolulu HI 96816

LOOMIS, HERSCHEL HARE, JR, b Wilmington, Del, May 31, 34; m 57; c 2. ELECTRICAL ENGINEERING, COMPUTER ENGINEERING. *Educ:* Cornell Univ, BEE, 57; Univ Md, MS, 59; Mass Inst Technol, PhD(elec eng), 63. *Prof Exp:* Staff engr, Lincoln Lab, Mass Inst Technol, 60-61; from asst prof to assoc prof elec eng, Univ Calif, Davis, 62-74, chmn dept, 70-75, prof elec eng, 74-83; chair prof Navelex, 81-83; PROF ELEC & COMPUTER ENG, NAVAL POSTGRAD SCH, MONTEREY, CA, 83- *Concurrent Pos:* Consult, Lawrence Livermore Lab, Univ Calif, 63-88; NSF grants, 64 & 67-69; consult, Signal Sci, Inc, Santa Clara, CA, 81- *Mem:* Inst Elec & Electronics Engrs; Asn Comput Mach. *Res:* Theory, design and applications of digital computers; digital design automation; digital signal processing systems. *Mailing Add:* 4086 Pine Meadows Way Pebble Beach CA 93953

LOOMIS, RICHARD BIGGAR, b Lincoln, Nebr, June 18, 25; m 47; c 3. ACAROLOGY, HERPETOLOGY. *Educ:* Univ Nebr, BSc, 48; Univ Kans, PhD(zool), 55. *Prof Exp:* Lab asst biol & vert zool, Univ Nebr, 47-48; asst, US Navy Chigger Proj, Univ Kans, 48-53, asst instr biol & comp anat, 53, asst, Sch Med, 53-55; PROF BIOL, CALIF STATE UNIV, LONG BEACH, 55- *Concurrent Pos:* Prin investr, USPHS Grant, 60-74; res assoc, Los Angeles County Mus Natural Hist; mem bd govs, Calif Desert Studies Consortium; NSF grant, 80-82. *Mem:* Am Soc Mammal; Am Soc Ichthyol & Herpet; Am Soc Parasitol; Am Soc Acarologists. *Res:* Systematics, life histories and ecology of parasitic acarines, especially trombiculid mites and their vertebrate hosts in North America; medical acarology. *Mailing Add:* Dept Biol Calif State Univ Long Beach CA 90840

LOOMIS, ROBERT HENRY, b Atlanta, Ga, Nov 9, 23; m 45; c 4. ZOOLOGY, LIMNOLOGY. *Educ:* Univ Ga, BS, 47; Okla State Univ, MS, 51, PhD(zool), 56. *Prof Exp:* Instr biol, Piedmont Col, 48, Cent State Col, Okla, 51-52 & Jimma Agr Sch, Ethiopia, 52-54; asst prof, Cent State Col, Okla, 54-55; from asst prof to prof, Northeastern State Col, 55-63; prof, Parsons Col, 63-68; prof & chmn div sci, Pikeville Col, 68-75; prof life sci, Sacramento City Col, 75-78; mem fac dept biol, Calif State Univ, 78-89; dept life sci, Consumnes River Col; RETIRED. *Mem:* AAAS; Am Inst Biol Sci. *Res:* Watershed conditions on fish populations; food habits of mesopelagic fishes; identification of photosynthetic active components of phytoplankton communities; temperature acclimation in crayfish populations. *Mailing Add:* 7056 Pine Cone Dr Pollock Pines CA 95726

LOOMIS, ROBERT MORGAN, b Mauston, Wis, Aug 31, 22; m 48; c 6. FORESTRY. *Educ:* Univ Mich, BS; Univ Mo, MS, 65. *Prof Exp:* Forester, Ochoco Nat Forest, Ore, 48-51; adminr, Ottawa Nat Forest, Mich, 51-56 & Mo Nat Forests, 56-57; res forester, Cent States Forest Exp Sta, Columbia, Mo, 57-66, NCent Forest Exp Sta, Columbia, Mo, 66-71 & E Lansing, Mich, 71-80; RETIRED. *Mem:* Soc Am Foresters. *Res:* Forest fire effects, fuels and danger rating. *Mailing Add:* 104 Redwood Ct Pine Knoll Shores NC 28512

LOOMIS, ROBERT SIMPSON, b Ames, Iowa, Oct 11, 28; m 51; c 3. PLANT PHYSIOLOGY. *Educ:* Iowa State Univ, BS, 49; Univ Wis, MS, 51, PhD(bot), 56. *Prof Exp:* Instr agron & jr agronomist, 56-58, from asst prof & asst agronomist to assoc prof & assoc agronomist, 58-68, dir, Inst Ecol, 69-72, assoc dean environ studies, 70-72, PROF AGRON & AGRONOMIST, UNIV CALIF, DAVIS, 68- *Concurrent Pos:* NIH spec fel, Harvard Univ, 63-64; NZ Nat Res Adv Coun res fel, 71; vis scientist, Agr Univ, Wagenigen, 79; vis prof, Melbourne Univ, 85. *Mem:* Fel Am Soc Plant Physiol (secy, 65-67); Am Soc Sugar Beet Technol; fel Am Soc Agron; Agr Hist Soc. *Res:* Physiology and ecology of field crops including growth and development; integrative physiology with emphasis on system simulation. *Mailing Add:* Dept Agron Univ Calif Davis CA 95616

LOOMIS, STEPHEN HENRY, b Flint, Mich, Oct 3, 52; m 80; c 2. COMPARATIVE PHYSIOLOGY. *Educ:* Univ Calif, Davis, BS, 74, PhD(zool), 79. *Prof Exp:* Res assoc, Rice Univ, 79-80; asst prof, 80-86, ASSOC PROF COMPARATIVE PHYSIOL & INVERTEBRATE ZOOL, CONN COL, 86- *Mem:* Am Soc Zoologists; AAAS; Soc Cryobiol; Sigma Xi. *Res:* Freezing tolerance of intertidal invertebrates. *Mailing Add:* Dept Zool Conn Col Box 5496 New London CT 06320

LOOMIS, TED ALBERT, b Spokane, Wash, Apr 24, 17; m; c 2. PHARMACOLOGY, TOXICOLOGY. *Educ:* Univ Wash, BS, 39; Univ Buffalo, MS, 41, PhD(pharmacol), 43; Yale Univ, MD, 46. *Prof Exp:* Intern, US Marine Hosp, 46-47; assoc prof, 47-59, PROF PHARMACOL & TOXICOL, SCH MED, UNIV WASH, 59- *Concurrent Pos:* State toxicologist, Wash, 55-77. *Res:* Pesticide and insecticide toxicology; anticoagulant agents; alcohol research; toxicological methods; mechanisms of drug action and action of toxic chemicals. *Mailing Add:* Drill Friess Hays Loomis & Shaffer Inc 2707 E Becker Rd Clinton WA 98236

LOOMIS, THOMAS CHARLES, analytical chemistry, for more information see previous edition

LOOMIS, TIMOTHY PATRICK, b Alhambra, Calif, May 25, 46. PETROLOGY, REACTION KINETICS. *Educ:* Univ Calif, Davis, BS, 67; Princeton Univ, PhD(geol), 71. *Prof Exp:* J W Gibbs instr geol, Yale Univ, 71-73; adj asst prof, Univ Calif, Los Angeles, 73-74; asst prof, 74-76, ASSOC PROF GEOL, UNIV ARIZ, 76- *Mem:* Geol Soc Am; Am Geophys Union. *Res:* Heat and mass transfer and reaction kinetics in chemical processes. *Mailing Add:* Dept Geosci Univ Ariz Gould Simpson Bldg 77 Tucson AZ 85721

LOOMIS, WALTER DAVID, b Fayetteville, Ark, Mar 2, 26; m 52. BIOCHEMISTRY. *Educ:* Iowa State Univ, BS, 48; Univ Calif, PhD(comp biochem), 53. *Prof Exp:* Instr biochem, 53-54, from asst prof to assoc prof, 54-68, PROF BIOCHEM, ORE STATE UNIV, 68- *Concurrent Pos:* USPHS res career develop award, 61-67; vis researcher, Univ Col Wales, 65-66. *Mem:* Am Chem Soc; Am Soc Plant Physiol; Am Soc Biol Chem; Phytochem Soc NAm; Can Soc Plant Physiol. *Res:* Plant enzymes and proteins; terpene metabolism. *Mailing Add:* Dept of Biochem & Biophysics Ore State Univ Corvallis OR 97331

LOOMIS, WILLIAM FARNSWORTH, JR, b Boston, Mass, Sept 17, 40; m 62; c 2. DEVELOPMENTAL BIOLOGY. *Educ:* Harvard Univ, BS, 62; Mass Inst Technol, PhD(microbiol), 65. *Prof Exp:* NIH fel, Brandeis Univ, 65-66; from asst prof to assoc prof, 66-79, PROF BIOL, UNIV CALIF, SAN DIEGO, 79- *Mem:* Soc Develop Biol (pres); Am Soc Biol Chemists. *Res:* Cellular interactions involved in the biochemical differentiation in Dictyostelium discoideum; genetics of slime molds; complex processes of cellular interaction can be dissected by molecular genetics; concepts generated in one system can often be applied to others. *Mailing Add:* Dept Biol Univ Calif San Diego La Jolla CA 92093

LOONEY, NORMAN E, b Adrian, Ore, May 31, 38; m 57, 83; c 3. POMOLOGY, PLANT GROWTH REGULATION. *Educ:* Wash State Univ, BS, 60, PhD(hort), 66. *Prof Exp:* Sr exp aid hort, Wash State Univ, 60-62, res asst post-harvest hort, 62-66; POMOLOGIST & PLANT PHYSIOLOGIST, 66-, HEAD POMOLOGY & VITICULTURE SECT, AGR CAN RES STA, 87- *Concurrent Pos:* Vis scientist, CSIRO, Sydney, Australia, 71-72, East Malling Res Sta, Maidstone, Eng, 81-82 & Lincoln Univ, NZ, 90-91. *Mem:* Fel Am Soc Hort Sci; Can Soc Hort Sci; Int Soc Hort Sci. *Res:* Physiology of growth, development and ripening of fruits; investigations of flowering physiology and plant growth regulator effects. *Mailing Add:* Pomology & Viticulture Sect Agr Can Res Sta Summerland BC V0H 1Z0 Can

LOONEY, RALPH WILLIAM, b Spencer, WVa, June 30, 31; m 61; c 3. PHYSICAL CHEMISTRY, POLYMER CHEMISTRY. *Educ:* WVa Univ, BS, 53, MS, 54; Univ Wis, PhD(phys chem), 60. *Prof Exp:* Res chemist, Esso Res & Eng Co, 60-63, sr chemist, 63-66, sect head new chem intermediates, 66-72, res assoc chem intermediates, Elastomers Tech Div, 72-76, res assoc chem intermediates, Tech Div, 76-78, RES ASSOC, SPECIALTIES TECH DIV, EXXON CHEM CO, 78- *Mem:* Am Chem Soc. *Res:* Polymerization catalysts; kinetics of polymerization; polymer physics and oxonation of olefins. *Mailing Add:* 3735 Lake Laberge Ct Baton Rouge LA 70816

LOONEY, STEPHEN WARWICK, b Atlanta, Ga, Sept 6, 52; m 80. MULTIVARIATE ANALYSIS, QUALITY CONTROL. *Educ:* Univ Ga, BS, 74, MS, 76, PhD(statist), 80. *Prof Exp:* Chief statist analyst, Northeast Ga Health Dist, 79-80; vis fel, Health & Welfare Can, 80-81; vis biostatist, Upjohn Co, 87-88; asst prof, 81-86, ASSOC PROF QUANT BUS ANAL, LA STATE UNIV, 86-, MBA DIR, 89- *Mem:* Am Statist Asn; Am Soc Qual Control; Int Asn Statist Comput. *Res:* Research activity in applied statistics and how it can be applied in business research; quality control. *Mailing Add:* 1034 Briarrose Dr Baton Rouge LA 70810

LOONEY, WILLIAM BOYD, b South Clinchfield, Va, Mar 18, 22; m 55; c 2. RADIOBIOLOGY, BIOPHYSICS. *Educ:* Emory & Henry Col, BS, 44; Med Col Va, MD, 48; Cambridge Univ, PhD(radiobiol, biophys), 60. *Hon Degrees:* DSc, Emory & Henry Col, 78. *Prof Exp:* Intern, Presby Hosp, Chicago, 48-49, asst resident, 49-50; asst prof radiol, Johns Hopkins Univ, 59-60; from asst prof to assoc prof, 61-68, PROF RADIOBIOL & BIOPHYS & DIR DIV, UNIV VA, 68- *Concurrent Pos:* Mem interdisciplinary prog biophys, Univ Va, 66-; vis fel med oncol, Mem Sloan-Kettering Cancer Ctr, 78. *Mem:* AAAS; Am Asn Cancer Res; Am Soc Cell Biol; Biophys Soc; Radiation Res Soc; Sigma Xi. *Res:* Cancer; mathematical evaluation of tumor growth curves; cell cycle and cell kinetics studies in experimental tumors; modification of tumor growth rates and cell kinetics by radiation, alone or in combination with different chemotherapeutic agents; host-tumor interaction. *Mailing Add:* Div of Radiobiol & Biophys Univ of Va Hosp Charlottesville VA 22908

LOOP, JOHN WICKWIRE, radiology, medicine; deceased, see previous edition for last biography

LOOP, MICHAEL STUART, b Pittsburgh, Pa, Feb 28, 46; c 4. VISION, HERPETOLOGY. *Educ:* Fla State Univ, BS, 68, MS, 71, PhD(psychobiol), 72. *Prof Exp:* NIH fel neurol surg, Univ Va, 72-74, Sloane Found fel physiol, 74-75; vis asst prof physiol & biophys, Univ Ill, 75-78; asst prof, 78-81, ASSOC PROF PHYSIOL OPTICS, UNIV ALA, BIRMINGHAM, 81- *Mem:* Soc Neurosci. *Res:* Vertebrate visual system psychophysics; comparative animal behavior. *Mailing Add:* Med Ctr/Sch of Optom Univ Sta Birmingham AL 35294

LOOR, RUEYMING, immunodiagnostics of cancer, for more information see previous edition

LOOS, HENDRICUS G, b Amsterdam, Neth, Dec 18, 25; nat US; m 52; c 2. ENVIRONMENTAL PHYSICS, NEURAL NETWORKS. *Educ:* Univ Amsterdam, Drs(math), 51; Univ Delft, ScD, 52. *Prof Exp:* Res engr, Nat Aeronaut Res Inst, Neth, 46-52; res fel, Calif Inst Technol, 52-55; sr engr, Propulsion Res Corp, 55-57; sr physicist, Giannini Sci Corp, 57-66; mem staff, Douglas Advan Res Lab, 66-70, sr staff scientist, McDonnell-Douglas Astronaut Co, 70-71; prof math, Cleveland State Univ, 71-74; DIR, LAGUNA RES LAB, 74- *Concurrent Pos:* Lectr, Univ Calif, Riverside, 63-64, assoc prof in residence, 64-70, adj prof, 70-76. *Mem:* Am Phys Soc; Soc Photo-Optical Instrumentation Engrs; Am Optical Soc; Int Neural Network Soc. *Res:* Gauge theory; atmospheric physics; fluid mechanics; general relativity; neural networks. *Mailing Add:* 3015 Rainbow Glen Fallbrook CA 92028-9765

LOOS, JAMES STAVERT, b Grafton, NDak, May 24, 40; m 61; c 2. PHYSICS. *Educ:* Univ NDak, BS, 62; Univ Ill, MS, 63, PhD(physics), 68. *Prof Exp:* Res assoc high energy physics, Stanford Linear Accelerator Ctr, 68-72; asst prof physics, Duke Univ, 72-77; RES PHYSICIST HIGH ENERGY PHYSICS, ARGONNE NAT LAB, 77- *Mem:* Am Phys Soc. *Res:* Experimental high energy physics; electron-positron collisions at high energies; high energy particle detectors and techniques. *Mailing Add:* AT & T Bell Labs 600 Mountain Ave 2D320 Murray Hill NJ 07974

LOOS, KARL RUDOLF, b New York, NY, July 10, 39; m 65; c 3. PHYSICAL CHEMISTRY, ENVIRONMENTAL ANALYSIS. *Educ:* Rensselaer Polytech Inst, BS, 60; Mass Inst Technol, PhD(phys chem), 65. *Prof Exp:* Res assoc, Inst Phys Chem, Swiss Fed Inst Technol, 65-66; STAFF RES CHEMIST, SHELL DEVELOP CO, 67- *Mem:* Am Chem Soc. *Res:* Environmental air analysis; source emissions; ambient air; trace organic determinations. *Mailing Add:* Shell Develop Co Westhollow Res Ctr PO Box 1380 Houston TX 77001

LOOSE, LELAND DAVID, b Reading, Pa, Jan 25, 40; m 71; c 3. PHYSIOLOGY, IMMUNOLOGY. *Educ:* Tenn Wesleyan Col, BS, 63; ETenn State Univ, MA, 65; Univ Mo, Columbia, PhD(physiol), 70. *Prof Exp:* Instr physiol, Lees-McRae Col, 65-67; asst prof, Sch Med, Tulane Univ, 70-74; asst prof physiol, Dept Physiol & Inst Exp Path & Toxicol, Albany Med Col, 74-75, assoc prof, 75-80; proj leader Immunother, 80-85, asst dir clin res, 85-88, ASSOC DIR CLIN RES, PFIZER INC, 88- *Mem:* Am Physiol Soc; Am Soc Trop Med & Hyg; NY Acad Sci; Am Soc Zool; Sigma Xi; Am Rheumatic Assoc; Am Soc Microbiol; Am Asn Immunol. *Res:* Physiological control mechanisms of immune responses; influence of environmental chemicals on immune responses; differentiation of lymphoid tissue with special reference to hormonal effects; macrophage antigen processing; calcium alterations in shock; pharmacological control of inflammation. *Mailing Add:* Pfizer Inc Eastern Point Rd Groton CT 06340

LOOSLI, JOHN KASPER, b Clarkston, Utah, May 16, 09; m 36; c 3. ANIMAL NUTRITION. *Educ:* Utah State Univ, BS, 31; Colo State Univ, MS, 32; Cornell Univ, PhD(animal nutrit), 38. *Prof Exp:* Instr agr, Col Southern Utah, 33-35; asst animal nutrit, Cornell Univ, 35-38; agent, Bur Biol Surv, USDA, 38-39; from asst prof to prof, 39-74, head, dept animal sci, 63-71, EMER PROF ANIMAL NUTRIT, CORNELL UNIV, 74- *Concurrent Pos:* Collabr, US Fish & Wildlife Serv, 39-56; vis prof, Univ Philippines, 53-54 & 66 & Univ Ibadan, 72-74; consult, US Army Vet Grad Sch, 54 & USAID, Nigeria, 61; ed, J Animal Sci, 55-58; Fulbright lectr, Univ Queensland, 60; mem comt animal nutrit, Agr Bd, Nat Res Coun; vis prof, Univ Fla, 74-80, actg chmn, dept animal sci, 75-76, actg dean for res, 77, adj prof, 80-85. *Honors & Awards:* Am Feed Mfrs Award Nutrit, 50; Borden Award Dairy Prod, 51; Morrison Award, 56. *Mem:* Fel Am Soc Animal Sci (vpres, 59, pres, 60); Am Dairy Sci Asn (pres, 70-71); fel Am Inst Nutrit; Brit Soc Animal Prod. *Res:* Fat metabolism and requirements; vitamin requirements; lactation; mineral requirements; feed composition. *Mailing Add:* 406 SW 40th St Gainesville FL 32607-2749

LOOV, ROBERT EDMUND, b Wetaskiwin, Alta, Oct 29, 33; m 79; c 2. STRUCTURAL ENGINEERING. *Educ:* Univ Alta, BSc, 58; Stanford Univ, MS, 59; Univ Cambridge, DPhil, 73. *Prof Exp:* Sales engr, Con-Force Prod Ltd, 59-61, chief engr, 61-63; from asst prof to assoc prof, Univ Calgary, 63-74, asst to vpres, 70-73, actg head civil eng, 80-81, head, 84-89, PROF CIVIL ENG, UNIV CALGARY, 74- *Concurrent Pos:* Nat Res Coun Can res grants, 64-67 & 69-; on leave, Churchill Col, Eng, 67-69; vis prof, Univ NSW, Australia, 83; mem, five Can Standards Asn Comts on Concrete Design & Construct; chmn, Can Standards Asn Comt, 90- *Honors & Awards:* Fel Can Soc Civil Eng, 88. *Mem:* Am Concrete Inst; Eng Inst Can; Can Soc Civil Engrs; Prestressed Concrete Inst. *Res:* Strength and behavior of precast connections; optimum design of reinforced and prestressed concrete; bond strength of reinforced and prestressed concrete; high strength concrete; generalized concrete stress-strain curves. *Mailing Add:* Dept Civil Eng Univ Calgary Calgary AB T2N 1N4 Can

LOOYENGA, ROBERT WILLIAM, b NDak, Oct 21, 39; m 63; c 4. ANALYTICAL CHEMISTRY. *Educ:* Hope Col, AB, 61; Wayne State Univ, PhD(anal chem), 69. *Prof Exp:* Fel chem, Univ Wis-Milwaukee, 70; res chemist, Printing Develop Inc, 70-72; from asst prof to assoc prof, 72-87, PROF CHEM, SDAK SCH MINES & TECHNOL, 87- *Concurrent Pos:* Chemist, SDak Racing Comn, 75-78; consult, SDak Law Enforcement Agencies. *Mem:* Am Chem Soc; Sigma Xi. *Res:* Analytical research and analysis of trace metals and organics in municipal and natural waters, of new chemical deicers and of abused drugs; analytical separations and methods development. *Mailing Add:* Dept Chem SDak Sch Mines & Technol Rapid City SD 57701

LOPARDO, VINCENT JOSEPH, b Pittsburgh, Pa, Dec 1, 25; m 50; c 4. MECHANICAL ENGINEERING. *Educ:* Univ Pittsburgh, BSME, 48, MSME, 51; Cath Univ Am, PhD(mech eng), 68. *Prof Exp:* Design engr, Peth & Reed Engrs, 48-49 & Hunting, Larsen & Dunnells Engrs, 51; from instr to asst prof mech eng, Univ Pittsburgh, 51-60; assoc prof, 60-68, chmn dept, 76-80, PROF MECH ENG, US NAVAL ACAD, 68- *Concurrent Pos:* Design engr, Hunting, Larsen & Dunnells Engrs, 51-55; consult, Charles M Wellons Consult Engrs, 55-60; res prof eng res div, Univ Pittsburgh, 51-53; sr assoc, Trident Eng Assocs, 61-; fac fel, Nat Sci Found, 66; Naval Acad Res Coun grant, US Naval Acad, 68-69; Nat Bur Standards grant, Naval Ship Res & Develop Ctr, 81. *Mem:* Soc Exp Stress Analysis; Am Soc Eng Educ; Am Soc Mech Engr. *Res:* Stress analysis; stress and strains in large deformations of polyurethanes using photoelasticity and moire; exergy and the second law analyses of power systems; computer simulation of gas turbine engines. *Mailing Add:* Dept Mech Eng US Naval Acad Annapolis MD 21402

LOPATIN, DENNIS EDWARD, b Chicago, Ill, Oct 26, 48. MICROBIOLOGY. *Educ:* Univ Ill, PhD(microbiol), 74. *Prof Exp:* From asst prof to assoc prof dent, 78-90, res scientist, 86- 90, PROF DENT, UNIV MICH, 90- *Mem:* Am Asn Immunologists; Am Asn Dent Res; Am Asn Microbiol. *Res:* Studies host immunological response to members of the oral flora; primary interest in immunology of periodontal diseases. *Mailing Add:* Univ Mich 300 N Ingalls Bldg Box 0402 Ann Arbor MI 48109-0402

LOPATIN, WILLIAM, b Brooklyn, NY, July 20, 46; m 67; c 2. BIOCHEMISTRY, BIO-ORGANIC CHEMISTRY. *Educ:* Univ Fla, BS, 67; Univ SFla, MA, 71, PhD(chem), 77. *Prof Exp:* Teacher chem, Hillsborough Co, Fla Bd Pub Instr, 69-73; res assoc biochem, Univ Tex, 77-80. *Concurrent Pos:* Chmn sci dept, Blake High Sch, Tampa, 70-71. *Mem:* Sigma Xi; AAAS; Am Chem Soc. *Res:* Application of physical organic techniques to the study of enzyme reaction mechanisms. *Mailing Add:* 50900 Mercury Dr Granger IL 46530-9795

LOPER, CARL R(ICHARD), JR, b Wauwatosa, Wis, July 3, 32; m 56; c 2. METALLURGICAL ENGINEERING, ENVIRONMENTAL SCIENCE. *Educ:* Univ Wis, BS, 55, MS, 58, PhD(metall eng), 61. *Prof Exp:* Metall engr, Pelton Steel Castings Co, 55-56; instr metall eng, 56-58, res proj asst, 58-60, from asst prof to assoc prof, 61-69, assoc dir, Univ-Indust Res Prog & assoc chmn, Dept Metall & Mineral Eng, 79-81, PROF METALL ENG & ENVIRON STUDIES, UNIV WIS-MADISON, 69- *Concurrent Pos:* Consult, Gray & Ductile Iron Founders Soc, 62-, Gen Motors Corp & Brillion Iron Works, 66-, Oil City iron Works, 73- & Sperry-New Holland, 74-; res metallurgist, Allis Chalmers Mfg Co, 61-; invited lectr, Korea Foundry Soc & Korea Inst Sci & Technol, Seoul, 68, Disamatic Convention, Copenhagen, Denmark, 69, Korea Inst Sci & Technol & Korea Inst Advan Series, 74, Kyushu Univ, Japan, 74, Waseda Univ, Japan, 81, Zhejiang Univ, China, 81 & Brazilian Soc Metals, Santa Catapiura, 81; pres, Int Comn Compacted Graphite Cast Iron, 77-; hon lectr, Antioquia Univ, Medellin, Colombia, 83 & Univ Nacional de Colombia, 84. *Honors & Awards:* Adams Mem Award, Am Welding Soc, 64; H F Taylor Award, Am Foundrymen's Soc, 67, John A Penton Gold Medal, 72. *Mem:* Am Foundrymen's Soc; Am Welding Soc; fel Am Soc Metals; Sigma Xi; Am Inst Metall Eng; Foundry Educ Found. *Res:* Solidification and process control of cast irons; solidification and property relationships in aluminum and copper base alloys; fracture toughness of cast components; welding metallurgy; failure analysis; recycling of metallic solid wastes. *Mailing Add:* Dept Mats Sci & Eng 1509 University Ave Madison WI 53706

LOPER, DAVID ERIC, b Oswego, NY, Feb 14, 40; m 66; c 4. MAGNETOHYDRODYNAMICS, APPLIED MATHEMATICS. *Educ:* Carnegie Inst Technol, BS, 61; Case Inst Technol, MS, 64, PhD(mech eng), 65. *Prof Exp:* Sr scientist, Douglas Aircraft Corp, 65-68; from asst prof to assoc prof, 68-77, PROF MATH, FLA STATE UNIV, 77- *Concurrent Pos:* Nat Ctr Atmospheric Res fel, 67-68; sr vis fel, Univ Newcastle-upon-Tyne, Eng, 74-75. *Mem:* Am Phys Soc; Am Geophys Union; Soc Indust & Appl Math; Sigma Xi. *Res:* Boundary layers in rotating, stably stratified, electrically conducting fluids; evolution of the earth's core including stratification, heat transfer, solidification and particle precipitation. *Mailing Add:* Dept Math Fla State Univ Tallahassee FL 32306

LOPER, GERALD D, b Brooklyn, NY, May 4, 37; m 60; c 1. NUCLEAR PHYSICS. *Educ:* Univ Wichita, AB, 59; Okla State Univ, MS, 62, PhD(physics), 64. *Prof Exp:* Asst prof, 64-67, chmn dept, 66-78, asst dean grad studies, 86-87, ASSOC PROF PHYSICS, WICHITA STATE UNIV, 67-, ASSOC DEAN LIBERAL ARTS & SCI, 87- *Mem:* Am Phys Soc; Sigma Xi. *Res:* Measurement of positron lifetimes in solids; nuclear spectroscopy; internal conversion. *Mailing Add:* Dept of Physics Wichita State Univ Wichita KS 67208

LOPER, GERALD MILTON, b Sykesville, Md, Jan 7, 36; m 62; c 2. AGRONOMY, BIOCHEMISTRY. *Educ:* Univ Md, Bsc, 58; Univ Wis, MSc, 60, PhD(agron), 61. *Prof Exp:* Res agronomist, USDA, SDak, 62-67; assoc prof, 69-74, PROF AGRON & PLANT GENETICS, UNIV ARIZ, 74-; RES PLANT PHYSIOLOGIST, FED HONEY BEE LAB, 67- *Mem:* Bot Soc Am; Am Soc Agron. *Res:* Effect of environment and infective organisms on the chemical composition of forages in relation to animal nutrition; attractiveness of forage legumes to honey bees; pollination physiology; seed production and crop physiology investigations. *Mailing Add:* Fed Honey Bee Lab 2000 E Allen Rd Tucson AZ 85719

LOPER, JOHN C, b Hadley, Pa, June 21, 31; m 56; c 3. ENVIRONMENTAL TOXICOLOGY, BIODEGRADATION. *Educ:* Western Md Col, BA, 52; Emory Univ, MS, 53; Johns Hopkins Univ, PhD(biol), 60. *Prof Exp:* From instr to asst prof pharmacol, Sch Med, St Louis Univ, 60-63; from asst prof to assoc prof, 63-74, PROF MICROBIOL, COL MED, UNIV CINCINNATI, 74-, PROF ENVIRON HEALTH, 79- *Concurrent Pos:* NIH res grants, 62-78; NIH spec vis fel genetics, Res Sch Biol Sci, Australian Nat Univ, 70-71; Environ Protection Agency grants, 76-; mem biol comt, Argonne Nat Lab-Argonne Univ Asn, 70-73; mem subcomt toxicol, Safe Drinking Water Comt, Nat Res Coun, 78-79; assoc dir, Dept Molecular Genetics, Univ Cincinnati, 88-; co-prin investr, Nat Inst Environ Health Sci Superfund basic res prog grant, 88- *Mem:* Am Soc Microbiol; Genetics Soc Am; Environ Mutagen Soc; Am Chem Soc. *Res:* Molecular genetics of cytochrome PHSO systems in yeasts; microbial pathways of detoxication and degradation of xenobiotic compounds; genetics of antifungal resistance. *Mailing Add:* Dept Molecular Genetics Univ Cincinnati Col Med Cincinnati OH 45267-0524

LOPER, WILLARD H(EWITT), b Alden, NY, Apr 30, 26; m 50; c 4. AGRICULTURAL ENGINEERING. *Educ:* Cornell Univ, BSA, 53. *Prof Exp:* Sales & serv rep, Holz Col, 53-54; design & prod engr, Cochran Equip Co, 54-55; asst prof, 55-63, ASSOC PROF AGR ENG, CALIF POLYTECH STATE UNIV, SAN LUIS OBISPO, 63- *Concurrent Pos:* Civil engr, Bur Reclamation, US Dept Interior & State Div Hwys, 57 & 58; tech leader, Foreign Agr Serv, USDA, 59. *Mem:* Am Soc Agr Engrs. *Res:* Agricultural crop harvest mechanization. *Mailing Add:* 266 Luneta Dr San Luis Obispo CA 93401

LOPEZ, ANTHONY, b Chile, SAm, May 13, 19; US citizen; m 47; c 3. FOOD SCIENCE. *Educ:* Catholic Univ, Chile, BS, 42; Univ Mass, PhD(food tech), 47. *Prof Exp:* Chemist, SA Organa, Chile, 42-45; tech dir, Indust de Productos Alimenticios, 48-52; assoc res prof food technol, Univ Mass, 52-53; assoc prof, Univ Ga, 53-54; prof food sci & technol, Va Polytech Inst & State Univ, 54-88; RETIRED. *Concurrent Pos:* Instr, UN Latin Am Fisheries Training Ctr, Chile, 52; lectr, Ministry Commerce, Spain, 60; consult food processing, Govt Spain, 62, 63; consult food technol, UN Food & Agr Orgn, Chile, 66 & Brazil, 69, 72, 75 & 79; Orgn Am States in Mex, 70-74; tech ed, Food Prod Mgt, 71-87; UN Food & Agr Orgn, Arg, 80, Chile, 84, Mex 89-90 & PR, 90-91. *Mem:* Am Chem Soc; fel Inst Food Technologists; Chilean Soc Nutrit. *Res:* Processing and nutritive value of fish; composition of fresh fruits and vegetables; processing of fruits and vegetables; chemical changes in processed foods during storage; food packaging; microwave irradiation of foods; effect of processing on nutritive value of foods. *Mailing Add:* 721 Hutcheson Dr Blacksburg VA 24060-3209

LOPEZ, ANTONIO VINCENT, b Montgomery, Ala, Apr 24, 38. PHARMACEUTICAL CHEMISTRY, PHARMACOGNOSY. *Educ:* Auburn Univ, BS, 59, MS, 61; Univ Miss, PhD(pharm chem), 66. *Prof Exp:* Chmn, Dept Pharmaceut Chem, 66-76, from asst dean to assoc dean, 78-85, CHMN DIV NATURAL SCI, SOUTHERN SCH PHARM, MERCER UNIV, 76-, DIR STUDENT AFFAIRS, 85- *Mem:* AAAS; Am Col; Am Asn Col Pharm. *Res:* Central nervous system drugs. *Mailing Add:* Southern Sch of Pharm Mercer Univ 345 Boulevard NE Atlanta GA 30312

LOPEZ, CARLOS, b Ponce, PR, Jan 15, 42; m 70; c 1. IMMUNOLOGY, VIROLOGY. *Educ:* Univ Minn, BS, 65, MS, 66, PhD(pub health), 70. *Prof Exp:* Res fel, Univ Minn, 70-72, asst prof path, 72-73; assoc mem, Sloan-Kettering Cancer Ctr & assoc prof biol, Sloan-Kettering Div, Cornell Univ Sch med, 73-87; DIR BIOL RES, ELI LILLY, 87- *Concurrent Pos:* NIH fel, 70-71; fel, Nat Thoracic & Respiratory Dis Asn, 71-73. *Mem:* Am Asn Immunologists; Am Asn Exp Pathologists; Am Soc Microbiol; AAAS. *Res:* Immunological resistance to virus infections; immunologic response to virus induced tumors. *Mailing Add:* Eli Lilly Lilly Corp Ctr Indianapolis IN 46285

LOPEZ, DIANA MONTES DE OCA, b Havana, Cuba, Aug 26, 37; US citizen; m 58; c 3. MICROBIOLOGY, IMMUNOLOGY. *Educ:* Univ Havana, BS, 60; Univ Miami, MS, 68, PhD(microbiol), 70. *Prof Exp:* Res assoc, 70-71, from instr to assoc prof, 71-83, PROF MICROBIOL, SCH MED, UNIV MIAMI, 83- *Concurrent Pos:* Sect leader tumor immunol, Sylvester Comprehensive Cancer Ctr, State Fla, 80- *Mem:* Am Soc Microbiol; Tissue Cult Asn; Sigma Xi; Am Asn Immunologists; NY Acad Sci; Int Asn Breast Cancer Res (pres-elect, 85, pres, 87-89). *Res:* Tumor immunology; viral oncogenesis; cell kinetics. *Mailing Add:* Dept Microbiol & Immunol Sch Med R-138 Univ Miami PO Box 016960 Miami FL 33101

LOPEZ, GENARO, b Brownsville, Tex, Jan 24, 47; m 72; c 2. ECONOMIC ENTOMOLOGY. *Educ:* Tex Tech Univ, BS, 70; Cornell Univ, PhD(econ entom), 75. *Prof Exp:* Res asst entom, Cornell Univ, 70-75; entomologist, Tex Agr Exten Serv, Tex A&M Univ, 75-76; ASST PROF BIOL, TEX SOUTHMOST COL, 76-; INSTR BIOL, PAN AM UNIV, BROWNSVILLE, 76- *Mem:* Entom Soc Am; Acaralogical Soc Am. *Res:* Bionomics, ecology and control of insects affecting man's home environment; teaching biology to the bicultural/bilingual student at the college level. *Mailing Add:* Dept of Biol Tex Southmost Col Brownsville TX 78520

LOPEZ, JORGE ALBERTO, b Monterrey, Mex, Jan 23, 55; m 79; c 2. HEAVY ION REACTIONS, COMPUTATIONAL PHYSICS. *Educ:* Tex A&M Univ, PhD(physics), 86. *Prof Exp:* Postdoctoral researcher, Niels Bohr Inst, Denmark, 85-87 & Lawrence Berkeley Lab, 87-89; assoc prof, Calpoly State Univ, San Luis Obispo, 89-90; ASSOC PROF PHYSICS, UNIV TEX, EL PASO, 90- *Mem:* Am Phys Soc. *Res:* Nuclear physics; heavy ion physics; computational physics. *Mailing Add:* Physics Dept Univ Tex El Paso TX 79968-0515

LOPEZ, JOSE MANUEL, b San Juan, PR, Jan 7, 50; m 81; c 4. ENVIRONMENTAL CHEMISTRY, CHEMICAL OCEANOGRAPHY. *Educ:* Univ PR, BS, 71; Univ Wis-Madison, MS, 73; Univ Tex, PhD(environ chem), 76. *Prof Exp:* Res scientist marine chem, 76-81, head, Marine Econ Div, 81-85, SR SCIENTIST, RES & DEVELOP CTR, CTR ENERGY & ENVIRON RES, 85- *Concurrent Pos:* Pres Sci Teachers Asn, 79-80; consult, indust & govt; asst prof marine chem, Univ PR. *Mem:* AAAS; Am Chem Soc; Am Soc Limnol & Oceanog; Am Bot Soc; Water Pollution Control Fedn. *Res:* Sources, fate and significance of chemicals in aquatic ecosystems; biological availability of contaminants to aquatic organisms; nutrient dynamics; mangroves ecology. *Mailing Add:* Res & Develop Ctr Ctr Energy & Environ Res College Station Mayaguez PR 00708

LOPEZ, LEONARD ANTHONY, b Waltham, Mass, Dec 27, 40; m 61; c 3. ENGINEERING SOFTWARE SYSTEMS. *Educ:* Tufts Univ, BS, 62; Univ Ill, MS, 63, PhD(civil eng), 66. *Prof Exp:* Asst prof civil eng, Lehigh Univ, 66-67; PROF CIVIL ENG, UNIV ILL, URBANA, 67- *Mem:* Am Soc Civil Engrs. *Res:* Digital simulation; numerical methods; mechanics of nonlinear solids; computer system. *Mailing Add:* Dept Civil Eng Univ Ill Urbana IL 61801

LOPEZ, MARIA DEL CARMEN, b Buenos Aires, Arg, Aug 3, 54. IMMUNOLOGY. *Educ:* Univ Buenos Aires, BS, 80, PhD(biochem), 87. *Prof Exp:* Res fac aide, Nat Res Coun Arg, 88-89; res fac aide, 89-91, RES ASST PROF, DEPT FAMILY & COMMUNITY MED, UNIV ARIZ, 91- *Mem:* Am Asn Immunol; Soc Mucosal Immunol. *Mailing Add:* Dept Family & Community Med Univ Ariz 1501 N Campbell Ave Rm 4307 Tucson AZ 85724

LOPEZ, R C GERALD, b London, Eng, Mar 12, 57; m 80; c 3. AGRICULTURAL FORMULATIONS RESEARCH. *Educ:* Oxford Univ, BA, 80, PhD(org chem), 82. *Prof Exp:* Sr scientist, 80-88, MGR AGR FORMULATIONS RES, ROHM & HAAS CO, 88- *Res:* Discovery and optimization of formulations for new and existing agricultural chemicals, including fungicides, herbicides and insecticides. *Mailing Add:* Rohm & Haas Co 727 Norristown Rd Spring House PA 19477

LOPEZ, RAFAEL, b Dominican Republic, Dec 15, 29; m 56; c 2. PEDIATRICS, HEMATOLOGY. *Educ:* Seton Hall Univ, BSc, 52; Univ PR, MD, 56. *Prof Exp:* assoc prof pediat, New York Med Col, Flower & Fifth Ave Hosp, 65-80, ASSOC PROF PEDIAT, NEW YORK MED COL, MISERICORDIA HOSP MED CTR, 80- *Mem:* Soc Study Blood; Int Soc Hemat; Am Soc Hemat; NY Acad Sci. *Res:* Glutathione reductase as a tool for diagnosis of riboflavin deficiency in infants, children, adolescents; malabsorption syndromes and the effect of phototherapy upon this vitamin in the newborn. *Mailing Add:* Dept Pediat Our Lady of Mercy Med Ctr Bronx NY 10466

LOPEZ-BERESTEIN, GABRIEL, b La Habana, Cuba, Aug 13, 47; US citizen; c 1. ONCOLOGY, IMMUNOLOGY. *Educ:* Univ PR, San Juan, BA, 70; Univ Navarre, Spain, MD, 76. *Prof Exp:* asst prof & asst internist, 81-84, ASSOC PROF MED, UNIV TEX, M D ANDERSON CANCER CTR & ASSOC PROF PHARM, MED SCH, 84- *Concurrent Pos:* Assoc prof, biomed sci, Univ Tex Health Sci Ctr, 84-; mem, Biomed Sci Study Sect, NIH, 88- *Honors & Awards:* Stohlman Award, 90. *Mem:* Am Asn Cancer Res; AMA; Am Soc Clin Oncol; AAAS; Am Soc Immunologists. *Res:* Oncology; immunology. *Mailing Add:* M D Anderson Cancer Ctr Sect Immunobiol & Drug Carriers Univ Tex Box 41 1515 Holcombe Blvd Houston TX 77030

LOPEZ-ESCOBAR, EDGAR GEORGE KENNETH, b Buenos Aires, Arg, Jan 7, 37. MATHEMATICS. *Educ:* Cambridge Univ, BS, 58, MA, 71; Univ Calif, Berkeley, MA, 61, PhD(math), 65. *Prof Exp:* PROF MATH, UNIV MED, 66- *Mem:* Asn Symbolic Logic; soc Exact Philos; Am Math Soc. *Res:* Computer application as applied to mathematical logic. *Mailing Add:* 2703 Ogleton Rd Annapolis MD 21403

LOPEZ-MAJANO, VINCENT, b Madrid, Spain, Apr 3, 21; US citizen; m 52; c 2. NUCLEAR MEDICINE, PULMONARY MEDICINE. *Educ:* Inst Cardenal Cisneros, BA & BS, 39; Univ Madrid, MD, 45, PhD, 51. *Prof Exp:* Resident, Gen Hosp, Madrid, 45-51; physician, Sanatorium Carlos Duran, Costa Rica, 51-56 & Tuberc Sanatorium, Md, 55-60; chief, Pulmonary Function Lab, Vet Admin Hosp, Baltimore, 60-70; asst prof environ med, Johns Hopkins Inst, Baltimore, 68-70; clin assoc prof med, Loyola Stritch Sch Med, 70-73; dir nuclear med, Gottlieb Mem Hosp, 73-77; assoc prof med, Chicago Med Sch, 80-87; CHMN NUCLEAR MED, COOK COUNTY HOSP, CHICAGO, 77- *Concurrent Pos:* Ed, Respiration, 70-71 & J Nuclear Med & Allied Sci, 82; chief training sect nuclear med, Vet Admin Hosp, Hine, Ill, 70-74; vis prof, Nat Univ Mex, 73; vis scientist, Nat Acad Sci, 81-84; vis scholar, Nat Cancer Inst, 85. *Mem:* Nuclear Med Soc; Physiol Soc; Mex Nuclear Med Soc; Am Fed Clin Res. *Res:* Inflammatory diseases of the lungs; regional lung function; staging of neoplasms; studies of cardiac function with radionuclides. *Mailing Add:* 3100 N Sheridan Rd Chicago IL 60657

LOPEZ-SANTOLINO, ALFREDO, b Salamanca, Spain, July 23, 31; m 62; c 2. MEDICINE, BIOCHEMISTRY. *Educ:* Inst Ensenanza Media, Salamanca, BS, 49; Lit Univ Salamanca, MD, 55, PhD(med sci), 58; Tulane Univ, PhD(biochem), 63. *Prof Exp:* Asst prof physiol med, Sch Med, Lit Univ Salamanca, 56-58; instr biochem, Cali Univ Sch Med, 58-59; asst prof internal med, Col Med & biochemist, Clin Res Ctr, Univ Iowa, 64-67; assoc prof, 67-74, PROF INTERNAL MED, MED SCH, LA STATE UNIV MED CTR, NEW ORLEANS, 74- *Concurrent Pos:* Mem coun atherosclerosis, Am Heart Asn. *Mem:* AAAS; Am Oil Chem Soc; Soc Nutrit Educ; Am Inst Nutrit; Am Soc Clin Nutrit. *Res:* Nutrition and metabolic diseases; metabolism of lipids and steroid hormones. *Mailing Add:* Dept Med La State Univ Sch Med 1542 Tulane Ave New Orleans LA 70112

LOPINA, ROBERT F(ERGUSON), b Jamestown, NY, May 13, 36; m 58; c 3. AERONAUTICAL ENGINEERING, AVIONICS. *Educ:* Purdue Univ, Lafayette, BS, 57; Mass Inst Technol, MSc, 65, ME, 66, PhD(mech eng), 67. *Prof Exp:* US Air Force, 57-83, assoc prof aeronaut, US Air Force Acad, 67-74, chief scientist, Europ Off Aerospace Res & Develop, 74-76, chief, Flight Control Div, Air Force Flight Dynamics Lab, 77-78, comdr & dir, Air Force Avionics Lab, 78-80, dep eng, Aeronaut Systs Div, 80-82, dep reconnaissance strike & extreme width, 82-83, dep Aeronaut Systs Div, 83-87; vpres eng & prog dir T-46, Fairchild Repub Co, Farmingdale, NY, 83-87; dir, Advan Develop, Ford Aerospace Corp, Detroit, Mich, 87- 88; VPRES, ADVAN PROGS OFF, FORD AEROSPACE/LORAL AERONUTRONIC DIV, NEWPORT BEACH, CALIF, 88- *Mem:* Am Inst Aeronaut & Astronaut; Am Soc Mech Engrs; Sigma Xi; Air Force Asn; Nat Mgt Asn; Asn Old Crows. *Res:* Swirl flow heat transfer; computer applications in aeronautical education; night attack systems development; trainer aircraft development and production; integrated circuits for radio frequency applications. *Mailing Add:* Seven Calle Aqua San Clemente CA 92672

LO PINTO, RICHARD WILLIAM, b New York, NY, Nov 7, 42; m 70; c 2. MARINE BIOLOGY, BIOMONITORING & AQUATIC TOXICOLOGY. *Educ:* Iona Col, BS, 63; Fordham Univ, MS, 65, PhD(physiol ecol), 72. *Prof Exp:* Res asst water pollution, Osborne Lab Marine Sci, 67-68, consult, develop biol testing, Org Econ Coop Develop, 84-85; PROF BIOL, FAIRLEIGH DICKINSON UNIV, TEANECK, 70- *Concurrent Pos:* Cosult, Hackensack Meadowlands Develop Comn, 71-, Hart Mountain Indust Inc, 76-77; dir, Marine Biol Prog, Fairleigh Dickinson Univ, Rutherford, NJ, 72-; asst dir, Meadowlands Regional Study Ctr, 74-75; assoc, Seminar Pollution & Water Resources, Columbia Univ, 75; chmn tech adv comt, N NJ Water Quality Prog, 77-; assoc ed, Bulletin of the NJ Acad Sci, 78; consult, US EPA, 84; consult, Orgn Econ Coop & Develop, Paris, 84-85. *Mem:* Sigma Xi. *Res:* Bioassay development for marine and fresh water organisms; aquatic toxicology; phytoplankton physiology; microbial ecology; mariculture of plankton feeders. *Mailing Add:* Dept Biol Scis Fairleigh Dickinson Univ Teaneck NJ 07666

LOPO, ALINA C, b Havana, Cuba, June 14, 51; US citizen; m 82; c 1. FERTILIZATION, TRANSLATIONAL CONTROL. *Educ:* Univ Miami, Coral Gables, BS, 72, MS, 74; Univ Calif, Davis, PhD(cell biol), 79. *Prof Exp:* Teaching asst, Univ Miami, Coral Gables, 72-74; instr gen biol, Miami-Dade Community Col, 72-75; teaching asst, Univ Calif, Davis, 75-77, AAUW predoctoral fel, 77-78; res asst, Univ Calif, San Diego, 78-79; fel, Rockefeller Found, Univ Calif, San Francisco, 79-80, NIH, 80-81, Univ Calif Sch Med, Davis, 81-85; asst prof biomed sci, Univ Calif, Riverside, 85- *Concurrent Pos:* Res asst, Univ Miami, Coral Gables, 73-75; lectr, Sch Optom, Univ Calif, Berkeley, 81. *Mem:* Am Soc Biochem & Molecular Biol; Am Soc Cell Biol; Soc Develop Biol. *Res:* Activation of protein synthesis at fertilizataion and its regulation during early development; molecular and biochemical basis of cystic fibrosis, with a focus on protein phosphorylation. *Mailing Add:* 6136 Hawarden Dr Riverside CA 92506

LOPPNOW, HARALD, Ger citizen. CELLULAR IMMUNOLOGY, VASCULAR CYTOKINES. *Educ:* Kiel Univ, dipl microbiol, 83, PhD(immunol), 86. *Prof Exp:* Res assoc immunol, Sclavo Res Ctr, Siena I, 87 & Tufts Univ, Boston, 88-90; lab instr microbiol, 81-83, RES ASSOC IMMUNOL, 83-88 & 90- *Concurrent Pos:* Travel award, IV Int Conf Immunopharmacol, 88; lectr, Kiel Univ, 90. *Mem:* Ger Soc Immunol; Am Asn Immunologists. *Res:* Vascular and immune cell responses to pathophysiologically relevant stimuli such as bacterial lipopolysaccharide or cytokines; determine proliferation, cytokine production, or adhesion of cells; biochemical, cell biological, immunological and molecular biological methods. *Mailing Add:* Dept Biochem Forschungsinst Borstel Parkallee 22 Borstel D-2061 Germany

LOPREST, FRANK JAMES, b New York, NY, Jan 8, 29; m 60; c 5. PHYSICAL CHEMISTRY. *Educ:* St John's Univ, NY, BS, 50; NY Univ, MS, 52, PhD, 54. *Prof Exp:* Res chemist, Oak Ridge Nat Lab, 54-56; sr res chemist & supvr adv res, Reaction Motors Div, Thiokol Chem Corp, 56-65; tech assoc, Res & Develop Div, GAF Corp, 65-67, sect mgr new imaging processes res, 67-69, mgr appl chem, Res & Develop & Res Serv, Indust Photo Div, 69-77; dir basic sci, Princeton Res Ctr, Am Can Co, 77-83; ASSOC DIR BASIC RES, COLGATE-PALMOLIVE CO, 83- *Mem:* AAAS; Am Chem Soc; Sigma Xi; Licensing Execs Soc. *Res:* Heterogeneous equilibria; kinetics of liquid solid reactions; high temperature materials; physical chemistry of liquid and solid propellants; adhesion phenomena; cellulose and paper science; surface and colloid chemistry, detergency. *Mailing Add:* 220 Hampton Dr Langhorne PA 19047

LOPRESTI, PHILIP V(INCENT), b Johnstown, Pa, Sept 27, 32; m 59; c 3. ELECTRICAL ENGINEERING. *Educ:* Univ Notre Dame, BSEE, 54, MSEE, 58; Purdue Univ, Lafayette, PhD(elec eng), 63. *Prof Exp:* Asst prof elec eng, Ill Inst Technol, 64-67 & Northwestern Univ, 67-70; CONSULT MEM RES STAFF, ENG RES CTR, AT & T, 70-,. *Concurrent Pos:* Instr, Univ Notre Dame, 58-60 & Purdue Univ, Lafayette, 60-63; consult, Ill Inst Technol, 64-70. *Honors & Awards:* Darlington Prize, Inst Elec & Electronics Engrs, Circuits & Systs Soc, 78. *Mem:* Inst Elec & Electronics Engrs; Sigma Xi. *Res:* Automatic control theory; digital signal processing; digital integrated circuits; hybrid integrated circuits. *Mailing Add:* Eng Res Ctr AT & T PO Box 900 Princeton NJ 08540

LOPUSHINSKY, THEODORE, b Brooklyn, NY, Oct 25, 37. GENERAL BIOLOGY, ECOLOGY. *Educ:* Pa State Univ, BS, 59; Univ Tenn, Knoxville, MS, 61; Mich State Univ, PhD(ecol, path), 69. *Prof Exp:* Asst prof natural sci, Mich State Univ, 69-70; prog rep, Mich Asn Regional Med Progs, 70-71, actg dir, 72, dir prog develop, 72-73; asst prof proj develop, Col Human Med, 73-75, asst prof, 75-82, ASSOC PROF, CTR INTEGRATIVE STUDIES SCI, MICH STATE UNIV, 82- *Concurrent Pos:* Archivist, Soc Col Sci Teachers, 81-86; nat mem chmn, Soc Col Sci Teachers, 84-87. *Mem:* Sigma Xi; Nat Sci Teachers Asn; Soc Col Sci Teachers (pres-elect, 87-89, pres, 89-91). *Res:* Parasitism and disease pathologies in wildlife populations; general education science; science-humanities relationships. *Mailing Add:* Rm 100 N Kedzie Lab Mich State Univ East Lansing MI 48824

LOPUSHINSKY, WILLIAM, b Rome, NY, July 25, 30; m 60; c 3. PLANT PHYSIOLOGY. *Educ:* State Univ NY, BS, 53, MS, 54; Duke Univ, PhD(plant physiol), 60. *Prof Exp:* Asst plant physiol, Duke Univ, 57-60, res assoc bot, 60-61; PLANT PHYSIOLOGIST, FORESTRY SCI LAB, USDA, 62- *Mem:* Am Soc Plant Physiol. *Res:* Plant water relations. *Mailing Add:* 1133 N Western Ave Wenatchee WA 98801

LORANCE, ELMER DONALD, b Tupelo, Okla, Jan 18, 40; m 69; c 2. SCIENCE EDUCATION. *Educ:* Okla State Univ, BA, 62, PhD(bio-org chem), 77; Kans State Univ, MS, 67. *Prof Exp:* From asst prof to assoc prof, 70-80, PROF CHEM, SOUTHERN CALIF COL, 80-, CHMN DIV NATURAL SCI & MATH, 85- *Concurrent Pos:* Adj prof, Calif State Univ, Fullerton, 77. *Mem:* Am Chem Soc; AAAS; Am Sci Affil. *Res:* Isolation and structure elucidation of compounds from marine organisms; organic synthesis; chemical taxonomy of desert plants. *Mailing Add:* Div Natural Sci & Math Southern Calif Col 55 Fair Dr Costa Mesa CA 92626

LORAND, JOHN PETER, b Wilmington, Del, Dec 6, 36; m 64; c 3. PHYSICAL ORGANIC CHEMISTRY. *Educ:* Brown Univ, ScB, 58; Harvard Univ, PhD(org chem), 64. *Prof Exp:* NSF scientist, Univ Calif, Los Angeles, 64-65; asst prof org chem, Boston Univ, 65-71; from asst prof to assoc prof, 71-77, PROF ORG CHEM, CENT MICH UNIV, 77- *Concurrent Pos:* Vis prof, Univ Groningen, Neth, 77-78; vis assoc prof, Rutgers State Univ NJ, 86-88. *Mem:* Am Chem Soc; Sigma Xi. *Res:* Free radicals; C-H hydrogen bonding; charge-transfer complexes. *Mailing Add:* Dept of Chem Cent Mich Univ Mt Pleasant MI 48859

LORAND, LASZLO, b Gyor, Hungary, Mar 23, 23; nat US; m 53; c 1. BIOCHEMISTRY, PHYSIOLOGY. *Educ:* Leeds Univ, PhD(biomolecular struct), 51; Budapest Univ, absolutorium med, 48. *Prof Exp:* Demonstr biochem, Budapest Univ, 46-48; asst biomolecular struct, Leeds Univ, 48-52; res assoc physiol & pharmacol, Wayne State Univ, 52-53, asst prof, 53-55; from asst prof to prof chem, 55-74, prof biochem & molec biol, 74-81, PROF BIOCHEM, MOLECULAR & CELL BIOLOGY, NORTHWESTERN UNIV, 81- *Concurrent Pos:* Beit Mem fel, Eng, 52; Lalor fac award, 57; USPHS career award, 62; mem corp, Marine Biol Lab, Woods Hole, Mass; dep dir basic sci, Northwestern Univ Cancer Ctr, 90-91; lectr, Japan Soc Promotion Sci, 90- *Honors & Awards:* James F Mitchell Found Award, Heart & Vascular Res, 73. *Mem:* AAAS; Am Soc Biol Chem; Soc Exp Biol & Med; Am Physiol Soc; Brit Biochem Soc; Am Chem Soc. *Res:* Blood proteins; coagulation of blood; muscle chemistry; protein and enzyme chemistry. *Mailing Add:* Dept Biochem & Molecular Biol Northwestern Univ Evanston IL 60201

LORANGER, WILLIAM FARRAND, b Detroit, Mich, Nov 6, 25. XERORADIOGRAPHY. *Educ:* Denison Univ, BA, 47; Univ Ill, MS, 50, PhD(chem & x-ray diffraction), 52. *Prof Exp:* Asst, Anal Div, Ill State Geol Surv, 47-49; proj scientist, Wright Air Develop Div, US Air Force, Ohio, 51-54, instr physics & chem, US Mil Acad, 54-56; sales engr, X-Ray Dept, Gen Elec Co, 56-57; asst prof, Univ Fla, 57-58; tech adv indust sales, X-Ray Dept, Gen Elec Co, 58-61; prod mgr x-ray & electron optics, Picker X-Ray Corp, NY, 62-70, mkt mgr, Indust Div, Picker Corp, 70-72; new mkt res mgr, Xerox Corp, 72-73, dir educ, xeroradiography, 73-80, consult, xeroxmed systs, 80-84; RETIRED. *Mem:* AAAS; Am Chem Soc; Sigma Xi; Am Crystallog Asn. *Res:* X-ray diffraction and emission; optical methods of instrumental analysis; instrumental chemical analysis; radiography; diseases of the breast; diagnostic ultrasound; applied x-rays. *Mailing Add:* 130 Exeter Ct Plantation Walk Hendersonville NC 28739

LORBEER, JAMES W, b Oxnard, Calif, Oct 30, 31; m 64. PLANT PATHOLOGY, MYCOLOGY. *Educ:* Pomona Col, BA, 53; Univ Wash, MS, 55; Univ Calif, Berkeley, PhD(plant path), 60. *Prof Exp:* Asst bot, Univ Wash, 53-55; asst plant path, Univ Calif, Berkeley, 55-60; from asst prof to assoc prof, 60-72, PROF PLANT PATH, CORNELL UNIV, 72- *Mem:* Mycol Soc Am; Am Phytopath Soc; NY Acad Sci; Brit Mycol Soc. *Res:* Diseases of vegetable crops; epidemiology; plant disease control; biology of Botrytis; fungal genetics. *Mailing Add:* Cornell Univ Main Campus 334 Plant Sci Bldg Plant Path Ithaca NY 14853

LORBER, HERBERT WILLIAM, b Indianapolis, Ind, July 12, 29; m 62; c 2. ELECTRONIC WARFARE, DECISION ANALYSIS. *Educ:* Purdue Univ, BS, 51; Rutgers Univ, MSc, 55; Univ Pa, PhD(elec eng), 62. *Prof Exp:* Engr, Signal Corp Eng Labs, 51 & 53-54; mem tech staff, RCA Labs, 55-62; sr sci specialist, Edgerton Germeshausen & Grier, Inc, 62-71; electron res specialist, Teledyne Ryan Aeronaut, 72-76; mem staff, Los Alamos Nat Lab, 76-82; SR STAFF SPECIALIST, LOCKHEED AERONAUT SYSTS CO, 82- *Concurrent Pos:* Consult, N J Damaskos, Inc, Los Alamos Tech Assocs, Inc & Convair Div Gen Dynamics, 82. *Mem:* Inst Elec & Electronic Engrs; Oper Res Soc Am; AAAS; Sigma Xi. *Res:* Interaction of spacecraft and military vehicles with radar systems; quantitative space-system concept assessment; applications of utility theory to management decision-making; analysis of military and business operations. *Mailing Add:* 3205 Deer Creek Dr Canton GA 30114

LORBER, MORTIMER, b New York, NY, Aug 30, 26; m 56; c 2. PHYSIOLOGY, HEMATOLOGY. *Educ:* NY Univ, BS, 45; Harvard Univ, DMD, 50, MD, 52. *Prof Exp:* Rotating intern, Univ Chicago Clins, 52-53; resident hemat, Mt Sinai Hosp, NY, 53-54, asst resident med, 57; med officer hemat res, Naval Med Res Inst, 55-56; sr asst resident med, Univ Hosp, 58, from instr to asst prof, 59-68, ASSOC PROF PHYSIOL, SCH MED, GEORGETOWN UNIV, 68- *Concurrent Pos:* Lederle Med Fac Award, Georgetown Univ, 60-63, USPHS res career develop award, 63-70. *Mem:* Am Soc Hemat; Int Soc Hemat; Am Soc Cell Biol; Int Asn Dent Res; Am Physiol Soc; Asn Res Vision Ophthal. *Res:* Splenic function; iron metabolism in Gaucher's disease; organ regeneration, particularly of mammalian submandibular salivary glands following removal of parenchyma; exocrin gland structure. *Mailing Add:* Dept Physiol & Biophys Georgetown Univ Sch Med Washington DC 20007

LORBER, VICTOR, b Cleveland, Ohio, Apr 22, 12; m 37; c 3. PHYSIOLOGY. *Educ:* Univ Chicago, BS, 33; Univ Ill, MD, 37; Univ Minn, PhD(physiol), 43. *Prof Exp:* From instr to asst prof, Med Sch, Univ Minn, 41-46; from assoc prof biochem to prof, Case Western Reserve Univ, 46-51; prof, 52-80, EMER PROF PHYSIOL, SCH MED, UNIV MINN, MINNEAPOLIS, 80- *Concurrent Pos:* Career investr, Am Heart Asn, 51. *Mem:* Am Physiol Soc; Soc Exp Biol & Med. *Res:* Cardiac metabolism; ionic fluxes in heart muscle. *Mailing Add:* 3707 Modena Way Santa Barbara CA 93105

LORCH, EDGAR RAYMOND, b Nyon, Switz, July 22, 07; nat US; m 37, 56; c 5. MATHEMATICS. *Educ:* Columbia Univ, AB, 28, PhD(math), 33. *Prof Exp:* Asst math, Columbia Univ, 28-30, instr, 31-33; Nat Res Coun fel, Harvard Univ, 33-34; Cutting traveling fel, Univ Szeged, 34-35; instr math, Columbia Univ, 35-41, from asst prof to assoc prof, 41-48, prof, 48-74, chmn dept, 68-72, Adrain prof, 74-76; CONSULT, 74- *Concurrent Pos:* Ed, Am Math Soc Bull, 41-46 & Ind Univ Math J, 66-75; res mathematician, Nat Defense Res Coun, 43-45; sci adv to chief of staff, US Army, 48-49; vis prof, Carnegie Inst Technol, 49, Univ Rome, 53-54, 66 & 82, Col de France, 58, Stanford Univ, 63, Mid East Tech Univ, Ankara, 65 & Univ Florence, 75; Fulbright lectr, Italy, 53-54 & France, 58; mem, Sec Sch Math Curric Improvement Study, 66-72; vis lectr, Fordham Univ, 66-73; ed, La Qualita, Sem de Venezia, 74; Fulbright lectr, Colombia, 77; co-chmn, univ sem, comput, man & soc. *Honors & Awards:* Fulbright lectr, Italy 53 & 54, France, 58, & Colombia, 77. *Mem:* Am Math Soc; Math Asn Am; Math Soc France; Austrian Math Soc; Math Union Italy. *Res:* Linear spaces; Banach algebras; theory of convex bodies and integration; point set topology; infinite permutation groups. *Mailing Add:* Dept Math Columbia Univ New York NY 10027

LORCH, JOAN, b Offenbach, Germany, June 13, 23; m 52; c 2. CELL BIOLOGY, PROTOZOOLOGY. *Educ:* Univ Birmingham, BSc, 45; Univ London, PhD(physiol), 48. *Prof Exp:* Nuffield fel, King's Col, Univ London, 49-52; res assoc cell biol, Ctr Theoret Biol, State Univ NY Buffalo, 63-68, res asst prof, 68-72; lectr, 71-72, from asst prof to assoc prof, 72-84, chair, Biol Dept, 81-84, PROF BIOL, CANISIUS COL, 84- *Concurrent Pos:* Vis prof for women, NSF, 84-85. *Mem:* AAAS; Hastings Ctr. *Res:* Nuclear-cytoplasmic relationships; species specificity; protozoa; bio-ethics; symbiosis. *Mailing Add:* Dept Biol Canisius Col Buffalo NY 14208

LORCH, LEE (ALEXANDER), b New York, NY, Sept 20, 15; m 43; c 1. MATHEMATICS. *Educ:* Cornell Univ, BA, 35; Univ Cincinnati, MA, 36, PhD(math), 41. *Hon Degrees:* LHD, City Univ NY, 90. *Prof Exp:* Asst mathematician, Nat Adv Comt Aeronaut, 42-43; instr math, City Col New York, 46-49; asst prof, Pa State Univ, 49-50; assoc prof & chmn dept, Fisk Univ, 50-53, prof & chmn dept, 53-55; prof & chmn dept, Philander Smith Col, 55-58; vis lectr, Wesleyan Univ, 58-69; from assoc prof to prof, Univ Alta, 59-68; prof, 68-85, EMER PROF MATH, YORK UNIV, 85- *Mem:* Am Math Soc; Can Math Soc; Asn Women Math; Nat Asn Mathematicians; fel Royal Soc Can. *Res:* Fourier series; special functions; summability; ordinary differential equations. *Mailing Add:* Math Dept York Univ North York ON M3J 1P3 Can

LORCH, STEVEN KALMAN, b New York, NY, Aug 21, 44; m 67; c 3. FORENSIC SCIENCE, MANAGEMENT. *Educ:* City Col New York, BS, 66; State Univ NY Binghamton, MA, 70; Univ Md, PhD(plant physiol), 72. *Prof Exp:* Res assoc, Mich State Univ-AEC Plant Res Lab, 72-73; crime lab scientist, 73-75, chief drug identification unit, 75-77, Div Crime Detection, Mich Dept Pub Health; supvr, narcotics & dangerous drug unit, East Lansing Sci Lab, 77-78, Madison Heights Sci Lab, 78-82, ASST LAB DIR, NORTHVILLE SCI LAB, FORENSIC SCI DIV, MICH STATE POLICE, 82- *Concurrent Pos:* Mich State Police rep, Sci Adv Comn, Mich Bd Pharm, 83- *Mem:* Am Chem Soc; Am Acad Forensic Sci. *Res:* Identification of controlled and prescription drugs; gas chromatographic-mass spectrometry; forensic plant identification; crime scene investigation, clandestine laboratories; development of latent fingerprints; automated fingerprint identification systems; major disaster victim identification. *Mailing Add:* Northville Forensic Sci Lab Michigan State Police 42145 W Seven Mile Rd Northville MI 48167

LORD, ARTHUR E, JR, b Buffalo, NY, Apr 7, 35; m 62; c 2. PHYSICS, GEOPHYSICS. *Educ:* Purdue Univ, BSc, 57, MSc, 59; PhD(metall), Columbia Univ, 64. *Prof Exp:* Res assoc appl math, Brown Univ, 64-66, asst res prof physics, 66-68; assoc prof, 68-75, PROF PHYSICS, DREXEL UNIV, 75- *Concurrent Pos:* Fel, Columbia Univ, 64; mem, NASA Electromagnetic Containerless Processing Task Team, 77; consult, acoust, House Comt Kennedy & King Assasinations, 78. *Honors & Awards:* IR-100 Award, Indust Res & Develop Mag, 77. *Mem:* Am Phys Soc; Acoustic Emission Working Group; Int Geotextile Soc. *Res:* Acoustic emission studies in soils and magnetic materials; nondestructive testing techniques in geotechnical problems; geomembranes and geotextiles; centrifuge modelling in geotechnical areas. *Mailing Add:* Dept Physics Drexel Univ 32nd & Chestnut St Philadelphia PA 19104

LORD, ARTHUR N(ELSON), b Los Angeles, Calif, May 7, 32; m 61; c 2. PHYSICAL METALLURGY. *Educ:* Stanford Univ, BS, 53, MS, 55, PhD(creep of aluminum), 60. *Prof Exp:* Physical metallurgist, Adv Tech Labs, 58-65, METALLURGIST, KNOLLS ATOMIC POWER LAB, GEN ELEC CO, 65- *Mem:* Am Inst Mining, Metall & Petrol Engrs; Am Phys Soc; Am Soc Metals. *Res:* Transport properties of solids; effects of radiation damage in metals. *Mailing Add:* Seven Spring Rd Scotia NY 12302-2614

LORD, EDITH M, b Kingman, Kans. IMMUNOLOGY. *Educ:* Univ Kans, BA, 70; Univ Calif, PhD(biol), 75. *Prof Exp:* Res immunologist, Univ Calif, San Francisco, 75-76; sr instr, 76-77, ASST PROF ONCOL, UNIV ROCHESTER, 77- *Mem:* Am Asn Immunologists; Radiation Res Soc. *Res:* Interaction between host immune cells and tumor cells; modulation of these interactions for therapeutic advantage. *Mailing Add:* Dept Microbiol & Cancer Ctr Univ Rochester 601 Elmwood Ave Box 704 Rochester NY 14642

LORD, ELIZABETH MARY, b Baltimore, Md, July 2, 49; m 84. GENERAL BIOLOGY. *Educ:* Univ Mass, BA, 72; Univ Calif, Berkeley, PhD(bot), 78. *Prof Exp:* From asst prof to assoc prof, 78-89, PROF BOT, UNIV CALIF, RIVERSIDE, 89- *Concurrent Pos:* Mem, develop biol panel, NSF, 83-86. *Honors & Awards:* Pelton Award, Bot Soc Am. *Mem:* Am Soc Plant Physiologists; Sigma Xi; Asn Women Sci; Bot Soc Am. *Res:* Use of comparative development data as a tool to elucidate sequence of events leading to a mature floral form; pollination processes in flowering plants. *Mailing Add:* Dept Bot & Plant Sci Univ Calif 900 University Ave Riverside CA 92521

LORD, GARY EVANS, applied mathematics, for more information see previous edition

LORD, GEOFFREY HAVERTON, pathology, for more information see previous edition

LORD, HAROLD WESLEY, b Clayton, Mich, July 12, 31; m 60; c 3. MECHANICS OF COMPOSITE MATERIALS, INDUSTRIAL NOISE CONTROL. *Educ:* Univ Mich, BSE, 60, MSE, 61; Northwestern Univ, PhD(mech eng, astronaut sci), 66. *Prof Exp:* Test engr systs div, Bendix Corp, 60; instr mech eng, Univ Maine, 61-63; assoc engr res div, Gen Am Transp Corp, 65-66; asst prof eng sci, Univ Western Ont, 66-67; assoc prof, 67-74, prof eng mech, chmn dept, 80-86, PROF ENG MECH, MICH TECHNOL UNIV, 74- *Concurrent Pos:* Lectr, Univ Western Ont, 67-68; NSF initiation grant, 68-69; vis fel, Inst Sound & Vibration Res, Southampton Univ, 76-77; forging indust noise control res proj, NSF-Forging Indust Educ & Res Found, 73-80; vis scientist, US Army Cold Regions Res & Eng Lab, Hanover Nat Health, 86-87. *Mem:* Inst Noise Control Eng; Am Soc Eng Educ; Am Acad Mech. *Res:* Transient thermoelastic phenomena; vibration and stress analysis; noise control; mechanics of composite materials. *Mailing Add:* 37 Peepsock Rd Houghton MI 49931

LORD, HAROLD WILBUR, b Eureka, Calif, Aug 20, 05; m 28; c 4. MEASUREMENT OF VOLTAGE TRANSIENTS. *Educ:* Calif Inst Technol, BS, 26. *Prof Exp:* Elec engr, Gen Elec Co, 26-66; CONSULT ELEC ENGR, 66- *Concurrent Pos:* Chmn, sci & electronic comt, Inst Elec & Electronics Engrs, 62-63. *Honors & Awards:* Centennial Award, Inst Elec & Electronics Engrs, 84. *Mem:* Inst Elec & Electronics Engrs (tech vpres, 62). *Res:* Development of and design procedures for electromagnetic devices in electronics circuits; voltage transients due to switching. *Mailing Add:* 1565 Golf Course Dr Rohnert Park CA 94928

LORD, HARRY CHESTER, III, b Utica, NY, May 28, 39; m 61, 72; c 5. PHYSICAL CHEMISTRY, ANALYTICAL CHEMISTRY. *Educ:* Tufts Univ, BS, 61; Univ Calif, San Diego, PhD(chem), 67. *Prof Exp:* Sr scientist, Jet Propulsion Lab, 67-69, vpres, 69-77; pres, Environ Data Corp, 77-81; pres, Syconex Corp, 80-88; PRES, AIR INSTRUMENTS & MEASUREMENTS, INC, 88- *Concurrent Pos:* chmn, Energy Technol & Control Ltd, 86-88; dir, Dosibi Environ Corp, 89-90. *Honors & Awards:* Gold Medal, Am Inst Chemists, 61. *Mem:* Am Chem Soc; Air Pollution Control Asn; Combustion Inst; Sigma Xi; Instrument Soc Am. *Res:* Modification of combustion, improved control techniques; hardware to increase efficiency and to reduce pollutant emissions; development of state-of-the-art sensors to monitor environmental emissions of toxic and reactive gases. *Mailing Add:* 1400 Edge Cliff Lane Pasadena CA 91107

LORD, JERE JOHNS, b Portland, Ore, Jan 3, 22; m 47; c 3. PHYSICS. *Educ:* Reed Col, AB, 43; Univ Chicago, MS, 48, PhD(physics), 50. *Prof Exp:* Civilian with Radiation Lab, Univ Calif, 42-46; res assoc physics, Univ Chicago, 50-52; instr, 52-62, PROF PHYSICS, UNIV WASH, 62- *Mem:* Fel Am Phys Soc; Am Asn Phys Teachers; fel AAAS. *Res:* Cosmic ray and high energy physics. *Mailing Add:* Dept Physics Univ Wash Seattle WA 98195

LORD, JERE WILLIAMS, JR, b Baltimore, Md, Oct 12, 10; m 41, 71; c 3. SURGERY. *Educ:* Princeton Univ, AB, 33; Johns Hopkins Univ, MD, 37; Am Bd Surg, dipl. *Prof Exp:* From intern to resident surgeon, NY Hosp, 37-44; PROF CLIN SURG, POSTGRAD SCH MED, MED CTR, NY UNIV, 53- *Concurrent Pos:* Consult surgeon, Univ Hosp, Bellevue Hosp, Fourth Div Med Bd & Doctors Hosp & Hackensack Hosp, NJ, St Luke's Hosp, Newburgh, NY, Norwalk Hosp, Conn, 50-, Cent Suffolk Hosp, Riverhead, NY, 51-, Elizabeth Horton Mem Hosp, Middletown, NY, 54-, St Agnes Hosp, White Plains, NY, 55-, Paterson Gen Hosp, NJ, 58 & Univ Hosp; chief, Vascular Surg, Columbus Hosp, NY, 66- *Mem:* Am Col Surg; Am Surg Asn; James IV Asn Surg (secy, 67-75); Am Heart Asn (secy, 53-55); Int Cardiovasc Soc (treas, 53-60, vpres, 61-63). *Res:* Cardiovascular surgery, especially atherosclerosis; gastrointestinal surgery, particularly portal hypertension and intestinal obstruction. *Mailing Add:* 50 Sutton Pl S New York NY 10022

LORD, NORMAN W, computer sciences, nuclear power, for more information see previous edition

LORD, PETER REEVES, b Ruckinge, Eng, Feb 10, 23; m 47; c 3. ENGINEERING, TEXTILE TECHNOLOGY. *Educ:* Battersea Polytech, Eng, BSc, 50; Univ London, PhD(eng), 66, DSc(eng), 76. *Prof Exp:* Res asst heat transfer, Delaney-Gallay Ltd, Eng, 45-46; draughtsman, Fairey Aviation Co Ltd, 46-47; sect leader eng, Vacuum Oil Co Ltd, 47-51; sr test engr, Vickers Armstrongs Ltd, 51-58; lectr textile technol, Univ Manchester, 58-69; from assoc prof to prof, 69-75, Abel C Lineberger prof textiles, 75-90, EMER PROF, ABEL C LINEBERGER, NC STATE UNIV, 90- *Concurrent Pos:* Alexander von Humboldt US sr scientist award, 80. *Honors & Awards:* Harold DeWitt Smith Award, Am Soc Testing & Mat, 79. *Mem:* Fel Brit Inst Mech Eng; fel Brit Textile Inst; Am Soc Mech Engrs; Am Fiber Soc; Sigma Xi. *Res:* Modern methods of yarn formation; open-end spinning; fabric forming systems; sliver and yarn monitoring systems; design of textile machinery; physics of fibrous assemblies. *Mailing Add:* 3116 Monticello Dr Raleigh NC 27612

LORD, SAMUEL SMITH, JR, b Rockland, Maine, Apr 10, 27; m 48; c 5. ANALYTICAL CHEMISTRY. *Educ:* Tufts Col, BS, 47; Mass Inst Technol, PhD(anal chem), 52. *Prof Exp:* Res chemist, Fabrics & Finishes Dept, E I du Pont de Nemours & Co, Inc, 47-49, res chemist, Org Chem Dept, 52-57, res supvr, Elastomer Chem Dept, 57-59, div head, 59-65, supt qual control, 65-67, supt monomer area, 67-70, gen prod supt, 70-71, asst works dir, Maydown Works, Du Pont Co (UK) Ltd, 71-75, WORKS MGR, BEAUMONT WORKS, E I DU PONT DE NEMOURS & CO, INC, 75- *Mem:* AAAS; Am Chem Soc. *Res:* Polarography; coulometry; infrared and ultraviolet spectrophotometry; urethane chemistry. *Mailing Add:* 1240 Nottingham Lane Beaumont TX 77706-4316

LORD, WILLIAM B, b Omaha, Nebr, Jan 2, 29; m 51; c 3. POLICY ANALYSIS, INSTITUTIONAL ANALYSIS. *Educ:* Univ Mich, BS, 51, MF, 58; Univ Wis, MS, 59; Univ Mich, PhD(forestry), 64. *Prof Exp:* Res forester, Lake States Forest Exp Sta US Forest Serv, 54-57; asst prof res, dept agr econ, Univ Wis, 59-65, dir, Ctr Res Policy Studies, 67-72; res assoc, Inst Behav Sci, Univ Colo, 74-77; pres, Policy Sci Assocs, 77-85; dir, Water Resources Res Ctr, 85-90, PROF AGR ECON, UNIV ARIZ, 90- *Concurrent Pos:* Economist policy anal, Off Secy, US Dept Army, 65-67; res Mex, Resources for Future, Inc, 72-74; tech asst expert, UN Develop Prog, 76; mem, Task Force on Oil Shale & Tar Sands, Nat Res Coun, 78-80; & comt groundwater resources & coal mining, 79-81; vis sr scientist, Nat Ctr Atmospheric Res, 79-80; vis prof econ, Univ Colo, 81. *Mem:* Am Water Resources Asn; Am Agr Econ Asn; Asn Evolutionary Econ; Asn Environ & Resources Economists; Nat Asn Environ Profs. *Res:* Natural resource policy and institutions; water conflict resolution; integrated water management. *Mailing Add:* Dept Agr Econ Univ Ariz Econ Bldg 23 Tucson AZ 85721

LORD, WILLIAM JOHN, b Farmington, NH, Nov 3, 21; m 47; c 1. POMOLOGY. *Educ:* Univ NH, BS, 43, MS, 53; Pa State Univ, PhD(hort), 55. *Prof Exp:* EXTEN PROF POMOL, AGR EXTEN SERV, UNIV MASS, AMHERST, 55- *Mem:* Am Soc Hort Sci. *Res:* Weed control; nutrition; growth regulators. *Mailing Add:* Plant & Soil Sci Dept Univ Mass French Hall Amherst MA 01003

LORDI, NICHOLAS GEORGE, b Orange, NJ, Mar 25, 30; m 61; c 3. PHARMACY. *Educ:* Rutgers Univ, BSc, 52 & MSc, 53; Purdue Univ, PhD(pharmaceut chem), 55. *Prof Exp:* From asst prof to assoc prof, 57-64, chmn dept, 77-82, PROF PHARM, RUTGERS UNIV, 64-, ASST DEAN, 81- *Mem:* Am Chem Soc; AAAS; Am Pharmaceut Asn; Acad Pharmaceut Soc. *Res:* Physical stability pharmaceutical systems; pharmaceutical technology. *Mailing Add:* Dept Pharm Rutgers Univ New Brunswick NJ 08903

LORDS, JAMES LAFAYETTE, b Salt Lake City, Utah, Apr 5, 28; m 55; c 2. PHYSIOLOGY. *Educ:* Univ Utah, BS, 50, MS, 51, PhD(plant physiol), 60. *Prof Exp:* Asst bot, Univ Utah, 56-58, instr biol, 58-59; proj assoc plant path, Univ Wis, 60-62; from asst prof to assoc prof, 62-75, PROF MOLECULAR & GENETIC BIOL, UNIV UTAH, 75-, PROF BIOL, 75- *Res:* Microwave interactions with biological systems. *Mailing Add:* Dept of Biol Univ of Utah Salt Lake City UT 84112

LORE, JOHN M, JR, b New York, NY, July 26, 21; m; c 4. OTOLARYNGOLOGY, SURGERY. *Educ:* Col Holy Cross, BS, 44; NY Univ, MD, 45; Am Bd Otolaryngol, dipl, 54; Am Bd Surg, dipl, 56. *Prof Exp:* Intern, St Vincent's Hosp, New York, 45-46, resident otolaryngol & head & neck surg, 48-50; asst resident gen surg, St Clare's Hosp, 50-52, sr resident, 54-55; asst resident surg & radiation, Mem Cancer Ctr, 52-53; asst clin prof surg & asst attend surgeon, NY Med Col, Flower & Fifth Ave Hosps, 64-66; prof otolaryngol & chmn dept, Sch Med, State Univ NY, Buffalo, 66-90; RETIRED. *Concurrent Pos:* Fel exp surg, St Clare's Hosp, 53-54; asst vis surgeon, Metrop Hosp Ctr, New York, 64-66; dir surg, Good Samaritan Hosp, Suffern, attend surgeon, St Clare's Hosp, New York & consult surgeon, Tuxedo Mem Hosp, 65-66; head dept otolaryngol & chief combined head & neck serv, Buffalo Gen Hosp & Buffalo Children's Hosp, 66-; head dept otolaryngol & chief combined head & neck serv, Eric County Med Ctr, 66-; consult, Buffalo Vet Admin Hosp, 66-; chmn dept otolaryngol, Sisters of Charity Hosp, 75-90; vis prof, Col Med, Baylor Univ, 67; consult, Roswell Park Mem Inst, 68-; clin consult, NY State Dept Health, 68-; vis prof, Denver Med Ctr, 69 & Dept Otolaryngol, Bethesda Naval Med Ctr, Md, 71; consult, Deaconess Hosp, Buffalo, NY & Buffalo Hearing & Speech. *Honors & Awards:* Hektoen Gold Medal, AMA, 52. *Mem:* Fel Am Col Surg; Am Cancer Soc; fel Am Acad Ophthal & Otolaryngol; AMA; James Ewing Soc. *Res:* General surgery, including maxillofacial surgery and plastic surgery of the head and neck. *Mailing Add:* Suite 208 Seton Prof Bldg 2121 Main St Buffalo NY 14214

LOREE, THOMAS ROBERT, b Seattle, Wash, Feb 1, 36; m 58; c 3. LASERS, SOLID STATE PHYSICS. *Educ:* Willamette Univ, BA, 57; Univ Wis, MS, 60, PhD(solid state physics), 62. *Prof Exp:* MEM RES STAFF, LOS ALAMOS NAT LABS, 62- *Concurrent Pos:* Consult, Particle Technol,

Inc, 73-76, Laser Photonics, 83-84, Laser Technics, 83-84, Spektrale Identifikationssysterme, 85-, GV Medical, 86-8, Site Microsurgical, 87-, J&J Cardiology, 88-89, Amoco Laser, 88-, Sci Med, 89-, Pfizar, 89-90, Aria, 89-, J&J, 89- *Mem:* Optical Soc Am. *Res:* Electronic properties of metals and semiconductors; shock waves in metals; excimer laser research; laser system design; non-linear optics research and design; photochemistry and spectroscopy; medical applications of lasers. *Mailing Add:* Los Alamos Nat Labs Box 1663 MS J564 Los Alamos NM 87545

LORENCE, MATTHEW C, EXPERIMENTAL BIOLOGY. *Prof Exp:* ASSOC PROD LINE MGR, BIO-RAD LABS, 90- *Mailing Add:* Bio-Rad Labs 15111 San Pablo Ave Richmond CA 94806

LORENSEN, LYMAN EDWARD, b Lincoln, Nebr, Sept, 26, 23; m 50; c 3. ORGANIC POLYMER CHEMISTRY. *Educ:* Univ Nebr, BS, 47; Cornell Univ, PhD(chem), 52. *Prof Exp:* Jr chemist, Bristol Labs, 47-48; asst org chem, Cornell Univ, 50-52; chemist, Shell Develop Co, 52-58 & 60-64, mem staff, Mfg Res Dept, Shell Oil Co, 58-60; mem staff, composites & polymer technol & actg technol leader, Lawrence Livermore Nat Lab, Univ Calif, 64-88; CONSULT, 88- *Mem:* Am Chem Soc; Sigma Xi. *Res:* High temperature polymers; polymers for geothermal applications; unsaturated glycols; possible precursors in biosynthesis of rubber; lubricating oil additives; silicone and epoxy polymers; filled polymers; foams; coatings. *Mailing Add:* 9 Broadview Terr Orinda CA 94563

LORENTE DE NO, RAFAEL, neurophysiology; deceased, see previous edition for last biography

LORENTS, ALDEN C, b Bagley, Minn, Apr 29, 37; m 60; c 2. DATABASE, COMPUTER AIDED SOFTWARE ENGINEERING. *Educ:* Concordia Col Minn, BSBA, 60; Univ Minn, MBA, 62, PhD(acct), 71. *Prof Exp:* Programmer & analyst, Honeywell, 60-66; dir comput, Bemidji State Univ, 66-71; PROF COMPUTER INFO SYSTS, NORTHERN ARIZ UNIV, 71- *Concurrent Pos:* Consult, Ariz Guid Ctr, 73-83 & Univ Kuwait, 83; internship, Lawrence Livermore Labs, 81, Sandia Labs, 83 & Ariz Pub Serv, 87-89. *Mem:* Soc Info Mgt; Data Processing Mgt Asn; Decision Sci Inst. *Res:* Software engineering; re-engineering; repository development; database development; computer aided software engineering. *Mailing Add:* Northern Ariz Univ Box 15066 Flagstaff AZ 86011

LORENTS, DONALD C, b Minn, Mar 26, 29; m 52; c 2. LASER PHYSICS, ATOMIC CLUSTER PHYSICS. *Educ:* Concordia Col, Moorhead, Minn, BA, 51; Univ Nebr, MA, 54, PhD(physics), 58. *Prof Exp:* Res physicist, Westinghouse Res Labs, 58-59; physicist, 59-63, chmn, Dept Molecular Physics, 63-67, head, Atomic & Molecular Collisions Sect, 67-68, physicist, 69-70, sr physicist, 70-75, assoc dir, 75-79, dir, Molecular Physics Lab, 80-84, dir, Chem Physics Lab, 84-90, SCI DIR, MOLECULAR PHYSICS LAB, SRI INT, 90- *Concurrent Pos:* Vis res physicist, Inst Physics, Aarhus Univ, 68-69, 70, 87. *Mem:* AAAS; fel Am Phys Soc. *Res:* Atomic and molecular collision processes with emphasis on scattering, charge transfer and excitation in ion-atom or ion-molecule collisions; kinetic processes in electronically excited dense gases; molecular spectroscopy; cluster physics. *Mailing Add:* SRI Int 333 Ravenswood Menlo Park CA 94025

LORENTZ, GEORGE G, b St Petersburg, Russia, Feb 25, 10; m 42; c 5. MATHEMATICAL ANALYSIS. *Educ:* Univ Leningrad, Cand, 35; Univ Tuebingen, Dr rer nat(math), 44. *Hon Degrees:* Dr, Univ Tuebingen. *Prof Exp:* Lectr math, Univ Leningrad, 36-42 & Univ Frankfurt, 46-48; prof, Univ Tubingen, 48-49; from asst to asst prof, Univ Toronto, 49-53; prof, Wayne State Univ, 53-58; prof, Syracuse Univ, 58-69; prof, 69-80, EMER PROF MATH, UNIV TEX, AUSTIN, 80- *Concurrent Pos:* Res grants, NSF & Off Sci Res. *Honors & Awards:* Humboldt Prize, A von Humboldt Stiftung, 73. *Mem:* Am Math Soc; Math Asn Am; Ger Math Soc. *Res:* Mathematical analysis, especially approximations and expansions; summability; Birkhoff interpolation; functional analysis, especially Banach function spaces; interpolation theorems for operators; published several monographs on Approximation and Interpolation. *Mailing Add:* 8804 Siverwood Court Austin TX 78759

LORENTZEN, KEITH EDEN, b Heber City, Utah, Apr 13, 21; m 47, 80; c 6. PHYSICAL ORGANIC CHEMISTRY. *Educ:* Univ Utah, BA, 42, MS, 47; Pa State Univ, PhD(chem), 51. *Prof Exp:* Chemist, Standard Oil Co (Ind), 51-62; from asst prof to assoc prof chem, Ind Univ NW, 63-88, from asst chmn to chmn dept, 66-88, chmn, dept physics & astron, 77-88; RETIRED. *Mem:* Am Chem Soc; fel Am Inst Chemists; Am Asn Univ Professors. *Res:* Conductivity measurements; chemistry of lubricating oils and additives; organic analytical chemistry; chromatography; polarography; Friedel-Crafts methylation of xylenes; aromatic deuteration of methylbenzenes. *Mailing Add:* 1505 Melbrook Dr Munster IN 46321

LORENZ, CARL EDWARD, b New York, NY, Aug 22, 33; m 56; c 3. ORGANIC CHEMISTRY. *Educ:* NY Univ, BA, 53, PhD(chem), 57. *Prof Exp:* Asst chem, NY Univ, 53-57; chemist, Plastics Dept, Exp Sta, 57-63, sr res chemist, 63-68, supvr, 68, sr supvr, 68-69, lab supt, 69-70, res lab mgr, Sabine River Works, 70-72, res mgr, Wilmington, 72-74, asst dir, Int Dept, 74-76, prod mgr, 76-78, dir, Feedstocks Div, 78-79, dir Res Div, Cent Res & Develop, 79-81, dir res & develop, Polymer Prod Dept, 81-83, dir, Ethylene Polymers Div, 83-85, dir, Chemicals & Pigments Dept, 85-90, VPRES RES & DEVELOP, DU PONT CHEMICALS, E I DU PONT DE NEMOURS & CO, INC, 90- *Honors & Awards:* Award, Am Inst Chem, 53. *Mem:* Am Chem Soc; Am Inst Chem; The Chem Soc. *Res:* Fluorocarbon monomer syntheses and polymerizations; high pressure hydrocarbon syntheses; heterogeneous catalysis; chemistry of anionic and radical polymerizations. *Mailing Add:* 103 Bellant Circle Wilmington DE 19807

LORENZ, DONALD H, b Brooklyn, NY, Oct 18, 36; m 62. ORGANIC CHEMISTRY, POLYMER CHEMISTRY. *Educ:* Polytech Inst Brooklyn, BS, 58, PhD(org chem), 63. *Prof Exp:* Asst org chem, Polytech Inst Brooklyn, 58-59, organometallics, 59-62; asst scientist chem eng res div, NY Univ, 62-63; sr polymer chemist, Tex-US Chem Co, 63-65; explor polymer chemist, Gen Aniline & Film Co, 65-70, group leader polymer synthesis, GAF Corp, 70-74, mgr vinyl polymer res, 74-80; dir res & develop, 80-89, EXEC VPRES, HYDROMER INC, 89- *Mem:* Am Chem Soc. *Res:* Organometallic chemistry; elastomers; resins; adhesives; polymers of vinyl ethers and vinyl amides; polyurethanes; fire retardants; ultraviolet and electron beam curable resins; coatings for medical devices; drug delivery systems; wound dressings. *Mailing Add:* 12 Radel Pl Basking Ridge NJ 07920

LORENZ, EDWARD NORTON, b West Hartford, Conn, May 23, 17; m 48; c 3. METEOROLOGY. *Educ:* Dartmouth Col, AB, 38; Harvard Univ, AM, 40; Mass Inst Technol, SM, 43, ScD(meteorol), 48. *Hon Degrees:* DSc, McGill Univ, 83, Univ Ariz, 88. *Prof Exp:* Asst meteorol, Mass Inst Technol, 46-48, mem staff, 48-54; vis assoc prof, Univ Calif, Los Angeles, 54-55; res staff, 48-54, from asst prof to assoc prof, 55-62, head dept, 77-81, PROF METEOROL, MASS INST TECHNOL, 62- *Concurrent Pos:* Vis scientist, Lowell Observ, 51, Norweg Meteorol Inst, 62; sr postdoctoral assoc, Nat Ctr Atmospheric Res, 73-74; vis sr scientist, Univ Oslo, 82. *Honors & Awards:* Clarence Leroy Meisinger Award, Am Meteorol Soc, 63, Carl Gustaf Rossby Res Med, 69; Symons Mem Gold Medal, Royal Meteorol Soc, 73; Holger & Anna-Greta Crafoord Prize, Royal Swed Acad Sci, 83; Elliott Creson Medal, Franklin Inst, 89. *Mem:* Nat Acad Sci; fel Am Acad Arts & Sci; hon mem Am Meteorol Soc; hon mem Royal Meteorol Soc. *Res:* General circulation of the atmosphere; dynamical and statistical weather prediction; chaotic dynamical systems. *Mailing Add:* Dept Earth Atmospheric & Planetary Sci Mass Inst Technol Cambridge MA 02139

LORENZ, JOHN DOUGLAS, b Talmage, Nebr, July 2, 42; m 67; c 1. MANUFACTURING SYSTEMS DESIGN, ASSEMBLY SYSTEMS DESIGN. *Educ:* Univ Nebr, Lincoln, BS, 65, MS, 67, PhD(indust eng), 73. *Prof Exp:* From asst prof to assoc prof, GMI Eng & Mgt Inst, 73-78, dept head indust eng, 84-87, asst dean acad serv, 86-87, asst dean res & grad studies, 87-88, PROF INDUST ENG, GMI ENG & MGT INST, 78-, PROVOST & DEAN FAC, 88- *Concurrent Pos:* Dir, Flint Area Sci Fair, 89-; Robotic Indusrs Asn, 89- & Jr Eng Tech Soc, 91-; Richard L Terrell prof acad leadership, GMI Eng & Mgt Inst, 90- *Mem:* Soc Mfg Engrs; Soc Automotive Engrs; Inst Indust Engrs; Am Soc Eng Educr. *Res:* Manufacturing systems design; assembly systems design; computer assisted assembly line balancing. *Mailing Add:* 3122 Beech Tree Lane Flushing MI 48433

LORENZ, KLAUS J, b Berlin, Ger, June 22, 36; US citizen; m 60; c 3. CEREAL CHEMISTRY. *Educ:* Northwestern Univ, Ill, PhB, 68; Kans State Univ, MS, 69, PhD(food sci), 70. *Prof Exp:* Baking technologist, Am Inst Baking, 61-65; food technologist, Nat Dairy Prod Corp, 65-68; asst prof, 70-74, assoc prof, 74-78, PROF FOOD SCI & NUTRIT, COLO STATE UNIV, 78- *Mem:* Am Asn Cereal Chem; Inst Food Technologists; Swiss Soc Food Sci & Technol. *Res:* Cereal chemistry and technology; carbohydrate chemistry. *Mailing Add:* Dept Food Sci & Human Nutrit Colo State Univ Ft Collins CO 80523

LORENZ, KONRAD ZACHARIAS, comparative anatomy & animal psychology; deceased, see previous edition for last biography

LORENZ, MAX R(UDOLPH), b Detroit, Mich, June 25, 30; m 55; c 3. PHYSICAL CHEMISTRY, PHYSICS. *Educ:* Rensselaer Polytech Inst, BChE, 57, PhD(phys chem), 60. *Prof Exp:* Res assoc, Gen Elec Res Lab, 60-63; res staff mem, Thomas J Watson Res Ctr, NY, 63-73; head dept inorg mat, Res Div, IBM Corp, San Jose, Calif, 73-82, consult prod assurance, IBM Corp, Maiz, WGer, 83-87, RES STAFF MEM, ALMADEN RES CTR, IBM CORP, SAN JOSE, CALIF, 87- *Mem:* Fel Am Phys Soc; Electrochem Soc; NY Acad Sci; fel Am Inst Chem; sr mem Inst Elec & Electronics Engrs. *Res:* Physics and chemistry of semiconductors, especially the role and control of defects and electrical and optical properties; magnetic disk coating technology and recording physics. *Mailing Add:* Almaden Res Ctr Dept K61-802 IBM 650 Harry Rd San Jose CA 95120

LORENZ, OSCAR ANTHONY, b Colorado Springs, Colo, Dec 5, 14; m 47. VEGETABLE CROPS, PLANT NUTRITION. *Educ:* Colo State Col, BS, 36; Cornell Univ, PhD(veg crops), 41. *Prof Exp:* Asst hort, Colo State Col, 36-37; asst veg crops, Cornell Univ, 37-41; from instr to assoc prof, 41-55, vchmn dept, 55-64, chmn dept, 64-70, PROF VEG CROPS, COL AGR, UNIV CALIF, DAVIS, 55- *Concurrent Pos:* Mem, Agr Comn to Bermuda, 39. *Honors & Awards:* Vaughn Award, 42. *Mem:* Fel Am Soc Hort Sci; fel Am Soc Plant Physiol; fel Am Soc Agron; fel Am Potato Asn. *Res:* Mineral nutrition of vegetable crops; boron deficiency in table beets; soils and plant nutrient relationships; environmental factors affecting vegetable production. *Mailing Add:* 44163 Lakeview Dr El Macero CA 95618

LORENZ, PATRICIA ANN, b New York, NY, Jan 31, 38; m 62; c 2. INFORMATION SCIENCE, ANALYTICAL CHEMISTRY. *Educ:* Marymount Manhattan Col, BS, 59; Polytech Inst Brooklyn, PhD(anal chem), 65. *Prof Exp:* Info chemist, Exxon Res & Eng Co, 65-67; consult info sci, 67-77; group head, Anal & Info Div, 78-83, sect head, Comput & Info Support Div, 83-87, RES ASSOC, INFO RES ANAL, EXXON RES & ENG CO, 87- *Mem:* Am Chem Soc; Am Soc Info Sci. *Res:* Mechanism of acid-base reactions in benzene. *Mailing Add:* Exxon Res & Eng Co PO Box 121 Linden NJ 07036

LORENZ, PHILIP BOALT, b Dayton, Ohio, Aug 14, 20; m 46; c 3. PHYSICAL CHEMISTRY, PETROLEUM PRODUCTION. *Educ:* Swarthmore Col, AB, 41; Harvard Univ, MA, 46, PhD(chem), 49. *Prof Exp:* Asst biol, Princeton Univ, 42-43; asst, Phys Chem, SAM Labs, Columbia Univ, 44-45; phys chemist surface chem, Petrol Res Ctr, US Bur Mines, 49-71, res chemist, Petrol Prod & Environ Res, 71-75, res chemist, Bartlesville

Energy Technol Ctr, US Dept Energy, 75-83; sci adv, Nat Inst Petrol & Energy Res, 83-85; CONSULT, 85- Mem: Am Chem Soc; Sigma Xi; Soc Petrol Engrs. Res: Surface chemistry; electrochemistry; petroleum engineering. Mailing Add: 1541 Keeler Ave Bartlesville OK 74003-5723

LORENZ, PHILIP JACK, JR, b Atlanta, Ga, Apr 15, 24; m 70; c 2. ATMOSPHERIC PHYSICS. Educ: Oglethorpe Univ, BS, 49; Vanderbilt Univ, MS, 52. Prof Exp: Lab asst, Oglethorpe Univ, 48; qual control tech, Transparent Package Co, 50-51; asst prof physics, Lemoyne Col, 52-54, Ky Wesleyan Col, 54-56 & Upper Iowa Univ, 56-61; res assoc, Syracuse Univ, 63-65, vis instr, 65-66; chmn dept, 66-74, assoc prof, 66-82, PROF PHYSICS, UNIV OF THE SOUTH, 82-, DIR OBSERV, 87- Concurrent Pos: Lab asst, Vanderbilt Univ, 52; consult physicist, Empirical Explor Co, Ky, 56; univ fel, Syracuse Univ, 58-59, Nat Sci Found fac fel, 61-63; textbook consult, J B Lippincott Co, 71-72. Mem: Sigma Xi; Am Phys Soc; Am Asn Physics Teachers; Hist Sci Soc. Res: Atmospheric electricity in fair and foggy weather; geophysics of environmental radioactivity at sandstone sinkhole sites; history of medieval Persian and Arabic science; history of astronomy; designing laboratory experiments for premedical physics, optics and introductory astronomy; history of science in the post-bellum south. Mailing Add: Dept of Physics Univ of the South Sewanee TN 37375

LORENZ, RALPH WILLIAM, b Waseca, Minn, Aug 19, 07; m 39; c 2. FORESTRY. Educ: Univ Minn, BS, 30, PhD(plant Physiol), 38. Prof Exp: Asst plant physiol, Univ Minn, 30-33; jr forester, US Forest Serv, 33-35; instr forestry, Univ Farm, Univ Minn, 35-38; assoc, Univ Ill, Urbana, 38-41, asst chief, 41-44, assoc chief, 44-47, assoc prof forest res, 47-55, prof, 55-73, actg head dept, 65-66, EMER PROF FORESTRY, 73-; RETIRED. Mem: Fel Soc Am Foresters. Res: Forest research in planting; silviculture; dendrology; regeneration and forest management. Mailing Add: Dept Forestry 110 Mumford Hall 101 W Windor Rd No 6212 Urbana IL 61881

LORENZ, RICHARD ARNOLD, b Fond du Lac, Wis, Mar 7, 42; m 65; c 4. WEAPON EFFECTS, SHOCK DYNAMICS. Educ: Marquette Univ, BS, 65, MS, 68. Prof Exp: Physicist, Naval Ord Lab, 67-70; res physicist, 70-74; res physicist appl physics, Naval Surface Weapons Ctr, 74-85; prin engr, Boeing Mil Airplanes, 85-90, PRIN ENGR, MISSLES & SPACE DIV, BOEING CO, 90- Concurrent Pos: Consult, Advan Reactors Br, Nuclear Regulatory Comn, 75-78. Mem: Am Phys Soc. Res: Weapon effects and explosion effects; shock waves; atmospheric sound focusing; data reduction techniques; hydrodynamics; computer simulations. Mailing Add: Missles & Space Div Boeing Co 2002 19th Dr NE Auburn WA 98002

LORENZ, ROMAN R, b Breslau, Ger, July 15, 35; US citizen; m 60; c 3. ORGANIC CHEMISTRY. Educ: Rensselaer Polytech Inst, BS, 58; Univ Mich, MS, 60, PhD(med chem), 62. Prof Exp: Res org chemist, Sterling-Winthrop Res Inst, 62-69, sr res chemist & sect head, 69-74, sr res assoc & sect head, 74-76, dir chem develop, 77-90, EXEC DIR CHEM DEVELOP, STERLING RES GROUP, 90- Mem: Am Chem Soc. Res: Synthesis of organic and medicinal compounds. Mailing Add: Sterling Res Group Rensselaer NY 12144

LORENZEN, CARL JULIUS, biological oceanography, food chain dynamics; deceased, see previous edition for last biography

LORENZEN, HOWARD O(TTO), b Atlantic, Iowa, June 24, 12; m 36; c 1. ELECTRONICS. Educ: Iowa State Col, BSEE, 35. Prof Exp: Develop engr electronics, Colonial Radio Corp, 35-39 & Zenith Radio Corp, 39-40; head, Electronic Countermeasures Br, Naval Res Lab, 40-66, supt, Electronic Warfare Div, 66-70, supt, Space Systs Div, 70-80; RETIRED. Concurrent Pos: Tech adv, Chief Naval Opers Off, Off Secy Defense, Joint Chiefs Staff, Dir Naval Intel & to Dir Naval Labs. Honors & Awards: Dexter Conrad Award, 72; Space Pioneer Award, US Navy, 85. Mem: Fel Inst Elec & Electronics Engrs. Res: Signal indication and analysis; direction finding; propagation; electronic countermeasures; satellite design; space data reduction. Mailing Add: 9000 Lake Washington Blvd NE Bellevue WA 98004

LORENZEN, JANICE R, b Chicago, Ill, May 29, 50. ENDOCRINOLOGY. Educ: Valparaiso Univ, BS, 72; Albany Med Col, PhD(physiol), 79; Univ Ill, MD, 86. Prof Exp: Asst prof, Dept Biol Sci, Western Mich Univ, 81-82; PHYSICIAN, ROCKFORD CLIN, COL MED, UNIV ILL, 86- Mem: Endocrine Soc; Am Col Physicians. Mailing Add: Rockford Clin Col Med Univ Ill 2300 N Rockton Ave Rockford IL 61108

LORENZEN, JERRY ALAN, b Grand Island, Nebr, Oct 3, 44; m 67; c 2. SURFACE CHEMISTRY, APPLIED STATISTICS. Educ: Midland Lutheran Col, BS, 66; Okla State Univ, PhD(chem), 70. Prof Exp: Instr chem, Okla State Univ, 69-70; SR CHEMIST, IBM CORP, 70- Mem: Am Chem Soc. Res: Environmental chemistry; gas-solid interaction; engineering statistics. Mailing Add: Rd N02 Box 52H Stone Ridge NY 12484-9802

LORENZETTI, OLE J, b Chicago, Ill, Oct 25, 36; m 62; c 3. PHARMACOLOGY, BIOCHEMISTRY. Educ: Univ Ill, Chicago, BS, 58; Ohio State Univ, MS, 62, PhD(pharmacol & toxicol), 65. Prof Exp: Asst chief pharmacist, West Suburban Hosp, 58; instr pharm, Univ Ill, Chicago, 58-59; asst instr pharmacol, Ohio State Univ, 59-62; from res pharmacologist to sr res pharmacologist, Therapeut Res Labs, Dome Chem Inc Div, Miles Labs, Inc, Ind, 64-69; sr res scientist, Alcon Labs Inc, 69-71, dir immunol & biochem res, 71-72, mgr biol res, 72-74, dir dermatol res & develop, 79-81; vpres, 67-70, PRES, PHARMACEUT CONSULT, INC, 70-; DIR OPHTHAL/SURG, ALCON LABS INC, 82- Concurrent Pos: Assoc prof pharmacol, Univ Tex Health Sci Ctr, Dallas, 70-; adj prof, Tex Christian Univ, 72- Mem: AAAS; Am Chem Soc; Soc Cosmetic Chem; Am Acad Clin Toxicol; Am Soc Pharmacol & Except Ther; Am Acad Ophthal; Am Intraocular Implant Soc; Am Pharmacol Asn. Res: Pharmacodynamics; evaluations of analgesic, anti-inflammatory agents and antiglaucoma agents; development of drug screening programs; autonomic and biochemical

pharmacology; topical pharmacology and toxicology of eye and skin; ophthalmology; ophthalmic surgical devices; toxicology; immunology; drug metabolism; pharmacokinetics. Mailing Add: Alcon Labs Inc 6201 S Freeway Ft Worth TX 76134

LORENZO, ANTONIO V, b Vigo, Spain, July 23, 28; US citizen; m 58; c 2. NEUROPHARMACOLOGY, NEUROCHEMISTRY. Educ: Univ Chicago, BA, 56, BS, 58, PhD(pharmacol), 66. Prof Exp: From asst to assoc neurol, Children's Hosp Med Ctr, 64-68, instr pharmacol, 66-68; asst prof, 69-71, ASSOC PROF PHARMACOL, HARVARD MED SCH, 71-; DIR NEUROSURG, CHILDREN'S HOSP MED CTR, 68- Concurrent Pos: Epilepsy Found Am fel, Children's Hosp Med Ctr, 65 & dir neurol res, 69-78; Nat Inst Neurol Dis & Stroke proj grant, 71-73; NIH career develop award, Harvard Med Sch, 70-75. Mem: AAAS; Am Soc Pharmacol & Exp Therapeut; NY Acad Sci; Am Soc Neurochem; Am Acad Neurol. Res: Pathophysiology of the blood, role of brain barrier, putative transmitter seizures; cerebrospinal fluid transport phenomena; cerebrospinal fluid dynamics. Mailing Add: Neurosurg Children's Hosp Med Ctr 300 Longwood Ave Boston MA 02115

LORENZO, GEORGE ALBERT, b Rochester, NY, Apr 18, 43; m 66; c 2. ELECTROCHEMICAL PROCESSES, ION EXCHANGE PROCESSES. Educ: John Fisher Col, BS, 65; Syracuse Univ, PhD(org chem), 70. Prof Exp: Res postdoc liquid crystal chem, State Univ NY Binghamton, 70-71; res chemist liquid crystals, Olivetti Corp, 71-72; dir, Lab Environ Health, Health Dept, Monroe Co, 72-74; dir appl res pollution control, CPAC, Inc, 74-89; environ consult, Sear-Brown Group, 90-91; NAT IMAGING CONSULTS, 91- Concurrent Pos: Prin investr, Wright Patterson AFB Res Contract, 74-76. Mem: Soc Photog Scientists & Engrs. Res: Res & develop to provide chemical recycle and pollution control processes, especially for photofinishing operations which include electrolysis, ion exchange, ozone and mechanic processes. Mailing Add: 2124 Dutch Hollow Rd Avon NY 14414

LORETZ, CHRISTOPHER ALAN, b Santa Monica, Calif, Apr 28, 51. ENDOCRINOLOGY. Educ: Univ Wash, BS, 72; Univ Calif, Los Angeles, MA, 74, PhD(comp physiol), 78. Prof Exp: Fel, Dept Zool & Cancer Res Lab, Univ Calif, Berkeley, 78-81; asst prof, 81-87, ASSOC PROF, DEPT BIOL SCI, STATE UNIV NY BUFFALO, 87- Mem: Am Soc Zoologists; NY Acad Sci; AAAS. Res: Osmoregulation in aquatic vertebrates; hormonal control of epithelial ion transport. Mailing Add: Dept Biol Sci 109 Cooke Hall State Univ NY Buffalo NY 14260

LORETZ, THOMAS J, b Oceanside, NY, Mar 19, 51; m 83; c 2. FIBEROPTIC IMAGING & COMMUNICATION, COMPUTER AIDED MANUFACTURING. Educ: State Univ NY, BS, 73, MS, 78. Prof Exp: Proj engr glass res & develop, Schott Optical Glass, Duryea, Pa, 74-78; dir res & develop mat & electro-optics, Galileo Electro-Optics, Sturbridge Mass, 78-82; sr scientist fiberoptic med develop, Johnson & Johnson, Southbridge, Mass, 82-85; DIR RES & DEVELOP ELECTRO-OPTIC GLASSES, DETECTOR TECHNOL, BROOKFIELD, MASS, 85- Concurrent Pos: Consult & inventor, Buffalo Med Specialties, Fla, 76-90; prin investr, NASA, SBIR Progs Advan Space Telescope, 85-; consult, NIH Spec Comt Fiberoptics Med, 86-87; secy bd, Bd Dirs, Charlton Credit Union, Mass, 90- Honors & Awards: IR 100 Award, Res & Develop Mag, 80. Mem: Am Ceramic Soc; Nat Inst Ceramic Engrs; Soc Photog Instrumentation Engrs. Res: solid state, electron multiplication; glasses and geometries to enhance lifetime and characteristics of continuous dynode single channel and microchannel plate devices. Mailing Add: Detector Technol PO Box K-300 Brookfield MA 01506

LOREY, FRANK WILLIAM, b Staten Island, NY, May 7, 29; m 51; c 3. PAPER CHEMISTRY. Educ: State Univ New York Col Forestry, Syracuse, BS, 51, MS, 52. Prof Exp: Res engr, Mead Corp, Ohio, 52-54; assoc prof pulp & paper chem & pilot plant group leader, State Univ New York Col Forestry, Syracuse, 54-66; asst to gen mgr, 66-67, corp tech dir, 67-75, vpres res, 75-86, SR VPRES, GARDEN STATE PAPER CO, GARFIELD, 86- Concurrent Pos: Develop consult, AB Kamyr, Sweden, 65. Mem: Tech Asn Pulp & Paper Indust; Int Asn Sci Papermakers. Res: Improved methods in pulping of wood and use of chemicals for influencing paper properties; development of processes and design of systems for deinking of waste papers. Mailing Add: Garden State Paper Co Inc 669 River Dr Ctr Two Elmwood Park NJ 07407

LORIA, EDWARD ALBERT, b Pittsburgh, Pa, Apr 29, 19; m 54; c 3. METALLURGY & PHYSICAL METALLURGICAL ENGINEERING. Educ: Carnegie Inst Technol, BS, 44, MS, 66. Prof Exp: Asst metallurgist, Res Lab, Carnegie-Ill Steel Div, US Steel Corp, Pittsburgh, Pa, 44-46; fel, Mellon Inst Indust Res, Pittsburgh, Pa, 46-48, sr fel, 48-50; sr engr metall, Res & Eng Ctr, Carborundum Co, Niagara Falls, NY, 50-52; staff metall engr alloy & stainless steels, Cent Metall Dept, Crucible Steel Co Am, Pittsburgh, Pa, 53-59, prod metall engr titanium & superalloys, 57-59; mgr alloy & stainless steels & superalloys develop, Climax Molybdenum Co, Div Am Metal Climax, NY, 59-63; supv res metallurgist, Reno Metall Res Ctr, US Dept Interior, Bur Mines, 63-64; supvr mat & processes, Res & Develop Eng Div, Nat Steel Corp, Weirton, WVa, 64-75; tech dir, Roll Mfrs Inst, Pittsburgh, Pa, 75-77; div prod metallurgist, Universal-Cyclops Specialty Steel Div, Cyclops Corp, Pittsburgh, Pa, 77-84, consult, 84-86; CONSULT, NIOBIUM PROD CO, SUBSID CBMM, PITTSBURGH, PA, 84- Concurrent Pos: Mem, Advan Res Metal Comt, Am Soc Metals, 75-77, Subcomt Roll Res, AISE-RMI, 75-77 & Comt Gen Res, Am Iron & Steel Inst, 81-84. Honors & Awards: Charles H Herty Award, Am Inst Mining & Metall Engrs, 67; Edgar C Bain Award, Am Soc Metals, 82, Andrew Carnegie Lectr, 84, William Hunt Eisenman Award, 91. Mem: Am Inst Mining & Metall Engrs; Am Soc Metals. Res: Practical application of metallurgy for an unusually wide range of metals and alloys; author of over 125 publications. Mailing Add: 1828 Taper Dr Pittsburgh PA 15241

LORIA, ROGER MOSHE, b Antwerp, Belgium, Apr 19, 40; US citizen; m 78; c 3. VIROLOGY, IMMUNOLOGY. *Educ:* Bar-Ilan Univ, Israel, BS, 65; State Univ NY, Buffalo, MS, 68; Boston Univ, PhD(microvirol), 72. *Prof Exp:* Asst prof biochem, Mass Col Optom, 69-70; asst virol, Sch Med, Boston Univ, 68-72, instr microbiol, 72-74; asst prof, 74-78, ASSOC PROF MICROBIOL & PATH, MED COL VA, 78- *Concurrent Pos:* Mass Heart Asn fel, 72-74; res assoc, Sch Med, Boston Univ, 74; NIH res grants, Arthritis & Metab Dis, 74, 78 & 79, Heart & Lung Div, 75; Young investr develop award, Am Diabetes Asn, 75-77; asst prof acad path, Sch Med, Harvard Univ, 80-82; instr pediat, Childrens Hosp, Boston, Mass, 80-81; adv, Consol Labs Commonwealth Va, 82-86; vpres, Va Commonwealth Chapt, Am Asn Univ Prof, 89, pres, 90. *Mem:* Am Soc Microbiol; AAAS; Am Fedn Clin Res; Reticuloendothelial Syst Soc; Am Diabetes Asn; Am Soc Virol; Am Asn Univ Professors; Int Soc for Anti-viral Res; Am Inst Nutrit; Am Soc Clin Nutrit; fel Am Acad Biol. *Res:* Investigation on the role of group B coxsackieviruses in diabetes, atherosclerosis and cardiovascular disease in experimental animal models; general aspects of host-virus interaction; viral infection by the oral route; nutritional hypercholesteremia; effects on host resistance; rapid viral diagnosis; immune-up regulations; publication of 51 manuscripts and 64 abstracts, 2 US patents. *Mailing Add:* Dept Microbiol & Immunol Med Col Va Box 678 Richmond VA 23298

LORIAUX, D LYNN, b Bartlesville, Okla, Apr 29, 40; m 60, 86; c 4. GROWTH & DEVELOPMENT. *Educ:* Baylor Med Sch, MD, 67, PhD(biochem), 68. *Prof Exp:* CLIN DIR, NAT INST CHILD HEALTH & HUMAN DEVELOP, NIH, 76-, CHIEF, DEVELOP ENDOCRINOL BD, 80- *Mem:* The Endocrine Soc. *Res:* The endocrinology of growth and development. *Mailing Add:* Clin Ctr NIH Bethesda MD 20892

LORIMER, GEORGE HUNTLY, b Eng, Oct 14, 42; m 70; c 2. CARBON METABOLISM, ENZYMOLOGY. *Educ:* Mich State Univ, PhD(biochem), 72. *Prof Exp:* RES LEADER, CENT RES & DEVELOP DEPT, EI DU PONT DE NEMOURS & CO, 78- *Mem:* Royal Soc; Am Soc Biol Chemists. *Mailing Add:* E I du Pont de Nemours & Co Exp Sta E402-2239 Cent Res & Develop Wilmington DE 19898

LORIMER, JOHN WILLIAM, b Oshawa, Ont, Apr 16, 29; m 54; c 3. PHYSICAL CHEMISTRY. *Educ:* Univ Toronto, BA, 51, MA, 52, PhD(phys chem), 54. *Prof Exp:* Asst phys chem, Univ Leiden, Netherlands, 54-56; asst res officer, Atlantic Regional Lab, Nat Res Coun Can, 56-61, assoc res officer, 61; from asst prof to assoc prof, 61-79, PROF PHYS CHEM, UNIV WESTERN ONT, 79- *Concurrent Pos:* Vis prof, Univ Southampton, 70-71; Murdoch Univ, Perth, Australia, 83 & Glasgow Univ, 83-84; mem, Comn V-8, Int Union Pure & Appl Chem, 79- & Comn V, 84-86, chmn, 87-91. *Mem:* fel Chem Inst Can. *Res:* Thermodynamics of liquids; transport in membranes; irreversible thermodynamics; electrochemistry. *Mailing Add:* Dept Chem Univ Western Ont London ON N6A 5B7 Can

LORIMER, NANCY L, b Mishawaka, Ind, Feb 8, 47; m 72; c 3. INSECT GENETICS. *Educ:* Ind Univ, AB, 69; Univ Notre Dame, PhD(biol), 75. *Prof Exp:* Fel genetic control, Int Centre Insect Ecol & Physiol, 74-75; RES ENTOMOLOGIST, NCENT FOREST EXP STA, FOREST SERV, USDA, 75-; ADJ ASST PROF, DEPT ENTOMOL, FISH WILDLIFE, UNIV MINN, 79- *Concurrent Pos:* Consult, WHO, 73; assoc ed, Am Midland Nat, 80- *Mem:* Entom Soc Am; Asn Women Sci; Genetics Soc Am; Sigma Xi. *Res:* Assessment of genetic variation in forest insect populations and how these variations interact with other factors to influence population dynamics. *Mailing Add:* 1433 Raymond Ave St Paul MN 55108

LORINCZ, ALLAN LEVENTE, b Chicago, Ill, Oct 31, 24; m 52; c 3. DERMATOLOGY. *Educ:* Univ Chicago, SB, 45, MD, 47. *Prof Exp:* Res fel dermat, Cancer Clin, 50-51, from instr to assoc prof, 51-67, PROF DERMAT, UNIV CHICAGO, 67- *Concurrent Pos:* Mem dermat training grants comt, USPHS, 61-64; mem comt cutaneous syst, Div Med Sci, Nat Res Coun, 62-65; nat consult to Surgeon Gen, US Air Force, 62-; mem dermat adv comt, Food & Drug Admin, 71-72. *Mem:* Soc Invest Dermat; Soc Exp Biol & Med; Am Soc Dermatopath; Am Dermat Asn; Am Fedn Clin Res; Sigma Xi. *Res:* Psoriasis; cutaneous fungus infections; biochemistry and physiology of the skin, especially melanin chemistry and sebaceous gland control by endocrine factors; immunology. *Mailing Add:* 9905 S Kilbourn Ave Oak Lawn IL 60453

LORINCZ, ANDREW ENDRE, b Chicago, Ill, May 17, 26; m 65. PEDIATRICS, BIOCHEMISTRY. *Educ:* Univ Chicago, PhB, 48, BS, 50, MD, 52. *Prof Exp:* From intern to jr asst resident pediat, Univ Chicago Clin, 52-54, jr asst resident fel, Rosenthal Clin, 54-55, instr, Sch Med, 56-59; from asst prof to assoc prof, Sch Med, Univ Fla, 59-68; assoc prof biochem & dir Ctr Develop & Learning Disorders, Med Ctr, 68-80, PROF PEDIAT, MED CTR, 68-, PROF, SCH PUB HEALTH, UNIV ALA, BIRMINGHAM, 84- *Concurrent Pos:* Res fel, Univ Chicago, 54-55 & Arthritis & Rheumatism Found res fel, 55-58; instr, La Rabida Inst, 57-59; sci adv comt, Nat Tay-Sachs & Allied Dis Asn, 79-; mem, Nat Coalition on Prev Mental Retardation, 85- *Mem:* Am Chem Soc; Soc Pediat Res; fel Am Acad Pediat; Soc Invest Dermat; fel Am Acad Cerebral Palsy & Develop Med. *Res:* Heritable disorders of connective tissue acid mucopolysaccharides; inborn errors of metabolism; mental retardation; biophysical cytochemistry. *Mailing Add:* BMSB Rm 124 Univ Ala Birmingham Univ Sta CBB Birmingham AL 35294

LORING, ARTHUR PAUL, b New York, NY, May 22, 36; m 63; c 3. GEOLOGY, ENVIRONMENTAL GEOLOGY. *Educ:* Columbia Univ, AB, 58; Pa State Univ, MS, 61; NY Univ, PhD(geol), 66. *Prof Exp:* Lectr geol, Brooklyn Col, 62-65; instr, 66-67; asst prof, Upsala Col, 67; asst prof, 67-73, ASSOC PROF & COORDR GEOL, YORK COL, NY, 73 - *Concurrent Pos:* Consult geol & environ, Rock Soil Water Int, Inc, 85 - *Mem:* Fel Geol Soc Am; Am Soc Photogram; Asn Eng Geol; Sigma Xi. *Res:* General geologic field mapping in areas of folded and faulted sediments; environmental geology on ground water hydrology. *Mailing Add:* Dept Geol York Col Jamaica NY 11451

LORING, BLAKE M(ARSHALL), b Belmont, NH, Mar 21, 14; m 74; c 3. PHYSICAL METALLURGY. *Educ:* Mass Inst Technol, SB, 37, ScD(metall), 40; George Washington Univ, MA, 45. *Prof Exp:* Asst x-ray metallog, Mass Inst Technol, 37-40; chief nonferrous br, US Naval Res Lab, 40-50; tech officer, US Naval Ord Lab, 51-53; CONSULT METALL ENGR, 53- *Concurrent Pos:* From instr to assoc prof, Univ Md, 47-53; US tech adv, First Geneva Conf Peaceful Uses Atomic Energy, 55; Chmn Conserv Comn, Belmont, NH. *Mem:* Am Soc Metals; Sigma Xi; Am Inst Mining, Metall & Petrol Engrs. *Res:* Metal processing; alloy and copper steels; high strength aluminum and bronze alloys for special Navy use. *Mailing Add:* 25889 Lancia St Moreno Valley CA 92388

LORING, DAVID WILLIAM, b Richmond, Ind, July 13, 56; m 88; c 1. NEUROPSYCHOLOGY. *Educ:* Wittenberg Univ, BA, 78; Univ Houston, MA, 80, PhD(clin neuropsychol), 82. *Prof Exp:* Fel, Baylor Col Med, 82-83; res assoc, Univ Tex Med Br, 83-84, instr, 84-85; asst prof, 85-89, ASSOC PROF, DEPT NEUROL, MED COL GA, 89- *Mem:* AAAS; NY Acad Sci; Int Neuropsychol Soc; Am Psychol Asn; Soc Psychol Res; Soc Philos & Psychol. *Res:* Electrophysiological measures of human hippocampus; neurochemical manipulation of human hippocampal responses; memory function in patients with mesial temporal lobe damage. *Mailing Add:* Dept Neurol Sect Behav Neurol Med Col Ga Augusta GA 30912-3275

LORING, DOUGLAS HOWARD, b Concord, NH, July 25, 34; Can citizen; m 61; c 3. MARINE GEOCHEMISTRY. *Educ:* Acadia Univ, BSc, 54, MSc, 56; Univ Manchester, PhD(geochem), 60. *Prof Exp:* Tech officer, Geol Surv Can, 54-55; res fel geochem, Univ Manchester, 57-60; RES SCIENTIST, BEDFORD INST, 60- *Concurrent Pos:* Spec lectr, Dalhousie Univ, 62-68. *Mem:* Mineral Asn Can; fel Geol Asn Can; Geochem Soc. *Res:* Geochemistry of ancient and modern marine sediments; marine geology of the Gulf of St Lawrence. *Mailing Add:* Atlantic Oceanog Lab Bedford Inst Box 1006 Dartmouth NS B2Y 4A2 Can

LORING, ROGER FREDERIC, b Berkeley, Calif, Sept 14, 58; m 90. NONEQUILIBRIUM STATISTICAL MECHANICS, THEORY OF MOLECULAR SPECTROSCOPY. *Educ:* Univ Calif, Davis, BS, 80; Stanford Univ, PhD(phys chem), 84. *Prof Exp:* Res assoc chem, Univ Rochester, 84-87; ASST PROF CHEM, CORNELL UNIV, 87- *Res:* Dynamics of molecular electronic and vibrational excited states in condensed phases; solvation effects in electronic spectroscopy; theory of nonlinear spectroscopy; structure and dynamics of polymer fluids. *Mailing Add:* Baker Lab Cornell Univ Ithaca NY 14853

LORING, STEPHEN H, b Boston, Mass, July 9, 46. PHYSIOLOGY. *Educ:* Amherst Col, BS, 68; Dartmouth Med Sch, BMS, 70; Harvard Med Sch, MD, 73. *Prof Exp:* Intern med, Univ Hosp, Boston, 73-74; asst prof, 74-85, ASSOC PROF PHYSIOL, DEPT ENVIRON HEALTH, HARVARD SCH PUB HEALTH, 85- *Mem:* Am Phys Soc; Am Thoracic Soc. *Mailing Add:* Dept Environ Health Harvard Sch Pub Health 665 Huntington Ave Boston MA 02115

LORING, WILLIAM BACHELLER, b Haileybury, Ont, Mar 4, 15; m 45; c 2. ECONOMIC GEOLOGY. *Educ:* Mich Col Min, BS, 40; Univ Ariz, MS, 47, PhD, 59. *Prof Exp:* Field geologist, Noranda Mines Co, Can, 41-42; inspector, US Dept Eng, 42-43; field engr, US Bur Mines, Mich, 43-44; party chief, Nfld Geol Surv, 44; party chief, Mining Geophys Co, Can, 44-45; mine mgr, Discovery Yellowknife Gold Mine, 46; geologist, Great Northern Explor Co, Ariz, 46-48; geologist, Eagle-Picher Mining & Smelting Co, 49-55; chief geologist, Big Indian Dist, Hidden Splendor Mining Co, 55-62; staff geologist, Atlas Minerals, 62-66; mine geologist, US Smelting, Ref & Mining Co, NMex, 66-67; dist geologist, Cities Serv Minerals Corp, Wyo, 67-71, staff geologist, 71-78; consult, 78-82; RETIRED. *Mem:* Am Inst Mining, Metall & Petrol Eng; Soc Econ Geol; Can Inst Mining & Metall; Int Asn Genesis Ore Deposits. *Res:* Ore deposits, especially controlling structures and surface indications. *Mailing Add:* 2058 11th St Douglas AZ 85607

LORING, WILLIAM ELLSWORTH, pathology; deceased, see previous edition for last biography

LORIO, PETER LEONCE, JR, b New Orleans, La, Apr 10, 27; m 57; c 6. FOREST SOILS, TREE PHYSIOLOGY. *Educ:* La State Univ, BS, 53; Duke Univ, MF, 54; Iowa State Univ, PhD(forestry-soils), 62. *Prof Exp:* Soil scientist, Standard Fruit & Steamship Co, 54-58, chief soil scientist, 58-59; soil scientist, 62-68, prin soil scientist, 68-76, SUPVRY SOIL SCIENTIST, FOREST INSECT RES PROJ, SOUTHERN FOREST EXP STA, US FOREST SERV, 76- *Honors & Awards:* Superior Serv Award, USDA, 84. *Mem:* Am Soc Agron; Int Soc Trop Foresters; Sigma Xi; Soil Sci Soc Am; Int Soc Soil Sci; Soc Am Foresters. *Res:* Soil, tree, and stand factors affecting pine susceptibility to bark beetles; tree physiology; soil water; tree rooting; stand composition, age, density. *Mailing Add:* 2500 Shreveport Hwy Pineville LA 71360

LORRAIN, PAUL, b Montreal, Que, Sept 8, 16; m 44; c 4. SPACE MAGNETOHYDRODYNAMICS. *Educ:* Univ Ottawa, BA, 37; McGill Univ, BSc, 40, MSc, 41, PhD(physics), 47. *Prof Exp:* Lectr physics, Sir George Williams Col, 42-43, Univ Laval, 43-46 & Inst Physics, Univ Montreal, 46; res assoc, Lab Nuclear Studies, Cornell Univ, 47-49; head dept, 57-66, PROF PHYSICS, UNIV MONTREAL, 49- *Concurrent Pos:* Vis prof fac sci, Univ Grenoble, France, 61-62 & Univ Madrid, 68-69; mem, Nat Res Coun, 60-66; vis fel, Oxford Univ, 81; vis prof, six Chinese univs, 85, Univ Murcia, 86-88; visitor, Inst Physics, Globe Paris, 89 & 90. *Mem:* Royal Soc Can; Am Phys Soc; Can Asn Physicists (pres, 64-65). *Res:* Space magnetohydrodynamics. *Mailing Add:* Dept Earth Sci McGill Univ 3450 University Montreal PQ H3A 2A7 Can

LORSCH, HAROLD G, b Frankfurt, Ger, Aug 25, 19. AIR CONDITIONING. *Educ:* Mass Inst Technol, SM 42, Columbia Univ, PhD(applied mech), 53. *Prof Exp:* Instr, civil eng, New York Univ, 48-50; asst prof, City Col of New York, 50-57; head, stress anal, Curtiss-Wright Corp, 57-60; mgr, aerospace, Gen Elec Co, 60-70; res assoc, solar energy, Univ Pa, 71-73; mgr, energy, Franklin Res Ctr, 74-83; LECTR, MECH ENG, DREXEL UNIV, 83- *Mem:* Am Soc Heating, Refrigerating & Air Conditioning Engrs; Int Solar Energy Soc; Am Solar Energy Soc; Am Soc Civil Eng. *Res:* Primary research in area of energy, energy conservation in buildings through improved insulation and the use of renewable energy sources. *Mailing Add:* Dept Mech Eng Mech Drexel Univ 32 & Chestnut St Philadelphia PA 19104

LORSCHEIDER, FRITZ LOUIS, b Rochester, NY, Aug 27, 39; m 67; c 4. PHYSIOLOGY, ENDOCRINOLOGY. *Educ:* Univ Wis, BS, 63; Mich State Univ, MS, 67, PhD(physiol, endocrinol), 70. *Prof Exp:* Res asst endocrinol, Radioisotope Unit, Med Col Wis, 63-64; from asst prof to assoc prof, 70-80, PROF MED PHYSIOL, FAC MED, UNIV CALGARY, 80- *Concurrent Pos:* NIH fel, Mich State Univ, 70. *Mem:* Am Physiol Soc; Am Soc Biochem & Molecular Biol; Can Physiol Soc; Can Soc Clin Invest; AAAS; Int Soc Oncodevelop Biol & Med; NY Acad Sci. *Res:* Reproductive and fetal physiology; chemistry and physiology of onco-fetal proteins; fetal macroglobulin and steroid metabolism; metabolism of mercury released from dental amalgam fillings. *Mailing Add:* Dept Med Physiol Univ Calgary Fac Med Calgary AB T2N 4N1 Can

LORY, HENRY JAMES, b Baltimore, Md, Mar 3, 36; m 60; c 3. ELECTRICAL ENGINEERING. *Educ:* Johns Hopkins Univ, BES, 58, PhD(elec eng), 63. *Prof Exp:* Asst, Air Res & Develop Command Contract Proj, Johns Hopkins Univ, 57-58, mem res staff, Radiation Lab, 61-63; MEM TECH STAFF, BELL TEL LABS, 63- *Mem:* Sigma Xi. *Res:* Development of Schottky barrier devices, especially analysis of high temperature failure mechanisms; design of linear integrated circuits. *Mailing Add:* 3221 Stoudts Ferry Bridge Rd Riverview Park Reading PA 19605

LOS, MARINUS, b Ridderkerk, Netherlands, Sept 18, 33; m 57; c 4. CHEMISTRY. *Educ:* Univ Edinburgh, BSc, 55, PhD(chem), 57. *Prof Exp:* Res fel, Nat Res Coun Can, 58-60; res chemist, Am Cynamid Co, 60-71, group leader, Organic Synthesis, 71-84, Herbicide Discovery, 84-86, mgr, Crop Protection Chemicals Discovery, 86-88, ASSOC DIR, CROP SCI, AM CYANAMID CO, 88- *Concurrent Pos:* Sr res fel, Dept Pharmacol, Univ Edinburgh, 69-70. *Mem:* Am Chem Soc; Plant Growth Regulator Soc Am; AAAS. *Res:* Aliphatic and aromatic chemistry, especially nitrogen heterocycles; natural products, especially alkaloids and terpenes; screening for mode of action of and field testing of herbicides. *Mailing Add:* Agr Div Am Cyanamid Co PO Box 400 Princeton NJ 08540

LOSCALZO, ANNE GRACE, b New York, NY, Sept 2, 17; m 40; c 1. MICROCHEMISTRY, ANALYTICAL CHEMISTRY. *Educ:* NY Univ, BA, 37, MS, 41, PhD(chem), 43. *Prof Exp:* Asst instr chem, Wash Square Col, NY Univ, 41-43, instr, 43-46; lectr, City Col New York, 53-58; from asst prof to assoc prof, 58-71, PROF CHEM, LONG ISLAND UNIV, 71- *Mem:* Am Chem Soc. *Res:* Educational projects to improve learning abilities of students in chemistry. *Mailing Add:* 3078 38th St Long Island City NY 11103

LOSCALZO, JOSEPH, b Camden, NJ, Oct 26, 51; m 74; c 2. THROMBOSIS, FIBRINOLYSIS. *Educ:* Univ Pa, AB, 72, PhD(biochem), 77, MD, 77. *Prof Exp:* Res fel biochem, Univ Pa, 78; clin fel med, 78-81, clin fel cardiol, 81-83, instr med, 83-85, asst prof med, 85-89, ASSOC PROF MED, HARVARD UNIV, 89- *Concurrent Pos:* Res fel med, Harvard Univ, 81-83; resident physician, Brigham & Women's Hosp, 78-81, chief resident physician, 83-84, assoc physician, 83- & dir, Ctr Res Thrombolysis, 87-; consult, Cardiol Dept, Children's Hosp, Boston, 87-; chief, Cardiol Sect, Va Med Ctr, W Roxbury, 87-; prin investr, NIH grants & Nat Heart, Lung, Blood Inst grants, 88-; asst ed, J Vascular Med & Biol; Res Career Develop Award, NIH, 89-94. *Honors & Awards:* Clin Scientist Award, Am Heart Asn, 83. *Mem:* Fel Am Col Cardiol; Fel Am Col Physicians; AAAS; Am Heart Asn; Biophys Soc; Am Soc Biol Chemists; Am Soc Hemat; Am Soc Clin Invest. *Res:* The relationship of thrombosis to cardiovascular disease; the role of platelets and the fibrinolytic system in thrombotic events; the interactions between thrombosis and otherosclerosis. *Mailing Add:* Dept Med Brigham & Womens Hosp 75 Francis St Boston MA 02115

LOSCHER, ROBERT A, b Philadelphia, Pa, May 2, 30. COMPUTERIZED MANAGEMENT OF INFORMATION, COMPUTER NETWORKING. *Educ:* Univ Pa, BS, 58, MS, 60. *Prof Exp:* Asst prof chem eng, Univ Pa, 59-60; res & develop engr, Selas Corp Am, 60-63 & Dupont Co, Chambers Works, 63-71; dir data processing, 71-75, dir MIS, 75-87, DIR TELECOMMUN, GLASSBORO STATE COL, 87- *Concurrent Pos:* Adj prof, Glassboro State Col & Del County Community Col. *Mem:* Am Inst Chem Engrs; AAAS; Asn Comput Mach; NY Acad Sci. *Res:* Computerization of gas chromatography and IR spectrometry; computer control of chemical manufacturing plants; digital process control; combustion control of lehrs and furnaces; computerization of information management; conversion from batch computer shops to on-line multi-station telecommunication systems. *Mailing Add:* Mem Hall Glassboro State Col Glassboro NJ 08028

LOSCHIAVO, SAMUEL RALPH, b Transcona, Man, June 28, 24; m 50; c 2. INSECT PHYSIOLOGY. *Educ:* Univ Man, BSc, 46, MSc, 50, PhD, 64. *Prof Exp:* Chemist, Man Sugar Co, 48; res scientist, Can Dept Agr, 49-87; RETIRED. *Concurrent Pos:* Hon prof, Univ Man; res assoc, Univ Wisc, 61; vis prof, Univ Hawaii, 76. *Mem:* hon mem Entom Soc Can. *Res:* Biology, behavior and control of insects associated with stored grain and milled cereal products; Canadian and US patents. *Mailing Add:* 112 Linacre Rd Winnipeg MB R3T 3G6 Can

LOSECCO, JOHN M, b New York, NY, Oct 21, 50; m 86; c 1. WEAK INTERACTIONS, COLLIDER PHYSICS. *Educ:* Cooper Union, BS, 72; Harvard, AM, 73, PhD(physics), 76. *Prof Exp:* Res assoc physics, Harvard Univ, 76-79; asst res scientist physics, Univ Mich, 79-81; asst prof physics, Calif Inst Technol, 81-85; PROF PHYSICS, UNIV NOTRE DAME DU LAC, 85- *Honors & Awards:* Bruno Rossi Prize, Am Astron Soc, 89. *Mem:* Am Phys Soc. *Res:* Studying extensions to the standard model of elementary particle; applications of particle physics to astrophysics and cosmology. *Mailing Add:* Physics Dept Univ Notre Dame Notre Dame IN 46556

LOSEE, DAVID LAWRENCE, b Mineola, NY, July 19, 39; m 63; c 2. SOLID STATE PHYSICS, SEMICONDUCTORS. *Educ:* Cornell Univ, BEng, 62, MS, 63; Univ Ill, PhD(solid state physics), 67. *Prof Exp:* RES ASSOC, EASTMAN KODAK CO, 67- *Mem:* Am Phys Soc; Electrochem Soc; Sigma Xi. *Res:* Physics of the noble gas solids; physics of semiconductors and semiconductor devices. *Mailing Add:* 100 W Church St Fairport NY 14450

LOSEE, FERRIL A, b Lehi, Utah, June 5, 28; m 53; c 9. ELECTRICAL ENGINEERING. *Educ:* Univ Utah, BSEE, 53; Univ Southern Calif, MSEE, 57. *Prof Exp:* Elec engr, Hughes Aircraft Co, 53-59 & Aeronutronic Div, Philco Corp, 59-65; prof elec eng & chmn dept, Brigham Young Univ, 65-83; engr, SRS Technologies, 84-89; ENGR, EG&G SP, 89- *Res:* Communication; electronic countermeasures; systems engineering; radar engineering. *Mailing Add:* 3145 Bannock Dr Provo UT 84604

LOSEKAMP, BERNARD FRANCIS, b Cincinnati, Ohio, July 16, 36; wid; c 4. POLYMER CHEMISTRY, ORGANIC CHEMISTRY. *Educ:* Xavier Univ, Ohio, BS, 58, MS, 61; Univ Akron, PhD(polymer chem), 66. *Hon Degrees:* LLD, Univ Akron, 90. *Prof Exp:* Res asst, Wm S Merrell Co, Ohio, 61; res chemist, Inst Polymer Sci, Univ Akron, 61-64; from asst ed to assoc ed, Chem Abstr Serv, 64-69, sr indexer, 69-71, group leader, 71-72, SR ED, CHEM ABSTR SERV, COLUMBUS, OHIO, 72- *Mem:* Am Chem Soc. *Res:* Acenaphthe arsenicals; synthesis and characterization of polymers; polymer nomenclature; thermal polymerization; information science. *Mailing Add:* 2011 Chelsea Rd Columbus OH 43212-1945

LOSEY, GEORGE SPAHR, JR, b Louisville, Ky, June 30, 42; m 67; c 2. MARINE ZOOLOGY, ETHOLOGY. *Educ:* Univ Miami, BS, 64; Scripps Inst Oceanog, Univ Calif, PhD(marine biol), 68. *Prof Exp:* NIH res fel fish behav, Hawaii Inst Marine Biol, 68-70; from asst prof to assoc prof zool, 70-80, assoc dir, 80-90, PROF ZOOL, HAWAII INST MARINE BIOL, UNIV HAWAII, 80-, CHAIR, 90- *Concurrent Pos:* Res fel, Univ Calif, Berkeley, 78; vis researcher, Univ Leiden, 87-88. *Honors & Awards:* Stoye Award, Am Soc Ichthyol & Herpet, 67. *Mem:* Animal Behav Soc; Am Soc Ichthyol & Herpet. *Res:* Ethology and ecology of fish; symbiotic cleaner fish and mimicry in Blenniidae fish; behavioral ecology of herbivorous fish; development of aggression; computerized data acquisition; learning and modification of species-typical behavior. *Mailing Add:* Dept Zool Univ Hawaii 2500 Campus Rd Honolulu HI 96822

LOSEY, GERALD OTIS, b Detroit, Mich, Nov 13, 30; m 63. ALGEBRA. *Educ:* Univ Mich, BS, 52, MS, 53, PhD(math), 58. *Prof Exp:* Res instr math, Princeton Univ, 57-58; instr, Univ Wis, 58-61, asst prof, 61-64; from assoc prof to prof math, 64-84, PROF COMPUT SCI, UNIV MAN, 84- *Mem:* Asn Comput Mach. *Res:* Group theory; ring theory. *Mailing Add:* Dept Computer Sci Univ Man Winnipeg MB R3T 2N2 Can

LOSICK, RICHARD MARC, b Jersey City, NJ, July 27, 43; m 70. MOLECULAR BIOLOGY. *Educ:* Princeton Univ, AB, 65; Mass Inst Technol, PhD(biochem), 69. *Prof Exp:* Harvard Soc fels jr fel biochem, 68-71, asst prof, 71-74, assoc prof biochem, 74-77, PROF BIOL, HARVARD UNIV, 77- *Honors & Awards:* Camille & Henry Dreyfus Award, Camille & Henry Dreyfus Found, 73. *Mem:* Am Soc Biol Chemists; Am Soc Microbiol. *Res:* Bacterial sporulation; regulatory subunits of RNA polymers. *Mailing Add:* Biol Labs Harvard Univ Cambridge MA 02138

LOSIN, EDWARD THOMAS, b Racine, Wis, July 9, 23; m 50; c 2. PHYSICAL ORGANIC CHEMISTRY, ENERGY CONVERSION. *Educ:* Univ Ill, BS, 48; Columbia Univ, AM, 50, PhD(chem), 54. *Prof Exp:* Res assoc, Eng Res Inst, Univ Mich, 54-57; res chemist, Union Carbide Corp, 57-61; chem dept mgr, Isomet Corp, 61-63; sr res scientist, Allis-Chalmers Corp, 63-71, mgr non-metallic mat, 71-73, sr res scientist, 73-88; CHEM CONSULT, 88- *Mem:* Am Chem Soc; The Chem Soc; NY Acad Sci; AAAS; Sigma Xi. *Res:* Reaction mechanisms of organic, stereospecific and free radical gas-phase reactions; electrical insulation materials and systems for various applications; epoxy technology; high temperature fuel gas cleanup; coal combustion of pulverized fuel in entrained-bed combustors; coal-fired cement and iron ore pelletizing systems; coal water slurry fuels technology. *Mailing Add:* 10000 Sheridan Rd Mequon WI 53092

LOSPALLUTO, JOSEPH JOHN, b New York, NY, Nov 8, 25. BIOCHEMISTRY. *Educ:* City Col New York, BS, 45; NY Univ, PhD, 53. *Prof Exp:* Chemist, Fleischman Labs, Stand Brands, Inc, 45-47; instr biochem, NY Univ, 53-58; from asst prof to assoc prof, 58-72, PROF BIOCHEM, UNIV TEX HEALTH SCI CTR, DALLAS, 72- *Concurrent Pos:* Res fel, Arthritis & Rheumatism Found, 56-58; res fel, Whitney Found, 58-61; sr investr, Arthritis Found, 61-66. *Mem:* Fel AAAS; Am Asn Immunol; Am Rheumatism Asn; Am Soc Biol Chemists; Sigma Xi. *Res:* Proteins, especially chemistry and immunology; antibodies and connective tissue chemistry. *Mailing Add:* Dept Biochem Univ Tex Health Med Sch Dallas TX 75235-9038

LOSS, FRANK J, b Homestead, Pa, May 14, 36; m 66; c 2. FRACTURE MECHANICS, FAILURE ANALYSIS. *Educ:* Carnegie Mellon Univ, BS, 58, MS, 59, PhD(mech eng), 61. *Prof Exp:* Engr, Westinghouse Bettis Atomic Power Lab, Pittsburgh, Pa, 61-62; first lieutenant, US Army Corps Engrs, 62-64; head, mech of mat br, US Naval Res Lab, Washington, DC, 64-82; EXEC VPRES, MAT ENG ASSOCS, LANHAM, MD, 82- *Mem:* Am

Nuclear Soc; Am Soc Mech Engrs; Am Soc Testing & Mat. *Res:* Structural technology development; fracture mechanics of structural steels; corrosion fatigue; failure analysis and radiation embrittlement. *Mailing Add:* Mat Eng Assocs 9700-B King Hwy Lanham MD 20706-1837

LOSSING, FREDERICK PETTIT, b Norwich, Ont, Aug 4, 15; m 38; c 3. CHEMICAL PHYSICS, ION CHEMISTRY. *Educ:* Univ Western Ont, BA, 38, MA, 40; McGill Univ, PhD(phys chem), 42. *Prof Exp:* Res chemist, Shawinigan Chem, Ltd, 42-46; prin res officer, Div Chem, Nat Res Coun Can, 46-49 & 77-80, asst dir, 69-77; HON SR RES PROF, DEPT CHEM, UNIV OTTAWA, 80- *Mem:* Fel Royal Soc Can; Royal Astron Soc Can; Am Soc Mass Spectrometry. *Res:* Mass spectrometry; chemical kinetics; photochemistry; heats of formation of organic cations and free radicals; ionization processes. *Mailing Add:* Dept Chem Univ Ottawa Ottawa ON K1N 6N5 Can

LOSSINSKY, ALBERT S, b Passaic, NJ, May 17, 46; m 81; c 4. EXPERIMENTAL NEUROPATHOLOGY, MICRO BLOOD VESSEL PATHOLOGY. *Educ:* Kans Wesleyan Univ, BA, 69; Empire State Univ, MS, 74. *Prof Exp:* Chief res immunolpathol, Johns Hopkins Univ, 73-76; assoc res neuropath, Univ Md Sch Med, 76-79; res scientist pathol, 79-81, res scientist second neurobiol, 81-85, res scientist neurobiol, INST BASIC RES DEVELOP DISABILITIES, 85- *Concurrent Pos:* Asst supvr, Clin Electron Microscopy Lab, Inst Basic Res Develop Disabilities, 83- *Mem:* Am Soc Cell Biol; Soc Neuroscience; Am Asn Neuropathologists; NY Acad Sci. *Res:* Investigation using animal modes of mechanisms of macromolecular and inflammatory cell transport across the altered blood-brain barrier of mammals; electron microscopic analysis of human tissue biopsy material for clinical diagnosis. *Mailing Add:* Dept Pathol Neurobiol NYS Inst Basic Res Develop Disabilities 1050 Forest Hill Rd Staten Island NY 10314

LOSURDO, ANTONIO, b Spadafora, Italy, Jan 1, 43; US citizen. PHYSICAL CHEMISTRY. *Educ:* Syracuse Univ, BA, 65, PhD(chem), 70. *Prof Exp:* Res asst chem, Syracuse Univ, 65-69; NIH fel, Rutgers Univ, New Brunswick, 69-70, instr, 70-71; mem vis fac chem, Syracuse Univ, 71-72; res assoc, Ohio State Univ, 72-73, lectr chem, 73-74; res assoc chem, Clark Univ, 74-75; chief chemist, Cambridge Instrument Co, 75-76; res asst prof chem oceanog, Univ Miami, 77-79, res assoc prof, 79-81; chief chemist & gas chromatography/mass spectrometry & qual assurance/qual control group leader, O'Brien & Gere Engrs, Inc, 82-87. *Mem:* Am Chem Soc; NY Acad Sci; Sigma Xi; AAAS. *Res:* Physical chemistry of multicomponent electrolyte solutions and seawater; thermochemistry and thermodynamics of solutions; solute-solvent and solute-solute interactions; transport properties of hydrophobic electrolytes; electroanalytical chemistry; trace organics analyses; gas chromatography and mass spectrometry of priority pollutants, polychlorinated dibenzo-p-dioxins and dibenzofurans; polynuclear aromatic hydrocarbons in several matrices of environmental concern. *Mailing Add:* GSA Raritan Depot Woodbridge Ave Bldg 20A Annex Edison NJ 08837

LOTAN, JAMES, b Mich, Mar 20, 31; m 51; c 5. SILVICULTURE, FIRE ECOLOGY. *Educ:* La State Univ, BSF, 59; Univ Mich, MF, 61, PhD, 70. *Prof Exp:* Forestry technician, Southern Forest Exp Sta, US Forest Serv, La, 57-59, fire control, Deerlodge Nat Forest, Mont, 59; asst forest res, Univ Mich, 60; res forester, 61-65, proj leader forest sci res, 65-74, prog mgr, Fire Mgt Res & Develop Prog, 74-79, Northern Forest Fire Lab, 79-84, res forester, Forestry Sci Lab, Forest Serv, USDA, 84-87; ADJ PROF, DEPT FOREST RESOURCES, UNIV IDAHO, MOSCOW 87- *Mem:* Am Forestry Asn; Soc Am Foresters. *Res:* Silviculture and ecology of Pinus contorta, Pinus ponderosa & Pseudotsuga menziesli; effects of fire on forests and rangelands of the northern Rocky Mountains; fire management RD & A program; fire effects research and development program. *Mailing Add:* 1550 Mid Burnt Fork Rd Stevensville MT 59870

LOTAN, REUBEN, b Sumarkand, USSR, Mar 19, 46. BIOCHEMISTRY. *Educ:* Tel Aviv Univ, MD, 71; Weizmann Inst, PhD(physics), 76. *Prof Exp:* Assoc prof biophys, Weizmann Inst, 80-84; PROF & CHMN, DEPT TUMOR BIOL, UNIV TEX, M D ANDERSON CANCER CTR, 84- *Concurrent Pos:* Vis asst prof develop biol, Univ Calif, Irvine, 78-80. *Mem:* Am Asn Cancer Res; Am Soc Cell Biol; Soc Complex Carbohydrates. *Mailing Add:* Tumor Biol Dept Univ Tex M D Anderson Cancer Ctr 1515 Holcombe Blvd Houston TX 77030

LOTH, JOHN LODEWYK, b Hague, Neth, Sept 14, 33; c 3. AERODYNAMICS. *Educ:* Univ Toronto, BASc, 57, MASc, 58, PhD(mech eng), 62. *Prof Exp:* French Govt fel aeronaut eng, Nat Ctr Sci Res, Ministry Ed, France, 58-59; lectr mech eng, Univ Toronto, 60-62; asst prof aeronaut eng, Univ Ill, Urbana, 62-67; assoc prof, 67-71, PROF AEROSPACE ENG, WVA UNIV, 71- *Concurrent Pos:* Consult, Ellard Wilson Assocs, Ont, 57-61, Air Force & ARO Inc, 63-66, Off Naval Res, 68-72 & Dept Energy, 73-; pres, Dynamic Flow Inc, 72-; prin investr, Lockheed Ga & US Dept Energy, chmn Allegheney sect, Am Inst Aeronaut & Astronaut. *Mem:* Assoc fel Am Inst Aeronaut & Astronaut; Sigma Xi; Am Soc Engr Educ. *Res:* Low speed aerodynamics; aerodynamic mixing and supersonics; combustion; steel mill coating and cooling processes. *Mailing Add:* Dept Aerospace Eng Eng Sci Bldg WVa Univ Morgantown WV 26506

LOTHSTEIN, LEONARD, b Newark, NJ, Aug 21, 54. PHARMACOLOGY. *Educ:* Bowdin Col, BA, 76; Vanderbilt Univ, PhD(molecular biol), 83. *Prof Exp:* Res assoc, Sloan Kettering Cancer Inst, 82-83; res assoc & postdoctoral fel, Albert Einstein Col Med, 83-88; ASST PROF, DEPT PHARMACOL, COL MED, UNIV TENN, 88- *Mem:* Am Asn Cancer Res; NY Acad Sci; Sigma Xi; Am Soc Cell Biol. *Mailing Add:* Dept Pharmacol Col Med Univ Tenn Memphis TN 38163

LOTLIKAR, PRABHAKAR DATTARAM, b Shirali, India, May 21, 28; US citizen; m 60; c 1. BIOCHEMISTRY, PHARMACOLOGY. *Educ:* Univ Bombay, BS, 50, MS, 54; Ore State Univ, PhD(biochem, pharmacol, bact), 60. *Prof Exp:* Asst chemist, Raptakos Brett & Co, Ltd, India, 50-55; proj assoc, McArdle Lab Cancer Res, Univ Wis, 63-65, instr, 65-66; res instr, 67-68, asst prof, 68-75, ASSOC PROF BIOCHEM, FELS RES INST, SCH MED, TEMPLE UNIV, 75-, INVESTR, 67- *Concurrent Pos:* Res fel oncol, McArdle Lab Cancer Res, Univ Wis, 60-63; res career develop award, USPHS, Nat Cancer Inst, 69-73. *Mem:* AAAS; Am Chem Soc; Am Asn Cancer Res; Am Soc Biol Chem; NY Acad Sci; Sigma Xi; Soc Exp Biol & Med; Biohem Soc (London). *Res:* Mechanisms of chemical carcinogenesis. *Mailing Add:* Fels Res Inst Temple Univ Sch Med Philadelphia PA 19140

LOTRICH, VICTOR ARTHUR, b Pueblo, Colo, July 10, 34; m 55; c 3. POPULATION ECOLOGY. *Educ:* Northern Colo Univ, BA, 56, MA, 60; Univ Ky, PhD(biol), 69. *Prof Exp:* ASSOC PROF ECOL, UNIV DEL, 69- *Res:* Population dynamics of tide marsh fish and tide marsh estuarine interactions. *Mailing Add:* Dept Life Scis Univ Del Newark DE 19716

LOTSPEICH, FREDERICK BENJAMIN, b Konawa, Okla, Jan 10, 14; m 48; c 4. ENVIRONMENTAL LAND CLASSIFICATION, GEOCHEMISTRY OF BIOTIC SYSTEMS. *Educ:* Wash State Univ, BS, 50, MS, 52, PhD(soil sci), 56. *Prof Exp:* Soil scientist, US Geol Surv, 54-57, Agr Res Serv, USDA, 57-66 & Environ Protection Agency, 66-79; VOL RES PHYS SCIENTIST, NORTHWEST FOREST & RANGE EXP LAB, US FOREST SERV, 79- *Honors & Awards:* Boggess Award, Am Water Resources Asn, 80. *Mem:* AAAS; Ecol Soc Am. *Res:* Geochemistry of soils and mineral exploration; physical properties of soils and soil compaction; geohydrology of groundwater, sediments and water quality; environmental land systems classification; geochemistry of nutrient supply and productivity of fresh water fishery. *Mailing Add:* 2006 Sterling Creek Rd Jacksonville OR 97530

LOTSPEICH, FREDERICK JACKSON, b Keyser, WVa, Mar 12, 25; m 48; c 1. BIO-ORGANIC CHEMISTRY. *Educ:* WVa Univ, BS, 48, MS, 51; Purdue Univ, PhD(chem), 55. *Prof Exp:* Res chemist, E I du Pont de Nemours & Co, 48-50; asst org chem, WVa Univ, 50-52; asst, Purdue Univ, 52-53; asst prof chem, Simpson Col, 54-56; from asst prof to prof biochem, Med Ctr, WVa Univ, 66-78; PROF BIOCHEM, MED SCH, MARSHALL UNIV, 78- *Mem:* Am Chem Soc; Sigma Xi; Am Asn Cancer Res; Am Soc Biochem & Molecular Biol. *Res:* The effect of B-carotene & vitamin A on carcinogenesis. *Mailing Add:* 2139 Easlow Blvd Huntington WV 25701

LOTSPEICH, JAMES FULTON, b Cincinnati, Ohio, Oct 22, 22; m 60. OPTICS. *Educ:* Princeton Univ, BA, 43; Univ Cincinnati, MS, 49; Columbia Univ, PhD(physics), 58. *Prof Exp:* Lab instr gen physics, Univ Cincinnati, 47-48; asst, Columbia Univ, 51-56; RES PHYSICIST, LABS, HUGHES AIRCRAFT CO, 56- *Mem:* AAAS; Am Phys Soc; Sigma Xi; NY Acad Sci; fel Optical Soc Am. *Res:* Microwave spectroscopy and molecular structure; electrooptic techniques; applied laser technology; photodetection techniques; integrated optics. *Mailing Add:* 25346 Malibu Rd Malibu CA 90265

LOTT, FRED WILBUR, JR, b Ohio, Oct 8, 17; m 41; c 3. MATHEMATICS, MATHEMATICAL STATISTICS. *Educ:* Cedarville Col, AB, 39; Univ Mich, MA, 46, PhD(math), 55. *Prof Exp:* From asst prof to prof, 49-84, asst vpres acad affairs, 71-84, EMER PROF MATH, UNIV NORTHERN IOWA, 84- *Concurrent Pos:* Opers analyst, US Air Force, 55-64. *Mem:* Am Math Soc; Math Asn Am; Inst Math Statist; Sigma Xi. *Mailing Add:* 1934 Merner Ave Cedar Falls IA 50613-3556

LOTT, FRED WILBUR, III, b San Mateo, Calif, Aug 21, 43; m 84. ELEMENTARY PARTICLE PHYSICS. *Educ:* Carleton Col, BA, 64; Univ Calif-Berkeley, MA, 66, PhD(physics), 78. *Prof Exp:* Comput programmer, ATC Med Technol, Inc, 79-80; instr physics, San Mateo Col, 80-81; asst prof, Wilkes Col, 82-84; asst prof, Weber State Col, 84-86; asst prof, Fla A & M Univ, 86-87; index ed, Munic Code Corp, 89-90; ASST PROF, JACKSON STATE UNIV, 90- *Mem:* Am Phys Soc; Am Asn Physics Teachers. *Res:* The interface between elementary particle physics and general relativity. *Mailing Add:* 809 N State St No 401 Jackson MS 39202

LOTT, JAMES ROBERT, b Houston, Tex, Jan 16, 24; m 42; c 4. PHYSIOLOGY, BIOPHYSICS. *Educ:* Univ Tex, BA, 49, MA, 51, PhD(physiol, bact), 56. *Prof Exp:* Med bacteriologist, Brackenridge Hosp, Austin, Tex, 55; lectr zool, Univ Tex, 55-56, res scientist, Radiobiol Lab, Balcones Res Inst, 56; instr physiol, Sch Med, Emory Univ, 56-57; from asst prof to assoc prof, 57-64, PROF BIOL, NTEX STATE UNIV, 64- *Concurrent Pos:* Sr res investr, AEC, 58-; NSF grant, 63-64. *Mem:* Am Physiol Soc; Alcohol Res Soc; Radiation Res Soc; Int Soc Biometeorol; Int Soc Bioelec; Am Col Sports Med; Am Coun Alcohol Res. *Res:* Neurophysiology; effects of electric fields on the nervous system; effects of electric fields on cancer growth; endocrinology; effects of ethanol on the response to stress; vasospasms in coronary arteries; effects of exertion on the heart using telemetric methods; effects of alcohol on the nervous system. *Mailing Add:* 1907 Locksley Lane Denton TX 76201

LOTT, JAMES STEWART, b Sarnia, Ont, Apr 10, 20; m 50; c 4. MEDICINE, RADIOLOGY. *Educ:* Univ Western Ont, BA, 43, MD, 46; Royal Col Physicians & Surgeons, dipl med radiother, 52 & specialist therapeut radiol, 54. *Prof Exp:* Instr radiol, Univ Western Ont, 52-62, assoc prof radiother & actg head dept, 62-63; assoc prof radiol, Sch Med, Johns Hopkins Univ & head div radiother, Hosp, 64-71; dir, Ont Cancer Found, Kingston Clin, 71-85; RETIRED. *Concurrent Pos:* Fel histol, Univ Western Ont, 46-47, fel path, 47-48, fel radiol, 48-49; Can Cancer Soc fel radiother, 50-51; Brit Empire Cancer Campaign exchange fel, 51-52; consult, Westminster Vet Hosp, 56-63 & St Joseph's Hosp, London, Can, 56-63; consult & radiologist-in-chg ther, Hackley Hosp, Muskegon, Mich, 63-64; prof therapeut radiol & chmn dept, Queen's Univ, Ont, 71- *Mem:* AMA; Can Med Asn; Can Asn Radiol. *Res:* Radiobiology applied to radiotherapy; clinical radiotherapy applied to cancer. *Mailing Add:* 142 Seaforth Kingston ON K7L 2V7 Can

LOTT, JOHN ALFRED, b Ger, Oct 30, 36; US citizen; m 63; c 1. CLINICAL CHEMISTRY. *Educ:* Rutgers Univ, BS, 59, MS, 61, PhD(anal chem), 65. *Prof Exp:* Instr chem, Rutgers Univ, 64-65; asst prof, Flint Col, Univ Mich, 65-68; from asst prof to assoc prof, 68-79, PROF PATH, OHIO STATE UNIV, 79- & DIR, CLINIC CHEM LAB, OHIO STATE UNIV HOSP, 79- *Concurrent Pos:* Expert witness, clin chem. *Honors & Awards:* Katchman Award, Am Asn Clin Chem, 79, Outstanding Contrib Educ Award, 87; Presidential Award, Nat Acad Biochemist, 83. *Mem:* Am Assoc Clin Chem; Nat Acad Clin Biochemists, (treas, 78-79, pres, 81-82); Am Chem Soc; Asn Clin Scientists. *Res:* Instrumentation; methodology development; enzymology; specific-ion electrodes. *Mailing Add:* Starling Loving M-368 Ohio State Univ Med Ctr Columbus OH 43210-1240

LOTT, JOHN NORMAN ARTHUR, b Summerland, BC, Jan 20, 43; m 66; c 2. PLANT ANATOMY, PLANT PHYSIOLOGY. *Educ:* Univ BC, BSc, 65; Univ Calif, Davis, MSc, 67, PhD(bot), 69. *Prof Exp:* Res asst bot, Univ Calif, Davis, 65-69; from asst prof to assoc prof, 69-81, PROF BIOL, MCMASTER UNIV, 81- *Mem:* Can Bot Asn; Am Soc Plant Physiol; Bot Soc Am; Micros Soc Can; Electron Micros Soc Am. *Res:* Ultrastructure and physiological studies of developing and germinating seeds, with special emphasis on protein bodies; mineral nutrient storage in seeds. *Mailing Add:* Dept Biol McMaster Univ Hamilton ON L8S 4K1 Can

LOTT, LAYMAN AUSTIN, b Ft Collins, Colo, Sept 21, 37; m 58; c 4. PHYSICS. *Educ:* Colo State Univ, BS, 59, MS, 61; Iowa State Univ, PhD(physics), 65. *Prof Exp:* Res physicist, Rocky Flats Div, Dow Chem USA, 65-71, sr res physicist, 71-73; SR ENG SPECIALIST, IDAHO NAT ENG LAB, 73- *Mem:* Am Phys Soc; Am Soc Nondestructive Test; Sigma Xi. *Res:* Solid state physics; physical properties of materials; nondestructive testing; development of advanced nondestructive testing methods. *Mailing Add:* 701 9th St Idaho Falls ID 83404

LOTT, PETER F, b Berlin, Germany, Mar 26, 27; nat US; m 56; c 2. PHYSICAL CHEMISTRY, ANALYTICAL CHEMISTRY. *Educ:* St Lawrence Univ, BS, 49, MS, 50; Univ Conn, PhD(chem), 56. *Prof Exp:* Asst instr chem, Univ Conn, 54-56; res chemist, E I du Pont de Nemours & Co, 56; assoc prof, Univ Mo, 56-59; chemist, Pure Carbon Co, 59-60; assoc prof chem, St John's Univ, NY, 60-64; PROF CHEM, UNIV MO-KANSAS CITY, 64- *Honors & Awards:* Benedetti-Pichler Award, Am Microchem Soc. *Mem:* Am Chem Soc; Am Microchem Soc; Royal Soc Chem. *Res:* Analytical methods development; trace and instrumental analysis; chemical kinetics; chemical microscopy; physical measurements; organic reagents; forensic chemistry; asbestos analysis. *Mailing Add:* Dept Chem Univ Mo Kansas City MO 64110

LOTT, SAM HOUSTON, JR, b New Orleans, La, Sept 22, 36; m 59. PHYSICS, HEALTH PHYSICS. *Educ:* La State Univ, BS, 58; Vanderbilt Univ, MS, 60, PhD(physics), 65. *Prof Exp:* Res assoc physics, 65-66, dir radiation safety off, Vanderbilt Univ, 66-76; consult, Nat Cancer Inst, 73-76; head, Health Physics Dept, King Faisal Specialist Hosp & Res Ctr, 76-85. *Concurrent Pos:* Consult, 66-76. *Mem:* Am Phys Soc; Health Phys Soc; Am Asn Physicists Med. *Res:* Three-color photometric study of variable stars; Zeeman & Faraday effects in high pulsed magnetic fields; calibration techniques for diagnostic and therapeutic machines. *Mailing Add:* 45 Vaugn Gap Rd Nashville TN 37205

LOTTES, P(AUL) A(LBERT), b Wilkinsburg, Pa, Aug 2, 26; m 47; c 3. MECHANICAL ENGINEERING. *Educ:* Purdue Univ, PhD(mech eng), 50. *Prof Exp:* Assoc mech engr, 50-60, SR MECH ENGR, ARGONNE NAT LAB, 60- *Mem:* Fel Am Soc Mech Engrs; fel Am Nuclear Soc. *Res:* Heat transfer and pressure drop in boiling; nuclear reactor safety. *Mailing Add:* Argonne Nat Lab RE 207 9700 S Cass Ave Argonne IL 60439

LOTTI, VICTOR J, b Trenton, NJ, Jan 6, 38. NEUROPHARMACOLOGY. *Educ:* Univ Conn, BS, 59; Univ Mo, MS, 61; Univ Calif, Los Angeles, PhD(pharmacol), 65. *Prof Exp:* Sr res pharmacologist, Dept Pharmacol, Merck, Sharp & Dohme Res Labs, 67-69, res fel neuropharmacol, 69-73, dir neuropsychopharmacol, 73-75, sr dir res coordr, 75-77, sr dir, Dept Pharmacol, Chibret, France, 77-79, SR SCIENTIST, DEPT PHARMACOL, MERCK, SHARP & DOHME RES LABS, 79- *Mem:* Am Soc Pharmacol & Exp Therapeut; Am Chem Soc. *Mailing Add:* Dept Pharmacol Merck Sharp & Dohme Res Labs West Point PA 19486

LOTTMAN, ROBERT P(OWELL), b Brooklyn, NY, Sept 24, 33; m 56. CIVIL ENGINEERING. *Educ:* Polytech Inst Brooklyn, BCE, 54; Purdue Univ, MSCE, 56; Ohio State Univ, PhD, 65. *Prof Exp:* Proj engr, Struct Appln Sect, Grumman Aircraft Eng Corp, 56-57; supvr, Asphalt Tech Serv Lab, Standard Oil Co, Ohio, 57-59; res supvr hwy mat, Transp Eng Ctr, Ohio State Univ, 59-65, asst prof civil eng, 65-66; PROF CIVIL ENG, UNIV IDAHO, 66- *Concurrent Pos:* Instnl & comt mem, Hwy Res Bd, Nat Acad Sci-Nat Res Coun, 60- *Mem:* Asn Asphalt Paving Technol; Am Soc Testing & Mat. *Res:* Study and evaluation of physical and chemical properties of construction materials to determine mechanical behavior under various loading and environmental conditions. *Mailing Add:* Dept Civil Eng Univ Idaho Moscow ID 83843

LOTTS, ADOLPHUS LLOYD, b Buchanan, Va, June 10, 34; m 54; c 4. NUCLEAR SAFETY, ISOTOPE SEPARATION. *Educ:* Va Polytech Inst, BS, 55, MS, 57. *Prof Exp:* Instr metall eng, Va Polytech Inst, 56-57; assoc scientist, Atomic Energy Div, Babcock & Wilcox Co, 58-59; assoc metallurgist, Metals & Ceramics Div, Oak Ridge Nat Lab, 59-61, group leader Fuel Cycle Technol, 61-66, head Fuel Cycle Technol Oper, Metals & Ceramics Div, 66-70, assoc dir Gas-cooled Reactor & Thorium Utilization Progs, 70-78, dir Nuclear Waste Prog, 78-81, dir, Nuclear Regulatory Comn Prog, 81-83, chmn long range planning group, 69-83; dir, Atomic Vapor Laser Isotope Separation Div, Martin Marietta Energy Systs, Inc, 83-90; div dir res reactors, Oak Ridge Nat Lab; RETIRED. *Honors & Awards:* E O Lawrence Mem Award, US Dept Energy, 76. *Mem:* Fel Am Nuclear Soc; Am Soc

Metals. *Res:* Nuclear fuel processing technology; economics and properties of nuclear fuel; materials for reactor systems; radioactive and toxic waste management; nuclear reactor and safety technology; laser isotope separation technology. *Mailing Add:* 801 Chateaugay Rd Knoxville TN 37923

LOTZ, W GREGORY, ENDOCRINOLOGY, THERMAL PHYSIOLOGY. *Educ:* Univ Rochester, PhD(biophysics), 77. *Prof Exp:* CHIEF, ENVIRON PHYSIOL DIV, NAVAL AEROSPACE MED RES LAB, 76- *Res:* Physiological affects of nonionizing radiation. *Mailing Add:* Naval Aerospace Med Res Lab Naval Air Sta Code 23 Pensacola FL 32508-5700

LOTZE, MICHAEL T, b Pasadena, Calif, July 11, 52; m 77; c 4. LYMPHOKINE RESEARCH, T-CELL IMMUNOBIOLOGY. *Educ:* Northwestern Univ, BS, 73, MD 74. *Prof Exp:* Jr med fel surg, Md Anderson Tumor Inst, 75; Intern surg, Univ Rochester, 75-82; med officer, Nat Health Serv Corp, 77-78; SR INVESTR SURG, NAT CANCER INST, 82-; ASST PROF SURG, UNIFORMED SERV UNIV HEALTH SCI, 83- *Mem:* Am Col Surgeons; Am Asn Immunol; Am Soc Clin Oncol; Am Asn Cancer Res; Soc Surg Oncol; Soc Univ Surgeons. *Res:* Tumor immunology and developmental therapeutics; lymphokine function and T-cell immunobiology; surgical treatment of primary and metastatic liver tumors and melanoma. *Mailing Add:* Dept Surg & Pittsburgh Cancer Inst Univ Pittsburgh 497 Scaife Hall Pittsburgh PA 15261

LOTZOVÁ, EVA, b Prague, Czech; US citizen; m. SURGERY. *Educ:* Charles Univ, Czech, MS, 65, PhD(immunol), 68. *Prof Exp:* Postdoctoral fel, Inst Exp Biol & Genetics, Czech Acad Sci, 68; res assoc, Dept Genetics, Albert Einstein Col Med, Yeshiva Univ, NY, 69; res instr, Dept Path, State Univ NY, Buffalo, 69-72, asst prof, Dept Path, Sch Med, 72-73; asst prof, Div Exp Biol, Baylor Col Med, Houston, Tex, 73-76; asst prof & asst immunologist, Dept Develop Therapeut, Univ Tex Syst Cancer Ctr, M D Anderson Hosp & Tumor Inst, 76-78, assoc prof & assoc immunologist, 78-84, prof & immunologist, Dept Gen Surg, 84-88; from asst prof to assoc prof, 76-84, PROF, GRAD SCH BIOMED SCI, UNIV TEX HEALTH SCI CTR, HOUSTON, 84-, PROF & IMMUNOLOGIST, M D ANDERSON CANCER CTR, 88-, FLORENCE MAUDE THOMAS CANCER RES PROF, UNIV TEX SYST CANCER CTR, M D ANDERSON HOSP & TUMOR INST, 88- *Concurrent Pos:* Dir, Lab Immunogenetics, M D Anderson Cancer Ctr, Univ Tex, 78-89, chief, Sect Natural Immunity, Dept Gen Surg, 89-; consult, TNI Pharmaceut, Inc, Tulsa, Okla, 85-; sci adv, Cellular Immunother, Inc, Tucson, Ariz, 88-90. *Mem:* Am Asn Immunologists; Am Asn Cancer Res; Transplantation Soc; Int Soc Exp Hemat; NY Acad Sci; AAAS; Asn Gnotobiotics; Am Soc Hemat; Clin Immunol Soc. *Mailing Add:* Univ Tex Dept Gen Surg M D Anderson Cancer Ctr 1515 Holcombe Blvd Box 18 Houston TX 77030

LOU, ALEX YIH-CHUNG, b Chungking, China, Nov 10, 38; US citizen; m 69; c 2. ENGINEERING MATERIALS, MECHANICS. *Educ:* Nat Taiwan Univ, BS, 60; Purdue Univ MS, 65, PhD(solid mech), 69. *Prof Exp:* Sr engr composites, The Boeing Co, 69-70; res scientist, 70-76, sr res scientist mat, Firestone Tire & Rubber Co, 76-81; PHILLIPS PETROL, 81- *Mem:* Am Soc Mech Engr. *Res:* Characterization and evolution of composite; polymer materials for engineering applications; tire mechanics such as rolling resistance. *Mailing Add:* 2564 Georgetown Dr Bartlesville OK 74006

LOU, DAVID YEONG-SUEI, b Yuncom, China, Nov 12, 37; m 64; c 2. MECHANICAL ENGINEERING. *Educ:* Taiwan Univ, BS, 59; Mass Inst Technol, MS, 63, MechE, 66, ScD(mech eng), 67. *Prof Exp:* Res asst mech eng, Mass Inst Technol, 61-63; thermodyn engr, Jackson & Moreland Consult Co, 63; asst thermionic energy conversion, Mass Inst Technol, 63-64, asst mech eng, 64-65, asst molecular beams, 65-67; from asst prof to prof mech eng, Univ Del, 67-79; PROF MECH ENG & CHMN DEPT, UNIV TEX, ARLINGTON, 79- *Mem:* Am Phys Soc; Am Inst Aeronaut & Astronaut; Am Soc Mech Engrs. *Res:* Solar energy; kinetic theory of gases; molecular beams; thermodynamics; fluid mechanics; heat transfer; biomedical engineering. *Mailing Add:* Dept Mech Eng Univ Tex Arlington TX 76019

LOU, KINGDON, b Stockton, Calif, Aug 3, 22; m 45; c 2. MICROBIOLOGY. *Educ:* Stanford Univ, AB, 52, AM, 56; Am Bd Bioanal, dipl. *Prof Exp:* Dir, immunol dept, res div, Hyland Labs, Baxter, 57-67; sr immunochemist, Res Div, Hoffmann-La Roche, 67-68; dir immunol, Kallestad Labs, 68-69; vpres & dir immunol res, ICL Sci, 70-81; CONSULT, IMMUNOASSAY TECHNOL, INC, 81- *Mem:* Am Soc Microbiol; Am Asn Clin Chemists; NY Acad Sci; AAAS. *Res:* Immunochemical diagnostic reagents; hybridoria and monoclonal antibodies; immunoassays. *Mailing Add:* PO Box 1849 Tustin CA 92681-1849

LOU, PETER LOUIS, b Shanghai, China, Dec 9, 45; US citizen; c 2. MOLECULAR BIOLOGY, OPHTHALMOLOGY. *Educ:* Univ Ottawa, BSc, 67, MD, 74; McMaster Univ, Can, MSc, 70, Univ Toronto, dipl ophthal, 77. *Prof Exp:* Instr biol, McMaster Univ, 67-70; resident ophthal, Univ Toronto, 74-77; INSTR OPHTHAL, HARVARD MED SCH, 79- *Concurrent Pos:* Retina fel, Retina Assoc, Mass Eye & Ear Infirmary, Boston, 77-78; consult, 82- *Mem:* Asn Res Vision & Ophthal; fel Am Acad Ophthal & Otolaryngol; Vitreous Soc; AMA. *Res:* Effect of near ultraviolet light on aphakic retina metabolism; diabetic retinopathy; pathophysiology of vitreous and retina. *Mailing Add:* 75 Blossom Ct Boston MA 02114

LOUCK, JAMES DONALD, b Grand Rapids, Mich, Dec 13, 28; m 60; c 3. MATHEMATICAL PHYSICS. *Educ:* Ala Polytech Inst, BS, 50; Ohio State Univ, MS, 52, PhD(physics), 58. *Prof Exp:* Staff mem, Los Alamos Nat Lab, 58-60; assoc res prof physics, Auburn Univ, 60-63; STAFF MEM, LOS ALAMOS NAT LAB, UNIV CALIF, 63- *Mem:* Am Phys Soc; AAAS; Int Asn Math Physicists. *Res:* Application and development of group theoretical methods in physics. *Mailing Add:* Los Alamos Nat Lab Univ Calif Los Alamos NM 87545

LOUCKS, DANIEL PETER, b Chambersburg, Pa, June 4, 32; m 67; c 2. SYSTEMS ANALYSIS. *Educ:* Pa State Univ, BS, 54; Yale Univ, MS, 55; Cornell Univ, PhD(systs eng, econ), 65. *Prof Exp:* Asst prof water resources eng, Col Eng, Cornell Univ, 65-70, assoc prof environ eng, 70-75, prof environ eng & chmn dept, 76-80, assoc dean res & grad study, 80-81; CONSULT, 81- *Concurrent Pos:* Prin investr, NSF, Environ Protection Agency, Nato, Ford Found, Resources for the Future & US Dept Interior Res Grants, 67-; sem assoc, Columbia Univ, 67-80; res fel, Harvard Univ, 68; consult, UN Develop Prog, WHO, Food & Agr Orgn, NATO, UN & IRBD; economist, World Bank, 72-73; vis prof, Mass Inst Technol, 77-78; res scholar, Int Inst Appl Systs Anal, Austria, 81-82. *Honors & Awards:* Res Award, Am Soc Civil Engrs, 70 & 86. *Mem:* Nat Acad Eng; AAAS; Opers Res Soc Am; Int Mgt Sci; Am Geophys Union; fel Am Soc Civil Engrs; Asn Comput Mach; Int Hydraul Res Asn. *Res:* Applications of operations research to problems in environmental and water resources engineering; public policy analysis; interactive modelling and computer graphics for decision aids. *Mailing Add:* Hollister Hall Cornell Univ Ithaca NY 14853

LOUCKS, ORIE LIPTON, b Minden, Ont, Oct 2, 31; m 55; c 3. BIOLOGY, ECOLOGY. *Educ:* Univ Toronto, BSc, 53, MSc, 55; Univ Wis, PhD(bot), 60. *Prof Exp:* Forest ecologist, Dept Forestry, Can Govt, 55-62; from asst prof to prof bot, Univ Wis-Madison, 62-78; sci dir Inst Ecol, 78-82; dir, Holcomb Res Inst, 83-89; OHIO EMINENT SCHOLAR, MIAMI UNIV, 89- *Concurrent Pos:* Univ Wis rep, State Bd Preserv Sci Areas, 64-78; coordr environ mgt progs, US/Int Biol Prog, Univ Tex, 73; co-chmn, NRC/RSC comt Great Lakes Water Quality Agreement, 84-85; chmn, Nature Conservancy Bd Gov Sect, 90-; pres, Asn Ecosyst Res Centers, 90- *Honors & Awards:* George Mercer Award, Ecol Soc Am, 64. *Mem:* Fel AAAS; Soc Am Foresters; Ecol Soc Am; Am Inst Biol Sci; Am Soc Limnol Oceanog; Int Ecol Asn. *Res:* Forest ecology and ecosystem dynamics; lake ecosystem modeling and analysis; watershed and water quality systems studies; air pollution effects on ecosystems; dynamics and trends in biological divesity; US/China research on global environmental change. *Mailing Add:* Zool Dept Miami Univ Oxford OH 45056

LOUD, ALDEN VICKERY, b Boston, Mass, Apr 6, 25; m 50; c 4. CELL BIOLOGY, BIOPHYSICS. *Educ:* Mass Inst Technol, BS & MS, 51, PhD(biophys), 55. *Prof Exp:* Res assoc, Detroit Inst Cancer Res & asst prof biophys, Col Med, Wayne State Univ, 57-65; asst prof path, Col Physicians & Surgeons, Columbia Univ, 65-68; assoc prof, 68-80, PROF PATH, NY MED COL, 80- *Concurrent Pos:* Res fel med, Mass Gen Hosp, 51-57. *Mem:* Electron Micros Soc Am; Am Soc Cell Biol; Int Soc Stereology; Royal Micros Soc; Am Heart Asn. *Res:* Stereologic morphometry; quantitative electron microscopy and methods of ultrastructure research; correlation of cellular ultrastructure with metabolic function. *Mailing Add:* Dept of Path NY Med Col Valhalla NY 10595

LOUD, OLIVER SCHULE, b Vernal, Utah, Jan 16, 11; m 35; c 2. HISTORY & PHILOSOPHY OF SCIENCE. *Educ:* Harvard Univ, AB, 29; Columbia Univ, AM, 40, EdD, 43. *Prof Exp:* Master, Nichols Sch, NY, 29-32; instr high sch, Ohio, 32-36; teacher gen sci, Sarah Lawrence Col, 36-40; res assoc, Bur Educ Res Sci, Columbia, 39-43; asst prof physics, Antioch Col, 43-44; instr, Ohio State Univ, 44; tech supvr, Tenn Eastman Corp, Tenn, 44-45; from assoc prof to prof phys sci, Antioch Col, 45-78, distinguished univ prof, 78-81; CONSULT, 81- *Concurrent Pos:* Ford Found fel, Harvard Univ, 52-53; mem staff fac develop prog, Great Lakes Cols Asn, Ann Arbor, Mich; mem staff, Wilmington Col, Ohio & Proj Talents, Lebanon Correctional Inst, Lebanon, Ohio. *Res:* Science in general education; suggestions for teaching problems of good land use. *Mailing Add:* 1430 Meadow Lane Yellow Springs OH 45387

LOUD, WARREN SIMMS, b Boston, Mass, Sept 13, 21; m 47; c 3. MATHEMATICS. *Educ:* Mass Inst Technol, SB, 42, PhD(math), 46. *Prof Exp:* Instr math, Mass Inst Technol, 43-47; from asst prof to assoc prof, 47-59, PROF MATH, UNIV MINN, MINNEAPOLIS, 59- *Concurrent Pos:* Res engr, Mass Inst Technol, 45-47, vis fel, 55-56; guest prof, Darmstadt Tech Univ, 64-65; vis prof, Kyoto Univ, Japan, 74-75 & Univ Florence & Univ Trento, Italy, 81-82. *Mem:* AAAS; Am Math Soc; Soc Indust & Appl Math; Math Asn Am. *Res:* Theory of differential equations; numerical methods of solution of differential equations; stationary solutions of Van der Pol's equation with a forcing term; nonlinear mechanics. *Mailing Add:* Sch Math Univ Minn 206 Church St SE Minneapolis MN 55455

LOUDA, SVATA MARY, b Prague, Czech; US citizen. PLANT-ANIMAL INTERACTIONS, PLANT DEMOGRAPHY. *Educ:* Pomona Col, BA, 65; Univ Wash, Seattle, BS, 68; Univ Calif, Santa Barbara, MS, 72; Univ Calif, Riverside & San Diego State Univ, PhD(ecol), 78. *Prof Exp:* Asst economist, Pac Northwest Bell Tel, 65-66 & Syst Develop Corp, 68-69; postdoctoral fel bot, Yale Univ, 79-81; res asst prof, Marine Lab, Beaufort & res scientist, Bot Dept, Durham, Duke Univ, 81-83; asst prof, 83-87, ASSOC PROF BIOL, UNIV NEBR, LINCOLN, 87- *Concurrent Pos:* Sr scientist, Rocky Mountain Biol Lab, 79-90; researcher & prin investr, Cedar Pt Biol Sta, 83-; assoc ed, Oecologia, 90-; coun mem, Am Inst Biol Sci, 91- *Honors & Awards:* George Mercer Award, Ecol Soc Am, 82. *Mem:* Ecol Soc Am; Am Inst Biol Sci; Entom Soc Am; Bot Soc Am; AAAS; Sigma Xi; Soc Ecol Res. *Res:* Plant population; community ecology; interaction of plants with insects; biological control of weeds in a variety of ecosystems in the central and western US. *Mailing Add:* Sch Biol Sci Univ Nebr Lincoln NE 68588-0343

LOUDEN, L RICHARD, b Monroe, Wash, July 8, 33; m 63. GEOCHEMISTRY, SATELLITE COMMUNICATIONS. *Educ:* Univ Wurzburg, PhD(geochem), 63. *Prof Exp:* Assoc prof geochem, Univ Houston, 63-64; geologist, Magnet Cove Barium Corp, 64-65, supvr, X-ray Dept, 65-67, mgr anal sect, 67-69, tech adv, 69-71, spec proj engr, 71-72, develop mgr, Dresser Pollution, Dresser Oilfield Prod Div, 72-73, prod mgr, 73-76, mkt mgr, Dresser-Swaco, 76-78; exec vpres res, eng, construct & mfg, The Analysts Inc, 78-80; vpres, satellite commun, Drilling Info Serv Co, 81-82; PRES, L-R RESOURCE DEVELOP CORP, 80-; VPRES ECCO INC, ENVIRON, 90- *Concurrent Pos:* Co-worker, NASA grant, Univ Houston, 63-64. *Mem:* AAAS; Marine Tech Soc; Clay Minerals Soc; Ger Geol Asn; Nat Oilfield Equip Mfrs & Distribr Soc. *Res:* Organic geochemistry, oceanography, clay mineralogy, and x-ray analysis; new and novel equipment and chemicals for oilwell and other drilling practices; geotechnical services; project management; geophysical analysis. *Mailing Add:* 8011 Highmeadow Houston TX 77063

LOUDON, CATHERINE, b Chanute, Kans, Nov 1, 58; m 84. PHYSIOLOGICAL ECOLOGY, INVERTEBRATE BIOMECHANICS. *Educ:* Brown Univ, ScB, 80; Duke Univ, PhD(zool), 86. *Prof Exp:* Postdoctoral res asst, Univ Minn, 86-88; NSF postdoctoral fel, Cornell Univ, 89-90; asst prof, Ithaca Col, 88-90; POSTDOCTORAL RES ASSOC, UNIV CALIF, BERKELEY, 90- *Mem:* AAAS; Entom Soc Am; Am Soc Zoologists; Sigma Xi. *Res:* Physiology; physiological ecology; invertebrate biomechanics. *Mailing Add:* Dept Biol Integrative Ithaca Col Univ Calif-Berkeley 345 Mulford Hall Berkeley CA 94720

LOUDON, GORDON MARCUS, b Baton Rouge, La, Oct 10, 42; m 64; c 2. BIOCHEMISTRY, ORGANIC CHEMISTRY. *Educ:* La State Univ, Baton Rouge, BS, 64; Univ Calif, Berkeley, PhD(org chem), 68. *Prof Exp:* USPHS fel, Univ Calif, Berkeley, 69-70, lectr biochem, 70; from asst prof to assoc prof chem, Cornell Univ, 70-77; assoc prof, 77-83, PROF MED CHEM, PURDUE UNIV, 83-, ASSOC DEAN SCH PHARM, 87- *Mem:* Am Soc Biol Chemists; Am Chem Soc; AAAS; Peptide Soc; Am Asn Cols Pharm. *Res:* Peptide chemistry; enzyme model systems; bioanalytical methods. *Mailing Add:* Dept Med Chem & Pharmacog Purdue Univ Sch Pharm West Lafayette IN 47907

LOUDON, ROBERT G, b Edinburgh, Scotland, June 27, 25; US citizen; m 55; c 3. INTERNAL MEDICINE. *Educ:* Univ Edinburgh, MB & ChB, 47. *Prof Exp:* House physician gen med, Western Gen Hosp, Edinburgh, Scotland, 47-48; sr house physician tuberc wards, City Hosp, 49-50; asst med officer, Tor-na-Dee Sanatorium, Aberdeen, 50-51; house physician, Chest Hosp, Brompton Hosp, London, Eng, 51-52; clin tutor gen med, Royal Infirmary, Edinburgh, 53-54; staff physician, South-East Kans Tuberc Hosp, Chanute, 56-60, supt, 60-61; from asst prof to assoc prof internal med, Univ Tex Southwestern Med Sch Dallas, 61-69; assoc prof med, Sch Med, George Washington Univ, 69-71; PROF INTERNAL MED, MED CTR & DIR PULMONARY DIS DIV, COL MED, UNIV CINCINNATI, 71- *Concurrent Pos:* Assoc med, Univ Kans, 57-61; staff physician, Woodlawn Hosp, Dallas, 61-69; chief res in respiratory dis, Vet Admin Cent Off, Washington, DC, 69-71. *Mem:* Am Thoracic Soc; AMA. *Res:* Chest diseases; tuberculosis; aerobiology. *Mailing Add:* Dept Med Pulmonary Dis Univ Cincinnati Col Med 231 Bethesda Ave Cincinnati OH 45267

LOUGEAY, RAY LEONARD, b Medford, Ore, Feb 9, 44; m 68. PHYSICAL GEOGRAPHY, REMOTE SENSING. *Educ:* Rutgers Univ, AB, 66; Univ Mich, MS, 69, PhD(phys geog), 71. *Prof Exp:* Lectr phys geog, Univ Mich, 69-70; asst prof, 71-79, ASSOC PROF GEOG & DIR ENVIRON STUDIES, STATE UNIV NY COL GENESEO, 79- *Mem:* Asn Am Geogr; AAAS; Am Meteorol Soc; Am Soc Photogram. *Res:* Remote sensing; applied climatology and environmental modification as a function of radiative energy balances and hydrologic water balances; Alpine periglacial environments. *Mailing Add:* Dept Geog State Univ NY Col at Geneseo Geneseo NY 14454

LOUGH, JOHN WILLIAM, JR, b St Louis, Mo, Apr 2, 43; m 68; c 3. ANATOMY, CELL BIOLOGY. *Educ:* St Louis Univ, BS, 65, MS, 68; Wash Univ, St Louis, PhD(cell biol & anat), 75. *Prof Exp:* Res assoc biol, Mass Inst Technol, 75-77; asst prof anat, 77-83, ASSOC PROF ANAT & CELLULAR BIOL, MED COL WIS, 83- *Mem:* Am Soc Cell Biol; Am Asn Anatomists; Am Heart Asn. *Res:* Muscle differentiation in cell culture; changes in chromosomal proteins during myoblast differentiation. *Mailing Add:* Dept of Anat Med Col Wis 8701 Watertown Plank Rd Milwaukee WI 53226

LOUGHHEED, THOMAS CROSSLEY, b Sherbrooke, Que, Oct 22, 29; m 53; c 3. QUALITY SYSTEMS. *Educ:* Bishop's Univ, BSc, 49; McGill Univ, MSc, 54; Univ London, PhD(microbiol); Imp Col, dipl, 58. *Prof Exp:* Res officer biochem, Can Dept Agr, 53-63; Res Officer Anal Biochem, 63-86, MGR TECH SERV, JOHN LABATT LTD, 86- *Mem:* Am Chem Soc; Micros Soc Can; Royal Micros Soc; Am Asn Cereal Chemists; Am Soc Qual Control. *Res:* Applications of analytical chemistry and microscopy in food science. *Mailing Add:* Dept Tech Serv John Labatt Ltd PO Box 5050 London ON N6A 4M3 Can

LOUGHLIN, BERNARD D, broadcast color tv systems, am stereo systems; deceased, see previous edition for last biography

LOUGHLIN, JAMES FRANCIS, glycerol, ethyl butyl & aceton alcohol; deceased, see previous edition for last biography

LOUGHLIN, THOMAS RICHARD, b Santa Monica, Calif, July 19, 43; m 71; c 2. MARINE MAMMALOGY, BEHAVIORAL ECOLOGY. *Educ:* Univ Calif, Santa Barbara, BA, 72; Humboldt State Univ, MA, 74; Univ Calif, Los Angeles, PhD(biol), 77. *Prof Exp:* MARINE MAMMAL RES SPECIALIST, NAT MARINE FISHERIES SERV, 77- *Concurrent Pos:* Biol consult, TerraScan, Inc, Environ Consults, 72-74; recipient res funds, Univ Calif, 75 & US Marine Mammal Comn, 75-77; vis scientist, Smithsonian Inst & US Dept Com alt mem, US Endangered Species Sci Authority, 77-80; Assoc Prof, Ore State Univ, 85- *Mem:* AAAS; Am Asn Biol Sci; Am Soc Mammalogists; Animal Behav Soc; Soc Mar Mamm. *Res:* Natural history, including physiological and behavioral ecology of marine mammals and the impact of man caused perturbations on them; recovery of endangered species; phylogenetic relationship between marine mammals; general oceanography. *Mailing Add:* Nat Marine Mammals Lab Fisheries Serv 7600 Sand Point Way NE Bldg 4 Seattle WA 98115

LOUGHLIN, TIMOTHY ARTHUR, b Bay Shore, NY, Nov 16, 42; m 65; c 4. APPLIED MATHEMATICS. *Educ:* State Univ NY Stony Brook, BS, 64; Rensselaer Polytech Inst, MS, 66, PhD(math), 69. *Prof Exp:* Asst prof math, Union Col, NY, 69-76; ASSOC PROF MATH, NEW YORK INST TECHNOL, 76- *Mem:* Math Asn Am. *Res:* Network theory; realization of matrices as impedance and admittance matrices. *Mailing Add:* Dept Math NY Inst Technol Central Islip NY 11722

LOUGHMAN, BARBARA ELLEN EVERS, b Frankford, Ind, Oct 26, 40; m 62; c 2. IMMUNOBIOLOGY. *Educ:* Univ Ill, BS, 62; Univ Notre Dame, PhD(microbiol & immunol), 72. *Prof Exp:* From asst res microbiologist to assoc res microbiologist, Ames Res Lab, Miles Labs Inc, 62-71, res scientist immunol, 71-72; staff fel immunol, Nat Inst Child Health & Human Develop, 72-74; res scientist, Hypersensitivity Dis Res, Upjohn Co, 74-79, res head immunol, 79- 84, res mgr, 84-85; dir immunol res, Monsanto Co, 85-88; DIR & CONSULT PROJ MGT, RORER CENT RES, 88- *Mem:* AAAS; Asn Women Sci; Am Asn Immunologists. *Res:* Cellular immunology; regulatory mechanisms in cells using controlled in vitro and in vivo systems as models for specific intervention in an immume response; clinical research immunobiology of transplantation and blood dyscrasia; management; strategic planning; project and portfolio management. *Mailing Add:* Rorer Cent Res 800 Business Center Dr Horsham PA 19044

LOUGHRAN, EDWARD DAN, b Canton, Ohio, June 2, 28; m 59; c 3. ANALYTICAL CHEMISTRY. *Educ:* Ohio State Univ, BS, 50; MS, 53, PhD(chem), 55. *Prof Exp:* Asst chem, Res Found, Ohio State Univ, 53-55; mem staff, Los Alamos Nat Lab, 55-90, assoc group leader, 81-86, sect leader, 86-90; RETIRED. *Concurrent Pos:* Lab assoc, Los Alamos Nat Lab, 91- *Mem:* Am Soc Mass Spectrometry. *Res:* Analytical mass spectrometry; surveillance and compatibility studies of plastic-bonded explosives; physical properties, modes of decomposition and radiation chemistry of organic explosives. *Mailing Add:* Box 1663 MS 920 Los Alamos NM 87545

LOUGHRAN, GERARD ANDREW, SR, b Mt Vernon, NY, Sept 10, 18; m 45; c 4. ORGANIC CHEMISTRY. *Educ:* Fordham Univ, BS, 41; NY Univ, MS, 48. *Prof Exp:* Analytical chemist, NY Quinine & Chem Works, 41-43; asst chem, Fordham Univ, 43-44; chemist, Am Cyanamid Co, 46-56; chemist, R T Vanderbilt Co, 56-59; chemist, 60-86, PROJ SCIENTIST, MAT LAB, WRIGHT AERONAUT LABS, USAF, 86-; CONSULT, 86- *Mem:* Fel Am Inst Chem; Am Chem Soc; AAAS; NY Acad Sci. *Res:* Petroleum and rubber chemicals; polymer chemistry; high temperature materials; elastomers; organic synthesis. *Mailing Add:* 4575 Irelan St Kettering OH 45440

LOUGHRIDGE, MICHAEL SAMUEL, b Jacksonville, Tex, Aug 27, 36; m 61; c 1. MARINE GEOLOGY. *Educ:* Rice Univ, BA, 58; Harvard Univ, MA, 61, PhD(geol), 67. *Prof Exp:* Grad res geologist II, Marine Phys Lab, Scripps Inst, Calif, 61-63, postgrad res geologist II, 63-64, postgrad res geologist III, 64-67, asst res geologist, 67-68; sci staff asst, Oceanog Surv Dept, US Naval Oceanog Off, 68-78; SUPVRY OCEANOGR, NAT GEOPHYS & SOLAR TERRESTRIAL DATA CTR, 78- *Mem:* AAAS; Geol Soc Am; Am Geophys Union; assoc mem Soc Explor Geophys. *Res:* Studies of specialized techniques of echo sounding and the micro-topography of the sea floor; studies of fine scale magnetics of the sea floor; instrumentation for marine geology; seismic profiling; quantitative geomorphology; stream hydraulics; relationships between archaeology and geology. *Mailing Add:* 2630 Iliff St Boulder CO 80303

LOUGHRY, FRANK GLADE, b Marion Center, Pa, Apr 16, 10; m 44. SOIL CONSERVATION. *Educ:* Pa State Univ, BS, 31, PhD(agron, soils), 60; Ohio State Univ, MS, 34. *Prof Exp:* Asst agron, Ohio Agr Exp Sta, 31-33; soil scientist, USDA Soil Conserv Serv, 34-35, asst regional soil scientist, Northeastern US, 36-45, state soil scientist, Pa, 45-66; soil scientist, Pa Dept Health, 66-70; chief, Soil Sci Unit, Pa Dept Environ Resources, 71-77, consult soil scientist, 77-84; RETIRED. *Mem:* Fel AAAS; Am Soc Agron; Int Soc Soil Sci; fel Soil Conserv Soc Am. *Res:* Relation of soil morphology to aeration; soil factors affecting renovation of waste; interpretation of soil data for environmental protection; use of soil surveys in environmental programs. *Mailing Add:* Brethern Village Box 5093 Lancaster PA 17601

LOUGHTON, ARTHUR, b Wisbech, Eng, May 25, 31; Can citizen; m 55; c 2. HORTICULTURE. *Educ:* Univ Nottingham, Eng, BSc, 54, MSc, 60. *Prof Exp:* Hort officer res, Stockbridge House Exp Hort Sta, Ministry Agr, Fisheries & Food, Yorkshire, Eng, 54-62, dep dir, 62-67; res scientist veg res, Hort Res Inst Ont, Vineland Sta, 67-75; dir hort res, Hort Exp Sta, 75-86, MGR, TRANSITION CROP TEAM ONT MINISTRY AGRI & FOOD, SIMCOE, 86- *Mem:* Can Soc Hort Sci; Agr Inst Can. *Mailing Add:* Ont Ministry Agr & Food Hort Exp Sta Box 587 Simcoe ON N3Y 4N5 Can

LOUI, MICHAEL CONRAD, b Philadelphia, Pa, June 1, 55; m 83; c 1. COMPUTER SCIENCE & ENGINEERING. *Educ:* Yale Univ, BS, 75; Mass Inst Technol, MS, 77, PhD(comput sci), 80. *Prof Exp:* Res asst prof, 82-86, RES ASSOC PROF, COORD SCI LAB & ASSOC PROF ELEC & COMPUT ENG, UNIV ILL, URBANA, 86- *Concurrent Pos:* Vis res asst prof & vis asst prof elec eng, Univ Ill, Urbana, 81-82; category ed, Comput Reviews, 87-; prog dir, NSF, Washington, DC, 90-91. *Mem:* Asn Comput Mach; Inst Elec & Electronic Engrs; Am Soc Eng Educ; Soc Indust & Appl Math. *Res:* Computational complexity theory; parallel and distributed computation. *Mailing Add:* Beckman Inst Univ Ill 405 N Mathews Ave Urbana IL 61801

LOUIE, DEXTER STEPHEN, b San Francisco, Calif. GASTROENTEROLOGY. *Educ:* Univ Calif, Berkeley, AB, 74, BS, 76, PhD(nutrit), 82. *Prof Exp:* Res fel gastroenterol, Univ Mich, Ann Arbor, 85-87, res investr, 87-90, asst res scientist, 90-91; ASST PROF NUTRIT, UNIV NC, CHAPEL HILL, 91- *Concurrent Pos:* Investr, Gastrointestinal Peptide Res Ctr, Univ Mich, Ann Arbor, 88-91, adj asst prof nutrit, 90-91. *Mem:* Am Gastroenterol Asn; Am Inst Nutrit; Am Pancreatic Asn; Am Physiol Soc. *Res:* Neurohormonal control of exocrine pancreatic secretion; intracellular messenger mechanisms. *Mailing Add:* Dept Nutrit CB No 7400 Univ NC Chapel Hill NC 27599

LOUIE, MING, b Canton, China, Dec 8, 48; US citizen; m; c 1. POLYMER IN ELECTRONIC APPLICATION, ELECTROCHEMISTRY. *Educ:* Univ Conn, BA, 71; Pa State Univ, MS(polymer sci) & MS(phys chem), 75. *Prof Exp:* Chemist, Zapata Industs Inc, 78-85; res chemist, West Co, 85-89; TECH STAFF, ELASTOMERIC TECHNOLOGIES INC, 89- *Mem:* Am Chem Soc; Soc Plastics Engrs; Int Inst Connector & Interconnection Technol Inc; Int Soc Hybrid Microelectronics. *Res:* Formulation of conductive silicone rubber for electronic application such as elastomeric connectors; conduct testing programs on the connectors for electronics applications; development of fine-line electronic connection technologies; granted one patent; author of several publications. *Mailing Add:* Elastomeric Technologies Inc 2940 Turnpike Dr Hatboro PA 19040

LOUIE, RAYMOND, b Canton, China, June 22, 36; US citizen; m 62; c 1. PLANT PATHOLOGY. *Educ:* Univ Calif, Berkeley, BS, 59; Cornell Univ, MS, 65, PhD(plant path), 68. *Prof Exp:* ASSOC PROF VIROL, OHIO STATE UNIV & RES PLANT PATHOLOGIST, OHIO AGR RES & DEVELOP CTR, USDA, 67- *Mem:* Am Phytopath Soc. *Res:* Epiphytology of plant viruses; virus vector relationships; mechanical transmission of plant viruses. *Mailing Add:* Dept Plant Path Oardc Crops Res Div Wooster OH 44691

LOUIE, ROBERT EUGENE, b Oakland, Calif, Aug 2, 29; m 62; c 1. VIROLOGY. *Educ:* Univ Calif, Berkeley, BA, 51, MA, 53, PhD(bacteriol), 63. *Prof Exp:* Res asst virol, Ft Detrick, Md, 54-55; res microbiologist virol, 61-77, MGR VIROL RES DEPT, CUTTER LABS, 77- *Mem:* Am Soc Microbiol; Sigma Xi. *Res:* Development of viral vaccines for human use; viral chemotherapy; virus-cell relationships. *Mailing Add:* Cutter Labs Fourth & Parker Sts Berkeley CA 94701

LOUIE, STEVEN GWON SHENG, b Canton, China, Mar 26, 49; US citizen; m 75; c 3. THEORETICAL SOLID STATE PHYSICS. *Educ:* Univ Calif, AB, 72, PhD(physics), 76. *Prof Exp:* NSF fel, dept physics, Univ Calif, Berkeley, 76-77; fel theoret solid state physics, T J Watson Res Ctr, IBM Corp, 77-79; asst prof physics, Univ Penn, 79-80; assoc prof, 80-84, PROF PHYSICS, UNIV CALIF, BERKELEY, 84- *Concurrent Pos:* A P Sloan fel, 80-82; prof, Miller Inst Basic Res Sci, 86-87; vis scholar, Univ Tokyo, 89; J S Guggenheim fel, 89-90; vis prof, Fourier Univ, Grenoble, France, 90. *Mem:* Fel Am Phys Soc; Am Vacuum Soc. *Res:* Theoretical solid state physics; electronic properties of solids and of solid surfaces and interfaces; many-body effects in solids. *Mailing Add:* Dept Physics Univ Calif Berkeley CA 94720

LOUIS, JEAN FRANCOIS, fluid mechanics; deceased, see previous edition for last biography

LOUIS, JOHN, b Chicago, Ill, June 21, 24; m 75. HEMATOLOGY, CLINICAL PHARMACOLOGY. *Educ:* Univ Ill, BS, 48, MS & MD, 50. *Prof Exp:* Instr med, Col Med, Univ Ill, 51-65; asst prof, Stritch Sch Med, Loyola Univ, Chicago, 65-70; Prof med, Chicago Med Sch, 75; chief hematol sect, Vet Admin Hosp, Downey, Ill, 75; assoc dir, Div Hematol & Oncol, Chicago Med Sch, 75; CONSULT HEMAT & ONCOL, 70- *Concurrent Pos:* Consult to various hosps & Chicago State TB Sanatorium, 58-; chmn leukemia criteria comt, NIH, 61-65, leukemia task force, 62-65; US deleg, Eighth Int Cancer Cong, 62. *Mem:* Am Soc Hemat; Am Soc Clin Oncol; Am Col Physicians; Am Soc Clin Path; emer mem Cent Soc Clin Res; Int Soc Hemat. *Res:* Clinical pharmacology of drugs relating to hematology and cancer. *Mailing Add:* 347 Circle Lane Lake Forest IL 60045

LOUIS, KWOK TOY, b Shanghai, China, Jan 22, 27; m 54; c 3. TEXTILE CHEMISTRY. *Educ:* Tex Tech Col, BS, 51. *Prof Exp:* Lab dir, Otto Goedecke, Inc, Tex, 53-54; develop chemist, Burlington Indust, Inc, NC, 55-56; chief chemist, United Piece Dye Works, SC, 57-61; applns chems, Ciba Chem & Dye Co, 61-62, group leader appln res & qual control, 63-64, admin mgr res & appln, Tech Appln Prod, 64-68, mgr cent lab, 68-71; dir tech dept, Dyes & Chem Div, Crompton & Knowles Corp, NJ, 71-76; tech dir, 76-77, VPRES, APEX CHEM CO, INC, 78- *Mem:* Am Asn Textile Chemists & Colorists; AAAS; NY Acad Sci. *Mailing Add:* 442 Ellis Place Wyckoff NJ 07481

LOUIS, LAWRENCE HUA-HSIEN, b Canton, China, Apr 23, 08; nat US; m 42; c 4. BIOCHEMISTRY. *Educ:* Univ Mich, BS, 32, MS, 33, ScD, 37. *Prof Exp:* Res org chem, Univ Berlin, Ger, 37-39; fel physiol, Univ Pa, 40-41; asst internal med, 41-46, instr biochem, 46-48, from asst prof to assoc prof, 48-69, prof biochem, Univ Mich, Ann Arbor, 70-78; RETIRED. *Mem:* AAAS; Am Chem Soc; Am Soc Biol Chem. *Res:* Endocrinology and metabolism. *Mailing Add:* 2302 Manchester Rd Ann Arbor MI 48104

LOUIS, THOMAS MICHAEL, b Pensacola, Fla, Dec 27, 44; m 69; c 2. REPRODUCTIVE ENDOCRINOLOGY. *Educ:* Va Polytech Inst & State Univ, BS, 68, MS, 71; Mich State Univ, PhD(sci), 75. *Prof Exp:* Lalor res fel reproductive endocrinol, Univ Oxford, 75-76; from asst prof to assoc prof, 76-85, PROF ANAT, SCH MED, EAST CAROLINA UNIV, 85- *Honors & Awards:* Richard Hoyte Res Prize, Am Dairy Sci Asn, 75. *Mem:* AAAS; Soc Gynec Invest; Sigma Xi; Soc Study Endocrinol; Am Asn Anatomists. *Res:* Chronic effects of alcohol, nicotine, and the nervous system on pregnancy and parturition; studies include endocrinology of parturition, fetal endocrinology, effects of fetal asphyxia on the neonate and endocrine control of the hypothalamus and pituitary. *Mailing Add:* Dept of Anat Sch of Med East Carolina Univ Greenville NC 27834

LOUIS-FERDINAND, ROBERT T, PHARMACOLOGY. *Prof Exp:* PROF, DEPT PHARMACOL, WAYNE STATE UNIV, 76- *Mailing Add:* Pharmaceut Sci Wayne State Univ Shapero Hall Detroit MI 48202

LOULLIS, COSTAS CHRISTOU, b Nicosia, Cyprus, Jan 5, 50. NEUROBIOLOGY, NEUROCHEMISTRY. *Educ:* Fairfield Univ, BS, 74; Syracuse Univ, MA, 75, PhD(biopsychol), 78. *Prof Exp:* Teaching asst biopsychol, Dept Psychol, Syracuse Univ, 74-78; NIMH fel neurochem &

behavior, Dept Psychiat, Sch Med, Ind Univ, 78-80; SCIENTIST, MED RES DIV, AM CYANAMID CO, 80- *Mem:* Soc Neurosci; NY Acad Sci; AAAS. *Res:* Psychopharmacology; CNS lesions; limbic system; schedule induced polydipsia; taste aversion; operant behavior; eating and drinking behaviors; neurotransmitters; aging; calmedulin; calcium; cyclic neucleotides. *Mailing Add:* 454 Sierra Vista Ln Valley Cottage NY 10989

LOULOU, RICHARD JACQUES, b Relizane, Algeria, Apr 19, 44; Can citizen; m 67; c 2. OPERATIONS RESEARCH, PROBABILITY. *Educ:* Sch Polytech, Paris, BSc, 66; Univ Calif, Berkeley, MSc, 68, PhD(opers res), 71. *Hon Degrees:* DSc, Univ Grenoble, France, 78. *Prof Exp:* From asst prof to assoc prof, 70-83, PROF OPERS RES, McGILL UNIV, 83- *Concurrent Pos:* Consult, Archer, Seaden & Assocs, 72-73, Can Ministry Energy, Mines & Resources, 84-85. *Mem:* Inst Mgt Sci; Opers Res Soc Am; Can Opers Res Soc; Soc Indust & Appl Math. *Res:* Queueing theory; congested service systems; stochastic processes simulation; heuristics in optimization. *Mailing Add:* Dept Mgt McGill Univ Sherbrooke St W Montreal PQ H3A 2M5 Can

LOUNIBOS, LEON PHILIP, b Petaluma, Calif, Aug 19, 47; div; c 1. INSECT ECOLOGY, INSECT BEHAVIOR. *Educ:* Univ Notre Dame, BS, 69; Harvard Univ MS, 70, PhD(biol), 74. *Prof Exp:* Res scientist & head, Int Ctr Insect Physiol & Ecol, Coastal Res Sta, 74-77; entomologist III, Fla Med Entom Lab, 77-83; ASSOC PROF ENTOM, UNIV FLA, 83- *Concurrent Pos:* NIH fel, 69-77, 88-89. *Mem:* AAAS; Sigma Xi; Entom Soc Am; Animal Behav Soc; Ecol Soc Am. *Res:* Insect ecology: seasonality, diapause strategies, predator-prey relationships, community organization; insect behavior: building, predatory, oviposition behaviors. *Mailing Add:* Fla Med Entom Lab 200 9th St SE Vero Beach FL 32962

LOUNSBURY, FRANKLIN, b Chicago, Ill, May 6, 12; m 41; c 3. MEDICINE. *Educ:* Univ Wis, AB, 34; Northwestern Univ, MD, 39, MS, 48. *Prof Exp:* Asst prof, 54-65, ASSOC PROF SURG, NORTHWESTERN UNIV, CHICAGO, 65- *Concurrent Pos:* Attend physician, Northwestern Mem Hosp, 46-80, emer physician, 80- *Mem:* Am Col Surg; Cent Surg Asn; Soc Surgery Alimentary Tract. *Res:* Abdominal surgery, especially of the biliary tract. *Mailing Add:* 165 N Kenilworth Oak Park IL 60301

LOUNSBURY, JOHN BALDWIN, b Urbana, Ill, Jan 30, 36; m 63; c 3. PHOTOLITHOGRAPHIC TECHNOLOGY, HIGH BANDWIDTH COMMUNICATION SYSTEMS. *Educ:* Univ Vt, BA, 57; Columbia Univ, MA, 58; Ill Inst Technol, PhD(phys chem), 66. *Prof Exp:* SR ENGR & MGR, IBM CORP, ARMONK, NY, 58- *Concurrent Pos:* Res assoc physics, Armour Res Found, Chicago, Ill, 59-62. *Res:* Career research activities include quantum chemistry of molecular structure, plasma chemistry, physics and materials/processes of photolithography for semiconductor fabrication; one patent and 20 publications. *Mailing Add:* PO Box 593 Billings NY 12510

LOURENCO, RUY VALENTIM, b Lisbon, Portugal, Mar 25, 29; US citizen; m 60; c 2. MEDICINE, PHYSIOLOGY. *Educ:* Univ Lisbon, BSc, 46, MD, 51. *Prof Exp:* Intern, Lisbon City Hosps, 52, resident internal med, 53-55; instr med, Sch Med, Lisbon, 56-59; from asst prof to assoc prof, NJ Col Med, 63-67; assoc prof, 67-69, dir pulmonary sect, Dept Med, 70-77, PROF MED & PHYSIOL, ABRAHAM LINCOLN SCH MED, UNIV ILL COL MED, 69-, CHMN DEPT MED, 77-, FOLEY PROF MED, 78- *Concurrent Pos:* Attend physician, Nat Cancer Inst, Lisbon, Portugal, 55-61, Lisbon Univ Hosp, 56-61, Jersey City Med Ctr & VA Hosp, NJ & Newark City Hosp, 63-67, Univ Ill Hosp, Cook County Hosp & W Side VA Hosp, 67-; fel med, Cologne Univ, 57 & Columbia-Presby Med Ctr, 59-63; dir, Respiratory Physiol Lab, Univ Lisbon Med Sch, Portugal, 57-61, Respiratory Physiol, dept med, NJ Col Med, 63-67, Respiratory Res, Hektoen Inst Med Res, Chicago, 67-71, Respiratory Physiol Lab, Cook County Hosp, 67-69, dept pulmonary med, 69-70 & Pulmonary Sect & Labs, Univ Ill Med Ctr, 70-77; Lederle int fel, 59-60; Polachek Found fel, 61-63; consult physician, Vet Admin Hosps, 65-; mem cardio-pulmonary coun, Am Heart Asn; mem task force sci basis respiratory therapeut, Nat Heart & Lung Inst, 71-72; mem study sect, NIH, 72-76; consult, Career Develop Prog, Vet Admin, 72-; chmn sci assembly, Am Thoracic Soc, 74-75; vis prof, Cardiothoracic Inst, Brompton Hosp, Univ London, 75-76; physician-in-chief, Univ Ill Hosps, 77-; mem, inhalation toxicol comt, Nat Ctr Toxicol Res, 77-, Asn Prog Dirs Internal Med, 77-, comt smoking & health, Am Lung Asn, 81- & comt int affairs, Am Col Chest Physicians, 84-; reviewer, var physiol, respiratory & clin journals; lectr, var univs in US, Brazil, Portugal & Spain. *Mem:* Am Physiol Soc; Am Fedn Clin Res; Am Thoracic Soc; Am Soc Clin Invest; Soc Exp Biol & Med; Int Soc Aerosols Med; Asn Profs Med. *Res:* Internal medicine; chest diseases; respiratory physiology and biochemistry; regulation of ventilation; muscles of breathing; pulmonary defense mechanisms. *Mailing Add:* NJ Med Sch 185 S Orange Ave Newark NJ 07103

LOURIA, DONALD BRUCE, b New York, NY, July 11, 28; m 55; c 3. INTERNAL MEDICINE, MICROBIOLOGY. *Educ:* Harvard Univ, BS, 49, MD, 53. *Prof Exp:* From instr to assoc prof med, Col Med, Cornell Univ, 58-69; PROF PREV MED & COMMUNITY HEALTH & CHMN DEPT, NJ MED SCH, COL MED & DENT NJ, 69- *Concurrent Pos:* Pres, NY State Coun Drug Addiction, 65-73. *Mem:* Am Soc Clin Invest; Am Fedn Clin Res; Am Soc Microbiol; Am Col Physicians. *Res:* Mycology, especially fungal toxins and the pathogenesis of Candida infections; prevention programs for adults; health education; health manpower; cancer epidemiology; health problems of the aging. *Mailing Add:* 100 Bergen Newark NJ 07103

LOURIE, ALAN DAVID, b Boston, Mass, Jan 13, 35; m 59; c 2. ORGANIC CHEMISTRY. *Educ:* Harvard Univ, AB, 56; Univ Wis, MS, 58; Univ Pa, PhD(org chem), 65; Temple Univ, JD, 70. *Prof Exp:* Res chemist, Monsanto Co, 57-59; res chemist, Wyeth Labs, 59-60, lit chemist, 60-62, patent chemist, 62-64; patent agent chem, Smith Kline & French Labs, 64-70, patent atty, 70-71, assoc patent coun, 71-74, asst dir, patent dept, 74-76, vpres corp patents, Smithkline Corp, 76-; JUDGE, COURT APPEALS, FED CIRCUIT. *Mem:* Am Chem Soc. *Res:* Synthesis of heterocyclic compounds; medicinal chemistry. *Mailing Add:* Court Appeals Fed Circuit 717 Madison Pl NW Nat Court Bldg Washington DC 20439

LOURIE, HERBERT, neurosurgery; deceased, see previous edition for last biography

LOUSTAUNAU, JOAQUIN, b San Louis Potosi, Mex, Sept 17, 36; m 66. MATHEMATICS. *Educ:* Okla State Univ, BS, 58, MS, 60; Univ Ill, PhD(math), 65. *Prof Exp:* Instr math, Inst Tech & Higher Educ, Monterrey, Mex, 60-61; ASST PROF MATH, NMEX STATE UNIV, 65- *Mem:* Math Asn Am; Am Math Soc. *Res:* Functional analysis. *Mailing Add:* Dept of Math NMex State Univ Las Cruces NM 88003

LOUTFY, RAFIK OMAR, b Cairo, Egypt, Nov, 43; Can citizen; m 65; c 2. PHOTOCHEMISTRY, PHYSICAL CHEMISTRY. *Educ:* Ain Shams Univ, Cairo, BSc, 64, MSc, 67; Univ Western Ont, PhD(photochem), 72; Univ Toronto, MBA, 85. *Prof Exp:* Fel laser flash photolysis, Nat Res Coun Can, 72-74; fel photochem, Univ Toronto, 74; mem sci staff, 74-80, AREA MGR, XEROX RES CTR CAN, 80- *Concurrent Pos:* Adj prof, Univ Western Ont, 79- *Mem:* Am Chem Soc; Chem Inst Can; Inter-Am Photochem Soc; Europ Photochem Soc; Soc Photographic Sci & Eng. *Res:* Photophysics of small molecules and polymers; solar energy conversion using organic semiconductors; dye sensitization of semiconductors; electrochemistry and spectroscopy of organic molecules and dyes; photo conductors. *Mailing Add:* Xerox Corp 800 Long Ridge Rd 2-2-B PO Box 1600 Stamford CT 06904-1600

LOUTTIT, RICHARD TALCOTT, b Bloomington, Ind, Dec 5, 32; m 54; c 2. BEHAVIORAL SCIENCES, NEUROSCIENCE. *Educ:* DePauw Univ, AB, 54; Univ Mich, MA, 59, PhD(psychol), 61. *Prof Exp:* Asst prof psychol, Univ Pac, 61-64; health sci adminr, NIH, 64-70; prof & head, Dept Psychol, Univ Mass, Amherst, 70-75; DIV DIR, DIV BEHAV & NEURAL SCI, NSF, 75- *Concurrent Pos:* Staff dir, President's Biomed Res Panel, 75. *Mem:* Fel AAAS; Am Psychol Soc. *Mailing Add:* Behav & Neural Sci Div NSF 1800 G St NW Washington DC 20550

LOUTTIT, ROBERT IRVING, b Honolulu, Hawaii, July 23, 29; m 54, 86; c 3. EXPERIMENTAL HIGH ENERGY PHYSICS. *Educ:* Univ NH, BS, 52; Wash Univ, PhD(physics), 58. *Prof Exp:* From asst physicist to physicist, 58-84, head accelerator develop br, 84-86, sr physicist, Brookhaven Nat Lab, 84-86; RETIRED. *Concurrent Pos:* Physicist, Nuclear Res Ctr, Saclay, France, 63-64. *Mem:* AAAS; Am Phys Soc; Sigma Xi. *Res:* Bubble chamber development; neutrino interactions. *Mailing Add:* PO Box 1418 Sarasota FL 34230

LOVAGLIA, ANTHONY RICHARD, b San Jose, Calif, Jan 25, 23; m 44; c 3. MATHEMATICS. *Educ:* Univ Calif, Los Angeles, AB, 45, PhD(math), 51; Stanford Univ, MS, 48; Univ Calif, Berkeley, PhD(math), 51. *Prof Exp:* From asst prof to assoc prof, 51-60, PROF MATH, SAN JOSE STATE UNIV, 60- *Mem:* Math Asn Am. *Res:* Analysis. *Mailing Add:* 278 Anchor Ct Boulder Creek CA 95006

LOVALD, ROGER ALLEN, b Marshall, Minn, Aug 8, 38; m 57; c 2. ORGANIC POLYMER CHEMISTRY. *Educ:* Univ Minn, BChem, 60; Univ Wis, PhD(org chem), 65. *Prof Exp:* Chemist, Spring Res Lab, Rohm & Haas Co, 65-67; cent res, 67-71, sect leader, Resin Develop, 71-75, TECH DIR RESINS, GEN MILLS CHEM, INC, 75- *Mem:* Am Chem Soc. *Res:* Heteroaliphatic and organic chemistry; addition and condensation polymerization; acrylics; polyamides; polyesters; polyurethanes. *Mailing Add:* 1738 Wildwood Lane Darien IL 60559-4920

LOVALLO, WILLIAM ROBERT, b Newark, NJ, Nov 16, 46. BEHAVIORAL MEDICINE, PSYCHOPHARMACOLOGY. *Educ:* Univ Calif, Los Angeles, BA, 68; Univ Okla Health Sci Ctr, PhD(biol & psychol), 78. *Prof Exp:* Asst prof, 80-85, ASSOC PROF, PSYCHIAT & BEHAVIORAL SCI, UNIV OKLA HEALTH SCI CTR, 85- *Concurrent Pos:* Prin investr, Vet Admin grant, Cardiovasc Dis, 78-; NIH grant, Caffeine Effects, 85-; dir, Behavior Sci Labs, Okla City Vet Admin Med Ctr, 86-; mem, Behavior Med Study Sect, NIH, 88-; asst ed, Int J Psychophysiol, 88- *Mem:* Am Heart Asn; AAAS; Am Psychol Asn; Soc Psychophysiol Res; Soc Behavioral Med. *Res:* Psychological and behavioral stress; the role of stress on the development of cardiovascular diseases. *Mailing Add:* Vet Admin Med Ctr 151A 921 NE 13th St Oklahoma City OK 73104

LOVAS, FRANCIS JOHN, b Cleveland, Ohio, July 29, 41; m 70; c 1. MOLECULAR SPECTROSCOPY, RADIO ASTRONOMY. *Educ:* Univ Detroit, BS, 63; Univ Calif, Berkeley, PhD(phys chem), 67. *Prof Exp:* Res grant, Lawrence Radiation Lab, Univ Calif, Berkeley, 67-68; NATO fel, Phys Inst, Free Univ Berlin, 68-70; Assoc, Nat Res Coun-Nat Bur Standards, 70-72, DIR, MOLECULAR SPECTRA DATA CTR, NAT INST STANDARDS & TECHNOL, 72- *Honors & Awards:* Gold Medal, Dept Com, 77. *Mem:* AAAS; Am Phys Soc. *Res:* Properties of diatomic molecules by high temperature microwave adsorption and molecular beam electric resonance techniques; microwave spectroscopy of transient molecules and molecular radio astronomy; critical evaluation of microwave spectroscopic data. *Mailing Add:* Molecular Physics Div 545 Nat Inst Standards & Technol Gaithersburg MD 20899

LOVASS-NAGY, VICTOR, b Debrecen, Hungary, Apr 25, 23; m 51; c 2. APPLIED MATHEMATICS. *Educ:* Budapest Tech Univ, dipl, 47, PhD(math), 49. *Prof Exp:* Instr math, Budapest Tech Univ, 47-49, from asst prof to assoc prof, 49-58; consult engr, Ganz Elec Works, Hungary, 60-64; reader eng math, Univ Khartoum, 64-66; PROF MATH, CLARKSON COL TECHNOL, 66-, PROF COMPUT SCI, 77- *Mem:* Soc Indust & Appl Math; Am Math Soc; Math Asn Am; Tensor Soc; sr mem Inst Elec & Electronics Engrs. *Res:* Matrix theory; numerical analysis; network theory. *Mailing Add:* Dept Elec & Comput Eng Clarkson Univ Potsdam NY 13699-5720

LOVATT, CAROL JEAN, b Kansas City, Mo, May 14, 47; div; c 2. METABOLIC REGULATION. *Educ:* Univ Mass, BA, 73; Univ RI, MS, 76, PhD(bot), 80. *Prof Exp:* Res assoc, 80, asst prof plant physiol & asst plant physiologist, 80-87, ASSOC PROF PLANT PHYSIOL & ASSOC PLANT PHYSIOLOGIST, DEPT BOT & PLANT SCI, UNIV CALIF, RIVERSIDE, 87- *Honors & Awards:* Fruit Publ Award, Am Soc Hort Sci, 87, Cross-Commodity Publ Award, 88. *Mem:* Am Soc Plant Physiologists; AAAS; Am Women Sci; Sigma Xi; Am Soc Hort Sci. *Res:* Metabolic regulation of nucleotide metabolism and arginine biosynthesis/urea cycle; citrus physiology: regulation of flowering, fruit set, and fruit growth; role of boron in plant metabolism. *Mailing Add:* Dept Bot & Plant Sci Univ Calif Riverside CA 92521

LOVE, ALLAN WALTER, b Toronto, Ont, May 28, 16; US citizen; m 46; c 3. ELECTROMAGNETISM, ANTENNAS & MICROWAVES. *Educ:* Univ Toronto, BA, 38, MA, 39, PhD(microwave physics), 51. *Prof Exp:* Res officer, Radiophysics Lab, Commonwealth Sci & Indust Res Orgn, Australia, 46-48; demonstr asst, Physics Lab, Univ Toronto, 48-51; chief instrumentation, Newmont Explor Ltd, Conn & Ariz, 51-57; staff scientist, Giannini Res Lab, Wiley Electronics Co, Ariz, 57-62, mgr, Physics Lab, Calif, 62-63; area mgr, Nat Eng Sci Co, 63-65; group scientist, Antenna Lab, Autonetics Div, NAm Aviation Inc, 63, group scientist theoret anal, 65-71, mem tech staff, Space Div, NAm Rockwell Corp, 71-73, prog mgr, 73-76, prin scientist, Satellite Systs Div, Rockwell Int, 76-90; RETIRED. *Mem:* Fel Inst Elec & Electronic Engrs. *Res:* Microwave and millimeter wave physics; antenna theory and design; development of spacecraft antenna systems. *Mailing Add:* 518 Rockford Pl Corona Del Mar CA 92625-2721

LOVE, CALVIN MILES, b Chicago, Ill, Mar 2, 37; m 60; c 3. PYROTECHNICS, RADIOCHEMISTRY. *Educ:* Ill Inst Technol, BS, 59; Mich State Univ, PhD(inorg chem), 64. *Prof Exp:* Res specialist, 64-80, SR RES SPECIALIST, EG&G MOUND APPL TECHNOLOGIES, INC, 80- *Mem:* AAAS; Am Chem Soc; NAm Thermal Anal Soc. *Res:* Kinetics and mechanisms of inorganic oxidation-reduction reactions; plutonium separation and recovery; polonium process development; metal distillation; metal hydrides; radiation damage; thermal analysis; hydrides for hydrogen storage; chemistry of pyrotechnics; chemistry of explosives. *Mailing Add:* 7601 Eagle Creek Dr Dayton OH 45459

LOVE, CARL G(EORGE), b Warsaw, NY, Sept 20, 40; m 71; c 2. SYSTEMS ANALYSIS. *Educ:* Rochester Inst Technol, BS, 63; Carnegie Inst Technol, MS, 65, PhD(elec eng), 67. *Prof Exp:* Coop student, Rochester Gas & Elec Corp, 58-60; coop student, Delco Appl Div, Gen Motors Corp, 60-63, proj engr, 63; sr engr, 67-71, fel engr, 72, MGR SYST PLANNING & TECH ASSESSMENT, WESTINGHOUSE RES & DEVELOP CTR, 72-, DIR CORP VENTURE PROGS, CORP PLANNING, 84- *Concurrent Pos:* Sr consult, Corp Planning, Westinghouse Elec Corp, 82-84. *Mem:* Inst Elec & Electronics Engrs; Inst Mgt Sci; Opers Res Soc Am. *Res:* Technology forecasting; business analysis; energy analysis. *Mailing Add:* Westinghouse Elec Corp Corp Venture Progs Parkway Ctr Bldg Five G F/r Pittsburgh PA 15220

LOVE, DAVID VAUGHAN, b St John, NB, Aug 25, 19; m 43; c 3. FOREST MANAGEMENT. *Educ:* Univ NB, BSc, 41; Univ Mich, MF, 46. *Prof Exp:* From lectr to prof, Univ Toronto, 46-72, asst dean, 72-76, assoc dean, 77-83, dean, 84-85. *Concurrent Pos:* Vpres, Conservation Coun Ont, 59-64, pres, 74-75; vchmn, Can Coun on Rural Develop, 75-79; rep, Can Forestry Asn, 73. *Mem:* Soc Am Foresters; Can Pulp & Paper Asn; Can Inst Forestry (secy-mgr, 48-54, pres, 65-66); Ont Forestry Asn (vpres, 70-71, pres, 72-73); Can Forestry Asn (vpres, 74 & 75, pres, 75). *Res:* Land use; forests; acid precipitation. *Mailing Add:* 16 Marchwood Dr Downsview ON M3H 1J8 Can

LOVE, DAVID WAXHAM, b Laramie, Wyo, Nov 1, 46; m. QUATERNARY STRATIGRAPHY. *Educ:* Beloit Col, BA, 69; Univ NMex, MS, 71, PhD(geol), 80. *Prof Exp:* Asst prof geol, Wash State Univ, 76-78; ENVIRON GEOLOGIST, NMEX BUR MINES & MINERAL RESOURCES, 80- *Mem:* Geol Soc Am; Sigma Xi; Soc Archeol Sci. *Res:* Geomorphic processes and stratigraphy of surficial deposits in New Mexico and adjacent areas for assessing natural hazards and for determining stability of land forms for siting industrial plants or for storing hazardous materials. *Mailing Add:* NMex Bur Mines & Mineral Resources Socorro NM 87801

LOVE, GEORGE M, b Lima, Ohio, Oct 5, 44; m 72. ORGANIC CHEMISTRY. *Educ:* DePauw Univ, BA, 66; Wake Forest Univ, MA, 68; Mich State Univ, PhD(org chem), 72. *Prof Exp:* Fel org chem, Rutgers Univ, 72-73; sr res chemist, 73-80, RES FEL, MERCK INC, 80-; ASSOC DIR, CHEM PROCESS DEVELOP, SCHERING-PLOUGH. *Mem:* Sigma Xi; Am Chem Soc. *Res:* Process research in organic chemistry. *Mailing Add:* Chem Process Develop Schering Plough 1011 Morris Ave Union NJ 07083-9977

LOVE, GORDON ROSS, b Cleveland, Ohio, July 31, 37; m 62; c 1. MATERIALS SCIENCE. *Educ:* Case Inst Technol, BS, 58; Carnegie Inst Technol, MS, 61, PhD(metall), 63. *Prof Exp:* Metallurgist, Oak Ridge Nat Lab, 62-64, group leader superconducting mat, 64-70; asst mgr technol, Mat Syst Div, Union Carbide Corp, 70-78; staff mem, Sprague Elec, 78-87; TECH DIR CERAMICS, ALCOA TECH CTR, ALCOA, 87- *Concurrent Pos:* Lectr, Univ Tenn, 66-67; ed, Trans Comp Hyb Mfrs Tech, Inst Elec Electronics Engrs, 81-85. *Mem:* Sigma Xi; Inst Elec & Electronics Engrs. *Res:* Diffusion; superconductivity; statistical process control; powder technology; surface and interface properties; ceramic dielectric materials. *Mailing Add:* Aluminium Co Am Alcoa Tech Ctr 100 Technical Dr Alcoa Center PA 15069-0001

LOVE, HARRY SCHROEDER, JR, b Idabel, Okla, Aug 20, 27; m 52; c 2. BOTANY, ECOLOGY. *Educ:* Okla State Univ, BS, 52, MS, 58, PhD(bot), 71. *Prof Exp:* assoc prof, 67-80, PROF BIOL, EAST CENT OKLA STATE UNIV, 80- *Mem:* AAAS; Am Inst Biol Sci; Nat Asn Biol Teachers. *Res:* Terrestrial plant ecology, especially clonal and root-graft relationships. *Mailing Add:* Dept of Biol E Cent State Univ Ada OK 74820

LOVE, HUGH MORRISON, b Northern Ireland, Aug 21, 26. PHYSICS. *Educ:* Queen's Univ, Belfast, BSc & PhD(physics), 50. *Prof Exp:* Asst lectr, Queen's Univ, Belfast, 46-50; lectr, Univ Toronto, 50-52; from asst prof to assoc prof, 52-65, PROF PHYSICS, QUEEN'S UNIV, ONT, 65-, VPRIN, 76- *Mem:* Am Phys Soc. *Res:* Solid state physics; surface physics. *Mailing Add:* Dept Physics Queen's Univ Stirling Hall Kingston ON K7I 3N6 Can

LOVE, JIM, b Bathgate, Scotland, Oct 21, 38; m 62; c 2. ORGANIC CHEMISTRY. *Educ:* Univ Edingurgh, BSc, 60, PhD(carbohydrate chem), 63. *Prof Exp:* Fel, Scripps Inst, Univ Calif, 63-64 & Ohio State Univ, 64-65; res chemist, Dow Chem Co Mich, 65-67, Western Div, 67-74, res specialist, 74-77, group leader, 77-79, res mgr, Western Div, 79-83, lab dir, Pittsburg, Calif, 83-88, LAB DIR, DOW CHEM CO, MIDLAND, WI, 88- *Mem:* Am Chem Soc; The Chem Soc. *Res:* Carbohydrate chemistry, particularly polysaccharide and mucopolysaccharide structural determination and biological activity; synthesis and biological activity of heterocyclic compounds. *Mailing Add:* 3016 Scarborough Lane Midland MI 48640-6908

LOVE, JIMMY DWANE, b Plainview, Tex, Feb 2, 46; m 67; c 2. ANALYTICAL CHEMISTRY, THERMODYNAMICS & MATERIAL PROPERTIES. *Educ:* Stephen F Austin State Univ, Nacogdoches, Tex, BS, 69, MS, 76. *Prof Exp:* Bench chemist, Moore Bus Forms, 70-74, sr chemist, 74-78, lab mgr, 78-83, tech serv mgr, 80-83, qual assurance mgr, 83-85, carbonless paper prod mgr, Moore Bus Forms, 85-88; COM DEVELOP MGR, GA-PAC CORP, 88- *Mem:* Am Chem Soc; Am Soc Qual Control; Am Soc Testing & Mat; Tech Asn Pulp & Paper Indust; Paper Indust Mgt Asn. *Res:* Micro encapsulation chemistry and techniques; coatings technology; papermaking chemistry; applied technology. *Mailing Add:* 1022 Fairwood Ct Acworth GA 30101

LOVE, JOHN DAVID, b Riverton, Wyo, Apr 17, 13; m 40; c 4. GEOLOGY. *Educ:* Univ Wyo, BA, 33, MA, 34; Yale Univ, PhD(geol), 38. *Hon Degrees:* LLD, Univ Wyo, 61. *Prof Exp:* Asst geologist, Geol Surv Wyo, 33-37; field asst, US Geol Surv, 38; from asst geologist to geologist, Shell Oil Co Inc , 38-42; asst geologist, US Geol Surv, 42-43, from assoc geologist to prin geologist, 43-56, supvr heavy metals, Jackson Proj, 66-68, Northern Rocky Mts Br, 64-66, 67-69, in charge Wyo basins fuels proj, 43-56, STAFF GEOLOGIST, US GEOL SURV, 56-, SUPVR, LARAMIE OFF, REGIONAL GEOL BR, 69- *Concurrent Pos:* Distinguished lectr, Univ Wyo, 65, adj prof, 69-; instr & trustee, Teton Sci Sch, 65-85; affil prof geol, Univ Idaho, 74-; exten instr geol, Univ Calif, Davis, 77-; grad res adv, Univ Wash; affil, Univ Wash, 79-83 & Univ Minn, 81-83. *Honors & Awards:* Meritorious Serv Award, US Dept Interior, 77 & Distinguished Serv Award, 87. *Mem:* AAAS; fel Geol Soc Am; Am Asn Petrol Geol; Sigma Xi. *Res:* Geology of fuels; uranium, vanadium and gold investigations; stratigraphic and structural geology; author or coauthor of about 220 scientific publications. *Mailing Add:* US Geol Surv Box 3007 Univ Sta Laramie WY 82071

LOVE, JOSEPH E(UGENE), JR, b Chicago, Ill, Apr 9, 20; m 42; c 2. CIVIL ENGINEERING. *Educ:* Northwestern Univ, BS, 42, MS, 48, PhD(civil eng), 51. *Prof Exp:* Struct analyst, Curtiss-Wright Corp, 42-43; instr math & eng, Ripon Col, 43-45; from instr to asst prof civil eng, Northwestern Univ, 46-51; struct engr, Hanford Atomic Prod Oper, Gen Elec Co, 52-55, struct engr, Atomic Power Equip Dept, 55-66, mgr arrangements & struct design, 66-72, mgr advan eng, 72-75, mgr plant struct systs, nuclear energy div, 75-84; CONSULT, 85- *Concurrent Pos:* Contribr, 1st Int Conf Peaceful Uses of Atomic Energy; chmn working group on containment, Int Orgn Standardization, Technol Comt 85, subcomt 3, 76-84, mem, working group on containment ISO, 84-88. *Mem:* Sigma Xi; Am Soc Civil Engrs. *Res:* Plasticity effects in flexure; nuclear power plant design. *Mailing Add:* 15605 On Orbit Dr Saratoga CA 95070

LOVE, L J CLINE, b Richmond, Mo, Oct 1, 40; m 72; c 2. LUMINESCENCE, MICELLAR CHEMISTRY. *Educ:* Univ Mo-Columbia, BS, 62, MA, 65; Univ Ill, Urbana, PhD(chem), 69. *Prof Exp:* Fel chem, Univ Fla, 69-70; asst prof anal chem, Mich State Univ, 70-72; asst prof, 72-77, assoc prof, 77-82, PROF ANAL CHEM, SETON HALL UNIV, 82-, DIR CTR APPL SPECTROS, 85- *Concurrent Pos:* Prin investr, NIH, 81-83, Environ Protection Agency, 82-84, NSF, 83-86; assoc ed, Applied Spectroscopy J. *Honors & Awards:* Spectros Medal, NY Soc Appl Spectros, 85. *Mem:* Am Chem Soc; Soc Appl Spectros; Am Microchem Soc; Sigma Xi. *Res:* Development of new instrumentation and methodology in luminescence; analytical applications of micellar and cyclodextrin systems; fluorescence and phosphorescene spectroscopy; automation of chemical instrumentation; high performance liquid chromatography; computer factor analysis of data. *Mailing Add:* Dept Chem Seton Hall Univ 400 S Orange Ave South Orange NJ 07079

LOVE, LEON, b New York, NY, Sept 7, 23; m 56; c 3. RADIOLOGY. *Educ:* City Col New York, BS, 43; Chicago Med Sch, MD, 46; Am Bd Radiol, dipl, 51. *Prof Exp:* Radiologist, Cook County Hosp, Chicago, 56-61; assoc prof radiol, Chicago Med Sch, 58-67, clin prof, 67-69; PROF RADIOL & CHMN DEPT, MED CTR, LOYOLA UNIV CHICAGO, 69- *Concurrent Pos:* Consult, Dwight Vet Admin Hosp, 56-62; dir diag radiol, Cook County Hosp, Chicago, 61-69; consult, House of Correction, Chicago, 61- & WSide Vet Admin Hosp, 62- *Mem:* Am Col Radiol; Radiol Soc NAm. *Res:* Renal radiology; radiology of the gastro-intestinal tract. *Mailing Add:* 235 Hawthorn Glencoe IL 60022

LOVE, NORMAN DUANE, b Howell, Mich, Jan 1, 39; m 62; c 3. LOW TEMPERATURE PHYSICS. *Educ:* Albion Col, AB, 60; Western Mich Univ, MA, 62; Mich State Univ, PhD(physics), 67. *Prof Exp:* From asst prof to assoc prof, Maryville Col, 67-77, dir comput serv, 71-77; software specialist, 77-82, COMMUN CONSULT, DIGITAL EQUIP CORP, 82- *Concurrent Pos:* Nat Sci Found comput grant, 68-73. *Mem:* Am Phys Soc; Am Asn Physics Teachers. *Res:* Effect of magnons on transport of phonons; phase boundaries in an antiferro magnetic material using calorimetric techniques. *Mailing Add:* Digital Equip Corp 412 Executive Tower Dr, Suite 300 Knoxville TN 37923

LOVE, RAYMOND CHARLES, b Washington, DC, July 30, 53; m 76. CLINICAL PHARMACY. *Educ:* Univ Md, Baltimore, DrPharm, 77. *Prof Exp:* Dir, Area Health Educ Ctr, Cumberland, 77-78, ASST PROF CLIN PHARM, SCH PHARM, UNIV MD, 77- *Concurrent Pos:* Consult, Mem Hosp, Cumberland, Md, Sacred Heart Hosp, Thomas B Finan Ctr, & Memt Health Clin, Allegany Health Ctr, 77-; lectr, Squibb Pharmaceut, E R Squibb & Son, 78-; mem adv coun, Md High Blood Pressure Coord Coun, 78- *Mem:* Am Soc Hosp Pharmacists; Am Asn Cols Pharm. *Res:* Tardive dyskinesia; psychotherapeutic agents; hypertension; geriatric health care. *Mailing Add:* Dept Pharm Univ MD Sch Pharm 20 N Pine St Baltimore MD 21201

LOVE, RICHARD HARRISON, b Brooklyn, NY, Aug 23, 39; m 63; c 2. UNDERWATER ACOUSTICS. *Educ:* Univ Md, BS, 61, MS, 63; Cath Univ Am, PhD(mech eng), 76. *Prof Exp:* Res scientist fluid mech, Hydronautics, Inc, 63-65; mech engr, Naval Res Lab, 65-67; oceanographer acoustics, Naval Oceanog Off, 67-76; OCEANOGRAPHER ACOUSTICS, NAVAL OCEAN RES & DEVELOP ACTIV, NAT SPACE TECH LABS, 76- *Mem:* Acoust Soc Am. *Res:* Scattering and reflection of underwater acoustic energy from marine organisms and ocean boundaries. *Mailing Add:* Naval Ocean Atmospheric Res Lab Stennis Space Ctr MS 39529

LOVE, ROBERT MERTON, b Tantallon, Sask, Can, Jan 29, 09; nat US; m 36; c 3. RANGE SCIENCE, ECOLOGY. *Educ:* Univ Sask, BSc, 32, MSc, 33; McGill Univ, PhD(genetics), 35. *Hon Degrees:* LLD, Univ SDak, 85; DSc, McGill Univ, 85. *Prof Exp:* Instr, Univ Sask, 30-32, asst, 32-33; asst, McGill Univ, 33-34; asst agr scientist, Cereal Div, Cent Exp Farm, Can Dept Agr, 35-40; instr agron & jr agronomist, 40-41, from asst prof & asst agronomist to prof & agronomist, 41-85, chmn dept, 59-70, EMER PROF AGRON, EXP STA, UNIV CALIF, DAVIS, 85-, CHMN GRAD PROF ECOL, 71- *Concurrent Pos:* Spec lectr, McGill Univ, 38; Can Dept Agr del, Int Genetics Cong, Scotland, 39; organizer, Cytogenetic Lab, Brazilian Ministry Agr, Rio Grande do Sul, 48-49; Fulbright res scholar, NZ & Australia, 56-57, Greece, 67; chmn range improvement adv comt, State Bd Forestry, Calif, 54-67; vis prof, NC State Univ, 63, Univ Ghana, 70-71 & Univ BC & Univ Guelph, 71; Rockefeller Found travel grant, 64; consult, Int Coop Admin, Govt Spain, 60, Food & Agr Orgn, Greece, 70, Kenya Meat Comn, 71, Spain, 75, AID, Ivory Coast, Univ Chile, 80 & Ctr Arg Engrs Agron, 81; mem panel resource technol, Nat Acad Sci-Nat Res Coun; mem ecol adv comt, Sci Adv Bd, US Environ Protection Agency, 74-78. *Honors & Awards:* Calouste Gulbenakian Award, Portugal, 67; Stevenson Award, Am Soc Agron, 52, Agronomic Serv Award, 66; Medallion Award, Am Forage & Grassland Coun, 66. *Mem:* Fel AAAS; Bot Soc Am; fel Am Soc Agron; hon mem Biol Soc; Can Soc Agron; fel Crop Sci Soc Am. *Res:* Cytogenetics of range forage crops and species; effect of grazing treatment on range species; interspecific hybridization of grasses; range plant improvement; ecology of grasslands. *Mailing Add:* Dept Ecol Univ Calif Davis CA 95616-3621

LOVE, RUSSELL JACQUES, b Chicago, Ill, Jan 11, 31; m 61; c 2. SPEECH PATHOLOGY. *Educ:* Northwestern Univ, Ill, BS, 53, MA, 54, PhD(speech path), 62. *Prof Exp:* Speech & hearing therapist, Moody State Sch Cerebral Palsied Children, Tex, 54-56; staff clinician, Cerebral Palsy Speech Clin, Northwestern Univ, Ill, 58-61; audiologist, WSide Vet Admin Hosp, Chicago, Ill, 61-62; res speech pathologist, Vet Admin Hosp, Coral Gables, Fla, 62-64; assoc prof speech path, DePaul Univ, 64-67; from asst prof to assoc prof, 67-78, PROF SPEECH & LANG PATH, SCH MED, VANDERBILT UNIV, 78- *Concurrent Pos:* Consult speech pathologist, Michael Reese Hosp & Med Ctr, Chicago, Ill, 64-67; chief speech pathologist, Bill Wilkerson Hearing & Speech Ctr, Tenn, 67-71, consult & res speech pathologist, 71-; consult, Vet Admin Hosp, Murfreesboro, 70-89; mem, Nat Adv Comt Accessible Environ, 74-78. *Mem:* Am Speech, Lang & Hearing Asn; Am Cleft Palate Asn; fel Am Speech Lang Hearing Assoc. *Res:* Aphasia; Dyspraxia of speech; Dysarthria; childhood motor speech disability; rights of the handicapped; cerebral palsy; neurology of speech and language. *Mailing Add:* Hearing & Speech Sci Vanderbilt Univ Med Sch Nashville TN 37232-8700

LOVE, SYDNEY FRANCIS, b Winnipeg, Man, June 20, 23. MANAGEMENT SCIENCE, ELECTRONICS. *Educ:* Univ Toronto, BASc, 47, MA, 48; Univ Waterloo, MASc(systs design), 70. *Prof Exp:* Supvr appln, Can Gen Elec Co Ltd, 52-59; mgr TV & organ eng, Electrohome Ltd, 59-66; consult electronics, Sparton of Can, 66-68; PRES MGT SCI, DESIGNECTICS INT INC, 70-, PRES, ADVAN PROF DEVELOP INST, 74- *Concurrent Pos:* Fel, Imp Oil Ltd, 67-68 & Cent Mortgage & Housing Corp, 68-70; consult, Xerox Corp, 72-74 & Govt of Can, 74-77. *Mem:* Sr mem Inst Elec & Electronic Engrs; Proj Mgt Inst. *Res:* The application of engineering principles and models to the practice of management, especially to engineering design management and to project management; author. *Mailing Add:* 23022 Maraleste Rd Laguna Niguel CA 92677-2917

LOVE, TOM JAY, JR, b Jonesboro, Ark, Oct 2, 23; m 45; c 3. HEAT TRANSFER, BIOMEDICAL ENGINEERING. *Educ:* Univ Okla, BS, 48; Univ Kans, MS, 56; Purdue Univ, PhD(mech eng), 63. *Prof Exp:* Proj engr, Colgate Palmolive Co, 47-52; sr res engr, Midwest Res Inst, 52-56; from asst prof to assoc prof mech eng, Univ Okla, 56-65, dir sch, 63-72, prof aerospace & mech eng, 65-72, Halliburton Prof Eng, 72-88, George Lynn Gross Prof, 73-88, EMER GEORGE LYNN CROSS PROF, UNIV OKLA, 88- *Concurrent Pos:* Mem bd dirs, Sverdrup-ARO, Inc, 77-81 & mem, adv coun, Sverdrup Technol, Inc, 77- *Honors & Awards:* Thermophysics Award, Am Inst Aeronaut & Astronaut, 84; Heat Transfer Mem Award, Am Soc Mech Engrs. *Mem:* fel Am Inst Aeronaut & Astronaut; fel Am Soc Mech Engrs; Am Soc Eng Educ; Am Soc Testing & Mat; Am Acad Thermology. *Res:* Physiological heat transfer; radiative heat transfer; thermography. *Mailing Add:* 865 ASP Rm 200 Sch Aerospace Mech & Nuclear Eng Univ Okla Norman OK 73019

LOVE, WARNER EDWARDS, b Philadelphia, Pa, Dec 1, 22; m 45; c 2. BIOPHYSICS. *Educ:* Swarthmore Col, BA, 46; Univ Pa, PhD(physiol), 51. *Prof Exp:* Asst instr physiol, Univ Pa, 48-49, fel biophys, Johnson Found, 51-53, assoc, 53-55; res asst physics, Inst Cancer Res, 55-56, res assoc, 56-57; from asst prof to assoc prof, 57-65, chmn dept, 72-75 & 80-83, PROF BIOPHYS, JOHNS HOPKINS UNIV, 65- *Honors & Awards:* Phillips lectr, Haverford Col, 55. *Mem:* Am Physiol Soc; Biophys Soc; Am Crystallog Asn; Am Soc Biol Chemists. *Res:* Biological ultrastructural basis of functions; x-ray crystallography of macromolecules, hemoglobins, hemocyamins and histone. *Mailing Add:* 5900 Wilmary Lane Baltimore MD 21210

LOVE, WILLIAM ALFRED, b Pittsburgh, Pa, Aug 4, 32; m 57; c 2. PHYSICS, ELEMENTARY PARTICLES. *Educ:* Carnegie Inst Technol, BS, 54, MS, 55, PhD(physics), 58. *Prof Exp:* Res physicist, Carnegie Inst Technol, 58-59; fel, Nat Sci Found, European Orgn Nuclear Res, Switzerland, 59-60; from asst physicist to assoc physicist, 60-66, PHYSICIST, BROOKHAVEN NAT LAB, 66- *Mem:* Fel Am Phys Soc. *Res:* Particle physics. *Mailing Add:* Brookhaven Nat Lab Bldg 510-A Upton NY 11973

LOVE, WILLIAM F, b Houston, Tex, July 3, 25; m 51; c 3. SOLID STATE PHYSICS. *Educ:* Rice Inst, BS, 45, MA, 47, PhD, 49. *Prof Exp:* Instr physics, Randal Morgan Lab, Univ Pa, 49-52, asst prof, 52-54; from asst prof to assoc prof, 54-63, PROF PHYSICS, UNIV COLO, BOULDER, 63- *Mem:* Am Phys Soc; Am Asn Physics Teachers; AAAS; Sigma Xi. *Res:* Symmetry properties of crystals; galvanomagnetic properties of metals and semiconductors in high magnetic fields; electrical noise in semiconductors. *Mailing Add:* Dept of Physics Univ of Colo Boulder CO 80309

LOVE, WILLIAM GARY, b Meridian, Miss, Aug 16, 41; m 66; c 2. NUCLEAR PHYSICS. *Educ:* Univ Tenn, BS, 63, PhD(physics), 68. *Prof Exp:* Res assoc physics, Fla State Univ, 68-70; from asst prof to assoc prof, 70-80, PROF PHYSICS, UNIV GA, 80- *Mem:* Fel Am Phys Soc. *Res:* Study of the properties of the nucleon-nucleon interaction as they are manifested in many nucleon systems, for example, in scattering. *Mailing Add:* Dept of Physics & Astron Univ of Ga Athens GA 30602

LOVECCHIO, FRANK VITO, b Syracuse, NY, Apr 30, 43. ANALYTICAL CHEMISTRY. *Educ:* Syracuse Univ, AB, 65, PhD(chem), 70. *Prof Exp:* Fel, Ohio State Univ, 70-73; RES CHEMIST, EASTMAN KODAK CO, 73- *Mem:* Am Chem Soc; Soc Photog Scientists & Engrs. *Res:* Reactions and mechanisms of coordination compounds, including electron transfer reactions. *Mailing Add:* 1185 Hidden Valley Trail Webster NY 14580-9133

LOVEJOY, DAVID ARNOLD, b Nashua, NH, Dec 12, 43; m 69, 80; c 2. MAMMALIAN ECOLOGY, LOCAL FLORA. *Educ:* Univ Conn, BA, 65, PhD(zool, ecol), 70. *Prof Exp:* From asst prof to assoc prof, 70-85, PROF BIOL, WESTFIELD STATE COL, 85- *Mem:* Am Soc Mammal. *Res:* Ecology of small mammals; Siphonapteran parasites of mammals; flora of Massachusetts. *Mailing Add:* Dept of Biol Westfield State Col Westfield MA 01086

LOVEJOY, DEREK R, b London, Eng, Jan 19, 28; Can citizen; m 53; c 3. ENERGY PHYSICS. *Educ:* Univ London, BS, 50; Univ Toronto, MA, 52, PhD(physics), 54. *Prof Exp:* Assoc res officer, Appl Physics Div, Nat Res Coun Can, 54-66; proj officer, Res Div, 66-72, sr tech adv, Tech Adv Div, UN Develop Prog, 72-78, SR TECH ADV NEW SOURCES OF ENERGY, UN, 78- *Concurrent Pos:* Expert thermal metrol, Nat Phys Lab Metrol Proj, Cairo, United Arab Repub, UNESCO, 64-65. *Mem:* AAAS; US/Int Solar Energy Soc. *Res:* Liquid helium physics; temperature scales and measurements from very low to very high temperatures. *Mailing Add:* United Nations New York NY 10017

LOVEJOY, DONALD WALKER, b New York, NY, Mar 29, 31. GEOMORPHOLOGY & GLACIOLOGY, GENERAL EARTH SCIENCES. *Educ:* Harvard Col, AB, 53; Columbia Univ, AM, 56 & PhD(geol), 58. *Prof Exp:* Asst prof geol, Univ Calif Los Angeles, 57-58; dept chair geol, Rollins Col, Winter Park, 59-62; asst dean, Northeastern Univ, Boston, 62-69; fac dean, Mass Bay Community Col, 69-73; vpres, Nasson Col, Springvale, 73-75; ASSOC PROF OCEANOG, PALM BEACH ATLANTIC COL, 79- *Mem:* fel, Geol Soc Am. *Res:* Deglaciation history of Adirondack Mountains, Upper NY State. *Mailing Add:* Palm Beach Col PO Box 3353 West Palm Beach FL 33402-3353

LOVEJOY, OWEN, b Paducah, Ky, Feb 11, 43; m 69. HUMAN BIOLOGY, BIOMECHANICS. *Educ:* Western Reserve Univ, BA, 65; Case Inst Technol, MA, 67; Univ Mass, Amherst, PhD(human biol), 70. *Prof Exp:* Assoc prof phys anthrop, 69-77, PROF SOCIOL & ANTHROP, KENT STATE UNIV, 77-; ASST CLIN PROF, DIV ORTHOP SURG, SCH MED, CASE WESTERN RESERVE UNIV, 70- *Mem:* Brit Soc Study Human Biol; Am Asn Phys Anthrop; Am Eugenics Soc. *Res:* Primate anatomy, biomechanics and taxonomy; human palaeontology and palaeodemography; skeletal biology. *Mailing Add:* Dept of Sociol & Anthrop Kent State Univ Kent OH 44242-0001

LOVEJOY, ROLAND WILLIAM, b Portland, Ore, June 18, 31; m 59; c 2. PHYSICAL CHEMISTRY. *Educ:* Reed Col, BA, 55; Wash State Univ, PhD(chem), 60. *Prof Exp:* Fel chem, Univ Wash, 59-62; from asst prof to assoc prof, 62-76, PROF CHEM, LEHIGH UNIV, 76- *Mem:* Am Phys Soc. *Res:* Analysis of vibration-rotation spectra of inorganic and organic molecules using infrared and Raman spectroscopy. *Mailing Add:* Dept Chem Lehigh Univ Bethlehem PA 18015

LOVEJOY, THOMAS E, b New York, NY, Aug 22, 41; m 66; c 3. ECOLOGY. *Educ:* Yale Col, BS, 64; Yale Univ, PhD(biol), 71. *Prof Exp:* Res assoc, biol, Univ Pa, 71-74; exec asst to sci dir & asst to vpres resources & planning, Acad Natural Sci, 72-73; vpres sci, World Wildlife Fund-US, 73-; ASST SECY EXTERNAL AFFAIRS, SMITHSONIAN INST. *Concurrent Pos:* Vis lectr, trop ecol, Yale Sch Forestry & Environ Studies, 82; chmn, Wildlife Preservation Trust Int, 74-; treas, Int Coun Bird Preservation, 73-; res assoc ornithol, Acad Natural Sci, 71- *Mem:* AAAS; Am Inst Biol Sci; Am Ornithologists Union. *Res:* Tropical ecology; ornithology; problems of ecology theory relating to conservation and natural resource management. *Mailing Add:* Smithsonian Inst Castle Bldg Rm 317 Washington DC 20560

LOVELACE, ALAN MATHIESON, b St Petersburg, Fla, Sept 4, 29; m 52; c 2. CHEMISTRY. *Educ:* Univ Fla, BA, 51, MA, 52, PhD(chem), 54. *Prof Exp:* Mem staff, Air Force Mat Lab, Wright Patterson AFB, 54-72; dir sci & technol, Andrews AFB, Washington, DC, 72-73; prin dep asst secy, Air Force Res & Develop, 73-74; assoc admin, Aerospace Technol Off Aeronaut & Space Technol, 74-76; dep admin, Aerospace Technol, NASA, 76-81; vpres sci & eng, Gen Dynamics Corp, 81-82, corp vpres prod & qual assurance, 82-85, corp vpres & gen mgr, Space Syst Div, 85-91, CORP VPRES, GEN DYNAMICS CORP & CHMN, COM LAUNCH SERV, INC, 91- *Concurrent Pos:* Chmn, Adv Group Aerospace Res & Develop, NATO, 79- *Honors & Awards:* Von Karman Medal, 84; Goddard Astronaut Award, Am Inst Aeronaut & Astronaut, 89. *Mem:* Nat Acad Eng; fel Am Inst Aeronaut & Astronaut; Am Astronaut Soc; Nat Space Club; Air Force Asn; Sigma Xi. *Res:* High performance macromolecular materials. *Mailing Add:* Gen Dynamics Com Launch Serv PO Box 85911 MS BB-9690 San Diego CA 92186-5911

LOVELACE, C JAMES, b Holdenville, Okla, Sept 26, 34; div; c 3. PLANT PHYSIOLOGY, BIOCHEMISTRY. *Educ:* Harding Col, BS, 61; Utah State Univ, MS, 64, PhD(plant physiol), 66. *Prof Exp:* PROF BOT, HUMBOLDT STATE UNIV, 65- *Concurrent Pos:* Summer res assoc, Utah State Univ, Justus-Liebig Univ, Inst Plant Nutrit, Giessen, Ger, 87-88. *Mem:* Am Soc Plant Physiol; Int Soc Fluoride Res. *Res:* Plant mineral nutrition; fluoride research in relation to enzyme reactions within plants; heavy metal toxicants; iron metabolism in plants; chlorophyl biosynthesis. *Mailing Add:* Dept of Biol Humboldt State Univ Arcata CA 95521

LOVELACE, CLAUD WILLIAM VENTON, b London, Eng, Jan 16, 34. THEORETICAL PHYSICS. *Educ:* Univ Capetown, BS, 54. *Prof Exp:* Dept Sci & Indust Res res fel, Imp Col, Univ London, 61-62, lectr physics, 62-65; sr physicist, Europ Orgn Nuclear Res, Geneva, 65-71; PROF PHYSICS, RUTGERS UNIV, NEW BRUNSWICK, 70- *Res:* Theoretical particle physics; strong interactions; high energy phenomenology. *Mailing Add:* Dept Physics & Astron Rutgers Univ New Brunswick NJ 08903

LOVELACE, RICHARD VAN EVERA, b St Louis, Mo, Oct 16, 41; c 2. PLASMA PHYSICS, ASTROPHYSICS. *Educ:* Wash Univ, BS, 64; Cornell Univ, PhD(physics), 70. *Prof Exp:* Res assoc, Lab Plasma Studies, Cornell Univ, 70-73 & Plasma Physics Lab, Princeton Univ, 73-74; FROM ASST PROF TO PROF APPL PHYSICS, CORNELL UNIV, 75- *Concurrent Pos:* Vis res assoc, US Naval Res Lab, 70-71; consult, Lawrence Livermore Lab, 71- & Plasma Physics Lab, Princeton Univ, 74-75; consult, Los Alamos Nat Lab, 86-; Guggenheim fel, 90. *Mem:* Am Astron Soc; Am Phys Soc; Int Astron Union. *Res:* Plasma physics of controlled fusion systems; collective phenomena of galaxies and quasars; wave propagation through random media; relativistic magnetohydrodynamics of astrophysical and laboratory jets and beams. *Mailing Add:* Dept Appl Physics Cornell Univ 237 Clark Hall Ithaca NY 14853

LOVELAND, DONALD WILLIAM, b Rochester, NY, Dec 26, 34. MATHEMATICS, COMPUTER SCIENCE. *Educ:* Oberlin Col, AB, 56; Mass Inst Technol, SM, 58; NY Univ, PhD(math), 64. *Prof Exp:* Mathematician & programmer, Int Bus Mach Corp, 58-59; instr math, NY Univ, 63-64, asst prof, 64-67; from asst prof to assoc prof, Carnegie-Mellon Univ, 67-73; PROF COMPUT SCI, DUKE UNIV, 73- *Mem:* Asn Comput Mach; Asn Symbolic Logic; AAAS. *Res:* Artificial intelligence; theorem proving by computer; logic programming; fast approximation algorithms for computationally hard problems. *Mailing Add:* 3417 Cambridge Rd Durham NC 27707

LOVELAND, ROBERT EDWARD, b Camden, NJ, May 3, 38; m 62; c 3. BIOLOGY. *Educ:* Rutgers Univ, Camden, AB, 59; Harvard Univ, MA, 61, PhD(biol), 63. *Prof Exp:* Asst prof biol, Long Beach State Col, 63-64; asst prof zool, 64-70, ASSOC PROF ZOOL, RUTGERS UNIV, NEW BRUNSWICK, 70- *Concurrent Pos:* NSF sci fac fel, Univ BC, 71-72. *Mem:* AAAS; Atlantic Estuarine Res Soc. *Res:* Distribution of marine invertebrates; behavioral modelling; population models of biological systems. *Mailing Add:* Dept Biol Rutgers Univ New Brunswick NJ 08903

LOVELAND, WALTER (DAVID), b Chicago, Ill, Dec 23, 39; m 62. NUCLEAR CHEMISTRY. *Educ:* Mass Inst Technol, SB, 61; Univ Wash, PhD(chem), 66. *Prof Exp:* Res assoc chem, Argonne Nat Lab, 66-67; res asst prof, 67-68, from asst prof to assoc prof, 68-81, PROF CHEM, ORE STATE UNIV, 81- *Concurrent Pos:* US Dept Energy res grant, Ore State Univ, 68-; vis scientist, Lawrence Berkeley Lab, 76, 77 & 80; Tartar fel, Ore State Univ, 77. *Mem:* AAAS; Am Phys Soc; Am Chem Soc. *Res:* Nuclear reactions, especially heavy ion reactions and fission; activation analysis; use of computers for data acquisition; environmental chemistry. *Mailing Add:* Radiation Ctr Ore State Univ Corvallis OR 97331

LOVELESS, SCOTT E, MACROPHAGE ACTIVATION, ADOPTIVE IMMUNOTHERAPY. *Educ:* Va Commonwealth Univ, PhD(pharmacol), 80. *Prof Exp:* RES IMMUNOTHERAPIST & PHARMACOLOGIST, E I DU PONT DE NEMOURS & CO, INC, 84- *Mailing Add:* Immunotoxicol Cent Res Dept E I du Pont de Nemours & Co Inc Haskell Lab Elkton Rd Newark DE 19714

LOVELL, BERNARD WENTZEL, b Greenfield, Mass. COMPUTER SCIENCE, ELECTRICAL ENGINEERING. *Educ:* Mass Inst Technol, BS, 58, MS, 58, EE, 63; Univ Conn, PhD(comput sci), 69. *Prof Exp:* Electronic engr, US Naval Ord Lab, 54-59; instr elec eng, Mass Inst Technol, 59-63; asst prof, Univ Mass, 63-67; ASSOC PROF ELEC ENG, UNIV CONN, 69- *Mem:* Inst Elec & Electronics Engrs; Am Phys Soc; Asn Comput Mach. *Res:* Automata theory; operating systems. *Mailing Add:* Comput Ctr & Eng Univ Conn Main Campus U-155 260 Glennbrook Storrs CT 06268

LOVELL, CHARLES W(ILLIAM), JR, b Louisville, Ky, Nov 16, 22; m 48; c 2. CIVIL ENGINEERING, SOIL MECHANICS. *Educ:* Univ Louisville, BCE, 44; Purdue Univ, MSCE, 51, PhD(civil eng), 57. *Prof Exp:* Instr civil eng, Univ Louisville, 46-48; from res asst to res engr & instr, 48-57, from asst prof to assoc prof, 57-76, PROF CIVIL ENG, PURDUE UNIV, 76- *Concurrent Pos:* Vis assoc prof, Mass Inst Technol, 62-63; mem, Hwy Res Bd, Nat Acad Sci-Nat Res Coun. *Mem:* Nat Soc Prof Engrs; Am Soc Civil Engrs; Am Soc Eng Educ; Am Soc Testing & Mat. *Res:* Frost action; load-deformation characteristics of soils; subsurface exploration. *Mailing Add:* Dept Civil Eng Purdue Univ Main Campus West Lafayette IN 47907

LOVELL, EDWARD GEORGE, b Windsor, Ont, May 25, 39; US citizen; div; c 2. ENGINEERING, STRUCTURAL MECHANICS. *Educ:* Wayne State Univ, BSAE, 60, MSEM, 61; Univ Mich, Ann Arbor, PhD(eng mech), 67. *Prof Exp:* Instr eng mech, Univ Mich, Ann Arbor, 63-67; Nat Acad Sci-Nat Res Coun res associateship, Langley Res Ctr, NASA, 67-68; PROF ENG MECH, UNIV WIS-MADISON, 68- *Concurrent Pos:* Proj engr, Boeing Co, Wash, 62; design engr, Pratt & Whitney Aircraft, Conn, 70; NATO sr sci fel, Univ Manchester, Eng, 73; Fusion Technol Inst, Wis Ctr Appl Microelectronics, Univ Wis. *Mem:* Sigma Xi; Am Soc Mech Engrs. *Res:* Nonlinear vibrations of structures; structural instability; stress analysis; nuclear reactor structural mechanics; microelectromechanical systems. *Mailing Add:* Dept Eng Mech Univ Wis Madison WI 53706

LOVELL, HAROLD LEMUEL, b Bellwood, Pa, July 13, 22; m 44; c 2. FUEL SCIENCE, MINERAL ENGINEERING. *Educ:* Pa State Univ, BS, 43, MS, 45, PhD(fuel tech), 52. *Prof Exp:* Asst chem micros, Pa State Univ, 43-44, microchem, 44-45; res chemist, Mallinckrodt Chem Works, 45-47; asst fuel technol, 47-51, res assoc spectros, 51-52, asst prof, 52-58, mineral prep, 58-64, assoc prof, 64-71, actg head dept mineral prep, 64-68, assoc prof, 71-76, prof mineral eng, 77-82, dir mine drainage res sect, 68-82, EMER PROF MINERAL ENG, PA STATE UNIV, 82- *Concurrent Pos:* Consult, US Dept Com-Com Tech Adv Bd, 74-75. *Mem:* Am Chem Soc; Am Inst Mining, Metall & Petrol Eng. *Res:* Mineral preparation; analytical chemistry; absorption and emission; microchemistry; coal constitution chemistry; chemical utilization of coal; mine water pollution-treatment; coal preparation; physical processing of coal for synfuel feed stock management. *Mailing Add:* 120 W Mitchell Ave State Col PA 16803-3544

LOVELL, JAMES BYRON, b Fallentimber, Pa, Mar 19, 27; c 2. ENTOMOLOGY. *Educ:* Pa State Univ, BS, 50; Univ Ill, MS, 55, PhD(entom), 56. *Prof Exp:* Entomologist, US Army Chem Ctr, Md, 50-53; asst, Univ Ill, 53-56; PRIN SCIENTIST, AGR RES DIV, AM CYANAMID CO, 56- *Mem:* AAAS; Entom Soc Am. *Res:* Insect physiology and toxicology; mode of action of insecticides; mechanism of resistance in insects. *Mailing Add:* Agr Res Div Am Cyanamid Co PO Box 400 Princeton NJ 08540

LOVELL, RICHARD ARLINGTON, b Kentland, Ind, Aug 4, 30; m 65; c 6. NEUROCHEMISTRY. *Educ:* Xavier Univ, Ohio, BS, 52, MS, 53; St Louis Univ, Lic Philos, 59; McGill Univ, PhD(biochem), 63. *Prof Exp:* Proj assoc physiol, Epilepsy Res Ctr, Univ Wis, 64-65; USPHS Pharmacol Res Training Prog fel psychiat, Yale Univ, 65-66; from instr to asst prof neurochem in psychiat, Univ Chicago, 69-75; MGR BIOCHEM PHARMACOL, CIBA-GEIGY CORP, SUMMIT, NJ, 75- *Concurrent Pos:* Res fel, Schweppe Found, 68-71. *Mem:* AAAS; Am Chem Soc; Am Soc Neurochem; Am Epilepsy Soc; Soc Neurosci. *Res:* Neurochemistry; neuropharmacology; biochemical control mechanisms in the nervous system. *Mailing Add:* 479 Snyder Ave Berkeley Heights NJ 07922-1492

LOVELL, RICHARD THOMAS, b Lockesburg, Ark, Feb 21, 34; m 63; c 2. FISHERIES. *Educ:* Okla State Univ, BS, 56, MS, 58; La State Univ, PhD(nutrit, biochem), 63. *Prof Exp:* From asst prof to assoc prof food sci, La State Univ, 63-69; assoc prof, 69-75, PROF FISHERIES & ALLIED AQUACULT, AUBURN UNIV, 75- *Concurrent Pos:* Consult fish cult, US AID, 72-74; columnist, Aquacult Mag, 74-; mem, Comt Animal Nutrit, Nat Res Coun-Nat Acad Sci, 74-; assoc ed, Trans Am Fisheries Soc, 75-; chmn, Comt Fish Nutrit, Nat Res Coun-Nat Acad Sci, 89-90. *Mem:* Fel Am Inst Chemists; Am Fisheries Soc; Am Chem Soc; Inst Food Technologists; Am Inst Nutrit. *Res:* Fish nutrition, especially vitamin C requirements and energy metabolism of warm water fish cultured for food; environment-related off-flavors in intensively-cultured food fishes. *Mailing Add:* Dept Fisheries & Allied Aquacult Auburn Univ Auburn AL 36849

LOVELL, ROBERT EDMUND, b Ann Arbor, Mich, Aug 12, 21; m 47. SYSTEMS ENGINEERING. *Educ:* Univ Mich, BSE, 43; Univ Ariz, MS, 71, PhD(systems eng), 75. *Prof Exp:* Navigator & electronics officer, US Navy, 43-46; engr & mgr, Dyncorp, 49-71; assoc prof indust eng, Ariz State Univ, 72-76; assoc prof systs eng, Univ Petrol & Minerals, Saudi Arabia, 76-77; assoc prof indust eng & comput sci, Ariz State Univ, 77-81; prof, 81-86, EMER PROF ELEC & COMPUTER ENG, UNIV PAC, 86- *Mem:* Inst Elec & Electronic Engrs Computer Soc; Am Soc Eng Educ. *Res:* Computer simulation of large scale systems; co-author of one book. *Mailing Add:* PO Box 7152 Stockton CA 95267-0152

LOVELL, ROBERT GIBSON, b Ann Arbor, Mich, May 13, 20; m 48; c 5. INTERNAL MEDICINE, ALLERGY. *Educ:* Univ Mich, MD, 44, AB, 57. *Prof Exp:* From instr to asst prof, Univ Mich, Ann Arbor, 50-73, fac secy, 54-56, asst dean sch med, 57-59, clin prof internal med, 73-90; RETIRED. *Concurrent Pos:* Consult physician, US Vet Admin Hosp, 54-55; consult, President's Comm Vet Pensions, US Air Force, 55 & Wayne County Gen Hosp,Mich, 59-; chmn med ed comt, St Joseph Mercy Hosp, 63- *Mem:* AMA; assoc Am Col Chest Physicians; fel Am Acad Allergy. *Res:* Use of medications and aerosol preparations in treatment of bronchial asthma. *Mailing Add:* 3000 Geddes Ave Ann Arbor MI 48104

LOVELL, ROBERT R(OLAND), b Gladwin, Mich, Feb 22, 37. SPACE TECHNOLOGY. *Educ:* Univ Mich, BS & MS. *Prof Exp:* Var tech & tech mgt positions, Lewis Res Ctr, NASA, Cleveland, Ohio, 62-80, dir, Satellite Commun Advan Res & Develop Prog, NASA Hq, 80-87; CORP VPRES & PRES, SPACE SYSTS DIV, ORBITAL SCI CORP, 87- *Concurrent Pos:* Mgr, Pegasus Prog, Orbital Sci Corp. *Honors & Awards:* Nat Medal of Technol, 91; Yuri Gagarin Medal, USSR Acad Cosmonautics. *Mem:* Am Inst Aeronaut & Astronaut. *Res:* Development of unmanned spacecraft technology; communications systems; rocket propulsion; author of over 50 technical publications; holder of several patents. *Mailing Add:* Orbital Sci Corp 12500 Fair Lakes Circle Fairfax VA 22033

LOVELL, STUART ESTES, b Seattle, Wash, Oct 8, 28; m 55; c 2. COMPUTER SCIENCE. *Educ:* Univ Wash, BS, 53; Brown Univ, PhD(chem), 58. *Prof Exp:* Proj assoc chem, Univ Wis, 58-63, asst prof comput sci, 63-67; mgr computer serv, 65-75, SYSTEM ANALYST, KITT PEAK OBSERV, 75- *Mem:* Asn Comput Mach. *Res:* Systems programming; computer based systems. *Mailing Add:* 5221 E Rosewoods Tucson AZ 85711

LOVELOCK, DAVID, b Bromley, Eng. MATHEMATICS, THEORETICAL PHYSICS. *Educ:* Univ Natal, BSc, 59, Hons, 60, PhD(math), 62, DSc, 74. *Prof Exp:* Res asst math, Univ Natal, 60-61; jr fel, Bristol Univ, 62-63, lectr, 63-69; assoc prof, 69-74, prof appl math, Univ Waterloo, Ont, 74; PROF MATH, UNIV ARIZ, 74- *Concurrent Pos:* Nat Res Coun Can grant, Univ Waterloo, 69-, adj prof appl math, 74- *Mem:* Am Math Soc; Tensor Soc. *Res:* General relativity; calculus of variations; differential geometry. *Mailing Add:* Dept of Math Univ Ariz Tucson AZ 85721

LOVELY, RICHARD HERBERT, b Santa Monica, Calif, Sept 20, 41; m 88. NEUROBEHAVIORAL TOXICOLOGY, HEALTH PSYCHOLOGY. *Educ:* Calif State Univ, Northridge, BA, 65; Cent Wash Univ, MS, 67; Univ Wash, Seattle, PhD(psychol), 74. *Prof Exp:* Instr psychol, Yakima Valley Col, 67-68; lectr psychol, Univ Wash, 70-72, psychol & rehab med, 73-75, asst prof, 75-79; sr res scientist neurosci, Dept Biol, 79-89, EPIDEMIOL & BIOMET, BATTELLE MEM INST, 89- *Concurrent Pos:* Vis prof, Calif State Univ, Chico, 72-73; consult & US-USSR exchange scientist, Nat Inst Environ Health Sci, 76-82, Nat Coun Radiation Protection & Measurements, 78-85; bd dirs, Bioelectromagnetic Soc, 85-88. *Mem:* Soc Neurosci; Psychonomic Soc; Neurobehav Teratol Soc; Bioelectromagnetics Soc. *Res:* Biopsychological effects of electromagnetic radiation exposure; in utero determinants of adult behavior; neurobehavioral toxicology; neural substrates of learning and memory, limbic system functions and constraints on animal behavior; AIDS in adolescents. *Mailing Add:* Epidemiol & Biomet Battelle Mem Inst Seattle WA 98105

LOVENBERG, WALTER MCKAY, b Trenton, NJ, Aug 9, 34; m 58; c 2. BIOCHEMISTRY. *Educ:* Rutgers Univ, BS, 56, MS, 58; George Washington Univ, PhD(biochem), 62. *Prof Exp:* Biochemist, Nat Heart, Lung & Blood Inst, 59-72, trainee, 62-63, head sect biochem pharmacol, Hypertension-Endocrine Br, 72-85; dir biochem sci, Merrel Dow Res Inst, 85-86, vpres Strasbourg, France, 86-89, PRES, MARION MERRELL DOW RES INST, 89- *Concurrent Pos:* Exec ed, J Anal Biochem; US ed, J Neurochem Int. *Mem:* Am Soc Biol Chem; Am Soc Pharmacol & Exp Therapeut; Biochem Soc; Am Soc Neurochem; Am Col Neuropsychopharmacol. *Res:* Enzymatic mechanisms and the chemistry of proteins involved in neurohumoral amine biosynthesis. *Mailing Add:* Marion Merrel Dow Res Inst 2110 E Galbraith Rd Cincinnati OH 45215-6300

LO VERDE, PHILIP THOMAS, b Benton Harbor, MI, Oct 5, 46; m 65; c 3. PARASITOLOGY, MEDICAL MALACOLOGY. *Educ:* Univ Mich, BS, 68, MS, 71, MS, PhD(epidemiol sci), 76. *Prof Exp:* Mus asst, Zool Mus, Univ Mich, 68-69, NIH fel, 70-75, curatorial asst, 73-75, teaching fel, Dept Zool, 72-73; res assoc med malacol, Ain Shams Univ, Cairo, 74-75; asst prof parasitol, dept biol sci, Purdue Univ, 76-81; ASST PROF PARASITOL, DEPT MICRO, SCH MED, STATE UNIV NY, BUFFALO, 81- *Concurrent Pos:* Guest scientist, Naval Med Res Unit No 3, Cairo, Egypt, 74-75; Spec Study Sect, NIH, 83, Tropical Dis Unit Study Sect, 84, SBIR Study Sect, 86. *Honors & Awards:* Chester A Herrick Award, Eli Lilly & Co, 74. *Mem:* Am Soc Parasitologists; AAAS; Am Soc Trop Med Hyg. *Res:* Host-parasite interrelationships; invertebrate defense mechanisms; parasite immunology and molecular biology; parasitology; malacology; schistosomiasis. *Mailing Add:* Microbiol 203 Sherman State Univ NY Buffalo Sch Med 3435 Main St Buffalo NY 14214

LOVERING, EDWARD GILBERT, b Winnipeg, Man, Oct 15, 34; m 58; c 3. PHARMACEUTICAL CHEMISTRY. *Educ:* Univ Man, BSc, 57, MSc, 58; Univ Ottawa, PhD(chem), 61. *Prof Exp:* Sci officer radiation chem, Defense Res Bd, 58-59; Nat Res Coun Can fel, Oxford Univ, 61-63; res chemist, Polymer Corp Ltd, 63-69, assoc scientist, Polymer Corp, 69-71; RES SCIENTIST HEALTH & WELFARE OF CAN, 71- *Mem:* Am Chem Soc; Chem Inst Can; Sigma Xi. *Res:* Contaminants in drugs and cosmetics; drug raw material characterization; drug stability; pharmaceutical and cosmetic analysis. *Mailing Add:* Bur Drug Res Tunney's Pasture Ottawa ON K1A 0L2 Can

LOVERING, THOMAS SEWARD, economic geology; deceased, see previous edition for last biography

LOVESTEDT, STANLEY ALMER, b Iliff, Colo, June 7, 13; m 40; c 3. ORAL SURGERY. *Educ:* Univ Southern Calif, BS & DDS, 38; Univ Minn, MS, 45; Am Bd Oral Surg, dipl. *Prof Exp:* Resident oral surg, Mayo Grad Sch Med, 38-43, mem fac, 46-60, from instr to assoc prof, 60-69, prof clin dent, 67-73, prof dent, 73-78; consult, 43-62, head sect, 55-62, SR CONSULT, DEPT DENT & ORAL SURG, MAYO CLIN, 62-; EMER PROF DENT, MAYO MED SCH, 78- *Concurrent Pos:* Chief oral diag & roentgenology, US Army Dent Corps, Brooke Army Hosp, Ft Sam Houston, Tex, 53-55; mem, dent study sect, NIH, 64-68. *Mem:* Fel AAAS; Am Soc Oral Surgeons; Am Dent Asn; fel Am Acad Dent Radiol; fel Am Col Dent (vpres, 65-66, pres, 68-69); Sigma Xi; Am Asn Dent Schs; Am Acad Hist Dent; Am Acad Oral Path; Am Acad Hist Med; Am Cancer Soc; Am Acad Oral Med. *Res:* Radiology; oral medicine. *Mailing Add:* Mayo Clin 200 First St SW Rochester MN 55901

LOVETT, EDMUND J, III, EXPERIMENTAL BIOLOGY. *Prof Exp:* DIR & CHIEF EXEC OFFICER, MAINE CYTOMETRY RES INST, 86- *Mailing Add:* Maine Cytometry Res Inst 125 John Roberts Rd Suite 8 South Portland ME 04106

LOVETT, EVA G, b Orange, NJ, Aug 17, 40; m 63. ANALYTICAL & POLYMER CHEMISTRY. *Educ:* Douglass Col, Rutgers Univ, BA, 62; Univ Rochester, PhD(chem), 66. *Prof Exp:* Sr chemist, Merck, Sharp & Dohme Res Lab, 66-67; res assoc chem, Washington Univ, St Louis, 69-76; res chemist, Tretolite Div, Petrolite Corp, St Louis, 76-89; res assoc chem, Victoria Univ, Wellington, NZ, 89-90; INSTR CHEM, FOREST PARK COMMUNITY COL, 91- *Mem:* AAAS; Am Chem Soc. *Res:* Synthesis, degradation and mass spectroscopy of natural products, particularly purines, pyrimidines and related heterocyclic compounds; polymer synthesis and characterization. *Mailing Add:* 6807 Pershing Ave St Louis MO 63130

LOVETT, GARY MARTIN, b Albany, NY, July 6, 53. FOREST NUTRIENT CYCLING, ATMOSPHERE-FOREST INTERACTIONS. *Educ:* Union Col, BS, 75; Dartmouth Col, PhD(biol), 81. *Prof Exp:* Res assoc, Oak Ridge Nat Lab, 81-85; asst scientist, 85-89, ASSOC SCIENTIST, INST ECOSYST STUDIES, 89- *Concurrent Pos:* Assoc mem, Grad Prog Ecol, Rutgers Univ, 90- *Mem:* Ecol Soc Am; AAAS. *Res:* Forest nutrient cycling, in particular patterns and mechanisms of atmospheric deposition and atmosphere/canopy interactions; physiological ecology and effects of air pollution on trees. *Mailing Add:* Inst Ecosyst Studies Box AB Millbrook NY 12545

LOVETT, JACK R, acoustics, physical oceanography, for more information see previous edition

LOVETT, JAMES SATTERTHWAITE, b Fallsington, Pa, Aug 22, 25; m 46; c 1. BOTANY, BIOCHEMISTRY. *Educ:* Earlham Col, AB, 53; Mich State Univ, PhD, 59. *Prof Exp:* Fel bot, Mich State Univ, 59-60; from asst prof to assoc prof, 60-69, assoc head, Dept Biol Sci, 75-79, prof mycol, 69-87, EMER PROF BIOL SCI, PURDUE UNIV, 87- *Concurrent Pos:* Nat Sci Found sr fel, 66-67; Europ Molecular Biol Orgn fel, 71; assoc ed, Exp Mycol, 76-82, ed-in-chief, 82-87; mem, Microbial Physiol Study Sect, NIH, 79-81; co-chair, Gordon Res Conf, Fungal Metabol, 82. *Mem:* Am Asn Univ Prof. *Res:* Physiology of fungi; genetic and metabolic control of development in the lower fungi, principally the aquatic Phycomycetes. *Mailing Add:* Dept Biol Sci Purdue Univ West Lafayette IN 47907

LOVETT, JOHN ROBERT, b Norristown, Pa, June 17, 31; m 56; c 3. ORGANIC CHEMISTRY. *Educ:* Ursinus Col, BS, 53; Univ Del, MS, 55, PhD(chem), 57. *Prof Exp:* Res chemist polymer processes, Esso Res & Eng Co, 57-59, prof leader high energy propellants, 59-60, sr chemist, 60-61, sect head, 61-64, dir govt res lab, 65-68, dir petrol additives lab, 68-70, vpres paramins dept, Exxon Chem Co, 70-73, worldwide tech mgr, 73-76; vpres res, Air Prod & Chem Inc, 76-81, mem bd dir, 77-81; PRES & MEM BD DIR, AIR PROD EUROPE, INC, 81- *Concurrent Pos:* Bd trustees, Cedar Crest Col, 77-81; adv bd, US Dept Energy, 78-81; dir, Amersham Int Plc & Am Chamber Com, UK, 82- *Mem:* Am Chem Soc; Indust Res Inst; Mfg Chemists Asn; AAAS. *Res:* Polymers; chemical additives; industrial gases; catalysts; fossil energy technology. *Mailing Add:* 2830 Liberty St Allentown PA 18104-4748

LOVETT, JOSEPH, b Columbus Co, NC, Feb 24, 33; m 57; c 2. ENVIRONMENTAL HEALTH, PUBLIC HEALTH. *Educ:* Wake Forest Univ, BS, 56; Univ NC, Chapel Hill, MSPH, 60; Univ Minn, Minneapolis, MS, 65, PhD(environ health, microbiol), 71. *Prof Exp:* Asst supt water treat, City of Raleigh, NC, 57-58; regional consult, Interstate Carrier Prog Environ Eng & Food Protection, USPHS, 60-64, chief mycol sect, Food Microbiol Br, 66-77, asst br chief, bact physiol br. Div Microbiol, Bur Foods, Food & Drug Admin, 77-87, CHIEF, MICROBIOL HAZARDS EVALUATION GROUP, LAB QUALITY ASSURANCE BR, USPHS, 87- *Mem:* Int Asn Milk, Food & Environ Sanitarians; Am Soc Microbiol; Sigma Xi; Inst Food Technologists. *Res:* Toxic microbial metabolites in foods and the ecology of toxigenic and pathogenic microorganisms. *Mailing Add:* 1682 Vaquera Pl Cincinnati OH 45255

LOVETT, PAUL SCOTT, b Philadelphia, Pa, Dec 14, 40; m 82; c 2. MICROBIOLOGY. *Educ:* Delaware Valley Col, BS, 64; Temple Univ, PhD(microbiol), 68. *Prof Exp:* USPHS fel microbiol, Scripps Clin & Res Found, Calif, 68-70; from asst prof to assoc prof, 70-78, PROF BIOL SCI, UNIV MD, BALTIMORE COUNTY, 78- *Concurrent Pos:* USPHS career develop award, 76-81; mem gen biol study sect, NSF, 78-81. *Honors & Awards:* Distinguished Young Scientist Award, Md Acad Sci, 75. *Mem:* Am Soc Microbiol. *Res:* Microbial genetics; mechanisms of bacteriophage infection; bacillus plasmids; regulation of inducible cat genes. *Mailing Add:* Dept of Biol Sci Univ of Md Baltimore County Catonsville MD 21228

LOVICH, JEFFREY EDWARD, b Alexandria, Va; m; c 1. MORPHOMETRIC ANALYSIS, EVOLUTIONARY ECOLOGY. *Educ:* George Mason Univ, BS, 82, MS, 84; Univ Ga, PhD(ecol), 90. *Prof Exp:* Fac res assoc, Savannah River Ecol Lab, Univ Ga, 90-91; WILDLIFE BIOLOGIST, US BUR LAND MGT, 91- *Mem:* Am Soc Ichthyologists & Herpetologists; Soc Study Amphibians & Reptiles; Herpetologists' League. *Res:* Evolutionary ecology, systematics and conservation of North American and Southeast Asian turtles; analysis and comparison of populations using demographic analysis and state-of-the-art morphometric techniques. *Mailing Add:* Bur Land Mgt 6221 Box Springs Blvd Riverside CA 92507-0714

LOVICK, ROBERT CLYDE, b Atchison, Kans, Aug 25, 21; m 45; c 1. ELECTRICAL ENGINEERING. *Educ:* Univ Nebr, BSc, 44. *Prof Exp:* Sr tech assoc, Eastman Kodak Co, 44-78; PRES, IDEAS FOR INDUST, 78- *Concurrent Pos:* Soc Motion Picture & TV Eng fel, 63; lectr Creativity & Innovation. *Mem:* fel Soc Motion Picture & TV Eng. *Res:* Development of proximity fuses for naval ordnance; systems for silver sound records on reversal color films; development of magnetic prestriping on removable backing color films; co-inventor multi-layer digital magnetic recording media; establishment of electronic-optical image evaluation center. *Mailing Add:* Ideas for Indust 2608 Kanuga Pines Dr Hendersonville NC 28739-7014

LOVINGER, ANDREW JOSEPH, b Athens, Greece, May 15, 48; US citizen; m 76; c 2. POLYMER & MATERIALS SCIENCE, CHEMICAL ENGINEERING. *Educ:* Columbia Univ, BS, 70, MS, 71, ScD, 77. *Prof Exp:* Mem tech staff, 77-85, DISTINGUISHED MEM TECH STAFF, AT&T BELL LABS, 85-, HEAD, POLYMER CHEM RES DEPT, 85- *Concurrent Pos:* Adj asst prof, dept chem eng & appl chem, Columbia Univ, NY, 81-82, adj assoc prof, 82-83. *Honors & Awards:* Dillon Medal, Am Phys Soc, 85. *Mem:* Fel Am Phys Soc; Am Chem Soc; fel AAAS. *Res:* Structure and properties of polymeric materials; piezoelectric and ferroelectric polymers; high temperature and high strength polymers and silicon based polymers; morphology and phase transitions. *Mailing Add:* AT & T Bell Labs Murray Hill NJ 07974

LOVINGOOD, JUDSON ALLISON, b Birmingham, Ala, July 18, 36; m 55; c 4. MATHEMATICS, ELECTRICAL ENGINEERING. *Educ:* Univ Ala, BSEE, 58, PhD(math), 68; Univ Minn, MS, 63. *Prof Exp:* Assoc engr, Martin Co, 58-59; res engr, Honeywell Inc, 59-62; aerospace engr, Marshall Space Flight Ctr, NASA, 62-64, dep chief astrodyn guid theory div, 64-69, chief dynamics & control div, Aero-Astrodyn Lab, 69-74, dir, Systs Dynamics Lab, 74-79, dept mgr, Space Shuttle Prog & Space Shuttle Main Engine, 79-87; DIR, ENG & RES, THIOKOL CORP, 88- *Concurrent Pos:* Asst prof, Univ Ala, Huntsville, 68- *Mem:* Am Inst Aeronaut & Astronaut. *Res:* Optimal and adaptive control theory research applications to launch and space vehicles; mathematical research in guidance theory, control theory and celestial mechanics. *Mailing Add:* 117 Hickory Hills Rd Gurley AL 35748

LOVINS, ROBERT E, b Ashgrove, Mo, Sept 25, 35; m 56; c 1. MASS SPECTROMETRY, PROTEIN BIOCHEMISTRY. *Educ:* Univ Calif, Riverside, AB, 58; San Jose State Col, MS, 61; Univ Calif, Davis, PhD(chem), 63. *Prof Exp:* Lectr, chem & res chemist, Univ Calif, 63-65; res assoc chem, Mass Inst Technol, 65-66, asst dir, Mass Spectrometry Lab, 66-69; assoc prof biochem & dir, High Resolution Mass Spectrometry Ctr, Univ Ga, 69-76; assoc prof basic & clin immunol & microbiol, Med Univ SC, 76-; EARLY-CLAY LABS, INC. *Concurrent Pos:* NIH Res Career Develop Award, 71- *Mem:* Am Soc Biol Chem; Am Asn Immunologists; Am Soc Mass Spectrometry. *Res:* Application of mass spectrometry to the sequence analysis of heterogeneous proteins; structural heterogeneity of antibody proteins; tumor antigen structures. *Mailing Add:* Dept Microbiol & Immunol Med Univ SC Col Med 171 Ashely Ave Charleston SC 29425

LOVRIEN, REX EUGENE, b Eagle Grove, Iowa, Jan 25, 28; m 56; c 3. PHYSICAL BIOCHEMISTRY. *Educ:* Univ Minn, BS, 53; Univ Iowa, PhD, 58. *Prof Exp:* Res assoc phys chem, Yale Univ, 58-61; from asst prof, to asoc prof, 65-76, PROF BIOCHEM, UNIV MINN, 76- *Mem:* Am Chem Soc; Biophys Soc; Sigma Xi. *Res:* Macromolecular biochemistry; solution physical chemistry; light energy utilization; calorimetry enzymology; protein separation. *Mailing Add:* Dept Biochem Gortner Lab Col Biol Sci Univ Minn St Paul MN 55108

LOVSHIN, LEONARD LOUIS, JR, b Rochester, Minn, Mar 21, 42; m 73; c 2. AQUACULTURE, HATCHERY MANAGEMENT. *Educ:* Miami Univ, BA, 64; Univ Wis, MS, 66; Auburn Univ, PhD(fisheries), 72. *Prof Exp:* Asst prof, 72-78, assoc prof, 78-85, PROF, FISHERIES, AUBURN UNIV, 85- *Concurrent Pos:* USAID-Auburn Univ proj coordr, Tech Assistance Prog Fisheries Develop, Ctr Ichthyol Res, Fortaleza, Brazil, 72-79, proj coordr to Govt of Panama, small farmer aquaculture develop, 81-84. *Mem:* Am Fisheries Soc; World Aquaculture Soc. *Res:* Fish culture research dealing with Tilapias, all male hybrid tilapias, native species indigenous to Brazil, and the extension of research results to local fish farmers; integrated aquaculture development in rural, tropical Latin America, hatchery management. *Mailing Add:* Dept Fisheries Auburn Univ Main Campus Auburn AL 36849

LOW, BARBARA WHARTON, b Lancaster, Eng, Mar 23, 20; nat US; m 50. PROTEIN STRUCTURE & FUNCTION. *Educ:* Oxford Univ, BA, 42, MA, 46, DPhil(chem), 48. *Prof Exp:* Res assoc, Harvard Med Sch, 48, assoc phys chem, 48-50, asst prof, Harvard Univ, 50-56; from assoc prof to prof biochem, 56-85, prof biochem & molecular biophysics, 85-90, EMER PROF & SPEC LECTR, COL PHYSICIANS & SURGEONS, COLUMBIA UNIV, 90- *Concurrent Pos:* Rose Sidgwick Mem fel, Am Asn Univ Women, 46-47; Spec Rockefeller Found fel, 47; assoc mem, Lab Phys Chem, Harvard Univ, 50-54; sr res fel, NIH, 59-63, career develop award, 63-68, mem, Biophysics & Biophys Chem Study Sect, Div Res Grants, 66-69; consult, USPHS; vis prof, Univ Strasbourg, 65 & Tohoku Univ, 75; invited lectr, Chinese Acad Sci, 81, Acad Sci, USSR, 88. *Mem:* AAAS; Am Inst Physics; Am Soc Biol Chem; Am Crystallog Asn; Am Acad Arts & Sci; Biophys Soc; Harvey Soc; Int Soc Toxinology. *Res:* X-ray crystal structure of non-enzyme proteins and peptides, particularly snake venom post-synaptic neurotoxins, cytotoxins and membrane proteins; protein-protein interactions; prediction of protein conformation; curaremimetic toxins, interaction with acetylcholine receptors; choleic acids. *Mailing Add:* Col Physicians & Surgeons Columbia Univ 630 W 168th St New York NY 10032

LOW, BOBBI STIERS, b Louisville, Ky, Dec 4, 42. EVOLUTIONARY BIOLOGY, ECOLOGY. *Educ:* Univ Louisville, BA, 62; Univ Tex, Austin, MA, 64, PhD(evolutionary zool), 67. *Prof Exp:* Can Med Res Coun fel physiol, Univ BC, 67-69; Commonwealth Sci & Res Orgn res assoc ecol, Univ Melbourne & Univ S Australia, 69-72; asst prof, 72-75, ASSOC PROF RESOURCE ECOL, SCH NATURAL RESOURCES, UNIV MICH, ANN ARBOR, 75- *Mem:* AAAS; Am Soc Naturalists; Sigma Xi; Soc Study Evolution. *Res:* Evolution of life history strategies; herbivorous competition; reproductive ecology in arid environments. *Mailing Add:* Dept Biol Univ Mich Ann Arbor MI 48109-1048

LOW, BOON-CHYE, b Singapore, Feb 13, 46; m 71; c 1. PLASMA PHYSICS, FLUIDS. *Educ:* Univ London, UK, BSc, 68; Univ Chicago, MS, 69, PhD(physics), 72. *Prof Exp:* Res assoc, Enrico Fermi Inst, Univ Chicago, 72-73; vis scientist, High Altitude Observ, Nat Ctr Atmospheric Res, 73-74; Japan Soc Prom Sci fel, Tokyo Astron Observ, Tokyo Univ, 78-79; Nat Acad Sci-Nat Res Coun sr res assoc, NASA-Marshall Space Flight Ctr, 80-81; scientist, 81-87, head coronal interplanetary physics, 87- 90, actg dir, High Altitude Observ, 89-90, SR SCIENTIST, HIGH ALTITUDE OBSERV, NAT CTR ATMOSPHERIC RES, 87- *Mem:* Am Phys Soc; Am Astron Soc; Am Geophys Union. *Res:* Theoretical research in the fluid dynamics and magnetohydrodynamics of solar and astrophysical plasmas. *Mailing Add:* High Altitude Observ Nat Ctr Atmospheric Res PO Box 3000 Boulder CO 80307

LOW, CHOW-ENG, b Perak, Malaysia, May 31, 38; m 66; c 3. IMMUNOPATHOLOGY, ORGANIC ANALYTICAL CHEMISTRY. *Educ:* Chung Chi Col, Chinese Univ, BS, 62; Tex Southern Univ, MS, 66; Univ Tex, Austin, PhD(org chem), 70. *Prof Exp:* Vis asst prof, Dept Chem, La State Univ, 70-71; res fel, Ind Univ, Bloomington, 72-75; res assoc, Dept Human Biol Chem & Genetics, Univ Tex Med Br, 76-78; asst prof biochem, George Washington Univ Med Ctr, 78-84; chmn dept chem, 86-89, PROF CHEM, NAT CHENG KUNG UNIV, TAINAN, TAIWAN, 84- *Concurrent Pos:* Dir, Inst Chem, Nat Chen Kung Univ, Tainan, Taiwan, 86-89. *Mem:* Am Chem Soc; AAAS; Chem Soc London; Sigma Xi; Chinese Chem Soc. *Res:* Autoxidation of polyunsaturated fatty acids; lipoxygenase metabolities of polyunsaturated fatty acids; analysis of pollutants; Friedel-Crafts reaction mechanisms; catalytic transfer hydrogenation. *Mailing Add:* Dept Chem Nat Cheng Kung Univ Tainan 70101 Taiwan

LOW, EMMET FRANCIS, JR, b Peoria, Ill, June 10, 22; m 74. APPLIED MATHEMATICS. *Educ:* Stetson Univ, BS, 48; Univ Fla, MS, 50, PhD(math), 53. *Prof Exp:* Instr phys sci, Univ Fla, 50-51, physics, 51-54; aeronaut res scientist, Nat Adv Comt Aeronaut, 54-55; asst prof math, Univ Miami, 55-59; vis res scientist, Courant Inst Math Sci, NY Univ, 59-60; assoc prof math & chmn dept, Univ Miami, 60-66, actg dean col arts & sci, 66-67, prof math & assoc dean faculties, 68-72; dean of col & prof math, 72-86, CHMN, DEPT OF MATH SCI & PROF MATH, CLINCH VALLEY COL, UNIV VA, 86- *Mem:* AAAS; Am Math Soc; Soc Indust & Appl Math; Sigma Xi; Nat Coun Teachers Math. *Res:* Stress and functional analysis. *Mailing Add:* Box 3417 Wise VA 24293

LOW, FRANCIS EUGENE, b New York, NY, Oct 27, 21; m 48; c 3. THEORETICAL PHYSICS. *Educ:* Harvard Univ, BS, 42; Columbia Univ, AM, 47, PhD(physics), 49. *Prof Exp:* Instr physics, Columbia Univ, 49-50; mem, Inst Advan Study, 50-52; from asst prof to assoc prof physics, Univ Ill, 52-56; prof, Mass Inst Technol, 57-68, Karl Compton prof, 68-85, dir, Ctr Theoret Physics, 74-83, dir, Lab Nuclear Sci, 79-85, provost, 80-85, INST PROF, MASS INST TECHNOL, 85- *Concurrent Pos:* Consult, AEC, 55-; Fulbright fel, 61-62; Guggenheim fel, 61-62; nat coun, Nat Acad Sci, 86-89. *Honors & Awards:* Leob Lectr, Harvard Univ, 59. *Mem:* Nat Acad Sci; Am Acad Arts & Sci (vpres, 86-87); fel Am Phys Soc. *Res:* Theoretical, atomic and nuclear physics; field theory. *Mailing Add:* Dept Physics Mass Inst Technol Cambridge MA 02139

LOW, FRANK JAMES, b Mobile, Ala, Nov 23, 33; m 56; c 3. SOLID STATE PHYSICS. *Educ:* Yale Univ, BS, 55; Rice Univ, MA, 57, PhD(physics), 59. *Prof Exp:* Mem tech staff, Tex Instruments, Inc, 59-62; assoc scientist, Nat Radio Astron Observ, WVa, 62-65; res prof, Lunar & Planetary Lab, 65-79, RES PROF, STEWARD OBSERV, UNIV ARIZ, 71-, REGENTS RES PROF, 88- *Concurrent Pos:* Prof space sci, Rice Univ, 66-71, adj prof, 71-79; pres, Infrared Labs, Inc, Ariz, 67- *Honors & Awards:* Helen B Warner Prize, Am Astron Soc, 68; NASA Medal Except Sci Achievement, NASA, 84; Rumford Prize, AAAS, 86. *Mem:* Nat Acad Sci; Am Phys Soc; Am Astron Soc; Sigma Xi. *Res:* Infrared astronomy; infrared physics; solid state physics; low temperature physics. *Mailing Add:* Steward Observ Univ Ariz Tucson AZ 85721

LOW, FRANK NORMAN, b Brooklyn, NY, Feb 9, 11. ANATOMY. *Educ:* Cornell Univ, AB, 32, PhD(micros anat), 36. *Hon Degrees:* ScD, Univ NDak, 83. *Prof Exp:* From instr to asst prof, Univ NC, 37-45; assoc, Sch Med, Univ Md, 45; assoc prof, Sch Med, WVa Univ, 46; asst prof, Sch Med, Johns Hopkins Univ, 46-49; from assoc prof to prof, Sch Med, La State Univ, 49-64; Hill res prof, Univ NDak, 64-73, Chester Fritz Distinguished prof, 75-77, res prof, 73-81; RETIRED. *Concurrent Pos:* Charlton fel anat, Med Sch, Tufts Univ, 36-37; mem, Great Plains Regional Res Rev & Adv Comt, Am Heart Asn, 72-74; assoc ed, Am J Anat, 75-; vis prof anat, Sch Med, La State Univ, New Orleans, 81- *Honors & Awards:* Henry Gray Award, Am Asn Anat, 88. *Mem:* Electron Micros Soc Am; Am Asn Anat; Am Asn Hist Med; Am Soc Cell Biol; Sigma Xi. *Res:* Transmission and scanning electron microscopy; fine structure of lung; subarachnoid space; development of connective tissues; microdissection by ultrasonication; medical history. *Mailing Add:* 1901 Perdido St La State Med Ctr New Orleans LA 70112

LOW, HANS, b Vienna, Austria, Oct 22, 21; nat US; m 49; c 3. ORGANIC CHEMISTRY. *Educ:* Marietta Col, BS, 50; Purdue Univ, MS, 52; St Louis Univ, PhD(chem), 59; Univ Tex, MPH, 76. *Prof Exp:* Instr German & Latin, Marietta Col, 47-50; res chemist, 52-67, group leader lubricant additives, 67-72, staff technologist, 72-78, SR STAFF TECHNOLOGIST, HEALTH, SAFETY & ENVIRON, SHELL OIL CO, 78- *Mem:* Am Chem Soc, NY Acad Sci. *Res:* Petroleum solvents and lubricants; synthetic lubricants; lubricant additives; industrial toxicology; public health. *Mailing Add:* 7855 Cowles Mountain Ct San Diego CA 92119-2501

LOW, JAMES ALEXANDER, b Toronto, Ont, Sept 22, 25; m 52; c 3. MEDICINE. *Educ:* Univ Toronto, MD, 49; FRCS(C). *Prof Exp:* Clin teacher obstet & gynec, Univ Toronto, 55-65; head dept, 65-85, PROF OBSTET & GYNEC, QUEEN'S UNIV, ONT, 85- *Mem:* Soc Gynec Invest; Can Soc Clin Invest; Soc Obstet & Gynec Can; Am Gynec & Obstet Soc; Am Acad Cerebral Palsy & Develop Med. *Res:* Perinatal medicine; urodynamics. *Mailing Add:* Dept Obstet & Gynec Queen's Univ Kingston ON K7L 3N6 Can

LOW, JOHN R(OUTH), JR, metallurgy; deceased, see previous edition for last biography

LOW, KENNETH BROOKS, JR, b New Rochelle, NY, Jan 19, 36; m 60; c 2. GENETICS, DNA RECOMBINATION. *Educ:* Amherst Col, BA, 58; Univ Pa, MS, 60, PhD(molecular biol), 65. *Prof Exp:* Asst prof radiobiol, 68-71, asst prof radiobiol & microbiol, 71-73, assoc prof, 73-78, sr scientist radiobiol, 78-81, sr scientist radiobiol & biol, 81-84, PROF RES, YALE UNIV, 84- *Concurrent Pos:* USPHS fel, Med Ctr, NY Univ, 66-68; mem, Microbiol Genetics Study Sect, NIH, 78-82; consult comn to study antibiotic use in animal feeds, Nat Acad Sci, 79; mem, Prokaryotic Genetics Rev Panel, NSF, 86- *Mem:* Am Soc Microbiol. *Res:* Molecular genetics; genetic recombination and control. *Mailing Add:* Radiobiol Labs Yale Univ Med Sch New Haven CT 06510

LOW, LAWRENCE J(ACOB), b New York, NY, June 22, 21; m 51; c 1. MECHANICAL ENGINEERING, OPERATIONS RESEARCH. *Educ:* Stevens Inst Technol, ME, 42. *Prof Exp:* Aerodynamicist, Curtiss Wright Airplane Div, NY, 42-43; res aerodynamicist, Cornell Aeronaut Lab, 46-50; sr res engr, Stanford Res Inst, 55-65; dir, Naval Weapons Res Ctr, 65-76, STAFF SCIENTIST, SRI INT, 76- *Concurrent Pos:* Mem US Marine air defense eval group, Off Naval Res, DC, 57-58; chmn opers anal sect, Advan Surface Missile Assessment Group, 65. *Mem:* AAAS; Sigma Xi; Am Inst Aeronaut & Astronaut; Opers Res Soc Am; Am Ord Asn. *Res:* Aerodynamics; fluid mechanics; weapon systems analysis and evaluation. *Mailing Add:* SRI Int 333 Ravenswood Ave Menlo Park CA 94025

LOW, LEONE YARBOROUGH, b Cushing, Okla, Aug 27, 35; div; c 2. MATHEMATICAL STATISTICS, APPLIED STATISTICS. *Educ:* Okla State Univ, BS, 56, MS, 58, PhD(math), 61. *Prof Exp:* Instr math, Univ Ill, 60-64; asst prof, 64-68, ASSOC PROF MATH, WRIGHT STATE UNIV, 68- *Concurrent Pos:* Nat Res Coun res assoc, Wright-Patterson AFB, 67-68; consult, Systs Res Lab, 71-; vis assoc prof, Iowa State Univ, 80-81. *Mem:* Inst Math Statist; Am Statist Asn; Sigma Xi; fel Royal Statist Soc. *Res:* Variance component models in the analysis of variance; bootstrapping; experimental design; Taguchi design. *Mailing Add:* Dept Math & Stat Wright State Univ Dayton OH 45435

LOW, LOH-LEE, b Kuala Lumpur, Malaysia, Jan 15, 48; m 73; c 2. FISHERIES. *Educ:* Univ Wash, BS, 70, MS, 72, PhD(fisheries), 74. *Prof Exp:* Fishery biologist, Univ Wash, 74; FISHERY BIOLOGIST & OPERS RES ANALYST, NORTHWEST FISHERIES CTR, NAT MARINE FISHERIES SERV, 74- *Concurrent Pos:* Consult, Food & Agr Orgn, UN, 75; affil asst prof, Univ Wash, 78- *Res:* Fisheries population dynamics; computer modelling of fisheries systems; international fisheries management. *Mailing Add:* NW & Alaska Fish Ctr Nat Marine Fish Serv 7600 Sandpoint Way NE BINCI570 Seattle WA 98115-0070

LOW, M DAVID, b Lethbridge, Alta, Mar 25, 35; m 59, 84; c 4. NEUROPHYSIOLOGY. *Educ:* Queen's Univ, Ont, MD & CM, 60, MSc, 62; Baylor Univ, PhD(physiol), 66; FRCP(C), 73. *Prof Exp:* From instr to asst prof physiol, dept physiol, Baylor Col Med, 65-68; from assoc prof to prof, Div Neurol, Dept Med, Univ BC, 68-89, clin assoc dean, Fac Med, 74-76, actg assoc dean res & grad studies, 77-78, actg head, Div Neurol, 79-80, coordr health sci, 85-89; PRES, UNIV TEX HEALTH SCI CTR, HOUSTON, 89-, DIR, INST HEALTH POLICY & EDUC, 90- *Concurrent Pos:* Dir, Dept Diag Neurophysiol, Vancouver Gen Hosp, 68-87; serv staff consult, 87-89; consult med staff neurol, Univ Hosp, Univ BC Site, 70-89; attend staff, 89, dir, Evoked Potential Lab, 86-89; vis staff neurol EEG, Shaughnessy Hosp, Vancouver, 71-89; interim dir, Vancouver Gen Hosp Res Inst, 81-83, dir, 83-86; prof Dept Neurol, Med Sch, Univ Tex Health Sci Ctr, Houston, 89-; prof mgt & policy sci, Sch Pub Health, 89-, prof neural sci, Grad Sch Biomed Sci, 89-; consult, Health Protection Br, Health & Welfare Can; mem, Coun Univ Teaching Hosps, 85-, vchmn, 86-88, chmn, 89; mem bd dirs, Health Environ Inst, Univ Houston, 89-; mem med sci adv comt, US Info Agency, 91-; vis prof & lectr, numerous universities, associations & hosps. *Mem:* AMA; Am Coun Educ; Int Fedn Socs EEG & Clin Neurophysiol (secy, 81-85); fel Am EEG Soc; Can Soc Clin Neurophysiol (secy, 70-72, pres, 72-74); Can Soc Clin Invest; Can Asn Med Educ; Sigma Xi. *Res:* Electrophysiology of the central nervous system; cognitive neuroscience; sleep disorders; health policy research and development; health services. *Mailing Add:* Univ Tex Health Sci Ctr PO Box 20036 Houston TX 77225-0036

LOW, MANFRED JOSEF DOMINIK, b Karlsbad, Bohemia, June 18, 28; nat US; m 65. PHYSICAL CHEMISTRY. *Educ:* NY Univ, BA, 52, MS, 54, PhD(phys chem), 56. *Prof Exp:* Asst chem, NY Univ, 52-55; res chemist, Davison Chem Co Div, W R Grace & Co, 56-58; sr chemist, Texaco, Inc, 58-61; asst prof chem, Rutgers Univ, 61-67; assoc prof, 67-72, PROF CHEM, NY UNIV, 72- *Mem:* Am Chem Soc; Soc Appl Spectros; NY Acad Sci. *Res:* Chemisorption; heterogeneous catalysis; infrared spectra of surfaces; infrared emission spectroscopy; surface chemistry and physics; Fourier transform spectroscopy; photoacoustic spectroscopy. *Mailing Add:* Dept of Chem NY Univ 4 Washington Place New York NY 10003

LOW, MARC E, b Ada, Okla, Sept 25, 35; m 57; c 2. MATHEMATICS. *Educ:* Okla State Univ, BS, 58, MS, 60; Univ Ill, PhD(math), 65. *Prof Exp:* Instr math, 64-65, asst prof, 65-71, ASSOC PROF MATH, WRIGHT STATE UNIV, 71-, ASST DEAN, COL SCI & ENG, 73- *Mem:* Math Asn Am; Am Math Soc. *Res:* Elementary and analytic number theory. *Mailing Add:* 6361 Shadow Lake Trail Dayton OH 45459

LOW, MARY ALICE, b Warren, Ohio, May 15, 49; m 71; c 2. NUTRITION. *Educ:* Univ Mass, BS, 71; Univ Maine, MS, 72; Cornell Univ, PhD(nutrit), 75. *Prof Exp:* Clin instr nutrit, Syracuse Univ, 74-76, asst prof, 76-78; asst prof, 78-81, assoc prof nutrit, 81- ASSOC PROF FOOD NUTRIT & INST ADMIN, AT GRAD SCH UNIV MD. *Concurrent Pos:* Nutrit res consult, Loretto Geriatric Ctr, 76-78, USDA, 78; atty, John B Low, Oxen Hill, Md. *Mem:* Am Dietetic Asn; Nutrit Today Soc; Am Pub Health Asn; Soc Nutrit Educ. *Res:* Socio-cultural correlates of nutritional status. *Mailing Add:* 5425 Indian Head Hwy Oxen Hill MD 20745

LOW, MORTON DAVID, b Lethbridge, Alta, Mar 25, 35; m 59, 84; c 4. NEUROPHYSIOLOGY, HEALTH SCIENCES ADMINISTRATION. *Educ:* Queen's Univ, Ont, MD & CM, 60, MSc, 62; Baylor Univ, PhD(physiol), 66; FRCP(C), 73. *Prof Exp:* From instr to asst prof physiol, Baylor Univ Col Med, 65-68; dir, Vancouver Gen Hosp Res Inst, 81-85, dir diag neurophysiol, Vancouver Gen Hosp, 68-87; from assoc prof to prof med, Univ BC, 68-89, coordr health sci, 85-89; biomed indust liaison, 87-89; PRES, UNIV TEX HEALTH SCI CTR, HOUSTON, 89-, PROF NEUROL MGT, POLICY SCI & NEURAL SCI. *Concurrent Pos:* Med Res Coun Can grants, Univ BC, 68-80; Mr & Mrs P A Woodward's Found grants, Vancouver Gen Hosp, 69-72, Univ Hosp, 85, Univ BC, 87-88 & Vancouver Found grants, 87; MRC-INSERM Can-France exchange scientist, 78-79; dir, Evoked Potential Lab, Univ Hosp, Univ BC, 85-89. *Mem:* AAAS; Am EEG Soc; Can Soc EEG (secy, 70-72, pres, 72-74); Am Epilepsy Soc; Int Fedn EEG Socs (secy, 81-85). *Res:* Neural basis of perception and performance; brain mechanisms in maintenance and disorders of consciousness; sleep; epilepsy; health policy. *Mailing Add:* 1100 Holcombe Blvd PO Box 20036 Houston TX 77225

LOW, NIELS LEO, b Copenhagen, Denmark, Dec 16, 16; nat US; m 43; c 2. MEDICINE. *Educ:* Med Col SC, MD, 40. *Prof Exp:* Clin instr pediat, Marquette Univ, 46-53; res assoc neurol, Univ Ill, 54; assoc res prof pediat, Univ Utah, 56-58; asst prof neurol, 60-67, assoc prof clin neurol, 67-75, PROF CLIN NEUROL & CLIN PEDIAT, COL PHYSICIANS & SURGEONS, COLUMBIA UNIV, 75-; DIR PEDIAT, BLYTHEDALE CHILDREN'S HOSP, VALHALLA, NY, 67- *Concurrent Pos:* Fel, Columbia Univ, 55 & 58-59; consult, NIH. *Mem:* Am EEG Soc; fel Am Acad Neurol; fel Am Acad Pediat; Am Epilepsy Soc; Int Child Neurol Asn (pres, 75-). *Res:* Pediatric neurology; metabolic disease affecting brain of children. *Mailing Add:* Blythedale Children's Hosp Valhalla NY 10595

LOW, PAUL R, ELECTRICAL ENGINEERING. *Educ:* Univ Vt, BS, 55, MS, 57; Stanford Univ, PhD(elec eng), 63. *Prof Exp:* Asst res, develop logic systs, Poughkeepsie, 57-69, lab dir, East Fishkill, 69-70, gen mgr mfg, IBM Burlington, 70-71, gen mgr, 71-74, dir, systs prod div, East Fishkill & Poughkeepsie, 74-75, vpres, develop & mfg, 75-79, dir logic, develop lab, East Fishkill, 79-80, gen mgr, 80-81, vpres, Gen Technol Div, 81-83, pres, 83-84, PRES, GEN PRODS DIV, IBM, 87- *Concurrent Pos:* Bd counr, Sch Eng, Univ Southern Calif. *Mem:* Fel Inst Elec & Electronics Engrs; Sigma Xi. *Mailing Add:* Gen Tech Div IBM Corp Rte 100 Somer NY 10589

LOW, PHILIP FUNK, b Carmangay, Alta, Oct 15, 21; nat US; m 42; c 6. COLLOID CHEMISTRY. *Educ:* Brigham Young Univ, BS, 43; Calif Inst Technol, MS, 44; Iowa State Univ, PhD(soil chem), 49. *Prof Exp:* Soil scientist, USDA, 49; from asst prof to assoc prof, 49-55, PROF SOIL CHEM, PURDUE UNIV, 55- *Concurrent Pos:* Distinguished vis award to Australia, Fulbright Educ Exchange Prog, 68; consult, Exxon Prod Res Lab, US Army Cold Regions Res & Eng Lab, Pac Northwest Labs, Battelle Mem Inst; guest prof & Thurburn vis fel, Univ Sydney, Australia, 83; hon prof, Zhejiang Agr Univ, People's Repub China, 87. *Honors & Awards:* Sigma Xi Res Award, Purdue Univ, 60 & Herbert Newby McCoy Award, 80; Soil Sci Award, Am Soc Agron, 63; Bouyoucos Distinguished Soil Sci Career Award, Soil Sci Soc Am, 84. *Mem:* Fel Soil Sci Soc Am (pres-elect, 71-72, pres, 72-73); fel Am Soc Agron; Clay Minerals Soc; Int Soc Soil Sci; Int Asn Study Clays. *Res:* Physical and colloidal chemistry of soils. *Mailing Add:* Dept Agron Purdue Univ West Lafayette IN 47907

LOW, PHILIP STEWART, b Ames, Iowa, Aug 8, 47; m 69; c 5. BIOCHEMISTRY. *Educ:* Brigham Young Univ, BS, 71; Univ Calif, San Diego, PhD(biochem), 75. *Prof Exp:* Res assoc, Dept Chem, Univ Mass, 75-76; from asst prof to assoc prof, 76-86, PROF BIOCHEM, DEPT CHEM, PURDUE UNIV, 86- *Honors & Awards:* Fel Int Union Against Cancer. *Mem:* Sigma Xi; Am Soc Biochem & Molecular Biol; Am Soc Cell Biol. *Res:* Biochemistry and physical chemistry of biological membranes. *Mailing Add:* Dept of Chem Purdue Univ West Lafayette IN 47907

LOW, ROBERT BURNHAM, b Greenfield, Mass, Sept 19, 40; m 67; c 2. PHYSIOLOGY. *Educ:* Princeton Univ, AB, 63; Univ Chicago, PhD(physiol), 68. *Prof Exp:* NIH fel biol, Mass Inst Technol, 68-70; asst prof physiol, 70-74, assoc prof, 74-79, PROF PHYSIOL & BIOPHYS & ASSOC DEAN RES, COL MED, UNIV VT, 79- *Concurrent Pos:* NIH & Muscular Dystrophy res grants, Univ Vt; Sr Fogarty Int fel, 79, Univ scholar, 88. *Mem:* Am Soc Cell Biol; Am Thoracic Soc. *Res:* Mammalian protein turnover; physiology and biochemistry of muscle; cytoskeleton; tissue and cell remodeling; lung epithelial cells; smooth muscle. *Mailing Add:* Dept Physiol & Biophys Given Bldg E-211 Univ Vt Burlington VT 05405

LOW, TERESA LINGCHUN KAO, b Hankow, China, Feb 17, 41; US citizen; m 66; c 3. PROTEIN CHEMISTRY, THYMIC HORMONES. *Educ:* Tunghai Univ, Taiwan, BS, 62; Tex Woman's Univ, MSc, 66; Univ Tex, Austin, PhD(biochem), 70. *Prof Exp:* Sci res specialist, Dept Biochem, La State Univ, 70-71; res assoc, Dept Zool, Ind Univ, 72-75; fel, Dept Human Biol Chem & Genetics, Univ Tex Med Br, 76-77, instr protein chem, 77-78; from asst prof to assoc prof protein chem, Dept Biochem, Med Sch, George Washington Univ, 81-84; chmn dept, 84-90, PROF, DEPT BIOCHEM, COL MED, NAT CHENG KUNG UNIV, 90- *Mem:* Sigma Xi; NY Acad Sci; Am Soc Biochem & Molecular Biol; Protein Soc. *Res:* Chemical and biological characterization of thymosin, a family of hormones derived from the thymus gland and demonstrated to have potent immunomodulating properties. *Mailing Add:* Dept Biochem Col Med Nat Cheng Kung Univ Tainan 70101 Taiwan

LOW, WALTER CHENEY, b Madera, Calif, May 11, 50; m 83; c 2. NEUROPHYSIOLOGY, NERVE REGENERATION & TRANSPLATION. *Educ:* Univ Calif, Santa Barbara, BS, 72; Univ Mich, MS, 74, PhD(bioeng), 79. *Prof Exp:* NIH-Nat Inst Gen Med Sci fel bioeng, Univ Mich, 75-78, res assoc neurophysiol, 78-79; res fel neurophysiol, Cambridge Univ, 79-80; res fel neurophysiol, Univ Vt, 80-83; from asst prof to assoc prof, Ind Med Sch, Ind Univ, 83-90, dir, Grad Prog Physiol, 85-88; ASSOC PROF,

UNIV MINN, 90- *Concurrent Pos:* NIH-Nat Neurol & Commun Disorders & Stroke fel, Univ Mich, 79; NSF/NATO fel, Cambridge Univ, 79-80; fel, Univ Vt, 80-81, NIH-Nat Heart, Lung & Blood Inst, 81-82; prin investr, NIH, 84-85, NIA, 85-87, Alzheimer's Dis Asn, 88-89, Am Heart Asn, 87-, Nat Inst Neurol Cardiol Dis Soc, 87-91, Minn Med Found, 90-91. *Honors & Awards:* Nat Res Serv Award, NIH, 79-80 & 81-83. *Mem:* AAAS; Soc Neurosci; Am Physiol Soc; NY Acad Sci; Am Parkinson Dis Asn; Am Heart Asn. *Res:* Central nervous system physiology; neural transplantation and the recovery of function; Parkinson's Disease; Alzheimer's Disease; stroke and cerebral ischemia; neural regeneration. *Mailing Add:* Dept Neurosurg Univ Minn Med Sch 420 Delaware St SE Box 96 UMHC Minneapolis MN 55455

LOW, WILLIAM, b Vienna, Austria, Apr 25, 22; nat Can; m 48, 70; c 9. PHYSICS. *Educ:* Queen's Univ, Ont, BA, 46; Columbia Univ, MA, 47, PhD, 50. *Prof Exp:* Tutor physics, Queen's Univ, Ont, 45-46; asst, Columbia Univ, 46-50; lectr, Hebrew Univ, Israel, 50-55, sr lectr, 55-59; res assoc & vis scholar, Univ Chicago, 55-56; assoc prof, 59-61, chmn Res & Develop, 63-65; prof physics, Columbia Univ, 85; PROF PHYSICS, HEBREW UNIV, ISRAEL, 61- *Concurrent Pos:* Vis scholar, Oxford Univ, 54 & Univ Chicago, 55-56; vis prof, Technion, Israel, 58-61, Inst Technol, 60-64, Mass Inst Technol, 65-66, 81-82 & Columbia Univ, 85-86; vis scientist, Nat Res Coun, Ottowa, Can & Lincoln Lab, 60; Guggenheim fel, 63-64; ed, Physics Letters; chmn Israel comt, Int Union Radio Sci; pres & rector, Jerusalem Col Technol, Israel, 69-81, chmn bd gov, 85; prof, Univ Toronto, 90. *Honors & Awards:* Morrison Award, NY Acad Sci, 56; Israel Prize Exact Sci, 61; Rothschild Prize Physics, 64. *Mem:* Am Phys Soc; NY Acad Sci; Phys Soc Israel (vpres, 58-60, pres, 60-61 & 70-72); Europ Phys Soc; Int Union Pure & Appl Physics. *Res:* Paramagnetic resonance in solids; microwave spectroscopy in gases; quantum electronics; electron density behind shock waves; light scattering from macromolecules. *Mailing Add:* Microwave Div Inst Physics Hebrew Univ Jerusalem Israel

LOWANCE, FRANKLIN ELTA, b Monroe Co, WVa, Dec 29, 07; m 31; c 1. PHYSICS. *Educ:* Roanoke Col, BS, 27; Duke Univ, MA, 31, PhD(physics), 35. *Prof Exp:* Prof eng & physics, Edinburg Col, 33-35; assoc prof math & astron, Wofford Col, 35-38; head dept physics & eng, Centenary Col, 38-42; assoc prof physics, Ga Inst Tech, 41-43, prof, 45-49; res assoc, Harvard Univ, 44; mem staff, Radiation Lab, Mass Inst Technol, 45; tech dir, Naval Civil Eng Lab, 49-53; assoc tech dir, Naval Ord Test Sta, 53-54; dir res & vpres, Westinghouse Air Brake Co, 55-58; vpres eng, Crosley Div, Avco Corp, 58-60; pres, Adv Tech Corp, 60-62; CONSULT & VPRES, MERCO CORP, 62- *Mem:* Am Phys Soc; Nat Soc Prof Eng; Am Asn Physics Teachers. *Res:* Microwave propagation and beacons; acoustics; ferromagnetism; magnetothermoelectricity. *Mailing Add:* 5822 Williamsburg Landing Dr Williamsburg VA 23185-3777

LOWDEN, J ALEXANDER, b Toronto, Ont, Feb 21, 33; m 56; c 4. BIOCHEMICAL GENETICS. *Educ:* Univ Toronto, MD, 57; McGill Univ, PhD(biochem), 64; CCFMG, 83. *Prof Exp:* Resident pediat, Hosp Sick Children, 58-60; res assoc, Univ Toronto, 65-67, assoc pediat, 67-80; assoc scientist, 64-74, ASSOC DIR, RES INST, HOSP SICK CHILDREN, 75-; PROF PEDIAT & CLIN BIOCHEM, UNIV TORONTO, 80-; PRES, HSC RES DEVELOP CORP, 82- *Concurrent Pos:* Fel neurochem, Montreal Neurol Inst, 61-64; Helen Hay Whitney Found fel, 63-66. *Mem:* AAAS; Can Biochem Soc; Am Soc Pediat; Soc Pediat Res. *Res:* Inborn errors of metabolism, especially lysosomal storage disease. *Mailing Add:* Dept Pediat Univ Toronto Fac Med One Kings College Circle Toronto ON M5S 1A8 Can

LOWDEN, RICHARD MAX, b Columbus, Ohio, Sept 27, 43; m 70; c 3. PLANT SYSTEMATICS. *Educ:* Ohio State Univ, BA, 64, MSc, 67, PhD, 71. *Prof Exp:* Asst prof bot, Ohio State Univ, 71; from asst prof to assoc prof, 71-85, PROF BIOL & BOT, CATH UNIV, SANTIAGO, 85-, DIR, MOSCOSO HERBARIUM, 73- *Mem:* Sigma Xi; Int Asn Plant Taxon; Asn Trop Biol; Asn Aquatic Vascular Plant Biologists; Acad Ciencias Republica Dominicana. *Res:* Aquatic freshwater vascular flora of Hispaniola; Latin American botany; botanical collectors; international index compiler. *Mailing Add:* Univ Cath Madre y Maestra Moscoso Herbarium Santiago de los Caballeros Dominican Republic

LOWDER, J ELBERT, b Pinedale, Wyo, Mar 18, 40; m 64; c 3. APPLIED PHYSICS. *Educ:* Univ Calif, Berkeley, BS, 63, MS, 65; Univ Calif, San Diego, PhD(eng physics), 71. *Prof Exp:* Flight test engr, Northrop Aircraft Corp, 63-64; proj engr, Aeronutronic Div, Philco-Ford Corp, 65-68; mem staff, Lincoln Lab, Mass Inst Technol, 71-75, assoc group leader appl physics, 75-80; VPRES, SPARTA INC, 80- *Mem:* Optical Soc Am; Am Inst Aeronaut & Astronaut; Soc Photo-Optical Instrumentation Engrs. *Res:* Effects of atmospheric aerosols on propagation of laser radiation; interaction of high power laser radiation with solid surfaces; laser radar applications; passive infrared detection systems. *Mailing Add:* Sparta Inc 21 Worthen Rd Lexington MA 02173

LOWDER, JAMES N, b Cleveland, Ohio, Aug 26, 50. HEMATOLOGY, ONCOLOGY. *Educ:* Case Western Reserve Univ, BS, 73, MD, 78. *Prof Exp:* Staff, Cleveland Clin Found, 86-88; assoc med dir, Immunocytometry Systs, 88-89, CORP MED DIR, ADVAN DIAGNOSTICS, BECTON DICKINSON, 89- *Mem:* Am Soc Hemat; AAAS; Am Soc Clin Chemists; Am Soc Clin Oncol; Am Soc Microbiol; Am Soc Immunol. *Mailing Add:* Dept Med Becton Dickinson Advan Diag 225 International Circle Cockeysville MD 21030

LOWDER, WAYNE MORRIS, b Chicago, Ill, Jan 6, 33; div; c 2. RADIATION PHYSICS, DOSIMETRY. *Educ:* Harvard Univ, AB, 54; Int Sch Nuclear Sci & Eng, Argonne Nat Lab, cert, 55. *Prof Exp:* Prog mgr, Off Health & Environment Res, 77-86, PHYSICIST ENVIRON MEASUREMENTS LAB, 55-, DIR RADIATION PHYSICS DIV, US DEPT ENERGY, 87- *Concurrent Pos:* Mem Sci Comt, Nat Coun Radiation Protection & Measurement, 73-; consult, UN Sci Comt, Effects Atomic Radiation, 77-; co-chmn, Radon Workgroup, Fed Comm Indoor Air Qual, 83-87; co-organizer, Nat Rad Environ Symp, 63, 72, 78, 87, 91. *Mem:* Am Phys Soc. *Res:* Measurement of ionizing radiation from natural and manmade environmental radionuclides and in the space environment and the assessment of dose to man from these sources. *Mailing Add:* US Dept of Energy 376 Hudson St New York NY 10014

LOWDIN, PER-OLOV, b Uppsala, Sweden, Oct 28, 16; m 60; c 2. THEORETICAL PHYSICS, QUANTUM BIOLOGY. *Educ:* Univ Uppsala, Fil Kand, 37, Fil Mag, 39, Fil Lic, 42, Fil Dr(theoret physics), 48. *Hon Degrees:* Dr, Univ Gent, Belgium, 75. *Prof Exp:* Lectr math & physics, Univ Uppsala, 42-48, asst prof theoret physics, 48-55, assoc prof, 55-60, prof & head dept, 60-83, emer prof quantum chem, Univ Uppsala, 83-; leader Fla Quantum Theory Proj, 60-82, GRAD RES PROF CHEM & PHYSICS, UNIV FLA, 60- *Concurrent Pos:* Fel, Swiss Fed Inst Technol, 46; H H Wells Phys Lab, Univ Bristol, 49; vis prof & consult, Duke Univ, Univ Chicago, Mass Inst Technol & Calif Inst Technol, 50-60; ed-in-chief, Advan in Quantum Chem, 64-, Int J Quantum Chem, 67-; mem Nobel Comt Physics, Swedish Royal Acad Sci, 72-84; foreign mem Sci Coun Inst Molecular Sci, Okazaki, Japan, 83-86; mem adv bd, Max Planck Soc, Carbon Res, Ruhr, WGer; dir, Uppsala-Fla Exchange Proj, quantum sci, 60-, dir Fla-Latinamerican & Caribbean Basin Exchange Proj, 85- *Mem:* Swed Royal Soc Arts & Sci; Swed Royal Soc Sci; Norweg Acad Sci & Letters; Int Soc Quantum Biol (pres, 71-72); Int Acad Quantum Molecular Sci (vpres, (vpres, 68-); Am Chem Soc. *Res:* Theoretical physics and chemistry; quantum theory of atoms, molecules and solid state; quantum genetics and pharmacology. *Mailing Add:* QTP 365 William Hall Univ Fla Gainesville FL 32611-2085

LOWE, A(RTHUR) L(EE), JR, b Boyce, Va, Jan 25, 27; m 53; c 3. METALLURGICAL & MATERIALS ENGINEERING. *Educ:* Va Polytech Inst, BS, 51; Lehigh Univ, MS, 55; Lynchburg Col, MBA, 73. *Prof Exp:* Proj engr, Richmond Eng Co, 51-53; res asst, Lehigh Univ, 53-54; welding engr, Metals Joining Div, Battelle Mem Inst, 54-57; group supvr liquid metal fuel reactor exp mat, 57-59, adv reactor concepts mat, 59-63, supvr metall eng group, Nuclear Develop Ctr, 63-65, staff specialist, Mat Processes Sect, 65-67, sr mat engr, Nuclear Power Generation Dept, 67-72, prin mat engr, 72-81, ADV ENGR MAT, NUCLEAR POWER GENERATION DIV, BABCOCK & WILCOX CO, 81- *Concurrent Pos:* Pvt consult engr, 63- *Mem:* Am Soc Metals; Am Inst Mining, Metall & Petrol Engrs; Brit Inst Metals; Am Soc Testing & Mat. *Res:* Nuclear materials applications; metal corrosion and fabrication problems; general materials application problems; forensic engineering-failure analysis and accident reconstruction; technical problems. *Mailing Add:* PO Box 10935 Lynchburg VA 24506-0935

LOWE, CARL CLIFFORD, b West Salem, Ohio, Jan 1, 19; m 42; c 3. PLANT BREEDING. *Educ:* Colo Agr & Mech Col, BS, 48; Cornell Univ, MS, 50, PhD(plant breeding), 52. *Prof Exp:* From asst prof to assoc prof, 52-62, PROF PLANT BREEDING, NY STATE COL AGR & LIFE SCI, CORNELL UNIV, 62-, PROF BIOMETRY, 70- *Mem:* Am Soc Agron. *Res:* Forage crops breeding. *Mailing Add:* 1517 Slaterville Rd Ithaca NY 14851

LOWE, CHARLES HERBERT, JR, b Los Angeles, Calif, Apr 16, 20; m 44; c 2. ZOOLOGY. *Educ:* Univ Calif, Los Angeles, AB, 43, PhD(zool), 50. *Prof Exp:* Consult, AEC, 47-50; instr, 50-53, from asst prof to assoc prof, 53-64, prof zool, 64-80, PROF ECOL & EVOLUTION, UNIV ARIZ, 80-, CUR AMPHIBIANS & REPTILES, 77- *Mem:* AAAS; Am Soc Ichthyol & Herpet; Ecol Soc Am; Soc Study Evolution; Soc Syst Zool. *Res:* Animal and plant ecology; systematics; evolution; vertebrate zoology. *Mailing Add:* Dept Ecol Univ Ariz Tucson AZ 85721

LOWE, CHARLES UPTON, b Pelham, NY, Aug 24, 21; m 55; c 4. PEDIATRICS. *Educ:* Harvard Univ, BS, 42; Yale Univ, MD, 45. *Prof Exp:* From intern to asst resident pediat, Children's Hosp, Boston, 45-46; resident, Mass Gen Hosp, 47; assoc prof pediat, Sch Med, State Univ NY Buffalo, 51-55, res prof, 55-65; prof, Col Med, Univ Fla, 65-68, dir human develop ctr, 66-68; sci dir, Nat Inst Child Health & Human Develop, 68-74; exec dir, Nat Comn Protection Human Subjects Biomed & Behav Res, HEW, 74-77, spec asst child health affair, Off of Asst Secy Health, 74-79; actg assoc dir, med appln res, 80-82, SPEC ASST TO THE DIR, NAT INST CHILD HEALTH & DEVELOP, NIH, 83- *Concurrent Pos:* Nat Res Coun fel, Med Sch, Univ Minn, 48-51; Buswell fel, Sch Med, State Univ NY Buffalo, 55; ed-in-chief, Pediat Res, 66-74; exec dir, President's Biomed Res Panel, 74-76, mem, President's Reorgn Proj Food & Nutrit Study, 78. *Honors & Awards:* John F Kennedy Mem Lectr, 66; Clifford G Grulee Award, Am Acad Pediat, 71; Special Recognition Award, NIH, 88; Grover Powers Mem Lectr, 69. *Mem:* Soc Pediat Res; Soc Exp Biol & Med; Am Soc Exp Path; Am Pediat Soc; Am Soc Clin Invest; Sigma Xi. *Res:* Clinical and laboratory study of nutritional disease, including celiac and cystic fibrosis of the pancreas; relationship between adrenocortical steroids and nucleic acid metabolism; inborn errors of metabolism and parenteral fluid therapy. *Mailing Add:* NIH NICHD Bldg 31 Rm 2A20 Bethesda MD 20892

LOWE, DONALD RAY, b Sacramento, Calif, Sept 22, 42; m 64; c 2. SEDIMENTOLOGY. *Educ:* Stanford Univ, BS, 64; Univ Ill, Urbana, PhD(geol), 67. *Prof Exp:* Instr geol, Univ Ill, Urbana, 67-68; res assoc, US Geol Surv, Calif, 68-70; asst prof, 70-73, assoc prof, 73-78, prof geol, La State Univ, Baton Rouge, 78-88; PROF GEOL, STANFORD UNIV, 88- *Concurrent Pos:* Prin investr, NSF, 77- *Mem:* Soc Econ Paleont & Mineral; Int Asn Sedimentologists; Int Asn Sedimentologists; Geol Soc Am; Int Soc Study Origin Life. *Res:* Archean sedimentology and the application of sedimentological principles to interpreting surface conditions on the early earth; composition of the early ocean and atmosphere; paleoecology of Archean life; deep-sea sedimentation and transport systems; Archean sedimentology. *Mailing Add:* Dept Geol Stanford Univ Stanford CA 94305

LOWE, FORREST GILBERT, b Gilman City, Mo, Mar 27, 27; m 48. GENERAL MATHEMATICS, ELECTRICAL ENGINEERING. *Educ:* Northwest Mo State Univ, BS(sec educ) & BS(physics & math), 51; Tex Christian Univ, MS, 62; Nova Univ, EdD, 89. *Prof Exp:* Teacher physics & math, Kansas City Sch Dist, 53-56; nuclear engr radiation effects, Convair Div, Gen Dynamics, Ft Worth, Tex, 56-59; instr physics, Kansas City Jr Col, 59-64 & Metrop Community Col Dist, 64-69; INSTR & CHMN, DIV PHYSICS & ENG, LONGVIEW COMMUNITY COL, 69- *Concurrent Pos:* Instr eng, Univ Mo, Kansas City, 83-91. *Mem:* Am Asn Physics Teachers; Nat Soc Prof Engrs; Soc Mfg Engrs; Am Soc Eng Educ; Am Inst Physics; Comput & Automated Systs Asn; Robotics Int; Math Asn Am; Nat Asn Indust Technol. *Res:* Nuclear radiation effects; physics and engineering curriculum. *Mailing Add:* 8412 E 49th St Kansas City MO 64129

LOWE, HARRY J, b Nogales, Ariz, Dec 21, 19; m 47; c 5. ANESTHESIOLOGY. *Educ:* Univ Ariz, BS, 44; Johns Hopkins Univ, SM, 45, MD, 49. *Prof Exp:* Assoc prof biochem, Univ Tex, Southwest Med Sch, 53-56; prin res scientist, Roswell Park Mem Inst, NY, 58-62; resident anesthesiol & dir hyperbaric med, Millard Fillmore Buffalo, 62-66; prof anesthesiol & chmn dept, Pritzker Sch Med, Univ Chicago, 66-73; prof anesthesiol, Univ Southern Calif, 73-78; prof anesthesiol, Univ Ala, Birmingham, 79-80; DIR, DEPT ANESTHESIOL, CITY OF HOPE MED CTR, 80- *Concurrent Pos:* Am Cancer Soc fel, Johns Hopkins Univ, 49-52. *Mem:* AAAS; AMA; Am Chem Soc; Am Anesthesiol Soc; Sigma Xi. *Res:* Quantitative automated administration of volatile anesthetics in closed circuit systems; acid-base regulation of physiological ventilation during anesthesia. *Mailing Add:* 614 N Old Ranch Rd Arcadia CA 91007

LOWE, IRVING J, b Woonsocket, RI, Jan 4, 29; m 53, 87; c 3. SOLID STATE PHYSICS, NUCLEAR MAGNETIC RESONANCE IMAGING. *Educ:* Cooper Union, BEE, 51; Washington Univ, St Louis, PhD(physics), 57. *Prof Exp:* Fel, Sloan Found & res assoc physics, Washington Univ, St Louis, 56-58; asst prof, Univ Minn, 58-62; assoc prof, 62-66, PROF PHYSICS, UNIV PITTSBURGH, 66- *Mem:* Am Phys Soc. *Res:* Experimental and theoretical studies of the structure and behavior of solids and biological systems using nuclear magnetic resonance techniques; nuclear magnetic resonance imaging; magnetic resonance in medicine. *Mailing Add:* Dept of Physics Univ of Pittsburgh Pittsburgh PA 15260

LOWE, JACK IRA, b Fairmount, Ga, Dec 8, 27; m 57. MARINE ECOLOGY, TOXICOLOGY. *Educ:* Berea Col, AB, 50; Univ Ga, MS, 55. *Prof Exp:* Biologist, US Fish & Wildlife Serv, 57-61 & US Bur Com Fisheries, 61-70; aquatic biologist, Environ Protection Agency, 70-71, dep lab dir, 71-75, assoc dir tech assistance, 75-76, chief exp environ br, Environ Res Lab, 76-85; RETIRED. *Mem:* Am Fisheries Soc; Nat Shellfisheries Asn; Gulf Estuarine Res Soc. *Res:* Estuarine and coastal ecology; effects of pollutants on marine organisms and their environment. *Mailing Add:* 4461 Sound Side Dr Gulf Breeze FL 32561

LOWE, JAMES EDWARD, b Brunswick, Ga, Dec 27, 46; m 69; c 2. THORACIC SURGERY, ELECTROPHYSIOLOGY. *Educ:* Stanford Univ, BA, 69; Univ Calif, Los Angeles, MD, 73. *Prof Exp:* From asst prof to assoc prof surg & path, 86-90, PROF SURG, DUKE UNIV, 91- *Concurrent Pos:* Investr, Am Heart Asn, 81-86; dir, Surg Electrophysiol Serv, 83- *Mem:* Am Col Surgeons; Am Col Cardiol; Am Col Chest Physicians; Am Heart Asn; Am Asn Thoracic Surg; Soc Thoracic Surgeons. *Res:* Etiology of cardiac arrhythmias; basic pathogenesis of global myocardial ischemic injury. *Mailing Add:* Duke Hosp Box 3954 Durham NC 27710

LOWE, JAMES HARRY, JR, b Vonore, Tenn, Mar 15, 31; m 55; c 3. ENTOMOLOGY. *Educ:* Univ Tenn, BA, 55; Ohio State Univ, MSc, 57; Yale Univ, PhD(forest entom), 66. *Prof Exp:* Res entomologist, Northeastern Forest Exp Sta, USDA, Conn, 59-62, insect res ecologist, 63-65; ASSOC PROF FORESTRY & ZOOL, UNIV MONT, 65- *Mem:* Entom Soc Am; Entom Soc Can. *Res:* Ecology of insects in forest communities; insect dispersal and distribution; alpine entomology; behavioral and meteorological aspects of flight of insects. *Mailing Add:* Sch of Forestry Univ of Mont Missoula MT 59812

LOWE, JAMES N, b Grand Forks, NDak, May 3, 36; m 61; c 3. ORGANIC CHEMISTRY. *Educ:* Antioch Col, BS, 59; Stanford Univ, PhD(chem), 64. *Prof Exp:* Asst prof chem, Smith Col, 63-65; from asst prof to assoc prof, 71-78, PROF CHEM, UNIV OF THE SOUTH, 78- *Concurrent Pos:* Am Chem Soc Petrol Res Fund grant, 64-66 & 67-69; fel, Univ Calif, Davis, 70-71; fel Univ Ill, 77-78; res corp grant, 80. *Mem:* Am Chem Soc; AAAS; Sigma Xi. *Res:* Coenzyme mechanisms. *Mailing Add:* Box 1225 Sewanee TN 37375

LOWE, JAMES URBAN, II, b Durham, NC, June 30, 21; m; c 4. PHYSICAL ORGANIC CHEMISTRY. *Educ:* Va State Col, BS, 42, MS, 46; Howard Univ, PhD, 63; Tenn State Univ, MPA, 82. *Prof Exp:* Asst prof chem, Tenn State Col, 47-52 & Ft Valley State Col, 52-56; fel, Howard Univ, 56-59, instr, 59-60; res chemist, US Govt, Md, 60-68; assoc dean admin, Sch Med, Meharry Med Col, 69-81, asst dean admis, 81-82, assoc prof biochem, 68-87, dir, Off Instnl Res, 81-84, Sch Med, Instnl Res, 84-87; RETIRED. *Concurrent Pos:* Co-Founder, pres, Lophelps, Inc; consulting, Info Serv; interim dir, Acad Dev & Support Serv Div, Grad Sch, Meharry, 89- *Mem:* Am Chem Soc; Sigma Xi. *Res:* Synthesis of 0-nitrobenzoates; aryloxyaliphatic acids; nitroguanidines; physical studies of beta diketones; nuclear magnetic resonance, ultraviolet, infrared spectroscopy of guanidines and perfluoroaromatics; longitudinal study of scholastic performance of Meharry medical students. *Mailing Add:* 4230 Eatons Creek Rd Nashville TN 37218

LOWE, JANET MARIE, b Ellensburg, Wash, Jan 13, 24. MICROBIOLOGY, EMBRYOLOGY. *Educ:* Univ Wash, BS, 45; Univ Chicago, SM, 47. *Prof Exp:* Res assoc bact, Univ Chicago, 47-49; instr biol, Cent Wash State Col, 49-54, from asst prof to assoc prof zool, 54-75, prof biol & dir Allied Health Sci Prog, 74-90; RETIRED. *Concurrent Pos:* NSF res grant, 58-60. *Mem:* AAAS; Am Soc Microbiol; Am Inst Biol Sci. *Res:* Chick embryology; bacteriology. *Mailing Add:* 7124 47th St SW No 203 Seattle WA 98136

LOWE, JOHN, III, b New York, NY, Mar 14, 16; m 43; c 3. DAM ENGINEERING, GEOTECHNICAL ENGINEERING. *Educ:* City Col New York, BS, 36; Mass Inst Technol, MS, 37. *Prof Exp:* Instr civil eng, Univ Md & Mass Inst Technol, 37-44; physicist, David Taylor Model Basin, US Navy, 45; head soil & rock eng dept, Tippetts-Abbett-McCarthy-Stratton, 45-56, assoc partner, 56-62, partner, 62-83; INDEPENDENT CONSULT, DAM ENGINEERING, 84- *Honors & Awards:* Eighth Terzaghi Lect, Soc Soil Mech Mex, 71, Fourth Nabor Carrillo Lectr, 78; Second Ann USCOLD Lectr, US Comt on Large Dams of Int Comn on Large Dams, 82; Martin Kapp Lectr, 86. *Mem:* Nat Acad Eng; fel Am Soc Civil Engrs; Int Soc Soil Mech & Found Eng; Int Soc Rock Mech; US Comn Large Dams. *Res:* Chapters in four engineering handbooks and author of more than 30 technical papers; redam and geotechnical engineering. *Mailing Add:* 26 GrandView Blvd Yonkers NY 10710

LOWE, JOHN EDWARD, b Newark, NJ, May 20, 35; m 57; c 2. VETERINARY SURGERY. *Educ:* Cornell Univ, DVM, 59, MS, 63. *Prof Exp:* Intern vet surg, 59-60, resident, 60-61, instr vet path, 61-63, asst prof vet surg, 63-68, ASSOC PROF VET SURG, NY STATE COL VET MED, CORNELL UNIV, 68-, COORD MGR, EQUINE RES PARK, 74-, ASSOC PROF, NY STATE COL AGR & LIFE SCI, 68- *Mem:* Am Vet Med Asn; Am Asn Equine Practitioners. *Res:* Endocrine control of the equine skeletal system; effect of nutrition on equine bone and joint disease; equine gastrointestinal surgery. *Mailing Add:* NY State Col Vet Med 205 Diag Lab Cornell Univ Ithaca NY 14853

LOWE, JOHN PHILIP, b Rochester, NY, Aug 28, 36; m 59; c 2. QUANTUM CHEMISTRY. *Educ:* Univ Rochester, BS, 58; Johns Hopkins Univ, MAT, 59; Northwestern Univ, PhD(quantum chem), 64. *Prof Exp:* Teacher high sch, NJ, 59-60; NIH fel theoret chem, Johns Hopkins Univ, 64-66; from asst prof to assoc prof, 66-86, PROF CHEM, PA STATE UNIV, UNIVERSITY PARK, 86- *Concurrent Pos:* Petrol Res Fund starter grant, 66-68, type AC grant, 69-71. *Mem:* AAAS; Am Chem Soc; Am Phys Soc. *Res:* Chemical carcinogenicity; chemical reactivities; relations between Huckel and ab initio calculations; quantum chemistry of solids. *Mailing Add:* Dept Chem Pa State Univ University Park PA 16802

LOWE, JOSIAH L(INCOLN), b Hopewell, NJ, Feb 13, 05; m 33; c 1. TAXONOMY OF FLOWERING PLANTS. *Educ:* Col Forestry, Syracuse, NY, BS, 27; Univ Mich, PhD(mycol), 38. *Prof Exp:* Asst, Bot Lab, 33-38, from asst prof to prof, 42-52, res prof, 59-75, EMER PROF MYCOL, COL FORESTRY, SYRACUSE, NY, 75- *Mem:* AAAS; Mycol Soc Am (pres, 61). *Res:* Taxonomy of polyporaceae, regional to continental. *Mailing Add:* 1202 Broad St Syracuse NY 13224

LOWE, KURT EMIL, b Munich, Ger, Nov 21, 05; nat US; m 40; c 1. PETROLOGY. *Educ:* City Col New York, BS, 33; Columbia Univ, MA, 37, PhD(petrol), 47; Asn Prof Geol Scientists, cert. *Hon Degrees:* DSc, Jersey City State Col, 81. *Prof Exp:* Lab asst & tutor geol, Eve Session, 33-42, tutor, 46-47, instr, 47-50, from asst to assoc prof, 50-64, chmn dept, 57-68, prof, 65-72, EMER PROF GEOL, CITY COL NEW YORK, 72- *Concurrent Pos:* Asst, Columbia Univ, 40-42; consult, NY, 36-42 & 47-; consult, NY Trap Rock Corp, 53-73. *Honors & Awards:* Neil Miner Award, Nat Asn Geol Teachers, 68. *Mem:* Fel AAAS; fel Am Geol Soc; fel Mineral Soc Am; fel NY Acad Sci; Nat Asn Geol Teachers (pres, 51-52). *Res:* Mineragraphy; optical mineralogy; structural petrology of granites; Storm King granite at Bear Mountain, New York; structure of the Palisades of Rockland County, New York. *Mailing Add:* 49-01 Francis Lewis Blvd Bayside NY 11364

LOWE, LAWRENCE E, b Toronto, Ont, Mar 29, 33; m 57; c 3. SOIL CHEMISTRY. *Educ:* Oxford Univ, BA, 54, MA, 61; McGill Univ, MSc, 60, PhD(agr chem), 63. *Prof Exp:* Soil chemist, Res Coun Alta, 63-66; from asst to assoc prof, 66-75, assoc dean agr sci, 85, PROF SOILS, UNIV BC, 75- *Mem:* Can Soc Soil Sci; Int Soc Soil Sci; Soil Sci Soc Am. *Res:* Soil organic matter; sulphur in soil. *Mailing Add:* Dept Soil Sci Univ Bril Col 2075 Westbrook Pl Vancouver BC V6T 1W5 Can

LOWE, REX LOREN, b Marshalltown, Iowa, Dec 28, 43; m 64; c 2. PHYCOLOGY. *Educ:* Iowa State Univ, BS, 66, PhD(phycol), 70. *Prof Exp:* Asst bot, Iowa State Univ, 66-69; asst prof, 70-74, PROF BIOL, BOWLING GREEN STATE UNIV, 80- *Concurrent Pos:* Consult, Icthyol Assocs, 71-; vis prof, Univ Mich Biol Sta, 74-85 & Va Polytech Inst & State Univ, 82-83; collabr, US Nat Park Serv, 75-76. *Mem:* Phycol Soc Am. *Res:* Periphyton ecology; diatom taxonomy. *Mailing Add:* Dept Biol Bowling Green State Univ Bowling Green OH 43403

LOWE, RICHIE HOWARD, b Huff, Ky, Apr 9, 35; m 58; c 2. PLANT PHYSIOLOGY, BIOCHEMISTRY. *Educ:* Univ Ky, BS, 58, MS, 59; Ore State Univ, PhD(plant physiol), 63. *Prof Exp:* Res plant physiologist, 63-74, PLANT PHYSIOLOGIST, AGR RES SERV, USDA, 74- *Mem:* Am Soc Plant Physiol. *Res:* Enzymatic activity and biochemical changes associated with plant senescence and post harvest physiology; inorganic nitrogen and phosphorous metabolism. *Mailing Add:* 1077 Spurlock Lane Nicholasville KY 40356

LOWE, ROBERT FRANKLIN, JR, b Chicago, Ill, Nov 14, 41. CARDIOVASCULAR PHYSIOLOGY. *Educ:* Univ Wis, BS, 64, PhD(physiol), 69. *Prof Exp:* ASST PROF PHYSIOL, SCH MED, TULANE UNIV, 70- *Concurrent Pos:* NIH fel, Univ Wis-Madison, 69-70. *Mem:* Am Physiol Soc; Am Heart Asn; Am Fedn Clin Res. *Res:* Autonomic pharmacology. *Mailing Add:* Dept Physiol Tulane Univ Sch Med 1430 Tulane Ave New Orleans LA 70112

LOWE, ROBERT PETER, b Cambridge, Eng, July 8, 35; Can citizen. AERONOMY, INFRARED ASTRONOMY. *Educ:* Univ Western Ont, BSc, 57, PhD(atomic physics), 67. *Prof Exp:* Sci officer, Defense Res Bd, Can, 56-68; from asst prof to assoc prof, 68-80, PROF PHYSICS, 80-, MEM, INST SPACE & TERRESTRIAL SCI, UNIV WESTERN ONT, 87- *Concurrent*

Pos: Vis prof elect eng, Utah State Univ, 82. *Mem:* Am Geophys Union; Can Asn Physicists. *Res:* Infrared airglow; stratospheric composition; infrared spectroscopy of HII regions and planetary nebulae; electronic, vibrational and rotational excitation in ion-molecular collisions. *Mailing Add:* Dept Physics Univ Western Ont London ON N6A 3K7 Can

LOWE, RONALD EDSEL, b Terre Haute, Ind, Jan 8, 35; m 55; c 6. PUBLIC HEALTH & EPIDEMIOLOGY, MENTAL HEALTH. *Educ:* Ohio State Univ, BSc, 62; Purdue Univ, PhD(entom), 67. *Prof Exp:* Res asst entom, Purdue Univ, 62-66; res entomologist, Cent Am Res, Int Progs Div, Sci & Educ Admin, USDA, 66-75, proj leader, 75-79; dir, Plant Opers, Animal & Plant Health Inspection Serv, Vet Serv, USDA, Tuxtla Gutierrez, Mex, 79-84; ENVIRON MGT OFFICER, USDA, ANIMAL & PLANT HEALTH INSPECTION SERV, VS, NAT PROG PLANNING STAFF, HYATTSVILLE, MD, 84-; REHAB COUNR, HUD, NEW PORT RICHEY. *Concurrent Pos:* Courtesy prof, Univ Fla, 66- *Mem:* AAAS; Entom Soc Am; Am Mosquito Control Asn; Soc Invert Path; Sigma Xi. *Res:* Population dynamics of sterile-male release programs; growth regulation compounds for control of medically important insects; pathogenic microorganisms for biological control programs; ecology and epidemiology; agricultural research administration; environmental management; public health. *Mailing Add:* 16011 Glen Haven Dr Tampa FL 33618

LOWE, TERRY CURTIS, b Spokane, Wash, Nov 10, 55; m 80. THEORETICAL MECHANICAL METALLURGY. *Educ:* Univ Calif, Davis, BS, 78; Stanford Univ, MS, 79, PhD(mat sci), 83. *Prof Exp:* mem tech staff, Sandia Nat Labs, Livermore, Calif, US Dept Energy, 82-; MEM STAFF & GROUP LEADER, MAT RES PROCESSING SCI, LOS ALAMOS NAT LABS. *Concurrent Pos:* Lectr, Div Eng, San Francisco State Univ, 80-81; vis scholar, Dept Mat Sci, Stanford Univ, 83-86; chmn, Interagency Metal Forming Work Group, US Dept Energy, 85- *Mem:* Am Soc Metals; Am Inst Mining & Metall Engrs; Am Soc Mech Engrs; Sigma Xi. *Res:* Mathematical modeling of mechanical and physical processes of metals; crystal plasticity modeling of large strain plasticity; finite element analysis of metal forming processes. *Mailing Add:* 1207 Calleton Arias Santa Fe NM 87501-8920

LOWE, WARREN, b San Francisco, Calif, June 4, 22; m 56. ORGANIC CHEMISTRY. *Educ:* Univ Calif, Berkeley, BS, 45. *Prof Exp:* Res asst, Manhattan Proj, US AEC, Univ Calif Radiation Lab, 43-45; from res chemist to sr res chemist, 45-74, SR RES ASSOC CHEM, CHEVRON RES CO, 74- *Mem:* AAAS; Am Chem Soc; Sigma Xi; fel Am Inst Chem. *Res:* Exploratory research of petroleum products and chemicals. *Mailing Add:* Calif Res Corp 576 Standard Ave Point Richmond CA 94807

LOWE, WILLIAM WEBB, b Bartlesville, Okla, Dec 18, 20. CHEMICAL ENGINEERING. *Educ:* Purdue Univ, BS, 47. *Prof Exp:* Staff mem radiochem, Los Alamos Sci Lab, 44-48; chief nuclear eng sect, USAEC, 48-54; nuclear engr, Bath Iron Works, Maine, 54-56; partner, Pickard, Lowe & Garrick Inc, 56-86; CONSULT MGT & APPL SCI, 87- *Concurrent Pos:* Co-ed, Power Reactor Technol, AEC, 60. *Mem:* Am Chem Soc; Am Nuclear Soc. *Res:* Nuclear engineering; engineering economics. *Mailing Add:* 300 N Pitt Alexandria VA 22314

LOWE-KRENTZ, LINDA JEAN, b Milwaukee, Wis, 1953; m 76; c 2. BIOCHEMISTRY. *Educ:* Northwestern Univ, PhD(biochem), 80. *Prof Exp:* Res asst prof biochem, Chicago Med Sch, Univ Health Sci, 85-86; ASST PROF CHEM, LEHIGH UNIV, 86- *Mem:* AAAS; Am Soc Chem Biologists. *Res:* Endothelial heparan sulfate proteoglycans and their receptors. *Mailing Add:* Lehigh Univ Mountaintop Campus Bldg A Bethlehem PA 18015

LOWELL, A(RTHUR) I(RWIN), b New York, NY, Nov 9, 25; m 54; c 3. POLYMER CHEMISTRY, EMULSION POLYMERS. *Educ:* Brooklyn Col, AB, 45; Univ Pa, MS, 48, PhD(chem), 51. *Prof Exp:* Res chemist, Air Reduction Co, Inc, 51-57, sect head, 57-58; res assoc, Lucidol Div, Wallace & Tiernan, Inc, 59-60, supvr appln res, 61-62; res assoc, Berkeley Chem Corp, 62 & Heyden Newport Chem Co, 63; sr res chemist, Mobil Chem Co, 64-66, group leader, 66-68, sect leader, 69-73; sci teacher pub schs, Edison, NJ, 73-76; group leader, Norton & Son, 77-78; SR RES CHEMIST, SUN CHEM CORP, 78- *Mem:* Am Chem Soc. *Res:* Polymerization kinetics; organic peroxide initiators; polymer process and product development; emulsion polymerization; coatings; inks in offset printing. *Mailing Add:* 11 Chandler Rd Edison NJ 08820-2642

LOWELL, GARY RICHARD, b Modesto, Calif, Sept 26, 42; m 68; c 2. SKARN PETROLOGY, GREISEN GEOCHEMISTRY. *Educ:* San Jose State Col, BS, 65; NMex Inst Mining & Technol, PhD(geol), 70. *Prof Exp:* From asst prof to assoc prof, 69-81, PROF GEOL, SOUTHEAST MO STATE UNIV, 81- *Concurrent Pos:* Vis prof geol, Univ Fed do Pará, Brazil, 78-80; consult, Houston Int Minerals Corp, Alaska, 80 & 81, Newmont Explor Ltd, 85 & 87. *Mem:* Mineral Asn Can. *Res:* Igneous and metamorphic petrology; tin and tungsten ore deposits. *Mailing Add:* Dept of Earth Sci Southeast Mo State Univ Cape Girardeau MO 63701

LOWELL, JAMES DILLER, b Lincoln, Nebr, Aug 17, 33; m 57; c 4. PETROLEUM GEOLOGY, STRUCTURAL GEOLOGY. *Educ:* Univ Nebr, BSc, 55; Columbia Univ, MA, 57, PhD(geol), 58. *Prof Exp:* Geologist, Am Overseas Petrol Ltd, 58-65; asst prof geol, Washington & Lee Univ, 65-66; sr res specialist, Esso Prod Res Co, 66-73; explor geologist, Exxon Co, USA, 73-74; mgr geol, Northwest Explor Co, 74-76; PRES, COLEXCON, INC, 78- *Concurrent Pos:* Assoc, Oil & Gas Consults Int Inc, 76-; consult geologist, 76- *Mem:* Fel Geol Soc Am; Am Asn Petrol Geologists; Explorer's Club. *Res:* Structural geology of sedimentary rocks. *Mailing Add:* 5836 S Colorow Dr Morrison CO 80465

LOWELL, PHILIP S(IVERLY), b Manila, Philippines, July 9, 31; US citizen; m 59, 74; c 3. CHEMICAL ENGINEERING. *Educ:* Univ Tex, BS, 54, MS, 63, PhD, 66. *Prof Exp:* Process engr, Jefferson Chem Co, Inc, 54-55; process engr, C F Braun & Co, 55-59, sr process engr, 59-60; sr engr, Tex Res Assocs, Inc, 60-64; asst dir chem res, Tracor, Inc, 64-69; vpres, Radian Corp, 69-77; pres, P S Lowell & Co, Inc, 77-87; PRIN ENGR, RADIAN CORP, 88- *Concurrent Pos:* Adj prof, Univ Tex, Austin, 74 & 86; proj group mem, Environ Protection Coop Effort, US-USSR, 74-79. *Mem:* Am Inst Chem Engrs; Am Chem Soc. *Res:* Process engineering of chemical plants and refineries; application of thermodynamics to practical problems; research in air pollution control processes. *Mailing Add:* Radian Corp PO Box 201088 Austin TX 78720-1088

LOWELL, ROBERT PAUL, b Chicago, Ill, Apr 10, 43; m 81; c 4. GEOPHYSICS. *Educ:* Loyola Univ Chicago, BS, 65; Ore State Univ, MS, 67, PhD(geophys), 72. *Prof Exp:* Asst prof, 71-78, ASSOC PROF GEOPHYS, GA INST TECHNOL, 78- *Mem:* Am Geophys Union; Sigma Xi; Geol Soc Am. *Res:* Thermal geophysics; modeling of magmatic processes hydrothermal ore deposits; geothermal energy; regional tectonics. *Mailing Add:* Sch Geophys Sci Ga Inst Technol Atlanta GA 30332

LOWELL, SHERMAN CABOT, b Olean, NY, Aug 15, 18; m 41; c 2. PHYSICS, THEORETICAL NUMERICAL ANALYSIS. *Educ:* Univ Chicago, BS, 40; NY Univ, PhD(math), 49. *Prof Exp:* Sci liaison officer math sci, Office Naval Res, London, Eng, 49-51; from asst to assoc prof math, NY Univ, 51-57; prof math, head dept & dir grad progs math & appl sci, Adelphi Univ, 57-62; prof math & info sci, 62-66, PROF PHYSICS & COMPUT SCI, WASH STATE UNIV, 66-, MATHEMATICIAN, COMPUT CTR, 62- *Concurrent Pos:* Asst to sci dir, Inst Math Sci, NY Univ, 53-57; vis scientist, Lab Physics of Solids, Paris, 68 & Nat Ctr Very Low Temp, Grenoble, France, 69; consult serv bur, Int Bus Mach Corp, 59-62 & Lawrence Livermore Lab, Univ Calif, 63- *Mem:* AAAS; Am Meteorol Soc; Soc Indust & Appl Math; Am Phys Soc. *Res:* Lattice dynamics; wave propagation; numerical analysis. *Mailing Add:* Dept Physics Wash State Univ Pullman WA 99163

LOWEN, GERARD G, b Munich, Ger, Oct 52; 21; US citizen; m; c 3. MECHANICAL ENGINEERING. *Educ:* City Col New York, BME, 54; Columbia Univ, MSME, 58; Munich Tech Univ, Dr Ing, 63. *Prof Exp:* prof mech eng, 54-87, DEPT HEAD, ASSOC DEAN GRAD STUDIES & H KAYSER PROF MECH ENG, CITY COL NY, 87- *Concurrent Pos:* Consult, var indust, 63-; NSF grants, Army Res Off; expert witness, Army Armament Res & Develop Command res grants. *Honors & Awards:* Mechanism Comt Award, Am Soc Mech Engrs, 84, Mach Design Award, 87. *Mem:* AAAS; fel Am Soc Mech Engrs; Am Soc Eng Educ; NY Acad Sci; Soc Mfg Engrs. *Res:* Dynamics of high speed machinery; rigid and elastic body behavior of linkages and mechanisms; kinematic synthesis and analysis; stress and vibration analysis. *Mailing Add:* Dept Mech Eng Sch Eng Convent Ave at 138th St New York NY 10031

LOWEN, W(ALTER), b Cologne, Ger, May 17, 21; nat US; m 43; c 2. SYSTEMS SCIENCES, MECHANICAL ENGINEERING. *Educ:* NC State Univ, BME, 43, MS, 47; Swiss Fed Inst Technol, DrSc(nuclear eng), 62. *Prof Exp:* Instr mech eng, NC State Col, 43-47; prof, Union Col, NY, 47-67, actg chmn dept, 59 & 67, chmn div eng, 56-59 & 66-67; dir sch advan technol, 67-68, dean, 68-77, prof systs sci, 67-90, EMER PROF, THOMAS J WATSON SCH, STATE UNIV NY, BINGHAMTON, 91- *Concurrent Pos:* Consult, Alco Prod, Inc, 52-54, 56, Oak Ridge Nat Lab, 54-57 & Gen Elec Co, 60; consult inst appl technol, Nat Bur Standards, 65-66; vis prof, Swiss Fed Inst Technol, 65-66; dir, Vols for Int Tech Assistance, Inc, 66-69, mem charter bd, 69-; guest sabbaticant, IBM Systs Res Inst, 78 & 79; acad guest, Swiss Fed Inst Technol, Zurich, Switz, 82; fel, NSF; auth, 82. *Honors & Awards:* Sigma Xi. *Mem:* Am Soc Mech Engrs; Am Nuclear Soc; Am Soc Eng Educ; NY Acad Sci; World Acad Arts & Sci; Sigma Xi. *Res:* Cognitive models and visual perception; human factors research. *Mailing Add:* T J Watson Sch State Univ NY Binghamton NY 13901

LOWENGRUB, MORTON, b Newark, NJ, Mar 31, 35; m 61; c 1. MATHEMATICS. *Educ:* NY Univ, BA, 56; Calif Inst Technol, MS, 58; Duke Univ, PhD(math), 61. *Prof Exp:* Instr math, Duke Univ, 60-61; asst prof, NC State Col, 61-62; Leverhulme res fel, Glasgow Univ, 62-63; asst prof, Wesleyan Univ, 63-66; NSF fel, Glasgow Univ, 66-67; assoc prof, 67-72, chmn dept, 77-80, PROF MATH, IND UNIV, BLOOMINGTON, 72- *Concurrent Pos:* Sr res fel, Sci Res Coun, Gt Brit, 73-74; ed, Ind Math J, 77-81 & Math Reviews, 81- *Mem:* Math Asn Am; Am Math Soc; Soc Indust & Appl Math; Am Math Soc. *Res:* Mathematical theory of elasticity. *Mailing Add:* Ind Univ Col Arts & Sci Kirkwood Hall 104 Bloomington IN 47405

LOWENHAUPT, BENJAMIN, b St Louis, Mo, July 15, 18; m 50; c 3. BIOPHYSICS, PLANT PHYSIOLOGY. *Educ:* Iowa State Col, BS, 40; Univ Chicago, MS, 41; Univ Calif, Berkeley, PhD(plant physiol), 54. *Prof Exp:* Res assoc, Univ Calif, Berkeley, 54-55; res assoc, Rockefeller Inst, 56-60, NIH fel, 60-62; sr res assoc physiol, Col Med, Univ Cincinnati, 62-67; PROF BIOL, EDINBORO STATE COL, 67- *Concurrent Pos:* Sabbatical, Flinders Univ SAustralia, 74-75. *Mem:* AAAS; Am Chem Soc; Am Soc Plant Physiol. *Res:* Biochemistry; role of inorganic ions in biology; ion transport and excitation. *Mailing Add:* Dept Biol Edinboro State Col Edinboro PA 16444

LOWENSOHN, HOWARD STANLEY, b Columbus, Ohio, Jan 23, 31; m 53; c 1. CORONARY PHYSIOLOGY, EXERCISE. *Educ:* Franklin & Marshall Col, BS, 56; Univ Southern Calif, MS, 62; Univ Md, PhD(physiol), 72. *Prof Exp:* RES PHYSIOLOGIST, WALTER REED ARMY INST RES, 63-; ASSOC PROF PHYSIOL, UNIFORMED SERV UNIV HEALTH SCI, 80- *Concurrent Pos:* Consult, Johns Hopkins Univ, 77-85; res adv, Nat Res Coun, 80-; mem, ACE Comt & Grip Comt, Am Physiol Soc, 88- *Mem:* Am Physiol Soc; NY Acad Sci; AAAS; Am Heart Res Coun. *Res:* Hemodynamics of coronary blood flow in chronic conscious dogs at rest, during exercise and with varying degrees of ischemia; hypertrophied hearts, including the initial chronic studies of phasic coronary artery blood flow in the right heart in normal and hypertrophied and dilated hearts; cardiovascular and respiratory research director for pre- clinical drug development, including cooperative efforts with WHO. *Mailing Add:* 9105 Louis Ave Silver Springs MD 20910-2129

LOWENSTAM, HEINZ ADOLF, b Siemianowitz, Ger, Oct 9, 12; nat US; m 37; c 3. PALEOECOLOGY. *Educ:* Univ Chicago, PhD(paleont), 39. *Hon Degrees:* Dr, Ludwig Maximilians Univ, Munich, Ger, 81. *Prof Exp:* Cur paleont, Ill State Mus, 40-43; assoc geologist, State Geol Surv, Ill, 43-49, geologist, 49-50; res assoc, Univ Chicago, 48-50, assoc prof, 50-52; prof, 52-83, EMER PROF PALEOECOL, CALIF INST TECHNOL, 83- *Concurrent Pos:* Spec staff to aid war effort, Coal & Oil Develop, State Geol Surv, Ill, 43-45; res assoc, Invert Paleont, Nat Hist Mus, Los Angeles County, 83- *Honors & Awards:* Medal of the Paleont Soc, 86. *Mem:* Nat Acad Sci; AAAS; fel Geol Soc Am; hon mem Paleont Soc; Am Acad Arts & Sci; hon mem Soc Econ Paleontologists & Mineralogists; Japanese Soc Biol Chem. *Res:* Paleoecology; biogeochemistry; paleo-temperatures; evolution of reef ecology; impact of the evolution of life on chemical and physical processes in the oceans; minerals in hard tissue precipitates of marine invertebrates. *Mailing Add:* Div Geol & Planetary Sci Calif Inst Technol Pasadena CA 91125

LOWENSTEIN, CARL DAVID, b New York, NY, Sept 3, 34; m 65; c 1. APPLIED PHYSICS. *Educ:* Kent State Univ, BA, 55; Harvard Univ, SM, 56, PhD(physics), 63. *Prof Exp:* Res fel, Harvard Univ, 63-64; asst res physicist, 64-69, ASSOC SPECIALIST, MARINE PHYSICS LAB, UNIV CALIF, SAN DIEGO, 69- *Concurrent Pos:* Mem sensors comt, US Navy Deep Submergence Syst Prog, 64- *Mem:* Acoust Soc Am; Audio Eng Soc; Inst Elec & Electronics Engrs. *Res:* Synthesis of directive arrays; signal processing; underwater acoustics; computer applications. *Mailing Add:* Marine Physics Lab Univ Calif San Diego 0704 9500 Gillman Dr La Jolla CA 92093-0704

LOWENSTEIN, DEREK IRVING, b Hampton Court, Eng, Apr 26, 43; US citizen; m 68; c 2. HIGH ENERGY PHYSICS. *Educ:* City Col New York, BS, 64; Univ Pa, MS, 65, PhD(physics), 69. *Prof Exp:* Res assoc, Univ Pa, 69-70 & Univ Pittsburgh, 70-73; asst physicist, 73-75, assoc physicist, 75-77, physicist & head, exp planning & support div, 77-84, sr physicist, 83, dep chmn, accelerator dept, 81-84, CHMN, AGS DEPT, BROOKHAVEN NAT LAB, 84 - *Concurrent Pos:* Mem, US/USSR Joint Coord Comt Fundamental Properties Matter, Dept Energy, US/Japan Comt High Energy Physics. *Mem:* Fel Am Phys Soc; AAAS; NY Acad Sci. *Res:* Experimental high energy physics; accelerator operations. *Mailing Add:* AGS Dept Brookhaven Nat Lab Upton NY 11973

LOWENSTEIN, EDWARD, b Duisburg, Ger, May 29, 34; US citizen; m 59; c 3. ANESTHESIOLOGY, CARDIOPULMONARY PHYSIOLOGY. *Educ:* Univ Mich, MS, 59; Am Bd Anesthesiol, dipl,Harvard Univ,MA, 81. *Prof Exp:* Assoc anesthesia, 68-70, from asst prof to assoc prof, 70-81, PROF ANESTHESIA, HARVARD MED SCH, 81- *Concurrent Pos:* Assoc anesthetist, Mass Gen Hosp, 68-71, anesthetist, 71- *Honors & Awards:* Distinguished Lectr Physiol, Am Col Chest Phys, 86. *Mem:* Am Soc Anesthesiol; Am Physiol Soc; Soc Critical Care Med. *Res:* Physiological effects of cardiac and pulmonary disease; cardiac anesthesia. *Mailing Add:* Anesthetist-in-Chief Beth Israel Hosp 330 Brookline Ave Boston MA 02115

LOWENSTEIN, J(ACK) G(ERT), b Frankfurt, Ger, Mar 19, 27; US citizen; m 50; c 3. CHEMICAL ENGINEERING, TECHNICAL MANAGEMENT. *Educ:* Pratt Inst, BChE, 50; Univ Md, MSChE, 58. *Prof Exp:* Asst prod mgr, Gen Gummed Prod, Inc, 50-51; chem engr, Army Chem Ctr, US Army, 53-56; res engr, Org Div, FMC Corp, 56-60, eng supvr, 60-64, process evaluator, Inorg Div, 64-66, asst dir res & develop, 66-72, asst dir res & develop, Niagara Chem Div, FMC Corp, 72-76, tech dir mfg, Agr Chem Group, 76-80, venture mgr, Chem Technol Dept, 80-82, tech dir, 82-84, dir admin, Princeton Res & Develop Ctr, 84-90, MGR PUB AFFAIRS, AGR CHEM GROUP, FMC CORP, 91- *Concurrent Pos:* Lectr, Sch Continuing Educ, NY Univ, 68-; dir, Am Nat Metric Coun, 86- *Mem:* Am Inst Chem Engrs; Am Chem Soc; Am Mgt Asn; Am Inst Chemists. *Res:* Study of research management and management science; chemical engineering research, particularly in the chemical separation unit operations; technical aspects of manufacturing, including quality assurance. *Mailing Add:* FMC Corp PO Box 8 Princeton NJ 08543

LOWENSTEIN, JEROLD MARVIN, b Danville, Va, Feb 11, 26; m 81; c 3. NUCLEAR MEDICINE. *Educ:* Columbia Univ, BS, 46, MD, 53. *Prof Exp:* Physicist, Los Alamos Sci Lab, 46-48; instr med & radiol, Sch Med, Stanford Univ, 57-58; asst clin prof, 63-68, assoc clin prof, 68-81, CLIN PROF MED THYROID RES, MED CTR, UNIV CALIF, SAN FRANCISCO, 81- *Concurrent Pos:* Nat Found fel radiobiol, 55-56; NIH res grant 60-66, 61-66 & 62-66; dir nuclear med, Presby Med Ctr, San Francisco, 59-; partic, Galapagos Int Sci Proj, 64. *Mem:* AMA; Soc Nuclear Med; fel AAAS. *Res:* Applications of physics to medicine, especially medical uses of radioactive isotopes; molecular evolution. *Mailing Add:* 2203 Scott San Francisco CA 94115

LOWENSTEIN, JOHN HOOD, b Newark, NJ, Mar 15, 41; m 67; c 2. QUANTUM FIELD THEORY, NONLINEAR DYNAMICS. *Educ:* Harvard Univ, AB, 62; Univ Ill, Urbana, MS, 63, PhD(physics), 66. *Prof Exp:* Res assoc physics, Univ Minn, 66-68; vis asst prof, Univ Sao Paulo, 68-70; res assoc, Univ Pittsburgh, 70-72; res asst prof, 72-74, assoc prof, 74-81, chmn dept, 85-88, PROF PHYSICS, NY UNIV, 81- *Mem:* Am Phys Soc. *Res:* Quantum field theory, with emphasis on renormalized perturbation theory and soluble two-dimensional models; nonlinear dynamics. *Mailing Add:* Dept Physics NY Univ Four Washington Pl New York NY 10003

LOWENSTEIN, JOHN MARTIN, b Berlin, Ger, Oct 28, 26; m 54. BIOCHEMISTRY. *Educ:* Univ Edinburgh, BSc, 50; Univ London, PhD, 53. *Prof Exp:* Demonstr chem & biochem, Med Sch, St Thomas' Hosp, Eng, 50-53; res assoc biochem, Med Sch, Univ Wis, 53-55; Beit mem fel med res, Oxford Univ, 55-58; prof, 59-77, HELENA RUBINSTEIN PROF BIOCHEM, BRANDEIS UNIV, 77- *Concurrent Pos:* Ed, Methods in Enzymol, Archives Biochem & Biophysics, 67-72, J Lipid Res, 79-, J Biol Chem, 79-; mem adv comt, Med Found Res Comt, 74-77, Biochem Study Sect, 77-81; lectr, Indian Dept Sci & Indust Res, 80. *Mem:* AAAS; Am Chem Soc; Am Soc Biol Chem; Brit Biochem Soc. *Res:* Regulated enzymes; integration and control of metabolism pathways. *Mailing Add:* Dept Biochem Brandeis Univ Waltham MA 02254

LOWENSTEIN, MICHAEL ZIMMER, b Hornell, NY, Oct 4, 38; m 62; c 2. SOLAR ENERGY. *Educ:* Oberlin Col, AB, 60; Ariz State Univ, MS, 62, PhD(x-ray crystallog), 65. *Prof Exp:* Asst prof chem, 64-71, prof chem, Adams State Col, 71-78; educ proj mgr, Joint US-Saudi Prog, Solar Energy Res Inst, 78-82, prog mgr, Biofuels Prog Off, 82-90; MGR, POWER QUAL SYSTS, TRANS-COIL, INC, 90- *Concurrent Pos:* AEC fac res assoc, Ariz State Univ, 70-71; consult, Citizen's Workshop, Energy Res & Develop Agency, 74-76, Dept Energy, 76-78; vis prof, Solar Energy Appln Lab, Colo State Univ, Ft Collins, 75-76; prof chem, Adams State Col, Alamosa, 71-77; dir, Solar Energy Div, Navarro Col, Tex, 77-78. *Mem:* Am Chem Soc. *Res:* Biomass energy systems; solar energy education; energy problems; instrumentation. *Mailing Add:* Trans-Coil, Inc 7878 N 86th St Milwaukee WI 53224

LOWENTHAL, DENNIS DAVID, b Yakima, Wash, Nov 10, 42; m 66; c 2. PLASMA PHYSICS, ELEMENTARY PARTICLE PHYSICS. *Educ:* Calif State Univ, Northridge, BS, 65; Univ Calif, Los Angeles, MS, 66; Univ Calif, Irvine, PhD(physics), 75. *Prof Exp:* Res & develop engr, Aeronutronic Div, Philco-Ford Corp, 66-75; physicist, Math Sci Northwest, 75-80. *Mem:* Am Phys Soc; Optical Soc Am; AAAS. *Res:* Experimental search for the double beta decay of selenium 82; geometrical and wave optics; plasma physics diagnostics. *Mailing Add:* Spectra Technol Inc 2755 Northup Way Bellevue WA 98004

LOWENTHAL, DOUGLAS H, b New York, NY, Dec 19, 48; m 82. RECEPTOR MODELLING, ANALYTIC CHEMISTRY. *Educ:* Tufts Univ, BA, 70; Univ RI, MS, 76, PhD(oceanog), 86. *Prof Exp:* Teaching asst org chem, 74, res asst, 74-80 & RES SPECIALIST, UNIV RI, 81- *Mem:* Air Pollution Control Asn. *Res:* Determination of regional sources of air pollution in the Arctic and eastern US; statistical source receptor modelling. *Mailing Add:* Desert Res Inst PO Box 60220 Reno NV 89506

LOWENTHAL, WERNER, b Krefeld, Ger, Dec 20, 30; US citizen; m 85; c 2. EDUCATION. *Educ:* Albany Col Pharm, Union Univ, NY, BS, 53; Univ Mich, PhD(pharmaceut chem), 58; VA Commonwealth Univ, MEd, 78. *Prof Exp:* Asst, Univ Mich, 53-55; res pharmacist, Abbott Labs, 57-61; asst prof pharm, 61-66, assoc prof, 66-71, PROF PHARM, SCH PHARM, MED COL VA, VA COMMONWEALTH UNIV, 71-, PROF EDUC PLANNING & DEVELOP, 74-, DIR CONTINUING EDUC, 80- *Concurrent Pos:* Mem US Pharmacopoeia Rev Comt, 75-80; chmn coun fac, Am Asn Cols Pharm, 80-81. *Mem:* Asn Psychol Types; Am Asn Cols Pharm; Am Educ Res Asn; Sigma Xi. *Res:* Pharmaceutical product development; drug absorption; programmed instruction; continuing education; curriculum development; ethics; Myers Briggs personality preferences. *Mailing Add:* Sch Pharm Med Col Va Va Commonwealth Univ Richmond VA 23298-0581

LOWER, RICHARD ROWLAND, thoracic surgery, cardiovascular surgery, for more information see previous edition

LOWER, STEPHEN K, b Oakland, Calif, Sept 8, 33; m 63. PHYSICAL CHEMISTRY. *Educ:* Univ Calif, Berkeley, BA, 55; Ore State Univ, MS, 58; Univ BC, MSc, 60, PhD(phys chem), 63. *Prof Exp:* Fel phys chem, Polytech Inst Brooklyn, 63-64 & Univ Calif, Los Angeles, 64-65; ASSOC PROF PHYS CHEM, SIMON FRASER UNIV, 65- *Concurrent Pos:* Nat Res Coun Can grants, 65-71, mem panel on comput assisted instruction lang, 70- *Mem:* Am Chem Soc. *Res:* Instructional systems design; computer-assisted instruction and instructional technology applied to college science teaching. *Mailing Add:* Dept Chem Simon Fraser Univ Burnaby ON V5A 1S6 Can

LOWER, WILLIAM RUSSELL, b La Junta, Colo, Oct 28, 30; m 71; c 2. GENETICS, ENVIRONMENTAL HEALTH. *Educ:* Univ Calif, Los Angeles, BA, 53; Univ Calif, Berkeley, PhD(genetics), 65. *Prof Exp:* Res assoc genetics of nematodes, Kaiser Found Res Inst, 64-66; res assoc, Clin Pharmacol Res Inst, 67-69; res assoc biol monitoring, Environ Health Surveillance Ctr, 70-72, ASSOC PROF COMMUNITY HEALTH & MED PRACT & BIOL, UNIV MO-COLUMBIA, 72-, GROUP LEADER, ENVIRON TRACE SUBSTANCES RES CTR, 72- *Concurrent Pos:* Fel, Univ Mo, 69-70. *Mem:* Genetics Soc Am; Environ Mutagen Soc; Soc Toxicol; Soc Environ Geochem & Health; Soc Environ Toxicol & Chem; Air Pollution Control Asn; Am Biol Safety Asn. *Res:* Research and monitoring of genetic, biochemical, and physiological effects of airborne, terrestrial and fresh water environmental pollutants in situ in the real world and under controlled laboratory conditions; mutagenesis of environmental contaminants; development and standardization of new bioassays. *Mailing Add:* Environ Trace Res Ctr Rr 3 Univ Mo Columbia MO 65203

LOWERY, LEE LEON, JR, b Corpus Christi, Tex, Dec 26, 38; m 60; c 2. STRUCTURAL ENGINEERING, STRUCTURAL FAILURES. *Educ:* Tex A&M Univ, BS, 60, MS, 61, PhD(struct eng), 67. *Prof Exp:* From asst prof to assoc prof, 61-71, PROF ENG, TEX A&M UNIV, 71- *Concurrent Pos:* Res engr, Albritton Eng Corp, 63-66, Tex Transp Inst, 67-; consult engr, Esso Prod Res Corp, 66-68, Shell Oil Corp, 76-78 & Marathon Oil Corp, 77-78; prof construct, Sch Archit, Tex A&M Univ, 65-69, prof struct, Dept Aerospace Eng, 67-70; failure analyst, Eng Consult, Inc, 70-76, prod failure analyst, 76-77. *Mem:* Am Soc Exp Stress Analysis; Am Soc Civil Engrs; Soc Marine Technol; Am Soc Eng Educ; Nat Soc Prof Engrs. *Res:* Basic research, engineering structures and products; applied research in areas of design and analysis of coastal, offshore structures; product failure analysis, consumer protection; engineering applications of computer analysis. *Mailing Add:* Dept of Civil Eng Tex A&M Univ College Station TX 77843-3136

LOWERY, R(ICHARD) L, b Haven, Kans, July 25, 35; m 59; c 2. MECHANICAL ENGINEERING. *Educ:* Tex Tech Col, BS, 56; Okla State Univ, MS, 57; Purdue Univ, PhD(mech eng), 61. *Prof Exp:* Instr mech eng, Tex Tech Col, 57-58; from asst prof to prof, 61-72, HALLIBURTON PROF MECH ENG & DIR CTR TEACHING, OKLA STATE UNIV, 72- *Concurrent Pos:* Consult, Fed Aviation Agency, 64-65. *Mem:* Acoust Soc Am; Am Soc Eng Educ. *Res:* Acoustics; sonic boom research; ultrasonics; vibrations; instrumentation. *Mailing Add:* Dept Mech Eng Okla State Univ Main Campus Stillwater OK 74078

LOWERY, THOMAS J, cytology, physiology; deceased, see previous edition for last biography

LOWES, BRIAN EDWARD, b Harrow, Eng, Sept 21, 35; Can citizen; m 66; c 2. GEOLOGY. *Educ:* Imperial Col, London Univ, BSc, 57; Queen's Univ, Ont, MSc, 63; Univ Wash, Seattle, PhD(geol), 72. *Prof Exp:* Mine geologist asst, Opemiska Copper Ventures Ltd, 57-59; explor geologist, Hollinger Consol Gold Mines, 61-62; tech asst, Can Geol Surv, 63-64; asst prof, 68-75, assoc prof, 75-82, CHMN EARTH SCI DEPT, 77-, PROF, 82- *Mem:* Geol Soc Am; Geol Asn Can; Mineral Asn Can. *Res:* Structural geology and metamorphic petrology of crustal basement rocks in Pacific Northwest. *Mailing Add:* Dept Earth Sci Pac Lutheran Univ Tacoma WA 98447

LOWEY, SUSAN, b Vienna, Austria, Jan 22, 33; nat US. PROTEIN CHEMISTRY, PHYSICAL CHEMISTRY. *Educ:* Columbia Univ, BA, 54; Yale Univ, PhD(chem), 58. *Prof Exp:* Res fel biol, Harvard Univ, 57-59; assoc prof biochem, 72-74, PROF BIOCHEM, BRANDEIS UNIV, 74-, MEM STAFF, ROSENSTIEL BASIC MED SCI RES CTR, 72- *Concurrent Pos:* Res assoc, Children's Cancer Res Found, 59-72. *Mem:* Am Chem Soc. *Res:* Physical chemistry of muscle proteins. *Mailing Add:* Rosenstiel Ctr Brandeis Univ 415 South St Waltham MA 02154-2700

LOWI, ALVIN, JR, b Gadsden, Ala, July 21, 29; m 53; c 4. THERMAL ENGINEERING & HEAT TRANSFER, ENGINE POWER & FUELS. *Educ:* Ga Inst Technol, BME, 51, MSME, 56. *Prof Exp:* Res asst eng exp pract, Ga Inst Technol, 54-56; design engr, Air Res Div, Garrett Corp, 56-58; mem tech staff eng, TRW Aerospace Corp, 58-66; pres, Terraqua, Inc, 59-76; CONSULT ENG, ALVIN LOURI & ASSOCS, 66-; VPRES, DAECO FUELS & ENG CO, INC, 76- *Concurrent Pos:* Lectr econ, Free Enterprise Inst, 60-70; fel, Inst Human Studies, 66-72; res assoc, Heather Found, 66-; vis lectr eng, Univ Pa, 72-74; dir, Southern Calif Tissue Bank, 83-; prin investr, Gas Res Inst, 86-; mem, Reactivity Adv Panel, Calif Air Resources Bd, 89- *Mem:* Am Soc Mech Engrs; Soc Automotive Engrs; Nat Soc Prof Engrs; Soc Am Inventors. *Res:* Simultaneous heat and mass transfer applied to novel cooling and distillation apparatus; supplementary fueling of diesel engines by fumigation of alternative volatile fuels; dissolution of natural gas in liquified petroleum materials for application to compact vehicular fuel storage; fire retardation of cellulose fibers for thermal insulation. *Mailing Add:* 2146 Toscanini Dr San Pedro CA 90732

LOWIG, HENRY FRANCIS JOSEPH, b Prague, Czech, Oct 29, 04; m 49; c 2. PURE MATHEMATICS. *Educ:* Ger Univ, Prague, Dr rer nat, 28; Univ Tasmania, DSc, 51. *Prof Exp:* Privatdozent math, Ger Univ, Prague, 35-38; lectr, Univ Tasmania, 48-51, sr lectr, 51-57; from assoc prof to prof, 57-70, EMER PROF MATH, UNIV ALTA, 70- *Concurrent Pos:* Vis fel, Res Sch Phys Sci, Australian Nat Univ, 66-67. *Mem:* Am Math Soc; Can Math Soc. *Res:* Functional analysis; lattice theory; universal algebra. *Mailing Add:* 15212 81st Ave Edmonton AB T5R 3P1 Can

LOWINGER, PAUL, b Chicago; m 48; c 3. PSYCHOTHERAPY & CLINICAL PHARMACOLOGY, PRISON & FORENSIC PSYCHIATRY. *Educ:* Northwestern Univ, BS, 45; State Univ Iowa, MD, 49, MS, 53; Am Bd Psychiat & Neurol, dipl, 56. *Prof Exp:* Clin instr, psychiat, Tulane Univ Sch Med, 53-55, instr, 55-59, asst prof, 59-62, assoc prof, 62-71, adj assoc prof, 71-74, assoc clin prof, 74-85; MED DIR, OCCUPATIONAL STRESS CLIN, INST LABOR & MENTAL HEALTH, CALIF, 82-; CLIN PROF, DEPT PSYCHIAT & DIV AMBULATORY & COMMUNITY MED, SCH MED, UNIV CALIF, SAN FRANCISCO, 85- *Concurrent Pos:* Psychiat consult, San Francisco Co Jail, 80, Superior Court, Alameda Co, Oakland, 80-86, Prisoners Legal Serv, State Prison, Salem, Ore, 81-82, mental patients rights, dept mental health, San Francisco, 82 & 85, Fulton Co Jail, Legal Aid, Atlanta, 88, public defender, city & co of San Francisco, 83-; psychiat consult, Calif Dept Corrections, Brobeck, Phleger & Harrison & Prison Law Off, 85, 89-; numerous hosp staff mem & consultancies, 53-; res grants, NIMH, 57-62 & 80-82, teaching grant, 59-71, Demonstration Proj grant, Health Educ & Welfare, 74-79. *Mem:* NY Acad Sci; fel Am Psychiat Asn; Sigma Xi; Am Psychosomat Soc; fel Am Orthopsychiat Asn; Am Asn Advance Sci; Am Asn Univ Profs; Am Psychopath Asn; Nat Med Asn; Am Pub Health Asn; Acad Psychoanal. *Res:* Psychotherapy; psychosomatic medicine; psychosis; clinical pharmacology; legal medicine and social psychiatry. *Mailing Add:* 77 Belgrave Ave San Francisco CA 94117

LOWITZ, DAVID AARON, b Newark, NJ, Dec 18, 28; m 53; c 4. CHEMICAL PHYSICS. *Educ:* Rutgers Univ, BA, 50; Pa State Univ, MS, 53, PhD(physics), 55. *Prof Exp:* Asst physics, 50-55, res assoc, Pa State Univ, 55-56; physicist, Gulf Res & Develop Co, 56-64; res assoc & head cent res physics sect, Lord Corp, 64-67; mgr, Physics Div, 67-79, technol planning coordr appl res, 79-82, asst to dir appl res, 82-87, SR SCI, PHILIP MORRIS RES CTR, 87- *Concurrent Pos:* Am Petrol Inst fel, 52-56. *Honors & Awards:* IR 100 Award, 72. *Mem:* Am Phys Soc; Int Soc Quantum Biol; Sigma Xi. *Res:* Microwave scattering; quantum mechanics; high pressure liquid viscosity; electromagnetic wave propagation; dielectrics; electron optics; electro-optic technology. *Mailing Add:* 4312 W Franklin St Richmond VA 23221

LOWMAN, BERTHA PAULINE, b Newton, NC, Mar 17, 29. MATHEMATICS. *Educ:* Lenoir-Rhyne Col, BS, 51; Univ Ala, MA, 52; George Peabody Col, PhD(math), 76. *Prof Exp:* Instr math, Campbell Col, 52-53; instr sci & math, Anderson Col, 53-54; asst, Univ NC, 55; asst prof math, Hardin-Simmons Univ, 55-59, ECarolina Col, 59-60 & Elon Col, 60-62; asst prof, 62-78, PROF MATH, WESTERN KY UNIV, 78- *Mem:* Math Asn Am; Nat Coun Teachers Math. *Res:* Number theory and algebra; geometry and history of mathematics; linear algebra. *Mailing Add:* 1025 Roselawn Way Bowling Green KY 42104

LOWMAN, PAUL DANIEL, JR, b Elizabeth, NJ, Sept 26, 31; m 58. ASTROGEOLOGY, PHOTOGEOLOGY. *Educ:* Rutgers Univ, BS, 53; Univ Colo, PhD(geol), 63. *Prof Exp:* Staff, Goddard Space Ctr, 59-87, GEOPHYSICIST, GEOPHYS BR, LAB TERRESTRIAL PHYSICS, GODDARD SPACE FLIGHT CTR, NASA, 87- *Concurrent Pos:* Vis lectr, US Air Force Inst Technol, 63-64; lectr, Cath Univ, 63-66 & Univ Calif, Santa Barbara, 70. *Honors & Awards:* John C Lindsay Mem Award, NASA, 74. *Mem:* Geol Soc Am; AAAS; Am Geophys Union. *Res:* Planetology; lunar geology; geologic application of orbital photography; remote sensing; comparative planetology. *Mailing Add:* Code 921 Goddard Flight Ctr Greenbelt MD 20771

LOWMAN, ROBERT MORRIS, b Baltimore, Md, Dec 31, 12; m 37; c 2. RADIOLOGY. *Educ:* Harvard Univ, AB, 32; Univ Md, MD, 36. *Hon Degrees:* MA, Yale Univ, 65. *Prof Exp:* Instr radiol, Grad Sch Med, Univ Pa, 36-38, asst dir dept radiol, Grad Hosp, 40-45; asst, Sch Med, Boston Univ, 38-40; assoc prof, 55-62, actg chmn dept, 73, PROF RADIOL, MED SCH, YALE UNIV, 62-; DIR MEM UNIT, YALE-NEW HAVEN HOSP, 62- *Concurrent Pos:* Angiol Res Found honors achievement award, 64-65; attend physician, Grace-New Haven Community Hosp, 45-, dir dept radiol, Mem Unit, 45-; consult, W Haven Vet Hosp, 62-; pres, New Eng Roentgen Ray Soc, 71-72, mem exec comt & chmn exec bd, 72-74; fel, Davenport Col, Yale Univ; staff radiologist, Emer Prof, West Haven, Va. *Honors & Awards:* Sigma Xi. *Mem:* Fel Am Col Radiol; Radiol Soc NAm; Sigma Xi. *Res:* Thoracic lymphatics; embryology of the bladder; cardiac kymography; experimental coronary arteriography; development of animal model with atherosclerotic change in the heart, kidney and brain. *Mailing Add:* Mem Unit Box 1001 Yale-New Haven Med Ctr New Haven CT 06504

LOWN, BERNARD, b Utena, Lithuania, June 7, 21; US citizen; m 46; c 3. ARRHYTHMOLOGY. *Educ:* Univ Maine, BS, 42; Johns Hopkins Univ, MD, 45. *Hon Degrees:* DSc, Univ Maine, 82, Worcester State Col, 83, Charles Univ, 87, State Univ NY, 88, Bowdoin Col, 88; LLD, Bates Col, 83, Queens Univ, 85; LHD, Colby Col, 86, Thomas Jefferson Univ, 88; DPhil, Univ Buenos Aires, 86. *Prof Exp:* Dir, Samuel Levine Cardiovasc Res Lab, Peter Bent Brigham Hosp, 56-58; DIR, CARDIOVASC RES LAB, HARVARD SCH PUB HEALTH, 61-, PROF CARDIOL, 74- *Concurrent Pos:* From asst prof to assoc prof cardiol, Harvard Univ Sch Pub Health, 61-74; sr assoc, Brigham & Womens Hosp, 63-70, physician, 70-84, sr physician, 84-; vis scientist, Mass Inst Technol, 87-; co-pres, Int Physicians Prev Nuclear War, 80-; consult cardiol, Beth Israel Hosp, 63-, Childrens Hosp, 64- *Honors & Awards:* Nobel Peace Prize, 85. *Mem:* Inst Med; Nat Acad Sci Hungary; Am Acad Arts & Sci; Am Col Cardiol. *Res:* The problem of sudden cardiac death, identified potential victims and evolved programs for their protection; investigated the role of neural and psychologic factors provoking life threatening disturbances of heart rhythm. *Mailing Add:* 21 Longwood Ave Brookline MA 02146

LOWN, JAMES WILLIAM, b Blyth, Eng, Dec 19, 34; m 62. BIOORGANIC CHEMISTRY. *Educ:* Univ London, BSc, 56, PhD(org chem) & dipl, Imp Col, 59. *Prof Exp:* Asst lectr chem, Imp Col, Univ London, 59-61; fel, Univ Alta, 61-62, asst prof, 62-63; res chemist, Walter Reed Army Inst Res, DC, 62-63; from asst to assoc prof, 64-74, PROF CHEM, UNIV ALTA, 74-; MEM, NAT CANCER INST CAN, 77- *Concurrent Pos:* Mem UN Educ Sci & Cultural Organ Global Network for Molecular & Cellular Biol, 89- *Mem:* Am Chem Soc; The Chem Soc; Sigma Xi; Am Asn Cancer Res. *Res:* Organic reaction mechanisms; heterocyclic synthesis; antibiotics; cancer and viral chemotherapy. *Mailing Add:* Dept Chem Univ Alta Edmonton AB T6G 2G2 Can

LOWNDES, DOUGLAS H, JR, b Pasadena, Calif, Jan 3, 40; m 61; c 2. SEMICONDUCTORS, PHOTOVOLTAIC CELL RESEARCH. *Educ:* Stanford Univ, BS, 61; Univ Colo, PhD(physics), 69. *Prof Exp:* Res asst solid state physics, Hewlett-Packard Assocs, Calif, 62-63; NSF fel physics, Sch Math & Phys Sci, Univ Sussex, 68-70; from asst prof to prof physics, Univ Ore, 70-79; SR RES STAFF MEM, SOLID STATE DIV, OAK RIDGE NAT LAB, 79- *Concurrent Pos:* Assoc, Solar Energy Ctr, 74-79; guest prof physics, Univ Nijemegen, 76-77; prof mat sci & eng, Univ Tenn, 86- *Mem:* Fel Am Phys Soc; Int Solar Energy Soc; sr mem Inst Elec & Electronic Engrs; Mat Res Soc. *Res:* Photochemical thin film growth; laser interactions with semiconductors; solar cells; nanosecond and piosecond laser measurements; pulsed laser annealing; superconductivity and magnetism; electronic materials. *Mailing Add:* 1101 W Outer Dr Oak Ridge TN 37830-6056

LOWNDES, HERBERT EDWARD, b Barrie, Ont, July 12, 43; m 66; c 3. NEUROTOXICOLOGY, NEUROPHARMACOLOGY. *Educ:* Univ Sask, BA, 64, MSc, 70; Cornell Univ, PhD(pharmacol), 72. *Prof Exp:* Fel pharmacol, Univ Western Ont, 72-73; from asst prof to prof pharmacol, Col Med & Dent, NJ Med Sch, 73-81; DISTINGUISHED PROF, COL PHARM, RUTGERS UNIV, 85- *Concurrent Pos:* Vis prof, Univ Paul Sabatier, Toulouse, France, 66-; consult, Toxicol Data Bank, Nat Libr Med, 81-85, Health Res, Effects Grants Rev Panel, Environ Protection Agency, 84-, Toxicol Study Sect, 80-84, 89-93, Safety & Occup Health Study Sect, NIH, 85- *Mem:* Am Soc Pharmacol & Exp Therapeut; NY Acad Sci; Soc Toxicol; Soc Neurosci; Am Asn Neuropathologists. *Res:* Neurotoxicology and neuropharmacology of central and peripheral nervous sytem, particularly electrophysiological, histochemical and morphological correlates. *Mailing Add:* Dept Pharmacol & Toxicol Col Pharm Rutgers Univ PO Box 789 Piscataway NJ 08854

LOWNDES, JOSEPH M, b Duluth, Minn, Feb 28, 55. BIOCHEMISTRY, MOLECULAR BIOLOGY. *Educ:* Univ Notre Dame, BS, 77; Univ Wis-Madison, MS, 83, PhD(biochem), 88. *Prof Exp:* Postdoctoral fel, Dept Pediat, Nat Jewish Ctr Immunol & Resp Med, 89-91; SR SCIENTIST, DEPT RES & DEVELOP, FIVE PRIME THREE PRIME, 91- *Mem:* Am Soc Biochem & Molecular Biol; AAAS; Am Chem Soc; Sigma Xi. *Mailing Add:* Dept Res & Develop Five Prime Three Prime 5603 Arapahoe Boulder CO 80303

LOWNDES, ROBERT P, b Derby, Eng, Dec 11, 39. PHYSICS. *Educ:* Univ London, BSc, 62, Queen Mary Col, PhD(exp solid state physics), 67; Northeastern Univ, MBA, 76. *Prof Exp:* Res assoc physics, Mass Inst Technol, 67-68; asst prof, 68-72, assoc prof, 72-78, PROF PHYSICS,

NORTHEASTERN UNIV, 78-, CHMN DEPT, 81- *Mem:* Am Inst Physics; Brit Inst Physics; fel Sci Res Coun; fel Am Coun Educ; Am Phys Soc. *Res:* High pressure dielectric and far infrared spectroscopic studies of solids. *Mailing Add:* Dept Physics Northeastern Univ Huntington Ave Boston MA 02115

LOWNEY, EDMUND DILLAHUNTY, b Port Arthur, Tex, Nov 8, 31; m 58; c 2. DERMATOLOGY. *Educ:* Univ Tex, BA, 53; Yale Univ, PhD(psychol), 57; Univ Pa, MD, 60. *Prof Exp:* From instr to asst prof dermat, Univ Mich, Ann Arbor, 64-67; assoc prof, Med Col Va, 67-69; PROF DERMAT, UNIV HOSP, COL MED, OHIO STATE UNIV, 69- *Mem:* Soc Invest Dermat; Am Dermatol Asn. *Res:* Immunology. *Mailing Add:* 907 Singing Hills Lane Worthington OH 43085

LOWNEY, JEREMIAH RALPH, b Fall River, Mass, Dec 16, 46; m 80. SEMICONDUCTOR ELECTRONICS, SEMICONDUCTOR PHYSICS. *Educ:* Mass Inst Technol, Cambridge, BS, 67, MS, 68, PhD(elec eng), 75. *Prof Exp:* Physicist, Naval Ord Lab, 68-72 & Naval Surface Weapons Ctr, 75-79; PHYSICIST, NAT INST STANDARDS & TECHNOL, 79- *Mem:* Am Phys Soc; Inst Elec & Electronics Engrs; AAAS; Sigma Xi. *Res:* Electronic properties of semiconducting materials, such as band structure, mobility, lifetime, deep-level spectroscopy and impact ionization in silicon and compound semiconductors. *Mailing Add:* Bldg 225 Rm A-305 Nat Inst Standards & Technol Gaithersburg MD 20899

LOWNIE, H(AROLD) W(ILLIAM), JR, b Buffalo, NY, July 11, 18; m 76; c 2. METALLURGY, ENGINEERING. *Educ:* Purdue Univ, BS, 39; Univ Pittsburgh, MS, 44; Ohio State Univ, MBA, 72. *Prof Exp:* Foundry engr, Westinghouse Elec Corp, Pa, 39-45; asst supvr, Columbus Div, Battelle Mem Inst, 45-50, chief process metall, 50-81, res leader, 71-83; RETIRED. *Concurrent Pos:* Consult, 83- *Honors & Awards:* Whiting Gold Medal, Foundrymens Soc, 59; Merit Award, Am Soc Testing & Mat, 66. *Mem:* Am Soc Metals; Foundrymens Soc; Am Inst Mining, Metall & Petrol Engrs; fel Am Soc Testing & Mat. *Res:* Metallurgy and inoculation of gray cast iron; use of foundry coke; cupola operation; blast-furnace practice; general foundry practice; research administration; direct reduction of iron ore; chemical metallurgy; economics of metallurgy processes. *Mailing Add:* 2902 Halstead Rd Columbus OH 43221

LOWNSBERY, BENJAMIN FERRIS, b Wilmington, Del, July 28, 20; m 50; c 1. PLANT NEMATOLOGY. *Educ:* Univ Del, BA, 42; Cornell Univ, PhD(plant path), 50. *Prof Exp:* Chemist explosives div, E I du Pont de Nemours & Co, 42-45; asst plant path, Cornell Univ, 45-50; asst plant pathologist, Conn Agr Exp Sta, 51-53; from asst nematologist to nematologist, Exp Sta, 54-83, lectr nematol, 60-70, prof, 70-83, EMER PROF NEMATOL, UNIV CALIF, DAVIS, 83- *Concurrent Pos:* Mem subcomt nematodes, Agr Bd, Nat Acad Sci-Nat Res Coun, 66-68; sr ed, J Nematol, 77-78, ed-in-chief, 78-81. *Honors & Awards:* Stark Award, Am Soc Hort Sci, 70. *Mem:* Soc Nematol; Am Phytopath Soc. *Res:* Forest and agricultural nematology. *Mailing Add:* Dept Nematol Univ Calif Davis CA 95616

LOWRANCE, EDWARD WALTON, b Ogden, Utah, June 17, 08; m 35; c 2. ANATOMY. *Educ:* Univ Utah, AB, 30, AM, 32; Stanford Univ, PhD(biol), 37. *Prof Exp:* Asst zool, Stanford Univ, 32-34, Rockefeller asst exp embryol, 34-36 & 37-38; from instr to assoc prof zool, Univ Nev, 38-49; asst prof anat, Sch Med, Univ SDak, 49-50; from assoc prof to prof, 50-78, EMER PROF ANAT, SCH MED, UNIV MO-COLUMBIA, 78- *Concurrent Pos:* Actg assoc prof, Sch Med, Univ Kans, 44-46; State secy, Mo State Anat Bd, 69-78. *Mem:* AAAS; Am Asn Anat; Am Micros Soc; NY Acad Sci. *Res:* Statistical treatment of weights and linear measurements of selected dimensions of bones of sub-adult and mature opossum and adult man; tendon growth and associated bone growth in postnatal rabbit. *Mailing Add:* Six Miller Dr Columbia MO 65201-5420

LOWRANCE, WILLIAM WILSON, JR, b El Paso, Tex, May 8, 43. SCIENCE POLICY, RISK ASSESSMENT. *Educ:* Univ NC, Chapel Hill, AB, 65; Rockefeller Univ, PhD(biochem), 70. *Prof Exp:* Res chemist, Tenn Eastman Co, Kingsport, 70-71; res consult, NC Dept Educ, Raleigh, 71-72; asst exec ed, J Cell Biol, New York, 72-73; resident fel, Nat Acad Sci, Washington, DC, 73-75; res fel, Prog Sci & Int Affairs, Harvard Univ, 75-76; spec asst to US Secy State, Washington, DC, 77-78; vis assoc prof human biol, Stanford Univ, 78-80; SR FEL & DIR, LIFE SCI & PUB POLICY PROG, ROCKEFELLER UNIV, NEW YORK, 80- *Mem:* AAAS. *Res:* National and international science policy; decisions regarding public health risks; ethical responsibilities of technical people; nuclear proliferation; synthetic and mechanistic organic photochemistry. *Mailing Add:* 460 E 63rd St New York NY 10021-6399

LOWREY, CHARLES BOYCE, b New Orleans, La, Mar 15, 41; m 61; c 3. PHYSICAL ORGANIC CHEMISTRY. *Educ:* Centenary Col, BS, 63; Univ Houston, PhD(heterocyclic chem), 68. *Prof Exp:* Teaching asst chem, Univ Houston, 63-66; from asst prof to assoc prof chem, Centenary Col La, 73-77, asst dean col, 74-77; gen mgr opers & prod, Petrol Assocs of Lafayette, Inc, 77-79; asst gen mgr & tech mgr, Port Arthur, Tex Facil, Chem Water Mgt, Inc, 79-81; consult hazardous waste disposal, Price-Curtis & Assoc Inc, 81-82; SOUTHERN REGIONAL VPRES SALES, CHEM WASTE MGT, 82- *Concurrent Pos:* Consult, Baifield Industs, La, 66-70; water pollution consult, Ford Battery Plant, Shreveport, 68-73 & Gould Battery Plant, Shreveport, 73-75. *Mem:* Am Chem Soc; Soc Petrol Engrs. *Res:* Synthesis and study of electronic effects in substituted benzo(b) furans and benzo(b) thiophenes. *Mailing Add:* 18035 Rolling Creek Houston TX 77090-1125

LOWREY, GEORGE HARRISON, pediatrics, for more information see previous edition

LOWRIE, ALLEN, b Washington, DC, Dec 30, 37; div; c 1. MARINE GEOLOGY, CONTINENTAL MARGINS. *Educ:* Columbia Univ, BA, 62. *Prof Exp:* Res asst marine geol, Lamont Geol Observ, 63-68; oceanogr marine geol, Naval Oceanog Off, 68-76 & 78-81 & Naval Ocean Res & Develop Act, 76-78; explorationist, Mobil Oil Corp, 81; MEM STAFF, NAVAL SPACE TECH LAB, NASA. *Concurrent Pos:* Consult geologist, Seagull Int Explor, Houston, Tex & Int Inc, Kenner, La; invited lectr, Catholic Univ Am, Washington, DC, 72-73 & Universidad de Los Andes, Bogota, Colombia, 78-; guest lectr oceanog & ecol, Calverton Sch, Huntington, Md, 74-76; consult, St Stanislaus Col, Bay St Louis, Miss, 76-; instr, Tulane Univ, New Orleans, La. *Mem:* Soc Econ Paleont & Mineral; Am Asn Petrol Geologists; NY Acad Sci; Am Inst Prof Geologists; Sigma Xi. *Res:* Interaction along subduction zones of Western North and South America, i.e., Chile, Isthmus of Panama, and Western Colombia; sediment type and thickness and acoustic response in ocean basins; evolution of passive margins. *Mailing Add:* Dept Navy US Naval Oceanog Off Stennis Space Ctr MS 39522-5001

LOWRIE, HARMAN SMITH, organic chemistry; deceased, see previous edition for last biography

LOWRIGHT, RICHARD HENRY, b Bethlehem, Pa, Aug 31, 40; m 66. SEDIMENTOLOGY. *Educ:* Franklin & Marshall Col, AB, 62; Pa State Univ, PhD(geol), 71. *Prof Exp:* Teacher pub sch, NY, 64-66; asst prof, 71-78, ASSOC PROF GEOL, SUSQUEHANNA UNIV, 78- *Concurrent Pos:* Consult geol, 73- *Mem:* Nat Water Well Asn; Soc Econ Paleontologists & Mineralogists. *Res:* Quantity and quality of ground water in Snyder County, Pennsylvania. *Mailing Add:* Dept Geol & Environ Sci Susquehanna Univ Selinsgrove PA 17870

LOWRY, BRIGHT ANDERSON, b Newberry, SC, Apr 6, 36; m 65; c 2. ASTRONOMY, COMPUTER INTERFACING. *Educ:* Mass Inst Technol, SB, 58; Univ Chicago, PhD(phys chem), 65. *Prof Exp:* Res assoc, Dartmouth Col, 63-64 & Univ NC, Chapel Hill, 64-66; from asst prof to assoc prof chem, Southern Methodist Univ, 66-74; PROF CHEM, ERSKINE COL, 74- *Mem:* Am Chem Soc; Am Phys Soc. *Res:* Physical properties of liquid crystals. *Mailing Add:* Dept of Chem Erskine Col Box 535 Due West SC 29639

LOWRY, ERIC G, b Berlin, Ger, Nov 23, 16; US citizen; m 54; c 1. PHYSICAL CHEMISTRY. *Educ:* Univ Geneva, PhD(phys chem), 43. *Prof Exp:* Res chemist fluorochem, Gen Chem Div, Allied Chem Corp, 47-49; res chemist photog, Remington-Rand Div, Sperry Rand Corp, 51-58; res chemist lithography, Polychrome Corp, 59; res chemist, Addressograph-Multigraph Corp, 59-65, chief chemist reprography, 65-77, sect supvr, Charles Bruning Co Div, 77-81; RETIRED. *Mem:* AAAS; Am Chem Soc; Soc Photog Sci & Eng; Tech Asn Pulp & Paper Indust. *Res:* Reprography. *Mailing Add:* 73 Lewis St Middleton CT 06457-5226

LOWRY, GEORGE GORDON, b Chico, Calif, Jan 12, 29; m 53; c 4. PHYSICAL CHEMISTRY. *Educ:* Chico State Col, AB, 50; Stanford Univ, MS, 52; Mich State Univ, PhD(phys chem), 63. *Prof Exp:* Res asst, Stanford Res Inst, 51; res chemist, Dow Chem Co, 51-62; NSF fel, 62-63; from asst prof to assoc prof chem, Claremont Men's Col, 63-68; assoc prof, 68-75, PROF CHEM, WESTERN MICH UNIV, 75- *Concurrent Pos:* Independent consult, Environ Safety & Health. *Mem:* Am Chem Soc; Sigma Xi. *Res:* Polymerization kinetics and processes; copolymerization; statistical theory of kinetic chain processes; hazardous materials; safety and health. *Mailing Add:* Dept Chem Western Mich Univ Kalamazoo MI 49008-3842

LOWRY, GERALD LAFAYETTE, b Harrisburg, Pa, Sept 12, 28; m 49; c 3. FORESTRY, SOIL SCIENCE. *Educ:* Pa State Univ, BS, 53; Ore State Univ, MS, 55; Mich State Univ, PhD(forestry), 61. *Prof Exp:* Asst, Ore State Univ, 53-55; instr stripmine reclamation, Ohio Agr Exp Sta, Wooster, 55-61; res forester, Pulp & Paper Res Inst Can, 61-72; assoc prof, 72-76, PROF, STEPHEN F AUSTIN STATE UNIV, 76- *Concurrent Pos:* Asst prof, Ohio State Univ, 57-58; spec res asst, Mich State Univ, 58-59; vchmn forestry comt, Coun Fertilizer Appln, 61-63, chmn, 63-65. *Mem:* Soc Am Foresters; Soil Sci Soc Am. *Res:* Forest soil-site relationships; rehabilitation of burned and cutover lands; coal stripmine reclamation; soil chemistry, physics and fertility; tree physiology and silviculture. *Mailing Add:* Sch of Forestry Stephen F Austin State Univ Nacogdoches TX 75962-6109

LOWRY, JAMES LEE, b Birmingham, Ala, Feb 19, 31; m 56; c 3. ELECTRICAL ENGINEERING. *Educ:* Auburn Univ, BEE, 55, MS, 57; Univ Fla, PhD(elec eng), 63. *Prof Exp:* From instr to asst prof elec eng, Auburn Univ, 55-59; teaching assoc, Univ Fla, 62-63; assoc prof, 63-65, PROF ELEC ENG, AUBURN UNIV, 65- *Concurrent Pos:* Consult, Ala Power Co. *Mem:* Sr mem Inst Elec & Electronics Engrs; Am Soc Eng Educ; Nat Soc Prof Engrs; Sigma Xi. *Res:* Circuit analysis and synthesis; power systems. *Mailing Add:* Dept Elec Eng Auburn Univ 200 Broun Hall Auburn AL 36849

LOWRY, JEAN, b Indianapolis, Ind, Feb 7, 21. GEOLOGY. *Educ:* Pa State Univ, BS, 42; Yale Univ, PhD(geol), 51. *Prof Exp:* Jr economist, Off Price Admin, 42-43; jr geologist, US Geol Surv, 43-46, asst geologist, 46-49; dist geologist, State Geol Surv, Va, 49-57; from asst prof to prof geol, E Carolina Univ, 58-83; RETIRED. *Concurrent Pos:* Vis prof, Concepcion Univ, Chile, 62-63. *Res:* Stratigraphy and structure of southern Appalachians; caves. *Mailing Add:* 211 S Eastern St Greenville NC 27858

LOWRY, JERALD FRANK, b Listie, Pa, Oct 22, 39; m 61; c 4. EXPERIMENTAL PHYSICS. *Educ:* Univ Pittsburgh, BS, 61; Cornell Univ, MS, 63. *Prof Exp:* Jr engr, Testing Reactor, Westinghouse Elec Corp, 61; teaching asst physics, Cornell Univ, 61-63; sr engr appl physics, 63-80, SR RES SCIENTIST, WESTINGHOUSE SCI & TECHNOL CTR, 81- *Mem:* Am Phys Soc; AAAS. *Res:* Low pressure plasmas; fluorescent lamp discharges; generation of high power electron beams; measurement of power density distribution and beam radiance; gas discharge lasers, electron-beam sustained discharges; superconductivity. *Mailing Add:* 1730 Yorktown Pl Pittsburgh PA 15235

LOWRY, LEWIS ROY, JR, b Little Falls, NY, Dec 3, 28; m 50; c 4. PHYSICS, ELECTRICAL ENGINEERING. *Educ:* Miami Univ, AB, 49, MS, 54; Ohio State Univ, PhD(physics), 67. *Prof Exp:* Design engr, Aerospace Elec Div, Westinghouse Res & Develop Ctr, 51-61, supvr engr, 61-63, fel engr, 63-68, mgr, appl res, 68-72, mgr device technol, 72-79, semiconductor res, 79-85, mgr, Semiconductor Specialties Develop, 85-87, mgr, Sci Specialties, 87-90, PROG MGR, SILICON CARBIDE SEMICONDUCTOR DEVICES, WESTINGHOUSE SCI & TECHNOL CTR, 90- *Mem:* Am Phys Soc; AAAS; Inst Elec & Electronic Engrs. *Res:* Investigations into processing and fabrication of semiconductor devices; recent concentration has been in silicon carbide semiconductors for hostile environments. *Mailing Add:* Westinghouse Sci & Technol Ctr 1310 Beulah Rd Pittsburgh PA 15235

LOWRY, NANCY, b Newburgh, NY, Sept 4, 38; m 61; c 3. PHYSICAL ORGANIC CHEMISTRY. *Educ:* Smith Col, AB, 60; Mass Inst Technol, PhD(chem), 65. *Prof Exp:* Res assoc chem, Mass Inst Technol, 65-66 & Amherst Col, 66-67; lectr, Smith Col, 67-69, res assoc, 69-70; from asst prof to assoc prof, 70-84, PROF CHEM, HAMPSHIRE COL, 84-, DEAN, NAT SCI, 89- *Mem:* AAAS; Associational Women Sci; Inst Soc, Ethics, & Life Sci. *Res:* Women and science; science education. *Mailing Add:* Sch Nat Sci & Math Hampshire Col Amherst MA 01002

LOWRY, OLIVER HOWE, b Chicago, Ill, July 18, 10; m 35; c 5. PHARMACOLOGY, BIOCHEMISTRY. *Educ:* Northwestern Univ, BS, 32; Univ Chicago, MD & PhD(biochem), 37. *Hon Degrees:* DSc, Wash Univ, 81. *Prof Exp:* Instr biochem, Harvard Med Sch, 37-42; mem staff, Pub Health Res Inst, NY, 42-44, assoc chief, Div Physiol & Nutrit, 44-47; prof pharmacol, 47-79, head dept, 47-76, dean, 55-58, DISTINGUISHED EMER PROF PHARMACOL, SCH MED, WASH UNIV, 79- *Concurrent Pos:* Commonwealth Found fel, Carlsberg Lab, Copenhagen Univ, 39. *Honors & Awards:* Midwest Award, Am Chem Soc, 62, Scott Award, 63; Borden Award, Asn Am Med Cols, 66. *Mem:* Nat Acad Sci; Am Soc Pharmacol & Exp Therapeut; Am Soc Biol Chem; Am Chem Soc; Histochem Soc. *Res:* Tissue electrolytes; chemistry of aging; nutrition and detection of nutritional deficiency; histochemistry; neurochemistry. *Mailing Add:* Dept Pharmacol Sch Med Wash Univ St Louis MO 63110

LOWRY, PHILIP HOLT, b New York, NY, Feb 20, 18; m 45; c 2. OPERATIONS RESEARCH. *Educ:* Princeton Univ, AB, 39; Yale Univ, MA, 42, PhD(int rels), 49. *Prof Exp:* Meteorologist, Brookhaven Nat Lab, 47-51; opers analyst, Opers Res Off, Johns Hopkins Univ, 51-61; opers analyst, Res Anal Corp, 61-72; opers analyst, Gen Res Corp, 72-80; consult, 80-88; RETIRED. *Mem:* Opers Res Soc Am; Am Meteorol Soc; Am Astron Soc. *Res:* Military operations research; impact of technology on international relations; nuclear policy and strategy. *Mailing Add:* 8701 Georgetown Pike McLean VA 22102

LOWRY, RALPH A(DDISON), b Clay County, Mo, Aug 9, 26; m 47; c 4. ENGINEERING, PHYSICS. *Educ:* Iowa State Univ, BS, 49, PhD(physics), 55. *Prof Exp:* Sr scientist, Res Labs Eng Sci, 55-62, from assoc prof to prof aerospace eng, 62-77, chmn dept aerospace eng & eng physics, 65-72, dean, Sch Eng & Appl Sci, 73-74, PROF NUCLEAR ENG & ENG PHYSICS, UNIV VA, 77-, JOHN LLOYD NEWCOMB PROF ENG & APPL SCI, 78- *Concurrent Pos:* Assoc dean, Sch Eng & Appl Sci, Univ Va, 86- *Mem:* Am Phys Soc; Am Soc Eng Educ; Sigma Xi. *Res:* Atomic and molecular physics; isotope separation; gas centrifuges; fluid mechanics. *Mailing Add:* Univ Va Thornton Hall Charlottesville VA 22901

LOWRY, ROBERT JAMES, b Chelsea, Mich, Aug 26, 12; m 34; c 1. BOTANY. *Educ:* Univ Mich, BS, 40, MS, 41, PhD(bot), 47. *Prof Exp:* Res assoc, Univ Mich, 42-45; asst prof bot, Mich State Univ, 46-48; from asst prof to prof, 48-81, EMER PROF BOT, UNIV MICH, ANN ARBOR, 81- *Mem:* AAAS. *Res:* Cytotaxonomy; electron microscopy. *Mailing Add:* 630 Hampstead Lane Ann Arbor MI 48103

LOWRY, STEPHEN FREDERICK, b Colombus, Ohio, Nov 1, 1947; c 3. SURGERY. *Educ:* Ohio, Wesleyan Univ, BA, 69; Univ Mich Sch Med, MD, 73. *Prof Exp:* Asst prof surg, 82-87, DIR HYPERALIMENTATION UNIT, NY HOSP-CORNELL MED CTR, 82-, ASSOC PROF SURG, 87- *Concurrent Pos:* Dir lab surg metab, NY Hosp Cornell Med Ctr, 82-; vis assoc physician, Rockefeller Univ, 82; asst attend surgeon, gastric & mixed tumor serv, 82 & nutrit, mem, Sloan-Kettering Cancer Ctr, 85-; traveling fel, James IV Asn Surgeons, 87. *Mem:* Am Col Surgeons; Asn Acad Surg; Soc Univ Surgeons; Fed Am Soc Exp Biol; Soc Surg Oncol; Int Soc Surg. *Res:* Identifications mechanisms inducing hypermetabolisms, protein regulation and tissue in trauma, sepsis and cancer; method for restoration of protein homeostasis by nutritional support. *Mailing Add:* Cornell Med Ctr NY Hosp 525 E 68th St Rm F2014 New York NY 10021

LOWRY, THOMAS HASTINGS, b New York, NY, June 16, 38; m 61; c 3. ORGANIC CHEMISTRY. *Educ:* Princeton Univ, AB, 60; Harvard Univ, PhD(chem), 65. *Prof Exp:* NIH fel chem, Mass Inst Technol, 64-65, res assoc, 65-66; from asst prof to assoc prof, 66-81, PROF CHEM, SMITH COL, 81- *Mem:* Am Chem Soc. *Res:* Physical organic chemistry. *Mailing Add:* Dept Chem Smith Col Northampton MA 01063

LOWRY, WALLACE DEAN, b Medford, Ore, Oct 5, 17; m 42. GEOLOGY. *Educ:* Ore State Univ, BS, 39, MS, 40; Univ Rochester, PhD(geol), 43. *Prof Exp:* Geologist, Ore Dept Geol & Mineral Indust, 42-47, Texaco, Inc, 47-49; assoc prof, 49-58, PROF GEOL, VA POLYTECH INST & STATE UNIV, 58- *Mem:* Fel Geol Soc Am; Am Asn Petrol Geol. *Res:* Late Cenozoic stratigraphy of the lower Columbia River basin; ferruginous bauxite deposits of Northwestern Oregon; silica sands of Western Virginia; porosity of sandstone reservoir rocks; role of Tertiary volcanism in tectonism; relation of silicification and dolomitization; geology of the Blue Mountains, Oregon; mechanics of Appalachian thrusting; North American geosynclines. *Mailing Add:* Dept of Geol Sci Va Polytech Inst & State Univ Blacksburg VA 24061

LOWRY, WILLIAM THOMAS, b Hobbs, NMex, Dec 11, 42; m 65; c 2. OCCUPATIONAL SAFETY & HEALTH. *Educ:* E Tex State Univ, BS, 65, MS, 67; Colo State Univ, PhD(natural prod chem), 71; Am Inst Chemists, cert, 75; Am Bd Forensic Toxicol, cert, 76. *Prof Exp:* Chemist, Fed Bur Invest, 65 & spec agent, 72-73; res assoc biochem, Va Polytech Inst & State Univ, 71-72; toxicologist, Southwestern Inst Forensic Sci, 73-85; from asst prof to assoc prof toxicol, Grad Sch Biomed Sci, Univ Tex Health Sci Ctr, 77-85. *Concurrent Pos:* Assoc consult, attend staff toxicol, Parkland Mem Hosp, 73 -; instr path, Univ Tex Southwestern Med Sch, 73-75, instr path & forensic sci, 75-77, asst prof path, 77 -; adj asst prof chem, E Tex State Univ, 76-77, adj assoc prof, 77-80; adj asst prof civil eng, Univ Tex, Arlington, 82 - *Mem:* Am Acad Clin Toxicol; Am Acad Forensic Sci; Am Chem Soc; Am Inst Chemists; Am Soc Pharmacog; Sigma Xi. *Res:* Environmental toxicology; biodegradation of toxic substances; utilizing bacteria; combustion and pulmonary toxicology. *Mailing Add:* Fielder Prof Pk 733 B N Fielder Rd Arlington TX 76012

LOWTHER, FRANK EUGENE, b Orrville, Ohio, Feb 3, 29; m 51; c 4. PETROLEUM ENGINEERING. *Educ:* Ohio State Univ, BS, 52. *Prof Exp:* Sr engr, Raytheon Mfg Co, 52-57; consult, Gen Elec Co, 57-65; founder dir & vpres, Purification Sci, Inc, 65-75; sr eng assoc, Union Carbide, 75-79; sr res scientist, 80-82, chief scientist, Energy Conversion & Mat Lab, 82-83, prin scientist, 83-85, RES ADV, ATLANTIC RICHFIELD CO, 85- *Mem:* Am Inst Aeronaut & Astronaut; Inst Elec & Electronics Engrs. *Res:* Ozone chemistry; plasma generators; solid state power devices; internal combustion engines; electrodesorption; thermoelectrics; virus and bacteria disinfection systems. *Mailing Add:* 2928 Clear Springs Plano TX 75075

LOWTHER, JAMES DAVID, b Jackson, Miss, June 22, 39; m 61; c 3. MECHANICAL ENGINEERING. *Educ:* Miss State Univ, BS, 61, MS, 62; Univ Tex, Austin, PhD(mech eng), 68. *Prof Exp:* Mech engr, Baton Rouge refinery, Humble Oil & Refining Co, 62-63; from asst prof to assoc prof, 63-73, UNIV DISTINGUISHED PROF MECH ENG, LA TECH UNIV, 73- *Concurrent Pos:* Prin investr, NSF res grant, 70-71, Naval Weapons Eng Suport Activ, 76-77, US Dept Energy, 80-81, Energy Anal & Diag Ctr, US Dept Energy, 84-85, La Dept Nat Resources, 86-89; consult, 73- *Mem:* Am Soc Mech Engrs; Am Soc Eng Educ; Sigma Xi. *Res:* Heat transfer; thermodynamics; energy conservation; computer-based measurement. *Mailing Add:* Dept of Mech Eng La Tech Univ Ruston LA 71272-0046

LOWTHER, JOHN LINCOLN, b Burlington, Iowa, Sept 5, 43. COMPUTER SCIENCE. *Educ:* Univ Iowa, BA, 65, MS, 67, PhD(comput sci), 75. *Prof Exp:* Instr math, Southwest State Univ, 67-71; from instr to asst prof, 74-77, ASSOC PROF COMPUT SCI, MICH TECHNOL UNIV, 77- *Mem:* Asn Comput Mach; Math Asn Am; Sigma Xi; Inst Elec & Electronics Engrs; Am Asn Artificial Intel. *Res:* Artificial intelligence; programming languages; computer graphics. *Mailing Add:* 404 Bridge St Houghton MI 49931-1295

LOWTHER, JOHN STEWART, b Cochrane, Ont, July 31, 25; m 53, 80. PALEONTOLOGY, PALEOBOTANY. *Educ:* McGill Univ, BSc, 49, MSc, 50; Univ Mich, PhD(geol), 57. *Prof Exp:* From instr to assoc prof, 56-80, PROF GEOL, UNIV PUGET SOUND, 80- *Res:* Sedimentology; Mesozoic paleobotany and stratigraphy; pollen microstructure; palynology. *Mailing Add:* Dept Geol Univ Puget Sound Tacoma WA 98416

LOWY, BERNARD, b New York, NY, Feb 29, 16; m 50; c 2. MYCOLOGY. *Educ:* Long Island Univ, BS, 38; Univ Iowa, MS, 49, PhD(bot), 51. *Prof Exp:* Tech asst biol, Long Island Univ, 38-42, instr, 46-48; instr bot, Univ Iowa, 49-51; from asst to assoc prof, 51-62, prof, 62-80, EMER PROF BOT & CUR MYCOL HERBARIUM, LA STATE UNIV, BATON ROUGE, 80- *Concurrent Pos:* Fulbright scholar, Peru, 58-59 & 72 & Brazil, 65-66; vis prof, Univ Tucuman, Argentina, 59; Am Philos Soc grant, Mex, 62; Sigma Xi grant, Guatemala, 63; res partic, Orgn Trop Studies, Costa Rica, 64; mem numerous mycol expeds, Mex, Cent Am, SAm & West Indies, 50-78; consult ed, Revista Interam, Interam Univ, PR, 71-; chmn, Ethnomycol Sect, Int Mycol Cong, 75-77; mem, ed bd Mycologia, 72-87, Proj Flora Amazonica, Brazil, 80-86. *Mem:* Mycol Soc Am; Bot Soc Am; Am Bryol & Lichenological Soc; hon mem Mex Soc Mycol; Int Asn Plant Taxon. *Res:* Taxonomy and phylogeny of neotropical tremellaceous fungi; ethnomycology of Central America. *Mailing Add:* Dept Bot La State Univ Baton Rouge LA 70803-1705

LOWY, DOUGLAS R, ONCOLOGY. *Prof Exp:* MEM STAFF, CELLULAR ONCOL LAB, NAT CANCER INST. *Mailing Add:* Nat Cancer Inst Cellular Oncology Lab Bldg 37 Rm 1B-26 Bethesda MD 20892

LOWY, PETER HERMAN, b Vienna, Jan 3, 14; nat US; m 40; c 3. ORGANIC CHEMISTRY. *Educ:* Univ Vienna, Dr(chem), 36. *Prof Exp:* Food chemist, Rochester, NY, 40-45; res asst, Calif Inst Technol, 46-49, from res fel chem to sr res fel, 49-72, from res assoc biol to sr res assoc, 72-85; RETIRED. *Mem:* AAAS; Am Chem Soc; Fedn Am Socs Exp Biol; Am Soc Hemat. *Res:* Organic chemical synthesis, particularly of radioactive compounds; isolation and structure determination of bio-organic substances. *Mailing Add:* 188 S Meridith Ave Pasadena CA 91106

LOWY, R JOEL, b Pittsburgh, Pa, Aug 24, 56. PHYSIOLOGY, CELL BIOLOGY. *Educ:* Col William & Mary, BS, 74; Va Inst Marine Sci, MA, 77; Ore State Univ, PhD(zool & biochem), 82. *Prof Exp:* Sr staff fel, NIH, 87-88; RES PHYSIOLOGIST, DEPT PHYSIOL, ARMED FORCES RADIOBIOL RES INST, 88- *Concurrent Pos:* Nat res serv award, NIH, 85-87. *Mem:* Sigma Xi; AAAS; Am Physiol Soc; Am Soc Cell Biol. *Mailing Add:* Dept Physiol Armed Forces Radiobiol Res Inst Bethesda MD 20814

LOWY, STANLEY H(OWARD), b New York, NY, Mar 10, 22; m 45; c 2. AEROSPACE ENGINEERING. *Educ:* Purdue Univ, BS, 43; Univ Minn, MS, 47. *Prof Exp:* Test engr, Allison Div, Gen Motors Corp, 47; instr mech eng, Ore State Col, 47-50; struct design engr, Willamette Iron & Steel Co, 50-51; standards engr, Hughes Aircraft Co, 52; chief engr, Peters Co, 52-53; chief engr, A Young & Son Iron Works, 53-56; consult engr, Stan H Lowy

& Assocs, 56-58; assoc prof aerospace eng, Univ Okla, 58-64; assoc prof, 64-77, assoc dir, Proj Themis, Res Found, 69-71, prof aerospace eng, 77-86, asst dean eng, 80-86, EMER PROF AEROSPACE ENG, TEX A&M UNIV, 86- *Concurrent Pos:* Proj dir space shuttle wind tunnel tests & analysis, Manned Spacecraft Ctr, NASA, 69-72. *Mem:* Am Soc Eng Educr; Am Inst Aeronaut & Astronaut; Am Helicopter Soc; Sigma Xi. *Res:* Aircraft design; aircraft power plants; orbital mechanics. *Mailing Add:* 1016 Walton Dr College Station TX 77840

LOXLEY, THOMAS EDWARD, b Beaver, Pa, Jan 20, 40; div. ENGINEERING. *Educ:* Case Western Univ, Cleveland, BS, 61. *Prof Exp:* Mech engr, US Naval Weapons Lab, 61-65 & US Army Watervliet Arsenal, 65-68; syst engr, Int Hydrodynamics, Ltd, 68-69; pres, Manned Submersible Syst Co, 69-71; mech engr, US Naval Surface Weapons Ctr, 71-75; asst prof tech resources, Va Polytech Inst & State Univ, 75-78; FOUNDER, INVERTED CAVE EDUC, 78- *Concurrent Pos:* Mem, Int Coun Bldg Res, W67 Working Comn, World Cong, Wash, 86, World Cong, Paris, 89; lectr, Nat Bur Standards, Denver, 80; Am Sol Energy Soc, Houston, 82; Royal Inst Tech, Stockholm, 87; N Sun Conf, Borlange, 88; Tech Univ, Vienna, 88. *Mem:* Am Soc Heating Refrig & Air Conditioning Engrs; Int Platform Asn; Int Coun Bldg Res Studies & Doc. *Res:* Geological aspect of global warming and inverted cave phenomenon for earth-coupling conventional looking low-rise buildings; such buildings use the ground under them as an energy resource and storage device for space heating and cooling. *Mailing Add:* 103 Baughman Lane Suite 138 Frederick MD 21702

LOY, JAMES BRENT, b Borger, Tex, Feb 28, 41; div; c 1. PLANT BREEDING, DEVELOPMENTAL GENETICS. *Educ:* Okla State Univ, BS, 63; Colo State Univ, MS, 65, PhD(genetics), 67. *Prof Exp:* from asst prof to assoc prof, 67-81, PROF PLANT SCI, UNIV NH, 81- *Concurrent Pos:* Vis scholar bot, Univ Calif, Berkeley, 74-75. *Mem:* Am Soc Hort Sci; Soc Econ Bot; Nat Agr Plastics Asn. *Res:* Cucurbit breeding; hormonal and genetic regulation of sex expression in Cucumis melo; morpho-physiological investigation of seed and fruit field in Cucurbito species. *Mailing Add:* Dept Plant Sci Univ NH Durham NH 03824

LOY, MICHAEL MING-TAK, b China, Jan 12, 45; US citizen; m 70; c 2. TECHNICAL MANAGEMENT. *Educ:* Univ Calif, Berkeley, BS, 66, PhD(physics), 71. *Prof Exp:* Res staff mem, 71-77, mgr, 78-86, tech planning staff, 86-87, DEPT MGR, THOMAS WATSON RES CTR, IBM CORP, 87- *Concurrent Pos:* Prin investr, Off Naval Res, 78-89; mem steering comt, Laser Sci Topical Group, Am Phys Soc, 91- *Mem:* Am Phys Soc. *Res:* Laser science; surface science; dynamic properties at or near surfaces; nonlinear optical study techniques. *Mailing Add:* Thomas Watson Res Ctr IBM Corp Yorktown Heights NY 10598

LOY, REBEKAH, b Berkeley, Calif, Dec 30, 47; m 78; c 4. NEURAL ANATOMY, DEVELOPMENT & PLASTICITY. *Educ:* Univ Calif, Irvine, BS, 71, PhD(psychobiol), 75. *Prof Exp:* Fel, Univ Calif, San Diego, 75-78, asst prof neurosci, 78-83; assoc prof, Dept Anat, 83-88, SCIENTIST, DEPT NEUROL & SURG, UNIV ROCHESTER, 88- *Concurrent Pos:* Prin investr, Nat Inst Neurol & Commun Dis & Stroke, 78-; panel mem, Neurobiol Prog, Subpanel Integrative & Motor Processes, NSF, 82-84. *Mem:* AAAS; Soc Neurosci; Int Soc Develop Neurosci; Am Asn Anatomists. *Res:* Neuronal reorganization in response to brain injury; sex differences in brain function, development and repair; control of synaptic specificity and plasticity in development, after injury and in response to chronic drug treatment; Alzheimer's disease. *Mailing Add:* Dept Neurol Univ Rochester 435 E Henrietta Rd Rochester NY 14642

LOY, ROBERT GRAVES, b Prescott, Ariz, Feb 7, 24; m 51; c 5. ANIMAL PHYSIOLOGY. *Educ:* Ariz State Univ, BS, 55; Univ Wis, MS, 56, PhD(physiol of reprod) 59. *Prof Exp:* Instr genetics, Univ Wis, 56-59; asst prof animal husb, Univ Calif, Davis, 59-66; from asst to assoc prof vet sci, Univ Ky, 66-71; agr consult, 71-74; from assoc prof to prof, 74-87, EMER PROF VET SCI, UNIV KY, 87- *Concurrent Pos:* Consult, Eguine Reproduction, 87- *Mem:* Am Soc Animal Sci. *Res:* Physiology and endocrinology of reproduction in horses. *Mailing Add:* Hagyarg Davidson McGee 848V Nandino Blvd Lexington KY 40510

LOYALKA, SUDARSHAN KUMAR, b Pilani, India, Apr 11, 43. NUCLEAR & MECHANICAL ENGINEERING. *Educ:* Univ Rajasthan, BEMech, 64; Stanford Univ, MS, 65, PhD(nuclear eng), 67. *Prof Exp:* From asst prof to assoc prof, 67-77, PROF NUCLEAR ENG & CUR PROF, DEPT NUCLEAR ENG, UNIV MO-COLUMBIA, 89- *Concurrent Pos:* Vis scientist, Max Planck Inst Aerodyn, Gottingen, 69-71; Huber O Croft chair eng, Univ Mo-Columbia, 83- *Honors & Awards:* Fel, Am Phys Soc; fel, Am Nuclear Soc. *Mem:* Sigma Xi; Am Nuclear Soc; Am Phys Soc; Am Chem Soc. *Res:* Kinetic theory of gases; neutron transport theory and reactor physics; nuclear reactor safety analysis; mechanics of aerosols. *Mailing Add:* Dept Nuclear Eng Univ Mo Columbia MO 65212

LOYD, DAVID HERON, b Shreveport, La, July 3, 41; m 60; c 2. ATOMIC PHYSICS, NUCLEAR PHYSICS. *Educ:* Univ Tex, Austin, BS, 63, MA, 64; Univ Wis-Madison, PhD(physics), 70. *Prof Exp:* ASST PROF PHYSICS, ANGELO STATE UNIV, 69- *Mem:* Am Phys Soc. *Res:* Atomic collisions. *Mailing Add:* Dept Physics Angelo State Univ 2601 West Ave N San Angelo TX 76909

LOYNACHAN, THOMAS EUGENE, b Oskaloosa, Iowa, Nov 18, 45; m 67; c 3. SOIL MICROBIOLOGY, SOIL FERTILITY. *Educ:* Iowa State Univ, BS, 68, MS, 72; NC State Univ, PhD(soil sci), 75. *Prof Exp:* Asst prof agron, Univ Alaska, 75-78; MEM TEACHING STAFF SOIL SCI, IOWA STATE UNIV, 78-, MEM RES STAFF FIXATION & SOIL ECOL, 78- *Mem:* AAAS; Soil Sci Soc Am; Am Soc Agron; Coun Agr Sci & Technol. *Res:* Nitrification inhibitors; oil degradation in Arctic soils; nitrogen fixation of legumes. *Mailing Add:* Dept Agron Iowa State Univ Ames IA 50011

LOZANO, EDGARDO A, b Tampico, Mex, Nov 20, 24; m 49; c 3. BACTERIOLOGY. *Educ:* Univ Tex, BA, 48; Univ Wis, MS, 54; Mont State Univ, PhD(microbiol), 65. *Prof Exp:* Bacteriologist vaccine prod, Agr Res Serv, 48-50; res, Am Sci Labs, 54-55; dept head prod & develop, Corn States Labs, 55-59; dir bio-prod, Philips Roxane Inc, 63-64; asst prof bact, 65-68, ASSOC PROF BACT, VET RES LAB, MONT STATE UNIV, 68-, ASSOC PROF MICROBIOL, 80- *Res:* Bacteriological antigens and their purification; bacterial toxins; electrophoresis; telemetry of domestic animals. *Mailing Add:* 924 1/2 Sourdough Rd Bozeman MT 59715

LOZERON, HOMER A, b Grande Prairie, Alta, July 24, 34; m 67; c 2. BIOCHEMISTRY. *Educ:* Univ Alta, BSc, 56, MSc, 59; Univ Wash, PhD(biochem), 64. *Prof Exp:* Proj assoc, McArdle Lab Cancer Res, 65-67, instr, 67-72; asst prof, 72-77, ASSOC PROF BIOCHEM, SCH MED, ST LOUIS UNIV, 77- *Mem:* Am Soc Biol Chemists; Am Soc Microbiol. *Res:* RNA processing pathways and regulation of gene expression in bacterial virus systems. *Mailing Add:* Dept Biochem St Louis Univ 1402 S Grand Blvd St Louis MO 63104

LOZIER, DANIEL WILLIAM, b Portland, Ore, Apr 10, 41; m 66; c 1. NUMERICAL ANALYSIS, MATHEMATICAL SOFTWARE. *Educ:* Ore State Univ, BA, 62; Am Univ, MA, 69; Univ Md, PhD(appl math), 79. *Prof Exp:* Mathematician, US Army Eng Res & Develop Lab, Ft Belvoir, Va, 63-69; MATHEMATICIAN, NAT BUR STANDARDS, US DEPT COM, 69- *Concurrent Pos:* Adj prof, Inst Phys Sci & Technol, Univ Maryland, College Park. *Mem:* Soc Indust Appl Math; Math Asn Am; Asn Comput Mach; Sigma Xi. *Res:* Numerical analysis and mathematical software; computation of special functions; forward and backward recurrence methods; floating-point and level-index computer arithmetic; numerical aspects of programming languages; computational fluid dynamics. *Mailing Add:* Appl & Computational Math Div Nat Inst Standards & Technol Gaithersburg MD 20899

LOZZIO, CARMEN BERTUCCI, b Buenos Aires, Arg, Dec 20, 31; US citizen; m 55; c 1. MEDICAL GENETICS, CELL BIOLOGY. *Educ:* Univ Buenos Aires, physician, 55, MD, 60. *Prof Exp:* Physician in chg cytol, Rivadavia Hosp, Buenos Aires, 56-60; instr genetics, Univ Buenos Aires, 60-65; from res assoc to asst res prof, 65-72, assoc res prof med genetics, 72-78, DIR, BIRTH DEFECTS CTR, MEM RES CTR & HOSP, UNIV TENN, KNOXVILLE, 66-, PROF MED BIOL, CTR HEALTH SCI, UNIV TENN, KNOXVILLE, 78- *Concurrent Pos:* Arg Asn Prog Sci Millet fel & Arg Nat Res Coun fel radiation res, Rivadavia Hosp & Arg AEC, 57-60; grants, Arg Nat Res Coun, Univ Buenos Aires, 61-65, Pan Am Union, Biol Div, Oak Ridge Nat Lab, 64, Am Cancer Soc, Univ Tenn, Knoxville, 66-71, Nat Found-March of Dimes, 66-80, Physicians Med Educ & Res Found, 69-70, NIH, 69-71 & 75-81, US Dept Health, Educ & Welfare, 70-74, Tenn Dept Human Serv, 74-, Tenn Dept Ment Health, 74- & Tenn Dept Pub Health, 78- *Honors & Awards:* Honor Cert, World Cong Obstet & Gynec & Int Cong Internal Med, 64. *Mem:* Genetics Soc Am; Genetics Soc Can; Am Asn Ment Deficiency; Am Soc Human Genetics; NY Acad Sci; Sigma Xi. *Res:* Studies on human genetics and cytogenetics; genetic counseling and prenatal diagnosis of hereditary disorders; experimental studies on cell culture of human diploid strains with genetic markers and the effect of antimetabolites on mammalian cell cultures. *Mailing Add:* Mem Res Ctr Knoxville TN 37920

LU, ADOLPH, b Chengtu, China, Feb 19, 42; Can citizen. HIGH ENERGY PHYSICS. *Educ:* Queen's Univ, BSc, 64; Univ Toronto, MA, 65; Univ Calif, Berkeley, PhD(physics), 73. *Prof Exp:* Researcher, Univ D'Orsay, Paris, 73-75; ASSOC RES PHYSICIST HIGH ENERGY PHYSICS, UNIV CALIF, SANTA BARBARA, 76- *Mem:* Am Phys Soc. *Res:* Bubble chamber physics; proton storage ring studies: high point events; photon cross sections; two-photon physics. *Mailing Add:* Physics Dept Univ Calif Santa Barbara CA 93106

LU, ANTHONY Y H, b Hupei, China, Jan 12, 37; m 65; c 1. BIOCHEMISTRY. *Educ:* Nat Taiwan Univ, BS, 58; Univ NC, Chapel Hill, PhD(biochem), 66. *Prof Exp:* Fel inst sci & technol, Univ Mich, Ann Arbor, 66-70; sr biochemist, Res Div, Hoffmann-La Roche Inc, 70-74, res fel, 74-78; SR INVESTR, RES LABS, MERCK SHARP & DOHME LABS, 78- *Mem:* AAAS; Am Chem Soc; Am Soc Pharmacol & Exp Therapeut; Am Soc Biol Chemists; NY Acad Sci. *Res:* Basic research in biochemistry and biochemical pharmacology. *Mailing Add:* Animal Drug Metab Merck Sharp & Dohme Res Lab Rahway NJ 07065

LU, BENJAMIN C(HIH) Y(EU), b Peking, China, Oct 20, 26; m 52; c 3. CHEMICAL ENGINEERING. *Educ:* Nat Cent Univ, China, BASc, 47; Univ Toronto, MASc, 51, PhD(chem eng), 54. *Prof Exp:* Asst engr, Chinese Petrol Corp, China, 47-50; res assoc, Ont Res Found, Can, 54-55; lectr, Univ Toronto, 55-56; from asst prof to assoc prof, 56-62, actg chmn dept, 60, chmn dept, 61-76, vdean eng, Fac Sci & Eng, 69-76, PROF CHEM ENG, UNIV OTTAWA, 62- *Concurrent Pos:* Mem, Grant Selection Comt, Nat Res Coun Can, 69-72 & Nat Comt Deans Eng & Appl Sci, 69-76; exchange scientist, Inst Chem Process Fundamentals, Czech Acad Sci, 75; vis prof, Univ Pittsburgh, 76, Nihon Univ, Japan, 90; exchange scientist, Japan Soc for Promo Sci, 77; UNESCO consult, Univ Zulia, Venezuela, 78; mem, Hazardous Prod Bd Rev, Can Govt, 80-82; chmn, Comm Grad Studies Sci, Univ Ottawa, 89-92; assoc ed, Can J Chem Eng, 90-; hon prof, Beijing Inst Chem Technol, China, 82, Nanjing Inst Chem Technol, China, 85. *Honors & Awards:* R S Jane Mem Lectr, Can Soc Chem Engrs, 90. *Mem:* Fel Chem Inst Can; Am Inst Chem Engrs; Am Chem Soc; Can Soc Chem Engrs; Sigma Xi. *Res:* Phase equilibria; thermodynamic properties of solutions; cryogenic research; energy engineering; supercritical fluid extraction. *Mailing Add:* Dept Chem Eng Univ Ottawa Ottawa ON K1N 9B4 Can

LU, BENJAMIN CHI-KO, b Changchow, China, Mar 9, 32; m 62; c 2. GENETICS, CELL BIOLOGY. *Educ:* Taiwan Univ, BS, 55; Univ Alta, MS, 62, PhD(bot, genetics), 65. *Prof Exp:* Instr bot, Taiwan Univ, 58-60; fel fungal genetics, Cambridge Univ, 65-67; vis fel, Copenhagen Univ, 66; asst prof, 67-

70, assoc prof, 70-79, PROF GENETICS, UNIV GUELPH, 79- *Concurrent Pos:* Nat Res Coun Can overseas fel, 65-67, res grant, Rask-Orsted Found fel & Carlsberg Found grant, 66-67; Nat Res Coun grant, 68-78; res assoc, Univ Calif, Berkeley, 73-74, Univ NC, Chapel Hill, 83-84; Natural Sci & Eng Res Coun, Can grant, 79-; mem grant comt (cell biol & genetics), Nat Sci & Eng Res Coun Can, 85-88. *Mem:* Genetics Soc Can. *Res:* Meiosis-specific nucleases and their genes; cellular programs in meiosis; genetic recombination. *Mailing Add:* Dept Molecular Biol & Genetics Univ Guelph Guelph ON N1G 2W1 Can

LU, CHIH YUAN, b Taiwan, Repub China, August 13, 50; m 78; c 2. SEMICONDUCTOR, INTEGRATED CIRCUIT. *Educ:* Nat Taiwan Univ, BS, 72; Columbia Univ, MA, 74, PhM, 75, PhD(physics), 77. *Prof Exp:* From assoc prof to prof electronics, Nat Chiao-Tung Univ, 78-83; MEM TECH STAFF, AT&T BELL LABS, 84- *Concurrent Pos:* Exec bd mem, Phys Soc ROC, 81-84; res mem, Sci & Tech Adv Group, ROC, 81-84; vis assoc prof elec eng, NC State Univ, 83-84. *Mem:* Sr mem Inst Elec & Electronics Engrs; Sr mem Electron Device Soc; Life mem Am Phys Soc; Life mem Phys Soc Repub China; Life mem Chinese Inst Eng. *Res:* Very large scale integration semiconductor technology; high voltage IC and submicrometer CMOS process integration and device technology. *Mailing Add:* 5371 Andrea Dr Wescosville PA 18106

LU, CHRISTOPHER D, b Taipei, Taiwan, Repub China, Aug 30, 51; US citizen; m; c 1. RUMINANT NUTRITION, INTERNATIONAL AFFAIR. *Educ:* Nat Taiwan Univ, BS, 74; Univ Wis-Madison, MS, 78, PhD(dairy sci & biochem), 81. *Prof Exp:* From res asst to res assoc ruminant nutrit, Univ Wis-Madison, 78-82; scientist biochem & nutrit, Int Harvester Co, 82; res scientist ruminant nutrit, Prairie View A&M Univ, 82-85; prof & dir, Goat Res Inst, 85-89, PROF & DIR INT PROGS, LANGSTON UNIV, 89- *Concurrent Pos:* Prin rep, Div Int Affairs, Nat Asn State Univs & Land Grant Cols, 88-; trustee, Bd of Southeast Consort Int Develop, 89-; chairperson, Livestock Comt Goats, Am Soc Animal Sci, 89-90; mem, Mgt Award Comt, Am Soc Animal Sci, 90- *Mem:* Am Dairy Sci Asn; Am Inst Nutrit; Nutrit Soc UK; Am Soc Animal Sci; Asn Dirs Int Progs. *Res:* Nutrient requirements for lactation, growth, pregnancy and fiber production in goats; energy and protein utilization; ruminant nutrition; metabolism and physiology. *Mailing Add:* Dir Int Progs Langston Univ PO Box 730 Langston OK 73050

LU, FRANK CHAO, b Hupeh, China, Mar 9, 15; nat US; m 39; c 3. PHARMACOLOGY, SCIENCE COMMUNICATION. *Educ:* Cheeloo Univ, MD, 39. *Prof Exp:* Assoc ed, Coun on Pub, Chinese Med Asn, 40-42; sr asst pharmacol, Cheeloo Univ, 42-44, lectr, 45-47; lectr, WChina Union Univ, 44-45; pharmacologist, Food & Drug Labs, Can Dept Nat Health & Welfare, 51-60, head pharmacol & toxicol sect, 60-65; chief food additives, WHO, 65-76; clin prof pharmacol, Sch Med, Univ Miami, 77-79; CONSULT TOXICOL, 79- *Concurrent Pos:* Res fel exp surg, McGill Univ, 47-48, med res fel pharmacol, 48-51; spec lectr pharmacol, Univ Ottawa, 59-62, spec lectr toxicol, Univ Toronto, 59-62, lectr, Joint China-WHO Toxicol Course, 82; vis prof toxicol, Shanghai Med Univ, 85; managing ed, Biomed & Environ Sci, Academic Press, Inc. *Honors & Awards:* Int Achievement Award, Int Soc Regulatory Toxicol, 87. *Mem:* Am Col Toxicol; Am Soc Pharmacol & Exp Therapeut; Soc Toxicol; Europ Soc Toxicol; Can Pharmacol Soc; Int Acad Environ Safety; Int Soc Regulatory Toxicol. *Res:* Physiology and pharmacology of coronary circulation; bioassay of drugs; cardiac glycosides; blood dyscrasias; toxicology of drugs, food additives, pesticides and contaminants; principles and procedures for toxicological evaluation of chemicals; assessment of the safety of chemicals, on the basis of toxicological data, by the use of the acceptable daily intake approach. *Mailing Add:* 7452 SW 143rd Ave Miami FL 33183-2919

LU, FRANK KERPING, b Taipei, Taiwan, Oct 17, 54; US citizen; m 83; c 1. EXPERIMENTAL TECHNIQUES, TURBULENT FLOWS. *Educ:* Cambridge Univ, BA, 76; Princeton Univ, MSE, 83; Pa State Univ, PHD(mech eng), 88. *Hon Degrees:* MA, Cambridge Univ, 80. *Prof Exp:* Eng Officer, Singapore Armed Forces, 76-79; admin asst, Singapore Civil Serv, 79; res asst, Princeton Univ, 79-82; proj engr, ICOS Corp of Am, 82-83; res asst, Pa State Univ, 84-87; ASST PROF AEROSPACE ENG, UNIV TEX ARLINGTON, 87- *Concurrent Pos:* Prin investr, NASA, 88-91. *Mem:* Am Inst Aeronaut & Astronaut; Inst Elec & Electronic Engrs; Am Phys Soc; Sigma Xi; Am Acad Mech. *Res:* Experimental supersonic and hypersonic aerodynamics; gas dynamics; turbomachinery and internal flows; unsteady flows and flow-induced vibrations; turbulence. *Mailing Add:* Aerospace Eng Dept Univ Tex Arlington TX 76019

LU, GRANT, b Ottawa, Ont, May 9, 56; US citizen. DIAMOND FILM, OPTICAL FIBERS. *Educ:* Univ Manchester, Eng, BSc, 76; Rutgers Univ, MS, 80, PhD(ceramic eng), 83. *Prof Exp:* Mat scientist, Naval Res Lab, Washington, DC, 83-88; SR RES ENGR, NORTON CO, NORTHBOROUGH, MA, 88- *Concurrent Pos:* Ed, Am Ceramic Soc, 85-87. *Honors & Awards:* Mat Res Soc Award, 81. *Mem:* Am Ceramic Soc; Int Soc Optical Eng. *Res:* Thermal and optical applications of diamond film. *Mailing Add:* Norton Co Goddard Rd Northboro MA 01532

LU, GUO-WEI, b Gaixian, China, Feb 10, 32; m 57; c 2. SPINAL PROJECTION NEURONS, PLASTICITY OF CENTRAL NERVOUS SYSTEM. *Educ:* China Med Univ, MD, 55. *Prof Exp:* Asst prof pathophysiol, Beijing Med Univ, 55-60; int res fel neurosci, Fogarty Int Ctr, NIH, 80-82; asst prof physiol, 60-72, assoc prof neurophysiol & chmn dept, 72-80, PROF NEUROBIOL, CHMN DEPT & DIR, INST EXP MED, CAPITAL INST MED, 83- *Concurrent Pos:* Vis prof neurol, Univ Wis-Madison, 87-88. *Mem:* Am Physiol Soc; Soc Neurosci; Int Asn Study Pain; Int Brain Res Orgn. *Res:* Pain physiology and antinociception; anatomico-physiological basis of acupuncture; singly and doubly projecting spinal systems; spinal injuries and stroke; developmental neurobiol of spinal cord and brain; adaptation to and plasticity of hypoxia and pain. *Mailing Add:* Dept Neurobiol Inst Exp Med Capital Inst Med You An Men St Beijing 100054 China

LU, HSIENG S, b Taiwan, China, July 28, 47. PROTEIN STRUCTURE. *Educ:* Nat Taiwan Univ, BS, 70, MS, 75; NTex State Univ, PhD(biochem), 81. *Prof Exp:* Res asst, Pharmacol & Microbiol Group, Panlabs, Inc, Taipei, 70-72; teaching asst biochem, Inst Biochem Sci, Nat Taiwan Univ, Taipei, 72-76, res asst prof, Dept Chem, 83-84; res scientist, Protein Develop & Microsequencing Group, 84-88, SR RES SCIENTIST PROTEIN STRUCT, AMGEN, 88- *Concurrent Pos:* Robert A Welch Found fel, 82-83. *Mem:* Am Soc Biochem & Molecular Biol; Protein Soc; AAAS. *Res:* Protein therapeutics; exploration and initial characterization of new therapeutic proteins; protein recovery process and structure-function studies; development of protein analytical methods, QC tests; extensive characterization of therapeutic proteins. *Mailing Add:* Dept Protein Struct Amgen 1900 Oak Terrace Lane Thousand Oaks CA 91320

LU, JOHN KUEW-HSIUNG, b Miaoli, Taiwan, China, Sept 16, 37; US citizen; m 69; c 2. ENDOCRINOLOGY, NEUROENDOCRINOLOGY. *Educ:* Nat Taiwan Normal Univ, BSc, 61; Nat Taiwan Univ, MSc, 67; Mich State Univ, PhD(physiol), 72. *Prof Exp:* Teacher biol, Hsinchu Sr High Sch, Taiwan, 61-62; instr biol, Nat Taiwan Normal Univ, 63-65; res asst physiol, Nat Taiwan Univ, 65-67; res asst endocrinol, Purdue Univ, 67-68; teaching asst physiol, Mich State Univ, 68-72; postdoctoral fel reprod endocrin, Univ Pittsburgh, 72-74; res assoc, Mich State Univ, 74-75; asst prof endocrin, Univ Calif, San Diego, 75-77; from asst to assoc prof, 77-88, PROF OBSTET, GYNEC & ANAT, CELL BIOL, UNIV CALIF, LOS ANGELES, 88- *Concurrent Pos:* Prin investr res grants, Nat Inst Aging, 80-91 & 84-92; mem biochem endocrinol study sect, NIH, 89-94. *Mem:* Soc Gynec Invest; Am Physiol Soc; NY Acad Sci; Endocrine Soc; Soc Study Reproduction; Sigma Xi. *Res:* Investigations to understand the biological and physiological changes during aging which are responsible for the progressive cessation of regular ovulatory cycles and decrease in reproductive function in mammals. *Mailing Add:* Dept Ostet & Gynec Div Reprod Endocrinol Sch Med Univ Calif CHS-22-177 10833 Le Conte Ave Los Angeles CA 90024-1740

LU, KAU U, b Canton, China, July 10, 39; US citizen; m 68; c 1. APPLIED MATHEMATICS, NUMBER THEORY. *Educ:* Nat Taiwan Univ, BS, 61; Calif Inst Technol, PhD(math), 68. *Prof Exp:* From asst prof to assoc prof, 68-79, PROF MATH, CALIF STATE UNIV, LONG BEACH, 79- *Concurrent Pos:* Consult, Tridea Electronics, 69-70; res assoc, Univ Calif, Berkeley, 81- *Mem:* Am Math Soc; Soc Indust & Appl Math; Planetary Soc. *Res:* Applied mathematics; theory of spiral galaxy; dynamics of earth, cyclone and pulsar; solar physics and sunspots; general relativistic bianary system; Riemann hypothesis. *Mailing Add:* Dept Math Calif State Univ Long Beach CA 90840

LU, KUO CHIN, b Singapore, Dec 26, 17; US citizen; m 58; c 1. SOIL MICROBIOLOGY, PLANT PATHOLOGY. *Educ:* Nanking Univ, BS, 37; Ore State Univ, PhD(microbiol), 53. *Prof Exp:* Jr bacteriologist, Ore State Univ, 53-57; asst soil microbiologist, Cornell Univ, 57-59; res scientist, US Army Biol Warfare Lab, Md, 59-60; soil microbiologist, 60-67, PRIN MICROBIOLOGIST, FORESTRY SCI LAB, USDA, 67-; ASSOC PROF SOIL MICROBIOL, ORE STATE UNIV, 67- *Mem:* AAAS; Am Soc Microbiol; Am Phytopath Soc; Mycol Soc Am; Soil Sci Soc Am. *Res:* Antagonistic organisms against root-rot pathogens; biological control of forest diseases; rhizosphere association of mycorrhizal roots; influence of characteristic carbon-nitrogen ratio in decomposition of forest litters; biochemistry. *Mailing Add:* 150 NW 35th St Corvallis OR 97330

LU, KUO HWA, b Antung, China, Jan 7, 23; US citizen; m 56; c 4. BIOSTATISTICS, GENETICS. *Educ:* Nat Cent Univ, China, BS, 45; Univ Minn, MS, 48, PhD(genetics), 51. *Prof Exp:* Agr adv, Continental Develop Found, 52-53; assoc prof appl statist, Utah State Univ, 56-60; assoc prof, 60-63, PROF BIOSTATIST, DENT SCH, ORE HEALTH SCI UNIV, 63-, PROF MED PSYCHOL, MED SCH, 71-, ADJ PROF MED GENETICS, 79- *Concurrent Pos:* Eli Lilly fel, Univ Minn, 53-56; sta statistician, Utah Agr Exp Sta, 56-60; consult, Lab Nuclear Med & Radiation Biol, Univ Calif, Los Angeles, 63-, NIH fel, vis prof, 66-67; consult, Appl Math Assoc, Inc, 63-; adj prof, Portland State Univ, 67-; consult, Tempo, Gen Elec, 78-80 & Procter & Gamble Co. *Mem:* AAAS; Biomet Soc; Am Math Soc; Am Statist Asn; Int Asn Dent Res. *Res:* Development and application of statistical methodology in biomedical research; statistical methods; dental public health; actuarial investigations in medical and dental insurance programs; simulation of oral diseases. *Mailing Add:* 11780 SW Terra Linda St Beaverton OR 97005

LU, LE-WU, b Shanghai, China, June 5, 33; m 63; c 2. STRUCTURAL ENGINEERING. *Educ:* Nat Taiwan Univ, BS, 54; Iowa State Univ, MS, 56; Lehigh Univ, PhD(civil eng), 60. *Prof Exp:* Res asst civil eng, Lehigh Univ, 58-59, res assoc, 59-61, res asst prof, 61-65, res assoc prof, 65-67, assoc prof, 67-69, PROF CIVIL ENG, LEHIGH UNIV, 69- *Concurrent Pos:* USSR Fulbright-Hays lectureship, Int Coun Exchange Scholars, 75; hon prof, Harbin Civil Eng Inst, 80. *Honors & Awards:* Leon Moisseiff Award, Am Soc Civil Engrs, 67. *Mem:* Assoc Am Soc Civil Engrs; Am Concrete Inst; Int Asn Bridge & Struct Engrs; Am Soc Eng Educ; Earthquake Eng Res Inst; Int Asn Struct Safety & Reliability. *Res:* Behavior of building frames and their components in the elastic and inelastic range; planning and design of tall buildings; response of steel and reinforced concrete building structures to earthquake ground motion. *Mailing Add:* Fritz Eng Lab No 13 Lehigh Univ Bethlehem PA 18015

LU, MARY KWANG-RUEY CHAO, b Liao-ning, China, Sept 6, 35; US citizen; m 61; c 2. ORGANIC CHEMISTRY, MATHEMATICS. *Educ:* Notre Dame Col, Ohio, BS, 59; Univ Detroit, MS, 61; Univ Tenn, Knoxville, PhD(org chem), 68. *Prof Exp:* Technician, Chem Lab, NY Hosp, New York, 59; chemist, US Testing Co, Inc, 61-63; asst prof chem, Morris Col, SC, 63-64; prof chem & math, Lincoln Mem Univ, 68-78; PROF CHEM, WALTERS STATE COMMUNITY COL, MORRISTOWN, TENN, 78- *Concurrent Pos:* US Dept Energy res grant. *Mem:* AAAS; Am Chem Soc. *Res:* Organometallic chemistry; silicon solar cells. *Mailing Add:* Dept Math & Sci Walters State Comm Col 500 S Davey Crockett Morristown TN 37813

LU, MATTHIAS CHI-HWA, b Fukien, China, Jan 3, 40; m; c 2. PHARMACY, MEDICINAL CHEMISTRY. *Educ:* Kaohsiung Med Col, Taiwan, BSc, 63; Ohio State Univ, PhD(med chem), 69. *Prof Exp:* Res asst med chem, Univ Iowa, 64-67 & Ohio State Univ, 67-69; res assoc, Col Pharm, Univ Mich, Ann Arbor, 69-71, instr, 71-72, asst prof, 72-73; asst prof, 73-78, ASSOC PROF MED CHEM, COL PHARM, UNIV ILL, CHICAGO, 78 - *Concurrent Pos:* Vis assoc prof, Grad Inst Pharmaceut Sci, Kaohsiung Col, Kaohsiung, Taiwan, 90, adj prof, Sch Pharm & Sch Chem, 90- *Mem:* Am Chem Soc; NAm Taiwanese Prof Asn. *Res:* Steroidogenesis and metabolisms; drug design; site-directed receptor-based design of radiopharmaceuticals; molecular structures as probes for cholinergic receptors; stereochemistry. *Mailing Add:* Dept Med Chem Col Pharm Univ Ill-Chicago 833 S Wood Chicago IL 60680

LU, NANCY CHAO, b Sian, China, May 29, 41; US citizen; m 66; c 1. NUTRITION, FOOD SCIENCE & TECHNOLOGY. *Educ:* Nat Taiwan Univ, Taipei, Taiwan, BS, 63; Univ Wyoming, Laramie, MS, 65; Univ Calif, Berkeley, PhD(nutrit), 73. *Prof Exp:* Res biochemist metab res, Highland Hosp, Oakland, Calif, 65-66; res biochemist cardiovasc res, Mt Zion Hosp, San Francisco, Calif, 66-68; NIH postdoctoral fel folic acid, Univ Calif, Berkeley, 73-75, res assoc nematode nutrit, 78-80; lectr nutrit & metab, 80-82, assoc prof, 82-87, PROF NEMATODE NUTRIT & NUTRIT METAB, SAN JOSE STATE UNIV, CALIF, 87- *Mem:* Am Inst Nutrit; Am Dietetic Asn; Inst Food Technologist; Soc Exp Biol & Med; Soc Nematol. *Res:* Developing nematodes as a model for food and nutritional research; nutritional requirement of vitamins, minerals and growth factors of nematodes; nematode as a screening organism for testing food additives; food toxins. *Mailing Add:* Dept Nutrit & Food Sci San Jose State Univ San Jose CA 95192

LU, PAU-CHANG, b Kiangsu, China, Apr 11, 30; m 63. MECHANICAL ENGINEERING, AEROSPACE SCIENCE. *Educ:* Nat Taiwan Univ, BS, 54; Kans State Univ, MS, 59; Case Western Reserve Univ, PhD, 63. *Prof Exp:* Mech engr, Taiwan Power Co, 54-56; asst eng, Cheng Kung Univ, Taiwan, 56-57, asst eng, Kans State Univ, 57-59 & Case Western Reserve Univ, 59-62, res assoc, 62-63, asst prof, 63-68; assoc prof, 68-72, PROF MECH ENG, UNIV NEBR, LINCOLN, 72- *Mem:* Am Soc Mech Engrs; Am Soc Eng Educ. *Res:* Viscous flow; magneto-fluid-mechanics; heat exchangers; free convection; integral transforms and other branches of applied mathematics. *Mailing Add:* 1930 Dover Court Lincoln NE 68506

LU, PHILLIP KEHWA, b Anhui, China, Oct 11, 32; m 59; c 3. ASTRONOMY, PHYSICS. *Educ:* Maritime Col, Taiwan, BS, 60; Welsleyan Univ, MA, 65; Columbia Univ, MPhil, PhD(astron & sci educ), 70. *Prof Exp:* Math analyst inst math, Chinese Acad Sci, 60-63; instr comput sci, Jefferson Prof Inst, 65-67; res assoc astron observ, Yale Univ, 67-70; asst prof earth & space sci, 70-77, chem dept, 73-74, assoc prof astron, 77-81, PROF ASTRON, WESTERN CONN STATE UNIV, 81- *Concurrent Pos:* Consult, Bd Educ, New York, 74-75; sci educ scholar, NSF, 74-75; vis prof & consult, Nat Cent Univ, Taiwan; Carnegie-Mellon fel astron, Yale Univ, 83- *Mem:* Fel Royal Astron Soc; Am Astron Soc; Am Phys Soc; Sigma Xi. *Res:* Primodial helium and stellar chemical abundance of halo and high velocity stars using speckle interferometry; missing mass problem of Milky Way Galaxy using stellar kinematics of faint F-stars to one kiloparsec; photometry and spectroscopy. *Mailing Add:* 33 Garella Rd Bethel CT 06801

LU, PONZY, b Shanghai, China, Oct 7, 42; US citizen. MOLECULAR BIOLOGY. *Educ:* Calif Inst Technol, BS, 64; Mass Inst Technol, PhD(biophys), 70. *Prof Exp:* Arthritis Found fel biophys, Max Planck Inst Biophys Chem, 70-73; Europ Molecular Biol Orgn fel genetics, Univ Geneva, 73; PROF CHEM, UNIV PA, 73- *Concurrent Pos:* NIH Biophys Chem study sect, 82-86; Univ Space Res Asn, NASA biotechnol discipline working group, 86- *Mem:* AAAS; Biophys Soc; Sigma Xi; Am Soc Biol Chemists. *Res:* Molecular components involved in the regulation of gene expression. *Mailing Add:* Dept of Chem Univ of Pa Philadelphia PA 19104

LU, RENNE CHEN, b China, Feb 13, 44; m 71; c 2. PROTEIN STRUCTURE, CHEMICAL MODIFICATION. *Educ:* Univ Calif, San Diego, PhD(biochem), 70. *Prof Exp:* PRIN STAFF SCIENTIST, BOSTON BIOMED RES INST, 71- *Mem:* Am Soc Biochem & Molecular Biol; Am Soc Cell Biol; Biophys Soc; Protein Soc. *Mailing Add:* Dept Muscle Res Boston Biomed Res Inst 20 Staniford St Boston MA 02114

LU, SHIH-LAI, b Fukien, China, Nov 1, 46; m 71; c 2. ORGANIC CHEMISTRY, POLYMER CHEMISTRY. *Educ:* Fu Jen Univ, BS, 68; Wright State Univ, MS, 71; Iowa State Univ, PhD(org chem), 75. *Prof Exp:* Res assoc acad res, Iowa State Univ, 75-76 & Univ Chicago, 77-78; SR RES SPECIALIST, 3M CO, 78- *Mem:* Am Chem Soc. *Res:* Synthesis and thermoxidative degradation studies of polymers; mechanistic studies of organic reactions involving carbonium, radical and carbanion intermediates; process research; total synthesis of natural products; new products developments; polymer characterization; adhesives; radiation curable systems. *Mailing Add:* 1273 Shannon Dr St Paul MN 55125

LU, TOH-MING, b Sibu, Malaysia, June 28, 43. CONDENSED MATTER PHYSICS. *Educ:* Cheng Kung Univ, BS, 68; Worchester Polytech Inst, 71; Univ Wis, Madison, PhD(physics), 76. *Prof Exp:* Guest scientist, US Nat Bur Standards, 79-80; res assoc, Univ Wis, Madison, 80-82; asst prof, 82-86, ASSOC PROF, RENSSELAER POLYTECH INST, 86- *Concurrent Pos:* Teacher, Catholic High Sch, Malaysia, 77-79. *Mem:* Am Phys Soc; Am Vet Soc; Mat Res Soc. *Res:* Thin film, interfaces and surfaces. *Mailing Add:* Ctr Integrated Electron CII Rensselaer Polytech Inst Troy NY 12180

LU, WEI-KAO, b Kiangsu, China, Apr 6, 33; m 64. METALLURGY, PHYSICAL CHEMISTRY. *Educ:* Cheng Kung Univ, Taiwan, BS, 57; Univ Minn, PhD(metall), 64. *Prof Exp:* Fel, Univ Minn, 64-65; from asst prof to assoc prof metall, 65-73, STELCO PROF METALL, MCMASTER UNIV, 73- *Mem:* Am Inst Mining, Metall & Petrol Engrs; Iron & Steel Inst Japan; Can Inst Mining & Metall. *Res:* Theoretical and experimental study of chemical kinetics of gas-solid and slag-metal reactions; heterogeneous kinetics of iron and steelmaking reactions; iron ore agglomeration; coke and carbonization. *Mailing Add:* Dept Metall & Mat McMaster Univ 1280 Main St W Hamilton ON L8S 4L8 Can

LU, WEI-YANG, b Taiwan, Repub China, Mar 24, 50; US citizen; m 81. PLASTICITY, EXPERIMENTAL STRESS ANALYSIS. *Educ:* Nat Taiwan Univ, BS, 72; Univ NMex, MS, 76; Yale Univ, PhD(eng & appl sci), 81. *Prof Exp:* Asst prof, 81-87, ASSOC PROF MECH, UNIV KY, 87- *Concurrent Pos:* Consult, Sandia Nat Labs, 85-; prin investr, NSF, 86-; chmn ed comt, Soc Exp Mech, 88-; reviewer, NSF, Acta Mech, Exp Mech, J Eng Mat & Technol. *Mem:* Soc Exp Mech; Am Soc Mech Engrs; Sigma Xi. *Res:* The effects of inelastic deformation on materials, its application on manufacturing such as machining and forming, and on ultrasonic nondestructive material characterization. *Mailing Add:* Dept Eng Mech Univ Ky Lexington KY 40506-0046

LU, YEH-PEI, US citizen. MECHANICAL ENGINEERING. *Educ:* Nat Taiwan Univ, BS, 58; Univ Houston, MS, 64, PhD(mech eng), 67. *Prof Exp:* Second Lieutenant eng, Chinese Air Force, 58-60; customer engr, IBM Corp, Taiwan, 61; teaching & res fel, Univ Houston, 62-67; mech engr, 68-72, SR PROJ ENGR, DAVID W TAYLOR NAVAL SHIP RES & DEVELOP CTR, 72- *Mem:* Am Soc Mech Engrs; Acoust Soc Am; Sigma Xi. *Res:* Vibration and acoustics; structural dynamics; fluid-structural interaction; numerical analyses. *Mailing Add:* 12712 Haven Lane Bowie MD 20716

LUBAN, MARSHALL, b Seattle, Wash, May 29, 36; m 82; c 2. PHYSICS. *Educ:* Yeshiva Univ, AB, 57; Univ Chicago, SM, 58, PhD(theoret physics), 62. *Prof Exp:* Mem, Inst Advan Study, 62-63; asst prof physics, Univ Pa, 63-66; Guggenheim Mem Found fel, Bar-Ilan Univ, Israel, 66-67, chmn dept, 67-70, dean fac natural sci, 69-71; assoc prof, 67-74, PROF PHYSICS, IOWA STATE UNIV 74- *Concurrent Pos:* Mem, Israel Coun Res & Develop, 70-73; mem bd trustees & exec coun, Bar-Ilan Univ, 71-74 & 79-81; mem bd trustees, Jerusalem Inst Technol, 79-81; vis prof, Wash Univ, Mo, 81-82. *Mem:* Israel Phys Soc (vpres, 78-79 & pres, 79-82); Am Phys Soc. *Res:* Theoretical condensed matter physics. *Mailing Add:* Dept Physics Iowa State Univ Ames IA 50011

LUBAR, JOEL F, b Washington, DC, Nov 16, 38; m 61; c 2. NEUROSCIENCES, PSYCHOPHYSIOLOGY. *Educ:* Univ Chicago, BS, 60, PhD(biopsychol), 63. *Prof Exp:* Asst prof psychol, Univ Rochester, 63-67; assoc prof, 67-71, PROF PSYCHOL, UNIV TENN, 71- *Concurrent Pos:* NIH grant, 65-73, prog dir, 70-75; vis lectr, Inst Physiol, Univ Bergen, Norway, 72; NSF fel, Sch Med, Univ Calif, Los Angeles, 75-76; regional ed, Physiol & Behav J, 70-; psychol consult, Vet Admin Hosp, 72-; co-dir, Southeastern Biofeedback Inst, 76-80, dir, 80- *Mem:* Am Psychol Asn; Sigma Xi; Soc Neurosci; Biofeedback Soc Am; fel NY Acad Sci. *Res:* Operant control of electroencephalographic and electrophysiological responses with special emphasis on epilepsy, hyperkinesis, learning disabilities and psychophysiological disorders; neuroanatomical substrates of emotional and motivational behavior. *Mailing Add:* Dept Psychol 310 AP Knoxville TN 37916

LUBAROFF, DAVID MARTIN, b Philadelphia, Pa, Feb 1, 38; m 61; c 3. IMMUNOLOGY. *Educ:* Philadelphia Col Pharm & Sci, BS, 61; Georgetown Univ, MS, 64; Yale Univ, PhD(microbiol), 67. *Prof Exp:* Assoc, Univ Pa, 69-70, asst prof, 70-73; from asst prof to assoc prof, 73-82, PROF UROL & MICROBIOL, 82-, DIR, UROL RES, UNIV IOWA, 82- *Concurrent Pos:* USPHS fel, Univ Pa, 67-69; assoc res career scientist, Vet Admin, 85- *Mem:* Am Asn Immunologists; Transplant Soc; AAAS; Int Soc Prev Oncol; Am Asn Cancer Res; Am Urol Asn. *Res:* Delayed hypersensitivity reactions; transplantation immunology; tumor immunology; lymphocyte membrane antigens. *Mailing Add:* Dept Urol Univ Iowa Iowa City IA 52242

LUBATTI, HENRY JOSEPH, b Oakland, Calif, Mar 16, 37; m 68; c 3. PHYSICS. *Educ:* Univ Calif, Berkeley, AB, 60, PhD(physics), 66; Univ Ill, Urbana, MS, 63. *Prof Exp:* Physicist, Boeing Co, Wash, 60-61; res assoc physics, Linear Accelerator Lab, Univ Paris, 66-68; asst prof, Mass Inst Technol, 68-69; assoc prof, 69-74, PROF PHYSICS, UNIV WASH, 74-, SCI DIR VISUAL TECH LAB, 69- *Concurrent Pos:* Vis lectr, Int Sch Physics, Erice, Sicily, 68, Herceg-Novi Int Sch, Yugoslavia, 69 & XII Cracow Sch Theoret Physics, Zacopane, Poland, 72; vis scientist, CERN, Geneva, Switzerland, 80-81; consult & collabr, Los Alamos Nat Lab, 82-86; mem ed adv comt, World Sci Publ Co, Ltd, 82- *Mem:* AAAS; fel Am Phys Soc. *Res:* Elementary particle physics, experimentalist; deep inelastic muon scattering and rare K-decay experiments. *Mailing Add:* Visual Tech Lab Dept Physics FM-15 Univ Wash Seattle WA 98195

LUBAWY, WILLIAM CHARLES, b South Bend, Ind, Nov 30, 44; m 71; c 3. PHARMACOLOGY. *Educ:* Butler Univ, BS, 67; Ohio State Univ, MS, 69, PhD(pharmacol), 72. *Prof Exp:* Asst prof, 72-77, assoc prof pharmacol & Grad Ctr Toxicol, 77-82, PROF PHARMACOL & TOXICOL & ASSOC DEAN ACAD AFFAIRS, COL PHARM, UNIV KY, 83- *Mem:* Am Asn Col Pharm. *Res:* Metabolism of tobacco smoke components by isolated organ systems; isolated lung synthesis of prostaglandins. *Mailing Add:* Col Pharm Univ Ky Lexington KY 40536-0082

LUBBERTS, GERRIT, b Oldemarkt, The Netherlands, Sept 15, 35; US citizen; m 59; c 2. SOLID STATE ELECTRONICS. *Educ:* Univ Rochester, 62, MS, 67, PhD, 71. *Prof Exp:* Technician, Case-Hoyt Corp, 56-58; technician, 58-62, res physicist, 62-71, sr res physicist, 71-78, RES ASSOC, EASTMAN KODAK CO, 78- *Mem:* Inst Elec & Electronics Engrs; Am Phys Soc. *Res:* Semiconductor physics; surface barrier photodetectors, charge coupled devices, high Tc superconducting thin films. *Mailing Add:* Eastman Kodak Co Res Lab 1669 Lake Ave Rochester NY 14650

LUBCHENCO, JANE, b Denver, Colo, Dec 4, 47; m 71; c 2. MARINE PLANT HERBIVORE INTERACTIONS, CHEMICAL ECOLOGY. *Educ:* Colo Col, BA, 69; Univ Wash, MS, 71; Harvard Univ, PhD(ecol), 75. *Prof Exp:* Asst prof ecol, Harvard Univ, 75-77; from asst prof to assoc prof, 78-88, PROF ZOOL, ORE STATE UNIV, 88-; RES ASSOC, SMITHSONIAN INST, 78- *Concurrent Pos:* Prin investr, NSF, 76-, mem adv panel, long term ecol res prog, 77; vis asst prof, Discovery Bay Marine Lab, 76; sci adv, Ocean Trust Found, 78-84 & West Quoddy Marine Sta, 81-88; vis assoc prof, Univ Antofagasta, Chile, 85; Inst Oceanol, Qingdao, Peoples Repub China, 87; nat lectr, Phycological Soc Am, 87-89. *Honors & Awards:* George Mercer Award, Ecol Soc Am, 79. *Mem:* Ecol Soc Am (coun mem, 82-84, awards comt chair, 83-86, nominating comt, 86); Phycol Soc Am; Am Soc Naturalists; Am Soc Zoologists; Am Inst Biol Sci. *Res:* Population and community ecology; plant-herbivore and predator-prey interactions; competition; marine ecology; algal ecology; algal life histories; biogeography, chemical ecology. *Mailing Add:* Dept Zool Ore State Univ Corvallis OR 97331-2914

LUBCKE, HARRY RAYMOND, b Alameda, Ca, Aug 25, 05. PHYSICS. *Educ:* Univ Calif, BS, 29. *Prof Exp:* Pres, Soc TV Engr, 42; Pres acad TV Arts & Sci, 49; RETIRED. *Mem:* Fel AAAS; Soc Motion Picture & TV Engrs; Sigma Xi; Acad TV Arts & Sci (pres, 49). *Mailing Add:* 2443 Creston Way Hollywood CA 90028

LUBECK, MICHAEL D, IMMUNOLOGY. *Prof Exp:* ASSOC DIR, WYETH-AYERST LABS INC, 85- *Mailing Add:* Wyeth-Ayerst Labs Inc PO Box 8299 Philadelphia PA 19101

LUBEGA, SETH GASUZA, b Mubende, Uganda, Dec 24, 36; m 71; c 2. EMBRYOLOGY, GENETICS. *Educ:* Oakwood Col, BA, 67; Howard Univ, MS, 69, PhD(zool), 75. *Prof Exp:* Instr biol, Oakwood Col, 71-72; asst prof, Ft Valley State Col, 75-76; ASSOC PROF BIOL, OAKWOOD COL, 76- *Mem:* Genetic Soc Am; Nat Inst Sci. *Res:* Isoenzymes of octanol dehydrogenase in populations of Drosophila species, developmental stages, and specific organs. *Mailing Add:* Dept Biol Oakwood Col Huntsville AL 35896

LUBELL, DAVID, b Brooklyn, NY, Apr 1, 32; m 60; c 3. MATHEMATICS. *Educ:* Columbia Univ, BS, 56; NY Univ, PhD(math), 60. *Prof Exp:* Benjamin Peirce instr math, Harvard Univ, 60-61; res instr, NY Univ, 61-62; sr mathematician, Systs Res Group Inc, 62-66; asst prof math, NY Univ, 66-70; assoc prof, 70-74, PROF MATH, ADELPHI UNIV, 75- *Concurrent Pos:* Consult, Systs Res Group Inc, 66-67 & US Air Force, 67-68; math adv, Nassau County Med Ctr, 69-72. *Mem:* Am Math Soc. *Res:* Combinatorics; biomathematics. *Mailing Add:* Dept of Math Adelphi Univ Garden City NY 11530

LUBELL, JERRY IRA, b New York, NY, Oct 19, 43; m 66; c 2. ELECTRONICS, NUCLEAR ENGINEERING. *Educ:* Univ Wash, BS, 66, MS, 68. *Prof Exp:* Mem tech staff, TRW Systs Group, 69-73, head, Response Analysis Sect, 73-75, asst mgr, Electronic Syst & Technol Dept, 75-77; res scientist, Kaman Sci Corp, 77-78, mgr radiation & electromagnetics, 78-80; from asst to pres, Systs Hardening, 80-87, VPRES, ELECTROMAGNETIC & RADIATION EFFECTS SECTOR, MISSION RES CORP, 87- *Concurrent Pos:* Prof nuclear engr, State Calif, 77- *Mem:* Am Nuclear Soc. *Res:* Nuclear weapon effects on electronic systems, subsystems and piece parts; electromagnetic pulse, system generated electromagnetic pulse and transient radiation effects causing both temporary and permanent damage. *Mailing Add:* 1975 Oak Hills Dr Colorado Springs CO 80919

LUBELL, MARTIN S, b New York, NY, June 5, 32; m 62; c 1. SOLID STATE PHYSICS. *Educ:* Mass Inst Technol, SB, 54; Univ Calif, Berkeley, MA, 56. *Prof Exp:* Asst, Univ Calif, 55-56; res physicist, Res Labs, Westinghouse Elec Corp, 56-67; res physicist, 67-73, asst dept mgr, 74-76, SECT HEAD, OAK RIDGE NAT LAB, 76- *Concurrent Pos:* Pres, Appl Superconductors Conf Inc, 80; chmn, standing comt fusion-technol, Inst Elec & Electronic Engrs, Nuclear & Plasma Sci Soc, 87; adv ed, Cryogenics; dep prog mgr, LCP, 83-88. *Mem:* Am Phys Soc; Inst Elec & Electronic Engrs. *Res:* Low temperature physics; superconductivity; fusion reactor technology; magnets. *Mailing Add:* Oak Ridge Nat Lab Bldg 9105 Fusion Energy Oak Ridge TN 38731-8040

LUBELL, MICHAEL S, b New York, NY, Mar 25, 43; m 69; c 1. ELECTRON AND POSITRON INTERACTIONS. *Educ:* Columbia Univ, BA, 63; Yale Univ, MS, 65, PhD(physics), 69. *Prof Exp:* AEC fel physics, Yale Univ, 70-71, from instr to assoc prof, 71-80, Sloan Found fel, 79-80; assoc prof, 80-82, Sloan Found fel, 80-83, vis sci, Brookhaven Nat Lab, 86-87, PROF PHYSICS, CITY COL, CITY UNIV NEW YORK, 83- *Concurrent Pos:* Prin investr, NSF, Dept Energy & Off Naval Res, 74-; sci & technol adv to US Sen Christopher J Dodd, Conn, 80-; steering comt, Nat Res Coun army res, 81-83, vchmn, 87-88, chmn, 88-90, past chmn, 90-91; mem panel pub affairs, Am Phys Soc, 84-85; prog comt, div electron atomic physics, Am Physics Soc, 78; org comt, div meeting dir electron, atomic physics, Am Physics Soc, 81; exec comt, International Conf Physics Electronic Atomic Collisions, 83-, cochmn, org comt, 84-89; org comt, fifth topical Am Physics Soc Conf atomic process in high temperature plasmas, 84-85; mem comt pub info, Am Inst Physics, 88-; vis lectr, Univ Tex, Austin, 90. *Mem:* AAAS; Am Phys Soc; NY Acad Sci; Sigma Xi. *Res:* Lepton-atom collisions; laser-atom interactions; polarized particle beams; electro-weak interactions at medium energy; high energy physics with Lepton beams. *Mailing Add:* 171 Bayberry Lane Westport CT 06880

LUBENSKY, TOM C, b Kansas City, Mo, May 7, 43; m 68; c 1. THEORETICAL CONDENSED MATTER PHYSICS. *Educ:* Calif Inst Technol, BS, 64; Harvard Univ, MA, 65, PhD(physics), 69. *Prof Exp:* NSF fel physics, Fac Sci, Orsay, France, 69-70; res asst, Brown Univ, 70-71; from asst prof to assoc prof, 71-80, PROF PHYSICS, UNIV PA, 80- *Concurrent Pos:* Sloan Found fel, 75; Guggenheim fel, 81-82; prof, Univ de Paris VI,

81-82. *Mem:* Am Phys Soc. *Res:* Liquid crystals, quasicrystals phase transitions, complex fluids cooperative phenomena in random systems and applications of the Wilson renormalization group. *Mailing Add:* Dept Physics Univ Pa Philadelphia PA 19104

LUBEROFF, BENJAMIN JOSEPH, b Philadelphia, Pa, Apr 17, 25; m 44; c 3. INDUSTRIAL CHEMISTRY. *Educ:* Cooper Union, BChE, 49; Columbia Univ, AM, 50, PhD(phys org chem), 53. *Prof Exp:* Statutory fel chem, Columbia Univ, 49-51; instr, Cooper Union, 51-53; chemist, high pressure lab, Am Cyanamid Co, 53-57; head gen chem res sect, Stauffer Chem Co, 57-62; mgr process res dept, Lummus Co, 62-70; CONSULT & ED, CHEMTECH, AM CHEM SOC, 70- *Concurrent Pos:* Actg dir, Continuing Sci Educ, Rutgers Univ, 76-78. *Honors & Awards:* Cooper Medal, 49. *Mem:* Am Chem Soc; Am Inst Chem; Am Inst Chem Eng; Am Soc Magazine Ed. *Res:* Research and development management; applied physical chemistry; high temperature and pressure processes; petrochemicals; catalysis; pesticides; analytical chemistry; technical journalism. *Mailing Add:* 19 Brantwood Dr Summit NJ 07901

LUBET, RONALD A, b New York, NY, July 7, 46; m; c 3. CANCER. *Educ:* Univ Tenn, Knoxville, BS, 69, MS, 73; Univ Tex Health Sci Ctr, PhD(radiation biol), 77. *Prof Exp:* Fel, Dept Biochem, Univ Tex Health Sci Ctr, 77-78; fel lab immunodiag, Nat Cancer Inst, 78-79; asst proj dir, Microbiol Assoc, 79-83, proj dir, 83-87; NAT CANCER INST SPEC EXPERT, LAB COMP CARCINOGENESIS, NAT CANCER INST, FREDERICK CANCER RES & DEVELOP CTR, 87- *Concurrent Pos:* Reviewer, Arch Environ Toxicol & Contamination, Arch Biochem & Biophys, Carcinogenesis, Biochem Pharm & J Nat Cancer Inst. *Mem:* Am Col Toxicol; Am Soc Pharm & Exp Therapeut; Am Asn Cancer Res. *Res:* Mechanisms of tumor promotion; induction of cytochrome P-450 by various xenobiotics; mechanisms of chemical carcinogenesis; mutagenicity of chemical carcinogens and chemotherapeutic compounds; metabolism of a variety of xenobiotics including carcinogens and chemotherapeutic agents; numerous publications. *Mailing Add:* Frederick Cancer Res Facil Nat Cancer Inst Frederick MD 21701-1013

LUBIC, RUTH WATSON, b Bucks County, Pa, Jan 18, 27; m 55. NURSE-MIDWIFERY. *Educ:* Sch Nursing Hosp Univ Pa, Dipl, 55; Teachers Col, Columbia Univ, BS, 59, MA, 61, EdD, 79. *Hon Degrees:* Univ Pa, LLD, 85; Univ Med & Dentistry, DSc, 86. *Prof Exp:* From staff nurse to head nurse, Mem Ctr Cancer & Allied Dis, 55-58; instr maternal nursing, Sch Nursing, Flower & Fifth Ave Hosp, 61; instr clin nurse-midwifery, Downstate Med Ctr, 62, parent educ & consult, 63-67, GEN DIR, MATERNITY CTR ASN, 70- *Concurrent Pos:* Mem, First Off Am Med Deleg to People's Repub of China, 73. *Honors & Awards:* Hattie Hemschemeyer Award, Am Col Nurse-Midwives, 83. *Mem:* Inst Med-Nat Acad Sci; fel AAAS; fel Am Acad Nurses; Am Col Nurse-Midwives; Am Pub Health Asn; fel Soc Appl Anthrop; Nat Asn Childbearing Ctrs. *Res:* Barriers and conflict in maternity care innovation. *Mailing Add:* 48 E 92nd St New York NY 10128-1397

LUBIN, ARTHUR RICHARD, b Newark, NJ, Mar 24, 47. MATHEMATICAL ANALYSIS. *Educ:* Mich State Univ, BS, 67; Univ Wis, MA, 68, PhD(math), 72. *Prof Exp:* Asst prof math, Tulane Univ, 72-73 & Northwestern Univ, 73-75; asst prof, 75-80, ASSOC PROF MATH ILL INST TECHNOL, 81- *Mem:* Am Math Soc. *Res:* Operator theory; functional analysis; Hardy spaces. *Mailing Add:* Dept Math Ill Inst Technol 3300 Federal St Chicago IL 60616

LUBIN, BERNARD, b Washington, DC, Oct 15, 23; m 58. CLINICAL PSYCHOLOGY, UNIVERSITY TEACHING. *Educ:* George Washington Univ, BA, 52, MA, 53; Pa State Univ, PhD(clin psychol), 58; Am Bd Prof Psychol, dipl, Am Bd Examr Psychol Hypnosis, dipl, exp hypnosis. *Prof Exp:* Fel psychother, Univ Wis Sch Med, 58; behav sci intern prog, NTL Inst, 60; assoc prof psychol, Ind Univ Med Ctr, 60-67; dir, Greater Kansas City Mental Health Found, 67-74; dir clin prog, dept psychol, Univ Houston, 74-76; PROF PSYCHOL, UNIV MO, KANSAS CITY, 76- *Concurrent Pos:* Dir, Div Res & Training, Ind Dept Mental Health, 63-67; chairperson, dept psychol, Univ Mo, Kansas City, 76-83; bd dir, Nat Training Labs, 86-; clin prof, dept psychiat, Univ Kans Med Ctr, 87-; chmn sponsor approval comt, Am Phychol Asn, 82-83, mem coun rep, 85-87. *Honors & Awards:* NT Veatch Award, 81. *Mem:* Am Psychol Asn. *Res:* The measurement of affect or mood; the small group in treatment and training; psychological test development and validation. *Mailing Add:* 5305 Holmes St Kansas City MO 64110

LUBIN, CLARENCE ISAAC, mathematical analysis; deceased, see previous edition for last biography

LUBIN, JONATHAN DARBY, b Staten Island, NY, Aug 10, 36. NUMBER THEORY, ALGEBRAIC GEOMETRY. *Educ:* Columbia Univ, AB, 57; Harvard Univ, AM, 58, PhD(math), 63. *Prof Exp:* Instr math, Bowdoin Col, 62-63, from asst to assoc prof, 63-67; assoc prof, 67-70, PROF MATH, BROWN UNIV, 70- *Concurrent Pos:* Assoc prof, Inst Henri Poincare, Univ Paris, 68-69; lectr, Math Inst, Copenhagen Inst, 74-75. *Mem:* Am Math Soc. *Res:* Algebraic geometry; number theory. *Mailing Add:* 24 John St Providence RI 02906

LUBIN, MARTIN, b NY, Mar 30, 23; m 42; c 4. CELL BIOLOGY. *Educ:* Harvard Univ, AB, 42, MD, 45; Mass Inst Technol, PhD(biophys), 54. *Prof Exp:* Res assoc biol, Mass Inst Technol, 53-54; assoc pharmacol, Harvard Med Sch, 54-57; asst prof, 57-68; PROF MICROBIOL, DARTMOUTH MED SCH, 68- *Concurrent Pos:* USPHS sr res fel, 56-61; Lalor Found fel, 57-59; Guggenheim fel & Commonwealth Fund fel, Lab Molecular Biol, Cambridge Univ, 65-67. *Mem:* Am Soc Microbiol; Am Soc Microbiol; Biophys Soc; Am Soc Cell Biol; Soc Gen Physiol. *Res:* Regulation of cell proliferation. *Mailing Add:* Dept Microbiol Dartmouth Med Sch Hanover NH 03755

LUBINIECKI, ANTHONY STANLEY, b Greensburg, Pa, Oct 4, 46; m 68; c 1. PROCESS DEVELOPMENT FOR BIOTECHNOLOGY PRODUCTS, MAMMALIAN CELL CULTURE. *Educ:* Carnegie Inst Technol, BS, 68; Univ Pittsburgh, ScD, 72. *Prof Exp:* Res asst microbiol, Grad Sch Pub Health, Univ Pittsburgh, 71-72, asst res prof, 72-74; prin scientist immunol & virol, Meloy Labs Inc, 74-79, managing dir, 79-80; tech div biol prod, Flow Labs Inc, 80-82; mgr, Cell Cult Oper, Genentech Inc, 82-83, dir res & demonstration, 83-86, dir process transfer, 86-88; VPRES BIOPHARM MFG & DEVELOP, SMITH KLINE BECKMAN, 88- *Concurrent Pos:* Mem, Dengue Task Force, US Army Med Res & Develop Command, 71-74, prin investr contract, 73-74; prin investr, Nat Inst Allergy & Infectious Dis grant, 73-74; prin investr contract, Nat Cancer Inst, 74-82, Nat Inst Child Health & Human Develop, 75-77; chmn, process technol comt, Biol & Biotechnol Sect, Pharmaceut Mfrs Asn, biotechnol adv comt. *Mem:* Am Soc Microbiol; AAAS; Soc Exp Biol & Med; NY Acad Sci; Europ Soc Animal Cell Technol; Parenteral Drug Asn. *Res:* Cell biology models of human genetic diseases and cancer; interferon; genetic mutants of mammalian cells and their viruses; infectious disease models; carcinogenesis; process development for recombinant; DNA pharmaceuticals and monoclonal antibodies. *Mailing Add:* 11681 Bennington Woods Rd Reston VA 22094

LUBINSKI, ARTHUR, b Antwerp, Belg, Mar 30, 10; nat US; m 35; c 2. MECHANICS. *Educ:* Univ Brussels, Cand Ing, 31, Ing CM & E, 34. *Hon Degrees:* PhD, NMex Inst Mining & Technol, 89. *Prof Exp:* Res engr, Barnsdall Res Corp, 47-50; spec res assoc, Marine & Arctic Opers Group, Res Ctr, Amoco Prod Co, 50-75; PVT TECH CONSULT, 75- *Concurrent Pos:* Mem, panel drill tech, Mohole Proj Phase I, AMSOComt, Nat Acad Sci; consult, Mohole Proj Phase II, Brown & Root; lectr, Univ Tex, 60 & 67; mem panel energy & resources, comt ocean eng, Nat Acad Eng, 67-; tech eval bd & tech assistance bd, Deep Sea Drilling Proj, Scripps Inst Oceanog, 67-70. *Honors & Awards:* Distinguished Achievement Award, Offshore Technol Conf, 76. *Mem:* Nat Acad Eng; fel Am Soc Mech Engrs; Soc Petrol Engrs; Am Petrol Inst; Belg Eng Asn; Belg Fedn Eng Asns. *Res:* Elastic stability of strings of pipe subjected to distributed weight and pressures; elasticity of porous bodies; hydraulics; shocks; wave propagation; applied mechanics; offshore structures; marine and arctic technology. *Mailing Add:* 4469 S Gary Ave Tulsa OK 74105

LUBITZ, CECIL ROBERT, b Brooklyn, NY, Mar 18, 25; m 46; c 3. NUCLEAR PHYSICS, NEUTRON CROSS SECTIONS. *Educ:* US Naval Acad, BS, 45; Univ Mich, MSEE, 49, PhD(physics), 60. *Prof Exp:* Res assoc elec eng, Res Inst, Univ Mich, 49-54; PHYSICIST, KNOLLS ATOMIC POWER LAB, GEN ELEC CO, 60- *Mem:* Am Nuclear Soc; Am Phys Soc. *Res:* Neutron cross sections for technological applications. *Mailing Add:* Knolls Atomic Power Lab Gen Elec Co Schenectady NY 12301

LUBKER, ROBERT A(LFRED), b Puyallup, Wash, May 19, 20; m 45; c 2. METALLURGICAL ENGINEERING. *Educ:* Univ Wash, BS, 42; Carnegie Mellon Univ, MS, 46. *Prof Exp:* Metall engr, Westinghouse Elec Corp, Pa, 42-46; supvr nonferrous metals, Metals Res Dept, Armour Res Found, Ill Inst Technol, 46-47, asst chmn, 47-51, assoc mgr, 51-53, mgr, 53-58; dir res & develop, Alan Wood Steel Co, 58-61, vpres res & develop, 61-67; dir res & develop, CF&I Steel Corp, 67-70 & Gen Cable Corp, 70-72; vpres technol, Assoc Metals & Minerals Corp, 72-74; vpres, AVA Steel Prod Int, Inc, 74-80, exec vpres, AVA-Toshin Corp, 76-80; pres, M&R Refractory Metals, Inc, Winslow, NJ, 80-83; RETIRED. *Mem:* Am Inst Mining, Metall & Petrol Engrs; Am Soc Metals; Am Iron & Steel Inst. *Res:* General physical metallurgy; welding; foundry; powder metallurgy; extractive metallurgy; mechanical metallurgy; copper, aluminum, titanium, molybdenum, tungsten and alloy steels; supervision and direction of research; engineering problems; wire and cable; steelmaking research and development. *Mailing Add:* 3150 Timberlake Point Ponte Vedra Beach FL 32082

LUBKIN, ELIHU, b Brooklyn, NY, Oct 25, 33; m 62; c 2. THEORETICAL PHYSICS. *Educ:* Columbia Univ, AB, 54, AM, 57, PhD(physics), 60. *Prof Exp:* Asst theoret physics radiation lab, Univ Calif, Berkeley, 59-61; res assoc high energy group, Brown Univ, 61-63, res asst prof theoret physics, 63-66; ASSOC PROF PHYSICS, UNIV WIS-MILWAUKEE, 66- *Mem:* Am Phys Soc. *Res:* Differential geometry used to interpret the old and for new constructions in physics; interpretation of quantum mechanics; quantum measurement theory; quantum psychology; thermodynamics. *Mailing Add:* Dept Physics Univ Wis PO Box 413 Milwaukee WI 53201

LUBKIN, GLORIA BECKER, b Philadelphia, Pa, May 16, 33; div; c 2. PHYSICS. *Educ:* Temple Univ, AB, 53; Boston Univ, MA, 57. *Prof Exp:* Mathematician aircraft div, Fairchild Stratos Corp, 54 & Letterkenny Ord Depot, US Defense Dept, 55-56; physicist tech res group, Control Data Corp, 56-58; actg chmn dept physics, Sarah Lawrence Col, 61-62; vpres, Lubkin Assocs, 62-63; assoc ed, 63-69, sr ed, 70-84, ED, PHYSICS TODAY, AM INST PHYSICS, 85- *Concurrent Pos:* Consult ctr for hist & philos of physics, Am Inst Physics, 66-67; Nieman fel, Harvard Univ, 74-75; mem exec comn, Forum Physics & Soc, Am Phys Soc, 77-78, exec comt, Hist Physics Div, 83-; mem, Nieman Adv Comt, Harvard Univ, 78-82; co-chair adv comn, Theoret Physics Inst, Univ Minn, 87-88, co-chair, Oversight Comt, 89- *Mem:* NY Acad Sci; fel Am Phys Soc; Nat Asn Sci Writers; fel AAAS. *Res:* Nuclear physics; science policy; physics reporting, writing and editing. *Mailing Add:* 160 S Middle Neck Rd Great Neck NY 11021

LUBKIN, JAMES LEIGH, b New York, NY, Mar 5, 25; m 48; c 2. STRUCTURAL ENGINEERING, ENGINEERING EDUCATION. *Educ:* Columbia Univ, BS, 44, MS, 47, PhD(appl mech), 50. *Prof Exp:* Consult, Appl Mech & Eng Probs, Mergenthaler Linotype Co, 49-50; sr proj analyst, Appl Physics Div, Midwest Res Inst, 50-56; sr res engr & head theoret anal group, Cent Res Lab, Am Mach & Foundry Co, Conn, 56-63; PROF CIVIL & SANIT ENG, MICH STATE UNIV, 63- *Concurrent Pos:* Fac fel, Ford Motor Co, 72-73. *Mem:* Am Soc Mech Engrs; Soc Exp Stress Analysis; Am Soc Eng Educ; Sigma Xi; Am Asn Univ Prof. *Res:* Computer-assisted testing and homework; individualized instruction; computer-aided design in engineering; computer applications in engineering education; vibration of vehicles; database management of traffic accident records. *Mailing Add:* Dept of Civil Eng Mich State Univ East Lansing MI 48824

LUBLIN, FRED D, b Philadelphia, Pa, Sept 28, 46; m 69. NEUROLOGY, BIOCHEMISTRY. *Educ:* Temple Univ, Philadelphia, Pa, AB, 68; Jefferson Med Col, Philadelphia, Pa, MD, 72; Am Bd Med Examr, dipl, 73, Am Bd Psychiat & Neurol, 77. *Prof Exp:* Instr neurol, Cornell Med Col, 75-76; from instr to assoc prof neurol, 76-86, res assoc biochem, 76-78, ASSOC PROF BIOCHEM & MOLECULAR BIOL, JEFFERSON MED COL, THOMAS JEFFERSON UNIV, 83-, PROF NEUROL, 86-, VCHMN, DEPT NEUROL, 87- *Concurrent Pos:* Asst neurologist, NY Hosp, 73-75, neurologist, 75-76; attend neurologist, Thomas Jefferson Univ Hosp, 76-; prin investr, Basic Res Support Grant, NIH, 76-77, Nat Multiple Sclerosis Soc res grant, 81-93; adj asst prof biochem, Jefferson Med Col, Thomas Jefferson Univ, 78-83, dir, Div neuroimmunol, Dept Neurol, 87-; consult neurologist, Coatesville Vet Admin Hosp, Wilmington Vet Admin Hosp & Wills Eye Hosp; teacher investr develop award, Nat Inst Neurol & Commun Dis & Stroke, 78-83; mem, Comt Drug Develop, Nat Multiple Sclerosis Soc, 83-, chmn, Computer Database Comt, 86-; co-dir, Multiple Sclerosis Comprehensive Clin Ctr, 84-; examr neurol, Am Bd Psychiat & Neuro, 86; co-investr, Triton Biosci Inc, 86-90; mem, Neurol Dis Prog Proj B Comt, Nat Inst Neurol Dis & Stroke, 90- *Honors & Awards:* Roche Award, Jefferson Med Col, 70; Henry M Phillips Prize, 72; William Potter Mem Prize, 72. *Mem:* Am Neurol Asn; Am Acad Neurol; Soc Neurosci; Asn Res Nervous & Ment Dis; Sigma Xi; NY Acad Sci; AAAS; Alpers Soc Clin Neurol; Int Brain Res Orgn; Am Asn Immunologists. *Res:* Author of various publications. *Mailing Add:* Neurol Dept Jefferson Med Col 1025 Walnut St Suite 511 Philadelphia PA 19107-5083

LUBLIN, PAUL, b New York, NY, Sept 8, 24; m 52; c 3. PHYSICAL CHEMISTRY. *Educ:* NY Univ, BA, 48; Purdue Univ, MS, 49. *Prof Exp:* Res chemist, Pigment Div, Am Cyanamid Co, 51-53; asst res staff mem, Res Div, Raytheon Mfg Co, 53-54; appln engr, Instrument Div, Philips Electronics, 54-56; sr eng, Gen Tel & Electronics Labs, Inc, Waltham, 56-59, res engr, 59-61, adv res engr, 61-63, engspecialist, 63-67, mem tech staff, 67-78, mgr mat eval, 78-84; RETIRED. *Concurrent Pos:* Consult, 85- *Mem:* Am Crystallog Asn; Soc Appl Spectros; Sigma Xi; Microbeam Anal Soc; fel Am Inst Chem. *Res:* Materials analysis, applications of x-ray diffraction and spectroscopy to structure and chemical identification of materials; electron probe and scanning electron microscopy as applied to electronic materials. *Mailing Add:* 16 Montgomery Dr Framingham MA 01701

LUBLINER, J(ACOB), b Lodz, Poland, May 5, 35; US citizen; m 60; c 3. MECHANICS, BIOPHYSICS. *Educ:* Calif Inst Technol, BS, 57; Columbia Univ, MS, 58, PhD(eng mech), 60. *Prof Exp:* Mem tech staff appl mech, Bell Tel Labs, 60; NSF fel, Polytech Sch, Paris, 60-61; preceptor civil eng, Columbia Univ, 61-62, asst prof, 62-63; from asst prof to assoc prof, 63-68, assoc prof eng sci, 68-73, PROF ENG SCI, UNIV CALIF, BERKELEY, 73- *Concurrent Pos:* NIH spec fel, Weizmann Inst Sci, Israel, 69-70; vis prof, Univ Andes, Bogota, Colombia, 77, Univ Costa Rica, San Jose, Costa Rica, 84, Univ Politec Catalunya, Barcelona, Spain, 86. *Mem:* Am Acad Mech. *Res:* Thermomechanics of viscoelastic, viscoplastic and plastic materials; wave propagation in solids; thermodynamics; mechanochemistry; wave propagation in biological systems; high frequency structural dynamics; segmented telescope design; modeling of concrete. *Mailing Add:* Dept Civil Eng Univ Calif Berkeley CA 94720

LUBMAN, DAVID, b Chicago, Ill, Aug 3, 34; wid; c 1. ACOUSTICS, ELECTRICAL ENGINEERING. *Educ:* Ill Inst Technol, BS, 60; Univ Southern Calif, MS, 62. *Prof Exp:* Sr scientist, LTV Corp Res Ctr, Anaheim, 67-68 & Bolt Beranek & Newman Inc, Van Nuys, 68-69; staff engr underwater acoust, 60-67, SR STAFF ENGR, GROUND SYSTS GROUP, HUGHES AIRCRAFT CO, 76- *Concurrent Pos:* Vis prof math, Chapman Col, Orange, 63-68; consult, D Lubman & Assocs, 69-; mem working group, Am Nat Standards Inst, 70-74; consult, Off Naval Res, Washington, DC, 69-76, Aircraft Engine Group, Gen Elec Co, 71-73, Nat Bur Standards, Washington, DC, 73-74 & Dept Archit & Construct, State of Calif, 76-; vis prof acoust, Calif State Univ, Los Angeles, 76-; vis lectr, Univ Calif, Santa Barbara, 76-; mem, Nat Coun Acoust Consult; chmn, Orange County Regional Chap, Acoust Soc Am, 89- *Mem:* Fel Acoust Soc Am; Inst Noise Control Eng; Am Soc Testing & Mat; Sigma Xi. *Res:* Architectural and underwater acoustics; characterization and measurement of the statistics of sound fields over space, time and frequency; reverberation chambers; measurement of sound power; noise quality; speech intelligibility. *Mailing Add:* Hughes Aircraft Co PO Box 3310 Bldg 675 MSE R315 Fullerton CA 92631

LUBMAN, DAVID MITCHELL, b Brooklyn, NY, April 23, 54; m 84; c 3. LASER SPECTROSCOPY, MASS SPECTROMETRY. *Educ:* Cornell Univ, AB, 75; Columbia Univ, MA, 76; Stanford Univ, PhD(phys chem), 79. *Prof Exp:* Staff scientist chem, Quanta-Ray, Inc, 80-83; asst prof, 83-87, ASSOC PROF CHEM, UNIV MICH, 87- *Concurrent Pos:* Vis scientist chem, Weizman Inst Sci, 82-83; Alfred P Sloan fel, 87-89, Eli Lilly Teaching fel, 84. *Mem:* Optical Soc Am; Am Phys Soc; Am Chem Soc; Am Soc Mass Spectrometry; Soc Appl Spectros. *Res:* Laser-induced selective ionization of small biologicals and peptides for supersonic beam spectroscopy studies and mass spectrometry. *Mailing Add:* Dept Chem Univ Mich Ann Arbor MI 48109

LUBORSKY, FRED EVERETT, b Philadelphia, Pa, May 14, 23; m 46; c 3. PHYSICAL CHEMISTRY. *Educ:* Univ Pa, BS, 47; Ill Inst Technol, PhD(phys chem), 52. *Prof Exp:* Asst chemist, Ill Inst Technol, 47-51; res assoc res lab, 51-52, phys chemist instrument dept, 52-55, physicist appl physics unit, 55-58, PHYS CHEMIST, RES & DEVELOP CTR, GEN ELEC CO, 58- *Concurrent Pos:* Mem div eng & indust res, Nat Acad Sci, 52-55; co-chmn, Tech Prog Comt, Int Conf Magnetism, 67; ed-in-chief, IEEE Trans Magnetics, 72-75; pres, Magnetics Soc, Inst Elec & Electronics Engrs, 75-77; gen chmn, Second Joint Int Conf Magnetics, Magnetism & Magnetic Mat Conf, 79; chmn adv comt, Conf Magnetism & Magnetic Mat, 80; Coolidge fel, Res & Develop, Gen Elec Co. *Honors & Awards:* Distinguished lectr, Inst Elec & Electronic Engrs, 79; Centennial Medal, Inst Elec & Electronics

Engrs, 84. *Mem:* Nat Acad Eng; Am Chem Soc; Am Phys Soc; fel Inst Elec & Electronics Engrs; NY Acad Sci; Am Inst Chemists; Materials Res Soc; fel Brit Sci Res Coun; AAAS. *Res:* Nucleation and growth of sub-micron size particles; development of single domain particle permanent magnetic materials; electrochemistry; magnetism; magnetic thin films; amorphous magnetic materials; magnetic separation; magnetic-optic materials; preparation and properties of magnetic materials; superconducting materials. *Mailing Add:* Res & Develop Ctr Gen Elec Co Box 8 Schenectady NY 12301

LUBORSKY, JUDITH LEE, MEMBRANE RECEPTORS, CELL FUNCTIONS. *Educ:* State Univ NY, Albany, PhD(biol), 75. *Prof Exp:* ASST PROF OBSTET & GYNEC, YALE UNIV, 76- *Res:* Cellular endocrinology. *Mailing Add:* Dept Obstet & Gynec Yale Med Sch 333 Cedar St New Haven CT 06510

LUBORSKY, SAMUEL WILLIAM, b Philadelphia, Pa, Jan 18, 31; m 53; c 3. MOLECULAR BIOLOGY. *Educ:* Univ Mich, BS, 52; Northwestern Univ, PhD, 57. *Prof Exp:* Fel, 57-58, BIOCHEMIST, NIH, 58- *Mem:* Am Chem Soc. *Res:* Properties of new components (antigens) which appear after cell transformation; biochemistry. *Mailing Add:* 7815 Custer Rd Bethesda MD 20814

LUBOWE, ANTHONY G(ARNER), b New York, NY, Dec 21, 37; m 59; c 2. ELECTRONIC PACKAGING, CONNECTORS. *Educ:* Columbia Univ, AB, 57, BS, 58, MS, 59, EngScD(eng mech), 61. *Prof Exp:* Res asst, Sch Eng, Columbia Univ, 60-61; mem tech staff, 61-73, SUPVR, INTERCONNECTION TECHNOL LAB, AT&T BELL LABS, 73- *Concurrent Pos:* Dir, Int Electronics Packaging Soc; corp dir, Int Inst Connector & Interconnection Technol. *Mem:* Am Soc Mech Engrs; Int Electronics Packaging Soc. *Res:* Elasticity; orbit prediction; electronic assembly; electronic packaging. *Mailing Add:* Interconnection Technol Lab AT&T Bell Labs Whippany NJ 07981

LUBOWSKY, JACK, b Brooklyn, NY, July 11, 40. BIOMATHEMATICS. *Educ:* City Col New York, BEE, 62; Polytech Inst Brooklyn, MSEE, 66, PhD(elec eng), 73. *Prof Exp:* Engr, Brookhaven Nat Labs, 61-62; proj engr, Airborne Instruments Lab, 62-66; res assoc, 66-67, instr med comput sci, 67-70, assoc proc comput sci & biophysics, 70-72, Dept Neurol, 72-73, DIR, SCI COMPUT CTR, STATE UNIV NY HEALTH SCI CTR, BROOKLYN, 73- *Concurrent Pos:* Co-investr, Spec Res Resources Div Biomath Comput Ctr, NIH, 72-73, prin investr, 73-75; asst prof, Dept Neurol, Down State Med Ctr, 73-78, assoc prof, Dept Neurol, 78-; assoc prof, Dept Biophysics, 80-; congressional sci fel, AAAS & Inst Elec & Electronics Engrs, 83; sci adv, Sen Subcomt on Energy. *Mem:* Sr mem Inst Elec & Electronics Engrs; AAAS; Sigma Xi. *Res:* Application of computers to biomedical research; investigation of adaptive and optimal search techniques to the determination of recognition properties of visual system neurons. *Mailing Add:* Sci Comput Ctr Box #7 State Univ NY Health Sci Ctr at Brooklyn Brooklyn NY 11203

LUBRAN, MYER MICHAEL, b London, Eng, Mar 9, 15; m 44; c 3. CLINICAL PATHOLOGY. *Educ:* Univ London, MB, BS, 38, BSc, 43, PhD(chem), 55; FRCPath. *Prof Exp:* Lectr physiol, biochem & clin chem, Guy's Med Sch, London, 38-43; lectr physiol & biochem, Med Sch, Univ Birmingham, 43-44; asst clin pathologist, Emergency Health Serv, Eng, 46-48; consult pathologist, Nat Health Serv, WMiddlesex Hosp, 48-64; prof path & dir clin chem, Sch Med, Univ Chicago, 64-70; PROF PATH, UNIV CALIF, LOS ANGELES, 70-; CHIEF CLIN PATH, HARBOR GEN HOSP, 70- *Concurrent Pos:* Mem hosp mgt comt & chmn med staff comt, WMiddlesex Hosp, 56-58; dir cent sterile supply dept & chmn cross-infection comt, 58-64; mem exam bd, Int Med Lab Technol, Eng; examr, Royal Col Path. *Mem:* AAAS; Asn Clin Path; Asn Clin Chem (pres, 74-75); Am Asn Clin Chem; NY Acad Sci. *Res:* Clinical pathology; trace metals. *Mailing Add:* Harbor UCLA Med Ctr Torrance CA 90509

LUBY, ELLIOT DONALD, b Detroit, Mich, Apr 3, 24; m 50; c 3. PSYCHIATRY, LAW. *Educ:* Univ Mo-Columbia, BS, 47; Wash Univ, MD, 49; Am Bd Psychiat & Neurol, dipl, 57. *Prof Exp:* Resident psychiat, Menninger Found, 50-51; sr asst surgeon, USPHS, 51-52; resident psychiat, Yale Univ, 52-54; chief adult inpatient sect, Lafayette Clin, Detroit, 57-62, assoc dir in chg clin serv, 62; prof law, 62-76, PROF PSYCHIAT, WAYNE STATE UNIV, 65-; CHIEF PSYCHIAT, HARPER HOSP, 78- *Concurrent Pos:* Prof law, Wayne State Univ, 80- *Honors & Awards:* Gold Medal Award, Am Acad Psychosom Med, 62. *Mem:* NY Acad Sci; AMA; Am Psychiat Asn; Am Psychosom Soc; fel Am Col Psychiat. *Res:* Psychopharmacology; drug induced model psychoses and sleep deprivation; law and psychiatry; schizophrenia. *Mailing Add:* Dept Psychiat Wayne State Univ Detroit MI 48202

LUBY, PATRICK JOSEPH, b Zanesville, Ohio, May 20, 30; m 56; c 4. AGRICULTURAL ECONOMICS. *Educ:* Univ Dayton, BA, 52; Purdue Univ, MS, 54, PhD(agr econ), 56. *Prof Exp:* Instr agr econ, Purdue Univ, 54-56, asst prof, 56-58; economist, 58-66, gen mgr provisions, 66-71, GEN MGR PROVISIONS & PROCUREMENT, OSCAR MAYER & CO, 71-, VPRES, 72-, CORP ECONOMIST, 74- *Mem:* Am Agr Econ Asn. *Res:* Use of statistical methods to analyze and forecast meat and livestock supplies and prices; efficient marketing of livestock and meats. *Mailing Add:* Oscar Mayer & Co PO Box 7188 Madison WI 53707

LUBY, ROBERT JAMES, b Kansas City, Mo, Apr 13, 28; m 51; c 8. OBSTETRICS & GYNECOLOGY. *Educ:* Rockhurst Col, BS, 48; Creighton Univ, MD, 52, MS, 59. *Prof Exp:* Intern obstet & gynec, Creighton Mem St Joseph Hosp, Omaha, 52-53, resident, 55-58; assoc prof, Col Med, Univ Nebr, 68-69; assoc dir obstet & gynec, 69-72, chmn dept, 72-77, PROF OBSTET & GYNEC, CREIGHTON UNIV, 69- *Mem:* AMA; Am Col Obstet & Gynec; Am Col Surg. *Res:* Nutritional aspects of infectious perinatal morbidity and mortality. *Mailing Add:* 7831 Chicago St Omaha NE 68114

LUCANSKY, TERRY WAYNE, b Massillon, Ohio, Aug 21, 42; m 66; c 1. BOTANY, PLANT MORPHOLOGY. *Educ:* Univ SC, BS, 64, MS, 67; Duke Univ, PhD(bot), 71. *Prof Exp:* ASSOC PROF BOT, UNIV FLA, 71- *Mem:* Bot Soc Am; Am Fern Soc; Am Inst Biol Sci; Sigma Xi; Torrey Bot Club. *Res:* Comparative anatomical and morphological studies of tropical pteridophytes; anatomical studies of aquatic plants and vines in relation to their habit and habitat; pteridology. *Mailing Add:* Dept Bot Rm 3175 McCarty Hall Univ of Fla Gainesville FL 32611-2009

LUCANTONI, DAVID MICHAEL, b Baltimore Md, Aug 31, 54. COMPUTATIONAL PROBABILITY. *Educ:* Towson State Univ, BS, 76; Univ Del, MS, 78, PhD(opers res), 82. *Prof Exp:* MEM TECH STAFF, BELL LABS, 81- *Mem:* Opers Res Soc Am; Math Asn Am; Am Math Soc. *Res:* Computationally stable algorithms for the solution of complex stochastic models such as those arising in the theory of queues. *Mailing Add:* Ten Oak Tree Lane Wayside NJ 07712

LUCAS, ALEXANDER RALPH, b Vienna, Austria, July 30, 31; US citizen; m 56; c 4. CHILD PSYCHIATRY. *Educ:* Mich State Univ, BS, 53; Univ Mich, MD, 57. *Prof Exp:* Rotating intern, Univ Mich Hosp, 57-58; resident child psychiat, Hawthorn Ctr, Northville, Mich, 58-59 & 61-62, from staff child psychiatrist to sr psychiatrist, 62-67; from asst prof to assoc prof psychiat, Wayne State Univ, 67-71; assoc prof, 73-76, PROF PSYCHIAT, MAYO MED SCH, 76-; head, 71-81, CONSULT SECT CHILD & ADOLESCENT PSYCHIAT, MAYO CLIN, 71- *Concurrent Pos:* Res child psychiatrist & res coordr, Lafayette Clin, Detroit, 67-71; consult, State of Minn Dept Pub Welfare, 72-80 & NIMH, 74-77. *Mem:* Am Orthop Asn; Am Psychiat Asn; Am Acad Child Psychiat; Soc Prof Child Psychiat; Soc Biol Psychiat. *Res:* Biologic aspects of child psychiatry; eating disorders. *Mailing Add:* Mayo Clin Rochester MN 55905

LUCAS, CAROL N, b Aberdeen, SDak, Feb 13, 40; m 61; c 2. BIOMEDICAL MATHEMATICS. *Educ:* Dakota Wesleyan Univ, BA, 61; Univ Ariz, MS, 67; Univ NC, Chapel Hill, PhD(biomed math, eng), 73. *Prof Exp:* Jr systs analyst, Cargill, Inc, Minneapolis, 62-65; res asst biomed math & eng, Univ NC, Chapel Hill, 67-68, res assoc, Div Cardiothoracic Surg, 72, lectr biomed math & eng, 76- 77, from asst prof to assoc prof, 77-89, PROF SURGERY & BIOMED MATH & ENG, UNIV NC, CHAPEL HILL, 89-, ACTG CHAIR, 90- *Concurrent Pos:* High sch teacher, US Army Educ Ctr, Furth, Ger, 61-62; teaching asst, Dept Math & Eng, Univ Ariz, 65-67, comput lab asst, 66-67. *Mem:* Am Heart Asn; Sigma Xi; Biomed Eng Soc; Inst Elec & Electronics Engrs. *Res:* Mathematical modelling and computer simulation of physiological systems; digital processing of dynamic physiological data. *Mailing Add:* Dept Surg Univ NC Chapel Hill NC 27599-7065

LUCAS, COLIN ROBERT, b Toronto, Can, Oct 11, 43; m 69; c 3. ORGANOMETALLIC CHEMISTRY. *Educ:* Acadia Univ, BSc, 68, MSc, 69; Oxford Univ, PhD(organometallic chem), 72. *Prof Exp:* Res fel, Univ Alberta, 73-74; asst prof chem, 74-79, ASSOC PROF CHEM, MEM UNIV, NFLD, 79- *Mem:* Fel Chem Inst Can. *Res:* Synthesis and properties of organometallic and coordination compounds containing sulphur; catalysis. *Mailing Add:* Dept Chem Mem Univ St Johns NF A1B 3X7 Can

LUCAS, DAVID OWEN, b Orange, Calif, Oct 19, 42; m 84; c 3. IMMUNOLOGY. *Educ:* Duke Univ, BA, 64, PhD(microbiol, immunol), 69. *Prof Exp:* From asst prof to assoc prof microbiol & immunol, Col Med, Univ Ariz, 70-86; vpres, Protein Technol, Petaluma, 86-90; DIR, TECHNOL DEVELOP NETWORK, SEBASTOPOL, 90- *Concurrent Pos:* Res fel immunol, Children's Hosp Med Ctr, Harvard Med Sch, 68-70, actg dept head, 77-79. *Mem:* AAAS; Am Asn Immunologists; Am Soc Microbiol; Reticuloendothelial Soc. *Res:* Cellular immunology; lymphocyte metabolism; interferon; hybridomas. *Mailing Add:* 825 Jewel Ave Sebastopol CA 95472

LUCAS, DOUGLAS M, b Windsor, Ont, May 5, 29; m 53; c 5. FORENSIC SCIENCE. *Educ:* Univ Toronto, BSc, 53, MSc, 57. *Prof Exp:* Instr pharmacol chem, Univ Toronto, 53-57; chemist, Atty Gen's Lab, 57-67; DIR, CTR FORENSIC SCI, 67- *Concurrent Pos:* Chmn, Comt Alcohol & Drugs, Nat Safety Coun, 77-79; vpres, Int Comt Alcohol, Drugs & Traffic Safety, 84-86. *Honors & Awards:* Adelaide Medal, Int Asn Forensic Sci, 90. *Mem:* Can Soc Forensic Sci (pres, 68-69); Am Acad Forensic Sci (pres, 72-73); Int Asn Forensic Sci (pres, 67-69); Am Soc Crime Lab Dirs (pres, 77-78). *Res:* Alcohol, drugs and traffic safety; investigation of fires and explosions; forensic science. *Mailing Add:* Ctr Forensic Sci 25 Grosvenor St Toronto ON M7A 2G8 Can

LUCAS, EDGAR ARTHUR, b Franklin, Ind, Oct 28, 33; m 60; c 2. ANATOMY, NEUROPHYSIOLOGY. *Educ:* Ball State Univ, BA, 61, MS, 65; Univ Calif, PhD(anat), 72. *Prof Exp:* Teacher, Sch, Town of Griffith, 61-62; planner admin, Rocketdyne Div, NAm Rockwell Corp, 62-63; assoc res engr, 64-65; from instr to asst prof anat, Univ Ark Med Ctr, Little Rock, 72-84; DIR, SLEEP DISORDERS CTR, ALL SAINTS EPISCOPAL HOSP, 84- *Concurrent Pos:* Mem comt Polysomnography Asn Sleep Disorder Ctrs, 76-87. *Mem:* Inc Soc Chronobiol; Asn Psychophysiol Study Sleep; Am Asn Anat; Sigma Xi; Clin Sleep Soc; Asn Prof Sleep Socs. *Res:* Biological rhythms; sleep; neurosciences. *Mailing Add:* Sleep Dis Ctr All Sts Episcopal Hosp PO Box 31 Ft Worth TX 76101-0031

LUCAS, FRED VANCE, b Grand Junction, Colo, Feb 7, 22; m 48; c 2. PATHOLOGY. *Educ:* Univ Calif, AB, 42; Univ Rochester, MD, 50. *Prof Exp:* From asst to asst prof path, Med Sch, Univ Rochester, 51-55; assoc prof, Col Physicians & Surgeons, Columbia Univ, 55-60; prof & chmn dept, Sch Med, Univ Mo-Columbia, 60-77, res assoc, Space Sci Res Ctr, 64-77; dir med serv, Univ Hosp & assoc dean & prof path, Sch Med, Vanderbilt Univ, 77-79, assoc vpres med affairs, 79-89; RETIRED. *Concurrent Pos:* Vet fel path, Med Sch, Univ Rochester, 50-51; Gleeson fel, 51-52; Lilly fel, 52-53; Lederle med fac award, 54; from asst resident to chief resident, Strong Mem Hosp, Rochester, 51-54; consult, Highland Hosp, Rochester, 54-55; assoc attend pathologist, Presby Hosp, New York, 55-; consult, NIH, 66- & Vietnam med educ proj, US AID-AMA, 67- *Mem:* Am Soc Exp Path; Harvey Soc; Am Asn Path & Bact; Am Soc Clin Path; Col Am Path. *Res:* Oxidative enzymes in proliferating tissue; plasma proteins studies employing C-14; hemoglobin; activation and inactivation of human chromosomes; ultrastructure of normal and abnormal human endometrium. *Mailing Add:* 3901 Harding Rd Apt 302 Nashville TN 37205

LUCAS, FREDERICK VANCE, JR, b Rochester, NY, Nov 27, 49; m 75; c 2. BLOOD COAGULATION & THROMBOSIS, HEMATOLOGY. *Educ:* Amherst Col, BA, 71; Univ Mo, MD, 75. *Prof Exp:* Res path, 75-79, staff physician, Lab Hemat, 79-88, CHMN DEPT PATH & LAB MED, CLEVELAND CLIN FOUND, 88- *Concurrent Pos:* Dir Sch Med Technol, Cleveland Clin Found, 79-83; prin investr, Am Heart Asn, 83; coun hemat, Am Soc Clin Pathologist, 82-87; coagulation resource comt, Col Am Pathologist, 83-85; clin asst prof, Case Western Reserve Univ, Sch Med, 84-; co investr, NIH, 87- *Mem:* Am Heart Asn; Int Acad Path; Am Soc Hemat; AAAS. *Res:* Clinical coagulation and role of soluble fibrin in cardiovascular disease, specifically the contribution of tissue transglutamivase to crosslinking fibrin; fibrinogen in blood of patients with vascular occlusion; development of flowing whole blood model of clotlysis. *Mailing Add:* Cleveland Clin Fla 3000 W Cypress Creek Rd Ft Lauderdale FL 33309

LUCAS, GENE ALLAN, b Des Moines, Iowa, Oct 15, 28; m 48; c 3. GENETICS. *Educ:* Drake Univ, BA, 54, MA, 58; Iowa State Univ, PhD(genetics), 68. *Prof Exp:* Lab instr biol, Drake Univ, 54-59, instr, 60-67; asst genetics, Iowa State Univ, 61-66; asst prof biol, 68-74, ASSOC PROF BIOL, DRAKE UNIV, 74- *Mem:* AAAS; Genetics Soc Am; Int Oceanog Found. *Res:* Pigmentation, especially of aquarium fish; pigment genetics of Siamese fighting fish; application of biological principles to world problems; race and population problems; teaching; biology and behavior of Siamese fighting fish. *Mailing Add:* Dept of Biol Drake Univ 25th Univ Ave Des Moines IA 50311

LUCAS, GEORGE BLANCHARD, b Philipsburg, Pa, Mar 8, 15; m 40, 55; c 7. PLANT PATHOLOGY. *Educ:* Pa State Col, BS, 40; La State Univ, MS, 42, PhD(plant path), 46. *Prof Exp:* From asst prof to assoc prof, 46-63, PROF PLANT PATH, NC STATE UNIV, 63- *Mem:* Bot Soc Am; Mycol Soc Am; Am Phytopath Soc. *Res:* Tobacco diseases; genetics of fungi. *Mailing Add:* 3040 Churchill Rd Raleigh NC 27607

LUCAS, GEORGE BOND, b New Orleans, La, Dec 21, 24; m 62; c 1. GEOCHEMISTRY. *Educ:* Tulane Univ, BS, 48; Iowa State Univ, PhD(chem), 52. *Prof Exp:* Postdoctoral fel, Northwestern Univ, 52-53; res chemist, Red Stong Arsenal Res Div, Rohm & Haas Co, 53-56; from asst prof to prof, 56-88, EMER PROF, COLO SCH MINES, 88- *Mem:* Am Chem Soc; Sigma Xi. *Res:* Reaction mechanisms organic particularly free radical mechanisms. *Mailing Add:* Dept Chem & Geochem Colo Sch Mines Golden CO 80401

LUCAS, GLENN E, b Los Angeles, Calif, Mar 8, 51; m 72; c 3. RADIATION EFFECTS, MECHANICAL METALLURGY. *Educ:* Univ Calif, Santa Barbara, BS, 73; Mass Inst Technol, MS, 75, ScD(nuclear eng), 77. *Prof Exp:* Engr, Exxon Nuclear, Inc, 78; from asst prof to assoc prof, 78-87, PROF NUCLEAR ENG, UNIV CALIF, SANTA BARBARA, 87- *Concurrent Pos:* Prin investr, numerous res contracts, 78-; consult, govt & private bus, 78-; vis res prof, Tokyo Univ, 85, Hokkaido Univ, 85. *Honors & Awards:* Young Eng Achievement Award, Am Nuclear Soc, 91. *Mem:* Am Nuclear Soc; Am Soc Metals; AAAS; Sigma Xi; Mat Res Soc; Am Soc Testing and Mat. *Res:* Effects of radiation and environment on microstructural evolution; mechanical properties of structural steels and composite materials. *Mailing Add:* 529 Dorset Goleta CA 93117

LUCAS, GLENNARD RALPH, b Marissa, Ill, Feb 22, 16; m 41; c 2. ORGANIC POLYMER CHEMISTRY. *Educ:* Monmouth Col, BS, 38; Columbia Univ, PhD(phys org chem), 42. *Prof Exp:* Asst chem, Monmouth Col, 36-38 & Columbia Univ, 38-41; res chemist, Gen Elec Co, Mass, 42-52, process engr, NY, 52-54, supvr process eng, 54-56, mgr adv proj develop, Mass, 56-58; res dir, Signode Corp, 58-81; RETIRED. *Mem:* AAAS; Am Chem Soc; Am Mgt Asn; Soc Plastics Eng. *Res:* Mechanisms of organic reactions; polymer studies of styrene and silicone resins; plastic and steel strapping materials; high speed paint cure. *Mailing Add:* 1011 Hunter Rd Glenview IL 60025

LUCAS, HENRY C, JR, b Omaha, Nebr, Sept 4, 44; m 68; c 2. INFORMATION TECHNOLOGY. *Educ:* Yale Univ, BS, 66; Mass Inst Technol, MS, 68, PhD, 70. *Prof Exp:* Consult, Arthur D Little, Inc, Cambridge, Mass, 66-70; asst prof computer & info systs, Grad Sch Bus, Stanford Univ, 70-74; assoc prof computer applications & info systs, Schs Bus, NY Univ, 74-78, prof & chmn, 78-84, prof info systs, Grad Sch Bus Admin, 84-88, RES PROF INFO SYSTS, LEONARD N STERN SCH BUS, NY UNIV, 88- *Concurrent Pos:* Ed, Sloan Mgt Rev, 67-68, Performance Eval Rev, 72-73 & Macmillan Series Info Systs; Chmn, Working Group Interaction of Info Systs & the Orgn, Int Fedn Info Processing, 75-80; assoc ed, Mis Quarterly, 78-83, Mgt Sci, 85-87, Asn Comput Mach Trans on Off, Info & Mgt; on leave, Europ Systs Res Inst, IBM, La Hulpe, Belgium, 81; vis prof, INSEAD, Fontainebleau, France, 85; vis researcher, Bell Commun Res, NJ, 91. *Mem:* Asn Comput Mach; Inst Mgt Sci; Inst Elec & Electronics Engrs. *Res:* Information technology and organizations; information systems implementation; expert systems; value of technology; impact of technology; systems analysis and design. *Mailing Add:* 18 Portland Rd Summit NJ 07901

LUCAS, J B, b Scottdale, Pa, May 3, 29; m 52; c 2. MINING & MINERAL ENGINEERING. *Educ:* Waynesburg Col, BS, 51; WVa Univ, BS, 52; Univ Pittsburgh, MS, 54; Columbia Univ, PhD(mining eng), 65. *Prof Exp:* Miner, Crucible Steel Co Am, 48-52; field engr, Joy Mfg Co, 52-54; mem fac mining eng, Ohio State Univ, 54-56, head div, 57-61; head dept, Va Polytech Inst & State Univ, 61-71, head, Div Minerals Eng, 71-76, head, Dept Mining & Minerals Eng, 76-87, MASSEY PROF MINING & MINERALS ENG, VA POLYTECH INST & STATE UNIV, 87- *Concurrent Pos:* Dir, US Off Coal Res Proj, Va Polytech Inst & State Univ, 62-; actg asst dir, Va Eng Exp Sta, 63-64; mem secy's res adv coun coal miner's health, HEW, 70-73; consult & reviewer, Prog Comn, Mining Safety & Health Admin, Dept Labor, 69-, NSF, 73-74, 76-77 & Ad-Hoc Panel Coal Mining Technol, Nat Res Coun, 75-78; mem, Joint Comt Coal Mining Health, Safety & Res, Mining Safety & Health Admin, US Dept Labor & Bur Mines, US Dept Interior, 70-, rev comt, Fel Prog Mining & Minerals Eng & Conserv, Off Educ, HEW, Wash DC, 79, prog comt, State Mine Recovery Competition, Div Mines, Va Mining Inst, 78-, ad hoc comt coal mine safety, Dept Indust & Resources, 79, Coal Conversion Fac, 79, exec comt, Va Mining & Mineral Resources Res Inst, 79-; coordr, Va Ctr Coal & Energy Res, Coal Inst, 77-; dir, Generic Mineral Technol Ctr, Mine Systs Design & Ground Control, 82- *Mem:* Am Inst Mining, Metall & Petrol Engrs; Am Soc Eng Educ; Am Mining Cong; Nat Soc Prof Engrs; AAAS. *Res:* Mining systems engineering; mineral property evaluation; mining design and layout; computer applications in underground coal mining systems; coal mining safety research; methane from coal seams; underground coal-mining research. *Mailing Add:* Dept Mining & Minerals Eng Va Polytech Inst & State Univ Blacksburg VA 24061

LUCAS, JAMES M, b Philipsburg, Pa, July 21, 41. STATISTICAL METHODOLOGY. *Educ:* Pa State Univ, BS, 63; Yale Univ, MS, 65; Tex A&M Univ, PhD(statist), 72. *Prof Exp:* SR CONSULT STATIST, E I DU PONT DE NEMOURS & CO, INC, 65- *Concurrent Pos:* Adj prof, Univ Del, 72-; assoc ed, Technometrics, 81-, J Qual Technol, 83-, Chemometrics & Intelligent Lab Systs, 88; past pres, Del chap, Am Statist Asn. *Honors & Awards:* Brumbaugh Award, Am Soc Qual Control, 76. *Mem:* Fel Am Statist Asn; Am Soc Qual Control. *Res:* Methods for the control and improvement of industrial processes; cumulative sum techniques; experimental designs. *Mailing Add:* DuPont Qual Mgt & Technol Ctr P O Box 6091 Newark DE 19714-6091

LUCAS, JAMES ROBERT, b Mankato, Minn, Apr 26, 47. GEOLOGY. *Educ:* Mankato State Univ, BA, 69; Univ Iowa, MA, 73, PhD(geol), 77. *Prof Exp:* Instr earth sci, Providence Sch, South St Paul, Minn, 69-70; res geologist, Iowa Geol Surv, 75-76; appln scientist water resources, Earth Resources Observ Systs Data Ctr, SDak, 76-80, prin applns scientist geol, 80-81; vpres, Centaur Explor, Inc, Amarillo, Tex, 81-83; gen partner, Orion, Ltd, Midland, Tex, 83-87, owner, 87-90; MGR, SPATIAL ANALYSIS LAB, LOCKHEED ENG & SCI CO, LAS VEGAS, NEV, 90- *Concurrent Pos:* Adj instr geol, Univ Iowa, 75-76; prin investr, NASA contract, 75-76; adj fac, Univ Tex Permian Basin, 84-85, Univ Nev, Las Vegas, 91- *Mem:* Am Soc Photogrammetry; Sigma Xi. *Res:* Remote sensing techniques applied to hydrocarbon, minerals and water resources exploration; Landsat digital image processing for natural resources analysis and interpretation; photographic enhancement of Landsat imagery for geological applications; land classification of SE Iowa from computer enhanced Landsat images; glacial geomorphology of NW Iowa; semi-quantitative analysis of clay minerals by x-ray diffraction. *Mailing Add:* Lockheed Eng & Sci Co 1050 E Flamingo Rd Las Vegas NV 89119

LUCAS, JOE NATHAN, b Lake Providence, La, Dec 18, 45; c 1. CHROMOSOMAL FLOW CYTOMETRY. *Educ:* Univ Calif, Los Angeles, BS, 70, MS, 72, PhD(biophysics), 77. *Prof Exp:* Sci instr math & sci, Mill Col, 72-76; scientist & engr, Lawrence Berkeley Nat Lab, Univ Calif, 77-78; instr physics, Calif State Univ, 78-79; SR SCIENTIST, LAWRENCE LIVERMORE NAT LAB, UNIV CALIF, 78- *Concurrent Pos:* Chmn bd, The Lucas Educ Found, Inc, 78-; Fulbright Scholar, 77. *Mem:* Soc Anal Cytometry; Radiation Res Soc. *Res:* Biological dosimetry: rapid image analysis to quantify chromosome aberrations (translocations) in man using chromosome specific probes; slit-scan and fringe-scan flow cytometers to measure the distribution of flourescent dye(s) along isolated chromosomes; flow cytometric devices and procedures for rapid, quantitative classification of chromosomes according to shape and/or flourescent band patterns. *Mailing Add:* Lawrence Livermore Nat Lab PO Box 5507 L-452 Livermore CA 94551

LUCAS, JOHN J, BIOCHEMISTRY, MOLECULAR BIOLOGY. *Prof Exp:* PROF & DIR RES, HEALTH SCI CTR, STATE UNIV NY, 89- *Mailing Add:* Biochem & Molecular Biol Health Sci Ctr State Univ NY 766 Irving Ave Syracuse NY 13210

LUCAS, JOHN PAUL, b Youngstown, Ohio, Nov 16, 45; m 68. MICROBIOLOGY. *Educ:* Univ Pittsburgh, BS, 67, MS, 69, ScD(microbiol), 73. *Prof Exp:* Res assoc virol, Grad Sch Pub Health, Univ Pittsburgh, 74; MICROBIOLOGIST, FOOD & DRUG ADMIN, 74- *Mem:* Am Soc Microbiol; Sigma Xi. *Res:* Develop growing area standards for shellfish. *Mailing Add:* 18908 Bluewillow Ln Gaithersburg MD 20879

LUCAS, JOHN W, b Pomona, Calif, Mar 14, 23; m 53; c 3. MECHANICAL ENGINEERING, HEAT TRANSFER. *Educ:* Univ Calif, Berkeley, BS, 48; Univ Calif, Los Angeles, MS, 49, PhD(mech eng), 53. *Prof Exp:* sr res engr, Jet Propulsion Lab, Calif Inst Technol, 54-59, group supvr, 59-65, res rep eng mech, 66-70, mgr res & planetary quarantine, 70-74, exec asst to dir, 74-76, mgr point focus distributed receiver solar energy technol proj, 76-85; RETIRED. *Mem:* Am Inst Aeronaut & Astronaut. *Res:* Ice nucleation in lemons; spacecraft advanced propulsion; radiation, conduction and convection heat transfer as related to spacecraft thermal control in space, on the moon and planets, and in solar thermal energy. *Mailing Add:* 865 Canterbury Rd San Marino CA 91108

LUCAS, KENNETH ROSS, b Bradford, Pa, June 4, 39; m 61; c 4. ANALYTICAL CHEMISTRY, POLYMER PHYSICS. *Educ:* Univ Pittsburgh, BS, 61; Univ Ill, MS, 64, PhD(anal chem), 66. *Prof Exp:* Res chemist, 66-71, sr res chemist, 71-81, ASSOC SCIENTIST, FIRESTONE TIRE & RUBBER CO, 81- *Mem:* Polymer Chem Div, Am Chem Soc, Rubber Div; Electron Micros Soc Am. *Res:* Molten salt and organic electrochemistry; x-ray diffraction; polymer physics; electro-organic synthesis; polymer morphology analysis; radiothermoluminescence; scanning electron microscopy; ESCA/Auger spectroscopy. *Mailing Add:* 314 Silver Ridge Akron OH 44321

LUCAS, LEON THOMAS, b Halifax, NC, July 30, 42; m 64; c 1. PLANT PATHOLOGY, MICROBIOLOGY. *Educ:* NC State Univ, BS, 64; Univ Calif, Davis, PhD(plant path), 68. *Prof Exp:* Res asst plant path, Univ Calif, Davis, 64-68; asst prof, 68-76, assoc prof, 76-80, PROF PLANT PATH, NC STATE UNIV, 80- *Mem:* Am Phytopath Soc. *Res:* Diseases of turfgrasses and forage crops in North Carolina; bacterial diseases of plants. *Mailing Add:* 7616 Plant Pathol NC State Univ Raleigh NC 27695-7616

LUCAS, MYRON CRAN, b Cincinnati, Ohio, Nov 15, 46. BIOCHEMICAL GENETICS. *Educ:* Lewis & Clark Col, BS, 69; Wash State Univ, PhD(genetics), 74. *Prof Exp:* Res assoc bot, Univ Ill, Urbana, 73-75; res assoc genetics, Univ Ga, 75-77; res assoc biochem, Univ Idaho, 77-78; from asst prof to assoc prof, 78-88, PROF BIOL, LA STATE UNIV, SHREVEPORT, 88- *Concurrent Pos:* Adj asst prof biol, Fla State Univ, 77. *Mem:* Genetics Soc Am; Am Soc Microbiol; NY Acad Sci; AAAS. *Res:* Biochemical genetics of Neurospora crassa; structure and function of low molecular weight RNA; gene regulation and synthesis of messenger RNA; characterization of egg jelly glycoproteins in salamanders. *Mailing Add:* Dept Biol Sci La State Univ Shreveport LA 71115

LUCAS, OSCAR NESTOR, b Resistencia, Arg, Aug 6, 32. HEMATOLOGY, PHYSIOLOGY. *Educ:* Univ Buenos Aires, Dentist, 58, DDS, 59; Univ Sask, PhD(physiol), 65. *Prof Exp:* Res assoc physiol, Med Sch, Univ Sask, 63-64; from asst prof to assoc prof, Med & Dent Sch, Univ Alta, 65-68; assoc prof, 68-70, PROF ORAL BIOL, DENT SCH, UNIV ORE, 70-, PROF DENT, MED SCH, 70- *Concurrent Pos:* Univ Buenos Aires fel, Jefferson Med Hosp, Philadelphia, Pa, 59-60, Cardeza Found fel, 60-63; affil, Div Hemat, Med Sch, Univ Ore, 71- *Mem:* Int Soc Hemat; Int Asn Dent Res. *Res:* Fibrinolysis; mast cell and connective tissue reparative process. *Mailing Add:* 375 Carolyn Dr Eugene OR 97404

LUCAS, ROBERT ALAN, b Allentown, Pa, June 13, 35; m 57; c 4. MECHANICAL ENGINEERING. *Educ:* Lehigh Univ, BS, 57, MS, 59, PhD(mech eng), 64. *Prof Exp:* Design engr, Air Prod & Chem, Inc, Pa, 57-58; from asst to asst prof, 58-69, ASSOC PROF MECH ENG, LEHIGH UNIV, 69- *Concurrent Pos:* Nat Res Coun-Naval Res Lab resident res assoc, Naval Res Lab, DC, 65-66. *Mem:* Am Soc Mech Engrs; Sigma Xi. *Res:* Machine system simulation and analysis; expert systems; optimization; applied mathematics; dynamics; computer aided design; computer aided instruction; vibrations. *Mailing Add:* Dept of Mech Eng & Mech Packard Lab Bldg 19 Lehigh Univ Bethlehem PA 18015

LUCAS, ROBERT ELMER, b Malolos, Philippines, June 27, 16; m 41; c 5. AGRONOMY, HORTICULTURE. *Educ:* Purdue Univ, BSA, 39, MS, 41; Mich State Col, PhD(soil sci), 47. *Prof Exp:* Asst soils, Va Truck Exp Sta, 41-43 & Mich State Col, 45-46; agronomist, Wm Gehring, Inc, Ind, 46-51, 77-78; from assoc prof to prof, 51-77, exten specialist, 53-77, EMER PROF SOIL SCI, MICH STATE UNIV, 77- *Concurrent Pos:* Vis prof, Univ Fla, 79-80. *Mem:* Fel Soil Sci Soc Am; fel Am Soc Agron; Int Peat Soc. *Res:* Micronutrients in crop production; soil organic matter dynamics and models; physical and chemical properties of organic soils (histosols); comparison of four management systems in vegetable production; plant nutrient requirements. *Mailing Add:* Dept Crop & Soil Sci Mich State Univ E Lansing MI 48824

LUCAS, ROBERT GILLEM, nuclear engineering, mechanical engineering, for more information see previous edition

LUCAS, RUSSELL VAIL, JR, b Des Moines, Iowa, Nov 2, 28; m 51; c 4. PEDIATRIC CARDIOLOGY. *Educ:* Macalester Col, BA, 50; Wash Univ, MD, 54. *Prof Exp:* From intern to resident pediat, Univ Hosp, Univ Minn, Minneapolis, 54-56, resident, 58-59; from asst prof to assoc prof pediat, Med Ctr, WVa Univ, 61-66; assoc prof, 66-69, PROF PEDIAT, UNIV MINN, MINNEAPOLIS, 69 - *Concurrent Pos:* NIH fel pediat cardiol, Univ Hosps, Univ Minn, Minneapolis, 59-61; NIH res career develop award, WVa Univ, 63-66. *Honors & Awards:* Distinguished Achievement Award, Am Heart Asn, 66. *Mem:* Soc Pediat Res; Am Acad Pediat; Asn Am Med Cols; Am Fedn Clin Res; AMA. *Res:* Physiology of ventricular function; pathology, physiology and natural history of congenital cardiac defects. *Mailing Add:* Pediat 284 Vchh Univ Minn Minneapolis MN 55455

LUCAS, THOMAS RAMSEY, b Tampa, Fla, June 9, 39; m 70; c 2. MATHEMATICS. *Educ:* Univ Fla, BS, 61; Univ Mich, Ann Arbor, MS, 62; Ga Inst Technol, PhD(math), 70. *Prof Exp:* Sr engr, Martin Co, 62-65; from instr to assoc prof, 69-85, PROF MATH UNIV NC, CHARLOTTE, 85- *Mem:* Am Math Soc; Soc Indust & Appl Math. *Res:* Numerical analysis; approximation theory; spline theory. *Mailing Add:* Dept Math Univ NC Charlotte NC 28223

LUCAS, WILLIAM FRANKLIN, b Detroit, Mich, Apr 21, 33; m 57; c 4. OPERATIONS RESEARCH, APPLIED MATHEMATICS. *Educ:* Univ Detroit, BS, 54, MA, 56, MS, 58; Univ Mich, PhD(math), 63. *Prof Exp:* Instr math, Univ Detroit, 56-58 & 61-62, asst prof, 62-63; res instr, Princeton Univ, 63-65; Fulbright fel & vis assoc prof econ & statist, Mid East Tech Univ, Ankara, 65-66; vis assoc prof, Math Res Ctr, Univ Wis-Madison, 66-67; mathematician, Rand Corp, 67-69; assoc prof opers res & appl math, 69-70, dir ctr appl math, 71-74, prof opers res & appl math & math, Cornell Univ, 70-84; PROF MATH, CLAREMONT GRAD SCH, 84- *Concurrent Pos:* Consult, Rand Corp, 69-77 & Educ Develop Ctr, 75-79; sci exchange with USSR, US Nat Acad Sci, 76 & 83; Chautauqua lectr, AAAS, 75-79. *Mem:* Am Math Soc; Math Asn Am; Soc Indust & Appl Math; Opers Res Soc Am; Asn Women Math. *Res:* Elasticity; applied mathematics; game theory. *Mailing Add:* Dept Math Claremont Grad Sch Claremont CA 91711

LUCAS, WILLIAM JOHN, b Adelaide, S Australia, Feb 23, 45; m 67; c 4. PLANT PHYSIOLOGY, PLANT BIOPHYSICS. *Educ:* Univ Adelaide, BSc, 71, PhD(plant physiol), 75, DSc(plant physiol), 90. *Prof Exp:* Res assoc, Dept Bot, Univ Toronto, 75-77; from asst prof to assoc prof plant physiol, 77-83, PROF BOT, UNIV CALIF, DAVIS, 83- *Concurrent Pos:* Guest prof, Univ Göttingen, WGer, 84-85; Nat Sci Found grant, 78-93; mem, Int Comt for Phloem Physiol, 84-90. *Mem:* Am Soc Plant Physiologists; Australian Soc Plant Physiologists; Can Soc Plant Physiologists; Bot Soc Am; Soc Exp Biol. *Res:* Biophysical and physiological aspects of transport across plant membranes, in particular the plasmalemma; cell-to-cell communication via plasmodesmata. *Mailing Add:* Dept Bot Univ Calif Davis CA 95616

LUCAS, WILLIAM R(AY), b Newbern, Tenn, Mar 1, 22; m 48; c 3. INORGANIC CHEMISTRY, MATERIALS SCIENCE ENGINEERING. *Educ:* Memphis State Univ, BS, 43; Vanderbilt Univ, MS, 50, PhD(chem, metall), 52. *Hon Degrees:* DHL, Mobile Col, 77; DSc, Southeastern Inst Technol, 80 & Univ Ala, Huntsville, 81. *Prof Exp:* Instr chem, Memphis State Univ, 46-48; chemist, guided missile develop div, Redstone Arsenal, 52-54, chief chem sect, 54-55; chief eng mat sect, Army Ballistic Missile Agency, 55-56, chief eng mat br, 56-60; chief Eng Mat Br, 60-63, chief Mat Div, 63-66, dir Propulsion & Vehicle Eng Lab, 66-68, dir prog develop, 68-71, dep dir, 71-74, dir, Marshall Space Flight Ctr, NASA, 74-86; CONSULT, 86- *Honors & Awards:* Except Sci Achievement Medal, NASA, 64; Oberth Award, Am Inst Aeronaut & Astronaut, 65, Holger N Toftoy Award, 76; Space Flight Award, Am Astronaut Soc, 82; Vet Foreign Wars Space Award, 83; Elmer A Speery Award, Am Instit Aeronaut & Astronautics, 86. *Mem:* Nat Acad Eng; fel Am Inst Aeronaut & Astronaut; fel Am Astronaut Soc; fel Am Soc Metals; Sigma Xi; Am Chem Soc. *Res:* Materials engineering, metallurgy and inorganic chemistry; environmental effects on materials, especially space. *Mailing Add:* 6805 Criner Rd Huntsville AL 35802

LUCAS-LENARD, JEAN MARIAN, b Bridgeport, Conn, July 17, 37; m 64. MOLECULAR BIOLOGY. *Educ:* Bryn Mawr Col, AB, 59; Yale Univ, PhD(protein synthesis), 63. *Prof Exp:* USPHS fel enzymol, Inst Physiochem Biol, Paris, 63-64; guest investr protein synthesis, Rockefeller Univ, 64-65, res assoc, 65-68, asst prof, 68-70; assoc prof, 70-76, PROF BIOL, UNIV CONN, 76- *Concurrent Pos:* Estab investr, Am Heart Asn, 70-71; NIH career develop award, 71-76. *Mem:* Am Soc Biol Chemists; Am Soc Microbiol; Am Soc Virol. *Res:* Mechanism of protein biosynthesis in eukaryotes; translational control mechanisms in virus infected cells. *Mailing Add:* Dept Molecular & Cell Biol Univ Conn Storrs CT 06269-3125

LUCAST, DONALD HURRELL, b Minneapolis, Minn, July 11, 46; m 75; c 3. ORGANIC CHEMISTRY. *Educ:* Univ Minn, BS, 68, PhD(org chem), 76. *Prof Exp:* Res chemist, Ethyl Corp, 77-83; RES CHEMIST, 3M, 83- *Concurrent Pos:* Fel, Univ Detroit, 75-76 & Wayne State Univ, 77. *Mem:* Am Chem Soc; Tech Asn Pulp & Paper Indust. *Res:* Organic synthesis; reaction mechanisms; polymer chemistry. *Mailing Add:* 2504 Skillman Circle North St Paul MN 55109

LUCATORTO, THOMAS B, b New York, NY, May 9, 37; m 79; c 2. LASERS. *Educ:* City Univ NY, BS, 60; Columbia Univ, MA, 64, PhD(physics), 68. *Prof Exp:* Res assoc physics, Columbia Univ, 68-69; RES PHYSICIST, NAT INST STANDARDS & TECHNOL, 69- *Honors & Awards:* IR-100 Award, Nat Bur Standards, 80 & 84. *Mem:* Fel Am Phys Soc; Am Optical Soc. *Res:* Atomic photoabsorption in the vacuum ultraviolet; laser-excitation of plasmas and laser ionization; multiphoton ionization. *Mailing Add:* 3600 Van Ness St NW Washington DC 20008-3129

LUCCA, JOHN J, b Brooklyn, NY, July 12, 21; m 46; c 6. DENTISTRY. *Educ:* NY Univ, AB, 41; Columbia Univ, DDS, 47; Am Bd Prosthodontics, dipl. *Prof Exp:* From instr to assoc prof dent, 47-64, PROF PROSTHODONTICS & DIR DIV, SCH DENT & ORAL SURG, 64-, EMER PROF, PROSTHODONTICS, COLUMBIA UNIV, 87. *Concurrent Pos:* Consult, Vet Admin & USPHS; hon police surgeon & consult, New York Police Dept, 64-; attend, Presby Hosp & Westchester County Med Ctr; consult ed prosthodont, Progreso-Odonto-Stomatologique. *Honors & Awards:* Ewell Medal, 47. *Mem:* Fel Am Col Dent; fel Am Col Prosthodont; fel Int Col Dent; fel Int Col Prosthodont. *Res:* Precision attachment; partial dentures. *Mailing Add:* Div of Prosthodontics Columbia Univ New York NY 10032

LUCCHESI, BENEDICT ROBERT, CARDIAC ARRHYTHMIAS, MYOCARDIAL ISCHEMIA. *Educ:* Univ Mich, MD, 64. *Prof Exp:* PROF PHARMACOL, MED SCH, UNIV MICH, 68- *Res:* Cardiovascular pharmacology. *Mailing Add:* Dept Pharmacol Univ Mich Med Sci Bldg 1M6322 Ann Arbor MI 48109

LUCCHESI, CLAUDE A, b Chicago, Ill, Apr 20, 29; m 54; c 2. ANALYTICAL CHEMISTRY, PHYSICAL CHEMISTRY. *Educ:* Univ Ill, BS, 50; Northwestern Univ, PhD, 54. *Prof Exp:* Asst, Northwestern Univ, 50-54; spectros group leader, Shell Develop Co, Tex, 54-56; dir anal res dept, Sherwin-Williams Co, 56-61; mgr anal & phys chem dept, Mobil Chem Co, 61-67, mgr cent coatings lab, 67-68; SR LECTR CHEM & DIR ANALYSIS SERV, NORTHWESTERN UNIV, 68- *Concurrent Pos:* Consult coatings, healthcare & instrument co; ed, Bull Anal Lab Mgr Asn, 84-87; contrib ed, Analyt Chem, 74-80. *Mem:* Am Chem Soc; Soc Appl Spectros; Instrument Soc Am; Analyt Lab Mgr Asn. *Res:* General applied spectroscopy; NMR spectroscopy; chelate chemistry; plastics and coating characterization and analysis. *Mailing Add:* Dept Chem Northwestern Univ Evanston IL 60208

LUCCHESI, JOHN CHARLES, b Cairo, Egypt, Sept 3, 34; US citizen; m 55; c 2. GENETICS. *Educ:* La Grange Col, AB, 55; Univ Ga, MS, 58; Univ Calif, Berkeley, PhD(zool), 63. *Prof Exp:* NIH res assoc biol, Univ Ore, 63-65; from asst prof to prof zool & genetics, 65-80, PROF BIOL & GENETICS, UNIV NC, CHAPEL HILL, 80- *Concurrent Pos:* Vis investr, Max Planck Inst Biol, Tübingen, Germany, 69 & dept genetics, Univ Cambridge, Eng, 84; NIH res career develop award, 70-75; vis Kenan prof, Dept Genetics, Univ Calif, Berkeley, 78; adj prof genetics, Duke Univ, Durham, NC, 80-; Carry C Boshamer prof biol, 82-90; overseas fel, Churchill Col, Eng, 84-; chair, Genetics Study Sect, Div Res Grants, NIH, 87-90; Aza G Chandler prof biol, 90- *Mem:* Genetics Soc Am (vpres, 90, pres, 91); Am Soc Cell Biol; Soc Develop Biol. *Res:* Molecular genetics; biochemistry of development; sex differentiation and dosage compensation in Drosophila. *Mailing Add:* Emory Univ Atlanta GA 30322

LUCCHESI, PETER J, b New York, NY, Sept 23, 26; m 49; c 2. PHYSICAL CHEMISTRY. *Educ:* NY Univ, AB, 49, MS, 53, PhD(chem), 54. *Prof Exp:* Instr chem, Adelphi Col, 52, NY Univ, 53-54 & Ill Inst Technol, 54-55; res chemist, Exxon Res & Eng Co, 55-68, dir, Corp Res Lab, 68-75, vpres corp res, 75-85; RETIRED. *Mem:* Am Chem Soc; AAAS. *Res:* Radiation chemistry; heterogeneous catalysis; crystal growth and dissolution. *Mailing Add:* 24 Brearly Rd Princeton NJ 08540-6766

LUCCHITTA, BAERBEL KOESTERS, b Muenster, Ger, Oct 2, 38; US citizen; m 64; c 1. PLANETARY GEOLOGY, GEOMORPHOLOGY. *Educ:* Kent State Univ, BS, 61; Pa State Univ, MS, 63, PhD(geol), 66. *Prof Exp:* GEOLOGIST, BR ASTROGEOL, US GEOL SURV, 68-, ASSOC CHIEF, 86- *Concurrent Pos:* Prin investr, three lunar projs, NASA, 74-78, guest investr, Viking Lander Imaging Team, 76, prin investr, four martian projs, 78-, mem, Planetary Geol Rev Panel, 80-82, coordr, Galilean Satellite Geol Mapping Prog, 80-; assoc ed, J Geophys Res, 80-84; proj chief Antarctica, 82-; secy/treas, Planetary Geol Div, Geol Soc Am, 87-89, 2nd vchmn, 89-90, 1st vchmn, 90-91; planet cartog working group, 89; lectr, Sigma Xi, 90-91. *Honors & Awards:* Spec Recognition Award, NASA, 79. *Mem:* Asn Women Geoscientists; Asn Women Sci; Geol Soc Am; Planetary Soc; Am Geophys Union; Int Glaciol Soc. *Res:* Dark mantles, secondary craters, basin formation, plains formation, scarps and ridges, northside and Apollo 17-site geological map of the moon; erosion, landform development, map of Ismenius Lacus, canyons and scarps, landslides, channels, glacial and periglacial features, Valles Marineris geology and structure of Mars; geomorphology and structural geology of earth; geologic map of Jupiter Satellite Europa; structure of Ganymede; Antarctic investigations with Landsat images, Antarctic coastal changes and glacier velocities. *Mailing Add:* Br Astrogeol US Geol Surv 2255 N Gemini Dr Flagstaff AZ 86001

LUCCHITTA, IVO, b Budweis, Czech, June 17, 37; US citizen; m 64; c 1. GEOLOGY. *Educ:* Calif Inst Technol, BSc, 61; Pa State Univ, PhD(geol), 67. *Prof Exp:* Geologist, proj chief & coordr Apollo geol methods, US Geol Surv, 66-70, geologist & proj chief earth resources technol satellite appln & anal, 70-73, geologist nat landslide overview map, 73-74, dep asst chief geologist, 85-87, GEOLOGIST & PROJ CHIEF WARIZ TECTONICS, US GEOL SURV, 73-, GEOLOGIST, SHIVWITS-GRAND WASH WILDERNESS AREA, 80-, PROJ CHIEF QUATERNARY GRAND CANYON, 90- *Concurrent Pos:* Penrose Bequest grant, Geol Soc Am, 63; Museum N Ariz grants, 63, 64; adj prof, Northern Ariz Univ, 74-; res fel, Univ Rome, Italy, 84; res grant, Nat Geog Soc, 84. *Honors & Awards:* Spec commendation, geol training astronauts, Geol Soc Am, 70; Group Achievement Award, Earth Resources Technol Satellite geol anal & image processing, NASA, 75; Super Serv Award, US Dept Int, 89. *Mem:* Fel Geol Soc Am. *Res:* Tectonic history of southwestern Colorado plateau and of plateau basin and range transition; basement control of structure; tectonic heredity; history of Colorado River and Grand Canyon; Cenozoic continental rocks; structure and tectonics of Cordilleran core complexes; Quaternary of the Grand Canyon. *Mailing Add:* US Geol Surv 2255 N Gemini Dr Flagstaff AZ 86001

LUCCI, ROBERT DOMINICK, b Norwalk, Conn, July 11, 50; m 71; c 2. ANALYTICAL CHEMISTRY. *Educ:* Univ Conn, BA, 72; Cornell Univ, PhD(org chem), 77. *Prof Exp:* SR SCIENTIST, HOFFMANN-LAROCHE, INC, 77- *Mem:* Am Chem Soc; Sigma Xi. *Res:* Safe, economic and environmentally sound industrial chemical processes from research synthesis. *Mailing Add:* 4 Parton Ct Lincolnshire IL 60045

LUCE, JAMES EDWARD, b Toronto, Ont, Aug 24, 35. PAPER CHEMISTRY, PHYSICS. *Educ:* Univ Toronto, BASc, 56; McGill Univ, PhD(chem), 60; NY Inst Tech, MBA, 80. *Prof Exp:* Asst mgr basic res, CIP Res Ltd, 60-71; sci admin officer, Atomic Energy Can, Ltd, 71; sr mgr oper systs develop, 72-81, assoc dir advan develop, 81-84, mgr papermaking technol, 84-87, MGR PAPER SCI & TECHNOL, INT PAPER CO, 87- *Honors & Awards:* Fel, Tech Asn Pulp & Paper Indust. *Mem:* Tech Asn Pulp & Paper Indust; Can Pulp & Paper Asn. *Res:* Application of modern instrumental techniques to control of pulp and paper processes; development of papermaking processes. *Mailing Add:* Int Paper Co Long Meadow Rd Tuxedo Park NY 10987

LUCE, R(OBERT) DUNCAN, b Scranton, Pa, May 16, 25; div; c 1. MATHEMATICAL PSYCHOLOGY, THEORY MEASUREMENT. *Educ:* Mass Inst Technol, BS, 45, PhD(math), 50. *Hon Degrees:* MS, Harvard Univ, 76. *Prof Exp:* Mem staff, Res Lab Electronics, Mass Inst Technol, 50-53; asst prof sociol & math statist, Columbia Univ, 54-57; prof psychol, Univ Pa, 59-67, Benjamin Franklin prof, 67-68; vis prof social sci, Inst Advan Study, Princeton, 69-72; prof soc sci, Univ Calif, Irvine, 72-75; Alfred North Whitehead prof psychol & math psychol, 76-81, Victor S Thomas prof psychol, 83-88, EMER VICTOR S THOMAS PROF, HARVARD UNIV, 88-; DISTINGUISHED PROF & DIR IRVINE RES UNIT MATH SOCIAL SCI, UNIV CALIF, IRVINE, 88- *Concurrent Pos:* Lectr social rels, Harvard Univ, 57-59; vis prof psychol, Catholic Univ Rio de Janeiro, 68-69; managing dir, Behav Models Proj, Columbia Univ, 53-57, fel, Ctr Advan Study Behav Sci, 54-55, 66-67 & 87-88; Guggenheim Found fel, 80-81. *Honors & Awards:* Distinguished Sci Contrib Award, Am Psychol Asn, 70. *Mem:* Nat Acad Sci; Am Acad Arts & Sci; Soc Math Psychol (pres, 79); Psychometric Soc (pres, 76-77); AAAS; Am Math Soc; fel Am Psychol Asn; Found Behav Psychol & Cognitive Sci (pres, 88-90). *Res:* Theoretical work on measurement and structures, especially conjoint and utility ones; theoretical and experimental work in psychophysics, including absolute identification, detection and recognition, magnitude estimation and reaction time. *Mailing Add:* Social Sci Tower Univ Calif Irvine CA 92717

LUCE, ROBERT JAMES, b Boston, Mass, Aug 7, 29; m 81. ROCK MAGNETISM. *Educ:* Drexel Univ, BS, 73; Univ Pittsburgh, MS, 75, PhD(geophysics), 80. *Prof Exp:* ASSOC PROF PHYSICS & GEOL, WASHINGTON & JEFFERSON COL, 80- *Mem:* Am Phys Soc; Am Asn Physics Teachers. *Res:* Theoretical models of hadronic atoms. *Mailing Add:* Dept Physics Washington & Jefferson Col Washington PA 15301

LUCE, WILLIAM GLENN, b Beaver Dam, Ky, Mar 21, 36; m 70; c 2. ANIMAL NUTRITION. *Educ:* Univ Ky, BS, 58; Univ Nebr, MS, 64, PhD(animal nutrit), 65. *Prof Exp:* Mgt trainee grocery & meat merchandising, Kroger Co, Ky, 58- 60, co-mgr grocery & meat merchandising, 60-62; asst nutrit res, Univ Nebr, 62-65; asst prof swine exten, Univ Ga, 65-68; PROF SWINE EXTEN, OKLA STATE UNIV, 68- *Honors & Awards:* Extrusion Award, Am Soc Animal Sci. *Mem:* Sigma Xi; Am Soc Animal Sci; Am Registry Prof Animal Scientists; Coun Agr Sci & Technol. *Res:* Swine nutrition; cereal grain utilization and amino acid requirements. *Mailing Add:* Dept Animal Sci Okla State Univ Stillwater OK 74078

LUCEY, CAROL ANN, b Johnstown, NY, Sept 16, 43; m 64; c 1. THEORETICAL PHYSICS, PHILOSOPHY OF SCIENCE. *Educ:* Harpur Col, BA, 65; State Univ NY Binghamton, MA, 68; Brown Univ, PhD(physics), 72. *Prof Exp:* Actg assoc dean instr, 76-78, PROF PHYSICS, JAMESTOWN COMMUNITY COL, 73- *Concurrent Pos:* Nat Endowment Humanities fel, 79-80. *Mem:* Am Phys Soc; Philos Sci Asn. *Res:* Study of cosmological implications for elementary particle physics; gauge theories and general relativity; scientific methodology. *Mailing Add:* 15 Clyde Ave Jamestown NY 14701

LUCEY, EDGAR C, b Feb 27, 45. PULMONARY. *Educ:* Morningside Col, Iowa, BS, 67; Idaho State Univ, Pocatello, MS, 71, PhD(physiol), 75. *Prof Exp:* Lectr physiol & instr human physiol & advan physiol, Humboldt State Univ, 74-76; instr respiratory syst, Dept Physiol, 76-78, asst res prof med, 79-87, ASSOC RES PROF, SCH MED, BOSTON UNIV, 87-; RES PHYSIOLOGIST, BOSTON VET ADMIN MED CTR, 79- *Concurrent Pos:* Mem, Animal Studies Subcomt, Vet Admin Med Ctr, 86- & chair, Res Safety Subcomt, 91- *Mem:* Am Physiol Soc; NY Acad Sci. *Res:* Pathogeneses of pulmonary emphysema and airway secretory cell metaplasia; relationship of neutrophils and neutrophil products to lung elastin and emphysema; interaction of lung macrophages, neutrophils and their secretory products; author of various publications. *Mailing Add:* Boston Vet Admin Med Ctr 150 S Huntington Ave Boston MA 02130

LUCEY, JEROLD FRANCIS, b Holyoke, Mass, Mar 26, 26; m 50; c 3. PEDIATRICS. *Educ:* Dartmouth Col, AB, 48; NY Univ, MD, 52. *Prof Exp:* Intern pediat, Bellevue Hosp, New York, 52-53; asst resident, Columbia-Presby Med Ctr, 53-55; from instr to assoc prof, 56-66, PROF PEDIAT, COL MED, UNIV VT, 66- *Concurrent Pos:* Bowen Brooks scholar, NY Acad Med, Bellevue Hosp, New York, 54; Mead Johnson fel, Columbia-Presby Med Ctr, 54-55; Nat Found Infantile Paralysis res fel, Harvard Med Sch, 55-56; Markle scholar, 59-64; res fel biochem, Harvard Med Sch, 60-61; consult, Vt State Health Dept, 56-81; chmn, Nat Bd Med Exam, 68-72; mem, Am Bd Pediat Exam, 70; ed-in-chief, Pediatrics, 73-; Humboldt Found fel, 78; Litchfield lectr, Oxford Univ, 78. *Honors & Awards:* Goulee Award, Am Asn Pediatrics, 81; United Cerebral Palsy Prize, 84; McDonald Award, 90. *Mem:* Soc Pediat Res; fel Am Acad Pediat; Am Pediat Soc; Am Soc Photobiol; Royal Soc Med; Sigma Xi. *Res:* Neonatal physiology; transcutaneous oxygen. bilirubin metabolism; surfactant. *Mailing Add:* Med Ctr Hosp Vt 52 Overlake Park Burlington VT 05401

LUCEY, JOHN WILLIAM, b Winthrop, Mass, Aug 21, 35; m 57; c 4. NUCLEAR ENGINEERING. *Educ:* Univ Notre Dame, BS, 57; Mass Inst Technol, SM, 63, PhD(nuclear eng), 65. *Prof Exp:* Asst prof, 65-68, ASSOC PROF NUCLEAR ENG, UNIV NOTRE DAME, 68- *Concurrent Pos:* Dir, Ind Civil Defense Prof Adv Serv, 69-73; dir, Notre Dame Energy Anal & Diag Ctr, 90- *Mem:* AAAS; Am Nuclear Soc; Am Soc Eng Educ; Health Physics Soc; Sigma Xi; Soc Radiol Protection. *Res:* Numerical methods for nuclear reactor calculations; radiation shielding; transport calculations. *Mailing Add:* Dept Aerospace & Mech Eng Univ Notre Dame Notre Dame IN 46556

LUCEY, JULIANA MARGARET, b Santa Monica, Calif. NUMERICAL ANALYSIS. *Educ:* Univ Wash, AM, 62; Univ Ariz, MS, 72; St Louis Univ, PhD(math), 75. *Prof Exp:* Teacher high sch math & chmn dept, Sisters of the Holy Names of Jesus & Mary, 65; asst prof math, Holy Names Col, 65-69; teaching asst, Univ Ariz, 69-71; teaching fel, St Louis Univ, 73-74; asst prof, Holy Names Col, 74-75; instr math, Wayne State Univ, 75-76; asst prof math, Marquette Univ, 76-79; vis asst prof math, Univ Alaska, 79-82, asst prof, San Jose State Univ, 82-85, MATHEMATICIAN, INTEGRATED SYSTEMS ANALYSTS, 85- *Concurrent Pos:* Marine Sci Cruise, Bering Sea, 83, Bering Sea & Gulf Alaska, 84. *Mem:* Am Math Soc; Am Statist Asn; Biometric Soc; Amer Asn Advancement Science; NY Acad Sci. *Res:* Solutions of stiff ordinary differential equations by a fifth order composite multistep method; index and Lefschetz number in the structure of gratings; ecosystems. *Mailing Add:* 3343 Willow Crescent Dr No T2 Fairfax VA 22030

LUCEY, ROBERT FRANCIS, b Worcester, Mass, Mar 13, 26; m 52; c 7. AGRONOMY. *Educ:* Univ Mass, BVA, 50; Univ Md, MS, 54; Mich State Univ, PhD(field crops), 59. *Prof Exp:* Asst prof agron, Univ NH, 57-61; from asst to assoc prof field crops, 61-70, PROF FIELD CROPS, NY STATE COL AGR & LIFE SCI, CORNELL UNIV, 70-, CHMN DEPT AGRON, 75- *Mem:* Am Soc Agron. *Res:* Production of field crops, especially crop-climate relationships; adaptability; plant competition. *Mailing Add:* 236 Emerson Ithaca NY 14850

LUCHER, LYNNE ANNETTE, b Houston, Tex, June 18, 54. VIROLOGY, PROTEIN CHEMISTRY. *Educ:* Lindenwood Cols, St Charles, Mo, BA, 76; Rice Univ, Houston, PhD(biochem), 83. *Prof Exp:* Res assoc, Med Sch, St Louis Univ, 82-85; ASST PROF VIROL & MICROBIOL, ILL STATE UNIV, NORMAL, 85- *Mem:* Am Soc Microbiol; Am Soc Virol; AAAS. *Res:* Biochemistry of adenovirus interaction with a host cell; reactions which determine whether a lytic infection or transformation occurs. *Mailing Add:* Dept Biol Sci Ill State Univ Normal IL 61761-6901

LUCHINS, EDITH HIRSCH, b Poland, Dec 21, 21; nat US; m 42; c 5. MATHEMATICS. Educ: Brooklyn Col, BA, 42; NY Univ, MS, 44; Univ Ore, PhD(math), 57. Prof Exp: Govt inspector anti-aircraft dirs, Sperry Gyroscope Co, NY, 42-44; instr math, Brooklyn Univ, 44-46 & 48-49; asst appl math lab, NY Univ, 46; Am Asn Univ Women res fel & res assoc math, Univ Ore, 57-58; from res assoc to assoc prof math, Univ Miami, 59-62; assoc prof, 62-70, PROF MATH, RENSSELAER POLYTECH INST, 70- Mem: Math Asn Am; Am Math Soc; Soc Indust & Appl Math. Res: Banach algebras; functional analysis; mathematical psychology. Mailing Add: 53 Fordham Ct Albany NY 12209

LUCHSINGER, WAYNE WESLEY, b Milaca, Minn, May 8, 24; m 43; c 4. BIOCHEMISTRY. Educ: Univ Minn, BS, 51, MS, 54, PhD(biochem), 56. Prof Exp: Asst biochem, Univ Minn, 51-55; sr chemist, Kurth Malting Co, 56-58, asst dir res, 58-60; assoc prof biochem, WVa Univ, 60-66; from assoc prof to prof chem, 66-84, EMER PROF CHEM, ARIZ STATE UNIV, 84- Mem: AAAS; Am Chem Soc; Am Soc Brewing Chem; Am Asn Cereal Chem. Res: Enzymes; barley carbohydrates; chemistry and mechanism of action of enzymes; carbohydrate structure. Mailing Add: 3329 S Stanley Pl Tempe AZ 85282

LUCHTEL, DANIEL LEE, b Carroll, Iowa, Jan 13, 42. ELECTRON MICROSCOPY, CELL BIOLOGY. Educ: St Benedict's Col, Kans, BS, 63; Univ Wash, PhD(zool), 69. Prof Exp: NIH fel, 69-71, res assoc biol struct, 71-73, res assoc environ health, 73-75, asst prof, 75-82, ASSOC PROF ENVIRON HEALTH, UNIV WASH, 82- Concurrent Pos: Res fel, Hubrecht Lab, Utrecht, Neth, 72. Mem: AAAS; Sigma Xi; Am Soc Cell Biol; Am Inst Biol Sci; Electron Micros Soc Am. Res: Lung ultrastructure and effects of gaseous and particulate air pollutants; respiratory tract mucus and mechanisms of mucous cell secretion; lung development; mechanisms of pulmonary edema; tracheal organ cultures. Mailing Add: Environ Health SC-34 Univ Wash Seattle WA 98195

LUCID, MICHAEL FRANCIS, b Indianapolis, Ind, Feb 23, 37; m 67; c 3. INORGANIC CHEMISTRY. Educ: Ind Univ, Bloomington, BS, 61; Purdue Univ, Lafayette, MS, 65. Prof Exp: Res chemist, Kerr McGee Corp, 65-67, sr res chemist, 67-75, res proj chemist, 75-78; staff engr, 78-80, mgr mining develop, 80-83, staff mining engr, 83-84, proj mgr, 85-87, SR STAFF MINING ENGR, SHELL MINING CO, 87- Mem: Am Chem Soc. Res: Hydrometallurgy; solvent extraction; ion exchange; solution chemistry; geochemistry; solution mining, uranium, vanadium, copper, gold; oil shale. Mailing Add: Shell Mining Co PO Box 2906 Houston TX 77252

LUCIER, GEORGE W, b Southbridge, Mass, June 23, 43. TOXICOLOGY. Educ: Univ Md, PhD(entom), 70. Prof Exp: CHIEF, LAB BIOCHEM RISK ANAL, NAT INST ENVIRON HEALTH SCI, 83-, ED, ENVIRON HEALTH PERSPECTIVES, 73- Mem: Am Soc Toxicol; Edocrine Soc; Teratology Soc. Res: Applications of biochemical data to human risk assessment. Mailing Add: Nat Inst Environ Health Sci NIH PO Box 12233 Research Triangle Park NC 27709

LUCIER, JOHN J, b Detroit, Mich, Aug 10, 17. ORGANIC CHEMISTRY. Educ: Univ Dayton, BS, 37; Western Reserve Univ, MS, 50, PhD(org chem), 51. Prof Exp: Instr chem, 45-47 & 51-52, from asst prof to assoc prof, 52-63, chmn dept, 64-79, PROF CHEM, UNIV DAYTON, 63- Concurrent Pos: Distinguished serv prof, Univ Dayton, 88. Mem: AAAS; Am Chem Soc; Soc Appl Spectros; NY Acad Sci; Chem Soc. Res: Organic synthesis; infrared spectroscopy; history of science. Mailing Add: Dept Chem Univ Dayton Dayton OH 45469-2357

LUCIS, OJARS JANIS, b Latvia, Apr 2, 24; Can citizen; m 49; c 2. ENDOCRINOLOGY, CLINICAL PHARMACOLOGY. Educ: Sir George Williams Univ, BSc, 54; McGill Univ, MSc, 57, PhD(invest med), 59, MD, CM, 61, cert clin chem, 74. Prof Exp: Res asst invest med, McGill Univ, 56-60; asst prof endocrinol, Dalhousie Univ, 65-71; med officer, Health & Welfare Can, 71-90, div chief endocrinol & Metab, 74-90. Concurrent Pos: Med Res Coun Can fel, McGill Univ, 62-63, res scholar, 63-65; Med Res Coun Can scholar steroid biochem, Dalhousie Univ, 65-68; asst pathologist, Prov NS Dept Pub Health, 66-68, assoc pathologist, 68-71. Res: Biosynthesis and metabolism of hormones; immunochemical assays of hormones; interaction of trace elements with the cells and the mammalian organism; biosynthesis and isolation of cadmium binding proteins; pharmacology and toxicology of drugs. Mailing Add: 1512 Caverly Ottawa ON K1G 0Y1 Can

LUCIS, RUTA, b Rujiena, Latvia, Apr 9, 25; Can citizen; m 49; c 2. COMPARATIVE ENDOCRINOLOGY. Educ: Sir George Williams Univ, BSc, 49; McGill Univ, MS, 64, PhD(invest med), 66. Prof Exp: Res asst endocrinol, McGill Univ, 62-65; res asst, Path Inst, 66-71; clin chemist, Animal Res Inst, Ottawa, 72-90; RETIRED. Mem: NY Acad Sci. Res: Biochemistry of steroids; immunochemical assays and metabolism of hormones; environmental health. Mailing Add: 1512 Caverley St Ottawa ON K1G 0Y1 Can

LUCK, DAVID JONATHAN LEWIS, b Milwaukee, Wis, Jan 7, 29. CYTOLOGY. Educ: Univ Chicago, SB, 49; Harvard Med Sch, MD, 53; Rockefeller Univ, PhD, 62. Prof Exp: From asst prof to assoc prof, 64-68, PROF CELL BIOL, ROCKEFELLER UNIV, 68- Concurrent Pos: Teaching fel, Harvard Med Sch, 57-59; fel, Rockefeller Univ, 59-64; res physician, Mass Gen Hosp, Boston, 57-59. Mem: Nat Acad Sci; Am Soc Cell Biol; Am Soc Biol Chemist. Res: Biochemical cytology; cell structure; biochemical function. Mailing Add: Rockefeller Univ New York NY 10021

LUCK, DENNIS NOEL, b Durban, SAfrica, Dec 8, 39; m 69; c 1. MOLECULAR BIOLOGY. Educ: Univ Natal, BSc, 61, MSc, 63; Oxford Univ, DPhil(molecular biol), 66. Prof Exp: Lectr biochem, Univ Natal, 66-68; vis asst prof pharmacol, Baylor Col Med, 69; asst prof zool, Univ Tex, Austin, 70-72; from asst prof to assoc prof, 72-82, PROF BIOL, OBERLIN COL, 82- Concurrent Pos: Eleanor Roosevelt Int Cancer fel, Univ Oxford, 78-79;

foreign expert, Shanxi Agr Univ, Taigu, Shanxi, People's Repub China, 82; consult biochemist, Gilford Instrument Lab, Oberlin, Ohio, 80-82; vis prof, Dept Biochem, Univ BC, Vancouver, Can, 84-85, 86-87 & 90-91. Mem: Brit Biochem Soc; Am Soc Develop Biol; Am Soc Cell Biol; Sigma Xi; Endocrine Soc; Am Soc Microbiol. Res: Regulation of pituitary gene expression; structure-function studies on growth hormone and prolactin. Mailing Add: Dept Biol Oberlin Col Oberlin OH 44074-1413

LUCK, JOHN VIRGIL, b Chalmers, Ind, Jan 20, 26; m 45; c 3. MICROBIOLOGY. Educ: Purdue Univ, BS, 49, MS, 51, PhD(microbiol, biochem), 54. Prof Exp: Dir beer fermentation res, Pabst Brewing Co, 53-55; proj chem res, Gen Foods Corp, NY, 55-58; head biol chem dept, Armour & Co, 58-60; dir res & develop, Durkee Famous Foods, Glidden Co, Ill, 60-70; SR VPRES & TECH DIR, GEN MILLS, INC, MINNEAPOLIS, 70- Concurrent Pos: Mem, Res & Develop Assocs. Mem: AAAS; Am Chem Soc; Am Oil Chem Soc; Inst Food Technol; Soc Indust Microbiol; Int Life Sci Inst-Nutrit Fedn. Res: Food chemistry; fats; starch; proteins emulsifiers; enzymology. Mailing Add: Gen Mills Inc One Marsh Bird Lane Savannah GA 31411-1602

LUCK, LEON D(AN), b Spokane, Wash, Apr 25, 21; m 41; c 2. CIVIL ENGINEERING. Educ: Wash State Univ, BS, 43; Univ Minn, MS, 51; Stanford Univ, CE, 60. Prof Exp: Mine engr, Pend Oreille Mines & Metals Co, 43 & 46-47; from instr to assoc prof civil eng, Wash State Univ, 47-57; lectr, Stanford Univ, 57-59; from assoc prof to prof, 59-83, chmn dept civil & environ eng, 72-76, EMER PROF CIVIL ENG, WASH STATE UNIV, 83- Concurrent Pos: Consult engr, Potlatch Forests, Inc, 56-60; Fulbright lectr, Chungbuk Nat Univ, Rep of Korea, 83-84. Honors & Awards: Western Elec Award, Am Soc Eng Educ, 82. Mem: Am Soc Civil Engrs; Am Soc Eng Educ; Nat Soc Prof Engrs. Res: Shear characteristics of Palouse clay; seepage flow through porous soil media; rigid frame analysis by matrix methods with the aid of a digital computer. Mailing Add: SE 920 High St Pullman WA 99163

LUCK, RICHARD EARLE, b Roanoke, Va, Mar 9, 50; m 78. ASTROPHYSICS, ASTRONOMY. Educ: Univ Va, BA, 72; Univ Tex, MA, 75; Univ Tex, PhD(astron), 77. Prof Exp: RES ASSOC, DEPT PHYSICS & ASTRON, LA STATE UNIV, 77- Mem: Am Astron Soc; Royal Astron Soc. Res: Chemical composition of late-type stars to determine the effects of stellar and galactic chemical evolution on such objects. Mailing Add: Dept Astron Case Western Reserve Univ University Circle Cleveland OH 44106

LUCK, RUSSELL M, b Reading, Pa, May 11, 26; m 63; c 2. POLYMER CHEMISTRY, ORGANIC CHEMISTRY. Educ: Albright Col, BSc, 47; Bucknell Univ, MSc, 48. Prof Exp: Asst chem, Bucknell Univ, 47-48; asst prod mgr, Wyomissing Glazed Papers, Inc, 48-51; engr, Mat Eng Dept, Westinghouse Elec Corp, 53-60, sr engr, Res & Develop Ctr, 60-71, fel scientist, 71-83, adv scientist, Res & Develop Ctr, 83-87; RETIRED. Mem: Am Chem Soc. Res: Organic and inorganic polymers for application as lubricants and electrical insulations with high temperature capabilities. Mailing Add: 1241 Harvest Dr Monroeville PA 15146

LUCKE, ROBERT LANCASTER, b Norfolk, Va, July 22, 45. ASTROPHYSICS. Educ: Johns Hopkins Univ, BA, 68, MA, 72, PhD(physics), 75. Prof Exp: Assoc res scientist physics, Johns Hopkins Univ, 75-76; Nat Res Coun fel, Goddard Space Flight Ctr, NASA, 76-78; ASST PROF PHYSICS & ASTRON, UNIV TOLEDO, OHIO, 79- Mem: Am Astron Soc. Res: Far ultraviolet albedo of the moon; coronal line emission in supernova remnants; x-ray astronomy; astronomical instrumentation. Mailing Add: 10608 Ridge Rd Clinton MD 20735

LUCKE, WILLIAM E, b Grand Island, Nebr, July 31, 36; m 59; c 5. ANALYTICAL CHEMISTRY. Educ: Univ Nebr, BS, 58; Ohio State Univ, PhD(chem), 63. Prof Exp: Res chemist, Olympic Res Div, Rayonier Inc, 63-69; res assoc, Cincinnati Milling Mach Co, 69-71, supvr, Cimcool Customer Lab Serv, Cincinnati Milacron Inc, 71-74, SR ANAL CHEMIST, CIMCOOL DIV, CINCINNATI, MILACRON INC, 74- Mem: AAAS; Am Chem Soc. Res: Carbohydrate, cellulose and wood chemistry; analytical chemistry of industrial metal working products. Mailing Add: 3447 Burch Cincinnati OH 45208-2003

LUCKEN, KARL ALLEN, crop breeding; deceased, see previous edition for last biography

LUCKENBILL-EDDS, LOUISE, b Lebanon, Pa, Nov 19, 36; m 71, 86. DEVELOPMENTAL BIOLOGY, NEUROBIOLOGY. Educ: Oberlin Col, BA, 58; Brown Univ, PhD(biol), 64. Prof Exp: Arthritis Found res fel arthritis & connective tissue dis, Sch Med, Boston Univ, 65-66, instr res dermat, 66-68; sci fel, Hubrecht Lab, Royal Netherlands Acad Sci & Letters, 68-69; asst prof biol sci, Smith Col, 69-75; instr, Dept Neuropath, Harvard Med Sch, 75-77; ASSOC PROF ZOOL & BIOMED SCI, OHIO UNIV, ATHENS, 77- Concurrent Pos: Guest scientist, Nat Inst Dent Res, NIH, Bethesda, Md, 85-86; sci fel, Max Planck Inst Biochem, Ger, 89-90; sr int fel, NIH, Fogarty Ctr, 89-90. Mem: AAAS; Am Soc Zool; Soc Develop Biol; Soc Neurosci; Sigma Xi. Res: laminin-mediated neurite outgrowth; histogenesis of sympathetic neurons; migration and differentiation of neural crestcells. Mailing Add: Col Osteop Med Ohio Univ Athens OH 45701

LUCKENS, MARK MANFRED, b Kiev, Russia, Apr 7, 12; US citizen; m 43; c 2. PHARMACOLOGY, TOXICOLOGY. Educ: Columbia Univ, BS, 35; NY Univ, MS, 50; Univ Conn, PhD(pharmacol, toxicol), 54; Polytech Inst New York, MSES, 72; Am Bd Indust Hyg, dipl. Prof Exp: Jr chemist, Wilkow Food Prod, 28-33, chemist, 33-36; chief chemist, Technichem Labs, 37-41; inspector, Chem Warfare Serv, 41-43; dir, Emmet Tech Assocs, 48-54; toxicologist, Conn State Dept Health, 54-61; from asst prof to assoc prof toxicol & pharmacol, Col Pharm, Univ Ky, 61-77, dir, Inst Environ Toxicol & Occup Hyg, 62-77, mem fac & co-dir interdisciplinary grad prog toxicol, 73-77; RETIRED. Concurrent Pos: Consult, Ky State Dept Human

Resources, 61-, Lexington-Fayette County Dept of Health, Ky Poison Info & Environ Health Control Prog, 61-, Lab Serv, Childrens's Hosp, Louisville, Ky, 63-, Spindletop Res Ctr, 65- & Nat Inst Occup Health & Safety, 77; Fulbright travel grant, 65-66; award, Partners-in-the-Americas, 65-66; mem, adv comt pesticides, Ky Dept Agr vis prof, Polytech Inst of Guayaquil; vis dir, Oceano vis prof, Polytech Inst of Guayaquil; dir, Hemispheric Prog Poison Info & Control; pvt pract, 77- *Mem:* Fel AAAS; fel Am Inst Chem; fel Am Acad Indust Hyg; fel Am Acad Forensic Sci; Am Chem Soc; Sigma Xi. *Res:* Toxicodynamics; comparative toxicology and pharmacology; environmental, occupational, clinical, analytical, food and forensic toxicology; chemical pathology; drug action in hibernation; biorhythms; effects of psychosocial parameters on toxicity and pharmacologic action. *Mailing Add:* Emmet Res Inst 664 Sheridan Dr Lexington KY 40503

LUCKERT, H(ANS) J(OACHIM), b Ger, Aug 26, 05; nat Can; m 53; c 1. AERODYNAMICS, APPLIED MATHEMATICS. *Educ:* Harvard Univ, AM, 29; Univ Berlin, Dr Phil, 33. *Prof Exp:* Asst to prof math, Mining Acad Freiberg, Ger, 29-34; aerodynamicist, Henschel Aircraft Co, 35-37; sr group leader aerodyn, Arado Aircraft Co, 37-45; scientist transl & aero res, Brit Ministry Supply, 45-47; consult aerodyn, Control Comn for Ger, 47-52; engr, Canadair, Ltd, 52-54, design specialist, 54-57, chief tech sect, Missiles & Systs Div, 57-63, sect chief missiles & space res, 63-64, staff scientist res & develop, 64-65; chief aerodynamicist, Space Res Inst, McGill Univ, 65-68; chief aerodynamicist, Space Res Inst, Inc, 68-69, Space Res Corp, 69-80; consult, Potton Tech Indust, Inc, 80-81, Phoenix Eng, Inc, Newport, Vt, 82-83, Space Res Corp, 83-86; RETIRED. *Concurrent Pos:* Chmn, Nat Res Coun Res Coord Group, Upper Atmosphere Res Vehicles, 64-65, mem assoc comt aerodyn, 63-66, mem assoc comt space res, 64-67; hon res assoc, McGill Univ, 67-85. *Mem:* Assoc fel Am Inst Aeronaut & Astronaut; fel Can Aeronaut & Space Inst; Ger Soc Aeronaut & Astronaut. *Res:* Aerodynamics and physics; astronautics; aircraft and missiles. *Mailing Add:* 197 - 58th Ave Laval des Rapides PQ H7V 2A5 Can

LUCKETT, WINTER PATRICK, b Atlanta, Ga, Mar 23, 37. ANATOMY, EMBRYOLOGY. *Educ:* Univ Mo, AB, 61, MA, 63; Univ Wis-Madison, PhD(anat), 67. *Prof Exp:* Instr, Col Physicians & Surgeons, Columbia Univ, 68-69, asst prof anat, 69-75; assoc prof anat, Sch Med, Creighton Univ, 75-; AT DEPT ANAT, UNIV PR, SAN JUAN. *Mem:* AAAS; Am Asn Anat; Soc Study Reproduction; Int Primatol Soc. *Res:* Comparative morphogenesis of the placenta and fetal membranes; comparative structure of the ovary; endocrinology of reproduction; evolution of primates. *Mailing Add:* Dept Anat Univ PR Med Sci GPO Box 5067 San Juan PR 00936

LUCKEY, EGBERT HUGH, internal medicine; deceased, see previous edition for last biography

LUCKEY, GEORGE WILLIAM, b Dayton, Ohio, Apr 17, 25; m 58; c 3. PHYSICAL CHEMISTRY. *Educ:* Oberlin Col, BA, 47; Rochester Univ, PhD(chem), 50. *Prof Exp:* Mem staff, Photog Theory Dept, Eastman Kodak Co, 50-56, Appl Photog Div, 56-60 & Spec Res Dept, 61-77, res fel & lab head, Spec Res Lab, 77-86; RETIRED. *Mem:* Am Chem Soc; Am Phys Soc; Royal Soc Chem; Soc Photog Scientists & Engrs; Electrochem Soc. *Res:* Photochemistry; photographic theory; luminescence; processing chemistry; photographic and radiographic systems. *Mailing Add:* 240 Weymouth Dr Rochester NY 14625

LUCKEY, PAUL DAVID, JR, b Pittsburgh, Pa, May 18, 28; wid; c 3. ELECTROMAGNETISM. *Educ:* Carnegie Inst Technol, BS, 49; Cornell Univ, PhD(physics), 54. *Prof Exp:* Res assoc physics, Cornell Univ, 53-56; mem sci res staff, 56-70, SR RES SCIENTIST PHYSICS, MASS INST TECHNOL, 70 - *Mem:* Am Phys Soc. *Res:* Meson physics; photoproduction of Pi mesons; electron synchrotrons. *Mailing Add:* CERN DP Div 1211 Geneva 23 Switzerland

LUCKEY, THOMAS DONNELL, b Casper, Wyo, May 15, 19; m 43; c 3. BIOCHEMISTRY, NUTRITION. *Educ:* Colo Agr Col, BS, 41; Univ Wis, MS, 44, PhD(biochem), 46. *Hon Degrees:* Hon prof, The Free Univ, Herborn, 82. *Prof Exp:* Asst, Agr & Mech Col, Tex, 41-42 & Univ Wis, 42-46; asst res prof biochem, Univ Notre Dame, 46-54; prof biochem, Sch Med, Univ Mo-Columbia, 54-84; RETIRED. *Concurrent Pos:* NSF traveling fel, Paris Nutrit Cong, 57; Univ Mo fel, Stockholm Microbiol Cong, 58; Commonwealth res fel, 61-62; Am Inst Nutrit traveling fel, Cong, 63; dir, WCent States Biochem Conf, 64-; moderator symp gnotobiol, Int Meeting Microbiol, Moscow, 66; mem subcomt interaction of infection & nutrit, Nat Acad Sci, 72-74; nutrit consult, NASA Johnson Space Ctr, Houston; consult, McDonnell Aircraft Corp, Mygrodol Prod Inc & Gen Elec Co; vis prof, Univ Qatar, 83; chief exec officer, Oralu Corp, 91- *Honors & Awards:* Av Humboldt Sr Sci Award, 78-80; Knighted, Greifenstein Castle, 84. *Mem:* AAAS; Am Chem Soc; Soc Exp Biol & Med; Am Soc Microbiol; Am Inst Nutrit; Sigma Xi. *Res:* Nutrition and metabolism of germ-free vertebrates; folic acid and related compounds in chick nutrition; comparative nutrition; modes of action of antibiotics; gnotobiology; thymic hormones; hormesis; low level radiation effects; biochemistry and nutrition; author. *Mailing Add:* 1009 Sitka Ct Loveland CO 80538

LUCKHAM, DAVID COMPTOM, b Kingston, Jamaica, Sept 7, 36. COMPUTER SCIENCE. *Educ:* Univ London, BSc, 56, MSc, 57; Mass Inst Technol, PhD(math logic), 63. *Prof Exp:* Res assoc comput sci, Mass Inst Technol, 63-65; lectr math, Univ Manchester, 65-68; res assoc comput sci, Stanford Univ, 68-70; from asst prof to assoc prof, Univ Calif, Los Angeles, 70-72; res comput scientist, 72-76, sr res assoc, 76-78, ADJ PROF ELEC ENG, STANFORD UNIV, 78- *Concurrent Pos:* Consult, Bolt, Beranek & Newman Inc, 63-65, Jet Propulsion Lab, 71- & Systs Control Inc, 78-; Sci Coun res grant, Univ Manchester, 65-68; lectr, Ctr Comput & Automation, Imp Col, Univ London, 67-68; Hayes sr fel, Harvard Univ, 76-77. *Mem:* Am Math Soc; Asn Comput Mach; Asn Symbolic Logic. *Res:* Theory of computation; automated proof procedures and applications to computer-aided programming, verification of programs; semantics of programming languages; parallel programs; microprocessor systems; artificial intelligence. *Mailing Add:* Dept Elec Eng Stanford Univ Stanford CA 94305

LUCKMANN, WILLIAM HENRY, b Cape Girardeau, Mo, Jan 15, 26; m 49; c 5. ENTOMOLOGY. *Educ:* Univ Mo, BS, 49; Univ Ill, MS, 51, PhD, 56. *Prof Exp:* Asst entomologist, State Natural Hist Surv, Ill, 51-53 & tech develop, Shell Chem Corp, Colo, 53-54; assoc entomologist, 54-59, entomologist, 59-65, ENTOMOLOGIST & HEAD SECT ECON ENTOM, STATE NATURAL HIST SURV, ILL, 65-; PROF & HEAD, OFF AGR ENTOM, COL AGR, UNIV ILL, 65- *Mem:* Entom Soc Am. *Res:* Ecology; biology; applied control. *Mailing Add:* State Natural Hist Surv 607 Peabody Champaign IL 61820

LUCKOCK, ARLENE SUZANNE, b Oakland, Calif, Nov 23, 48; m 76; c 2. NEUROPHYSIOLOGY, ENDOCRINOLOGY. *Educ:* Univ Calif, Berkeley, BA, 69, PhD(physiol), 74. *Prof Exp:* Fel, Dept Psychiat, Med Sch, Stanford Univ, 74-76 & Dept Genetics, 76-78; instr physiol, West Valley Col, Saratoga, Calif, 78-79; ASSOC PROF PHYSIOL, PALMER COL CHIROPRACTIC-W, 79- *Res:* Effects of thyroid hormones on mammalian brain development; genetic differences in testosterone synthesis in two strains of mice; genetic polymorphisms in testosterone-estradiol binding globulin in human populations. *Mailing Add:* Palmer Col Chiropractic W 1095 Dunford Way Sunnyvale CA 94087

LUCKRING, R(ICHARD) M(ICHAEL), b Canton, Ohio, Feb 3, 17; m 54; c 6. TECHNICAL MANAGEMENT. *Educ:* Heidelberg Col, BS, 40; Lehigh Univ, BSChE, 42. *Prof Exp:* Field engr, Eng Dept, E I Du Pont de Nemours & Co, 42-52, res engr, Pigments Dept, 52-53, res supvr, 53-55, res mgr, 55-71, tech mgr inorg fibers, 71-75, environ mgr, 75-78, planning assoc, Chem, Dyes & Pigments Dept, 78-81; RETIRED. *Mem:* Am Chem Soc; Am Inst Chem Engrs. *Res:* Process development; extractive metallurgy; refractory metals; titanate and titanium dioxide products and processes. *Mailing Add:* 108 Meriden Dr Canterbury Hills Hockessin DE 19707

LUCKY, GEORGE W(ILLIAM), b Dallas, Tex, Nov 7, 23; m 48; c 1. ELECTRICAL ENGINEERING. *Educ:* Okla State Univ, BSEE, 44, MS, 60, PhD(eng), 65. *Prof Exp:* Asst engr, Southwestern Bell Tel Co, 46-52, sr engr, 52-56; asst prof, Okla State Univ, 56-64; from asst prof to assoc prof, 64-69, PROF ELEC ENG, N MEX STATE UNIV, 69- *Concurrent Pos:* Consult, J B Payne Assocs, Inc, 60- *Mem:* Am Soc Eng Educ; Inst Elec & Electronics Engrs. *Res:* Computer characterization of electric networks; network synthesis. *Mailing Add:* Dept of Elec Eng NMex State Univ University Park NM 88003

LUCKY, ROBERT W, b Pittsburgh, Pa, Jan 9, 36; m 61; c 2. ELECTRICAL ENGINEERING. *Educ:* Purdue Univ, BSEE, 57, MSEE, 59, PhD(elec eng), 61. *Hon Degrees:* DEng, Purdue Univ, 88; DSc, NJ Inst Technol, 91. *Prof Exp:* Mem tech staff, Holmdel, NJ, 61-64, supvr signal theory group, 64-65, head data theory dept, 65-76, dept head digital-switching, processing res dept, 76-77, asst dir elec comput, 77-78, dir elec & comput systs, 78-82, EXEC DIR, RES COMMUN SCI DIV, BELL LABS, 82- *Concurrent Pos:* Asst ed, Trans Commun, Inst Elec & Electronics Engrs, 70-73, assoc ed, Trans Info Theory, 71-74, ed, Proc Inst Elec & Electronics Engrs, 74-76, consult ed, J Telecommun Networks, 78-86; vchmn, Sci Adv Bd, USAF, 83-86, chmn, 86-89; mem, Computer Sci & Technol Bd, Nat Res Coun, 86-, Strategic Defense Initiative Adv Comt, 86-89, Off Sci & Technol Policy Nat Crit Technol Panel, 90- & vis comt advan technol, Nat Inst Standards & Technol, 91-92. *Honors & Awards:* Edwin Armstrong Award, Inst Elec & Electronic Engrs, Commun Soc, 75; Centennial Medal, Inst Elec & Electronics Engrs, 84. *Mem:* Nat Acad Eng; fel Inst Elec & Electronics Engrs (vpres, 78-79 & 81-82); Inst Elec & Electronics Engrs Commun Soc (vpres, 78, pres, 78-79); Sigma Xi. *Res:* Communication theory; information theory; data transmission; author of over 50 publications; awarded 11 patents. *Mailing Add:* AT&T Bell Labs Rm 4E-605 Holmdel NJ 07733-1988

LUCOVSKY, GERALD, b New York, NY, Feb 28, 35; m 57; c 5. SOLID STATE PHYSICS. *Educ:* Univ Rochester, BS, 56, MA, 58; Temple Univ, PhD(physics), 60. *Prof Exp:* Mem staff solid state physics, Philco Corp, Pa, 58-65; sr scientist, Xerox Corp, 65-67; assoc prof eng, Case Western Reserve Univ, 67-68; mgr, Photoconductor Res Br, Xerox Corp, 68-69, solid state res br, 69-70, solid state sci br, Palo Alto Res Ctr, 70-73, assoc lab mgr, Gen Sci Lab, 73-74, sr res fel, Gen Sci Lab, Palo Alto Res Ctr, 74-80; UNIV PROF PHYSICS, NC STATE UNIV, RALEIGH, 80- *Mem:* Fel Am Phys Soc. *Res:* Optical properties of solids; lattice dynamics; amorphous semiconductors. *Mailing Add:* Dept Physics NC State Univ Box 8202 Raleigh NC 27695-8202

LUDDEN, GERALD D, b Quincy, Ill, Sept 6, 37; m 61; c 3. MATHEMATICS. *Educ:* St Ambrose Col, BA, 59; Univ Notre Dame, MS, 61, PhD(math), 66. *Prof Exp:* Lectr math, Ind Univ, 65-66; from asst prof to assoc prof, 66-77, PROF MATH, MICH STATE UNIV, 77- *Mem:* Math Asn Am; Am Math Soc; Tensor Soc. *Res:* Hypersurfaces of manifolds with an f-structure; submanifolds of real and complex space forms. *Mailing Add:* Dept of Math Mich State Univ East Lansing MI 48824

LUDDEN, PAUL W, b Omaha, Nebr, Nov 7, 50; m 74. BIOCHEMISTRY. *Educ:* Univ Nebr, Lincoln, BS, 72; Univ Wis-Madison, PhD(biochem), 77. *Prof Exp:* Res asst, Univ Wis-Madison, 72-77; res assoc, Mich State Univ, 77-78; asst prof biochem & asst biochemist, Univ Calif, Riverside, 78-81; asst prof, 81-85, ASSOC PROF BIOCHEM, UNIV WIS-MADISON, 85- *Concurrent Pos:* Fel, Rockefeller Found, 77-78. *Mem:* Am Soc Plant Physiol; Am Soc Microbiol. *Res:* Plant biochemistry; nitrogen metabolism in plants and bacteria; carbon monoxide oxidation. *Mailing Add:* Dept Biochem Univ Wis 420 Henry Mall Madison WI 53706

LUDDEN, THOMAS MARCELLUS, b Kansas City, Mo, Jan 16, 46; m 67, 79; c 4. BIOPHARMACEUTICS, DRUG METABOLISM. *Educ:* Univ Mo-Kansas City, BS, 69, PhD(pharmacol), 73. *Prof Exp:* Vis res assoc pharmaceut, Ohio State Univ, 74-75; from asst prof to assoc prof, 75-85, PROF PHARMACOL, UNIV TEX, AUSTIN, 85- *Concurrent Pos:* Tech consult, Audie Murphy Vet Hosp, San Antonio, 77-; from asst prof to assoc prof, Univ Tex Health Sci Ctr, San Antonio, 76-85, prof pharmacol, 85-;

Southwestern Drug Centennial fel pharm, 85- *Mem:* Am Pharmaceut Asn; Am Asn Pharmaceut Sci; Sigma Xi; NY Acad Sci; Fel Am Col Clin Pharm. *Res:* Applied pharmacokinetics and new drug development. *Mailing Add:* 7703 Floyd Curl Dr San Antonio TX 78284

LUDEKE, CARL ARTHUR, b Cincinnati, Ohio, Sept 26, 14. PHYSICS, OCEANOGRAPHY. *Educ:* Univ Cincinnati, AB, 35, PhD(physics), 38. *Prof Exp:* Instr math, John Carroll Univ, 38-40; instr, Univ Cincinnati, 40-43, from asst prof to assoc prof mech, 42-54, prof physics, 54-72; prof, Phys Oceanog Lab, NY Inst Technol at Nova Univ, 72-75, sr scientist, 75-79; RETIRED. *Concurrent Pos:* Consult, Gen Elec Co, 56-70. *Mem:* Int Asn Analog Comput. *Res:* Nonlinear mechanics; vibration analysis; shock mounts; mathematical physics; energy from the sun, sea and atmosphere. *Mailing Add:* PO Box 21682 Ft Lauderdale FL 33335

LUDEKE, RUDOLF, b Hannover, Ger, May 6, 37; m 64; c 2. SOLID STATE PHYSICS, MATERIAL SCIENCE. *Educ:* Univ Cincinnati, BS, 61; Harvard Univ, MA, 62, PhD(appl physics), 68. *Prof Exp:* RES STAFF MEM, T J WATSON RES CTR, IBM CORP, 68- *Concurrent Pos:* Vis scientist, Max Planck Inst, Stuttgart, Ger, 77-78; Alexander Von Humboldt Found fel, 77; assoc ed, J Vacuum Sci & Tech, 82-84; vis scholar, Dept Physics, Univ Utah, 86-87; prog chmn & conf chmn, Conf on Physics & Chem Semi-conductor Interfaces, 87; comt mem, Elec Mat & Processing Div, Am Vacuum Soc, 86- *Mem:* Am Phys Soc; Sigma Xi; Mat Res Soc, Am Vacuum Soc. *Res:* Semiconductor physics; surface and interface physics; thin film technology; structure and electronic properties of semiconductor interfaces, growth and characterization of semiconductor thin films and structures by molecular beum epitary, co-investor of the man-made semiconductor super lattice, optial properties of solids. *Mailing Add:* IBM T J Watson Res Ctr PO Box 218 Yorktown Heights NY 10598

LUDEL, JACQUELINE, b Boston, Mass, Mar 17, 45. BIOPSYCHOLOGY. *Educ:* Queens Col, NY, BA, 66; Ind Univ, PhD(psychol), 71. *Prof Exp:* Asst prof psychol, Jacksonville Univ, 71-73 & Stockton State Col, 73-76; PROF BIOL & PSYCHOL, GUILFORD COL, 76- *Concurrent Pos:* Grad fel, Nat Sci Found, 66-71; assoc instr, Ind Univ, 67-71; Danforth Assoc, 79-85; trustee, Marine Mammal Stranding Ctr, 79-86; Kenan grant, Guilford Col, 77- *Mem:* Psychol Social Responsibility. *Res:* Sensory anatomy and physiology; stranded and beached cetaceans. *Mailing Add:* Depts of Biol & Psychol Guilford Col Greensboro NC 27410

LUDEMA, KENNETH C, b Dorr, Mich, Apr 30, 28; m 55; c 5. MECHANICAL ENGINEERING, SURFACE PHYSICS. *Educ:* Calvin Col, BS, 55; Univ Mich, BS, 55, MS, 56, PhD(mech eng), 63; Cambridge Univ, PhD(physics), 65. *Prof Exp:* Instr mech eng, Univ Mich, 55-62; Ford Found & Univ Mich Inst Sci & Technol fac develop grant, Cambridge Univ, 62-64; from asst prof to assoc prof, 64-72, PROF MECH ENG, UNIV MICH, ANN ARBOR, 72- *Mem:* Am Soc Mech Engrs; Am Soc Testing & Mat. *Res:* Sliding friction and wear behavior of solids, steels, plastics and rubbers; fundamental adhesion mechanisms between dissimilar materials; skid resistance properties of tires and roads. *Mailing Add:* Dept Mech Eng Univ Mich Main Campus Ann Arbor MI 48109-2125

LUDEMANN, CARL ARNOLD, b Brooklyn, NY, June 21, 34; m 56; c 2. ACCELERATOR DESIGN & CONTROL. *Educ:* Brooklyn Col, BS, 56; Univ Md, PhD(nuclear physics, elec eng), 64. *Prof Exp:* Res assoc physics, Univ Md, 64-65; vis scientist, 64-65; physicist, Electronuclear Div, 65-71; PHYSICIST, OAK RIDGE NAT LAB, 71- *Mem:* Am Phys Soc; Am Asn Physics Teachers. *Res:* Neutron threshold measurements; gamma ray spectroscopy; angular correlation and nuclear reaction mechanism; nuclear structure studies; accelerator control, accelerator design. *Mailing Add:* Physics Div Oak Ridge Nat Lab Bldg 600 MS 6368 Oak Ridge TN 37831

LUDERS, RICHARD CHRISTIAN, b Staten Island, NY, July 23, 34; m 57; c 2. ANALYTICAL CHEMISTRY. *Educ:* Wagner Col, BS, 56. *Prof Exp:* Chemist, S B Penick & Co, 56-57; chemist, Ciba Pharmaceut Co, 58-66, group supvr anal res, 66-67, supvr, 67-70, head bioanal studies, Drug Metab Sect, 70-72, SR CHEMIST, DRUG METAB DIV, CIBA-GEIGY CORP, 72- *Mem:* Am Chem Soc. *Res:* Gas liquid chromatographic analysis and methods development for pharmaceutical compounds, preparations and raw materials; blood level determinations of pharmaceutical compounds. *Mailing Add:* RD 2 Katonah NY 10536-9802

LUDIN, ROGER LOUIS, b Jersey City, NJ, June 13, 44; m 66; c 2. NUCLEAR PHYSICS. *Educ:* Brown Univ, ScB, 66; Worcester Polytech Inst, MS, 68, PhD(physics), 69. *Prof Exp:* Fel, Worcester Polytech Inst, 69-71; prof physics, Burlington County Col, NJ, 71-86; LECTR PHYSICS, CALIF POLYTECH STATE UNIV, SAN LOUIS OBISPO, 84- *Mem:* AAAS; Am Phys Soc; Am Asn Physics Teachers. *Res:* Neutron-deuteron scattering. *Mailing Add:* Dept Physics Calif Polytech State Univ San Louis Obispo CA 93407

LUDINGTON, MARTIN A, b Detroit, Mich, Mar 7, 43; m 79; c 2. NUCLEAR PHYSICS. *Educ:* Albion Col, AB, 64; Univ Mich, MS, 65, PhD(physics), 69. *Prof Exp:* PROF PHYSICS, ALBION COL, 69-, CHMN DEPT, 80-83, 89- *Mem:* Am Phys Soc; Am Asn Physics Teachers. *Res:* Nuclear spectroscopy; prompt-gamma activation analysis. *Mailing Add:* Dept of Physics Albion Col Albion MI 49224

LUDKE, JAMES LARRY, b Vicksburg, Miss, Jan 11, 42; m 65. ENVIRONMENTAL BIOLOGY. *Educ:* Millsaps Col, BS, 64; Miss State Univ, MS, 67, PhD(physiol), 70. *Prof Exp:* Res asst physiol, Miss State Univ, 70-71; RES PHYSIOLOGIST, PATUXENT WILDLIFE RES CTR, US FISH & WILDLIFE SERV, 71-, DEP CHIEF, OFF RES SUPPORT. *Mem:* Am Soc Zoologists; Sigma Xi; AAAS. *Res:* Study of the chronic or lethal effects of pollutants on nontarget species; emphasis on fate of chemicals, diagnostic methods and chemical interactions. *Mailing Add:* Nat Fisheries Res Ctr Leetown Box 700 Kearneysville WV 25430

LUDLAM, WILLIAM MYRTON, b Teaneck, NJ, Mar 31, 31; m 54; c 4. OPTOMETRY, PHYSIOLOGICAL OPTICS. *Educ:* Columbia Univ, BS, 53, MS, 54; Mass Col Optom, OD, 63. *Prof Exp:* Dir, Vision Res Lab, Optom Ctr NY, 61-73; assoc prof physiol optics & optom, Col Optom, State Univ NY, 71-73; assoc prof, 74-80, PROF PHYSIOL OPTICS & OPTOM, COL OPTOM, PAC UNIV, 80- *Concurrent Pos:* Res grants, Am Optom Found, 55-56, NY Acad Optom, 57, Optom Ctr Res Fund, 60-61 & NIH, 63-74. *Honors & Awards:* Skeffington Award, 77. *Mem:* Fel AAAS; fel Am Acad Optom; fel Optical Soc Am; fel NY Acad Sci. *Res:* Ocular dioptric components; pathophysiology of strabismus and its remediation; ametropia and its etiology; vision and learning. *Mailing Add:* Col of Optom Pac Univ 2043 College Way Forest Grove OR 97116

LUDLOW, CHRISTY L, b Montreal, Que, June 7, 44; m 68. SPEECH PATHOLOGY, AUDIOLOGY. *Educ:* McGill Univ, BSc, 65, MSc, 67; NY Univ, PhD(psycholing, speech path), 73. *Prof Exp:* Res asst, McGill Univ, 66-67; res speech pathologist, Med Ctr, NY Univ, 67-70, W A Anderson fel, 70-72; vis lectr speech & hearing sci, Univ Md, 73-74; proj mgr, Am Speech & Hearing Asn, 73-74; RES SPEECH PATHOLOGIST, NAT INST DEAFNESS & OTHER COMMUN DIS, 74- *Concurrent Pos:* Liaison rep, AAAS, 77-81; ed consult, J Speech & Hearing Dis, 77- & J Speech & Hearing Res, 77-; chief, speech path unit intramural res prog, Nat Inst Neurol & Commun Dis & Stroke, NIH; consult, Vietnam Head Injury Study, Walter Reed Army Med Ctr, 81-85. *Honors & Awards:* Dir Award, NIH, 77. *Mem:* Int Neuropsychol Soc; Acoust Soc Am; Soc Neurosci; AAAS; Asn Res Otolaryngol; fel Am Speech-Lang-Hearing Asn; Acad Aphasia. *Res:* Speech science; neurolinguistics; aphasia; developmental language disorders; neuropharmacology; vocal pathologies; neurological disorders affecting speech and language functioning; neurophysiology of laryngeal movement control during speech in disorders of spasmodic dysphonia and stuttering. *Mailing Add:* Bldg 10 Rm 5038 Voice & Speech Sect VSLB NIDCD 9000 Rockville Pike Bethesda MD 20892

LUDLUM, DAVID BLODGETT, b Brooklyn, NY, Sept 30, 29; m 52; c 2. ENVIRONMENTAL HEALTH. *Educ:* Cornell Univ, BA, 51; Univ Wis-Madison, PhD(chem), 54; NY Univ, MD, 62. *Prof Exp:* Res chemist, E I du Pont de Nemours & Co, Inc, Del, 54-58; intern, 3rd & 4th Med Divs, Bellevue Hosp, 62-63; asst prof pharmacol & Am Cancer Soc fac res assoc, Sch Med, Yale Univ, 63-68; from assoc prof to prof pharmacol, Sch Med, Univ Md, Baltimore City, 70-76; prof pharmacol, Albany Med Col, 76-86 chmn dept, 76-80, prof med, 80-86; PROF PHARMACOL & MED, UNIV MASS SCH MED, 86- *Concurrent Pos:* Markle scholar acad med, Yale Univ & Univ Md, 67-72; Nat Inst Gen Med Sci career develop award, Yale Univ, 68; assoc ed, Cancer Res, 80- *Mem:* Am Chem Soc; Am Soc Pharmacol & Exp Therapeut; Am Soc Biol Chem; Am Asn Cancer Res; Am Soc Clin Pharmacol & Therapeut. *Res:* Pharmacology of antineoplastic agents; cancer chemotherapy; mutagenesis and carcinogenesis; molecular and clinical pharmacology. *Mailing Add:* Dept of Pharmacology Univ of Mass Med Sch 55 Lake Ave N Worcester MA 01655

LUDLUM, JOHN CHARLES, b Chevy Chase, Md, Feb 2, 13; m 40. GEOLOGY. *Educ:* Lafayette Col, BS, 35; Cornell Univ, MS, 39, PhD(struct geol), 42. *Prof Exp:* Mem staff, Socony Vacuum Oil Co, 35-37 & Amerada Petrol Corp, 37; from asst instr to instr geol, Cornell Univ, 37-42; from asst prof to prof, 46-72, dir ctr resource develop, 62-63, dir off res & develop, Appalachian Ctr, 63-66, from asst dean to dean grad sch, 66-72, EMER PROF GEOL, WVA UNIV, 72- *Concurrent Pos:* Consult, 46-62; coop econ geologist, State Geol Surv, 46-62. *Mem:* Fel Geol Soc Am; Soc Econ Geol; Am Asn Petrol Geol; Am Inst Mining, Metall & Petrol Eng. *Res:* Structural and economic geology of West Virginia; natural and human resources research applied toward improvement of the economy and life in West Virginia and the Appalachian highlands. *Mailing Add:* 612 Callen Ave Morgantown WV 26505

LUDLUM, KENNETH HILLS, b Albany, NY, Nov 16, 29; m 53; c 4. PHYSICAL CHEMISTRY. *Educ:* Col Educ Albany, BA, 51, MA, 52; Rensselaer Polytech Inst, PhD(phys chem), 61. *Prof Exp:* Chemist, Beacon Res Lab, 61-62, sr chemist, 62-65, res chemist, 65-73, sr res chemist, 73-80, RES ASSOC, TEXACO INC, 80-, COORDR, TEXACO ENV CONS & TOXICOL, 87- *Mem:* Am Chem Soc; Catalysis Soc; Air Pollution Control Asn. *Res:* Reaction kinetics; air pollution studies and related environmental science; catalysis and surface chemistry. *Mailing Add:* 117 N Elm St Beacon NY 12508

LUDMAN, ALLAN, b Brooklyn, NY, Mar 7, 43. GEOLOGY, PETROLOGY. *Educ:* Brooklyn Col, BS, 63; Ind Univ, Bloomington, AM, 65; Univ Pa, PhD(geol), 69. *Prof Exp:* Asst prof geol, Smith Col, 69-75; from asst to assoc prof, 75-82, PROF EARTH & ENVIRON SCI, QUEEN COL, NY, 82- *Concurrent Pos:* Field geologist, Maine Geol Surv, 66- *Mem:* Geol Soc Am; Sigma Xi; AAAS; Geol Asn Can. *Res:* Regional geologic mapping in central and eastern Maine; low-temperature metamorphism of pelitic and calcareous rocks; tectonic evolution of northeastern New England. *Mailing Add:* Dept Geol Queens Col Flushing NY 11367-0904

LUDMAN, JACQUES ERNEST, b Chicago, Ill, Nov 26, 34; m 70; c 1. SOLID STATE PHYSICS. *Educ:* Middlebury Col, BA, 56; Northeastern Univ, PhD(solid state physics), 73. *Prof Exp:* Res physicist, Air Force Cambridge Res Lab, 59-75, chief, Optical Processing Sect, 75-89; PRES, NORTHEAST PHOTOSCI, INC, 90- *Mem:* Sigma Xi. *Res:* Injection laser development; radiation damage effects on semiconductor devices; infrared sensor physics. *Mailing Add:* 18 Flagg Rd Hollis NH 03049

LUDOVICI, PETER PAUL, b Pittsburgh, Pa, Aug 9, 20; m 45; c 5. BACTERIOLOGY. *Educ:* Washington & Jefferson Col, BS, 42; Univ Pittsburgh, MS, 49, PhD(bact), 51. *Prof Exp:* Res bacteriologist immunol, West Penn Hosp, 49-51; res assoc, Univ Pittsburgh, 51; res assoc obstet & gynec, Univ Mich, 51-54; from instr to asst prof, 54-63, microbiol, obstet & gynec, 63-64, microbiol & cent tissue cult facil, 64-65, from assoc prof to prof,

65-87, EMER PROF MICROBIOL, UNIV ARIZ, 87- *Mem:* AAAS; Am Soc Microbiol; Tissue Cult Asn; Soc Exp Biol & Med. *Res:* Tissue culture; cancer; virology; cell transformations. *Mailing Add:* 5425 E Rosewood Ave Tucson AZ 85711

LUDOWIEG, JULIO, biochemistry, for more information see previous edition

LUDUENA, RICHARD FROILAN, b San Francisco, Calif, Feb 9, 46; m 81; c 1. BIOCHEMISTRY. *Educ:* Harvard Univ, BA, 67; Stanford Univ, PhD(biol), 73. *Prof Exp:* Fel pharmacol, Sch Med, Stanford Univ, 73-75, fel genetics, 75-76; from asst prof to assoc prof, 76-88, PROF BIOCHEM, UNIV TEX HEALTH SCI CTR, SAN ANTONIO, 88- *Concurrent Pos:* Jane Coffin Childs Mem Fund Med Res fel, 73-75. *Mem:* Am Soc Cell Biol; Int Soc Neurochem; Am Soc Biochem & Molecular Biol. *Res:* Regulation of microtubule assembly; structure and evolution of tubulin; pharmacology of microtubule proteins. *Mailing Add:* Dept Biochem Univ Tex Health Sci Ctr San Antonio TX 78284

LUDVIGSEN, CARL W, JR, b Palo Alto, Calif. PATHOLOGY. *Educ:* Univ Colo, Boulder, BA, 74; Wash Univ, St Louis, Mo, MD & PhD, 80; Am Bd Path, cert, 85. *Prof Exp:* Teaching asst, Dept Anat, Sch Med, Wash Univ, St Louis, Mo, 76, tutor physiol, path & pharmacol, 76-77; clin path resident, Univ Minn, Minneapolis, 80-82, clin chem fel, 82-83; dir chem, spec chem & toxicol, Med Ctr, Univ Nebr, 83-86; emergency rm physician, Lutheran Hosp, Omaha, Nebr & Bryan Mem Hosp, Lincoln, Nebr, 86-88; dir, Emergency Rm & Labs, Sandstone Area Hosp, Minn, 88-89; CHIEF PATHOLOGIST & SR VPRES, HOME OFF RES LAB, LENEXA, KANS, 89-; CLIN ASSOC PROF, DEPT PATH, SCH MED, MED CTR, UNIV KANS, KANSAS CITY, 90- *Concurrent Pos:* Young investr award, Acad Clin Lab Physicians & Scientists, Seattle, Wash, 82; mem, Prod Eval & Standardization Comt, 83-85; consult, Ariel Answer Prizes, 86-; assoc med dir, Bus Mens Assurance Co, Kansas City, Mo, 89- *Mem:* Fel Am Col Legal Med; Am Med Asn; fel Clin Asn Path; Am Soc Clin Pathologists; Am Soc Law & Med; AAAS; Am Asn Clin Chemists; Am Diabetes Asn; Acad Clin Lab Physicians & Scientists; Asn Clin Scientists. *Res:* Validation, verification and correlation of various toxicologic measurement modalities; aspects of lipid measurements as related to coronary artery dosage risk; general population studies; development of alcohol abuse markers suitable for population screening; author of various publications. *Mailing Add:* Dept Path Home Off Ref Lab PO Box 2035 Shawnee Mission KS 66201

LUDWICK, ADRIANE GURAK, b Passaic, NJ, June 16, 41; m 68; c 2. ORGANIC CHEMISTRY. *Educ:* Rutgers Univ, New Brunswick, AB, 63; Univ Ill, Urbana, MS, 65, PhD(chem), 67. *Prof Exp:* Asst prof chem, Tuskegee Univ, 67-68; vis asst prof & res assoc, Univ Ill, Urbana, 68-69; from asst prof to assoc prof, 69-77, PROF CHEM, TUSKEGEE UNIV, 77- *Concurrent Pos:* Res assoc, Environ Sci Div, Oak Ridge Nat Lab, 74, Chem Div, Lawrence Livermore Lab, 78 & AT&T Bell Labs, 82-85; NIH fac fel, Macromolecular Res Ctr, Univ Mich, 78-79. Mem; NSF fac fel, Macromolecular Res Ctr, Univ Mich, 79-80. *Mem:* AAAS; Am Chem Soc; Sigma Xi. *Res:* Synthetic macromolecules and simpler organic molecules with potential biological activity; polycarbosilanes, fluoroepoxies. *Mailing Add:* Dept Chem Tuskegee Univ Tuskegee AL 36088

LUDWICK, JOHN CALVIN, JR, b Berkeley, Calif, Apr 25, 22; m 50, 57. MARINE GEOLOGY. *Educ:* Univ Calif, Los Angeles, AB, 47; Scripps Inst Oceanog, MS, 49, PhD(oceanog), 51. *Prof Exp:* Res asst, Scripps Inst Oceanog, 49-50; sedimentationist, Gulf Res & Develop Co, Tex, 50, group leader, 51-52, party chief, 53-59, Pa, 59-61, sr res geologist, 61-63, sect suvpr, 63-68; SLOVER PROF OCEANOG, OLD DOMINION UNIV, 68- *Concurrent Pos:* Old Dominion Univ eminent scholar, 74- *Mem:* Soc Econ Paleont & Mineral; fel Geol Soc Am; Am Geol Inst; AAAS. *Res:* Marine sedimentation; mechanics of marine sediment transport; physical oceanography; environmental interpretation of ancient sediments; shoal construction; tidal current analysis; beach processes. *Mailing Add:* 8565-L Tidewater Dr Norfolk VA 23503

LUDWICK, LARRY MARTIN, b Jamestown, NY, Oct 15, 41; m 68; c 2. INORGANIC CHEMISTRY. *Educ:* Mt Union Col, BS, 63; Univ Melbourne, BSc, 65; Univ Ill, Urbana, MS, 67, PhD(inorg chem), 69. *Prof Exp:* Res chemist, PPG Industs, 65; asst prof, 69-73, assoc prof, 74-76, PROF CHEM, TUSKEGEE UNIV, 76- *Concurrent Pos:* NIH fel, Biophys Res Div, Univ Mich, 78-80. *Mem:* AAAS; Am Chem Soc; Sigma Xi; Nat Sci Teachers Asn. *Res:* Metal binding studies; copper and zinc binding constants using superoxide dismutase. *Mailing Add:* Dept Chem Tuskegee Inst Tuskegee AL 36088

LUDWICK, THOMAS MURRELL, b Cox's Creek, Ky, Aug 2, 15. ANIMAL PHYSIOLOGY. *Educ:* Eastern Ky Teachers Col, BS, 36; Univ Ky, MS, 39; Univ Minn, PhD(dairy sci, animal genetics), 42; Univ Chicago, dipl, 42; Univ Va, dipl, 43. *Prof Exp:* Dairy & tobacco farmer, Ky, 25-39; asst cattle breeding & physiol, Univ Minn, 39-42; asst prof dairy sci, Univ Ky, 46-48; from assoc prof to prof dairy sci, 48-83, EMER PROF DAIRY SCI, OHIO STATE UNIV, 83- *Concurrent Pos:* Teacher high sch, Ky, 37-38; dir, Ohio Regional Dairy Cattle Breeding Proj, 48-; instr meteorol & weather forecasting, US Air Force, 42-47. *Honors & Awards:* Am Dairy Sci Award, Ohio State Univ. *Mem:* Am Soc Animal Sci; Am Dairy Sci Asn. *Res:* Physiology of reproduction and milk secretion; artificial insemination; animal breeding and biometeorology. *Mailing Add:* 7614 Stafford Rd Greenfield OH 45123

LUDWIG, ALLEN CLARENCE, SR, b San Antonio, Tex, Nov 3, 38; m 60; c 4. ASPHALT-EMULSION TECHNOLOGY, PLASTICS TECHNIQUE. *Educ:* Tex A&M Univ, BS, 60. *Prof Exp:* Chem engr, Tech Serv Div, Monsanto Chem Co, 60; nuclear res chemist, ASAF, 60-63; res develop, Southwest Res Inst, 63-69, sr res engr, Systs Develop, Div Automotive Res, 69-74, Process Res & Eng Dept Vehicle & Traffic Safety, 74-82, Process Res & Eng, Dept Energy Conversion & Combustion Technol, Fuels & Lubricants Res Div, 82- 86; OWNER, ALLEN C LUDWIG, PE CONSULT, 86- *Honors & Awards:* IR 100 Award, 79. *Mem:* Am Chem Soc. *Res:* Sulfur product and process development have been principal areas of interest; numerous US and foreign patents; many technical publications. *Mailing Add:* 5914 Brenda Lane San Antonio TX 78240

LUDWIG, CHARLES HEBERLE, b Minneapolis, Minn, May 1, 20; m 56; c 2. WOOD CHEMISTRY. *Educ:* Macalester Col, BA, 42; Univ Wash, PhD(chem), 61. *Prof Exp:* Chemist, D A Dodd, Mfg Chemist, 47-55 & Univ Wash, 56-61; mem res staff, Ga Pac Corp, 61-82; RETIRED. *Mem:* Am Chem Soc; Sigma Xi. *Res:* Nuclear magnetic resonance spectroscopy of lignins and lignin models; chemistry of lignosulfonates and other lignins; nuclear magnetic resonance studies of lignin; model compounds and related materials; product development of lignosulfonates. *Mailing Add:* Waldron WA 98297

LUDWIG, CLAUS BERTHOLD, b Berlin, Ger, Nov 18, 24; m 54; c 2. MOLECULAR SPECTROSCOPY, ENVIRONMENTAL PHYSICS. *Educ:* Aachen Tech Univ, MS, 51, PhD(physics), 53. *Prof Exp:* Design analyst, Eng Dept, Int Harvester Corp, 53-58; sr staff scientist, Space Sci Lab, Gen Dynamics-Convair, 58-72; scientist, Sci Appln, Inc, 72-77; SR SCIENTIST & VPRES, PHOTON RES ASSOCS, 77- *Mem:* Am Phys Soc; Optical Soc Am; assoc fel Am Inst Aeronaut & Astronaut Engrs. *Res:* Molecular physics; high temperature molecular spectroscopy; infrared phenomena; radiative energy transfer; optical properties of small solid particles; guiding research and development in remote sensing of air pollution; development of air pollution monitors based on optical methods. *Mailing Add:* 5218 Cassandra Lane San Diego CA 92109

LUDWIG, DONALD A, b New York, NY, Nov 14, 33; m 53; c 2. MATHEMATICS, APPLIED MATHEMATICS. *Educ:* NY Univ, BA, 54, MS, 57, PhD(math), 59. *Prof Exp:* Res assoc math, Inst Math Sci, NY Univ, 59-60; Fine instr, Princeton Univ, 60-61; asst prof, Univ Calif, Berkeley, 61-64; from assoc prof to prof, NY Univ, 64-74; PROF MATH, UNIV BC, 74- *Concurrent Pos:* Guggenheim fel, Tel Aviv, Rehovot, Dundee, 70-71. *Mem:* Am Math Soc; Soc Indust & Appl Math; fel Royal Soc Can. *Res:* Partial differential equations; mathematical methods for population biology. *Mailing Add:* Dept Math Univ BC Vancouver BC V6T 1W5 Can

LUDWIG, EDWARD JAMES, b New York, NY, Apr 13, 37; m 58; c 4. NUCLEAR PHYSICS. *Educ:* Fordham Univ, BS, 58; Ind Univ, MS, 60, PhD(physics), 63. *Prof Exp:* Res fel physics, Rutgers Univ, 63-66; from asst prof to assoc prof, 66-76, PROF PHYSICS, UNIV NC, CHAPEL HILL, 76- *Concurrent Pos:* Prin investr, DOE res contract; vis prof, Univ Birmingham, 73, Lawrence Berkeley Lab, 80, Univ Munich, 89. *Mem:* Am Phys Soc. *Res:* Nuclear reactions and scattering cross sections and polarization effects; reaction mechanisms; resonance studies. *Mailing Add:* Dept of Physics Univ of NC Chapel Hill NC 27599-3255

LUDWIG, FRANK ARNO, b West Reading, Pa, Jan 17, 31; m; c 7. ELECTROCHEMICAL OR THERMAL REGENERATIVE FUEL CELLS. *Educ:* Calif Inst Technol, BS, 53; Case Western Reserve Univ, MS, 65, PhD(phys chem), 68. *Prof Exp:* Proj engr, Carter Labs, Inc, 53-56; res engr, Hughes Aircraft Co, 56-57; vpres, Tech Commun, Inc, 55-58; dept mgr fuel cells, thermogalvanics, Electro-Optical Systs, Inc, 58-62; dept mgr org electrolyte batteries, electrochem trace gas sensors, Whittaker Corp, 68-69; supvr, Res Lab, Ford Motor Co, 69-78; mgr, Near-Term Elec Vehicle Battery Contracts, Argonne Nat Lab, 78-79; prin engr corrosion, mat develop, Ford Aerospace & Commun Corp, 79-82; CHIEF SCIENTIST, MAT TECHNOL LAB, HUGHES AIRCRAFT CO, 82- *Honors & Awards:* Hughes Electro-optical & Data Syst Group Pat Award. *Mem:* Electrochem Soc; Am Chem Soc; Sigma Xi; Am Electroplaters & Surface Finishers Soc. *Res:* Materials, corrosion, chemical and electrochemical kinetics, ac impedance techniques; development of new batteries and fuel cells; electroanalytical device inventions; improvements in batteries for electric vehicles; energy storage and conversion device inventions; electrochemical sensors inventions; thermodynamics; electroplating and surface finishing innovations. *Mailing Add:* 29443 Whitley Collins Dr Ranch Palos Verdes CA 90274

LUDWIG, FREDERIC C, b Bad Nauheim, WGer, Jan 22, 24; US citizen; m 58; c 4. EXPERIMENTAL PATHOLOGY. *Educ:* Univ T bingen, MD, 49; Univ Paris, ScD(radiobiol), 58. *Prof Exp:* Sect chief radiation path, AEC, France, 55-59; assoc res pathologist, Med Ctr, Univ Calif, San Francisco, 58-62, lectr path, 62-65, assoc prof in residence, 65-71; PROF PATH & RADIOL SCI, COL MED, UNIV CALIF, IRVINE, 71- *Concurrent Pos:* Consult, Stanford Res Inst, 65-; vis fel, St John's Col, Cambridge. *Honors & Awards:* Award, Nat Inst Hyg, France, 59. *Mem:* Radiation Res Soc Am; Am Soc Exp Path; NY Acad Sci; Fr Asn Anat; Ger Path Soc. *Res:* Abscopal effects of radiation; radiation injury in blood forming organs; pathogenesis of radiation leukemia; homeostasis of white blood cells; gerontology. *Mailing Add:* Dept Path & Radiol Sci Univ Calif Col Med Irvine CA 92717

LUDWIG, FREDERICK JOHN, SR, b St Louis, Mo, June 20, 28; m 56; c 2. ANALYTICAL CHEMISTRY, ORGANIC CHEMISTRY. *Educ:* Washington Univ, AB, 50; St Louis Univ, PhD(chem), 53. *Prof Exp:* Lab asst chem, St Louis Univ, 50-53; res chemist, Uranium Div, Mallinckrodt Chem Corp, 55-59; group leader, Petrolite Corp, 59-73, RES SCIENTIST, TRETOLITE DIV, PETROLITE CORP, 73- *Mem:* Am Chem Soc; Sigma Xi. *Res:* Gas-liquid and liquid-solid chromatography; infrared spectroscopy; wax-polymers; water-treatment chemicals; nuclear magnetic resonance spectroscopy. *Mailing Add:* Res Lab Petrolite Corp 369 Marshall Ave St Louis MO 63119

LUDWIG, GARRY (GERHARD ADOLF), b Mannheim, Ger, Sept 4, 40; Can & German citizen; m 68. GENERAL RELATIVITY. *Educ:* Univ Toronto, BSc, 62; Brown Univ, PhD(physics), 66. *Prof Exp:* From asst prof to assoc prof, 66-82, PROF MATH, UNIV ALTA, 82- *Concurrent Pos:* Nat Res Coun grants, 67-93. *Mem:* Am Math Soc; Am Phys Soc; Can Math Soc;

Can Appl Math Soc; Int Soc Gen Relativity & Gravitation. *Res:* General relativity and gravitation; asymptotically flat spacetimes, H-space, exact solutions, spin-coefficient formalism. *Mailing Add:* Dept of Math Univ of Alta Edmonton AB T6G 2G1 Can

LUDWIG, GEORGE H, b Johnson Co, Iowa, Nov 13, 27; m 50; c 4. SPACE SYSTEMS DESIGN, SPACE SCIENCES. *Educ:* Univ Iowa, BA, 56, MS, 59, PhD(elec eng), 60. *Prof Exp:* Res assoc space res, Univ Iowa, 60; head instrumentation sect, Goddard Space Flight Ctr, NASA, 60-65, chief info processing div, 65-71, assoc dir data opers, 71-72; dir systs integration, Nat Earth Satellite Serv, Nat Oceanic & Atmospheric Admin, 72-75, dir opers, 75-80, tech dir, 80, sr scientist, Environ Res Labs, 80-81, dir, 81-83; asst chief scientist, NASA, 83-84; SR RES ASSOC, UNIV COLO, 85-; VIS SR SCIENTIST, CALIF INST TECHNOL, 89- *Concurrent Pos:* Consult, space systems, 84- *Honors & Awards:* Golden Plate Award, Acad Achievement, 62; NOAA Prog & Mgt Award, Nat Oceanic & Atmospheric Admin, 77; Except Sci Achievement Medal, NASA, 84. *Mem:* Am Meteorol Soc; Am Geophys Union; Inst Elec & Electronic Engrs; AAAS. *Res:* Cosmic rays; development of space instrumentation; on board and ground data processing; co-discovery and investigation of Van Allen radiation belts; atmospheric, oceanic, hydrologic remote sensing and forecasting; direction of space, atmospheric and oceanic environmental research. *Mailing Add:* HC33 Box 641 Winchester VA 22601

LUDWIG, GERALD W, b New York, NY, Jan 7, 30; m 51; c 3. POWER ELECTRONICS, SIMULATION. *Educ:* Harvard Univ, AB, 50, AM, 51, PhD(chem physics), 55; Rensselaer Polytech Inst, MA, 87. *Prof Exp:* Physicist, Res & Develop Ctr, Gen Elec Corp, 55-63, liaison scientist, 63-65, physicist, 65-71, mgr, Integrated Circuits Br, 71-74. *Mem:* Fel Am Phys Soc; sr mem Inst Elec & Electronics Engrs; Electrochem Soc. *Res:* Transport properties of semiconductors; electron paramagnetic resonance; Gunn effect; x-ray and cathode ray phosphors; semiconductor materials and processing; charge transfer devices; integrated circuits; modelling and simulation of power electronic circuits and systems. *Mailing Add:* Res & Develop Ctr Gen Elec Co PO Box 8 Schenectady NY 12301

LUDWIG, HARVEY F, b Saskatoon, Sask, Can, Dec 4, 16. ENVIRONMENTAL & SANITARY ENGINEERING. *Educ:* Univ Calif, Berkeley, BS, 38, MS, 42; Clemsen Univ, DEng, 65. *Prof Exp:* Sanit engr, USPHS, 43-46; assoc prof eng, Univ Calif, Berkeley, 49-51; chmn & pres, Eng Sci, Inc, 56-72; PRES SOUTHEAST ASIA TECHNOL, BANGKOK, THAILAND, 73- *Mem:* Nat Acad Eng; Sigma Xi. *Mailing Add:* 43 Alston Pl Santa Barbara CA 93108

LUDWIG, HOWARD C, b Beaver Falls, Pa, July 31, 16; m 41; c 1. CHEMICAL PHYSICS, PLASMA PHYSICS. *Educ:* Geneva Col, BS, 41. *Prof Exp:* Chem analyst, Armstrong Cork Co, 41-42; spectroscopist, Propeller Div, Curtiss-Wright Corp, 42-46; res engr, Res Labs, Westinghouse Elec Corp, 46-59, fel scientist, Res & Develop Ctr, 59-76; consult plasma physics, 76-80; RETIRED. *Honors & Awards:* IR 100 Award, 63; Lincoln Gold Medal, Am Welding Soc, 56. *Res:* Research and development of high and low pressure plasmas. *Mailing Add:* 159 Roberta Dr Pittsburgh PA 15221

LUDWIG, HUBERT JOSEPH, b Lincoln, Ill, July 27, 34; m 65; c 2. MATHEMATICS. *Educ:* Univ Ill, Urbana, BS, 56; St Louis Univ, MS, 64, PhD(math), 68. *Prof Exp:* Instr math, chem & eng mech, Springfield Col, Ill, 56-65; teaching asst math, St Louis Univ, 65-68; from asst prof to assoc prof, 68-81, PROF MATH, BALL STATE UNIV, 81- *Mem:* Math Asn Am; Nat Coun Teachers Math. *Res:* Chaotic dynamics; 2-metric spaces; logo and fractals. *Mailing Add:* 3209 W Twickingham Dr Muncie IN 47304

LUDWIG, JOHN HOWARD, b Burlington, Vt, Mar 7, 13; m 46; c 2. ENVIRONMENTAL SCIENCES. *Educ:* Univ Calif, Berkeley, BS, 34; Univ Colo, MS, 41; Harvard Univ, MS, 56, ScD(indust hyg), 58; Environ Eng Intersoc Bd, dipl. *Prof Exp:* Design engr, US Bur Reclamation, Colo, 36-39; design engr, Corps Engrs, Ore, 39-43, chief dams design sect, Calif, 46-48; consult engr, Ludwig Bros, Engrs, 48-51; chief tech opers br, Div Civilian Health Requirements, USPHS, 51-53, spec asst to chief, Div Water Supply & Pollution Control, 53-55 & Div Air Pollution, 55-62, chief lab eng & phys sci, 62-67; assoc dir control tech res & develop, Nat Ctr Air Pollution Control, 67-68; assoc comnr, Nat Air Pollution Control Admin, 68-70; dir tech coordr, Off Air Progs, Environ Protection Agency, 70-72; CONSULT, 72- *Concurrent Pos:* Vis lectr, Harvard Univ, 64-74; mem expert adv panel air pollution, WHO, 65-72; US deleg, Orgn Econ Coop & Develop & Econ Comn Europe, 68-72. *Mem:* Nat Acad Eng; AAAS. *Res:* Air pollution in physical sciences and engineering controls. *Mailing Add:* 43 Alston Pl Santa Barbara CA 93108

LUDWIG, MARTHA LOUISE, b Pittsburgh, Pa, Aug 16, 31; m 61. BIOCHEMISTRY. *Educ:* Cornell Univ, BA, 52, PhD(biochem), 56; Univ Calif, Berkeley, MA, 55. *Prof Exp:* Res fel biochem, Harvard Med Sch, 56-59; res assoc biol, Mass Inst Technol, 59-62; res fel chem, Harvard Univ, 62-67; from asst prof to assoc prof, 67-75, PROF BIOL CHEM, UNIV MICH, ANN ARBOR, & RES BIOPHYSICIST, BIOPHYS RES DIV, 75- *Mem:* Am Chem Soc; Am Soc Biol Chemists; Biophys Soc; Am Crystallog Asn. *Res:* Protein crystallography; protein structure and function. *Mailing Add:* 3455 Woodland Rd Ann Arbor MI 48104-4257

LUDWIG, OLIVER GEORGE, b Philadelphia, Pa, Nov 15, 35. PHYSICAL CHEMISTRY. *Educ:* Villanova Univ, BS, 57; Carnegie Inst Technol, MS, 60, PhD(quantum chem), 61. *Prof Exp:* Mem math lab & sr res worker theoret chem, Cambridge Univ, 61-63; asst prof chem & fac assoc, Comput Ctr, Univ Notre Dame, 63-68; ASSOC PROF CHEM, VILLANOVA UNIV, 68- *Concurrent Pos:* NSF fel, 61-63; actg chmn, Dept Chem, Villanova Univ, 69-70. *Mem:* Am Chem Soc; Am Phys Soc; Asn Comput Mach. *Res:* Quantum chemistry; chemical applications of digital computers; development of methods for scientific computing. *Mailing Add:* Dept of Chem Villanova Univ Villanova PA 19805

LUDWIG, THEODORE FREDERICK, b Castlewood, SDak, July 8, 24; m 45; c 1. PROSTHODONTICS. *Educ:* Cent Col, Iowa, AB, 45; Ohio State DDS, 59, MSc, 63. *Prof Exp:* Asst prof dent, Sch Dent, WVa Univ, 63-67; asst prof, Sch Dent, Univ Iowa, 67-69; ASSOC PROF PROSTHODONTICS, COL DENT MED, MED UNIV SC, 69- *Concurrent Pos:* NIH grant, 62-63; mem, Carl O Boucher Prosthodontic Conf, 66- *Mem:* Am Dent Asn. *Res:* Esthetics in complete dentures; design and metals in removable partial dentures. *Mailing Add:* Dept Prosthodontics Col Dent Med Univ SC 171 Ashley Ave Charleston SC 29425

LUDWIN, ISADORE, b Malden, Mass, Feb 23, 15; m 49; c 5. ANIMAL GENETICS, PHYSIOLOGY. *Educ:* Univ Mass, BS, 37; Univ Wis, MS, 39; Harvard Univ, PhD(genetics), 48. *Prof Exp:* Statist analyst, US Dept Navy, DC, 41-42; asst prof biol, Univ Mass, 46-48; fel, Tufts Col, 48; assoc cancer biologist, Roswell Park Mem Inst, 49-51; prof sci, Calvin Coolidge Col, 57-62; prof biol, Cambridge Jr Col, 57-67; lectr, Northeastern Univ, 67-68; pvt res & develop, 68-79; RETIRED. *Mem:* AAAS; Asn Advan Med Instrumentation. *Mailing Add:* 1073 Centre St Newton MA 02159

LUEBBE, RAY HENRY, JR, b Schenectady, NY, Mar 31, 31; m 59; c 3. PHYSICAL CHEMISTRY. *Educ:* Dartmouth Col, AB, 53; Univ Wis, PhD(phys chem), 58. *Prof Exp:* Asst phys chem, Univ Wis, 53-55; chemist, Photo Prod Dept, E I du Pont de Nemours & Co, 58-64; scientist, Xerox Corp, 64-79; unit mgr, Qwip Systs, Exxon Enterprises, 79-82; dir process technol, Environ Technol Inc, 82-85; Sr process develop engr, Harris Graphics Co, 85-86, MGR, ELECTROPHOTOGRAPHIC MAT & PROCESS DEVELOP, AM GRAPHICS, AM INT, 86- *Mem:* Am Chem Soc; Soc Photo Sci & Eng. *Res:* Hot atom and photo chemistry; photographic science; photopolymerization; electrophotography. *Mailing Add:* 305 Blackstone Dr Centerville OH 45459

LUEBBERS, RALPH H(ENRY), b Burlington, Iowa, Mar 24, 06; m 35; c 3. CHEMICAL ENGINEERING. *Educ:* Iowa State Col, BS, 27, MS, 32, PhD(chem eng, sanit bact), 35. *Prof Exp:* Plant chemist, Universal Gypsum Co, Iowa, 27; plant chemist & chem engr, Des Moines Water Works, 28; jr engr, Int Combustion Eng Corp, NY, 28-29; develop engr, Dorr Co, Inc, 29-31; chem & sanit engr, US Army Dept, Kans, 35-37; from instr to prof, 38-72, EMER PROF CHEM ENG, UNIV MO-COLUMBIA, 72- *Concurrent Pos:* Eng consult. *Mem:* Am Chem Soc; Am Soc Eng Educ; Am Water Works Asn; Am Inst Chem Eng; Nat Soc Prof Engrs; Sigma Xi. *Res:* Mixing of dry powders and liquids; heat transfer in packed columns; biological oxidation processes; fluid flow of suspensions. *Mailing Add:* 532 W Rte K Columbia MO 65203

LUEBKE, EMMETH AUGUST, b Manitowoc, Wis, Aug 1, 15; m 37; c 2. PHYSICS. *Educ:* Ripon Col, BA, 36; Univ Ill, PhD(physics), 41. *Prof Exp:* Asst physics, Univ Ill, 36-41; group leader, Radiation Lab, Mass Inst Technol, 41-45; res assoc, Res Lab, Gen Elec Co, 45-50, mgr reactor eval, Knolls Atomic Power Lab, 50-55, gen physicist, Missile & Space Vehicle Dept, 55-58 & Gen Eng Lab, 58-63, physicist, Tempo, 63-72; admin judge, US Nuclear Regulatory Comn, 72-87; RETIRED. *Concurrent Pos:* Mem, Joint Liquid Metals Comt, US Navy AEC, 50-55; presiding tech mem, Atomic Safety & Licensing Bd. *Honors & Awards:* Presidential Cert Merit. *Mem:* Fel Am Phys Soc; Am Nuclear Soc. *Res:* Linear accelerator; velocity spectrometer measurement of neutron cross section; microwave radar components; system design; liquid metal heat transfer; design and evaluation of reactor power plants; breeder; submarine propulsion; central station types; environmental controls. *Mailing Add:* 5500 Friendship Blvd Apt 1923 N Chevy Chase MD 20815

LUEBS, RALPH EDWARD, b Wood River, Nebr, Mar 21, 22; m 51; c 4. SOIL & WATER MANAGEMENT, SOIL FERTILITY. *Educ:* Univ Nebr, BS, 48, MS, 52; Iowa State Univ, PhD(soil fertil), 54. *Prof Exp:* Asst agron, Univ Nebr, 48-49; soil scientist, Agr Res Serv, Univ Nebr, USDA, 55-56, Ft Hays Exp Sta, Kans, 56-59 & Univ Calif, Riverside, 59-75; chief, agron div, Woodward-Clyde Consults, 75-81, sr proj scientist, Environ Systs Div, 81-82. *Concurrent Pos:* Int consult agronomist, 82- *Mem:* Am Soc Agron; Soil Sci Soc Am; Sigma Xi; Soil Conserv Soc Am. *Res:* Nitrogen availability and rainfall use efficiency for dryland crops; mined land reclamation; diagnosis of low crop production. *Mailing Add:* 13347 W Exposition Dr Lakewood CO 80228

LUECK, CHARLES HENRY, b St Paul, Minn, Oct 1, 28; m 55; c 6. ANALYTICAL CHEMISTRY. *Educ:* Col St Thomas, BS, 50; Univ Detroit, MS, 53; Wayne State Univ, PhD, 56. *Prof Exp:* Res assoc & anal res supvr, Textile Fibers Dept, E I du Pont de Nemours & Co, 56-85; CHEM INSTR, BEAUFORT COMMUNITY COL, WASH, NC. *Mem:* Am Chem Soc; Am Soc Qual Control. *Res:* Spectrophotometric analysis; chemical degradation studies; quality systems; test method uniformity & control. *Mailing Add:* PO Box 789 Chocowinity NC 27817

LUECKE, DONALD H, b St Paul, Minn, Aug 29, 36; m; c 2. MEDICAL RESEARCH. *Educ:* Macalester Col, BA, 59; Univ Ill, MS, 61; Univ NDak, BS, 73; Mich State Univ, MD, 75. *Prof Exp:* Teaching asst, Dept Microbiol, Med Ctr, Univ Ill, 59-62; microbiologist, Dept Health, NDak State, 62-64; virologist-res supvr, Dept Microbiol, Univ NDak, 65-69; virologist-dep proj dir, Spec Virus Cancer Prog, Nat Cancer Inst, 69-73; grad teaching & res asst, Dept Human Develop, Col Human Med, Mich State Univ, 74-75; med officer & head, Clin Studies Sect, Collab Res Br, Viral Oncol Prog, Div Cancer Cause & Prev, Nat Cancer Inst, NIH, 76, med officer, Prog Admin & actg head, Physiol Sci Sect, Physiol & BiomedEng Prog, Nat Inst Gen Med Sci, 77-78, med officer & chief, Spec Prog Br, Div Cancer Cause & Prev, Nat Cancer Inst, 79-81, med officer & dep dir, Stroke & Trauma Prog, Nat Inst Neurol & Commun Dis & Stroke, 81-82, med officer & dep dir, Extramural Activ Prog, 82-87, MED OFFICER & DEP DIR, DIV RES GRANTS, NIH, 87- *Mem:* Am Soc Microbiol; AMA; NY Acad Sci. *Res:* Medicine; cancer. *Mailing Add:* Div Res Grants Off Dir Westwood Bldg Rm 448 5333 Westbard Ave Bethesda MD 20892

LUECKE, GLENN RICHARD, b Bryan, Tex, May 19, 44; m 67; c 2. MATHEMATICAL ANALYSIS. *Educ:* Mich State Univ, BS, 66; Calif Inst Technol, PhD(math), 70. *Prof Exp:* Asst prof, 69-74, assoc prof, 74-80, PROF MATH, IOWA STATE UNIV, 80- *Mem:* Am Math Soc; Soc Indust & Appl Math; Asn Comput Mach. *Res:* Numerical solution of integral equations. *Mailing Add:* Comput Ctr Iowa State Univ Ames IA 50011

LUECKE, RICHARD H, b Cincinnati, Ohio, Mar 27, 30; m 53; c 5. CHEMICAL ENGINEERING, MATHEMATICAL MODELLING. *Educ:* Univ Cincinnati, BChE, 53; Univ Okla, MChE, 63, PhD(chem eng), 66. *Prof Exp:* Engr, E I du Pont de Nemours & Co, Inc, 53-62; res engr, Monsanto Co, 66-67; assoc prof, 67-80, PROF CHEM ENG, UNIV MO-COLUMBIA, 80- *Concurrent Pos:* Consult, Chemshare Corp, Okla, 69- *Mem:* Am Inst Chem Eng. *Res:* Process control; optimization; mathematical methods; bioengineering. *Mailing Add:* Chem Engr 1030 Engr Bldg Univ Mo 1030 Engr Bldg Columbia MO 65211

LUECKE, RICHARD WILLIAM, b St Paul, Minn, July 12, 17; m 41; c 3. BIOCHEMISTRY, NUTRITION. *Educ:* Macalester Col, BA, 39; Univ Minn, MS, 41, PhD(biochem), 43. *Prof Exp:* prof biochem, Mich State Univ, 45-87; RETIRED. *Concurrent Pos:* Assoc prof biochem, Tex A&M Univ, 43-45; consult, Armour Res Labs, Chicago, 55-66; mem comt on animal nutrit, Nat Res Coun, 55-65; mem food & nutrit bd, Food & Agr Orgn, UN, 60-65; consult, Merck Sharp & Dohme Res Labs, 62-69. *Mem:* Am Chem Soc; Am Inst Nutrit; Am Soc Biol Chemists. *Res:* Trace element metabolism in animals. *Mailing Add:* 1893 Birchwood Dr Okemos MI 48864-2766

LUEDECKE, LLOYD O, b Hamilton, Mont, July 28, 34; m 57; c 2. DAIRY BACTERIOLOGY. *Educ:* Mont State Col, BS, 56; Mich State Univ, MS, 58, PhD(food sci), 62. *Prof Exp:* Asst prof dairy sci, 62-70, assoc prof & assoc dairy scientist, 70-73, assoc prof, 73-77, PROF FOOD SCI, WASH STATE UNIV, 77- *Mem:* Am Dairy Sci Asn; Inst Food Technol. *Res:* Heat resistance of psychrophiles; bacteriological aspects of mastitis. *Mailing Add:* Food Sci Wash State Univ Pullman WA 99164-6330

LUEDEMAN, JOHN KEITH, b Ft Wayne, Ind, Apr 27, 41; div; c 4. ALGEBRA, MATHEMATICS EDUCATION. *Educ:* Valparaiso Univ, BA, 63; Southern Ill Univ, Carbondale, MA, 65; State Univ NY, Buffalo, PhD(math), 69. *Prof Exp:* Instr math, State Univ NY, Buffalo, 67-68; from asst prof to assoc prof, 68-80, PROF MATH, CLEMSON UNIV, 80-, PROF EDUC, 88- *Concurrent Pos:* Consult math, Oconee County Sch Syst, SC, 74-; dir, Ctr Ex Math Sci Educ, Clemson Univ, 84- *Mem:* Am Math Soc; Math Asn Am; Sigma Xi. *Res:* Mathematics education; mathematical biology; semigroups; graph theory; computing on graphs. *Mailing Add:* Dept Math Clemson Univ 0-14 Martin Hall Clemson SC 29634-1907

LUEDER, ERNST H, b Schiltach, Ger, Feb 20, 32; c 2. OPTIMIZATION OF SYSTEMS, REALIZATION OF FLAT PANEL DISPLAYS. *Educ:* Univ Stuttgart, Dipl Ing, 58, Dr Ing(elec commun), 62, Dr Ing habil, 66. *Prof Exp:* Mem tech staff, Bell Tel Labs, NJ, 68-71; privat-dozent, 66-68, FULL PROF & DIR, INST NETWORK & SYSTS THEORY, UNIV STUTTGART, 71- *Concurrent Pos:* Consult, Fed Ministry Sci & Technol, Bonn & Ger Res Commun, Bonn, 72-; chmn sci adv group, Heinrich-Hertz-Inst, Berlin, 84-87, mem sci adv group, 90- *Mem:* Fel Inst Elec & Electronic Engrs; Soc Info Display; Int Soc Hybrid Microelectronics; Int Soc Optical Eng. *Res:* Design of passive, re-active, switched capacitor, digital and saw filters; digital and optical signal processing; optimization of systems; realization of flat panel displays. *Mailing Add:* Univ Stuttgart Pfaffenwaldring 47 7000 Stuttgart 80 Germany

LUEG, RUSSELL E, b Chicago, Ill, Nov 24, 29; m 56; c 5. ELECTRICAL ENGINEERING. *Educ:* Univ Ark, BS, 51; Univ Tex, MS, 56, PhD(elec eng), 61. *Prof Exp:* Prog engr, Gen Elec Co, NY, 53-54; radio engr & instr elec eng, Univ Tex, 54-60; assoc prof, 60-64, actg head dept, 66-68, PROF ELEC ENG, UNIV ALA, 64-, ASSOC DEAN ADMIN, 88- *Concurrent Pos:* Consult, Army Missile Command, Ala, 64-65. *Mem:* Inst Elec & Electronics Engrs; Am Soc Eng Educ. *Res:* Nonlinear control systems. *Mailing Add:* Box 870200 Tuscaloosa AL 35487-0286

LUE-HING, CECIL, b Jamaica, WI, Nov 3, 30; m 52; c 2. CIVIL & ENVIRONMENTAL ENGINEERING. *Educ:* Marquette Univ, BCE, 61; Case Inst Technol, MS, 63; Washington Univ, St Louis, DSc(sanit eng), 66. *Prof Exp:* Chief technician, Col Med, Univ WI, 50-55; instr histol & cytol chem & lab supvr, Sch Med Technol, Mt Sinai Hosp, Wis, 55-61; res assoc clin biochem, Huron Rd Hosp, Ohio, 61-63; res assoc environ eng, Washington Univ, St Louis, 63-65, asst prof, 65-66; assoc, Ryckman, Edgerley, Tomlinson & Assocs, 66-68, sr assoc, 68-77; DIR RES & DEVELOP, METROP SANIT DIST, CHICAGO, 77- *Concurrent Pos:* Fel, Washington Univ, Mo. *Mem:* AAAS; Am Soc Civil Engrs; Am Pub Health Asn; Water Pollution Control Fedn; Am Water Works Asn. *Res:* Pesticide pollution of water supplies; significance of enzyme response in pesticide detection in water supplies; phosphorus and nutrient removal from water supplies; industrial wastes detoxification and biodegradation. *Mailing Add:* Water Reclaimation Dist 100 E Erie St Chicago IL 60611

LUEHR, CHARLES POLING, b Plentywood, Mont, Sept 27, 30. APPLIED MATHEMATICS, MATHEMATICAL PHYSICS. *Educ:* Ore State Univ, BS, 53, MS, 56; Univ Calif, Berkeley, PhD(appl math), 62. *Prof Exp:* Mem prof staff, Gen Elec Co, Tempo, Calif, 62-68; fel, Univ Fla 68-70, from asst prof to assoc prof math, 70-82; SR RES SCIENTIST, NMEX ENG RES INST, UNIV NMEX, 85- *Concurrent Pos:* Res visitor, Inst Nuclear Sci, Nat Univ Mex, 73-; vis assoc prof math, Ore State Univ, 80-82; res scholar, Air Force Weapons Lab, Kirtland AFB, 83-84; intergovt personnel act, Phillips Lab, Kirtland AFB, 90- *Mem:* Math Asn Am; Am Math Soc; Am Phys Soc; Soc Indust & Appl Math; Sigma Xi. *Res:* Methods of mathematical physics; tensor analysis; abstract theory of spinors with applications in quantum mechanics and relativity theory; modern differential geometry applied to general relativity; scientific programming. *Mailing Add:* 920 Continental Lp SE Apt 27 Albuquerque NM 87108

LUEHRMANN, ARTHUR WILLETT, JR, b New Orleans, La, Mar 8, 31; m 61; c 2. COMPUTER SCIENCE, SCIENCE EDUCATION. *Educ:* Univ Chicago, AB, 55, SB, 57, SM, 61, PhD(physics), 66. *Prof Exp:* From instr to asst prof, 65-70, adj assoc prof physics & dir, Off Acad Comput, Dartmouth Col, 70-77; assoc dir, Lawrence Hall Sci, Univ Calif, Berkeley, 77-80; PARTNER, COMPUTER LITERACY, 80- *Concurrent Pos:* Consult, NSF Off Comput Activities, 68-, res grant, 69-78. *Honors & Awards:* Fulbright lect, Fulbright Comn, Colombia, 69. *Mem:* Am Asn Physics Teachers; Am Phys Soc; Asn Computing Machines; Inst Elec & Electronics Engrs. *Res:* Solid state theory; band structure; computational physics; computer graphics; computer-based instruction; solid state physics. *Mailing Add:* Computer Literacy 1466 Grizzly Peak Blvd Berkeley CA 94708

LUEHRS, DEAN C, b Fremont, Nebr, Apr 20, 39; m 69; c 1. INORGANIC CHEMISTRY. *Educ:* Mich State Univ, BS, 61; Univ Kans, PhD(chem), 65. *Prof Exp:* Asst prof, 65-69, ASSOC PROF CHEM, MICH STATE TECHNOL UNIV, 69- *Mem:* Am Chem Soc. *Res:* Nonaqueous solvents; electrochemistry. *Mailing Add:* Dept Chem Mich Technol Univ Houghton MI 49931

LUEKING, DONALD ROBERT, b Cincinnati, Ohio, Nov 24, 46; m 73. MICROBIAL BIOCHEMISTRY. *Educ:* Ind Univ, Bloomington, BS, 69, PhD(microbiol), 73. *Prof Exp:* Trainee microbiol, Univ Pa, 73-74, fel, 74-75; fel microbiol, Univ Ill, Urbana, 75-78; ASST PROF MICROBIOL, TEX A&M UNIV, 78-, assoc prof Dept Biol Sci Mich Technol Univ. *Mem:* Am Soc Microbiol; AAAS; Sigma Xi. *Res:* The use of the photosynthetic bacteria as a model system for the study of the factors involved in the regulation of membrane biosynthesis and differentiation. *Mailing Add:* Dept Biol Sci Mich Technol Univ Houghton MI 49931

LUENBERGER, DAVID GILBERT, b Los Angeles, Calif, Sept 16, 37; m 62; c 4. SYSTEMS ENGINEERING. *Educ:* Calif Inst Technol, BS, 59; Stanford Univ, MS, 61, PhD(elec eng), 63. *Prof Exp:* Engr, Westinghouse Elec Corp, 61-63; asst prof elec eng, 63-67, assoc prof eng-econ systs & elec eng, 67-71, PROF ENG-ECON SYSTS & ELEC ENG, STANFORD UNIV, 71-, CHMN, ENG-ECON SYSTS, 80- *Concurrent Pos:* Consult, Stanford Res Inst, 66-, Intasa, Inc, 70-72, Systs Control Inc, 74-83, Time & Space Processing, 81 & Optimization Technol Inc, 83-; tech asst to dir, Off Sci & Technol, Exec Off of the President, 71-72; vis prof, Mass Inst Technol, 76 & Tech Univ Denmark, 86. *Honors & Awards:* Hendrik W Bode Lectr Prize, Control Systs Soc, Inst Elec & Electronic Engr, 90. *Mem:* Inst Mgt Sci; Am Asn Univ Profs; Inst Elec & Electronics Engrs; Econometric Soc; Am Soc Eng Educ; Am Finance Asn; Soc Econ Dynamics & Control (pres, 87-); Soc Prom Econ Theory; Math Prog Soc; Soc Advan Econ Theory. *Res:* Control systems, particularly multivariable systems; optimization, including control, operations research and estimation; economic systems; finance. *Mailing Add:* Dept Eng-Econ Systs Terman Eng Ctr 306 Stanford Univ Stanford CA 94305

LUER, CARL A, b St Louis, Mo, Nov 9, 48. BIOCHEMISTRY, IMMUNOLOGY. *Educ:* Duke Univ, BA, 70; Univ SFla, MS, 74; Univ Kans, PhD(biochem), 78. *Prof Exp:* Staff scientist, 79-85, SR SCIENTIST, MOTE MARINE LAB, 85- *Concurrent Pos:* Adj asst prof, Dept Med, Brown Univ, 91- *Mem:* Am Soc Biochem & Molecular Biol; Sigma Xi. *Mailing Add:* Mote Marine Lab 1600 Thompson Pkwy Sarasota FL 34236

LUERSSEN, FRANK W, b Reading, Pa, Aug 14, 27; m 50; c 5. METALLURGY, PHYSICAL CHEMISTRY. *Educ:* Pa State Univ, BS, 50; Lehigh Univ, MS, 51. *Hon Degrees:* LLD, Calumet Col. *Prof Exp:* Jr res engr, Bethlehem Steel Corp, 51-52; metallurgist, Inland Steel Co, 52-54, chief reduction & ref, 54-57, chief res engr, 57-61, asst mgr, Res Dept, 62-63, assoc mgr, 63-64, mgr, 64-68, vpres res, 68-77, vpres steel mfg, 77-78, exec vpres & pres, 78-82, pres & chief exec officer, 82-83, CHMN & CHIEF EXEC OFFICER, INLAND STEEL CO, 83- *Concurrent Pos:* Trustee, Northwestern Univ; chmn, Phys Chem Steelmaking Group, Am Iron & Steel Inst, Gen Res Comt & Comt Mfg; mem, Indust Policy Adv Comt, US Dept Com, 88 & bd trustees, Mus Sci & Indust. *Honors & Awards:* Howe Mem Lectr, Am Inst Mining, Metall & Petrol Engrs, 84, Benjamin Fairless Award, 85. *Mem:* Nat Acad Eng; Brit Inst Metals; Am Iron & Steel Inst; fel Am Soc Metals; Am Inst Mining, Metall & Petrol Engrs. *Res:* Physical chemistry of slag metal systems in steel refining; process research in ironmaking and steelmaking; physical metallurgy of iron base alloy systems. *Mailing Add:* Inland Steel Industs Inc 30 W Monroe St Chicago IL 60603

LUESCHEN, WILLIAM EVERETT, b Springfield, Ill, Jan 29, 42; m 65; c 2. AGRONOMY. *Educ:* Southern Ill Univ, BS, 64; Univ Ill, MS, 66, PhD(agron), 68. *Prof Exp:* PROF AGRON & AGRONOMIST, SOUTHERN EXP STA, UNIV MINN, 68- *Mem:* Am Soc Agron; Crop Sci Soc Am; Weed Sci Soc Am; Coun Agr Sci & Tech; Am Reg Cert Prof Agron Crops & Soils; Am Forage & Grassland Coun. *Res:* Crop production, management and weed science. *Mailing Add:* Southern Exp Sta Univ of Minn Waseca MN 56093

LUESSENHOP, ALFRED JOHN, b Chicago, Ill, Feb 6, 26; m 52; c 4. MEDICINE, NEUROSURGERY. *Educ:* Yale Univ, BS, 49; Harvard Med Sch, MD, 52. *Prof Exp:* Intern surg, Univ Chicago, 52-53; resident neurosurg, Mass Gen Hosp, 53-58; vis scientist, Nat Inst Neurol Dis & Blindness, 59-60; from instr to assoc prof neurosurg, 60-73, PROF SURG, SCH MED, GEORGETOWN UNIV, 73-, CHIEF DIV NEUROSURG, 65- *Concurrent Pos:* Teaching fel, Harvard Med Sch, 57-58; res fel neurosurg, Harvard Med Sch, 53-54; res consult, Nat Inst Neurol Dis & Stroke, 60-65, clin consult, 65-; clin consult, Vet Admin Hosp, 65-; consult, Fed Aviation Agency, 67 & Nat Naval Med Ctr, 67- *Mem:* Cong Neurol Surg; Am Asn Neurol Surg; Soc Neurol Surg; Am Acad Neurosurg. *Res:* Cerebrovascular disease. *Mailing Add:* Georgetown Univ Hosp 3800 Reservoir Rd Washington DC 20007

LUETZELSCHWAB, JOHN WILLIAM, b Hammond, Ind, Sept 8, 40; m 63; c 2. HEALTH PHYSICS. *Educ:* Earlham Col, AB, 62; Wash Univ, MA & PhD(physics), 68; cert Am Bd Health Physics. *Prof Exp:* From asst prof to assoc prof, 68-83, PROF PHYSICS, DICKINSON COL, 83- *Mem:* Am Asn Physics Teachers; Health Physics Soc. *Res:* Environmental radioactivity; radon in the environment and in homes. *Mailing Add:* Dept Physics & Astron Dickinson Col Carlisle PA 17013-2896

LUFKIN, DANIEL HARLOW, b Philadelphia, Pa, Sept 26, 30; m 51; c 3. SOLAR PHYSICS. *Educ:* Mass Inst Technol, BS, 52, MS, 58; Univ Stockholm, Fil lic meteorol, 64. *Prof Exp:* Meteorol officer, Air Weather Serv, US Air Force, DC, 53-69, dir solar forecast facility, 69-73; asst prof astron, Hood Col, 75-87; MGR, INSTRUMENT DIV, FAIRCHILD SPACE CO, 88- *Concurrent Pos:* Dir, Off Systs & Advan Technol, Nat Oceanic & Atmospheric Admin, 76-86; consult, Solar Energy Sci Serv,74-87. *Mem:* Am Meteorol Soc. *Res:* Optical instrumentation for satellite remote sensing. *Mailing Add:* 303 W College Terr Frederick MD 21701

LUFT, HAROLD STEPHEN, b Newark, NJ, Jan 6, 47; m 70; c 2. HEALTH ECONOMICS, HEALTH SERVICES RESEARCH. *Educ:* Harvard Univ, AB, 68, MA, 70, PhD(econ), 73. *Prof Exp:* Instr, econ, Tufts Univ, 72-73; asst prof, health econ, Stanford Univ, 73-78; assoc prof, 78-82, PROF HEALTH ECON, INST HEALTH POLICY STUDIES, UNIV CALIF, SAN FRANCISCO, 82-, ASSOC DIR, 86- *Concurrent Pos:* Postdoctoral fel, Harvard Ctr Community Health & Med Care, Harvard Univ, 72-73; mem, Health Servs Study Sect, Nat Ctr Health Scis Res, 81-83; mem, Comt Design Strategy Qual Rev & Assurance Medicare, Inst Med, 88-90, Health Adv Comt, US Gen Acct Off, 88- & coun, Inst Med, 89-; res assoc, Nat Bur Econ Res, 82-; Flinn distinguished scholar healthcare mgt & policy, Univ Ariz & Ariz State Univ, 86; co-prin investr, Nat Ctr Health Serv Res, 86-91, prin investr, Agency Health Care Policy Res, 90-92 & 90-95; fel, Ctr Advan Study Behav Sci, 88-89; consult numerous orgns. *Honors & Awards:* Distinguished Article of the Year Award, Asn Health Servs Res, 88. *Mem:* Inst Med Nat Acad Sci; Am Pub Health Asn; Asn Social Scis Health; Asn Health Servs Res; Am Econ Asn. *Res:* Topics in health economics: health maintenance organizations, biased selection in health insurance, hospital competition, relation between volume and outcome in hospital care, applicability of incentive systems in health care; author of numerous publications. *Mailing Add:* Inst Health Policy Studies Sch Med Univ Calif 1388 Sutter St 11th Floor Palo Alto CA 94301

LUFT, JOHN HERMAN, b Portland, Ore, Feb 6, 27; m 49; c 3. HISTOLOGY. *Educ:* Univ Wash, BS, 49, MD, 53. *Prof Exp:* Intern, Peter Bent Brigham Hosp, Boston, Mass, 53-54; from asst prof anat to assoc prof, 56-67, PROF BIOL STRUCT, MED MED SCH, UNIV WASH, 67- *Concurrent Pos:* Nat Res Coun Rockefeller fel, Harvard Med Sch, 54-56; USPHS sr fel, 57-65. *Mem:* AAAS. *Res:* Microscopy and electron microscopy; fixatives; basic cellular structure and function; external cell coats; ultrastructure; biomechanics. *Mailing Add:* Dept Biol Struct Univ Wash Med Sch SM-20 Seattle WA 98195

LUFT, LUDWIG, b Lvov, Poland, Nov 9, 26; nat US; m 52; c 2. PHYSICAL CHEMISTRY. *Educ:* Univ Frankfurt, Dipl, 51; Univ Kans, PhD(phys chem), 56. *Prof Exp:* Asst, Univ Kans, 52-55; asst prof chem, Univ Miami, 55-57; res supvr, MSA Res Corp, 57-58; tech & managerial mem staff, Gen Elec Co, 58-62; sr scientist, Allied Res Assocs, 62-63; dir res, Instrumentation Lab Inc, 63; PRES, LUFT INSTRUMENTS, INC, 63- *Concurrent Pos:* Lectr & adj prof chem eng, Tufts Univ, Medford, Mass, 82-85. *Mem:* AAAS; Am Chem Soc; Instrument Soc Am. *Res:* Chemical engineering; automatic controls; methods development; electrochemistry. *Mailing Add:* 3 Hillside Rd Lincoln MA 01773-0214

LUFT, STANLEY JEREMIE, b Turin, Italy, Sept 26, 27; US citizen; wid; c 4. GEOLOGY. *Educ:* Syracuse Univ, AB, 49; Pa State Col, MS, 51. *Prof Exp:* Asst geol, Pa State Col, 49-51; explor geologist, NJ Zinc Co, 51-54; geologist mineral deposits, US Geol Surv, 54-56; geologist, Northern Pac Rwy Co, 56-58; prof geol & mineral & head dept, Oriente Univ, 59-60; proj geologist, Callahan Mining Corp, 61; geologist, US Geol Surv, 61-; RETIRED. *Honors & Awards:* Gerard Gilbert Award. *Mem:* Geol Soc Am; Soc Econ Geol. *Res:* Petrography and petrology of volcanic rocks; stratigraphy; drainage evolution of Kentucky; geology of uranium in Tertiary intermontaine basins; basin analysis; Northern Powder River Basin. *Mailing Add:* 870 S Miller Ct 870 S Miller Court Lakewood CO 80226

LUFT, ULRICH CAMERON, b Berlin, Ger, Apr 25, 10; nat US; m 41; c 1. HUMAN PHYSIOLOGY. *Educ:* Univ Berlin, MD, 37. *Prof Exp:* Chief high altitude physiol, Aeromed Res Inst, Univ Berlin, 37-45, actg dir dept physiol, Univ, 46-47; res physiologist & assoc prof physiol, Sch Aviation Med, Air Univ, Randolph AFB, Tex, 47-54; head, Dept Physiol, Lovelace Found, 54-80; RETIRED. *Concurrent Pos:* Consult human factors group, Comt Space Technol, Nat Adv Comt Aeronaut, 58; consult, Nat Acad Sci, 58 & Air Res & Develop Command, 58; mem adv bd, Manned Space Flight, NASA, 65-81; assoc physiol, Univ NMex. *Mem:* Fel AAAS; Soc Exp Biol & Med; Am Physiol Soc; fel Aerospace Med Asn; fel Am Col Chest Physicians; fel Am Col Sports Med. *Res:* Physiology of respiration and circulation; aviation medicine; acclimatization to high altitudes; cold and hot climates; physical exercise; clinical physiology. *Mailing Add:* 1900 Ridgecrest Dr SE Albuquerque NM 87108

LUFTIG, RONALD BERNARD, b Brooklyn, NY, Dec 8, 39; m 61; c 4. MICROBIOLOGY, BIOPHYSICS. *Educ:* City Col New York, BS, 60; NY Univ, MS, 62; Univ Chicago, PhD(biophys), 67. *Prof Exp:* Asst prof microbiol, Med Ctr, Duke Univ, 69-73; sr scientist, Worcester Found Exp Biol, 74-79; prof microbiol, Med Sch, Univ SC, 79-83; PROF MICROBIOL & HEAD DEPT, MED SCH, LA STATE UNIV, 83- *Concurrent Pos:* NSF fel, Calif Inst Technol, 67-69; NIH res grant, Med Ctr, Duke Univ, 70-73; res grants, Worcester Found, 74-79 & Univ SC Med Sch, 79- *Mem:* AAAS; Am Soc Biol Chem; Am Soc Microbiol; Am Soc Cell Biol; Sigma Xi. *Res:* Viral and membrane ultrastructure; leukemia virus morphogenesis; microtubule function. *Mailing Add:* Dept Microbiol La State Univ Med Ctr 1901 Perdido St New Orleans LA 70112

LUGAR, RICHARD CHARLES, b Philadelphia, Pa; div; c 1. COMPUTERS IN CHEMISTRY. *Educ:* Univ Pa, BS, 62, PhD(chem), 69. *Prof Exp:* PROF ORG CHEM, DELAWARE VALLEY COL, 67- *Mem:* Am Chem Soc. *Res:* Conformational analysis of alicyclic systems. *Mailing Add:* Dept Chem Delaware Valley Col Doylestown PA 18901

LUGASSY, ARMAND AMRAM, b Kenitra, Morocco, July 23, 33; m 66; c 2. MATERIALS SCIENCE, PROSTHODONTICS. *Educ:* Toulouse Fac Med & Pharm, France, Chirurgien-Dentiste, 59; Univ Pa, DDS, 62, PhD(metall, mat sci), 68. *Prof Exp:* Monitor oper dent, Toulouse Fac Med & Pharm, France, 58-59; instr, Sch Dent Med, Univ Pa, 62-63; asst prof biol mat, Dent-Med Sch, Northwestern Univ, 68-71; assoc prof, 71-77, PROF FIXED PROSTHODONTICS, SCH DENT, UNIV OF THE PAC, 77- *Concurrent Pos:* Nat Inst Dent Res traineeship, Sch Metall & Mat Sci, Univ Pa, 63-68; consult, USPHS Hosp, San Francisco, Calif, 71- *Mem:* Am Soc Metals; Int Asn Dent Res. *Res:* Physical properties of calcified tissues; behavior of materials and devices in clinical applications. *Mailing Add:* Dept Fixed Prosthodontics Univ Pac Sch Dent 2155 Webster St San Francisco CA 94115

LUGAY, JOAQUIN CASTRO, b Manila, Philippines, Apr 3, 38; US citizen; m 62; c 3. BIOCHEMISTRY, FOOD SCIENCE. *Educ:* Univ Santo Thomas, Manila, BS, 60; State Univ NY, PhD(chem), 69. *Prof Exp:* Chemist brewing, San Miguel Breweny Inc, 60-62; chemist rice lipids, Int Rice Res Inst, 62-63; sr chemist biotechnol, 69-71, res specialist, 71-73, sr res specialist protein, 73-77, SR LAB MGR PROTEIN BIOTECHNOL COFFEE, GEN FOODS CORP, 78- *Mem:* AAAS; Am Chem Soc; Sigma Xi. *Res:* Isolation and characterization of enzymes; utilization of enzymes in foods; protein texturization; meat analogs; pet food palatability; functional properties of proteins; protein modification; alternate sources of proteins. *Mailing Add:* Gen Foods Corp 555 S Broadway Tarrytown NY 10591

LUGER, GEORGE F, b Spokane, Wash, Dec 1, 40; m 69; c 3. COMPUTER SCIENCE, INTELLIGENT SYSTEMS. *Educ:* Gonzaga Univ, MS, 66; Univ Notre Dame, MS, 69; Univ Pa, PhD(artificial intel), 73. *Prof Exp:* PROF COMPUTER SCI & PSYCHOL, DEPT COMPUTER SCI, UNIV NMEX, ALBUQUERQUE, 79- *Concurrent Pos:* Consult, Learning Tree Int, 84- *Mem:* Inst Elec & Electronics Engrs; Am Asn Artificial Intel; Asn Comput Mach. *Res:* Aritificial intelligence, especially related to modelling human problem solving, machine learning and expert system design; cognitive science. *Mailing Add:* Dept Comput Sci Univ NMex Albuquerque NM 87106

LUGINBUHL, GERALDINE HOBSON, b Los Angeles, Calif, Feb 27, 44; m 65; c 1. MICROBIOLOGY. *Educ:* Stanford Univ, BA, 65; Univ NC, Chapel Hill, PhD(bact, immunol), 71. *Prof Exp:* NIH fel bact, Duke Univ, 71-74; asst prof, 74-80, ASSOC PROF MICROBIOL, NC STATE UNIV, 80- *Mem:* Am Soc Microbiol; Sigma Xi. *Res:* Genetics and physiology of virulence; alcaligenes. *Mailing Add:* Microbiol Box 7615 NC State Univ Main Campus Raleigh NC 27695-7615

LUGINBUHL, WILLIAM HOSSFELD, b Des Moines, Iowa, Mar 11, 29; m 55; c 5. PATHOLOGY. *Educ:* Iowa State Univ, BS, 49; Northwestern Univ, MD, 53. *Prof Exp:* Intern, Wesley Mem Hosp, Chicago, Ill, 53-54; resident path, Children's Mem Hosp, 54-55; resident, Univ Hosps Cleveland, Ohio, 55-57; from asst prof to assoc prof, 60-67, assoc dean col, 67-70, DEAN HEALTH SCI & COL, COL MED, UNIV VT, 70-, PROF PATH, 67- *Concurrent Pos:* Fel, Col Med, Univ Vt, 59-60. *Mem:* Col Am Path; Am Soc Clin Path; Sigma Xi. *Res:* Gynecologic and obstetrical pathology; endometrial anatomy and physiology. *Mailing Add:* Deans Off Col Med Univ Vt Burlington VT 05405

LUGMAIR, GUENTER WILHELM, b Wels, Austria, Feb 5, 40; m 65; c 2. COSMOCHEMISTRY, GEOCHRONOLOGY. *Educ:* Univ Vienna, Austria, PhD(physics), 68. *Prof Exp:* Fel nuclear physics, Max Planck Inst, Mainz, Ger, 65-68; chemist, 68-71, asst res chemist, 71-77, assoc res chem, 77-84, assoc res geochem, Scripps Inst Oceanog, 79-84, RES CHEMIST, UNIV CALIF, SAN DIEGO, 84- *Concurrent Pos:* Consult, Jet Propulsion Lab, Calif Inst Technol, 69-71; co-investr, Lunar & Planetary Sci Prog, NASA, 69-81, mem-consult, Rev Panel, 78-80; prin investr, NASA & NSF, 81-; assoc ed, J Geophys Res, 81-84; lunar & planetary sci rev panel, NASA, 85-87. *Honors & Awards:* G P Merrill Award, Nat Acad Sci, 87. *Mem:* Am Geophys Union; fel Meteoritical Soc; AAAS. *Res:* Origin and history of the solar system; nucleosynthesis; extinct radioactivities; geo dating of terrestial and extraterrestial materials; cosmic ray effects. *Mailing Add:* SIO-GRD 0212 Univ of Calif San Diego 9500 Gilman Dr La Jolla CA 92093-0212

LUGO, HERMINIO LUGO, b San German, PR, June 6, 18; m 41; c 2. PLANT PHYSIOLOGY. *Educ:* Polytech Inst, PR, BA, 39; Cornell Univ, MS, 48, PhD, 54. *Prof Exp:* Teacher pub sch, PR, 41-46; instr biol & bot, Polytech Inst PR, 46-47; from asst prof to prof biol, bot & plant physiol, Col Agr, Mayaguez, 48-60; prof biol, 60-69, asst dean studies, 60-66, acad coord, Rio Piedras Campus, 66-69, PROF ECOL, UNIV PR, 69-, DIR PREMED STUDIES, 71-, DIR CAYEY UNIV COL, 78- *Concurrent Pos:* Fel, Inst Ecol, Univ Ga, 68-69. *Mem:* Bot Soc Am; Am Soc Agr Sci; Sigma Xi. *Res:* Germination of vanilla seeds. *Mailing Add:* Dept Biol/Environ Sci Inter American Univ Box 1293 Hato Rey PR 00919

LUGO, TRACY GROSS, SOMATIC CELL GENETICS, GENE TRANSFER. *Educ:* Mass Inst Technol, PhD(biol), 82. *Prof Exp:* FEL, NORRIS CANCER HOSP & RES INST, 82- *Mailing Add:* 4648 Feather River Rd Corona CA 91720-9408

LUGO-LOPEZ, MIGUEL ANGEL, b Mayaguez, PR, July 21, 21; m 45; c 2. SOIL SCIENCE. *Educ:* Univ PR, BSA, 43; Cornell Univ, MS, 45, PhD(soil sci), 50. *Prof Exp:* Asst, Fed Exp Sta, PR, 43-44; asst prof agron, Univ PR, 46-48, asst scientist & assoc soil scientist, Agr Exp Sta, 48-57, assoc soil scientist & soil scientist chg, Gurabo Substa, 57-60; asst dir chg, Univ PR, Mayaguez, 60-61, asst dir, Agr Exp Sta, 61-64, actg dir, 64-66, assoc dir, 66-69, dir off progs & plans & assoc dean, Col Agr Sci, 69-72, dean students, 72-

74, prof soil sci & soil scientist, 74-76, EMER PROF, UNIV PR, MAYAGUEZ, 89- Concurrent Pos: Consult, Cornell Univ-AID, 74-78, Tech Servs Caribbean, 80-85, Agr Exp Sta, Univ PR, 80-. Secy State Agr, Dominican Repub, Ministry Agr & Livestock, Costa Rica, Ministry Agr, Bolivia, Coop Agr, Norte, Uruguay. Mem: Soil Sci Soc Am; Am Soc Agron; Am Soc Agr Sci; Int Soc Soils; Caribbean Food Crops Soc. Res: Tropical soils, fertility, management; physical properties of tropical soils. Mailing Add: Box 506 Isabela Agr Exp Sta Isabela PR 00662

LUGT, HANS JOSEF, b Bonn, Ger, Sept 12, 30; US citizen; m 57; c 2. FLUID DYNAMICS. Educ: Univ Bonn, Vordiplom, 52; Aachen Tech Univ, Diplom, 54; Stuttgart Tech Univ, PhD(eng), 60. Prof Exp: Asst hydraul, Ruhrgas AG, Essen, Ger, 54-57, head physics lab, 57-60; res physicist hydrodyn, US Naval Weapons Lab, Va, 60-66; scientist consult, 67-74, head, Numerical Mech Div, 74-78, SR RES SCIENTIST, DAVID TAYLOR RES CTR, 78- Concurrent Pos: Lectr, Am Univ, 62-66; prof lectr, 68-69, George Washington Univ, 88-; Alexander von Humboldt US sr scientist award, Ger Govt, 81. Honors & Awards: David W Taylor Award 1974, US Navy, 75; Sigma Xi Club Award, 80; Distinguished Civilian Serv Award, USN, 82. Mem: Fel Am Phys Soc; Sigma Xi; Asn Appl Math & Mech Ger; Am Hist Soc. Res: Mathematical fluid dynamics; vortex motion; rotating fluids; numerical solution of Navier-Stokes equations. Mailing Add: Code 1205 David Taylor Res Ctr Bethesda MD 20084-5000

LUGTHART, GARRIT JOHN, JR, b Los Angeles, Calif, Feb 11, 23; m 55; c 3. ENTOMOLOGY, GENETICS. Educ: Mich State Univ, BS, 50, MS, 51; Univ Wis, PhD(entom), 59. Prof Exp: Assoc prof biol, Adrian Col, 56-61; ASSOC PROF BIOL, LE MOYNE COL, NY, 61-, CHMN DEPT, 79- Mem: AAAS; Entom Soc Am; Am Genetic Asn; Sigma Xi. Res: Biology and control of insects injurious to man. Mailing Add: Dept Biol Le Moyne Col Syracuse NY 13214-1399

LUH, BOR SHIUN, b Shanghai, China, Jan 13, 16; m 40; c 1. FOOD SCIENCE. Educ: Chiao Tung Univ, BS, 38; Univ Calif, MS, 48, PhD(agr chem), 52. Prof Exp: Instr, Chiao Tung Univ, 38-41; chemist, Ma Ling Canned Foods Co, Ltd, 41-46; asst, 48-51, jr specialist, Dept Food Technol, 52-56, from jr food technologist to assoc food technologist, 56-69, FOOD TECHNOLOGIST, UNIV CALIF, DAVIS, 69-, LECTR FOOD TECHNOL, 57- Concurrent Pos: Consult Food & Agr Orgn, UN, 68 & 88, UN Indust Develop Orgn, UN, Vienna, Austria, 68, 72, 80 & 82. Mem: AAAS; Am Chem Soc; fel Inst Food Technol; Am Oil Chem Soc; Am Asn Cereal Chemists. Res: Chemistry of foods; food processing; biochemistry. Mailing Add: Div Food Technol Univ Calif Davis CA 95616

LUH, JIANG, b Haining, Chekiang, China, June 24, 32; m 56; c 3. ALGEBRA. Educ: Taiwan Normal Univ, BS, 56; Univ Nebr, MS, 59; Univ Mich, PhD(math), 63. Prof Exp: Assoc prof math, Ind State Univ, 63-66 & Wright State Campus, Miami-Ohio State Univ, 66-68; assoc prof, 68-71, PROF MATH, NC STATE UNIV, 71- Mem: Am Math Soc; Math Asn Am. Res: Ring theory; semi-group theory; linear algebra. Mailing Add: NC State Univ Raleigh NC 27650

LUH, JOHNSON YANG-SENG, b Shanghai, China, Apr 9, 25; US citizen; m 57; c 2. ELECTRICAL ENGINEERING, APPLIED MATHEMATICS. Educ: Utopia Univ, China, BS, 47; Harvard Univ, MS, 50; Univ Minn, PhD(elec eng), 63. Prof Exp: Teaching fel elec eng, Harvard Univ, 50-51; engr, Nat Pneumatic Co, 51-56 & Curtiss-Wright Corp, 56-57; assoc engr, Int Bus Mach Corp, 57-58; staff engr, 62-63; instr elec eng, Univ Minn, 58-60; sr res scientist, Honeywell, Inc, 63-65; assoc engr, 65-71, PROF ELEC ENG, PURDUE UNIV, 71- Concurrent Pos: Lectr, Univ Minn, 63-65; prin investr, NASA res grant, Jet Propulsion Lab, 65- Mem: Soc Indust & Appl Math; sr mem Inst Elec & Electronics Engrs; sr mem Am Astronaut Soc; Sigma Xi. Res: Control and information systems and computer aided engineering design, especially bounded-state, stochastic control, learning and communication, and data reduction systems. Mailing Add: Dept Elect & Comput Eng Clemson Univ Clemson SC 29634

LUH, YUHSHI, b Kaohsiung, Taiwan, Feb 14, 49; m; c 2. SPECIALTY CHEMICALS & MATERIALS, ADHESIVES. Educ: Nat Taiwan Univ, BS, 71; Rice Univ, PhD(chem), 76. Prof Exp: Proj investr, M D Anderson Hosp & Tumor Inst, 77-78; NIH postdoctoral fel, Mass Inst Technol, 78-79; res chemist, Mobil Oil Corp, Mobil Res & Develop Co, 79-81; res scientist, Gulf Oil Co, Gulf Sci & Technol Co, 81-83; res chemist, Mine Safety & Appliances Co, 83-84; SR RES CHEMIST, AM CYANAMID CO, 84- Concurrent Pos: Spec lectr, Univ New Haven, 89-90. Honors & Awards: Welch Fund Award, 74. Mem: Am Chem Soc. Res: Biomaterials; ultraviolet stabilizers; photochromics; adhesives; crosslinking chemistry; specialty chemicals; enhanced oil recovery; chemotherapy; pharmaceuticals; petrochemicals; process development; organic and polymer synthesis; product formulation and testing; product development; structure-property relationships. Mailing Add: 948 Red Fox Rd Orange CT 06477

LUHAN, JOSEPH ANTON, neurology; deceased, see previous edition for last biography

LUHBY, ADRIAN LEONARD, b New York, NY, Dec 21, 16; m 67; c 1. HEMATOLOGY, PEDIATRICS. Educ: Columbia Univ, AB, 38; NY Univ, MD, 43. Prof Exp: Intern path & bact, Mt Sinai Hosp, New York, 44-45, intern med & surg, 45-46; res assoc immunol, Children's Hosp, Ohio State Univ, 48-49, asst resident pediat, Hosp & instr, Univ, 49-50; from instr to assoc prof, 50-59, PROF PEDIAT, NEW YORK MED COL, 59- Concurrent Pos: Fel hemat, Children's Hosp, Boston, Mass, 46-48; pres, Am Bd Nutrit, 76- Mem: Am Asn Cancer Res; Am Physiol Soc; Am Inst Nutrit; Am Soc Clin Nutrit; Am Hemat Soc. Res: Morphologic hematology; oncology; nutrition; megaloblastic anemias; physiology, metabolism, biochemistry and nutrition of folic acid, vitamin B-12 and vitamin B-6. Mailing Add: Dept Pediat NY Med Col 2792 Webb Ave Bronx NY 10468

LUI, YIU-KWAN, b Hong Kong, Mar 24, 37; US citizen; m 67; c 1. PHYSICAL CHEMISTRY. Educ: Chung Chi Col, Hong Kong, BS, 59; Lehigh Univ, MS, 61, PhD(phys chem), 66. Prof Exp: Res chemist, Titanium Pigment Div, NL Indust, 66-75; res chemist, Indust Chem Div, 75; res chemist, 76-78, sr res chemist, 78-81, RES ASSOC, ENGELHARD INDUST DIV, ENGELHARD CORP, 81- Mem: Am Chem Soc. Res: Heterogeneous catalysis; preparation and characterization of precious metal catalysts; colloid and surface properties of silica and alumina; physical properties of rheological additives; dispersion stability; physical and surface properties of titanium dioxide pigments. Mailing Add: Three Paprota Ct Parlin NJ 08859

LUIBRAND, RICHARD THOMAS, b Detroit, Mich, Apr 13, 45. ORGANIC CHEMISTRY. Educ: Wayne State Univ, BS, 66; Univ Wis, PhD(org chem), 71. Prof Exp: Fel, Alexander von Humboldt Found, WGer, 71-72; from asst prof to assoc prof, 72-81, PROF ORG CHEM, CALIF STATE UNIV, HAYWARD, 81- Concurrent Pos: Cottrell res grant, Res Corp, 73. Mem: Am Chem Soc; AAAS. Res: Reaction mechanisms in organic chemistry; natural products chemistry. Mailing Add: Dept of Chem Calif State Univ Hayward CA 94542

LUINE, VICTORIA NALL, b Pine Bluff, Ark, Apr 22, 45; c 1. NEUROCHEMISTRY. Educ: Allegheny Col, BS, 67; State Univ NY, Buffalo, PhD(pharmacol), 71. Prof Exp: Res assoc, Rockefeller Univ, 72-75, asst prof neurochem, 75-77, mem fac, dept physiol, 77-87; PROF, DEPT PSYCHOL, HUNTER COL & PROG BIOPSYCHOL & BIOL, CITY UNIV NY, 87- Concurrent Pos: Adj prof, Rockefeller Univ, 87- Mem: AAAS; Soc Neurosci; Endocrine Soc. Res: Steroid hormone regulation of central neurotransmitters and their role in behavior, memory and aging. Mailing Add: Dept Psychol Hunter Col 695 Park Ave New York NY 10021

LUISADA, ALDO AUGUSTO, medicine, physiology; deceased, see previous edition for last biography

LUK, GORDON DAVID, b Shanghai, China, Nov 15, 50; US citizen; m 73; c 2. POLYAMINES, GASTROENTEROLOGY. Educ: Univ Pa, BA, 71; Harvard Med Sch, MD, 75; Am Bd Internal Med, cert med, 78, cert gastroenterol, 79. Prof Exp: Resident med, 75-77, fel gastroenterol, 77-79, PHYSICIAN, JOHNS HOPKINS HOSP, 79-, ASST PROF MED & ONCOL, JOHNS HOPKINS UNIV, 80- Concurrent Pos: Instr med, John Hopkins Univ, 79-80. Mem: Am Fedn Clin Res; Am Col Physician; Am Gastroenterol Asn; Am Soc Gastrointestinal Endoscopy; Am Asn Study Liver Dis. Res: Cell proliferation and differentiation with special emphasis on the potential regulatory role of polyamines; diseases of gastrointestinal epithelia and neoplastic diseases. Mailing Add: Dept Gastroenterol Wayne State Univ Sch Med Harper Hosp 3990 John R St Detroit MI 48201

LUK, KING SING, b Canton, China, Sept 1, 32; US citizen; m 57; c 4. STRUCTURAL ENGINEERING. Educ: Los Angeles State Col, BS, 57; Univ Southern Calif, MSCE, 60; Univ Calif, Los Angeles, PhD(dynamics, soils & struct eng), 71. Prof Exp: Chief engr, R E Rule, Inc, Calif, 58-60; from asst prof to assoc prof, 60-65, from assoc prof to chmn dept, 66-72, prof, 70-83, EMER PROF CIVIL ENG, CALIF STATE UNIV, LOS ANGELES, 83- Concurrent Pos: Consult, King S Luk & Assoc, Calif, 60-; pres, Cathay Pac Inc, 74-; comnr, Calif Seismic Safety Comn, 80-84; dir, Mech Nat Bank, 82- Mem: Fel Am Soc Civil Engrs. Res: Engineering education; structural and earthquake engineering in design and practice of reinforced concrete and steel structures; foundations; time dependent soil and foundation engineering; reinforced concretes. Mailing Add: 1825 Alpha Ave South Pasadena CA 91030

LUKACH, CARL ANDREW, b Wilkes-Barre, Pa, Dec 18, 30; m 53; c 3. ORGANIC CHEMISTRY, POLYMER CHEMISTRY. Educ: Lehigh Univ, BS, 52, MS, 53; Univ Notre Dame, PhD(org chem), 57. Prof Exp: Res chemist, Hercules Inc, 56-69, res supvr, 69-73, res mgr, Org Div, 73-78, mgr, Chem Sci Div, 78-79, proj mgr cellulose derivatives, 80-81, petrol recovery appln, 82-85; MGR, RES & DEVELOP, AQUALON. Mem: Am Chem Soc; Sigma Xi; Soc Petrol Engrs; Am Petrol Inst. Res: Polymerization and copolymerization of olefins and olefin oxides; polymerization kinetics; conformational analysis; reverse osmosis; cross-linking agents; paper chemistry; cellulose chemistry; oil and gas drilling fluids, fracturing fluids and completion fluids; casing cement. Mailing Add: Aqualom c/o Deb Van Mater 2711 Centerville Rd L/F/C/1 Wilmington DE 19850-5417

LUKACS, EUGENE, b Szombathely, Hungary, Aug 14, 06; US citizen; m 35. MATHEMATICS. Educ: Univ Vienna, PhD(math), 30. Prof Exp: Mathematicianm US Naval Ord Test Sta, Calif, 48-50 & Nat Bur Standards, 50-53; head statist br, Off Naval Res, 53-55; prof math & dir statist lab, Cath Univ, 55-72; prof math, Bowling Green State Univ, 72-75; vis prof, Tech Univ Vienna, 75-77; RETIRED. Concurrent Pos: From lectr to adj prof, Am Univ, 54-56; vis prof, Sorbonne, 61, 66, Swiss Fed Inst Technol, 62, Inst Technol Austria, 70 & Univ Hull, 71. Mem: Fel AAAS; fel Inst Math Statist; Am Math Soc; fel Am Statist Asn; Biomet Soc. Res: Probability theory; mathematical statistics. Mailing Add: 3727 Van Ness St NW Washington DC 20016

LUKAS, GEORGE, b Budapest, Hungary, Mar 16, 31; m 56; c 2. INDUSTRIAL PHARMACY. Educ: Univ Budapest, BS, 54; Polytech Inst Brooklyn, MS, 60; Mass Inst Technol, PhD(org chem), 63; NY Univ, MBA, 72. Prof Exp: Develop engr, United Pharmaceut Works, Hungary, 54-56; chemist, Avery Industs, Calif, 57; develop engr, Chas Pfizer & Co, 57-59; NIH fel, Inst Chem Natural Substances, Gif-Sur-Yvette, France, 63-64; res chemist, Ciba-Geigy Corp, 64-65, res biochemist, 65-67, group leader, Biochem Dept, 67-71, mgr drug metabol, 71-80, assoc dir, 80-81, dir pharmaceut & pharm technol, 81-90, EXEC DIR, PHARMACEUT & PHARM TECHNOL, CIBA-GEIGY CORP, 91- Concurrent Pos: adj assoc prof, Dept Pharmacol, NY Med Col, 80-90, adj prof, Col Pharm, Univ RI, 89- Mem: Am Asn Pharmaceut Sci. Res: Chemistry of natural products; pharmacodynamics; absorption and disposition of drugs. Mailing Add: Pharmaceuticals Div CIBA- Geigy Corp 556 Morris Ave Summit NJ 07901

LUKAS, JOAN DONALDSON, b New Haven, Conn, June 19, 42; m 63, 90; c 2. MATHEMATICS. *Educ:* Columbia Univ, AB, 63; Mass Inst Technol, PhD(math), 67. *Prof Exp:* Asst prof, 67-74, ASSOC PROF MATH, UNIV MASS, BOSTON, 74- *Concurrent Pos:* Vis lectr, Brandeis Univ, 71 & 79; consult, 81- *Mem:* Am Math Soc; Math Asn Am; Asn Symbolic Logic; Am Col Med; Inst Elec & Electronic Engrs Computer Soc. *Res:* Mathematical logic; recursive function theory; compiler optimization for parallel architectures. *Mailing Add:* Dept Math Univ Mass Harbor Campus Boston MA 02125

LUKAS, RONALD JOHN, b Syracuse, NY, Aug 22, 49; m 72; c 1. NEUROCHEMISTRY. *Educ:* State Univ NY, Cortland, BS, 71; State Univ NY Downstate Med Ctr, PhD(biophysics), 76. *Prof Exp:* Fel, Univ Calif, Berkeley, 76-78; res assoc, Lab Chem Biodynamics, 78-79; fel neurobiol, Stanford Univ, 79-80; NEUROCHEMIST NEUROPHARMACOL, BARROW NEUROL INST, PHOENIX, ARIZ, 80- *Concurrent Pos:* Res asst prof pharmacol, Univ Ariz, Tucson, 80-88, res assoc prof pharmacol, 89-, adj prof chem, Arizona State Univ, Tempe, 88. *Mem:* Soc Neurosci; Am Soc Neurochem; Biophys Soc; Sigma Xi. *Res:* Neurotransmitters, neurotoxins and synaptic receptors, nervous system hormone, tropic factors and molecular aspects of developmental neurobiology. *Mailing Add:* Div Neurobiol Barrow Neurol Inst 350 W Thomas Rd Phoenix AZ 85013

LUKASEWYCZ, OMELAN ALEXANDER, US citizen; m 68; c 2. IMMUNOBIOLOGY, ACADEMIC ADMINISTRATION. *Educ:* St Joseph's Col, Pa, AB, 64; Villanova Univ, MS, 68; Bryn Mawr Col, PhD(microbiol), 72. *Prof Exp:* Res asst microbiol, Univ Tex, Austin, 70-72; lectr microbiol, Med Sch, Univ Mich, Ann Arbor, 73-75; asst prof, 75-78, ASSOC PROF MED MICROBIOL & IMMUNOL, SCH MED, UNIV MINN, DULUTH, 78-, ASST DEAN CURRICULAR AFFAIRS, 77- *Concurrent Pos:* Res scholar tumor immunol, Med Sch, Univ Mich, Ann Arbor, 73-75; fel, Bush Found, Minn, 83. *Mem:* Am Soc Microbiol; AAAS; Am Asn Immunologists; Fedn Am Socs Exp Med; Sigma Xi. *Res:* Evaluation of immunocompetent cell populations in immune mechanisms of leukemia; contribution of B and T cell subsets; role of macrophage; role of histocompatibility antigens; role of copper in the immune response; effects of copper deficiency on tumor immunity. *Mailing Add:* Dept Curric Affairs Univ Minn Ten University Dr Duluth MN 55812

LUKASIEWICZ, JULIUS, b Warsaw, Poland, Nov 7, 19; US citizen; m 41; c 2. AEROSPACE ENGINEERING. *Educ:* Univ London, BSc, 43, DIC, 45, DSc(eng), 66; Polish Tech Univ, Eng, dipl, 44. *Prof Exp:* Sr sci officer, Aerodyn Dept, Royal Aircraft Estab, Eng, 45- 48; head high speed aerodyn lab, Nat Res Coun Can, 49-57; chief Von Karman Gas Dynamics Facil, Arnold Eng Develop Ctr, ARO, Inc, Tenn, 58-68; prof aerospace eng & assoc dean grad studies & res, Col Eng, Va Polytech Inst & State Univ, 68-70, Whittemore prof eng, 70-71; PROF ENG, CARLETON UNIV, 71- *Concurrent Pos:* Chmn, Aeroballistic Range Asn, 61-62 & Supersonic Tunnel Asn, 61-62; mem, Adv Group Aeronaut Res & Develop, NATO, 62-68; consult adv comt, US Air Force Systs Command, Nat Acad Sci, 69-71; mgr transp study, Sci Coun Can, 77-78. *Mem:* Fel Am Inst Aeronaut & Astronaut; fel Can Aeronaut & Space Inst; fel Brit Inst Mech Engrs; NY Acad Sci. *Res:* High speed aerodynamics; test facilities; energy and transportation; technology-society interaction. *Mailing Add:* 46 Whippoorwill Dr Ottawa ON K1J 7H9 Can

LUKASIK, STEPHEN JOSEPH, b Staten Island, NY, Mar 19, 31; m 83; c 6. PHYSICS. *Educ:* Rensselaer Polytech Inst, BS, 51; Mass Inst Technol, SM, 53, PhD(physics), 56. *Hon Degrees:* DEng, Stevens Inst Technol, 87. *Prof Exp:* Asst physics, Mass Inst Technol, 51-55; scientist, Westinghouse Elec Corp, 55-57; chief, Fluid Physics Div, Davidson Lab, Stevens Inst Technol, 57-66, assoc res prof physics, 59-66; dir nuclear test detection, Advan Res Projs Agency, 66-68, dept dir, 68-71, dir, 71-74; vpres, Systs Develop Div, Xerox Corp, 75-76; vpres nat security res, Rand Corp, 77-78, chief scientist, 78-79; chief scientist, Fed Commun Comn, 79-82; vpres & mgr, Northrop Res & Technol Ctr, 82-85, CORP VPRES TECHNOL, NORTHROP CORP, 85- *Concurrent Pos:* Acoust engr, Bolt, Beranek & Newman Co, 52-55; consult, Vitro labs, Vitro Corp Am, 59-66; mem, Bd Trustees, Stevens Inst Technol, 75-, Harvey Mudd Col, 87-; mem, Comput Sci Adv Comt, Stanford Univ, 76-82; ed, Info Soc, 78-; mem, bd dirs, Software Productivity Consortium, 85-; mem bd trustees, Nat Security Indust Asn, 86- *Honors & Awards:* Ottens Res Award, 63. *Mem:* AAAS; Am Phys Soc; Sigma Xi. *Res:* Relaxation processes in gases and liquids; viscous boundary layer phenomena; energy dissipation processes in water waves; interaction of explosives with magnetic fields. *Mailing Add:* 1714 Stone Canyon Rd Los Angeles CA 90077

LUKASKI, HENRY CHARLES, b Dearborn, Mich, Sept 28, 47; m 77; c 2. TRACE ELEMENT METABOLISM, BODY COMPOSITION ASSESSMENT. *Educ:* Eastern Mich Univ, BS, 73; Pa State Univ, MS, 76, PhD(physiol & nutrit), 79. *Prof Exp:* Res collabr, Brookhaven Nat Lab, 78-79; USDA Agr Res Serv postdoctoral fel nutrit, Grand Forks Human Nutrit Ctr, Univ NDak, 79-80, biologist, 80-82, res physiologist, 83-90, INSTR MED, UNIV NDAK SCH MED, 88-, RES LEADER, GRAND FORKS HUMAN NUTRIT CTR, AGR RES SERV, USDA, 90- *Concurrent Pos:* Consult, NIH, 87- & Nat Sci & Eng Res Coun, Can, 89- *Mem:* Am Physiol Soc; Am Inst Nutrit; Am Soc Clin Nutrit; fel Am Col Sports Med; fel Human Biol Coun; NY Acad Sci. *Res:* Human nutritional requirements for trace minerals and the physiologic and functional effects of graded trace element deficiencies in humans; human body composition assessment. *Mailing Add:* Grand Forks Human Nutrit Res Ctr Agr Res Serv USDA Box 7166 Univ Sta Grand Forks ND 58202

LUKE, HERBERT HODGES, b Pavo, Ga, Feb 2, 23; m 46; c 2. PLANT PATHOLOGY. *Educ:* Univ Ga, BS, 50; La State Univ, MS, 52, PhD, 54. *Prof Exp:* Plant pathologist, Delta Br Exp Sta, USDA, Miss, 54-55, PLANT PATHOLOGIST, AGR EXP STA, UNIV FLA, USDA, 55-, PROF PLANT PATH, 70- *Mem:* Am Phytopath Soc. *Res:* Chemical nature of disease resistance in plants, particularly isolation and identification of host metabolites that inhibit pathogenesis of pathogen; chemical and genetic control of small grain diseases. *Mailing Add:* Dept Plant Path 2557 Hs/Pp Univ Fla Gainesville FL 32603

LUKE, JAMES LINDSAY, b Cleveland, Ohio, Aug 29, 32; m 57; c 3. PATHOLOGY. *Educ:* Columbia Univ, BS, 56; Western Reserve Univ, MD, 60. *Prof Exp:* Intern Path, Yale-New Haven Hosp, 60-61; chief resident, Inst Path, Western Reserve Univ, 61-63; staff researcher, Lab Exp Path, Nat Inst Arthritis & Metab Dis, 63-65; assoc med examr forensic path, Off Chief Med Examr, New York, 65-67; prof forensic path, Sch Med, Univ Okla, & chief med examr, 71-83; chief med examr, Washington, DC, 71-83; DIST SCIENTIST, ARMED FORCES INST PATH; CLIN PROF PATH & LAB MED, UNIV CONN HEALTH CTR, 88- *Concurrent Pos:* State med examr, Okla, 67-71; clin prof path, Georgetown Univ, George Washington Univ & Howard Univ, 71-83. *Mem:* Fel Am Acad Forensic Sci. *Res:* Epidemiological research in legal medicine; pathology of strangulation, hanging and sudden natural death; aspects of forensic pathology as related to pediatrics; experimental pathology of quantitated blunt force injury. *Mailing Add:* Forensic Path Behav Sci Invest Support Unit FBI FBI Acad Quantico VA 22135

LUKE, JON CHRISTIAN, b Minneapolis, Minn, Aug 10, 40; m 83; c 2. APPLIED MATHEMATICS. *Educ:* Mass Inst Technol, SB, 62, SM, 63; Calif Inst Technol, PhD(appl math), 66. *Prof Exp:* NSF fel, 66-68; asst prof math, Univ Calif, San Diego, 68-73; postdoctoral assoc, Univ Minn, 73-74; vis assoc, Calif Inst Technol, 74-75; asst prof, 75-79, ASSOC PROF MATH SCI, IND UNIV-PURDUE UNIV, INDIANAPOLIS, 79- *Mem:* Sigma Xi; Soc Indust & Appl Math. *Res:* Nonlinear methods in applied mathematics; applications in nonlinear wave problems, geomorphology, economics biophysics, and acoustics. *Mailing Add:* Dept Math Ind Univ-Purdue Univ Indianapolis 1125 E 38th St Indianapolis IN 46205

LUKE, ROBERT A, b Rigby, Idaho, Jan 5, 38; m 64; c 6. PARTICLE PHYSICS. *Educ:* Utah State Univ, BS, 62, MS, 66, PhD(physics), 68. *Prof Exp:* From asst prof to assoc prof, 68-77, PROF PHYSICS, BOISE STATE UNIV, 77-, CHMN DEPT, 83- *Mem:* Am Asn Physics Teachers; Am Nuclear Soc. *Res:* X-ray investigation of clay mixtures; multi-pion production in pion proton interactions. *Mailing Add:* 9121 Pattie Dr Boise ID 83704

LUKE, STANLEY D, b Sialkot, WPakistan, Jan 1, 28; m 52; c 5. MATHEMATICS. *Educ:* Univ Panjab, WPakistan, BA, 47, MA, 49; Carnegie-Mellon Univ, MS, 54; Univ Pittsburgh, PhD(math), 68. *Prof Exp:* Prof math, Gordon Col, WPakistan, 49-64; instr, Univ Pittsburgh, 67-68; prof math, Nebr Wesleyan Univ, 68-; AT DEPT MATH, SEATTLE PAC COL, WASH. *Mem:* Math Asn Am. *Res:* Mathematical analysis with special interest in summability. *Mailing Add:* Dept Natural & Math Sci Seattle Pac Univ Nickerson Seattle WA 98119

LUKE, YUDELL LEO, mathematics; deceased, see previous edition for last biography

LUKEHART, CHARLES MARTIN, b DuBois, Pa, Dec 21, 46; m 73; c 3. ORGANOMETALLIC CHEMISTRY. *Educ:* Pa State Univ, BS, 68; Mass Inst Technol, PhD(inorg chem), 72. *Prof Exp:* Res assoc chem, Tex A&M Univ, 72-73; from asst prof to assoc prof, 73-82, PROF CHEM, VANDERBILT UNIV, 82- *Concurrent Pos:* Alfred P Sloan res fel, 79-83. *Mem:* Am Chem Soc. *Res:* Synthesis, characterization and chemical reactivity of organometallic and coordination complexes containing transition metals. *Mailing Add:* Dept of Chem Vanderbilt Univ Nashville TN 37235

LUKENS, HERBERT RICHARD, JR, b Coquille, Ore, May 19, 21; m 45; c 2. CHEMISTRY, PSYCHOPHYSIOLOGY. *Educ:* Univ Calif, Berkeley, BA, 45; US Int Univ, San Diego, MA, 75, PhD(human behav), 78. *Prof Exp:* Chemist, Albers Milling Co, 45-46, Consumers Yeast Co, 46-48, Tracerlab Inc, 48-55, Shell Develop Co, 55-62 & Gen Atomic, 62-73; CHEMIST, IRT CORP, 73- *Concurrent Pos:* Family counsr, San Diego Youth Serv, 75-77, consult, 77- *Mem:* Am Chem Soc. *Res:* Anxiety, its psychophysiology and existential aspects; biochemistry, immunochemical applications; nucleonics, nuclear fuel cycle. *Mailing Add:* 5616 Abalone Pl La Jolla CA 92037

LUKENS, LEWIS NELSON, b Philadelphia, Pa, Jan 21, 27; m 64; c 4. BIOCHEMISTRY. *Educ:* Harvard Univ, AB, 49; Univ Pa, PhD(biochem), 58. *Prof Exp:* Instr biochem, Mass Inst Technol, 56-58; Nat Res Coun res fel chem, Columbia Univ, 58-59, USPHS res fel, 59-60; asst prof biochem, Yale Univ, 64-66; assoc prof, 66-73, chmn biol dept, 78-81, PROF BIOCHEM, WESLEYAN UNIV, 73- *Mem:* Am Soc Biol Chem. *Res:* Protein synthesis and its control in eukaryotes, especially collagen. *Mailing Add:* Dept Molecular Biol & Biochem Wesleyan Univ Hall Atwater Lab Lawn Ave Middletown CT 06457

LUKENS, PAUL W, JR, b Hibbing, Minn, Apr 24, 28; m 60; c 2. MAMMALOGY. *Educ:* Univ Minn, BS, 52, PhD(zool), 63; Tex A&M Univ, MS, 56. *Prof Exp:* From instr to assoc prof, 61-70, PROF ZOOL, UNIV WIS-SUPERIOR, 70- *Concurrent Pos:* Bd regents res grant, Univ Wis, 65-66. *Mem:* Am Soc Mammal. *Res:* Identification, interpretation and paleoecology of vertebrate faunas from archaeological sites; paleozoology; environmental conservation. *Mailing Add:* Dept of Biol Univ of Wis-Superior Superior WI 54880

LUKENS, RAYMOND JAMES, b Beverly, NJ, Feb 25, 30; m 54; c 5. PLANT PATHOLOGY. *Educ:* Rutgers Univ, BS, 54, MS, 55; Univ Md, PhD(bot), 58. *Prof Exp:* Asst plant pathologist, Conn Agr Exp Sta, 57-60, assoc plant pathologist, 60-69, plant pathologist, 70-75; SR PLANT PATHOLOGIST, ORTHO DIV, CHEVRON CHEM CO, 75- *Concurrent Pos:* Lectr plant path, Univ Calif, Berkeley, 77-78. *Mem:* Soc Indust Microbiol; Am Phytopath Soc; Bot Soc Am; Sigma Xi. *Res:* Chemistry of fungicides; correlation between structure and activity of fungicides; fungicide screening and plant disease control. *Mailing Add:* 2009 Westview Ct Modesto CA 95351

LUKER, WILLIAM DEAN, b Yazoo City, Miss, Sept 21, 20. ANALYTICAL CHEMISTRY, CHEMICAL ENGINEERING. *Educ:* La State Univ, BS, 41; Univ Wis, PhD(chem), 55. *Prof Exp:* Process engr, Union-Camp Corp, Ga, 41-52; technologist, E I du Pont de Nemours & Co, Inc, 55-58; res chemist, Miss State Chem Lab, Miss State Univ, 58-65, assoc prof chem eng, 58-65, asst state chemist, 65-67, assoc prof chem, 65- 79, assoc state chemist, 67-79; RETIRED. *Mem:* Am Chem Soc; assoc mem Am Inst Chem Eng. *Res:* Fats and oils, including studies of unsaponifiable matter. *Mailing Add:* PO Box 1083 Mississippi State MS 39762

LUKERT, MICHAEL T, b Kansas City, Mo, June 28, 37; m 61; c 3. GEOLOGY, GEOCHEMISTRY. *Educ:* Univ Ill, BS, 60; Northern Ill Univ, MS, 62; Case Western Reserve Univ, PhD(geol), 73. *Prof Exp:* Instr geol, Northern Ill Univ, 62-64; from asst prof to assoc prof, 67-74, PROF GEOL, EDINBORO UNIV PA, 74- *Concurrent Pos:* Consult, Pa Geol Surv, 75; adj prof, Thiel Col, 75-76 & Mercyhurst Col, 79 & 81; geologist C, Va Div Mineral Resources, 76-77. *Mem:* Geol Soc Am. *Res:* Geochronology; igneous and metamorphic petrology; Precambrian geology; geostatistics. *Mailing Add:* Dept of Geosci Edinboro Univ Pa Edinboro PA 16444

LUKERT, PHIL DEAN, b Topeka, Kans, Nov 1, 31; m 56; c 4. MICROBIOLOGY. *Educ:* Kans State Univ, BS, 53, DVM, 60, MS, 61; Iowa State Univ, PhD(microbiol), 67. *Prof Exp:* Res assoc microbiol, Kans State Univ, 60-61; res vet, Nat Animal Dis Lab, Agr Res Serv, USDA, Iowa, 61-67; MEM STAFF, COL VET MED, UNIV GA, 67- *Mem:* Am Vet Med Asn; Am Soc Microbiol; Am Asn Avian Path. *Res:* Animal virology, particularly pathogenesis of viral infections, identification of new pathogenic viruses and the development of new diagnostic methods for viral diseases. *Mailing Add:* Dept Med Microbiol Col Vet Med Univ Ga Athens GA 30602

LUKES, ROBERT MICHAEL, b San Francisco, Calif, Mar 27, 23; m 49; c 6. ORGANIC CHEMISTRY. *Educ:* Univ San Francisco, BS, 43; Univ Calif, MS, 47; Univ Notre Dame, PhD(org chem), 49. *Prof Exp:* Res chemist, Merck & Co, Inc, 49-53; res assoc, Res Labs, Gen Elec Co, 54-58, supvr, Insulation Lab, Locomotive & Car Equip Dept, 58-64, MGR FINISH SYSTS LAB, MAJOR APPLIANCE LABS, GEN ELEC CO, 64- *Mem:* Am Chem Soc; Am Electroplaters Soc; fel Am Inst Chemists; Fedn Socs Paint Technol. *Res:* Hydrogenation; steroid synthesis; plastics; resins; electrical insulation; surface coatings; paint; surface chemistry; electroless plating. *Mailing Add:* 223 Bramton Rd Louisville KY 40207-3419

LUKES, THOMAS MARK, b San Jose, Calif, Mar 28, 20; m 52; c 4. FOOD SCIENCE. *Educ:* San Jose State Col, BS, 47; Univ Calif, Berkeley, MS, 49. *Prof Exp:* Microbiologist, Real Gold Citrus, Mutual Orange Distributor, 49-51; head lab qual control, Gentry Div, Consol Food Corp, 51-62; assoc prof, 62-73, PROF FOOD PROCESSING & HEAD DEPT FOOD INDUST, CALIF POLYTECH STATE UNIV, SAN LUIS OBISPO, 73- *Mem:* AAAS; Am Chem Soc; Inst Food Technol. *Res:* Application of evolutionary operations to the food processing industry; development of chemical methods of flavor evaluation and application of new developments in food dehydration to the industrial scale. *Mailing Add:* 176 Del Norte San Luis Obispo CA 94301-1508

LUKEZIC, FELIX LEE, b Florence, Colo, May 27, 33; m 55; c 2. PLANT PATHOLOGY. *Educ:* Colo State Univ, BS, 56, MS, 58; Univ Calif, PhD(plant path), 63. *Prof Exp:* Asst plant path, Colo State Univ, 56-58; lab technician, Univ Calif, 58-63; plant pathologist, Div Trop Res, United Fruit Co, Honduras, 63-65; from asst prof to assoc prof, 65-75, PROF PLANT PATH, PA STATE UNIV, 75- *Mem:* Am Phytopath Soc; Am Soc Microbiol. *Res:* Physiology of plant parasitism, especially bacterial caused diseases. *Mailing Add:* 211 Buckhout Labs Pa State Univ University Park PA 16802

LUKIN, L(ARISSA) S(KVORTSOV), b Lvov, Poland, Aug 30, 25; nat US; m 45; c 1. PHYSIOLOGY. *Educ:* Col Women, Poland, BA, 44; Univ Heidelberg, Cand med, 49; Columbia Univ, PhD, 55; Univ of the Pacific, MA, 79. *Prof Exp:* Asst physiol, Columbia Univ, 51-55, instr, 55-56; asst prof, Ohio State Univ, 56-63; sr scientist, Hamilton Standard, 63-64; asst res physiologist, Biomech Lab, Med Ctr, Univ Calif, San Francisco, 64-66, assoc res physiologist, 66-70; ASSOC PROF PHYSIOL, SCH DENT, UNIV OF THE PAC, 70- *Concurrent Pos:* Lectr physiol, Sch Med, Univ Calif, San Francisco, 74- *Mem:* Am Physiol Soc. *Res:* Cardiovascular physiology; blood volumes; cardiac outputs; energy metabolism in exercise; cardiopulmonary physiology. *Mailing Add:* Univ the Pac Sch Dent 2115 Webster St San Francisco CA 94115

LUKIN, MARVIN, b Cleveland, Ohio, Feb 12, 28; m 62; c 2. ORGANIC CHEMISTRY, BIOCHEMISTRY. *Educ:* Ohio Univ, BS, 49; Case Western Reserve Univ, MS, 54, PhD(org chem), 56. *Prof Exp:* Fel org synthesis, Mellon Inst, 56-57; fel protein chem, Albert Einstein Col Med, 57-61; res assoc immunochem, St Lukes Hosp, 61-63; staff asst, Cleveland Clin, 63-65; fel antibiotics, Case Western Reserve Univ, 66-67; from asst prof to assoc prof, 75-83, PROF CHEM, YOUNGSTOWN STATE UNIV, 83- *Mem:* Am Chem Soc; Sigma Xi. *Res:* Organic synthesis; peptide synthesis. *Mailing Add:* Dept Chem Youngstown State Univ Youngstown OH 44503

LUKOW, ODEAN MICHELIN, b Winnipeg, Man. CEREAL CHEMISTRY, CEREAL QUALITY. *Educ:* Univ Man, BSc, 74, MSC, 77, PhD(cereal chem), 82. *Prof Exp:* Lectr microbiol, Univ Man, 77-78; RES SCIENTIST CEREAL CHEM, AGR CAN, 82- *Mem:* Am Asn Cereal Chemists; Inst Food Technologists; Sigma Xi. *Res:* Annual cereal quality evaluation of Western Canadian breeders' lines of wheat; the biochemical basis of cereal quality as related to the protein component. *Mailing Add:* Res Sta Agr Can 195 Dafoe Rd Winnipeg MB R3T 2M9 Can

LUKOWIAK, KENNETH DANIEL, b Newark, NJ, Jan 10, 47. NEUROPHYSIOLOGY, NEUROETHOLOGY. *Educ:* Iona Col, BSc, 69; State Univ NY, Albany, PhD(neurophysiol), 73. *Prof Exp:* Fel neurophysiol, Univ Ky, 73-75; asst prof physiol, McGill Univ, 75-78; from asst prof to assoc

prof, 78-85, PROF MED PHYSIOL, UNIV CALGARY, 85- *Concurrent Pos:* NIH fel, 73-75; Med Res Coun Can grant, 75-78, 78-84 & 84-; vis prof med, Tribuhvan Univ, Kathmandu, Nepal, 81- *Mem:* Am Physiol Soc; Am Soc Zoologists; Can Physiol Soc; Sigma Xi; AAAS; Soc Neurosci. *Res:* Neural and peptidergic mechanisms of adaptive behavior including associative learning in invertebrates; interactions between the central and peripheral nervous systems in the mediation of habituation, sensitization and dishabituation; central pattern generations and behavior. *Mailing Add:* Dept Med Physiol Fac Med Univ Calgary Calgary AB T2N 4N1 Can

LULL, DAVID B, b Rochester, NY, Feb 21, 23; m 49; c 4. CHEMICAL ENGINEERING. *Educ:* Mass Inst Technol, BS, 47, Univ Mich, MS, 49. *Prof Exp:* Res engr, Arthur D Little Inc, 49-62; prin scientist, Appl Sci Lab, GCA Tech Div, 62-71; gen engr, US Naval Weapons Lab, 71-76, gen engr, 76-80, CHEM ENGR, NAVAL SURFACE WEAPONS CTR, 80- *Res:* Generation and assessment of aerosols; propagation and suppression of dust explosions; fracture and propulsion of solids and liquids by high explosives; effectiveness of spaced armor. *Mailing Add:* 16621 Cashell Rd Rockville MD 20853

LULLA, JACK D, FLEXIBLE DIELECTRICS. *Educ:* City Col NY, BS, 50. *Prof Exp:* SR VPRES, TECHNICAL TAPE, 86- *Mem:* Am Chem Soc. *Res:* Polymer coatings and adhesives. *Mailing Add:* 40 E 88th St Apt 11D New York NY 10028

LULLA, KOTUSINGH, atomic physics; deceased, see previous edition for last biography

LUM, BERT KWAN BUCK, b Honolulu, Hawaii, May 9, 29; m 52; c 4. PHARMACOLOGY. *Educ:* Univ Mich, BS, 51, PhD(pharmacol), 56; Univ Kans, MD, 60. *Prof Exp:* From instr to asst prof pharmacol, Med Ctr, Univ Kans, 56-62; from asst prof to prof, Sch Med, Marquette Univ, 62-69, asst chmn dept, 64-69; PROF PHARMACOL & CHMN DEPT, SCH MED, UNIV HAWAII, MANOA, 69- *Mem:* Am Soc Pharmacol & Exp Therapeut; Cardiac Muscle Soc; Asn Med Sch Pharmacol. *Res:* Cardiovascular and autonomic pharmacology. *Mailing Add:* Dept Pharmacol Sch Med Univ Hawaii Honolulu HI 96822

LUM, KIN K, b Ipoh, Malaya, Sept 4, 40; US citizen; m 65; c 2. PHOTOGRAPHIC CHEMISTRY. *Educ:* Hong Kong Baptist Col, BSc, 62; Baylor Univ, PhD(org chem), 66. *Prof Exp:* Res fel, Utah State Univ, 66-68; SR RES ASSOC, EASTMAN KODAK CO RES LABS, 68- *Mem:* Am Chem Soc; Soc Photog Sci & Eng. *Res:* Application of novel imaging chemistry into color image transfer systems. *Mailing Add:* 633 Chalelaine Dr Webster NY 14580

LUM, LAWRENCE, EXPERIMENTAL BIOLOGY. *Prof Exp:* PROF INTERNAL MED, WAYNE STATE UNIV, 89- *Mailing Add:* Hemat & Oncol Div Wayne State Univ PO Box 02188 Detroit MI 48201-1998

LUM, PATRICK TUNG MOON, b Honolulu, Hawaii, Nov 6, 28; div; c 2. ENTOMOLOGY. *Educ:* Earlham Col, BA, 50; Univ Ill, MS, 52, PhD(entom), 56. *Prof Exp:* Asst entom, Univ Ill, 54-56, res assoc, 56, USPHS res fel, 57; res biologist, Entom Res Ctr, Fla State Bd Health, 57-65; res entomologist, Stored Prod Insect Res & Develop, USDA, 65-; RETIRED. *Mem:* Int Mgt Coun; Entom Soc Am. *Res:* Pathogenecity of micro-organisms to insects; photoperiodism and circadian rhythms in insects; physiology, behavior, and morphology of reproduction in moths. *Mailing Add:* 318 E 49th St Savannah GA 31405

LUM, VINCENT YU-SUN, b China, Sept 26, 33; US citizen; m 60; c 3. COMPUTER SCIENCE, ELECTRICAL ENGINEERING. *Educ:* Univ Toronto, BAS, 60; Univ Wash, MS, 61; Univ Ill, Urbana, PhD(elec eng), 66. *Prof Exp:* Assoc engr, IBM Develop Lab, IBM, 62-63, res staff mem, IBM Res Lab, 66-82, proj mgr, 74-82, proj mgr, IBM Heidelberg Sci Ctr, Germany, 82-85; chmn, 85-88, PROF, COMPUT SCI DEPT, NAVAL POSTGRAD SCH, 85- *Concurrent Pos:* Adj asst prof, Loyola Col, Fordham Univ, 67-68. *Mem:* Inst Elec & Electronics Engrs; Asn Comput Mach; Comput Soc; Sigma Xi. *Res:* Data base application and research; office automation; information and data organization and management; computer system application, analysis and design. *Mailing Add:* 15248 Janor Ct Mt Sereno CA 95030

LUMB, ETHEL SUE, b Huntsville, Mo, Dec 21, 16. BIOLOGY. *Educ:* Wash Univ, BS, MA & PhD(zool), 41. *Prof Exp:* Asst prof zool, Univ Rochester, NY, 49-54; prof zool & biol, Vassar Col, 54-82; RETIRED. *Mem:* Soc Develop Biol; Sigma Xi; Soc Cell Biol. *Mailing Add:* Five Lenoir Ct Columbia MO 65201

LUMB, GEORGE DENNETT, b London, Eng, Jan 26, 17; nat US; m 45; c 1. PATHOLOGY. *Educ:* Univ London, MB & BS, 39, MD, 46; MRCP, Eng, 54; FRC(path), 55; FCAP, 56. *Prof Exp:* Assoc prof path, Univ London, 53-57; prof, Univ Tenn, 57-59; dir clin labs & pathologist, James Walker Mem Hosp, 59-65; dir, Warner-Lambert Res Inst Can, 65-69, vpres & dir, Pharmaceut Co, Fla, 69-71; dir med serv & res & develop, 71-73; vpres med affairs, Synapse Commun Serv Inc, 73-77; vpres prod safety assessment, Searle Labs, 77-80; PROF PATH, HAHNEMANN MED COL & HOSP, 80- *Concurrent Pos:* Traveling res fel, Westminster Hosp, London, consult pathologist, 48-57; vis assoc prof health affairs, Sch Med, Univ NC, 60-; assoc prof, Univ Toronto, 69-71. *Mem:* Am Soc Exp Path; Am Asn Path & Bact; Col Am Path; Int Acad Path; Path Soc Gt Brit & Ireland. *Res:* Cardiac research; conduction of specialized muscle pathways in hogs; experimental production of infarcts in canine hearts. *Mailing Add:* c/o Cathy Ramsey Duke Univ Hosp PO Box 3712 Durham NC 27710

LUMB, JUDITH RAE H, b Bridgeport, Conn, Mar 19, 43; m 64; c 2. IMMUNOLOGY. *Educ:* Univ Kans, BA, 65, MA, 66; Stanford Univ, PhD(med microbiol), 69. *Prof Exp:* From asst prof to assoc prof, 69-83, actg chmn, 83-85, PROF BIOL, ATLANTA UNIV, 83- *Concurrent Pos:* NIH career develop award, 75-80. *Mem:* AAAS; Reticuloendothelial Soc; Am Soc

Microbiol; Am Soc Cell Biol; Cellular Kinetics Soc. *Res:* Biochemistry of alkaline phosphatase of C57BL lymphomas; derepression of embryo functions in C57BL lymphomas; computer simulation of the development of the thymus; early lymphocyte differentiation. *Mailing Add:* Caye Caulker Belize City Belize

LUMB, RALPH F, b Worcester, Mass, May 27, 21; m 41; c 8. PHYSICAL CHEMISTRY, NUCLEAR SCIENCES. *Educ:* Clark Univ, AB, 47, PhD(phys chem), 51. *Prof Exp:* Instr chem, Assumption Col, 47-48 & Northeastern Univ, 49-51; chief, Chem-Physics Br, Div Nuclear Mat Mgt, US AEC, 51-56; proj leader, Quantum Inc, 56-59, vpres, 59-60; dir, Western NY Nuclear Res Ctr, Inc, 60-68; pres, Advan Technol Consult Corp, 68-71; pres, Nusac Inc, 71-84; sr consult, Wackenhut Advan Technol, 84-86; PROP, RALPH LUMB ASSOCS, 86- *Concurrent Pos:* Secy, adv comt uranium standards, AEC, 53-56; mem, adv comt safeguarding spec nuclear mat, Atomic Indust Environ, 67-69 & Safeguards steering Group, 66-; consult, Univ Buffalo Nuclear Reactor Proj, 56-60 & Safeguards Br, Int Atomic Energy Agency, 63- *Mem:* Fel AAAS; fel Am Inst Chemists; Am Nuclear Soc; fel Inst Nuclear Mat Mgt. *Res:* Applications of nuclear energy; nuclear research; reactor design, operation and utilization. *Mailing Add:* 63 Maple St Somersville CT 06072

LUMB, ROGER H, b Union, NJ, June 29, 40; m 62; c 3. BIOCHEMISTRY. *Educ:* Alfred Univ, AB, 62; Univ SC, MS, 65, PhD(biol), 67. *Prof Exp:* Instr biol, Univ SC, 65-67; from asst prof to assoc prof, 67-74, PROF BIOL, WESTERN CAROLINA UNIV, 74- *Concurrent Pos:* Damon Runyon fel, 71-73; researcher, Utrecht, Neth, 75-76. *Mem:* Sigma Xi. *Res:* Lipid metabolism in lung; lipid metabolism in cancer cells; membrane biochemistry. *Mailing Add:* Box 1197 Cullowhee NC 28723

LUMB, WILLIAM VALJEAN, b Sioux City, Iowa, Nov 26, 21; m 49; c 1. VETERINARY MEDICINE. *Educ:* Kans State Univ, DVM, 43; Tex A&M Univ, MS, 53; Univ Minn, PhD(vet med), 57; Am Col Vet Anethesiologists, Dipl; Am Col Vet Surgeons, Dipl. *Prof Exp:* From intern to resident, Angell Mem Animal Hosp, Boston, 46-48; from instr to assoc prof med & surg, Tex A&M Univ, 49-52; assoc prof clin & surg, Colo State Univ, 54-58 & surg & med, Mich State Univ, 58-60; assoc prof med, 60-63, dir surg lab, 63-79, prof surg, 63-81, EMER PROF, COL VET MED, COLO STATE UNIV, 81- *Honors & Awards:* Gaines Award, 65; Ralston-Purina Res Award, 80; Jakob Markowitz Award, Am Acad Surg Res, 87. *Mem:* Nat Acad Sci; NY Acad Sci; Am Asn Vet Clinicians; Am Col Vet Surg; AAAS; Am Vet Med Asn; Am Col Vet Anesthesiologists. *Res:* Experimental surgery and anesthesiology. *Mailing Add:* Vet Teaching Hosp Colo State Univ Ft Collins CO 80521

LUMBERS, SYDNEY BLAKE, b Toronto, Ont, Aug 6, 33; m 82. GEOLOGY, PETROLOGY. *Educ:* McMaster Univ, BSc, 58; Univ BC, MSc, 60; Princeton Univ, PhD(geol), 67. *Prof Exp:* Geologist, Ont Div Mines, Ministry Natural Resources, 62-73; CUR GEOL, ROYAL ONT MUS, 73-, CUR-IN-CHARGE, 80- *Concurrent Pos:* Mem comt study of solid earth sci Can, Sci Coun Can, 68-69; corresp, Subcomt Precambrian Stratig, Int Union Geol Sci, 72- *Mem:* Geol Soc Am; Mineral Asn Can; Geol Asn Can; Sigma Xi. *Res:* Precambrian geology; evolution of Grenville Province of Canadian Precambrian Shield; metamorphism; petrogenesis of anorthosite suite rocks and alkalic rocks; geochronology; relationship of mineral deposits to stratigraphy, metamorphism and plutonism. *Mailing Add:* 1042 Rouge Valley Dr Pickering ON L1V 4N6 Can

LUMENG, LAWRENCE, b Manila, Philippines, Aug 10, 39; US citizen; m 66; c 2. MEDICINE, BIOCHEMISTRY. *Educ:* Ind Univ, Bloomington, BS, 60; Ind Univ, Indianapolis, MD, 64, MS, 69; Am Bd Internal Med, dipl, 70. *Prof Exp:* asst prof, 71-74, assoc prof, 74-79, PROF MED BIOCHEM, SCH MED, IND UNIV, INDIANAPOLIS, 79- *Concurrent Pos:* Res & educ associateship, Vet Admin Hosp, Indianapolis, 71-73, clin investr, 73-76, chief in gastroenterol, 77-; dir gastroenterol & hepatol, Ind Univ Med Ctr, 84- *Mem:* Am Soc Clin Investr; Am Col Phys; Am Soc Biol Chemists; Am Gastroenterol Asn; Am Asn Study Liver Dis; Res Soc Alcoholism. *Res:* Regulation of metabolic pathways; ethanol metabolism; pyridoxine and thiamine metabolism; clinical liver diseases. *Mailing Add:* Dept Med Ind Univ Med Ctr Indianapolis IN 46202

LUMLEY, JOHN L(EASK), b Detroit, Mich, Nov 4, 30; m 53; c 3. FLUID MECHANICS, TURBULENCE. *Educ:* Harvard Univ, AB, 52; Johns Hopkins Univ, MSE, 54, PhD(aeronaut), 57. *Hon Degrees:* Haute Distinction Honoris Causa, Ecole Centrale Lyon, 87. *Prof Exp:* Asst & jr instr mech eng, Johns Hopkins Univ, 53-54, asst aeronaut, 54-57, res assoc mech eng, 57-59; from asst prof to assoc prof eng res, Pa State Univ, 59-61, from assoc prof to prof aerospace eng, 61-74, Evan Pugh prof, 74-77; WILLIS H CARRIER PROF ENG, SIBLEY SCH MECH & AEROSPACE ENG, CORNELL UNIV, 77- *Concurrent Pos:* Instr, McCoy Col, 56-59; courtesy fel, Johns Hopkins Univ, 57-58, fel, 58-59; exchange prof, Univ Aix-Marseille, 66-67; vis prof, Univ Louvain-la-Neuve, Belg & Fulbright sr lectr, Univ Liege, 73-74; Guggenheim fels, Mech Fluids Lab, Sch Cent Lyon & Inst Mech Statist Turbulence, Univ d'Aix-Marseille II, France, 73-74; coordr, Grad Exchange Prog Cornell Univ/Ecole Centrale de Lyon, 84-; mem adv comt, Stanford/NASA Ames Ctr Turbulence Res, 89, chmn, 90 & 91. *Honors & Awards:* Medallion, Univ Liege, 71; Fluid & Plasmadynamics Prize, Am Inst Aeronaut & Astronaut, 82; Fluid Dynamics Prize, Am Phys Soc, 90. *Mem:* Nat Acad Eng; Soc Natural Philos; NY Acad Sci; fel Am Acad Mech; Am Inst Aeronaut & Astronaut; fel Am Acad Arts & Sci; fel Am Phys Soc. *Res:* Turbulence; stochastic processes; electronic instrumentation. *Mailing Add:* 256 Upson Hall Cornell Univ Ithaca NY 14853

LUMMA, WILLIAM CARL, JR, b Detroit, Mich, Apr 21, 41; m 75; c 2. ORGANIC CHEMISTRY, MEDICINAL CHEMISTRY. *Educ:* Wayne State Univ, BS, 63; Mass Inst Technol, PhD(org chem), 66. *Prof Exp:* Asst prof chem, St Louis Univ, 66-70; sr res chemist process develop, Rahway, NJ, 70-72, res fel, 72-81, SR RES FEL MED CHEM, MERCK, SHARP & DOHME RES LABS, WEST POINT, PA, 81- *Mem:* Am Chem Soc. *Res:* Heterocyclic and organic synthetic chemistry. *Mailing Add:* 25 Newman Rd MR1 Pennsburg PA 18073-1925

LUMPKIN, LEE ROY, b Oklahoma City, Okla, Sept 6, 25; m 53; c 5. DERMATOLOGY, PATHOLOGY. *Educ:* Univ Okla, BA, 49, MD, 53; Am Bd Dermat, dipl. *Prof Exp:* Intern, Tripler Gen Hosp, US Air Force, Honolulu, Hawaii, 53-54, resident dermat, Walter Reed Gen Hosp, Washington, DC, 58-61, chief dermat serv & clins, 3070th Air Force Hosp, Torrejon AFB, Spain, 61-64, chief dermat & clins, Air Force Hosp, Carswell AFB, Tex, 65-67, chief dermat serv, Wilford Hall Air Force Med Ctr, Lackland AFB, 67-72, dir residency training, 69-72; prof, 72-80, CLIN PROF DERMAT, ALBANY MED COL, 80- *Concurrent Pos:* Fel dermal-path, Armed Forces Inst Path, 64-65; vis lectr, US Air Force Sch Aerospace Med; clin assoc prof, Univ Tex Med Sch, San Antonio; assoc mem comn of cutaneous dis, Armed Forces Epidemiol Bd; US Air Force rep, Nat Prog Dermat; pres, NY State Soc Dermat, 81-83. *Honors & Awards:* Cert of Appreciation, Strategic Air Command, 64 & Surgeon Gen Air Force, 69; James Clarke White Award, 71. *Mem:* Fel Am Col Physicians; fel Am Acad Dermat; Soc Air Force Physician (pres-elect, 71). *Mailing Add:* 22 New Scotland Ave Albany NY 12208

LUMPKIN, MICHAEL DIRKSEN, b Dallas, Tex, Feb 2, 53. NEUROENDOCRINOLOGY, NEUROIMMUNOLOGY. *Educ:* Univ Tex, Austin, BA, 75; Univ Tex Health Sci Ctr, Dallas, PhD(physiol), 81. *Prof Exp:* Teaching asst physiol, Univ Tex Health Sci Ctr, Dallas, 75-76, NIH fel, 76-81, res assoc neuroendocrinol, 81-83; asst prof physiol, 84-89, ASSOC PROF, MED SCH, GEORGETOWN UNIV, 89- *Concurrent Pos:* Lectr, Univ Tex Health Sci Ctr, Dallas, 78-82; prin investr, NIH grants, 86-94; consult, Adamha, 88-, NSF, 88-, VA, 88- & March of Dimes, 88-; lectureship award, Univ Modena, Italy, 90. *Mem:* Endocrine Soc; Soc Neurosci. *Res:* Role of neuropeptides in the control of anterior pituitary gland function; role of hypothalamic and pituitary hormones in the regulation of male and female gonadal function; cytokine regulation of neuroendocrine function. *Mailing Add:* Dept Physiol Sch Med Georgetown Univ 3900 Reservoir Rd NW Washington DC 20007

LUMRY, RUFUS WORTH, II, b Bismarck, NDak, Nov 3, 20; div; c 3. BIOCHEMISTRY. *Educ:* Harvard Univ, AB, 42, MS, 48, PhD(chem physics), 48. *Prof Exp:* Res assoc, Div Eight, Nat Defense Res Comt, 42-45; Merck fel, Univ Utah, 48-50, asst prof phys chem, 50-53, asst res prof biochem, 51-53; assoc prof, 53-57, PROF PHYS CHEM, UNIV MINN, MINNEAPOLIS, 57-, dir lab biophys chem, 63-86. *Concurrent Pos:* NSF sr fel & vis prof, Lab Carlsberg, Copenhagen, 59-60; vis prof, Inst Protein Res, Osaka, Japan, 61, Inst Biol Chem Rome, 63, Univ Calif, San Diego, 77-78 & Univ Granada, Spain, 85. *Mem:* Am Chem Soc; Soc Biol Chem; Sigma Xi; Biophys Soc. *Res:* Biophysical chemistry; enzymes, proteins; fast reactions; water and water solutions. *Mailing Add:* Chem Dept Sch Chem Univ Minn 207 Pleasant St SE Minneapolis MN 55455

LUMSDAINE, EDWARD, b Hong Kong, Sept 30, 37; US citizen; m 59; c 4. SOLAR ENERGY, MECHANICAL ENGINEERING. *Educ:* NMex State Univ, BSME, 63, MSME, 64, DSc(mech eng), 66. *Prof Exp:* Res engr, Boeing Co, Wash, 66-67; assoc prof mech eng, SDak State Univ, 67-72; prof mech & aerospace eng, Univ Tenn, Knoxville, 72-77; prof mech eng & res engr, Phys Sci Lab, N Mex State Univ, 77-78, dir, N Mex Solar Energy Inst & prof mech eng, 78-; DEAN ENG, UNIV MICH, DEARBORN. *Concurrent Pos:* Lectr, Seattle Univ, 66-67; consult, Tenn Valley Authority, 78- & Phys Sci Lab, NMex State Univ, 78-; prin investr of many res grants, NSF & US Dept Energy. *Mem:* Am Soc Mech Engrs; Am Soc Eng Educ; Int Solar Energy Soc; Am Soc Heating, Refrig & Air-Conditioning Engrs; Sigma Xi. *Res:* Compressor noise; unsteady transonic flow; boundary layer; solar energy applications in photovoltaics, desalination and irrigation; collector testing; passive solar building design; high-temperature solar industrial applications; development of Egyptian village with renewable resources. *Mailing Add:* Dean Eng Univ Toledo 2801-W Bancroft Toledo OH 43606

LUMSDAINE, EDWARD, b Hong Kong, Sept 30, 37; US citizen; m 59; c 4. SOLAR ENERGY, PRODUCT QUALITY. *Educ:* NMex State Univ, BSME, 63, MSME, 64, ScD(eng), 66. *Prof Exp:* Res engr, Boeing Co, 66-67; from asst prof to assoc prof mech eng, SDak State Univ, 67-72; from assoc prof to prof fluid flow & aeroacoust, Univ Tenn, Knoxville, 72-77; sr res engr, Phys Sci Lab, NMex State Univ, 77-78, prof mech eng, 77-81; dir, NMex Solar Energy Inst, 78-81; dir & prof mech & aerospace eng, Univ Tenn, Knoxville, 81-83; dean & prof mech eng, Univ Mich, Dearborn, 82-88; DEAN & PROF MECH ENG, UNIV TOLEDO, 88- *Concurrent Pos:* Vis prof, Cairo Univ, Egypt, 74, Tatung Inst Technol, Taipei, Repub of China, 78 & Qatar Univ, Doha, 83; UNESCO expert consult, Cairo Univ, 79-80; lectr, US Info Serv, 81; consult, Ford Motor Co, 84-; bd mem, Am Solar Energy Soc, 85-89; consult & bd mem, Am Supplies Inst, 86-; prin investr, NSF, Am Soc Heating, Refrig & Air Conditioning Engrs, NASA, Dept Energy, HEW and others. *Mem:* Fel Am Soc Mech Engrs; Am Soc Eng Educ; assoc fel Am Inst Aeronaut & Astronaut; Am Soc Testing & Mat; Am Solar Energy Soc. *Res:* Heat transfer; fluid mechanics; turbomachinery; aeroacoustics; solar energy; energy conversation; teaching with microcomputers; product quality; noise-harshness vibrations; author on software in engineering mathematics and creativity and problem solving. *Mailing Add:* Univ Toledo 2801 W Bancroft St Toledo OH 43606-3390

LUMSDEN, CHARLES JOHN, b Hamilton, Ont, Can, Apr 9, 49. THEORETICAL PHYSICS. *Educ:* Univ Toronto, BSc, 72, MSc, 74, PhD(theoret physics), PhD, 78. *Prof Exp:* Spec lectr biophys, Univ Toronto, 75-78; fel, Dept Biol, Harvard Univ, 79-82; assoc prof, 83-90, PROF DEPT MED, UNIV TORONTO, 91- *Concurrent Pos:* Scholar, Med Res Coun Can, 83-88, scientist, 88. *Honors & Awards:* Sir John Cunningham McLennan Award Physics, 72; E C Stevens Award Physics, 77. *Mem:* Am Phys Soc; Soc Math Biol; Biophys Soc. *Res:* Published numerous articles on sociobiology, physiology, statistical mechanics & mathematical biology. *Mailing Add:* Clin Sci Div Univ Toronto Rm 7313 Med Sci Bldg Toronto ON M5S 1A8 Can

LUMSDEN, DAVID NORMAN, b Buffalo, NY, Aug 29, 35; m 63; c 2. GEOLOGY. *Educ:* State Univ NY, Buffalo, BA, 58, MA, 60; Univ Ill, PhD(geol), 65. *Prof Exp:* Res engr, Carborundum Co, 60-62; sr geologist, Pan Am Petrol Corp, 65-67; from asst prof to assoc prof, 67-77, PROF GEOL, MEMPHIS STATE UNIV, 77- *Mem:* Geol Soc Am; Am Asn Petrol Geol; Soc Econ Paleont & Mineral; Sigma Xi. *Res:* Study of carbonate and quartzose sedimentary rocks. *Mailing Add:* Dept of Geol Memphis State Univ Memphis TN 38152

LUMSDEN, RICHARD, b New Orleans, La, Apr 6, 38; div; c 1. CELL BIOLOGY, PARASITOLOGY. *Educ:* Tulane Univ, BSc, 60, MSc, 62; Rice Univ, PhD(biol), 65. *Prof Exp:* From asst prof to assoc prof, Tulane Univ, 65-73, prof biol, 73-87, trop med, 74-87, anat, 75-87; vpres res & develop, Compumed, Inc, 88-90; PROF & CHMN BIOL, INST CREATION RES, SANTEE, CALIF, 90- *Concurrent Pos:* Am Cancer Soc res grant, 65-66; NIH res grants, 65-68, 69-73 & 72-77, career develop award, 69-74; NSF res grant, 68-72; ed consult, J Parasitol, 75-86; res contract, US Food & Drug Admin, 74-76; dean, Grad Sch, 76-78; training grant, Saudi Arabia Educ Mission, 82-85. *Mem:* AAAS; Am Soc Parasitol; Am Soc Zoologists. *Res:* Cytology and biochemistry of parasitic helminths and host-parasite relationships; pharmacology; nutrition. *Mailing Add:* 109 Ave E Metairie LA 70005

LUMSDEN, ROBERT DOUGLAS, b Washington, DC, June 21, 38; m 60; c 2. PLANT PATHOLOGY. *Educ:* NC State Univ, BS, 61, MS, 63; Cornell Univ, PhD(plant path), 67. *Prof Exp:* RES PLANT PATHOLOGIST, BIOCONTROL OF PLANT DIS LAB, PLANT SCI INST, AGR RES CTR W, BELTSVILLE AGR RES CTR, USDA, 66- *Concurrent Pos:* Fel, USDA Agr Res Serv, 87. *Mem:* Am Phytopath Soc. *Res:* Physiology of plant diseases, including the physiology of pathogenesis and disease resistance; pathology and biological control of plant pathogens, especially soilborne plant pathogens; soil ecology; mechanism of action of biological control agents. *Mailing Add:* 262 Biosci Bldg Agr Res Ctr W USDA Beltsville MD 20705

LUMSDEN, WILLIAM WATT, JR, b Dallas, Tex, Dec 21, 20; m 45; c 2. GEOLOGY, PALEONTOLOGY. *Educ:* Univ Calif, Los Angeles, AB, 55, PhD, 64. *Prof Exp:* Asst geol, Univ Calif, Los Angeles, 55-58; from asst prof to assoc prof, Calif State Univ, Long Beach, 58-70, chmn dept, 58-74, prof, 70-83; RETIRED. *Concurrent Pos:* Leverhulme fel før Gt Brit, Aberdeen Univ. *Mem:* AAAS; Paleont Soc; Soc Econ Paleont & Mineral; Am Asn Petrol Geol. *Res:* Invertebrate paleontology; field geology; stratigraphy. *Mailing Add:* PO Box 1556 Idyllwild CA 92549

LUNA, ELIZABETH J, b Poplar Bluff, Mo, Oct 18, 51; m 74. CELL BIOLOGY, MEMBRANE BIOCHEMISTRY. *Educ:* Southern Ill Univ-Carbondale, BA, 72; Stanford Univ, PhD(chem), 77. *Prof Exp:* Res assoc, dept cell biol & develop biol, Harvard Univ, 77-81; asst prof biol, Princeton Univ, 81-88; SR SCIENTIST, WORCESTER FOUND EXP BIOL, 88- *Concurrent Pos:* FRA, Am Chem Soc, 84-89. *Mem:* AAAS; Am Chem Soc; Am Soc Cell Biol; Biophys Soc. *Res:* Cytoskeleton membrane interaction and regulation. *Mailing Add:* Worcester Found Exp Biol 222 Maple Ave Shrewsbury MA 01545

LUNARDINI, VIRGIL J(OSEPH), JR, b Holyoke, Mass, May 10, 35; m 60; c 3. MECHANICAL ENGINEERING, HEAT TRANSFER. *Educ:* Univ Notre Dame, BS, 57; Ohio State Univ, MS, 60, PhD(mech eng), 63. *Prof Exp:* Instr eng, Ohio State Univ, 58-63; asst prof, Clarkson Col Technol, 63-66; assoc prof, State Univ NY Buffalo, 66-69; from assoc prof to prof mech eng, Univ Ottawa, 76-79; RES ENGR, US COLD REGIONS LAB, 79- *Concurrent Pos:* Consult, NASA Lewis Labs, Ohio, 64, Pratt & Whitney Aircraft Div, United Aircraft Corp, 66-67, Chisolm-Ryder, NY, 68- & Govt Can, 78-79, Gulf Interstate, 83, Pace Consults, 83-85; State Univ NY Buffalo fac fel, 67; adj prof Thayer Sch, Dartmouth, 79- *Honors & Awards:* Eugene Jacob Award, Petroleum Div, Am Soc Mech Engrs, 81; Ralph James Award, 85; Am Soc Mech Engrs Award, 87. *Mem:* Am Soc Mech Engrs; Sigma Xi. *Res:* Permafrost heat transfer; cold regions engineering; energy conservation; radiative heat transfer. *Mailing Add:* US Cold Regions Lab 72 Lyme Rd Hanover NH 03755-1290

LUNCHICK, CURT, b New York, NY, June 28, 53; m 85; c 2. FARM WORKER PROTECTION, HUMAN EXPOSURE TO PESTICIDES. *Educ:* Univ Md, College Park, BS, 74; NC State Univ, MS, 77. *Prof Exp:* Sect supvr, Toxicol Lab, Litton Bionetics, Inc, 78-80 & Dynamac Corp, 81-84; chemist, 84-91, SUPVRY CHEMIST, OCCUP & RESIDENTIAL EXPOSURE, OFF PESTICIDE PROGS, US ENVIRON PROTECTION AGENCY, 91- *Res:* Exposure of humans to pesticides applied in residential settings; use of immunoassay and protective clothing to protect agricultural workers from pesticides; regulatory guidelines concerning human exposure to pesticides and hazard reduction options. *Mailing Add:* OPP-HED-OREB H7509C US Environ Protection Agency 401 M St SW Washington DC 20460

LUND, ANDERS EDWARD, b Luverne, Minn, Sept 26, 28. WOOD SCIENCE. *Educ:* Colo State Univ, BS, 55; Duke Univ, MF, 56, DF(wood sci, bus mgt), 64. *Prof Exp:* Forest prod technician, US Forest Prod Lab, 56-58; sr scientist, Koppers Co, Inc, 58-66; assoc prof wood sci, Clemson Univ, 66-67; head admin, Tex Forest Prod Lab, Tex A&M Univ, 67-73; dir, Inst Wood Res, Mich Technol Univ, 74-84; CONSULT, ANDERS E LUND, INC, 84- *Concurrent Pos:* Consult, Forest Prod Co, 66-; prof, Tex A&M Univ, 67-73; consult, Cent States Energy Res Comn, 77, Sci & Educ Admin, USDA, 78 & Mich Energy Admin, 78-84. *Mem:* Int Asn Wood Anatomists; Int Res Group Wood Preservation; Am Wood Preservers Asn; Forest Prod Res Soc; Soc Wood Sci & Technol; fel Inst Wood Sci; Am Soc Testing & Mat; Rwy Tie Asn. *Res:* Wood deterioration and prevention; composite wood products; research administration; wood preservation; fire behavior of wood; physical, mechanical and chemical properties of wood. *Mailing Add:* Anders E Lund Inc PO Box 2534 Farmington Hills MI 48333-2534

LUND, CHARLES EDWARD, b Fremont, Mich, Apr, 4, 46; div; c 1. STRUCTURAL & THERMAL ANALYSIS, FINITE ELEMENT METHODS. *Educ:* Univ Mich, BSE, 68, MSE, 70; Stanford Univ. *Prof Exp:* Assoc engr, McDonnell-Douglas Astro Co, 68-69; structures engr, Lockheed Missiles & Space Co, 70-77; STRESS ENGR, NWL CONTROL SYSTS, PNEUMO-ABEX CORP, IC, INDUSTS, 77- *Concurrent Pos:* Lectr, dept mech eng, Western Mich Univ, 84 & 85. *Mem:* Am Inst Aeronaut & Astronaut. *Res:* Carbon fiber-epoxy composite hydraulic actuaters. *Mailing Add:* 3090 Sturgeon Bay Kalamazoo MI 49002

LUND, DARYL B, b San Bernardino, Calif, Nov 4, 41; m 63; c 2. FOOD ENGINEERING, FOOD PROCESSING. *Educ:* Univ Wis, Madison, BS, 63, MS, 65, PhD(food sci, chem eng), 68. *Prof Exp:* From instr to assoc prof food sci, Univ Wis-Madison, 67-77, prof food sci & agr eng, 77-87; PROF CHMN FOOD SCI, ASSOC DIR NJ AGR EXP STA, RUTGERS UNIV, 88- *Concurrent Pos:* Invited vis prof, Agr Univ, Wageningen, Holland, 79. *Honors & Awards:* Food Eng Award, Dairy & Food Industs Supply Asn/Am Soc Agr Eng, 87; Fel Inst Food Technol. *Mem:* Inst Food Technol; Am Soc Agr Eng; Am Inst Chem Engrs; Sigma Xi. *Res:* Food engineering; fouling of heat exchangers; nutrient retention in processing; starch gelatinization; water movement in foods; microwave heat transfer. *Mailing Add:* Dept Food Sci Rutgers Univ New Brunswick NJ 08903

LUND, DONALD S, b Evanston, Ill, Sept 23, 32. PHYSICS. *Educ:* Northwestern Univ, BS, 54; Univ NMex, MA, 61. *Prof Exp:* Res asst, Univ NMex, 59-61; physicist, Nat Bur Standards, 61-62; res asst, High Altitude Observ, 62-65; physicist, Wave Propagation Lab, Nat Oceanic & Atmospheric Admin, 65-77; SR STAFF ENGR, MARTIN-MARIETTA AEROSPACE DIV, 77- *Mem:* Sigma Xi; Am Geophys Union. *Res:* Aeronomy; radio physics and astronomy. *Mailing Add:* PO Box 1664 Boulder CO 80306

LUND, DOUGLAS E, b Newcastle, Nebr, Dec 12, 33; m 58; c 2. GENETICS, EMBRYOLOGY. *Educ:* Nebr Wesleyan Univ, BA, 58; Univ Nebr, MS, 60, PhD(zool), 62. *Prof Exp:* Asst prof zool, 62-68, PROF BIOL, KEARNEY STATE COL, 68- *Mem:* AAAS; Sigma Xi. *Res:* Temperature effects on early developmental stages of mammalian embryos; carbon dioxide sensitivity in Drosophila. *Mailing Add:* Dept of Biol Kearney State Col Kearney NE 68847

LUND, FREDERICK H(ENRY), b Seattle, Wash, June 2, 29; m 50; c 4. AEROSPACE & ELECTRONICS ENGINEERING. *Educ:* Univ Wash, BSEE, 51; Mass Inst Technol, SM, 57. *Prof Exp:* Proj engr, US Naval Air Missile Test Ctr, Calif, 53-57; sr proj engr, 57-58, head systs employ br, 58-61, plans & analysis group officer, 61-65; sr res engr, Stanford Res Inst, 65-69; sr staff engr, 69-78, MEM PROF STAFF, MARTIN MARIETTA CORP, ORLANDO, 78- *Concurrent Pos:* Mem exec comt, Mil Opers Res Symp, Off Naval Res, 62-66. *Mem:* Inst Elec & Electronics Engrs; Sigma Xi; Asn Old Crows; Am Inst Aeronaut & Astronaut. *Res:* Conduct of system analyses; operations research studies; analysis of electronic and optical countermeasures systems; development, test and evaluation of missile weapon systems; development of electronic instrumentation for guided missile systems; military requirements analyses; ballistic missile defense studies; strategic defense initiative architecture studies; pershing II/intermediate range nuclear force (INF) studies. *Mailing Add:* Martin Marietta Missile Systs PO Box 5837 Mp 81 Orlando FL 32855

LUND, HARTVIG ROALD, b Fargo, NDak, May 15, 33; m 57; c 4. AGRONOMY. *Educ:* NDak State Univ, BS, 55, MS, 58; Purdue Univ, PhD(agron, plant breeding), 65. *Prof Exp:* Res asst agron, NDak State Univ, 55-58, asst prof, 59-62; res asst, Purdue Univ, 62-65; assoc prof, 65-74, assoc dean, Col Agr & assoc dir, 74-79, PROF AGRON, NDAK STATE UNIV, 74-, DEAN, COL AGR & DIR, AGR EXP STA, 79- *Concurrent Pos:* Asst dean, Col Agr & asst dir, Agr Exp Sta, NDak State l)niv, 71-74. *Mem:* Am Soc Agron; Crop Sci Soc Am. *Res:* Rust genetics of durum wheat; chemical mutagenesis in corn; corn breeding and corn endosperm genetics. *Mailing Add:* Col of Agr Morrill Hall NDak State Univ Fargo ND 58105

LUND, J KENNETH, b Brooklyn, NY, Feb 11, 33; m 60; c 2. CHEMICAL ENGINEERING. *Educ:* Polytech Inst Brooklyn, BChE, 55; Princeton Univ, MSE, 58, PhD(chem eng), 63. *Prof Exp:* Sr res engr, Plastics Div, Monsanto Co, Mass, 61-65, res group leader, Hydrocarbons & Polymers Div, Tex, 65-69, prod develop mgr, NJ, 69-71; dir res & develop, Polyester Div, Olin Corp, 71-73; asst to pres, Occidental Res Corp, 74-78; SR VPRES, EXEC TECH RES, VARO CORP, 78- *Mem:* Am Chem Soc; Soc Plastics Engrs; Am Inst Chem Engrs. *Res:* Polymer melt rheology; shear degradation of polymer melts; polymer fatigue failure; high speed tensile studies; extrusion processing of polyolefins and foamed polystyrene polyolefins. *Mailing Add:* Lund & Assoc Inc Suite 160 1520 W Camern Ave West Covina CA 91790-2711

LUND, JOHN EDWARD, b Detroit, Mich, Mar 16, 39; m 59; c 2. VETERINARY PATHOLOGY. *Educ:* Mich State Univ, BS, 62, MS & DVM, 64; Wash State Univ, PhD(vet sci), 69; Am Cl Vet Pathologists, dipl. *Prof Exp:* Instr path, Med Sch, Stanford Univ, 68-70; from asst prof to assoc prof, Sch Vet Sci & Med, Purdue Univ, 70-73; sr scientist, Battelle Northwest Labs, Wash, 73-74, mgr & res assoc, Exp Path Sect, Biol Dept, 74-77; sr res scientist, 77-80, res head, 80-85, ASSOC DIR, UPJOHN CO, 85- *Concurrent Pos:* Consult & vet pathologist, Inst Chem Biol, Univ San Francisco, 69-72. *Mem:* Am Vet Med Asn; AAAS; Am Soc Vet Clin Path; Int Acad Path. *Res:* Hematologic diseases of animals; neutrophil kinetics in blood and bone marrow; chemical carcinogenesis. *Mailing Add:* Upjohn Pharmaceut Ltd 17th Floor Greentower Bldg 14-1 6-Chome Shinjuku-Ku Tokyo 160 Japan

LUND, JOHN TURNER, b Brooklyn, NY, Nov 3, 29; m 55; c 2. PHYSICAL CHEMISTRY, INTELLIGENT SYSTEMS. *Educ:* Brown Univ, AB, 51; Univ Wash, PhD(phys chem), 54; Univ Del, MS, 85. *Prof Exp:* Fel, Univ Wash, 54-55; res chemist, 55-86, EXEC DIR EDUC AID, E I DU PONT DE NEMOURS & CO, INC, 86- *Mem:* Am Chem Soc; Am Phys Soc. *Res:* Industrial research on textile fibers; intelligent computer systems for problem solving. *Mailing Add:* External Affairs Dept E I du Pont De Nemours & Co Inc Wilmington DE 19898

LUND, LANNY JACK, b Dalton, Nebr, May 1, 43; m 64; c 2. SOIL MORPHOLOGY. *Educ:* Univ Nebr, BS, 65, MS, 68; Purdue Univ, PhD, 71. *Prof Exp:* Asst prof & asst soil scientist, 71-77, assoc prof & assoc soil scientist, 77-83, chmn, 85-90, PROF SOIL SCI & SOIL SCIENTIST, UNIV CALIF, RIVERSIDE, 83-, ASSOC DEAN, AGR EXP STA, 90- *Mem:* Am Soc Agron; Soil Sci Soc Am. *Res:* Soil morphology, genesis and classification; soil and the environment. *Mailing Add:* Dept Soil & Environ Sci Univ Calif Riverside CA 92521

LUND, LOUIS HAROLD, b Jefferson City, Mo, Mar 17, 19; m 42; c 2. CHEMICAL PHYSICS. *Educ:* Kans Wesleyan Univ, AB, 40; Univ Mo, AM, 43, PhD(physics), 48. *Prof Exp:* Instr physics, Univ Mo, 43-44; physicist, Lucas-Harold Corp, 44-45; instr math, 45-47, from asst prof to prof, 48-80, EMER PROF PHYSICS, UNIV MO-ROLLA, 80- *Mem:* Am Phys Soc. *Mailing Add:* 36 McFarland Dr Rolla MO 65401

LUND, MARK WYLIE, b Santa Rosa, Calif, Sept 9, 52; m 77; c 5. X-RAY OPTICS, LENS DESIGN. *Educ:* Brigham Young Univ, BS, 77; San Diego State Univ, MS, 79; Ariz State Univ, PhD(physics), 89. *Prof Exp:* Mem tech staff, Hughes Aircraft Co, 79-81; optical physicist, Night Vision Div, Litton Industs, 81-83; sr engr, Govt Electronics Div, Motorola, 83-87; prin optical engr, Optical Disk Div, Honeywell, 87-90; DIR X-RAY OPTICS, MULTILAYER OPTICS & X-RAY TECHNOL, INC, 90- *Mem:* Optical Soc Am; Soc Photo-Optical Instrumentation Engrs. *Res:* Optical engineering for far infrared through x-rays including display systems, night vision, thermal infrared, optical data storage and multilayer x-ray optics; lens design and system analysis. *Mailing Add:* 8318 W Donald Dr Peoria AZ 85382

LUND, MELVIN ROBERT, b Siren, Wis, Oct 17, 22; m 46; c 3. DENTISTRY. *Educ:* Univ Ore, DMD, 46; Univ Mich, MS, 54. *Prof Exp:* From instr to prof restorative dent, Loma Linda Univ, 53-71; PROF OPER DENT & CHMN DEPT, IND UNIV-PURDUE UNIV, INDIANAPOLIS, 71- *Concurrent Pos:* Fel, Claremont Grad Sch, 69-70. *Mem:* Int Asn Dent Res; Acad Oper Dent; Am Acad Gold Foil Opers (pres, 79-80). *Res:* Physical research in dental materials; biologic research in dental procedures. *Mailing Add:* 14706 Little Eagle Creek Ave Zionsville IN 46077

LUND, PAULINE KAY, b Golborne, Lancashire, April 20, 55; m 80. GROWTH FACTORS, GASTROENTEROLOGY. *Educ:* Univ Newcastle, UK, BSc Hons, 75, PhD(gastrointestinal endocrinol), 79. *Prof Exp:* Demonstr physiol, Univ Newcastle, UK, 77-79; res fel, Lab Molecular Endocrinol, Harvard Med Sch, Mass Gen Hosp, 79-88; ASSOC PROF PHYSIOL, DEPT PHYSIOL, UNIV NC, 88-, ASSOC PROF PEDIAT, DEPT PEDIAT, 89-, ASSOC DIR, CTR GASTROINTESTINAL BIOL & DIS, 89- *Concurrent Pos:* Mem Endocrinol Study Sect, NIH, 89- *Mem:* Endocrine soc. *Res:* Insulin-like growth factors biosynthesis and action; role of growth factorsand peptide hormones in intestinal growth and development; molecular correlation of neuronal function; ubiquitin biosynthesis. *Mailing Add:* Dept Physiol Univ NC Chapel Hill NC 27514

LUND, RICHARD, b New York, NY, Sept 17, 39; m 65, 78; c 3. VERTEBRATE PALEONTOLOGY. *Educ:* Univ Mich, Ann Arbor, BS, 61, MS, 63; Columbia Univ, PhD(zool), 68. *Prof Exp:* Asst cur fossil fish, Sect Vert Fossils, Carnegie Mus, 66-69; asst prof earth & plant sci, Univ Pittsburgh, 69-74; from asst prof to assoc prof, 74-81, PROF BIOL, ADELPHI UNIV, 81- *Concurrent Pos:* Pittsburgh Found fel, Carnegie Mus, 67-69, res assoc, 69-; Pittsburgh Found fel, Univ Pittsburgh, 71-74; res assoc, WVa Geol Surv, 74; prin investr, NSF, Mississippian fishes from Montana, 74-85; res assoc Ichthyology, Am Mus Natural Hist, 82- *Mem:* AAAS; Soc Vert Paleont; Am Soc Icthyol & Herpet; Am Soc Zoologists; Ecol Soc Am; Am Elasmobranch Soc (secy, 84). *Res:* Fossil fish; late Paleozoic biostratigraphy; morphology and relationship of early osteichthyan and chondrichthyan fishes with emphasis on the fishes of the Mississippian Bear Gulch Limestone. *Mailing Add:* Dept of Biol Adelphi Univ Garden City NY 11530

LUND, STEVE, b Wis, Dec 3, 23; m 46; c 5. AGRONOMY. *Educ:* Clemson Col, BS, 49; Univ Wis, MS, 51, PhD(agron), 53. *Prof Exp:* Exten agronomist, Clemson Col, 53-54; from asst res specialist to assoc res specialist farm crops, Rutgers Univ, New Brunswick, 54-62, res prof, 62-75; supt & prof, Columbia Basin Agr Res Ctr, 75-85, EMER PROF, ORE STATE UNIV, 85-; EMER PROF, RUTGERS UNIV, 75- *Concurrent Pos:* Chmn dept soils & crops, Rutgers Univ, New Brunswick, 71-75. *Mem:* Am Soc Agron; Crop Sci Soc Am. *Res:* Cereal breeding. *Mailing Add:* 1201 SW 23rd St Pendleton OR 97801

LUND, WILLIAM ALBERT, JR, ichthyology; deceased, see previous edition for last biography

LUNDBERG, GEORGE DAVID, b Pensacola, Fla, Mar 21, 33; m 56, 83; c 5. PATHOLOGY. *Educ:* Univ Ala, BS, 52; Med Col Ala, MD, 57; Am Bd Path, dipl anat & clin path, 62; Baylor Univ, MS, 64. *Hon Degrees:* ScD, State Univ NY, Syracuse, 88. *Prof Exp:* Intern, Tripler Gen Hosp, Honolulu, Hawaii, 57-58; resident path, Brooke Gen Hosp, San Antonio, Tex, 58-62; chief anat path, Letterman Gen Hosp, San Francisco, Calif, 62-63, res officer, 63-64; chief path, William Beaumont Gen Hosp, El Paso, Tex, 64-67; from assoc prof to prof path, Sch Med, Univ Southern Calif, 67-77; from asst dir to assoc dir labs, Los Angeles County/Univ Southern Calif Med Ctr, 68-77; prof & chair path, Sch Med, Univ Calif, Davis & dir path & labs, Med Ctr, Sacramento, 77-82; CLIN PROF PATH, GEORGETOWN & NORTHWESTERN, 82- *Concurrent Pos:* US Army Med Res & Develop Command res grants, 63-64 & 65-67; vis prof forensic med, Lund Univ, Sweden, 76 & Univ London, 76-77; vpres sci info & ed-in-chief jour, AMA, 82-; mem bd dirs, Am Soc Clin Path. *Honors & Awards:* Distinguished Serv Award, Asn Path Chairmen, 90. *Mem:* Col Am Path; Am Soc Clin Path (pres); Am Asn Path & Bact; Int Acad Path; Am Acad Forensic Sci. *Res:* Laboratory computer applications; diseases produced by drugs; toxicology; drug abuse; laboratory management; boxing and brain damage; methods of education. *Mailing Add:* Am Med Asn 515 N State St Chicago IL 60610

LUNDBERG, GUSTAVE HAROLD, b Fremont, Nebr, Sept 5, 01; m 35. APPLIED MATHEMATICS. *Educ:* Midland Col, BS, 24; Colo State Col, MA, 37; Vanderbilt Univ, MA, 32; George Peabody Col, PhD(math), 51. *Prof Exp:* Prof math & sci, Dana Col, 24-29; teacher high sch, Colo, 29-34; instr, Allen Acad, 34-41; prof appl math, Vanderbilt Univ, 42-67; prof, Austin Peay State Univ, Tenn, 67-72, emer prof math, 72-88; emer prof appl math, Vanderbilt Univ, 67-88; RETIRED. *Concurrent Pos:* Res partic, Oak Ridge Nat Lab, 57; vis prof, George Peabody Col, 60 & 61; ed, J Tenn Acad Sci, 63-66, pres, Tenn Acad Sci, 69. *Mem:* Sigma Xi; Math Asn Am. *Res:* Engineering mathematics. *Mailing Add:* Kahler Hotel 20th Ave SW No 548 Rochester MN 55902

LUNDBERG, JOHN L(AUREN), b St Paul, Minn, Oct 8, 24; m 55; c 4. POLYMER SCIENCE, TEXTILE ENGINEERING. *Educ:* Univ Minn, BChE, 48; Univ Calif, PhD(chem), 52. *Prof Exp:* Mem tech staff, Bell Tel Labs, Inc, 52-68; assoc prof textile chem, Clemson Univ, 68-71, chmn dept, 70-71; CALLAWAY PROF TEXTILE CHEM, GA INST TECHNOL, 72- *Concurrent Pos:* Vis assoc prof, Polytech Inst Brooklyn, 61-62, lectr, 63, adj prof, 64-68. *Mem:* AAAS; Am Asn Textile Chem & Colorists; Am Asn Textile Technol; Am Chem Soc; Am Inst Chem; Sigma Xi. *Res:* Physical chemistry and physics of polymers, fibers and textiles; solution chemistry, diffusion and physical properties of polymer solutions; light scattering by fibers, liquids and solutions. *Mailing Add:* 3306 Stratfield Dr Atlanta GA 30319

LUNDBERG, ROBERT DEAN, b Valley City, NDak, May 30, 28; m 53; c 2. POLYMER CHEMISTRY. *Educ:* Harvard Univ, BA, 52, MA & PhD(phys chem), 57. *Prof Exp:* Chemist, Eastman Kodak Co, 52-53; chemist, Union Carbide Corp, 57-62, group leader, Res & Develop Dept, Union Carbide Chem & Plastic Co, 62-69; vpres res, Inter-Polymer Res Corp, 69-70; res assoc, Exxon Chem Co, 70-71, sr res assoc, 71-76, sci adv, Exxon Res & Eng Res Lab, 76-84, chief scientist, 84-90; RETIRED. *Concurrent Pos:* Consult, Exxon Res & Eng Co, 90- *Honors & Awards:* Chem Pioneer, Am Inst Chem, 86. *Mem:* Am Chem Soc; NY Acad Sci. *Res:* Synthesis of synthetic polypeptides; polymer interactions; ionic polymers; block copolymers; thermoplastic elastomers; polymer blends; dilute polymer solution behavior. *Mailing Add:* Corp Res Exxon Res & Eng Co Annandale NJ 08801

LUNDBLAD, ROGER LAUREN, b San Francisco, Calif, Oct 31, 39; div. BIOCHEMISTRY, HEMATOLOGY. *Educ:* Pac Lutheran Univ, BS, 61; Univ Wash, PhD(biochem), 65. *Prof Exp:* Res assoc biochem, Univ Wash, 65-66; res assoc, Rockefeller Univ, 66-68; asst prof, 68-71, assoc prof, 71-77, PROF PATH & BIOCHEM, DENT RES CTR, UNIV NC, CHAPEL HILL, 77- *Concurrent Pos:* Mem coun basic sci & coun thrombosis, Am Heart Asn. *Honors & Awards:* Distinguished Career Award, Univ NC Thrombosis Ctr, 87. *Mem:* AAAS; Am Chem Soc; Am Soc Biol Chem; Am Soc Microbiol. *Res:* Mechanism of blood coagulation; protein chemistry; salivary proteins; oral microbiology. *Mailing Add:* Dept of Path CB #7455 Dent Res Ctr Univ NC Chapel Hill NC 27599-7455

LUNDE, BARBARA KEGERREIS, (BK), b Oak Park, Ill, Aug 10, 37; div; c 2. ELECTRICAL ENGINEERING, RADIO ENGINEERING. *Educ:* Northwestern Univ, Ill, BA, 57, MS, 59; Iowa State Univ, PhD(solid state physics), 70. *Prof Exp:* Res engr, Charles Stark Draper Lab, Mass Inst Technol, 59-61; aerospace engr, Goddard Space Flight Ctr, NASA, 61-65; vpres & chief engr, Radio Sta KEZT, 67-75; asst prof food & nutrit, Iowa State Univ, 70-71, assoc biophys, 71-72, asst prof aerospace eng, 71-76; assoc engr, Ames Lab, 74-76, assoc prof civil eng, 80-81; solar & elec engr, Brooks Borg & Skiles, Engrs-Architects, 76-80; vpres & chief engr, Radio Sta KJJY, 76-80; asst prof physics, Iowa State Univ, 86-87, assoc prof archit, 87-88; ENGR, US WEST, 80-; PRES, SILVER LINING, 87- *Concurrent Pos:* Mem, Nat Adv Comn Foods, Food & Drug Admin, 71-72; mem, Iowa Building Code Coun, 84- *Mem:* Asn Energy Engrs; Am Soc Heating, Refrig & Air Conditioning Engrs; Am Inst Aeronaut & Astronaut; Nat Soc Prof Engrs; Inst Elec & Electronic Engrs; Sigma Xi. *Res:* Nuclear magnetic resonance of metals; solar energy for heating and cooling; electrical design of buildings; energy management in buildings; protection of communication circuits; planning communication networks; protection of people from electromagnetic fields. *Mailing Add:* 2209 SW Park Ave Des Moines IA 50321-1503

LUNDE, KENNETH E(VAN), b Great Falls, Mont, Mar 6, 18; m 43; c 3. CHEMICAL ENGINEERING. *Educ:* Mont State Col, BS, 40; Univ Wash, MS, 41. *Prof Exp:* Process engr, Permanente Metals Corp, 41-42, chem engr, 43-44; chem engr, Henry J Kaiser Co, 42-43; sr engr, Kaiser Aluminum & Chem Corp, 46-47; mech engr, Ralph M Parsons Co, 47-48; head chem eng, Stanford Res Inst, 48-55, mgr indust air res, 55-59; sr engr, Yuba Consol Industs, 59; mgr, Carad Chem Corp, 59-60, vpres, 60-62; mgr process econ, Stanford Res Inst, 62-69, dir chem indust econ, 69-77, DIR SPEC PROJS, SRI INT, 77- *Concurrent Pos:* Lectr, Stanford Univ, 53. *Mem:* Am Chem Soc; Am Inst Chem Engrs. *Res:* Dust collection; gas absorption; drying; heat transfer; cost estimation. *Mailing Add:* 720 Russett Terr Sunnyvale CA 94087

LUNDE, MILFORD NORMAN, b Dodgeville, Wis, Apr 17, 24; m 50; c 2. PARASITOLOGY. *Educ:* Luther Col, AB, 47; Univ NC, MPH, 48. *Prof Exp:* Bacteriologist, WVa State Hyg Lab, 48-51; parasitologist, Inst Trop Med, Bowman-Gray Sch Med, 51-52 & Am Found Trop Med, 52-53; bacteriologist, Army Med Ctr, Ft Detrick, Md, 53-55; RES PARASITOLOGIST, LAB PARASITIC DIS, NAT INST ALLERGY & INFECTIOUS DIS, NIH, 55- *Mem:* Am Soc Parasitologists; Am Soc Trop Med & Hyg. *Res:* Immunodiagnosis of parasitic diseases; application of enzyme immunoassay for detection of antigens and characterization of antibodies; toxoplasmosis; amebiasis and schistosomiasis. *Mailing Add:* Lab Parasitic Dis/NIH 9000 Rockville Pike Bldg 5 Rm 110 Bethesda MD 20892

LUNDE, PETER J, b New York, NY, June 8, 31; m 57; c 3. CHEMICAL & SOLAR ENGINEERING. *Educ:* Pa State Univ, BS, 53, MS, 60, PhD(chem eng), 62. *Prof Exp:* Instrument engr, Union Carbide Plastics Co, 56-57; chem engr process design, Chevron Res Corp, 61-63; sr chem engr, Res Div, Carrier Corp, 63-67; sr chem engr, Hamilton Standard Div, United Aircraft Corp, 67-69, head advan design & develop, Space Life Support Systs, 69-73, head chem process anal, 71-73; sr res scientist, Ctr Environ & Man, Inc, 73-77; solar eng consult, 77-81; PRES, NEW ENERGY RESOURCES, INC, 81- *Concurrent Pos:* Vis prof, Univ Conn, 74-75, 77, 83-86; adj prof, Hartford Grad Ctr, 75-82. *Mem:* Am Inst Chem Engrs; Am Chem Soc; Am Soc Heating, Refrig & Air Conditioning Engrs; Int Solar Energy Soc; Sigma Xi. *Res:* Reaction kinetics, adsorption and catalysis; solar system performance prediction; solar air-conditioning. *Mailing Add:* 4 Daniel Lane West Simsbury CT 06092

LUNDEEN, ALLAN JAY, b New York, NY, Aug 24, 32; m 54; c 4. ORGANIC CHEMISTRY. *Educ:* Southwestern Col, Kans, AB, 54; Rice Univ, PhD(chem), 57. *Prof Exp:* Res chemist org chem, 57-60, sr res chemist, 60-62, res group leader, 62-70, dir explor res, 70-78, DIR PLASTICS RES, CONTINENTAL OIL CO, 78- *Mem:* Am Chem Soc. *Res:* Chemistry of mustard oil glucosides; reactions of carbonium ions; heterogenous catalysis; chemistry of organoaluminum compounds; hydrocarbon oxidation; polymer chemistry. *Mailing Add:* 15203 Park E States Lane Houston TX 77062-3657

LUNDEEN, CARL VICTOR, JR, b Baltimore, Md, Jan 20, 43; m 65; c 3. BIOCHEMISTRY. *Educ:* Univ NC, Chapel Hill, AB, 65; Rockefeller Univ, PhD(life sci), 72. *Prof Exp:* Res assoc plant biol, Rockefeller Univ, 71-72; asst prof chem, 72-74, asst prof biol, 74-77, ASSOC PROF BIOL, UNIV NC, WILMINGTON, 77- *Mem:* Sigma Xi. *Res:* Attempting to elucidate the mechanisms by which autonomous cells attain the capability for rapid growth. *Mailing Add:* Biol Sci Univ NC 601 S Coll Rd Wilmington NC 28403-3297

LUNDEEN, GLEN ALFRED, b Sterling, Colo, June 7, 22; m 48; c 6. FOOD SCIENCE, AGRICULTURE & FOOD CHEMISTRY. *Educ:* Univ Calif, BS, 47; Ore State Univ, MS, 49, PhD(food technol), 52. *Prof Exp:* In chg qual control, Smith Canning & Freezing Co, Ore, 50; asst prof hort & asst horticulturist, Univ Ariz, 51-52; prof subtrop hort, Am Univ, Beirut, 52-54, prof & head dept food technol, 54-57; asst prof hort, Mich State Univ, 57-60 & food sci, 60-64; res chemist, Nutrit Div, Wyeth Lab, Inc, 64; assoc prof, Fresno State Col, 64-68; DIR FOOD SCI RES CTR, 68- *Mem:* AAAS; Am Chem Soc; Am Soc Hort Sci; Inst Food Technol. *Res:* Food antioxidants; oxidation-reduction potentials in food products; enzymes in fruits and vegetables; compounding nutrition foods; milk chemistry; preservation of fruit and vegetable products; world food problems; human nutrition problems. *Mailing Add:* 3451 E Bellaire Way Fresno CA 93726

LUNDEGARD, ROBERT JAMES, b Youngstown, Ohio, Feb 22, 27; m 51; c 1. SCIENCE POLICY, TECHNICAL MANAGEMENT. *Educ:* Ohio Univ, BS, 50; Purdue Univ, MS, 52, PhD(math), 56. *Prof Exp:* Dir, Math Sci Div, Dept Naval Reserve, 68-78, dep dir, 78-81, dir, naval cost anal, Dept Navy, 81-87; CHIEF, STATIST ENG DIV, NAT INST STANDARDS & TECHNOL, 87- *Honors & Awards:* Fleming award, US Govt, 67. *Mem:* Fel Am Statist Asn. *Res:* Develop statistical methods for engineering, with emphasis on achieving quality goals through the design of processes and products. *Mailing Add:* Nat Inst Standards & Technol Gaithersburg MD 20899

LUNDELIUS, ERNEST LUTHER, JR, b Austin, Tex, Dec 2, 27; m 53; c 2. VERTEBRATE PALEONTOLOGY. *Educ:* Univ Tex, BS, 50; Univ Chicago, PhD(paleozool), 54. *Prof Exp:* Fulbright scholar vert paleont, Univ Western Australia, 54-55 & 76; res fel paleoecol, Calif Inst Technol, 56-57; from asst prof to assoc prof, 57-69, PROF GEOL, UNIV TEX, AUSTIN, 69-, JOHN A WILSON PROF VERT PALEONT, 78- *Mem:* Soc Vert Paleont (secy-treas, 75-, pres, 81); Soc Study Evolution; Am Soc Mammalogists; Geol Soc Am; Am Soc Naturalists. *Res:* Pleistocene vertebrates; paleoecology; adaptive morphology; Australian marsupials. *Mailing Add:* Dept of Geol Sci Univ of Tex Austin TX 78712

LUNDELL, ALBERT THOMAS, b Riverside, Calif, Dec 23, 31; m 52; c 3. MATHEMATICS. *Educ:* Univ Utah, AB, 52, AM, 55; Brown Univ, PhD(math), 60. *Prof Exp:* Instr math, Brown Univ, 59-60; lectr, Univ Calif, Berkeley, 60-62; asst prof, Purdue Univ, 62-66; assoc prof, 66-69, chmn dept, 70-72, PROF MATH, UNIV COLO, BOULDER, 70- *Mem:* Am Math Soc. *Res:* Algebraic topology. *Mailing Add:* Dept Math Univ Colo Box 426 Boulder CO 80309

LUNDELL, FREDERICK WALDEMAR, b Revelstoke, BC, Jan 31, 24; Canadian citizen; m 50; c 5. PSYCHIATRY, CHILD PSYCHIATRY. *Educ:* Univ BC, BA, 47; McGill Univ MDCM, 51. *Hon Degrees:* Dipl, McGill Univ, 56. *Prof Exp:* Asst resident psychiat, Montreal Gen Hosp, 55, clin asst psychiat, 56, dir psychiat res, 63-67; asst psychiat, Montreal Children Hosp, 57, dir mental assessment & guidance clin, 61-64; asst prof, 64-67, ASSOC PROF DEPT PSYCHIAT, MCGILL UNIV, 67- *Concurrent Pos:* Consult Psychiat, St Anne Mil Hosp, 56; lectr psychiat, Dept Psychiat, Sch Occup & Physiotherapy, McGill Univ, 59 & 61; coordr psychiat res, Queen Mary Vet & Ste Ann Mil Hosp, 63-76. *Honors & Awards:* Nobel Peace Prize. *Mem:* Fel Royal Col of Physicians & Surgeons Can; Am Psychiat Asn; Am Geriat Soc fel; Royal Soc Health fel; fel AAAS; NY Acad Sci; Can Gerontol Soc; Am Orthopsychiat Asn. *Res:* Active research and publications on substance abuse in adolescents; post-traumatic stress disorders in Canadian Vietnam veterans. *Mailing Add:* Suite 1019 1538 Sherbrooke Montreal PQ H3G 1L5

LUNDELL, O ROBERT, b Revelstoke, BC, Nov 7, 31; m 56; c 2. PHYSICAL CHEMISTRY. *Educ:* Queen's Univ, Ont, BA, 54; Mass Inst Technol, PhD(phys chem), 58. *Prof Exp:* Lectr chem, Royal Mil Col, Ont, 58-61; from asst prof to assoc prof, 61-71, actg chmn dept biol, 67-68, assoc dean, 68-74, PROF CHEM, YORK UNIV, 71-, DEAN, FAC SCI, 74- *Concurrent Pos:* Res assoc, Mass Inst Technol, 60-62. *Mem:* Chem Inst Can. *Res:* Calorimetry and kinetics of gas phase reactions. *Mailing Add:* Dept Chem York Univ 4700 Keele St Downsview ON M3J 2R3 Can

LUNDEN, ALLYN OSCAR, b Toronto, SDak, Feb 5, 31; m 55; c 3. PLANT BREEDING, PLANT GENETICS. *Educ:* SDak State Col, BS, 52, MS, 56; Univ Fla, PhD(plant genetics), 60. *Prof Exp:* Asst agronomist, SDak State Col, 55-56; asst scientist plant genetics, Univ Tenn-AEC Agr Res Lab, 59-62, assoc prof agron, 62-64; assoc prof agron, SDak State Univ, 64-76, head, Seed Lab, 76-80, seed researcher, 81-; RETIRED. *Mem:* Am Soc Agron; Asn Off Seed Anal. *Res:* Irradiation sensitivity of plant tissues; genetic effects of ionizing and ultraviolet irradiation of plant tissues; seed testing techniques; seed vigor testing; seed germination; seed technology; seed storage research. *Mailing Add:* 614 Seventh Ave S Brookings SD 57006

LUNDERGAN, CHARLES DONALD, b Washington, Ind, Sept 24, 23. SYSTEMS RESEARCH. *Educ:* Univ Notre Dame, BSc, 47, MSc, 51. *Prof Exp:* Instr math, St Louis Univ, 51-54, acting dir aeronaut eng, 52-54; instr physics, Agr & Mech Col, Tex, 54-56; physicist mat sci, Sandia Lab, 56-61, sect supvr, 61-62, div supvr, 62-67, mem staff, Mat Res, 67-73, Reactor Safety Res, 73-75 & Mgt Staff, 75-78, mem staff, Systs Res, 78-89; RETIRED. *Concurrent Pos:* Res consult George Mallinckrodt Res, 53-54 & Ohio State, Wright-Patterson AFB, 62-63. *Res:* Equations of state of solids, propagation of shock waves in solids, dynamic stress-strain relations of metals; dynamic behavior of composites; effects of nuclear explosions; remote detection of nuclear effects. *Mailing Add:* 4409 Kellia Lane Albuquerque NM 87111

LUNDGREN, CLAES ERIK GUNNAR, b Stockholm, Sweden, Jan 21, 31. PHYSIOLOGY. *Educ:* Univ Lund, MD, 59, PhD(physiol), 67. *Hon Degrees:* Docent Aviation & Navl Med, Univ Lund, 67. *Prof Exp:* Assoc prof physiol, Fac Med, Univ Lund, Sweden, 67-77; PROF PHYSIOL, SCH MED, STATE UNIV NY, BUFFALO, 77- *Concurrent Pos:* Consult aviation med, Royal Swed Air Force, 59-77; vis assoc prof physiol, Univ Lund, Sweden, 74-77; dir, Ctr Res Spec Environ, State Univ NY, Buffalo, 85- *Mailing Add:* Ctr Res Special Environ State Univ NY 124 Sherman Hall Buffalo NY 14214

LUNDGREN, DALE A(LLEN), b Duluth, Minn, Apr 26, 32; m 54; c 6. AIR POLLUTION, INDUSTRIAL HYGIENE. *Educ:* Univ Minn, BS, 58, MS, 62, PhD(environ health), 73. *Prof Exp:* Engr, Link-Belt Co, Minn, 55-58; asst mech eng, Univ Minn, 58-61; scientist, Electronics Div, Gen Mills, Inc, 61-63; prin scientist, Appl Sci Div, Litton Industs, Inc, 63-65; head air & particle anal lab, Ctr Air Environ Studies & instr mech eng & air pollution, Pa State Univ, 65-67; specialist & head aerosol lab, Statewide Air Pollution Res Ctr, Univ Calif, Riverside, 67-69; chief engr-dir, Air Pollution Control Equipment Sect, Environ Res Corp, St Paul, Minn, 69-72; PROF, ENVIRON ENG DEPT, UNIV FLA, 72- *Concurrent Pos:* Consult, Dale A Lundgren Assoc, 72- & various indust. *Mem:* Air Pollution Control Asn; Am Indust Hyg Asn; Am Soc Mech Engrs; Am Asn Aerosol Res; Soc Aerosol Res Germany. *Res:* Aerosol physics; air pollution; industrial hygiene; air sampling instrumentation; air pollution control equipment. *Mailing Add:* Environ Eng Sci Univ Fla 410 Blk Bldg Gainesville FL 32601

LUNDGREN, DAVID L(EE), b Aberdeen, Wash, Sept 28, 31; wid; c 5. RADIOBIOLOGY, INHALATION TOXICOLOGY. *Educ:* Ore State Univ, BS, 54; Univ Utah, MS, 61, PhD(microbiol), 68. *Prof Exp:* Bacteriologist, Univ Utah, 54-59, chief epizool diag lab, 59-62, chief infectious disease lab, 61-64, microbiologist, Biol Div, Dugway Proving Ground, 64-66; RADIOBIOLOGIST, LOVELACE INHALATION TOXICOL RES INST, 66- *Mem:* Radiation Res Soc; Am Soc Microbiol; Health Physics Soc; Soc Exp Biol & Med; Sigma Xi; AAAS; Am Pub Health Asn. *Res:* Toxicity of inhaled radionuclides from nuclear energy generation; biological effects of internally deposited radionuclides in experimental animals; extrapolation of data from laboratory animals to man. *Mailing Add:* Lovelace Inhalation Toxicol Res Inst PO Box 5890 Albuquerque NM 87185

LUNDGREN, HARRY RICHARD, b Chicago, Ill, May 2, 28; m 55. STRUCTURAL ENGINEERING. *Educ:* Purdue Univ, BSCE, 50; Ariz State Univ, MSE, 62; Okla State Univ, Phd(struct eng), 67. *Prof Exp:* Proj engr, Kawneer Co, Mich, 53-58; vpres eng, R B Feffer & Sons, Ariz, 58-59; sr civil engr, Salt River Proj, 59-61; from instr to assoc prof eng, 62-73, PROF STRUCT ENG, ARIZ STATE UNIV, 73- *Mem:* Am Soc Civil Engrs; Sigma Xi. *Res:* Finite element applications to structural engineering problems; structural stability; light gauge steel structures; wind engineering; software development. *Mailing Add:* Eng Ctr Ariz State Univ Tempe AZ 85287

LUNDGREN, J RICHARD, b Springfield, Mass, Oct 1, 42; m 64; c 2. APPLICATIONS OF GRAPH THEORY. *Educ:* Worcester Polytech Inst, BS, 64; Ohio State Univ, MS, 69, PhD(math), 71. *Prof Exp:* Proj engr, New Eng Tel, 64-67; asst prof math, Allegheny Col, 71-77, assoc prof math, 77-81; assoc prof, 81-86, chmn dept, 84-90, PROF MATH, UNIV COLO, DENVER, 86- *Mem:* Am Math Soc; Math Asn Am; Soc Indust & Appl Math; fel Inst Combinatorics & Its Appln. *Res:* Applications of graphs and matrices; mathematical modeling. *Mailing Add:* 1413 S Ward St Lakewood CO 80228

LUNDGREN, LAWRENCE WILLIAM, JR, b Attleboro, Mass, Mar 17, 32; m 81; c 3. ENVIRONMENTAL GEOLOGY. *Educ:* Brown Univ, AB, 53; Yale Univ, PhD(geol), 58. *Prof Exp:* From instr to assoc prof geol, 56-67, chmn dept geol sci, 71-74 & 76-86 PROF GEOL, UNIV ROCHESTER, 67- *Concurrent Pos:* Fulbright lectr, Finland, 67-68; NSF fac fel geog & environ eng, Johns Hopkins Univ, 76; mem staff, US Geol Surv, Menlo Park, 77. *Mem:* AAAS; Geol Soc Am; Sigma Xi; Am Geophys Union. *Res:* Impact of Chernobyl on Sweden; geology and public policy. *Mailing Add:* Dept Geol Univ Rochester Rochester NY 14627

LUNDHOLM, J(OSEPH) G(IDEON), JR, b Emporia, Kans, Feb 19, 25; m 56; c 2. APPLIED PHYSICS, ENGINEERING PHYSICS. *Educ:* Kans State Univ, BS, 46, MS, 48; NC State Univ, PhD(eng physics), 56. *Prof Exp:* Instr math, Kans State Univ, 47-48; instrumentation develop engr, Oak Ridge Nat Lab, 48-52; res assoc physics, NC State Univ, 52-56, supvr, Raleigh Res Reactor, 52-57; staff res specialist, Reactor Develop Dept, Atomics Int Div, NAm Aviation, Inc, 57-59, supvr syst control & safety, Compact Power Plants, 59-60; mem tech staff, Res & Adv Develop Div, Avco Co, Mass, 60-

62, proj mgr, Adv Space Systs, 62-64; dir adv res & tech, Space Systs Div, Fairchild-Hiller Corp, 64-65; mgr exp prog, Skylab Prog, Hq, 65-74, res prog mgr, Off Aeronaut & Space Technol Res Div, 74-81, adv technol mgr, adv land observations systs off, 81- 83, study mgr, Adv Missions Anal Off, Goddard Space Flight Ctr, NASA, 83-86; CONSULT, 86- Concurrent Pos: Mem, Comt on Radioactive Waste Mgt, Nat Acad Sci, 76-78; res assoc, Mat Sci Dept, Univ Md, 78-79. Mem: Assoc fel Am Inst Aeronaut & Astronaut; sr mem Am Astronaut Soc. Res: Space research and technology, especially space payloads, laser systems, advance energy conversion methods, ultra low temperature coolers; nuclear systems technology and safety; ultra high pressure research; instrumentation and control systems; earth remote sensing technology. Mailing Add: 8106 Post Oak Rd Rockville MD 20854

LUNDIN, BRUCE T(HEODORE), b Alameda, Calif, Dec 28, 19; m 46; c 3. MECHANICAL ENGINEERING. Educ: Univ Calif, BS, 42. Prof Exp: Design engr, Standard Oil Co, Calif, 42-43; res engr, Aircraft Engine Res Lab, NASA, 44-45, head, Jet Propulsion Res Sect, Lewis Flight Propulsion Lab, 46-49, chief, Engine Res Div, 50-58, asst dir, Lewis Res Ctr, Ohio, 58-61, assoc dir, 61-68, dep assoc adminstr, Off Advan Res & Technol, Washington, DC, 68-69, dir, Lewis Res Ctr, 69-77; RETIRED. Concurrent Pos: Staff dir, Pres's Comn on Three Mile Island. Mem: Nat Acad Eng; AAAS; Am Inst Aeronaut & Astronaut. Res: High energy chemical, nuclear and electric propulsion systems; direct and indirect energy conversion systems for space vehicles, including processes, components and systems research; launch vehicle development and operation. Mailing Add: 5859 Columbia Rd North Olmsted OH 44070

LUNDIN, CARL D, b Yonkers, NY, Dec 16, 34; m 57; c 3. PHYSICAL METALLURGY. Educ: Rensselaer Polytech Inst, BMetEng, 57, PhD(mat sci), 66. Prof Exp: Res asst metall, Rensselaer Polytech Inst, 60-62, from instr to asst prof, 62-68; from assoc prof to prof metall & dir welding res, Univ Tenn, 68-75; welding sect mgr, Babcock & Wilcox Co, 75-77; MAGNOVOX PROF ENG, TENN TOMORROW PROF & DIR WELDING RES, UNIV TENN, KNOXVILLE, 77- Concurrent Pos: Supvr welding res mat div, Rensselaer Polytech Inst, 60-68; consult, Oak Ridge Nat Labs, 67-; NSF res initiation grant, 67-68; mem, Welding Res Coun & Pressure Vessel Res Comt. Honors & Awards: Adams Mem Award, Am Welding Soc, 68 & 73, Sparager Award, 78; McKay-Helm Award, 81; Adams Mem lectr, Am Welding Soc, 81. Mem: Am Welding Soc; Am Soc Metals. Res: Physical metallurgy associated with welding and joining--solid state transformations, solidification, diffusion, fissuring, arc physics; process development in welding industry. Mailing Add: Dept Mat Sci Univ Tenn Knoxville TN 37996

LUNDIN, FRANK E, JR, b Chicago, Ill, Aug 25, 28; m 49, 79; c 4. EPIDEMIOLOGY. Educ: Manchester Col, BA, 49; Ind Univ, MD, 53; Johns Hopkins Univ, MPH, 59, DrPH, 62. Prof Exp: USPHS, 53-, intern, Hosp, Norfolk, Va, 53-54, staff physician, Hosp, Carville, La, 54-56, epidemiologist, Cancer Invest, Nat Cancer Inst, Univ Tenn, 56-58, instr, Johns Hopkins Univ, 60-61, res assoc, 61-62, head special studies section, Epidemiol Br, Nat Cancer Inst, 62-67, sr epidemiologist, Occup Studies, Nat Inst Environ Health Sci, NIH, 67-71, sr epidemiologist, Epidemiol Br, Epidemiol & Biomet Div, Nat Inst Child Health & Human Develop, 71-74, dep chief, 74-75, chief, 75-80, SR EPIDEMIOLOGIST, EPIDEMIOL STUDIES BR, BUR RADIOL HEALTH, FOOD & DRUGS ADMIN, USPHS, 80- Mem: Soc Epidemiol Res; Am Pub Health Asn; Soc Occup & Environ Health; Am Med Asn. Res: Epidemiology of cancer, especially of the cervix; lung cancer; leukemia and lymphoma; occupational cancer, infant and fetal mortality and parental smoking; health effects of radiation. Mailing Add: 5600 Fishers Lane Rm 1542 Rockville MD 20857

LUNDIN, ROBERT ENOR, b Boston, Mass, Mar 19, 27; m 52; c 2. NUCLEAR MAGNETIC RESONANCE. Educ: Harvard Univ, AB, 50; Univ Calif, Berkeley, PhD(chem), 55. Prof Exp: Res chemist, Res Ctr, Texaco, Inc, 55-58; res chemist, 58-81, res leader, 81-87, COLLAB, WESTERN REGIONAL RES CTR, USDA, 87- Mem: Am Chem Soc; Soc Appl Spectros. Res: High resolution nuclear magnetic resonance spectroscopy; catalysis; radiation chemistry; gaseous thermodynamics. Mailing Add: Western Regional Res Ctr USDA Berkeley CA 94710

LUNDIN, ROBERT FOLKE, b Rockford, Ill, July 20, 36; m 58. GEOLOGY, PALEONTOLOGY. Educ: Augustana Col, Ill, AB, 58; Univ Ill, MS, 61, PhD(geol), 62. Prof Exp: From asst prof to assoc prof, 62-74, res comt res grants, 66-67, 70-74 & 76, PROF GEOL, ARIZ STATE UNIV, 74-, ASSOC CHMN DEPT, 78- Concurrent Pos: Petrol Res Fund res grants, 63-65, 66-68 & 70-72; Res Corp res grant, 70; guest scientist, Univ Uppsala, 70, distinguished vis prof, 73-74 & 81; Swed Natural Sci Res Coun res grant, 73-74 & 81; co-ed, J Paleont, 74-80. Mem: Soc Econ Paleont & Mineral; Am Asn Petrol Geol; Geol Soc Am; Paleont Soc; Int Paleont Asn. Res: Siluro, Devonian and Mississippi ostracodes, conodonts and stratigraphy; Cenozoic stratigraphy; freshwater ostracodes. Mailing Add: Dept of Geol Ariz State Univ Tempe AZ 85287

LUNDQUIST, BURTON RUSSELL, b Chicago, Ill, June 29, 27; m 50; c 4. FOOD SCIENCE, FOOD CHEMISTRY. Educ: Univ Ill, Urbana, BS, 50. Prof Exp: Chemist & bacteriologist, Dairy Div, Borden Co, 50-53; anal chemist, Res & Develop Ctr, Swift & Co, 53-54, packaging engr, 54-63, mgr packaging develop, 63-67; dir mkt, Champion Packaging Co, Champion Int, 67-71; MGR PACKAGING RES & DEVELOP, RES CTR, ARMOUR FOOD CO, 71- Concurrent Pos: Mem packaging coun, Am Mgt Asn, 76- Mem: Am Soc Testing & Methods; Tech Asn Pulp & Paper Indust. Res: Food preservation using package designing and systems as primary tool to maintain quality throughout merchandising cycle and until consumed; food science bacteriology. Mailing Add: Res & Develop Ctr Armour Swift Eckrich Downers Grove IL 60515

LUNDQUIST, CHARLES ARTHUR, b Webster, SDak, Mar 26, 28; m 51; c 5. SPACE SCIENCES. Educ: SDak State Univ, BS, 49; Univ Kans, PhD(physics), 53. Hon Degrees: DSc, SDak State Univ, Brookings, 79. Prof Exp: Asst prof eng res, Pa State Univ, 53-54; physicist, Tech Feasibility Study Off, Redstone Arsenal, 54-56, chief physics & astrophys sect, Army Ballistic Missile Agency, 56-60; chief, Physics & Astrophys Div, Marshall Space Flight Ctr, NASA, 60-62; asst dir sci, Smithsonian Astrophys Observ, 62-73; dir, Space Sci Lab, Marshall Space Flight Ctr, NASA, 73-82; dir res, 82-90, ASSOC VPRES RES, UNIV ALA, HUNTSVILLE, 90- Concurrent Pos: Assoc, Harvard Col Observ, 62-73; vis prof physics, Univ Ala, Huntsville, 73-81; dir, Consortium Mat Develop in Space, 85- Honors & Awards: Herman Oberth Award, Am Inst Aeronaut & Astronaut, 78. Mem: AAAS; Int Astron Union; Am Astron Soc; Am Geophys Union; Am Phys Soc; NY Acad Sci; Nat Speleol Soc; Meteoritic Soc; Sigma Xi. Res: Spacecraft orbital mechanics and orbit determination; space technology; classical mechanics; radiative transfer. Mailing Add: 214 Jones Valley Dr Huntsville AL 35802

LUNDQUIST, MARJORIE ANN, b Newport News, Va, Aug 17, 38. INDUSTRIAL HYGIENE. Educ: Randolph-Macon Woman's Col, AB, 59; Univ Va, MS, 62, PhD(physics), 65. Prof Exp: Res fel, Dept Phys & Inorg Chem, Univ Adelaide, 65-66; physicist mat sci, 67, tech prog planner lead-acid battery eng, 67-72, supvr monitoring, anal & compliance occup safety & health, 72-74, MGR INDUST HYG, GLOBE-UNION, INC, 74- Concurrent Pos: Adj asst prof, Dept Energetics, Col Eng & Appl Sci, Univ Wis-Milwaukee, 74-75, mem bd dirs, Occup Health Inst, 79-81. Mem: Am Indust Hyg Asn; Soc Occup & Environ Health; Air Pollution Control Asn; Am Soc Testing & Mat. Res: Cigarette smoking as a source of occupational exposure to lead and other metals; biological monitoring of lead-exposed employees; correlation between airborne lead exposure and lead absorption of employee populations. Mailing Add: PO Box 11831 Milwaukee WI 53211

LUNDSAGER, C(HRISTIAN) BENT, b Denmark, Feb 27, 25; nat US; m 47; c 4. ENGINEERING, PLASTICS. Educ: Tech Univ Denmark, MSc, 50. Prof Exp: Engr, Tech Univ Denmark, 47-52 & E I du Pont de Nemours & Co, Inc, 52-62; res assoc, res div, W R Grace & Co, Columbia, Md, 62-90; CONSULT, 90- Mem: Soc Plastics Engrs; Am Ceramic Soc. Res: Thermoplastics processing; concept development of novel products and processes including ceramics. Mailing Add: 1308 Patuxent Dr Ashton MD 20861-9759

LUNDSTROM, LOUIS C, b Tekamah, Nebr, June 7, 15; m 40; c 4. AUTOMOTIVE ENGINEERING, HIGHWAY SAFETY. Educ: Univ Nebr, BS & MS, 39, PhD(eng), 62. Prof Exp: Dir proving ground, Gen Motors Corp, 56-65, dir auto safety, 65-73, exec dir environ activ, 73-80; RETIRED. Concurrent Pos: Chmn, Dept Transp Motor Vehicle Safety Adv Coun. Mem: Nat Acad Eng; fel Soc Automotive Engrs. Res: Vehicle and highway safety; vehicle and highway noise. Mailing Add: 66 W Ranch Trail Morrison CO 80465

LUNDSTROM, MARK STEVEN, b Alexandria, MN, June 8, 51; m 72; c 2. SEMICONDUCTOR DEVICES, COMPUTER SIMULATION. Educ: Univ Minn, BEE, 73, MSEE, 74; Purdue Univ, PhD(elec eng), 80. Prof Exp: Mem tech staff, Hewlett-Packard Corp, 74-77; from asst prof to assoc prof, 80-88, PROF ELEC ENG, PURDUE UNIV, 88-, DIR, OPTO ELECTRONICS RES CTR, 89-, ASST DEAN ENG, 91- Mem: Inst Elec & Electronics Engrs; Am Inst Physics; Sigma Xi. Res: Physics and computer simulation of heterostructure semiconductor devices; semiconductor materials properties which influence device performance. Mailing Add: Sch Elec Eng Purdue Univ W Lafayette IN 47907

LUNDSTROM, RONALD CHARLES, b Lynn, Mass, Mar 15, 52; m 77; c 2. BIOTECHNOLOGY, SEAFOOD TECHNOLOGY. Educ: Northeastern Univ, BA, 75. Prof Exp: RES FOOD TECHNOLOGIST, NAT MARINE FISHERIES SERV, 75- Concurrent Pos: Assoc referee, Asn Off Analytical Chemists, 78- Honors & Awards: Silver Medal, US Dept Com, 89. Mem: Inst Food Technologists; Asn Off Anal Chemists; Electrophoresis Soc. Res: Seafood quality and safety; development of biochemical species identification methods; development of monoclonal antibody based immunoassay methods for fishery biology and technology based applications. Mailing Add: 21 Madison Ave Beverly MA 01915

LUNDVALL, RICHARD, b Boxholm, Iowa, Dec 10, 20; m 41; c 3. VETERINARY MEDICINE. Educ: Iowa State Univ, DVM, 44, MS, 56. Prof Exp: From instr to assoc prof, 44-71, PROF VET MED & SURG, IOWA STATE UNIV, 71- Res: Large animal surgery; ophthalmology. Mailing Add: Vet Med Clin Iowa State Univ Ames IA 50011

LUNDY, JOHN KENT, b Vancouver, Wash, Jan 21, 46; m 68; c 1. FORENSIC ANTHROPOLOGY. Educ: Western Wash Univ, Bellingham, BA, 76, MA, 77; Univ Witwatersrand, Johannesburg, SAfrica, PhD(anat), 84; Am Bd Forensic Anthropol, dipl, 88. Prof Exp: Asst lectr anat, Univ Witwatersrand, 80; asst prof, Nat Col Naturopathic Med, 81-82; med examr & forensic anthropologist, Multnomah County Med Examr, Portland, Ore, 82-86; FORENSIC ANTHROPOLOGIST, CENT IDENTIFICATION LAB, US ARMY, HAWAII, 86- Concurrent Pos: Consult forensic anthrop, 82-; adj asst prof, dept anthrop, Portland State Univ, 82-; adj asst prof & former dir, forensic studies, Dent Sch, Ore Health Sci Univ, 84-86; identification consult to US Navy, 87. Mem: Am Acad Forensic Sci; Am Asn Phys Anthropologists; Am Anthrop Asn; Sigma Xi. Res: Human variation and evolution; morphometric analysis; forensic anthropology; physical anthropology of Southern Africa; Pacific Northwest. Mailing Add: Anthrop Prog Clark Col 1800 E McLoughlin Blvd Vancouver WA 98663

LUNDY, RICHARD ALAN, b Sullivan, Ind, Aug 20, 34; m 60; c 2. HIGH ENERGY PHYSICS. Educ: Univ Chicago, PhD(physics), 62. Prof Exp: Assoc dir, Fermi Nat Accelerator Lab; RETIRED. Honors & Awards: Nat Medal Technol, US Govt, 89. Res: High energy physics; large superconducting magnet systems. Mailing Add: PO Box 506 White Salmon WA 98672

LUNDY, TED SADLER, b Sumner Co, Tenn, Apr 24, 33; m 55; c 1. RESEARCH ADMINISTRATION, METALS & CERAMICS. *Educ:* Univ Tenn, BS, 54, MS, 57, PhD(metall), 64; Oak Ridge Sch Reactor Technol, Dr Pile Eng, 58. *Prof Exp:* Instr eng drawing, 55-57; metallurgist, Metals & Ceramics Div, Oak Ridge Nat Lab, 57-59, group leader diffusion studies, 59-71, supvr corrosion res, 71-72, group leader diffusion studies, 72-75, Prog Planning & Anal, 75-76, nat prog mgr, Building Thermal Envelope Systs & Insulating Mats, 77-85, mgr, Energy Conversion & Utilization Technologies, 85-88; ASSOC PROF MECH ENG, TENN TECHNOL UNIV, 88-, DIR, MFG CTR, 90- *Concurrent Pos:* Lectr, Univ Tenn, 66-; mem, Knox County Ct; mem bd dirs, Knoxville Urban League; consultative coun, Nat Inst Bldg Sci. *Mem:* Sigma Xi; Am Inst Mining, Metall & Petrol Engrs; Am Soc Metals; Am Soc Testing & Mats; Am Soc Heating Refrigerating & Air-Conditioning Engrs; Nat Asn Civil Engrs. *Res:* Solid state reactions; diffusion in metals and ceramics; building sciences; heat transfer and moisture flow; materials sciences. *Mailing Add:* 2875 Seven Springs Rd Cookeville TN 38501

LUNER, PHILIP, b Vilno, Poland, June 1, 25; US citizen; m 51; c 2. PHYSICAL CHEMISTRY. *Educ:* Loyola Col, BSc, 47; McGill Univ, PhD(phys chem), 51. *Prof Exp:* Res chemist, Pulp & Paper Res Inst Can, 51-54; group leader, Sulfite Pulp Mfrs League, 54-57; from res assoc to assoc prof, 57-64, PROF PULP & PAPER RES, STATE UNIV NY COL ENVIRON SCI & FORESTRY, 64-, SR RES ASSOC, EMPIRE STATE PAPER RES INST, 77- *Mem:* Am Chem Soc; Tech Asn Pulp & Paper Indust; Can Pulp & Paper Asn; Sigma Xi. *Res:* Diffusion and penetration studies of pulping; chromophores in model lignin compounds; mechanical properties of fibers and paper; surface chemical properties of wood polymers. *Mailing Add:* SUNY Col Environ Sci for Espri Syracuse NY 13210

LUNER, STEPHEN JAY, b New York, NY, Oct 2, 40; m 65; c 3. BIOPHYSICS. *Educ:* Calif Inst Technol, BS, 61; Univ Calif, Los Angeles, PhD(biophys), 69. *Prof Exp:* Res biophysicist, Univ Calif, Los Angeles, 68-71, asst res biophysicist, Biophys Lab, 71, asst prof in residence pediat, 72-77; ASST PROF PATH, DALHOUSIE UNIV, 77- *Concurrent Pos:* NIMH trainee, Univ Calif, Los Angeles, 69-71. *Mem:* Am Soc Cell Biol. *Res:* Biophysics of the cell surface; electrophoresis; cell surface antigens; effects of enzymes on cell interactions; immunohematology. *Mailing Add:* Dept Path Dalhousie Univ Tupper Bldg Halifax NS B3H 4H7 Can

LUNGSTROM, LEON, b Lindsborg, Kans, July 22, 15; m 65; c 2. MEDICAL ENTOMOLOGY. *Educ:* Bethany Col, Kans, BS, 40; Kans State Univ, MS, 46, PhD(med entom), 50. *Prof Exp:* Entomologist, USPHS, Commun Dis Ctr, 49-52; biologist, 52-73, prof biol & head dept, 52-80, EMER PROF, BETHANY COL, KANS, 81- *Concurrent Pos:* NSF fac fel, Stanford Univ, 59-60, Univ Okla, 65, Ariz State Univ, 62, Tulane Univ, 70 & Ft Hays Kans State Univ, 77; co dir, McPherson County Old Mill Mus & Park, 81-84; munic mosquito control, 89-91. *Mem:* Emer Mem Am Mosquito Control Asn; emer mem Sigma Xi. *Res:* Mosquitoes. *Mailing Add:* 518 E Lincoln Lindsborg KS 67456

LUNIN, MARTIN, b New York, NY, Aug 31, 17; m 47. PATHOLOGY. *Educ:* Okla Agr & Mech Col, BS, 38; Wash Univ, DDS, 50; Columbia Univ, MPH, 52. *Prof Exp:* Assoc prof path, Univ Tex Dent Br, 59-64; asst dean curriculum affairs, 69-71, assoc dean acad affairs, 71-74, prof path & head Sch Dent, 64-85, EMER PROF, SCH DENT, UNIV MD, BALTIMORE, 85- *Concurrent Pos:* Sr consult, Univ Tex M D Anderson Hosp & Tumor Inst, 60-64; consult, Vet Admin Hosp, 62- & Children's & Lutheran Hosps, Baltimore, Md, 64- *Mem:* AAAS; Am Dent Asn; Am Acad Oral Path; Int Asn Dent Res. *Res:* Diseases of the soft and hard tissues of the head and neck. *Mailing Add:* 666 W Baltimore St Baltimore MD 21201

LUNINE, JONATHAN IRVING, b New York, NY, June 26, 59; m. PLANETARY SCIENCES. *Educ:* Univ Rochester, BS, 80; Calif Inst Technol, MS, 83, PhD(planetary sci), 85. *Prof Exp:* Res assoc, 84-86, asst prof, 86-90, ASSOC PROF PLANETARY SCI, UNIV ARIZ, 90- *Concurrent Pos:* Vis asst prof, Univ Calif, Los Angeles, 86. *Honors & Awards:* Harold C Urey Prize, Div Planetary Sci, Am Astron Soc, 88; Zeldovich Prize, Comt Space Res, Int Coun Sci Unions, 90. *Mem:* Sigma Xi; Am Astron Soc; Am Geophys Union. *Res:* Theoretical studies of outer solar system satellites, comets, their present nature and evolution, emphasizing physical chemistry of ices and volatiles; modeling of the evolution of substellar mass objects, brown dwarfs; terrestrial photochemical processes. *Mailing Add:* Lunar & Planetary Lab Univ Ariz Tucson AZ 85721

LUNK, WILLIAM ALLAN, b Johnstown, Pa, May 6, 19; m 47; c 4. ORNITHOLOGY. *Educ:* Univ WVa, AB, 41, MS, 46; Univ Mich, PhD(zool), 55. *Prof Exp:* Instr biol, Univ WVa, 46-47; preparator, Exhibit Mus, Univ Mich, Ann Arbor, 49-59, assoc cur exhibits & lectr zool, actg dir, 85-88, cur exhibits, 64-89; RETIRED. *Concurrent Pos:* Consult, Kalamazoo Nature Ctr, Mich, 63-77. *Mem:* Cooper Ornith Soc; Wilson Ornith Soc; assoc Am Ornith Union. *Res:* Ornithological life history; taxonomy and distribution; fossil birds; exhibit techniques. *Mailing Add:* 865 N Wagner Rd Ann Arbor MI 48103

LUNN, ANTHONY CROWTHER, b Huddersfield, Eng, Sept 25, 46, US citizen; m 72. BIOMATERIALS, POLYMER PHYSICS. *Educ:* Cambridge Univ, Eng, BA, 67; Harvard Univ, MS, 68; Mass Inst Technol, ScD, 72. *Prof Exp:* Res scientist, Pioneering Res Lab, Du Pont Co, 69; res assoc, Mass Inst Technol, 72-73; proj leader, Chem Res Div, Am Cyanamid Co, 73-81; SECT MGR, ETHICON INC, JOHNSON & JOHNSON CO, 81- *Mem:* Am Chem Soc; Fiber Soc; Soc Plastics Engrs; Sigma Xi. *Res:* Development of novel polymeric products for use in surgery, tissue repair and wound closure; implantable devices and sutures for surgical use, including bioabsorbable polymers. *Mailing Add:* J&J Interventional Systs PO Box 4917 Warren NJ 07060

LUNNEY, DAVID CLYDE, b Charleston, SC. ANALYTICAL CHEMISTRY. *Educ:* Univ SC, BS, 59, PhD(phys chem), 65. *Prof Exp:* NIH fel, Duke Univ, 66-68; from asst prof to assoc prof, 68-80, PROF CHEM, 80 & DIR, SCI INST DISABLED, E CAROLINA UNIV, 86- *Mem:* Audio Eng Soc; Am Chem Soc; AAAS; Found Sci & Handicapped. *Res:* Laboratory computerization and automation; laboratory aids for disabled scientists and engineers. *Mailing Add:* Dept Chem E Carolina Univ Greenville NC 27834

LUNNEY, JOAN K, b Philadelphia, Pa, July 19, 46; m 79. ANIMAL INFECTIOUS DISEASES, ANIMAL GENOME. *Educ:* Chestnut Hill Col, Philadelphia, Pa, BS, 68; Johns Hopkins Univ, Baltimore, Md, PhD(biochem), 76. *Prof Exp:* Chemist cell surface receptors, Lab Biochem Pharmacol, Nat Inst Arthritis, Metab Digestive Dis, NIH, 73-76, postdoctoral fel swine immunogenetics & biochem, Immunol Br, Nat Cancer Inst, 76-79, sr staff fel, 79-83; RES IMMUNOLOGIST SWINE IMMUNOGENETICS, HELMINTHIC DIS LAB, AGR RES SERV, USDA, 83- *Concurrent Pos:* Panel mem, Cellular Physiol Panel, NSF, 83-87; adv coun mem, Portuguese NSF, 87-90; comt mem, Comt Res Animal Genome, US Exp Stas Comt Orgn & Policy, 90; chairperson, Swine CP Workshop, Int Union Immunol Sci, 90-92; mem, Vet Immunol Comt, Am Asn Immunologists, 90-93; comt mem, Nat Animal Genetic Resources Comt, USDA, 91- *Mem:* Am Asn Immunologists; Transplantation Soc; Asn Women Sci; Int Soc Animal Genetics; Am Asn Vet Parisitologists; Am Asn Vet Immunologists. *Res:* Analyses of immunologic mechanisms and genetic control of swine responses to infectious diseases; Understanding of basic swine immune responses and of complexity of swine genome. *Mailing Add:* LPSI Helminthic Dis Lab Bldg 1040 Rm 2 Agr res Serv USDA Beltsville MD 20705

LUNSFORD, CARL DALTON, b Richmond, Va, Feb 11, 27; m 47; c 3. PHARMACEUTICAL CHEMISTRY. *Educ:* Univ Richmond, BS, 49, MS, 50; Univ Va, PhD(chem), 53. *Prof Exp:* Instr chem, Univ Va, 52-53; res chemist, 53-57, assoc dir chem res, 58, dir, 59-64, dir labs, 62-64, dir res, 64-66, asst vpres, 66-74, vpres, 73-80, SR VPRES, A H ROBINS CO, INC, 80- *Mem:* AAAS; Am Chem Soc; Am Inst Chemists. *Res:* Medicinal and organic chemistry and development. *Mailing Add:* 1807 Poplar Green Dr Richmond VA 23233-4171

LUNSFORD, JACK HORNER, b Houston, Tex, Feb 6, 36; m 60; c 7. HETEROGENEOUS CATALYSIS. *Educ:* Tex A&M Univ, BS, 57; Rice Univ, PhD(chem eng), 62. *Prof Exp:* Asst prof chem eng, Univ Idaho, 61-62; asst prof chem, Sam Houston State Col, 65-66; from asst prof to assoc prof, 66-71, PROF CHEM, TEX A&M UNIV, 71- *Honors & Awards:* Paul H Emmett Award, Catalysis Soc, 75; Catalysis Soc Metropolitan NY Award, Excellence Catalysis, 86. *Mem:* Am Chem Soc. *Res:* Surface chemistry and heterogeneous catalysis, using modern spectroscopic techniques. *Mailing Add:* Dept Chem Tex A&M Univ College Station TX 77843-5000

LUNSFORD, JESSE V(ERNON), b Ninnekah, Okla, Sept 4, 23; m 48; c 5. CIVIL & SANITARY ENGINEERING. *Educ:* Univ NMex, BS, 53; Univ Calif, MS, 54. *Prof Exp:* Asst prof & asst res engr, Wash State Univ, 54-57; assoc prof, Rensselaer Polytech Inst, 57-58; PROF CIVIL ENG, N MEX STATE UNIV, 58- *Mem:* Am Soc Civil Engrs; Nat Soc Prof Engrs; Am Soc Eng Educ; Am Pub Health Asn; Am Water Works Asn. *Res:* Anaerobic digestion; stream sanitation; algae production; water reclamation and utilization. *Mailing Add:* Civil Agr & Geol Eng Dept 3CE Box 0001 NMex State Univ Las Cruces NM 88003

LUNSFORD, RALPH D, b Ninety Six, SC, Jan 7, 34; m 57; c 1. AUDIO-VIDEO EQUIPMENT & TAPES REGARDING CONSUMER PRODUCTS, LINEAR SOLID STATE APPLICATIONS ENGINEERING. *Educ:* Clemson Univ, BS, 56. *Prof Exp:* Elec engr audio electronics acoustics, RCA Corp, 56-66; proj engr audio & acoust, CBS TV Network, 66-68; vpres eng, Audio Tape Duplication, Nat Tape Serv, Inc, 68-71; proj mgr, Audio & FM Receivers & Amplifiers, Brit Industs, Div of Avnet, 71-73; sr proj engr, Audio Amplifiers Design, Dynaco, Inc, Div of Tyco Labs, 73-76; sr engr audio for auto radios, Ford Motor Co, 76-79; MGR PROD ENG & QUAL OR SOLID STATE APPLN ENGR, THOMSON CONSUMER ELECTRONICS, INC, 79- *Concurrent Pos:* Liaison officer UL/CSA matters, Thomson Consumer Electronics, Inc, 85-; mem, Camcorder Battery Standards Comt, Electronic Industs Asn, 90- *Mem:* Audio Eng Soc; Am Inst Physics. *Res:* Microphones used in space program; designer of audio electronic systems used in Ford auto radio; developed state of the art audio tape duplicating system; developed state of the art battery cycling equipment. *Mailing Add:* Thomson Consumer Electronics 200 Clements Bridge Rd Deptford NJ 08096

LUNT, HARRY EDWARD, b New York, NY, Apr 30, 24; m 50; c 5. FAILURE ANALYSIS, STANDARDS DEVELOPMENT. *Educ:* Syracuse Univ, AB, 48; Iowa State Univ, MS, 53. *Prof Exp:* Res asst, Ames Lab, US Atomic Energy Comn, 50-53; develop metallurgist, US Steel Corp, 53-63; sr engr, Westinghouse Res Labs, 63-66; corp metallurgist, Worthington Corp, 67-74; CORP CONSULT ENGR, BURNS & ROE ENTERPRISES, INC, 74- *Concurrent Pos:* Chmn, Comt A-1 Steel, Am Soc Testing & Mat, 86-91, mem bd dirs, 91- *Honors & Awards:* Award of Merit, Am Soc Testing & Mat, 81; Robert J Painter Award, Standards Eng Soc, 89. *Mem:* Fel Am Soc Testing & Mat; fel Am Soc Metals; Nat Asn Corrosion Engrs; Am Welding Soc. *Res:* Development of standards for steel and liason among national and international standards organizations; metallurgy and failure analysis, particularly for power generation equipment. *Mailing Add:* 13 Brockden Dr Mendham NJ 07945

LUNT, OWEN RAYNAL, b El Paso, Tex, Apr 8, 21; m 53; c 3. SOIL FERTILITY. *Educ:* Brigham Young Univ, AB, 47; NC State Univ, PhD(agron), 51. *Prof Exp:* Lectr soil chem, NC State Univ, 50; from instr to assoc prof soil sci, 51-63, actg dir lab nuclear med & radiation biol, 65-68, actg chmn dept biophys, 65-70, PROF BIOL, UNIV CALIF, LOS ANGELES, 63-, DIR LAB NUCLEAR MED & RADIATION BIOL, 68- *Concurrent Pos:* Tech Expert, Int Atomic Energy Agency, Columbia, 71, Kenya, 83,

Malaysia, 85. *Mem:* Fel Soil Sci Soc Am; fel Am Soc Agron; Am Soc Hort Sci; Am Nuclear Soc. *Res:* Soil chemistry; environmental pollution. *Mailing Add:* Dir Lab Nuclear Med/Radiol Biol 900 Veteran Ave Los Angeles CA 90024

LUNT, STEELE RAY, b Mammoth, Utah, Jan 5, 35; m 59; c 5. MEDICAL ENTOMOLOGY. *Educ:* Univ Utah, BS, 57, MS, 59, PhD(entom), 64. *Prof Exp:* From asst prof to assoc prof, 64-74, PROF BIOL, UNIV NEBR AT OMAHA, 74- *Concurrent Pos:* Mem bd dirs, Am Mosquito Control Asn, 75-76; ed bd, Mosquito Systs, 84-89. *Mem:* Am Inst Biol Sci; Am Mosquito Control Asn; Entom Soc Am. *Res:* Control, systematics, ecology, and medical importance of mosquitoes; ecology. *Mailing Add:* 3853 N 100th Ave Omaha NE 68134

LUNTZ, MYRON, b New York, NY, Jan 16, 40; m 64; c 2. RADIATION PHYSICS. *Educ:* City Col New York, BS, 62; Univ Conn, MS, 64, PhD(physics), 68. *Prof Exp:* Res asst physics, Univ Conn, 64-68, fel, 68; vis scientist, Inst Physics, Univ Aarhus, 68-69; from asst prof to assoc prof, 69-82, chmn dept, 78-84, PROF PHYSICS, STATE UNIV NY COL FREDONIA, 82- *Concurrent Pos:* Vis assoc prof physics, Univ Del, 75-76. *Mem:* Am Phys Soc; Am Asn Physics Teachers; Sigma Xi; Soc Physics Students. *Res:* Theoretical study of the penetration of matter by energetic charged particles, with emphasis on effects associated with the spatial distribution of energy deposition about particle tracks; experimental study of surface alteration of metal substrates by ion beam irradiation. *Mailing Add:* Dept Physics State Univ NY Col Fredonia NY 14063

LUO, PEILIN, b Tianjin, China, Dec 30, 13; m 41; c 3. ELECTRONICS CIRCUIT & SYSTEM, MANUFACTURE METHODS. *Educ:* Nat Chiaotung Univ, BS, 35; Calif Inst Technol, PhD(elec eng, physics & math), 52. *Prof Exp:* Var tech positions, Chinese factories, 35-48; dept dir, Technol Dept, Admin Telecom Indust, 50-53; chief engr, NChina Combine Radio & Component Mfg, 53-56; dep chief engr, Admin Electronics Indust, 56-62; dep dir, Sci & Tech Admin, 93-82 & Sci & Tech Comt, Ministry Electronics Indust, 80-88, CONSULT, SCI & TECH ADV COMT, MINISTRY MACH & ELECTRONICS INDUST, 88- *Concurrent Pos:* Mem, Nat Natural Sci Award Comt, 65-90; guest prof, Peking Univ, Chinese Electronic Sci & Univ, Xi-Dian Electronics Sci & Tech Univ; hon prof, Beijing Inst Technol. *Honors & Awards:* Centennial medal, Inst Elec & Electronic Engrs, 84. *Mem:* Sigma Xi; fel Inst Elec & Electronic Engrs; fel Chinese Inst Electronics. *Res:* Electronic circuit, transmitters and receivers, radar system and decision theory; computer arithmatics; policy of science and especially electronics development; application of mathematics to national economics. *Mailing Add:* No Two Second Entrance Bldg 11 Nanshagou Sanlihe Beijing 100045 China

LUOMA, ERNIE VICTOR, b Sault Ste Marie, Mich, Sept 1, 32; m 54; c 5. INDUSTRIAL CHEMISTRY. *Educ:* Mich Technol Univ, BS, 54; Univ Calif, Berkeley, MS, 56; Mich State Univ, PhD(phys inorg chem), 66. *Prof Exp:* Instr chem, Mich Technol Univ, 56-57; chemist, 57-62, group leader, 62-70, res mgr, 70-77, tech dir, 77-78, dir, Anal Labs, 78-80, DIR ANALYTICAL SCI, CORP RES & DEVELOP, DOW CHEM CO, 80- *Mem:* Am Chem Soc; Am Inst Chem Engrs. *Res:* Industrial research. *Mailing Add:* Lubrizol Corp 29400 Lakeland Blvd Wickliffe OH 44092-2298

LUOMA, JOHN ROBERT VINCENT, b Huntingdon, Pa, June 3, 38; m 61; c 3. PHYSICAL CHEMISTRY. *Educ:* Ohio Univ, BA & BS, 61; Purdue Univ, Lafayette, PhD(phys chem), 66. *Prof Exp:* Asst prof chem, NDak State Univ, 66-69; asst prof, 69-74, ASSOC PROF CHEM, CLEVELAND STATE UNIV, 74- *Mem:* Am Chem Soc. *Res:* Chemical education; chemical demonstrations. *Mailing Add:* Dept of Chem Cleveland State Univ Euclid Ave E 24th St Cleveland OH 44115

LUONGO, CESAR AUGUSTO, b Montevideo, Uruguay, Oct 5, 54; m 85; c 2. SUPERCONDUCTING MAGNETS, THERMAL-FLUID SCIENCES. *Educ:* Univ Uruguay, Montevideo, Ing, 79; Stanford Univ, Palo Alto, MS, 81, PhD(mech eng), 85. *Prof Exp:* Sr engr, Res & Develop Div, Bechtel, 86-88; MGR, DEVELOP GAS PIPELINES, STONER ASSOCS INC, 88- *Concurrent Pos:* Consult, Superconducting Supercollider Lab, 90. *Mem:* Am Soc Mech Engrs; Inst Elec & Electronic Engrs. *Res:* Superconducting magnetic energy storage; magnet design; thermal and fluid dynamics analyses; electromagnetics; system studies; optimization applied to simulation of compressible flow in pipelines. *Mailing Add:* 3334 21st St San Francisco CA 94110-2317

LUPAN, DAVID MARTIN, b Cleveland, Ohio, Oct 23, 45; m 68; c 2. MEDICAL MYCOLOGY. *Educ:* Univ Ariz, BS, 67; Univ Iowa, MS, 70, PhD(microbiol), 73. *Prof Exp:* From asst prof to assoc prof, 73-87, PROF MICROBIOL, SCH MED SCI, UNIV NEV, RENO, 87- *Mem:* Sigma Xi; Am Soc Microbiol; Int Soc Human & Animal Mycol; Med Mycol Soc of the Americas. *Res:* The mechanism of pathogenesis of fungi. *Mailing Add:* Sch Med Sci Univ of Nev Reno NV 89557-0046

LUPASH, LAWRENCE O, b Bucharest, Romania, May 29, 42; US citizen; m 75. KALMAN FILTER, NUMERICAL METHODS IN CONTROL THEORY. *Educ:* Polytech Inst Bucharest, Romania, MSc, 65, PhD(control & comput eng), 72. *Prof Exp:* Researcher-engr, control eng, Inst Automation, Bucharest, Romania, 65-68; sr researcher, 71-72; sr analyst & sr researcher control & comput appln, Univ Bucharest Comput Ctr, 72-79; SR ANALYST, INTERMETRICS, INC, 80 - *Concurrent Pos:* Asst prof, Fac Automation, Polytech Inst Bucharest, 66-68 & 71-72; lectr informatics & math, Univ Bucharest, Romania, 73-78; vis lectr, Univ Tirana, Albania, 73. *Mem:* Inst Elec & Electronics Engrs; Asn Comput Mach; Soc Indust & Appl Math. *Res:* Numerical techniques for software applications in technical-scientific problems; applied mathematics; optimization; numerical methods in control theory; estimation; stability and control of multivariable systems. *Mailing Add:* Intermetrics Inc 5312 Bolsa Ave Huntington Beach CA 92649

LUPINSKI, JOHN HENRY, b Schenectady, NY, Feb 28, 27; m 54; c 3. POLYMER CHEMISTRY. *Educ:* State Univ Leyden, BS, 49, MS, 53, PhD(chem), 59. *Prof Exp:* Res chemist, Res & Develop Ctr, 60-72, proj mgr, Corp Res & Develop Ctr, 72-79, UNIT MGR CORP RES & DEVELOP, GEN ELEC CO, 79- *Mem:* AAAS; Am Chem Soc; Fedn Am Scientist. *Res:* Organic conductors; polymer electro-chemistry; electrical insulation and polymer application processes; electrostatics. *Mailing Add:* 26 Country Fair Lane Scotia NY 12302-3712

LUPLOW, WAYNE CHARLES, b Milwaukee, Wis, Jan 16, 40; m 60; c 4. DEVELOPMENT OF HIGH DEFINITION TELEVISION FOR TERRESTRIAL BROADCAST, QUALITY & RELIABILITY OF SEMICONDUCTOR COMPONENTS. *Educ:* Univ Wis, BSEE, 62; Univ Pa, MSE, 64. *Prof Exp:* Engr television res, RCA, 62-64; engr & leader television res, 64-74, eng dir components & reliability, 74-87, EXEC DIR RES & DEVELOP HDTV, ZENITH ELECTRONICS CORP, 87- *Concurrent Pos:* Publ chmn, Consumer Electronics Soc, Inst Elec & Electronic Engrs, 76-, Transactions ed, 80-; admin comt, Consumer Electronics Soc, 75- *Mem:* Sr mem Inst Elec & Electronic Engrs. *Res:* High definition television and other comunications systems relating to consumer electronics. *Mailing Add:* Zenith Electronics Corp 1000 Milwaukee Ave Glenview IL 60025

LUPSKI, JAMES RICHARD, b Rockville Center, NY, Feb 22, 57; m 86. HUMAN GENETICS, GENETIC ENGINEERING. *Educ:* NY Univ, BA, 79, MS, 83, PhD(molecular biol), 84; NY Univ Med Ctr, MD, 85. *Prof Exp:* Res asst prof biochem, dept biochem, NY Univ Med Ctr, 85-86; res asst prof pediat, 86-87, RES ASST PROF, INST MOLECULAR GENETICS, BAYLOR COL MED, 87- *Concurrent Pos:* Guest prof, Ctr Advan Molecular Biol, Punjab Univ, Lahore, Pakistan, 86. *Mem:* AMA; AAAS; NY Acad Sci. *Res:* Regulation of complex gene systems in E coli and mechanisms of DNA rearrangements. *Mailing Add:* 11102 Ashcroft Dr Houston TX 77096

LUPTON, CHARLES HAMILTON, JR, b Norfolk, Va, July 17, 19; m 45; c 3. MEDICINE, PATHOLOGY. *Educ:* Univ Va, BA, 42, MD, 44; Am Bd Path, dipl, 51. *Prof Exp:* Asst prof path, Sch Med, Univ Va, 51-53; from assoc prof to prof, 55-83, chmn dept, 61-74, EMER PROF PATH, MED CTR, UNIV ALA, BIRMINGHAM, 83- *Concurrent Pos:* Consult, Vet Admin, Birmingham, Tuskeegee. *Mem:* AAAS; Am Asn Path; Int Acad Path; AMA; Col Am Path. *Res:* Cardiovascular-renal diseases, especially the kidney as studied by simpler histochemical techniques. *Mailing Add:* Dept Path Volker Hall G025 Univ Ala Birmingham AL 35294

LUPTON, JOHN EDWARD, b Bakersfield, Calif, July 30, 44; m. CHEMICAL OCEANOGRAPHY, ISOTOPE GEOLOGY. *Educ:* Princeton Univ, BA, 66; Calif Inst Technol, PhD(physics), 72. *Prof Exp:* Asst res physicist, Scripps Inst Oceanog, 73-81; assoc res oceanogr, 81-89, RES OCEANOGR, MARINE SCI INST, UNIV CALIF, SANTA BARBARA, 89- *Concurrent Pos:* Cruise coordr res vessel, Melville Vulcan Exped, Scripps Inst Oceanog, 80-81; adj assoc prof, 81-89, adj prof geol, Univ Calif, Santa Barbara, 89- *Mem:* Am Geophys Union. *Res:* Application of helium and rare gas isotopes to ocean circulation studies; geothermal and volcanic gases; outgassing of mantle volatiles; numerical modeling of ocean tracer distributions. *Mailing Add:* Marine Sci Inst Univ Calif Santa Barbara CA 93106

LUPTON, WILLIAM HAMILTON, b Charlottesville, Va, July 25, 30. PLASMA PHYSICS. *Educ:* Univ Va, BA, 50; Univ Md, PhD(physics), 60. *Prof Exp:* Physicist, Radio Div, Nat Bur Stand, 52-55; physicist, Plasma Physics Div, US Naval Res Lab, 60-85; SR SCIENTIST, JAYCOR, 85- *Mem:* Am Phys Soc; Inst Elec & Electronics Engrs; Sigma Xi. *Res:* Plasma spectroscopy; high voltage and high current pulse technology; high power laser development. *Mailing Add:* 16509 Montecrest Lane Gaithersburg MD 20878-2163

LUQI, , b Shanghai, China, May 4, 49; m 85; c 1. RAPID PROTOTYPING, REAL-TIME SYSTEMS. *Educ:* Jilin Univ, China, BS, 75; Univ Minn, MS, 84, PhD(computer sci), 86. *Prof Exp:* Asst researcher, Sci Acad China, Peking, 75-80; Teaching res & proj asst, Computer Sci Dept, Univ Minn, 81-86; asst prof, 86-90, ASSOC PROF, COMPUTER SCI DEPT, NAVAL POSTGRAD SCH, 90- *Concurrent Pos:* Mem tech staff, Honeywell Systs, Inc, 84-85; adj asst prof, Computer Sci Dept, Univ Minn, 86-90; assoc ed, Inst Elec & Electronic Engrs Expert, J Systs Integration; mem, prog comt Syst Design & Network Conf, Inst Elec & Electronic Engrs, 89 & 90, first & second Int Conf Systs Integration, Inst Elec & Electronic Engrs/Asn Comput Mach; tech consult, Int Software Systs, Inc & Honeywell Res Ctr; prin investr, NSF, Dept Navy & Rome Air Force Develop Ctr, 87-93; NSF presidential young investr award, 90. *Mem:* Inst Elec & Electronic Engrs; Inst Elec & Electronic Engrs, Computer Soc; Asn Comput Mach. *Res:* Computer aided software engineering; designs computer languages and computer software tools which support software automation; rapid prototyping methodology and computer aided prototyping tools in terms of user interface, prototyping language specifications, software base and design database; numerical analysis; author of numerous publications and books. *Mailing Add:* Dept Computer Sci Naval Postgrad Sch Monterey CA 93943

LURA, RICHARD DEAN, b Kenosha, Wis, Aug 21, 45; m 68. PHYSICAL ORGANIC CHEMISTRY. *Educ:* Univ Wis, BS, 67; Iowa State Univ, PhD(chem), 71. *Prof Exp:* asst prof, 71-80, ASSOC PROF CHEM, MILLIGAN COL, 80- *Concurrent Pos:* Consult, R I Schattner Co, 72-79. *Res:* Research and devlepoment of germicidal and sporicidal solutions for hospital and home use. *Mailing Add:* Dept Chem Milligan Col Milligan TN 37682

LURAIN, JOHN ROBERT, III, b Princeton, Ill, Oct 27, 46; m 69; c 2. GYNECOLOGIC ONCOLOGY. *Educ:* Oberlin Col, AB, 68; Univ NC, MD, 72. *Prof Exp:* Resident obstet-gynec, Magee Womens Hosp, Univ Pittsburgh, 72-75; lieutenant commander, med corps, US Naval Reserve, Naval Regional

Med Ctr, Portsmouth, Va, 75-77; fel gynec oncol, Roswell Park Mem Inst, 77-79; from asst prof to assoc prof, 79-88, PROF OBSTET-GYNEC, SCH MED, NORTHWESTERN UNIV, 89- *Concurrent Pos:* Galloway fel, Mem Sloan Kettering Cancer Ctr, 75; clin fel, Am Cancer Soc, 77-79; jr fac fel, 80-83; dir, div gynec oncol, Prentice Women's Hosp, Sch Med, Northwestern Univ, 85-; assoc dir, John I Brewer Trophoblastic Dis Ctr, 79- *Honors & Awards:* Purdue-Frederick Award, Am Col Obstet Gynec, 83. *Mem:* Soc Gynec Oncologists; Am Soc Clin Oncol; Am Soc Colposcopy & Cervical Path; Am Col Obstetricians & Gynecologists. *Res:* Gestational trophoblasic disease; endometrial cancer; cervical cancer; laser therapy; hormone receptors; chemotherapy and surgery; ovarian cancer. *Mailing Add:* Prentice Women's Hosp 333 E Superior St Chicago IL 60611

LURCH, E(DWARD) NORMAN, b Morristown, NJ, Dec 23, 19; m 41; c 4. ELECTRICAL ENGINEERING. *Educ:* NY Univ, BEE, 40, MEE, 43. *Prof Exp:* Tutor elec eng, City Col New York, 41-43; instr, Manhattan Col, 43-47; asst prof, Univ Fla, 47; asst prof, Clarkson Col Technol, 48-49; assoc prof electronics, State Univ NY Agr & Tech Col, Farmingdale, 49-60; chief engr, Chemtronics, Inc, 60-61; aerospace technologist, Goddard Space Flight Ctr, NASA, 62-65; assoc prof, 65-66, PROF ELECTRONICS, STATE UNIV NY AGR & TECH COL, FARMINGDALE, 66- *Concurrent Pos:* Lectr grad div, State Univ NY Col, New Paltz, 58-61; consult engr, Oil Heat Inst, Long Island, 57-59. *Res:* Fundamentals of electronics; electric circuits. *Mailing Add:* 11 Black Duck Dr Stony Brook NY 11790

LURIA, S(AUL) M(ARTIN), b Athol, Mass, Dec 24, 29; m 63; c 2. PHYSIOLOGICAL PSYCHOLOGY. *Educ:* Univ Richmond, BS, 49; Univ Pa, MA, 51, PhD(psychol), 55. *Prof Exp:* Res psychologist, 57-83, HEAD, VISION DEPT, US NAVAL SUBMARINE MED CTR, 83- *Concurrent Pos:* Lectr, Univ RI, 66-, Univ Conn, 68-70 & Univ New Haven, 71- 90. *Mem:* Fel AAAS; fel Am Psychol Asn; fel Optical Soc Am; Psychonomic Soc; fel NY Acad Sci. *Res:* Vision. *Mailing Add:* 35 Beacon Hill Dr Waterford CT 06385

LURIA, SALVADOR EDWARD, bacteriology; deceased, see previous edition for last biography

LURIE, ALAN GORDON, b Los Angeles, Calif, Apr 23, 46; m 69; c 2. RADIATION BIOLOGY, CARCINOGENESIS. *Educ:* Univ Calif, Los Angeles, DDS, 70; PhD(radiation biol, biophys), 74. *Prof Exp:* Asst prof oral radiol, 73-77, asst prof oral diag, 77, ASSOC PROF ORAL DIAG, UNIV CONN HEALTH CTR, 77- *Concurrent Pos:* Consult oral radiol, Newington Vet Admin Hosp, 75-; HEW/NIH grants, Am Cancer Soc, 75- *Honors & Awards:* E H Hatton Award, Int Asn Dent Res, 69. *Mem:* Am Asn Cancer Res; Radiation Res Soc; Int Asn Dent Res. *Res:* Radiation pathophysiology; radiation carcinogenesis and cocarcinogenesis at low doses; chemical carcinogenesis; mechanistic roles of vascular changes during carcinogenesis. *Mailing Add:* Dept Oral Radiol Univ Conn Health Ctr Farmington CT 06032

LURIE, ARNOLD PAUL, b Brooklyn, NY, July 22, 32; m 54; c 3. ORGANIC CHEMISTRY. *Educ:* NY Univ, BA, 54; Purdue Univ, PhD(org chem), 58. *Prof Exp:* Lab asst org chem, Purdue Univ, 54-56, fel, 58; res chemist, Eastman Kodak Co, 58-61, sr res chemist, 61-66, info scientist, Res Lab, 65-89, res assoc, 66-89; RETIRED. *Mem:* Am Chem Soc. *Res:* Synthetic and theoretical organic chemistry related to photographic systems; computerized handling of information. *Mailing Add:* 4380 Camrose Lane West Palm Beach FL 33417

LURIE, FRED MARCUS, b Boston, Mass, Nov 16, 30. PHYSICS. *Educ:* Univ NC, Chapel Hill, BS, 52; Univ Ill, Urbana, MS, 57, PhD, 63. *Prof Exp:* Teaching asst physics, Univ Ill, Urbana, 57-59; from instr to asst prof, Univ Pa, 63-67; asst prof, 67-70, ASSOC PROF PHYSICS, IND UNIV, BLOOMINGTON, 70- *Mem:* Am Phys Soc. *Mailing Add:* Dept Physics Ind Univ Bloomington IN 47401

LURIE, HAROLD, b Durban, SAfrica, Mar 28, 19; nat US; div; c 2. ENGINEERING. *Educ:* Univ Natal, BSc, 40, MSc, 46; Calif Inst Technol, PhD(aeronaut), 50; Northeastern Univ, JD, 89. *Prof Exp:* Lectr aeronaut, Calif Inst Technol, 48-50; head weapons effectiveness group, Rand Corp, 50-52; from asst prof appl mech to prof eng sci & assoc dean grad studies, Calif Inst Technol, 53-71; dir res & develop, New Eng Elec Syst, 71-79; dean eng, Polytech Inst NY, 79-81; dean eng, Northeastern Univ, Boston, 81-86; ASSOC DIR, CTR LAW, SCI & TECHNOL, UNIV WASH, 86- *Concurrent Pos:* Sr develop engr, Oak Ridge Nat Lab, 56-57; consult, Yankee Atomic Elec Co, 70-71; actg dir, Advan Systs Dept, Elec Power Res Inst, 74-75. *Mem:* Am Nuclear Soc; Am Soc Eng Educ; assoc fel Am Inst Aeronaut & Astronaut; assoc fel Royal Aeronaut Soc. *Res:* Energy conversion; nuclear and aerospace engineering; structural mechanics; law and technology. *Mailing Add:* Sch Law JB 20 Univ Wash Seattle WA 98105

LURIE, JOAN B, b New York, NY, Jan 21, 41; m 61; c 2. THEORETICAL SOLID STATE PHYSICS. *Educ:* Brooklyn Col, BS, 61; Rutgers Univ, MS, 62, PhD(physics), 67. *Prof Exp:* Mem tech staff physics res, RCA Labs, 62-66; fel appl math, Univ Col, Univ London, 67-68; syst programmer comput sci, Appl Data Res, 69-70; fel solid state physics, Rutgers Univ, 70-72; from asst prof to assoc prof physics, Rider Col, 72-81; mem res staff, IDA, NJ, 81-84; dept mgr, Hughes, 84-89; prin scientist, Mitre Corp, 89-90; ADVAN SYSTS MGR, TRW, 91- *Concurrent Pos:* Am Phys Soc indust fel, Colgate Palmolive Res Lab. *Mem:* Am Phys Soc; Inst Elec & Electronics Engrs. *Res:* Theoretical research in lattice dynamics of solid state of rare gases; computer assisted instruction, particularly in physics and mathematics; image analysis. *Mailing Add:* TRW M/S R2/2170 One Space Park Redondo Beach CA 90277

LURIE, NORMAN A(LAN), b Detroit, Mich, Dec 2, 40; m 67; c 2. NUCLEAR PHYSICS & ENGINEERING, ELECTROOPTICS. *Educ:* Univ Mich, BSE, 63, MSE, 65, PhD(nuclear eng), 69. *Prof Exp:* Fel physics, Univ Mo-Columbia, 69-71; sr res assoc, Brandeis Univ, 71-74; sr physicist, IRT Corp, 74-75, staff physicist, 75-76, prog mgr res, Nuclear Systs Div, 78-81, prin physicist, 76-86, mgr tech pers, 81-86; SR SCIENTIST, SCI APPLICATIONS INT CORP, 87-, DEPT DIR MGR, 89- *Concurrent Pos:* Res collabr, Brookhaven Nat Lab, 71-74. *Mem:* Sigma Xi. *Res:* Applied nuclear physics; electrooptics. *Mailing Add:* 10436 El Comal Dr San Diego CA 92124

LURIE, ROBERT M(ANDEL), b Boston, Mass, Feb 24, 31; m 53; c 3. CHEMICAL ENGINEERING, COLLOIDAL CHEMISTRY. *Educ:* Mass Inst Technol, SB, 52, ScD(chem eng), 55. *Prof Exp:* Chem engr & prod res mgr, Dewey & Almy Chem Co Div, W R Grace & Co, 55-60; sr chem engr, Ionics, Inc, 60-63; mgr mat develop, Res & Adv Develop Div, Avco Corp, 63-65; dir mats, Systs Div, 65-70; PRES, NYACOL PROD INC, 70- *Mem:* Soc Plastics Engrs; Am Inst Aeronaut & Astronaut; Am Chem Soc; Am Inst Chem Engrs; Fire Retardant Chem Asn. *Res:* Polymer synthesis; adhesion of polymers; unit operations of polymer manufacture and polymer fabrications; electrochemistry; fuel cells; ablation phenomena; physics of reinforced plastics; reentry vehicle design; organic dyes; colloidal chemicals. *Mailing Add:* 4 Tufts Rd Lexington MA 02173

LURIX, PAUL LESLIE, JR, b Bridgeport, Conn, Apr 6, 49; m 70; c 3. INFRARED SPECTROSCOPY, DATA BASE APPLICATIONS. *Educ:* Drew Univ, BA, 71; Purdue Univ, MS, 73. *Prof Exp:* Tech dir, Analysts Inc, 76-77; chief chemist, Caleb Brett USA, Inc, 77-80; vpres, Tex Labs, Inc, 80-82; PRES, LURIX CORP, 82- *Concurrent Pos:* Consult, 77-82, Phillips 66, 86- & Conoco, Inc, 87-; vpres, Diesel King Corp, 80-82. *Mem:* Fel Am Inst Chemists; Am Chem Soc; Am Soc Testing & Mat; Soc Appl Spectros; NY Acad Sci; AAAS. *Res:* Design and implementation of multi-user information systems; studies of liquid structure through infrared spectroscopy. *Mailing Add:* PO Box 148 Fulshear TX 77441

LURKIS, ALEXANDER, b New York, NY, Oct 1, 08; m 30; c 1. ENGINEERING EXPERT IN POWER & LIGHTING. *Educ:* Cooper Union, BSEE, 30; NY Univ, BSEE, 34; Univ State NY, Tech Teacher Cert, 36. *Prof Exp:* Jr elec engr, New York City Bd Transp & Transit Authority, 30-40; sr elec engr, 42-58; elec engr, F R Harris, Inc, consult engrs, 40-41; chief eng, Bur Gas & Elec, New York City Dept Water, Gas & Elec, 59-64, act comm, 61; pres consult engrs, 64-90; CONSULT ENGR, ALEXANDER LURKIS, PE, 91- *Concurrent Pos:* Vpres, Peak Tech Asn, 51-59; arbitrator, Am Arbitration Asn, 64-; chmn, Illuminating Eng Soc Energy Comt, 74-77; secy-treas, Glimmer Security Systs, 76-81; pres, Icare Press, 81-85. *Honors & Awards:* Design Excellence, Fifth Biennial Am Iron & Steel Inst, 72; Design Excellence, Fifth Biennial HUD Award, 72. *Mem:* Fel Illuminating Eng Soc; sr mem Inst Elec & Electronic Engrs; fel NY Acad Sci. *Res:* Ten utility standard US patents; 1 Canadian standard patent; 1 traffic signal US patent; 1 US patent for museum security. *Mailing Add:* Alexander Lurkis PE 193-12 Nero Ave Holliswood NY 11423

LURYI, SERGE, b Leningrad, USSR, Oct 9, 47; Can citizen; m 82; c 3. PHYSICS OF EXPLORATORY SEMICONDUCTOR DEVICES, PHYSICS OF ELECTRONICS MATERIALS. *Educ:* Univ Leningrad, USSR, dipl physics, 71; Univ Toronto, Can, MSc, 75, PhD(physics), 78. *Prof Exp:* Res engr, VNIIG, Leningrad, 71-73; Postdoctoral fel, Univ Toronto, 78-80; mem tech staff, 80-85, supvr, 85-90, DISTINGUISHED MEM TECH STAFF, AT&T BELL LABS, MURRAY HILL, 90- *Concurrent Pos:* Ed, Inst Elec & Electronic Engrs Elec Devices, 86-90. *Mem:* Fel Inst Elec & Electronic Engrs; Inst Elec & Electronic Engrs, Electron Devices Soc; Am Phys Soc. *Res:* Physics of exploratory semiconductor devices; inventing novel devices and principles, electronic and optoelectronic. *Mailing Add:* AT&T Bell Labs 600 Mountain Ave Rm 2d-230 Murray Hill NJ 07974

LUSAS, EDMUND W, b Woodbury, Conn, Nov 25, 31; m 57; c 3. FOOD SCIENCE, FOOD TECHNOLOGY. *Educ:* Univ Conn, BS, 54; Iowa State Univ, MS, 55; Univ Wis, PhD(food technol), 58; Univ Chicago, MBA, 72. *Prof Exp:* Proj leader, Res Labs, Quaker Oats Co, 58-64, mgr canned foods res, 64-66, mgr pet foods res, 66-72, mgr sci serv, 72-77; DIR, FOOD PROTEIN RES & DEVELOP CTR, TEX A&M UNIV, 78- *Mem:* Sigma Xi; Inst Food Technol; Am Chem Soc; Am Oil Chemists Soc; Am Cereal Chem Asn; Am Soc Agr Engr. *Res:* Protein and oil utilization from cottonseed, peanuts, soy, sunflower and sesame; development of processes for converting crops into food, feed and industrial ingredients; human and pet food development; research and development administration; technical staff services management. *Mailing Add:* Food Protein Res & Develop Ctr FM 183 Tex A&M Univ College Station TX 77843-2476

LUSBY, WILLIAM ROBERT, b Washington, DC, June 23, 40. MASS SPECTROMETRY, STRUCTURAL ELUCIDATION. *Educ:* Univ Md, BS, 67. *Prof Exp:* Chemist, Meat & Carcass Lab, 67-73, Anal Chem Lab, 73-75, Pesticide Degradation Lab, 75-81, RES CHEMIST, INSECT & NEMATODE HORMONE LAB, USDA, BELTSVILLE, MD, 81- *Mem:* Am Soc for Mass Spectrometry; Am Chem Soc; NY Acad Sci. *Res:* Development of mass spectrometric methods of analysis for agricultural and natural products; structural elucidation of natural products. *Mailing Add:* USDA ARS Insect Neurobiol & Hormone Lab Bldg 467 Barc East Beltsville MD 20705

LUSCHER, ULRICH, b Oftringen, Switz, July 18, 32; m 62, 83; c 2. ARCTIC ENGINEERING, HAZARDOUS WAST ENGINEERING. *Educ:* Swiss Fed Inst Technol, BS, 56; Mass Inst Technol, SM, 59, ScD(civil eng & soil mech), 63. *Prof Exp:* Designer, Vevey Metal Works, Switz, 57 & Stone & Webster Eng Corp, 58-59; res engr, Mass Inst Technol, 59-60, asst prof civil eng, 63-67; mem staff, 67-74, PRIN MEM, WOODWARD-CLYDE CONSULTS, 75- *Mem:* Am Soc Civil Engrs; Int Soc Soil Mech & Found Eng; Am Consult Engrs. *Res:* Soil mechanics and foundation engineering; research in soil mechanics; underground structures; permafrost and arctic engineering. *Mailing Add:* Woodward-Clyde Consults 500 12th St Oakland CA 94607-4014

LUSCOMBE, HERBERT ALFRED, b Johnstown, Pa, Aug 9, 16; m 42; c 3. DERMATOLOGY. *Educ:* St Vincent Col, BSc, 36; Jefferson Med Col, MD, 40. *Prof Exp:* prof, 59-, EMER PROF DERMAT, JEFFERSON MED COL. *Mem:* Sigma Xi. *Mailing Add:* 111 S 11th St Suite 4265 Philadelphia PA 19107

LUSHBAUGH, CLARENCE CHANCELUM, b Covington, Ky, Mar 15, 16; m 42, 63; c 3. PATHOLOGY. *Educ:* Univ Chicago, BS, 38, PhD(path), 42, MD, 48. *Prof Exp:* Asst path, Univ Chicago, 39-42, from instr to asst prof, 42-49, pathologist, Toxicity Lab, 41-49; mem staff, Los Alamos Sci Lab, Univ Calif, 49-63; chief scientist appl radiobiol, 63-75, NASA Total Body Irradiation Proj, 64-75, chmn, Med & Health Sci Div, 75-83, CHIEF, RADIATION MED, OAK RIDGE ASSOC UNIVS, 84-, PATHOLOGIST, RADIATION ASSISTANCE CTR-TRAINING SITE, 74- *Concurrent Pos:* Pathologist, Los Alamos Med Ctr, NMex, 49-63; mem path study sect, NIH, 61-64; mem radiobiol adv panel, Space Sci Bd, Nat Acad Sci-Nat Res Coun, 66-72; mem adv comt space radiation effects lab, Col William & Mary, 68-70. *Mem:* Am Soc Exp Path; Soc Exp Biol & Med; Health Physics Soc; Radiation Res Soc; AAAS. *Res:* Pathology of obstetric shock; chemotherapy of cancer; mitotic poisons; radiation damage; diagnostic radioisotopology; human radiobiology; electronic clinical pathology; primate pathology. *Mailing Add:* Dist Scientist Med Sci Div Oak Ridge Assoc Univ PO Box 117 Oak Ridge TN 37831-0117

LUSHBOUGH, CHANNING HARDEN, b Watertown, SDak, Aug 11, 29; m 52; c 4. NUTRITION, RESOURCE MANAGEMENT. *Educ:* Univ Chicago, AB, 48, AM, 52, PhD(nutrit, biochem), 56. *Prof Exp:* Res chemist, Res Lab, Carnation Co, Wis, 50-51; assoc biochemist & actg chief, Div Biochem & Nutrit, Am Meat Inst Found, Ill, 56-59; dir prod info, Res Ctr, Mead Johnson & Co, Ind, 59-67; vpres planning & develop, Blue Cross, NY, 67-71; assoc dir, Consumers Union US, 71-73; dir & exec secy, Citizens Comn on Science, Law & Food Supply, Rockefeller Univ, 73-75; vpres qual assurance, Kraft, Inc, 75-81; mkt rep, Tweedy, Browne Inc, 81-84; MKT REP, ROUND HILL ASSET MGT INC, 86- *Concurrent Pos:* Instr grad nutrit, Ill Inst Technol, 56; lectr, Univ Chicago, 57-59 & Northwestern Univ, 58-59. *Mem:* Am Inst Nutrit; AAAS; Gt Brit Nutrit Soc; Inst Food Technologists; NY Acad Sci; Sigma Xi. *Res:* Nutritional quality of natural proteins; effects of processing on vitamin retention; relations of dietary fat, protein and carbohydrate to atherosclerosis. *Mailing Add:* 420 Elm St Glenview IL 60025-4949

LUSIGNAN, BRUCE BURR, b San Francisco, Calif, Dec 22, 36; m 58; c 3. ELECTRICAL ENGINEERING. *Educ:* Stanford Univ, BS, 58, MS, 59, PhD(elec eng), 63. *Prof Exp:* Instr, 62-63, res asst, 62-64, actg asst prof, 63-65, asst prof, 65-68, ASSOC PROF ELEC ENG, STANFORD UNIV, 68-, DIR, COMMUN SATELLITE PLANNING CTR, 68- *Res:* Applications of satellite, radio and digital technology to communications; transfer of planning and manufacturing knowledge to developing countries. *Mailing Add:* Commun Satellite Planning Ctr Stanford Univ Bldg ERL Rm 202 Stanford CA 94305-4053

LUSIS, ALDONS JEKABS, b Esslingen, Ger, June 22, 47; US citizen. MOLECULAR BIOLOGY. *Educ:* Wash State Univ, BS, 69; Ore State Univ, PhD(biochem), 73. *Prof Exp:* Res assoc molecular biol, Roswell Park Mem Inst, 73-80; AT DEPT MICROBIOL, UNIV CALIF, LOS ANGELES. *Concurrent Pos:* NIH fel, 74- *Mem:* Sigma Xi. *Res:* Mechanisms controlling developmental expression of enzymes in mammals; processing of mouse lysosomal enzymes. *Mailing Add:* 1901 S Bentley Los Angeles CA 90025

LUSK, JOAN EDITH, b Teaneck, NJ, July 29, 42. BIOCHEMISTRY. *Educ:* Radcliffe Col, BA, 64; Harvard Univ, PhD(biol chem), 70. *Prof Exp:* Nat Cystic Fibrosis Res Found fel biol, Mass Inst Technol, 70-71, NIH fel biol, 71-72; asst prof, 72-77, ASSOC PROF CHEM, BROWN UNIV, 77- *Concurrent Pos:* Prin investr, NIH res grant, 73- & NSF grant, 74-; NIH career develop award, 76. *Mem:* Am Soc Microbiol; AAAS. *Res:* Membrane structure and function; colicin action; transport. *Mailing Add:* Assoc Dean Grad Sch Box 1867 Brown Univ Providence RI 02912

LUSKIN, LEO SAMUEL, b Buffalo, NY, Feb 1, 14; m 39; c 3. ORGANIC CHEMISTRY, POLYMER CHEMISTRY. *Educ:* Univ Mich, BSc, 36, MSc, 42. *Prof Exp:* Chemist coal tar prod, Barrett Div, Allied Chem Corp, 43-44; chemist org chem & polymers, Rohm & Haas Co, 44-62, head tech writing sect, Spec Prod Dept, 62-68, promotion mgr, indust chem, plastics intermediates, 68-76, sr tech writer, Advert Dept, 76-84; RETIRED. *Concurrent Pos:* Consult, Rohm & Haas Co & A I Paul Lefton Co. *Mem:* Am Chem Soc. *Res:* Synthesis of organic chemicals; polymers; monomers. *Mailing Add:* 2317 Spruce St Philadelphia PA 19103

LUSKIN, MITCHELL B, b Pasadena, Calif, Nov 13, 51; m 76; c 3. NUMERICAL ANALYSIS, PARTIAL DIFFERENTIAL EQUATIONS. *Educ:* Yale Univ, BS, 73; Univ Chicago, MS, 74, PhD(math), 77. *Prof Exp:* Hildebrandt asst prof math, Univ Mich, 77-79, asst prof, 79-81; prof appl math, Calif Inst Technol, 89-90; from asst prof to prof, 81-89, PROF MATH, UNIV MINN, 90- *Concurrent Pos:* Vis prof, Ecole Polytechnique Federale, Lausanne, Switz, 80; vis mem, Courant Inst, NY Univ, 80-81; ed, Soc Indust & Appl Math J Numerical Anal, 82-90, managing ed, 90-; fel, Minn Supercomputer Inst, 85-, mem grad fac, Dept Aerospace Eng & Mech, 87-; ed, Dynamics & Differential Equations, 88- *Mem:* Am Math Soc; Soc Indust & Appl Math. *Res:* Scientific computing; numerical analysis; applied mathematics; partial differential equations; computational mechanics. *Mailing Add:* Sch Math Univ Minn 206 Church St SE Minneapolis MN 55455

LUSS, DAN, b Tel Aviv, Israel, May 5, 38; m 66; c 3. CHEMICAL REACTION ENGINEERING. *Educ:* Israel Inst Technol, BSc, 60, MSc, 63; Univ Minn, Minneapolis, PhD(chem eng), 66. *Prof Exp:* Asst prof chem eng, Univ Minn, 66-67; from asst prof to assoc prof, 67-72, PROF CHEM ENG, UNIV HOUSTON, 72-, CHMN DEPT, 75- *Concurrent Pos:* Dir, Am Inst Chem Engrs, 86-88. *Honors & Awards:* Honor Scroll, Indust & Eng Div, Am Chem Soc, 68; A P Colburn Award, 72; Prof Progress Award, Am Inst Chem Engrs, 78 & Wilhem Award, 86. *Mem:* Nat Acad Eng; Am Chem Soc; Am Inst Chem Engrs. *Res:* Stability of chemical reactors; diffusional effects in catalysts; lumping of complex reactions networks; synthesis of superconducting ceramics. *Mailing Add:* Dept Chem Eng Univ Houston Houston TX 77004

LUSSIER, ANDRE (JOSEPH ALFRED), b Sherbrooke, Que, May 27, 33; m 61; c 3. INTERNAL MEDICINE, RHEUMATOLOGY. *Educ:* Univ Montreal, BA, 54, MD, 59; Col Med, Que, cert internal med, 64 & cert rheumatology, 70; Royal Col Physicians & Surgeons, cert internal med, 65; FCRCP(C), 72. *Prof Exp:* Assoc clin res, Can Arthritis Soc, 69-79; dir, Rheumatic Dis Unit, 69-84, assoc prof med, 69-75, dir, Clin Res Ctr, 80-84, PROF MED, UNIV SHERBROOKE & UNIV HOSP CTR, 75- *Concurrent Pos:* Mem, Med Res Coun Can, 88- *Honors & Awards:* Basic Res Prize, Asn Fr Lang Physicians Can. *Mem:* Am Rheumatism Asn; Royal Col Med; Can Rheumatism Asn (secy, 81-84 & vpres, 84-86, pres, 86-88); Pan Am League Against Rheumatism (vpres, 82-86); Can Soc Clin Invest; hon mem Fr Soc Rheumatology. *Res:* Etiopathogenesis and pathological ossification of the spine; mechanism of microcrystal arthritides; normalization of terminology used in semiology of musculoskeletal system; efficacy and toxicity of non steroidal antiinflammatory drugs, especially of the consecutive gastro-intestinal microbleeding; hyperostotic disease: basic and clin research (pathogenesis). *Mailing Add:* Dept Med Rheumatol Univ Sherbrooke Fac Med Sherbrooke PQ J1H 5N4 Can

LUSSIER, GILLES L, b St Charles, Que, Oct 5, 34; m 57; c 2. VETERINARY PATHOLOGY, VIROLOGY. *Educ:* Univ Montreal, BA, 55, DVM, 59; Univ Toronto, MSc, 61, Univ Sask, PhD(path), 73. *Prof Exp:* From res asst to res assoc, Armand-Frappier Inst, 61-74, prof, 74-75, head virol, Res Ctr, 75-83, PROF PATH & VIROL, ARMAND-FRAPPIER INST, 83- *Mem:* Can Vet Med Asn; Can Soc Microbiologists; Can Asn Lab Animal Sci; Can Asn Vet Pathologists; Am Asn Lab Animal Sci. *Res:* Comparative pathology of viral diseases and viral vaccines; diagnosis of viral diseases of laboratory animals. *Mailing Add:* Inst Armand-Frappier PO Box 100 Laval PQ H7N 4Z3 Can

LUSSIER, ROGER JEAN, b Newport, RI, Apr 29, 43; m 66; c 2. SYNTHETIC INORGANIC & ORGANOMETALLIC CHEMISTRY. *Educ:* Univ Mass, Amherst, BS, 65; Brown Univ, PhD(inorg chem), 69; Johns Hopkins Univ, MA, 75. *Prof Exp:* NSF grant, Cath Univ Am, 69-70; res chemist, 70-80, sr res chemist, 80-82, res assoc, 81-84, ENG SPECIALIST, DAVISON DIV, W R GRACE & CO, 85- *Res:* Heterogeneous catalysis; reaction mechanisms; homogeneous catalysis; transition metal chemistry; surface chemistry; mineral synthesis and chemistry; zeolite synthesis and characterization. *Mailing Add:* Davison Tech Ctr W R Grace & Co 5603 Chemical Rd Baltimore MD 21226

LUSSKIN, ROBERT MILLER, b Dec 14, 21; m 47; c 2. ORGANIC CHEMISTRY. *Educ:* Harvard Univ, AB, 43; NY Univ, MS, 46, PhD(chem), 50. *Prof Exp:* Chemist, Spencer Kellogg & Sons, 43 & Grosvenor Labs, 45-46; with Trubek Labs, 47-55, res dir, 56-60; dir chem res, UOP Chem Co, Universal Oil Prod Co, 60-67; supt nonwoven lab, Kimberly-Clark Corp, Wis, 67-68, mgr basic & explor res, 68-72, mgr new concepts res, 72-75; tech dir, Resource Planning Assocs, 75-77; dir tech serv, Raltech Sci Serv, 77-82; dir, Cent Res & Develop, Ralston Purina Co, 82-86; adj prof chem, Univ Mo, St Louis, 87; RETIRED. *Mem:* AAAS; Am Chem Soc. *Res:* Business strategy development; consumer new products; polymer and fiber research; chemical intermediates; energy and materials management; analytical and environmental chemistry management. *Mailing Add:* 12856 Hawthicket Lane Des Peres MO 63131

LUSTBADER, EDWARD DAVID, b Baltimore, Md, June 15, 46; m 69; c 2. BIOSTATISTICS. *Educ:* Case Inst Technol, BS, 67; Univ Pa, PhD(statist), 72. *Prof Exp:* STATISTICIAN, FOX CHASE CANCER CTR, 72- *Concurrent Pos:* Res assoc, Mgt Sci Ctr, Univ Pa, 70-72. *Mem:* Am Statist Asn; Biometric Soc; AAAS. *Res:* Model building for survival studies; statistical computing; relation of diet and growth to cancer. *Mailing Add:* Fox Chase Cancer Ctr 7701 Burholme Ave Philadelphia PA 19111

LUSTED, LEE BROWNING, b Mason City, Iowa, May 22, 22; m 43; c 2. RADIOLOGY. *Educ:* Cornell Col, BA, 43; Harvard Med Sch, MD, 50; Am Bd Radiol, dipl. *Hon Degrees:* DSc, Cornell Col, 63. *Prof Exp:* Spec res assoc, Radio Res Lab, Harvard Univ, 43-46; from instr to asst prof radiol, Med Sch, Univ Calif, San Francisco, 55-57; asst radiologist, NIH, 57-58; from asst prof to assoc prof radiol, Sch Med, Univ Rochester, 58-60, prof biomed eng, 60-62; prof radiol, Med Sch, Univ Ore & sr scientist, Ore Primate Res Ctr, 62-68; prof radiol & chmn dept, Stritch Sch Med, Loyola Univ Chicago, 68-69; prof radiol & vchmn dept, Univ Chicago, 69-78; radiologist, Southern Calif Permanente Med Group, 78-89; RETIRED. *Concurrent Pos:* Chmn, Comt Comput Biol & Med, Nat Acad Sci-Nat Res Coun, 58-59; consult, Strong Mem Hosp, 58-62; chmn, Adv Comt Comput Res, NIH, 60-64; assoc dean prof affairs & chief of staff, Loyola Univ Hosp, Chicago, 68-69; clin prof radiol, Univ Calif, San Diego, 78-; ed-in-chief, Int J Med Decision Making, 78-85; adj distinguished clin mem, Scripps Clin & Res Found, La Jolla, 78- *Mem:* Fel AAAS; fel Am Col Radiol; fel Inst Elec & Electronics Engrs; Roentgen Ray Soc; Radiol Soc NAm; fel Am Col Med Informatics. *Res:* Study of medical decision making; application of signal detection theory to assess system and observer performance in radiographic diagnosis. *Mailing Add:* 323 Concord Dr Menlo Park CA 94025

LUSTER, MICHAEL I, b Malden, Mass, Sept 18, 47. IMMUNOTOXICOLOGY. *Educ:* Loyola Univ, PhD(microbiol & immunol), 74. *Prof Exp:* GROUP LEADER, IMMUNOTOXICOL GROUP, NAT INST ENVIRON HEALTH, NIH, 76- *Mem:* Soc Toxicol; Am Asn Immunol; Int Soc Immunopharmacol. *Mailing Add:* Nat Inst Environ Health NIH PO Box 12233 Research Triangle Park NC 27709

LUSTGARTEN, RONALD KRISSES, b New York, NY, Feb 24, 42. SOLVOLYSIS, PATENTS. *Educ:* Columbia Univ, AB, 62; Pa State Univ, PhD(chem), 66. *Prof Exp:* NIH fel, Univ Calif, Los Angeles, 66-68; res assoc & Mellon fel chem, Carnegie-Mellon Univ, 68-72; RES/STAFF SCIENTIST, UPJOHN CO, 72- *Concurrent Pos:* NSF fel, 65; secy, Am Chem Soc, Kalamazoo Sect, 83-85. *Mem:* AAAS; Am Chem Soc; Fedn Am Scientists; Sigma Xi. *Res:* Organic mechanisms; reactive intermediates; kinetics. *Mailing Add:* Upjohn Co 7298-24-2 Kalamazoo MI 49001-3298

LUSTICK, SHELDON IRVING, b Syracuse, NY, Aug 16, 34; m 70. ENVIRONMENTAL PHYSIOLOGY, VERTEBRATE ZOOLOGY. *Educ:* San Fernando Valley State Col, BA, 63; Syracuse Univ, MS, 65; Univ Calif, Los Angeles, PhD(zool), 68. *Prof Exp:* From asst prof to assoc prof, 68-74, PROF ZOOL, OHIO STATE UNIV, 76-, DIR, ENVIRON BIOL GRAD PROG, 77- *Concurrent Pos:* Dept of Interior res grant, 69-72 & 73-75; NSF grant, 76-78 & 80-83; Air Force Off Sci Res grant, 78-80. *Mem:* AAAS; Cooper Ornith Soc; Am Ornith Soc; Ecol Soc Am; Sigma Xi. *Res:* How animals adapt physiologically to environmental stress. *Mailing Add:* Dept Zool Ohio State Univ 1735 Neil Ave Columbus OH 43210

LUSTIG, BERNARD, b Kolomea, Austria, Dec 21, 02; nat US; m 38; c 2. BIOCHEMISTRY. *Educ:* Univ Vienna, PhD(chem), 25. *Prof Exp:* Chemist, Rudolf Hosp, Vienna, 26-32; chief biochemist, Pearson Cancer Found, 33-38 & West London Hosp, 38-40; biochemist, Lawrence R Bruce Inc, 40-44, dir res, 45-58; vpres chg res, Clairol Inc, 58-68, vpres & dir basic res, 68-70; RETIRED. *Honors & Awards:* Prize, Asn Chocolate Mfrs, 30. *Mem:* Fel AAAS; Am Chem Soc; Soc Exp Biol & Med; fel Textile Inst (Eng); fel NY Acad Sci. *Res:* Chemistry and biochemistry of proteins and lipids; biochemistry of cancer; chemistry and technology of keratin fibers; author of over 100 publications. *Mailing Add:* 38 Chester St Stamford CT 06905

LUSTIG, HARRY, b Vienna, Austria, Sept 23, 25; nat US; m 80; c 2. SOLAR ENERGY. *Educ:* City Col N Y, BS, 48; Univ Ill, MS, 49, PhD(physics), 53. *Prof Exp:* Asst physics, Univ Ill, 49-53; from instr to prof, 53-86, chmn dept, 65-70, exec officer PhD prog physics, 68-70, assoc dean sci, 72-75, dean col lib arts, 73-74, dean sci, 75-82, provost & vpres acad affairs, 82-85, RESIDENT PROF PHYSICS, CITY COL NY, 86- *Concurrent Pos:* Prin scientist, Nuclear Develop Corp Am, 56-61; vis res asst prof, Univ Ill, 59-60; fel, Colo Inst Theoret Physics, 60; Fulbright lectr, Univ Dublin, 64-65; vis prof, Univ Colo, 66 & Univ Wash, 67 & 69; sr officer, UNESCO, Paris, 70-72; consult, UNESCO, Paris, 72-75 & 79, US Int Commun Agency, 78, Univ SDak, 80, Univ SFla, 81, Univ Mass, 83, NJ State Dept Higher Educ, 83-; educ adv comt, NY Acad Sci, 82-84; mem, gov bd, NY Acad Sci, 82-84, Am Inst Physics, 85-; mem, US Nat Comt, Int Union Pure & appl Physics. *Mem:* Fel Am Phys Soc (treas, 85-); AAAS; fel NY Acad Sci (vpres, 84); Sigma Xi; Am Asn Physics Teachers; Am Asn Univ Prof. *Res:* Theoretical nuclear physics; Mossbauer effect; solar energy; science education. *Mailing Add:* 54 Riverside Dr New York NY 10024

LUSTIG, HOWARD E(RIC), b Vienna, Austria, Oct 23, 25; US citizen; m 50; c 3. ELECTRONICS, SYSTEMS ENGINEERING. *Educ:* Columbia Univ, BS, 49, MSEE, 51, EE, 56. *Prof Exp:* Instr electronics, Sch Eng, Cooper Union, 49-51; proj supvr electronic eng, Ford Instrument Co, Sperry Rand Corp, 51-59; prod area mgr eng mgt, Radio Receptor Div, Gen Instrument Corp, 59-67; corp dir eng, Superior Mfg & Instrument Corp, 67-70; vpres eng, Am Comput Commun Co, Inc, 70-71 & Phonplex Corp, 71-74; asst vpres, Citibank, 74-76; prog mgr, 76-80, VPRES, MGT INFO SYSTS TELEPHONICS CORP, 80- *Mem:* Sr mem Inst Elec & Electronics Engrs; Am Soc Photogram; Marine Technol Soc. *Res:* Military reconnaissance systems; digital interface and processing systems; oceanographic sensors; engineering management. *Mailing Add:* Helus Electronics Ltd 196-35 53rd Ave Flushing NY 11365

LUSTIG, MAX, b Chicago, Ill, Apr 9, 32; m 54; c 1. INORGANIC CHEMISTRY, AIR POLLUTION. *Educ:* Univ Calif, Los Angeles, BS, 57; Univ Wash, PhD(inorg chem), 62. *Prof Exp:* Chemist, Olin Mathieson Chem Corp, 57-58; res chemist, Redstone Arsenal Res Div, Rohm and Haas Co, Ala, 62-68; asst prof chem, Memphis State Univ, 68-73; res chemist, IIT Res Inst, 73-78, consult environ effects & chem hazards, 78-80; CONSULT, 80- *Concurrent Pos:* Eve instr, Univ Ala, 63-68. *Mem:* Am Chem Soc; fel Am Inst Chemists. *Res:* Physical and chemical studies of boron hydrides; chemistry of non-metal compounds with oxygen and fluorine, especially peroxides and hypofluorites; free radical chemistry; organometallic compounds; air pollution studies; high vacuum techniques; reaction kinetics involving air pollutants in the troposphere and stratosphere. *Mailing Add:* 8303 Steven Lane Canoga Park CA 91304

LUSTIG, STANLEY, b Brooklyn, NY, Feb 23, 33; m 60; c 2. PLASTICS ENGINEERING. *Educ:* Univ Toledo, BS, 58. *Prof Exp:* Chemist, Save Elec Corp, 58-59; res chemist, Union Carbide Corp 59-65, res proj leader, 65-70, group leader, 70-74, tech mgr, Films-Pkg Div, 74-86; res & develop mgr, 86-90, DIR, RES & DEVELOP, VISKASE CORP, 90- *Mem:* Am Chem Soc; Soc Plastics Engrs; Inst Packaging. *Res:* Plastic products and processes. *Mailing Add:* 561 Lakewood Blvd Park Forest IL 60466

LUSTMAN, BENJAMIN, b Pittsburgh, Pa, Oct 31, 14; m 46; c 3. METALLURGY, NUCLEAR MATERIALS. *Educ:* Carnegie-Mellon Univ, BS, 36, MS, 38, DSc(metall), 40. *Prof Exp:* Metallurgist, Standard Steel Spring Co, 39-41, Metals Res Lab, 41-43 & Int Minerals & Chem Corp, 43-44; metallurgist res lab, Bettis Atomic Power Lab, Westinghouse Corp, 44-49, metallurgist, 49-79; RETIRED. *Honors & Awards:* Order of Merit, Westinghouse Elec Corp, 56; Achievement Award, Am Nuclear Soc, 68; Kroll Mem Award, Colo Sch Mines, 78. *Mem:* Nat Acad Eng; Am Soc Metals; Am Nuclear Soc; Am Inst Mining Engrs. *Res:* Metallic corrosion; surface reactions; nuclear metallurgy; fuel element development. *Mailing Add:* 5253 1/2 Forbes St Pittsburgh PA 15217

LUSTY, CAROL JEAN, b Chicago, Ill, Sept 25, 36. RECOMBINANT DNA. *Educ:* Univ Mich, BS, 58; Wayne State Univ, PhD(biochem), 63. *Prof Exp:* Assoc, Pub Health Res Inst, 68-78, assoc mem, Dept Biochem, 78-81, assoc mem, Dept Molecular Genetics, 81-86, MEM DEPT MOLECULAR GENETICS, PUB HEALTH RES INST, 86- *Concurrent Pos:* Prin investr, 74-, mem Nat Adv Comt, PIR, 85-88, mem Phys Biochem study sect, 90- *Mem:* Am Chem Soc; Am Soc Biol Chemists; AAAS; NY Acad Sci; Harvey Soc. *Res:* Gene structure and evolution of carbamyl phosphate synthetases; regulatory mechanisms of mammalian arginine biosynthesis; protein structure and function. *Mailing Add:* Pub Health Res Inst 455 First Ave New York NY 10016

LUSZTIG, GEORGE, b May 20, 46; nat US citizen; m 72; c 2. THEORY OF GROUP REPRESENTATIONS. *Educ:* Princeton Univ, MA, 71, PhD(math), 71; dipl lic, math, Bucharest Univ, 68. *Prof Exp:* From res fel to prof math, Univ Warwick, UK, 71-77; PROF MATH, MASS INST TECHNOL, 78- *Concurrent Pos:* Mem, Inst Advan Study, Princeton, NJ, 69-71. *Honors & Awards:* Cole Prize, Am Math Soc, 85. *Mem:* Am Math Soc; London Math Soc; Fel Royal Soc. *Mailing Add:* Dept Math, MIT Rm 2-276 Cambridge MA 02139

LUTES, CHARLENE MCCLANAHAN, b Grundy, Va, Feb 4, 38; div. GENETICS, DEVELOPMENTAL BIOLOGY. *Educ:* Radford Col, BS, 59; Ohio State Univ, MSc, 62, PhD(genetics), 68. *Prof Exp:* From instr to prof, Radford Col, 64-80, chmn dept biol, 81-87, PROF BIOL, RADFORD UNIV, 80-, DEAN COL ARTS & SCI, 87- *Mem:* AAAS; Am Soc Zoologists; Am Inst Biol Sci. *Res:* Developmental genetics of wing venation patterns in Drosophila melanogaster. *Mailing Add:* Radford Univ Box 5724 Radford VA 24142

LUTES, DALLAS D, b St Louis, Mo, July 12, 25; m 45; c 2. PLANT PATHOLOGY. *Educ:* La Polytech Inst, BS, 49; Univ Mo, PhD(bot), 54. *Prof Exp:* Instr bot, ETex State Col, 54-55; from assoc prof to prof bot, 55-74, head dept, 63-73, PROF BOT & BACT, LA TECH UNIV, 74- *Mem:* AAAS. *Res:* Disease resistance by breeding; virus transmission; seed germination affected by light; mistletoe seed germination; fern taxonomy and distribution. *Mailing Add:* Dept Bot La Tech Univ Ruston LA 71272

LUTES, LOREN DANIEL, b Stapleton, Nebr, Dec 1, 39; m 82; c 4. ENGINEERING MECHANICS. *Educ:* Univ Nebr, BSc, 60, MSc, 61; Calif Inst Technol, PhD(appl mech), 67. *Prof Exp:* Res engr, Jet Propulsion Lab, 67; from asst prof to prof civil eng, Rice Univ, 67-87; PROF CIVIL ENG, TEX A&M UNIV, 88- *Concurrent Pos:* Vis prof, Univ Chile, 71; vis assoc prof civil eng, Univ Waterloo, 74-75. *Honors & Awards:* Wason Res Medal, Am Concrete Inst, 64; State-of-the-Art Award, Am Soc Civil Engrs, 83. *Mem:* Am Soc Civil Engrs. *Res:* Response of linear and nonlinear systems to random excitations; first-passage probabilities for stochastic processes; fatigue damage caused by stochastic loadings. *Mailing Add:* Dept Civil Eng Tex A&M Univ Col Sta TX 77843-3136

LUTES, OLIN S, b Faribault, Minn, Apr 29, 22; m 48, 71; c 3. ELECTROMAGNETICS, PHYSICS OF MAGNETIC FILMS. *Educ:* Carnegie Inst Technol, BS, 44; Columbia Univ, MA, 50; Univ Md, PhD(physics), 56. *Prof Exp:* Physicist, Sinclair Ref Co, 46-48 & Nat Bur Stand, 51-56; sr prin scientist, Honeywell, Inc, 56-83. *Concurrent Pos:* Consult scientist, 83- *Mem:* Am Phys Soc; Inst Elec & Electronics Engrs. *Res:* Superconductivity; low temperature thermometry; magnetic alloys; thin film devices for computer memories and integrated sensors. *Mailing Add:* 10413 Nicollet Circle Bloomington MN 55420

LUTEYN, JAMES LEONARD, b Kalamazoo, Mich, June 23, 48. SYSTEMATIC BOTANY. *Educ:* Western Mich Univ, BA, 70; Duke Univ, MA, 72, PhD(bot), 75. *Prof Exp:* Assoc cur, 75-81, CUR BOT, NEW YORK BOT GARDENS, 81- *Concurrent Pos:* Assoc ed, BRITTONIA, 76-81; ed, Proceedings Int Rhododendron Conf, 78; assoc ed,-Flora Neotropica 80-83, co-ed, 84- *Mem:* Am Soc Plant Taxonomists; Int Asn Plant Taxon; Bot Soc Am. *Res:* Evolution and systematics of the neotropical Ericaceae, Plumbaginaceae, and Companulaceae-Lobelioideae. *Mailing Add:* 58 Columbia Ave Hartsdale NY 10530

LUTH, WILLIAM CLAIR, b Winterset, Iowa, June 28, 34; m 53; c 3. GEOLOGY, GEOCHEMISTRY. *Educ:* Univ Iowa, BA, 58, MS, 60; Pa State Univ, PhD(geochem), 63. *Prof Exp:* Res assoc geochem, Pa State Univ, 63-65; asst prof, Mass Inst Technol, 65-68; from assoc prof to prof geol, Stanford Univ, 68-79; supvr, Geophys Res Div, Sandia Nat Labs, 79-82, mgr, geosci dept, 82-89, mgr, environ technol dept, 89-90; MGR, GEOSCI RES PROG, US DEPT ENERGY, 90- *Concurrent Pos:* Alfred P Sloan Found res fel, Mass Inst Technol, 66-67; geoscientist, Off Basic Energy Sci, Dept Energy, Washington, DC, 76-78; vis staff mem, Los Alamos Nat Lab, 78. *Mem:* Am Geophys Union; Geol Soc Am; Mineral Soc Am; Geochem Soc; Sigma Xi. *Res:* Experimental petrology; physical chemistry of the igneous and metamorphic rocks; phase equilibria in silicate-volatile systems at high pressure and temperature; disposal radioactive wastes. *Mailing Add:* 7516 Miller Fall Rd Rockville MD 20855

LUTHER, EDWARD TURNER, b Nashville, Tenn, Feb 11, 28; m 55; c 2. GEOLOGY. *Educ:* Vanderbilt Univ, BA, 50, MS, 51. *Prof Exp:* From geologist to chief geologist, 51-77, ASST STATE GEOLOGIST, TENN DIV GEOL, 67- *Concurrent Pos:* Instr, Univ Tenn, Nashville, 55-57 & 76-78; fuels engr, Tenn Valley Authority, 57. *Mem:* Fel Geol Soc Am; Sigma Xi. *Res:* Areal and economic geology of various areas in Tennessee, particularly the stratigraphy and structural geology of the Cumberland Plateau; coal resources, particularly in Eastern United States. *Mailing Add:* 823 Summerly Dr Nashville TN 37209

LUTHER, GEORGE WILLIAM, III, b Philadelphia, Pa, Feb 17, 47; m 71; c 2. INORGANIC & MARINE CHEMISTRY. *Educ:* LaSalle Col, BA, 68; Univ Pittsburgh, PhD(chem), 72. *Prof Exp:* From asst prof to prof chem, Kean Col, NJ, 72-86 & chmn, dept physics, 76-84; assoc dean, Col Marine Studies, 86-88, PROF MARINE CHEM, UNIV DEL, 86- *Concurrent Pos:* Investr, Nat Oceanic & Atmospheric Admin grant, 76-80, Investr, NSF, 83-; chmn, North Jersey Am Chem Soc. *Mem:* Am Chem Soc; AAAS; Microbeam Analysis Soc; Sigma Xi; Am Geophys Union; Am Soc Limnol & Oceanog. *Res:* Sulphur, iodine, metal speriation in seawater and sediments; mineral dissolution and formation; x-ray microanalysis of particulates; chemical oceanography-ocean, estuaries and anoxic basins. *Mailing Add:* Col Marine Studies Univ Del Lewes DE 19958

LUTHER, HERBERT GEORGE, b Brooklyn, NY, Oct 1, 14; m 38; c 4. BIOENGINEERING & BIOMEDICAL ENGINEERING. *Educ:* Cooper Union, New York, BChE, 40; NY Univ, MS, 44; Polytech Inst Brooklyn, DChE, 57. *Prof Exp:* With Sunshine Biscuit Co, 34-41, dir biochem labs, 41-44; asst dir tech serv, Chas Pfizer & Co, 45-52, dir agr res & develop, 52-59, sci dir agr, 59-69; dir animal health res, Hoffman-La-Roche Inc, Nutley, NJ, 74-82; PRES, LUTHER ASSOCS INC, 69-74 & 82- *Concurrent Pos:* Expert, Comn Food Additives, WHO/Food Agr Orgn; consult res & develop. *Mem:* Am Chem Soc; Am Inst Chem Engrs; Am Asn Animal Sci; Poultry Sci Asn; Am Inst Chemists; Am Asn Agr Eng; Asn Consult Chemists & Chem Engrs; Am Asn Indust Vet; Inst Food Technol; NY Acad Sci; Math Asn Am; Animal Nutrit Res Coun; Sigma Xi; Int Union Pure & Appl Chem; AAAS. *Res:* Antibiotics; antibacterials; vitamins; steroids; tranquilizers; unidentified growth factors; enzymes; antioxidants; nutrition; animal health; pharmacokinetics; operations research; food and feed technology; bioengineering; biotechnology; agricultural engineering. *Mailing Add:* The Mill Head of the River Smithtown NY 11787

LUTHER, HOLGER MARTIN, b Gdynia, Poland, Feb 4, 40; US citizen; m 69. MASS SPECTROMETRY, ELECTRON OPTICS. *Educ:* Marietta Col, BScL, 63; Pa State Univ, MS, 66, PhD(physics), 70. *Prof Exp:* Sr res physicist, CBS Labs, 69-75, EPSCO Labs, 75-76 & Electron Sci & Tech Ctr, Div Carson Alexiou Corp, 76-77; staff mem, Avco Everett Corp, 77-80; staff mem, 80-87, CHIEF SCIENTIST, SENSOR SYSTS GROUP, C S DRAPER LABS, 87- *Concurrent Pos:* Fac mem, Bridgeport Eng Inst, 71-77. *Mem:* Soc Photo-Optical Instumentation Engrs. *Res:* Electron-optical and elctro-optical instrumentation; compact radio frequency mass spectrometers; electron beam recorders and storage tubes; high speed tracking cameras for charged particle beams and ultra high resolution angle sensors. *Mailing Add:* 150 Bear Hill Rd Waltham MA 02154

LUTHER, LESTER CHARLES, b Joliet, Ill, Apr 19, 31; m 54; c 4. INDUSTRIAL ENGINEERING. *Educ:* Univ Ill, Urbana, BS, 53 & 58; Univ Nebr, Lincoln, MS, 60; Ariz State Univ, PhD(indust eng), 68. *Prof Exp:* Instr mech eng, Univ Nebr, 58-61; indust engr, Reynolds Metals Co, Ariz, 61-62; qual assurance engr, Motorola, Inc, 62-68; assoc prof, 68-72, PROF MECH ENG, CALIF STATE UNIV, SACRAMENTO, 72- *Concurrent Pos:* Indust engr, Cushman Motor Works, Nebr, 59-61 & McClellan AFB, Calif, 69-70; Nat Sci Found fel, Sacramento State Col, 71-72. *Mem:* Am Inst Indust Engr; Am Soc Eng Educ. *Res:* Economic interactions between quality assurance and inventory control. *Mailing Add:* Dept Mech Eng Calif State Univ 6000 J St Sacramento CA 95819

LUTHER, MARVIN L, b Waterloo, Iowa, Nov 16, 34; m 59; c 4. ATOMIC PHYSICS, NUCLEAR PHYSICS. *Educ:* Macalester Col, BA, 57; Univ Fla, MS, 60; Va Polytech Inst, PhD(physics), 67. *Prof Exp:* Asst prof physics, Randolph-Macon Men's Col, 60-63; asst prof, 66-67, ASSOC PROF PHYSICS, ILL STATE UNIV, 67- *Mem:* Am Asn Physics Teachers; Sigma Xi. *Res:* Atomic spectroscopy using collisionally excited beams provided by an accelerator. *Mailing Add:* 718 S Catalina Ave Four Redondo Beach CA 90227

LUTHER, NORMAN Y, b Palo Alto, Calif, June 3, 36; m 58; c 4. MATHEMATICAL DEMOGRAPHY. *Educ:* Stanford Univ, BS, 58; Univ Iowa, MS, 60, PhD(math), 63. *Prof Exp:* Instr math, Univ Iowa, 63; NSF fel, 63-64; from asst prof to assoc prof math, Wash State Univ, 64-87; PROF MATH, HAWAII PAC COL, 87- *Concurrent Pos:* Assoc prof, Albany State Col, Ga, 71-72; mem staff, East-West Ctr, 78-; vis assoc prof, Univ Hawaii, 82-83; Danforth Assoc, 72- *Mem:* Am Math Soc; Math Asn Am; Pop Asn Am. *Res:* Probability and statistics; measure theory; mathematical demography. *Mailing Add:* Dept Math Hawaii Pacific Col Honolulu HI 96813

LUTHERER, LORENZ O, b Cleveland, Ohio, Jan 20, 36; m. PHYSIOLOGY, INTERNAL MEDICINE. *Educ:* Haverford Col, AB, 58; Univ Iowa, MS, 64; Univ Fla, PhD(physiol), 69; Tex Tech Univ, MD, 77. *Prof Exp:* Grad res asst, Col Med, Univ Iowa, Iowa City, 61-62, grad teaching asst, 62-63, NIH trainee fel, 63-64; res physiologist, US Army Res Inst Environ Med, Natick, Mass, 64-65; NIH trainee fel, Col Med, Univ Fla, Gainesville, 66-69, postdoctoral fel, Dept Physiol, 69-71, Div Genetics, Endocrinol & Med, 71-72; from asst prof to assoc prof, Dept Physiol, 72-86, asst prof, Dept Internal Med, 81-86, PROF, DEPT PHYSIOL, HEALTH SCI CTR, SCH MED, TEX TECH UNIV, LUBBOCK & PROF DEPT INTERNAL MED, 86- *Concurrent Pos:* Res assoc, Arctic Inst NAm, Point Barrow, Alaska, 61; mem, bd dirs, Lubbock Chap, Tex Affil, Am Heart Asn, 75-79, 85-88, Prof Educ Comt, 75-76, Hypertension Screening Comt, 76-77, chmn, Prog Comt, 78 & 91, Hypertension Task Force, 85-87; coordr curric, Health Sci Ctr, Sch Med, Tex Tech Univ, Lubbock, 78-79, asst dean curric, 79-81, coordr, Grad Prog, Dept Physiol, 82-86, chmn, Instnl Animal Care & Use Comt, 85-89; mem, Radiation Safety Comt, Health Sci Ctr, Sch Med, Tex Tech Univ, 81-; secy & mem bd sci dirs, Tex Soc Biomed Res, 89- *Mem:* Am Physiol Soc; Shock Soc. *Res:* Author of various publications. *Mailing Add:* Physiol & Internal Med Dept-Sch Med Tex Tech Univ Health Sci Ctr Lubbock TX 79430

LUTHEY, JOE LEE, b Winslow, Ariz, Sept 21, 43. SPACE PHYSICS. *Educ:* Univ Calif, Berkeley, AB, 65; Univ Kans, Lawrence, PhD(physics), 70. *Prof Exp:* Res assoc space physics, Univ Iowa, Iowa City, 70-73; resident res assoc space physics, 73-75, consult radiation physics, 75-77, MEM TECH STAFF, NEW EARTH PROBE, JET PROPULSION LAB, CALIF INST TECHNOL, 77- *Concurrent Pos:* Consult, Physics Dept, Univ Iowa, 73-74; resident res assoc, Nat Res Coun, Jet Propulsion Lab. 73-75. *Mem:* Am Geophys Union; Am phys Soc. *Res:* Test/create Jovian radiation belt models; determine x-ray and gamma-ray emission from natural and artificial satellites in the Jovian trapped electron proton belts. *Mailing Add:* 80 N Marion Ave four Victor Sq Pasadena CA 91106

LUTHRA, HARVINDER SINGH, b Amritsar, India, Mar 14, 45; m 75; c 3. RHEUMATOLOGY, IMMUNOLOGY. *Educ:* Christian Med Col, India, MB & BS, 67; Am Bd Internal Med, 73; Am Bd Internal Med & Rheumatology, 74. *Prof Exp:* Intern, Christian Med Col, India, 67, resident, 68; intern, Middlesex Gen Hosp, NB, 69; resident internal med, Mt Sinai Hosp, Chicago, Ill, 70-72; trainee rheumatology, Mayo Grad Sch, 72-74, assoc consult, 74-75, CONSULT RHEUMATOLOGY, MAYO CLIN, ROCHESTER, MINN, 75- *Mem:* Int Med Acad Sci; fel Am Rheumatism Asn; Sigma Xi; Am Fedn Clin Res; AAAS; fel Am Col Physicians; Am Asn Immunologists. *Res:* Trying to understand genetic mechanisms involved in development of rheumatoid arthritis and potential new treatments; internal medicine. *Mailing Add:* Div Rheumatology & Internal Med Mayo Clin & Mayo Found 200 First St SW Rochester MN 55905-0002

LUTHRA, KRISHAN LAL, b Jaipur, India, Sept 28, 49. HIGH TEMPERATURE & METALLURGICAL CHEMISTRY. *Educ:* Univ Rajasthan, BEng, 70; Indian Inst Technol, Kanpur, MTech, 72; Univ Pa, PhD(metall & mat sci), 76. *Prof Exp:* METALLURGIST, CORP RES & DEVELOP, GEN ELEC CO, 76- *Concurrent Pos:* Sr res asst, Dept Metall Eng, Indian Inst Technol, Kanpur, 71-72; res fel, Dept Metall & Mat Sci, Univ Pa, 72-76. *Mem:* Electrochem Soc; Metall Soc; Am Inst Mining, Metall & Petrol Engrs. *Res:* Thermodynamic and kinetics of high temperature reactions; corrosion at elevated temperatures; gas-liquid-solid reactions; high temperature materials. *Mailing Add:* Gen Elec Res & Develop Lab Bldg K1 Rm 3B4 PO Box 8 Schenectady NY 12301

LUTHY, JAKOB WILHELM, b Staefa, Switz, Jan 31, 19; nat US; m 48; c 3. CHEMISTRY. *Educ:* Swiss Fed Inst Technol, MS, 44, DSc(org chem), 47. *Prof Exp:* Asst prof org technol, Swiss Fed Inst Technol, 46-47; chemist, Gen Aniline & Film Corp, NJ, 47-48; chemist, Sandoz Chem, 48-51; head appln lab, Chem Div, Sandoz, Inc, 51-54, dir appln & promotion, 54-58, tech mgr, Dyestuff Div, 58-84, exec vpres, 64-84, pres, Colors & Chem Div & dir, 67-84; RETIRED. *Concurrent Pos:* Dir, Toms River Chem Corp, 58-81. *Mem:* Emer fel Am Chem Soc; fel Swiss Chem Soc. *Res:* Dyestuffs. *Mailing Add:* 643 Mountain Rd Kinnelon NJ 07405

LUTHY, RICHARD GODFREY, b June 11, 45; m 69; c 3. ENVIRONMENTAL & CIVIL ENGINEERING. *Educ:* Univ Calif, Berkeley, BS, 67, MS, 74, PhD(civil eng), 76; Univ Hawaii, MS, 69. *Prof Exp:* Res asst, Dept Civil Eng, Univ Hawaii, 68-69; res proj officer, Naval Civil Eng Lab, Civil Eng Corps, US Navy, 70-71, asst officer-in-chg underwater construct team, 71-72; res asst, Div Sanit Eng, Univ Calif, Berkeley, 73-75; actg head, 85, from asst prof to assoc prof, 75-83, PROF, DEPT CIVIL ENG, CARNEGIE INST TECHNOL, 83-, ASSOC DEAN, 86-, HEAD DEPT, 89- *Concurrent Pos:* Consult, Allied-Signal, Environ Res & Technol Inc, IT Corp, Baker Chem Co, Koppers Co, Inc, US Steel, Alcoa, SmithKline-Beckman, FMC Corp, Exxon, Aetna Casualty & Ins, Remediation Technol Inc, US Dept Energy & US Environ Protection Agency; chmn, Asn Environ Eng Prof, 88; deleg, Water Sci & Technol Bd, Nat Acad Eng, Wash, Beijing, 88. *Honors & Awards:* G Tallman Ladd Award, Carnegie Inst Technol, 77; Nalco Award, Asn Environ Eng Prof, 78 & 82; Eddy Medal, Water Pollution Control Fedn, 80; Founders Award, US Nat Comn Int Asn Water Purification Res & Control, 84; Eng Sci Award, Asn Environ Eng Prof, 88. *Mem:* Water Pollution Control Fedn; Am Soc Civil Engrs; Am Chem Soc; Int Asn Water Pollution Res; Asn Environ Eng Prof (vpres, 86-87, pres, 87-88); Am Water Works Asn; Am Acad Environ Engrs. *Res:* Hazardous substances in wastewaters and ground waters; Wastewater treatment and industrial wastewater treatment; chemistry of dilute aqueous systems; treatment of wastewaters from petroleum refining, chemical manufacturing, coal conversion, and iron and steel making. *Mailing Add:* Dept Civil Eng Carnegie Mellon Univ Pittsburgh PA 15213-3890

LUTON, EDGAR FRANK, b Memphis, Tenn, Mar 3, 21; m 44; c 3. INTERNAL MEDICINE. *Educ:* Univ Tenn, Memphis, MD, 44; Am Bd Internal Med, dipl & cert nephrol, 74. *Prof Exp:* Staff physician neuropsychiat serv med teaching group, Vet Admin Hosp, Memphis, 48-49, resident internal med, 49-51, staff physician med serv, 51-59, sect chief internal med & allergy, 59-67, sect chief allergy & nephrology, 67-77; from asst to assoc prof, 61-74, PROF MED, CTR HEALTH SCI, UNIV TENN, MEMPHIS, 74- *Mem:* Fel Am Col Physicians; Am Soc Nephrology; Int Soc Nephrology. *Res:* Nephrology. *Mailing Add:* 2242 Tidmington Dr Cordora TN 38018

LUTRICK, MONROE CORNELIUS, b Grayson, La, July 22, 27; m 52; c 4. AGRONOMY, SOIL CHEMISTRY. *Educ:* La State Univ, BS, 51, MS, 53; Ohio State Univ, PhD(agron), 56. *Prof Exp:* Asst agron, La State Univ, 51-53; asst agronomist, 56-67, assoc soil chemist, 67-77, SOIL CHEMIST, AGR RES EDUC CTR, UNIV FLA, 77- *Mem:* Am Soc Agron; Soil Sci Soc Am. *Res:* Soil chemistry and maximum production of field crops; utilization of liquid digested sludge on agricultural lands; micronutrient status of field crops grown in North Florida. *Mailing Add:* Agr Res & Educ Ctr Rte 3 Box 575 Jay FL 32565-9524

LUTS, HEINO ALFRED, medicinal chemistry, for more information see previous edition

LUTSCH, EDWARD F, b Chicago, Ill, Nov 23, 30; m 65; c 2. ZOOLOGY. *Educ:* Northern Ill Univ, BS, 52; Northwestern Univ, MS, 57, PhD(biol), 62. *Prof Exp:* Asst prof zool, Univ Ill, Chicago, 62-68; from asst prof to assoc prof, 68-74, PROF BIOL, NORTHEASTERN ILL UNIV, 74- *Concurrent Pos:* Lectr, Northwestern Univ, 69- *Mem:* AAAS; Am Inst Biol Sci; Am Soc Zoologists. *Res:* Biological rhythms and clocks; rhythmic response of animals to pharmacological drugs; comparative physiology; animal behavior. *Mailing Add:* Dept Biol Northeastern Ill Univ 5500 N St Louis Ave Chicago IL 60625

LUTSKY, IRVING, b Paterson, NJ, June 12, 26; m 48; c 4. LABORATORY ANIMAL MEDICINE. *Educ:* Rutgers Univ, BS, 48; Purdue Univ, MS, 51; Univ Pa, VMD, 55; Am Col Lab Animal Med, dipl, 65. *Prof Exp:* Res asst poultry diseases, Purdue Univ, 49-51; staff vet, Fromm Labs, 55-58; asst prof vet sci, Med Col Wis, 60-66, assoc prof comp med, 66-72, adminr surg res lab, Allen Bradley Med Sci Lab, 60-72; PROF & CHMN, DEPT COMP MED, SCH MED, HEBREW UNIV, 72- *Concurrent Pos:* Vis prof comp med, Sch Med, Hebrew Univ, 71-72. *Honors & Awards:* Res Award, Am Asn Lab Animal Sci, 77. *Mem:* Am Soc Microbiol; Am Asn Lab Animal Sci; Am Vet Med Asn; Asn Gnotobiotics; Am Soc Lab Animal Pract; fel Am Acad Allergy & Immunol. *Res:* Infectious diseases; natural disease resistance; applied gnotobiology; occupational allergies; occupationally related hypersensitivity; lung disease in laboratory animal workers, veterinarians and poultry workers. *Mailing Add:* Dept Comp Med Sch Med Hebrew Univ POB 1172 Jerusalem 91010 Israel

LUTT, CARL J, b Guthrie Co, Iowa, Feb 10, 21; m 45; c 2. ANATOMY, PHYSIOLOGY. *Educ:* Creighton Univ, BSM, 42, MD, 45. *Prof Exp:* Dir, Student Health Serv, Calif State Univ, 60-65, Hayward, 60-65, prof, Biol Health & Kinesbiol, 60-87, asst dir student health serv, 73-85, dir sports med, 75-87; EMER PROF BIOL SCI, KINESIOL & PHYS EDUC, 87- *Mem:* Am Col Sports Med. *Mailing Add:* 20964 Woodside Way Groveland CA 95321-9410

LUTTER, LEONARD C, MOLECULAR BIOLOGY. *Prof Exp:* DIR, DEPT MOLECULAR BIOL, HENRY FORD HOSP, 88- *Mailing Add:* Dept Molecular Biol Henry Ford Hosp 7025 Educ & Res Bldg 2799 W Grand Detroit MI 48202

LUTTGES, MARVIN WAYNE, b Chico, Calif, Feb 3, 41; m 69. NEUROBIOLOGY, BIOENGINEERING. *Educ:* Univ Ore, Eugene, BSc, 62; Univ Calif, Irvine, PhD(biol sci), 68. *Prof Exp:* Res asst dept psychol, Univ Ore, 62-64; teaching asst psychobiol, Univ Calif, Irvine, 64-68; USPHS fel neurochem, Med Sch, Northwestern Univ, 68-69; asst prof to assoc prof, 69-79, PROF BIOENG, AEROSPACE ENG SCI, UNIV COLO, 80- *Concurrent Pos:* Noise consult, City of Boulder, Colo, 76-78; Am Eng Soc res award, 79. *Mem:* Soc Neurosci; AAAS; Sigma Xi; Biophys Soc. *Res:* Neurobiological basis of learning and memory; nervous system degeneration and regeneration; biological and physical acoustics; comparative studies of brain structure and function. *Mailing Add:* 3800 Pleasant Ridge Rd Boulder CO 80301

LUTTINGER, JOAQUIN MAZDAK, b New York, NY, Dec 2, 23. THEORETICAL PHYSICS. *Educ:* Mass Inst Technol, BS, 44, PhD(physics), 47. *Prof Exp:* Swiss-Am exchange fel, 47-48; Nat Res Coun fel, 48-49; Jewett fel, Inst Advan Study, 49-50; from asst prof to assoc prof physics, Univ Wis, 50-53; assoc prof, Univ Mich, 53-57 & Ecole Normale Superieure, Paris, 57-58; prof, Univ Pa, 58-60; chmn dept, 77-81, PROF PHYSICS, COLUMBIA UNIV, 60- *Mem:* Nat Acad Sci; fel Am Phys Soc; Am Acad Arts & Sci. *Res:* Theoretical magnetism; quantum field theory; statistical mechanics; theory of solids; condensed matter physics. *Mailing Add:* Dept Physics 826 Pupin Lab Columbia Univ New York NY 10027

LUTTMANN, FREDERICK WILLIAM, b New Brunswick, NJ, Aug 9, 40; div. MATHEMATICS. *Educ:* Amherst Col, AB, 61; Stanford Univ, MS, 64; Univ Ariz, PhD(math), 67. *Prof Exp:* Assoc, Univ Ariz, 63-67; assoc prof math, Alaska Methodist Univ, 67-70; asst prof, 70-74, assoc prof, 74-81, PROF MATH, SONOMA STATE UNIV, 81- *Mem:* Math Asn Am. *Res:* Steiner symmetrization of convex bodies; polynomial interpolation. *Mailing Add:* Dept of Math Sonoma State Univ 1801 E Cotati Ave Rohnert Park CA 94928

LUTTON, JOHN D, b Sioux City, Iowa, Feb 3, 37; c 2. EXPERIMENTAL HEMATOLOGY. *Educ:* Univ Nebr, BS, 61, MS, 63; NY Univ, PhD(cell biol & physiol), 69. *Prof Exp:* Instr gen physiol, Dept Biol, NY Univ, 66-68, asst prof, 70-71, instr & res scientist, Dept Cell Biol, Med Sch, 71-76; asst prof physiol & hemat, Dept Physiol, Mt Sinai Sch Med, 76-77; asst prof hemat, Downstate Med Ctr, State Univ NY, 77; assoc prof med, 83-89, RES ASSOC PROF MED & ANAT, NY MED COL, 77-, PROF MED, 89-, ASSOC PROF MICROBIOL & IMMUNOL, 90- *Concurrent Pos:* Adj asst prof, Dept Biol, City Col, City Univ New York, 71-73, Baruch Col, City Univ New York, 81-85. *Mem:* Am Soc Hemat; Int Soc Exp Hemat; Reticulo Endothelial Soc; AAAS. *Res:* Growth factors and the regulation of hematopoiesis: bone marrow growth and in vitro aspects on the regulation of erythropoiesis including regulatory aspects of hemebiosynthesis and degradation; in vitro characteristics of disorders such as anemia, polycythemins, neoplastic states and disorders of iron metabolism; granulopoiesis; differentiation of leukemic cells. *Mailing Add:* Dept Med NY Med Col Munger Pavilion Valhalla NY 10595

LUTTON, JOHN KAZUO, b Tokyo, Japan, July 11, 49; US citizen; m 71; c 1. NEUROPHARMACOLOGY, ENZYMOLOGY. *Educ:* Pac Lutheran Univ, BS, 71; Purdue Univ, PhD(biochem), 76. *Prof Exp:* Grad student, Dept Biochem, Purdue Univ, 71-76 & chem anal, Ind State Chem Off, 72-73; res assoc pharmacol, Med Sch, Univ Colo, 76-77 & Univ NC, 77-80; ASST PROF, CHEM DEPT, KENYON COL, 80- *Mem:* Am Chem Soc; AAAS; Sigma Xi. *Res:* Molecular mechanisms of hormone action especially the role of cyclic nucleotides in brain function and cell growth; enzymatic mechanisms of redox enzymes especially flavin-containing dehydrogenases. *Mailing Add:* Dept Chem Kenyon Col Gambier OH 43022

LUTTRELL, ERIC MARTIN, b Wheeling, WVa, May 12, 41; m 63; c 2. PETROLEUM GEOLOGY. *Educ:* Univ Wis-Madison, BS, 62, MS, 65; Princeton Univ, PhD(geol), 68. *Prof Exp:* Geologist, Producing Dept, Texaco Inc, 68-69, sr geologist, 69-73, res geologist, Res & Tech Dept, 73-76, asst supvr geol res, 76-79, consult explor geologist, 79-80; regional geologist, 80-82, asst explor mgr, 82-84, proj mgr, Anadarko, 84-86, ONSHORE EXPLOR MGR, STANDARD OIL PROF CO, 86- *Mem:* Geol Soc Am; Soc Econ Paleontologists & Mineralogists; Am Asn Petrol Geologists. *Res:* Applications of seismic stratigraphy clastic sedimentology; organic geochemistry and geologic thermometry to petroleum exploration. *Mailing Add:* 3215 W Autumn Run Circle Sugar Land TX 77479

LUTTRELL, GEORGE HOWARD, b Glendale, Calif, Dec 23, 41; m 64. ANALYTICAL CHEMISTRY. *Educ:* Univ Tex, BS, 65; Southern Methodist Univ, MS, 69; Univ Ga, PhD(chem), 75. *Prof Exp:* Res chemist anal, Alcon Labs, 69-72, res chemist anal, Ctr Labs, 75-77, MEM STAFF, ALCON LABS PR, 77- *Mem:* Am Chem Soc; Sigma Xi. *Res:* Preconcentration of trace metal cations and oxyanions for analysis by x-ray fluorescence using immobilized complexing and chelating reagents. *Mailing Add:* Alcon Labs PO Box 6600 Ft Worth TX 76115

LUTTS, JOHN A, b Baltimore, Md, Feb 26, 32; m 67; c 7. COMPUTER SCIENCE. *Educ:* Spring Hill Col, BS, 57; Univ Pa, MA, 59, PhD(math), 61; Woodstock Col, Md, STL, 65. *Prof Exp:* From instr to asst prof math, Loyola Col, Md, 65-66; asst prof, 66-70, fac growth fel, 67, fac res grant, 70-71, ASSOC PROF MATH, UNIV MASS, HARBOR CAMPUS, 70- *Mem:* Math Asn Am. *Res:* Cultural history of mathematics; approximation theory; computer languages. *Mailing Add:* Dept of Math/Computer Sci Univ of Mass Harbor Campus Boston MA 02125-3393

LUTWAK, ERWIN, b USSR, Feb 9, 46; US citizen; m 68. MATHEMATICS. *Educ:* Polytech Inst Brooklyn, BS, 68, MS, 72; Polytech Inst NY, PhD(math), 74. *Prof Exp:* Asst prof math, Col Pharmaceut Sci, Columbia Univ, 70-75; from asst prof to assoc prof, 75-86, PROF MATH, POLYTECH UNIV, 86- *Concurrent Pos:* Chair, Math Sect NY Acad Sci, 88-90. *Mem:* Am Math Soc; London Math Soc; Math Asn Am; Sigma Xi; NY Acad Sci. *Res:* Convexity; integral geometry; analytic and geometric inequalities. *Mailing Add:* Polytech Univ NY 333 Jay St Brooklyn NY 11201

LUTWAK, LEO, b New York, NY, Mar 27, 28; m 50, 76; c 8. ENDOCRINOLOGY, NUTRITION. *Educ:* City Col New York, BS, 45; Univ Wis, MS, 46; Univ Mich, PhD(biochem), 50; Yale Univ, MD, 56. *Prof Exp:* Biochemist med, Brookhaven Nat Lab, 50-52; clin assoc metab, Metab Dis Br, Nat Inst Arthritis & Metab Dis, 57-59, sr investr, 60-63; Jameson prof clin nutrit, Grad Sch Nutrit, Cornell Univ, 63-72; prof med, Univ Calif, Los Angeles, 72-76, prof nutrit, Sch Pub Health, 73-76; sect chief metab, Vet Admin Hosp, Sepulveda, 72-76; prof med, Northeastern Ohio Univ, 76-84, prof nutrit & prog chief, 76-84; CONSULT, 84- *Concurrent Pos:* NSF sr NASA fel, Ames Res Lab, Moffett Field, Calif, 70-71; prin investr, NASA, 63-80; consult, Div Res Grants, NIH, 64-69, NASA, 80-; consult, Tompkins County Hosp, Ithaca, NY, 64-69; vis prof, Sch Med, Stanford Univ, 70-71; chmn dept med, Akron City Hosp, Ohio, 76-78; mem, Am Bd Clin Nutrit, pres, 81-82; pvt pract, endocrinol, Huntsville, Ala, 85- *Mem:* AAAS; Endocrine Soc; Am Inst Nutrit; fel Am Col Physicians; fel Am Col Nutrit. *Res:* Isotope kinetics in metabolic bone disease; calcium, phosphorus and magnesium in human nutrition; effect of space flight on bone and muscle metabolism; obesity control; diabetes and electrolyte metabolism; hospital malnutrition. *Mailing Add:* 5827 Jones Valley Dr Huntsville AL 35802-1916

LUTY, FRITZ, b Essen, Ger, Apr 12, 28; m 60; c 2. SOLID STATE PHYSICS. *Educ:* Univ Gottingen, dipl physics, 53; Stuttgart Univ, Dr rer nat(physics), 56. *Prof Exp:* Asst physics, Stuttgart Univ, 53-62, dozent, Physics Inst, 64-65; vis assoc prof, Univ Ill, Urbana, 63; PROF PHYSICS, UNIV UTAH, 65- *Concurrent Pos:* Vis prof, Soc Advan Sci, Japan, 73; distinguished res award, Univ Utah, 84. *Mem:* Fel Am Phys Soc; Ger Phys Soc. *Res:* Defects in ionic crystals; radiation damage; absorption and emission spectroscopy; field emission; magneto-optics, paraelectric and paraelastic effects; low temperature dielectric and electro-caloric studies; Raman-scattering; phase transitions; material development for tunable laser application. *Mailing Add:* Dept of Physics Univ of Utah Salt Lake City UT 84112

LUTZ, ALBERT WILLIAM, b Baltimore, Md, Sept 26, 24; m 51; c 2. AGRICULTURAL CHEMISTRY. *Educ:* Johns Hopkins Univ, AB, 49, MA, 50, PhD(chem), 53. *Prof Exp:* Assoc prof chem, Col William & Mary, 53-56; res chemist, Chemagro Corp, 56-57; res chemist, 57-59, sr res chemist, 59-69, group leader herbicides, 69-85, GROUP LEADER CHEM DISCOVERY, AGR DIV, AM CYANAMID CO, 85- *Honors & Awards:* J Shelton Horsley Award. *Mem:* Am Chem Soc. *Res:* Pesticides, particularly growth regulants and herbicides. *Mailing Add:* Am Cyanamid Co PO Box 400 Princeton NJ 08540

LUTZ, ARTHUR LEROY, b Louisville, Ohio, Oct 22, 08; m 37; c 2. NUCLEAR PHYSICS. *Educ:* Capital Univ, BS, 31; Ohio State Univ, MS, 36, PhD(physics), 43. *Prof Exp:* High sch teacher, Ohio, 31-40; asst physics, Ohio State Univ, 40-43; prof, 43-75, EMER PROF PHYSICS, WITTENBERG UNIV, 75- *Concurrent Pos:* Fac fel, NSF, 60-61. *Mem:* Sigma Xi; Am Phys Soc; Am Asn Physics Teachers. *Res:* Radioactive isotopes; internal conversion and K-capture in the radioactive isotopes of lead and bismuth. *Mailing Add:* 1500 Villa Rd No 105 Wittenberg Univ Springfield OH 45503

LUTZ, BARRY LAFEAN, b Windsor, Pa, Jan 2, 44; m 81. PLANETARY ATMOSPHERES, MOLECULAR SPECTROSCOPY. *Educ:* Lebanon Valley Col, BS, 65; Princeton Univ, AM, 67, PhD(astrophys sci), 68. *Prof Exp:* Fel physics, Nat Res Coun Can, 68-70; res astronr, Lick Observ, Univ Calif, 70-71; from adj asst prof to adj assoc prof, State Univ NY Stony Brook, 73-77, sr res assoc, 71-78; ASTRONOMER, LOWELL OBSERV, 77- *Concurrent Pos:* Vis astron, Observ Paris, 79 & Univ Dijon, 80; adj assoc prof, Ariz State Univ, 81-83, adj prof, 83-86, Northern Ariz Univ, 87-; prin investr,

NSF, 72-88, NASA, 75-; consult, Kitt Peak Nat Observ, 73, NASA Planetary Astron Mgt Oper Working Group, 85- *Mem:* Int Astron Union; Am Astron Soc; Sigma Xi; Am Geophys Union. *Res:* High resolution spectroscopy of the interstellar medium, of comets and of stellar and planetary atmospheres; laboratory astrophysics; intensity measurements and long path length planetary atmospheres simulations; absolute spectrophotometry of planetary atmospheres and of comets and narrow band photopolarimetric imaging of planets. *Mailing Add:* Lowell Observ Mars Hill Rd 1400 W Flagstaff AZ 86001

LUTZ, BRUCE CHARLES, b London, Ont, May 16, 20; m 45; c 3. PHYSICS. *Educ:* Western Ont Univ, BA, 42, MA, 44; Johns Hopkins Univ, PhD, 54. *Prof Exp:* Instr electronics & radio, Western Ont Univ, 41-44; lectr electronics & physics, Univ Man, 45-47; instr electronics, 47-57, assoc prof elec eng, 57-62, prof elec eng, Univ Del, 62-, actg chmn dept, 73-; RETIRED. *Mem:* Am Inst Aeronaut & Astronaut (treas, Rocket Soc, 59-60); Inst Elec & Electronics Eng. *Res:* Nuclear reactor physics and engineering; plasma-microwave interaction; signal analysis. *Mailing Add:* 560 Terrapin Lane Newark DE 19711

LUTZ, DONALD ALEXANDER, b Syracuse, NY, Apr 2, 40. MATHEMATICS. *Educ:* Syracuse Univ, BS, 61, MS, 63, PhD(math), 65. *Prof Exp:* Instr math, Syracuse Univ, 65; from asst prof to prof math, Univ Wis-Milwaukee, 65-86; PROF MATH, SAN DIEGO STATE UNIV, 86- *Concurrent Pos:* Lectr, Univ Md, 67-69; vis asst prof, Math Res Ctr, Univ Wis-Madison, 69-70; vis assoc prof math, Univ Southern Calif, 73; Humboldt fel, Univ Ulm, WGer, 75-76. *Mem:* Am Math Soc; German Math Union. *Res:* Systems of linear ordinary differential equations with meromorphic coefficients; systems of linear difference equations. *Mailing Add:* Dept Math San Diego State Univ San Diego CA 92182

LUTZ, GEORGE JOHN, b New England, NDak, May 9, 33; m 65; c 2. PHYSICAL CHEMISTRY. *Educ:* Augustana Col, BA, 53; Iowa State Univ, PhD(phys chem), 62. *Prof Exp:* Jr chemist, Ames Lab, Univ Iowa, 55-58; resident res assoc, Argonne Nat Lab, 62-64; CHEMIST, NAT BUR STANDARDS, 64- *Honors & Awards:* Alexander von Humboldt Found, Sr US Scientist award, Bonn, WGer, 74. *Mem:* Am Chem Soc; Am Nuclear Soc; Am Soc Metals. *Res:* Activation analysis; applications of radioactive isotopes. *Mailing Add:* Nat Bur Standards Chem Bldg Rm 402 Washington DC 20234

LUTZ, HAROLD JOHN, b Saline, Mich, Aug 11, 00; m 26; c 2. FORESTRY. *Educ:* Mich State Col, BS, 24; Yale Univ, MF, 27, PhD(forestry), 33. *Prof Exp:* Tech asst, US Forest Serv, 24-26, assoc silviculturist, Allegheny Forest Exp Sta, 28-29; asst forester, Conn Exp Sta, 27-28; asst prof forestry, Pa State Col, 29-31; from asst prof to prof, Yale Univ, 33-48, Morris K Jesup prof silvicult, 48-65, Oastler prof forest ecol, 65-68, OASTLER EMER PROF FOREST ECOL, YALE UNIV, 68- *Concurrent Pos:* Walker-Ames prof, Univ Wash, 59; H R MacMillan lectr, Univ BC, 59; summers, US Forest Serv, 49-52 & 57 & vis prof, Univ Colo, 64-69. *Honors & Awards:* Soc Am Foresters Award, 57. *Mem:* Fel Soc Am Foresters. *Res:* Forest ecology and soils. *Mailing Add:* 440 W Russell Saline MI 48176

LUTZ, HARRY FRANK, b Philadelphia, Pa, Jan 30, 36; m 60; c 2. NUCLEAR PHYSICS. *Educ:* Univ Pa, AB, 57; Mass Inst Technol, PhD(physics), 61. *Prof Exp:* PHYSICIST, LAWRENCE LIVERMORE LAB, 61- *Mem:* Am Phys Soc. *Res:* Nuclear reactions and nuclear spectroscopy. *Mailing Add:* L-390 LLNL PO Box 808-L531 Livermore CA 94550

LUTZ, JULIE HAYNES, b Mt Vernon, Ohio, Dec 17, 44; m 66; c 2. ASTRONOMY. *Educ:* San Diego State Univ, BA, 65; Univ Ill, MS, 68, PhD(astron), 72. *Prof Exp:* From asst prof to assoc prof, 72-84, PROF ASTRON, WASH STATE UNIV, 84- *Concurrent Pos:* Asst dean sci, Wash State Univ, 78-79, assoc provost, 81-82; dir, Div Astron Sci, NSF, 90- *Mem:* Int Astron Union; Royal Astron Soc; Am Astron Soc; Astron Soc Pac (pres, 90-92). *Res:* Planetary nebulae; stellar evolution. *Mailing Add:* Prog Astron Wash State Univ Pullman WA 99164-2930

LUTZ, PAUL E, b Hickory, NC, June 25, 34; m 78, 86; c 1. INVERTEBRATE ZOOLOGY, ECOLOGY. *Educ:* Lenoir-Rhyne Col, AB, 56; Univ Miami, MS, 58; Univ NC, PhD(zool), 62. *Hon Degrees:* LHD, Lenoir-Rhyne Col, 83. *Prof Exp:* Asst zool, Univ Miami, 56-58 & Univ NC, 58-61; from instr to assoc prof, 61-70, PROF BIOL, UNIV NC, GREENSBORO, 70- *Concurrent Pos:* Am Philos Soc grant, 64; NSF grants, 65-67 & 69-71. *Mem:* AAAS; Ecol Soc Am; Sigma Xi; Am Inst Biol Sci. *Res:* Ecology and physiology of aquatic insects, especially effects of temperature and photoperiod as they affect seasonal regulation of developmental patterns in the Odonata. *Mailing Add:* Dept Biol Univ NC Greensboro NC 27412-5001

LUTZ, PETER LOUIS, b Glasgow, Scotland, Sept 29, 39. RESPIRATION, OSMOREGULATION. *Educ:* Glasgow Univ, Scotland, BSc, 64, PhD(zool), 70. *Prof Exp:* Lectr physiol, Univ Ife, Nigeria, 64-66, biol, Univ Glasgow, Scotland, 69-70; asst prof biol, Duke Univ, NC, 70-72; lectr biol, Bath Univ, Eng, 72-76; assoc prof, Miami Univ, Fla, 76-82, prof physiol & chmn, Marine Sch, 82-90; MCGINTY CHAIR MARINE BIOL, FLA ATLANTIC UNIV, 90- *Mem:* Soc Exp Biol; Am Soc Zoologists; Am Physiol Soc. *Res:* Animal physiology, particularly respiration and osmoregulation; anerobic brain metabolism; applied physiology of aquaculture of crustaceans; pollution; oil and plastics. *Mailing Add:* 561 Satinwood Dr Key Biscayne FL 33149

LUTZ, RAYMOND, b Oak Park, Ill, Feb 27, 35; m 58. INDUSTRIAL ENGINEERING, ENGINEERING ECONOMICS. *Educ:* Univ NMex, BS, 58, MBA, 62; Iowa State Univ, PhD(eng valuation), 64. *Prof Exp:* Instr mech eng, Univ NMex, 58-61; asst indust eng, Iowa State Univ, 61-64; asst prof mech eng, NMex State Univ, 64-67; from assoc prof to prof indust eng, Univ Okla, 68-72; dean, Sch Mgt, 73-78, PROF, 73-, EXECUTIVE DEAN GRAD STUDIES & RES, UNIV TEX, DALLAS, 79- *Concurrent Pos:* Ed, Eng Economist, 72-77, Indust Mgt, 83-87; bd dirs, Sigma Xi. *Honors & Awards:* E L Grant Award, Am Soc Eng Educ, 72; Fred Crane Distinguished Serv Award, Inst Indust Engrs, 87. *Mem:* AAAS; fel Am Inst Indust Engrs; Opers Res Soc Am; Inst Sci; Sigma Xi. *Res:* Operations management; industrial management; shipbuilding technology. *Mailing Add:* 10275 Hollow Way Dallas TX 75229

LUTZ, RAYMOND PAUL, b Cleveland, Ohio, May 31, 32. PHYSICAL ORGANIC CHEMISTRY. *Educ:* Univ Fla, BS, 53, MS, 55; Calif Inst Technol, PhD(org chem), 62. *Prof Exp:* Res chemist, E I du Pont de Nemours & Co, Ky & Mich, 55-57; instr chem, Harvard Univ, 61-64, lectr, 64-65; asst prof, Univ Ill, Chicago, 65-68; from asst prof to assoc prof, 68-83, PROF CHEM, PORTLAND STATE UNIV, 83- *Mem:* Am Chem Soc. *Res:* Reaction mechanisms, including displacement reactions and thermal isomerizations. *Mailing Add:* Dept Chem Portland State Univ Portland OR 97207-0751

LUTZ, RICHARD ARTHUR, b New York, NY, June 8, 49; m 81; c 3. BIOLOGICAL OCEANOGRAPHY, MARINE ECOLOGY. *Educ:* Univ Va, BA, 71; Univ Maine, PhD(oceanog), 75. *Prof Exp:* Res asst, dept oceanog, Univ Maine, 71-75, res assoc, Darling Ctr, 75-78; res assoc, dept geol & geophys, Yale Univ, 77-79; asst prof, 79-84, PROF, DEPT MARINE & COASTAL SCI, RUTGERS UNIV, 87-, DIR, FISH & AQUACULT TEX CTR, 86- *Concurrent Pos:* Print investr, Nat Oceanic & Atmospheric Admin sea grants, 75-78; biol consult, Blue Gold Sea Farms, 76-; assoc investr, Nat Oceanic & Atmospheric Admin sea grant, Yale Univ, 78-; co-prin investr NSF grants, Univ Calif, Santa Barbara, 78-; prin investr NSF grant, Rutgers Univ, 81- *Honors & Awards:* Thurlow C Nelson Award, Nat Shellfisheries Asn, 73. *Mem:* World Mariculture Soc; Nat Shellfisheries Asn (vpres, 81-); Am Soc Zoologists; Estuarine Res Fedn; AAAS; Sigma Xi. *Res:* Shellfish biology; molluscan shell structure and mineralogy; shellfish aquaculture; bivalve larval ecology; marine ecology and paleoecology; malacology; waste heat, especially power plant effluent, utilization; paleoclimatology; deep-sea hydrothermal vents, ecology. *Mailing Add:* Fish & Aquacult Tex Ctr Rutgers Univ Cook Campus PO Box 231 New Brunswick NJ 08903

LUTZ, ROBERT WILLIAM, b Mason City, Iowa, Sept 14, 37; m 56; c 4. CHEMICAL PHYSICS, COMPUTER SCIENCE. *Educ:* Drake Univ, BA, 62; Univ NMex, MS, 66; Ill Inst Technol, PhD(physics), 69. *Prof Exp:* Res asst physics, Los Alamos Sci Lab, 62-64, staff mem, 64-66; asst prof, 69-73, dir comput serv, 74-86, ASSOC PROF PHYSICS, DRAKE UNIV, 73-, DIR COMPUT & TELECOM, 86- *Concurrent Pos:* Chair, ACM-SIGUCCS, 83-85. *Mem:* Combustion Inst; Sigma Xi; Asn Comput Mach. *Res:* Computer assisted instruction; computers in undergraduate curriculum. *Mailing Add:* Comput Ctr Drake Univ Des Moines IA 50311

LUTZ, THOMAS EDWARD, b Teaneck, NJ, Nov 20, 40; m 66; c 2. ASTRONOMY. *Educ:* Manhattan Col, BME, 62; Univ Ill, MS, 65, PhD(astron), 69. *Prof Exp:* From asst prof to assoc prof, 69-81, PROF ASTRON, WASH STATE UNIV, 81-, DIR PROG ASTRON, 80- *Concurrent Pos:* Vis asst prof, Univ Wash, 71; vis astronr, Royal Greenwich Observ, 76-77; hon res fel, Univ Col, London, 82-83; vis resident Astromr, Cerro Tololo Interamerican Observ, 86 & 88-89. *Mem:* Am Astron Soc; Int Astron Union; Royal Astron Soc; AAAS; Sigma Xi; Astron Soc Pacific. *Res:* Calibration of luminosity criteria; observational astrophysics; application of statistical techniques to astronomy. *Mailing Add:* Prog Astron Wash State Univ Pullman WA 99164-3113

LUTZ, WILSON BOYD, b Mogadore, Ohio, May 12, 27; m 50; c 2. BIOCHEMISTRY, ORGANIC CHEMISTRY. *Educ:* Manchester Col, BA, 50; Ohio State Univ, PhD(org chem), 55. *Prof Exp:* Fel biochem, Med Col, Cornell Univ, 55-57; scientist, Warner-Lambert Res Inst, 57-60, sr scientist, 60-62; from asst prof to assoc prof, 62-72, PROF CHEM, MANCHESTER COL, 72- *Concurrent Pos:* Consult, Warner-Lambert Res Inst, 63-66; guest worker, NIH, 71; res assoc & dir, Inst Biomed Res, Univ Tex, Austin, 81; consult, Miles Lab, Elkhart, Ind, 82. *Mem:* Am Chem Soc. *Res:* Synthesis of new derivatives of hydroxylamine and substances of biological interest including melanogenic indoles. *Mailing Add:* Dept Chem Manchester Col North Manchester IN 46962-1299

LUTZE, FREDERICK HENRY, JR, b Brooklyn, NY, Nov 27, 37; m 65. AEROSPACE ENGINEERING. *Educ:* Worcester Polytech Inst, BS, 59; Univ Ariz, MS, 64, PhD(aerospace eng), 67. *Prof Exp:* Mech engr, Eclipse Pioneer Div, Bendix Corp, 59-60; instr aerospace eng, Univ Ariz, 65-66; from asst prof to assoc prof, 66-81, PROF AEROSPACE ENG, VA POLYTECH INST & STATE UNIV, 81- *Concurrent Pos:* Mem staff, Boeing Corp, 63 & NAm, 64, Adv Technol, Inc, 86-89; consult, EG&G, Inc, 81-89, Optimization Inc, 86-89. *Mem:* Assoc fel Inst Aeronaut & Astronaut; Am Astronaut Soc. *Res:* Trajectory optimization; flight mechanics; aircraft stability and control; orbital mechanics. *Mailing Add:* Dept Aerospace Eng Va Polytech Inst & State Univ Blacksburg VA 24061

LUTZER, DAVID JOHN, b Sioux Falls, SDak, Mar 27, 43; m 82; c 4. MATHEMATICS. *Educ:* Creighton Univ, Omaha, NE, BS, 64, Oxford Univ, Eng, Dpl Advan math, 66; Univ Wash, Seattle, PhD(math), 70. *Prof Exp:* From asst prof to assoc prof math, Univ Pittsburgh, 70-78; prof math, Tex Tech Univ, 76-82; prof & chair, Miami Univ, Oxford Ohio, 82-87; DEAN, DEPT MATH, COL WILLIAM & MARY, 87- *Mem:* Am Math Soc; Math Asn Am; Soc Indust & Appl Math. *Res:* Self-theoretic topology in ordered spaces and function spaces. *Mailing Add:* Dean Arts & Sci Col William & Mary Willamsburg VA 23185

LUTZKER, EDYTHE, b Berlin, Ger, June 25, 04; c 3. HISTORY OF SCIENCE, HISTORY OF MEDICINE. *Educ:* City Col New York, BA, 54; Columbia Univ, MA, 59. *Prof Exp:* Res asst hist of sci, City Col New York, 52-55; RES & WRITING, 55- *Concurrent Pos:* Am Philos Soc Johnson Fund Grant, 64; Penrose Fund Grant, 65; NIH grant, 66 & 68-74. *Mem:* AAAS; Am Asn Hist Med; Am Soc Microbiol; Int Soc Hist Med; fel Royal Soc Med. *Res:* Social history; participation by women in science and medicine; pioneers of 19th century medicine in British Empire and India. *Mailing Add:* 201 W 89th St New York NY 10024

LUUS, R(EIN), b Tartu, Estonia, Mar 8, 39; Can citizen; m 73. CHEMICAL ENGINEERING. *Educ:* Univ Toronto, BASc, 61, MASc, 62; Princeton Univ, AM, 63, PhD(chem eng), 64. *Prof Exp:* Fel optimal control, Princeton Univ, 64-65; from asst prof to assoc prof, 65-74, PROF CHEM ENG, UNIV TORONTO, 74- *Concurrent Pos:* Consult, Can Gen Elec Co, Ltd, 65-66, Shell Oil Co Can, 66-70 & 78-79, Milltronics Ltd, 67-71 & Imperial Oil Ltd, 74-77; dir, Chem Eng Res Consults Ltd, 66-; Nat Res Coun Can sr indust fel, 72-73; vis assoc, Calif Inst Technol, 79-80. *Honors & Awards:* Steacie Prize, Nat Res Coun Can, 76; ERCO Award, Can Soc Chem Eng, 80. *Mem:* Can Soc Chem Eng (secy, 67-68, vchmn, 68-69, chmn 69-70, past chmn, 70-71); fel Chem Inst Can. *Res:* Development of optimization procedures suitable for optimal and suboptimal control of nonlinear systems; nonlinear analysis; optimal control of time delay systems; parameter estimation; model reduction. *Mailing Add:* Dept of Chem Eng Univ of Toronto Toronto ON M5S 1A4 Can

LUVALLE, JAMES ELLIS, b San Antonio, Tex, Nov 10, 12; m 46; c 3. PHYSICAL CHEMISTRY, PHOTOGRAPHIC CHEMISTRY. *Educ:* Univ Calif, Los Angeles, AB, 36, AM, 37; Calif Inst Technol, PhD(chem), 40. *Prof Exp:* Instr chem, Fisk Univ, 40-41; res chemist, Res Labs, Eastman Kodak Co, 41-42, phys chemist, Kodak Res Labs, 43-53; phys chemist, Nat Defense Res Comt, Chicago, 42 & Calif Inst Technol, 42; proj dir, Tech Opers, Inc, 53-59; dir basic res, Fairchild Camera & Instrument Corp, NY, 59-63, dir res, Ill, 63-68; tech dir microstatics lab, SCM Corp, 68-69; dir physics & chem res, Smith Corona Marchant Labs, 69-70; sci coordr, Res & Develop Labs, SCM Bus Equip Div, Calif, 70-75; lab adminr, chem dept, Stanford Univ, 75-84; RETIRED. *Concurrent Pos:* Vis lectr, Brandeis Univ, 57-59; vis scholar, Stanford Univ, 71-75; independent consult, 75- *Mem:* AAAS; Am Chem Soc; Am Phys Soc; Soc Photog Scientists & Engrs; fel Royal Soc Chem. *Res:* Photochemistry; electron diffraction; magnetic susceptibility; reaction kinetics and mechanisms; photographic theory; magnetic resonance; solid state physics; neurochemistry, chemistry of memory and learning. *Mailing Add:* 3580 Evergreen Dr Palo Alto CA 94303

LUX, SAMUEL E, IV, ONCOLOGY. *Prof Exp:* CHIEF, DIV HEMAT & ONCOL, CHILDREN'S HOSP, 85- *Mailing Add:* Div Hemat & Oncol Children's Hosp 300 Longwood Ave Boston MA 02115

LUXEMBURG, WILHELMUS ANTHONIUS JOSEPHUS, b Delft, Neth, Apr 11, 29; m 55; c 2. MATHEMATICAL ANALYSIS. *Educ:* State Univ Leiden, BSc, 50, MSc, 53; Delft Univ Technol, PhD, 55. *Prof Exp:* Fel math, Queen's Univ, Can, 55-56; asst prof, Univ Toronto, 56-58; from asst prof to assoc prof, 58-62, exec officer, 70-85, PROF MATH, CALIF INST TECHNOL, 62- *Concurrent Pos:* Humboldt award, 80. *Mem:* Am Math Soc; Can Math Cong; Neth Math Soc; corresp mem Royal Acad Sci Amsterdam. *Res:* Functional analysis, particularly measure and integration theory, Banach function space theory and theory of locally convex spaces; Riesz spaces; nonstandard analysis. *Mailing Add:* 817 S El Molino Ave Pasadena CA 91106

LUXENBERG, HAROLD RICHARD, b Chicago, Ill, Feb 2, 21; m 42; c 3. APPLIED MATHEMATICS. *Educ:* Univ Calif, Los Angeles, BA, 42, MA, 48, PhD(math), 50. *Prof Exp:* Mathematician, Nat Bur Stand, 50-51; res physicist, Hughes Res & Develop Labs, 51-53; consult engr, Remington Rand, Inc, 53-55; proj consult, Litton Industs, 56-58; mgr display dept, Thompson-Ramo-Wooldridge Corp, 59-60; vpres eng & asst gen mgr, Houston Fearless Corp, 61-63; consult, Lux Assocs, 64-70; PROF COMPUT SCI, CALIF STATE UNIV, CHICO, 70- *Concurrent Pos:* Lectr & instr, Univ Calif, Los Angeles, 52-69. *Mem:* Sr mem Inst Elec & Electronics Engrs; Sigma Xi. *Res:* Data display; document storage and retrieval; photo-optical systems; digital computers in command and control applications. *Mailing Add:* Dept Comput Sci Calif State Univ Chico CA 95929

LUXHOJ, JAMES THOMAS, b Staten Island, NY, Jan 13, 56; m 87; c 2. LOGISTICS, DECISION SUPPORT SYSTEMS. *Educ:* Va Polytech Inst & State Univ, BS, 84, MS, 85, PhD(indust eng & opers res), 86. *Prof Exp:* ASST PROF INDUST ENG, RUTGERS STATE UNIV NJ, 86- *Concurrent Pos:* Co-investr, USDA, 87-89, Hackensack Water Co, 88 & NSF, 91-92; chief fac adv, Rutgers Univ Chap Inst Indust Engrs, 87-; prin investr, NASA Langley Res Ctr, 88; co-prin investr, Fed Aviation Admin, 89; assoc ed, Inst Indust Engrs Trans, 89-; treas-secy, Eng Econ Div, Am Soc Eng Educ, 89, vchmn & prog chmn, 90-91, chmn, 91-92. *Honors & Awards:* Ralph R Teetor Award Eng Educ Excellence, Soc Automotive Engrs, 89. *Mem:* Sigma Xi; sr mem Inst Indust Engrs; sr mem Soc Logistics Engrs; Am Soc Eng Educ. *Res:* Production and operations management; logistics; decision support and expert systems. *Mailing Add:* Dept Indust Eng Rutgers Univ PO Box 909 Piscataway NJ 08855-0909

LUXMOORE, ROBERT JOHN, b Adelaide, Australia, Nov 7, 40; m 75. SOIL PHYSICS, WHOLE PLANT PHYSIOLOGY. *Educ:* Univ Adelaide, BAgSc, 62, BAgSc Hons, 63; Univ Calif, Riverside, PhD(soil physics), 69. *Prof Exp:* Agronomist, Dept Agr, SAustralia, 63-66; res asst, Univ Calif, Riverside, 66-69; res assoc, Univ Ill, 69-70; fel, Univ Calif, Riverside, 70-71; res assoc, Univ Wis, Madison, 71-72; SOIL & PLANT SCIENTIST, OAK RIDGE NAT LAB, 73- *Concurrent Pos:* Vis scientist, Commonwealth Sci & Indust Res Orgn, Australia, 76; consult, Ctr Law & Social Policy, Washington, DC, 79; mem, Rural Abandoned Mines Prog, Tenn, 80-81. *Mem:* Am Soc Agron; fel Soil Sci Soc Am, 88; Crop Sci Soc Am; Am Geophys Union; fel AAAS, 85; Int Soc Soil Sci. *Res:* Experimental and computer modeling research on the relationships between environmental variables and whole plant physiological processes including disruptions induced by pollutant stress and soil variability effects on hydrologic transport by pollutant processes; the relationships between environmental variables and whole plant physiological processes including disruptions induced by pollutant stress; soil variability effects on hydrologic transport processes. *Mailing Add:* Environ Sci Div Oak Ridge Nat Lab PO Box 2008 Oak Ridge TN 37831-6038

LUXON, BRUCE ARLIE, b Ft Dodge, Iowa, Mar 8, 55; m 85. ANIMAL PHYSIOLOGY, MEDICINE. *Educ:* Univ Iowa, BS, 76; Univ Mo-Columbia, PhD(math), 83, MD, 85. *Prof Exp:* Asst prof, Dept Med, Univ Mo, 85-89; FEL GASTROENTEROL, UNIV CALIF, SAN FRANCISCO, 89- *Mem:* Soc Indust & Appl Math; Soc Math Biol. *Res:* Biomathematics of hepatic transport; biophysics of bile formation; hepatic drug metabolism; removal of albumin-bound substances by liver cells. *Mailing Add:* 1368 Plymouth Ave San Francisco CA 94112

LUXON, JAMES THOMAS, b Norwalk, Ohio, Nov 12, 34; m 61; c 2. LASER SURFACE MODIFICATION, BEAM PROPAGATION. *Educ:* Wabash Col, BA, 58; Mich State Univ, MS, 64, PhD(eng), 69. *Prof Exp:* Anal engr, New Departure Div, Gen Motors Corp, 59-61; instr physics, Gen Motors Inst, 61-67, assoc prof elec eng, 69-80, prof mat sci, 81-85, dept head sci math, 88-89, ALLIED DISTINGUISHED PROF & DIR LASER LAB, ENG & MGT INST, GEN MOTORS INST, 81-, ASSOC DEAN, GRAD STUDIES, EXT SERVS & RES. *Concurrent Pos:* Vis prof laser, Univ Luleå, Sweden, 81; Rhodes prof, Eng & Mgt Inst, Gen Motors Inst alumni grant, 83; laser mat processing consult. *Mem:* Optical Soc Am; Laser Inst Am (pres, 86); Soc Mfg Engrs. *Res:* Mathematical description of focusing and beam propagation for high-order mode laser beams; laser surface modification. *Mailing Add:* GMI Eng & Mgt Inst 1700 W Th Ave Flint MI 48502

LUYBEN, WILLIAM LANDES, b Omaha, Nebr, Oct 17, 33; m 63; c 2. CHEMICAL ENGINEERING. *Educ:* Pa State Univ, BS, 55; Rutgers Univ, MBA, 58; Univ Del, MSChE, 62, PhD(chem eng), 63. *Prof Exp:* Process engr, Humble Oil & Refining Co, 55-58 & Iranian Oil & Refining Co, 58-60; tech serv engr, E I du Pont de Nemours & Co, Inc, 63-67; assoc prof, 67-73, PROF CHEM ENG, LEHIGH UNIV, 73- *Concurrent Pos:* Lectr, Univ Del, 63-66; consult, E I du Pont de Nemours & Co, Inc & Sun Oil Co, 67- *Mem:* Am Inst Chem Eng. *Res:* Process dynamics, control and simulation, particularly in distillation columns and chemical reactors. *Mailing Add:* Dept Chem Eng Lehigh Univ Mountaintop Campus Bldg A Bethlehem PA 18015

LUYENDYK, BRUCE PETER, b Freeport, NY, Feb 23, 43; m 67. MARINE GEOPHYSICS. *Educ:* San Diego State Col, BS, 65; Scripps Inst Oceanog, Univ Calif, San Diego, PhD(oceanog), 69. *Prof Exp:* Geophysicist, US Navy Electronics Lab, 65-66; res asst oceanog, Scripps Inst Oceanog, Univ Calif, San Diego, 65-69; fel, Woods Hole Oceanog Inst, 69-70, asst scientist, 70-73; asst prof, 73-75, assoc prof, 75-81, PROF GEOL SCI, UNIV CALIF, SANTA BARBARA, 81- *Concurrent Pos:* Mem working group marine geophys data, Comn Oceanog, Nat Acad Sci, 71; mem working group Mid-Atlantic Ridge, US Geodyn Comn, 71; ed adv, Geol Mag, 74- *Mem:* AAAS; Am Geophys Union; fel Geol Soc Am. *Res:* Geotectonics; paleomagnetism; paleoceanography. *Mailing Add:* Dept Geol Sci Univ Calif Santa Barbara CA 93106

LUYKX, PETER (VAN OOSTERZEE), b Detroit, Mich, Dec 14, 37; m 78; c 4. CYTOGENETICS, HUMAN GENETICS. *Educ:* Harvard Univ, AB, 59; Univ Calif, Berkeley, PhD(zool), 64. *Prof Exp:* Asst prof zool, Univ Minn, Minneapolis, 64-67; asst prof, 67-74, assoc prof, 74-82, PROF BIOL, UNIV MIAMI, 82- *Concurrent Pos:* NIH res grants, 65-73, NSF res grants, 78-88. *Mem:* AAAS; Genetics Soc Am; Am Soc Cell Biol; Entom Soc Am; Am Soc Human Genetics. *Res:* Meiosis and mitosis; cytogenetics of termites; sex chromosome evolution. *Mailing Add:* Dept Biol Univ Miami PO Box 249118 Coral Gables FL 33124

LUYTEN, JAMES REINDERT, b Minneapolis, Minn, Dec 26, 41; m 67; c 3. PHYSICAL OCEANOGRAPHY. *Educ:* Reed Col, AB, 63; Harvard Univ, AM, 65, PhD(chem physics), 69. *Prof Exp:* Res fel geophys fluid dynamics, Harvard Univ, 69-71; asst scientist, 71-75, assoc scientist, 75-86, SR SCIENTIST, WOODS HOLE OCEANOG INST, 86-, DEPT CHMN, 89- *Concurrent Pos:* Vis sci, NCAR, 83-84. *Mem:* Am Geophys Union. *Res:* Theoretical and observational study of the dynamics of low frequency variability of ocean circulation; moored current meter arrays; observations of the Gulf Stream system; equatorial current systems in Pacific and Indian Oceans. *Mailing Add:* Woods Hole Oceanog Inst Woods Hole MA 02543

LUYTEN, WILLEM JACOB, b Semarang, Dutch E Indies, Mar 7, 99; nat US; m 30; c 3. ASTRONOMY. *Educ:* Univ Amsterdam, BA, 18; State Univ Leiden, PhD(astron), 21. *Hon Degrees:* DSc, Case Western Reserve Univ, 67 & Univ St Andrews, 71. *Prof Exp:* Asst, Observ, State Univ Leiden, 20-21; fel, Lick Observ, Univ Calif, 21-22, Kellogg fel, 22-23; astronr, Harvard Observ, 23-27, asst prof astron, Univ, 27-30; from asst prof to prof, 31-75, EMER PROF ASTRON, UNIV MINN, MINNEAPOLIS, 67- *Concurrent Pos:* Mem, Lick Observ Calif Eclipse Exped, Ensenada, Baja, Calif, 23 & Hamburg Observ Eclipse Exped, Jokkmokk, Lapland, 27; Guggenheim fel, 28-30 & 37-38. *Honors & Awards:* Watson Medal, Nat Acad Sci, 64; Bruce Medal, Astron Soc of Pac, 68. *Mem:* Nat Acad Sci; AAAS; Am Astron Soc; Am Asn Variable Star Observers; Int Astron Union. *Res:* Stellar motions; nearby stars; white dwarfs; origin of the solar system. *Mailing Add:* Space Sci Ctr Univ of Minn Minneapolis MN 55455

LUZZI, THEODORE E, JR, b Floral Park, NY, June 15, 27; m 55; c 2. ENGINEERING SCIENCE. *Educ:* Stevens Inst Technol, ME, 51; Mass Inst Technol, MS, 53; Columbia Univ, Eng ScD, 63. *Prof Exp:* Engr, M W Kellogg Co, NY, 53-58; res engr, Grumman Aerospace Corp, 58-61, res scientist plasma physics, 63-73, staff scientist, 73-82, sr staff scientist, 82-90; RETIRED. *Mem:* Am Phys Soc; Am Inst Aeronaut & Astronaut; Am Soc Mech Eng; Sigma Xi. *Res:* Gas dynamics; plasma physics; heat transfer. *Mailing Add:* 4489 Terra Lane St Joseph MI 49085

LUZZIO, ANTHONY JOSEPH, b Lawrence, Mass, Oct 13, 24; m 52; c 4. IMMUNOLOGY. *Educ:* Univ Mass, BS, 47; Kans State Col, MS, 50, PhD(microbiol), 55. *Prof Exp:* Bacteriologist, Wyo State Vet Lab, 47-49; asst, Univ Kans, 50-52; chemist, Hercules Powder Co, 52-54; chief, Immunol Br, US Army Med Res Lab, 55-71, res immunologist, Blood Transfusion Res Div, 71-74, immunologist, Letterman Army Inst Res, 74-80; CONSULT, 80-

Concurrent Pos: Lectr, Univ Louisville, 55-74. Mem: AAAS; Am Soc Microbiol; Am Asn Immunol; Radiation Res Soc. Res: Effects of ionizing radiation on immune mechanisms; effects of arctic climates on immunity; protein degradation and alterations in antigenic specificity by exposure to ionizing rays; immune mechanisms in leishmaniasis. Mailing Add: 2167 Bay St San Francisco CA 94123

LUZZIO, FREDERICK ANTHONY, b Lawrence, Mass, Sept 17, 53. SYNTHETIC ORGANIC CHEMISTRY. Educ: Vanderbilt Univ, BS, 76; Tufts Univ, MS, 79, PhD(chem), 82. Prof Exp: Res chemist, Arthur D Little, Inc, 76-78; postdoctoral fel, Harvard Univ, 82-85; sr develop chemist, E I duPont de Nemours, 85-88; ASST PROF, UNIV LOUISVILLE, 88- Concurrent Pos: Consult, Arthur D Little, Inc, 78-85. Mem: Am Chem Soc. Res: Synthetic organic chemistry; synthesis of natural products, nucleosides, carbohydrates; synthetic methods, chiral oxidation, ultrasound-promoted reactions; isolation and structural elucidation of marine natural products. Mailing Add: Dept Chem Univ Louisville Louisville KY 40292

LWOWSKI, WALTER WILHELM GUSTAV, b Garmisch, Ger, Dec 28, 28; US citizen. ORGANIC CHEMISTRY. Educ: Univ Heidelberg, dipl, 54, Dr rer nat, 55. Prof Exp: Fel, Univ Calif, Los Angeles, 55-57; asst, Univ Heidelberg, 57-59; res fel chem, Harvard Univ, 59-60; asst prof, Yale Univ, 60-66; RES PROF CHEM, NMEX STATE UNIV, 66- Concurrent Pos: Mem bd dirs, Boehringer-Mannheim Corp, Indianapolis. Mem: Fel AAAS; Am Chem Soc; fel NY Acad Sci; Ger Chem Soc; Royal Soc Chem. Res: Reactions mechanisms; electron-deficient nitrogen intermediates; photochemistry; heterocyclic chemistry; heteroatom rearrangements. Mailing Add: Dept of Chem NMex State Univ PO Box 30001 Las Cruces NM 88003-0001

LYBECK, A(LVIN) H(IGGINS), b Trenton, NJ, Feb 28, 19; m 45; c 1. INDUSTRIAL CHEMISTRY. Educ: Polytech Inst, Brooklyn, BS, 41. Prof Exp: Asst develop engr, US Rubber Co, 41-44; res chemist, Gen Cable Co, 44-47; sr res chemist, Congoleum-Nairn, Inc, 47-50; lab dir, William Brand & Co, Inc, 50-59; develop mgr, Brand-Rex Div, Am Enka Corp, Conn, 60-67; TECH MGR, CERRO WIRE & CABLE CO, 67- Mem: Am Chem Soc; Inst Elec & Electronics Eng; Am Inst Chem. Res: Development of electrical insulation systems for wire and cables. Mailing Add: 31 Surrey Lane Branford CT 06405

LYBRAND, TERRY PAUL, b Augusta, Ga, Oct 8, 57; m 84; c 2. MOLECULAR BIOPHYSICS, MOLECULAR SIMULATION & MODELING. Educ: Univ SC, BS, 80; Univ Calif, San Francisco, PhD(pharmaceut chem), 84. Prof Exp: Postdoctoral molecular biophys, Univ Houston, 85-87; asst prof med chem, Univ Minn, 88-90; ASST PROF BIOENG, UNIV WASH, 90- Concurrent Pos: NSF presidential young investr, 87-92; Searle scholar, 88. Mem: Am Chem Soc; Protein Soc; AAAS. Res: Computer simulation of biological molecules to gain an understanding of their properties and behavior in atomic detail; atomic motions in large biological molecules and the relationship of these motions to biological function. Mailing Add: Ctr Bioeng BF-10 Univ Wash Seattle WA 98195

LYCETTE, R(ICHARD) (MILTON), b Houlton, Maine, Sept 20, 26; m 52; c 5. PHYSIOLOGY, MICROBIOLOGY. Educ: Univ Maine, Orono, BS, 50; Ill Inst Technol, MS, 63, PhD(physiol), 68. Prof Exp: Food technologist bacteriol & foods chem, Gen Foods Corp, Albion, NY, 50-52; res scientist, Continental Can Co, Chicago, 62-67; res assoc blood physiol, Presby St Luke's Hosp & Med Sch, Univ Ill, Chicago, 62-69; dir blood prod res & develop, Parke-Davis Co, Detroit, 73-74; sr chemist polymers, Fuller/OBrien Corp, South Bend, Ind, 74-76; res assoc physiol, Med Sch, Wayne State Univ, 76-77; lab dir, World Wide Chem Corp, 80-82; lab dir & consult, SNP Chem Co, Saginaw, Mich, 82-84; LAB DIR & VPRES RES & DEVELOP, AFFIL MARINE PROD CO, BIOMED SYST CO, MAINE, 84- Concurrent Pos: NIH fels, Nat Heart Inst & Off Surgeon Gen, US Army, 62-69; consult, Ind Biomed Systs Co, Mich, 69-; vis lectr & prof, Univ Maine, Augusta, 69-73; sci adv to gov, Off Res & Develop, Maine, 70-73; consult biochemist, Togus Vet Admin Hosp, Maine, 71-72; dir white cell res sect, Blood Res Ctr, Am Nat Red Cross, Bethesda, Md, 72-73; adj & res liaison, Detroit Polymer Inst & Indust Labs, 80-83; trustee, CCEE Soc, NY, 90. Mem: Am Soc Microbiol; Am Chem Soc; fel Royal Microbiol Soc. Res: Cell physiology and microbiology; influence of cell membranes and lipids on aggregation; bioenergetics in cancer; degradation polymers; blood coagulation process; immunology; Limulus substances for wide scale bacterial/viral diagnosis cures including AIDS, herpes, pioneering new device for study live membranes at angstrom levels. Mailing Add: Biomed Syst Co Box 7 New Limerick ME 04761

LYDA, STUART D, b Bridger, Mont, June 6, 30; m 53; c 5. PLANT PATHOLOGY. Educ: Mont State Col, BS, 56, MS, 58; Univ Calif, PhD(plant path), 63. Prof Exp: Lab technician, Univ Calif, 59-62; assoc prof plant path, Univ Nev, Reno, 62-67; assoc prof, 67-77, PROF PLANT PATH, TEX A&M UNIV, 77- Mem: AAAS; Mycol Soc Am; Am Phytopath Soc. Res: Fungus and plant physiology; mycology. Mailing Add: Dept of Plant Path & Microbiol Tex A&M Univ College Station TX 77843

LYDING, ARTHUR R, b New York, NY, May 12, 25; m 57; c 1. RHEOLOGICAL ADDITIVES, POLYURETHANES. Educ: Cornell Univ, BA, 45; Univ Pa, MS, 48, PhD(chem), 51. Prof Exp: Instr, Cornell Univ, 44-45; control chemist, Gen Baking Co, 46; res chemist, Heyden Chem Corp, 50-52; sr res chemist, Olin Industs, 52-56; group leader polymers div, Olin Mathieson Chem Corp, Conn, 57-64; tech asst to vpres res & develop, Pkg Div, 64-69; sr res scientist, FMC Corp, 69-75; SECT LEADER NL CHEM, NL INDUSTS, INC, HIGHTSTOWN, 75- Concurrent Pos: Asst prof, Southern Conn State Col, 64-69; sci Ger translr, 79- Mem: Am Chem Soc; Sigma Xi. Res: Agricultural chemicals; vinyl monomers and polymers; polyurethanes; oil additives; cellulose chemistry; textile stain repellents and flame retardants; fluorochemicals; emulsion polymerization; synthesis of polymers and plastics additives; coatings; rheological additives. Mailing Add: 24 Broadripple Dr Princeton NJ 08540

LYDY, DAVID LEE, b Elwood, Ind, Apr 27, 36; m 59; c 3. SKIN RESEARCH, DENTAL RESEARCH. Educ: Ind Univ, AB, 58; Univ Ill, PhD(inorg chem), 63. Prof Exp: Res chemist, 63-68, SECT HEAD, MIAMI VALLEY LABS, PROCTER & GAMBLE CO, 68- Mem: Am Chem Soc. Res: New opportunities research. Mailing Add: 3617 Pamaisra Dr Oxford OH 45056

LYE, ROBERT J, b St Paul, Minn, May 30, 55. GENETICS. Educ: Johns Hopkins Univ, BA, 77; Univ Colo, Boulder, PhD(cell biol), 89. Prof Exp: POSTDOCTORAL RES FEL, SCH MED, WASH UNIV, 89- Mem: Genetics Soc Am; Am Soc Cell Biol. Mailing Add: Genetics Dept Sch Med Wash Univ 4566 Scott Ave Box 8232 St Louis MO 63110

LYERLA, JO ANN HARDING, b Long Beach, Calif, Sept 28, 40; m 64; c 1. BIOLOGY, ECOLOGICAL GENETICS. Educ: Univ Calif, Davis, BS, 62; San Diego State Univ, MA, 67; Clark Univ, PhD(biol), 78. Prof Exp: Lab technician, Univ Calif, Davis, 62-63; Gen Atomics Div, Gen Dynamics Corp, 63-64; Rockefeller Univ, 66-67 & Pa State Univ, 67-70; ASSOC PROF BIOL, BECKER JR COL, LEICESTER, 76- Mem: Genetics Soc; Am Soc Zoologists. Res: Ecological genetics of terrestrial isopods; isozyme studies in animal population. Mailing Add: Dept Biol Sci Becker Jr Col Three Paxton St Leicester MA 01524

LYERLA, TIMOTHY ARDEN, b Long Beach, Calif, Mar 5, 40; m 64; c 1. DEVELOPMENTAL GENETICS. Educ: Univ Calif, Davis, BA, 63; San Diego State Col, MA, 67; Pa State Univ, University Park, PhD(zool), 70. Prof Exp: NIH fel, Northwestern Univ, Ill, 70-71; from asst prof to prof biol, Clark Univ, 71-88; ASSOC BIOCHEM, SHRIVER CTR MENT RETARDATION, WALTHAM, MASS, 80- Concurrent Pos: NSF sci fac fel, 78-79. Mem: Fel AAAS; Am Soc Zoologists; Soc Develop Biol; Sigma Xi; NY Acad Sci. Res: Pigment genetics in vertebrates; cell differentiation in amphibian development; lysosomal storage diseases in humans. Mailing Add: Dept Biol Clark Univ Worcester MA 01610

LYERLY, HERBERT KIM, b San Diego, Calif, Aug 26, 58. HIV ASSOCIATED MALIGNANCIES, GENE THERAPY. Educ: Univ Calif, Riverside, BA, 80; Univ Calif, Los Angeles, MD, 83. Prof Exp: ASST PROF SURG, DUKE UNIV MED CTR, 90-, ASST PROF PATH, 91- Concurrent Pos: Investr, Ctr AIDS Res & mem, Comprehensive Cancer Ctr, Duke Univ Med Ctr, 91-; mem, Sci Adv Comt, Am Found AIDS Res, 91- Honors & Awards: Achievement Award, Am Col Surgeons, 89. Mem: Sigma Xi. Res: Surgical oncology. Mailing Add: Duke Univ Med Ctr Box 3551 Durham NC 27710

LYFORD, JOHN H, JR, b Chicago, Ill, July 10, 28; div; c 6. ECOLOGY. Educ: Carleton Col, BA, 50; Ore State Univ, MS, 62, PhD(bot), 66. Prof Exp: Pub sch teacher, Wash, 55-62; res biologist, Ore State Game Comn, 63-65; asst prof, 65-72, ASSOC PROF BIOL, ORE STATE UNIV, 72- Mem: AAAS; Ecol Soc Am; Am Bryol Soc. Res: Trophic structure of aquatic communities; ecology and distribution of mosses. Mailing Add: Dept of Gen Sci Ore State Univ Corvallis OR 97331

LYFORD, SIDNEY JOHN, JR, b Exeter, NH, Jan 20, 37; m 61; c 3. ANIMAL NUTRITION, BIOCHEMISTRY. Educ: Univ NH, BS, 58; NC State Univ, MS, 60, PhD(animal nutrit), 64. Prof Exp: PROF ANIMAL NUTRIT, UNIV MASS, AMHERST, 63- Mem: Am Dairy Sci Asn; Am Soc Animal Sci; Sigma Xi. Res: Mechanism of action of certain natural inhibitors and of volatile fatty acid absorption from the ruminant stomach; pectin degradation; nutritive evaluation of byproduct materials as animal feedstuffs. Mailing Add: Dept of Vet & Animal Sci Stockbridge Hall Univ of Mass Amherst MA 01003

LYGRE, DAVID GERALD, b Minot, NDak, Aug 10, 42; m 66; c 2. BIOCHEMISTRY. Educ: Concordia Col, Moorhead, Minn, BA, 64; Univ NDak, PhD(biochem), 68. Prof Exp: Am Cancer Soc fel, Case Western Reserve Univ, 68-70; from asst prof to assoc prof, res corp grant, Cent Wash Univ, 70-79, asst dean, 80-83, assoc dean, 83-89, PROF CHEM, CENT WASH UNIV, 79- Concurrent Pos: Lectr, Am Inst Chem Engrs. Mem: Am Chem Soc; Sigma Xi; AAAS. Res: Enzymology of carbohydrate metabolism; biochemistry of aging; writing chemistry textbooks. Mailing Add: Dept Chem Cent Wash Univ Ellensburg WA 98926

LYJAK, ROBERT FRED, b Detroit, Mich. MATHEMATICS. Educ: Wayne State Univ, BS, 51; Univ Mich, Ann Arbor, MA, 53, PhD(math), 60. Prof Exp: Assoc mathematician, Res Inst, Univ Mich, 53-56, instr math, Univ, 56-58, mathematician, Res Inst, 59-62; res mathematician, Conduction Corp, 62-63 & Res Inst, Univ Mich, 63-66; assoc prof, 66-69, chmn dept, 67-70, PROF MATH, UNIV MICH-DEARBORN, 69- Mem: Am Math Soc. Res: Transformation groups; mathematical models of stochastic systems. Mailing Add: 2222 Needham Ann Arbor MI 48104

LYKE, EDWARD BONSTEEL, b Boston, Mass, Nov 9, 37; m 62; c 2. CYTOLOGY, INVERTEBRATE ZOOLOGY. Educ: Miami Univ, BA, 59; Univ Wis-Madison, MS, 62, PhD(zool), 65. Prof Exp: From asst prof to assoc prof, 65-73, PROF BIOL SCI, CALIF STATE UNIV, HAYWARD, 73- Mem: AAAS; Am Soc Zoologists; Am Inst Biol Sci; Marine Biol Asn UK; Sigma Xi. Res: Invertebrate cytology and histology; spermatogenesis and oogenesis; ecology of estuarine invertebrates. Mailing Add: Dept of Biol Sci Calif State Univ Hayward CA 94542

LYKKEN, DAVID THORESON, b Minneapolis, Minn, June 18, 28; m 52; c 3. GENETICS, PSYCHIATRY. Educ: Univ Minn, BA, 49, MA, 52, PhD(psychol), 55. Prof Exp: From asst prof to assoc prof, 57-65, PROF PSYCHOL, UNIV MINN, 65- Concurrent Pos: Fel, Ctr Advan Study Behav Sci, 59-60. Honors & Awards: Distinguished Contrib to Psychol Award, Am Psychol Asn, 90. Mem: Fel Am Psychol Asn; fel AAAS; Soc Psychophysiol Res (pres), 80; Behav Genetics Soc. Res: Psychological study of twins, reared together or apart, & their families; study of emergenic traits which are genetic but do not run in families; studies of polygraphic interogation (lie detection). Mailing Add: Dept Psychol Univ Minn Minneapolis MN 55455

LYKKEN, GLENN IRVEN, b Grafton, NDak, Jan 27, 39; m 64; c 4. PHYSICS, NUTRITION. *Educ:* Univ NDak, BSEE, 61; Univ NC, MS, 64, PhD(physics), 66. *Prof Exp:* Asst physics, Univ NDak, 61-62 & Univ NC, 62-65; from asst prof to assoc prof, 65-76, PROF PHYSICS, UNIV NDAK, 76- *Concurrent Pos:* Vis prof, Univ NC, 69-70; res physicist, Grand Forks Human Nutrit Res Ctr, USDA, Grand Forks, NDak, 77-88, health physicist, 88- *Mem:* Sigma Xi; Am Phys Soc; Am Inst Nutrit. *Res:* Whole body counting of low level gamma emissions from humans; environmental radon uptake and distribution in the body; health effects in humans. *Mailing Add:* PO Box 8008 Grand Forks ND 58202

LYKOS, PETER GEORGE, b Chicago, Ill, Jan 22, 27; m 50; c 3. PHYSICAL CHEMISTRY. *Educ:* Northwestern Univ, BS, 50; Carnegie Inst Technol, PhD(chem), 55. *Prof Exp:* Instr chem, Carnegie Inst Technol, 54-55; from instr to assoc prof chem, 55-64, dir comput ctr & comput sci dept, 64-71, PROF CHEM, ILL INST TECHNOL, 64- *Concurrent Pos:* Consult, Solid State Sci Div, Argonne Nat Lab, 58-67; consult, Dept Radiation Ther, Michael Reese Hosp, 66-70; pres, Four Pi, Inc, 66-; mem-at-large & chmn comt comput in chem, Nat Acad Sci-Nat Res Coun, 68-74; prog dir, Off Comput Activities, NSF, 71-73; originator series int conferences comput in chem res & educ, Ill, 71, Yugoslavia, 73, Venezuela, 76, USSR, 78, Japan, 80, Washington, DC, 82, WGer, FRG, 85, & Beijing, 87, Italy, 89; co-chmn, Nat Resource Comput in Chem Proposal Develop Team, Argonne Univs Asn-Argonne Nat Lab, 74-77; chmn comput in chem div, Am Chem Soc, 73-77, mem comt prof training, 77- & adv comt, Chem & Eng News, 77-80; dir, Interactive Instr TV Network, 76-78; mem bd, Asn Media-based Continuing Eng Educ, 76-78; Sci consult, Video Satellite Delivery, Nat Tech Univ. *Mem:* Asn Comput Mach; Am Chem Soc; Sigma Xi. *Res:* Quantum chemistry; computational chemistry; computers in chemical education. *Mailing Add:* Dept Chem Ill Inst Technol Chicago IL 60616

LYKOUDIS, PAUL S, m 53; c 1. AERONAUTICAL ENGINEERING. *Educ:* Nat Tech Univ, Greece, Mech & Elec Engr, 50; Purdue Univ, MS, 54, PhD, 56. *Prof Exp:* From asst prof to assoc prof, Purdue Univ, 56-60, dir aerospace sci lab, 68-73, head dept nuclear eng, 73-85, PROF AEROSPACE, ASTRONAUTICS & ENG SCI, PURDUE UNIV, 60-, PROF NUCLEAR ENG, 85- *Concurrent Pos:* Consult, Rand Corp, 60-; Nat Sci Found grant, 60- *Honors & Awards:* Res Award, Sigma Xi, 87. *Mem:* Assoc fel Am Inst Aeronaut & Astronaut; Am Phys Soc; Am Astron Soc; Am Nuclear Soc; Sigma Xi. *Res:* Contributor of numerous papers in field of fluid mechanics, magneto-fluid-mechanics, astrophysics, and fluid mechanics of physiological systems. *Mailing Add:* Sch Nuclear Eng Purdue Univ Lafayette IN 47907

LYLE, BENJAMIN FRANKLIN, b Johnson City, Tenn, Aug 14, 33; m 57; c 3. INDUSTRIAL ENGINEERING, SYSTEMS ANALYSIS. *Educ:* Univ Tenn, Knoxville, BS, 55, MS, 56; E Tenn State Univ, MA, 62; NMex State Univ, ScD(indust eng), 69. *Prof Exp:* Engr artist, Fisher Body Div, Gen Motors Corp, Mich, 55; pres, Lyle Furniture Co, Tenn, 55-61; instr math, E Tenn State Univ, 61-66; from instr to asst prof indust eng, NMex State Univ, 66-70; assoc prof, 70-76, PROF MATH, E TENN STATE UNIV, 76- *Mem:* Am Inst Indust Eng; Am Soc Eng Educ; Nat Soc Prof Engrs. *Res:* Decision theory; economic evaluation; mathematical modeling. *Mailing Add:* Dept Technol East Tenn State Univ Johnson City TN 37614

LYLE, EVERETT SAMUEL, JR, b Dyersburg, Tenn, Mar 17, 27; m 47; c 2. FORESTRY, SOIL SCIENCE. *Educ:* Univ Ga, BSF, 51; Duke Univ, MF, 52; Auburn Univ, PhD(soil sci), 69. *Prof Exp:* Staff asst, Union Camp Corp, 52-57; researcher, 57-86, LAND RECLAMATION CONSULT, AUBURN UNIV, 86- *Concurrent Pos:* Researcher, Ala Surface Mine Reclamation Coun, 73-77; state comnr, Ala Surface Mining Reclamation Comn, 76-80. *Mem:* Soc Am Foresters; Am Soc Agron; Soil Sci Soc Am; Can Land Reclamation Asn. *Res:* Coal surface mine reclamation; tree nutrition; forest soils. *Mailing Add:* Rte 10 Box 371-B Jasper AL 35501

LYLE, GLORIA GILBERT, b Atlanta, Ga, Aug 7, 23; m 47. ORGANIC CHEMISTRY. *Educ:* Vanderbilt Univ, BA, 45; Emory Univ, MS, 46; Univ NH, PhD, 58. *Prof Exp:* Instr chem, Hollins Col, 46-47; res assoc, McArdle Lab Cancer Res, Univ Wis, 47-49; from instr to prof chem, Univ NH, 51-77; asst prof pharmacol, Tex Col Osteop Med, 77-80; ADJ PROF, UNIV TEX, SAN ANTONIO, 81- *Concurrent Pos:* USPHS res fel, 58-59; vis assoc prof, Univ Va, 73-74. *Honors & Awards:* Harry & Carol Mosher Nat Award, Am Chem Soc, 86. *Mem:* Am Chem Soc; NY Acad Sci; Royal Soc Chem; Sigma Xi. *Res:* Organic synthesis; natural products; optical rotatory dispersion and circular dichroism; stereochemistry. *Mailing Add:* 12814 Kings Forest San Antonio TX 78230

LYLE, JAMES ALBERT, b Lexington, Ky, Sept 19, 16; m 48; c 1. BOTANY. *Educ:* Univ Ky, BS, 40; NC State Col, MS, 46; Univ Minn, PhD(plant path), 53. *Prof Exp:* Jr plant pathologist, Exp Sta, Univ Hawaii, 46-47; from asst plant pathologist to assoc plant pathologist, 47-54, prof bot & microbiol & head dept, 54-79, EMER PROF, BOT, PLANT PATH & MICROBIOL DEPT, AUBURN UNIV, 79- *Mem:* Am Phytopath Soc; Am Asn Adranament Sci; Coun Agr Sci & Technol; Am Inst Biol Sci. *Res:* Fungus ecology and plant disease control, especially peanuts; field and forage crops. *Mailing Add:* 1678 Millbranch Dr Auburn AL 36830-7008

LYLE, LEON RICHARDS, b Ottumwa, Iowa, Nov 28, 41; m 72; c 2. NUCLEAR MEDICINE. *Educ:* Drake Univ, BA, 63, MA, 67; Mont State Univ, PhD(microbiol), 69. *Prof Exp:* Postdoctoral fel immunol, Sch Med, Wash Univ, 70-73; chemist immunol, Mallinckrodt Med Inc, 73-75, group leader immunol, 75-80, asst dir, Hybridoma Lab, 80-85, assoc dir nuclear med, 85-89, assoc dir tech planning, 89-91, DIR TECH PLANNING, MALLINKRODT MED INC, 91- *Concurrent Pos:* Indust rep, Immunol Devices Adv Panel, Off Med Devices, Food & Drug Admin, 81-88; sci prog chmn, Nat Meeting St Louis, Clin Ligand Assay Soc, 86-87. *Mem:* Am Asn Immunologists; Soc Nuclear Med; Am Soc Microbiol; Am Asn Clin Chem; AAAS. *Res:* Identification, implementation and administration of extramural research programs in nuclear medicine, magnetic resonance, ultrasound and x-ray diagnostic imaging. *Mailing Add:* Mallinckrodt Med Inc 675 McDonnell Blvd PO Box 5840 St Louis MO 63134

LYLE, ROBERT EDWARD, JR, b Atlanta, Ga, Jan 26, 26; m 47. PHARMACEUTICAL CHEMISTRY, STRUCTURAL CHEMISTRY. *Educ:* Emory Univ, BA, 45, MS, 46; Univ Wis-Madison, PhD(org chem), 49. *Prof Exp:* Asst prof chem, Oberlin Col, 49-51; prof chem, Univ NH, 51-76; prof & chair chem, Univ NTex, 76-79; VPRES CHEM & CHEM ENG, SOUTHWEST RES INST, 79- *Concurrent Pos:* Adj prof, Univ Tex, San Antonio, 82- *Honors & Awards:* Harry-Canl Mosher Award, Am Chem Soc, 87. *Mem:* Am Chem Soc; Royal Soc Chem; AAAS; Am Asn Cancer Res; Soc Nuclear Med. *Res:* Stereochemistry of nitrogen heterocycles; organic synthesis of heterocyclic compounds; microencapsulation and drug delivery systems. *Mailing Add:* 12814 Kings Forest San Antonio TX 78230

LYLE, WILLIAM MONTGOMERY, b Summerside, PEI, Oct 4, 13; m 56; c 3. OPTOMETRY. *Educ:* Col Optom Ont, dipl, 38, OD, 58; Ind Univ, Bloomington, MS, 63, PhD(physiol optics), 65. *Prof Exp:* Pvt pract optom, 38-60; res assoc physiol optics, Ind Univ, 60-62, lectr, 62-65; asst prof optom, Col Optom Ont, 65-67; chief path sect, Univ Waterloo, 67-74, from assoc prof to prof optom, 67-84, dir clins, 74-77, adj prof, 84-89, EMER PROF OPTOM, UNIV WATERLOO, 89- *Concurrent Pos:* Pres, Asn Schs Optom Can, 71-73; ed, Am J Optom & Physiol Optics, 79-88; ed, Optom & Vision Sci, 89- *Honors & Awards:* President's Award, Can Asn Optom. *Mem:* Can Asn Optom (pres, 55-57); AAAS; Am Acad Optom; Am Soc Human Genetics; Am Optom Asn; Sigma Xi. *Res:* Side effects of drugs; inheritance of astigmatism; intraracial differences in refraction. *Mailing Add:* Sch Optom Univ Waterloo Waterloo ON N2L 3G1 Can

LYLES, LEON, b Wetumka, Okla, Jan 18, 32; m 51; c 3. AGRICULTURAL ENGINEERING, RESEARCH ADMINISTRATION. *Educ:* Okla State Univ, BS, 55; Kans State Univ, MS, 59, PhD(mech eng), 70. *Prof Exp:* Agr engr erosion res, Agr Res Serv, USDA, 57-60, water mgt res, 60-64, wind erosion res, 64-75, res leader, Wind Erosion Res Unit, 75-88; RETIRED. *Honors & Awards:* Distinguished Serv Award, USDA, 84. *Mem:* Am Soc Agr Engrs; Soil Sci Soc Am. *Res:* Wind erosion and water management (dryland) research. *Mailing Add:* Kans State Univ E Waters Hall Rm 147 Manhattan KS 66506

LYLES, SANDERS TRUMAN, b Reeves, La, May 24, 07; m 46; c 4. BACTERIOLOGY. *Educ:* Rice Univ, BA, 30, MA, 31; Southwestern Baptist Sem, ThM, 36, ThD, 49; Univ Tex, PhD(bact, biochem), 55. *Prof Exp:* From instr to prof biol, Tex Christian Univ, 46-77, res scientist, 52-54; EMER CONSULT ECOL & ENVIRON, 70- *Mem:* Am Soc Microbiol; Sigma Xi. *Res:* Epidemiology and antibiotic resistance of staphylococcus; biochemical studies of blood serum. *Mailing Add:* 3901 Stadium Dr Ft Worth TX 76109-3714

LYMAN, BEVERLY ANN, b Philadelphia, Pa, Aug 22, 56; m 81; c 1. BIOCHEMISTRY, TOXICOLOGY. *Educ:* Thomas Jefferson Univ, BS, 78; Univ Pa, MS, 82; Hahnemann Med Col, PhD(biochem), 86. *Prof Exp:* Postdoctoral fel, Chem Indust Inst Toxicol, 87-88; ASST PROF CLIN LAB SCI & MED CHEM, UNIV TENN, MEMPHIS, 89- *Mem:* Am Soc Biochem & Molecular Biol; AAAS; Sigma Xi; Am Soc Med Technol; Soc Toxicol. *Res:* Biochemical toxicology of anticancer drugs; biochemistry of surface-active antithrombotic agents for prostheses. *Mailing Add:* Dept Clin Lab Sci Univ Tenn 800 Madison Ave Memphis TN 38163

LYMAN, CHARLES PEIRSON, b Brookline, Mass, Sept 23, 12; m 41; c 5. BIOLOGY. *Educ:* Harvard Univ, AB, 36, MA, 39, PhD(biol), 42. *Prof Exp:* Asst physiol, 42, asst cur, Mus Comp Zool, 45-50, fel anat, Med Sch, 46-48, res assoc, 48-62, assoc cur, 50-57, res assoc, 58-68, from asst prof to assoc prof anat, 62-76, cur mammal, 68-81, prof, 76-81, EMER PROF BOIL, HARVARD UNIV, 81- *Mem:* AAAS; Am Physiol Soc; Am Soc Zoologists; Am Acad Arts & Sci; Sigma Xi. *Res:* Hibernation and temperature regulation in mammals; physiological ecology. *Mailing Add:* 105 Elm St Canton MA 02021

LYMAN, DONALD JOSEPH, b Chicago, Ill, Nov 5, 26; m 78; c 2. POLYMER CHEMISTRY, BIOMATERIALS. *Educ:* Univ Nev, BS, 49; Univ Del, MS, 51, PhD(chem), 52. *Prof Exp:* Asst chem, Univ Del, 50-52; res chemist high polymers, E I du Pont de Nemours & Co, 52-61; sr polymer chemist, Stanford Res Int, 61-64, head biomed polymer res, 64-69; PROF MAT SCI & RES ASSOC PROF SURG, UNIV UTAH, 69-, PROF BIOENG, 74-; pres, Vascular Int Inc, 83-86. *Concurrent Pos:* Lectr, Dept Mat Sci, Stanford Univ, 64-68; chmn, Gordon Conf Sci & Technol Biomat; mem, comt surv mat sci & eng, Nat Acad Sci; mem ,eval panel polymer div, Nat Bur Standards, 73-76. *Honors & Awards:* Am Soc Artificial Internal Organs Award, 69; Clemson Award Basic Res, Soc Biomaterials, 82; Distinguished Res Award, Univ Utah, 82. *Mem:* AAAS; Am Chem Soc; Am Soc Artificial Internal Organs; Soc Biomat; Int Soc Artificial Organs. *Res:* Synthetic polymers and polymer intermediates; mechanisms of polymerization; structure-property relationships of polymers; biomedical polymers; implants for artificial organs and reconstruction surgery. *Mailing Add:* Dept Bioeng Salt Lake City UT 84112

LYMAN, ERNEST MCINTOSH, physics; deceased, see previous edition for last biography

LYMAN, FRANK LEWIS, b Springfield, Ill, Nov 6, 21; m 47; c 6. TOXICOLOGY. *Educ:* Swarthmore Col, AB, 43; Hahnemann Med Col, MD, 46; Bd Toxicol Sci, dipl. *Prof Exp:* Intern, WJersey Hosp, Camden, NJ, 47; physician, coach & instr biol, William Penn Col, 47-48; pvt pract, Iowa, 48-55; staff pediatrician, US Naval Hosp, Beaufort, SC, 55-57; assoc med dir, Mead Johnson & Co, 57-60; assoc dir med dept, Geigy Chem Corp, 60-61, asst to med dir, 61-63; dir indust med, Ciba-Geigy Corp, 63-76; CONSULT TOXICOL, 76- *Concurrent Pos:* Instr, Seton Hall Col, 60-62; assoc prof, Sch Med, Temple Univ, 77-; mem various comts, Nat Acad Sci, 75-81. *Mem:* Am Col Toxicol; Soc Toxicol; fel Am Acad Clin Toxicol. *Res:* Dietary management of phenylketonuria; toxicology of fluorescent whitening agents; pesticide toxicology. *Mailing Add:* 68F Long Beach Blvd North Beach NJ 08008

LYMAN, FREDERIC A, b Syracuse, NY, Sept 4, 34; m 54; c 3. MECHANICAL ENGINEERING. *Educ:* Syracuse Univ, BME, 55, MME, 57; Rensselaer Polytech Inst, PhD(eng mech), 61. *Prof Exp:* Preceptor eng mech, Columbia Univ, 61-62; aerospace res engr, Lewis Res Ctr, NASA, 62-66, head plasma flow sect, 66-67; assoc prof eng, Case Western Reserve Univ, 67-70; assoc prof, 70-78, PROF MECH & AEROSPACE ENG, SYRACUSE UNIV, 78- *Concurrent Pos:* Vis res engr, Princeton Univ, 77-78; prin res engr, Case WRU/NASA Lewis Inst Computational Mech Propulsion, 85-86. *Mem:* AAAS; Am Phys Soc; Am Soc Mech Engrs. *Res:* Fluid mechanics; heat transfer; plasma dynamics; acoustics; combustion. *Mailing Add:* 323 Scott Ave Syracuse NY 13224-1725

LYMAN, GARY HERBERT, b Buffalo, NY, Feb 24, 46; m 78; c 3. MEDICAL ONCOLOGY, HEMATOLOGY. *Educ:* State Univ NY, BA, 68, MD, 72; Harvard, MPH, 82. *Prof Exp:* PROF MED & BIOSTATIST, UNIV S FLA, 77-, CHIEF MED ONCOL, INTERNAL MED, 79-, CHIEF MED, H LEE MOFFITT CANCER CTR & RES INST, 85- *Mem:* AAAS; Am Soc Hemat; Am Soc Clin Oncol; Am Asn Cancer Res; Am Col Physicians. *Res:* Cancer epidemiology; mathematical modeling of complex systems. *Mailing Add:* H Lee Moffitt Cancer Ctr & Res Inst 12902 Magnolia Dr Tampa FL 33612

LYMAN, HARVARD, b San Francisco, Calif, Sept 25, 31. PLANT PHYSIOLOGY, MOLECULAR BIOLOGY. *Educ:* Univ Calif, Berkeley, BA, 53; Univ Wash, MS, 57; Brandeis Univ, PhD(biol), 60. *Prof Exp:* Asst biol, Univ Wash, 55-57; instr, Brooklyn Col, 60-62; vis scientist biochem, Brookhaven Nat Lab, 62-63; asst prof biol, Brooklyn Col, 63-65; asst scientist microbiol, Brookhaven Nat Lab, NY, 65-67, assoc scientist, Med Dept, 67-68; ASSOC PROF BIOL, STATE UNIV NY STONY BROOK, 68- *Concurrent Pos:* NIH res grant, 63-65; NSF travel grant, 64, grant, 70-72. *Mem:* AAAS; Am Soc Plant Physiologists; Soc Protozoologists; Am Soc Cell Biologists; Biophys Soc. *Res:* Biosynthesis and inheritance of cellular organelles; development, physiology and differentiation of algae and fleshy and unicellular fungi. *Mailing Add:* Dept of Biol State Univ of NY Stony Brook NY 11794

LYMAN, JOHN (HENRY), b Santa Barbara, Calif, May 29, 21; wid; c 2. SELF ORGANIZING SYSTEMS. *Educ:* Univ Calif Los Angeles, BA, 43, MS, 50, PhD(exp psychol), 51. *Prof Exp:* Res technician math, Lockheed Aircraft Corp, 40-44; PROF ENG & PSYCHOL, UNIV CALIF-LOS ANGELES, 51- *Concurrent Pos:* Managing ed, Human Factors Soc, 58-63, Ann Biomed Eng, Biomed Eng Soc, 71-76; head, Biotechnol Lab, Sch Eng & Appl Sci, Univ Calif-Los Angeles, 58-84; consult, Vet Admin, Los Angeles, 62 & 66-72, Human Factors Soc, 76-; vis prof bioeng, Technol Inst, Delft, Neth, 65; spec consult, Comt Prosthetics Res & Develop, Nat Acad Sci, 73. *Honors & Awards:* Paul M Fitts Award, Human Factors Soc, 71. *Mem:* Human Factors Soc (pres, 67-68); Biomed Eng Soc (pres, 80-81); fel Am Psychol Asn; Systs Man & Cybernetics Soc; fel Am Psychol Soc; Inst Elec & Electronic Engrs Med & Biol Soc. *Res:* Functional optimization of human machine environment design interfaces; applications to neuromuscularly handicapped, teleoperations, robotic; systems and human-computer shared functions in normal and stressful environments. *Mailing Add:* 3512 Beverly Ridge Dr Sherman Oaks CA 91423-4505

LYMAN, JOHN L, b Delta, Utah, June 16, 44; m 68; c 5. LASER PHOTOCHEMISTRY. *Educ:* Brigham Young Univ, BS, 68, PhD(phys chem), 73. *Prof Exp:* Phys scientist asst, US Army Dugway Proving Ground, 69-70; mem staff, 73-81, asst group leader, 81-82, FEL, LOS ALAMOS NAT LAB, 83- *Concurrent Pos:* Guest prof, Ctr Interdisciplinary Res, Univ Bielefeld, WGer, 80; vis scientist, Max Planck Inst Quantum-Optics, Garching, WGer, 82, 84 & 90; vis prof, Univ NewSWales, Kensington, Australia, 87. *Mem:* Am Chem Soc; Optical Soc Am; fel Am Inst Chem. *Res:* Chemical kinetics; interaction of laser radiation with polyatomic molecules, including laser photochemistry, laser istope separation, infrared excitation of polyatomic molecules, and vibrational energy dynamics. *Mailing Add:* 270 Donna Ave Los Alamos NM 87544

LYMAN, JOHN TOMPKINS, b Berkeley, Calif, May 25, 32; m 80; c 3. BIOPHYSICS. *Educ:* Univ Calif, AB, 54 & 58, PhD(biophys), 65. *Prof Exp:* Res asst, 59-65, biophysicist, 65-79, STAFF SR SCIENTIST, LAWRENCE BERKELEY LAB, 79- *Mem:* Am Asn Physicists in Med; AAAS; Sigma Xi. *Res:* Radiation physics; radiation therapy; radiobiology; heavy charged-particle radiation dosimetry; radiobiology and radiotherapy. *Mailing Add:* Lawrence Berkeley Lab MS 55-121 Berkeley CA 94720

LYMAN, ONA RUFUS, b Jamaica, Vt, Nov 18, 30; m 54; c 3. PHYSICS. *Educ:* Univ Vt, BA, 52. *Prof Exp:* Jr engr, Sprague Elec Co, 52-54; PHYSICIST, TERMINAL BALLISTICS LAB, BALLISTICS RES LAB, ABERDEEN PROVING GROUND, 56- *Res:* Neutron shielding; combustion; interaction of laser beams with materials; blast and fragment protection for industrial workers; initiation mechanisms of explosives; explosive safety in storage and transport; vulnerability of gun propellants to hostile threats. *Mailing Add:* 303 Carter St Aberdeen MD 21001

LYMAN, W(ILKES) STUART, b Mt Vernon, SDak, Apr 13, 24; m 48; c 3. PHYSICAL METALLURGY. *Educ:* Univ Notre Dame, BS, 44; Univ Calif, MS, 52. *Prof Exp:* Jr metallurgist, Nat Adv Comt Aeronaut, Ohio, 44; head adv planning unit, Off Chief Engr, US Forces Frankfurt, Ger, 46-47, engr, Spec Assignment, Heidelberg, 48-49; asst, Univ Calif, 49-50, res engr, Inst Eng Res, 50-51; staff metallurgist, Mat Adv Bd, Nat Res Coun, 51-54; asst dept consult, Battelle Mem Inst, 55-57, div consult ferrous metall, 57-62, div chief, 62-64; mgr tech serv & mkt res, 64-79, vpres, 79-81, SR VPRES, COPPER DEVELOP ASN INC, 81- *Mem:* Am Soc Metals Inst; Metall Soc; Am Soc Mech Engrs; Am Soc Testing & Mat; Soc Automotive Engrs; Sigma Xi; Inst Mat. *Res:* Metal fabrication; materials application; alloy selection. *Mailing Add:* 15 N Bridge Terrace Mt Kisco NY 10549

LYMAN, WILLIAM RAY, b Stratton, Vt, May 30, 20; m 44; c 4. CHEMISTRY, RADIOISOTOPES IN RESEARCH. *Educ:* Univ Vt, BS, 41; Mass Inst Technol, PhD(org chem), 47; Columbia Univ, AM, 47. *Prof Exp:* Asst chem, Columbia Univ, 41-44; jr chemist, Tenn Eastman Corp Div, Eastman Kodak Co, 44-46; res chemist, Resinous Prod & Chem Co, 47-48; res chemist, 48-66, lab head, 66-73, proj leader, 73-81, res sect mgr, 81-84, spec assignment, 84-85, consult, Rohm & Haas Co, 85-87; RETIRED. *Mem:* Am Chem Soc. *Res:* Pesticide residue analysis; fate of pesticides in plant and animal systems and in the environment. *Mailing Add:* 2125 Guernsey Ave Abington PA 19001

LYMANGROVER, JOHN R, b Ft Wayne, Ind, July 24, 44. ENDOCRINOLOGY, ELECTROPHYISOLOGY. *Educ:* Xavier Univ, BS, 66; Univ Ky, MS, 68; Univ Cincinnati, PhD(physiol), 72. *Prof Exp:* Fel res, Dept Biochem, Med Col Ohio, 72-75; asst prof, Dept Physiol, Tulane Univ, 75-80; ASSOC PROF, DEPT PHYSIOL & PHARMACOL & DIR MED PHYSIOL TEACHING, BOWMAN-GRAY MED SCH, 80- *Concurrent Pos:* Adj assoc prof, Dept Elec Eng, Tulane Univ, 80-; consult, grants reviewer, NIH, 79-81; mem, Basic Sci & High Blood Pressure Coun, Am Heart Asn. *Mem:* AAAS; Sigma Xi; Bioelectromagnetics Soc; Endocrine Soc. *Res:* Neuroendocrinology; biological effects of electric fields; mechanism of peptide hormone action on adrenal cortical hormone release; role of endogenous opioids on adrenal steroid secretion and regulation of blood pressure. *Mailing Add:* Nat Ctr Res Resources NIH Westwood Bldg Rm 10A16 Bethesda MD 20892

LYMN, RICHARD WESLEY, b Flushing, NY, July 26, 44; m 70; c 2. BIOPHYSICS, BIOCHEMISTRY. *Educ:* Johns Hopkins Univ, BA, 64; Univ Chicago, PhD(biophys), 70. *Prof Exp:* USPHS fel biophys, Univ Chicago, 70-71; Brit-Am fel, Am Heart Asn, MRC Lab Molecular Biol, Cambridge, Eng, 71-74; sr staff fel biophys, Phys Biol Lab, Nat Inst Arthritis Metab & Digestive Dis, 74-78, grants assoc, Div Res Grants, 78-79, asst assoc dir, Arthritis, Musculoskeletal & Skin Dis, Nat Inst Arthritis, Diabetes, Digestive & Kidney Dis, 79-83, DIR MUSCLE BIOL PROG, NIH, 84- *Concurrent Pos:* Treas, Biophys Soc, 82-87; Publ Comt, Biophys Soc, 87- *Mem:* Biophys Soc; Am Soc Biol Chemists; AAAS. *Res:* Enzyme kinetics; cellular and morphological movement; mathematical modelling; molecular mechanism of muscle contraction and tension development; science administration. *Mailing Add:* Muscle Biol NIAMS NIH Westwood Bldg Rm 403 Bethesda MD 20892

LYNCH, BENJAMIN LEO, b Omaha, Nebr, Dec 29, 23; m 56; c 6. ORAL SURGERY. *Educ:* Creighton Univ, BS, 45, DDS, 47, MA, 53; Northwestern Univ, MSD, 54; Am Bd Oral Surg, dipl. *Prof Exp:* From asst instr to assoc prof oral surg, Creighton Univ, 48-57, dean, Sch Dent, 54-61, dir, Oral Surg Dept, 54-67, coordr, Dent Sch Grad & Post-grad Prog, 67, PROF ORAL SURG, CREIGHTON UNIV, 57- *Concurrent Pos:* Pres dent staff, Children's Mem Hosp, 52-53 & 59-60; fac mem, San Antonio Jr Col, 55, Med Field Sch, Ft Sam Houston, 55-56, Walter Reed Army Post-grad Sch Med, 56-57, guest lectr, 57-58; mem, Omaha-Douglas County Health Bd, 66-68, vpres, 67, pres, 68; mem, bd dir, Nebr Blue Cross-Blue Shield, 68-, exect comt, 73-81; mem bd dirs, Nebr Dental Serv Corp, 72-78, pres, 74-78; treas, Children's Mem Hosp Med-Dent Staff, 79-81, staff mem, 50-; consult, Vet Hosp & Strategic Air Command Hq, Omaha, Nebr & Jenny Edmundson Hosp, Council Bluffs, Iowa; staff mem & exec comt, Omaha Surg Ctr, 79-83, secy staff, 81-83. *Mem:* Am Soc Oral & Maxillofacial Surgeons; Am Dent Asn; fel Am Col Dent. *Res:* Dental education. *Mailing Add:* 509 Happy Hollow Blvd Omaha NE 68106

LYNCH, BRIAN MAURICE, b Melbourne, Australia, Jan 20, 30; m 56; c 2. PHYSICAL ORGANIC CHEMISTRY. *Educ:* Univ Melbourne, BSc, 52, MSc, 54, PhD(chem), 56. *Prof Exp:* Fel & vis prof cancer chemother, NMex Highlands Univ, 56-57; asst prof org chem, St Francis Xavier Univ, Can, 57-58; res officer chem, Div Coal Res, Commonwealth Sci & Indust Res Orgn, Australia, 58-59; asst prof phys chem, Mem Univ Nfld, 59-62; assoc prof, 62-68, chmn dept, 72-79, & 80-86, PROF ORG CHEM, ST FRANCIS XAVIER UNIV, 68- *Concurrent Pos:* Nat Res Coun Can sr res fel, Australian Nat Univ, 68-69; Natural Sci & Eng Res Coun sr indust fel, NS Res Found, 81-82; proj Seraphim fel, Eastern Mich Univ, 88. *Mem:* Am Chem Soc; fel Can Soc Chem; fel Royal Soc Chem London; Soc Appl Spectros; fel Chem Inst Can. *Res:* Nuclear magnetic resonance; infrared spectra by Fourier transform techniques. *Mailing Add:* Campus PO Box 53 St Francis Xavier Univ Antigonish NS B2G 1C0 Can

LYNCH, CAROL BECKER, b New York, NY, Dec 3, 42; m 67. BEHAVIORAL GENETICS, EVOLUTIONARY GENETICS. *Educ:* Mt Holyoke Col, AB, 64; Univ Mich, MA, 65; Univ Iowa, PhD(zool), 71. *Prof Exp:* NSF fel, Inst Behav Genetics, Univ Colo, 72-73; from asst prof to prof, 73-85, PROF BIOL, WESLEYAN UNIV, 85-, DEAN SCI, 88- *Mem:* Behav Genetics Asn; Genetics Soc Am; Animal Behav Soc; Soc Study Evolution; Am Soc Naturalists. *Res:* Genetic and environmental influences on behavioral and physiological thermoregulation in mice; empirical tests of quantitative; genetic theory; genetic influence on circadian rhythms. *Mailing Add:* Dept of Biol Wesleyan Univ Middletown CT 06457

LYNCH, CHARLES ANDREW, b Brooklyn, NY, Jan 6, 35; m 60; c 2. SYNTHETIC FAT SUBSTITUTES, SYNTHETIC LUBRICANTS. *Educ:* Manhattan Col, BS, 56; Univ Notre Dame, PhD(org chem), 60. *Prof Exp:* Res chemist, Esso Res & Eng Co, 60-65; mgr org appln res, FMC Corp, 65-74; exec vpres & tech dir, Am Oil & Supply Co, 74-80; tech dir, dir sales & mkt/ new bus develop, Hatco Corp, 81-90; VPRES TECHNOL, HATCO ADVAN TECHNOL CORP, 91- *Mem:* Am Chem Soc; Am Oil Chemists Soc; Soc Tribologists & Lubrication Engrs; Soc Automotive Engrs. *Res:* Advanced high temperature liquid lubricant development; advanced dielectric fluids; fat substitutes. *Mailing Add:* Hatco Corp King George Post Rd Fords NJ 08863

LYNCH, DAN K, b San Francisco, Calif, Aug 6, 20; m 53; c 1. INDUSTRIAL CHEMISTRY. *Educ:* Principia Col, BS, 42; Stanford Univ, MA, 44. *Prof Exp:* Instr, Principia Col, 44-48; anal chemist, Monsanto Co, Mo, 48-53, res chemist, Org Chem Div, 53-63, res specialist, 63-71; sr group leader plant process technol, Wm G Krummrich Plant, Monsanto Chem Intermediates Co, 71-85; RETIRED. *Mem:* Am Chem Soc; fel Am Inst Chemists; AAAS. *Res:* Plant process improvement and maintenance research. *Mailing Add:* 12 Lemp Rd St Louis MO 63122

LYNCH, DANIEL MATTHEW, b Detroit, Mich, June 28, 21. PLANT ECOLOGY. *Educ:* Univ Detroit, AB, 43; Mich State Univ, MS, 48; Wash State Univ, PhD(bot), 52. *Prof Exp:* Asst bot, Mich State Univ, 47-48 & Wash State Univ, 48-52; from instr to assoc prof, 54-65, PROF BIOL, ST EDWARD'S UNIV, 65- *Mem:* AAAS; Ecol Soc Am; Bot Soc Am. *Res:* Ecology of the southwestern grasslands and woodlands. *Mailing Add:* St Edward's Univ Austin TX 78704

LYNCH, DARREL LUVENE, b Dewey, Okla, Feb 6, 21; m 49; c 4. ORGANIC CHEMISTRY, SOIL MICROBIOLOGY. *Educ:* Univ Ill, PhD(agron), 53; Univ Del, MS, 57. *Prof Exp:* Instr & asst soil biol, Univ Ill, 48-52; asst prof agron, Univ Del, 52-58; asst prof soil sci, Univ Alta, 58-60; assoc prof chem, Ga Southern Col, 60-62; from assoc prof to prof, 66-82, EMER PROF BIOL SCI, NORTHERN ILL UNIV, 82- *Mem:* Am Soc Microbiol. *Res:* Nitrogen fixation of Rhizobia and nodulation; soil organic matter; soil polysaccharides; morphology and nutrition studies with algae; ultrastructure studies with the Actinoplanaceae; pigment production in bacteria. *Mailing Add:* 306 Dresser Rd DeKalb IL 60115

LYNCH, DAVID DEXTER, b Brooklyn, NY, May 22, 34; m 54; c 4. THEORY OF INERTIAL INSTRUMENTS, MODELING & SIMULATION OF AUTOMOTIVE CRASH SENSORS. *Educ:* Tufts Univ, BS, 56; Harvard Univ, Am, 57, PhD(theoret physics), 67; Univ Calif, Santa Barbara, BA, 86. *Prof Exp:* Instr physics, Tufts Univ, 59-63; engr, Delco Systs Opers, Delco Electronics Corp, 63-67, head physics group, 67-69, head Advan Instrument Technol Sect, 69-80, STAFF ENGR, DELCO SYSTS OPERS, DELCO ELECTRONICS CORP, 80- *Concurrent Pos:* Mem, Task Group on Design Marine Risers, Am Petrol Inst, 74-75, Tech Working Group 4-B (inertial) organized by Inst Defense Anal for Dept Defense, 87-89 & Gyro & Accelerometer Panel, Inst Elec & Electronics Engrs Aerospace & Electronic Systs Soc, 86- *Mem:* Optical Soc Am. *Res:* Theory and design analysis of inertial instruments including the laser gyroscope, the fiber-optic gyroscope, and the hemispherical-resonator gyroscope; math modeling and simulation of crash sensors for automotive air-cushion restraint systems. *Mailing Add:* 5442 Berkeley Rd Santa Barbara CA 93111

LYNCH, DAVID H, b San Francisco, Calif, Aug 5, 50. IMMUNOLOGY. *Educ:* Univ Calif, Santa Cruz, BS, 74; Univ Utah, PhD(exp path), 79. *Prof Exp:* Postdoctoral fel, Dept Path, Sch Med, Univ Utah, 79-81, res instr, Dept Obstet-Gynec, 81-82, res asst prof, 85-88; IPA investr, Immunol Br, Nat Cancer Inst, 82-85; STAFF SCIENTIST IMMUNOL, IMMUNEX CORP, 88- *Mem:* Sigma Xi; Am Asn Immunologists. *Mailing Add:* Immunex Corp 51 University St Seattle WA 98101

LYNCH, DAVID WILLIAM, b Rochester, NY, July 14, 32; wid; c 3. SOLID STATE PHYSICS. *Educ:* Rensselaer Polytech Inst, BS, 54; Univ Ill, MS, 55, PhD(physics), 58. *Prof Exp:* Fulbright fel, Pavia, Italy, 58-59; from asst prof to prof, 59-85, chmn dept, 85-90, DISTINGUISHED PROF PHYSICS, IOWA STATE UNIV, 85- *Concurrent Pos:* Sr physicist, Ames Lab, US Dept Energy, 66-; vis prof, Univ Hamburg, 74; actg assoc dir, Synchrotron Radiation Lab, Stoughton, Wis, 84; chmn dept, Iowa State Univ, 85-90. *Mem:* AAAS; fel Am Phys Soc; Optical Soc Am. *Res:* Optical properties of solids, including use of synchrotron radiation and modulation-spectroscopy; photoelectron spectroscopy. *Mailing Add:* Dept Physics Iowa State Univ Ames IA 50011

LYNCH, DENIS PATRICK, b Kansas City, Kans, Oct 5, 51; m 73; c 2. ORAL PATHOLOGY, ORAL MEDICINE. *Educ:* Univ Calif, San Francisco, DDS, 76. Univ Ala, PhD, 86. *Prof Exp:* Assoc dean, Acad Affairs, 87-89, ASST PROF PATH, DENT BR, UNIV TEX, 81-, ASSOC PROF DIAG SCI & EXEC ASSOC DEAN, 89- *Concurrent Pos:* Mem, Comn Accreditation, Am Dent Asn, 75-79; adj asst prof path & lab med, Col Med, Tex A&M Univ, 83-89, adj assoc prof, 89-; adj assoc prof community med, Baylor Col Med, 89-; chmn, sect path, Am Asn Dent Schs, 84-85; mem, Janssen Res Coun, Janssen Pharmaceut, 86-; mem, Pres Task Force on AIDS, Univ Tex Health Sci Ctr, Houston, 86; vpres dent res & develop, Pearce Sci & Tech Assocs, 86-; consult, Bering Clin, 87-, curric consult, Comn Dental Accreditation, 90- *Honors & Awards:* Gabbs Award, Am Acad Oral Path Award, & Am Acad Oral Med Award, Univ Calif, San Francisco, 76; Golden Pen Award, Int Col Dent, 85. *Mem:* Am Asn Dent Schs; Int Asn Dent Res (vpres, 90); Am Acad Oral Path; Am Asn Dent Res; Am Dent Asn. *Res:* Oral manifestations of acquired immune deficiency syndrome; opportunistic fungal infections; oral candidasis, mucocutaneous disease; recurrent oral ulcerations. *Mailing Add:* Dent Br Univ Tex PO Box 20068 Houston TX 77225

LYNCH, DERMOT ROBORG, b Johannesburg, SAfrica, Feb 9, 40; Can citizen; m 65; c 2. PLANT BREEDING, PLANT PHYSIOLOGY. *Educ:* Univ Natal, SAfrica, BSc, 63, MSc, 69; Univ Guelph, Can, PhD(plant physiol), 74. *Prof Exp:* Crop specialist, Tech Servs, Dept Agr, 65-66, res scientist potato mgt & physiol, 68-71; res scientist, McCain Foods Ltd, 74-75; asst prof potato and vegetable crops, NS Agr Col, RES SCIENTIST POTATO BREEDING, LETHBRIDGE RES STA, AGR CAN, 78- *Mem:* Agr Inst Can; Potato Asn Am; Europ Asn Potato Res. *Res:* Potato breeding; physiology of the potato and development of superior management options. *Mailing Add:* Agr Can Res Sta Lethbridge AB T1J 4B1 Can

LYNCH, DON MURL, b Delano, Calif, Feb 19, 34. ORGANIC CHEMISTRY. *Educ:* Fresno State Col, AB, 60; Univ Calif, Berkeley, PhD(org chem), 64. *Prof Exp:* Sr res chemist, Abbott Labs, 64-67; sr res chemist, Cutter Labs, 67-68; sr res chemist, 68-78, sr clin res assoc, 78-81, clin monitor, 81-84, CLIN PROJ MGR, ABBOTT LABS, 84- *Mem:* Am Chem Soc. *Res:* Synthesis of potential pharmaceuticals and agricultural chemicals; isolation, structure and synthesis of natural products; preparation of clinical protocols and monitoring clinical investigations. *Mailing Add:* Abbott Labs Abbott Park IL 60064

LYNCH, EDWARD CONOVER, b Fayette, Mo, Feb 24, 33; m 55; c 4. INTERNAL MEDICINE, HEMATOLOGY. *Educ:* Wash Univ, BA, 53, MD, 56. *Prof Exp:* From intern to asst resident med, Barnes Hosp, St Louis, Mo, 56-58; from assoc resident to chief resident, Strong Mem Hosp, Rochester, NY, 58-60; from instr to assoc prof, 62-72, assoc dean, Med Sch, 71-74, dean student affairs, 74-76, PROF MED, BAYLOR COL MED, 72-, ASSOC CHMN DEPT, 77- *Concurrent Pos:* Adj assoc prof biomed eng, Rice Univ, 71-73; adj prof, 73- *Mem:* Am Col Physicians; Am Fedn Clin Res; Am Soc Hemat. *Res:* Effects of physical forces on erythrocytes and blood rheology; internal distribution of iron in various anemias. *Mailing Add:* 311 Wilchester Houston TX 77079

LYNCH, EUGENE DARREL, b Danville, Ill, Sept 4, 21; m 43; c 3. CERAMICS. *Educ:* Univ Ill, BS, 43, MS, 45, PhD, 55. *Prof Exp:* Spec asst, Univ Ill, 43-45; ceramic engr & gen mgr, Kentuckiana Pottery Co, Ky, 45; asst develop mineral, Univ Tex, 45-46; asst high temperature ceramics, Air Materiel Command proj, Univ Ill, 46-47, from asst prof to prof ceramic eng, 47-58; assoc ceramist, Argonne Nat Lab, 58-66; mgr, Mat Lab, Lynchburg Res Ctr, Babcock & Wilcox Co, 66-72, mgr, Mat & Chem Lab, 72-84; RETIRED. *Mem:* Emer mem & fel Am Ceramic Soc. *Res:* High temperature materials; ceramic nuclear fuels. *Mailing Add:* 2217 Oriole Pl Lynchburg VA 24503

LYNCH, FRANK W, b San Francisco, Calif, Nov 26, 21; m 50; c 2. ENGINEERING, ELECTRONICS. *Prof Exp:* Res lab analyst, Boeing Airplane Co, Seattle, 48-50; res engr, Northrop Aircraft, Inc, 50, supvr dynamic anal, 50-51, gen supvr component develop, 51-52, asst chief guid & controls, 52-54 & 55-57, chief flight controls, 54-55; vpres eng, Hallamore Electronics, 57-59; vpres & mgr, Electro-Mech Div, 69-75, sr vpres opers, 76-79, SR VPRES, TACTICAL & ELECTRONIC SYSTS GROUP, NORTHROP CORP, 79- *Mem:* Aerospace Indust Asn Am; sr mem Inst Elec & Electronics Engrs; Am Ord Asn. *Res:* Analog computers and flight simulators; autopilots; guidance sensor systems; astroinertial guidance systems. *Mailing Add:* Northrop Corp 1840 Century Park E Los Angeles CA 90067

LYNCH, GEORGE ROBERT, b Pittsburgh, Pa, Oct 5, 41; m 67. PHYSIOLOGICAL ECOLOGY, COMPARATIVE PHYSIOLOGY. *Educ:* Grove City Col, BS, 64; Univ Mich, MS, 66; Univ Iowa, PhD(zool), 72. *Prof Exp:* Instr zool, Ohio Wesleyan Univ, 66-67; instr biol, Augustana Col, Ill, 67-69; asst prof zool, Univ Maine, Orono, 73-74; asst prof, 74-86, PROF BIOL, WESLEYAN UNIV, 86- *Concurrent Pos:* NIH fel, Inst Behav Genetics, Univ Colo, 72-73; Von Humbodlt fel, 84. *Mem:* Am Soc Zoologists; Ecol Soc Am. *Res:* The role of circadian clocks in photoperiod-induced adjustments in mammalian reproduction and temperature regulation. *Mailing Add:* Dept Biol Wesleyan Univ Lawn Ave Middletown CT 06457

LYNCH, GERARD FRANCIS, b Glascow, Scotland, Oct 10, 45; Can citizen; m 67; c 2. PHYSICS, ENGINEERING PHYSICS. *Educ:* Glasgow Univ, BSc, 67; Queen's Univ, Can, PhD(physics), 71. *Prof Exp:* Lectr physics, Queen's Univ, 71-73; scientist, 73-81, head, Electronics Br, 81-84, exec asst, 84-85, GEN MGR, LOCAL ENERGY SYSTS BUS UNIT, CHALK RIVER NUCLEAR LABS, ATOMIC ENERGY CAN LTD, 85- *Concurrent Pos:* Fel, Queen's Univ, 71, spec lectr, 79-; mem working group, Int Electrotech Comn, 76- *Mem:* Instrument Soc Am. *Res:* Instrumentation development for nuclear reactor applications; infrared spectroscopy; radiation detection and measurement, and analytical techniques. *Mailing Add:* Chalk River Nuclear Labs Atomic Energy Can Ltd Chalk River ON K0J 1J0 Can

LYNCH, HARRY JAMES, b Glenfield, Pa, Jan 18, 29; m 63. NEUROENDOCRINOLOGY. *Educ:* Geneva Col, BS, 57; Univ Pittsburgh, PhD(biol), 71. *Prof Exp:* Clin chemist, Western Pa Hosp, Pittsburgh, 55-66; sr tech fel res asst, Univ Pittsburgh, 66-71; NIH fel, 71-73, res assoc, Lab Richard Wurtman, 74-75, RES SCIENTIST, LAB NEUROENDOCRINE REGULATION, DEPT APPL BIOL SCI, MASS INST TECHNOL, 81- *Mem:* Endocrine Soc; Am Asn Clin Chemists; AAAS; Am Soc Zoologists. *Res:* Neuroendocrine regulation exemplified by the pineal gland of vertebrate animals; pineal gland function as evidenced by melatonin biosynthesis and excretion; physiological, pharmacological, and environmental factors that influence pineal function. *Mailing Add:* Dept Brain & Cognitive Sci Mass Inst Technol E25-604 Lab Neuroendocrine Reg Cambridge MA 02139

LYNCH, HENRY T, b Lawrence, Mass, Jan 4, 28; m 51; c 3. MEDICAL GENETICS, MEDICAL ONCOLOGY. *Educ:* Univ Okla, BS, 51; Univ Denver, MA, 52; Univ Tex, MD, 60. *Prof Exp:* Intern, St Mary's Hosp, Evansville, Ind, 60-61; resident internal med, Col Med, Univ Nebr, 61-64; asst prof biol & asst internist, M D Anderson Hosp & Tumor Inst, Tex, 66-67; assoc prof, 67-71, PROF PREV MED & PUB HEALTH, SCH MED, CREIGHTON UNIV, 71-, CHMN DEPT, 67- *Concurrent Pos:* USPHS sr clin cancer trainee, Eppley Cancer Inst, Nebr, 64-66. *Honors & Awards:* Billings Silver Medal, AMA, 66; Ungerman-Lubin lectr Cancer Res, 87. *Mem:* Am Soc Clin Oncol; Am Asn Cancer Res. *Res:* Cancer genetics. *Mailing Add:* Dept of Prev Med & Pub Health Creighton Univ Sch of Med Calif 24th St Omaha NE 68178

LYNCH, JAMES CARLYLE, b Clifton Hill, Mo, Mar 1, 42; m 65; c 2. NEUROPHYSIOLOGY, NEUROANATOMY. *Educ:* Univ Mo, AB, 64; Stanford Univ, MA & PhD(neurol sci), 71. *Prof Exp:* Instr physiol, Sch Med, Johns Hopkins Univ, 74-76; asst prof physiol, Mayo Med Sch, 76-81; asst prof, 81-82, ASSOC PROF ANAT, UNIV MISS MED CTR, 82-, ASST PROF RES OPHTHAL, 87- *Concurrent Pos:* Nat Inst Neurol Dis & Stroke neurophysiol training grant, Dept Physiol, Sch Med, Johns Hopkins Univ, 71-73; assoc consult physiol, Mayo Found, 76-81; Andrew Mellon career develop award, 74. *Mem:* AAAS; Europ Neurosci Asn; Soc Neurosci; Asn Res Vision & Ophthal. *Res:* Central neural mechanisms of sensation, perception and motor control. *Mailing Add:* Dept Anat Univ Miss Med Ctr Jackson MS 39216

LYNCH, JOHN AUGUST, b Jan 29, 47; US citizen; m 71, 82. ANALYTICAL CHEMISTRY. *Educ:* St Peter's Col, NJ, BS, 70; Pa State Univ, PhD(chem), 76. *Prof Exp:* NSF teaching asst chem, Pa State Univ, 70-71, PHS res fel, 71-75; asst prof, 75-80, ASSOC PROF CHEM, UNIV TENN, CHATTANOOGA, 80- *Concurrent Pos:* Univ Chattanooga Found grant, 76-78; Res Corp grant, 77-; NSF/URP grants, 80, 81; fel, Univ Chattanooga Found, 84 & 88. *Mem:* Am Chem Soc. *Res:* Thermometric methods of analysis used in conjunction with computer interpretation of data; automated titrations; development of instructional microcomputer software. *Mailing Add:* Dept of Chem Univ of Tenn Chattanooga TN 37403

LYNCH, JOHN BROWN, b Akron, Ohio, Feb 5, 29; m 50; c 2. PLASTIC SURGERY. *Educ:* Vanderbilt Univ, BS, 49; Univ Tenn, MD, 52; Am Bd Surg & Am Bd Plastic Surg, dipl. *Prof Exp:* Internship, John Gaston Hosp, Tenn, 53-54; resident surg, Univ Tex Med Br, Galveston, 56-59, res plastic surg, 59-62, from instr to assoc prof, 62-73; PROF PLASTIC SURG, VANDERBILT UNIV SCH MED, 73-, CHMN DEPT, 73- *Concurrent Pos:* Nat consult plastic surg to Surgeon Gen, USAF, 74-; mem, Food & Drug Admin Adv Panel, HHS, Gen Surg & Plastic Surg Devices, 74- *Mem:* AMA; Am Soc Plastic & Reconstructive Surgeons; Am Asn Plastic Surg; fel Am Col Surg; Plastic Surg Res Coun. *Res:* Pathophysiological aspects of burns and laboratory projects related to congenital anomalies. *Mailing Add:* Dept Plastic Surgery Vanderbilt Univ Nashville TN 37240

LYNCH, JOHN DOUGLAS, b Collins, Iowa, July 30, 42; m 64; c 2. ZOOLOGY, HERPETOLOGY. *Educ:* Univ Ill, Urbana, BA, 64, MS, 65; Univ Kans, PhD(zool), 69. *Prof Exp:* Asst prof, 69-73, assoc prof zool, 73-80, PROF LIFE SCI, UNIV NEBR, LINCOLN, 80- *Concurrent Pos:* Vis prof, Nat Univ Columbia, Inst Natural Sci, 85. *Mem:* Am Soc Ichthyol & Herpet; Soc Systs Zool; Soc Study Amphibians & Reptiles; Herpetologists' League. *Res:* Systematics and zoogeography of leptodactyloid frogs especially of neotropical genus Eleutherodactylus; evolution in tropical ecosystems; conservation biology of cyprinodont fishes. *Mailing Add:* Sch Biol Sci Univ of Nebr Lincoln NE 68588

LYNCH, JOHN EDWARD, b Taunton, Mass, Feb 3, 23; m 46; c 2. BACTERIOLOGY, PARASITOLOGY. *Educ:* Providence Col, BS, 49; Mich State Univ, MS, 50, PhD(bact), 52; Am Bd Med Microbiol, dipl. *Prof Exp:* Res bacteriologist, Chas Pfizer & Co, Inc, 52-56, head parasitol lab, 57-60; res microbiologist, Hoffmann-La Roche, Inc, 60-61; res virologist, Chas Pfizer & Co, Inc, 61-63, mgr dept bact & parasitol, 63-70, asst dir dept pharmacol, 71-76, dir, bact & parasitol res, Pfizer, Inc, 76-84; RETIRED. *Mem:* Am Soc Microbiol; Soc Exp Biol & Med; fel Am Acad Microbiol; fel Royal Soc Trop Med & Hyg. *Res:* Chemotherapy of infectious diseases; parasitology. *Mailing Add:* Chas Pfizer Co Med Res Lab Groton CT 06340

LYNCH, JOHN THOMAS, b Washington, DC, Mar 21, 38; m 59, 80; c 2. SPACE PLASMA PHYSICS. *Educ:* Va Polytech Inst, BS, 63; Univ Wis, MS, 65; PhD(physics), 72. *Prof Exp:* Res assoc, Univ Wis-Madison, 72-75, asst scientist, 75-78, assoc scientist physics, 78-79; staff mem, Los Alamos Nat Lab, 79-81; prog scientist, NASA Hq, 81-85; PROG DIR, NSF, 85- *Concurrent Pos:* Lectr physics, Univ Wis-Madison, 72-78; vis staff mem, Los Alamos Nat Lab, 78-79. *Mem:* Am Geophys Union; AAAS. *Res:* Polar aeronomy and astrophysics. *Mailing Add:* NSF DPP 1800 G St NW Washington DC 20550

LYNCH, JOSEPH J, JR, b Baltimore, Md, Apr 18, 56. CARDIOVASCULAR PHARMACOLOGY. *Educ:* Loyola Col, BA, 78; Ohio State Univ, PhD(pharmacol), 82. *Prof Exp:* Sr res pharmacologist, 88-91, ASSOC DIR, DEPT PHARMACOL, MERCK, SHARP & DOHME RES LABS, 91- *Mem:* Am Soc Therapeut. *Mailing Add:* Dept Pharmacol Merck Sharp & Dohme Res Labs WP26-265 West Point PA 19486

LYNCH, MAURICE PATRICK, b Boston, Mass, Feb 24, 36; m 65; c 2. BIOLOGICAL OCEANOGRAPHY, PHYSIOLOGICAL ECOLOGY. *Educ:* Harvard Col, AB, 57; Col William & Mary, MA, 65, PhD(marine sci), 72. *Prof Exp:* Assoc marine scientist, Col William & Mary, 71-73, sr marine scientist & head, dept spec progs, 73-75, asst dir & head, Div Biol Oceanogr, 75-77; asst dir & head, Div Spec Progs & Sci Serv, Va Inst Marine Sci,77-82; asst dir & head, Div Marine Resource Mgt, Col William & Mary, Univ Va, 81-86, assoc prof, 76-79, assoc grad dean, Sch Marine Sci, 87-89, PROF, SCH MARINE SCI & HEAD, OFF SPEC PROGS, SCH MARINE SCI, VA INST MARINE SCI, COL WILLIAM & MARY, VA UNIV, 75- *Concurrent Pos:* Adj prof Earth Sci, Va State Col, 74; US Navy, 57-88; vpres, Coastal Eviron Assoc, Inc, 74-; dir, Va Sea Grant Prog & Chesapeake Res Chesapeake Res Consortium, 84-88; chmn, Sci & Tech Adv Comt, Chesapeake Bay Prog, 85-89. *Mem:* Am Inst Biol Sci; Marine Technol Soc; Am Soc Zoologists; Am Fisheries Soc; Am Soc Limnol & Oceanog; Coastal Soc (pres-elect, 81-83, pres, 83-85); Atlantic Estuarine Res Soc; AAAS. *Res:* Management of marine and estuarine resources with special emphasis on management-research interactions and communications; physiology of marine and estuarine organisms with special emphasis on development of physiological condition indices. *Mailing Add:* Col William & Mary Gloucester Point VA 23062

LYNCH, PETER JOHN, b Minneapolis, Minn, Oct 22, 36; m 64; c 2. DERMATOLOGY. *Educ:* Univ Minn, Minneapolis, BS, 59, MD, 61. *Prof Exp:* Clin instr dermat, Univ Minn, Minneapolis, 65-66; from asst prof to assoc prof, Univ Mich, Ann Arbor, 70-73; assoc prof, 73-75, PROF DERMAT & CHIEF DIV, UNIV ARIZ, 75-, ASSOC HEAD, DEPT INTERNAL MED, 77- *Concurrent Pos:* Consult, Wayne County Gen Hosp, Eloise, Mich, 68-73; Vet Admin Hosp, Ann Arbor, Mich, 71-73, Vet Admin Hosp, Tucson, Ariz & Kino Community Hosp, Tucson, Ariz, 74- *Mem:* AAAS; Am Acad Dermat; Soc Invest Dermat; Am Dermat Asn; Asn Am Med Cols. *Res:* Clinical subjects in diseases of the skin. *Mailing Add:* 420 Delaware St SE Derm Box 98 Minneapolis MN 55455

LYNCH, PETER ROBIN, b Philadelphia, Pa, July 18, 27; m 53; c 3. PHYSIOLOGY. *Educ:* Univ Miami, BS, 50; Temple Univ, MS, 54, PhD(physiol), 58. *Prof Exp:* From instr to assoc prof physiol, 58-70, PROF INTERNAL MED, 86-, CHMN, PHYSIOL DEPT, TEMPLE UNIV SCH MED, 87- *Concurrent Pos:* Prof Physiol & Radiol, Temple Univ Sch Med, 70-; adj prof, Druckkhammerlaboratorium, Kantonsppital, Zurich, Switz, 77-78. *Mem:* Am Physiol Soc; Sigma Xi; Am Heart Asn; AAAS; NAm Soc Cardiac Radiol. *Res:* Cardiovascular and radiologic physiology; rheology. *Mailing Add:* Dept Physiol Temple Univ Med Sch Philadelphia PA 19140

LYNCH, RICHARD G, b Apr 9, 34. PATHOLOGY. *Educ:* Univ Mo, BA, 61; Univ Rochester, MD, 66. *Prof Exp:* Path resident, Wash Univ, St Louis, Mo, 66-69, postdoctoral immunol res fel, 69-72, from asst prof to assoc prof path, 72-80, dir, NIH Training Prog Membranes & Immunol, Sch Med, 80-81; PROF & HEAD, DEPT PATH, COL MED, UNIV IOWA, IOWA CITY, 81-, PROF MICROBIOL, 82- *Concurrent Pos:* Chmn, Path B Study Sect, NIH, 83-86, mem, 82 & Bd Sci Counr, Div Cancer Biol & Diag, Nat Cancer Inst, 87-91; block chmn, Tumor Immunol Prog, 85 & Meetings, Am Asn Immunologists, 86. *Res:* Pathology; immunology. *Mailing Add:* Dept Path Col Med Univ Iowa Iowa City IA 52242

LYNCH, RICHARD VANCE, III, b Philadelphia, Pa, Feb 24, 44; m 68; c 2. MARINE BIOLOGY, BIOPHYSICS. *Educ:* Yale Univ, BS, 66; Univ Pittsburgh, PhD(biophys), 71. *Prof Exp:* Fel org chem, Tokyo Kyoiku Univ, 72-73; res biologist marine biol, Naval Res Lab, Dept Defense US, 83; RES BIOLOGIST MARINE BIOL, NAVAL OCEAN RES & DEVELOP ACTIVITY, DEPT DEFENSE US, 83- *Mem:* Am Soc Photobiol; Sigma Xi. *Res:* Bioluminescence. *Mailing Add:* Lockheed Missiles & Space Co 3251 Hanover St Palo Alto CA 94304-1191

LYNCH, RICHARD WALLACE, b Ft Leavenworth, Kans, June 17, 39; m 62; c 3. CHEMICAL PHYSICS, CHEMICAL ENGINEERING. *Educ:* Univ Calif, Berkeley, BS, 62; Univ Ill, MS, 64, PhD(chem eng), 66. *Prof Exp:* Tech staff mem chem physics, 66 & 68-71, supvr appl mat sci div, 71-73, supvr chem technol div, 73-76, mgr, Waste Mgt & Environ Progs, 76-83, DIR, NUCLEAR WASTE MGT & TRANSP, SANDIA LABS, 83- *Mem:* AAAS; Am Inst Chem Engrs; Am Phys Soc; Sigma Xi. *Res:* Nuclear waste solidification; geologic isolation of nuclear wastes. *Mailing Add:* 7500 Osuna Rd NE Albuquerque NM 87109

LYNCH, ROBERT EARL, b Luxora, Ark, Oct 4, 43; m 61; c 2. ENTOMOLOGY. *Educ:* Ark State Univ, BSE, 65; Iowa State Univ, MS, 69, PhD(entom), 74. *Prof Exp:* Agr res technician, 66-68, entomologist, 68-69, res entomologist, 69-83, SUPVRY RES, ENTOMOLOGIST, USDA, 83- *Concurrent Pos:* Prin investr, US-Aid Res Grant, Burkina Faso, West Africa. *Mem:* Entom Soc Am; Am Peanut Res & Educ Asn. *Res:* Population distributions and economic thresholds of insects on forage grasses and peanuts; resistance in peanuts and forage grasses to insects. *Mailing Add:* Coastal Plain Exp Sta USDA ARS PO Box 748 Tifton GA 31793-0748

LYNCH, ROBERT EMMETT, b Chicago, Ill, Feb 5, 32; m 55; c 3. APPLIED MATHEMATICS. *Educ:* Cornell Univ, BEngPhys, 54; Harvard Univ, MA, 59, PhD(appl math), 63. *Prof Exp:* Sr res mathematician, Res Labs, Gen Motors Corp, 61-64; asst prof math & res mathematician, Univ Tex, Austin, 64-66, assoc prof, 66-67; assoc prof, 67-84, PROF COMPUT SCI & MATH, PURDUE UNIV, 85- *Mem:* Am Math Soc; Math Asn Am; Soc Indust & Appl Math; Sigma Xi. *Res:* Numerical analysis, particularly numerical solution of partial differential equations and applied mathematics. *Mailing Add:* Dept Comput Sci Purdue Univ West Lafayette IN 47907

LYNCH, ROBERT MICHAEL, b Brooklyn, NY, May 30, 44; m 69; c 2. STATISTICS, COMPUTER SCIENCE. *Educ:* State Univ NY, Brockport, BSc, 66; Univ Northern Colo, PhD(statist), 71. *Prof Exp:* Asst prof mgt, Eastern Ill Univ, 71-73; PROF STATIST, UNIV NORTHERN COLO, 73-, ASSOC DEAN, 84- *Concurrent Pos:* Fel WIE, Inst Educ Leadership, George Washington Univ, 72-73; ed collabr & reviewer, current index to statist, J Computing Reviews, 73-; consult ed, J Exp Educ, 78-82; Fulbright prof, Thammasat Univ, Bangkok, 78-79; vis prof, Col VI, 81-82; consult, Weiss & Assoc, Aurora Co, 82-; labor & policies group, Oak Ridge Assoc Univs, Tenn, 82-; vis prof, Info Systs, Cowen Univ, Perth, Australia, 86-87; reviewer, Australian Computer J, 87- *Mem:* Royal Statist Soc; Am Statist Asn; Asn Comput Mach. *Res:* Linear models; data base management systems. *Mailing Add:* Assoc Dean-COBA Univ Northern Colo Greeley CO 80639

LYNCH, STEVEN PAUL, b Los Angeles, Calif, Aug 19, 46; m 67; c 1. SYSTEMATIC BOTANY, POLLINATION ECOLOGY. *Educ:* Calif Polytech State Univ, San Luis Obispo, BS, 69, MA, 71; Univ Calif, Davis, PhD(bot), 77. *Prof Exp:* ASST PROF BIOL, LA STATE UNIV, 77- *Concurrent Pos:* Researcher, Univ Calif, Davis, 77; environ consult, Demopulos & Ferguson Inc Assoc Engrs, 78- *Mem:* Bot Soc Am; Am Soc Plant Taxonomists; Int Soc Plant Taxonomists; Sigma Xi. *Res:* Plant-animal coevolution; floral biology of Asclepias; Monarch Butterfly migratory and feeding behavior; systematics of the Asclepiadaceae and Euphorbiaceae; pollen morphology and Angiosperm Phylogeny; scanning electron microscopy techniques. *Mailing Add:* Dept Biol La State Univ 8515 Youree Dr Shreveport LA 71115

LYNCH, T(HOMAS) E(LWIN), b Mexico, Maine, Aug 7, 14; m 44. ENGINEERING. *Educ:* Univ Maine, BS, 38. *Prof Exp:* Engr, Brush Develop Co, 39-43, head, Dept Electronics Eng, 43-52 & Clevite Corp, 52-57, gen mgt, Ord Prod Div, 57-59, vpres, 65-69; vpres, Gould Inc, 69-75; CHMN, CLEVELAND CRYSTALS INC, 72-; CHMN, DESIGN & MFG CO, 82- *Mem:* Audio Eng Soc; Am Defense Prep Asn; Nat Security Indust Asn; Inst Elec & Electronics Engrs. *Res:* Underwater sound; disc and magnetic recording; underwater ordnance; government contracting; energy technology. *Mailing Add:* Old Mill Rd Gates Mills OH 44040

LYNCH, THOMAS JOHN, b Quincy, Mass, Mar 3, 41; m 69. POLYMER PROCESSING, APPLICATION OF ENGINEERING POLYMERS. *Educ:* Boston Col, BS, 62; Mass Inst Tech, PhD(org chem), 66. *Prof Exp:* Res chemist, Gulf Res & Develop Co, 66-71, sr res chemist, 71-75 & Gulf Oil Chem Co, 75-78, res assoc, 78, mgr polystyrene prod res, 78-83; dir, Environ Progs, 83-87, Prog Mgt, 87-90, DIR, POLYMER PROCESS TECHNOL, AMP INC, 90- *Concurrent Pos:* Woodrow Wilson fel, 58-62; fel NSF, 62-66. *Mem:* AAAS; Am Chem Soc; The Chem Soc. *Res:* Application and modification of engineering polymer for electrical/electronic connectors; solid state polymerization and polymer recycling; program and production management. *Mailing Add:* Amp Inc MS 106-08 PO Box 3608 Harrisburg PA 17105

LYNCH, VINCENT DE PAUL, b Niagara Falls, NY, May 27, 27; m 54; c 4. PHARMACOLOGY. *Educ:* Niagara Univ, BS, 50; St John's Univ, NY, BS, 54; Univ Conn, MS, 56, PhD(pharmacol), 59. *Prof Exp:* Asst pharmacol, Univ Conn, 56-58; from asst prof to assoc prof, 58-66, chmn dept pharmacog, pharmacol & allied sci, 61-73, PROF PHARMACOL, ST JOHN'S UNIV, NY, 66-, DIR DIV TOXICOL, 69-, CHMN DEPT PHARMACEUT SCI, 73- *Concurrent Pos:* Res fel pharmacol, Univ Conn, 54-56; consult, NY State Off Drug Abuse Serv & NY State Assembly Ment Health Comt Drug Abuse Adv Coun. *Mem:* Soc Forensic Toxicol; Int Soc Psychoneuroendocrinol; Int Narcotic Enforcement Off Asn; Sigma Xi. *Res:* Neuropharmacology; toxicology; drug abuse. *Mailing Add:* 420 Convent Ave New York NY 10031

LYNCH, WESLEY CLYDE, b Vancouver, Wash, Feb 28, 44; m 65. NEUROPSYCHOLOGY. *Educ:* Univ Hawaii, BA, 67; Hollins Col, MA, 68; Univ NMex, PhD(exp psychol), 72. *Prof Exp:* Fel physiol psychol, Rockefeller Univ, 71-73, asst prof, 73-75; vis asst fel physiol psychol, John B Pierce Found Lab, 75-80; ASST PROF PSYCHOL, MONT STATE UNIV, 80- *Concurrent Pos:* Adj asst prof, Rockefeller Univ, 75-; res assoc psychol, Yale Univ, 75- *Mem:* Am Psychol Asn; AAAS; Sigma Xi. *Res:* Psychological and physiological bases of motivation, reward and learning. *Mailing Add:* Dept Psychol Mont State Univ Bozeman MT 59717

LYNCH, WILLIAM C, b Cleveland, Ohio, Apr 27, 37; div; c 4. MATHEMATICS, COMPUTER SCIENCE. *Educ:* Case Univ, BS, 59; Univ Wis, MS, 60, PhD(math), 63. *Prof Exp:* Actg instr numerical anal, Univ Wis, 62-63, asst prof, 63; from asst prof to assoc prof comput eng, Case Western Reserve Univ, 63-76; PRIN SCIENTIST, XEROX CORP, 76- *Concurrent Pos:* Vis prof, Comput Lab, Univ Newcastle, 70-71; vis prof, Univ Fed Rio de Janeiro, 75. *Mem:* AAAS; Asn Comput Mach; Am Math Soc; Sigma Xi. *Res:* Mathematical linguistics; design, construction, measurement and modelling of operating systems. *Mailing Add:* 3331 Thomas Dr Palo Alto CA 94303

LYND, JULIAN QUENTIN, b Joplin, Mo, Feb 11, 22; m 43; c 2. SOIL SCIENCE. *Educ:* Univ Ark, BS, 43; Mich State Univ, MS, 47, PhD(soil sci), 48. *Prof Exp:* Asst prof soil sci, Mich State Univ, 48-51; assoc prof, 52-57, PROF AGRON, OKLA STATE UNIV, 57- *Mem:* Fel Am Soc Agron; Soil Sci Soc Am; Int Soc Soil Sci; Am Soc Microbiol; Mycol Soc Am. *Res:* Soil microbiology; induced antibiosis to carcinogenic mycotoxins and biopathway of biotoxin degradation. *Mailing Add:* Dept Agron Okla State Univ Stillwater OK 74074

LYND, LANGTRY EMMETT, b Can, Feb 8, 19; nat US; m 42; c 2. EXTRACTIVE METALLURGY, INORGANIC CHEMISTRY. *Educ:* Univ Man, BSc, 41; Rutgers Univ, MS, 55, PhD(geol), 57. *Prof Exp:* Res scientist & mgr, raw mat sect, res & develop dept, Titanium Pigment Div, Sayreville, NJ, N L Industs Inc, 48-72, sr res scientist, Cent Res Lab, Hightstown, NJ, 72-75; phys scientist & titanium specialist, Bur Mines, US Dept Interior, 77-91; RETIRED. *Mem:* Geol Soc Am; Am Inst Mining, Metall & Petrol Eng. *Res:* Preparation and evaluation of concentrates for titanium dioxide pigment processes; utilization of titaniferous magnetite; treatment of industrial plant wastes; titanium geology and mineralogy; petrography and mineragraphy. *Mailing Add:* 10213 Raider Lane Fairfax VA 22030-1908

LYNDE, RICHARD ARTHUR, b Orange, NJ, Apr 12, 42; m 61; c 2. INORGANIC CHEMISTRY. *Educ:* Hamilton Col, AB, 64; Iowa State Univ, PhD(inorg chem), 70. *Prof Exp:* Asst prof, 70-75, chmn, Dept Chem, 73-76, assoc prof, 75-80, actg dean, 76-80, PROF & DEAN, SCH MATH & NATURAL SCI, MONTCLAIR STATE COL, 80- *Mem:* Am Chem Soc; AAAS; Sigma Xi. *Res:* Elucidation of the stoichiometry, structure and bonding of compounds formed by the post-transition and transition metals in unusual oxidation states. *Mailing Add:* 138 Thackeray Dr Basking Ridge NJ 07920

LYNDON, ROGER CONANT, mathematics; deceased, see previous edition for last biography

LYNDS, BEVERLY T, b Shreveport, La, Aug 19, 29; wid; c 1. ASTRONOMY. *Educ:* Centenary Col, BS, 49; Univ Calif, PhD(astron), 55. *Prof Exp:* Res assoc astron, Nat Radio Astron Observ, Green Bank, WVa, 60-62; from asst prof & asst astronomer to assoc prof astron & assoc astronomer, Steward Observ, Univ Ariz, 62-71; asst dir, Kitt Peak Nat Observ, 71-78, astronomer, 74-86; RETIRED. *Concurrent Pos:* Consult, Astron Adv Panel, NSF, 75-77 & NSF Sci & Technol Policy Off, Adv Group Sci Progs, 75-77; Asn Univ Res Astronomers,87 & Univ Hawaii, 88. *Mem:* Am Astron Soc; Int Astron Union. *Res:* Interstellar medium; galactic structure; composition of galaxies. *Mailing Add:* 3244 6th St Boulder CO 80302

LYNDS, CLARENCE ROGER, b Kirkwood, Mo, July 28, 28; m 54; c 1. OBSERVATION COSMOLOGY. *Educ:* Univ Calif, AB, 52, PhD(astron), 55. *Prof Exp:* Asst, Lick Observ, 52; astronr, Univ Calif, 53-54; jr res ASTRONR & assoc astronr, 55-58; Nat Res Coun Can fel, Dom Astrophys Observ, Can, 58-59; asst astronr, Nat Radio Astron Observ, 59-61; from asst to assoc astronr, 61-68, ASTRONR, KITT PEAK NAT OBSERV, 68- *Mem:*

Nat Acad Sci; Am Astron Soc; Royal Astron Soc; Int Astron Union. *Res:* Photometry and spectroscopy of quasi-stellar objects and galaxies; observational cosmology; optical interferometry. *Mailing Add:* Kitt Peak Nat Observ PO Box 26732 Tucson AZ 85726

LYNE, LEONARD MURRAY, SR, b Riverhurst, Sask, Aug 20, 19; m 46; c 1. PAPER CHEMISTRY. *Educ:* Queen's Univ Ont, BSc, 42, MSc, 46. *Prof Exp:* Chemist, Int Nickel Co, 42-43; res chemist, Dom Plywoods, Ltd, 45-46; res chemist, E B Eddy Co, 46-56, res mgr, 56-62; head printability, Pulp & Paper Res Inst, 62-65; asst res dir pulp & paper, Ont Paper Co, 65-67, dir qual assurance, 67-83; RETIRED. *Mem:* Tech Asn Pulp & Paper Indust; Can Pulp & Paper Asn. *Res:* Fundamental and applied research of pulp and paper. *Mailing Add:* 128 William St Box 402 Niagara-on-the-Lake ON L0S 1J0 Can

LYNK, EDGAR THOMAS, b Kansas City, Mo, Aug 26, 41. LASERS. *Educ:* Yale Univ, BS, 63, MS, 65, PhD(physics), 70. *Prof Exp:* Assoc prof physics, Southern Univ, 69-74; STAFF PHYSICIST, RES & DEVELOP CTR, GEN ELEC CO, 74- *Mem:* AAAS; Am Phys Soc; Inst Elec & Electronics Engrs. *Res:* Atomic excitation cross sections; computerized tomography; ultrasound for medical imaging. *Mailing Add:* 70 Park Terr Apt 2G New York NY 10034

LYNN, DENIS HEWARD, b Kingston, Ont, Apr 20, 47; m 73; c 2. CILIATOLOGY, ELECTRON MICROSCOPY. *Educ:* Univ Guelph, BSc, 69; Univ Toronto, PhD(zool), 75. *Prof Exp:* Res assoc protozool, Dept Zool Univ Md, College Park, 72-73; fel cell biol, Dept Zool, Univ St Andrews, Scotland, 75-77; asst prof, 77-85, ASSOC PROF ZOOL & PROTISTOL, UNIV GUELPH, 85- *Mem:* Am Micros Soc; Soc Protozoologists; Can Soc Cell Biologists; Int Soc Evolutionary Protistology. *Res:* Form and function of ciliated protists as unicellular organisms using techniques of light and electron microscopy; ecology and systematics of protists, especially ciliates; using techniques of cytology, molecular biology (electrophoresis, DNA) and numerical taxonomy. *Mailing Add:* Dept Zool Univ Guelph Guelph ON N1G 2W1 Can

LYNN, HUGH BAILEY, b Verona, NJ, Aug 13, 14; m 40; c 3. SURGERY. *Educ:* Princeton Univ, AB, 36; Columbia Univ, MD, 40. *Prof Exp:* Assoc surg, Newark Babies Hosp, 52-53; assoc prof surg & chief sect pediat surg, Sch Med, Univ Louisville, 53-60; head sect pediat surg, Mayo Clin, 61-78, prof surg, Mayo Grad Sch Med, Univ Minn, 71-78; PROF SURG, UNIV ALA, BIRMINGHAM, 78- *Concurrent Pos:* Teaching fel, Harvard Univ, 51-52; surgeon-in-chief, Children's Hosp, Louisville, Ky, 53-60. *Mem:* Fel Am Col Surg; Am Acad Pediat. *Mailing Add:* Stonehedge Farm PO Box 1040 Middleburg VA 22117

LYNN, JEFFREY WHIDDEN, b Hackensack, NJ, Mar 2, 47; m 64; c 2. SOLID STATE PHYSICS. *Educ:* Ga Inst Technol, BS, 69, MS, 70, PhD(physics), 74. *Prof Exp:* Res asst physics, Oak Ridge Nat Lab, 72-74; res assoc, Brookhaven Nat Lab, 74-76; from asst prof to assoc prof, 76-86, actg dir, Ctr Superconductivity Res, 88-89, PROF PHYSICS, UNIV MD, 86- *Concurrent Pos:* Brookhaven Nat Lab fel, 74-76; NSF grant, 76-; consult, Nat Inst Standards & Technol, 76-; Inst Laue Langevin, Grenoble, France, 83-84; Res Corp grant, 77-80. *Mem:* Am Phys Soc; Am Inst Physics; AAAS. *Res:* Neutron scattering-solid state research; magnetic properties of solids; spin dynamics; magnetic and structural phase transitions; structurally amorphous solids; magnetic superconductors; fundamental physics of neutrons. *Mailing Add:* Dept Physics Univ Md College Park MD 20742

LYNN, JOHN R, b Dallas, Tex, Mar 8, 30; m 54; c 5. OPHTHALMOLOGY. *Educ:* Rice Univ, BA, 51; Univ Tex, MD, 55. *Prof Exp:* Res assoc, Univ Iowa Hosps, 61-63; from asst prof to assoc prof, 63-70, PROF SURG, UNIV TEX HEALTH SCI CTR DALLAS, 70-, CHMN DEPT OPHTHAL, 63- *Concurrent Pos:* Nat Inst Neurol Dis & Blindness spec fel, Univ Iowa Hosps, 61-63 & Eye Clin, Univ T bingen, 62-63. *Mem:* AMA; Am Acad Ophthal & Otolaryngol; Asn Res Vision & Ophthal. *Res:* Methods of clinical perimetry; acute visual function effects by raising the intraocular pressure; threshold, summation and visual acuity of accentric scotomatous areas during phototopic, mesopic and scotopic adaptations. *Mailing Add:* 7150 Greenville Ave Suite 300 Dallas TX 75231

LYNN, JOHN WENDELL, b New York, NY, Mar 23, 25; m 46; c 3. ORGANIC CHEMISTRY. *Educ:* Yale Univ, BS, 48, PhD(chem), 51. *Prof Exp:* Res chemist & proj leader, Org Chem Res Dept, Union Carbide Corp, 51-55, group leader, 55-60, res assoc, 60-61, asst dir res & develop, 61-69, mgr new mkt develop, 69-70, dir technol, Fibers & Fabrics, 69-72, new venture mgr, Chem & Plastics, 72-73, assoc dir res & develop, chem & plastics, 73-85; CONSULT, 85- *Mem:* Am Chem Soc; Electrochem Soc; AAAS. *Res:* Nitrogenous substances; vinyl monomers; organic synthesis; synthetic fibers; vinyl fabrics; nonwovens; thermoplastic M & E resins; phenolic resins, water-soluble polymers. *Mailing Add:* 14 Rolling Hills Ct Seven Lakes, Box 646 West End NC 27376

LYNN, KELVIN G, b Rapid City, SD, Feb 2, 48. SOLID STATE PHYSICS, MATERIALS SCIENCE. *Educ:* Univ Utah, BS, 71, BS, 72, PhD(mat sci), 74. *Prof Exp:* Res assoc, Dept Mat Sci, Univ Utah, 73-74; PHYSICIST, BROOKHAVEN NAT LAB, 74- *Concurrent Pos:* Res vis, Bell Labs, 74-77; vis prof, State Univ NY, Stony Brook, 77-; adj prof, Univ Guelph, Ont. *Mem:* Am Phys Soc; Am Inst Metall Engrs; Am Soc Metals. *Mailing Add:* 510B Brookhaven Nat Lab Upton NY 11973-5000

LYNN, MERRILL, b New Columbia, Pa, Nov 20, 30; m 57; c 2. POLYMER CHEMISTRY. *Educ:* Bucknell Univ, BS, 56; Univ Fla, PhD(chem), 61. *Prof Exp:* Res chemist, Esso Res & Eng Co, 61-69; DEVELOP ASSOC CHEM, CORNING GLASS WORKS, 70- *Mem:* Am Chem Soc; Am Inst Chem; Am Ceramic Soc. *Res:* Bonding to glass surfaces; glass reinforced plastics; polymer modifications; coating resins; immobilized enzymes; ceramic binders. *Mailing Add:* 2920 Olcott Rd Big Flats NY 14814

LYNN, R(ALPH) EMERSON, b Elkhart, Ind, Mar 17, 20; m 46. CHEMICAL ENGINEERING. *Educ:* Purdue Univ, BS, 42; Univ Tex, MS, 49, PhD(chem eng), 53. *Prof Exp:* Tech serv supvr, US Rubber Co, 43-46; sr res engr, B F Goodrich Co, 52-56, res scientist, 56, mgr chem eng res, 56-60, prog planning, B F Goodrich Chem Co Div, 60-66, mgr, E P Rubber Develop, B F Goodrich Chem Co, Ohio, 66-67; alcoa prof chem eng, Ohio State Univ, 67-82; RETIRED. *Mem:* AAAS; Am Chem Soc; fel Am Inst Chem Engrs; Soc Plastics Engrs; Am Soc Eng Educ. *Res:* Economics; thermodynamics; kinetics; polymerization and polymer processing. *Mailing Add:* 467 Zacapa Ave Venice FL 34292

LYNN, RALPH BEVERLEY, b Penetanguishene, Ont, Aug 24, 21; m 44; c 4. SURGERY. *Educ:* Queen's Univ, Ont, MD, CM, 45; FRCS(E), 48; FRCS, 49; Royal Col Physicians & Surgeons Can, cert, 57; FRCS, 58; FRCS(C), 65. *Prof Exp:* Jr intern, Kingston Gen Hosp, 44-46; sr intern surg, Royal Victoria Hosp, Montreal, Que, 46-47; sr registr, Post-Grad Med Sch, Univ London, 47-48; clin tutor, Royal Infirmary, Edinburgh, Scotland, 48-49; asst lectr surg, Post-Grad Med Sch, Univ London, 49-50 & 52-54; sr registr, Southampton Chest Hosp, Eng, 54-55; from asst prof to assoc prof surg, Univ Sask, 55-58; assoc prof, 58-62, PROF SURG, SCH MED, QUEEN'S UNIV, ONT, 62-; HEAD CARDIOTHORACIC UNIT, KINGSTON GEN HOSP, 58-, EMER PROF, 80- *Concurrent Pos:* Nat Res Coun Can scholar, Western Reserve Univ, 50-51; traveling fel, Post-Grad Med Fedn, Johns Hopkins Univ, 51-52; Markle scholar, Univ Sask, 55-57; surgeon, Cleveland City Hosp, Ohio, 50-51; consult, Hotel Dieu & Can Forces Hosp, 58- & Dept Vet Affairs, 58-; fel coun clin cardiol, Am Heart Asn, 65. *Mem:* Asn Thoracic Surg; fel Am Col Surg; fel Am Col Chest Physicians; NY Acad Sci; Can Thoracic Soc; Royal Soc Med. *Res:* Thoracic, cardiovascular and peripheral vascular surgery. *Mailing Add:* Dept Surg Queen's Univ Sch Med Kingston ON K7L 3N6 Can

LYNN, RAYMOND J, b Bitner, Pa, Oct 23, 28; m 58; c 3. MEDICAL MICROBIOLOGY, HOST-PARASITE INTERACTION. *Educ:* Univ Pittsburgh, BS, 52, MS, 53; Univ Pa, PhD(med microbiol), 56. *Prof Exp:* Asst biol, Univ Pittsburgh, 52-53; res investr microbiol, Univ Pa, 53-56, res microbiologist, 56-57; res assoc microbiol, Sch Med, Univ Pittsburgh, 58-60, instr, 60-61; from asst prof to assoc prof, 61-70, PROF MICROBIOL, SCH MED, UNIV SDAK, VERMILLION, 70-, ASSOC DEAN, 83- *Concurrent Pos:* Secy-treas, SDak Bd Examr in Basic Sci, 71-79; rep Dak Affil, regional rev comt, Am Heart Asn, 75-78. *Mem:* AAAS; Am Pub Health Asn; Am Soc Microbiol; Soc Exp Biol & Med; NY Acad Sci. *Res:* Immunology of the Mycoplasmataceae; role of L-forms in sequelae disease states; immunochemistry of streptococcal L-forms and relation of such antigens to rheumatic fever and acute glomerular nephritis; cell-wall defective microorganisms as agents of immunoregulation. *Mailing Add:* Dept of Microbiol Univ of SDak Sch of Med Vermillion SD 57069

LYNN, ROBERT K, b Ky, Oct 29, 47. DRUG METABOLISM. *Educ:* Murray State Univ, BA, 69; Australian Nat Univ, PhD(chem), 74. *Prof Exp:* Res asst, Dept Pharmacol, Sch Med, Vanderbilt Univ, Nashville, 69-71; res scholar, Med Chem Group, Australian Nat Univ, Canberra, 71-74; res assoc, Dept Pharmacol, Sch Med, Ore Health Sci Univ, Portland, 74-77, res instr, Clin Pharmacol Div, 77-78, res asst prof, 78-82; asst dir drug metab, Smith Kline & French Labs, Philadelphia, 82-86, dir, King of Prussia, Pa, 86-87, group dir drug metab & pharmacokinetics, Welwyn, Eng, 87-89, group dir drug metab, SmithKline Beecham Pharmaceuticals, King of Prussia, Pa, 89-90, DIR & VPRES DRUG METAB & PHARMACOKINETICS, SMITHKLINE BEECHAM PHARMACEUTICALS, WELWYN, ENG, 91- *Concurrent Pos:* Prin investr, Nat Inst Environ Health Sci, 77-83, young environ health scientist award, 78-81. *Mem:* Am Soc Pharmacol & Exp Therapeut; Am Soc Mass Spectrometry; Am Chem Soc; AAAS. *Res:* Drug metabolism; pharmacokinetics; analytical chemistry; environmental chemical metabolism. *Mailing Add:* SmithKline Beecham Pharmaceuticals PO Box 1539 Hartfordshire England

LYNN, ROBERT THOMAS, b Coleman, Tex, Jan 15, 31; m 54; c 2. ANIMAL BEHAVIOR, ECOLOGY. *Educ:* Fla State Univ, BA, 56, MA, 57; Univ Okla, PhD(zool), 63. *Prof Exp:* Instr biol, Austin Col, 57-59; asst prof, Emory & Henry Col, 63-64; assoc prof, Presby Col, SC, 64-67; PROF BIOL SCI, SOUTHWESTERN OKLA STATE UNIV, 67- *Mem:* AAAS; Ecol Soc Am; Am Inst Biol Sci; Am Ornith Union; Wilson Ornith Soc. *Res:* Ecology and behavior of birds and lizards. *Mailing Add:* 1208 N Indiana Weatherford OK 73096

LYNN, ROGER YEN SHEN, b Shanghai, China, Jan 18, 41; m; c 1. COMPUTER SCIENCE, OPERATIONS RESEARCH. *Educ:* Cheng Kung Univ, Taiwan, BS, 61; Brown Univ, MS, 64; Courant, NY Univ, PhD(math), 68. *Prof Exp:* Lectr math, Univ Ind, Bloomington, 68-69, asst prof, 69-71; ASST & ASSOC PROF MATH, VILLANOVA UNIV, SC, 71- *Mem:* Am Math Soc; Soc Indust & Appl Math; NY Acad Sci; Asn Comp Mach. *Res:* Asymptotic solutions of differential equations; operations research; computer graphics. *Mailing Add:* Dept of Math Sci Villanova Univ Villanova PA 19085

LYNN, SCOTT, b Iola, Kans, June 18, 28; m 54; c 4. CHEMICAL ENGINEERING. *Educ:* Calif Inst Technol, BS, 50, MS, 51, PhD(chem eng), 54. *Prof Exp:* Asst, Tech Hogesch, Holland, 53-54; res engr, Dow Chem Co, Calif, 54-67; actg prof, 67-69, PROF CHEM ENG, UNIV CALIF, BERKELEY, 69- *Concurrent Pos:* Ed, Indust Electrolytic Div, J Electrochem Soc. *Mem:* Am Chem Soc; fel Am Inst Chem Engrs; Electrochem Soc. *Res:* Separation processes; gas absorption; electrochemistry and electrochemical engineering; process synthesis and development. *Mailing Add:* 2646 San Antonio Dr Walnut Creek CA 94598

LYNN, THOMAS NEIL, JR, b Ft Worth, Tex, Feb 14, 30; m 52; c 3. MEDICINE, PREVENTIVE MEDICINE. *Educ:* Univ Okla, BS, 51, MD, 55. *Prof Exp:* From intern to asst resident med, Barnes Hosp, St Louis, 55-57; clin assoc, Nat Heart Inst, Md, 57-59; chief res, Med Ctr, Univ Okla, 59-61, instr, 61-63, asst prof prev med, 61-64, asst prof med, 63-69, assoc prof prev med & pub health, 64-69, vchmn dept, 63-69, prof family pract, community med & dent & chmn dept, 69-80, actg dean, Col Med, 74-76, dean, 76-80; VPRES, BAPTIST MED CTR, OKLAHOMA CITY, 80- *Mem:* AAAS; Asn Teachers Prev Med; AMA; Am Fedn Clin Res; Asn Am Med Cols. *Res:* Epidemiology of coronary artery disease; psycho-social aspects of dependence and rehabilitation; ballistocardiography and electrocardiography. *Mailing Add:* 3300 Northwest Exp Oklahoma City OK 75112

LYNN, WALTER R(OYAL), b New York, NY, Oct 1, 28; m 60; c 1. CIVIL & ENVIRONMENTAL ENGINEERING. *Educ:* Univ Miami, Fla, BSCE, 50; Univ NC, MSSE, 54; Northwestern Univ, PhD, 63. *Prof Exp:* Asst prof civil eng, Univ Miami, 54-58, assoc prof, 58-61; assoc prof Cornell Univ, 61-67, dir, Ctr Environ Qual Mgt, 66-76, Sch Civil & Environ Eng, 70-78, Sci Tech & Soc Prog, 80- 88, PROF CIVIL & ENVIRON ENG, CORNELL UNIV, 67-, DEAN UNIV FAC, 88- *Concurrent Pos:* Dir res, Ralph B Carter Co, 55-57; consult, Reeder & Lynn, Consult Engrs, 57-61, Rockefeller Fdn, 76-80 & WHO, 69-; adj prof pub health, Med Col, Cornell Univ, 71-80, dir, Prog Sci, Tech & Soc, 80-; mem bd dir, Cornell Res Found, 78-, bd trustees, Cornell Univ, 80-85; assoc ed, J Oper Res, 68-76, J Environ Econs & Mgt, 78-; chmn, Water Sci & Technol bd, Nat Res Coun, 82-85, Comn Water Res, 87-91; chmn NY State Water Res Plng Coun, 86- *Mem:* AAAS; Am Soc Civil Engrs; Sigma Xi. *Res:* Systems analysis and operations research applications in civil and environmental engineering and public health; environmental control; science, technology policy, science and technology for development. *Mailing Add:* Sch of Civil & Environ Eng Cornell Univ Ithaca NY 14853

LYNN, WARREN CLARK, b Satanta, Kans, Dec 4, 35; m 60; c 3. SOIL SCIENCE. *Educ:* Kans State Univ, BS, 57, MS, 58; Univ Calif, PhD(soil sci), 64. *Prof Exp:* SOIL SCIENTIST, NAT SOIL SURV LAB, USDA, 63- *Mem:* Int Soc Soil Sci; Soil Sci Soc Am; Clay Minerals Soc; Int Peat Soc. *Res:* Properties of cat clays or acid sulfate soils; clay minerals in relation to soil properties; organic soils. *Mailing Add:* 2820 Leonard Lincoln NE 68507

LYNN, WILLIAM GARDNER, b Washington, DC, Dec 26, 05; m 33; c 2. ZOOLOGY. *Educ:* Johns Hopkins Univ, AB, 28, PhD(zool), 31. *Prof Exp:* From asst to instr to assoc zool, Johns Hopkins Univ, 28-42; from assoc prof to prof, 42-74, head dept, 58-63, EMER PROF BIOL, CATH UNIV AM, 74- *Concurrent Pos:* Fel Rockefeller Found, Yale Univ, 39-40; Fulbright scholar, Univ Col, WIndies, 52-53. *Mem:* Am Soc Naturalists. *Res:* Anatomy of reptiles; amphibian metamorphosis. *Mailing Add:* 2908 Harvard Dr Wilmington NC 28403

LYNN, WILLIAM SANFORD, b June 14, 22. MEDICINE. *Educ:* Ala Polytech Inst, BS, 43; Columbia Univ, MD, 46. *Prof Exp:* Fel pulmonary physiol, Trudeau Sanitarium, Saranac Lake, NY, 46-49; fel, Dept Endocrinol, Med Ctr, Duke Univ, 50, resident, Dept Med, 51-52; resident physician, Raybrook, NY State Tuberc Sanitarium, NY, 53; fel, Dept Biochem, Univ Pa, 53-55; assoc med & biochem, Med Ctr, Duke Univ, 55-72, prof med & assoc prof biochem, 72-87; RES PROF, DIV HUMAN NUTRIT, MED BR, UNIV TEX, 87- *Concurrent Pos:* Dir, Diabetic Clin, 55-58, Toxicol Prog, 81-86; Markle scholar, Duke Univ, 56-61; ed, Arch Environ Health; chief, Pulmonary Serv, Durham Vet Admin Hosp, 74-82; mem, Duke Comprehensive Cancer Ctr. *Mem:* Am Soc Biol Chemists; Am Soc Clin Investigators; Biophys Soc; Am Thoracic Soc; Asn Am Physicians; Soc Toxicol. *Res:* Pathogenesis of HIV, TNF and dexamethasone and cotton dust; role of lean beef in lipid metabolism in man; mechanism of cytotoxicity of oxidized sterols; cholesterol and cell growth; numerous publications. *Mailing Add:* Prev Med Dept Univ Tex Med Bd Galveston TX 77550

LYNN, YEN-MOW, b Shanghai, China, Jan 17, 35; m 64; c 3. APPLIED MATHEMATICS. *Educ:* Nat Taiwan Univ, BS, 55; Calif Inst Technol, MS, 57, PhD, 61. *Prof Exp:* From asst res scientist to assoc res scientist, Courant Inst Math Sci, NY Univ, 60-64; assoc prof, Ill Inst Technol, 64-67; assoc prof, 67-72, chmn dept, 76-82, PROF MATH, UNIV MD, BALTIMORE COUNTY, 72- *Concurrent Pos:* Consult, Ames Res Ctr, NASA, 66; consult, Ballistic Res Lab, US Army, 69-75. *Mem:* Am Math Soc; Soc Indust & Appl Math; Am Phys Soc. *Res:* Magneto-gasdynamics; plasma physics; partial differential equations; rotating fluids. *Mailing Add:* Dept of Math Univ Md Baltimore County Baltimore MD 21228

LYNNE-DAVIES, PATRICIA, b Swansa, Wales, July 4, 33. RESPIRATORY PHYSIOLOGY. *Educ:* Conjoint Bd, London, MRCS-LRCP, 61; McGill Univ, PhD(physiol), 69. *Prof Exp:* Asst prof, Dept Med, Univ Alta, 69-74; assoc prof, Stanford Univ, 74-80; PROF, DEPT INTERNAL MED, WAYNE STATE UNIV, 80- *Mem:* Am Fedn Clin Res; Am Physiol Soc. *Mailing Add:* Res & Develop ISI Vet Admin Med Ctr Allen Park MI 48101

LYNTON, ERNEST ALBERT, b Berlin, Ger, July 17, 26; nat US; m 53; c 2. ACADEMIC ADMINISTRATION, LOW TEMPERATURE PHYSICS. *Educ:* Carnegie Inst Technol, BS, 47, MS, 48; Yale Univ, PhD(physics), 51. *Prof Exp:* Asst, Off Naval Res, Yale Univ, 48-50; AEC fel, Univ Leiden, 51-52; from asst prof to prof physics, Rutgers Univ, 52-74, dean Livingston Col, 65-74; sr vpres acad affairs, 74-80, COMMONWEALTH PROF PHYSICS, UNIV MASS, 74- *Concurrent Pos:* Vis prof, Univ Grenoble, 59-60; mem, Comn Higher Educ, Mid States Asn, 70-75. *Mem:* Fel Am Phys Soc; Sigma Xi; Am Asn Higher Educ; Am Educ Res Asn. *Res:* Low temperature physics helium 3 and helium 4 mixtures; superconductors; dilute metallic alloys; thermal conductivity; policy in higher education emphasis on changing mission of universities and collaboration with industry; professional preparation and continuing education. *Mailing Add:* 14 Allerton St Brookline MA 02146

LYNTS, GEORGE WILLARD, b Edgerton, Wis, July 26, 36; m 59; c 2. GEOLOGY, PALEONTOLOGY. *Educ:* Univ Wis, BS, 59, MS, 61, PhD(geol), 64. *Prof Exp:* USPHS fel, Columbia Univ, 64-65; from asst prof to assoc prof geol, Duke Univ, 65-82; GEOL SPECIALIST, SAUDI ARAMCO, 90- *Concurrent Pos:* Mem, Cushman Found Foraminifera Res. *Mem:* AAAS; Soc Econ Paleont & Mineral; Paleont Soc; Protozool Soc. *Res:*

Biology and ecology of the Foraminifera; application of quantitative techniques to the solution of geological and biological problems; micropaleontology and paleoecology of the oceans. *Mailing Add:* Saudi ARAMCO PO Box 11250 Dhahran Saudia Arabia

LYO, SUNGKWUN KENNETH, b Pyongnam, Korea, July 3, 41; m 71; c 3. SEMICONDUCTOR PHYSICS, QUANTUM TRANSPORT & MANY-BODY THEORY. *Educ:* Seoul Nat Univ, Korea, BA, 64; Univ Calif Los Angeles, PhD(physics), 72. *Prof Exp:* Asst res physicist, Univ Calif, Los Angeles, 72-73; res assoc physics, Univ Chicago, 73-74; adj asst prof, Univ Calif, Los Angeles, 74-77; mem tech staff, 77-88, SR MEM TECH STAFF, PHYSICS, SANDIA NAT LABS, 88- *Concurrent Pos:* Vis prof, Korea Advan Inst Sci, 80. *Mem:* Am Phys Soc; Mat Res Soc. *Res:* Ferromagnetic Hall effect; spin-lattice relaxation; hopping transport in disordered solids; quantum transport and many-body effects in metals, semiconductors and organic conductors; optical properties of quantum wells. *Mailing Add:* Sandia Nat Labs PO Box 5800 Albuquerque NM 87185-5800

LYON, CAMERON KIRBY, b Islampur, India, July 23, 23; US citizen; m 48; c 3. ORGANIC CHEMISTRY. *Educ:* Col Wooster, BA, 47; Northwestern Univ, PhD(chem), 52. *Prof Exp:* Chemist, Jackson Lab, E I du Pont de Nemours & Co, 51-59; chemist, Western Regional Res Lab, USDA, Albany, 59-86; RETIRED. *Mem:* Am Chem Soc; Am Oil Chem Soc. *Res:* Polymers; urethanes; fats and oils; oilseed and leaf proteins. *Mailing Add:* Five North Lane Orinda CA 94563-2204

LYON, DAVID LOUIS, b Oshkosh, Wis, Jan 20, 35; m 57; c 3. ECOLOGY, ORNITHOLOGY. *Educ:* Beloit Col, BA, 56; Univ Mo, MA, 59; Iowa State Univ, PhD(wildlife ecol), 65. *Prof Exp:* Wildlife biologist, Nebr Game & Parks Comn, 59-61; asst prof, 65-73, ASSOC PROF BIOL, CORNELL COL, 73- *Mem:* AAAS; Ecol Soc Am; Am Ornithologists' Union. *Res:* Competition ecology, particularly territoriality and its relation to resource utilization; pollination ecology. *Mailing Add:* Dept of Biol Cornell Col Mt Vernon IA 52314

LYON, DAVID N, b Altoona, Kans, Apr 15, 19; m 42; c 2. PHYSICAL CHEMISTRY. *Educ:* Univ Mo, MA, 42; Univ Calif, PhD(chem), 48. *Prof Exp:* Res assoc, 48-51, asst res chemist, 51-53, assoc res chemist, 53-59, res chem engr, 59-65, lectr chem eng, 57-65, asst dean, Col Chem, 69-72, PROF CHEM ENG, COL CHEM, UNIV CALIF, BERKELEY, 65- *Mem:* NY Acad Sci; AAAS; Am Chem Soc; Am Inst Chem Eng; Sigma Xi. *Res:* Chemical thermodynamics; cryogenic engineering; chemical process design. *Mailing Add:* 266 Corliss Dr Moraga CA 94556

LYON, DONALD WILKINSON, b Manchester, Eng, Aug 6, 16; nat US; m 42; c 3. INORGANIC CHEMISTRY. *Educ:* Ohio Wesleyan Univ, BA, 37; Ohio State Univ, PhD(inorg chem), 41. *Prof Exp:* Res chemist, E I du Pont de Nemours & Co, Inc, 41-54, tech supvr, 54-62, admin supvr, Pigments Dept, 62-77, personnel coordr, Chem Dyes & Pigments Dept, 77-81; RETIRED. *Honors & Awards:* Borman Award, Am Soc Eng Educ, 81. *Mem:* Am Chem Soc; Am Soc Eng Educ. *Res:* Titanium dioxide. *Mailing Add:* 110 Banbury Dr Windsor Hills Wilmington DE 19803

LYON, DUANE EDGAR, b Muskegon, Mich, Mar 12, 39; m 61; c 2. FOREST PRODUCTS. *Educ:* Univ Mich, BS, 62, MS, 63; Univ Calif, Berkeley, PhD(forest prod), 75. *Prof Exp:* Asst technologist, Dept Wood Technol, Wash State Univ, 63-66; asst specialist, Forest Prod Lab, Univ Calif, 66-73; asst prof, 73-78, ASSOC PROF FOREST PROD, MISS FOREST PROD LAB, MISS STATE UNIV, 78- *Mem:* Forest Prod Res Soc; Soc Wood Sci & Technol. *Res:* Development and characterization of composite engineering materials made wholly or in part from wood. *Mailing Add:* Drawer Fr-Forestry Miss State Univ Mississippi State MS 39762

LYON, EDWARD SPAFFORD, b Chicago, Ill, Feb 26, 26; m 51; c 12. GENITOURINARY SURGERY. *Educ:* Univ Chicago, PhB, 48, SB, 50, MD, 53. *Prof Exp:* Intern, Univ Hosps, 53-54, resident surg, 54-56, resident urol, 56-59, asst prof, Univ, 59-65, assoc prof, 65-83, PROF UROL, UNIV CHICAGO, 84- *Mem:* Am Urol Asn; Soc Univ Urologists; Int Soc Urol Endoscopy; Endourology Soc. *Res:* Urolithiasis. *Mailing Add:* Dept Surg Univ Chicago Med Ctr Box 403 Chicago IL 60637

LYON, GORDON EDWARD, b New London, Wis, June 8, 42; m 71; c 2. COMPUTER SCIENCE. *Educ:* Mich Technol Univ, BS, 64; Univ Mich, MS, 66 & 67, PhD(comput sci), 72. *Prof Exp:* Mathematician, Comput Sci Dept, Gen Motors Res Labs, 67-68; res assoc, dept psychiat, Ment Health Res Inst, Univ Mich, 70-72; COMPUT SCIENTIST, NAT BUR STANDARDS, 72- *Concurrent Pos:* Assoc prof lectr, Dept Elec Eng & Comput Sci, George Wash Univ, 78-79; adj prof lectr, dept decision sci, George Mason Univ, 84-85. *Honors & Awards:* Silver Medal Award, US Dept Com, 78. *Mem:* Asn Comput Mach; Soc Indust Appl Math; Comput Soc Inst Elec & Electronics Engrs. *Res:* Algorithm design; programming techniques; parallel computation (performance). *Mailing Add:* Nat Inst Standards & Technol Gaithersburg MD 20899

LYON, GORDON FREDERICK, b London, Eng, May 10, 22; Can citizen; m 43; c 1. PHYSICS. *Educ:* Univ Sask, BA, 56, MA, 58, PhD(physics), 61. *Prof Exp:* Instr physics, Univ Sask, 56-62; from asst prof to assoc prof,Univ Western Ont, 62-69, prof physics, 69-88; CONSULT, 88- *Concurrent Pos:* Mem comn 6, Int Union Geod & Geophys-Int Asn Geomag & Aeronomy, 63-; mem subcomt aeronomy, Nat Res Coun Can, 66- *Mem:* Am Geophys Union; Am Asn Physics Teachers; Can Asn Physicists. *Res:* Radio physics of the upper atmosphere; scattering of radio waves by ionospheric inhomogeneities; ionospheric absorption; travelling ionospheric disturbances; ionospheric electron content utilizing beacon satellites; associated geophysical phenomena; aurora. *Mailing Add:* 634 Middlewoods Dr London ON N6G 1W8 Can

LYON, HARVEY WILLIAM, dental research; deceased, see previous edition for last biography

LYON, IRVING, b Los Angeles, Calif, May 10, 21; m 48; c 3. HEPATIC GLUTATHIONE HOMEOSTASIS. *Educ:* Univ Calif, Los Angeles, AB, 42, MA, 49; Univ Calif, Berkeley, PhD(physiol), 52. *Prof Exp:* Lab & teaching asst mammalian anat & gen embryol, Univ Southern Calif, Los Angeles, 47-49, researcher & gen lab asst mammalian physiol, Univ Calif, Berkeley, 49-52; res biochemist physiol & biochem skin, med dept, Toni Co, Chicago, 54-58; res assoc physiol & biochem bone, orthop surg, Presby-St Luke's Hosp, Chicago, 58-62; asst prof biol chem, Univ Ill, Chicago, 58-62; assoc prof biochem, Chicago Med Sch, Ill, 62-67; prof biol, sci fac, Bennington Col, Vt, 67-72; sr visitor, Inst Biol Chem A, Univ Copenhagen, Denmark, 72-74; spec consult energy resources & conserv, Calif State Comn, Los Angeles, 75; res physiologist tumor-lipid biochem, Univ Calif, Los Angeles, 79-81; RES BIOCHEMIST HEPATOL, US VET ADMIN WADSWORTH HOSP CTR, LOS ANGELES, 81- *Concurrent Pos:* Res & teaching fel, Rockefeller Found-Med Sci Dept Nutrit, Harvard Sch Pub Health, 52-54; lectr, Soc Gen Physiologists, Woods Hole, Mass, 63, Inst Med Res, Putnam Mem Hosp, Bennington, Vt, dept biochem, physiol & oncol, Univ Wis-Madison, Will Rogers Mem Hosp, Saranac Lake, NY & lab pharmacol, Baltimore Cancer Res Ctr, Nat Cancer Inst, 68; NSF res fel, dept physiol & biophys, Univ Ill, Urbana, 70; vis investr, Jackson Lab, Bar Harbor, Maine, 71; consult, environ health & nutrit, 75- *Mem:* Fel AAAS; fel Int Col Appl Nutrit; Am Physiol Soc; NY Acad Sci. *Res:* Liver transplantation studies; prevention of post-ischemic injury; hepatic enzymes and oxidant injury. *Mailing Add:* 3529 Greenfield Ave Los Angeles CA 90034

LYON, JEFFREY A, m; c 2. IMMUNOLOGY. *Educ:* Va Mil Inst, BS, 70; Univ SC, PhD(biochem), 74. *Prof Exp:* Res chemist, Div Biochem, 75-80, RES CHEMIST, DEPT IMMUNOL, WALTER REED ARMY INST RES, 80- *Mem:* Am Asn Immunologists; Am Soc Trop Med & Hyg; Sigma Xi. *Mailing Add:* Dept Immunol Walter Reed Army Inst Res Washington DC 20307-5100

LYON, JOHN B(ENNETT), b Washington, DC, Mar 13, 27; m 57; c 1. CHEMICAL ENGINEERING. *Educ:* Catholic Univ, BChE, 50; Univ Del, PhD(chem eng), 53. *Prof Exp:* Res engr, Polychem Dept, 53-57, tech investr, Film Dept, 57-58, res engr, 58-59, engr res supvr, Yerkes Res Lab, 60-62, process develop supvr, Clinton Film Plant, Iowa, 62-65 & Spruance Film Plant, Richmond, Va, 65-76, sr engr, 76-80, staff engr, Sabine River Works, 80-85, SR TECH ASSOC, E I DU PONT DE NEMOURS & CO, INC, 86 - *Mem:* Am Chem Soc; Am Inst Chem Engrs; Sigma Xi. *Res:* Heat and mass transfer; application of reaction kinetics. *Mailing Add:* 1824 Linderwood Dr Orange TX 77630

LYON, JOHN BLAKESLEE, JR, b Auburn, NY, Mar 17, 25; m 48; c 2. BIOCHEMISTRY. *Educ:* Hamilton Col, AB, 50; Brown Univ, ScM, 52, PhD(biol), 54. *Prof Exp:* Asst biol, Brown Univ, 50-52; Life Ins Med Res Fund fel biochem, 54-56, from instr to assoc prof, 56-70, PROF BIOCHEM, EMORY UNIV, 70- *Concurrent Pos:* Lederle Med Fac award, Emory Univ, 56-59, USPHS sr res fel, 59. *Mem:* Am Soc Biol Chem. *Res:* Regulatory mechanisms of metabolism; glycogen metabolism; vitamin B-6. *Mailing Add:* Dept Biochem Emory Univ Atlanta GA 30322

LYON, K(ENNETH) C(ASSINGHAM), b La Harpe, Ill, Jan 22, 08; m 33; c 3. CERAMICS ENGINEERING, GLASS TECHNOLOGY. *Educ:* Univ Ill, BS, 31, MS, 33, PhD(ceramic eng), 36. *Prof Exp:* Asst, Eng Exp Sta, Univ Ill, 31-33; ceramic engr, Glass Tech Lab, Gen Elec Co, 35-40; asst chief chemist, Armstrong Cork Co, NJ, 40-42, chief chemist, 42-46, asst mgr glass res, 46-54; tech mgr, Ind Glass Co, 54-55; mgr glass res, Ball Bros Res Corp, 55-67, mgr glass process develop, Ball Bros Co, 67-71; res assoc & glass container mfgrs inst fel, Nat Bur Standards, Md, 72-73; RETIRED. *Concurrent Pos:* Mem div chem & technol, Nat Res Coun, 53-56. *Mem:* Fel Am Ceramic Soc (vpres, 55-56); fel Am Inst Chem; NY Acad Sci. *Res:* Quantitative relationship composition to physical and chemical properties of soda-lime glasses; surface tension of glass; electric melting of glass. *Mailing Add:* 317 E Washington St Dunkirk IN 47336

LYON, LEONARD JACK, b Sterling, Colo, Oct 31, 29; m 56; c 2. WILDLIFE ECOLOGY, FOREST ECOLOGY. *Educ:* Colo State Univ, BS, 51, MS, 53; Univ Mich, PhD(wildlife mgt), 60. *Prof Exp:* Res biologist & proj leader pheasant habitat, Colo Game & Fish Dept, 55-62; WILDLIFE BIOLOGIST & PROJ LEADER FOREST WILDLIFE HABITAT, FORESTRY SCI LAB, INTERMOUNTAIN RES STA, US FOREST SERV, 62- *Concurrent Pos:* Res assoc, Univ Mont, 65-, Univ Idaho, 89- *Mem:* Wildlife Soc; Am Inst Biol Sci; Ecol Soc Am. *Res:* Forest seral ecology; wildlife habitat. *Mailing Add:* Forestry Sci Lab US Forest Serv Intermountain Res Sta Missoula MT 59807-8089

LYON, RICHARD HALE, b Marquette, Mich, Nov 15, 20; c 3. MICROBIOLOGY, BIOCHEMISTRY. *Educ:* Univ Minn, BA, 47, MS, 62, PhD(microbiol, biochem), 65. *Prof Exp:* City bacteriologist, Sioux City Dept Health, Iowa, 48-49; bacteriologist, Vet Admin Ctr, Sioux Falls, SDak, 49-54; res microbiologist, Bact Res Lab, Vet Admin Hosp, Minneapolis, 54-77; res microbiologist, Mastitis Res, Col Vet Med, Univ Minn, 77-79; DIR QUAL CONTROL, PABST MEAT SUPPLY INC, INVER GROVE HEIGHTS, MINN, 79- *Mem:* Am Soc Microbiol; fel Am Inst Chem; Am Thoracic Soc; Inst Food Technologists; Sigma Xi. *Res:* Microbial physiology, specifically as it pertains to metabolic differences in the mycobacteria and to the relationship of these differences to drug susceptibility, taxonomy and virulence. *Mailing Add:* 8567 134 St W St Paul MN 55124-6781

LYON, RICHARD KENNETH, b Cleveland, Ohio, Dec 22, 33; m 68. PHYSICAL CHEMISTRY. *Educ:* Col William & Mary, BS, 55; Harvard Univ, PhD(phys chem), 60. *Prof Exp:* Chemist, 60-64, sr chemist, Cent Basic Res Lab, 64-67, res assoc, 67-75, sr res assoc, 75-80, SCI ADV, EXXON RES & ENG CO, 80- *Honors & Awards:* Indust Res 100 Award; Chem Award, Am Chem Soc. *Mem:* Am Chem Soc; Combustion Inst. *Res:* Chemical reaction kinetics; combustion science; cage effect in solution and gas phase; gas phase detonations and shock waves; radiation and high pressure chemistry; laser isotope separation. *Mailing Add:* Rd 3 Box 267 Finn Rd Pittstown NJ 08867-9445

LYON, RICHARD NORTON, nuclear engineering; deceased, see previous edition for last biography

LYON, ROBERT LYNDON, b Dolgeville, NY, Apr 17, 27; m 84; c 7. FOREST ENTOMOLOGY, INSECT TOXICOLOGY. *Educ:* Syracuse Univ, BS, 53, MS, 54; Univ Calif, Berkeley, PhD(insect toxicol), 61. *Prof Exp:* Res entomologist, 53-72, supvry res entomologist & proj leader, Insecticide Eval Proj, Pac Southwest Forest & Range Exp Sta, 72-76, MEM NAT STAFF, FOREST INSECT & DIS RES, US FOREST SERV, 76-, STAFF RES FOREST ENTOMOLOGIST, 76- *Mem:* Entom Soc Am. *Res:* Development of safe, selective, nonpersistent and effective chemical insecticides and techniques to manage forest insect populations and protect forest resource values with minimal adverse effects on the environment. *Mailing Add:* USDA Forest Serv Box 96090 Washington DC 20090-6090

LYON, RONALD JAMES PEARSON, b Northam, WAustralia, Jan 15, 28; US citizen; m 61; c 4. GEOLOGY, MINERALOGY. *Educ:* Univ Western Australia, BS, 48, Hons, 49; Univ Calif, Berkeley, PhD(geol), 54. *Prof Exp:* Geologist, Lake George Mines, Captains Flat, NSW, 49-51; Goewey res fel geol, Univ Calif, Berkeley, 51-54; res off mining, Commonwealth Sci Res Orgn, Australia, 54-56; geochemist, Kennecott Res Ctr, Utah, 56-59; sr geochemist, Stanford Res Inst, 59-63; Nat Acad Sci sr fel geol, Ames Res Ctr, NASA, 63-65; assoc prof, 65-71 PROF MINERAL EXPLOR, STANFORD UNIV, 72- *Concurrent Pos:* Fulbright travel grant, 51-54 & 78-79; chmn geol panel, Nat Acad Sci, Woods Hole, Mass, 67-69; consult planetary atmospheres, NASA, 68-70; consult & prin assoc, Earth Satellite Corp, 70-; mem remote sensing group, Int Hydrol Decade, Nat Acad Sci, 72- *Honors & Awards:* Photog Interpretation Award, Am Soc Photogram, 72. *Mem:* AAAS; Soc Econ Geol; Am Soc Photogram. *Res:* Use of airborne geophysical techniques and remote sensing in exploration for mineral deposits; recognition of rock and soil materials using land satellite and Skylab spectral data; airborne scanners. *Mailing Add:* Dept Earth Sci Stanford Univ Stanford CA 94305

LYON, WALDO (KAMPMEIER), b Los Angeles, Calif, May 19, 14; m 37; c 2. PHYSICS. *Educ:* Univ Calif, Los Angeles, AB, 36, MA, 37, PhD(physics), 41. *Prof Exp:* Asst physics, Univ Calif, Los Angeles, 40-41; chief scientist, Arctic Submarine Res, US Navy Electronics Lab, 41-66, dir, Arctic Submarine Lab, 66-84, CHIEF SCIENTIST, NAVAL OCEAN SYSTS CTR, 84- *Concurrent Pos:* Sr scientist, Wave Measurement Group, Bikini atom bomb tests, 46; lectr, Univ Calif, Los Angeles, 48-49; physicist, Submarine Opers, US Navy-Byrd Antarctic exped, 46-47; chief scientist, US-Can Aleutian exped, 49, Beauford Sea expeds, 51-54; sr scientist, Transpolar Submarine Exped, 57-82. *Honors & Awards:* Am Soc Naval Engrs Gold Medal Award, 59; Bronze Medal, Royal Inst Navigation, 85; Silver Medal, Société De Géographie, Paris, 83. *Mem:* Fel Am Phys Soc; fel Acoust Soc Am; fel Arctic Inst NAm; Am Soc Naval Eng; Am Geophys Union; Int Glaciological Soc. *Res:* Ocean-cryology and physics of sea ice; underice acoustics. *Mailing Add:* 1330 Alexandria Dr San Diego CA 92107

LYON, WILLIAM FRANCIS, b Mt Gilead, Ohio, Jan 24, 37; m 62; c 5. ECONOMIC ENTOMOLOGY. *Educ:* Ohio State Univ, BSc, 59, MSc, 62, PhD(entom), 69. *Prof Exp:* County exten agent, Ohio Coop Exten Serv, 59-61; exten entomologist, 66-72; surv entomologist, Ohio Agr Res & Develop Ctr, 62-64; asst prof entom & plant protection entomologist, Makerere Univ, Uganda, 72-73; asst prof & pest mgt entomologist, Univ Nairobi, Kenya, 73-74; assoc prof, Afgoi Agr Res Sta, Mogdiscio, Somalia, 76-78; asst & assoc, 74-76, PROF ENTOM, OHIO STATE UNIV, 82- *Concurrent Pos:* USAID entom consult, Guinea Bissau, 88; res fel, Ohio State Univ Ctr African Studies. *Mem:* Entom Soc Am; Am Inst Biol Sci; E African Acad. *Res:* Identification and control of household/structural, livestock, poultry & pet pests, mosquito insects; 4-H youth projects. *Mailing Add:* Dept Entom Ohio State Univ 1991 Kenny Rd Columbus OH 43210-1090

LYON, WILLIAM GRAHAM, b Chelsea, Mass, Apr 29, 44; m 65; c 2. PHYSICAL CHEMISTRY. *Educ:* Univ Mich, BS, 66, MS, 68, PhD(chem), 73. *Prof Exp:* Fel phys chem, Univ Mich, 73-74; fel phys chem, Argonne Nat Lab, 74-76; fel phys chem, Phillips Petroleum Co, 76-88; ENVIRON PHYS CHEMIST, MAN TECH ENVIRON TECHNOL INC, 88- *Mem:* Am Chem Soc; Sigma Xi. *Res:* Environmental geochemistry. *Mailing Add:* Man Tech Environ Technol Inc PO Box 1198 Ada OK 74821-1198

LYON, WILLIAM SOUTHERN, JR, b Pulaski, Va, Jan 25, 22; m 46; c 2. RADIOCHEMISTRY. *Educ:* Univ Va, BS, 43; Univ Tenn, MS, 68. *Prof Exp:* Chemist, E I du Pont de Nemours & Co, WVa, 43-44 & Wash, 44-45; lab foreman, Tenn Eastman Corp, 45-47; chemist, Oak Ridge Nat Lab, 47-62, group leader radiochem, 62-77, head, anal methodol, 77-85; CONSULT, 85- *Concurrent Pos:* Consult, Thai Atomic Energy for Peace Inst, Bangkok, 66-; mem sci comt 25, Nat Coun Radiation Protection, 67-; assoc ed, Radiochem-Radioanal Lett, 70-; regional ed, J Radioanal Chem, 71- *Honors & Awards:* Radiation Indust Award, Am Nuclear Soc, 80; Hevesy Medal, 81. *Mem:* Am Chem Soc; Am Nuclear Soc. *Res:* Trace element analysis; new energy sources; nuclear decay schemes; specialized radioactivity measurements; scientometrics. *Mailing Add:* Anal Chem Div Oak Ridge Nat Lab PO Box X Oak Ridge TN 37831-6142

LYON, CARL J(OHN), b Chicago, Ill, Apr 20, 24; m 47; c 3. CHEMICAL ENGINEERING. *Educ:* Pa State Univ, BSc, 47; Ohio State Univ, MSc, 50. *Prof Exp:* From asst div chief fuels & phys chem to mgr biol, environ & chem dept, 47-73, assoc dir, Res Opers, 73-76, assoc dir, proj mgt & prog develop, 76-80, ASSOC DIR, INTELLECTUAL PROPERTY DEVELOP, BATTELLE COLUMBUS LABS, 80- *Mem:* AAAS; Am Chem Soc; Am Ord Asn. *Res:* Physical chemistry of fuel reactions; surface chemistry; environmental effects of combustion; solid waste technology; application of physical sciences to medical sciences; circulating fluid bed boiler. *Mailing Add:* Battelle Mem Inst 505 King Ave Columbus OH 43201

LYONS, DONALD HERBERT, b Buffalo, NY, Feb 28, 29; m 51; c 3. PHYSICS. *Educ:* Univ Buffalo, BA, 49; Univ Pa, MA, 51, PhD(physics), 54. *Prof Exp:* Staff scientist, Lincoln Lab, Mass Inst Technol, 56-61 & Sperry Rand Res Ctr, 61-63; res prof physics, Inst Solid State Physics, Univ Tokyo, 63-64; staff scientist, Sperry Rand Res Ctr, Mass, 64-66; assoc prof, 66-68, chmn dept, 67-68 & 70-72, PROF PHYSICS, UNIV MASS, BOSTON, 68- *Concurrent Pos:* Fulbright grant, 63-64. *Mem:* Am Phys Soc. *Res:* Theoretical magnetism; communication theory; theoretical nuclear physics. *Mailing Add:* Dept Physics Harbor Campus Univ Mass Boston MA 02125

LYONS, EDWARD ARTHUR, b Halifax, NS, Mar 15, 43; m 67; c 1. RADIOLOGY, ULTRASOUND. *Educ:* Univ Man, BSc, 63, MD, 68; FRCP(C), 73. *Hon Degrees:* FACR, 83. *Prof Exp:* DIR, ULTRASOUND SECT, HEALTH SCI CTR, WINNIPEG, 73-, ST BONIFACE HOSP, 73-; ASSOC PROF MED, UNIV MAN, 76- *Mem:* Am Inst Ultrasound in Med; Can Asn Radiologists; Soc Radiol Ultrasound; Am Col Radiol. *Res:* Long term effects of ultrasound; immunological effects of ultrasound. *Mailing Add:* Sect Diag Ultrasound Health Sci Ctr 700 William Ave Winnipeg MB R3E 0Z3 Can

LYONS, EUGENE T, b Yankton, SDak, May 6, 31. PARASITOLOGY. *Educ:* SDak State Univ, BS, 56; Kans State Univ, MS, 58; Colo State Univ, PhD(parasitol), 63. *Prof Exp:* Asst prof, 58-60 & 63-70, assoc prof, 70-77, PROF PARASITOL, UNIV KY, 77- *Mem:* Am Soc Parasitol; Wildlife Dis Asn. *Res:* Parasites of jackrabbits, fur seals, horses, sheep and cattle. *Mailing Add:* 1149 E Cooper Dr Lexington KY 40502

LYONS, GEORGE D, b New Orleans, La, Jan 19, 28; m 54; c 5. OTOLARYNGOLOGY. *Educ:* Southeastern La Col, BS, 50; La State Univ, New Orleans, MD, 54. *Prof Exp:* From clin instr to clin assoc prof, 58-70, assoc prof, 70-71, PROF OTOLARYNGOL & HEAD DEPT, SCH MED, LA STATE UNIV, NEW ORLEANS, 71-, PROF BIOCOMMUN, 77- *Concurrent Pos:* Mem, Soc Acad Chmn Otolaryngol, 72- *Honors & Awards:* Recognition Award, AMA. *Mem:* Fel Am Laryngol, Rhinol & Otol Soc; fel Am Acad Facial Plastic & Reconstruct Surg; fel Am Col Surg; fel Pan-Am Soc Otolaryngol. *Res:* Regional plastic surgery; otology. *Mailing Add:* La State Univ Med Ctr 2020 Gravier Suite A New Orleans LA 70112-2234

LYONS, HAROLD, b New York, NY, Mar 27, 19; m 41; c 3. ENVIRONMENTAL SCIENCE, MOLECULAR PATHOLOGY. *Educ:* City Col New York, BS, 45; Okla State Univ, MS, 49, PhD(chem), 51. *Prof Exp:* Chemist, Climax Rubber Co, 37-41; res chemist, Ruberoid Co, 45-48; sr res chemist, Gen Elec Co, 51-52 & Pa Salt Mfg Co, 52-55; lab mgr, Koppers Co, Inc, 55-58; prof path, Med Units Univ Tenn, Memphis, 63-85; assoc prof, 58-60, PROF CHEM, RHODES COL, 60- *Mem:* AAAS; Am Chem Soc. *Res:* Instrumental analysis; forensic toxicology; analytical biochemistry; biochemistry. *Mailing Add:* Dept Chem Rhodes Col Memphis TN 38112

LYONS, JAMES EDWARD, b Montpelier, Vt, Oct 20, 37; m 63; c 2. ORGANIC CHEMISTRY, ORGANOMETALLIC CHEMISTRY. *Educ:* Boston Col, BS, 59; Purdue Univ, MS, 61; Univ Calif, Davis, PhD(org chem), 68. *Prof Exp:* Chemist, Res & Develop Ctr, Gen Elec Co, 62-64; res chemist, 68-74, sr res chemist, 74-77, GROUP LEADER, SUN CO, 77- *Mem:* AAAS; Am Chem Soc; NY Acad Sci. *Res:* Mechanisms and synthetic applications of transition metal catalyzed reactions in organic and organometallic systems. *Mailing Add:* 211 Cooper Dr Wallingford PA 19086-6827

LYONS, JAMES MARTIN, b Livermore, Calif, Oct 9, 29; m 56; c 2. PLANT PHYSIOLOGY. *Educ:* Univ Calif, Berkeley, BS, 51; Univ Calif, Davis, MS, 58, PhD(plant physiol), 62. *Prof Exp:* Asst plant physiologist, Univ Calif, Riverside, 62-66; vchmn dept veg crops, 64-66, asst prof, 65-66, assoc prof, chmn dept & assoc plant physiologist, 66-70; chmn, Dept Veg Crops, 70-73, assoc dean, Col Agr & Environ, 73-81, PROF VEG CROPS & PLANT PHYSIOLOGIST, DAVIS, 70-, ASST DIR, EXP STA, UNIV CALIF, 81- *Honors & Awards:* Campbell Award, Am Inst Biol Sci, 71. *Mem:* Am Soc Hort Sci; Am Soc Plant Physiol; Int Soc Hort Sci. *Res:* Biochemistry and physiology of fruit ripening; low temperature biology and chilling injury in vegetable crops. *Mailing Add:* Dept Vegetable Crops Univ Calif Davis CA 95616

LYONS, JOHN BARTHOLOMEW, b Quincy, Mass, Nov 22, 16; m 45; c 5. GEOLOGY. *Educ:* Harvard Univ, AB, 38, AM, 39, PhD(geol), 42. *Prof Exp:* Geologist, US Geol Surv, 41-45; from asst prof to prof, 46-87, EMER PROF GEOL, DARTMOUTH COL, 87- *Mem:* Fel Geol Soc Am; Mineral Soc Am; Arctic Inst NAm. *Res:* Petrology; structural geology; glaciology. *Mailing Add:* Dept of Earth Sci Dartmouth Col Hanover NH 03755

LYONS, JOHN WINSHIP, b Reading, Mass, Nov 5, 30; m 53; c 4. PHYSICAL CHEMISTRY. *Educ:* Harvard Univ, AB, 52; Washington Univ, AM, 63, PhD(phys chem), 64. *Prof Exp:* Prof chemist, Monsanto Co, 55-73; dir, Ctr Fire Res, Nat Bur Standards, 73-77; dir, Nat Eng Lab, 77-90; DIR, NAT INST STANDARDS & TECHNOL, 90- *Concurrent Pos:* Chmn, Prod Res Comt, 74-79; mem adv comt eng, NSF & adv coun, Col Eng, Univ Md, 79-90; mem bd dir, Nat Fire Protection Agency, 78-84. *Honors & Awards:* Gold Medal Award, US Dept Com, 77. *Mem:* Nat Acad Engr; fel AAAS; Am Chem Soc; Am Inst Chem Engrs; Sigma Xi. *Res:* Phosphorus compounds; rheology; fire and fire retardants; surface chemistry; polyelectrolytes; solution behavior of DNA. *Mailing Add:* Nat Inst Standards & Technol Admin A-1134 Gaithersburg MD 20899

LYONS, JOSEPH F, b Wappingers Falls, NY, Nov 27, 20; m 46; c 6. PETROLEUM CHEMISTRY, ALTERNATE ENERGY. *Educ:* Fordham Univ, BS, 41; Purdue Univ, MS, 48, PhD(chem), 50. *Prof Exp:* Chemist, Texaco, Inc, 41-46 & 50-53, group leader, 53-60, res supvr, 60-73, asst mgr, 73-82; RETIRED. *Mem:* Am Chem Soc; Sigma Xi. *Mailing Add:* 17 Broadview Rd Poughkeepsie NY 12603

LYONS, JOSEPH PAUL, b Ardmore, Pa, Dec 9, 47; m 70; c 1. OPERATIONS RESEARCH, PUBLIC HEALTH. *Educ:* Bloomsburg State Col, BA, 70; Johns Hopkins Univ, ScD, 75. *Prof Exp:* Syst analyst ment health, Pa Off Ment Health, 70-71; SCIENTIST ALCOHOLISM, RES INST ALCOHOLISM, 75- *Concurrent Pos:* Nat Inst Ment Health trainee, Johns Hopkins Univ, 71-75; assoc consult, Elliott Assocs, 71-74; admin consult, Md Dept Ment Hyg, 73; asst clin prof, Dept Psychiat, Sch Med & adj asst prof, Dept Indust Eng, Sch Eng, State Univ NY Buffalo, 75- *Mem:* Oper Res Soc Am; AAAS; Asn Ment Health Admin. *Res:* Problem oriented record and its application to alcoholism service delivery; treatment planning in both in-patient and out-patient settings and systems design for delivery of alcoholism services. *Mailing Add:* Dept Indust Eng Bell Hall SUNY at Buffalo North Campus Buffalo NY 14260

LYONS, KENNETH BRENT, b St Louis, Mo, Aug 31, 46; m 68; c 2. SOLID STATE PHYSICS. *Educ:* Univ Okla, BS, 68; Univ Colo, MS, 69, PhD(physics), 73. *Prof Exp:* Mem res staff, 73-87, DISTINGUISHED MEM STAFF, AT&T BELL LABS, 87- *Mem:* Am Phys Soc. *Res:* Raman and Brillouin light scattering in solids, with emphasis on non-equilibrium phenomena, phase transitions, and magnetic scattering in oxide superconductors; magnetooptics. *Mailing Add:* 1A126 Bell Tel Labs 600 Mountain Ave Murray Hill NJ 07974

LYONS, NANCY I, b Akron, Ohio, Sept 17, 46. ECOLOGICAL STATISTICS. *Educ:* Kent State Univ, BS, 68, MA, 70; NC State Univ, PhD(statist), 75. *Prof Exp:* Statistician, Res Triangle Inst, 74-75; asst prof, 75-81, ASSOC PROF STATIST, UNIV GA, 81- *Mem:* Am Statist Asn; Biomet Soc. *Res:* Statistical inference with applications to ecology; computer simulation techniques; sample surveys. *Mailing Add:* Dept Statist Univ Ga Athens GA 30602

LYONS, PAUL CHRISTOPHER, b Cambridge, Mass, Oct 1, 38; m 63; c 5. MINERALOGY, PETROLOGY. *Educ:* Boston Univ, AB, 63, AM, 64, PhD(geol), 69. *Prof Exp:* Pub sch teacher, Mass, 64-68; instr, Boston Univ, 68-69, asst prof phys sci, Boston Univ, 69-75; ADJ PROF CHEM, UNIV PITTSBURGH, 85- *Concurrent Pos:* Res grants, Boston Univ, 71-72 & Mineral Soc Gt Brit; lectr, Lowell Technol Inst, 72 & Boston Univ Metrop Col, 72-73. *Honors & Awards:* Fel, Geol Soc Am. *Mem:* AAAS; Geol Soc Am; Am Asn Petrol Geologist. *Res:* Geology of granites, eastern Massachusetts and Rhode Island; Pennsylvanian stratigraphy, coal geology. *Mailing Add:* 11904 Escalante Ct Reston VA 22091-1834

LYONS, PETER BRUCE, b Hammond, Ind, Feb 23, 43; m 63; c 3. PLASMA PHYSICS. *Educ:* Univ Ariz, BS, 64; Calif Inst Technol, PhD(physics), 69. *Prof Exp:* Staff mem, 69-76, assoc group leader, 76-77, alt group leader, 77-79, group leader, 79-84, prog mgr, 84-85, DEP ASSOC DIR, LOS ALAMOS NAT LAB, 85- *Mem:* Am Phys Soc; Optical Soc Am; Inst Elec & Electronics Eng. *Res:* X-ray interactions and dosimetry; high intensity monoenergetic x-ray generation; x-ray and nuclear detectors and instrumentation; low energy nuclear physics; fiber optic technology; plasma diagnostics; accelerator technology; plastic scintillators. *Mailing Add:* Group ADDRA MS-A110 Los Alamos Nat Lab Los Alamos NM 87545

LYONS, PETER FRANCIS, b Philadelphia, Pa, Nov 29, 42; m 68; c 3. PHYSICAL CHEMISTRY, POLYMER SCIENCE. *Educ:* Villanova Univ, BS, 64; Princeton Univ, MA, 67, PhD(chem), 70. *Prof Exp:* Res chemist, 68-71, sr res chemist, 71-73, mkt rep, 73-78, mkt supvr, 78-80, bus strategist, 80-81, sr planning consult, 82-83, tech marketing mgr, 84, bus develop mgr, 84-86, int marketing mgr, 87-89, STRATEGIC PLANNING MGR, E I DU PONT DE NEMOURS & CO, INC, 89- *Mem:* Am Chem Soc. *Res:* Physical chemistry of polymeric systems including work on degradation, strength mechanisms and viscosity theory. *Mailing Add:* Dupont Fibers E I du Pont de Nemours & Co Inc Wilmington DE 19898

LYONS, PHILIP AUGUSTINE, b Lancashire, Eng, May 26, 16; US citizen; m 49; c 4. PHYSICAL CHEMISTRY. *Educ:* La Salle Col, BA, 37; Univ Wis, PhD(chem), 48. *Prof Exp:* From instr to assoc prof, 48-65, PROF CHEM, YALE UNIV, 65- *Concurrent Pos:* Consult, Audiotape Corp; vis prof, Univ Islamabad, WPakistan, 71. *Mem:* Am Chem Soc. *Res:* Raman spectra; nonaqueous solutions; diffusion in liquids; Soret effect; critical solution phenomena. *Mailing Add:* Ridgewood Terrace N Haven CT 06473

LYONS, RUSSELL DAVID, b Stoneham, Mass, Sept 6, 57. TREES, MEASURES. *Educ:* Case Western Reserve Univ, BA, 79; Univ Mich, PhD(math), 83. *Prof Exp:* Asst prof math, Stanford Univ, 85-90; ASSOC PROF MATH, INDIANA UNIV, 90- *Mem:* Am Math Soc; Math Asn Am. *Res:* Research combines harmonic analysis, functional analysis, erdotic theory and probability factors; random walks and percolation. *Mailing Add:* Dept Math Ind Univ Bloomington IN 47405

LYONS, RUSSETTE M, b Smithtown, NY, Feb 13, 53. CELL BIOLOGY. *Educ:* State Univ NY, BA, 75; Univ Nebr, MS, 78, PhD(life sci), 85. *Prof Exp:* Postdoctoral cancer biol, Vanderbilt Med Ctr, 85-88, asst res prof, Dept Cell Biol, 88-90; CELL BIOL GROUP LEADER & RES SCIENTIST, GENETIC THER, INC, 90- *Mem:* Am Asn Cancer Res; Am Soc Cell Biol. *Mailing Add:* Genetic Ther Inc 19 Firstfield Rd Gaithersburg MD 20878

LYONS, WILLIAM BERRY, b Gainesville, Fla, Feb 8, 47. GEOCHEMISTRY. *Educ:* Brown Univ, BA, 69; Univ Conn, MSc, 72 & PhD(oceanog), 79. *Prof Exp:* Postdoctoral, geochem, 76-79, res scientist, 79-80, asst prof, 80-85, ASSOC PROF GEOCHEMISTRY, UNIV NH, 85- *Mem:* Am Geophys Union; Geochem Soc; Am Soc Immunol & Oceanog; Soc Econ Paleontologists & Mineralogists; Int Glaciol Soc; Int Asn Geochemistry & Cosmochemistry. *Res:* Chemistry of glacial ice and snow; geochemistry of lakes and lacustrine sediments as well as paleoclimatic studies. *Mailing Add:* Hydrol/Hdrogeol Prog Univ Nev Mackay Sch Mines Reno NV 89557

LYRENE, PAUL MAGNUS, b Ala, Apr 16, 46. PLANT BREEDING. *Educ:* Auburn Univ, BS, 68; Univ Wis, MS, 70, PhD(plant breeding), 74. *Prof Exp:* Asst prof agron, 74-77, ASST PROF HORT, UNIV FLA EXP STA, 77- *Mem:* Am Soc Hort Sci. *Res:* Blueberry variety improvement; blueberry interspecific hybridization; blueberry cytogenetics and polyploidy; Zizyphus (Chinese date) investigations. *Mailing Add:* Dept Hort Univ Fla Gainesville FL 32611

LYS, JEREMY EION ALLEYNE, b Dannevirke, NZ, Apr 17, 38; m 68; c 2. HIGH ENERGY PHYSICS. *Educ:* Univ Canterbury, BSc, 58, MSc, 60; Oxford Univ, PhD(physics), 64; Mitchell Col, New South Wales, DipEd, 74. *Prof Exp:* Res fel physics, Univ Liverpool, 63-65; res assoc physics, Univ Mich, 66-72; sch teacher sci, De la Salle Col, New South Wales, 72-74; res assoc physics, Fermi Nat Accelerator Lab, 75-77; physicist, Lawrence Berkeley Lab, 77-81; PHYSICIST, UNIV CALIF, BERKELEY, 82- *Mem:* Am Phys Soc. *Res:* High energy physics. *Mailing Add:* Bldg 50B-5239 Lawrence Berkeley Lab Berkeley CA 94720

LYSAK, ROBERT LOUIS, b Chicago, Ill, Jan 18, 55. SPACE PLASMA PHYSICS, MAGNETOSPHERIC PHYSICS. *Educ:* Mich State Univ, BS, 75; Univ Calif, Berkeley, PhD(physics), 80. *Prof Exp:* Teaching asst physics, Univ Calif, 75-77, res asst, 76-80,asst researcher, 80-82; asst prof, 82-87, ASSOC PROF PHYSICS, UNIV MINN, 87- *Concurrent Pos:* Stipendiat, Max-Planch Inst fur Extraterrestrische Physik, 81. *Mem:* Am Geophys Union; Am Phys Soc. *Res:* Theoretical and numerical investigations of auroral current dynamics, particle accelerations, plasma instabilities, magnetic reconnection and MHD waves and turbulence. *Mailing Add:* Sch Physics & Astron Univ Minn 116 Church St SE Minneapolis MN 55455

LYSEN, JOHN C, b Benson, Minn, Sept 2, 31; m 58; c 2. MECHANICAL ENGINEERING. *Educ:* St Olaf Col, BA, 53; Iowa State Univ, BS, 58, PhD(mech & aerospace eng), 62. *Prof Exp:* Instr mech eng, Iowa State Univ, 58-60, asst prof, 60-63; assoc prof, 63-66, res coord, Col Eng, 66-68, prof mech & aerospace eng, 68-76, DIR ENG EXP STA, UNIV MO-COLUMBIA, 76- *Mem:* AAAS; Am Soc Mech Eng; Am Soc Eng Educ. *Res:* Flow characteristics in converging passages, particularly axially-symmetric annular passages with rotating center bodies. *Mailing Add:* Elec Engr Bldg Rm 349 Univ Mo Columbia MO 65211

LYSER, KATHERINE MAY, b Berkeley, Calif, May 11, 33; m 65. NEUROEMBRYOLOGY. *Educ:* Oberlin Col, AB, 55, Radcliffe Col, MA, 57, PhD(biol), 60. *Prof Exp:* Instr zool, Oberlin Col, 57-58; NSF fel exp embryol, Col France, 60-61; res fel, Med Col, Cornell Univ, 61-62, instr anat, 62-64; asst prof, Sch Med & Dent, Georgetown Univ, 64-65; from asst prof to assoc prof, 65-76, PROF BIOL SCI, HUNTER COL CITY UNIV NY, 76- *Concurrent Pos:* Part-time fac mem, Sarah Lawrence Col, 61-62; USPHS res grants, Med Col, Cornell Univ, 63 & Hunter Col, 65-70; United Cerebral Palsy Res & Educ Found grant, Cornell Univ & Georgetown Univ, 64-65; guest investr, P A Weiss Lab, Rockefeller Univ, 67-70; fac res award, City Univ New York, 76-82, 87-89. *Mem:* Int Soc Develop Neurosci; Am Soc Zool; Soc Neurosci; Soc Develop Biol. *Res:* Factors controlling development in the embryonic nervous system, especially cytological differentiation and cellular morphogenesis in retinal and other neurons and in neuronal tumors. *Mailing Add:* Dept Biol Sci Hunter Col 695 Park Ave New York NY 10021

LYSIAK, RICHARD JOHN, b Chicago, Ill, Dec 29, 28; m 53; c 2. PHYSICS. *Educ:* Aeronaut Univ, Chicago, BSAE, 50; Tex Christian Univ, BA, 59, MA, 60, PhD(physics), 63. *Prof Exp:* Sr aerosyst engr, Gen Dynamics/Ft Worth, 54-61; teaching fel, 61-63, asst prof, 63-67, ASSOC PROF PHYSICS, TEX CHRISTIAN UNIV, 67-, CHMN DEPT, 69-, DIR PRE-ENG PROG, 77- *Mem:* Am Phys Soc. *Res:* Quantum electronics; optics; random noise theory. *Mailing Add:* 1700 Ems Rd W Ft Worth TX 76116

LYSMER, JOHN, b Copenhagen, Denmark, Aug 18, 31; US citizen. EARTHQUAKE ENGINEERING, SOIL DYNAMICS. *Educ:* Tech Univ Denmark, MSc, 54; Univ Mich, PhD(civil eng), 65. *Prof Exp:* Civil engr, Ove Arup & Partners, London, 55-61; PROF SOIL MECH, UNIV CALIF, BERKELEY, 65- *Concurrent Pos:* Eng consult, 65-; Thomas Middlebrooks Award, Am Soc Civil Engrs, 67, Walter Huber Civil Eng Prize, 76. *Mem:* Am Soc Civil Engrs; Seismol Soc Am; Earthquake Eng Res Inst. *Res:* Theoretical soil mechanics and dynamics; developed computer codes for seismic response analysis of earth dams and soil-structure interaction analysis. *Mailing Add:* Dept Civil Eng Univ Calif Berkeley CA 94720

LYSNE, PETER C, b Milwaukee, Wis, July 20, 39; m 62; c 2. APPLIED PHYSICS. *Educ:* Grinnell Col, BA, 61; Ariz State Univ, PhD(physics), 66. *Prof Exp:* Staff mem, Shock Physics Res, 66-77, staff mem, Geothermal Res, 77-89, DISTINGUISHED MEM TECH STAFF, GEOTHERMAL RES, GEOSCI RES DRILLING OFF, SANDIA NAT LABS, 89- *Mem:* Am Phys Soc; Am Geophys Union. *Res:* Soc Prof Well Log Analysts. *Res:* Thermodynamics and its relation to shock physics; shock propagation in solid, liquid and porous media; shock-wave induced depolarization of ferroelectrics; neutron log analysis. *Mailing Add:* 1000 Upland Ct NE Albuquerque NM 87112

LYSTER, MARK ALLAN, b Kalamazoo, Mich, Jan 5, 53; m 75; c 4. ORGANIC CHEMISTRY. *Educ:* Albion Col, BA, 75; Univ Calif, Los Angeles, PhD(org chem), 79. *Prof Exp:* RES CHEMIST, UPJOHN CO, 79- *Mem:* Am Chem Soc. *Res:* Developing processes to produce bulk quantities of prospective new drugs. *Mailing Add:* Upjohn Co 1510-91-1 Kalamazoo MI 49001

LYSYJ, IHOR, b Tarnow, Poland, Apr 13, 29; nat US; m 57; c 2. WASTE TREATMENT, ENVIRONMENTAL TECHNOLOGY. *Educ:* Ukrainian Tech Inst, Ger, MS, 50. *Prof Exp:* Anal chemist, Ex-Lax, Inc, NY, 52-54; dir res, Gaston Johnston Corp, 54-56; anal chemist, Cent Res Lab, Food Mach & Chem Corp, 56-60; res scientist, Ethicon, Inc, 60-61; prin scientist, Rocketdyne Div, Rockwell Int Corp, 61-84; prog mgr, combustion, 84-89; SR SCIENTIST, FURGO-MCCLELLAND, 89- *Mem:* Am Chem Soc. *Res:*

Waste water and hazardous materials treatment; chemical detection and sensing technology; environmental quality monitoring systems and networks; regulatory analysis and pollution assessments; environmental engineering. *Mailing Add:* 8485 Carla Lane Canoga Park CA 91304

LYTHCOTT, GEORGE I, b New York, NY, Apr 29, 18. PEDIATRICS. *Educ:* Boston Univ, MD, 43. *Prof Exp:* Intern, Boston City Hosp, 43-44, from asst resident to resident pediat, 44-46; USAF, 51-53; dean, Med Sch, City Univ New York, 53-91; CONSULT, 91- *Concurrent Pos:* Assoc dean, Urban Community Health Affairs, Univ Wis, 69-74, assoc vchancellor health sci & prof pediat, 74-77. *Mem:* Inst Med-Nat Acad Sci; Am Pub Health. *Mailing Add:* Five Van Wardt Pl Tappan NY 10983

LYTLE, CARL DAVID, b Millersburg, Ohio, Jan 28, 41; c 2. BIOPHYSICS. *Educ:* Kent State Univ, BS, 63; Cornell Univ, MS, 65; Pa State Univ, PhD(biophys), 68. *Prof Exp:* Res biophysicist, Bur Radiol Health, USPHS, 68-70, chief path studies sect, 70, chief path studies sect, Environ Protection Agency, 70-71, chief multi environ stresses br, 71-74, res biophysicist, 74-85, dir, Div Life Sci, 85-89, BIOPHYSICIST CTR DEVICES & RADIOL HEALTH, FOOD & DRUG ADMIN, USPHS, 89- *Concurrent Pos:* Adj prof, George Washington Univ, 71-; assoc ed, Photochem & Photobiol, 78-83. *Mem:* AAAS; Am Soc Photobiol. *Res:* Radiation virology; photodynamic virus inactivation; virus penetration of barrier materials. *Mailing Add:* Ctr Devices & Radiol Health FDA 5600 Fishers Lane Rockville MD 20857

LYTLE, CHARLES FRANKLIN, b Crawfordsville, Ind, May 13, 32; m 55; c 5. INVERTEBRATE ZOOLOGY. *Educ:* Wabash Col, AB, 53; Ind Univ, MA, 58, PhD(zool), 59. *Prof Exp:* Asst zool, Ind Univ, 53-55, 57-58, res assoc, 59-60; asst prof, Tulane Univ, 60-62; res analyst, US Govt, 62-64; from asst prof to assoc prof zool, Pa State Univ, 64-69; assoc prof, 69-72, PROF ZOOL, NC STATE UNIV, 72-, COORDR BIOL SCI PROG, 69- *Concurrent Pos:* Fel embryol, Ind Univ, 59; consult, US Dept Army, 62-63, Educ Testing Serv, 69- & Col Bd, 80-; res assoc, NC Mus of Natural Hist, 77-; vis prof, Duke Univ, Univ Ala, Fla Atlantic Univ; exec dir, NC Student Acad Sci, 85-90; pres sci, NC State Univ, 89- *Mem:* Fel AAAS; Am Soc Zool; Am Inst Biol Sci; Sigma Xi; Soc Col Sci Teachers; Nat Asn Sci Teachers; Nat Asn Biol Teachers. *Res:* Invertebrate zoology; cell biology; cellular structure and function in invertebrate development; differentiation and regulation of cellular organelles; systematics and ecology of Hydrozoa; biological education; instructional television; academic computing. *Mailing Add:* Dept Biol Sci Raleigh NC 27695-7611

LYTLE, DEAN WINTON, b Long Beach, Calif, May 23, 27; m 55; c 4. ELECTRICAL ENGINEERING. *Educ:* Univ Calif, BS, 50; Stanford Univ, MS, 54, PhD(elec eng), 57. *Prof Exp:* Electronic scientist, US Navy Electronics Lab, Calif, 50-53; asst elec eng, Stanford Univ, 53-57; asst prof, Robert Col, Turkey, 57-58; assoc prof, 58-69, PROF ELEC ENG, UNIV WASH, 69- *Concurrent Pos:* Consult, Aerospace Div, Boeing Co, 59- & Seattle Develop Lab, Honeywell, Inc, 63- *Mem:* Inst Elec & Electronics Engrs. *Res:* Information and communication theory. *Mailing Add:* Dept EE Ft-10 Univ of Wash Seattle WA 98195

LYTLE, FARREL WAYNE, b Cedar City, Utah, Nov 10, 34; m 54; c 4. SOLID STATE PHYSICS, STRUCTURAL CHEMISTRY. *Educ:* Univ Nev, BS, 56, MS, 58. *Prof Exp:* Chemist, Univ Bur Mines, 55-58; sr basic res scientist, Boeing Sci Res Labs, 60-74, PRIN RES SCIENTIST, BOEING CO, 74-; PRES, EXAFS CO, 74- *Concurrent Pos:* Grad Study, Univ Wash, 60-63. *Honors & Awards:* Warren Diffraction Physics Award, Am Crystallog Asn, 79. *Mem:* AAAS; fel Am Phys Soc; Am Inst Chem; Am Chem Soc; Mat Res Soc. *Res:* X-ray physics, x-ray absorbtion fine structure spectroscopy and diffraction; materials science; structural inorganic chemistry; amorphous structures; structure of catalysts. *Mailing Add:* 10815 24th Ave S Seattle WA 98168-1808

LYTLE, FRED EDWARD, b Lewisburg, Pa, Jan 13, 43; m 67; c 2. CHEMISTRY. *Educ:* Juniata Col, BS, 64; Mass Inst Technol, PhD(chem), 68. *Prof Exp:* From asst prof to assoc prof, 68-74, PROF CHEM, PURDUE UNIV, WEST LAFAYETTE, 79- *Honors & Awards:* Merck Co Found Fac Develop Award, 69; Am Chem Instrumentation Award, 86; Am Chem Soc Award, 88. *Mem:* Am Chem Soc; Soc Appl Spectros. *Res:* Time resolved spectroscopy; trace analysis; use of lasers in applied spectroscopy. *Mailing Add:* Dept of Chem Purdue Univ 1393 Brwn Bldg West Lafayette IN 47907-1393

LYTLE, LOY DENHAM, b Glendale, Calif, Apr 8, 43; m 74; c 2. PSYCHOPHARMACOLOGY, NEUROSCIENCES. *Educ:* Univ Calif, Santa Barbara, BA, 66; Princeton Univ, PhD(psychol), 70. *Prof Exp:* NIMH fel neuropharmacol, Mass Inst Technol, 70-72, asst prof psychopharmacol, 72-77; ASSOC PROF PSYCHOPHARMACOL, UNIV CALIF, SANTA BARBARA, 77- *Concurrent Pos:* Alfred P Sloan fel neurosci, 75. *Mem:* Am Soc Pharmacol & Exp Therapeut; Nutrit Soc; Int Soc Develop Psychobiol; Neurosci Soc; AAAS; Sigma Xi. *Res:* Effects of drugs on physiological and behavioral development; diet and drug induced changes in behavior; effects of drugs on brain and peripheral neurotransmitters. *Mailing Add:* Dept Psychol Univ Calif Lab Psychopharmacol Santa Barbara CA 93106

LYTLE, RAYMOND ALFRED, b Spartanburg, SC, Sept 23, 19; m 44; c 4. MATHEMATICS. *Educ:* Wofford Col, BS, 40; Univ Va, MA, 46; Univ Ga, PhD, 55. *Prof Exp:* Instr math, Univ Va, 42-46; RETIRED. *Concurrent Pos:* Researcher, Univ Ga, 52-; emer prof math, Univ SC, 86- *Mem:* Am Math Soc; Math Asn Am; Sigma Xi. *Res:* Topology. *Mailing Add:* 1301 Brentwood Dr Columbia SC 29206-2803

LYTTON, BERNARD, b London, Eng, June 28, 26; c 4. UROLOGY. *Educ:* Univ London, MB, BS, 48; FRCS, 55. *Prof Exp:* House officer med & surg, London Hosp, 55-61; from asst prof to assoc prof, 62-71, PROF UROL, SCH MED, YALE UNIV, 71-, CHIEF SECT UROL, 67- *Concurrent Pos:* Brit Empire Cancer res fel surg, Univ Hosp, King's Col, Univ London, 61-62; USPHS grant; resident surg, Royal Victoria Hosp, McGill Univ, 57-58; attend, Yale-New Haven Hosp, 62-; consult, West Haven Vet Admin Hosp, 62- & Hartford Hosp & Hosp of St Raphael, 68- *Honors & Awards:* Hugh H Young Award, Am Urol Asn, 85. *Mem:* AAAS; fel Am Col Surg; Soc Pelvic Surg; Am Asn Genito-Urinary Surg; Clin Soc Genito-Urinary Surgeons. *Res:* Immunologic aspects of cancer; delayed hypersensitivity response to autogenous tumor extracts; problems of renal ischemia; renal responses to alterations in bladder pressure; compensatory renal growth in parabiotic animals and effects of hemodialysis; endoscopic treatment of urinary calculi. *Mailing Add:* 70 High St New Haven CT 06511-6643

LYTTON, JACK L(ESTER), b Los Angeles, Calif, Aug 4, 33; m 54; c 4. MATERIALS SCIENCE, METALLURGY. *Educ:* Univ Calif, Berkeley, BS, 56, MS, 57; Stanford Univ, PhD(mat sci), 62. *Prof Exp:* Res engr, Inst Eng Res, Univ Calif, Berkeley, 56-57; res scientist, Lockheed Missiles & Space Co, 60-65; PROF METALL ENG, VA POLYTECH INST & STATE UNIV, 65- *Mem:* Am Soc Metals; Am Inst Mining, Metall & Petrol Engrs. *Res:* Mechanical behavior of solids, recovery and creep at high temperatures; plastic flow and fracture; failure analysis; structure-property relationships; electronmicroscopy. *Mailing Add:* Dept Metall & Mat Va Polytech Inst & State Univ Blacksburg VA 24061

LYTTON, ROBERT LEONARD, b Port Arthur, Tex, Oct 23, 37; m 61; c 3. PAVEMENTS, EXPANSIVE SOILS. *Educ:* Univ Tex, Austin, BS, 60, MS, 61, PhD(civil eng), 67. *Prof Exp:* Civil engr, Naval Civil Eng Lab, Calif, 60; engr officer, 35th Eng Construct Group, US Army, 61-63; assoc, Dannenbaum Eng Corp, 63-65; asst prof mat & soils, Univ Tex, Austin, 67-68; fel NSF, Australian Commonwealth Sci Inst Res Orgn, 63-65; assoc prof, 71-76, PROF SOILS & PAVEMENTS, TEX A&M UNIV, 76- *Concurrent Pos:* Mem pub adv bd, Int J Analysis & Numerical Methods Geo Mech, 77-; mem tech adv bd, Post-Tensioning Inst, 78-; dir, Meyer, Lytton, Allen, Whitaker, Inc, 80- & ERES Consults, Inc, 81-; bd consult, US Army Corps Engrs, 84-87; consult, Strategic Hwy Res Prog, 85-87; US Rep, Comt TC-6, Int Soc Soil Mech & Found Engrs, 87- *Honors & Awards:* John B Hawley Award, Tex Sect, Am Soc Civil Engrs, 66; Everite Bursary Award, Coun Sci & Indust Res, SAfrica, 84- *Mem:* Am Soc Civil Engrs; Trans Res Bd; Asn Asphalt Paving Technologists; Post-Tensioning Inst; Am Concrete Inst; Int Soc Soil Mech & Foundations Eng. *Res:* Nondestructive testing of pavements; analysis and design of pavement evaluation, foundations and pavements on expansive clays; fracture mechanics; probabilistic design; operations research; pavement network optimization; climatic and environmental effects. *Mailing Add:* Texas A&M Univ 508G CE/TTI Bldg Col Sta TX 77843

LYUBSKY, SERGEY, b Moscow, USSR, June 2, 45; US citizen; m; c 1. PATHOLOGY. *Educ:* Moscow Univ, MD, 68; Inst Human Morphol, Moscow, PhD(cell biol), 75; Am Bd Path, dipl, 85. *Prof Exp:* Postdoctoral fel cytogenetics, Soviet Nat Cancer Inst, Moscow, 68-70; prin investr, Lab Cell Biol, Inst Human Morphol, Moscow, 75-78; vis assoc, Lab Path, Nat Cancer Inst, NIH, 79; resident anat & clin path, George Washington Univ, 80-83; chief resident path, Yale-New Haven Hosp, Conn, 84; ASST PROF PATH, STATE UNIV NY, STONY BROOK, 85-; ASSOC DIR, ELECTRON MICROS LAB, VET ADMIN MED CTR, NORTHPORT, NY, 85-, STAFF PATHOLOGIST, 85- *Concurrent Pos:* Mem, Res & Develop Comt, Cancer Comt & Tumor Bd, Vet Admin Hosp. *Res:* Author of numerous papers and abstracts. *Mailing Add:* Vet Admin Hosp Middleville Rd Northport NY 11768